2014 IEEE 27th International Conference on Micro Electro Mechanical Systems

(MEMS 2014)

San Francisco, California, USA
26 – 30 January 2014

Pages 1-643

IEEE Catalog Number: CFP14MEM-POD
ISBN: 978-1-4799-3510-9

Copyright © 2014 by the Institute of Electrical and Electronic Engineers, Inc
All Rights Reserved

Copyright and Reprint Permissions: Abstracting is permitted with credit to the source. Libraries are permitted to photocopy beyond the limit of U.S. copyright law for private use of patrons those articles in this volume that carry a code at the bottom of the first page, provided the per-copy fee indicated in the code is paid through Copyright Clearance Center, 222 Rosewood Drive, Danvers, MA 01923.

For other copying, reprint or republication permission, write to IEEE Copyrights Manager, IEEE Service Center, 445 Hoes Lane, Piscataway, NJ 08854. All rights reserved.

***This publication is a representation of what appears in the IEEE Digital Libraries. Some format issues inherent in the e-media version may also appear in this print version.**

IEEE Catalog Number: CFP14MEM-POD
ISBN 13:   978-1-4799-3510-9

Additional Copies of This Publication Are Available From:

Curran Associates, Inc
57 Morehouse Lane
Red Hook, NY 12571 USA
Phone: (845) 758-0400
Fax: (845) 758-2633
E-mail: curran@proceedings.com
Web: www.proceedings.com

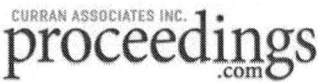

MEMS 2014 Program Schedule

Scroll to the title and select a **Blue** link to open a paper. After viewing the paper, use the bookmarks to the left to return to the beginning of the Table of Contents.

Sunday, January 26

17:00 - **Registration and Wine & Cheese Welcome Reception**
19:00

Monday, January 27

08:00 **Opening and Welcome Address**
Farrokh Ayazi, *Georgia Institute of Technology, USA*
Chang-Jin "CJ" Kim, *University of California, Los Angeles, USA*

Plenary Speaker I
Session Chairs:
F. Ayazi, *Georgia Institute of Technology, USA*
CJ Kim, *University of California, Los Angeles, USA*

08:20 **MICROFABRICATED IMPLANTABLE WIRELESS MICROSYSTEMS: PERMANENT AND BIODEGRADABLE IMPLEMENTATIONS** 1
Mark G. Allen
University of Pennsylvania, USA

Session I – Biomedical Microdevices
Session Chairs:
K. Peterson, *Profusa, USA*
J. Lammertyn, *KU Leuven, BELGIUM*

09:00 **AN INTEGRATED MICROFLUIDIC SYSTEM FOR RAPID ISOLATION AND DETECTION OF LIVE BACTERIA IN PERIPROSTHETIC JOINT INFECTIONS** 5
W.H. Chang[1], C.H. Wang[1], S.Y. Yang[2], Y.C. Lin[2], J.J. Wu[3], M.S. Lee[4], and G.B. Lee[1]
[1]National Tsing Hua University (NTHU), TAIWAN, [2]Jabil Circuit Inc., Ltd, TAIWAN, [3]National Cheng Kung University, TAIWAN, and [4]Chia-Yi Chang Gung Memorial Hospital, TAIWAN

An integrated microfluidic system was presented in this work, which could distinguish the existence of live bacteria within 1 hour. This is the first time that a microfluidic platform was reported to detect live bacteria in periprosthetic joint infection samples. The results demonstrated that the proposed system can detect live bacteria successfully in the micro-environment of clinical samples. The proposed system can be a promising tool for the clinicians with timely medical decisions.

09:15 **A CABLE-TIE-TYPE PARYLENE CUFF ELECTRODE FOR PERIPHERAL NERVE INTERFACES** 9
H.Q. Yu, W.J. Xiong, H.Z. Zhang, W. Wang, and Z.H. Li
Peking University, CHINA

We design, fabricate and characterize a cable-tie-type parylene cuff electrode for peripheral nerve interfaces, whose diameter is adjustable to accommodate the nerve properly during implantation. Cuffs made of thin and flexible parylene minimize mechanical damage to surrounding tissues after implantation. Moreover, the integrated parylene cable and pads facilitate connection with external circuits through wired or wireless interfaces. The acute in vivo rat experiments were performed to verify the ability for the neural recording and selective stimulation of different nerve fascicles.

09:30   **A SILICON ELECTRO-MECHANO TISSUE ASSAY SURGICAL TWEEZER** 13
P.-C. Chen[1], C. Wu[2], F. Michelassi[1], and A. Lal[1]
[1]Cornell University, USA and [2]University of Pennsylvania, USA

This paper reports a first-ever silicon surgical tweezer for characterizing electromechanical properties of tissue. Unlike the other probe-like tissue stiffness tactile sensors, the tweezer structure provides a platform for the clinical use during surgery. We chose to pursue an all-silicon tweezer, instead of attaching sensors to existing tweezers, for repeatable tissue assessment across surgeries without external calibration.

09:45   **HIGHLY PACKED LIPOSOME ASSEMBLIES TOWARD SYNTHETIC TISSUE** 17
H. Hamano[1], T. Tonooka[1], T. Osaki[1,2], and S. Takeuchi[1,2]
[1]University of Tokyo, JAPAN and [2]Kanagawa Academy of Science and Technology, JAPAN

We present a highly-packed liposome assembly that implements lipid bilayer-lipid bilayer contact at the interfaces for mimicking cell-cell connection on living tissues. The closely packed liposomes, based on our previous technique producing a monodisperse liposome array, facilitate easy modification in size and components of the model structures as well as long-term observation of their interfaces. We believe that the assembly technique would help providing a synthetic tissue model.

10:00   **Break & Exhibit Inspection**

Session II – Gyros & Accelerometers
Session Chairs:
D. Horsley, *University of California at Davis, USA*
H. Külah, *Middle East Technical University, TURKEY*

10:45   **WHOLE-ANGLE-MODE MICROMACHINED FUSED-SILICA**
       BIRDBATH RESONATOR GYROSCOPE (WA-BRG) 20
J.-K. Woo, J.Y. Cho, C.W. Boyd, and K. Najafi
University of Michigan, USA

We report the fused-silica birdbath resonator gyroscope (BRG) with a large angular gain, controlled in the whole angle (WA) mode. The BRG is fabricated using the micro-blow-torching process and exhibits good mechanical symmetry, which is ideal for WA mode operation. We adopted the control algorithm for the hemispherical resonator gyroscope (HRG). We report a large bandwidth and full-scale range of 700 deg/s with a large angular gain (A_g = 0.27).

11:00   **100K Q-FACTOR TOROIDAL RING GYROSCOPE IMPLEMENTED IN**
       WAFER-LEVEL EPITAXIAL SILICON ENCAPSULATION PROCESS 24
D. Senkal[1], S. Askari[1], M.J. Ahamed[1], E.J. Ng[2], V. Hong[2], Y. Yang[2],
C.H. Ahn[2], T.W. Kenny[2], and A.M. Shkel[1]
[1]University of California, Irvine, USA and [2]Stanford University, USA

This paper reports a new type of degenerate mode gyroscope with measured Q-factor of > 100,000 on both modes at a compact size of 1760 μm diameter. The toroidal ring gyroscope consists of an outer anchor ring, concentric rings nested inside the anchor ring and an electrode assembly at the inner core. Devices were fabricated using high-temperature, ultra-clean epitaxial silicon encapsulation (Epi-Seal) process.

11:15   **SINGLE PROOF-MASS TRI-AXIAL PENDULUM**
       ACCELEROMETERS OPERATING IN VACUUM 28
D.E.Serrano[1,2], Y. Jeong[1], V. Keesara[2], W.K. Sung[2], and F. Ayazi[1,2]
[1]Georgia Institute of Technology, USA and [2]Qualtré, USA

This paper reports on the design, fabrication and characterization of single proof-mass tri-axial capacitive accelerometers coexisting in a wafer-level packaged (WLP) low-pressure environment with high-frequency gyroscopes, for the implementation of monolithic 6-degree-of-freedom (6-DOF) inertial measurement units (IMUs). The accelerometers are designed to operate as quasi-static devices (i.e. non-resonant sensors) in high vacuum levels (1 – 10 Torr) by increasing squeeze-film air damping through the use of capacitive nano-gaps (< 300 nm).

11:30   **SIMULTANEOUS DETECTION OF LINEAR AND CORIOLIS ACCELERATIONS**
       ON A MODE-MATCHED MEMS GYROSCOPE 32
S. Sonmezoglu, H.D. Gavcar, K. Azgin, S.E. Alper, and T. Akin
Middle East Technical University (METU), TURKEY

This paper presents a novel "in operation acceleration sensing and compensation method" for a single-mass mode-matched MEMS gyroscope. In this method, the amplitudes of the sustained residual quadrature signals on the differential sense-mode electrodes are compared to measure the linear acceleration acting on the sense-axis of the gyroscope. Measuring the acceleration along the sense-axis, the sensitivity of the gyroscope output to linear accelerations along this axis is eliminated without using a dedicated accelerometer.

11:45   **Lunch & Exhibit Inspection**

13:00   **Poster/Oral Session I**

Session III– Materials and Process Characterization
Session Chairs:
M. Despont, *CSEM, SA, SWITZERLAND*
H. Toshiyoshi, *University of Tokyo, JAPAN*

15:00   **A STATIC CAPACITANCE PROBE STRUCTURE FOR RESOLVING THE SIDEWALL SKEW ANGLE OF SILICON DEEP REACTIVE-ION ETCHING** 36
K. Jia, A. Geisberger, A. Dickens, R. Steimle, D.C. Chang, P. Winebarger, L. Liu, and A. McNeil
Freescale Semiconductor Inc., USA

This work presents a design that enables quantitative characterization of the silicon Deep Reactive-Ion Etching sidewall skew angle using static LCR prober at ambient pressure, which provides an easy, accurate and batch solution to the long existing challenge of resolving such process features in an industrial manufacturing environment.

15:15   **INVAR-36 MICRO HEMISPHERICAL SHELL RESONATORS** 40
N. Mehanathan, V. Tavassoli, P. Shao, L. Sorenson, and F. Ayazi
Georgia Institute of Technology, USA

We report the successful fabrication as well as operational characterization of electroplated Invar micro-hemispherical shell resonators. Additionally, the heat treatment of the samples and its effect on the quality factor of the resonators is studied. We show that thermal annealing shifts the coefficient of thermal expansion (CTE) of the alloy towards its minimum, as a result of which the Q increases at least 3 times and reaches ~7500. An annealed electroplated Invar µHSR shows Q of 7500, where unannealed electroplated Invar µHSRs have Qs in the range 2000-3000.

15:30   **IMPLEMENTATION OF SINGLE/MULTI-LAYER MAGNETIC-ANISOTROPY MAGNETIC POLYMER COMPOSITES FOR MAGNETIC PROPERTY MODULATION** 44
F.-M. Hsu, W.-C. Chen, W.-M. Lai, Y.-C. Sun, and W. Fang
National Tsing Hua University (NTHU), TAIWAN

This study extends the two-stage solidification technology to fabricate the isotropic/anisotropic magnetic polymer composites (MPC, polymer with magnetic particles). Multilayer magnetic-anisotropy/isotropic MPC film can also be implemented using the two-stage solidification process layer by layer. Merits of proposed technology: (1) material properties of magnetic-anisotropy MPC layer is realized using the two stage solidification and anisotropic-magnetization processes, and (2) film of various magnetic properties can also be implemented using the different combination of multilayer magnetic-anisotropy MPCs. In applications, the multilayer polymer-NdFeB magnetic composites are realized in silicon substrate and further integrate with MEMS structures. The 1-4 layers of different magnetic-anisotropy MPC are demonstrated. Performances enhancement of magnetic-anisotropy 30wt%-NdFeB MPC (vs isotropic MPC) are: coercivity force (3.4%), remanence (304%), and saturation magnetization (268%). Anisotropic magnetostatic shielding effect (reduce from 0.45Telsa to 0.3-0.35Telsa) is achieved. Moreover, change of magnetic field distributions after stacking of different magnetic-anisotropy MPC layers is also demonstrated.

15:45   **CNT BUNDLES GROWTH ON MICROHOTPLATES FOR DIRECT MEASUREMENT OF THEIR THERMAL PROPERTIES** 48
C. Silvestri, B. Morana, G. Fiorentino, S. Vollebregt, G. Pandraud, F.Santagata,
G.Q. Zhang, and P.M. Sarro
Delft University of Technology, THE NETHERLANDS

Vertically aligned Carbon Nanotubes (CNT) arrays were successfully grown on top of freestanding microheaters. This was made to investigate the thermal dissipation properties of CNTs bundles and their applicability as heat exchanger. The 70µm high bundles have a diameter of 20 and 200µm. A Platinum thin film microheater, integrated on a freestanding SiN membrane, is used as heat source and as temperature sensor. The power consumption of the micro-heaters with different CNTs patterns, is measured in air. At 300 °C a power increase up to 31% was recorded for the microheaters equipped with the CNTs.

16:00   **Break & Exhibit Inspection**

Session IV– Fabrication
Session Chairs:
J. Kim, *Pohang University of Science & Technology, SOUTH KOREA*
B. Pruitt, *Stanford University, USA*

**16:30   3D ICE PRINTING AS A FABRICATION TECHNOLOGY OF
MICROFLUIDICS WITH PRE-SEALED REAGENTS** 52
H.Z. Zhang, H. Li, M.X. Wu, H.Q. Yu, W. Wang, and Z.H. Li
Peking University, CHINA

We propose a innovative and inexpensive method to fabricate 3D structure featured by ice printing.This "bottom-up" 3D fabrication method is achieved by printing water onto the cold substrate and turning into ice structure layer by layer.Through this method, fluid with reagents, such as drugs and nanoparticles, is sealed into microfluidics during fabrication which can be used for drug delivery or other medical care applications.Moreover, a complex microchannels is easily fabricated using 3D ice structure as soft lithography mould, which can be used for microfluid-mixer, three dimensional flow focusing ect.

**16:45   LIQUID-FILLED SEALED MEMS CAPSULES FABRICATED
BY FLUIDIC SELF-ASSEMBLY** 56
M. Mastrangeli, L. Jacot-Descombes, M.R. Gullo, and J. Brugger
Ecole Polytechnique Federale de Lausanne (EPFL), SWITZERLAND

We present an innovative method based on fluidic self-assembly for the encapsulation of functional liquids into sealed picoliter MEMS capsules. Capsules self-assembly and liquid co-encapsulation are achieved through the interplay of global fluidic stirring and local capillary forces ensuing from a selectively-precipitated insoluble polymeric phase. Our encapsulation method is massively parallel, scalable and compatible with batch MEMS fabrication. It can address a large variety of applications, including distributed MEMS sensors, self-healing materials, fragrance release and drug delivery.

**17:00   3D MASK MODULES USING TWO-PHOTON DIRECT LASER WRITING
TECHNOLOGY FOR CONTINUOUS LITHOGRAPHY PROCESS ON FIBERS** 60
M. Hayashi[1,2], Y. Zhang[1], M. Hayase[2], T. Itoh[1], and R. Maeda[1]
[1]National Institute of Advanced Industrial Science and Technology (AIST), JAPAN and
[2]Tokyo University of Science, JAPAN

In this paper, we report a new fabrication method of 3D mask modules using two-photon direct laser writing technology in 140 µm-diameter half-pipe structures on quartz substrates. For the first time, the two-photon direct laser writing technology is utilized for the high resolution patterning process on a curved surface. The minimum feature sizes of about 2 µm line and space are successfully fabricated across the whole 140 µm-diameter half-pipe structures. Using the new 3D mask modules, fine metal patterns are prepared on 125 µm-diameter fiber.

**17:15   FABRICATION OF A MONOLITHIC MICRODISCHARGE-BASED
PRESSURE SENSOR FOR HARSH ENVIRONMENTS** 64
X. Luo, C.K. Eun, and Y.B. Gianchandani
University of Michigan, USA

We present a 6-mask monolithic fabrication process for a pressure sensor that uses a differential microdischarge signal to sense diaphragm deflection. Microdischarge-based transduction is advantageous for harsh environments because of its immunity to temperature and inherently large signals. This work reports the first monolithic fabrication process that successfully addresses a number of challenges for microdischarge-based pressure sensors. Compared to prior work, it results in a ≈30× smaller exterior volume (0.05mm3), a ≈30× wider pressure range (40MPa), and backside terminals for appropriate packages.

17:30   Adjourn for the day

Tuesday, January 28

08:00 ANNOUNCEMENTS

Plenary Speaker II
Session Chairs:
H. Toshiyoshi, *University of Tokyo, JAPAN*
X. Wang, Tsinghua *University, CHINA*

08:05 BIONIC SKINS USING FLEXIBLE ORGANIC DEVICES 68
Takeo Someya[1,2] and T. Sekitani[1,2]
[1]*University of Tokyo, JAPAN and* [2]*Japan Science and Technology Agency (JST), JAPAN*

We have fabricated ultrathin, ultra-lightweight, ultraflexible, organic devices, such as organic thin-film transistors (TFTs), organic photovoltaic (OPV) cells, and organic light-emitting diodes (OLEDs) on polymeric films with the thickness of only 1 μm. The ultrathin organic devices are utilized to fabricate human-machine interfaces such as a touch sensor and wearable electronic systems such as an electromyogram measurement sheet with a two-dimensional array of organic amplifiers. The transistor films exhibit extraordinarily tough mechanical robustness such as minimum bending radius of 5 μm for organic TFTs.

Session V – Optical & Magnetic Microdevices
Session Chairs:
N. Miki, *Keio University, JAPAN*
H. Zappe, *University of Freiburg, GERMANY*

08:45 DEVELOPMENT OF MICRO VARIABLE OPTICS ARRAY 72
Y. Kwon[1], Y. Choi[1], K. Choi[1], Y. Kim[1], S. Choi[2], J. Lee[2], and J. Bae[1],
[1]*Samsung Electronics Co., Ltd., SOUTH KOREA and* [2]*Seoul National University, SOUTH KOREA*

We develop a micro variable optics array which modulates the direction of a light beam. Each pixel of the device has an interface of two immiscible liquids at which the lights are deflected and the interface is actuated on the four separated wall electrodes of the pixel by electrowetting. The four separated electrodes enable the every single pixel to work independently with multiple degrees of freedom, e.g. various tilting angles in every direction for prism or a large number of curvatures for lens mode.

09:00 MINIATURIZED MAGNETOELASTIC TAGS USING FRAME-SUSPENDED
HEXAGONAL RESONATORS 76
J. Tang, S.R. Green, and Y.B. Gianchandani
University of Michigan, USA

Magnetoelastic resonators are of considerable interest for passive wireless interrogation and detection. This paper presents miniaturized magnetoelastic tags using hexagonal resonators with an overall size of about ⌀1.3mm X 27μm, and a resonant frequency as high as 2.13MHz. The tags are 100X smaller than typical commercial tags. The frame-suspension results in ≈75X improvement in signal amplitude of hexagonal tags compared to that of non-suspended disc tags. This paper also demonstrates that the signal amplitude can be boosted by utilizing signal superposition of an ensemble of tags.

09:15 SINGLE-STRUCTURE 3-AXIS LORENTZ FORCE MAGNETOMETER
WITH SUB-30 nT√Hz RESOLUTION 80
M. Li[1], E.J. Ng[2], V.A. Hong[2], C.H. Ahn[2], Y. Yang[2], T.W. Kenny[2], and D.A. Horsley[1]
[1]*University of California, Davis, USA and* [2]*Stanford University, USA*

This work demonstrates a 3-axis Lorentz force magnetometer for electronic compass purposes. The magnetometer measures magnetic flux in 3 axes using a single structure with sub-30 nT/√Hz resolution. Assuming 10 μT Earth's field, the magnetometer has an angular resolution of 0.17 deg/√Hz with 1 mW power consumption. Compared to the 3-axis Hall sensors currently used in smartphones, the 3-axis magnetometer shown here has the advantages of 10× lower noise floor and the ability to be co-fabricated with MEMS inertial sensors.

09:30 DESIGN, FABRICATION AND CHARACTERIZATION OF TUNABLE
PERFECT ABSORBER ON FLEXIBLE SUBSTRATE 84
X. Zhao[1], K. Fan[1], J. Zhang[1], H.R Seren[1], G.D. Metcalfe[2], M. Wraback[2], R.D. Averitt[1], and X. Zhang[1]
[1]*Boston University, USA and* [2]*U.S. Army Research Laboratory, USA*

This paper reports our recent progress on a highly flexible actively tunable metamaterial (MM) perfect absorber at terahertz frequencies. The MM array on GaAs thin-film was patterned on 5μm polyimide substrate via transfer printing technique, and the backside of the substrate was coated with gold. THz time-domain-spectroscopy measurements show that the absorptivity at resonance frequency of 1.59THz can be tuned up to 60% by photo-excitation of free carriers in GaAs patches. Our flexible tunable MM perfect absorber has potential applications in energy harvesting, and imaging.

09:45   **TUNABLE META-FLUIDIC-MATERIALS BASE ON MULTILAYERED MICROFLUIDIC SYSTEM** 88

W.M. Zhu[1], B. Dong[1], Q.H. Song[2], W. Zhang[1], R.F. Huang[3], S.K. Ting[3], and A.Q. Liu[1,2]
[1]*Nanyang Technological University, SINGAPORE,* [2]*Xi'an Jiao Tong University, CHINA, and* [3]*Temasek Laboratories, SINGAPORE*

We demonstrate a multilayered microfluidic system with a flexible substrate, which has tunable optical chirality within THz spectrum range. The optical properties of the multilayered microfluidic system can be tuned by either changing the liquid pumped into each layer or stretching the flexible substrate. It is feasible for the multilayered microfluidic structure to be integrated to an optofluidic system, where strong or tunable optical chirality are needed, which not only can be used as traditional optic components such as THz polarizers and filters but also has potential applications on imaging and sensor of bio-materials.

10:00   **Break & Exhibit Inspection**

Session VI – Fluidic Microdevices
Session Chairs:
G.-B. Lee, *National Tsing Hua University (NTHU), TAIWAN*
A. Dietzel, *Technische Universität Braunschweig, GERMANY*

10:45   **MULTIPLE SIZE-ORIENTED PASSIVE DROPLET SORTING DEVICE AND BASIC APPROACH FOR DIGITAL CHEMICAL SYNTHESIS** 92

S. Numakunai, A. Jamsaid, D.H. Yoon, T. Sekiguchi, and S. Shoji
Waseda University, JAPAN

This paper presents a multiple size-oriented passive droplet sorting utilizing a balance between surface free energy and flow force. We propose a multi-stage sorting structure and passive five different-sized droplets sorting of about 100 droplets/sec is achieved without any active elements. Also, we fabricated a prototype of the integrated micro fluidic system with droplet generation, merging and sorting for digital chemical synthesis.

11:00   **SURFACE ENERGY MICROPATTERN INHERITANCE FROM MOLD TO REPLICA** 96

G. Pardon, T. Haraldsson, and W. van der Wijngaart
KTH Royal Institute of Technology, SWEDEN

We report a novel surface energy patterning phenomenon, in which a novel polymer composition inherits the surface energy of the medium it is in contact with during polymerization. This process occurs via spontaneous alignment of hydrophilic and hydrophobic monomers contained in the prepolymer. This single-step method for simultaneous structuring and surface energy micropatterning of polymer structures is potentially more robust and lower cost than state-of-the-art. We further demonstrate the self-assembly of a liquid droplet array on the replicated polymer surfaces.

11:15   **ELECTROSPRAY DEPOSITION FROM AFM PROBES WITH NANOSCALE APERTURES** 100

J. Geerlings[1], E. Sarajlic[1,2], J.W. Berenschot[1], R.G.P. Sanders[1], L. Abelmann[1,3], and N.R. Tas[1]
[1]*MESA+, University of Twente, THE NETHERLANDS,* [2]*SmartTip B.V., THE NETHERLANDS, and* [3]*Korea Institute of Science and Technology (KIST) – Europe, GERMANY*

In this contribution we present for the first time extraction of liquid from nano-sized apertures in fountain pen AFM probes by means of electrospray. This technique allows for contactless deposition and we show that droplets with radii in the order of one micrometer can be deposited. The required onset voltage for electrospray as function of gap spacing and applied pressure is studied and a simple model is presented which is in qualitative agreement with our measurements.

11:30   **A MICROBUBBLE PRESSURE TRANSDUCER WITH BUBBLE NUCLEATION CORE** 104

L. Yu and E. Meng
University of Southern California, USA

We present a microchannel-based microbubble (μB) pressure transducer (μBPT) with μB nucleation core for characterization of μB dynamics and pressure transduction in wet environments with low power consumption. The transducer leverages electrochemical impedance-based measurement to monitor the instantaneous response of μB size induced by hydrostatic pressure changes. We demonstrated on-demand μB nucleation and real-time pressure tracking (0-350 mmHg). Biocompatible construction and liquid-based operation of μBPTs are ideal for *in vivo* pressure monitoring.

11:45 MICROFLUIDIC ELECTROCHEMILUMINESCENCE (ECL) INTEGRATED
CELL FOR PORTABLE FLUORESCENCE DETECTION 108
M. Tsuwaki[1], J. Mizuno[1], T. Kasahara[1], T. Edura[2], E. Kunisawa[2], R. Ishimatsu[2],
S. Matsunami[2], T. Imato[2], C. Adachi[2], and S. Shoji[1]
[1]Waseda University, JAPAN and [2]Kyushu University, JAPAN

We propose a portable electrochemiluminescence (ECL)-induced fluorescence chip which consists of flow channels for fluorescence sample and multi-color emitting ECL excitation source. A prototype ECL-induced fluorescence chip was fabricated by conventional photolithography and bonding technique. Device performance was evaluated using ECL of rubrene as excitation source and resorufin as fluorescent dye. Fluorescence of 500 μM resorufin (600 nm) was successfully detected using 10 mM rubrene solution (560 nm) under the applied voltage of 4 V. The proposed principle is applicable for portable and on-demand multi fluorescence detection device using its freedom of choice for combination of the ECL light source.

12:00 **A MONOLITHIC KNUDSEN PUMP WITH 20 sccm FLOW RATE
USING THROUGH-WAFER ONO CHANNELS** 112
S. An, Y. Qin, and Y.B. Gianchandani
University of Michigan, USA

We report a lithographically microfabricated Knudsen pump for high gas flow. Knudsen pumps operate by thermal transpiration and require no moving parts. To achieve high gas flow, high-density arrays of microchannels are used in parallel (with over 4000 channels/mm^2). These vertically oriented microchannels have 2×120 μm^2 openings surrounded by 0.1 μm-thick silicon oxide-nitride-oxide (ONO) sidewalls. The thin ONO sidewalls provide thermal isolation between a heat sink formed within the Si substrate, and a Cr/Pt thin film heater formed above the microchannels that provides a temperature bias for thermal transpiration. The Knudsen pump is monolithically microfabricated on a single SOI wafer using a four-mask process. It has a total footprint of 8×10 mm^2. It produces a measured air flow of 20 sccm, with typical response times of 0.1-0.4 sec.

12:15 **MEMS 2015 Announcement**

12:30 **Lunch on Own**

14:00 **Poster/Oral Session II**

Session VII– Resonant Microdevices & Sensors
Session Chairs:
M. Rais-Zadeh, *University of Michigan, USA*
L. Buchaillot, *IEMN, FRANCE*

16:00 **SUSPENDED NANOCHANNEL RESONATORS AT ATTOGRAM PRECISION** 116
S. Olcum[1], N. Cermak[1], S.C. Wasserman[1], K. Payer[1], W. Shen[2], J. Lee[3], and S.R. Manalis[1]
[1]*Massachusetts Institute of Technology, USA*, [2]*Innovative Micro Technology, USA,* and
[3]*Sogang University, SOUTH KOREA*

We developed a nanomechanical resonator that can directly measure the mass of individual nanoparticles down to 10 nm in solution at room temperature with single-attogram precision, also enabling access to many of the engineered nanoparticles used in nanomedicine, most of the virions like HIV, HCV, and natural sub-cellular structures like exosomes. To achieve this, we demonstrate an oscillator system with frequency stability down to 4 ppb, approaching the fundamental limit imposed by intrinsic thermomechanical fluctuations of the resonator.

16:15 **DUAL-MODE VERTICAL MEMBRANE RESONANT PRESSURE SENSOR** 120
R. Tabrizian and F. Ayazi
Georgia Institute of Technology, USA

We present a novel dual-mode resonant pressure sensor operating based on mass loading of air molecules on transversely vibrating vertical silicon membranes. Identical piezoelectrically-transduced silicon bulk acoustic resonators are acoustically coupled through thin vertical membranes, resulting in two high-Q resonance modes with small frequency split, but large difference in pressure sensitivity. Being proportional to the flexural resonance frequency of the thin membranes, the small beat frequency (fb) extracted from subtraction of the two coupled modes shows amplified pressure sensitivity. A proof-of-concept device implemented on silicon substrate and transduced by aluminum nitride film shows an fb of 370 kHz with a linear pressure sensitivity of 280 ppm/kPa.

16:30 **HIGHLY RESPONSIVE CURVED ALUMINUM NITRIDE PMUT** 124
S. Akhbari[1], F. Sammoura[1,2], S. Shelton[3], C.Yang[1], D. Horsley[3], and L. Lin[1]
[1]*University of California, Berkeley, USA,* [2]*Masdar Institute of Science and Technology, UAE, and*
[3]*University of California, Davis, USA*

We have successfully demonstrated highly responsive, curved piezoelectric micromachined ultrasonic transducers (PMUT) based on a CMOS-compatible fabrication process using AlN as the transduction material for the first time. A prototype device using a 2μm-thick AlN layer on a curved diaphragm surface with a radius of curvature of 1065μm and physical size of 140μm in diameter has shown a measured resonant frequency at 2.19MHz. The DC response has been experimentally measured as 1.1nm/V, which is 50X higher than that of a planar device with same size and operation conditions. As such, this new class of curved PMUT could replace the state-of-art, planar PMUT to achieve high electromechanical coupling for various ultrasonic transduction applications, including gesture recognition and medical imaging.

16:45 **ENCASED CANTILEVERS FOR LOW-NOISE FORCE AND MASS SENSING IN LIQUIDS** 128
D. Ziegler[1], A. Klaassen[2], D. Bahri[1], D. Chmielewski[1], A. Nievergelt[1], F. Mugele[2],
J.E. Sader[3], and P.D. Ashby[1]
[1]*Lawrence Berkeley National Laboratory, USA,* [2]*MESA+, University of Twente, THE NETHERLANDS, and*
[3]*University of Melbourne, AUSTRALIA*

Viscous damping severely limits the performance of cantilever based sensing in liquids. Encased cantilevers achieve low damping in liquids by keeping the resonator dry. This is achieved by fabricating a hydrophobic encasement from which only few microns of the sensing tip protrude into the liquid. We achieve Q-factors and associated noise levels as if operating in air. We discuss fabrication of these devices and demonstrate successful application for low-noise mass sensing and gentle AFM imaging of soft matter in liquids.

17:00 **RESONANT MICRO-SENSOR PLATFORM FOR CONTACT-FREE**
CHARACTERIZATION OF LIQUID PROPERTIES 132
J.M. Gonzales and R. Abdolvand
Oklahoma State University, USA

This paper presents a novel resonant microsensor platform which maintains high quality factors(Q) when measuring ultrasonic properties of liquid samples such as blood. By avoiding the direct contact of the liquid with the resonator, significant losses due to liquid loading are mitigated and the physical properties of various fluids, including viscous samples, can be determined without adversely affecting the resonator performance. Devices have been fabricated and tested, achieving quality factors up to 6000 in air and the results show that the output signals measured from the device are sensitive to the properties of the liquid under test.

17:15 **NOVEL IN-PLANE GAP CLOSING CMOS-MEMS MICROPHONE WITH NO BACK-PLATE** 136
C.-I. Chang[1], S.-C. Lo[1], C. Wang[2], Y.-C. Sun[1], and W. Fang[1,2]
[1]*National Tsing Hua University (NTHU), TAIWAN and* [2]*MotionsTek Inc., TAIWAN*

The stacking of metal/tungsten layers as the sensing electrodes for CMOS-MEMS microphone without the back-plate has been proposed and demonstrated for the first time (Fig.1a). The acoustic pressure will deform the spring-diaphragm structure and further cause the in-plane gap-closing between sensing electrodes (Fig.1b). Thus, acoustic pressure and dynamic response of spring-suspension can be determined by the sensing capacitance changes. Such design has the following merits: (1) no back-plate is required, (2) bias voltage to pull diaphragm close to back-plate is not required, (3) in-use pull-in and process stiction between diaphragm and back-plate is also prevented, (4) easy integration with sensing circuits [1]. The design was implemented using the standard TSMC CMOS process. Typical microphone with 200μm-diameter diaphragm and 48-pairs sensing electrodes has been realized. Measurements show the sensitivity of microphone is -67.17dBV/Pa at 1KHz.

17:30 **Adjourn for the day**

Wednesday, January 29

08:00 **ANNOUNCEMENTS**

Plenary III
Session Chairs:
J. Brugger, *Ecole Polytechnique Federale de Lausanne (EPFL), SWITZERLAND*
W. van der Wijngaart, *KTH – Royal Institute of Technology, SWEDEN*

08:05 **CAVITY QUANTUM OPTOMECHANICS:**
COUPLING LIGHT AND MICROMECHANICAL OSCILLATORS 140
E. Verhagen, S. Deléglise, S. Weis, A. Schliesser, and **Tobias J. Kippenberg**
Ecole Polytechnique Federale de Lausanne (EPFL), SWITZERLAND

Session VIII – Nanodevices
Session Chairs:
L. Lin, *University of California, Berkeley, USA*
Y.-J. Yang, *National Taiwan University, TAIWAN*

08:45 **AMORPHOUS CARBON ACTIVE CONTACT LAYER FOR RELIABLE**
NANOELECTROMECHANICAL SWITCHES 143
D. Grogg[1], C.L. Ayala[1], U. Drechsler[1], A. Sebastian[1], W.W. Koelmans[1], S.J. Bleiker[2],
M. Fernandez-Bolanos[1], C. Hagleitner[1], M. Despont[1], and U.T. Duerig[1]
[1]IBM Research – Zurich, SWITZERLAND and [2]KTH Royal Institute of Technology, SWEDEN

This paper reports an amorphous carbon (a-C) contact coating for ultra-low-power curved nanoelectromechanical (NEM) switches. a-C addresses important problems in miniaturization and low-power operation of mechanical relays: i) the surface energy is lower than that of metals, ii) active formation of highly localized a-C conducting filaments offers a way to form nano-scale contacts, and iii) high reliability is achieved through the excellent wear properties of a-C, demonstrated in this paper with more than 100 million hot switching cycles.

09:00 **NEAR INFRARED PHOTO-DETECTOR USING SELF-ASSEMBLED FORMATION**
OF ORGANIC CRYSTALLINE NANOPILLAR ARRAYS 147
Y. Ajiki[1,4], T. Kan[2], M. Yahiro[3], A.Hamada[3], J. Adachi[3], C. Adachi[3], K. Matsumoto[2], and I. Shimoyama[2]
[1]Olympus Corporation, JAPAN, [2]University of Tokyo, JAPAN, [3]Kyushu University, JAPAN, and
[4]NMEMS Technology Research Organization, JAPAN

We proposed a near infrared photo-detector (NIR-PD) using self-assembled formation of organic crystalline arrays, which were formed on an n-type silicon (n-Si) substrate and covered with an Au film. These structures act as antennas for near infrared (NIR) light, resulted in an enhancement of the light absorption on the Au film. In this paper, the fabrication process of the NIR-PDs and the estimation results of photo-responsivity are described. The maximum value of the responsivity to NIR light (wavelength = 1.2 µm) was 1.79 mA/W without applying forward bias. This value is 10 times larger than that of a conventional Au/n-Si typed Schottky diode, which is fabricated as a reference.

09:15 **ULTRASENSITIVE SI NANOWIRE PROBE FOR MAGNETIC RESONANCE DETECTION** 151
Y.J. Seo, M. Toda, Y. Kawai, and T. Ono
Tohoku University, JAPAN

We have fabricated and evaluated an atto-newton-sensitive Si nanowire probe with a Nd-Fe-B magnet for magnetic resonance force microscopy. The width, thickness and length of the nanowire are 210 nm, 200 nm and 32 µm, respectively. The nanowire probe has a resonance frequency f0 of 11.256 kHz and a Q factor of order 12000. Then, we have demonstrated the measurement of force mapping based on electron spin resonance for three-dimensional imaging of radicals.

09:30 **A VERTICALLY INTEGRATED NANOSCALE TIPPED**
MICROPROBE INTRACELLULAR ELECTRODE ARRAY 155
Y. Kubota, H. Oi, H. Sawahata, A. Goryu, Y. Ando, R. Numano, M. Ishida, and T. Kawano
Toyohashi University of Technology, JAPAN

Here we report an integration of vertical 120-µm-long nanoscale tipped microprobe electrode (NTE) array and the intracellular recordings using a gastrocnemius muscle of a mouse. The tip diameter of the NTE was < 200 nm, with the height of 4 µm exposed from the parylene-shell. The impedance of the NTE exhibited 3.1 MΩ at 1 kHz in saline, with the output/input signal amplitude ratio of 50% for intracellular recordings. The penetrated NTE into the muscle of a mouse detected the residual potentials with the amplitude of ~ -200 mV, confirming the intracellular recording capability of the NTE.

09:45 **Break & Exhibit Inspection**

Session IX – Energy Harvesting & Power
Session Chairs:
Z. Li, *Peking University, CHINA*
J. Judy, *University of Florida, USA*

10:30 **ENERGY HARVESTING USING UNIAXIALLY ALIGNED CARDIOMYOCYTES** 159
X. Liu[1,2], X. Wang[1], S. Li[2] and L. Lin[2]
[1]Tsinghua University, CHINA and [2]University of California, Berkeley, USA

This study presents the concept of energy harvesting from uniaxially-aligned cardiomyocytes on a flexible substrate for the first time. Experimentally, synchronously contracting neonatal rat ventricular cardiomyocytes (NRVCs) at 0.5Hz have been found to cause the mechanical straining of a piezoelectric energy harvester to produce 87.5nA and 92.3mV of peak current and voltage, respectively. This work presents a successful step toward mechanical energy harvesting via living biological cells and tissues.

10:45 **DIFFUSION REFUELING BIOFUEL CELL MOUNTABLE ON INSECT** 163
K. Shoji[1], Y. Akiyama[1], M. Suzuki[2], N. Nakamura[2], H. Ohno[2], and K. Morishima[1]
[1]Osaka University, JAPAN and [2]Tokyo University of Agriculture and Technology, JAPAN

This paper reports an insect-mountable biofuel cell (imBFC) using trehalose, main sugar of insect hemolymph. The imBFC is refueled trehalose by diffusion from insect hemolymph automatically and generates electric power by oxidizing glucose which is obtained by hydrolyzing trehalose enzymatically. We fabricated the imBFC consisted of a connector, two chambers separated by a dialysis membrane and electrodes and succeeded in driving a light-emitting diode by the imBFC. The results have shown a potentially to be applied for a battery of novel ubiquitous robots such as insect cyborgs.

11:00 **ALD RUTHENIUM OXIDE-CARBON NANOTUBE ELECTRODES
FOR SUPERCAPACITOR APPLICATIONS** 167
R. Warren[1], F. Sammoura[1,2], A. Kozinda[1], and L. Lin[1]
[1]University of California, Berkeley, USA and [2]Masdar Institute of Science and Technology, UAE

This work presents the first demonstration of atomic layer deposition (ALD) ruthenium oxide (RuO_2) and its conformal coating onto vertically aligned carbon nanotube (CNT) forest as supercapacitor electrodes. The ALD method allows precise control over the RuO_2 layer thickness and composition without the use of binder molecules. The ALD RuO_2 coated CNTs achieve a specific capacitance of 100 mF/cm^2 and retain their high-performance over repeated cycling.

11:15 **SUB 3-MICRON GAP MICROPLASMA FET WITH 50 V TURN-ON VOLTAGE** 171
P. Pai and M. Tabib-Azar
University of Utah, USA

We report the smallest microplasma transistor reported till date that operates with a low turn-on voltage of 50V dc. The device achieves more than 3x reduction in the turn-on voltage and 100x reduction in size compared to devices reported by other groups in the past. Earlier work reported by our group used plasma from an external source to operate the transistor. Our recent work successfully generated direct current plasma within the device with a turn-on voltage of 180V. This paper reports gate field-effect characterization results performed under dc and rf excitation and draws a comparison.

11:30 **Lunch on Own**

13:00 **Poster/Oral Session III**

Session X– Microdevices for Cell Manipulation
Session Chairs:
I. Park, *Korea Advanced Institute of Science and Technology (KAIST), SOUTH KOREA*
Y. Sun, *University of Toronto, CANADA*

15:00 **FORMATION OF CROSS-SHAPED ESCHERICHIA COLI** 175
K. Hirayama, Y.J. Heo, and S. Takeuchi
University of Tokyo, JAPAN

We develop a method to regulate the shapes of bacteria by confining a single cell of bacteria into micro chamber. Escherichia coli has a cell wall structure which determines the rod-shape. We removed the cell wall of E. coli and then confined the spheroplasts into cross-shaped microchambers. If the bacteria re-synthesize its cell wall within the microchamber, we could obtain the cross-shaped bacteria. By analyzing the behavior of proteins and the cross-shaped bacteria, we believe that we could contribute to the comprehensive understanding of the shape regulation mechanism of E. coli.

15:15 **SINGLE-CELL 3D BIO-MEMS ENVIRONMENT WITH ENGINEERED GEOMETRY AND PHYSIOLOGICALLY RELEVANT STIFFNESSES** 177
M. Marelli[1], N. Gadhari[2], G. Boero[1], M. Chiquet[2], and J. Brugger[1]
[1]Ecole Polytechnique Federale de Lausanne (EPFL), SWITZERLAND and
[2]University of Bern, SWITZERLAND

We present a 3D microenvionment for on-chip cell culture, made of stress-bent cantilevers and designed to mimic essential physical properties of the in vivo environment, at the single-cell scale and with a high degree of parallelization. In particular, we report on a combinatorial fabrication approach bringing to the realization of a palette of devices with constant sizes (fitting a single-cell), but with stiffnesses spanning two orders of magnitude and matching physiologically relevant values.

15:30 **INTEGRATED MICRO CULTURE DEVICE FOR FULLY AUTOMATED CLOSED CULTURE EXPERIMENT OF EMBRYONIC BODY** 181
A. Yasukawa[1], T. Nishijima[2], M. Ikeuchi[1], and K. Ikuta[1]
[1]University of Tokyo, JAPAN and [2]Nagoya University, JAPAN

This paper reports formation, differentiation and analysis of embryonic bodies (EBs) from human iPS cells in a palm-size device "PASMA (Pressure Actuated Shapable Microwell Array)". By incorporating a transparent heat film and $CO2$ concentration adjusting system, PASMA realized miniaturization of the whole process of EB experiment in one chip. Moreover, the fully automated closed culture system can eliminate the risks of contamination due to manual operation. EBs from human iPS cells were successfully fabricated using this system.

15:45 **MECHANICAL CELL PAIRING SYSTEM BY SLIDING PARYLENE RAILS** 185
Y. Abe[1,2], K. Kamiya[1,3], T. Osaki[1,4], R. Kawano[1], N. Miki[1,2], and S. Takeuchi[1,4]
[1]Kanagawa Academy of Science and Technology, JAPAN, [2]Keio University, JAPAN,
[3]Japan Science and Technology Agency (JST), JAPAN and [4]University of Tokyo, JAPAN

This paper proposes a cell pairing system that is capable of defining the number and the position of trapped cells by mechanically sliding the parylene rail films (PRF). This system allows us to control the number as well as the order of lined-up cells. We successfully demonstrated lining up of three different cells in a designated order. The proposed system is readily applicable to study the cell-cell interactions using the single cell pairing.

16:00 **Break & Exhibit Inspection**

Session XI– Bio-Inspired Microactuators
Session Chairs:
Y. Yamanishi, *Shibaura Institute of Technology, JAPAN*
T.-H. Wang, *The Johns Hopkins University, USA*

16:30 **TWO-DIMENSIONALLY STEERING MICROSWIMMER PROPELLED BY OSCILLATING BUBBLES** 188
J. Feng and S.K. Cho
University of Pittsburgh, USA

An acoustically excited and oscillated bubble can generate a propelling force in micro scale. We develop a simple and efficient method that allows a bubble-propelled micro swimmer to propel and steer two dimensionally as wirelessly and remotely commanded.

16:45 **A MICROFABRICATED, BIOHYBRID, SOFT ROBOTICS FLAGELLUM** 192
B.J. Williams[1], S.V. Anand[1], J. Rajagopalan[2], and M.T.A. Saif[1]
[1]University of Illinois, Urbana-Champaign, USA and [2]Arizona State University, USA

We develop a microfabricated soft robotics biohybrid swimmer utilizing the contractions of one to several cardiomyocytes to provide on-board actuation to a thin, deformable, polydimethylsiloxane (PDMS) filament. The actuated filament deforms passively in response to fluid drag, producing a time-irreversible deformation pattern and net propulsive force at low Reynolds number. We utilize an elastohydrodynamic model to determine appropriate filament parameters and realize a functional swimmer.

17:00 **EVALUATION AND OPTICAL CONTROL OF SOMATIC MUSCLE MICRO BIOACTUATOR OF CHANNELRHODOPSIN TRANSGENIC DROSOPHILA MELANOGASTER** 196
M. Hirooka[1], S.P. Beh[1], T. Asano[1], Y. Akiyama[1], T. Hoshino[2], K. Hoshino[3], H. Tsujimura[3], K. Iwabuchi[3], and K. Morishima[1]
[1]Osaka University, JAPAN, [2]University of Tokyo, JAPAN, and
[3]Tokyo University of Agriculture and Technology, JAPAN

In this research, we first developed light-activated somatic muscle of transgenic Drosophila melanogaster expressing a blue light sensitive cation channel, channelrhodopsin-2, and incorporated into a micro device. We successfully demonstrated that optogonetic stimulation using light pulses was able to control the contractile activity with a given temporal pattern. The contractile force was evaluated with varying light intensities and pulse widths. These results have shown that mechanical systems powered by muscles will be controlled more accurately under our strategy.

17:15 **FERROFLUID-ASSISTED MICRO ROTARY MOTOR FOR MINIMALLY INVASIVE ENDOSCOPY APPLICATIONS** 200
B. Assadsangabi, M.H. Tee, S. Wu, and K. Takahata
University of British Columbia, CANADA

Micro-scale rotary motors have a vast range of potential applications in broad areas. One promising area is medical applications. Minimally-invasive endoscopic catheters are an excellent example. The micromotors used in these catheters, however, generally have large axial lengths (e.g., >1 cm). This paper presents the first rotary micromotor enabled by ferrofluid bearing that drastically shortens the axial length of micromotors, suitable for medical catheter applications, owing to the extremely simple and reliable bearing mechanism.

17:30 **Adjourn for the day**

18:30 - **Conference Banquet**
21:30

Thursday, January 30

08:00 ANNOUNCEMENTS

Plenary IV
Session Chairs:
F. Ayazi, *Georgia Institute of Technology, USA*
CJ Kim, *University of California, Los Angeles, USA*

08:05 COMMERCIALIZATION OF WORLD'S FIRST PIEZOMEMS RESONATORS FOR HIGH PERFORMANCE TIMING APPLICATIONS 204
Harmeet Bhugra, S. Lee, W. Pan, M. Pai, and D. Lei
Integrated Device Technology, Inc. (IDT), USA

Session XII – Resonators & RF MEMS
Session Chairs:
A. Seshia, *Cambridge University, UK*
K. Böhringer, *University of Washington, USA*

08:45 MEMS-RECONFIGURABLE WAVEGUIDE IRIS FOR SWITCHABLE V-BAND CAVITY RESONATORS 206
Z. Baghchehsaraei and J. Oberhammer
KTH Royal Institute of Technology, SWEDEN

We present, for the first time, a reconfigurable waveguide iris based on a MEMS-reconfigurable surface integrated into a WR-12 rectangular waveguide (60–90 GHz, 3.099 mm × 1.549 mm). The reconfigurable surface is only 30 μm thick and incorporates 252 simultaneously switched contact points for activating or deactivating an inductive iris. The switchable irises can be utilized to implement components such as reconfigurable filters and cavity resonators. We also present for the first time a reconfigurable cavity resonator based on the novel MEMS-reconfigurable iris.

09:00 A MICROMECHANICAL PARAMETRIC OSCILLATOR FOR FREQUENCY DIVISION AND PHASE NOISE REDUCTION 210
T.O. Rocheleau, R. Liu, J. Naghsh Nilchi, T.L. Naing, and C.T.-C. Nguyen
University of California, Berkeley, USA

A capacitive-gap RF MEMS resonator array demonstrates a first on-chip MEMS frequency divider with 61-MHz output generated from a 121-MHz electrical drive through use of a parametric oscillation effect. This provides not only the expected 6dB reduction in close-to-carrier phase noise, but also a remarkable 23dB reduction in far-from-carrier noise due to filtering by the high-Q mechanical resonator. In contrast with traditional frequency division, the parametric oscillator here requires no active devices, adds no noise to the signal and has essentially zero power consumption, limited in principle only by MEMS resonator loss.

09:15 TEMPERATURE-COMPENSATED PIEZOELECTRICALLY ACTUATED LAMÉ-MODE RESONATORS 214
V.A. Thakar and M. Rais-Zadeh
University of Michigan, USA

In this work, we present a passive compensation strategy for Lamé-mode resonators using silicon dioxide refilled islands within the resonator body. With this technique we achieve compensation of the first-order TCF and further demonstrate that the turnover temperature (TOT) can be tuned across a wide range from -40°C to +120°C by optimizing the placement of oxide refilled islands.

09:30 A MEMS AUTONOMOUS SWITCHED OSCILLATOR 218
G.B. Torri[1,2], J. Bienstman[2], X. Rottenberg[1], H.A.C. Tilmans[2], C. Van Hoof[1,2], and R. Puers[1,2]
[1]*imec, BELGIUM and* [2]*KU Leuven, BELGIUM*

We design and measure an autonomous electrostatic MEMS oscillator that exhibits periodic and aperiodic behavior. The system consists of an electrostatic MEM, a dc voltage source, and a displacement dependent resistive circuit. The applied voltage above the pull-in and the position dependent circuit are responsible for sustaining the oscillations. Electronic and non-electronic information can be input or retrieved from the oscillator system. Applications for such device range from chaotic signals for communications, sensitive mass sensors, and signal processing.

10:00 Poster/Oral Session IV

Session XIII – Chemical Sensors & Systems

Session Chairs:
A. Llobera, *Centre Nacional de Microelectronica, SPAIN*
H. Moon, *University of Texas, Arlington, USA*

**12:00 SURFACE TENSION MONITORING IN A "SOFT" INTERFACE
USING ELECTROWETTING ON DIELECTRIC** 222
S. Choi[1], Y. Kwon[2], E. Jeon[1], S. Kim[1], and J. Lee[1]
[1]Seoul National University, SOUTH KOREA and [2]Samsung Electronics Co., Ltd., SOUTH KOREA

We report the first demonstration of interfacial tension monitoring across two immiscible liquids using electrowetting on dielectric (EWOD). Impedance measurement during EWOD reveals the variation of surfactant concentration at the liquid-liquid interface in real time. We also show such approach can be used for label-free monitoring of DNA hybridization. Our approach opens a new horizon of EWOD used as a molecular sensing mechanism across a "soft" interface between liquids.

12:15 MICROFLUIDIC BUBBLE-BASED GAS SENSOR 226
A. Bulbul, H.-C. Hsieh, and H. Kim
University of Utah, USA

We report a new class of a gas sensor that utilizes the variations in bubble sizes, when different gases are introduced into a liquid flow, to identify gas types and even quantify the amounts. To verify the feasibility of the concept, the sizes of discretely formed bubbles were optically monitored and analyzed through a custom-developed MATLAB software.

12:30 A FLEXIBLE GRAPHENE FET GAS SENSOR USING POLYMER GATE DIELECTRICS 230
Y. Liu, J. Chang, and L. Lin
University of California, Berkeley, USA

We have successfully demonstrated a Graphene FET (GFET) gas sensor on a flexible plastic substrate for the first time with a sensitivity of 0.00428ppm^{-1} ($\Delta R/R_0$) for ammonia detection. Compared with the state-of-art technologies, four distinctive advancements have been achieved: (1) first demonstration of a flexible graphene FET gas sensor; (2) a new fabrication process to achieve embedded-gate GFET on a flexible substrate; (3) proof of utilizing polymeric materials of parylene and polyethylenimine (PEI) as the gate dielectrics and channel dopant for graphene FET, respectively; and (4) validation of a real-time gas sensing mechanism utilizing n-type doping of graphene induced by ammonia exposure.

12:45 Awards Ceremony

13:00 Conference Adjourns

Poster/Oral Presentations

M – Monday (13:00 - 15:00)
T – Tuesday (14:00 - 16:00)

W – Wednesday (13:00 - 15:00)
Th – Thursday (09:45 - 11:45)

Bio MEMS (Bio)

M-001 A CONTINUOUS OPTICALLY-INDUCED CELL ELECTROPORATION DEVICE WITH ON-CHIP MEDIUM EXCHANGE MECHANISMS 234
C.-J. Chang, M.-Y. Lu, and G.-B. Lee
National Tsing Hua University (NTHU), TAIWAN

We present a novel design of continuously optically-induced electroporation (OIE) device capable of replacing culture medium and electroporation buffer in a seamless fashion. With this approach, the entire process for continuous cell electroporation could be performed automatically without human intervention. Furthermore, the survival rate of the cells could be greatly improved due to this fast, automatic procedure.

T-002 A HIGH-THROUGHPUT PERMEABILITY ASSAY PLATFORM FOR SHEAR STRESS CHARACTERIZATION OF ENDOTHELIAL CELLS 238
R. Booth, S. Noh, and H. Kim
University of Utah, USA

Here we present the first permeability assay platform to enable rapid characterization of shear stress effects fully spanning the physiologically relevant spectrum (1-60dyn/cm2) for endothelial cells in vitro. The structure comprised 4 parallel channels to enable independent permeability assays, and generated 15x shear stress range indicated by simulation and an integrated micro-flow sensor array. Endothelial cells exhibited decreased permeability of fluorescent tracers with shear stress, as well as increased elongation and cell alignment with increases in shear stress.

W-003 A MICROFLUIDIC DEVICE FOR ISOLATION OF CELL-TARGETING APTAMERS 242
J. Zhu[1], T. Olsen[1], R. Pei[2], M. Stojanovic[1], and Q. Lin[1]
[1]Columbia University, USA and [2]Chinese Academy of Sciences, CHINA

This paper presents a microfluidic device for synthetically isolating cell-targeting aptamers from a randomized single-strand DNA (ssDNA) library. The device integrates cell culturing with affinity selection of cell-binding ssDNA, which is then amplified by bead-based polymerase chain reaction (PCR). Coupling of the selection and amplification using pressure-driven flow realizes multi-round aptamer isolation on a single chip.

Th-004 A NANOWIRE-INTEGRATED MICROFLUIDIC DEVICE FOR HYDRODYNAMIC TRAPPING AND ANCHORING OF BACTERIAL CELLS 246
D. Kwon, J. Kim, S. Chung, and I. Park
Korea Advanced Institute of Science and Technology (KAIST), SOUTH KOREA

We develop nanowire-integrated microfluidic devices for hydrodynamic trapping and anchoring of bacterial cells and demonstrate that the mesh-like cages formed by integrating ZnO nanowires are effective in trapping and anchoring of Escherichia coli as a model bacterium. We present two anchoring modes, impaling and wedging, followed by irreversible damage or reversible deformation of the cell wall, respectively.

M-005 AN INTEGRATED MICROFLUIDIC SYSTEM USING FIELD-EFFECT TRANSISTORS FOR CRP DETECTION 250
C.L. Lin, Y.W. Kang, K.W. Chang, W.H. Chang, Y.L. Wang, and G.B. Lee
National Tsing Hua University (NTHU), TAIWAN

Cardiovascular diseases are responsible for 25-million deaths worldwide on a yearly basis. Timely diagnosis of the disease is therefore an important research area. Toward this end, C-reactive protein (CRP) is a general biomarker for inflammation and infection, and has become a good marker for evaluating risks of cardiovascular diseases. Previously, a CRP-specific aptamer (nucleic acid-based antibody, in a nutshell) with high sensitivity and specificity was used to detect CRP in microfluidic system, which was capable of performing the detection in an automated fashion while consuming tiny volumes of reagents and samples. In parallel, field-effect transistors (FET) have emerged as sensors to detect small molecules, proteins and even viruses, and have demonstrated rapid and highly sensitive detection in a compact system. In this work, an integrated device that combined the advantages of microfluidics, aptamers and FET-based sensors was developed to achieve rapid, sensitive and specific CRP detection. The developed integrated microfluidic system with FET sensor and CRP-specific aptamer can be promising for fast, sensitive and specific CRP detection.

T-006 ASSEMBLY OF NEURAL CELL-LADEN MICROPLATES ON A MICROFABRICATED BREADBOARD 254
S. Yoshida[1], K. Sato[1,2], and S. Takeuchi[1,2]
[1]University of Tokyo, JAPAN and [2]Japan Science and Technology Agency (JST), JAPAN

We developed a method to form an arbitrary network of neural cell-laden microplates using a breadboard-like microelectrode array (MEA). We cultured single neural cells on perforated microplates with control of their morphologies, then mounted the plates on the micro-sockets attached on the breadboard. Through a pore punched in the center of the microplate, the electrical activities of cells would be accessible with gold electrodes patterned on the MEA.

W-007 CD4 CELL ISOLATION FROM BLOOD USING FINGER-ACTUATED ON-CHIP MAGNETOPHORESIS FOR RAPID HIV/AIDS DIAGNOSTICS 256
M. Glynn, D.J. Kinahan, D. Chung, and J. Ducrée
Dublin City University, IRELAND

We have developed a microfluidic based disposable device, actuated solely by finger pressure from the operator, to repeatedly pass micro quantities of blood past a magnetic based cell capture structure to allow CD4 cell based HIV diagnostics in less than 1 minute following sample application. This is aimed for deployment in resource-poor regions where HIV is endemic but diagnostics is a challenge.

Th-008 CENTRIFUGALLY AUTOMATED SOLID-PHASE PURIFICATION OF RNA 260
N. Dimov, J. Gaughran, D. McAuley, D. Boyle, D.J. Kinahan, and J. Ducrée
Dublin City University, IRELAND

We report for the first time on a fully centrifugally automated solid-phase purification of RNA on an integrated microfluidic disc with sequential release of on-board reagents.

M-009 FABRICATION METHOD TO A HIGH RESOLUTION CONTROL IN THE SPACE OF CELL CULTURING ENVIRONMENT WITH MICROFLUIDIC SYSTEM 264
T. Hiraiwa[1], T. Kimura[1], Y. Takenaka[1], R. Tanamoto[1], H. Ota[3], H. Kimura[2], Y. Taguchi[1], N. Miki[1], Y. Matsumoto[1], K. Oka[1], A. Funahashi[1], and N. Hiroi[1]
[1]Keio University, JAPAN, [2]Tokai University, JAPAN, and [3]University of California, Berkeley, USA

We develop a reusable Cell Culturing Device, designed for single cells and cellular network analysis. This is the first success of combination of Microcontact Printing (μCP) and Vacuum Device. This device has following advantages: (1) cells stay within the micropatterns more than 12hrs, (2) cell culture environment is regulated precisely using laminar flow, (3) this device is reusable for further experiments.

T-010 INDIVIDUALLY ADDRESSABLE MULTI-CHAMBER ELECTROPORATION PLATFORM WITH DIELECTROPHORESIS-ASSISTED CELL POSITIONING 268
S. Park and G. Yossifon
Technion–Israel Institute of Technology, ISRAEL

We developed a novel, multifunctional platform with an array for high throughput screening which integrates DEP trapping/positioning and electroporation mechanisms in a multi-chamber array where each chamber may be individually activated. As a proof of concept, platforms with an array of 40×2 cell microchambers were fabricated and tested to demonstrate the feasibility of integrating the two functionalities: 1) the manipulation of the particles/cells using DEP, and 2) gradient electroporation using by producing a spatial gradient of electroporation parameters to optimize electroporation efficiency.

W-011 INTEGRATION OF HEAT-TRANSFER RESISTANCE MEASUREMENTS ONTO A DIGITAL MICROFLUIDIC PLATFORM TOWARDS THE MINIATURIZED AND AUTOMATED LABEL-FREE DETECTION OF BIOMOLECULAR INTERACTIONS 272
E. Perez-Ruiz[1], T. Vandenryt[2], D. Witters[1], D. Decrop[1], B. van Grinsven[2], D. Spasic[1], P. Wagner[2], and J. Lammertyn[1]
[1]KU Leuven, BELGIUM and [2]University of Hasselt, BELGIUM

In this work the successful integration of heat-transfer resistance measurements with a digital microfluidic chip is shown. The integrated miniaturized platform allows the automated label-free detection of biomolecular interactions. To immobilize biomolecules on the hydrophobic chip surface, hydrophilic gold sensing patches are created by means of a novel dry lift-off technique that leaves the chip surface unaffected. In order to validate the integrated device, DNA melting analysis was performed in the set-up.

Th-012 NANOFLUIDIC SENSING AT NORMAL PHYSIOLOGICAL CONDITION
BY COUPLING ION CONCENTRATION POLARIZATION WITH A
LIMIT OF DETECTION OF ONE FEMTOMOLE 276
W. Ouyang[1], W. Wang[1,2], H. Zhang[1], W. Wu[1,2], and Z. Li[1,2]
[1]*Peking University, CHINA and*
[2]*National Key Laboratory of Science and Technology on Micro/Nano Fabrication, CHINA*

Traditional nanofluidic sensing devices lose function at normal physiological condition (e.g. 0.1 M). This work enables nanofluidic sensing of Biotin at normal physiological condition by coupling ion concentration polarization (ICP) in a nanofluidic crystal (NFC) device. The enrichment effect of ICP was utilized for target molecule preconcentration for lower limit of detection, while the depletion effect of ICP was utilized for creating a nanofluidic regime in high ionic concentration buffer. A limit of detection of 1 fM was realized in our work.

M-013 MASS-PRODUCTION AND PROLONGED UNDIFFERENTIATED STATE OF EMBRYONIC
BODIES BY USING A SEMI-PERMEABLE TAPERED MICROWELL ARRAY 280
M. Ikeuchi[1], S. Hayashi[2], M. Osonoi[2], and K. Ikuta[1]
[1]*University of Tokyo, JAPAN and*
[2]*The Foundation for Promotion of State of the Art in Medicine and Health Care, JAPAN*

We have demonstrated embryonic body (EB) formation and prolongation of undifferentiated state by using an improved tapered microwell array equipped with semipermeable bottom. The array had hydrophilic surface to prevent cell adhesion, and thus, the cells seeded onto the array aggregated into EBs in each microwell. The semipermeable bottom only permitted liquid to go through, and promoted exchange of culture medium in each microwell, resulting in prolonged term of undifferentiated state of EBs.

T-014 ON CHIP PRODUCT PURIFICATION FOR COMPLETE
MICROFLUIDIC RADIOTRACER SYNTHESIS 284
S. Chen, A.A. Dooraghi, M. Lazari, R.M. van Dam, A.F. Chatziioannou, and C.J. Kim
University of California, Los Angeles, USA

We developed on-chip removal of excess radioactive fluoride to follow the radiolabeling of a neurotransmitter with an electrowetting-on-dielectric (EWOD) device. Solid phase extraction of fluoride was achieved by adding alumina particles in the radiolabeled droplet and filtering them out by passing the droplet through pillars in the device. Purification was analyzed both on chip with Cerenkov radiation imaging and off chip with radio-thin layer chromatography.

W-015 RECONSTITUTION OF FUNCTIONAL MEMBRANE PROTEINS INTO
ASYMMETRIC GIANT LIPOSOMES BY USING A PULSED JET FLOW 288
K. Kamiya[1], R. Kawano[1], T. Osaki[1,2], and S. Takeuchi[1,2]
[1]*Kanagawa Academy of Science and Technology, JAPAN and* [2]*University of Tokyo, JAPAN*

We develop the reconstitution of functional membrane proteins (flippases) into asymmetric lipid giant liposomes that were prepared by deforming a planar asymmetric bilayer using a microfluidic pulsed jet flow. we observed the translocation of phosphatidylserine from the extracellular leaflet to the cytoplasmic leaflet which was catalyzed by flippases.

Th-016 TRACTION FORCE OF SMOOTH MUSCLE CELL
DURING GROWTH ON A RIGID SUBSTRATE... 290
U.G. Jung, T. Tsukagoshi, H. Takahashi, T. Kan, K. Matsumoto, and I. Shimoyama
University of Tokyo, JAPAN

This paper reports on the traction force of smooth muscle cell during cell growth on the rigid substrate, specially designed for measuring the x and z axis forces. The proposed piezoresistive force sensor is characterized by three points: 1) a rigid substrate, 2) high force sensitivity (10 nN force resolution), 3) small gaps between sensor pads. We measured the traction force of the smooth muscle cells during culture at 37°C.

Chemical Sensors and Systems (CSS)

M-017 A MICRO GAS CHROMATOGRAPH WITH INTEGRATED
BI-DIRECTIONAL PUMP FOR QUANTITATIVE ANALYSES 294
Y. Qin and .B. Gianchandani
University of Michigan, USA

We report a micro gas chromatography (μGC) system that comprises a Knudsen pump with bi-directional flow capability (KP2), a two-stage preconcentrator-focuser (PCF2) and a separation column. In this valveless system, the bi-directionality of the pump allows flow reversal in the multi-stage preconcentrator. The KP2, PCF2, and separation column are arranged in a 4.3 cm^3 stack, and used with a commercial flame ionization detector. In preliminary experiments, the μGC system demonstrated quantitative separations of benzene, toluene, and xylene (BTX) with concentrations of 43-328 mg/m^3. The separations were completed in 80 sec using room air as the carrier gas.

T-018   A MICRO TRACE HEAVY METAL SENSOR BASED ON DIRECT PROTOTYPING MESOPOROUS CARBON ELECTRODE 298

F. Teng[1], X.H. Wang[1,2], and C.W. Shen[1]
[1]Tsinghua University, CHINA and [2]Chinese Academy of Sciences, CHINA

We present a micro trace heavy metal sensor based on bismuth-modified mesoporous carbon electrodes that are direct prototyped on silicon wafer. The proposed device features a great electrochemical sensing platform for voltammetric analysis because of the thicker mesoporous carbon electrode has high surface area, high electric conductivity, and can be integrated into microsystems. The novel sensor achieves excellent sensing performance, the limits of detection are an order of magnitude lower than other reported and the peak current also exhibits well linear response.

W-019   A NOVEL QUANTITATIVE DESIGN MODELING ON GAS SENSING PARAMETER OF NANO-MATERIALS BASED ON MICRO-GRAVIMETRIC THERMO-DYNAMIC EXPERIMENTS 302

P.C. Xu and X.X. Li
Chinese Academy of Sciences, CHINA

The study aims to build a novel quantitative adsorbing/sensing model for chemical gas sensing-materials, with which various key sensing-parameters can be comprehensively evaluated and optimally designed. Gravimetric resonant-cantilevers are used in experiment to real-time record sensing curves at different temperatures, which are further used to extract adsorbing/sensing performance of the specific materials for quantitative evaluation and optimal sensor design. The model is well validated by choosing the best trimethylamine (TMA) sensing-material among three similar mesoporous-silica nano-particles (MSNs).

Th-020   A POLYMER-BASED MEMS DIFFERENTIAL SCANNING CALORIMETER 306

Y. Jia, B. Wang, J. Zhu, and Q. Lin
Columbia University, USA

We present a new MEMS differential scanning calorimeter (DSC). The device, using polymer calorimetric microstructures inexpensively fabricated on a polymer substrate, is mechanically flexible and highly robust, and well suited to disposable use for measurement of biomolecular energetics. We demonstrate this polymer MEMS DSC device with the characterization of lysozyme unfolding in a 1 micro liter volume at low concentrations of practically useful levels.

M-021   DESIGN AND MOTION CONTROL OF SELF-PROPELLED DROPLETS 310

A. Suzuki[1], S. Maeda[2], Y. Hara[3], and S. Hashimoto[1]
[1]Waseda University, JAPAN, [2]Shibaura Institute of Technology, JAPAN, and
[3]National Institute of Advanced Industrial Science and Technology (AIST), JAPAN

We report a new oil droplet system that is autonomously driven by the energy of oil-water interactions and its motion control. This droplet moves by ejecting polymers in alkaline water and displays large driving force. The oil droplet can move along a flow channel by deforming in an amoeboid motion on the micro- to milli- scale, so we specifically demonstrated its motion in a micro fluidic channel. Such droplets offer considerable potential for the application as a transportable actuator.

T-022   DETECTION OF MUTATIONS IN THE BINDING DOMAIN OF TAU PROTEIN BY KINESIN– MICROTUBULE GLIDING ASSAY 314

S.P. Subramaniyan[1], M.C. Tarhan[2], S.L. Karsten[3], H. Fujita[2], H. Shintaku[1], H. Kotera[1], and R.Yokokawa[1]
[1]Kyoto University, JAPAN, [2]Tokyo University, JAPAN, and [3]NeuroInDx Inc., USA

We have studied the kinesin- microtubule (MTs) based gliding assay for its application as a diagnostic tool indetecting neuronal marker - tau protein. In this paper we report our findings; that the landing rate and density of MTs have depicted the type of tau protein decorated on them, we have discussed five major tau mutants located in thebinding domain of tau protein in addition to its isoforms, and we have also demonstrated a micro device to detect MTsby their landing rate and gliding density.

W-023   FROG EGG-ARRAY DEVICE INTEGRATED WITH FLUIDIC CHANNEL AND MICROELECTRODES FOR CHEMICAL SENSING 318

M. Tomida, Y. Murakami, and N. Misawa
Toyohashi University of Technology, JAPAN

This study describes a membrane-protein based chemical sensor device consisting of microfluidic channels and Xenopus laevis oocytes. The fluidic device has Si-based microfabricated electrodes to measure of the oocyte's response to each chemical by two-electrode voltage clamping method. After cell trapping, the fluidic device can be separated to each fluidic channel that can measure an individual oocyte membrane potential change. We successfully placed each oocyte into the device and detected individual Xenopus oocyte responses to chemical stimulus.

Th-024 INTEGRATION OF DIAMOND MICROELECTRODES ON CMOS-BASED AMPEROMETRIC BIOSENSOR ARRAY BY FILM TRANSFER TECHNOLOGY 322
T. Hayasaka, S. Yoshida, K.Y. Inoue, M. Nakano, T. Ishikawa, T. Matsue, M. Esashi, and S. Tanaka
Tohoku University, JAPAN

This study reports on integration of boron-doped diamond (BDD) microelectrodes on CMOS-based 20×20 amperometric biosensor array. The BDD electrodes are once formed on a Si wafer at 800°C, and then transferred to a 0.18 μm CMOS wafer with a benzocyclobutene (BCB) bonding interlayer. The fully-integrated device successfully detects biomolecules such as histamine owing to a wide potential window of the BDD electrode, and offers 2-dimensional real-time imaging of histamine diffusion in a solution. This type biosensor promises sensing and imaging applications of various biological materials which cannot be detected by conventional sensors.

M-025 IONIC LIQUID-GATED GRAPHENE FET ARRAY WITH ENHANCED SELECTIVITY FOR ELECTRONIC NOSE 326
A. Inaba, Y. Takei, K. Matsumoto, and I. Shimoyama
University of Tokyo, JAPAN

Graphene has high sensitivity to gases, but has poor selectivity. Therefore, graphene is hard to apply to electronic nose. Because the gas absorbability of ionic liquid (IL) depends on its type, graphene FETs (GFETs) gated by different ILs have different gas responses. The response pattern of the IL-gated GFETs (ILGFETs) enables gas species detection, i.e. ILs provide gas selectivity to graphene. We assembled three ILGFETs with three kinds of ILs into an ILGFET array. The response patterns to several gases were measured to demonstrate the feasibility of graphene electronic nose.

T-026 LoC SENSOR ARRAY PLATFORM FOR REAL-TIME COAGULATION MEASUREMENTS 330
O. Cakmak[1], N. Kilinc[1,2], E. Ermek[1], A. Mostafazadeh[1], C. Elbuken[1], G.G. Yaralioglu[3], and H. Urey[1]
[1]Koc University, TURKEY, [2]Gebze Institute of Technology, TURKEY, and [3]Ozyegin University, TURKEY

We report the first demonstration of MEMS-based sensor array enabling multiple tests in one disposable microfluidic cartridge using plasma. The LoC (Lab-on-Chip) platform technology is versatile and demonstrated here for real-time coagulation and clot-time tests (activated Partial Thrompoblastin Time (aPTT) and Prothrombin Time (PT)). The start and the end of fibrin generation during coagulation can be clearly seen in real-time for both of the tests. Magnetic actuation and optical read-out is used. Hence no electrical connection to the MEMS chip is required. This makes the system convenient for point-of-care tests.

W-027 PARALLELIZATION OF FISSION AND FUSION+OPERATIONS FOR HIGH THROUGHPUT GENERATION OF COMBINATORIAL DROPLETS 334
H.C. Zec, T.D. Rane, P. Ma, and T.H. Wang
Johns Hopkins University, USA

In this paper, we present a parallelized droplet-based platform for on-demand, combinatorial generation of nanoliter droplets. By parallelizing fission and fusion modules, throughput is increased by two orders ofmagnitude. With 32 Hz droplet generation, the projected throughput of this parallelized design is nearly 3 million sample-probe droplets per day on a single device (with 4 replicates of 750 thousand different mixtures). This translates to 240 unique sample-probe mixtures with 4 replicates per minute each.

Th-028 TOWARDS ON-CHIP CHEMICAL REACTION MONITORING BY EWOD IMPEDANCE MEASUREMENT 338
X. Ma, S. Chen, C.J. Kim, and R.M. van Dam
University of California, Los Angeles, USA

We develop an EWOD impedance measurement system for in situ chemical reaction monitoring to maximize the advantage of microscale chemical synthesis. As a demonstration, we measure the droplet impedance on EWOD at various stages of an acid-base titration, showing its capability of detecting the equivalence point of neutralization.

Energy Harvesting and Power MEMS (EHPM)

M-029 3D SOLID-STATE SUPERCAPACITORS OBTAINED BY ALD COATING OF HIGH-DENSITY CARBON NANOTUBES BUNDLES 342
G. Fiorentino, S. Vollebregt, F.D. Tichelaar, R. Ishihara, and P.M. Sarro
Delft University of Technology, THE NETHERLANDS

A three-dimensional solid-state miniaturized supercapacitor based on double conformal coating of Multiwalled Carbon Nanotubes (MWCNTs) bundles is presented. Atomic Layer Deposition (ALD) is used to deposit Al_2O_3 as dielectric layer and TiN as high aspect-ratio conformal counter-electrode on 2μm long MWCNTs bundles. The devices are realized using an IC wafer-scale manufacturing process and show a remarkable volumetric capacitance density value of 12mF/cm^3 with high reproducibility ($\leq$0.3E-12F deviation). The small footprint (100μm^2 to 625μm^2), a thickness of only 2μm, the extremely high capacitance density and the novel and easy-to-integrate fabrication process make it possible to realize high performance energy storage micro-devices.

T-030 A HIGH PERFORMANCE TRIBOELECTRIC GENERATOR FOR HARVESTING LOW FREQUENCY AMBIENT VIBRATION ENERGY 346
B. Meng, W. Tang, X.S. Zhang, M.D. Han, X.M. Sun, W. Liu, and H.X. Zhang
Peking University, CHINA

We present a novel triboelectric generator for vibration energy harvesting based on the mass production manufacture of flexible printed circuit (FPC). An elastic zigzag-shaped structure was employed as a natural spring, making the generator simple and easy to be stimulated. The use of FPC manufacture makes the fabrication efficient and low-cost. Low resonant frequency of 16 Hz and wide bandwidth of 37 Hz was achieved. The maximum effective output power of 77.3 μW was obtained at 16 Hz.

W-031 A GAP-VARYING ELECTROSTATIC TRANSDUCER UTILIZING FERROFLUID-BASED ACTUATION FOR MOTION HARVESTING 350
T. Galchev, D. Barutçu, and O. Paul
University of Freiburg – IMTEK, GERMANY

This paper provides electrical characterization of the gap-varying electrostatic springless proximity inertial harvester (SPIH). This is a new type of harvester for converting multi-dimensional motion from low-frequency sources such as humans or other environment. Each 2-mm-diameter transducer is capable of producing between 0.05-4.2 nJ of energy per actuation cycle at bias voltages of 10-100 V under controlled experiments.

Th-032 A HYBRID SUPERCAPACITOR USING VERTICALLY ALINGED CNT-POLYPYRROLE NANOCOMPOSITE 354
F. Sammoura[1,3], K.S. Teh[2], A. Kozinda[3], X. Zang[3], and L. Lin[3]
[1]*Masdar Institute of Science and Technology, UAE,* [2]*San Francisco State University, USA, and* [3]*University of California, Berkeley, USA*

We have successfully demonstrated, for the first time, the fabrication of vertically aligned carbon nanotube (VACNT)-polypyrrole (PPY) nanocomposites as a "hybrid supercapacitor material directly integrated on silicon-based electrodes. In contrast to previous works, three distinctive achievements are accomplished: (1) a "hybrid supercapacitor" using VACNT forest with electroplated PPY and dodecylbenzenesulfonate (DBS) as a dopant in acetonitrile, (2) realizing 500% higher capacitance as compared to the capacitance of electrodes made of VACNT or DBS-doped PPY alone, and (3) highly reversible cycling between -1 V and +1 V with improved knee frequency at 797Hz. As such, this hybrid nanocomposite could become a new class of material for future supercapacitors.

M-033 A MEMS-ENABLED BIODEGRADABLE BATTERY FOR POWERING TRANSIENT IMPLANTABLE DEVICES 358
M. Tsang[1], A. Armutlulu[1], A. Martinez[1], F. Herrault[1], S.A. Bidstrup[1], and M.G. Allen[1,2]
[1]*Georgia Institute of Technology, USA and* [2]*University of Pennsylvania, USA*

A series of MEMS-enabled biodegradable batteries, composed of a Mg anode and Fe cathode in a 0.1M MgCl2 electrolyte, were developed to power transient implantable medical devices (IMD). Biodegradable energy sources would enable active devices for the monitoring and treatment of transient disease states, such as bone fracture healing and drug delivery. The anode was fabricated by electroplating Mg from a non-aqueous solution and passivated with either polycaprolactone or polyglycerol sebacate. The batteries demonstrated a capacity and power of up to 1.1mAh and 22uW, respectively, which are sufficient for a typical IMD.

T-034 A MICRO-SCALE MICROBIAL SUPERCAPACITOR 362
H. Ren[1], H. Tian[2], T.L. Ren[2], and J. Chae[1]
[1]*Arizona State University, USA and* [2]*Tsinghua University, CHINA*

We report, for the first time, a micro-scale microbial supercapacitor to substantially enhance the current and power density, aiming for a carbon-neutral renewable miniaturized electrochemical power converter. Current and power density of 501.5A/m2, and 251.4W/m2 are achieved, which is more than 18 and 32 folds of the previous records, yielding the supercapacitor an attractive alternative to existing energy conversion and storage device.

W-035 A THREE-DIMENSIONAL ELECTROSTATIC/ELECTRET MICRO POWER GENERATOR FOR LOW ACCELERATION AND LOW FREQUENCY VIBRATION ENERGY HARVESTING 366
K. Tao, S.W. Liu, J.M. Miao, and S.W. Lye
Nanyang Technological University, SINGAPORE

In this work, a three-dimensional (3D) multimodal electret-based micro power generator is developed for scavenging energy from low acceleration (<0.05g) and low frequency (<100Hz) vibrations, which are ubiquitous existence and readily available in our daily life.

Th-036 A WEARABLE SYSTEM OF MICROMACHINED PIEZOELECTRIC CANTILEVERS COUPLED TO A ROTATIONAL OSCILLATING MASS FOR ON-BODY ENERGY HARVESTING 370

R. Lockhart, P. Janphuang, D. Briand, and N.F. de Rooij
École Polytechnique Fédérale de Lausanne (EPFL), SWITZERLAND

We present a compact, wearable piezoelectric on-body harvesting system that uses a small eccentric mass from a common watch movement to mechanically deflect a set of micromachined piezoelectric cantilevers when excited by the low frequency movements of the human body. The piezoelectric cantilevers are directly coupled to the rotating mass via a set of pins located near its rotational center. The energy produced by each pluck of a single cantilever is 545 nJ, corresponding to a maximum output power of 11 µW for continuous plucking. This concept could be used to power on-body electronics.

M-037 ALL-POLYMER PIEZOELECTRET ENERGY HARVESTER WITH EMBEDDED PEDOT ELECTRODE 374

Y. Feng and Y. Suzuki
University of Tokyo, JAPAN

We develop a novel all-polymer high-aspect-ratio (HAR) piezoelectret energy harvester with embedded PEDOT electrode, and demonstrate its performance for low-resonant-frequency in-plane vibration energy harvesting. Butterfly-shaped stop valves is devised to control the PEDOT capillary flow inside the parylene channels. With the present early prototype, 5.2 V open circuit voltage and 53 nW output power have been obtained at the low resonant frequency of 205 Hz with 3 g acceleration.

T-038 ELECTROMAGNETIC ENERGY HARVESTER WITH HIGH EFFICIENCY USING MICRO-MACHINING SI SPRINGS 378

T. Shirai, Y. Wakasa, T. Nakagawa, K. Nomura, and H. Yagyu
Panasonic Corporation, JAPAN

We developed electromagnetic vibration energy harvester with quite high generating efficiency over 40%. This high efficiency was realized by using the 1mm-thick uniquely-structured Si springs in the harvester, which was created using micromachining process.

W-039 ELECTROMAGNETIC ENERGY HARVESTER WITH AN IN-PHASE VIBRATION BANDWIDTH BROADENING TECHNIQUE 382

S.J. Chen, J.Y. Wu, and S.Y. Liu
National Central University, TAIWAN

This paper develops a duo-mode vibration structure for increasing usable bandwidth in a micromachined electromagnetic energy harvester. Compared to a pure cantilever harvester, the proposed cantilever-spiral coupled energy harvester has lower resonant frequencies and larger bandwidth.

Th-040 ELECTROSTATIC GENERATOR WITH FREE MICRO-BALL AND ELASTIC STOPPERS FOR LOW-FREQUENCY VIBRATION HARVESTING 385

F. Cottone[1], P. Basset[2], F. Marty[2], D. Galayko[3], L. Gammaitoni[1], and T. Bourouina[2]
[1]University of Perugia, ITALY, [2]Université Paris-Est, FRANCE, and
[3]UPMC-Sorbonne Universités, FRANCE

We present a novel MEMS electrostatic vibration harvester based on frequency amplification through multiple-mass impacts in combination with elastic stoppers. The harvester proof mass hosts a tungsten micro-ball free to travel along the vibration direction. At low frequencies (10-60 Hz) the micro-ball impacts with the oscillating mass of the generator transferring kinetic energy to the gap-closing comb transducer which in turn resonates at 92 Hz. In addition, the elastic stoppers amplify the proof mass and ball velocity by collision with the rigid frame. Output power between 0.25 and 0.45 µW is achieved at 0.3-g amplitude and only 15 V bias in the range of 10-60 Hz with a -3db bandwidth of 50 Hz

M-041 FLEXIBLE MICRO-SUPERCAPACITORS FROM PHOTORESIST-DERIVED CARBON ELECTRODES ON FLEXIBLE SUBSTRATES 389

M.-S. Kim, B. Hsia, C. Carraro, and R. Maboudian
University of California, Berkeley, USA

We demonstrate a simple and scalable technique for the fabrication of flexible micro-supercapacitors. The flexible high surface area electrodes are fabricated via a photoresist pyrolysis process followed by transfer for the electrodes to a flexible substrate. An energy density of 1 mWh/cm^3 is measured and the mechanical stability of the device is demonstrated through mechanical cycling tests.

T-042   GRAPHENE ELECTRODES ENHANCE PERFORMANCE FOR MICROLITER-SCALE MICROBIAL FUEL CELLS 393
V. Jayaprakash, R.D. Sochol, R. Warren, K. Iwai, and L. Lin
University of California, Berkeley, USA

In this work, graphene electrodes are employed to increase the power output of a microliter-scale microbial fuel cell (µMFC) for the first time. Previously, researchers have predominantly used Au/Cr electrodes in µMFCs, and have operated these fuel cells under controlled anodic conditions to attain high current densities and columbic efficiencies. At present, relatively low power outputs and open circuit potentials have limited such fuel cells from implementation in practical applications. To improve such performance, here we introduce a graphene-based µMFC (G-MFC) that utilizes graphene electrodes.

W-043   HIGH PERFORMANCE NONLINEAR MICRO ENERGY HARVESTER INTEGRATED WITH (K,Na) NbO$_3$/Si COMPOSITE QUAD-CANTILEVER 397
L.V. Minh, M. Hara, and H. Kuwano
Tohoku University, JAPAN

We developed a lead-free (K,Na)NbO3 (KNN) nonlinear microenergy harvester. The harvester was densely integrated with a quatrefoil-shaped proof mass and quad cantilevers using bulk micromachining. The KNN/Si composite cantilever with two-separated KNN capacitors was to effectively collect charge. Clamped-clamped beam design was also adopted for wide band operation. The experimental results showed that the wide bandwidth of 253 Hz and the highest power density of 1623 uW/cm3 among the developed piezoelectric nonlinear MEMS harvesters were achieved at the low acceleration of 6 m/s2.

Th-044   HIGH-ENERGY-DENSITY ON-CHIP LI-ION CAPACITORS 401
S. Li[1] and X. Wang[1,2]
[1]Tsinghua University, CHINA and [2]Chinese Academy of Sciences, CHINA

This paper presents a new on-chip Li-ion capacitor featured by higher energy density than that of supercapacitor under the same level of charge/discharge rate. Activated carbon (AC), a supercapacitor material, is used as positive electrode, while graphite, an anode material of Li-ion battery, is used as negative electrode and the electrolyte used in Li-ion battery serves as electrolyte. The prototype with 100-µm-thick electrodes shows a capacity of 175µAh/cm^2 and an energy density of about 1550mJ/cm^2 under a charge/discharge current of 0.5mA/cm^2, and a cell voltage of 3.4V.

M-045   HIGH-ENERGY-DENSITY ON-CHIP SUPERCAPACITORS USING MANGANESE DIOXIDE-DECORATED DIRECT-PROTOTYPED POROUS CARBON ELECTRODES 405
S. Li[1], X. Wang[1,2], and C. Shen[1]
[1]Tsinghua University, CHINA and [2]Chinese Academy of Sciences, CHINA

This paper presents the high performance on-chip micro supercapacitors using manganese dioxide (MnO$_2$) decorated into direct-prototyped porous carbon electrodes. By a new method of incorporating MnO$_2$ into carbon framework, both electric double layer capacitance (EDLC) and pseudocapacitance contribute to total capacity. Therefore, about 4-time increase in volumetric capacitance (0.8mF/(cm^2 µm) vs. 2.9mF/(cm^2 µm) under the scan rate of 50mV/s) is achieved. The procedure makes such devices potentially to be integrated into multi-function microsystems.

T-046   MEMS VIBRATION ELECTRET ENERGY HARVESTER WITH COMBINED ELECTRODES 409
Q.Y. Fu and Y. Suzuki
University of Tokyo, JAPAN

We have developed a novel in-plane MEMS electret energy harvester with overlapping-area change and gap-closing electrodes that provides large output power both at low and high vibration accelerations. An early prototype has been successfully microfabricated with the single layer silicon-on-insulator process. Soft X-ray charging is employed to establish uniform surface potential around 60 V on vertical electrets on the sidewall of the comb fingers. Up to 1.6 µW output power has been obtained, which corresponds to the effective as high as 57%.

W-047   MICRO PATTERN OF CHARGE IN PTFE ELECTRET FOR ENERGY HARVESTERS 413
W. Bian, X. Wu, and X. Wang
Tsinghua University, CHINA

This paper presents a novel fabrication process to pattern the charge in electret film for vibration energy harvester (VEH) Applications. Compared with previously reported techniques, PTFE electret material, which is inexpensive and has highSurface potential, is used. The line width of the charge pattern is determined by photolithography process. Experiment results show that the surface potential on the patterned charge zone of PTFE is higher than -200V when line width reaches 20um. A demoed VEH is built by using the pattern technique and tested.

**Th-048 NANO-POROUS SIO2 ELECTRET WITH HIGH SURFACE
POTENTIAL AND HIGH THERMAL RESISTANCE** 417

M. Suzuki[1], T. Wada[1], T. Takahashi[1], T. Nishida[2], Y. Yoshikawa[2], and S. Aoyagi[1]
[1]Kansai University, JAPAN and [2]ROHM Co., Ltd., JAPAN

This study proposes a new electret with high surface potential and high thermal resistance, which is made of nano-porous SiO2. Electrical charge density in the nano-porous SiO2 is higher than that in a normal SiO2 because the interfaces between void and SiO2 trap electrical charges. Thermal stability of the nano-porous SiO2 electret is better than that of a polymer electret. Output power generated by vibration energy harvesting using the nano-porous SiO2 electret is also larger than that using the normal SiO2 or CYTOP electret.

**M-049 NANOFLUIDIC REVERSE ELECTRODIALYSIS PLATFORM USING CONTROLLED
ASSEMBLY OF NANOPARTICLES FOR HIGH POWER ENERGY GENERATION** 421

E. Choi, K. Kwon, D. Kim, and J. Park
Sogang University, SOUTH KOREA

This paper presents a novel microplatform for high power energy generation based on reverse electrodialysis. The ideal cation-selective membrane for power generations is realized using geometrically controlled in-situ self-assembled nanoparticles and it can be constructed with simple and cost effective process using microdroplet control containing nanoparticles in microchannel. Another advantage in our system is that maximum powers and the energy conversion efficiency can be improved by changing the geometry of microchannel and proper selection of size and materials in nanoparticles

**T-050 SPRINGLESS CUBIC HARVESTER FOR CONVERTING
THREE DIMENSIONAL VIBRATION ENERGY** 425

M.D. Han, W. Liu, B. Meng, X.S. Zhang, X.M. Sun, and H.X. Zhang
Peking University, CHINA

This paper reports the design, fabrication and measurement of a springless cubic energy harvester. Coils are fabricated onto polyimide substrate and folded to form a cubic box. Output performance of the device is theoretically and experimentally investigated. Vibration in all dimensions can be effectively harvested and the maximum output is achieved at low frequencies with a wide bandwidth. Moreover, the device can be placed on a backpack or wrist to harvest vibration energy from daily life.

**W-051 ON THE OPTIMIZATION AND PERFORMANCES OF A COMPACT
PIEZOELECTRIC IMPACT MEMS ENERGY HARVESTER** 429

P. Janphuang, R. Lockhart, D. Briand, and N.F. de Rooij
École Polytechnique Fédérale de Lausanne (EPFL), SWITZERLAND

This paper presents the development of a compact energy harvesting configuration to convert low frequency, mechanical oscillations into usable electrical energy using AFM-like MEMS piezoelectric cantilevers coupled to a rotating gear. The harvester, with an active device volume of 3.5 mm3 (3×5×0.23 mm3), is able to produce an average output power of 12 µW measured across an optimal resistive load of 4.7 kΩ at a rotational speed of 19 rps, demonstrating the potential of the compact MEMS piezoelectric micro-power generator.

Fabrication Technologies (FAB)

**Th-052 "ASSIST-FREE" ASSEMBLY TECHNIQUE OF STANDING OPTICAL
DEVICES ON SOFT SPRING ACTUATOR STAGES** 433

Y. Oka, R. Shinozaki, K. Terao, T. Suzuki, F. Shimokawa, F. Oohira, and H. Takao
Kagawa University, JAPAN

In this study, a new assembly technique of independently fabricated optical devices on fragile MEMS actuator stages has been developed to realize novel functional optical-MEMS devices. This technique realizes the "assist-free" alignment and fixing of vertically mounted optical devices by combination of "micro spring slider" and "trapezoidal alignment slit". In the experiments, micro mirrors were attached on electrostatic linear actuators and rotational actuators using this assembly technique, and a small average value of relative-angle error around 4/100° was successfully obtained."

**M-053 3D NANOFABRICATION ON COMPLEX SEED
SHAPES USING GLANCING ANGLE DEPOSITION** 437

H.-H. Jeong[1], A.G. Mark[1], J.G. Gibbs[1], T. Reindl[2], U. Waizmann[2], J. Weis[2], and P. Fischer[1,3]
*[1]Max Planck Institute for Intelligent Systems, GERMANY, [2]Max Planck Institute for Solid State Research,
GERMANY, and [3]University of Stuttgart, GERMANY*

We report a 3D fabrication scheme that combines two existing techniques, electron beam lithography (EBL) and glancing angle deposition (GLAD), to fabricate nanostructures with complex 3D shapes both parallel and perpendicular to the growth direction. GLAD is a physical vapor deposition (PVD) technique where evaporant is delivered to a substrate at a high angle of incidence. Local shadowing and azimuthal manipulation of the substrate allows a rich variety of complex 3D structures to be grown down to the nanoscale. Herein, we use EBL to write custom seed layers with complex shapes, and do so at the resolution limit for the GLAD technique.

**T-054   A FLEXIBLE TACTILE AND SHEAR SENSING ARRAY FABRICATED
BY A NOVEL BUCKYPAPER PATTERNING TECHNIQUE** 441
C.-W. Ma, L.-S. Hsu, J.-C. Kuo, and Y.-J. Yang
National Taiwan University, TAIWAN

In this work, we present a flexible tactile and shear sensing array utilizing patterned buckypaper as the sensing elements. A novel fabrication process for realizing patterned buckypaper with high aspect ratio was proposed. The fabricated sensing device possesses the advantages such as anisotropic sensing capability, flexibility, simple fabrication process, and low cost. Measured results show excellent sensitivity and repeatability. In addition, the anisotropic sensing capability, which can be employed for better shear sensing, was also observed and discussed.

**W-055   A TECHNOLOGY FOR MONOLITHIC MEMS-CMOS INTEGRATION AND ITS
APPLICATION TO THE REALIZATION OF AN ACTIVE-MATRIX TACTILE SENSOR** 445
F. Zeng and M. Wong
Hong Kong University of Science and Technology, HONG KONG

A scheme of MEMS-CMOS integration based on the surface-migration of silicon is presently described. A cavity sealed with a cover-diaphragm is first formed, the electronic devices are next fabricated, and the suspended mechanical components are finally realized using the cover-diaphragm, without a sacrificial layer etch. With this scheme, the material and process incompatibility issues inherent in the existing integration techniques are largely eliminated. As a demonstration, a 16x16 active-matrix tactile sensor integrating a total of 256 force-sensing diaphragms, 512 transistors and 512 piezoresistors was designed, realized and characterized.

Th-056   ALD HONEYCOMB PLATES ENABLING ROBUST ULTRATHIN MEMS 449
K. Davami, L. Zhao, and I. Bargatin
University of Pennsylvania, USA

We report rigid MEMS structures made of ALD films, with a thickness of the order of 10 nanometers and patterned in the shape of a 3D honeycomb. Unlike planar ALD films, the 3D honeycomb plates do not warp due to fabrication stress gradients and are promising for a number of applications. For example, honeycombs made from refractive metals, such as ALD tungsten can be used to create thermionic energy converters with a well-controlled gap. Honeycomb cantilevers or beams can be used as resonant gas sensors, thanks to their low thickness and high stiffness, which lead to high resonance frequencies and high sensitivity to surface adsorbates. The transparency of the alumina ALD films may even enable their use as support films in electron microscopy.

**M-057   ARRAYS OF MICRO PENNING-MALMBERG TRAPS:
AN APPROACH TO FABRICATE VERY HIGH ASPECT RATIOS** 453
A. Narimannezhad, J. Jennings, M.H. Weber, and K.G. Lynn
Washington State University, USA

This paper reports on the progress of fabrication of very high aspect ratio (1000:1) micro-Penning-Malmberg trap arrays designed to store antimatter. The structure consists of thousands of 100μm diameter tubes etched by deep reactive ion etching through Si wafers. Cycles of thermal oxidation and wet etching in buffered oxide etch (BOE) minimized the sidewalls roughness and ensured a complete coating during gold sputtering. The wafers were then aligned and stacked in order to create the microtubes. Uniform plating with mean roughness of R_a=600nm was achieved by tuning the electroplating parameters.

**T-058   BATCH RELEASE OF MONODISPERSE LIPOSOMES TRIGGERED
BY PULSED VOLTAGE STIMULATION** 457
T. Osaki[1], K. Kamiya[1], R. Kawano[1], and S. Takeuchi[2]
[1]Kanagawa Academy of Science and Technology, JAPAN and [2]University of Tokyo, JAPAN

We present a batch release technique for monodisperse liposomes immobilized on a substrate. A single pulsed voltage to the substrate induced detachment of the arrayed liposomes previously developed. Simultaneous release was observed shortly after the electrical stimulation. The release technique produced monodisperse and solvent-free liposomes freely suspended on the substrate, and allowed manipulation of the liposomes.

**W-059   CLARITAS™ – A UNIQUE AND ROBUST ENDPOINT TECHNOLOGY FOR
SILICON DRIE PROCESSES WITH OPEN AREA DOWN TO 0.05%** 459
O. Ansell[1], R. Barnett[1], T. Haase[2], L. Xie[3], S. Vargo[1], and D. Thomas[1]
*[1]SPTS Technologies Limited, UK, [2]Fraunhofer Institute for Photonic Microsystems (IPMS,) GERMANY,
and [3]Harvard University, USA*

SPTS' Claritas is an enhanced method of OES endpoint detection for the Bosch process. It will be shown that Claritas has the capability to endpoint very low open area patterns (<1%), Bosch process recipes with high process pressures and show potential use of Claritas with other process solutions, including vapour phase etching.

Th-060 CONCURRENT REACTIVE ION ETCHING EMPLOYING MICROMACHINED IONIC LIQUID ION SOURCE ARRAY 463
R. Yoshida, M. Hara, H. Oguchi, T. Suzuki, and H. Kuwano
Tohoku University, JAPAN

This paper describes concurrent reactive ion etching using micro ionic liquid ion source (ILIS) array. The ILIS array was fabricated using bulk micromachining and consists of micro needle emitters and a reservoir for the ionic liquid (IL) of 1-ethyl-3-methylimidazolium tetrafluoroborate. The ion beam etching of a (100) silicon substrate using the fabricated ILIS array was demonstrated. Monitoring mass spectra during the etching, the peaks of $SiF+$, $SiF2+$, and $SiF3+$ could be observed. These peaks indicate the chemical reaction between the silicon and fluorine based ions from the IL.

M-061 DEVELOPMENT OF MEMS PIERCE-TYPE NANOCRYSTALLINE SI ELECTRON-EMITTER ARRAY FOR MASSIVELY PARALLEL ELECTRON BEAM DIRECT WRITING 467
H. Nishino[1], S. Yoshida[1], A. Kojima[2], N. Ikegami[3], N. Koshida[3], S. Tanaka[1], and M. Esashi[1]
[1]Tohoku University, JAPAN, [2]Crestec Corporation, JAPAN, and
[3]Tokyo University of Agriculture and Technology, JAPAN

This study reports on development of the fabrication process for 100×100 Pierce-type nanocrystalline Si electron-emitter array for massively-parallel electron-beam (EB) direct writing system based on active-matrix operation of large-scaled-integrated circuit (LSI). The 100-μm-pitch emitter array with each diameter of ~40 μm is prototyped and successfully demonstrates EB resist patterning by 1:1 projection exposure at CMOS-compatible operation voltages. This study also successfully establishes the integration process of the emitter array on a CMOS-LSI wafer. This achievement is a giant step for realizing the novel EB lithography system.

T-062 DIRECT LASER WRITING OF 3D PROTEIN STRUCTURES WITH NANOSCALE FEATURE SIZES 471
D. Serien[1,2] and S. Takeuchi[1,3]
[1]University of Tokyo, JAPAN, [2]CIRMM-IIS, JAPAN, and [3]Japan Science and Technology Agency (JST), JAPAN

We report the fabrication of three-dimensional (3D) protein structures with nanoscale feature sizes by two-photon direct laser writing (DLW). For this fabrication technology, we combine the established DLW technology with previously reported 3D protein structure fabrication by photosensitized crosslinking. We demonstrate the fabrication of 2D and 3D protein structures with nm-sized features.

W-063 FABRICATION OF ANISOTROPIC AND HIERARCHICAL UNDULATIONS BY BENCHTOP SURFACE WRINKLING 474
K. Wei and Y. Zhao
Ohio State University, USA

This paper describes a benchtop wrinkling process where highly ordered sinusoidal wrinkles with tailored wavelength and amplitude are created atop a PDMS foundation by atmospheric electric discharge. The method is used to fabricate hierarchical and anisotropic wrinkle-on-wrinkle and wrinkle-on-microstructure surface patterns. Its accessibility in general wet lab environments and simplicity to create multi-scale roughness are believed to facilitate applications in optical gratings, topography guidance for cell alignment, and micro/nanofluidics.

Th-064 FABRICATION OF THIN STENCIL WITH BUFFER RESERVOIR UTILIZING THE COMBINATION OF AZ4620 AND SU-8 ELECTROPLATING MOLDS 478
P.H. Chen[1], C.W. Huang[2], and C.H. Lin[1]
[1]National Sun Yat-sen University, TAIWAN and [2]Metal Industries Research and Development Center, TAIWAN

This work develops a novel process for fabricating ultra-thin stencil with buffer reservoir utilizing the combination of AZ4620 positive photoresist (PR) and SU-8 negative PR as the electroplating molds. A 5 um thick AZ4620 layer is used to precisely define the printing patterns while a 3 um thick of nickel layer is electroplated. A SU-8 layer of the thickness 50 um is patterned as the second electroplating mold. The high transparency of SU-8 PR makes it easy to align the two PR plating molds. A 30 um thick nickel layer is then electroplated onto the first nickel layer. The developed stencil can be used to printing ultra fine line and thin film pattern.

M-065 FREE-STANDING SUBWAVELENGTH GRID INFRARED REJECTION FILTER OF 90 MM DIAMETER FOR LPP EUV LIGHT SOURCE 482
Y. Suzuki, K. Totsu, M. Moriyama, M. Esashi, and S. Tanaka
Tohoku University, JAPAN

A subwavelength grid infrared filter as large as 90 mm in diameter was fabricated and tested on a 6 inch Si wafer for a laser-produced plasma extreme ultraviolet (EUV) light source used in the next generation lithography tools. The IR filter has a free-standing Mo-coated Si honeycomb grid structure with a thickness of 5 μm, a wire width of only 0.35 μm and a pitch of 4.5 μm, showing 99.7% rejection for 10.6 μm IR light. Such a large-size free-standing microstructure was successfully fabricated by carefully balancing film stress at each process step.

T-066   GRAPHENE SYNTHESIS VIA DROPLET CVD AND ITS PHOTONIC APPLICATIONS 486
X.N. Zang and L. Lin
University of California, Berkeley, USA

The process of "droplet CVD" for the synthesis of graphene sheets and its photonic applications have been demonstrated for the first time. Metal (Cu or Ni) droplets are naturally transformed from thin films in a high temperature furnace and utilized to grow graphene sheets via the chemical vapor deposition (CVD) process. As such, this new class of fabrication process could open up various graphene-based device/system applications, including photonic sensors."

**W-067   HIGH ASPECT RATIO, LARGE AREA SILICON-BASED
GRATINGS FOR X-RAY PHASE CONTRAST IMAGING** 490
J.J. Baborowski, V. Revol, C. Kotler, R. Kaufman, P. Niedermann, F. Cardot,
A. Dommann, A. Neels, and M. Despont
CSEM SA, SWITZERLAND

The presented work reports on the latest developments in the manufacturing of high aspect ratio silicon-based gratings used for X-ray phase contrast imaging (XPCI. XPCI reveals subtle changes in the microstructure of the samples, such as micro-cracks in composite materials or micro-calcifications in breast tissues. In fields as diverse as medicine, non-destructive testing or security, the gained information of this technique allows early diagnostic or detection of defects, tumors or explosives. The range of opportunities offered by depends highly on the achievable gratings parameters, such as periodicity, depth, duty cycle and aspect ratio. We have developed large (100x100mm2) Au-Si-Au-Air gratings with a periodicity of down to 2 μm and a depth of up to 100 μm with extremely low defect density (<1 defect/cm2). The fabricated gratings have been implemented on a XPCI set-up and used to demonstrate unprecedented imaging quality in material quality control.

**Th-068   HIGH RESOLUTION MICRO ULTRASONIC MACHINING (HR-μUSM) FOR
POST-FABRICATION TRIMMING OF FUSED SILICA 3-D MICROSTRUCTURES** 494
A. Viswanath, T. Li, and Y.B. Gianchandani
University of Michigan, USA

Post-fabrication trimming is interesting for devices such as inertial sensors, timing references, and mass-balance resonators to adjust stiffness, mass, and potentially damping. The trimming process should be capable of micro machining brittle materials, without inducing stress or subsurface cracks. We have developed and evaluated a subtractive trimming technique based on micro ultrasonic machining (μUSM), for high-resolution trimming of complex 3D microstructures made from fused silica. Machining rates as low as 10 nm/sec and surface roughness as low as 30 nm have been achieved.

**M-069   FABRICATION OF CARBON NANOFIBROUS MICROELECTRODE ARRAY (CNF-MEA)
USING NANOFIBER IMMERSION PHOTOLITHOGRAPHY** 498
P.F. Jao[1], E. Franca[1], S.P. Fang[1], J. Yoon[1], K. Cho[1,3], D.E. Senior[2], G.J. Kim[1],
B. Wheeler[1], and Y.K. Yoon[1]
*[1]University of Florida, USA, [2]Universidad Tecnologica de Bolivar, COLOMBIA, and
[3]Korea Basic Science Institute, SOUTH KOREA*

Microelectrode arrays are used for stimulating and receiving neural electrical signals in vitro neural study. This work demonstrates the fabrication process of nanofibrous 3D microelectrodes using immersion lithography. Oil immersion negates the diffraction effects intrinsic in the photopatterning of electrospun nanofibers to give higher aspect ratio. Nanofiber electrode resistivity is characterized and its performance compared to that of carbon thin film. In vitro testing of electrodes are performed using E18 cortical neurons and cell density and cell viability analyzed.

**T-070   INDUCTIVELY COUPLED PLASMA ETCHING OF BULK TUNGSTEN
FOR MEMS APPLICATIONS** 502
L. Song, N. Li, S. Zhang, J. Luo, J. Hu, Y. Zhang, S. Chen, and J. Chen
Peking University, CHINA

Tungsten based MEMS devices have the potential to be used for many applications, such as tools for micro electrical discharge machining and ultrasonic machining, or mold for inject molding. For the first time, bulk tungsten ICP etching was developed and characterized, which is capable of producing high aspect ratio (>13) structures with feature size below 3μm. Etching depth of 230μm has been achieved at an etch rate up-to 2.2μm/min. This technology offers big opportunities for MEMS applications.

W-071   LARGE ARRAYS OF INKJET-PRINTED MEMS MICROBRIDGES ON FOIL 506
F. Molina-Lopez, D. Briand, and N.F. de Rooij
Ecole Polytechnique Fédérale de Lausanne (EPFL), SWITZERLAND

This works describes the fabrication of an array of printed MEMS microbridges on polymeric foil in only four easy steps.Each functional material was deposited exclusively by inkjet-printing technique, compatible with large-area fabrication. Thearray consists of 60 to 80 individual microbridges, occupying an area of 2 mm x 2 mm, and displaying a total capacitancevalue of 1 - 1.7 pF when connected in parallel.

Th-072 LIGHTWEIGHT MICRO LATTICES WITH NANOSCALE FEATURES FABRICATED FROM PROJECTION MICROSTEREOLITHOGRAPHY 510

X. Zheng[1], J. Deotte[1], J. Vericella[1], M. Shusteff[1], T. Weisgraber[1], H. Lee[2], N. Fang[2], and C.M. Spadaccini[1]

[1]*Lawrence Livermore National Laboratory, USA and* [2]*Massachusetts Institute of Technology, USA*

We demonstrate the utility of three-dimensional Projection Microstereolithography manufacturing system by producing a variety of microstructures with complex geometries and explored the potential of using the system to build meso-scale structures with micro-scale architecture and nano-scale features. These achievements pave the way for large scale micro- and nano- manufacturing that extends the current state-of-the-art of three-dimensional fabrication technologies.

M-073 FABRICATION OF MICRO-HEATERS EMBEDDED IN PDMS USING A DRY PEEL-OFF PROCESS 514

I. Byun, R. Ueno, and B.J. Kim

University of Tokyo, JAPAN

We shows a reliable fabrication method of micro-heaters embedded in polydimethylsiloxane (PDMS), and shows characterization of the micro-heaters. Metallization of PDMS is achieved using a dry peel-off process which involves modifying the surface properties of the substrate and metal patterns through self-assembled monolayer and manually peeling off the PDMS with embedded metal layers. Thus, micro-heaters can be fabricated by a simpler and easier way compared to conventional methods.

T-074 MICROPATTERNING OF BACTERIAL CELLULOSE AS DEGRADABLE SUBSTRATE FOR CELL CULTURE 518

Y. Karita, K. Hirayama, H. Onoe, and S. Takeuchi

University of Tokyo, JAPAN

This paper describes microfabrication of bacterial cellulose membrane, which is a nanofibrous cellulosic material produced by a bacterium, Acetobacter xylinum. We successfully micropatterned bacterial cellulose membrane by applying MEMS process and this patterned bacterial cellulose was confirmed to work as a scaffold for mouse embryonic fibroblast cells. Moreover, formation of cell cluster was observed by the treatment of cellulose degrading enzyme. We believe that this micropatterned cellulose plate would be useful in degradable microscaffolds for cell culture.

W-075 MICROSCALE MAGNETIC PATTERNING OF HARD MAGNETIC FILMS USING MICROFABRICATED MAGNETIZING MASKS 520

A. Garraud[1], O.D. Oniku[1], W.C. Patterson[1], E. Shorman[1], D. Le Roy[2], N.M. Dempsey[2], and D.P. Arnold[1]

[1]*University of Florida, USA and* [2]*University of Grenoble-Alpes, FRANCE*

We present a batch-fabrication process to imprint microscale magnetic pole patterns (perpendicular north/south poles) into hard magnetic films using field-shaping, soft magnetic "magnetizing masks". Using 7-μm-thick, electroplated Fe–Co magnetizing masks, magnetic stripes with widths down to 50 μm have been imprinted into both 15-μm-thick Co–Pt films and 5-μm-thick Nd–Fe–B films. These patterned films exhibit a sinusoidal stray magnetic field pattern with ~4 and ~7 mT_{pk-pk} variations and corresponding field gradients of 80 and 140 T/m, respectively. We also demonstrate the ability to transfer more complex patterns by showing magnetization of various geometric shapes.

Th-076 MONOLITHIC PIEZOELECTRIC IN-PLANE MOTION STAGE LOW-CROSS-AXIS-COUPLING 524

S. Nadig, S. Ardanuc, and A. Lal

Cornell University, USA

In this work we present a rotary dither stage which can provide rotation stimulus with high dynamic range of 1800-deg/s, and parts-per-thousand cross axis actuation, and is planar compatible with in-package inertial sensor calibration. We use bulk PZT-4 beams, laser cut out from plates, to achieve monolithic integration of lateral actuators and flexures. This process enables high-aspect ratio beams (500um thick x150um wide) resulting in parts-per-thousand in-plane to out-of-plane motion coupling

M-077 POLYMER MICROMACHINING BASED ON CU ON POLYIMIDE SUBSTRATE AND ITS APPLICATION TO FLEXIBLE MEMS SENSOR 528

Y. Niimi, S. Shibata, and M. Shikida

Nagoya University, JAPAN

MEMS technologies have produced various types of MEMS sensors on a Si or Silicone On Insulator (SOI) wafers. To realize MEMS sensors in the flexible fashion, we newly proposed to apply a Cu On Polyimide (COP) substrate as a starting material, and introduced a sacrificial etching for producing a cavity and an electrical feed through structures on the COP substrate. Finally, a flexible thermal MEMS sensor was fabricated on COP substrate.

T-078 PRINTING AND ENCAPSULATION OF ELECTRICAL CONDUCTORS ON POLYLACTIC ACID (PLA) FOR SENSING APPLICATIONS 532
A. Vásquez Quintero, N. Frolet, D. Märki, G. Mattana, A. Marette, D. Briand, and N.F. de Rooij
Ecole Polytechnique Fédérale de Lausanne (EPFL), SWITZERLAND

This paper presents the printing of resistive and capacitive devices for temperature and humidity sensing applications, respectively, on biodegradable polylactic acid (PLA) substrates. Inkjet and gravure printing were assessed as direct silver-based nanoparticles inks transfer methods. Flash photonic ink sintering methodologies were optimized due to the low PLA glass transition temperature (58 °C) and maintain its mechanical integrity. An encapsulation method for electrical conductive structures is proposed by means of laminating PLA sheets at relatively low temperatures (< 60 °C). These fabricated structures are now exposed for long periods (months) in compost and high humidity environments to evaluate their degradation.

W-079 RAPID PROTOTYPING OF RESISTIVE MEMS SENSING DEVICES ON PAPER SUBSTRATES 536
T. Meiss[1], R. Werthschützky[1], and B. Stoeber[2]
[1]Technische Universität Darmstadt, GERMANY and [2]University of British Columbia, CANADA

We have developed an inexpensive inkjet printing process to rapidly fabricate resistive sensor devices on paper substrates. Since we use a commercial inexpensive inkjet printer and the design can be modified and tested within minutes, the process is especially useful to easily develop and test MEMS sensor models. Additional applications encompass disposable medical sensors, sensors for paper packaging, as well as very low cost strain sensing.

Th-080 REAL-TIME DYNAMICALLY RECONFIGURABLE LIQUID METAL BASED PHOTOLITHOGRAPHY 540
D. Kim[1], J.H. Yoo[1], W. Choi[1], K. Yoo[2], and J.B. Lee[1]
[1]University of Texas at Dallas, USA and [2]Hanbat National University, SOUTH KOREA

We report real-time dynamically reconfigurable photolithography technique using liquid metal Galinstan as UV opaque material and PDMS as UV transparent material. We demonstrated dynamically reconfigured on-demand patterning of single digit numbers in positive photoresists along with various patterns with minimum feature size of 10 μm. To the best of our knowledge, this is the first demonstration of true real-time reconfigurable photolithography in UV wavelengths.

M-081 RELEASE AND TRANSFER OF LARGE-AREA ULTRA-THIN PDMS 544
J. Gao, D. Guo, S. Santhanam, Y.J. Yu, A.J.H. McGaughey, S.C. Yao, and G.K. Fedder
Carnegie Mellon University, USA

This paper reports on a successful fabrication of ultra-thin 10μm PDMS films with embedded Au electrodes, as well as the releasing and transferring of large area films. The motivation for this work is in development of a miniature pump actuator for moving working fluid in an electrocaloric microcooler. It opens a promising route for fabricating ultra-thin and compliant MEMS and electronic devices.

T-082 SOLID STATE MEMS DEVICES ON FLEXIBLE AND SEMI-TRANSPARENT SILICON (100) PLATFORM 548
S.M. Ahmed, A.M. Hussain, J.P. Rojas, and M.M. Hussain
King Abdullah University of Science and Technology, SAUDI ARABIA

We report fabrication of MEMS thermal actuators on flexible and semi-transparent bulk silicon <100> substrate. The fabricated thermal actuators exhibit similar performance before and after bending. We fabricate the devices first and then release the top portion of the silicon (≈ 19 μm) which is flexible and semi-transparent. Then we perform chemical mechanical polishing to reuse the remaining wafer. Prior demonstrations on flexible MEMS devices had limited thermal budget (<150 °C) compatibility and they did not have cost-saving wafer recycling process.

W-083 WAFER-SCALE FABRICATION OF HIGHLY INTEGRATED RUBIDIUM VAPOR CELLS 552
T. Overstolz, J. Haesler, G. Bergonzi. A. Pezous, P.-A. Clerc, S. Ischer, J. Kaufmann, and M. Despont
CSEM SA, SWITZERLAND

CSEM is developing a highly integrated chip scale atomic clock based on coherent population trapping (CPT) of 87-Rb atoms which are confined in a vapor cell. The vapor cells are batch fabricated, based on pipetting dissolved RbN3 into cell cavities etched into a silicon wafer, closing the cavities by anodic bonding, and UV decomposition of recrystallized RbN3 deposits into Rb and N2. The vapour cells are equipped with resistive heaters, temperature sensors, and Helmholtz coils integrated on both sides of the cell windows.

Th-084 WAFER-SCALE FLEXIBLE GRAPHENE LOUDSPEAKERS 556
H. Tian, Y.L. Cui, Y. Yang, D. Xie, and T.L. Ren
Tsinghua University, CHINA

Wafer-scale flexible graphene loudspeakers are fabricated in one-step laser scribing technology. Current fabrication process for graphene devices is mainly based on CVD graphene, which needs several hours' graphene growth, transfer and patterning. By using this new laser scribing technology, wafer-scale graphene patterns can be obtained in 25 minutes. The loudspeaker is demonstrated to be high performance with wide-band sound generation from 1~50 kHz. Our results show that the laser scribed graphene could be widely used in integrating wafer-scale graphene-based electroacoustic devices.

Industry

M-085 A POLYMER MEMS MIRROR FOR ON-DEMAND LIGHT DISTRIBUTION FABRICATED BY INJECTION MOLDING AND TRANSFER OF PRINTED LAYERS 560

K. Kurihara[1], O. Nagumo[2], H. Takagi[1], and R. Maeda[1]
[1]*National Institute of Advanced Industrial Science and Technology (AIST), JAPAN and*
[2]*Designtech Co., Ltd, JAPAN*

A low-cost and wide mirror area polymer MEMS scanner for on-demand light distribution was fabricated by combined process of the injection molding and layer transfer of screen-printed patterns on a film. This fabrication process realizes low cost polymer MEMS. It is expected that the low-cost MEMS can be a killer application for new industry field.

T-086 BI-CHAMBER ELECTROMAGNETIC FLUIDIC PUMP 564

C.S. Gudeman[1], P.J. Rubel[1], and J.S. Foster[2]
[1]*Innovative Micro Technology, USA and* [2]*Owl Biomedical Inc., USA*

We describe the design, fabrication and performance of a MEMS pump that is actuated electromagnetically and is capable of pumping very high viscosity liquids. Valve motion relative to that of the pumping piston is described in detail.

W-087 IMPROVED MECHANICAL RELIABILITY OF MEMS PIEZOELECTRIC VIBRATION ENERGY HARVESTERS FOR AUTOMOTIVE APPLICATIONS 568

M. Renaud, Z. Wang, M. Jambunathan, S. Matova, R. Elfrink, M. Rovers, M. Goedbloed,
C. de Nooijer, R.J.M. Vullers, and R. van Schaijk
Holst Centre-Imec, THE NETHERLANDS

We present a comprehensive approach to address the issue of the mechanical reliability of MEMS piezoelectric vibration harvesters. These harvesters generate sufficient electrical power for powering a tire pressure monitoring system. However, their reliability, particularly in terms of shock resilience, has to be optimized for in-tire applications. This paper showcases experimentally verified improvements of the mechanical reliability, which is achieved by optimizing both the package design and the wafer processing.

Th-088 LOW-COST MICROBOLOMETER WITH NANO-SCALED PLASMONIC ABSORBERS FOR FAR INFRARED THERMAL IMAGING APPLICATIONS 572

F. Utermöhlen[1], D. Etter[2], D. Borowsky[1], I. Herrmann[1], C. Schelling[1], F. Hutter[2],
S.H. Sun[2], and J. Burtghartz[2]
[1]*Robert Bosch GmbH, GERMANY and* [2]*Institut für Mikroelektronik Stuttgart (IMS CHIPS), GERMANY*

We have developed a scalable low-cost microbolometer which can be used for automotive nightvision as well as consumer applications with a broad variety of requirements regarding image resolution and sensitivity. In contrast to state-of-the-art microbolometers which are based on a standard CMOS ASIC process with CMOS-compatible MEMS post-processing on the same wafer, we use a MEMS wafer with the microbolometer pixels and a standard CMOS wafer with the read-out ASIC which are mechanically, electrically and hermetically connected. This concept allows for significant cost reduction since the two dedicated technologies for the MEMS and the ASIC can be optimized and fabricated independently and because the ASIC chip can serve as a hermetic package for the MEMS.

Materials and Device Characterization (MDC)

M-089 2D PHOTONIC-CRYSTALS FOR HIGH SPECTRAL CONVERSION EFFICIENCY IN SOLAR THERMOPHOTOVOLTAICS 576

A. Lenert[1], V. Rinnerbauer[1], D.M. Bierman[1], Y. Nam[1,2], I. Celanovic[1], M. Soljacic[1], and E.N. Wang[1]
[1]*Massachusetts Institute of Technology, USA and* [2]*Kyung Hee University, SOUTH KOREA*

We present a high-efficiency 2D photonic crystal based solar thermophotovoltaic (STPV) device operating at high temperatures (~1300 K) under moderate solar concentration (~100 Suns). These results were only possible by tailoring the spectral properties of the absorber-emitter through surface nanostructuring of tantalum and minimizing parasitic thermal losses through an innovative vacuum-enclosed experimental setup.

T-090 AN OPTICAL IN-PLANE DISPLACEMENT MEASUREMENT TECHNIQUE WITH SUB-NANOMETER ACCURACY BASED ON CURVE FITTING 580

J. Kokorian[1], F. Buja[1], U. Staufer[1], and W.M. van Spengen[1,2]
[1]*Delft University of Technology, THE NETHERLANDS and* [2]*Falco Systems BV, THE NETHERLANDS*

We will show a technique, based on plain optical microscopy and curve fitting, for measuring in-plane displacements in MEMS applications. We modeled and experimentally verified how the measurement accuracy is influenced by quantization noise, photon shot noise, optical magnification, camera resolution and pixel binning. We found that when the noise figure was dominated by shot noise, the measurement error was lowered into the deep-subnanometer range.

W-091 CHARACTERIZATION OF IMPROVED CAPACITIVE MICROMACHINED ULTRASONIC TRANSDUCERS (CMUTS) USING ALD HIGH-K DIELECTRIC ISOLATION 584

T. Xu, C. Tekes, and F.L. Degertekin
Georgia Institute of Technology, USA

We show the advantages of high-k dielectric, ALD HfO2 over traditional PECVD silicon nitride isolation for Capacitive Micromachined Ultrasonic Transducers (CMUTs) fabricated by a low temperature, CMOS compatible, sacrificial release method. ALD HfO2 dielectric properties are characterized to optimize CMUT design in transmit and receive mode. Performances of the two different dielectric isolation devices are evaluated through parallel plate modeling and experimentally measured pressure outputs and receive sensitivities.

Th-092 CHARACTERIZATION OF STICTION FORCES IN ULTRA-CLEAN ENCAPSULATED MEMS DEVICES 588

D.B. Heinz, V.A. Hong, E.J. Ng, C.H. Ahn, Y. Yang, and T.W. Kenny
Stanford University, USA

We show that stiction in contact between encapsulated MEMS devices and the surrounding sidewalls generally results in a reversible adhesion with a consistent adhesion force. This force is small enough (25 μN) to be overcome by the restoring force of the springs in inertial sensors with resonant frequency above 4 kHz. Therefore, it should be possible to design and build stiction-free inertial sensors in this process – a significant advantage over approaches that rely on deposition, tuning and maintenance of chemical coatings for inertial sensors.

M-093 COMPOSITE OF THERMALLY RESPONSIVE SOLUTION AND LUBRICATING MICRO BEADS AS SEALING MATERIAL FOR PISTON-CYLINDER ACTUATOR 592

T. Chishiro, S. Honda, and S. Konishi
Ritsumeikan University, JAPAN

This paper proposes a novel sealing technique for miniaturized piston-cylinder actuator. The sliding part of a piston actuator requires both high sealing and excellent lubrication. This paper will show a smart sealing material composed of thermally responsive solution and lubricating micro beads. Micro beads are expected to contribute to provide lubricant between a sliding part and an inner wall when thermally responsive solution gels by heating. We will present the concept, design, and characterization of proposed sealing technique for a piston actuator.

T-094 CONTINUOUS DYNAMIC TIMING MEASUREMENTS TO MONITOR SPRING AND SURFACE FORCES IN MEMS SWITCH RELIABILITY 596

C. Kosla[1], P. Fitzgerald[2], and M. Hill[1]
[1]Cork Institute of Technology, IRELAND and [2]Analog Devices, IRELAND

We demonstrate an automatic reliability detection/prediction system for industry manufactured MEMS switches based on dynamic time measurements, allowing for non-invasive and continuous device monitoring. The developed method highlights the influence of both restoring and surface forces evolution on switch reliability and for the first time allows identification of an imminent device failure due to its continuous monitoring. Additionally we present the scalability of this approach by testing it on different switch types and on a large number of samples.

W-095 DIRECT MEASUREMENT OF SHEAR PIEZORESISTANCE COEFFICIENT ON SINGLE CRYSTAL SILICON NANOWIRE BY ASYMMETRICAL FOUR-POINT BENDING TEST 600

T. Kimura, N. Saito, T. Takeshita, K. Sugano, and Y. Isono
Kobe University, JAPAN

This research evaluated the shear piezoresistance property of p-type single crystal silicon nanowire (SiNW) by the asymmetrical four-point bending (AFPB) testing proposed by the authors. We fabricated the p-type SiNW on the AFPB specimen with "V" shaped notches made of single crystal silicon. Bending the specimen by the asymmetrical four point-supports, simple shear stress can be produced at the center of the specimen. Consequently, we have succeeded in evaluating the shear piezoresistance coefficient of SiNW directly."

Th-096 DISSOLVABLE MATERIAL FOR HIGH-ASPECT-RATIO FLEXIBLE SILICON-MICROWIRE PENETRATIONS 604

S. Yagi, S. Yamagiwa, T. Imashioya, H. Oi, Y. Kubota, M. Ishida, and T. Kawano
Toyohashi University of Technology, JAPAN

For realization of low invasive electrode penetrations into biological tissue, here we improved the penetration capability of high-aspect-ratio flexible silicon-microwires by coating a dissolving material of silk fibroin. The silk fibroin was coated over vertically vapor-liquid-solid (VLS) grown silicon-microwires. The 420-μm-long silicon-wire with a ~200-μm-thick silk film exited the stiffness of 4.03 N/m, which is 72% improved value compared to that of the silicon-wires without silk (2.34 N/m). The effects of the silk support on the wire penetration were observed by demonstrating the gelatin penetrations. These results suggest that the numerous high-aspect-ratio flexible bioprobes can be penetrated by using the silk support.

M-097 DOME-DISC: DIFFRACTIVE OPTICS METROLOGY ENABLED DITHERING INERTIAL SENSOR CALIBRATION 608

S. Nadig[1], S. Ardanuç[1], B. Clark[2], and A. Lal[1]
[1]Cornell University, USA and [2]Analog Devices Inc., USA

We demonstrate ~100-ppm accurate scale-factor and bias calibration of a commercial gyroscope, in which the typical un-calibrated scale factor variations are 100,000-ppm. In this paper, we present a Diffractive Optical Metrology Enabled Dithering Inertial Sensor Calibration consisting of a novel piezoelectric dither stage, the motion of which is measured by imaging the diffraction pattern off the stage of a long-term stable wavelength laser. The architecture presented here illustrates how atomically stable lasers and CMOS imagers can be combined to form a miniature atomically stable self-calibrated inertial sensor platform.

T-098 ELECTRICAL CHARACTERIZATION OF ALD COATED SILICON DIOXIDE MICRO-HEMISPHERICAL SHELL RESONATORS 612

P. Shao, V. Tavassoli, C.-S. Liu, L.D. Sorenson, and F. Ayazi
Georgia Institute of Technology, USA

A micro-hemispherical shell resonator (µHSR) is the beating heart of a micro-scale hemispherical resonator gyroscope (µHRG). Small damping and high symmetry are two essential requirements for µHRGs. Damping can be quantified by mechanical quality factor (Q) of the resonance, and structural symmetry can be quantified by the frequency split between two degenerate resonance modes. This paper reports on important electrical characterizations of Q and frequency split of ALD coated thermally-grown silicon-dioxide µHSRs, and analysis on how the performance will change with fabrication and measurement parameters.

W-099 FABRICATION AND DEGRADATION CHARACTERISTIC OF SPUTTERED IRIDIUM OXIDE NEURAL MICROELECTRODES FOR FES APPLICATION 616

X.-Y. Kang, J.-Q. Liu, H.-C. Tian, J.-C. Du, B. Yang, H.-Y. Zhu, Y. NuLi, and C.-S. Yang
Shanghai Jiao Tong University, CHINA

We have fabricated reactively sputtered iridium oxide film (SIROF) microelectrodes under different oxygen flows and the stimulus-evoked degradation properties are also tested. The SIROF microelectrodes prepared under 25 sccm oxygen flow shows the least degradation from continuous electrical stimulation. That the charge storage capacity (CSC) is only 9.6 % lost and the 1 kHz impedance is only 4.23% increase. Hence, the 25 sccm one can be an ideal microelectrode modification material for electrical stimulation with the least degradation.

Th-100 FABRICATION AND TESTING OF PIEZOELECTRIC HYBRID PAPER FOR MEMS APPLICATIONS 620

S.K. Mahadeva, K. Walus, and B. Stoeber
University of British Columbia, CANADA

We have developed a new inexpensive functional paper based material that can be used as a piezoelectric substrate for sensing applications. In our simple method, nanostructured $BaTiO_3$ is embedded onto the fibers prior to forming paper sheet, which involves immersion of wood fibers in aqueous solution of poly(diallyldimethylammonium chloride) PDDA and poly(sodium 4-styrenesulfonate) and once again in PDDA, and results in the creation of a positively charged surface on wood fiber. The treated wood fibers are then immersed in a $BaTiO_3$ suspension, leading to the electrostatic binding of $BaTiO_3$. The hybrid paper showed the highest d33 of 4.8±0.4 pC/N.

M-101 GRAPHENE WOVEN FABRIC AS HIGH-RESOLUTION SENSING ELEMENT OF CONTACT-LENS TONOMETER 624

Y. Zhang[1], T. Man[1], X. Li[2], H. Zhu[2], and Z. Li[1]
[1]Peking University, CHINA and [2]Tsinghua University, CHINA

In our work, the graphene woven fabrics (GWFs), the combination of highly sensitive strain sensing and transparency, is investigated as the sensing element of the contact-lens tonometer, which enables precisely monitor IOP. The relationship between the current changes when keeping the voltage constant and effective IOP increasing has been obtained.

T-102 INCREASED THERMAL CONDUCTIVITY POLYCRYSTALLINE DIAMOND FOR LOW-DISSIPATION MICROMECHANICAL RESONATORS 628

H. Najar[1], A. Thron[1], C. Yang[2], S. Fung[1], K. van Benthem[1], L. Lin[2], and D.A. Horsley[1]
[1]University of California, Davis, USA, and [2]University of California, Berkeley, USA

We report an investigation of microcrystalline diamond (MCD) films deposited under different conditions to increase thermal conductivity and therefore mechanical Q in MEMS resonators. Here, through a study of different deposition conditions, we demonstrate a three-fold increase in thermal conductivity (i.e. k = 100W/mK) and therefore Q-TED. We further present a study of the unique microstructure of hot filament CVD diamond films and relate growth conditions to observed microstructural defects.

W-103   INTERFACE LOSSES IN MULTIMATERIAL RESONATORS 632
L.G. Villanueva[1,3], B. Amato[1], T. Larsen[1], and S. Schmid[1]
[1]Denmark Technical University, DENMARK and
[2]Ecole Polytechnique Federale de Lausanne (EPFL), SWITZERLAND

We present an extensive study shedding light on the role of surface and bulk losses in micromechanical resonators. We fabricate a set of Si3N4 square membranes with different lateral dimensions, thickness and thickness of metal on top and characterize the 81 lowest flexural modes, obtaining more than 3000 experimental points to eventually quantify the importance of interface losses in multimaterial resonators.

Th-104   INVESTIGATION OF DOMINANT FACTORS TO CONTROL C-AXIS
TILT ANGLE OF ALN THIN FILMS FOR EFFICIENT ENERGY HARVESTING 636
Q. Wang, H. Oguchi, M. Hara, and H. Kuwano
Tohoku University, JAPAN

We investigated growth conditions to enhance the c-axis inclination of aluminum nitride (AlN) thin films grown on silicon substrates using the electron cyclotron sputtering. Higher substrate tilt angles, lower substrate temperature, and rougher buffer layer surface resulted in higher c-axis tilt angle, mainly due to decrease in ad-atom mobility on the surface. This study deepens the understanding of how to control c-axis inclination of AlN thin film to control the electro-mechanical coupling coefficient for larger output power of the AlN-based energy harvesters.

M-105   INVESTIGATION OF THE FATIGUE ORIGIN AND PROPAGATION
IN SUBMICROMETRIC SILICON PIEZORESISTIVE LAYERS 640
G. Langfelder[1], S. Dellea[1], P. Rey[2], A. Berthelot[2], and A.F. Longoni[1]
[1]Politecnico di Milano, ITALY and [2]CEA - LETI – Minatec, FRANCE

We present the study performed on structures designed and tested for the analysis of long-term reliability and fatigue of 250-nm-thick crystalline Silicon that can be used as piezoresistive sensing layer in low-power 10-axis inertial measurement units. With a specimen surface-to-volume ratio 100 times smaller than previous literature, this work extends to the nanometric domain the debate data previously published about the origin and propagation of fatigue in Silicon at the micro scale.

T-106   NANOFIBER FORESTS WITH HIGH INFRARED ABSORPTANCE 644
H.Y. Mao[1,3], C. Lei[1,4], Y.J. Chen[1,4], Z.J. Chen[2], W. Ou[1,3], W.G. Wu[2], A.J. Ming[1,3], and D.P. Chen[1,3]
[1]Chinese Academy of Sciences, CHINA, [2]Peking University, CHINA,
[3]Jiangsu R&D Center for Internet of Things, CHINA, and [4]North University of China, CHINA

Nanofiber forests with high infrared(IR) absorptance are reported in this work. In wavelength range from 1.5 to 5 μm, the absorptance of the nanofiber forests reaches a minimum of 96%, which is much higher than that of Si3N4-based IR absorbers and the polymer coatings from which the nanofibers are obtained. Such nanofiber forests are fabricated by using a plasma-stripping-of-polymer technique, which is fast, high-yield, and compatible with micro-fabrication. By introducing the nanofiber forests in MEMS IR devices, improved performance of the devices is expected to be acquired.

W-107   NANOSTRUCTURED SILICON FLAPPING WING WITH HIGHER STRENGTH
AND LOW REFLECTIVITY FOR SOLAR POWERED MEMS AIRCRAFT 648
K. Kashyap[1], A. Kumar[1], C.-N. Chen[1], M.T. Hou[2], and J.A. Yeh[1,3]
[1]National Tsing Hua University (NTHU), TAIWAN, [2]National United University, TAIWAN, and
[3]National Applied Research Laboratories, TAIWAN

We develop a novel way of higher strength silicon flapping wings design for MEMS aircraft achieved by silicon nanostructures, which breaks the limitation of silicon as a fragile material. Silicon flapping wings were designed for MEMS aircraft which increases the bending strength of wings by 6 times and reduces the reflectance to 2%. Both the benefits simultaneously were achieved from nanostructure surface texturing by low cost wet chemical etching.

Th-108   POSSIBILITY OF CEMENTED CARBIDE AS STRUCTURAL MATERIAL FOR MEMS 652
T. Morikaku[1], T. Fujii[1], K. Kuroda[2], Y. Takami[2], S. Inoue[1], and T. Namazu[1,3]
[1]University of Hyogo, JAPAN, [2]Silveralloy Co., Ltd. JAPAN, and
[3]Japan Science and Technology Agency (JST), JAPAN

We present the possibility of WC-Co cemented carbide as mechanical elements in MEMS. The cemented carbide is typically used as material for working tool because it has superior characteristics, such as very high Young's modulus, excellent rigidity, good chemical inertness, and good thermal stability. These are also very attractive as structural material in MEMS. We investigated the influences of specimen size and WC-Co composition ratio on mechanical properties of FIB-fabricated WC-Co cemented carbide nanowires by means of on-chip uniaxial tensile testing.

M-109 THIN-FILM MAGNESIUM AS A SACRIFICIAL AND BIODEGRADABLE MATERIAL 656
Y. Liu, J. Park, J.H.-C. Chang, and Y.-C. Tai
California Institute of Technology, USA

Magnesium (Mg) and magnesium alloys have drawn great attention as biodegradable materials. It means that magnesium could be an interesting dual "sacrificial and biodegradable MEMS material". This work then reports the first etching tests of the dual properties of ebeam-deposited thin-film Mg (i.e., 0.3 and 1.0 micron thick). Here we have tested etchants including diluted hydrochloric acid, saline, and culture medium. Data are fitted by "First-and-Second order" model. The initial results do show that thin-film Mg indeed is a promising dual sacrificial and biodegradable material."

T-110 THREE-DIMENSIONAL (3-D) RESHAPING TECHNIQUE IN MEMS DEVICES BY
SOLELY ELECTRICAL CONTROL WITH ULTRAFINE TUNING RESOLUTION 660
Y.H. Yoon, C.H. Han, and J.-B. Yoon
Korea Advanced Institute of Science and Technology (KAIST), SOUTH KOREA

We propose an innovative and simple three-dimensional (3-D) reshaping (plastic deformation) technique in MEMS devices by solely electrical control with ultrafine tuning resolution. While voltage input induces stress on the device, Joule heating is applied to make plastic deformation in the device, where the tuning resolution was demonstrated at a sub-100nm level. The proposed technique is expected to be favorably used in many integrated MEMS devices where reshaping feature is required avoiding any external instruments.

W-111 TUNABLE THZ FILTER BASED ON RANDOM ACCESS
METAMATERIAL WITH LIQUID METAL DROPLETS 664
Q.H. Song[1], W.M. Zhu[2], W. Zhang[2], E.M. Chia[2], M. Ren[2], and A.Q. Liu[1]
[1]Xi'an Jiao Tong University, CHINA and [2]Nanyang Technological University, SINGAPORE

We report a tunable THz filter based on random access metamaterial with liquid metal droplet, which is tuned by controlled electrowetting effects. The random access metamaterial consists of micro droplets formed by lotus effect. In experiment, it measures a near 0.01-THz frequency shift of the dipole resonance spectrum induced by changing of the droplets shape via electrowetting effect. The random access metamaterial is flexible in tuning and easy in fabrication, which has potential application on tunable filters, controllable beam steering and flat lens.

Th-112 WETTING DYNAMICS STUDY OF UNDERWATER SUPERHYDROPHOBIC
SURFACES THROUGH DIRECT MENISCUS VISUALIZATION 668
M. Xu, G. Sun, and C.J. Kim
University of California, Los Angeles, USA

We report the study of underwater wetting transition of superhydrophobic (SHPo) surfaces from dewetted (Cassie) to wetted (Wenzel) state through direct and continuous meniscus visualization. The result confirmed two meniscus states of pinning and wetting, the latter leading to the Wenzel state. Furthermore, the result revealed that the Cassie state can (or cannot) be indefinite if (or unless) the water is saturated with air and the hydrostatic pressure is low enough.

Mechanical Sensors and Systems (MECH)

M-113 3-D HEMISPHERICAL MICRO GLASS-SHELL RESONATOR WITH INTEGRATED
ELECTROSTATIC EXCITATION AND CAPACITIVE DETECTION TRANSDUCERS 672
M.M. Rahman, Y. Xie, C. Mastrangelo, and H. Kim
University of Utah, USA

This paper reports the development and performance of a 3D hemispherical micro glass-shell resonator with integrated electrostatic excitation and capacitive detection transducers. This paper presents the first performance results of the 3D shell resonator with integrated micro fabricated excitation and sensing units that produced the first vibration mode of resonance at 5.843 kHz with a quality factor of 730 at atmosphere with the time decay constant of 39.78ms.

T-114 A CMOS MEMS PIRANI VACUUM GAUGE WITH COMPLEMENTARY
BUMP HEAT SINK AND CAVITY HEATER 676
Y.-C. Sun[1], K.-C. Liang[1,2], C.-L. Cheng[1], and W. Fang[1]
[1]National Tsing Hua University (NTHU), TAIWAN and
[2]Taiwan Semiconductor Manufactury Company Ltd., TAIWAN

We design and manufacturing a new CMOS-MEMS Pirani vacuum gauge with complementary bump heat-sink and cavity heater. By using the bump heat-sink and cavity heater design, the active area of heat-sink and heater can be increased without changing device footprint size. In addition, the cavity in heater reduces the thermal mass for low power operation. The proposed design have larger dynamic range, higher sensitivity and lower power consumption as compare to the typical type.

W-115 A LARGE RANGE MULTI-AXIS CAPACITIVE FORCE/TORQUE SENSOR REALIZED IN A SINGLE SOI WAFER 680
D. Alveringh, R.A. Brookhuis, R.J. Wiegerink, and G.J.M. Krijnen
University of Twente, THE NETHERLANDS

A miniature silicon capacitive force/torque sensor is designed and realized to be used for biomechanical applications. The sensor is capable of measuring 5 degrees of freedom with a force range of 2 N in shear and normal direction and a torque range of 6 Nmm. The fabrication of the sensor requires only two masks, making the sensor cost-effective to fabricate. This is the first 5 degrees of freedom force/torque sensor in this force range made in a single SOI wafer.

Th-116 A NOVEL ELECTRET ROTATIONAL SPEED SENSOR 684
W. Bian, X. Wu, and X. Wang
Tsinghua University, CHINA

In this paper, a novel rotational speed sensor based on electrostatic variation is presented, which is fabricated by typical micro fabrication processes. Compared to the other rotational sensors, the merits of the presented sensor are its simple configuration, small size, and low cost.

M-117 ALN-BASED PIEZOELECTRIC RESONATOR FOR INFRARED SENSING APPLICATION 688
W.C. Ang[1,2], P. Kropelnicki[1], H. Campanella[1], Y. Zhu[1,2], A.B. Randles[1],
H. Cai[1], Y.A. Gu[1], K.C. Leong[3], and C.S. Tan[2]
[1]Agency for Science, Technology and Research (A*STAR), SINGAPORE, [2]Nanyang Technological University, SINGAPORE, and [3]GLOBALFOUNDRIES Singapore Pte Ltd, SINGAPORE

We develop a highly sensitive AlN-based resonant uncooled infrared (IR) detector utilizing the photoresponse and piezoelectric properties of polycrystalline AlN. The design, fabrication, and IR sensing characterization of the device are presented. Different from other reported works, photoresponse mechanism was proposed in this paper instead of thermal effect. Without the need of vacuum, AlN-based IR detector brings the great advantage in device packaging and thus further reduces the manufacturing and operation cost.

T-118 AN ALL OPTICAL SHOCK SENSOR BASED ON BUCKLED DOUBLY-CLAMPED SILICON BEAM 692
B. Dong[1], J.G. Huang[3], H. Cai[2], P. Kropelnicki[2], A.B. Randles[2], Y.D. Gu[2], and A.Q. Liu[1]
[1]Nanyang Technological University, SINGAPORE, [2]Agency for Science, Technology and Research (A*STAR), SINGAPORE, and [3]Xi'an Jiao Tong University, SINGAPORE

An all optical shock sensor is designed, fabricated and experimentally demonstrated. Fabricated with CMOS compatible process, this optical shock sensor can be easily integrated with other photonic devices. The opto-mechanical shock sensor can be potentially used at hash environment like in oil industry, or military usage in a complex electromagnetic environment. It also has potential applications such as inertial sensor, optical switch and other optomechanical devices.

W-119 AN ANGULAR ACCELERATION SENSOR INSPIRED BY THE VESTIBULAR SYSTEM WITH A FULLY CIRCULAR FLUID-CHANNEL AND THERMAL READ-OUT 696
J. Groenesteijn, H. Droogendijk, M.J. de Boer, R.G.P. Sanders, R.J. Wiegerink, and G.J.M. Krijnen
University of Twente, THE NETHERLANDS

We report on an angular accelerometer based on the semicircular channels of the vestibular system. The accelerometer consists of a water-filled circular tube, wherein the fluid flow velocity is measured thermally as a representative for the external angular acceleration. Measurements show a linear response for angular acceleration amplitudes up to 2×10^5 °s^{-2}.

Th-120 AN EFFICIENT EARTH MAGNETIC FIELD MEMS SENSOR: MODELLING AND EXPERIMENTAL RESULTS 700
M. Bagherinia[1], A. Corigliano[1], S. Mariani[1], D.A. Horsley[2], M. Li[2], and E. Lasalandra[3]
[1]Politecnico di Milano, ITALY, [2]University of California, Davis, USA, and [3]STMicroelectronics, ITALY

We present the experimental results and performance indexes of a new z-axis Lorentz force MEMS magnetometer with reduced dimensions and high efficiency and exploit an ad-hoc formulated multi-physics approach and its solutions to model the sensor dynamics. The obtained sensor has a good resolution for earth magnetic field detection and navigation, and is very efficient in terms of exciting current, surface area and bandwidth.

M-121 AN ELECTRET-BIASED RESONANT RADIATION SENSOR 704
S.S. Lee, C.K. Yoon, S.H. Song, and B. Ziaie
Purdue University, USA

In this work, an electret-biased resonant radiation sensor capable of measuring accumulated radiation dosage is presented. The sensor consists of a positive corona-charged Teflon electret placed underneath a ZnO piezoelectric cantilever. As ionizing radiation passes through the ambient air surrounding the electret, ions are generated in the air and drift toward the Teflon substrate. These ions neutralize the electret's surface charges and thus reduce the electrostatic force. The force reductions result in the cantilever's resonant frequency back to its natural frequency.

T-122 AN SOI TACTILE SENSOR WITH A QUAD SEESAW ELECTRODE FOR 3-AXIS COMPLETE DIFFERENTIAL DETECTION 709

Y. Hata[1], Y. Nonomura[1], H. Funabashi[1], T. Akashi[1], M. Fujiyoshi[1], Y. Omura[1], T. Nakayama[2],
U. Yamaguchi[2], H. Yamada[2], S. Tanaka[3], H. Fukushi[3], M. Muroyama[3], M. Makihata[3], and M. Esashi[3]
[1]Toyota Central R&D Labs., Inc., JAPAN, [2]Toyota Motor Corp., JAPAN, and [3]Tohoku University, JAPAN

This paper presents an SOI capacitive tactile sensor with a quad-seesaw electrode for 3-axis differential detection. For differentially detecting 3-axis forces, we propose a novel seesaw-electrode structure composed of four rotating plates individually suspended by torsion beams. We successfully fabricated the test device that integrates an SOI with seesaw electrodes and an LTCC with fixed electrodes. The test results demonstrated that the proposed sensor differentially detects 3-axis forces.

W-123 DIELECTRICAL LIQUID-BASED TACTILE SENSING ARRAY WITH ADJUSTABLE SENSING RANGES AND SENSITIVITY 713

K.W. Liao[1], M.T. Hou[2], H. Fujita[3], and J.A. Yeh[1]
*[1]National Tsing Hua University (NTHU), TAIWAN, [2]National United University, TAIWAN, and
[3]University of Tokyo, JAPAN*

We present a novel tactile sensing array with adjustable sensing ranges. Each sensing element contains a low dielectric constant droplet covered with high liquid. We controlled the contact angle of the droplet by controlling the electric flux passing through the element. Then, the sensing ranges and sensitivity were also adjusted due to the variation of the droplet shape. The results show the sensor's the sensing range is easily adjusted from 0.04N ~ 0.60N to 0.33N ~ 1.05N. The sensitivity increases 1.9 times in at small force range from 1.47pF/N to 2.90.pF/N.

Th-124 ELECTRET-BASED LOW POWER RESONATOR FOR ROBUST PRESSURE SENSOR 717

H. Mitsuya[1], H. Ashizawa[1], T. Sugiyama[2], M. Kumemura[3], M. Ataka[3], H. Fujita[3], and G. Hashiguchi[2]
[1]Saginomiya Seisakusho, Inc., JAPAN, [2]Shizuoka University, JAPAN, and [3]University of Tokyo, JAPAN

We have developed a membrane-less pressure sensor based on squeeze-film damping in a 2-μm driving gap of a silicon ring-shape resonator. Its sensing range is from sub-atmospheric to over 1MPa; very wide-range pressure measurement is possible with one sensor element. An electret film having the 200-V-bias voltage was incorporated to the resonator; this allows the excitation of the resonator at very low AC voltage. This membrane-less pressure sensor has robust and low power consumption (nW-range) characteristics.

M-125 ELECTRIC GRADIENT FORCE DRIVE MECHANISM FOR NOVEL MICRO-SCALE ALL DIELECTRIC GYROSCOPE 721

R. Perahia, J.J. Lake, S.S. Iyer, D.J. Kirby, H.D. Nguyen, T.J. Boden, R.J. Joyce, L.X. Huang,
L.D. Sorenson, and D.T. Chang
HRL Laboratories, USA

This paper reports a novel drive mechanism used to excite a cylindrical, all-dielectric micro-shell gyroscope structure. The drive mechanism operates by generating a gradient electric-field force from a set of interdigitated electrodes placed adjacent to the gyroscope structure. This novel transduction mechanism enables mechanical actuation of a pristine dielectric structure without the need for direct metallization which could otherwise degrade mechanical performance. Design, fabrication, and experimental demonstration are presented.

T-126 ELECTROMECHANICAL DAMPING IN MEMS ACCELEROMETERS: A WAY TOWARDS SINGLE-CHIP GYROMETER-ACCELEROMETER CO INTEGRATION 725

Y. Deimerly[1], P. Rey[1], P. Robert[1], T. Bourouina[2], and G. Jourdan[1]
[1] CEA - LETI – Minatec, FRANCE and [2]Université Paris-Est, FRANCE

This work proposes a method for controlling mechanical damping in MEMS devices. By capacitively coupling a micro mechanical sensor to an electrical resistance, mechanical energy is dissipated by an additional damping sink. In this study, the damping rate of a MEMS accelerometer has been tuned under vacuum, in compliance with a simple electromechanical model that will be further detailed. Using this phenomenon, this presentation will discuss the possibility of co integrating accelerometers with gyrometers on a single chip inside a same cavity to form compact System In Package.

W-127 EXPERIMENTALLY VALIDATED ALUMINUM NITRIDE BASED PRESSURE, TEMPERATURE AND 3-AXIS ACCELERATION SENSORS INTEGRATED ON A SINGLE CHIP 729

F. Goericke[1], K. Mansukhani[1], K. Yamamoto[2], and A. Pisano[3]
*[1]University of California, Berkeley, USA, [2]Murata Manufacturing Co., Ltd, JAPAN, and
[3]University of California, San Diego, USA*

This paper reports a unified fabrication process used to build multiple aluminum nitride (AlN) based micro-electromechanical system (MEMS) sensors on a single chip. A fully functional AlN-based sensor cluster has been demonstrated and is presented in this paper. This sensor cluster is a "five degree-of-freedom" cluster; it measures 3-axis acceleration, temperature and pressure fabricated on a 1 cm x 1 cm die. In addition to utilizing AlN as both the structural and active layer of the sensors, this work is novel because all sensors are fabricated in the same fabrication run.

Th-128 FABRICATION AND CHARACTERIZATION OF ALL HYDROGEL CANTILEVERS FOR ATOMIC FORCE MICROSCOPY APPLICATIONS 733

I. Lee and J. Lee

Sogang University, SOUTH KOREA

We develop a novel method for fast and simple fabrication of hydrogel microcantilevers for atomic force microscopy applications. Fabricated hydrogel microcantilevers exhibit imaging performance comparable to that of commercial silicon microcantilevers in case of non-contact mode operation.

M-129 FULLY PRINTED, LARGE-SCALE, HIGH SENSITIVE STRAIN SENSOR ARRAY FOR STRESS MONITORING OF INFRASTRUCTURES 737

S. Harada, W. Honda, T. Arie, S. Akita, and K. Takei

Osaka Prefecture University, JAPAN

We demonstrated a macroscale sensor sheet by fabricating the fully printed, large-scale, and high sensitive strain sensor array on flexible substrates, which can cover any surfaces, for the application of real-time secure infrastructure maintenance as a proof-of-concept. Printed strain sensor array exhibits that the impressively high gauge factor ~106 and successfully detects the small deformation <20µm distributions.

T-130 HARBOR SEAL INSPIRED MEMS ARTIFICIAL MICRO-WHISKER SENSOR 741

A.G.P. Kottapalli[1], M. Asadnia[1], H. Hans[1], J.M. Miao[1], and M. Triantafyllou[2]

[1]Nanyang Technological University, SINGAPORE and [2]Massachusetts Institute of Technology, USA

Harbor seal whiskers possess a unique geometry along the length of the whisker which is believed to perform vortex induced vibrations (VIV) in frontal flows. The geometry of the whisker appears to be well-tuned to offer maximum allowable sensitivity for sensing by minimizing the self-induced vibrations until an upstream stimulus is encountered. In this work we develop artificial MEMS versions of seal whiskers using stereolithography. These artificial sensors demonstrate a threshold velocity detection limit as low as 193µm/s which rivals the abilities of the Harbor seal's real whisker. Experiments conducted in water tunnel reveal VIV suppression by the whisker structure.

W-131 HIGH FREQUENCY PIEZOELECTRIC MICROMACHINED ULTRASONIC TRANSDUCER ARRAY FOR INTRAVASCULAR ULTRASOUND IMAGING 745

Y. Lu, A. Heidari, S. Shelton, A. Guedes, and D.A. Horsley

University of California, Davis, USA

This paper presents a 1.2 mm diameter high fill-factor array of 1,261 piezoelectric micromachined ultrasonic transducers (PMUTs) operating at 18.6MHz for medical imaging applications. This process incorporates a sacrificial polysilicon release pit that precisely defines the PMUT diameter, thereby enabling 10× smaller device spacing and eliminating the need for through-wafer etching. Measurements show a large voltage response of 2.5nm/V and good frequency matching in air, a high center frequency 18.6MHz and wide bandwidth 4.9MHz when immersed in fluid, and phased array simulations based on measured PMUT parameters show high output pressure of the focused acoustic beam.

Th-132 IMPACT OF GYROSCOPE OPERATION ABOVE THE CRITICAL BIFURCATION THRESHOLD ON SCALE FACTOR AND BIAS INSTABILITY 749

S.H. Nitzan[1], T.-H. Su[1], C.H. Ahn[2], E.J. Ng[2], V.A. Hong[2], Y. Yang[2], T.W. Kenny[2], and D.A. Horsley[1]

[1]University of California, Davis, USA and [2]Stanford University, USA

We investigate the impact of operating a vibratory rate gyro (VRG) at large oscillation amplitude where the VRG's driven becomes nonlinear. Nonlinearities arising at large amplitudes cause the resonator's amplitude-frequency response to become multi-valued above a level known as the critical bifurcation threshold, xc. Open-loop resonators operating at amplitudes above xc are subject to large amplitude instabilities. We demonstrate using closed-loop operation, that scale-factor and bias instability are not affected by operation above xc and angle random walk is reduced.

M-133 IMPROVED ACOUSTIC COUPLING OF AIR-COUPLED MICROMACHINED ULTRASONIC TRANSDUCERS 753

S. Shelton[1], O. Rozen[1], A. Guedes[1], R. Przybyla[2], B. Boser[2], and D. Horsley[1]

[1]University of California, Davis, USA and [2]University of California, Berkeley, USA

A micromachined ultrasonic transducer (MUT) achieves maximum acoustic coupling when its radius approaches the acoustic wavelength. Previously, this fact posed a critical limitation on size for MUTs operating in air. We present a new approach to increase the acoustic coupling and bandwidth of MUTs using a resonant cavity etched beneath the MUT. The result is a 4x increase in sound pressure level for MUTs having radius equal to one-eighth the acoustic wavelength and an 8x improvement in the bandwidth, thereby enabling much smaller transducers.

T-134 MECHANICAL FORCE-DISPLACEMENT TRANSDUCTION STRUCTURE FOR PERFORMANCE ENHANCEMENT OF CMOS-MEMS PRESSURE SENSOR 757

C.-L. Cheng[1], H.-C. Chang[1], C.-I. Chang[1], Y.-T. Tuan[2], and W. Fang[1]

[1]National Tsing Hua University (NTHU), TAIWAN and [2]National Nano Device Laboratories, TAIWAN

This study implements a mechanical force-displacement transduction structure using the TSMC 0.18um 1P6M CMOS process to improve CMOS-MEMS capacitive pressure sensor. The membrane will be deformed by pressure and cause the sensing-gap change between undeformed movable-electrode and fixed-electrode. Feature of this study is CMOS-MEMS deformed membrane and undeformed movable-electrode to enable the parallel-plate gap-closing pressure detection. Thus, the performance of pressure sensor can be improved and stabilized.

W-135 LOW NOISE VACUUM MEMS CLOSED-LOOP ACCELEROMETER USING SIXTH-ORDER MULTI-FEEDBACK LOOPS AND LOCAL RESONATOR SIGMA DELTA MODULATOR 761

F. Chen[1], W.Z. Yuan[1], H.L. Chang[1], I. Zeimpekis[2], and M. Kraft[3]

[1]Northwestern Polytechnical University, CHINA, [2]University of Southampton, UK, and
[3]University of Duisburg-Essen, GERMANY

We report a novel sixth-order sigma-delta modulator MEMS closed-loop accelerometer with extended bandwidth operating in a vacuum environment, which can coexist on a single die with other sensors requiring vacuum packaging. The sensing element was fabricated on a common SOI substrate, four electronic integrators with local resonators are cascaded with the sensing element form high-order noise shaping and notch to suppress the total in-band quantization noise. The feedback voltage signal was applied to the proof-mass to artificially damp the system, which guarantees stable operation in vacuum.

Th-136 MICRO LIQUID-BASED THERMO-ACOUSTIC TRANSMITTER FOR EMITTING ULTRASOUND IN LIQUID MEDIUM 765

D. Hoang-Giang[1], N. Thanh-Vinh[1], K. Noda[1], P. Hoang-Phuong[2], N. Binh-Khiem[1], T. Takahata[1], K. Matsumoto[1], and I. Shimoyama[1]

[1]University of Tokyo, JAPAN and [2]Griffith University, AUSTRALIA

We proposed a thermo-acoustic transmitter using a nanometer thickness metal layer encapsulated with a micrometer thickness liquid layer on thermal-insulator substrate for emitting ultrasound in liquid medium. To improve energy efficiency we take advantage of low specific heat capacity liquid which is encapsulated physically and thermally in small volume by a thin parylene film to fabricate the device. The experiment results demonstrated that by using silicone oil (HIVAC-F5) encapsulated on glass composite, we can obtain ultrasound with sound pressure 3Pa in water medium.

M-137 MULTI-AXIS FORCE SENSOR WITH DYNAMIC RANGE UP TO ULTRASONIC 769

P. Quang-Khang[1], N. Minh-Dung[1], N. Binh-Khiem[1], H.P. Phan[2], K. Matsumoto[1], and I. Shimoyama[1]

[1]University of Tokyo, JAPAN and [2]Griffith University, AUSTRALIA

We proposed a multi-axis force sensor that has a dynamic range up to ultrasonic. The sensor utilizes multilayer structure of elastomer/polymer/viscous liquid to conduct forces and acoustic vibrations to four piezoresistive cantilevers. Experiment results showed that the sensor was capable of measuring normal and lateral forces with high linearity in the range up to 40kPa. Moreover, the dynamic range of the sensor covers ultrasonic frequencies, with the first resonant frequency located at 170kHz.

T-138 MULTIFUNCTIONAL INTEGRATED SENSOR IN A 2X2 MM EPITAXIAL SEALED CHIP OPERATING IN A WIRELESS SENSOR NODE 773

C.L. Roozeboom[1], V.A. Hong[1], C.H. Ahn[1], E.J. Ng[1], Y. Yang[1], B.E. Hill[1], M.A. Hopcroft[2], and B.L. Pruitt[1]

[1]Stanford University, USA and [2]Hewlett-Packard Labs, USA

We present multifunctional integrated sensors that combine temperature, humidity, pressure, air speed, chemical gas, magnetic, and acceleration sensing on a single 2x2 mm chip. We fabricate the multi-sensor in a wafer scale encapsulation process to hermetically seal the sensor functions with moving parts at low vacuum, and then surface micromachine the environmental sensors on top of the sealed layer. We demonstrate the multi-sensor in a wireless sensor node that combines energy harvesting, power management, and low power electronics to transmit data using a cloud-based service.

W-139 OUT-OF-PLANE MICRO TRIPLE-HOT-WIRE ANEMOMETER BASED ON PYREX BUBBLE FOR AIRFLOW SENSING 777

S.W. Liu[1], S.S. Pan[1], F. Xue[1], N. Lin[1], H.B. Liu[1], J.M. Miao[1], L.K. Norford[2], and H.B. Lim[1]

[1]Nanyang Technological University, SINGAPORE and [2]Massachusetts Institute of Technology, USA

We report novel design and fabrication of out-of-pane micro airflow sensors based on the hot-wire sensing principle, i.e. gas cooling of electrically-heated hot-wires. With three micro Cr/Au/Cu hot-wire components fabricated on a Pyrex bubble, the anemometer has demonstrated the ability to detect velocity (<10m/s) and to determine flow direction with an error less than ±8° when the velocity is 10m/s.

Th-140 A PAPER-BASED PIEZOELECTRIC TOUCH PADS INTEGRATING ZINC OXIDE NANOWIRES 781

Y.H. Wang, X. Li, C. Zhao, and X.Y. Liu
McGill University, CANADA

We report a new type of paper-based piezoelectric touch pads integrating zinc-oxide nanowires (ZnO-NWs) as the sensing component. We directly grew ZnO-NWs on cellulose paper using a simple hydrothermal approach, and fabricated single-layer piezoelectric touch pads from ZnO-NW-coated paper. The presented piezoelectric touch pads are inexpensive, easy-to-fabricate, ultra-thin, lightweight and disposable, and will further enrich the tool set of paper electronics.

M-141 PIEZORESISTIVITY OF AG NWS-PDMS NANOCOMPOSITE 785

M. Amjadi, A. Pichitpajongkit, S. Ryu, and I. Park
Korea Advanced Institute of Science and Technology (KAIST), SOUTH KOREA

In this work, we developed a conductive silver nanowire (AgNW)-PDMS composite thin film for a flexible strain sensing application. The piezoresistivity of Ag NWs-PDMS nanocomposite thin film was experimentally investigated and analyzed by a computational model. Finally, a finger motion detection device was developed by using Ag NWs-PDMS nanocomposite thin film as a highly stretchable, flexible and sensitive strain sensor.

T-142 SINGLE-CHIP ATOMIC FORCE MICROSCOPE WITH INTEGRATED Q-ENHANCEMENT AND ISOTHERMAL SCANNING 789

N. Sarkar[1,2] and R.R. Mansour[1,2]
[1]University of Waterloo, CANADA and [2]ICSPI Corp., CANADA

We report on the design, fabrication, and imaging performance of a single-chip Atomic Force Microscope (AFM) that does not require any off-chip scanning or sensing hardware. The first AM-AFM images obtained with such a device reveal that 90nm vertical features (on an AFM calibration standard) can be resolved. The design comprises improved lateral and vertical actuators, an isothermal electrothermal scanner design that maintains constant tip-temperature while traversing a 50um x 10um area, and a Q-enhancement mechanism that improves the force resolution of the instrument.

W-143 SMART-CUT 6H-SILICON CARBIDE (SIC) MICRODISK TORSIONAL RESONATORS WITH SENSITIVE PHOTON RADIATION DETECTION 793

R. Yang[1], K. Ladhane[2], Z. Wang[1], J. Lee[1], D. Young[2], and P.X.-L. Feng[1]
[1]Case Western Reserve University, USA and [2]University of Utah, USA

We report on experimental demonstration of a new type of microdisk torsional resonators based on a smart-cut 6H-silicon carbide (6H-SiC) technology. We carefully calibrate these torsional mode resonances by employing highly sensitive multi-wavelength laser interferometric techniques. To utilize these first 6H-SiC torsional resonators, we further demonstrate sensitive detection of radiations from both blue and infrared (IR) photons. Toward force detection applications which are well suited for torsional resonators, our calibration measurements demonstrate impressive intrinsic force resolutions in these SiC torsional resonators.

Th-144 SUB-0.05° PRECISION OPTOFLUIDIC DUAL-AXIS INCLINOMETER 797

S. Wahl, F. Marty, N. Pavy, B. Mercier, and D.E. Angelescu
Université Paris-Est, FRANCE

We present a low-power bi-axial miniaturized inclinometer based on a mobile mass (spherical ball or fluidic droplet) positioned on a precision curved surface that is generated using a novel MEMS process. The detection of the mobile mass was implemented through an external optical system, using a quadrant photodetector. Nanotopography and chemical treatment of the curved surface have been implemented to increase accuracy when using a fluidic mobile mass, by tailoring wetting properties and minimizing contact angle hysteresis. Fluidic damping was also implemented to render the sensor less sensitive to vibrations.

M-145 TUNING OF NONLINEARITIES AND QUALITY FACTOR IN A MODE-MATCHED GYROSCOPE 801

E. Tatar, T. Mukherjee, and G.K. Fedder
Carnegie Mellon University, USA

This paper examines methods to electrically tune cubic nonlinearity and quality factor (Q) of a mode-matched MEMS gyroscope by changing the DC voltages across specially shaped combs. The gyroscope includes traditional combs for drive-sense and dedicated shaped combs for cubic nonlinearity and frequency tuning. In addition to nonlinearity, Q can be tuned by understanding the nature of the losses with the appropriate model. The electrical loss components are added to the electromechanical resonator model to account for the electrical losses which depend on the applied voltages.

Medical Microsystems (MEDM)

T-146   2D RESONANT MICROSCANNER FOR DUAL AXES CONFOCAL FLUORESCENCE ENDOMICROSCOPE 805
H. Li, Z. Qiu, X. Duan, K. Oldham, K. Kurabayashi, and T.D. Wang
University of Michigan, USA

We demonstrate a parametrically-excited 2D microscanner for a miniature dual axes confocal fluorescence endomicroscope. The scanner has a compact and robust gimbal structure which can perform resonant scanning with large tilting angle at high speed. A single-wafer based SOI process has been developed for improving the quality and the yield of the device. Ex vivo imaging on mouse colon is performed using the fabricated endomicroscope, and the near infrared fluorescence en-face image of dysplasia crypts over a large field-of-view of 800µm×400µm with subcellular resolution is obtained.

W-147   A CAPACITIVE IMMUNOSENSOR USING ON-CHIP ELECTROLYTIC PUMPING AND MAGNETIC WASHING TECHNIQUES FOR POINT-OF-CARE APPLICATIONS 809
J.-C. Kuo, P.-H. Kuo, H.-T. Hsueh, C.-W. Ma, C.-T. Lin, S.-S. Lu, and Y.-J. Yang
National Taiwan University, TAIWAN

This work presents a capacitive immunosensor using on-chip electrolytic pumping and magnetic washing techniques. The proposed device possesses the advantages such as simple operation, low power consumption, and portability. The proposed device was fabricated using typical micromachining process, and is suitable for mass-production. We also demonstrated the detection of N-Terminal pro-brain-Type natriuretic peptide (NT-proBNP) using the fabricated device integrated with a CMOS capacitance sensing chip. The proposed device potentially can be used as a portable system for point-of-care applications.

Th-148   A WIRELESS SLANTED OPTRODE ARRAY WITH INTEGRATED MICRO LEDS FOR OPTOGENETICS 813
K. Kwon[1], H. Lee[2], M. Ghovanloo[2], A. Weber[1], and W. Li[1]
[1]Michigan State University, USA and [2]Georgia Institute of Technology, USA

We develop a wireless-enabled, flexible optrode array with multichannel micro-LEDs for selective optical stimulation of cortical neurons and simultaneous recording of light-evoked neural activity. The array integrates wirelessly addressable micro-LED chips with slanted polymer waveguides for precise light delivery to multiple cortical layers simultaneously. A droplet backside exposure (DBE) method was developed to monolithically fabricate varying-length optrodes on a single polymer platform.

M-149   AN ELECTROPORATION CHIP BASED ON FLEXIBLE MICRONEEDLE ARRAY FOR IN VIVO NUCLEIC ACID DELIVERY 817
Z. Wei[1], R. Wang[2], S. Zheng[3], Z. Liang[3], and Z. Li[3]
[1]National Center for Nanoscience and Technology, CHINA, [2]North University of China, CHINA, and [3]Peking University, CHINA

We reports a flexible microneedle array electroporation chip for in vivo nucleic acid delivery. Silicon MNA is proposed to penetrate the high-resistant stratum corneum, while flexible parylene substrate is used to fit the natural shape of electroporated objects. Using the proposed chip, we successfully achieved plasmid DNA expression and siRNA delivery in living tissue with low voltage (30-40V), neither physical nor biological harm to skin was observed.

T-150   AN INTEGRATED MICROFLUIDIC SYSTEM FOR DIAGNOSIS OF QUINOLONES RESISTANCE OF *HELICOBACTER PYLORI* 821
C.Y. Chao[1], C.H. Wang[1], Y.J. Che[1], C.Y. Kao[2], J.J. Wu[2], and G.B. Lee[1]
[1]National Tsing Hua University (NTHU), TAIWAN and [2]National Cheng Kung University, TAIWAN

Helicobacter pylori play a crucial role in gastric diseases. The incidence rate of duodenal ulcer and gastric ulcer from H. pylori infected patients were found to be about 90-100% and 60-100%.Recently, some point mutations were found in gyrase genes against Quinolones. In this study a new method was therefore developed to perform molecular diagnostic techniques of SNP-PCR on an integrated microfluidic system to detect the Quinolones resistance of H. pylori.

W-151   ANNEALING EFFECTS ON FLEXIBLE MULTI-LAYERED PARYLENE-BASED SENSORS 825
B.J. Kim[1], E.P. Washabaugh IV[2], and E. Meng[1]
[1]University of Southern California, USA and [2]University of Michigan, USA

The mechanical and electrochemical properties and sensing performance of untreated and annealed Parylene-platinum electrochemical impedance-based force sensors were compared. Annealing reduced the height and increased the stiffness of the Parylene structure, and smoothed electrode surfaces, affecting sensor performance. Our results indicate that annealing effects cannot be ignored for Parylene-metal device systems and that mechanical and electrochemical properties and performance must be determined after heat treatment, such as annealing and sterilization.

**Th-152 AUTOMATED VITRIFICATION OF MAMMALIAN
EMBRYOS ON A DIGITAL MICROFLUIDIC DEVICE** 829
D.G. Pyne[1], J. Liu[1], M. Abdelgawad[2], and Y. Sun[1]
[1]University of Toronto, CANADA and [2]Assiut University, EGYPT

We present, for the first time, the development of a digital microfluidic device to achieve automated vitrification of mammalian embryos for clinical in vitro fertilization (IVF) applications. Micro drops are used as vessels to move an embryo and subject it to a series of cyroprotectants of different concentrations, as required by the IVF vitrification protocols.

**M-153 CHARACTERIZATION OF RED BLOOD CELL DEFORMABILITY
CHANGE DURING BLOOD STORAGE** 833
Y. Zheng[1], J. Chen[1], T. Cui[2], N. Shehata[3], C. Wang[3], and Y. Sun[1]
[1]University of Toronto, CANADA, [2]University of Minnesota, USA, and [3]Mount Sinai Hospital, CANADA

Deformability change of stored red blood cells over an 8 weeks' storage period was measured using a microfluidic device and high-speed imaging. Multiple parameters including deformation index (DI), time constant, and RBC circularity were quantified. Compared to previous RBC deformability studies, our results include a significantly higher number of cells (>1,000 cells/sample vs. a few to tens of cells/sample) and, for the first time, reveal deformation changes of stored RBCs when traveling through human-capillary-like microchannels.

**T-154 DETERMINATION OF MULTIDRUG RESISTANCE LEVEL IN K562 LEUKEMIA
CELLS BY 3D-ELECTRODE CONTACTLESS DIELECTROPHORESIS** 837
Y. Demircan, M. Erdem, E. Özgür, U. Gündüz, and H. Külah
Middle East Technical University, TURKEY

We designed, fabricated and tested a MEMS based cell identification 3D-electrode contactless dielectrophoresis system. As an application for this system, the determination of multidrug resistance degree of K562 cells was presented in this study.

**W-155 FLEXIBLE MEA FOR ADULT ZEBRAFISH ECG RECORDING
COVERING BOTH VENTRICLE AND ATRIUM** 841
X. Zhang[1], J. Tai[2], J. Park[1], and Y.C. Tai[1]
[1]California Institute of Technology, USA and [2]Tufts University, USA

We develop a parylene based MEA to monitor adult zebrafish ECG, for the first time, in both ventricle and atrium viewing angles, during its heart regeneration post injury. It is a novel tool to allow the discovery of fine bio-electrical activities in the entire heart.

Th-156 MEASUREMENT OF MECHANOMYOGRAM 845
T. Kaneko, N. Minh-Dung, R. Aoki, T. Takahata, K. Matsumoto, and I. Shimoyama
University of Tokyo, JAPAN

We proposed an approach for measuring mechanomyogram (MMG) by taking advantage of the acoustic impedance matching between liquid and human skin to convey the pressure signal of MMG to a piezo-resistive cantilever.In experiments, the sensor was placed on the skin surface above bicepcs brachii. The MMG signal, the frequency of which was 10-15 Hz, was able to be detected using silicone oil as the propagating medium, while it was not using air as the medium.Experiment results also indicated that the proposed sensor was able to detect the vascular oscillations.

**M-157 MEASURING FLOW VELOCITY OF SWALLOWED LIQUID IN THE HUMAN PHARYNX
BY TONGUE PRESSURE SENSOR AND SWALLOWING SOUND SENSOR** 849
Y. Takei, T. Kaneko, K. Noda, K. Matsumoto, and I. Shimoyama
University of Tokyo, JAPAN

We measured flow velocity of swallowed liquid passing through pharynx. We put pressure sensor on palate and two acoustic sensors on the neck skin. From the output of these three sensors, we can know the timing of the liquid passing through each sensor points and can calculate the flow velocity of the swallowed liquid at the pharynx. In this paper, we compare the flow velocity between two swallowing positions, "sit straight position" and "look upward position." As a result, we found that the flow speed of the "look upward position" was 2.5 times faster than that of "sit straight position."

**T-158 MEMS NEURAL PROBE ARRAY FOR MULTIPLE-SITE OPTICAL STIMULATION WITH
LOW-LOSS OPTICAL WAVEGUIDE BY USING THICK GLASS CLADDING LAYER** 853
Y. Son[1,2], H.J. Lee[1], D. Kim[1], Y.K. Kim[1], E.-S. Yoon[1], J.Y. Kang[1], N. Choi[1], T.G. Kim[2], and I.-J. Cho[1]
*[1]Korea Institute of Korea Institute of Science and Technology (KIST), SOUTH KOREA and
[2]Korea University, SOUTH KOREA*

We present a MEMS neural probe array for multiple-site optical stimulation with low-loss SU-8 optical waveguides. The 20-μm-thick cladding layer was formed by glass reflow process and no additional thickness was required due to embedded structure. Furthermore, the low-loss optical waveguide enables multiple-site stimulation with the two-step optical splitter. We also demonstrate a successful in-vivo optical stimulation and recording of neural signals of a transgenic ChR2-YFP mouse. Recorded neural signals are synchronized with light pulses which confirm that neurons were successfully stimulated and recorded.

W-159 MICRO-ELECTRODE ARRAYS FOR MULTI-CHANNEL MOTOR UNIT EMG RECORDING 857

S. Yamagiwa, H. Sawahata, M. Ishida, and T. Kawano
Toyohashi University of Technology, JAPAN

We report an array of micro-electrodes, which can record motor unit (MU) electromyogram (EMG) signals. As a basic structure of the electrode, we prepared 200-μm-square Si-pyramids with the height of 200 μm by Tetramethylammonium hydroxide (TMAH), resulting in robust MU-EMG recordings without conductive gel. Platinum (Pt) was used as an electrode material and parylene-C was deposited as an insulator. Fabricated μEMG electrodes connected to a recording system clearly detected MU-EMG action potentials from a human forearm. In addition, different MU-EMG signals between μEMG electrodes were detected by crooking fingers. These results indicate that the μEMG array device becomes a powerful tool for medical applications including myoelectric prosthetic technologies.

Th-160 MICRO-WING AND PORE DESIGN IN AN IMPLANTABLE FPC-BASED NEURAL STIMULATION PROBE FOR MINIMALLY INVASIVE SURGERY 861

Y.-H. Wang[1], D. Tsai[1,2], B.-A. Chen[1], Y.-Y. Chen[1], C.-C. Huang[1], P.-C. Huang[1],
C.-Y. Lin[1], J. Yu[1], W.-P. Shih[1], C.-W. Lin[1], and H.-J. Sheen[1]
[1]National Taiwan University, TAIWAN and [2]University of California, San Diego, CA

A bipolar porous probe for an implantable nerve stimulation treatment utilizing minimally invasive surgery is presented. The flexible printed circuit (FPC) probe features micro-wings, which can increase fixation after implantation, and contains porous structures for cell growth to promote permanence in the body. Two recording pairs detect whether or not cells grow into the pores, and one pair of stimulating pads stimulates the target nerve. This probe is composed of two SU-8 layers and one FPC layer, to form a 3-D porous structure.

M-161 NANOELECTROPORATION AND CONTROLLABLE INTRACELLULAR DELIVERY INTO LOCALIZED SINGLE CELL WITH HIGH TRANSFECTION AND CELL VIABILITY 865

T.S. Santra, J. Borana, P.-C. Wang and F.-G. Tseng
National Tsing Hua University (NTHU), TAIWAN

Here we demonstrate controllable nano-electroporation platform for HeLa and human Caucasian Gastric Adenocarcinoma (AGS) cell to achieve high efficient bimolecular delivery with high cell viability.

T-162 OPTO-MECHANICAL MICROBRIDLES FOR THE DETERMINATION OF STRUCTURAL AND FUNCTIONAL PROPERTIES OF SMALL RESISTANCE ARTERIES 869

R. Rodríguez-Rodríguez[1], J.A. Plaza[2], V. Matchkov[3], U. Simonsen[3], M.D. Herrera[1],
S. Büttgenbach[4], A. Llobera[2], and X. Munoz-Berbel[2]
*[1]University of Seville, SPAIN, [2]Centro Nacional de Microelectrónica (CNM), SPAIN,
[3]University of Aarhus, DENMARK, and [4]Technische Universität Braunschweig, GERMANY*

We develop and optimize an opto-mechanical system for monitoring diameter of arterial segments in vitro.

W-163 PDMS MICROCHANNEL SCAFFOLDS FOR NEURAL INTERFACES WITH THE PERIPHERAL NERVOUS SYSTEM 873

Y. Choi[1], S. Park[2], Y. Chung[3], R.K. Gore[4], A.W. English[4], and R.V. Bellamkonda[2]
*[1]University of Texas – Pan American, USA, [2]Georgia Institute of Technology, USA,
[3]University of Rhode Island, USA, and [4]Emory University, USA*

Neural interfaces with the peripheral nervous system have been developed to provide a direct communication pathway between peripheral nerves and prosthetic limbs. This paper reports a regenerated peripheral nervous system which can control the reinnervated muscles and interpret neurological signals. The acquired bioelectrical signals can be used for the interpretation of mind which will be used to monitor prosthetic limbs. Transected nerves were regenerated through PDMS scaffolds and transferred signals through embedded microwires and acquisition systems.

Th-164 SELECTIVE RF HEATING OF RESONANT STENT TOWARD WIRELESS ENDOHYPERTHERMIA FOR RESTENOSIS INHIBITION 877

Y. Luo, M. Dahmardeh, X. Chen, and K. Takahata
University of British Columbia, CANADA

Stents have served as a critical device for minimally invasive treatment of cardiovascular disease, the leading cause of death in North America. Artery renarrowing (known as restenosis) often occurs after stent implantation due to excess growth of vessel tissue, blood clot(thrombus) formation, and/or other factors. This paper presents, for the first time, a novel active stent that serves as a resonant heater with high frequency selectivity controlled using external RF fields, offering a new therapeutic path to wireless endohyperthermia for in-stent restenosis.

M-165 **TUNABLE MEMS FIBER SCANNER FOR CONFOCAL MICROSCOPY** 881
N. Weber, T. Meinert, H. Zappe, and A. Seifert
University of Freiburg – IMTEK, GERMANY

We present an endoscopic probe with forward-looking piezoelectric fiber scanner for confocal optical imaging with a very short length of 13.1 mm. The system is based on Si bench technology with integrated fluidics for realizing tunable liquid-filled membrane-lenses. The tunability in focal length allows confocal depth scanning up to 100 μm without any movable optics or stages. The lateral and axial resolution were demonstrated to be 2 μm and 20 μm, respectively.

T-166 **ULTRASOUND-ASSISTED MICRO-KNIFE FOR CELLULAR SCALE SURGERY** 885
H. Jeong[1], T. Li[2], Y. Gianchandani[2], and J. Park[1]
[1]*Pohang University of Science and Technology (POSTECH), SOUTH KOREA and*
[2]*University of Michigan, USA*

We developed a microknife for cellular scalce surgery.The work includes modeling, fabrication, measurement and actual cell-cutting. The result showed that developed knife can cut cell monolayer successfully with 2 micro cut line width by utilizing the ultra-sharp edge and ultrasound assiat.

W-167 **WRINKLE CELLOMICS: SCREENING BLADDER CANCER**
CELLS USING AN ULTRA-THIN SILICONE MEMBRANE 889
J.H. Appel[1], L.Y. Sin[2], J.C. Liao[2], and J. Chae[1]
[1]*Arizona State University, USA and* [2]*Stanford University, USA*

We report a visualization platform comprised of an ultra-thin silicone membrane to differentiate between the biophysical properties of cancerous and healthy cells. Cancerous cells adhere to and spread on the membrane inducing deformation, termed 'membrane wrinkling', while healthy cells do not generate wrinkle patterns on the membrane. Quantitative measurement of wrinkling represents a powerful, non-invasive diagnostic tool for common cancers such as bladder cancer.

Th-168 **AN X-RAY DETECTABLE PRESSURE MICROSENSOR FOR**
MONITORING CORONARY IN-STENT RESTENOSIS 893
M.N. Gulari, M. Ghannad-Rezaie, P. Novelli, N. Chronis, and T.C. Marentis
University of Michigan, USA

We present a novel implantable X-ray-addressable MEMS Blood Pressure sensor, the X-BP, for the non-invasive and cost-effective surveillance of coronary in-stent restenosis. We successfully fabricated and tested the X-BP sensor and its pressure response curve. We placed the X-BP sensor in a coronary stent and prove adequate visibility in a clinically realistic scenario.

Micro-Actuators (ACT)

M-169 **A NOVEL ACTUATOR FOR ENERGY HARVESTING USING**
AN ACOUSTICALLY OSCILLATING LIQUID DROPLET 897
Y.R. Lee, J.H. Shin, I.S. Park, and S.K. Chung
Myongji University, SOUTH KOREA

This paper presents a novel actuator for energy harvesting from ambient acoustic noise using acoustically oscillating droplets. When a water droplet sitting on a piezocantilever is excited by an acoustic wave around its natural frequency, it oscillates and simultaneously bends the piezocantilever by the reaction of the droplet oscillation, resulting in electric power generation from the piezocantilever. The envisioned energy harvesting system can extract mechanical power from acoustic noise in a wide range of frequencies using liquid droplets in different sizes and natural frequencies and convert the mechanical power to electrical power for wireless electronic devices. This new type of actuation technique is a simple but useful tool not only for the energy harvesting system but also potential acoustic wave sensors and actuators.

T-170 **A NOVEL ON-CHIP MICROMANIPULATION METHOD USING A MICROBUBBLE**
FOR SINGLE CELL MANIPULATION AND CHARACTERIZATION 901
J.H. Shin, Y.R. Lee, I.S. Park, and S.K. Chung
Myongji University, SOUTH KOREA

This paper presents a novel on-chip micromanipulation method using a microbubble actuated by optical and acoustical excitation for single cell manipulation and characterization in a microfluidic chip, along with the experimental verification of bubble manipulation (generating and transporting operations) and micro-object manipulation (capturing, carrying, and releasing operations).

W-171 A THERMOTROPIC LIQUID CRYSTAL ELASTOMER MICRO-ACTUATOR WITH INTEGRATED DEFORMABLE MICRO-HEATER 905

S. Petsch[1], R. Rix[2], P. Reith[1], B. Khatri[1], S. Schuhladen[1], D. Ruh[1], R. Zentel[2], and H. Zappe[1]

[1]University of Freiburg – IMTEK, GERMANY and [2]University of Mainz, GERMANY

We present a large-stroke thermal actuator with an integrated, MEMS fabricated, deformable heater based on the phase transition in a thermotropic liquid crystal elastomer (LCE) material. The transition from nematic to isotropic phase in the LCE causes a contraction of 28% (1.15mm) when the integrated system is heated to 120°C. With the heater buried in the LCE, full contraction is reached after 19.7s at 320mW when heated from room temperature. Complete back actuation is achieved in 5.6s.

Th-172 A TUNABLE LIQUID LENS DRIVEN BY A CONCENTRIC ANNULAR ELECTROACTIVE ACTUATOR 909

K. Wei, N. Domicone, and Y. Zhao

Ohio State University, USA

We present a membrane-enveloped fluidic lens hydrostatically coupled to a concentric annular electroactive elastomer. Electrical activation deforms the annular elastomer, which induces fluid transmission between the lens part and the actuation part. The lens changes the shape and thereby the focal length from 12.5 mm to 105.2 mm within 1.0 kV. Compared to existing fluidic lenses driven by electroactive polymer, this lens implements a larger focusing range at a lower voltage. It finds applications in miniaturized optical components where adaptive focalization is at a premium.

M-173 LONG STROKE OUT-OF-PLANE ACTUATOR USING COMBINATION OF ELECTROSTATIC AND PNEUMATIC FORCES 913

T.K. Kan, A.I. Isozaki, H.T. Takahashi, K.M. Matsumoto, and I.S. Shimoyama

University of Tokyo, JAPAN

We propose an out-of-plane MEMS actuator with a large stroke length comparable to the size of the actuator itself. The proposed device is actuated by the combined forces of the electrostatic and the pneumatic forces. This combination of two independent forces enlarges a stable area during the actuation, resulting in a large stroke which is difficult to be achieved with only either force. The 3D profiling by the laser scanning microscopy confirmed the largest stroke of 103 μm was obtained with the 150-μm-diameter actuator area with the electric field of 5.8×10^5 V/m (330V between the electrodes) and the air pressure of 2.0 kPa.

T-174 MECHANO-ACTIVE TISSUE SCAFFOLD SYSTEM BASED ON A MAGNETIC NANOPARTICLE EMBEDDED NANOFIBROUS MEMBRANE 917

S.P. Fang[1], H. Shang[1], P.F. Jao[1], K.T. Kim[1], G.J. Kim[1], J.H. Yoon[1], K. Cho[1,2], A.J. Katz,[1] and Y.K. Yoon[1]

University of Florida, USA and Korea Basic Science Institute, SOUTH KOREA

A mechano-active nanofibrous scaffold system consisting of iron oxide nanoparticle embedded electrospun nanofibers, a membrane holder and an electro magnet, is designed and demonstrated. The scaffold provides mechanical stress on culturing cells by external AC magnetic fields. The mechanical properties of the nanoporous membrane including the density, the porosity, and the effective Young's modulus are characterized. Cell viability with and without magnetic nanoparticle embedded has been tested.

W-175 NANO-SCALE BIOMECHANICAL ANALYZER FOR STUDYING STIMULUS DEPENDENT SELF-ASSEMBLY OF ACTIN FILAMENT 921

N. Shimada[1], M. Ikeuchi[1,2], and K. Ikuta[1]

[1]University of Tokyo, JAPAN and [2]Japan Science and Technology Agency, JAPAN

We have developed nano-scale mechanical analysis system by using "optically driven nano-beam" to measure elasticity of self-assembled actin filament under dynamic mechanical stimulus. In this report, we worked on developing a new nano-beam to specifically capture actin on its surface. By using the new nano-beam, we have successfully measured elasticity of self-assembled actin filament in water. The nano-mechanical analysis system unravels cell life phenomenon which can't be dealt with through conventional methodologies."

Th-176 PINCHING AND RELEASING OF CELLULAR AGGREGATE BY MICROFINGERS USING PDMS PNEUMATIC BALLOON ACTUATORS 925

S. Shimomura[1], Y. Teramachi[1], Y. Muramatsu[1], S. Tajima[2], Y. Tabata[2], and S. Konishi[1]

[1]Ritsumeikan University, JAPAN and [2]Kyoto University, JAPAN

This paper proposes microfingers for manipulation of spherical cellular aggregates (φ200μm). The microfinger is driven by pneumatic balloon actuator "PBA" to pinch and release a spherical cellular aggregate directly. The paper presents the design, operation principle, fabrication, and characterization. The pinching force of developed fingers was estimated with the aim of evaluating the damage to the cellular aggregate. A series of operation of a real cellular aggregate by developed microfingers will be successfully demonstrated.

M-177 PNEUMATICALLY ACTUATED BIOMIMETIC PARTICLE TRANSPORTER 927
A. Rockenbach[1], C. Brücker[2], and U. Schnakenberg[1]
[1]RWTH Aachen University, GERMANY and [2]Technical University Bergakademie Freiberg, GERMANY

To prevent the adhesion of particles at surfaces by transporting them along the surface this paper reports on a pneumatically actuated new type of biomimetic particle transporter. Rows of flaps are positioned asymmetrically on movable membranes. Each flap row can be deflected separately by an induced pneumatic force. This membrane movement converts to a large deflection of the flaps in x-direction (lateral). Due to the high aspect ratio of the flaps the angle rotation results in a fluid movement parallel to the surface which prevents the particle deposition.

**T-178 QUANTITATIVE ANALYSIS OF SURFACE TEXTURES CREATED BY
MEMS TACTILE DISPLAY USING MICROFABRICATED TACTILE SAMPLES** 931
Y. Kosemura[1], S. Hasegawa[1], H. Ishikawa[1], J. Watanabe[1], and N. Miki[1,2]
[1]Keio University, JAPAN and [2]Japan Science and Technology Agency (JST), JAPAN

This paper discusses characterization of the surface textures created by MEMS tactile display with a large displacement MEMS actuator array. The actuator consists of piezoelectric actuators and a hydraulic displacement amplification mechanism to achieve large enough displacement to stimulate human tactile receptors. In our prior work, we successfully displayed smooth and rough surfaces using MEMS tactile display by controlling the vibration frequency and the driving voltage of the actuators. In this paper, we propose "sample comparison method" to further characterize the virtually created surface textures, where microfabricated tactile samples are used. In this method, by requesting the subjects to select samples that they felt most similar to the displayed surfaces, the control parameters of the MEMS tactile display were successfully correlated with the surface properties of the samples.

Micro-Fluidic Components and Systems (μFLUIDIC)

**W-179 A NOVEL CONSTANT FLOW REGULATION PRINCIPLE
FOR COMPACT BREATH DIAGNOSTICS** 935
S.B. Johansson, G. Stemme, and N. Roxhed
KTH Royal Institute of Technology, SWEDEN

Our work reports on a passive compact flow regulator designed to maintain a steady flow during breath diagnostics using a novel flow regulation principle. The fabricated prototype consists of two 3D-printed plastic parts with an integrated cantilever aligned in the direction of the flow, to control comparatively large air flows in the 50 ml/s regime suitable for asthma diagnostics.

**Th-180 APPARENT SIZE CORRELATION: A SIMPLE METHOD TO DETERMINE
VERTICAL POSITIONS OF PARTICLES USING CONVENTIONAL MICROSCOPY** 939
M.H. Winer, A. Ahmadi, and K.C. Cheung
University of British Columbia, CANADA

We have developed and implemented a simple three-dimensional (3D) particle tracking method for use in particle focusing applications. Using conventional fluorescence microscopy and a multi-step image post-processing algorithm based on particle defocusing principles, this technique was experimentally verified with results comparable to theoretical predictions of (1) gravitational settling and (2) inertial focusing. Our technique determines particle positions to micron accuracy in microfluidic systems for Re < 100.

**M-181 CELL-NICHE-ON-CHIP: PAIRED SINGLE CELL CO-CULTURE PLATFORMS
USING IMMISCIBLE LIQUID ISOLATION AND SEMI-PERMEABLE MEMBRANES** 943
Y.-C. Chen, Y. Cheng, and E. Yoon
University of Michigan, USA

The fundamental difficulty in single cell co-culture is to provide a controlled microenvironment. The cell culture chamber must be isolated for secreted cytokines to be accumulated inside the chamber over time. However, in an isolated environment nutrition factors will deplete. It is important to find a way to continuously supply nutrition factors while isolating cells sectional view of the proposed co-culture chip and its operation. We placed a semi-permeable membrane between the cell culture chamber and the media exchange channel. Nutrition can be supplied to the cells through the membrane, but the secreted cytokines are accumulated inside the chamber because their molecule sizes are too large to escape. The preliminary result demonstrated the capability of studying interaction between two cells and its potential to investigate modeling of more complicated cell niches.

**T-182 DEVELOPMENT OF VACUUM ASSISTED MICROFLUIDIC CELL TRAPPING DEVICE
FOR REPOSITIONING OF OOCYTE INTRACELLULAR CHROMOSOMES** 947
J. Hong, P. Purwar, S. Lee, N. Verma, and J. Lee
Seoul National University, SOUTH KOREA

We report the design and fabrication of vacuum assisted microfluidic trapping device for the capture of single cell such as an oocyte. We also suggest an application of such device for an intracellular monitoring of a cell that has an interaction with external environments. Real time monitoring is enabled through the fabrication on a silicon-on-glass substrate, offering excellent optical imaging window. We demonstrate the single cell capture event and monitoring of chromosome activity. This result will provide a powerful tool for investigating the physiological and pathological cellular functions.

W-183 DIELECTROPHORETIC (DEP) SEPARATION OF LIVE/DEAD CELLS ON A GLASS SLIDE FUNCTIONALIZED WITH INTERDIGITATED 3D SILICON RING MICROELECTRODES 951

X. Xing and L. Yobas
Hong Kong University of Science and Technology, HONG KONG

An elegant device with 3D interdigitated silicon ring electrodes is developed here for DEP-activated cell sorting. The integration of transparent glass substrate makes the device cost-effective and aids the coupling of DIC microscopy. The self-aligned lateral rings form multiple flow lines thus enhancing the throughput of the whole device. A capture efficiency of live mammalian cancer cells approaching 100% is achieved in separating them from a dead group with high flow rate.

Th-184 DROPLET DISPENSING AND SPLITTING BY ELECTROWETTING ON DIELECTRIC DIGITAL MICROFLUIDICS 955

N.Y.J.B. Nikapitiya[1], S.M. You[2], and H. Moon[1]
[1]University of Texas at Arlington, USA and [2]University of Texas at Dallas, USA

This paper reports an experimental study of two essential capabilities of electrowetting-on-dielectric (EWOD) digital microfluidics (DMF) − 1) high precision and consistency in volume of unit nanodrop dispensed from a reservoir, and 2) reduction of time to dispense and split drops. These capabilities are sought in applications that need tiny but accurate volume of liquid delivery at high flow rate.

M-185 ENZYME-DOPED POLYESTER THREAD COATED WITH PVC MEMBRANE FOR ON-SITE UREA AND GLUCOSE DETECTION ON A THREAD-BASED MICROFLUIDIC SYSTEM 959

Y.-A. Yang[1], W.-C. Kuo[2], and C.-H. Lin[1]
[1]National Sun Yat-sen University, TAIWAN and
[2]National Kaohsiung First University of Science and Technology, TAIWAN

This study presents a novel enzyme-doped thread with PVC membrane coating for on-site urea and glucose detection on a thread-based microfluidic device. The enzyme can be directly applied on the thread without delicate pretreatment or surface modification process. The passing biomolecules are digested by the enzymes and then electrochemically detected downstream. With this approach, CE-EC detection with on-site bio-reaction can be simply achieved. A thin layer of PVC membrane is coated on the enzyme-doped thread to further fix the applied enzyme and to prevent from the rapid evaporation of the running buffer due to the Joule heating effect. In addition, the PVC coated thread can be operated at a higher separation electric field of 500 V/cm due to the reducing buffer evaporation. Successfully on-site enzyme digestion, CE separation and EC detection of urea and glucose samples in a single test run is demonstrated with the enzyme-doped microfluidic system. Results also indicate that the developed system exhibits good linear dynamic range for detecting urea and glucose sample in concentrations from 0.1 mM – 10.0 mM (R2=0.9850) and 0.1 mM – 13.0 mM (R2=0.9668), which is suitable for adoption in detecting the BUN concentration in serum (1.78~7.12 mM) and the standard glucose fasting measuring range (3.89~6.11 mM).

T-186 FABRICATION OF HIGH ASPECT RATIO INSULATING NOZZLE ARRAY USING GLASS REFLOW PROCESS AND ITS ELECTROHYDRODYNAMIC PRINTING CHARACTERISTICS 963

K.I. Lee[1], B. Lim[1], S.W. Oh[1], S.H. Kim[1], C.S. Lee[1], J.W. Cho[1], and Y. Hong[2]
[1]Korea Electronics Technology Institute, SOUTH KOREA and
[2]Seoul National University, SOUTH KOREA

We develop micromachining process including glass filling to fabricate high aspect ratio glass nozzles which is more appropriate for electrohydrodynamic inkjet printing head. With this nozzle array, we print very narrow lines of various materials which is not obtained by conventional piezoelectric or thermal inkjet printing systems.

W-187 GALLIUM-BASED LIQUID METAL INKJET PRINTING 967

D. Kim[1], J.H. Yoo[1], Y. Lee[1], W. Choi[1], K. Yoo[2], and J.B. Lee[1]
[1]University of Texas at Dallas, USA and [2]Hanbat National University, SOUTH KOREA

We report clog-free and oxide-free metal inkjet printing using gallium-based liquid metal. Unlike typical metal nanoparticles or metal alloys, gallium-based liquid metal alloys are in liquid-phase at room temperature. Therefore, there is no need for heating or dispersing in solvent for inkjet printing. Another distinctive benefit is it maintains liquid-phase after printing if the substrate stays at around room temperatures. This is extremely useful to create 3D freeform rapid prototyping of metallic patterns that can conform to virtually any dynamic deformation of substrates.

Th-188 IN-PLANE CAPACITIVE MEMS FLOW SENSOR FOR LOW-COST METERING OF FLOW VELOCITY IN NATURAL GAS PIPELINES 971

S.D. Nguyen[1], I. Paprotny[2], P.K. Wright[1], and R.M. White[1]
[1]University of California, Berkeley, USA and [2]University of Illinois, Chicago, USA

This paper presents the design, fabrication, and experimental results of an in-plane capacitive MEMS flow sensor that uses the displacement of a micro-fabricated paddle caused by dynamic pressure for measuring the velocity of the flow of surrounding gas. Simplicity of fabrication, combined with insensitivity to variations in ambient temperature makes this sensor ideal for widespread deployment in natural gas pipelines.

M-189 INTEGRATED MULTI-PARAMETER FLOW MEASUREMENT SYSTEM 975

J.C. Lotters[1], E. van der Wouden[1], J. Groenesteijn[2], W. Sparreboom[1], T.S. Lammerink[2], and R.J. Wiegerink[2]
[1]*Bronkhorst High-Tech BV, THE NETHERLANDS and*
[2]*MESA+ - University of Twente, THE NETHERLANDS*

We have designed and realised an integrated multi-parameter flow measurement system, consisting of an integrated Coriolis and thermal flow sensor and a pressure sensor. The integrated system enables on-chip measurement, analysis and determination of flow and several physical properties of both gases and liquids. With the system, we demonstrated the feasibility to measure the flow rate, density, viscosity and heat capacity of hydrogen, helium, nitrogen, air, argon and water.

T-190 INTERACTION FORCES DURING THE SLIDING OF A
WATER DROPLET ON A TEXTURED SURFACE 979

N. Thanh-Vinh, H. Takahashi, K. Matsumoto, and I. Shimoyama
University of Tokyo, JAPAN

Using a MEMS 2-axis force sensor array, we have directly measured the pressure and shear force during the sliding of a water droplet on a Su-10 micropillar array. The measurement results showed a fluctuation in the interaction forces when the micropillar was close to the trailing edge or leading edge of the droplet. Meanwhile, in the inner region of the contact line, both the normal and lateral interaction forces were relatively stable. These results indicate that the interaction forces at the edges of the droplet are important factors controlling the sliding motion of the droplet.

W-191 LIQUID DROPLET MICRO-BEARINGS ON
DIRECTIONAL CIRCULAR SURFACE RATCHETS 983

C. Varel and K.F. Bohringer
University of Washington, USA

This paper presents de-ionized water droplets used as torque-generating micro-bearings between a glass plate and a micromachined Si substrate. The pattern on the micromachined Si substrate includes circular tracks, which allow droplet motion in a single direction. When vertical vibration is applied to the system, a rotation in the transverse plane is triggered. The system can be tailored to respond to a specific vibration frequency, from 36.5 to 83 Hz by droplet volumes from 13 to 1 µL.

Th-192 LOW GAS PERMEABLE AND NON-ABSORBENT RUBBERY
OSTE+ FOR PNEUMATIC MICROVALVES 987

J. Hansson, J.M. Karlsson, C.F. Carlborg, W. van der Wijngaart, and T. Haraldsson
KTH Royal Institute of Technology, SWEDEN

We present an elastomeric, low gas permeable off-stoichiometric thiol–ene-epoxy (OSTE+) polymer fully compatible with standard micro-molding manufacturing and demonstrate its use in pneumatic pinch microvalves for lab-on-chip. The polymer is shown to have rubbery properties (similar to PDMS), low permeability to gases, low absorption of molecules from liquid samples, and the ability to bond layers in room temperature without the need for adhesives or plasma treatment.

M-193 FREE SURFACE PROPULSION BY
ELECTROWETTING-ASSISTED 'CHEERIOS EFFECT' 991

J. Yuan and S.K. Cho
University of Pittsburgh, USA

We combine electrowetting principle with Cheerios effect in order to manipulate floating objects in centimeter and millimeter scales. By turning electrowetting on/off, we attract or repel floating objects. Using an array of electrowetting electrodes, we generates translationally and rotationally continuous motions on floating objects.

T-194 MESOPOROUS-SILICA NANO-CHANNELS INTEGRATED IN MICRO-FLUIDIC
CHIP FOR FAST LIQUID MICRO-EXTRACTION OF PESTICIDE RESIDUAL 995

P.C. Xu, C.Z. Chen, H. Yu, and X.X. Li
Shanghai Institute of Microsystem and Information Technology (SIMIT), CHINA

The paper reports a micro-chip with nano-channels integrated as extraction-reservoir for quickly extracting analyt from aqueous solution to water-soluble organic solvent. Using this novel technology, trace-level residual of organophosphorus pesticide in water-solution can be micro-extracted to a common organic-solvent (e.g. ethanol) and, thereafter, quantitatively detected by GC-MS (Gas Chromatography-Mass Spectrometry) analysis.

W-195 MINIATURISED PRANDTL TUBE WITH INTEGRATED PRESSURE SENSORS FOR MICRO-THRUSTER PLUME CHARACTERISATION 999
M. Dijkstram, K. Ma, M.J. de Boer, J. Groenesteijn, J.C. Lötters, and R.J. Wiegerink
MESA+ - University of Twente, THE NETHERLANDS

Micro chemical propulsion systems (μCPS) have been identified by ESA as emerging compact propulsion system. Within the PRECISE project a MEMS-based monopropellant propulsion system applying catalytic decomposition of hydrazine is being developed. Investigation of the micro-thruster rarefied plume flow as well as direct simulation Monte Carlo (DSMC) validation is of great importance for nozzle design and performance evaluation. A novel 6 mm long 40 μm diameter micro Pitot tube with integrated pressure sensors for characterisation of rarefied plume flow during hot-firing test has therefor been developed.

Th-196 MICROFLUIDIC-BASED DROPLET MERGING DEVICE WITH A NON-CONTACT DROPLET PAIRING METHOD 1003
S. Lee, H. Kim, and J. Kim
Pohang University of Science and Technology (POSTECH), SOUTH KOREA

We developed a novel droplet merging method based on the deformability characteristic of a droplet in pressure-driven shear flow using only fluid flow control by a unique Laplace trap that performs a multi-step 'trapping–releasing–non-contact pairing–washing–and–merging' process. Using the unique Laplace trap array, parallel merging was successfully performed within a short time variation in non-contact pairing (SD ±4.3 s) compared with conventional contact pairing (SD ±136.4 s).

M-197 MINIATURE CIRCULATORY COLUMN SYSTEM FOR GAS CHROMATOGRAPHY 1007
H.-C. Hsieh and H. Kim
University of Utah, USA

We develop the first micro-scale circulatory column system for functioning gas chromatography and the resultant highest separation capacity demonstrated by any commercial and non-commercial GC column systems beyond the current state-of-art, by enabling the extension of the effective column length through the circulatory loop without increasing the device volume.

T-198 MIRRORED ANODIZED DIELECTRICS FOR RELIABLE ELECTROWETTING 1011
S. Chen and C. Kim
University of California, Los Angeles, USA

Anodized metal oxides are an attractive dielectric material for electrowetting-on-dielectric (EWOD) devices because of their ability to limit current leakage, high dielectric constants and low cost fabrication. However, the reliability is for only one actuation polarity because of their rectifying effect. To overcome this limitation, we developed parallel-plate EWOD devices using anodized aluminum on both plates so that one is always under the correct bias to limit the leakage current. Lifetime and current leakage were tested across a range of actuation biases.

W-199 NANOPARTICLES SORTING AND ASSEMBLY BASED ON DOUBLE-AXICON IN AN OPTOFLUIDIC CHIP 1015
Y.Z. Shi[1,2], S. Xiong[2], L.K. Chin[2], M. Ren[2], and A.Q. Liu[1]
[1]Xi'an Jiao Tong University, CHINA and [2]Nanyang Technological University, SINGAPORE

We present a novel optofluidic system of sorting and assembly of nanoparticles by tunable interference patterns generated from injecting a Gaussian beam through a double-axicon. The tightly confined (several micrometers) Bessel beam is used to sort the 100-nm gold, 200-nm and 500-nm polystyrene nanoparticles massively and simultaneously by controlling the flow rate and the laser power (from 300 mW to 500 mW). In addition, the 500-nm polystyrene particles are assembled into a 2D array by the discrete interference pattern.

Th-200 NANOSLIT MEMBRANE INTEGRATED FLUIDIC CHIP FOR MICRO/NANO PARTICLE TRAPPING AND SEPARATION 1019
Y. Koh[1], H.M. Kang[1], J.H. Kim[2], Y.S. Lee[1], and Y.K. Kim[1]
[1]Seoul National University, SOUTH KOREA and [2]Hanyang University, SOUTH KOREA

We propose a nanoslit fluidic chip that has a large number of nanoslit array membrane (Nanoslit-Chip) for trapping and concentrating particles of a desired size. The proposed Nanoslit-Chip has several benefits such as low flow resistance and little non-specific nanoparticle clogging for the separation of nanoparticles.

M-201 ON-CHIP CONTROL OF PNEUMATIC-BASED BISTABLE VALVE SWITCH 1023
A. Chen and T. Pan
University of California, Davis, USA

We present pneumatic-based, bistable valve (BSV) switches for immediate on-chip fluid-flow manipulation without the requirement of external microcontroller circuitries. The applicability of the on-chip controller is demonstrated in a 4-to-1 microfluidic multiplexor and its clinical relevance is further supported in a point-of-care ABO blood-typing diagnostic chip.

T-202 RAPID MICROFLUIDIC PROTOTYPING OF SOPHISTICATED PROTEIN ANALYSIS PLATFORMS USING GRAYSCALE PHOTOPATTERNING 1027

T.A. Duncombe[1], K. Maurer[2], and A.E. Herr[1]

[1]University of California, Berkeley/University of California, San Francisco Joint Graduate Group in Bioengineering, USA and [2]ETH Zurich, USA

We introduce, characterize, and demonstrate a novel grayscale fabrication technique for rapid prototyping of complex spatially varied hydrogels as lab-on-a-chip devices optimized to address important protein measurement questions. Our technique utilizes hydrogel photopatterning via grayscale masks to define non-uniform pore-size distributions from a single UV exposure and precursor solution. Using this method we realize two workhorse analytical electrophoresis platforms: (1) a 24-plex electrophoresis screening assay and (2) a 96-plex gradient gel-based protein sizing assay.

W-203 REALIZATION OF 240 NANOMETER RESOLUTION OF CELL POSITIONING BY A VIRTUAL FLOW REDUCTION MECHANISM 1031

S. Sakuma[1], K. Kuroda[1], M. Kaneko[1], and F. Arai[2]

[1]Osaka University, JAPAN and [2]Nagoya University, JAPAN

For cell manipulation in a microchannel, it is often required to control the cell as accurate as possible. However, the issue for using a syringe pump is that the flow rate is geometrically amplified in microchannel. Therefore, we propose a virtual flow reduction mechanism in this paper. By using elastic feature of the PDMS chip, we designed and developed the total system for cell manipulation. Through experiments, we confirmed that the cell positioning resolution is 240 nm with the frequency up to 20 Hz.

Th-204 SINGLE CELL SEPARATION BY USING ACCESSIBLE MICROFLUIDIC CHIP 1035

T. Hayakawa[1], T. Fukuhara[2], K. Ito[1], and F. Arai[1]

[1]Nagoya University, JAPAN and [2]Tokyo University of Pharmacy and Life Science, JAPAN

In this paper, we proposed novel single cell separation method that used accessible microfluidic channel. Various single cell separation method by using microfluidic chip have been proposed. However, the biggest problem is that those microfluidic chip are closed and separated cells are tend to missed at interface of the chip to outer world. Therefore, we used cover opened microfluidic chip that can be accessible in order to collect the separated cell. And also, we proposed single cell pick-up tool to collect the separated single cell.

M-205 STUDY OF HOTSPOT COOLING USING ELECTROWETTING ON DIELECTRIC DIGITAL MICROFLUIDIC SYSTEM 1039

G.S. Bindiganavale[1], S.M. You[2], and H. Moon[1]

[1]University of Texas at Arlington, USA and [2]University of Texas at Dallas, USA

This paper presents a novel digital microfluidic (DMF) cooling system using electrowetting on dielectric (EWOD) developed for demonstrating and studying hotspot cooling for applications in electronics thermal management. The merits of this cooling system lies in the fact that no mechanically moving parts such as valves, pumps and fans are required to achieve hotspot cooling, thus having smaller form factor than bulky heatpipes and other conventional cooling systems. This study reveals close profiles of temperature change during coolant drop motion over hotspot as well as importance of phase change in the proposed cooling system.

T-206 SURFACE-ACOUSTIC-WAVE DRIVEN POINT SOURCE ATOMIZER INTEGRATED WITH PICOLITER MICRO PUMPS FOR POLYMERIC NANOPARTICLES SYNTHESIS 1043

S. Sugimoto, M. Hara, H. Oguchi, A. Yabe, and H. Kuwano

Tohoku University, JAPAN

We developed a surface acoustic wave (SAW) driven atomizer integrated with picoliter micro pumps for polymeric nanoparticles synthesis. The pumps consisted of the reservoir and a pair of interdigital transducer (IDT). The atomizer also consisted of arc-shaped IDT for focusing the SAW energy into the liquid. As an experimental result using water, when applying the burst signal to the IDT, discharge in which the rate was 0.3 pl a burst could be observed. Moreover, we succeeded in ejection of narrow mist spray from the atomizer.

W-207 TEFLON WETTING AND DEWETTING ON EWOD DEVICE FOR CHEMILUMINESCENCE DETECTOR 1047

X.Y. Zeng[1], K.D. Zhang[1], G.W. Tao[1], S.K. Fan[2], and J. Zhou[1]

[1]Fudan University, CHINA, and [2]National Taiwan University, TAIWAN

We develop a hydrophobicity recoverable EWOD (electrowetting-on-dielectric) based chemiluminescence detector with an integrated signal and heater electrode. A series of experiments and X-ray photoelectron spectroscopic analysis are used to reveal the wetting and dewetting mechanism of Teflon in the EWOD device, and get the recovery relationships between the recovered contact angle, the recovery threshold time and heating temperature.

Th-208 TRANSIENT INERTIAL FLOWS: A NEW DEGREE OF FREEDOM
FOR PARTICLE FOCUSING IN MICROFLUIDIC CHANNELS 1051
M.H. Winer, A. Ahmadi, and K.C. Cheung
University of British Columbia, CANADA

We have investigated the unique effect of transient flow rate on inertial particle focusing in microfluidic systems. A comparative analysis was conducted using both constant and transient flow rates on polystyrene (PS) beads in various channel geometries. Results show that particle focusing equilibrium positions are affected by the use of a transient (changing) flow rate. Transient inertial flows provide a new degree of freedom for manipulation of particle positioning in microfluidic channels.

Nano-Electro-Mechanical Devices and Systems (NANO)

M-209 BIAXIAL STRAIN IN SUSPENDED GRAPHENE
MEMBRANES FOR PIEZORESISTIVE SENSING 1055
A.D. Smith[1], F. Niklaus[1], S. Vaziri[1], A.C. Fischer[1], M. Sterner[1],
F. Forsberg[1], S. Schröder[1], M. Östling[1], and M.C. Lemme[2]
[1]KTH Royal Institute of Technology, SWEDEN and [2]University of Siegen, GERMNAY

This work compares through both theory and experiment the effect of cavity shape and size on the sensitivity of piezoresistive pressure sensors based on suspended graphene membranes. Further, the paper analyzes the effect of both biaxial and uniaxial strain on the membranes.

T-210 FABRICATION OF GOLD NANOPARTICLE-EMBEDDED NANOCHANNELS
FOR SURFACE-ENHANCED RAMAN SPECTROSCOPY 1059
K. Suekuni, T. Takeshita, K. Sugano, and Y. Isono
Kobe University, JAPAN

A micro/nanofluidic device including linearly-arranged gold nanoparticles embedded into nanochannels was developed for highly-sensitive Surface-Enhanced Raman Spectroscopy (SERS) analysis. The nanochannels array was fabricated by a "photo" lithography-based process without costly and time-consuming process such as EB lithography. Then particles with diameters of 100 nm are arranged into the nanochannels by a nanotrench-guided self-assembly process. The device was successfully fabricated and it was active for SERS analysis with 4,4'-bypiridine as a target molecule."

W-211 FINFET WITH FULLY PH-RESPONSIVE HFO$_2$ AS
HIGHLY STABLE BIOCHEMICAL SENSOR 1063
S. Rigante[1], M. Wipf[2], A. Bazigos[1], K. Bedener[3], D. Bouvet[1], and A.M. Ionescu[1]
[1]Ecole Polytechnique Federale de Lausanne (EPFL), SWITZERLAND, [2]University of Basel, SWITZERLAND, and [3]Paul Scherrer Institute, SWITZERLAND

We present a sensing platform based on high-stability low-power n-channel fully depleted FinFETs on Si-bulk. Efficient chemical and biological label-free sensing has been demonstrated, paving the way towards non-invasive simultaneous monitoring of human physiological signals such as pH and proteins. In contrast to other SiNW-based sensors, the use of scalable high-k dielectric FinFETs for both applications is in accordance with the material constraints which come along Moore's Law of scaling.

Th-212 FREQUENCY DEPENDENT AC ELECTROOSMOTIC FLOW IN NANOCHANNELS 1067
W.T.E. van den Beld, W. Sparreboom, A. van den Berg, and J.C.T. Eijkel
MESA+ - University of Twente, THE NETHERLANDS

We report frequency-dependent bidirectional AC electroosmotic flow (AC-EOF) in a nanochannel with double layer overlap. Simulations confirm the observed bidirectionality. By this frequency-dependent bidirectional pumping, nanochannel AC-EOF behaves in fundamentally different way than microchannel AC-EOF. The results are of importance for the understanding of ion and liquid transport in nanoconfinement.

M-213 FULLY MONOLITHIC AND ULTRA-COMPACT NEMS-CMOS SELF-OSCILLATOR BASED
ON SINGLE-CRYSTAL SILICON RESONATORS AND LOW-COST CMOS CIRCUITRY 1071
J. Philippe, G. Arndt, E. Colinet, M. Savoye, T. Ernst, E. Ollier, and J. Arcamone
CEA - LETI – Minatec, FRANCE

This work reports on the first experimental demonstration of a self-oscillator based on a single crystal silicon NEMS resonator monolithically co-integrated with a simple electronic circuitry manufactured with a very low-cost 0.35μm CMOS technology. This NEMS-CMOS self-oscillator pixel is as small as 50x70 μm^2 (pads excluded).

T-214 A GRAPHENE NANOSENSOR FOR DETECTION OF SMALL MOLECULES 1075

C. Wang[1,2], J. Kim[1], J. Zhu[1], R. Pei[3], G. Liu[2], J. Hone[1], M. Stojanovic[1], and Q. Lin[1]
[1]Columbia University, USA, [2]Nankai University, CHINA, and [3]Chinese Academy of Sciences, CHINA

We developed a graphene field effect transistor (GFET) biosensor for detection of an important small molecular hormone DHEA-S. In view of the low charged small biomolecules can't excite sufficient electrical response of GFET, we proposed a competitive dehybridization strategy based on the aptamer-target specific association. We experimentally demonstrated that on the graphene surface, aptamer dehybridization caused by DHEA-S specific association provides strong response of GFET. And the concentration of target DHEA-S can be quantitatively detected by observing the "half time period" of aptamer dehybridization kinetic process.

**W-215 INTERROGATING CONTACT-MODE SILICON CARBIDE (SiC)
NANOELECTROMECHANICAL SWITCHING DYNAMICS BY
ULTRASENSITIVE LASER INTERFEROMETRY** 1079

T. He, J. Lee, Z. Wang, and P.X.-L. Feng
Case Western Reserve University, USA

We report the experimental demonstration of probing the dynamics of nanoscale contacts in robust nanoelectromechanical switches based on silicon carbide (SiC) nanocantilevers. For the first time, we measure the dynamical behavior of contact-mode SiC nanoelectromechanical switches, in both frequency- and time-domain, by directly probing the tips of the SiC nanocantilevers, using ultrasensitive laser interferometric techniques.

Th-216 MATRIX INDEPENDENT LABEL-FREE NANOELECTRONIC BIOSENSOR 1083

R. Esfandyarpour[1,2], M. Javanmard[2], Z. Koochak[1], J.S. Harris[1], and R.W. Davis[2]
[1]Stanford University, USA and [2]Stanford Genome Technology Center, USA

We fabricated a novel, label free and real time electrical impedance biosensor, referred to as the nanoneedle biosensor. The nanoneedle is an ultrasensitive and localized device, which has the ability to directly measure biomolecular binding as a function of time (real-time). The utility of this sensor in affinity biosensing was demonstrated. As a practical example with clinical relevance, we demonstrated the detection of Vascular Endothelial Growth Factor (VEGF) for cancer diagnosis. Our demonstration of label-free and real-time detection of VEGF with this sensor can be envisioned to allow for one-step point-of-care cancer diagnosis. This work provides a strong starting point for a new class of electronic biosensing devices with the capability of rapid direct large-scale integration.

M-217 MECHANICAL PROPERTIES OF FEW LAYER GRAPHENE CANTILEVER 1087

K. Matsui, A. Inaba, Y. Oshidari, Y. Takei, H. Takahashi, T. Takahata, R. Kometani,
K. Matsumoto, and I. Shimoyama
University of Tokyo, JAPAN

We report the spring constant measurement of few-layer (1-, 2-, and 3-layer) graphene (FLG) cantilevers by optical heterodyne interferometry. We fabricated FLG cantilever with a weight of diamond-like carbon using focused ion beam. The effective spring constants were obtained from the measured resonant frequency and the mass of the weight, and were calculated to be about 2.7×10^{-3} N/m. This result indicates FLG cantilever structure is more rigid than that predicted from the literature data.

**T-218 NANO-OPTO-MECHANICAL MEMORY BASED ON
OPTICAL GRADIENT FORCE INDUCED BISTABILITY** 1091

B. Dong[1,2], J.G. Huang[3], H. Cai[1], P. Kropelnicki[2], A.B. Randles[2], Y.D. Gu[2], and A.Q. Liu[1]
*[1]Nanyang Technological University, SINGAPORE, [2]Agency for Science, Technology and Research (A*STAR),
SINGAPORE, and [3]Xi'an Jiao Tong University, CHINA*

A bistable nano-opto-mechanical memory is designed, fabricated and experimentally demonstrated.Fabricated with CMOS compatible process, this optical memory can be easily packaged and integrated with otherphotonic devices. The nano-size of the memory enable for large scale integration, high speed operation and low powerconsumption. It has other potential applications such as optical switch, logic gate and actuator.

**W-219 SUBMICRON THREE-TERMINAL SIGE-BASED
ELECTROMECHANICAL OHMIC RELAY** 1095

M. Ramezani[1,2], S. Cosemans[1], J. De Coster[1], X. Rottenberg[1], V. Rochus[1],
H. Osman[1], H.A.C. Tilmans[1], S. Severi[1], and K. De Meyer[1,2]
[1]imec, BELGIUM and [2]KU Leuven, BELGIUM

We demonstrate functional cantilever switches based on a CMOS-compatible low-T(400ºC) CVD SiGe process flow. Devices with dimensions in the micrometer range, thickness and gap smaller than 100nm were successfully fabricated and electrically characterized. Typical switches characteristics such as high I_{on}/I_{off} ratio, sharp sub-threshold slope and zero off-state leakage current were observed. A minimum of 1000 cycles device lifetime was demonstrated. The maximum current which can flow through the device without causing stiction due Joule-heating was investigated.

Th-220 PIEZOELECTRIC BUCKLING-BASED NEMS RELAYS FOR MILLIVOLT MECHANICAL LOGIC 1099

U. Zaghloul[1,2] and G. Piazza[1]

[1]*Carnegie Mellon University, USA and* [2]*Electronics Research Institute, EGYPT*

We report on the design, fabrication, characterization, and scaling analysis of novel NEMS relays that use, for the first time, buckling piezoelectric actuators. The fabricated switches exhibit low actuation voltage (< 2 V) and reduced threshold voltage (~110 mV). Also, hysteresis in the switching process was observed and limits the minimum swing voltage to ~250 mV. A scaling analysis highlights the possibility of achieving milliVolt switching at aggressively scaled device footprints.

M-221 NANOELECTROMECHANICAL TUNNELING SWITCHES BASED ON SELF-ASSEMBLED MOLECULAR LAYERS 1103

F. Niroui[1]. P.B. Deotare[1], E.M. Sletten[1], A.I. Wang[1], E. Yablonovitch[2], T.M. Swager[1], J.H. Lang[1], and V. Bulovic[1]

[1]*Massachusetts Institute of Technology, USA and* [2]*University of California, Berkeley, USA*

We propose and experimentally investigate nanoelectromechanical switches that operate via electromechanical modulation of tunneling current through compressible molecular films. This approach utilizes self-assembled molecular layers to define few nanometer-thick switching gaps, and has the potential to enable low-voltage operation while simultaneously mitigating device failure due to stiction.

T-222 TRANSITION OF Q-DOT DISTRIBUTION ON MICROTUBULE ARRAY ENCLOSED BY PDMS SEALING FOR AXONAL TRANSPORT MODEL 1107

K. Fujimoto[1], H. Shintaku[1], H. Kotera[1], and R. Yokokawa[2,1]

[1]*Kyoto University, JAPAN and* [2]*Japan Science and Technology Agency (JST), JAPAN*

We developed an experimental system which enables kinesin driven transport on arrayed microtubules in enclosed micro channels. To avoid an expected difficulty of exchanging solution, surface fabricated micro tracks were encapsulated after reagents introduction using flow cells with deformable PDMS chip at its top. After enclosed micro channels were formed, directed transport and continuous accumulation of fluorescent labeled kinesin molecules were observed. These results indicate a possibility of application as in vitro model of intracellular transport as seen in axons.

W-223 WAFER-SCALE FABRICATION OF SCANNING THERMAL PROBES WITH INTEGRATED METAL NANOWIRE RESISTIVE ELEMENTS FOR SENSING AND HEATING 1111

K. Hatakeyama[1], E. Sarajlic[2], M.H. Siekman[1], L. Jalabert[3], H. Fujita[3], N. Tas[1], and L. Abelmann[4]

[1]*MESA+ - University of Twente, THE NETHERLANDS,* [2]*SmartTip B.V., THE NETHERLANDS,* [3]*University of Tokyo, JAPAN, and* [4]*Korea Institute of Science and Technology (KIST), GERMANY*

We present a novel scanning resistive probe aimed for thermal imaging and localized thermal analysis. The probe features an AFM cantilever with a sharp pyramidal tip. Metal nanowires are integrated at the inner edges of the pyramidal tip forming an electrical cross-junction at the apex. The cross-junction can be utilized both as a local temperature sensor and a heater.

Packaging Technologies (PCK)

Th-224 CAPILLARY EFFECT BASED TSV FILLING METHOD 1115

J. Gu, X. Jiang, H. Yang, and X. Li

Shanghai Institute of Microsystem and Information Technology (SIMIT), CHINA

We explore a capillary liquid solder through-hole filling method, which utilizes liquid bridge pinch-off effect. The filling is completed by first pushing solder into via holes from a solder pool through nozzle orifices, and then followed by cutting the solder pillars in the via holes off from the solder pool in the nozzle orifices. The whole TSV filling process can be completed by a cycle of pressure change. In addition, 'wafer sandwich' structure is utilized to neutralize pressure differential, which causes wafer breakage.

M-225 CHARACTERISATION AND SIMULATION OF LOW TEMPERATURE Si-Si-DIRECT BONDING THROUGH VELCRO-LIKE SURFACES BASED ON POROUS SILICON 1119

S. Keshavarzi[1,2], U. Mescheder[1], and H. Reinecke[2]

[1]*Furtwangen University - IAF, GERMANY and* [2]*University of Freiburg – IMTEK, GERMANY*

We develop, characterize and model a new bonding technique based on Pours Silicon (PS) technology. PS allows strong permanent bonding between needle like surfaces as well as multiple bonding and un-bonding of chips similar to Velcro principle. This approach provides low temperature Si-Si direct bonding, a fully CMOS compatible approach suitable in system integration using the Si-motherboard concept.

T-226 FROM CHIPS TO DUST: THE MEMS SHATTER SECURE CHIP 1123
N. Banerjee, Y. Xie, M. Rahman, H. Kim, and C.H. Mastrangelo
University of Utah, USA

This paper presents the implementation of transience silicon microchips through post-processing microfabrication and micropackaging steps that transform almost any electronic, optical or MEMS substrate chips into transient ones. When transience is activated the chip mechanically shatters, and it is literally reduced to a heap of silicon dust. The massive cleavage action is achieved by the triggered release of mechanical energy stored within the silicon substrate in expandable microparticles.

W-227 LONG TERM GLASS-ENCAPSULATED PACKAGING FOR IMPLANT ELECTRONICS 1127
J.H. Chang, Y. Liu, and Y.C. Tai
California Institute of Technology, USA

This paper studies a new long-term packaging scheme for implant electronics using glass encapsulation featuring a controlled failure mode from fast diffusion to slow undercut. The experimental results show that this packagingscheme can easily survive for more than 10 years by accelerated "active" lifetime soaking test (i.e. with electric field applied) in 0.9 wt.% saline solution. This method provides advantages of easy employment, controllable long life time, and enhanced heat dissipation."

**Th-228 LOW-TEMPERATURE GOLD-GOLD BONDING USING SELECTIVE FORMATION
OF NANOPOROUS POWDERS FOR BUMP INTERCONNECTS** 1131
H. Mimatsu[1], J. Mizuno[1], T. Kasahara[1], M. Saito[1], S. Shoji[1], and H. Nishikawa[2]
[1]Waseda University, JAPAN and [2]Osaka University, JAPAN

We proposed low-temperature Au-Au bonding using nanoporous Au-Ag powders as an electrical connective adhesion between bump interconnects. The nano-porous powders were formed by de-alloying Au-Ag alloy sheet with Au:Ag. The influence of the annealing temperature on the porous structures was investigated. Selective formation of the powders on bumps was achieved by stamping. Bonding strength of about 2.4 MPa was achieved by using nanoporous Au-Ag powders at 150 °C. This result indicates that the proposed powder is a useful material for low-temperature Au-Au bonding.

**M-229 MICRO DEVICES INTEGRATION WITH LARGE-AREA 2D CHIP-NETWORK
USING STRETCHABLE ELECTROPLATING COPPER SPRING** 1135
W.L. Sung[1], W.C. Lai[1], C.C. Chen[2], K. Huang[2], and W. Fang[1]
[1]National Tsing Hua University (NTHU), TAIWAN and [2]imec Taiwan Inc., TAIWAN

This study presents a large-area multi-devices integration scheme using stretchable electroplating copper spring. Advantages of this approach: (1) using the existing process technologies and materials for semiconductor in large-area applications; (2) stretchable electroplating-copper spring with large maximum strain acts as mechanically and electrically connection; (3) Si-node acts as a hub for devices implementation and integration; and (4) the chip-network can apply to curved surfaces. The proposed expand network using stretchable spring integrated multi devices has been implemented and tested.

**T-230 SOLID-STATE ISFET FLOW METER FABRICATED WITH A PLANAR PACKAGING
PROCESS FOR INTEGRATING MICROFLUIDIC CHANNEL WITH CMOS IC CHIP** 1139
J.J. Wang[1], C.F. Lin[2], Y.Z. Juang[2], H.H. Tsai[2], H.H. Liao[2], and C.H. Lin[1]
[1]National Sun Yat-sen University, TAIWAN and [2]National Applied Research Laboratories, TAIWAN

We presents a solid-state ISFET flow meter fabricated with an innovative planar packaging process. The developed method provides a simple yet efficient method to integrate CMOS IC chip with microfluidic systems and the whole packaging process can be achieved in 40 min. The sealed ISFET chip is used for measuring the flow rate of non-ionic solutions including acetone, ethanol and glycerol of slow flow rate. And the flow rate measurement exhibited good reproducibility in the flow rate ranging from 66 to 1700 µm/s.

Physical MEMS (PHYS)

W-231 A TUNABLE LASER BASED ON NANO-OPTO-MECHANICAL SYSTEM 1143
M. Ren[1], H. Cai[2], Y.D. Gu[2], P. Kropelnicki[2], A.B. Randles[2], and A.Q. Liu[1]
*[1]Nanyang Technological University, SINGAPORE and
[2]Agency for Science, Technology and Research (A*STAR), SINGAPORE*

A tunable laser based on nano-opto-mechanical system is presented in this paper. A novel tuning approach is demonstrated which applies optical force to adjust the cavity mode via controlling the mechanical displacement of the silicon waveguide. In the experiments, a 24-nm wavelength tuning is realized due to a deflection of 14-nm. The optomechanical wavelength tuning coefficient is 214 GHz/nm. The demonstrated device has potential applications for optical communication system, pulse trapping/release, and chemical sensing, with easy on-chip integration on a silicon platform.

Th-232 **A TUNABLE OPTICAL IRIS BASED ON ELECTROMAGNETIC ACTUATION FOR A HIGH-PERFORMANCE MINI/MICRO CAMERA** 1147

H.W. Seo[1], J.B. Chae[1], S.J. Hong[1], I.U. Shin[1], K. Rhee[1], J.-H. Chang[2], and S.K. Chung[1]
[1]*Myongji University, SOUTH KOREA and*
[2]*Samsung Advanced Institute of Technology (SAIT), SOUTH KOREA*

This paper presents a tunable optical iris based on electromagnetic actuation for a high-performance mini/micro camera. In optics, an iris, an aperture stop, is placed in the light path of a lens or objective and regulates the amount of light that passes through the lens by controlling the size of the aperture, an opening at it center. The iris not only controls light flux, field of view, depth of field (DOF), but also blocks scattered light and improves image quality by limiting spherical aberration. Hence, the iris is an indispensable element in most optical systems. However, the conventional mechanical iris, consisting of movable sliding blades, requires a complicated sliding rotary mechanism that has to be operated by bulky motors and is therefore difficult to miniaturize. We develop a variable optical iris operated by electromagnetic actuation. According to electromagnetic induction, when an electrical current flows in an electric coil, a magnetic field is generated in its surroundings. In this work, the magnetic field is used to actuate or pull an optically opaque ferrofluid initially filled inside the sub-channel of the iris to the center of the main channel, resulting in controlling the diameter of an aperture.

M-233 **CALORIMETRIC DEVICE FOR NON-DESTRUCTIVE MEASUREMENT OF THE THERMAL DIFFUSIVITY DEPENDENCY BY PHASE DELAY** 1151

T. Suzuki, Y. Ichikawa, T. Takahata, K. Matsumoto, and I. Shimoyama
University of Tokyo, JAPAN

We developed a device for measuring thermal diffusivity dependency of the contacted surface layer non-destructively. The device was based on the principle that temperature phase delay between a heater and a resistance temperature detector (RTD) is affected by thermal diffusivity of the contacted surface layer. The device consisted of an Au wire, as oscillating heat source and a piezoresistance as a RTD. We exerted the simulation and the experiment for the device, and found that the phase delay decreased as thermal diffusivity increased.

T-234 **CAPACITIVE FEEDBACK CONTROLLED PZT MICRO MIRROR ARRAYS FOR WAVELENGTH SELECTIVE SWITCH** 1155

R. Uchino, T. Misaki, T. Fujimura, and O. Torayashiki
Sumitomo Precision Products Co., Ltd., JAPAN

We develop a single-axis mechanical micro mirror array used for gridless wavelength selective switch (WSS). The mirrors are driven by lead zirconate titanate (PZT) unimorph actuators, which is adequate for low-voltage actuation and low interference with adjacent mirrors in operation. In addition, the mirror tilt angle is feedback controlled using comb-shaped capacitance in order to realize high control resolution. We fabricated a prototype of the mirror array, and evaluated its basic performance.

W-235 **CARBON SP^2-SP^3 TECHNOLOGY: GRAPHENE-ON-DIAMOND THIN FILM UV DETECTOR** 1159

K. Yao[1], C. Yang[1], X. Zang[1], F. Feng[2], and L. Lin[1]
[1]*University of California, Berkeley, USA and* [2]*Chinese Academy of Sciences, CHINA*

We for the first time demonstrates the graphene-diamond-metal (GDM) vertical sandwich structure as a thin film UV detector. New scientific and engineering breakthroughs are: (1) first experimental investigation of the carbon-based sp2-sp3 junctions; (2) a peel-and-stick fabrication process to make flexible diamond films; and (3) first GDM vertical UV sensors. As such, the proposed detector/architecture can open up a new class of scheme to build diamond-based optoelectronic systems.

Th-236 **CLOSE-PACKED LIQUID-FILLED TUNABLE MICROLENS ARRAY** 1163

Y. Iimura[1], H. Onoe[1,2], and S.Takeuchi[1,2]
[1]*University of Tokyo, JAPAN and* [2]*ERATO Takeuchi Biohybrid Innovation Project,*

We develop close-packed liquid-filled tunable microlens arrays for optical devices such as integral imaging systems. These lenses are simply composed of poly(dimethylsiloxane)(PDMS) microchannels and applied pressure deforms the top membrane of microchannels to become convex lenses. These lenses have three advantages: (i)Uniform deformation by pressure-driven actuation, (ii)Adjustable optical characteristics without patterned electrode, (iii)High-density integration of tunable microlenses. We fabricated three types of lenses based on closed-packed structure and showed that the Spiderweb type packing is the most suitable for closer packing.

M-237 **COMPACT TUNABLE HYPERSPECTRAL IMAGING SYSTEM** 1167

P.-H. Cu-Nguyen[1], A. Grewe[2], C. Endrödy[2], S. Sinzinger[2], H. Zappe[1], and A. Seifert[1]
[1]*University of Freiburg – IMTEK, GERMANY and* [2]*Ilmenau University of Technology, GERMANY*

We demonstrate a compact tunable hyperchromatic lens system for imaging an object with highly resolved spectral information. This hybrid device is composed of a diffractive optical element, a tunable concave liquid-filled membrane lens, and an integrated magnetic actuator for hydraulically tuning the focal length of the refractive lens. The lens system can generate a hyperspectral datacube in the visible wavelength range, 400 – 730 nm, proved here with a spectral sampling interval of 2.4 nm.

T-238 CYLINDRICAL LENS WITH INTEGRATED PIEZO ACTUATION FOR FOCAL LENGTH TUNING AND LATERAL SCANNING 1171
M. Stuermer, A. Schatz, and U. Wallrabe
University of Freiburg – IMTEK, GERMANY

We present a cylindrical lens which features integrated piezo bending actuators for focal length tuning. The design is based on a PDMS membrane which encloses an optical liquid. We optimize the shape of the actuators for good cylindricity and show a process for prototype fabrication. The lens provides a large usable aperture of ca. 4 x 10 mm, a tuning range of more than 20 dpt, and the possibility to move the lens vertex along one axis. Therefore, it enables scanning of the line focus.

W-239 FRESNEL LENS BASED ON SILICON NANOWIRES 1175
Y.-S. Lu, J. Fernandes, H. Liu, and H. Jiang
University of Wisconsin-Madison, USA

We demonstrate silicon-based Fresnel lenses by photolithography techniques and metal assisted chemical etching, where the opaque zones are composed of 2 µm-tall silicon nanowires formed directly in silicon. The reflective Fresnel lens showed a high-contrast light intensity distribution between the bright and dark zones, leading to a focused spot with strong contrast above the lens. The lens has the potential to be integrated with dye-sensitive solar cells by reflecting and focusing light onto the photosensitive dye to improve their light absorption efficiency and photocurrent.

Th-240 ENHANCED WAVELENGTH SELECTIVE INFRARED EMISSION USING SURFACE PLASMON POLARITON AND THERMAL ENERGY CONFINED IN MICRO-HEATER 1179
T. Sawada[1], K. Masuno[2], S. Kumagai[1], M. Ishii[2], S. Uematsu[2], and M. Sasaki[1]
[1]Toyota Technological Institute, JAPAN and [2]Yazaki Corporation, JAPAN

A new surface plasmon polariton (SPP) based wavelength selective IR emitter is combined with microheater. IR emitted from the microheater is basically confined except SPP propagation on the metal grating carrying IR energy to the outside. The limited condition for SPP excitation realizes the narrow wavelength filtering. SPP related emission is obtained having the peak width similar order compared with the bandwidth of gas absorption. Since the microheater can minimize the thermal conduction loss, the high efficiently is expected at SPP related wavelength.

M-241 EFFECT OF NEEDLE SHAPE ON PERFORMANCE OF NEEDLE-TYPE ELECTRO TACTILE DISPLAY 1183
N. Kitamura[1], J. Chim[1], and N. Miki[1,2]
[1]Keio University, JAPAN and [2]Japan Science and Technology Agency (JST), JAPAN

In our prior work, we revealed that a needle-type electrotactile display that penetrates through a stratum corneum of a finger skin can display tactile information at 20 times as low voltage as that with flat electrodes. We discovered that the needle-tip shapes greatly affected the performance of the display. In this work, we experimentally deduced the optimum shape of the needle tip using titanium micro-needles patterned by electrochemical etching. The needles can be readily applicable to efficient electrotactile displays.

T-242 INCLINATION-INDEPENDENT TRANSFORMATION OF LIGHT BEAMS USING HIGH-THROUGHPUT UNIQUELY-CURVED MICROMIRRORS 1185
Y.M. Sabry[1,2], D. Khalil[2,3], B. Saadany[2], and T. Bourouina[1,2]
[1]Université Paris-Est, FRANCE, [2]Si-Ware Systems, EGYPT, and [3]Ain-Shams University, EGYPT

This paper reports a novel class of deeply-etched, specifically-designed curved micromirrors enabling phase-transformation of light beams independent of the inclination angle of the incident light with respect to the mirror surface. The micromirrors were fabricated on silicon by deep reactive ion etching technology. The profile of the specifically-designed mirrors' surfaces was controlled precisely, thanks to the photolithographic process. High optical throughput micromirrors exhibiting submillimeter focal lengths were fabricated with depth larger than 300 µm. Optical measurements show stable dimensions for the optical beam spot with less than ± 5% dependence on the inclination angle up to 60 degrees.

W-243 MAGNETOSTRICTIVE TYPE TACTILE SENSOR BASED ON METAL EMBEDDED POLYMER ARCHITECTURE 1189
H.-C. Chang[1], W.-L. Sung[1], H.-S. Hsieh[1], J.-H. Wen[1], C.-C. Fu[1], S.-C. Liao[1],
C.-H. Lai[1], W.-C. Lai[2], C.-H. Chang[2], C.-P. Chang[2], C.-H. Chen[2], and W. Fang[1]
[1]National Tsing Hua University (NTHU), TAIWAN and [2]WinMEMS Technologies Co., Ltd., TAIWAN

This study presents new process scheme to fabricate polymer structure with embedded metal on silicon substrate. The primary merit of presented process scheme is: simple approach for the integration of 3D structures with different materials (e.g. metal, glass, polymer) on substrate. To demonstrate the feasibility, a tactile sensor design consisting of polymer structure with embedded 3D Ni inductor is demonstrated. As the polymer diaphragm deformed by tactile force, the magnetostriction effect of 3D Ni inductor will induce the permeability change. Thus, the permeability change as well as the tactile force can be detected by the inductance difference of embedded 3D Ni inductor.

Th-244 A MULTI-MATERIAL Q-BOOSTED LOW PHASE NOISE OPTOMECHANICAL OSCILLATOR 1193

T. Beyazoglu, T.O. Rocheleau, K.E. Grutter, A.J. Grine, M.C. Wu, and C.T.-C. Nguyen
University of California, Berkeley, USA

We present a multi material Radiation Pressure driven Optomechanical Oscillator (RP-OMO) with simultaneously high mechanical Q_m >22,000 and optical Q_o >190,000 achieving best-to-date phase noise performance of -125 dBc/Hz at 5 kHz offset from its 52-MHz carrier, which is 12 dB better than the previous best RP-OMO constructed of silicon nitride alone. The device not only reduces phase noise, but does so with a lower input laser power of only 3.6 mW. The key to achieving this performance is the addition of polysilicon material as an inner ring that boosts the overall mechanical Q_m of the total structure. The addition of polysilicon further provides a mechanism for voltage-controlled electrical stiffness tuning of the oscillation frequency.

M-245 NOVEL TUNABLE OPTICAL MODULATION LENS USING MAGNETHORHEOLOGICAL EFFECT 1197

F.-M. Hsu, R. Chen, and W. Fang
National Tsing Hua University (NTHU), TAIWAN

This study extends the fluid dispensing and sealing technology to realize a novel MR-fluid lens (MR-fluid: liquid polymer with magnetic particles) for light intensity modulation. Merits of the device: Optical transmittance of lens is controlled by (1) weight fraction of magnetic powder, and (2) orientation of columnar particles controlled by magnetic field. In applications, the MR-fluid lens is realized on glass substrate and suspended MEMS structures. The light intensity modulation of MR-fluid lens (diameter: 2000µm) by magnetic field is demonstrated. Measurements show the NdFeB- liquid polymer (10wt%) has a 40% dark area change and 290% laser transmittance difference after applying magnetic field.

T-246 OPTICAL CONTROL AND TUNING OF THERMAL-PIEZORESISTIVE SELF-SUSTAINED OSCILLATORS 1201

H.J. Hall[1], L. Wang[1], J.S. Bunch[1], S. Pourkamali[2], and V.M. Bright[1]
[1]University of Colorado, Boulder, USA and [2]University of Texas, Dallas, USA

We experimentally demonstrate the ability to frequency tune and provide on/off control of electrically driven thermal-piezoresistive self-sustained oscillators through the direct application of HeNe laser illumination at the device surface. These phenomena, which are unique to this class of oscillator, are explained by photoexcitation of charge carriers in the device's single crystal silicon structure inducing changes to the effective electrical resistivity and piezoresistivity.

W-247 PHOTOTHERMAL PROBING OF PLASMONIC HOTSPOTS WITH NANOMECHANICAL RESONATOR 1205

S. Schmid, K. Wu, T. Rindzevicius, and A. Boisen
Technical University of Denmark, DENMARK

We present a novel technique to probe and image plasmonic structures with nanoscale resolution by measuring the photothermally induced frequency detuning of highly temperature sensitive nanomechanical resonators. We employ the high temperature sensitivity of a nanomechanical string resonator to directly probe the heating pattern produced by a gold nanoslit illuminated by a scanning laser beam. The experimental approach allows a sensitive heat mapping of single localized surface plasmons, thereby helping to shed light on the underlying thermal effects in hotspots.

Th-248 RADIATION-PRESSURE ENHANCED OPTO-ACOUSTIC OSCILLATOR 1209

M.J. Storey, S. Tallur, and S.A. Bhave
Cornell University, USA

We present a driving scheme for integrated chip-scale opto-acoustic oscillators (OAO) in silicon with improved phase noise performance. Through simultaneous incorporation of radiation-pressure (RP) and RF feedback oscillating mechanisms, we have demonstrated a silicon RP enhanced OAO with a 10dB close-to-carrier phase noise improvement and thereby 10dB improvement in the oscillator's figure of merit.

M-249 THERMOPILE INFRARED ARRAY SENSOR FOR HUMAN DETECTOR APPLICATION 1213

J. Tanaka, M. Shiozaki, F. Aita, T. Seki, and M. Oba
OMRON Corporation, JAPAN

This paper reports the design of thermopile infrared sensor for human detector application. Sensitivity and response time of thermopile infrared sensor element are important for human detector application. In order to fulfill the specification, we developed S-shaped structure for thermopile infrared sensor element and fabrication process of chip scale vacuum package for mass production of the thermopile infrared sensors. As the result, 140V/W sensitivity and 17msec response time of the thermopile infrared sensor element are achieved.

T-250 TRANSFER-PRINTED COMPOSITE MEMBRANES FOR ELECTRICALLY-TUNABLE ORGANIC OPTICAL MICROCAVITIES 1217

A. Wang, W. Chang, A. Murarka, J. Lang, and V. Bulovic
Massachusetts Institute of Technology, USA

We present a method for fabricating organic optical microcavities using a transfer-printed composite membrane which can be electrostatically actuated for dynamic tuning of the cavity emission spectra. Electrical actuation and optical characterization of a completed device show cavity mode tuning greater than 20 nm. The device structure and transfer technique is easily applicable to large area fabrication of electrostatically tunable organic lasers, and potentially allows single-point contactless-readout for large area pressure sensor arrays.

W-251 TUNABLE METAMATERIALS BY CONTROLLING SUB-MICRON GAP FOR THE THZ RANGE 1221

A. Isozaki[1], T. Kan[1], H. Takahashi[1], N. Kanda[2], N. Nemoto[1], K. Konishi[1],
M. Kuwata-Gonokami[1], K. Matsumoto[1], and I. Shimoyama[1]
[1]University of Tokyo, JAPAN and [2]Institute of Physical and Chemical Research (RIKEN), JAPAN

We propose a tunable metamaterial actuated by pneumatic force. The tunable metamaterial has a double of sprit-ring-resonators (SRRs) whose gap between each other is controllable in sub-micron-order. Results of a terahertz (THz) spectroscopy confirmed that controlling gap in sub-micron-order or a few micron order was suitable for tuning resonant frequency of a metamaterial compared with that in 10 micron-order.

Th-252 UNCOOLED MULTI-BAND IR IMAGING USING BIMATERIAL CANTILEVER FPA 1225

W. Ma[1], S. Wang[1], Y. Wen[1], Y. Zhao[2], L. Dong[2], M. Liu[2], X. Liu[2], and X. Yu[1]
[1]Peking University, CHINA and [2]Institute of Beijing Technology, CHINA

A 256×256 bimaterial cantilever focal plane array (FPA), which is able to work in the three infrared atmospheric windows simultaneously, is fabricated and characterized. The FPA employs a silicon-framed structure by selectively etching away the substrate with Deep Reactive Ion Etching technique, and can be conveniently readout by an optical system. By combining the Chromium nano-films with silicon nitride as the multi-band IR absorber, the images of short wavelength, middle wavelength and long wavelength infrared are captured successfully with the same FPA.

M-253 VERY LOW POWER CONSUMPTION MEMS SCANNER WITH ALKALI ELECTRET COMB DRIVE 1229

T. Sugiyama, M. Aoyama, K. Kawai, and G. Hashiguchi
Shizuoka University, JAPAN

We developed the very low power consumption MEMS scanner that utilizes the electrostatic field generated by alkali-ion electret. The alkali-ion electret formed on comb electrodes of the scanner provides built-in potential for the electro-static actuator so that no bias voltage is necessary. The power consumption of prototype MEMS scanner was 0.57 μW (bias voltage: DC 0 V, driving voltage: AC 9 V_{pp}, deflection angle: 12°, resonance frequency: 1.4 kHz).

RF MEMS (RF)

T-254 A LOW-LOSS RF MEMS SILICON SWITCH USING REFLOWED GLASS STRUCTURE 1233

J. Hwang, S.-H. Hwang, Y.-S. Lee, and Y.-K. Kim
Seoul National University, SOUTH KOREA

This paper firstly reports on a low-loss RF MEMS switch that contains a reflowed glass structure beneath a contact metal. The reflowed glass structure is employed to reduce the electromagnetic wave loss brought about by the conductive silicon bulk underneath the contact metal. RF MEMS switch totally made of silicon is used as a reference model and the insertion loss is reduced as much as 0.26 dB for the proposed model in the frequency range of 5 to 30 GHz.

W-255 A UHF 4TH-ORDER BAND-PASS FILTER BASED ON CONTOUR-MODE PZT-ON-SILICON RESONATORS 1237

H. Yagubizade, M. Darvishi, M. Elwenspoek, and N.R. Tas
MESA+, University of Twente, THE NETHERLANDS

A novel RF-MEMS filter configuration around 700 MHz is proposed. It is based on a differential read-out of two in-phase actuated contour-mode resonators with slightly different resonance frequencies. The resonators are actuated independently in-phase and the outputs of the resonators are subtracted. This method is effective for improving the stopband rejection by canceling the feed-through signal. The BPF is presented using 50Ω termination with bandwidth of approximately 28.6 MHz and 35 dB stopband rejection. The ultimate rejection of the filter is improved by more than 20 dB compared to the individual resonators.

Th-256 AN 880MHZ LADDER FILTER FORMED BY ARRAYS OF LATERALLY VIBRATING THIN FILM LITHIUM NIOBATE RESONATORS 1241

S. Gong[1], and G. Piazza[2]

[1]*University of Illinois, Urbana Champaign, USA and* [2]*Carnegie Mellon University, USA*

This paper reports on the first implementation of a ladder filter using Lithium Niobate (LN) based laterally vibrating resonator (LVR) arrays. This demonstration is made possible by engineering the device orientation and using a distributed configuration of resonator arrays to simultaneously reduce spurious vibrations and insertion loss in a low impedance RF system. An almost spurious-free filter with < 3.5 dB of IL at 880 MHz was demonstrated by arraying properly sized devices into a ladder configuration. A total of 37 resonators were used for this demonstration. This work sets an important milestone in the development of a thin film LN technology platform for wide-band and frequency-agile RF filtering.

M-257 CAPACITIVE SILICON RESONATOR STRUCTURE WITH MOVABLE ELECTRODES TO REDUCE CAPACITIVE GAP WIDTHS BASED ON ELECTROSTATIC PARALLEL PLATE ACTUATION 1245

N.V. Toan and T. Ono

Tohoku University, JAPAN

This paper presents the design and fabrication of a capacitive silicon resonator with movable electrodes to obtain smaller capacitive gap widths, which results in smaller motional resistance and lower insertion loss. It also helps to increase tuning frequency range for compensation of temperature drift of silicon resonator.

T-258 COMBINED ELECTRICAL AND MECHANICAL COUPLING FOR MODE-RECONFIGURABLE CMOS-MEMS FILTERS 1249

C.-Y. Chen, M.-H. Li, C.-H. Chin, C.-S. Li, and S.-S.Li

National Tsing Hua University (NTHU), TAIWAN

This work presents a novel filter scheme which combines electrical and mechanical coupling implemented in a CMOS-MEMS filter to simultaneously attain narrow bandwidth and decent stopband rejection. As compared to the parallel-class filters and mechanically-coupled filters, the proposed filter structure features flexible electrical routing and non-conductive mechanical couplers, hence enabling single-ended to differential (SIDO) and differential to single-ended (DISO) reconfigurable modes in a single device. The proposed 8.6-MHz CMOS-MEMS filter was successfully demonstrated with narrow passband of 0.41% bandwidth and stopband rejection more than 20dB under proper termination.

W-259 DYNAMIC CHARACTERIZATION OF TUNABLE RF MEMS PRODUCTS 1253

D. DeReus, S. Cunningham, S. Natarajan, A. Morris, and J. Hilbert

WiSpry, Inc., USA

We present the dynamic characterization of tunable capacitors for RF MEMS products. The dynamic measurements have been made electrically by laser doppler vibrometry, and correlated with conventional finite element and high-order, parametric finite element models.

Th-260 ETCH-HOLE-ASSISTED ENERGY DISPERSION FOR ENHANCING QUALITY FACTOR IN SILICON BULK ACOUSTIC RESONATORS 1257

C. Tu and J.E.-Y. Lee

City University of Hong Kong, HONG KONG

This paper empirically demonstrates how the quality factor (Q) of a width-extensional mode silicon bulk-acoustic-resonator (SiBAR) can be enhanced by three times through strategic placement of holes on the structure. The holes serve to disperse the strain energy field concentrated around the nodal lines, ultimately re-distributing strain energy away from the anchors. This in turn reduces anchor loss and thus enhances Q. These results agree well with our finite-element (FE) simulations and we envisage the concepts herein to be transferable to other higher performance resonators like piezoelectric-AlN CMRs.

M-261 NANO-OPTO-ELECTRO-MECHANICAL (NOEM) OSCILLATOR WITH CONTROLLABLE NON-LINEAR DYNAMICS 1261

B. Dong[1,2], J.G. Huang[3], H. Cai[2], P. Kropelnicki[2], A.B. Randles[2], Y.D. Gu[2], and A.Q. Liu[1]

[1]*Nanyang Technological University, SINGAPORE,* [2]*Agency for Science, Technology and Research (A*STAR), SINGAPORE, and* [3]*Xi'an Jiao Tong University, CHINA*

An opto-mechanical oscillator with controllable non-linear dynamics is designed, fabricated and experimentally demonstrated. Fabricated with CMOS compatible process, this opto-mechanical oscillator can be easily packaged and integrated with other photonic devices. It has potential applications such as optical resonator type gyroscope, accelerometer and optical communication devices.

T-262 ON/OFF SWITCHABLE HIGH-Q CAPACITIVE-PIEZOELECTRIC ALN RESONATORS 1265
R.A. Schneider and C.T.-C. Nguyen
University of California, Berkeley, USA

AlN disk resonators having suspended (non-contacting) electrodes are demonstrated to have quality factors as high as 8,850 at 300 MHz, show no spurious modes, have single disk motional impedances of 3.0kOhm, and to possess an electrode collapse based off/on switching capability that operates via the application and subsequent removal of a strong bias voltage.

**W-263 PARAMETRIC FILTERING SURPASSES RESONATOR NOISE
IN ALN CONTOUR-MODE OSCILLATORS** 1269
C. Cassella, N. Miller, J. Segovia-Fernandez, and G.Piazza
Carnegie Mellon University, USA

We developed a new method for lowering the phase noise of oscillators where the intrinsic resonator frequency fluctuations represent the dominant noise source. We called this technique "parametric filtering".This method has been applied to a 227 MHz aluminum nitride contour-mode MEMS resonator that shows high level of intrinsic noise which limits the oscillator phase noise. By using this new approach we have obtained an improvement of more than 20 db and 26 db respectively at 1 khz and 10 khz offset. This resulted in the lowest phase noise level ever measured for any MEMS based oscillator.

**Th-264 POLYCIDE CONTACT INTERFACE TO SUPPRESS SQUEGGING
IN MICROMECHANICAL RESOSWITCHES** 1273
Y. Lin, R. Liu, W. Li, and C.T.C. Nguyen
University of California, Berkeley, USA

The use of a Pt-silicide-based contact interface has greatly reduced impact-induced energy loss in comb-driven resonant micromechanical switches (a.k.a., resoswitches) to the point where squegging phenomena (whereby impacts do not occur on every cycle) are eliminated, so no longer constrain the clock frequency of recently demonstrated mechanical charge pumps. This opens the application range of such charge pumps to higher power converters capable of delivering currents much higher than those of previously demonstrated version, which targeted low current-draw MEMS dc-biasing applications.

**M-265 STABLE CHARGE-BIASED CAPACITIVE RESONATORS
WITH ENCAPSULATED SWITCHES** 1277
E.J. Ng, K.L. Harrison, C.L. Everhart, V.A. Hong, Y. Yang, C.H. Ahn, D.B. Heinz, R.T. Howe, and T.W. Kenny
Stanford University, USA

We show that an electrically isolated silicon resonator within an epi-seal polysilicon encapsulation can retain a charge for prolonged periods of time with no noticeable leakage, even at elevated temperature. The charge is applied using a silicon contact switch that operates within the epi-seal cavity to isolate the resonator from the environment.

T-266 STABLE PULL-IN ELECTRODES FOR NARROW GAP ACTUATION 1281
E.J. Ng[1], Y. Yang[1], V.A. Hong[1], C.H. Ahn[1], D.L. Christensen[1], B.A. Gibson[2],
K.R. Qalandar[2], K.L. Turner[2], and T.W. Kenny[1]
[1]Stanford University, USA and [2]University of California, Santa Barbara, USA

We report on the use of pull-in electrodes for achieving narrower gaps than lithography/etch capabilities. Resonant devices with sub-ppm stability are demonstrated within the epi-seal polysilicon encapsulation process using pulled-in electrodes. The pull-in effect is reversible and can be made permanent by welding.

**W-267 ULTRA-STABLE NONLINEAR THIN-FILM PIEZOELECTRIC-ON-SUBSTRATE
OSCILLATORS OPERATING AT BIFURCATION** 1285
H. Fatemi, M. Shahmohammadi, and R. Abdolvand
Oklahoma State University, USA

Presented is a ~27MHz oscillator incorporating a thin-film piezoelectric-on-silicon (TPoS) resonator with a phase noise (PN) of -139 dBc/Hz at 1kHz and -157 dBc/Hz at 1MHz from the carrier. The close-to-carrier PN is equivalent to -148 dBc/Hz when normalized to 10MHz and is the lowest reported to date for MEMS oscillators. Additionally, it is experimentally proven that the PN significantly improves when the resonator is driven at or beyond the bifurcation point in the closed-loop oscillator circuit.

**Th-268 VARIABLE CAPACITOR WITH SWITCHING MECHANISM FOR
WIDE TUNING RANGE AND LOW POWER CONSUMPTION** 1289
D. Baek, Y. Eun, D.S. Kwon, M.O. Kim, T. Chung, and J. Kim
Yonsei University, SOUTH KOREA

We developed a variable capacitor with mechanical switching mechanism and reversible mechanical latching system to enhance tuning ratio and reduce power consumption. The switching mechanism could connect four sets of capacitors arranged in parallel sequentially by controlling the displacement of a microactuator for abrupt and coarse tuning of total capacitance. Continuous and fine tuning was also achieved by gap-closing mode of comb-finger type capacitors. The resultant maximum tuning ratio was 5.71 by combining coarse and fine tuning.

LETTER FROM THE CHAIRS

Welcome to the 27th IEEE International Conference on Micro Electro Mechanical Systems (MEMS 2014) in San Francisco, California! The IEEE MEMS Conference series originated in 1987, and has been known as the IEEE International Conference on Micro Electro Mechanical Systems since 1999. Over the last three decades, the MEMS community has experienced immense growth in science and technology of miniaturization, as well as commercialization. Many companies have developed and launched new products in various areas, including the recent success in consumer electronics, a cost competitive market that has been difficult for MEMS to crack into. Today, a large number of MEMS devices can be batch fabricated on 8" silicon wafers with very high yield and competitive cost at a number of foundries across the globe.

It is with great pleasure to announce that we have received 909 abstracts, a record number of submission for Americas! A total of 324 papers were carefully selected by the 30 experts of the Technical Program Committee (TPC). The presentations are arranged in a single-session format that includes 4 invited keynote presentations, 56 oral presentations and 268 posters. Poster presentations, which have always been a highlight of this Conference, are divided into four sessions to facilitate interaction with the authors, who will have their posters on display for two days. This year, for the first time, a new submission category was devised on industry research and development to encourage participation and presentation of our industry colleagues. Also new this year is the addition of an industrial keynote presentation to the program, scheduled for Thursday morning.

We will continue with the outstanding paper award program that was started last year in Taipei. The awards are to recognize excellence amongst the papers presented by students. All student papers presented at the conference, in either oral or poster sessions, will be eligible for award selection. We will recognize the award finalists at the conference banquet on Wednesday evening. The outstanding paper awardees will be announced just prior to adjourning the conference, early afternoon on Thursday.

We would like to express our sincerest gratitude to all the authors of the submitted abstracts. Their high quality work serves as the foundation for the success of this Conference. The papers were selected by the TPC made up with equal representation from three regional divisions: America, Europe & Africa, and Asia & Oceania. Five sub-committees were formed in order to facilitate a careful review of the large number of submitted abstracts. Each abstract has been evaluated and rated by six expert members of the TPC. The committee recommendations on the acceptance or rejection of papers were taken as binding. We are grateful to all TPC members who volunteered their valuable time, including participation in a two-day on-site meeting in Atlanta, GA, for paper selection. We are also indebted to the International Steering Committee for generously sharing their experience. We gratefully acknowledge the industrial support groups, exhibitors and benefactors for their involvement in this Conference, and the IEEE Robotics and Automation Society for their continued support of this meeting. The dedicated and relentless effort of Ms. Katharine Cline and her team at PMMI in managing this Conference is highly appreciated.

In closing, we hope you enjoy the technical presentations, exhibition booths and events of the Conference this week in San Francisco! The Embarcadero waterfront anchors the world's most majestic skylines. San Francisco's attractions such as Chinatown, Fisherman's Wharf, the vibrant Union Square and the Embarcadero Center are added bonuses, or board the cable car conveniently located outside of the conference hotel for a riding tour of the city.

Farrokh Ayazi
General Chair

Chang-Jin "CJ" Kim
General Chair

ORGANIZING COMMITTEE

General Co-Chairs

Farrokh Ayazi
Georgia Institute of Technology, USA

Chang-Jin "CJ" Kim
University of California, Los Angeles, USA

International Steering Committee

Co-Chairs

Gwo-Bin "Vincent" Lee
National Cheng Kung University, TAIWAN

Toshiyiki Tsuchiya
Kyoto University, JAPAN

Members

Farrokh Ayazi
Georgia Institute of Technology, USA

Karl Böhringer
University of Washington, USA

Jürgen Brugger
Ecole Polytechnique Federale de Lausanne (EPFL), SWITZERLAND

Lionel Buchaillot
IEMN, FRANCE

CJ Kim
University of California, Los Angeles, USA

Gwo-Bin "Vincent" Lee
National Cheng Kung University, TAIWAN

Liwei Lin
University of California, Berkeley, USA

Hiroshi Toshiyoshi
University of Tokyo, JAPAN

Toshiyiki Tsuchiya
Kyoto University, JAPAN

Xiaohong Wang
Tsinghua University, CHINA

Wouter van der Wijngaart
KTH - Royal Institute of Technology, SWEDEN

Hans Zappe
University of Freiburg, GERMANY

Technical Program Committee

Farrokh Ayazi
Georgia Institute of Technology, USA

Jürgen Brugger
EPFL, SWITZERLAND

Michel Despont
CSEM, SA, SWITZERLAND

Andreas Dietzel
Technische Universität Braunschweig, GERMANY

David Horsley
University of California, Davis, USA

Jack Judy
University of Florida, USA

Chang-Jin "CJ" Kim
University of California, Los Angeles, USA

Joonwon Kim
Pohang University of Science & Technology, SOUTH KOREA

Haluk Kulah
Middle East Technical University, TURKEY

Jeroen Lammertyn
KU Leuven, BELGIUM

Gwo-Bin "Vincent" Lee
National Tsing Hua University, TAIWAN

Junghoon Lee
Seoul National University, SOUTH KOREA

Zhihong Li
Peking University, CHINA

Andreu Llobera
Centre Nacional de Microelectronica, SPAIN

Norihisa Miki
Keio University, JAPAN

Hyejin Moon
University of Texas, Arlington, USA

Kurt Petersen
Profusa, USA

Beth Pruitt
Stanford University, USA

Mina Rais-Zadeh
University of Michigan, USA

Ashwin Seshia
Cambridge University, UK

Yu Sun
University of Toronto, CANADA

Hiroshi Toshiyoshi
University of Tokyo, JAPAN

Toshiyuki Tsuchiya
Kyoto University, JAPAN

Deepak Uttamchandani
University of Strathclyde, UK

Ruud Vullers
IMEC, THE NETHERLANDS

Jeff Tza-Huei Wang
Johns Hopkins University, USA

Xiaohong Wang
Tsinghua University, CHINA

Wouter van der Wijngaart
KTH - Royal Institute of Technology, SWEDEN

Yoko Yamanishi
Shibaura Institute of Technology, JAPAN

Yao-Joe Yang
National Taiwan University, TAIWAN

MEMS 2014 TECHNICAL PROGRAM COMMITTEE

1 – Jeroen Lammertyn
2 – Haluk Kulah
3 – Xiaohong Wang
4 – Yoko Yamanishi
5 – Gwo-Bin "Vincent" Lee
6 – Hyejin Moon
7 – Farrokh Ayazi
8 – Kurt Petersen
9 – CJ Kim
10 – Hiroshi Toshiyoshi
11 – Jeff Tza-Huei Wang
12 – Joonwon Kim
13 – Yu Sun
14 – Jack Judy
15 – David Horsley

16 – Yao-Joe Yang
17 – Junghoon Lee
18 – Andreu Llobera
19 – Toshiyuki Tsuchiya
20 – Zhihong Li
21 – Deepak Uttamchandani
22 – Mina Rais-Zadeh
23 – Norihisa Miki
24 – Ashwin Seshia
25 – Andreas Dietzel
26 – Ruud Vullers
27 – Wouter van der Wijngaart
28 – Beth Pruitt
29 – Michel Despont
30 – Jürgen Brugger

ACKNOWLEDGEMENTS

We gratefully acknowledge the support and involvement of this Conference from the following companies and institutions as of the printing of January 3, 2014:

AMMT
Berkeley Sensor & Actuator Center (BSAC)
Coventor, Inc.
DJ DevCorp
Femtotools
FRT of America
Georgia Tech - Institute for Electronics and Nanotechnology
Heidelberg Instruments Inc.
Institution of Engineering and Technology (IET)
IntelliSense
IOP Publishing
iX-factory GmbH
LPE S.p.A.
Lurie Nanofabrication Facility
MEMS Exchange
MEMS Industry Group
MEMS Journal
memsstar Ltd.
MEMStaff Inc.
MicroChem Corp.
Micronarc
Muegge GmbH
OAI
Oxford Instruments Plasma Technology
Polytec, Inc.
Qualtré, Inc.
scia Systems GmbH
SemiProbe Inc.
Si-Ware Systems
Silex Microsystems AB
SoftMEMS LLC
SPTS Technologies
SSEC
Tousimis
Tystar Corporation
Ulvac Technologies
University of Louisville Micro/Nano Technology Center
WIMS^2, Wireless Integrated Microsensing and Systems
Yield Engineering Systems, Inc.
Zurich Instruments AG

MICROFABRICATED IMPLANTABLE WIRELESS MICROSYSTEMS: PERMANENT AND BIODEGRADABLE IMPLEMENTATIONS

Mark G. Allen

Department of Electrical and Systems Engineering

School of Engineering and Applied Science

University of Pennsylvania, Philadelphia, PA USA

ABSTRACT

The tremendous technological convergence of microfabrication technology, wireless communication technology, and low-power circuitry has opened the possibility of widespread use of microfabricated implantable wireless microsystems. A typical operational mode for these microsystems is to transduce a physiological parameter relevant to a disease state of interest, and wirelessly communicate this parameter external to the body to guide therapy. For chronic disease states, long-term, permanent sensors are of interest; while for acute disease states, biodegradable wireless microsystems may be of interest. Two microsystem examples, permanent pressure sensors for chronic monitoring of patients with congestive heart failure, and biodegradable pressure sensors for acute monitoring of patients with transient conditions, are given.

INTRODUCTION

Implantable microsystems are small-scale systems that can be implanted in the human body and can perform structural or transduction tasks. An example of such a task is the measurement of a relevant physiological parameter of interest within the body, and the transmission of this measurement wirelessly to an external reader for guiding disease treatment. Such systems have been discussed for nearly six decades; an early example is the "endoradiosonde", a swallowable device that could monitor gastrointestinal pressure and wirelessly report the results to external electronics [1]. Since that time, many researchers have designed, fabricated, and tested an incredible variety of such microsystems both *in-vitro* and *in-vivo*.

In the early part of the twenty-first century a tremendous technology convergence has taken place. Advances in micro- and nanofabrication technologies have enabled the practical and functional realization of devices with size scales that match well with the size scales of many key portions of the human body. Advances in wireless communication technology have not only made real the ubiquitous communication that we all take for granted today; but have also enabled the ability to transmit and receive extremely weak signals from deep within electrically lossy media (such as the human body). Finally, advances in power sources, energy harvesting, and energy utilization (e.g., low power integrated circuits) have enabled devices that can operate autonomously over time scales that range from days to years.

Exploiting this convergence, research on a large number of microsystems intended for both acute and chronic monitoring has been performed by many groups around the world. Unfortunately, a comprehensive review of this extensive literature cannot be accommodated in this paper. Instead, this paper will attempt to focus on illustrating key principles of chronic and acute implants by means of selected examples from our own research group.

CHRONIC CONDITIONS: PERMANENTLY IMPLANTABLE DEVICES

Over the past thirty years, the medical community has had tremendous success in ameliorating acute mortality from cardiac events [2]. However, the resultant incidence of patients living with impaired heart function has concomitantly increased. One way of thinking about this problem is that an acute condition has been transformed into a chronic one. Although clearly the avoidance of sudden death from a coronary event is highly desirable progress, the medical community must now turn its attention to chronic management of impaired heart function, or *heart failure*. The prevalence of heart failure has increased over the past thirty years due to this shift in disease modality. Approximately half of the people diagnosed with heart failure will die within five years of diagnosis [2], and the costs of heart failure in the United States are estimated at $32 billion USD each year [3]. A typical manifestation of heart failure is *congestive heart failure*, in which the insufficient pumping capacity of the damaged heart causes buildup of fluid in the lungs. Such so-called decompensation events are typically treated by hospitalization, in which the patient is stabilized and a cocktail of drugs are titrated to restore as best as possible heart functionality.

Analogously to the manner in which the lack of control of the level of blood sugar is an indicator of diabetes, it has been postulated that lack of control of the level of blood pressure deep within the heart is associated with congestive heart failure [4]. This leads to the hypothesis that if the pressure within the heart can be measured on a continuous basis, decompensation and the resultant hospitalization can be prevented by titrating medication on a continuous or semicontinuous basis, relying on measurement of this pressure output to guide this titration. The current approach to pressure measurement is a Swan-Ganz catheterization of the pulmonary artery [5], which is typically performed in a hospital setting. A wirelessly interrogatable implanted microsystem placed in the pulmonary artery would be a

978-1-4799-3510-9/14 $31.00 © 2014 IEEE

tremendous improvement in monitoring since on-demand data without the need for hospitalization would be available to guide therapy.

Previously we reported on micromachined implantable pressure sensors for the application of monitoring the pressure in the excluded space of stent-graft-repaired abdominal aortic aneurysms [6,7]. These devices are passive and wireless, with no need for an internal power source or circuitry. As reported in [6], the sensor comprises flexible plates bearing inductor windings (along with associated distributed capacitances) that bound a hermetically-sealed reference cavity. The inductor windings serve two purposes: to form a resonant electrical circuit with the capacitor and to magnetically couple with an external loop. A change in the pressure surrounding the sensor changes the position of the plates, thereby changing the capacitance and resonant frequency of the sensor. This resonant frequency change can be monitored by external electronics and through appropriate calibration be translated to the pressure sensed by the sensor.

Sensor readout is accomplished by a 'ping-and-listen' approach that exploits phase correlation to exclude ambient noise or interfering signals and enables detection of sensors embedded in lossy media at significant distances. As discussed in [8], the sensor is energized with a low duty cycle, gated burst of radio frequency energy having a predetermined frequency or set of frequencies and a predetermined amplitude. The energizing signal is coupled to the sensor via magnetic coupling and induces a current in the sensor that oscillates at the resonant frequency of the sensor. The system receives the response of the sensor via magnetic coupling and does a phase comparison of the signal to determine whether the energizing signal frequency matches the resonant frequency of the sensor. If not, the energizing frequency is adjusted using a phase locked loop and the process is repeated until the sensor is optimally energized at resonance.

As further described in [6], sensor fabrication is based on fused silica MEMS. A lower substrate of fused silica has a lower inductor electrodeposited on it using a plate-through photoresist mold approach. An upper substrate of fused silica has a recess etched into it using isotropic wet etching and an upper inductor is electrodeposited into the recess using the same plate-through-mold approach. The two substrates are then fusion-bonded. Typical resonant frequencies of sensors are in the 30-40 MHz range with Q-factors of approximately fifty and readout distances of 8-12 inches. Later iterations of sensors used in heart failure trials (described below) have had a variety of winding fabrication approaches and configurations but the basic sensor configuration is as described above.

The hypothesis of sensor-guided treatment resulting in reduced heart-failure-related hospitalization was tested in the so-called CHAMPION clinical trial [9,10]. In that trial, 550 patients suffering from heart failure had sensors implanted in branches of their pulmonary arteries using a catheter-based delivery approach (after right heart

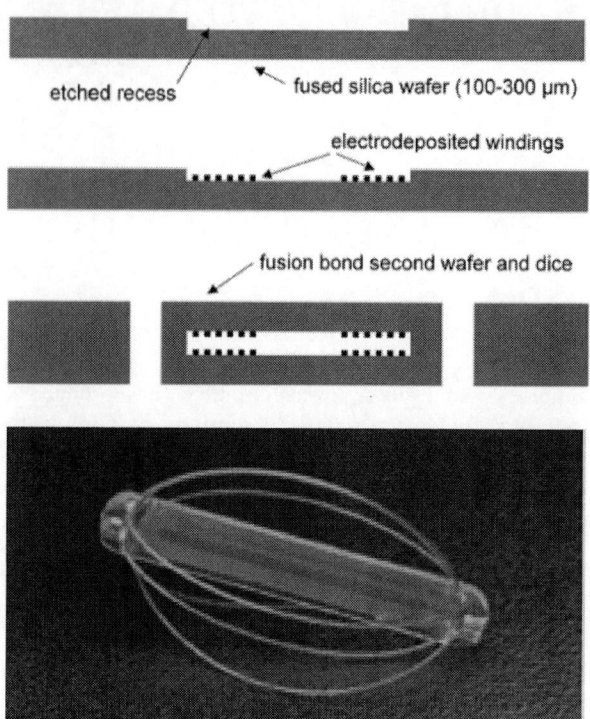

Figure 1. (top) Simplified fabrication sequence (side view) of sensor; (bottom) a fabricated sensor. The windings are on the interior of the sensor; the external wires surrounding the sensor have no electrical functionality but act to keep the sensor fixed in place in-vivo. Multiple configurations of sensor (size, fixation, and winding type) have been used; the devices used in heart failure testing are approximately 15 mm long, 3.5 mm wide, and 2 mm thick.

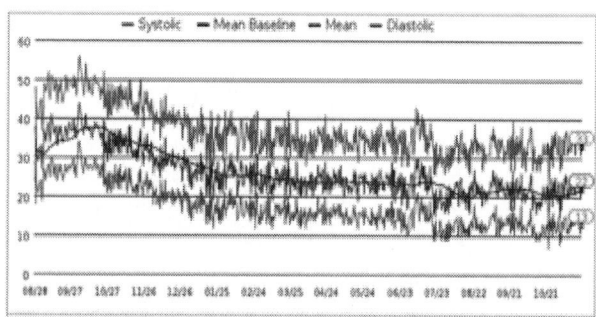

Figure 2. Pulmonary artery (PA) pressure data typical of that obtained from a patient during the CHAMPION trial. The upper red curves are systolic pressure; the lower green curves are diastolic pressure; and the intermediate blue curves are mean pressure, respectively. The x-axis represents over a year of daily-collected data. As can be seen, after a 3-4 month period, the patient's mean PA pressure has been stabilized within the physiologically desirable 10-25 mm Hg range.

catheterization took place, mean delivery time for the sensor was 7 minutes). The implanted patients were divided approximately evenly into a control arm and a treatment arm. Data from all patients were collected; however, in the control arm, patients received traditional heart failure disease management (including drug regimens) regardless of data supplied by the sensor, while in the treatment arm, patients' therapy and drug regimens were also guided by sensor data.

Zero sensor failures were observed over the six month duration of the trial. Further, patients in the treatment arm enjoyed an approximately 30% reduction in heart failure related hospitalizations at 6 months, and an annualized 38% reduction in heart failure related hospitalizations for the entire randomized followup.

ACUTE CONDITIONS: BIODEGRADABLE DEVICES

Many disease states manifest themselves as acute rather than chronic conditions. Examples include healing of wounds or fractured bones; therapeutic body or tissue remodeling, such as arterial angioplasty or cosmetic procedures; and timed diagnostic procedures, such as imaging. In such cases, if implants are indicated, it may be desirable to consider *biodegradable* devices.

Biodegradable structural devices such as biodegradable stents [11] or biodegradable microneedles for transdermal drug delivery [12] have been demonstrated; however, the development of biodegradable wireless *sensors* is more complex. Passive wireless devices such as those described in the previous section are amenable to biodegradable adaptation due to their simplicity. Several groups have investigated approaches to passive wireless resonators for stimulation [13] or readout characterization [14].

Recently, we demonstrated a passive wireless pressure sensor made completely of biodegradable materials [15]. Philosophically the device is similar to that described in Figure 1; however, the fused silica dielectric has been replaced by biodegradable poly(L-lactic acid) (PLLA) and polycaprolactone (PCL); and the windings have been replaced with biodegradable Zn/Fe bilayers, the degradation time of which can be tailored by exploiting galvanic corrosion between these materials.

Since these biodegradable materials are not nearly as robust as typical MEMS materials, unconventional fabrication processes may be required for their realization. Figure 3 shows the fabrication process exploited for the pressure sensors described here. Conventional MEMS processing (lithography and electroplating) is utilized to fabricate Zn/Fe bilayer conductors comprising inductor coils and capacitor plates on a polyimide substrate. Separately, a PLLA sheet is prepared by solvent casting. The conductors are embossed into the PLLA sheet, the polyimide is peeled off, and the seed layer is polished, leaving the all-biodegradable device layers realized with no additional solvent or water-based processing. After incorporation of a spacer to define the cavity, the structure is then folded and

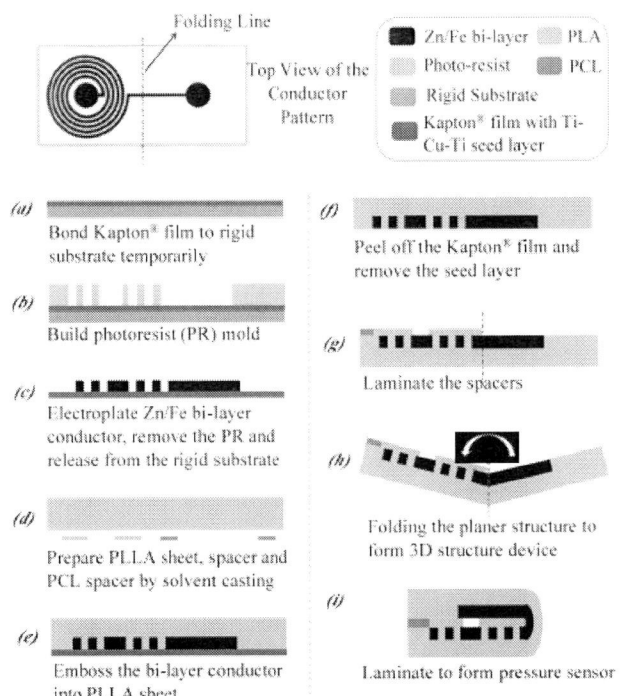

Figure 3. Biodegradable pressure sensor fabrication process.

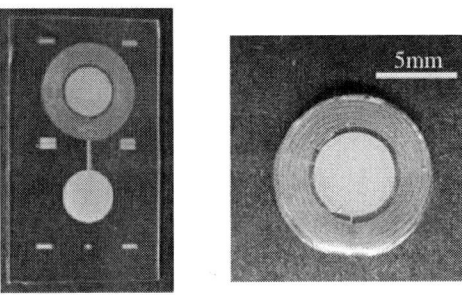

Figure 4. Biodegradable pressure sensors. (Left) prior to spacer incorporation, folding, and lamination; (Right) after lamination and trimming.

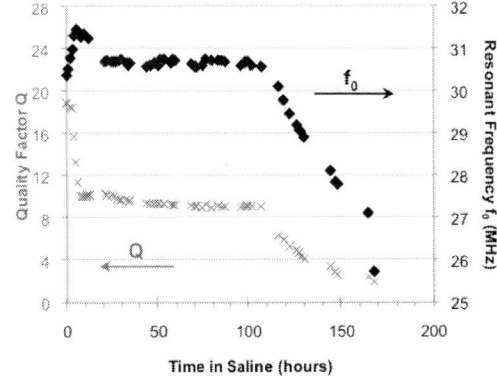

Figure 5. Resonant frequency and quality factor of biodegradable sensors over time. The sensors were stable with frequency sensitivity of approximately 55 kHz/kPa and functional lifetime of approximately 4 days.

laminated together to form the sensor. Figure 4 shows photographs of fabricated sensors.

Sensors were operated in both air and saline environments and after an initial stabilization period exhibited a frequency-shift-based pressure sensitivity of approximately 55 kHz/kPa. The functional lifetime of the sensors was approximately four days (Figure 5); the functional lifetime can be tailored by the choice of polymer encapsulation and area ratio of the bilayer galvanic couple. Other choices for biodegradable conductors, such as Mg, are also under consideration [16].

FUTURE DIRECTIONS:
ACTIVE BIODEGRADABLE MEMS AND UBIQUITOUS IMPLANTABLE SENSING

The devices discussed above rely on passive mechanisms with no batteries or integrated circuits. Such devices do not exploit the third technological convergence of low power circuitry; and by necessity the functionality of passive wireless sensors is limited. It is therefore useful to consider future directions in biodegradable microsystems that incorporate active elements.

To realize active biodegradable microsystems, transient silicon and transient power sources (or energy harvesters) should be considered. Transient degradable silicon has been demonstrated [17], as have batteries made from biodegradable materials with sufficient stored energy to operate microsystems [18-20]. Given that many of the fundamental technological hurdles to the realization of these systems have been overcome, it is only a matter of time before both active biodegradable and permanent implantable microsystems become widespread for the treatment of disease and improvement of the human condition.

ACKNOWLEDGEMENTS

The author would like to acknowledge the significant contributions of members past and present of the Microsensors and Microactuators Research Group, as well as the team at CardioMEMS, Inc., of Atlanta, GA, to the work described in this paper.

REFERENCES

[1] R.S. Mackay and B. Jacobson, "Endoradiosonde," *Nature*, vol. 179, p. 1239-1240 (1957)

[2] A.S. Go *et al.*, "Heart Disease and Stroke Statistics – 2013 Update," *Circulation*, 127:e6-e245 (2013)

[3] P.A. Heidenreich *et al.*, "Forecasting the Future of Cardiovascular Disease in the United States," *Circulation*, 123(8):933-44 (2011)

[4] F.M. Merchant *et al.*, "Advances in Arrhythmia and Electrophysiology," *Circulation: Arrhythmia and Electrophysiology*, 3:657-667 (2010)

[5] H.J. Swan *et al.*, "Catheterization of the heart in man with use of a flow-directed balloon-tipped catheter," *New England Journal of Medicine*, 283:447 (1970)

[6] M.G. Allen, "Micromachined Endovascularly Implantable Wireless Pressure Sensors: From Concept to Clinic," *Proc. Transducers '05*, p 275-278 (2005)

[7] M.A. Fonseca, J. Kroh, J. White, and M.G. Allen, "Flexible wireless passive pressure sensors for biomedical applications," *Solid-State Sensor, Actuator, and Microsystems Workshop (Hilton Head)* (2006)

[8] J. Joy *et al.*, "Communicating with Implanted Wireless Sensor," U.S. Patent 7,245,117 (2007)

[9] W.T. Abraham and P.B. Adamson, "Primary Results of the CardioMEMS Heart Sensor Allows Monitoring of Pressure to Improve Outcomes in Class III Heart Failure Patients, Presentation to the 2010 European Society of Cardiology Heart Failure Congress, Berlin, Germany. Slides are accessible from: http://www.escardio.org/congresses/hf2010/slides-trials /documents/hf2010-champion-abraham.pdf (2010)

[10] W.T. Abraham *et al.*, "Wireless Pulmonary Artery Haemodynamic Monitoring in Chronic Heart Failure: A Randomized Controlled Trial," *The Lancet*, vol. 377, no. 9766, pp 658-666 (2011)

[11] R. Waksman, "Biodegradable stents: they do their job and disappear," *J. Inv. Card.*, vol. 18, pp.70-74, (2006)

[12] J.H. Park, M.G. Allen, and M.R. Prausnitz, "Biodegradable Polymer Microneedles: Fabrication, Mechanics, and Transdermal Drug Delivery," *J. Controlled Release*, vol. 104, p. 51-66 (2005)

[13] B. Finamore, G. Fedder *et al.*, "Development of an Implantable Biodegradable Electrical Stimulator for Bone Repair," *Proc. Biomed Eng Soc Fall Mtg* (2009)

[14] C. M. Boutry, C. Hierold *et al.*, "Characterization of miniaturized RLC resonators made of biodegradable materials for wireless implant applications," *Sensors Actuat. A-Phys.*, vol. 189, pp.344-355 (2013)

[15] M. Luo, C. Song, F. Herrault, and M.G. Allen, "A Microfabricated Wireless RF Pressure Sensor Made Completely of Biodegradable Materials," *Solid-State Sensor, Actuator, and Microsystems Workshop (Hilton Head)* (2012); expanded version to appear in *J. Microelectromechanical Systems* (in press)

[16] M. Tsang, F. Herrault, R. Shafer, and M.G. Allen, "Methods for the Microfabrication of Magnesium," *Proc. IEEE MEMS Conference*, p. 247-250 (2013)

[17] J. Rogers et al., "A Physically Transient Form of Silicon Electronics," *Science*, vol. 337, no. 6102, p. 1640-1644 (2012)

[18] K.B. Lee, "Urine-Activated Paper Batteries for Biosystems," *J. Micromechanics Microeng.*, vol. 15, p. S210-S214 (2005)

[19] M. Zdeblick, "How Wireless Health Will Change Health Care Delivery," *Solid-State Sensor, Actuator, and Microsystems Workshop (Hilton Head)* (2012)

[20] M. Tsang et al., "A MEMS-Enabled Biodegradable Battery For Powering Transient Implantable Devices," *Proc. IEEE MEMS Conference* (2014)

CONTACT

M. Allen, tel: +1-215-898-5901; mallen@upenn.edu

AN INTEGRATED MICROFLUIDIC SYSTEM FOR RAPID ISOLATION AND DETECTION OF LIVE BACTERIA IN PERIPROSTHETIC JOINT INFECTIONS

Wen-Hsin Chang[1], Chih-Hung Wang[1], Sung-Yi Yang[4], Yi-Cheng Lin[4], Jiunn-Jong Wu[5], Mel S Lee[6]* and Gwo-Bin Lee[1,2,3]*

[1]Department of Power Mechanical Engineering, [2]Institute of Biomedical Engineering, [3]Institute of NanoEngineering and Microsystems, National Tsing Hua University, Hsinchu, Taiwan, [4]Medical R&D, Jabil Circuit Inc., Ltd., Taichung, Taiwan, [5]Department of Medical Laboratory Science and Biotechnology, National Cheng Kung University, Tainan, Taiwan, [6]Department of Orthopaedic Surgery, Chia-Yi Chang Gung Memorial Hospital, Taiwan

ABSTRACT

Periprosthetic joint infection (PJI) is difficult to treat and the incidence is between 1% and 2% in primary arthroplasties. Implant-associated infections usually arise via either primary infections from bacterial invasion at the time of implant surgery or secondary infections from hematogenous sources. The two-stage re-implantation protocol that consists of extensive debridement at the first stage followed by delayed re-implantation is currently the standard process for chronic PJI with a success rate between 82% to 95%. Furthermore, re-implantation arthroplasty should be only performed after ensuring the complete eradication of bacterial infection to avoid devastating complications. However, it is still a challenge in clinical practice to accurately determine the eradication of infections before or during implantation. Conventional diagnostic methods such as measurements of serum C-reactive protein or interleukin-6 levels, culture of joint aspirates, and microscopic examination of tissue biopsy are either non-specific or relatively time-consuming. For critical decision-making before or during the re-implantation surgery, a quick method with high sensitivity and specificity is therefore of great need. Previous studies reported bacterial ribosomal ribonucleic acids (rRNAs) as a target for the diagnosis of infections since rRNAs are highly conserved among bacterial species and abundant in amount. By using universal primers, the presence of bacterial rRNA could be amplified by using reverse-transcription polymerase chain reaction (RT-PCR). Currently the RT-PCR method for detection of bacterial rRNA is highly sensitive with a limit of detection (LOD) as low as a pictogram level. However, RT-PCR signals could only indirectly distinguish live from dead bacteria based on the degradation of rRNA in the tissue. Furthermore, the whole detection procedure of 16s rRNA RT-PCR is labor-intensive. Therefore, an integrated microfluidic system was presented in this work, which could distinguish the existence of live bacteria within 1 hour with a LOD of 10^4 colony formation unit (CFU). In this study, the fabrication of the microfluidic chip was improved so that the consistency of the transported liquid volume was increased. Moreover, by using an ethidium monoazide (EMA) assay, the cumbersome pre-treatment process of rRNA in live bacteria can be alleviated. This is the first time that a microfluidic platform was reported to detect live bacteria successfully in PJI samples.

INTRODUCTION

Prosthetic joint replacement surgery has been used extensively to improve the life quality of patients suffering from immobility and pain. It is estimated that the demand for the primary total hip arthroplasties may reach 572,000 cases and the demand for primary total knee arthroplasties will be even higher (about 3.48 million case) by the year of 2030 [1]. However, PJI occurs at the surgical site prevents arthroplasty from being a most perfect medical treatment. PJI needs to be treated by implant removal and extensive debridement of the joint followed by delayed reimplantation of prosthesis after complete eradication of infection. It is likely to implant a prosthesis into a joint where PJI has not been completely eradicated with residual microorganisms. For critical decision-making before or during the arthroplasty, a rapid method for detection of bacterial infections with high sensitivity and specificity would be extremely valuable.

Criteria for diagnosing PJI are challenging and relied on systemic or local infection signs in conjunction with laboratory tests of serum C-reactive protein (CRP) level, histopathological examination [2], microbiological culture, Gram staining, white blood cell counts, erythrocyte sedimentation rate, radiograph and nuclear bone scans [3]. However, laboratory tests are non-specific and may fluctuate with underlying medical conditions. Image studies such as radiograph or radionuclide scans can only provide additional information in controversial cases [4]. At present, microbiological culture is taken as the gold standard of PJI diagnosis. Unfortunately, many cases of established PJI that were inadequately treated with antibiotics could not have a positive culture result with average culture negative rates of 11% [5]. Moreover, it usually takes 3-7 days to get the definite culture results and antibiotic-sensitivity tests.

After the emergence of genetic mapping, molecular diagnosis such as polymerase chain reaction (PCR) has been applied to many different fields including PJI diagnosis. Previous studies used bacterial ribosomal RNAs (rRNAs) as a target for the diagnosis of infections [6]. The bacterial rRNAs are highly conserved among bacterial species and abundant in amount but absent in human. By using universal primers, the presence of rRNAs can be amplified by RT-PCR. Currently the detection limit of RT-PCR for bacterial rRNA is highly sensitive and can reach the pictogram level. rRNAs will be degraded rapidly after bacteria death. Hence, using rRNA as amplification templates can provide information of live bacteria only so that the false-positive results of dead bacteria can be avoided. However, the handling and storage processes of rRNA are very difficult because the easy degradation of rRNA. Therefore, a live bacteria detection system for PJI is of great need.

To meet the demand for clinicians, a novel microfluidic system which can perform the entire detection process for live bacteria from PJI samples was purposed in current study. Universal probes for bacteria identification, introduction of EMA for differentiating live and dead bacteria, universal 16s rRNA primers, a microfluidic chip and an integrated control system for PJI diagnosis were established. To waive the complicated procedure of RNA handling, a DNA staining dye, EMA, was employed to distinguish live and dead bacteria because EMA can only penetrate into dead bacteria with broken cell walls and then intercalate into DNA so that the denature of DNA was prevented even under 95°C [7]. Vancomycin, an antibiotic, was then coated on the surface of magnetic beads and was used as universal probes for bacteria isolation due to its ability to bind on the peptidoglycans of bacterial cell walls [8]. Note that it can both bind on Gram-positive and Gram-negative bacteria [9-10]. The experimental results showed that the developed microfluidic system was capable of performing live bacterial detection in PJI clinical samples. The developed system is promising for PJI diagnosis and can be a useful tool for clinical applications in the near future.

MATERIALS AND METHODS
Experimental procedures

Figure 1 shows a schematic illustration of the experimental process for the bacteria isolation, the on-chip PCR and fluorescence detection. First, the clinical PJI sample was pre-treated with EMA and exposed under visible light (Figure 1(a)). At the same time, the entire bacteria were then captured by Vancomycin-coated magnetic beads which were used as a universal probe for bacteria isolation because they could capture various species of bacteria (Figure 1(b)). After incubation for 10 minutes, a magnet was placed under the microfluidic chip automatically by the integrated control system [11] (Figure 1(c)). Wash buffer was then transported into a reaction chamber to wash out the unbound materials (Figure 1(d)). After the washing process, the magnet underneath the microfluidic chip was removed to enhance the contact area of PCR reagents and bacteria bound on the Vancomycin-coated magnetic beads. Next, PCR reagents were transported into the reaction chamber and the pre-programmed temperature cycles for PCR reaction was executed by the temperature control module (Figure 1(e)). The fluorescence detection was carried out on the microfluidic chip at the end of PCR reaction and the diagnostic results were shown on the personal computer of the integrated control system (Figure 1(f)). Note that the whole process could be completed within 1 hour automatically.

(a) EMA addition and visible light exposure

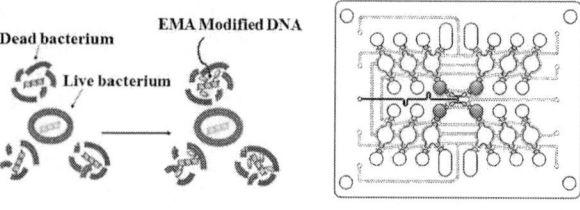

(b) Bacteria captured by magnetic beads coated with Vancomycin

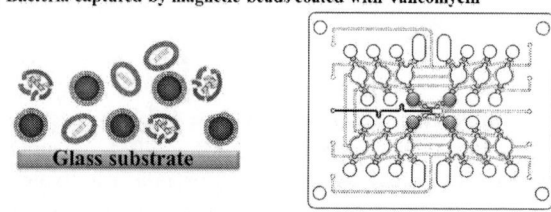

(c) Bacteria-magnetic beads complexes collection by the magnet

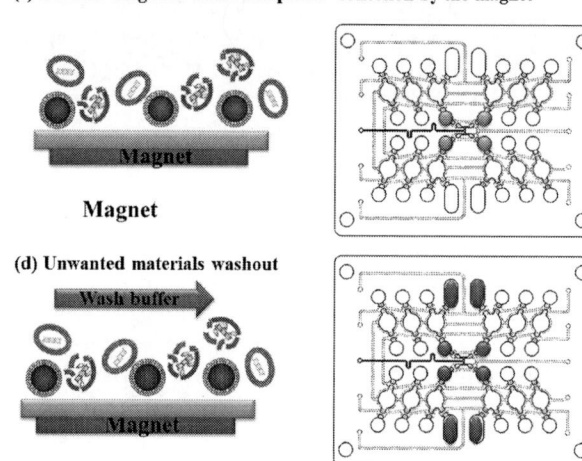

(d) Unwanted materials washout

(e) PCR reaction

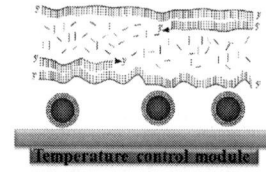

(f) On-chip fluorescence detection

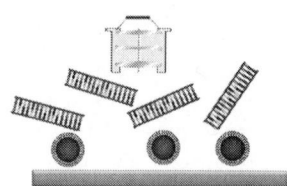

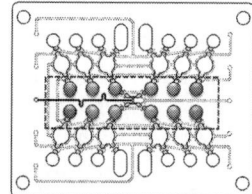

Figure 1: Schematic illustration of the experimental procedure. (a) EMA addition; (b) Bacteria capture; (c) Bacteria-magnetic beads complexes collection; (d) Unwanted materials washout; (e) PCR reaction and (f) Fluorescence detection.

Chip design

The microfluidic chip used in this study is an improved version adopted from our previous work [11]. Four identical diagnostic sets including reaction chambers, positive chambers, negative chambers, and micro-components for liquid automatic handling were integrated in the chip. Furthermore, micro pillars were specially designed under the thin membranes in order to make the pumping volume more accurate. Hot embossing fabrication process was used to mass-fabricate the chips.

Figure 2 shows the photographs of the microfluidic chip. The details of the improved parts were demonstrated in Fig. 3. For instance, quantification pillars (Fig 3(a)) under the thin membranes were designed to control the pumping volume each time precisely. Furthermore, the micro-structures on the normally-closed micro-valves (Fig 3(b)) which simplified the fabrication process were demonstrated..

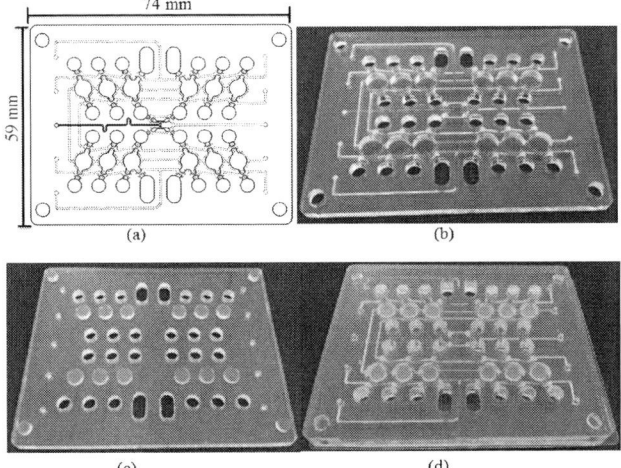

(a)   (b)

(c)   (d)

Figure 2: Photographs of (a) mass-produced chips, (b) the bottom layer, (c) the upper layer and (d) assembled microfluidic chip. The chip was measured to be 75.0 mm x 59.0 mm x 6.7 mm.

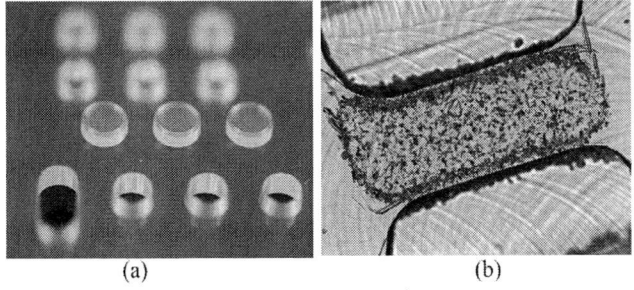

(a)   (b)

Figure 3: Photographs of (a) the quantification pillars and (b) the micro-valve of the hot-embossing microfluidic chip.

RESULTS AND DISCUSSION

The pumping rate of the integrated microfluidic chip

The pumping rates of the micropumps on the microfluidic chip were tested to verify that the use of cylinders under the thin membranes can improve the consistence of the amount of transported liquid. Table 1 indicated that the consistence of the pumping volume each time of different chambers was improved because of the newly designed micro pillars.

Table 1: Consistence of pumping volumes of different chambers on the microfluidic chip

Chamber no.	#1		#2		#3		#4	
Pumping times	Ave	Std	Ave	Std	Ave	Std	Ave	Std
1	17.9	0.26	16.8	0.23	17.7	0.10	17.5	0.40
2	35.4	0.35	34.7	0.64	34.4	0.45	34.6	0.44
3	53.6	0.45	52.9	0.15	52.6	0.35	52.3	0.38

Determination of EMA dosage

To determine the dosage of EMA applicable to the developed bacteria detection process, eight bacteria commonly seen in PJI [2,3] were tested under the same experimental procedure except being treated with EMA of different concentrations. The results of EMA dosage were demonstrated in Figure 4. All PCR reactions were performed successfully when there was no EMA and hence live and dead bacteria cannot be distinguished. Live and dead bacteria were well distinguished only while the dosage of EMA reached 1 mg/mL. Therefore, 1 mg/mL was chosen as the EMA optimal dosage for the subsequent experiments.

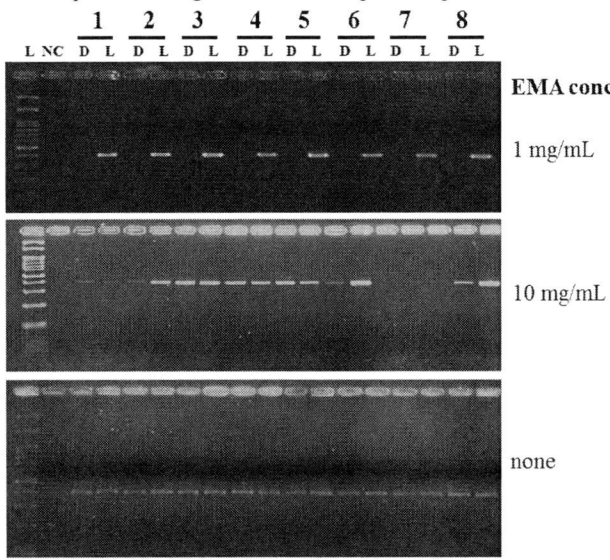

Figure 4: EMA dosage determination; L: 100-bp DNA ladder, NC: negative control using distilled water, 1. E. coli, 2. Enterococcus sp., 3. Coagulase-negative staphylococci, 4. Streptococcus pyogenes, 5. Klebsiella pneumonia, 6. methicillin-resistant Staphylococcus aureus, 7. Staphylococcus aureus, 8. Pseudomonas syringae. D: dead bacteria L: live bacteria.

LOD of the proposed microfluidic system

Figure 5 shows the limit of detection of the proposed microfluidic system. An *E. Coli* culture of 10^6 CFU was 10-fold diluted to single CFU and all dilutes were subjected to the proposed bacteria detection process. Figure 5(a) shows the results from gel electrophoresis. The 16s rRNA PCR products with an initial concentration as low as 10^4 CFU could be visualized under UV exposure. The fluorescent signals of 16s rRNA PCR products were obtained from the integrated control system. The results demonstrated that the fluorescent signal for bacterial concentration as low as 10^4 CFU could be distinguished from negative control (3.5±0.27

V vs. 1.8±0.12 V), as shown in Figure 5(b). These results also demonstrated that the performance of the integrated control system was comparable with the traditional gel electrophoresis.

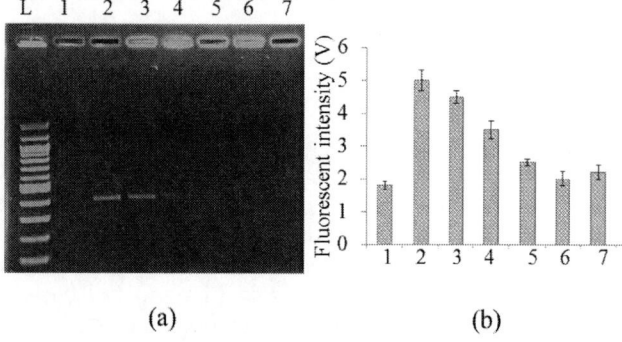

(a)   (b)

Figure 5: LOD of the integrated microfluidic system; (a) The results from gel electrophoresis; (b) Fluorescent signals of PCR products obtained from different amounts of E. coli. Lane 1: negative control using distilled water, lane 2-7: 10-fold serial dilutions of E. coli starting with 10^6 CFU.

The feasibility of Vancomycin-coated magnetic beads and16s rRNA PCR in PJI clinical samples spiking with bacteria

In order to validate if the developed system could be used in PJI clinical samples, *E. Coli* was first spiked into a PJI clinical sample without bacterial infections and subjected to the above-mentioned experimental procedure. Figure 6 shows the results of the PJI sample-spiking test. The 16s rRNA PCR products of *E. Coli* only and the PJI clinical sample spiked with *E. Coli* could be both detected by using the traditional gel electrophoresis (Figure 6(a)) and the fluorescent signals measured using the integrated control system (Figure 6(b)). The data demonstrated that the proposed experimental scheme could be feasible in the microenvironment of the PJI clinical sample.

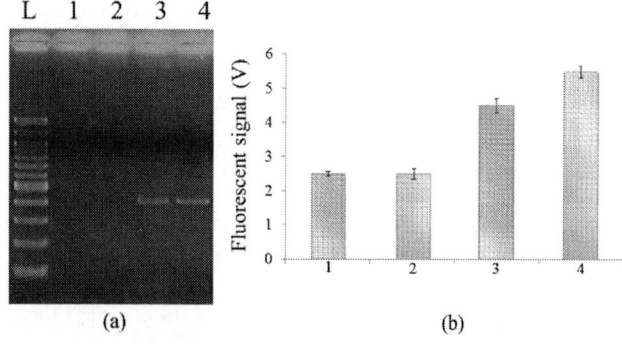

(a)   (b)

Figure 6: Bacteria detection in spiked clinical samples using the integrated microfluidic system; (a) The results from gel electrophoresis; (b) Fluorescent signals of RT-PCR products obtained from different amounts of total RNA extracted from E. coli. Lane 1: negative control using distilled water ,lane 2:aspetic PJI clinical sample, lane 3: positive control using E. Coli, lane 4: aseptic PJI clinical sample spiked with E. Coli.

CONCLUSION

In this study, an automatic microfluidic system capable of carrying out the detection of bacteria from the PJI clinical sample has been demonstrated. The entire process including bacteria isolation, on-chip PCR and fluorescence detection can be completed within 60 minutes. The proposed experimental procedure has a detection limit of 10^4 CFU. This is the first time that an integrated microfluidic system could be used in the PJI clinical sample for bacteria detection and would be a useful tool during the primary or reimplantation arthroplasty.

ACKNOWLEDGEMENTS

The authors would like to thank the Chang Gung Memorial Hospital in Taiwan for the financial support. Partial financial support from National Science Council is greatly appreciated.

REFERENCES

[1] S. Kurtz, K. Ong, E. Lau, F. Mowat and M. Halpern, *J. Bone. Joint. Surg. Am.*, 89-A (2007), pp. 780-785.

[2] W. Zimmerli, A. Trampuz and P.E. Ochsner, *N. Engl. J. Med.*, 351(2004), pp. 1645-1654.

[3] B.D. Mariani and R.S. Tuan, *Mol. Med. Today.*, 98 (1998), pp. 207-214.

[4] T.W. Bauer, J. Parvizi, N. Kobayashi and V. Krebs, *J. Bone. Joint. Surg. Am.*, 88-A (2006), pp.869-881.

[5] E.F Berbari, C. Marculescu, I. Sia, B.D. Lahr, A.D. Hanssen, J.M. Steckelberg, R. Gullerud, D.R. Osmon, *Clin. Infect. Dis.*, 45 (2007), pp.1113–1119.

[6] P.F. Bergin, J.D. Doppelt, W.G. Hamilton, G.E. Mirick, A.E. Jones, S. Sritulanondha, J.M. Helm and R.S. Tuan, *J. Bone. Joint. Surg. Am.*, 92-A (2010), pp. 654-663.

[7] Y.H. Liu, C.H. Wang, J.J. Wu and G.B. Lee, *Biomicrofluidics.*, 6 (2012), pp. 034119 -1-034119-11.

[8] G. M. de Tejada, S. Sanchez-Gomez, I. Razquin-Olazaran, I. Kowalski, Y. Kaconis, L. Heinbockel, J. Andra, T. Schuerholz, M. Hornef, A. Dupont, P. Garidel, K. Lohner, T. Gutsmann, S.A. David and K. Brandenburg, *Curr. Cancer. Drug. Targets.*,13 (2012), pp.1121-1130.

[9] D.H. William and B. Bardsley, *Angew. Chem. Int. Ed.*, 38 (1999), pp.1172-1193.

[10] A. J. Kell, G. Stewart, S. Ryan, R. Peytavi, M. Boissinot, A. Huletsky, M. G. Bergeron and B. Simard, *ACS. NANO.*, 2 (2008), pp.1777-1788.

[11] W.H. Chang, S.Y. Yang, C.H. Wang, M.A. Tsai, P.C. Wang, T.Y. Chen, S.C. Chen and G.B. Lee, *Sens. Actuator. B-Chem.*, 180 (2013), pp. 96-106.

CONTACT

* Gwo-Bin Lee, tel:+886-3-5715131 ext. 33765; gwobin@mail.ncku.edu.tw

* Mel S Lee, tel:+886-975-365-587; mellee@adm.cgmh.org.tw

A CABLE-TIE-TYPE PARYLENE CUFF ELECTRODE FOR PERIPHERAL NERVE INTERFACES

Huaiqiang Yu, Wenjie Xiong, Hongze Zhang, Wei Wang and Zhihong Li
National Key Laboratory of Science and Technology on Micro/Nano Fabrication,
Institute of Microelectronics, Peking University, Beijing 100871, China

ABSTRACT

We have designed, fabricated and characterized a cable-tie-type parylene cuff electrode for peripheral nerve interfaces, whose diameter is adjustable to accommodate the nerve properly during implantation. Cuffs made of thin and flexible parylene minimize mechanical damage to surrounding tissues after implantation. Moreover, the integrated parylene cable and pads facilitate the connection with external circuits through wired or wireless interfaces. Using the fabricated cable-tie-type parylene cuff electrodes, acute neural recoding and stimulation tests were performed on the rat sciatic nerve to verify the capabilities of recording the neural activity and selectively stimulating different nerve fascicles.

INTRODUCTION

A large number of neuroprostheses, developed to link the peripheral nervous system (PNS) with an external device via neuro-electrode interfaces, have significant potential to treat or assist patients with neurological disorders such as spinal cord injuries (SCI) and stroke [1]. Recently, there has been the development of functional electrical stimulation (FES) systems based upon nerve-based electrode technology to substitute the brain control and directly apply current into the intact peripheral nerves, with intent of producing selective and controlled activation of different muscles [2]. In such devices, one key component is the electrode that allows electrical stimulation of different portions of the nerve and/or recording neural activity as sensory feedback information. Among the most successful nerve-based electrodes, cuff electrodes have been investigated over decades and have extensively been used for both basic research and clinical practice [3], [4] due to the high biocompatibility and simplicity of handling and implantation.

The cuff electrode consists of an insulating sheath encircling a nerve, a number of electrode sites arranged on the inside walls and lead wires. To minimize passively-induced neural damage, the cuff electrode itself should be made as small and flexible as possible. Among the factors related to safety, it is cuff nerve diameter ratio (CNR) that requires the most attention. In general, a small CNR prevents the flow of blood, nutrients and other critical fluids into nerves [5], while a large CNR leads to poor contact between electrode and nerve tissue [6]. The Association for the Advancement of Medical Instrumentation (AAMI) recommends that CNR should be larger than 1.5 for safe implantation. Thus, CNR is important for safe implantation and effective stimulation or high-quality recording.

A typical cuff electrode is the split-cylinder that has a longitudinal slit along the silicon rubber tube. The slit has to be closed by sutures, wax, or silicon rubber flaps. The split-cylinder cuff electrode that is commercially available from MicroProbes for Life Sciences (Gaithersburg, MD) has been used for many years by Loeb's group. Another widely used type is the spiral cuff, which consists of electrode sites embedded within a self-curling sheet [7]. The cuff is fabricated by bonding a pre-stretched polymer sheet to an un-stretched polymer sheet. The spiral cuff electrode is expandable so that it can be sized to fit snugly around a nerve or adjust for neural swelling. However, Loeb and Peck [8] reported that it is difficult to stretch and place the tightly wrapped spiral cuff around the nerve. There exist many variants of the split-cylinder type and the spiral type. In most cases, the method of fabrication is based on the assembly of discrete components, and thus the device has a bulky size and a limited number of electrode sites, which may limit its use in clinical practice. In contrast, the microelectromechanical system (MEMS) techniques offer great capability in minimization and system integration. Recently, microfabricated, polyimide-based spiral cuff electrodes have been developed and characterized [9]. To achieve a cylinder spiral shape, a thermal treatment is required, after the flat cuff electrode is released from substrate and rolled using a designed tool.

This paper proposed a cuff electrode, called the cable-tie-type cuff electrode, which is fabricated by electroplating and parylene deposition. In contrast to the conventional cuff electrodes, the cuff diameter of this type can be intra-operatively adjusted to fit the nerve properly by using the locking structure including paired ratchet teeth and a locking loop. The cuff is made out of thin, flexible and biocompatible parylene, which can minimize the likelihood of mechanically-induced neural damage after implantation. The integrated parylene ribbon cable facilitates connection with external circuits through wired or wireless interfaces. Further, the acute neural recoding and stimulation tests were performed on the rat sciatic nerve using the fabricated cable-tie-type parylene cuff electrodes.

DESIGN CONCEPT

The design of the cable-tie-type parylene cuff electrode is illustrated in Fig. 1. The device consists of a parylene strip with a guide tongue at the end and gold ratchet teeth along the edges, platinum MEAs, a gold locking loop, a parylene ribbon cable and pads for external connection (or coil for telemetry). Similar to the common cable tie, after the guide tongue is inserted into the locking loop, the parylene strip is carefully pulled until the paired ratchet teeth are caught. The parylene strip can only move along one direction and be locked at any required place. The increment of cuff diameter depends on distance between adjacent teeth. An electroplated gold layer (10 μm thick) is locally deposited on the teeth and the locking loop to guarantee locking strength. Due to the flexibility of parylene and the malleability of gold, the closed cuff electrode can also be released for re-use by careful open-up procedure.

In this work the cable-tie parylene cuff electrode is designed to fit the rat sciatic nerves (1.0-1.5 mm in diameter), illustrating the flexibility of the approach in its capability to produce customized electrodes to fit various sized nerves. The cuff diameter can be adjusted from 1.5 to 2.0 mm with an increment of 0.25 mm. The width and length of the opening in the locking loop are 70 μm and 8 mm, respectively. The width of the parylene strip is also 8 mm, while the length of the projection portion of the ratchet teeth is 100 μm. It should be noted that a long projection will increase the difficulty for the ratchet teeth in passing through the locking loop, while the locking loop will fail to catch the ratchet teeth if it is too short. A tripolar configuration is used to reduce the noise caused by sources outside the cuff. The microelectrode array is composed of two outer reference electrode sites (250 μm wide, 3850 μm long) and six central working electrode sites (300 μm in diameter). The longitudinal distance (center to center) between the reference electrode and working electrode is 3 mm. The pads and through holes in the parylene connection sheet are designed to match the position and size of the pins of the 1.27 mm pitch double row female header, which can be connected to some conventional male connectors such as the Omnetics connectors.

EXPERIMENTAL METHODS

Microfabrication

Surface micromachining processes have been used to fabricate the cable-tie-type parylene cuff electrode with an integrated flexible parylene ribbon cable. Parylene C was chosen as the substrate material for its biocompatibility, mechanical strength, electrical insulation properties, chemical resistance and low permeability to gas and moisture. Gold electroplating technique was used to create the locking structure for its simplicity and low cost. Furthermore, gold is biocompatible, ductile and malleable.

A 1-μm-thick aluminum layer and an 8-μm-thick parylene layer were deposited on the silicon wafer, respectively. The platinum electrode sites, interconnects and contact pads were patterned by lift-off on a parylene substrate. The sputtered metal layer was composed of 15 nm titanium, 150 nm gold and 50 nm platinum (Figure 2b). Subsequently, a parylene insulation layer (2 μm) was deposited. To construct the ratchet teeth and locking loop, a 10 μm thick gold layer was electroplated using a thick photoresist mould (Figure 2c). Opening for electrode sites and contact pads, separation of devices and etching through holes were simultaneously accomplished by using oxygen plasma etching. Finally, devices were released from the silicon wafer by removing the sacrificial aluminum layer (Figure 2d).

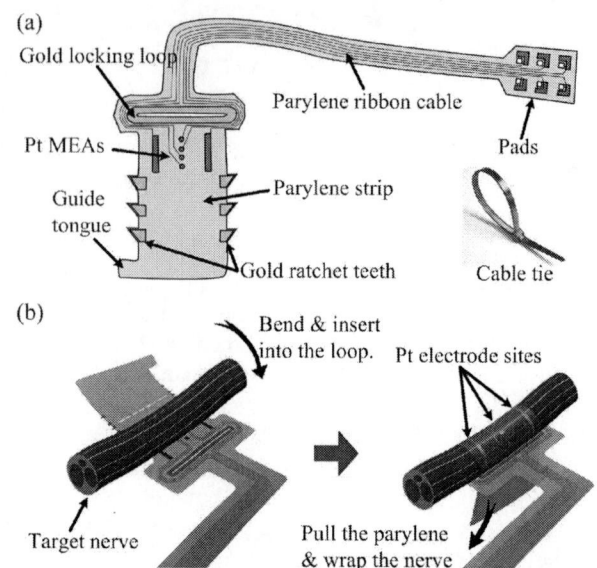

Figure 1: Schematic view of (a) proposed cable-tie-type cuff electrode and (b) its implantation on a nerve.

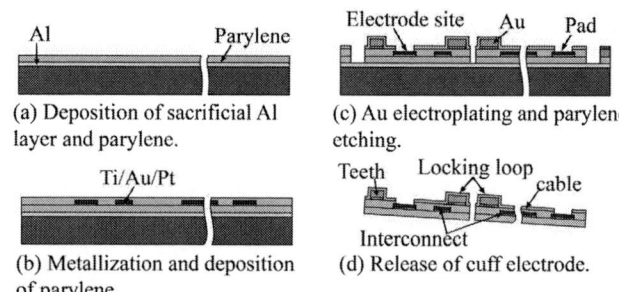

Figure 2: Abbreviated process flow for cable-tie-type cuff electrode fabrication.

Electrode Assembly

Commercial 40-pin 1.27 mm pitch double row female headers were used as an adaptor allowing the flexible cuff electrodes to be connected to some male connectors such as Omnetics connectors. The connection sheet with contact pads was carefully assembled along the pins of female header and subsequently the electrical connection was created by using conductive glue (Epo-Tek H20E, Epoxy Technology Inc, Billerica, MA). The similar packaging approach has been previously presented in Huang's work [10].

Experimental Setup

Acute *in vivo* tests were performed in the male Sprague-Dawley rat (200–250 g) under urethane anesthesia (1.4 g/kg, i.p.). The sciatic nerve was carefully dissected from the surrounding tissues over ~10 mm. The nerve diameter was then measured with a sterile ruler. A cable-tie-type parylene cuff electrode with a CNR of 1.5 was implanted around the nerve (Figure 3). The rat sciatic nerve bifurcates into the common peroneal nerve that innervates the tibialis anterior (TA) and the tibial nerve that innervates the gastrocnemius medialis (GM).

978-1-4799-3510-9/14 $31.00 © 2014 IEEE   10

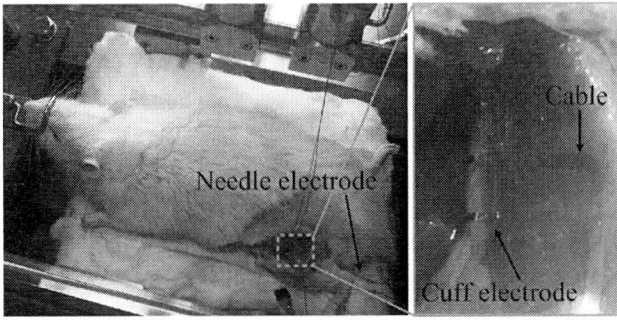

Figure 3: Acute rat experiment setup and the close-up view of one cable-tie-type parylene cuff electrode implanted around the sciatic nerve at the thigh. The needle electrodes are used for CMAPs recording.

For recording test, the compound action potentials (CAPs) of the rat sciatic nerve were detected differentially between the six shorted central electrodes and two shorted outer electrodes. The bipolar silver hook electrode was placed on the common peroneal nerve and connected to an external stimulator. Monophasic, squarewave pulses with intensity of 4-500 µA, pulse width of 100 µs and train length of 5 pulses were used as stimulus source. The evoked CAPs in the sciatic nerve were recorded and analyzed as the stimulus intensity was gradually increased above absolute threshold.

For stimulation test, the stimulus was applied to each one of the six central working electrodes (namely placed at 0°, 45°, 90°, 135°, 180° and 225°) in the cuff, using central electrodes as cathode and outer electrodes as anodes. The motor responses were monitored from compound muscle action potentials (CMAPs) of the gastrocnemius medialis (GM) and the tibialis anterior (TA), recorded by monopolar needle electrodes (Figure 3). Selective fascicular stimulation of the rat sciatic nerve was demonstrated by Navarro et al. using polyimide spiral cuff electrodes [9].

RESULTS

Fabrication results

Figure 4 shows the optical and SEM pictures of the fabricated cable-tie-type parylene cuff electrode. The flexible cuff electrode with a 35-mm-long 2-mm-wide parylene cable has a weight of 8 mg (Figure 4a). The measured impedance of 300-µm-diameter working electrode is 8.1kΩ at 1 kHz. The cuff is closed to form a 1.7-mm-diameter, 15-µm-thick and 8-mm-long cylinder. The silver epoxy was used to make the electrical connection. The packaged cuff electrode with a weight of 163 mg can be conveniently and reliably connected with external instruments through male connectors (Figure 4b).

It shows a very good adhesion between the parylene layers and electroplated gold layer. The electroplated gold structure was conformally coated with parylene. The slit in locking loop has a width of 70 µm ensuring a nearly closed cuff to reduce unwanted electrical noise.

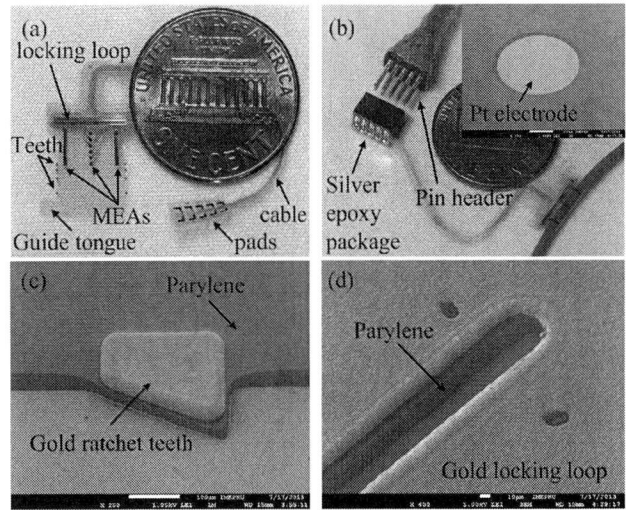

Figure 4: Photograph of (a) cuff electrode released from substrate and (b) packaged cuff electrode wrapped around a 1.3-mm-diameter hook-up wire; (c) (d) SEM of the ratchet teeth and locking loop with 70-µm-wide and 8-mm-long slit.

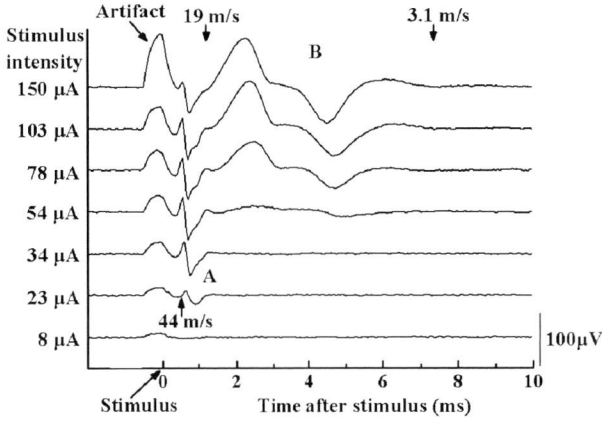

Figure 5: Series of mean CAPs recorded in the sciatic nerve cuff electrode following electrical stimulation of common peroneal nerve.

Acute Recording Test

The evoked CAP, which is the algebraic summation of all the action potentials produced by all the fibers within a nerve excited by the stimulus, was recorded using a cuff electrode wrapped around the nerve trunk. With increasing stimulus intensity, two waves representing nerve fiber groups with different conduction velocities were observed (Figure 5). The threshold current value was about 20 µA, which agrees with the reported values of 20-50 µA in Loeb's study [8].The latencies of A- and B-components are approximately 0.5-1.2 ms and 1.3-7.1 ms, respectively. At a distance of 23mm between the stimulation and recording electrodes, these latencies correspond to conduction velocities of 44-19 m/s and 18-3.1 m/s, respectively. The peek-to-peek amplitude of noise was 8 µV.

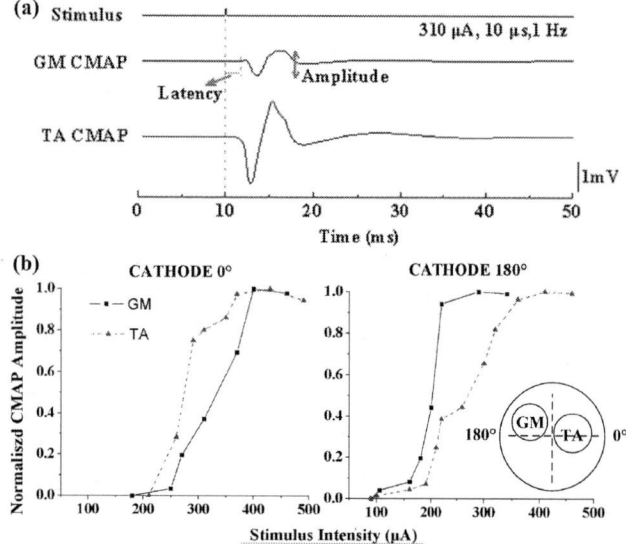

Figure 6: Selective fascicular stimulation of the rat sciatic nerve using tripolar cuff electrode; (a) sample recordings of GM and TA CMAPs (peek to peek) evoked by stimulating through cathode 0°; (b) recruitment curves of the GM and TA CMAPs and hypothetical location of the fascicles innervating GM and TA muscles in a cross-section.

Acute Stimulation Test

The stimulation tests after acute implantation proved that the proposed multipolar cuff electrode placed on a nerve trunk could selectively activate different fascicles within the nerve. Figure 6a shows sample recordings of GM and TA CMAPs evoked by stimulating through cathode at 0° with intensity of 310 µA. Figure 6b shows the recruitment curves of GM and TA normalized CMAPs with increasing stimulus intensity for two working electrodes (placed at 0° and 180°). For cathode at 0°, when the stimulus intensity was 290µA, the GM and TA CMAPs were 28.7% and 75.2%, respectively, showing preferential activation of TA. On the contrary, for cathode 180°, when the stimulus intensity was 220µA, the GM and TA CMAPs were 94.1% and 38.5%, respectively, showing preferential activation of GM. It should be pointed out that selective activation of one muscle results in recording the CMAP of the non-activated muscle, which may be explained by the cross-contamination between GM and TA muscle recording channels.

The nerve fascicle to the GM should be near cathode at 180°, while the nerve fascicle to the TA should be near cathode at 0°. It could be roughly inferred the location of the fascicles innervating GM and TA muscles in the sciatic nerve cross-section (Figure 6b, lower cell of right picture).

CONCLUSIONS

We designed, fabricated and tested a cable-tie-type parylene cuff electrode for peripheral nerve interfaces, whose diameter can be adjusted to fit the nerve properly during implantation. We demonstrated the ability to record the evoked CAPs of the rat sciatic nerve with a noise floor of 8µV. From the recordings, we classified axons within the nerve into two types of fibers according to their conduction velocities. We also proved the feasibility of using the proposed cuff electrode to produce selective activation of two different muscles.

ACKNOWLEDGEMENTS

The authors would like to thank the stuff in the National Key Laboratory of Science and Technology on Micro/Nano Fabrication in Peking University for their valuable assistance. This work was supported by the National Basic Research Program of China (No. 2011CB707505).

REFERENCES

[1] X. Navarro, T. B. Krueger, N. Lago, S. Micera, T. Stieglitz, P. Dario, "A Critical Review of Interfaces with the Peripheral Nervous System for the Control of Neuroprostheses and Hybrid Bionic Systems", *J. Peripher. Nerv. Syst.*, vol. 10, pp. 229-258, 2005.

[2] K. H. Polasek, H. A. Hoyen, M. W. Keith, R. F. Kirsch, D. J. Tyler, "Stimulation Stability and Selectivity of Chronically Implanted Multicontact Nerve Cuff Electrodes in the Human Upper Extremity", *IEEE Trans. Neural Syst. Rehabil. Eng.*, vol. 17, pp. 428-437, 2009.

[3] J. A. Hoffer, G. E. Loeb, C. A., Pratt, "Single Unit Conduction Velocities from Averaged Nerve Cuff Electrode Records in Freely Moving Cats", *J. Neurosci. Methods*, vol. 4, pp. 211-225, 1981.

[4] N. D. Engineer, J. R. Riley, J. D. Seale, W. A. Vrana, J. A. Shetake, S. P. Sudanagunta, M. S. Borland, M. P. Kilgard, "Reversing Pathological Neural Activity Using Targeted Plasticity", *Nature*, vol. 470, pp. 101-104, 2011.

[5] F. A. Jr. Cuoco, D. M. Durand, "Measurement of External Pressures Generated by Nerve Cuff Electrodes", *IEEE Trans. Rehabil. Eng.*, vol. 8, pp. 35-41, 2000.

[6] N. S. Korivi, P. K. Ajmera, "Clip-on Micro-cuff Electrode for Neural Stimulation and Recording", *Sens. Actuator B-Chem.*, vol. 160, pp. 1514-1519, 2011.

[7] G. G. Naples, J. T. Mortimer, A. Scheiner, J. D. Sweeney, "A Spiral Nerve Cuff Electrode for Peripheral Nerve Stimulation", *IEEE Trans. Biomed. Eng.*, vol. 35, pp. 905-916, 1988.

[8] G. E. Loeb, R. A. Peck, "Cuff Electrodes for Chronic Stimulation and Recording of Peripheral Nerve Activity", *J. Neurosci. Methods*, vol. 64, pp. 95-103, 1996.

[9] X. Navarro, E. Valderrama, T. Stieglitz, M. Schüttler, "Selective Fascicular Stimulation of the Rat Sciatic Nerve with Multipolar Polyimide Cuff Electrodes", *Restor. Neurol. Neurosci.*, vol. 18, pp. 9-21, 2001.

[10] R. Huang, C. Pang, Y. C. Tai, J. Emken, C. Ustun, R. Andersen and J. Burdick, "Integrated Parylene-cabled Silicon Probes for Neural Prosthetics", in *Proc. MEMS'08,* Tucson, AZ, USA, January 13-17, 2008, pp. 240-243.

CONTACT

*Z.H. Li, tel: +86-10-62766581; zhhli@pku.edu.cn

A SILICON ELECTRO-MECHANO TISSUE ASSAY SURGICAL TWEEZER

Po-Cheng Chen[1], Connie Wu[2], Fabrizio Michelassi[3], and Amit Lal[1]

[1]SonicMEMS Lab, School of Electrical and Computer Engineering, Cornell University, Ithaca, NY, USA
[2]Department of Electrical Engineering, University of Pennsylvania, Philadelphia, PA, USA
[3]Weill Cornell Medical College, Cornell University, New York, NY, USA

ABSTRACT

Surgeons make decisions on the use of different surgical tools providing a spectrum of contact forces to cut and manipulate tissue. These decisions are mostly made without quantitative data about the mechanical integrity and mechanical properties of the tissue. Here we report on an instrumented silicon tweezer for characterizing the electromechanical properties of tissue that is being tweezed by the device. The tweezer is designed for characterizing tissue during surgical procedures. This silicon tweezer was designed with a spring constant of 9 N/m, and maximum silicon stress of 80.7 MPa during the tweezing motion to prevent silicon fracture. Multiple thin-film sensors are integrated along with the silicon tweezer, four sets of strain gauges, two sets of permittivity sensors and sixteen platinum bio-potential recording electrodes. Therefore, insertion force, permittivity and electrical properties of tissue can be monitored simultaneously at different locations provide fast information in time critical surgeries. A set of piezoelectric transducers is attached on the legs of the tweezer for gap monitoring with 20 μm displacement resolution. The tissue stiffness can then be estimated by the measured through applied force and distance variation. This device addresses a key problem during intestinal anastomoses surgical operation where stapling devices are used to seal tissue.

INTRODUCTION

Tissue properties can be measured using different physical modalities. These include electrical and mechanical characterization at frequencies ranging from DC to microwave.

Admittivity: In a recent study [1-2] malignant and benign prostate tissue was investigated using a tetrapolar electrode at frequencies up to 100 kHz. Discriminatory power of admittivity properties of cancer and other tissues was proven by ROC analysis for in-vitro tissue studies. This study however was on in-vitro tissue, which can be different than in-vivo tissue. Experiments could not be done on in-vivo experiments because the probe and electronics are not small enough for in-vivo measurements. From the experimental values of conductivity and permittivity, a data model is used to get parameters from the complex admittance $\sigma_f + j\varepsilon_f = \sigma_\infty + [\Delta\sigma/1+(jf/f_c)^\alpha]$. These parameters include σ_∞ which is the conductivity at very high frequencies, f_c is the characteristic frequency at the apex point of the admittance curve, f is the sample frequency, and $\Delta\sigma$ is the difference in conductivity between the high and low frequencies, while α is a fitting parameter.

Probe-insertion force: In [3] a needle with a centimeter-scale force gauge was used to characterize insertion force into prostate tissue. It was intended to be used in robotic brachytherapy where a needle is inserted in the tissue, force-torque data collected and results show that cancer tissue is harder, prostate density and PSA have significant effects on mean forces. In [4] the measurement of prostate visco-elastic properties indicates the normal tissue Young's modulus is 15.9 ± 5.9 kPa, while cancerous tissue is 40.4 ± 15.7 kPa. Our own previous work has shown that the penetration force in tissue can be used to image fine vessels and measure morphology of tissue at 25-50 micron resolution [5]. Hence, measurement of insertion force and mechanical modulus could differentiate healthy from unhealthy tissue.

Resonance properties: Studies have also shown that resonance properties of piezoelectrically driven actuators in tissue have significantly different resonance frequencies, and loss-factors [6]. Moreover, the viscosity of cancer tissue was found higher than normal tissue [7].

Given that different approaches can be used to characterize tissue, it would be beneficial to the surgeon to have a tweezer that can measure the various properties during surgery with minimal effect on the surgery time or procedure. In particular, during intestinal anastomoses (the joining of two intestinal loops after removal of a diseased intestinal segment) are commonly performed with stapling devices. To compensate for the difference in intestinal wall thickness due to individual patient variability and different pathologic conditions, stapling devices come loaded with staples of different height, between 1.0 mm and 2.0 mm. Despite the high reliability of stapling devices, intestinal anastomosis fail to heal appropriately in about 1-7% of cases. The failure is called a "dehiscence". Dehiscences are catastrophic events for patients in terms of additional morbidity, the need for additional interventions, increased length of hospital stay and recovery, occasional mortality, and overall increased cost to the health care system [8]. There are many reasons for the non-healing of intestinal anastomosis. Two of them are a) the mismatch between the size of the staplers and the thickness of the intestine, and b) a decrease in compliance and pliability of the intestinal wall. In the first case, the intestinal wall may be too thick even for the largest staple; in the second case, the intestinal wall may be too rigid to accept a staple without being fractured. A device which could accurately measure both intestinal wall thickness and compliance could help surgeons to choose the appropriate size staples or to identify situations where alternative methods to perform an anastomosis should be used, for example hand sewn anastomosis rather than stapled anastomosis.

978-1-4799-3510-9/14 $31.00 © 2014 IEEE

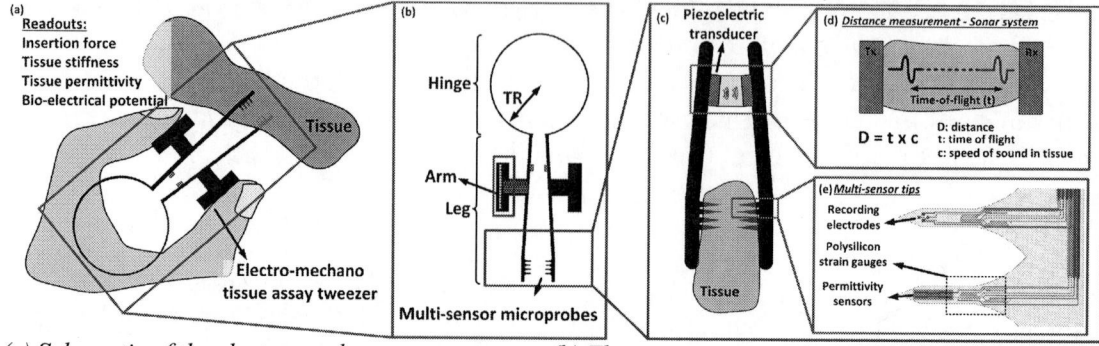

Figure 1. (a) Schematic of the electro-mechano assay tweezer. (b) The tweezer consist four different parts, hinge, arm, leg and microprobes. (c) The probes part will be inserted into tissue during the tweezing motion for measuring different tissue properties (d). The distance between two probe legs can be measured through the time of flight signal transmitted and received from the piezoelectric transducers with known speed of sound in tissue. (e) Tissue stiffness can be calculated from the distance variation verse the force variation. Tissue permittivity properties can also be monitored.

DEVICE ARCHITECTURE

This paper reports a silicon surgical tweezer for characterizing electromechanical properties of tissue (Figure 1), specifically addresses a key problem during intestinal anastomoses surgical operation where stapling devices are used to seal tissue. Unlike the other probe-like tissue stiffness tactile sensors [9-10], the tweezer structure provides a platform for the clinical use during surgery. We chose to pursue an all-silicon tweezer, instead of attaching sensors to existing tweezers, for repeatable tissue assessment across surgeries without external calibration. Lithographic precision of a few microns over a 10-cm tweezer provides ~1-10 part-per-million repeatability of spring constants and force sensitivity. The all-silicon structure also enables extensive embedded sensor integration, potentially alongside CMOS circuits, for a highly functional surgical tweezer. With integrated permittivity and strain gauge sensors, this tool-set has the potential of revolutionizing the choice of the anastomotic technique and technology used by providing measurements of various tissue characteristics, for example, compliance, electrical potential and permittivity (Figure 1c-e) which better inform the surgeon at the time of the surgical procedure. Specifically, the measurement of tissue stiffness and thickness will allow the proper determination of the tissue suturing technique. By knowing if the tissue is too stiff or too thick, hand suturing maybe warranted. Within the appropriate tissue thickness, the measurement will indicate what size staples are needed for optimum tissue apposition and healing. These "live" measurements will help the surgeon to make better tissue-suturing decisions and reduce the 1-7% postoperative dehiscence rate. The reduction of the dehiscence rate has the potential to spare thousands of patients this catastrophic complication, save millions of dollars to healthcare and reduce the cost of medical malpractice to hospitals and physicians, surgery safer and saving lives, while providing higher confidence during training for younger inexperienced surgeons. The fabrication process of the silicon tweezer is similar to one used previously [11] for neural probes. The fabricated silicon tweezer is shown in Figure 2.

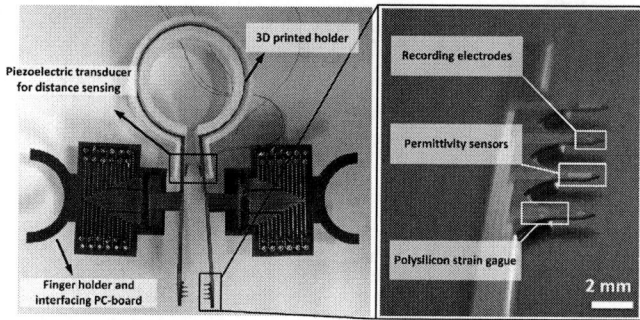

Figure 2. (a) Optical images of the fabricated tweezer. (b) Zoom in picture of the multi-sensor probes.

PERFORMANCE ANALYSIS

In order to measure the elastic properties of the tissue, applied force and deformation needs to be measured simultaneously. In our design, the force can be measured through the integrated strain gauge, and the distance of deformation can be measured through a pair of ultrasonic transducer.

Table 1. Young's modulus of different soft tissues. [12]

Tissue	Average Young's modulus (kPa)
Liver	~950
Arteriovenous	~3600
Breast tissue	~8
Muscle	~7
Spinal cord	~3

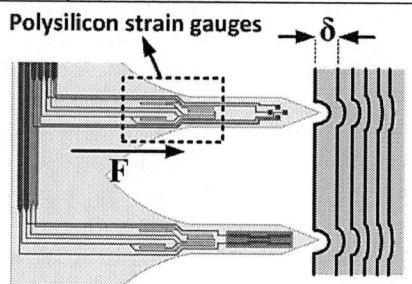

Figure 3. Soft tissue deformation during tweezing motion.

Force measurement: The microprobe will indent tissue longitudinally during tweezing motion (Figure 3). The force versus displacement for a rigid indenter can be estimated as

$$F = \frac{2}{\pi}\tan\alpha\,\frac{E_{tissue}}{1-v^2}\delta^2, [12]$$

where E_{tissue} is the tissue Young's modulus, δ is the displacement, v is the Poisson's ratio, α is the half angle opening in the indentation. The Young's modulus of various tissues are given in Table 1. Taking muscle as an example, the Young's modulus is ~8kPa, Poisson ratio of soft tissue is ~0.5[13], and assuming half angle opening is 15 degree, and assume the displacement is 10% of the 1 centimeter tissue thickness. The force can then be estimated as ~1.19 mN. The strain from the strain gauge on the silicon probe can then be calculated as 1.68×10^{-7}. The voltage obtain from the Polysilicon strain gauge with gauge factor of 20 and 5 V applied voltage ($V_{applied}$) can be estimated as 16.8 µV. With amplification, this signal is sufficient for tissue characterization with high signal to noise ratio.

Distance measurement: A Time-of-Flight (TOF) based system is used for measuring the distance between the tweezer legs. Two PZT transducers are placed on the leg with 2% of gelatin gel as acoustic impedance matching layer in between. A 20 MHz pulse is emitted from one of the transducer through the gelatin gel and pick up by the other one. The distance between the tweezer legs (D) can be estimated by the time delay (T_f) between the pulse emission and receiving as $D = C \times T_f$, where C is the propagation velocity of acoustic in the medium. The measured distance resolution is limited by the SNR and bandwidth of the receive amplifier to capture the received wave.

TWEEZER DESIGN

The all-silicon tweezer was designed with four different parts:

Hinge: The silicon tweezing radius-of-curvature (TR=Turn Radius) was designed for taking less than the maximum silicon stress of 1-3 GPa to avoid silicon fracture. COMSOL simulation shown that with tweezer structure of 8000 µm hinge radius, 40000 µm leg length, and 1 cm in inter-probe distance, the highest stress point is at the top of the circular hinge and it has a maximum stress of 80.7 MPa (Figure 4).

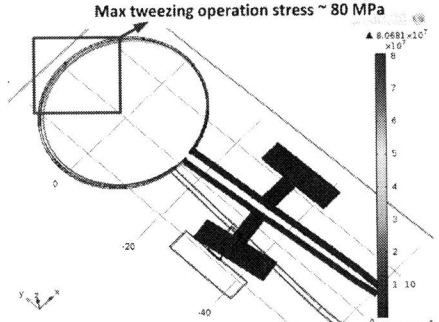

Figure 4. COMSOL simulation indicated that the maximum stress occurred at the top of the hinge during the tweezing movement.

Leg and arm design: The leg carried interconnects from the mircoprobes to the clamping-arm for electronic interfaces. At the same time, two piezoelectric PZT (1.7x3.5x0.5 mm) transducers were adhesively attached on legs near the arm to measure the distance between the legs transmitting sonic pulses at a frequency of 20-MHz, edge-to-edge of the two legs to measure distance between the two legs, with 15 µm resolutions. The pulses will be received on one hand of the leg and filtered with passive LC components and rectified to obtain a received pulse modified in amplitude and time-of-arrival, corresponding to the gap.

Microprobes design: Quad 3-mm long with 300 µm width and 140 µm thick multi-sensor microprobes are integrated at the end of each leg to provide measurements along the tissue length. Simultaneous measurements of tissue stiffness at different points provide fast measurement of stiffness gradients in time critical surgeries. Polysilicon strain gauges were integrated at the junction of the leg and the microprobe to measure the longitudinal and flexural strain due to tissue contact. Permittivity sensor was implemented by measuring capacitance versus frequency across two electrodes. Total of sixteen platinum recording electrodes were also co-fabricated to provide the capability of bio-potential measurement.

EXPERIMENTAL RESULTS

Spring constant: The tweezer spring constant was measured with a commercial force gauge with one arm fixed. The measurement results yielded a spring constant of 9 N/m, which is sufficient for human tweezing motion (Figure 5).

Integrated stain gauge signal: For testing the strain gauge, four probes were inserted into gelatin mixtures at varying speeds. The voltage difference before and after insertion with different gelatin samples under different speed were measured. The insertion force increases with insertion speed by 15% for every 1000um/s increase in speed. Surprisingly, the gelatin concentration from 9% to 18% decreases the insertion force by an average of 28% (Figure 6).

Sonar system – distance measurement: The tweezer speed and gap versus time can be monitored by the PZT pulse-echo displacement versus time measurement (Figure 7). The distance between two probe legs can be measured through the time of flight signal transmitted and received from the piezoelectric transducers with known speed of sound in tissue.

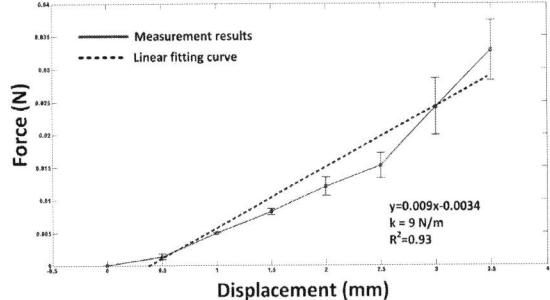

Figure 5: Spring constant of tweezer structure.

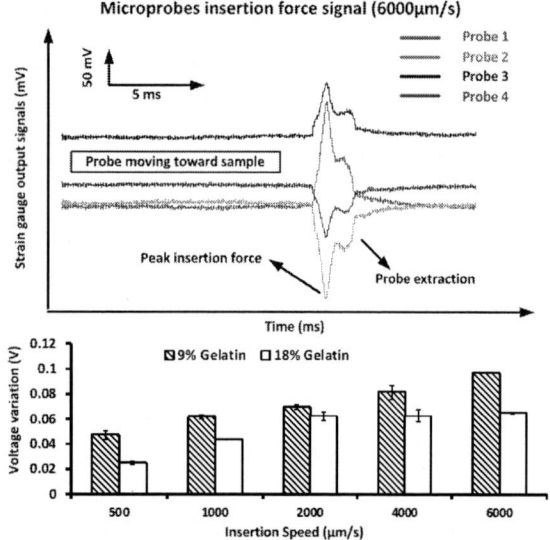

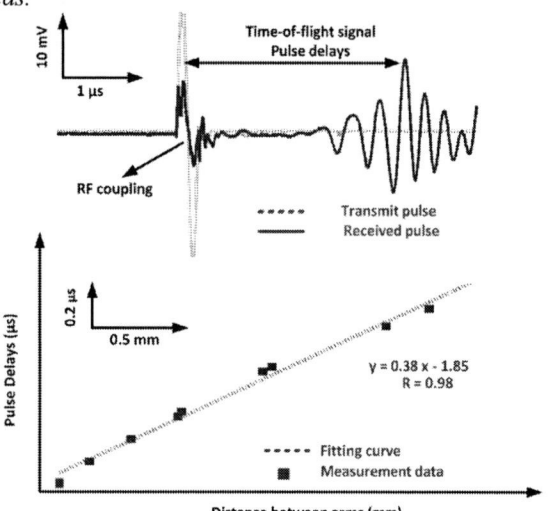

Figure 6. Measured strain gauge signal (a) Time dependent signal from strain gauges. (b) Voltage variations of two different percentage of gelatin gel under different insertion speeds.

Figure 7. The ToF is measured between the highest amplitude of emission and receiving pulses. (a) Piezoelectric transducer time-of-flight signal measurement. (b) Characterization of pulse delays verse different distances between two legs.

CONCLUSIONS

A silicon surgical tweezer for characterizing electromechanical properties of tissue was demonstrated in this paper. The silicon tweezer structure can perform tweezing motion without silicon fracture and spring constant of the tweezer was characterized as 9 N/m. Insertion forces for two types of Gelatin samples under various insertion speeds were monitored through the integrated stain gauge. The gap during the tweezing motion was monitored by a set of piezoelectric transducer attached on the leg from pulse-echo displacement versus time measurement. We demonstrated a multi-function silicon tweezer for characterizing tissue properties. This device can potentially

provide more information for surgeons during the surgery operation. Future work and challenge include integrated with CMOS integrated circuit for signal amplification and wireless transmission.

ACKNOWLEDGEMENTS

This work was sponsored by National Nanotechnology Infrastructure Network (NNIN) Research Experience for Undergraduates (REU) program. We would like to thank the Cornell Nanofabrication Facility (CNF) at Cornell University for fabrication of the tweezers.

REFERENCES

[1] R. J. Halter, et al., "Electrical properties of prostatic tissues: I. Single frequency admittivity properties", J. Urol., 182, pp. 1600-1607, 2009.

[2] R. J. Halter, et al., "Electrical properties of prostatic tissues: II. Spectral admittivity properties", J. Urol., 182, pp. 1608-1613, 2009.

[3] K. Yan, et al., "A real-time prostate cancer detection technique using needle insertion force and patient-specific criteria during percuatenous intervention", Med. Phy., 36(7), pp. 3356-3362, 2009.

[4] M. Zhang, et al., "Quantitative characterization of viscoelastic properties of human prostate correlated with histology", Ultrasound Med. Biol., 34(7), pp. 1033-1042, 2008.

[5] A. Ramkumar, et al., "An ultrasonically actuated silicon-microprobe-based testicular tubule assay", IEEE Trans. Biomed. Eng, 56(11), pp. 2666-267, 2009.

[6] V. Jalkanen, "Hand-held resonance sensor for tissue stiffness measurements- a theoretical and experimental analysis", Meas. Sci. Technol., 21, 055801, 2010.

[7] K. Hoyt, et al., "Tissue elasticity properties of biomarkers for prostate cancer", Cancer Biomarkers, 4, pp. 213-225, 2008.

[8] D. A. Telem, et al., "Risk Factors for Anastomotic Leak Following Colorectal Surgery", Arch. surg., 145(4), pp. 371-376, 2010

[9] P. Peng, et al, "Novel MEMS stiffness sensor for force and elasticity measurements", Sens. Actuator A, 158(1), pp. 10-17, 2010.

[10] Y. Zhang, Y. Mukaibo, et al., "A Multi-purpose Tactile Sensor Inspired by Human Finger for Texture and Tissue Stiffness Detection", Proc. IEEE Int. Conf. Robotics and Biomimetics, pp. 159-184, 2006.

[11] P-C. Chen, et al., "Ultrasonically Actuated Inserted Neural Probes for Increased Recording Reliability," in Digest Tech. Papers Transducers '13 Conference, Barcelona, pp. 872-875 ,June, 2013.

[12] C. T. Mckee, et al., "Indentation Versus Tensile Measurments of Young's Modulus for Soft Biological Tissues", Tissue Eng. Part B, 17(3), pp. 155-164, 2011.

[13] T. Glozman, et al., "A Method for Characterization of Tissue Elastic Properties Combining Ultrasonic Computed Tomography With Elastography", J. Ultrasound Med., 29, pp.387-398, 2010.

HIGHLY PACKED LIPOSOME ASSEMBLIES
TOWARD SYNTHETIC TISSUE

Hiroshige Hamano[1], Taishi Tonooka[1], Toshihisa Osaki[1, 2], and Shoji Takeuchi[1, 2]
[1]Institute of Industrial Science, The University of Tokyo, JAPAN
[2]Kanagawa Academy of Science and Technology, JAPAN

ABSTRACT

This paper presented an approach for preparation of a highly-packed liposome assembly that implemented lipid bilayer–lipid bilayer contact at the interfaces to mimic a cell–cell connection on living tissues. Cell-sized liposomes were closely packed using our previous technique that allowed monodisperse liposomes arrayed on a substrate. We explored the lipid patterning conditions that would provide a packed structure of the liposomes. By further works, we believe that the assembled structure would be useful for a tissue model.

INTRODUCTION

Cells *in vivo*, which is merely a diameter of about 10 µm, have gained surprisingly significant and various functions [1]. Reconstruction of those functions *in vitro* within a cellular size has been a great challenge both in science and engineering, and a part of them, e.g. DNA replication and protein synthesis, was successfully mimicked. On the one hand, model cell studies using aqueous micro-droplets or liposomes aim to reproduce and encapsulate the functions in a confined system. There have been numbers of studies reported those model systems presenting metabolic reactions, growth of a cell model, division cycles , and even Darwinian evolution [2-4].

A next challenge beyond the replication of single model cells will be the reconstruction of living-tissue structures and functions. Recently, a few works provided tissue models using aqueous droplets or liposomes connected each other [5, 6]. However, the connection manners of the model cells in those works did not follow the realistic tissue systems. The model cells were contacted with a single lipid bilayer or connected by thin lipid tubes, while living cells are faced with lipid bilayer–lipid bilayer contacts through membrane proteins or extracellular matrices [1].

We consider that an array of giant liposomes is the prerequisite platform to develop a tissue-like liposome assembly consisting of lipid bilayer–lipid bilayer contacts. Commonly, giant liposomes are produced by a gentle hydration or an electroformation method, yet these methods have difficulty in controlling the size, density and position of the formed liposomes; therefore, it is hardly feasible to form a liposome array by these methods. In MEMS 2011, we developed a method that enabled a precise lipid patterning by the integration of an electrospray deposition (ESD) technique and a micro-fabrication process, as shown in Figure 1 [7,8]: By the ESD of lipids on a conductive/non-conductive patterned substrate, the sprayed lipids were electrically led only to the conductive regions. With a simple hydration

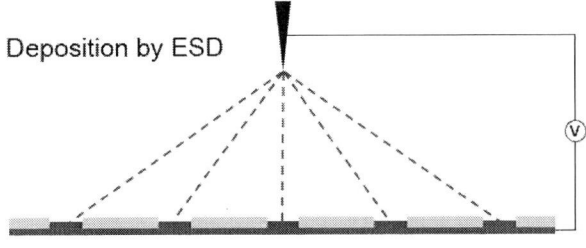
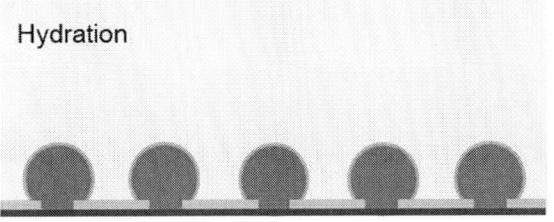

Figure 1: Schematic illustration of the liposome array formation method previously developed. (Top) ESD of lipids on a conductive/non-conductive pattern produced a lipid pattern on the conductive area. (Bottom) Arrayed liposomes were formed by hydration of the lipid pattern, with a controlled size, position, and density.

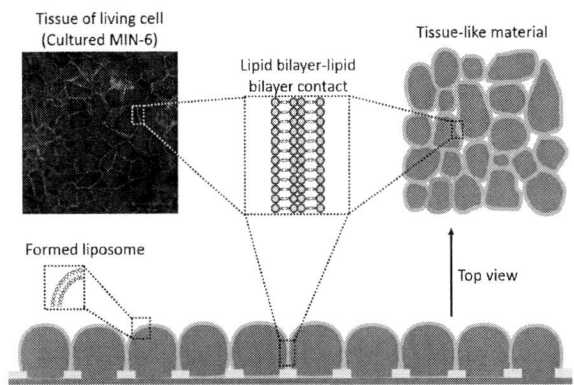

Figure 2: Conceptual diagram of a highly-packed cell-sized liposome assembly system, replicating a living tissue structure. The structure presents lipid bilayer–lipid bilayer contact at the interfaces, mimicking a cell–cell connection on living tissues.

process of these dried lipid patterns, we succeeded in the formation of giant liposomes in an array format with a narrow range of the size distribution, and moreover obtained the desired sizes of liposomes by changing the sizes of lipid patterns, close to common cellular sizes. In this paper, we are motivated to produce a tissue-like liposome structure, which implements the lipid bilayer–lipid bilayer contact, based on the liposome array technique (Figure 2).

978-1-4799-3510-9/14 $31.00 © 2014 IEEE

EXPERIMENTAL

Conductive/non-conductive substrate

We first fabricated a conductive/non-conductive patterns on a substrate using a common photolithography technique. As shown in Figure 3a, a thin polymer film of parylene C was deposited on an indium-tin-oxide (ITO)-coated glass slide. Then, an Al-layer and a photoresist (S1818) were coated on the substrate by using a vacuum vapor deposition and a spin-coating, respectively. The photoresist was patterned by UV lithography, and the Al-layer was then patterned by an Al etchant. Finally, the polymer layer was patterned by O_2 plasma, and the residual Al and photoresist was removed. The ITO area was exposed after the fabrication process as shown in Figure 3b. In this work, the conductive ITO pattern was closely located to decrease the space between the liposomes formed on the substrate (diameter: 20 μm, center-to-center spacing: 28 μm).

Electrospray deposition of lipids

The ESD setup was shown in Figure 4. As a result of the electrification of the solution, the sprayed lipid was deposited on the conductive pattern on the substrate. In this paper, we used a mixture of lipids (DOPC and DOPG) dispersed in a solvent consisting of chloroform and methanol for the electrospray. A rhodamine-labeled lipid was also mixed for fluorescent observation (Rhod-DPPE, 1 wt%). The tip of the glass capillary was between 10 and 15 μm in diameter. After the deposition, the samples were kept in vacuum before use. Lipids were purchased from Avanti Polar Lipids, AL, USA. All chemicals of this experiment were used without further purification.

RESULTS & DISCUSSION

Lipid amount and the ESD condition

To produce a tissue-like structure by using an array of cell-sized liposomes, the position and density of liposomes have to be adjusted. First, we examined the dried lipid amount on the substrate pattern by changing the ESD condition. Figure 5 shows the relationship between the spray time and the fluorescence intensity of the lipid pattern that is the index of the lipid amount. The spray time was set at 6, 10, 14, and 18 min, respectively. It was observed that the lipid amount inside each pattern increased with the spray time. We found that the lipid deposition occurred also on the non-conductive polymer area with increasing the spray time. This result indicated that the thick deposited lipids decreased the conductivity of ITO-electrode over time and hindered the precision of electrospray patterning.

Liposome–liposome contact

Liposomes were immediately formed after hydration of the patterned lipids (Figure 6a). The liposomes became larger with the increase of the spray time, and over 14 min of the spray we observed the liposomes contacting one another, similar to living tissues. The liposomes shape also changed from spherical to hexagonal by the contact. Figure 6b and 6c respectively represent the area of the formed liposomes and

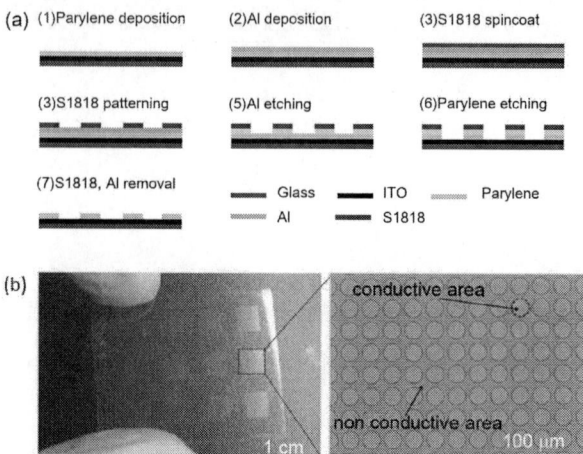

Figure 3: (a) Schematic of the fabrication process of a conductive/non-conductive substrate using a thin polymer film coated on an ITO-glass substrate. (b) Photo of the fabricated substrate and its close-up image representing circular conductive micro-patterns, diameter of 20 μm.

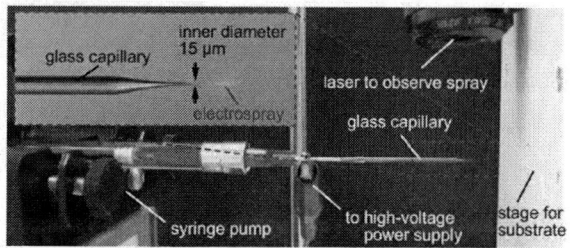

Figure 4: Electrospray setup for lipid micro-patterning at the conductive area of the substrate.

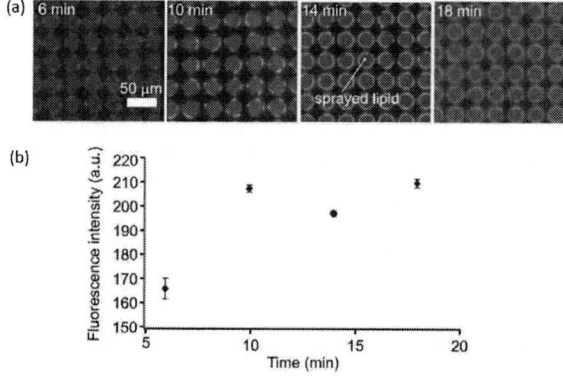

Figure 5: Lipid patterning vs. the time of electrospray deposition. (a) Fluorescent microscopic images of lipid patterned-substrates with different spray times. Red fluorescent was from rhodamine-labeled PE lipid. (b) Spray time-dependent fluorescence intensity at the conductive circular area evaluated from Figure 5a.

the occupancy rate of the liposomes over the surface depending on the spray time. The area proportionally increased with the spray time although the occupancy dropped due to detachment of the liposomes from the substrate at 10 min. Over 14 min, the occupancy rate saturated. We also confirmed that coalescence of liposomes increased at longer spray; there might be feasible spray duration for producing a tissue-like assembly.

CONCLUSIONS

In this work, we presented a highly-packed liposome structure oriented to synthetic tissue studies. By adjusting the micro-patterned lipid amount, the formed liposomes were contacted via lipid bilayer–lipid bilayer interfaces, which would be applied to mimicked living tissues. For future studies, we propose biochemical connections between the liposomes to construct a self-organizing structure or to express membrane proteins to reproduce communication functions between model cells.

ACKNOWLEDGEMENTS

The authors thank Maiko Onuki and Utae Nose for their technical assistance. This work was partly supported by JSPS (Grant-in-Aid for JSPS Fellowship and for Young Scientists A; 25706015) and MEXT (Platform for Dynamic Approaches to Living System), Japan.

REFERENCES

[1] B. Alberts, A. Johnson, J. Lewis, M. Raff, K. Roberts, P. Walter, *Molecular Biology of the Cell 5E: Reference Edition*, Garland Science, 2008.

[2] K. Nishimura, T. Matsuura, T. Sunami, H. Suzuki, T. Yomo, "Cell-free protein synthesis inside giant unilamellar vesicles analyzed by flow cytometry", *Langmuir*, vol. 28, pp. 8426-8432, 28, 2012.

[3] K. Kurihara, M. Tamura, K. Shohda, T. Toyota, K. Suzuki, T. Sugawara, "Self-reproduction of supramolecular giant vesicles combined with the amplification of encapsulated DNA", *Nat. Chem.*, vol. 3, pp. 775-781, 2011.

[4] N. Ichihashi, K. Usui, Y. Kazuta, T. Sunami, T. Matsuura, T. Yomo, "Darwinian evolution in a translation-coupled RNA replication system within a cell-like compartment", *Nat. Commun.*, vol. 4, p. 2494, 2013.

[5] A. Karlsson, R. Karlsson, M. Karlsson, A-S. Cans, A. Strömberg, F. Ryttsén, O. Orwar, "Networks of nanotubes and containers", *Nature*, vol. 409, pp. 150-152, 2001.

[6] G. Villar, A. D. Graham, H. Bayley, "A tissue-like printed material", *Science*, vol. 340, pp. 48-52, 2013.

[7] T. Osaki, K. Kuribayashi-Shigetomi, R. Kawano, H. Sasaki, S. Takeuchi, "Uniform-sized liposome array formation with gentle hydration", *Proc. IEEE MEMS 2011*, Cancun, pp. 103-106, 2011.

[8] T. Osaki, K. Kamiya, R. Kawano, H. Sasaki, S. Takeuchi, "Towards artificial cell array system: Encapsulation and hydration technologies integrated in liposome array", *Proc. IEEE MEMS 2012*, Paris, pp. 333-336, 2012.

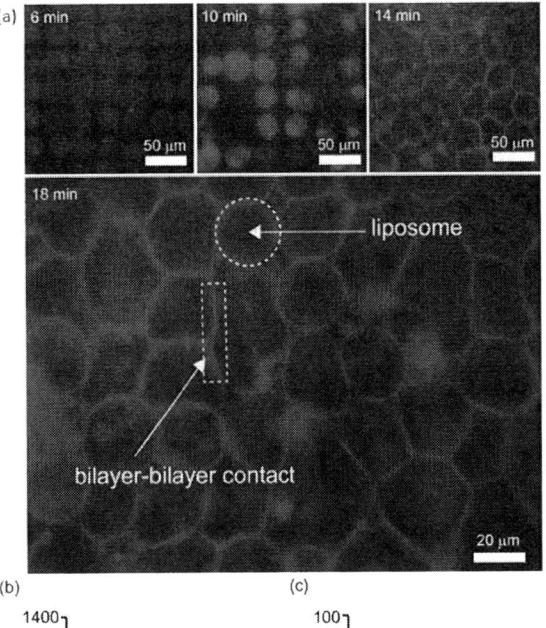

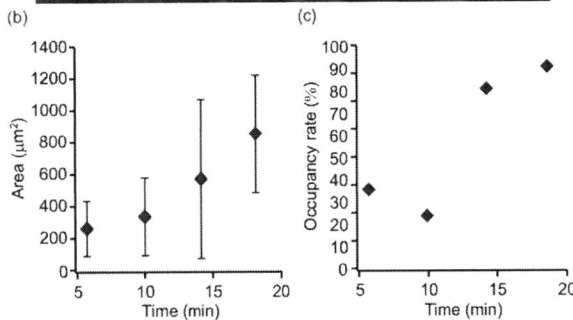

Figure 6: Liposome assemblies after hydration of the lipid patterned substrates. (a) Fluorescent microscopic images of liposome structures formed from the different times of electrospray deposition. (b) Cross-sectional area of single liposomes on the substrate vs. the spray time. Error bars show standard deviation. (c) Occupancy rate of the tonal numbers of formed liposomes within each image over the surface of the substrate vs. the spray time.

CONTACT

*H. Hamano, Institute of Industrial Science, The University of Tokyo, 4-6-1 Komaba, Meguro, Tokyo 153-8505, Japan; Tel: +81-3-5452-6650; Fax: +81-3-5452-6649; Email: hamano@iis.u-tokyo.ac.jp

WHOLE-ANGLE-MODE MICROMACHINED FUSED-SILICA BIRDBATH RESONATOR GYROSCOPE (WA-BRG)

Jong-Kwan Woo, Jae Yoong Cho, Christopher Boyd, and Khalil Najafi
Center for Wireless Integrated MicroSensing and Systems (WIMS[2])
University of Michigan, USA

ABSTRACT

We present the fused-silica micromachined birdbath resonator gyroscope (μ-BRG) operating in the whole-angle (WA) mode. The key advantages of the whole angle mode operation is rotation angle measurement, large bandwidth, and full-scale range which is needed in detecting the motion of fast-moving objects. The μ -BRG is made with fused silica using a micro blow-torching process and has $n = 2$ wineglass modes at 10.46 kHz with a small frequency mismatch ($\Delta f = 10$ Hz) and a decay time (τ) of 2.2s. The WA-BRG achieves a stable angular gain (A_g) and a large full scale range (700 °/s).

INTRODUCTION

Micro vibratory gyroscopes have been adopted in a large variety of consumer, industry, and military applications due to their significant performance improvement over the past two decades. Micro-gyroscopes can measure either rotation rates (Ω) or rotation angles (Θ). Depending on whether they measure Ω and Θ, they are categorized as the rate gyroscope (RG) and the rate-integrating gyroscope (RIG), respectively.

RIGs operate on the same principle as the Foucault pendulum by tracking the angle (θ) between the direction of the vibrating wave and a reference point. θ is equal to the Θ multiplied by a scale factor, called the angular gain (A_g). A_g is a constant determined solely by the sensor geometry. RIGs have several advantages over RGs. First, they can measure angles directly. While RGs can also calculate angles by integrating Ω over time, the calculation is less accurate than Θ measured by the RIG because of the errors generated from the integration of thermal noise in the Ω data, so called the angle random walk (*ARW*). Second, RIGs have unlimited mechanical detection bandwidth (*BW*) and full-scale range (*FSR*). This is because the wave stays at a constant angle in the inertial reference regardless of the speed and the frequency of the applied rotations. The direction of the RG's wave changes in the inertial coordinates as the sensor is rotated; however, due to the long decay time constants (τ) of its resonance modes, the direction cannot change quickly under fast rotation speed. For this reason, RGs have limited *BW* and *FSR*. Third, the A_g is constant regardless of vibration amplitude.

An important challenge for the RIG, however, is that its bias drift ($\dot{\theta}$) is a direct function of the Δf and $\Delta \tau^{-1}$ of its resonance modes as well as θ. Therefore, it is necessary for the RIG to have very good mechanical symmetry. We recently introduced the micro birdbath resonator gyroscope

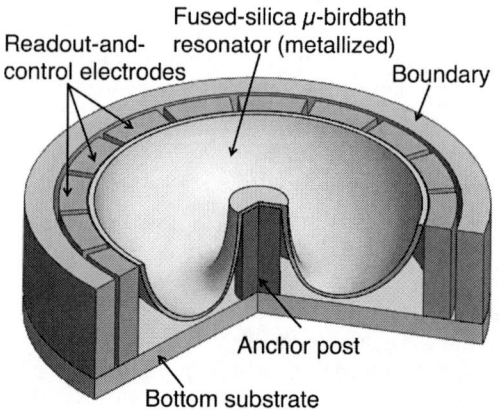

Figure 1: Architecture of the micromachined fused-silica birdbath resonator gyroscope (μ-BRG).

(μ-BRG) operating in the force-rebalance mode [1]. The gyro demonstrated 1 °/hr bias stability with >400 °/s *FSR* when its mode frequencies were completely matched. In this study, we demonstrate the μ-BRG operating in the whole-angle (WA) mode for a better dynamic range. The Si cylindrical rate-integrating gyroscope (CING) was reported in [2], and a control algorithm to improve its performance by actively compensating for Δf and $\Delta \tau^{-1}$ was demonstrated [3]; however, CING performance was limited by its low A_g (=0.013). Other researchers have also reported the WA-mode quad-mass tuning-fork gyroscope [4] with good accuracy; however, their algorithm did not contain the control loops to sustain the vibration amplitude or correct quadrature errors, thus it cannot be applied to gyros having short or intermediate τ.

In this paper, we will discuss the architecture and the fabrication process of the μ-BRG, WA-mode control system architecture, and the measurement results of the WA-BRG.

DEVICE ARCHITECTURE

The μ-BRG (Fig. 1) consists of a fused-silica birdbath resonator and sixteen readout and control electrodes surrounding its outer perimeter. The μ-BRG has several attractive features. First, the shell can be fabricated with precision on the micro scale, because the anchor and the rest of the resonator are self-aligned. Second, using the three-dimensional (3D) micro-blow-torching process [6], the BB resonator can be fabricated to have a large aspect ratio [Height (*H*) / Outer Radius (*R*)], which leads to a large angular gain ($A_g = 0.25$) [7]. Third, the shell has a large frequency difference between its n=2 wineglass modes and the parasitic modes, which potentially enables the sensor to achieve good vibration stability over a wide *BW*.

978-1-4799-3510-9/14 $31.00 © 2014 IEEE

FABRICATION PROCESS

The fabrication process steps of the μ-BRG are shown in Fig. 2. First, 1.6-mm-deep trenches are defined on a p-type 2-mm-thick Si (ρ < 5 mili-Ω-cm) using deep-reactive-ion-etching (DRIE) (Fig. 2a). A 4-μm-thick Al protection layer is evaporated to cover the top surface, sidewall, and the bottom surface of the trenches. The Al layer is wet-etch patterned to define the regions that will be released in a later step. The Si wafer is bonded face-down to a 500-μm-thick Pyrex wafer with 3-μm deep recesses in locations where Si will be released in a later step. The wafers are bonded at 400 °C and a voltage of 1500 V (Step 2b). Si is removed with DRIE from the topside, leaving Si electrodes and an anchor post (Step 2c). The Al layer is etched away in dilute hydrochloric acid (HCl : H₂O = 1 : 3), releasing the Si pieces located between the anchor post and the electrodes (Step 2d). The resonator is sputter coated with Cr/Au = 50/1000 Å and attached face up to the anchor post using polymer adhesive (Crystalbond™ 509, SPI Supplies, West Chester, PA, USA). (Step 2e). Fig. 3 shows the photograph of μ-BRG [dimension: 8 mm (width) × 8 mm (length) × 2.5 mm (height)]. The average electrode-to-shell gap (g_{avg}) is measured to be ~13.2 μm.

WHOLE ANGLE MODE CONTROL

Gyro Dynamics

The WA gyroscope dynamics can be expressed using the Lagrange's equations of motion for both $n=2$ wineglass modes. The equations of motion for an ideal WA-mode gyro with perfect f and τ symmetry is expressed in terms of the generalized displacement along the two axes (q_1, q_2), effective mass (M), Coriolis mass ($\gamma = 2MA_g$), yaw-axis rotation rate (Ω_z), and spring constants ($k = 4\pi^2 f^2 M$) by [7]:

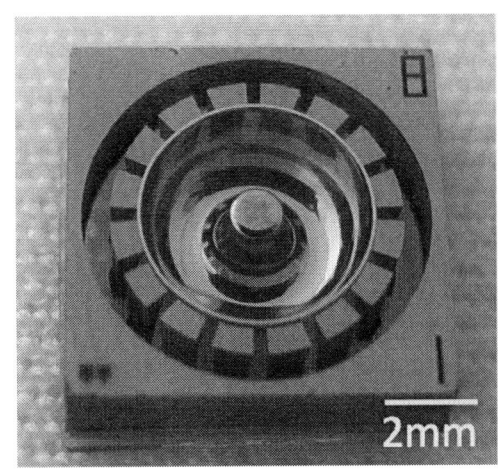

Figure 3: Micro Birdbath Resonator Gyro (BRG) [Resonator size: radius (R) = 2.5 mm, anchor radius (AR) = 0.5 mm, height (H) = 1.55 mm. Device size: 8 (width) × 8 (length) × 2.5 mm (height)].

$$\begin{bmatrix} M & 0 \\ 0 & M \end{bmatrix}\begin{bmatrix} \ddot{q}_1 \\ \ddot{q}_2 \end{bmatrix} + \begin{bmatrix} 0 & -2\gamma\Omega_z \\ 2\gamma\Omega_z & 0 \end{bmatrix}\begin{bmatrix} \dot{q}_1 \\ \dot{q}_2 \end{bmatrix} \quad (1)$$

$$+ \begin{bmatrix} k & 0 \\ 0 & k \end{bmatrix}\begin{bmatrix} q_1 \\ q_2 \end{bmatrix} = \begin{bmatrix} 0 \\ 0 \end{bmatrix}$$

Note that (1) is valid for $\Omega_z \ll 2\pi f$. The solution to (1) is:

$$q_1 = a\cos(\theta_o - \frac{\gamma}{M}\int\Omega_z dt)\cos(\omega' t + \phi_o) \quad (2a)$$

$$q_2 = a\sin(\theta_o - \frac{\gamma}{M}\int\Omega_z dt)\cos(\omega' t + \phi_o) \quad (2b)$$

$$\omega' \approx \sqrt{k/M} \quad (2c)$$

where a is the in-axis vibration amplitude, ω' is the resonance frequency, and ϕ_o is an offset phase. The angle of rotation, Θ, can be simply found from $\tan^{-1}(q_2/q_1)/2A_g$.

In reality, however, all gyros have asymmetric f and τ, and stiffness and damping asymmetry introduces different types of $\dot\theta$. First, f asymmetry introduces sinusoidal errors in q_1 and q_2 that have 90° phase difference from the original signals in (2a) and (2b). These errors are called the quadrature errors. The quadrature error amplitudes increase as the stiffness mismatch increases, and it prevents the vibrating wave from having a stable scale factor [8]. Second, the damping asymmetry generates sinusoidal errors in q_1 and q_2 that have same phase with the original signals in (2a) and (2b). These errors drive the wave to align at the direction of the lowest damping even though the sensor is not rotated. The $\dot\theta$ is a sum of the angular change due to real rotation ($-2A_g\Omega$) and the errors due to mismatches in the inverse decay time constants ($\Delta\tau^{-1}$) and resonant frequencies ($\Delta\omega$, $\omega = 2\pi f$) as well as errors due the phase mismatch between the vibrating wave and the driving signals ($\delta\phi$) and the amplitudes of in-phase and quadrature control forces (F_a, F_q) [9]:

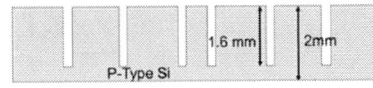

(a) Define 1.6 mm-deep DRIE trenches.

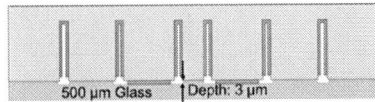

(b) Evaporate and pattern 4 μm thick Al protection layer. Anodically bond to borosilicate wafer with 3μm recess patterns.

(c) Etch the remaining thickness of Si.

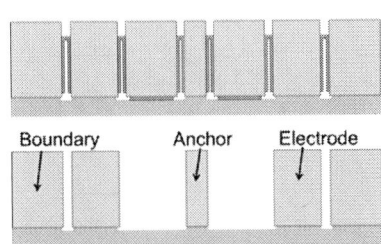

(d) Dissolve protection layer.

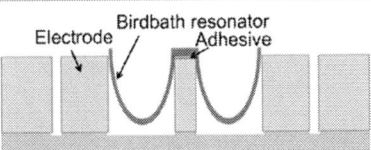

(e) Metallize birdbath resonator. Assemble resonator to the electrode substrate.

Figure 2: Fabrication process flow.

$$\dot{\theta} = -2A_g\Omega + \frac{1}{2}\Delta\left(\frac{1}{\tau}\right)\sin 2(\theta - \theta_\tau)\frac{E}{\sqrt{E^2 - Q_u^2}}$$

$$+ \frac{1}{2}\Delta\omega\cos 2(\theta - \theta_\omega)\frac{Q_u}{\sqrt{E^2 - Q_u^2}} \qquad (3)$$

$$+ \Re\left[\frac{i}{2\omega}e^{-i\delta\phi}\frac{(F_q a - iF_a q_u)}{\sqrt{E^2 - Q_u^2}}\right]$$

In Eq. (3), E is the overall motional energy, Q_u is the energy proportional to the quadrature error, θ_τ is the angle of the principal damping axes in the sensor coordinates, θ_ω is the angle of the principal elastic axes in the sensor coordinates, and q_u is the quadrature vibration amplitude.

The quadrature error is suppressible, because the error amplitude is distinguishable from the Coriolis force using the phase-sensitive demodulation method. The f can be also tuned to less than its BW using electronic tuning. The error due to damping mismatch, however, has the same phase as the Coriolis force. In order to reduce this error, the sensor has to have a very long and symmetric τ. In addition, although not demonstrated in this work, $\dot{\theta}$ due to damping mismatch is a sinusoidal function of θ, so the error due to damping mismatch can be compensated by calibrating $\dot{\theta}$ at zero Ω for all θ and subtracting their time integrals from the measured θ.

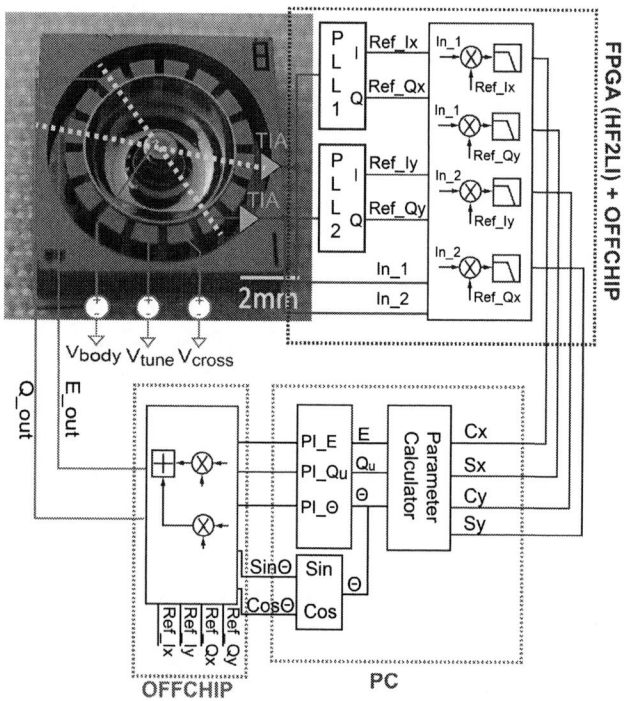

Figure 4: Control architecture of the BRG.

Control Architecture

The whole-angle-control algorithm is similar to that of the HRG [9]. The whole-angle-mode gyroscope interface circuitry (Fig. 4) consists of two phase-locked-loops (PLL),

a demodulator block, a parameter calculator, PI controllers, and a modulator block. These blocks are configured using a combination of FPGA (Zurich Instrument HF2LI lock-in amplifier), off-chip circuits, and Labview. The PLLs lock to the $f_{n=2}$ Axis 1 and Axis 2 and create in-phase and quadrature reference signals. The in-phase and quadrature amplitudes of both channels (c_x and s_x for Axis 1, c_y and s_y for Axis 2) are calculated by demodulating the output signals from both channels with the reference signals.

The parameter calculator block calculates the E, Q_u, and θ of the wave following the equations derived in [9]. The signal amplitudes for maintaining E, nulling Q_u, and maintaining θ (during force-rebalance-mode operation) are calculated using proportional-integral (PI) controllers. The control amplitudes are modulated by the in-phase and quadrature reference signals and applied to the drive electrodes for both axes.

TEST RESULT

The μ-BRG is evaluated on an Ideal Aerosmith® Aero900 rotation table at <1 mTorr vacuum pressure and uncontrolled temperature. The resonator has $n=2$ wineglass modes at 10.454 and 10.464 kHz and has τ of 2.22 and 2.06s ($Q = 72.87k$, 67.69k) (Fig. 5). The frequencies are exactly matched using electrical tuning. Fig. 6 demonstrates the whole-angle-mode operation of the μ-BRG by measuring increments using θ controller. Fig. 6a demonstrates the θ change under rotation rates of 100, 200, 300, and 400 °/s over a duration of 160s. Fig. 6b and 6c show the close-up plots of the measured θ at 100 and -100 °/s rotation rates, respectively. The A_g is calculated from the average $\theta/2\Theta$ to be ~0.27. This value is close to the theoretical calculation (0.25) in [7]. The nonlinearity in the measured θ is believed to be due to the anisotropy in damping and differences in shell-to-electrode capacitances. One of the PLLs also temporarily loses the lock state at θ near 0° or 90°. Fig. 7 shows the dependency of A_g over input rotation speeds from 100 to 700 °/s in clockwise and counterclockwise directions. The A_g is found to be stable at ~0.27 for these rates. The testing at higher speeds was not possible due to the latency of the control hardware. Nevertheless, we were able to prove the concept of whole-angle-mode BRG with a stable A_g over a wide dynamic range.

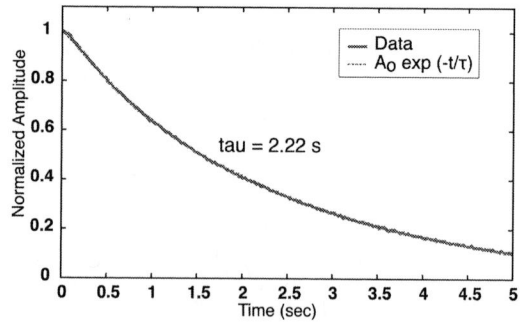

Figure 5: Ring down plot for one of n=2 wineglass modes (f = 10.454 kHz, τ = 2.22s, Q = 72.87k). Another mode has f = 10.464 kHz, τ = 2.06s, and Q = 67.69k.

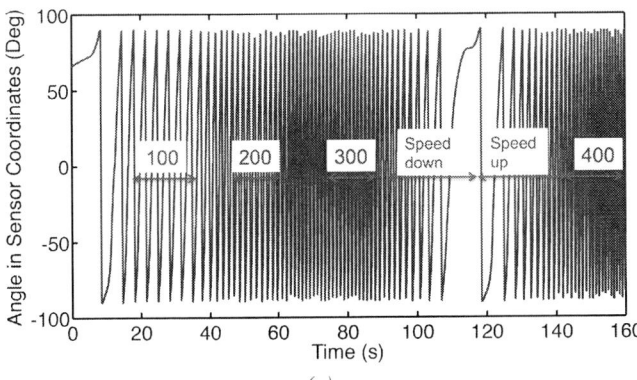

(a)

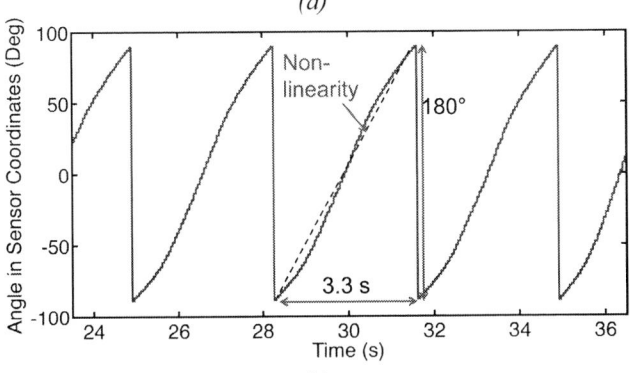

(b)

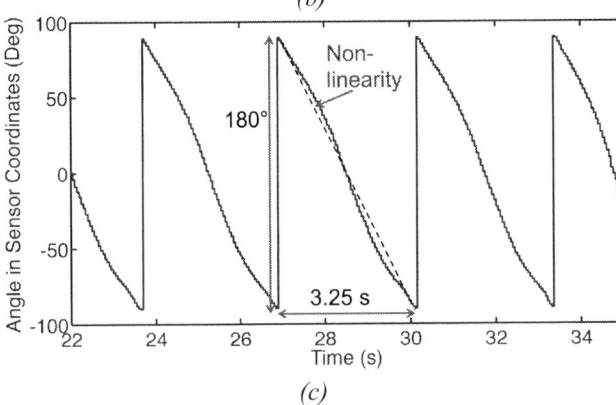

Figure 6: (a) Angular position(θ) of the wineglass modes in whole-angle mode under of 100, 200, 300, and 400 °/s rotation rate (Ω). (b) Close-up view of θ under Ω of 100°/s. (c) Close-up view of θ under Ω of -100 °/s.

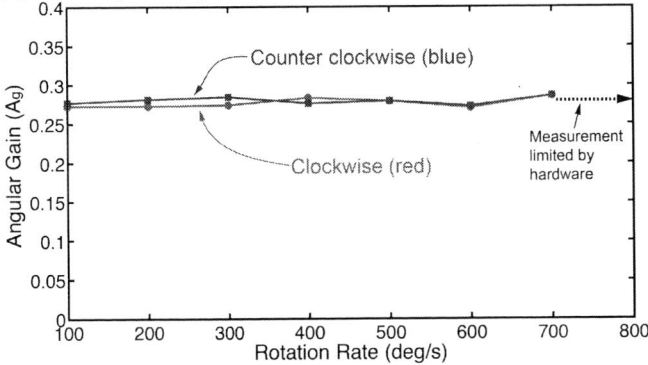

Figure 7: Angular gain (A$_g$) for Ω of 100 to 700 °/s and -100 to -700 °/s.

SUMMARY

We presented architecture and testing result of the WA-BRG. The fused silica μ-BRG is fabricated using a micro blow-torching process and has the $n=2$ wineglass modes at 10.46 kHz with a small Δf (=10 Hz) and a τ of 2.2s. We controlled the BRG in the whole-angle mode and achieved a full-scale range of 700 °/s. The gyro performance will be further improved by improving the τ of the resonator and more advanced control algorithms.

AKNOWLEGEMENT

This work is supported by DARPA MRIG award #W31P4Q-11-1-0002. We thank Dr. Radheshyam Tewari and Professor Craig Friedrich at Michigan Technological University for machining of graphite molds, and Mr. Robert Gordenker for his help in testing the BRG. Portions of this work were done in the Lurie Nanofabrication Facility (LNF), a site of the National Nanotechnology Infrastructure Network (NNIN), which is supported in part by the National Science Foundation (NSF).

REFERENCE

[1] J. Cho *et al.*, "A high-Q birdbath resonator gyroscope (BRG)," in *IEEE TRANSDUCERS'13*, June 2013, pp. 1847-1850.

[2] J. Cho *et al.*, "High-Q, 3kHz single-crystal-silicon cylindrical rate-integrating gyro (CING)," in *IEEE MEMS'12*, Jan. 2012, pp. 172-175.

[3] J. A. Gregory *et al.*, "Novel Mismatch Compensation Methods for Rate-Integrating Gyroscopes", in IEEE/ION PLANS 2012, May 2012, pp. 252-258.

[4] I. P. Prikhodko *et al.*, "Foucault pendulum on a chip: angle measuring silicon MEMS gyroscope," in *IEEE MEMS'11,* Jan. 2011, pp. 161-164.

[5] L. D. Sorenson *et al.*, "3-D micromachined hemispherical shell resonators with integrated capacitive transducers," in *IEEE MEMS'12*, Jan. 2012, pp. 168-171.

[6] J. Cho *et al.*, "High-Q fused silica birdbath and hemispherical 3-D resonators made by blow torch molding," in IEEE MEMS'13, Jan. 2013, pp. 177-180.

[7] J. Cho *et al.*, "Fused-Silica Micro Birdbath Resonator Gyroscope (μ-BRG)," *IEEE JMEMS*, to be published.

[8] C.C. Painter and A. M. Shkel, "Active structural error suppression in MEMS vibratory rate integrating gyroscopes," *IEEE Sensors Journal*, vol. 3, no. 5, pp. 595-606, Oct 2003.

[9] D. D. Lynch, "Vibratory gyro analysis by the method of averaging," in *Proc. 2nd Saint Petersburg Int. Conf. on Gyroscopic Technology and Navigation*, May 1995, pp. 26–34.

CONTACT

Khalil Najafi; najafi@umich.edu

100K Q-FACTOR TOROIDAL RING GYROSCOPE IMPLEMENTED IN WAFER-LEVEL EPITAXIAL SILICON ENCAPSULATION PROCESS

D. Senkal[1], S. Askari[1], M.J. Ahamed[1], E.J. Ng[2], V. Hong[2], Y. Yang[2], C.H. Ahn[2], T.W. Kenny[2], A.M. Shkel[1]

[1]University of California, Irvine, California, USA
[2]Stanford University, Palo Alto, California, USA

ABSTRACT

This paper reports a new type of degenerate mode gyroscope with measured Q-factor of > 100,000 on both modes at a compact size of 1760 μm diameter. The toroidal ring gyroscope consists of an outer anchor ring, concentric rings nested inside the anchor ring and an electrode assembly at the inner core. Current implementation uses n = 3 wineglass mode, which is inherently robust to fabrication asymmetries. Devices were fabricated using high-temperature, ultra-clean epitaxial silicon encapsulation (EpiSeal) process. Over the 4 devices tested, lowest as fabricated frequency split was found to be 8.5 Hz (122 ppm) with a mean of 21 Hz ($\Delta f/f$ = 300 ppm). Further electrostatic tuning brought the frequency split below 100 mHz (< 2 ppm). Whole angle mechanization and pattern angle was demonstrated using a high speed DSP control system. Characterization of the gyro performance using force-rebalance mechanization revealed ARW of 0.047°/√hr and an in-run bias stability of 0.65 deg/hr. Due to the high Q-factor and robust support structure, the device can potentially be instrumented in whole angle mechanization for applications which require high rate sensitivity and robustness to g-forces.

INTRODUCTION

Coriolis Vibratory Gyroscopes (CVGs) can be divided into two broad categories [1] based on the gyroscope's mechanical element: Degenerate mode gyroscopes which have x-y symmetry (Δf = 0 Hz ideal) and non-degenerate mode gyroscopes which are designed intentionally to be asymmetric in x and y modes ($\Delta f \neq 0$ Hz). Degenerate mode CVGs have potential advantages over non-degenerate mode CVGs in terms of rate sensitivity, signal to noise ratio, power consumption, and potential to implement whole angle mechanization [2]. However, mechanical elements with high-Q factor and very good frequency symmetry are required to utilize these advantages.

Realizing this potential, many MEMS degenerate mode gyroscopes emerged in the recent years. For example, high aspect ratio ring gyroscopes have been demonstrated in [3]. A cylindrical rate integrating gyroscope with a Q-factor of ~21,800 at 2.5 mm diameter was demonstrated in [4], a high frequency poly-silicon disk resonator gyroscope (DRG) [5] with a Q-factor of ~50k at 264 kHz and 600 μm diameter was presented in [6], later a crystalline-silicon version of similar geometry was demonstrated with Q-factor ~100k in [7]. Q-factors as high as 1 million was demonstrated on a quadruple mass gyroscope (QMG) [8], however the device had a 9 mm x 9 mm footprint. Despite these successful implementations

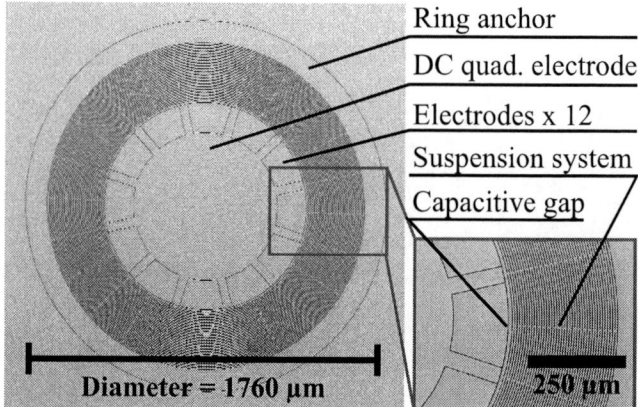

Figure 1: Toroidal ring gyroscope consists of an outer ring anchor, distributed suspension system and inner electrodes.

of degenerate mode operation, obtaining a high-Q factor in a compact volume remains to be a challenge due to factors such as support losses, thermo-elastic dissipation, and viscous damping.

In this paper, we present an epitaxial silicon encapsulated [9] toroidal ring gyroscope with a robust outer perimeter anchor and a distributed suspension system [10]. In contrast to axi-symmetric designs with central support structures, such as [3-7], we explore an alternative support structure for anchor loss minimization. The vibrational energy in the introduced design is concentrated towards the innermost ring, and the device is anchored at the outer perimeter. The distributed support structure prevents vibrational motion propagating to the outer anchor, which helps trap the vibrational energy within the gyroscope providing a Q-factor of > 100,000 at a compact size of 1760 μm.

The toroidal ring gyroscope was fabricated using a wafer-level epitaxial silicon encapsulation process (EpiSeal) [9]. EpiSeal process utilizes epitaxially grown silicon to seal the device layer at extremely high temperatures, which results in an ultra-clean wafer-level seal. This results in high vacuum levels (as low as 1 Pa [9]) without the need for getter materials for absorption of sealing by-products. Due to the high Q-factor and robust support structure, the device can potentially be instrumented in high-g environments that require high angular rate sensitivity.

In the next section, we will present design of toroidal ring gyroscope. This will be followed by experimental characterization of the mechanical element and implementation of force rebalance as well as whole angle mechanizations. The paper concludes with a discussion of the results.

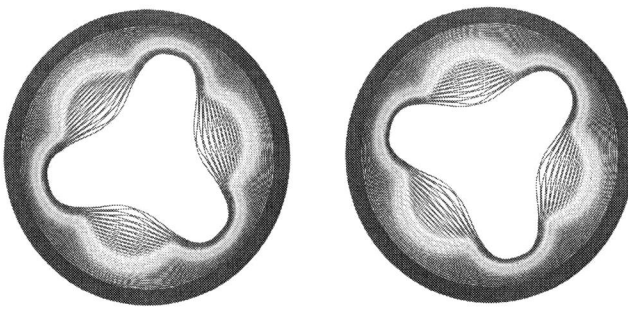

Vibration amplitude (nm) for 200 nm peak displacement

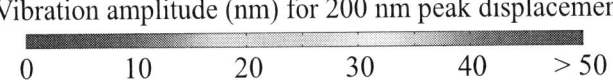

0   10   20   30   40   > 50

Figure 2: Due to the distributed suspension system vibrational energy is trapped within the gyroscope.

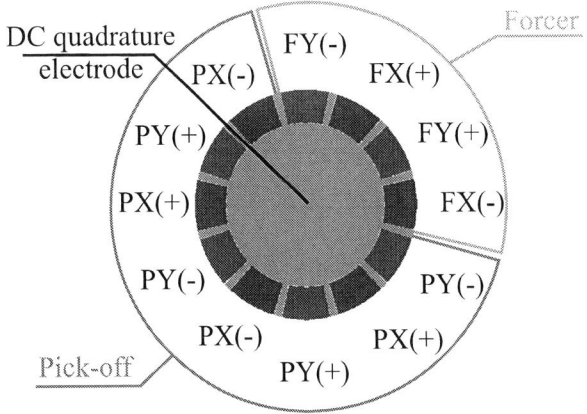

Figure 3:Electrode assembly of the toroidal ring gyroscope, showing pick-off (P) and forcer electrodes (F).

DESIGN

Toroidal ring gyroscope consists of an outer anchor that encircles the device, a distributed ring support structure and an inner electrode assembly, Fig. 1. As opposed to axi-symmetric devices with central support structures [3-7], vibration energy is concentrated at the innermost ring, Fig. 2. The distributed support structure [10] decouples the vibrational motion from the substrate, Fig. 2. This decoupling mitigates anchor losses into the substrate and prevents die/package stresses from propagating into the vibratory structure.

The device was fabricated on single-crystal <100> silicon wafers. For this reason device was designed to operate in n = 3 wineglass modes instead of the more commonly used lower order n = 2 wineglass modes. This eliminates frequency split induced by anisotropic modulus of elasticity of crystalline silicon and makes the frequency splits (Δf) insensitive to misalignment errors in crystalline orientation of the silicon wafer. The draw-back of operation in n = 3 modes are slightly lower angular gain factor, higher resonance frequency and smaller amplitude of motion.

Electrode assembly is located at the center of the gyroscope and consists of 12 discrete electrodes and 1 central electrode, Fig. 3. Discrete electrodes are distributed in groups of 6 onto each degenerate mode. 2 electrodes were used as a forcer and 4 as a pick-off for each mode, giving a total of 4 forcer and 8 pick-off electrodes across the gyro.

The central electrode has two primary functions: (1) it acts as a shield by sinking parasitic currents between adjacent discrete electrodes, (2) it can function as a DC quadrature null electrode as it forms a capacitive gap with the gyroscope at 12 points 15 deg. offset from the discrete electrodes. This eliminates the need for any of the 12 discrete electrodes to be assigned for quadrature null electrodes and hence allows all 12 discrete electrodes to be dedicated as forcer or pick-off.

The device was fabricated on a 2 x 2 mm die, the mechanical element has an outer diameter of 1760 µm and was fabricated on a device layer thickness of 40 µm, Table 1. The suspension system consists of 44 concentric rings. The rings are connected to each other using 12 spokes between the rings, the spokes are interleaved with an offset of 15 deg. between two consecutive rings. The suspension rings have a thickness of 5 µm. The innermost ring is designed to have a slightly higher ring thickness of 8.5 µm, this mitigates the effect of spokes on the overall mode shape and helps retain a truer wineglass shape at the electrode interface.

Table 1: Summary of geometric parameters.

Device diameter (µm)	1760
Device layer thickness (µm)	40
Capacitive gaps (µm)	1.5
Ring thickness (µm)	5
Innermost ring thickness (µm)	8.5
Number of rings	44

RESULTS

Devices were wirebonded to ceramic Leadless Chip Carriers (LCCs) and istrumented with discrete electronics. Electromechanical Amplitude Modulation (EAM) at 1 MHz was used to mitigate the effects of parasitic feed-through on the pick-off electronics. DC bias voltage of 1 V on both modes and AC voltage of 20 mV was used for initial characterization. Frequency response characterization of the fabricated gyroscopes revealed a Q-factor of > 100,000 on both n = 3 modes at ~70 kHz center frequency, Fig. 4.

As-fabricated frequency split (Δf) of 4 devices were characterized. Lowest frequency split observed was at 8.5 Hz, Fig. 3, ($\Delta f/f$ = 122 ppm) with a mean frequency split of 21 Hz ($\Delta f/f$ = 300 ppm) across 4 devices, Table 2. Low frequency split is attributed to robustness of the high order (n=3) wineglass mode to fabrication imperfections and the ultra-clean EpiSeal process.

Table 2: As-fabricated frequency symmetry of 4 devices.

Device	Δf (Hz)	f (kHz)	$\Delta f/f$ (ppm)
#1	8.5	69.75	122
#2	11	69.69	158
#3	25	71.29	350
#4	40	69.4	576

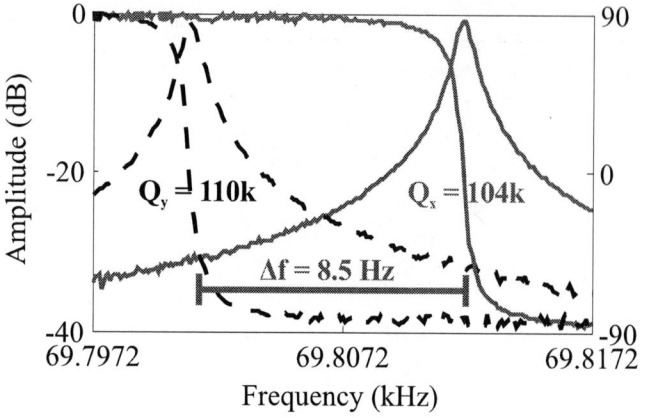

Figure 4: Freq. sweep showing the n=3 wineglass modes with Q-factor above 100k at central freq. of 69.8 kHz.

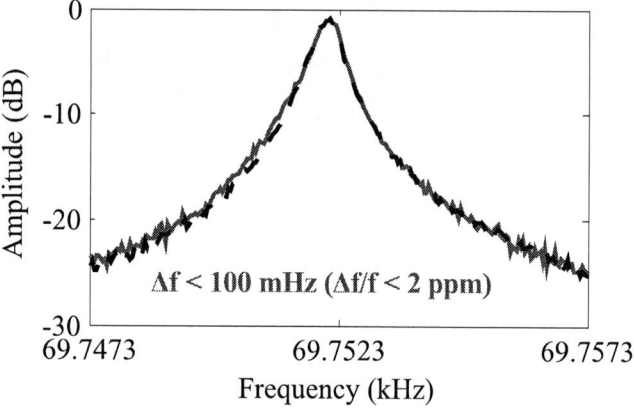

Figure 5: Electrostatic tuning with 3.26 V and 0.5 V resulted in Δf < 100 mHz (Δf/f < 2ppm @ 69.75 kHz).

After initial characterization the frequency split was further reduced using electrostatic tuning of DC bias on forcer and quadrature null electrodes. DC bias voltages of 3.26 V and 0.5 V was sufficient to reduce the frequency split to < 100 mHz (Δf/f < 2 ppm) on Device #1, Fig. 5.

Control system

A DSP-based control system was developed for rate and rate integrating operation. Pattern angle control and whole angle operation is enabled by the use of pendulum variables defined by [11], Fig. 6. The key component of this approach is a PLL loop that tracks the gyro motion at any arbitrary pattern angle as opposed to locking onto one of the primary gyro axis. Once the PLL is implemented key system parameters such as amplitude (E), quadrature error (Q), and pattern angle (θ) can be extracted. Three closed loops act on these parameters: Amplitude gain control (AGC) to keep the drive amplitude stable, quadrature null loop to suppress the quadrature component of the vibratory motion, and force rebalance (FRB) loop to control the pattern angle. Once the required drive vector is determined from the AGC, quadrature null and FRB loops a coordinate transformation takes place to align the force vector with the current pattern angle, which later is modulated at the gyro reference frequency and injected into the respective X and Y forcer

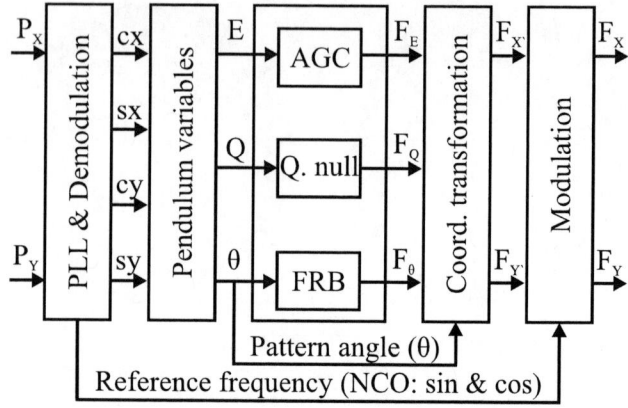

Figure 6: Whole angle mechanization implemented in DSP, allowing rate-integrating operation and control of pattern angle.

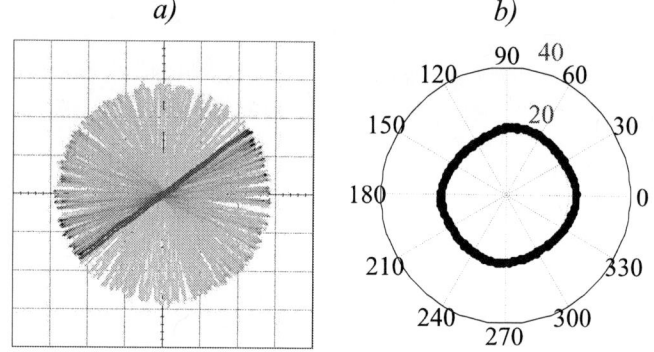

Figure 7: Demonstration of WA operation: real-time capture from carouseling (a), AGC output (mV) vs pattern angle (b).

electrodes. In this control scheme, disabling the FRB loop allows free precession of the pattern angle, also known as whole angle operation, Fig. 7 (a).

All control algorithms were implemented on an ADAU1442 DSP board from Analog Devices. A 24 bit audio codec (AD1938) operating at 192 kHz sampling rate was used for forcer and pick-off signals. An Arduino Due micro-controller board was interfaced with the DSP board over I2S protocol, which was used to down-sample the gyro output and transmit over RS232 protocol for data acquisition.

Pattern angle dependent behaviour of AGC and quadrature null loops as well as error in pattern angle estimation were evaluated through virtual carouseling. Amplitude of the standing wave was kept constant and AGC output was recorded with respect to pattern angle. Due to mismatches in force and pick-off gains a 484 μV std. deviation in AGC output was observed for a mean forcer output of 21.5 mV, Fig. 7(b). For DC quadrature control the output of the quadrature control loop was amplified and fed to the star shaped quadrature null electrode at a slowly varying rate. Instantaneous changes in quadrature error were compensated by using AC quadrature control. Maximum quadrature error was observed at ~45 deg. electrical pattern angle as expected from [11] with a mean value of 512 μV AC, Fig. 8(b).

Accuracy of pattern angle detection was evaluated by

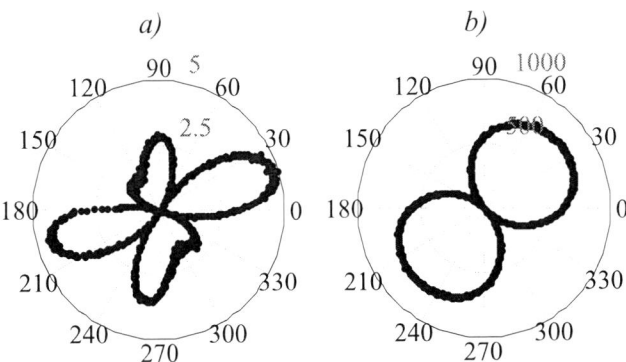

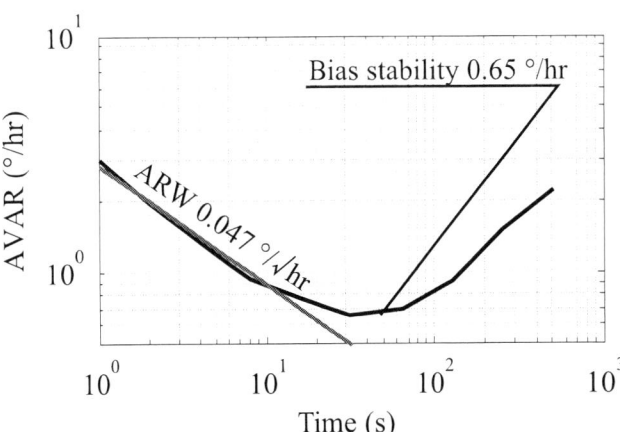

Figure 8: Pattern angle error (deg) vs pattern angle (a) and output of the quadrature null loop (μV) vs pattern angle (b).

Figure 9: Allan deviation of gyroscope force-rebalance output, showing bias stability of 0.65 deg/hr.

disabling the FRB loop and virtual carouseling the gyro. An average of 2.33 deg. offset was found between estimated pattern angle and drive vector, a 2θ dependence on pattern angle was also observed, Fig. 8(a). To evaluate rate gyro performance the pattern angle was locked to X axis by enabling the FRB loop. Alan variance analysis of FRB output revealed ARW of 0.047°/√hr.and an in-run bias stability of 0.65 deg/hr at 32 s integration time, Fig. 9.

CONCLUSIONS

A toroidal ring gyroscope with Q-factor above 100,000 at central frequency of ~70 kHz was designed and implemented in wafer-level epitaxial silicon encapsulation process. The gyroscope consists of a robust ring anchor and a distributed suspension system. In contrast to axi-symmetric designs with central support structures, the vibrational energy in the introduced design is concentrated towards the innermost ring, and the device is anchored at the outer perimeter. The distributed support structure prevents vibrational motion propagating to the outer anchor, which helps trap the vibrational energy within the gyroscope. Combined with the ultra-clean EpiSeal process Q-factors above > 100,000 were obtained at ~70 kHz and a compact size of 1760 μm without using getter materials.

Frequency splits (Δf) as low as 8.5 Hz ($\Delta f/f$ = 122 ppm) and a mean frequency split of 21 Hz ($\Delta f/f$ = 300 ppm) were observed across 4 devices. Low frequency split is attributed to robustness of the high order (n = 3) wineglass mode to fabrication imperfections and the ultra-clean EpiSeal process.

Whole angle and pattern angle control loops were implemented on a custom DSP system. Characterization of the gyro performance using force-rebalance mechanization revealed ARW of 0.047°/√hr and an in-run bias stability of 0.65 deg/hr. Due to the high Q-factor, degenerate mode operation on n = 3 modes and robust support structure, the device can potentially be instrumented for high-g environments that require high angular rate sensitivity.

ACKNOWLEDGEMENTS

Design and characterization was done in UCI Microsystems Laboratory. Devices were fabricated at Stanford Nanofabrication Facility.

REFERENCES

[1] A. M. Shkel, "Type I and Type II Micromachined Vibratory Gyroscopes," *IEEE/ION PLANS*, pp. 586–593, 2006.

[2] D. M. Rozelle, "The hemispherical resonator gyro: From wineglass to the planets", *Proc. AAS/AIAA Space Flight Mechanics Meeting*, pp. 1157–1178, 2009.

[3] F. Ayazi, K. Najafi, "Design and fabrication of high-performance polysilicon vibrating ring gyroscope," *IEEE MEMS*, pp. 621-626, 1998.

[4] J. Cho, J. Gregory, K. Najafi, "Single-crystal-silicon vibratory cylinderical rate integrating gyroscope (CING)," *TRANSDUCERS*, pp. 2813–2816, 2011.

[5] R.L. Kubena, D.T. Chang, "Disc resonator gyroscopes," *US Patent 7,581,443*, 2009.

[6] S. Nitzan, C. H. Ahn, T.-H. Su, M. Li, E. J. Ng, S. Wang, Z. M. Yang, G. O'Brien, B. E. Boser, T. W. Kenny, D. A. Horsley, "Epitaxially-encapsulated polysilicon disk resonator gyroscope," *IEEE MEMS*, pp. 625–628, 2013.

[7] C. H. Ahn, E. J. Ng, V. A. Hong, Y. Yang, B. J. Lee, M. W. Ward, T. W. Kenny, "Geometric compensation of (100) single crystal silicon disk resonating gyroscope for mode-matching," *TRANSDUCERS*, pp. 1723–1726, 2013.

[8] A. A. Trusov, A. R. Schofield, A. M. Shkel, *U.S. Patent 8,322,213*, 2012.

[9] R. N. Candler, M. A. Hopcroft, B. Kim, W-T. Park, R. Melamud, M. Agarwal, G. Yama, A. Partridge, M. Lutz, T. W. Kenny, "Long-Term and Accelerated Life Testing of a Novel Single-Wafer Vacuum Encapsulation for MEMS Resonators," *JMEMS*, 15, (6), pp. 1446-1456, 2006.

[10] A. M. Shkel, R. Howe, "Micro-machined angle-measuring gyroscope," *US Patent 6,481,285*, 2002.

[11] D. D. Lynch, "Vibratory gyro analysis by the method of averaging," *St. Petersburg Int. Conf. on Gyroscopic Technology and Navigation*, pp. 26-34, 1995.

CONTACT

*D. Senkal, tel: +1-949-689-3370; dsenkal@uci.edu.

SINGLE PROOF-MASS TRI-AXIAL PENDULUM ACCELEROMETERS OPERATING IN VACUUM

D. E. Serrano[1,2], Y. Jeong[1], V. Keesara[2], W. K. Sung[2], and F. Ayazi[1,2]
[1]Georgia Institute of Technology, Atlanta, GA, USA
[2]Qualtré Inc., Marlborough, MA, USA

ABSTRACT

This paper reports on the design, fabrication and characterization of single proof-mass tri-axial capacitive accelerometers coexisting in a low-pressure environment with high-frequency gyroscopes, for the implementation of monolithic 6-degree-of-freedom inertial measurement units. The accelerometers are designed to operate as quasi-static devices (i.e. non-resonant sensors) in mid vacuum levels (1 – 10 Torr) by increasing squeeze-film air damping through the use of capacitive nano-gaps (< 300 nm). Reduced die area is achieved utilizing a pendulum-like structure composed of a 450x450x40 μm³ proof-mass anchored to the substrate by a cross-shaped polysilicon spring. The small capacitive gaps, allow for the design of devices with high resonance frequency (~ 15 kHz) that provide large shock and vibration immunity.

INTRODUCTION

Micromachined accelerometers have played a critical role in the commercialization of almost all MEMS devices [1]. Their success in both the automotive industry and consumer electronics motivated the development of fabrication and integration techniques necessary for the implementation of other types of micromachined structures, such as resonators, microphones and gyroscopes, among others. Today—after almost twenty years since the first commercial product—sales projections for standalone MEMS accelerometers in the consumer space are starting to show a slow decline in market share due to the demand of highly integrated multi-degree-of-freedom sensors [2]. This new breed of devices incorporate accelerometers, gyroscopes and magnetometers on the same package, indicating that for further reductions in size, monolithic single-die implementations of inertial sensors will be necessary [3].

Conventionally, accelerometers and vibratory gyroscopes required different pressure levels to attain optimal operation conditions. Gyroscopes are implemented under high vacuum environments to achieve increased quality factors (Q), which are essential for large scale-factor and low noise performance. On the other hand, accelerometers are packaged at levels close to atmospheric pressure (760 Torr) in order to increase squeeze-film damping; this prevents the devices from experiencing overshoot and long settling times associated with a high Q response. The critical difference in pressures necessary for each type of sensor indicates that, in order for them to coexist in the same package, a paradigm shift in the design methodology of either accelerometers or resonant gyroscopes is necessary.

Methods to operate MEMS accelerometers by monitoring the change in their resonance frequency as a function of acceleration have been suggested as a possible solution [4]. If implemented as resonant sensors rather than quasi-static devices, accelerometers can then be integrated in the same pressure levels with gyroscopes. However, this comes at the expense of higher power consumption (required to operate them in an oscillator configuration), strong dependency on temperature (due to the temperature coefficient of frequency), and increased system complexity.

As an alternative, the use of deep-submicron capacitive gaps has been proposed and demonstrated as a technique for the reduction of Q values in accelerometers operating at low pressures [5]. In this paper, this technique is utilized for the implementation of single proof-mass tri-axial capacitive accelerometers co-integrated in the same wafer-level package with bulk-acoustic wave (BAW) tri-axial gyroscopes in a moderate vacuum environment (1 – 10 Torr). The pressure range is optimized to achieve maximum Q for the high-frequency gyros without compromising accelerometer performance. Acceleration sensors with proof-mass areas of 450x450 μm² were implemented on 40 μm-thick SOI wafers using the high-aspect-ratio poly and single-crystal silicon (HARPSS™) process [6]. Out-of-plane capacitive nano-gaps (< 300 nm) provide high scale-factor and increased squeeze film damping (SFD) to guarantee stable behavior. The devices have been interfaced with front-end electronics to characterize their performance.

ACCELEROMETER DESIGN

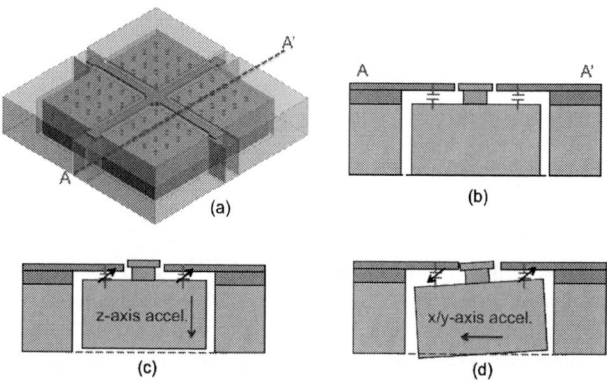

Figure 1. (a) Single-proof-mass tri-axial accelerometer composed of SCS mass supported by cross-shaped polysilicon tethers. (b) Fixed top electrodes form capacitive gaps with mass. (c) Mass translates under z-axis acceleration and (d) tilts with x/y-axis acceleration.

A schematic diagram of the proposed tri-axial accelerometer is shown in Fig. 1. The design consists of a pendulum-like structure composed of a 450x450x40 μm³ single-crystal silicon proof-mass anchored to the substrate by a cross-shaped polysilicon spring (Fig 2) [7]. The tethers that compose the spring are attached to the mass using a self-aligned process that prevents offsets in the center-of-mass, which would result in cross-axis sensitivity. Four pick-off electrodes placed on top of the moving structure are multiplexed to read out

978-1-4799-3510-9/14 $31.00 © 2014 IEEE

changes in capacitance generated by the x-, y- and z-axis acceleration components. In presence of acceleration along the x-axis, the tethers act as torsional springs, allowing the mass to tilt. This causes a differential change in capacitance $\Delta C_x = (C_1+C_2)-(C_3+C_4)$, where C_1 through C_4 correspond to the individual capacitances between the proof-mass and fixed electrodes (Fig. 2). Similarly, acceleration along the y-axis causes a differential change of $\Delta C_y = (C_1+C_4)-(C_2+C_3)$. Lastly, z-axis acceleration produces out-of-plane translation of the proof-mass, causing an effective capacitance variation $\Delta C_z = (C_1+C_2+C_3+C_4)-4C_0$, where C_0 is the zero-input acceleration rest capacitance. Figure 3 shows the simulated displacement response in the presence of acceleration along each individual axis.

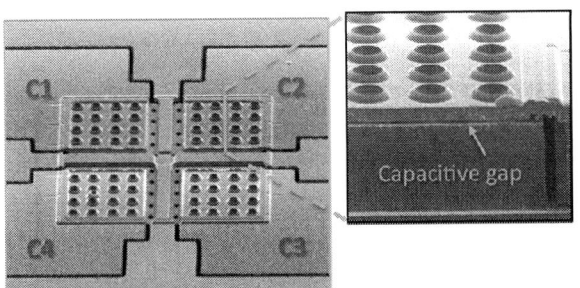

Figure 2. Top and side SEM view of single-proof mass tri-axial accelerometer. Electrodes over pendulum mass are multiplexed to detect acceleration along all 3 axes.

In a parallel-plate capacitive accelerometer, the overall scale-factor can be expressed as:

$$SF = \frac{\Delta C}{a_{in}} \approx \frac{1}{\omega_0^2} \frac{n \varepsilon A_{elec}}{g_0^2} \qquad (1)$$

where ω_0 is the resonance frequency of the structure and n, A_{elec} and g_0 are the number of electrodes, the electrode area and rest gap of the parallel plate capacitors, respectively. Therefore, the use of narrow capacitive gaps ($\sim$ 300 nm) between the top electrodes and the pendulum proof-mass provides large electromechanical coupling. This allows for the design of much smaller structures with larger resonance frequencies ($\sim$ 15 kHz) as compared to other commercial accelerometers (1 – 6 kHz), making them less prone to stiction during fabrication and more immune to shock and vibration. For instance, it takes up to 8,000 g to bring the proof-mass in contact with the electrodes separated by a 300 nm gap.

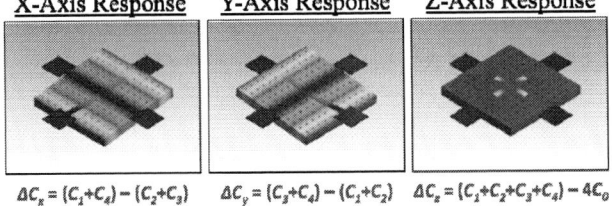

Figure 3. Displacement simulation response to x-, y- and z-axis input acceleration in ANSYS (electrodes not shown and displacements greatly exaggerated for clearer visualization).

Capacitive nano-gaps also provide increased squeeze-film air damping to avoid high quality factors that could cause overshoot and ringing in the accelerometer response. In a parallel-plate moving capacitor, the damping coefficient b, which is inversely proportional to Q, can be expressed as [8]:

$$\frac{1}{Q} \propto b = n \mu_{eff} l \left(\frac{w}{g_0} \right)^3 \qquad (2)$$

where l and w are the electrode length and width ($A_{elec} = w*l$) and μ_{eff} is the effective viscosity of the gas inside the package. The strong dependency of b with respect to g_0 indicates that for gaps of only a few hundreds of nanometers, Q can be reduced by 3 to 4 orders of magnitude in comparison with devices implemented with conventional capacitive gaps of 2 to 5 μm in the same pressure environment.

Equation (2) serves as a starting guideline for the initial selection of the required capacitive gaps. However, a more complete expression should be utilized to capture the effects of fluid-wall interactions, presence of release holes and non-trivial boundary conditions [8-10]. For the particular case of non-uniform wall displacements—such in the case of the x- and y-axis acceleration tilt response (Fig. 3)—close-form expressions are challenging to derive, thus finite element analysis (FEA) tools provide a much better estimation of Q. Figure 4 shows the pressure distribution in the electrode gaps for the x-axis and z-axis response. Q values of 0.5 and 0.6 were extracted at 10 Torr for the x/y- and z-axis response, respectively.

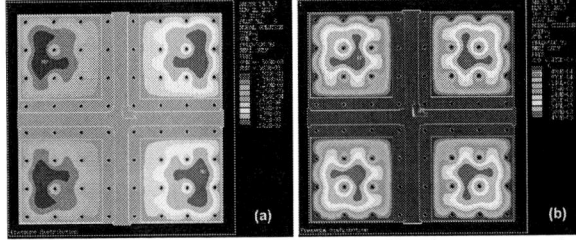

Figure 4. Pressure distribution of squeeze-film damping simulation for (a) x-axis and (b) z-axis acceleration response. $Q_{xy} = 0.5$, $Q_z = 0.6$ at 10 Torr, and $Q_{xy} = 4$, $Q_z = 5$ at 1 Torr.

FABRICATION

The single-proof mass tri-axial accelerometers were batch-fabricated with bulk-acoustic wave (BAW) tri-axial gyroscopes on a 40 μm-thick silicon-on-insulator (SOI) substrate utilizing a modified version of the HARPSS^{TM} process [11]. Lateral trenches were etched through DRIE to define the proof-mass and electrodes in the device layer by the use of a thermal oxide mask. A 300 nm layer of sacrificial oxide was grown to define the lateral gaps required for the gyros. Trenches were filled with polysilicon and TEOS oxide to implement side electrodes. By growing an additional 300 nm thermal oxide layer, out-of-plane gaps were defined, followed by the deposition and patterning of polysilicon for top electrodes. The devices were then fully released in hydrofluoric acid (HF).

A capping wafer, which is processed independently, was bonded to the base wafer in order to provide hermetic wafer-level packaging (WLP). Figure 5 shows a visual image of a 6-degree-of-freedom inertial sensor consisting of the single proof-mass accelerometer co-integrated with high-frequency tri-axial gyroscopes [11, 12].

978-1-4799-3510-9/14 $31.00 © 2014 IEEE

Figure 5. Wafer-level packaged (WLP) 6 degree-of-freedom (6-DOF) inertial measurement unit (IMU) containing single proof-mass 3-axis accelerometer in low vacuum environment.

MEASUREMENT RESULT

The resonance response of the tri-axial accelerometer was first measured by probing an uncapped wafer inside a vacuum probe station with adjustable pressure. Electrodes C1 and C4 were tied to excite the device, whereas C2 and C3 were connected together to the input of a network analyzer; the proof mass was biased at the same potential as the electrodes, thus the measured peak frequencies are twice the value of the device resonances. This is attributed to the quadratic relation between the drive voltage and the generated excitation force. In this configuration, the x-axis and z-axis modes are primarily excited (Fig. 6), but small capacitance mismatches cause a minor response of y-axis mode. Measured values of 12 kHz and 14 kHz are in good agreement with designed frequencies of 13 kHz and 16 kHz for the x/y- and z-axis modes, respectively.

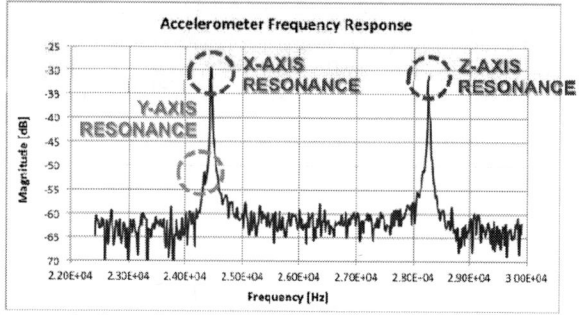

Figure 6. Open-loop measured frequency response of tri-axial accelerometer. Excitation signal applied on electrodes C1 and C4, signal read from C2 and C3. Frequency peak of y-axis is cross axis excitation due to small capacitance mismatches.

Figure 7 shows the Q values for the z-axis response at different vacuum levels. It is seen that for pressures above 200 mTorr, the device starts approaching a desired over-damped condition, which guarantees stable operation at the WLP pressure levels in the range of 1 to 10 Torr.

Accelerometers were mounted on an evaluation board and interfaced with a front-end switched-capacitor integrated circuit to verify functionality. Since mismatch between the parasitic capacitances translates into offset that limits or saturates the operation range of the electronics, additional compensation circuitry was implemented with discrete components to calibrate the system. A programmable voltage V_{cal} is switched between amplification phases to cancel the DC differential charge generated by the static capacitance offset (Fig. 8) [13].

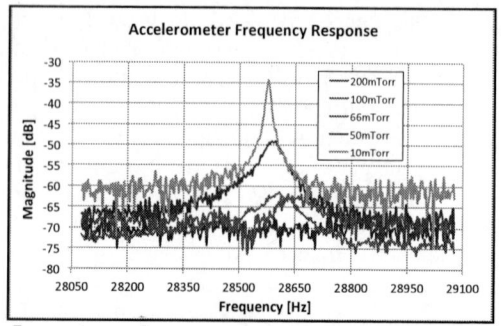

Figure 7. z-axis mode resonance response at different pressure levels. For 200 mTorr and below, Q becomes low enough to guarantee stable behavior at WLP vacuum levels (1 to 10 Torr).

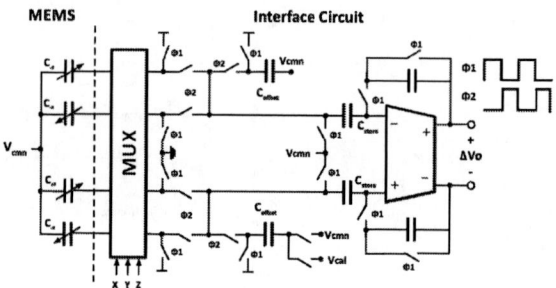

Figure 8. Front-end switched-capacitor interface circuit based on [5]. Multiplexer switches input capacitors to measure x, y and z-axis acceleration in different time slots. Additional calibration phase used to cancel DC offset charge generated by mismatch.

Acceleration sensitivities of 5 mV/g, 6 mV/g and 11 mV/g measured for the x-, y- and z-axis response, respectively (Fig 9). Higher z-axis sensitivity is observed due to larger change in capacitance during out-of-plane translation as compared to in-plane tilting. Differences between x and y responses are attributed to alignment inaccuracies between the evaluation board and the shaker table. Higher sensitivity can be achieved by having larger gain through circuit optimization. Similarly, cross-axis sensitivity levels in the order of 1 to 3% shown in Fig. 10 can be further reduced by proper alignment methods the measurement setup in order to reach the simulated cross-axis value of < 0.1% (Table 1).

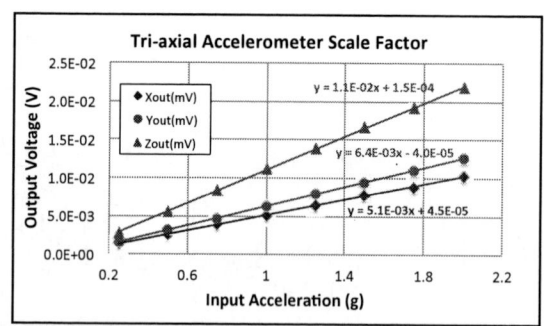

Figure 9. Measured scale factor for x-, y- and z-axis acceleration. Differences between x and y response attributed to setup alignment.

The measured output noise of the system (MEMS+IC) is in the order of 3 to 6 mg/√Hz, with a bias drift of 20 mg. This is attributed to the discrete electronics added for the offset calibration scheme (Fig. 11). Adding the calibration on-chip and further optimization will make the response Brownian noise limited, which is designed to be 30 µg/√Hz at 10 Torr.

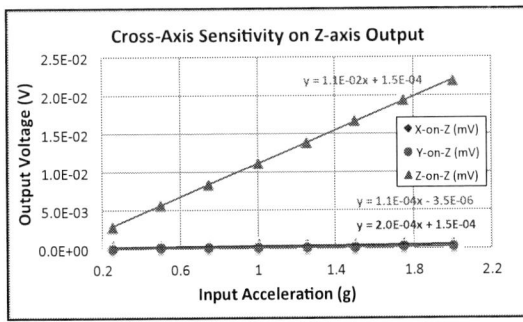

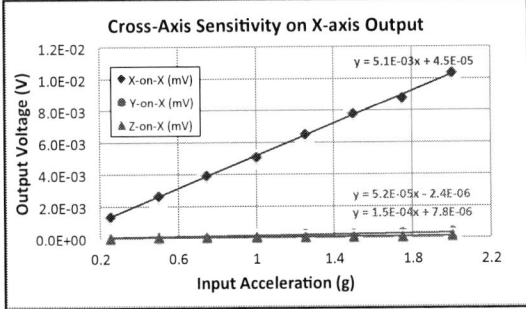

Figure 10. (top) Cross-axis sensitivity of z-axis output (x-axis input: 1.8%, y-axis input: 1%). (bottom) Cross-axis response of x-axis ouput (y-axis input: 3%, z-axis input: 1%). Response of y-axis output is similar to x-axis and attributed to mounting alignment.

CONCLUSIONS

A new single proof-mass tri-axial capacitive accelerometer, co-integrated with high-frequency BAW gyroscopes, was designed and fully characterized. Capacitive nano-gaps were used to achieve increased sensitivities using small form-factor structures, which provide high immunity to shock and vibration. Additionally, the small gaps provide optimal squeeze-film damping to stabilize accelerometer response in low-pressure environments, preventing the device from undergoing overshoot and long settling times.

Table 1. MEMS and system specifications

Accelerometer Design Parameters		
Dimensions	450 μm x 450 μm x 40μm (proof mass)	
Brownian noise	13 μg/√Hz (x/y-axis), 30 μg/√Hz (z-axis) @ 10 Torr (worst case)	
Resonance Frequency	12 kHz (x/y-axis), 14 kHz (z-axis)	
Bandwidth	6 kHz (x/y-axis), 10 kHz (z-axis) @ 10 Torr (worst case)	
Sensitivity	3.0 fF/g (x/y-axis) 8.0 fF/g (z-axis)	
Cross-axis Sensitivity (% of FS)	Simulation @ 20 g S_{XY}: 0.001% S_{XZ}: 0.09% S_{ZX}: 0.08% S_{ZY}: 0.08%	Measured @ 6 g S_{XY}: 3.0% S_{XZ}: 1.1% S_{ZX}: 1.0% S_{ZY}: 1.8%
Nonlinearity (% of FS)	0.5% (x/y-axis), 1.0% (z-axis) Measured @ 6 g	
MEMS + Interface IC		
C/V gain	~ 2 mV/fF	
Sensitivity	6 mV/g (x), 5 mV/g (y), 11 mV/g (z)	
Noise floor	-90 dBVrms/√Hz @ 1 Hz	

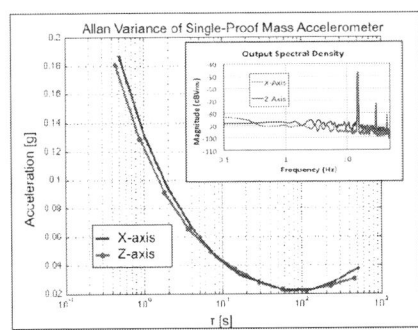

Figure 11. Allan variance of x- and z-axis response. Measured bias drift of 20 mg. (inset) Noise density of ~ -90 dBVrms at 1 Hz (3 to 6 mg/√Hz) attributed to external circuitry.

REFERENCES

[1] N. Yazdi, *et al.*, "Micromachined inertial sensors," *Proceedings of the IEEE*, vol. 86, pp. 1640-1659, 1998.

[2] "Inertial sensors in mobile products: Market and technology trends, adoption of combo solutions, sensor fusion," Yole Développement, 2012.

[3] F. Ayazi, "Multi-DOF inertial MEMS: from gaming to dead-reckoning," in *Tech. Digest of the 16th International Conference on Solid-State Sensors, Actuators and Microsystems (Transducers'11)*, Beijin, China, 2011, pp. 2805-2808.

[4] C. Comi, *et al.*, "A new biaxial silicon resonant micro accelerometer," in *Proc. 24th IEEE International Conference on Micro Electro Mechanical Systems (MEMS '11)*, 2011, pp. 529-532.

[5] Y. Jeong, *et al.*, "Wafer-level vacuum-packaged triaxial accelerometer with nano airgaps," in *Proc. 26th IEEE International Conference on Micro Electro Mechanical Systems (MEMS '13)*, 2013, pp. 33-36.

[6] F. Ayazi, *et al.*, "High aspect-ratio polysilicon micromachining technology," in *Proc. 10th Int. Conf. Solid-State Sensors and Actuators (Transducers '99)*, 1999, pp. 320-323.

[7] M. Mehregany, "Three-Axis Accelerometers," US Patent US 7,578,189 B1, Aug. 25, 2009.

[8] M. Bao, *et al.*, "Squeeze film air damping in MEMS," *Sensors and Actuators A: Physical*, vol. 136, pp. 3-27, 2007.

[9] M. Bao, *et al.*, "Energy transfer model for squeeze-film air damping in low vacuum," *Journal of Micromechanics and Microengineering*, vol. 12, pp. 341-346, 2002.

[10] T. Veijola, *et al.*, "Extending the validity of squeezed-film damper models with elongations of surface dimensions," *Journal of Micromechanics and Microengineering*, vol. 15, pp. 1624-1636, 2005.

[11] W. K. Sung, *et al.*, "A Mode-Matched 0.9 MHz Single Proof-Mass Dual-Axis Gyroscope," in *Tech. Digest of the 16th International Conference on Solid-State Sensors, Actuators and Microsystems (Transducers '11)*, 2011.

[12] H. Johari, *et al.*, "High-Frequency Capacitive Disk Gyroscopes in (100) and (111) Silicon," in *Proc. 20th IEEE International Conference on Micro Electro Mechanical Systems (MEMS '07)*, 2007, pp. 47-50.

[13] Y. Jeong, *et al.*, "A Novel Offset Calibration Method to suppress Capacitive Mismatch in MEMS Accelerometer," in *Samsung ElectroMechanics 9th 'Inside Edge'*, 2013.

CONTACT:

Diego Emilio Serrano: dserrano@qualtre.com

SIMULTANEOUS DETECTION OF LINEAR AND CORIOLIS ACCELERATIONS ON A MODE-MATCHED MEMS GYROSCOPE

S. Sonmezoglu[1], H. D. Gavcar[1], K. Azgin[3], S. E. Alper[1], and T. Akin[1,2]

[1]METU-MEMS Research Center, Middle East Technical University, Ankara, TURKEY
[2]Dept. of Electrical and Electronics Eng., Middle East Technical University, Ankara, TURKEY
[3]Dept. of Mechanical Eng., Middle East Technical University, Ankara, TURKEY

ABSTRACT

This paper presents a novel "in operation acceleration sensing and compensation method" for a single-mass mode-matched MEMS gyroscope. In this method, the amplitudes of the sustained residual quadrature signals on the differential sense-mode electrodes are compared to measure the linear acceleration acting on the sense-axis of the gyroscope. By measuring the acceleration acting along the sense-axis, the g-sensitivity of the gyroscope output to these accelerations is mitigated without using a dedicated accelerometer. It has been experimentally demonstrated that the g-sensitivity of the studied gyroscope is substantially reduced from 1.08°/s/g to 0.04°/s/g, and the effect of the linear acceleration on the gyroscope output is highly-suppressed (by 96%) with the use of the compensation method proposed in this work.

INTRODUCTION

Over the last decade, significant performance improvements took place in the operation of vibratory MEMS gyroscopes, either by matching the resonance mode frequencies or by minimizing the undesired mechanical cross-talk, called quadrature error, between the resonance modes of the gyroscope [1, 2]. However, most of the high performance gyroscopes suffer from the effects of external accelerations on the gyroscope output. The external acceleration acting on the sense-axis of the gyroscope causes fluctuations in the output bias and shifts the scale factor, which directly degrade the reliability of the gyroscope output [3]. Therefore, for high reliability, various methods have been introduced in the literature to reduce the acceleration effect on the gyroscope output, or shortly its g-sensitivity. A common method to reduce the effect of the acceleration or external vibration is to use the tuning-fork type vibrating gyroscopes [4]. In this type of gyroscopes, the common mode signals such as linear acceleration are mitigated by the differential reading of the sense-mode signals coming from the two mechanically-connected differential masses. However, the effect of the acceleration on the gyroscope output cannot be completely eliminated in the tuning-fork type gyroscopes due to the fabrication imperfections [5]. A compensation method can also be used by calibrating the gyroscope output using an external accelerometer [6]. However, this method requires at least one additional accelerometer, and the measured acceleration cannot always be the acceleration sensed by the gyroscope due to their physical locations. Implementing a micro-mechanical shock absorber under the sensor structure is

another method to eliminate the effect of random high-frequency accelerations or vibrations [7]. Although this method provides a solution for the vibration sensitivity of a gyroscope, it does not eliminate the effects of static or low frequency accelerations.

This paper proposes a novel "in operation acceleration sensing" method which utilizes the amplitude difference information between the residual quadrature signals on the differential sense-mode electrodes. In this study, the fluctuations in the gyroscope output caused by the linear accelerations have also been compensated by using the acceleration output of the gyroscope. Furthermore, it has been experimentally shown that the g-sensitivity of the studied gyroscope is reduced by 96% (from 1.08°/s/g to 0.04°/s/g) with the use of the compensation method proposed in this work.

PROPOSED ACCELERATION SENSING AND COMPENSATION METHOD

Figure 1 shows the simplified structural view of the single-mass fully-decoupled gyroscope. The main idea behind the gyroscope operation is the Coriolis coupling, which enables energy transfer from the drive to sense mode in the presence of an angular rate. The drive and sense frames are restricted to movement in only one direction, whereas the proof mass frame vibrates along two orthogonal directions to accomplish the Coriolis coupling between the operating modes (drive and sense). Although the coupling between the operating modes is totally eliminated in the mechanical design, unavoidable fabrication tolerances always cause a quadrature error in the system. The quadrature signal resulting from the quadrature error is adjusted to a level of about 10°/s with the help of the closed-loop quadrature control electronics [2]. Note that the sense-mode electrodes of the gyroscope are designed to be differential. The amplitude information of these sustained quadrature signals on the sense-mode electrodes are used to sense the linear accelerations acting on the sense-axis of the gyroscope.

Figure 2 shows the conceptual view of the differential sense-mode electrodes. During the gyroscope operation, in the absence of external acceleration, the sense mode frame is kept stationary thanks to the sense-mode force-feedback electronics, acting along the sense-axis. Therefore, the amplitudes and phases of the sustained quadrature signals (I_{SP+} and I_{SP-}) on the differential sense-mode electrodes are identical and opposite. However, it is seen from Figure 2-(b) that when the net force resulting from an applied quasi-static

978-1-4799-3510-9/14 $31.00 © 2014 IEEE

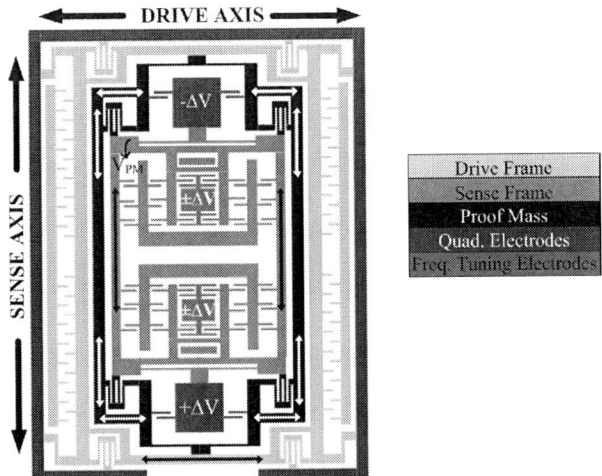

Figure 1: Simplified structural view of the single-mass fully-decoupled gyroscope.

acceleration on the movable sense frame is different than zero, the capacitive gap between the differential sense-mode electrodes decreases on one side and increases on the other, resulting in a shift at the sense-mode vibration axis. This shift causes a variation at the amplitudes of the quadrature signals on the differential sense-mode electrodes, i.e., the amplitude of the quadrature signal increases (or decreases) with a decreasing (or increasing) capacitive gap under the condition of an applied static acceleration condition compared to the case when there is no acceleration.

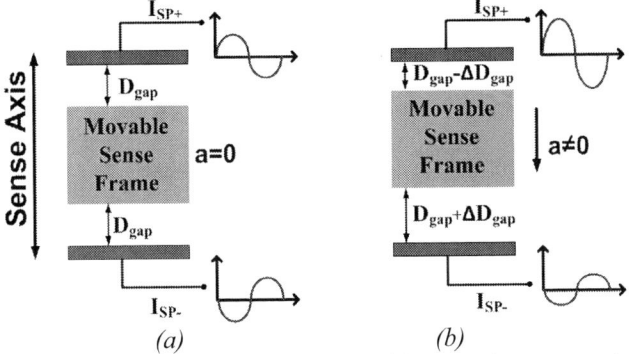

Figure 2: Conceptual view of the differential sense-mode electrodes under the conditions of (a) zero and (b) static accelerations.

The main motivation behind the acceleration sensing and compensation method is to sense the linear acceleration acting along the sense-axis and eliminate the effect of the acceleration on the gyroscope output. The operation principle of the acceleration sensing mechanism is mainly based on the comparison of the amplitudes of the residual quadrature signals, directly picked from the differential sense-mode electrodes. Figure 3 shows the block diagram of the sense-mode electronics of the gyroscope with the acceleration sensing electronics. In Figure 3, the differential sense-mode channel outputs (SP) are fed to the summing amplifier to obtain the amplitude difference information

between the quadrature signals. The output of the summing amplifier is then demodulated with the drive-mode channel output (DP), and the resulting rectified signal is passed through a low-pass filter (LPF) to get the amplitude information of the acceleration. Furthermore, the sense-mode force-feedback electronics is used only to generate the rate output by cancelling the movement caused by an angular rate in the sense mode of the gyroscope. It should be mentioned that the amplitude of the sustained quadrature signals of the gyroscope is adjusted to be about 10°/s to set the scale factor of the acceleration output of the gyroscope to a known level and to prevent saturation at the force-feedback electronics.

Linear acceleration acting along the sense-axis of the gyroscope affects the gyroscope output in three distinct ways. Firstly, shifting the sense-mode vibration axis does not equally affect the amplitude of the differential sense-mode channel outputs due to the nonlinearity in the capacitive detection. Consequently, in a differential reading scheme, the difference of positive and negative sense-mode channel outputs fluctuates, causing a variation at the rate output. Secondly, the capacitive gap of the force-feedback electrodes changes with an applied linear acceleration. The change in these gaps affects the force generated by these electrodes. As a result, the scale factor of the gyroscope shifts along with the bias in the rate output. Furthermore, the linear acceleration slightly alters the level of the residual quadrature signal by disturbing the quadrature cancellation electrodes. A portion of the quadrature signal (~5%) leaks to the rate output of the gyroscope due to the phase error of about 3° coming from the electronics. Hence, any change in the quadrature signal is reflected to the rate output.

Although the linear acceleration acting along the sense-axis affects the rate output of the gyroscope in different ways, it is possible to extract a relation between the rate output and acceleration experimentally. For small amplitudes of acceleration, the change in the capacitive electrode gaps would also be small. Thus, a linear relation can be assumed between the rate output variation of the gyroscope and the applied acceleration. The same assumption is also valid for the acceleration output. Under these assumptions, the rate and acceleration outputs of the gyroscope can be represented as follows;

$$V_{rate} = A.a + B.\Omega + V_{rateoffset} \qquad (1)$$

$$V_{acce} = C.a + D.\Omega + V_{acce.offset} \qquad (2)$$

where "a" is the external acceleration and "Ω" is the input angular rate. A is the g-sensitivity of the gyroscope, B is the gyro scale factor, C is the accelerometer scale factor and D is the rate sensitivity of the accelerometer output. According to these expressions, the rate and acceleration outputs are functions of both the angular rate and the linear acceleration. The reason of the change in the acceleration output with the angular rate is the unavoidable phase error in the electronics. These acceleration and rate constants for a specific gyroscope are determined by the static acceleration and angular velocity tests.

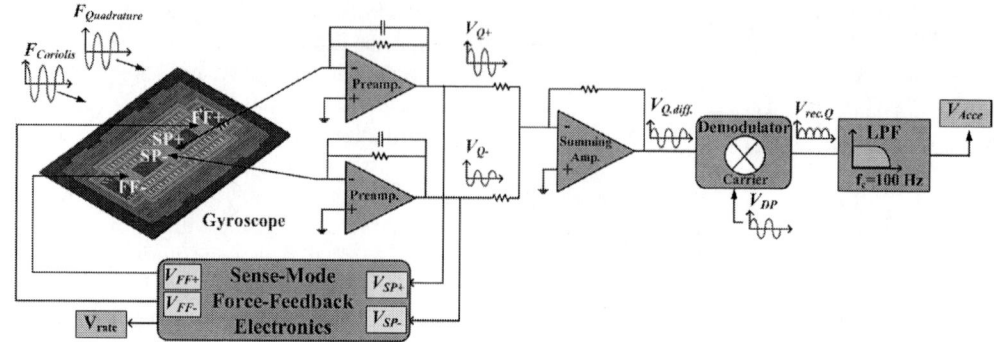

Figure 3: Block diagram of the sense-mode electronics of the gyroscope with the acceleration sensing electronics.

TEST RESULTS
Static Acceleration Tests

A static acceleration test is performed by rotating the sensor about its out-of-plane axis, in order to measure the amplitude of the static acceleration and to determine its effect on the rate output of the gyroscope. Figure 4 shows (a) the schematic and (b) actual views of the test setup for the static acceleration test. During the test, the rate and acceleration data are collected with a step angle of $10°$ under the gravitational ($1g$) acceleration condition. The g-acceleration (a) acting on the gyro sense-axis can be directly computed by $a=g*sin\Theta$, where Θ is the rotation angle.

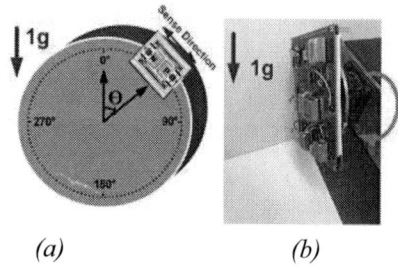

(a)　　　　　　　　(b)

Figure 4: (a) Schematic and (b) actual views of the test setup for the static acceleration test.

Figure 5 presents the simultaneously measured acceleration and rate outputs of the gyroscope as a function of the rotation angle. As observed from Figure 5, both the rate and acceleration outputs vary linearly with the applied static acceleration. The measured acceleration and rate output variations for $1g$-acceleration equals to 0.45mV and 2.6mV, respectively. The change in the rate output of the gyroscope corresponds to the g-sensitivity of $1.08°/s/g$.

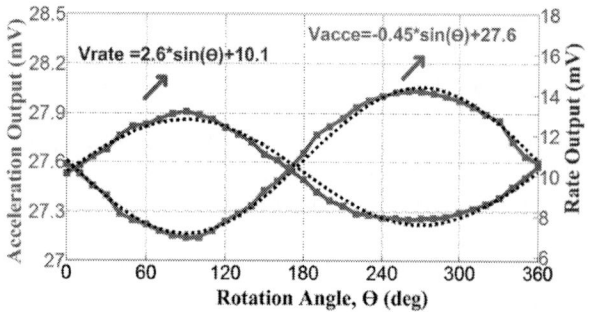

Figure 5: Simultaneously measured acceleration and rate outputs of the gyroscope as a function of the rotation angle.

By using (1) and (2), the rate output is compensated to eliminate the effect of the acceleration. Figure 6 shows the measured and compensated rate outputs as a function of the rotation angle during the static acceleration test. The compensated rate output variation for $1g$-acceleration equals to 0.1mV that corresponds to $0.04°/s/g$. It has been experimentally shown that the g-sensitivity of the rate output is reduced from $1.08°/s/g$ to $0.04°/s/g$, which directly indicates that the g-sensitivity of the gyroscope is highly-suppressed by up to 96% with the aid of the compensation method.

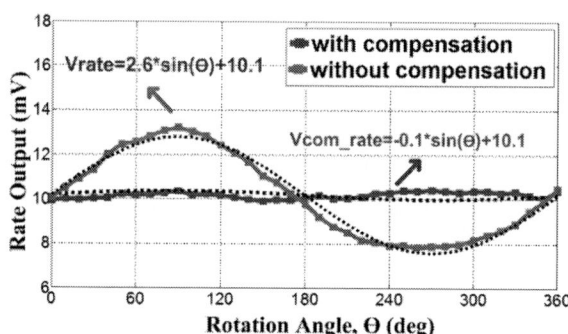

Figure 6: Measured and compensated rate outputs as a function of the rotation angle during the static acceleration test. With the compensation, the g-sensitivity of the rate output is highly-suppressed by up to 96%.

Centrifugal Acceleration Tests

In this test, the angular velocity is applied to the sensor in two different positions: centered and off-centered. Figure 7 shows (a) the schematic and (b) actual views of the test setup for the centrifugal acceleration test. The sense-axis of the sensor is purely subjected to the Coriolis force in the centered position, whereas both the centrifugal and Coriolis forces act on this axis in the off-centered position. In the off-centered position, the sensor is placed 2 inches (5 cm) far away from the rotation axis. The centrifugal acceleration (a) acting on the gyroscope is dependent on the angular rate (w) and radius (r) by $a=w^2r$.

Figure 8 shows the rate and acceleration output responses to the angular rate in the centered and off-centered positions on the rate table. In this test, the acceleration output also changes with the angular rate due to the phase error of about $3°$ coming from the electronics.

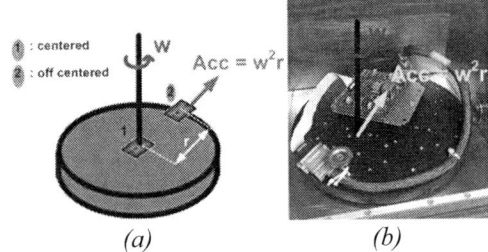

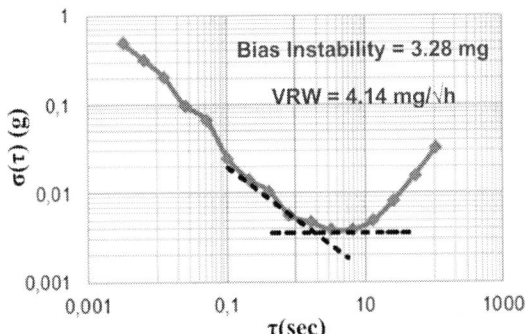

| (a) | (b) |

Figure 7: (a) Schematic and (b) actual views of the test setup for the centrifugal acceleration test.

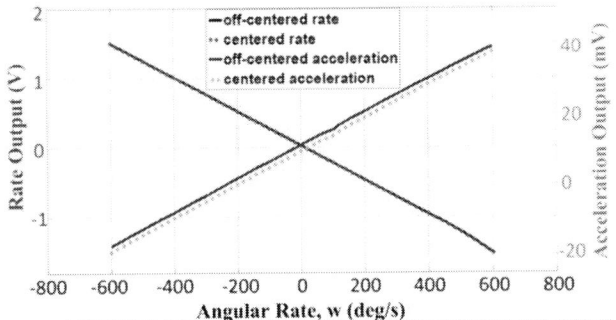

Figure 8: Rate and acceleration output responses of the gyroscope to the angular rate in the centered and off-centered positions on the rate table.

In order to determine the effect of the acceleration on the rate and acceleration outputs of the gyroscope under the angular velocity, the difference of the measured outputs in the centered and off-centered positions are obtained by eliminating the effect of the Coriolis acceleration on these outputs. Figure 9 shows the measured differences in the rate and acceleration outputs of the gyroscope with respect to the responses to the angular rate in the centered and off-centered positions during the centrifugal acceleration tests. The non-zero change in the rate output demonstrates that the scale factor of the gyroscope changes with the centrifugal acceleration. The rate output is compensated by using (1) and (2) to eliminate the effect of the acceleration on the scale factor. As shown in Figure 9, the rate output is significantly improved with the compensation.

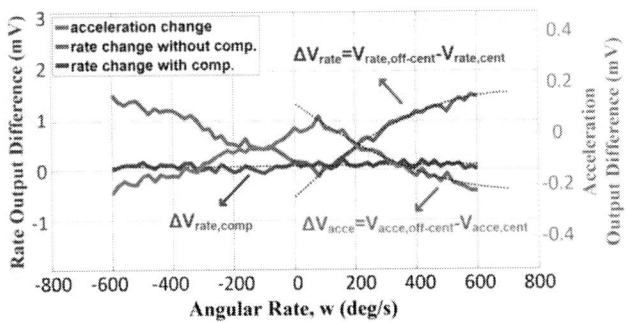

Figure 9: Measured differences in the rate and acceleration outputs of the gyroscope with respect to the responses to the angular rate in the centered and off-centered positions during the centrifugal acceleration test, and the compensated voltage difference between the rate outputs measured in these two positions.

Performance Result of the Acceleration Output

Figure 10 shows the Allan variance graph of the acceleration output of the gyroscope with a bias instability of 3.28 mg and a velocity-random walk (VRW) of 4.14 mg/√h.

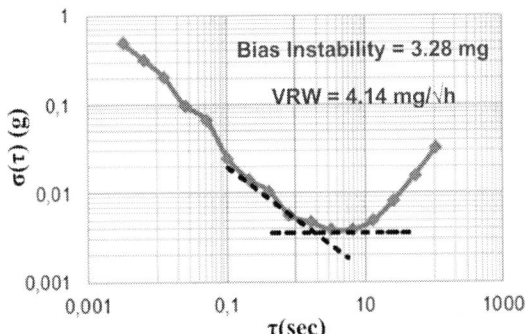

Figure 10: Allan variance graph of the acceleration output of the gyroscope with a bias instability of 3.28 mg and a VRW of 4.14 mg/√h.

CONCLUSION

This paper investigates the linear acceleration sensitivity and proposes a novel "in operation acceleration sensing and compensation method" for a single-mass mode-matched MEMS gyroscope. This sensing method uses the amplitude difference information between the residual quadrature signals on the differential sense-mode electrodes. By using relation between the rate and acceleration outputs, the effect of the acceleration on the gyroscope output is suppressed by 96%. It has been experimentally shown that the acceleration sensitivity of the studied MEMS gyroscope is reduced from 1.08°/s/g to 0.04°/s/g, and the effect of the linear acceleration on the gyroscope output is highly-suppressed up to 26 times with the use of the compensation method proposed in this work.

REFERENCES

[1] S. Sonmezoglu, S.E. Alper, and T. Akın, "An Automatically Mode-Matched MEMS Gyroscope with 50 Hz Bandwidth," *IEEE MEMS 2012*, pp. 523-526, Feb. 2012.

[2] E. Tatar, S.E. Alper, and T. Akin, "Effect of Quadrature Error on the Performance of a Fully-Decoupled MEMS Gyroscope," *IEEE MEMS 2011*, pp. 569-572, Jan. 2011.

[3] K. Azgin, "High Performance MEMS Gyroscopes," *M.Sc. Thesis*, Middle East Technical University, Feb. 2007.

[4] S. W. Yoon, S.W. Lee, and K. Najafi, "Vibration-Induced Errors in MEMS Tuning Fork Gyroscopes," *Sensor and Actuators A: Physical*, Vol. 180, pp. 32-34, Jun. 2012.

[5] A. R. Schofield, A. A. Trusov, and A.M. Shkel, "Multi-degree of Freedom Tuning Fork Gyroscope Demonstrating Shock Rejection," *IEEE Sensors*, pp.120-123, Oct. 2007.

[6] H. Weinberg, "Gyro Mechanical Performance: The Most Important Parameter," *Tech. art., Analog Devices*, Sep. 2011.

[7] S. W. Yoon, "Vibration Isolation and Shock Protection for MEMS," *Ph.D. Dissertation*, Uni. of Michigan, 2009.

CONTACT

*: H. D. Gavcar, tel: +90-312-2104409; hgavcar@mems.metu.edu.tr

A STATIC CAPACITANCE PROBE STRUCTURE FOR RESOLVING THE SIDEWALL SKEW ANGLE OF SILICON DEEP REACTIVE-ION ETCHING

Kemiao Jia, Aaron Geisberger, Andrew Dickens, Robert Steimle, David C. Chang, Paul Winebarger, Lianjun Liu, and Andrew McNeil

Freescale Semiconductor, Inc. Tempe, Arizona, USA

ABSTRACT

This work presents a static capacitive probe structure that enables quantitative characterization of the effective sidewall skew angle of the Silicon Deep-Reactive-Ion-Etching (DRIE) using static LCR prober at ambient environment. The design is capable of resolving sidewall skew angles around both in-plane axes independently and simultaneously with the same sensitivity. The measured distributions of the sidewall skew angle across 8-inch wafers conform to empirical expectation and correlate tightly with quadrature error distributions measured gyroscopes from the same wafers. This work provides an easy, accurate and batch solution to the long existing challenge of resolving such process features in an industrial manufacturing environment.

INTRODUCTION

Silicon DRIE is one of the critical process technologies used in the manufacturing of nearly all Micro-electro-mechanical Systems (MEMS) products. The most common DRIE method, also known as the Bosch Process, uses an interplay of etching and passivation steps, which makes it possible to achieve high aspect ratio Silicon structures with relatively smooth side walls. [1-3]

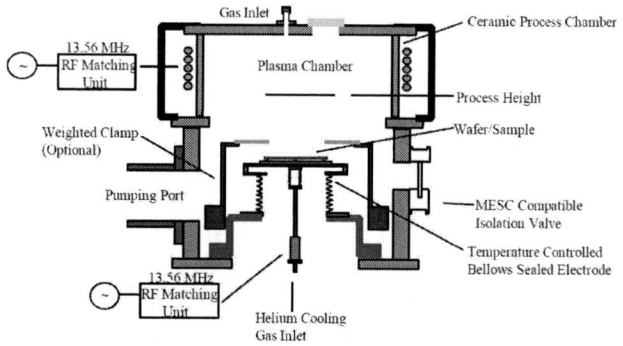

Figure 1: Schematic of a modern DRIE etcher system [1]

The typical schematic of a modern DRIE etcher is shown in Fig. 1 [1]. The top of the etcher is the inlet of the gas etchant. The RF coils, usually coupled inductively on the side of the etch chamber, creates alternating magnetic field that excites the etchant gas into a plasma. The plasma is drawn toward the bottom of the etch chamber by a second RF bias applied to the wafer substrate, which is clamped down to the holding chuck with Helium cooling inlet connect to the back. The power of the top RF coils generally determines the number of ions reaching the substrate, while the bias applied to the substrate generally controls the incident energy of the ions [4].

Although the Silicon DRIE technology has matured greatly over the years, it is also known to have inherent varying responses across wafer on a number of process features, including etch rate, depth uniformity, loss of critical dimension (CD), footing at dielectric interface, and the overall sidewall profile [3]. In particular, the variation and non-uniformity of the plasma density can lead to off-axis ion incident trajectory that results in asymmetric tilting of the overall etch profile, progressively worse towards edge of the wafer [5]. A good example of such tilted profile is shown in [6]. This type of etch profile induces an angular skew of the primary inertial axis of the etched silicon beams, which is the basic building element of most of the MEMS devices. In this paper, we refer to this skew angle as the side wall skew angle, or SSA. Although other factors, such as the formation of asymmetric footing, can also contribute to the SSA, we believe that the primary cause of the SSA is the off-axis trajectory of the ion. Owing to the powerful compensation capability of the application specific integrated circuit (ASIC), most MEMS products can tolerate some degree of SSA, however, MEMS gyroscopes are extremely sensitive to this type of process imperfection because it is the main driving mechanism of the gyroscope's quadrature error [2, 5, 6].

SSA AND GYROSCOPE QUADRATURE

The impact of SSA on the gyroscope quadrature error can be understood using the simplest lateral-axis gyroscope as an example. Fig. 2(b) shows the cross section of a spring tether with SSA, compared to ideally etched spring tether shown in Fig. 2(a). Assuming the gyroscope drives in direction x_d, and senses in direction z_s, the SSA is denoted by θ_{SSA}, and the actual drive and sense directions are denoted by x_d' and z_s'.

The existence of θ_{SSA} introduces a cross axis stiffness term which can be expressed by [6]:

$$k_{xz} = k_{zx} = \frac{k_{z_s} - k_{x_d}}{2} \cdot \sin(2\theta_{SSA}) \qquad (1)$$

in which k_{xd} denotes the ideal drive spring stiffness, and k_{zs} denotes the ideal sense spring stiffness. For the simplest gyroscope, drive and sense modes share the same spring and mass; therefore their stiffness relation can be expressed as:

$$\frac{k_{z_s}}{k_{x_d}} = \frac{t^2}{w^2} \qquad (2)$$

in which t stands for the thickness of the beam tether and w

stands for the width of the beam tether. For most state-of-the-art gyroscopes based on bulk silicon/polysilicon, it is true to assume $t^2 >> w^2$, therefore the force coupling ratio from the drive to the sense direction can be expressed as:

$$\alpha_{ds} = \frac{k_{xz}}{k_{x_d}} \approx \frac{k_{z_s}}{k_{x_d}} = \frac{t^2}{w^2} \cdot \sin(\theta_{SSA}) \qquad (3)$$

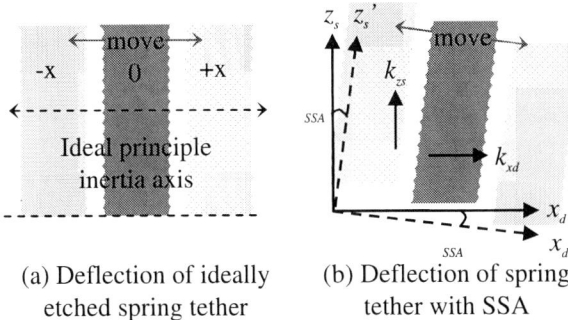

(a) Deflection of ideally etched spring tether

(b) Deflection of spring tether with SSA

Figure 2: (a) ideally etched beam tether; (b) Beam tether with SSA

Using this ratio, one can translate the force coupled from drive to sense to equivalent rate signal by equating this force to Coriolis force [7], which renders:

$$\Omega_{quad} = \alpha_{ds} \cdot \frac{\omega_d}{2} \approx \frac{\omega_d}{2} \cdot \frac{t^2}{w^2} \cdot \sin(\theta_{SSA}) \qquad (4)$$

in which ω_d is the drive resonance frequency.

Ω_{quad} is in-phase with the drive displacement therefore in quadrature with the desired rate signal, hence called the quadrature error. Based on Eq. (4), one can estimate that, assuming t/w=2-5, and ω_d=10-20 kHz, the magnitude of Ω_{quad} can be as high as 100000dps even with θ_{SSA} as small as 0.1°. Although in theory Ω_{quad} can be eliminated by phase demodulation, it is difficult in real practice because small phase instability of the ASIC will result in unacceptable bias drift. Researchers have long pointed out the necessity of modeling such process imperfections for quadratue error and gyroscope performance in general [5-7], yet it is very difficult to measure the SSA on real silicon and thus prevent designers from making realistic assumptions for modeling and performing validation of the model after silicon returns. On the manufacturing side, an efficient and accurate method of characterizing such process feature is also highly desired for tool evaluation and in-line process control. Traditional techniques, such as FIB/SEM cross sectioning, are time consuming and lack the resolution to resolve sub-degree level SSA. Directly probing the quadrature error on finished gyroscope wafers is one solution, however, it provides little help in the design phase. Also, dynamic quadrature error probing requires costly and sophisticated toolset, and the wafer has to complete full cycle of fabrication which results in long learning cycle. The above described issues and challenges call for an easy, accurate and batch solution of characterizing such a process feature, which motivated the work in this paper.

DESIGN OF THE SSA TEST STRUCTURE

One reason that the dynamic quadrature error probing is possible is that the gyroscope is capped under vacuum and thus can be excited into resonance, amplifying the static quadrature response by the system quality factor. The SSA test structure design in this paper implements a mechanical amplification feature that enhances signal level statically, making this measurement feasible by routine DC LCR probing at ambient pressure.

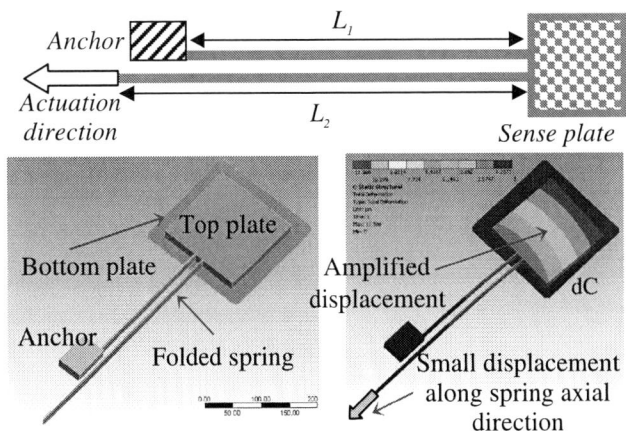

Figure 3: Mechanical amplification: Small displacement in the axial direction of the folded beam generates a large displacement at the end of the spring fold.

The concept of the mechanical amplification is shown in Fig. 3. A single amplification and detection arm consists of a folded beam, a movable plate, and a fixed electrode. A small displacement applied along the axial direction of the beam can generate an amplified in-plane displacement on the movable plate. With the presence of SSA, this in-plane displacement converts to out-of-plane displacement, which is detected by the bottom electrode. The amplification factor is highly dependent on the width W_1 and length L_1 of the first beam, while relatively immune to the width W_2 and length L_2 of the second beam. Using the dimensions shown in Table 1, finite element analysis shows that for an axially applied displacement of 1.0μm, the maximum displacement on the movable plate is ~12.9μm, an amplification factor of ~12.9×. With 0.1° SSA applied to the spring tether, the displacement of the movable plate in the out-of-plane direction is around ~22nm, resulting in a capacitance change of 5.6fF. Connecting two such detection arms opposing each other, as shown in Fig. 4(a) can achieve differential sensing, as well as push-pull actuation capability, therefore further amplifies the delta capacitance by 4×.

Table 1: Design parameters of one detection arm

Parameters	Dimensions	Unit
L_1	250	μm
W_1	2.5	μm
L_2	430	μm
W_2	2.5	μm
Plate size	160×160	μm²

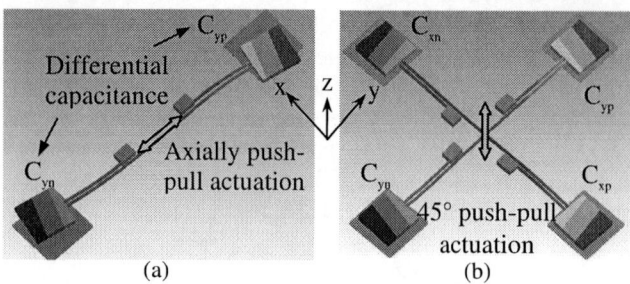

Figure 4: Placing detection arms orthogonally and opposing each other achieves further amplification as well as dual axis resolvability

Furthermore, connecting four such detection arms orthogonally and actuating it along the 45° direction, as shown in Fig. 4(b), can realize dual-axis x`resolvability simultaneously. Connecting four arms together increases the stiffness of the structure significantly. This needs to be taken into account while designing the actuators so that the voltage required to pull-in the structure is sufficiently low for the prober. To verify the independent resolvability of the 45° actuation method, FEA model was built based on Fig. 4(b). First the SSA applied to x axis is kept at 0.2°, while the SSA applied to y axis is swept from -0.2° to 0.2° as the center of the spring tether connection point is pulled along the 45° direction by 1μm, which is equivalent to 0.7μm displacement in both x and y direction. Similarly, the simulation is repeated with y axis SSA fixed at 0.2° while sweeping the x axis SSA from -0.2° to 0.2°. The result of the delta capacitance change in both simulations is shown in Fig. 5(a) and (b), respectively. It is evident from this simulation result that as the SSA is swept around x/y axis, only the corresponding electrode pair showed a differential signal output while the other electrode pair remains undisturbed.

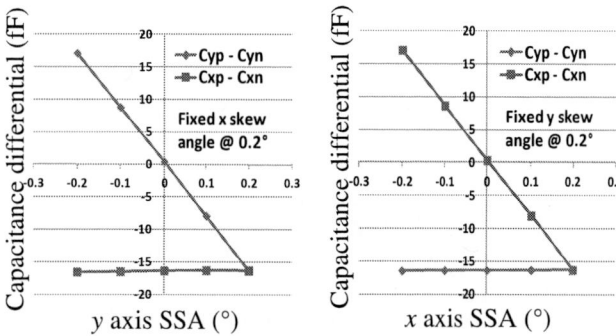

Figure 5: Simulated differential capacitance delta as function of SSA, showing independent resolving capability for both in-plane axes

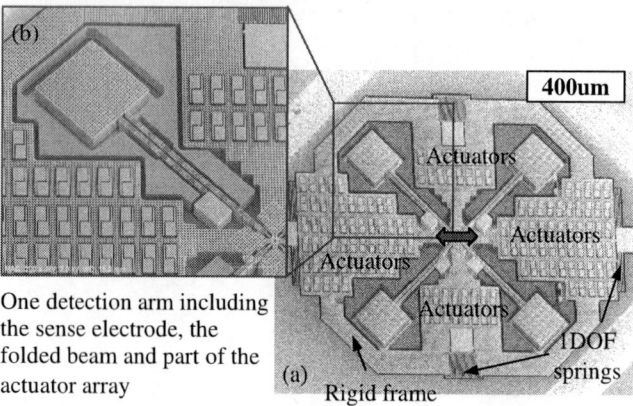

Figure 6: SEM images of the fabricated design using a 25um MEMS process. (a) Overview; (b) One detection arm

RESULT AND DISCUSSION

The design can be fabricated using a 25um polysilicon-MEMS process. The SEM images of a fabricated device are shown in Fig. 6. Besides the four detection arms, additional structures are needed in order to achieve the actuation in the desired direction at sufficiently low voltage. As shown in Fig. 6, these additional structures include four electrostatic actuator arrays for actuation, a rigid frame that can be actuated by the actuator arrays, and four springs that support the rigid frame while limiting the degrees of freedom of the rigid frame to the 45° direction. To prevent the rigid frame from touching the actuator electrodes in the pull-in event, several travel stops are added to the anchors of the spring, limiting the maximum travel distance of the rigid frame to be exactly 1.0μm. Finally, shield structures were added between the sense elements and the rest of structures to make sure that the measured signal is clean. Considering all design and process parameters, the final sensitivity of the SSA test structure is estimated to be 165.4fF/°, with a pull-in voltage around 13V.

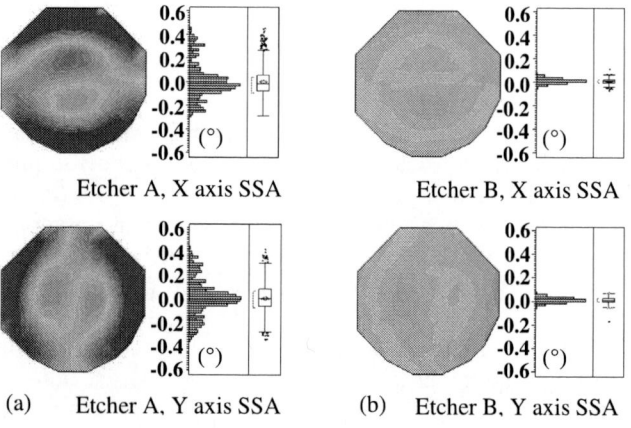

Figure 7: Measured SSA distributions on 5 wafers using Etcher A and 16 wafers using Etcher B

Using this test structure, SSA measurements were done across multiple wafers etched by two different Etchers. Roughly 100 data points were taken on each wafer. For each Etcher, the result shows very consistent cross wafer distribution of the SSA for both axes. Fig. 7(a) and (b) show the measured SSAs for Etcher A and B, respectively. The results agree well with the empirical expectations in that the center of wafer is close to ideal (0°) while the far edges in the x direction show the worst SSA around the y axis while far edges in the y direction show the worst SSA around the x axis. Comparing the two etchers, Etcher B shows a significantly better etching result, achieving ~±0.08° worst case SSA with a standard deviation of σ=0.02° for both axes across 16 wafers and 1595 data points, while Etcher A gives ~±0.45° worst case SSA with a standard deviation of σ=0.15° across 5 wafers and 470 data points.

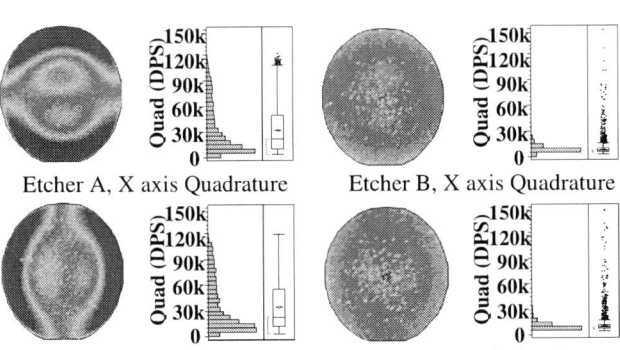

Etcher A, X axis Quadrature   Etcher B, X axis Quadrature

Etcher A, Y axis Quadrature   Etcher B, Y axis Quadrature

Figure 8: Measured quadrature error, in degrees per second (DPS), distributions on gyroscope wafers etched by Etcher A and Etcher B.

To further verify the result, we measured the quadrature error of actual gyroscope dies on wafers from each Etcher using dynamic probing. As expected, the wafer from Etcher B shows significantly lower quadrature level and tighter quadrature error distribution compared to Etcher A, as shown in Fig. 8. The dynamic probing method only captures the magnitude of the quadrature error, therefore the result in Fig. 8 does not reflect the polarity of the SSA, however, it confirms that the dies toward the far edges in the x direction have worse quadrature error for y axis gyroscope, while the dies toward the far edges in the y direction have worse quadrature error for x axis gyroscope. This distribution is consistent with the result shown in Fig. 7, which proves that the direct measurement of the SSA using the presented design also provides a strong prediction of the gyroscope quadrature level and its distribution across wafer. The measurement result on the SSA provides valuable process information for future gyroscope design and modeling.

It should be noted that several factors were not fully considered in the measurement and calculation of the SSA. First of all, a constant sensitivity of 165.4fF/° is assumed across wafer, however, this is not necessarily true given the fact that a number of factors can cause the sensitivity to vary

which include the loss of CD on spring tether, variation of the structural polysilicon thickness, and the variation of the sense gap, etc. Secondly, the loss of CD on the travel stops also impacts how much the rigid frame can travel, therefore further complicates the variation of sensitivity across wafer. Finally, the impact of the residual stress of the thick polysilicon film on the measured SSA is not considered. With the above mentioned variation factors taken into consideration, the result in Fig. 7 can be further improved.

The differences in observed SSA between Etcher A and Etcher B may be related to differences in the construction and configuration of each chamber. Both chambers are inductively coupled plasma etch chambers. Differences are found in the design of the top coil, the volume and layout of the chamber, the positions in the chamber where process gasses are introduced, the distance between the wafer and the top coil, and the types of chamber furniture used.

Some of these parameters, such as the chamber furniture and the spacing between the wafer and the plasma source, can be adjusted by changing hardware components. One advantage of this SSA test structure is the ability to quickly observe the effect on products after changing the configurations of these components, making it suitable for inline process control and monitoring.

REFERENCES

[1] J. Bhardwaj, H. Ashraf, A. McQuarrie, "Dry Silicon Etching for MEMS", in *The Symposium on Microstructures and Microfabricated Systems*, Montreal, Quebec, Canada. May 4-9, 1997.

[2] F. Laermer, A. Urban, "Challenges, Developments and Applications of Silicon Deep Reactive Ion Etching", *Microelectronic Engineering*, 67-68 2003, pp. 349-355.

[3] M. Wasilik, N. Chen, "Deep Reactive Ion Etch Conditioning Recipe", in *Proc. of SPIE*, Vol. 5342, pp. 103-110.

[4] M. D. Henry, "ICP Etching of Silicon for Micro and Nanoscale Devices", *Thesis*, California Institute of Technology, 2010.

[5] P. Merz, W. Pilz, F. Senger, K. Reimer, M. Grouchko, T. Pandhumsoporn, W. Bosch, A. Cofer, S. Lassig, "Impact of Si DRIE on Vibratory MEMS Gyroscope Performance", in *Proc. of Transducer 2007*, pp. 1187-1190.

[6] A. Shkel, R. T. Howe, and R. Horowitz, "Modeling and Simulation of Micromachined Gyroscopes in the Presence of Imperfections", *in Proc. International Conference on Modeling and Simulation of Microsystems*, 1999, pp. 605-608.

[7] M. S. Weinberg, A. Kourepenis, "Error Sources in In-Plane Silicon Tuning Fork MEMS Gyroscopes", *J. Microelectromech. Syst.*, vol. 15, pp. 479-491, 2006.

CONTACT

*Kemiao Jia, #:1-480-413-5205; kemiao.jia@freescale.com

INVAR-36 MICRO HEMISPHERICAL SHELL RESONATORS

Nishanth Mehanathan, Vahid Tavassoli, Peng Shao, Logan Sorenson and Farrokh Ayazi
Georgia Institute of Technology, USA

ABSTRACT

We report, for the first time, on the successful fabrication and operational characterization of electroplated Invar Micro-Hemispherical Shell Resonators (μHSR). The heat treatment of the samples and its effect on the quality factor (Q) of the resonators is studied. We show that thermal annealing shifts the coefficient of thermal expansion (CTE) of the alloy towards its minimum of ~2ppm/°C, as a result of which the Q of a 29kHz μHSR with diameter of 780 μm increases at least 3 times and reaches 7500 in vacuum.

INTRODUCTION

The widespread success of macro-scale Hemispherical Resonator Gyros (HRG) [1] has been an inspiration and a motivation towards design and implementation of their micro-scale counterparts (μHRG). Although micro-scale freestanding hemispherical shells have been fabricated three decades ago for inertial fusion experiments [2], recent efforts are particularly aimed at their damping and resonant behavior [3,4]. High Q μHSRs can lead to the implementation of high performance resonant gyroscopes that can directly measure angle of rotation instead of angular velocity, or rate of rotation, which most commercial gyroscopes measure today. A range of materials is reported for μHSRs including poly-silicon [3] and silicon dioxide [4], but a potentially high Q metallic μHSR can be advantageous.

Nickel-Iron alloys, a class of metallic materials with exceptional thermal properties are interesting candidates for microelectromechanical systems (MEMS). Elinvar for example, is an Iron-Nickel-Chromium alloy with a zero temperature coefficient of elasticity (TCE). Super-Invar, a less studied variation, contains 64% Fe, 31 % Ni, and 5% Co, which provides a CTE very close to zero. Invar36 is a Nickel (36%) Iron (64%) alloy with a very low CTE comparable to that of fused quartz and silica [5]. The small CTE, an anomalous property called the Invar effect [6], can translate into low thermoelastic damping (TED), which makes Invar an attractive material choice for high-Q 3D MEMS resonators. Furthermore, being a conductor gives Invar a significant advantage over other low CTE materials that are mostly insulator as it makes capacitive actuation and sensing possible without requiring additional coating material, which can often reduce Q.

In this work, results from electrodeposition and sputtering of Invar are presented and compared. A process flow for fabrication of Invar μHSRs is introduced; thermal annealing of Invar shells is explored and experimental results related to the effect of annealing on CTE and Q of Invar microshell resonators are provided.

Figure 1: An electroplated Invar μHSR with assembled electrodes suitable for actuation and detection of m=2 wineglass modes.

FABRICATION OF INVAR MICROSHELL RESONATORS

Invar μHSRs with capacitive electrodes are successfully fabricated and tested (Figure 1). Two different structural material deposition techniques are examined: electrodeposition and sputtering. Figure 2 demonstrates a sputtered super-Invar μHSR (left), which shows poor structural integrity with visible folding of the shell at the rim, while electroplated Invar microshell (right) shows excellent structural integrity. Higher deposition rate compared to sputtering and scalability to larger film thickness are other advantages of the electroplating process over sputtering.

An optimized fabrication process for electroplated Invar μHSRs has been developed, the details of which are shown in Figure 3. First, the backside of a low resistivity Silicon wafer is patterned to define electrode holes, followed by anisotropic etching of silicon using the Bosch process (a). Next, the front side oxide mask is patterned to form the mold openings (b). Isotropic etching in SF_6 plasma is done

Figure 2: (left) A sputtered super-Invar μshell with poor structural integrity, particularly, the rim is folded towards the center. (right) An electroplated Invar μshell with superior structural integrity compared to sputtered Invar.

978-1-4799-3510-9/14 $31.00 © 2014 IEEE

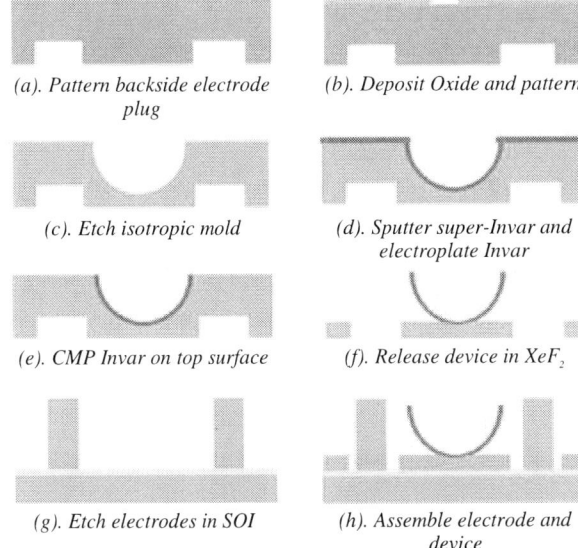

(a). Pattern backside electrode plug

(b). Deposit Oxide and pattern

(c). Etch isotropic mold

(d). Sputter super-Invar and electroplate Invar

(e). CMP Invar on top surface

(f). Release device in XeF$_2$

(g). Etch electrodes in SOI

(h). Assemble electrode and device

Figure 3: Invar hemispherical shell resonator fabrication process (cross-sectional view)

on the exposed silicon to form the hemispherical molds (c). A 0.1 μm thick Titanium adhesion layer followed by a 0.5 μm thick Invar or super-Invar as seed layer is sputtered onto the sample. Invar is then electroplated onto the seed layer (d). The surface Invar is polished using a Chemical-Mechanical Planarization (CMP) process leaving invar only in the mold, while the structural invar shell is protected by a photoresist (PR) layer (e). The structure is finally released using XeF$_2$ gas which etches the silicon around the shell as well as the exposed Titanium layer (f). On a separate wafer, electrodes are fabricated by etching electrode patterns on a 700um/2um/700um SOI (g). Finally the structure and the electrodes are assembled to form the device (h).

Electrodeposition of Invar

Table 1 shows the recipe for electroplating Invar, which is a slightly modified version of what is given in [7]. Particularly, the Ferrous Sulfate concentration is increased from 0.07 M/L to 0.533 M/L. The solution is stirred at 60RPM during the process to maintain uniformity of composition across the sample.

The anode is made of Invar and the cathode houses the sample to be electroplated. The current source provides 15-30 mA/cm^2 pulsed current (50 ms ON and 30 ms OFF).

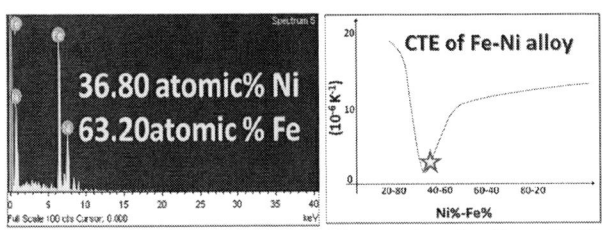

Figure 4: EDS analysis of electroplated alloy (left), showing a composition close to the Fe-Ni Invar (right)

Table 1: The electrolyte recipe

Nickel Sulfamate	0.51 M/L
NiBr2	0.04 M/L
FeSO4	0.533 M/L
H3B03	24 g/l
Sodium Saccharin	4 g/l
Ascorbic acid	1 g/l
Wetting Agent	0.01 g/l
pH	3.0
Temperature	40 (°C)
Current density	20-25 mA/cm2

Sodium Saccharin is an essential element in the bath to reduce film stress, but should not exceed a limit, above which it will leave carbon residue on the film. Pulsed current is used as a technique to reduce the stress even more, to prevent film delamination, which occurs around the invar composition.

After each deposition, the composition is verified by Energy-Dispersive X-ray Spectroscopy (EDS) as shown in Figure 4 to confirm 36% Nickel and 64% Iron composition. Using our current plating setup, the electroplated film shows 9% composition variation across the surface, which reduces the fabrication yield.

HEAT TREATMENT OF INVAR

Heat treatment of the alloy is needed in addition to accurate composition to reach the invar effect. The necessity of annealing is sometimes linked to the lattice allotropy of the atoms. The Invar effect only develops if the material is in γ-phase (Face-Centered Cubic: FCC), and not in α-phase (Body-Centered Cubic: BCC), and heat treatment is needed to transform the BCC units to FCC units [8]. The importance of annealing is also identified in association

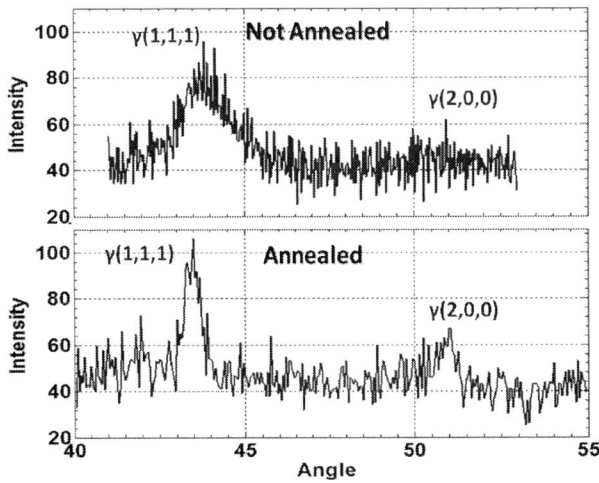

Figure 5: XRD intensity graph for (top) non-annealed and (bellow) annealed Invar electrodeposited films with only FCC peaks present. The annealed film has a sharper peak implying a larger grain size as derived from the Scherrer formula

978-1-4799-3510-9/14 $31.00 © 2014 IEEE

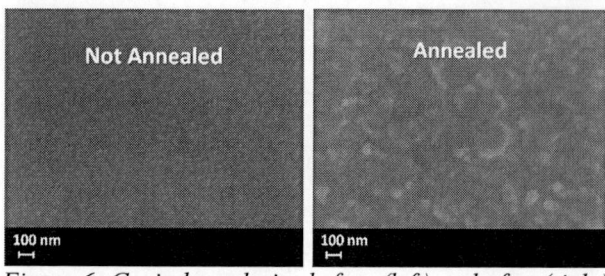

Figure 6: Grain boundaries before (left) and after (right) annealing the Invar films, showing significant grain growth caused by the heat treatment

with the fact that the CTE is low only for the crystallites of the alloy and not the grain boundaries. Although Invar can be in FCC phase as electroplated, annealing is still needed to enforce grain growth and therefore to reduce the CTE to its minimum [9].

The annealing is performed by placing the samples in an air tight chamber, which is purged for 5 minutes. The temperature is ramped up to the annealing temperature (620 °C) in 300 seconds and held for 5-30 seconds. After the samples are annealed, the temperature is ramped down slowly to ambient.

The electroplated invar exists as pure FCC (γ-phase) directly after deposition, which can be inferred from the graphs of Figure 5. The top graph shows XRD pattern of the electroplated Invar film right after deposition only with peaks corresponding to γ-phase; the left peak with Miler index (1,1,1) and the right peak (2,0,0). The lower graph of Figure 5 shows that the width of γ(1,1,1) peak has decreased after annealing, which based on Scherrer formula translates into bigger grain size [9]. To confirm the XRD results, more accurate measurements of the grain size are done directly from the SEMs shown in Figure 6 with visible grain boundaries. The average grain sizes before and after annealing are measured as ~20 nm and ~70 nm respectively.

Both XRD analysis and direct grain size measurements confirm that the grains of electroplated Invar films, already in FCC phase, grow significantly as a result of annealing. Therefore, the annealed films are expected to show a lower CTE.

CTE Characterization

To measure the CTE of electroplated Invar before and after annealing, Vernier microgauge structures [10] are designed and fabricated. As shown in Figure 7, the microfabricated Vernier-like structure measures expansions or contractions of the test beams caused by residual or thermal strain. They bend the slope arm and create an angle, which is magnified by the indicator beam and can be visually observed at the Vernier gauge.

Similar to the μHSR, Invar is electroplated on sputtered titanium and super-Invar layers, and after release some samples are annealed to be compared to non annealed ones. The deflections of the Vernier are measured at two temperatures, namely -20 °C and 50 °C and are compared to calculate the CTE of electroplated Invar before and after

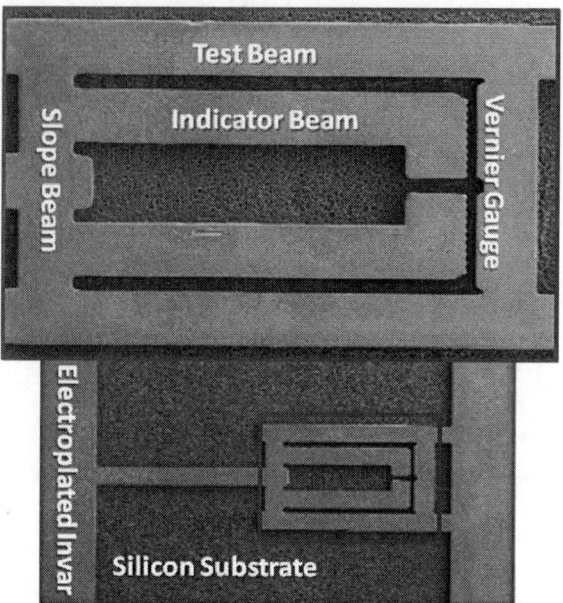

Figure 7: SEM of Vernier-like structure to measure length changes in response to heating or cooling

annealing using analytic models given in [10]. The average measured CTE of electroplated Invar before annealing is 5.9075 ppm/°C. After annealing at 620°C for 5-6 seconds, the CTE of the film was reduced. The average CTE of annealed Invar was found to be 2.975 ppm/°C. The lowest CTE was observed as 2.09 ppm/°C for an Invar film which was annealed at 620 °C for 6 seconds.

RESONATOR MEASUREMENT

Several Electroplated Invar μHSR were fabricated and electrically tested. Devices were capacitively actuated and sensed using low resistivity capacitive silicon electrodes. The dimensions of the tested shells, their measured m=2 frequencies and Q factors are given in table 2. As shown in Figure 8a, an annealed electroplated Invar μHSR with a frequency of 29kHz shows a Q of 7500. Measurements of non-annealed electroplated Invar μHSRs show Qs in the range 1500-3000; an example is shown in Figure 8b. Sputtered super-Invar μHSRs demonstrated a lower Q of ~1300 as shown on Figure 8c.

As the trend shows, annealed Invar μHSRs have the best performance in terms of energy loss. Also, a small frequency split of 27 Hz was measured from an annealed Invar microshell between the two degenerate m=2 modes.

CONCLUSION

Electrodeposition and sputtering of Invar36 for fabrication of 3D μHSRs were presented and compared. It was shown that electroplated Invar is a suitable material for fabrication of 3D curved microstructures with high quality factor. The effect of annealing on CTE was studied and was linked to grain size growth after annealing. Experimental data on CTE were provided using microfabricated Vernier-like structures. Thermal annealing shifts the CTE of the

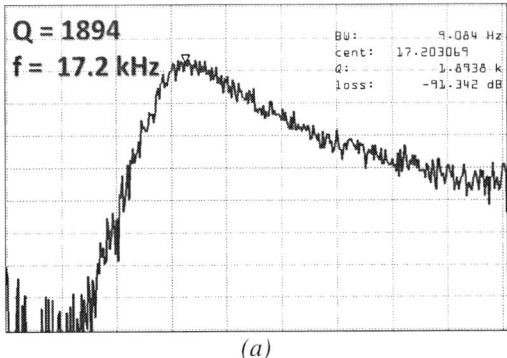

(a)

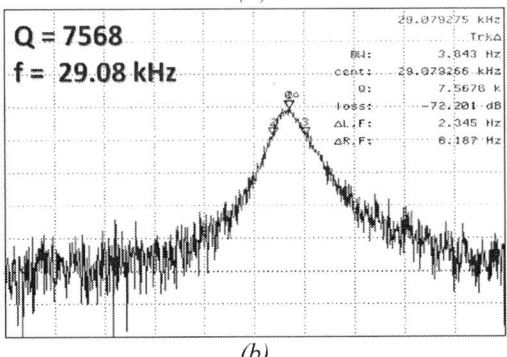

(b)

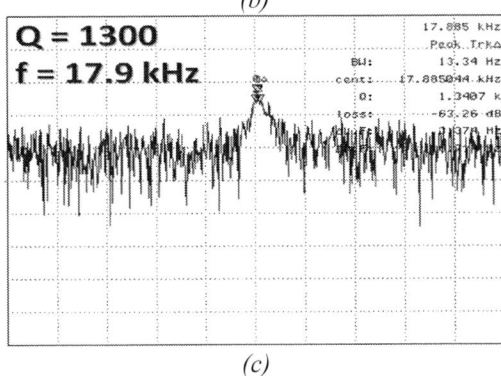

(c)

Figure 8: The m=2 frequency response of a non-annealed (a), an annealed (b) electroplated Invar μHSR with thickness of 5.2μm and diameters of 920 μm and 780μm respectively. The m=2 frequency response of a sputtered super-Invar shell is shown in (c)

alloy towards its minimum of ~2ppm/°C, as a result of which the Q of a 29kHz μHSR with diameter of 780 μm increases at least 3 times and reaches 7500.

ACKNOWLEDGEMENTS

This work was supported by the DARPA Microsystems Technology Office, Microscale Rate Integrating Gyroscope (MRIG) program under contract #HR0011-00-C-0032 led by Northrop Grumman. The authors would like to thank the cleanroom staff at Georgia Tech's Institute for Electronics and Nanotechnology for processing assistance.

REFERENCES

[1] D.M. Rozelle, "The hemispherical resonator gyro: From wineglass to the planets", *19th AAS/AIAA Space Flight Mechanics Meeting*, 2009, pp. 1157-1178.

Table 2: Measured Q-factors of electroplated Invar μHSRs with different dimensions. The annealed device shows the highest Q

Annealing	Diameter (μm)	Thickness (μm)	$f_{m=2}$ (kHz)	Q factor
No	910	3.5	11.64	2,000
No	910	5.2	19.70	1,862
No	910	5.2	19.70	2,000
No	910	5.2	18.3	3,196
No	780	5.6	36.1	1,500
Yes	780	5.2	29.08	7,567

[2] K. D. Wise, M. G. Robinson, and W. J. Hillegas, "Solid-State Processes to Produce Hemispherical Components for Inertial Fusion Targets", *J. Vac. Sci. Technol.*, 18, April 1981, pp. 1179-1182.

[3] L. Sorenson, X. Gao, and F. Ayazi, "3-D Micromachined Hemispherical Shell Resonators with Integrated Capacitive Transducers", *IEEE International Conference on Micro Electro Mechanical Systems (MEMS 2012)*, Paris, France, Jan. 2012, pp. 168-171.

[4] L. Sorenson, P. Shao, and F. Ayazi, "Effect of thickness anisotropy on degenerate modes in oxide micro-hemispherical shell resonators", *IEEE International Conference on Micro Electro Mechanical Systems (MEMS 2013)*, Taibei, Taiwan, Jan. 2013, pp. 169-172.

[5] A.K. Dokania, B. Kocdemir, U. Herr, "Thermal Expansion of Invar Thin Film for MEMS Application", *International Symposium of Research Students on Materials Science and Engineering*, 20-22, pp. 1-6, 2004.

[6] M.V. Schilfgaarde, I.A. Abrikosov, B. Johansson, "Origin of the Invar effect in iron-nickel alloys", *NATURE*, 400, pp. 46-49,1999

[7] T. Hirano, L.S. Fan, "Invar electrodeposition for MEMS application", *in SPIE Proceedings: Micromachining and Microfabrication Process Technology II*,Vol. 2879, Austin, TX, pp. 252-259, SPIE 1996.
Int. Soc. Opt. Instrumentation Eng.

[8] A.K. Dokania, et. al. "alpha to gamma phase transformation in electrodeposited Invar film by short pulse laser treatment", *Materials Science & Engineering A,* 15 May 2007, vol.456, no.1-2, pp. 64-71.

[9] Y. Liu, L. Liu, B. Shen, W. Hu, "A study of thermal stability in electrodeposited nanocrystalline Fe–Ni invar alloy", *Materials Science and Engineering: A*, 2011, 528(18), 5701-5705.

[10] W. H. Chuang, *MEMS-based Silicon Nitride Thin Film Materials and Devices at Cryogenic Temperatures for Space Applications,* Ph.D. Thesis, University of Maryland, College Park, MD, March 2005.

CONTACT

Nishanth Mehanathan, nmehanathan3@gatech.edu
Farrokh Ayazi, ayazi@gatech.edu

IMPLEMENTATION OF SINGLE/MULTI-LAYER MAGNETIC-ANISOTROPY MAGNETIC POLYMER COMPOSITES FOR MAGNETIC PROPERTY MODULATION

Fu-Ming Hsu[1], Wen-Chien Chen[1], Wei-Ming Lai[1], Yi-Chiang Sun[1], and Weileun Fang[1,2]
[1]Power Mechanical Engineering dept., [2]NEMS Inst., National Tsing Hua University, HsinChu, TAIWAN

ABSTRACT

This study exploits the two stage solidification technology to fabricate the isotropic/anisotropic magnetic polymer composites (MPC, polymer with magnetic particles). Multilayer magnetic isotropic/anisotropic MPC film can also be implemented using the two stage solidification process layer by layer. Merits of proposed technology: (1) material properties of magnetic-anisotropy MPC layer is realized using the two stage solidification and anisotropic magnetization processes, and (2) film with various magnetic properties can also be implemented using different composition of multilayer magnetic anisotropy MPCs. In applications, the multilayer polymer NdFeB magnetic composites are realized in silicon substrate and further integrate with MEMS structures. The 1-3 layers of different magnetic anisotropy MPC are demonstrated. Performances enhancement of magnetic anisotropy 30wt% NdFeB MPC (vs isotropic MPC) are: coercivity force (3.4%), remanence (304%), and saturation magnetization (268%). Anisotropic magnetostatic shielding effect that reduced from 0.45Telsa to 0.3-0.35Telsa) is also achieved. Moreover, changing of magnetic field distributions after different magnetic anisotropy MPC layers stacking is also demonstrated.

INTRODUCTION

Permanent magnets can find various applications for bidirectional driven microactuators (e.g., scanning mirror, pumps, relays, etc.). Many approaches have been reported to deposit permanent magnet materials on MEMS device [1-4]. Sputter deposition is a popular approach to prepare hard magnet films. The magnetic moment of sputtered films can be aligned by annealing. However, the film thickness and annealing temperature (600°C) are two concerns [1,10]. Such high temperature process is usually not compatible with many existing MEMS devices. For low temperature screen printing [2] and electroplating [3] process, control of magnetization direction is a major concern [4].

Magnetic particles spread in fluid would be redistributed to form the chain-like structure by external magnetic field. This phenomenon is so called the magnetorheological effect [5]. The shape of chain-like structure is affected by the dipole motion and attraction between magnetic powders. The material properties, such as mechanical, magnetic, and thermal, could be changed from isotropic to anisotropic by the magnetorheological effect [6]. For instance, this effect has been employed in [7] to change the material properties of solid magnetic polymer. As discussed in [8], the characteristics of magnetorheological effect could be influenced by the carrier fluid viscosity, size/shape of magnetic particles, volume fraction of particles in MPC, and magnetic field intensity.

This study further exploits the two stage solidification technologies [9] to present the multilayer tunable anisotropy magnetization of MPC film by screen print process. In application, the suspend MEMS structures with multilayer magnetic polymer composites (MPC, polymer with magnetic powders) is integrated and tested. According to the direction of magnetic field, the distribution of NdFeB powders would change and further enhance the performances of MPC. Polarity, flux density (flux meter), magnetostatic shielding effect, and M-H curve (VSM, vibrating sample magnetometer) testing on fabricated anisotropic NdFeB MPC film are characterized. These could show the influence of NdFeB fraction of powders and applied magnetic field during presented solidification process.

DESIGN AND FABRICATION

Fig.1 shows the two stage solidification approach to form multilayer micro magnetic anisotropy MPC. In Fig.1a, the liquid phase MPC was prepared using UV

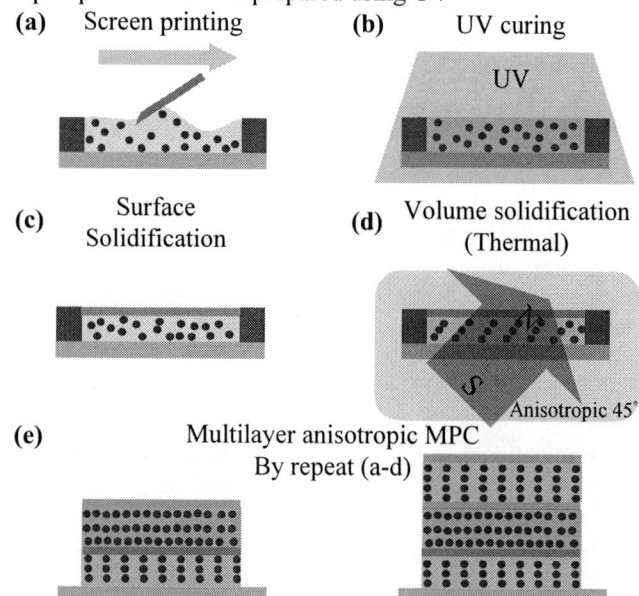

Fig.1 (a) Screen printing for liquid MPC, (b-c) 1st stage: Surface solidification by UV, (d) 2nd stage: Volume solidification by thermal heating and anisotropic magnetization, and (e) repeat (a-d) to form multilayer magnetic anisotropy MPC

curable polymer with specified wt% magnetic powders (1μm NdFeB particles in this study). The wt% NdFeB could change magnetic properties of MPC. The liquid phase MPC was patterned by screen printing. In Fig.1b-c, the 1st stage UV curing was for MPC surface solidification. As in Fig.1d, the 2nd stage thermal curing was for MPC volume solidification, and the magnetic anisotropy of MPC was defined by applying magneticfield during curing. As in Fig.1e, by repeating the above processes, structure with multilayer MPC of different magnetic anisotropy was achieved.

Fig.2 shows the fabrication and integration processes for multilayer MPC. The 1μm anisotropic NdFeB magnetic particles and SOI wafer (device layer 200μm, handle layer 500μm) are employed in this study. In Figs.2a-d, the back side cavity and front side device pattern were defined by photolithography and the following DRIE etching. After that, the back side cavity was covered with photoresist. Finally, the front side photoresist on MEMS structure was removed by O_2 plasma. Fig.2e-g shows the integration of multilayer anisotropy NdFeB MPC with MEMS structure using screen printing through two stage solidification processes. The pattern and thickness of each NdFeB MPC layer was defined by screen print plate. The 1st stage solidification (surface solidification) of MPC was realized by UV curing (60mW/cm^2, 30sec). The 2nd stage solidification (volume solidification) of MPC was achieved by thermal

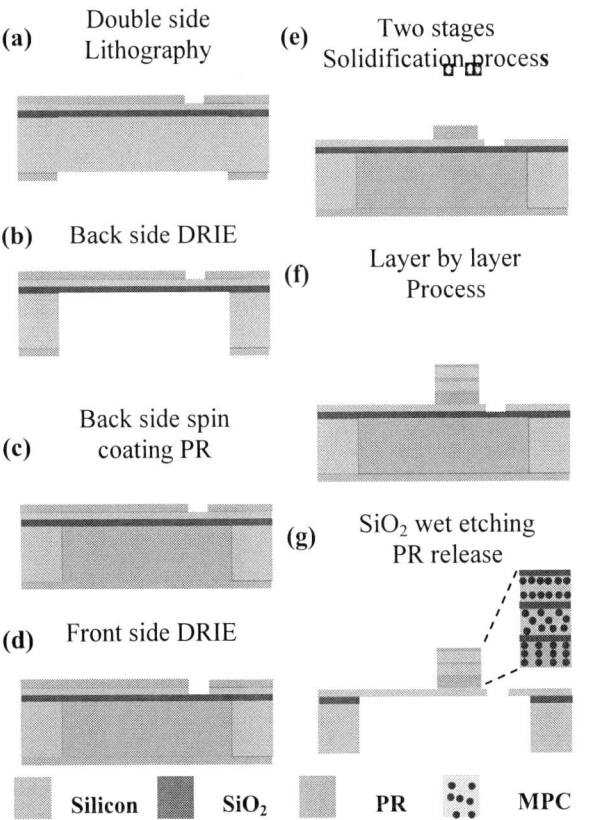

Fig.2 Process steps for the fabrication and integration of multilayer MPC on MEMS structure

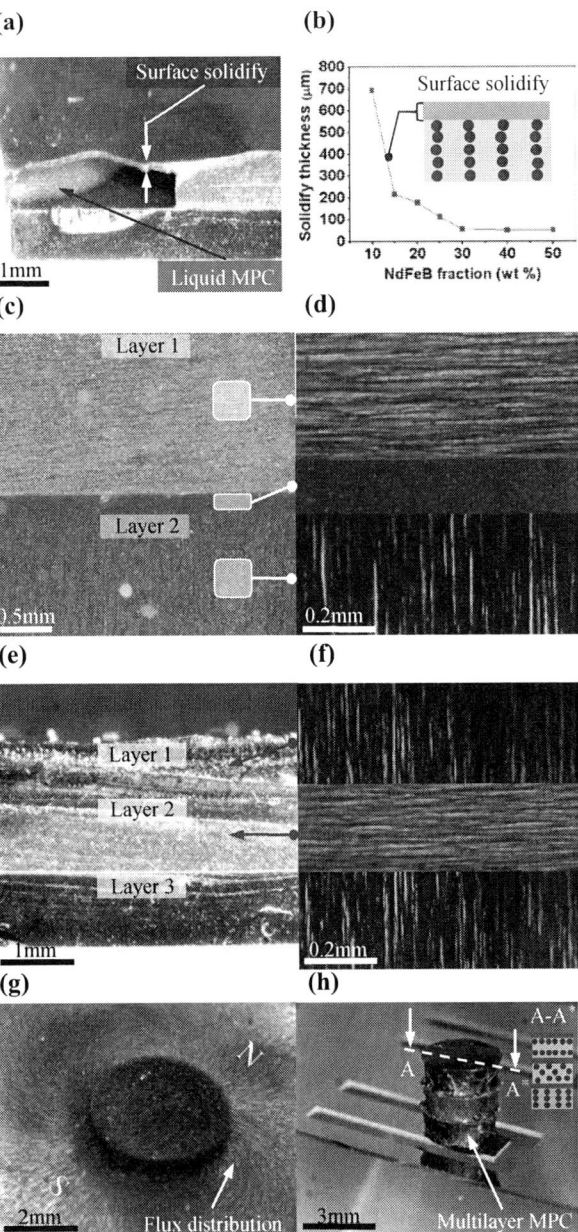

Fig.3 (a) Cross section view of 2 layer MPC after surface solidification, (b) solidification thickness for different wt% of NdFeB, (c-f) zoom in micrographs of 2 layer and 3 layer MPC with different magnetic anisotropy, (g) flux distribution of 1 layer anisotropic MPC, and (h) integration of 3 layers magnetic anisotropy MPC with MEMS structure

curing (at 80°C for 30minutes, or at room temperature for 24hours). Meanwhile, the magnetization is performed during the volume solidification of each MPC layer. Thus, the multilayer magnetic anisotropic MPC was achieved after the repeating of solidification processes for different MPC layers. After that, the exposed SiO_2 layer was etched in buffered oxide etchant (BOE) solution for 15minutes. The device was finally released by acetone solution[2].

Fig.3a shows the cross section micrograph of 2 layers MPC film (1 layer thickness: 500μm, 50 wt% of NdFeBpowders) after surface solidification (solidification thickness is 54μm) by UV curing. Fig.3b indicates the solidification thickness for different weight fraction of 10, 30 and 50wt% of NdFeB powders. Zoom in micrographs in Fig.3c-f indicate that the multilayer MPC film of NdFeB powders were orderly distributed on solidified surface where is properly aligned by magnetic field inside the surface. Fg.3g shows fabricated 1 layer NdFeB MPC film (thickness: 500μm, diameter: 3000μm) with anisotropic magnetization attracted each magnetic particles by magnetic force. Fig.3h shows the integration of multilayer anisotropic NdFeB MPC film on suspended MEMS structure for magnetic driving.

MEASUREMENT RESULT

This study performed the polarity, flux density, and M-H curve testing on the fabricated anisotropic single multilayer NdFeB MPC film to show the magnetic property with NdFeB fraction of powders after applying magnetic field (3400kGauss) solidification. Fig.4a shows the shielding effect tests on both in-plane and out-of-plane anisotropic magnetized MPC layers. Moreover, the MPC layers of different NdFeB weight fraction (10, 30 and 50wt%) are characterized. Measurements in Fig.4b show a 22~34% flux decay for solidified MPC film of 10~50wt% NdFeB after anisotropic (in-plane or out-of-plane) magnetization in Fig.1. Figs.5a-b shows the increment of coercivity force (3.4%), remanence (304%), and saturation magnetization (268%), for solidified MPC film (single layer) of 30wt% NdFeB. In short, the change of MPC layer from isotropic to anisotropic has contributed to the variation of magnetic properties [8-9]

Measurements in Fig.6 show the distributions of magnetic field for different stacking of magnetic anisotropy MPC layers (30% NdFeB). As marked with

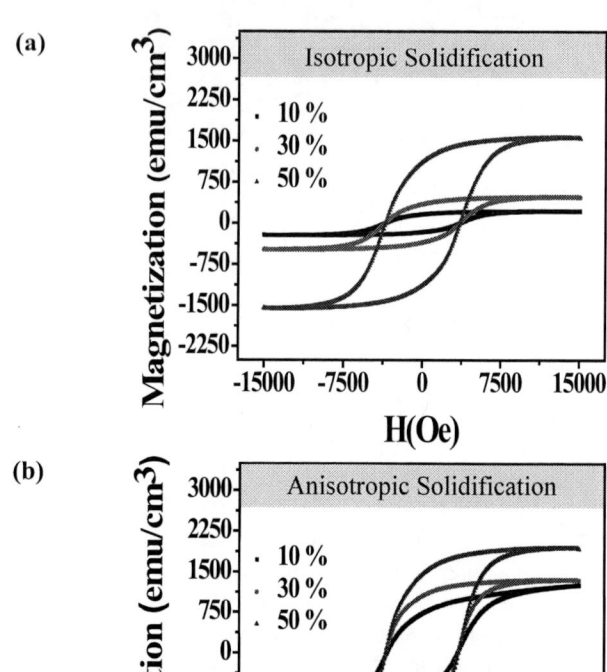

Fig.5 Measurements setups and results of the M-H characteristics for 1 layer magnetic isotropic and anisotropy MPC

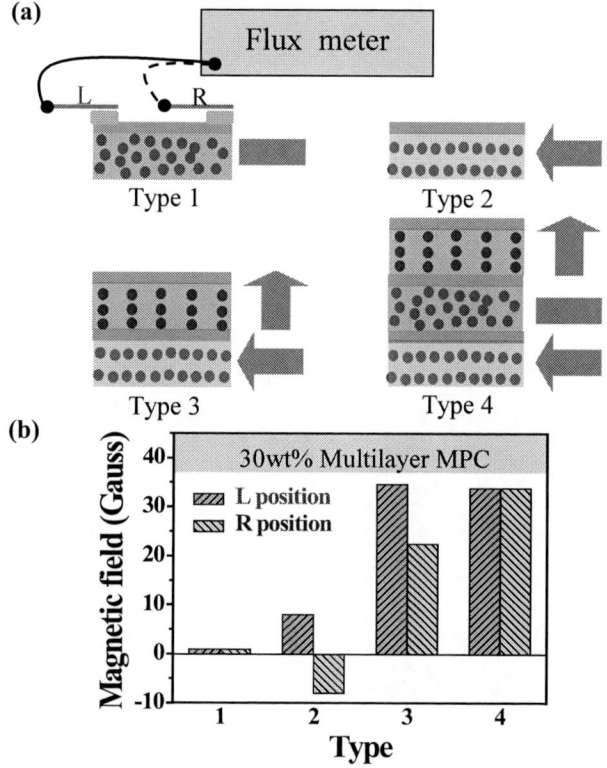

Fig.6 Measurements steps and results of polarity and flux density for MPC with different layers and anisotropy

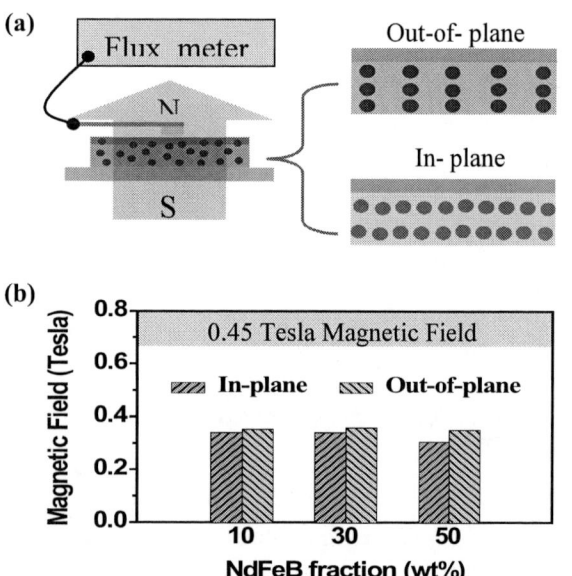

Fig.4 Measurements setups and results of the shielding effect for 1 layer in-plane and out-of-plane magnetic-anisotropy MPCs

(a)

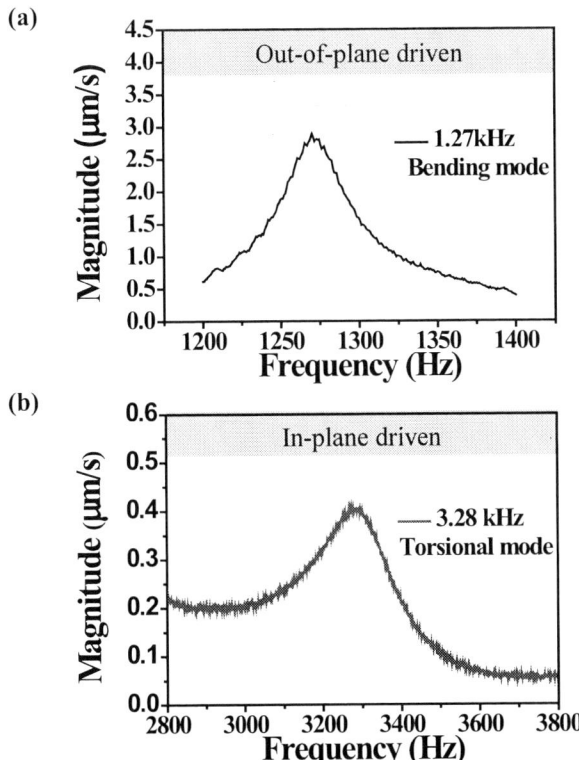

(b)

Fig.7 Magnetic field driving tests on MEMS structures with 3 layer magnetic-anisotropy MPC (in Fig.3h); the dynamic response excited by (a) out-of-plane, and (b) in-plane AC magnetic field

"L" and "R" in Fig.6a, the magnetic field intensity on samples was characterized by gauss meter at two different locations. Measurements in Fig.6b show the magnetic distribution of four samples (Type1~4) of different layer stacking. It demonstrates the distribution of magnetic field could be modulated using the stacking of MPC layers. This study also performed the magnetic driving of MEMS structure with MPC layers. Fig.7a-b respectively shows the dynamic responses of micro cantilever with three MPC layers (Fig.3h) after the excitations of out-of-plane and in-plane AC magnetic fields. Due to the magnetic field distribution of three layer MPC, the bending and torsional modes are excited respectively.

CONCLUSIONS

In summary, this study has successfully implemented two stage solidification technology to realize multilayer anisotropic magnetic polymer composites films, and the integration of multilayer anisotropic MPC film on MEMS structures for magnetic driving has also been demonstrated. Various types of NdFeB MPC films by anisotropic magnetization have been achieved. This study has characterized the performances enhancement of NdFeB MPC (30wt%) by anisotropic magnetization during the proposed two-stage solidification. Measurement shows that anisotropic magnetostatic shielding effect is reduced from

0.45Telsa to 0.3-0.35Telsa. The improvement of the coercivity force, remanence, and saturation magnetization of NdFeB MPC are 3.4%, 304%, and 268%, respectively. Integration of MPC on MEMS structures and the following driving tests have also been demonstrated.

ACKNOWLEDGMENTS

This paper was partially supported by National Science Council of Taiwan under grant number NSC 102-2221-E-007-027-MY3 and Brain Research Center at the National Tsing Hua University, Taiwan, under contract 102A0129JA. The authors would like to express his appreciation to the Nano Science and Technology Center of National Tsing Hua University, and Nano Facility Center of National Chiao Tung University in providing the fabrication facilities.

REFERENCES

[1] C. Prados, et al., "Magnetic and structural properties of high coercivity Sm(Co, Ni, Cu) sputtered thin films", *J. of Applied Physics*, vol. 83, No. 11, pp. 6253-6255, 1998.

[2] L. K. Lagorce, et al., "Magnetic Microactuators Based on Polymer Magnets", *J. of Microelectromechanical Systems*, 83, pp. 2-9, 1998.

[3] F. Herrault, et al., "Fabrication and Performance of Silicon-Embedded Permanent-Magnet Microgenerators", *J. of Microelectromechanical Systems*, 19, pp. 4-13, 2010.

[4] D. P. Arnold, et al., "Permanent Magnets for MEMS ", *J. of Microelectromechanical Systems*, 18, pp. 1255-1266, 2009.

[5] A. G. Olabi, et al., "Design and application of magneto-rheological fluid", *Materials and Design*, 28 pp. 2658-2664, 2007.

[6] G. Filipcsei, et al., "Magnetic Field-Responsive Smart Polymer Composites" , *Adv Polym Sci*, 206, pp. 137-189, 2007.

[7] Z. Varga, et al., "Smart composites with controlled anisotropy" *Polymer*, 46, pp. 7779-7787, 2005.

[8] M. R. Jolly, et al., "A model of the behaviour of magnetorheological materials", *Smart Mater. Struct.*, 5, pp. 607-614. 1996.

[9] F.-M. Hsu, et al., "Formation and integration of tunable anisotropic magnetic polymer composites by two stages solidification process", *IEEE Conference Transducers*, pp. 2672-2675, 2013.

[10] N. M.Dempsy, et al., "High performance hard magnetic NdFeB thick films for integration into Micro-Electro-Mechanical-Systems", *Applied Physics Letters*,90(2007)092509.

CONTACT

*Weileun Fang, Department of PME/Institute of NEMS, National Tsing Hua University, Hsinchu, 30013, Taiwan, Tel: +886-3-574-2923; Fax: +886-3-573-9372; E-mail: fang@pme.nthu.edu.tw

CNT BUNDLES GROWTH ON MICROHOTPLATES FOR DIRECT MEASUREMENT OF THEIR THERMAL PROPERTIES

Cinzia Silvestri*, Bruno Morana, Giuseppe Fiorentino, Sten Vollebregt, Gregory Pandraud, Fabio Santagata, Guo Qi Zhang and Pasqualina M. Sarro
*Delft University of Technology, ECTM-DIMES, Delft, The Netherlands

ABSTRACT

Vertically aligned Carbon Nanotubes (CNT) arrays were successfully grown on top of a freestanding microhotplate, to investigate the thermal dissipation properties of CNT bundles and their applicability as heat exchanger. Two CNT configurations are employed: a group of six bundles, each with a diameter of 20 μm, and a single CNT bundle with a diameter of 200 μm. In both configurations the bundles are 70 μm high. The microhotplate consists of a platinum thin film microheater integrated on a freestanding silicon nitride membrane. The microhotplate is used as heat source and as temperature sensor. Results show that at 300 °C, 20% and 31% of power can be saved with the circular six and single bundle configurations, respectively

INTRODUCTION

Due to the increasing performance requirements, system downscaling, and features integration, thermal management has become a critical issue in electronic design. It is therefore crucial to efficiently drain off the heat generated within the integrated electronics, in order to avoid damage or malfunctioning of the components.

To increase the effectiveness of thermal management components, new nanomaterials and advanced micro-scale features are explored [1]. Due to their prominent thermal properties, vertical aligned CNTs have been suggested as promising material for thermal management [2]. Numerous studies, mostly theoretical, have been published to evaluate the thermal performance of CNTs and their applicability for heat removal in integrated system devices. However, to estimate the thermal performance of nanotubes is still a big challenge, because conventional thermal measurement techniques do not have enough spatial resolution to accurately detect temperature drop across micrometer length scales [1]. Most data published on CNTs thermal performance, are extracted employing complex techniques, like photothermal metrology [3], time-resolved infrared pyrometry tests [4] and the three-ω method [5]. However, none of these techniques is adopted for CNT structures with high aspect ratio. Moreover, they require a top metallization after the growth process, which can affect CNT quality.

Here we propose a novel approach to evaluate the thermal properties of CNT bundles based on a direct measurement technique to assess their suitability for thermal management applications such as thermal interface material (TIM) [6] or as external heat-sink [7]. This technique has been successfully employed for the thermal characterization of thin films [8] and single silicon nanowire [9].

The test structure used in this study consists of an array of vertical CNT bundles, directly synthetized on top of a freestanding microheater (Fig 1). In this way the thermal isolation necessary to detect the cooling effect of the CNT bundles is achieved. No further process steps are required after the CNTs growth, thus keeping intact their intrinsic qualities. By means of this microfabricated device, we want to understand the average CNT bundle performance instead of the single nanotube, in order to demonstrate the feasible application as heat exchanger material and, more generally, as sensing element. Two different CNT configurations are studied: a group of six symmetrically distributed bundles each with a diameter of 20 μm, and a single bundle with a diameter of 200 μm. The power dissipated through the air by CNTs is then measured to extract their thermal properties.

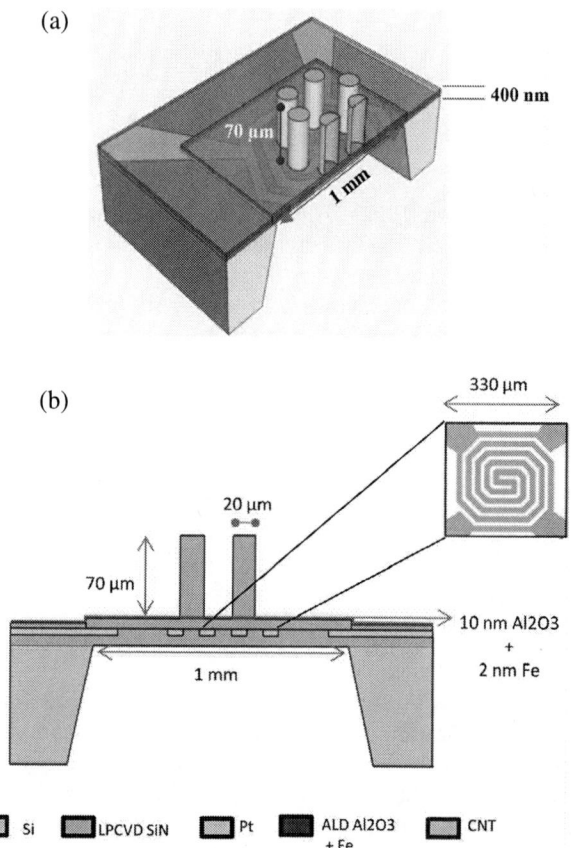

Figure 1: The proposed test structure and the main geometrical parameters: (a) 3D sketch, and (b) detailed schematic cross-section (not to scale). The heater side is 1/3 of the membrane width. The spiral layout is designed to guarantee a uniform temperature distribution.

978-1-4799-3510-9/14 $31.00 © 2014 IEEE   48

DEVICE DESIGN

The microhotplate consists of a squared-shaped low-stress SiN$_x$ membrane, with a side length of 1 mm and a thickness of 400 nm, and a spiral Pt microheater. The microheater is a 3.3 mm long and 18 μm wide spiral, set in the middle of the membrane, covering an area of 330 μm x 330 μm. It is equipped with four contact pads in order to measure the average resistance of the heated part with the four-point method. The spiral layout is designed to guarantee uniform temperature distribution. The CNT bundles are directly grown in the center of the suspended microhotplate. A 3D sketch and a detailed schematic cross-section of the proposed test structure are shown in Fig 1. In order to perform reliable measurements, three different configurations are processed on the same wafer: a bare sample without CNT bundles to be used as reference device; a group of six bundles, having a diameter of 20 μm each, uniformly distributed in a circular pattern (Fig 2(a)); a single bundle with a diameter of 200 μm (Fig 2(b)).

DEVICE FABRICATION

The starting material is a single side polished silicon (Si) wafer, with a thickness of 525 μm and 100 mm diameter. First, a 200 nm thick low stress LPCVD SiN$_x$ layer is deposited to provide electrical insulation. This step is followed by the evaporation of 160 nm of Pt on an adhesion layer of 20 nm of Tantalum (Ta). A lift-off process is used to pattern the Ta/Pt layer into an 18 μm wide resistive spiral. A second layer of 200 nm of low-stress LPCVD SiN$_x$ is deposited to seal the heater and to create a symmetrical thermal configuration. The final value of the spiral sheet resistance is equal to 1.76 Ω/□. Contact openings to the Pt contact pads are realized by dry etching of the SiN$_x$ to allow electrical characterization of the microheaters. In order to release the 1 x 1 mm^2 membrane, the backside is patterned and the Si substrate is removed in a 33wt% KOH solution at 85°C, using SiN$_x$ as hard mask.

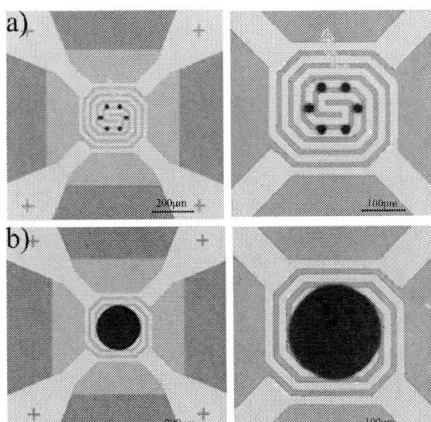

Figure 2: Optical microscope images of the fabricated devices. The CNT bundles directly grown on a suspended microheater are visible as black spots. The geometries fabricated are: a) six circular CNT bundles (diameter 20 μm each); b) a single large CNT bundle (diameter 200 μm).

The microhotplates are now ready for the CNTs growth. First, a 10 nm layer of Aluminum Oxide (Al$_2$O$_3$) is deposited by Atomic Layer Deposition (ALD). The Al$_2$O$_3$ layer is employed not only as buffer layer, but also to promote the formation of nanosized Iron (Fe) particles suitable for CNT growth. A 1.5 nm of Fe, catalyst layer for CNTs growth, is evaporated and patterned by lift-off, to define circular features of 20 μm and 200 μm. The microheater samples with the Al$_2$O$_3$/Fe catalyst are then loaded into a commercial system (Black Magic II, Aixtron) for CNT synthesis. The CNTs are grown using low-pressure CVD at 600°C. The CNTs height can be tuned by varying the process time. In our case, after five minutes, instead of the expected 150 μm high CNTs, a height of about 70 μm is achieved (Fig 3). The growth rate on a suspended membrane, although successful, is almost half of what is achieved on the bulk silicon substrate. This can be caused by a different thermal transfer in the released structure. Further, investigation is needed to identify all the possible causes and to optimize the growth process.

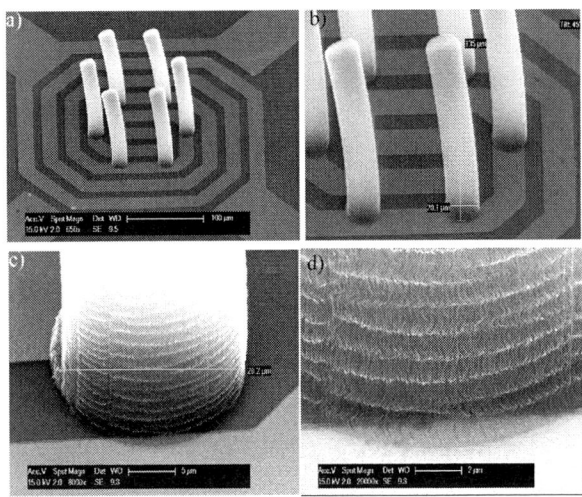

Figure 3: SEM images of the active area of the device: (a) Global view of the six CNT bundles on the microheater; (b) close-up of the CNT bundles. Each feature is 20 μm large and ~70 μm long; (c-d) close ups of the CNT base showing the good adhesion to the substrate;

MEASUREMENT PRINCIPLE

The purpose of this work is to measure the thermal properties of vertically aligned CNT bundles, using a microfabricated device. In our microhotplate without CNTs, the total heating power is composed by the effective power, defined as the amount of power necessary to raise the temperature of the test device, and the power loss. Under steady-state operating conditions, the electrical power applied to the microhotplate, P_{heat}, is dissipated into the surrounding area through natural convection, thermal radiation and conduction within the membrane material. The described thermal loss paths are defined in terms of power losses, P_{loss}. In first approximation, the heating power of our

978-1-4799-3510-9/14 $31.00 © 2014 IEEE

heater fully embedded in a membrane, can be expressed as:

$$P_{heat} = R_{heat} * I^2_{heat} = R_0(1 + \alpha\,\Delta T) * I^2_{heat} \quad (1)$$

were I_{heat} is the current applied through the terminals and R_{heat} is the Pt resistance in function of the heater temperature. The resistance of the platinum spiral changes linearly with the temperature, where, α corresponds to the temperature coefficient of resistance (TCR) and R_0 is the resistance value at ambient temperature.

To estimate the temperature of the heater, the Pt layer of a bare microhotplate is characterized to determine resistance and temperature coefficient of resistance (TCR). The reference sample experienced the same processing steps of the sample equipped with CNTs bundles, as they are processed on the same wafer. The extracted values are obtained performing a four-probe measurement.

The TCR, extracted from the relative change of the heater resistance with the temperature, corresponds to 2300 ppm K^{-1} with a standard deviation of 40 ppm K^{-1}. The TCR value is based on previously calibrated hotplates with the same Ta/Pt layer. Once the heater calibration curve is obtained, the heater can also be used as sensing element.

In the case of the microhotplate equipped with CNT bundles, additional thermal loss paths are introduced. In fact, the CNT bundles extend the surface area in contact with the environment, increasing the radiative and convective contributions (Fig 4). As the power losses increase, we need to supply more power to the entire system to reach the same temperature value of the reference microhotplate.

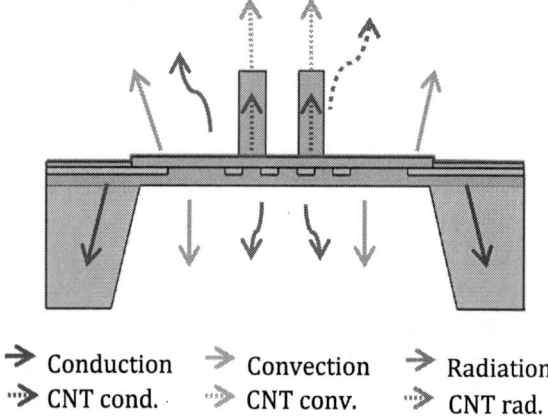

Figure 4: Schematic cross-section with possible thermal paths after the surface extension due to the CNTs growth. The additional contribution appears in terms of convective and radiative losses (dashed arrows).

DEVICE CHARACTERIZATION

Device characterization was performed using an Agilent 4156C Semiconductor Parameter Analyzer and a Cascade probe station with a temperature-controlled chuck.

As mentioned before, the microhotplate is connected to the contact pads by four Pt tracks, allowing four-probe resistance measurements. The measurements are performed in air. The recorded value of the resistance is representative of the average temperature of the microhotplate. The microhotplate was electrically heated up from room temperature to 400°C by supplying a stepwise current with steps of 2 s. In order to preserve the structural integrity of the device under test, the maximum input current value used is 10 mA.

RESULTS AND DISCUSSION

Our preliminary study is carried out on two types of CNT patterns: six circular CNT bundles, with a diameter of 20 µm each; a single big bundle, 200 µm in diameter. The six bundles occupy about 2% of the total microheater area, and the single bundle covers around 29% of the total area. The thermal performances of CNT geometries are extracted by comparison with a reference microhotplate without CNTs. Applying current between contact pads, Joule heating will heat the testing structure. The applied current and the corresponding voltage across the heating spiral are recorded.

In Figure 5, the normalized resistance (R/R_0) as function of the applied current for a six bundle and a big single bundle samples are shown. The normalized resistance of the microheater with CNT bundles is lower compared to the reference one. This is due to the additional CNT thermal contribution and related surface extension. In particular, at 10 mA up to 9% of resistance decrease between the reference and the single bundle, is recorded.

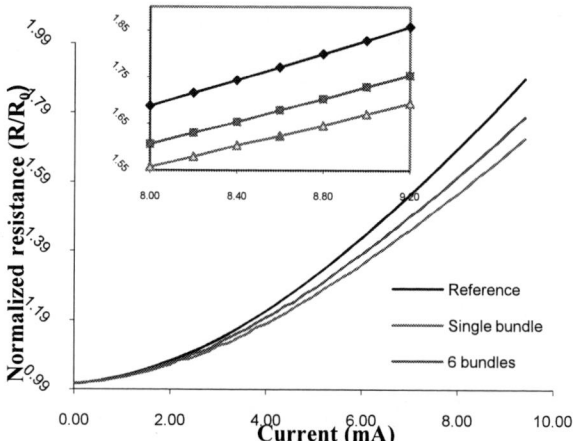

Figure 5: Normalized resistance (R/R_0) versus applied current. For a given current the resistance of the microheaters equipped with the CNTs is lower compared to the reference one (up to 9% at 10mA). The recorded data between 8mA and 9.2mA are shown in the inset.

Figure 6 shows the heating power of both reference and test structures versus temperature. As expected the power required to achieve a certain temperature value is higher for the microheater equipped with CNTs. For example, setting the temperature at 300°C the power surplus is 20% for the microhotplate equipped with 6 bundles and 31% for the single bundle. Results are summarized in Table 1.

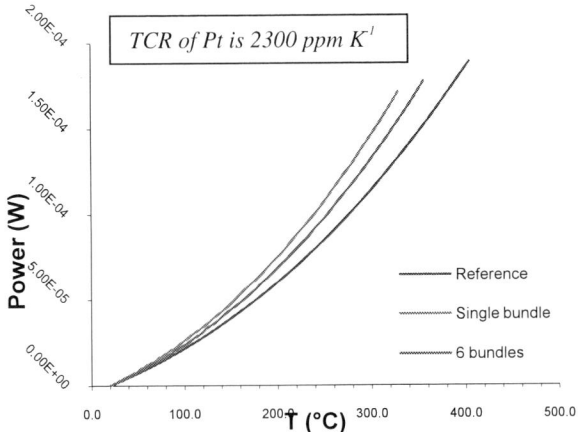

Figure 6: Steady-state power consumption versus temperature measured for a microhotplate with and without CNT bundles. The power required to achieve a certain temperature is higher for the microheaters equipped with the CNTs. At 300°C the power increase is 20% and 31% for the six bundles and the single bundle respectively.

Table 1: Summary of the preliminary results.

Sample	Resistance at $T = 300°C$	Power diss. at $T = 300°C$	Dissipation percentage
Reference	324 Ω	1,17 E-4	/
6 bundles	588Ω	1,34 E-4	~ 20%
Single bundle	570 Ω	1,54 E-4	~ 31%

As the measurement results show, the single bundle configuration is more effective then the six bundles one. Indeed, the total surface are of the single bundle is exactly 1.6 times larger than the total surface area of the six bundles added up together. This proportion is reflected also on the power percentage reported.

The recorded values confirm the effectiveness of the proposed measurement technique and the possibility of using the CNTs bundles as heat exchangers.

CONCLUSIONS

A simple method of measuring the dissipative power of vertically aligned CNTs has been proposed. A Pt/SiN$_x$ microhoplate device has been used as test structure. CNT bundles as high as 70 μm were grown directly on top of the suspended microhotplate. No further processing steps are required after CNTs growth, leaving intact their properties. Using a differential measurement method, it is possible to emphasize only the CNTs contribution, shadowing the basic thermal effects of the bare membrane.

Heat characterization was performed measuring the dissipative power of two types of CNT patterns. When fixing the temperature at 300°C, 20% and 31% of power can be saved with the six bundles and single bundle geometries, respectively.

The described results demonstrate the dissipative power of the investigated CNT bundle geometries and the effectiveness of the proposed test microstructure for thermal performance measurement.

In order to quantify the magnitude of the thermal loss contributions, further investigation is required. In particular, a comparison between measurements recorded both in air and vacuum environment can help to extract relevant data such as the average thermal conductivity of the CNT bundle.

ACKNOWLEDGEMENTS

The authors gratefully acknowledge the technical support and advice of the staff at the Dimes Technology Centre. This work was performed in the framework of NanoNextNL Project.

REFERENCES

[1]   X. Tong, *Advanced Materials for Thermal Management of Electronic Packaging*, Springer S. 2011.

[2]   J. Che, T. Cagin, and W. G. III, "Thermal conductivity of carbon nanotubes," *Nanotechnology*, vol. 65, 2000.

[3]   Y. Xu, Y. Zhang, E. Suhir, and X. Wang, "Thermal properties of carbon nanotube array used for integrated circuit cooling," *J. Appl. Phys.*, vol. 100, no. 7, p. 074302, 2006.

[4]   M. Gaillard, H. Mbitsi, A. Petit, E. Amin-Chalhoub, C. Boulmer-Leborgne, N. Semmar, E. Millon, J. Mathias, and S. Kouassi, "Electrical and thermal characterization of carbon nanotube films," *J. Vac. Sci. Technol. B Microelectron. Nanom. Struct.*, vol. 29, no. 4, p. 041805, 2011.

[5]   S. Vollebregt, S. Banerjee, K. Beenakker, and R. Ishihara, "Thermal conductivity of low temperature grown vertical carbon nanotube bundles measured using the three-ω method," *Appl. Phys. Lett.*, vol. 102, no. 19, p. 191909, 2013.

[6]   A. J. McNamara, Y. Joshi, and Z. M. Zhang, "Characterization of nanostructured thermal interface materials – A review," *Int. J. Therm. Sci.*, vol. 62, pp. 2–11, Dec. 2012.

[7]   F. Santagata, G. Almanno, S. Vollebregt, C. Silvestri, G. Q. Zhang, and P. M. Sarro, "Carbon Nanotube based heat-sink for solid state lighting," *8th Annu. IEEE Int. Conf. Nano/Micro Eng. Mol. Syst.*, vol. 1, pp. 1214–1217, Apr. 2013.

[8]   A. Roncaglia, F. Mancarella, M. Sanmartin, I. Elmi, G. C. Cardinali, and M. Severi, "Measurement of Thermal Conductivity on Thin Films," pp. 20–23, 2006.

[9]   D. Li, Y. Wu, P. Kim, L. Shi, P. Yang, and A. Majumdar, "Thermal conductivity of individual silicon nanowires," *Appl. Phys. Lett.*, vol. 83, no. 14, p. 2934, 2003.

CONTACT

*C. Silvestri, tel: +31630187549; C.Silvestri@tudelft.nl

3D ICE PRINTING AS A FABRICATION TECHNOLOGY OF MICROFLUIDICS WITH PRE-SEALED REAGENTS

Hongze Zhang, Hui Li, Mengxi Wu, Huaiqiang Yu, Wei Wang and Zhihong Li*

National Key Laboratory of Science and Technology on Micro/Nano Fabrication,
Institute of Microelectronics, Peking University, China
Beijing 100871, People's Republic of China.

ABSTRACT

This paper proposes a fast and inexpensive method to fabricate 3D structure by "ice printing". This "bottom-up" 3D fabrication method is achieved by printing water onto the cold surface and turning into ice structure layer by layer. Through this method, fluid with reagents, such as drugs and nanoparticles, is sealed into microfluidics during fabrication, which is available for drug delivery or other medical care applications. Moreover, complex micro-channels are easily fabricated using 3D ice structure as soft lithography mould, which can be used for microfluidic mixer, three dimensional flow focusing, etc.

INTRODUCTION

Water is the most important solvent for drugs and plays irreplaceable role in microfabrication. Recently, because ice is contamination-free and easy to fabricate, it has been used in micro- or nano- fabrications, such as a "photoresist" [1] in electron beam lithography or a sacrificial layer in packaging application [2]. In addition, a microfluidic channel was achieved in ice block by laser [3]. In their works, the complex and expensive electron beam or laser system is used to pattern the ice film or block, or the hydrophobic and hydrophilic regions should be defined by complex lithography and plasma treatments. Compared with these "top-down" technologies, an innovative and inexpensive "bottom-up" method is proposed to fabricate 3D structure featured by ice printing.

The ice printing is quite similar to other 3D printing technologies. This bottom-up 3D fabrication method is achieved by printing water onto the cold surface then freezing into ice structure layer by layer. A tiny liquid droplet can be printed separately from the nozzle and will be frozen into solid ice as soon as attaching to the cold surface under ice point. Repeating this step with automatically moving printing head or stage, a complex 3D ice structure with the fine feature size can be constructed, as shown in the schematic view of Figure 1.

As water is an excellent solvent, various liquids with different reagents, such as drugs and nanoparticles, can be used for ice printing.

This ice printing method can be used on a variety of substrates, such as glass, metal, silicon and polymer film. Moreover, theoretically any substrate which is cold enough is valid for ice printing. So this method is not selective to the substrates, which could widely extend its application.

MEMS drug delivering micro-devices are highly promising for the biomedical applications [4]. These micro

fabricated MEMS drug delivering devices can be very tiny and precisely controlled. While due to the limitation of conventional microfabrication, drugs are not easy to be packaged during the fabrication process. Thus drugs should be injected into the assembled reservoir individually [5, 6], which limits the integration and mass production of MEMS drug delivering devices. With the ice printing technology, the drug solution can be printed as a reservoir and ready for packaging during the fabrication process. Thus the ice printing method may be available for drug delivering application.

The ice printed 3D structure can also be used as soft lithography mould for complex microfluidic channels with varied heights, which are difficult or even infeasible for conventional photolithography method.

ICE PRINTING SYSTEM

Here a prototype of ice printing system has been made to demonstrate the 3D ice printing method. As shown in the photo of Figure 1, a commonly used commercial 4-color inkjet printer (EPSON®, L111) is modified to build up the ice printing system. Recently, the desktop printer with inkjet printing technology has been rebuilt to utilize in cell printing [7] and biochemical applications [8], because it is simple and versatile. The one we choose is a piezoelectric printer with 90×2 micro-nozzles. Moreover, a Peltier cooler with a cooling water circulator and a temperature controller is used

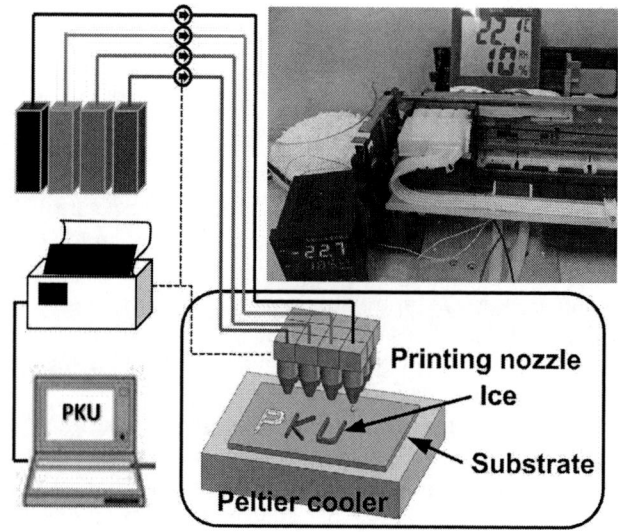

Figure 1: The schematic view and the photo of the ice printing system.

to keep the substrate in the low temperature (-22℃ lowest). Then the whole system is put inside a glove box filled with pure nitrogen to isolate from the outside environment. Inside the glove box drying agent is utilized to keep the inside atmosphere dry enough in order to avoid frosting on the cold surface. A digital thermometer and hygrometer are used to monitor the inside environment. The hygrometer has a minimum detectable limit of 10% RH (relative humidity), which means the humidity of the inside environment could be much lower than that.

Controlled by a computer, the modified inkjet printer can print 4 different kinds of liquid on a 30mm×200mm region as designed. Here 4 different dyes are used to substitute different reagents to demonstrate the system feasibility. Then 4-colored liquids representing 4 kind liquids with different chemical reagents participate the ice printing process.

EXPERIMENTS AND RESULTS

3D Ice Printing Ability

For simplicity, the test pattern for ice printing is designed though Microsoft Office [TM] software, which means the ice printing process is quite similar to that printing a picture on a paper. Thus the precision of ice printing is almost the same to that of printing ink on the paper. In Microsoft Office software, the thinnest line that can be drawn is 100 μm. Therefore, the thinnest line that we ice-printed is approximately 100μm wide. The Epson micro piezoelectric print head has 3-picoliter Ultra Micro Dot [TM] print technology, which means the printed droplet could be smaller than 20μm in diameter. However, limited by the poor mechanical system (repeatability of the moving dimension for the printing head is over ±100 μm) and the complicated control system of the inkjet printer, the potential of the printer head has not been shown yet.

With a professional mechanical and control system, the precision of ice printing could be improved greatly. For example, with a commercial picoliter inkjet print head and an inexpensive mechanical x-y stage with ±5μm accuracy, an ice printing system with accuracy of at least 20μm could be built up. That accuracy is available for majority of microfluidic applications. Moreover, the improvements of inkjet printing technology make sub –femtoliter printing possible [9]. Hence, with the sub –femtoliter method, an ice printing system with accuracy of 1μm could be achieved.

As shown in Figure 2(a), a 20×20 ice pillar array is printed on a silicon substrate with a penny on top. While in Figure 2(c), the ice pattern is printed on glass slide. In addition, ice printing is also achieved on parylene-C thin film attached on silicon, which is widely used in implantable devices. These results show the ice printing capability on various substrates and topographies. Thus this ice printing method could be apply on curved, rough surface or even that with big step, which is normally difficult or infeasible for conventional photolithography method .

Considering different thickness and thermal conductivity of different substrates, the Peltier cooler should be set at different temperatures. Some thermal conductive silicone could be used to improve the thermal conduction between the substrate and the Peltier cooler.

In Figure 2(b), an ice micro-pillar array with different heights is printed on a silicon surface. The pillar size is 400μm×300μm, and heights increase gradually from about 150 microns to 2 millimeters. The different heights can be achieved by different printing circles. The 2 mm ice pillar is achieved after 210 circles, which means during each circle the pillar increase approximately 10μm.

Figure 2(c) shows ice printed double-layered mesas with different sizes, where the first layer is colorless and the second layer is dyed. The inset of Figure 2(c) is a close-up view, in which the size of colorless layer is 5mm×5mm,

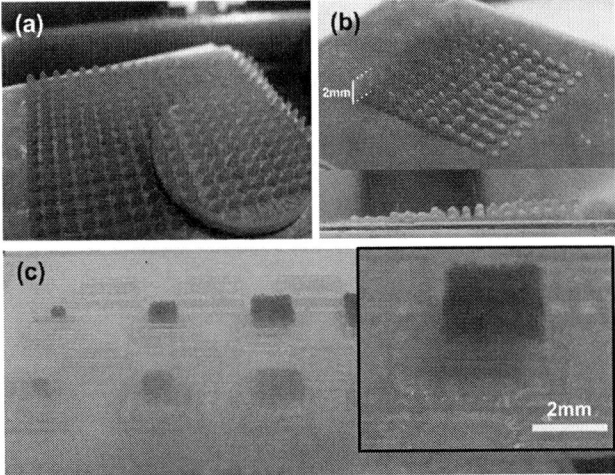

Figure 2: Photos of printed ice structures: (a) array of micro-pillars printed on a surface of coin-silicon stack, (b) pillars with varied heights, (c) layered mesas.

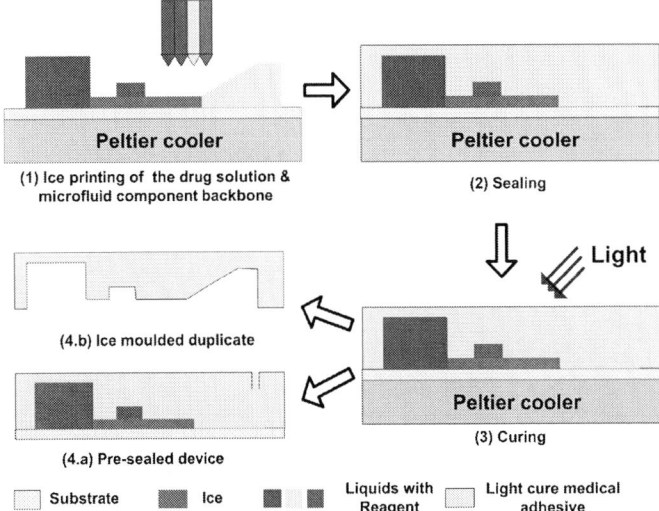

Figure 3: The schematic view of the fabrication process of pre-sealed drug delivery chip or ice moulded duplicate.

while that of the upper pink layer is 2mm×2mm. The colorless layer is firstly printed, and then the second pink and yellow mesas are printed simultaneously. Meanwhile the size of the pink and yellow foursquare vary from 1mm to 4mm.

Process and Method

The processes to fabricate the pre-sealed device and the ice moulded duplicate are similar, but pre-treatment is needed for ice moulded duplicate. The mold-release agent is spun on the substrate to make the remolding process easier. The fabrication process for pre-sealed device and ice moulded duplicate has been shown in Figure 3. Firstly, the water or solution is printed and frozen to form the reservoir; multiple drug solutions can be printed on demand. Meanwhile, other microfluidic components are ice-printed on the cold substrate, as shown in Figure 3.1. Then a visible-light cure medical device adhesive (Loctite® 3311™) is poured onto the ice structure to get the 3D ice moulded structure or capsulation sealing the drugs, as shown in Figure 3.2. The Loctite® 3311™ light cure medical device adhesive, which is qualified to ISO 10993 Protocol, has been widely used in many biomedical applications [10, 11]. After leveling, the medical device adhesive is cured by light below ice point when the 3D ice structure maintains, as shown in Figure 4.3. Then the drug-release hole can be punched right away or just before use, shown in Figure 4.4a. Alternatively, the cured adhesive can be remolded to get the ice moulded duplicate, as shown in Figure 4.4b. As the 3D ice printed structure could have varied heights, the microfluidic component with varied depths could be achieved via the process mentioned above.

Ice Moulded Duplicate

Microfluidic channels with varied heights can be achieved using ice printed 3D structure as mould. Figure 4 (a) and (b) show SEM photos of a microchannel and its cross section. A microchannel with varied depth is shown in Figure

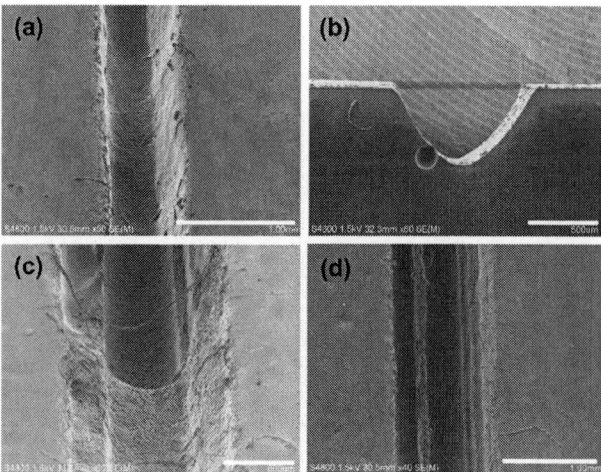

Figure 4: SEM of printed ice moulded light cure medical adhesive structures: (a) a micro channel, (b) the cross section, (c) micro channel with varied depth, and (d) paralleled channels with a micro fence.

4(c), where the upper section is deeper. It can be used for microfluid mixer or three dimensional flow focusing [12]. Figure 3(d) shows paralleled channels with a micro-fence, which can be used for particle or blood cell separation [13].

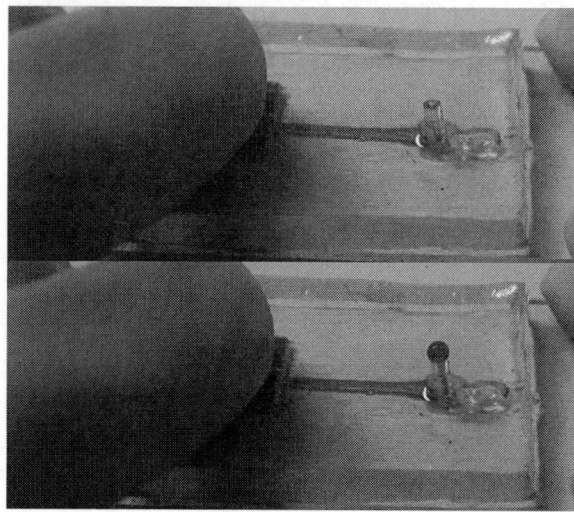

Figure 5: The photo of pre-sealed drug delivery chip. Drug is printed and sealed in the capsule, and delivered by finger press.

Pre-sealed Devices

Here two prototypical pre-sealed devices are fabricated to validate the technology. A simplified pre-sealed drug delivery chip, where the drug is delivered by finger press is shown in Figure 5. The substrate for this chip is glass slide, and the size of the drug reservoir is 10mm×10mm×2mm (0.2ml). Furthermore, different agents are pre-sealed in different reservoir and connected with ice channel, as shown in Figure 6. Thus, the pre-sealed reagents can be preserved

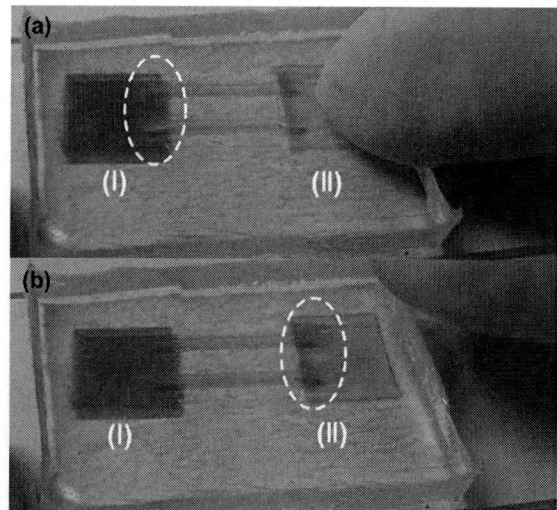

Figure 6: Primary demonstration of pre-sealed mixing reactor: (a) reagent in reservoir I flows into reservoir II driven by finger press, (b) the mixed solution flow back to reservoir II after the finger release.

without reaction below ice point. After its melting, the different reagents can be released and mixed, and the reaction can be triggered; subsequently the reaction products can be used for following medical use.

A prototype of the pre-sealed mixing reactor chip with two reagents is fabricated, and also triggered by finger press, as shown in Figure 6. Two reservoirs are printed with different colored liquids which represent different drugs. Meanwhile they are connected with ice printed microchannels. Under the finger press, the yellow liquid in reservoir II flow into the reservoir I. After the finger release, the mixed solution flow back to reservoir II, as shown in Figure 6(a) and (b). After several-circle finger pressings, two different liquids mix evenly.

CONCLUSIONS

We propose a 3D ice printing technology, which is a simple and inexpensive bottom-up 3D fabrication method. A prototypical ice printing system is built up with a modified desktop inkjet printer to validate this technology. Using 3D ice printing structure as a mould, complex microfluidic components can be achieved. With the ice printing technology, the drug solution can be printed as a reservoir and packaged during the fabrication process. Moreover, two prototypical pre-sealed chips are fabricated to demonstrate this idea. The ice printing method shows promising potential for drug delivery applications.

REFERENCES

[1] Anpan Han, Dimitar Vlassarev, Jenny Wang, Jene A. Golovchenko and Daniel Branton "Ice lithography for nanodevices." *Nano letters*, vol.10(12), pp.5056-5059, 2010.

[2] Sha Li, Li-Wei Pan, and Liwei Lin. "Frozen water for MEMS fabrication and packaging applications." *IEEE The Sixteenth Annual International Conference on Micro Electro Mechanical Systems 2003,* Kyoto, January, 2003, pp. 650-653.

[3] A.M. Bossi, M. Vareijka, E.V. Piletska, A.P.F. Turner, I. Meglinski and S.A. Piletsky, "Ice matrix in reconfigurable microfluidic systems." *Laser Phys.*, vol. 23, 075605, 2013.

[4] N.M. Elman, Y. Patta, A.W. Scott, B. Masi, H.L. Ho Duc and M.J. Cima, "The next generation of drug-delivery microdevices." *Clinical Pharmacology & Therapeutics*, vol.85, No.5, pp.544-547, 2009.

[5] F.N. Pirmoradi, John K. Jackson, Helen M. Burtb and Mu Chiao "A magnetically controlled MEMS device for drug delivery: design, fabrication, and testing." *Lab Chip*, vol.11(18), pp. 3072-3080, 2011.

[6] N.M. Elman, H.L. Ho Duc and M.J. Cima, "An implantable MEMS drug delivery device for rapid delivery in ambulatory emergency care." *Biomed. Microdevices*, vol.11(3), pp.625-631, 2009.

[7] Paul Calvert, "Printing cells." *Science*, vol.318, pp. 208-209, 2007.

[8] H. Li, Y. Huang, H.Q. Yu, Y. Ma, C.Y. Tang, Z.W. Wei, Z.C. Liang, W. Wang, Z.J. Yang and Z.H. Li, "High throughput synthesis of oligonucleotide utilizing inkjet printing and micro-reactor array filled with robust opal." *The 17th International Conference on Miniaturized Systems for Chemistry and Life Sciences (MicroTAS 2013)*, Freiburg ,Germany, October 27-31, 2013, pp.1242-1244.

[9] Tsuyoshi Sekitani, Yoshiaki Noguchi, Ute Zschieschang, Hagen Klauk and Takao Someya "Organic transistors manufactured using inkjet technology with subfemtoliter accuracy. " *Proceedings of the National Academy of Sciences*, vol.105(13), pp. 4976-4980, 2008.

[10] M. L. Fugoso, K. M. Rowean, M. E. Fourmont, C. T. Brahana, S. M. Schwab and M. E. Minas "Balloon attachment at catheter tip. " U.S. Patent No. 5,964,778. 12 Oct. 1999

[11] Yuya Morimoto, Kaori Kuribayashi-Shigetomi, and Shoji Takeuchi "Biohybrid muscle fibers integrated in a three-dimensional cellular construct." *Proc. of The 16th International Conference on Miniaturized Systems for Chemistry and Life Science*, Okinawa, Japan, 2012, pp.1645-1647.

[12] Yanrui Ju, Jian Song, Zhaoxin Geng, Hongze Zhang, Wei Wang, Lide Xie, Weijuan Yao and Zhihong Li, "A microfluidics cytometer for mice anemia detection." *Lab Chip*, vol. 12, pp.4355-4362, 2012.

[13] Zhaoxin Geng, Yanrui Ju, Wei Wang, Zhihong Li, "Continuous blood separation utilizing spiral filtration microchannel with gradually varied width and micro-pillar array." *Sensors and Actuators B: Chemical*, vol.180, pp.122-129, 2013.

CONTACT

* Zhihong Li, Peking University, Beijing, 100871, China, Phone: 86-10-62766581; Fax: 86-10-62751789; E-mail: zhhli@pku.edu.cn

LIQUID-FILLED SEALED MEMS CAPSULES FABRICATED BY FLUIDIC SELF-ASSEMBLY

Massimo Mastrangeli*, Loïc Jacot-Descombes, Maurizio R. Gullo, Juergen Brugger
Microsystems Laboratory (LMIS1), École Polytechnique Fédérale de Lausanne (EPFL),
Station 2, 1015 Lausanne, Switzerland

ABSTRACT

We present a new method to encapsulate functional liquids into sealed MEMS capsules by fluidic self-assembly. Self-assembly of 200 µm SU-8 cargos and picoliter liquid co-encapsulation are driven by the interplay of global fluidic drag and short-range capillary forces. The latter ensues from the localized surface-selective precipitation of a photopolymerizable adhesive onto the capsules' rim. Assembly yield higher than 50% is achieved, and can be improved by optimized agitation and shape matching. The method is massively parallel, scalable and compatible with batch MEMS fabrication. It can address a variety of applications, including distributed MEMS, cell encapsulation and drug delivery.

INTRODUCTION

Liquid-filled, mechanically-rugged MEMS containers affording controlled content release are uniquely suited for important technological applications such as, *e.g.*, targeted drug delivery [1], cell encapsulation [2] and MEMS networks [3]. Computing and sensing capabilities can be embedded in such barely visible cargos [4], whose collective potential is tackled in large distributed swarms [3]. In this perspective, optimal and cost-effective resource utilization require massive parallelism and scalability for both fabrication and assembly of microcapsules. For the former, wafer-scale photolithography offers unmatched yield. For the latter, dexterous servoed yet serial pick-and-place assembly of micrometric parts is hardly efficient due to stiction effects expected to worsen in presence of liquid interfaces [5]; and wafer-level bonding/sealing is challenged by liquid-related issues, as well. Conversely, the intrinsically parallel pairwise fluidic self-assembly (SA) [6] of half-capsules into full liquid-filled capsules would yield naturally to the task, if stable sealing of capsules be integrated in the process. In this respect, Whitesides' [7] and Parviz' [8] groups pioneered the fluidic SA of microcomponents mediated by the

precipitation of an insoluble phase from a polar solution; and Nemani *et al.* evidenced the mechanical stability of SU-8-based microcapsules bonded through adhesives [2].

We hereby demonstrate seamless photo-activated sealing and liquid co-encapsulation in fluidically self-assembled MEMS capsules. Our research builds on and extends the aforementioned techniques by their joint application to high-throughput encapsulation into mechanically-stable MEMS containers. This work improves our prior findings on pairwise fluidic SA mediated by hydrophobic interactions [9] by: 1) achieving significantly higher assembly selectivity, yield and throughput, and 2) enabling the mechanical sealing of the self-assembled microcapsules. As an alternative to capillary self-folding and sealing of polyhedral microcontainers [10], our SA method affords a simpler fabrication process still compatible with the embedding of functionalities (*e.g.* microchip [4]) within the microcapsules.

MATERIALS AND METHODS

The fluidic SA procedure and fabrication process of the microcapsules are illustrated in Figs. 1 and 2-3, respectively, and described in the following sections.

Fabrication of half microcapsules

A 100 mm Si wafer is masked by photolithographically-patterned posts of positive tone resist (Fig. 2a) and structured by a fluorine-based, Bosch-like deep reactive ion etching (DRIE) step to obtain arrays of thousands of 50 µm-high triangular pillars with edgelength varying from 20 to 40 µm (Fig. 2b). Cylindrical half capsules with diameter of 100 µm and flat bases are then correspondingly patterned on top of the pillars out of a single, 100 µm-thick spun SU-8 layer (GM 1075, Gersteltec, CH; Fig. 2c). The concentric Si pillars constitute the positive for the cavities embossed in the half capsules (Fig. 3 I-II). A 300 nm-thick layer of SiO_2 is subsequently deposited over the substrate by oxygen-rich

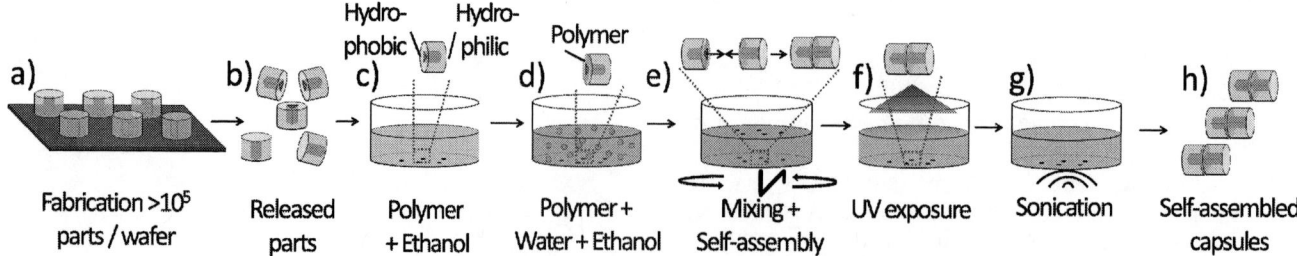

Figure 1: *The fluidic process for the self-assembly and sealing of liquid-filled MEMS capsules. Self-assembly and sealing are mediated by an insoluble polymeric phase precipitated from a polar solvent. Controlled orbital agitation and sonication of the assembly solution are pivotal to the yield and throughput of the process.*

RF sputtering (Pfeiffer Spider 600) to render the exterior surface of the half capsules fully and stably hydrophilic (Figs. 2d and 3b). The half capsules are finally released from the substrate by immersion in an aqueous, heated KOH bath (1 M, 80 °C) for up to 2 hours (Fig. 2e). The open (*i.e.* bottom) sides of the microcapsules are geometrically shielded during the SiO$_2$ sputtering step. They preserve the rather hydrophobic character of the native SU-8 surface [9] after release. The ensuing surface energy difference between the sidewalls and the open base of the half capsules is pivotal for the selective precipitation of the polymeric phase, described in the next section.

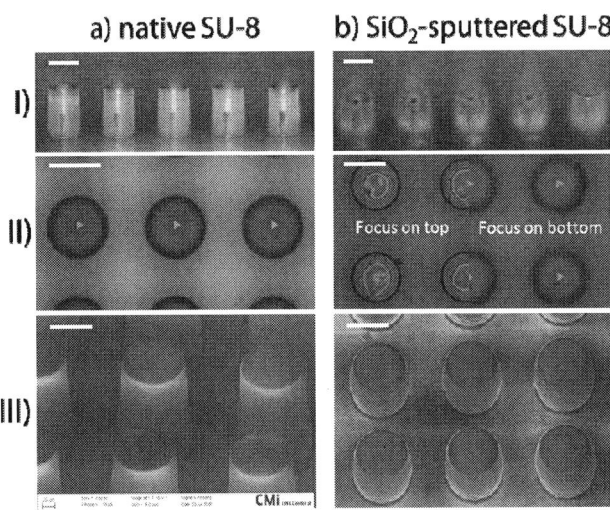

a) native SU-8   b) SiO$_2$-sputtered SU-8

Figure 3: Optical images of the lithographical patterning (a) and SiO$_2$ coating (b) of the SU-8 half microcapsules. I) side views: inner Si pillars can be seen in transparency; II) top views, showing the triangular cross-section of inner pillars; III) SEM micrographs. Conformality to SU-8 and surface smoothness of the sputtered SiO$_2$ layer increase with its thickness. Scale bars are 100 μm.

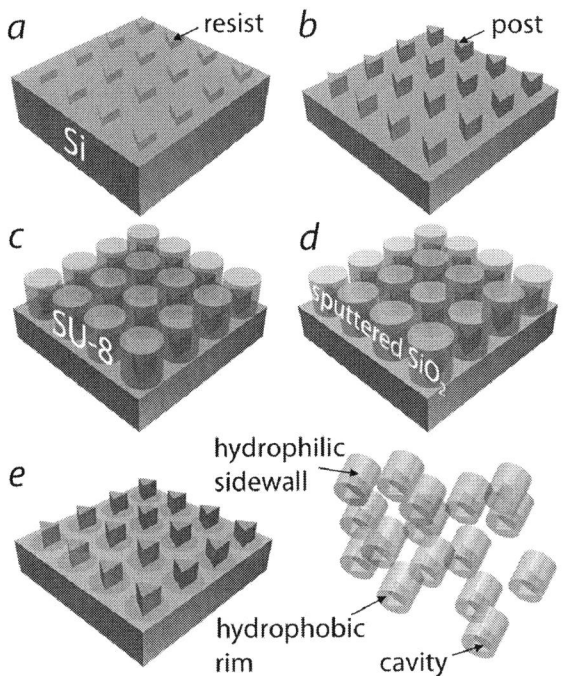

Figure 2: Fabrication of the half microcapsules. a) Lithographical resist patterning over Si substrate, b) DRIE of 50 μm-high triangular pillars, c) patterning of the half capsules with concentric cavities out of a single, 100 μm-thick SU-8 layer, d) sputtering of 300 nm-thick SiO$_2$ over the substrate, and e) release of the half capsules in aqueous heated KOH bath (1 M, 80 °C, up to 2 hours).

Fluidic self-assembly procedure

After release, the half capsules are transferred into an ethanolic solution of a photo-crosslinkable, acrylate-based polymer (0.5 % wt.) at the bottom of a clean glass beaker (Schott-Duran, 10 mL) (Fig. 1c). The solution is then diluted with 5 parts of water, and 3 mL are orbitally stirred (Labnet Orbit 1000) at 100 rpm for 30' to allow for uniform mixture and selective precipitation of the polymer phase from the polar solvent onto the hydrophobic annular rim of the half capsules' open side (Fig. 1d). The consequent formation of localized thin films of hydrophobic adhesive is a result of the tailored minimization of the excess surface energy of the system [8]. Driven by the interplay of global fluidic drag and local capillary forces, the fluidic SA of full

microcapsules and concurrent liquid encapsulation take place during subsequent 30' of orbital agitation at 150 rpm (Fig. 1e). After further water dilution to remove excess polymer, the assembly solution is exposed to UV radiation for 5' to cross-link the interstitial adhesive films, thus enforcing the sealing of self-assembled microcapsules (Fig. 1f). The assembly solution is then sonicated by dipping the beaker in an ultrasonic water bath at 30 kHz for 5' (Fig. 1g). The disposable microcapsules can finally be stored in water or in low surface tension liquids such as ethanol (Fig. 1h).

RESULTS

Photopatterning of the single SU-8 layer on top of the structured Si substrate results in axisymmetric half microcapsules with slightly sloped sidewalls (Fig. 3). This is attributed to 1) dispersion and pillars-induced scattering of UV light inside the thick SU-8 layer during masked exposure, and 2) the relaxation of the cylindrical structures during SU-8 development and thermal curing. Sputtering of SiO$_2$ produces a smooth, conformal coating which does not affect the shape nor, importantly, the release of the half capsules. The detachment of the half capsules is in fact promoted by the permeation of diluted KOH through the SU-8/Si interfaces, in turn eased by the poor adhesion between the two materials. Half capsules with large cavities detach by slipping off the Si pillars, while small pillars remain embedded inside the half capsules after breaking off from the substrate, and are thereby dissolved by the etching bath.

Optical snapshots of a typical progression of the fluidic SA process are shown in Fig. 4. We used batches of about 500 equal half capsules per experiment. Effective low-speed orbital shaking and concurrent selective precipitation of the

978-1-4799-3510-9/14 $31.00 © 2014 IEEE

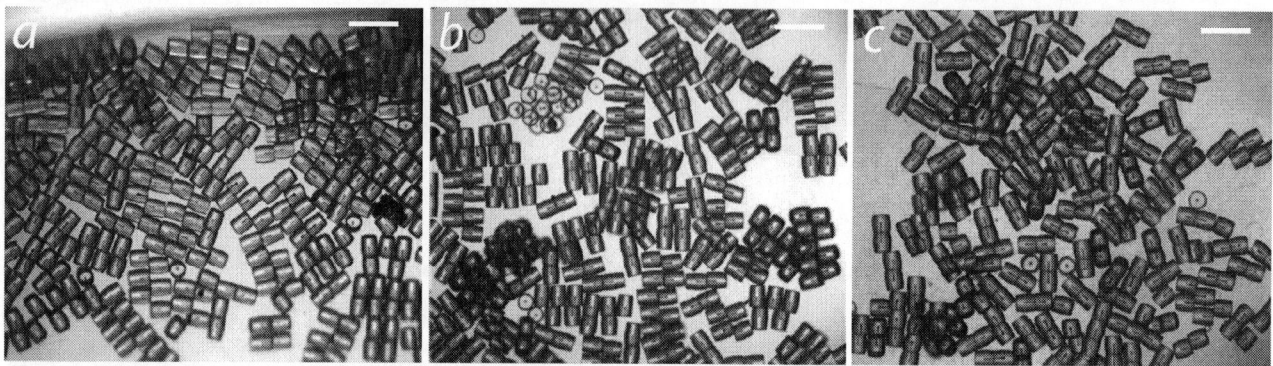

Figure 4: Optical snapshots from a typical fluidic SA run with 500 half capsules: a) after initial low-speed mixing and 10' of agitation, b) after 30' of agitation and UV exposure, c) after subsequent 5' of sonication. Scale bars are 200 μm.

insoluble phase from the assembly solution cause an initial, diffusion-limited aggregation of almost all the half capsules into bridging bi-layered structures (Fig. 4a). Within these extended aggregates the half capsules assume the required, face-selective orientation, yet their pair-wise alignment is mostly skewed and only partial. Upon increase of the orbital agitation speed, larger fluidic shear stresses are applied to the aggregates and can overcome the local capillary forces of the fluid menisci interposed between partially overlapping half capsule pairs. Elongated multimeric aggregates are consequently broken up, and the capillary self-alignment of facing half capsules is enhanced (Fig. 4b). After photopolymerization of the adhesive, the acousto-fluidic forces of the ultrasonic field further increase the singulation of the self-assembled full microcapsules while keeping them intact (Fig. 4c). Our open-loop fluidic SA procedure routinely achieves yield in excess of 50% in about 1 hour, as measured by the relative number of full capsules assembled out of the initial half capsules. The mechanical strength of the thin, conformal polymeric sealing is proved by 1) its resistance to ultrasonic irradiation, and 2) leakage-free encapsulation of colored ink as an instance of functional liquid (Fig. 5). Self-assembled full capsules showed no apparent degradation or leakage upon storage in water for 1 week.

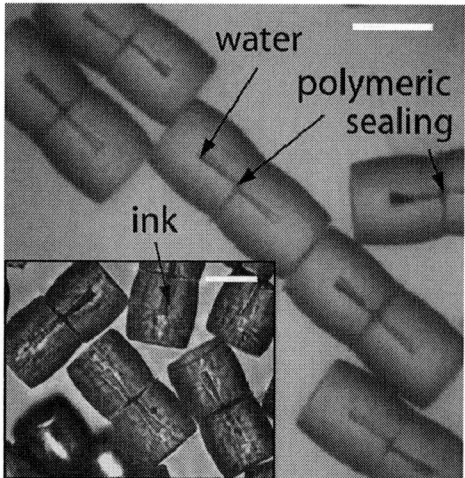

Figure 5: Sealed self-assembled full microcapsules containing water and red ink (inset). Scale bars are 100 μm.

DISCUSSION

Through iterative experimental optimization we elucidated three factors as key for the assembly throughput of our process: face-selective precipitation of the immiscible adhesive, control of fluidic drag forces by orbital agitation, and active use of ultrasonic energy for capsule singulation.

The spatial self-coordination of the half capsules is informed by the controlled precipitation of the immiscible adhesive onto the annular rim of their open sides. The surface selectivity is enforced by the sharp chemical contrast between the native and SiO_2-coated SU-8 surfaces of the half capsules, quantified by a water contact angle difference of ~ 90°. The SiO_2 coating of SU-8 preserves its full hydrophilicity upon prolonged exposure to diluted KOH, as well. It is therefore preferable to SU-8 hydrophilization by oxygen plasma [9], besides being more stable. The energetic interface between the adhesive films and the polar solvent drives the capillary adhesion between approaching and correctly-oriented half capsules. The thickness of the precipitated film increases with time and can be controlled by the initial adhesive concentration [8]. The film thickness can easily be larger than the roughness and bowing of the SU-8 surface. Thus, related issues concerning short-range interactions [9] are avoided by the use of the liquid bridge [7].

Efficient SA is achieved by balancing the attractive forces of the localized capillary menisci with the disjoining fluidic shear stresses enacted by global orbital agitation. The magnitude of fluidic drag forces can be tuned by choosing the orbital speed in relation to: 1) the dimension of the beaker relative to the radius of orbital stirring, and 2) the volume-dependent height of the solution in the beaker [11]. Observations through an inverted microscope evidenced that, in our experimental conditions, the drag induced at 150 rpm by the laminar flow at the bottom of the beaker was effective to partially disperse large multimers—thus overcoming capillary forces of the order of 100 μN [7]—while crowding the half capsules toward the center of the beaker. Conversely, half capsules motion at 100 rpm was barely noticeable. Capsules mobility depends also on their height-to-diameter aspect ratio (AR), and is hindered by friction for AR < 1. Final, active use of ultrasonic energy proved significant to overcome residual local minima in the energy land-

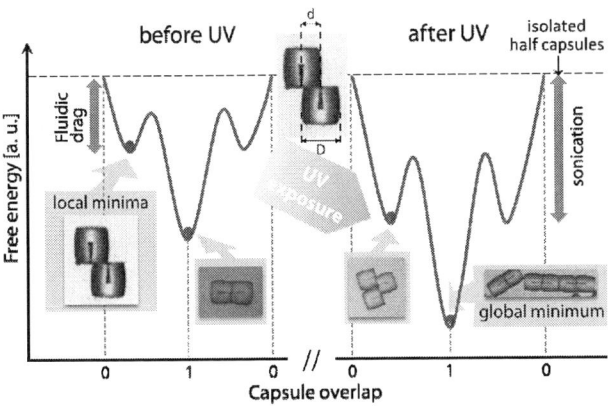

Figure 6: Sketch of the energy landscape as function of the overlap $O = 1 - |d|/D$ of facing half capsules pairs (see inset; D half capsules diameter, d distance between centers of open faces). Clusters of partially overlapping capsules (i.e. local energy minima) are disassembled by fluidic drag forces and ultrasonic field forces respectively before and after exposure of the solution to UV radiation, actively improving the yield of the process.

scape, *i.e.* to break apart remaining aggregates and increase the fraction of correct dimers after UV exposure. The consistent contributions of fluidic and ultrasonic forces to assembly yield are depicted in Fig. 6.

The process' main failure modes originate from: 1) incomplete adhesive coating of the half capsules' rims, and 2) flat geometry of the open sides of the capsules. The former hinders the complete self-alignment of capsule pairs, as they get stuck in partial overlap for lack of a continuous, hydrophobic lubricating film. This issue is eased by a longer precipitation stage prior to SA. The latter mode entails the initial formation of elongated bi-layered aggregates out of the skewed overlap of tens of half capsules. Breaking apart such aggregates, which may extend out of the horizontal plane (as seen in Fig. 4b), significantly extends the duration of the SA process. Enhanced throughput can therefore be achieved by fabricating half capsules with geometrically complementary (*i.e.* lock & key) open sides, as improved shape matching can constrain exclusively dimeric capsule SA.

CONCLUSIONS

We presented an innovative fluidic SA process for the fabrication of sealed MEMS cargos encapsulating functional liquids. The process achieves an assembly yield higher than 50% for cylindrical half capsules—a figure expected to further improve through the use of more sophisticated geometries. The massively parallel nature of SA makes our fluidic encapsulation method compliant with batch MEMS fabrication and scalable to high volume manufacturing provided a consistent scaling of fluidic stirring [11]. Our biofriendly SA process is compatible with most organic liquids as it takes place in a polar solvent. It can easily be adapted to biocompatible materials other than SU-8. Particularly, the use of bioerodible polymers, such as polylactide, could allow triggerable content release and superior disposability in advanced medical and environmental applications [1].

ACKNOWLEDGEMENTS

This work was sponsored by the Nano-Tera.ch research initiative within the framework of the SelfSys project. The authors thank the great staff of EPFL's Center for Micronanotechnology for constant availability and support, Arnaud Bertsch of EPFL's LMIS4 for providing and helping with the preparation of the adhesive, and Victor J. Cadarso for insightful discussions.

REFERENCES

[1] A. P. Esser-Khan, S. A. Odom, N. R. Sottos, S. R. White, J. S. Moore, "Triggered release from polymer capsules", *Macromolecules*, vol. 44, pp. 5539-5553, 2011.

[2] K. Nemani, J. Kwon, K. Trivedi, W. Hu, J.-B. Lee, B. Gimi, "Biofrendly bonding processes for nanoporous implantable SU-8 microcapsules for encapsulated cell therapy", *J. Microencapsul.*, vol. 28, pp. 771-782, 2011.

[3] J. Bourgeois, S. C. Goldstein, "Distributed intelligent MEMS: Progresses and Perspectives", in *ICT Innovations 2011, AISC*, vol. 150, pp. 15-25, 2012

[4] L. Jacot-Descombes, M. R. Gullo, M. Mastrangeli, J. Brugger, "Inkjet printed SU-8 hemispherical microcapsules and silicon chip embedding" *Micro Nano Lett.*, vol. 8, pp. 633-636, 2013

[5] P. Lambert, *Capillary forces in microassembly*, Springer, 2007

[6] N. B. Crane, O. Onen, J. Carballo, Q. Ni, R. Guldiken, "Fluidic assembly at the microscale: progress and prospects", *Microfluid. Nanofluid.*, vol. 14, pp. 383-419, 2013

[7] A. Terfort, N. Bowden and G. M. Whitesides, "Three-dimensional self-assembly of millimeter scale components", *Nature*, vol. 386, pp. 162-4, 1997

[8] C. J. Morris, H. Ho and B. A. Parviz, "Liquid polymer deposition on free-standing microfabricated parts for self-assembly", *J. Microelectromech. Syst.*, vol. 15, pp. 1795-1804, 2006

[9] L. Jacot-Descombes, C. Martin-Olmos, M. R. Gullo, V. J. Cadarso, G. Mermoud, L. G. Villanueva, M. Mastrangeli, A. Martinoli, J. Brugger, "Fluid-mediated parallel self-assembly of polymeric micro-capsules for liquid encapsulation and release", *Soft Matter*, vol. 9, pp. 9931-9938, 2013

[10] R. Fernandes, D. H. Gracias, "Self-folding polymeric containers for encapsulation and delivery of drugs", *Adv. Drug Deliver. Rev.* vol. 64, pp. 1579-1589, 2012

[11] W. Weheliye, M. Yianneskis, A. Ducci, "On the Fluid Dynamics of Shaken Bioreactors-Flow Characterization and Transition", *AIChE J.*, vol. 59, pp. 334-344, 2013

CONTACT

*Massimo Mastrangeli. Current Address: Bio, Electro And Mechanical Systems (BEAMS), École Polytechnique, Université Libre de Bruxelles, 1050 Bruxelles, Belgium.
tel.: + 32 2 650 47 66, massimo.mastrangeli@ulb.ac.be

3D MASK MODULES USING TWO-PHOTON DIRECT LASER WRITING TECHNOLOGY FOR CONTINUOUS LITHOGRAPHY PROCESS ON FIBERS

Masahiro Hayashi[1,2], Yi Zhang[1], Masanori Hayase[2], Toshihiro Itoh[1], and Ryutaro Maeda[1]

[1]National Institute of Advanced Industrial Science and Technology (AIST), Tsukuba, Japan
[2]Tokyo University of Science, Noda, Japan

ABSTRACT

In this paper, we report a new fabrication method of three dimensional (3D) mask modules using two-photon direct laser writing technology in 140 μm-diameter half-pipe structure on quartz substrates. For the first time, the two-photon direct laser writing technology is successfully utilized for high-resolution patterning process on a curved surface. The minimum feature sizes of about 2 μm are successfully fabricated in the 140 μm-diameter half-pipe structure. Using the new 3D mask modules, fine metal patterns are prepared on 125 μm-diameter fiber. The results demonstrated that the preparation of the 3D mask modules became more feasible by using the new method, and thus could be expected in the practical application of 3D photolithography process on fibers.

INTRODUCTION

Electronic textiles (E-textiles) and other flexible sheet-type electronic devices are attractive in flexible display, health-care, security, medical and life technology, energy and other applications. Textile fibers are cheap, easy-to-fabrication, and can be made of many materials including metals, ceramics and polymers. E-textiles can be prepared through embedding of conventional electronic devices and batteries into woven sheets, or directly integrating electronics into textile fibers. In order to realize more functionality and applications, it is necessary to directly integrate electronic, computational and other functionality into textile fibers. Therefore, there are strong academic and industry interests in developing high resolution micromachining technology on fiber substrates [1-4]. As a matter of facts, practical applications cannot be expected unless the aforementioned micromachining technology can be utilized for mass production because the fiber substrates are not compatible with batch processes.

In our previous works [5-7], we have suggested a 3D mask module technology for the high resolution micromachining process on the fibers. Figure 1 shows schematic of the 3D mask module technology. The fibers are coated with thin resist film and loaded into the 3D mask modules. The print gap between the fiber and the half-pipe structure is very narrow so that complicated optical system is not necessary. A stepping-forward mode is adapted so that a roll-to-roll process and mass production can be possible. In Fig. 1, lift-off process is used for the formation of metal patterns. A roll-to-roll 3D exposure system has been already introduced in our previous work [6]. As a result, mass production of fine patterns can be possible on the fibers in the stepping-forward mode. The key challenge is how to fabricate the 3D mask modules with high resolution.

The 3D mask module technology is involved of surface micromachining process on curved surfaces. Projection photolithography method is utilized in our previous work [5-6]. The state-of-the-art of projection lithography has the resolution of 6 ~ 7 μm and the depth of focus of about 50 ~ 70 μm. Fine patterns with the line and space of better than 10 μm have been successfully prepared in the half-pipe structures with the diameter of about 140 μm. However, textile fibers have verified shapes and dimensions so that the 3D mask modules would be of complicated shapes instead of half-circular ones. In addition, the cross-sectional sizes of the fibers are within from several tens to hundreds μm, the projection lithography technology cannot meet the requirements. New fabrication method is required for the 3D mask modules.

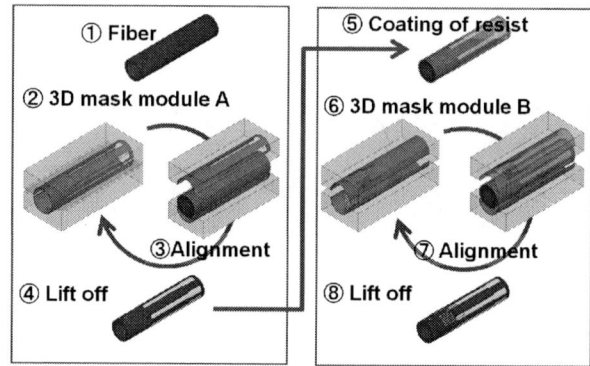

Fig. 1 Mass production of fine metal patterns on fibers using the 3D mask modules. Metal patterns are formed using lift-off process. Positive photoresist is used. Mass production can be possible in a stepping-forward mode using a roll-to-roll system.

Recently, two-photon laser writing technology has matured and been already commercialized [8-9]. Arbitrary 3D structures with the lateral resolution as better as 100 nm can be fabricated in a volume of $300 \times 300 \times 300$ μm and even larger. It is thus inspired us that the two-photon laser writing technology could be utilized for preparing finer patterns in the 3D mask modules. Unfortunately, there are few reports on its application on either the curved or high topography surfaces. In this work, we would develop a new fabrication method of the 3D mask modules using the two-photon laser writing technology.

EXPERIMENTAL

A commercial two-photon laser writing system

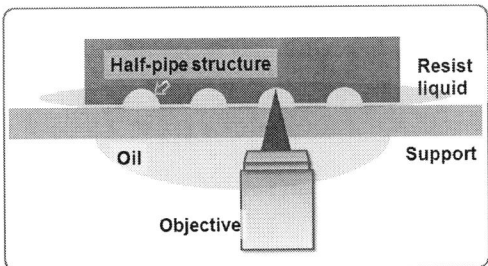

Figure 2: Schematic of writing set-up using the quartz substrate with half-pipe substrates in the two-photon laser direct writing system. Slide glass is used as support.

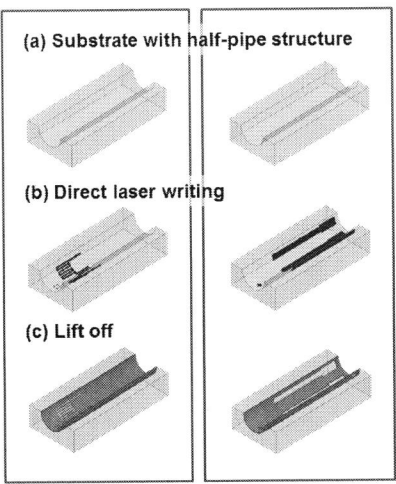

Figure 3: Fabrication sequence of the 3D mask modules. The half-pipe structures are about 140 µm in diameter.

(Photonic Professional PP-AI/HP-001, Nanoscribe GMbH) was used. Figure 2 is schematic of writing set-up using oil immersion lens (×100, NA=1.4). A 70 µm-thick slide glass was used as support. In order to examine the feasibility of the two-photon laser writing technology on curved surface, the metal patterns in Figure 1 would be prepared on 125 µm-diameter fiber in this work. The patterns are a typical MEMS structure that consists of wiring and sensing parts. The wiring structure was 20 µm wide and 250 µm long. The sensing structure has the minimum feature size of 2 µm, which is enough for most MEMS structures. Figure 3 is fabrication schematic of the 3D mask modules. Half-pipe structures with the radius of about 70 µm were prepared in quartz substrate using wet-etching process, which was described in more details in our previous report [7]. IP-L negative resist (Nanoscribe GMbH) was used in the two-photon laser writing process. Extra attentions were made during the preparation of the IP-L resist on the half-pipe substrate because the resist liquid was subject to air bubbles on curved surfaces. After the laser writing, the sample was developed in SU-8 developer. Then it was rinsed with isopropyl alcohol (IPA) and purified water. Thin Cr film of about 100 nm thick was sputtered and patterned by removing the photoresist layer. The prepared 3D mask modules were then loaded into the home-made 3D exposure system.

125 µm-diameter quartz fiber was coated with an about 3 µm-thick photoresist film (Shipley 1830, Shipley Company LLC) using a home-made sprayer coater [7]. Then the photoresist-coated fiber was exposed using the 3D mask modules and the 3D exposure system. After developed in MF-319 solution and rinsed with purified water, the photoresist patterns were annealed at 120°C for 20 min. The fiber sample was coated with 100 nm-thick Cr film and residual photoresist layer was removed using acetone.

RESULTS AND DISCUSSIONS

Figure 4 are SEM image of the prepared resist patterns in the half-pipe structures. The patterns are about 20 µm wide and 250 µm long. Figure 4 (a) shows that the resist patterns had peeled off although the substrate surface had been carefully modified. It is difficult to achieve good metal patterns because metal atoms can bypass the resist patterns during the sputtering process. The peel-off phenomenon might be related to residual stress in the photoresist patterns occurred during the laser writing, which would become more severe on the curved surface. In order to improve the adhesion force of the resist patterns to the curved surface, we have tried to divide the pattern into several parts but with thin connection structures. Figure 4 (b) shows the SEM image of the prepared resist patterns. It indicated that the thin connection structure released the residual stress and depressed the occurring of the peel-off defects.

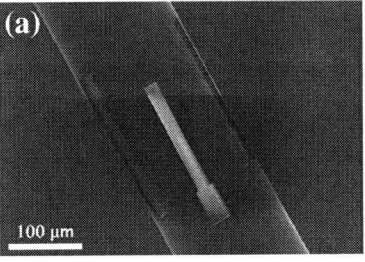

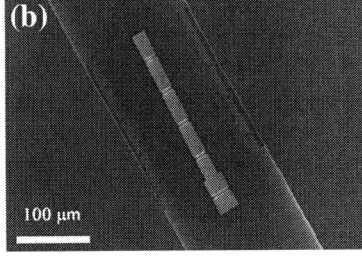

Figure 4: SEM image of IP-L photoresist patterns in the 140 µm-diameter half-pipe structure.

Figure 5 are optical photography of the prepared Cr patterns in the half-pipe structures. The photography was taken from the backside of the half-pipe structure. The prepared patterns are black under optical microscope. The mask patterns are well prepared as designed. The wiring structure was about 250 µm long and 20 µm wide. The alignment mark of the cross pattern was about 4 µm. Figure 6

978-1-4799-3510-9/14 $31.00 © 2014 IEEE

are corresponding SEM images of the prepared mask patterns in the half-pipe structure. The minimum feature sizes are about 2 µm. Some distortions are visible in the prepared patterns, particularly at the corners. It indicated that the drawing parameters should be further improved.

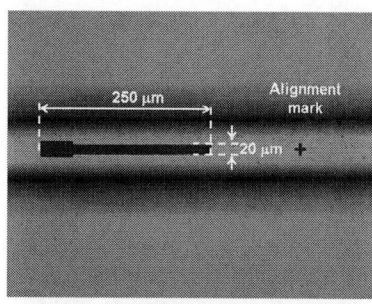

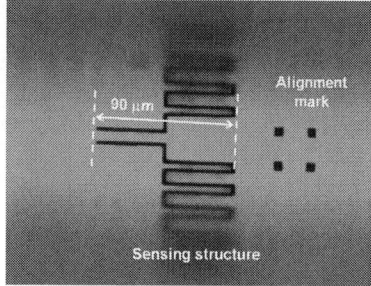

Figure 5: Optical photography of the prepared Cr film patterns in the half-pipe structures. The photography was taken from the back side of the 3D module so that the patterns are in black because the lights went through.

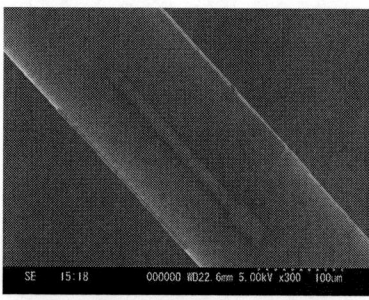

Figure 6: SEM image of the prepared Cr patterns in the half-pipe structure.

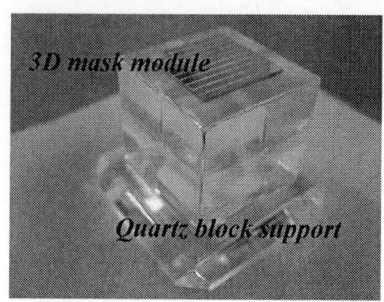

Figure 7: Optical photography of the prepared 3D mask module on quartz block support.

Figure 7 are photography of the prepared 3D mask modules on quartz block support. Then the module assemblies were loaded into the home-made exposure system for performing photolithography process on the 125 µm-diameter optical fibers. Figure 8 are optical images of the achieved resist patterns on the 125 µm-diameter optical fibers. It could be the seen that fine patterns have been formed on the 125 µm-diameter optical fibers. The alignment marks were also successfully prepared without noticeable distortions. Considering that the roll-to-roll 3D exposure system is not involved of complicated optical parts, it could be expected that the minimum feature size could be reduced to 4 µm and below after the improvement of the photolithography process. The patterns of 2 µm line and space were also achieved but there are visible distortions and other defects. It is mainly related to the mismatch between the optical fiber and the half-pipe structure, which was prepared using wet-etching method [7].

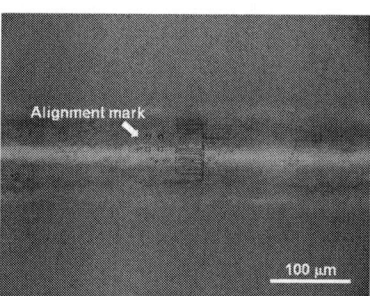

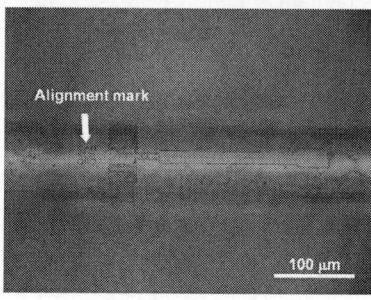

Figure 8: Optical Photography of the prepared resist patterns on the 125 µm-diameter fiber.

Figure 9 is photography of the prepared metal patterns by using lift-off process of Cr films. The achieved patterns had visible distortions. It is mainly related to that although a lift-off process had been utilized, there had been not any extra treatments on the photoresist film of the optical fibers. Although the prepared patterns are not satisfied relative to those matured technology, the minimum features of 2 ~ 4 µm are prepared on the 125 µm-diameter fiber for the first time by using the 3D mask module.

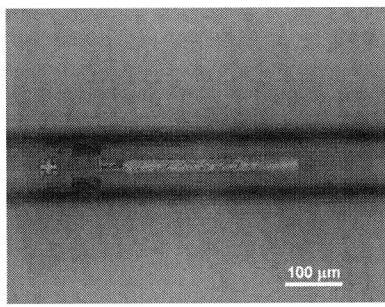

Figure 9: Optical photography of the prepared metal patterns on the 125 µm-diameter fiber.

CONCLUSIONS

In this work, two-photon laser direct writing technology was successfully utilized for the preparation of fine patterns on the half-pipe substrates. Preliminary knowledge and process of the two-photon laser writing technology have successfully established for the direct writing process on curved surfaces. The minimum feature sizes were about 2 µm. The photolithography process of the 125 µm-diameter fiber could be improved further through reducing the mismatch between the fibers and the half-pipe structures. Therefore, with the improvement of the half-pipe structures, i.e. to the reducing of size deviations, better resolution could be expected using the 3D mask modules. In addition, owing to that the two-photon laser writing technology has larger capabilities including larger writing depth and better resolutions, high resolution 3D mask modules can be also expected for the fibers with different shapes and dimensions.

REFERENCES

[1] http://www.beanspj.org/kyoten/macro/index.html.

[2] G. R. Fox, N. Setter, H. G. Limberger, "Fabrication and structural analysis of ZnO coated fiber optical phase modulators", *J. Mater. Res.* 11(8) 1996, pp. 2051-2061.

[3] T. Katoh, N. Nishi, M. Fukagawa, H. Ueno, S. Sugiyama, "Direct writing for three-dimensional microfabrication using synchrotron radiation etching", *Sensors and Actuators A 89 (1-2), 2001*, pp. 10-15.

[4] D. C. Abeysinghe, S. Dasgupta, H. E. Jackson, J. T. Boyd, "Novel MEMS pressure and temperature sensors fabricated on optical fibers", *J. Micromech. Microeng. 12, 2002*, pp. 229-235.

[5] Y. Zhang, J. Lu, A. Mimura, S. Matsumoto and T. Itoh, "MEMS-based exposure module for continuous lithography process on fiber substrate", *Proc. MEMS 2010*, pp. 380-383.

[6] Y. Zhang, J. Lu, A. Ohtomo, H. Mekaru and T. Itoh, "Continuous photolithography system and technology for fiber substrate", *Tech. Digest Transducers 2011*, pp. 370-373

[7] Y. Lu, Y. Zhang, J. Lu, A. Mimura, S. Matsumoto and T. Itoh, "Three-dimensional photolithography technology for a fiber substrate using a microfabricated exposure module", J. Micromech. Microeng, *20, 2010*, 125013 (10pp).

[8] M. Wegener, "3D photonic metamaterials and invisibility cloaks: the making of", *Proc. MEMS 2011*, pp. 1-4.

[9] M. Kurihara, Y. J. Heo, K. Kuribayashi-shigetomi and S. Takekuchi, "3D laser lithography combined with parylene coating for the rapid fabrication f 3D microstructure", *Proc. MEMS 2012*, pp. 196-199.

CONTACT

*Y. Zhang, Tel: +81-29-8617297; yi.zhang@aist.go.jp

FABRICATION OF A MONOLITHIC MICRODISCHARGE-BASED PRESSURE SENSOR FOR HARSH ENVIRONMENTS

Xin Luo, Christine K. Eun, and Yogesh B. Gianchandani
Center for Wireless Integrated MicroSensing and Systems (WIMS²)
University of Michigan, Ann Arbor, USA

ABSTRACT

This paper presents a 6-mask monolithic fabrication process for a pressure sensor that uses a differential microdischarge signal to sense diaphragm deflection. Microdischarge-based transduction is potentially advantageous for device miniaturization and harsh environments because of inherently large signals and immunity to temperature. This work reports a monolithic fabrication process that successfully addresses a number of challenges for microdischarge-based pressure sensors, such as three-dimensional (3D) electrical connection by electroplating laser-drilled through-glass vias (TGVs), and backside terminals for appropriate packages. The device has an exterior volume of 585×540×200 μm³ (0.05 mm³). Preliminary results show an estimated average sensitivity equivalent to 9,800 ppm/MPa over 0-40 MPa pressure range.

INTRODUCTION

A variety of microscale pressure sensing solutions have been explored in the past five decades, of which the most commonly used are piezoresistive and capacitive pressure sensors [1]. Although piezoresistive and capacitive pressure sensors with diaphragm diameter of ≈1 mm have been reported, further miniaturization has been a challenge for both approaches, but for different reasons. For piezoresistive pressure sensors, reducing the diaphragm diameter presents a challenge in localizing the resistor. If the resistor extends too far from the edge toward the center of the diaphragm, it loses signal due to stress averaging. Making the resistor smaller is a challenge as well. Smaller resistors demand more current to generate a measurable voltage, and are relatively imprecise, which affects calibration and yield. Capacitive pressure sensors present a scaling challenge because the capacitance decreases in proportion to the area of the diaphragm. This scaling puts the burden of detection on the interface circuit, which must not only be precise, but must also be located in the immediate vicinity of the sensor in order to prevent the signal from leaking into parasitic capacitance.

A new transduction principle based on microdischarges was recently reported for pressure sensing [2-3]. Microdischarge-based pressure sensors operate by measuring the change with pressure in spatial current distribution of a confined plasma. Microdischarges are localized glow discharge plasmas or arcs created in gaseous media, which, due to their size, have characteristics different from larger scale discharges [4]. Microdischarges can be used in a variety of microsensors, including micro total analysis systems that use optical emission spectroscopy for chemical sensing [5-6], radiation detectors, sputter ion pumps, etc. [7]. Devices incorporating microdischarges are very appealing for miniaturization. This is because the inherent signals are large compared to both piezoresistive and capacitive devices, eliminating the need for a proximal interface circuit and amplification. Additionally, microdischarge-based devices are suited for high temperature operation as electron temperatures are typically many eV (1eV=11,600K) and so are not significantly perturbed by a high ambient temperature [8]. Microdischarge-based pressure sensors have been reported for high temperature operation as high as 1000°C [3].

DEVICE CONCEPT

The new device structure is illustrated in Figure 1. It primarily consists of a glass substrate with copper filled through-glass vias (TGVs), a silicon diaphragm, one anode, and two competing cathodes. A microdischarge chamber is formed by the glass substrate, silicon diaphragm, and a Au-In eutectic bond ring. All three electrodes are made of thin-film Ni. The anode (*A*) and reference cathode (*K1*) are located on the glass side facing the microdischarge chamber, where the terminal contacts are located on the exterior surface of the glass substrate.

The sensing cathode (*K2*) is located on the silicon diaphragm, and is electrically connected to the exterior

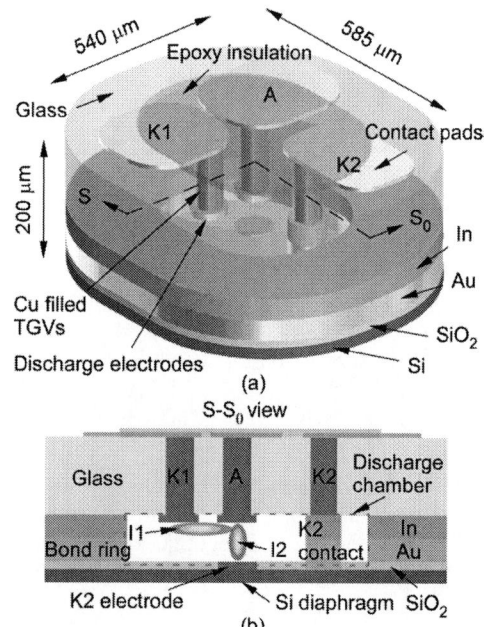

Figure 1: Concept of the microdischarge-based pressure sensor. (a) 3D model of the pressure sensor. (b) S-S₀ view. I1 and I2 are discharge currents from two paths.

978-1-4799-3510-9/14 $31.00 © 2014 IEEE

contact pad through a doped silicon layer and the *K2* contact, which is a sandwich of Au and In layers in the interior of the chamber that mates with a TGV. All the electrical connections from within the chamber are routed to the exterior of the glass substrate through copper filled TGVs.

In this three-electrode configuration, two discharge current paths *I1* and *I2* are established when high voltage (≈500V) pulses break down the chamber gas. As the diaphragm deflects due to external pressure, the spacing between the anode and the sensing cathode (*AK2*) decreases, but the spacing between the anode and the reference cathode (*AK1*) is essentially unaffected. This change of interelectrode spacing redistributes the currents: *I1* and *I2*. Differential current $(I1-I2)/(I1+I2)$, is used to indicate the external pressure. This ratio minimizes the contributions of common mode cross sensitivities, and pulse-to-pulse variation.

The exterior dimensions of the device are shown in Figure 1. The total volume of the sensor in this design is $585 \times 540 \times 200$ μm^3 (0.05 mm^3), whereas the discharge chamber is 2.2×10^{-4} mm^3.

FABRICATION

The fabrication process requires 6 masks: three for glass processing (Figure 2) and three for silicon processing (Figure 3). The glass wafer processing includes the laser-drilling of the TGVs, followed by the filling of the vias by copper electroplating. The next steps include the patterning of the contact pads on the exterior side of the wafer and the indium bond ring on the interior of the discharge chamber. Finally, the Ni electrodes are patterned on the interior side of the glass wafer. The silicon processing includes the deposition and patterning of an insulating oxide on the Si device layer of a silicon-on-insulator (SOI) wafer. This is followed by the patterning of the Au bond ring and *K2* electrode. Next, the glass and silicon chips are aligned and attached using a Au-In eutectic bonding method. Post-bonding, the Si diaphragm is released from the handle wafer by a deep reactive ion etching (DRIE) process, using the buried oxide layer as the etch stop.

Glass Wafer Processing

The glass wafer processing uses 300 μm-thick Schott Borofloat® glass wafers. In order to provide electrical contact from the pressure sensor electrodes to the contact pads, vias are drilled (Precision Microfab, Severna Park, MD) using a 193 nm ArF excimer laser. This machining process has a depth control of approximately ±5 μm, a lateral precision of 1-2 μm and a profile taper of 87°. The profile of fabricated TGVs is illustrated in Figure 4.

A variety of methods can be used for achieving an electrical connection through the glass vias, including thin-film deposition, solder reflow, and electroplating. The high aspect ratios of the TGV structures make it impractical to achieve sufficient sidewall coverage for reliable electrical contacts using thin film deposition. The use of solder particles yields limited success because of inconsistent reflow when heated to the melting temperature (183°C for

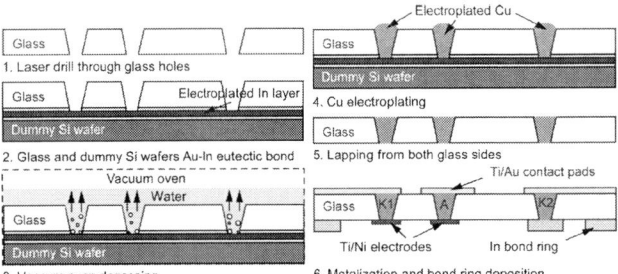

Figure 2: Process sequence for the glass wafer. 1) Through-holes are laser drilled. 2) Glass wafer is attached to a dummy Si wafer by eutectic bonding. 3) Degassing is performed in vacuum oven to remove trapped bubbles. 4) Through-holes are filled by Cu electroplating. 5) A lapping step removes excess Cu and the dummy wafer. 6) Deposition and patterning of Au contact pads, Ni electrodes and In bond ring.

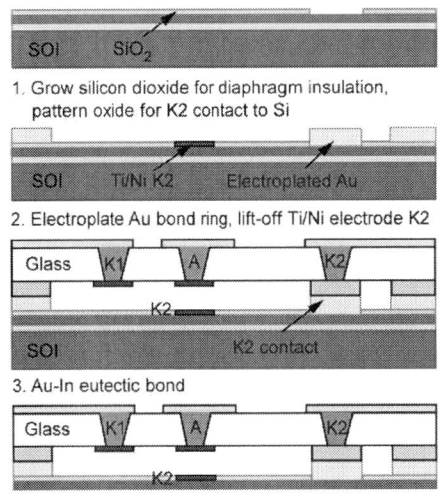

Figure 3: Process sequence for the Si wafer. 1) Oxide growth and patterning. 2) Plating Au bond ring and lift-off patterning of K2 electrodes. 3) Eutectic bonding between glass and silicon wafers to form the microdischarge chamber. 4) Handle wafer release by DRIE.

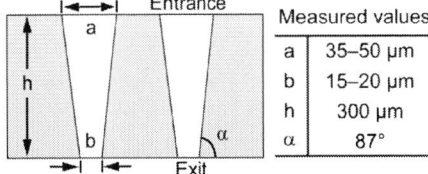

Figure 4: Profile of laser drilled TGVs.

37Pb/63Sn) and beyond (up to 280°C), which is possibly related to the large ratio of surface area to volume.

Electroplating provides the consistency and scalability for filling the TGVs. Although a variety of plating metals are available, In and Cu are attractive candidates for this application. Indium has a low reflow temperature (156°C), which allows temperature cycling post-plating in order to remove pinholes or voids. Copper offers lower resistivity and a higher plating rate. The higher re-melting temperature can also accommodate a higher operating temperature for

978-1-4799-3510-9/14 $31.00 © 2014 IEEE

the pressure sensor. Both metals were successfully plated in experiments.

Before electroplating, the glass wafer is attached to a dummy Si wafer coated with the appropriate metal seed layer (e.g. Ti/Au) for electroplating. Maintaining close contact and minimizing movement between the glass wafer and seed layer are critical. To ensure this, the dummy Si wafer is bonded to the glass wafer using Au-In eutectic bonding. Following a degassing step in vacuum oven (55°C, 30mins) to remove bubbles from the vias, Cu plating is performed (Enthone Cuprostar® CVF1) at 24°C. Pulse plating with periodic reversal of polarity is used to provide uniform plating across the TGVs on the wafer. The effective current density is 15–20 mA/cm^2. After the plating, the stacked structure is lapped from the front to remove excessive metal build-up and planarize the surface, and from the back to grind off the dummy Si wafer. The measured resistance of the TGV is <5 Ω. An scanning electron microscopy (SEM) image of plated TGVs is shown in Figure 5a.

The next processing steps involve patterning the Ti/Au (30 nm/300 nm) contact pads located on the exterior of the glass chip, followed by the patterning of the indium bond ring (4 μm-thick) and Ti/Ni (20 nm/200 nm) electrodes by lift-off as shown in Figure 5b.

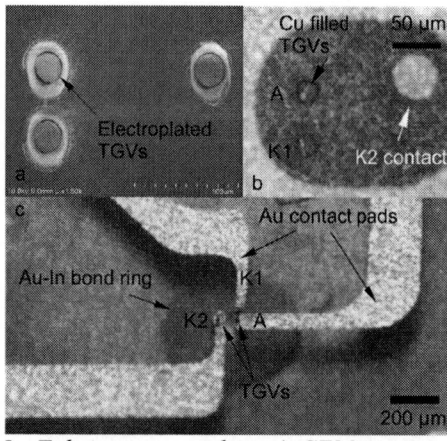

Figure 5: Fabrication results: a) SEM image of plated TGVs. b) Optical photos of discharge chamber and c) final assembled device.

Silicon Wafer Processing

The process utilizes SOI wafers with a 5 μm-thick device layer, a 2 μm-thick buried silicon dioxide layer, and a 500 μm-thick Si handle wafer. The Si device layer includes As doping for low resistivity (<0.005 Ω-cm). When the glass and Si wafers are bonded, the K2 contact on the Si wafer electrically connects to the K2 contact on the glass wafer. This connection is then routed through the TGV to the contact pad for K2. The buried oxide layer provides a well-defined etch stop, which can later facilitate the final diaphragm release of the Si device layer by a backside dry etch of the handle wafer.

Silicon dioxide is grown (100 nm-thick, by dry oxidation at 1000 °C) and then deposited (900 nm-thick low

temperature oxide) for a total thickness of 1 μm on the Si device layer to provide electrical isolation of the bond ring from the K2 electrode and contact. The oxide is then patterned to expose the doped device layer for the K2 contact. Next, an 8 μm-thick Au bond ring and K2 contact are electroplated. In a following step, the oxide is removed in the region of the K2 electrode, which is then formed by sputtering and lift-off of Ti/ Ni (20 nm/200 nm).

Eutectic Bonding

The transient liquid phase bonding technique [9] has been used to bond the fabricated glass and silicon wafers (Figure 3). The bonding was performed in a vacuum oven at 200 °C with an applied pressure >1 MPa for 90–120 mins. Figure 6a and 6b show the cross-section of the bond ring structure captured using SEM. Electron dispersive spectroscopy (EDS) was used to evaluate the composition of the bond ring (Figure 6c). The EDS shows diffusion of the Au and In layers that form the intermetallic compounds.

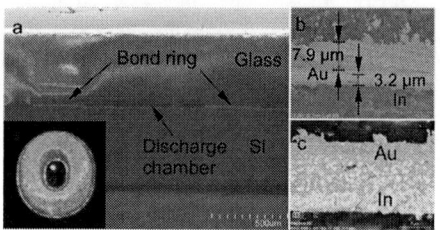

Figure 6: Bonding results: a) SEM image of bonding cross-section with inset showing the top view of a bonded ring, and b) details. c) Electro dispersive spectroscopy shows intermetallic compounds formation.

EXPERIMENTAL RESULTS

To demonstrate the pressure sensing and to evaluate the copper filled TGVs under high voltage conditions, preliminary experiments were conducted on an assembled device, in which the SOI wafer was thinned to 100 μm, but the handle wafer was not completely removed (Figure 5c). In the test structure, a glass chip and a SOI chip were held together with a porous epoxy bond instead of an eutectic bond. Tests were conducted in an argon environment. The experimental setup is illustrated in Figure 7. An external micromanipulator was used to apply a force at the center of the assembled chip to induce a diaphragm deflection, which emulates a large external pressure. Voltage pulses (480V, 1 ms) were applied to the anode. Multiple microdischarge pulses were produced during each voltage pulse. A ballast resistor 20 MΩ was utilized, while the currents going through two competing cathodes I1 and I2, were captured as voltages across 1 kΩ resistors.

A representative waveform of a single microdischarge pulse is shown in Figure 8. The typical duration is several hundred nanoseconds with decaying oscillation. Possible sources of parasitic capacitance, to which the oscillation may be attributed, include the oscilloscope probes connected to K1 and K2. When a voltage pulse is applied to the anode, it also charges the parasitic capacitance on the anode, which can potentially contribute to the peak transient discharge currents.

The relation between differential current and estimated equivalent pressure is plotted in Figure 9. The force applied to the diaphragm was converted to equivalent pressure using finite element analysis. In the experiments, *AK2* spacing was changed from 30 μm to 25 μm, and corresponding estimated pressure was 0–40 MPa. For a chamber pressure of 770 Torr, the operating voltage was 480 V and the differential currents ranged from -0.35 to -0.5, with an estimated average sensitivity equivalent to 9,800 ppm/MPa. It was observed that copper filled TGVs survive after hundreds of high voltage pulses without noticeable damage.

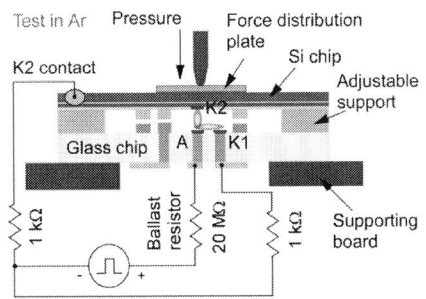

Figure 7: Test setup. A micro-manipulator applies force to deflect the Si chip and change the AK2 spacing.

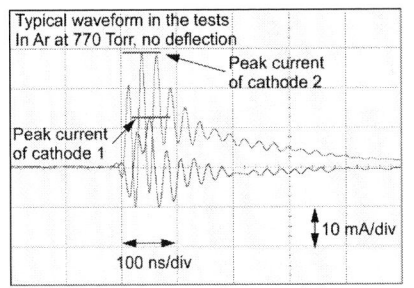

Figure 8: Representative waveforms of the microdischarges collected at the cathodes obtained during the tests.

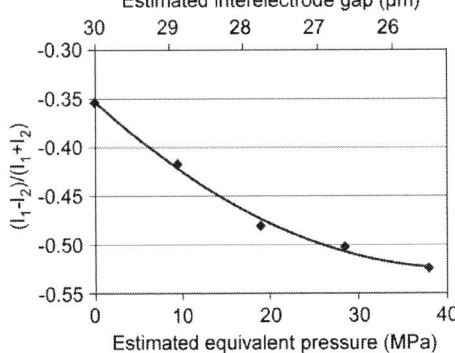

Figure 9: Differential currents from test results in response to equivalent external pressure and inter-electrode gap changes which are estimated from FEA based on a 5 μm thick Si diaphragm. I1 and I2 are peak values of envelope curves for AK1 and AK2 discharge waveforms, respectively. Every data point is the average of 5–8 measurements.

CONCLUSION

A 6-mask fabrication process was investigated for a microdischarge-based pressure sensor. A number of challenges, including 3D electrical connections using laser drilled and electroplated TGVs, are addressed. The differential cathodes arrangement that was investigated – with anode (*A*) and reference cathode (*K1*) on the glass substrate, and the sensing cathode (*K2*) on the diaphragm – was demonstrated to produce a differential output current that was a function of diaphragm deflection. The exterior volume of the device is $585 \times 540 \times 200 \ \mu m^3$ (0.05 mm^3).

ACKNOWLEDGEMENTS

This work is supported in part by a contract from the Advance Energy Consortium (AEC). Facilities used for this research include the Lurie Nanofabrication Facility (LNF), the University of Michigan, Ann Arbor.

REFERENCES

[1] Y.B. Gianchandani, C. Wilson, J.-S. Park, "Micromachined Pressure Sensors: Devices, Interface Circuits, and Performance Limits," *The MEMS Handbook*, ed: M. Gad-el-Hak, CRC Press, 2006.

[2] S.A. Wright, H.Z. Harvey, Y.B. Gianchandani, "A Microdischarge-Based Deflecting-Cathode Pressure Sensor in a Ceramic Package," *J. Microelectromech. Syst.,* 22, pp. 80-86, 2013.

[3] S.A. Wright and Y.B. Gianchandani, "Discharge-Based Pressure Sensors for High-Temperature Applications Using Three-Dimensional and Planar Microstructures," *J. Microelectromech. Syst.,* 18, pp. 736-743, June 2009.

[4] R. Foest, M. Schmidt, K. Becker, "Microplasmas, an emerging field of low-temperature plasma science and technology," *Int. J. Mass Spectrometry,* 248, pp. 87-102, Feb 15 2006.

[5] B. Mitra, Y.B. Gianchandani, "The detection of chemical vapors in air using optical emission spectroscopy of pulsed microdischarges from two- and three-electrode microstructures," *IEEE Sensors Journal,* 8, pp. 1445-1454, Jul-Aug 2008.

[6] C.G. Wilson, Y.B. Gianchandani, "Spectral detection of metal contaminants in water using an on-chip microglow discharge," *IEEE Transactions on Electron Devices,* 49, pp. 2317-2322, Dec 2002.

[7] C.K. Eun, Y.B. Gianchandani, "Microdischarge-Based Sensors and Actuators for Portable Microsystems: Selected Examples," *IEEE Journal of Quantum Electronics,* 48, pp. 814-826, Jun 2012.

[8] M.J. Kushner, "Modelling of microdischarge devices: plasma and gas dynamics," *Journal of Physics D-Applied Physics,* 38, pp. 1633-1643, Jun 7 2005.

[9] E.E. Aktakka, H. Kim, K. Najafi, "Wafer level fabrication of high performance MEMS using bonded and thinned bulk piezoelectric substrates," *IEEE Int. Conf. Solid-State Sensors, Actuators and Microsystems (Transducers),* pp. 849-852, 21-25 June 2009.

CONTACT

*X. Luo, tel:+1-734-7647428; xinluo@umich.edu

 Y. Gianchandani, tel:+1-734-6156407; yogesh@umich.edu

BIONIC SKINS USING FLEXIBLE ORGANIC DEVICES

Takao Someya[1,2] and Tsuyoshi Sekitani[1,2]

[1] Department of Electrical and Electronic Engineering and Information Systems, University of Tokyo, Tokyo, Japan.

[2] Exploratory Research for Advanced Technology (ERATO), Japan Science and Technology Agency (JST), Tokyo, Japan

ABSTRACT

We have fabricated ultrathin, ultra-lightweight, ultraflexible, organic devices, such as organic thin-film transistors (TFTs), organic photovoltaic (OPV) cells, and organic light-emitting diodes (OLEDs) on polymeric films with the thickness of only 1 μm. The ultrathin organic devices are utilized to fabricate human-machine interfaces such as a touch sensor and wearable electronic systems such as an electromyogram measurement sheet with a two-dimensional array of organic amplifiers. The transistor films exhibit extraordinarily tough mechanical robustness such as minimum bending radius of 5 μm for organic TFTs.

INTRODUCTION

Flexible and stretchable electronic devices [1] are expected to open up a new class of applications ranging from flexible displays [2], wearable sensors, flexible RFID [3], to flexible large-area sensors [4,5] and actuators [6]. As one of the promising applications of flexible and stretchable electronics, biomedical sensors [7-10] have attracted much attention recently. Sensors and electronic circuits for healthcare and medical applications have been fabricated using silicon and other rigid electronic materials. In order to minimize the discomfort of wearing rigid sensors, it is highly desirable to use soft electronic materials particularly for devices that come directly into contact with the skin and/or biological tissues [10]. In this regard, electronics manufactured on thin polymeric films are very attractive: in general, a thinner substrate provides better mechanical flexibility. However, directly manufacturing sensors or electronic circuits on ultrathin polymeric films with thicknesses of several micrometers or less is a difficult task when conventional semiconductor processes are used.

In this paper, we report on the recent progresses of ultrathin, ultra-lightweight, ultraflexible, organic devices, such as organic thin-film transistor (TFT) integrated circuits, organic photovoltaic (OPV) cells, and organic light-emitting diodes (OLEDs) on polymeric films with a thickness of only 1 μm. The ultrathin organic devices are used to fabricate human-machine interfaces such as touch sensors and wearable electronic systems such as an electromyogram (EMG) measurement sheet with a two-dimensional array of organic amplifiers. The transistor films exhibit extraordinarily tough mechanical robustness such as minimum bending radius of 10 μm or less. Ultrathin integrated circuits are applied to the surface EMG measurement sheet. In order to improve the performance uniformity of organic transistors for analog circuits, the feasibility of organic transistors with floating gate structures has been demonstrated.

ULTRATHIN ORGANIC TFT

We have manufactured organic TFTs on 1.2-μm-thick polyethylenenaphthalate (PEN) films (Fig. 1). Their weight is 3 g/m^2, which is approximately 1/30 of that of the standard weight office paper, and their total thickness is 2 μm, which is approximately 1/5 of that of plastic kitchen wrap. This was possible because of a novel technique to form a high-quality 19-nm-thick insulating layer on the rough surface of the 1.2-μm-thick polymeric film [10]. First, Al gate layers are deposited on the base film by the vacuum system with a shadow mask. The gate dielectric layers comprise anodic aluminum oxide and a phosphonic acid self-assembled monolayer (SAM) [11]. A 30-nm-thick dinaphtho[2,3-b:29,39-f]thieno[3,2-b]thiophene (DNTT) layer is deposited as the air-stable organic semiconductor by vacuum evaporation. Finally, Au layers are deposited as source/drain electrodes by vacuum evaporation. The channel length and width are 40 μm and 500 μm, respectively. The transistors exhibit a saturation mobility of 3 cm^2/ Vs (Fig. 2).

The organic transistor ICs exhibit extraordinary robustness in spite of being super-thin [10]. Indeed, the electrical properties and mechanical performance of the transistor ICs were practically unchanged, even when shrunk

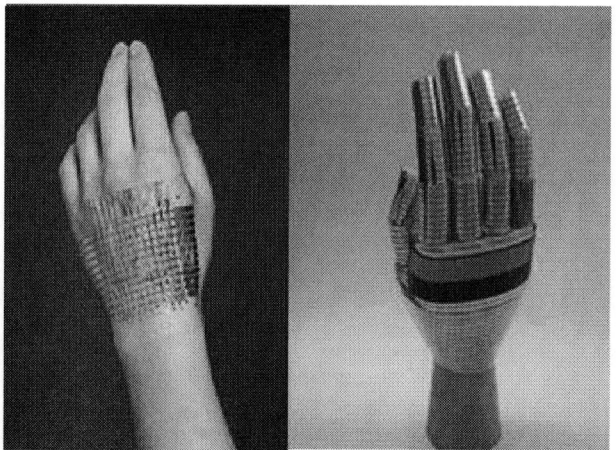

Figure 1: Images of the flexible organic transistor integrated circuits. Thanks to inherent softness of the organic materials, organic transistors are mechanically flexible and used to realize e-skins, which are applied on a robot hand (right) as a tactile sensor and human hand (left) as a healthcare sensor.

978-1-4799-3510-9/14 $31.00 © 2014 IEEE

to reduce the bending radius to 5 μm, crumpled like paper, or dropped from a height of a meter. Furthermore, the high durability was demonstrated in the study: after immersing the ICs in physiological saline (with components that are the same as in the bodily fluids or sweat) for more than two weeks, no obvious deterioration in the electrical properties was observed. Furthermore, the electric and mechanical performances of the organic transistor ICs were practically unchanged even when stretched by up to 233%.

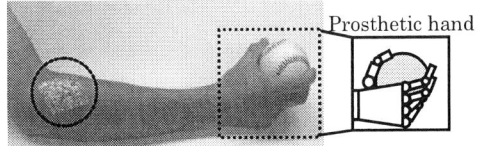

Figure 3: The 64 channel surface electromyogram (EMG) measurement sheet with 2V organic transistors is manufactured on a 1.2 μm thick polymeric film for prosthetic hand control.

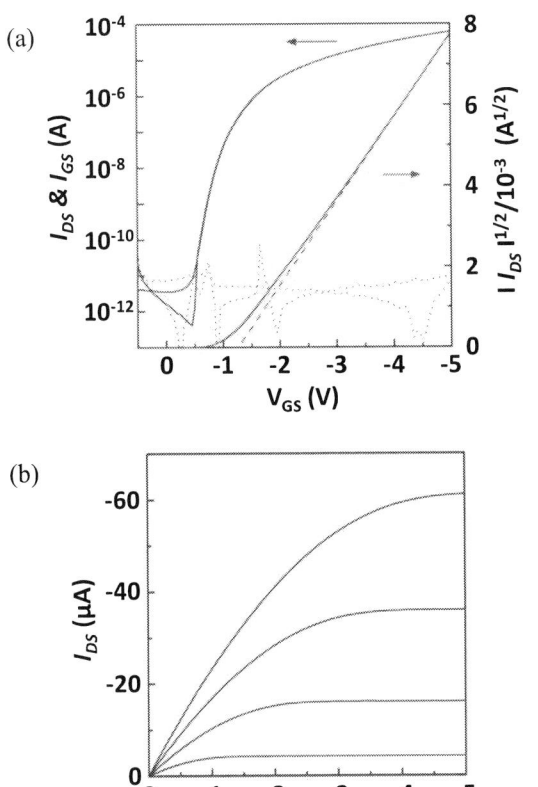

Figure 2: Organic transistors are manufactured on 1.2 μm thick polymeric films. The channel layer is formed with dinaphtho[2,3-b:2',3'-f]thieno[3,2-b]thiophene (DNTT) with channel length of 40 μm. The transistors exhibit a saturation mobility of 3 cm²/ Vs.

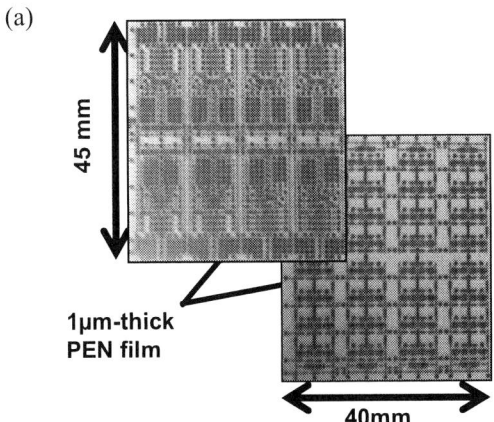

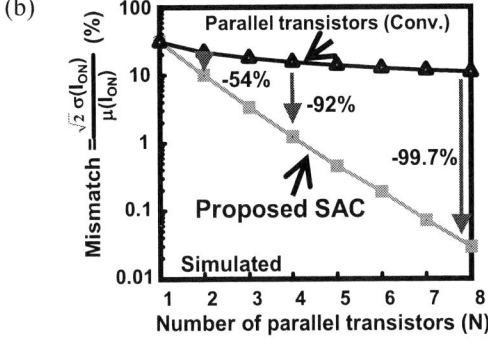

Figure 4: (a) The sheet with 8 × 2 amplifier array and the sheet with 8 × 8 EMG electrode array are laminated. (b) When the number of parallel transistors (redundancy) increases, the transistor mismatch and the power are reduced by 92% and 56%, respectively, compared with the conventional parallel transistors.

ELECTROMYOGRAM MEASUREMENT SHEET

In order to demonstrate the feasibility of ultrathin organic devices, we have manufactured 64-channel surface EMG measurement sheet with 2-V organic transistors on an ultraflexible PEN film for prosthetic hand control [12]. The distributed and shared amplifier architecture enables an in-situ amplification of the myoelectric signal with a four-time increase in the EMG electrode density. Postfabrication select-and-connect method reduces the transistor mismatch and the power by 92% and 56%, respectively, when compared with the conventional parallel transistors (Figs. 3 and 4).

ORGANIC AMPLIFIER SYSTEM

We present another approach using organic transistors with a floating gate structure to avoid the redundancy in circuit designs. In this development (Fig. 5), we have manufactured a large-area flexible strain-sensing system using a 2-D array of organic self-bias-feedback amplifier with a signal gain over 200 [13]. The amplifier system consists of three layers: a SAM capacitor matrix, a 2-D array of organic pseudo-CMOS inverters with a floating-gate structure using SAM gate dielectric [14], and an active matrix of organic thin-film transistors. The amplifier sheet comprises 8 × 8 amplifier cells. When the threshold voltage

978-1-4799-3510-9/14 $31.00 © 2014 IEEE

is controlled by accumulating appropriate charges in each floating gate, the distribution in the switching voltage of amplifier cells is suppressed from 400 to 20 mV (Fig. 6).

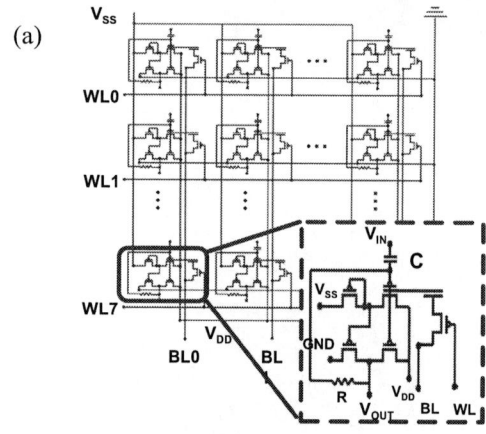

(a)

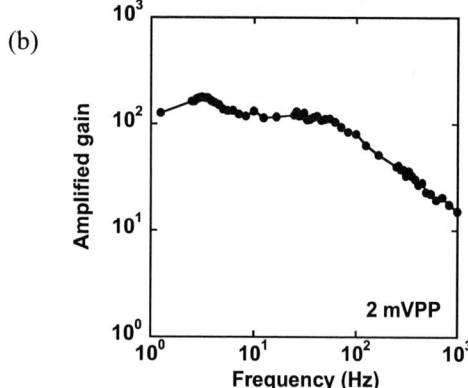

(b)

Figure 5: (a) Circuit diagram of an organic amplifier system consisting of 8 × 8 amplifier cells. Each amplifier cell has one capacitor, a pseudo-CMOS inverter with a floating-gate structure, and an organic transistor. (b) The gain exceeds 200 when the channel length is set to be 6 μm.

ULTRATHIN ORGANIC LED

We have succeeded in developing the world's lightest (3 g/m^2) and, simultaneously, the world's thinnest (2 μm) mechanically flexible ultrathin OLEDs (Fig. 7) [15]. We have demonstrated a very unique technique in forming organic semiconductor materials. The new ultraflexible OLED devices were manufactured on 1.4 μm polyethylene terephthalate (PET) foil substrates by combining highly flexible conducting polymers, semiconducting polymers, and thin metal layers.

The key to this success is the low temperature processing that enables OLEDs to be fabricated on rough 1.4-μm-thick polymeric film without damaging its mechanical properties. Instead of using indium tin oxide (ITO) transparent electrode that demands high temperature and high energy process, we have adopted poly(3,4-ethylenedioxythiophene):poly (styrenesulphonate) (PEDOT:PSS) formulation, as a

conductive polymer, which can be fabricated at a low temperature with a low-loss anode. A poly(p-phenylene-ethynylene)-alt-poly(p-phenylene-vinylene) (PPE-PPV) derivative with a statistical distribution of linear octyloxy and branched 2-ethylhexyloxy side groups (AnE-PVstat) is used as a semiconducting, red light-emitting polymer.

These ultrathin OLEDs can be operated as free-standing ultrathin films and allow crumpling during device operation. They are shown to be extremely flexible, with the radii of curvature below 10 μm, with display operational luminance of 100 cd/m^2. These LEDs could be as flexible as rubber, when OLEDs were pasted onto soft and stretchable rubber.

These soft, ultrathin and ultralight OLEDs can be very useful in various fields: for instance, they could become OLED lightings and OLED displays that may be applied to almost any surface and unique light sources for imperceptible healthcare sensors.

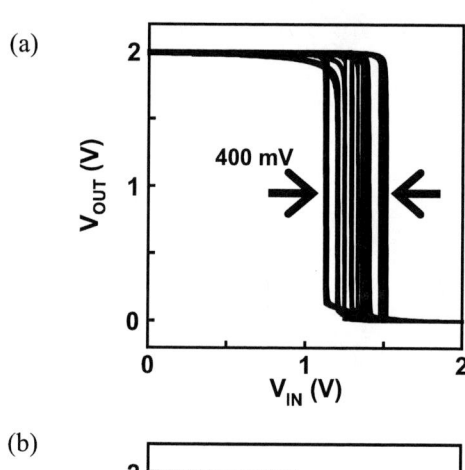

(a)

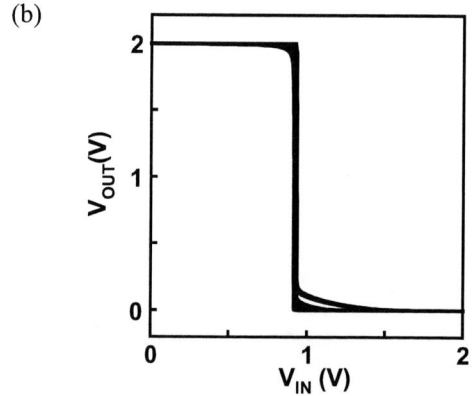

(b)

Figure 6: Performance variations of ten pseudo-CMOS inverters with floating gates. (a) Input–output characteristics of pseudo-CMOS inverters before programming. (b) Input–output characteristics of pseudo-CMOS inverters after programming. The variation in the switching voltage is reduced from 400 to 20 mV.

ULTRATHIN ORGANIC PHOTOVOLTAIC

We have demonstrated polymer-based photovoltaic devices on a plastic foil substrate less than 2-μm thick, with equal power conversion efficiency (4.2%) to glass-based counterparts (Fig. 8) [16]. We employ an ITO-free OPV architecture with high conductivity PEDOT:PSS as the transparent electrode, P3HT:PCBM as the photoactive layer, and Ca/Ag as the metal cathode. Our OPV devices constructed on 1.4-μm PET have a per-area mass of 4 g/m^2 and 4.2% efficiency, providing 10 W/g. This is the largest value reported for any solar cell technology (organic and inorganic) to date and among the highest in a broader picture of weight-specific power.

These ultrathin OPVs can reversibly withstand extreme mechanical deformation. Indeed, they can be wrapped firmly around a human hair without causing mechanical damages. Furthermore, they become stretchable when attached to a prestretched, highly optically transparent acrylic elastomer. Upon relaxation, an irregular structure of wrinkles and folds is formed, with the wrinkle wavelengths below 100 μm and bending radii as small as 10 μm.

The ultrathin OPV structures can be used as organic photodetectors (OPDs) and have the process compatibility with OTFTs and OLEDs; thereby, the integration of all the ultranthin organic devices, including OLEDs as light sources, OPDs as detectors, and OTFTs as readout circuits, will create a novel wearable photonic systems for heal-monitoring and many other applications.

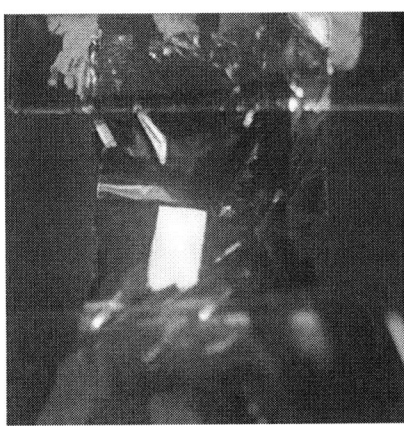

Figure 7: An ultrathin organic LED that exhibit excellent mechanical robustness.

CONCLUSIONS

Imperceptible electronics, namely, extremely thin, lightweight electronics whose presence cannot be perceived when worn, will open up a wide range of new applications in the fields ranging from healthcare and biomedicine to welfare. Many new applications will emerge including wearable healthcare sensor systems, stress-free input units for welfare machines such as smart wheelchairs, sensors for medical electronic equipment, and tough sensors for sports usage.

Figure8: A stretchable organic photovoltaic cell, which is manufactured on a 1.4-μm PET film and laminated with a rubber substrate.

ACKNOWLEDGEMENTS

A part of this research was carried out in collaboration with Professors Siegfried Bauer and Niyazi Serdar Sariciftci, Johannes Kepler Universität in Linz, Austria. We greatly appreciate our collaborators: Martin Kaltenbrunner, Matthew S. White, Jonathan Reeder, Tomoyuki Yokota, Kazunori Kuribara, Takeyoshi Tokuhara, Michael Drack, Reinhard Schwoediauer, Ingrid Graz, Simona Bauer-Gogonea, Eric D. Głowacki, Kateryna Gutnichenko, Gerald Kettlgruber, Safae Aazou, Christoph Ulbricht, Daniel A. M. Egbe, Matei C. Miron, Zoltan Major, and Markus C. Scharber.

REFERENCES

[1] J. A. Rogers, et al., Science, 327, 1603 (2010).
[2] J. A. Rogers, et al., PNAS, 98, 4835 (2001).
[3] E. Cantatore, IEEE Journal of Solid-State Circuits, 42, 84 (2007).
[4] T. Someya, et al., PNAS, 101, 9966 (2004)
[5] T. Someya, et al., PNAS, 102, 12321 (2005)
[6] Y. Kato, et al., IEEE Transactions on Electron Devices, 54, 202 (2007).
[7] D. H. Kim, et al. Science, 333, 838 (2011).
[8] J. Viventi, et al., Nature Neuroscience, 14, 1599 (2011).
[9] G. Schwartz, et al., Nature Communications, 4, 1859 (2013).
[10] M. Kaltenbrunner, et al., Nature, 499, 458 (2013).
[11] H. Klauk, et al., Nature, 445, 745 (2007).
[12] H. Fuketa, et al., IEEE ISSCC, #6.4 (2013).
[13] T. Yokota, et al., IEEE Transactions on Electron Devices, 59, 3434 (2012).
[14] T. Sekitani, et al., Science, 326, 1516 (2009).
[15] Matthew S. White, et al., Nature Photonics 7, 811 (2013).
[16] M. Kaltenbrunner, et al., Nature Communications 3, Article number: 770 (2013).

CONTACT

*T. Someya, tel: +81(Japan)-3-5841-0411; someya@ee.t.u-tokyo.ac.jp

DEVELOPMENT OF MICRO VARIABLE OPTICS ARRAY

Yongjoo Kwon[1], Yoonsun Choi[1], Kyuhwan Choi[1], Yunhee Kim[1], Seungyul Choi[2], Junghoon Lee[2], and Jungmok Bae[1]

[1]Samsung Advanced Institute of Technology, Samsung Electronics Co., Ltd., KOREA
[2]Nano/Micro Systems Laboratory, School of Mechanical and Aerospace Engineering
Seoul National University, Seoul KOREA

ABSTRACT

This research is on the development of a micro variable optics array which employs electrowetting as the working principle. The single pixel of the array has four separated electrodes and each of them is controlled independently giving the device multi-degree of freedom. The separated electrodes are fabricated using a thick photoresist and electroplating. Several formulas showing the relation among the radius of curvature, the prism angle, and electrowetting parameters are provided. The prism angles are measured to be ±30° and compared to the calculated values. The measurement of the radius of curvature is also presented showing that the various radiuses of curvature are achievable from concave to convex.

INTRODUCTION

As optical applications are diversifying, the requirement of the optical devices which are able to modulate a variety of lights properties is increasing. The modulation of the direction of lights is important in the optical applications such as a stereoscopic display with an eye-tracking scheme, integral photography, holography with a tracked viewing window technology, optical communication, and laser radar [1-4]. An early approach is using liquid crystals to steer beam with a sawtooth phase profile [1]. It showed limitation in steering angle and low efficiency by fringing field effects. There has been tries to use two immiscible liquids to form a continuous refractive interface and electrowetting on dielectric (EWOD) to control the interface efficiently in small scale. Berge and Peseux reported a millimeter-scaled liquid lens with variable focal length showing the reliable performance [5]. With the micro fabrication technology, a micro-sized variable prism array for display has been initiated by Hou and Heikenfeld [6]. The device has only two electrodes working independently in a pixel due to the difficulty in patterning the electrode on the vertical wall.

Here we propose the micro variable optics array with four separated electrodes. The great benefit of having the four separated electrodes is that the device works with multi-degree of freedom. In other words, the device can function as a variable optics array every individual pixel of which can have its own prism angle and focal length simultaneously. The relations of the prism angle and the radius of curvature to the liquid contact angle which is controlled by electrowetting on each wall are prepared. The fabrication process for the separation of the four electrodes is developed and a series of experimental results are provided.

WORKING PRINCIPLE

The unit pixel of the proposed optics array has cubic shape and consists of four separated electrodes on each wall and two immiscible and transparent liquids one of which is conductive. The schematic diagram of the single pixel is shown in figure 1. The interface of the two liquids is formed on the sidewall of the electrodes and controlled by electrowetting to have various curved or flat surfaces. Electrowetting is the phenomenon that a wetting property of a liquid on a solid surface is changed from its initial state by an external electric field. The wetting property is represented by contact angle of a liquid on a solid surface. Particularly when the phenomenon takes place on a dielectric solid layer (EWOD), the relationship among the contact angle and the external voltage is given as,

$$\cos\theta_{YL} = \cos\theta_Y + \frac{c}{2\gamma}V^2 \qquad (1)$$

where, θ_{YL} is the contact angle modified by the external electric potential, V, θ_Y the contact angle without external electric potential, c the capacitance of the unit area the dielectric layer makes, γ the surface tension of the liquid. Here, the electric potential is the difference of the potential across the dielectric layer and it makes no difference whether DC or AC as far as electrowetting is concerned. However, when DC is used, it should be considered that the surface charging and dielectric breakdown arise more easily than AC. On the other hand, in the usage of AC electric field, the conductive liquid with higher conductivity should be taken to avoid the considerable voltage drop in the liquid due to the AC current.

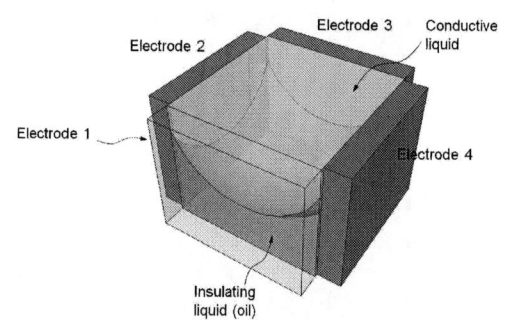

Figure 1: The single pixel of the optics array

The interface on the each wall of a unit pixel can have an independent contact angle according to the independent electrical potential. Depending on the one set of the four contact angles, the unit pixel can have a variety of interfacial shapes.

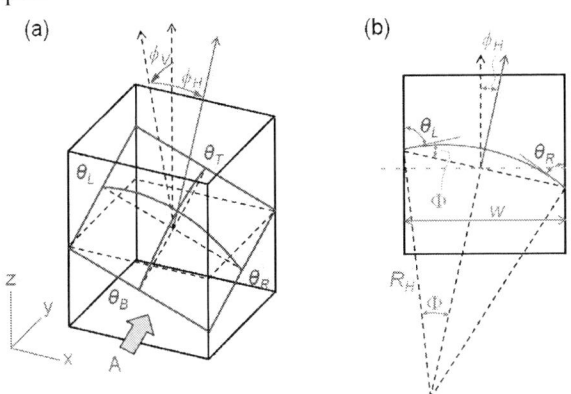

Figure 2: Geometry of the unit pixel. Parallel projection view of the unit pixel with arbitrary interfacial shape (a), Sectional view of the unit pixel at the view direction A (b).

Figure 2 (a) shows an arbitrary shape of the interface with four different contact angles on the each wall. The shape of the interface is regarded as the lens the axis of which is tilted to certain direction. Therefore, the geometrical feature of the shape can be represented by radius of curvature and prism angle. A pair of the facing electrodes shapes the interface in its direction dominantly and has less effect on the other perpendicular direction. Thus, the two directions can be dealt separately, i.e. horizontal and vertical directions. Figure 2 (b) shows the cross section of the geometry in the horizontal direction. Through the geometric reasoning, it is obvious that the two-dimensional shape for the section is determined by the two contact angles on the two facing walls. Likewise, for the vertical direction, the same inference is applicable. The relational expressions of the contact angles and the geometrical features are given

$$R_{H/V} = \frac{w}{\cos\theta_{L/B} + \cos\theta_{R/T}} \quad (2)$$

$$\phi_{H/V} = \frac{\theta_{L/B} - \theta_{R/T}}{2} \quad (3)$$

where, R is the radius of curvature, w the width of the pixel, θ the contact angle, ϕ the prism angle, and the subscript H represents the horizontal direction, V the vertical direction, L, R, B, and T the left, right, bottom, and top respectively. Merging the relation formulas, we obtain the formulas for the contact angle on the each wall in terms of the radius of curvature and the prism angle.

$$\theta_{L/R} = \cos^{-1}\left(\frac{w}{2R_H\cos(\phi_H)}\right) \pm \phi_H \quad (4)$$

$$\theta_{B/T} = \cos^{-1}\left(\frac{w}{2R_V\cos(\phi_V)}\right) \pm \phi_V \quad (5)$$

Given a radius of curvature and a prism angle, the contact angles on the four walls are determined by the equation (4) and (5), and consequently, the voltage each wall should have to make the contact angle is determined by the

equation (1) as well. Those geometrical features of the prism angle and the radius of curvature give the device the optical specifications of beam steering angle and focal length in combination with the refractive index of the materials.

FABRICATION

The schematic configuration of the micro variable optics array is shown in Figure 3. The four electrodes of each pixel comprise the four walls of the pixel, which are electrically separated. The size of the single pixel is 200 μm x 200 μm. In each pixel, two immiscible liquids, e.g. aqueous electrolyte and oil are filled. The interface of the two liquids is placed on the surface of the wall electrodes. The brief fabrication process flow appears in Figure 4. We formed the wall electrodes using electroplating with a mold patterned by using a thick photoresist, and the gaps in between the wall electrodes were filled with the thick photoresist again. For uniform coating of the photoresist on the surface with the vertical structure in micrometer scale, capillary force is used to avoid bubble trapping. Figure 5 shows the scanning electron microscope (SEM) images of the fabricated device. The four electrodes of the single pixel are well separated and the gaps are filled with the photoresist without defects such as air bubble trapped.

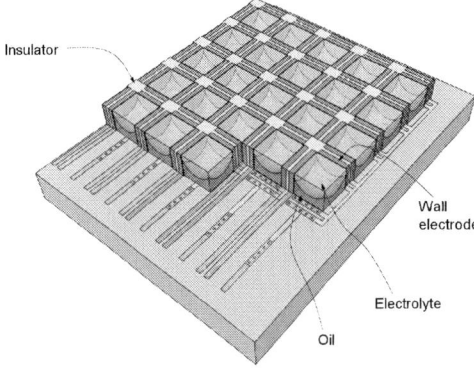

Figure 3: Schematic configuration of the micro variable optics array

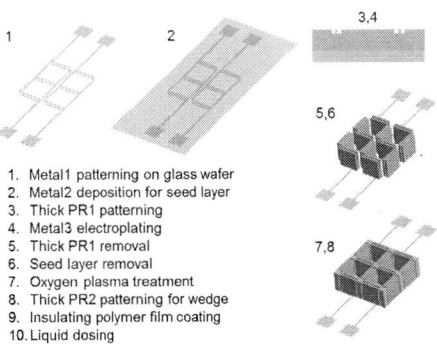

1. Metal1 patterning on glass wafer
2. Metal2 deposition for seed layer
3. Thick PR1 patterning
4. Metal3 electroplating
5. Thick PR1 removal
6. Seed layer removal
7. Oxygen plasma treatment
8. Thick PR2 patterning for wedge
9. Insulating polymer film coating
10. Liquid dosing

▒ Metal1 ▒ Metal2 ▒ Thick PR1 ▒ Metal3 ▒ Thick PR2

Figure 4: Fabrication process flow for the micro variable optics array.

Figure 5: SEM images of the fabricated device.

OPERATION AND MEASUREMENT

We demonstrated a couple of operation modes in the device by controlling the wall electrodes. First, prism mode is demonstrated using the 5x5 array of the pixels. The operation schematics are illustrated in the insets of Figure 6. Left, right, top, and bottom electrodes in all pixels were linked up respectively and the independent voltage sources are connected to each linkage. The results of the operations are shown in Figure 6. The interfacial shape of the liquids changes according to the combination of the voltage assigned to each electrode linkage to e.g. flat (b) or tilted (c, d, e, f, and g) shape while concave (a) shape without voltage application.

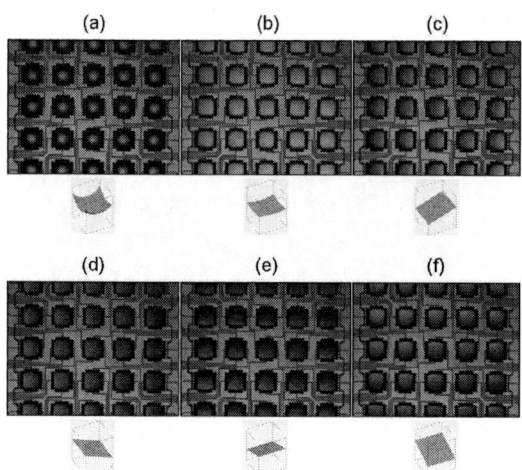

Figure 6: Movie clips of the 5x5 array pixels under working for flat(b) and tilted shape(c,d,e, and, f)

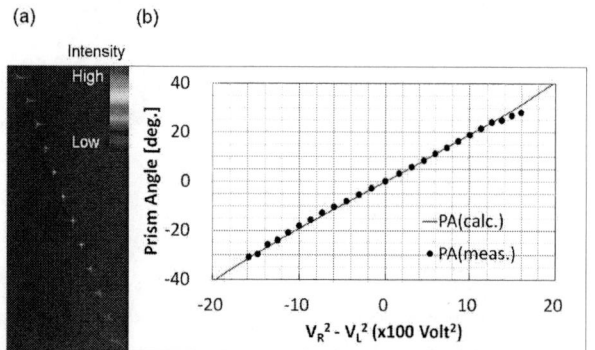

Figure 7: Beam intensity on each steered position (a), Beam steering angle according to the difference of the voltage-square between the facing electrodes (b)

The angle of the prism is determined by the voltage applied to the each wall electrode of the pixel. Using the equations given above, we reach another formula which shows the relation between the prism angle and the voltage each of the facing electrodes has. The relation is drawn as,

$$\sin\phi_{H/V} = \frac{c}{4\gamma}\left(V_{R/B}{}^2 - V_{L/T}{}^2\right) \qquad (6)$$

Here, it is noticeable that the prism angle is determined by the difference of the voltage-squares of the two facing electrodes rather than the voltages each electrode takes.

The prism angle is measured indirectly by measuring the profile of the laser beam passing through the single pixel and an optical system. The position of the detection is on the angular plan of the optical system and the change of the lateral position of the beam on that plan reflects the change of the direction of the beam steered by the device [7]. Figure 7 (a) shows the intensity profiles of the laser beam as a color map. The beam is steered according to the angle of the interface up to ±10° in horizontal direction. Assessing based on the angle of the beam, we obtain the prism angle of ±30°. The graph in figure 7 (b) shows the measured prism angles and the curve drawn by the calculation corresponding to the voltages that the two facing electrodes have. The measured prism angles are in a good agreement with the calculated one.

The device can also work in a lens mode with various radiuses of curvature according to the amplitude of the voltage applied. The relation between the radius of curvature and the voltage is, again, obtained using the formulas given above as,

$$R_{H/V} = \frac{w}{2\cos\theta_Y + \frac{c}{2\gamma}\left(V_{R/B}{}^2 + V_{L/T}{}^2\right)} \qquad (7)$$

Likewise, we note that the radius of curvature has a stronger relation to the summation of the voltage-squares of the two facing electrodes than the individual voltage.

We find the radius of curvature of the interface in a pixel by measuring the focal length. For the measurement, the distance that the device moves along with the optic axis to make the maximum peak of the beam intensity is taken for the focal length. Then the focal length is converted to the radius of curvature by calculation. The insets of figure 8 show the profiles of the beam which is, first, broadened by the lens effect of the single pixel and then focused with the movement of the device to the new position of focal point matching. The graph in figure 8 indicates the radiuses of curvature measured and calculated corresponding to the voltages applied to the walls. Because every parameter such as the refractive index of the glass for packaging is not taken into account of the calculation for simplification, the measured values diverge from the calculated one. However, it is worth noting that not only is the formula still valid with the strict consideration of all the parameters but the tendencies in both are in a good accordance. The radius of curvature for a certain point diverges to infinity. At this point, the fallowing relation is satisfied and the interface becomes

978-1-4799-3510-9/14 $31.00 © 2014 IEEE

flat.

$$V_{L/B}^2 + V_{R/T}^2 = -4\frac{\gamma}{c}\cos\theta_Y \quad (8)$$

On the other hand, when the voltage approaches to zero or goes to infinity, the radius of curvature gets closer to the half the aperture size, $w/2$.

We demonstrate further performance of the device using the independent voltage source for each electrode. Figure 9 (a) and (b) show Fresnel lens modes of the device. As shown in the figure, every single pixel with various prism angles makes together the concave (a) or convex (b) shapes of Fresnel lens. Figure 9 (c) and (d) also show the lens mode of the device with several radius of curvatures. In accordance with the amplitude of the voltage, the every single pixel has its own curvatures from concave to convex.

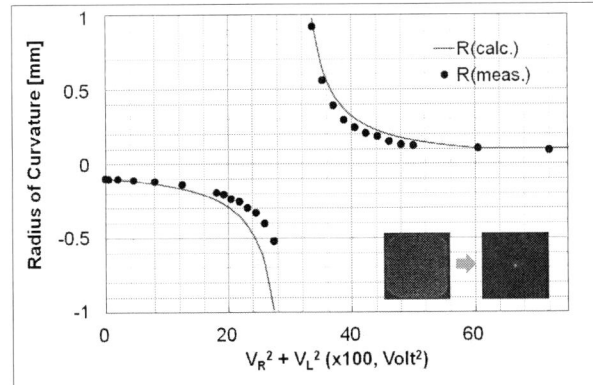

Figure 8: Radius of curvature plots according to the summation of the voltage-square between the facing electrodes for the calculation and measurement.

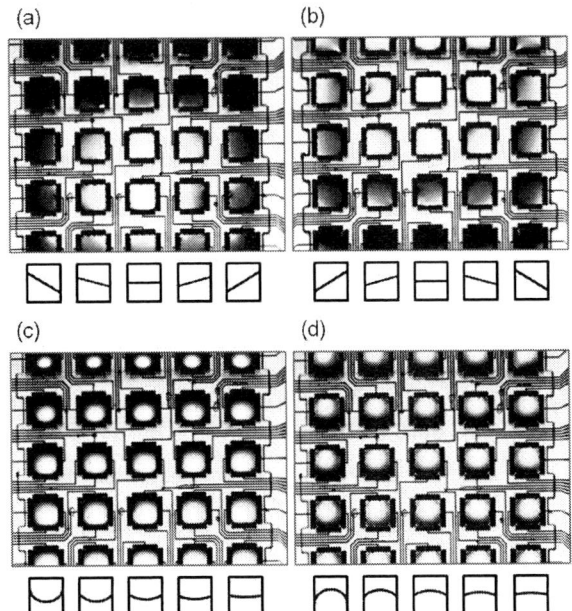

Figure 9: Results of the experiment with the independent voltage source for every single electrode.

CONCLUTION

We developed an electrowetting-based micro variable optics array with the fabrication process of thick-photoresist patterning and electroplating to make the separated vertical electrodes with the height in hundreds of micron scale for the cubic pixel. In order to characterize the single pixel of the array, a couple of formulas are derived from the consideration of the pixel geometry and electrowetting. The formulas clearly show the relations between the geometrical features, prism angle and radius of curvature which are the main parameters for optical application, and the required electrowetting parameters such as voltage and interfacial tension. Through the measurement of the prism angle the device modulates, it is shown that the prism angle of $\pm30°$ is achieved and the formula for the prism angle reflects the measured one. We also carried out the measurement of radius of curvature and showed that the various radiuses of curvature are achievable and the formula for the radius of curvature can be used to determine the electrowetting voltage for a given radius of curvature, especially even flat. Finally, we made demonstrations of the device in Fresnel mode and lens mode with various radiuses of curvature showing the versatility of the device in the modulation of light direction.

REFERENCES

[1] McManamon, Paul F., et al. "A review of phased array steering for narrow-band electrooptical systems", *Proceedings of the IEEE* Vol. 97, No. 6 (2009): 1078-1096.

[2] Shi, Lei, Paul F., et al., "Liquid crystal optical phase plate with a variable in-plane gradient", *Journal of Applied Physics* 104.3 (2008): 033109-033109.

[3] Reichelt, Stephan, et al., "Large holographic 3D displays for tomorrow's TV and monitors-solutions, challenges, and prospects" *IEEE Lasers and Electro-Optics Society*, 2008. 21st Annual Meeting of the. IEEE, 2008.

[4] Okano, Fumio, et al., "Three-dimensional video system based on integral photography" *Optical Engineering* 38.6 (1999): 1072-1077.

[5] B. Berge and J. Peseux, "Variable focal lens controlled by an external voltage: An application of electrowetting", *Eur. Phys. J. E.* 3, 159-163 (2000).

[6] Hou, L., et al., "A full description of a scalable microfabrication process for arrayed electrowetting microprisms" *Journal of Micromechanics and Microengineering*, 20(1), 015044.

[7] Kim, Yunhee, et al., "Measurement of the optical characteristics of electrowetting prism array for three-dimensional display", *Proceedings of the SPIE*, Volume 8643, article id. 864305, 6 pp. (2013)

MINIATURIZED MAGNETOELASTIC TAGS USING FRAME-SUSPENDED HEXAGONAL RESONATORS

Jun Tang, Scott R. Green, and Yogesh B. Gianchandani
Center for Wireless Integrated MicroSensing and Systems (WIMS²), University of Michigan,
Ann Arbor, Michigan, USA

ABSTRACT

Magnetoelastic tags – also referred to as acousto-magnetic or magnetomechanical tags – are used in wireless detection systems that electromagnetically query the resonant response of the tags. This paper presents miniaturized magnetoelastic tags using hexagonal resonators with an overall size of about ø1.3 mm X 27 µm and a resonant frequency as high as 2.13 MHz. The tags are 100X smaller than typical commercial tags. A unique feature is the frame-suspension, which results in ≈75X improvement in signal amplitude compared to that of non-suspended disc tags with similar size and frequency. This paper also demonstrates that the signal amplitude can be boosted by utilizing signal superposition of an ensemble of tags.

INTRODUCTION

Magnetoelastic resonators have been widely utilized in electronic article surveillance (EAS) systems used in the retail sector for theft prevention. The coupling between an applied magnetic field and the resulting resonant response due to magnetoelastic material strain allows wireless interrogating and sensing [1-2]. Further, magnetoelastic resonators typically operate at a specific frequency, which limits interference from spurious sources. Magnetoelastic resonators are also passive devices – no power sources or circuits are required with the tags – so that with proper packaging, they can be easily adapted for applications in a variety of environments [3-4]. These advantages – wireless operation, signal isolation, and passivity – make magnetoelastic resonators very attractive for remote detection applications, such as wireless detection of a blockage or leakage in piping systems or sophisticated machines, tagging of surgical instruments, and implantable sensors.

Many applications would benefit from miniaturization of magnetoelastic tags. For example, a smaller tag would be less conspicuous for anti-theft systems. A number of medical applications can be envisioned as well. Despite progress in increasing signal strength and detection range, miniaturization remains a challenge due to the signal loss resulting from smaller resonator size and due to higher sensitivity to fabrication tolerances [5]. Signal strength is related to the effective volume of the magnetoelastic material. The typical commercial magnetoelastic tags operating at 58 kHz are about 38 mm long, 12.7 mm or 6 mm wide and 27 µm thick [6]. Smaller tags operating at 120 kHz, with adequate signal strength for commercial use, still have a length of about 20 mm and a width of 6 mm [7]. These magnetoelastic tags are usually strips or ribbons and the length:width ratio is normally larger than 3:1. As length is reduced while maintaining a large aspect ratio, the width diminishes below hundreds of microns, which reduces signal strength.

This paper addresses some of the challenges in miniaturization of magnetoelastic tags. It presents ≈1 mm diameter magnetoelastic tags with frame-suspensions that allow strong signal amplitude for such small tags. Signal superposition for arrayed hexagonal tags is also demonstrated.

DESIGN AND MODELING

Concept

A typical magnetoelastic tag detection system includes a transmit coil, a receive coil, magnetoelastic tags and DC bias magnets (Fig. 1). When biased by a DC magnetic field and excited by an AC magnetic field, the magnetoelastic tags resonate and generate a magnetic flux that can be detected by the receive coil. The AC magnetic field is generated by the transmit coil and the DC bias can be provided by magnets packaged with the tags.

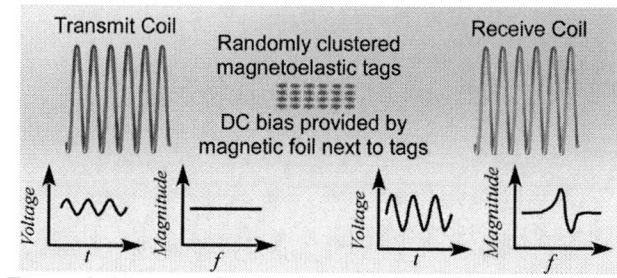

Figure 1: Magnetoelastic tags resonate under an applied AC magnetic field generated by a transmit coil; simultaneously, resonating tags produce a magnetic flux. This, along with the transmit signal, is detected by a receive coil. The resonant response indicates the presence of the tags.

Modeling

A custom magneto-mechanical harmonic finite element technique [8] is used to estimate displacements, mode shapes, and resonant frequencies for the magnetoelastic tags. Although magnetoelastic materials are generally non-linear, it is appropriate to use linearized constitutive equations describing the coupling between flux, field strength, stress, and strain in a magnetostrictive material:

$$\vec{\sigma} = [C]\vec{\varepsilon} - \frac{[C][d]^T}{\mu_0\mu_r}\vec{B} \tag{1}$$

$$\vec{H} = -\frac{[d][C]}{\mu_0\mu_r}\vec{\varepsilon} + \frac{1}{\mu_0\mu_r}\vec{B} \tag{2}$$

where σ is the stress vector, C is the stiffness matrix, ε is the strain, d is the magnetostrictivity matrix, B is the magnetic flux density vector, H is the field strength vector, μ_0 is the permeability of free space, and μ_r is the relative permeability. The operating point around which the equations are linearized is set by the DC biasing field. Equations (1) and (2) are implemented in this work utilizing COMSOL Multiphysics and coupled modes for time-harmonic induction current and stress-strain frequency response. Details of the finite element analysis (FEA) implementation for magnetostrictive materials are presented in [9].

Figure 2 shows the mode shape and displacement of a hexagonal tag at resonant frequency of 2.09 MHz. The desired mode shape, which exhibits both longitudinal and transverse motion, generates an oscillating magnetic field with one major response component that is orthogonal to the

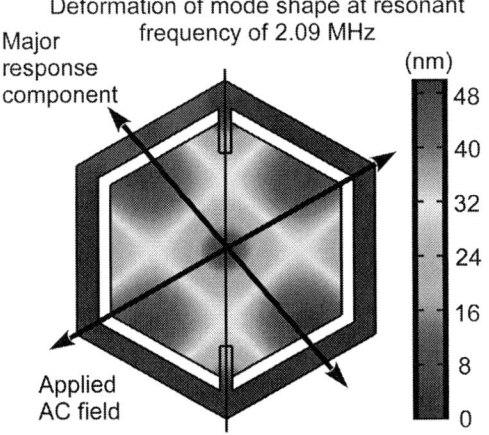

Figure 2: FEA simulation was carried out utilizing a custom magneto-mechanical coupled model in COMSOL. The results show a hexagonal resonator has a resonant frequency of 2.09 MHz for in-plane mode shape.

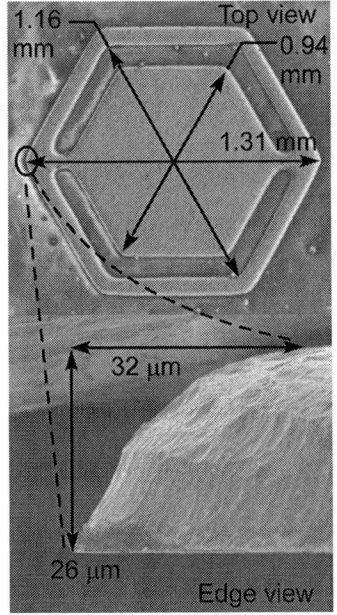

Figure 3: SEM images of the geometry and sidewall of hexagonal tags fabricated utilizing PCM.

transmitted magnetic field, facilitating decoupling of the transmit signal from the receive signal by placing the transmit coil and receive coil orthogonally.

FABRICATION

Magnetoelastic tags are batch patterned from a ≈27 μm thick foil of as-cast Metglas™ 2826MB, an amorphous NiFeMoB alloy [10], utilizing a "tabless" photochemical machining (PCM) process [11]. Normally, PCM fabricated devices have tabs, which keep the devices connected to the foil after etching. However, the "tabless" process is utilized in this work because it allows hundreds of tags to "drop" from the Metglas™ foil automatically during the etching process, eliminating the extra time, cost, and geometrical variability resulting from an additional tab cutting process. As shown in Fig. 3, the lateral undercut for sidewalls of a hexagonal tag is 32 μm, which is small compared to the size of the tag, so predictability and consistency in the resonant frequency across a batch of tags is expected.

EXPERIMENTAL

Experimental methods

In this work, an experimental configuration included a network analyzer, an RF amplifier, DC Helmholtz coils, a transmit coil and a receive coil (Fig. 4.). The transmit coil and the receive coil were configured orthogonally. The symmetry of the resonators and the combined longitudinal and transverse motion of the desired mode shape allow the oscillating magnetic field developed in response to the interrogation to be orthogonal to the direction of the transmitted oscillating magnetic field. This arrangement of coils and the symmetrical design of resonators contribute to decoupling of the transmit signal from the received signal, reducing the signal feedthrough and revealing the resonant response of the tags. For all data presented here, the baseline signal feedthrough (measured without tags present) has been subtracted.

Single tag

The measured signal amplitude of a frame-suspended hexagonal tag is 75X that of a disc-shaped tag without a suspension that was measured for comparison. The typical resonant response of a frame-suspended hexagonal tag

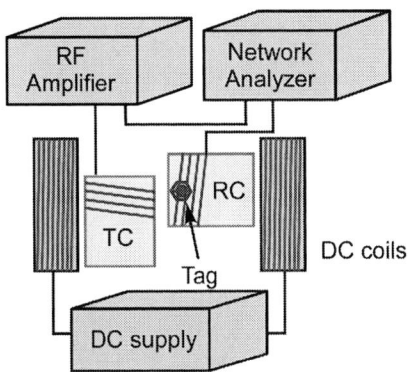

Figure 4: In preliminary experiments, a transmit coil and a receive coil are placed next to each other orthogonally. A DC magnetic bias is provided by DC coils.

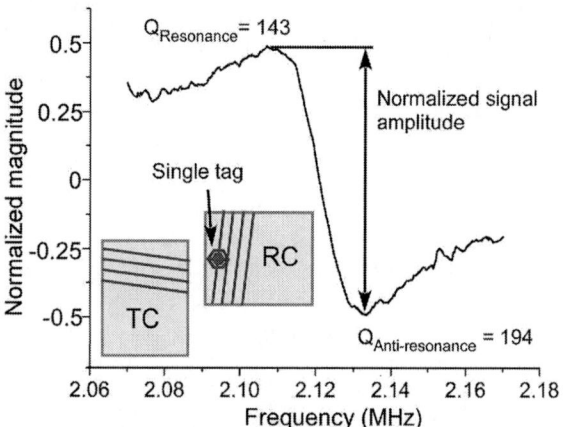

Figure 5: Typical frequency response of a single tag with strong signal and high quality factor.

shows a quality factor as high as 100-200 (Fig. 5). Since the signal amplitudes of tags vary dramatically with different experimental setups and the measuring conditions, signal amplitudes in this paper are normalized to the typical signal amplitude of a single frame-suspended hexagonal tag with a

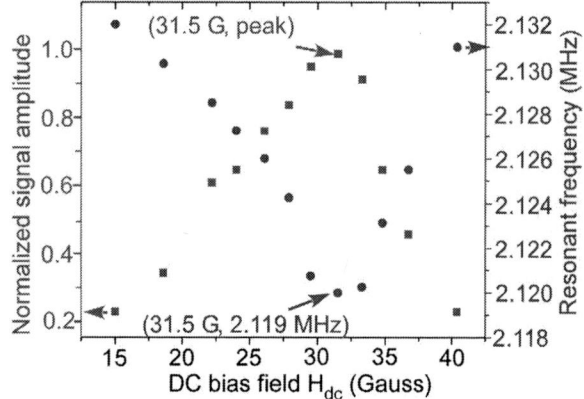

Figure 6: Resonant frequency and signal amplitude as a function of DC bias.

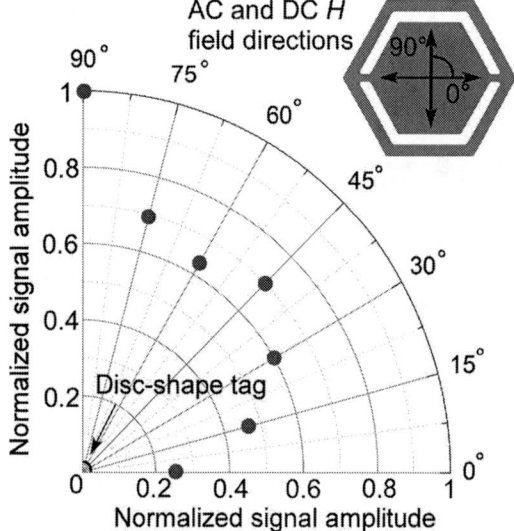

Figure 7: Signal amplitude as a function of orientations of applied AC magnetic field.

preferred DC bias.

The DC bias field shifts the operating point of the material to where the strain is most sensitive to the applied AC magnetic field. Figure 6 shows the measured signal amplitude and resonant frequency of a hexagonal tag as a function of DC bias. The signal amplitude reaches a maximum and the resonant frequency reaches a minimum when a preferred 31.5 Oe DC bias is applied.

Compared to azimuthally symmetric disc-shaped tags, hexagonal tags with frame-suspensions might exhibit variation of signal amplitudes with different azimuthal orientations of the AC magnetic field. The signal amplitude as a function of orientation of applied AC magnetic field is plotted in Fig. 7. Although signal amplitude does indeed exhibit some variation with different interrogating orientations, the signal amplitude is larger in every orientation compared to that of disc tags without a frame-suspension.

Signal superposition

In order to get insight into the concept of signal superposition, small quantities of tags (up to 10) were

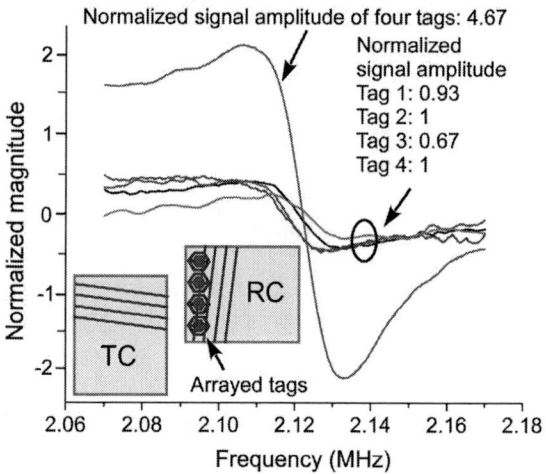

Figure 8: Frequency response of a collection of four tags and that of each individual tag, demonstrating signal superposition.

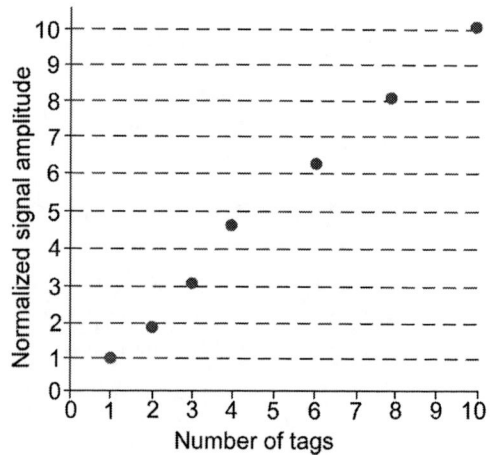

Figure 9: Signal amplitude as a function of number of tags, demonstrating signal superposition for at least 10 tags.

placed in an array, with each tag in the orientation that provided the largest signal amplitude. Signal superposition was demonstrated with this approach, for example by an array of 4 tags (Fig. 8). Although there are small variations of resonant frequencies of these tags, the cumulative resonant response for four tags has a standard response that is similar in form to that of a single tag. The overall normalized signal amplitude from four tags is 4.67, which is larger than the sum of the normalized signal amplitude from each tag (0.93, 1, 0.67 and 1).

As shown in Fig. 9, the signal strength increases linearly with the number of arrayed tags for modest counts. The experimental results of frame-suspended hexagonal tags are compared with those of disc-shaped tags (without frame suspensions) of similar size and resonant frequency, indicating approximately $\approx 75X$ improvement in signal amplitudes in Table I.

Table 1: Comparison of resonant response of hexagon and disc magnetoelastic tags, indicating $\approx 75X$ improvement.

PCM'D TAG SHAPE	RES. FREQ.	NORMALIZED SIGNAL AMPLITUDE	
Hexagon	2.13 MHz	1 tag	4 tags
		75X	350X
Disc (no suspension)	2.18 MHz	1 tag	4 tags
		1X	4.5X
Improvement		75X	78X

CONCLUSION

This work demonstrates the potential utility of miniaturized magnetoelastic frame-suspended hexagonal tags. The frame-suspension of these tags contributes to the significant improvement of signal amplitude. Tags with overall size of about ø1.3 mm X 27 μm are approximately 100X smaller than commercial tags. The preferred DC field bias for these tags is ≈ 31.5 Oe. At this bias value, the frame-suspension of hexagonal tags provides $\approx 75X$ improvement in signal amplitude compared to that of non-suspended disc tags with similar size and frequency. The typical frequency response of a hexagonal tag shows quality factors of 100-200. The signal amplitude of a hexagonal tag is a function of the orientation of the applied AC magnetic field. For 1-10 tags, the signal amplitudes of arrayed tags are at least the sum of the amplitude of each tag.

The applications of magnetoelastic tags are not limited to EAS systems. Magnetoelastic tags of such a small scale are very attractive for applications like surgical instrument tagging, or implantable sensors.

ACKNOWLEDGEMENTS

The authors acknowledge Metglas, Inc. for the foil samples provided for this project. This work was supported in part by a contract from a corporation.

REFERENCES

[1] Calkins, A.B. Flatau, M.J. Dapin, "Overview of magnetostrictive sensor technology," *Journal of Intelligent Material Systems and Structures*, vol. 18, pp. 1057-1066, Oct 2007

[2] A.S. Belyakov, "Magnetoelastic sensors and geophones for vector measurements in geoacoustics," *Acoustical Physics*, vol. 51, pp. S43-S53, 2005.

[3] S.R. Green, Y.B. Gianchandani, "Tailored magnetoelastic sensor geometry for advanced functionality in wireless biliary stent monitoring systems," *J. Micromechanics and Microengineering*, vol. 20, pp. 75040-75053, Jul 2010.

[4] E.L. Tan, B.D. Pereles, B. Horton, R. Shao, M. Zourob, K.G. Ong, "Implantable Biosensors for Real-time Strain and Pressure Monitoring," *Sensors*, vol. 8, pp. 6396-6406, Oct 2008.

[5] G. Herzer, "Magneto-Acoutic Marker for Electronic Article Surveillance Having Reduced Size and High Signal Amplitude," U.S. Patent, 6359563 B1, Mar 19, 2002.

[6] P.M. Anderson, G.R. Bretts, J.E. Kearney, "Surveillance System Having Magnetomechanical Marker," U.S. Patent 4510489, Apr 9, 1985.

[7] P.M. Anderson, G.E. Fish, "Miniature Magnetomechanical Marker for Electronic Article Surveillance System," U.S. Patent, 7075440 B2, Jul 11, 2006.

[8] S.R. Green and Y.B. Gianchandani, "Wireless Magnetoelastic Monitoring of Biliary Stents," *J. Microelectromechanical Systems*, vol. 18, pp. 64-78, Feb 2009.

[9] J. Benatar, *FEM Implementations of Magnetostrictive-Based Applications*, MS Thesis, Univ. of Maryland, 2005

[10] Metglas, Inc. *Magnetic Alloy 2826MB (nickel-based) Technical Bulletin.* [Online]. Available: http://www.metglas.com.

[11] *ASM Handbook*, 16, ASM International, 1989.

CONTACT

J.Tang, juntang@umich.edu
S.R. Green, greensr@umich.edu
Y.B. Gianchandani, yogesh@umich.edu

SINGLE-STRUCTURE 3-AXIS LORENTZ FORCE MAGNETOMETER WITH SUB-30 nT/√HZ RESOLUTION

Mo Li[1], Eldwin J. Ng[2], Vu A. Hong[2], Chae H. Ahn[2], Yushi Yang[2],
Thomas W. Kenny[2] and David A. Horsley[1]
[1]University of California, Davis, USA
[2]Stanford University, USA

ABSTRACT

This work demonstrates a 3-axis Lorentz force magnetometer for electronic compass purposes. The magnetometer measures magnetic flux in 3 axes using a single structure. With 1 mW power consumption, the sensor achieves sub-30 nT/√Hz resolution in each of the 3 axes. Compared to the 3-axis Hall sensors currently used in smartphones, the 3-axis magnetometer shown here has the advantages of 10× lower noise floor and the ability to be co-fabricated with MEMS inertial sensors.

INTRODUCTION

Unlike Hall-effect or magnetoresistive sensors, a Lorentz force magnetometer is a force sensor and typically operates at its mechanical resonance, f_n. The sensitivity of the sensor is directly proportional to the bias current flowing through the structure and to the length through which the current travels ($F_L = Li \times B$). One type of Lorentz force magnetometer modulates a low-frequency magnetic field up to f_n [1-3], thus the motion resulting from the Lorentz force is amplified by the mechanical quality factor (Q). Closed-loop operation of this type of sensor has also been demonstrated [4]. Another type of Lorentz force magnetometer has a frequency output; magnetic field is measured by monitoring the change in f_n as Lorentz force changes the stiffness of the resonator. In both types of magnetometers, it is desirable to have a larger bias current and longer current-carrying beam to improve the sensitivity and resolution of the sensor. Moreover, having high Q is always desirable for resonant sensors, as it improves the signal to noise ratio (SNR).

Multi-axis sensing and sensor integration is the trend of MEMS inertial sensors as they allow chip size and fabrication cost to be reduced while maintaining the same performance and reliability. Single-structure 3-axis gyroscopes are currently in full production and have been widely used in smart phones and portable devices [5]. Single-structure 3-axis accelerometers have also been demonstrated [6] whereas single-structure 3-axis magnetometers have not. Moreover, the lowest noise floor achieved in a MEMS Lorentz force magnetometer, fabricated in a process that requires electrically-isolated metal layers on top of the silicon MEMS structure, was reported as 70 nT/√Hz for z-axis field and 10 nT√Hz for x-/y- axis field [7]. Most other Lorentz force magnetometers have noise levels greater than 200 nT/√Hz. Here, we demonstrate a single-structure 3-axis Lorentz force magnetometer for electronic compass purposes, which has a noise floor below 30 nT/√Hz. The magnetometer is fabricated in the epi-seal process developed at Stanford University [8]. The process is currently used for high-volume manufacturing of micromechanical oscillators for timing applications by SiTime Inc. and provides a very low operating pressure (1 Pa). Pressure sensors, thermometers, accelerometers, gyroscopes [9] and magnetometers have been demonstrated using the same fabrication process. The magnetometer presented in this work is fully compatible with production MEMS processes such as the STMicroelectronics THELMA process and the Bosch's surface micromachining process, and can be fabricated on the same die with accelerometers and gyroscopes.

DESIGN

The 3-axis magnetometer consists of a 1.8×1.8 mm^2 resonator fabricated in 40 μm thick (100) single-crystal silicon and wafer-level vacuum sealed at less than 1 Pa to reduce damping and increase quality factor (Q). An n-type wafer with doping level of 6E19 cm^{-3} (resistivity 1mΩ·cm) is used to reduce the resistance of the current-carrying flexure. Figure 1 shows the schematic view and SEM image (inset) of the magnetometer. The structure is long to maximize Lorentz force, while the area is optimized to only 0.72 mm^2 (18% of the 2×2 mm^2 die area). During operation, an ac bias current at a frequency close to f_n is injected through the structure which generates Lorentz force and therefore movement of the structure in the presence of magnetic field.

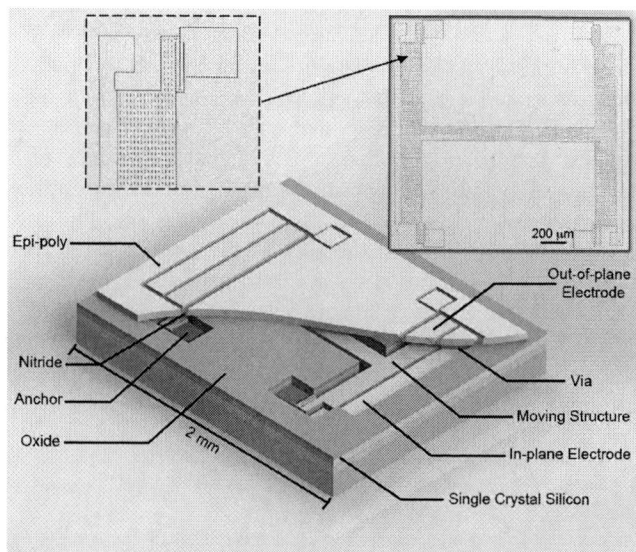

Figure 1: 3-axis Lorentz force magnetometer. Inset: SEM image of the magnetometer before epitaxial sealing process.

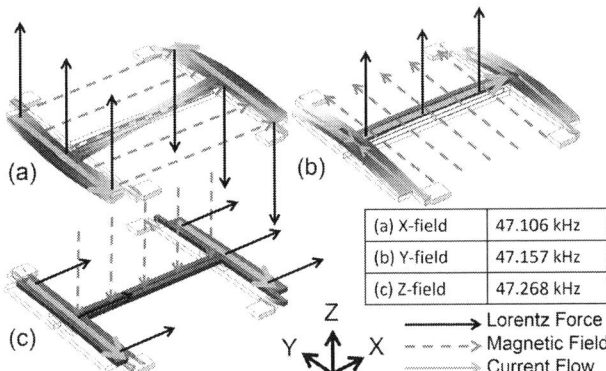

(a) X-field	47.106 kHz
(b) Y-field	47.157 kHz
(c) Z-field	47.268 kHz

——— Lorentz Force
- - - → Magnetic Field
········→ Current Flow

Figure 2: FEM simulation of the magnetometer. The excitation current (green arrows) is modulated at the resonance frequency, and low frequency magnetic field (blue dashed arrows) generates Lorentz force (black arrows) near the resonance frequency.

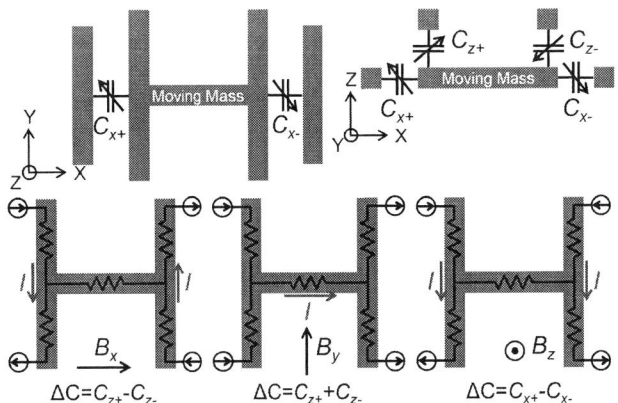

Figure 3: Capacitive pick-off placement for in-plane motion sensing (top left) and out-of-plane motion sensing (top right). Bottom schematics show the electrode configurations for x, y, and z axis magnetic field sensing.

Multi-Axis Resonator

Figure 2 shows the FEM simulation of the three resonant modes excited by x, y and z field, with a detailed schematic of current flow and Lorentz force direction. Time-multiplexing is used to measure each axis of magnetic field in sequence by switching the direction of current flow. While time-multiplexing reduces the measurement bandwidth, cross-axis coupling is greatly reduced.

The three resonant frequencies were designed to be close to each other, so that when operating in open loop, only one frequency reference is required for 3-axis sensing. With the trade-off of reducing sensitivity, the bandwidth can be greatly enhanced by operating at a frequency slightly off-resonance [10]. The magnetometer can also be operated in closed-loop, providing the maximum sensitivity and at the same time better stability over temperature. The closed-loop operation also reduces the effect of self-heating due to the bias current, which would otherwise reduce the sensitivity because of the change in the natural frequency.

Capacitive Sensing

Capacitive pick-offs are used for displacement sensing, as illustrated in Figure 3. In-plane motion (z-axis field) is measured with a pair of differential parallel plates (C_{x+} and C_{x-}), while out-of-plane motion (x- or y- axis field) is measured by taking the difference (x-axis field) or sum (y-axis field) of the signals from two out-of-plane sense pick-offs (C_{z+} and C_{z-}) located above the moving structure. The electrode configurations for each axis are different, further reducing cross-axis coupling.

When designing the capacitive pick-offs, damping has to be considered. For typical micromechanical resonators with capacitive sensing, squeeze film gas damping and slide film gas damping dominate the damping for pressure levels down to 10-100 Pa [11]. Large sensing capacitance requires more die area and increases both damping and thermomechanical noise, however too little sensing capacitance causes electronic noise to dominate. Comb-finer and parallel plate capacitors are the most commonly used capacitive sensing types. Compared to comb-finger, parallel plate capacitive sensing gives larger sensitivity for the same area but also introduces transduction nonlinearity and electrostatic nonlinearity. However, for the full-range of Earth's field detection, the displacement resulting from Lorentz force is on the order of 1 nm and the nonlinearity from the parallel plate can be neglected. We considered device size, capacitance sensitivity, sensor's resolution and quality factor (Q) during the sensor design. The designed displacement-to-capacitance sensitivity is ~1 µF/m in each axis.

RESULTS

Lorentz force bias current generating circuitry and capacitive sensing circuitry are implemented with discrete components on a PCB. We currently use 4 switches to manually select the sensing axis by setting the current flow direction, however this switching can be automated in the future. During operation, a 5.8 mA$_{rms}$ bias current is injected through the MEMS sensor. This Lorentz force bias current passes through a resistance of ~30 Ω in each axis, which results in a power consumption of 1 mW. A dc bias voltage is also applied to the moving mass at the same time for capacitive sensing. The Lorentz force results in a change in capacitance that is measured by a transimpedance amplifier. The measured electronic noise floor is 0.03 aF/√Hz with a 4 V DC bias. A digital lock-in amplifier (Zurich Instruments HF2-LI) is used to generate the bias current at sensor's resonant frequency and demodulate the signal. Table I summarizes the magnetometer parameters for each axis.

Frequency Response and Sensitivity

Figure 4 shows measured frequency responses (left) and capacitance change vs. input magnetic field (right). Note here that the measured z-axis frequency response (in-plane resonating mode) shows that the resonance frequency is 2 kHz higher than the expected value from FEM simulation,

978-1-4799-3510-9/14 $31.00 © 2014 IEEE

Table 1. Sensor Parameters

Parameter	X-field	Y-field	Z-field
Natural Frequency	47.43 kHz	47.29 kHz	49.10 kHz
Q	7100	3200	12700
Bandwidth	3.3 Hz	7.4 Hz	1.9 Hz
Power	1 mW	1 mW	1 mW
Sense Cap.	1947 fF	1947 fF	990 fF
Sense Gap	2 μm	2 μm	1 μm
Sensitivity	17.44 pF/T	9.28 pF/T	24.2 pF/T
Resolution	19 nT/√Hz	32 nT/√Hz	17 nT/√Hz

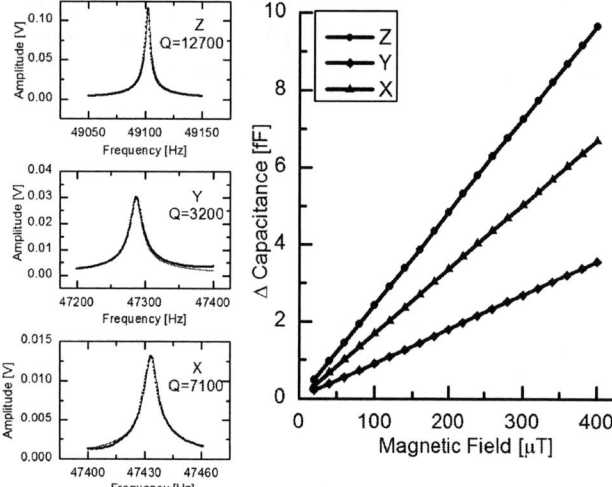

Figure 4: Frequency responses (left) and measured capacitance change vs. input magnetic field (right). The magnetometer is linear up to 400 μT in all three axes.

mainly due to the overcompensation for over-etch in the device layout. Although all three axes have similar 1 μF/m displacement-to-capacitance sensitivities, they have different field-to-capacitance sensitivities due to different Q, an effect which can be compensated by the detection electronics. The difference in Q also results in slightly different thermomechanical noise-limited resolution. The magnetometer is linear (<1% nonlinearity) up to 400 μT in all three axes limited by the measurement setup.

Earth's Field Measurement

We use a commercial geomagnetometer (Integrity Deisgn & Research IDR-321) to measure the Earth's field in our lab (Davis, California) as a reference (13 μT). To demonstrate that the MEMS magnetometer is capable of measuring Earth's field, the device is rotated 360° about its y-axis and the horizontal component of Earth's field is measured from the sensor's x- and z-axis outputs, showing

the expected sin(θ) and cos(θ) dependency in Figure 5. This measurement was performed with the magnetometer excited slightly off-resonance (Δf_x = 7.2 Hz, Δf_z = 4.2 Hz), trading sensitivity for increased bandwidth (BW_x = 12 Hz, BW_z = 6.8 Hz) and reduced sensitivity to temperature, similar to an approach recently proposed for gyroscopes [12] . Since the Brownian (thermomechanical) noise is at least 10× larger than the electronic noise floor, the resolution is not sacrificed by operating off-resonance [10].

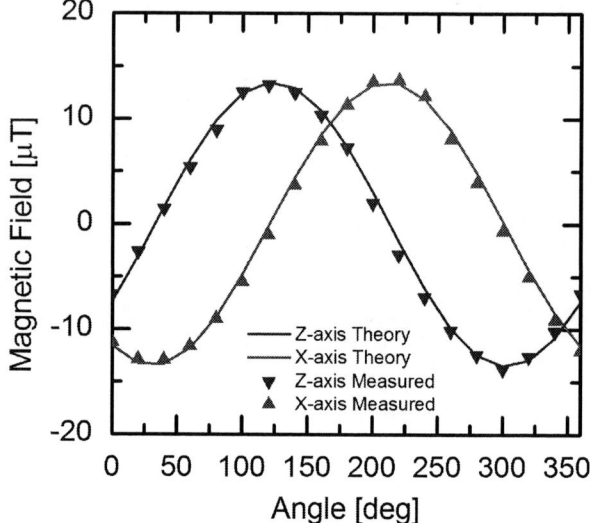

Figure 5: Magnetometer's response to the horizontal component of Earth's field (13 μT).

Cross-axis Sensitivity

The cross-axis sensitivity is measured by mounting the magnetometer inside a 2-axis Helmholtz coil. The accuracy of the measurement is limited by the accuracy of the alignment in our measurement setup (tens to hundreds of millidegrees). Note that a one-degree misalignment between the coil and the magnetometer introduces a -35 dB error in the cross-axis sensitivity.

Table 2 shows the measured cross-axis sensitivity. The low cross-axis sensitivity results from using (a) three different resonating modes, (b) time-multiplexing and (c) three different capacitive sensing configurations. The y-axis output shows larger coupling from the x-axis input compared to other combinations. This is possibly because a voltage source rather than a current source is used to provide the Lorentz force bias current. A resistance-mismatch in the current-carrying flexure due to fabrication variation results in a mismatch of the bias current. The mismatched bias current is orthogonal with the original bias current, which increases

Table 2. Measured Cross-axis Sensitivity

Input	X-OUT	Y-OUT	Z-OUT
X	0 dB	-29 dB	-65 dB
Y	-37 dB	0 dB	-63 dB
Z	-43 dB	-36 dB	0 dB

the cross-axis sensitivity. This effect can be suppressed by using a current source instead of voltage source.

Figure 6 compares the performance of recently reported MEMS Lorentz force magnetometers which are capacitively transduced. For both sensitivity and resolution, the bias current for Lorentz force generation is normalized to 1 mA for comparison. Kyynäräinen's [7] torsional magnetometer and double-ended-tuning-fork magnetometer keep the record for sensitivity and resolution thanks to the electrically-isolated metal layer on the silicon MEMS structure, which increases the effective length of the current carrying beam by a factor of 5X-65X.

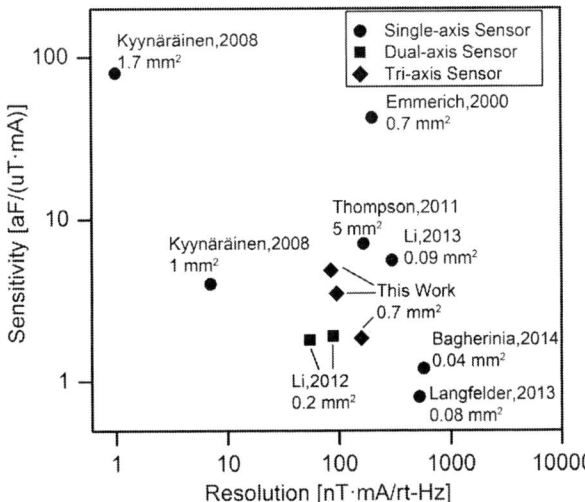

Figure 6: Comparison of capacitance sensitivity versus resolution of recently reported MEMS Lorentz force magnetometers.

CONCLUSION

We demonstrate a 3-axis Lorentz force magnetometer in a single structure. By optimizing the mechanical design, it achieves sub-30 nT/√Hz resolution for all three axes. Experimental results verify that the device is capable of measuring Earth's field. The cross-axis sensitivity is reduced by both mechanical design and time-multiplexing operation. Compared to previous single-axis and dual-axis magnetometers, the 3-axis magnetometer demonstrated in this work provides very good performance considering resolution, sensitivity and sensor size.

ACKNOWLEDGEMENTS

This work was supported by the Defense Advanced Research Projects Agency (DARPA) Precision Navigation and Timing program (PNT) managed by Dr. Andrei Shkel and Dr. Robert Lutwak under contract # N66001-12-1-4260 and the National Science Foundation under award number CMMI-0846379. The work was performed in part at the Stanford Nanofabrication Facility (SNF) which is supported by National Science Foundation through the NNIN under Grant ECS-9731293. The authors would also like to thank the SNF staff, particularly M. M. Stevens for the timely

assistance with the epitaxial reactor.

REFERENCES

[1] L. M. Miller, J. A. Podosek, E. Kruglick, *et al.*, "A µ-magnetometer based on electron tunneling," in *IEEE MEMS*, 1996, pp. 467-472.

[2] G. Langfelder, C. Buffa, A. Frangi, *et al.*, "Z-Axis Magnetometers for MEMS Inertial Measurement Units Using an Industrial Process," *Industrial Electronics, IEEE Transactions on,* vol. 60, pp. 3983-3990, 2013.

[3] M. Li, V. T. Rouf, M. J. Thompson, *et al.*, "Three-Axis Lorentz-Force Magnetic Sensor for Electronic Compass Applications," *Microelectromechanical Systems, Journal of,* vol. 21, pp. 1002-1010, 2012.

[4] M. Li, E. J. Ng, V. A. Hong, *et al.*, "Lorentz force magnetometer using a micromechanical oscillator," *Applied Physics Letters,* vol. 103, 173504, 2013.

[5] B. Vigna, "It Makes Sense: How Extreme Analog and Sensing Will Change the World," in *Tech. Digest 2012 Solid-State Sensors, Actuators and Microsystems Workshop*, Hilton Head, SC, 2012, pp. 58-65.

[6] M. A. Lemkin, B. E. Boser, D. Auslander, *et al.*, "A 3-axis force balanced accelerometer using a single proof-mass," in *Solid State Sensors and Actuators*, , 1997, vol. 2, pp. 1185-1188.

[7] J. Kyynäräinen, J. Saarilahti, H. Kattelus, *et al.*, "A 3D micromechanical compass," *Sensors and Actuators A: Physical,* vol. 142, pp. 561-568, 2008.

[8] R. N. Candler, M. A. Hopcroft, B. Kim, *et al.*, "Long-Term and Accelerated Life Testing of a Novel Single-Wafer Vacuum Encapsulation for MEMS Resonators," *Microelectromechanical Systems, Journal of,* vol. 15, pp. 1446-1456, 2006.

[9] S. Nitzan, C. H. Ahn, T. H. Su, *et al.*, "Epitaxially-encapsulated polysilicon disk resonator gyroscope," in *Micro Electro Mechanical Systems (MEMS), 2013 IEEE 26th International Conference on*, 2013, pp. 625-628.

[10] G. Langfelder and A. Tocchio, "On the operation of Lorentz-force MEMS magnetometers with a frequency offset between driving current and mechanical resonance," *Magnetics, IEEE Transactions on,* vol. PP, pp. 1-6, 2013.

[11] M.-H. Bao, *Analysis and design principles of MEMS devices* (Amsterdam: Elsevier).

[12] M. W. Judy, J. A. Geen, and H. Johari-Galle, "Non-Degenerate Mode MEMS Gyroscope," US Patent 20120137774, 2012.

CONTACT

*Mo Li, tel: +1-530-752-5180; moxli@ucdavis.edu

DESIGN, FABRICATION AND CHARACTERIZATION OF TUNABLE PERFECT ABSORBER ON FLEXIBLE SUBSTRATE

X. Zhao[1], K. Fan[1], J. Zhang[1], H. R. Seren[1], G. D. Metcalfe[2], M. Wraback[2], R. D. Averitt[1] and X. Zhang[1]

[1] Boston University, Boston, MA, USA
[2] U.S. Army Research Laboratory, Adelphi, MD, USA

ABSTRACT

This paper reports our recent progress on a highly flexible dynamic perfect absorber at terahertz (THz) frequencies. Metamaterial unit cells were patterned on thin GaAs patches, which were fashioned in an array on a 5μm polyimide substrate via transfer printing technique, and the backside of the substrate was coated with gold film as ground plane. Optical-pump THz-probe reflection measurements show that the absorptivity at resonance frequency of 1.58THz can be tuned up to 57% through photo-excitation of free carriers in GaAs layers in presence of 800nm pump beam. Our flexible tunable MM perfect absorber exhibits potential applications in energy harvesting, imaging and stealth coating.

INTRODUCTION

During the past decade, electromagnetic (EM) metamaterials (MMs) have attracted considerable interest due to their unusual EM response and promising applications [1]. MM perfect absorber (MPA) has been considered as one particular and important branch of MMs enabling near-unity absorption in a thin slab with thickness much smaller than λ/4 [1, 2]. Since 2008, great numbers of examples have been demonstrated in different frequency regimes, including microwave [3], terahertz (THz) [4], infrared [5], mid-infrared [6], near-infrared [7] and visible frequency [8].

Tunable metamaterials have shown the ability in manipulating EM waves to build modulators, sensors, detectors, and many other devices [9]. Routes to realize tunable MM include photo-excitation [10], electrical [11, 12], phase change [13], and mechanical reconfiguration [14]. In our previous work [15], we demonstrated the fabrication and characterization of optically tunable metamaterials on thin and flexible substrate. The gold MM structure on GaAs patch was successfully transferred to 4μm thick polyimide. The modulation depth is 60% at 0.98THz with the pump beam power of 8mW.

Introducing dynamics into MPAs provides a new path towards exotic devices thus extending applications of MPAs for imaging, energy harvesting, even camouflage of IR emission from detection. Furthermore, integration of tunable MPAs on ultrathin and flexible substrates enables high adaptability to wrap and fit on arbitrary surfaces for practical applications.

Here, we report our progress on a novel flexible tunable perfect absorber in THz regime. We used the transfer printing method to pattern the MMs and GaAs patches on polyimide and coated the backside with gold thin film to construct the flexible tunable perfect absorber. It was characterized by the

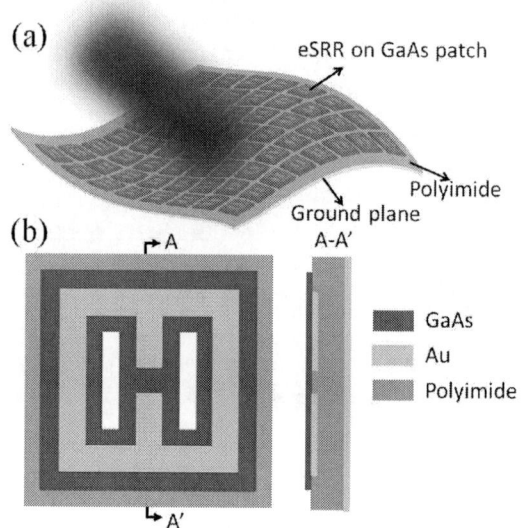

Figure 1: (a) The illustration of flexible tunable MM perfect absorber illuminated by an 800nm beam; (b) unit cell of the MM perfect absorber (top view and cross section view).

optical pump THz-probe spectroscopy. The modulation depth is about 57% at 1.58THz when the pump beam power is 10mW. Our flexible perfect absorber can be used for applications including camouflage coating, THz modulation and switching, energy harvesting, and chemical/biological sensors.

DESIGN AND FABRICATION

Fig.1 (a) schematically shows the tunable flexible perfect absorber photo-excited by an 800nm pump beam. The unit cell of the absorber as shown in Fig.1 (b) is composed of GaAs patch, gold electric split-ring resonator (eSRR), polyimide spacer and gold ground plane. The array of GaAs patches and eSRRs constructs tunable MM, sandwiching the polyimide spacer along with the ground plane. The windows in each unit are intentionally fabricated to etch the sacrificial layer during the releasing of the flexible absorber.

The metallic eSRRs and ground plane together determine the effective electromagnetic properties (i.e. permittivity-ε_{eff} and permeability-μ_{eff}) of the perfect absorber. When the electric field of the incident THz wave is perpendicular to the capacitive gap in the eSRR structure, it will couple to the eSRR and excite the LC resonance that arise an effective permittivity. At the same time, the magnetic field generates circulating currents between the two metallic layers as shown in Fig.2 (a) and (b), resulting in effective permeability. With appropriate geometric design, the

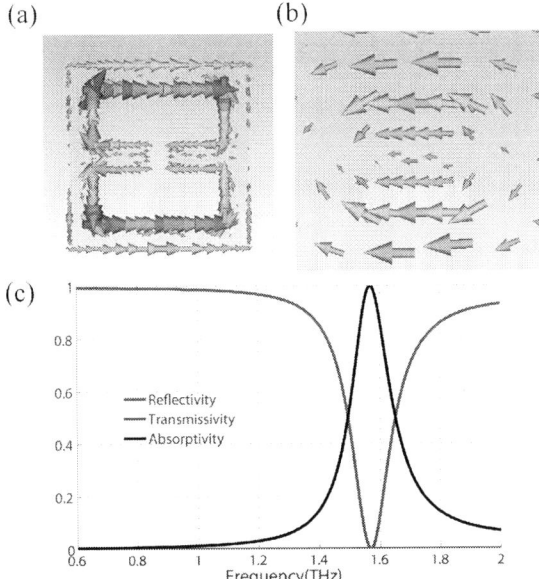

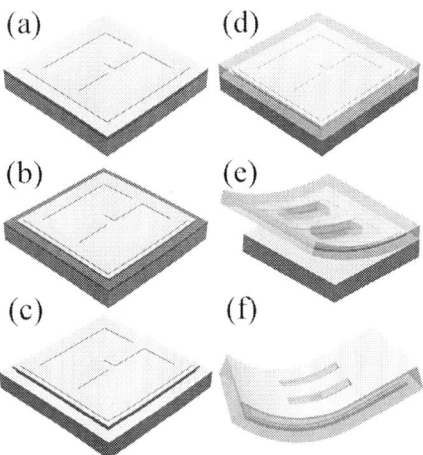

Figure 2: (a) Simulated current distribution on the eSRR structure; (b) simulated current distribution on the ground plane; (c) simulated reflectivity (R: black curve), transmissivity (T: green curve) and absorptivity (A: black curve) for the tunable MM perfect absorber.

impedance ($Z = \sqrt{\mu/\varepsilon}$) of the material can be matched with free-space around the resonant frequencies to eliminate the reflection; meanwhile, the ground plane ensures that the transmission is negligibly small. As a result, the near-unity absorption can be achieved at the LC resonance as shown in Fig.2 (c).

MPAs provide a new access to construct absorbers, in which the permittivity and permeability can be engineered independently. That is, the absorption response can be easily tailored by modifying either the permittivity or permeability. As a simple way, changing the capacitance in the eSRRs' gap can modify the resonant frequency, amplitude and phase of the reflection, thus altering the permittivity ε_{eff}. If an optical pump beam is incident on our MM perfect absorber, the free carriers will be generated in GaAs patches due to photo-excitation, which can in turn lead to increase of their conductivity. This will alter the ε_{eff} and cause mismatch of impedance and higher reflection. Consequently, the absorption can be modulated.

Fig.3 shows the fabrication process of the flexible perfect absorber. At first, 300nm sacrificial layer of $Al_{0.95}Ga_{0.05}As$ and 300nm semi-insulating GaAs were epitaxially grown on the 2'' SI-GaAs wafer. Then, a 15×15 mm^2 array of 200nm-thick gold eSRRs was patterned on the epitaxial SI-GaAs layer by lift-off process (Fig.3 (a)). The epilayer was wet etched with citric acid:H_2O_2 (10:1) solution to form the GaAs patch that is concentric with the eSRRs after photoresist S1813 was spun on and patterned (Fig.3 (b)). The following step was to dip the wafer into diluted HF solution to remove the uncovered sacrificial layer (Fig.3 (c)).

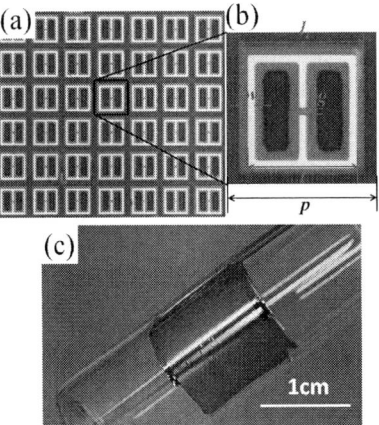

Figure 3: Fabrication process of the flexible tunable MM perfect absorber.

The sacrificial layer was etched in isotropic manner, yielding an undercut about 500nm underneath GaAs layer. Then, a 5μm thick polyimide (PI5878G, HD Microsystems) layer was spun on and cured at 275°C (Fig.3 (d)). The undercut volume was filled with polyimide. Next, the polyimide was etched by RIE with oxygen gas to define the etching windows, following with wet etching of GaAs and soaking the wafer in diluted HF solution (Fig.3 (e)). After all of the sacrificial layer was etched, the polyimide was released from the substrate and the GaAs patch and eSRR were transferred to the polyimide successfully. Finally, the backside of the released polyimide was coated with 200nm thick gold film via e-beam evaporator to form the ground plane (Fig.3 (f)). It should be emphasized that this fabrication process can be applied to other semiconductors (e.g. silicon and InAs) under which the sacrificial layer can be grown.

Figure 4: Images of the perfect absorber after releasing. (a) Microscopic images of the eSRRs; (b) closed-up view of one eSRR unit, li = 36μm, l = 30μm, w = 4μm, g = 4μm, p = 42μm; (c) the flexible tunable MM perfect absorber wrapped on a plastic bottle.

978-1-4799-3510-9/14 $31.00 © 2014 IEEE   85

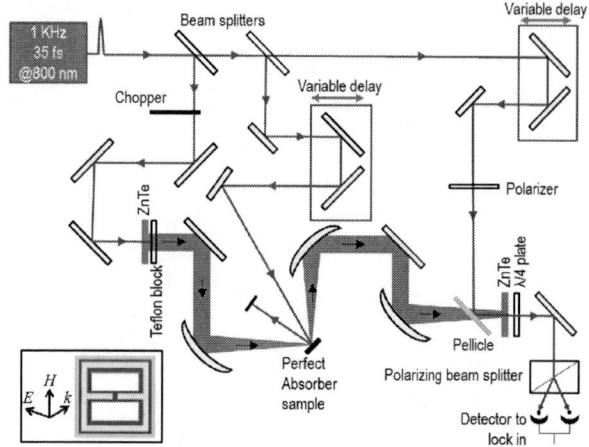

Figure 5: THz-TDS reflection measurement setup.

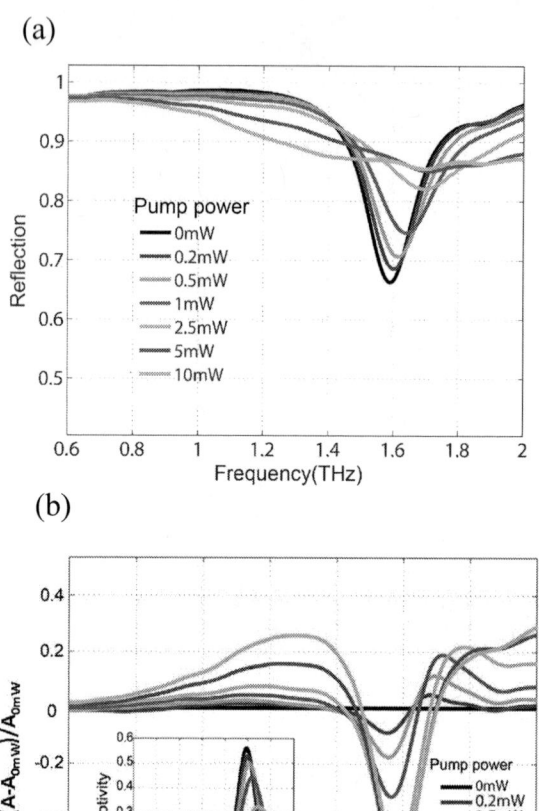

Figure 6: Experimental results. (a) Reflection coefficient at different pump power; (b) the measured modulation depth $(A-A_{0mW})/A_{0mW}$ with the pump power, which shows the tunability of the perfect absorber; inset: The measured absorptivity of the flexible MM perfect absorber at different pump power.

Fig.4 (a) and (b) are showing the eSRR array of the flexible perfect absorber and the dimensions of the fabricated eSRR unit, respectively. The fabricated perfect absorber is highly flexible and can be wrapped on a plastic bottle as in Fig.4 (c).

EXPERIMENT AND RESULTS

The reflection response of the MPA was characterized by optical pump THz-probe (OPTP) spectroscopy, as shown in Fig.5. The experiment was conducted in a humidity controlled environment where the humidity is less than 0.1% at room temperature. The flexible absorber sample was mounted on a holder so that the transverse magnetic (TM) THz pulses were 45° incidence to the sample, as well as a gold coated reflector that is used as reference. The OPTP spectroscopy system delivered amplified 35fs laser pulses at center wavelength of 800nm and a repetition rate of 1 kHz to excite the free carriers in the GaAs patches. Because the lifetime of the photo-excited carrier in GaAs is about 1ns [15], the THz probe pulse was fixed to arrive on the sample 10ps after the 800nm pump pulse to ensure that the steady accumulation of the carriers is established.

The experimental results of the reflection spectrum at different pump power are shown in Fig.6 (a). With the absence of 800nm pump beam, the LC resonance frequency of the perfect absorber is 1.58THz with reflection amplitude of 56% (the black curve). The resonance strength becomes weaker and amplitude increases to 86% when the pump power increases from 0mW to 2.5mW. If the pump power is high enough, such as higher than 5mW (equivalent to a fluence of $25\mu J/cm^2$), the LC resonance is totally damped since the pump beam excites sufficient photo-carriers in the GaAs. Because the backside of the polyimide is coated with gold film, the transmissivity (T) is negligible according to the simulation. Hence, the absorptivity (A) can be estimated by the measured reflection coefficient (r) by $A = 1-|r|^2$ as shown in the inset of Fig.6 (b). The maximum modulation depth of absorptivity of 57% is acquired at 1.58THz with pump power increasing from 0mW to 10mW as illustrated by Fig.6 (b).

SIMULATION AND DISCUSSION

To understand the nature of the modulation of the absorptivity, we performed the finite-difference simulation by CST Microwave Studio. The dimensions of the perfect absorber in our model are defined based on the fabricated samples. The gold layer was modeled as lossy metal with conductivity of $\sigma_{gold} = 4.5 \times 10^5$ $(\Omega\text{-cm})^{-1}$, while the 5 μm polyimide was modeled as lossy dielectric material with dielectric constant of 2.88 and a tangent delta of 0.03 [15]. The 300nm thick semi-insulating (SI) GaAs layer was modeled as Drude model, in which the plasma frequency $\omega_p \propto \sqrt{n}$ where n is the carrier density. The initial carrier density in GaAs was set to N = 1.0×10^{14} cm^{-3}. The carriers were generated in the SI GaAs layer when the pump beam is incident on the sample. The unit cell boundary condition is applied in our simulation and the THz source was obliquely incident on the sample at 45°.

978-1-4799-3510-9/14 $31.00 © 2014 IEEE   86

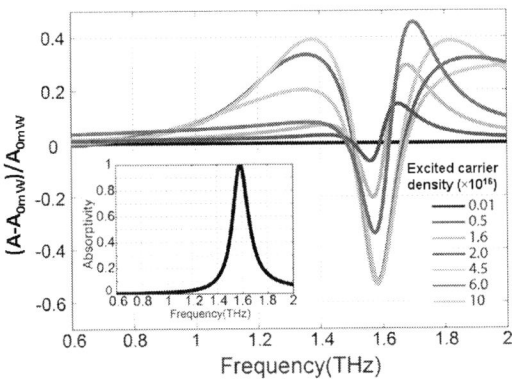

Figure 7: The simulated results of modulation depth with different excited carrier densities in GaAs; inset: the simulated absorptivity of the perfect absorber without pumped beam.

The simulated results of modulation depth of absorptivity as a function of carrier density in SI GaAs layer were shown in Fig.7. With the carrier density increase from $N=1.0 \times 10^{14} cm^{-3}$ to $10 \times 10^{16} cm^{-3}$, the variation in simulated modulation depth we found shows a similar trend with the experimental results (Fig.5 (b)). A significant difference between the measured and simulated results is seen at the initial state where maximum absorptivity is almost unity in the simulation, while it is only about 58% experimentally. This is because of the imperfections during the fabrication, especially in the thickness of the polyimide thickness on which the absorptivity is highly dependent [1]. In general, the simulated modulation depth in absorptivity agrees well with the experimental results.

CONCLUSIONS

A novel flexible tunable perfect absorber was fabricated by transfer printing and characterized by OPTP spectroscopy. The experimental results show that the 800nm ultrafast laser beam with sufficient power on GaAs patch can modulate the absorptivity with 57% at 1.58THz. The numerical simulation at different GaAs carried density was performed to understand the nature of the tunability and the results agreed well with the experiments.

ACKNOWLEDGEMENTS

This work was supported in part by the National Science Foundation under Contract ECCS 1309835, the Air Force Office of Scientific Research under Contract FA9550-09-1-0708, and DTRA under Contract W911NF-06-2-0040 administered by the Army Research Laboratory. The authors would like to thank Boston University Photonics Center for technical support.

REFERENCES

[1] C. M. Watts, X. Liu, W. J. Padilla, "Metamaterial Electromagnetic Waver Absorbers", *Advanced Material*, vol. 24, pp. 98-120, 2012.

[2] L. Huang, H. T. Chen, "A brief review on terahertz metamaterial perfect absorbers", *Terahertz Science and Technology*, vol. 6, pp. 26-39, 2013.

[3] N. I. Landy, S. Sajuyigbe, J. J. Mock, D. R. Smith, W. J. Padilla, "Perfect Metamaterial Absorber", *Physical Review Letters,* vol. 100, 207402, 2008.

[4] H. Tao, N. I. Landy, C. M. Bingham, X. Zhang, R. D. Averitt, W. J. Padilla, "A metamaterial absorber for the terahertz regime: Design, fabrication and characterization", *Optical Express,* vol. 16, pp. 7181-7188, 2008.

[5] Y. Avitzour, Y. A. Urzhumov, G. Shvets, "Wide-angle infrared absorber based on a negative-index plasmonic metamaterial", *Physical Review B*, vol. 79, 045131, 2009.

[6] X. Liu, T. Starr, A. F. Starr, W. J. Padilla, "Infrared Spatial and Frequency Selective Metamaterial with Near-Unity Absorbance", *Physical Review Letter*, vol. 104, 208403, 2010

[7] J. Hao, J. Wang, X. Liu, W. J. Padilla, L. Zhou, M. Qiu, "High performance optical absorber based on a plasmonic metamaterial", *Applied Physics Letters*, vol. 96, 251104, 2010.

[8] K. Aydin, V. E. Ferry, R. M. Briggs, H. A. Atwater, "Broadband polarization-independent resonant light absorption using ultrathin plasmonic super absorbers", *Nature Communications*, vol. 2, 2011.

[9] N. I. Zheludev, Y. S. Kivshar, "From metamaterials to metadevices", *Nature Materials*, vol. 11, pp. 917-924, 2012

[10] K. Fan, A. C. Strikwerda, X. Zhang, R. D. Averitt, "Three-dimensional broadband tunable terahertz metamaterials", *Physical Review B*, vol. 87, 161104, 2013.

[11] H.-T. Chen, W. J. Padilla, M. J. Cich, A. K. Azad, , R. D. Averitt, and A. J. Taylor, "A metamaterials solid-state terahertz phase modulator," *Nature Photon.* vol. 3, pp. 148-151, 2009.

[12] D. Shrekenhamer, W. C. Chen, W. J. Padilla, "Liquid Crystal Tunable Metamaterial Perfect Absorber", *Physical Review Letter*, vol. 110, 177403, 2013.

[13] T. Driscoll, S. Palit, M. M. Qazilbash, D. R. Smith, D. N. Basov, et al. "Dynamic tuning of an infrared hybrid-metamaterial resonance using vanadium dioxide," *Applied Physics Letter*, vol. 93, 024101, 2008.

[14] X. Liu, W. J. Padilla, "Dynamic Manipulation of Infrared Radiation with MEMS Metamaterials", *Advanced Optical Material*, vol. 1, pp. 559-562, 2013.

[15] K. Fan, X. Zhao, J. Zhang, K. Geng, G. R. Keiser, H. R. Seren, G. D. Metcalfe, M. Wraback, X. Zhang, R. D. Averitt, "Optically Tunable Terahertz Metamaterials on Highly Flexible Substrates", *IEEE Transaction on Terahertz Science and Technology*, vol. 3, pp. 702-708, 2013.

CONTACT

*X. Zhang, tel: +1- 617-3582702; xinz@bu.edu

TUNABLE META-FLUIDIC-MATERIALS BASE ON MULTILAYERED MICROFLUIDIC SYSTEM

W. M. Zhu[1], B. Dong[1], Q. H. Song[2], W. Zhang[1], R. F. Huang[3], S. K. Ting[3] and A. Q. Liu[1, 2†]

[1]School of Electrical and Electronic Engineering, Nanyang Technological University
50 Nanyang Avenue, Singapore 639798
[2]School of Mechanical Engineering, Xi'an Jiaotong University, Xi'an 710049, China
[3]Temasek Laboratories, 5A Engineering Drive 1, Singapore 117411

ABSTRACT

We demonstrate a multilayered microfluidic system with a flexible substrate, which has tunable optical chirality within THz spectrum range. The optical properties of the multilayered microfluidic system can be tuned by either changing the liquid pumped into each layer or stretching the flexible substrate. In experiment, the polarization rotation angle is tuned from zero (non-chiral structure) to 16.9° (strong-chiral structure). Furthermore, the tuning resolution can be well controlled due to the fine refractive index change of the liquid with different concentrations. It is feasible for the multilayered microfluidic structure to be integrated to an optofluidic system, where strong or tunable optical chirality are needed, which not only can be used as traditional optical components such as THz polarizers and filters but also has potential applications on imaging and sensor of bio-materials.

INTRODUCTION

The concept of artificial materials, such as metamaterials and photonic crystals, has created a platform of new functionalities of devices, which can manipulate light across the entire electromagnetic spectrum. The extraordinary optical properties of artificial materials, such as negative index [1-3], zero epsilon [4] and giant chirality [5] arise from their sub-wavelength elements, which are both designable and controllable for different applications. Artifical materials, such as metamaterials, are now widely studied for the enhanced nonlinear switching [6, 7] and light emission performance over conventional active materials. Furthermore, the possibilities on waveform manipulation and cloaking have promising potential in vast applications, which cannot be realized using materials already existed in nature. Those thrilling technological prospects have stimulated a wide search for developing artificial materials with tunable and switchable properties using MEMS, phase change media, liquid crystal, magnetic media and superconductors [8-12]. However, the tunabilites of the artificial materials are still limited by the nature materials, which are used to fabricate the sub-wavlength elements. Pervious works on structural reconfigurable metamaterials, manage to gain the tunablties from varying the near field coupling of the metal structures [13-16]. However, the tuning range is still limited since the metal structure cannot be changed once fabricated.

Microfludics systems have been intensively studied for inkjet print heads, DNA chips, lab-on-a-chip technology and micro-propulsion applications. The concept of digital microfluidics, the formation and control of nano litter droplets, offers great opportunities in particle Lab-on-a-chip applications, such as bio-imaging, sensor and particle sorting. Although the demands for new functional materials of above mentioned applications are increasing, the possibilities of developing functional materials using microfluidics systems have never been explored, which is mainly due to the low contrast ratio of the fluidics refractive index. The liquid metal, on the other hand, can be used to construct functional materials when patterned and controlled by using the microfluidic systems. Different from the solid state metal structures, the geometry of the sub-wavelength liquid metal elements can be redefined by the valves and pumps via microfluidic channels, which offers more flexibility in the dynamic control of the functional materials properties.

MICROFLUIDIC SYSTEM DESIGN

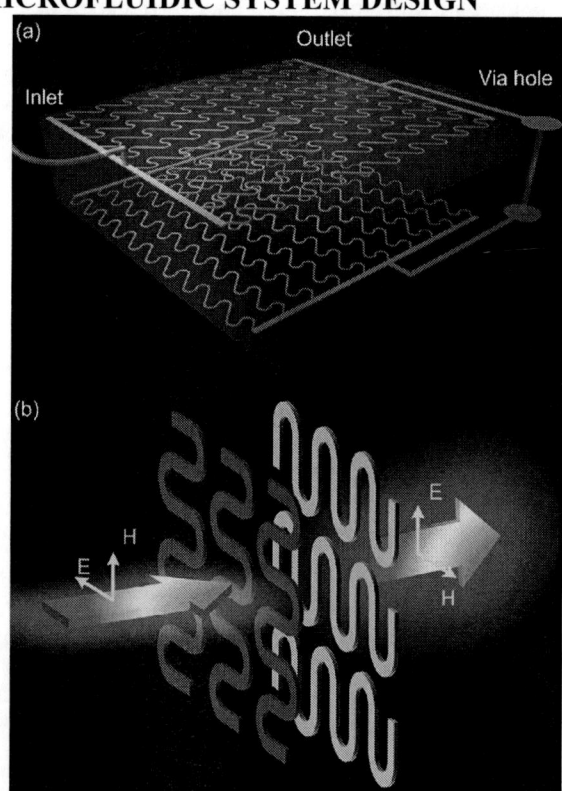

Figure 1: Schematics of .the microfluidic system. (a) the working principle of the tunable optical chirality of the multilayered microfluidic system. (b) the polarization rotation due to the change of the liquid in the micro channels (purple).

Figure 1 shows the schematic and working principle of the microfluidic system, which consists of two layers of "S" shaped microfluidic channels separated by a 100-µm Polydimethylsiloxane (PDMS) layer. This microfluidic system is designed as a THz chiral material with tunable polarization rotator function. The microfluidic channels are all connected together at the inlet and outlet points, where the liquid can be pumped in and out, respectively. A via hole can be used to connect the outlet of the top layer and inlet of the bottom layer if both layers are designed to be pumped with identical liquid, as shown in Fig. 1(a). In this paper, we don't use the via hole since only the liquid in the top layer need to be changed during the tuning of the polarization rotator. The S-shaped microchannels on the top layer and the bottom layer are identical but perpendicular to each other.

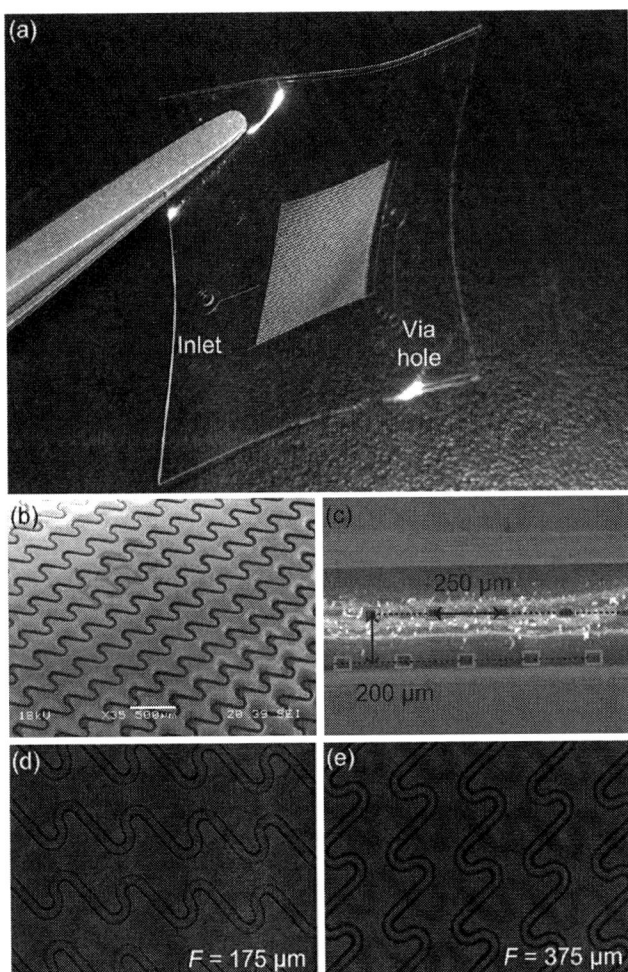

Figure 2: Photographs of microfluidic systems. (a) The over view graph of the microfluidic system with flexible substrate. (b) and (c) The SEM graph of the over view and cross section of the system. (b) shows the overview SEM graph of the top layer microfluidic channels. (c) shows the cross section of the bi-layer microfluidic system (d) and (e) the microscope graph of the polarizer taken at different focus distance F. (d) shows the top layer graph (F = 175 µm). (e) shows the bottom layer graph (F = 375 µm).

Once pumped with liquid metal, each single layer of the microfluidic systems are planar metamaterials, which cannot be superimposed on its in-plane mirror image. However, these planar metamaterials still possess mirror symmetry for the mirror parallel to the plane and require tilted incidence to achieve the chiroptical effects. Although the out-of-plane mirror symmetry can be broken by adding a substrate to the microfluidic system, the achieved chirality is inevitably low in these metamaterials due to the weak magnetic responses. Therefore, another layer of microfluidic channels is added to the system to achieve strong chiroptical effects with the out-of-plane mirror symmetry broken by the 90-degree rotated microchannels at different layers. However, the microfluidic system is a non-chiral structure with plain mirror symmetry when pumped with identical liquid. The symmetry can be broken by pumping the top and bottom layers with different liquid. Furthermore, the out-of-plane symmetry of the microfluidic system can be changed by pumping the liquid with different refractive indexes to the front layer, which can be used for the tuning of the optical chirality.

The fabrication of the microfluidic system is based on the polymer soft lithography technique. The designs of the microfluidic channels of each PDMS layers are drawn using a computer-aided design (CAD) program (L-Edit). A dark field clear feature plastic mask is fabricated based on the CAD drawing. A 6-inch silicon is cleaned using a piranha solution ($H_2SO_4 + H_2O_2$) and spin coated with a 50-µm SU-8 photoresist (MicroChem, SU-8 50), which is achieved by spinning coat the photoresist at 2000 rpm for 30 s using the spin coater (CEE, 200). The silicon substrate is soft baked using a hot plate at 65 °C for 6 mins and 95 °C for 20 mins after the spin coating. Then, the substrate is exposed to UV light for 30 s under the plastic mask using the mask aligner (OAI, J500-IR/VIS). The post expose bake is performed at 65 °C for 1 min and 95 °C for 5 mins after the exposure. The SU-8 layer, which is used as the master of the PDMS channels, is developed using the SU-8 developer (MicroChem) for 6 mins. The PDMS channels is fabricated using the replica molding, which is the casting of PDMS prepolymer against a master and obtaining the negative replica of the master. Three masters with different patterns are fabricated for metamolecules array layer, air pumping channels layer and the control panel layer, respectively. There are altogether 3 PDMS layers. Two layers with microchannels and one layer without any pattern as the substrate. The microfluidic system is fabricated by plasma boning the three PDMS layers.

Figure 2 shows the graphs of the multilayered microfluidic system with flexible substrate. The micro channels are 20-µm wide and the total thickness of the microfluidic system is approximately 500 µm while the foot print of the microchannels region is 1 cm × 1 cm. Fig. 2(b) and Fig. 2(c) show the SEM graphs of top view and cross section of the microfluidic system, respectively. The cross section view is taken by cutting the microfluidic system into two pieces from the microchannel region. The microchannels are highlighted with pink color. The channel

978-1-4799-3510-9/14 $31.00 © 2014 IEEE

spacing within and between the layers is 250 μm and 200 μm, respectively, as shown in Fig. 2(c). Photographs of the microfluidic system with different focus lengths show that the top and bottom layer are perpendicular to each other. Fig. 2(d) and 2(e) shows the graph of the top and bottom layer, which are 175 μm and 375 μm from a reference plane. Therefore, the vertical spacing is approximately 200 μm between the top and bottom PDMS channels.

RESULTS AND DISCUSSIONS

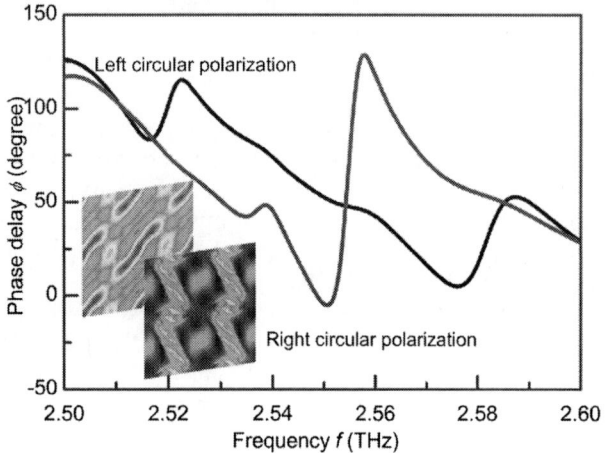

Figure 3: Numerical simulation of the transmittances phase delay under different polarization states. The insert shows the electric field concentration at the top and bottom layers of the multilayered metamaterial, which shows the polarization rotation due to the rotation of the resonance modes.

Figure 3 shows the numerical analysis of the transmittance phase delay under different polarization states for the microfluidic system. The magenta and navy blue lines represent the phase delay of right and left circular polarized lights, respectively, which show large difference at 2.56 THz.

The optical activity is defined as polarization azimuth rotation of elliptically polarized light

$$\theta = (\Phi_L - \Phi_R)/2 \qquad (1)$$

where Φ_L and Φ_R are the phase change of the left and right circular transmittance, respectively.

The insert shows the electric field distribution at the top and bottom layers of the microfluidic system when the microchannels of the top and bottom layer are filled with air and mercury, respectively. The rotation of the resonance modes indicates the rotation of the polarization states.

The circular transmittance, Φ_L and Φ_R, is difficult to measure due to the lack of corresponding polarizer in THz region. In experiment the phase and amplitude of linear polarized light are measured using TeraView Spectra 3000 and the results are converted to the circular polarized light using the method as shown in [17]. Dry air is supplied in the measuring chamber to dispel water in atmosphere, which has large absorption in terahertz regime.

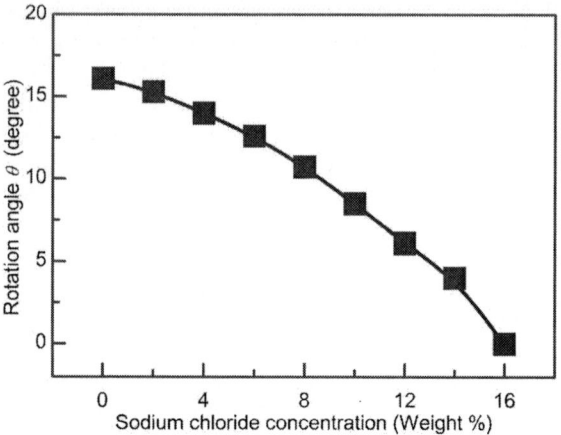

Figure 4: Experimental results showing the change of the rotation angle due to the pumping of different liquid into the micro channels of the top layer of the metamaterial. The polarization rotation decreases when the concentration of the sodium chloride solution is increasing.

Figure 4 shows the experimental results of the rotation angle θ as the function of the concentration of sodium chloride solution within the top layer. The left circular polarized incident light is at 2.56 THz. The bottom layer is pumped with mercury. The polarization rotation angle is tuned from 16.9° to 0°.

CONCLUSIONS

In conclusion, a microfluidic system with tunable optical chirality is designed, fabricated and demonstrated. The microfluidic system is 500 μm in thickness, which is approximately 2.5 times of the working wavelength. In experiment, it measures a large tunability in optical chirality, which has potential applications not only on traditional optic components such as THz polarizers and filters but also on imaging devices and sensor of bio-materials.

ACKNOWLEDGEMENTS

The work is main supported by the Environmental and Water Industry Development Council of Singapore (EWI), RPC programme (Grant No. 1102-IRIS-05-01 and 1102-IRIS-05-02)

REFERENCES

[1] R. A. Shelby, D. R. Smith and S. Schultz, "Experimental Verification of a Negative Index of Refraction," *Science*, Vol. 292, pp. 77, 2001.

[2] J. B. Pendry, "Negative Refraction Makes a Perfect Lens," *Phys. Rev. Lett.*, Vol. 85, pp. 3966, 2000.

[3] Y. Poo, R. X. Wu, G. H. He, P. Chen, J. Xu and R. F. Chen, "Experimental verification of a tunable left-handed material by bias magnetic fields," *Appl. Phys. Lett.*, vol. 96, 161902, 2010.

[4] M. Silveirinha and N. Engheta, "Tunneling of electromagnetic energy through subwavelength channels and bends using epsilon-Near-Zero Materials," *Phys. Rev. Lett.*, Vol. 97, 157403, 2001.

[5] A. V. Rogacheva, V. A. Fedotov, A. S. Schwanecke and N. I. Zheludev, "Giant gyrotropy due to electromagnetic-field coupling in a bilayered chiral structure," *Phys. Rev. Lett.*, Vol. 97, 177401, 2006.

[6] M. Ren, J. Huang, H. Cai, J. M. Tsai, J. Zhou, Z. Liu, Z. Suo, and A. Q. Liu, "Nano-optomechanical actuator and pull-back instability," *ACS Nano*, Vol 7, pp.1676 - 1681, 2013.

[7] B. Dong, H. Cai, G. I. Ng, P. Kropelnicki, J. M. Tsai, A. B. Randles, M. Tang, Y. D. Gu, Z. G. Suo and A. Q. Liu, "A nanoelectromechanical systems actuator driven and controlled by Q-factor attenuation of ring resonator," *Appl. Phys. Lett.*, Vol 103, 181105, 2013.

[8] J. Y. Ou, E. Plum, L. Jiang and N. I. Zheludev, "Reconfigurable photonic metamaterials," *Nano Lett.*, vol. 11, pp. 2142-2144, 2011.

[9] F. Zhang, Q. Zhao, W. Zhang, J. Sun, J. Zhou and D. Lippens, "Voltage tunable short wire-pair type of metamaterial infiltrated by nematic liquid crystal," *Appl. Phys. Lett.*, vol. 97, 134103, 2010.

[10] A. Minovich, D. N. Neshev, D. A. Powell, I. V. Shadrivov and Y. S. Kivshar, "Tunable fishnet metamaterials infiltrated by liquid crystals," *Appl. Phys. Lett.*, vol. 96, 193103, 2010.

[11] W. M. Zhu, A. Q. Liu, X. M. Zhang, D. P. Tsai, T. Bourouina, J. H. Teng, X. H. Zhang, H. C. Guo, H. Tanoto, T. Mei, G. Q. Lo and D. L. Kwong, "Switchable magnetic metamaterials using micromachining processes," *Adv. Mat.*, vol. 23, pp. 1792-1796, 2011.

[12] Y. H. Fu, A. Q. Liu, W. M. Zhu, X. M. Zhang, D. P. Tsai, J. B. Zhang, T. Mei, J. F. Tao, H. C. Guo, X. H. Zhang, J. H. Teng, N. I. Zheludev, G. Q. Lo, and D. L. Kwong, "A micromachined reconfigurable metamaterial via reconfiguration of asymmetric split-ring resonators," *Adv. Func. Mat.*, vol 21, pp. 3589-3594, 2011.

[13] W. Zhang, W. M. Zhu, H. Cai, M. L. J. Tsai, G. Q. Lo, D. P. Tsai, H. Tanoto, J. H. Teng, X. H. Zhang, D. L. Kwong and A. Q. Liu, "Resonance switchable metamaterials using MEMS fabrications," *IEEE Journal of Selected Topics in Quantum Electronics*, Vol 19, 4700306, 2013.

[14] W. M. Zhu, A. Q. Liu, T. Bourouina, D. P. Tsai, J. H. Teng, X. H. Zhang, G. Q. Lo, D. L. Kwong and N. I. Zheludev, "Microelectromechanical Maltese-cross metamaterial with tunable terahertz anisotropy," *Nature Communications*, Vol 3, 1274, 2012.

[15] W. Zhang, W. M. Zhu , E. P. Li , H. Tanoto , Q. Y. Wu , J. H. Teng , X. H. Zhang , J. M. Tsai , G. Q. Lo , D. L. Kwong and A. Q. Liu, "Micromachined Switchable Metamaterial with Dual Resonance," *Appl. Phys. Lett.*, Vol 101, 151902 2012.

[16] W. M. Zhu, A. Q. Liu, W. Zhang, J. F. Tao, T. Bourouina, J. H. Teng, X. H. Zhang, Q. Y. Wu, H. Tanoto, H. C. Guo, G. Q. Lo and D. L. Kwong, "Polarization dependent state to polarization independent state change in THz metamaterials," *Appl. Phys. Lett.*, Vol 99, 221102, 2011.

[17] G. M. Kuwata, N. Saito, Y. Ino, M. Kauranen, K. Jefimovs, T. Vallius, J. Turunen and Y. Svirko, "Giant Optical Activity in Quasi-Two-Dimensional Planar Nanostructures;" *Phys. Rev. Lett.*, Vol. 95 227401, 2005.

CONTACT

*A. Q. Liu, Tel: +65-67904336; eaqliu@ntu.edu.sg

MULTIPLE SIZE-ORIENTED PASSIVE DROPLET SORTING DEVICE AND BASIC APPROACH FOR DIGITAL CHEMICAL SYNTHESIS

Satoshi Numakunai[1], Afshan Jamsaid[1], Dong Hyun Yoon[1], Tetsushi Sekiguchi[2] and Shuichi Shoji[1]

[1]Major in Nanoscience and Nanoengineering, Waseda University, Japan
[2]Institute for Nanoscience and Nanotechnology, Waseda University, Japan

ABSTRACT

This paper presents a multiple size-oriented passive droplet sorting utilizing a balance between surface free energy and flow force. We proposed a simple multi-stage sorting structure which consists of a microchannel and guide grooves formed in parallel. Passive five different-sized droplets sorting of up to 200 droplets/s is achieved without any active elements. Also, we fabricated a prototype of the integrated micro fluidic system for droplet generation, merging and sorting for digital chemical synthesis.

INTRODUCTION

Droplet-based microfluidics has been attracted because it enables to control very small amount of reagents or samples as droplets. For instance, monodispersed emulsions in microfluidic channels can also be used as miniaturized reactors in which the reaction time could be rigidly controlled [1, 2]. Therefore droplet-based microfluidics is used in various applications like chemical, bio-chemical [3] and optical devices [4] due to its advantages in high throughput sample handling and contamination free. Recently, micro-capsule techniques for single cell analysis using droplet-based microfluidics are also reported [5, 6].

Droplet-based microfluidics requires some droplet manipulations such as generation, mixing, fusion and sorting [7-9]. In actual applications, high throughput and uniform size microdroplet generation is requested. Droplet size deviation is sometimes occurred due to the instability of injection pumps and unexpected bubble generation in the channel. This causes misoperation in droplet-based microfluidics detection or synthesis systems. Precise and high throughput droplet size selection by sorting is required in practical uses.

Various sorting methods using electric field [7] and laser [8] are reported. However, these methods are active sorting which requires complex operation and large external flow control equipments. In addition, since active controls utilizing electric or laser force can cause damage to samples, these controls are not suitable for biological applications. On the other hand, passive droplet sorting method using flow focusing [9] is also reported. Passive droplet sorting is achieved with minimal external equipments, easy operation and little damage to samples. Therefore, passive method suits to biological assay. However, the accuracy of droplets sorting is not enough under unstable flow condition because previous passive droplet sorting systems strongly depend on external flow control.

In this work, we propose a passive five different-sized droplet sorting device consisting of a simple channel structure without precise flow control. A prototype of droplet sorting device integrated with droplet merging system for digital chemical and bio-chemical analysis is also developed.

CONCEPT

In our previous work, we achieved a size-oriented passive droplet sorting with two parallel grooves using balancing between surface free energy trapping force and flow force [10]. In this study, three different-sized droplet sorting structure having three parallel grooves as shown in Figure 1 is used by utilizing droplet manipulation concept the same as the previous study. The structure consists of a center narrow groove along with the flow channel and a wide upper and lower grooves located different positions at a sorting area in parallel (Figure 1(a)). Flows toward the upper and lower groove exist due to the low flow resistance of the wide grooves. Downward flow force is stronger than upward flow force because the lower groove is formed closer to the center groove than the upper groove. Hence, large droplets move up and sift to the upper groove along with upward flow due to lower trapping force to the center groove (Figure 1(b)). Similarly, medium droplets shift to the lower groove (Figure 1(c)) while small droplets flow along the center groove (Figure 1(d)).

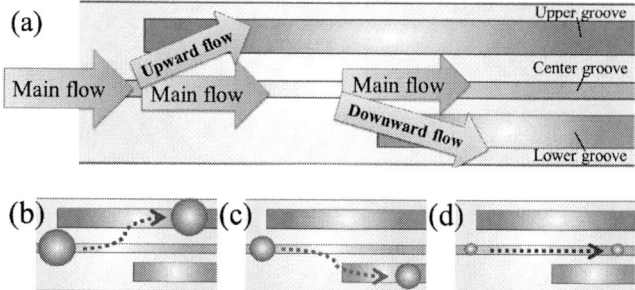

Figure 1: (a) Illustration of the fluid flow at the sorting area. Movement of (b) large droplet, (c) medium droplet and (d) small droplet around the sorting area.

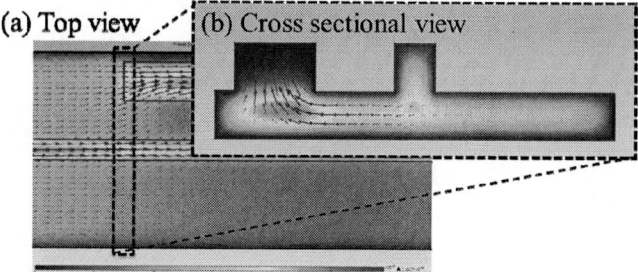

Figure 2: Result of CFD simulation. (a) Top view and (b) cross sectional view of the microchannel with two grooves.

978-1-4799-3510-9/14 $31.00 © 2014 IEEE

CFD (Computational Fluid Dynamics) simulation was applied for the microchannel having two parallel grooves and the flow toward the upper groove is confirmed. The flow velocity distributions in microchannel are shown in Figure 2.

FIVE DIFFERENT-SIZED DROPLET SORTING DEVICE

Principle and Device Design

Figure 3(a) explains a five different-sized droplets moving behavior around the sorting area. The proposed device has two sorting areas by connecting the three grooves structure in series. Five parallel guide grooves are formed from the sorting area to each outlet. Five different-sized droplets can be sorted step by step depending on their size.

Figure 3(b) shows an overall view of five different-sized droplets sorting device. A cross sectional view of the sorting area and the width of guide grooves are shown in Figure 3(c). The distance from each guide groove to the center groove is shown in Table 1. The height of flow channel and the depth of grooves are 50 μm.

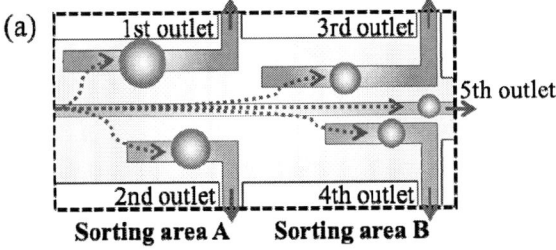

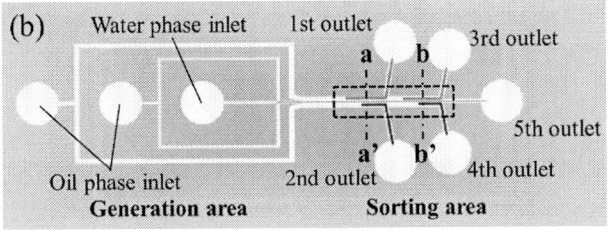

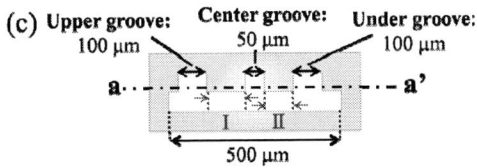

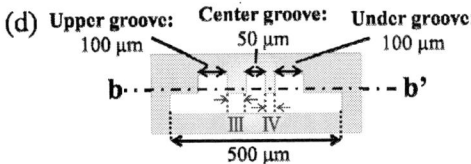

Figure 3: (a) Principle of five different-sized droplet sorting. (b) Overall view of the proposed device. (c) Cross sectional view of the sorting area A and (d) the sorting area B.

Table 1: Groove pitch of the sorting area A and B.

I [μm]	II [μm]	III [μm]	IV [μm]
100	75	50	25

Fabrication

The proposed device is fabricated by PDMS (poly-dimethylsiloxane, Dow Coming Toray SILPOT 184) molding with SU-8 two-stage mold as shown in Figure 4. The SU-8 mold is made by photolithography process. SU-8 3050 (MicroChem) is spin-coated on the silicon substrate and exposed in order to the fabricate channel structure (Figure 4(a)). This process is performed again for the groove structure (Figure 4(b)) and the mold is developed with SU-8 developer as shown in Figure 4(c). The PDMS base resin and curing agent are mixed in 10:1 ratio and poured onto the SU-8 mold. After degassing process, they are cured at 75 ℃ for 45 min (Figure 4(d)). The molded PDMS microchannel chip is bonded to a PDMS which is coated on a glass substrate after O_2 plasma treatment (Figure 4(e, f)).

Figure 5 shows SEM (Scanning Electron Microscope, KEYENCE VE-7800) images of the fabricated PDMS replicas in each area. Figure 5(a) is the SEM image of droplet generation area, and Figure 5(b) is the SEM image of droplet sorting area A. As shown in these SEM images, the structure of the proposed device is fabricated precisely.

Experimental setup

For fluidic experiments, syringe pumps (KDS210, KD Scientific) are used to control volumetric flow rates of water and oil phase fluid. Deionized water in which Methylene Blue is solved and plant oil are introduced into the each inlet as water and oil phase fluids. In order to evaluate the generation rate and the diameter of the droplet, the high speed CCD camera (FASTCAM-NEO, Photoron) is employed. Using this system, the droplet movement is observed and memorized as an image data. The high intensive light source of a halogen lamp is used to obtain clear images of the droplet generation and sorting.

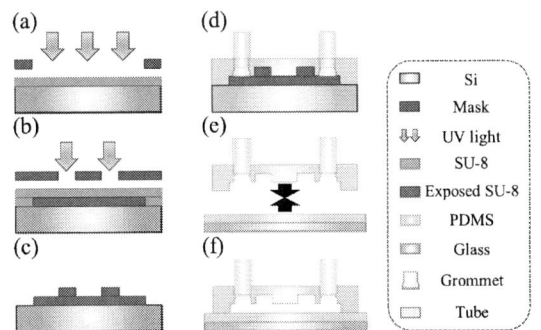

Figure 4: (a, b) spin coat and exposure of SU-8, (c) development, (d) PDMS molding and (e, f) plasma bonding.

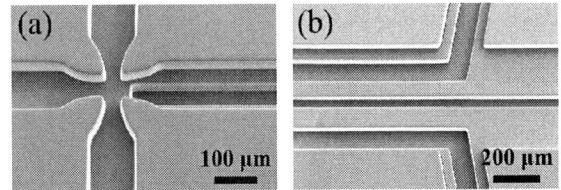

Figure 5: Device fabrication results of five different-sized droplet sorting device. SEM image of (a) the droplet generation area and (b) the droplet sorting area A.

Results

Figure 6 shows the movement of five different-sized droplets around the sorting area. Droplets of about 90 μm, 80 μm, 75 μm, 60 μm and 55 μm in radius were separated step by step from the 1st outlet to the 5th outlet, respectively. Five different-sized droplet sorting of up to 200 droplets/s was successfully performed. Figure 7 shows the radius ranges of droplet flowing into each outlet. The droplets are collected from each outlet as follows: 1st outlet; > 88 μm, 2nd outlet; 77-85 μm, 3rd outlet; 73-77 μm, 4th outlet; 63-72 μm and 5th outlet; < 62.3 μm in radius.

(a) To 1st outlet

(b) To 2nd outlet

(c) To 3rd outlet

(d) To 4th outlet

(e) To 5th outlet

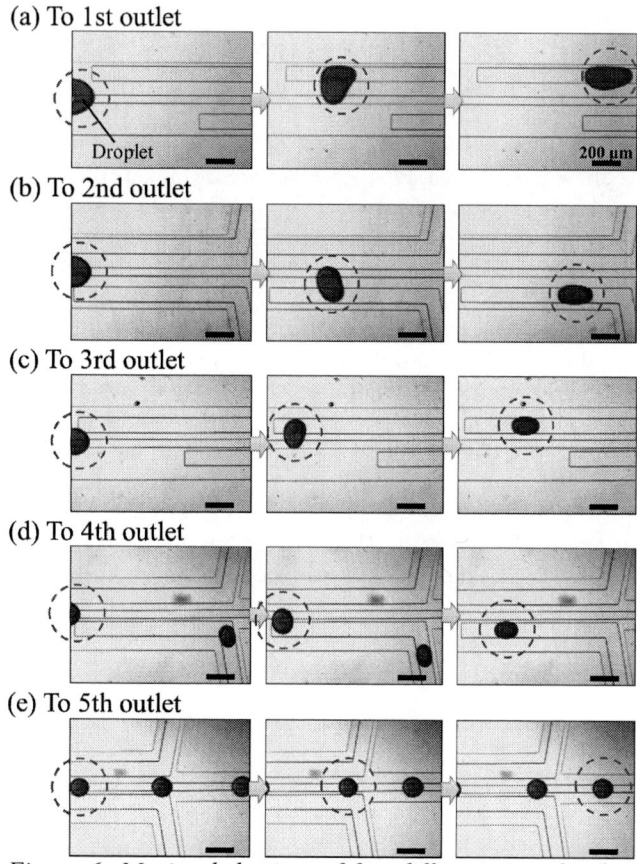

Figure 6: Moving behaviors of five different-sizes droplets around the (a-b) sorting area A and (c-e) sorting area B.

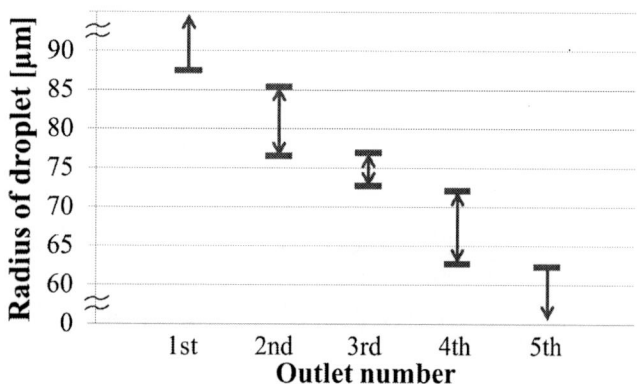

Figure 7: Result of five different-sized droplets sorting.

INTEGRATED MICRO FLUIDIC DEVICE FOR DROPLET SORTING AND MERGING

Principle and Device Design

The illustration of droplet merging area (our previous work [11]) is shown in Figure 8(a). Droplets can be merged at the merging area using pneumatic actuators. Based on the principle shown in Figure 1(b), droplets without merging can be removed from the center outlet and merged droplets are collected to the lower outlet (Figure 8(b)).

Figure 9(a) shows an overall view of droplet sorting device integrated with droplet merging system. Droplets of different colors are generated at the T-junction. The merging area consists of main channel, pillars and side channel. Horizontal pneumatic micro actuators are formed beside the two channels. Detailed dimensions of merging area are shown in Figure 9(b). Figure 10 is the SEM image of the fabricated droplet merging area.

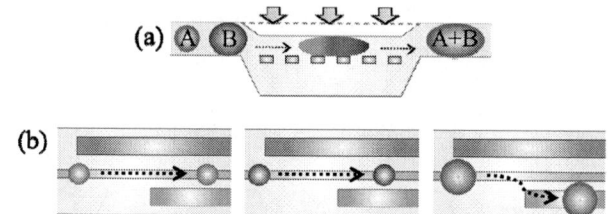

Figure 8: Principle of droplet merging and sorting. (a) Illustration of droplet merging and (b) merged and non-merged droplet sorting.

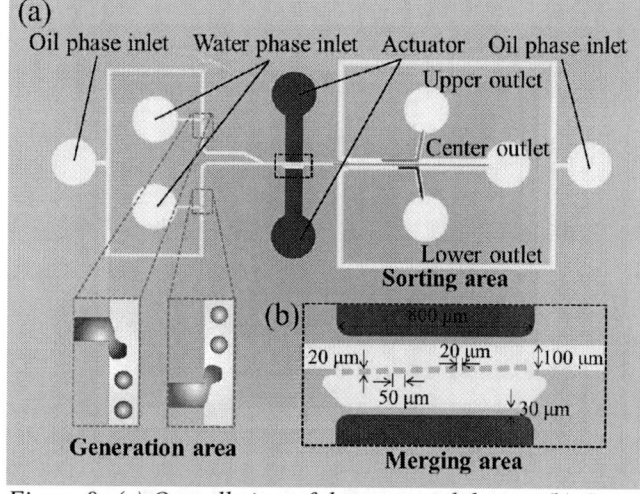

Figure 9: (a) Overall view of the proposed device. (b) Cross sectional view of the sorting area.

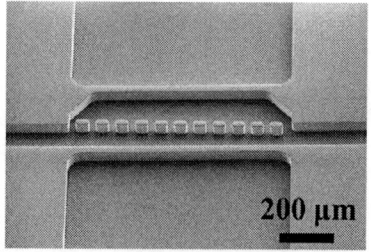

Figure 10: SEM image of droplet merging area.

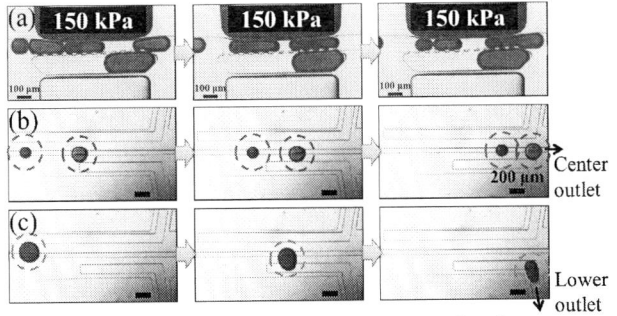

Figure 11: (a) Experimental result of two droplet merging. Movements of (b) blue droplet A and red droplet B and (c) merged A+B droplet at the sorting area.

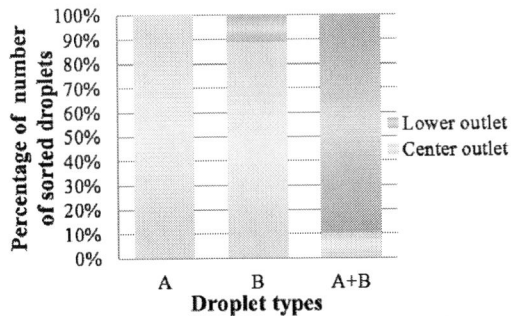

Figure 12: Result of droplet separation of the proposed sorting device integrated with merging system.

Results

Two droplets merging was achieved as shown in Figure 11(a) when flow resistance of the main channel was controlled by applying pressure of 150 kPa. Blue droplet A (Methylene Blue solved in water) of 55 μm in radius and red droplet B (Rhodamine B solved in water) of 65 μm in radius on average were generated. These droplets were merged into purple droplet of about 76 μm in radius. Also, droplets without merging and merged droplets were sorted successfully as shown in Figure 11 (b) and (c). Figure 12 shows the result of droplet A, B and merged A+B droplet sorting. Collection of merged droplet of about 90% in accuracy was achieved.

CONCLUSION

We successfully achieved multiple size-oriented passive droplet sorting and the fabricated prototype of sorting device including droplet generation and merging systems. The proposed sorting device is one of the key elements for practical digital chemical synthesis systems. In order to realize a reliable total flow system, optimization of connection structure between the flow elements is an important issue.

ACKNOWLEDGEMENTS

This work is partly supported by Japan Ministry of Education, Culture, Sports Science & Technology (MEXT) Grant-in-Aid for Scientific Basic Research (S) No. 23226010. The authors thank for MEXT Nanotechnology Platform Support Project of Waseda University.

REFERENCES

[1] Z. T. Cygan, J. T. Cabral, K. L. Beers, E. J. Amis, "Microfluidic Platform for the Generation of Organic-Phase Microreactors", *Langmuir*, vol. 21, 2005, pp. 3629-3634.

[2] E. M. Chan, A. P. Alivisatos, R. A. Mathies, "High-Temperature Microfluidic Synthesis of CdSe Nanocrystals in Nanoliter Droplets", *J. Am. Chem. Soc.*, vol. 127, 2005, pp. 13854-13861.

[3] B. Zheng, L. S. Roach, R. F. Ismagilov, "Screening of Protein Crystallization Conditions on a Microfluidic Chip Using Nanoliter-Size Droplets", *J. Am. Chem. Soc.*, vol. 125, 2003, pp. 11170-11171.

[4] A. Fernández-Nieves, D. R. Link, D. Rudhardt, D. A.Weitz, "Electro-Optics of Bipolar Nematic Liquid Crystal Droplets" *Phys. Rev. Lett.*, vol.92, 2004, pp. 105503.

[5] M. He, J. S. Edgar, G. D. M. Jeffries, R. M. Lorenz, J. P. Shelby, D. T. Chiu, "Selective Encapsulation of Single Cells and Subcellular Organelles into Picoliter- and Femtoliter-Volume Droplets", *Anal. Chem.*, vol.77, 2005, pp. 1539-1544.

[6] E. Brouzesa, N. Medkova, N. Savenelli, D. Marran, M. Twardowski, J. B. Hutchison, J. M. Rothberg, D. R. Link, N. Perrimon, M. L. Samuels, "Droplet Microfluidic Technology for Single-Cell High-Throughput Screening" *Proc. Natl. Acad. Sci. USA*, vol.106, 2009, pp. 14195-14200.

[7] F. Guo, X.-H. Ji, K. Liu, R.-X. He, L.-B. Zhao, Z.-X. Guo, W. Liu, S.-S. Guo, X.-Z. Zhaoa, "Droplet Electric Separator Microfluidic Device for Cell Sorting", *Appl. Phys. Lett.*, vol.96, 2010, pp. 193701.

[8] C. N. Baroud, J.-P. Delville, F. Gallaire, R. Wunenburger, "Thermocapillary Valve for Droplet Production and Sorting", *Phys. Rev. E*, vol.75, 2007, pp. 046302.

[9] H. Maenaka, M. Yamada, M. Yasuda, M. Seki, "Continuous and Size-Dependent Sorting of Emulsion Droplets Using Hydrodynamics in Pinched Microchannels", *Langmuir*, vol.24, 2008, pp. 4405-4410.

[10] S. Numakunai, Y. Harada, D. H. Yoon, T. Sekiguchi, S. Shoji, "High Throughput Size-Oriented Passive Droplet Sorting Device Using Surface Free Energy with Simple Parallel Guide Grooves", in *Digest Papers Microprocesses and Nanotechnology Conference'12*, Kobe, October 30- November 2, 2012, 2B-8-3.

[11] D. H. Yoon, S. Numakunai, J. Ito, T. Fukuda, T. Sekiguchi, S. Shoji, "Number Controllable Active Microdroplet Merging with Wide Range of Volume for Digital Chemical Synthesis", in *Digest Tech. Papers Transducers'13 Conference*, Barcelona, June 16-20, 2013, pp. 2660-2663.

CONTACT

*S. Numakunai, tel: +81-3-5286-3384; numakunai@shoji.comm.waseda.ac.jp

SURFACE ENERGY MICROPATTERN INHERITANCE FROM MOLD TO REPLICA

Gaspard Pardon[1], Tommy Haraldsson[1]* and Wouter van der Wijngaart[1]
[1] KTH Royal Institute of Technology, Sweden

ABSTRACT

We report a novel surface-energy patterning phenomenon, in which a novel polymer composition inherits the surface energy of the medium it is in contact with during polymerization. This surface property mimicking process occurs via spontaneous selective molecular alignment of hydrophilic and hydrophobic monomers mixed into an off-stoichiometry thiol-ene (OSTE) formulation. This single-step method for simultaneous structuring and surface energy micropatterning of polymer structures is potentially more robust and lower cost than state-of-the-art processes requiring post-processing surface modification steps. We further demonstrate the self-assembly of a liquid droplet array on the replicated polymer surfaces.

INTRODUCTION

Most polymer surface modification protocols are performed in back-end operations, i.e. *after* device structuring, using plasma activation and/or chemical reactions [1]-[7]. However, these methods suffer from limited reliability, lifetime or spatial control [8] and greatly increase the complexity and cost of microfluidic device production [9]. Two solutions for integrating surface modifications during device structuring are surface topography nanostructuring [10] and addition of a modifier to the primary prepolymer [11], [12]. The former suffers from lack of robustness, i.e. fouling of the surface nanostructures, or from complex manufacturing; and the latter failed so far to offer spatial control, i.e. micropatterning, of the surface energy.

Here, we report a novel surface-energy patterning phenomenon, in which the surface energy of a novel polymer composition mimics that of the phase it is in contact with during polymerization. The phenomenon occurs for specifically designed prepolymer mixtures that contain hydrophilic and hydrophobic monomers. The mimicking process is thought to occur via selective enrichment of hydrophilic and hydrophobic monomers at a surface with a matching surface energy. Unlike state-of-the-art methods relying on complex, unreliable and time-consuming post-processes for surface modification, here we demonstrate a novel, single-step manufacturing method for simultaneous structuring and surface energy micropatterning of polymer structures that is potentially more robust and lower cost. We further demonstrate the self-assembly of a liquid droplet array on the replicated polymer surfaces.

MATERIALS AND METHODS
Materials

We designed a thiol-ene polymer of the OSTE polymer family, which has shown a number of promising features for lab-on-chip applications, such as tunable mechanical properties, robust sealing via UV-activated covalent bonding, ease of surface modification and excellent photolithographic capabilities. [13], [14]

An OSTE-Thiol 80 prepolymer was prepared using the following monomers: Tetraallyloxyethane (TAOE) (Tokyo Chemical Industry Co., Ltd., Japan) and Pentaerythritol tetrakis(2-mercaptoacetate) (PETMA) (Sigma-Aldrich Co., USA). These two monomers were mixed in a ratio 24.55 %wt and 74.15 %wt, to obtain 80 % excess of thiol functional groups. 0.3 %wt of Lucirin TPO-L (BASF Corp., USA) was added as photoinitiator.

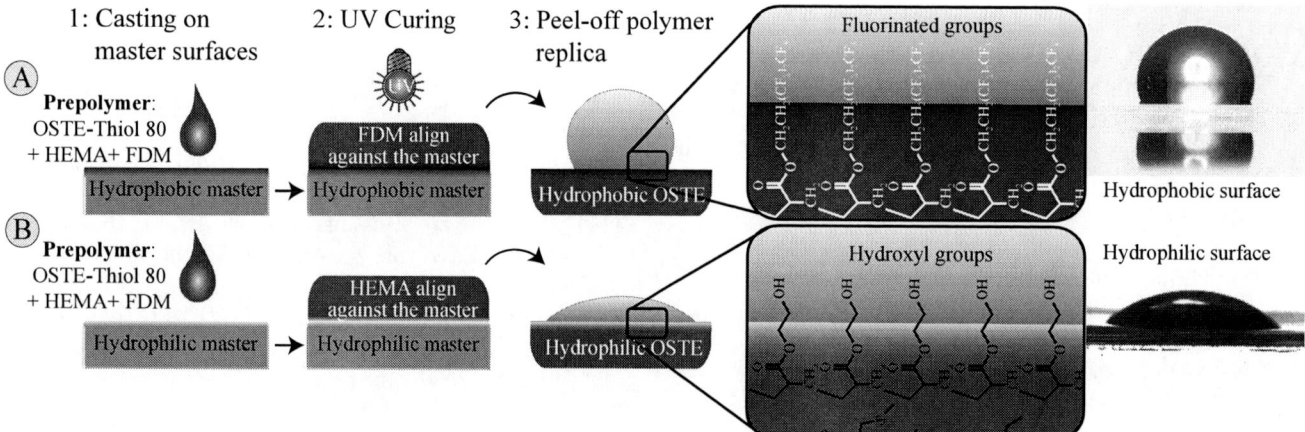

Figure 1: Description of the phenomenon. When the prepolymer, loaded with fluorinated and hydroxylated methacrylates, is UV-cured, it inherits the surface energy of the master surface, i.e. it becomes hydrophilic or hydrophobic when cured against a hydrophilic or hydrophobic master surface, respectively.

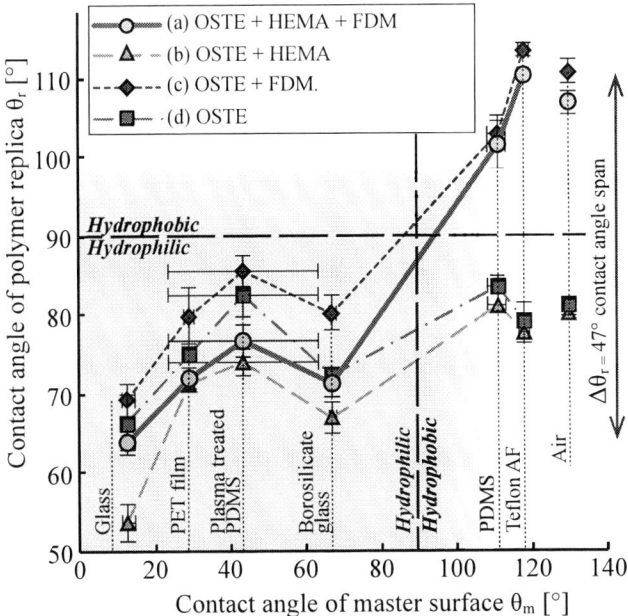

Figure 2: Measurements of the contact angle of the polymer replica versus that of the master surface in contact during photopolymerization. A total contact angle span of 47° is achieved. Note the large error bar for the plasma treated PDMS master surface due to short-lived surface treatment

The following functionalized methacrylates were used: 3,3,4,4,5,5,6,6,7,7,8,8,9,9,10,10,-heptadecafluorodecyl methacrylate (FDM) (Sigma-Aldrich Co., USA) 20 %wt in toluene and 2-hydroxyethyl methacrylate (HEMA) (Sigma-Aldrich Co., USA) 5 %wt in toluene.

One prepolymer formulation was specifically designed for dual energy patterning (a) and three prepolymer formulation, (b), c), and (d), were used for evaluation purposes (Table 1).

Table 1: Prepolymer formulations prepared for this work.

	OSTE-Thiol 80	FDM	HEMA
(a)	Base polymer	0.5 %wt	0.5 %wt
(b)	Base polymer	-	0.5 %wt
(c)	Base polymer	0.5 %wt	-
(d)	Base polymer	-	-

Solvents used were isopropyl alcohol (IPA), for cleaning and rinsing of molds and cured OSTE-Thiol 80 polymer surfaces, and toluene (Merck KGaA, Germany), for dissolving the methacrylate monomers in the OSTE prepolymer.

Master surfaces were prepared as follows. Glass substrates were microscopy slides (VWR, Sweden). Borosilicate glass was microscope cover slips glass (VWR, Sweden). PET films were provided by AGFA (Belgium). PDMS substrates were prepared in-house within an hour before the experiments. Plasma treatment was performed using an H_2O plasma oven (Femto 1, Diener electronic GmbH, Germany) at 0.3 mbar, 40 w power, 720 cm^3/min flow, for 15 s. Teflon surfaces were prepared on glass slides

by spinning Teflon AF (DuPont, USA) 0.15 %wt in Teflon AF solvent at 1200 rpm for 60 s, followed by 10 min at 110° annealing process. The thickness of the resulting Teflon AF film was measured to be 8 nm, using ellipsometry.

Planar surface replica manufacturing and testing

The polymer replicas were prepared as shown in Figure 1. Each of the four liquid prepolymers, (a), (b), (c) or (d), was sandwiched between a supporting substrate and one of the master surfaces, with 500 µm interspacing. One sample was also cured without master surface, exposed to the ambient air. The prepolymer was cured under UV light. The curing was performed using a collimated 12 mW cm^{-2} near-UV mercury lamp (OAI, Milpitas, USA) with 8 s exposure time. The master surfaces were carefully removed to expose flat replica surfaces that were subsequently rinsed using IPA and dried using nitrogen gas, before proceeding with contact angle measurements.

Contact angles were measured using 5 µl droplets using the *Contact angle* plugin of the ImageJ software. A minimum of 6 droplets per sample were measured. Master surface contact angles were measured before and after being used to cure the polymers to ensure that material transfer was not the main mechanism behind the observed phenomena. Results are reported in Figure 2 and described below.

Patterned replica manufacturing and testing

We also developed a novel single-step fabrication method for simultaneous microstructuring and surface energy micropatterning. A glass master was pattern-coated with a Teflon-AF layer, leaving a microarray of 200 µm diameter circular hydrophilic patches exposed with a pitch of 1 mm. The hydrophobic-hydrophilic patterns were prepared on a glass slide using the soft-lithography MIMIC process, [15] using the Teflon AF solution defined above (Fig. 3A). For this process, the micropattern was photolithographically structured in an SU8 layer on a silicon substrate. PDMS was then cast on this mold, to obtain a PDMS layer containing a fluidic network with an inverted pattern layout. The PDMS layer was soft-bonded to a glass slide and the fluidic network was filled with the Teflon AF solution. The Teflon was subsequently cured in an oven. The PDMS layer was removed, leaving the area in contact with the PDMS unchanged and the area in contact with the Teflon modified. The pattern consisted of 200 µm diameter hydrophilic patches surrounded by a hydrophobic surface. Results are shown in Figure 3B and described below.

For testing, the chip was submerged in a liquid, which was capilarilly withdrawn. Both high surface tension (DI water) and low surface energy blue dye solution (Alphazurine A, Sigma-Aldrich Co., USA), were tested.

RESULTS
Surface energy mimicking

Figure 2 plots the measured contact angles of replica surfaces, θ_r, versus master surfaces, θ_m, clearly illustrating

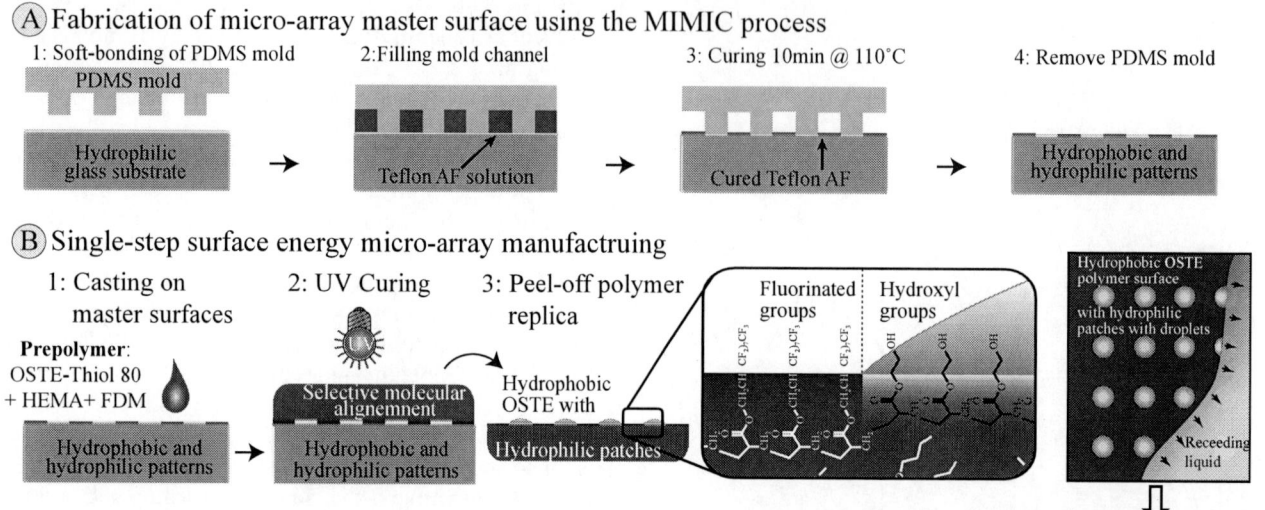

(A) Fabrication of micro-array master surface using the MIMIC process

1: Soft-bonding of PDMS mold → 2: Filling mold channel → 3: Curing 10min @ 110°C → 4: Remove PDMS mold

(B) Single-step surface energy micro-array manufactruing

1: Casting on master surfaces → 2: UV Curing → 3: Peel-off polymer replica

Figure 3: A: Fabrication of the micro-array master surface using the MIMIC process [15]; B: Single-step manufacturing of a surface-energy micropattern for the self-assembly of liquid droplets.

the mimicking effect: the designed polymer (a) is hydrophilic ($\theta_{r,a,glass} = 63°$) when cured against a hydrophilic glass master ($\theta_{m,glass} = 13°$) and hydrophobic ($\theta_{r,a,Teflon} = 110°$) when cured against a hydrophobic Teflon-AF master ($\theta_{m,Teflon} = 118°$), resulting in a contact angle span of $\Delta\theta_{r,a} = 47°$. In comparison, the intrinsic contact angle span for unmodified OSTE-Thiol 80 (d) was only $\Delta\theta_{r,d} < 12°$.

Polymer (a) follows the same trend as polymer (c) against hydrophobic master surfaces, while it follows that of polymer (b) against hydrophilic master surfaces, showing that selectivity between hydrophobic or hydrophilic expression is obtained with little interference between the two added methacrylates. However, it seems more difficult to obtain significantly different surface energy compared to the original OSTE (d) on the hydrophilic end of the master surface types, probably because of the comparatively lower number of functional groups on the hydrophilic methacrylate compared to the hydrophobic methacrylate used in this work.

Surface energy micropatterning

Figure 4 shows a stereomicroscope image of the fabricated polymer chip while the liquid was capillarily withdrawn, and where microdroplets are left on the hydrophilic patches in a self-assembly process.

Because of the high surface tension of water, it was not possible to detach microdroplets from the bulk to generate the microdroplet array. Hence, IPA was added to water in a 1:3 ratio to lower the surface tension. The volume of the deposited droplets was calculated to be in the picoliter range. Using the MIMIC process or other microfabrication processes, it is possible to simultaneously structure and surface-energy micropattern the chip.

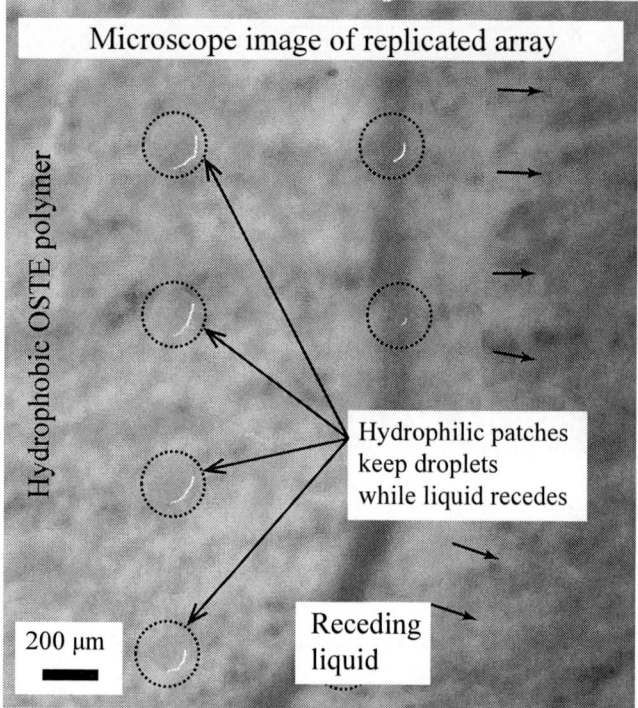

Figure 4: Stereomicroscope image of the single-step fabricated surface-energy micropattern, showing the self-assembly of liquid droplet on the hydrophilic patches when the bulk liquid recedes.

DISCUSSION

We hypothesize that, in an energy minimization process, hydrophobic functional groups preferentially self-align at a prepolymer liquid surface in contact with a hydrophobic phase, while hydrophilic functional groups are repelled from such interface, and vice versa. This results in that the functional group of the methacrylates are either expressed or repelled at the surface of the polymer replica. The methacrylate monomers are subsequently reacted with the thiols and enes during polymerization, resulting in a surface energy mimicking of the hydrophilic and

978-1-4799-3510-9/14 $31.00 © 2014 IEEE

hydrophobic regions of the mold in the finished polymer piece.

Two observations support the hypothesis that the contact angle is determined by a molecular restructuring of the replica polymer, and not by a material transfer from the master to the replica. First, a clear differentiation of the resulting contact angle is observed for the different polymers (a) to (d) on the different master surfaces. Second, polymerizing in contact with air or with Teflon results in comparable liquid contact angles, and this both for modified (a), $\theta_{r,a,Teflon} \approx \theta_{r,a,air}$, and unmodified (d) polymers, $\theta_{r,d,Teflon} \approx \theta_{r,d,air}$.

Since the functional methacrylate monomers are polymerized with the bulk polymer via covalent bonds, the replica surface modifications are robust and experimentally verified to last for at least 2 months, which is the time since the experiments were conducted.

CONCLUSION

The method presented enables simple, reliable, durable and cost-efficient simultaneous structuring and surface energy micropatterning using a polymer system designed for lab-on-chip applications. The process can be performed on structured molds to produce both microstructures and surface energy patterns in a single step. Since the pattering does not involve lateral surface diffusion, which is the case with most post-processing method, it can potentially be used for nanopattern generation. The generated surface energy patterns on the polymer surface can potentially be used for selective biofunctionalization and improved spotting resolution.

REFERENCES

[1] L. E. Locascio, A. C. Henry, T. J. Johnson, and D. Ross, "Lab-on-a-chip: Miniaturized Systems for (bio)chemical Analysis and Synthesis - Google Books," *Lab Chip*, 2003.

[2] S. A. Soper, A. C. Henry, B. Vaidya, M. Galloway, M. Wabuyele, and R. L. McCarley, "Surface modification of polymer-based microfluidic devices," *Analytica Chimica Acta*, vol. 470, no. 1, pp. 87–99, 2002.

[3] A. C. Henry, T. J. Tutt, M. M. Galloway, Y. Y. Davidson, C. S. McWhorter, S. A. Soper, and R. L. McCarley, "Surface modification of poly(methyl methacrylate) used in the fabrication of microanalytical devices.," *Anal. Chem*, vol. 72, no. 21, pp. 5331–5337, Oct. 2000.

[4] Seguin, McLachlan, Norton, "Surface modification of poly(dimethylsiloxane) for microfluidic assay applications," *Applied Surface Science*, vol. 256, no. 8, pp. 8–8, Jan. 2010.

[5] J. P. Lafleur, R. Kwapiszewski, T. G. Jensen, and J. P. Kutter, "Rapid photochemical surface patterning of proteins in thiol–ene based microfluidic devices," *Analyst*, vol. 138, no. 3, p. 845, 2013.

[6] J. Zhou, D. A. Khodakov, A. V. Ellis, and N. H. Voelcker, "Surface modification for PDMS-based microfluidic devices," *Electrophoresis*, vol. 33, no. 1, pp. 89–104, Jan. 2012.

[7] C. F. Carlborg, F. Moraga, F. Saharil, W. van der Wijngaart, and T. Haraldsson, "Rapid Permanent Hydrophilic and Hydrophobic Patterning of Polymer Surfaces via Off-Stoichiometry Thiol-Ene (OSTE) Photographting," presented at the 16th International Conference on Miniaturized Systems for Chemistry and Life Sciences, 2012, pp. 677–679.

[8] B. Levaché, A. Azioune, M. Bourrel, V. Studer, and D. Bartolo, "Engineering the surface properties of microfluidic stickers.," *Lab Chip*, vol. 12, no. 17, pp. 3028–3031, Sep. 2012.

[9] H. Becker and C. Gärtner, "Polymer microfabrication technologies for microfluidic systems.," *Analytical and bioanalytical chemistry*, vol. 390, no. 1, pp. 89–111, Dec. 2007.

[10] K. Tsougeni, D. Papageorgiou, A. Tserepi, and E. Gogolides, "'Smart' polymeric microfluidics fabricated by plasma processing: controlled wetting, capillary filling and hydrophobic valving.," *Lab Chip*, vol. 10, no. 4, pp. 462–469, Feb. 2010.

[11] J. Wang, A. Muck Jr, M. P. Chatrathi, G. Chen, N. Mittal, S. D. Spillman, and S. Obeidat, "Bulk modification of polymeric microfluidic devices.," *Lab Chip*, vol. 5, no. 2, pp. 226–230, Jan. 2005.

[12] Y. Xiao, X.-D. Yu, J.-J. Xu, and H.-Y. Chen, "Bulk modification of PDMS microchips by an amphiphilic copolymer.," *Electrophoresis*, vol. 28, no. 18, pp. 3302–3307, Aug. 2007.

[13] C. F. Carlborg, T. Haraldsson, K. Öberg, M. Malkoch, and W. van der Wijngaart, "Beyond PDMS: off-stoichiometry thiol-ene (OSTE) based soft lithography for rapid prototyping of microfluidic devices.," *Lab Chip*, vol. 11, no. 18, pp. 3136–3147, Sep. 2011.

[14] G. Pardon, F. Saharil, J. M. Karlsson, O. Supekar, C. F. Karlborg, T. Haraldsson, and W. van der Wijngaart, "Rapid mold-free manufacturing of microfluidic devices with robust and spatially directed surface modifications," *Microfluidics and Nanofluidics*.

[15] Y. Xia, E. Kim, and G. M. Whitesides, "Micromolding of Polymers in Capillaries: Applications in Microfabrication," *Chem. Mater*, vol. 8, no. 7, pp. 1558–1567, Jan. 1996.

CONTACT

*T. Haraldsson, KTH Royal Institute of Technology, Micro and Nanosystems, Osquldas väg 10, 10044 - Stockholm, Sweden; tel: +4687909059; tommyhar@kth.se

ELECTROSPRAY DEPOSITION FROM AFM PROBES WITH NANOSCALE APERTURES

J. Geerlings[1], E. Sarajlic[1,2], J.W. Berenschot[1], R.G.P. Sanders[1], L. Abelmann[1,3] and N.R. Tas[1]

[1]MESA+, University of Twente, Enschede, THE NETHERLANDS
[2]SmartTip B.V., Enschede, THE NETHERLANDS
[3]Korea Institute of Science and Technology, Saarbrücken, GERMANY

ABSTRACT

Electrospray deposition utilizes a high electric field to extract liquid droplets from a capillary nozzle. In this contribution we demonstrate non-contact droplet deposition by electrospray from atomic force microscopy (AFM) probes with a fully integrated microfluidic system, so called FluidFM probes. Electrospray experiments were performed using probes with a pyramidal tip with a sub-micron size aperture in a dedicated setup. The onset voltage as function of the gap between the probe tip and the substrate was measured and compared with a numerical model. Onset voltages in the range 360-410 V were found at 8.5 μm gap height. We observed a reduction in onset voltage with an increase in external pressure. Wetting of the outside of the tip could be reduced by applying a fluorocarbon coating.

INTRODUCTION

Liquid deposition on surfaces by means of AFM techniques has promising application in biomedicine and technology [1-4]. A well-known example is dip-pen lithography [1]. An important shortcoming of the dip-pen lithography is the necessity to regularly recoat and realign the tip during the pattering. In order to extent the writing, the nanofountain pen concept was introduced [2]. It consist of a microfluidic channel to deliver ink from a relatively large liquid reservoir located at the probe holder to a small aperture at the probe tip. The first working fountain pen probe was presented by our group [3] and further developed as FluidFM probe [4]. A typical application is surface modification by deposition of liquids or suspended particles.

Up to now, deposition has been achieved by bringing the tip of the AFM probe in contact with the substrate, which limits the tip/liquid/substrate combinations. In this contribution we present extraction of liquid from nanosized apertures in fountain pen AFM probes by means of electrospray. Not only does this technique allow contactless deposition, we also show that droplets with radii in the order of one micrometer can be deposited.

Electrospray is a well-known technique for the dispersion of liquid into droplets and also commonly used for direct deposition, however, performed from glass capillaries (for example in [5]). Electric field assisted deposition in an AFM setup has been demonstrated before by Kaisei et al. [6]. However, in their experiments they used traditional AFM probes which were modified by focused ion beam milling. No fluidic channel was present in these probes which prevents continuous operation. We extend on this work by electrospray experiments from batch

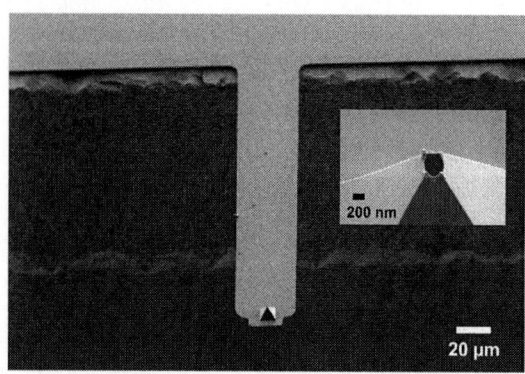

Figure 1: Scanning electron microscope (SEM) image of a FluidFM probe with pyramidal shaped tip. Inset shows the aperture at the apex of the pyramid, which is approximately 300 nm (fabricated using corner lithography).

Table 1: Dimensions of the FluidFM probes.

	Cantilever	Channel
Length	200 μm	1500 μm
Width	36 μm	28 μm
Height	2.2 μm	1 μm
Wall thickness	600 nm	---

fabricated AFM probes with integrated microchannel and liquid reservoir. Apertures of the probes, fabricated using corner lithography [7], were in the order of 300 nm (shown in Figure 1). Typical dimensions are given in Table 1.

THEORY

In Figure 2 the setup required for electrospray is shown. A voltage is applied between the liquid inside the reservoir of the probe and a conductive substrate. The fluidic channel inside the probe can be modeled as an electric resistance for conductive liquids. Based on the ionic concentrations and the 1 μm channel height, we expect the surface conduction to be small compared to the bulk conductivity. Filled with DI water (and carbonic acid formed from carbon dioxide), the resistance of the channel is estimated to be ~100 GΩ.

Electric field

With use of COMSOL Multiphysics the relation between potential at the meniscus and the electric field close to the meniscus was simulated. A simplified model of the probe, consisted of a nitride cantilever with uniform electrode on top and a droplet at the same potential. The half-spherical meniscus and pyramidal shape tip, give rise to a field enhancement (γ_E), ranging from 10.1 for a 5 μm gap to 29.7 for a 30 μm gap according to the definition

$\gamma_E = E_{local} d_{meniscus}/V_{tip}$. The gap height ($d_{gap}$) is defined as the distance between ground plane and the center of the aperture. The planar field is calculated using $d_{meniscus}$, which is the distance between the ground plane and the edge of the meniscus ($d_{meniscus} = d_{gap} - r_{aperture}$). The simulated values of E_{local} were determined close to the meniscus where the fields are maximal.

Expected onset voltage

For electrospray to occur, the electrostatic pressure acting on the meniscus must overcome the Laplace pressure (which is the pressure difference over the liquid meniscus due to the curvature of the meniscus). The minimal voltage at which electrospray begins, is called the onset voltage. If a half spherical meniscus shape is assumed with the two orthogonal radii of curvature equal to the aperture radius ($r_{aperture}$), the Laplace pressure can be found using

$$P_{Laplace} = \frac{2\gamma}{r_{aperture}}, \qquad (1)$$

where γ is the surface tension in J/m^2. The electrostatic pressure acting on a conducting surface is given by the Maxwell stress

$$P_{Maxwell} = \frac{1}{2}\varepsilon_0 E_s^2. \qquad (2)$$

In this equation E_s is the electric field in V/m perpendicular to the surface and ε_0 the permittivity of vacuum. The electric field is determined from the COMSOL calculations described in the previous section. From Equation 1 and 2, the onset electric field can be found. By applying an external pressure (P_a) to the liquid, the onset field can be reduced. The onset electric field including external pressure is given by

$$E_{onset} = \pm\sqrt{\frac{2}{\varepsilon_0}\left(\frac{2\gamma}{r_{aperture}} - P_a\right)}. \qquad (3)$$

Under the assumption that the liquid meniscus remains half spherical. Equation 3 shows that the onset field is dependent on the surface tension of the liquid, however, not on the conductivity.

The silicon nitride walls of the channel give rise to an electric double layer when they are in contact with aqueous solutions. When a voltage is applied over the channel, the resulting electric field will give rise to an electric osmotic flow (EOF). This EOF supports the liquid flow to the tip during positive mode electrospray. However, because of the relatively high channel (1 µm), the pressure generated by the EOF is negligible compared to the Laplace pressure. This was confirmed by COMSOL simulations.

Using the surface tension of water (72 mJ/m^2), the onset electric field for the probes with 300 nm aperture is 4.6×10^8 V/m (Equation 3). These fields can be translated to tip-substrate voltages using the COMSOL field calculations, and will be compared to measurements in the Results section. The cantilever stiffness of the FluidFM probes was estimated to be in the order of 3 N/m. Calculations showed

that electrostatic pull-in of the cantilever occurs at voltages higher than the expected onset voltages.

EXPERIMENTAL

We constructed a dedicated measurement setup, on top of a Leica inverted microscope (DMI 5000 M), Figure 2.

The probes were glued to a polycarbonate block which was connected to a plastic tube. By feeding pressurized nitrogen through the tube, pressure could be applied on the liquid inside the probe reservoir. The applied pressure was measured by a Panasonic DP-100 pressure sensor.

The polycarbonate block with probe was connected to a linear translation motor, which could vary the height of the probe (with respect to the substrate) with a resolution of 1 µm. A glass plate coated with indium tin oxide (ITO) was used as conductive (transparent) substrate. The angle between cantilever and substrate was approximately 10 degrees. A positioning meter (with a resolution of 1 µm) was in contact with the objective stage of the microscope. The gap height was measured by focusing successively on the topside of the ITO substrate and the end of the cantilever (taking the height difference with the tip into account).

The resistance between the ITO substrate and the source measurement unit (SMU) ground wire was smaller than 50 Ω. The current through the setup was limited by the compliance setting of the SMU. An additional 100 GΩ series resistance (Ohmite) was inserted between the high voltage connection of the SMU and the setup when measurements with high conductive liquids were performed. A N$_2$-fumehood was installed over the setup to reduce the leakage currents (by reduction of the relative humidity).

Typical properties of the liquids that were used for the experiments are given in Table 2. All liquids were filtered through a 0.2 µm PTFE syringe filter before use. A mixture of water, alcohol and acid is often used for electrospray experiments. The alcohol lowers the surface tension, while the added acid increases the conductivity. Pure DI water is a desirable solvent choice for electrospray mass spectrometry of biomolecules because water is the natural solvent of most biomolecules.

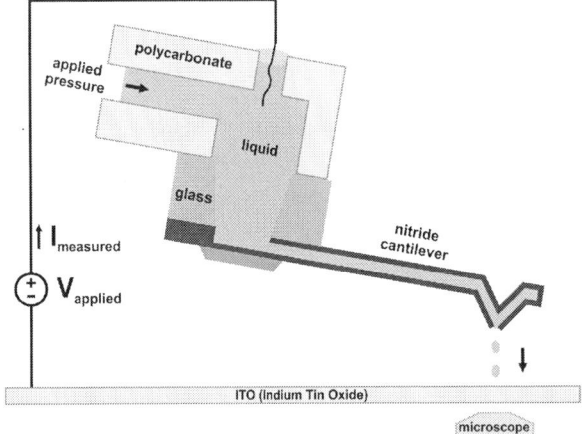

Figure 2: Schematic of the experimental setup with a FluidFM probe with pyramidal tip.

Table 2: Typical properties of the liquids used (DI = deionized water, IPA = isopropyl alcohol (Merck KGaA), AA = acetic acid (Merck KGaA), v%=volume%, m%=massa%).

	DI	DI/AA	DI/IPA/AA
Composition	100%	98v%/2v%	50v%/50v% + 5 m%
Resistivity ρ	9.8 kΩ·m	13 Ω·m	80 Ω·m
Surface tension γ	72 mJ/m^2	67 mJ/m^2	25 mJ/m^2 (estimation)

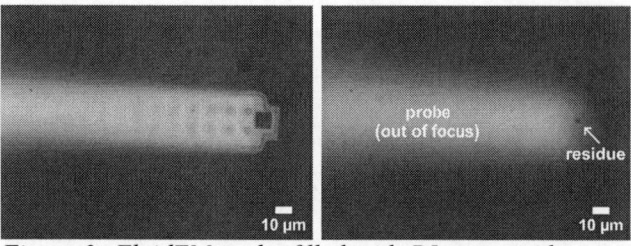

Figure 3: FluidFM probe filled with DI water with acetic acid (8.5 µm gap). Drying residue on the substrate after the experiment (see arrow) indicate spraying from the aperture of the probe (residue diameter was approximately 2 µm).

RESULTS

During spraying liquid was deposited practically always on the substrate directly under the aperture. Spot sizes depended on the spray time, applied voltage and applied external pressure. Typical drying residue diameters were in the order of 10 µm. Minimal spots with a diameter of 2 µm were observed, shown, for example, in Figure 3. A voltage of 300 V was applied in that case.

For small gap heights (8.5-13.5 µm) and voltages around 400 V, a deflection of the cantilever end in the order of 1-2 µm was observed. This deflection is in agreement with the estimation of the spring constant (3 N/m) which gave similar calculated deflections. Based on the observed gap height and the spot sizes, it is reasonable to assume that contactless deposition was achieved.

A pulsating effect in the deposition was observed even when the compliance setting was not reached. Frequency of the pulsation was in the order of one to several Hz. We expect that the specific electrospray spray mode or the limited supply of liquid or charge to the tip are possible causes for this effect.

Onset voltage versus gap

For the electric current measurements, the applied voltage was increased with steps of 10 V every 20 seconds. For each voltage, 50 current measurements were taken. The voltage, after which the current remains larger than 0.01 nA, was considered as the onset voltage. At this voltage, liquid deposition was visible. For several gaps the onset voltage is plotted in Figure 4. The series resistance of 100 GΩ was inserted for experiments with DI/AA. A logarithmic function was fitted to the experimentally obtained onset voltages for DI water in Figure 4, because of the logarithmic

relation between the electric field at the tip and the gap. The root mean square error (RMSE) is depicted by the gray bar.

Onset voltage versus applied external pressure

In Figure 5 the onset voltage as function of the applied external pressure is plotted. The relation between pressure and onset voltage, according to the simulation model, can be approximated by a linear function (in the range between 0 and 2 bar) based on a Taylor expansion. Therefore, the measured onset voltages (for DI) were fitted to a linear function of the pressure. A decreasing trend for increasing pressure is observed as expected (based on Equation 3). The simulated line falls within the RMSE margin of the linear best fit, for pressures between 1 and 2 bar.

Fluorocarbon coating

During the measurements liquid on the pyramidal tip and part of the cantilever could be observed. By coating the

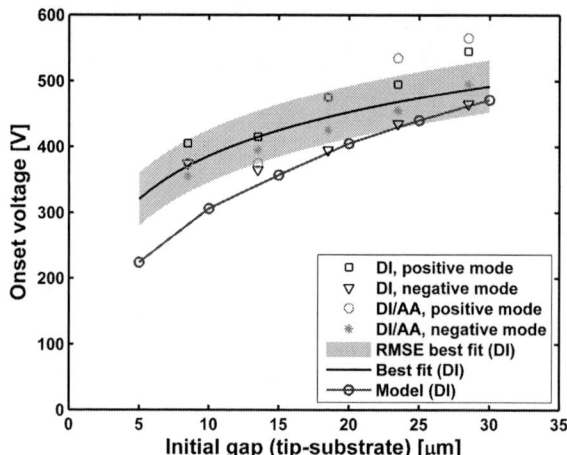

Figure 4: Onset voltage as function of the gap (no applied external pressure). A logarithmic function was fitted to the measurements with DI water.

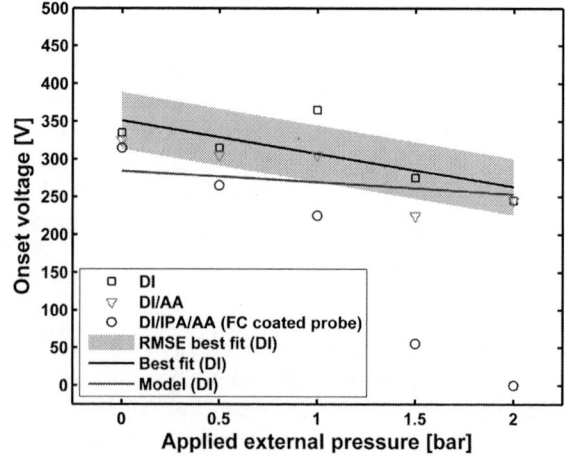

Figure 5: Onset voltage (positive mode) as function of the applied external pressure for a gap of 8.5 µm. A linear best fit to the measurements with DI water is shown.

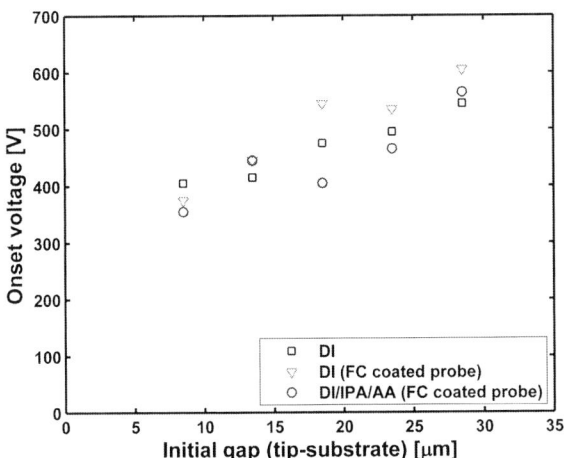

Figure 6: The effect of fluorocarbon coating and surface tension reduction on the onset voltage (positive mode) as function of the gap (no applied external pressure).

probe with a 40 nm thick layer of fluorocarbon (FC), the wetting for DI could be prevented. The conformal coating also led to a reduction in aperture size. Onset voltage as function of gap is plotted in Figure 6. Furthermore, measurements conducted with DI/IPA/AA (with 100 GΩ series resistance) are shown. A reduction in onset voltage is observed, although smaller than expected based on Equation 3. When in addition an external pressure is applied, a much stronger reduction in onset voltage can be obtained (shown in Figure 5). In the presence of the FC coating, a slight wetting during the DI/IPA/AA measurements was observed.

DISCUSSION

The experimentally determined onset voltages are in the same order of magnitude as our relatively simple model. For smaller gaps the deviation from the model is larger. In the model a half-spherical meniscus is assumed. This might not be realistic because the meniscus is dependent on the actual aperture shape and might not be half-spherical for small gaps. (Before spraying the meniscus can further deform into a Taylor cone.) Furthermore, the applied external pressure could deform the meniscus. The expected onset voltages are lower than the experimentally obtained values. We expect that the experimental definition of the point at which electrospray occurs explains the difference.

Based on the high onset voltages (and micrometer sized gaps), electric breakdown and corona discharge (combined with electrospray) require further investigation. However, in [8] experiments with electrospraying of water indicated that electrospray occurs before corona discharge. Furthermore, electric discharge will be suppressed by the large resistance of the channel (or series resistance) and the compliance setting of the SMU. By using liquids with low surface tension and applying external pressure, the onset voltage can be lowered, reducing the chance of breakdown.

In the present setup only z-actuation is possible. Application in a commercial AFM setup is pursued. This gives the possibility for a more accurate gap measurement and, with x-y translation, writing would become possible.

CONCLUSIONS

We present a new technique for liquid deposition on the nanoscale by means of electrospray, making use of cantilevers with microfluidic channels connected to a micrometer sized pyramid with 300 nm aperture (FluidFM probes). Liquid deposition was observed and resulting spots with a diameter in the order of 2 μm could be obtained. The onset voltage ranges from 360 V to 570 V for tip substrate distances between 8.5 and 28.5 μm. A reduction in onset voltage is observed for an increasing applied pressure and can be further reduced by liquids with lower surface tension. The resistivity of the liquid did not have a significant influence on the onset voltage as expected. Further reduction of the onset voltages is anticipated when scaling down the gap height into the sub-micrometer regime.

These promising initial results open up the way towards AFM controlled contactless nanoscale deposition.

ACKNOWLEDGEMENTS

The authors would like to thank Mark Smithers for taking the SEM images and SmartTip and Cytosurge (Switzerland) for providing the FluidFM probes. This work is carried out within the FunTips-project, funded by STW/NWO through a vidi grant.

REFERENCES

[1] R. D. Piner et al., "Dip-pen nanolithography", *Science*, vol. 283, pp. 661-663, 1999.

[2] K. –H. Kim et al. "Massively parallel multi-tip nanoscale writer with fluidic capabilities - fountain pen nanolithography (FPN)", in *Proc. 4th Int. Symp. MEMS and Nanotechnology*, 2003, pp. 235-238.

[3] S. Deladi et al., "Micromachined fountain pen for atomic force microscope-based nanopatterning", *Appl. Phys. Lett.*, vol. 85, pp. 5361-5363, 2004.

[4] A. Meister et al., "FluidFM: Combining atomic force microscopy and nanofluidics in a universal liquid delivery system for single cell applications and beyond", *Nano Lett.*, vol. 9, pp. 2501-2507, 2009.

[5] J. -U. Park et al., "High-resolution electrohydrodynamic jet printing", *Nat. Mater.*, vol. 6, pp. 782-789, 2007.

[6] K. Kaisei et al., "Nanoscale liquid droplet deposition using the ultrasmall aperture on a dynamic mode AFM tip", *Nanotechnology*, vol. 22, pp. 1-5, 2011.

[7] E. Sarajlic et al., "Fabrication of 3D nanowire frames by conventional micromachining technology", in *Digest Tech. Papers Transducers '05*, 2005, pp. 27-30.

[8] J. M. López-Herrera et al., "An experimental study of the electrospraying of water in air at atmospheric pressure", *J. Amer. Soc. Mass Spectrometry*, vol. 15, pp. 253-259, 2004.

CONTACT

* J. Geerlings, j.geerlings@utwente.nl

A MICROBUBBLE PRESSURE TRANSDUCER WITH BUBBLE NUCLEATION CORE

Lawrence Yu and Ellis Meng

University of Southern California, Los Angeles, USA

ABSTRACT

A microchannel-based microbubble (μB) transducer (μBPT) having a μB nucleation core (μBNC) was developed to achieve low power operation in wet environments. In this work, we investigate μB dynamics within the transducer structure and pressure transduction. The transducer leverages electrochemical impedance (EI)-based measurement to monitor the instantaneous response of μB size changes induced by hydrostatic pressure changes. We demonstrated on-demand μB nucleation by electrolysis and real-time pressure tracking (-93 Ω/mmHg over 0-350 mmHg). Repeatable, efficient electrolytic generation of stable microbubbles (< 1.5 nL with < 2% size variation) was achieved using a μBNC structure attached centrally to the microchannel. Biocompatible construction (only Parylene and Pt), small footprint, low power consumption (< 60 μW), and liquid-based operation of μBPTs are ideal for *in vivo* pressure monitoring applications.

INTRODUCTION

Typically MEMS pressure transducers consist of a flexible diaphragm affixed to a stiffer supporting structure. Such a structure conveniently utilizes capacitive or piezoelectric transduction mechanisms. Here, we employ an alternative approach which exploits the ability of a microbubble to respond instantaneously to external pressure variations [1]. By combining localization of a microbubble with electrochemical transduction, our pressure transduction approach obviates the need for hermetic packaging required of conventional pressure sensor structures when used for *in vivo* monitoring and thereby significantly reduces the overall sensor footprint.

DESIGN

Transduction Principle

Previously, pressure response of μBs generated within silicon microfluidic channels was investigated [2] but with poor control over the repeatability during electrolytic bubble nucleation (> 6% size variation). We developed a μBPT based on a circular Parylene chamber and flexible substrate but achieved limited sensitivity due to the short EI measurement path. Also, the electrolysis/measurement electrodes were shared and thus switching was required for sensor operation [3]. Electrode spacing for optimal electrolytic nucleation and EI measurement have opposing requirements: increased spacing between electrodes results in a larger measurement range while closely spaced electrodes enables localized bubble formation. Therefore, to improve sensitivity for *in vivo* applications, the use of distinct pairs of smaller area electrodes for electrolysis residing in a separate structure (a μBNC) and larger area electrodes for high precision EI measurement was investigated (Fig. 1).

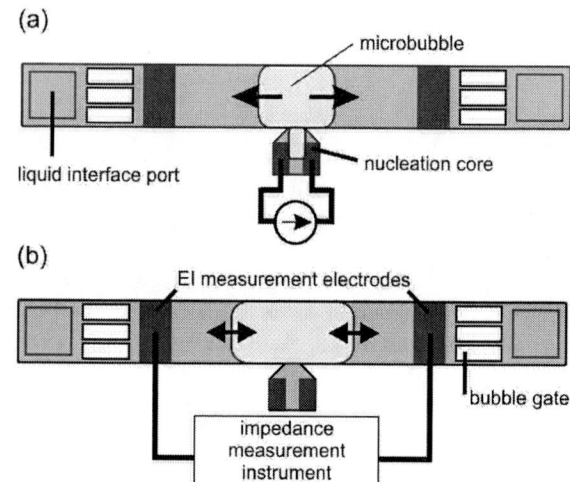

Figure 1: Top-down illustration of pressure transduction principle and layout of pressure transducer. (a) Microbubbles are electrolytically generated within a nucleation core and coalesce into a single bubble. The growing bubble extends outwards and fills measurement region of the channel. (b) The μB then detaches from the μBNC and remains in the measurement channel where it can respond to local pressure changes transmitted via the liquid interface ports.

In order to perform a pressure measurement, first a μB is formed via electrolysis in the nucleation core and enters the measurement channel as it grows (Fig. 1a), forming capillaries of fluid in the channel corners (Fig. 2). Once a sufficient size is reached, electrolysis current is terminated and the μB detaches from the nucleation core. The electrically insulating bubble now resides in the measurement channel region and is available for pressure transduction (Fig 1b). A low power, high frequency alternating current applied across the EI measurement electrodes permits monitoring of the volumetric conductive path (solution resistance R_s in Fig. 3) [4]. External pressure of the liquid media is transferred through the interface ports (Fig. 1a), which directly influences bubble size. These pressure variations induce μB size changes that are reflected in the measured impedance, Z:

$$|Z|_{f \geq 10kHz} \approx R_s = \frac{\rho \ell}{S_{Tot}} \qquad (1)$$

where R_s is solution resistance, ρ is conductivity of

978-1-4799-3510-9/14 $31.00 © 2014 IEEE

electrolyte, ℓ is the length of the fluid capillary formed by a µB residing in the microchannel, and

$$S_{Tot} = 4\frac{r^2}{R^2}\left[\frac{\cos\theta}{\sin\pi/4}\cos(\pi/4+\theta)-(\pi/4-\theta)\right] \quad (2)$$

R is the inscribed inner radius of the channel, r the radius of curvature of the corner meniscus, and θ the contact angle of the µB with the channel wall (Fig. 3) [2].

This conductive path geometry improves sensitivity over our previous design by leveraging the large changes in impedance observed in submicron fluid capillaries [5].

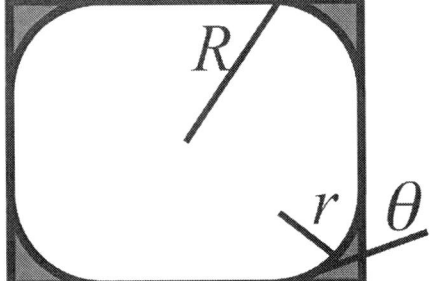

Figure 2: Cross sectional view of bubble in microchannel in measurement region. Capillaries of fluid (shown in blue) reside in corners between the bubble boundary and channel walls. Total cross sectional area (S_{Tot}) of conductive fluid capillary path is dependent on contact angle of channel sidewall material.

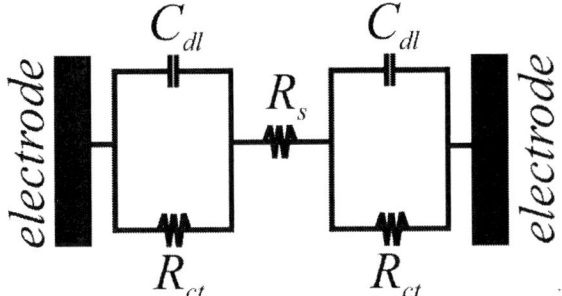

Figure 3: Impedance between measurement electrodes modeled by the Randles circuit is proportional to length of microbubble in measurement channel.

Nucleation Core

The µBNC consists of a pair of closely-spaced microelectrodes (50 µm apart) enclosed within a tapered cavity shaped so as to direct bubbles into the adjacent microchannel. Typically, nucleation via electrolysis creates µBs within natural microcavities randomly distributed on the electrode surface [6]; the cavity structure of the µBNC forces µBs to coalesce before entering the microchannel. When the µB spans the width of the microchannel, surface energy of the hydrophobic Parylene sidewall causes the bubble to adhere, and termination of electrolysis current results in µB detachment from the µBNC. Thus, a single bubble is generated and localized in the measurement area.

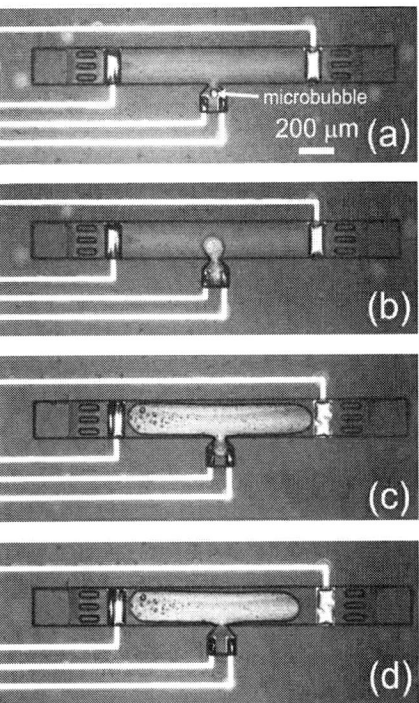

Figure 4: Bubble nucleation and formation process. (a) Nucleation via electrolysis in µB nucleation core. (b) µB enters measurement channel. (c) Continued growth fills microchannel. (d) Detachment of µB from µBNC and localization in the measurement channel.

Microchannel Measurement Area

A microchannel serves to confine the µB and provide an area for EI measurements. Gating structures in the microchannel keep µBs within the measurement area and prevent escape through the open-ended channels. Situating the µBNC between and equidistant from the measurement microelectrodes located at the ends of the microchannel maximizes area for µB growth, thus maximizing measurement range.

METHODS
Fabrication

The µBPT was fabricated (Fig. 5) on a flexible Parylene C substrate with thin film Pt electrodes based on previously reported techniques [7]. Platinum was deposited (2000 Å thick) and patterned on a Parylene coated (10 µm thick) silicon wafer. Following deposition of a 2 µm Parylene insulation layer, sacrificial photoresist was patterned to establish the microchannel and µBNC structures. A final 2 µm thick layer of Parylene was used to enclose the structure, and access ports were patterned and opened with oxygen plasma.

Devices were released in an acetone bath, and an electrical connection was established with a ZIF connector, using previously reported methods [8]. The microchannel was first soaked in isopropyl alcohol to facilitate filling with 1X PBS, the measurement solution.

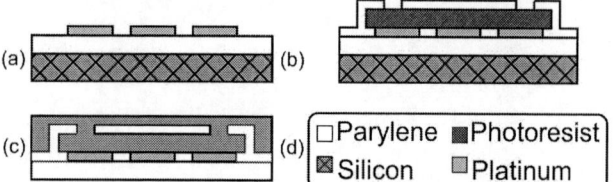

Parylene ■ Photoresist
⊠ Silicon □ Platinum

Figure 5: Overview of transducer microfabrication process. (a) Deposit 1st Parylene and perform Pt lift-off. (b) Pattern sacrificial photoresist, deposit 2nd Parylene layer, and etch interface ports. (c) Release device from silicon substrate and soak in electrolyte to fill channel.

Experimental Setup

Microbubbles were electrolytically generated (Fig. 4) in 1X PBS under LabVIEW control over 1-4 µA for 5-25 s at the µBNC and consistently exited into the measurement channel. During current injection, multiplexed impedance measurements were taken (Fig. 6) with a precision LCR meter (Agilent E4980A). Bubble size was quantified optically and with EI measurement. Electrochemical impedance spectroscopy (EIS) yielded 10 kHz as the optimum frequency (minimum system phase) for maximizing the solution resistance component of the impedance response (Fig. 7). Hydrostatic pressure measurement used a calibrated pressure source attached to a custom test fixture housing the µBPT. A µB was generated at 0 mmHg and subsequent pressure oscillations were tracked in real-time (Fig. 9).

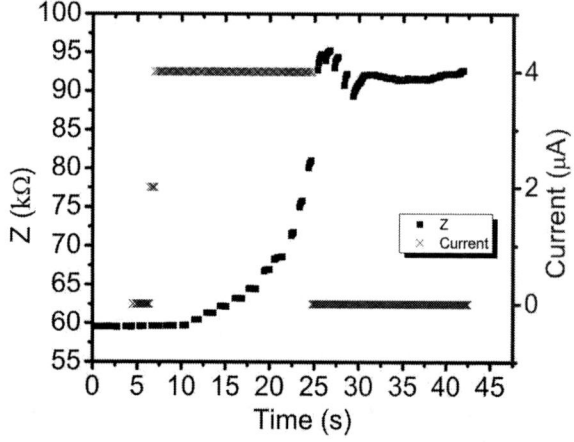

Figure 6: Impedance tracking of bubble nucleation in 1X PBS superimposed on the injected current pulse.

RESULTS AND DISCUSSION

Localized and metered bubble nucleation was repeatable. Power draw was ~1 nW for EI measurement ($1V_{p-p}$) and < 60 µW for electrolytic bubble generation. Impedance monitoring during µB nucleation enabled precise control of bubble formation, and this rate is hypothesized to be correlated to the hydrostatic pressure [9].

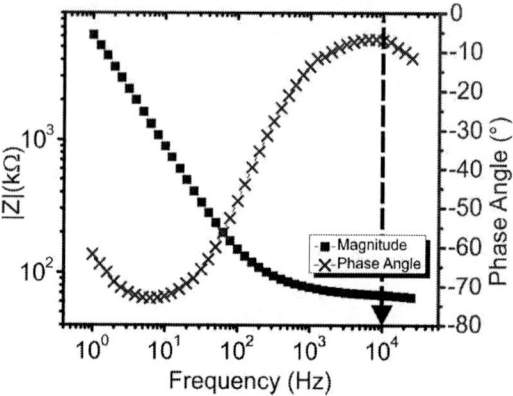

Figure 7: Magnitude and phase for measurement electrodes in 1X PBS. To eliminate capacitive effects, measurement frequency was selected where phase ~0° (10 kHz).

Three types of µBs were observed (Fig. 8): type I µBs were confined in the µBNC (Fig. 4a, < 16 µC); type II (Fig. 4b) spanned both the µBNC and microchannel (16-40 µC); type III (Fig. 4c) resided in the microchannel, detached (Fig. 4d) from the µBNC (> 40 µC). Bubble type can be categorized based on the total injected charge during electrolysis. Pressure transduction was performed only with type III µBs. The gas evolved from electrolysis of PBS is nearly all hydrogen [10], resulting in the short lifetime (< 5 s) of type I and II µBs due to H_2 contact with Pt electrodes, which catalyze recombination. Type III µBs maintained constant size (> 15 min) due to diffusion-limited dissolution (Fig. 6; partial time course shown). High precision EI measurement (SD < 2%) of bubble size was achieved.

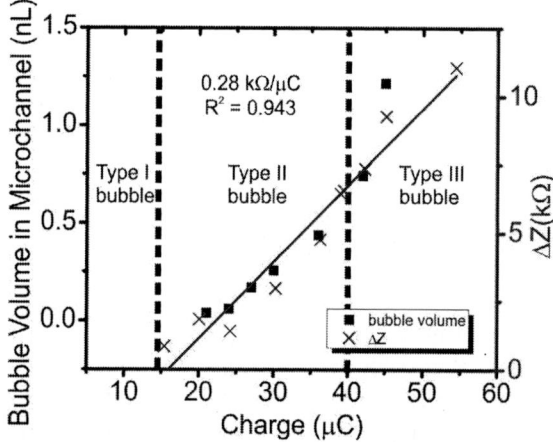

Figure 8: Metered current injection for µB nucleation correlates to a change in impedance. Dashed lines distinguish bubble morphologies based on injected charge.

A linear trend between EI and applied pressure was observed (Fig. 9), suggesting bubble length is proportional to applied pressure as described in Eq 1. For the same measurement range, sensitivity was improved three orders of magnitude compared to the previous circular chamber design (0.22Ω/mmHg); this improvement is attributed to the high impedance response of the submicron fluid capillaries.

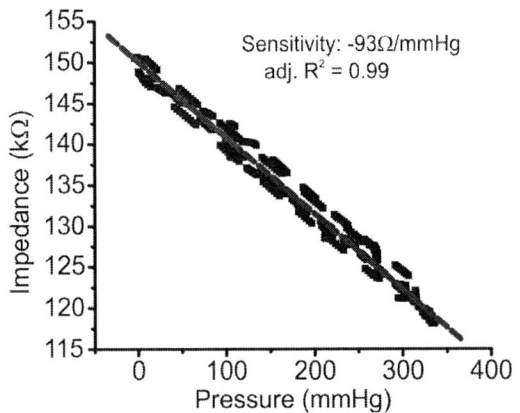

Figure 9: Impedance-pressure correlation for Type III μBs. Obtained highly linear response with improved sensitivity compared to previous chamber design.

On a type III μB, pressure oscillations were tracked in real-time (Fig. 10). Dissolution of the bubble was relatively slow and dependent on mass transport and interfacial kinetics, enabling stable pressure tracking measurement for an adequate time period (> 15 min). Bubble detachment from the μBNC also favors the slow diffusion-limited process over the Pt-catalyzed recombination reaction.

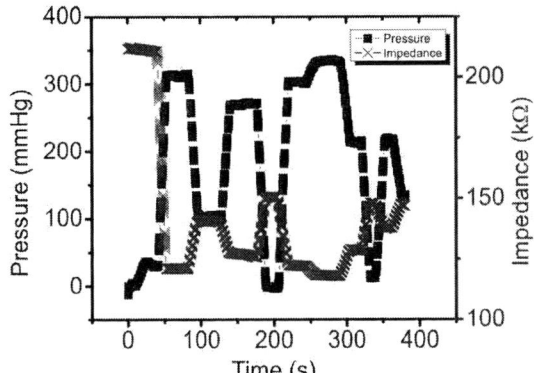

Figure 10: Real-time pressure tracking with type III microbubble.

CONCLUSIONS

Precise control of μB formation is essential in pressure transduction based on μB response to pressure variations. The use of a μBNC structure for high precision μB nucleation and pressure tracking using the localized μB in an adjacent microchannel was demonstrated. Generated μBs were stable (> 15 min) and the new sensor geometry allowed improved sensitivity over the previous circular chamber design. Further sensor characterization under varying environments (electrolyte composition, temperature, and pH) is underway.

ACKNOWLEDGEMENTS

This work was funded by the NSF under award number ECCS-1231994. The authors would like to thank Mr. J. Nishida, Mr. A. Jibodu, and the members of the USC Biomedical Microsystems Laboratory for their assistance.

REFERENCES

[1] P. S. Epstein and M. S. Plesset, "On the Stability of Gas Bubbles in Liquid-Gas Solutions," *Journal of Chemical Physics,* vol. 18, pp. 1505-1509, 1950.

[2] D. A. Ateya, A. A. Shah, and S. Z. Hua, "Impedance-based response of an electrolytic gas bubble to pressure in microfluidic channels," *Sensors and Actuators, A: Physical,* vol. 122, pp. 235-241, 2005.

[3] C. A. Gutierrez and E. Meng, "A subnanowatt microbubble pressure sensor based on electrochemical impedance transduction in a flexible all-parylene package," in *24th IEEE International Conference on Micro Electro Mechanical Systems, MEMS 2011, January 23, 2011 - January 27, 2011*, Cancun, Mexico, 2011, pp. 549-552.

[4] J. E. B. Randles, "Kinetics of Rapid Electrode Reactions," *Discussions of the Faraday Society,* vol. 1, pp. 11-19, 1947.

[5] D. A. Ateya, A. A. Shah, S. Z. Hua, and F. Sachs, "Bubble based microfluidic sensors," in *2004 ASME International Mechanical Engineering Congress and Exposition, IMECE 2004, November 13, 2004 - November 19, 2004*, Anaheim, CA, United states, 2004, pp. 437-441.

[6] A. Volanschi, D. Oudejans, W. Olthuis, and P. Bergveld, "Gas phase nucleation core electrodes for the electrolytical method of measuring the dynamic surface tension in aqueous solutions," in *Sixth International Meeting on Chemical Sensors, 22-25 July 1996*, Switzerland, 1996, pp. 73-9.

[7] C. A. Gutierrez, C. McCarty, B. Kim, M. Pahwa, and E. Meng, "An Implantable All-Parylene Liquid-Impedance Based MEMS Force Sensor," in *MEMS 2010: 23rd IEEE International Conference on Micro Electro Mechanical Systems, Technical Digest*, ed New York: IEEE, 2010, pp. 600-603.

[8] B. J. Kim, C. A. Gutierrez, G. A. Gerhardt, and E. Meng, "Parylene-based electrochemical-MEMS force sensor array for assessing neural probe insertion mechanics," in *Micro Electro Mechanical Systems (MEMS), 2012 IEEE 25th International Conference on*, 2012, pp. 124-127.

[9] N. P. Brandon and G. H. Kelsall, "Growth kinetics of bubbles electrogenerated at microelectrodes," *Journal of Applied Electrochemistry,* vol. 15, pp. 475-484, 1985/07/01 1985.

[10] S. Z. Hua, F. Sachs, D. X. Yang, and H. D. Chopra, "Microfluidic Actuation Using Electrochemically Generated Bubbles," *Analytical Chemistry,* vol. 74, pp. 6392-6396, 2002/12/01 2002.

CONTACT

*E. Meng, tel: +1-213-8213949; ellis.meng@usc.edu

MICROFLUIDIC ELECTROCHEMILUMINESCENCE (ECL) INTEGRATED FLOW CELL FOR PORTABLE FLUORESCENCE DETECTION

Miho Tsuwaki[1], Jun Mizuno[1], Takashi Kasahara[1], Tomohiko Edura[2], Eri Kunisawa[2], Ryoichi Ishimatsu[2], Shigeyuki Matsunami[2], Toshihiko Imato[2], Chihaya Adachi[2], and Shuichi Shoji[1]

[1]Waseda University, Japan
[2]Kyushu University, Japan

ABSTRACT

We propose a portable electrochemiluminescence (ECL)-induced fluorescence chip which consists of flow channels for fluorescence sample and multi-color emitting ECL excitation source. A prototype ECL-induced fluorescence chip was fabricated by conventional photolithography and bonding technique. Device performance was evaluated using ECL of rubrene as excitation source and resorufin as fluorescent dye. Fluorescence of 500 µM resorufin (600 nm) was successfully detected using 10 mM rubrene solution (560 nm) under the applied voltage of 4 V. The proposed principle is applicable for portable and on-demand multi fluorescence detection device using its freedom of choice for combination of the ECL light source.

INTRODUCTION

In the field of molecular biology and biochemical researches, detection methods for proteins, secretion from cells and DNA are essential for pathological analysis. Fluorescent staining and fluorescent labels have been widely used in these detection methods because of their stability and high sensitivity [1]. At present, micro total analysis (MicroTAS) and lab-on-a-chip have attracted attentions because they can provide advantages such as simple, fast multiprocessing and high accuracy [2]. Microfluidic fluorescent detection systems are typically composed of laser light as excitation source, microfluidic chip for fluorescent-labeled and/or -stained samples, and a detector [3]. In spite of the development for miniaturizations of the systems, however, these kinds of detection systems still remains to be bulky and complicated. To realize the portable and low-cost detection system, several studies reported on microfluidic devices integrated with light emitting diode (LED) [4] or organic light emitting diode (OLED) [5] as the excitation sources. In our previous research, we proposed multi-color microfluidic electrochemiluminescence (ECL) device [6]. The fabricated device exhibited various ECL light-emissions (blue, yellow, and red) from ECL solutions in the microchannels. ECL is a luminous phenomenon which occurs by collision of anion and cation radicals of emitting molecules when an appropriate voltage is applied [7]. Compared to LED and OLED, ECL device has unique properties as follows: 1) Emitting light wavelength can be controlled on-demand by injecting chosen ECL solutions. 2) ECL solution is easy to be integrated to microfluidic device because ECL occurs in liquid form and change its shape easily to adapt to any structure. In this study, we propose

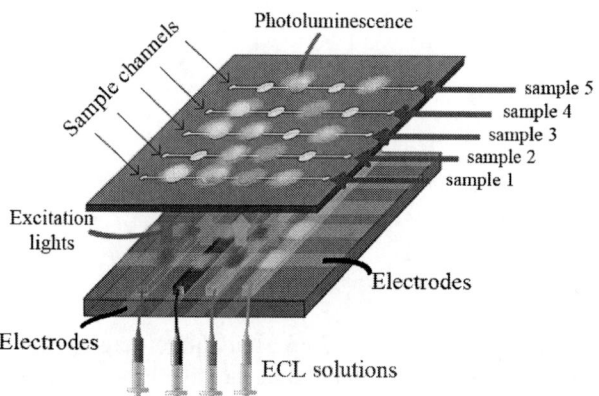

Figure 1: Concept of multi-color ECL-induced fluorescence chip.

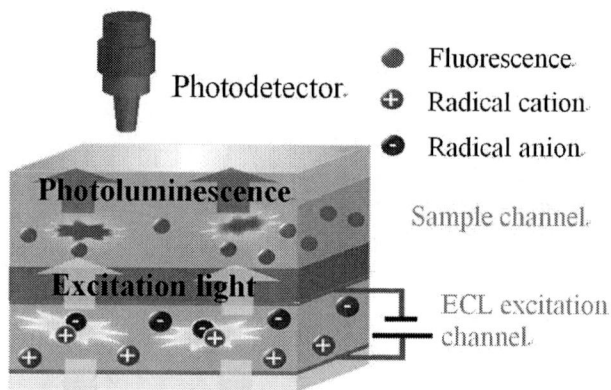

Figure 2: Principle of exciting fluorescence by bottom placed ECL.

microfluidic ECL integrated flow channels for portable ECL-induced fluorescence chip.

CONCEPT AND PRINCIPLE

The concept of portable ECL-induced fluorescence chip is shown in figure 1. Flow channels for fluorescent-dye samples are formed vertically on multi-color microfluidic ECL excitation source. Microfluidic ECL excitation part has multi-color light emitting pixels which are realized at cross point of electrodes matrix. Fluorescent sample is injected to the flow channels while certain ECL solutions which can excite target fluorescence samples are injected into pixels of the ECL microchannels. Figure 2 shows fluorescence excitation principle of the proposed device. When proper voltage is applied to the ECL solution, the radical anions and cations of light emitting molecules are generated at the

978-1-4799-3510-9/14 $31.00 © 2014 IEEE

surfaces of cathode and anode, respectively. ECL derives from the collision of radical anions and cations. The ECL (excitation light) is irradiated above the flow channels, and excites fluorescence. Thereby fluorescence signal (photoluminescence (PL)) is generated and detected by an appropriate photodiode.

EXPERIMENT

Device design

The prototype design of the integrated chip is shown in figure 3. ECL-induced fluorescence chip is composed of microfluidic ECL part [8] at the bottom and flow channels for fluorescence sample at the top. On the microfluidic ECL part, three SU-8 microchannels are sandwiched by a 3×3 matrix of the indium tin oxide (ITO) anode and cathode. Microchannel depth is chosen about 6 μm, while widths are all 1000 μm. Emitting pixels are at the cross points of ITO anodes and cathodes, and each emitting area is 2×1 mm^2. Emitting light is extracted from glass substrate. Inlet and outlet ports for ECL solutions were located on polyethylene naphthalate (PEN) film. On the flow channel part, polydimethylsiloxane (PDMS)-based three flow channels with inlet and outlet ports are formed on a glass side of microfluidic ECL part. Flow channels for fluorescence sample are placed vertically to the ECL microchannels. Depths of the flow channels are 1000 μm.

Device fabrication

Microfluidic ECL part and flow channel for fluorescence sample were individually fabricated, and finally they were bonded. The fabrication of flow channel for fluorescence sample is made by conventional single step soft-lithography process of PDMS using a silicon mold. Figure 4 shows the fabrication process of the microfluidic ECL part [8]. The anode and cathode substrates were separately fabricated and subsequently bonded using epoxy- and amine self-assembled monolayers (SAMs) [9].

For the cathode substrate, an ITO-coated PEN film substrate was used. The ITO cathodes were fabricated by the screen printing techniques with wet-etching. Dilute aqua regia solution of HCl: HNO3: H2O in the ratio of 5:1:6 was used as ITO etchant (figure 4 (a)). The inlet and outlet for ECL solutions were mechanically punched out using a sharpened needle tip (figure 4 (b)). For the anode substrate, an ITO-coated glass substrate was used. ITO anodes were fabricated by the conventional photolithography with wet-etching (figure 4(c)). The SU-8 3005 (MicroChem Co.) was spin-coated on a patterned-ITO-coated glass substrate and 6-μm-thick open microchannels were fabricated by photolithography.

Before SAM formation processes, the anode and cathode substrates were pre-treated by VUV/O$_3$ using the Xe$_2$* excimer lamp source (UER20-172 from Ushio Inc.) (Figure 4(e)). Epoxy-terminated and amine-terminated SAMs of 3-glycidodyloxyproplyltrimethoxysilane (GOPTS) and 3-aminopropyltriethoxysilane (APTES) were utilized (figure 4 (f)) [9]. The anode and cathode substrates were immersed into 1% (v/v) GOPTS and 5 % (v/v) APTES solutions. The anode substrate and cathode substrate were both rinsed with ethanol. Finally, the surfaces were bonded under contact pressure of 1.5 MPa at 140 °C for 5 min with SB6e from SUSS (figure 4 (g)). Finally, PDMS flow channels were aligned and stuck to glass side of microfluidic ELC part.

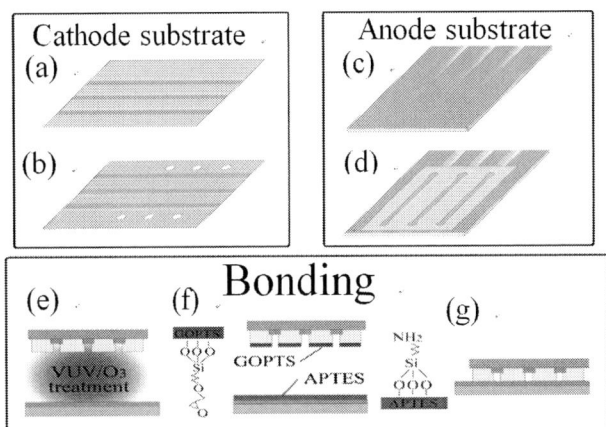

Figure 4: Fabrication process of the microfluidic ECL device.

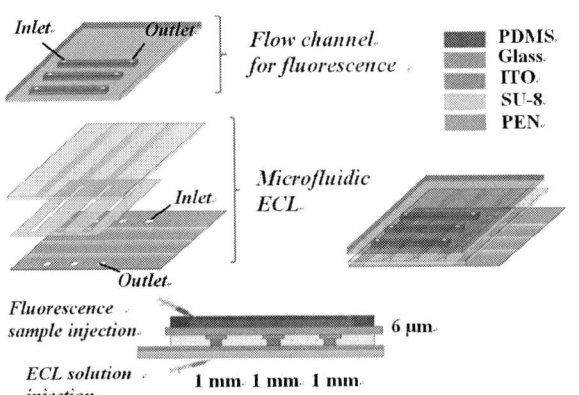

Figure 3: Design of ECL-induced fluorescence chip.

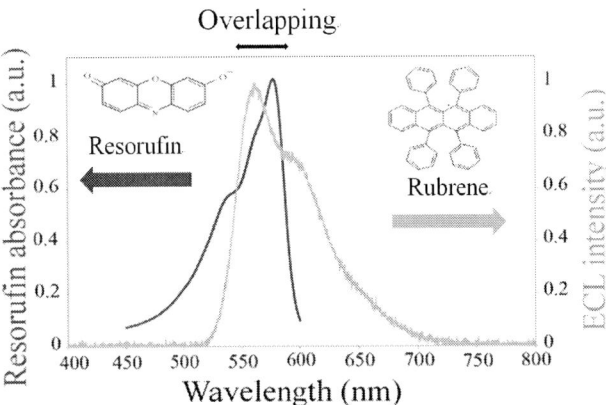

Figure 5: Resorufin absorption spectra and ECL of rubrene spectra.

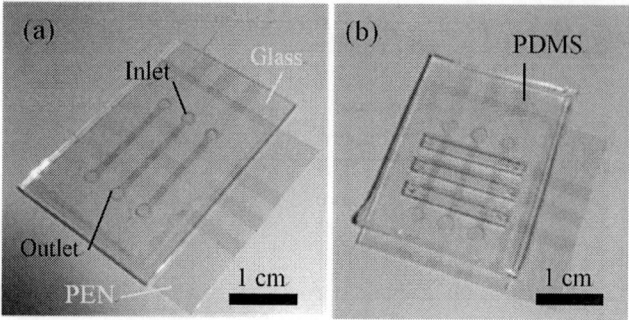

Figure 6: Photographs of (a) fabricated microfluidic ECL device, (b) fabricated ECL-induced fluorescence chip.

Materials

We used resorufin (pink and highly fluorescence) for fluorescence and ECL of rubrene for excitation source because ECL spectrum of rubrene overlaps with absorbance spectrum of resorufin as shown in figure 5. Resorufin sample was prepared in 1 mM and 500 μM concentration by dissolving resorufin in solution [water and ethanol= 4:1 (v/v)]. Rubrene solution was prepared in the same method as the previous work [6-7].

Experimental set-up

Performance of the fabricated microfluidic ECL device with rubrene solution was evaluated by current density-voltage-luminance (J-V-L) measurement using Keithley 2400 source meter and Konica-Minolta LS-110 luminance meter.

Exciting resorufin experiment was carried out. A DC power supply was connected to the anode and cathode of microfluidic ECL device to produce ECL. A spectrometer (USB 4000 from Ocean Optics) was placed on the flow channels for fluorescence samples to detect fluorescence of resorufin.

RESULTS AND DISCUSSION
Fabricated ECL-induced fluorescence chip

The fabricated microfluidic ECL device is shown in figure 6(a). Three microchannels were enclosed successfully by bonding techniques using epoxy- and amine-SAMs. In addition, no voids were observed at interface between anode and cathode substrates. Figure 6(b) shows the fabricated ECL-induced fluorescence chip. PDMS-based flow channels for fluorescence sample were placed and stuck to glass side of the microfluidic ECL device.

Evaluation of microfluidic ECL device

Figure 7 shows the J-V-L characteristics of the one pixel of the microfluidic ECL device with 10 mM rubrene solution. Current density increased with applied voltage. The luminance was observed above 2 V at the 6 μm-deep microchannel. Luminance also increased with applied voltage, and the device exhibited maximum luminance of about 3.8 cd/m^2 at 7.5 V. Figure 8 shows the ECL spectra of microfluidic ECL device with 10 mM rubrene solution. The same ECL spectra with peak wavelength of 560 nm were

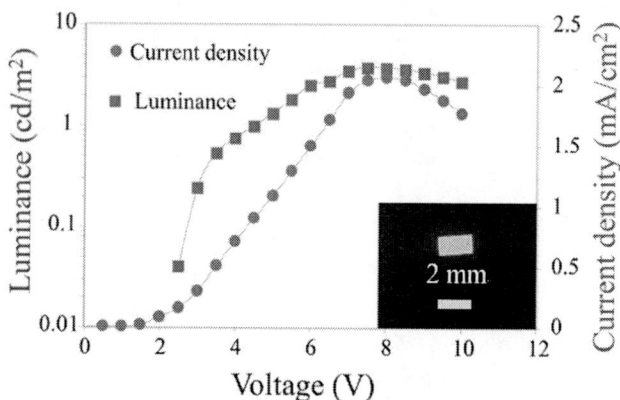

Figure 7: J-V-L characteristics of one pixel of microfluidic ECL device.

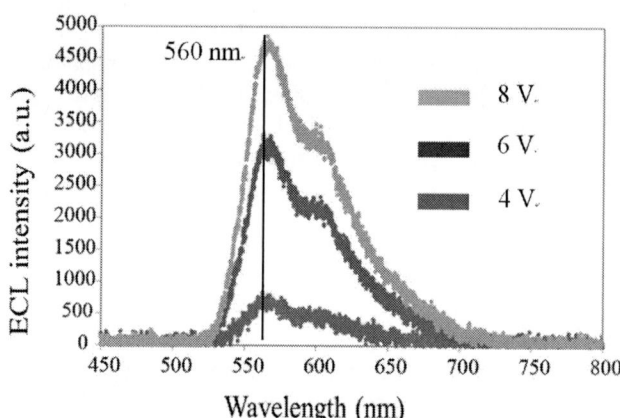

Figure 8: ECL spectra of rubrene in the microfluidic ECL device at different voltages.

obtained under different voltages of 4, 6, and 8 V. This indicates that the fabricated microfluidic ECL device works as stable light source in this voltage of this range.

ECL-induced fluorescence

Figure 9 shows the photographs of demonstration of fabricated ECL-induced fluorescence chip without and with 500 μM resorufin sample. Only yellow ECL of rubrene were extracted from the flow channel side when without resorufin (figure 9(a)). On the other hand, when with injecting resrorufin into the flow channel, different color emitting was observed from flow channel side (figure 9(b)). This result indicated the ECL of rubrene was irradiated to resorufin sample in the flow channel, and its fluorescence was excited. Figure 10 shows the emission spectra of 1 mM and 500 μM resorufin excited by ECL of rubrene driven at 4 V. Different spectra were observed from 1 mM and 500 μM resorufin in comparison with the ECL of rubrene.

CONCLUSION

We proposed and fabricated a prototype ECL-induced fluorescence chip consisting of microfluidic ECL device

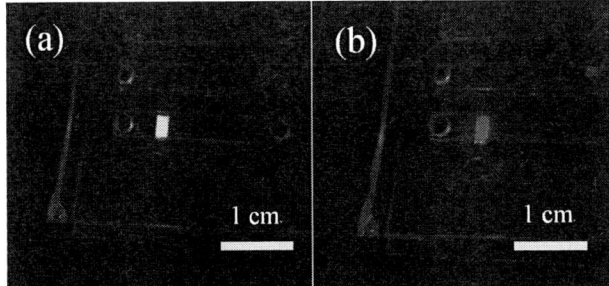

Figure 9: Demonstration of fabricated ECL-induced fluorescence chip without and with injecting 500 μM resorufin.

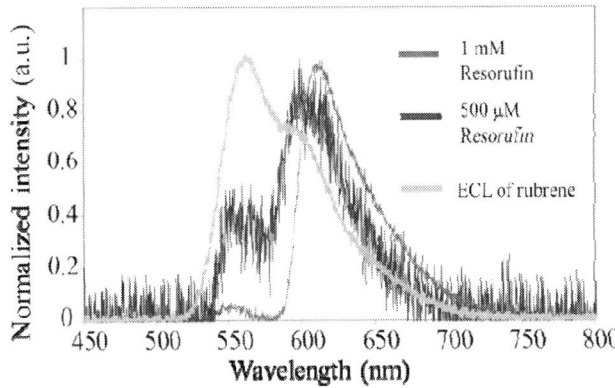

Figure 10: Emission spectra of 1 mM and 500 μM resorufin excited by ECL of rubrenen driven at 4 V.

integrated PDMS flow channel for fluorescence sample. The performance of microfluidic ECL device using rubrene solution was evaluated. In excitation test, 1 mM and 500 μM resorufin were excited by microfluidic ECL device using 10 mM rubrene solution, and their fluorescence signals were successfully detected. The proposed concept of ECL-induced fluorescence chip enhances technical impact for future portable and on-demand fluorescence chips. The proposed device characteristics are still primitive compared with those of the well-established LED- and/or OLED-induced fluorescence chip. Further investigation of optimizing ECL light source is necessary. Integration of detector is also key issue to realize complete portable systems.

ACKNOWLEDGEMENTS

This work was supported in part by the Funding Program for World-Leading Innovative R&D on Science and Technology (FIRST) and the International Institute for Carbon Neutral Energy Research (WPI-I2CNER) sponsored by MEXT and Scientific Basic Research (S) No. 23226010. The authors thank for Nanotechnology Support Project of Waseda University for their technical advices.

REFERENCES

[1] M. Calcerrada, P. Roy, C. García-Ruiz, M. González-Herráez, "Photonic crystal fibres as efficient separation component in capillary electrophoresis", *Sens. Actuators B*, vol.191, pp. 264-269, 2014.

[2] B. Tossanaitada, T. Masadome, T. Imato, "Sequential injection analysis of nitrate ions using a microfluidic polymer chip with an embedded ion-selective electrode", *Anal. Methods*, vol. 4, pp. 4384-4388, 2010.

[3] Y.M. Hsu, C.C. Chang, "Linearity improvement of Cy5 fluorescence concentration detection using series photodetector frequency circuit system with ITO–PolySi–metal–semiconductor–metal photodetector and its application for DNA detection and bacterial identification", *Optik*, 124 , pp. 4963-4970, 2013.

[4] H, Ji, M, Li, L, Guo, H, Yuan, C. Wang, D, Xiao, "Design and evaluation of capillary coupled with optical fiber light-emitting diode induced fluorescence detection for capillary electrophoresis", *Electrophoresis* 34, pp. 2546-2552, 2013.

[5] F. Lefèvre, A., Chalifour, L. Yu, V. Chodavarapu, P. Juneau, R. Izquierdo, "Algal fluorescence sensor integrated into a microfluidic chip for water pollutant detection", *Lab. Chip*, vol. 12, pp. 787-793, 2012..

[6] T. Kasahara, J. Mizuno, S. Matsunami, T. Edura, M. Tsuwaki, J. Oshima, C. Adachi, S. Shoji, "Multi-color microfluidic organic light emitting device using electroluminescence and electrochemiluminescence" *Proc. IEEE Int. Conf. Micro Electro Mechanical Systems 2013*, pp. 1133-1136

[7] K. Nishimura, Y. Hamada, T. Tsujioka, S. Matsuta, K. Shibata, T. Fuyuki, "Solution electrochemiluminescent cell with a high luminance using an ion conductive assistant dopant", *Jpn. J. Appl. Phys*, vol. 40, pp. L1323-L1326, 2001.

[8] T. Kasahara, S. Matsunami, T. Edura, J. Oshima, C. Adachi, S. Shoji. J. Mizuno, "Fabrication and performance evaluation of microfluidic organic light emitting diode", *Sens. Actuators A,* vol. 195, pp. 219-223, 2013.

[9] N.Y. Lee and B.H. Chung, "Novel Poly(dimethylsiloxane) Bonding Strategy via Room Temperature "Chemical Gluing"", *Langmuir*, vol. 25, pp. 3861-3866, 2009.

CONTACT

*M. Tsuwaki, tel: +81-3-5286-3384;
tsuwaki@shoji.comm.waseda.ac.jp
J. Mizuno, tel: +81-3-3205-3181; mizuno@waseda.jp

A MONOLITHIC KNUDSEN PUMP WITH 20 sccm FLOW RATE USING THROUGH-WAFER ONO CHANNELS

Seungdo An, Yutao Qin, and Yogesh B. Gianchandani
Center for Wireless Integrated MicroSensing and Systems (WIMS²)
University of Michigan, Ann Arbor, MI 48109, USA

ABSTRACT

This paper describes a lithographically microfabricated Knudsen pump for high gas flow. Knudsen pumps operate by thermal transpiration and require no moving parts. To achieve high gas flow, high-density arrays of microchannels (with over 4000 channels/mm²) are used in parallel. These vertically oriented microchannels have 2×120 μm² openings surrounded by 0.1 μm-thick silicon oxide-nitride-oxide (ONO) sidewalls. The thin ONO sidewalls provide thermal isolation between a heat sink formed within the Si substrate, and a Cr/Pt heater that provides a temperature bias for thermal transpiration. The Knudsen pump is monolithically microfabricated on a single wafer using a four-mask process. It has a footprint of 8×10 mm². It produces a measured air flow of 20 sccm (*i.e.*, 0.8 sccm/mm²), with typical response times of 0.1-0.4 sec.

INTRODUCTION

Knudsen pumps generate a pressure difference and gas flow based on thermal transpiration in narrow channels that have hydraulic diameter similar to or less than the mean free path of gas molecules. With a temperature gradient applied along the narrow channel, gas molecules flow from the cold end to the hot end [1].

Compared to mechanical gas micropumps, which typically use electrostatic [2-4], piezoelectric [5-6], or electromagnetic [7] actuation to drive flexible membranes, Knudsen pumps have no moving parts and consequently promise high reliability and long lifetimes. A Knudsen pump has been reported to operate continuously for ≈11750 hours without showing degradation [8]. In addition, unlike mechanical gas pumps that typically rely on AC driving voltages, Knudsen pumps only require DC voltages to provide Joule heating. Therefore, Knudsen pumps operate with simple electronic control circuitry.

Two basic approaches for creating Knudsen pumps are microfabricating thermal transpiration channels and employing porous media. In the former approach, the previously reported Knudsen pumps were microfabricated with in-plane channels (i.e., channels are in the plane of the wafer) cascaded into multiple stages for high-vacuum generation. A 162-stage Knudsen pump achieved a vacuum pressure of 0.9 Torr from an atmospheric ambient pressure [9], but its flow rate was limited to ~10⁻⁶ sccm due to the small number of parallel channels. In the latter approach, the porous media have numerous nano/micro channels which permit thermal transpiration to occur in parallel, in order to generate high gas flow. Various porous media, such as naturally occurring zeolite [8] and mixed cellulose

ester (MCE) membranes [10], have been used. The MCE-based Knudsen pump generated an air flow rate of ~1 sccm, the highest among previously reported Knudsen pumps. The integration of thermoelectric elements with the MCE-based Knudsen pump [11], and the application of the MCE-based Knudsen pump in micro gas chromatography systems [12] have also been demonstrated. The approach of using porous media is low-cost and facile, but the generated gas flow rate was relatively small compared to theoretical estimates. The limitation of flow rate can be explained by the defect-induced leakage in the porous media and the inefficient temperature gradient applied by the assembly [10].

This paper describes a method of lithographically microfabricating thermal transpiration channels with uniform and controllable dimensions. In contrast to the prior efforts directed at the lithographic fabrication of Knudsen vacuum pumps [9, 13], this effort is directed at providing high flow instead of high pressure. Consequently, the channels described in this paper are vertically oriented, allowing dense integration and parallel operation.

DESIGN

The high-gas-flow Knudsen pump contains three regions: the active pumping area, the thermal isolation zone, and the Si rim (Fig. 1). The active pumping area has arrays of rectangular, vertical channels ($a \approx 2$ μm, $b \approx 120$ μm, $l \approx 20$ μm) oriented such that a, b, and l are aligned to x, y, and z directions, respectively. The arrays have density >4000 channels/mm². The channel sidewalls are silicon oxide-nitride-oxide (ONO) layers of ≈0.1 μm total thickness. On the channel array is a double-ONO layer (the upper ONO layer) of ≈0.72 μm thickness, which has rectangular openings along the x direction. The upper ONO layer provides mechanical support for the ONO sidewalls from

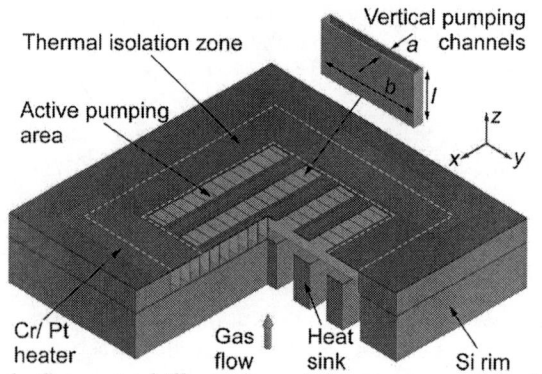

Fig. 1: Conceptual illustration of the structure of the high-gas-flow Knudsen pump.

978-1-4799-3510-9/14 $31.00 © 2014 IEEE

above, whereas the openings serve as the gas outlet windows for the vertical channels. A metal layer (Cr/Pt ≈25/100 nm) on the upper ONO pattern serves as the heater. Beneath the channel array is the Si heat sink with grid-patterned openings. The Si heat sink dissipates Joule heat laterally to the outer Si rim, while the grid-patterned openings serve as the gas inlet windows.

Surrounding the active pumping area is a region that serves as a thermal isolation zone. This region comprises the same vertical channels as the active pumping area, but the channels are fully covered by the upper ONO layer. The region is designed to provide lateral thermal isolation between the heated active pumping area and the cooler Si rim, thus maintaining a relatively uniform temperature in the active pumping area. The Si rim is also used for physical handling as well as for attachment to the external setup.

FABRICATION

The Knudsen pump is fabricated in a single silicon-on-insulator (SOI) wafer (with device layer thickness ≈20 μm, buried oxide thickness ≈0.5 μm, and handle layer thickness ≈380 μm) using a 4-mask lithographic process (Fig. 2). The SOI wafer is processed from the front side to construct the vertical channels in the device layer and the heater on top of the device layer. The handle layer and the buried oxide layer are micromachined from the back side to form the heat sink.

First, arrays of channels (with width and separation ≈2 μm) are etched through the SOI device layer by deep reactive-ion etching (DRIE, Mask #1). The buried oxide layer is used as the etch-stop (Fig. 2(a)). Second, an ONO layer (of thickness ≈42/17/42 nm, total thickness ≈0.1 μm) is deposited by low-pressure chemical vapor deposition (LPCVD) for constructing the channel sidewalls (Fig. 2(b)). Third, polysilicon (thickness ≈2 μm) is deposited by LPCVD to refill the channels. A subsequent $SF_6/O_2/Ar$ plasma is used to dry polish the polysilicon surface until the ONO sidewalls are exposed (Fig. 2(c)). Chemical-mechanical polishing may alternatively be used for this step.

Fourth, a double-ONO layer (i.e., ONO–ONO, 84/51/84–210/83/210 nm, total thickness ≈0.72 μm) is deposited by LPCVD to construct the upper ONO layer, which is then patterned by $CF_4/SF_6/Ar$ plasma (Mask #2). The ONO thicknesses are selected to provide a mild tensile stress level of ≈42 MPa. Next, a Cr/Pt layer (thickness ≈25/100 nm) is deposited by evaporation and patterned by lift-off on top of the upper ONO layer (Mask #3), forming the heater (Fig. 2(d)).

Fifth, the SOI handle layer is micromachined from the back side to form the heat sink. During the previous deposition steps, layers of ONO and polysilicon have covered the back side. These layers are lithographically patterned by corresponding plasmas in a reversed order, followed by the DRIE of the SOI handle layer and buried oxide layer. All these etching steps use the same photoresist pattern as the mask (Mask #4). Next, a layer of Al_2O_3 (thickness ≈10 nm) is deposited by atomic layer deposition

(ALD) and anisotropically etched by $CF_4/SF_6/Ar$ plasma, leaving Al_2O_3 only on the sidewalls of the heat sink (Fig. 2(e)). This Al_2O_3 layer is intended to protect the heat sink from the final XeF_2 etching (similar to the SCREAM process [14]). During this ALD step, the front side of the SOI wafer is protected by photoresist from being deposited with Al_2O_3. Finally, after wafer dicing, the sacrificial silicon and polysilicon in the channel region are etched by XeF_2 dry gas, which releases the vertical ONO channel sidewalls (Fig. 2(f)). The scanning electron microscope (SEM) images and the photograph of the fabricated devices are shown in Fig. 3 and Fig. 4, respectively.

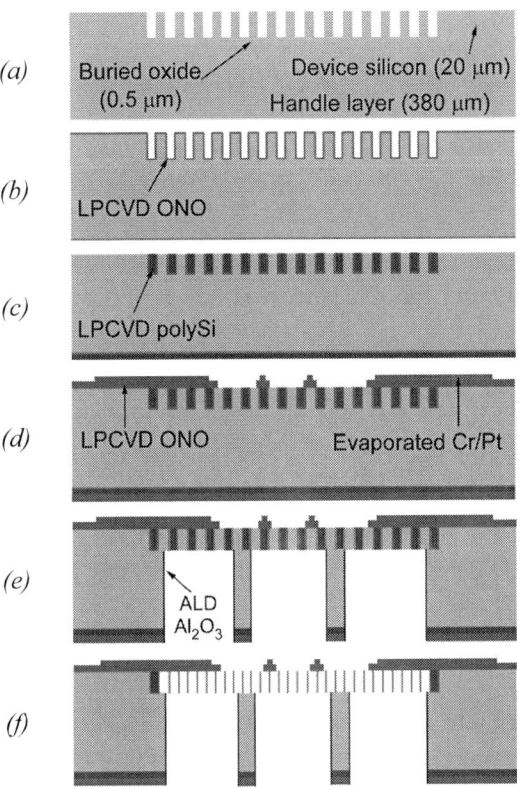

Fig. 2: Fabrication process. (a) DRIE to create channels. (b) LPCVD 0.1 μm-thick ONO. (c) LPCVD channel refill and plasma etching. (d) Deposition and patterning of the upper ONO layer and the Cr/Pt heater. (e) DRIE to create the heat sink. (f) XeF_2 etching of sacrificial Si.

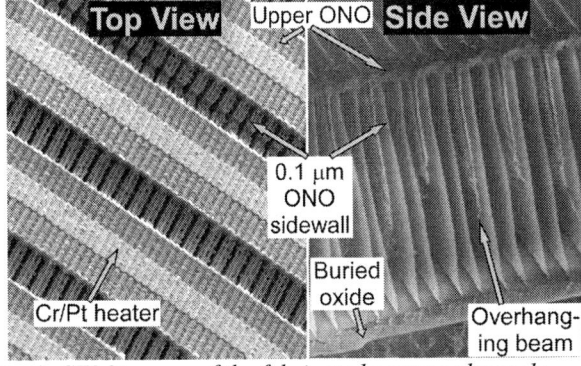

Fig. 3: SEM images of the fabricated narrow channels.

Fig. 4: Image of a fabricated Knudsen pump.

EXPERIMENTAL EVALUATION

To test the flow generation, the Knudsen pump was mounted on an Al plate, which had perforations for providing fluidic connection to the gas inlet of the Knudsen pump. The Si rim was in good thermal contact with the Al block, and epoxy was applied around the perimeter to prevent leakage. A commercial pressure sensor (Model # MPX5010DP, Freescale Semiconductor, Inc., AZ) and flow meter (Model # MW-1SLPM-D, Alicat Scientific, Inc., AZ) were used to provide measurement from the Knudsen pump inlet. Tubes with various internal diameters were used to alter the fluidic loading to the Knudsen pump.

The steady state flow rate (Q) and back pressure (ΔP) of the fabricated Knudsen pumps supplied with various input power levels were measured (Fig. 5). For each input power level, the data points were linearly fitted, forming a Q-ΔP line. The highest measured air flow rate provided by the Knudsen pump was $\approx$20 sccm at ΔP $\approx$39 Pa. The maximum equilibrium back pressures ΔP_{eq} was $\approx$270 Pa.

The transient flow rate and back pressure responses of the fabricated Knudsen pumps were also measured by turning the input power on and off (Fig. 6). The transient flow response was obtained under the maximum-flow condition, *i.e.*, of no external loading except that of the flow meter. The transient back pressure response was obtained under the maximum-ΔP condition, *i.e.*, where air flow was fully blocked. The Knudsen pump showed 0.1-0.4 sec response time for reaching 90% of the steady state values, which is ~20x faster than comparable prior work [10-11].

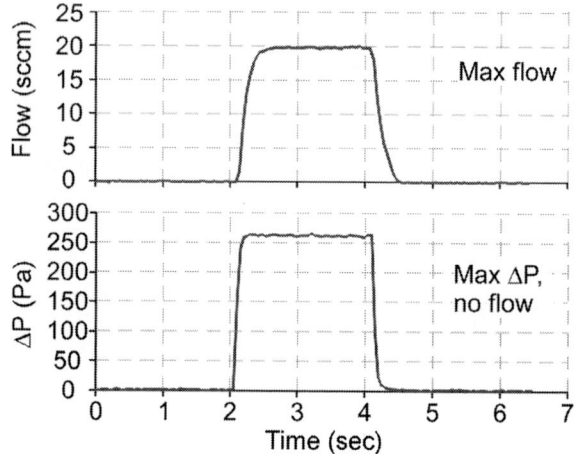

Fig. 6: Transient back pressure and flow rate response in the ambient air created by the high-gas-flow Knudsen pump.

DISCUSSION AND CONCLUSION

Various gas micropump technologies are benchmarked in terms of flow per unit area (Fig. 7). The Knudsen pump in this paper demonstrates a measured air flow per unit area that is among the highest, and 35-200x higher than prior reports on Knudsen pumps [10-11]. Additionally, it demonstrates ~20x faster response than prior reports on Knudsen pumps [10-11] because of its low thermal mass in the heating zone (*i.e.*, the Cr/Pt heater and the upper ONO layer).

One result of the fabrication that is not ideal is the formation of ONO overhanging beams in every other

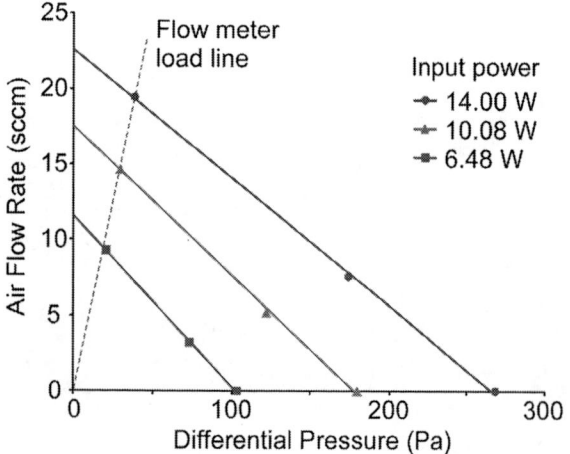

Fig. 5: Steady state pressure drop and flow rate in the ambient air created by the high-gas-flow Knudsen pump.

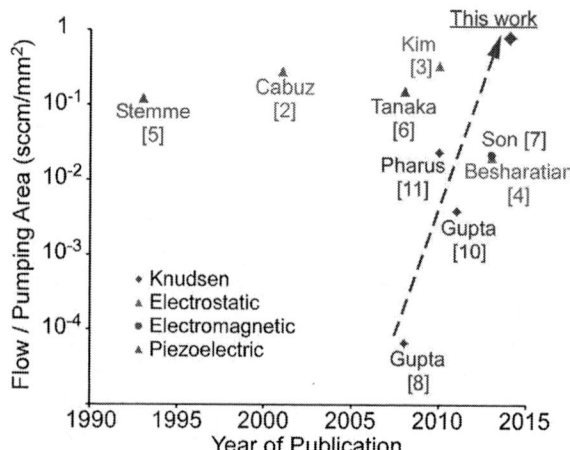

Fig. 7: Benchmarking of flow rate in gas micropumps.

vertical channel (Fig. 3). These beams are caused by the combined effect of several process steps. During the polysilicon refill step, voids exist in the refilled channels. These voids are enlarged into grooves during the dry polishing step. During the subsequent LPCVD of the upper ONO layer, the ONO fills the grooves and remains after the final XeF_2 etching, forming the overhanging beams. (This is the reason why the upper ONO layer is composed of a double-ONO layer; if a 0.72 μm-thick single-ONO layer were used, the grooves would be filled only with oxide, resulting in unbalanced stress in the overhanging beams.) These beams cause additional heat dissipation and thus higher power consumption, as well as decrease the flow rate. To prevent the formation of these overhanging beams, plasma-enhanced chemical vapor deposition (PECVD) can be used for the deposition of the upper ONO layer, as PECVD provides a less conformal deposition profile than LPCVD.

In summary, this effort explores a monolithic, fully lithographic microfabrication approach for making a Knudsen pump for high-gas-flow generation. Vertical channels are constructed in dense arrays for parallel pumping, and channel sidewalls are made of ≈0.1 μm-thick ONO layers to minimize power consumption. The experimentally measured air flow rate has reached ≈20 sccm (*i.e.*, ≈0.8 sccm/mm^2). Future work includes tailoring the fabrication process for achieving both high flow and high back pressure.

ACKNOWLEDGEMENTS

The study was supported in part by the Microsystems Technology Office of the Defense Advanced Research Projects Agency High-Vacuum Program (DARPA Contract #W31P4Q-09-1-0011). Facilities used for this research include the Lurie Nanofabrication Facility (LNF) operated by the Solid-State Electronics Laboratory (SSEL) and the University of Michigan. The authors thank SSEL and WIMS^2 staff.

REFERENCES

[1] M. Knudsen, "Eine Revision der Gleichgewichtsbedingung der Gase. Thermische Molekularstromung," *Annalen der Physik, Leipzig,* 336, pp. 205-229, 1909. (in German).

[2] C. Cabuz, W.R. Herb, E.I. Cabuz, S.T. Lu, "The dual diaphragm pump," *IEEE Int. Conf. on Micro Electro Mechanical Systems (MEMS)*, Interlaken, Switzerland, 2001, pp. 519-522.

[3] H. Kim, K. Najafi, L.P. Bernal, "Helmholtz Resonance Based Micro Electrostatic Actuators for Compressible Gas Control: a Microjet Generator and a Gas Micro Pump," *J. Microelectronics and Electronic Packaging*, 7, pp. 1-9, 2010.

[4] A. Besharatian, K. Kumar, R. L. Peterson, L. P. Bernal, K. Najafi, "Valve-Only Pumping in Mechanical Gas Micropumps," *IEEE Int. Conf. on Solid-State Sensors, Actuators and Microsystems*

(Transducers), Barcelona, Spain, June 2013.

[5] E. Stemme, G. Stemme, "A valveless diffuser/nozzle-based fluid pump," *Sensors and Actuators A (Physical)*, A39, pp. 159-67, 1993.

[6] K. Tanaka, V.T. Dau, R. Sakamoto, T.X. Dinh, D.V. Dao, S. Sugiyama, "Fabrication and basic characterization of a piezoelectric valveless micro jet pump," *Japanese J. Applied Physics*, 47, pp. 8615-8618, 2008.

[7] J. Son, H. Kim, H. Kim, "Pneumatic-less high-speed vacuum meso-pump driven by programmable hydraulics," *IEEE Int. Conf. on Micro Electro Mechanical Systems (MEMS)*, Taipei, Taiwan, Jan. 2013, pp. 584-587.

[8] N.K. Gupta, Y.B. Gianchandani, "Porous Ceramics for Multistage Knudsen Micropumps-modeling Approach and Experimental Evaluation," *J. Micromechanics and Microengineering*, 21, paper 095029, pp. 1-14, 2011.

[9] S. An, N.K. Gupta, Y.B. Gianchandani, "A Si-Micromachined 162-Stage Two-Part Knudsen Pump for On-Chip Vacuum," *J. Microelectromechanical Systems*, JMEMS-2013-0139, in press 2013.

[10] N.K. Gupta, Y.B. Gianchandani, "Thermal transpiration in mixed cellulose ester membranes: Enabling miniature, motionless gas pumps," *Microporous and Mesoporous Materials*, 142, pp. 535-541, 2011.

[11] K. Pharas, S. McNamara, "Knudsen pump driven by a thermoelectric material," *J. Micromechanics and Microengineering*, 20, paper 125032, pp. 1-7, 2010.

[12] J. Liu, N.K. Gupta, K.D. Wise, Y.B. Gianchandani, X. Fan, "Demonstration of motionless Knudsen pump based micro-gas chromatography featuring micro-fabricated columns and on-column detectors," *Lab on a Chip*, 11, pp. 3487-3492, 2011.

[13] S. McNamara, Y.B. Gianchandani, "On-chip vacuum generated by a micromachined Knudsen pump," *J. Microelectromechanical Systems*, 14, pp. 741-746, 1993.

[14] Z.L. Zhang, N.C. MacDonald, "Fabrication of submicron high-aspect-ratio GaAs actuators," *J. Microelectromechanical Systems*, 2, pp. 66-73, 1993.

CONTACTS

S. An, sdan@umich.edu
Y. Qin, yutaoqin@umich.edu
Y.B. Gianchandani, yogesh@umich.edu

SUSPENDED NANOCHANNEL RESONATORS AT ATTOGRAM PRECISION

Selim Olcum[1], Nathan Cermak[1], Steven C. Wasserman[1], Kris Payer[1], Wenjiang Shen[2], Jungchul Lee[3], and Scott R. Manalis[1]

[1] Massachusetts Institute of Technology, Cambridge, MA, 02139 USA
[2] Innovative Micro Technology, Santa Barbara, CA, 93117 USA
[3] Sogang University, Seoul 121-742 Republic of Korea

ABSTRACT

Nanomechanical resonators can quantify individual particles down to a single atom; however the applications are limited due to their degraded performance in solution. Suspended micro- and nanochannel resonators can achieve vacuum level performances for samples in solution since the target analyte flows through an integrated channel within the resonator. Here we report on a new generation suspended nanochannel resonator (SNR) that operates at approximately 2 MHz with quality factors between 10,000-20,000. The SNR is measured to have a mass sensitivity of 8.2 mHz/attogram. With an optimized oscillator system, we show that the resonator can be oscillated with a mass equivalent frequency stability of 0.85 attogram (4 parts-per-billion) at 1 kHz bandwidth, which is 1.8 times the calculated stability imposed by the thermal noise. We demonstrate the use of this mass resolution by quantifying the mass and concentration of nanoparticles down to 10 nm in solution.

INTRODUCTION

Established nanoparticle characterization methods in solution are limited for diameters below about 50 nm, particularly for heterogeneous samples. Nanomechanical resonators can detect masses as low as single atoms due to their very small sizes and high quality factors. However their performance degrades dramatically in solution because of the fluid loading. Suspended micro- and nanochannel resonators (SMRs and SNRs) [1], [2] have opened up the possibility of achieving high-Q mass sensing in the aqueous environment by integrating a fluidic channel inside a cantilever. Previously, the performance of nanomechanical resonators in vacuum has been studied extensively. However, SNRs received only theoretical approaches so far [3] to study their limits of detection. Previously as a proof of concept for nanoparticle detection, SNRs achieved a limit of detection higher than 100 attograms at 1 kHz bandwidth [2], far above the limit imposed by the thermal noise at room temperature.

Here we report on the device and system level advancements that have enabled the mass equivalent stability down to 0.85 attogram at 1 kHz, which is 1.8 times the calculated thermomechanical noise limit. We first discuss the resonator and oscillator design and demonstrate an analysis on the stability of the developed system. Finally, we demonstrate the use of this mass resolution by measuring the mass and concentration of nanoparticles down to 10 nm in solution.

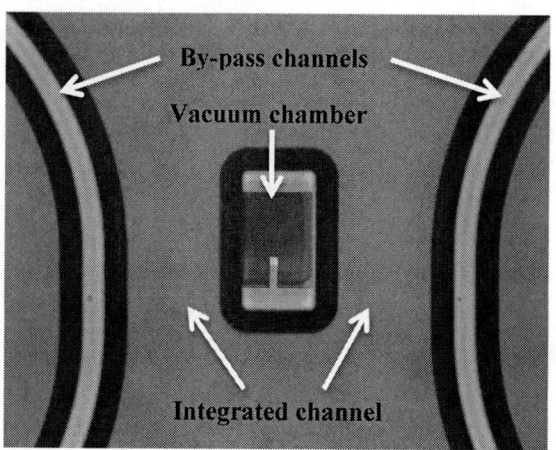

Figure 1: *A micrograph of the fabricated SNR. The cantilever, which is seen in the vacuum cavity, is 27 microns in length. Larger microfluidic channels at the sides are the by-pass channels for delivering and discarding the sample. 1 μm wide integrated channel, which is difficult to see in this micrograph, runs between the by-pass channels.*

RESONATOR DESIGN

The mass resolution of a mechanical mass sensor improves with higher mass sensitivity and/or better frequency stability. The mass sensitivity increases with increased resonant frequency and decreased resonator mass. For achieving sub-attogram level precision, we designed (Table 1) and fabricated (Fig. 1) 8-fold more sensitive cantilevers to changes in mass than the previous designs [2].

SNRs are manufactured by a previously described micro-fabrication process [2]. The process enables the cantilevers to vibrate in a vacuum cavity. Quality factors up to 20,000 are measured on the fabricated devices. An on-chip getter maintains the vacuum environment required for high-Q operation. There are four fluidic ports drilled on the top glass wafer to access the two bypass channels (50 μm X 20 μm) separated 285 μm apart at each side of the cantilever (Fig. 1). The U-shaped channel in the cantilever is connected to these bigger bypass channels by 140 μm-long channels with the same cross-section that is in the cantilever.

OSCILLATOR DESIGN

SNRs are operated in self-oscillation mode in a positive feedback loop. The oscillator consists of a photodetector as the vibration detector, a FPGA for phase shifting and signal processing and a piezo-ceramic for actuation.

Table 1: *Design dimensions and theoretically calculated properties of SNRs. Properties were calculated assuming the cantilevers are filled with water.*

Type	Length (μm)	Thickness (μm)	Width (μm)	Channel Height (nm)	Channel Width (μm)	Resonant Frequency (MHz)	Stiffness (N/m)	Mass (pg)	Sensitivity (mHz/ag)
SNR in [2]	50	1.3	10	700	2	0.76	7.4	1328	- 1.19
new SNR	27	1	7.5	400	1	1.99	16.9	443	-9.3

We implemented an optical lever setup for detecting the cantilever vibration. The vibration of the cantilever causes the angle of the cantilever tip to change, which is converted to the deflection of a laser beam bounced off of the tip. We used an ultra-low noise diode lab laser module (Coherent 635 nm, 5 mW) as the laser source. We implemented a spatial filter (0.5NA, 13.86 mm aspherical lens, 20 μm pinhole) to clean the laser beam. The beam is then collimated using a spherical lens (f = 30 mm) to achieve a ~3 mm-diameter filtered laser beam. For tuning the laser power hitting the cantilever, we used a film polarizer on a rotational stage at the output of the spatial filter since excessive laser power increases the frequency noise of the cantilever in closed loop operation. We focused the laser on the cantilever tip using a 20X objective (Nikon LU Plan ELWD 0.4NA). The reflected laser beam is then focused on a custom-made photodetector by an achromatic doublet lens (Thorlabs AC254-30-A).

For achieving the best signal-to-noise ratio we used a fast, split photo-diode (Hamamatsu S4204) as the photodetector with ultra-low-noise transimpedance amplifiers (OPA847) with 5 MHz 3 dB-bandwidth. The differential optical signals from the two channels of the split photodiode are converted to a single-ended voltage signal using a high speed instrumentation amplifier (AD8130). Finally, an automatic gain control (AGC) stage maintains a fixed amplitude at the output, which can be tuned by a DC input voltage.

The noise of the optical lever and photodetector system is dominated by the intrinsic thermomechanical fluctuations of the cantilever around its resonant frequency (Fig. 2). The voltage noise power spectrum density of the photodetector output, S_V^{pd}(V^2/Hz) is measured using a spectrum analyzer (HP4395A) when the only driving force on a water-filled cantilever is generated by the thermal noise due to the room temperature, T (~300 K). To convert the measured voltage spectrum density to displacement noise spectral density, S_x^{th}, we used the derivation given in [4]:

$$S_x^{th}(\omega) = \frac{4\omega_0^3 k_B T}{kQ} \frac{1}{(\omega_0^2 - \omega^2)^2 + (\omega\omega_0/Q)^2} \quad (1)$$

where, k_B, k, Q and ω_0 are Boltzmann constant and stiffness, quality factor and resonant frequency of the cantilever. We can use (1) evaluated at the resonant frequency ($4k_B TQ/\omega_0 k$) to calculate the responsivity (V/m) of the vibration detection as follows:

$$R = \sqrt{\frac{S_V^{pd}(\omega_0) - S_V^{pd}(\omega \neq \omega_0)}{4k_B TQ/\omega_0 k}} \quad (2)$$

The quality factor and the resonant frequency is obtained by an open loop frequency scan. The stiffness is calculated by using the measured mass sensitivity of the cantilever using:

$$\frac{\delta f}{\delta m} = -\frac{f_0}{2m^*} = -\frac{2\pi^2 f_0^3}{k} \quad (3)$$

where m^* is the effective mass of the cantilever. Measured displacement power spectral density of a SNR is given in Fig. 2. The red circles are the spectrum analyzer measurements referred to displacement domain performed at 3 Hz resolution bandwidth with 100 averages. The dashed black curve is calculated using equation (1) and the dashed green line is the noise density of the photodetector referred to the displacement domain. We measured the detection noise floor as 100 $fm/\sqrt{Hz}$. It is seen in Fig. 2 that the noise at the photodetector output is dominated by the thermomechanical vibrations of the cantilever at the vicinity of its resonant frequency and by the detection noise floor at frequencies away from the resonant frequency.

The detected vibration signal at the output of the photodetector circuit is digitized (14 bits Terasic AD/DA at 100 MSPS) and passed to a FPGA (Altera Cyclone IV on DE2-115). The optimal delay required in the loop for stable self-oscillation is controlled by the FPGA with sub-clock precision (39 ps) using numerical interpolation. The FPGA and the conversion board run on an external 100 MHz oven controlled crystal oscillator clock (Abracon AOCJY2-E-H1C). The delayed signal is then converted back to analog by the conversion board to be used by the motion actuator.

To operate and oscillator with the best frequency stability, the amplitude of oscillation should be increased

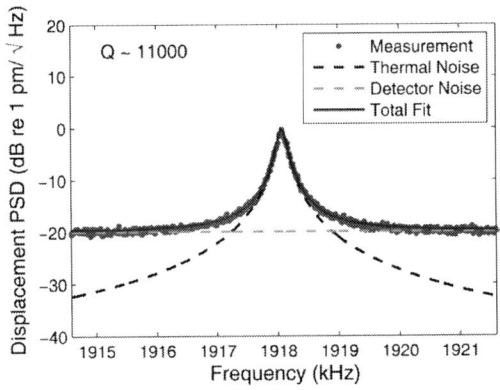

Figure 2: *Displacement noise density of a SNR due to thermal noise. The red circles are the spectrum analyzer measurements; the black line is the theoretical thermal noise and the green line is the photodetector noise (100 fm/Hz$^{1/2}$).*

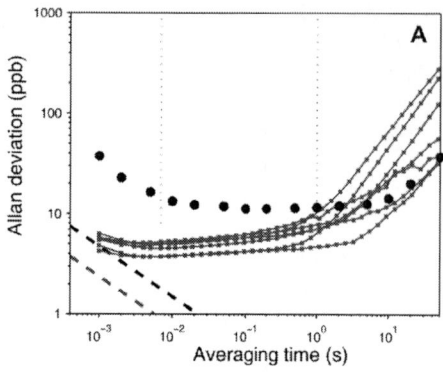 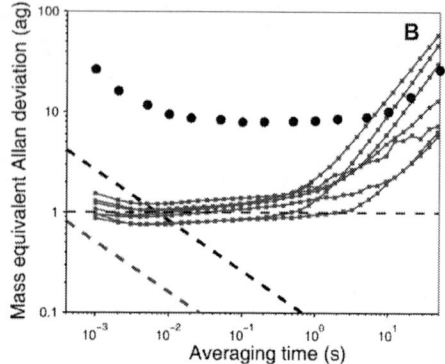

Figure 3: *A. Allan deviation is plotted as a function of averaging (gate) time for new generation SNRs (red) and for previous studies [2] (black). Ultimate noise limits imposed by thermal energy at room temperature are shown as colored dashed lines.* ***B***, *Mass precision of SNRs, defined as the mass equivalent Allan deviation.*

until Duffing type mechanical nonlinearity is observed [5]. For achieving high mechanical actuation amplitudes at 2 MHz, we used a single layer, hard piezo-ceramic (APC841, 7x7x0.2 mm) integrated underneath the SNR chip. The piezo-ceramic is driven by an amplifier utilizing a high current opamp (LT1210) output stage.

The frequency of oscillation is measured by period counting at 100 MHz using a digital heterodyne mixer and a low-pass filter coded in the FPGA. The input signal is mixed with a reference sinusoid generated by a numerically-controlled oscillator (NCO). The heterodyned signal is then filtered by a low pass filter and subsequently up-sampled (1st-order hold) back to 100 MHz. The period of oscillation is then determined by counting the clock ticks between the zero-crossings of the filtered signal.

The oscillator setup is controlled by a control computer which communicates with the FPGA through the Ethernet. The frequency measurements is monitored by the computer and recorded for post-processing.

OSCILLATOR STABILITY

To access the frequency stability of the fabricated SNRs in closed loop, we calculated the Allan deviation of the measured oscillation frequency as a function of averaging time, when the cantilevers were filled with water. The measured frequency stability (red dots in Fig. 3A) ranges from 4 to 7 parts-per-billion at room temperature (without temperature control) using measurement bandwidths of 5-1,000 Hz. The red curves in Fig. 3A correspond to various tested SNR chips. Compared to the previous studies (black dots) the stability of the oscillator system is improved 10-fold at 1 kHz. When converted to mass (Fig. 3B) using the mass sensitivities given in Table 1, the improvement is more than 30-fold, which translates to a stability better than 1 attogram (10^{-21} kg) at 1 kHz.

To investigate the potential for further improvements in the stability and the mass precision, we calculated the ultimate limit of frequency stability imposed by the thermal energy due to room temperature. The power spectral density of the phase fluctuations of the oscillation, $S_\varphi^{th}(\omega)$, is

calculated using the power spectral density of the random displacements of the cantilever due to thermal noise, $S_x^{th}(\omega)$, which was given in (1) as follows:

$$ S_\varphi^{th}(\omega) = \frac{S_x^{th}(\omega)}{\langle x_c \rangle^2} \qquad (4) $$

For very high Q oscillators and as a function of the baseband modulation frequency, $\omega_m = \omega - \omega_0$, (4) can be approximated [4] as:

$$ S_\varphi^{th}(\omega_m) \approx \frac{\omega_0 k_B T}{k \langle x_c \rangle^2 Q} \frac{1}{\omega_m^2} \qquad (5) $$

The Allan variance of the oscillation frequency due to the phase noise density is calculated as given in [6]:

$$ \sigma_A^2(\tau) = 2 \left(\frac{2}{\omega_0 \tau} \right)^2 \int_0^\infty S_\varphi(\omega) \, sin^4(\omega\tau/2) \, d\omega \qquad (6) $$

where τ is the averaging duration. If we evaluate the above integral using the phase noise density in (5), we get:

$$ \sigma_A^{th}(\tau) = \sqrt{\frac{\pi k_B T}{\tau \, k \langle x_c \rangle^2 \, Q \, \omega_0}} \qquad (7) $$

The SNRs used in this work are driven at their onsets of nonlinearity. Therefore we use the following approximate displacement expression of a cantilever given by Arlett et. al. [3] converted to RMS at the onset of nonlinearity in (7).

$$ \langle x_c \rangle = 5.46 \frac{L}{\sqrt{2Q}} \qquad (8) $$

where L is the length of the cantilever. So the thermal noise limited Allan deviation at the onset of nonlinearity is:

$$ \sigma_{A,non}^{th}(\tau) = \frac{1}{5.46L} \sqrt{\frac{k_B T}{\tau \, k \, f_0}} \qquad (9) $$

which is independent of the quality factor. Therefore, as long as an oscillator is driven at its onset of nonlinearity, the lowest Allan deviation that can be achieved does not depend on the quality factor of the cantilever. The ultimate limit of stability for SNRs is calculated and plotted in Fig. 3A and B using the same color codes. Measured frequency stability values at 1 ms averaging time are 1.8- to 3-fold (5-10 dB) above the thermomechanical noise limits.

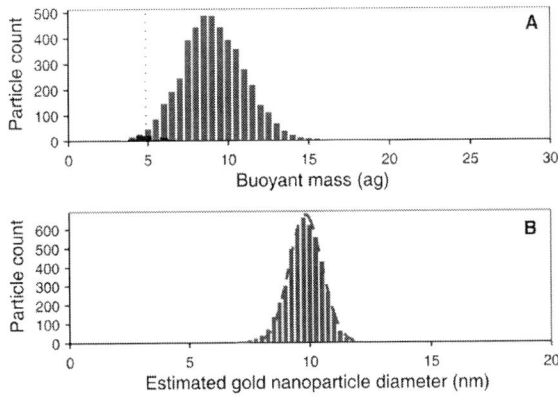

Figure 4: *A, Buoyant mass histogram (red) of 4,700 particles detected during a 70-minute experiment using a population of 10 nm gold nanoparticles in DI water. **B**, Diameter histogram is calculated assuming the particles are spheres of density 19.3 g/cm³.*

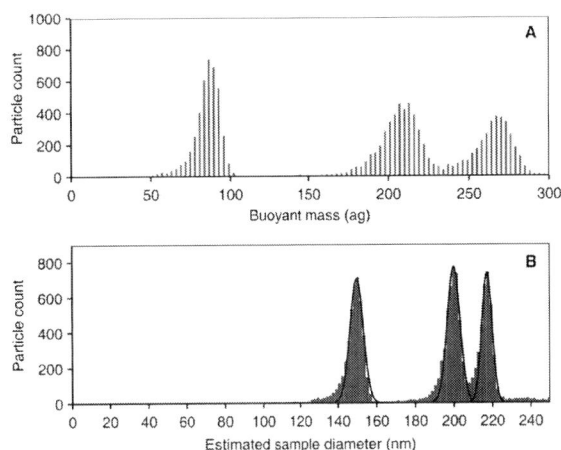

Figure 5: *A, Buoyant mass histogram of more than 12,500 detected particles in less than 30 minutes in a mixture of 150 nm, 200 nm and 220 nm polystyrene beads. **B**, Diameter histogram calculated assuming particles are spheres of density 1.05 g/cm³.*

NANOPARTICLE ANALYSIS

The mass distribution and absolute particle concentration are important measures for analyzing nanoparticle populations. We sought to demonstrate the mass resolution of our system using size-calibrated 10 nm gold nanoparticles (NIST RM-8011). Before analyzing the sample, we calibrated the mass sensitivity (Hertz/gram) of the employed cantilever using size-calibrated 30 nm gold nanoparticles (NIST RM-8012), which are approximately 25 times heavier than the 10 nm particles. We determined that the mass sensitivity of the cantilever used is 8.2 mHz/ag slightly lower than the expected calculation. In a 70-minute experiment on 10 nm gold suspension, we measured 4,700 individual particles. Our results show a distinct population (Fig. 4A) around 9 attograms. The overlaid black histogram includes the detected peaks during a measurement with the same duration but without the particles as a control. We then estimated the diameter of each nanoparticle (Fig. 4B), assuming each particle is spherical and equally dense (the density of gold is 19.3 g/cm³). The estimated mean size is 9.83 nm with a coefficient of variance of 6.9%, which agrees well with the manufacturer specification of 9.9 nm. The detection and estimation algorithm used can successfully estimate the transit time of a particle. The transit time information combined with the dimensions of the buried microfluidic channel enables the calculation of the particle concentration. We estimate the concentration of the nanoparticles as 1.95×10^9 particles/ml compared to the concentration of 1.90×10^9 particles/ml calculated using the information in the datasheet.

We also tested the dynamic range of the SNRs by weighing larger particles (150 nm, 200 nm and 220 nm polystyrene beads) using the same operation, detection and estimation conditions. The resulting mass histograms are given in Fig. 5 indicating the detection of each sub-population in the mixture successfully.

OUTLOOK

The range of buoyant mass that can only be reached by this new generation of SNRs includes many of the engineered nanoparticles used in nanomedicine, most of the virions like HIV, HCV, and natural sub-cellular structures such as exosomes. We envision this device will be useful for accurately analyzing nanoparticles in solution for a wide range of fields

REFERENCES

[1] T. P. Burg, M. Godin, S. M. Knudsen, W. Shen, G. Carlson, J. S. Foster, K. Babcock, and S. R. Manalis, "Weighing of biomolecules, single cells and single nanoparticles in fluid," *Nature*, vol. 446, no. 7139, pp. 1066–1069, Apr. 2007.

[2] J. Lee, W. Shen, K. Payer, T. P. Burg, and S. R. Manalis, "Toward Attogram Mass Measurements in Solution with Suspended Nanochannel Resonators," *Nano Lett.*, vol. 10, no. 7, pp. 2537–2542, Jul. 2010.

[3] J. L. Arlett and M. L. Roukes, "Ultimate and practical limits of fluid-based mass detection with suspended microchannel resonators," *J. Appl. Phys.*, vol. 108, no. 8, Oct. 2010.

[4] T. R. Albrecht, P. Grütter, D. Horne, and D. Rugar, "Frequency modulation detection using high-Q cantilevers for enhanced force microscope sensitivity," *J. Appl. Phys.*, vol. 69, no. 2, pp. 668–673, Jan. 1991.

[5] D. Greywall, B. Yurke, P. Busch, A. Pargellis, and R. Willett, "Evading Amplifier Noise in Nonlinear Oscillators," *Phys. Rev. Lett.*, vol. 72, no. 19, pp. 2992–2995, May 1994.

[6] A. N. Cleland and M. L. Roukes, "Noise processes in nanomechanical resonators," *J. Appl. Phys.*, vol. 92, no. 5, pp. 2758–2769, Sep. 2002.

DUAL-MODE VERTICAL MEMBRANE RESONANT PRESSURE SENSOR

Roozbeh Tabrizian and Farrokh Ayazi

Georgia Institute of Technology, Atlanta, USA

ABSTRACT

This paper presents a novel dual-mode resonant pressure sensor operating based on mass loading of air molecules on transversely resonating vertical silicon membranes. Two silicon bulk acoustic resonators (SiBAR) are acoustically coupled through thin vertical membranes, resulting in two high-Q resonance modes with small frequency split, but large difference in pressure sensitivity. The membranes are designed to couple 180° out-of-phase vibrations of piezoelectrically-transduced SiBARs through pressure-insensitive extensional Lamb waves and without changing their resonance frequency. The in-phase vibrations, on the other hand, induce a high-order pressure-sensitive transverse flexural resonance in vertical membranes while slightly changing the resonance frequency of SiBAR due to stiffness and mass loading. A combinatorial of the two modes is used as a pressure sensor with an amplified sensitivity. A proof-of-concept device implemented on a 20μm silicon substrate and activated by a thin aluminum nitride film shows a combinatorial beat frequency (f_b) of 1.3 MHz with a linear pressure sensitivity of 346 ppm/kPa over 0-100kPa range.

INTRODUCTION

The majority of MEMS pressure sensors require low-pressure hermetic encapsulation to exploit the pressure difference between inside and outside of a package as an input force deflecting a thin membrane [1]. While realization of such encapsulation adds into the overall cost of the device, the membrane may experience large stress in operation, which can degrade their performance; hence necessitating co-integration of stress sensors on the same die for extraction and consequent compensation of packaging-induced effects [2]. Furthermore, additional calibration steps may be required to determine the reference pressure level inside the encapsulated volume. Therefore, absolute pressure sensing techniques which obviate the need for hermetic packaging are desirable.

Besides several advantages offered by miniaturized resonant sensors, acoustic engineering of microstructures can further provide promising opportunities for realization of in-situ physical relative references. This can be accomplished by formation of synthesized resonant modes with largely different sensitivities to physical/environmental signals in a single structure. A combinatorial of these modes can result in a frequency with an amplified sensitivity to the physical input of interest. The authors have recently shown an application of such techniques for highly-sensitive resonant temperature sensing [3].

In this paper we report, for the first time, on a combinatorial resonant pressure sensor using acoustically coupled microresonators (Figure 1). Inertial loading of gas molecules [4] on thin vertical membranes driven by silicon bulk acoustic resonators (SiBAR) into their transverse flexural resonance are used in contrast with pressure-insensitive extensional modes to realize absolute air pressure measurement without a need for a reference pressure.

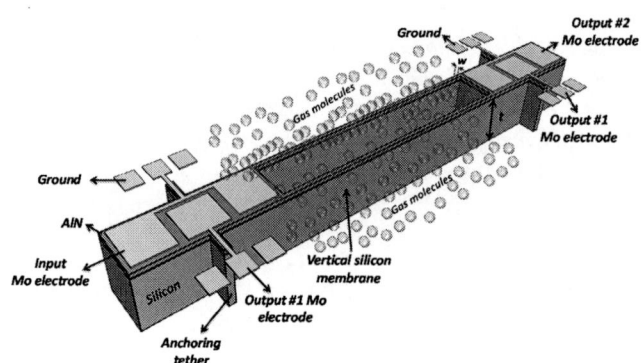

Figure 1: Schematic view of the dual-mode vertical membrane resonant pressure sensor with aluminum nitride (AlN) film transduction.

RESONANT PRESSURE SENSING

The acoustic interaction between solid resonant structure and its ambient fluid molecules at their interface gives rise into energy dissipation [5, 6] and inertial-loading [6, 7] of the resonator. Such effects will be amplified as surface to volume ratio of the resonant structure increases. Furthermore, depending on the particle polarization at solid-fluid interface and frequency of vibration, different resonance modes of a single structure experience different Q and f_0 loading when operated in fluids: while transversely polarized resonance modes interact efficiently with surrounding molecules resulting in radiation of compression waves in fluid, shear-based vibrations do not couple into propagating waves and their frequency remains nearly constant.

Pressure Sensitive Flexural Membranes

Thin membranes resonating in their transverse flexural modes are specifically sensitive to their surrounding fluid. Such sensitivity can be used to implement simple resonant pressure sensors. For a rectangular membrane operating in n^{th} flexural mode, the resonance frequency when operating in an arbitrary fluid of finite density can be estimated as [6]:

$$f_n \approx f_{n,vac} \cdot \left(1 - \frac{1}{2} \cdot \frac{\mu}{\rho_0 A}\right) \tag{1}$$

Here $f_{n,vac}$ is the resonance frequency in vacuum; A is the surface area of membrane lateral cross-section and ρ_0 is the mass density of the membrane; and μ is the added mass per unit length as a result of inertial loading of fluid molecules carried along by membrane vibration which can be estimated from [4]:

$$\mu \approx \frac{\pi}{4} \cdot \rho_f \cdot t^2 \qquad (2)$$

Here ρ_f is the fluid density and t is the vertical membrane depth. For a membrane resonating in air, (2) can be re-written as a function of ambient pressure and temperature using the ideal gas law:

$$\mu_{air} \approx \frac{\pi}{4} \cdot t^2 \cdot \frac{P}{R_{specific} \cdot T} \qquad (3)$$

where P and T are absolute pressure and temperature; and $R_{specific}$ is the specific gas constant of air. Using (1-3), the pressure coefficient of frequency (PCF) of the flexural modes can be written as:

$$\frac{1}{f_n} \cdot \frac{\partial f_n}{\partial P} \approx -\frac{\pi}{4} \cdot \frac{1}{\rho_0 \cdot R_{specific} \cdot T} \cdot \left(\frac{t}{w}\right) \qquad (4)$$

It can be concluded from (4) that PCF of the flexural modes can be improved by opting for thin membranes (i.e. small w) with large depth (t). This, on the other hand, results in placement of lower-order flexural modes in low-frequencies resulting in significant air damping [8, 9] especially for high-pressure applications. This is undesirable since realization of any accurate resonant-based sensor requires implementation of low phase-noise oscillator, which in turn needs high Q of the resonance mode, consistent over the entire sensing range. On the other hand, although the Q of higher-order flexural modes are less affected by air damping, their PCF cannot be approximated accurately using (4) since Euler-Bernoulli beam theory is not valid for modeling of flexural resonance of beams/membranes with small aspect ratios ($= \frac{l}{nw}$).

Dual-Mode Pressure Sensor

In this paper, acoustic coupling of vertical membranes with bulk acoustic wave resonators is used to generate two high Q coupled resonance modes ($CM_{1,2}$) with small frequency split ($\Delta f_{res} = f_{res,1} - f_{res,2}$) but large PCF difference (ΔPCF). A combinatorial beat frequency generated from the two modes shows an amplified PCF:

$$PCF_b \approx \frac{f_{res,1}}{\Delta f_{res}} \cdot \Delta PCF \qquad (5)$$

The large (ΔPCF) of the coupled modes is due to different polarization of the vibrations induced in vertical membranes when coupling SiBARs in-phase and 180°-out-of-phase. While the former excites a high-order transverse flexure in membranes, the latter results in pressure-insensitive extensional Lamb waves making the

corresponding coupled mode a good relative pressure reference. Furthermore, although thin membranes coupling bulk acoustic resonators are operating in a high-order tone with reduced pressure sensitivity compared to fundamental transverse flexural mode, the PCF_b can be considerably increased by proper design for frequency split reduction.

RESONATOR DESIGN

Two identical SiBARs with characteristic length of L are connected through thin vertical membranes with a length of $2L$, resulting in creation two coupled resonance modes when SiBARs are resonating in their 3rd length-extensional mode (Figure 2).

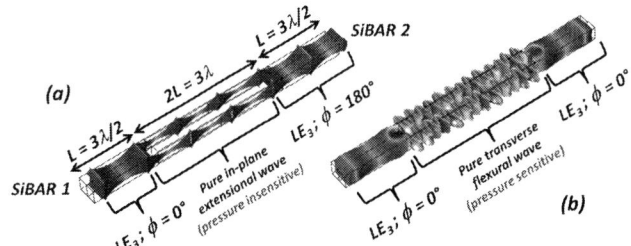

Figure 2: Mode shapes of (a) pressure insensitive CM_1; and (b) pressure sensitive CM_2.

The 180° out-of-phase coupling of SiBARs excites an extensional Lamb wave with a wavelength of $\lambda = 2L/3$ at the same frequency in thin membranes of length 3λ resulting in mode CM_1 (Figure 2a). In this case, vertical membranes do not load the equivalent stiffness/mass of the SiBARs.

On the other hand, since the in-phase coupling of LE_3 modes in SiBARs results in anti-periodic stress at two terminations of the membranes, extensional wave cannot be excited since the length of the membrane is an integer number of λ. Therefore, the vibration of the beam will be purely flexural, and load the resonance of SiBARs resulting in slight shift in their frequency (mode CM_2, figure 2b). While the longitudinal polarization of the membrane in CM_1 makes this mode insensitive to the gas mass loading, the transverse vibration of the membrane in CM_2 makes it pressure sensitive. Figure 3 shows a simplified electrical equivalent circuit for the single-input, double-output electromechanical transduction configuration (Figure 1).

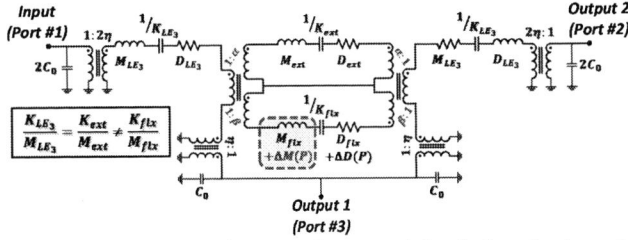

Figure 3: Electrical equivalent model of the dual-mode resonant pressure sensor.

Here M_i, K_i and D_i ($i \in \{LE_3, ext, flx\}$) are the equivalent mass, spring and damping of LE_3 mode in SiBARs and extensional and flexural modes in vertical membranes. η is the piezoelectric transduction coefficient per single finger electrode (Figure 1), and α and β are the acoustic coupling efficiency between LE_3 mode in SiBARs mode with extensional and flexural modes in vertical membranes respectively. Pressure sensitive fluid mass and damping loading are shown by $\Delta M(P)$ and $\Delta D(P)$ in the model. Proper acoustic engineering guarantees $\frac{K_{LE_3}}{M_{LE_3}} = \frac{K_{ext}}{M_{ext}} \neq \frac{K_{flx}}{M_{flx}}$. Therefore, f_b can be formulated as:

$$f_b \approx \frac{1}{2\pi} \left| \sqrt{\frac{K_{LE_3}+\alpha^2 \cdot K_{flx}}{M_{LE_3}+\alpha^2 \cdot (M_{flx}+\Delta M(P))}} - \sqrt{\frac{K_{LE_3}}{M_{LE_3}}} \right| \quad (6)$$

It's worth to note the absolute sign in (6). The frequency sequence of the two coupled modes depends on the membrane thickness, resulting in placement of transverse flexural resonance frequency before or after that of extensional mode. Also considering the electrical equivalent model, since the main portion of acoustic energy is concentrated in high-frequency SiBARs (i.e. small α and β), the effect of air damping ($\Delta D(P)$) on overall Q of the modes remains negligible over the entire pressure range from vacuum to atmospheric level.

Figure 4 shows the COMSOL simulated pressure characteristic of the resonance frequency for the two coupled modes at 30°C.

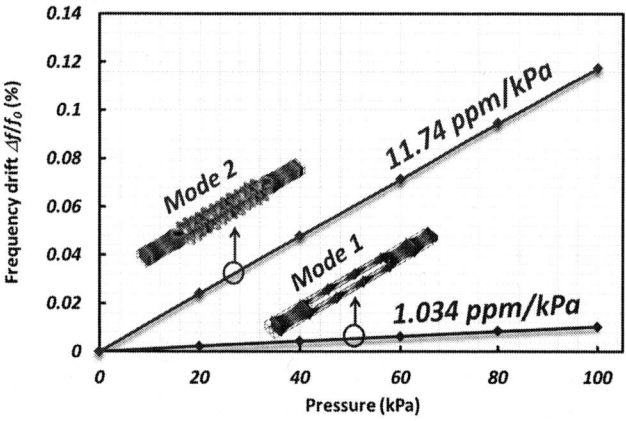

Figure 4: Simulated pressure characteristic of the two resonance modes of figure 2.

An amplified pressure sensitivity can be realized by generating a combinatorial beat frequency from subtraction of reference (CM_1) from pressure sensing mode (CM_2) using a system with a block diagram schematically shown in figure 5a. Figure 5b shows the pressure characteristic of f_b extracted from coupled modes in figure 4. The small frequency split of 453 kHz and large ΔPCF has resulted in a large pressure sensitivity of ~2900 ppm/kPa. Considering (4), since the mass-loading effect of fluid molecules linearly increases with the depth of the vertical membranes, thick silicon substrates can be used to further increase the pressure sensitivity.

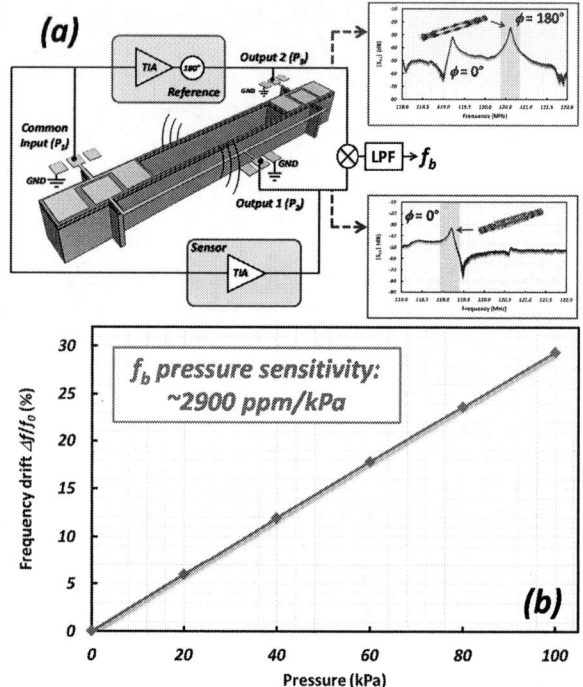

Figure 5: (a) Block diagram required for extraction of f_b. (b) Simulated pressure sensitivity of f_b.

DEVICE FABRICATION

A proof-of-concept design consisting of high frequency SiBARs operating in their LE_3 at ~120 MHz and coupled through two vertical membranes of 2μm thickness, has been implemented on AlN-on-Si acoustic platform using TPoS fabrication process [10] and a 20μm SOI wafer. A 0.5μm AlN film sandwiched between metal electrodes used to provide efficient and selective electromechanical transduction for the two modes. Figure 6 shows the SEM image of the device. A single-input double-output configuration of top electrodes has been used to facilitate implementation of two electrically-isolated oscillators locking separately into the two modes.

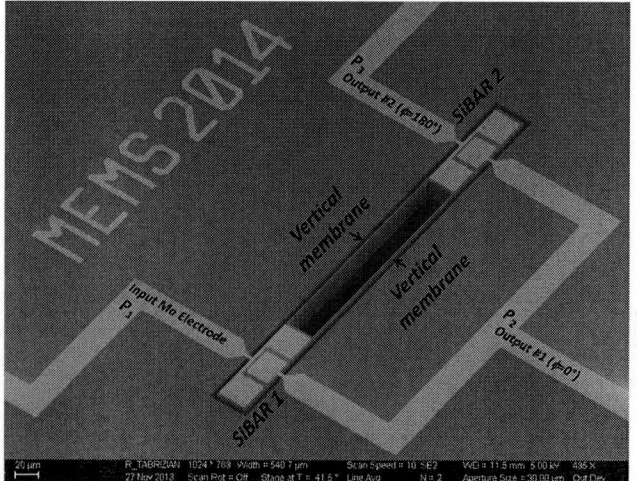

Figure 6: SEM image of the dual-mode vertical membrane resonant pressure sensor.

978-1-4799-3510-9/14 $31.00 © 2014 IEEE

DEVICE CHARACTERIZATION

Figure 7 shows the measured transmission frequency response for both S_{21} and S_{31} configurations.

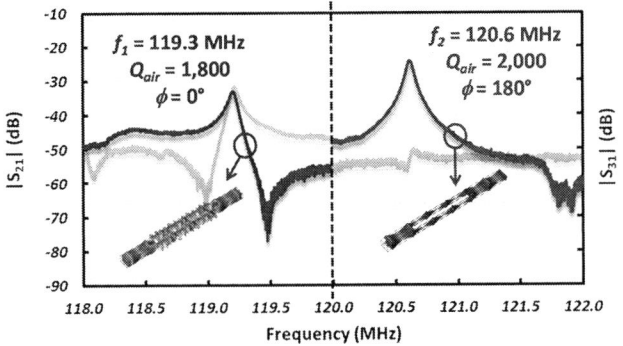

Figure 7: Measured frequency response of the dual-mode resonant pressure sensor.

An f_b of ~1.3 MHz extracted from subtraction of the two modes shows a linear pressure sensitivity of ~346 ppm/kPa over the pressure range of 0-100 kPa (Figure 8).

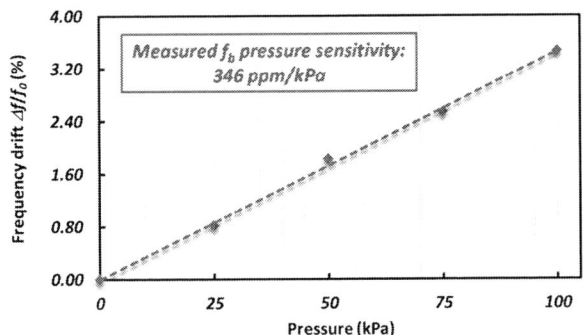

Figure 8: Pressure characteristics of f_b measured at 30°C, over 0-100 kPa pressure range.

The smaller measured pressure sensitivity compared to simulated value (Figure 5b) is due to the larger f_b of the device resulted from fabrication processing imperfections. Figure 9 shows the pressure characteristic of the Q for the two coupled modes. The Q of the two modes remains consistent across the entire pressure range with a maximum drop of ~9% for the pressure sensitive mode CM_2, highlighting the role of SiBARs to store the main portion of acoustic energy, and hence minimizing the effect of air damping on resonance Q.

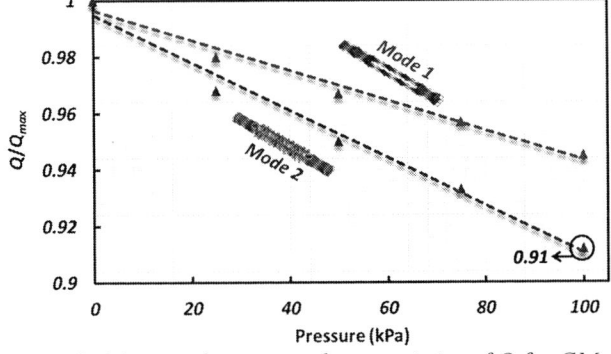

Figure 9: Measured pressure characteristics of Q for $CM_{1,2}$.

CONCLUSION

A dual-mode resonant pressure sensor has been implemented based on gas mass loading of vertical thin membranes operating in their transverse flexural mode. The membranes are driven using two identical SiBARs operating in 3rd length-extensional bulk acoustic mode. SiBAR and membrane dimensions are designed to provide two acoustically coupled resonance modes with small frequency split but large difference in their pressure sensitivity. This facilitated extraction of a combinatorial beat frequency with amplified pressure sensitivity. A high pressure sensitivity of 346 ppm/kPa has been measured over the vacuum to atmospheric pressure range, while the Q of f_b constituent modes remained consistently high over the entire pressure range.

REFERENCES

[1] W. P. Eaton, et. al., "Micromachined pressure sensors: review and recent developments," Smart Materials and Structures 6.5 (1997), pp. 530-539.

[2] C. F. Chiang, et. al., "Resonant pressure sensor with on-chip temperature and strain sensors for error correction," in Proc. IEEE Int. Conf. MEMS, 2013, pp. 45-48.

[3] R. Tabrizian and F. Ayazi, "Acoustically-Engineered multi-port AlN-on-Silicon resonators for accurate temperature sensing," to be presented at IEEE International Electron Devices Meeting (IEDM 2013), Washington, DC, Dec. 2013.

[4] W. K. Blake, "The radiation from free-free beams in air and in water," Journal of Sound and Vibration, vol. 33. No. 4, 1974, pp. 427-450.

[5] Y.-H. Cho, A. P. Pisano, and R. T. Howe, "Viscous damping model for laterally oscillating microstructures", J. Microelectromech. Syst., vol. 3, no. 2, pp.81 -87 1994.

[6] M. Christen, "Air and gas damping of quartz tuning forks," Sensors and Actuators, vol. 4, No. 4, 1983, pp. 555-564.

[7] L. Chen, and M. Tabib-Azar, "Air and gas damping of quartz tuning forks," IEEE Sensors 2011, pp. 740-742.

[8] W. E. Newell, "Miniaturization of tuning forks," Science 161, no. 3848, 1968, pp. 1320-1326.

[9] W. Zhang and K. Turner, "Frequency dependent fluid damping of micro/nano flexural resonators: Experiment, model and analysis," Sensors and Actuators A: Physical 134, no. 2, 2007, pp. 594-599.

[10] W. Pan and F. Ayazi, "Thin-Film Piezoelectric-on-Substrate Resonators with Q Enhancement and TCF Reduction," IEEE International Conference on Micro Electro Mechanical Systems (MEMS 2010), Hong Kong, 2010, pp. 104-107.

CONTACT

*R. Tabrizian, tel: +1-404-259-7322; roozbeh@gatech.edu

HIGHLY RESPONSIVE CURVED ALUMINUM NITRIDE PMUT

Sina Akhbari[1], Firas Sammoura[1,2], Stefon Shelton[3], Chen Yang[1], David Horsley[3], and Liwei Lin[1]
[1]Department of Mechanical Engineering, University of California, Berkeley, USA
[2]Department of Electrical Engineering and Computer Science, Masdar Institute of Science and Technology, Abu Dhabi, UAE
[3]Department of Mechanical Engineering, University of California, Davis, USA

ABSTRACT

We have successfully demonstrated highly responsive, curved piezoelectric micromachined ultrasonic transducers (pMUTs) based on a CMOS-compatible fabrication process using AlN (aluminum nitride) as the transduction material. Micro fabrication techniques have been used to control the radius of curvature of working diaphragms from 400~2000 μm and theoretical analysis have been developed for the optimal dimensions of the transducers to boost the electromechanical coupling and acoustic pressure. A prototype device made of a 2μm-thick AlN on a curved diaphragm with a nominal size of 140μm in diameter and a radius of curvature of 1065μm has been fabricated. The measured resonant frequency is 2.19MHz and DC response is 1.1nm/V, which is 50X higher than that of a planar device with the same nominal diameter. As such, this new class of curved pMUTs could dramatically enhance the responses of the state-of-art, planar pMUTs with high electromechanical coupling for various ultrasonic transduction applications, such as gesture recognition and medical imaging.

INTRODUCTION

Ultrasonic imaging is one of the most important and widely used medical imaging techniques, which uses high-frequency sound waves to take images of soft tissues such as muscles, internal organs as well as blood flows in blood vessels [1]. The advancements of microelectromechanical systems (MEMS) have produced ultrasonic transducers based on plate flexural mode with good improvements in bandwidth, cost, and yield over the conventional large scale, thickness-mode PZT sensors [2]. In the past two decades, micro fabrication technologies have been utilized to realize both capacitive (cMUTs) and piezoelectric (pMUTs) micromachined ultrasonic transducers with mechanical impedances closely matched to those of the imaging media, resulting in improved bandwidth and system efficiency. Although cMUTs are constrained by high DC polarization voltage and small gap requirements, they have better electromechanical coupling than pMUTs [3]. In order to improve the electromechanical coupling in pMUTs, efforts by means of "dome-shape" piezoelectric transducers and actuators have been proposed based on the intuition that dome-shape structures can achieve "better efficiency in converting in-plane strain to volumetric deflection" for sensing and actuation applications. For instance, Feng *et al.* have developed a fabrication process using spherical balls to make wax molds which are coated with parylene as the dome-shape transducers based on ZnO [4]. However, the

process is not IC compatible and not suitable for large array fabrication. Morris *et al.* have analyzed and measured the actuation characteristics of piezoelectric diaphragms with an initial curvature that is induced by a static pressure [5]. Hajati *et al.* have reported a dome-shaped pMUT device with 45% coupling using PZT (lead zirconate titanate) but it is not compatible with IC process [6]. This work presents the concept of "curved pMUT" with several unique accomplishments: (1) a CMOS-compatible AlN fabrication process for curved pMUTs; (2) analyses and experimental validations of optimal designs for curved pMUTs with high coupling and acoustic pressure; and (3) the demonstration of more than one-order of higher DC displacements compared with planar pMUTs of similar geometry.

CONCEPT

Fig. 1a shows the schematic diagram of a curved pMUT and Fig. 1b is the typical state-of-art planar pMUT structure. The generated piezoelectric moment is expressed:

$$M^P = Y'_0 d'_{31} ZV \qquad (1)$$

where Y'_0 is the modified Young's modulus, d'_{31} is the modified piezoelectric charge constant, Z is the distance of the piezoelectric layer to neutral axis, and V is the applied voltage [7]. In order to excite a planar pMUT, the Laplacian of the piezoelectric moment about the neutral axis of the structure should be nonzero. As a result, an additional structural layer, silicon is shown here as an example, is needed to generate a non-zero piezoelectric moment.

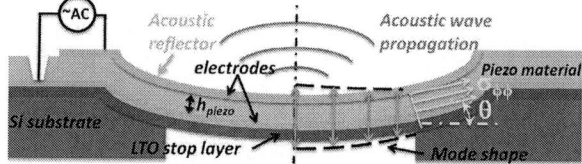

Figure 1a: The schematic cross-sectional view of a curved pMUT, which promotes the conversion of in-plain stress '$\sigma_{\phi\phi}$' to vertical mechanical forcing function.

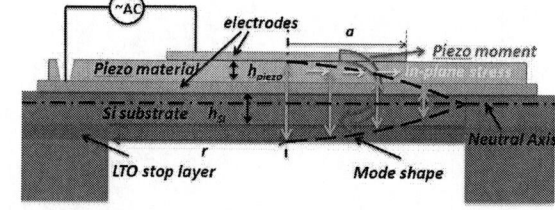

Figure 1b: The schematic cross-sectional view of a planar pMUT with the silicon structural layer with height of 'h_{Si}', diaphragm radius of 'r', top electrode radius of 'a', and piezoelectric material thickness of 'h_{piezo}'.

In contrast to a planar pMUT which relies on the excessive plain strain due to the d_{31} effect to induce vertical deformation, the induced piezoelectric in-plane strain has a vertical component, which is in the direction of the normal motion, as illustrated in Fig. 1(a). Hence, the curved-shape diaphragm promotes the conversion of in-plane strain to vertical mechanical motion for higher electromechanical coupling and acoustic pressure. A simplified formula is derived for the uniform normal pressure generated on a piezoelectric hemispherical shell under applied voltage V:

$$p_{piezo} = 2Y_0' d_{31}' V / R_c \qquad (2)$$

where R_c is the radius of curvature of the diaphragm. The normal driving force helps to eliminate the necessity of the additional structural layer to generate a moment about the neutral axis [8]. As such, the piezoelectric layer alone can also serve as the structural layer. It is noted that this term goes to zero as the radius of curvature goes to infinity for a planar pMUT. Moreover, curved pMUTs can be constructed based on a CMOS compatible process as illustrated in Fig. 2. A concave diaphragm, with a radius of curvature of R_c is chosen as the demonstration example by etching a cavity into the silicon substrate. The nominal diaphragm size is determined by the backside through-hole etching process, with an opening radius, r, and the rest of the curved surface serves as acoustic reflector/concentrator which could further enhance the transduction performances.

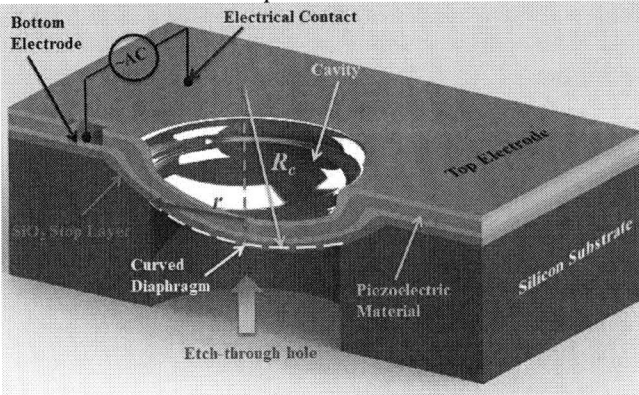

Figure 2: 3D schematic of the curved pMUT with a radius of curvature of 'R_c' and diaphragm radius of 'r', defined by the backside etching opening, based on a CMOS-compatible fabrication process using Aluminum Nitride as the piezoelectric material.

FABRICATION

Process Flow

The fabrication process flow begins with silicon wet etching using HNA (hydrofluoric, nitric, acetic) to form the curved device structural base using a 1.2 μm low stress LPCVD nitride as a hard mask (Fig. 3a). A 1.1 μm-thick LPCVD low temperature oxide (LTO) is then grown to form the backside etching stop layer as shown in Fig. 3b, which is followed by the sputtering of the active stack of Mo/AlN/Mo with thicknesses of 100 nm, 2 μm and 100 nm, respectively. The Molybdenum layers constitute the top and bottom electrodes, while the AlN layer serves the purpose of the piezoelectric and main structural layer, simultaneously. Figure 3c shows the via opening to the bottom electrode, which is made by SF_6 plasma etching of the top Mo and followed by a combination of plasma dry etching in chlorine based gases and MF-319 developer wet etching of AlN. The last process step is to release the diaphragm using backside deep reactive ion etching (DRIE) and the nominal diameter of the released membrane is defined by the backside etch opening process, as illustrated in Fig. 3d. In order to target medical imaging frequency for deep structures (e.g. kidneys, liver, etc.), curved pMUTs with controlled radii of curvature within 400~2000 μm have been fabricated with nominal diameters of 120 μm to 190 μm in the prototype devices for characterizations and investigations.

Fabrication Results

An SEM cross-sectional view photo showing a curved pMUT is in Fig. 4a, where the cavity opening is 320μm in diameter and the curved pUMT diaphragm is 170μm in diameter. Figure 4b is the close-up view of Fig. 4a near the anchor area where the undercut phenomenon during the DRIE process is clearly observed. Figure 4c is a high-resolution SEM photo revealing that desirable polarization direction of AlN - perpendicular to the curvature of the diaphragm, has been achieved.

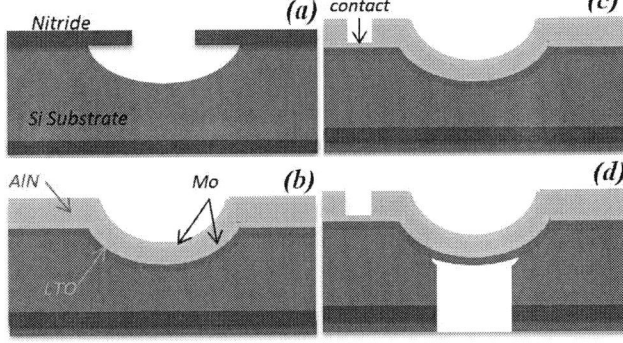

Figure 3: Process flow for the curved pMUT: (a) silicon wet etching using nitride as the mask for cavity formation; (b) LTO deposition as the stop layer and Mo/ALN/Mo deposition as the piezoelectric structure layer; (c) contact opening; (d) backside DRIE silicon etching.

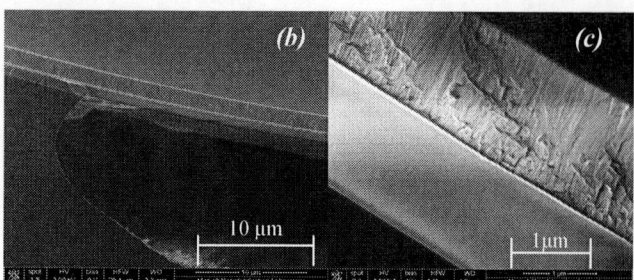

Figure 4: SEM micrographs: (a) a cleaved device showing the cross sectional view of a fabricated pMUT; (b) close-up view of (a) showing the undercut due to excessive DRIE etching; (c) the close view SEM photo showing that the polarization direction of the AlN crystalline structure is perpendicular to the curvature of the diaphragm.

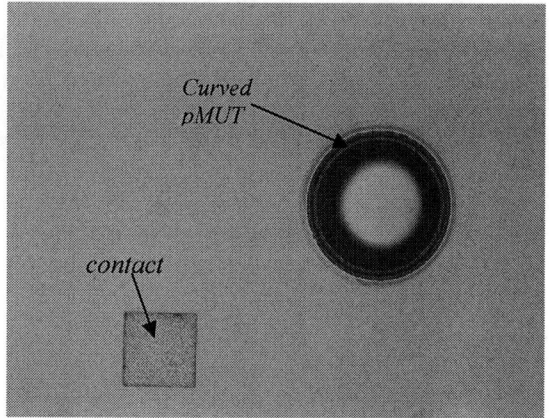

Figure 5: a top view optical image of a curved pMUT

Figure 5 is a top-view optical image of the same curved pMUT shown in Fig 4a with the reflection of light from the smooth concave surface of the curved pMUT and the via contact opening to the bottom electrode.

RESULTS AND DISCUSSIONS

In order to investigate the effect of the curvature of the structure, both an analytical model [8] and finite element analyses using COMSOL Multiphysics have been built with good consistency. The solid curve in Fig. 6b is the predicted DC displacement of a curved pMUT with 2-μm thick AlN, 140μm in nominal diameter with respect to different radii of curvature. It is noted that the DC displacement starts to increase as the radius of curvature increases; reaches an optimum point; and starts to decrease with further increase of the curvature. Clearly, a curved pMUT has a higher resonant frequency than a flat pMUT with the same nominal diameter. As a result the curved pMUT will have a higher volumetric velocity (the product of higher frequency and higher displacement) than a planar pMUT to generate a higher acoustic pressure. The symbols in Fig. 6b are the experimental data measured from the prototype devices with different radii of curvature. The data from a planar pMUT is from a similar process run with 1 μm-thick AlN, 3 μm-thick silicon as the structural layer and 70% top electrode

coverage. The experimental results for curved pMUTs show good consistency with the simulation results both in terms of the center displacement and resonant frequency. It is noted that more than one order of magnitude higher DC displacement is achieved from the curved pMUT as compared with the planar pMUT.

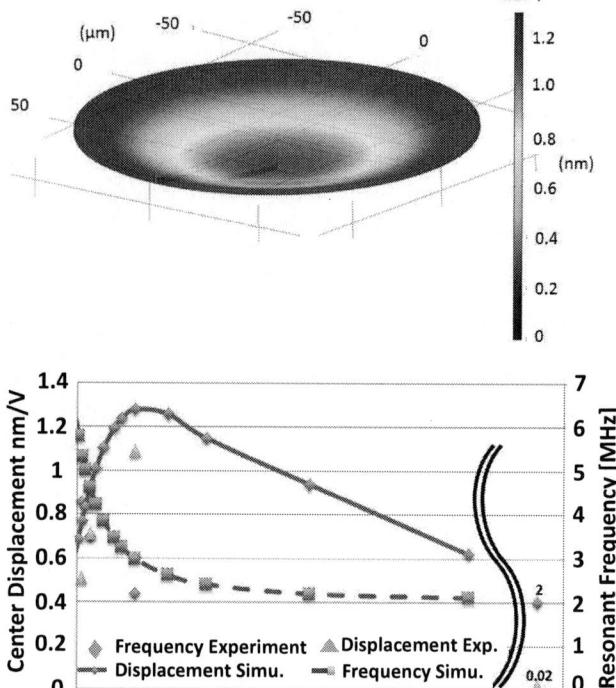

Figure 6: (a, top) Simulated operation mode shape and DC displacement of a curved pMUT. (b, bottom) Good consistency between simulations (line) and experiments (symbols) with 140μm in diameters curved PMUT and different radii of curvature. Experimental results from a planar PMUT are also plotted.

Acoustic-piezoelectric frequency domain simulations have also been conducted in COMSOL Multiphysics to estimate the frequency response of the diaphragms. Figure 7a shows center displacement (nm/V) versus frequency plot for a curved AlN pMUT with total thickness of 2.5 μm (including the metal electrode and bottom oxide layer), 190 μm nominal diameter and 1065 μm in radius of curvature operated in air. The simulated center DC displacement is 1.4 nm/V and the displacement at its resonance is 75 nm/V at 2.45 MHz. The measurements were conducted using a Laser Doppler Vibrometer (LDV) in air as shown in Fig. 7b of a prototype device with 190 μm in nominal diameter and 1065 μm in radius of curvature. It has measured resonant frequency at 2.19 MHz, 1.1nm/V DC vertical displacement and 45 nm/V of center displacement at resonance. It is noted that LDV can only measure the velocity/displacement when the target is under motion such that the DC displacement is measured at very low frequencies away from the resonance. These results show good consistency between the simulation and the experiment. Experimental results show slightly lower displacement per volt and lower resonant frequency probably due to geometry mismatches (e.g. the fabricated diaphragm might not be perfect spherical

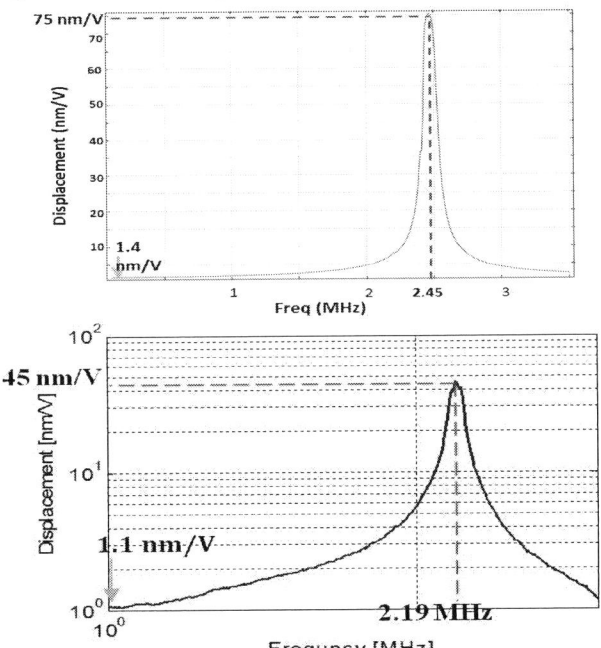

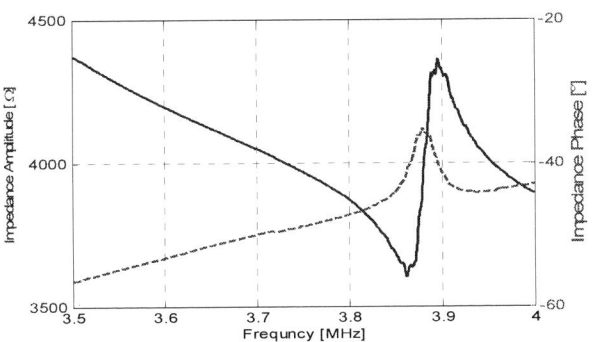

Figure 7: Displacement versus frequency plots of a curved pMUT with 190µm in diameter and 1065µm in radius of curvature. (a, top) simulation and (b, bottom), experiment.

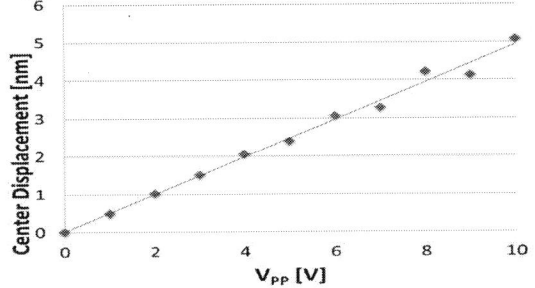

Figure 8: Impedance versus frequency measurements (in air) of another curved pMUT with 120 µm in diameter and 550µm in radius of curvature.

shape), additional loss mechanisms in air damping, anchor and boundaries.

Fig. 8 is the impedance measurement in air for another curved pMUT with 120 µm in nominal diameter and 550 µm in radius of curvature which has a resonant frequency of 3.86MHz. The electromechanical coupling coefficient is calculated as 2.1% in air from Fig. 8 using resonant and anti-resonant frequencies [3]. Although the transducer is designed for operation in water, which has higher acoustic impedance, the measured coupling is as good as other state-of-art planar pMUTs operated in air. Fig. 9 shows excellent measured linearity response between center displacement and the input voltage from a curved pMUT operated at 500 kHz – another desirable characteristic. As such, curved pMUTs could provide high sensitivity, miniaturization, and low manufacturing cost for broad applications.

ACKNOWLEDGEMENTS

The authors would like to thank Jeffery Clarkson, Ryan Rivers and Joseph Donnelly from the UC Berkeley Marvell nanolab and Mr. Amir Heidari and Yipeng Lu from ME Department of UC Davis for valuable discussions.

REFERENCES

[1] G. S. Kino, *Acoustic Waves: Device, Imaging, and Analog Signal Processing*, Prentice Hall, 1987.

[2] R. Pryzbyla et al., "An Ultrasonic Rangefinder Based on an AlN Piezoelectric Micromachined Ultrasound Transducers," in *Proc. IEEE Sensors 2010*, pp. 2417-242, Nov. 2010.

[3] A. S. Ergun et al., "Capacitive Micromachined Ultrasonic Transducers: Theory and Technology," *Journal of Aerospace Engineering*, vol. 16, pp. 76-84, 2003.

[4] G. Feng et al., "Fabrication of MEMS ZnO Dome Shaped-diaphragm Transducers for High-frequency Ultrasonic Imaging," *Journal of Micromechanics and Microengineering*, Vol. 15, pp. 586-590, 2005

[5] D. J. Morris et al., "Enhanced actuation and acoustic transduction by pressurization of micromachined piezoelectric diaphragms," *Sensors and Actuators A*, vol. 161, pp. 164-172, 2010.

[6] Hajati et al., "Three-dimensional micro electromechanical system piezoelectric ultrasound transducer," *Applied Physics Letters*, vol. 101 (253101), pp. 1-5, 2012.

[7] F. Sammoura et al., "An Analytical Analysis of the Sensitivity of Circular Piezoelectric Micromachined Ultrasonic Transducers to Residual Stress," in *Proc. 2012 IEEE Int. Ultrason. Symp. (IUS)*, pp. 580-583, Dresden, Germany, October 7-10, 2012.

[8] F. Sammoura et al., "An Analytical Solution for Unimorph Curved Piezoelectric Micromachined Ultrasonic Transducers with Spherical–shape Diaphragms," *IEEE Transactions on Ultrasonics, Ferroelectrics, and Frequency Control*, submitted.

CONTACT

*S. Akhbari, tel: +1-510-9267150;
sina.akhbari@berkeley.edu

Figure 9: Center displacement vs. the input voltage of a curved pMUT with 140µm in diameter and 1065µm in radius of curvature operated at 500 kHz.

ENCASED CANTILEVERS FOR LOW-NOISE FORCE AND MASS SENSING IN LIQUIDS

D. Ziegler[1], A. Klaassen[2], D. Bahri[1], D. Chmielewski[1], A. Nievergelt[1],
F. Mugele[2], J.E. Sader[3], and P.D. Ashby[1]*

[1]Lawrence Berkeley National Laboratory, Berkeley, USA
[2] University of Twente, Twente, THE NETHERLANDS
[3] University of Melbourne, Melbourne, AUSTRALIA

ABSTRACT

Viscous damping severely limits the performance of resonator based sensing in liquids. We present encased cantilevers that overcome this limitation with a transparent and hydrophobic encasement built around the resonator. Only a few micrometers of the cantilever probe protrude from the encasement and water does not enter the encasement. This maintains high Q-factors and reduces the thermo-mechanical noise levels by over one order of magnitude and reaches minimal detectable forces of 12 fN/•Hz in liquids. These probes expand the frontiers of cantilever based sensing. We discuss their design and fabrication with special focus on squeeze film damping and demonstrate their successful application for quantitative mass sensing of single nanoparticles and gentle Atomic Force Microscopy imaging of soft matter in liquids.

INTRODUCTION

One of the most prominent applications of mechanical resonators is Atomic Force Microscopy (AFM)[1], which takes a topographic image by scanning a sharp tip attached to a cantilever over a surface while measuring the interaction forces through the change in the cantilever's deflection, resonance frequency, or oscillation amplitude. Beyond height information it enables visualization of materials properties such as magnetism[2], specific chemical interactions[3], [4], or electrical properties at high resolution [5]. Its resolution, label-free detection, and extended observation time without damage or bleaching, lift AFM to arguably the best tool to study the structure and behavior of biomolecules at surfaces[6], [7], [8]. Another application of mechanical resonators, which emerged in the mid-1990s, is chemical or bio-sensing[9] and is today a well-known technique for cheap, portable, analysis systems [10],[11]. The minimum detectable forces or mass changes are associated with losses due to damping of the resonator arising from the exchange of energy between the resonator and its surrounding environment [12], [13]. The smallest measurable force is given by $F_n=(2kk_bTB/\pi Qf_0)^{1/2}$, where k is the spring constant, k_bT is the thermal energy, B the measurement bandwidth, Q the mechanical quality factor, and f_0 the resonance frequency. Upon immersion in water, a typical cantilever's resonance frequency drops about by a factor of three and its Q-factor by a factor of 50. Compared to operation in air, the resulting smallest detectable force is increased by more than one order of magnitude.

Researchers have been attempting to decrease viscous damping by reducing the size of the cantilever[14], [15] or

keep the resonator in air such that only the sensing tip is immersed in solution. Practically, the latter has been difficult to achieve. The few attempts so far include bell jar designs for length extensional resonators[16], tuning forks[17], or optical fiber probes[18]. But in all these designs the resonator and their enclosures are so large, that their size offsets gains due to lower viscosity, furthermore, they all demand specialized instrumentation and cannot be used with standard AFM instruments. Here, we present a solution, where a transparent encasement is directly fabricated around the cantilever with only the tip protruding from the encasement (see Figure 1a). As the encasement is transparent these probes can be used on any commercially available AFM without modification of the instrument.

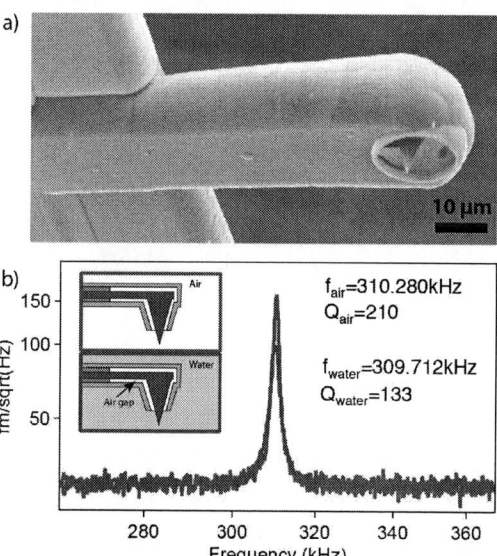

Figure 1: a) Electron microscope image of an encased cantilever b) Thermal noise spectra recorded in air (red curve) and water (blue curve) show that exceptionally high Q-factor and resonance frequency are mainained while operating in liquids. Note that without the encasement immersion in water would lower the resonance frequency by about a factor three and the Q-factor would drop by a factor of 50.

Once immersed in liquid, the hydrophobic interior of the encasement pins the air-water interface at the orifice preventing water from entering, and the cantilever effectively vibrates in air. The thermal spectra shown in figure 1b) clearly reveal that a high resonance frequency and Q factor are maintained while submerged in water. The resulting smallest detectable force in liquid of 12 fN/√Hz outperforms regular non-encased cantilevers (150 fN/√Hz) by more than one order of magnitude.

FABRICATION

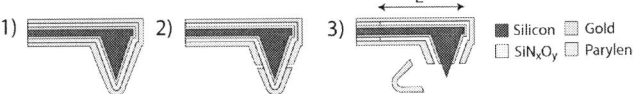

Figure 2: Simplified illustration of the mask-less process to fabricate encased cantilevers. 1) Deposition of a thick sacrificial silicon oxynitride (gap) and Parylene layer (encasement) on commercially available gold-coated silicon cantilevers. 2) Focused ion beam opening of the Parylene layer around the apex. 3) A hydrofluoric acid wet etch releases the cantilever and defines its final length (L).

A simplified fabrication process of the encasement is illustrated in figure 2. In the first step (1) regular gold coated silicon cantilevers with a nominal width of 35 μm, thickness of 1 μm, and a length of 90,110, and 130 μm (NSC36, Mikromasch) are coated with a sacrificial layer of silicon oxynitride using plasma enhanced chemical vapor deposition (PECVD, Plasmalab80 Plus, Oxford Instruments). Deposition using 1182 sccm of 1%SiH$_4$/Ar and 710 sccm of N$_2$O, at 350ºC and a pressure of 1000 mbar results in a rate of about 140 nm/min. The resulting low stress oxynitride layer reduces bending of the cantilever and its low quality gives conveniently high etch rates in the subsequent release process. The encasement itself consists of a uniformly deposited 2 μm thick poly(p-xylylene) layer (Parylene C, PDS2010, SCS). In the next step (2) we use focused ion beam milling (FIB, 1540ESB, Zeiss) to cut a roughly 0.5 μm wide opening around the cantilever's tip. This exposes the sacrificial layer and enables releasing of the cantilever by a hydrofluoric acid etch (>40% HF) (3). Note that the length of the final cantilever (L) is adjusted by the time of this etch step. The cap falls off during etching and the tip apex protrudes from the encasement. The protecting gold layer can be selectively removed from the apex only (wet gold etch, G1814, Transene). Transmission electron microscopy (TEM) and using AFM confirmed that the fabrication process does not affect probe sharpness and a radius smaller than 10 nm is preserved.

SQUEEZE FILM DAMPING

The thickness of the sacrificial layer is an important design parameter because it defines the gap size, which determines how much the tip protrudes. But the gap size also influences squeeze film damping as discussed in this section. Figure 3 a) shows a cross-section of an encased cantilever, where the red arrows indicate the flow of air inside the encasement when the cantilever moves upwards. The built-up pressure inside the encasement leads to a special type of viscous damping in confined spaces known as squeeze film damping (SFD). This phenomenon has been studied to optimize MEMS resonators[19], [20], but these designs commonly experience SFD on only one side of the

resonator. In encased cantilevers, however, the effect is two-sided and thus it is unknown for which gap size SFD starts deteriorating the device's performance. Modeling squeezed films involves coupling the modified Reynolds equation with the equations for the structural displacements which remains a challenging mathematical problem until today[21]. Hence, we used finite elements to model the hydrostatic pressures and fluid/solid interactions inside the encasement (FLUID80 model, ANSYS).

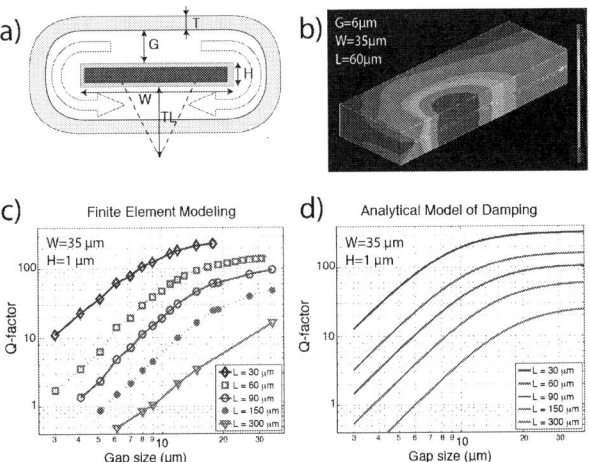

Figure 3: Squeeze film damping in encased cantilevers a) Cross-section of an encased cantilever (not to scale) the gap size (G), encasement thickness (T), cantilever thickness (H), width (W) and tip length (TL) are indicated. b) Pressure distribution inside cavity c) Q-factors for different gap size (G) and cantilever length (L) extracted by FEM simulations. d) Analytical models of squeeze film damping using Sader equations.

Figure 3b) shows a longitudinal cross-section of the air volume inside the gap. The color scale shows the absolute value of the pressure upon motion of the cantilever. The pressure is highest near the free end of the cantilever where displacement is greatest. The high pressure extends to the encasement surface pushing the air around the side of the cantilever and increasing damping. By fitting a simple harmonic oscillator model to the power spectrum of the calculated cantilever thermal motion, we extract resonance frequency and damping of the cantilever. Figure 3c) shows the Q-factor as a function of gap size for five different cantilever lengths (L) computed using this finite element model.

The damping cantilevers can be calculated analytically in the extreme cases of being completely dominated by squeeze film damping (G<<W/2) [22],[23] and when completely unbounded (G>>W) [13]. For encased cantilever we assume a doubling in the squeeze film damping and we approximate the transition between the two extremes with the function A/(1+B/(G/W)^3) where A is the unbounded Q factor and B is the coefficient matching the behavior of the doubly bounded system at small gap (see figure 3d). Our findings from the finite element analysis and hydrodynamic models are in good qualitative agreement and clearly show that SFD only becomes significant for G<<W/2. For most commercially available cantilevers this is difficult to achieve, because as illustrated in Figure 3a), the tip length (TL) is required to be larger than G+T for the tip to protrude into the liquid (see Figure 3a).

APPLICATIONS

Gentle Imaging of Soft Matter

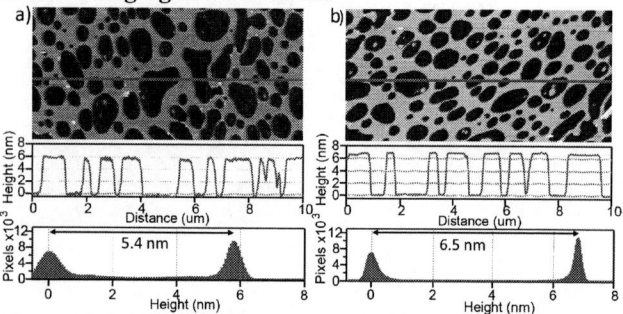

Figure 4: DPPC lipid bilayers supported on mica measured using regular silicon nitride cantilevers (a)) and encased cantilevers (b)). The cross-sections and histogram analysis clearly reveals a higher measured thickness of the bilayer when measured with encased cantilevers, which demonstrates that soft samples deform less.

High viscous damping has limited the AFM's utility for imaging soft materials in solution at high resolution without deforming or even damaging the samples[25]. As lipid bilayers readily deform under the force applied by the tip, it is a useful sample to demonstrate the gentleness of encased cantilevers. Supported DPPC lipid bilayers (L-α-dipalmitoyl-phosphatidylcholine) are prepared using a Langmuir-Blodgett trough, transferred onto mica and imaged in aqueous buffer. While the softest possible imaging with conventional silicon nitride cantilevers (Hydra Cantilevers, Applied Nanosciences Inc., Santa Clara CA, USA) results in a height of 5.4 nm (figure 2a) we obtain a thickness of 6.5 nm when using encased cantilevers (figure 2b). This exceeds the highest previously reported thickness using AFM [26], and clearly demonstrates gentle imaging enabled by the low damping.

Quantitative Mass Sensing

Cantilever based mass sensors detect minute amounts of specific chemicals or biomolecules and may significantly impact biological and chemical diagnostics. The resonance frequency shift of a cantilever is a direct measure of the added mass, but the challenge of measuring resonance frequency in a high damping/low Q-factor environment leads many sensor incarnations to indirectly measure adsorbed material using static deflection caused by surface stress[27]. The low damping of encased cantilevers, however, enables in-situ mass sensing in liquids that surpasses conventional cantilever based mass sensing, where the uncertainty of the location of the added mass, or the indirect measurement of stress induced bending require various assumptions or models in order to extract the exact value of the added mass [28],[29]. For encased cantilevers,

functionalization and binding of an analyte only occur at the tip (figure 5a). Compared to a uniform loading, this confined location of the added mass results in better responsivity, R_p, Responsivity takes into account a normalizing effective mass ($m^* = n_e\,m_c$, with $n_e \approx 0.24$) (see Eq.1).

$$R_p = \frac{\Delta f}{\Delta m} = -2\pi^2 \frac{f_0^3}{k_c} = -\frac{1}{2n_e}\frac{f_0}{m_c}. \qquad (1)$$

As the mass of the cantilever (m_c) is known, the added mass (Δm) can be extracted quantitatively from the measured frequency shift (Δf) [30]. Combined with a factor 20 of improvement from maintaining high resonance frequency f_0 and Q-factor a 80-fold reduction of the smallest detectable mass (δm) is achieved compared to a non-encased cantilever.

$$\delta m = -\frac{1}{\pi A_c}\sqrt{\frac{2k_B T B n_e m_c}{f_0^3 Q}}. \qquad (2)$$

where A_c is the oscillation amplitude. The concept of quantitative mass sensing is confirmed by measuring the mass of gold particles.

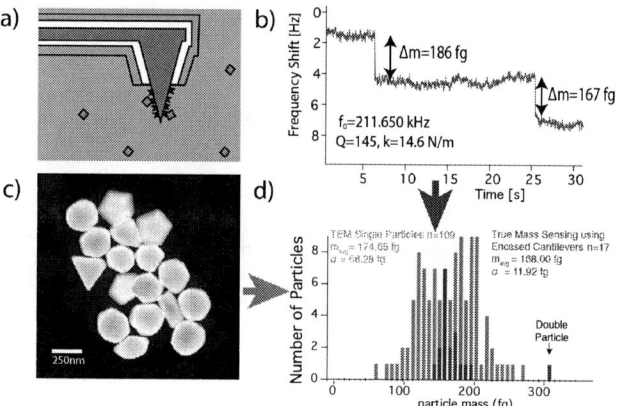

Figure 5: Encased cantilevers for mass sensing in liquids, a) local functionalization and attachment of the analyte at the tip b) the frequency shifts upon attachment of 250 nm gold particles and a corresponding added average mass of 169 fg was found c) TEM image of the gold particles, d) histogram of the estimated mass using TEM (green) and encased cantilevers as mass sensors (red).

Figure 5b shows distinct steps in the frequency shift resulting from consecutive attachment of 250 nm gold particles (BBI Solutions, EMGC250), using cantilevers etched back to a length of 135 µm, with a thickness of 3.5 µm. This results in f_0=211.65 kHz and Q=150 in liquid. Including the tip mass, m_c is 29.7 ng and mass responsivity is -15mHz/fg. We find an average mass of 168±12 fg which agrees will with the expected value 175±68 fg based on TEM characterization of the particles. (See red histogram in Figure 5d). We confirm their real size using TEM imaging (JEOL, 2100-F). Figure 5c) shows a high-angle annular dark field image used to estimate the volume from the projected areas by assuming a spherical form factor. Only considering the particles with no occlusions we find an average estimated mass of 174.6±68.3 fg. The large deviation as shown in figure 5d) is partly due to the different geometries deviating from perfect spheres. These data derive from cantilevers, which were not optimized for detecting exceedingly small masses but with the first flexural eigenmode and an amplitude Ac of 100 nm it results in a smallest detectable mass of 80 attogram/•Hz (See Eq. 1). Building the encasement around stiffer levers with shorter geometries has the potential to extend the frontier of biosensing in liquids down to the zeptogram range.

CONCLUSIONS

We have described successful fabrication of encasements around mechanical resonators the encasement traps an air bubble and thereby greatly reduces viscous damping. Improvements in performance by more than one order of magnitude for force sensing and nearly two orders of magnitude for mass sensing show that encased cantilevers revolutionize cantilever based sensing in liquids. Finite elements modeling and analytical expressions of hydrodynamic damping have been used to establish design rules for low squeeze film damping. The superior performance of encased cantilevers has been demonstrated imaging supported lipid bilayers and measuring the mass of gold nanoparticles.

ACKNOWLEDGEMENTS

The authors specifically thank E. Wong for fast and high-quality technical support, Virgina Altoe for her support in TEM imaging. We would like to thank Dr. Deirdre Olynick and Dr. Stefano Cabrini from the Nanofabrication Facility for fruitful discussions and their support in clean room processing. Work at Molecular Foundry was supported by the Office of Science, Office of Basic Energy Sciences, of the US Department of Energy under contract No. DE-AC02-05CH11231.

REFERENCES

[1] G. Binnig, C. F. Quate, and C. Gerber, "Atomic Force Microscope," *Physical Review Letters*, vol. 56, no. 9, pp. 930+, 1986.

[2] Y. Martin and H. Wickramasinghe, "Magnetic imaging by "force microscopy"with 1000 Å resolution," *Appl. Phys. Lett*, vol. 50, no. 20, pp. 1–3, 1987.

[3] W. J. Cho, A. Jeremic, and B. P. Jena, "Size of Supramolecular SNARE Complex:☐ Membrane-Directed Self-Assembly," *J. Am. Chem. Soc.*, vol. 127, no. 29, pp. 10156–10157, 2005/07/01 2005.

[4] A. Noy, D. V. Vezenov, and C. M. Lieber, "Chemical force microscopy," *Annual Review of Materials Science*, vol. 27, no. 1, pp. 381–421, 1997.

[5] D. Ziegler and A. Stemmer, "Force gradient sensitive detection in lift-mode Kelvin probe force microscopy," *Nanotechnology*, vol. 22, p. 075501, Jan. 2011.

[6] D. A. Walters, J. P. Cleveland, N. H. Thomson, P. K. Hansma, M. A. Wendman, G. Gurley, and V. Elings, "Short cantilevers for atomic force microscopy," *Review of Scientific Instruments*, vol. 67, no. 10, pp. 3583–3590, 1996.

[7] P. Hansma, V. Elings, O. Marti, and C. Bracker, "Scanning tunneling microscopy and atomic force microscopy: application to biology and technology," *Science*, vol. 242, no. 4876, pp. 209–216, Oct. 1988.

[8] H. G. Hansma and J. H. Hoh, "Biomolecular Imaging with the Atomic Force Microscope," *Annual Review of Biophysics and Biomolecular Structure*, vol. 23, no. 1, pp. 115–140, 1994.

[9] J. Fritz, "Cantilever biosensors," *Analyst*, vol. 133, no. 7, p. 855, Jan. 2013.

[10] J. L. Arlett, E. B. Myers, and M. L. Roukes, "Comparative advantages of mechanical biosensors," *Nature Nanotechnology*, vol. 6, no. 4, pp. 203–215, Mar. 2011.

[11] A. Boisen, S. Dohn, S. S. Keller, S. Schmid, and M. Tenje, "Cantilever-like micromechanical sensors," *Rep. Prog. Phys.*, vol. 74, no. 3, p. 036101, 2011

[12] H. Butt and M. Jaschke, "Calculation of thermal noise in atomic force microscopy," *Nanotechnology*, vol. 6, no. 1, pp. 1–7, 1995.

[13] J. Sader, "Frequency response of cantilever beams immersed in viscous fluids with applications to the atomic force microscope," *Journal of Applied Physics*, 1998.

[14] B. Sanii and P. D. Ashby, "High Sensitivity Deflection Detection of Nanowires," *Physical Review Letters*, vol. 104, no. 14, Apr. 2010.

[15] T. E. Schaeffer, M. Viani, D. A. Walters, B. Drake, E. K. Runge, J. P. Cleveland, M. A. Wendman, and P. K. Hansma, "An atomic force microscope for small cantilevers," *Proc. SPIE*, vol. 3009, p. 48, 1997.

[16] S. Torbrügge, O. Schaff, and J. Rychen, "Application of the KolibriSensor to combined atomic-resolution scanning tunneling microscopy and noncontact atomic-force microscopy imaging," *Journal of Vacuum Science \& Technology B: Microelectronics and Nanometer Structures*, vol. 28, p. C4E12, 2010.

[17] T. van Zanten, M. Lopez-Bosque, and M. Garcia-Parajo, "Imaging Individual Proteins and Nanodomains on Intact Cell Membranes with a Probe-Based Optical Antenna," *Small*, vol. 6, no. 2, pp. 270–275, 2010.

[18] J. M. LeDue, M. Lopez-Ayon, S. A. Burke, Y. Miyahara, and P. Gruetter, "High Q optical fiber tips for NC-AFM in liquid," *Nanotechnology*, vol. 20, no. 26, p. 264018, Jun. 2009.

[19] M. Bao and H. Yang, "Squeeze film air damping in MEMS," *Sensors and Actuators A: Physical*, vol. 136, no. 1, pp. 3–27, May 2007.

[20] T. Veijola, "Compact models for squeezed-film dampers with inertial and rarefied gas effects," *J. Micromech. Microeng.*, vol. 14, p. 1109, 2004.

[21] C. R. Doering, "The 3D Navier-Stokes Problem," *Annu. Rev. Fluid Mech.*, vol. 41, no. 1, pp. 109–128, Jan. 2009.

[22] C. P. Green and J. E. Sader, "Frequency response of cantilever beams immersed in viscous fluids near a solid surface with applications to the atomic force microscope," *Journal of Applied Physics*, 2005.

[23] T. Naik, E. K. Longmire, and S. C. Mantell, "Dynamic response of a cantilever in liquid near a solid wall," *Sensors and Actuators A: Physical*, vol. 102, no. 3, pp. 240–254, Dec. 2002.

[24] J. E. Sader, I. Larson, P. Mulvaney, and L. R. White, "Method for the calibration of atomic force microscope cantilevers," *Review of Scientific Instruments*, vol. 66, no. 7, pp. 3789–3798, 1995.

[25] A. Sanpaulo, "High-Resolution Imaging of Antibodies by Tapping-Mode Atomic Force Microscopy: Attractive and Repulsive Tip-Sample Interaction Regimes," *Biophysical Journal*, vol. 78, no. 3, pp. 1599–1605, Mar. 2000.

[26] D. Ebeling and H. Holscher, "Analysis of the constant-excitation mode in frequency-modulation atomic force microscopy with active Q-Control applied in ambient conditions and liquids," *Journal of Applied Physics*, vol. 102, no. 11, p. 114310, 2007.

[27] J. Fritz, M. Baller, H. P. Lang, H. Rothuizen, P. Vettiger, E. Meyer, H. J. Guntherodt, C. Gerber, and J. K. Gimzewski, "Translating Biomolecular Recognition into Nanomechanics," *Science*, vol. 288, no. 5464, pp. 316–318, Apr. 2000.

[28] A. Boisen, "Nanoelectromechanical systems: Mass spec goes nanomechanical," *Nature Nanotechnology*, vol. 4, no. 7, pp. 404–405, Jul. 2009.

[29] S. Dohn, S. Schmid, F. Amiot, and A. Boisen, "Position and mass determination of multiple particles using cantilever based mass sensors," *Appl. Phys. Lett*, vol. 97, no. 4, p. 044103, 2010.

[30] U. Rabe, J. Turner, and W. Arnold, "Analysis of the high-frequency response of atomic force microscope cantilevers," *Appl Phys A*, vol. 66, no. 7, pp. S277–S282, Mar. 1998.

CONTACT

*P.D. Ashby, Tel: +1-510-486-7081; pdashby@lbl.gov

RESONANT MICRO-SENSOR PLATFORM FOR CONTACT-FREE CHARACTERIZATION OF LIQUID PROPERTIES

Jonathan Gonzales and Reza Abdolvand
Oklahoma State University, Tulsa, USA

ABSTRACT

This paper presents a novel resonant micro-sensor platform which is designed to measure the acoustic properties of liquid samples by quantification of ultrasonic waves reflected from and transmitted through the liquid sample. By avoiding the direct contact of the liquid with the resonator, significant losses due to liquid loading are mitigated and the physical properties of various fluids, including viscous samples, can be determined without adversely affecting the resonator performance. Devices have been fabricated and tested, achieving quality factors up to 6000 in air at 112MHz and the results show that the output signals measured from the device (i.e. the resonator Q/vibration amplitude and the amplitude of the signal passing through the liquid) are sensitive to the properties of the liquid under test.

INTRODUCTION

Sensors based on acoustics are commonly used to measure physical, chemical, and biological quantities and are important to many industries such as: medical, chemical, and petroleum [1,2,3]. Resonant sensors such as quartz crystal microbalances (QCM), bulk acoustic wave (BAW) and film bulk acoustic wave resonators (FBAR) have been used for such measurements and offer many advantages such as quasi-digital outputs, simple electronic interfacing, and high sensitivities [4]. For these sensors, the measured samples are in direct contact with the resonant body and the change in the quantity of interest will impact the amplitude and the frequency of the resonant structure due to mass loading.

In the case of measuring in liquid environments, the employed resonator suffers from high radiation losses (damping), especially for devices whose displacements are normal to the sensing surface. These losses lead to performance degradations of the device and limit the sensor applicability, especially for measuring viscose samples such as blood.

In this paper, we demonstrate an alternative measurement platform in which an extensional mode resonator is used for liquid sensing using a contact-free method of sample interrogation. Since the resonator is not directly exposed to the liquid environment, excessive signal attenuation is avoided. The principle of operation, design, fabrication, and experimental results of the sensor are discussed.

PRINCIPLE

Anchor Loss

The amplitude of a resonator's vibration in response to a constant excitation force at a natural resonance frequency of the structure is a function of the resonator quality factor (Q) which is defined as 2π times the ratio of its maximum energy stored to the energy dissipated per vibration cycle. The structural shape and the material of the resonator limits the maximum Q, which is governed by physical loss phenomena such as air damping [5], thermoelastic dissipation (TED) [6], and anchor/tether loss [7].

Anchor loss, also referred to as attachment or clamping loss, is the quantification of the elastic energy leaving the resonator through the structural elements connecting it to a frame. As a resonator vibrates, it exerts forces on the tether, causing acoustic energy to travel down the tether and to subsequently be transmitted into the substrate and lost.

TPoS Resonator

Thin-film piezoelectric-on-substrate (TPoS) resonators [8] built upon a silicon-on-insulator (SOI) substrate are used as the sensing element in this work. The resonator is composed of the main resonant body (silicon) which is suspended at the ends and connected to the substrate through small tethers. A piezoelectric film (aluminum nitride) is sandwiched between two metallic electrodes (molybdenum) and stacked on top of the resonant structure. The use of a low-loss substrate (e.g. single crystal silicon) provides for high quality factor values in air.

Acoustic Reflectors

Our group has previously shown the effectiveness of in-plane acoustic reflectors to increase the confinement of the radiated acoustic energy. Consequently, the quality factor can be greatly improved by etching a pair of properly shaped trenches into the substrate at a fixed distance from the resonator [9]. And when the same trench is filled with a fluid, it was suggested that the quality factor of the resonator (or the resonator vibration amplitude in response to a fixed excitation power) and the acoustic energy that is transmitted through the liquid are functions of the acoustic properties of the fluid [10].

SENSOR DESIGN

A schematic of the sensor is shown in Fig 1. The system is composed of two main components: liquid delivery and acoustic interrogation. The liquid which is to be sampled is transported from a main reservoir and passively drawn into the microfluidic channels by capillary forces into an interrogation area (reflector site). A TPoS resonator is used as a generator of bulk acoustic waves which propagate into the substrate due to the anchor losses. The ultrasonic waves reach the fluidic channel where they interact with the sample fluid and the acoustic energy is

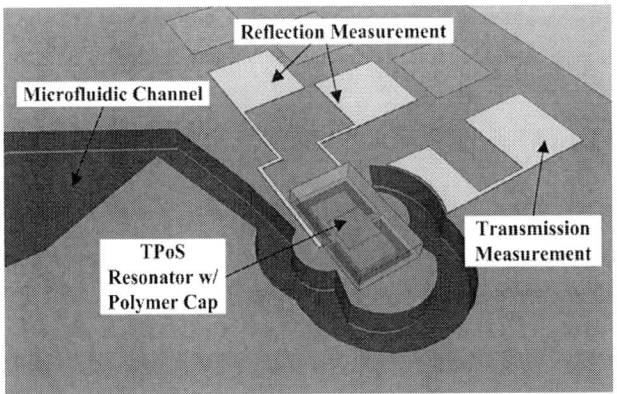

Figure 1: Schematic diagram of the TPoS-based ultrasonic sensor illustrating the different components involved in operation.

either transmitted through or reflected from the fluidic channel.

Electrical signals are generated by the waves via piezoelectric transduction which are used to determine the physical attributes of the liquid sample. The output signals of the sensor are taken from the transmission electrode on the periphery of the fluidic channel and from the resonator electrodes using a one-port or two-port configuration. The resonance frequency is largely determined by the width of the resonator and is 112MHz in this work.

FABRICATION

The sensor platform is fabricated using both surface and bulk micromachining techniques. A stack of aluminum nitride (1um) and molybdenum (100nm) are sputtered onto an SOI wafer (20um device layer) for this work. The top metal resonator fingers and the I/O electrodes are defined first by dry etching the molybdenum in RIE. The AlN is then wet etched in TMAH in order to access the bottom molybdenum electrode. Then, the resonator body and microfluidic channels are defined by consecutively etching with chlorine using an inductively-coupled plasma (ICP) etcher and in SF6/C4F8 plasmas using a multiplexed BOSCH process. The handle silicon is then etched from the backside using a BOSCH recipe up to the BOX layer. Next, the device is dipped into a buffered oxide etch (BOE) bath to etch the buried oxide and release the structure.

The devices can be tested at this point using open channels. However, to guide the liquid while avoiding excessive evaporation for applications where small sample sizes are used or to cover the resonator from direct contact for applications where the sensor is submerged in liquid, a simple polymeric capping technique can be utilized. Figure 2 shows the two different capped designs In one, the microfluidic channel is capped from the top using an insoluble polymer (parylene). To achieve such structures, the fluidic channels of the devices are filled with photoresist and then a film of parylene-C is deposited on top and patterned in O2 plasma. Acetone is then used to dissolve the photoresist from underneath the parylene:

leaving hollow, covered channels (Fig. 2, left). For the second design, the resonator is encapsulated. Photoresist is spun and patterned over the device and a parylene film is deposited afterwards. These are then submerged in an acetone bath to remove the photoresist from the resonator but leaving the parylene covering untouched (Fig. 2, right).

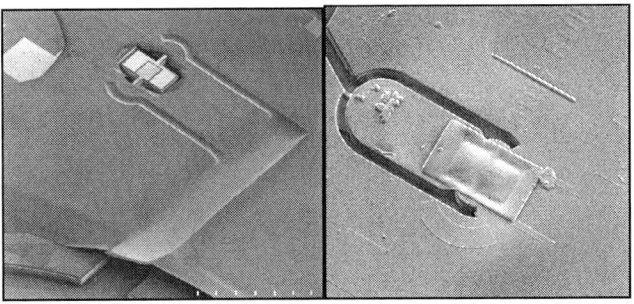

Figure 2: SEM of fabricated resonator with parylene-C thin film deposited over the fluidic channel (left) and a device which has been encapsulated with a parylene package (right).

EXPERIMENT AND RESULTS

Dynamic Response

A simple test set-up is developed to verify that the sample liquids travel through the fluidic channels and that the transmission signal and the quality factor of the resonator are functions of the acoustic impedance of the liquid. Our set up includes a microliter syringe with a fine needle head that is installed onto a precision manipulator. Photoresist (Shipley 1827 positive resist) is used to test the channels' filling capability and the sensor's dynamic response since it is a relatively viscous liquid and also solidifies over time resulting in a change in sensor output over time. Several devices of varying channel designs are used in this work and the device used in the following preliminary measurements is shown in figure 3.

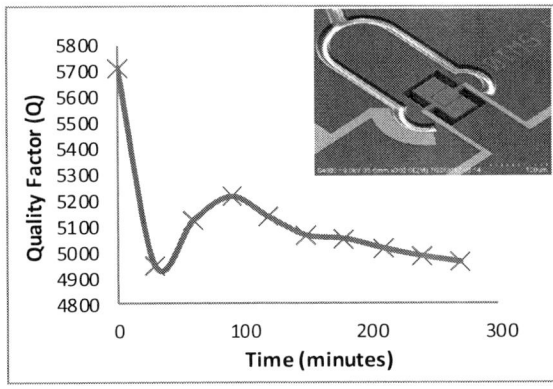

Figure 3: The quality factor of the resonator vs. time as the injected photoresist dries. SEM of dried resist in channels (inset).

978-1-4799-3510-9/14 $31.00 © 2014 IEEE

The device response is measured using a network analyzer and GSG RF probes. The initial frequency response is collected and then the photoresist is injected into the microfluidic channels while the frequency response is continuously observed. The liquid is observed to uniformly and quickly fill both channels as shown in the inset SEM of figure 3 and the quality factor is immediately reduced.

The change in the quality factor of the device is plotted as a function of time in figure 3. This plot is a significant experimental proof of concept for our proposed sensing techniques that validates two claims: first, the quality factor of the resonator is a function of the channel filling although the liquid is not directly in contact with the resonator; second, the dynamic change in the properties of the liquid can be detected.

The large initial change in the Q is believed to be a result of the high initial drying rate and the variation in the volume of the resist during drying. The quality factor of the resonator is further reduced to 4300 after baking the resist. It should be noted that the resonator is not optimized for maximum Q, therefore, the initial Q and Q sensitivity can be potentially increased.

Liquid Differentiation

The next experiment that we conducted was to demonstrate the capability of our sensor to differentiate between different liquid fillings. We chose water, photoresist, and whole blood as our different assays. The test was started by filling the channels with DI water. The change in the quality factor immediately after filling the channel was recorded and the test was repeated several times after the water was evaporated from the channel. The drop in the Q was very consistent in each measurement and the Q of the device after evaporation would return to the initial value.

Next, the same process was repeated with photoresist (Shipley 1813 positive resist). The only difference was that after each measurement the sample was cleaned with acetone, isopropanol, and DI water. Lastly, blood was injected into the channels and cleaned with water several times. The results of this experiment are compiled in figure 4. The change in the Q shown in this figure is the average values measured from all applications. This plot signifies adequate sensitivity of the sensor to distinguish between the liquid fillings with different properties. The resolution of the sensor can be further improved by optimization of the structure.

Signal Transmission

Our final experiment was performed in an attempt to show that the acoustic wave transmitted through the liquid in the channel can be measured and corresponds to the properties of the liquid. The phase of the transmitted signal can be used to extract the acoustic velocity in the fluid and the intensity will be required to estimate the acoustic attenuation and scattering in the liquid under test.

To perform this experiment, the resonator was excited

using the network analyzer through one of the probes and the second probe was contacted to the pad that is connected to the electrode beyond the fluidic channel.

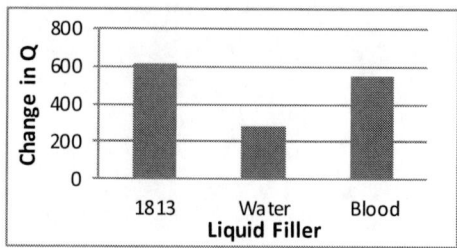

Figure 4: The change in quality factor of the sensor with various liquid fillers.

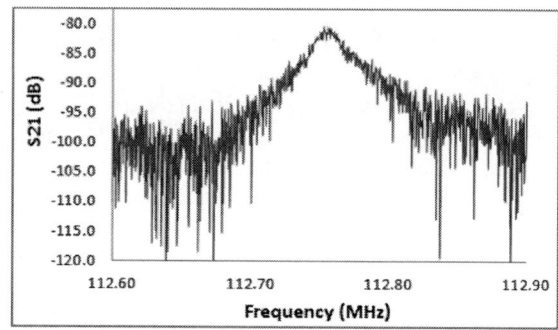

Figure 5: Example of the transmitted signal across the fluid-filled trench.~20dB above noise floor.

Before filling the channel, the weak signal that was measured at the resonance frequency was calibrated out. Next, the photoresist was injected into the channel while the signal was continuously monitored. Immediately after the channel is filled with the resist, the resonance peak is clearly detected. The signal level increased as the resist dried out and the signal intensity reached a maximum of ~20dB above the noise level over time (Fig. 5).

Another signal transmission test was carried out using ethylene glycol/water solutions for the liquid filler. Different concentrations of glycol were mixed and injected into the fluidic channels and the resulting quality factor and transmission signals were collected and plotted (Fig. 6).

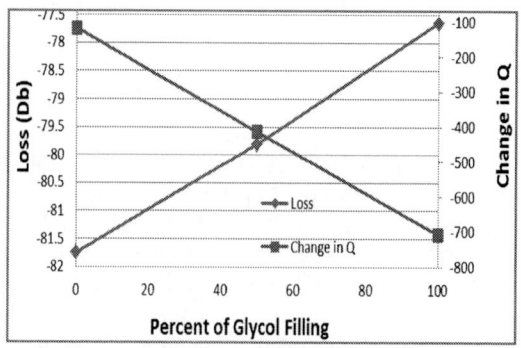

Figure 6: Transmission signal and Q of the sensor with various glycol concentrations.

The higher concentrations of glycol result in more energy being lost from the resonator (lowered Q) and a higher signal detected across the filled trench (higher signal level or lower loss) due to glycol's higher acoustic impedance match to silicon compared to water. The difference of ~550 in Q corresponds to a difference in acoustic impedance of ~360x10^3 Rayl.

Finally, the design from figure 2 was tested for survivability while being fully submerged in a liquid environment. This device was capped with a parylene enclosure as described previously and wirebonded to nearby electrodes for probing. The frequency response was initially measured and then photoresist was injected onto the wafer surface to fully cover the resonator area (Fig. 7). The device continued to operate once submerged and showed a change in Q of ~470 between empty channels and resist-filled channels.

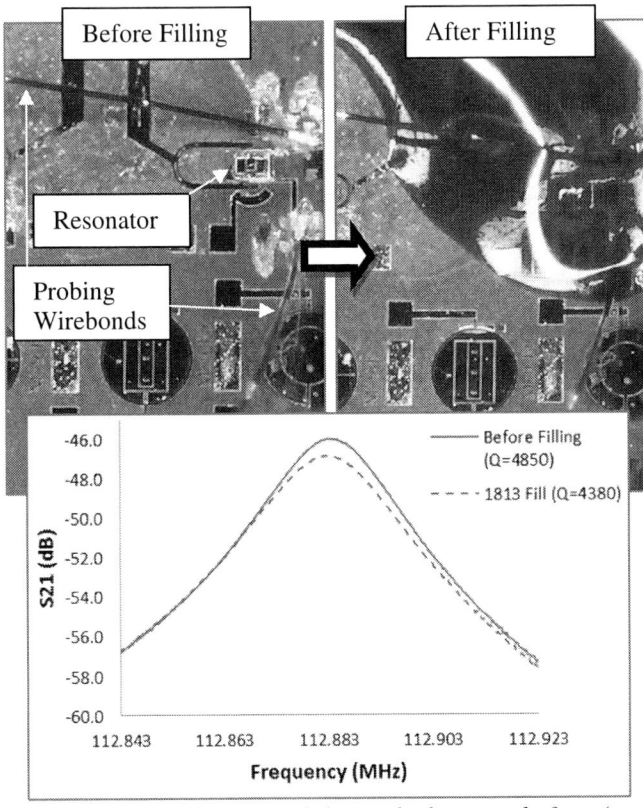

Figure 7: Optical image of the probed sensor before (top left) and after (top right) the substrate is covered with photoresist. The resulting resonator performance is also shown (bottom). Note that the Parylene cap is transparent.

CONCLUSION

In this paper, a contact-free resonant microsensor platform for measuring liquids was presented which can operate in a submerged environment. Several sensors based on TPoS resonator technology were designed and fabricated and used to measure different liquids while maintaining high quality factor(Q). The sensor shows a change in Q of ~700 when measuring photoresist and is sensitive to its

dynamic changes in physical properties. The devices have also been shown to consistently respond to water and photoresist with a difference of ~300 between the two liquids. And glycol measurements cause a change of 550 Q with a variation in impedance of 360x10^3 Rayl. With these data, both the reflection and transmission signals can be used in tandem to characterize a liquid under test.

REFERENCES

[1] W. Xu, J. Appel, J. Chae, "Real-time monitoring of whole blood coagulation using a microfabricated contour-mode film bulk acoustic resonator", J. Microelectromechanical Systems, vol. 21, no.2, April 2012.

[2] W. Pang, H. Zhao, E.S. Kim, H. Zhang, H. Yu, X. Hu, "Piezoelectric microelectromechanical resonant sensors for chemical and biological detection", Lab Chip, Vol. 12, p. 29-44, 2012.

[3] D. S. Ballantine, R. M. White, S. J. Martin, A. J. Ricco, E. T. Zellers, G. C. Frye, H. Wohltjen, "Acoustic Wave Sensors: Theory, Design, and Physico-Chemical Applications", Academic Press, San Diego, 1997.

[4] F. Lucklum, B. Jakoby, "Novel magnetic-acoustic resonator sensros for remote liquid phase measurement and mass detection", Sensors and Actuators A, Vols. 145-146, p. 44-51, 2008.

[5] W. Ye, X. Wang, W. Hemmert, D. Freeman, and J. White, "Air damping in laterally oscillating microresonators: a numerical and experimental study," J. Microelectromechanical Systems, vol. 12, 2003, p. 557–566.

[6] R. Abdolvand, G.K. Ho, A. Erbil, and F. Ayazi, "Thermoelastic damping in trench-refilled polysilicon resonators," Transducers, Solid-State Sensors, Actuators and Microsystems, 12th International Conference on, 2003, 2003, p. 324–327.

[7] J.A. Judge, D.M. Photiadis, J.F. Vignola, B.H. Houston, and J. Jarzynski, "Attachment loss of micromechanical and nanomechanical resonators in the limits of thick and thin support structures," J. Applied Physics, vol. 101, 2007, p. 013521.

[8] R. Abdolvand, H. Mirilavasani, G.K.Ho, F. Ayazi, "Thin film piezoelectric-on-silicon resonators for high-frequency reference oscillator applications," IEEE Ultrasonics, Ferroelectrics, and Frequency Control, vol. 55, no. 12 p. 2596-2606, Dec. 2008.

[9] B. P. Harrington, R. Abdolvand, "In-plane acoustic reflectors for reducing effective anchor loss in lateral-extensional MEMS resonators", J. Micromech. Microeng. 21, 2011.

[10] J.M. Gonzales, M. Shahmohammadi, R. Abdolvand, "Sensing acoustic properties of materials using piezoelectric lateral-mode resonators", IEEE Ultrasonics Symposium, p.200-203. 2011.

CONTACT

*J. Gonzales; jonathan.gonzales@okstate.edu

NOVEL IN-PLANE GAP CLOSING CMOS-MEMS MICROPHONE WITH NO BACK-PLATE

Chun-I Chang[1], Sung-Cheng Lo[2], Chuanwei Wang[3], Yi-Chiang Sun[2], and Weileun Fang[1,2]
[1]Institute of NanoEngineering and MicroSystems, National Tsing Hua University, Hsinchu, TAIWAN
[2]Power Mechanical Engineering, National Tsing Hua University, Hsinchu, TAIWAN
[3]MotionsTek Inc., Taipei, TAIWAN

ABSTRACT

The stacking of metal/tungsten layers as the *sensing electrodes* for *CMOS-MEMS microphone* without the back-plate has been proposed and demonstrated for the first time. The acoustic pressure will deform the spring-diaphragm structure and further cause the in-plane gap-closing between sensing electrodes. Thus, acoustic pressure and dynamic response of spring-suspension can be determined by the sensing capacitance changes. Such design has the following merits: (1) no back-plate is required, (2) bias voltage to pull diaphragm close to back-plate is not required, (3) in-use pull-in and process stiction between diaphragm and back-plate is also prevented, (4) easy integration with sensing circuits. The design was implemented using the standard TSMC CMOS process. Typical microphone with 200μm-diameter diaphragm and 48-pairs sensing electrodes has been realized. Measurements show the sensitivity of microphone is -64.78dBV/Pa at 1kHz.

INTRODUCTION

The microphones converting acoustic input to electrical output are one of the key devices for various multimedia applications such as mobile phone, computer, personal digital assistants, etc. The MEMS condenser microphones consisting of a diaphragm and a back-plate have been extensively reported [1-11]. The well-known surface micromachining process consisted of two poly-Si structure layers has successfully been employed to realize a commercial microphone [1]. In general, the back plate is employed to serve as a stationary reference electrode for capacitive sensing MEMS microphone [1, 8]. Thus, the stiffness of back plate is an important design consideration. The air damping of microphone will also be influenced by the design of vent holes on back plate [11]. In addition, more process steps are required to implement the back-plate design [1, 11].

The CMOS-MEMS sensors have the advantages of using existing foundry service, electrical routing compatibility, and IC/MEMS monolithic integration [12]. Thus, the CMOS-MEMS microphones have also been investigated in [9-10]. The deposition of polymer film is required for the diaphragm of the microphone in [9]. Moreover, the condenser microphone in [10] employs a corrugated metal diaphragm to improve its performance. However, non-standard CMOS processes are required to realize the polymer and corrugated metal diaphragm. The time control deep Si etching is employed to define the thickness of back-plate. Such process may introduce thickness variation of back-plate.

This study takes the advantages of standard CMOS process to develop a novel CMOS-MEMS microphone with in-plane gap closing sensing electrodes to detect the acoustic pressure. In addition, the proposed microphone design has no back-plate. The proposed microphone is designed and implemented based on the standard TSMC 0.35μm 2P4M process. Note the proposed design concept can be further extended to realize microphone array on a single unit [13]. Monolithic integration of proposed microphone with other MEMS sensors can also be achieved through standard CMOS process.

DESIGN CONCEPT

Fig.1 shows schematic illustrations of the proposed CMOS-MEMS capacitive microphone design, including diaphragm, springs, movable electrodes (on diaphragm), and fixed electrodes (on anchor). The movable electrodes on diaphragm and the stationary electrodes on anchor form the in-plane gap-closing sensing capacitor. The cross-section illustration shows the sensing electrodes and the diaphragm consisting of the stacked metal, dielectric, and tungsten layers for CMOS process.

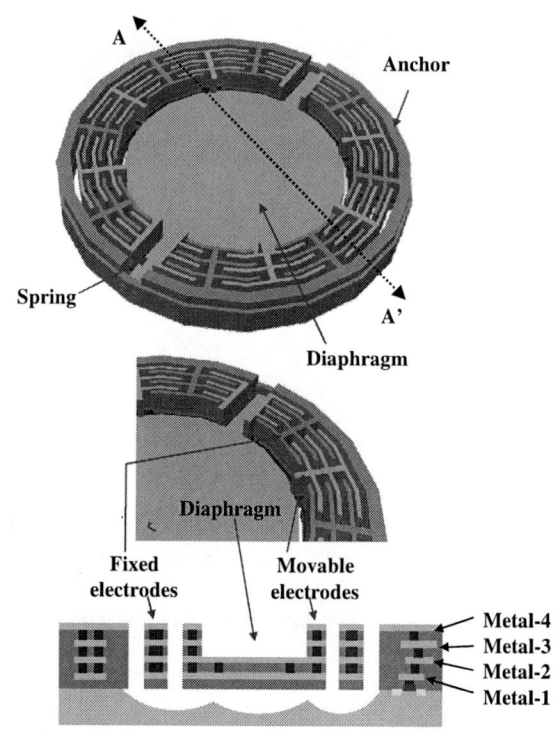

Figure 1: (a) Design concept of microphone and zoom-in configuration of sensing electrodes, and (b) the diaphragm under the 1Pa acoustic pressure.

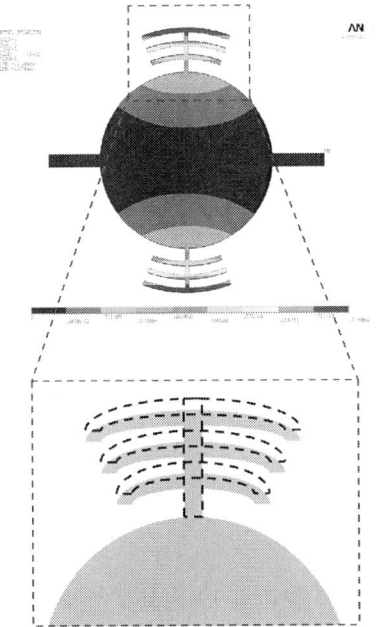

Figure 2: FEM mode shape for the diaphragm under the 1Pa acoustic pressure using ANSYS software.

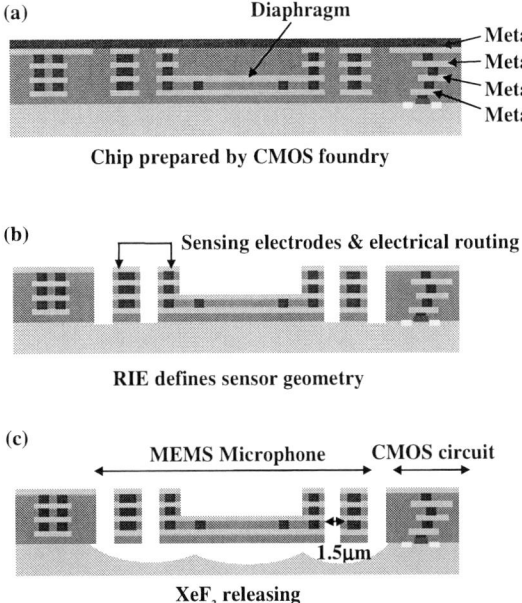

Figure 3: Process steps, (a) chip prepared by TSMC, and (b-c) in-house post-CMOS RIE and structure releasing.

The device is designed and implemented based on the standard TSMC 0.35μm 2-poly 4metals CMOS process. Briefly, this standard process provides four metal (named Metal-1~Metal-4), dielectric, and tungsten layers for design. This study exploited the Metal-1 and Metal-2 layers to fabricate the diaphragm. According to the test keys, such layer stacking has smaller diaphragm bending deformation by the thin film residual stresses. Moreover, the movable and fixed electrodes have four metal layers to increase the initial capacitance and sensing area.

Simulation in Fig.2 shows the deformation of spring-diaphragm by acoustic pressure. The sound pressure on the microphone will cause the deformation of the flexible diaphragm. As indicated by the inset, the bending of spring and diaphragm will cause the movable electrodes an in-plane displacement. Thus, as diaphragm deformed by acoustic pressure, the movable-electrodes will approach the fix-electrodes on anchor, so as to cause the in-plane gap-closing between sensing electrodes. Thus, the dynamic response of spring-diaphragm and the acoustic pressure are detected by the gap-closing capacitance change.

FABRICATION AND RESULTS

Fig.3 shows the process steps to fabricate the microphone. Fig.3a shows the chip prepared by standard TSMC 0.35μm 2P4M CMOS process. In Fig.3b, the reactive ion etching (RIE) was used to remove the silicon dioxide, and the top metal layer acted as the etching mask. The 1.5μm in-plane sensing gap was defined by the minimum line width of metal layer for post-CMOS process. Finally, the Si substrate was etched isotropically by XeF$_2$ to suspend MEMS structures (Fig.3c). Fig.4 shows typical fabrication results. The micrograph in Fig.4a shows a

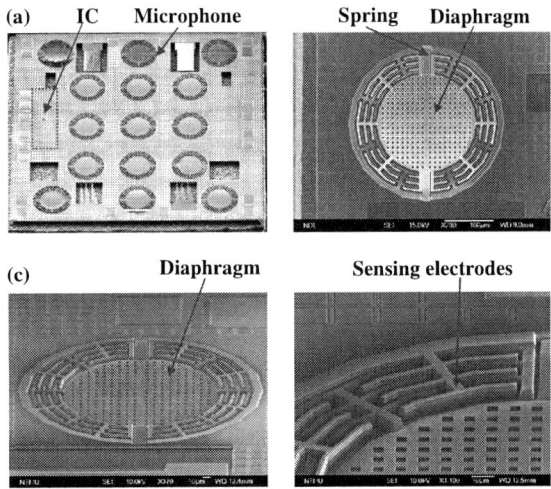

Figure 4: SEM images of (a) typical fabricated CMOS chip consisting of microphone structure and sensing circuit, (b-c) the mechanical structures, diaphragm and springs, and (d) the sensing electrodes.

CMOS chip consisting of various microphone structure designs and sensing circuits. Moreover, this CMOS-MEMS chip shows the potential to realize not only microphone but also microphone array [13], and also monolithically integrated sensing IC [14]. The number and distribution of microphone array will be arranged based on the acoustic design. Micrographs in Fig.4b-c depict the microphone for testing. The microphone has a

978-1-4799-3510-9/14 $31.00 © 2014 IEEE

200μm-diameter diaphragm, two 60μm×15μm springs, and 48-pairs sensing electrodes with 1.5μm sensing-gap. Zoom-in micrographs in Fig.4d depict the flexible diaphragm consisted of Metal-1~Metal-2 layers, and the sensing electrodes (movable and fixed electrodes) consisted of Metal-1~Metal-4 layers.

MEASUREMENT RESULTS

The initial deformation of the diaphragm was characterized using the commercial optical interferometer. The top-view color contour in Fig.5a displays the out-of-plane bending deformation of diaphragm. The surface profile of diaphragm is observed by colors. Measurement results in Fig.5b further show the deformation profile of the diaphragm along the AA' cross section indicated in Fig.5a. The measurement shows the diaphragm has a radius of curvature (ROC) of 1.25mm. Fig.6a shows the sensing chip after packaging. The measurement setup in Fig.6b was established to characterize the performance of the microphone. The packaged microphone was placed inside a semi-anechoic chamber for test. During the test, the loudspeaker was used to specify a sound pressure to excite the microphone, and the sound pressure level was monitored by a commercial reference microphone (PCB 130E20). The output signals from the capacitance-to-voltage converter were recorded by spectrum. Fig.7 shows the typical measured mechanical frequency response of the microphone. The first two resonant frequencies are 44.87kHz (1st mode) and 74.34kHz (2nd mode). Fig.8 shows the typical measured frequency response of microphone. The measurements depict the sensitivity of microphone is -64.78dBV/Pa at 1kHz. Table1 summarizes detail measurement results.

(a)

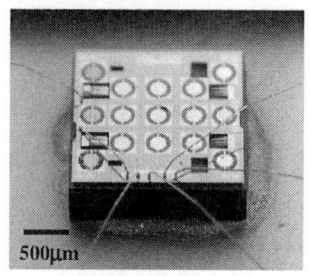

(b)

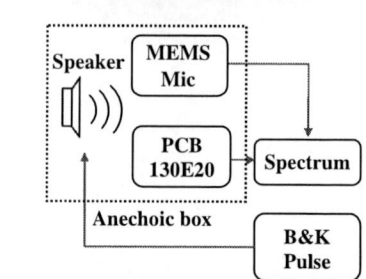

Figure 6: (a) Monolithic CMOS-MEMS microphone after packaging, and (b) measurement setup for testing.

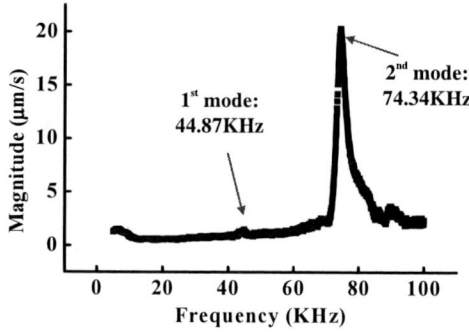

Figure 7: The measured mechanical frequency response of microphone.

(a)

(b)

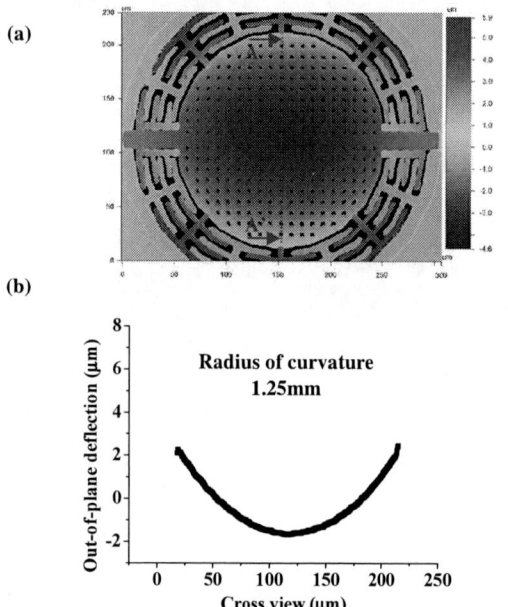

Figure 5: Measurements of structure deformations at diaphragm cross section (a) top view, and (b) suface profile of diaphragm along the cross section AA'.

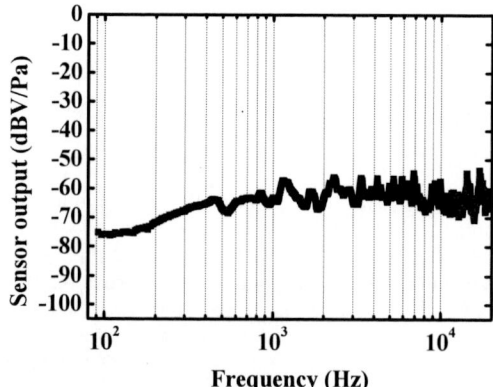

Figure 8: The measured frequency response of microphone.

Table 1: Summary and comparison of presented microphone.

	This study	Ref. [7]	Ref. [10]
Diaphragm size	Diameter 0.2mm	Square 0.5x0.5mm	Diameter 0.8mm
Diaphragm thickness	3.5μm	3μm	1.1μm
Back-plate	No	Yes	Yes
Sensing gap	1.5μm (in-plane)	1μm (out-of-plane)	4.2μm (out-of-plane)
Bias voltage	0V	105V	1.65-3.6V
Sensitivity	-64.78dBV/Pa	-69.5dBV/Pa	-42dBV/Pa

Note: measurements in the low frequency range (0~1kHz) need to be improved. The improvement can be achieved by reducing the etching release holes on diaphragm. Moreover, the present approach could realize the microphone array as indicated in Fig.4a. Thus, the acoustic performance can be further improved.

CONCLUSIONS

In this study, the stacking of metal/tungsten layers as the sensing structures and electrodes for a novel no back-plate CMOS-MEMS microphone has been proposed and demonstrated based on the standard TSMC 0.35μm 2P4M process. The acoustic pressure will deform the spring-diaphragm structure and further cause the in-plane gap-closing between sensing electrodes. Thus, acoustic pressure and dynamic response of spring-diaphragm can be determined by the sensing capacitance changes. The design was implemented using the standard TSMC CMOS process and in-house post-CMOS release process. Typical microphone with diaphragm of 200μm in diameter and 48-pairs sensing electrodes has been realized. Measurements show the sensitivity of microphone is -64.78dBV/Pa at 1kHz.

ACKNOWLEDGEMENTS

This research was sponsored in part by the National Science Council of Taiwan under grant of NSC-102-2221-E-007-027-MY3, NSC-102-2622-E-007-014-MY3, and NSC-102-2218-E-007-003-MY3. The authors wish to appreciate the TSMC and the National Chip Implementation Center (CIC), Taiwan, for the supporting of CMOS chip manufacturing. The authors would also like to thank the National Center for High-Performance Computing for support of simulation tools. The authors also would like to appreciate the Center for Nanotechnology, Materials Science and Microsystems of National Tsing Hua University and the National Nano Device Lab. in providing fabrication facilities.

REFERENCES

[1] P. V. Loeppert, and S. B. Lee, "SiSonicTM – The first commercialized MEMS microphone," *Solid-State Sensors, Actuators, and Microsystems Workshop*, Hilton Head Island, SC, June, 2006, pp. 27-30.

[2] H. Guckel, J. J. Sniegowski, and T. R. Christenson, "Fabrication of micromechanical devices from polysilicon films with smooth surfaces," *Sens. Actuators A*, vol. 20, pp. 117-122, 1989.

[3] X. Zhang, T.-Y. Zhang, M. Wong, and Y. Zohar, "Residual-stress relaxation in polysilicon thin films by high-temperature rapid thermal annealing," *Sens. Actuators A*, vol. 64, pp. 109-115, 1998.

[4] J. Yang, H. Kahn, A.-Q. He, S. M. Phillips, S. Member, and A. H. Heuer, "A new technique for producing large-area as-deposited zero-stress LPCVD polysilicon films: the multiPoly process," *J. Microelectromech. Syst.*, vol. 9, pp. 485-494, 2000.

[5] P. R. Scheeper, W. Olthuis, and P. Bergveld, "The design, fabrication and testing of corrugated silicon nitride diaphragms," *J. Microelectromech. Syst.*, vol. 9, pp. 36-42, 1994.

[6] J. Miao, R. Lin, L. Chen, Q. Zou, S. Y. Lim, and S. H. Seah, "Design considerations in micromachined silicon microphones," *Microelectronics J.*, vol. 33, pp. 21-28, 2002.

[7] B. A. Ganjia and B. Y. Majlis, "Design and fabrication of a new MEMS capacitive microphone using a perforated aluminum diaphragm," *Sens. Actuators A*, vol. 149, pp. 29-37, 2002.

[8] J. W. Weigold, T. J. Brosnihan, J. Bergeron, and X. Zhang, "A MEMS condenser microphone for consumer applications," *Proc. 19th Int. Conf. on Micro Electro Mechanical System,* Istanbul, Turkey, Jan. 22-26, pp. 86-89, 2006.

[9] J. J. Neumann Jr., and K. J. Gabriel, "CMOS-MEMS membrane for audio-frequency acoustic actuation," *Sens. Actuators A*, vol. 95, pp. 175-182, 2002.

[10] C.-H. Huang, C.-H. Lee, T.-M. Hsieh, L.-C. Tsao, S. Wu, J.-C. Liou, M.-Y. Wang, L.-C. Chen, M.-C. Yip and W. Fang, "Implementation of the CMOS MEMS condenser microphone with corrugated metal diaphragm and silicon back-plate," *Sensors*, vol. 11, pp. 6257-6269, 2011.

[11] Q. Zou, Z. Li, and L. Liu, "Theoretical and experimental studies of single-chip-processed miniature silicon condenser microphone with corrugated diaphragm," *Sens. Actuators A*, vol. 63, pp. 209-215, 1997.

[12] G. K. Fedder, R. T. Howe, T.-J. King Liu, E. P. Quevy, "Technologies for co-fabricating MEMS and electronics," *Proceedings of IEEE*, vol. 96, pp. 306-322, 2008.

[13] W. A. Ryan, M. Abry, and P. V. Loeppert, "Microphone having multiple transducer elements," U.S. Pat. 8,170,244, 2012.

[14] M.-H. Tsai, C.-M. Sun, Y.-C. Liu, C. Wang, and W. Fang, "Design and fabrication of a metal wet-etching post-process for the improvement of CMOS-MEMS capacitive sensors," *J. Micromech. Microeng.*, vol. 19, 105017, 2009.

CONTACT

* W. Fang, Tel: +886-3-5742923; fang@pme.nthu.edu.tw

CAVITY QUANTUM OPTOMECHANICS: COUPLING LIGHT AND MICROMECHANICAL OSCILLATORS

Ewold Verhagen, Samuel Deléglise, Stefan Weis, Albert Schliesser, Tobias J. Kippenberg
[1]École Polytechnique Fédérale de Lausanne (EPFL), CH1015 Lausanne, Switzerland

ABSTRACT

Cavity optomechanics[1] is a new research field that has seen spectacular advances in recent years. Optomechanics combines advances in nano- and electromechanical systems with radiation pressure enabled control. The radiation pressure backaction enables to readout mechanical motion of micro- and nanoscale mechanical oscillators with an imprecision at the standard quantum limit, enables to amplify[2] mechanical motion – enabling coherent mechanical oscillators. Likewise the cooling[3,4] of mechanical oscillators has enabled to access the quantum regime of optomechanical systems. Likewise mechanical degrees of freedom provide new ways to control the propagation of light via the phenomenon of optomechanically induced transparency[5], which can e.g. enable switching, slowing or advancing of electromagnetic pulses[6].

Cavity optomechanical systems also have reached the quantum regime of mechanical oscillators, which has been long anticipated. As one example of the possible range of optomechanical phenomena, we review an optomechanical microresonator in which optical and mechanical degrees of freedom exchange energy at a rate exceeding the relevant decoherence rates in the system, enabling quantum control of a mechanical oscillator with light. Such quantum coherent coupling provided a quantum coherent link[7] between engineered microscale oscillators and the light field.

INTRODUCTION

Optical laser fields have been widely used to achieve quantum control over the motional and internal degrees of freedom of atoms, ions and molecules enabling novel quantum states to be generated and enabling the most precise atomic clocks[8-10]. A route to controlling the quantum states of *microfabricated nano- or microscale mechanical oscillators* in a similar fashion is to exploit the parametric coupling between optical and mechanical degrees of freedom through radiation pressure in suitably engineered optical cavities [11] and as first theoretically analyzed by Braginsky in the context of gravity wave detection[12]. If the optomechanical coupling is 'quantum coherent', i.e., if the coherent coupling rate exceeds both the optical and the mechanical decoherence rate, quantum states are transferred from the optical field to the mechanical oscillator and vice versa, allowing control of the mechanical oscillator state using the wide range of available quantum optical techniques. Previously such quantum coherent coupling has only been achieved in the microwave domain[13,14].

DEMONSTRATION OF QUANTUM COHERENT COUPLING

We demonstrate quantum-coherent coupling between optical photons and a micromechanical oscillator [15]. Simultaneously, coupling to the cold photon bath cools the mechanical oscillator to an average occupancy of 1.7±0.1 motional quanta. Pulsed optical excitation reveals the exchange of energy between the optical light field and the micromechanical oscillator in the time domain.

The studied system is a silica microcavity depicted in Fig. 1(a), which simultaneously exhibits optical whispering gallery modes and a mechanical radial breathing mode with a frequency of $\Omega_m/2\pi = 78$ MHz (Fig. 1(b)). The optical and mechanical degrees of freedom are parametrically coupled through radiation pressure. In the resolved sideband regime

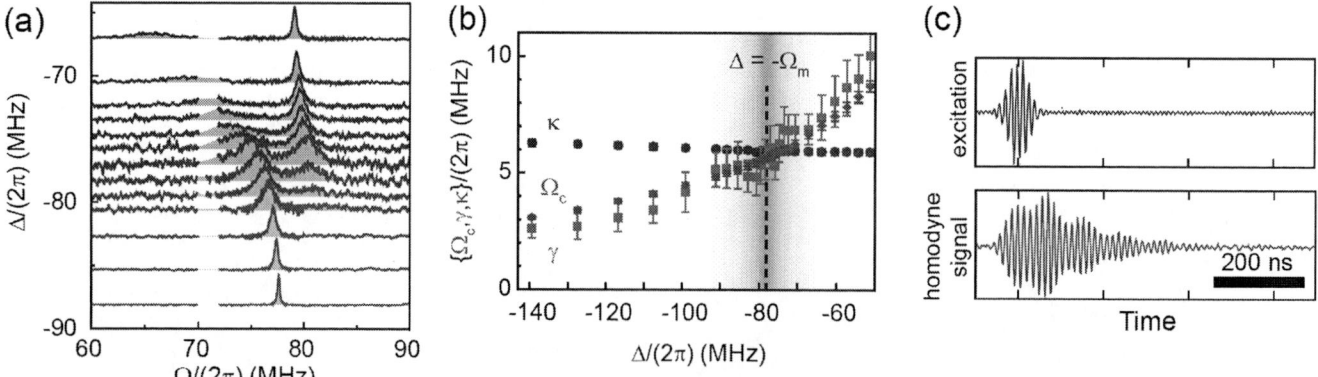

Fig. 2 Quantum-coherent coupling. (a) Mechanical noise spectra as a function of the laser's detuning Δ from the cavity resonance frequency. Normal mode splitting occurs when $\Omega_c \geq \kappa$. (b) For the laser detuned to the lower mechanical sideband ($\Delta = -\Omega_m$), the coherent coupling rate Ω_c is comparable to the optical and mechanical decoherence rates κ and γ. (c) Swapping of a weak pulsed coherent excitation between the optical and mechanical mode is evidenced by oscillations in the homodyne signal envelope at frequency Ω_c.

978-1-4799-3510-9/14 $31.00 © 2014 IEEE

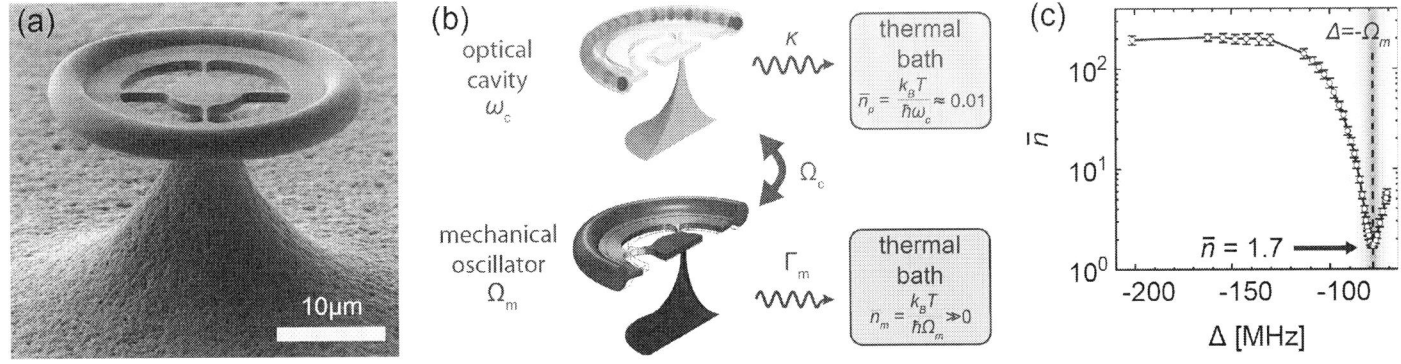

Fig. 1 (a) Silica toroidal optomechanical resonator. (b) Sketch of the optical cavity mode and the calculated displacement of the mechanical radial breathing mode. Quantum-coherent coupling occurs when $\Omega_c > (\kappa, n_m \Gamma_m)$. (c) The oscillator's average phonon occupancy is reduced to 1.7±0.1 when the laser is detuned to $\Delta = -\Omega_m$.

(where Ω_m exceeds the cavity decay rate κ), with an intense laser tuned to the lower optomechanical sideband, the coupling can be described by the interaction Hamiltonian $H = \hbar\Omega_c (\hat{a}\hat{b}^\dagger + \hat{a}^\dagger\hat{b})/2$. Here, $\hat{a}$ and $\hat{b}$ are the photon and phonon annihilation operators (displaced by their steady state values), and $\Omega_c = 2\sqrt{\bar{n}_c}\, g_0$ denotes the coherent coupling rate, which is proportional to the square root of the average number of photons in the cavity $\bar{n}_c$ and the vacuum optomechanical coupling rate g_0. The evolution described by this Hamiltonian corresponds to swapping of the (displaced) optical and mechanical states at a rate Ω_c. In practice, this unitary evolution is however compromised by the coupling of both degrees of freedom to their respective environments. Hence, it is important for Ω_c to exceed both the optical decoherence rate κ and the mechanical decoherence rate γ, defined here as the inverse time needed for a single excitation to be lost into the environment. Because of the relatively small mechanical frequency, the mechanical mode is coupled to an environment with a large thermal occupation $\bar{n}_m = k_B T / \hbar\Omega_m$, the mechanical *decoherence* rate $\gamma = \Gamma_m(\bar{n}_m + 1)$ is much larger than the *dissipation* rate Γ_m, which is relevant for high-frequency resonantly coupled systems. For parametrically coupled systems, the relevant regime in which quantum-coherent dynamics can occur is defined by $\Omega_c > (\kappa, \gamma)$, which we denote as 'quantum-coherent' coupling.

The use of an optimized spoke-anchored microtoroid design (Fig. 2(a)) isolates the micromechanical motion from the supports to reduce mechanical clamping losses, thereby effectively allowing to increase the optomechanical coupling strength g_0 through a reduction of cavity radius and resonator mass. To reduce the mechanical decoherence rate, the microcavity is embedded in a helium-3 cryostat (T_{min} = 650 mK; $\bar{n}_m \approx 200$). A quantum-limited continuous-wave Ti:Sapphire laser is coupled to the microcavity and the weak phase fluctuations imprinted on the outgoing field by mechanical displacement fluctuations are measured using balanced homodyne detection.

Both the coherent coupling rate Ω_c as well as the decoherence rates γ and κ can be accurately determined from a combination of coherent response measurements and measurements of the random thermal fluctuations of the resonator. Moreover, these simultaneously allow to extract the effective temperature of the resonator mode, which is strongly reduced when the laser is tuned to the lower mechanical sideband. In the case of weak coupling ($\Omega_c < \kappa$), the optical decay is faster than the swapping between the vacuum in the displaced optical field and the thermal state in the mechanical oscillator. In this case, the mechanical oscillator is coupled to an effective optical bath at near-zero thermal occupancy, leading to a pronounced cooling of the mechanical mode. Figure 1(d) shows how the mode's average occupancy is reduced to a value of $\bar{n} = 1.7 \pm 0.1$, corresponding to 37% ground state occupation.

Increasing the coupling strength to reach $\Omega_c \approx \kappa$, the signature of normal mode splitting can be observed in the fluctuation spectra (Fig. 2(a)) [16,17], which show an anti-crossing of the optical and mechanical modes as the laser is detuned across the lower mechanical sideband. When the laser is tuned to the lower mechanical sideband, the coupling rate Ω_c becomes equal to both the optical and mechanical decoherence rates (Fig. 2(b)), thus bringing the system into the regime of quantum-coherent coupling. The dynamical exchange of weak coherent excitations between the optical and mechanical degrees of freedom can be demonstrated in a classical proof of principle experiment by sending a weak pulse resonant with the cavity in the presence of the strong detuned pump field. The envelope of the resulting homodyne signal shows several oscillations before decaying, due to the cyclic energy exchange with the mechanical oscillator (cf. Fig. 2(c)), which can also be detected for pulses containing less than one quantum on average.

The reported experiments demonstrate the possibility to couple a mechanical oscillator and an optical cavity mode quantum-coherently, thus setting the stage for the creation of nonclassical states of motion by swapping nonclassical states of the light field to the mechanical oscillator, as well as employing micromechanical oscillators as quantum state transducers between optical fields and other degrees of freedom [18].

SUMMARY

The realization that optical micro-resonators combine optical and mechanical degree of freedom provides the bases for novel opto-mechanical devices that have been in recent year lead to a plethora of opto-mechanical phenomena to be explored, ranging from cooling, amplification or quantum limited readout of mechanical motion – to slowing advancing or storing of light pulses using micro- and nanomechanical oscillators. The range of optomechanical systems extends from macro-scale mirrors to nanoscale photonic crystal or whispering gallery mode resonators. The radiation pressure enabled coupling can be harnessed for sensors of force, charge, acceleration or displacement with unprecedented sensitivity. Moreover, optomechanical coupling enables the preparation of mechanical systems in the quantum regime, by removing thermal quanta of the mechanical oscillator by optomechanical sideband cooling.

References

1 Kippenberg, T. J. & Vahala, K. J. Cavity Optomechanics: Back-Action at the Mesoscale. *Science* 321, 1172-1176, doi:10.1126/science.1156032 (2008).

2 Kippenberg, T., Rokhsari, H., Carmon, T., Scherer, A. & Vahala, K. Analysis of Radiation-Pressure Induced Mechanical Oscillation of an Optical Microcavity. *Physical Review Letters* 95, doi:10.1103/PhysRevLett.95.033901 (2005).

3 Schliesser, A., Arcizet, O., Rivière, R., Anetsberger, G. & Kippenberg, T. J. Resolved-sideband cooling and position measurement of a micromechanical oscillator close to the Heisenberg uncertainty limit. *Nature Physics* 5, 509-514, doi:10.1038/nphys1304 (2009).

4 Schliesser, A., Del'Haye, P., Nooshi, N., Vahala, K. & Kippenberg, T. Radiation Pressure Cooling of a Micromechanical Oscillator Using Dynamical Backaction. *Physical Review Letters* 97, doi:10.1103/PhysRevLett.97.243905 (2006).

5 Weis, S. *et al.* Optomechanically induced transparency. *Science* 330, 1520-1523, doi:10.1126/science.1195596 (2010).

6 Zhou, X. *et al.* Slowing, advancing and switching of microwave signals using circuit nanoelectromechanics. *Nature Physics* 9, 179-184, doi:10.1038/nphys2527 (2013).

7 Verhagen, E., Deleglise, S., Weis, S., Schliesser, A. & Kippenberg, T. J. Quantum-coherent coupling of a mechanical oscillator to an optical cavity mode. *Nature* 482, 63-67, doi:10.1038/nature10787 (2012).

8 Brown, K. R. *et al.* Coupled quantized mechanical oscillators. *Nature* 471, 196-199, doi:10.1038/nature09721 (2011).

9 Diedrich, F., Bergquist, J., Itano, W. & Wineland, D. Laser Cooling to the Zero-Point Energy of Motion. *Physical Review Letters* 62, 403-406, doi:10.1103/PhysRevLett.62.403 (1989).

10 Schmidt, P. O. *et al.* Spectroscopy using quantum logic. *Science* 309, 749-752, doi:10.1126/science.1114375 (2005).

11 Kippenberg, T. J. & Vahala, K. J. Cavity Optomechanics: Backaction at the mesoscale. *Science* 321, 1172 (2008).

12 <Investigation of dissipative pondermotive effects of electromagnetic radiation Braginsky 1970 Soviet Physics Letters.pdf>.

13 Teufel, J. D. *et al.* Sideband cooling of micromechanical motion to the quantum ground state. *Nature* 475, 359-363, doi:10.1038/nature10261 (2011).

14 O'Connell, A. D. *et al.* Quantum ground state and single-phonon control of a mechanical resonator. *Nature* 464, 697-703, doi:10.1038/nature08967 (2010).

15 Verhagen, E., Deléglise, S., Weis, S., Schliesser, A. & Kippenberg, T. J. Quantum-coherent coupling of a mechanical oscillator to an optical cavity mode. *arXiv:1107.3761* (2011).

16 Groblacher, S., Hammerer, K., Vanner, M. R. & Aspelmeyer, M. Observation of strong coupling between a micromechanical resonator and an optical cavity field. *Nature* 460, 724-727, doi:http://www.nature.com/nature/journal/v460/n7256/suppinfo/nature08171_S1.html (2009).

17 Teufel, J. D. *et al.* Circuit cavity electromechanics in the strong-coupling regime. *Nature* 471, 204-208 (2011).

18 Stannigel, K. *et al.* Optomechanical Transducers for Long-Distance Quantum Communication. *Phys. Rev. Lett.* 105, 220501 (2010).

AMORPHOUS CARBON ACTIVE CONTACT LAYER FOR RELIABLE NANOELECTROMECHANICAL SWITCHES

Daniel Grogg[1], Christopher L. Ayala[1], Ute Drechsler[1], Abu Sebastian[1], Wabe W. Koelmans[1],
Simon J. Bleiker[2], Montserrat Fernandez-Bolanos[1], Christoph Hagleitner[1],
Michel Despont[3] and Urs T. Duerig[1]

[1]IBM Research – Zurich, Switzerland
[2]KTH Royal Institute of Technology, Sweden
[3]IBM Research – Zurich, now at CSEM, Switzerland

ABSTRACT

This paper reports an amorphous carbon (a-C) contact coating for ultra-low-power curved nanoelectromechanical (NEM) switches. a-C addresses important problems in miniaturization and low-power operation of mechanical relays: i) the surface energy is lower than that of metals, ii) active formation of highly localized a-C conducting filaments offers a way to form nanoscale contacts, and iii) high reliability is achieved through the excellent wear properties of a-C, demonstrated in this paper with more than 100 million hot switching cycles. Finally, a full inverter using a-C contacts is fabricated to demonstrate the viability of the concept.

INTRODUCTION

Mechanical switches, with their abrupt (on-off) *I-V* characteristic offer, a distinct feature not found in other electronic devices. Recent research results highlight the importance of the contact characteristics for the operation of electromechanical relays. In particular, the switching energy of a NEM switch is limited by the size and the nature of its contact [1]. This means that to construct truly low-power NEM switches contact adhesion forces in the nN range must be achieved.

Miniaturization of the contact reduces the adhesion forces in a metallic contact [2]; however, the size of a single contact asperity and the tendency of metals to form a neck impose practical limits. Contacts using non-metallic materials, such as graphene [3], carbon nanotubes [4] and metal oxides [5, 6], have been explored as a means to achieve high reliability and low adhesion forces in NEM relays. Nonetheless, finding a suitable trade-off between good electrical contact and a low adhesion force remains a challenging task.

In this paper, we investigate a-C as contact material for NEM switches. The a-C used as a coating has a low surface energy, similar to other carbon-based materials, and it can be deposited with a simple physical sputter-deposition process suitable for full wafer integration. Moreover, Joule-heating-induced clustering of sp^2 hybridized carbon atoms [7] enables current conduction in a localized filament without creating excessive adhesion force, thus enabling NEM relays to operate with a restoring force well below 100 nN.

NEM SWITCH

Device Design

The silicon NEM switch shown in Figure 1a is designed using the curved cantilever approach described in [8]. The motion of the free-standing cantilever can be described as a rotation around the hinge, which is close to the source. The hinge region is used to design the equivalent stiffness of the switch, which is 1.7 N/m for this design. The contact of these devices is shown in detail in Figure 1b and 1c in the open and closed state, respectively. In the open state, the actuation gap is about 60 nm wide, as is also the gap at the tip of the switch. When the tip comes into contact, the actuation gap remains open and is designed to remain half the width of the initial gap.

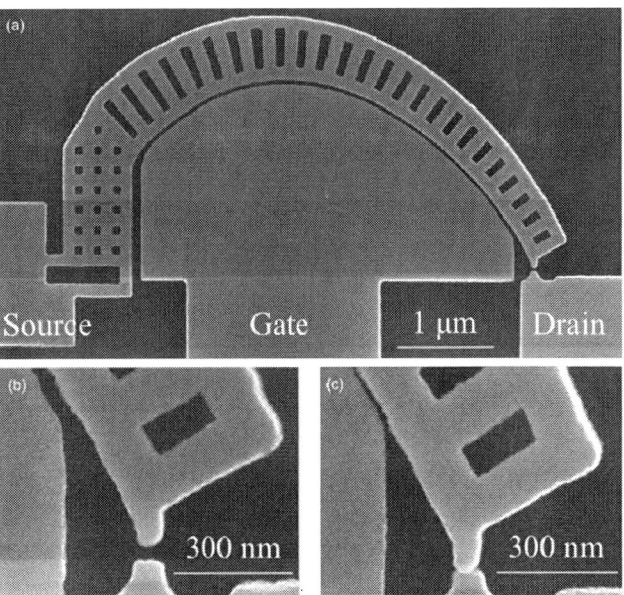

Figure 1: a) SEM micrograph of NEM switch fabricated. b) Contact region in the open and c) the closed state.

The effective benefit of the curved switch design is shown in Figure 2. The curvature helps avoid the excessive field strength often observed at the edge of a straight cantilever, which causes breakdown at low voltages. To confirm the advantages of the curved design experimentally, we fabricated both straight and curved cantilevers, as depicted in Figure 2a, on the same substrate next to each other. Figure 2b reports the lowest and highest values for the pull-in and pull-out voltages as well as the breakdown voltages (destructive test) of three devices of each design. The curved and straight switches have similar pull-in and pull-out voltages, as expected from simulation. However, to confirm the influence of the curvature on these two characteristics a larger number of devices would be necessary. In contrast, the increase in breakdown voltage can be distinguished even on this small sample of devices and confirms a breakdown voltage of > 13.5 V. This is more than twice the pull-in voltage in the worst case. Moreover, it clearly increases the margin between pull-in and breakdown compared with a straight cantilever.

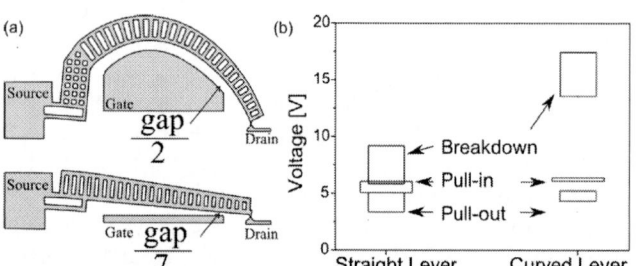

Figure 2: Comparison of switch designs: (a) Curved cantilever has a larger gap in the closed state, resulting in (b) higher measured breakdown voltages and smaller hysteresis windows.

Fabrication

The curved NEM switch shown in Figure 1 was fabricated using electron beam lithography on a silicon-on-insulator (SOI) substrate and the fabrication process described in Figure 3. The silicon device layer is 220 nm thick and etched using an ICP coupled HBr plasma to create vertical and smooth surfaces (step 1). The 60-nm air-gap is directly etched in the same process step, followed by sacrificial etching of the underlying silicon dioxide (step 2). A conduction layer is evaporated onto the device, enabling a low series resistance along the suspended cantilever of the switch (step 3). Good results have been achieved using platinum as conduction layer, but the inherent stress in the platinum layer creates a strong out-of-plane bending of the suspended parts of the NEMS switch. Therefore, a thin gold layer is used to relax the stress in the platinum layer. A low stress with nearly no visible out-of-plane bending is achieved using subsequent evaporation of the two materials while maintaining the harder platinum closer to the contact surface. Finally, a 10-nm-thin a-C layer is deposited over all devices using sputter deposition from a graphite target (step 4). The deposition is done in argon atmosphere, optimized for low stress in the as-deposited a-C layer. The main drawback of this process comes from potential electrical shortcuts created by the a-C layer. However, this drawback can be eliminated rather simply by using an additional photolithography step in the future.

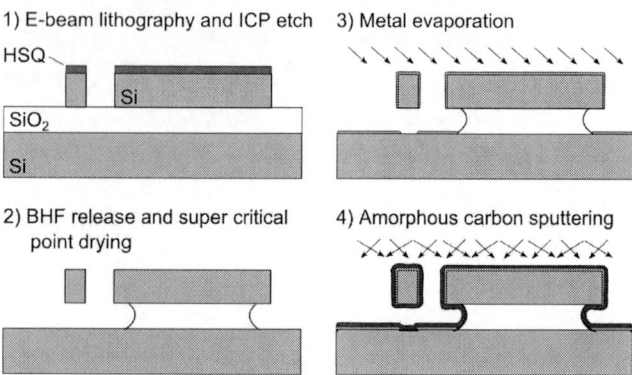

Figure 3: NEM switch fabrication process with a-C sputtering as the final step.

CONTACT MATERIAL

Static Characteristics

A schematic cross section through the contact region is shown in Figure 4a. The contact tip fabricated is 30 nm wide and has a slightly curved shape because of fabrication-related effects. The resistivity of the as-deposited a-C carbon is high, and the measured I-V characteristic fits a modified Poole–Frenkel model that accounts for the non-isolated nature of the traps arising from the high trap-density intrinsic to a-C (Figure 4b).

A conduction filament forms in the a-C upon application of a sufficiently high voltage; currently voltages in the range of 1.5 to 2.5 V are necessary. This filament formation process is an active process, i.e., the filament is created post-fabrication during the first contact events. After this process, the filament remains in the a-C layer even when the switch is opened. Detailed investigations of the filament-formation process in a-C at the nanoscale, including the field/temperature dependence, had been performed using a conductive-mode AFM [7]. Once the filament has been formed, there is a marked difference in the conduction behavior: metallic transport is observed, and the total resistance of the switch reduces to 15 kΩ. This resistance depends on several factors, including the effective contact area, the layer thickness and the current flow (heat) during the formation process.

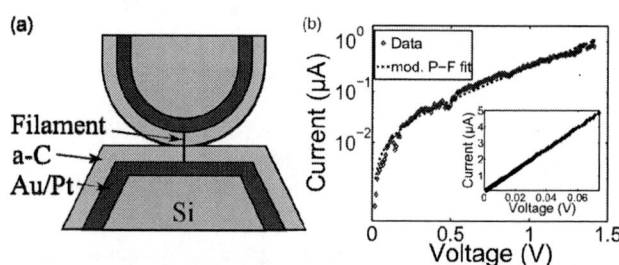

Figure 4: (a) Schematic cross section of contact region. (b) Experimental I-V characteristics of the drain-source contact after fabrication fit a modified Poole–Frenkel (P–F) conduction model. Inset: Nearly perfect metallic conduction after formation of a filament in the a-C.

978-1-4799-3510-9/14 $31.00 © 2014 IEEE   144

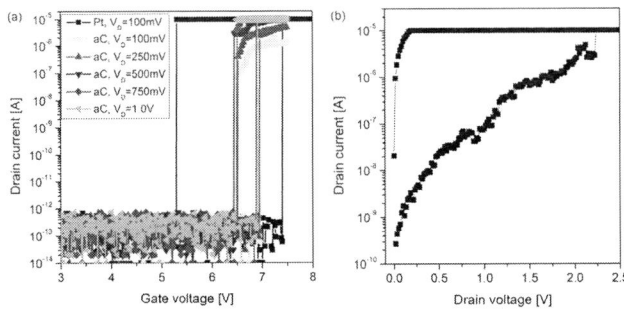

Figure 5: (a) I_D-V_G characteristic of a NEM relay: The larger hysteresis of the Pt contact (black) considerably reduces when using a-C coating. (b) I_D-V_D characteristic: Filament formation in a fresh a-C layer occurs at 2.2 V.

The benefit of the a-C coating is evident in the I_D–V_G characteristic (Figure 5a): the hysteresis window of the switch reduces from > 2 V for a platinum-platinum contact to 0.5 V for the a-C contact. Based on results from electromechanical simulations of the NEM switch, these hysteresis windows correspond to adhesion forces of about 50 nN for the platinum contact and only 10 nN for the a-C contact. Moreover, the resistance of the contact directly influences the maximum voltage the contact can handle during a measurement before stiction, neck formation and microwelding are observed. Using the a-C contact layer, static measurements with a current compliance of 10 μA are possible up to 2.5 V. Figure 5b shows the I_D-V_D characteristic of a fresh a-C layer with filament formation at 2.2 V.

Reliability

Devices have been subjected to a long-term reliability test to assess the lifetime of nanoscale a-C contacts. The devices were operated using a gate voltage pulse of –10 V at 50 kHz. The switching transient ramps have been 200 ns long, i.e., much slower than the simulated switching time of ≤ 42 ns.

A two-level drain voltage is used to create a filament in the a-C layer at each contact cycle, if necessary, and to extract the total resistance at a lower voltage. The drain voltage levels are set to V_1 and $V_1/10$, to extend the measurement range of the linear transimpedance amplifier used as an output. In addition, this voltage pulse also mimics the voltage conditions a NEM switch sees under capacitive load: a high voltage when closing the contact and a low voltage when opening the switch.

Figure 6 reports the extracted total ON-resistance (R_{ON}) for each switching cycle under hot switching conditions. One experiment was run with a drain voltage V_1 of 1.6 V (0.16 V for the extraction of R_{ON}, plotted in red) and the other with $V_1 = 2.0$ V (0.2 V for the extraction of R_{ON}, plotted in blue). These cycling tests with more than 100 million cycles at 1.6 V and 2.0 V prove the high reliability of the a-C.

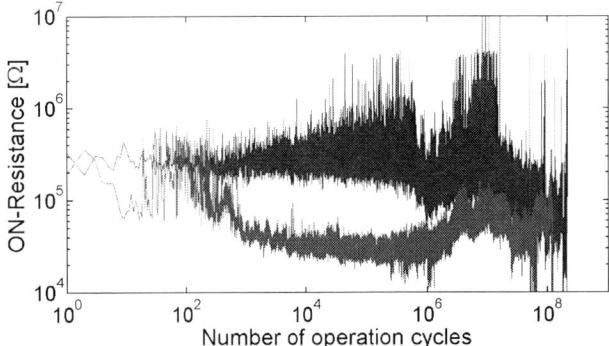

Figure 6: Lifetime test results over more than 100 million hot switching cycles with 1.6 V (blue) and 2.0 V (red). Filament formation mainly occurs in the first switching cycles (higher resistance) and for higher voltages.

Detailed analysis of the measurement data reveals additional information. R_{ON} is often higher in the beginning of an experiment and then decreases after some time. We attribute this to the formation of multiple filaments in the contact tip, caused by the probability of the tip to contact the previously formed filament in the exact same location. Therefore, the conduction behavior becomes stable only after a number of events that seems to be linked also to the voltage drop over the contact. This is observed in the experiment run at $V_1 = 2.0$ V, where the initial contact resistance is about 200 kΩ, dropping to < 50 kΩ within a few hundred cycles. Closing the switches with a lower V_1 of 1.6 V shows a more gradual decrease of the extracted contact resistance starting from the same initial value. This is an indication that the filament formed at 1.6 V is either less stable or it does not contribute to the current conduction of the subsequent switching cycle as much as a filament formed at higher voltage.

BUILDING BLOCK: INVERTER

The new NEM switch proposed in this work meets all requirements for logic circuits: a reliable contact, small hysteresis, low pull-in voltage, high breakdown voltage, small size, and the capability to sustain high voltage at the output. The inverter is the classical building block to demonstrate these attributes in a single device.

Figure 7 shows an SEM micrograph of a NEMS-based inverter. The left side shows two NEM switches in an inverter configuration that drive the gate of an output switch (right side). Decoupling the inverter from the measurement setup by means of an output switch reduces the capacitive load from the measurement setup. These parasitic loads from the measurement setup are a major limitation when measuring devices with a high drain voltage as they lead to high surge currents and welding of the contact tip.

Figure 8 shows the measurement results and the measurement setup of the inverter presented in Figure 7. Figure 8a shows the measurement results of the NEM-switch-based inverter with $V_{DD} = 4$ V and $V_{SS} = –4$ V. Cycling the input signal (blue) between the two supply voltages results in a corresponding oscillation of the output signal (V_{Output}). To limit and control the voltage on the output switch, the virtual ground of the transimpedance amplifier is set to 1.75 V, resulting in a drain-to-source voltage drop of 2.25 V on the output switch.

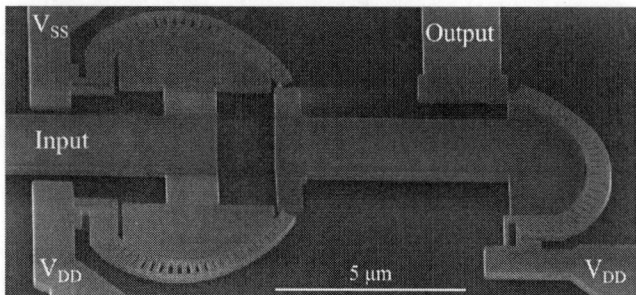

Figure 7: SEM image of an inverter driving a single output NEM switch that decouples the inverter from the measurement setup.

Figure 8b shows a schematic of the measurement setup including the inverter, the output switch and the transimpedance amplifier at the output. The gain of the amplifier and the steady current in all NEM switches are limited by using discrete resistors. The measured output signal is the result of three inversions by the total setup and switches between the open state (V_{Output} = 1.75 V) and the closed state ($V_{Output} \sim 0$ to 0.5 V) of the output switch. However, the voltage of the internal gate must cycle through the full voltage span to create this characteristic. Therefore, we demonstrate with this inverter setup that carbon-coated NEM switches can drive additional devices over the entire supply voltage range.

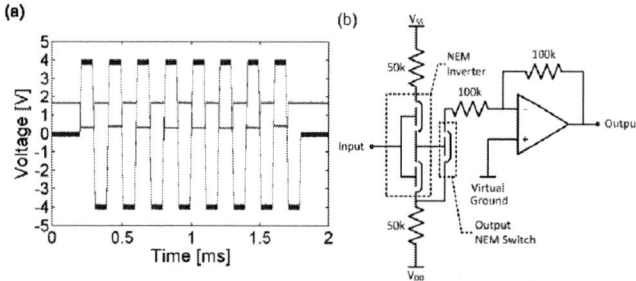

Figure 8: (a) Measurement of an inverter as shown in Figure 7, with the output connected to the virtual ground of a transimpedance amplifier (setup in (b)). The input (blue) actuates the inverter; the output (red) is measured from the inverting amplifier. The virtual ground (V_{Output}) is set to 1.75 V. (V_{DD} = +4 V and V_{SS} = –4 V).

CONCLUSION

A first evaluation of amorphous carbon (a-C) as a contact material for nanoelectromechanical (NEM) switches is presented in this paper. a-C has a low surface energy and offers an elegant way of creating nanoscale contacts by forming filaments in the a-C layer. Combining these advantages with a curved NEM switch design, we demonstrate all attributes necessary for nanomechanical

logic: a reliable contact (R_{ON} = 15kΩ, >10^8 switching cycles), small hysteresis, low pull-in voltage, high breakdown voltage, small size, and the capability to sustain the supply voltage at the output. In addition, this work provides the characteristics of an inverter used as basic building block for logic circuits, revealing a way towards achieving more complex logic systems.

ACKNOWLEDGEMENTS

The authors gratefully acknowledge the support of Kevin Lister, Meinrad Tschudy, Richard Stutz and the whole team working at the IBM Binnig and Rohrer Nanotechnology Center.

The research leading to these results has received funding from the European Union's Seventh Framework Programme (FP7/2007-2011) under grant agreement No. 288670.

REFERENCES

[1] A. Knoll *et al.*, "Fundamental scaling properties of electro-mechanical switches", New J. Phys. 14 (2012) 123007.

[2] J. Yaung *et al.*, "Adhesive force characterization for MEM logic relays with sub-micron contacting regions", J. Microelectromech. Systems (2013), DOI: 10.1109/JMEMS.2013.2269995

[3] S.M. Kim *et al.*, "Suspended few-layer graphene beam electromechanical switch with abrupt on-off characteristics and minimal leakage current", Appl. Phys. Lett. 99 (2011) 023103.

[4] O. Loh *et al.*, "Carbon-Carbon Contacts for Robust Nanoelectromechanical Switches", Adv. Mater. 24 (2012) 2463.

[5] M.P. de Boer *et al.*, "Design, fabrication, performance and reliability of Pt- and RuO$_2$-coated microrelays tested in ultra-high purity gas environments", J. Micromech. Microeng. 22 (2012) 105027.

[6] H. Kam *et al.*, "Design and reliability of a micro-relay technology for zero-standby-power digital logic applications", in Proc. IEEE IEDM, 2009, p. 809.

[7] A. Sebastian *et al.*, "Resistance switching at the nanometre scale in amorphous carbon", New J. Phys. 13 (2011) 013020.

[8] D. Grogg *et al.*, "Curved in-plane electromechanical relay for low power logic applications", J. Micromech. Microeng. 23 (2013) 025024.

CONTACT

*D. Grogg, dgr@zurich.ibm.com
*U.T. Duerig, drg@zurich.ibm.com

NEAR INFRARED PHOTO-DETECTOR USING SELF-ASSEMBLED FORMATION OF ORGANIC CRYSTALLINE NANOPILLAR ARRAYS

Yoshiharu Ajiki[1,5], Tetsuo Kan[2], Masayuki Yahiro[3], Akiko Hamada[3], Junji Adachi[4], Chihaya Adachi[3], Kiyoshi Matsumoto[2] and Isao Shimoyama[2]

[1]Microtechnology R&D Division, Olympus Corporation, Japan
[2]The University of Tokyo, Japan
[3]Center for Organic Photonics and Electronics Research (OPERA), Kyushu University, Japan
[4]Office for Strategic Research Planning, Kyushu University, Japan
[5]NMEMS Technology Research Organization, Japan

ABSTRACT

We proposed a near infrared photo-detector (NIR-PD) using self-assembled formation of organic crystalline arrays, which were formed on an n-type silicon (n-Si) substrate and covered with an Au film. These structures act as antennas for near infrared (NIR) light, resulting in an enhancement of the light absorption on the Au film. The NIR-PDs thus have higher photo-responsivity compared with that of an Au/n-Si typed Schottky diodes, which was fabricated as a reference. In this paper, the fabrication process of the NIR-PDs and the estimation results of photo-responsivity were described. The maximum responsivity to NIR light (wavelength = 1.2 μm) was 1.79 mA/W without applying forward bias. This value is 10 times larger than the responsivity of the Au/n-Si typed Schottky diode as a reference.

INTRODUCTION

Detection of near infrared (NIR) light is useful for many applications such as infrared imaging systems [1], optical telecommunication systems, and spectroscopy systems [2]. Si-based near infrared photo-detectors, which employ internal photoemission effect (IPE) are studied [3]. The cut-off wavelength is determined by a Schottky barrier height between a metal and a semiconductor so that the NIR light becomes detectable. The cut-off wavelength is about 1.55 [μm] in the case of gold and Si junction.

But photo-responsitivities of IPE based near infrared photo-detectors (NIR-PDs) are low due to poor absorption of the NIR light in a metal film. Knight *et al.* investigated NIR-PDs with Au nano-antenna structures [4]. In the study, nano-antenna structures were used for enhancing absorption of the NIR light incident on an Au film by Surface Plasmon Resonance (SPR). However, it is usually difficult to fabricate nano-antenna structures because it requires nano-patterning process.

We propose a near infrared photo detector (NIR-PD) using self-assemble formation of an organic crystalline pillar array. In this paper, the fabrication process of the NIR-PDs and the estimation results of photo-responsivity were described. The maximum value of the responsivity to NIR light (wavelength = 1.2 [μm]) was 1.79 [mA/W] without applying forward bias.

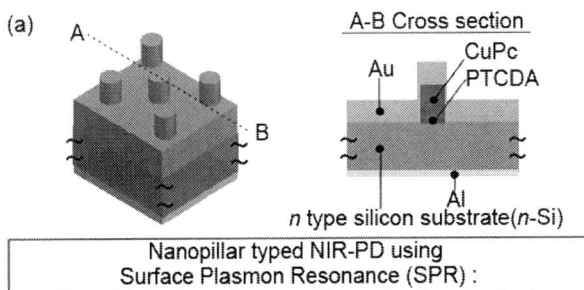

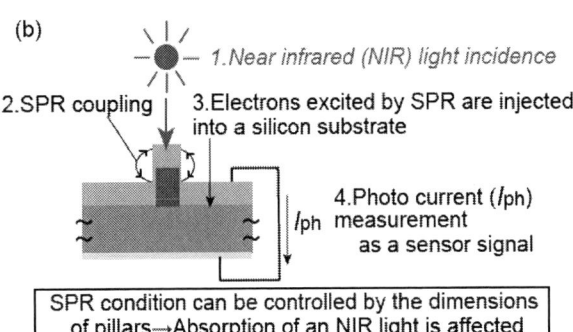

Figure1. (a) Device configuration of an NIR photo-detector (NIR-PD). Nano-pillars of the NIR-PD are fabricated by self assembled formation of organic semiconductors (PTCDA and CuPc). (b) Detection method of the NIR-PD.

This value is 10 times larger than that of an Au/n-Si typed Schottky diode, which was fabricated as a reference.

FABRICAION PROCESSES

The organic crystals were formed on an n-type silicon substrate and covered with an Au film as shown in Figure 1(a). In this configuration, copper phthalocyanine (CuPc) and 3,4,9,10-perylene-tetracarboxylic-dianhydride (PTCDA) were used as organic crystal materials. In a previous paper, it was reported that CuPc deposited on a PTCDA seed layer was used to form the self-assembled nano-pillars [5].

978-1-4799-3510-9/14 $31.00 © 2014 IEEE

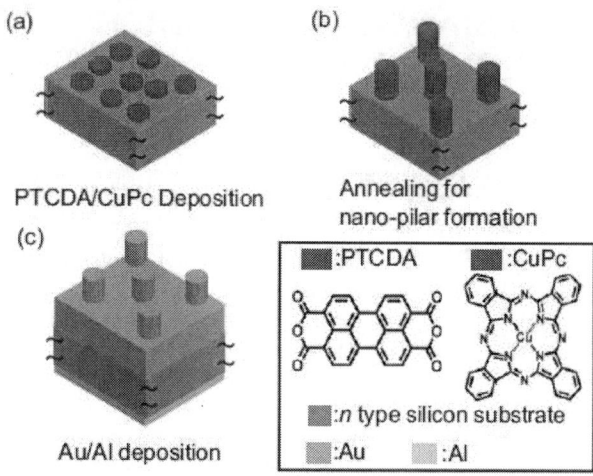

Figure 2. (a)-(c) Fabrication processes of the proposed NIR-PD.

The dimensions of the pillars can be controlled by changing the process parameters such as annealing temperature so that various nano-antenna structures can be obtained. The detection method is shown in Figure 1(b).

In this case, when the near infrared light is irradiated to the NIR-PD, an SPR coupling occurs on the surface of the nano-antennas. SPR coupling enhances absorption of the NIR light in a metal film so that a lot of electrons in the metal film are excited and injected to the Si substrates as the IPE phenomenon. These injected electrons become a photo-current, which is a sensor signal of the proposed NIR-PD.

A fabrication process of proposed NIR-PDs is shown in Figure 2. First, an *n*-type silicon substrate (<100>, ρ = 40 [Ωcm]) was dipped into a HF solution for removing a Si native oxide. After that, PTCDA and CuPc were deposited on the *n*-type silicon substrate. Next, the substrate was annealed in the nitrogen atmosphere for nano-pillar formation. Finally, a 50-nm-thick Au film was deposited onto the nano-pillar array by the EB evaporation. Also, Al film was deposited on to the backside of the *n*-type silicon substrate for an anode electrode of the proposed NIR-PD. In this configuration, the Au film acts as not only a component of the nano-antenna but also a cathode electrode.

As shown in Figure 3, SEM images of nano-pillar were obtained to confirm the nano-pillar formation.

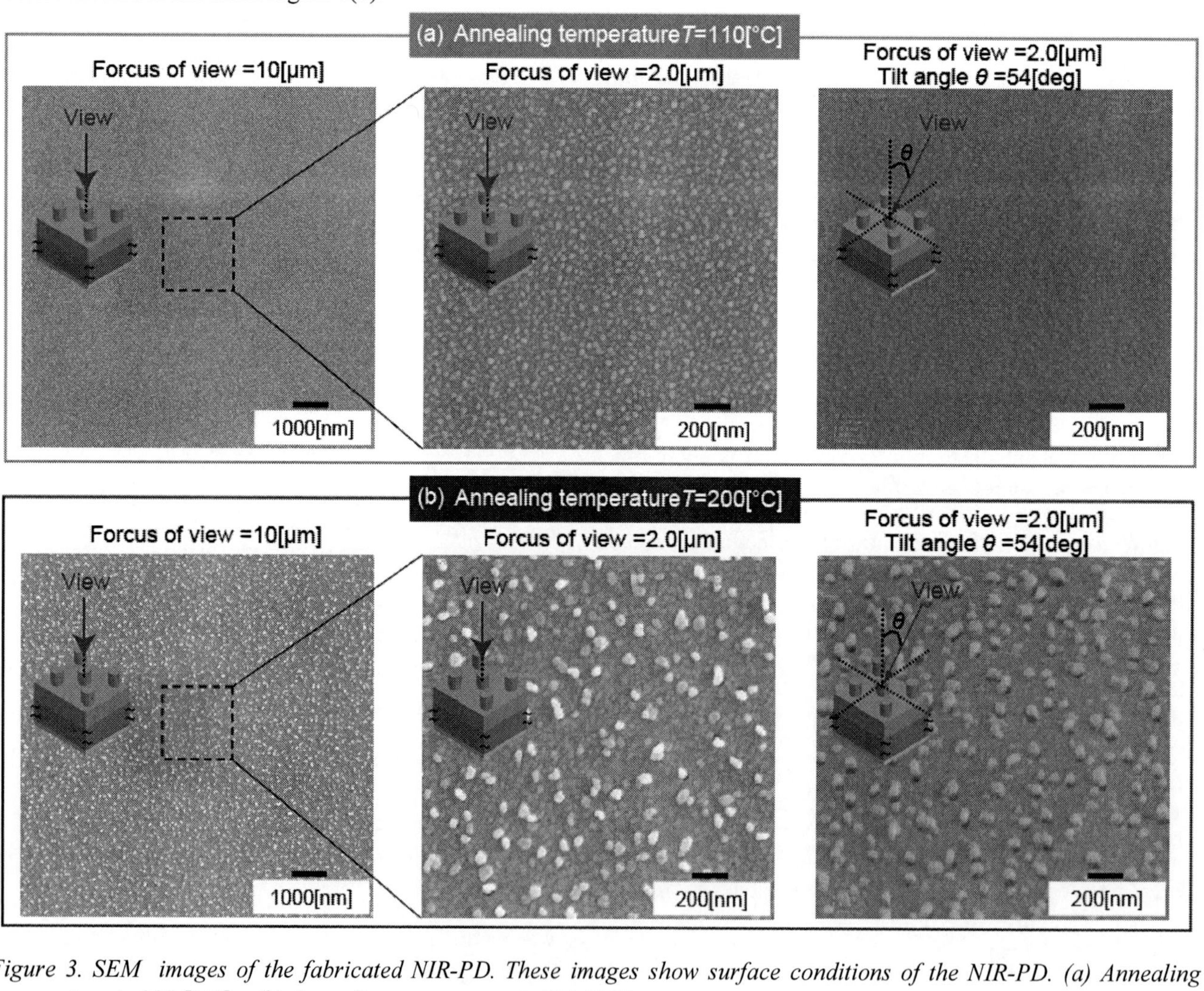

Figure 3. SEM images of the fabricated NIR-PD. These images show surface conditions of the NIR-PD. (a) Annealing temperature is 110 [° C]. (b) Annealing temperature is 200 [° C].

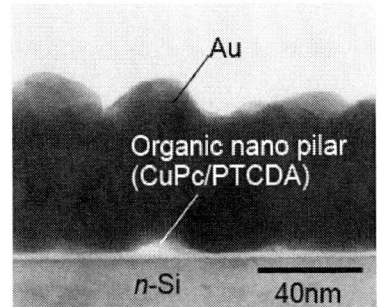

(a) Annealing temperature T=110[°C]

Au

Organic nano pilar (CuPc/PTCDA)

n-Si   40nm

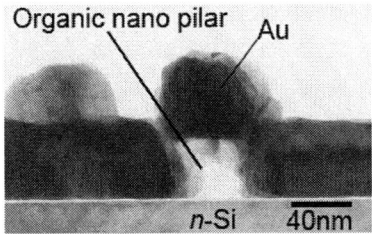

(b)Annealing temperature T=200[°C]

Organic nano pilar   Au

n-Si   40nm

Figure 4. TEM images of the fabricated NIR-PD. These images show cross section of the NIR-PD. (a) Annealing temperature is 110 [° C]. (b) Annealing temperature is 200 [° C]

The size of the nano-pillar tended to be larger as the annealing temperature is higher. Additionally, to obtain cross-section images of the NIR-PDs, TEM images were obtained as described in Figure 4. Average heights of nano-pillar were 15.8 [nm] and 41.6 [nm], in the annealing temperature were 110 [°C] and 200 [°C], respectively. Thus, we can say that the forms of organic crystalline nano-pillars were controlled by the annealing condition.

ELECTRIC CHARACTERISTICS OF THE FABRICATED NIR-PDS

To verify whether our fabricated devices can work as a diode, Current-Voltage (I-V) characteristics were obtained. NIR light was irradiated into the devices for measuring the photo-responsivity of our fabricated devices simultaneously. Figure 5 shows the experimental set up for verifying the photo-responsivity. The measurement set up for the photo-responsivity of the fabricated devices was consisted of three components; a light source of NIR light (Fianium SC450), a multi-meter (ADCMT 6242), and our fabricated devices. A semiconductor parameter analyzer (HP 4156B) was used for I-V characteristic measurements. These experiments were carried out in a photo dark room. The wavelength range of the irradiated NIR light could be changed from 1.1 to 1.3 [μm] so that we could measure the photo-responsitivity above the cut-off wavelength of Si-PD. The incident angle of the NIR light was 0 [deg] (normal incidence).

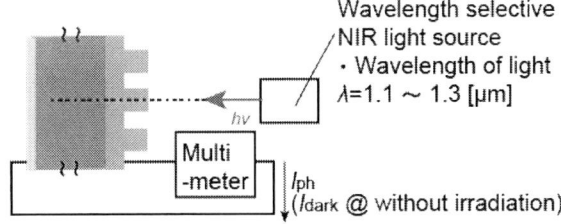

Wavelength selective NIR light source
· Wavelength of light
λ=1.1 ~ 1.3 [μm]

Multi-meter

I_{ph} (I_{dark} @ without irradiation)

Figure5. Experimental set up for verifying the photo-responsivities of fabricated diodes

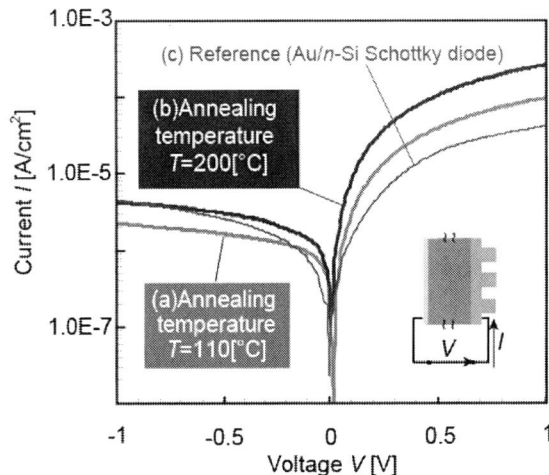

Figure 6. Current-Voltage characteristics of the fabricated diodes at dark condition .
(a) Annealing temperature is 110[° C].
(b) Annealing temperature is 200[°C].
(c) Au/n-Si Schottky diode fabricated as a reference.

The photo-responsivity is calculated by using equation (1),

$$R = \frac{(I_{ph} - I_{dark})}{P_{in}} , \qquad (1)$$

where P_{in} is the power of the incident light, I_{ph} is a short circuit current with the NIR light irradiation, and I_{dark} is a short circuit current without the NIR light irradiation.

Figure 6 shows the I-V characteristics of the NIR-PDs in cases of two annealing temperatures (T = 110 [°C] and 200 [°C]). The I-V characteristics of an Au/n-Si typed Schottky diode, which was fabricated as a reference is also shown. The obtained diode characteristics proved that a diode was formed between the Au film and an n-Si substrate even though the nano-pillars were inserted. Additionally, the on-state current of nano-pillars was higher than that of a reference Au/n-Si Schottky diode when the applied voltage was 1.0[V].

Figure 7 shows the relationship between the wavelength and the responsivity of the NIR-PD. From these relationships, responsivities of the fabricated devices were higher than that of a reference Au/n-Si Schottky diode.

978-1-4799-3510-9/14 $31.00 © 2014 IEEE   149

(a)

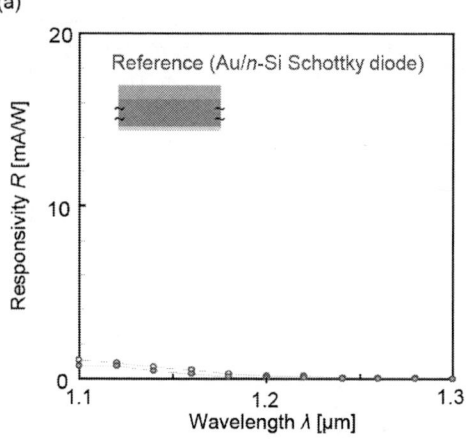

(b)

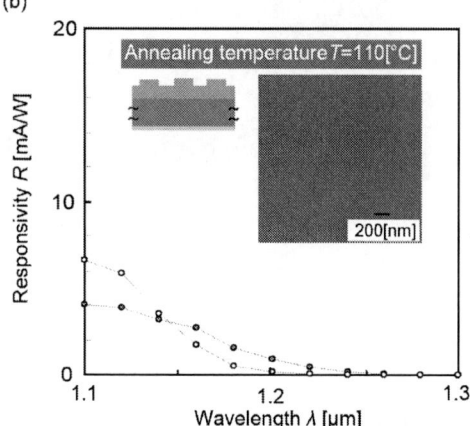

(c)

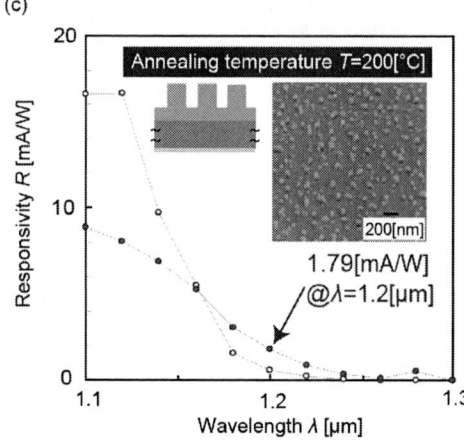

Figure 7. Photo responsivity of the fabricated NIR-PD (n=2).
(a) Au/n-Si Schottky diode as a reference.
(b) Annealing temperature is 110 [° C].
(c) Annealing temperature is 200 [° C].

As shown in Figure 7(c), the maximum photo-responsivity of the fabricated NIR-PD was 1.79[mA/W], when the wavelength of the NIR light was 1.2 [μm]. The responsivity is about 10 times larger than that of the reference Au/n-Si Schottky diode.

In addition, the responsivity changed according to the change of the configuration of the nano-pillars. As larger the nano-pillars were formed, as higher photo-responsivities were obtained. These phenomena imply that the plasmonic absorption occurred in the proposed NIR-PD.

CONCLUSION

We proposed a near infrared photo-detector (NIR-PD) using self-assembled formation of organic crystalline arrays, which were formed on an *n*-type silicon (*n*-Si) substrate and covered with an Au film. By using these structures, nano-antennas, which can enhance the NIR light absorption by SPR, were obtained on Si substrates. The formations of pillars could be controlled by annealing condition of these devices. The maximum photo-responsivity of the fabricated NIR-PD is about 10 times larger than that of an Au/*n*-Si typed Schottky diode, which was fabricated as a reference when the wavelength of the NIR light was 1.2 [μm]. In addition, the responsivity changed according to the configuration of the nano-pillars, implying that the plasmonic absorption occurred in the proposed NIR-PD.

ACKNOWLEDGEMENTS

This work is partly supported by New Energy and Industrial Technology Development Organization (NEDO). This work was partially conducted in Center for Nano Lithography & Analysis, The University of Tokyo, supported by the Ministry of Education, Culture, Sports, Science and Technology (MEXT), Japan.

REFERENCES

[1] W.F. Kosonocky, F.V. Shallcross, T.S. Villani, "160x244 Element PtSi Schottky Barrier IR-CCD Image Senser," *IEEE Trans. Electron. Dev.*, vol. ED-32, pp. 1564, 1985.

[2] A. Zwielly, S. Mordechai, I. Sinielnikov, E. Bogomolny, S. Argov, "Advanced statistical techniques applied to comprehensive FTIR spectra on human colonic tissues," *Medical Physic.*, vol. 37, pp. 1047-1055, 2010.

[3] M.Casalino, L.Sirleto, L.Moretti, L.Moretti, F.D.Corte, I.rendina, "Design of a silicon resonant cavity enhanced photodetector based on the internal photoemission effect at 1.55 μm," *Journal of Optics A:Pure and applied optics*, vol. 8, pp. 909-913, 2006.

[4] M.W. Knight, H. Sobhani, P. Nordlander, N.J. Halas, "Photodetection with Active Optical Antennas," *Science*, vol. 332, pp. 702-704, 2011.

[5] M. Hirade, H. Nakanotani, M. Yahiro, C. Adachi, "Formation of Organic Crystalline Nanopillar Arrays and Their Application to Organic Photovoltaic Cells," *Applied Materials & Interfaces*, vol. 3, pp. 80-83, 2010.

CONTACT

Ajiki Yoshiharu, Olympus Corporation
2-3 Kuboyama-cho, Hachioji-shi, Tokyo 192-8512, Japan.
E-mail:yoshiharu_ajiki@ot.olympus.co.jp

ULTRASENSITIVE SI NANOWIRE PROBE
FOR MAGNETIC RESONANCE DETECTION

Yong-Jun Seo[1], Masaya Toda[1,2], Yusuke Kawai[1], and Takahito Ono[1]
[1]Graduate School of Engineering, Tohoku University, Sendai, JAPAN
[2]Micro System Integration Center, Tohoku University, Sendai, JAPAN

ABSTRACT

In this study, we have fabricated a 210 nm-wide and 32 μm-long Si nanowire probe with a Si mirror from a silicon-on-insulator wafer. Additionally, a Nd-Fe-B magnet is mounteded at the end of the nanowire for magnetic force detection in MRFM measurements. The fabricated probe shows a resonance frequency f_0 of 11.256 kHz and a Q factor of 12000 after annealing at 800°C for 2 hours in forming gas. The probe exhibits atto-newton sensitivity, and the measurement of force mapping based on electron spin resonance is demonstrated for three-dimensional imaging of radical density. The detected force is approximately 8.5 aN at room temperature.

INTRODUCTION

Magnetic resonance imaging (MRI) is well known as a useful instrument for visualizing three-dimensional structure inside a human body. However, the sensitivity and spatial resolution of the conventional MRI is limited by several tens to hundreds of micrometer scales due to limitations of conventional inductive detection techniques [1–3]. Magnetic resonance force microscopy (MRFM) was proposed to improve the sensitivity and spatial resolution of MRI. [4–6]. The MRFM uses a probe to detect a force between a small magnet and spins in the sample while the spins are excited using magnetic resonance techniques. If sufficient spatial resolution can be achieved, MRFM could have an impact on many applications such as determining three-dimensional molecular structures of biological samples and imaging of dopant distributions in semiconductor materials [7]. In order to obtain the higher spatial resolution, larger magnetic field gradients and higher force sensitivity are required for MRFM [8]. In particular, the detection of single proton requires an ultrasensitive probe [9].

Recently, researchers made three-dimensional images of individual tobacco mosaic viruses with a resolution better than 10 nm using MRFM technique [10]. The smallest force using a freely vibrating probe is limited by thermo mechanical noise as given by

$$F_{min} = \sqrt{\frac{4kk_BTB}{f_0Q}}, \qquad (1)$$

where k is the spring constant, k_B is the Boltzmann constant, T is the temperature, B is the bandwidth, f_0 is the resonant frequency, and Q is the quality factor. The factor k/f_0 can be written in terms of the probe structure. The following spring constant and the resonance frequency of a rectangular probe determined by the dimension and the material properties are used.

$$\omega_0 = \frac{3.516}{2\pi}\left(\frac{E}{12\rho}\right)^{\frac{1}{2}}\frac{t}{l^2}, \qquad (2)$$

$$k = \frac{1.030}{4}E\frac{t^3w}{l^3}, \qquad (3)$$

where E is the Young's modulus, ρ is the density, l is the length, w is the width, and t is the thickness. From equations (1), (2) and (3), the minimum detectable force [11] for a simple rectangular probe can be expressed in terms of the probe dimensions as following equation,

$$F_{min} = t\left(\frac{w}{lQ}\right)^{\frac{1}{2}}(E\rho)^{\frac{1}{4}}(k_BTB)^{\frac{1}{2}}. \qquad (4)$$

According to this equation (4), the design of probes is suggested that a high Q factor, thin, narrow, and long probe structure such as a nanowire is desirable for detecting small force.

The goal of this study is to fabricate a nanowire probe with a magnet and evaluate the possibility for MRFM. We have fabricated a 210 nm-wide and 32 μm-long Si nanowire probe with a Si mirror. Additionally, a Nd-Fe-B magnet is mounted or Ni magnet pattan is formed at the end of the nanowire for magnetic force detection in MRFM measurements. Then, we have demonstrated the measurement of force mapping based on electron spin resonance using the fabricated probe.

EXPERIMENTS

The design of a Si nanowire probe with a mirror and a magnet support part for MRFM is shown in Fig 1(a). The probe beam with a total length of 32 μm and width of 300 nm is designed. We have fabricated two types of Si nanowire probe using different magnets: Ni and Nd-Fe-B magnets. The magnet is formed on the magnet support part with a diameter of 2 μm. In order to measure the deflection and vibration of the probe, an octagon-shaped mirror with 5 μm of the inradius is designed at the middle of the beam. This mass of the mirror reduces the resonant frequency of the probe because the resonant frequency is inversely proportional to the mass. Designing a smaller size of the mirror having low mass is efficient due to higher frequency and smaller detectable force. Then, the spring constant k is 3.58×10^{-2} N/m, as calculated from the two separated beam of 10 μm-length and mirror as spring. The resonant frequency of the nanowire probe without magnet is 146.8 kHz, as calculated using the equation of classical vibrational frequency: $f_0 = 1/2\pi\times(k/m^*)^{1/2}$. As shown in Fig. 1 (b), the

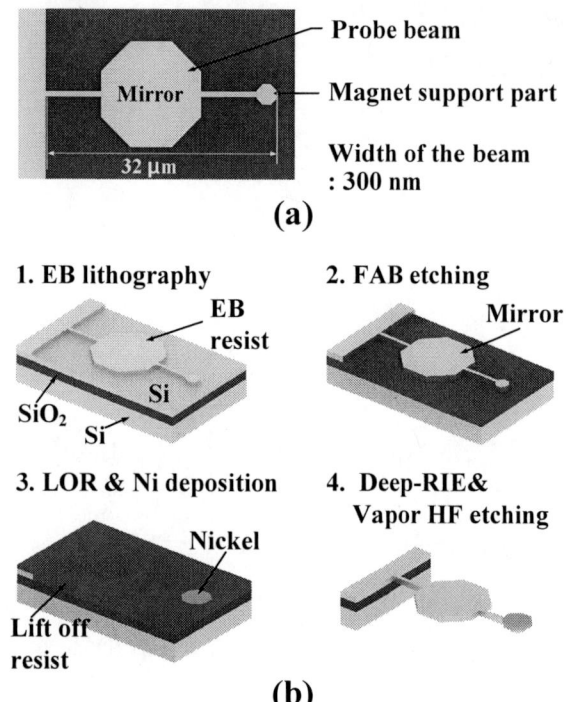

(a)

(b)

Figure 1: Schematic illustration of (a) design, (b) the fabrication process of Si nanowire probe with the Ni magnet.

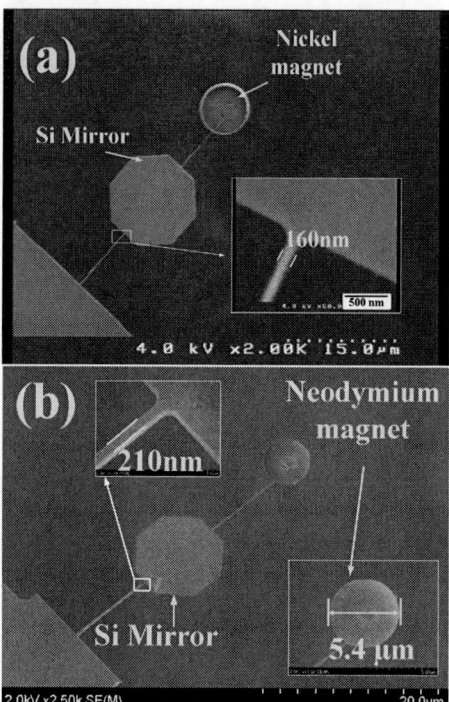

Figure 2: SEM image of the fabricated silicon nanowire probe with (a) the Ni magnet and (b)the Nd-Fe-B magnet. The width and length of the nanowire are 210 nm and 32 μm and the octagon mirror with 5 μm of the inradius is formed at the middle of the probe.

starting material for Si nanowire probe with the Ni magnet fabrication is a silicon on insulator wafer having a top silicon thickness of 200 nm. The Si nanowire probe is patterned by electron beam lithography [Fig. 1(b).1]. The top silicon layer is etched by fast atom beam using SF_6 gas for making the pattern of the nanowire probe [Fig.1(b).2]. In order to integrate a Ni magnet, nickel with a diameter of 5 μm was deposited on silicon using DC facing magnetron sputtering and patterned by a lift off technique [Fig.1(b).3]. Then, the handling silicon layer is etched from the bottom side using deep reactive ion etching. The buried oxide layer is removed, and the structure is released by a HF vapor-phase etching system [Fig.1(b).4]. This system is an effective technique to release fragile and suspended structures without stiction due to the surface tension during drying process. In order to obtain larger magnetic field gradients, we had also used the Nd-Fe-B magnet instead of the Ni magnet. The fabrication of the Si nanowire probe with the Nd-Fe-B magnet is basically same as the previous process, except for the nickel deposition and the lift-off process. Lastly, the Nd-Fe-B magnet of 5.4 μm-diameter is putted on the magnet support part of probe using a manipulator. Then, the fabricated silicon nanowire probe had ashed with O_2 plasma and annealed at 800°C for 2 hours in forming gas (3% hydrogen and 97 % argon) to clean its surface and increase its quality factor [12]. The Nd-Fe-B magnet on the probe was magnetized along to the vertical direction of the probe length by applying a magnetic field higher than 0.7 T. Figure 2 shows the fabricated Si nanowire probe with the Ni magnet and Nd-Fe-B magnet for MRFM. The width and length of the nanowire probe with the Nd-Fe-B magnet are 210 nm and 32 μm, respectively.

RESULTS AND DISCUSSIONS

The resonance frequency of the nanowire probe with the Nd-Fe-B is measured by a laser Doppler vibrometer at room temperature in vacuum of $\sim 10^{-3}$ Pa. Figure 3 shows the mechanical resonant spectrum of the probe with the Nd-Fe-B magnet at room temperature. The resonant frequency f_0 is 10.450 kHz before annealing and f_0 is 11.256 kHz after annealing. The Q factor is approximately 7000 before annealing and 12000 after annealing. The oxygen plasma ashing and annealing process could remove some absorbates, such as H_2O and some organics, on the probe surface and thus modifying the surface stress condition, changing the resonant frequency, and improving the Q factor.

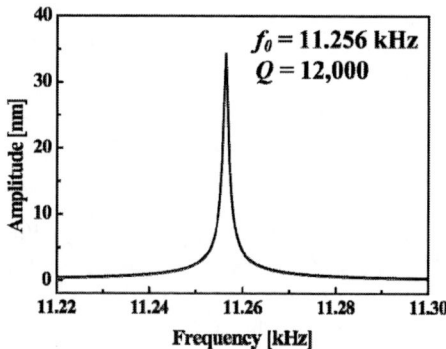

Figure 3: Mechanical resonant spectrum of the nanowire probe.

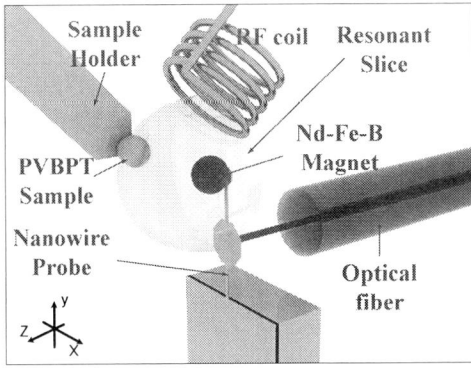

Figure 4: Schematic illustration of the MRFM experimental apparatus. The 6 μm-diameter PVBPT sample is placed at the end of the sample holder.

The theoretical force sensitivity improves from 1.12×10^{-15} N/√Hz for the non-annealed probe to 8.26×10^{-16} N/√Hz for the annealed probe at room temperature.

Schematic of the MRFM system used in our experiments is shown in Fig. 4. Electron spins in the sample are resonantly excited by rf magnetic field which is generated by an rf coil. The vibration of the probe caused by magnetic force due to electron spin resonance is detected by a fiber-optic interferometer using a laser diode with 1550 nm wavelength. In addition, a modulation coil has used for anharmonic modulation of resonant detection [13]. As a single 6 μm-diameter sample, the radical containing PVBPT (Poly-10-(4-vinylbenzyl)-10H-phenothiazine, [14]) is glued on the sample holder. The probe vibration amplitude from the fiber-optic interferometer is measured by a lock-in amplifier. The glued sample on the sample holder is moved by an *XYZ* scanner and the resonance force is mapped through the resonant slice. The resonant slice is paraboloidal shell, representing the region in which the field from the magnet tip matches the condition for magnetic resonance. The force signal is proportional to a number of spins at the resonant slice. In Fig. 5, the magnetic resonance force signals are plotted as a function of translation along *Z* axis, in other words, the distance between the magnet and the sample. The rf frequency f_{res} were 1.6, 1.7, 1.8 and 1.9 GHz. The positive peak corresponds to the resonance of the PVBPT particle. The changes of the signal sign are due to the phase shift in the phase sensitive detection of the lock-in amplifier [15]. The measured amplitude of the peak at f_{res} = 1.9 GHz was 0.0361 nm. The detected force can be calculated from the equation $F = A(k/Q)$, where A is amplitude of probe, k is spring constant, Q is quality factor. The detected force at the peak is calculated to be 8.5 aN. Through the resonance force position as a function of magnetic resonance frequency, the magnetic field distribution generated by the magnet can be estimated.

Figure 5(b) shows the resonant frequency of one-dimensional scan measurement as a function of the magnetic resonance position. By approximating the observed field variation to be linear, we determined magnetic field gradient ~15100 *T/m* around the resonant slice. The estimated number of electron spins can be calculated from the equation

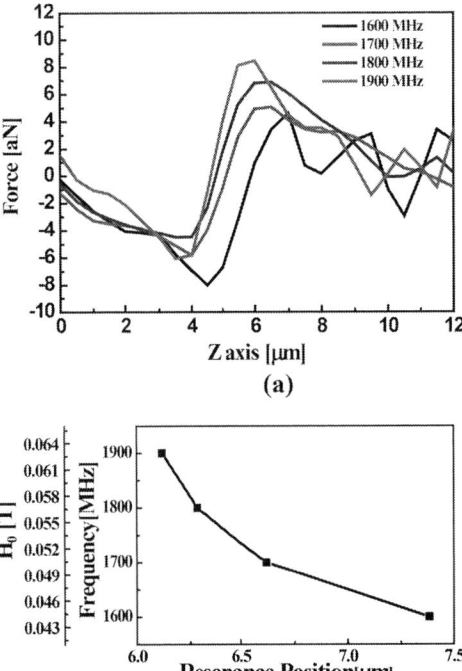

Figure 5: Magnetic resonance force is plotted as a function of translation along Z at magnetic fields with frequencies of 1.6, 1.7, 1.8 and 1.9 GHz. (b) The magnetic resonance frequency of one-dimensional scan measurement as a function of the resonant position.

$$F_Z = N_e \times \mu_B \, \frac{\partial H_Z}{\partial z}, \qquad (5)$$

The detected number of radicals (unpaired electron spins) can be calculated from the detected force. The number of radicals at f_{res}= 1900 MHz was 8.6×10.

The successive *XY* images of magnetic resonance force signal sliced at different depths *Z* are displayed in Fig. 6. This two-dimensional force map was taken for a single 6 μm-diameter PVBPT particle at f_{res} = 1.8 GHz (H_{res}= 64.2 mT) and the scan range was meshed into 50×50 pixels. The force map on *XY* plane shows circle-like image as the response of the single particle sample. In a *XY* plane at fixed distance z from the magnet tip, the spins are resonantly excited only when the sample is located on the circle-like intersection between the sample and paraboloid of resonant slice. The circle means that the magnetic resonance force signal of electron spins are detected. Red and black pixels indicate positive and negative force signals due to the phase shift in the phase sensitive detection of the lock-in amplifier, respectively. The force signal had been detected in a circle at $Z = Z_0 + 2$, 4 and 6 μm, where Z_0 is initial distance between sample and magnet tip. The circle diameter depends on the distance of the plane. The radius of the intersection decreases as the scan plane is moved away from the magnet tip until the apex of the resonant slice. Finally, the force signal of Z_4 disappeared because the sample position is outside of the resonant slice.

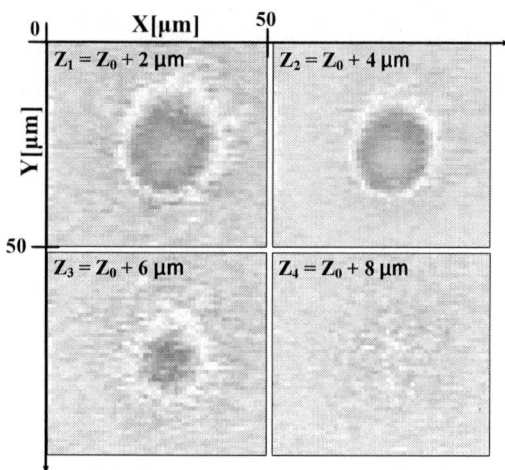

Figure 6: Some of two-dimensional force maps obtained at different Z positions for the PVBPT sample.

As a consequence, the fabricated Si nanowire probe with the magnet has a high sensitivity and applied to detection of tenth-order radicals by electron spin resonance at room temperature.

CONCLUSION

An atto-newton-sensitive Si nanowire probe with a mirror has been fabricated based on top-down fabrication process. A Nd-Fe-B magnet particle is integrated at the end of the probe. A scanning measurement of force map based on electron spin resonance for three-dimensional imaging of radicals has been demonstrated using the fabricated nanowire probe. We could achieved the detected force of 8.46 aN, and calculated number of radicals is 8.5×10. We believe that the silicon nanowire probe will be a good choice for magnetic resonance detection in MRFM applications.

ACKNOWLEDGEMENTS

Parts of this work were performed in the Micro/Nanomachining Research Education Center (MNC) and Micro System Integration Center (μSIC) of Tohoku University. This work was supported in part by Special Coordination Funds for Promoting Science and Technology, Formation of Innovation Center for Fusion of Advanced Technologies from the Japanese Ministry of Education, Culture, Sports, Science and Technology (MEXT), also supported in part by Grant-in Aid for Young Scientists from Japan Society for the Promotion of Science (JSPS).

REFERENCES

[1] S.C. Lee, K.S. Kim, J.H. Kim, S.C. Lee, J.H. Yi, S.W. Kim, K.S. Ha, C.J. Cheong, "One micrometer resolution NMR microscopy", *Journal of Magnetic Resonance*. Vol.150, pp.207, 2001.

[2] D.A. Seeber, L. Ciobanu, C.H. Pennington, "Advances toward MR microscopy of single biological cells", *Applied Magnetic Resonance*, Vol.22, pp. 139, 2002.

[3] K.R. Minard, R.A. Wind, "Picoliter ^{1}H NMR spectroscopy", *Journal of Magnetic Resonance*. Vol. 154, pp.336, 2002.

[4] J.A. Sidles, "Noninductive detection of single proton magnetic resonance", *Applied Physics Letters*, Vol.58, pp.2854-2856, 1991.

[5] J.A. Sidles, J.L. Garbini, G.P. Drobny, "The theory of oscillator-coupled magnetic resonance with potential applications to molecular imaging", *Review of Scientific Instruments*, Vol.63, pp.3881-3884, 1992.

[6] O. Züger, S. Hoen, C.S. Yannoni, D. Rugar, "Three dimensional imaging with a nuclear magnetic resonance force microscope", *Journal of Applied Physics*, Vol.79, pp.1881–1884, 1996.

[7] M. Poggio, C.L. Degen, "Force-detected nuclear magnetic resonance: Recent advances and future challenges", *Nanotechnology*, Vol.21, No.342001, pp.1–15, 2010.

[8] S. Tsuji, Y. Yoshinari, H. S. Park, and D. Shindo, "Three dimensional magnetic resonance imaging by magnetic resonance force microscopy with a sharp magnetic needle" *Journal of magnetic resonance*, Vol.178, No.2, pp. 325–328, 2006.

[9] J. Nichol, E. Hemesath, L. Lauhon, and R. Budakian, "Nanomechanical detection of nuclear magnetic resonance using a silicon nanowire oscillator", *Physical Review B*, Vol.85, No.5, pp. 054414, 2012.

[10] C.L. Degen, M. Poggio, H.J. Mamin, C.T. Rettner, D. Rugar, "Nanoscale magnetic resonance imaging", *Proceedings of the National Academy of Sciences of the United States of America*. Vol.106, pp.1313–1317, 2009.

[11] T.D. Stowe, K. Yasumura, T.W. Kenny, D. Botkin, K. Wago, D. Rugar, "Attonewton force detection using ultrathin silicon cantilevers", *Applied Physics Letters*, Vol.71, pp.288–290, 1997.

[12] S. Mouaziz G. Boero, G. Moresi, C. Degen, Q. Lin, B. Meier, J. Brugger, "Combined Al-protection and HF-vapor release process for ultrathin single crystal silicon cantilevers" *Microelectronic Engineering*, Vol.83, pp. 1306-1308, 2006.

[13] K.J. Bruland, J. Krzystek, J.L. Garbini, J.A. Sidles, "Anharmonic modulation for noise reduction in magnetic resonance force microscopy", *Review of Scientific Instruments*, Vol.66, pp.2853–2856,1995.

[14] A.A. Golriz, T. Kaule, J. Heller, M.B. Untch, P. Schattling, P. Theato, M. Toda, et al., "Redox active polymers with phenothiazine moieties for nanoscale patterning via conductive scanning force microscopy", *Nanoscale*, Vol.3, pp.5049–5058, 2011.

[15] S. Tsuji, T. Masumizu, Y. Yoshinari, "Magnetic resonance imaging of isolated single liposome by magnetic resonance force microscopy", *Journal of Magnetic Resonance*, Vol.167, pp.211–220, 2004.

CONTACT

*Yong-Jun Seo, tel: +81-22-795-5810;
E-mail: seo@nme.mech.tohoku.ac.jp

A VERTICALLY INTEGRATED NANOSCALE TIPPED MICROPROBE INTRACELLULAR ELECTRODE ARRAY

Yoshihiro Kubota[1], Hideo Oi[2], Hirohito Sawahata[1], Akihiro. Goryu[1], Yoriko Ando[2], Rika Numano[3], Makoto Ishida[1,2] and Takeshi Kawano[1]

[1]Department of Electrical and Electronic Information Engineering, Toyohashi University of Technology, Japan.
[2]Electronics-Inspired Interdisciplinary Research Institute (EIIRIS), Toyohashi University of Technology, Japan.
[3]Department of Environmental and Life Science Engineering, Toyohashi University of Technology, Japan.

ABSTRACT

Here we report integration of nanoscale tipped 120-μm-long vertical microprobe electrode (NTE) array and intracellular recordings using a gastrocnemius muscle of a mouse. The tip diameter and curvature radius of the NTE was < 200 nm, respectively, and the controlled height of the exposed tip section was 4 μm. The impedance of the fabricated NTE exhibited 3.1 MΩ at 1 kHz in saline, with the output/input signal amplitude ratio of 50%. The penetrated NTE into the muscle of a mouse detected the resting membrane potentials with the amplitude of ~ −200 mV, indicating that the NTE device detected intracellular signals from the mouse's muscle. Although we have demonstrated the intracellular recording capability using a muscle, such nanoscale electrodes with a high aspect ratio can be used for multisite intracellular recordings within numerous neuronal tissues including brain slice.

INTRODUCTION

Electrophysiological methodology is a way to understand the communication between neurons in a tissue. Especially, multi-channel intracellular recording within a tissue becomes a powerful methodology, in terms of the large amplitude (~100 mV) and the signal quality (synaptic potential measured) of individual neurons, compared to extracellular recording (< 100 μV)(Fig.1). Intracellular signals include action potentials, subthreshold excitatory- and inhibitory-postsynaptic potentials (EPSP, IPSP), and subthreshold membrane oscillations; monitoring these information will offer further understanding of the neuronal communication[1].

Nanoscale fabrication techniques have enabled the development of new class of intracellular devices, enhancing the recording performance by increasing number of the recording sites and/or improving the recording sensitivity of the electrodes [2, 3]. However these nanodevices cannot be used for the penetration deep within a thick biological tissue due to the short probe length of < 10 μm.

To realize the multi-channel intracellular recordings in a tissue, we have proposed an array of nanoscale tipped silicon-microwire with a high-aspect-ratio. The nanoitp microwire array can be fabricated by vapor-liquid-solid (VLS) growth of silicon-microwire and the nanotip formation by silicon chemical etching [2]. Here we report the integration of the multi-channel NTE array and the intracellular recording using a gastrocnemius muscle of a mouse.

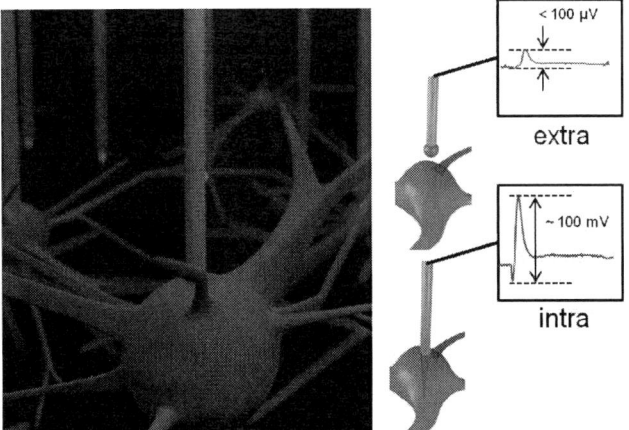

Fig. 1 Schematics for neural recordings. Intracellular recording offers large amplitude (~100 mV) and synaptic potentials, which cannot be obtained in extracellular recording (< 100 μV).

FABRICATION

The NTE arrays can be fabricated based on VLS growth of vertical silicon-microwires and the subsequent sharpening process [4, 5]. A SOI substrate, which consists of 2-μm-thick (111)-top-Si (n-type with a resistivity of < 0.02 Ω·cm)/5-μm-thick BOX/525-μm-thick (111)-Si substrate (n-type with a resistivity of < 0.02 Ω·cm), was used for the fabrication of the NTE array. An array of the top-Si islands was patented by reactive ion etching (RIE), in order for the electrical isolation between the wire-sites. After the island formation, these islands are covered with an insulating layer of SiO_2 by wet oxidation (Fig. 2a). The SiO_2 film was then exposed, where a catalytic-Au is placed by evaporation and lift-off (Fig. 2b). With the Au catalysts, vertical silicon-microwires are formed by VLS growth of silicon [7].

Silicon chemical etching of the tip section of the microwire results in nanoscale tipped silicon microwire array (Fig. 2c, d)[1]. We have demonstrated that the angle of the nanotip can be varied by changing the ratio of the etching solution (HF, HNO_3 and H_2O) and the etching time [4]. Both the wire-metallization and the device interconnection can be formed by a single step metal sputtering [Ir/Ti (120 nm) or Al-Si/Ti (170 nm)] (Fig. 2e). After the device metallization, the sidewall of the metal/silicon-wire and the device interconnection are insulated by forming Parylene (0.7 μm). The recording-site of the wire-tip and the bonding pad are exposed by O_2 plasma (Fig. 2f).

The height/area of the nanotip section exposed from the parylene-shell can be controlled by the process parameters of the O_2 plasma (Fig. 3). After spray coating of the photoresist over the wire, the photoresist at the tip section was removed by the O_2 plasma, and the underneath the Parylene was subsequently removed by the O_2 plasma. Finally, the photoresist is removed. Figure 4 shows the fabricated NTE array device. The length of the NTE was set at 120 μm, which value can be used for a thin biological samples including brain slice. intracellular potentials. The diameter and curvature radius of the nanotip exhibited < 200 nm and < 100 nm, respectively. The tip exposure process realized the parylene shelled Ir-nanotip with a height of 4 μm, which was controlled by the O_2 plasma for 27 min.

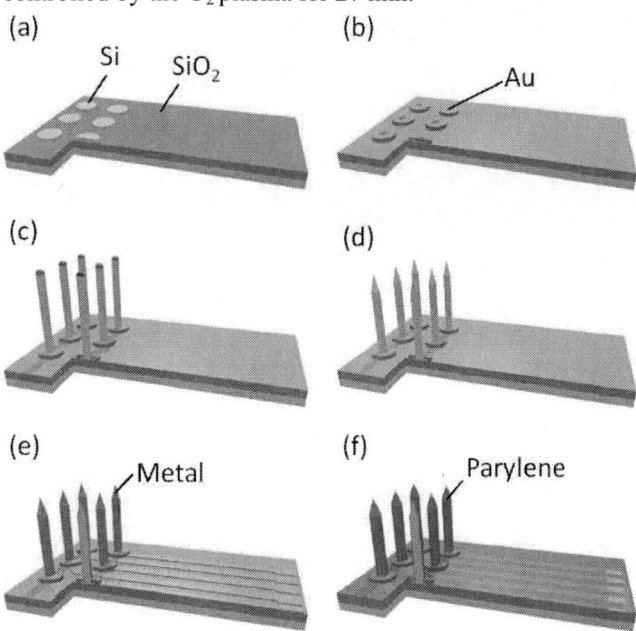

Fig. 2 Schematically illustrated fabrication process steps for the NTE array device: (a)Si-island formation, (b-c) Au-catalyzed VLS growth of silicon-microwires, (d) nanotip formation, (e) metallization, (f) parylene coating and exposing the wire-tip and bonding pads.

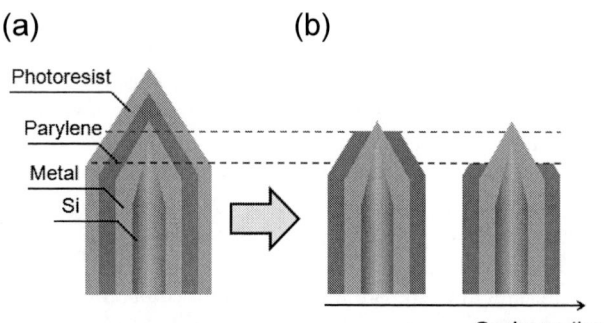

Fig. 3 Nanotip exposure: (a) Schematics of photoresist coated Parylene/metal/silicon-wire system, (b) Metal/silicon section exposed from the parylene-shell. The height of the nanotip section can be controlled by the process parameters (time) of the etching of photoresist/Parylene with O_2 plasma.

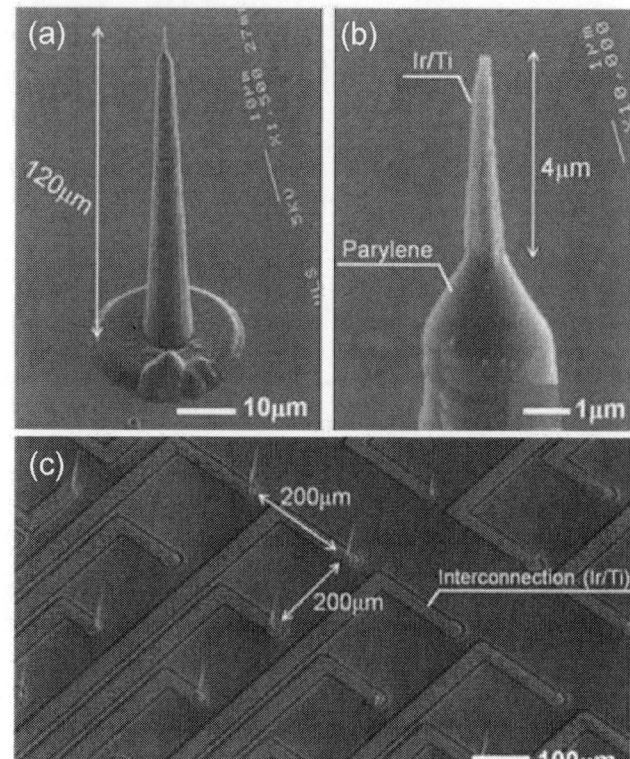

Fig. 4 SEM images of fabricated NTEs: (a) overall view of an NTE, (b) the nanotip section, and (c) an array of NTEs. Gap between NTEs is set at 200 μm.

ELECTRICAL PROPERTIES

The electrolyte/metal interfacial impedance of the electrode is an important characteristic of the intracellular electrode. The equivalent circuit of an individual NTE consists of an electrode impedance (Z_e) and a parasitic impedance associated with the interconnection (Z_l)(Fig. 5a). Z_e can be obtained by considering the total impedance of the device (Z) and Z_l. Figure 5b shows the device impedances (Z) taken from an Ir-NTE device, which was immersed in phosphate buffered saline (PBS, room temperature). Since the interconnection with Z_l=9.4 MΩ (1 kHz) was designed, Z_e was calculated to be 3.1 MΩ (1 kHz). The impedance value depends on the exposed area of the Ir-tip section (1.05 μm² from the SEM observation).

The output/input (O/I) signal amplitude ratio of the device was also measured by applying input test signals (1.6 mV$_{pp}$ sinusoidal waves at 1 kHz) to the PBS-bath. The O/I ratio during the test signal recording exhibited ~ 70% (Figs. 6a and b), which becomes 50% during intracellular recording by considering the grounded Z_l (Fig. 6c). Since the measured noise of the recording system was tens of microvolts, 50% attenuated intracellular potentials (>> noize level) can be obtained by the NTE. Note that Z_p is the cable impedance between the device and the amplifier (5.4 MΩ, measured value) and Z_a is the input impedance of the amplifier of the recording system (10^{14} Ω, RA16AC, RZ2 BioAmp Processor, Tucker-Davis Technologies).

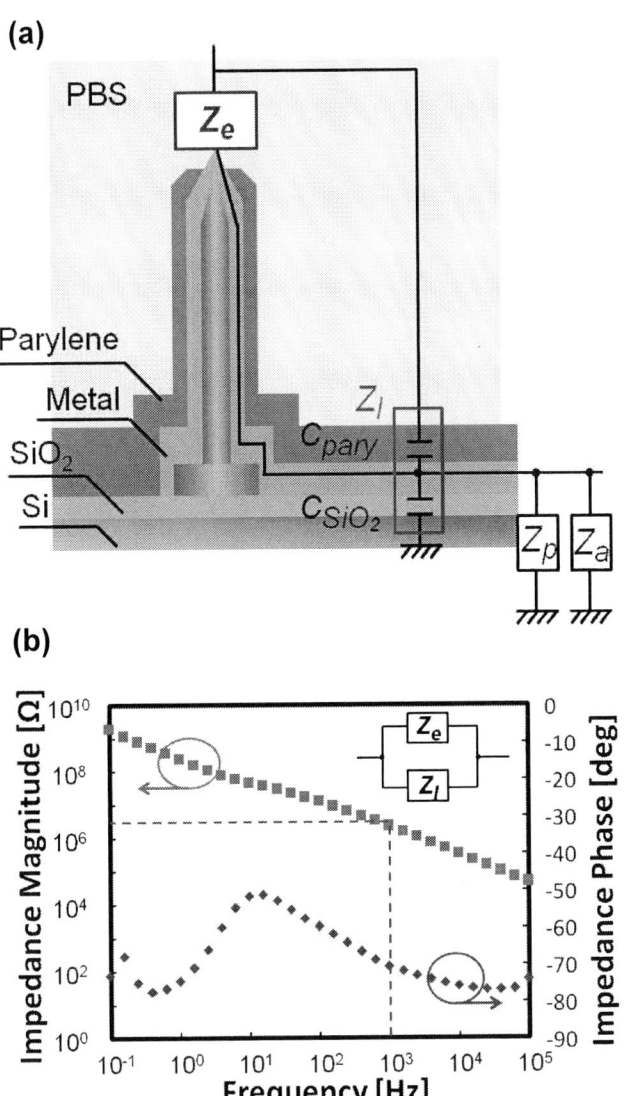

Fig. 5 Impedance measuremnts. (a) Shcematic showing the cross-sectional image of an individual NTE device in saline. The shcematic also consists of the equivalent cirucit model between the NTE and the amplifier of the recording system. (b) Magnitude and phase of the impedance of an NTE device meadued in PBS.

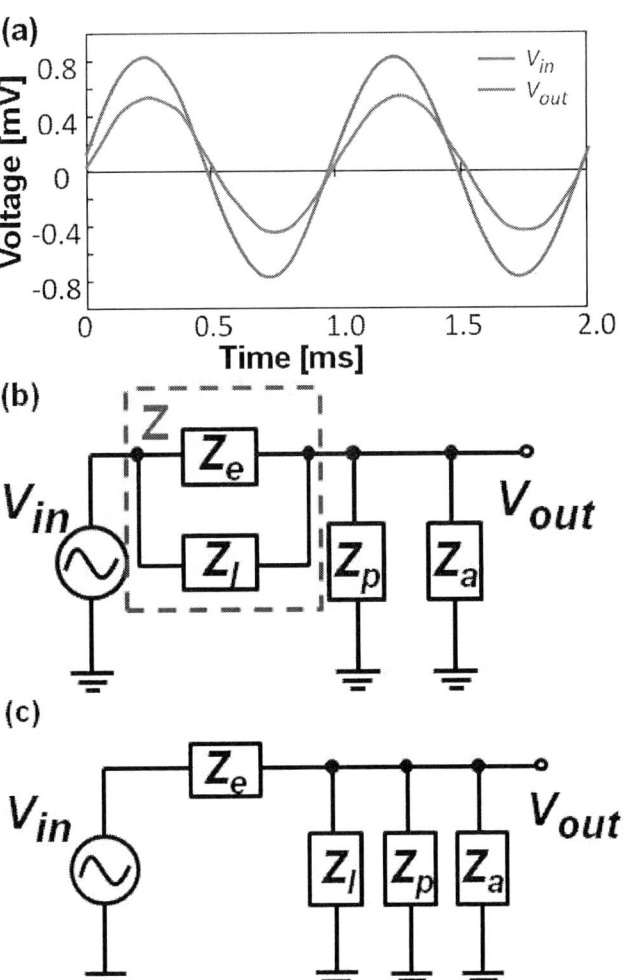

Fig. 6 O/I signal amplitude ratio. (a) O/I ratio of a NTE device taken in PBS. Red and blue lines are iput and ouput signals, respectively. (b) Equivalent circuit model showing the test signal recording (a). Impedances of NTE (Z_e) and device interconnection (Z_l) are connected in parallel as shown in Fig. 4a. (c) Model for intracellular recording with grounded Z_l.

INTRACELLULAR RECORDING

Figure 7 shows the intracellular recording using an Al-Si NTE array, where the gastrocnemius muscle of a mouse was placed in Ringer's solution. Once the NTEs were penetrated into the muscle, > 100 mV potential changes between the un-penetrated (I in Fig.8) and penetrated NTE (II in Fig.7) were observed. The results indicated that the penetrated NTE detected the resting membrane potentials (intracellular potential) of the muscle. Potential drift for the un-penetrated NTE (data for I in Fig.8) was due to the electrical characteristics of the electrolyte-electrode (Al-Si) interface. For stable recordings of intracellular signals, electrical properties of the electrolyte/Al-Si interface can be improved by using a recording metal with a low ionization such as platinum.

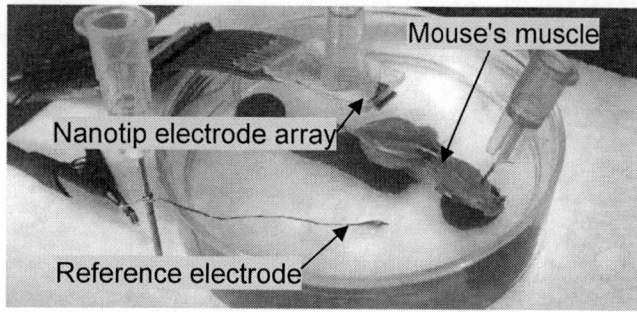

Fig. 7 Photograph of intracellular recording using a gastrocnemius muscle of a mouse.

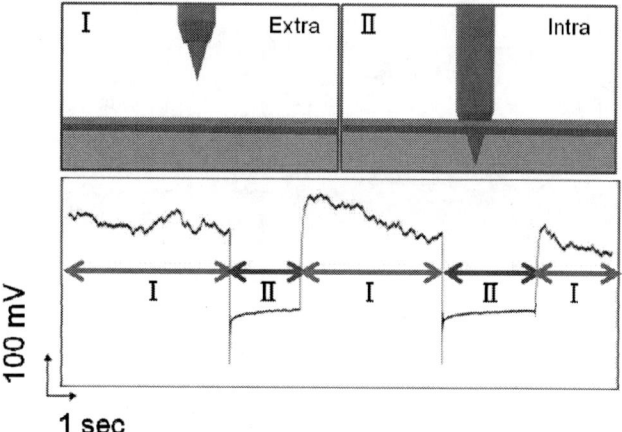

Fig. 8 Intracellular recording with a NTE. (I) un-penetrated, and (II) penetrated NTE into the muscle shown in Fig. 7.

CONCLUSIONS

We fabricated 120-μm-long vertical NTE arrays for recording intracellular potentials. With the developed integration process, we fabricated NTEs with the tip diameter of < 200 nm and the controlled tip height of 4 μm. Intracellular recordings with our NTE array was demonstrated using the gastrocnemius muscle of a mouse. The recording results indicated that the penetrated NTE detected the resting membrane potentials of the muscle.

Issues of the fabrication and properties of the NTE still remains to be done to enhance the device performance. The length of the integrated wires was 120 μm, which will be exceeded for numerous thick biological samples. The O/I ratio of the NTE device (~50%) was still attenuated due to the set of the NTE- and parasitic impedances. Improving the ratio will be important for investigating intracellular signals. However, we believe that to the NTE array device can be used for numerous biological samples including brain slice and perform multi-channel intracellular potential recordings deep within the tissue.

ACKNOWLEDGEMENTS

This work is supported by a Grant-in-Aid for Scientific Research (S), Young Scientists (B), the Global COE Program, the Strategic Research Program for Brain Sciences (SRPBS) from MEXT, and the PRESTO Program from JST.

A Goryu is a recipient of a JSPS fellowship. R Numano is supported by MEXT's 'Program to Foster Young Researchers in Cutting-Edge Interdisciplinary Research', by the TOYOAKI Scholarship Foundation and The Research Foundation for Opto-Science and Technology.

REFERENCES

[1] Micha E. Spira and Aviad Hai, "Multi-electrode array technologies for neuroscience and cardiology", *Nature Nanotechnology*, vol. 8, 83, 2013.

[2] Xiaojie Duan, *et al.*, "Intracellular recordings of action potentials by an extracellular field-effect transistor", *Nature Nanotechnology*, vol. 7, 174, 2012.

[3] Jacob T.Robinson, *et al.*, "Vertical nanowire electrode arrays as a scalable platform for intracellular interfacing to neuronal circuits", *Nature Nanotechnology*, vol. 7, 180, 2012.

[4] Akihiro Goryu, Akihito Ikedo, Makoto Ishida and Takeshi Kawano, "Nanoscale sharpening tips ofvapor–liquid–solid grown silicon microwire arrays", *Nanotechnology*, 21, 12, 125302, 2010

[5] Akihiro Goryu, Rika Numano, Makoto Ishida and Takeshi Kawano "Nanoprobe Array for Gene Transfer into Individual Cells", 2012 MRS Spring meeting, 2012.

[6] Akifumi Fujishiro, Hidekazu Kaneko, Takahiro Kawashima, Makoto Ishida and Takeshi Kawano, "A Penetrating Micro-scale Diameter Probe Array for in-vivo Neuron Spike Recordings", IEEE Micro Electro Mechanical Systems (IEEE-MEMS) Conference 2011, Cancun, Mexico, 2011.

[7] Akihito Ikedo, Takahiro Kawashima, Takeshi Kawano and Makoto Ishida, "Vertically Aligned Silicon Microwire Arrays of Various Lengths by Repeated Selective Vapor-Liquid-Solid Growth of n-type Silicon/n-type Silicon," *Applied Physics Letters*, vol. 95, 033502, 2009.

CONTACT

*Yoshihiro Kubota, kubota-y@int.ee.tut.ac.jp;
Tel: +81-532-44-6717

ENERGY HARVESTING USING UNIAXIALLY ALIGNED CARDIOMYOCYTES

Xia Liu[1,2,3], Xiaohong Wang[1], Song Li[2], Liwei Lin[3]

[1]Institute of Microelectronics, Tsinghua University, Beijing, China
[1]Tsinghua National Laboratory for Information Science and Technology, China
[2]Department of Bioengineering, University of California, Berkeley, USA
[3]Department of Mechanical Engineering, University of California, Berkeley, USA

ABSTRACT

This paper presents the concept of energy harvesting from uniaxially-aligned cardiomyocytes (CMs) on a flexible substrate for the first time. Experimentally, synchronously contracting neonatal rat ventricular cardiomyocytes (NRVCMs) at 0.5Hz have been found to cause the mechanical straining of a piezoelectric energy harvester to produce 87.5nA and 92.3mV of peak current and voltage, respectively. This work has been accomplished: (a) fabrication of a bio-hybrid energy harvester combining living cells, bio-compatible PDMS polymer substrate and piezoelectric PVDF films; (b) engineered living cell patterns on PDMS with uniaxially-aligned direction for enhanced mechanical actuation; and (c) up to one month of continuous synchronous contractions from NRVCMs for energy harvesting demonstration. This paper will detail the concept, design, fabrication, and experiments of the bio-hybrid energy harvester.

INTRODUCTION

With the combination and development of information technologies and medical sciences, more and more promising research on personal healthcare exists in implantable devices which enable for real-time in-body detection and treatment, diagnosis and therapy of major diseases, and organ transplantation and so on. Recent progress in microfabrication and bioengineering technologies are increasingly motivating the development of a variety of miniaturized implantable systems for sensing, health monitoring and deficiency treatments.

Among many diseases that may seriously impair health, some are extremely difficult to cure and only by means of implanted machines or the self-recovery mechanisms of the human body. Medical devices, some of which work in the body to help or replace the function of certain organs, are thus of great necessity. For example, a pacemaker is an electrical device that is implanted in a human body and issues regular electrical impulses to keep the heart beating. Nowadays it has been widely studied and used in clinical application [1]. Additionally, neural prosthesis [2], vascular grafts [3] and orthopedic devices [4], which are familiar to human life, are typically passive devices. The difference between passive devices and active devices is that latter ones need power supply.

Now the mainstream of power supply for implantable passive devices relies on primary or rechargeable lithium-ion batteries that may have a life of up to 5 to 7 years and after that should be replaced or recharged to continue to use [5]. In terms of applications in micro/nano devices, lithium-ion battery is hardly integrated to meet requirements of device sizes; in addition, there are a certain amount of battery safety and environmentally friendly issues. The optimized choice to feed implantable electronic devices is permanent power supplies or the batteries with a cycle life of no less than device lifetime. To achieve power supplies, many studies on the potential of view are carried out on energy conversion from chemistry or mechanical energy to electrical energy [6].

The current research focuses mainly on biofuel cells (BFCs), energy convertors and energy harvesters [7-8]. BFCs make use of in vivo biological substances (such as glucose) to achieve electrical output by electromechanical reaction [9-10]. D. Erickson et al. created a non-enzymatic glucose biofuel cell to achieve high power output, high output stability and great potential for integration [10]. But BFCs have strict requirements for the environment they exist, which causes challenges in packaging and design of fuel transportation. In the energy conversion field, Zhong Lin Wang et al. implanted a piezoelectric generator on a rat heart in 2010 for the first time and obtained steady power output [9], but such energy converter based on mechanical activity harvest vibration from the external environment. If the device works as long as needed, it is easier to make use of the in-body mechanical energy without invasive. To explore in-body usable mechanical vibrations, converting mechanical energy into electricity based on transducer materials is a good idea to power implantable electronic devices [11-12]. Kevin Kit Parker et al. conducted a fundamental study on CMs under the influence of the external environment like that single CM, cultured on various microstructures and shapes, can completely get access to the microstructural and shape constraints [13] and research on the corresponding frequencies of CMs [14]. They published their fancy work about microactuators based muscular thin film (MTF) in Science 2007. They achieved that in vitro experiments, the spatially ordered MTF can generate specific forces as high as 4 millinewtons per square millimeter, which inspires CMs practical applications in microactuators, microrobotics and microgenrators [15].

In this work we present the concept of energy harvesting from uniaxially aligned CMs on a flexible PDMS film. Based on sliding filament mechanism, CMs cultured in vitro convert biochemical energy into kinetic energy efficiently and beat themselves spontaneously. A bio-hybrid energy harvesting device combines living CMs, bio-compatible PDMS polymer substrate and piezoelectric PVDF films. The contractile force is transmitted to the piezoelectric film which is electromechanical coupling, and then the voltage and current is generated by piezoelectric effect.

DEVICE DESIGN AND FABRICATION

Working principle and device structure

Muscle cells are micro linear actuators driven by activation of actin-myosin motors, coordinated in space and time through excitation-contraction (EC) coupling. Previously, researchers have demonstrated artificial muscular mechanical actuators [14] and tissue-engineered jellyfish [16] using CMs on a flexible substrate. This work advances these results by showing the feasibility to construct an energy harvester using CMs as the power source.

The principle of the bio-hybrid energy harvester is that uniaxially aligned myocardial cell sheet can generate mechanical stress to drive the piezoelectric film beneath to bend periodically. Figure 1A illustrates the conceptual drawing of the energy harvester, mainly consisting of a PDMS cantilever structure on top of a PVDF thin film coated with gold electrodes on both sides. The uniaxially aligned CMs are grown on top of the PDMS film. The contraction and relaxation of cells in Figure 1B generate mechanical stress on PDMS film and electrical outputs are collected from the PVDF thin film (Figure 1C).

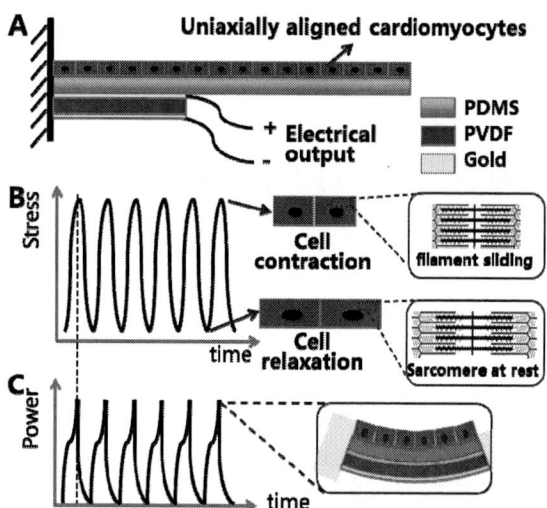

Figure 1: (A) Schematic of energy harvesting using uniaxially aligned cardiomyocytes, (B) stress of cell sheet varying periodically due to shortening of cardiomyocytes during synchronous contraction, (C) power generation collected from PVDF due to periodic mechanical strain from the contractions of cardiomyocytes.

Device fabrication

Figure 2 describes the fabrication process of the bio-hybrid device. Firstly, glass coverslips were cleaned by sonicating for 60min in 95% ethanol and nitrogen dried. Next a sacrificial film was spin coated for 60s at 6,000 RPM using a solution with 10 wt% poly(N-isopropylacrylamide) (PIPAAm) in 99.4% 1-butanol. Sylgard 184 (Dow Corning) polydimethylsiloxane (PDMS) elastomer was mixed at 10:1 base to curing agent ratio and spin coated on top of the

PIPAAm coated glass coverslip (step 2A). The PDMS film thickness is 28 um. A piece of PDMS film with the same shape and area as a metallized PVDF film (thick: 28um) was cut and replaced by the PVDF film (step 2B). Part of PDMS film was peeled off from step 2A and bonded on top of the device structure from step 2B (step 2C).

The next step is the key to the uniaxially aligned cell patterns. Microcontact printing (µCP) process is utilized to control the 2D cell patterns with alternating high- and low-density fibronectin (FN) lines for orientating cell bodies/sarcomeres and interconnecting adjacent cells, respectively. PDMS stamps were fabricated with 20um wide, 20um tall ridges separated by 20um spacing. Prior to use, the stamps were sonicated in 50% ethanol for 30min to sterilize and remove surface contaminants. Once dried, the stamps were inked with a droplet of 50ug/mL FN in DI water and incubated for 1 hour and then rinsed twice in DI water to remove excess protein and dried. High density FN lines were transferred from the stamps to the PDMS film by making conformal contact for 1min (step 2D). In order to create high and low density FN lines alternating, a droplet of 2.5ug/mL FN in DI water was spread over the patterned area and incubated on the PDMS surface for 15min (step 2E). Following the incubation period, the PDMS film was washed three times with DI water, air dried (step 2F). CMs were isolated from neonatal rats and seeded on the FN functionalized PDMS film with the concentration of 1 million per ml and cultured in a 37℃ & 5% CO2 incubator for 3-4 days (step 2G). Finally, desired shape was cut out and peeled off PDMS film with bonded piezoelectric film (step H). Then the bio-hybrid film was fixed in a culture dish with DMEM media and connected with measuring setup.

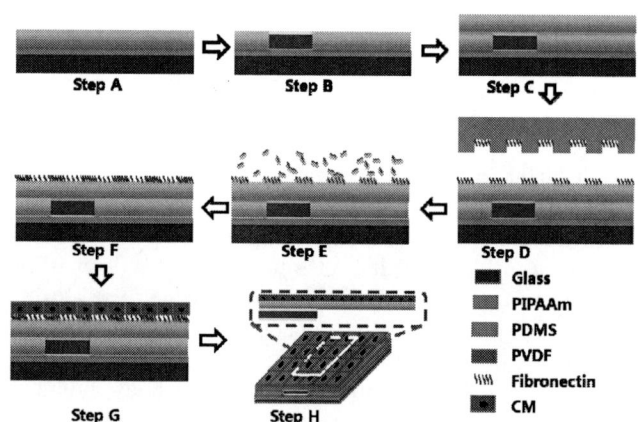

Figure 2: Bio-hybrid device design and fabrication: (step A-C) assemble piezoelectric film and PDMS film; (step D-F): fibronectin patterns using microcontact printing; (step G) isolate and seed neonatal rat ventricular cardiomyocytes; (step H) cut desired shape.

CELL ISOLATION AND CULTURE

NRVCMs are isolated from two-day-old neonatal Sprague-Dawley rats based on published protocols [17]. All procedures are conducted in accordance with the guidelines of the Institutional Animal Care and Use Committee at University of California, Berkeley. Ventricles were surgically isolated and homogenized by washing in Hank's balanced salt solution (HBSS) followed by digestion with 0.1% trypsin overnight at 4 ℃. After the supernatant discarded, the second digestion was started by adding digest enzyme (0.1% (v/v) collagenase type II in HBSS) at 37 ℃ water bath with stir bar for 5-8min and then was repeated for around eight times until the ventricle tissues became smaller. Subsequently, cells were re-suspended in DMEM culture medium supplemented with 10% (v/v) heat-inactivated fetal bovine serum (FBS), and 1% (v/v) Penicillin/Streptomycin. The isolated CMs were loaded on the aforementioned devices and incubated under standard conditions at 37 ℃ and 5% CO_2. After 24 hours incubation the devices were washed three times with HBSS to remove non-adherent cells and then covered with media. Subsequently, every other day media was changed with maintenance media.

Images in Figure 3 show the microstructures of cell sheets on PDMS substrates without and with the alternating high- and low-density FN line patterns. In contrast to Figure 3A, the 3rd-day cell image in Figure 3B shows grown and uniaxially-aligned cardiomyocytes. In a recorded video clip, uniaxial contraction of a PDMS film at a frequency of 1 Hz has been observed, which validates the successful construction of confluent and interconnected cell lines.

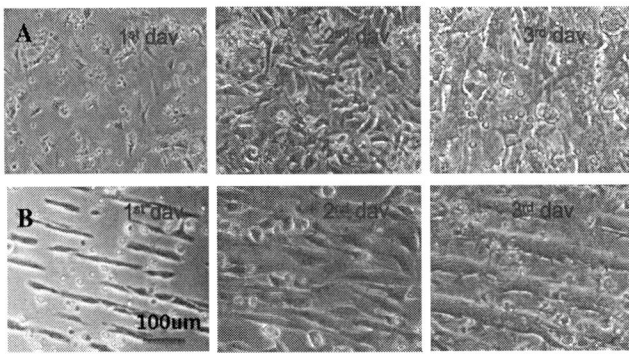

Figure 3: Microstructures of 2D cardiomyocyte sheets on FN functionalized PDMS thin film: (A) FN coatings without patterns and (B) FN coatings with alternating high- and low-density line patterns.

EXPERIMENTAL RESULTS

The fabricated bio-hybrid film is tested under an optical microscope in a culture dish with DMEM media for deformation and bending measurements as illustrated in Figure 4. In the prototyped device of 6mm in length, the bio-hybrid film is found to deform hundreds of micrometers at the tip of the device during one contraction operation of a second in Figure 4A. Figure 4C recorded the specific magnitudes of deformation and bending angle from Figure 4A.

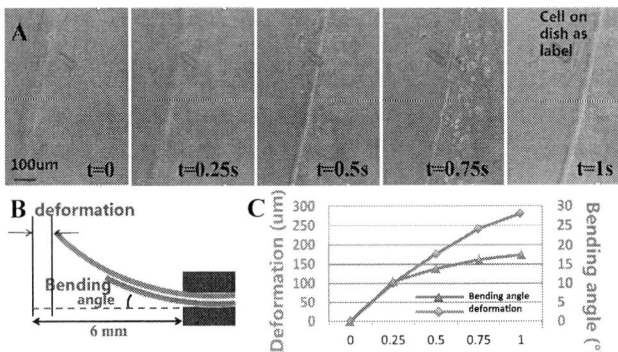

Figure 4: Deformation of bio-hybrid film: (A) aerial view of film bending during half period, (B) schematic of bio-hybrid film bending experiment, (C) analysis of deformation and bending angle.

The real-time periodic deformation and energy generation of the harvester is recorded in Figure 5. In this case, the beating frequency is about 0.5Hz and the resulting signals are much larger than the noise levels as shown in the figure. In our prototype tests, the peak output voltage is measured at 92.3mV, the peak current at 87.5nA, and the calculated power output is $14.6\mu W/cm^3$.

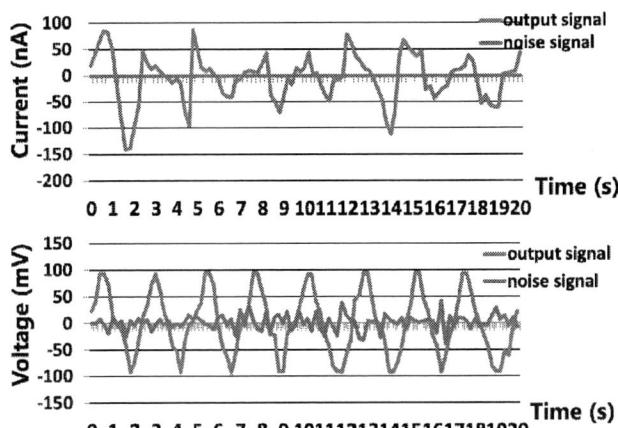

Figure 5: Experimentally measured output current and voltage with frequency of ~0.5Hz.

CONCLUSION

We succeed in electrical energy harvesting using bio energy of uniaxially aligned heart muscle cells for the first time. A bio-hybrid energy harvester combines living cardiomyocytes, bio-compatible PDMS polymer substrate and piezoelectric PVDF films, and was fabricated by patterning living cells on PDMS and PVDF films with uniaxially-aligned direction for enhanced mechanical actuation. Continuously synchronous contraction of NRVCMs can last stable for up to one month for energy harvesting applications. The peak output voltage was measured at 92.3mV, the peak current at 87.5nA and the calculated power output is $14.6\mu W/cm^3$.

These results show the energy conversion happens from CMs movements to electrical energy for applications in power supply. In the future, the prototyped energy harvester may be used to power miniaturized pacemakers or implanted micro/nano passive devices. The demonstrated device will probably be a stress sensor that the electrical power output can reflect the mechanical performance of CMs.

ACKNOWLEDGEMENTS

This work is support by grants from the National Natural Science Foundation of China (No. 60936003), National Basic Research Program (973 Program, No. 2009CB320304), and a fellowship from China Scholarship Council (CSC). The author would like to thank J. Chu at CTE lab of UC Berkeley, X. Qiu, C. W. Huang, D. Wang, W. Huang, J. Sia, R. D. Sochol and Y. Liu for their assistance and fruitful discussions.

REFERENCES

[1] C. L. Schmidt and P. M. Skarstad, "the Future of Lithium and Lithium-Ion Batteries in Implantable Medical Devices", *Journal of Power Sources*, vol. 97–98, pp.742–746, 2001.

[2] L. R. Hochberg, M. D. Serruya, G. M. Friehs, "Neuronal Ensemble Control of Prosthetic Devices by a Human with Tetraplegia", *Nature*, vol. 442, pp. 164-171, 2006.

[3] R. A. Roeder, G. C. Lantz, L. A. Geddes, "Mechanical Remodeling of Small-Iintestine Submucosa Small-Diameter Vascular Grafts: a Preliminary Report", *Biomedical Instrumentation and Technology*, vol. 35, pp.110-120, 2001.

[4] Q. Li, V. Naing, J. M. Donelan, "Development of a Biomechanical Energy Harvester", *Journal of Neuroengineering and Rehabilitation*, vol. 6, pp. 22-23, 2009.

[5] S. M. Kurtz et al., "Implantation Trends and Patient Profiles for Pacemakers and Implantable Cardioverter Defibrillators in the United States: 1993–2006", *Pacing and Clinical Electrophysiology*, vol. 33, pp. 705–711, 2010.

[6] U. Schröder, "from in Vitro to in Vivo—Biofuel Cells are Maturing", *Angewandte Chemie International Edition*, vol. 51, pp. 7370–7372, 2012.

[7] Z. Li, G. Zhu, R. Yang, A. C. Wang, Z. L. Wang, "Muscle-Driven in Vivo Nanogenerator", *Advanced Materials*, vol. 22, pp. 2534–2537, 2010.

[8] M. Falk, C. Villarrubia, "Biofuel Cells for Biomedical Applications, Colonizing the Animal Kingdom", *Chem. Phys. Chem.*", vol. 14, pp. 2045 – 2058, 2013.

[9] V. Oncescu, D. Erickson, "High Volumetric Power Density, Non-Enzymatic, Glucose Fuel Cells", *Scientific Reports*, vol. 3, 2013.

[10] Y. Qi, M. C. McAlpine, "Nanotechnology-Enabled Flexible and Biocompatible Energy Harvesting", *Energy & Environmental Science*, vol. 3, pp. 1275–1285, 2010.

[11] A. Khaligh, Z. Peng, Z. Cong, "Kinetic Energy Harvesting Using Piezoelectric and Electromagnetic Technologies—2014: State of the Art Industrial Electronics", *IEEE Transactions*, vol. 57, pp. 850–860, 2010.

[12] K. K. Parker, D. E. Ingber, "Extracellular Matrix, Mechanotransduction and Structural Hierarchies in Heart Tissue", *Philosophical Transactions: Biological Sciences*, vol. 362, pp. 1267-1279, 2007.

[13] M. A. Bray, S. P. Sheehy, K. K. Parker, "Sarcomere Alignment is Regulated by Myocyte Shape", *Cell Motility and the Cytoskeleton*, vol. 65, pp. 641–651, 2008.

[14] A. W. Feinberg et al., "Muscular Thin Films for Building Actuators and Powering Devices", *Science*, vol. 317, pp. 1366-1370, 2007.

[15] S. Cosnier, A. L. Goff, M. Holzinger, "Towards Glucose Biofuel Cells Implanted in Human Body for Powering Artificial Organs: Review", *electrochemistry communications*, DOI: 10.1016/j.elecom.2013.09.021, 2013.

[16] J. Nawroth et al., "A tissue-engineered jellyfish with biomimetic propulsion", *Nature Biotechnology*, vol. 30, pp. 792-797, 2012.

[17] W. J. Adams, T. Pong, N. A. Geisse, S. P. Sheehy, B. Diop-Frimpong, K. K. Parker, "Engineering design of a cardiac myocyte", *J. Comput. Aided. Mat. Des.*, vol. 14, pp. 19–29, 2007.

CONTACT

*X. Liu, tel: +86-10-62793062; bobo20074447@126.com

DIFFUSION REFUELING BIOFUEL CELL MOUNTABLE ON INSECT

Kan Shoji[1], Yoshitake Akiyama[1], Masato Suzuki[2],
Nobuhumi Nakamura[2], Hiroyuki Ohno[2] and Keisuke Morishima[1]
[1]Osaka University, Osaka, JAPAN
[2]Tokyo University of Agriculture and Technology, Tokyo, JAPAN

ABSTRACT

This paper reports an insect mountable biofuel cell (imBFC) which generates electric power from trehalose found in insect hemolymph and refuel trehalose from insect hemolymph by diffusion automatically. First, we designed and fabricated the imBFC consisted of a connector, two chambers, a dialysis membrane and electrodes. Then, we evaluated the power density and life-time of the imBFC. The maximum power of 50.2 μW (42.9 μW/cm^2) was obtained. The power output of the imBFC was maintained at more than 10 μW for 3 h. Finally, we connected a light-emitting diode (LED) device to the imBFC and succeeded to driving the LED device. The results indicate that the imBFC is a promising micro battery to power environmental monitoring micro-tools.

INTRODUCTION

Insects are extremely successful animals, living almost everywhere on the earth [1]. Insect cells have been used for recombinant protein production [2] and high performance bio-actuators [3] [4]. Furthermore, high performing micro-robots have been developed by mimicking functions and structures of insect [5]. In particular, insect cyborgs which are robots controlled by electrical stimulation of their muscles and neurons would be desirable exploration robots in disasters, for environmental monitoring. For example, Takeuchi et al. [6] and Hozer and Shimoyama [7] analyzed the locomotion reaction of an insect and controlled its walking by electrical stimulation of its neurons. Kanzaki et al. [8] utilized moth antennas as a sensitive sensor for their robot. Sato et al. [9] and Bozkurt et al. [10] succeeded in controlling the flight of insects by electrical stimulation to their muscles.

Coin batteries used currently for insect cyborgs not only interfere with the insect's movement owing to their size and weight but also have to be recharged or replaced to drive an integrated circuit for a long time. It is desirable to have a battery which is small and light enough to mount it on insects and that has a semi-permanent lifetime.

Recently, implantable biofuel cells (BFCs) which generate electric power from sugar found in body fluid have been attracted attention as a new power source. For example, Cinquin et al. [11] succeeded in generating an electric power by implanting a glucose BFC covered by dialysis membranes into a rat. Rasmussen et al. [12] developed a trehalose BFC immobilized trehalase and glucose oxidase and implanted the BFC into an insect. Szczupak et al. [13] succeeded in driving a motor by connecting three BFCs implanted in a clam in series. Halámkova et al. [14] implanted a glucose BFC into a snail and confirmed the possibility of refueling by feeding. Our group also developed a BFC using trehalose found in

insect hemolymph and succeeded in driving a melody IC and a piezo speaker by connecting five BFCs in series [15]. However, because of the size limitation for implanting into living organisms, the power outputs of the implanted BFCs are from several dozen nW to several μW and the outputs are not enough to drive general electric devices

Therefore, in this study, we demonstrate an insect-mountable BFC (imBFC) refueling trehalose from insect body by diffusion so as to generate electric power semi-permanently (Figure 1). Trehalose diffuses from insect hemolymph to a chamber of the imBFC. Then, the trehalose is decomposed to glucose by trehalase and mutarotase included in the chamber. The glucose is oxidized by glucose oxidase immobilized on an anode and then electrons are produced. On a cathode, oxygen is reduced by bilirubin oxidase immobilized on the electrode and electrochemical conversion of proton and oxygen into water occurs. As a result, electric power is generated.

In this paper, we fabricate the imBFC by a rapid prototyping and compare the power output and the stability of the imBFC with a glucose BFC. Then, we demonstrate the ability of the imBFC to drive a light-emitting diode (LED) device.

EXPERIMENTAL
Preparation of Electrodes

Ketjen black (KB) electrodes were prepared by casting a mixture (4:1, v/v) of 5 mg/mL KB in N methyl-2-pyrrolidone

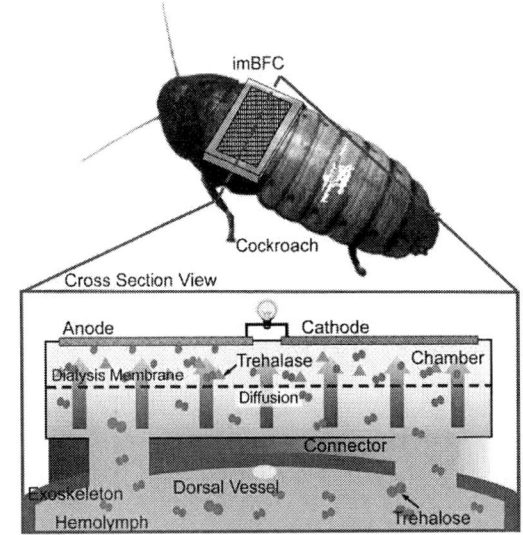

Figure 1: Schematic image of the imBFC. Trehalose diffuses from insect body to the chamber. Then, the trehalose is decomposed to glucose and oxidized enzymatically.

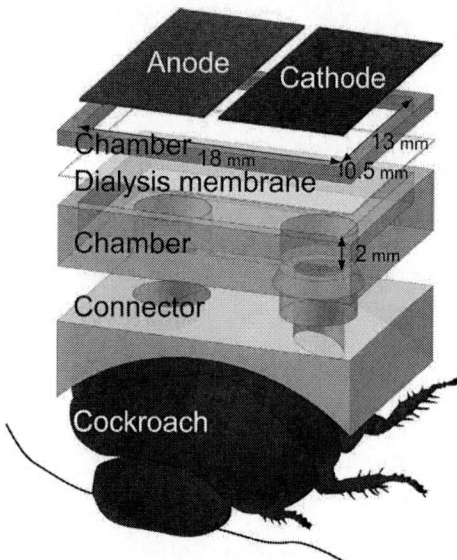

Figure 2: Image of the imBFC consisting of a connector, two chambers, a dialysis membrane and electrodes.

(NMP) and 10 wt. % poly (vinylidene fluoride) in NMP on a carbon paper. The anode was made by successively dropping 10 mM ferrocene in acetone and 50 mg/mL glucose oxidase (GOD) solution onto a KB electrode. The cathode was made by successively dropping 100 mM 2,2'-Azinobis (3-ethylbenzothiazoline-6-sulfonic Acid) solution and 50 mg/mL bilirubin oxidase solution onto a KB electrode. The electrodes were dried at room temperature after each solution was dropped. Then, in order to immobilize enzymes and mediators to electrode by forming the insoluble PLL-PSS complex film (poly ion complex), poly L-lysine (50 mM monomer unit) and poly sodium 4-styrenesulfonate (60 mM monomer unit) were casted on the electrodes.

Fabrication of imBFC

The imBFC consisted of a connector, two chambers (top: length, 18 mm; width, 13 mm; depth, 0.5 mm, bottom: length, 18 mm; width, 13 mm; depth, 2 mm) separated by dialysis membrane (molecular weight cut off (MWCO): 500-1000) and electrodes (length, 13 mm; width, 9 mm; surface area, 117 mm^2) (Figure 2). In order to protect the electrodes from adsorption of macromolecules in insect hemolymph, the chambers were separated by the dialysis membrane. The connector, fabricated by a rapid prototyping (RVS-S1, Real Vision Systems Inc.), was glued onto the exoskeleton by an epoxy adhesive (Araldite, Huntsman) (Figure 3(a)). Then, two holes 3 mm in diameter were made through the exoskeleton. The chambers were filled with cockroach saline (pH 6.5) [16] and fixed to the connector by the epoxy adhesive. The anode and the cathode were attached to the top chamber twelve hours after trehalase and mutarotase were added into the top chamber (Figure 3 (b)).

Electrochemical Measurements of imBFCs

We confirmed the electric power of the glucose BFC

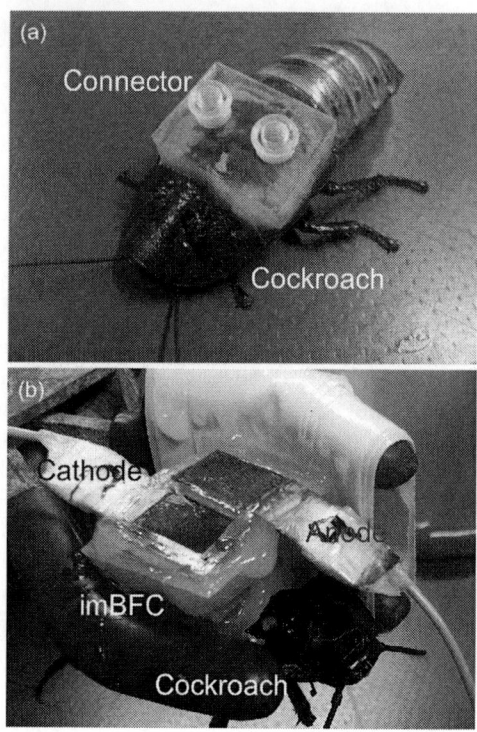

Figure 3: Photographs of Madagascar Cockroach with (a) the connector and (b) the imBFC.

with 100 mM glucose solution (0.1 M phosphate potassium buffer, pH 6.5) and the imBFC. The electric powers were measured by an electrochemical measurement system (SI 1280b, Solartron). The glucose BFC was fabricated by inserting the electrodes which were identical with the electrodes of the imBFC to the glucose solution.

The stability of the imBFC and the glucose BFC was evaluated by applying potentials of 0.4 V and 0.25 V for 5 h, respectively.

Flashing LED Experiment

A LED device consisted of a charge pump IC (input voltage, 0.3-3 V; output voltage, 1.8 V), a 10 μF electrolytic capacitor and a red LED. The output voltage of the imBFC was boosted to 1.8 V by the charge pump IC. Then, the electrical charge was stored to the capacitor. When the capacitor was charged full, electricity in the capacitor was discharged and the LED was flashing.

RESULT AND DISCUSSION
Power Generation by imBFC

Maximum power of 151 μW (129 μW/cm^2) and 50.2 μW (42.9 μW/cm^2) were obtained from the glucose BFC and the imBFC, respectively (Figure 4). Open-circuit voltages of the glucose BFC and the imBFC were 563 mV and 701 mV. Maximum currents of the glucose BFC and the imBFC were 791 μA (676μA/cm^2) and 199 μA (170 μA/cm^2). The power and the current of the imBFC was one third and quarter as low as the glucose BFC, while the open-circuit voltage of the imBFC was about 140 mV higher than the glucose BFC.

978-1-4799-3510-9/14 $31.00 © 2014 IEEE

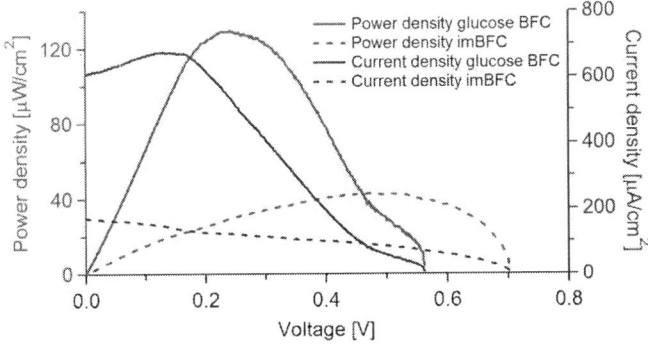

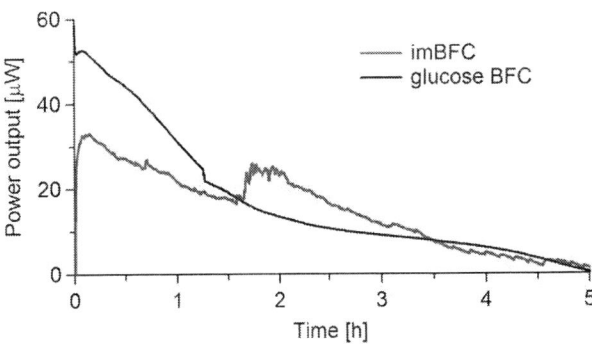

Figure 4: Performances of the imBFC and the glucose BFC. The perpendicular axes on the left and the right show power density and current density, respectively.

Figure 5: Time course power outputs of the glucose BFC (black line) and the imBFC (red line). The power outputs both of the imBFC and glucose BFC dropped to almost zero for 5 h.

The current of BFCs is determined by the oxidation current of sugar and the reduction current of oxygen by enzymes immobilized on the electrodes. In the imBFC, the top chamber and the bottom chamber were separated by the dialysis membrane and trehalose was diffused from the bottom chamber to the top chamber. Hence, the glucose concentration in the top chamber was lower than insect hemolymph. As a result, the oxidation current of the anode decreased and the current of the imBFC became lower than the glucose BFC. In addition, the anode of the imBFC has three-step enzymatic reactions, whereas the anode of the glucose BFC has just one-step reaction. Due to the energy losses in every enzymatic reaction, the power output of the imBFC decreased.

On the other hand, the open-circuit voltage is determined by the difference of redox potentials between the anode and cathode. The MWCO of the dialysis membrane used in the present study was 500–1000. In the imBFC, while macromolecules found in insect hemolymph were cut off by the dialysis membrane, many ions such as Na^+, K^+, Mg^{2+}, Ca^{2+}, Cl^- and organic acid ions found in insect hemolymph passed through the dialysis membrane. These ions might cause electrochemical reactions on the electrodes and the open-circuit voltage increased.

Rasmussen et al. also reported power generation from insect hemolymph by implanting a trehalose BFC [11]. The power output of the BFC was just 0.07 µW because the electrodes size implanted into insect was as small as 0.04-0.05 cm in diameter. In this paper, the BFC was not implanted in insect but mounted on insect. Thus, the larger electrodes were used in the imBFC and the power output of 50.2 µW was obtained.

Lifetime of imBFCs

The time course power outputs of the glucose BFC and the imBFC are shown in Figure 5. The initial power output of the imBFC was lower than that of the glucose BFC. On the other hand, the decreasing rate of the imBFC was lower than that of the glucose BFC. The power outputs of both the imBFC and glucose BFC maintained at 10 µW for 3 h and decreased to 0 µW within 5 h.

The power degradation of BFCs is caused by deactivation of enzymes immobilized onto electrodes. In the imBFC, the electrodes were protected by the dialysis membrane in order to prevent adsorption of macromolecules in insect hemolymph. As a result, the adsorption of macromolecules was prevented and the equivalent lifetime of the BFCs was obtained. The power output of the insect BFC reported by Rasnussen et al. decreased to 6.5 % after 2 h [11]. Our imBFC with the dialysis membrane is able to prevent the adsorption of macromolecules and power reduction for a long time.

We reported the gold nanoparticle-based BFC fabricated by sputtering gold onto the electrode and succeeded in

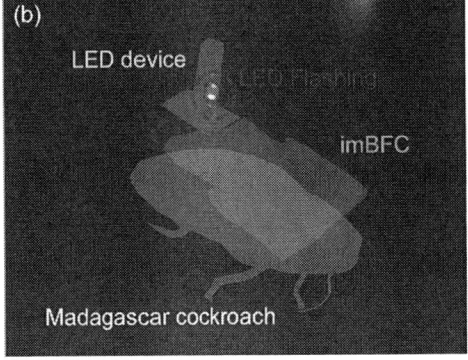

Figure 6: Photographs of (a) the LED device connected to the imBFC and (b) the flashing LED.

extending the lifetime compared with the enzymatically BFC [17]. By applying the gold nanoparticle-based BFC to the imBFC, the lifetime of the imBFC will be extended.

Driving LED Device with imBFC

The LED device connected to the imBFC was shown in Figure 6 (a). The LED flashed about every 5 s by the imBFC (Figure 6 (b)). The open-circuit voltage of the imBFC was about 0.7 V and the driving voltage of the LED was 1.5 V. The output voltage of the imBFC was boosted to 1.8 V by the charge pump IC, which enables us to flash the LED. In order to apply the imBFC to the micro battery for insect cyborgs, it needs to drive higher energy consumption electric devices such as a radio sensor device and a microcomputer that the driving voltages are around 3 V. Therefore, we will attempt to boost the output voltage of the imBFC to around 3 V by a two-stage boost converter and drive these electric devices.

CONCULUTION

In this paper, we proposed the imBFC refueling trehalose from insect body to the chamber by diffusion and fabricated the imBFC by using the rapid prototyping. Then, we generated the electric power using the imBFC and a maximum power of 50.2 μW (42.9 μW/cm^2) was obtained. The power output of the imBFC was maintained at more than 10 μW for 3 h. Furthermore, we demonstrated to drive the LED device consisted of the charge pump IC, the 10 μF electrolytic capacitor and the red LED powered by the imBFC and the LED flashed about every 5 s.

As a result, the imBFC had enough potential to be applied for a semi-permanent power source of novel ubiquitous robots such as insect cyborgs.

ACKNOWLEDGEMENTS

This work was partly supported by the Industrial Research Program of NEDO, Grants-in-Aid for Scientific Research from the Ministry of Education, Culture, Sports, Science and Technology in Japan (Nos. 20034017, 21676002, 21225007, 21111503, 23111705) and TEPCO Memorial Foundation.

REFERENCES

[1] P. Gullan, P. Cranston, *The Insects*, 3rd edition, Blackwell, 2005.

[2] B. Maiorella, D. Inlow, A. Shauger, D. Harano, "Large-Scale Insect Cell-Culture for Recombinant Protein Production", *Nat. Biotech.*, Vol. 6, pp. 1406-1410, 1988.

[3] Y. Akiyama, K. Iwabuchi, Y. Furukawa, K. Morishima, "Long-term and room temperature operable bioactuator powered by insect dorsal vessel tissue," *Lab Chip*, vol. 9, no. 1, pp. 140–144, 2009.

[4] Y. Akiyama, T. Sakuma, K. Funakoshi, T. Hoshino, K. Iwabuchi, K. Morishima, "Atmospheric-operable bioactuator powered by insect muscle packaged with medium," *Lab Chip*, vol. 13, no. 24, pp. 4870–4880, 2013.

[5] K. Y. Ma, P. Chirarattananon, S. B. Fuller R. J. Wood, "Controlled Flight of a Biologically Inspired, Insect-Scale Robot," *Science*, vol. 340, no. 6132, pp. 603–607, 2013.

[6] S. Takauchi, J. Ishihara, I. Shimoyama, H. Miura, "Hibrid Insect Robot," *Proc. The 13st Annual Conference of The Robotics Society of JAPAN,* vol. 13, no. 3, Kawasaki, 1995, pp. 1161-1162

[7] R. Holzer, I. Shimoyama, "Locomotion control of a bio-robotic system via electric stimulation," *Proc. IROS 1997*, Grenoble, 1997, pp. 1514-1519.

[8] Y. Kuwana, S. Nagasawa, I. Shimoyama, R. Kanzaki, "Synthesis of the pheromone-oriented behaviour of silkworm moths by a mobile robot with moth antennae as pheromone sensors," *Biosens Bioelectron,* vol. 14, no. 2, pp. 195–202, 1999.

[9] H. Sato, C. W. Berry, Y. Peeri, E. Baghoomian, B. E. Casey, G. Lavella, J. M. VandenBrooks, J. F. Harrison, M. M. Maharbiz, "Remote radio control of insect flight," *Front. Integr. Neurosci.*, vol. 3, pp. 1-11, 2009.

[10] A. Bozkurt, R. F. Gilmour, A. Lal, "Balloon-Assisted Flight of Radio-Controlled Insect Biobots," *IEEE Trans. Biomed. Eng.*, vol. 56, pp. 2304-2307, 2009.

[11] P. Cinquin, C. Gondran, F. Giroud, S. Mazabrard, A. Pellissier, F. Boucher, J.-P. Alcaraz, K. Gorgy, F. Lenouvel, S. Mathé, P. Porcu, S. Cosnier, "A Glucose BioFuel Cell Implanted in Rats," *PLoS ONE*, vol. 5, no. 5, p. e10476, 2010.

[12] M. Rasmussen, R. E. Ritzmann, I. Lee, A. J. Pollack, D. Scherson, "An Implantable Biofuel Cell for a Live Insect," *J. Am. Chem. Soc.*, vol. 134, no. 3, pp. 1458–1460, 2012.

[13] A. Szczupak, J. Halamek, L. Halámková, V. Bocharova, L. Alfonta, E. Katz, "Living Battery – Biofuel Cells Operating In Vivo in Clams," *Energy Environ. Sci.*, 2012.

[14] L. Halámková, J. Halámek, V. Bocharova, A. Szczupak, L. Alfonta, E. Katz, "Implanted Biofuel Cell Operating in a Living Snail," *J. Am. Chem. Soc.*, vol. 134, no. 11, pp. 5040–5043, 2012.

[15] K. Shoji, Y. Akiyama, M. Suzuki, T. Hoshino, N. Nakamura, H. Ohno, K. Morishima, "Insect biofuel cells using trehalose included in insect hemolymph leading to an insect-mountable biofuel cell," *Biomed. Microdevices*, vol. 14, no. 6, pp. 1063–1068, 2012.

[16] W. J. Bell, *The Laboratory Cockroach,* Chapman & Hall, 1982.

[17] K. Shoji, Y. Akiyama, M. Suzuki, N. Nakamura, H. Ohno, K. Morishima, "Gold nanoparticle-based biofuel cell using insect body fluid circulation," *Proc. Transducers 2013,* Barcelona, June, 2013, pp. 2811–2814.

CONTACT

*K. Morishima,
E-mail: morishima@mech.eng.osaka-u.ac.jp

ALD RUTHENIUM OXIDE-CARBON NANOTUBE ELECTRODES FOR SUPERCAPACITOR APPLICATIONS

Roseanne Warren[1], Firas Sammoura[1,2], Alina Kozinda[1], and Liwei Lin[1]
[1]University of California, Berkeley, USA
[2]Masdar Institute of Science and Technology, Abu Dhabi, UAE

ABSTRACT

This work presents the first demonstration of atomic layer deposition (ALD) ruthenium oxide (RuO_2) and its conformal coating onto vertically aligned carbon nanotube (CNT) forests as supercapacitor electrodes. Specific accomplishments include: (1) successful demonstration of ALD RuO_2 deposition, (2) uniform coating of RuO_2 on a vertically aligned CNT forest, and (3) an ultra-high specific capacitance of 100 mF/cm^2 from prototype electrodes with a scan rate of 100 mV/s. Advantages of the ALD method include precise control of the RuO_2 layer thickness and composition without the use of CNT-binder molecules. In addition to high capacitance, preliminary results indicate that the ALD RuO_2-CNTs have good stability over repeated cycling. Besides its use in supercapacitors, ALD-RuO_2 has potential NEMS applications: in biosensors and pH sensing [1], as a strong oxidative material in multiple chemical processes [2], and in catalytic reactions for photocatalytic systems [3].

INTRODUCTION

Supercapacitors are electrochemical energy-storage devices that store charge by: 1) reversible adsorption of ions onto high-surface area, porous materials (known as "electric double layer capacitors"), or 2) reversible surface reduction-oxidation (redox) reactions (known as "pseudo-capacitors"). With their high power density and long cycle stability, supercapacitors are well-suited to complement or replace batteries in a wide range of applications, including transportation, renewable energy, and portable electronics [4]. High-performance supercapacitors are characterized by: high specific capacitance, good stability over repeated cycling, and low series resistance.

RuO_2 is considered one of the best pseudo-capacitive materials in terms of charge storage ability and fast, reversible reaction kinetics [4]. Because of its high electrical conductivity (comparable to that of metals), RuO_2 supercapacitors can have low resistive energy losses [5]. In addition, RuO_2 has three available oxidation states as it cycles between its oxide and hydroxide states, enabling RuO_2 electrodes to store large quantities of charge [4]. CNTs are a good supporting electrode material for active RuO_2 pseudo-capacitive layers because CNTs have: 1) high surface area and porosity, 2) good electrical conductivity, and 3) excellent mechanical flexibility and strength [6].

While there have been numerous studies of RuO_2 - CNT composite materials for supercapacitor applications, there is a need for improved synthesis methods to maximize device performance. Current methods for depositing RuO_2 on CNTs include sputtering, electroplating, chemical vapor deposition (CVD), and pyrolysis. These methods are often time-consuming, can yield incomplete surface coverage, and result in poor control over critical properties such as chemical composition and thickness [7, 8]. Due to the high cost of RuO_2, supercapacitor devices should use only the minimum amount of RuO_2. Since pseudo-capacitors store charge by surface redox reactions, the RuO_2 layer in RuO_2-CNT composites can be just a few nanometers in thickness. In this work, we have found that the ALD method is well-suited for creating high-performance RuO_2-CNT supercapacitor electrodes with excellent conformality and precise control over film thickness and chemical composition.

ALD is a form of CVD in which reactant precursors are pulsed sequentially, enabling films to be deposited on the substrate one monolayer at a time. The ALD process enables angstrom-level control over film thickness, and can achieve uniform film coverage even when used with high aspect-ratio geometries. While the ALD process for Ru is well-established, the ALD process for RuO_2 is still poorly understood. In addition to demonstrating a high-performance RuO_2-CNT supercapacitor, the results of this study have potential wide-ranging applicability by furthering our understanding of the ALD RuO_2 process.

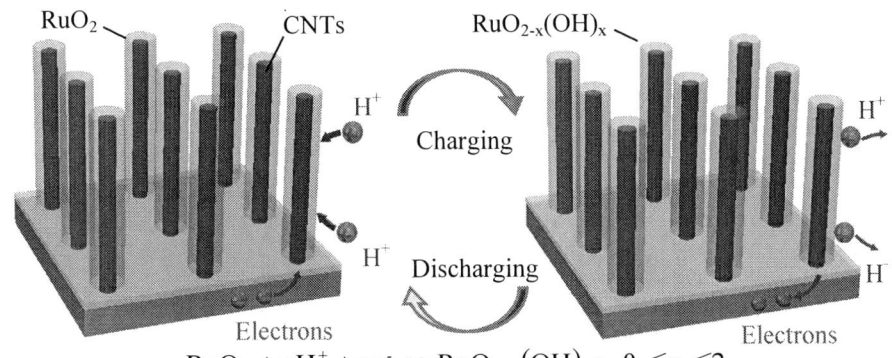

Figure 1: Conceptual illustration of the ALD RuO_2-coated CNTS on a conductive substrate. During charging, the RuO_2 surface adsorbs protons from the electrolyte and a current flows into the CNTs to create a $RuO_{2-x}(OH)_x$ surface layer, where $0 < x \leq 2$. During discharging, the reaction is reversed.

$$RuO_2 + xH^+ + xe^- \leftrightarrow RuO_{2-x}(OH)_x; \quad 0 \leqslant x \leqslant 2$$

978-1-4799-3510-9/14 $31.00 © 2014 IEEE

CONCEPT

Figure 1 shows a conceptual illustration of our device design. A forest of vertically-aligned CNTs is covered with a layer of RuO_2 several nanometers thick deposited by ALD. As the device charges, the active RuO_2 layer absorbs protons from the electrolyte and electrons flow into the device from an external circuit as the RuO_2 layer is converted to $Ru(OH)_2$ by a surface redox reaction. When discharging, stored electrons flow out of the device to an external circuit and protons are released into the electrolyte as $Ru(OH)_2$ is converted back to RuO_2 in the reverse redox reaction. In our device, the high strength and mechanical flexibility of the CNT forest is important for relieving mechanical stresses in the RuO_2 layer that arise from repeated cycling through redox states.

EXPERIMENT

Vertically-aligned, multi-walled CNTs are grown by chemical vapor deposition in a horizontal tube furnace with ethylene gas as the carbon-source. The CNTs are grown on a molybdenum- and oxide-coated silicon wafer, using aluminum and iron as catalyst layers [9]. Following the CNT growth, RuO_2 is deposited on the CNTs using a Cambridge Fiji F200 Plasma ALD with ruthenium bis(ethylcyclopentadienyl) $(Ru(EtCp)_2)$ and oxygen as precursors. In our experiment, the ALD deposition temperature was varied from $300^{\circ}C$ to $400^{\circ}C$.

Imaging of the RuO_2-coated CNTs was done using an FEI Nova NanoSEM 650 scanning electron microscope (SEM) and an FEI Technai 12 transmission electron microscope (TEM). Materials characterization was done using a Siemens D5000 X-ray diffractometer (XRD), and a PHI 5400 X-ray photoelectron spectrometer (XPS) was used for glancing incident XPS (GIXPS) measurements. Both XRD and GIXPS measurements were taken of thin-film ALD-RuO_2 samples deposited on silicon substrates under the same conditions as the ALD RuO_2-coated CNTs. Electrochemistry measurements were conducted using a three-electrode test set-up in 0.5 M H_2SO_4 electrolyte with an Ag/AgCl reference electrode, Pt counter electrode, and a Gamry Reference 600 potentiostat.

RESULTS AND DISCUSSION

A cross-sectional SEM image of the RuO_2-coated CNTs (Figure 2a) shows that the CNT forest is approximately 5 μm tall with large pore spaces available for the electrolyte to penetrate into the dense matrix of CNTs. A close-up SEM image (Figure 2b) shows that the CNTs are uniformly coated by the ALD process. A TEM image (Figure 2c) shows that the ALD coating is approximately 20 nm thick and has good adhesion to the CNT surface.

The composition of the ALD coating was investigated by GIXPS and XRD measurements. Figure 3a shows the GIXPS spectrum for ALD RuO_2 deposited at $350^{\circ}C$. The results indicate a near-stoichiometric composition (65% O, 35% Ru) within the GIXPS measurement range (top 10 nm of the film). Figure 3b is a plot of the oxygen content of the ALD films measured by GIXPS as a function of deposition temperature. In the range of $300^{\circ}C$ to $400^{\circ}C$, we have found that higher deposition temperatures correspond to increased oxygen content of the films.

The ALD process for RuO_2 is believed to occur via the accumulation of subsurface oxygen in a depositing Ru film [10]. In a recent study of ALD growth mechanisms for Ru vs. RuO_2, Methaapanon et al. found that several hundred deposition cycles of Ru were required before RuO_2 layers began to form [11]. The authors proposed that a certain thickness of Ru film is needed before there are enough defect sites to accumulate sufficient quantities of subsurface oxygen to form RuO_2. In our study, XRD measurements of ALD-RuO_2 films deposited at temperatures ranging from $300^{\circ}C$ to $400^{\circ}C$ show primarily Ru diffraction peaks (Figure 3c). Based on these results, it is likely that an underlying Ru layer is deposited during the ALD process before RuO_2 is deposited as a surface layer. If the surface RuO_2 layer is amorphous or very thin, its XRD signal would not be easily detected compared to the underlying Ru. Ru does not form a native oxide under ambient conditions [12], so we conclude that the surface oxide measured by GIXPS was deposited by ALD. For supercapacitor applications, only a surface layer of RuO_2 is needed for charge storage. The presence of an underlying Ru layer with good electrical conductivity could be beneficial to supercapacitor performance.

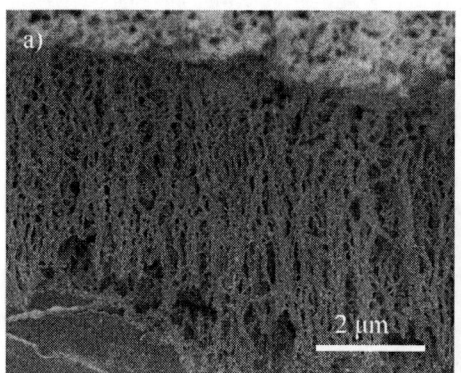 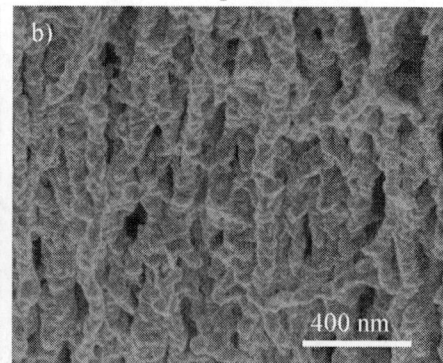 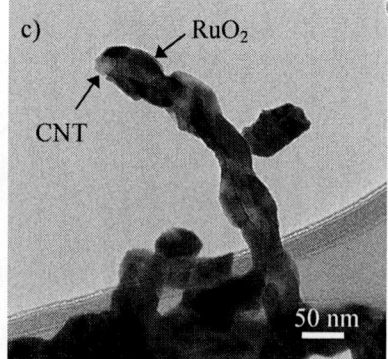

Figure 2: Electron microscope images of the as-fabricated RuO_2-coated CNTs. (a) Cross-sectional SEM image shows that the ALD coating covers the full length of the approximately 5 μm-tall CNTs, (b) Close-up SEM image of the ALD coated CNTs, (c) TEM image of a single RuO_2- coated CNT.

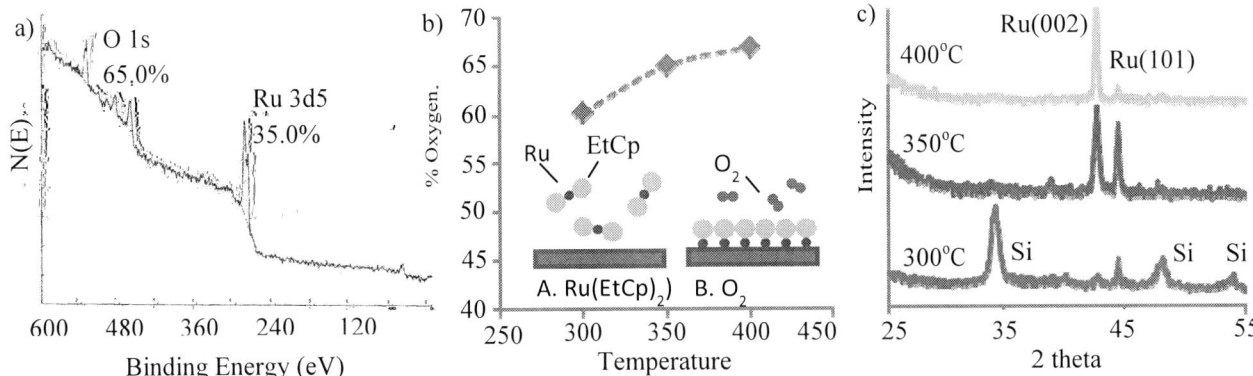

Figure 3: a) GIXPS measurement of ALD RuO_2 thin-film sample deposited at 350^oC, (b) Percent oxygen in ALD RuO_2 films as a function of temperature (measured with GIXPS) (Inset) Schematic of the ALD RuO_2 process with alternating $Ru(EtCp)_2$ and O_2 pulses, c) XRD measurement of ALD RuO_2 thin-film samples at 300^oC, 350^oC, and 400^oC deposition.

When tested as a supercapacitor electrode, the ALD RuO_2-coated CNTs demonstrate high capacitive energy storage capability. Figure 4 shows cyclic voltammetry (CV) measurements of the coated CNTs compared to an uncoated CNT sample. A capacitive current of 10 mA/cm^2 for the RuO_2-coated CNTs at a scan rate of 100 mV/s corresponds to a specific capacitance of 100 mF/cm^2, which represents one of the best values in the literature [13, 14]. The high specific capacitance of the RuO_2-coated CNTs can be attributed in part to the quality of the ALD coating. Previous studies have indicated the importance of a hydrous RuO_2 (RuO_2 · nH_2O) surface for achieving high supercapacitor performance [15]. From the CV measurements of the RuO_2-coated CNTs, we conclude that the ALD process is well-suited to fabricating such a surface layer. Besides the ALD RuO_2 quality, the high surface area and density of the CNT forest also contribute to the high specific capacitance of our device.

As shown in Figure 4, charge and discharge currents measured by CV for the coated CNTs are approximately fifty times that of the uncoated CNTs. For the uncoated CNT electrode, charge storage is by electric double layer capacitance. The dramatic improvement in supercapacitor performance with the ALD RuO_2-coating is due to the

pseudo-capacitive charge storage mechanism of the RuO_2 surface layer. We note that the shape of the CV curve for the RuO_2-coated CNTs in Figure 4 resembles that reported in other studies of RuO_2 supercapacitors [16].

Electrochemical impedance measurements of the RuO_2-coated CNTs are shown in the Nyquist plot in Figure 5. From the x-intercept, we can see that the device has an equivalent series resistance (ESR) of 7.1 Ω, a remarkably low value. This low ESR value can be attributed in part to the conformal nature of the ALD coating which minimizes contact resistance between the active RuO_2 layer and the CNTs [8]. Based on a Randles circuit model of cell impedance [17], the shape of the Nyquist plot suggests that the impedance of our device is dominated by diffusion (Warburg impedance) with low charge-transfer resistance and low double layer capacitance. A low charge-transfer resistance is characteristic of RuO_2 supercapacitors because of their fast reaction kinetics [4].

Repeating chronoamperometry was used to measure the device response to a step-change in applied potential. As shown in Figure 6, the ALD-RuO_2 coated CNTs display rapid charge-discharge characteristics that remain stable over time.

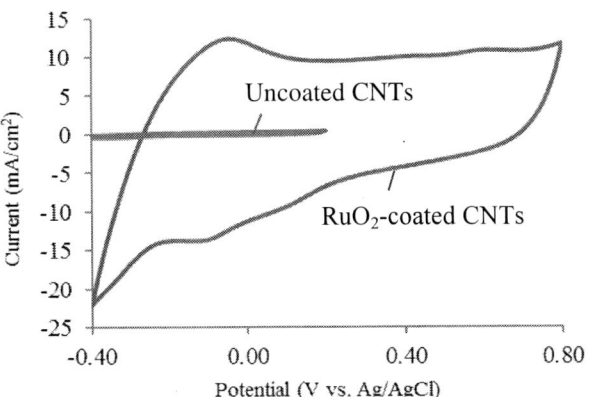

Figure 4: CV measurements of uncoated CNTs and RuO_2-coated CNTs at 100 mV/s scan rate in 0.5 M H_2SO_4 electrolyte with Ag/AgCl reference electrode and Pt counter electrode.

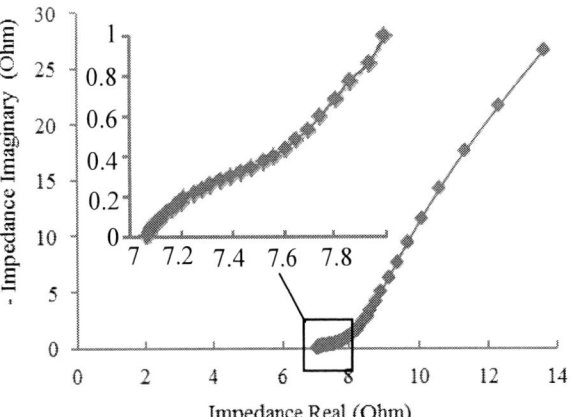

Figure 5: Nyquist plot of the RuO_2-coated CNTs showing a contact resistance (x-intercept) of 7.1 Ω.

978-1-4799-3510-9/14 $31.00 © 2014 IEEE

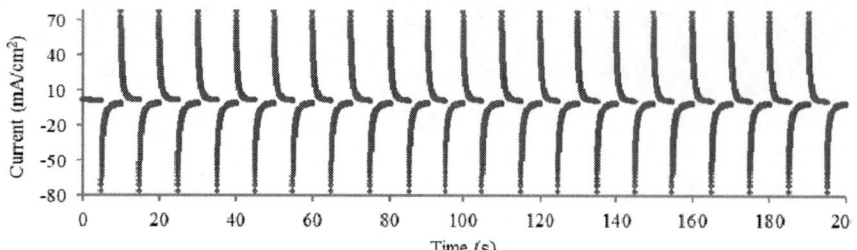

Figure 6: Charge-discharge measurements of the RuO$_2$-coated CNTs show rapid and consistent supercapacitor cycling as the input potential alternates between 0 and 0.5 V vs. Ag/AgCl.

CONCLUSIONS

In this work we have demonstrated for the first time the application of ALD RuO$_2$ to supercapacitor energy storage devices. Our results show that ALD RuO$_2$-coated CNTs can be used as high-performance supercapacitor electrodes. In addition to high specific capacitance, our device has low equivalent series resistance, rapid charge-discharge characteristics, and good stability over repeated cycling. The exceptionally high-performance of the ALD RuO$_2$-coated CNTs can be attributed to: 1) excellent conformal coverage of the CNTs by the ALD RuO$_2$ coating, 2) a high-quality RuO$_2$ surface layer capable of fast, reversible redox reactions, and 3) high surface area of the dense, vertically-aligned CNT forest. We expect that these results can be further improved by optimizing the stoichiometry and thickness of the ALD RuO$_2$ coating. With the ALD fabrication method, these properties can be controlled with angstrom-level precision to achieve exceptionally high-performance supercapacitor devices.

ACKNOWLEDGEMENTS

The authors would like to thank Ryan Rivers and Jason Chukes from the University of California Berkeley Marvell Nanofabrication Laboratory for their assistance with the ALD, Dr. Frank Ogletree and Dr. Adam Schwartzberg at Lawrence Berkeley National Lab for the XPS measurements, as well as all the members of the Liwei Lin Lab. This project is supported in part by the Advanced Technology Investment Company (ATIC, Abu Dhabi, United Arab Emirates) under the TwinLab project grant no. 12RAZB9.

REFERENCES

[1] J. Lenz, V. Trieu, R. Hempelmann, A. Kuhn, "Ordered Macroporous Ruthenium Oxide Electrodes for Potentiometric & Amperometric Sensing Applications", *Electroanalysis*, vol. 23, pp. 1186-1192, 2011.

[2] D. Rolison, P. Hagans, K. Swider, J. Long, "Role of Hydrous Ruthenium Oxide in Pt−Ru Direct Methanol Fuel Cell Anode Electrocatalysts:☐ The Importance of Mixed Electron/Proton Conductivity", *Langmuir*, vol. 15, pp. 775-779, 1999.

[3] S. Tilley, M. Schreier, J. Azevedo, M. Stefik, M. Graetzel, "Ruthenium Oxide Hydrogen Evolution Catalysis on Composite Cuprous Oxide Water-Splitting Photo-cathodes", *Adv. Funct. Mater.* vol. 23, 2013.

[4] P. Simon, Y. Gogotsi, "Materials for Electrochemical Capacitors", *Nature Mater.*, vol. 7, pp. 845-854, 2008.

[5] M. Steeves, "Electronic transport properties of ruthenium and ruthenium dioxide thin films", Ph.D. Thesis, University of Maine, 2011.

[6] A. Jorio, G. Dresselahus, M.S. Dresselhaus, *Carbon Nanotubes: Synthesis, Structure, Properties, and Applications*, Springer, 2008.

[7] G. Cui, L. Zhi, A. Thomas, I. Lieberwirth, U. Kolb, K. Müllen, "A Novel Approach Towards Carbon-Ru Electrodes with Mesoporosity for Supercapacitors", *ChemPhysChem*, vol. 8, pp. 1013-1015, 2007.

[8] S. Boukhalfa, K. Evanokff, and G. Yushin, "Atomic Layer Deposition of Vanadium Oxide on Carbon Nanotubes for High-Power Supercapacitor Electrodes", *Energy Environ. Sci.*, vol. 5, pp. 6872-6879, 2012.

[9] Y. Jiang, P. Wang, X. Zang, Y. Yang, A. Kozinda, L. Lin, "Uniformly Embedded Metal Oxide Nanoparticles in Vertically Aligned Carbon Nanotube Forests as Pseudocapacitor Electrodes for Enhanced Energy Storage," *Nanoletters* Vol. 13, pp. 3524-3530, 2013.

[10] T. Aaltonen, P. Alén, M. Ritala, M. Leskelä, "Ruthenium thin films grown by atomic layer deposition", *Chem. Vap. Dep.*, vol. 9, pp. 45-49, 2003.

[11] R. Methaapanon, S. Geyer, S. Brennan, S. Bent, "Size Dependent Effects in Nucleation of Ru and Ru Oxide Thin Films by Atomic Layer Deposition Measured By Synchrotron Radiation X-ray Diffraction", *Chem. Mater.*, 2013.

[12] B. Herd, J. Goritzka, H. Over, "Room Temperature Oxidation of Ruthenium", *J. Phys.Chem. C*, vol. 117, pp.15148-15154, 2013.

[13] J.-S. Ye, H. F. Cui, X. Liu, T. M. Lim, W.-D. Zhang, F.-S. Sheu, "Preparation and Characterization of Aligned Carbon Nanotube-Ruthenium Oxide Nanocomposites for Supercapacitors", *Small*, vol. 1, pp. 560-565, 2005.

[14] J. Zhang, J. Ma, L.L. Zhang, P. Guo, J. Jiang, X. S. Zhao, "Template Synthesis of Tubular Ruthenium Oxides for Supercapacitor Applications", *J. Phys. Chem. C*, vol. 114, pp. 13608-13613, 2010.

[15] M. T. Brumbach, T.M. Alam, P. G. Kotula, B.B. McKenzie, B.C. Bunker, "Nanostructured Ruthenium Oxide Electrodes via High-Temperature Molecular Templating for Use in Electrochemical Capacitors", *ACS Appl. Mater. Interfaces*, vol. 2, pp. 778-787, 2010.

[16] D. Rochefort, A.-L. Pont, "Pseudocapacitive Behavior of RuO$_2$ in a Proton Exchange Ionic Liquid", *Electrochem Commun*, vol. 8, pp. 1539-1543, 2006.

[17] A. Bard, L. Faulkner, *Electrochemical Method*, John Wiley & Sons, Inc., 2001.

SUB 3-MICRON GAP MICROPLASMA FET WITH 50 V TURN-ON VOLTAGE

Pradeep Pai and Massood Tabib-Azar

Department of Electrical and Computer Engineering, University of Utah, Salt Lake City, USA

ABSTRACT

This work reports the smallest microplasma field-effect transistor reported till date that operates with a low turn-on voltage of ~50V dc; a more than 3x reduction in the turn-on voltage compared to earlier reported work. Our previous work used plasma from an external source to operate the transistor while in the present work we use rf or dc voltage to directly generate plasma in the transistor channel and use dc gate field effect to control the device conduction. The reduction in turn-on voltage is achieved by a small plasma cavity of 1.5μm gap width.

INTRODUCTION

Microplasmas possess the same properties as large volume plasmas in addition to being spatially confined to small region, which improves their flexibility. Due to this, Micro-Plasma Devices (MPD) have found numerous applications in optical, chemical, electrical and biological devices. MPDs can operate under pressures ranging from a few milli-Torr to above atmospheric pressure. Use of higher pressure increases the density of plasma which is desirable in some applications.

Due to its low electrical resistance, plasma has been used as the conductive element in microplasma switches [1] and transistors [2-6]. The earliest reported microplasma transistors operated on the principles of a BJT [2-3]. These transistors had plasma cavities larger than 350μm that required plasma turn-on voltages in excess of 200V.

The motivation behind this work has been to reduce the operating voltages of microplasma transistors that would enable them to operate with regular on-board electronics. In the past, the authors' group has successfully demonstrated the working of a Micro-Plasma Field Effect Transistor (MOPFET) [4-6]. Although the earlier MOPFETs were similar in construction to the ones used in this work, they differ in their operation. The earlier devices operated on plasma that was created external to the device. The novelty of this work is in making the MOPFET operate independently by producing the plasma required for operation within the device. This approach significantly reduces the overall device package size.

WORKING PRINCIPLE

The MOPFET is similar in construction to a MOSFET, consisting of three terminals namely source, drain and gate as shown in fig 1. Two sets of source-drain fingers are shown in fig 1 to represent a generic structure. In practice, the number of fingers varies. Unlike the MOSFET that uses a semiconducting substrate as an active region for current conduction, the MOPFET uses plasma as the conducting

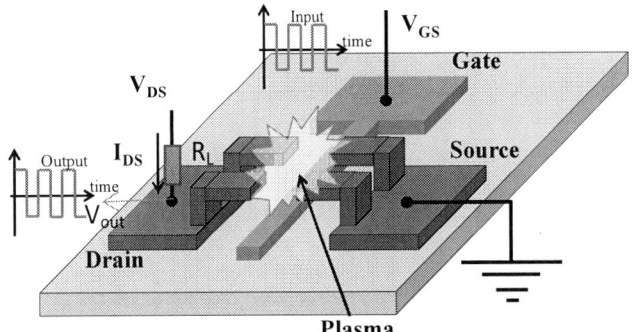

Fig 1: Schematic representation of the working principle of a MOPFET. The drain voltage breaks down the gas in the source-drain gap to create plasma. The gate voltage modulates the plasma current between drain and source through field-effect.

medium. Thus, the region between the source and drain is an air gap filled with a suitable noble gas (He in this work). Plasma is created between the source and drain electrodes by applying a voltage greater than the breakdown voltage of the gas. At the onset of breakdown, the resistance of the gas is significantly reduced which sets a steady current flow between the source-drain. Current flow can be due to electrons and ions. The insulated gate is capable of electrostatically interacting with the charges in the plasma, but does not conduct current. A voltage applied to the gate interacts with the plasma and modulates the drain-source current. The main difference between the MOSFET and MOPFET is in the "channel" turn-on mechanism. The MOSFET uses the gate bias to turn-on the channel, whereas the MOPFET uses drain voltage.

Sub-Paschen breakdown

Generation of plasma is governed by Townsend's breakdown criterion (eqn. 1), that mathematically defines the condition for the electrical breakdown of a gas [7].

$$\gamma_i \left[e^{Apd \, exp\left(-Bpd/V_b\right)} \right] = 1 \qquad (1)$$

where γ_i is the secondary emission coefficient for bombarding ions, A and B are empirical constants for a given gas, p is the pressure in Torr, d is the distance in cm and V_b is the breakdown voltage in volts.

Solution of this equation produces the well-known Paschen's curves for any given gas. Typically Paschen's curve shows an exponentially increasing trend in breakdown voltage for small values of (pd). In other words, the breakdown voltage of a gas is expected to increase exponentially for smaller electrode spacing under constant pressure. However, experimental works have shown a deviation from this behavior and a steady decrease in breakdown voltage at smaller gaps [8-10]. The main reason for this behavior is the ion-enhanced field-emission, which was not incorporated in Paschen's curve. Ion-enhanced field-emission is the phenomenon by which electrons are emitted from the cathode due to the localized electric field of an approaching positive ion.

Following these works, researchers have succeeded in theoretically modeling the modified Paschen's curve by incorporating the coefficient for ion-enhanced emission [11]. The resulting expression provides a unified model that is applicable to both Paschen and non-Paschen regime of gas behavior. Eqn. 2 expresses the modified Townsend's criterion mathematically as defined by [11]. The Paschen curve is compared against the true breakdown behavior of air and He (this work) at atmospheric pressure for small electrode spacing in fig 2. As shown in the figure, the breakdown voltage decreases almost linearly with gap for sub 8μm electrode spacing.

$$(\gamma_i + \gamma')\left[e^{Apd\ exp\left(-Bpd/V_b\right)}\right] = 1 \qquad (2)$$

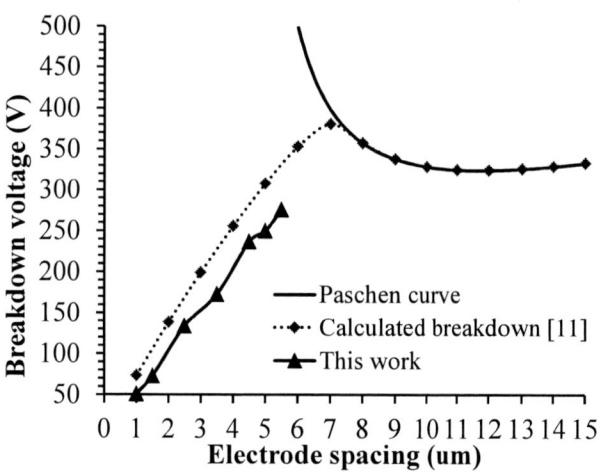

Fig 2: Dependence of breakdown voltage of air at atmospheric pressure on electrode spacing as initially determined by Paschen's curve (solid line) compared against the actual behavior as theoretically modelled by ref. [11] (dotted line). Experimental breakdown voltage in He from this work are included for gaps below 6μm.

FABRICATION

The fabrication process is straightforward with standard micromachining steps. Fig 3 shows a schematic representation of fabrication sequence. The gate and source-drain electrodes are made of 0.5μm thick sputtered TiW. The gate layer is deposited and patterned first (fig 3a), followed by a blanket deposition of 100nm Al_2O_3 for gate insulation (fig 3b). A sacrificial strip of 300nm thick poly-silicon is formed on top of the gate electrode to separate the source-drain from gate (fig 3c). The poly-silicon strip is etched away in XeF_2 gas after patterning the source-drain electrodes (fig 3e). Fig 4 shows SEM micrographs of one of the devices before sacrificial etch. The source-drain

electrode consists of 10 sets of 10μm-wide fingers with an inter-finger separation of 10μm.

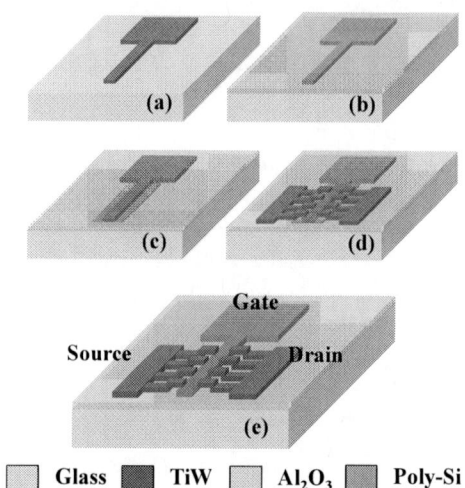

Fig 3: Schematic representation of the fabrication process flow. The source-drain and gate electrodes are made of 0.5 μm sputtered TiW. The source-drain electrodes are suspended over the gate electrodes by a stand-off distance defined by the sacrificial poly-Si layer.

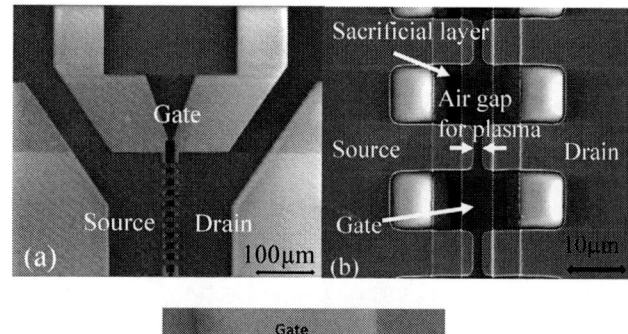

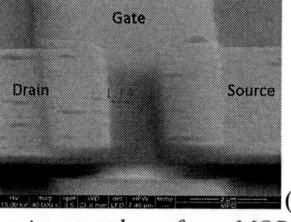

Fig 4: SEM micrographs of a MOPFET before the sacrificial etch. (b) shows a magnified view of the source-drain and gate electrodes of the device in (a). c) Tilted SEM of a self-aligned gate MOPFET.

EXPERIMENT

The MOPFETs are characterized for their current-voltage dependencies under both dc and rf excitations. The devices are wire bonded to a 40-pin hybrid package. The package is loosely sealed to maintain atmospheric pressure and supplied with a continuous flow of He. DC characterization is performed using 4156C semiconductor parameter analyzer. At the onset of plasma, the device can draw a large current due to the very small resistance of plasma. To prevent device damage, the current is limited by the compliance setting in 4156C. RF characterization is performed by an electro-optical setup. A high rf voltage is applied to the source-drain to ignite the plasma. The power transfer efficiency of the rf source is increased by using a tuning coil that reduces the power required for plasma ignition by greater than 10dB due to the Q of the coil and source-drain capacitance combination at the resonant frequency. To simplify the switching detection mechanism and to prevent interference with high rf voltage, the switching is optically detected using a photomultiplier tube. Fig 5 shows the experimental setup for dc and rf characterization along with snapshots of the device during operation.

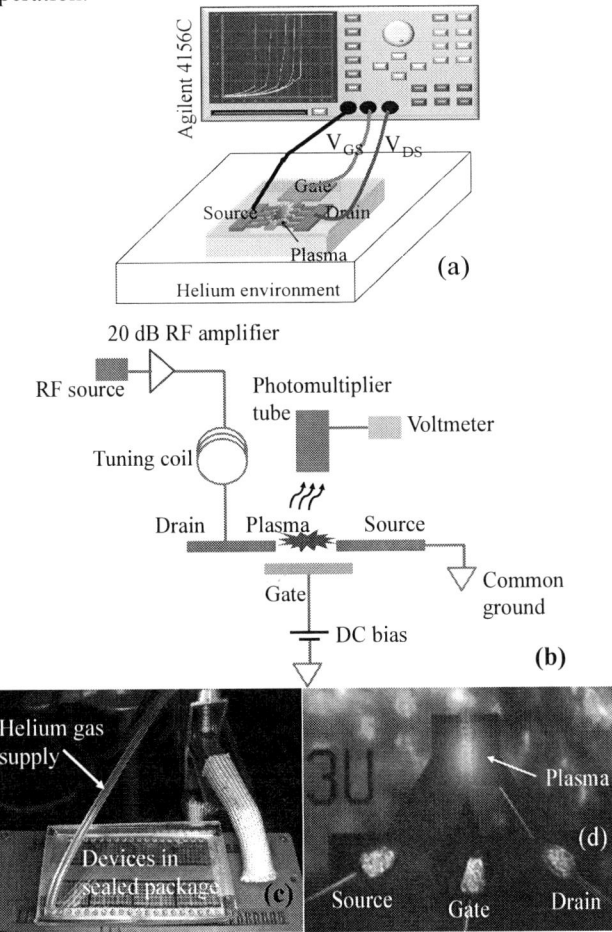

Fig 5: Experimental setup for dc and rf plasma gate control tests. (a) shows the setup for dc characterization and (b) is the setup for rf plasma characterization using optical detection. (c) shows a 40-pin hybrid package used to test devices in He environment. (d) shows a MOPFET operating with rf plasma at 511MHz.

RESULTS AND DISCUSSION

As mentioned earlier, plasma is ignited when the voltage (drain-source) exceeds a threshold value, leading to a constant current flow. The device is considered to be ON at this point. Due to the visual glow of the plasma, the switching can be alternately detected using an optical

sensor. Due to simplicity of the dc characterization setup, all measurements are performed electrically.

DC characterization

Under dc excitation, the current conduction is due to both electrons and ions. The ions cause sputtering of the cathode which can severely damage these devices due to their negligible thickness. To prevent device damage, the current is limited to 10nA and less to enable repeated measurements. At such small currents, the plasma does not radiate visible light due to insufficient excited electrons. A bidirectional voltage sweep shows a hysteresis in the I-V switching which conforms to earlier observation [6], confirming a true plasma switching as shown in fig 6(a). The theory behind this is explained in detail in [6]. The gate bias can assist or retard the breakdown mechanism and thus changing the breakdown voltage. The I-V curve of the MOPFET with gate bias (fig 6b) shows a symmetric and well controlled gate field-effect.

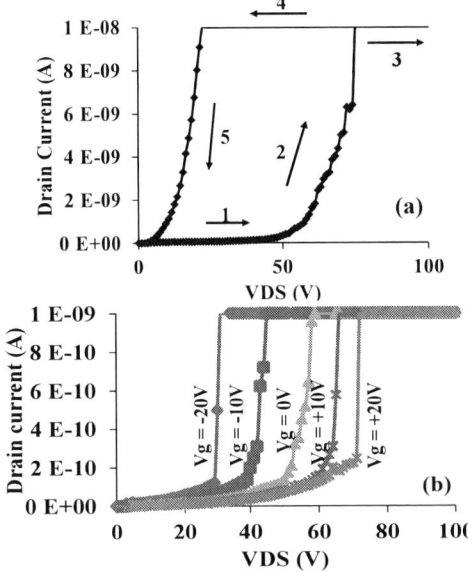

Fig 6: I-V characterization of the MOPFET under dc excitation without gate bias (a) and with gate bias (b). The switching voltage changes roughly by 1V for a 1V change in gate bias.

RF characterization

Current conduction in an rf plasma is almost entirely due to electrons since ions are heavy and cannot follow the rf oscillations. This results in significantly lower sputtering of the electrodes and thereby allowing operation at larger currents. The device suffered negligible damage with rf plasma for currents as high as 80μA when operated continuously for up to 20 hrs. In contrast, a dc current of only 10nA deteriorated the electrodes severely while in less than 5 minutes. Due to this the device is operated with a glowing plasma with rf excitation. This allows optical detection of the switching which also prevents interference of the detection circuitry with rf signal.

978-1-4799-3510-9/14 $31.00 © 2014 IEEE

Similar to dc plasma, the rf plasma also shows hysteresis in its switching behavior as shown in fig 7(a) due to the same phenomenon. The gate bias has similar effect with the rf plasma as shown in fig 7(b). The photomultiplier response in fig 7 is limited to the turn-ON value. In reality, the photo-voltage increased almost linearly after the device turned ON since there is nothing limiting the current through the device, unlike the compliance setting in dc testing.

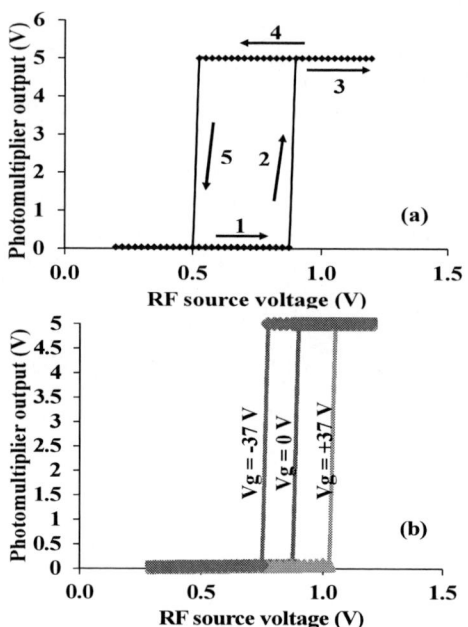

Fig 7: Switching characterization of the MOPFET under rf excitation without gate bias (a) and with gate bias (b). Switching is optically detected using a photomultiplier tube.

Further tests are required to test the ac response of gate field-effect. To test the switching response of plasma, the MOPFETs are tested at different frequencies ranging from 100MHz – 10GHz. The device responded at all of these frequencies with suitable frequency tuning. Fig 8 shows the power spectrum of MOPFET between 9 and 10GHz. Detailed discussion of the high frequency testing will be reported elsewhere.

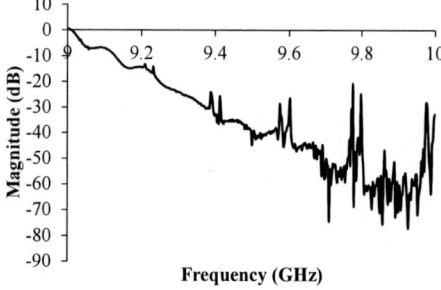

Fig 8: Power spectrum of MOPFET (S_{21}) for 9-10GHz input signal. The peaks indicate plasma ignition.

CONCLUSION

This work discussed miniaturization of microplasma transistor setting a new trend of scaling similar to MOSFETs. Gate field-effect was successfully shown for the first time. Sputtering problem associated with dc excitation was mitigated by using rf excitation that prolonged the life of the device. High frequency (10GHz) switching of microplasma was demonstrated for the first time.

ACKNOWLEDGMENT

This work is supported by the DARPA MPD program under Dr. Daniel Purdy. The authors appreciate Dr. Lingyao Chen's technical assistance.

REFERENCES

[1] H. Rahaman et al, "Switching Characteristics of Microplasmas in a Planar Electrode Gap", Applied Physics Letters, vol. 90, p. 131505, 2007.

[2] K. F. Chen and J. G. Eden, "The Plasma Transistor: A Microcavity Plasma Device Coupled with a Low Voltage, Controllable Electron Emitter", Applied Physics Letters, vol. 93, p. 161501, 2008.

[3] C. J. Wagner, P. A. Tchertchian and J. G. Eden, "Coupling Electron-Hole and Electron-Ion Plasmas: Realization of an npn Plasma Bipolar Junction Phototransistor", Applied Physics Letters, vol. 97, 134102, 2010.

[4] F. K. Chowdhury, Y. Zhang and M. Tabib-Azar, "Fabrication and Characterization of 3D Micro-Plasma Field Effect Transistors", IEEE Conf. on MEMS 2013, p. 669-672, 2013.

[5] M. Cai, F. K. Chowdhury and M. Tabib-Azar, "Micro-plasma field-effect transistors", IEEE Sensors 2012 Conf., 2012.

[6] Y. Zhang, P. Pai, F. K. Chowdhury and M. Tabib-Azar, "Operation principles of micro-plasma field effect transistor," Int'l conf. on Solid-State Sensors, Actuators and Microsystems 2013, p 578-81, 2013.

[7] Y. P. Raizer, "Gas Discharge Physics", Springer-Verlag, 1991.

[8] R. T. Lee, H. H. Chung and Y. C. Chiou, "Arc erosion behavior of silver electric contacts in a single arc discharge across a static gap," IEE Proceedings Science, Measurement and Technology, Vol. 149, No. 4, p 172-180, 2002.

[9] J. M. Albright, L. L. Raja, M. Manley, K. Ravi-Chandar, and S. Satapathy, "Studies of asperity-scale plasma discharge phenomena," IEEE Transactions On Plasma Science, Vol. 39, NO. 6, p 1560-1565, 2011.

[10] J. M. Torres and R. S. Dhariwal, "Electric field breakdown at micrometre separations," Nanotechnology, Vol. 10, p 102-107, 1999.

[11] R. Tirumala and D. B. Go, "An analytical formulation for the modified paschen's curve," Applied Physics Letters, vol. 97, p 151502, 2010.

CONTACT

Prof. Massood Tabib-Azar, tel. +1-(801)-5818775, azar.m@utah.edu.

FORMATION OF CROSS-SHAPED ESCHERICHIA COLI

Kayoko Hirayama[1], Yun Jung Heo[1,2], and Shoji Takeuchi[1,2]
[1]Institute of Industrial Science, The University of Tokyo, JAPAN
[2]ERATO Takeuchi Biohybrid Innovation Project, JAPAN

ABSTRACT

This paper describes preliminary results of a method to regulate the shapes of a single bacterial cell by confining the cell into a micro chamber. The cell wall of Escherichia coli determines the rod-shape (0.5 μm in width, 2 μm in length) of the bacteria and protects from outer stress. We removed its cell wall by enzyme treatment and confined the cell into cross-shaped microchambers formed on the surface of 2% agar contained the growth media. We believe that our method could contribute to the comprehensive understanding of the shape regulation mechanism of *E. coli*. This paper focuses on the fabrication of cross-shaped microchambers formed on agar and confinement of the bacteria into the microchambers.

INTRODUCTION

A rod-shaped bacterium divides into two rod-shaped cells and maintains its rod-shape (Figs. 1A and 2A) generation after generation. This shape retaining mechanism is extensively studied recently and several crucial proteins (MreB, MinCDE) are identified[1, 2]. However, the mechanism controlling the behavior of the proteins still remain unclear. Many biologists are trying to reveal the whole mechanism in the shape regulation of *E. coli* through conventional genetic engineering which often requires much time and efforts to obtain the expecting results. On the other hand, MEMS technology provides us with methods to regulate micro-geometries of various materials and enables us to take approaches completely different from the conventional methods; we applied physical stress to a single cell of *E. coli* to reveal the unknown shape regulation of *E. coli*.

When the cell wall is removed, *E. coli* cells turn into round shape, osmotically sensitive structure (Figs. 1A and 2B). This structure is called spheroplasts. Ranjit D et al reported that spheroplasts re-obtained its rod shape by culturing the spheroplasts on agar plate containing osmotically protective nutrient broth [3]. Then what would happen if spheroplasts re-synthesize its cell wall under physical pressure? (Fig. 1B-D)

Here we present a method to apply physical pressure to spheroplasts by confining the spheroplasts into the cross-shaped microchambers (Fig. 1B and C).

MATERIALS AND METHODS
Bacterial strain and growth conditions

E. coli strain DH5α was used in this study. The bacteria were routinely grown in Luria-Bertani (LB) liquid medium (10 g/L tryptone, 5g/L yeast extract, 5g/L NaCl, pH 7).

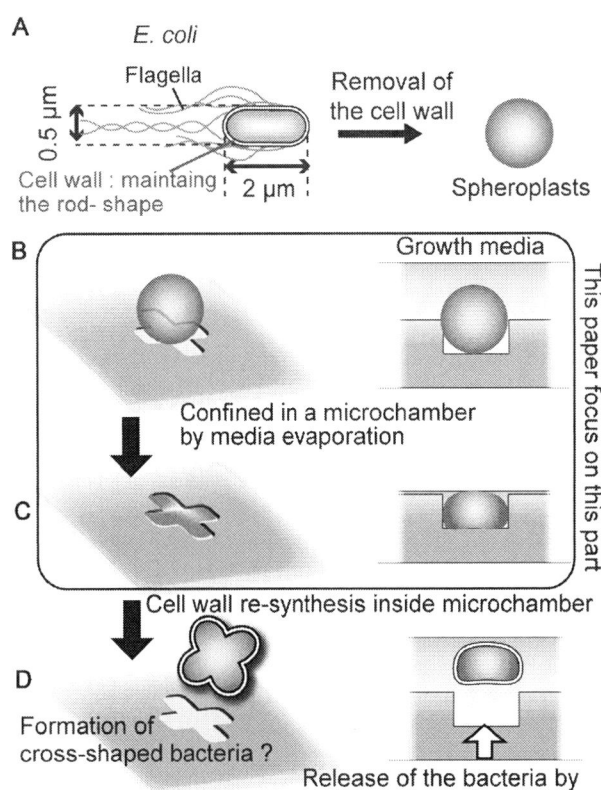

Fig.1 Conceptual illustration of our method to form cross-shaped E. coli. A) Illustration of E. coli and a spheroplast. B) We placed spheroplast suspension on the microchamber and wait 5 -10 min for the confinement of the spheroplasts. C) The spheroplasts were confined in cross-shaped microchamber as water evaporated. D) After cell wall formation, addition of a droplet of growth media would release the cross-shaped E. coli.

Formation of spheroplasts

E. coli were cultured in LB broth for about 18 hours at 37°C with shaking at 150 rpm. The cultured cells were pelleted at 120,000 rpm for 5 min. The pellet was washed twice by 1ml of 10 mM Tris-HCl (pH 8.0). Then the cells were resuspended in 2 ml of lysozyme solution (1.5 ml of 1M sucrose, 200 μl of 1M Tris-HCl (pH 8.0), 200 μg of lysozyme, 300 μl of water) and incubated for 30 minutes at 37°C. The cell suspension was then slowly diluted by 4 ml of iced 1.5 mM EDTA (pH 8.0) and 6 ml of sucrose-recovery medium. The solution was gently mixed by inverting the test tube for several times. The spheroplasts were collected (500 × g, 30 min) and gently resuspended in sucrose recovery medium.

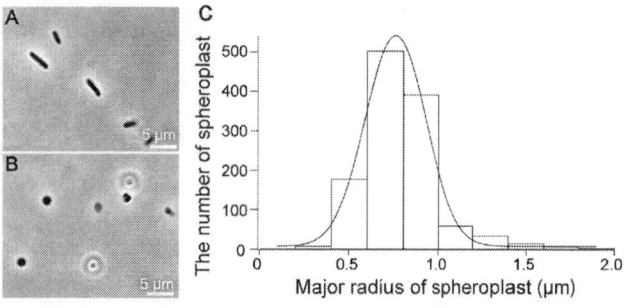

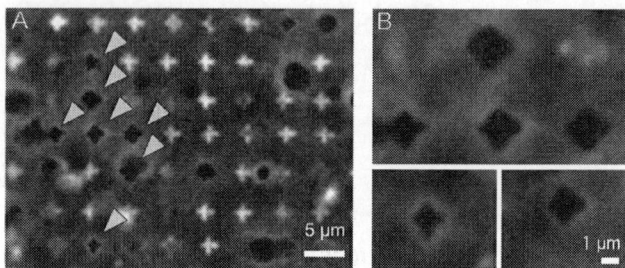

Fig. 2 A) A phase contrast image of rod-shaped E. coli. B) A phase contrast image of spheroplasts after the treatment of lysozyme. C) Distribution of major diameter of spheroplasts.

The major radius of spheroplasts were analyzed using Image J.

Microfabrication of cross-shaped structure array

First, we used a two-photon direct laser writing system (Photonic Professional, Nanoscribe GmbH equipped with a 100×, Numerical Aperture=1.4, oil immersion objective) to fabricate cross-shaped structures with UV-curable IP-L photoresist (Nanoscribe GmbH). The cross-shaped structures were fabricated on a glass slide. After the direct writing process, the IP-L photoresist was developed by SU-8 developer solution. We used this array of cross-shaped microstructures as an initial mold.

Next, 0.08g of parylene was deposited on the cross-shaped micostructures of IP-L. The cross-shaped microstructures were then transferred to polydimethylsiloxane (PDMS) molds. The PDMS molds were coated by CHF_3 before imprinted the array of cross-shaped microchambers to another PDMS molds. As a result, the array of cross-shaped microstructures were formed on PDMS surface. We used this PDMS microstructures as a template to fabricate microchambers on agar surface.

The pre-heated glass-bottom dishes were rinsed with hot sucrose recovery media contained 2% agar and dried out at 65 °C to make adhesive surface on the glass. Then a drop (about 1 μl) of the same agar media was placed on the glass and the cross-shaped microstructures of PDMS was stamped on the drop and allowed to gel for about 10 minutes and the PDMS mold was peeled away from agar containing sucrose recovery media.

Confinement of spheroplasts into microchambers

1 μl of E. coli spheroplasts suspended in sucrose recovery media was placed on the microchambers formed on nutrient agar layer. During 5-10 minutes, some spheroplasts were automatically trapped into the microchambers as the liquid culture media was evaporated.

Microscopy

The microchambers formed on agar surface and the confined spheroplasts were observed with Olympus IX71 microsope equipped with ×100 objective lens (Numerical Aperture=1.4, oil immersion objective).

Fig. 3 A) An overview photo of the cross-shaped microchambers. Yellow arrow heads show the cross shaped spheroplasts. B) Magnified images of cross-shaped spheroplasts.

RESULTS
Formation of spheroplasts

The distribution of major radius of the spheroplasts were shown in Fig. 2C. The average major radius of spheroplasts was 0.8 μm. If we assume the spheroplasts as spheres with radius of 0.8 μm, the average volume of spheroplasts was estimated to be 1.6 $μm^3$. Thus, we design the microchambers to have the volume of 1.6-2.0 $μm^3$.

Confinement of the spheroplasts

We fabricated a thin nutrient agar layer with the microchambers by applying hot stamping method using PDMS as master plate of the microchambers. Then we successfully confined the spheroplasts into cross-shaped microchambers (Fig. 3A). The cross-shaped spheroplasts in the microchambers (Fig.3B) remained stable at least for 2 hours.

CONCLUSION

We believe that by combining this method with genetic engineering techniques, our method could be powerful tools to analyze the shape regulation mechanism of the bacteria.

ACKNOWLEDGEMENTS

We thank Dr. K. Sato and Dr. H. Onoe at The University of Tokyo for fruitful discussions. We also thank Ms. M. Onuki for her technical assistance.

REFERENCES

[1] Gitai Z, Dye N, Reisenauer A, Wachi M, Shapiro L. *Cell.* 2005;120:329-41.
[2] Adams D, Errington J. *Nature reviews Microbiology.* 2009;7:642-53.
[3] Ranjit D, Young K. *Journal of bacteriology.* 2013;195:2452-62.

CONTACT

Prof. Shoji Takeuchi, Institute of Industrial Science, The University of Tokyo, 4-6-1, komaba Meguro-ku, Tokyo, JAPAN,Tel: +81-3-5452-6650; Fax: +81-3-5452-6649, E-mail: takeuchi@iis.u-tokyo.ac.jp

SINGLE-CELL 3D BIO-MEMS ENVIRONMENT WITH ENGINEERED GEOMETRY AND PHYSIOLOGICALLY RELEVANT STIFFNESSES

Mattia Marelli,[1]* Neha Gadhari,[2] Giovanni Boero,[1] Matthias Chiquet[2] and Jürgen Brugger[1†]

[1]École Polytechnique Fédérale de Lausanne (EPFL), Switzerland

[2]University of Bern, Switzerland

ABSTRACT

We present a three dimensional (3D) microenvironment for on-chip cell culture, with engineered geometrical and mechanical properties. The device, named μ-flower, is based on micorfabricated cantilever beams bent out of plane by the intrinsic stresses of a bilayer structure. The use of Ti-SiO$_2$ bilayers with various thicknesses allows spanning a large range of rigidities while keeping the size nearly constant. The geometrical and mechanical properties of the devices are thus decoupled, and the degrees of stiffness of several physiological tissues are matched. These characteristics make μ-flowers a microfabricated cell-culture substrate designed to mimic essential physical properties of the *in vivo* environment (dimensionality, shape and rigidity) in a precisely controlled way, at the single-cell scale, and with a high degree of parallelization.

INTRODUCTION

Geometrical and mechanical features of the cell milieu at the microscale are crucial regulators of various cell functions [1, 2]. Micro-engineered cell-culture substrates are thus becoming key tools in domains such as tissue engineering and pharmacological tests, where they can simulate physiological conditions more realistically than two dimensional (2D) Petri dishes [3]. MEMS technologies represent an opportunity to engineer the features of artificial cell microenvironments at the cellular or sub-cellular scale, enriching them with controlled topographies, patterns, mechanical structures and microfluidics [1, 2, 3]. Earlier results showed the possibility to use bent cantilevers as 3D cell culture substrate with potential applications in cell traction forces measurements [4]. Yet, these prototypical devices were limited to have one single value of stiffness, while the rigidity of the microenvironment has been shown to greatly affect cell functions, such as cytoskeletal organization and stem cell differentiation [2]. Here we report a substantial improvement in MEMS fabrication that allows for the stiffness to be tuned in a wide range of nearly two orders of magnitude, making it possible to mimic on chip a variety of physiological rigidities found *in vivo*.

FABRICATION

The devices are based on bent cantilevers, whose curvature is caused by the internal stress of the beams themselves. A pictorial view of the process flow is given in Figure 1.

A SiO$_2$ layer of 200 nm is first grown on 10 cm Si (100) wafers, with a wet oxidation process at 850 °C. The

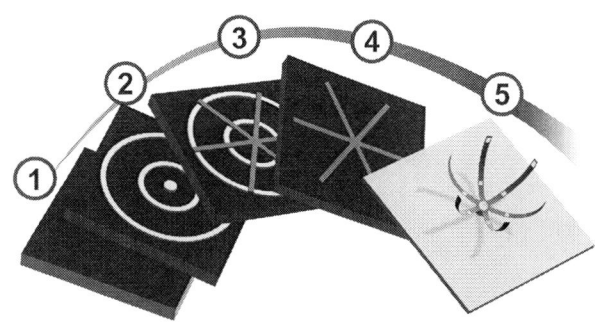

Figure 1: Process flow for the fabrication of μ-flowers. 1) growth of SiO$_2$ and deposition of Ti, for the formation of a bilayer; 2) Au patterns deposition via lift-off; 3) photolithography; 4) etching of unprotected Au, Ti and SiO$_2$, to shape the μ-flowers; 5) beams release by dry etching, and photoresist stripping.

relatively low oxidation temperature is chosen in order to reduce thermal relaxation of the intrinsic stress, and have oxide films retaining a higher compressive stress [5].

Later, the oxide film thickness is modified by etching in a buffered hydrofluoric acid solution (BHF). This step is used to prepare SiO$_2$ layers with various thicknesses (100, 125, 150, and 180 nm).

A Ti film is then deposited on top of the oxide, by means of physical vapor deposition (PVD). 4 Ti thicknesses have been tested (30, 60, 110, and 150 nm), combining them with each of the 4 silica thicknesses, for a total of 16 different combinations.

Circular Au patterns are deposited on top of the bilayer, with a lift-off process. A second photolithographic mask is then used to shape the μ-flowers.

Au and Ti are subsequently etched away from the wafer area that is unprotected by the resist. Au is etched with a solution containing KI (25 g/l) and I$_2$ (12 g/l), and Ti with an inductively coupled plasma (ICP) of Cl$_2$ and BCl$_3$ gases. At this stage the wafer is cut in 1.0 × 1.2 cm^2 chips. After cleaning in DI water, the chips are glued on a support wafer and the process is resumed by etching SiO$_2$ with an ICP of C$_2$F$_6$ gas.

Finally the Ti-SiO$_2$ beams with Au spots are released from their Si substrate by means of an ICP of SF$_6$ gas, while the resist is stripped with a gentle oxygen plasma (100 W) followed by wet chemical removal (Microposit 1165 remover for 30 min at 70 °C).

BILAYER BENDING MECHANICS

The bending moment M_{tot} in a pre-stressed beam is given by:

$$M_{\text{tot}} = \int m(z)\, dz = w \int z \cdot \sigma(z)\, dz \qquad (1)$$

$m(z)$ being the moment density, w the width of the beam, $\sigma(z)$ the intrinsic stress and z the distance from the neutral axis (as shown in Figure 2). In a SiO_2 single layer the

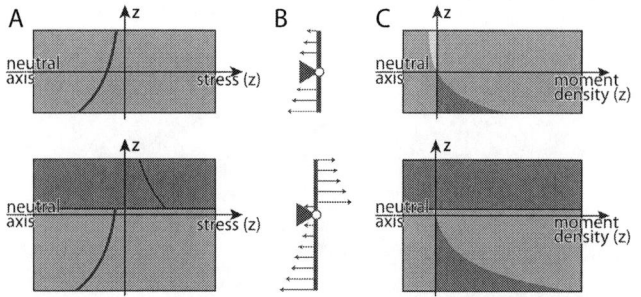

Figure 2: *Schematics of stress and bending moment profiles in single- and bi-layer beams. A: stress profiles for a mono- and a bi-layer. B: analogy with a leverage. C: moment density $m(z) = z \cdot \sigma(z)$.*

compressive stress $\sigma(z)$ is not constant through the depth z, and shows a decaying profile, due to thermal stress relaxation during the oxidation process [5] (Figure 2A). The generation of bending moment can be easily understood by considering the analogy with a leverage (Figure 2B). The case of the SiO_2 single layer is similar to a leverage whose two arms are loaded in the same direction with loads of different magnitudes. This yields a negative moment density $m(z)$ (yellow area in Figure 2C), which reduces M_{tot}. The use of a Ti top layer, characterized instead by a tensile stress, gives a positive $m(z)$, which sums to the one provided by SiO_2 and increases the resulting bending moment M_{tot}, allowing for relatively thick and stiff cantilevers to be bent

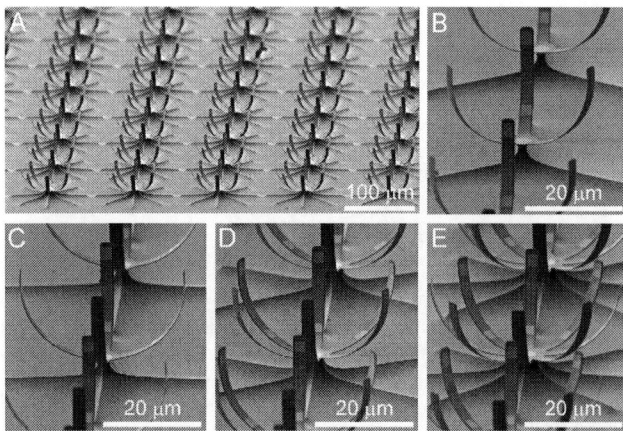

Figure 3: *Engineering μ-flower geometry. A: SEM micrograph showing an array of μ-flowers allowing for parallelized cell-culture tests. Chips with a density of about 7000 μ-flowers/cm^2 have been fabricated. B to D: close up views of devices featuring a variable number of cantilevers (3, 4, 5, 6 and 8, respectively). Au spots are deposited in the center of each μ-flower and onto cantilevers, and further define the geometry of the microenvironment by providing cells with adhesive patterns in 3D.*

up to a radius R of 30 μm. This size is suitable for the fabrication of μ-flowers fitting single-cell dimensions (Figure 3). The main problem when dealing with the intrinsic stress of thin films is the difficulty, or even the impossibility, to predict and model the stress profile $\sigma(z)$ at the nanometer scale [6], an thus to calculate the bending moment. This makes it challenging to design the bilayer thicknesses *a priori* in order to have a given bending radius R. For this reason a combinatorial empiric process has been chosen instead, aiming to investigate the attainable radii and spring constants k.

GEOMETRICAL AND MECHANICAL PROPERTIES

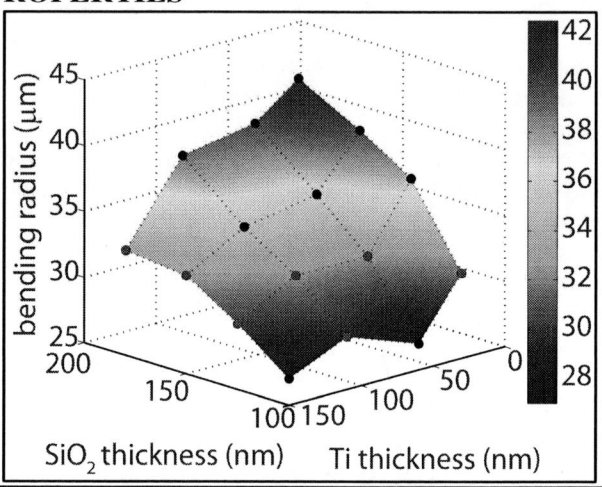

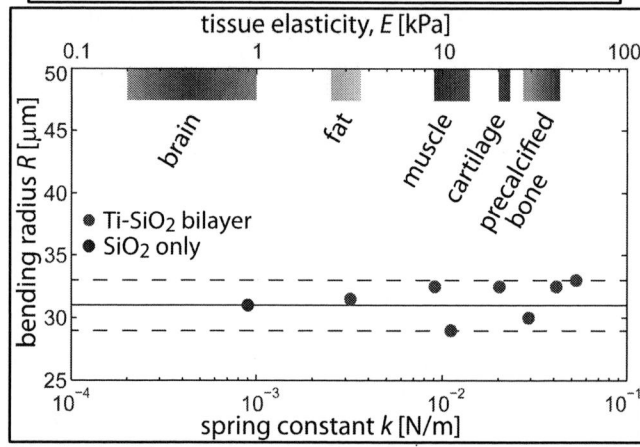

Figure 4: *Engineering μ-flower stiffness. Top: bending radii of bilayer cantilever beams as a function of Ti and SiO$_2$ thicknesses. Highlighted with red markers are seven Ti-SiO$_2$ combinations having values of R suitable for the fabrication of μ-flowers fitting the size of a single cell (29 μm < R < 33 μm). Bottom: the seven selected combinations span an extended range of spring constants k, while having all similar radii, allowing for the design of single-cell culture devices with decoupled geometrical and mechanical properties. Together with SiO$_2$ single layer beams, the attained spring constants cover a span of nearly two orders of magnitude, matching the rigidities of various physiological tissues [7].*

We performed a systematic study of the Ti-SiO$_2$ bilayer by combining four Ti thicknesses with four SiO$_2$ thicknesses (Figure 4). Seven combinations have been heuristically determined, which give values of R around 31 μm (29 μm < R < 33 μm), while spanning an extended range of stiffness. The selected radii are suitable for the fabrication of μ-flowers with the size of single cells (in this case mammalian fibroblasts, which can spread out to about 60-70 μm).

Together with SiO$_2$ single layer beams that we had previously fabricated [8], the spring constants k covers nearly two orders of magnitude (10^{-3} < k < 53 × 10^{-2} N/m), while R experiences only minor variations, achieving the decoupling of the mechanical properties from the geometrical ones.

Moreover, this stiffness range matches the rigidity of physiological tissues, as diverse as brain (0.2-1 kPa), fat (2.5-3.5 kPa), muscles (10-15 kPa), cartilage (20 kPa) and precalcified bone (30-40 kPa) [7] (Figure 4), making it possible for this kind of devices to mimic a variety of mechanical conditions found *in vivo* (a comparison between the elastic modulus E of continuous elastic media, such as tissues, and the spring constant k of discrete elements, such as cantilevers, is given by Ghibaudo *et al.* [9]).

A preliminary measurement of the spring constant has also been performed on the stiffer cantilevers, in order to cross check the theoretical values of k. A capacitive force sensor from FemtoTools [10] has been used to acquire force curves while probing the bent cantilevers horizontally. Figure 5 shows good agreement between the measured k (69 × 10^{-2} N/m) and the theoretical expectation (53 × 10^{-2} N/m).

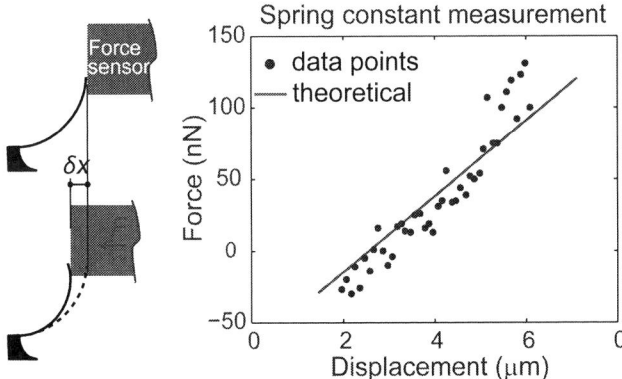

Figure 5: Left: schematics of the spring constant measurement performed with a force sensor from FemtoTools [10]. Right: force curve for one of the stiffest cantilevers, showing good agreement with the theoretical expectation.

CELL CULTURE

The device has been validated by culturing cells inside μ-flowers (Figure 6). Two cell lines have been tested: mouse embryo fibroblasts (MEF, also used elsewhere [4]), and rat embryo fibroblasts (REF), used for the first time in this kind of device. First, the Au spots deposited along the cantilevers are selectively functionalized with an adhesive

peptide (Figure 7), in order to facilitate the formation of cellular focal adhesions (FAs). The peptide used for this purpose has a cysteine amino acid, characterized by the thiol group of the side chain, which allows the peptide to be selectively adsorbed on the Au spots. Moreover, the peptide features the arginine-glycine-aspartic acid (RGD) sequence, which acts as a ligand for the integrin protein, responsible for cell adhesion. The remainder of the beam surface is constituted either by SiO$_2$ or by a layer of native TiO$_2$, which are both passivated against cell adhesion by means of poly(L-lysine)-*grafted*-poly(ethylene glycol) (PLL-*g*-PEG) [11]. The presence of Au adhesive patterns increases the

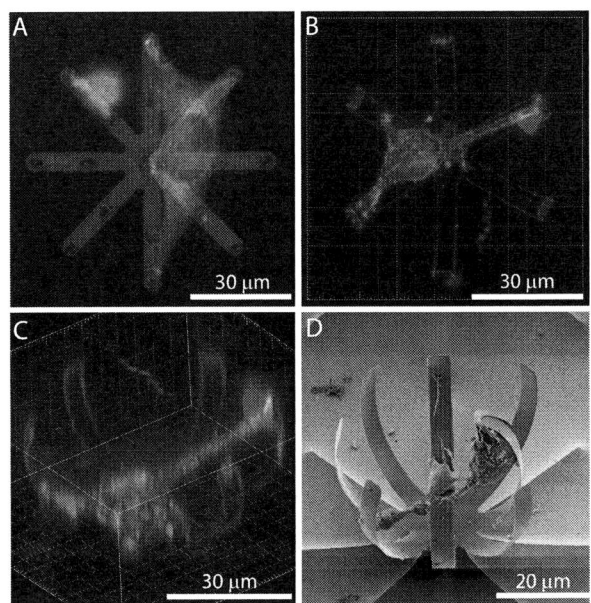

Figure 6: A: MEF (green) adhering selectively on functionalized Au spots (dark spots on the cantilever). Au spots are functionalized with a peptide to enhance cell adhesion, while the remainder of the beam is passivated with PLL-g-PEG (here stained in red) against cell adhesion. In this image the cantilevers are not released from the chip and kept 2D in order to facilitate imaging. B: Top view of a REF inside a released 3D μ-flower (confocal micrograph). C: Perspective view of B, showing the REF spread and suspended inside the μ-flower. D: SEM micrograph of a MEF spreading in a μ-flower and bending a cantilever beam (false color).

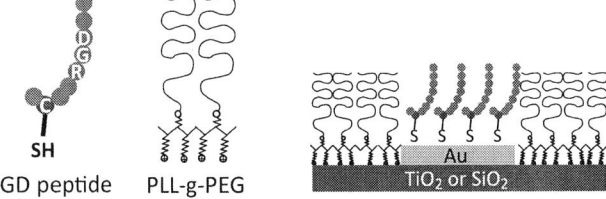

Figure 7: Schematics of the functionalization/passivation strategy. An integrin binding peptide is self-assembled onto Au surfaces via the thiol of a cysteine amino acid. The remainder of the oxidized surfaces is later passivated against cell adhesion by adsorbing PLL-g-PEG.

control over cell geometry and confines fibroblasts exclusively inside μ-flowers, where they spontaneously adhere. Typically, about 5000 (50%) μ-flowers per chip contain cells, after seeding them and leaving them spread overnight. This shows the device suitability for highly parallelized cell culture tests within 3D microenvironments with precisely engineered geometrical and mechanical properties.

CONCLUSION

Ti-SiO$_2$ bilayers have been used for the fabrication of 3D microenvironments based on bent cantilevers. The opposite intrinsic stresses of these two materials (tensile for Ti [12] and compressive for SiO$_2$ [5]) largely increased the bending moment with respect to previously tested SiO$_2$ single layers. In that case, the limited moment given by the intrinsic stress was able to bend only very soft beams (10^{-3} N/m) to the size required to hold a single cell [8]. .

A combinatorial approach was chosen to systematically investigate the attainable radii and spring constants of the bilayer cantilevers. It was found that the mechanical properties (k) can be decoupled from the geometrical ones (R). Moreover, the full range of spring constants achieved by single-cell μ-flowers corresponds to the typical rigidities of several physiological tissues, from nervous tissues to tendons and pre-calcified bones [7].

The geometry of the devices is also engineered, by changing the number of cantilevers in each μ-flower, and by creating cell-adhesive Au spots that can be deposited at arbitrary positions along the beams.

These features make μ-flowers a parallelized 3D single-cell culture substrate particularly suitable to mimic adult connective tissues (e.g. tendons). Indeed, in this kind of *in vivo* environments, the highly dense extra cellular matrix (ECM) confines the movements, position, shape, and attachment points of cells, similarly to what happens when culturing cells inside μ-flowers.

ACKNOWLEDGEMENTS

This work was supported by the Swiss National Science Foundation grants Nos. 125290 and 140623.

REFERENCES

[1] M. Nikkhah, F. Edalat, S. Manoucheri, A. Khademhosseini, "Engineering microscale topographies to control the cell-substrate interface", *Biomaterials*, vol. 33, pp. 5230-5246, 2012.

[2] J. Fu, Y. K. Wang, M. T. Yang, R. A. Desai, X. Yu, Z. Liu, C. S. Chen, "Mechanical regulation of cell function with geometrically modulated elastomeric substrates", *Nat. Meth.*, vol. 7, pp. 733-736, 2010.

[3] D. J. Beebe, D. E. Ingber, J. den Toonder, "Organs on Chips 2013", *Lab Chip*, vol. 13, pp. 3447-3448, 2013.

[4] M. Marelli, N. Gadhari, G. Boero, M. Chiquet, J. Brugger, "Cell force measurements in 3D microfabricated environments based on compliant cantilevers", *Lab on a Chip*, published online, DOI: 10.1039/C3LC51021B.

[5] A. Fargeix, G. Ghibaudo, "Dry oxidation of silicon: A new model of growth including relaxation of stress by viscous flow", *J. Appl. Phys.*, vol. 54, pp. 7153-7158, 1983.

[6] M. Ohring, *Materials science of thin films*, Academic press, 1993. (p. 425)

[7] D. E. Discher, P. Janmey, Y. Wang, "Tissue Cells Feel and Respond to the Stiffness of Their Substrate", *Science*, vol. 310, pp. 1139-1143, 2005.

[8] M. Marelli, N. Gadhari, P. Biro, M. Chiquet, J. Brugger, "Micro-beams with tunable stiffness and curvature for mechano-sensitive cell culture substrates", 37th International Conference on Micro and Nano Engineering (MNE), Berlin, September 19-23, 2011.

[9] M. Ghibaudo, A. Saez, L. Trichet, A. Xayaphoummine, J. Browaeys, P. Silberzan, A. Buguin, B. Ladoux, "Traction forces and rigidity sensing regulate cell functions", *Soft Matter*, vol. 4, pp. 1836-1843, 2008.

[10] www.femtotools.com

[11] N.-P. Huang, R. Michel, J. Voros, M. Textor, R. Hofer, A. Rossi, D. L. Elbert, J. A. Hubbell, N. D. Spencer, "Poly(L-lysine)-g-poly(ethylene glycol) layers on metal oxide surfaces:□ surface-analytical characterization and resistance to serum and fibrinogen adsorption", *Langmuir*, vol. 17, pp. 489-498, 2000.

[12] T. Tsuchiya, M. Hirata, N. Chiba, "Young's modulus, fracture strain, and tensile strength of sputtered titanium thin films", *Thin Solid Films*, vol. 484, pp. 245-250, 2005.

CONTACTS

*mattia.marelli@epfl.ch, tel. +41 21 693 67 37
†juergen.brugger@epfl.ch, tel. +41 21 693 65 73

INTEGRATED MICRO CULTURE DEVICE FOR FULLY AUTOMATED CLOSED CULTURE EXPERIMENT OF EMBRYONIC BODY

Akane Yasukawa[1], Takuya Nishijima[2], Masashi Ikeuchi[1], and Koji Ikuta[1]
[1]The University of Tokyo, Tokyo, JAPAN
[2]Nagoya University, Aichi, JAPAN

ABSTRACT

This paper reports the development of a palm-size device called "PASMA (Pressure Actuated Shapable Microwell Array)" for formation, differentiation and analysis of embryonic bodies (EBs). By incorporating a transparent heat film, sterilized water including cotton and CO_2 generation agents, PASMA realized miniaturization of the whole process of EB experiments in one chip without a conventional big incubator. EB fabrication from human adipose-derived stem cells (hASCs) and human induced pluripotent stem cells (hiPS cells) were successfully conducted using this system. The massively parallel analysis system "PASMA" should accelerate the regenerative medicine as a total EB research platform.

INTRODUCTION

Regenerative medicine has been paid attention to as a replacement of organ transplant. In regenerative medicine, a tissue or an organ made from patient's own cells is transplanted. Figure 1 shows the typical process of regenerative medicine. First, patient's cells are initiated to stem cells which has a capacity of changing to any kinds of cell types. This change is called differentiation. Before differentiation, a 3D multicellular spheroid of stem cells called an embryonic body (EB) is fabricated to promote differentiation [1]. Those differentiated cells are organized into a tissue or an organ and transplanted to that patient.

Usually, specific reagents are introduced in order to induce EB differentiation into a typical cell type. In addition, the size of EB is also an important factor for differentiation [2]. Therefore, development of massively parallel automated process is required for reagent introduction and size control

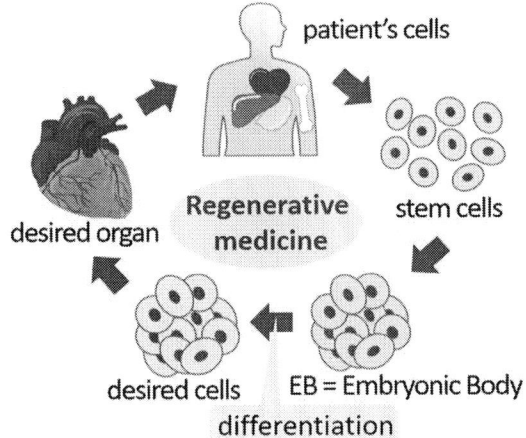

Figure 1: process of regenerative medicine

of EBs to promote regenerative medicine.

There are several conventional methods to make EBs: suspension culture, hanging-drop method [3] and so on. In suspension culture, big low-adhesive plates are used. It can make many EBs easily, but it is difficult to control the size of EBs, and the more, only one combination of reagents can be introduced to EBs. In the hanging drop method, cell including droplets are hanged from the lid of a plate. The size of EBs can be controlled, but it needs researchers' elaborate technical skills. In this method, reagents cannot be introduced during culture. Thus, time-consuming processes are needed to move all EBs to another space for differentiation. Therefore, an EB handling device which can be used to fabricate, differentiate and manipulate EBs efficiently is strongly required.

To achieve this goal, we developed a "Pressure Actuated Shapable Microwell Array (PASMA)" [4]. EB fabrication and differentiation were demonstrated in PASMA [5]. Our final goal is to develop the chip for fully automated closed culture experiments without a conventional big incubator. Here, we reported the development of PASMA equipped with the whole function of an incubator on a palm-size chip.

CENTRAL PART OF PASMA

PASMA was firstly developed as a combinatorial analysis device of EBs.

Figure 2(a) shows the central part of PASMA. It has 100 microwells all of which is arranged in 1cm-square area, and almost the same size of EBs are easily formed in each microwell (Figure 2(b)). Figure 2(c) shows the magnified view of the central part. There are 10 independent fluidic lines and 10 independent pressure actuation lines, making 10x10 orthogonal coordinate system. The bottom of fluidic lines are composed of a thin elastomer membrane, and pressure actuation lines are placed under the membrane. This design enables selective collection, differentiation induction to multiple cell types, and on-chip analysis of EBs.

Figure 2(d) shows the EB formation and collection process using PASMA. The membrane is bent downward to form microwells by applying negative pressure to pressure actuation lines. Then, cell suspension liquid is introduced into fluidic lines. The surface of fluidic lines are modified to be non-adhesive, so after cells fall into each microwell, they self-assemble into an EB. After forming EBs, reagent introduction for differentiation induction, EB stain for differentiation verification and other operation can be conducted. PASMA has a function of selective EB collection. Target EBs are collected by applying positive pressure to a pressure actuation line and then flowing medium into a

978-1-4799-3510-9/14 $31.00 © 2014 IEEE

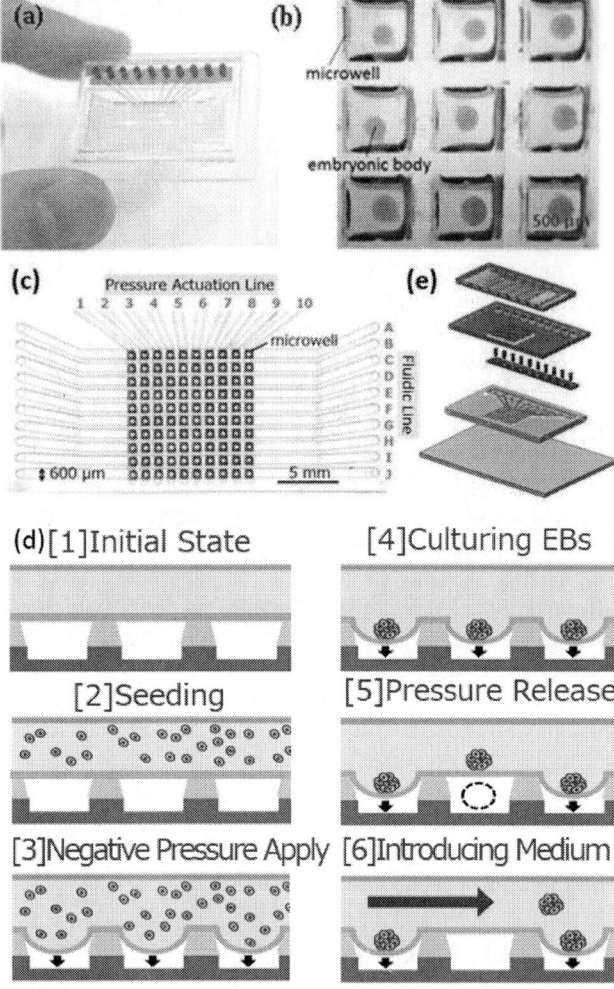

Figure 2: central part of PASMA (a) total figure (b) EB fabrication (c) magnified view (d) EB formation and collection process (e) structure

fluidic line. Applying negative pressure again, EB culture or collection can be continued.

The structure of PASMA is shown in Figure 2(e). It is made from three poly-dimethylsiloxane (PDMS) layers. The middle layer has through-holes arrayed 10x10 with a thin PDMS membrane attached on it. The bottom layer has 10 pressure actuation lines. The connecter part attached to the bottom layer is fabricated with photocurable resin. It is used for connecting PASMA to a vacuum pump through silicon tubes. This part is postcured in order to get biocompatibility and stiffness [6]. The top layer has 10 fluidic lines placed perpendicularly to the pressure actuation lines. These lines were modified hydrophilic with amphiphilic polymer (Anti-Link, Alvivo Inc.) before seeding cells to prevent cell adhesion to the thin membrane and promote self-assembling. The three layers are stacked and sealed together by using atmospheric-pressure plasma and uncured PDMS as an adhesive.

INCUBATOR FUNCTION EQUIPPMENT

Incubators have several functions for culturing cells: 1) temperature control (37°C), 2) humidity control (100%), 3) carbon dioxide (CO_2) concentration (5%) control, 4) maintenance of the clean environment.

The concept of a new PASMA equipped with cell incubation capability is shown in Figure 3. To keep the temperature, a transparent heater is placed on top of the fluidic lines. Cotton including sterilized water is set in a small chamber to keep humidity, and CO_2 generation agent is stored in another small chamber. The whole device is enclosed in a palm-sized outer case.

Heater

For efficient and precise temperature control of the medium, the heater had to be attached onto the top surface of PASMA. The heater needed to be transparent for EB observation under an optical microscope. It was also required not to become an obstacle for smooth operation during experiments. To meet these requirements, we applied a heat film made of ITO (Indium Tin Oxide) with transparency and electric conductivity.

The temperature of ITO film depends on the applied voltage. The relationship between applied voltage and the maximum temperature of an ITO film was measured and figure 4(a) shows the result. There was a proportional relationship between the voltage and the temperature of an ITO film.

Figure 4(b) shows the image of an ITO film. The size of the film is 20x17 mm. Both of the long sides of the film were coated with gold by sputtering, and connected with a wire by using conductive tape made from copper foil. It was put as the wires are parallel to the fluidic lines and they led to outside of the case from the side and connect to a power supply. Figure 4(c) shows the thermography image of an ITO film. The temperature was almost uniform because two wires existed parallel.

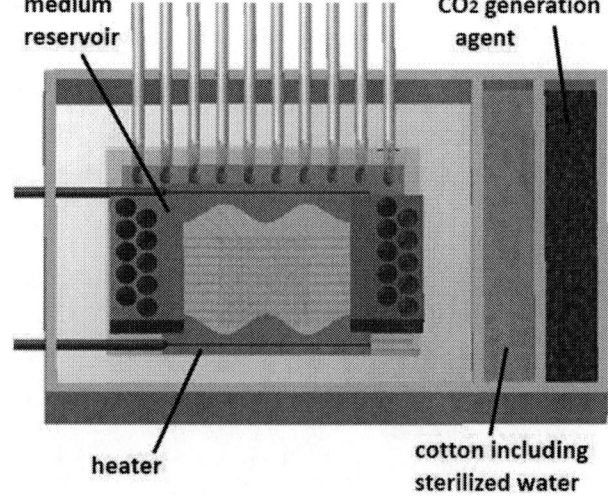

Figure 3: The concept of new PASMA equipped with cell incubation capability

978-1-4799-3510-9/14 $31.00 © 2014 IEEE

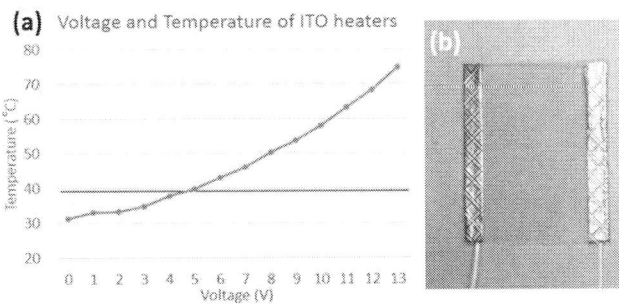

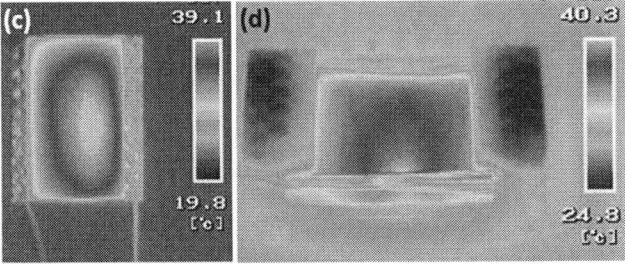

Figure 4: New PASMA equipped with the function of incubators (a) the relationship between applied voltage and maximum temperature of an ITO film (b) image of an ITO film (c) thermography image of an ITO film (c) thermography image of central part of PASMA for calibration (e) total figure of new PASMA

Each ITO film has different proportional constant between the voltage and the temperature, so calibration is required every time before culturing. We used a thermography for calibration. A typical thermography image for calibration is shown in Figure 4(d). Usually, the temperature is kept at 37 °C in cell culture incubators. In PASMA, however, the maximum temperature of the ITO heater was kept about 39 °C which was little high but not toxic to cells. This is because the medium was heated only from the upper side in PASMA, making heat diffuse easily.

Humidity

The volume of culture medium is only 20 µL in each fluidic line and PDMS has a good breathability. Also, a heater is put only on the surface of fluidic lines. These make all the medium evaporate easier.

To keep the humidity inside the PASMA, a cotton including sterilized water was set in a small chamber. Water evaporation was the fastest at the beginning of culture, because it was under saturated vapor pressure. In order to promote water evaporation from cotton at the beginning, the temperature of sterilized water was set at 45 °C.

Also, medium reservoirs were fabricated, and the both ends of fluidic lines were connected to the reservoirs in order to avoid evaporation of medium. The volume of the reservoir was equivalent to that of fluidic lines. It had 10 independent chambers for each line to keep the medium, so it also prevented mixing of medium in adjacent fluidic lines. It was designed to be stable on the central part of PASMA and it also worked in fixing the heater on the fluidic line layer.

CO_2 concentration

CO_2 concentration is 5 % in usual cell culture incubators to keep the freshness of medium. To control the CO_2 concentration, CO_2 generation agent was set in new PASMA. It is difficult to measure CO_2 concentration directly, but it can be checked by the color of medium due to the existence of phenol red. The color of medium with CO_2 generation agent was almost the same as the color of that in usual incubators and was different from the color of that put outside of incubators. This means CO_2 has been rightly provided to cells. For keeping CO_2 concentration, new PASMA should be closed completely and sealed. Sealing tape which both gas and water cannot pass through was used for sealing. It can also keep the humidity.

Fabrication

All of the culture system, except for a pump and a power supply, was packed in a palm-sized case (68*39*15 mm) shown in Figure 4(e). It has some holes for leading tubes and codes outside, so it is easy to put the lid.

DEMONSTRATION

We demonstrated EB fabrication for verifying the new PASMA with the function of an incubator.

EB fabrication from hASCs

First, human Adipose-derived Stem Cells (hASCs) capable of differentiation into several cell types were used [7]. Cell suspension liquid of hASCs were prepared at a density of 1.0×10^6 cell/mL, and 20µL of it was seeded to each fluidic lines of new PASMA.

1day after seeding, hASCs self-assembled into EBs shown in Figure 5(a). 1 EB was fabricated in all wells. The shape of fabricated EBs were almost orbicular. The diameter of EBs was almost the same and was about 200 µm. Collected EBs were stained by Calcein-AM/PI shown in figure 5(b). All cells were fluorescent to green indicating to be alive.

EB fabrication from hiPS cells

Next, human induced pluripotent stem cell (hiPS) were used [8]. They are fabricated from skin cells and can differentiate into any kinds of cell types , so it doesn't need ethical controversy. From these reasons, iPS cells attracted

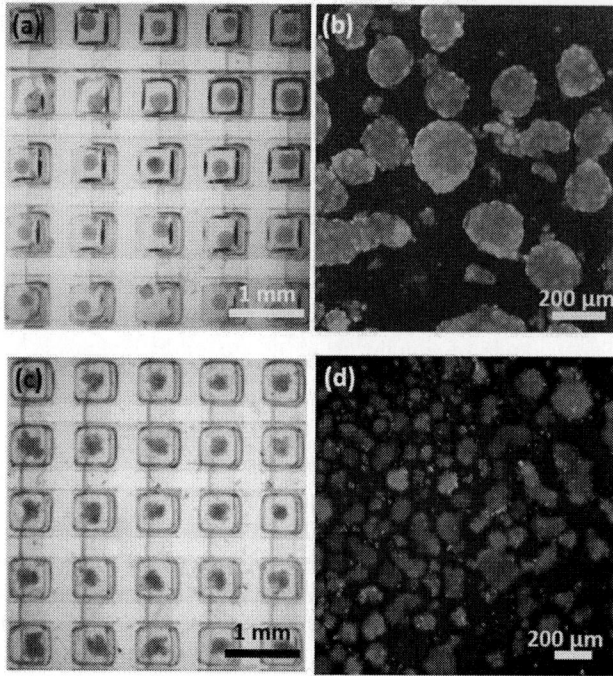

Figure 5: Demonstration of new PASMA (a) EBs fabricated from hASCs (b) Calcein-AM/PI stain of EBs fabricated from hASCs (c) EBs fabricated from hiPS cells (d) Calcein-AM/PI stain of EBs fabricated from hiPS cells

attentions and used in regenerative medicine. For future clinical applications, it is important to verify EB fabrication from iPS cells.

iPS cells are usually cultured on feeder cells in order to keep undifferentiation. However, when EBs were fabricated from on-feeder culture of iPS cells, feeder cells couldn't be completely removed and feeder cells disturbed fabrication and observation of EBs. So feeder-less culture of iPS cells were prepared for EB fabrication. Cell suspension liquid should be prepared in order to make the same size of EBs, but iPS cells lead to apoptosis when they become single cells. To prevent iPS cells from apoptosis at the time of cell suspension liquid, we applied rock inhibitor of Y-27632 to the culture medium [9].

One day after seeding, EBs were fabricated as shown in Figure 5(c). Figure 5(d) shows EBs in another PASMA collected and stained by Calcein-AM/PI stain. EBs were fluorescent to green and confirmed to be alive. It also shows that single cells or very small aggregation were fluorescent to red and confirmed to be dead. Apoptosis of single iPS cells couldn't be avoided even with Y-27632, but iPS cells forming EBs were kept alive. Thus, EB fabrication from hiPS cells were successfully conducted in new PASMA.

CONCLUSION

We developed a new EB culture device PASMA which has the function of incubators. Using an ITO film, cotton including sterilized water and CO_2 generation agent, we succeeded in realizing the same environment of a conventional big incubator in a palm-size chip. We also verified EB fabrication from hASCs and hiPS cells. The massively parallel analysis system "PASMA" should accelerate the regenerative medicine as a total EB research platform.

REFERENCES

[1] Lin R.Z. and Chang H.Y. "Recent advances in three-dimensional multi cellular spheroid culture for biomedical research", *Biotechnology Journal* 3 (2008), pp.1172-1184

[2] Hwang Y.S., Chung B.G., Ortmann D., Hattori N., Moeller H.C., and Khademhosseini A. "Microwell-mediated control of embryoid body size regulates embryonic stem cell fate via differential expression of WNT5a and WNT11", *Proceedings of the National Academy of Sciences* 106.40 (2009), pp.16978-16983

[3] Yoon B.S., Yoo S.J., Lee J.E., You S., Lee H.T. and Yoon H.S. "Enhanced differentiation of human embryonic stem cells into cardiomyocytes by combining hanging drop culture and 5-azacytidine treatment", *Differentiation* 74 (2006), pp.149-159

[4] Nishijima T., Ikeuchi M. and Ikuta K. "Pneumatically actuated spheroid culturing Lab-on-a-Chip for combinatorial analysis of embryonic body", *Micro Electro Mechanical Systems (MEMS), 2012 IEEE 25th International Conference on* (2012), pp. 92-95

[5] Yasukawa A., Ikeuchi M. and Ikuta K. "Combinatorial differentiation induction of embrionic bodies in" PASCL (Pneumatically Actuated Spheroids Culture Lab-on-chip)"", *Micro Electro Mechanical Systems (MEMS), 2013 IEEE 26th International Conference on* (2013), pp.931-934

[6] Inoue Y. and Ikuta K. "Cell culture biochemical IC chip with cell-level biocompatibility", *Micro Electro Mechanical Systems (MEMS), 2012 IEEE 25th International Conference on* (2012), pp.788-791

[7] Bunnell B.A., Flaat M., Gagliardi C., Patel B. and Ripoll C. "Adipose-derived stem cells: isolation, expansion and differentiation", *Methods* 45 (2008), pp.115-120

[8] Takahashi K., Tanabe K., Ohnuki M., Narita M, Ichisaka T., Tomoda K. and Yamanaka S. "Induction of pluripotent stem cells from adult human fibroblasts by defined factors", *cell* 131.5 (2007), pp.861-872

[9] Watanabe K., Ueno M., Kamiya D., Nishiyama A., Matsumura M., Wataya T. and Sasai Y. "A ROCK inhibitor permits survival of dissociated human embryonic stem cells", *Nature biotechnology* 25.6 (2007), pp.681-686

CONTACT

Koji Ikuta, RCAST, The University of Tokyo, Room 506, Building 4, 4-6-1 Komaba, Meguro-ku, Tokyo, 153-8904, Japan, ikuta@rcast.u-tokyo.ac.jp,
Tel: +81-3-5452-5162, Fax: +81-3-5452-5163

MECHANICAL CELL PAIRING SYSTEM BY SLIDING PARYLENE RAILS

Yuta Abe[1,2], Koki Kamiya[1,3], Toshihisa Osaki[1,4], Ryuji Kawano[1], Norihisa Miki[1,2] and Shoji Takeuchi[1,4]

[1] Kanagawa Academy of Science and Technology, Japan

[2] Keio University, Japan

[3] JST PRESTO, Japan

[4] Institute of Industrial Science, The University of Tokyo, Japan

ABSTRACT

This paper proposes a cell pairing system that is capable of defining the number and the position of trapped cells by mechanically sliding the parylene rail films (PRF). The device can control the area of trapping sites of cells by sliding PRF on a SU-8 comb layer. This mechanism allows us to control the number as well as the order of lined-up cells. We successfully demonstrated lining up of three different cells in a designated order. The proposed system is readily applicable to study the cell-cell interactions using the single cell pairing.

INTRODUCTION

Methodologies of pairing different types of cells can be powerful tools for various purposes such as the observation of cell-cell interaction [1], fusion of cells [2], and co-culture [3] for drug screening, tissue engineering and immunology. In recent years, two cells trap and contact method have been developed using microfluidic technologies. Skelley et al. developed a microfluidic device which contains a dense array of U-shaped structures [2]. With three-step loading, this device could trap thousands of proper two cells pairing with different cell types. Şen et al. proposed a device that uses positive dielectrophoresis for trapping cells and negative dielectrophoresis for contact within the microwells [4]. Our group has presented an improved meander-shaped dynamic microfluidic device for pairing different types of microobjects [5]. Given that the device allows sequential trapping, it will be useful for pairing expensive or rare samples in an array without losing them. However, these proposed devices require complex three-dimensional fabrication techniques or lack the throughput. In addition, it is impossible to trap three or more cells in contact because of the device characteristics.

In our previous study, we developed a device which could trap and bring into contact three or more cells in a line using a parylene membrane containing an array of micro-ordered rails. The number of cells could be controlled based on the rail lengths [6]. We also demonstrated that the direct dye transfer through the cell membranes was mediated by the membrane protein, connexin. However, the ordered cell pairing remained difficult to achieve, because cells trapped on rails depending on the concentration or spatial position. Therefore, the ordered pairing of three or more cells is challenging.

In this study, we developed a mechanical cell pairing system by sliding PRF on the SU-8 comb layer. The area of the trapping sites, where the gentle downward flow was generated, could be controlled by changing the open rail

lengths with tweezers. Here, we demonstrated that cells were captured at designated positions as well as the number of cells was controlled. By opening the trapping site area after trapping one cell, it is possible to capture another cell next to the cell in contact. Finally, by repeating this operation, we could pair the multiple cells in correct order on PRF, as shown in figure 1. We believe that the proposed device can accelerate the analysis of cell-cell interactions aiding not only drug discovery but also biology in general.

FABRICATION

The proposed device consists of a glass substrate, SU-8 comb layer, PRF, PDMS box, and PDMS cover. PDMS box works as the carrier of PRF. Figure 2a,b show the fabrication process of PDMS box and PDMS cover. Pre-cured PDMS was casted into a chamber of the acrylic mold and cured to get the PDMS box replica. The master molds for PDMS replicas were fabricated by milling the acrylic plates and

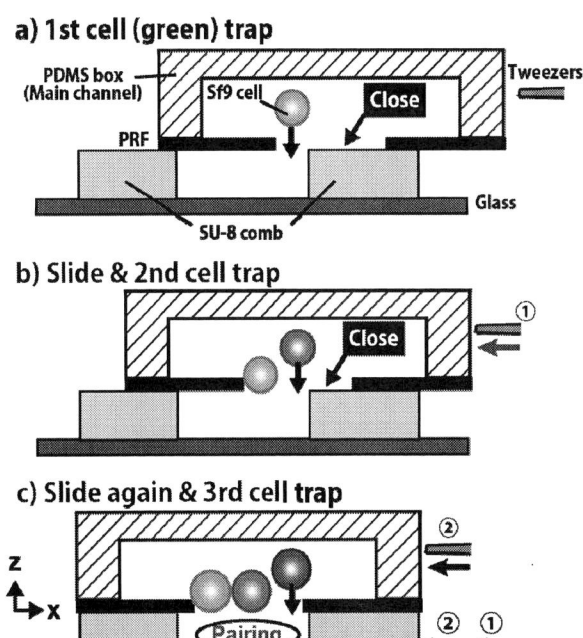

Figure 1: Conceptual illustration images of mechanical parylene slide system for ordered three cells pairing. a) At first, PRF on SU-8 layer was slide at open positions. First cells (green) are trapped on left side of rails. B) Next, PRF, which is attached to the PDMS box, slide a little by tweezers and second cells (red) are trapped at center of the rail. c) Finally, third cells (blue) are trapped on the opposite side (right) of the rails by the same way.

978-1-4799-3510-9/14 $31.00 © 2014 IEEE

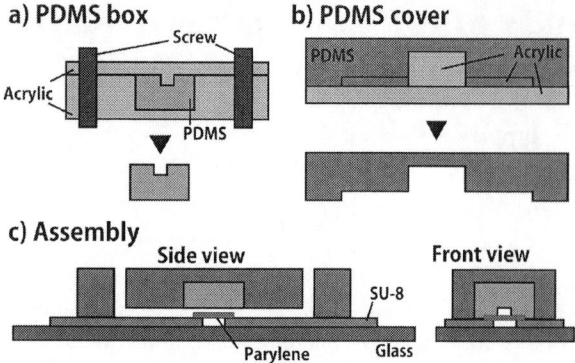

a) PDMS box

Screw

Acrylic

PDMS

b) PDMS cover

PDMS

Acrylic

c) Assembly

Side view

SU-8

Parylene

Glass

Front view

Figure 2: Fabrication process of the device. The device consists of six parts; glass, SU-8 comb layer, PRF, PDMS box, PDMS cover and acrylic holder. a, b) PDMS box and cover was molded by the acrylic structures. SU-8 comb layer and PRF were fabricated by standard lithography. c) After the parts fabrication, SU-8 layer and PDMS box were exposed to oxygen plasma and assembled.

assembled. For the introduction of different cells, the PDMSs have a straight microchannel connected to a syringe pump. The PRF and the SU-8 layer were fabricated by standard photolithography [1,6,7]. We patterned the micro rails on a parylene film, whose thickness was 5 μm, and the film was peeled off from Si wafer [8] in order to be inserted into PDMS box and the SU-8 layer. A rail length of PRF was 36 μm. The thickness of the SU-8 layer was 50 μm by spin-coat on a glass [9]. The surface of the SU-8 layer and PDMS box were exposed to oxygen plasma to make surfaces hydrophilic [10,11], which enables a smooth slide of the PDMS box on the SU-8 layer in water solution. Figure 2c shows the assembled image. There was a gap (~200 μm) between PDMS box and PDMS cover. By deforming the outer PDMS cover with tweezers, PDMS box with PRF can slide on the SU-8 layer. The alignment of PRF and SU-8 comb layer was carried out under a microscopy. Finally, a set of glass, SU-8 comb layer, PRF, PDMS box and PDMS cover was bind by acrylics to support the adhesion.

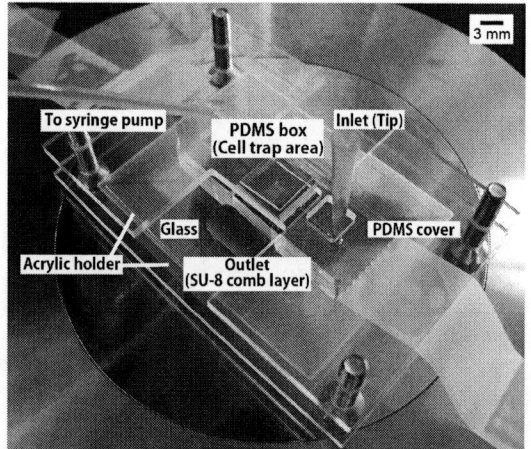

3 mm

To syringe pump

PDMS box
(Cell trap area)

Inlet (Tip)

Glass

PDMS cover

Acrylic holder

Outlet
(SU-8 comb layer)

Figure 3: Image of fabricated device. A set of glass, SU-8 layer, PRF, PDMS box and PDMS cover was bind by acrylics to support the adhesion.

EXPERIMENTAL
PRF Sliding on SU-8 Comb Layer

The device characteristic of sliding the PRF on SU-8 comb layer was confirmed. After the device fabrication, the device was filled with cell culture medium. PDMS cover was pushed and deformed by manual tweezers manipulation. The point of effort was near the center of PDMS box and the direction was perpendicular to the main microchannel. The experiments were conducted with or without oxygen plasma exposure to the surface of the SU-8 layer and PDMS box. The movement of PRF and PDMS box was observed under a microscopy.

Selective Cell Trap by Controlling the Area of Trapping Sites

For the multiple cells pairing in correct order, the selective trapping of target cell at designated position is important. To confirm the selective cell trap, Sf9 cell was used. The population of cell stock solution was about 3.0×10^5 cell/mL and 10 μL of cell stock solutions were injected into the inlet by a pipette. After 5 min, the exceeded cell solution was washed out with cell culture medium for 15 min. The flow rate of micro pump was 2.0 μL/min. The experiments were carried out when the left sides of the rails were full close, 5 μm open and 36 μm (full) open states. The position and the number of trapped cells were observed under a microscopy.

Triple Cell Pairing

Next, to confirm the device capability of multiple cells pairing in the correct order, the green, red and blue cells pairing was conducted. Sf9 cells were stained with calcein AM (green), calcein red-orange AM (red) or DAPI (blue). Each population of stained Sf9 cells was adjusted to be about 3.0×10^5 cell/mL and 10 μL of cell stock solutions were injected into the inlet by a pipette. At first, PRF on SU-8 layer was slid at 5 μm open positions and green cells are introduced for 5 min. The exceeded cell solution was washed out with cell culture medium for 15 min. After washing process, the PRF was slid at 20 μm open positions and red cells were introduced by same step. Finally, the PRF was slid at 36 μm open positions and blue cells were introduced. The flow rate of micro pump was 2.0 μL/min. By introducing cells in these ways, cells were trapped in the order of green, red and blue cells from left side of PRF. The ordered cell pairing was observed under a fluorescence microscopy.

RESULTS AND DISCUSSION
PRF Sliding on SU-8 Comb Layer

The movement of PRF on SU-8 comb layer was observed. In this experiment, we compared the movement of PDMS box with PRF with or without oxygen plasma exposure. As a result, when the oxygen plasma was exposed to the surface of the SU-8 layer and PDMS box, the PDMS box could slide on the SU-8 layer. On the other hand, without oxygen plasma exposure, the PDMS box could not move. These results suggest that wettability of the surfaces is

978-1-4799-3510-9/14 $31.00 © 2014 IEEE

important for the movement of PRF on the SU-8 layer. Because the exposure of oxygen plasma made the both surface hydrophilic [10,11], the liquid between PDMS and SU-8 acted as a lubricant which reduces the friction between PDMS box and the SU-8 layer. Therefore PDMS box with PRF could slide smoothly on the SU-8 layer.

Selective Cell Trap by Controlling the Area of Trapping Sites

After the device preparation, we used the Sf9 cell to demonstrate that the position and the number of trapped cells could be controlled by the trapping site area. As a result, when the rails were fully closed by the SU-8 layer, the cells were not trapped on PRF. On the other hand, when the rails were open at 5 μm and 36 μm (full open state), cells were trapped on the left and center side of PRF. The number of cells was 1.4 and 1.9 cells/rail for 5 μm and 36μm open states, respectively. These results suggest that the SU-8 layer and PRF were physical contact and the generation of gentle downward flow could be controlled.

Triple Cell Pairing

To confirm the device capability of three or more cells pairing in the correct order, we conducted the green, red and blue cells pairing. Figure 4 shows the microscopic image of trapped triple cells. At first, PRF on SU-8 layer was slid at 7 μm open positions and green cells were trapped on the left side of the rails. Next, PRF was slide at 23 μm open positions and red cells are trapped on the center of the rail. Finally, blue cells were trapped on the right side of the rails by fully opening the rails. As a result, we successfully carried out the ordered pairing of three cells as shown in figure 4.

CONCLUSION

We developed the mechanical cell pairing system by sliding the PRF on SU-8 comb layer. PRF was slid on the surface of SU-8 comb layer and the area of trapping sites was controlled with tweezers manipulation. We first demonstrated that the PRF could be slid on the SU-8 layer with oxygen plasma exposure. Next, the position and the number of cells were controlled with the area of trapping sites. Finally, we successfully demonstrated lining up of three different cells in a designated order. We believe that this mechanical cell pairing system will play an important role in the observation and analysis of cell-cell interaction, fusion or co-culture for drug discovery.

ACKNOWLEDGEMENTS

We thank Utae Nose, Yoshimi Nozaki, Maiko Uchida, and Yumi Kagamihara for technical assistance in device preparation and arranging experimental laboratory.

REFERENCES

[1] N. Ye, C. Bathany, and S. Z. Hua, "Assay for Molecular Transport across Gap Junction Channels in One-Dimensional Cell Arrays", *Lab Chip*, vol. 11, pp. 1096-1101, 2011.

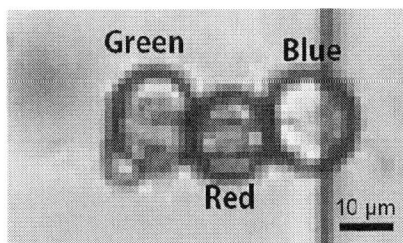

Figure 4: Microscopic image of green, red and blue cells pairing on one rail.

[2] A. M. Skelley, O. Kirak, H. Suh, R. Jaenisch, and J. Voldman, "Microfluidic Control of Cell Pairing and Fusion", *Nat. Method*, vol. 6, pp. 147-152, 2009.

[3] S. Hong, Q. Pan, and L. P. Lee, "Single-Cell Level Co-Culture Platform for Intercellular Communication", *Integr. Biol.*, vol. 4, pp. 374-380, 2012.

[4] M. Şen, K. Ino, J. Ramón-Azcón, H. Shiku, and T. Matsue, "Cell Pairing Using a Dielectrophoresis-Based Device with Interdigitated Array Electrodes", *Lab Chip*, vol. 13, pp. 3650-3652, 2013.

[5] T. Teshima, H. Ishihara, K. Iwai, A. Adachi, and S. Takeuchi, "A dynamic Microarray Device for Paired Bead-Based Analysis", *Lab Chip*, vol. 10, pp. 2443-2448, 2010..

[6] Y. Abe, K. Kamiya, T. Osaki, R. Kawano, K. Akiyoshi, N. Miki, and S. Takeuchi, "Mechanical Cell Contact System by a Parylene Rail Filter for Study of Cell-Cell Interaction Mediated by Connexin Gap Junction", in *Proc. MicroTAS'13 Conference*, Freiburg, Oct 27-31, 2013, pp. 407-409.

[7] W. Tan, and S. Takeuchi, "A Trap-and-Release Integrated Microfluidic System for Dynamic Microarray Applications", *PNAS*, vol. 104, pp. 1146-1151, 2007.

[8] R. Kawano, T. Osaki, H. Sasaki, and S. Takeuchi, "A Polymer-Based Nanopore-Integrated Microfluidic Device for Generating Stable Bilayer Lipid Membranes", *Small*, vol. 6, pp. 2100-2104, 2010.

[9] Y. Abe, K. Kamiya, T. Osaki, R. Kawano, N. Miki, and S. Takeuchi, "Confocal Laser Scanning Microscopic Observation of Deformation, Biological Reaction, and Contact of Cells Using Mechanical Trapping System with Parylene Micro Filter", *Proc. Transducers '13 Conference*, Barcelona, June 16-20, 2013, pp. 416-417.

[10] S. H. Tan, M.-T. Nguyen, Y. C. Chua, and T. G. Kang, "Oxygen Plasma Treatment for Reducing Hydrophobicity of a Sealed Polydimethylsiloxane Microchannel", *Biomicrofluidics*, vol. 4, pp. 032204, 2010.

[11] F. Walther, P. Davydovskaya, S. Zürcher, M. Kaiser, H. Herberg, A. M. Gigler, and R. W. Stark, "Stability of the Hydrophilic Behavior of Oxygen Plasma Activated SU-8", *J. Micromech. Microeng.*, vol. 17, pp. 524-531, 2007.

CONTACT

* Y. Abe, tel: +81-44-819-2037; yuta.a@a6.keio.jp

TWO-DIMENSIONALLY STEERING MICROSWIMMER PROPELLED BY OSCILLATING BUBBLES

Jian Feng and Sung Kwon Cho
University of Pittsburgh, Pittsburgh, Pennsylvania, USA

ABSTRACT

This paper presents how two-dimensional propulsion with steering capability is developed and realized for micro swimmers that may be used to navigate inside of human body. The propelling and steering forces are produced by oscillating multiple bubble columns, which are arranged in different directions on the micro swimmer. Since the resonant frequency of the bubbles highly depends on their length, only bubbles having the frequency-matching length can be selectively resonated by an acoustic excitation. This allows us to control which bubbles among many can be activated for propulsion, simply by adjusting the frequency. As a result, the controlling of propulsion and steering can be made by changing the frequency. This paper describes experimental results of a variety of controlled 2-D motions including back/forth translations, forward-and-backward rotations, steered propulsion on a 2-D plane, and carrying of objects in a T-junction microchannel.

INTRODUCTION

A number of attempts have been made to fabricate to and realize micro swimmers that possibly navigate inside of human body including the artificial magnetic bacteria flagella [1, 2], chemical micro swimmer [3] and natural organism based swimmer [4]. Micro bubble propulsion is also one of the promising methods. Thermally-actuated bubbles [5] and AC-electrowetting-actuated bubbles [6] propelled swimmers which used the flow field generated by periodical oscillation of the bubbles. Dijkink *et al.* [7] built the 'acoustic scallop' in millimeter scale and demonstrated the linear swimming. The advantage of acoustic excitation compared with other bubble driving methods is that the power is transferred wirelessly which makes in vivo applications easier. Later, Feng *et al.* [8] reduced the size of the acoustic bubble device down to micron scales by photolithography and showed a fast swimming speed (maximum ~ 80 body lengths per second). However, all of the previous acoustic bubble works is limited to 1-D linear motions. In this paper, the micro swimmers which propel and steer two-dimensionally are presented including translational and rotational motions.

Propulsion Concept

The propulsion concept is illustrated in Figure 1, including the conceptual and fabricated devices of 1-D linear propulsion and conceptual devices of 2-D translation and rotation. The propulsion physics is that when the bubble is excited by an acoustic field, it expands and shrinks periodically [9]. In particular, when the bubble column is trapped in a one-end-open micro tube, the meniscus of the bubble oscillates back and forth at the open end. As a result,

the fluid near the micro tube is drawn in and ejected out periodically. Importantly, the flow patterns in the intake and discharge cases [7] are asymmetric. This asymmetric flow patterns will generate a non-zero time-averaging streaming flow, which is responsible for propulsion. When the generated force is strong enough to overcome resistance including flow drag and friction with the tank bottom surface, the swimmer is pushed away. In order to realize propelling and steering on a 2-D plane, multiple micro bubbles should be arranged in different directions. In addition, the bubbles should be different in length so that they respond to different frequencies. As such, the bubbles will be able to be activated individually and selectively. This allows us to generate propulsion forces in any direction.

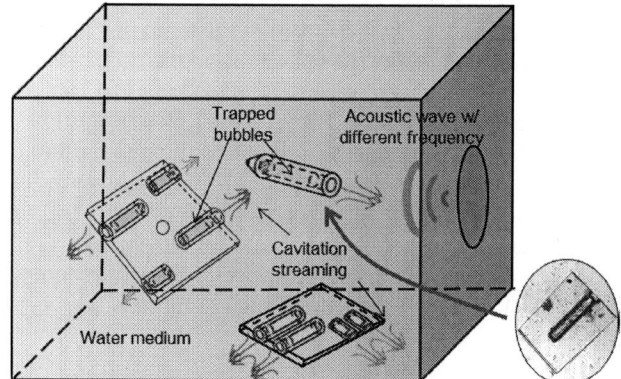

Figure 1: Concept of 2-D propulsion by oscillating bubble columns

MICROFABRICATION/EXPERIMENTAL SETUP

Microfabrication processes are used to build the micro tubes which are used to trap micro bubbles. Micro photolithography is used to accurately define the size, shape and arrangement of the micro tubes. The fabrication process is quite similar with that in the previous work [8]. The main difference is that an aluminum layer is deposited on the second parylene layer by a sputter as a protective mask before dry etching. This circumvents some issues caused by the similar RIE etching rates of photoresist and parylene in the oxygen/carbon tetrafluoride plasma. This aluminum layer is patterned with the second thin photoresist (AZP4210) layer. The advantage of using the aluminum layer over other types of metals is that it can be easily dissolved by an alkaline-based developer and does not need the traditional acid wet etching process which means both photoresist layer and aluminum layer are patterned at the same time. As a result, the entire process is simplified. The silicon wafer that is used as the substrate is separated from the parylene structure by simply being dipped into water after RIE etching.

978-1-4799-3510-9/14 $31.00 © 2014 IEEE

All the photoresist layers serve as sacrificial layers and are eventually removed by acetone. After dehydration, only the parylene structure is left in the end.

The bottom view of a single micro tube structure is shown in inset of Figure 1. The micro bubble is automatically trapped when the device is submerged into water. The tube is composed of two layers of parylene, and each layer is 7.5 μm thick. Below this thickness, the micro tube is not strong enough to serve as a rigid body. In other words, the tube would expand and shrink when an acoustic wave is applied. This reduces the amplitude of bubble oscillation and thus momentum transfer from the open end of the micro tubes. On the other hand, a thicker layer would bring more mass to the device and thus more friction between the device and tank bottom surface. The height of the micro tube opening is about 45 μm. Generally speaking, a larger height or larger cross sectional area would enhance the micro streaming flow at the open end and thus increase the propelling force. However, too large height will affect the cross sectional shape in exposure process. The length of micro tube is varied to get the different bubble resonance. There are several reasons for using parylene for the tube structure. First, parylene is hydrophobic, which means a micro bubble would be trapped inside of the micro tube automatically when submerged in water. The length of bubble is determined by the micro tubes, which allows us to control the bubble length relatively easily. Second, parylene is transparent such that the meniscus motion can be easily observed and the fluid pattern near the meniscus can be visualized by particle tracking. Third, parylene is flexible (but not much stretchable), which facilitates swimming inside of a micro channel and overcoming collisions. Last, the density of parylene is a little larger than that of water. So, the bubble trapped device will not easily float on the water but stay on the bottom surface with minimal normal forces and friction.

The external acoustic field is generated by a disk type ceramic piezoelectric actuator (dia. 27-mm, resonance freq. 4.6 +/-0.5 kHz). The actuator is glued to the side wall of a rectangular open acrylic water tank. The thickness of all the tank walls is 3 mm, and the dimensions of the tank are $11 \times 11 \times 5$ cm^3. The water depth is maintained at 3 cm for all the propulsion testing. The sinusoidal input signal to the actuator generated by a function generator and an amplifier is monitored by an oscilloscope. The frequency of the signal is incremented by 100 Hz from 4 kHz to observe the device swimming and meniscus behavior. The device is placed at the bottom of the testing tank which is mounted on an inverted microscope. The microscope is connected to a high speed camera to capture the device propulsion.

BUBBLE FREQUENCY RESPONSE
When the bubble was oscillated at large amplitudes, the generated force was strong in most of the previous experiments. The theoretical bubble resonance frequency was calculated based on the existing models [10, 11]. The acoustic wave frequency was selected as close to the theoretical resonance frequency as possible. Typically,

however, the overall oscillation of the bubble is determined not only by the bubble characteristics but also other elements including the resonance behaviors of the actuator and tank. In addition, the water depth would also significantly influence the bubble oscillation due to fluid-loaded structure coupled between water and testing tank [12]. On the other hand, in 2-D propulsion, the difference between the forces generated by different bubbles must be high enough to generate a clear propelling motion in a desired direction. This is a quite complex condition different from 1-D propulsion which requires only the large bubble oscillation. Therefore, it is important to know the detailed bubble oscillation and propelling over a wide range of the frequency.

The generated streaming flow speed for two different bubble lengths is measured by sweeping the frequency from 4 kHz to 12 kHz. To measure the flow speed, micro particles (20 μm dia. polystyrene) are seeded into water and tracked with the high speed camera. By measuring the displacement and frame rate, the streaming velocity can be measured. The streaming speed obtained by averaging the speed of multiple particles is shown in Figure 2.

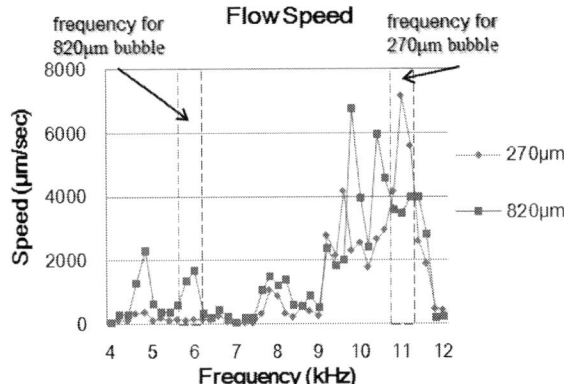

Figure 2: Propulsion speed vs. frequency for 820 μm and 270 μm long bubbles at 150 Vpp.

In Figure 2, it is clear that the shorter micro bubble (270 μm) generates a much stronger streaming flow at the frequency of 11 kHz, about twice larger than that of the longer bubble (820 μm) while the longer bubble does at the frequency of 6 kHz. Consequently, 6 kHz and 11 kHz are chosen for the operation frequency to selectively activate the 270 μm and 820 μm micro bubbles, respectively.

EXPERIMENT RESULTS
Back and Forth Propulsion
Switching of 1-D propelling direction is examined as shown in Figure 3. Six 270 μm and three 820 μm long bubbles are placed in parallel on the same device but the opening of the shorter bubbles are placed at the opposite side of those of the longer bubbles. When the acoustic input signal is set at 5.8 kHz (150 V$_{pp}$), the longer bubbles produce stronger micro streaming than that by the shorter bubbles as indicated by the black arrows in Fig. 3. This force by the longer bubbles is strong enough to overcome water drag,

friction with the tank surface and the force generated by the shorter bubbles in the opposite direction. As a result, the device is pushed up as indicated by the green arrow in Figure 3(a). On the contrary, when the frequency is switched to 11.2 kHz (95 V_{pp}), the shorter bubbles give a stronger pushing force, and the device moves downward in Figure 3(b). This experiment is repeated several times by alternatively using 5.8 kHz and 11.2 kHz signals. The result is always consistent, no matter how the device is placed (e.g., in vertical direction or horizontal direction). However, the propelling speed is a little slower than that of the 1-D uni-direction propelling in the previous publication [8]. The reason is that the resultant force is reduced due to the opposite bubble propulsion.

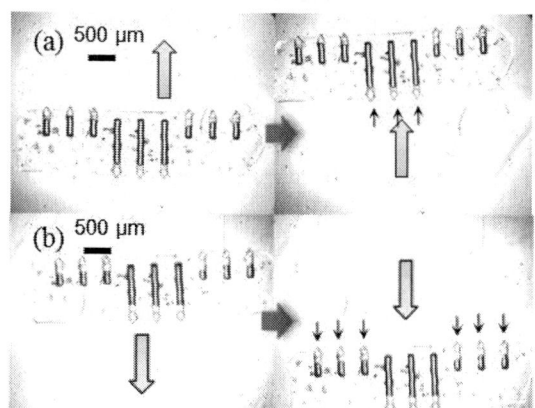

Figure 3: Back and forth propulsion (bottom view) by (a) 820 μm long bubbles at 5.8 kHz and (b) 270 μm short bubbles at 11.2 kHz.

Propulsion with 2-D Steering

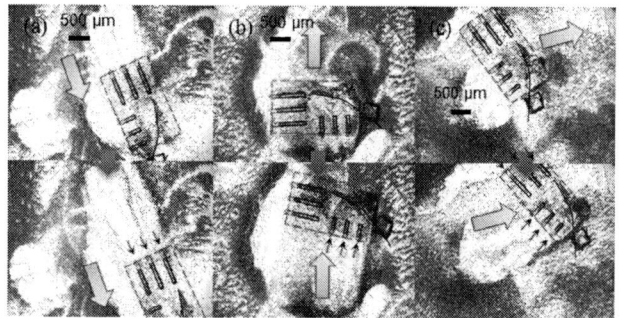

Figure 4: Figure 4: 2-D Motions: (a) the longer bubbles push; (b) the shorter push; (c) all the bubbles push under superposed acoustic field of the two frequencies (particles were seeded for flow visualization).

The similar device is also examined for 2-D steering propulsion as shown in Fig. 4. The major modification is that the number of 820 μm and 270 μm channels are both three and the alignment is orthogonal. A square shape plastic film load (~400 × 325 × 80 $μm^3$) is also attached to the device through a copper wire. Micro particles are seeded for fluid visualization. Figure 4(a) and (b) show propulsion by longer and shorter bubbles for the longitudinal and lateral motions, respectively. The direction of the movement is changed a little because of the non-uniform friction and the weak

oscillation of the other bubbles. For example, at 11.2 kHz, the shorter bubbles give a dominant propulsion force, however, the long bubbles also give a weaker force. When the superposed acoustic field is given, all the bubbles are oscillated and generates the motion in the direction of the resultant force as shown in Figure 4(c).

Rotation with Direction Reversing and Acoustic Motor

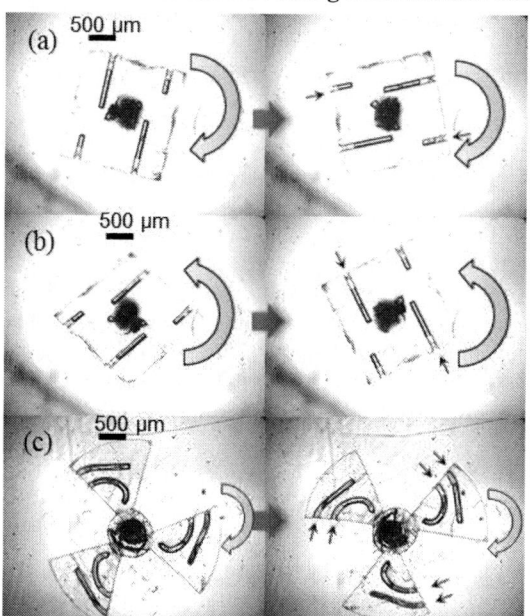

Figure 5: Rotation (bottom view) (a) clockwise rotation by 270 μm long bubbles at 11.0 kHz, (b) counterclockwise rotation by 820μm bubbles at 5.8 kHz and (c) bubble turbo-motor at 3.8 kHz.

The similar idea to 1-D back-and-forth motion is extended to rotation, as shown in Fig. 5. Two 820 μm and 270 μm micro channels are made on the same rectangular device. Each pair of channels could generate a rotation torque in the opposite directions to the other pair. In order to keep the device rotation always under the field of observation, the center of the device is punctured by an optical fiber. The bottom of this optical fiber anchors the device on the bottom plate of the testing tank as seen as the large black area in the center in Figure 5. As a result, the device can only rotate around this axis fiber instead of moving around.

When the acoustic field is applied at 11 kHz (170 V_{pp}), a torque generated by the shorter bubbles is larger than that by the longer bubbles, as shown by black arrows, and the device rotates clockwise as indicated by the blue arrow in Figure 5(a). Figure 5(b) shows switching of rotational direction (counterclockwise) by changing the frequency. Figure 5(c) shows the concept for acoustic bubble-turbomachinery. The fan-shaped device is fabricated to simulate a real motor. All six longer channels are placed in the same direction to get a faster rotation speed. The micro channels are made curved to reduce the device size. Radius of this device is about 1750 μm. Because all the bubble excitation could generate a torque in the same direction, as long as the oscillation amplitude is large enough, this motor would rotate. To examine this, another frequency, 3.8 kHz (480 V_{pp}) is used other than 5.8 kHz (310 V_{pp}) used in the previous tests. When the acoustic field is applied, this motor could rotate continuously at the speed about 75 rpm.

Carrying Load in T-Junction Microchannel

To simulate steering propulsion in a blood vessel branch, a micro channel ($\sim$ 2000 μm) with a T-junction is made. The previous orthogonally arrayed device is tested without the load first, as shown in Figure 6(a). The acoustic signal is set to be about 6 kHz (370 V_{pp}) at the beginning. With this signal, the three longer bubbles are working and pushing the device downward into the T-junction. Then, the acoustic field is set to be 11.0 kHz (210 V_{pp}). At this moment, the three shorter bubbles are activated and generate a torque to align the device with the horizontal channel. Then, frequency is set back to 6.0 kHz (370 V_{pp}) and the longer bubbles are excited again, give a pushing force to the device and swim out of the T-junction area. In Figure 6(b), a similar steering motion with a load ($\sim$400 × 325 × 80 μm^3) is shown. Finally, the device leaves the T-junction in the right direction.

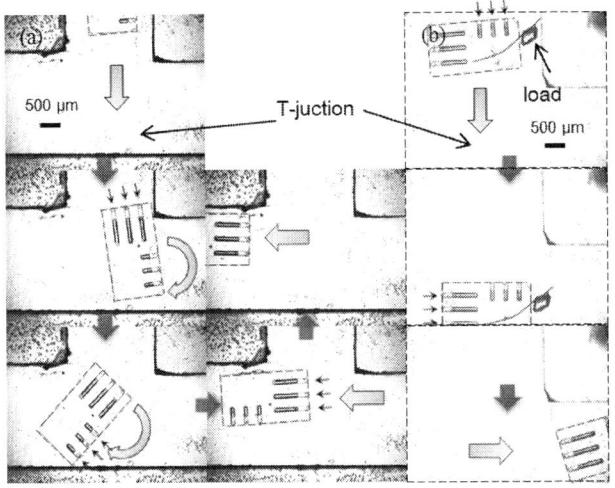

Figure 6: Steering capability in a T-junction of the microchannel (bottom view) by 820μm and 270μm bubbles (a) without load and (b) with load.

CONCLUSION AND FUTURE WORK

This paper presents 2-D steering micro propulsion that is achieved by selectively oscillating micro bubble columns. The selectivity in bubble oscillation is determined by the bubble resonance behavior which is mainly function of the bubble length and the frequency. By arranging multiple bubbles of different length on the same chip and carefully changing the acoustic excitation frequency, various 2-D motions are achieved including 1-D direction switching in translational and rotational motions and steering propelling on an open 2-D plane and in a T-junction of the microchannel. In addition, these motions can be repeated with a load attached to the device. All these motions can be controlled by simply changing the acoustic excitation frequency.

ACKNOWLEDGEMENTS

This research is supported by NSF Grant No. ECCS-1029318.

REFERENCES

[1] R. Dreyfus, J. Baudry, M. L. Roper, M. Fermigier, H. A. Stone, and J. Bibette1, "Microscopic artificial swimmers," *Nature,* vol. 437, pp. 862-865, 2005.

[2] D. J. Bell, S. Leutenegger, K. M. Hammar, L. X. Dong, and B. J. Nelson, "Flagella-like Propulsion for Microrobots Using a Nanocoil and a Rotating Electromagnetic Field," presented at the 2007 IEEE International Conference on Robotics and Automation, Roma, Italy.

[3] A. A. Solovev, Y. Mei, E. B. Ureña, G. Huang, and O. G. Schmidt, "Catalytic Microtubular Jet Engines Self-Propelled by Accumulated Gas Bubbles," *small,* vol. 5, pp. 1688-1692, 2009.

[4] B. Behkam and M. Sitti, "Bacterial flagella-based propulsion and on/off motion control of microscale objects," *Appl. Phys. Lett.,* vol. 90, pp. 023902-023902-3, 2007.

[5] J.-H. Tsai and L. Lin, "Active microfluidic mixer and gas bubble filter driven by thermal bubble micropump," *Sens. Actuators, A,* vol. 97-98, pp. 665-671, 2002.

[6] K. Ryu, J. Zueger, S. K. Chung, and S. K. Cho, "Underwater Propulsion Using AC-Electrowetting-Actuated Oscillating Bubbles for Swimming Robots," in *2010 IEEE 23rd International Conference on Micro Electro Mechanical Systems (MEMS),* Wanchai, Hong Kong, pp. 160-163.

[7] R. J. Dijkink, J. P. v. d. Dennen, C. D. Ohl, and A. Prosperetti, "The 'acoustic scallop': a bubble-powered actuator," *J. Micromech. Microeng.,* vol. 16, pp. 1653-1659, 2006.

[8] J. Feng and S. K. Cho, "Micro propulsion in liquid by oscillating bubbles," in *2013 IEEE 26th International Conference on Micro Electro Mechanical Systems (MEMS),* Taipei, pp. 63-66.

[9] M. S. Plesset and A. Prosperetti, "Bubble Dynamics and Cavitation," *Ann. Rev. Fluid Mech.,* vol. 9, pp. 145-185, 1977.

[10] X. M. Chen and A. Prosperetti, "Thermal processes in the oscillations of gas bubbles in tubes," *J. Acoust. Soc. Am.,* vol. 104, pp. 1389-1398, 1998.

[11] X. Geng, H. Yuan, H. N. Oğuz, and A. Prosperetti, "The oscillation of gas bubbles in tubes: Experimental results," *J. Acoust. Soc. Am.,* vol. 106, pp. 674-681, 1999.

[12] M. P. Norton and D. G. Karczub, *Fundamentals of Noise and Vibration Analysis for Engineers*: Cambridge Univeristy Press, 2003.

A MICROFABRICATED, BIOHYBRID, SOFT ROBOTICS FLAGELLUM

Brian J. Williams[1], Sandeep V. Anand[1], Jagannathan Rajagopalan[2], and M. Taher A. Saif[1]
[1]University of Illinois at Urbana-Champaign, Urbana, IL, USA
[2]Arizona State University, Tempe, AZ, USA

ABSTRACT

We present a microfabricated soft robotics flagellum powered by living cells that can generate propulsion at low Reynolds number (*Re*). The swimmer utilizes contractile cardiomyocytes to provide on-board actuation to a thin, deformable, polydimethylsiloxane (PDMS) filament. To enable propulsion at low *Re*, the filament is designed such that it deforms passively in response to fluid drag, producing a time irreversible cyclical deformation and a net propulsive force. This work provides a new paradigm by integrating microfabrication and biological cells to enable the realization of an independent, soft robotics actuator with micron-scale dimensions.

INTRODUCTION

Soft robotics is a rapidly developing field utilizing the deformations of compliant members to produce a mechanical output. The use of soft robotics instead of classical mechanical joints creates a robust, impact resistant design, facilitates the potential to exhibit complex, continuous deformation shapes, enables delicate handling of fragile payloads, and can be produced at exceptionally low costs. Common designs to date rely on pneumatic actuators and electroactive polymers to provide the deforming force [1-3]. The associated construction techniques and inclusion of such actuators require that the machine be on the macroscopic scale and be coupled to a suitable power source.

Here, we present a microfabricated soft robotics actuator consisting of a continuously deformable elastic filament powered by living cells. The filament can be engineered to produce the time irreversible deformations required to generate a net propulsive force to swim at low *Re*. The method of fabrication by photolithography enables the precise control over the geometry of the flagellum. The use of biological cells with a novel seeding technique permits the inclusion of a self-contained driving mechanism without the need for high precision tools for actuator placement or coupling to an external power source. The system geometry is designed by using a slender body elastohydrodynamic model to determine the appropriate physical parameters to achieve the required temporal dynamics under the driving power of cyclical contractions from one to several cardiomyocytes.

DESIGN

The proposed flagellum consists of a long PDMS filament with a rigid head and a compliant tail (Fig. 1a). Cardiomyocytes are cultured directly on preselected regions of the tail, near the head/tail junction, to provide the driving force. The filament parameters are selected such that it can be deflected by the 1-10 µN peak forces provided by the contractions of cardiomyocytes on the surface of the filament (Fig. 1b). Additionally, the structure must be sized to facilitate cell adhesion, but must be rigid enough to maintain its form under the influence of static (non-periodic) cell contractile forces. Finally, in order to generate propulsion at low *Re*, the filament must deform passively in response to fluid drag to provide a time irreversible deformation (Fig. 1c). This asymmetric deformation is required to produce a nonzero net reaction force against the suspending fluid and achieve net propulsion at low *Re* [4]. To produce a design that simultaneously satisfies all of these requirements, we first pick the filament cross-section such that the contractions of small clusters of 1-3 cardiomyocytes produce filament deflections of 15-30°, and then determine appropriate filament lengths by considering the appropriate fluid mechanics. We experimentally determine that a filament tail cross-section of 8 µm by 20 µm is suitable for the prescribed driving forces.

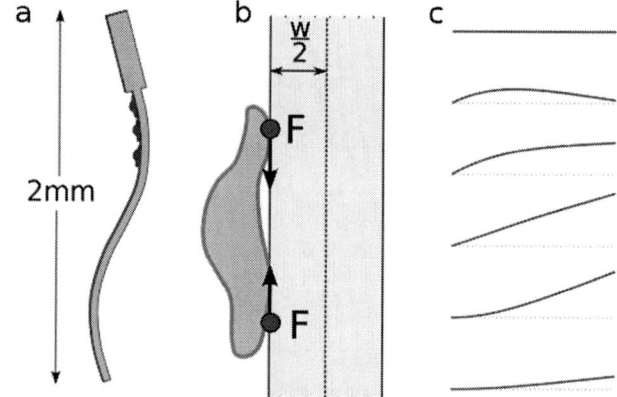

Figure 1: (a), Schematic of soft robotics flagellum, consisting of a microfabricated PDMS filament with cells seeded near the head/tail junction, (b), Actuation of flagellum achieved by contraction of cardiomyocytes on surface of filament, and (c), To achieve propulsion at low Re, the filament must exhibit a time irreversible deformation during its actuation and relaxation strokes.

Based on established theory [4-6], we consider a slender body elastohydrodynamic system actuated by a one-sided cyclical bending moment using the coordinate system defined in Fig. 2a. We consider the effects of fluid drag, $F_D = -\varsigma_\perp \frac{\partial y}{\partial t}$, and the filament elastic restoring force, $F_E = -A \frac{\partial^4 y}{\partial x^4}$, where $\varsigma_\perp$ is the normal drag coefficient, and A is the filament bending stiffness. We drive the system with

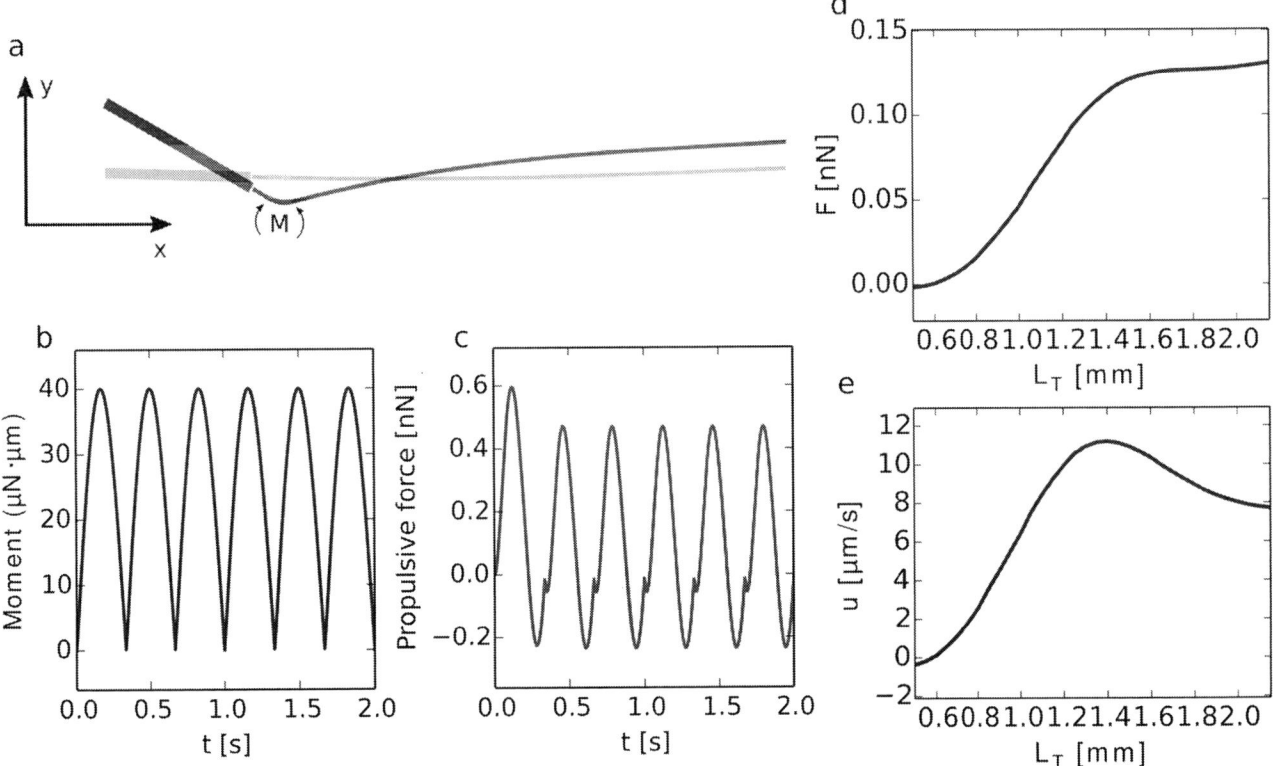

Figure 2: (a), Coordinate system for analysis of flagella, (b), driving function for actuation of filament, (c), resultant propulsive force generated by fluid drag, (d), Mean propulsive force as a function of tail length, and (e), Mean predicted swimming velocity as a function of tail length.

the normal force $F_{cell} = \frac{\partial^2 M}{\partial x^2}$, which is equivalent to the driving moment $M(x,t)$ generated by cell contractions on the region on which they are attached (Fig. 2b). Finally, using zero force and zero moment boundary conditions at the filament ends, we numerically solve the governing equation:

$$\varsigma_\perp \frac{\partial y}{\partial t} + A \frac{\partial^4 y}{\partial x^4} = \frac{\partial^2 M}{\partial x^2} \qquad (1)$$

We then calculate propulsive force by projecting the total fluid drag F_D on the x-axis (Fig. 2c), and calculate the resultant swimming velocity $U(t)$ by balancing the propulsive force against longitudinal drag $F_{prop} = \int_0^L \varsigma_\parallel U \, dx$, where L is the length of the filament and $\varsigma_\parallel$ is the longitudinal drag coefficient. We utilize the mean velocity $\bar{U}$ as an optimization metric, and pick filament parameters to optimize this output (Fig. 2d-e). Drag coefficients are experimentally measured, and the filament stiffness is computed as the product of the area moment of inertia I and the elastic modulus E, which is measured by nanoindentation. For the specified tail with a 8 μm wide, 20 μm deep cross-section, we determine that a 1.4mm long tail produces the optimal swimming velocity of 12 μm/sec, and construct our soft robotics flagellum to be on this size scale.

FABRICATION

Filament construction

PDMS filaments with rectangular cross sections are fabricated [7] to highly accurate, predetermined specifications determined by the previous analysis by etching the desired pattern in a silicon wafer by standard photolithography in a process amenable to batch fabrication (Fig. 3a-d). The wafer is coated in polytetrafluoroethylene (PTFE), and the channels are filled with PDMS (Sylgard 184, 4:1 base to cross-linker ratio) by capillary draw (Fig. 3e). The filaments are cured at 70°C, soaked in ethanol for 30 minutes, peeled out (Fig. 3f-g), and transferred to a petri dish.

Cell Culture

To selectively place cells near the head/tail junction, the filament is suspended between a pair of PDMS coated silicon chips mounted in a petri dish such that the head of the swimmer as well as the portion of the tail on which cell adhesion is desired is resting on one silicon pedestal. The remainder of the tail, on which no cell adhesion is desired, is suspended between the pedestals. A small droplet of Fibronectin (25 μg/mL) is deposited on the pedestal holding the head, which acts as a hydrophobic region of confinement for selective functionalization. The remainder of the system is treated with Pluronics F127 to deter

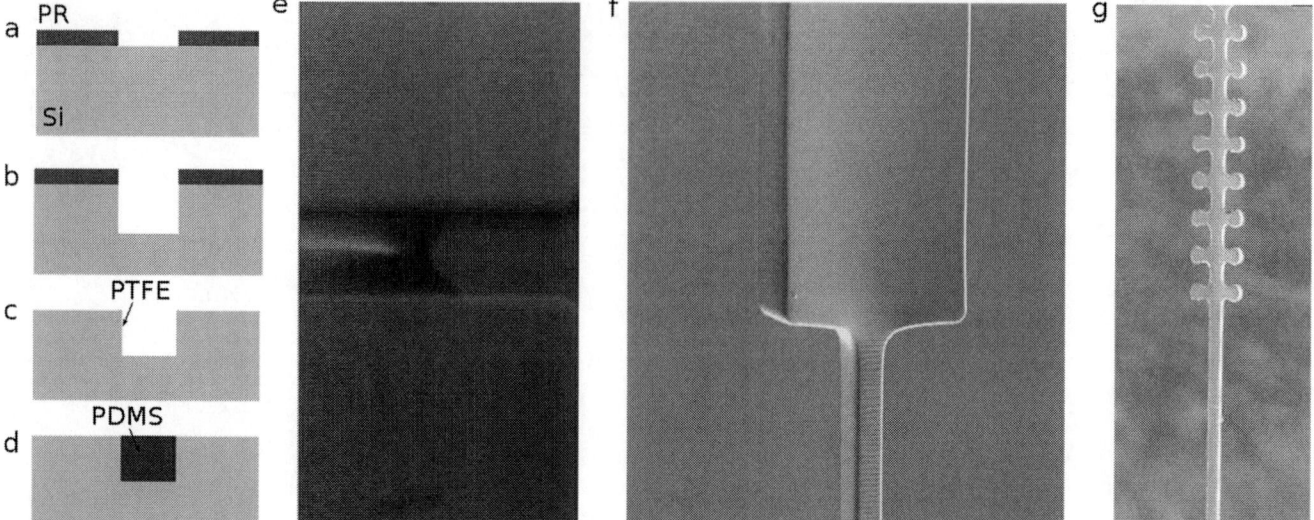

Figure 3: (a-c), The filament shape is etched into a Si wafer by standard photolithography, then coated with PTFE to minimize stiction. (d-e), The channel is then filled with PDMS by capillary draw and cured before removal, (f), SEM micrograph of cured filament, and (g), Complex shapes of constant depth can be fabricated by this process.

adhesion where not desired. Primary cardiomyocytes and fibroblasts are extracted from neonatal rat pups (2-5 days old) by standard procedures [8], and seeded directly on the suspended filaments. Sedimentation around the freestanding portions of the filament produces streamlines that minimize cell exposure to the tail, while cells are free to sediment on the silicon pedestals and adhere to the head and functionalized portion of the tail. The system is maintained at 37°C and 5% CO_2 in cell culture media.

RESULTS

1-2 days after culturing cells, the tail is cut free, the portion of the tail adjacent to the head is gently detached from the silicon pedestal, and the flagellum is moved to a region of the petri dish facilitating observation. The system is left in this unattached state for another 1-2 days, providing a highly compliant environment. We find that samples that develop in this unfixed configuration exhibit significantly higher deformation amplitudes. Adherent cardiomyocytes contract with a frequency of approximately 3 *Hz*, inducing peak angular deflections of over 30° (Fig. 4). As predicted by the elastohydrodynamic model, significant passive deformation in response to fluid drag is observed. As the cells contract in Fig. 4, the free end of the tail is driven downwards, providing an upwards curvature. As the cells relax, elastic restoring forces pull the tail back upwards, generating a downwards curvature. This asymmetry yields a traveling wave from head to tail and a resultant propulsive force towards the head.

CONCLUSIONS

Here, we demonstrated the use of common microfabrication techniques to construct molds for the fabrication of soft robotics flagella that can be designed to

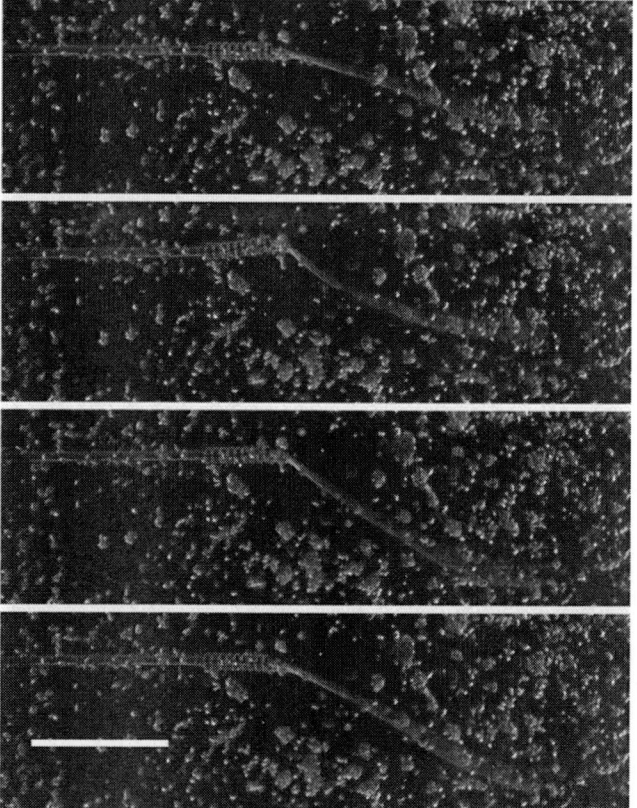

Figure 4: Set of displacement shapes achieved by the contraction of several cardiomyocytes on a soft robotics flagellum. Time irreversibility is clearly visible in the variation of curvature of the filament. Scale bar, 500μm.

produce propulsive force at low *Re* without relying on external power supplies. We provide a technique that can be scaled to produce large quantities of actuators (Fig. 5) or

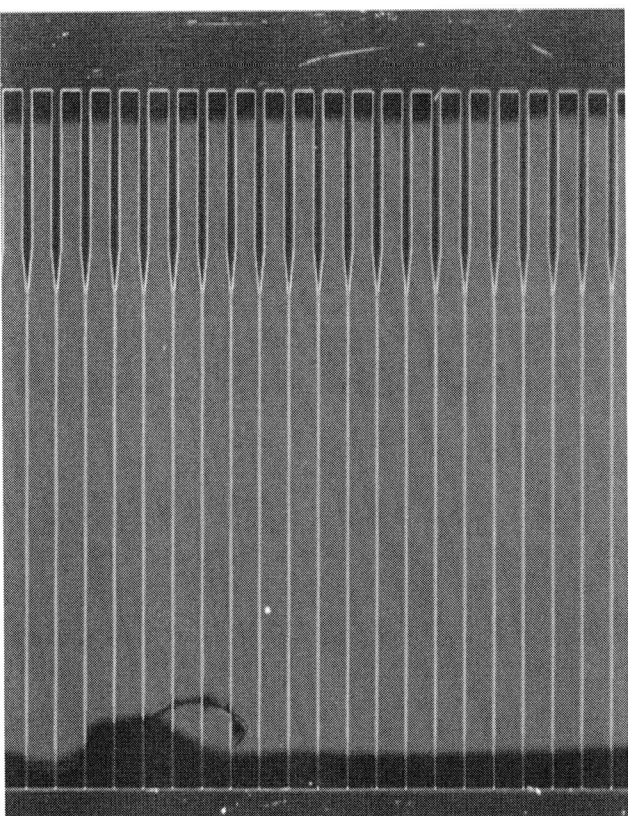

Figure 5: Bulk production of soft robotics actuators enabled by microfabrication technique.

be integrated into freestanding functional swimmers. Our approach facilitates the construction of microscale mechanisms that generate large scale deformations without relying on prohibitively expensive and fragile miniaturized classical joints.

ACKNOWLEDGEMENTS

This project was funded by the National Science Foundation (NSF), Science and Technology Center on Emergent Behaviors in Integrated Cellular Systems (EBICS) Grant CBET-0939511, a cooperative agreement that was awarded to UIUC and administered by the U.S. Army Medical Research & Material Command (USAMRMC) and the Telemedicine & Advanced Technology Research Center (TATRC), under Contract #: W81XWH0810701, and the National Institute of Health (NIH) grant: RO1 NS063405-01.

REFERENCES

[1] D. Trivedi, C. D. Rahn, W. M. Kier, and I. D. Walker, "Soft Robotics: Biological Inspiration, State of the Art, and Future Research", *Appl. Bionics Biomech.*, vol. 5, pp. 99-117, 2008.

[2] A. Albu-Schaffer et al, "Soft Robotics", *IEEE Robot. Autom. Mag.*, vol. 15, pp. 20-30, 2008.

[3] F. Ilievski et al, "Soft Robotics for Chemists", *Angewandte Chemie*, vol. 123, pp 1930-1935, 2011.

[4] E. Lauga and T.R. Powers, "The Hydrodynamics of Swimming Microorganisms", *Rep. Prog. Phys.*, vol. 72, 2009.

[5] C. H. Wiggins et al, "Trapping and Wiggling: Elastohydrodynamics of Driven Microfilaments", *Biophys. J.*, vol. 74, pp. 1043-1060, 1998.

[6] R. Dreyfus, et al. "Microscopic Artificial Swimmers", *Nature*, vol. 437, pp. 862-865, 2005.

[7] J. Rajagopalan, M. T. A. Saif, "Fabrication of Freestanding 1-D PDMS Microstructures Using Capillary Micromolding", *J. Microelectromech Syst.*, vol. 22, pp. 992-994, 2013.

[8] V. Chan et al, "Development of Miniaturized Walking Biological Machines", *Sci. Rep.*, vol. 2, 2012.

CONTACT

*M. T. A. Saif, tel: +1-217-333-8552; saif@illinois.edu

EVALUATION AND OPTICAL CONTROL OF SOMATIC MUSCLE MICRO BIOACTUATOR OF CHANNELRHODOPSIN TRANSGENIC DROSOPHILA MELANOGASTER

Masaya Hirooka[1], Sze Ping Beh[1], Toshifumi Asano[1], Yoshitake Akiyama[1], Takayuki Hoshino[2], Keita Hoshino[3], Hidenobu Tsujimura[3], Kikuo Iwabuchi[3] and Keisuke Morishima[1]

[1]Osaka University, Osaka, JAPAN
[2]The University of Tokyo, Tokyo, JAPAN
[3]Tokyo University of Agriculture and Technology, Tokyo, JAPAN

ABSTRACT

In this paper, we developed light-activated somatic muscle (body-wall muscle) from transgenic *Drosophila melanogaster* larvae expressing a blue light sensitive cation channel, channelrhodopsin-2, and incorporated it into a micro device. We successfully demonstrated that optogenetic stimulation using light pulses enabled control of contractile activity with a given temporal pattern. The contractile force was evaluated with varying light intensities and pulse widths. These results have shown that muscle-powered bioactuator system that combines light-activated muscle cells and microfabrication techniques is a useful model for the study of wet-robotics and muscle bioassays for investigating mechanical properties.

INTRODUCTION

Recently, soft robots incorporating biological components such as tissues and cells have raised much attention for the development of novel engineering devices. In particular, muscle plays a key role as an efficient actuator with consumes little energy, resources and space for soft robotic devices. Muscle-powered actuators have many advantages such energy conversion efficiency, self-organization/-renewal, plasticity and scalability, compared with conventional actuators. Contractile muscles have been incorporated into engineered bio-devices such as actuators [1, 2] and pump [3, 4]. Also, robots incorporating insect heart muscle tissue are generally environmentally robust compared to mammalian muscles, and can autonomously move at room temperature [5]. Muscle-powered actuators have been controlled by temperature stimulation [6], chemical stimulation [7] and electrical stimulation [8] to generate muscle contractile force. Although these methods are simple techniques to stimulate the muscle cell, many issues remain when it comes to precise control. Temperature and chemical stimulation are generally non-uniform and many unintended muscle cells are stimulated simultaneously. Although temporal pattern of activation is easy to control using electrical stimulation, the spatial resolution is low as with the above methods.

Optogenetic stimulation methods have raised much attention and several advantages over the other methods. They have high space and time resolution, allows parallel stimulations at multiple sites, wireless and low invasiveness. Optogenetics has become effective tool for investigation of neural function and networks [9] by using light sensitive cation channel, channelrhodopsin-2 (ChR2) [10], which was discovered from a unicellular green alga, *Chlamydomonas reinhardtii*. There are also potential bioengineering applications such as, wireless drive of muscle-powered actuators [11, 12]. On the other hand, optogenetic pacing using a nonviral strategy could be achieved in cardiomyocytes [13].

We have previously controlled insect heart muscle bioactuator by light stimulation [14, 15]. However, current optical stimulation methods are not able to precisely control the desired output (e.g. the muscular contractile force). In this study, to establish the stimulation condition which can precisely control the desired output for muscle-powered mechanical systems, we developed a light-activated somatic muscle (body-wall muscle) from *Drosophila melanogaster* (*Drosophila*) larvae expressing ChR2. A poly-dimethylsiloxane (PDMS) micro cantilever device was fabricated to quantitatively evaluate responses of optically regulated muscle contractions. We have derived control variables by evaluating transient responses of the light-activated muscle-cantilever.

MATERIALS AND METHODS

Development of Light-activated Insect Somatic Muscle

Drosophila is a common model organism employed in many biological researches, hence there is a wealth of genetic knowledge and technology regarding it. The GAL4/UAS system is a powerful gene manipulation method frequently employed in this field especially the original powerful gene manipulation method. The female adult expressing ChR2(H134R), which is a variant of the ChR2 with enhanced light responsiveness, were isolated and mated with male adults encoding Myocyte enhancer factor-2 (Mef2) proteins. The larvae were grown in the media contained all-trans retinal (Sigma-Aldrich) for inducing the functional expression of ChR2. After six days from mating, 3rd instar larvae expressing ChR2(H134R) in all muscle cells were obtained. Somatic muscle tissues were isolated from the larvae by cutting their anterior and posterior (Fig. 1a).

ChR2(H134R) was combined with fluorescent protein mCherry to indicate the expression of ChR2(H134R). When expressed in living cells, ChR2(H134R)–mCherry conjugate was identified under the conventional inverted-fluorescent microscopy (BZ-9000, KEYENCE). The *Drosophila* were used in this study were supplied from Bloomington Stock Center.

978-1-4799-3510-9/14 $31.00 © 2014 IEEE

(a)

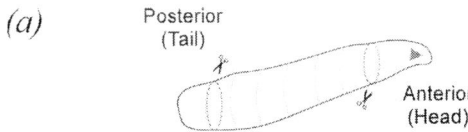

(b)

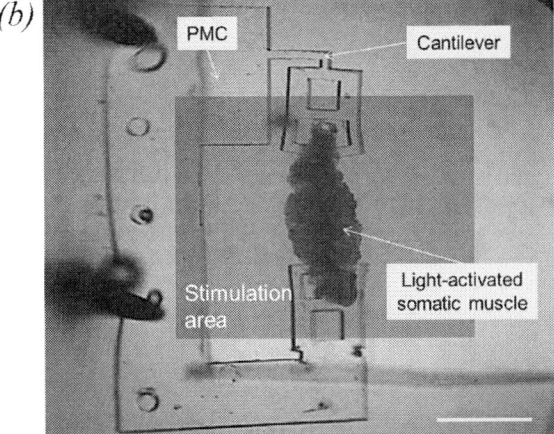

(c)

Fig. 1: Contractile force sensor actuated by photocontrollable muscle-cantilever (PMC). (a) Isolation of the somatic muscle tissue from Drosophila larva. (b) Microscopic image of the PDMS cantilever incorporated with the somatic muscle. (Scale bar: 1 mm). (c) Displacement of the point A by contraction force.

Device fabrication and PMC construction

A photocontrollable muscle-cantilever (PMC) made by PDMS was designed as shown in Fig. 1b. The PMC was fabricated by soft lithography. Briefly, SU-8 photoresist (Micro-chem) were spin-coated on the wafer to a thickness of 200μm, and were irradiated with ultraviolet (405nm) using the mask aligner (MA-10, MIKASA). Unexposed resist was removed with PM thinner. Then, a mixture (10:1, w/w) of uncured PDMS (SYIPOT 184, Dow Corning) and cross-linker was poured into the SU-8 micromold, and baked at 120 °C for 45 min.

The isolated somatic muscle was incorporated into the PMC (Fig. 1b). It was bending due to muscle contraction when the muscle was stimulated with blue light (Fig. 1c).

The PMC incorporating somatic muscle tissue was examined in insect medium (Schneider's Drosophila Medium "DAIGO", Wako) supplemented with 10% Fetal Bovine Serum (FBS, Gibco) and 1% Penicillin Streptomycin

(a)

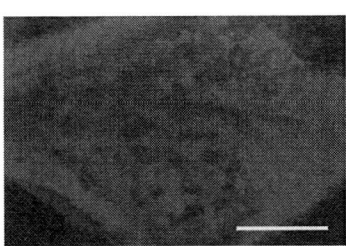

(b)

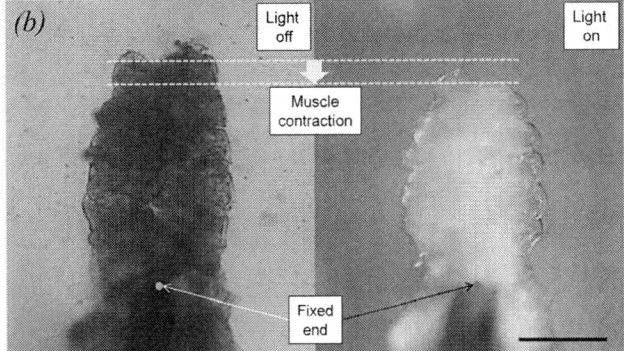

Fig. 2: Light-activated muscle tissues expressing ChR2(H134R). (a) Fluorescence images of the Drosophila somatic muscles. (b) Contractile motion of the light-activated somatic muscle tissue. (Scale bars: (a) and (b), 500 μm).

(Sigma-Aldrich) at room temperature immediately after tissue isolation.

Optical Stimulation and Motion Analysis Systems

The PMC incorporating light-activated muscles and its motion were observed with a microscope (Ti-U, Nikon). Optical stimulation patterns from the projector (LightCommander, LogicPD) were generated through the microscope by a lens system. The light source of the projector consists of a LED with light output wavelength of 466nm. Every input waveform to samples was a square wave. The power density of the blue light was measured with a power meter (PM100D, Thorlabs). The microscope images were captured with a CCD camera (FASTCAM SA3, Photron; 20 frames/s). The displacement of the micro PDMS device caused by muscle contraction was measured by tracking the point A as shown in Fig. 1c. All images were analyzed using image processing software (DIPP- Motion PRO, DITECT). The muscle contractile force F was calculated from the following equation,

$$F = 3EI\delta_A / L^3 \qquad (1)$$

where E is Young's modulus of PDMS, I is second moment of area for the beam of the cantilever device, δ_A is the displacement of the point A ,and L is length of the beam.

RESULTS AND DISCUSSION

Contractile Activity of Isolated Somatic Muscle

A transgenic Drosophila expressing red fluorescent protein (mCherry) in the somatic muscle tissue was

978-1-4799-3510-9/14 $31.00 © 2014 IEEE

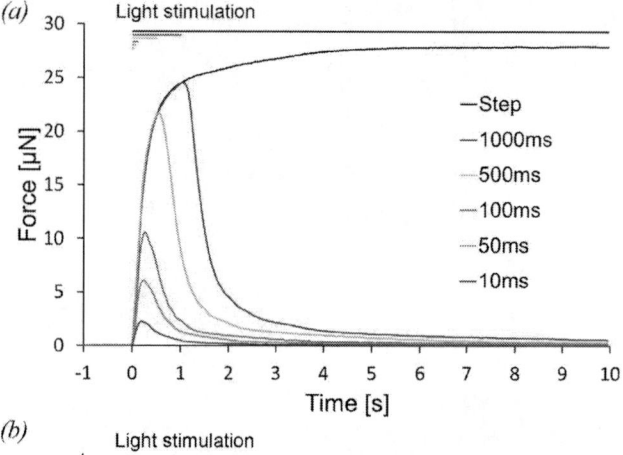

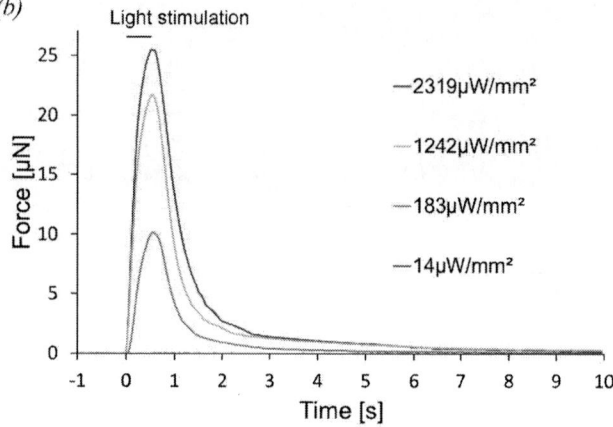

Fig. 3: The transient responses of the muscle-powered device to one pulse stimulation. (a) Pulse responses at constant pulse width. The pulse width is 500 ms. (b) Pulse and Step responses at constant light intensity. The light intensity is 1242 μW/mm².

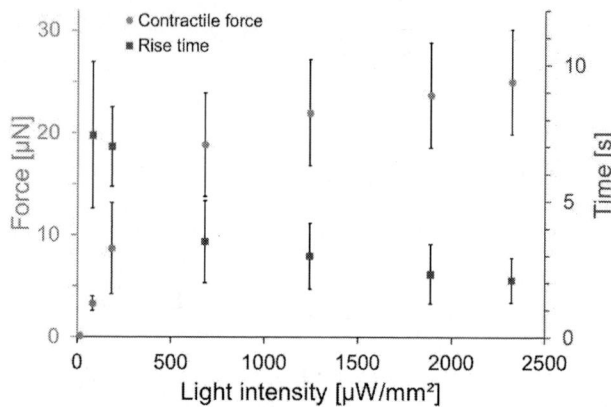

Fig. 4: Steady-state value of force and rise time of the step responses (n=3).

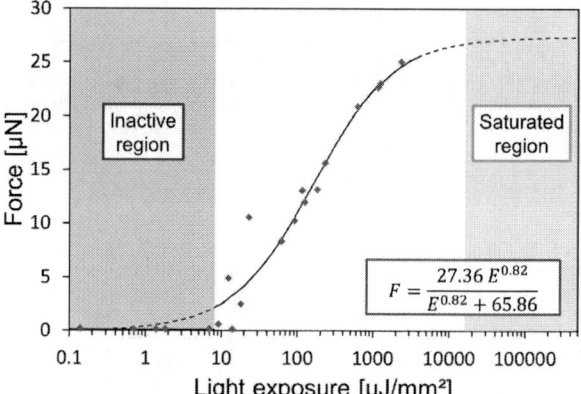

$$F = \frac{27.36\, E^{0.82}}{E^{0.82} + 65.86}$$

Fig. 5: The Maximum force of the pulse responses for the light exposure (n=3).

generated and the expression pattern was observed in third instar of development larvae (Fig. 2a). After the larvae somatic muscle tissue was isolated, it was examined by using optical stimulation system (1477 μW/mm², continued irradiation). The somatic muscle tissue contracted with contraction ratio of 19.3 ± 3.4% (natural length: 1342 ± 100 μm, contractile displacement: 257 ± 36 μm) (Fig. 2b). We have previously demonstrated the contractile performance of the dorsal vessel which is heart muscle tissue in *Drosophila* [14, 15]. However, heart muscle beats autonomously, so it was difficult to control contractile displacement and force with desired accuracy.

In contrast, somatic muscle produces robust contractions and intense contractile displacement without spontaneous beatings. These data shows that the somatic muscle is a stable driving source, which allows the precise control of the motion in muscle-powered actuators.

Optical Responses of PMC

After tissue assembly, the performance of PMC incorporating somatic muscle tissue was examined using optical stimulation with the various given conditions. The

light pulse had two parameters, light intensity (14, 183, 1242 and 2319 μW/mm²) and pulse width (10, 50, 100, 500, 1000 ms and continuous time). The force is determined from the displacement of the cantilever exerted during light-stimulated PMC. The magnitude of tension is calculated by measuring the cantilever deflection. The transient responses of the PMC with light stimulation were obtained by tracking point A, as shown in Fig. 1c. The generated force depended on the width of the light pulse during the constant intensity (Fig. 3a). Furthermore, the rate of contraction was almost identical even at different pulse widths. On the other hand, when stimulating with constant pulse width, the force was increased with increasing light intensity (Fig. 3b). Furthermore, peak times of the responses were almost similar at different intensities. The kinetics of contraction and relaxation in ChR2(H134R) expressing somatic muscle tissue was almost similar to the response with stimulation of the nervous system in *vivo*. These results indicated that the PMC was able to estimate the stimulation conditions which enable the control of the desired position and power output for muscle-powered mechanical systems.

Analysis of Transient Responses

We evaluated significant characteristics of step responses, and found that the somatic muscle generated large force and responded quickly at high light intensity (Fig. 4).

In responses to single pulse stimulation, the maximum force (peak value) of each response increased with increasing the light intensity, and was fitted to an approximated curve shown in Fig. 5. The light exposure was calculated from the product of light intensity and pulse width. The responses of the light-activated somatic muscle to single pulse stimulation have an inactive region and a linearly increasing region on the horizontal logarithmic axis. We estimated that they also have a saturated region, as shown in dot line (Fig. 5), because of the physical limitations of muscular contractile structure.

CONCLUSION

In this paper, we reported the development of light-activated insect somatic muscle tissue expressing ChR2, and evaluated the transient responses of the muscle using PMC system. Optogenetic stimulation was able to control their contractile activity with a given temporal pattern and intensity. In future, such a system that combines light-activated muscle tissues and microfabrication technique could be controlled more precisely. This technique would have many potential bioengineering and biorobotics applications such as wireless driving system of muscle-powered actuators/microdevices, muscle bioassays for biological and mechanical properties, and soft and wet robotic component.

ACKNOWLEDGEMENTS

This work was supported in part by the Industrial Research Program of NEDO and Grants-in-Aid for Scientific Research from the Ministry of Education, Culture, Sports, Science and Technology in Japan (No. 20034017, 21676002, 21225007, 21111503, 22860020, 23111705 and 23700557).

REFERENCES

[1] J. Xi, J.J. Schmidt, C.D. Montemagno, "Self-assembled Microdevices Driven by Muscle", *Nat. Mater.*, vol. 4 (2), pp. 180-184, 2005.

[2] A. W. Feinberg, A. Feigel, S. S. Shevkoplyas, S. Sheehy, G. M. Whitesides, K. K. Parker, "Muscular Thin Films for Building Actuators and Powering Devices", *Science*, vol. 317 (5843), pp. 1366–1370, 2007.

[3] Y. Tanaka, K. Morishima, T. Shimizu, A. Kikuchi, T. Okano, T. Kitamori, "An actuated pump on-chip powered by cultured cardiomyocytes", *Lab Chip*, vol. 6 (3), pp. 362-368, 2006.

[4] Y. Tanaka, K. Sato, T. Shimizu, M. Yamato, T. Okano, T. Kitamori, "A micro-spherical heart pump powered by cultured cardiomyocytes", *Lab Chip*, vol. 7 (2), pp. 207–212, 2007.

[5] Y. Akiyama, K. Odaira, K. Iwabuchi, K. Morishima, "Long-term and Room Temperature Operable Bio-Microrobot Powered by Insect Heart Tissue", *Proc.*

of MEMS 2011, Cancun, January, 2011, pp. 145-148.

[6] Y. Akiyama, K. Iwabuchi, Y. Furukawa, K. Morishima, "Fabrication Evaluation of Temperature-Tolerant Bioatuator Driven by Insect Heart Cells", *Proc. of Micro TAS 2008*, San Diego, October, 2008, pp. 1669-1671.

[7] Y. Akiyama, K. Iwabuchi, Y. Furukawa, K. Morishima, "Biological Contractile Regulation of Micropillar Actuator Driven by Insect Dorsal Vessel Tissue", *Proc. of BioRob 2008*, Arizona, October, 2008, pp. 501-505.

[8] J. C. Nawroth, H. Lee, A. W. Feinberg, C. M. Ripplinger, M. L. McCain, A. Grosberg, J. O. Dabiri, K. K. Parker, "A tissue-engineered jellyfish with biomimetic propulsion", *Nat. Biotechnol*, vol. 30 (8), pp. 792-797, 2012.

[9] E. S. Boyden, F. Zhang, E. Bamberg, G. Nagel, K. Deissrroth, "Millisecond-timescale, genetically targeted optical control of neural activity", *Nat. Neurosci*, vol. 8 (9), pp. 1263-1268, 2005.

[10] G. Nagel, T. Szellas, W. Huhn, S. Kateriya, N. Adeishvili, P. Berthold, D. Ollig, P. Hegemann, E. Bamberg, "Channelrhodopsin-2, a directly light-gated cation-selective membrane channel", *Proc. of Natl. Acad. Sci. USA*, vol. 100, pp. 13940-13945, 2003.

[11] M. S. Sakar, D. Neal, T. Boudou, M. A. Borochin, Y. Li, R. Weiss, R. D. Kamm, C. S. Chen, H. H. Asada, "Formation and optogenetic control of engineered 3D skeletal muscle bioactuators", *Lab Chip*, vol. 12 (23), pp. 4976-4985, 2012.

[12] T. Asano, T. Ishizua, H. Yawo, "Optically controlled contraction of photosensitive skeletal muscle cells", *Biotechnology and Bioengineering*, vol. 109 (1), pp. 199-204, 2012.

[13] Z. Jia, V. Valiunas, Z. Lu, H. Bien, H. Liu, H. -Z. Wang, B. Rosati, P. R. Brink, I. S. Cohen, E. Entcheva, "Stimulating cardiac muscle by light cardiac optogenetics by cell delivery", *Circ. Arrhythm. Electrophysiol.*, vol. 4 (5), pp. 756-760, 2011.

[14] K. Suzumura, K. Funakoshi, T. Hoshino, H. Tsujimura, K. Iwabuchi, Y. Akiyama, K. Morishima, "A light-regulated bio-micro-actuator powered by transgenic *Drosophila melanogaster* muscle tissue", *Proc. of MEMS 2011*, Cancun, January, 2011, pp. 149-152.

[15] S. P. Beh, M. Hirooka, T. Hoshino, K. Hoshino, Y. Akiyama, H. Tsujimura, K. Iwabuchi, K. morishima, "Visual servo of muscle-powered optogenetic bioactuator", *Proc. of Transducers 2013*, Barcelona, June, 2013, pp. 1444-1447.

[16] A. H. Brand, N. Perrimon, "Targeted gene expression as a means of altering cell fates and generating dominant phenotypes", *Development*, vol. 118 (2), pp. 401-415, 1993.

CONTACT

*K. Morishima,
E-mail: morishima@mech.eng.osaka-u.ac.jp

FERROFLUID-ASSISTED MICRO ROTARY MOTOR FOR MINIMALLY INVASIVE ENDOSCOPY APPLICATIONS

Babak Assadsangabi, Min Hian Tee, Simon Wu, and Kenichi Takahata
University of British Columbia, Vancouver, CANADA

ABSTRACT

This paper reports a micro rotary motor that is enabled with magnetic fluid called ferrofluid used as an extremely simple, miniaturized bearing material for microendoscopy applications. The ferrofluid bearing is magnetically sustained on the permanent magnet rotor that is levitated by the bearing layer inside a tubular substrate, an endoscope catheter. The levitated rotor is electromagnetically driven by two photo-defined meander-type coils formed around the outer walls of the catheter that enables 90°-step angular actuation of the rotor. The fabricated prototype with the rotor coupled with a 1-mm-sized prism mirror is revealed to provide both step-wise and continuous rotations with revolution rates up to 1875 rpm, verifying the effectiveness of the bearing and motor mechanism. The prototype device is operated to demonstrate its ability of endoscopic imaging in an experimental model.

INTRODUCTION

Micro-scale rotary motors have a vast range of potential applications in broad areas. One promising area is medical applications [1-3]. Microendoscopic catheter for minimally invasive medical imaging is an excellent application example. Advanced catheter devices that use rotary motors coupled with prism-shaped mirrors to perform full 360° circumferential scan for endoscopy have been widely investigated [4-7]. Piezoelectric rotary micromotors were developed for this type of application [4]. The use of magnetic actuation, however, has been a dominant approach in this area [5-7] for various reasons, including higher speeds, smaller sizes, and wider commercial availability compared with the piezoelectric type. One of the critical needs in the design of imaging endoscopes is the flexibility of the catheter. The magnetic micromotors used in these catheters, have large axial lengths (>1 cm [5-7]) in most cases, due to the need for a long cylindrical magnet, with a gear box to increase the output torque and compensate for large frictions between rotor and stator. The large axial size of the motors, rigid parts embedded in the catheters, significantly limit the flexibility and thus maneuverability of the catheters inside the body. Moreover, manufacturing of this type of micromotors requires high-precision assembly/packaging processes such as coil winding, which increase the overall production costs.

One fundamental challenge in the realization of practical micromotors is the development of a reliable rotor bearing. At present, most commercially available miniaturized motors utilize microballs (e.g., Maxon Motor, Switzerland) or sintered metals (e.g., Faulhaber, Germany) as the bearing components. The fabrication and packaging necessary to integrate microball bearings with the other motor components including guiding structures further raises the complexity and cost of motor manufacturing. Moreover, the size of microballs (the smallest being a few hundreds of microns reported in, e.g., [8]) limits the miniaturization of the bearings and motors. Sintered-metal bearings are structurally simple and lower cost; however, this type of bearings is not well suited for low-speed actuations (e.g., 240 rpm [5]) used for endoscopy imaging as the friction force increases at lower speeds. The wear of the bearing material is another critical concern in these types of bearings. Hydrodynamic lubrication is also not practical for low speeds that make it difficult to maintain thin lubrication film [9].

Ferrofluids belong to a category of smart fluids that can be manipulated using magnetic fields [10]. In the presence of a magnetic field gradient, this type of fluids flows toward the location with the highest magnetic flux density. As shown in our previous study [11], the levitation of permanent magnets can be achieved by the combinational use of ferrofluid that provides a reliable, low-friction bearing function highly suitable for micromotor applications. In the present study, the first rotary micromotor enabled with ferrofluid bearing is designed and fabricated. Due to the extremely simple and self-sustained bearing mechanism as will be described, drastically shorter axial length of micromotors, suitable for medical catheter applications, can be realized. In order to demonstrate the feasibility of the ferrofluid-enabled rotary micromotor for endoscopic imaging applications (Figure 1), a proof-of-concept device has been developed and characterized. The core functionality of endoscopic optical imaging through rotating mirror is demonstrated using the fabricated prototype.

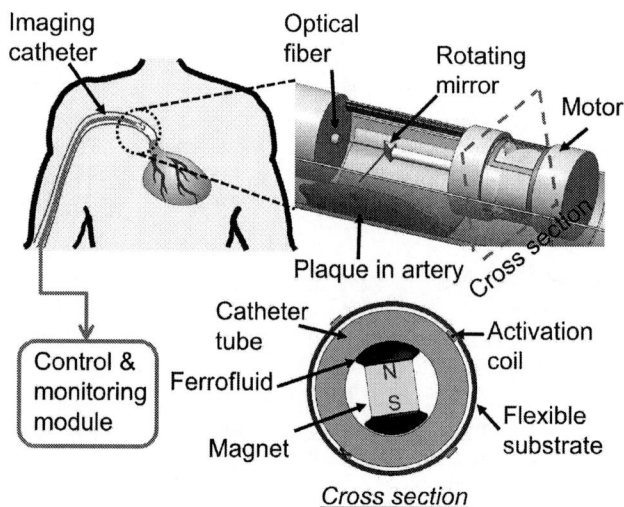

Figure 1: Conceptual diagram of the developed micro rotary motor and its application to microendoscopic catheters.

WORKING PRINCIPLE AND DESIGN

The developed micro rotary motor electromagnetically drives a permanent magnet as its rotor that is levitated by a ferrofluid layer inside a cylindrical hollow tube, or a catheter for the target application. When ferrofluid is applied onto a magnet, the fluid accumulates on the poles of the magnet due to the highest magnetic field gradient provided at those locations. As the ferrofluid layer is established on the magnet surfaces, the magnet is lifted up above the substrate surface due to a magnetic pressure generated in the fluid. Therefore, the ferrofluid layer not only acts as a lubricant for the rotor to reduce the friction when rotated but also physically supports the rotor to be levitated from surrounding inner surfaces of the tube without external pressurizing means. Moreover, the ferrofluid layer is attracted to and self-sustained on the magnet and follows the magnet as it rotates, eliminating the need for any precision alignment or special assembly to maintain the bearing layer in the motor construction.

The stator of the developed motor is composed of two meander-type coils, each of which is printed on a different side of the flexible polymer substrate. This flexible substrate is wrapped and bonded around the catheter tube to establish the stator circuit on the tube's outer surfaces (as presented in Figure 1). Figure 2 illustrates the design of the stator circuitry and its connection with the controller. The coils' patterns are designed to create four magnetic poles (at the middle locations between the vertical coil leads shown in Figure 2) with 90°-phase difference around the tube when currents pass through the coils. For continuous rotation, the coils are activated sequentially with specific current directions so that the magnet rotates 90° at each step. Figure 3 shows the logic steps of the coil activation that produces a counter-clockwise revolution of the ferrofluid-levitated magnet rotor. The direction of rotation can be easily switched by changing the current directions to the two coils at each step.

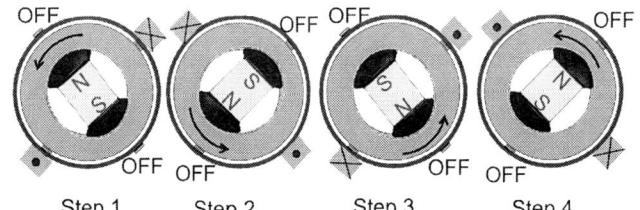

Figure 3: Activation logic showing the directions of the coil currents used to produce a 360° rotation.

FABRICATION

The flexible stator circuit is fabricated using double-sided Cu-clad polyimide (PI) film (Pyrlux AP8525R, DuPont, NC, USA). The PI film and its Cu-clad layers are 50-μm and 18-μm thick, respectively. The meander-shaped coils are fabricated by photo-patterning of dry-film photoresist (PM240, DuPont, NC, USA) laminated on both sides of the PI film, followed by wet etching of the Cu layer to shape the coils. The patterned circuit film is then bonded around the substrate tube using epoxy. For this proof-of-concept effort, glass tube with 2-mm inner diameter and 1-mm wall thickness is selected as the tube material. A NdFeB permanent magnet with 1.6-mm cubic shape (K&J Magnetics Inc., PA, USA) is used as the rotor of the device. This magnet is coupled with a steel shaft with 0.8-mm diameter, and a 45°-angled prism mirror with 1-mm size is bonded at the free end of the shaft (Figure 4(a)). A commercially available oil-based ferrofluid (EFH1, Ferrotec Co., NH, USA; relative permeability ~2.6) is applied onto the magnet to form a layer of the fluid between the magnet and the inner wall of the tube into which the magnet is inserted (Figure 4(b)). The tube is then capped by a 3D-printed plastic component with a hole that the rotor shaft passes through. Figure 4(c) shows a completed sample of the fabricated prototype.

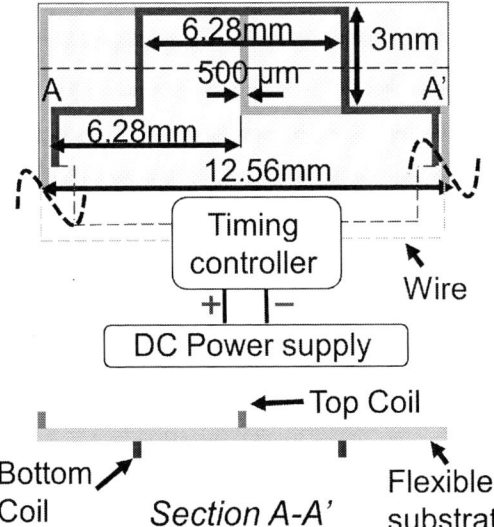

Figure 2: Design of the stator component (in its planar form before assembly), showing two driving coils patterned on opposing sides of the flexible substrate, and their connections with the current controller.

Figure 4: Optical images of (a) the rotor assembly with the magnet, prism mirror, and coupling shaft, (b) the ferrofluid-levitated magnet rotor in the stator tube, and (c) a side view of the completed proof-of-concept device.

EXPERIMENTAL RESULTS

The operation of the fabricated device was tested and characterized using the set-up shown in Figure 5. In order to demonstrate the rotation of the mirror-assembled motor and quantify its revolution speeds, a laser beam was directed to the rotating mirror from the axial direction, and the motion of the reflected beam was video recorded as well as captured by a photodiode connected to an oscilloscope that displayed photo-induced voltages due to the incident beam (Figure 5(a)). The illustration in Figure 5 also includes the set-up used for optical imaging of the inner wall of a test tube (Figure 5(b)), which will be discussed later.

The fabricated device was observed to provide stable rotation at different speeds using the control logic (Figure 3) by varying the switching time of the driving currents. This operation led to circumferential scanning of the reflected laser beam. Figure 6 shows the images of the reflected beam extracted from a video that recorded the rotation of the beam by 90° at each actuation step (with a rotation speed of 30 rpm). The rotation speed was characterized while shortening the switching time from the initial value of 500 ms. The coil current used for the initial condition was 700 mA, and this current level was raised, up to 2 A (limited by the controller), as needed to maintain a continuous rotation at a given switching time. The probable reason of the need for raising the current is that the torque, which is raised with the current, must be increased in order for the rotating rotor to catch up the speed of current switching. The measurement results (Figure 7) revealed that the fastest switching time to sustain the continuous rotations with the 2-A limit was 8 ms (i.e., 32 ms per rotation), corresponding to a rotation speed of 1875 rpm or a scanning rate of 31.3 Hz. This maximum level of revolution speed obtained is well enough for endoscopic imaging applications (e.g., 240 rpm [5]) and indicates a potential capability of the motor for real-time imaging (e.g., 30 fps [12]).

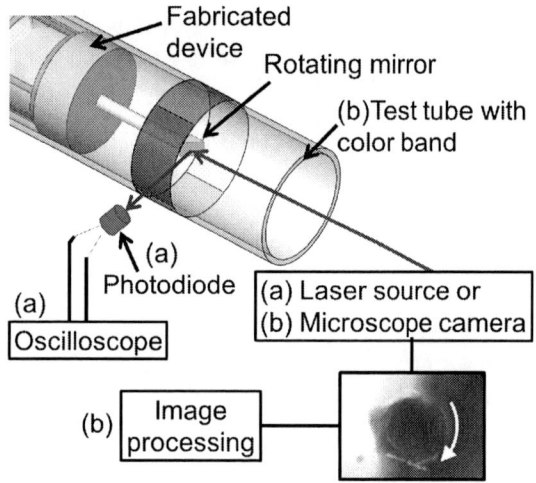

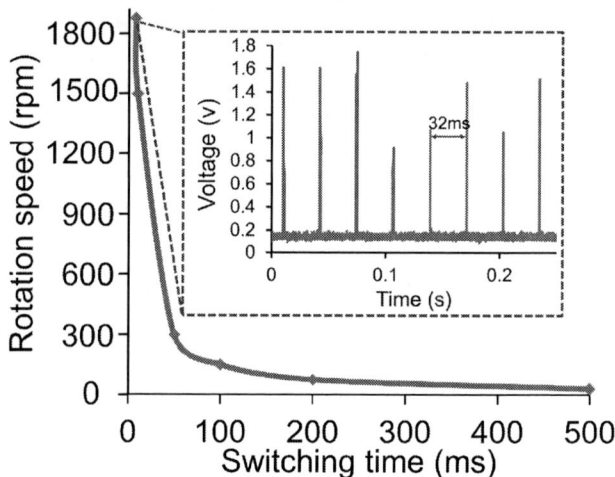

Figure 7: Rotation speed vs. switching time. Inset graph shows photodiode readings for the case of 8-ms switching time corresponding to the max speed recorded.

Figure 5: Experimental set-up used for (a) characterization of the fabricated rotary motor device and (b) preliminary test of endoscopic optical imaging through rotating mirror performed in a test tube with a color band. The image shown after the camera is an optical image of the actual mirror under rotation.

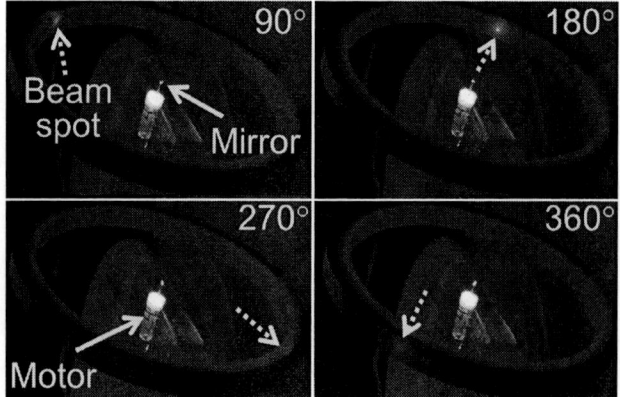

Figure 6: Angular scanning of laser beam (directed onto the inner wall of a plastic ring) by the rotating mirror showing the beam spot captured at each 90° step of one full rotation.

Another preliminary test was performed using an experimental model of body conduit/vessel to emulate the conditions involved in the endoscopic application. In this test, as illustrated in Figure 5(b), the device was placed inside the sample tube whose inner surface was circumferentially sectioned with four different colors (black, red, yellow, and green). The mirror was rotated with 1-s switching time while recording the optical image seen on the rotating mirror from the axial direction of the device/tube through a microscope. As shown in Figure 8(a), optical images of all the color sections were successfully acquired via the rotating mirror, which were then analyzed using the MATLAB® image processing toolbox to quantify a selected color component (in 8 bits) of the acquired optical signals as a function of time (Figure 8(b)). This demonstration verifies the feasibility of circumferential imaging through the fabricated micromotor device, an essential ability required for the targeted endoscopic application.

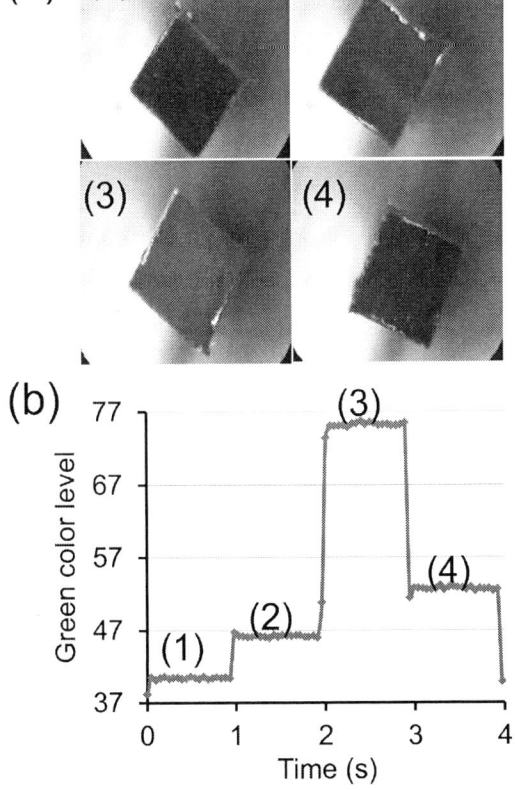

Figure 8: Endoscopic imaging test: (a) Rotating mirror at each 90° step detecting four different colors defined on the inner wall of the test tube; (b) detection of the green color component from the optical signals collected through the rotating mirror.

CONCLUSION

A novel micro rotary motor enabled with ferrofluid was designed, fabricated and characterized. Ferrofluid was used as a low-friction, self-sustained liquid bearing that levitated a 1.6-mm permanent magnet that served as the rotor driven in the stator tubing. The magnet was electromagnetically actuated with microfabricated coils established around the stator tube. The fabricated prototype of the motor was successfully operated to rotate the prism mirror connected to the motor with revolution rates as high as 31.3 Hz. The endoscopic imaging function was demonstrated with the device coupled with the test tube used as a vessel model by collecting optical images of the inner walls of the model through the rotating mirror. The ferrofluid-based bearing and levitation mechanism, with its simplicity and efficiency, offers potential advantages in the miniaturization with remarkably shortened axial sizes as well as in the production cost, enabling robust and low-cost rotary micromotors suitable for not only endoscopic imaging but also other applications in the medical area and beyond.

ACKNOWLEDGMENT

This work was partially supported by the Natural Sciences and Engineering Research Council of Canada, the Canada Foundation for Innovation, and the British Columbia Knowledge Development Fund. K. Takahata is supported by the Canada Research Chairs program.

REFERENCES

[1] A. M. Flynn, K. R. Udayakumar, D. S. Barrett, J. D. McLurkin, D. L. Franck, A. N. Shectman, "Tomorrow's Surgery: Micromotors and Microrobots for Minimally Invasive Procedures," *Min. lnvas. Ther. Allied. Technol.*, vol. 7, pp. 343-352, 1998.

[2] V. R. C. Kode, M. C. C□avus□og□lu, "Design and Characterization of a Novel Hybrid Actuator using Shape Memory Alloy and DC Micromotor for Minimally Invasive Surgery Applications," *IEEE/ASME Trans. Mechatron.*, vol. 12, pp. 455-464, 2007.

[3] J. Shang, D. P. Noonan, C. Payne, J. Clark, M. H. Sodergren, A. Darzi, G. Z. Yang, "An Articulated Universal Joint Based Flexible Access Robot for Minimally Invasive Surgery," in *Proc. IEEE Int. Conf. Robot. Autom.*, Shanghai, May 9-13, 2011, pp. 1147-1152.

[4] S. Chang, E. Murdock, Y. Mao, C. Flueraru, J. Disano, "Stationary-Fiber Rotary Probe with Unobstructed 360° View for Optical Coherence Tomography," *Opt. Lett.*, vol. 36, pp. 4392-4394, 2011.

[5] J. Yang, R. Chen, C. Favazza, J. Yao, C. Li, Z. Hu, Q. Zhou, K. Shung, L. Wang, "A 2.5-mm Diameter Probe for Photoacoustic and Ultrasonic Endoscopy," *Opt. Express*, vol. 20, pp. 23944-23953, 2012.

[6] J. M. Yang, C. Favazza, R. Chen, J. Yao, X. Cai, K. Maslov, Q. Zhou, K. K. Shung, L. V. Wang, "Simultaneous Functional Photoacoustic and Ultrasonic Endoscopy of Internal Organs In Vivo," *Nat. Med.*, vol. 18, pp. 1297-1302, 2012.

[7] J. M. Yang, K. Maslov, H. C. Yang, Q. Zhou, K. K. Shung, L. V. Wang, "Photoacoustic Endoscopy," *Opt. Lett.*, vol. 34, pp. 1591-1593, 2009.

[8] M. McCarthy, C. M. Waits, R. Ghodssi, "Dynamic Friction and Wear in a Planar-Contact Encapsulated Microball Bearing using an Integrated Microturbine," *J. Microelectromech. Syst.*, vol. 18, pp. 263-273, 2009.

[9] C.H. Wong, X. Zhang, S. A. Jacobson, A. H. Epstein, "Self-Acting Gas Thrust Bearing for High-Speed Microrotors," *J. Microelectromech. Syst.*, vol. 13, pp. 158-164, 2004.

[10] R. E. Rosensweig, *Ferrohydrodynamics*, Cambridge University Press, 1985.

[11] B. Assadsangabi, M. H. Tee, K. Takahata, "Ferrofluid-Assisted Levitation Mechanism for Micromotor Applications," in *Proc. IEEE Transducers 2013*, Barcelona, June 16-20, 2013, pp. 2720-2723.

[12] J. Su, J. Zhang, L. Yu, Z. Chen, "In Vivo Three-Dimensional Microelectromechanical Endoscopic Swept Source Optical Coherence Tomography," *Opt. Express*, vol. 15, pp.10390-10396, 2007.

CONTACT

*B. Assadsangabi, bobasad@interchange.ubc.ca

COMMERCIALIZATION OF WORLD'S FIRST PIEZOMEMS RESONATORS FOR HIGH PERFORMANCE TIMING APPLICATIONS

Harmeet Bhugra, Seungbae Lee, Wanling Pan, Minfan Pai, and Dino Lei

Integrated Device Technology, Inc. (IDT), San Jose, CA, USA

ABSTRACT

We present in this paper IDT's efforts and progress in the development and commercialization of world's first piezoelectric MEMS resonators for high performance timing applications. IDT's pMEMS™ resonators consist of a piezoelectric material on a single-crystal silicon layer substrate. Based on pMEMS™ technology, IDT is now shipping world's lowest jitter MEMS Oscillators with less than 0.3ps RMS jitter (12kHz to 20MHz). In addition, pMEMS™ oscillators have demonstrated great long term frequency stability and immunity to shock / vibration testing.

INTRODUCTION

Over the past several decades, quartz crystal resonators (XTALs) and oscillators (XOs) have been used for timing applications in almost all electronic systems due to their stability and superior electrical performance such as low phase noise and frequency stability. Several alternative technologies, including MEMS oscillators [1] have been trying to replace quartz crystals. Fig. 1 below outlines the primary reasons why system architects prefer MEMS oscillators.

Feature	Want?	MEMS Oscillators
Frequency	Higher	pMEMS™ resonators can cost effectively provide higher native frequencies that enable lower jitter (sub-ps).
Size	Smaller	MEMS enables sizes smaller that traditional XOs
Stability Long Term and Short Term	Better	MEMS demonstrates either comparable or better stability
Integrated Functions	More	Configurable PLL, multipliers, dividers, programmable, multiple Outputs
Power Supply	Lower	LVPECL, LVDS, 3.3V, 2.5V and lower...
Activity Dip	Absence	None
Lead Times	Short	Very short lead times, Customer Configurable
Inventory	Small	No Shortages >Semiconductor Level Availability
Reliability	Better	Higher Reliability >Silicon Level Reliability, Production Cost Savings
Cost	Competitive	Competitive costs due to semiconductor scaling and plastic packages
Operating Temperature ranges	Wider	Wide temperature ranges commercial/industrial

Figure 1: Why System designers pefer MEMS Oscillators?

In this paper, we present the design overview and performance characteristics of IDT MEMS resonators, as well as challenges we have faced during technology development.

STRUCTURE OF pMEMS™ RESONATORS

Two major transduction mechanisms, piezoelectric and electrostatic (capacitive) actuation, have been actively investigated by the industry for the commercialization of MEMS resonators. Typically, a capacitively-transduced MEMS resonator requires small gaps of a few hundred nanometers between the resonator body and the electrodes in addition to requiring a DC bias voltage for actuation. The existence of such gaps limits the insertion loss of capacitive resonators and poses challenges in their fabrication. In contrast, and similar to quartz, a piezoelectric MEMS resonator has its transduction layer in direct contact with the electrodes and does not need a DC bias to be operational, resulting in more flexibility in the implementation of piezo-electric MEMS resonators[2].

In implementing the piezoelectric principle, IDT's pMEMS™ platform uses a composite structure consisting of a piezoelectric material (*e.g.*, AlN) and a single crystal silicon (SCS) layer (Fig. 2). This structure takes advantage of the high electromechanical coupling of piezoelectric transduction and the low damping and strong stability of silicon.

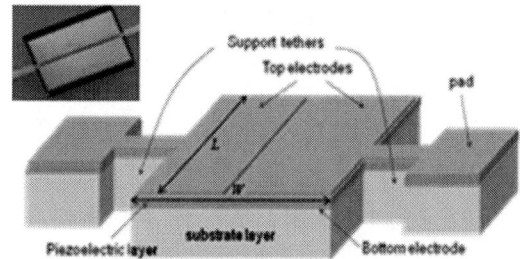

Figure 2: Schematic view of a pMEMS™ resonator.

FABRICATION OF pMEMS™ RESONATORS

IDT has developed a commercially viable CMOS-compatible process for hermetically sealed wafer-level packaged (WLP) pMEMS™ resonators. Fig. 3 shows the cross-sectional view of the resonator with WLP. The resonator itself has a ultra small size of $550 \times 450 \times 200 \ \mu m^3$.

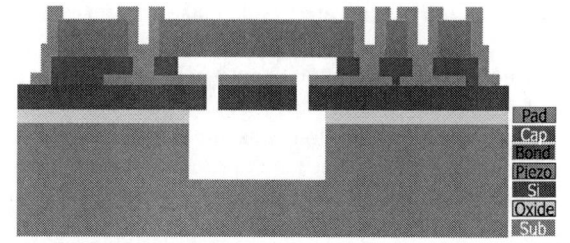

Figure 3. Cross-sectional schematic view of the pMEMS™ resonator with a microshell.

pMEMS™ RESONATOR PERFORMANCE

IDT currently provides MEMS oscillators operating at frequencies of up to 650 MHz based on resonators of different native frequencies. To our knowledge, IDT's pMEMS™

oscillators provide the best phase jitter performance among the MEMS and CMOS oscillators to this day (Fig. 4).

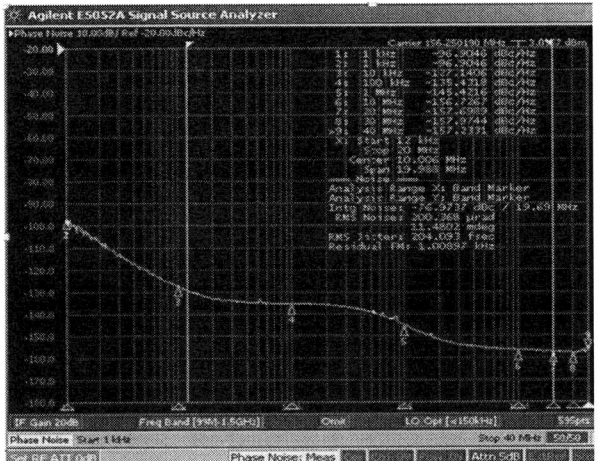

Figure 4 PN measurement plot shows 204fs of jitter from 12kHz to 20 MHz

Shock and Vibration Resistance

pMEMS Oscillators [3] easily survive more than 1500G of shock and 20G of vibration testing. This is a significant advantage over quartz crystals since quartz resonators can shatter easily under shock due to insufficient surface finish and acceleration. The small size of pMEMS™ resonators enables better reliability, *i.e.*, less mass results in better vibration/shock sensitivity. In addition to meeting military grade shock/vibration specifications pMEMS™ devices also remain functional after shock tests of 70,000G.

Long-Term Frequency Stability

Long-term frequency stability has also been demonstrated by subjecting these plastic QFN packaged devices to long term aging (frequency drift) tests. At 25°C, pMEMS™ devices have frequency variations of less than ±2 ppm measured over 2 years, which is comparable to the typical quartz performance of ±5 ppm (Fig.5).

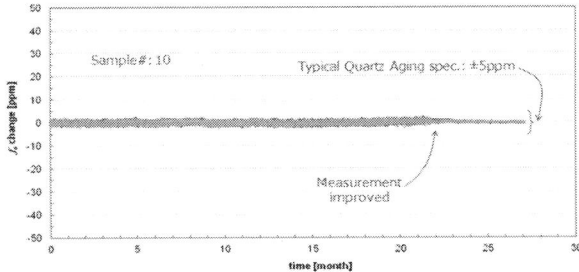

Figure 5: Aging measurements of ten pMEMS™ resonators over 2 years at 25°C (Measurement accuracy ±2ppm) .

Frequency Drift over Temperature

It is well known that the frequencies of mechanical resonators are subject to temperature change. Many teams have devised different mechanisms to reduce the temperature

coefficient of frequency (TCf) of MEMS resonators, including passive layer compensation such as introducing SiO_2 in the resonator body structure [4], Si doping [5] and crystal orientation change [6]. IDT's pMEMS™ resonators have used a combination of above mechanisms to implement reduced TCf resonators in each of successive three generations of resonators in production today (Fig 6).

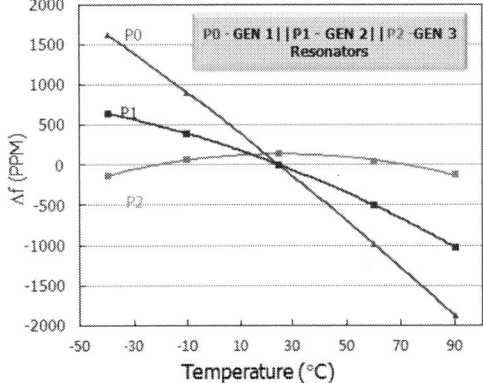

Figure 6: Native TCf improvement of three successive generations of pMEMS resonators.

CONCLUSIONS

A new class of pMEMS™ oscillators has been introduced for high frequency low phase noise (<0.3 ps) timing reference applications. The high performance, as well as compactness and stability of the pMEMS™ devices has proven this technology to be a cost-effective and more reliable replacement for quartz oscillators for the high-frequency frequency reference applications.

ACKNOWLEDGEMENTS

Big credit goes to the IDT MEMS Engineering, Circuit engineering, Product and Test engineering, Sales and Marketing to make these Timing products a reality.

REFERENCES

[1] W.-T. Hsu and A. R. Brown, "Frequency trimming for MEMS resonator oscillators," *IEEE Intl. Frequency Control Symposium*, pp. 1088 – 1091, 2007.

[2] S. Humad, R. Abdolvand, G. Ho, G. Piazza, and F. Ayazi, "High frequency micromechanical piezo-on-silicon block resonators," *IEEE Int. Electron Devices Meeting, Dec. 2003*, pp.39.3.1-39.3.4.

[3] H. Bhugra, S. Lee, W. Pan, D. Lei, "Introducing High Performance CrystalFree™ pMEMS Oscillators", *IEEE International Frequency Control Conference, 2012*.

[4] Roozbeh Tabrizian, Giorgio Casinovi, and Farrokh Ayazi, "Temperature-stable silicon oxide (SilOx) micromechanical resonators", *IEEE Transactions on Electron Devices,* vol. 60, no. 8, pp. 2656 – 2663, Aug. 2013

[5] Ashwin K. Samarao and Farrokh Ayazi, "Temperature compensation of silicon micromechanical resonators via degenerate doping", *IEEE International Electron Devices Meeting (IEDM), 2009*.

[6] M. Shahmohammadi, B. P. Harrington, J. Gonzales, and R. Abdolvand, "Temperature-compensated extensional-mode MEMS resonators on highly N-type doped silicon substrates", *Solid-State Sensors, Actuators, and Microsystems Workshop, 2012*.

MEMS-RECONFIGURABLE WAVGUIDE IRIS
FOR SWITCHABLE V-BAND CAVITY RESONATORS

Zargham Baghchehsaraei and Joachim Oberhammer

Micro and Nanosystems, KTH Royal Institute of Technology, Stockholm, Sweden

ABSTRACT

This paper presents for the first time a novel MEMS-reconfigurable inductive iris based on a 30-μm thick reconfigurable transmissive surface and reports on its application to create a switchable cavity resonator in a WR-12 rectangular waveguide (60-90 GHz). The reconfigurable surface incorporates 252 simultaneously switched contact points for activating (ON state) and deactivating (OFF state) the inductive iris by a 24 μm lateral displacement of two sets of distributed vertical cantilevers. In the ON state, these contact points are short-circuiting the electric field lines of the TE_{10} waveguide mode on the cross-sectional areas of a symmetric inductive waveguide iris, and are not interfering with the wave propagation in the OFF state. Thus, this novel concept allows for completely switching the inductive iris ON or OFF. The inductive iris has an insertion loss of better than 1.0 dB in the OFF state, of which 0.8 dB is attributed to the measurement setup alone. In the ON state the measured performance of the switchable iris is in good agreement with the simulation results. Furthermore, a novel, switchable cavity resonator was implemented based on such a MEMS-reconfigurable iris, and was characterized to a Q-factor of 186.13 at the resonance frequency of 68.87 GHz with the iris switched ON, and an OFF-state insertion loss of less than 2 dB (including the measurement setup) without any resonance, which is for the first time reported in this paper.

INTRODUCTION

Cavity resonator filters are of interest for millimeter-wave frequencies but also for lower frequencies when low insertion loss and high power handling capability are important. With the increasing complexity of wireless technology there is a desire to implement tunable/switchable components for frequency-agile and multi-frequency systems. The resonance frequency of a waveguide cavity is commonly tuned manually by using a screw, or by using either piezoelectric [1] or electrostatic radio frequency microelectromechanical systems (RF MEMS) actuators [2]. Resonators based on piezoelectric tuning mechanism are large and slow, while electrostatic RF MEMS actuators benefit from near-ideal signal handling behavior, ultra-low power consumption, faster tuning speed, and high Q-factor [3,4].

In this paper, we present a switchable inductive iris based on a MEMS-reconfigurable transmissive surface. The transmissive surface is integrated into a WR-12 rectangular waveguide (60-90 GHz, inner dimensions 3.1 mm × 1.55 mm) and can be switched between its activated (ON) state and deactivated (OFF) state by on-chip MEMS electrostatic comb-drive actuators. Previous attempts of MEMS-reconfigurable waveguide elements only provide limited

tuning functionality [5,6], but the novel concept presented here allows for completely switching ON and OFF the irises and thus enables new possibilities for MEMS-reconfigurable waveguide devices, such as resonators/filters which can be switched ON and OFF, as demonstrated by the very first results in this paper.

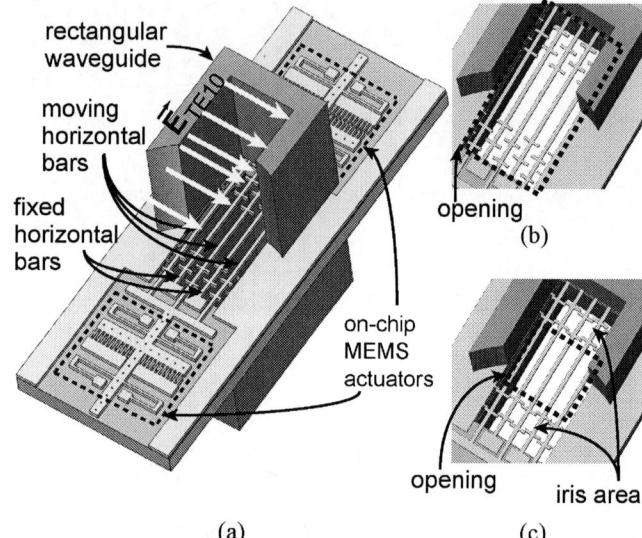

Figure 1: Simplified schematic view of presented MEMS-reconfigurable inductive iris: (a) chip with reconfigurable iris surface and on-chip actuators integrated into a rectangular waveguide; (b) iris surface in the deactivated/OFF state; (c) iris surface in the activated/ON state.

CONCEPT

A simplified schematic view of the micromachined chip with the integrated MEMS-reconfigurable inductive-iris surface, inserted into the WR-12 waveguide cross-section, is shown in Figure 1. The MEMS-reconfigurable surface, placed perpendicularly to the wave propagation, consists of distributed fully-metallized vertical cantilever sections, similar to a MEMS waveguide switch recently presented by the authors [7]. The distributed vertical cantilevers are placed on the two areas of a conventional, fixed waveguide iris. Furthermore, the cantilever sections are grouped into movable and fixed cantilevers and mechanically suspended through horizontal suspension bars which are transparent to the wave propagation. In the deactivate-iris state, the movable set and the fixed set of vertical cantilever sections are not in contact with each other and the reconfigurable surface is totally transmissive allowing the electromagnetic wave of the TE_{10} mode (dominant rectangular waveguide mode) to propagate freely through the surface

978-1-4799-3510-9/14 $31.00 © 2014 IEEE

(see Figure 1(b)). On-chip electrostatic comb-drive MEMS actuators, which are placed outside the narrow (E-plane) walls of the waveguide to avoid any disturbance of the wave propagation, can displace the movable set of vertical cantilevers laterally via the movable set of horizontal bars to bring them in contact with the fixed set to activate the iris (switched ON). As a result, the wave propagation will be partially blocked due to short-circuiting of the electric field lines of TE_{10} mode by vertical cantilevers in this configuration (see Figure 1(c)). The placement of vertical cantilevers toward the narrow wall of the waveguide results in localized short-circuiting of the field in that region similar to a symmetric inductive iris. Therefore, utilization of this novel reconfigurable iris in switchable waveguide components provides the possibility of completely deactivating (OFF state) or activating (ON state) the component.

FABRICATION

An overview of the two-mask micromachining fabrication flow of the chips is shown in Figure 2. The process starts with an silicon-on-insulator (SOI) wafer with device layer thickness of 30 μm, buried oxide (BOX) layer thickness of 2 μm and handle wafer thickness of 500 μm. Bosch deep reactive ion etching (DRIE) technique was used to etch the backside of the chip with a pocket as large as the waveguide inner dimensions to have only the device layer interacting with the wave propagation. Then, the MEMS structure is patterned on the device layer by DRIE. Afterwards, the moving elements of the MEMS structure were release-etched in hydrofluoric acid. Finally, 50-nm layer of titanium tungsten adhesion layer and 1-μm layer of gold is deposited on both sides of the chip by sputtering. The skin depth is 290 nm at 60 GHz and this metallization step (with the thickness equal to more than three time the skin depth) eliminates the metal losses due to skin depth. Furthermore, complete metallization of the chip eliminates any requirement of using a high resistivity SOI wafer to lower the substrate loss, thus an SOI wafer with resistivity of 1-20 Ω-cm is used here.

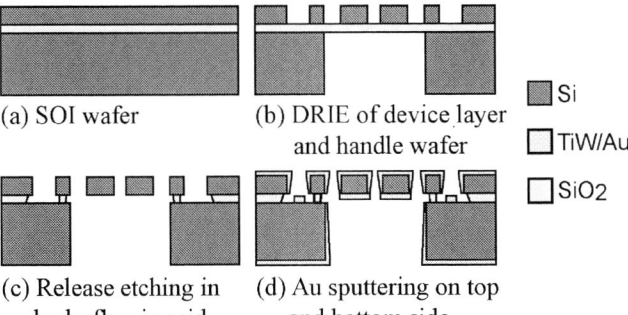

(a) SOI wafer (b) DRIE of device layer and handle wafer

(c) Release etching in hydrofluoric acid (d) Au sputtering on top and bottom side

☐ Si
☐ TiW/Au
☐ SiO2

Figure 2: Two-mask fabrication flow of the MEMS-reconfigurable iris chips: (a) SOI wafer; (b) Deep reactive ion etching of the handle wafer followed by the device layer; (c) Release etching of the device layer in hydrofluoric acid; (d) Double side gold sputtering.

Figure 3 shows a scanning electron microscope (SEM) image of the reconfigurable surface of a chip. The vertical cantilevers have 5 μm width with 5 μm overlap between the movable and the fixed pair. The horizontal suspension bar has 25 μm thickness. The gap between the edge of the movable cantilever (longer cantilevers) and the horizontal bar is 5 μm.

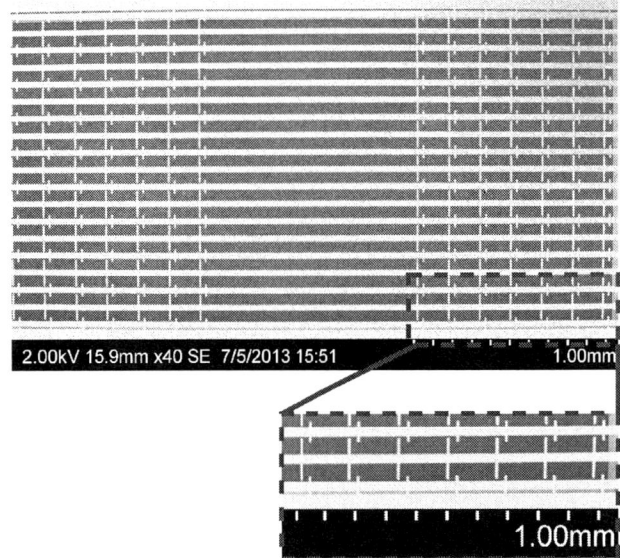

Figure 3: SEM image of the MEMS-reconfigurable surface of a switchable inductive iris.

ON-CHIP MEMS ACTUATORS

The on-chip electrostatic MEMS actuator is designed as a standard push-pull comb-drive actuator with folded-beam springs with the dimensions shown in Figure 4.

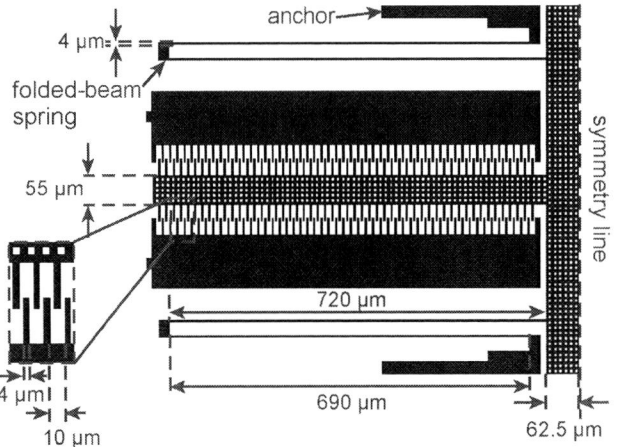

Figure 4: Dimensions of on-chip electrostatic comb-drive MEMS actuators.

The 24-μm displacement of the reconfigurable surface is achieved by engaging four comb-drive actuators for 12 μm active pushing or active pulling in opposite directions around the central rest position. The actuator shown in Figure 4, with 106 fingers on each actuator and with spring constant of 1.97 N/m, is mirrored along the E-plane and H-plane of the waveguide resulting in total eight comb-drive actuators. The average actuation voltage for full displacement in one direction was measured to 40.0 V with a reproducibility of 0.1 V.

As all structural elements are fabricated by bulk micromachining in silicon, all moving parts have a monocrystalline silicon core which is metal-cladded for RF purpose. Furthermore, the design comprises an all-metal design, i.e. no dielectric isolation layers are used which could be prone to reliability impact.

MEASUREMENT SETUP AND RESULTS
Reconfigurable inductive iris

The MEMS-reconfigurable chip is integrated into a WR-12 (60-90 GHz, inner dimensions 3.1 mm × 1.55 mm) rectangular waveguide by placing it in a recess of tailor-made waveguide flanges and by applying conductive polymer interposers between the top surface of the chip and wide walls of the waveguide [8]. The compliant conductive polymer layer, with a nominal resistivity of 8 mΩ-cm, avoids any discontinuity in the surface currents on the wide walls of the waveguide by filling the gap between the waveguide and the chip. The conductive polymer is only applied on the wide waveguide walls (H-plane) to allow the mechanical feed-throughs (horizontal suspension bars of the moving cantilevers) to be connected to on-chip MEMS actuators through the narrow walls (E-plane). The narrow walls are nearly current-free so that the mechanical feed-throughs are not disturbing the RF performance [8]. A schematic cross-section of the integration is shown in Figure 5.

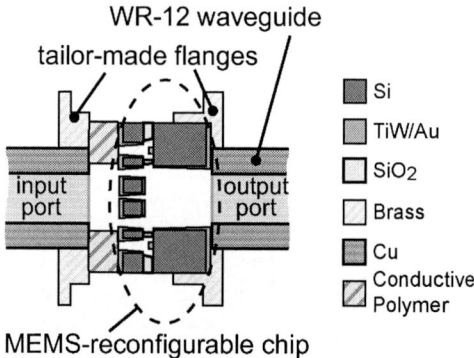

Figure 5: Cross-sectional view of integration of a chip with a MEMS-reconfigurable iris into WR-12 rectangular waveguide.

The results for measured performance and HFSS simulation of a reconfigurable inductive iris are shown in Figure 6. The reconfigurable surface of the iris, consisting of 252 simultaneously switched contact points, has 21 horizontal suspension bars and seven vertical cantilever columns on each side, resulting in an iris opening of 1.075 mm in the waveguide H-plane. In the activated-iris (ON) state, the simulation result is also compared to a

dummy chip with an ideal reference iris with completely filled metal areas. The measurements fit the simulations very well except for a higher insertion loss of 0.4-1.0 dB in the deactivated-iris (OFF) state, of which the majority of 0.3-0.8 dB is attributed to the flange assembly of the measurement setup, which could be demonstrated by the measurement of a reference chip. The reference chip is fabricated in the same process flow as the MEMS-reconfigurable chips but contains no MEMS elements; its device layer and handle wafer is etched through completely with pockets as large as the waveguide inner dimensions.

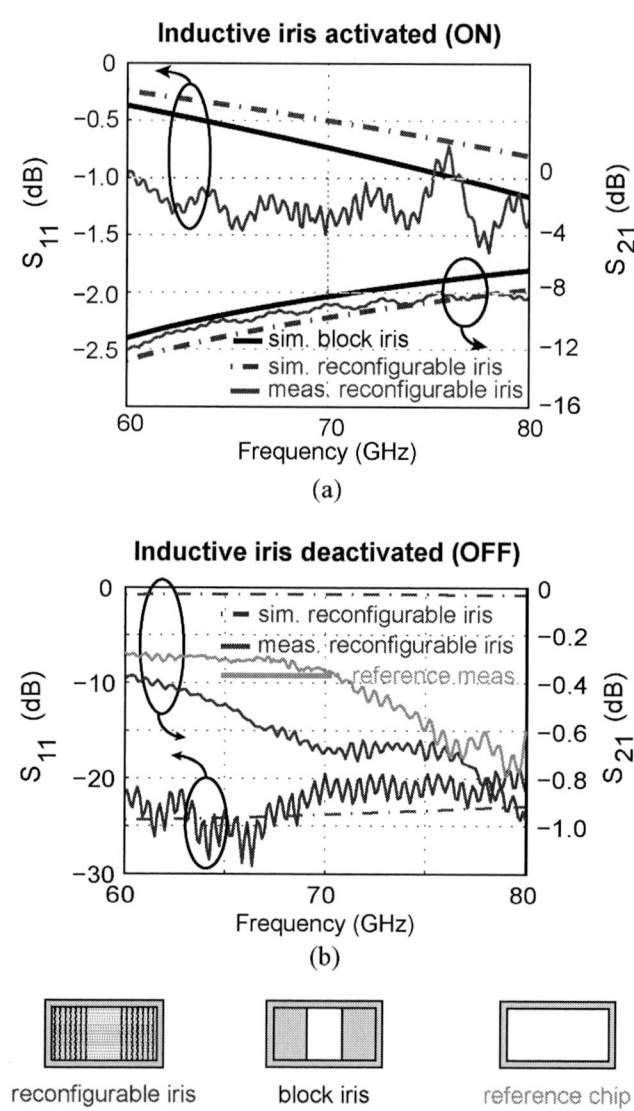

Figure 6: Measured and HFSS-simulated performance of a chip with MEMS-reconfigurable inductive iris compared to a block iris and a reference chip: (a) Inductive iris in the activated/ON state; (b) Inductive iris in the deactivated/OFF state; (c) Schematic cross-sectional view of the reconfigurable iris, the block iris, and the reference chip.

Switchable cavity resonator

A switchable cavity resonator was created by placing a MEMS-reconfigurable surface half a wavelength away from the back-shorted end of a WR-12 waveguide as shown in Figure 7. When the inductive iris is deactivated, i.e. configured as a complete transmissive surface, the configuration is similar to a standard back-shorted waveguide.

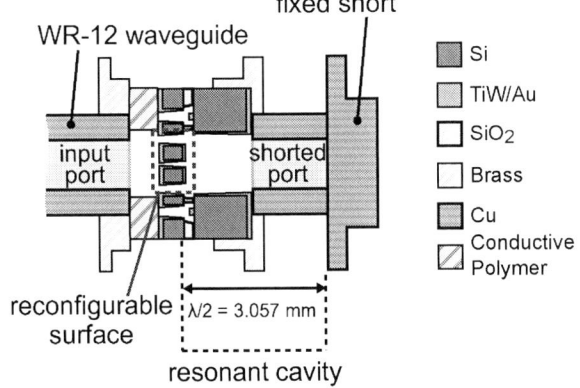

Figure 7: Cross-sectional view of integration of a chip with a MEMS-reconfigurable iris into a back-shorted WR-12 rectangular waveguide to implement a switchable cavity resonator.

The measurement results of the switchable cavity resonator, shown in Figure 8, confirm that the resonator can clearly be switched ON and OFF. The ON-state reflection coefficient is 14.5 dB at the sharp resonance frequency of 68.87 GHz with a Q-factor of 186.13. In contrast to that, the OFF-state configuration does not show any resonance behavior and has an insertion loss of less than 2 dB in the frequency range of interest. Also here, the OFF-state insertion loss is to a major extent contributed by the measurement setup.

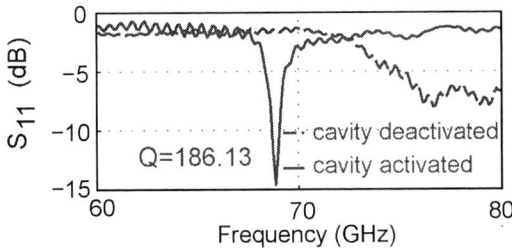

Figure 8: Measured cavity resonator (as in Figure 7) based on a single MEMS-switchable iris in ON and OFF states.

CONCLUSIONS

This paper reported on a switchable inductive waveguide iris with only 30-μm thick MEMS-reconfigurable surface for V-band (50-75 GHz) application. The measurement results of such an iris integrated into a WR-12 rectangular waveguide demonstrated the insertion loss is better than 1.0 dB for the iris in the deactivated (OFF) state, to the major part up to 0.8 dB attributed to the

assembly, and the ON-state measurements agree very well to idealized irises. Furthermore, the application of switchable inductive iris was shown by implementing a cavity resonator by integrating a MEMS-reconfigurable inductive iris half-wavelength from a back-short end of a waveguide. The switched resonator (ON-state) showed Q-factor of 186.13 at the resonance frequency of 68.87 GHz, and its OFF-state isolation is better than 2 dB (including the measurement setup).

ACKNOWLEDGEMENTS

The authors wish to thank Alan Cheshire and Gabriel Roupillard from Applied Materials for their help with the Centura Etch. The authors would also like to thank Jan Åberg from MicroComp Nordic AB for the fabrication of the customized flanges for the measurement setup.

REFERENCES

[1] H. Joshi, H. H. Sigmarsson, S. Moon, D. Peroulis, W. J. Chappell, "High-Q fully reconfigurable tunable bandpass filters", *IEEE Trans. Microw. Theory Tech.*, vol. 57, no. 12, pp. 3525-3533, Dec. 2009.

[2] S.-J. Park, I. Reines, C. Patel, G. M. Rebeiz, "High-Q RF-MEMS 4–6 GHz tunable evanescent-mode cavity filter", *IEEE Trans. Microw. Theory Tech.*, vol. 58, no. 2, pp. 381-389, Feb. 2010.

[3] J. Small, M.S. Arif, A. Fruehling, D. Peroulis, "A Tunable Miniaturized RF MEMS Resonator With Simultaneous High Q (500–735) and Fast Response Speed (< 10–60μs)", *J .Microelectromech. Syst.*, vol. 22, no. 2, pp. 395-405, April 2013.

[4] G. M. Rebeiz, *RF MEMS theory, design, and technology*, Wiley, 2003.

[5] P. Blondy, D. Peroulis, "Handling RF Power: The Latest Advances in RF-MEMS Tunable Filters", *IEEE Microwave Magazine*, vol. 14, no. 1, pp. 24-38, Jan.-Feb. 2013.

[6] D. Dancila, P. Ekkels, X. Rottenberg, I. Huynen, W. De Raedt, H. A. C. Tilmans, "A MEMS variable faraday cage as tuning element for integrated silicon micromachined cavity resonators", in *IEEE 23rd International Conference on Micro Electro Mechanical Systems*, Wanchai, Hong Kong, Jan. 24-28, 2010, pp. 723-726.

[7] Z. Baghchehsaraei, U. Shah, J. Aberg, G. Stemme, J. Oberhammer, "Millimeter-wave SPST waveguide switch based on reconfigurable MEMS surface", in *IEEE MTT-S Int. Microw. Symp. Dig.*, June 2-7, 2013, pp. 1–3.

[8] Z. Baghchehsaraei, M. Sterner, J. Aberg, J. Oberhammer, "Integration of microwave MEMS devices into rectangular waveguide with conductive polymer interposers", *Journal of Micromechanics and Microengineering*, vol. 23, no. 12, pp. 1-10, 2013.

CONTACT

*Z. Baghchehsaraei, tel: +46-762315809; zargham@kth.se

A MICROMECHANICAL PARAMETRIC OSCILLATOR FOR FREQUENCY DIVISION AND PHASE NOISE REDUCTION

Tristan O. Rocheleau, Ruonan Liu, Jalal Naghsh Nilchi, Thura Lin Naing, and Clark T.-C. Nguyen
University of California, Berkeley, USA

ABSTRACT

A capacitive-gap transduced micromechanical resonator array has demonstrated a first on-chip MEMS-based frequency divider with 61-MHz output generated from a 121-MHz electrical drive through use of a parametric oscillation effect that provides not only the 6 dB reduction in close-to-carrier phase noise expected for a frequency divide-by-two function, but also a remarkable 23 dB reduction in far-from-carrier noise provided by filtering with an extremely high mechanical Q of 91,500. Unlike conventional frequency dividers (i.e., prescalers), the parametric oscillator dispenses with active devices and their associated noise, and operates with close to zero power consumption, limited in principle only by the power required to overcome MEMS resonator loss, estimated here at 100 nW. With an output voltage swing of 450 mVpp generated from only 445 mVpp of input swing on a differential version of this MEMS divider, cascaded chains of fully passive dividers are possible, as needed for use in real-world phase-locked loops and frequency synthesizers.

INTRODUCTION

Frequency dividers have become essential components in phase-locked loops (PLL) and frequency synthesizers used a myriad of applications, from instrumentation to wireless handsets. In a typical frequency synthesizer application, frequency dividers often limit the achievable phase noise performance and contribute a large or even majority portion of the total power consumption. Common digital dividers offer good noise performance, but at the cost of power far in excess of that permissible for mobile applications and with poor scaling as frequency is increased, with 135 mW power consumption typical for a low-noise divide-by-16 at 3 GHz [1].

To alleviate this, injection-locked oscillator dividers [2] that lock a free running oscillator to an input signal at a harmonic of the oscillation frequency have emerged as lower power options at high frequencies. With operating power below 100 μW even at GHz frequencies [3], such dividers present a compelling alternative to traditional technologies, but come with performance limitations due to the active transistors used to sustain oscillation [4].

To overcome these limitations, this paper introduces a new frequency divider design using a parametric amplification effect in a capacitive-gap transduced MEMS device, such as that of Fig. 1, based on an array of wine-glass disk resonators [5], or Fig. 2, based on a stand-alone device operating in a differential mode. Each of these parametric oscillators works via modulation of a frequency-determining parameter of a resonator, in this case electrical stiffness [6], at twice the resonance frequency. This produces an effect analogous to a child on a swing, where swinging legs modulate frequency (or stiffness), providing gain and enhancement of in-phase resonant motion that with sufficient pump strength drives the resonator into self-sustained oscillation. While performing

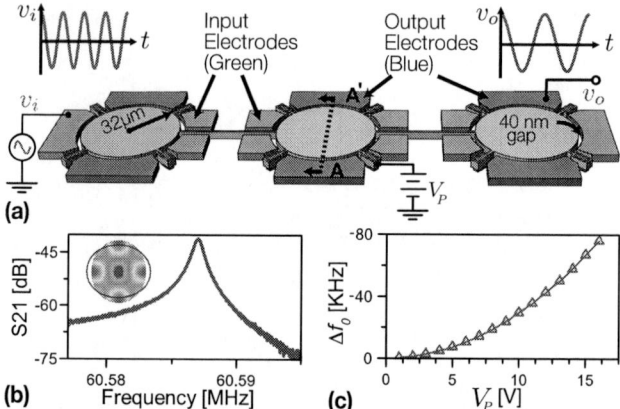

Fig. 1. Schematic view (a) of the wine-glass disk array parametric divider showing 121-MHz pump signal applied to the input electrodes (green) and resultant 60.6-MHz signal on the output electrodes (blue). (b) presents the high-Q frequency response of the resonator as measured by two-port transmission with a network analyzer, while (c) demonstrates the resonance frequency tuning required for parametric amplification through an electrical stiffness effect with applied voltage V_P

a frequency divide-by-two function, the single-ended version of Fig. 1 reduces phase noise by 6 dB at close-to-carrier frequencies and 23 dB far-from-carrier. Building upon this, the differential version of Fig. 2 enables an effective voltage gain, generating a measured output voltage swing of 450 mVpp, larger than the input swing of only 445 mVpp.

Though similar in operation to injection-locked frequency dividers, the parametric oscillator here requires no active devices, and thus, adds no additional noise sources beyond Brownian to the signal. Furthermore, with power consumption fundamentally limited only by the MEMS resonator loss, power usage can be nearly zero. Such MEMS dividers are expected to enable exceptionally low-power frequency synthesizers for applications where the operating frequency remains within the tuning range of the MEMS resonator, such as the Chip-Scale Atomic Clock (CSAC). Indeed, commercial CSACs presently consume a battery-unfriendly 120 mW

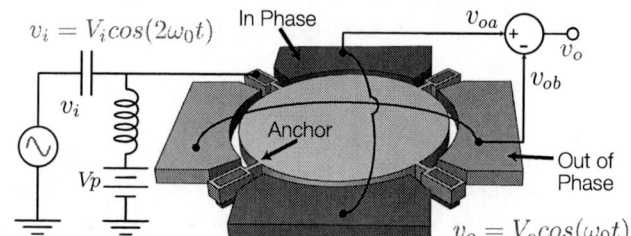

Fig. 2. Schematic of the differential frequency divider used on a single disk in order to boost divided output voltage swing and minimize common-mode harmonic feedthrough.

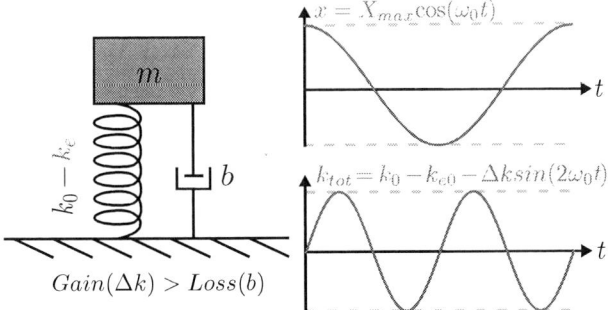

Fig. 3: Theoretical representation of the parametric oscillator divider consisting of a resonator (mass-spring-damper) undergoing motion at frequency ω_0 (blue curve). When stiffness, k, is modulated as in the red curve, restoring force is increased following x maxima and decreased near zero, leading to a parametric gain effect on motion.

[7], with a good portion of this burned by the frequency dividers in their synthesizers.

PARAMETRIC OSCILLATION

In a parametrically-driven device, an applied pump modulates a frequency-determining "parameter", realized here as the electrical stiffness k_e, arising from the applied voltage across the electrode-disk capacitive gaps [6]. This gives rise to the equations of motion

$$m\frac{\partial^2 r}{\partial t^2} + \frac{m\omega_0}{Q}\frac{\partial r}{\partial t} + [k_0 - k_e(t)]r = F(t) \qquad (1)$$

$$k_e(t) = \frac{1}{2}\frac{\partial^2 C_0}{\partial r^2}[V_g(t)]^2 \qquad (2)$$

where m is effective resonator mass, r is effective radial resonator displacement at a location of maximum displacement amplitude, ω_0 is the resonance frequency, k_0 is the effective mechanical spring constant of the resonator, $F(t)$ is the resonator driving force (here, comprised of random thermal noise only), C_0 is the resonator-to-electrode capacitance, and V_g is the total voltage applied across this capacitive gap.

Driving V_g with an ac signal v_i at twice the resonance frequency, together with dc bias voltage V_P, generates electrical stiffness $k_e(t) = \Delta k sin(2\omega_0 t) + k_{e0}$ composed of a static shift k_{e0} combined with a modulation component at twice the resonance frequency with $\Delta k \propto C_0 V_i V_P$. This modulation, illustrated in Fig. 3, leads to an increase in restoring force following max displacement and decrease near zero resulting in a phase-dependent parametric gain of resonant motion. For resonator motion at the correct phase relative to the parametric pump, this can be found [8] to give an amplitude gain of

$$G = \frac{1}{1 - Q\Delta k/2k_0} \approx \frac{1}{1 - \Delta f Q/f_0} \qquad (3)$$

where Δf is the amplitude of frequency shift induced by the modulated stiffness. Though produced entirely by modulation of the resonator frequency, this gain can be understood conceptually as equivalent to a traditional oscillator where resonator motion is amplified in a closed-loop configuration. Recognizing f_0/Q as the 3 dB bandwidth of the resonator and, therefore, equivalent to the resonator damping, makes clear the equivalence of amplifier gain with depth of the applied stiffness modulation. When driven with sufficient pump

strength, this gain overcomes resonator losses and amplifies the, initially Brownian, motion into steady-state oscillation at the fundamental resonator frequency, limited only by resonator nonlinearities. Eqn. (3) in fact can be seen as the classic positive feedback equation with loop gain equal to the ratio of frequency pull over 3 dB bandwidth, $A_l = \Delta f/(f_0/Q)$.

Given the above mechanism for parametric gain, the use of such a device as a frequency divider is clear: an input electrical drive at twice resonance frequency atop a dc-bias voltage V_P produces a parametric gain via modulation of the electrical stiffness of the capacitive gap. When sufficient input voltage swing at $2f_0$ is provided, the parametric gain drives the resonator into oscillation at f_0, the motion of which, when combined with the bias voltage, generates an output frequency-divided electrical signal.

As the parametric frequency modulation is produced by a time varying voltage applied across a pure capacitance, the power required for this drive can be, in principle, limited only by the energy transferred to the mechanical resonator. For a steady-state oscillator, this power transfer must be sufficient to balance resonator losses. With the extremely high Q achievable in MEMS resonators, this energy loss rate is tiny, requiring less 100 nW to sustain full oscillation amplitude for a typical Q of 90 k at 61 MHz.

DEVICE DESIGN AND OPERATION

To be useful in modern PLL applications, the MEMS resonators used must possess both high operating frequency and the capability to accurately define multiple unique frequencies on the same die, i.e., their frequencies should be definable via CAD layout. To this end, the wine-glass disk resonators depicted in Fig. 1(a) and Fig. 2 are quite suitable. These devices comprise 2 µm-thick, 32 µm-radius polysilicon disks supported by beams at quasi nodal points and coupled along their sidewalls to input-output electrodes by tiny 40 nm capacitive gaps. In the three-device array, coupling beams sized to correspond to half the acoustic wavelength force the individual resonators to move in-phase at a single resonance, allowing output current to add constructively to boost electromechanical coupling by the number of individual resonators [9]. To excite the composite resonator into motion, a bias voltage V_P on the disk structure combines with an ac drive voltage applied to all input electrodes to produce forces across the input electrode-to-resonator gaps that, at resonance, excite the wine-glass (i.e., compound (2, 1)) mode shape, shown in the inset of Fig. 1(b). The frequency of resonance is given by [10]:

$$f_{nom} = \frac{K}{R}\sqrt{\frac{E}{\rho(2+2\sigma)}}\left[1 - \frac{k_e}{k_0}\right]^{\frac{1}{2}} \qquad (4)$$

where R is the disk radius, $K = 0.373$ for polysilicon structural material, k_e is the electrical stiffness given by Eqn. (3), and E, σ, and ρ are the Young's modulus, Poisson ratio, and density of the structural material, respectively.

To exploit the parametric amplification effect for frequency division, the lateral 40 nm capacitive gap of the device of Fig. 1(a) produces both the strong voltage-dependent frequency tuning, cf. Fig. 1(c), required for parametric excitation and, with a bias voltage, an output current propor-

978-1-4799-3510-9/14 $31.00 © 2014 IEEE

	Q	Radius [μm]	Gap [nm]	h [μm]	m [ng]	k_0 [N/m]	C_0 [fF]	f_0 [MHz]	V_P [V]	v_i [mVpp]	v_o [mVpp]
Array Device	91 k	32	40	2	16.1	23.0×10^5	106.8	60.587	4	700	265
Differential Device	84 k	32	40	2	5.37	5.37×10^5	35.6	60.560	4	445	450

Table 1. Design and measured values for the fabricated MEMS dividers

tional to the resonator motion. By enhancing tuning, the small capacitive gaps enable parametric oscillation at UHF frequencies well beyond the kHz range of earlier parametrically amplified sensors [11], and further allow adjustment of the operating frequency of the divider, as needed for many real-world applications.

NOISE FILTERING

Because the MEMS resonator used here has such high mechanical Q, it responds with long time constants to changes in the input signal, and so effectively filters out noise signals not within its bandwidth. This means that when operated as a self-sustained parametric oscillator, this device removes oscillation perturbations at frequencies greater than the bandwidth of the resonator. Though similar in principle to injection-locked oscillators or PLLs where far-from-carrier phase noise is suppressed, the extremely high Q-factor of the MEMS resonator deployed here produces a pronounced filter cut-off even at frequency offsets below 1 kHz. Combined with the lack of active devices, hence lack of associated noise, the MEMS-based parametric oscillator provides not only divide-by-two frequency division with the expected 6 dB phase noise reduction, but also additional filtering of phase noise outside of the resonator bandwidth. It should be noted that while an effective narrow lock range of the oscillator provides additional filtering, the operating frequency can still be tuned over a much larger range through use of the voltage-controllable electrical stiffness, allowing this device to operate over appreciable bandwidths.

EXPERIMENTAL RESULTS

Using Eqn. (4), the single-disk differential device and 3-disk single-ended array device summarized in Table 1 were designed to provide divide-by-two functions with 120-MHz input and 60-MHz output. The fabrication process used for the devices was based on a surface micromachining process similar to that of [5], summarized in the cross sections of Fig. 4. The devices and electrical interconnect were fabricated from polysilicon deposited via Low-Pressure Chemical-Vapor Deposition (LPCVD) at 615°C, in-situ doped with phosphorus. A High-Temperature Oxide (HTO) deposition provided the high-quality sacrificial sidewall spacer that enabled 40nm electrode-to-resonator gaps. Two Chemical-Mechanical Polishing (CMP) steps, one before structural polysilicon deposition and patterning and the other after the electrode polysilicon deposition, provided the planar surfaces needed for precise lithography, as well as removed electrode-disk overhangs that can cause pull-in and device failure at low bias voltages. Fig. 5 presents SEMs of fabricated devices along with FEM simulations of their mode shapes.

Devices were tested in a Lakeshore model FWPX vacuum probe station capable of providing 10 μTorr vacuum.

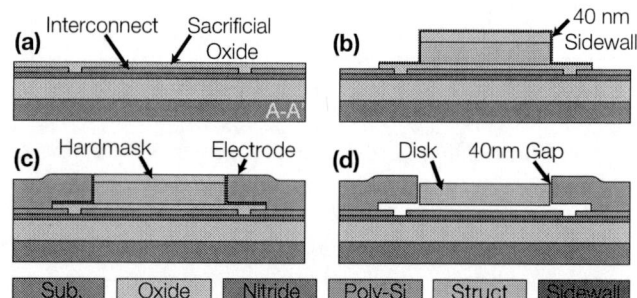

Fig. 4: Fabrication process consisting of repeated thin-film polysilicon and oxide deposition and etching on a silicon substrate, achieving, first, the electrical interconnect of (a). Subsequent depositions and etching yields the structure and thin oxide sidewall sacrificial of (b), after which a polysilicon deposition, patterning and CMP gives the planarized structure of (c). Finally, a wet-etch in 49% HF yields the released resonator structure of (d).

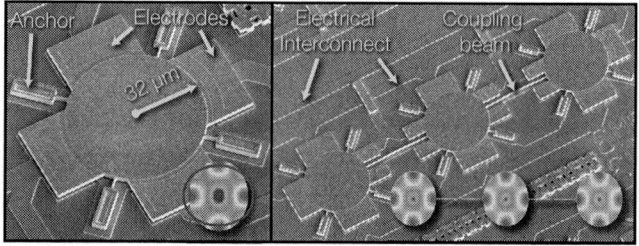

Fig. 5: SEMs of the released frequency-dividers measured in this work with (left) a single-disk, differential and (right) a 3-disk arrayed single-ended divider. Inset FEM simulations illustrate the resonance mode shapes.

A GGB Industries Picoprobe model 35, modified to operate in the Lakeshore vacuum system, provided accurate divider input and output voltage measurement with minimal signal loading. Phase noise was measured on an Agilent E5505A phase noise measurement system.

Fig. 6 presents the output of an arrayed device like that shown in Fig. 1(a), driven by 700 mVpp at 121.174 MHz. The resultant 60.587-MHz oscillation combined with a 4 V bias voltage generates a current at the output electrode that drives the modestly large 350 fF output capacitance (composed of bondpad plus Picoprobe capacitance) to a 265 mVpp swing. While the output-to-input voltage ratio is below unity in this configuration due to the large output capacitance, the generated output current would be sufficient to produce an output voltage swing equivalent to input when driving the much smaller 50 fF capacitance of an integrated 2nd MEMS division stage or on-chip buffer.

To achieve an output voltage of similar magnitude as the input even in the face of large bondpad capacitors, the balanced wine-glass disk configuration of Fig. 2 using a differential output was also investigated. Here, driving the pump

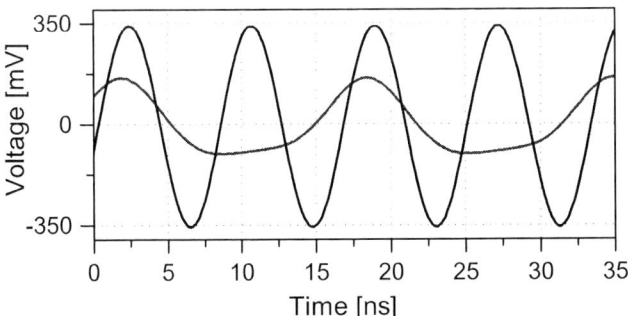

Fig. 6: Input 121.2-MHz pump (black) and resultant 60.6-MHz output waveform measured on the single-ended divider. Input and output signals spanned 700 mVpp and 265 mVpp, respectively. Distortion can be seen in the output waveform due to feedthrough of the pump signal.

signal on the disk itself not only enhances the electrical stiffness, which now derives from all electrodes, but also doubles the differential output swing, all while canceling common-mode feedthrough. The resultant output swing shown in Fig. 7 spans 450 mVpp using only a 445 mVpp pump. Thus, voltage gain is provided, as needed for a cascaded divider chain, without need for power-hungry active devices.

Fig. 8 presents measured phase noise of the single-ended divider output when driven by a custom-built VCO with mediocre phase noise, showing the 6 dB improvement for close-to-carrier phase noise expected due to frequency division. Perhaps more impressively, the high-Q resonator response provides filtering of noise past a 1 kHz offset, leading to a remarkable 23 dB decrease in far-from-carrier phase noise, limited only by the poor 50 nV/rtHz noise performance of the Picoprobe used to measure the output.

CONCLUSIONS

Through use of a parametric amplification effect enabled by the electrical stiffness of capacitive gap MEMS resonators, a new type of frequency divider has been demonstrated. This divider provides both the expected 6 dB phase noise improvement close to carrier for a divide-by-two function as well as additional noise filtering for offsets above 1 kHz due to the high-Q mechanical response function—a unique advantage of this MEMS technology. Consuming only 100 μm × 100 μm die space for a single 120-MHz divider, this MEMS-based approach further offers significant space savings over similar CMOS based dividers [12], where bulky inductors consume 750 μm × 320 μm for an operating frequency of 20 GHz, and offer no possibility of division down to typical reference frequencies, e.g., 10 MHz. Adding to the list of benefits, as frequencies increase, this MEMS-based frequency divider shrinks in size: A 3.4-GHz version would occupy under 30 μm × 30 μm of die area. Future efforts to design frequency-matched chains of such MEMS dividers would be expected to enable complete low-power PLL topologies at up to GHz frequencies. Clearly, this device adds a previously missing frequency divider capability to the MEMS toolbox.

Acknowledgement: This work was supported under the DARPA CSSA program.

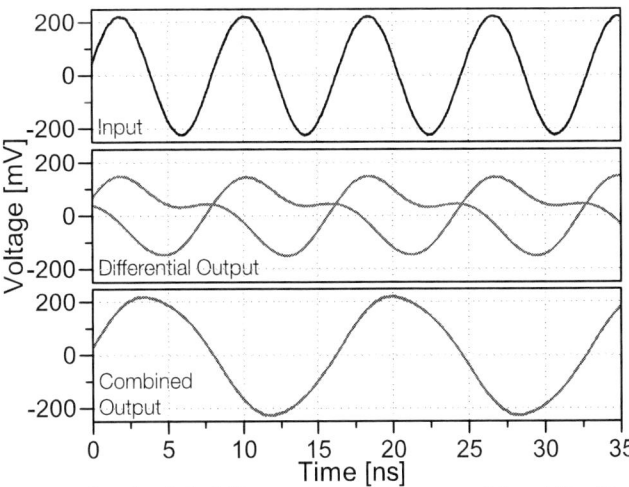

Fig. 7: Single-disk differential measurement with a 445mVpp input pump (top), separate differential outputs (middle) and combined differential output of 450mVpp (bottom).

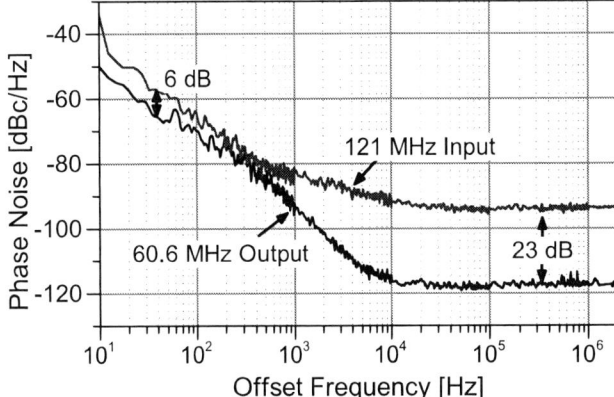

Fig. 8: Single side-band phase noise comparison of input pump signal from a low quality VCO operating at 121 MHz and resultant 60.6-MHz output of the single-ended divider. Phase noise shows both the expected 6 dB improvement, but also additional filtering from the high Q of the resonator.

REFERENCES

[1] ON Semiconductor, Divide by 16 Divider, Part: NB6N239S
[2] M. Tiebout, *JSSC*, vol. 39, pp. 1170-1174, 2004.
[3] P. Dubey and R. Agarwal, *Proceedings, VLSI Design*, 2013, pp. 158-162.
[4] S. Verma, *et al.*, *JSSC*, vol. 38, pp. 1015-1027, 2003.
[5] M. A. Abdelmoneum, *et al.*, *Proceedings, IEEE Int. Conf. on MEMS*, 2003, pp. 698-701.
[6] H. C. Nathanson, *et al.*, *IEEE Transactions on Electron Devices*, vol. 14, pp. 117-133, 1967.
[7] Symetricom CSAC model number SA.45
[8] D. Rugar and P. Grütter, *PRL*, vol. 67, p. 699, 1991.
[9] M. U. Demirci and C.-C. Nguyen, *JMEMS*, vol. 15, pp. 1419-1436, 2006.
[10] M. Onoe, *JASA*, vol. 28, p. 1158, 1956.
[11] M. J. Thompson and D. A. Horsley, *J-MEMS*, vol. 20, pp. 702-710, 2011.
[12] W. Lee and E. Afshari, *JSSC*, vol. 45, pp. 1834-1844, 2010.

Contact: Tristan O. Rocheleau, Tel: +001-607-232-2162; tristan@eecs.berkeley.edu

TEMPERATURE-COMPENSATED PIEZOELECTRICALLY ACTUATED LAMÉ-MODE RESONATORS

Vikram Thakar[1] and Mina Rais-Zadeh[1]
[1]University of Michigan, Ann Arbor, USA

ABSTRACT

Electrostatically actuated Lamé-mode resonators are known to offer high quality factors (*Q*) in the low MHz frequency range [1], [2] but require large bias voltages and suffer from low power handling. In this work, we utilize piezoelectric transduction to circumvent the limitations of electrostatic actuation. Silicon dioxide refilled islands, used to achieve temperature compensation, are shown to provide a 20× improvement in the total charge pick-up, enabling piezoelectric actuation of Lamé-mode resonators. By optimizing the placement of the oxide-refilled islands and without changing the total oxide volume, the turnover temperature (TOT) can be designed to occur across a wide range from -40 °C to +120 °C without any significant *Q* degradation. Using such an approach multiple piezoelectric resonators with different TOTs can be fabricated on a single wafer, enabling multi-resonator systems stable across a wide temperature range.

INTRODUCTION

Low-power timing references are typically implemented using resonators operating at frequencies less than 100 MHz. From the considerations of phase noise, which translates into timing jitter, it is critical that the resonator has a high *Q*, large power handling limit, and low motional impedance. The isochoric mode shape of the Lamé mode of resonance enables low thermoelastic damping (TED) and thus a high device *Q* in the low MHz frequency regime [3] and electrostatically actuated Lamé-mode resonators have been demonstrated with *Q*s over a million at 6 MHz (Figure 1) [1], [4]. However, the relatively large bias voltage requirements and poor power handling capability makes it difficult to realize precision timing references with such electrostatically actuated resonators [5].

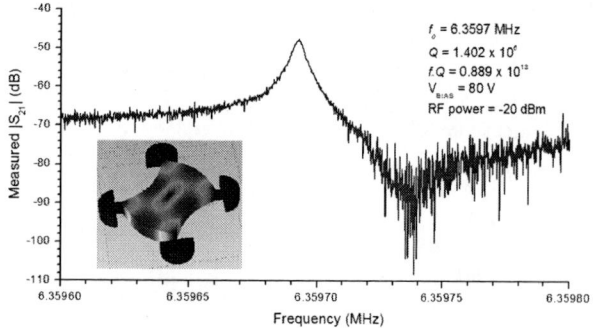

Figure 1: Measured frequency response of an electrostatically actuated Lamé-mode resonator showing a *Q* of 1.4 million. Inset shows the mode shape of the resonator.

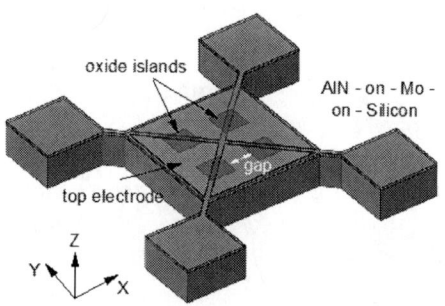

Figure 2: Schematic of a temperature-compensated Lamé-mode resonator. The location of the oxide-refilled trenches and the triangle shaped top electrode layout can be clearly seen.

Piezoelectric actuation obviates the DC bias requirements and ensures an improved power handling ability [6] and has been chosen as the transduction mechanism in this work. We compensate the temperature coefficient of frequency (TCF) of silicon using silicon dioxide refilled trenches positioned deliberately across the resonator body. Lamé-mode resonators actuated using piezoelectric transduction layer with Wurtzite symmetry (such as AlN and GaN) are susceptible to charge cancellation due to the equal magnitude and polarity of the d_{31} and d_{32} piezoelectric coefficients. We show that the presence of the oxide islands not only compensates the first-order TCF but also skews the strain profile across the resonator and provides a 20× improvement in the total charge pickup, enabling piezoelectric actuation of the Lamé-mode resonators.

RESONATOR DESIGN

Temperature Compensation

Silicon dioxide has been successfully used to overcome the large TCF of silicon [7]. The compensated composite resonators show a parabolic dependence of frequency with respect to the temperature with the overall frequency shift reduced to under 200 ppm across the industrial temperature range (as compared to ~3750 ppm for silicon only resonators) [7]. The TOT *i.e.* turnover temperature is defined as the inflection point of this parabola and determines the temperature at which the local TCF is zero.

For the compensation of bulk mode resonators, uniformly distributed oxide pillars have been used to limit the oxide thickness deposited [8]. In lieu of a uniform distribution of oxide, we utilize its location dependence on temperature compensation and place the oxide trenches around the resonator center, where strain energy is high [9], [10]. By careful control over the placement of these oxide-refilled trenches, it is possible to finely control the amount of compensation, which manifests as a change in the TOT without increasing any processing steps. This enables the realization of multiple resonators with different TOTs and can enable multi-resonator systems that are temperature stable across the industrial temperature range.

Figure 2 shows a schematic of a piezoelectrically actuated Lamé-mode resonator with the location of the oxide-refilled trenches. The size of the resonator plate is 100 μm × 100 μm, while the four oxide-refilled islands are 20 μm × 20 μm each in size. The resonance frequency of the first order Lamé-mode can be approximately written as

$$f_{Lame} = \frac{1}{2L}\sqrt{\frac{E}{(1+\nu)\rho}}, \qquad (1)$$

where L is the side length of the resonator and E, ν and ρ are the effective Young's modulus, Poisson's ratio and density of the composite resonator, respectively. Assuming a temperature independent ν and ρ, the TCF of a composite Lamé-mode resonator can be written as

$$TCF = -\frac{1}{2}TCE - \alpha_L, \qquad (2)$$

where α_L is the coefficient of thermal expansion and TCE is the temperature coefficient of Young's modulus. Typically, the effective E of composite resonators is estimated by averaging the volumetric contribution of the different materials within the resonator. However, it has been shown that the effective E and consequently the TCE is not only a function of the volumetric composition but also of the location of the different materials within the resonator body [9]. Here, we show that a modified modeling framework considering the ratio of strain energy in oxide and silicon to that in the whole resonator can more accurately predict the temperature behavior of a composite resonator:

$$TCF_{resonator} = TCF_{Si} \times k + TCF_{ox} \times (1-k), \qquad (3)$$

where k is ratio of strain energy in silicon to that of the whole resonator. Figure 3 (inset) shows the simulated strain energy gradient across a Lamé-mode resonator. From the energy distribution, it can be inferred that when the oxide islands are placed closer to the center their total strain energy content is significantly higher than when they are placed far away from the center. Figure 3 shows the comparison between the simulated TOT with the one analytically estimated using (3). For the analytical estimation, the magnitude of strain energy in the oxide and silicon is required. Table 1 summarizes the strain energy values taken from simulations and Table 2 details the material parameters used in the simulations.

Table 1: Simulated strain energy in silicon (E_{si}), total strain energy (E_{total}) and the calculated k. Gap is defined in Figure 3.

gap	14	16	18	20	22	24	26	28	30
E_{si} (×10^{12})	1.116	1.19	1.247	1.291	1.353	1.385	1.433	1.485	1.491
E_{total} (×10^{12})	1.676	1.731	1.764	1.78	1.824	1.829	1.859	1.897	1.879
k	0.666	0.687	0.707	0.725	0.742	0.757	0.771	0.783	0.793

Table 2: Material properties used in the estimation of the TOT using FEM and analytical formulation using (3).

Parameter	Silicon	Oxide	Unit
Young's modulus	169	71	GPa
First-order TCE	-64	187.5	ppm/K
Second-order TCE	-75	40	ppb/K^2

Piezoelectric actuation

As pointed out earlier, the crystal symmetry of AlN makes it challenging to actuate the fundamental Lamé-mode of resonance. Each point in the resonator undergoes similar expansion and contraction in the two in-plane orthogonal directions. Since the d_{31} and d_{32} piezoelectric coefficients of AlN have the same magnitude and polarity, the net piezoelectric charge pick up is very small. The presence of the oxide islands within the resonator volume skews the strain profile in the resonator and the temperature compensation strategy has an added benefit of improving the total charge pickup and helps improve the insertion loss of the resonator. Figure 4 plots the net strain across (a) an uncompensated and (b) a compensated Lamé-mode resonator across the AlN surface and clearly highlights the effect of including the oxide islands. To estimate the improvement in charge pickup of compensated resonator (compared to the uncompensated one), the strain gradient is integrated across the AlN layer and the results are summarized in Table 3. A 20× improvement in the total strain is seen due to the inclusion of oxide islands.

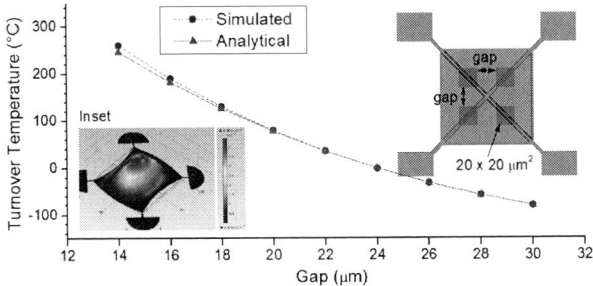

Figure 3: Comparison of the simulated and analytically estimated turnover temperature for the compensated Lamé-mode resonators. (Right) Top view showing the "gap" and location of the oxide-refilled islands. (Inset) Simulated strain energy across the resonator for the Lamé-mode shape. The gradient of energy, which is maximum at the center to its minima around the device edge, can be clearly seen.

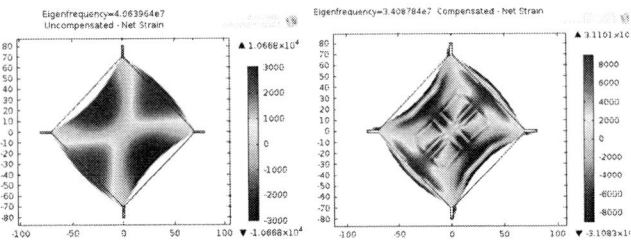

Figure 4: Net strain across the AlN surface for (a) uncompensated and (b) compensated Lamé-mode resonator. The presence of the oxide islands is seen to skew the strain profile. The color gradient plots the change in net strain across the surface, with red and blue representing positive and negative strains respectively.

Table 3: Estimated volume integral of strain in the AlN layer along the resonator in-plane axes showing the effect of silicon and oxide-refilled trenches on the effective charge pick-up.

Resonator	Strain in X (ε_X) (m^3)	Strain in Y (ε_Y) (m^3)	$\varepsilon_X - \varepsilon_Y$ (m^3)
AlN	3.95×10^{-15}	3.96×10^{-15}	2.51×10^{-18}
AlN-on-Si	4.03×10^{-15}	4.10×10^{-15}	6.74×10^{-17}
AlN-on-Si with oxide	2.03×10^{-15}	0.662×10^{-15}	1.37×10^{-15}

Based on the observed strain profile, four triangular shaped electrodes with the diametrically opposite electrodes connected together are used to achieve piezoelectric actuation. Figure 5 shows the simulated frequency response of a compensated and uncompensated resonator which highlights the improvement in insertion loss seen due to the presence of the oxide islands.

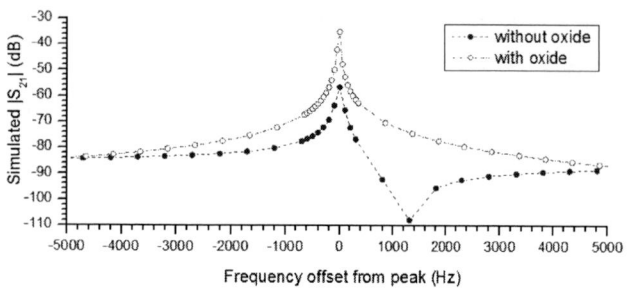

Figure 5: Simulated $|S_{21}|$ for a temperature compensated (with oxide) and uncompensated (without oxide) Lamé-mode resonator.

Tether Optimization for reduced Anchor dissipation

While Lamé-modes do not suffer from TED, they have been shown to be susceptible to anchor Q degradation [1]. The increased anchor dissipation has been shown to be caused by the flexural resonance modes of the tether, when matched to the frequency of the Lamé-mode [1]. With the inclusion of the oxide islands, the frequency of the fundamental Lamé-mode is seen to decrease (for a given resonator side length) and thus changes the optimum tether geometry. Using the approach presented in [1], the support tethers of the temperature-compensated Lamé-mode resonator are optimized to have a tether length of 10.5 μm and tether width of 2 μm.

FABRICATION

Devices are fabricated on a silicon-on-insulator (SOI) wafer with a 25 μm thick high-resistivity (>1000 Ω.cm) device layer. Figure 6 shows the schematic of the fabrication process and is similar to the one used in [9]. In order to obtain completely refilled trenches, the trench DRIE is optimized to provide a straight sidewall with an opening of 1.2 μm. The spacing between the trenches is set to be 0.8 μm. Figure 7 shows cross-section SEM images of the DRIE trenches and oxide-refilled islands. Figure 8 shows an SEM image of a fabricated Lamé-mode resonator.

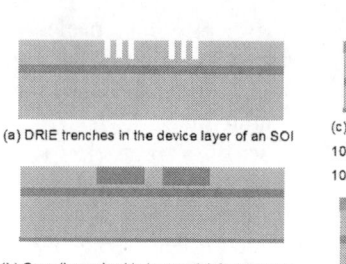
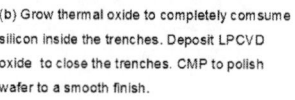
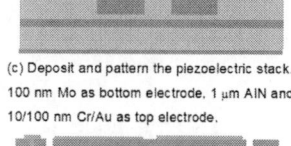

(a) DRIE trenches in the device layer of an SOI

(b) Grow thermal oxide to completely consume silicon inside the trenches. Deposit LPCVD oxide to close the trenches. CMP to polish wafer to a smooth finish.

(c) Deposit and pattern the piezoelectric stack. 100 nm Mo as bottom electrode, 1 μm AlN and 10/100 nm Cr/Au as top electrode.

(d) Etch the resonator contour using RIE. Backside DRIE to release the device. The BOX layer is etched using BHF.

Figure 6: Process flow used for the fabrication of temperature-compensated piezoelectrically actuated Lamé-mode resonators.

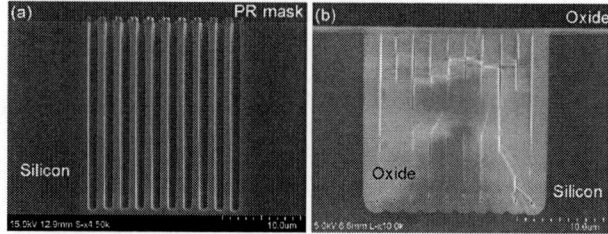

Figure 7: Cross-section SEM images of (a) the DRIE trenches to be oxidized, (b) an oxide-refilled island.

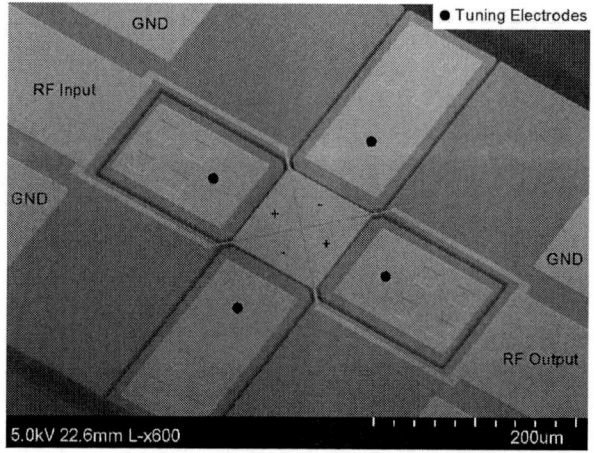

Figure 8: A SEM image of a fabricated Lamé-mode resonator. The top input and output ports are labeled with '+' and '-', respectively and resembles the net strain profile seen in Figure 4. The oxide-refilled trenches are embedded within the silicon body and are not visible through the AlN layer.

MEASURED RESULTS

In order to characterize the performance of the resonators, on wafer measurements are carried out in a temperature-controlled probe station at a pressure of ~100 μTorr. Figure 9 (a) and (b) plot the measured frequency response of a temperature-compensated and uncompensated piezoelectrically-actuated Lamé-mode resonator, respectively. As predicted, the insertion loss of the compensated resonator is lower and the signal to noise ratio is higher than the resonator without the compensating trenches. Also, note that the Qs of the two resonators are comparable.

978-1-4799-3510-9/14 $31.00 © 2014 IEEE

Figure 10 shows the measured frequency shift with temperature for three temperature-compensated Lamé-mode resonators. The three resonators have the same volume of oxide but different "gap" between the oxide islands. The measured results in Figure 10 compare well with the estimated TOT presented in Figure 3. There is a nominal TOT shift of approximately -80 °C for all designs which can be attributed to the piezoelectric stack. The TOT estimates given in Figure 3 ignore the presence of the bottom electrode, AlN, and the top electrode layers.

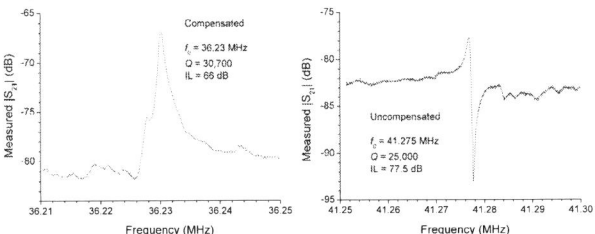

Figure 9: Measured frequency response of a (left) temperature compensated and (right) uncompensated Lamé-mode resonator, measured at room temperature and in vacuum.

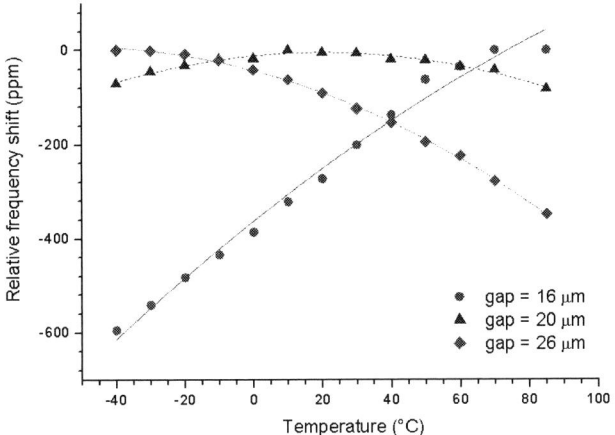

Figure 10: Measured peak frequency shift as a function of temperature for three different "gaps" between the oxide islands.

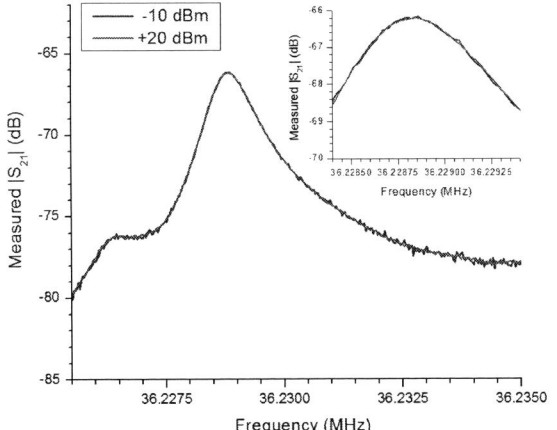

Figure 11: Measured frequency response of a Lamé-mode resonator at input RF power of -10 dBm and +20 dBm, showing no non-linearity in the response.

Figure 11 shows the measured frequency response of the compensated device shown in Figure 9(left) at -10 dBm and +20 dBm of input RF power. No non-linear behavior is visible, indicating the high power handling capability of these resonators.

CONCLUSIONS

Through the inclusion of oxide islands within the resonator volume, we achieved complete cancellation of the first-order TCF and demonstrated the ability to tune the turnover temperature through placement of the oxide islands within the resonator. The presence of the oxide islands was shown to improve the charge pickup by 20× and enabled piezoelectric actuation of temperature-compensated Lamé-mode resonators. Using this approach three compensated resonators with different TOT were demonstrated. Such resonators can be utilized in a multi-resonator system to improve the temperature stability of timing references across the entire industrial temperature range.

ACKNOWLEDGEMENTS

The authors acknowledge staff at the Lurie Nanofabrication Facility, a member of NSF NNIN, for their help with device fabrication. This work is supported by NASA under the Chip-Scale Precision Timing Unit project (Grant #NNX12AQ41G).

REFERENCES

[1] V. Thakar and M. Rais-Zadeh, "Optimization of tether geometry to achieve low anchor loss in Lamé-mode resonators," *IFCS '13*, Prague, CZ, July, 2013.

[2] J. Lee, J. Yan and A. Seshia, "Study of lateral mode SOI-MEMS resonators for reduced anchor loss," *J. Micromech. Microeng.*, vol. 21, pp. 045010, 2011.

[3] S. Chandorkar, *et al.*, "Limits of quality factor in bulk-mode micromechanical resonators," *MEMS 2008*, Jan. 2008.

[4] L Khine and M. Palaniapan, "High-Q bulk-mode SOI square resonators with straight-beam anchors," *J. Micromech. Microeng.*, vol. 19, pp. 015017, 2009.

[5] Y. Xu and J. E.-Y. Lee, "Mechanically coupled SOI Lamé-mode resonator-arrays: synchronized oscillations with high Q factors of 1 million," *IFCS 2013*, Prague, CZ, 21-25 Jul 2013.

[6] Z. Wu, A. Peczalski, V. Thakar, and M. Rais-Zadeh, "A low phase-noise Pierce oscillator using a piezoelectric-on-silica micromechanical resonator," *Transducers '13*, June, 2013.

[7] R. Melamud, *et al.*, "Temperature-compensated high-stability silicon resonators," *Appl. Phys. Lett.*, vol. 90, no. 24, pp. 244107, Jun. 2007.

[8] R. Tabrizian, G. Casinovi and F. Ayazi, "Temperature-stable high-Q AlN-on-silicon resonators with embedded array of oxide pillars," *Hilton Head '10*, Hilton Head Island, SC, pp. 100-101, June 2010.

[9] V. Thakar, Z. Wu, A. Peczalski, and M. Rais-Zadeh, "Piezoelectrically transduced temperature-compensated flexural-mode silicon resonators," *JMEMS*, Vol. 22, No. 3, pp. 819-823, 2013.

[10] M. Allah, *et al.*, "Temperature compensated solidly mounted bulk acoustic wave resonators with optimum piezoelectric coupling coefficient," *IEDM '09*, pp. 1-4, 7-9 Dec. 2009.

CONTACT

Vikram Thakar, tel: +1-734-3443480; thakar@umich.edu
M. Rais-Zadeh, tel: +1-734-7644249; minar@umich.edu

A MEMS AUTONOMOUS SWITCHED OSCILLATOR

G. B. Torri[1,2], J. Bienstman[1], X. Rottenberg[1], H. A. C. Tilmans[1], C. Van Hoof[1,2], and R. Puers[1,2]

[1]Imec, Kapeldreef 75, 3001 Heverlee, BELGIUM

[2]K.U.Leuven, Kastelpark Arenberg 10, 3001 Heverlee, BELGIUM

ABSTRACT

This work presents the model, design and characterization of an autonomous electrostatic MEMS oscillator for sensing application. The proposed system, consisting of a parallel-plate electrostatic MEMS device, a DC voltage source and a displacement-dependent resistive circuit, is capable of sustaining oscillations autonomously. It can exhibit periodic or aperiodic behavior, impacted by its working point and by environmental parameters, e.g. pressure, temperature. Electronic and non-electronic information can be submitted or retrieved from the oscillator system that present a very rich behavior sensitive to external stimuli what makes it a good candidate for threshold sensing.

INTRODUCTION

Electrostatic MEMS devices have been used as (linear) resonant transducers when used at small displacements and at frequencies close to a mechanical resonance frequency [1]. At large displacements, and including nonlinearities, the electrostatic autonomous MEMS impact resonator was demonstrated in periodic and aperiodic regimes [2]. Expanding on this, we report an autonomous oscillator with a displacement-dependent resistive switch that, in difference to the device from [3], does not require a periodic actuation. It works in autonomous vibration / self-switching. Applications for such a nonlinear oscillator range from chaotic signals [4], sensitive mass sensors [3], to arrays for computation and signal processing [5].

In this paper, we report on the design and characterization of an autonomous MEMS oscillator. This device changes its dynamic behavior as the pressure in the environment is varied. The above is analyzed using a numerical model with parameters extracted from measurements of basic device parameters. A prototype of a MEMS autonomous switched oscillator is presented alongside with measurement results. Simulations and measurements of the proposed electromechanical oscillator are in good agreement and have shown similarities with reported electronic oscillators [4,6].

CONCEPTUAL DESIGN

The electromechanical model and working principle of the device are illustrated in Figure 1. The mechanics of the device is represented as a lumped spring-mass-damper system. The moving mass forms an electrostatic gap to the substrate, which is actuated by a voltage source through a switched resistive divider. Applying a voltage V_{dc} above the pull-in voltage, the displacement u-dependent capacitance $C_a(u)$ is charged over R_C, building up the electrostatic force that drives the system into instability, i.e. pull-in. The u-dependent resistor $R_d(u)$, in this case modeled as a

switched resistor, separates the behavior of the system into a charge and a discharge phase. As u progresses and crosses u_{th}, the discharge phase activates. The switched resistor $R_d(u)$ reduces the charge on $C_a(t,u)$. The electrostatic force reduces and the device moves away from pull-in. Crossing u_{th} again, $R_d(u)$ opens initiating a new charging phase.

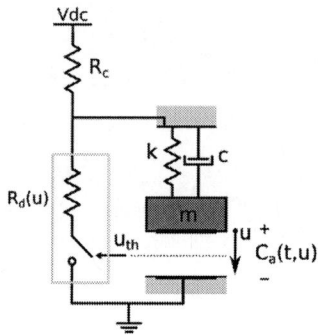

Figure 1: Electromechanical model of autonomous oscillator with position feedback to the driving signal.

The electromechanical model is represented in Equation (1). It consists of a lumped spring-mass-damper system with an extra equation for the charge stored on the capacitively actuated gap.

$$\dot{u} = v$$
$$m\,\dot{v} + cv + ku = \frac{\varepsilon_0 A}{2g_{eff}^2}\left(\frac{q_a}{C_a}\right)^2$$
$$\frac{V_{dc}}{R_c} = \frac{q_a}{C_a}\left(\frac{1}{R_c} + \frac{1}{R_d(u)}\right) + \dot{q}_a \qquad (1)$$
$$C_a = \frac{\varepsilon_0 A}{g_{eff}}$$
$$g_{eff} = d - u(t) + \frac{t_1}{\varepsilon_{r1}} + \frac{t_2}{\varepsilon_{r2}}$$

where the state variables u, v and q_a denote the plate displacement, plate velocity and charge stored on the actuation capacitance $C_a = C_a(u)$. The parameters m, c and k represent the lumped mass, viscous drag and lumped stiffness respectively. To better represent the electrostatic gap an effective actuation gap g_{eff} is defined between the plate area A and the counter electrode. It consists of the released gap d at vacuum permittivity ε_0, and two dielectric layers deposited over the counter electrode. The layer thicknesses and relative permittivities are t_1, t_2 and ε_{r1}, ε_{r2}, respectively A constant voltage source V_{dc}, a charging resistor R_C and a u-dependent discharge resistor $R_d(u)$ are used for actuation and feedback. Note that $u(t)$ modulates the gap d.

The value of the u-dependent discharge resistor $R_d(u)$ an the charge resistor R_C are defined so that the frequency of resonance f_0 is inversely proportional to the charge and

978-1-4799-3510-9/14 $31.00 © 2014 IEEE   218

discharge times constants. Assuming the charging starts at rest position and the discharge starts approximately when u reaches to 50% of the gap d, the resistors are defined as,

$$R_d(u) = \begin{cases} \infty & , u < u_{th} \\ R_c/2 & , u \geq u_{th} \end{cases} \quad (2)$$

In practice, a fraction of the calculated R_C value can be taken. This results in smaller electric time constants than the mechanical period, $T = 1/f_0$. In the simulations, R_C was taken as 1/3 of the value calculated with Equation (3).

$$f_0 = \frac{1}{R_c C_a(u=0)} \quad (3)$$

MODEL CALIBRATION

In this section the tested device is described, characterized, fitted to the model of which simulations are presented.

Device

Figure 2 presents the SiGe (Young's modulus: E=120GPa; mass density: ρ=4775kg/m^3) MEMS device used to setup the autonomous oscillator. It consists of a 200x200x4µm^3 square plate suspended 3µm above a dielectric coated electrode by four 130x10x4µm^3 legs. The counter electrode is coated by SiC (t_1=0.4µm; ε_{r1}=4.4) and SiO$_2$ (t_2=0.6µm; ε_{r2}=4.9). The pull-in was calculated with ANSYS at approximately 45V.

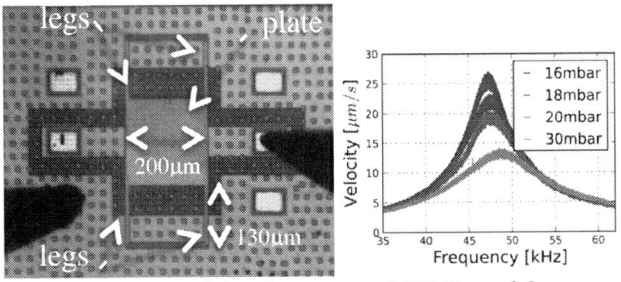

Figure 2: Image of the electrostatic MEMS used for parameter extraction and closed-loop measurements. Resonance peaks at different pressures. Fitted parameters: f_0=[47.3, 47.5, 47.8, 48.6] kHz; ζ=[0.044, 0.0501, 0.0514, 0.0679]; quality factor Q=[11.37, 9.97, 9.73, 7.36].

Parameter Extraction

Before implementing the MEMS device as autonomous closed-loop oscillator, we measured its mechanical properties under low pressure conditions. We extracted its resonance frequency f_0 and damping ratio ζ from optical measurements using a laser Doppler vibrometer (LDV) from Polytec GmbH. Applying a periodic chirp signal of 2V peak-to-peak superposed on a 3V DC offset, we obtained the resonance peak responses in Figure 2 for a few environment pressure conditions. We subsequently fitted these using Lorentzian shape functions. We further assume that the total lumped mass m is defined by the plate volume. As a result, the stiffness and viscous damping can be respectively estimated by $k = m(2\pi f_0)^2$ and. $c = \zeta 4\pi f_0 m$.

Model Simulation

With the extracted mechanical parameters, equations (1) and (2) were modeled in Modelica language and simulated with the OpenModelica simulation environment. Scripts, written in Python programing language, were used to steer the simulator and to post-process the results.

The dynamics of the model in Figure 1 were explored with the bifurcation diagram in Figure 4. A Poincaré section was placed at zero velocity. Hence, it captures the range of the displacement in the moments when the device changing its direction of movement. In these simulations the parameter V_{dc} is swept, revealing the various periodic oscillations and chaos. Parameters were extracted at 20mbar.

Beside the extracted and physical parameters, the value of the resistors were: R_c=60MΩ and $R_d(u)$=30MΩ if $u \geq u_{th}$ or open circuit otherwise. At the mechanical resonance frequency of 47.8kHz the period is 21µs. The electrical time constant $\tau_e = R_c C_a(u=0)$ is approximately 7µs. The u_{th} was set to 40% of the gap d (u_{th}=1.2µm).

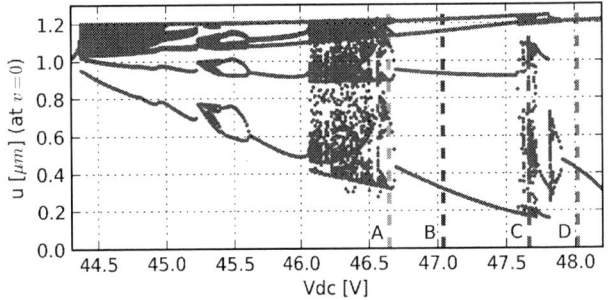

Figure 4: Simulated bifurcation diagram of model in Figure 1. Phase diagrams for lines A, B, C and D are in Figure 5. Parameters extracted from measurement at 20mbar.

Observing the transition from lines D to C, one can identify a bifurcation mechanism similar to the period-doubling cascade reported on electronic oscillators [4]. Several periodic and aperiodic trajectories can be observed.

In Figure 5, the trajectories marked in Figure 4 (A, B, C and D) are plotted as phase diagram showing the dynamic behavior at a fix value of V_{dc}. The period-4, period-2, chaos and period-1 are shown below.

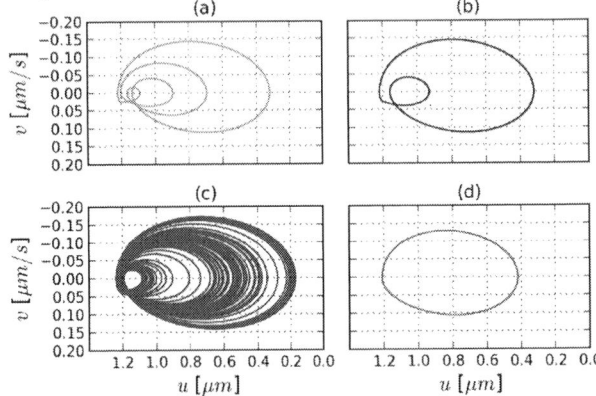

Figure 5: Phase diagrams of periodic and chaotic regimes. Lines as indicated on Figure 4: (a) A, period-4. (b) B, period-2. (c) C, chaotic attractor. (d) D, period-1.

To illustrate the dynamics of the system, the period-2 (line D, from Figure 4) is shown in Figure 6. Short time traces and the three-dimensional attractor for the state variables u, v, and q_a are plotted. The trajectory shows that the device is charging as it oscillates towards the substrate until it hits the switching plane. Then it is partially discharged to restart into the charging phase. This is a periodic orbit.

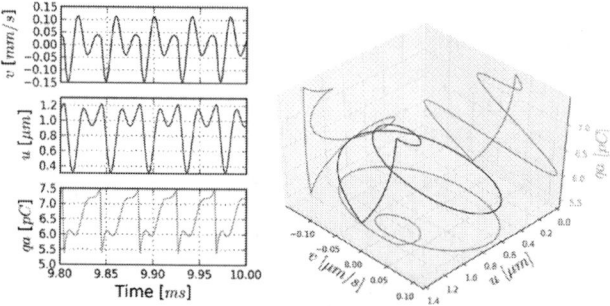

Figure 6: Periodic Orbit. State variables, simulated time traces and three-dimensional attractor for phase diagram in Figure 5 (b). Pressure of 20mbar and switching plate at 40% of the gap (1.2µm) are assumed.

EXPERIMENTAL RESULTS

Experimental Setup

To validate the working principle and above simulations, an experimental setup was built. Measurements were performed at three different pressures. At each pressure value the applied V_{dc} was swept and the response were recorded, analyzed and combined in bifurcation diagrams.

Figure 7 illustrates the block diagram of the experimental setup. With the laser beam pointed to the center of me moving plate, the LDV was set with the velocity decoder VD02 (125mm/s/V range). The velocity is then integrated into the instantaneous plate position. The plate position is compared to a reference position (voltage source). The resulting reset signal closes the loop controlling the actuation voltage by switching the resistive divider.

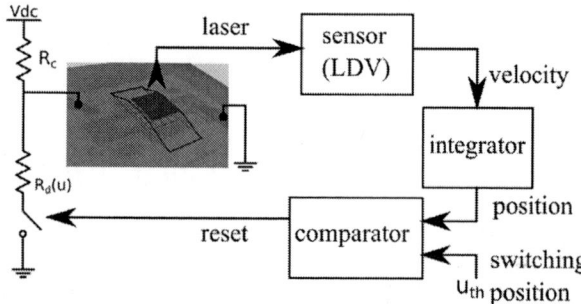

Figure 7: Block diagram of the implemented closed-loop system. A laser Doppler vibrometer (LDV) is used as velocity sensor. The plate velocity is integrated and compared to a switching position. The reset signal closes the loop switching the actuation voltage via the resistive divider.

Implementation

The system was implemented with off-the-shelf components soldered on a prototyping board, as in Figure 8. The high-pass filter, integrator and gain stages used high-bandwidth precision operational amplifiers (OPA228). A precision comparator (AD790) generates the reset signal. The reset is buffered to drive the bipolar transistor (ZTX653) as a switch. To improve the transistor turn-off switching time a Baker clamp circuit had to be used to avoid saturation and the storage time. Besides the LDV, the MEMS device was set on a pressure controlled (vacuum) chamber and electrically accessed via probing needles. Care was taken to avoid op-amp self-oscillations and common-impedance noise.

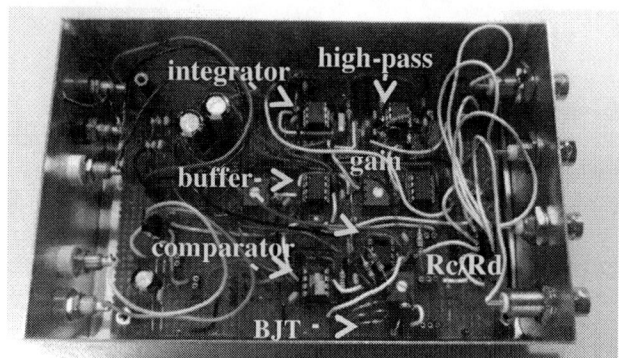

Figure 8: System prototype. It inputs velocity and outputs position, reset signal and drive voltage to be applied to the MEMS device.

Results

To calibrate the integrator and reset blocks a high-voltage (15V peak, 15V offset) sinusoid with a frequency of 7kHz was applied (below resonance). The displacement was monitored at the integrator output and compared with the digital displacement decoder DD500. The circuit was trimmed for the zero position and desired scale (µm/V).

For each 10mV step increase in V_{dc}, voltages were recorded during 4ms with a digital oscilloscope to a computer. Signals were post-processed with a Gaussian filter. The filtered data was further processed to extract the bifurcation diagram. A Poincaré section was placed at zero velocity.

The finite turn-on and turn-off times of the bipolar transistor and the resistive divider introduce electric time constants other than the ones seen with an ideal switch. The turn-on and turn-off time were measured to approximately 8µs for R_c=100kΩ and $R_d(u)$=51kΩ if $u \geq u_{th}$. This is close to 7µs electric time constant used on the simulations (at the cost of much larger resistors).

The switching position was set at 2V for all measurements. With the range of the position signal calibrated to $\approx$ 0.35µm/V the switching plane is at $u_{th} \approx 0.7$µm.

The experimental bifurcation diagrams at three different pressure values (16mbar, 18mbar, 20mbar) are shown in Figure 9. In each diagram transitions between periodic and aperiodic oscillations are observed. Fixing the value of V_{dc} one can see how the dynamics of the system change as a function of pressure variation.

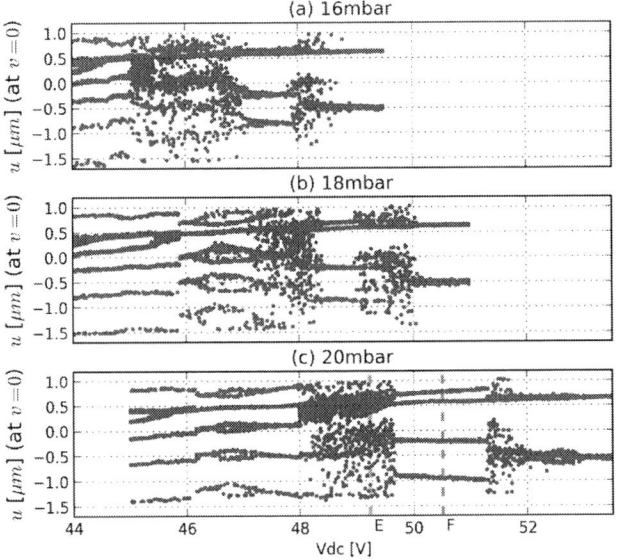

Figure 9: Experimental V_{dc} bifurcation diagrams. (a) Diagram at 16mbar. (b) Diagram at 18mbar (c) Diagram at 20mbar (lines E, F are plotted in Figures 10, 11).

A chaotic attractor is marked as line E in Figure 9 (c). In Figure 10 a segment of the time traces and the complete phase diagram are plotted to show the measured attractor.

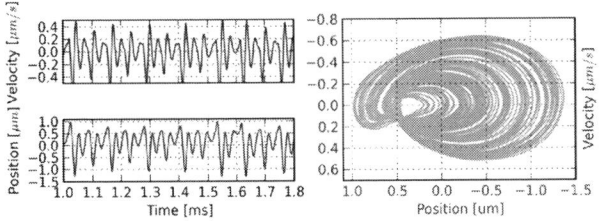

Figure 10: Chaotic attractor measured at 20mbar, V_{dc}=49.23V. Marked as line E in Figure 8 (c).

A period-2 trajectory is marked as line F in Figure 9 (c). In Figure 11 a part of the velocity and position time traces are plotted alongside with the phase diagram.

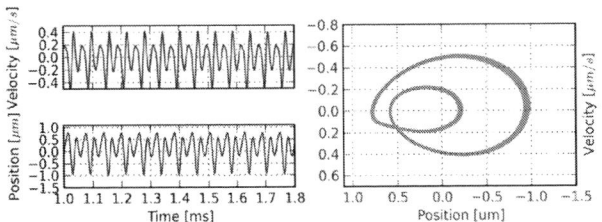

Figure 11: Period-2 trajectory measured at 20mbar, V_{dc}=50.51V. Marked as line F in Figure 8 (c).

Discussion

The simulated model is simplified to a rigid plate with a single mode of resonance. No spring stiffening or softening effects are accounted. Including the effective gap related to the deposited dielectrics helped to bring the pull-in closer to the values observed in the measurements. Even with the

performed calibration it still seems the measurements have a systematic error towards negative displacement.

Despite the mentioned limitations, the results are consistent across simulation and measurements. They describe a system that acts as parameter controlled oscillator, capable of exhibiting multiple well-defined periodic orbits (could be used for easy detection) as well as aperiodic regions (expected to have enhanced sensitivity).

CONCLUSIONS

In conclusion, the rich dynamical behavior of the proposed MEMS autonomous switched oscillator was simulated and recorded experimentally.

Despite non-idealities of the experimental prototype the periodic and aperiodic oscillations together with the intermittency between neighboring periods are clearly noted. The results resemble the bifurcation mechanism similar to the one described in switched electronic circuits.

Experiments demonstrated the sensitivity to pressure variations and applied voltage. Other sensitivities (switching position, mass deposition, acceleration) remain to be investigated. Owing to the observed chaotic nature of the oscillator we expect it to be highly sensitive to stimuli.

The possibility of submitting and retrieving electronic and non-electronic information are advantages of the proposed MEMS device. Applications ranging from sensitive sensors to coupled arrays with collective behavior for computation and signal processing are envisioned.

ACKNOWLEDGEMENTS

The authors wish to thank the EUROPRACTICE IC for the access to the MEMS MPW service, Simone Severi for overseeing the fabrication and Xiao Sun for providing us her samples.

REFERENCES

[1] H. A. C. Tilmans, "Equivalent circuit representation of electromechanical transducers: I. Lumped-parameter systems," *J. Micromech. Microeng.*, vol. 6, p. 157, 1996.

[2] J. Bienstman, R. Puers, and J. Vandewalle, "Periodic and Chaotic Behaviour of the Autonomous Impact Resonator," *IEEE MEMS*, pp. 562–567, 1998

[3] A. Seleim, S. Towfighian, E. Delande, E. Abdel-Rahman, and G. Heppler, "Dynamics of a close-loop controlled MEMS resonator," *Nonlinear Dynamics*, vol. 69, no. 1–2, pp. 615–633, Dec. 2011.

[4] G. M. Maggio el al., "Nonsmooth bifurcations in a piecewise-linear model of the Colpitts oscillator," *IEEE TCAS*, vol. 47, no. 8, pp. 1160–1177, 2000.

[5] F. C. Hoppensteadt and E. M. Izhikevich, "Synchronization of MEMS resonators and mechanical neurocomputing," *IEEE TCAS*, vol. 48, no. 2, pp. 133–138, 2001.

[6] T. Saito, "A chaos generator based on a quasi-harmonic oscillator," *IEEE TCAS*, vol. 32, no. 4, pp. 320–331, 1985.

SURFACE TENSION MONITORING IN A "SOFT" INTERFACE USING ELECTROWETTING ON DIELECTRIC

Seungyul Choi[1], Yongjoo Kwon[2], Eunyong Jeon[1], Sunggu Kim[1], and Junghoon Lee[1]
[1]Seoul National University, Seoul, South Korea
[2]Samsung Advanced Institute of Technology, Yongin, South Korea

ABSTRACT

We report the first demonstration of interfacial tension monitoring across two immiscible liquids using electrowetting on dielectric (EWOD). Impedance measurement during EWOD reveals the variation of surfactant concentration at the liquid-liquid interface in real time. The discrepancy with optical method was within 3.5 mN/m and the noise level was under 0.5 mN/m. We also show such approach can be used for label-free monitoring of DNA hybridization. Our approach opens a new horizon of EWOD used as a molecular sensing mechanism across a "soft" interface between liquids.

INTRODUCTION

Last decade has seen vivid applications in the smart use of surface tension on micro-scale. The surface tension becomes dominant in micro-scale devices due to low Bond number. In this sense, it of utmost importance to measure the surface tension in micro-device. EWOD is at the center of such showcase with its interesting capabilities in handling small amount of chemical/biological samples and liquid-based optical displays [1-3].

Many EWOD systems use dual-phase approach with two immiscible liquids such as an aqueous electrolyte and an oil that constitute conductive and insulating phases, respectively. Conducting liquid usually contains biomolecules, particles and/or additives like surfactant that strongly affects the surface tension or the interfacial tension between two phases. These changes in the interfacial tension need to be considered in the design and even operation processes for an accurate control of EWOD-based actuation. Furthermore, the monitoring of interfacial tension resulting from a molecular binding will lead to a new sensing mechanism with advantages in label-free and high sensitivity.

Currently, there is no technology that meets the needs for real time measurement of interfacial tension on a single chip. There are several methods known for experimental purposes such as pendant droplet tensiometry, Wilhelmy plate, and Du Noüy ring. However, these approaches use relatively large amount of sample, and require additional equipment such as bulky optical device and image processing system or force measurement system.

Here we demonstrate that the surface tension can be measured using electrowetting with impedance measurement [4]. We used simple patterned electrode device and modified Young-Lippmann's equation which represents energy balance between surface and electrical interfaces to calculate the surface tension via an electrical measurement.

SAMPLE PREPARATION FOR IMPEDANCE MEASUREMENT

We fabricated measurement samples on a glass substrate. Chromium (Cr) / gold (Au) / Cr layers were deposited on a glass wafer (Eagle XG, Corning Incorporated) with the thickness of 200 Å / 500 Å / 200 Å respectively. The metal layer was patterned by photolithography as shown in Figure 1. Patterned electrode was then covered with 5000 Å silicon nitride (SiN_x) by plasma enhanced chemical vapor deposition (PECVD) as a passivation layer. Finally, we coated CYTOP (CTL-809M, Asahi Glass) by spin coating to make the top surface hydrophobic.

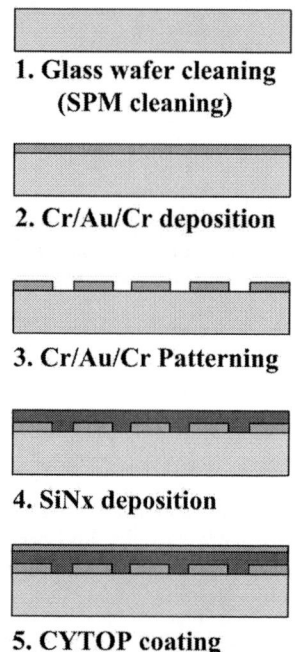

1. Glass wafer cleaning (SPM cleaning)

2. Cr/Au/Cr deposition

3. Cr/Au/Cr Patterning

4. SiNx deposition

5. CYTOP coating

Figure 1: Measurement sample fabrication process.

Our EWOD system is dual phase, so we need a reservoir to maintain a certain amount of conducting liquid on the measurement sample. We used PDMS as a material for the reservoir. PDMS was convenient to attach on and detach from any surface and had some hydrophobic property which kept conducting liquid from leaking to outside area. Sides of PDMS reservoir had to be transparent for optical observation. Thus, we used polished glass molds as in Figure 2. Conducting liquid was Na_2SO_4 solution with the conductivity of 19 mS/cm (Lutron YK-2014CD) which was close to 1×PBS buffer solution (~17.49 mS/cm). N-decane was used as an oil.

978-1-4799-3510-9/14 $31.00 © 2014 IEEE

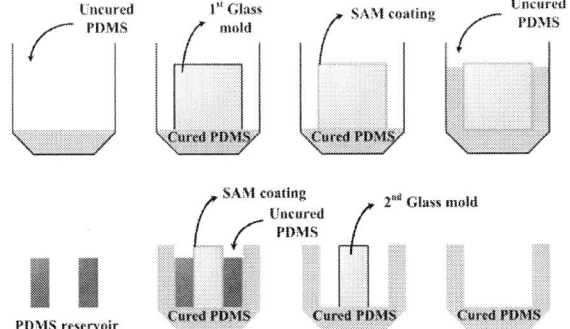

Figure 2: PDMS reservoir fabrication process. Two glass molds were required to achieve clear sides of the reservoir.

SURFACE TENSION CALCULATION

We modified Young-Lippmann's equation (1) to calculate surface tension.

$$\cos\theta_V - \cos\theta_Y = \frac{\varepsilon_0 \varepsilon_r}{2d\gamma_{OL}} V^2 \qquad (1)$$

where θ_V is contact angle at operation voltage V, θ_Y is Young's angle which is contact angle at zero operation voltage, ε_0 is vacuum permittivity, ε_r is a dielectric constant, d is the thickness of dielectric layer, and γ_{OL} is the surface tension value between two immiscible liquids.

We can consider V^2 as X value and $\cos\theta_V - \cos\theta_Y$ as Y value, then the term of $\varepsilon_0\varepsilon_r/2d\gamma_{OL}$ is the slope of X-Y graph. The surface tension can be derived from the slope term as in equation (2).

$$slope = \frac{\varepsilon_0 \varepsilon_r}{2d\gamma_{OL}} \rightarrow \gamma_{OL} = \frac{1}{2}\frac{\varepsilon_0 \varepsilon_r}{d}\frac{1}{slope} \qquad (2)$$

It is challenging to achieve an exact $\varepsilon_0\varepsilon_r/d$ value in (2), but we could calculate it accurately by using impedance measurement. Here, only the capacitance term is considered in impedance value because capacitance is dominant factor at the dielectric layer. We know the exact area of the patterned electrode contacting with Na_2SO_4 solution without oil (A_T) and its capacitance value (C_T) measured from LCR meter. Capacitance is simply expressed with the contact area and $\varepsilon_0\varepsilon_r/d$ term as $C_T = \varepsilon_0\varepsilon_r A_T/d$. This calculation implies that we can calculate accurate value of $\varepsilon_0\varepsilon_r/d$ from C_T and A_T, $\varepsilon_0\varepsilon_r/d = C_T/A_T$. Figure 3 shows the geometric information of our measurement system. There are two contact regions. A_{ls} is the contact area between conducting liquid and patterned electrode. A_{os} is the area between oil and the electrode. A_T is $A_{ls} + A_{os}$. We can derive contact angle equation with the function of A_{ls} and the volume of oil (Vol) using geometric information, and the results appear in (3) and (4). The contact area A_{ls} is calculated by measuring capacitance between the conducting liquid and electrode, C_{ls} as in the calculation of A_T from C_T value in the previous paragraph. A_{os} is $A_T - A_{ls}$ in (3) and (4), so if we know A_{ls}, A_{os} can be simply calculated.

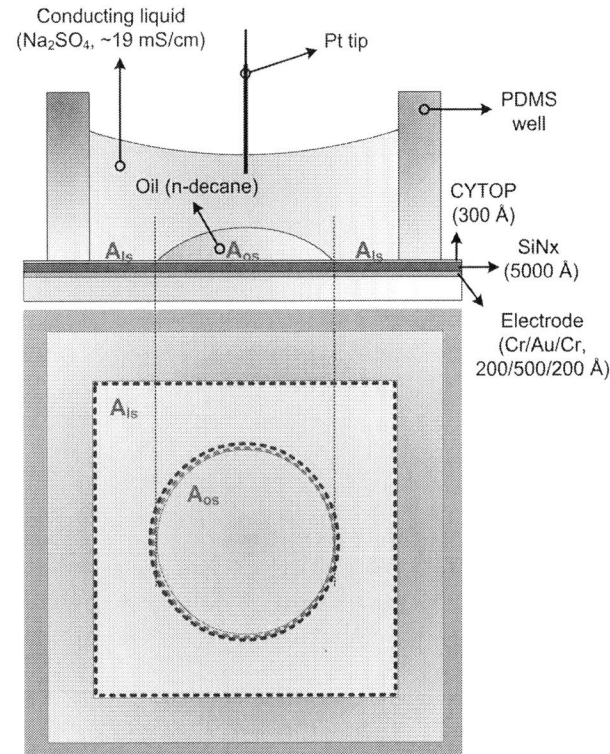

Figure 3: Geometric information of measurement setup.

Volume of oil is fixed and A_{ls} is defined by capacitance measured by LCR meter at each voltage. We can plot contact angle and voltage using (4). For instance, we apply V_1 and then the oil droplet has a certain contact angle which means we can obtain the value for the contact area between oil and electrode (A_{os}) from capacitance measurement. Using all information above, we can calculate the contact angle (θ_{V1}) at V_1. We choose several different operation voltages and calculate contact angle at each voltages using (4) and capacitance measurement.

MEASUREMENT SETUP, METHOD, AND RESULTS

Figure 4 shows the impedance measurement setup. PDMS reservoir was attached on the measurement sample. The reservoir was then filled with conducting liquid (Na_2SO_4 solution), and 1.5 µl n-decane oil droplet was dosed inside at the center of the sample surface. LCR meter (Agilent E4980A) was connected with Na_2SO_4 solution and patterned electrode through platinum (Pt) wire and BNC cable. LCR meter was used to apply operation voltage and measure the capacitance formed by hydrophobic and dielectric layer placed between conducting liquid and the electrode at the same time. Capacitance data measured by LCR meter is transferred to PC through GPIB (General Purpose Interface Bus) cable. We designed a Labview code to calculate surface tension from the data. All process was operated automatically. CCD camera captured droplet images that were required for image analysis for comparison.

978-1-4799-3510-9/14 $31.00 © 2014 IEEE

$$A_{OS} = \pi R^2 \sin^2 \theta_{V.cap} = \pi \left[\frac{3Vol}{\pi \left(2 + \cos \theta_{V.cap} \right) \left(1 - \cos \theta_{V.cap} \right)^2} \right]^{\frac{2}{3}} \sin^2 \theta_{V.cap} \qquad (3)$$

$$\theta_{V.cap} = f\left(A_{OS}, Vol \right) \ , where \quad A_{OS} = A_T - A_{ls} \qquad (4)$$

Measurement process consisted of three steps. First, 3, 6, 9, 12 and 15V was gradually applied between electrode and Na$_2$SO$_4$ solution through LCR meter. The voltage application was paused for 3 seconds at each step for capacitance measurement with LCR meter, followed by voltage changes to the next step. For example, LCR meter applied 3V and measured capacitance at the end of 3-second duration. In the second step, five data points were put into (4) for the calculation of contact angle at each voltage. Finally, Labview code plotted the graph for $cos\theta_V$ - $cos \theta_Y$ vs. square of voltage. The slope was calculated through linear fitting as shown in Figure 5. The surface tension between n-decane and Na$_2$SO$_4$ solution was derived from (2). This process comprised one cycle, and the program ran the same process continuously and display the values on a chart graph in real time. It is essential that the slope is totally independent from $cos\theta_Y$ which indicates initial contact angle. This implies that the results is not affected by initial state such as contact angle variation from hysteresis or surface absorption of proteins or surfactant.

Figure 6 shows the time course monitoring of interfacial tension as the surfactant at various concentrations were added. We used Tween 80 as a surfactant. It was added at a certain concentration, and we waited until the surface tension value measured by our system was saturated followed by the addition of the surfactant at a higher concentration. The surface tension decreased as the surfactant was introduced, dropping dramatically between 2.14×10^{-7}M and 5.19×10^{-7}M.

This indicates that critical micelle concentration is within this range. When the measured interfacial tension values were compared with the conventional image processing using CCD camera and Low Bond Axisymmetric Drop Shape Analysis, the discrepancy was found to be within ~3.5 mN/m.

BIO MOLECULAR DETECTION WITH SOFT INTERFACE

We further developed a protocol for molecular sensing by immobilizing receptors in the interface using binding chemistry for functionalized lipid, DSPE-PEG(2k)-MAL (Avanti lipid) with thiol modified single strand DNA (SH-GGTTGGTGTGGTTGG, SH-ssDNA) as shown in Figure 7. Conducting liquid for this experiment was 1×PBS in order to optimize the condition for DNA hybridization reaction. The conductivity was similar to Na$_2$SO$_4$ solution in the previous test. We used complementary DNA as a target molecule, and non-complementary DNA for control test. 1×PBS buffer with the lipid dissolved in it was placed into PDMS reservoir, and 1.5 µl n-decane was dosed as in Figure 7. The lipid molecules were self-aligned at the interface.

Lipid molecule was functionalized when SH-ssDNA was introduced into the reservoir. The functionalized lipid was finally formed at the interface and worked as receptors. Non-complementary ssDNA (nc-ssDNA) was added, followed by the addition of complementary ssDNA (c-ssDNA) at the concentration of 5 µM in both cases. Then we observed the surface tension change over time.

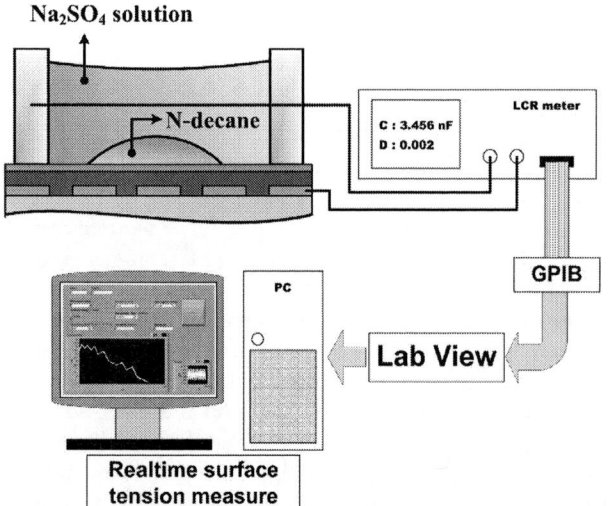

Figure 4: Impedance measurement setup schematic. Surface tension is measured automatically by LCR meter and Labview program.

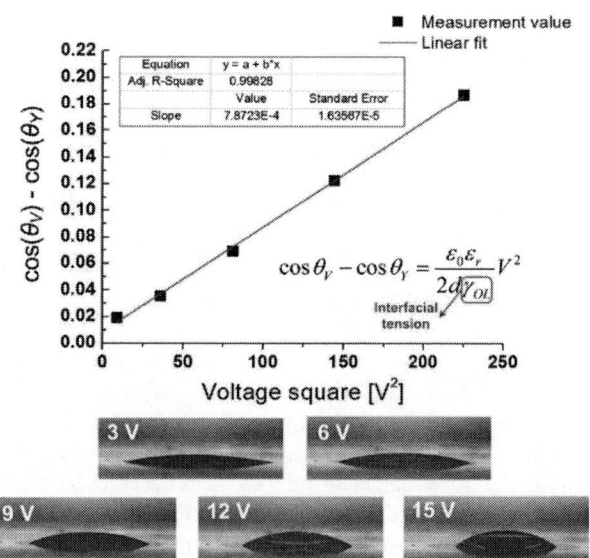

Figure 5: The graph between operation voltage and contact angle at each operation voltage.

978-1-4799-3510-9/14 $31.00 © 2014 IEEE   224

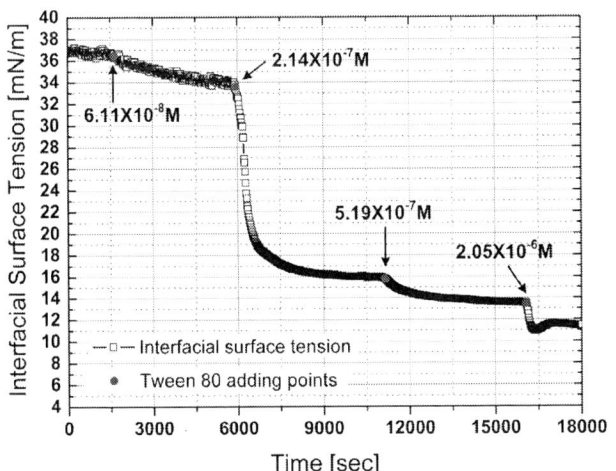

Figure 6: Monitoring of surface tension at prescribed surfactant (Tween 80) concentration.

Figure 8 shows the first EWOD-based monitoring of specific DNA binding event. There was hardly any response after nc-ssDNA. However, the addition of c-ssDNA resulted in the increase of the interfacial tension over time. The noise level of detection was ~0.5 mN/m.

CONCLUSION

Our measurements are in good agreement with results obtained through a conventional approach, and also show exciting potential for molecular sensing. With the possibility for integration with multi-droplet EWOD microfluidic platform, our first demonstration shows a new approach for a

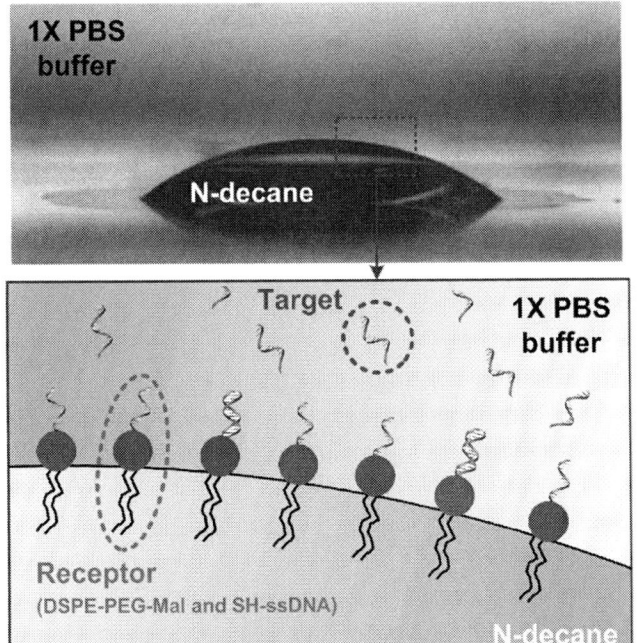

Figure 7: DNA hybridization sensing at the soft interface between n-decane and 1×PBS buffer.

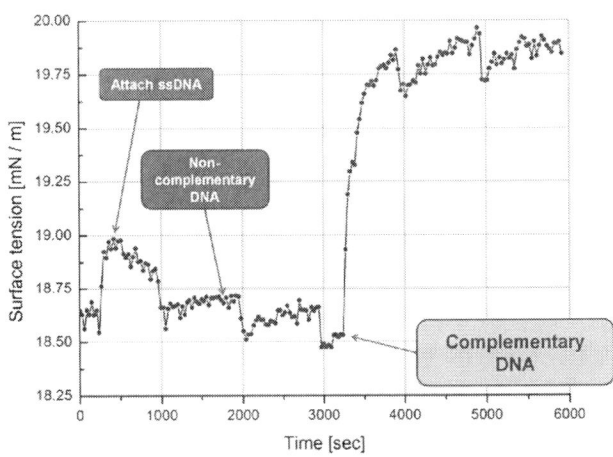

Figure 8: DNA hybridization sensing results. The surface tension changes after introducing complementary DNA.

surface energy-based label-free sensing using a "soft" interface since the development of chemo-mechanical transduction on a "hard" interface [5].

REFERENCE

[1]   J. Lee, H. Moon, J. Fowler, T. Schoellhammer, and C.-J. Kim, "Electrowetting and electrowetting-on-dielectric for microscale liquid handling," *Sensors and Actuators A: Physical,* vol. 95, pp. 259-268, 1/1/ 2002.

[2]   H. Moon, A. R. Wheeler, R. L. Garrell, J. A. Loo, and C. J. Kim, "An integrated digital microfluidic chip for multiplexed proteomic sample preparation and analysis by MALDI-MS," *Lab Chip,* vol. 6, pp. 1213-9, Sep 2006.

[3]   J. Heikenfeld, K. Zhou, E. Kreit, B. Raj, S. Yang, B. Sun, *et al.*, "Electrofluidic displays using Young–Laplace transposition of brilliant pigment dispersions," *Nature Photonics,* vol. 3, pp. 292-296, 2009.

[4]   S. Choi, Y. Kwon, Y. S. Choi, E. S. Kim, J. Bae, and J. Lee, "Improvement in the breakdown properties of electrowetting using polyelectrolyte ionic solution," *Langmuir,* vol. 29, pp. 501-9, Jan 8 2013.

[5]   J. Fritz, "Translating Biomolecular Recognition into Nanomechanics," *Science,* vol. 288, pp. 316-318, 2000.

MICROFLUIDIC BUBBLE-BASED GAS SENSOR

Ashrafuzzaman Bulbul, Hao-Chieh Hsieh, and Hanseup Kim
University of Utah, Salt Lake City, Utah, USA

ABSTRACT

This paper reports a new class of a gas sensor that utilizes the variation in bubble sizes when different gases are introduced into a liquid flow to identify gas types. The fabricated device enabled the detection of pentane (3μL) by producing the average bubble size of 130μm^3, which showed size reduction by 2.4μm (1.7%) from the background bubble sizes (132.4μm) of nitrogen. The measurement also showed a preliminary correlation between the injection volume and the bubble size: increase in the injection volume of pentane from 3μL to 5μL (66.7% increases) resulted in the bubble size reduction from 1.7% to 2.1% (24% decreases) and the time span increase from 5s to 10s (100% increases).

INTRODUCTION

Recent reports raised the importance of identifying gas types in various applications, ranging from environmental monitoring and homeland security to space exploration and analysis of natural gas products. These applications require reliable and small volume gas sensors that are able to detect a wide range of gas molecules in efficient and economical manner to assist in national, societal and personal decisions on surrounding atmospheric pollutions and variations.

Micro gas chromatography (μGC) system, which is the most widely utilized gas sensing system in analytical community, has been suggested as one of the best candidates for miniaturized gas sensors due to its capability of simultaneously detecting multiple numbers of gas compounds per analysis as well as their potentially small sizes, weights, rapid analysis tend low power consumption [1]. The μGC system enables the detection of a wide range of gas molecules by "pre-separating" them in space and time before detection utilizing a gas sensor component.

However, previous micro gas chromatography (μGC) systems up to date have suffered from some limitations caused by its sensor component: (1) degradation or non-repeatability of output signals over a span of time, (2) miniaturization difficulty in sensors due to the required ancillary instruments, and (3) fabrication complexity. The signal degradation was mainly caused by the chemistry-dependence of the gas-reactive films incorporated into some micro sensors, such as in chemiresistors [2], resonators [3], surface acoustic wave (SAW) sensors [4], metal oxide semiconductor (MOS) sensors [5] and Fabry-perot sensors [6]. These films easily become oxidized over time and cause significant changes in their properties. Miniaturization difficulty, caused by the needs for ancillary instruments for vacuum and laser, reduced the practical use of some other gas sensors for an integrated micro GC system, including optical fiber sensor [7], flame ionization detector [8], and mass-spectrometers [9]. Fabrication complexity still

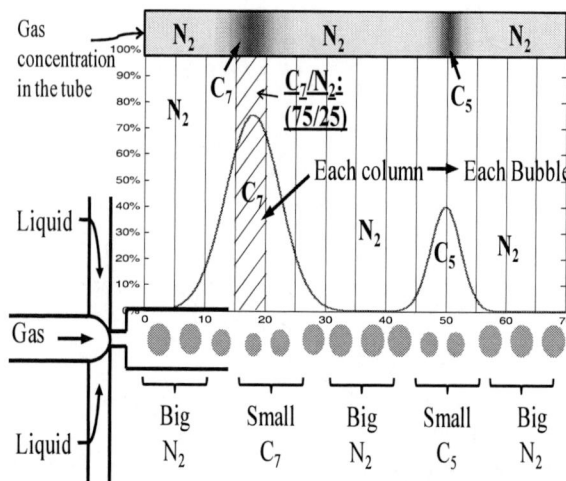

Figure 1: Concept of gas sensing: initially, a carrier/background gas (e.g. nitrogen (N_2)) flows into a liquid-filled micro channel forming a train of gas bubbles, of which the size is uniform. As a dose of different gas (e.g. pentane (C_5)) is injected, it causes the bubble sizes to become smaller. Thus, monitoring of the size changes enables the construction of a chromatogram for gas identification.

remains the challenge in constructing thermal conductivity detectors (TCD) [10] within micro channels.

To address such issues, a new concept of gas sensor has been developed, which is simple to fabricate, easy to miniaturize, provide time-invariant signals for a micro gas chromatography system. Particularly, this new class of a gas sensor utilizes the variation in bubble sizes when different gases are introduced into a liquid flow to identify gas types, thus eliminating the previous requirements for chemistry-dependent thin films or bulky ancillary components.

This paper reports the concept, fabrication and testing results of the new class of a gas sensor by providing two main results: the identification of the injected gas (C_5) by measuring the produced bubble sizes in liquid flow and the preliminary correlation between the span and size of such bubbles and varied injection amounts.

CONCEPT

Figure 1 illustrates the concept of the bubble-based gas sensor. When a gas (e.g. nitrogen, N_2) stream is introduced into a micro channel where there is a liquid flow, it forms a train of bubbles with a specific size depending on the flow rates, surface tension, dissolvability and viscosity. This gas, N_2, acts as a background or reference gas, while its resultant bubble size is uniform. When a dose of a different gas (e.g. pentane, C_5) is injected into the background nitrogen flow, the produced bubbles experience size changes due to the

978-1-4799-3510-9/14 $31.00 © 2014 IEEE

different gas properties aforementioned. Thus, by monitoring the size changes and the time span of the bubbles under size change, the device enables the identification of the gas types in comparison to the previously known gas retention times. Since the gradual size changes have been previously proven as proportional to the mixture ratio [11], it also becomes possible to determine precise mixture ratios at each bubble.

STRUCTURE AND FABRICATION

Structure

The bubble-based gas sensor primarily consists of a bubble generation nozzle, a bubble flow channel and inlets & outlet for both gas and liquid, all of which were constructed on a planar PDMS substrate. (Fig.2) The bubble generation nozzle combines the two-phase flows of both gas and liquid into a tiny gas-liquid co-flowing point and produces gas bubbles, as shown in Fig.2. At the co-flowing point, the gas stream is periodically cut by the liquid flow through the cross-channel configuration, as widely accepted [11]. The widths of the gas and liquid channels were 150 and 250µm, respectively, while the height of all channels was identically set as 100µm. The width of the nozzle was 40µm. The bubble flow channel forms a long and meander path for the bubbles to travel, allowing a wide range of optical measurement spots. The inlets and outlets of both gas and liquid flows were located at each end of the channel and connected through a flow controller to gas cylinders and to a liquid syringe pump, respectively. The diameters of both inlets and outlets were 600µm, and the footprint of the total sensor structure was $20\times5mm^2$.

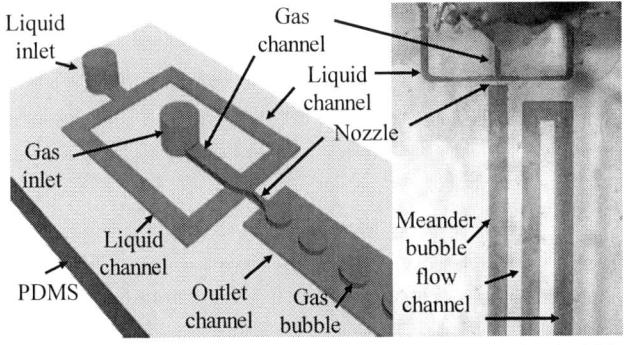

Figure 2: Device structure: (left) illustration of the bubble generation nozzle, the bubble flow channel and the fluidic connections. (right) photo of the fabricated microfluidic flow-focusing device.

Fabrication

Fabrication was performed mainly by molding two layers of PDMS substrates, including the channel layer and the cover layer, and stacking them with an oxygen plasma bonding technique (Fig.3). First, the channel layer was formed by pouring 10:1 PDMS-solvent mixture (Sylgard 186 Silicone Elastomer Kit) on top of a SU-8 (SU-8 2050) mold and curing the agent mixture at 65^0C for 6h. The SU-8

(SU-8 2050) mold had previously been constructed on a silicon wafer by standard photolithography technique, which provided a uniform height of ~100µm. This height of the SU-8 mold determines the height of a microfluidic channel. Next, the top layer was similarly constructed on a flat petri dish. Then, both layers were bonded to form a closed microfuidic channel by utilizing the oxygen plasma bonding. To ensure the stable formation of bubbles the hydrophilic nature of the oxidized PDMS walls inside the channel was preserved by immediate liquid flow after the sealing.

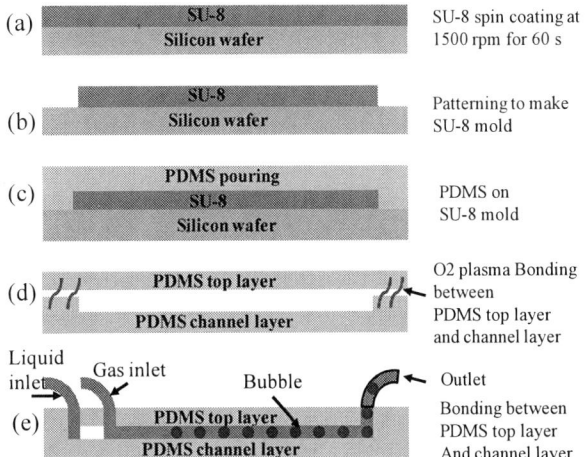

Figure 3: Fabrication steps of the microfluidic flow-focusing device.

TESTING METHODOLOGY

To validate the concept of the gas sensor, the microfluidic set-up was established to produce gas bubbles into a micro channel while the sizes of the produced bubbles were optically monitored utilizing a camera and a custom-built video processing software. First, the microfluidic set-up was established respectively for both gas and liquid flows. Gases of interests (e.g. nitrogen) flow from pressurized gas tanks through a pressure reduction valve and a commercial GC injector to the gas inlet of the device. The pressure reduction valve lowers the gas flow pressure to a moderate level for microfluidic application from 13.8GPa to 551.5kPa, while the GC selectively injects different types of gases in precisely defined amounts down to picograms. The GC injector controlled the gas flow rate at 0.3µL/minute. Liquid flow was manipulated by a syringe pump (KD scientific, KDS-210) that reliably set the flow rate at 1µL/s which is a moderate flow rate for a stable bubble generation. The liquid phase utilized DI water containing surfactant (Tween 20) to reduce surface tension of the liquid phase and avoid coalescence among the bubbles. Second, the optical measurement set-up was established by utilizing a microscope, a video recorder and the image processing software. The microfluidic device (footprint $20\times5mm^2$) was placed on the base of a microscope (Mitutoyo) that was connected to a video camera (Edmund Optics) to record the bubble generation process (Fig.4-(a)). The video was

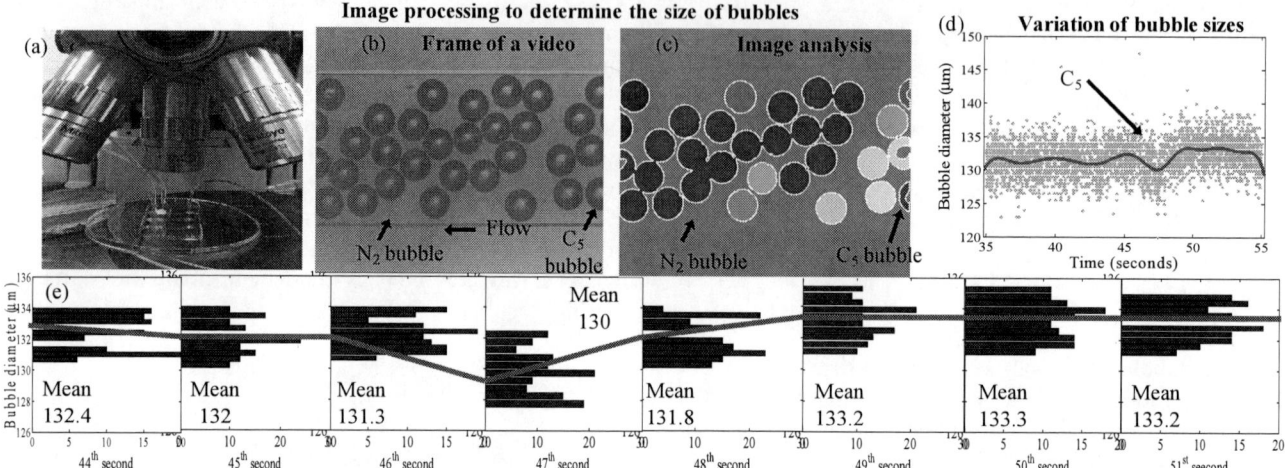

Figure 4. (a) Microfluidic channel under the microscope, (b) a frame from the recorded video using the microscope (c) each bubble is being encircled to determine the diameter, different color represents different zones of connected circles, (d) variation of bubble size after analyzing all of the frames over a certain period, (e) a detail second-by-second bar graph of bubbles between 44 and 51 seconds, indicating most of the bubbles within a second gets low in size when C_5 forms the bubbles.

recorded with 5X optical zoom at an 8-mm distance from the bubble generation nozzle to allow the stabilizing time for the bubble sizes. The span of video recording was set for 4 minutes starting from the injection of C_5 sample. The video was then analyzed to determine the bubble size by developing a customized software based on MATLAB. The developed software first divided the recorded video into frames (Fig.4-(b)). It converted the frames of a grayscale image (Fig.4-(b)) into those of a color image displaying the diameter sizes of each bubble (Fig.4-(c)). Such recorded diameters of all the bubbles in all frames of a video were then collected and plotted as in Fig.4-(d) for statistical analysis.

Gas Type Identification

To identify the presence of an injected gas, the bubble size changes were monitored while a dose of C_5 (J. T. Baker 98% Pentane) was injected into a stable N_2 gas flows. Monitoring of the bubble diameters through video analysis was performed over the 4.0 minute period of time, as shown in Fig.4-(d). Fitting into the collected data, a Gaussian polynomial curve was generated. Finally, to analyze the size variation, a statistical analysis was performed. The analysis period was set from 44th to 51st seconds (Fig.4-(e)). During the analysis, approximately 100-120 bubbles per second were contained.

Gas Amount vs. Signal Response

To observe any potential correlation between the bubble sizes and the amount of injected volume, two different amounts (3 and 5µL) of C_5 were injected while the bubble size changes were monitored. To minimize the broadening of the target gas before injection, the length of the capillary tube was kept minimum under our circumstances as 80cm.

Sensitivity, Detection Limit and Dynamic Range

To evaluate the performance, some performance parameters, such as sensitivity, minimum detection limit and dynamic range, were investigated. The sensitivity was defined as the slope between the output bubble and the input injection volumes, while dynamic range was the ratio between the maximum and the minimum detectable bubble sizes.

RESULTS AND DISCUSSION

The fabricated bubble-based gas sensor demonstrated a significant bubble size changes upon the injection of a target gas. It also showed a reasonable sensitivity, an excellent detection limit with a large dynamic range.

Gas Type Identification

Measurement results showed that the injection of 3µL of pentane into the nitrogen gas flow reduced the bubble size from 132.4µm to 130.0µm by 2.4µm on average, which is by 1.7% in comparison to the background nitrogen bubbles with an average size of 132.4µm, indicating the gas sensing capability utilizing microfluidic bubbles (Fig.4-(d)). Figure 4-(e) shows the gas bubble size distribution between 44th and 51st seconds from the injection at 1.0 second intervals. The average sizes of the produced bubbles noticeably decreased at 47th second: the bubble size span before injection was between 130.5 and 134.0µm while it dropped to between 127.5 and 132.0µm after injection, clearly indicating the existence of C_5 gases. The total number of the measured bubbles within the time period was 1324. The horizontal bar graph indicates ~100 bubbles/s in Fig.4-(e).

Gas Amount vs. Signal Response

Measurement results also showed that an increase in the

978-1-4799-3510-9/14 $31.00 © 2014 IEEE   228

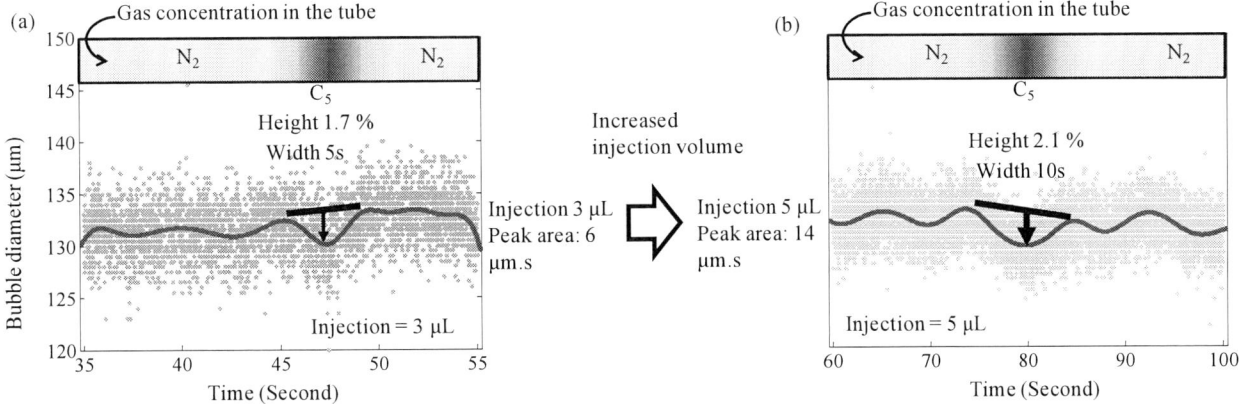

Figure 5. Plot of bubble diameters for a certain period of time. Each dot indicates each bubble's diameter. The data is fitted using Gaussian curve. It shows that, bubble size gets reduced due to C_5 sample in a flow of N_2 carrier gas. It also shows, an increase in injection volume from (a) 3 μL to (b) 5μL results in the decrease in areas.

injection volume from 3μL to 5μL caused (1) the further decrease in bubble sizes from 130.0 to 129.5μm on average by 24% and (2) increase in the time span of such reduced bubbles from 5s to 10s by 100% (Fig.5). Resultantly, the total area, which can be calculated as 0.5×time-span×bubble-size-variation on Figure 5, has increased from 6 to 14μm.s by 133.3%. Note that the average line was constructed based on the Gaussian fitted curves for each injection volumes of 3μL and 5μL in Fig.5-(a and b).

Sensitivity, Detection Limit and Dynamic Range

The preliminary sensitivity was approximated as 10^{-5} by comparing the calculated areas between 3μL and 5μL injection, despite the availability of only two data points as of yet. The minimum measurable volume of the bubbles was $10^3 \mu m^3/s$, which was currently limited by the speed of the video recording (22 fps). The results correspond to the minimum detectable weight of ~1 picogram of target gas molecules per second. The dynamic range under current set-up was measured as 10^7 with the minimum and maximum measurable volumes of 10^{-12} and 10^{-5}L.

CONCLUSION

We report a new class of a microfluidic bubble-based gas sensor that provides time invariant output signals, is simple to fabricate and easy to integrate into a micro-scale gas chromatography system. The fabricated device demonstrated the feasibility of the concept by enabling the detection of pentane (3μL) with the reduced average bubble size of 130μm by 2.4μm (1.7%) from the background bubble sizes (132.4μm) of nitrogen. The measurement also showed a preliminary correlation between the injection volume and the bubble size: increase in the injection volume of pentane from 3μL to 5μL (66.7% increases) resulted in the bubble size reduction from 1.7% to 2.1% (24% decreases) and the variation period elongation from 5s to 10s (100% increases).

ACKNOWLEDGEMENTS

This work was performed in the Utah Nanofabrication facility at the University of Utah in Salt Lake City.

REFERENCES

[1] H. Kim, et al., "Micropump-driven high-speed MEMS gas chromatography system," Transducers '07, Lyon, France, Jun. 10-14, 2007, pp. 1505-1508.

[2] S. Bedair, G. Fedder, "CMOS MEMS oscillator for gas chemical detection", *IEEE Sensors'04*, pp.955-958.

[3] M. Li, et al., "Nanoelectromechanical resonator arrays for ultrafast, gas phase chromatographic chemical analysis", *Nano Lett.*, vol. 10, pp. 3899-3903, 2010.

[4] P. Lewis, et al., "Recent advancement in the gas phase microchemlab", *J. Sensors*, v6, pp.784-795, 2006.

[5] S. Herberger, et al., "Detection of human effluents by a MOS gas sensor in correlation to VOC quantification byGC/MS",*Building&Environment*,v45,2430-39, 2010.

[6] J. Liu, et al., "Fabry-Perot Cavity Sensors for Multipoint On-Column Micro Gas Chromatography Detection", *Anal. Chem.*, v82, pp. 4370-4375, 2010.

[7] C. Ho, et al., "Review of Chemical Sensors for In-Situ Monitoring of Volatile Contaminants", *Sandia Report*, 2001.

[8] S. Zimmermann, et al., "Micro Flame Ionization Detector and Micro Flame Spectrometer", *SNA B*, v63, pp.159-166, 2000.

[9] K. Cheung, L. Velasquez-Garcia, A. Akinwande, "Chip-Scale Quadruple Mass Filters for Portable Mass Spectrometry", *JMEMS*, v19, pp.469-483, 2010.

[10] S. Narayanan, et al., "Two-Port Static Coated Micro Gas Chromatography Column with an Embedded Thermal Conductivity Detector", *J. Sensors*, v12, pp. 1893-1900, 2012.

[11] A. Bulbul and H. Kim, "Characterization of microbubbles in of multiple gases in microfluidic channels", *μTAS'13*, Germany, Oct.27-31, 2013.

CONTACT: ashrafuzzaman@utah.edu, hanseup@ece.utah.edu

A FLEXIBLE GRAPHENE FET GAS SENSOR USING POLYMER AS GATE DIELECTRICS

Yumeng Liu[1], Jiyoung Chang[2], and Liwei Lin[1]

[1]Berkeley Sensor and Actuator Center, University of California Berkeley, USA

[2] Department of Physics, University of California at Berkeley, USA

ABSTRACT

We have successfully demonstrated a Graphene Field Effect Transistor (GFET) gas sensor on a flexible plastic substrate with a sensitivity of $0.00428ppm^{-1}$ ($\Delta R/R_0$) for ammonia. Compared with the state-of-art technologies, four distinctive advancements have been achieved: (1) first demonstration of a flexible graphene FET gas sensor; (2) a new fabrication process to achieve embedded-gate GFET on a flexible substrate; (3) proof of utilizing polymeric materials of parylene and polyethylenimine (PEI) as gate dielectrics and channel dopant for graphene FETs, respectively; and (4) validation of a gas sensing mechanism utilizing the real-time, n-type graphene doping process induced by the exposure of graphene FET to the targeted gases. As such, the proposed sensing scheme/platform could open up a new class of research in graphene-based, flexible gas sensing systems.

INTRODUCTION

In recent years, graphene has been an attractive material for flexible electronics due to its ultrathin, 2D structure and strong mechanical properties. Multiple research groups have shown the feasibility to construct graphene FETs with inorganic gate dielectrics on flexible substrates [1,2] with superior performances in high mobility and high cut-off frequency [3,4]. The inorganic dielectric materials have been reported to generate cracks within the crystal domain in the applications of flexible substrates under large mechanical deformations [5]. It is also well known that the pristine graphene sheet is sensitive to water vapor doping and often results in p-type channel FETs with large gate voltage biases to reach their Dirac Points. In this work, organic parylene is proposed as the gate dielectric material and the physisorption of PEI is utilized as the chemical dopant which dramatically lowers the gate voltage for the Dirac Point in the prototype GFETs from 80V to 2.3V and considerably simplifies the device operations.

Graphene based sensing systems have also been under intensive investigations since every atom on the 2D graphene structure can react to the surrounding gas as a system with the largest possible surface area to volume ratio of any structure for highly-sensitive sensors. For example, ammonia has a lone pair of electrons such that it can act as temporary dopant to provide electron to graphene and lower the conductance by increasing the carrier concentration. It has been reported that the resistance of graphene is very sensitive to the exposure to ammonia and other gases [6], and the corresponding limit of detection (LOD) can be as good as one gas molecular under a special measurement setup [7]. In addition to sensors made from pristine graphene, chemically converted graphene from

graphene oxide has also been demonstrated as gas sensors with typically higher resistances and noises [8].

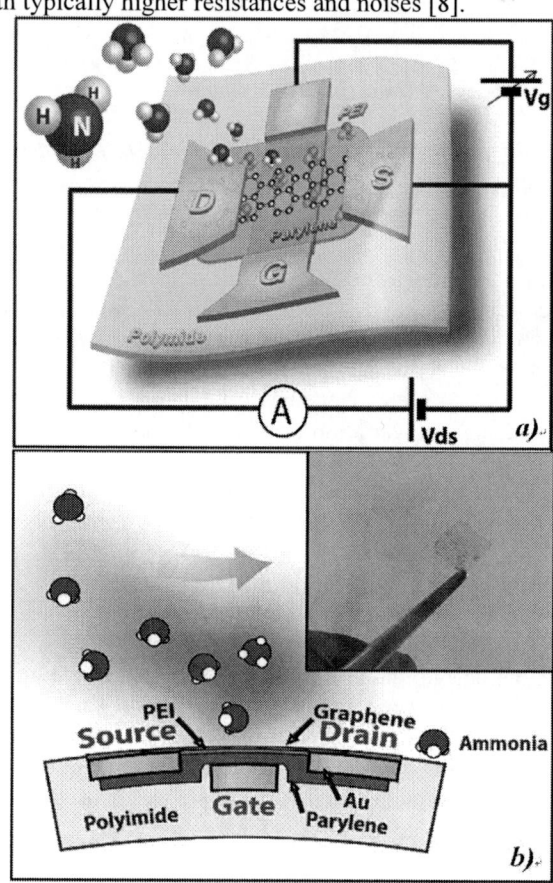

Figure 1, a) Schematic diagram of the flexible graphene FET gas sensor; b) cross-sectional view showing the graphene channel is open to the environment for gas sensing with 170nm-thick parylene and 70nm-thick PEI as gate dielectrics and channel dopant, respectively. Inset shows an array of 9x9 as-fabricated graphene gas sensors on a flexible polyimide substrate.

Although graphene devices have been built on flexible substrate and graphene gas sensors have been demonstrated, graphene based gas sensor on a flexible substrate has not been fabricated previously. Here we demonstrate a gas sensor utilizing graphene as the sensing material, as shown in Figure 1a, and implement it on a flexible polyimide substrate. By monitoring the channel current under a fixed V_{DS} and specific V_G, the real-time doping effect from ammonia to graphene could be detected as the gas sensing mechanism. Figure 1b shows the cross-section schematic of the flexible gas sensor, and the inset shows the as-fabricated 9x9 GFET array on a flexible substrate. In this paper, the characteristics of the GFET on both rigid and flexible substrates with polymer gate

978-1-4799-3510-9/14 $31.00 © 2014 IEEE

dielectrics have been experimentally characterized. The sensing responses under different gas concentrations and gate voltage biases have been recorded. Furthermore, observation of time-variant doping of graphene in FET sensors after their exposures to ammonia has been monitored as a potentially new sensing scheme.

FABRICATION

Firstly, the high quality single layer graphene sheet is synthesized via chemical vapor deposition (CVD) on a copper foil under $1000^{o}C$ and transferred onto a thermally grown 285nm-thick SiO_2 on a p-doped silicon wafer as described in our previous work [9]. The quality of the graphene sheet is verified using Raman spectroscopy. Figure2 illustrates the following fabrication process. The source and drain electrodes are deposited and patterned by Ti/Au (2nm/50nm) e-beam evaporation using a shadow mask (Fig. 2b). Then a 5um×5um graphene channel is patterned and etched by a 50W oxygen plasma process for 7s using the standard optical lithography process (Fig. 2c). The whole device is spin-coated with a 1um-thick, 50% w.t. PEI solution using H_2O as the solvent (Fig. 2d). Due to the physisorption, a thin layer of PEI is left on the graphene sheet after DI water rinse, and the residual PEI is utilized as the n-type dopant for the graphene sheet (Fig. 2e). A layer of 170nm-thick Parylene-C is then deposited using MVD (Special Coating System PDS2010) after a Gamma-MPS coating process (Fig. 2f) and then the parylene is patterned by a 100W oxygen plasma process for 60s using the standard optical lithography (Fig. 2g). The top gate (TG) is patterned by the Cr/Au (2nm/50nm) e-beam evaporation process using a shadow mask and the p-doped silicon functions as the back gate (BG) (Fig. 2h). Afterwareds, a 40um-thick polyimide (HD Systems, PI-2574) layer is spin-coated and cured at $180^{o}C$ (Fig. 2i). Finally, the flexible structure is released by using the 10:1 buffered oxide etch (BHF) (Fig. 2j). It is worth noting that a hydrophobic surface is required for a uniform parylene MVD deposition, such that the Gamma-MPS treatment is necessary after the PEI deposition process as PEI layer alters the surface contact angle from 85° of pristine graphene to 15°, as shown in Figure 3.

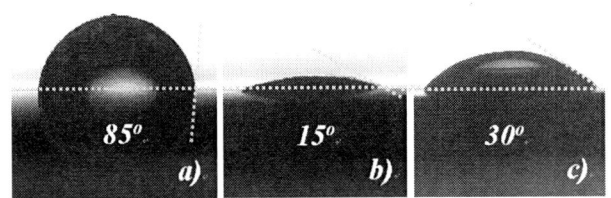

Figure 3, a) Contact angel of a) pristine graphene; b)graphene with PEI coating; c)graphene with PEI coating and a second coating of Gamma-MPS.

Figures 4a/4b show the optical microscopic pictures of a 3x3 device array before/after the transfer process from the rigid wafer onto a flexible plastic substrate, respectively and Figures 4c/4d shows enlarged views. After transferred onto the flexible substrate, graphene is directly exposed to the environment for sensing and the gate electrode is embedded in the polyimide substrate. The flexible transistor is made contact to the breadboard via conductive paste and fixed in the gas chamber. Not all GFET could be successfully transferred onto the polyimide substrate, and the major reason for failure is delamination of metal contact pads as shown in the upper left corner of Figure 4b.

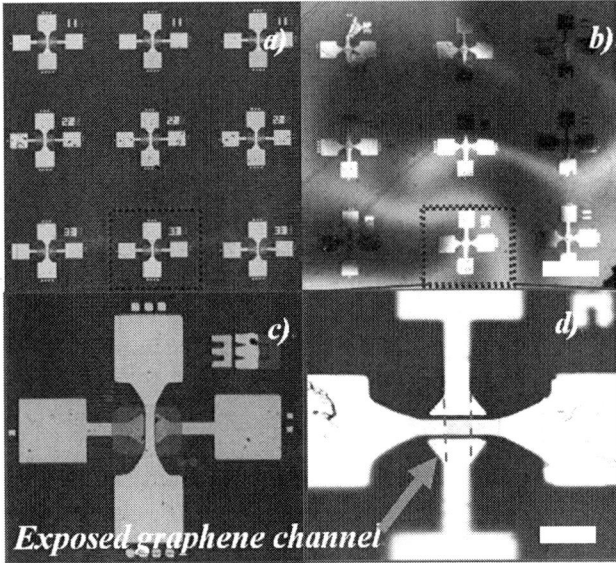

Figure 4, An array of 3x3 devices a) before polyimide coating and b) after polyimide coating and separation on a flexible substrate, scale bar 70um. c) and d) magnified views of the corresponding single transistor in a) and b), respectively, scale bar 7um.

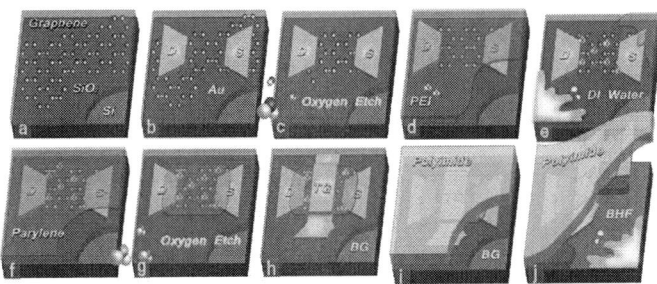

Figure2, The device fabrication process. (a) graphene transferred on to the silicon wafer (with 285nm-thick oxide), (b) formation of contract electrodes, (c) definition of channel areas, (d) PEI coating as graphene dopants, (e) DI water rinse, (f) paralyne coating as the gate dielectric material, (g) definition of paralyne by plasma etching, (h) gate electrode deposition and definition, (i) polyimide coating, and (j) peeling off the flexible substrate with devices on top.

Figure 5 shows the schematic of the gas sensor testing system. The flexible gas sensor is sealed in the chamber, and nitrogen is purged, and followed by a pumping down process to 10^{-3} torr. Afterwards, the target gas is introduced into the chamber by opening the control valve (V_2 in Figure 5). The amount of gas concentration inside the chamber is monitored by the pressure gauge. The source drain voltage of the GFET is fixed at $V_{DS}=0.1V$ throughout the experiment, and the channel current I_{DS} is sampled twice per second using Agilent digital multimeter 34401A.

978-1-4799-3510-9/14 $31.00 © 2014 IEEE   231

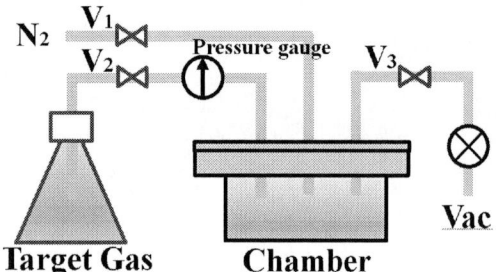

Figure 5, The schematic of the experimental setup. Valves with #1, #2, and #3 are used to control the flow of nitrogen, target gas, and vacuum pump, respectively.

RESULTS AND DISCUSSION

Experimentally, GFETs under dual-gate operations are characterized on solid substrate before the transfer process. Figure 6 shows the I_{DS}-V_{TG} curves (source-drain current versus top gate voltage) of a GFET under various V_{BG}. The majority carriers in graphene on the non-overlapping area (see inset in Fig. 6) between the top gate and source/drain electrodes are mainly controlled by the back gate bias, and weakly controlled by the top gate bias due to the weak fringing field effect. The graphene channel directly under the top gate is controlled by both of the top and back gate biases. Specifically, electrons are attracted to the graphene channel under positive bias and holes are attracted to the graphene channel under negative gate bias. Experimentally, under V_{BG}=50V, the majority carriers in the non-overlapping, overlapping, and non-overlapping graphene channel, (left to right in the inset of Figure 6) are electrons when the top gate voltage, V_{TG} changes from -10V to 90V. On the other hand, if V_{BG} is -50V, the majority carriers in the aforementioned regions switch to holes, electrons, and holes as V_{TG} changes from -10V to 90V. Therefore, it is possible to form a p-n-p junction in the graphene channel under the right biases of V_{TG} and V_{BG} and double local maximum resistances can be identified is in the I_{DS}-V_{TG} curves in Fig. 6, similar to the previous report [10].

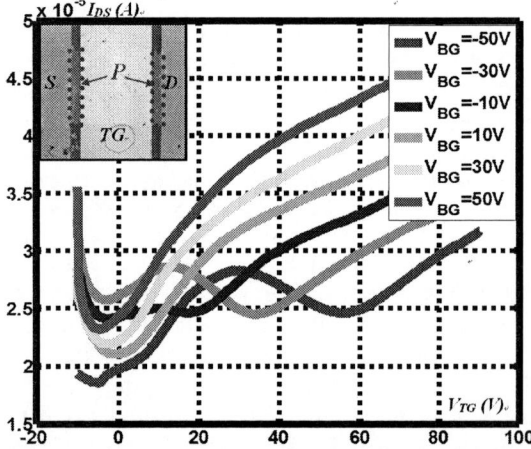

Figure 6, Dual gate operation of a GFET under V_{DS}=1V. The graphene channel can transition from pnp junction to pure n type when V_{BG} is increased from -50V to 10V and above. The inset shows the p type regions can be formed if V_{BG} is less than -10V. When V_{BG} is larger than 10V, n-type regions are formed as electrons will fill up the two p-type regions. scale bar - 3μm.

The I_{DS}-V_{DS} and I_{DS}-V_G curves of a flexible GFET under different gate voltage are measured and shown in Figure 7a and 7b respectively. In Figure 7a, a crossover of two curves (V_G=20V, V_G=0V) is observed after V_{DS}>3V. This is because the majority carrier of graphene channel switches from a suppression of holes to an accumulation of electrons as V_{DS} increases to alter the voltage distribution of graphene channel and flips the major carrier type. In Figure 7b, the hysteresis phenomenon is observed with forward and backward sweeping of the gate voltage at 1V/s, which has been known as the charge transfer process and a thermal annealing process of GFET at 300°C in N_2 has been demonstrated to eliminate this effect [11]. The Dirac Points changes from 0V to 15V as V_G sweeps forward and backward, respectively. Furthermore, under the n-type doping from the residual thin PEI layer, the maximum Dirac Point of the graphene channel is shifted from 80V of pristine graphene to 2.3V. A Dirac Point near zero helps the device operation to bias the GFET into either p-type or n-type region with low voltages without breaking down the gate dielectrics.

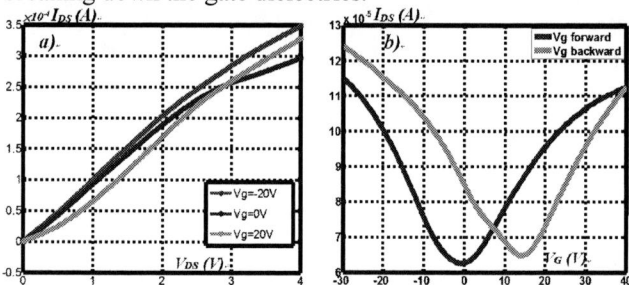

Figure 7, a) The I_{DS} -V_{DS} curves of a prototype flexible graphene FET under different gate voltages; b) The I_{DS} -V_G curves (V_{DS}=1V) showing the hysteresis of the Dirac Point at 0V to 15V as V_G sweeping forward and backward, respectively.

The sensing behavior is tested using the setup introduced above. The channel resistance of the flexible gas sensor responds to different concentrations of ammonia as shown in Figure 8, where the nitrogen purge process is used at each green arrow to remove ammonia under room temperature. The GFET is biased with V_G=-15V and V_{DS}=0.1V, and the majority carriers in the graphene channel are holes. During the adsorption phase, each ammonia molecule transfers one electron to the graphene channel, which decreases the major carrier density and increases the channel resistance. During desorption phase, nitrogen is purged into the chamber, and nearly half of the molecules can be detached from graphene surface quickly, while the rest of them follow a desorption process with a larger time constant. The sensitivity of the flexible graphene gas sensor is measured as 0.00428ppm^{-1}, which is calculated by using the term $\Delta R/R_0$, where ΔR is the difference between graphene channel resistance with and without the absorption of ammonia, and R_0 is the original graphene channel resistance without ammonia.

We also examine the gas sensor response under different gate voltages. Figure 9 illustrates the response of a flexible GFET gas sensor when exposed to 3500ppm of ammonia under different gate voltages.

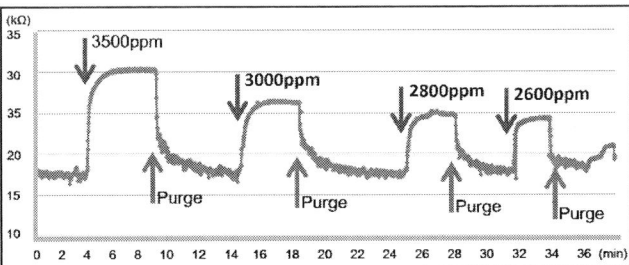

Figure 8, Transient responses of a flexible graphene gas sensor under various concentrations of ammonia. The chamber is purged with nitrogen in every cycle.

As shown in each figure, the intersection points between the R_{DS}-V_G curves and the vertical green line (working gate voltage) determine the channel resistance. As the R_{DS}-V_G curve shifts from the blue to red curve during the ammonia doping process, changes in the channel resistance (R_{DS}) are recorded and they behave differently under different gate voltages. For example, in Figure 9a, the graphene channel is biased with V_G=10V and the transition of channel resistance starts from the intersection of the blue and green line (blue point); follows the black vector on the blue line; and settles to a lower channel resistance at the red point. In Figures 8b and 8c, the black vector increases initially to pass the Dirac Point, then settles at a higher and lower magnitude as compared to the initial resistance, respectively. It is noted that passing the Dirac Point results in a resistance peak in the corresponding R_{DS}-time measurement plots, and provides direct observation of n-type doping of graphene passing through the Dirac Point shift in real time. Figure 8d shows that under a negative gate voltage V_G=-15V, the channel resistance increase with time during the measurements in the R_{DS}-time curve.

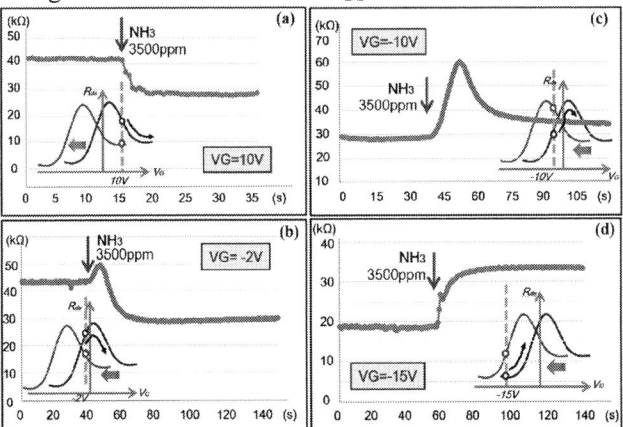

Figure 9, The graphene gas sensor responses under different gate voltages, showing channel resistance a) drops to a lower value; b) rises then drops to a lower value (passing the Dirac Point on the green working gate voltage); c) rises then drops to a higher value; and d) rises to a higher value, under different gate bias voltages.

CONCLUSION

We have successfully demonstrated a flexible GFET gas sensor with a sensitivity of 0.00428ppm^{-1} for sensing the ammonia. Organic materials, including gate dielectric (170nm-thick parylene C layer), channel dopant (PEI), and substrate (40μm-thick polyimide) are used to construct the flexible GFETs. Experimental sensing measurements show fast adsorption/desorption process of ammonia purged at room temperature under various gas concentrations. The doping effects of ammonia molecule to graphene channel are further investigated in four different working regimes and the R_{DS}-time curves agree well with the analyses. Under specific gate voltage bias, a peak of channel resistance appears in the transient response before R_{DS} settles at the final value, providing a direct observation of the transition of major carrier type at the Dirac Point due to the ammonia doping. Novel gas sensing application utilizing this phenomenon is under further investigation. This process is scalable for possible mass production of graphene based flexible sensors.

ACKNOWLEDGEMENTS

These devices have been fabricated in the Marvel Nanofabrication Lab at UC Berkeley. This project is supported in part by NSF grants CMII-10311749 and ECCS-0901864.

REFERENCES

[1] J. Lee, "25 GHz Embedded-Gate Graphene Transistors with High-K Dielectrics on Extremely Flexible Plastic Sheets", *ACS Nano, 2013, 7 (9), pp 7744–7750*

[2] C. Yan, "Graphene-based flexible and stretchable thin film transistors" *Nanoscale*, 4(2012), pp. 4870-4882

[3] C. Lu, **"High Mobility Flexible Graphene Field-Effect Transistors with Self-Healing Gate Dielectrics"**, *ACS Nano*, 2012, *6* (5), pp 4469–4474.

[4] W. Zhu "Graphene radio frequency devices on flexible substrate", *Appl. Phys. Lett*. 102, 233102 (2013)

[5] H. Chang, "High-Performance, Highly Bendable MoS$_2$ Transistors with High-K Dielectrics for Flexible Low-Power Systems", *ACS Nano*, 2013, 7 (6), pp 5446–5452

[6] W. Yuan, "Graphene-based gas sensors", *Journal of Material Chemistry A*, 2013, 1, 10078-10091

[7] F. Schedin, "Detection of individual gas molecules absorbed on graphene", *Nature Materials, 2007, 6*, 652 - 655 [8] H. Bai, "Functional Composite Materials Based on Chemically Converted Graphene" *Adv. Mater*., 2011, 23, 1089–1115

[9] J. Chang, Y. Liu, and L. Lin, "Direct-write n- and p-type graphene channel FETs", 2013 IEEE 26th International Conference on Micro Electro Mechanical Systems (MEMS), pp.201,204, 20-24 Jan. 2013.

[10] R. Gorbachev, "Conductance of p-n-p Graphene Structures with "Air-Bridge" Top Gates" *Nano Lett.*, 2008, *8* (7), pp 1995–1999

[11] H. Wang, "Hysteresis of Electronic Transport in Graphene Transistors", *ACS Nano*, 2010, *4* (12), pp 7221–7228

CONTACT

*Yumeng Liu,
tel: +1-510-520-0391;yumengliu@berkeley.edu

A CONTINUOUS OPTICALLY-INDUCED CELL ELECTROPORATION DEVICE WITH ON-CHIP MEDIUM EXCHANGE MECHANISMS

Chia-Jung Chang, Ming-Yu Lu and Gwo-Bin Lee*

Department of Power Mechanical Engineering, National Tsing Hua University, Hsinchu, Taiwan

ABSTRACT

We present a new design for continuous optically-induced electroporation (OIE) on a microfluidic device, which is capable of replacing culture medium and electroporation buffer in a seamless fashion. The seamless on-chip integration of medium exchange mechanisms and optically-induced electroporation device could avoid critical issues such as cell losses and cell damage during manual operation in the traditional centrifuge system, and is therefore suitable for handling small or rare cell population. Furthermore, the survival rate of the cells could be greatly improved due to this fast, automatic procedure.

INTRODUCTION

Cell electroporation has been widely utilized in various biological applications involving the delivery of drug, gene and other substances of interest, into the cytoplasm. However, the traditional methods of cell electroporation have short-comings such as the requirement of a high electric voltage, non-uniformity of the generated electric field and high heat generation during the electroporation process, resulting in a relatively low cell viability and low overall transfection efficiency in such systems. Therefore, a wide variety of microfluidic electroporation devices have been developed to tackle these issues [1].

Optically-induced electroporation (OIE), also known as light-induced electroporation, is one of the microfluidic approaches toward cell electroporation, with great flexibility with respect to other

fixed-metal-electrode based designs, making it possible for the operators to re-configure the electrode geometry immediately to perform different electroporation applications, without going through the entire fabrication process but only to change the optical pattern [2]. OIE was demonstrated to transfer substances of interest such as fluorescent dyes and genes into mammalian cells [3]. Nonetheless, in certain optimized electroporation protocols, the process took place in an electroporation buffer rather than the original cell culture medium. This indicates that the cells in the original medium have to be manually transferred into the electroporation buffer by centrifuging, removing the supernatant, and re-suspending into the suitable electroporation buffer. The traditional medium exchange process would inevitably lead to cell losses and was not suitable for the handling of small cell populations. Therefore, a seamless on-chip integration of medium exchange mechanisms and optically-induced electroporation was explored to tackle this problem in this study. With this approach, the entire process for continuous cell electroporation could be performed automatically without human intervention. Furthermore, the survival rate of the cells could be greatly improved due to this fast, automatic procedure performed on the integrated microfluidic chip.

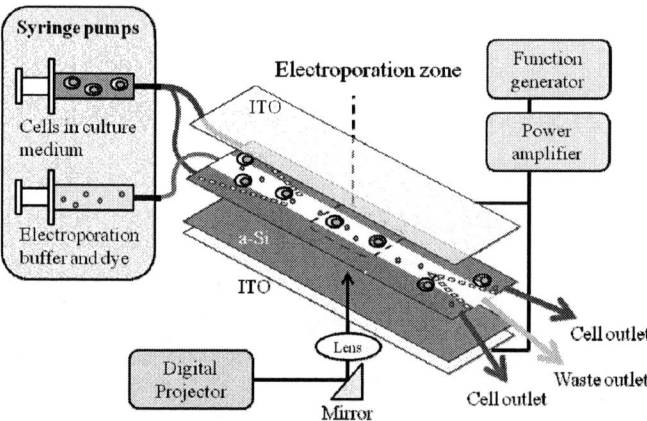

Figure 2: A schematic view of experiment setup for continuous cell electroporation.

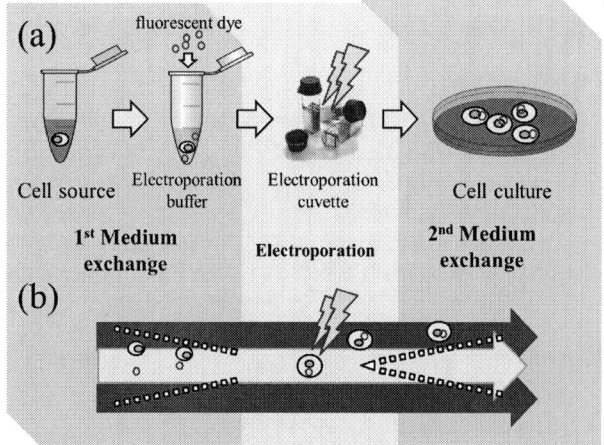

Figure 1: (a) Traditional cell electroporation procedure required to change medium manually. (b) The proposed device integrated μPAR and optically-induced electroporation (OIE) to realize continuous medium exchange and cell electroporation.

DESIGN

The experimental design and the corresponding traditional procedures are schematically shown in Figure 1. In order to transfer the cells from the original medium into the electroporation buffer, the cells in outer streams of the culture medium were guided into the central stream of electroporation buffer by using micropost array railing (μPar) structures [4]. Then the cells were electroporated in the buffer using OIE. As the cells flew downstream, a second μPar structure guided them back into the original streams of culture medium. A schematic view of experimental setup is shown in Figure 2. Streams of medium and buffer were injected by syringe pumps continuously between the ITO glass and amorphous (a-Si) layers (photoconductive layers). The PDMS μPar structures (about 20 μm wide, 10 μm gap and 1 ° Inclination angle) were located in the channel to guide the cells. A digital projector illuminated a light spot onto the a-Si layer in the electroporation zone, creating a "virtual electrode" on the a-Si surface while an alternating-current (AC) electric field was applied between the top and bottom ITO layers to electroporate the cells due to the fact that the transmembrane potential could be induced on the cell membrane.

technique [5] as our fabrication strategy. The SU-8 3050 (MicroChem, USA) was patterned onto the Si wafer by standard lithography process as the positive mold. The mold was treated with diluted commercial detergent "Salatt", the releasing agent, and rinsed with DI water and blown dry. Then the PDMS mold of A:B ratio of 10:1 was casted by the SU-8 mold. The PDMS mold was cured at 80°C in an oven for approximately 1 hour. Then the PDMS mold surface was treated with releasing agent of HPMC dissolved in DI water then rinsed with DI water and blown dry. A second PDMS of A:B ratio of 10:1 for about 100 μm as a final structure was spin-coated onto the PDMS mold and cured at 80°C in an oven for approximately 1 hour. A hole was then drilled for the electroporation zone. The thin PDMS film and the a-Si was treated with oxygen plasma and bonded with pressure on a hotplate. Then, the PDMS mold was peeled off, leaving the PDMS μPAR structure on the a-Si. An ITO-coated PET film with drilled via holes was then aligned and bonded onto the device by oxygen plasma surface treatment and bonded with pressure on a hotplate. A PDMS stab with drilled via holes was then aligned and bonded to the PET film with oxygen plasma treatment. A customized PMMA holder was made to fasten the OIE chip for microscopic observation and OIE operation.

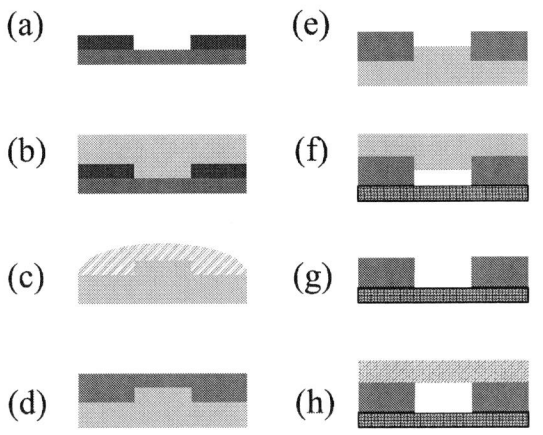

Figure 3: Device fabrication. (a) SU-8 patterned on Si by standard lithography; (b) Casted PDMS master mold; (c) Treat the surface with 5% detergent in ethanol and rinse again in ethanol; (d) Spin-coated PDMS film on PDMS master mold and cure; (e) Drill holes for electroporation zone; (f) Treat both PDMS and a-Si surface with oxygen plasma and then bond with pressure at room temperature; (g) Peel off the PDMS master mold; (h) Treat both PDMS and ITO surface with oxygen plasma and bond with pressure on a 85℃ hotplate .

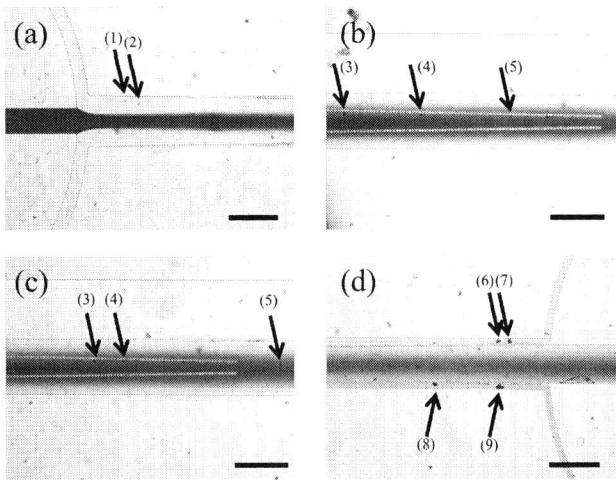

Figure 4: Medium exchange mechanism demonstrated on a PDMS-on-glass OIE chip. HEK 293T cells in DMEM were transferred into colored dye and returned to DMEM by μPAR structures. All Scale bars are 500 μm. (a) Cells (1~2) were injected from upper-left channel. (b)(c) Cells (3~5) were successfully guided from DMEM into the colored dye. (d) Cells (6~9) were transferred back into DMEM and went to outlet (to the right).

FABRICATION

The fabrication process of the device is schematically shown in Figure 3. We adopted a PDMS double casting

RESULTS AND DISCUSSION

Tests on separated modules had been first performed to ensure the successful integration of proposed

978-1-4799-3510-9/14 $31.00 © 2014 IEEE   235

device. The medium exchange mechanism was demonstrated to be capable of exchange mediums with cells, as shown in Figure 4. As shown in the figure, cells could be exchanged from the culture medium to the electroporation buffer, and then changed back to the culture medium afterwards.

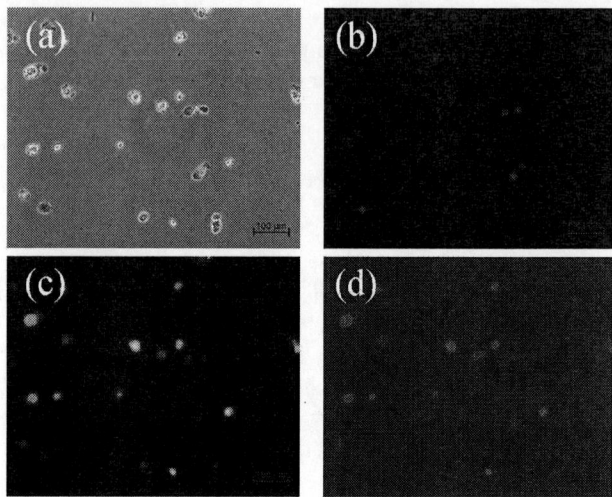

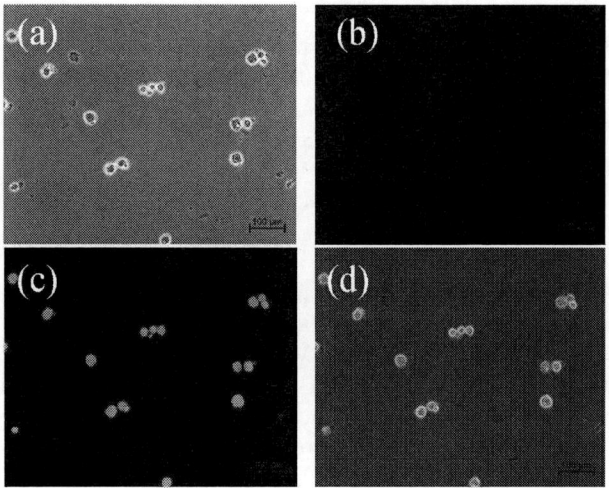

Figure 5: MEF OIE: Control Groups: nil. All Scale bars are 100 μm. (a) Bright field image; (b) Propidium iodide stained image; (c) Calcein Acetoxymethyl Ester stained image; (d) Merged image.

To evaluate the efficiency of OIE, propidium iodide (PI) and Calcein Acetoxymethyl Ester (CaAM) staining were conducted in the experiment. PI is a nucleic acid staining dye which is membrane impermeant. In the presence of DNA, the dye will bind to the nucleic acids and exhibit strong fluorescent red color. Therefore, successfully electroporated cells will uptake the PI and exhibit a strong red fluorescent signature. The viability of the cells is evaluated by CaAM. The dye is membrane permeable and will diffuse across the cell membrane. Once the dye is inside the cell, the CaAM molecule will be broken down by the enzymes in the cell cytosol, producing a membrane impermeable fluorescent green product. Under proper OIE conditions, the successfully electroporated cell will exhibit both red and green fluorescent colors, indicating the PI molecules were transferred into the cell without compromising the cell viability. In this study, propidium iodide (PI) dye has been successfully transferred into the MEF cells by the developed OIE device, while a light spot was illuminated on the cells. In Figure 5, as a negative control group of the experiment, the MEF cells were not treated by OIE and no PI dye was mixed into the solution. The CaAM Staining shown in Figure 5(c), showing that the cells exhibiting the light green color were alive.

Figure 6: MEF OIE: Control Groups: PI dye mixture. All Scale bars are 100 μm. (a) Bright field image; (b) Propidium iodide stained image; (c) Calcein Acetoxymethyl Ester stained image; (d) Merged image.

For another negative control group, MEF cells and PI dye were mixed, as shown in Figure 6. Still, there was no OIE treatment in this group. There were a few PI signals within the cells, suggesting that the PI was able to slightly penetrate through the cell membrane. The signals were due to the small population of dead cells in the solution.

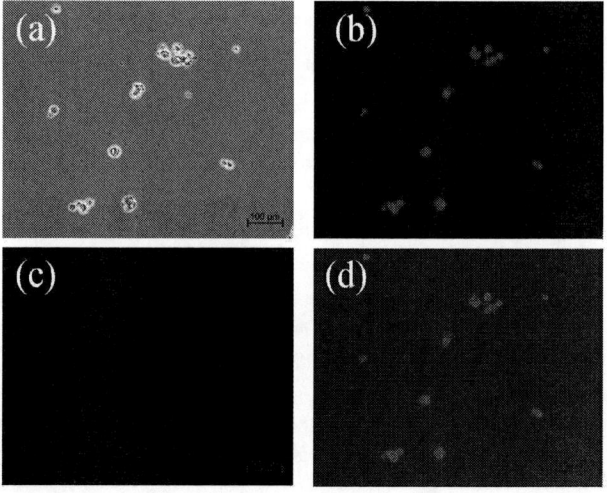

Figure 7: MEF OIE: Control Groups: PI dye and 95% Ethanol. All Scale bars are 100 μm. (a) Bright field image; (b) Propidium iodide stained image; (c) Calcein Acetoxymethyl Ester stained image; (d) Merged image.

For the positive control group, cells and PI dye, 95% ethanol were mixed and observed, as shown in Figure 7. The result shows that the PI dye has successfully penetrated into the cell membranes, as shown in Figure 7(b). However, the viability tests using CaAM show that the survival rate was significantly lower than the previous negative control groups.

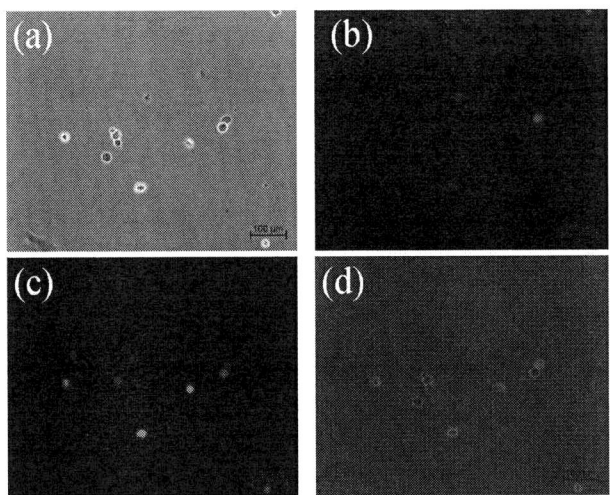

Figure 8: MEF OIE: samples. All Scale bars are 100 μm. (a) Bright field image; (b) Propidium iodide stained image; (c) Calcein Acetoxymethyl Ester stained image; (d) Merged image.

For the sample group, the MEF cells were mixed with PI dye solution and treated with OIE, as shown in Figure 8. The PI dye was successfully introduced into the cells by OIE (Figure 8(b)). The viability of the cells was evaluated by the CaAM staining (Figure 8(c)).

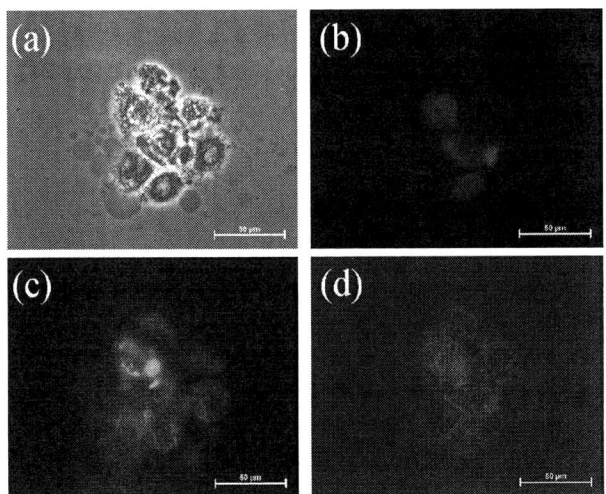

Figure 9: MEF OIE: Samples. All Scale bars are 50 μm. (a) Bright field image; (b) Propidium iodide stained image; (c) Calcein Acetoxymethyl Ester stained image; (d) Merged image.

Figures 9(a) to 9(d) were the closed-up pictures of the cells after OIE treatment. The merged image shows that the cells were not only successfully electroporated but also stayed alive after the OIE treatment (Figure 9(d)).

Table1: OIE electroporation rate and cell survival rate of each experiment.

	PI(+)
Nil	N.D
PI stained during OIE	27.5%
PI mixed (w/o OIE)	14.9%
95% Ethanol + PI stained	76.1%

The electroporation rate and the survival rate of the cells evaluated by PI and CaAM staining were summarized in Table 1. The cell electroporation occurred during the OIE treatment, resulting in the increase of the percentage of PI signals in the cells from 14.9% to 27.5%.

CONCLUSIONS

We presented a new design of a continuous optically-induced electroporation (OIE) microfluidic device with on-chip medium exchange functionality. The preliminary data showed that it is promising to realize a continuous medium exchanging and optically-induced electroporation device in a single microfluidic chip. And the system shows great potential in various electroporation related applications such as iPS cells, gene therapy etc.

ACKNOWLEDGEMENTS

The authors would like to thank the financial support from National Science Council in Taiwan.

REFERENCES

[1] S. Movahed, D. Li, "Microfluidics cell electroporation," *Microfluidics and Nanofluidics,* vol. 10, pp. 703-734, 2011.

[2] J. K. Valley, S. Neale, H.-Y. Hsu, A. T. Ohta, A. Jamshidi and M. C. Wu, "Parallel single-cell light-induced electroporation and dielectrophoretic manipulation," *Lab Chip*, vol. 9, Iss. 12, pp. 1714-1720, 2009.

[3] H.-T. Kuo, Y.-H. Lee, C.-H. Wang, C.-M. Chang and G.-B. Lee, "An optical-induced platform for gene transfection," in *Digest Tech. Papers MicroTAS'12 Conference*, Okinawa, October 28-November 12, 2012, pp. 1657-1659.

[4] R. D. Sochol, S. Li, L. P. Leebc and L. Lin, "Continuous flow multi-stage microfluidic reactors via hydrodynamic microparticle railing," *Lab Chip*, vol. 12, Iss. 20, pp. 4168-4177, 2009.

[5] L. Gitlin, P. Schulze and D. Belder, "Rapid replication of master structures by double casting with PDMS," *Lab Chip*, vol. 9, Iss. 20, pp. 3000-3002, 2009.

CONTACT

*Professor Gwo-Bin Lee, Tel: + 886-3-5715131-33765; E-mail: gwobin@pme.nthu.edu.tw

A HIGH-THROUGHPUT PERMEABILITY ASSAY PLATFORM FOR SHEAR STRESS CHARACTERIZATION OF ENDOTHELIAL CELLS

Ross Booth, Seungbeom Noh, and Hanseup Kim
University of Utah, Salt Lake City, UT, USA
Department of Electrical & Computer Engineering

ABSTRACT

We characterized the first high-throughput permeability assay platform enabling compound permeability assays, at the full spectrum (1-60dyn/cm^2) of shear stress, on endothelial cells. The platform comprises four parallel channels, with a porous membrane bonded between layers enabling permeability assays under four shear stresses per chip, ranging ~15x in magnitude. In the bEnd.3 brain endothelial cell line, decreased permeability was observed at rates of 4.06e^{-8} and 6.04e^{-8}cm/s per unit shear stress (dyn/cm^2) for FITC-Dextran and propidium iodide, respectively. Image analysis of cell stains indicated increased elongation and cell alignment with shear stress at rates of 9.15e^{-4} and 0.12° per dyn/cm^2, respectively.

INTRODUCTION

The study of Endothelial cells (ECs) is very important to understanding many diseases and their treatments, because their functional and morphological properties have significant affect on both pathology and treatment [1]. These properties are highly dependent on their vascular *in vivo* microenvironment, where they are exposed to a wide range of fluidic shear stress (1-60 dyn/cm^2) [2], resulting from significant variations in vessel geometries. It is known that shear stress causes a mechano-transductive response in ECs, inducing various phenotypic changes in protein expression and morphology relevant to barrier properties of the EC monolayer [3]. However, the quantitative relationship between EC physiological and morphological traits and the full range of *in vivo* shear stresses has not yet been effectively elucidated by existing *in vitro* models.

Microfluidic *in vitro* models, which are advantageous over *in vivo* models because they can generate well-defined shear stress in a controllable and repeatable manner, have been used to study shear stress effects on morphology [4] and permeability [5], but no previous system has managed to simultaneously induce the full-spectrum range of shear stress and investigate effects on permeability. Microfluidic systems have induced a shear stress range with geometric variation [6], but the non-uniform shear stress applied to the cell populations make such channels unfeasible for non-imaging measurement methods. Though recently a closed-loop braille-display device has applied distinctly different WSS to three parallel channels [7], it covered only a limited shear stress range of <12dyn/cm^2 due to limited flow rates, and did not allow trans-monolayer assays. In short, a parallel array structure with an integrated porous membrane is needed to allow trans-monolayer assays of shear stress effects in a high-throughput manner.

We previously introduced a parallel-array structure [8] allowing trans-endothelial electrical resistance (TEER) monitoring of microfluidic blood-brain barrier (μBBB) [9,10] cultures at various shear stress, but permeability assays were unfeasible due to the single I/O design. In order to resolve such issues, a high-throughput device allowing permeability assays at multiple simultaneous shear stress is necessary for rapid elucidation of shear stress effects. The presented structure (Fig. 1), comprised of 4 parallel channels of varying widths, enables application of simultaneous shear stress variation up to ~15x with independent outlets and allows testing of permeability at discrete shear stress rates in a rapid turn-around. First, in order to evaluate the effectiveness of the platform to produce the full range of shear stress, quantitative testing was done with (1) COMSOL simulation and (2) by experimental measurements with an integrated micro-flow sensor array. Second, the validity of the platform was tested on the brain microvascular endothelial cell line bEnd.3 (3) by assaying trans-monolayer permeability and (4) cell morphology (cell elongation and orientation).

Endothelial cells naturally experience a wide range of shear stress *in vivo* (~1-60 dyn/cm^2)

↓

Endothelial cells respond to shear stress through mechanotransduction, having effects on barrier properties

↓

To characterize these effects at the full range of shear stress *in vitro*, a high-throughput model is needed which allows permeability assays in a ***Parallel Array***

- Parallel channels from a common inlet have 4 distinct flowrates

Horizontal Cross Section of Parallel Array

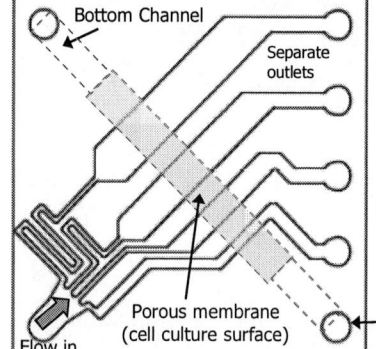

- *Fastest channel shear stress is approximately ~15x the slowest*

- *Thus the full physiological spectrum is testable in two experiments.*

Bottom channel crosses all 4 top channels, allowing permeability measurement

Figure 1: Multiple-outlet parallel array structure allows high-throughput assays of tracer permeability through endothelial cells at the full-spectrum of physiologically relevant shear stress.

METHODS

Structure & Fabrication

The parallel array platform (Fig.2A) consists of four top flow channels (0.73, 1.53, 2.33, 3.13mm widths) and one bottom flow channel (2mm), with a free-standing membrane culture surface. Channel heights were 200µm. To directly measure the shear stress in the channel, a 10µm-wide micro-flow sensor array was fixed in each channel based on the thermal conductivity detector configuration [11], with a fixed position at 70µm above the bottom wall.

The microfluidic device was fabricated as previously [2] with some modifications. Channel molds were lithographically constructed from SU-8 2075 (200µm thick) on a silicon substrate, and PDMS was cast (10:1 elastomer:curing agent) and heat-cured (110°C, 30m). High bond strength was achieved by bonding the polycarbonate membrane between the poly-dimethylsiloxane (PDMS) channel layers (200µm) with O_2 plasma following membrane treatment with 5% 3-aminopropyltriethoxysilane (APTES) at 80°C (30m) [12]. This bonding method significantly decreased occurrences of leaks under high flows compared to the standard method [13], with no observed losses of cell adhesion.

The micro-flow sensor array structure was fabricated as previously [11] with some modifications (Fig. 2B). LPCVD (1µm) silicon nitride was deposited on a silicon substrate, followed by sputter-deposition of a 200nm/10nm Pt/Ti layer, which was patterned to form sensor signal feed-through lines to the electrical pads. The metal was then deposited and patterned with a PECVD silicon nitride passivation layer (450nm), and etched to define the sensor structure with RIE, followed by DRIE and Xactive XeF_2 etching to freely suspend the sensor over the 70µm channel structure. The substrate was bonded to a 130µm PDMS channel structure, completing the final platform structure.

Simulation & Testing

The structure was analyzed with COMSOL simulations to predict the velocity fields and resultant shear stress distribution in the parallel array structure. The simulation used the following equation to calculate the shear stress

$$\tau = \mu \frac{dU}{dz} \quad (1)$$

where τ is shear stress at the wall, μ is dynamic viscosity, and dU/dz is the velocity gradient adjacent to the wall.

For fluidic testing, chips were fitted with marprene or silicone tubing and manipulated with a peristaltic pump. To calibrate the flow sensor, flow was reversed through each outlet to calibrate voltage/velocity relationship by applying 5V to a Wheatstone Bridge and measuring output voltage. Under forward flow, the output voltages were measured and fit to the calibration curves to find the uniform velocity. For correlation with simulations, the flow sensor flow measurements were used to calculate average shear stress τ_A

$$\tau_A = \frac{6Q\mu}{h^2 w} \quad (2)$$

where Q is the flow rate, h is the channel height, and w is the channel width.

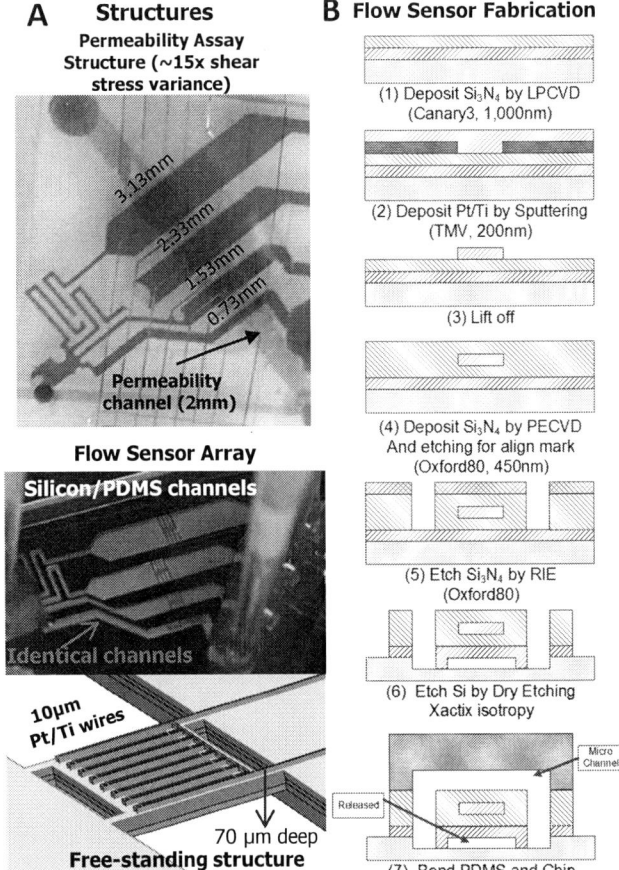

Figure 2: (A) Permeability structure and flow sensor structures were fabricated with identical channel design. (B) Sensor fabrication process for micro flow sensor array.

Sterilized culture devices (with 70% ethanol) were coated with Fibronectin and Collagen IV (100µg/mL each) overnight to facilitate cell adhesion, and b.End3 cells were seeded ($6e^4/cm^2$) for two hours. Media used was DMEM/F12 with 5% FBS, changed daily. Following three days at near-static flow, experimental shear stress was applied for 24h.

Permeability of fluorescent tracers FITC-dextran (4kD) and propidium iodide (PI) was measured in each channel by flowing tracer from each outlet and measuring tracer flux to the bottom channel. The permeability coefficients P were calculated using the following equation [14].

$$P = \frac{\Delta \cdot C_B \cdot V_S}{\Delta t \cdot A \cdot C_T} \quad (3)$$

where ΔC_B is the concentration change in the bottom, Δt is the assay time, V_S is the bottom perfusate sample volume, A is the membrane area, and C_T is the top concentration.

Following fixation with 4% paraformaldehyde and permeabilization with 0.1% Triton X-100, hematoxylin and eosin stain was applied to visualize cell dimensions. Cell orientation (0-90°) and shape index (0-1) were analyzed using CellProfiler software. Shape index SI is defined as

$$SI = \frac{4\pi A}{p^2} \quad (4)$$

where A is the cell area and P is the cell perimeter.

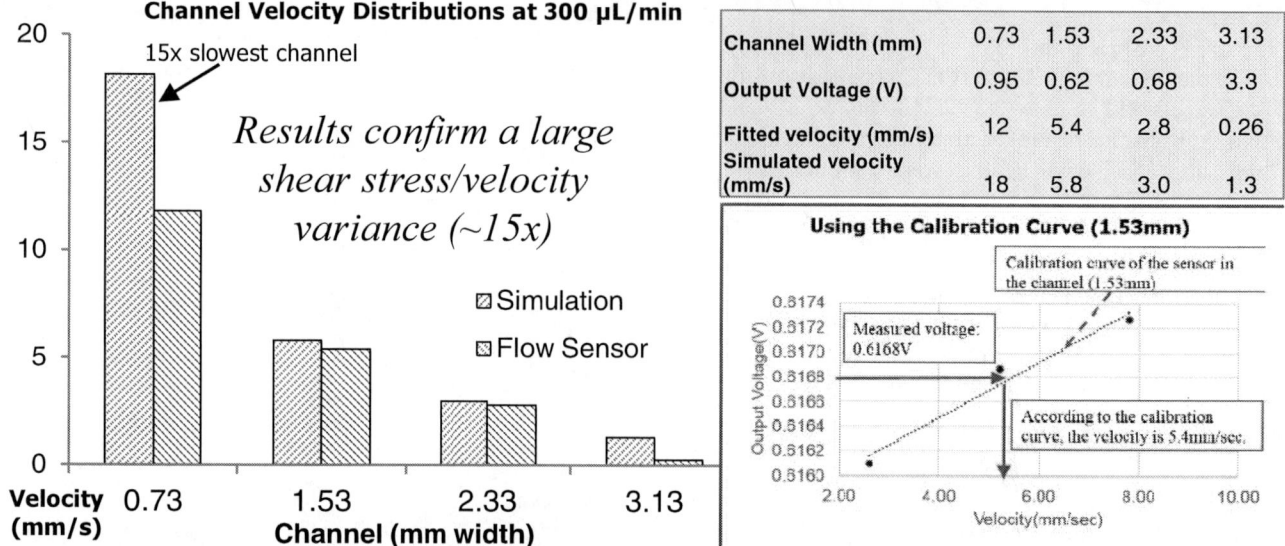

Channel Width (mm)	0.73	1.53	2.33	3.13
Output Voltage (V)	0.95	0.62	0.68	3.3
Fitted velocity (mm/s)	12	5.4	2.8	0.26
Simulated velocity (mm/s)	18	5.8	3.0	1.3

Figure 3: Distribution of velocity at 300 μL/min input flow. Results are compared for uniform velocity predicted by COMSOL simulation and measured by A micro-flow sensor array (70μm height). Also shown is a table of measured/simulated values and an example calibration curve (for 1.53mm channel). COMSOL simulations indicated a high on-chip shear stress range (~15x according to simulation), with close correlation for the two central channels.

RESULTS & DISCUSSION

Simulation results showed that distribution of velocity (Fig. 3) and resultant shear stress achieved a span ratio of ~15x relatively between the fastest and slowest channels. Velocities/shear stress matched within 10% error for the two center channels, indicating the validity of the COMSOL results, though the discrepancy became larger for the smallest (0.73mm) and largest (3.13mm) channel sizes by 22% and 66% of the simulated values, respectively. The discrepancy is under investigation. COMSOL-derived values for shear stress (eq. 1) were used for correlation of assay data.

The permeability results (Fig. 4) indicated decreased permeability with increasing shear stress for both tracers at a rate of $4.06e^{-8}$ and $6.04e^{-8}$cm/s per dyn/cm^2 for FITC-dextran and PI, respectively, ranging from <1dyn/cm^2 ($7.4e^{-6}$ and $2.3e^{-5}$cm/s) to 86dyn/cm^2 ($4.0e^{-6}$ and $1.9e^{-5}$cm/s). These decreases indicate an increase in barrier strength with shear stress. A notable reduction in standard deviation was observed at WSS above 20 dyn/cm^2. The permeability was consistently higher for propidium iodide than for FITC-dextran, as expected due to their differences in molecular weight (668D and 4kD).

The optical measurement results (Fig. 5) on cell shapes showed that bEnd.3 cells tend to elongate more at a shape index change rate of $9.15e^{-4}$ per dyn/cm^2 and align closer to the flow direction at a rate of 0.12° per dyn/cm^2, as WSS increases from static, (0 dyn/cm^2), to 86dyn/cm^2. The total shape index reduced, which indicates cell elongation, from 0.28 to 0.18, by a shape index change of 35.7%. The measurement results also showed that the average cell alignment angle from the flow direction also reduced from 47.9° to 39.9° by 16.7%. These results clearly indicate that shear stress has notable impacts on cell morphology, with increases in elongation and alignment with flow.

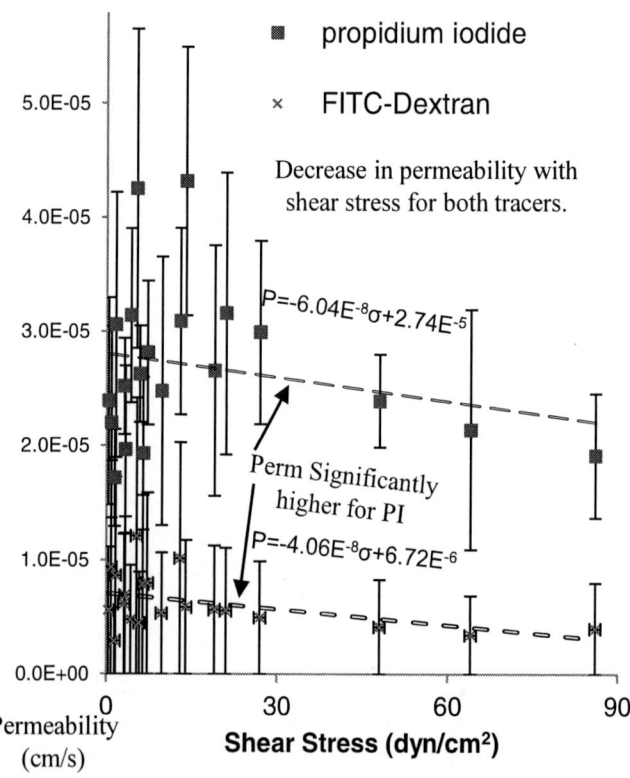

Figure 4: Permeability Analysis. Tracer permeabilities of FITC-conjugated Dextran (4kD) and propidium iodide were assayed through each channel following shear stress exposure across the full physiologically relevant spectrum.

978-1-4799-3510-9/14 $31.00 © 2014 IEEE

CONCLUSIONS

This study confirmed the production of well-defined shear stress at a wide on-chip range (~15x) with (1) velocity field simulation by COMSOL and (2) direct measurement with an integrated micro-flow sensor array in each channel. Measurement results also validated the effects of shear stress on the barrier and morphological characteristics of the bEnd.3 cell line by measurement of (3) permeability to tracers FITC-dextran 4kD and PI, as well as (4) cell morphology (cell elongation and orientation). With only a few iterations at different input flow rates, the testing conditions in this study exceeded the range of the physiologically relevant shear stress spectrum (1-60dyn/cm^2) for analyzing the quantitative effect on the bEnd.3 cells. Morphometric image analysis revealed increased elongation and cell alignment with increases in WSS at rates of 9.15e^{-4} and 0.12° per unit WSS (dyn/cm^2), respectively. Decreased permeability with increasing WSS was observed at rates of 4.06e^{-8} and 6.04e^{-8}cm/s per dyn/cm^2 for FITC-conjugated Dextran and propidium iodide, respectively. These results indicate that the bEnd.3 cell line responds to shear stress *in vitro* in a magnitude-dependent manner, providing practical insight into optimal flow conditions for dynamic EC culture models at the particular tested combination of cell type and culture conditions in future dynamic studies.

A

Quantifying Cell Traits

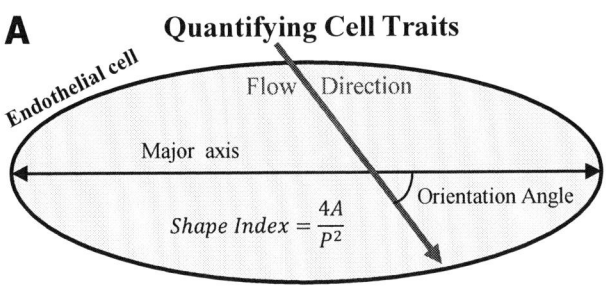

B

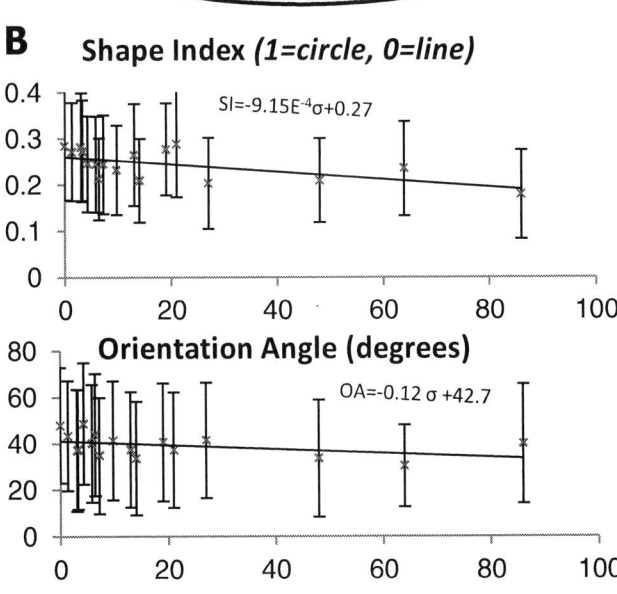

Figure 5: (A) Quantitative measurement of cell morphology to measure (B) shape index and orientation at a range of applied shear stresses in the presented structure.

REFERENCES

[1] Michiels, C., "Endothelial cell functions", *Journal of cellular physiology*, vol. 196, pp. 430-443.

[2] Loscalzo, J. and A.I. Schafer, *Thrombosis and hemorrhage*. 2003: Lippincott Williams & Wilkins.

[3] Siddharthan, V., et al., "Human astrocytes/astrocyte-conditioned medium and shear stress enhance the barrier properties of human brain microvascular endothelial cells", *Brain Res*, vol. 1147, pp. 39-50.

[4] van der Meer, A.D., et al., "Analyzing shear stress-induced alignment of actin filaments in endothelial cells with a microfluidic assay", *Biomicrofluidics*, vol. 4, pp. 11103.

[5] Shao, J., et al., "A microfluidic chip for permeability assays of endothelial monolayer", *Biomed Microdevices*, vol. 12, pp. 81-8.

[6] Wang, J., J. Heo, and S.Z. Hua, "Spatially resolved shear distribution in microfluidic chip for studying force transduction mechanisms in cells", *Lab Chip*, vol. 10, pp. 235-9.

[7] Song, J.W., et al., "Computer-controlled microcirculatory support system for endothelial cell culture and shearing", *Anal Chem*, vol. 77, pp. 3993-9.

[8] Booth, R.H. and H. Kim, "A Parallel Array Microfluidic Blood-Brain Barrier Model For High-Throughput Quantitation Of Shear Stress Effects", *MicroTAS Conference*, vol. pp. 491-493.

[9] Booth, R. and H. Kim, "Characterization of a microfluidic in vitro model of the blood-brain barrier (μBBB)", *Lab on a Chip*, vol. 12, pp. 1784-1792.

[10] Booth, R. and H. Kim, "A Multi-Layered Microfluidic Device For *In Vitro* Blood-Brain Barrier Permeability Studies", *MicroTAS Conference*, vol. pp. 1388-1390.

[11] Kuo, J.T., et al., "A microfluidic platform with integrated flow sensing for focal chemical stimulation of cells and tissue", *Sensors and Actuators B: Chemical*, vol. 152, pp. 267-276.

[12] Aran, K., et al., "Irreversible, direct bonding of nanoporous polymer membranes to PDMS or glass microdevices", *Lab Chip*, vol. 10, pp. 548-52.

[13] Chueh, B.H., et al., "Leakage-free bonding of porous membranes into layered microfluidic array systems", *Anal Chem*, vol. 79, pp. 3504-8.

[14] Yuan, W., et al., "Effect of surface charge of immortalized mouse cerebral endothelial cell monolayer on transport of charged solutes", *Ann Biomed Eng*, vol. 38, pp. 1463-72.

CONTACT

*R. Booth, tel: +1-208-8615597; Ross.Booth@utah.edu
*H. Kim, tel: +1-801-5879497; hanseup.kim@utah.edu

A MICROFLUIDIC DEVICE FOR ISOLATION OF CELL-TARGETING APTAMERS

Jing Zhu[1], Tim Olsen[1], Renjun Pei[3], Milan Stojanovic[2], and Qiao Lin[1]
[1]Department of Mechanical Engineering, [2]Department of Medicine,
Columbia University, New York, NY, USA
[3]Suzhou Institute of Nano-Tech and Nano-Bionics, Chinese Academy of Sciences, Suzhou, China

ABSTRACT

This paper presents a microfluidic device for affinity selection and amplification of cell membrane protein-binding strands from a randomized single-strand DNA (ssDNA) oligomer library, thereby isolating specific cell-targeting aptamers. The device consists of the selection and amplification microchambers situated on a temperature control chip. Affinity selection, integrated with cell culturing, of cell-binding ssDNA is performed in the selection chamber; the selected strands are then amplified by bead-based polymerase chain reaction (PCR) in the amplification chamber. Transfer between the selection and amplification microchambers using pressure-driven flow realizes multi-round aptamer isolation on a single chip. Experimental results demonstrate the feasibility of using this device to develop aptamers that specifically bind to target cells.

KEYWORDS

Aptamer, Cell Targeting, Cell Culture, SELEX, Selection, Amplification

INTRODUCTION

Aptamers are oligonucleotides that recognize biological targets by specific affinity binding and are synthetically isolated from a randomized oligonucleotide library via systematic evolution of ligands by exponential enrichment (SELEX). Such molecules, in particular cell-targeting aptamers (i.e., aptamers that bind to specific cell membrane proteins), have attractive applications in clinical diagnostics and therapeutics [1,2]. Conventional SELEX platforms for developing cell-targeting aptamers are labor-, time- and resource-intensive [3]. Microfluidic technology has been employed to improve the SELEX efficiency via process integration, but early attempts mostly focused on molecular targets. Limited work reported on microfluidic isolation of cell-targeting aptamers (cell-SELEX) has used bead-based cell immobilization in open wells [4] or electrokinetic DNA transfer [5], which may have drawbacks such as inefficient DNA manipulation and/or potential loss of cell viability during selection. We address these issues by presenting a microfluidic device incorporating hydrodynamic single strand DNA (ssDNA) transfer for cell-targeting aptamer development. The device integrates cell culturing with affinity selection of cell-binding ssDNA, and couples selection and bead-based amplification of the cell-binding ssDNA using pressure-driven flow to realize multi-round isolation of cell-targeting aptamers on a single chip.

PRINCIPLE AND DESIGN

Our microfluidic cell-targeting aptamer development approach integrates cell culture, affinity selection of cell-binding ssDNA, and bead-based polymerase chain reaction (PCR) into a single microfluidic device. Cells are first cultured in the selection chamber for a sufficiently long time to ensure cell attachment and surface biomarker regeneration. Affinity selection of ssDNA is then performed by introducing a random ssDNA library into the chamber (Fig. 1A), followed by multiple washes to remove weakly bound ssDNA (Fig. 1B). Next, primer-functionalized magnetic beads are introduced and held in the amplification chamber via external magnet. The remaining ssDNA that strongly bind to membrane proteins on cells are thermally eluted (Fig. 1C), hydrodynamically transferred to the amplification chamber (Fig. 1D), captured by the surface immobilized primers (Fig. 1E), and amplified via bead-based PCR (Fig. 1F). Afterward, the ssDNA strands are released from the bead surfaces (Fig. 1G) and are transported back to the selection chamber (Fig. 1H) as the process is repeated.

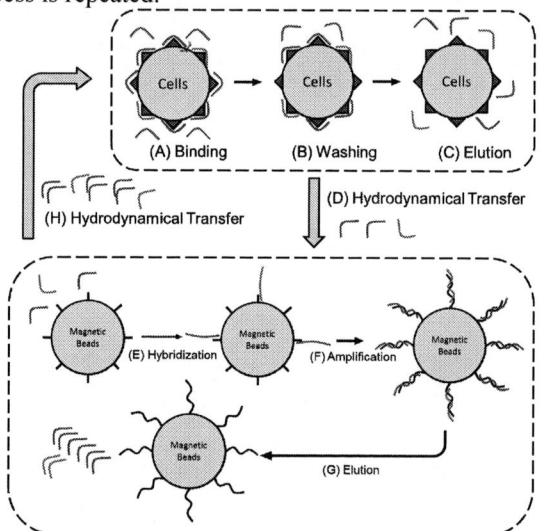

Figure 1: Principle of microfluidic aptamer development: (A) ssDNA with random sequence binds to cell membrane protein in the selection chamber; (B) weak binders are removed by washing; (C) strong binders are eluted and (D) transferred to the amplification chamber; (E) the strands are captured by magnetic beads with surface-immobilized reverse primers and (F) amplified through PCR; (G) the amplified single strands are released from bead surfaces and (H) transported back to the selection chamber for the next round.

The microfluidic device consists of two microchambers situated on a temperature control chip for selection and amplification (Fig. 2A). The surfaces of both microchambers (30 μm in height) with an approximately 1.5 μL volume are coated with Parylene C to prevent evaporative loss of reactants. A permanent magnet is placed beneath the amplification chamber to retain the streptavidin-coated magnetic beads (2.8 μm in diameter). The two channels that connect the microchambers have semi-circular profiles (20 μm in height) that can be completely sealed by elastomeric microvalves actuated by pressure-driven, oil-filled channels (30 μm in height) above (Fig. 2B). Integrated on the temperature control chip are two groups of temperature control units, each containing a serpentine-sharped resistive temperature sensor (linewidth: 25 μm) and a serpentine-shaped heater (linewidth: 300 μm) beneath the center of selection or amplification microchamber. The chamber temperatures can hence be controlled in a closed loop separately using corresponding integrated temperature sensor and heater.

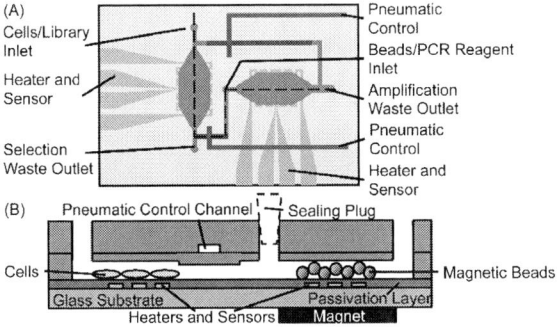

Figure 2: (A) Top view and (B) cross-sectional view along the dash line of the aptamer development microfluidic device.

EXPERIMENTAL

Materials

All chemicals were purchased from Sigma-Aldrich (St. Louis, MO) unless otherwise indicated. ssDNA random library (5' – GCC TGT TGT GAG CCT CCT GTC GAA – 40N – TTG AGC GTT TAT TCT TGT CTC CC – 3') and primers (Forward Primer: 5' – FAM – GCC TGT TGT GAG CCT CCT GTC GAA -3', and Reverse Primer: 5' – dual biotin – GG GAG ACA AGA ATA AAC GCT CAA – 3') were synthesized and purified by Integrated DNA Technologies (Coralville, IA). Deoxyribonucleotide triphosphates (dNTPs) and GoTaq Flexi DNA polymerase were obtained from Promega Corp. (Madison, WI). Minimum Essential Medium (MEM), fetal bovine serum (FBS), penicillin-streptomycin (P/S, penicillin 10,000 unit/mL, streptomycin 10,000 μg/mL), Dulbecco's phosphate-buffered saline (D-PBS), 0.25% Trypsin-EDTA and streptavidin coupled magnetic beads (Dynabeads® M-270 Streptavidin) were purchased from Invitrogen (Carlsbad, CA). MCF-7 cell line was obtained from the American Type Culture Collection (ATCC, Manassas, VA).

Microchip Fabrication

The temperature control chip was fabricated using standard microfabrication techniques. Briefly, gold (100 nm) and chrome (5 nm) thin films were thermally evaporated onto the glass substrate, patterned by photolithography and wet etched, which resulted in resistive temperature sensors and resistive heaters. Then, 1 μm of silicon dioxide was deposited using plasma-enhanced chemical vapor deposition (PECVD) to passivate the sensors and heaters, the contact regions for electrical connections to which were opened by etching the oxide layer using hydrofluoric acid (Fig. 3A).

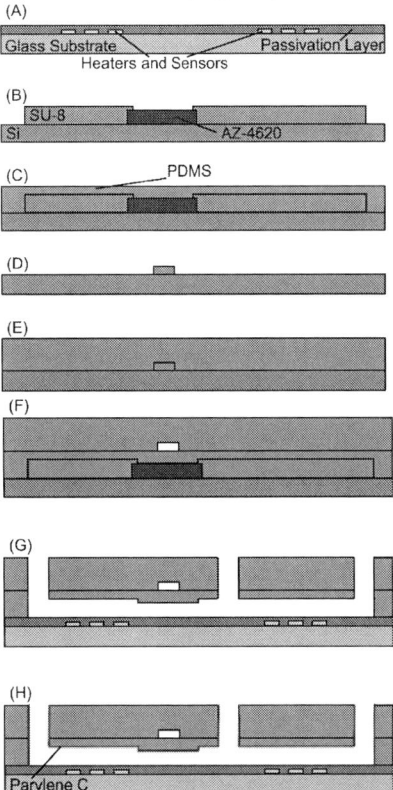

Figure 3: (A) Deposition, patterning and passivation of gold sensors and heaters. (B) Fabrication of fluidic channel mold including on-chip valves and microfluidic channels using AZ-4620 and SU-8. (C) PDMS spin-coating. (D) Fabrication of SU-8 mold for pneumatic controlled oil-filled valve actuation channels. (E) Casting of PDMS microfluidic channels. (F) Bonding of PDMS slab for pneumatic controlled oil-filled valve actuation channels to PDMS-coated mold for microfluidic channels. (G) Peeling off and bonding of PDMS sheet containing fluidic channels and pneumatic controlled oil-filled valve actuation channels to temperature control chip. (H) Deposition of Parylene C.

Subsequently, the microfluidic slab bearing microfluidic and pneumatic features was fabricated from PDMS using soft lithography techniques. A layer of AZ-4620 positive photoresist (20 μm, Clariant Corp. Somerville, NJ) was spin-coated on a silicon wafer (Silicon Quest International, Inc., San Jose, CA), exposed to ultraviolet light through photomasks, developed, and baked to form round-shaped flow channels that can be sealed completely. Then, a layer of SU-8 photoresist was patterned to finalize the mold defining microfluidic features (Fig. 3B).

Next, a PDMS prepolymer solution (base and curing agent mixed in a 10:1 ratio) was spin-coated onto the silicon wafer, and cured on a hotplate at 72 °C for 15 min (Fig. 3C). In parallel, a layer of SU-8 photoresist was patterned on another silicon wafer to establish pneumatically controlled oil-filled valve actuation channels (Fig. 3D). Another PDMS prepolymer solution was cast onto the mold and cured on a hotplate at 72 °C for 1 hour (Fig. 3E). The resulting PDMS slab was peeled off from the mold, punched to form pneumatic inlets, and bonded to the PDMS membrane on the silicon mold bearing the microfluidic features (Fig. 3F). Afterwards, the PDMS slab along with the thin PDMS membrane was peeled off, punched to establish fluidic inlets and outlets, and then bonded onto the temperature control chip irreversibly after another oxygen plasma treatment (Fig. 3G). Finally, the surface of SBE microchanmber was coated with a thin layer of Parylene C via chemical vapor deposition (Fig. 3H). A fabricated device is shown in Fig. 4.

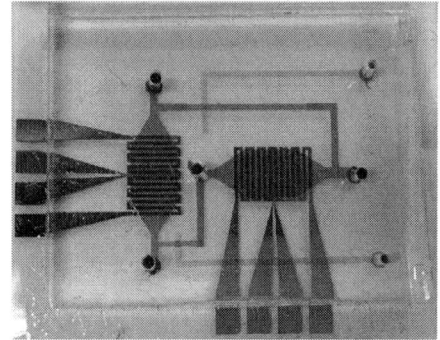

Figure 4: Image of a fabricated device.

Cell Culture and Preparation

MCF-7 cells were incubated with the complete culture media, including MEM supplemented with 10% FBS and 1% P/S, and were kept at 37 °C in a humidified incubator containing 5% CO2. Before microfluidic experiments, cells were treated in 3mL of trypsin-EDTA for 5 min to detach them from the substrate, and a triple volume of the complete culture media was added to stop trypsinization. Cells were then collected through centrifugation, resuspended at 1×10^8 cells/mL in complete culture media.

Experimental Procedure

Development of aptamers in the microfluidic device starts from culturing cells in the selection chamber for a sufficiently long time (>4 hours) to ensure cell attachment and surface biomarker regeneration. Selection of ssDNA is then performed by infusing a random ssDNA library through the chamber (1 μL/min), followed by multiple washes (10 μL/min) to remove weakly bound ssDNA. Next, primer-functionalized magnetic beads are introduced and held in the amplification chamber by an external magnet. The remaining strongly bound ssDNA are thermally eluted, hydrodynamically transferred to the amplification chamber (1 μL/min), and amplified via bead-based PCR. Afterward, the ssDNA are released from the bead surfaces at 95 °C and are transported back to the selection chamber (1 μL/min) as the process is repeated.

RESULTS AND DISCUSSION

Characterization of Cell-Binding ssDNA Selection

We first investigated the isolation of cell-binding ssDNA from a random library in the selection chamber. The temperature of selection chamber was kept at 37 °C for the whole procedure by using the temperature control unit beneath. 100 pmol of ssDNA library in 20 μL binding buffer was infused into the chamber at 1 μL/min for 20 min. Then, cells were washed with 9 aliquots of washing buffer at 10 μL/min, each for 3 min to remove undesired ssDNA. The eluents were amplified and analyzed with agarose gel electrophoresis (Fig. 5A). As the cells were rinsed, the amount of weakly bound ssDNA in each washing waste (identically amplified) declined, indicated by the monotonically decreased band intensities of lanes W1 to W9. In addition, the low band intensity of lane W9 suggests that nearly no ssDNA existed in the waste of final wash (Fig. 5B).

To demonstrate the thermal elution of strongly bound ssDNA after washing, the microchamber temperature was raised to 60 °C using the same temperature control unit. The cells were then rinsed with washing buffer (10 μL) at 1 μL/min. The high band intensity of Lane E indicates successful enrichment of strongly bound ssDNA (Fig. 5A&B).

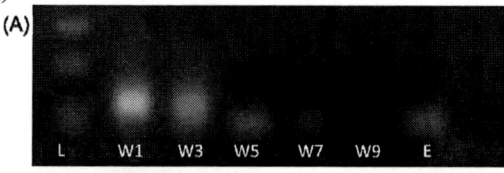

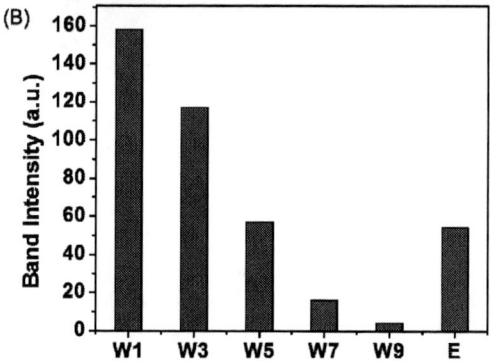

Figure 5. (A) Gel electropherogram of amplified eluents obtained during selection process. (B) Bar graph indicating band intensities for lanes W1-E. Lane L: Ladder; Lane W1: wash 1; Lane W3: wash 3; Lane W5: wash 5; Lane W7: wash 7; Lane W9: wash 9; Lane E: thermal elution.

Characterization of Cell-Binding ssDNA Amplification and Elution

To verify the on chip bead-based PCR, 10 μL of thermal eluent, followed by PCR reagents, was driven to the amplification chamber with primer-coated magnetic beads at 1 μL/min and at room temperature, and was then thermally cycled. The fluorescent bead intensities were then measured and compared with those without thermal cycling (Fig. 6A&B). Following amplification, the fluorescent bead intensity was 10-fold higher, indicating efficient ssDNA amplification through bead-based PCR (Fig. 6D).

The efficiency of ssDNA elution from magnetic bead surfaces was tested by rinsing at 95 °C and 1 µL/min for 10 min, followed by the measurement of the fluorescent intensity. The rinsed beads (Fig. 6C) showed an intensity that was 10% of the pre-elution intensity, and was only 2.6% higher than that of pre-thermal cycling intensity (Fig. 6D), indicating a highly efficient ssDNA dehybridization and separation.

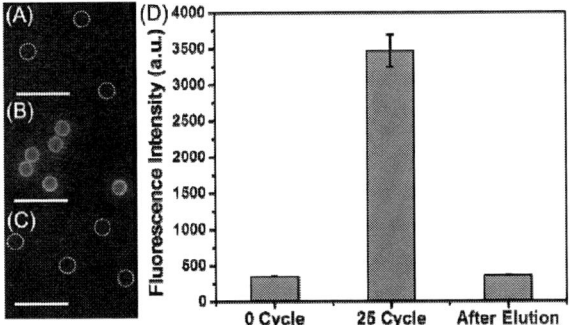

Figure 6: Verification of bead-based PCR and ssDNA elution. Fluorescent images of beads (A) before and (B) after 25 cycles of PCR, and (C) after thermally induced ssDNA elution. (D) Bar graph depicting the fluorescent intensities of the beads. Scale bar 10 µm.

Demonstration of Closed-Loop Cell-Targeting Aptamer Generation

To demonstrate the feasibility of multi-round, closed-loop cell specific aptamer generation, a three-round ssDNA selection, enrichment and amplification process was carried out. The weakly bound ssDNA of each wash for all three rounds, and the thermally eluted strongly bound ssDNA after the third round were collected from the selection chamber, amplified and analyzed with gel electrophoresis (Fig. 7A). In the first round, some ssDNA existed in the waste of the 9th wash. However, in the second and third round, there was nearly no ssDNA in the waste of the 9th wash. This indicates that the aptamer generation process was able to successfully increase the binding affinity of ssDNA pool to the target cells after each round. The high fluorescent intensity of the thermal elution lane in the gel image (Fig. 7B) suggests that ssDNA specific to the target cells were successfully isolated, enriched and amplified.

CONCLUSION

Cell-targeting aptamers are of great importance in a wide variety of fields, but their applications have been hindered by the limited availability of aptamers and labor-, time- and resource-intensive aptamer development processes. We developed a microfluidic device for synthetically isolating and enriching cell-targeting aptamers from a randomized ssDNA library. The device integrates cell culturing with affinity selection of cell-binding ssDNA, which is then amplified by bead-based PCR. Coupling of the selection and amplification processes using pressure-driven flow realizes multi-round aptamer isolation on a single chip. Experimental results have shown successful

cell-targeting ssDNA selection and amplification, as well as closed-loop multi-round iteration for aptamer generation.

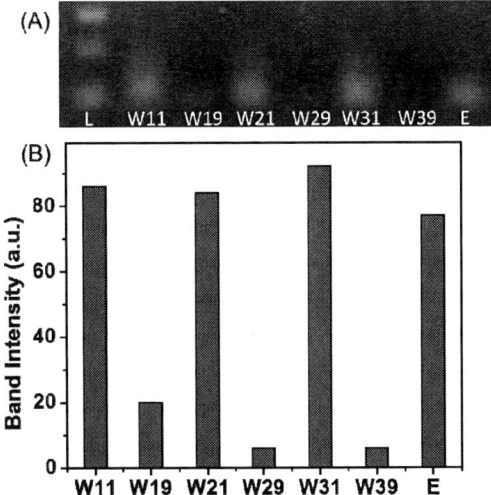

Figure 7. (A) Gel electropherogram of amplified eluents obtained during the three-round selection processes. (B) Bar graph indicating band intensities for lanes W11-E. Lane L: ladder; Lane W11: round 1, wash 1; Lane W19: round 1, wash 9; Lane W21: round 2, wash 1; Lane W29: round 2, wash 9; Lane W31: round 3, wash 1; Lane W39: round 3, wash 9; Lane E: thermal elution.

ACKNOWLEDGEMENTS

We gratefully acknowledge financial support from the National Institute of General Medical Sciences of the National Institutes of Health (grant number: 8R21GM104204), and the National Institute of Allergy and Infectious Diseases of the National Institutes of Health through the Center for High-Throughput Minimally Invasive Radiation Biodosimetry (grant number U19 AI067773).

REFERENCES

[1] E. Brody, M. Willis, et al., "The Use of Aptamers in Large Arrays for Molecular Diagnostics," *Mol Diagn*, 4: 381-8, 1999.

[2] S. Nimjee, C. Rusconi, and B. Sullenger, "Aptamers: An Emerging Class of Therapeutics," *Annu Rev Med*, 56: 555-583, 2005.

[3] M. Cho, Y. Xiao, et al., "Quantitative Selection of DNA Aptamers through Microfluidic Selection and High-Throughput Sequencing," *Proc Natl Acad Sci U S A* 107: 15373-15378, 2010.

[4] C. Weng, I. Hsieh, et al., "An Automatic Microfluidic System for Rapid Screening of Cancer Stem-Like Cell-Specific Aptamers," *Microfluid Nanofluid*, 14: 753-765, 2013.

[5] J. Kim, J. Hiltion, et al., "Electrokinetically Integrated Microfluidic Isolation and Amplification of Biomolecule- and Cell-Binding Nucleic Acids," *IEEE MEMS '13*, pp. 1007-1010, Taipei, Taiwan, 2013.

CONTACT

Jing Zhu, tel: +1-212-854-3221; jz2340@columbia.edu

A NANOWIRE-INTEGRATED MICROFLUIDIC DEVICE FOR HYDRODYNAMIC TRAPPING AND ANCHORING OF BACTERIAL CELLS

Donguk Kwon[1], Jung Kim[1], Soochan Chung[1], and Inkyu Park[1]
[1]Korea Advanced Institute of Science and Technology (KAIST), Daejeon, KOREA

ABSTRACT

In this work, we proposed a novel method for facile hydrodynamic trapping and anchoring of bacterial cells using nanowire array with fishnet-like structure in microfluidic channel. Vertically well-aligned ZnO nanowires were directly synthesized onto side walls of microslit structures by hydrothermal method to form mesh-like cage structures. We found that the mesh-like cages were effective in trapping and anchoring of *Escherichia coli* cells as model bacteria. In addition, we observed two anchoring modes; impaling and wedging, by electron microscopy and they resulted in irreversible and reversible damage to the anchored cells, respectively. We expected that the suggested bacterial cell trapping method can be used as a simple cell-manipulating platform for advanced microfluidic system.

INTRODUCTION

Trapping or retaining of biological cells at intended position is a key operation to develop an advanced microfluidic system for studying their biochemical reaction and behavior against external environment [1]. Trapping of biological targets in microfluidic device can provide controllable experimental environment for post-processes (e.g. monitoring [2], culturing [3], pairing [4], and cytotoxicity screening [5]). The trapping function also can be used as a means of detecting and analyzing platform itself [6].

Especially, in the case of bacterial cells, trapping the cells has some experimental difficulty due to the relatively small size of sub-micrometer dimension and mobility of the cells. Nevertheless, there have been many researches about the trapping of bacterial cells in microfluidic systems such as optical tweezing [7], dielectrophoretic trapping [8], magnetic trapping [9], and acoustic trapping [10]. However, these methods require expensive equipment or complicated setups to apply the trapping methods. Some researchers employed a hydrodynamic trapping method as an alternative, but this approach still has complexity in the flow control [11].

In this work, we developed nanowire-integrated microfluidic devices for facile hydrodynamic trapping and anchoring of bacterial cells and demonstrated that mesh-like cages formed by integrating ZnO nanowires are effective in trapping and anchoring of *Escherichia coli (E. coli)* as a model bacterium. Our method does not require expensive and complicate experimental instruments and delicate system control. It was shown that living bacteria were trapped both in the bare microslit structures and the nanowire-integrated structures with different efficiencies. We also observed two anchoring modes, impaling and wedging, resulted in irreversible damage or reversible deformation of the cell wall, respectively.

DESIGN AND TRAPPING MECHANISM

In Figure 1(a), the design of the microfluidic device is shown. Cell solution entered into the four parallel chambers that are identical with each other. The empty region is present before the cage arrays to make cells spread. The cage arrays were arranged in a series of zigzags as shown in Figure 1(c). Each cage has slit structures with width of 3~4 μm. A photograph of the device is shown in Figure 1(b).

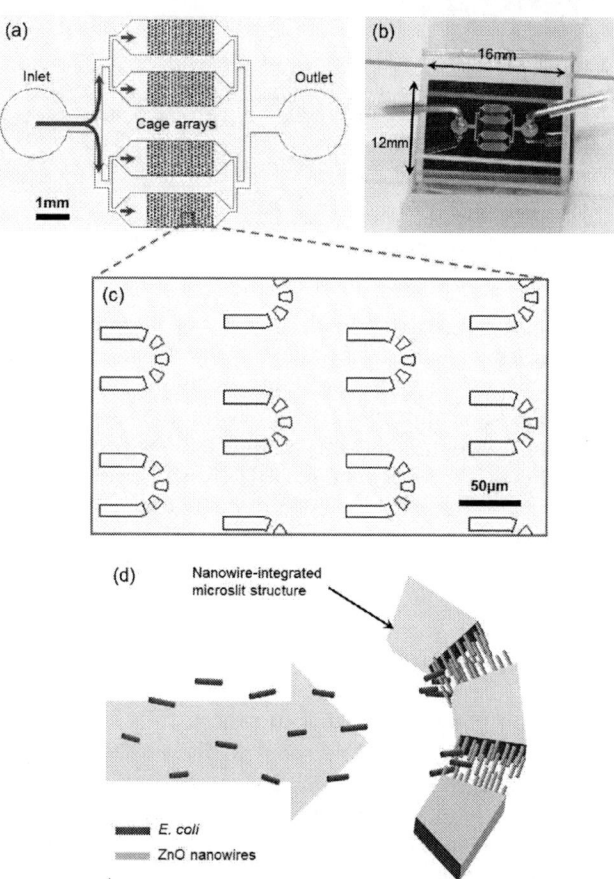

Figure 1: Design of the device and trapping mechanism. (a) Schematic illustration of the microfluidic device from top view. (b) A photograph of the microfluidic device. (c) Detailed view of cage array in a series of zigzags. (d) A schematic diagram of the mesh-like cage formed by integration of ZnO nanowires. E. coli cells are trapped at the cage while they travel along the microchannel.

A schematic diagram for trapping mechanism is described in Figure 1(d). ZnO nanowire array was employed to arrange dense nanostructures with shape of fishnet. Robust ZnO nanowires synthesized by hydrothermal method had high-aspect-ratio with diameter of 40~80 nm and length of few micrometers [12]. The ZnO nanowires were also well-aligned vertically on the substrate where they grew on. Moreover, the nanowire array was dense enough to form a fishnet for trapping of sub-micrometer sized bacterial cells. Therefore, we integrated ZnO nanowire array directly onto the microslit structure to form a mesh-like cage as shown in Figure 1. When *E. coli* cells meet the mesh-like cage during travel along the microchannel, they will be trapped at the cage.

DEVICE FABRICATION

Figure 2 shows the fabrication process of the device. First, chrome (Cr) etch mask was patterned on silicon substrate by photolithography and e-beam evaporation (Figure 2(a), (b)). Microslit structures were fabricated by deep reactive ion etching (DRIE) with the Cr etch mask (Figure 2(c)), which was removed from the substrate after DRIE (Figure 2(d)). Next, ZnO seed layer was deposited by sputtering directly onto the microslit structures (Figure 2(e)). Then, ZnO nanowires were synthesized hydrothermally in precursor solution at 95 □C for 2 hours (Figure 2(f)). After synthesis of ZnO nanowires, PDMS cover was capped by stamp-and-stick method to ensure the compact capping and visibility (Figure 2(g)) [13].

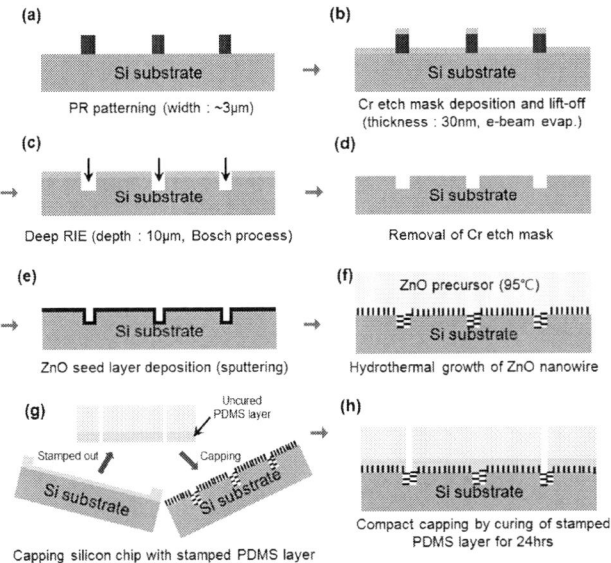

Figure 2: Fabrication process of the nanowire-integrated microfluidic device. First, microslit structures were patterned by photolithography and e-beam evaporation and fabricated by DRIE on silicon substrate. Next, ZnO nanowires were synthesized directly onto the side walls of microslit structures through hydrothermal method. Finally, PDMS cover is capped by stamp-and-stick method to ensure the compact capping.

In Figure 3, scanning electron microscope (SEM) images of microslit structures with ZnO nanowires are shown. Microslit structures without nanowire array were constructed in an array on silicon substrate with the slit width of 3~4 µm

and height of ~10 µm (Figure 3(a)). After hydrothermal synthesis of ZnO nanowire array, mesh-like cage structure was formed (Figure 3(b)). The length of ZnO nanowires was ~2 µm, so the slit could be filled with the nanowire array.

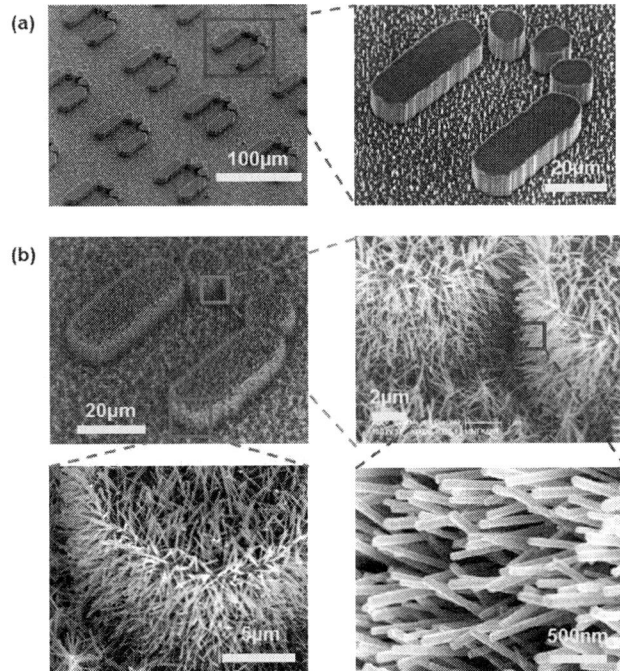

Figure 3: A group of SEM images for detailed view of microslit structures on silicon substrate. (a) SEM images of bare microslit structures before integration of ZnO nanowires. (b) SEM images of mesh-like cages after integration of ZnO nanowires into microslit structures.

EXPERIMENTAL SETUP FOR TRAPPING OF BACTERIAL CELLS

In Figure 4, the experimental setup for monitoring of trapping bacterial cells is shown. The setup simply consists of the nanowire-integrated device, fluorescent microscope and syringe pump. *E. coli* cells were genetically encoded with red fluorescent proteins (RFPs), so they express the red fluorescence with the fluorescent microscopy. *E. coli* cells in phosphate buffered saline (PBS) solution were simply forced to flow by syringe pump at the flow rate of 0.1 ml/hr. Total trapping period was 30 minutes.

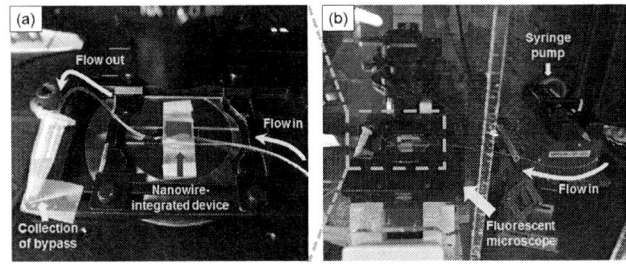

Figure 4: Experimental setup for monitoring of trapping bacterial cells.

HYDRODYNAMIC TRAPPING MODES

To understand how *E. coli* cells are trapped at the structures, we made *E. coli* cells flow through trapping chambers with bare microslit array. Originally, the microstructure of silicon material is not visible because it does not emit fluorescence. By introducing fluorescein isothiocyanate (FITC) into trapping chambers before injection of the cell solution, the silhouette of microslit structure could be discriminated clearly from surrounding area. Consequently, *E. coli* cells also could be detected as green fluorescent signals because the cell wall was coated with FITC (Figure 5). Although most cells just passed through the microslit structures, some cells which had relatively thicker diameter could be trapped at the microslit structures.

As shown in Figure 5, we observed two different trapping modes in bare microslit structures. First, bending occurs when a cell is trapped in transverse direction at the microslit (Figure 5(a)). Second, wedging happens when a cell burrows into microslit from the apical terminus of the cell in longitudinal direction (Figure 5(b)). The trapping process did not exert as much shear force to trapped cells as generating fracture of the cells.

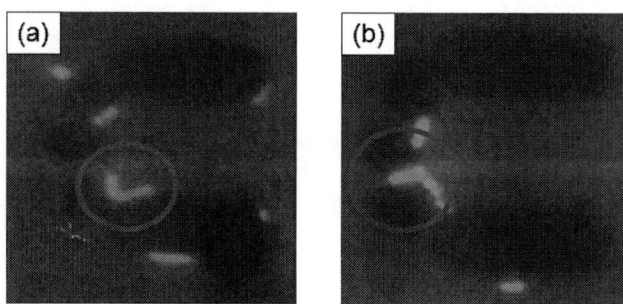

Figure 5: Fluorescent images of two different trapping modes of E. coli in bare microslit structures. (a) Bending (b) Wedging

TRAPPING OF BACTERIAL CELLS WITH NANOWIRE ARRAY

Results of trapping *E. coli* cells using mesh-like cages integrated with nanowire array are shown in Figure 6. *E. coli* cell solution whose concentration of 4×10^5 cells/ml was injected into trapping chambers to monitor the trapping progression with time at the mesh-like cages. Mesh-like cages were arranged as shown in Figure 6(a), and all the images of Figure 6 were taken at the same spot. For Figure 6(b) – (f), yellow mark array is overlapped to indicate position of cages. The number of *E. coli* cells trapped at the cages were increased and determined with red fluorescence. Most of cages were filled with *E. coli* cells within 10 minutes. After 20 minutes, *E. coli* cells started to clog even the bypassing channel area.

To examine the effectiveness of mesh-like shape of ZnO nanowires in cell trapping, we compared the bacterial cells trapping tendency in the bare microslit structures and nanowire-integrated microslit structures at the same flow rate.

After same trapping time (0.5 hour), a couple of *E. coli* cells were trapped in each unit of bare microslit structures as shown in Figure 7(a). On the other hand, dozens of *E. coli* cells were trapped and anchored in each unit of nanowire-integrated microslit structures as shown in Figure 7(b).

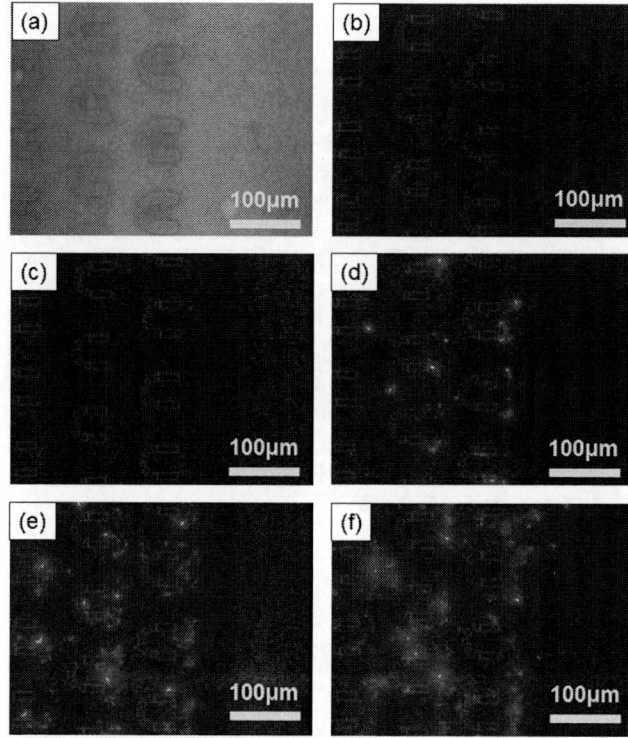

Figure 6: (a) Nanowire-integrated cage structures before injection of the cell solution. For the images (b) – (f), yellow mark arrays are overlapped artificially for visual help. E. coli cells were forced to flow during (b) 0 min. (c) 1 min. (d) 5 min. (e) 10 min. (f) 20 min. All the images were taken at the same spot.

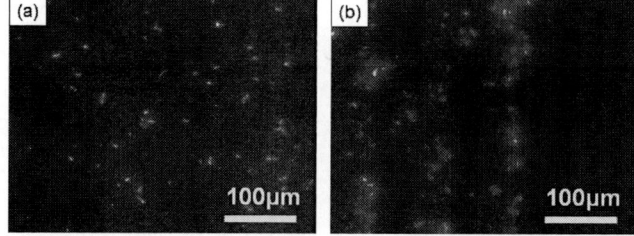

Figure 7: Fluorescent images of the trapping chambers after same trapping period of half an hour (a) Bare microslit structures (b) Nanowire-integrated microslit structures

HYDRODYNAMIC ANCHORING MODES

After trapping of *E. coli* cells, the cells were dried with the compact capping of PDMS cover. Then, we checked SEM images of the cells, and we observed two different anchoring modes. The first anchoring mode, impalement of *E. coli* on ZnO nanowires, is shown in Figure 8(a). *E. coli* cells were pierced by sharp tip of ZnO nanowires, and irreversible damage to cell wall occurred. The second mode, wedging, is shown in Figure 8(b). In the wedging mode, *E. coli* cells were deformed only in the reversible manner and stuck in the mesh-like cages.

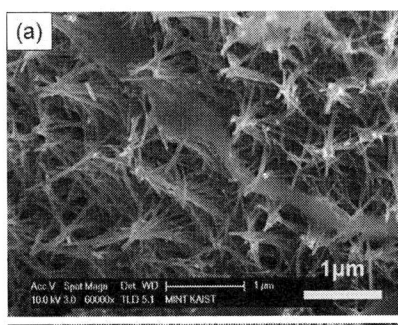
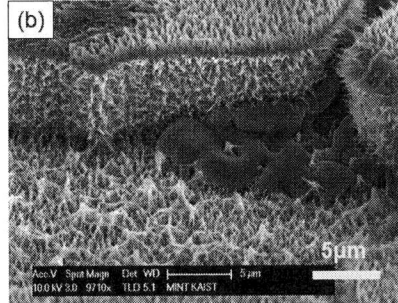

Figure 8: SEM images of two different anchoring modes of E. coli in nanowire-integrated microslit structures. (a) Impaling of E. coli on ZnO nanowires (b) Wedging into microslit structures integrated with ZnO nanowires

CONCLUSION

Trapping of biological targets at intended position is important to develop an advanced microfluidic system. In this work, novel hydrodynamic trapping and anchoring method of bacterial cells that utilizes the fishnet-like shape of nanowire array has been introduced. It was demonstrated that implementation of mesh-like cages integrated with ZnO nanowire array is effective in hydrodynamic trapping and anchoring of *E. coli* cells.

The proposed method is simpler than other trapping methods without complicated working principle, and has the potential for direct connection to many applications such as bacterial cell assay or inactivation.

ACKNOWLEDGEMENTS

This work was supported by the Center for Integrated Smart Sensors funded by the Ministry of Education, Science and Technology as Global Frontier Project (CISS-2012M3A6A6054201). Donguk Kwon would like to thank Prof. Je-Kyun Park for providing *Escherichia coli* cells and laboratory facilities.

REFERENCES

[1] J. Nilsson, M. Evander, B. Hammarström, and T. Laurell, "Review of Cell and Particle Trapping in Microfluidic Systems", *Anal. Chim. Acta.*, vol. 649, pp. 141-157, 2009.

[2] M. Yang, C. –W. Li, and J. Yang, "Cell Docking and On-Chip Monitoring of Cellular Reactions with a Controlled Concentration Gradient on a Microfluidic Device", *Anal. Chem.*, vol. 74, pp. 3991-4001, 2002

[3] D. Di Carlo, L. Y. Wu, and L. P. Lee, "Dynamic Single Cell Culture Array", *Lab Chip*, vol. 6, pp. 1445-1449, 2006.

[4] S. Cui, Y. Liu, W. Wang, Y. Sun, and Y. Fan, "A Microfluidic Chip for Highly Efficient Cell Capturing and Pairing", *Biomicrofluidics*, vol. 5, 032003, 2011

[5] Z. Wang, M. –C. Kim, M. Marquez, and T. Thorsen, "High-Density Microfluidic Arrays for Cell Cytotoxicity Analysis", *Lab Chip*, vol. 7, pp. 740-745, 2007.

[6] R. Davidsson, B. Johansson, V. Passoth, M. Bengtsson, T. Laurell, and J. Emneus, "Microfluidic Biosensing Systems. Part II. Monitoring the Dynamic Production of Glucose and Ethanol from Microchip-Immobilised Yeast Cells Using Enzymatic Chemiluminescent μ-biosensors", Lab Chip, vol. 4, pp. 488-494, 2004.

[7] U. Mirsaidov, W. Timp, K. Timp, M. Mir, P. Matsudaira and G. Timp, "Optimal Optical Trap for Bacterial Viability", *Phys. Rev. E*, vol. 78, 021910, 2008.

[8] B. H. Lapizco-Encinas, B. A. Simmons, E. B. Cummings, and Y. Fintschenko, "Dielectrophoretic Concentration and Separation of Live and Dead Bacteria in an Array of Insulators", *Anal. Chem.*, vol. 76, pp. 1571-1579, 2004.

[9] A. Krichevsky, M. J. Smith, L. J. Whitman, M. B. Johnson, T. W. Clinton, L. L. Perry, B. M. Applegate, K. O'Connor, and L. N. Csonka, "Trapping Motile Magnetotactic Bacteria with a Magnetic Recording Head", *J. Appl. Phys.*, vol. 101, 014701, 2007.

[10] B. Hammarström, T. Laurell, and J. Nilsson, "Seed Particle-enabled Acoustic Trapping of Bacteria and Nanoparticles in Continuous Flow Systems", *Lab Chip*, vol. 12, pp. 4296-4304, 2012.

[11] M. –C. Kim, B. C. Isenberg, J. Sutin, A. Meller, J. Y. Wong, and C. M. Klapperich, "Programmed Trapping of Individual Bacteria Using Micrometre-size Sieves", *Lab Chip*, vol. 11, pp. 1089-1095, 2011.

[12] L. E. Greene, B. D. Yuhas, M. Law, D. Zitoun, and P. Yang, "Solution-Grown Zinc Oxide Nanowires", *Inorg. Chem.*, vol. 45, pp. 7535-7543, 2006.

[13] S. Satyanarayana, R. N. Karnik, and A. Majumdar, "Stamp-and-Stick Room-Temperature Bonding Technique for Microdevices", *J. Microelectromech. Syst.*, vol. 14, pp. 392-399, 2005.

CONTACT

*D. Kwon, tel: +82-42-350-5240;
E-mail: dukwon8158@kaist.ac.kr

AN INTEGRATED MICROFLUIDIC SYSTEM USING FIELD-EFFECT TRANSISTORS FOR CRP DETECTION

Chih-Lin Lin[1†], Yen-Wen Kang[3†], Ko-Wei Chang[1],Wen-Hsin Chang[1], Yu-Lin Wang[3] and Gwo-Bin Lee[1,2,3*]*

[1]Department of Power Mechanical Engineering, [2]Institute of Biomedical Engineering,
[3]Institute of NanoEngineering and Microsystems, National Tsing Hua University, Hsinchu, Taiwan
[†]These authors contributed equally to this work; *Co-corresponding authors

ABSTRACT

Cardiovascular diseases are responsible for 25-million deaths worldwide on a yearly basis. Timely diagnosis of the disease is therefore an extensive research area. Toward this end, C-reactive protein (CRP) has become a reliable biomarker for evaluating risks of cardiovascular diseases. For commercial methods of CRP detection, the diagnosis of the enzyme-linked immunosorbent assay (ELISA) protocol and the high-sensitivity CRP (hs-CRP) were relatively time-consuming and the sensitivity and the detection limit of the above methods were not satisfactory. Recently, a CRP-specific aptamer (DNA-based) with high sensitivity and specificity was used to detect CRP in a microfluidic system, which was capable of performing the detection in an automated fashion while consuming tiny volumes of reagents and samples. Alternatively, AlGaN/GaN HEMT-based field-effect transistor (FET) sensors have emerged as promising biosensors to detect small molecules, proteins and even viruses, and have demonstrated rapid and highly sensitive detection in a compact system. To date, however, the combination of the microfluidic system, CRP-specific aptamer and high sensitivity FET sensors has not yet attempted. In this study, an integrated device that combined the advantages of microfluidics, CRP-specific aptamer and FET-based sensors was developed to achieve rapid, sensitive and specific CRP detection.

INTRODUCTION

Cardiovascular diseases present one of the most significant threats to human health, as they cause 25-million deaths in a year worldwide. Timely diagnosis of the disease is therefore important and could pose efficient treatment for these diseases. C-reactive protein (CRP) is an important biomarker for evaluating risks of cardiovascular diseases [1]. Specifically, the risks for cardiovascular diseases defined by the American Heart Association (AHA) and the Center for Disease Control and Prevention (CDC), are low for a CRP concentration below 1.0 mg/L (9.0 nM), moderate for a CRP ranging from 1.0 to 3.0 mg/L (9.0 nM to 27.0 nM), and high for concentrations over 3.0 mg/L (27.0 nM) [1]. Therefore, there is a critical need for development of detection methods that can not only detect CRP with high sensitivity and specificity, but also with a wide enough dynamic range to cover these concentration ranges.

Recently in biological application, the AlGaN/GaN HEMT-based sensors with high sensitivity were successfully demonstrated in detecting biomolecules such as glucose, specific antigen and single-strand DNA or RNA [2-6]. If they could be integrated with microfluidic systems for point-of-care diagnostics applications, they can not only reduce the consumption of expensive reagents and samples, but also improve reproducibility and throughput by automating fluid handling. Furthermore, the performance could be further improved with a significant reduction of the operating cost of conventional platforms. However, an integrated microfluidic system with FET sensors and CRP-specific aptamer to perform the entire procedure for CRP detection with a small sample volume in an automatic format had yet to be reported. In this study, a new microfluidic system with FET sensing devices for automatic measurement of CRP is therefore reported. The developed system achieved a high sensitivity of 2.9×10^{-2} mg/L (2.6×10^{-2} nM) in an automatic format within 10 minutes. The working range in this study from 2.9×10^{-2} mg/L (0.026 nM) to 2.9×10^{-2} mg/L (2600 nM) covered the range of the three categories of CRP concentrations as cardiovascular disease markers. Therefore, this integrated system may provide a promising tool to assess cardiovascular disease risk.

Materials and methods

Experimental procedures

The experimental process of the proposed microfluidic system integrated with FET sensors for CRP detection is schematically shown in Figure 1. First, the CRP-specific aptamer was immobilized on the gold surface of the gate region of the FET sensors for 24 hours. The concentration of the aptamer was 1 µM after optimization. Second, free (unbound) aptamer was removed with washing buffer by activating a micropump. Then, samples such as purified CRP or human whole blood with different concentrations of CRP were added. After that, unbound CRP was removed with washing buffer by activating the micropump. Finally, FET electric signals were measured.

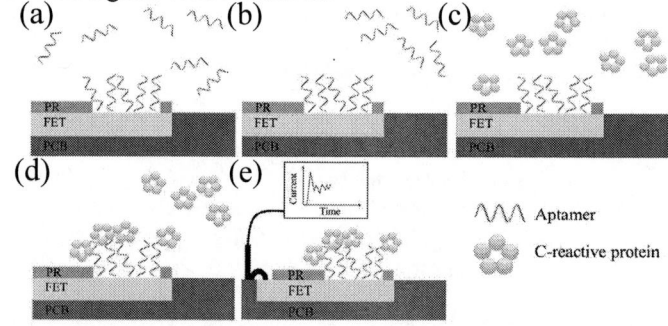

Figure 1: Experimental procedure for aptamer immobilization and CRP detection performed on the integrated microfluidic system. (a) Immobilization of aptamer; (b) Removal of free aptamer by pumping washing

buffer; (c) Addition of CRP; (d) Removal of unbound CRP by pumping washing buffer; (e) Detection of FET signals.

Integrated chip design

The integrated microfluidic FET chip is shown in Figure 2. Figure 2(a) is a photograph of the integrated microfluidic chip. The chip was composed of five layers as indicated by an exploded view of the microfluidic chip shown in Figure 2(b). Briefly, it was composed of two polydimethylsiloxane (PDMS) layers including a liquid channel layer (thick-film PDMS layer) and a pneumatic layer (thin-film PDMS layer), a double-sided tape layer binding with PDMS and a printed circuit board (PCB), AlGaN/GaN HEMT-based FET sensors and a PCB layer. The dimensions of the integrated microfluidic chip were measured to be 13 mm x 26 mm x 4.5 mm. Note that red dye indicates the liquid layer and blue dye indicates the air layer in the Figure 2(a). The PCB substrate (about 2 mm thick) was used for metal connection of FET sensors and to seal the microfluidic chip with PDMS layers by the double-sided tape. With this approach, AlGaN/GaN HEMT-based FET sensors can be integrated into the microfluidic chip such that a compact a semiconductor-based CRP analyzer could be feasible to detect the CRP with high sensitivity and specificity. Note that the proposed system can eliminate the manual calibration process through auto-sampling by using a pneumatic fluid control module and an automatic detection module so that the complicated steps of manual operation and setup of large-scale measuring instruments can be avoided.

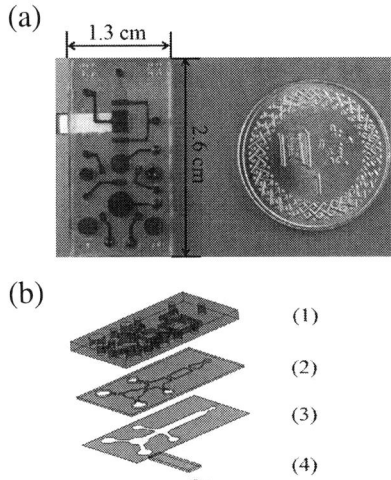

Figure 2: (a) A photograph of the integrated microfluidic chip. The dimensions of the integrated microfluidic chip were measured to be 13 mm × 26 mm × 4.5 mm. Red dye indicates the liquid layer and blue dye indicates the air layer. (b) An exploded view of the microfluidic chip. There are three layers in the microfluidic chip. Layer 1 is the air layer. Layers 2 and 3 are liquid layers. Layer 3 is a double-sided tape for bonding PDMS, the AlGaN/GaN HEMT FET and the PCB layer. The layer 4 is the FET sensor layer. The layer 5 is a PCB.

Microfluidic control module design

A new microfluidic system integrated with AlGaN/GaN HEMT-based FET sensors was developed in the present study. The integration of a sensitive electric signal detection module (FET) enabled direct measurement of changes in current after CRP binding to the CRP-specific aptamer by the semiconductor analyzer. As shown in Figure 3, two pneumatic micropumps [7], a pneumatic normally-open micromixer and normally-closed valves were integrated into the chip so that the integrated microfluidic chip with FET sensors for CRP detection can automatically perform the transportation, aptamer immobilization, mixing, washing and binding with CRP-specific aptamer and target. The entire process was controlled by a customer-made control system and electromagnetic valves (EMVs). Briefly, the experiment process started with loading the CRP-specific aptamer into the aptamer chamber and then immobilized it for 24 hours. Note that the transportation of these liquids was performed by the suction-type micropumps and microvalves. After that, the CRP-specific aptamer was immobilized on the gate area of the AlGaN/GaN HEMT FET. Simultaneously, the normally-closed valves were closed to confine the aptamer solution and the normally-open micromixer was activated to gently mix the aptamer solution for 24 hours to enhance the aptamer immobilization. Then, the CRP samples such as the purified CRP, the serum spiked with CRP or the clinical samples were loaded into the CRP sample chamber. Similarly, washing buffers were pre-loaded into each loading chamber. Next, the aptamer washing buffer was transported into the microchannel to wash out the free aptamer. After the aptamer washing process, the CRP sample was transported into the microchannel. Again, the normally-closed valves were closed to confine the CRP sample and the normally-open micromixer was activated to gently mix the CRP sample to facilitate efficient binding. Similarly, unbound CRP sample was removed by the CRP washing buffer. Finally, the FET electric signal was measured by a semiconductor analyzer (B1500A Semiconductor Device Analyzer, Agilent, USA).

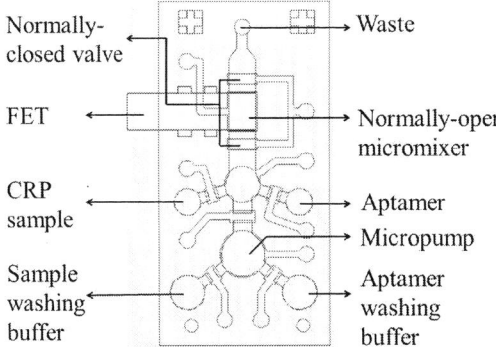

Figure 3: Schematic diagram of the microfluidic chip. There are four open chambers (included an aptamer chamber, a CRP sample chamber, a sample washing buffer chamber and an aptamer washing buffer chamber), two suction-type micropumps, six normally-closed valves and one normally-open micromixer.

AlGaN/GaN HEMT-based FET sensor chip design

AlGaN/GaN HEMT-based FET sensors with immobilized aptamers were developed in the present study. The schematic diagrams of the FET sensor are shown in Figure 4. Figure 4(a) is the cross-sectional schematic of the FET sensor, which was made of AlGaN/GaN HEMT materials. Specifically, AlGaN and GaN thin films were coated on a sapphire substrate so that the two-dimensional electron gas was formed between the AlGaN and GaN interface. The gate metal was gold and uncovered by a layer of photoresist so that the CRP-specific aptamer, which was modified with thiol on the 5' end, could be immobilized on the exposed gate surface area. If CRP binds to the aptamers, the electrical potential bias of gate, V_G, will be changed accordingly. The output current, I_D, will also change. The current I_D can be measured between the source (S) and the drain (D) by a semiconductor analyzer. Figure 4(b) is the top-view schematic of the FET sensor and the sensing region. Figure 4(c) is a photograph of the FET and the top-view microphotograph of the sensing region.

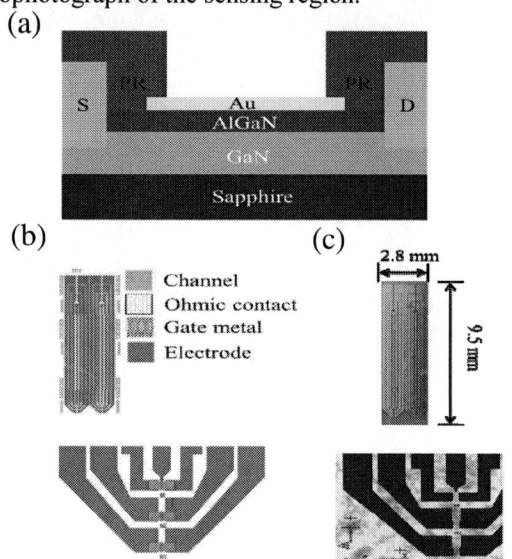

Figure. 4: (a) The cross-sectional schematic illustration of the AlGaN/GaN HEMT sensor. (b) and (c): The top-view and a photograph of the FET and the sensing region.

CRP-specific aptamer

The CRP-specific aptamer is single-strain DNA (ssDNA), which is specific to CRP and screened by a systematic evolution of ligands by exponential enrichment (SELEX) process [8]. The sequence of the CRP-specific DNA aptamer used in this microfluidic system screened by our team with another microfluidic chip is as follows: 5'-GGCAGGAAGACAAACACGATGGGGGGG TATGATTTGATGTGGTTGTTGCATGATCGTGGTCTG TGGTGCTGT-3'. Note that the aptamer is 72 bases in length [8] and 5' is modified with thiol such that they could be immobilized on gold surfaces. Previous work has demonstrated that this CRP-specific aptamer has high affinity for CRP binding and therefore is suitable for CRP measurements [9].

RESULTS AND DISCUSSION

Optimization of CRP-specific aptamer immobilization

To reduce the consumption of aptamer and to achieve good performance of the CRP sensor, optimization tests including aptamer immobilization time and concentration were first performed. Specifically, a fluorophore-tagged DNA probe complementary to part of the CRP aptamer was used to evaluate the optimal amount of aptamer immobilized on the FET sensing region. The result of optimization time is shown in Fig 5(a). Note that W indicated that the aptamer was immobilized on the gold surface, and W/O indicated bare gold surface only. Experimental data showed that 24-hour immobilization resulted in the greatest difference in fluorescence signals. In addition, the results of aptamer concentration optimization are shown in Fig 5(b). Optimization results revealed that good aptamer coverage could be obtained by immobilizing 1 µM aptamer on the surface for 24 hours.

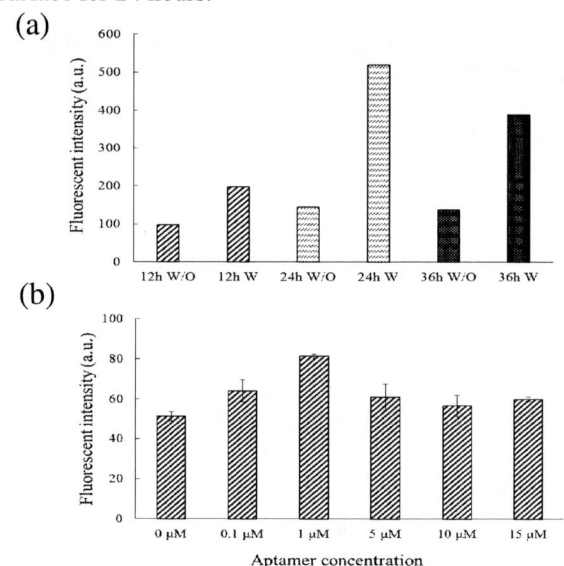

Figure 5: Optimization of CRP-specific aptamer for (a) immobilization time and (b) aptamer concentration.

Electric detection test with and without the microfluidic control chip

The microfluidic control chip improved the stability of the FET electrical signal. The sensitive electric signal was detected by the semiconductor analyzer after PBS was added. Figure 6(a) was the real-time detection of PBS with using the microfluidic control chip. With the microfluidic control chip, the signal fluctuation was observed to be about 0.2-0.5 µA. Conversely, Figure 6(b) was the real-time detection of PBS without using the microfluidic control chip. Without the microfluidic control chip, the signal fluctuation was measured to be about 0.5-0.9 µA. It indicated that electric signals became more stable because the microfluidic control chip would prevent the PBS from being evaporated or disturbed, thereby enabling more sensitive detection. It means this integrated chip could have a great potential to evaluate the CRP concentration more reliably while using the microfluidic control chip to reduce the disturbance from the air.

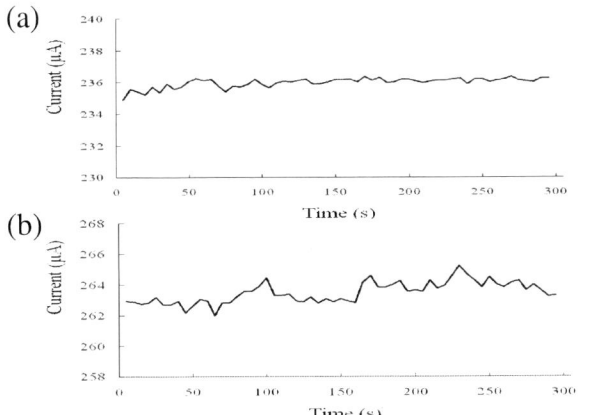

Figure 6: Real-time current detection of PBS (a) with using the microfluidic control chip and (b) without using the microfluidic control chip.

Sensitive electric detection of CRP

Sensitive electric detection of CRP has been demonstrated in Figure 7. Figure 7 (a) shows the real-time detection of serial dilutions of CRP ranging from 0.026 nM to 2600 nM. The current increased with higher concentration of CRP added and gradually saturated at 2600 nM. The current difference from real-time detection could be measured and the concentration of the target CRP would be determined by this curve, as shown in the dose-response curve in Figure 7(b). This result shows a linear correlation between the log CRP concentration (nM) ranging from 0.029 mg/L (0.026 nM) to 2900 mg/L (2600 nM). Note that the R-square of this curve is 0.9876. It means this developed chip could have potential to evaluate the CRP concentration by simply measuring the current difference from FET electric signals, especially in the human physiological concentrations. More tests using clinical samples are undergoing.

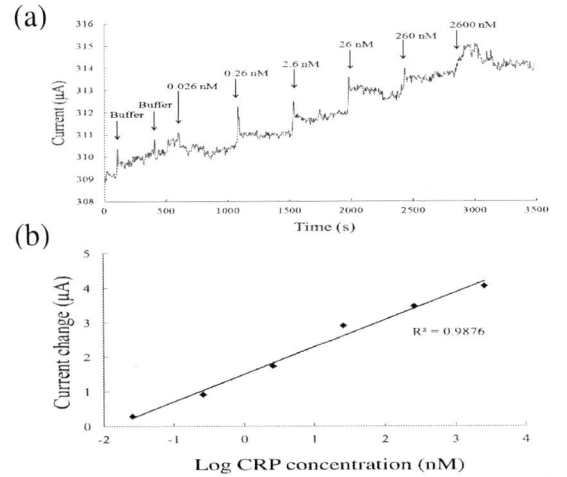

Fig 7: (a) Real-time detection of CRP and (b) dose-response curve of CRP detection.

CONCLUSION

In this work, an integrated device that combined the advantages of microfluidics, CRP-specific aptamer and FET-based sensors was developed to achieve rapid, sensitive and specific CRP detection. The working range spanned 0.029 mg/L (0.026 nM) to 2900 mg/L (2600 nM), which covers the entire range for the three categories of CRP concentrations as cardiovascular disease markers. The developed system therefore has the potential to assess the risks of cardiovascular diseases at the point of care in the future.

ACKNOWLEDGEMENTS

The authors would like to thank the financial support from National Science Council in Taiwan.

REFERENCES

[1] P. M. Ridker, et al., "C-reactive protein and other marjers of inflammation in the prediction of cardiovascular disease in women, " *New England journal of medicine*, 12(2000), 836-843

[2] B. S. Kang et al., " Electrical detection of deoxyribonucleic acid hybridization with AlGaN/GaN high electron mobility transistors," *Applied physics letters*, 89 (2006), 122102

[3] Y. W. Kang et al., " Human immunodeficiency virus drug development assisted with AlGaN/GaN high electron mobility transistors and binding-site models," *Applied physics letters*, 17 (102),173704

[4] B. W. Chu, et al., "Wireless Detection System for Glucose and pH Sensing in Exhaled Breath Condensate Using AlGaN/GaN High Electron Mobility Transistors," *IEEE sensors journal*, 10(2010), 64-70

[5] B. S. Kang, et al., "Prostate specific antigen detection using AlGaN/GaN high electron mobility transistors," *Applied physics letters*, 91 (2007), 112106

[6] Y. L. Wang, et al., "Fast detection of a protozoan pathogen, Perkinsus marinus, using AlGaN/GaN high electron mobility transistors," *Applied physics letters*, 94 (2009), 243901

[7] C. H. Weng, et al., "A suction-type, pneumatic microfluidic device for liquid transport and mixing," *Microfluidics and nanofluidics*, 2011 (10), 301-310

[8] C. J. Huang, et al., "Integrated microfluidic system for rapid screening of CRP aptamers utilizing systematic evolution of ligands by exponential enrichment (SELEX)" *Biosensors& Bioelectronics* , 25 (2010), 1761-1766

[9] W. B. Lee, et al., "Synthesis of Iron oxide Nanoparticles Using a Microfluidic System," *Biomedical Microdevices*, 11 (2009) , 161-171

CONTACT

*Dr. Gwo-Bin Lee, tel: +886-3-5715131 Ext.33765; gwobin@pme.nthu.edu.tw

ASSEMBLY OF NEURAL CELL-LADEN MICROPLATES ON A MICROFABRICATED BREADBOARD

Shotaro Yoshida [1], Koji Sato [1,2], Shoji Takeuchi [1,2]

[1]Institute of Industrial Science, The University of Tokyo, JAPAN
[2]ERATO Takeuchi Biohybrid Innovation Project, JST, JAPAN

ABSTRACT

This paper describes a method to form an arbitrary network of neural cell-laden microplates on a breadboard-like microelectrode array (MEA) for neural network studies. We cultured single neural cells on microplates with control of their morphologies, and mounted the plates on the breadboard-like MEA. Through a pore in the center of the microplate, the cells and gold electrodes patterned on the MEA would be electrically accessible. The device allows arrangement of morphological connectivity at single cell resolution with interfaced to single electrodes. Thus, the method will be adaptable to bottom-up neuronal circuit construction and electrical analysis of synaptic integrating mechanism.

INTRODUCTION

Multiple networks of neural cells connected by synapses are the core of nervous systems, which are capable of processing electrical signals based on action potentials. To understand how the neural ensemble processes the information, measuring action potentials as the output of network is essential. The microelectrode array (MEA) has been widely used in the field of neuroscience and for their performability in large-scale electrical measurements of neural networks [1-3]. Although conventional MEA allowed massively parallel recording/stimulation of neural cells on its electrodes, the cell-electrode arrangement was difficult due to natures of neural cells: non-relocatable adhesion and uncontrolled morphology. To measure electrical activities of multiple neural cells at individual cellular level, controlled location of single neural cells on MEA with defined morphology is needed.

In this study, we developed "perforated microplate" technique for single cell-electrode interfacing on MEA (*Fig. 1*). As previously reported [4-7], a microplate was single-cell sized, flat-topped substrate that allowed culturing single adherent cell on it for handling individually. We specifically designed circle/line patterns for controlling cell-body/neurite morphology, and a pore in the center of the circle to expose the cell body under the microplate. We also fabricated MEA that had micro-sockets for arranging the microplates. By mounting the single neural cell-laden microplates to the micro-socket on the MEA, the single cells and microelectrodes were precisely interfaced. The advantage of this method is that it is possible to design and achieve finely-tuned multiple cell-electrode configurations at single-cell resolution.

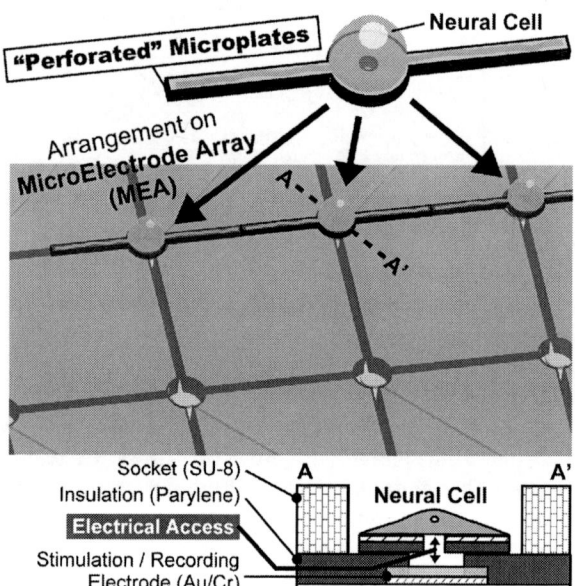

Figure 1. *Conceptual illustration of this study. Single neural cells were cultured on perforated microplates and morphologically controlled by their shapes. The cell-laden microplates were sequentially arranged on a microbreadboard (MEA with sockets) for constructing a network. Each cell was faced to a stimulation/recording electrode through the hole for electrical access.*

RESULTS

Fabricated Microplates and Microbreadboard

Microscopic pictures of fabricated microplates and a microbreadboard are shown in *Fig. 2*. Each microplate had a 50-μm-diameter center circle with a pore, two 10 x 100 μm branches, and arrayed by 50 μm and 250 μm apart for long and short axe, respectively (*Fig. 2(a)*). A microbreadboard had four electrodes with its circle area exposed and other area electrically insulated by parylene (*Fig. 2(b)*). SU-8 sockets were designed to be 5 μm larger than microplates.

Impedance of Microbreadboard

Measured impedance of the electrodes (*Fig. 3*) around 100 Hz to 1 kHz input voltage was approximately 1 MOhm. The result suggested the microbreadboards' suitability for electrophysiological experiments since typical requirements for MEA is 1 Ohm to 1 MOhm at the frequency of the range.

Single Neural Cells on Fabricated Microplates

Morphology of PC12 cells cultured on microplates were

controlled by the shape of the microplates (*Fig. 4(a)*). The neural cell bodies located on the circles covered the center pores. Thus, their cell membranes should be exposed below the microplates.

Assembly

We demonstrated arrangement of single neural cells onto the microbreadboard (*Fig. 4(b)*). The morphologically controlled single neural cells were picked up by a microtweezer, and mounted on the socket of the MEA (*Fig. 6(a, b)*). Fluorescent imaging showed that the mounted cells retained their controlled morphology after the cell handling process (*Fig. 6(c)*). Here, single neural cell-electrode interfacing was achieved.

CONCLUSION

We developed a method of arranging single neural cells to MEA using the microplate technique. We demonstrated neural cell culture on the perforated microplates, and arrangement of the morphologically controlled single neural cells to individual microelectrode of the MEA. In future works, we will assemble multiple single neurons and perform selective electrical stimulation and recording on the device to achieve micro-breadboards for the neural circuit assemblies.

ACKNOWLEDGEMENTS

The authors gratefully acknowledge Prof. Kazuhiko Ishihara at the University of Tokyo for providing the MPC polymer, and Asumi Yuzuriha at AOI Electronics. Co., Ltd. for providing microtweezers. The authors would like to thank Tetsuhiko Teshima, Kaori Kuribayashi-Shigetomi, and Midori Kato-Negishi for helpful discussions, and A. Hsiao for useful comments on the manuscript. This work was partly supported by Grant-in-Aid for Scientific Research on Innovative Areas "Bio Assembler" (23106008) from the Ministry of Education, Culture, Sports, Science and Technology of Japan. S. Yoshida is supported by Research Fellowship of the Japan Society for the Promotion of Science (JSPS) for Young Scientists, Japan.

REFERENCES

[1] M. E. Spira, and A. Hai, *Nature Nanotechnology*, 8, pp. 83-94, (2013).

[2] J. T. Robinson, et al., *Nature Nanotechnology*, 7, pp. 180-184, (2012).

[3] C. Py, M. Martina, et al., *Frontier in Pharmacology*, 2, 51, doi: 10.3389/fphar.2011.00051, (2011).

[4] H. Onoe, and S. Takeuchi, *Journal of Micromechics and Microengineering*, 18, 095003, (2008).

[5] K. Kuribayashi-Shigetomi, et al., *PLoS ONE*, 7(12), p. e51085, (2012).

[6] T. Teshima, et al., *Small*, online: doi: 10.1002/smll.2013 01993, (2013).

[7] S. Yoshida, et al., *in Proceedings of microTAS2013*, pp. 1986-1988, (2013) .

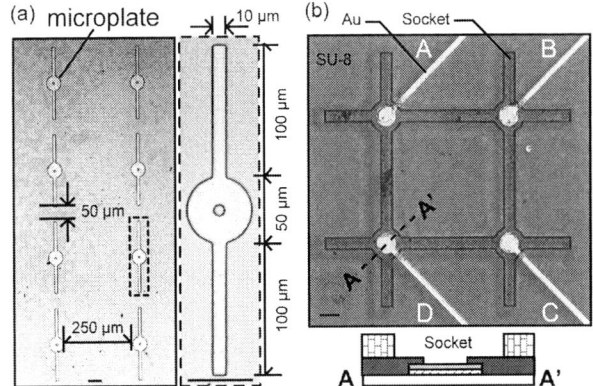

Figure 2. Fabricated microplates and microbreadboard. (a) Dimensions of microplate arrays. Diameter of the pore was 2 to 8 μm. (b) Top: picture of microbreadboard, and bottom: illustration of cross-section indicated by A-A'. Scale: 50 μm.

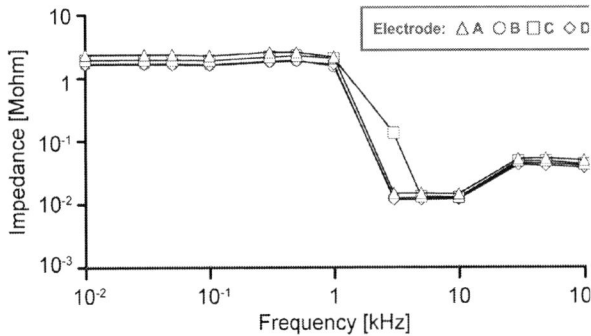

Figure 3. Impedance measurement of the fabricated MEA with micro-sockets. A, B, C, D indicate the four electrodes on a MEA device.

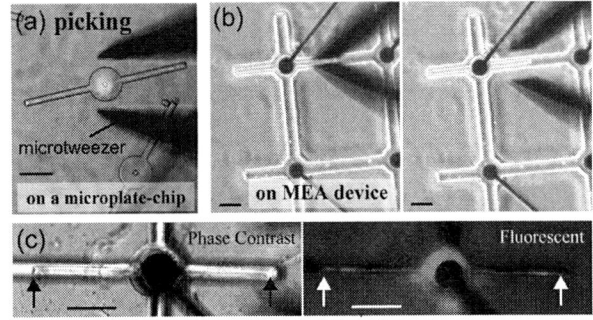

Figure 4. Assembly of single neural cell-laden microplates on a microbreadboard. (a) Picking up, and (b) placing procedure. (c) A phase contrast image of a single neural cell-laden microplate placed inside a socket (left), and its fluorescent image (right).

CONTACT

*S.Yoshida, tel: +81-3-5452-6650;
syoshida@iis.u-tokyo.ac.jp

CD4 CELL ISOLATION FROM BLOOD USING FINGER-ACTUATED ON-CHIP MAGNETOPHORESIS FOR RAPID HIV/AIDS DIAGNOSTICS

Macdara Glynn, David J. Kinahan, Danielle Chung, and Jens Ducrée

Biomedical Diagnostics Institute, School of Physical Sciences, Dublin City University, IRELAND

ABSTRACT

With timely diagnosis and correct treatment, people living with HIV/AIDS can consider the disease as a chronic rather than a terminal illness. Still, in regions were HIV is endemic, rapid diagnosis is a challenge due to the complexity of the instrumentation required, the poor infrastructure in these countries, as well as the technical expertise required to carry out the diagnosis. This paper presents a microfluidic chip based approach allowing semi-quantitative CD4+ cell counting on a cheap, rapid, highly portable and instrumentation-free Point-of-Care HIV diagnostic device. Flow is driven by finger-pressing a flexible reservoir, and the target cells are immobilized through magnetophoresis. The fluidic test completes within ca. 30 seconds of sample application to the chip.

INTRODUCTION

HIV (human immunodeficiency virus) infection and the subsequent development of the associated symptoms know as AIDS (acquired immunodeficiency syndrome) remains an ongoing disease of pandemic proportions since its identification in the early 1980s. Yet, application of modern molecular and cellular instrumentation to the diagnostics of the infection, as well as availability of [highly active] anti-retroviral therapy ([HA]ART) have brought the infection to low penetrance in countries where these technologies and interventions are readily available. However, epidemiological statistics show that low- to mid-income countries consistently maintain the largest viral pool of HIV (Fig. 1), with Sub-Saharan Africa being the most heavily affected (69% of the 34 million global cases in 2011) [1]. Access to rapid and reliable diagnosis in these regions is low due to limited infrastructure, costs and technical know-how. Alternatives to the expensive "gold standard" are currently investigated to allow reliable diagnostics to be deployed in such regions. These "Point-of-Care" (PoC) devices aim to be capable of operating at remote villages with minimal power and personnel. ART can then be immediately delivered to a patient presenting as positive for HIV in these areas.

The target of the HIV virus is the T-helper cells of the patient. Progressive loss of these cells results in the suppression of the immune system and the pathology of AIDS. T-helper cells have a specific protein marker on the cell surface called CD4, and hence CD4 cell counts are a key clinical determinant for diagnosis of HIV patients and the initiation of treatment. While healthy individuals present a CD4 cell count of > 1200 cells μl^{-1} in whole blood, common practice is that HIV positive patients begin ART treatment when this number falls below a specified threshold – currently set by the World Health Organization

(WHO) at 500 cells μl^{-1} [2]. The ability to identify HIV patients with CD4 cell counts around this threshold is therefore critical. A number of devices are on the market that performs at the required level, but the instruments and/or per-test cost can be large, and the number of tests-per-day can also limit the use of the devices [3].

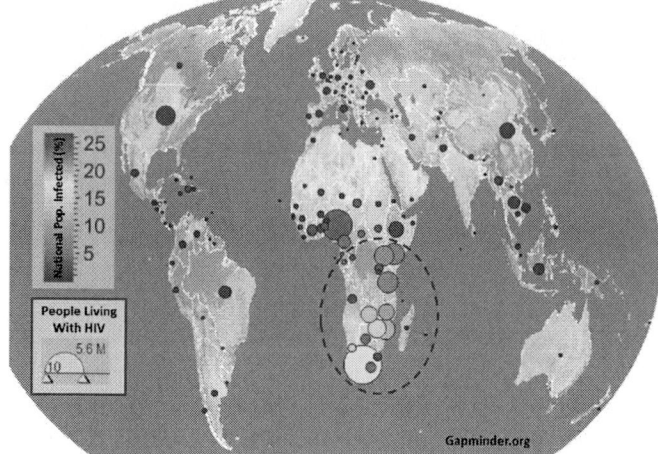

Figure 1: Global snapshot of the HIV pandemic in 2011. The diameters of the circles indicate the population numbers, while the color represents the percentage of a national population living with HIV. The area circled in red contains 50% of all people in the world living with HIV.

We present here a finger-press driven magnetophoretic CD4 cell isolation on an autonomous microfluidic chip for isolating CD4 cells from whole blood within 30 seconds of sample loading. The very simple, minimum operator-skill and instrument-free microfluidic procedure delivers accurate results and thus meets the criteria for deployment in resource-limited regions where HIV / AIDS is largely endemic, and monitoring remains a challenge. The strategy involves the incubation of whole blood with (super)para-magnetic beads that specifically bind to the CD4 epitope of T-helper cells. This sample is then introduced to a flui-dically primed microfluidic chip, and passaged a number of times past a magnetic capture chamber, specifically isolating the CD4 cells to a location removed from the bulk of the blood sample. Critically, the forces needed to drive both the loading of the sample, and the fluidic movements are generated by deflecting a fluidic reservoir connected to the main chamber *via* a flexible membrane. This can be achieved simply by depressing the membrane with the finger of the operator. Following isolation of the CD4 cells, they can be semi-quantitatively enumerated by visual inspection of the capture chamber.

SYSTEM DESIGN

The architecture of the chip displays the pneumatic fluidic reservoir (P1) at the base of the chip. P1 connects directly to the separation channel, a secondary reservoir (P2) is at the top of the separation channel and the sample input port is directly below P2. Midway down the separation channel is the CD4 capture locus. On the opposite side to the capture locus is a vent to allow any overflow to occur during the cell separation process (Fig. 2).

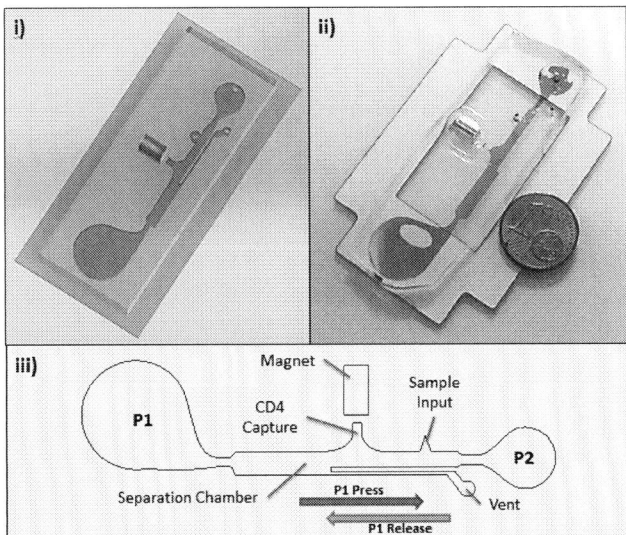

Figure 2: Finger-press actuated magnetophoretic CD4 isolation chip. i) 3D rendering of chip. ii) Image of a chip with the microfluidic features shown using green dye. iii) Schematic showing features of interest. Direction of flow when P1 is depressed or released is indicated with Blue and Green arrows respectively.

Microchannels (40 µm deep) are fabricated in PDMS [4] and adhered to a glass base. A permanent magnet is placed perpendicular to the pneumatically-driven flow of liquid (Fig. 2 iii), and the chamber is filled with priming buffer.

For biosafety reasons, we cannot handle HIV patient blood in our lab. Therefore finger-prick derived whole blood from healthy donors was first depleted of native CD4+ cells. It is then spiked with defined concentrations of fluorescently stained CD4-positive HL60 cells at medically relevant concentrations between 10 and 1000 cells µl^{-1}, and incubated for 3 min with paramagnetic beads specific to the CD4 epitope.

SYSTEM OPERATION

To load the chip, the pneumatic chamber (P1) is first depressed to reduce the internal volume of the chip. 4 µl of sample is applied to the loading inlet and P1 is released (Fig. 3 i-ii). The subsequent reconstitution of its initial volume generates a flow from the sample inlet, past the capture region, and into the P1 reservoir (Fig. 3 iii-iv). The actuation of the second reservoir (P2) compensates any

shortfall in liquid volume, insuring that the chamber always remains fully primed. The vast background of non-magnetically tagged cells follows the laminar flow towards P1; only the magnetically tagged, CD4-positive cells are deflected to the base of the flow-free capture structure (Fig. 4). In order to sequester remaining CD4 cells to the capture chamber, P1 is again pressed to drive the sample back towards P2 (Fig. 3 v-vi). By releasing P1 the cycle re-commences. By repeating these steps a second time, a dual-pass actuation will complete in approximately 30 seconds. Quantification of recovered cells was carried out by fluorescence and bright-field microscopy (Figs. 4-5), but the chip is amenable to detection using a handheld LED-based UV-fluorescence detector.

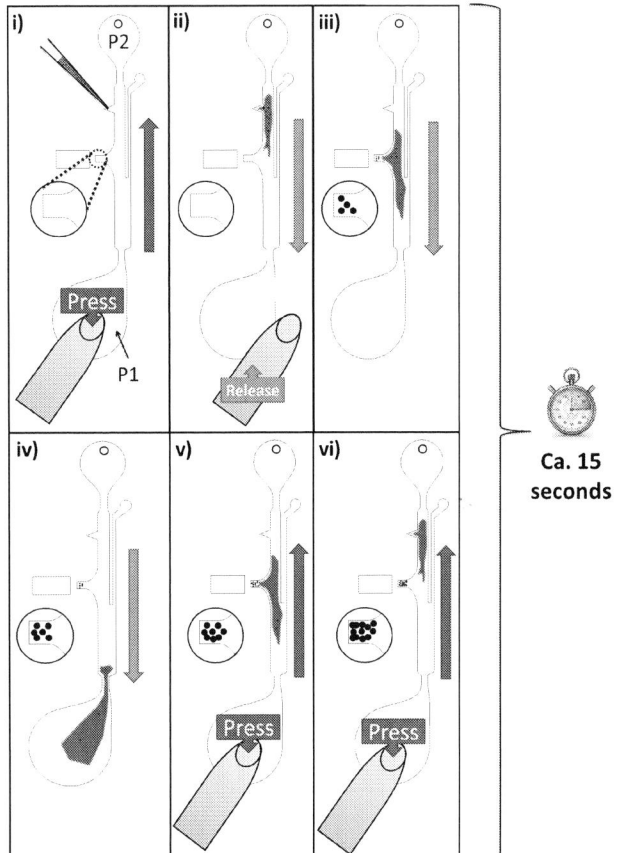

Figure 3: Operation of the magnetophoresis chip to isolate CD4 cells (shown as black circles) from whole blood. Direction of flow pressure is shown with Blue (towards P2) and Green (towards P1) arrows.

METHODOLOGY

Chip manufacture

The microfluidic chips used in this paper were formed from polydimethylsiloxane (PDMS; Dow Corning, MI) mixed at a ratio of 10:1 base and curing agent. The procedures for making a master and for securing the co-rotating magnet in the PDMS have been described in detail elsewhere [4-5]. Loading holes and vents were defined in the PDMS at appropriate locations using a dot punch. The

PDMS slab containing the microfluidic features was placed on a 25 × 60 mm glass coverslip and allowed to bond for 1 min. Finally, the glass / PDMS chip was mounted to a PMMA base. To prime the microchannels and structures, the chip was placed under vacuum for at least 1 hour, following which a large drop of priming buffer (phosphate-buffered saline [PBS] pH 7.4, 0.1% w/v bovine serum albumin [BSA], 1 mM EDTA) was immediately placed on the surface of the PDMS, covering both the sample port of the loading chamber, and the vent. Degas driven flow then primed the channels. Magnets used were NdFeB N45 cylindrical magnets, with a diameter of 3 mm and a height of 6 mm (Supermagnete, Germany). Magnets were placed in the molded cavities before the samples were loaded to the chip.

Blood Processing and Cell Culture

Blood was extracted directly from healthy donors *via* finger prick using 1.5 mm sterile lancets (BD Biosciences, NJ, USA). To prevent coagulation of the blood sample, 60 mM EDTA solution was immediately added to the sample to result in a final concentration of 6 mM EDTA in the whole blood. Blood was isolated and prepared fresh, directly before experimental use. Native CD4 cells were depleted using the T4 Quant Kit (Life Technologies, CA, USA) according to manufacturer's instructions.

HL60 cells (DSMZ, Braunschweig, Germany) were cultured in 75 cm^2 flasks in RPMI 1640 media, with 10% un-inactivated fetal bovine serum, 100 U mL^{-1} penicillin, and 100 µg mL^{-1} streptomycin. Cultures were maintained at 37 °C with 5% CO_2. Where indicated, cells were fluorescently stained with NucBlue Live Cell Stain (Life Technologies) according to the manufacturer's instructions.

Experimental samples composed of blood spiked with HL60 were treated with Dynabeads CD4 magnetic beads (Life Technologies). Incubations were carried out in a 2 mL Eppendorf tube, and final incubation volume was 200 µL per sample, composed of 140 µL incubation buffer (PBS pH 7.4, 0.1% w/v BSA, 1 mM EDTA), 50 µL experimental sample, and 10 µL Dynabead CD4. Incubation was performed at room temperature with rotation for 3 min. 4 µl of sample was introduced to the chip as described.

RESULTS

Following a dual-pass actuation of blood samples spiked with 4 concentrations of HL60 cells, CD4 concentrations as low as 10 cells µl^{-1} in whole blood were detected. Furthermore, in the medically relevant diagnostic range of 10-1000 CD4 cells µl^{-1} of whole blood, the system showed a linear fluorescent signal output corresponding to the concentration of cells. Over this range, the square of the Pearson correlation coefficient (RSQ) was calculated as 0.97 (Fig. 5-i). This suggests that the CD4 concentration in patient blood over diagnostic ranges lie within in the linear range of the chip performance – indicating that the strategy can lead to high predictability of CD4 count using a single data-point.

Furthermore, when the area of the space occupied with beads / cells is measured using the free image analysis software ImageJ (NIH, USA), a similar correlation to HL60 concentration (0.99) is observed (Fig. 5-ii). This suggests that imaging and quantification can be carried out using only bright-field optics.

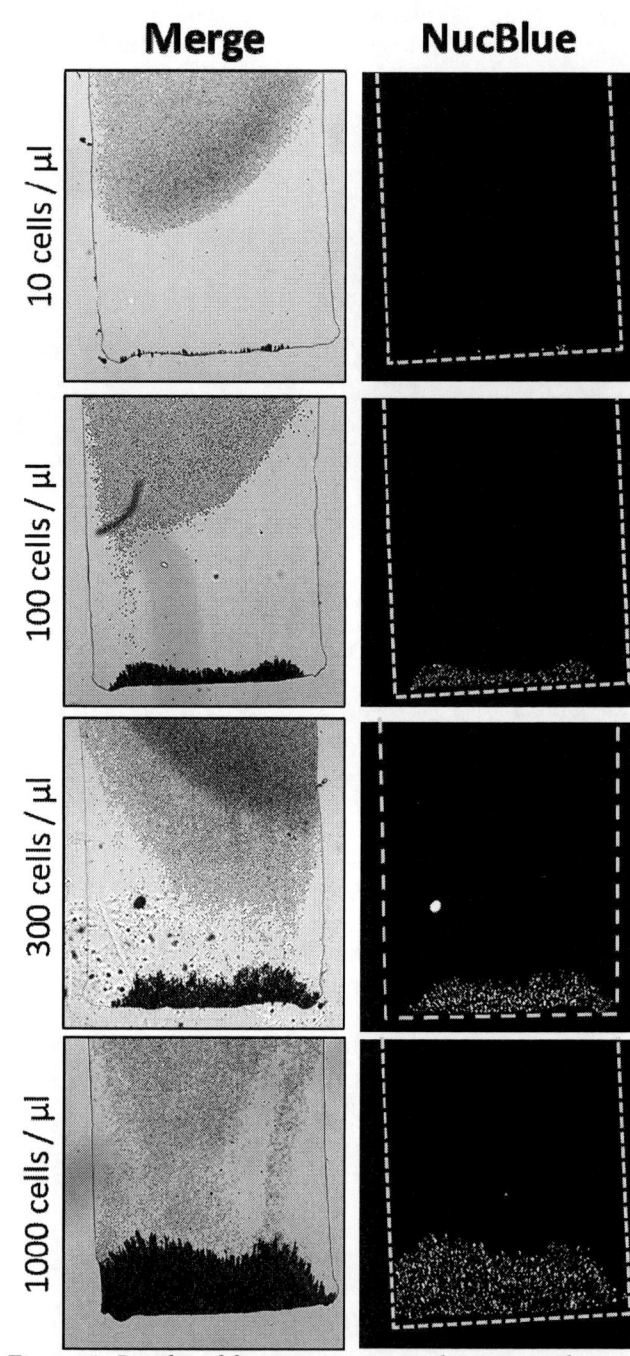

Figure 4: Results of finger-press pumped magneto-phoretic isolation of HL60 cells from CD4-depleted whole blood. Merged DIC / fluorescent images (left), raw fluorescent images (right). The edge of the capture chamber is shown as a yellow dashed line.

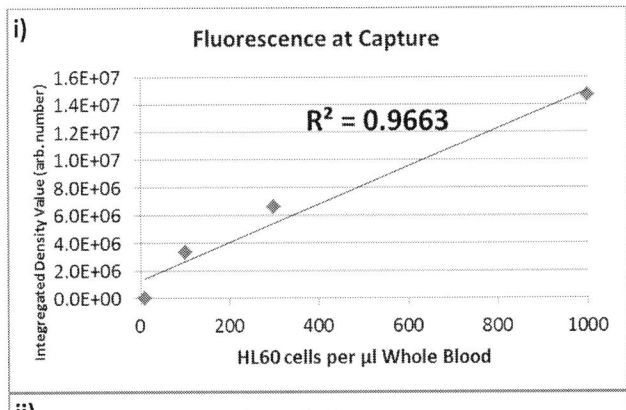

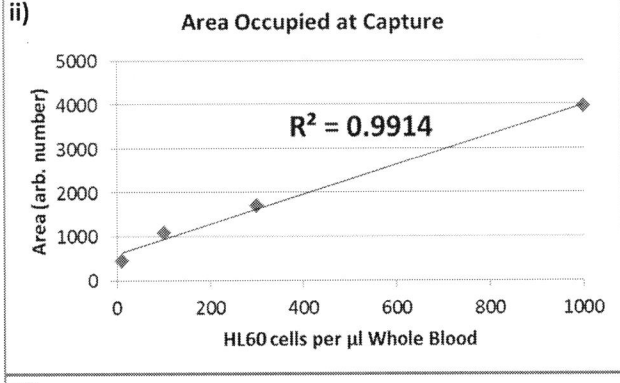

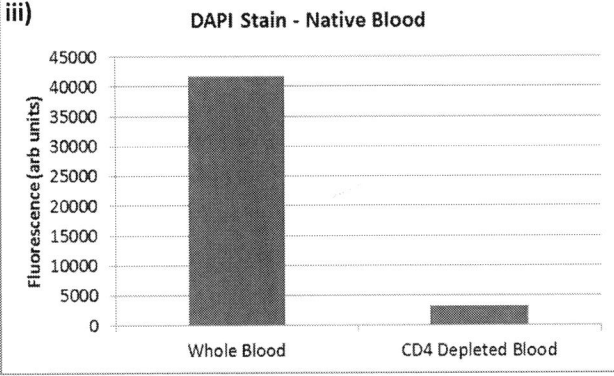

Figure 5: Fluorescent analysis of images from Fig. 4. i) The integrated density values (IDV) from identically sized region-of-interests (ROIs) from the images were calculated and shown relative to the number of HL60 cells spiked to the blood. ii) ROIs were generated on the brightfield images shown in Fig. 4. The area of these were compared to the number of HL60 cells spiked to the blood. RSQ was calculated on i) and ii) to show linearity of the data. iii) UV fluorescence was measured at the capture chambers of chips containing native- and CD4-depleted whole blood treated with DAPI.

In spite of the promising results using area calculation for the quantification of the data, a future analysis instrument may be more accurate using fluorescence imaging on a dark-field background. In this case, to keep the cost-per-test at a minimum, specific staining using expensive antibody based fluorescence markers should be avoided. However, as the chip will exclusively isolate CD4 cells using positive capture, when run under non-spiked native conditions a general UV based white blood stain will be sufficient to allow CD4 specific fluorescent readout without the need for expensive antibody binding reagents. To this end, we co-incubated native whole blood (both CD4-depleted and non-CD4 depleted) with DAPI (1 µg ml⁻¹) concurrently with the CD4 para-magnetic beads, and examined the resulting on-chip isolations under UV. A clear difference was observed in UV signal measured in the capture chamber between native and CD4-depleted blood (Fig. 5-iii).

DISCUSSION AND CONCLUSION

To enable rapid, cheap and simple HIV diagnostics based on CD4 cell count in a finger-prick volume of whole blood, we have developed an autonomous microfluidic chip capable of delivering a semi-quantitative diagnosis of HIV infection. Critically, this chip runs without instrumentation for flow control; and if deployed, would potentially require only a bright-field light source with an associated optical camera similar to that on mid-range mobile phones for readout. Compared to the complexity of devices currently on the market for PoC HIV diagnostics, the strategy presented here would be notably more affordable, albeit without a fully quantitative result. Nevertheless, in most cases, a medical practitioner on site needs only to know if a patient falls into a "treat" or "no-treat" category for the distribution of ART pharmaceuticals. Hence, given the simplicity of the presented device, the semi-quantitative nature of the data generated would be an acceptable caveat.

ACKNOWLEDGEMENTS

This work was supported by Enterprise Ireland under grant No. CF 2011 1317, the ERDF and the Science Foundation Ireland under grant No 10/CE/B1821.

REFERENCES

[1] UNAIDS, Global report: UNAIDS report on the global AIDS epidemic 2012, UNAIDS / JC2417E, WHO, Geneva, 2012.

[2] WHO, Consolidated guidelines on the use of antiretroviral drugs for treating and preventing HIV infection, ISBN: 978 92 4 150572 7, Geneva, 2013

[3] M. Glynn, D. J. Kinahan and J. Ducrée, *Lab Chip*, 13 (2013), pp. 2731-2748.

[4] Kirby, D.; Siegrist, J.; Kijanka, G. *et al.*, *Microfluid. Nanofluid.* 2012, 16, 1–10.

[5] M. Glynn, D. Kirby, D. Chung, D. J. Kinahan, G. Kijanka and J. Ducrée, *JALA*, 2013 – in press DOI: 10.1177/2211068213504759.

CONTACT

M. Glynn, tel: +353 1 7007899, e: macdara.glynn@dcu.ie
J. Ducrée, tel: +353 1 7007870, e: jens.ducree@dcu.ie

CENTRIFUGALLY AUTOMATED SOLID-PHASE PURIFICATION OF RNA

Nikolay Dimov, Jennifer Gaughran, Darren Mc Auley, David Boyle, David J. Kinahan, and Jens Ducrée
Biomedical Diagnostics Institute, Dublin City University, Dublin, IRELAND

ABSTRACT

The purity and integrity of RNA after extraction from a biosample like blood are critical for both downstream analysis and development of on-chip diagnostics. We report for the first time on a fully centrifugally automated solid-phase purification of RNA on an integrated microfluidic disc with sequential release of on-board reagents. Building on our earlier work the system successfully concatenates event-triggered valving and centrifugally controlled routing in order to collect purified RNA separately from the aqueous and organic wastes. In this article two-stage purification of RNA and sample conditioning through novel high-frequency release and hybrid event-triggered release valves is demonstrated.

INTRODUCTION

Effective point-of-care diagnostics require automated sample preparation strategies that are easily reproduced and efficiently controlled. Lab-on-a-disc (LOAD) microfluidic devices approach these criteria by minimizing the needs for auxiliary equipment and maximizing the efficiency of integrated flow control. Recent advances in on-disc valving have presented a dissolvable film (DF) based centrifugo-pneumatic technique [1]. This DF valving has been refined by Kinahan *et al.* to establish event-triggered fluidic networks, which can be run constant angular velocity [2]. This compound valving is built on the equilibrium between the centrifugally induced pressure acting on a liquid column and the compressed air restricted by a dissolvable film tab, actuation valve (AV). Dissolution of the AV causes the compressed air to escape while the liquid column contacts a second DF tab called release valve (RV). Thus, one liquid initiates the activation of another without external stimulus. In the current article we extend the functionality of the event-triggered release [2] by adding an external, second, control mechanism, the angular velocity. It allows short-term storage of liquid reagents over a wide range of frequencies of rotation.

Functionally different valves build the solvent selective routing of fluid flows [3], here at the pivot of RNA purification using acid-washed glass beads [4]. At high frequencies of rotation a hydrophobic membrane (HM) valve is permeated by the organic wash, carrying the remaining salts from the acid phenol extraction of RNA [5] on bench. A classical siphon valve restricts the organic; the aqueous solution passes through at low frequencies. Another DF valve downstream from the siphon covers a wide vertical channel and redirects the first aqueous flow in a side chamber. As the first flow has opened the DF valve, the second elution liquid flows over the siphon and is routed through the wide vertical channel in a separate chamber. In its initial version [3] this system allows for multiple organic

washes until the desired purity of RNA is achieved; however, we reduced the number of steps in the purification protocol and coupled the briefly described solvent selective router with the hybrid event-triggered release valves.

Downstream processing of the purified RNA often involves requires the addition of specific buffers and conditioning solutions prior to analysis. Typically, such amplification methods involve the addition of primers and / or enzymes. Here, we address this issue by engineering a separate chamber that introduces a conditioning buffer directly in the RNA collection chamber. By utilizing a novel high-frequency release valve we can precisely control the timing of events. Not only does the valve provide the possibility for tailored incubation times depending on the protocol, but also the sample RNA, purified with our system, can be directly transferred for analysis.

The integration of a novel high-frequency release valve, together with the improved hybrid event-triggered valves, and the router for solid-phase purification, extends the preceding designs and completes the automated sample preparation method on-disc.

SYSTEM DESIGN
High-frequency release valves on-disc

We utilized the geometry, orientation and dimensions of reservoirs and channels to achieve sophisticated levels of flow control on-disc.

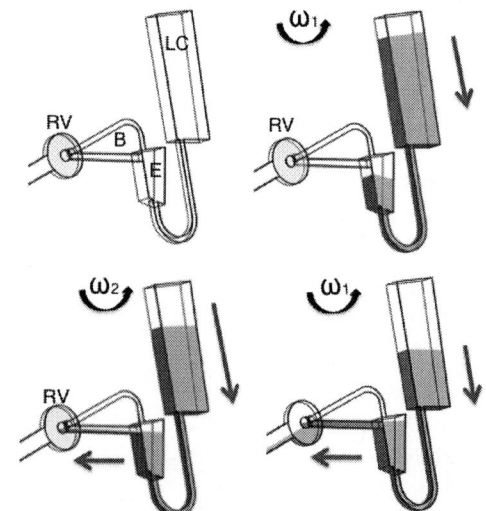

Figure 1: Schematic of the high-frequency release valve, controlled externally by increasing the angular velocity (ω_1 < ω_2) until the release valve (RV) is contacted and opened. The loading chamber (LC) is filled with aqueous buffer that flows in to the expansion reservoir (E) and at a critical angular velocity (ω_2) the liquid flows through the bypassed channel (B) and dissolves the RV.

978-1-4799-3510-9/14 $31.00 © 2014 IEEE

An aqueous buffer solution is introduced in the loading chamber (Fig. 1, LC) that is connected through a milled channel to an expansion chamber (E). Depending on the position of the solution in the LC and the height of the liquid column the liquid displacement in the expansion chamber (Fig. 1) can be calculated for a certain range of angular velocities. Furthermore, the range of velocities relates to the orientation of the bypassed channel (B) towards the expansion chamber (E). High frequency valving can be achieved when the channel is attached at a high point to E then the release of fluid requires large compression ratios; a low connection point allows the liquid to meter and contact the RV at low compression of air corresponding to low angular velocities. Liquid displacement in E correlates to the centrifugally induced pressure through the Boyle's law and has been derived earlier [6] for the development of pneumatic pumping on-disc.

The novelty in the system is the bypass (B), which prevents air trapping between the RV and the compressing liquid. In contrast to the centrifugo-pneumatic valves [1], here the RV is opened when the liquid is metered from the expansion chamber, and does not depend on the equilibrium between capillary effects and the centrifugal pseudo force. In such way, we improve the control over the consecutive flows on-disc and the reproducibility of valving.

Implementing the same bypassing strategy with the previously engineered tabs [2][3], the event-triggered release valves evolved into a hybrid valve with a bypass channel that provided dual (external and internal) control mechanism. The internal mechanism depends on the escape of compressed air through an actuation valve (Fig. 2) and acts as a primary control. Externally controlling the angular velocity determines whether the valve would open or remain closed and depends on the compression of air and metering of the advancing fluid. This is a secondary control lever as it acts after the dissolution of the AV. Altering the position of connecting bypassed channel can define valve functionality, a low- or high-frequency release. Moreover, the external control overrides the internal one, if it is activated before the AV dissolution. These finely tunable valves extend the LOAD toolbox for miniaturizing, hosting and automating multistep assays.

METHODOLOGY
Disc micromachining and assembly

The system consists of three polymethyl methacrylate (PMMA, 1.5 mm) layers, which are bonded through pressure sensitive adhesive sheets (PSA, 86 μm and 56 μm), shown in Figure 2. The top disc outlines (OD 120 mm, ID 15mm), inlets, outlets and two alignment-markers, through wholes were laser ablated (30W Laser (CO_2), Epilog, US) in the top layer. In the middle disc, through-cuts with the laser defined the chambers, vertical channels and the two alignment markers. On the surface of the bottom disc shallow air ducts were engraved and alignment markers were cut. Siphon valves, connecting channels and also valve beds were milled (MDX-40A, Roland, DG, UK) in the

middle PMMA disc. Channel depth was set to 500 μm for the siphon and 600 μm for the valve beds. Prior assembly, all PMMA components were cleaned for 30 min in an ultrasonic bath at 50°C using 2% aq. soln. of Micro 90 and consecutively were sonicated in DI water for 30 min longer; finally, the discs were dried at 80°C for 45 min.

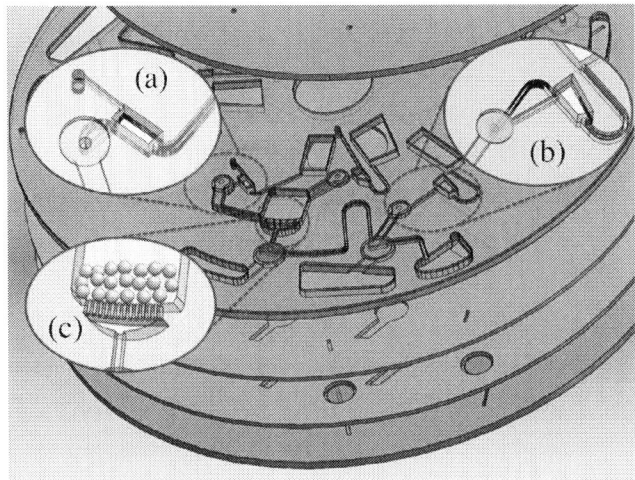

Figure 2: Exploded view of the six-layered disc: two 86 μm PSA layers (grey), one transparent 56 μm PSA and three PMMA discs (15 × 120 × 1.5mm). Magnified elements: a) hybrid event-triggered release valve and b) high frequency release valve both with bypass channels to avoid air trapping; c) glass beads inside the loading chamber. Color coded elements: hydrophobic membrane valve (green) and dissolvable film tabs (yellow). The hybrid event-triggered release valves comprise of RVs, small tabs in the middle disc, control internally the reagent discharge as the actuation valves (AVs), large yellow tabs, on the bottom disc are dissolved by the fluid flow.

The connective PSA layers were modified with a knife cutter (Graphtec, Yokohama, Japan). From the top PSA layer (56 μm) silhouettes were cutout following the outline of the loading, expansion chambers and the fluid channels around them. Thus, the top layer defined the bypass and also the air channels connecting the expansion chambers through vertical through channels to the actuation valves placed on the bottom PMMA. On the bottom PMMA disc a layer of 86 μm thick PSA sealed the laser-ablated air channels and also hosted the valve beds cutouts where the actuation valves were placed (Fig. 2). In order to close the gap between a release valve and the cutout valve bed a second PSA (86 μm) covered the valves; leaving an open slit for the liquid from a chamber to contact the DF of the AV. All features of the device were designed using SolidWorks 2011 (Dassault systems, US).

Our integrated device comprised of three identical, independent microfluidic structures. Each structure on the disc had three small valves (OD 4 mm, ID 1 mm) assembled on the topside of the middle disc to function as RVs. These were coupled to two AVs (OD 6 mm) sitting on the bottom PMMA disc. Routing was realized as reported earlier [3]

978-1-4799-3510-9/14 $31.00 © 2014 IEEE

through a hydrophobic membrane and dissolvable film (DF) valves with the same specifications as the actuation valve. All valves were assembled manually by applying pressure on the PSA periphery. Adhesion between the PMMA discs was secured with multiple runs through a mechanical roller press (Hot Roll Laminator, Chemsultant Int., US).

Bacterial culture and RNA extraction

The *E. coli* bacteria were collected in their log phase after reaching density of 6.4×10^6 colony forming units (CFU) per mL (based on a plate count). Duplicated samples were reduced to 25 µL by centrifugation at 2600 rpm for 10 min, corresponding to bacterial content 6.4×10^6 and 3.2×10^6 CFU. An acid phenol chloroform three-phase extraction was performed by adding 80 µL of Trizol reagent and 2 µL of 4-bromoanysole to each vial. The samples were centrifuged at 14000 rpm for 15 min at 4°C in order to reach phase separation. Bench top control samples were also extracted and later purified using previously reported 2-propanol precipitation with ethanol washes [3][5].

Protocol for purification of RNA on-disc

The aqueous phase, containing the total RNA (60 µL), was filled to 100 µL with water. This solution was introduced in the loading chamber of the disc on top of 50 mg acid-washed glass beads (150 - 200 µm). Preloaded on-disc were: 75 µL of 75% aq. soln. ethanol (EtOH) for the washing; 100 µL of RNAase free water for eluting the RNA; and for the conditioning step 85 µL of 50 mM Tris-Cl, pH 8.1 with 1 × EDTA buffer.

Bench top detection and analysis of the purified RNA

The concentration and integrity of the purified total RNA were determined by capillary electrophoresis (RNA 6000 Pico kit, Bioanalyzer 2100, Agilent technologies, US) according to the manufacturers instructions. The purified samples were also screened for small (less than 200 nt) RNA content (Small RNA kit, Bioanalyzer 2100, Agilent).

RESULTS

Microfluidic tests

As a pilot validation assay, the device was used for total RNA purification from the aqueous lysate of *E. coli*. Influenced by the chaotropic salt, the RNA was retained on the glass beads [4]. The depleted from RNA solution was driven over a siphon valve into an aqueous waste chamber while two dissolvable valves were opened: one of the router [3] and a second of the actuation valve coupled to an upstream release valve. The EtOH wash (Fig. 3, c) was triggered by the aqueous waste and, after washing over the beads, was collected in the organic waste chamber. Switching the angular velocity to 100 Hz released the conditioning solution into a large collection chamber, where a second actuation valve was opened (Fig. 3e). It triggered the aqueous elution buffer to dissolve the release valve. This final flow stripped the beads off the RNA and mixed it with the conditioning buffer. Purification was completed in less than 8 minutes after implementing the rotational protocol (Fig. 4). Afterwards, the samples were collected from the disc for further analysis.

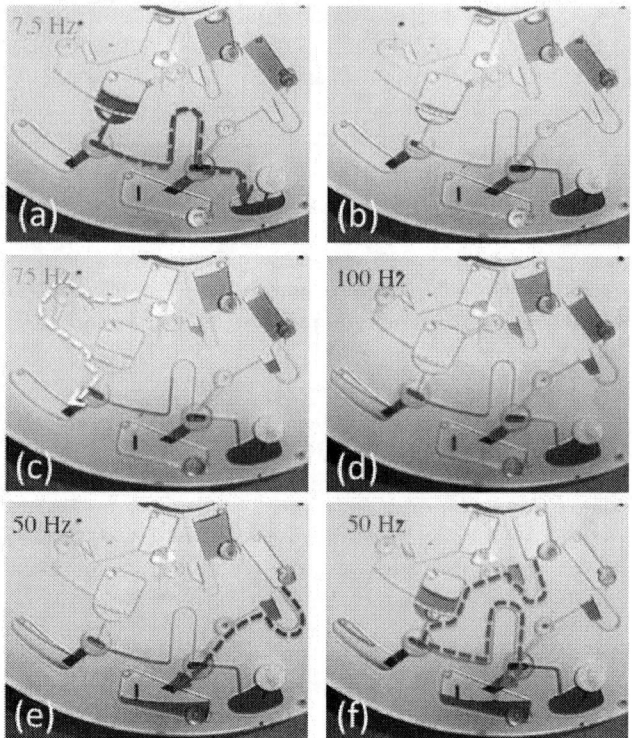

Figure 3: Complete sample preparation and conditioning on automated centrifugal microfluidic disc. The dashed lines follow the fluid flow on-disc: a), b) Crude extract (red) is collected in the aqueous waste chamber while the RNA is retained on glass beads (not shown in this fluidics experiment). c) Aqueous solution (white) of ethanol (75% aq.) is released over the beads and is driven into organic waste carrying the salts away. d) Conditioning solution contacts the high-frequency, externally controlled, release valve. e) It dissolves also the actuation valve that triggers the release of the elution buffer. f) Aqueous elution phase re-suspends the RNA and mixes with the conditioning solution. The pure sample is collected and analyzed (Bioanalyzer 2100) on-bench for RNA content.

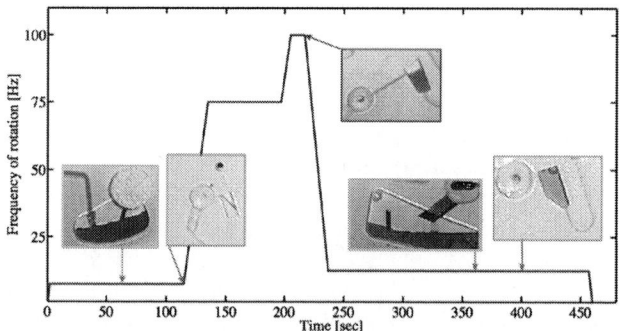

Figure 4: Frequency of rotation on disc versus experimental time. The arrows relate flow-triggered release of fluids with the timeline. Actuation valves (orange) open in contact with the fluid and allow the liquid to contact the release valves (green). A novel type of high-frequency hybrid valve opens at 100 Hz (red) as a bypass channel prevents air trapping.

Validation of the RNA purification on-disc

The purity and integrity of the RNA samples that were processed on-disc were found comparable to those extracted on bench. The electropherograms from the on-disc two-step purification process with conditioning resulted in lower total RNA (Fig. 5) and small RNA (Fig. 6) concentrations in comparison to their four-stage bench top counterparts.

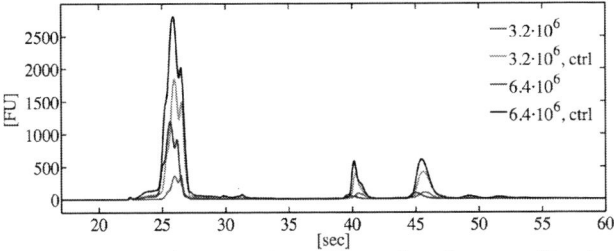

Figure 5: Total RNA from E. coli after purification. Integrity and purity of the RNA are comparable between bench-top and on-disc. Duplicate samples were analyzed for each assay, and after purification from 6.4×10^6 cells, on-disc were extracted 10.7 ± 2.1(blue) versus 27.6 ± 8.2 ng/µL (black) during bench top control; from 3.2×10^6 cells 4.9 ± 0.3 ng/µL (red) versus 20.7 ± 1.4 ng/µL (green) bench top.

In order to calculate the amount of total RNA extracted from each sample the concentration must be multiplied by the sample volume. For the LOAD processed RNA the volumes are the sum of the elution (100 µL) and the effective volume of conditioning buffer (70 µL). Effective volume is the nominal volume of the chamber (86 µL) that is corrected by the volume of the expansion chamber (16 µL), which is retained after the elution due to the orientation of the expansion chamber. The resulting RNA contents are 1.83 µg from disc versus 2.76 µg bench-top for the higher concentration, and 0.83 µg versus 2.07 µg from the lower concentration of cells (Fig. 5). Capture and elution efficiencies are thus between 40% and 66% from the bench-top yield.

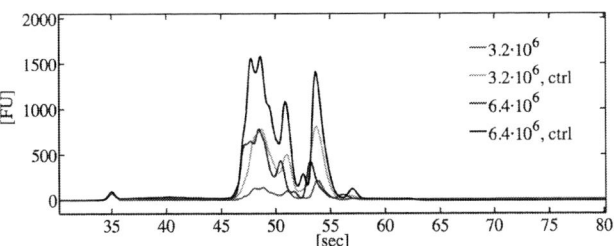

Figure 6: Comparison between small RNA from two-step purification with conditioning and four-stage bench-top purification without conditioning. The peaks showed identical footprint and resembling pattern after fewer purification steps and in less time.

DISCUSION AND CONCLUSSION

The integration of a novel high-frequency release valve together with event-triggered valves improves the solid-phase purification method that we implemented in our earlier router design [3]. The improved system for automated purification required significantly less time (8 min) in fewer purification steps, and provided conditioning of the eluted RNA. The higher extraction efficiency is a result of an exchange of the solid phase. In our previous work we used 30 mg of glass beads with diameter less than 106 µm [3] and recovered 12 % of the initial RNA. Further research will focus on the optimization of capture and elution efficiency by exploring the performance of the system with various concentrations of chaotropic salts in the aqueous extract. Despite the somewhat lower recovery, the quality of the total RNA is extremely good with no signs of degradation. After an optimization, our automated system with its high fidelity controlled valving could be directly implemented for complete sample preparation from aqueous cell lysate in biological RNA related studies.

ACKNOWLEDGEMENTS

This work was supported by the Science Foundation Ireland under grant 10/CE/B1821 and the BioAnalysis and Therapeutics Structured PhD Programme (Bio-AT) funded by HEA-PRTLI V.

REFERENCES

[1] R. Gorkin III, C.E. Nwankire, J. Gaughran, X. Zhang, G.G. Donohoe, M. Rook, R. O'Kennedy, J. Ducrée, "Centrifugo-pneumatic valving utilizing dissolvable films", *Lab Chip*, vol. 12, pp. 2894–2902, 2012.

[2] D.J. Kinahan, S. Kearney, J. Ducrée, "Auto-actuated Sequential Release Valves for Lab-on-a-Disc systems" in *Digest Papers Solid-State Sensors, Actuators and Microsystems Transducers & Eurosensors Conference: The 17th International Conference on Transducers'13*, Barcelona, June 16-21, 2013, pp. 2189-2192.

[3] J. Gaughran, N. Dimov, E. Clancy, T. Barry, T.J. Smith, J. Ducrée, "Multi-stage, Solvent-controlled Routing for Automated on-disc Extraction of Total RNA from Breast Cancer Cell Line Homogenate", in *Digest Papers Solid-State Sensors, Actuators and Microsystems Transducers & Eurosensors Conference: The 17th International Conference on Transducers'13*, Barcelona, June 16-21, 2013, pp. 305-308.

[4] R. Boom, C.J. Sol, M.M. Salimans, C.L. Jansen, P.M. Wertheim van Dillen, and J. van der Noordaa, "Rapid and simple method for purification of nucleic acids", *J. Clinical Microbiology*, vol. 28, pp. 495–503, 1990.

[5] P. Chomczynski, N. Sacchi, "Single-step Method of RNA Isolation by Acid Guanidinium Thiocyanate-phenol-chloroformextraction", *Analytical Biochemistry, vol.*162, pp. 156-159, 1987.

[6] R. Gorkin, L. Clime, M. Madou, H. Kido, "Pneumatic Pumping in Centrifugal Microfluidic Platforms", *Microfluid. Nanofluidics*, vol. 9, pp. 541–549, 2010.

CONTACT

*Jens Ducrée, jens.ducree@dcu.ie

FABRICATION METHOD TO A HIGH RESOLUTION CONTROL IN THE SPACE OF CELL CULTURING ENVIRONMENT WITH MICROFLUIDIC SYSTEM

Takumi Hiraiwa[1], Tadamasa Kimura[1], Yuma Takenaka[1], Ryo Tanamoto[1], Hiroki Ota[4], Hiroshi Kimura[3], Yoshihiro Taguchi[2], Norihisa Miki[2], Yoshinori Matsumoto[1], Kotaro Oka[1], Akira Funahashi[1], and Noriko Hiroi[1]

[1]School of Fundamental Science and Technology, Keio University, JAPAN
[2]School of Integrated Design Engineering, Keio University, JAPAN
[3]Department of Mechanical Engineering, Tokai University, JAPAN
[4]Electrical Engineering, University of California, Berkeley, USA

ABSTRACT

This paper describes the fabrication and the evaluation of a reusable Cell Culturing Device, designed for a single cell and cellular networks analysis. This is the first success of combination of Microcontact Printing (mCP) and Vacuum Device. This combination has following advantages; (1) cells stay within the micropatterns for long enough duration to achieve local activation of cells or cellular networks (more than 24 h), (2) the displacement distance of laminar flow from the interface of two fluids in Cell Culturing Device keeps smaller than the diameter of a cell, (3) all components of our device except micropatterned substrate are reusable for further analyses. The success of the combination of above techniques provides a controllable environment for the local activation of a single cell and cellular networks. Our device allows to exhibit the different responses induced with the various conditions in a single observation sight at exactly the same time point.

INTRODUCTION

Spatiotemporal changes in a single cell and cellular networks cause complex phenotypic behaviors of cells: differentiation [1], migration [2], and polarization [3]. These behaviors affect correct tissue formation and functions in individual. Analysis of these cellular behaviors requires methods that can perturb or manipulate spatiotemporal environment precisely. In previous work, perturbing and manipulating cellular spatiotemporal environment to elucidate the mechanisms of neural polarization, a glass bead coated with forskolin was manipulated into contact with a neurite of undifferentiated hippocampal neurons [3]. However, this stimulation method using a glass bead has two problems; (1) a glass bead method has limitation dependent on the molecular specificity (2) to control the location of the bead is difficult with this method in principle. To overcome these problems, fluidic stimuli has advantages. Therefore, analysis of cellular behaviors requires liquid manipulation method.

Microfluidics is a field of study that use devices with small channel to control and manipulate minute amounts of fluids. Microfluidic device were first developed to miniaturize chemical and biochemical analyses. Recently, microfluidic devices fabricated in poly-dimethylsiloxane (PDMS) using soft-lithography have found increasing applications in basic and applied biomedical research. Microfluidic device has several advantages in biology. First, high Reynolds number in micro channel let us to control fluid precisely and stably. Second, PDMS has low cytotoxicity and high air permeability. Using microfluidic device, users can localize fluidic stimuli at a single cell level and long-term culture for analysis of cell differentiation. However, previous works of functionally similar devices had a problem to keep preciseness of stimulation pattern in long-term culture, which is prevented by cell migration. To solve this problem, we applied mCP [4] which forces cells to stay within expected areas. Now the solution we chose became the next problem to solve; because we need to keep water-tight seal with the other methods except exposing the printed substrate to UV, otherwise the printed protein or peptide could lose its physiological structure and the desired function. We solved this problem by Vacuum Device [5], which can adhere to the substrate by negative pressure.

This work demonstrates that the Cell Culturing Device fabricated by Vacuum Device and mCP is useful to analyse spatiotemporal changes in a single cell and cellular networks by controlling micro-environment for long-term culture.

METHODS

Cell Culturing Device fabrication

The master mold of the Vacuum Device was designed to have two kinds of channel: liquid channels and vacuum channels (Figure 1A). Liquid channels consist of inlet channels (width 200 μm, height 100 μm), narrow channels (width 75 μm, height 100 μm) and main channel (width 400 μm, height 100 μm). Narrow channels have high resistance. Therefore, lager number of cells are maintained in the main channel. In addition, vacuum channels consist of vacuum port and vacuum channel (width 100 μm – 200 μm, height 100 μm).

The master mold of the device was made in a cleanroom photolithography facility. Glass wafer (MICRO SLIDE GLASS, S9111, MATSUNAMI) applied photoresist (SU-8 3050) was spun at 2000 rpm. After spincoat, we transferred the wafer to hotplate and performed softbake (100°C, 5 min). This process (spincoat - softbake) was repeated twice. After spincoat process, the wafer was exposed to UV light (250

978-1-4799-3510-9/14 $31.00 © 2014 IEEE

mJ/cm^2) through emulsion photomask. Exposed wafer was baked at 65°C for 1 min, 100°C for 2 min and developed by SU-8 developer. The mold applied PDMS (SILPOT, TORAY) mixed with curing agent at 10:1 ratio. The PDMS covered master was placed into a 65°C incubator and allowed curing for 2 h. After curing, device was cut out from PDMS by using razor blade. We carefully peeled the device off the master using tweezers.

Following processes of device fabrication method is shown Figure 1B. The Port3 of the PDMS replica were cut a hole with a biopsy punch (φ 1.5 mm). A needle (19 gauge) was inserted into the Port3. Furthermore, to fix the needle, we placed silicone tube (φ 3 × 5 mm) at the Port3 and poured PDMS into silicone tube. Reservoirs fabricated by PDMS were placed on the inlets of the PDMS replica and fixed by PDMS. After curing the PDMS, the Port1, 2 were cut holes with a biopsy punch (φ 1.5 mm). Then, an upper part of the device was placed on the reservoirs and fixed by PDMS. After device fabrication, fluid channels were coated with 2-methacryloyloxyethyl phosphorylcholine (MPC polymer, NOF Corporation) to prevent from non-specific adhesion of cells.

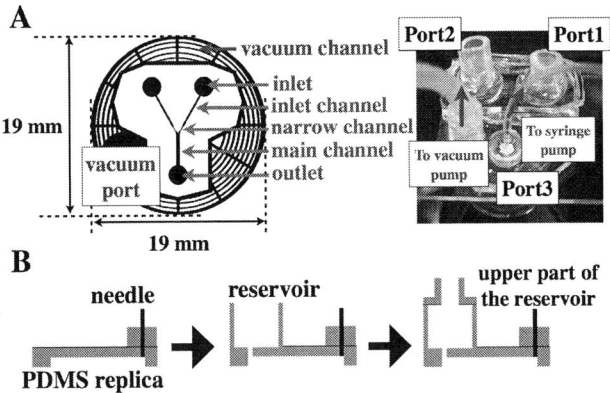

Figure 1: Design and fabrication method of the Cell Culturing Device (A) Design of the Vacuum Device and picture of the Cell Culturing Device, (B) fabrication method of the Cell Culturing Device.

Stamp fabrication and Microcontact Printing on a glass base dish

Molds for the stamps were produced with above mentioned lithography technique by exposure of UV to SU-8 3050 spun at 4000 rpm through a chrome photomask. After lithography, molds were applied PDMS mixed with curing agent at 10:1 ratio and cured for 2 h in 65°C incubator. After curing, we cut off the PDMS stamp.

The PDMS stamp was sonicated for 5 min in acetone, iso propanol and ethanol sequentially. After drying under a stream of nitrogen, the PDMS stamp was activated in the plasma chamber during 1 min and inked with laminin-fibronectin solution (laminin 200 µg/ml, Sigma-Aldrich, fibronectin conjugated with HiLyte488 20 µg/ml, Cytoskelton, in phosphate-buffered salt; PBS). The stamp was incubated for 45 min at room temperature. After 45 min, solution was removed and rinsed with PBS twice and deionized water twice, and then dried under a stream of nitrogen for 1 min. The stamp was immediately placed in contact with plasma treated glass base dish (IWAKI) and pressed slightly with tweezers. After 1 min, the stamp was peeled from the substrate carefully. The glass base dish was incubated on a hot plate at 140°C for 30 sec [6]. The stamp was cleaned for 1 h in deionized water, dried in ethanol and reused several times.

PLL-g-PEG backfilling after mCP

The glass base dish after mCP was incubated in PBS containing poly-L-lysine(20)-g[3.5]-poly-ethyleneglycol(2) (PLL-g-PEG, SurfaceSolutionS) 0.1 mg/ml for 1 h at room temperature. After PLL-g-PEG coating, PLL-g-PEG was removed. The dish was then rinsed with PBS once and ultra-deionized water once. The dish was stored in Hank's balanced salt solution (HBSS, Life Technologies) until experiment at 4°C or directly used for next steps.

Cell culture

SK-N-SH cells (Human Neuroblastoma Cell Line, obtained from RIKEN) can differentiate into nerve-like cell by stimulation of retinoic acid. Cells were maintained in alpha-minimum essential medium (α-MEM, WAKO) with 10% fetal bovine serum (FBS, Biowest) and were grown in a 5% CO_2 incubator at 37°C.

Attaching the Vacuum Device to micropatterned glass

The Vacuum Device is possible to attach the device body to various substrates by producing negative pressure in the vacuum channel (Figure 2). The device body was placed onto the glass base dish, and then vacuum pump (DAP-15, ULVAC KIKO) was turned on. The ultimate pressure in the vacuum channel is probably 39.9 kPa. Negative pressure in the vacuum channel enabled to seal the device and the substrate. Sealed device was incubated for 30 min at 37°C before cell culturing. To take off the device from the substrate, users simply needed to turn off the vacuum pump switch. Detached device is reusable for the further experiment.

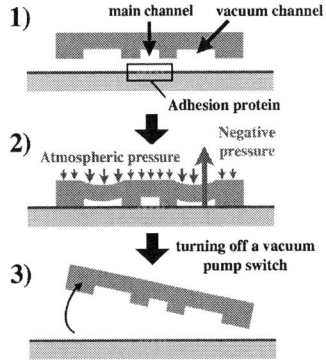

Figure 2: Illustration of attaching the Vacuum Device body to micropatterned substrate.

Operation of the Cell Culturing Device

After detaching the SK-N-SH cells by incubation with

0.25 w/v% trypsin solution (WAKO) at 37°C for 1 min, cells were resuspended with fresh α-MEM and centrifuged at 1000 rpm for 5 min. Finally, the concentration of the cell suspension was adjusted to 1×10^6 cells/ml. The injection of cell suspension to microchannel from the Port3 was performed by gravitationally using φ 3 × 5 mm silicone tube (Figure 3). After this process, the device was placed in Stage-Top Incubator (TOKAI HIT) at 5% CO_2, 37°C. After cell injection, the silicone tube was removed from the Port3 and the device was incubated until cell adhesion. After cell adhesion, until the height of medium in the Port1, 2 got higher than the height of medium in the Port3 (total volume in the Port1, 2 was approximately 200 – 300 μl), fresh medium was injected into Port1, 2 and the device was incubated at 5% CO_2, 37°C. After 30 min from fresh medium injection, the Port3 was connected to micro-syringe pump (NEXUS 3000, ISIS) with φ 1 × 2 mm silicone tube. Then, at every 10 min, medium in the device was withdrawn using micro-syringe pump at volumetric flow rate of 0.5 μl/min and cells in the device were cultured in a 5% CO_2 incubator at 37°C. Every 12 h, we injected fresh medium into the Port1, 2.

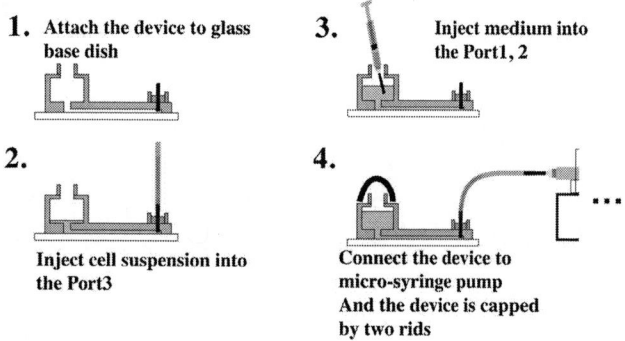

Figure 3: Illustration of cell injection into the Cell Culturing Device.

RESULTS

Observation of laminar flow in the main channel of the Cell Culturing Device

To confirm whether or not laminar flow has a potential to localize stimuli for a single cell, laminar flow interface of PBS and PBS containing 0.25 mg/ml fluorescein isothiocyanate (FITC) conjugated with dextran (mw = 9800, Sigma-Aldrich) in the main channel caused by withdrawing using micro-syringe pump at 2, 3, 4 μl/min was observed by fluorescence microscope (IX81, Olympus). And we measured displacement distance of the interface from the position of the start of measurement (observation at every 10 sec during 500 sec) at 500 μm downstream from the Y-junction of the device (Figure 4). We used ImageJ (NIH) to determine the interface of the two flows. Standard deviations of the displacement distance at 3 μl/min and 4 μl/min were ±5.14 μm and ±4.84 μm respectively. However, the displacement distance of less than 2 μl/min fluctuated intensely. This result indicates that laminar flow more than 3 μl/min has a potential to localize stimuli for a single cell.

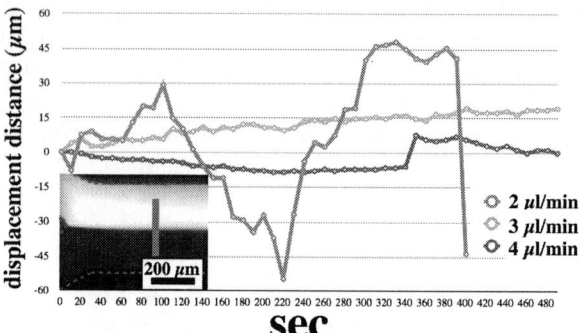

Figure 4: Displacement distance of the interface of the two flows at each volumetric flow rates.

Produce of heterogeneous environment at a single cell

SK-N-SH cells were injected into the Port3 and cultured in the device for 24 h (Figure 5). Figure 5A shows that the micropattern of laminin-fibronectin conjugated with HiLyte488 in the main channel at 12 h after cell injection. Figure 5B and Figure 5C show growing cells at 12 h and 24 h after cell injection. Right pictures of Figure 5 are produced by merging the image of predictive micropattern and the image of growing cells at 12 h and 24 h, respectively. We confirmed that SK-N-SH cells were kept on the same position more than 12 h (Figure 5, dotted red line).

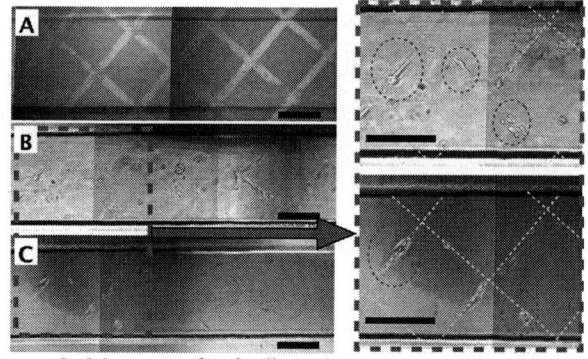

Figure 5: Micrograph of adhered SK-N-SH cells in the main channel (A) Micropattern of laminin-fibronectin conjugated with HiLyte488 at 12 h after cell injection, (B) growing SK-N-SH cells at 12 h after cell injection, (C) growing SK-N-SH cells at 24 h after cell injection. Right pictures are merged images that show predictive micropattern and growing cells at 12 h and 24 h, respectively. Scale bar = 200 μm.

After cell culturing in the device, to produce heterogeneous environment at a single SK-N-SH cell, we adjusted staining solution consisting of serum free α-MEM and Cell Tracker Orange (LONZA, 1 μM). This staining solution was injected into the Port2 and withdrawn at 4 μl/min. During the first 5 min after staining solution injection, the interface of the staining solution and α-MEM was fluctuated in the range indicated in Figure 6 (between dotted red and white line). However, after approximately 10 min, the interface became stable at the center in the main channel (Figure 6).

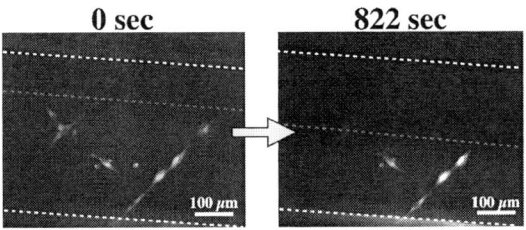

Figure 6: Predictive position of the interface of the two flows in the main channel at each elapsed time after staining solution injection.

The interface was stable for 30 min. We observed that a single cell in the main channel was stained locally by Cell Tracker Orange (Figure 7). We confirmed whether local staining at a single cell level was accomplished or not by detailed image analysis as follows. We set a Region Of Interest (ROI) in a single cell that adhered on the micropattern at center of the main channel and measured fluorescent intensity in the ROI (Figure 7, middle lane pictures). At 267 sec after staining solution injection, staining solution flowed at over half region in the main channel by impact of solution injection into the Port2. Therefore, a single cell or cells at center of the main channel were stained completely (Figure 7, middle left). However, at 1090 sec after staining solution injection, the interface of the two flows became stable. Therefore, a single cell or cells at center of the main channel were stained locally (Cell Tracker Orange molecules adhered on an upper region of a single cell were washed away by flow). By calculating the Intensity Ratio in the ROI, we confirmed that an upper region of a single cell at 1090 sec was not stained by Cell Tracker Orange.

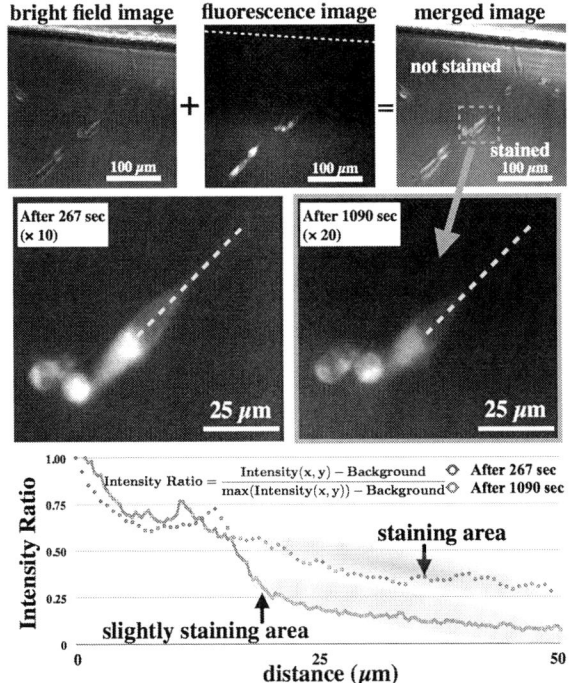

Figure 7: Local staining for a single cell using laminar flow in the main channel.

Furthermore, we evaluated the environment around a single cell at center of the main channel (Figure 8). We set a line ROI and measured fluorescent intensity. Then, gradient of intensity was observed in the ROI. Cell Tracker Orange molecule has low molecular weight (mw = 550). Therefore, Cell Tracker Orange may diffuse in the main channel toward α-MEM region. However, it was not clear that if the gradient has brought by diffusion. We doubted if there could be another reason to produce the gradient of Cell Tracker. Position of the cell was about 1 mm downstream from Y-junction. Diffusion time was about 1.6 sec and mean square displacement of 2D was about 3.3 μm (diffusion coefficient of Cell Tracker Green is 3.4×10^{-8} cm^2/sec [7], mw = 464). Therefore, the gradient was caused by not diffusion but other causes.

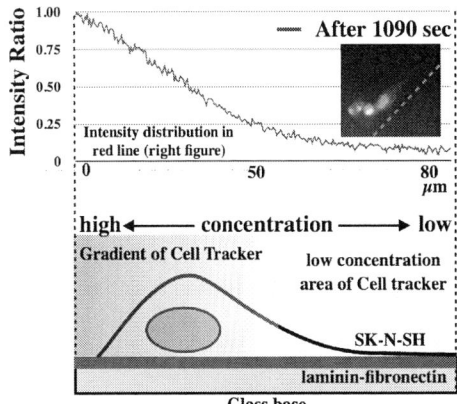

Figure 8: Environment around a single cell.

CONCLUSIONS

Cell Culturing Device is the first success of combination of mCP and Vacuum Device. This device has a potential to localize stimuli at a single cell, it was evaluated by the evidence of local staining of a single cell at 4 μl/min and stable laminar flow in the main channel. Adhered cells in the main channel were kept within the micropatterns more than 24 h. These results show that this device and method have a potential to analyze the molecular mechanisms of physiological morphology dynamics under spatially controlled signals, especially which take long-term to complete.

REFERENCES

[1] N Mochizuki et al., *Nature*, Vol. 411, pp. 1065-1068, 2001

[2] C Cecilia et al., *Nature Reviews Neuroscience*, Vol. 10, pp. 433-446, 2009

[3] M Shelly et al., *Neuron*, Vol. 71, pp. 433-446, 2011

[4] A Bernard et al., *Advanced Materials*, Vol. 12, pp. 1067-1070, 2000

[5] B G Chung et al., *BMC biotechnology*, Vol. 7, pp. 60, 2007

[6] J Fink et al., *Lab on a chip*, Vol. 7, pp. 672-680, 2007

[7] M Wartenberg et al., *FASEB J*, Vol. 15, pp 995-1005, 2001

INDIVIDUALLY ADDRESSABLE MULTI-CHAMBER ELECTROPORATION PLATFORM WITH DIELECTROPHORESIS-ASSISTED CELL POSITIONING

Sinwook Park[1], Gilad Yossifon[1]

[1] Technion–Israel Institute of Technology, Haifa, ISRAEL

ABSTRACT

We developed a multifunctional microfluidic platform which integrates electroporation and dielectrophoretic (DEP) trapping/positioning mechanisms. Its feasibility was demonstrated by characterizing and optimizing electroporation efficiency using on-chip spatial gradient of electroporation parameters. For simplicity, human dermal fibroblasts cells with a membrane impermeable fluorescent dye, PI, were used as to indicate electroporation. The results clearly show that the developed microfluidic platform has the potential to achieve high-throughput screening for electroporation with spatial control and uniformity.

INTRODUCTION

Electroporation (EP) has received much attention in the biological field as a powerful tool, with the aim to exploit the ability to physically introduce foreign molecules into the intra-cellular space with enhanced transfection efficiency [1]. Furthermore, the pores created in the cell membrane by impulse of high electric fields enables the introduction of membrane-impermeable exogenous molecules into the cells or to release intracellular components from the cells. The specific parameters of the applied electric field (i.e. pulse amplitude, pulse duration, numbers of pulse) determine whether, the exposed cell membranes can be resealed rapidly ensuring cell viability (reversible EP) or if they will be disrupted permanently resulting in cell death (irreversible EP). Both states of EP can be applicable to different fields, for example, the reversible EP is useful for delivery of gene, proteins, drugs while the irreversible EP in necessary for extracting cellular materials from the cells or tumor-ablation [2]. Therefore, determining the optimal field parameters is critical but also challenging due to the large number of parameters and strong dependence of the outcome on cell-type or environmental condition. The commonly used bulk EP system has drawbacks such as uncontrolled and nonuniform transfection efficiency and the requirement of high voltages (kV).

Microfluidic based EP systems can offer a solution to overcoming the problems associated with conventional EP systems owing to its unique characteristics of miniaturization [2]. A number of microfluidic platform which focus on viable electroporation, electrolysis, single cell level analysis and multi-cell arrays for high throughput screening have recently be developed[3]. In order to improve the processing rate or efficiency of electroporation, it is highly advantageous to maximize the benefits of microfluidic platforms by integration of multi-functionality, like the combination of DEP-assisted cell manipulation with electroporation [4] or the coupling of electroporation with capillary electrophoresis (CE) and electrochemical detection [5].

Here, we develop a novel, multifunctional platform with potential for high throughput screening which integrates dielectrophoretic trapping/positioning and electroporation mechanisms in a multi-chamber array where each chamber may be individually activated. The key advantage of our platform lies in its ability to produce a spatial gradient of electroporation parameters (i.e. pulse amplitude, duration, number) to optimize electroporation efficiency in conjunction with DEP-assisted cell trapping or positioning. Additionally, the use of transparent indium tin oxide (ITO) electrodes, rather than the more traditional metals which are opaque, enables visualization of cell transfection in real time. As a proof of concept, platforms with an array of 40×2 cell microchambers were fabricated and tested to demonstrate the feasibility of integrating the two functionalities: 1) the manipulation of the particles/cells using DEP, and 2) gradient electroporation.

METHODS

Design of microfluidic platform

The developed microfluidic platform consists of two ITO-patterned glass substrates sandwiched by a patterned SU8 structure and a thin film of PDMS (Fig.1a). The patterned SU8 structure defines the geometry of the channels and micro-chamber array and serves as a spacer between the top and bottom substrates. The corresponding patterned ITO electrode array enables the individual activation of each microchambers, exploiting the parallel plate design to introduce an electroporation gradient of uniform intensity within each of the microchambers. Additionally, by simply not applying a pulse at a single "control" chamber, we create a unique internal control which is under identical environmental conditions to the active cell chambers. Each microchamber and electrode is 400 μm in length, 200 μm in width and is separated from adjacent chamber by 150 μm via the SU8 wall. In order to prevent leakage of solution between the two plates, a thin patterned PDMS film was coated onto the top ITO electrode. The final space between top and bottom ITO electrodes is determined by the thickness of PDMS film (30 μm) and the SU8 structure (15 μm).

Due to the parallel plate electrode structure, the electric field is uniformly distributed inside the microchamber array, except the very edges of the chambers, which is well-demonstrated in simulated electric field distribution between parallel electrodes across the chamber using the finite elements program (Comsol™ 4.3) shown in Fig. 1b. In addition, this electrode design enables the dynamic manipulation of cells/and or particles using DEP forcing. The simulated electric in Fig. 1b indicates that high electric field gradients occur at the edge of the ITO electrodes, which coincides with the entrance of each microchamber

(Fig. 1b). Thus, the particles may be attracted to the corners of the chamber using attractive positive-DEP (P-DEP) forces or repulsed by negative DEP (N-DEP) as illustrated in Fig. 1b (right).

Fabrication of microfluidic platform

The layered view of the fabricated platform is illustrated in Fig. 1c. To fabricate the top and bottom electrode layer, Indium Tin Oxide (ITO) coated glass slides were pre-cleaned with Acetone/Methanol/ deionized water. Prior to patterning the bottom ITO electrode layer, metal of Au/Ti (200nm/ 30nm) was deposited onto the electrode pads only by lift-off process which enables easy alignment during the multi-step photolithography process. Afterwards, the ITO was patterned using standard photolithography and wet etching processes [6]. To create the walls of the channels and micro-chamber array, 15 μm-thick SU-8 negative photoresist (SU-8 2015, MicroChem) was then patterned on the ITO electrode array. For the top cover plate, an ITO coated glass was spin-coated with PDMS with a speed of 5000 rpm as an intermediate layer. In order to leave the top area of the chamber, channel and electric pads without PDMS film, a 1 mm-width rectangular scotch tape was simply attached at a predefined position before PDMS coating. After peeling off the tape, the 30 μm-thick PDMS film was cured on hotplate at 80 °C for 10 min, and 1.0 mm-diameter inlet holes were mechanically drilled. The fabricated top and bottom plates were then assembled and packaged using a mechanically screwed holder

Preparation of Cell Suspension

Normal human dermal fibroblasts (NHDF-Ne, Lonza Walkersville Inc, cc- 2509), Passage 8, were maintained in cell culture medium at 37°C and 5% CO2 incubator. The cells were consistently harvested at 80% confluence by using Trypsin and pelleted by centrifugation and the excess medium was removed.

DEP particle/cell manipulation

In order to examine the capability of DEP-assisted cell/particle manipulation, $1 \mu m$ fluorescent polystyrene beads were diluted to volume concentrations of 2×10^{-4} % in deionized (DI) water (σ_m=0.05 μS/cm), and were loaded into the platform. The motions of particles and/or cells through the microchannel/ chamber interface were recorded for different AC field frequencies (i.e. 20 KHz and 5 MHz) with 20 V_{pp} amplitude. In addition, the loaded cell suspensions with cell medium (σ_m=16 mS/cm) were activated at the condition of 5 MHz, 20 Vpp to observe the repelling motion of the N-DEP.

Electroporation

To optimize the electroporation parameters, PI and FDA were used to assess electroporation efficiency and cellular viability. For efficiency assays, PI with concentrations of 20μg/mL was mixed in cell suspension and loaded into the platform. Electroporation was simultaneously addressed to

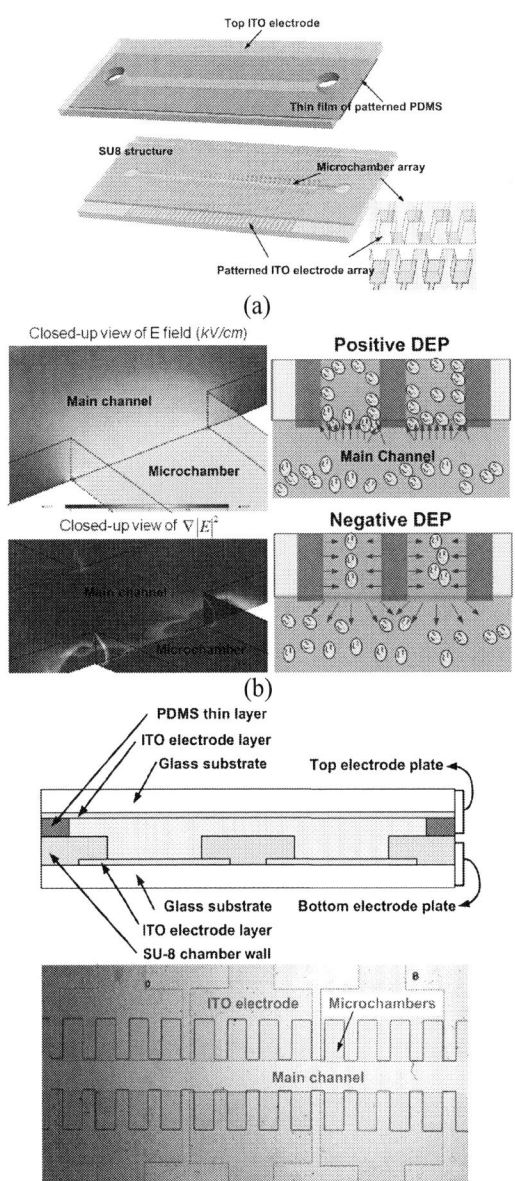

Figure 1: Design of microfluidic platform (a) Schematics of the combined DEP-electroporation platform; (b) 3D numerical simulation results of the electric field distribution indicating the uniformity of the electric field within microchamber excluding its edges where strong field gradient exist (left) and schematic illustration of cells' DEP response (right); (3) layered view of the platform and a micrograph of fabricated microchamber array.

the microchamber array with various sets of electric pulseparameters using an electrical voltage source (Keithley 2636). Thereafter, the microchambers are optically monitored for additional 10 minutes under an inverted microscope to observe the uptake of PI through the cell membrane. 40 minutes after electroporation, cells in the platform were further stained by 100μg/mL fluorescein diacetate (FDA) to assess cell viability. The captured images were converted to gray scale and further processed using image processing software (ImageJ, NIH). The resulting intensities of PI-stained cells can be normalized as follows:

$$I^* = \frac{I - I_0}{I_f - I_0} \qquad (1)$$

where, I^* is the normalized intensity; and I, I_0, and I_f are the intensities of a given PI-stained cell, non-stained cell, and fully PI-stained cell (i.e., dead cells), respectively.

RESULTS

Figure 2 shows typical motions of 1 μm particles and cells activated by DEP forcing. We see that the DEP forces produced at the chamber entrance and edges enables particle trapping by P-DEP at 20KHz, after which the trapped particles at the edges were repelled toward center of the chamber by N-DEP at 5MHz (Fig. 2a). This result corresponds well with the simulated electric field gradients. Similar behavior was observed with cells, wherein under application of N-DEP (5MHz with 20Vpp), cells located at the edges were moved toward the center of the microchamber or repelled to outer chamber (Fig. 2b). With this manipulation, more uniform and accurate electroporation experiments can be readily achieved. These results clearly demonstrate that our platform has a capability to produce a functionality of DEP-assisted cell manipulation which enhances the efficiency of electroporation.

The electroporation efficiency and cell viability were determined by varying the parameters of the applied electroporation pulse (i.e. a single pulse with 10 different amplitudes from 2V to 20V and, durations of 0.5, 1.0, 3.0 ms). The electroporation rate and viability of the cells were characterized by analyzing the number of stained cells which were subject to electroporation, symbolized by PI uptake and their viability demonstrated by FDA staining. Due to the membrane impermeable property of PI, the uptake with PI is an indication that exogenous molecules are diffused through the membrane by electric stimulation. To distinguish reversible and irreversible EP including cell death, viability assays using FDA are performed in which FDA-stained cells indicate viability.

Figure 3a shows the micrographs of the PI/ FDA stained cells in the chamber with various electric pulse parameters 40 minutes after electroporation. For visualization, three images with different coloring were merged, wherein the red and green colors represent PI and FDA stained cells respectively, while the gray background is obtained from images taken in bright field. The results illustrate that as pulse amplitude is increased, the number of PI-stained cells was increased, while cell viability decreased (Fig. 3a and b). At shorter pulse duration, more than 50% of PI-stained cells are observed at high pulse amplitude, while at higher duration, lower pulse amplitude is required to achieve that percent of PI-uptake.

The number of both PI and FDA stained cells indicates the electroporation efficiency as shown in Fig. 3c. The average maximum efficiency at 0.5, 1, and 3 ms are 50.1 ± 11.5 %, 52.2 ± 11.0 %, and 9.8 ± 9.4 %, respectively. From the results, it could be anticipated that shorter pulse duration result in better viable electroporation efficiency. This assessment is further quantified by the associated fluorescence intensity which reflects its free diffusion across the membrane (Fig. 3d).

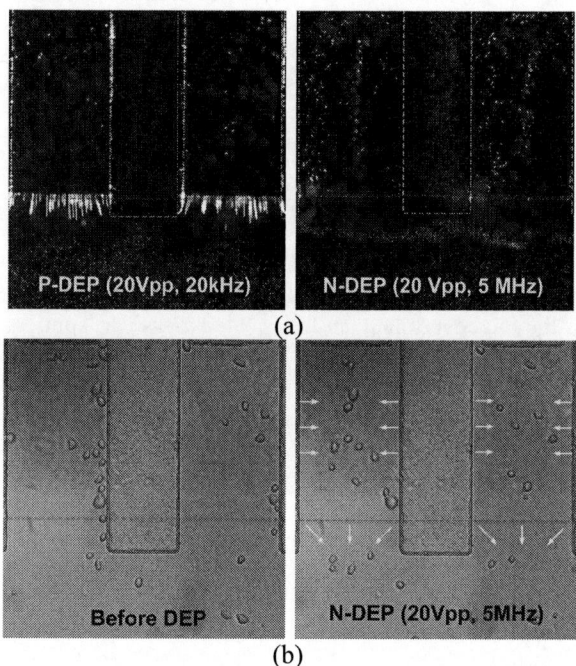

Figure 2: Manipulation of microparticles/cells using DEP Force: (a)1μm particles either trapped or repelled from the mcirochamber edges via p-DEP or n-DEP, respectively; (b) Similar n-DEP behavior experienced by cells

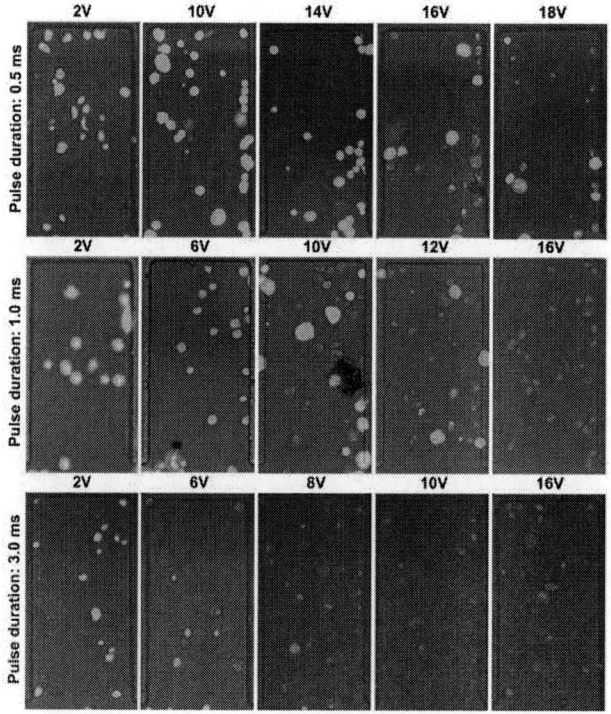

978-1-4799-3510-9/14 $31.00 © 2014 IEEE   270

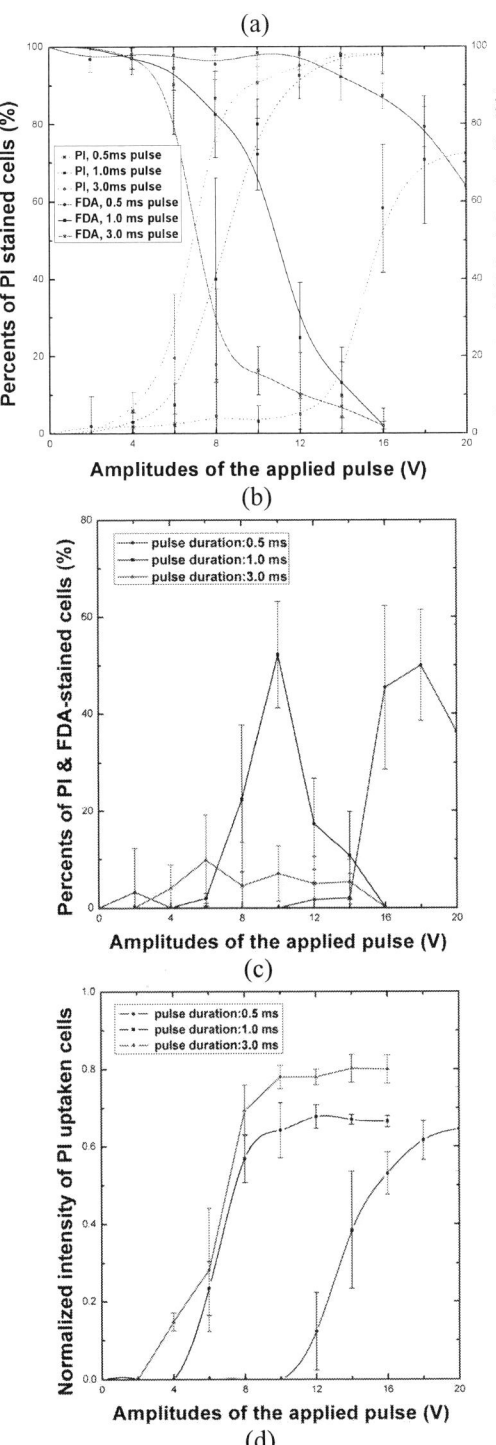

(a)

(b)

(c)

(d)

Figure 3: Electroporation and transport of PI into the Human dermal fibroblast cells. (a) Micrographs of the cells stained with PI (red) and FDA (green) after electroporation at increasing pulse amplitudes with single pulse of 0.5,1, 3ms pulse duration. (b) Effects of pulse amplitude on the transfection efficiency of PI and cell viability. (c) Effects of pulse amplitude on electroporation efficiency. (d) Normalized intensity of PI stained cells vs. pulse amplitude.

As expected, higher pulse amplitudes result in higher

intensities, irrespective of pulse duration. Finally, according to the results, we can distinguish that irreversible EP appears at higher normalized intensity ($I^* > 0.6$), non-activated EP at lower intensity ($I^* < 0.2$), and reversible EP occurs in range of $0.3 < I^* < 0.6$. Thus, we successfully valuate the optimal reversible electroporation parameter of HDNF cells and demonstrate the potential for our multi-functional, multi chamber electroporation platform.

CONCLUSIONS

A multi-functional microfluidic platform was fabricated to demonstrate the feasibility of on-chip electroporation integrated with DEP-assisted cell/particle manipulation. A spatial gradient of EP parameters were successfully produced on the chip so that the optimal reversible electroporation parameter of HDNF cells was obtained. As such, the preliminary results clearly indicate that the developed microfluidic platform has the potential to achieve high-throughput screening for electroporation with spatial control and uniformity, assisted by DEP manipulation/trapping of particles/cells into individual microchambers.

ACKNOWLEDGEMENTS

The research was supported by MOST – Tashtiyot grant #880011. The fabrication of the chip was possible through the financial and technical support of the Technion RBNI (Russell Berrie Nanotechnology Institute) and MNFU (Micro Nano Fabrication Unit).

REFERENCES

[1] J. C. Weaver and Y. A. Chizmadzhev, "Theory of electroporation: A review," *Bioelectrochem. Bioenerg.*, vol. 41, no. 2, pp. 135–160, 1996.

[2] S. Movahed and D. Li, "Microfluidics cell electroporation," *Microfluid. Nanofluidics*, vol. 10, no. 4, pp. 703–734, 2011.

[3] T. Geng and C. Lu, "Microfluidic electroporation for cellular analysis and delivery," *Lab. Chip*, vol. 13, no. 19, pp. 3803–3821, 2013.

[4] Y. Xu, H. Yao, L. Wang, W. Xing, and J. Cheng, "The construction of an individually addressable cell array for selective patterning and electroporation," *Lab. Chip*, vol. 11, no. 14, pp. 2417–2423, 2011.

[5] A. D. Hargis, J. P. Alarie, and J. M. Ramsey, "Characterization of cell lysis events on a microfluidic device for high-throughput single cell analysis," *ELECTROPHORESIS*, vol. 32, no. 22, pp. 3172–3179, 2011.

[6] S. Park, P. A. L. Wijethunga, H. Moon, and B. Han, "On-chip characterization of cryoprotective agent mixtures using an EWOD-based digital microfluidic device," *Lab. Chip*, vol. 11, no. 13, pp. 2212–2221, 2011.

CONTACT

*Gilad Yossifon, e-mail: yossifon@technion.ac.il

INTEGRATION OF HEAT-TRANSFER RESISTANCE MEASUREMENTS ONTO A DIGITAL MICROFLUIDIC PLATFORM TOWARDS THE MINIATURIZED AND AUTOMATED LABEL-FREE DETECTION OF BIOMOLECULAR INTERACTIONS

Elena Pérez-Ruiz[1], Thijs Vandenryt[2], Daan Witters[1], Deborah Decrop[1], Bart van Grinsven[2], Dragana Spasic[1], Patrick Wagner[2], and Jeroen Lammertyn[1,]*

[1]KU Leuven-University of Leuven, Belgium
[2] Hasselt University, IMO-IMOMEC, Belgium

ABSTRACT

In this paper the successful integration of heat-transfer resistance measurements with a digital microfluidic chip is shown. The integrated miniaturized platform allows the automated label-free detection of biomolecular interactions. To immobilize biomolecules on the hydrophobic chip surface, hydrophilic gold sensing patches are created by means of a recently described dry lift-off technique that leaves the chip surface unaffected. DNA melting analysis was performed for validating the integrated device.

INTRODUCTION

Microfluidic-based lab-on-a-chip (LOC) devices are able to integrate and scale down laboratory processes to a miniaturized chip format. In recent years, digital microfluidics (DMF) has been displacing traditional continuous flow microfluidics in the development of high-throughput, automatized LOC devices [1]. In DMF, nano- to microliter droplets are independently actuated on top of a hydrophobic surface by applying software-controlled electrical signals. Among all different droplet-actuation forces, electrowetting-on-dielectric (EWOD) principle [2], based on the modulation of the interfacial tension between a liquid droplet and an electrode coated with a hydrophobic dielectric layer by the application of a voltage, appears to be the most promising due to its ability to perform all the essential fluidic operations on-chip, such as droplet dispensing from a reservoir, splitting, merging, mixing and transporting, with high degree of reconfiguration and flexibility [3].

Because of the small sample volumes handled on DMF, diffusion distances are short, which results in little response time and fast analysis. These outstanding features, together with a great possibility of automation and multiplexing, next to cost-effective production, make DMF especially well-suited for analytical applications. Thus, recent research is focused on the integration of DMF platforms with different analysis techniques towards performing detection directly in the droplets on the chip. Numerous techniques, including fluorescence [4,5] and absorbance detection, surface plasmon resonance [6], voltammetry [7] and mass spectrometry [8], have been already coupled with DMF.

In this work we present the integration of DMF with a novel technique based on the measurement of heat-transfer resistance at solid-liquid interfaces [9]. This recently developed heat-transfer method (HTM) already showed its big potential for analysis of DNA mutations [9][10] and detection of cancer cells [11]. HTM based sensors offer the advantage of performing real-time label-free analysis with a simple and inexpensive instrumentation, since the readout requires no more than a controlled heat source and two temperature sensors. Moreover, the measurements are not sensitive to environmental and electronic conditions and any solid material, functionalized with an adequate receptor, can be used as sensing platform, provided that it does not inhibit the heat flow. Although this makes HTM flexible and potentially applicable to different fields, both sample preparation and measurements are difficult to automate in HTM based sensors and likewise they require rather high sample volumes. We propose the integration of HTM on a DMF platform to overcome these shortcomings, and allow for automated real-time label-free detection of biomolecular interactions in a miniaturized and compact manner.

As a proof-of-concept, denaturation of DNA was monitored in the integrated LOC platform and the obtained results were validated by traditional HTM.

MICROFABRICATION OF THE INTEGRATED DEVICE

The integrated set-up is based on a double-plated digital microfluidic device, as schematically depicted in figure 1A. A heating element positioned on a copper block is placed under the chip. The internal temperature of this block, T_1, is measured by a thermocouple and steered through a PID controller connected to a power resistor. Monitoring of heat-transfer resistance is realized through an ITO meander structure patterned on the top plate, aligned with the chromium electrodes, which move the droplet (figure 1B). The electric resistance, R, of the meander structure increases with increasing temperature (metallic behavior). This behavior allows us to use the meander structure as a temperature sensor which monitors the temperature inside the droplet, T_2. The heat-transfer resistance (°C/W) is then calculated by dividing the temperature gradient ($T_1 - T_2$) through the input power (P).

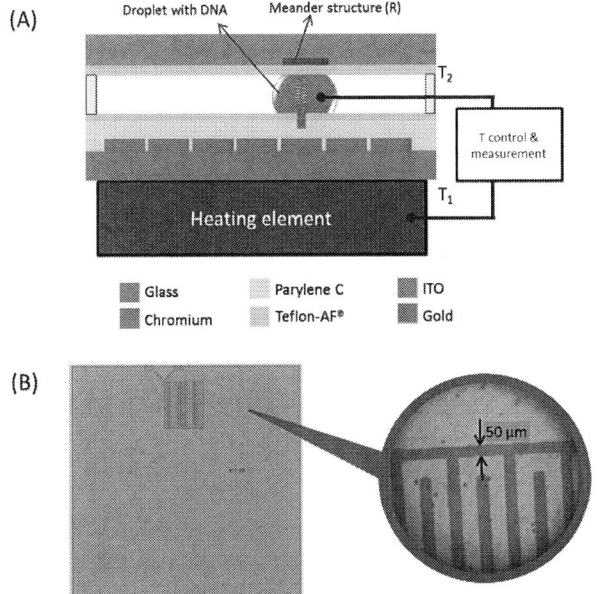

Figure 1: (A) Schematic side view of the integrated set up based on a double-plate microfluidic chip. (B) Top view of the top plate. The zoomed region shows a microscopy image of the patterned meander structure in the indium tin oxide (ITO).

The construction of the bottom plate is depicted in Figure 2. Microfabrication of the gold sensing patches is based on a dry lift-off technique recently developed, which provides a biocompatible way to pattern and biofunctionalize the DMF chip surface without affecting its hydrophobicity [12].

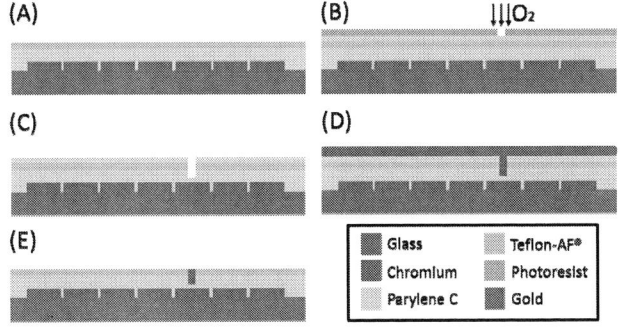

Figure 2: (A) Glass substrate with patterned chromium electrodes covered with parylene-C and Teflon-AF layers. (B) Deposition of parylene-C mask and coating and patterning of protective photoresist layer followed by oxygen plasma removal of underlying parylene-C and Teflon layers. (C) Removal of photoresist. (D) Sputtering of thin gold layer on top and (E) mechanical peel-off of the sacrificial parylene C mask.

This method uses a Parylene-C mask for the micropatterning process. In short, first, chromium electrodes

were patterned by standard photolithography on a clean glass substrate. After treating the surface with both O2-plasma and silane A174 in order to improve adhesion, the electrode layer was coated with both a dielectric Parylene-C (3.5 µm) and a hydrophobic Teflon-AF® (200 nm) layers. Subsequently, a Parylene-C mask (1 µm) was deposited over the Teflon-AF®, by chemical vapor deposition, and a thick protective SPR220-7 positive photoresist was spin-coated on top. The layer of photoresist was then patterned by exposing it to a UV light through a photomask bearing circular patches (500µm diameter). Next, the unprotected zones of the Parylene-C mask and the underlying Teflon-AF® and dielectric layers were selectively etched away using O_2-plasma and the photoresist was removed by rinsing the substrates with acetone and isopropyl alcohol. Finally, a thin layer of gold (50 nm) was sputtered on top and the sacrificial protective Parylene-C mask was peeled-off mechanically. In this way, hydrophilic gold round micropatches strategically localized on top of the measuring electrodes are patterned, with the hydrophobic Teflon-AF® on top of the microfluidic plate being unaffected.

DNA MELTING ANALYSIS

Label-free heat-transfer monitoring already proved to be a promising alternative to the currently used denaturation-based techniques for DNA characterization and the identification of new point mutations[9,10]. Therefore, to prove the integrated set-up, we performed DNA melting analysis on chip and compared the obtained results to the ones found when using a traditional HTM set-up.

Functionalization of Gold Sensing Patches with DNA

Gold sensing patches were functionalized with a thiol-terminated DNA probe (sequences can be found in Table 1), by incubating the chip substrates with a thiolated DNA solution (300 pmol DNA/ cm^2 of sensor surface) at RT overnight.

Table 1: Sequences of the probe DNA and its complementary strand. The probe DNA was extended with a spacer consisting of seven A bases to avoid steric hindering with the sensor surface.

	sequence
probe	5'-SH–AAAAAAA-CCCCTGCAGCCCATGTATACCCCCG AACC- 3'
complement	5'-GGTTCGGGGGTATACATGGGCTGCA GGGG-3'

Unbound probe DNA was rinsed off by washing the substrates with a 2xSSC (sodium clorihide/sodium citrate) buffer supplemented with 0.5% SDS (sodium dodecyl sulfate). The hybridization with the complementary DNA sequence (Table 1) was performed on-chip by incubating, on top of the measuring electrode, a droplet of 600 pmol DNA solution, prepared in 10x PCR homemade buffer, during 2 h at 30°C. To prevent evaporation of reagents during the hybridization reaction, an oil shell was created around the droplet and substrates were kept under a saturated water vapor atmosphere. After hybridization, unreacted target DNA was removed by two consecutive washing steps with 2xSSC/0.5%SDS and Phosphate Buffered Saline (PBS) standard buffer. Finally a droplet of PBS buffer, surrounded by an oil shell, was deposited on top of the measuring electrode.

On-Chip Heat-Transfer Measurement of DNA Denaturation

In order to verify the distinctive increase in heat-transfer resistance upon DNA denaturation, the temperature of the cupper heating element, T_1, was increased from 35°C to 70°C at a constant rate of 1°C/min and afterwards decreased back to 35°C at the same rate. In the meantime, the resistance (R) of the sample droplet, representing T_2, was monitored though the meander in the top plate. As a control, heat-transfer on a gold patch with DNA in the single strand state was measured. Figure 3 summarizes the obtained results. As can be derived from the graphs, an increase in heat transfer resistance was observed for the double stranded DNA (black line) through a decrease in T_2, while, when conducting the same experiment with DNA in the single stranded state, such a shift did not occur (blue line). This change in resistance indicates that DNA denaturation in the studied complex occurs at 61.7°C, which is in the range of the theoretically calculated melting temperature, 67.8 °C. However, the obtained value for the denaturation temperature is a bit lower than the predicted one, but the same effect was previously observed in heat-transfer resistance measurements without affecting the relative order of stability of different DNA complexes [9,10].

Validation with Traditional Heat-Transfer Method

To validate the results obtained in the integrated platform, traditional HTM was applied to study the denaturation of the same DNA complex. The used instrumental set up was described and employed previously by van Grisven *et al.* [9]

Gold sensor samples 1 x 1 cm^2 were prepared by sputtering a 50 nm gold layer onto a clean glass substrate. Functionalization with thiol-DNA probe and subsequent hybridization with the complementary DNA strand was performed following a protocol similar to the one used for functionalizing the gold patches on the chip substrates.

The results are presented in Figure 4. An expected stepwise increase in heat-transfer resistance, Rth, from 7.5 to almost 9 °C/W, is shown upon DNA denaturation at 61 °C. As can be seen, the melting temperature obtained experimentally by traditional HTM is similar to the one acquired in the integrated set-up.

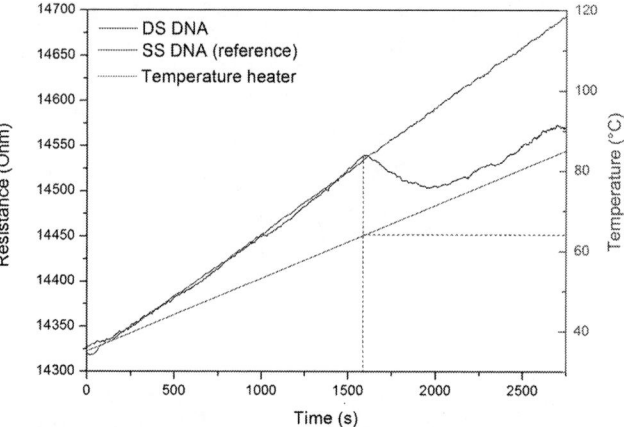

Figure 3:. Monitored temperature (red) and measured resistance for single stranded DNA (blue) and double stranded DNA (black), immobilized on the gold surface, represented as a function of the assay time.

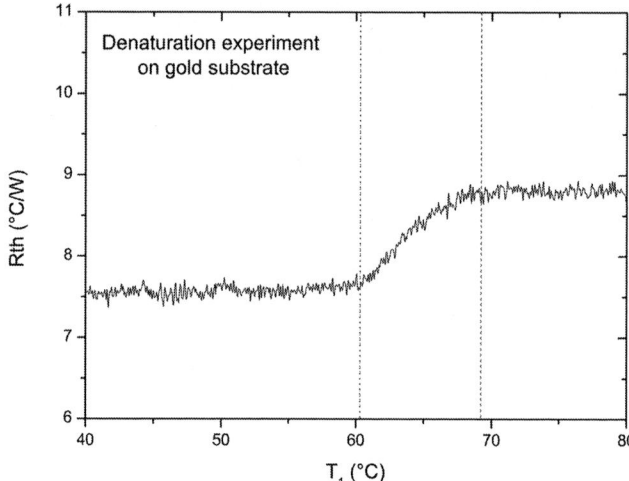

Figure 4: Heat-transfer resistance, Rth, as a function of temperature on a gold sensor with covalently attached DNA complex.

CONCLUSIONS

A two-plate digital microfluidic platform with integrated heat-transfer resistance measurements was developed. In order to integrate the heat-transfer resistance monitoring in a miniaturized fashion, a meander structure was patterned on the top plate and, to allow the immobilization of biomolecules on the chip surface, gold sensing patches were patterned on the hydrophobic chip surface by means of a dry lift-off technique based on a mechanically removable Parylene-C mask. The integrated platform was used to successfully measure DNA denaturation processes, giving comparable results as standard HTM measurements on gold electrodes.

From performed experiments it is evident that the combination of DMF and HTM results in a powerful system for the miniaturized detection of biomolecular interactions. Future work will be focused on optimizing the microfabrication of the platform and enabling sample multiplexing and automation. Also, other applications than DNA interactions will be investigated.

ACKNOWLEDGEMENTS

The research leading to these results has received funding from the European Commission's Seventh Framework Programme (FP7/2007-2013) under the grant agreement BIOMAX (project n° 264737), from the EFRO-INTERREG NanosensEU European project, from the KU Leuven and from the Flanders Fund for Scientific Research (FWO G.0997.11).

REFERENCES

[1] K. Choi, A. H. C. Ng, R. Fobel, A. R. Wheeler, "Digital Microfluidics.", *Annu. Rev. Anal. Chem.*, vol. 5, pp. 413–40, 2012.

[2] T. B. Jones, "On the Relationship of Dielectrophoresis and Electrowetting", *Langmuir*, vol. 18, pp. 4437–43, 2002.

[3] M. G. Pollack, A. D. Shenderov, R. B. Fair, "Electrowetting-based Actuation of Droplets for Integrated Microfluidics", *Lab Chip*, vol. 2, pp. 96–101, 2002.

[4] D. Witters, K. Knez, F. Ceyssens, R. Puers, J. Lammertyn, "Digital Microfluidics-enabled Single-Molecule Detection by Printing and Sealing Single Magnetic Beads in Femtoliter Droplets", *Lab Chip*, vol. 13, pp. 2047–54, 2013.

[5] R. Sista, Z. Hua, P. Thwar, A. Sudarsan, V. Srinivasan, A. Eckhardt, M. Pollack, V. Pamula, "Development of a Digital Microfluidic Platform for Point of Care Testing", *Lab Chip*, vol. 8, pp. 2091–104, 2008.

[6] L. Malic, T. Veres, M. Tabrizian, "Two-dimensional Droplet-based Surface Plasmon Resonance Imaging Using Electrowetting-on-dielectric Microfluidics", *Lab Chip*, vol. 9, pp. 473–5, 2009.

[7] M. D. M. Dryden, D. D. G. Rackus, M. H. Shamsi, A. R. Wheeler, "Integrated Digital Microfluidic Platform for Voltammetric Analysis", *Anal.Chem.* vol. 85, pp. 8809−16, 2013.

[8] A. E. Kirby, A. R. Wheeler, "Microfluidic Origami: a New Device Format for In-line Reaction Monitoring by Nanoelectrospray Ionization Mass Spectrometry", *Lab Chip*, vol. 13, pp. 2533–40, Jul. 2013.

[9] B. Van Grinsven, N. Vanden Bon, H. Strauven, L. Grieten, M. Murib, S. D. Janssens, K. Haenen, M. J. Schöning, K. L. Jiménez-Monroy, V. Vermeeren, M. Ameloot, L. Michiels, R. Thoelen, W. De Ceuninck, P. Wagner, "Heat-Transfer Resistance at Solid-Liquid Interfaces□: A Tool for the Detection of Single-Nucleotide Polymorphisms in DNA", *ACS Nano*, vol. 6, pp 2712–21, 2012.

[10] K. Bers, B. van Grinsven , T. Vandenryt, M. Murib,W. Janssen, B. Geerets, M. Ameloot, K. Haenen, L. Michiels, W. De Ceuninck, P. Wagner, "Implementing Heat-transfer Resistivity as a Key Element in a Nanocrystalline Diamond Based Single Nucleotide Polymorphism Detection Array", *Diamond Relat. Mater.*,vol.38, pp. 45–51, 2013.

[11] K. Eersels, B. Van Grinsven, A. Ethirajan, S. Timmermans, J. F. J. Bogie, S. Punniyakoti, T. Vandenryt, J. J. A. Hendriks, T. J. Cleij, M. J. A. P. Daemen, V. Somers, W. De Ceuninck, and P. Wagner, "Selective Identification of Macrophages and Cancer Cells Based on Thermal Transport through Surface-Imprinted Polymer Layers," *Appl. Mater. Interfaces*, vol. 5, pp 7258–67, 2013.

[12] D. Witters, N. Vergauwe, S. Vermeir, F. Ceyssens, S. Liekens, R. Puers, J. Lammertyn, "Biofunctionalization of Electrowetting-on-dielectric Digital Microfluidic Chips for Miniaturized Cell-based Applications." *Lab Chip*, pp. 2790–94, 2011.

CONTACT

*Jeroen Lammertyn; tel:+3216321459; jeroen.lammertyn@biw.kuleuven.be

NANOFLUIDIC SENSING AT NORMAL PHYSIOLOGICAL CONDITION BY COUPLING ION CONCENTRATION POLARIZATION WITH A LIMIT OF DETECTION OF ONE FEMTOMOLE

Wei Ouyang[1], Wei Wang[1, 2], Haixia Alice Zhang[1, 2], Wengang Wu[1, 2], and Zhihong Li[1, 2]

[1]Institute of Microelectronics, Peking University, Beijing, P.R. CHINA

[2]National Key Laboratory of Science and Technology on Micro/Nano Fabrication, Beijing, P.R. CHINA

ABSTRACT

Traditional nanofluidic devices lose function at normal physiological condition (e.g. 0.1 M), because the thickness of electrical-double-layer (EDL) λ_D (~1 nm) under this circumstance is much smaller than typical nanochannel size, which significantly limits their clinical applications. This work enables nanofluidic sensing of Biotin at normal physiological condition by coupling ion concentration polarization (ICP) with nanofluidic crystal (NFC), which utilizes the enrichment effect for target preconcentration and the depletion effect for buffer desalination, with a limit of detection of 1 fM.

INTRODUCTION

Nanofluidics has become a major research subject in the recent years, with wide applications in ion rectification [1], energy conversion [2], biomolecule separation [3] and sensing. Nanofluidic sensing attracts great attention with its potential of charge-sensitive label-free detection [4]. Nanofluidic devices usually depend on nanofabrication of nanochannels or nanopores, which considerably increases the cost and complexity of nanofluidic devices. We previously reported self-assembled nanofluidic crystal (NFC), as a facile and nanofabrication-free method of obtaining nanofluidic sensors [5, 6]. However, one bottleneck problem of conventional nanofluidic sensors is that, they all require extremely highly diluted buffer to ensure the dominating role of EDL in the system. In other words, they lose function at normal physiological condition, which significantly limits their clinical applications.

Nafion-induced ion concentration polarization (ICP) received renewed interest in the recent years [7, 8]. When a voltage is applied across a surface-patterned Nafion membrane, an ion depletion region would be induced near Nafion, and an enrichment plug forms right beside the depletion region due to electroosmosis and charge-repelling effect of the depletion region, so-called ICP. Recently, we utilized the depletion effect of ICP to temporarily form a low ionic concentration region in normal physiological concentration buffer (0.1 M PBS buffer), thereby realize nanofluidic sensing in the desalinated buffer [9]. In this work, we coupled both the enrichment and depletion effect of ICP with NFC, to further lower the limit of detection.

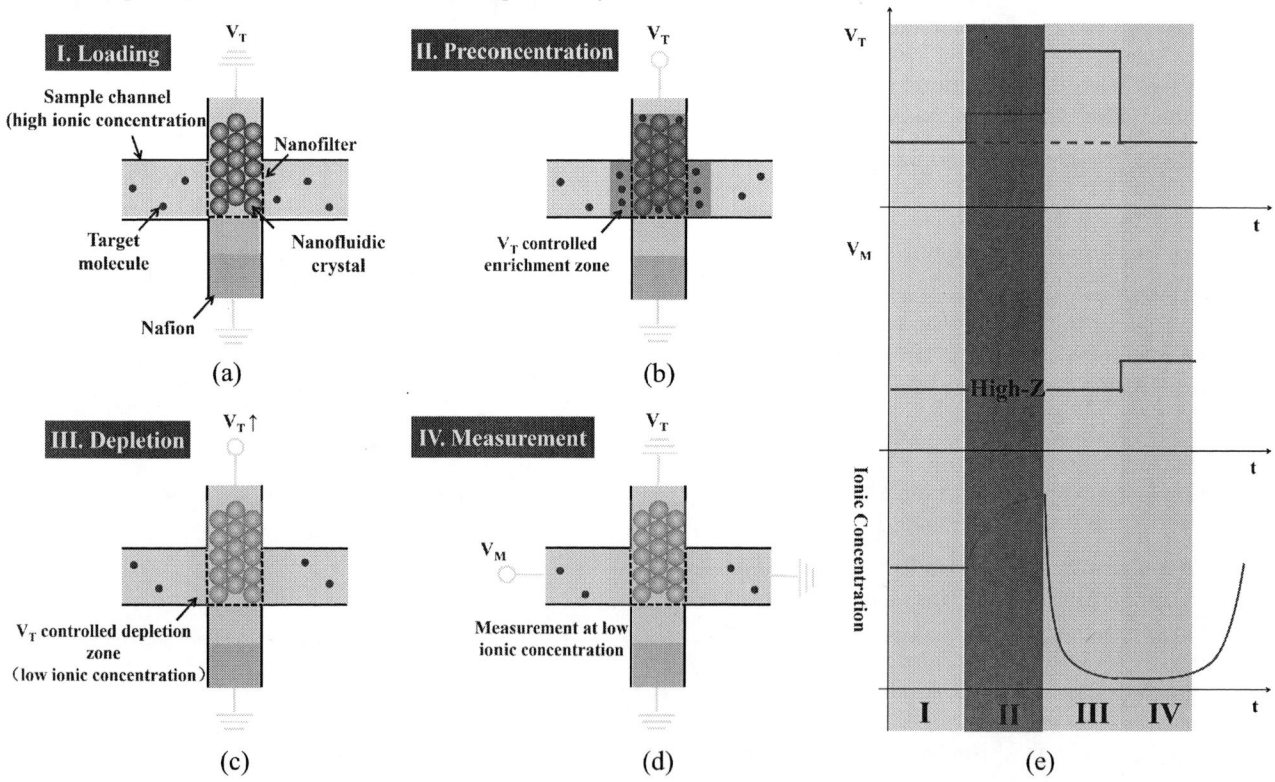

Figure 1: Principle and protocol of nanofluidic sensing at normal physiological condition by coupling ion concentration polarization and nanofluidic crystal. (a) Step I. Loading . (b) Step II. Preconcentration. (c) Step III. Depletion. (d) Step IV. Measurement. (e) Time sequence diagram of V_T and V_M, and ionic concentration variation in a sensing cycle.

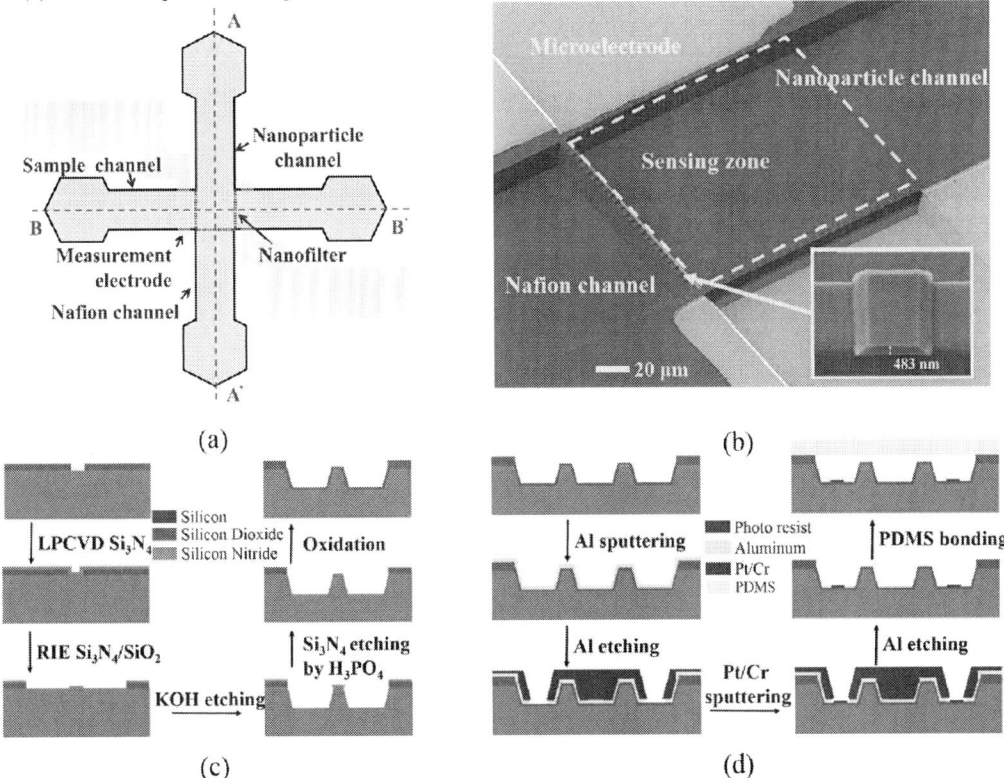

Figure 2: Chip design and fabrication. (a) Layout (b) SEM image of the central zone. Inset is magnified view of the nanofilter (c) Micro/Nanochannel fabrication process (A-A' cross-section) (d) Microelectrode fabrication process (B-B' cross-section).

PRINCIPLE AND PROTOCOL

Figure 1 shows the principle and protocol of the device. Before the sensing experiment, the device was prepared with Nafion membrane and NFC. The nanoparticles of the NFC were surface-modified with streptavidin. When the target molecule, Biotin, binds with streptavidin, the surface charge density of nanoparticles changes. The electrical conductance of NFC is a function of the surface charge density of nanoparticles at nanofluidic regime. Nanofluidic sensing is realized by measuring the electrical conductance of NFC.

The sensing experiment includes four steps: sample loading (I), preconcentration (II), depletion (III), and measurement (IV).

(I) Sample loading: Biotin in 0.1 M PBS buffer was pumped into the chip from the central channel, which allowed the reaction between Biotin and Streptavidin.

(II) Preconcentration: A DC voltage was applied across the Nafion membrane to generate an enrichment plug at the NFC region. The concentration of Biotin inside the enrichment plug was significantly amplified, which enhanced the binding efficiency of Biotin and Streptavidin.

(III) Depletion: After the reaction of Biotin and Streptavidin, the DC voltage was increased, so the depletion region would extend to cover the whole NFC structure. The ionic concentration of the depletion region was depleted to

less than 1 mM, at which surface charge density dominated the electrical conductance of NFC.

(IV) Measurement: The DC voltage for ICP was turned off. At the same time, electrical conductance of NFC at nanofluidic regime was measured by microelectrodes. The electrical measurement should be conducted immediately, before ions diffused back to the depletion region.

CHIP DESIGN AND FABRICATION

Our chip comprised of a cross-shaped channel network, including one nanoparticle channel, one Nafion channel, and two sample channels. The sensing zone was at the center and connected to channels by three nanofilter arrays (Figure 2(a)). The channels were 150 μm wide and 10 μm deep, while the nanofilter array was 480 nm deep (Figure 2(b)). The fabrication process is shown in Figure 2(c). The microchannels were fabricated by KOH etching. The nanofilter depth was determined by two oxidations. Pt/Cr electrodes were deposited on the microchannels by Al/photoresist dual-layer lift-off (Figure 2(d)). Finally, the chip was bonded with PDMS.

EXPERIMENT

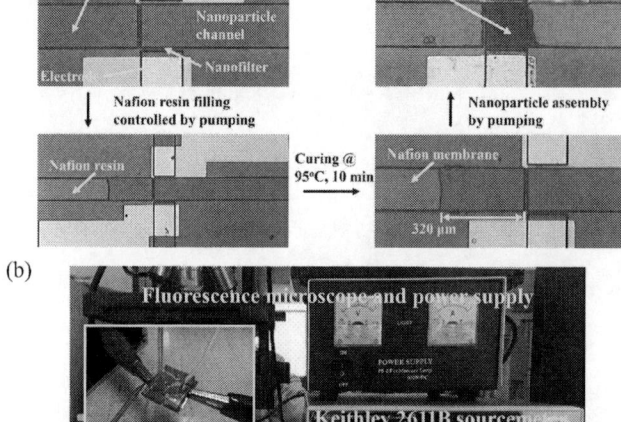

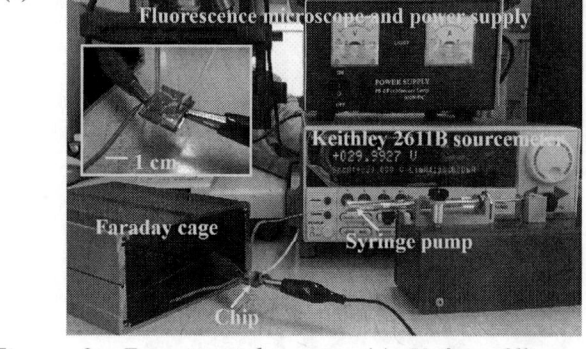

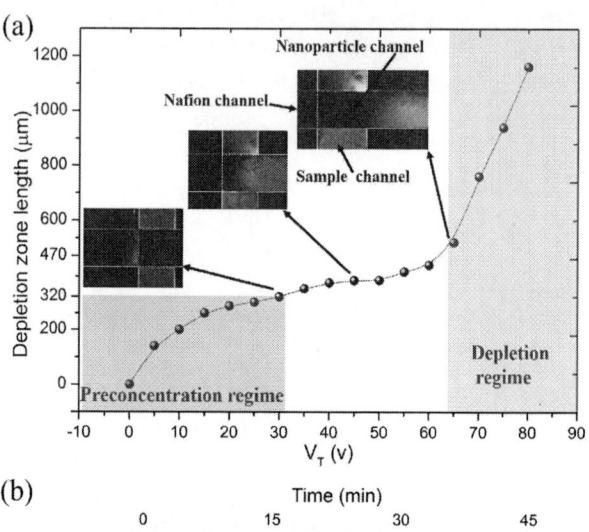

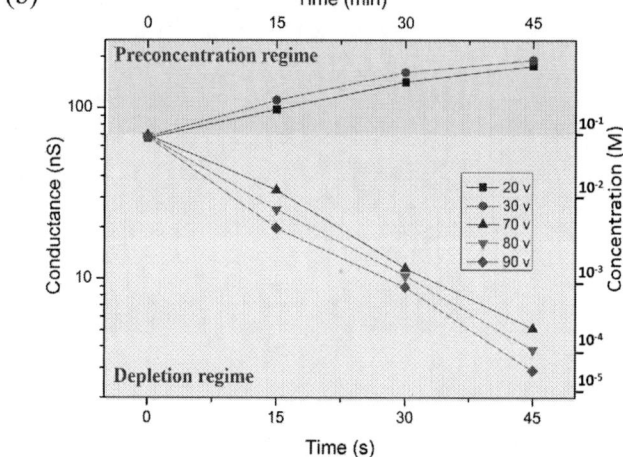

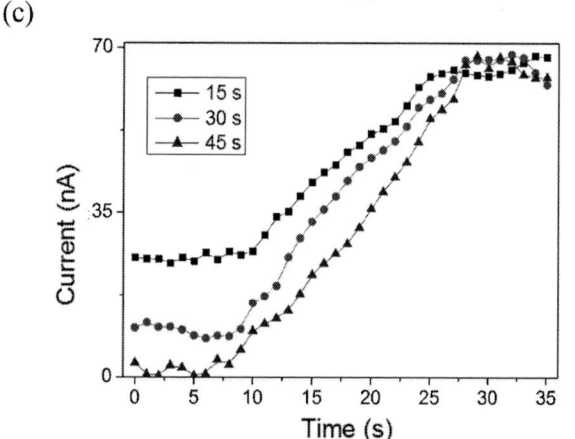

Figure 3: Experimental setup. (a) Nafion filling and nanoparticle packing process in the chip (b) Electrical testing system, inset is the photo of the chip.

Chip Preparation

Before the sensing experiment, we need to prepare the device with Nafion membrane and NFC. First, Nafion resin was filled in the Nafion channel by pump control. The front of the Nafion resin was 320 μm from the nanofilter, as shown in Figure 3(a). Then, Nafion resin was cured at 95°C for 10 min. Finally, monodispersed nanoparticle suspension (Bangs Laboratory Inc.) was pumped through the nanoparticle channel. The nanoparticles had a diameter of 540 nm, and were modified with streptavidin. Nanoparticles assembled into NFC due to nanofilters.

Characterization of Ion Concentration Polarization

Both fluorescence microscopy and electrical measurement were conducted to characterize ICP, including the relationship between the voltage applied and the length of depletion/enrichment region, the ionic concentration variation of the depletion/enrichment region. FITC-labeled aptamer was used for fluorescence microscopy. DC voltage supply and electrical current measurement were provided by a souremeter, Keithley 2611B (Figure 3(b)). The buffer used for ICP characterization was 0.1 M PBS buffer.

Biotin sensing experiment

The sample used was Biotin in 0.1 M PBS buffer. Biotin sensing experiment was conducted following the protocol in Figure 1 (I → II → III → IV). The gating voltage V_T=30 V for 30 min for preconcentration of Biotin, and V_T=90 V for 45 s for depletion of the buffer. Control experiments were also conducted, including depletion-only (I → III → IV) and no-ICP (I → IV).

RESULTS AND DISCUSSION

Figure 3: (a) Relationship between the gating voltage and the length of the depletion/enrichment zone. (b) Ionic concentration variation at preconcentration and depletion regime. (c) Current variation after depletion at 90 V for 15/30/45 s and turning off the gating voltage.

Ion Concentration Polarization

As Figure 4(a) shows, at V_T<35 V, the depletion region did not reach NFC, and NFC was bathed in the enrichment plug, which was suitable for target preconcentration. While

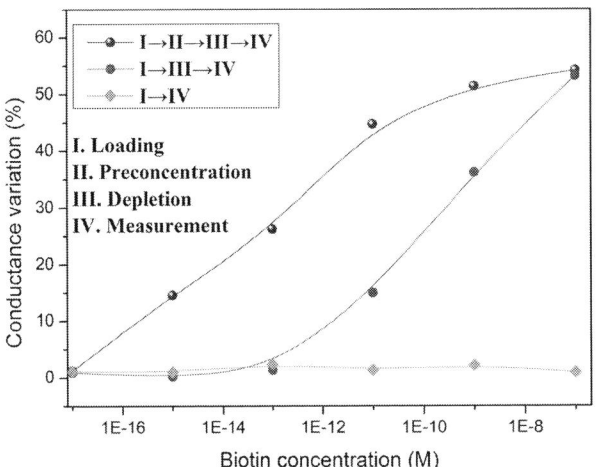

Figure 5. Biotin detection result following different protocols.

at V_T>65 V, the depletion region became long enough to cover NFC, which could be used for bulk ionic concentration depletion. The ionic concentration variation was also monitored by electrical measurement. In the depletion region, ionic concentration decreased dramatically, and reached its minimum value at about 45 s (Figure 4(b)). The minimum ionic concentration depended on the gating voltage. When V_T=90 V, it could reach about 10^{-4} M. After turning off the gating voltage, ions would spontaneously diffuse back to the depletion region, which was not desired for nanofluidic sensing. As shown in Figure 4(c), there was a time window of about 10 s before the back-diffusing ions affected the ionic concentration of the sensing zone.

Biotin sensing results

When ICP was not introduced into the sensing operations (I → IV), no significant difference of conductance could be observed. At buffer concentration of 0.1 M, the Debye Length (~1 nm) was much smaller than the interstices of nanoparticles, so surface charge did not affect the electrical conductance of NFC.

When only the depletion effect of ICP was coupled (I → III → IV), difference in conductance at different Biotin concentrations could be detected, with a limit of detection of 1 pM. At buffer concentration of about 0.1 mM, the Debye Length (~ 30 nm) became comparable to the interstices of nanoparticles, so surface charge dominated the conductance of NFC, which enabled electrical sensing.

When both the depletion and enrichment effect of ICP were coupled, a 1000 times lower limit of detection (1 fM) was obtained. The preconcentration of Biotin significantly increased the binding efficiency of Biotin and Streptavidin, which made the detection of Biotin of lower concentration possible.

CONCLUSION

In conclusion, we realized Biotin sensing with a limit of detection of 1 fM in nanofluidic crystal by utilizing the enrichment effect of ion concentration polarization for target preconcentration and the depletion effect of ion concentration polarization for lowering buffer ionic concentration. This work is promising for rapid biomolecule detection in clinical samples.

ACKNOWLEDGEMENTS

This work was financially supported by the Major State Basic Research Development Program (973 Program) (Grant Nos. 2009CB320300 and 2011CB309502), the National Natural Science Foundation of China (Grant No. 91023045) and the 985-III program (clinical applications) in Peking University.

REFERENCES

[1] I. Vlassiouk, S. Smirnov, and Z. Siwy, "Ionic selectivity of single nanochannels," *Nano letters,* vol. 8, pp. 1978-1985, 2008.

[2] W. Ouyang, W. Wang, H. Zhang, W. Wu, and Z. Li, "Nanofluidic crystal: a facile, high-efficiency and high-power-density scaling up scheme for energy harvesting based on nanofluidic reverse electrodialysis," *Nanotechnology,* vol. 24, p. 345401, 2013.

[3] J. Han and H. Craighead, "Separation of long DNA molecules in a microfabricated entropic trap array," *Science,* vol. 288, pp. 1026-1029, 2000.

[4] R. Karnik, K. Castelino, R. Fan, P. Yang, and A. Majumdar, "Effects of biological reactions and modifications on conductance of nanofluidic channels," *Nano letters,* vol. 5, pp. 1638-1642, 2005.

[5] Y. Lei, F. Xie, W. Wang, W. Wu, and Z. Li, "Suspended nanoparticle crystal (S-NPC): A nanofluidics-based, electrical read-out biosensor," *Lab on a Chip,* vol. 10, pp. 2338-2340, 2010.

[6] J. Sang, H. Du, W. Wang, M. Chu, Y. Wang, H. Li, *et al.*, "Protein sensing by nanofluidic crystal and its signal enhancement," *Biomicrofluidics,* vol. 7, p. 024112, 2013.

[7] J. H. Lee, Y.-A. Song, S. R. Tannenbaum, and J. Han, "Increase of reaction rate and sensitivity of low-abundance enzyme assay using micro/nanofluidic preconcentration chip," *Analytical chemistry,* vol. 80, pp. 3198-3204, 2008.

[8] S. J. Kim, S. H. Ko, K. H. Kang, and J. Han, "Direct seawater desalination by ion concentration polarization," *Nature Nanotechnology,* vol. 5, pp. 297-301, 2010.

[9] W. Ouyang, J. Sang, Y. Shi, W. Wang, M. Chu, Y. Wang, *et al.*, "Nanofluidic Crystal Sensing at Normal Physiological Condition by Coupling Ion Concentration Polarization," in *Proceedings of the 17th International Conference on Miniaturized Systems for Chemistry and Life Sciences (MicroTAS 2013),* Freiburg, Germany, Oct. 27-31, 2013, pp. 663-665.

CONTACT

*W. Wang, tel: +86-10-62769183; w.wang@pku.edu.cn

MASS-PRODUCTION AND PROLONGED UNDIFFERENTIATED STATE OF EMBRYONIC BODIES BY USING A SEMIPERMEABLE TAPERED MICROWELL ARRAY

Masashi Ikeuchi[1,2], Shuji Hayashi[3], Makoto Osonoi[3] and Koji Ikuta[1]

[1]The University of Tokyo, JAPAN

[2]PRESTO, Japan Science and Technology Agency, JAPAN

[3]The Foundation for Promotion of State of the Art in Medicine and Health Care, JAPAN

ABSTRACT

In this paper, we demonstrate embryonic body (EB) formation and prolongation of undifferentiated state by using an improved tapered microwell array equipped with semipermeable bottom. The permeability has promoted exchange of culture medium in each microwell, resulting in prolonged term of undifferentiated state of EBs. For rapid production of the microwell array, we have also developed a new process combining thermal imprinting and punching. The process realized precise imprinting of the tapered microwell structures and simultaneous punching of the through-holes at the bottom of microwells.

INTRODUCTION

Background

Organ transplant has been the only treatment for patients with an ailing organ through accident or illness. However, shortage of donors and rejection after transplantation are the longstanding issues. Recently, regenerative medicine has been attracting much attention as a new treatment instead of organ transplant (Fig1). Stem cell is the key to achieve the regenerative medicine. Stem cell is defined as a cell which has self-renewability - ability to replicate itself, and multiline age potential -ability to differentiate into multiple cell types. It is contemplated that stem cell can be differentiated into other cell types by applying single or combinatorial stimulation of genetic, chemical and physical cues. Over the years, embryonic stem (ES) cell has been the only cell type that can differentiate into any other cell types. Since ES cell

is derived from a fertilized egg, ethical issue has hampered the progress of regenerative medicine.

The invention of induced pluripotent stem cells (iPS cells) changed the situation. In the typical process of regenerative medicine using iPS cells, at first, body cells are collected from a patient who has ailing organs, and the cells are reprogrammed to pluripotent stem cells similar to ES cells. The iPS cells are then induced to differentiate into various types of body cells, and finally the patient is treated with the artificial tissue or organ composed of cells derived from him or herself. Since the tissues or organs are derived solely from the patient, the issues concerning the conventional organ transplant and the ES cells - shortage of donor, immune rejection and ethical issues, can be solved.

Embryonic Body

In the recent protocol of tissue regeneration, iPS cells are differentiated into typical cell types through culturing as 3-D multicellular spheroids called embryonic bodies (EBs) [1, 2]. There are mainly two conventional methods to fabricate EBs. In "hanging-drop" method [3], EBs are formed by manually placing cell suspension droplets on the reverse side of the top of a cell culture dish. The cells cultured in the hanging droplets have no solid surface to attach, and form an EB in each droplet by self-assembly. However, the time-consuming and elaborate manual deposition of droplets prevents large scale production of EBs. In the other method, a hydrophilically modified culture dish is used to culture stem

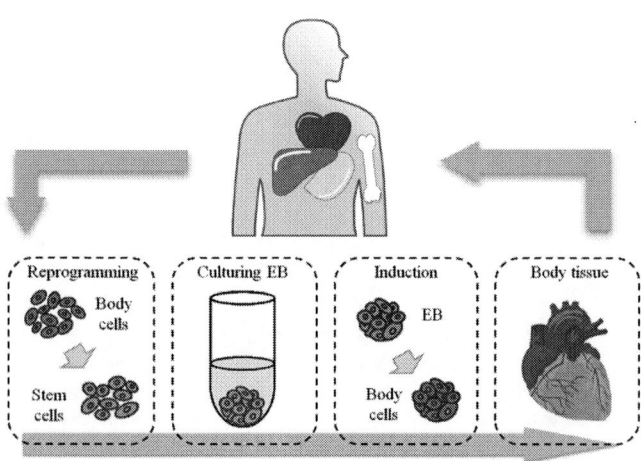

Fig.1 Typical flowchart of regenerative medicine

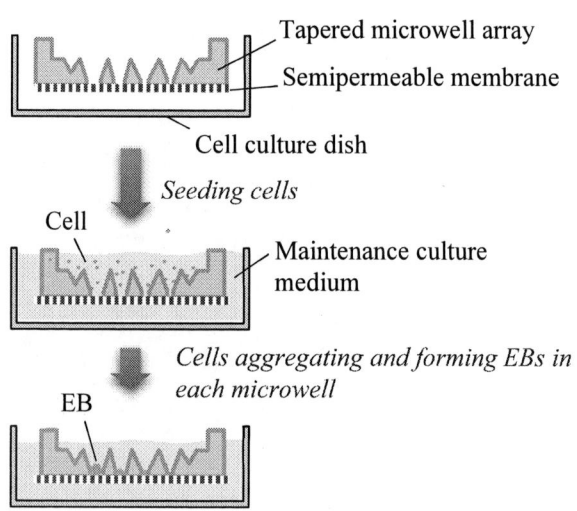

Fig.2 EB formation process using the microwell device

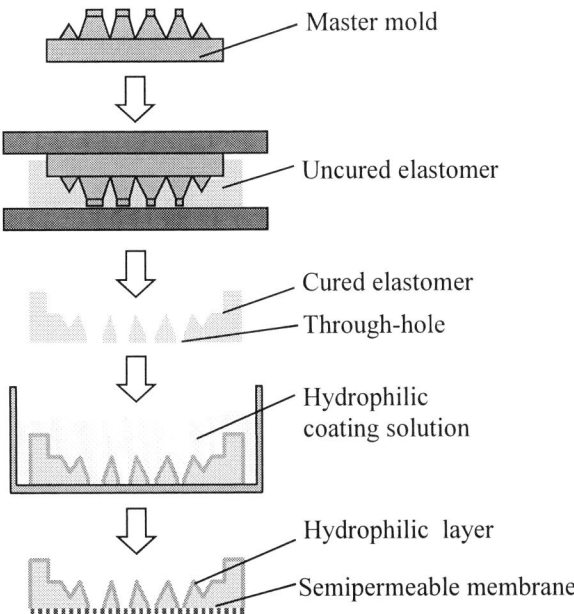

Fig.3 Fabrication of the device by simultaneous thermal imprinting and punching process

cells. Since cells cannot attach onto the hydrophilic surface, they randomly attach with adjacent cells to form a spherical aggregates. Many EBs can easily be produced at one time by using this method, however, it is difficult to control the size of EBs, and further purification process is required to obtain EBs of specified size.

In short, there are no conventional methods to efficiently mass-produce EBs with high uniformity. For rapid promotion of regenerative medicine, there is an urgent need to develop more efficient experimental system for large-scale and uniform production of EBs at low cost.

Tapered Stencil for Cluster Culture (TASCL)

To solve this problem, we have developed TASCL which is an array of tapered micro-apertures made of PDMS to enable mass production of EBs [4]. By placing a TASCL on a conventional low-attachment cell culture substrate, microwells are formed on the substrate. The surface of TASCL is also modified with hydrophilic polymer to prevent cell adhesion. Thus, the cells seeded onto the TASCL aggregates at the bottom of each microwell by gravity. By using TASCL, we verified mass-production of EBs of mouse iPS cells and pancreatic stem cells, which was proved to be more than ten times efficient compared to conventional hanging-drop method [4, 5].

However, the EBs could not maintain undifferentiated state more than 4days in TASCL because of poor exchange of culture medium at the bottom of microwells. Since the microwell is too small and densely-packed in TASCL, medium exchange of each microwell by using a conventional micropipette is time-consuming and impractical. Therefore, the culture medium in each microwells could not be completely exchanged.

SEMIPERMEABLE TAPERED MICROWELL ARRAY
Device Design

To overcome this issue of TASCL, here, we propose a new tapered microwell array which has semipermeable bottom (Fig.2). The device has an array of tapered holes optimized for EB formation, and the surface is modified with hydrophilic polymer to prevent cell adhesion. The semipermeable membrane at the bottom has $\phi 1\mu m$ pores, and only permits liquid and small molecules to go through. The device can be easily incorporated into conventional cell culture protocols by just hanging the device on a culture dish filled with culture medium. By immersing the device in culture medium, the microwells are filled with the medium from the bottom until the water level balances with outside. Then, cells are seeded onto the device and aggregate into EBs in each microwell due to low-attachment coating on the device. There is no flat surface between microwells so that each cell falls into any one of the microwells, and unexpected intrusion of cells after initial aggregation can be prevented. Therefore, EB formation can be initiated under precisely controlled conditions by just dropping cell suspension on the device. Different from the previous TASCL, medium exchange of each microwell in the array can be easily carried out at once by changing the medium in the culture dish

Fig.4 (a) Whole view of the fabricated microwell device (b) Magnified SEM image of the microwell

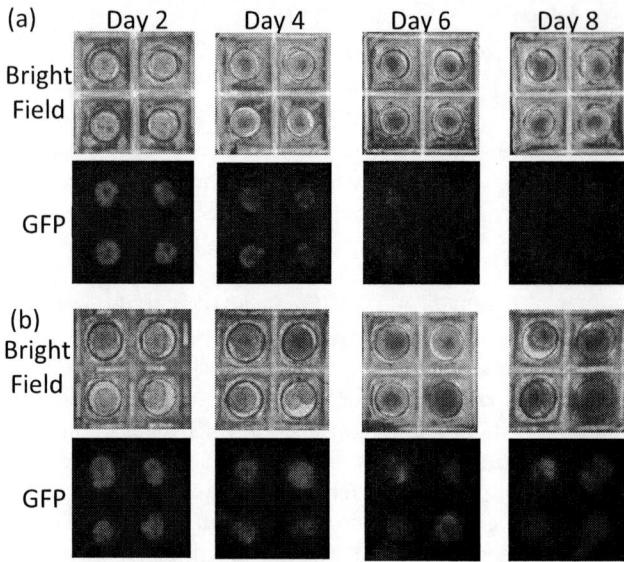

Fig.5 EBs of mouse iPS cells with GFP marker cultured in the previous non-permeable microwell device (a) and the new semipermeable microwell device (b)

through the gap between the dish and the device.

Fabrication Process

The upper part of the device is made of Polydimethylsiloxane (PDMS)-like elastomer and the bottom semipermeable membrane is made of polyethylene terephthalate. In our previously reported device TASCL, the device was made of PDMS, and the device was fabricated by conventional molding process [5]. Although the process was advantageous in high accuracy with inexpensive instruments, it was not suitable for low-cost mass-production due to manual time-consuming steps such as defoaming under vacuum and peel-off from the mold. These problems has limited the commercial use of microdevices made of PDMS, contrary to its wide-spread in academic researches.

For efficient production of our new device, we developed a new thermoplastic elastomer and its processing method combining thermal imprinting and punching (Fig.3) [6]. In this process, uncured PDMS-like elastomer was prepared on a substrate in advance, and then, imprinted and punched with an ultra-hard metal master mold. Then, it was cured at high temperature to make solid structures. The whole process took only a few minutes, being quite faster and simpler than conventional PDMS molding process taking several hours. After detaching from the master mold, the device was sealed with a semipermeable membrane insert (Greiner Bio-One, Germany).

The prototype device had 400 microwells in 10x10 mm^2, and each microwell was 500μm square at the top, ϕ300μm at the bottom, and 600μm deep (Fig.4a). The tapered microwell structure and the knife-edge structure between the microwells were precisely transferred onto the cured elastomer (Fig.4b).

EMBRYONIC BODY FORMATION AND CULTURE OF MOUSE IPS CELLS

Mouse iPS Cell

The iPS cell (iPS-MEF-Ng-20D-17) established by Prof. S. Yamanaka at Kyoto University were purchased from RIKEN Bio Resource Center (Institute of Physical and Chemical Research, Tsukuba, Japan). GFP gene was expressed by Nanog promoter in the cell line to monitor the undifferentiated state of the iPS cells. Therefore, GFP expression should be lost in accordance with differentiation of the iPS cells. The iPS cells were maintained at 37°C with 5% CO2 in stem cell culture medium on feeder layers of Mitomycin-treated MEF cells. The medium was changed every day, and passage was carried out every four days.

EB Formation

EB formation of mouse iPS cells was conducted on the previously reported non-permeable TASCL and the newly developed semipermeable device. Before use in cell culture experiment, the surface of both device was treated with an aqueous solution of polyethylene oxide (PEO) - polypropylene oxide - polyethylene oxide triblock copolymer (1wt% Anti-Link, Allvivo Vascular) for overnight at room temperature. The copolymer self-assembled on the hydrophobic surface of the device, and made a hydrophilic PEO layer which prevents cell adhesion. After rinsing three times with phosphate buffered saline, 2x10^3 cells were seeded in each microwells, and cultured in the medium adjusted for maintaining the undifferentiated state. The medium was changed every day. On both the previous device and the new device, iPS cells self-assembled into one EB in each microwell within two days after seeding (Fig.5).

The expression level of GFP was calculated by measuring the brightness of the EBs in the green fluorescent image. The brightness of each pixel was averaged over the whole areas of the EBs, and the averaged brightness of the 2nd day was set

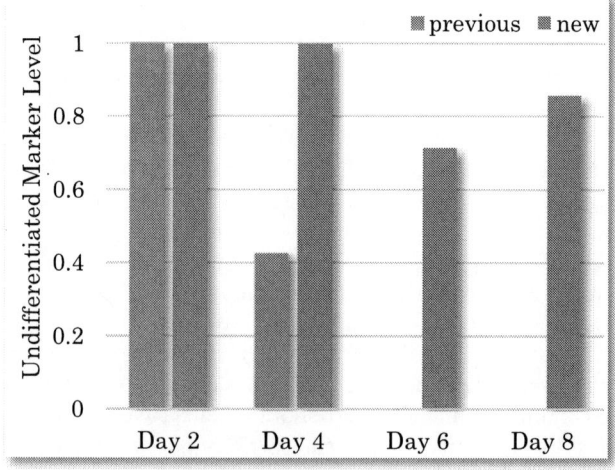

Fig.6 Levels of undifferentiated marker expressing in EBs cultured in the previous and the new microwell device

as 100% on each devices. While the GFP expression was almost lost in 4 days on the previous device, the GFP expression was kept over 60% for more than 8 days in the new device (Fig.6). This result indicates that the permeability at the bottom of the new device has promoted smooth exchange of medium in the microwell, and has made uniform distribution of maintenance culture medium around each EBs. Furthermore, the medium exchange procedure became much faster and safer because the hanging setup of the new device eliminated pipetting directly on the EBs.

CONCLUSION

In this report, we have developed a new device to culture a large number of EBs and maintain their undifferentiated state. Different from our previously reported device TASCL, the new device has liquid permeability at the bottom by using semipermeable membrane. EB formation of mouse iPS cells and prolongation of undifferentiated state were verified by using the device. And the more, from a commercial application perspective, a novel PDMS-like elastomer together with its micro-processing technology was developed for low-cost mass-production of the device. The new device should facilitate mass-production of EBs and long-time culture, which is highly necessary in stem cell research, drug screening and regenerative medicine.

ACKNOWLEDGEMENTS

This study was supported by JSPS Grant-in Aid for Scientific Research (S) 22220008 for which Koji Ikuta was the principal investigator and JST PRESTO "Development of emergent platform for tissue engineering based on membrane micromachining technology" for which Masashi Ikeuchi is the principal investigator.

REFERENCES

[1] YY. Choi, BG. Chung, DH. Lee, A. Khademhosseini, JH. Kim, SH. Lee, "Controlled-size embryoid body formation in concave microwell arrays" Biomaterials 31(15) (2010), pp.4296–4303

[2] Ruei-Zhen Lin, Hwan-You Chang, "Recent advances in three-dimensional multicellular spheroid culture for biomedical research", Biotechnology Journal 3 (2008), pp.1172-1184

[3] Byung Sun Yoon, Seung Jun Yoo, Jeoung Eun Lee, "Enhanced differentiation of human embryonic stemcells into cardiomyocytes by combing hanging drop culture and 5-azacytidinetreatment", Differentiation 74 (2006), pp.149–159

[4] M. Ikeuchi, K. Oishi, H. Noguchi, S. Hayashi, K. Ikuta, "Soft tapered stencil mask for combinatorial 3D cluster formation of stem cells", Proc. μTAS 2010, pp.641-643

[5] H. Yukawa, M. Ikeuchi, H. Noguchi, Y. Miyamoto, K. Ikuta, S. Hayashi, "Embryonic body formation using the tapered soft stencil for cluster culture device", Biomaterials 32(15) (2011), pp.3729-38

[6] M. Ikeuchi, R. Kibe, Y. Toyota, K. Ikuta, S. Hayashi, Regenerative Medicine, 2013, 12 sup: 218. (in Japanese)

CONTACT

*Masashi Ikeuchi, TEL: +81-3-54525162, FAX: +81-3-54525163, ikeuchi@micro.rcast.u-tokyo.ac.jp

ON-CHIP PRODUCT PURIFICATION FOR COMPLETE MICROFLUIDIC RADIOTRACER SYNTHESIS

Supin Chen[1], Alex A. Dooraghi[2,3], Mark Lazari[1,2,3], R. Michael van Dam[1,2,3], Arion F. Chatziioannou[2,3], and Chang-Jin "CJ" Kim[1,4]

[1]Bioengineering Department, [2]Crump Institute for Molecular Imaging, [3]Deparment of Molecular and Medical Pharmacology, [4]Mechanical and Aerospace Engineering Department,
University of California, Los Angeles (UCLA), USA

ABSTRACT

Solid phase extraction was incorporated into an electrowetting-on-dielectric chip for radiochemical purification of a positron emission tomography tracer that was radiolabeled on the same chip. The radiotracer droplet was mixed with alumina particles, and the alumina particles were filtered out from the droplet through a line of pillars, all by electrowetting droplet movement. Fluorination reaction and on-chip purification were analyzed with both Cerenkov imaging and off-chip radio-thin layer chromatography measurements. The measurements were compared to test the validity of the combined use of filtration on-chip and Cerenkov imaging as an alternative approach for monitoring reaction yield without the need to extract sample from the chip.

INTRODUCTION

Positron emission tomography (PET) is a medical imaging modality used in disease studies and treatment evaluation by detection of photons emitted from the radioactive decay of a positron emitting radiotracer (a biomolecule that has been labeled with radioactivity) [1]. It is a functional imaging modality because *in vivo* spatial and time distributions of an injected radiotracer can be recorded. The imaging function (such as glucose metabolism or immune response) depends on the radiotracer that is used. However, only a few radiotracers are commercially available, despite the development of over 1,600 unique radiotracers [2].

In order to increase availability, radiosynthesizers are being developed with the goal of producing a wide range of radiotracers at the imaging center site after delivery of the radionuclide from a radiopharmacy or cyclotron [3]. Previous work has demonstrated electrowetting-on-dielectric (EWOD) droplet manipulation as a promising approach for diverse synthesis of radiotracers [4]. Because it used electric potentials to move liquid, the EWOD chip did not require mechanical valves or pumps, and liquid movement was not restricted to pre-defined channels. Also, the open-to-air configuration enabled rapid drying and solvent exchange, which are critical steps for [18F]fluoride radiosynthesis. However, purification of the synthesized radiotracer to make the final product was always performed off chip, either using high-performance liquid chromatography (HPLC) or cartridges packed with solid phase extraction (SPE) resins.

In this work, solid phase extraction was incorporated on

chip and radiotracer in the droplet was separated from the purification materials using mechanical filtration on the same chip. [18F]Fallypride, a neurotransmitter analog, was radiolabeled for evaluation of the SPE and filtration. Both the fluorination reaction and purification were evaluated with the typical method of radio-thin layer chromatography (radio-TLC). Cerenkov imaging was conducted as an alternative to radio-TLC for on-chip quantification of the reaction yield without extracting any sample.

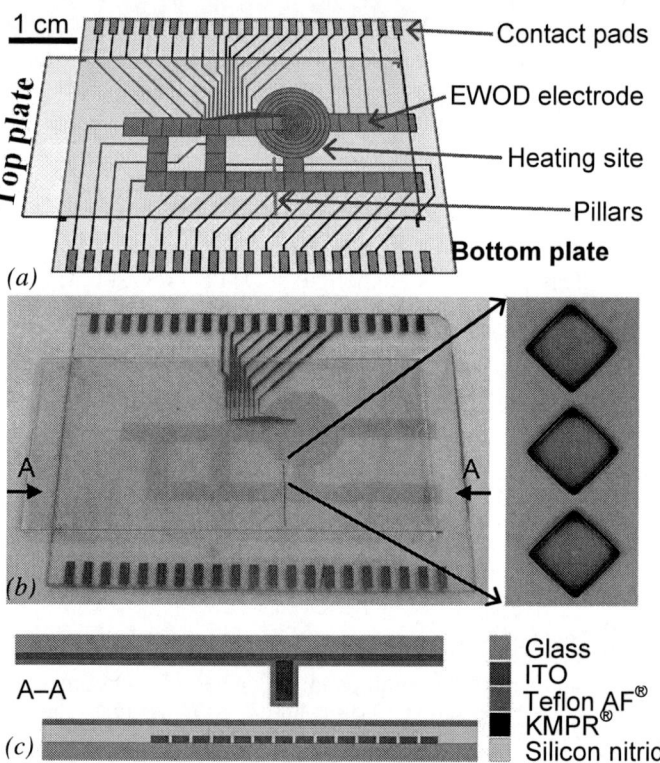

Figure 1: (a) Schematic of the chip showing the actuation glass plate patterned with EWOD electrodes (purple), the electric contact pads along the two edges, and the electrical ground glass plate assembled on top. (b) Assembled EWOD device with 12 mm circular reactor site (feedback-controlled resistive heaters) and filtration pillars on the cover plate. The zoomed in view shows pillars (140 μm tall, 60 μm gap between pillar edges). (c) Cross-sectional view of the chip.

RADIOSYNTHESIZER EWOD DEVICE
Design and fabrication

The EWOD chip has a two-parallel plate configuration

with actuation electrodes on the bottom glass plate and a ground electrode on the top glass plate (Fig. 1). Made of indium tin oxide (ITO), the electrodes are significantly transparent. The reactor site on the bottom plate was not changed from previous work [4]. It consisted of 4 concentric multifunctional electrodes made from ITO that could be used either for EWOD actuation or for feedback-controlled heating. Three droplet pathways of ITO square electrodes led to the reactor site, and a pathway loop was introduced for mixing reagents. Contact pads and connection lines to the multifunctional electrodes were patterned from gold and chrome using photolithography and wet etching. A silicon nitride layer was deposited by plasma-enhanced chemical vapor deposition to serve as a dielectric for the EWOD-driven microfluidics. Teflon AF 2400® was spin coated and annealed at 340°C to make the surface hydrophobic.

A mechanical filter was fabricated on the cover plate. It consisted of KMPR® pillars (140 μm height, 60 μm side-to-side gap between pillars) that were patterned into diamond-shapes by photolithography, similar to the SU-8 pillars fabricated on the bottom plate by Shertzer et al. [5]. In contrast to our previous work [4,6-8], no dielectric was added to the cover plate except for the thin Teflon® layer used as a hydrophobic coating.

Operation

EWOD actuation voltage was supplied from a 10 kHz signal (33220A waveform generator, Agilent Technologies) that was amplified to 100 V_{rms} (Model 601C, Trek). The voltage was applied selectively to desired electrodes by solid-state relays (AQW610EH PhotoMOS relay, Panasonic) that were controlled through a digital I/O device (NI USB-6509, National Instruments). Heating and feedback temperature control were performed by a multichannel heater driver that had been designed and built in house.

REAGENTS

No-carrier-added [^{18}F]fluoride ion was obtained by a cyclotron (RDS-112, Siemens, Knoxville, TN) at the UCLA Ahmanson Biomedical Cyclotron Facility. 2,3,-dimethyl-2-butanol (thexyl alcohol) and anhydrous acetonitrile (MeCN, 99.8%) were purchased from Fischer Scientific. Methanol, ethyl acetate, and triethylamine were purchased from Sigma Aldrich. Tetrabutylammonium bicarbonate solution (75 mM) (TBAHCO3), tosyl-fallypride (fallypride precursor), and neutral alumina were purchased from ABX Advanced Biochemical Compounds (Radeberg, Germany).

FALLYPRIDE SYNTHESIS

[^{18}F]Fallypride is a dopamine receptor antagonist, useful for studying Alzheimer's and Parkinson's disease. Despite its proven effectiveness, [^{18}F]fallypride is not commercially available. However, [^{18}F]fallypride can be synthesized on-chip by a single-step fluorination reaction (Fig. 2) [6].

The radioactive material was loaded onto chip as two

2.5 μL droplets of [^{18}F]fluoride in water for a total of 4.4 mCi of radioactivity. The droplets were loaded at the top plate edge by pipette, and then electrowetting was used to pull the droplets into the plate gap and move them to the reactor site. A 2.5 μL droplet of tetrabutylammonium bicarbonate (25 mM) was moved to the reactor site in a similar fashion and then mixed with the fluoride droplets using electrowetting.

Although [^{18}F]fluoride is obtained from the cyclotron in water, the water needs to be removed for no-carrier-added fluorination reactions, which are water sensitive. Water was removed from the loaded droplets by heating the reactor site to 105 °C for 1 minute.

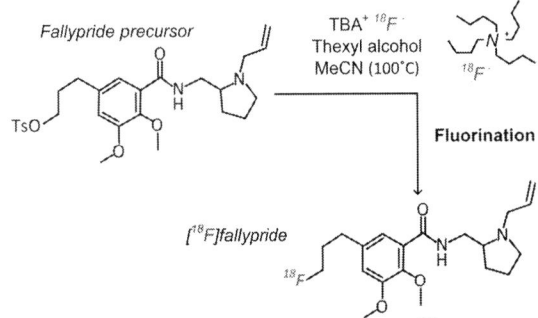

Figure 2: Scheme for one-step synthesis of [^{18}F]fallypride.

An azeotropic drying step was completed to ensure water removal. A 9 μL droplet of acetonitrile was loaded to the chip by pipette and then mixed with the dried fluoride and tetrabutylammonium bicarbonate using electrowetting. The mixture was then heated to 105 °C for 1 minute.

The material to be radiolabeled was added as a 4 μL droplet of tosyl-fallypride in thexyl alcohol (75 mM). It was moved to the reactor site by electrowetting and heated to 100 °C for 7 minutes to perform the fluorination reaction.

To measure fluorination yield, an aliquot of the reaction droplet was removed from the device with a capillary tube and spotted onto a silica gel plate (J.T. Baker, New Jersey). The silica gel plate was developed in a 1:1 mixture of methanol:ethyl acetate with a droplet of triethylamine and then analyzed in a radio-TLC scanner (MiniGITA star, Raytest) (Fig. 3).

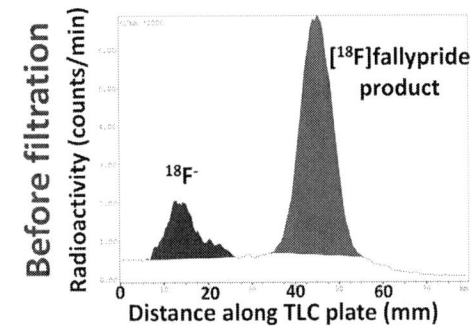

Figure 3: Radio-TLC result of crude mixture before purification. 84% of the radioactivity was labeled to the desired product [^{18}F]fallypride, and the other 16% was unreacted ^{18}F.

ON-CHIP SOLID PHASE EXTRACTION

Alumina particles (80 μm diameter) suspended in acetonitrile (0.5 mg alumina / μL acetonitrile) were pipetted onto the chip and mixed with the reaction droplet using electrowetting. The mixture droplet was then pulled across a line of pillars, which filtered alumina particles out of the droplet (Fig. 4).

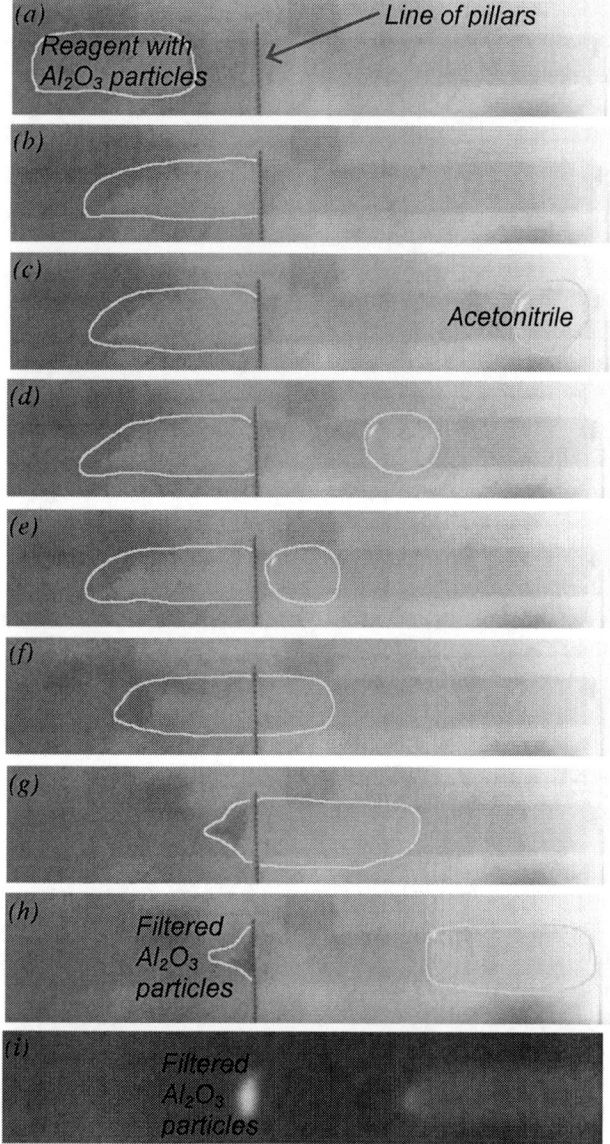

Figure 4: Filtration across line of pillars using electrowetting (from top down). (a-b) The reagent mixture droplet on the left side contains alumina particles and is moved to the filter. (c-g) A pure droplet of acetonitrile is loaded on the right side and brought to the filter site to help the reagent mixture droplet cross the filter. (h) After the droplet crosses the filter, particles remain on the left side. Outlines were added to the droplet images for clarity. (i) The last picture is a Cerenkov image, showing purified product droplet (right) and alumina-containing droplet (left).

The effectiveness of fluoride removal was evaluated using radio-TLC of an aliquot sampled from the [18F]fallypride droplet. A measurement of 99% ± 2% (n=4) showed near quantitative removal of [18F]fluoride (Fig. 5).

Typically SPE cartridges are packed with more than 10 mg of alumina to ensure enough surface interaction for trapping. However, with electrowetting mixing, only 1.5 mg of alumina was needed.

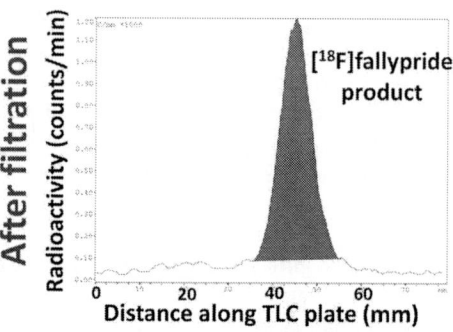

Figure 5: Radio-TLC result of sample from droplet after filtration. The single peak of [18F]fallypride at 45 mm from the left indicates a successful purification.

CERENKOV MEASUREMENT

Cerenkov radiation is emitted light in the UV and visual spectrum that is produced when a beta particle travels faster than the speed that light travels in the surrounding medium. Because the EWOD chip is transparent and the speed of light in the glass material is lower than the speed of a significant fraction of the beta particles emitted by the [18F]fluoride, Cerenkov imaging can both localize and quantify the on-chip radioactivity *in situ* [7].

Cerenkov images were taken of the EWOD chip in a light tight enclosure using a scientific cooled camera (QSI 540, Quantum Scientific Imaging, Poplarville, MS). Radioactivity within a region of interest was quantified from image intensity using calibration and correction factors with MATLAB as described in previous work [7]. After alumina had been filtered from the product droplet, a single Cerenkov image could substitute radio-TLC as a measurement of fluorination yield. The yield was measured by comparing the intensity of the Cerenkov radiation in the filtered alumina (and trapped [18F]fluoride) region with that of the [18F]fallypride droplet (Fig. 4i). The fluorination yield measured by Cerenkov imaging (82.0%) was consistent with the yield measured by radio-TLC (83.9%).

CONCLUSIONS

Addition of on-chip purification enables all four general stages of radiochemistry synthesis on one EWOD chip: concentration of cyclotron-produced [18F]fluoride ion, solvent exchange, reaction, and product isolation [9]. Although only alumina was used in the current demonstration, the same mixing and filtration method are expected to be valid for other SPE resins and impurity removal, opening the door for complete on-chip production of [18F]fallypride and additional PET radiotracers. Other SPE resins of interest are cation-exchange resin, ion-exchange resin, and C18, which in addition to alumina can be used to remove all the impurities for SPE purification of the most commonly used radiotracer, [18F]fluoro-2-deoxy-D-glucose ([18F]FDG) [8].

Filtration of alumina from the radiotracer droplet, which separates the radiofluorinated product from impurities (e.g. unreacted [18F]fluoride), enabled on-chip measurement of the fluorination reaction yield using Cerenkov imaging. Although more development is needed to demonstrate its reliability and validity as an analytical method, the technique holds potential as an on-chip measurement for synthesis development or quality control.

ACKNOWLEDGEMENTS

This work was supported in part by the UCLA Foundation from a donation made by Ralph & Majorie Crump for the UCLA Crump Institute for Molecular Imaging and the Department of Energy [DE-SC0005056]. We thank Prof. David Stout for use of radiolabeling facilities and Dr. Saman Sadeghi for production of radioactive fluoride.

REFERENCES

[1] M.E. Phelps, "Positron emission tomography provides molecular imaging of biological processes", *Proc. National Academy of Sciences*, vol. 97, pp. 9226-9233, 2000.

[2] C.H. Hsieh, *Positron Emission Tomography – Current Clinical and Research Aspects*, InTech, New York, United States, 2012, 153-182.

[3] C. Rensch, A. Jackson, S. Lindner, R. Salvamoser, V. Samper, S. Riese, P. Bartenstein, C. Wangler, and B. Wangler, "Microfluidics: a groundbreaking technology for PET tracer production", *Molecules*, vol. 18, pp. 7930-7956, 2013.

[4] S. Chen, R. Javed, J. Lei, H.-K. Kim, G. Flores, R.M. van Dam, P.Y. Keng, and C.-J. Kim, "Synthesis of diverse tracers on EWOD microdevice for positron emission tomography (PET)", in *Tech. Dig. Solid-State Sensor and Actuator workshop*, Hilton Head Island, SC., USA, June, 2012, 189-192.

[5] M.J. Schertzer, R. Ben-Mrad, and P.E. Sullivan, "Mechanical filtration of particles in electrowetting on dielectrid devices", *Journal of Micromechanical Systems*, vol. 20, pp. 1010-1015, 2011.

[6] M.R. Javed, S. Chen, J. Lei, J. Collins, M. Sergeev, H.-K. Kim, C.-J. Kim, M. van Dam, and P.Y. Keng, "High yield and high specific activity synthesis of [18F]fallypride in a batch microfluidic reactor for micro_PET imaging", *Chemical Communications*, 2013, Accepted.

[7] A.A. Dooraghi, P.Y. Keng, S. Chen, M.R. Javed, C.-J. Kim, A.F. Chatziioannou, and R.M. van Dam, "Optimization of microfluidic PET tracer synthesis with Cerenkov imaging", *Analyst*, vol. 138, pp. 5654-5664, 2013.

[8] P.Y. Keng, S. Chen, H.J. Ding, S. Sadeghi, G.J. Shah, A. Dooraghi, M.E. Phelps, N. Satyamurthy, A.F. Chatziioannou, C.-J. Kim, and R.M. van Dam, "Micro-chemical synthesis of molecular probes on an electronic microfluidic device", *Proc. National Academy of Sciences*, vol. 109, pp. 690-695, 2012.

[9] A.M. Elizarov, "Microreactors for radio-pharmaceutical synthesis", *Lab on a Chip*, vol. 9, pp. 1326-1333, 2009.

CONTACT

S. Chen, tel: +1-310-825-3977; supinchen@ucla.edu

RECONSTITUTION AND FUNCTION OF MEMBRANE PROTEINS INTO ASYMMETRIC GIANT LIPOSOMES BY USING A PULSED JET FLOW

Koki Kamiya,[1,2] Ryuji Kawano,[1] Toshihisa Osaki,[1,3] and Shoji Takeuchi[1,3]

[1]Kanagawa Academy of Science and Technology, Japan
[2]JST, PRESTO, Japan
[3]Institute of Industrial Science, The University of Tokyo, Japan

ABSTRACT

This report describes preliminary results of a reconstitution and function of membrane proteins (flippases) into asymmetric lipid giant liposomes (GLs) that were prepared by deforming a planar asymmetric bilayer using a pulsed jet flow. Frist, we confirmed the expression of flippases on the baculovirus membrane by a western blot analysis. Next, we successfully reconstituted flippases into the asymmetric GL of phosphatidylserine by a baculovirus-liposome fusion method. Finally, we observed the translocation of phosphatidylserine from the extracellular leaflet to the cytoplasmic leaflet which was catalyzed by flippases. The flippase-reconstituted asymmetric liposomes that emulate the function of cell membranes will be useful for the study of elementary lipid-lipid or lipid-protein interactions.

INTRODUCTION

Giant liposomes or giant vesicles are composed of phospholipid membranes similar to those commonly found in living cells. Giant vesicle is typically ~10 μm in diameter, which is sufficiently large for observation using an optical microscopy [1]. Therefore, such an artificial liposomal system has played an important role in the biochemical and biophysical studies such as the interactions of lipid membranes with cytoskeletal proteins, and the microencapsulation of gene expression systems [2]. It has been considered that an assay based on a single giant liposome may help to clarify biochemical properties that can be missed in the analysis of an ensemble of many conventional small liposomes, and this is known as a single giant liposomes assay. Unsymmetrical membrane structures have also been visualized on giant liposome, which suggests that they may be useful as platforms for the reconstitution of intricate cell membrane functions, and for the membrane deformation of fusion, fission and collapse by the ionic strength, the osmolality or the proteins. Recently, the giant vesicles have also been used in the study of membrane proteins function [3]. In the new membrane-protein-integrated giant vesicle method, membrane fusion between liposomes and glycoprotein 64 (gp64) displayed on recombinant budded viruses (BVs) of baculovirus (*Autographa californica* nuclear polyhedrosis viruses (AcNPV)) is analyzed under acidic conditions. The human nicotinic acetylcholine receptor a-subunit (AChRa); the connexin [4], which forms cellular gap junctions; the

adrenergic receptor [5]; and cadherin [6] were successfully prepared by using this method.

In our previous study, to effectively form uniformly sized giant vesicles and encapsulate biological molecules into giant vesicles, we created giant vesicles by the microfluidic-jetting-induced deformation of a planar lipid bilayer including an organic solvent. However, this method has limitations; for instance, the giant vesicle is a W/O/W emulsion having a thin shell of an organic solvent. Further, it is unstable and very large (over 300 μm in diameter). We improved double-well device wherein an acrylic separator with a small hole, was mounted between the wells. Cell-sized vesicles were stably formed using this improved double-well device. Moreover, we could successfully prepare asymmetric lipid vesicles to emulate asymmetric lipid leaflets of plasma membrane, and directly prepared a mimic intracellular vesicular traffic system of a cell-sized vesicle containing small vesicles using a triple-well device.

In this study, to closely emulate the membrane structure, we demonstrated the reconstitution of flippases into phosphatidylserine (PS)/ phosphatidylcoline (PC) asymmetric liposomes by a baculovirus-liposome fusion method. The activity of flippases was confirmed on PS/PC asymmetric liposome membranes (Fig.1).

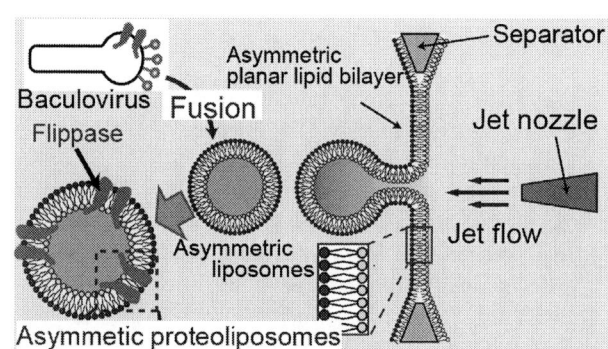

Figure 1: Illustration of membrane protein reconstition system using lipid asymmetirc liposomes using a device. Flippase utilizes energy from ATP hydrolysis to flip aminophospholipids from the exocytoplasmic to the cytoplasmic leaflet of cellular membranes. Consequently, the phospholipid flippases play an essential role in maintaining the asymmetricity of the phospholipid membrane on the cell surface.

EXPERIMENTAL

The double-well chamber was fabricated by machining a poly methacrylate (PMMA) plate using an automated CAD/CAM modeling machine. Fig.3a shows the preparation process of cell-sized liposomes by pulsed jet flow. A planar lipid bilayer membrane was formed in the double-well chamber, wherein a thin acrylic film, having a 150-μm-diameter hole, was placed between the chamber wells. To form a planar asymmetric lipid bilayer membrane, the lipid solution [1,2-dioleoyl-*sn*-glycero-3-phosphocholine (DOPC) and 1,2-dioleoyl-*sn*-glycero-3-phospho-L-serine (DOPS) dissolved in *n*-decane] was added to each well. Next, aqueous solution was added to each well and an asymmetric bilayer was easily formed. Asymmetric giant liposomes were formed by impinging the pulsed jet flow against the asymmetric lipid bilayer. The micro jet nozzle was fabricated by pulling a 1mm glass capillary tube with a micropipette puller. It was bent with a microforge to adjust its shape to access near the planar membrane. The pulsed jet flow was generated by opening an electromagnetic valve between a glass capillary nozzle and an air compressor.

Flippase BVs were harvested as follows: an insect cell line Sf9 derived from the fall armyworm *Spodoptera frugiperda* was cultured in the culture flasks (75 cm^3) containing Sf-900 III medium each until about 80% of each of the flasks was covered. A suspension of flippase BVs was spread in each flask at a multiplicity of infection (MOI) of 1. Flasks were incubated at 27°C for 94 h. Flippase BVs were purified by sucrose density gradient centrifugation. We observed the flippases-reconstituted liposomes by using confocal laser scanning microscopy (CLSM).

RESULTS

The sample from recombinant flippase budded viruses (BVs) contained a flippase band of an approximate molecular mass of 110 kDa that was not observed in wild type BVs (control) by using the western blotting analysis. To form asymmetric proteoliposomes, we generated DOPS/DOPC asymmetric GLs from the asymmetric lipid bilayer by applying a microfluidic jet flow. Flippase-reconstituted asymmetric GLs were prepared by fusion between flippase BVs and asymmetric GLs that had one leaflet comprising 100% DOPC and the other leaflet comprising DOPC/DOPS at a 1/1 molar ratio. Next, we examined the flippase activity on asymmetric GLs by fluorescence annexin V assay. The fluorescence of annexin V on flippase-reconstituted asymmetric GLs containing ATP decreased, indicating that the translocation of phosphatidylserine from extracellular leaflet to cytoplasmic leaflet was catalyzed by flippase activity (Fig.2).

CONCLUSION

We successfully prepared flippase-reconstituted asymmetric GLs using the baculovirus-liposome fusion method, which can more closely emulate the environment of asymmetric lipid leaflets of plasma membrane.

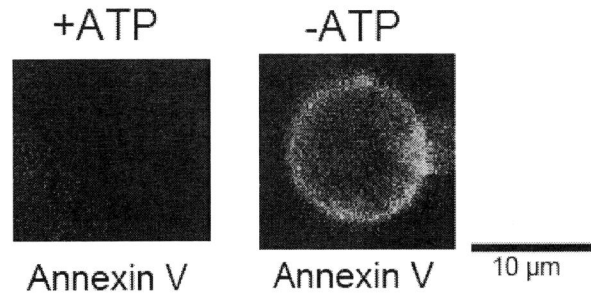

Figure 2: CLSM images of flippase activity assay.

ACKNOWLEDGEMENTS

The authors acknowledge the technical support provided by Maiko Uchida. This work was partly supported by the Grant-in-Aid from the Ministry of Education, Culture, Sports, Science and Technology of Japan (Challenging Exploratory Research: 25620138 (K.K.)).

REFERENCES

[1] F.M. Menger, M.I. Angelova, Giant vesicles: imitating the cytological processes of cell membranes, *Acc. Chem. Res.* 31 (1998) 789–797.

[2] D. Deamer, A giant step towards artificial life? Trends Biotechnol. 23 (2005) 336–338.

[3] J. Kriegsmann, I. Gregor, I. Hocht , J Klare,M Engelhard, J Enderlein, J. Fitter, Translational Diffusion and Interaction of a Photoreceptor and Its Cognate Transducer Observed in Giant Unilamellar Vesicles by Using Dual-Focus FCS, chembiochem. 10 (2009) 1823-1829.

[4] K. Kamiya, K. Tsumoto, S. Arakawa, S. Shimizu, I. Morita, T. Yoshimura, K. Akiyoshi, Preparation of Connexin43-Integrated Giant Liposomes by a Baculovirus Expression Liposome Fusion Method, *Biotechnol. Bioeng.* 107 (2010) 836-843.

[5] K. Kamiya, J. Kobayashi, T Yoshimura, K Tsumoto, Confocal microscopic observation of fusion between baculovirus budded virus envelopes and single giant unilamellar vesicles, *Biochim Biophys Acta Biomembranes* 1798 (2010) 1625-1631.

[6] K. Kamiya, K. Tsumoto, T.Yoshimura, K. Akiyoshi, Cadherin-integrated liposomes with potential application in a drug delivery system, *Biomaterials* 32 (2011) 9899-9907.

CONTACT

K. Kamiya, Kanagawa Academy of Science and Technology; KSP East 303, 3-2-1 Sakado, Takatsu, Kawasaki, Kanagawa 213-0012, JAPAN; Phone +81-44-819-2037; FAX +81-44-819-2092; Email kamiya@iis.u-tokyo.ac.jp

TRACTION FORCE OF SMOOTH MUSCLE CELL DURING GROWTH ON A RIGID SUBSTRATE

U. G. Jung, T. Tsukagoshi, H. Takahashi, T. Kan, K. Matsumoto and I. Shimoyama

The University of Tokyo, Japan

ABSTRACT

This paper reports on the sensor for traction force measurement of a smooth muscle cell during cell growth on a rigid substrate, specially designed for horizontal and vertical directional forces. For quantitative measurement, the cells are cultured only on the sensor pads using a cover chip. The length of the sensor is 1130μm. The size of the sensor pad is 125μm×15μm×5μm (length×width× thickness). The gaps between the sensor pads are 3μm. We confirmed that the cells spread on the two sensor pads at least. We measured the traction forces of bovine aortic smooth muscle cells (BAOSMCs, CAB35405) using the proposed sensor. When the three cells spread on the pads, the measured traction forces in x and z direction increased 30nN and 20nN for 8 min, respectively.

INTRODUCTION

Recently advanced technology is able to make "Bionic man" whose 70% of body is artificial one except brain and digestive organ [1]. Integration of mechanical equipment into a bionic environment is indispensable technology of our lives. For instance, missing and broken body parts can be replaced artificial ones by medical engineering. Commonly used materials of artificial organs such as plastic and polyethylene ($0.2×10^9$[Pa]) are rigid compared to muscle (10^4[Pa]) or blood vessel (10^6[Pa]). However, cell behaviors are different according to a stiffness of a substrate. When implanting the artificial organ into human body or developing an application using cells, it is important to know cell behaviors in those unnatural environments.

Traction force is a crucial key to understand cell behaviors [3-5]. Cell probes the substrate stiffness and regulates their processes such as growth, differentiation, migration, and tumor metastasis [2]. Cell generates forces on the substrate during spreading or motility. The traction force is also changed to cell shape and status [3-5].

The traction force on a rigid substrate is still unknown. In previous studies, the traction forces have been measured by observing a deformation of the flexible substrate with microscopy. However, the deformation of the rigid substrate is difficult to be observed. Therefore, we proposed the method for traction force measurement on the rigid substrate using a piezoresistive cantilever [6]. The spring constant of the cantilever was 0.9N/m which is 10 times larger than those of previous works [3][5].

In this paper, we upgraded the sensor to detect x and z multi directional traction forces. The sensor has sensor pads and sensor beams. The cover chip was designed for culturing cells only on the sensor pads. We measured the traction forces of smooth muscle cells at 37°C to measure a cell growth.

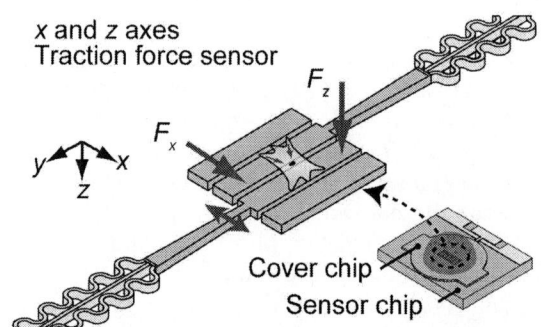

Figure 1: *Concept of a piezoresistive force sensor. Traction forces of smooth muscle cells are measured with the proposed sensor. The cover chip prevents cells from adhering on the sensor beams.*

BASIC DESIGN

Figure 1 illustrates the concept of the piezoresistive force sensor for the traction force measurement. Smooth muscle cells spread on the sensor pad, and then the sensor measures the traction forces. The sensor is divided into two parts (Figure 2). First part is a sensor pad for culturing cells (Figure 2(a)). Second part is the sensor beams for detecting forces (Figure 2(b)). The size of the sensor pad was designed 125μm×15μm×5μm (length×width×thickness). The diameter of a single cell is about 20μm so that cells are spreading at least on the two pads. The gap between the sensor pads is 4μm.

The sensor beams are composed of two resistors and one wiring. The piezoresistors were formed on the surface and the side wall of the beams. When x directional forces are applied on the sensor pad, a strain is concentrated on the side wall of the beam. The two resistances of the beams change in a reverse direction. On the other hand, when z directional forces are applied to the sensor, the two resistances decrease. Figure 3 shows the simulation of the sensor strain to x and z directional forces. The details of principle of the sensor for x and z directional force measurements using the side-wall doping was reported in a previous work [7].

The total length of the sensor and the width of each beam are 1130μm and 5μm, respectively. The spring constants of the proposed sensor in x direction and z direction were 0.4N/m and 0.06N/m, respectively.

Moreover, we fabricated a cover chip with a Si wafer. The cover chip had a through hole of the center. It prevents cells from adhering on the sensor beams because cells settle down on the sensor randomly in the dish. We assembled the cover chip with the sensor chip using guiding holes of the sensor chip as shown in Figure 2. Then, the cover chip and the sensor chip were glued with bees wax. The size of the

978-1-4799-3510-9/14 $31.00 © 2014 IEEE

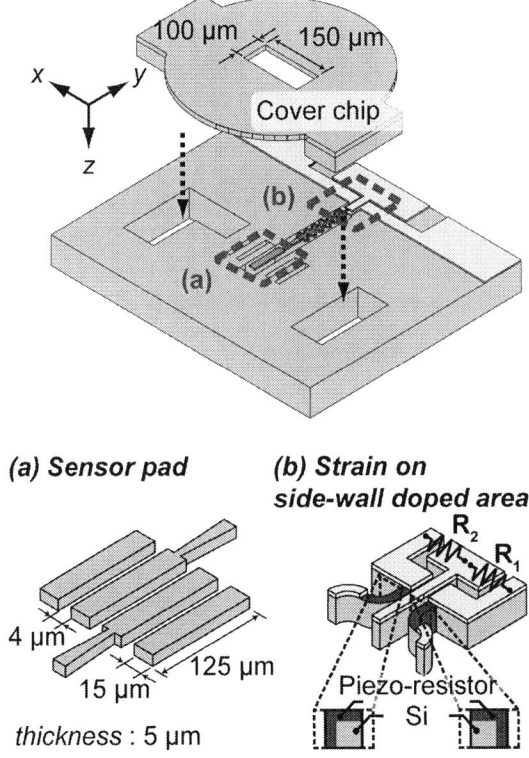

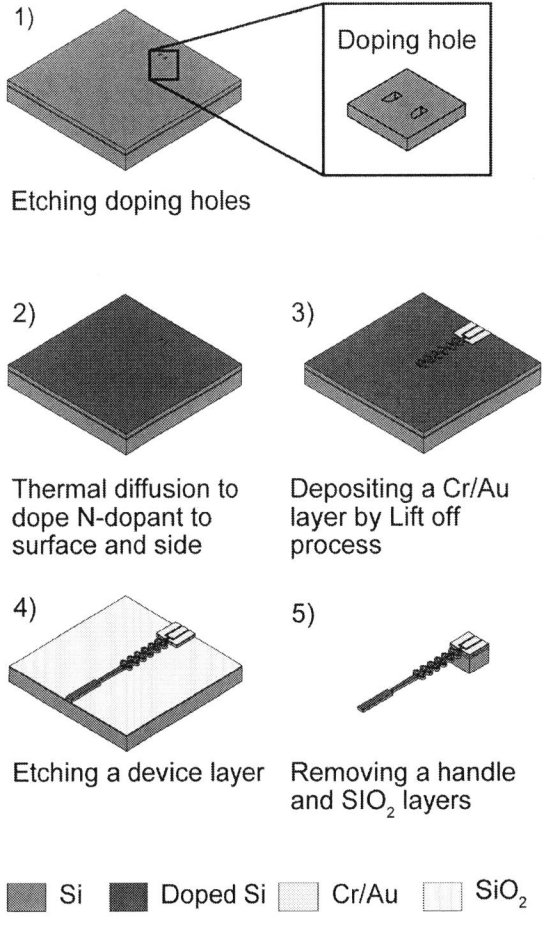

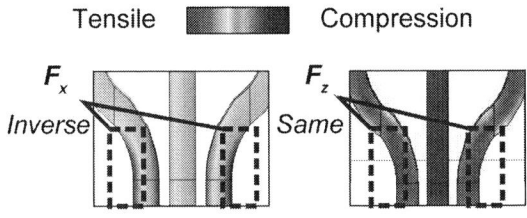

(a) Sensor pad   **(b) Strain on side-wall doped area**

4 µm
15 µm
125 µm
thickness : 5 µm

Piezo-resistor
Si

Figure 2: *Cell traction force measurement method. (a) Dimension of the sensor pads (b) Two piezo- resistors are formed on the surface and side-wall*

Tensile   Compression

F_x F_z

Inverse   Same

Figure 3: *Simulation of the sensor strain to x and z directional forces.*

Etching doping holes

Doping hole

2) Thermal diffusion to dope N-dopant to surface and side

3) Depositing a Cr/Au layer by Lift off process

4) Etching a device layer

5) Removing a handle and SIO₂ layers

Si   Doped Si   Cr/Au   SiO₂

Figure 4: *1) Etch the device Si layer 2) side-wall doping with rapid thermal diffusion 3) Depositing Cr/Au layer for wiring 4) Etch the device Si layer for sensor shape 5) Remove the SiO₂ layer and a handling Si layer.*

patterned by the lift off process. In step 4, the handling Si layer was etched by ICP-RIE. The oxide layer was removed with hydrogen fluoride vapor. In step 5, the sensor chip was coated with parylene-C (*thickness: 0.5~1µm*). Figure 5 shows a photograph of the sensor and an SEM image of the sensor pad. The actual gap between sensor pads was reduced down to 3µm due to the parylene coating.

SENSOR CHARACTERISTICS

The force sensitivity of the sensor was calibrated using a load cell and a reference cantilever. The spring constant of the reference cantilever was 2.0N/m. We calibrated fractional resistance change of the reference cantilever using the load cell. The x and z directional forces (µN) were applied to the reference cantilever. Then, the fractional resistance change of the proposed sensor was calculated. Figure 6 shows the relation between the resistance changes and the applied force. The force sensitivities in x directional forces of the sensor beams corresponding to R_1 and R_2 are $-7.4 \times 10^{-6} \mathrm{nN}^{-1}$ and $4.7 \times 10^{-6} \mathrm{nN}^{-1}$, respectively. The force sensitivities in z

hole on the cover chip was 100µm×150µm.

FABRICATION

The fabrication process of the sensor is shown in Figure. 4. The sensor was fabricated using a P-type SOI wafer (5/2/300µm). In step l, two holes were etched on the device Si layer by an inductive coupled plasma reactive ion etching (ICP-RIE). In step 2, the doping holes were filled with dopant liquid. Then, the surface and side-wall were doped with rapid thermal diffusion. In step 3, Cr/Au layers were deposited for electrical wiring by vacuum deposition equipment. The thicknesses of Cr and Au layer were 4 and 40nm, respectively. The wiring shape of device Si layer was

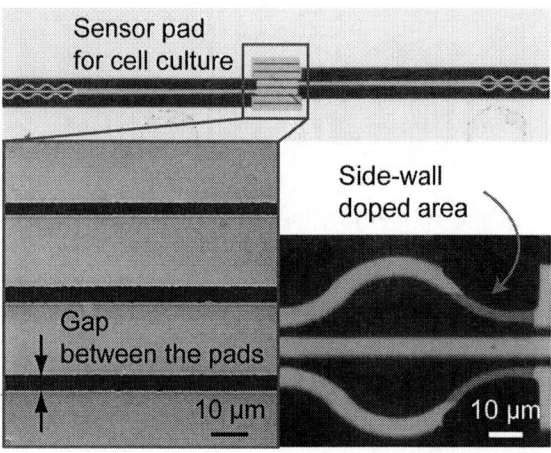

Figure 5: *Photograph of the fabricated sensor and SEM image of the sensor pads.*

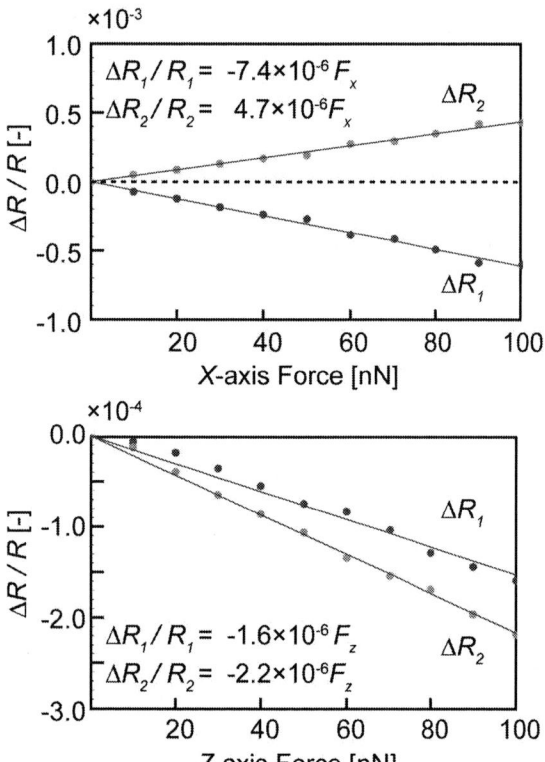

Figure 6: *The fractional resistance changes of two beams when x directional and z directional forces are applied.*

directional forces of the sensor beams are -2.0×10^{-6}nN^{-1} and -2.3×10^{-6}nN^{-1}, respectively. The characteristic matrix calculated from the fitting line is;

$$\begin{pmatrix} F_x \\ F_z \end{pmatrix} = 10^5 \times \begin{pmatrix} -0.9 & 0.8 \\ -1.8 & -2.8 \end{pmatrix} \begin{pmatrix} \Delta R_1 / R_1 \\ \Delta R_2 / R_2 \end{pmatrix}$$

The unit of force is nN.

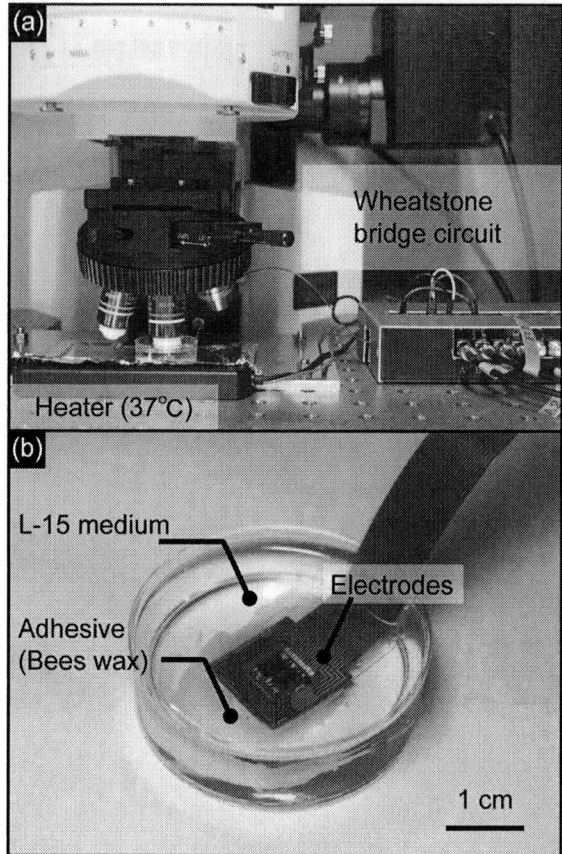

Figure 7: *Fibroblasts were cultured for 1 h in the incubator: photograph of (a) the experimental setup (b) sensor chip in the culture solution*

EXPERIMENTAL SETUP

Figure 7 shows an experimental setup. The sensor was fixed to the dish with bees wax and rinsed with 70% ethanol, PBS and Earle's balanced salt solution (EBSS). Then, the surface of the sensor was coated with fibronectin for 1 hour at room temperature. Smooth muscle cells (BAOSMCs, CAB35405) were suspended in the dish. After 1 hour, the medium was replaced by Leibovitz's L-15 medium with 20 mM HEPES pH 7.4. L-15 medium was filtered (φ:450μm) to prevent particles from touching the piezoresistors area. The hot plate was used to maintain the temperature of the medium at 37°C. The positions of cells were observed with a differential interference contrast microscope.

EXPERIMENT AND RESULTS

Figure 8 shows output signals of the sensor without cells in the L-15 medium at 37°C for 20 min. The sensor had a temperature compensator. The output signals were stable. The noise levels were corresponding to 10nN in x direction, 20nN in z direction, respectively.

Figure 9 shows photographs of cells ((a), (b)) and traction forces of BAOSMCs for 10 min. Three cells adhered

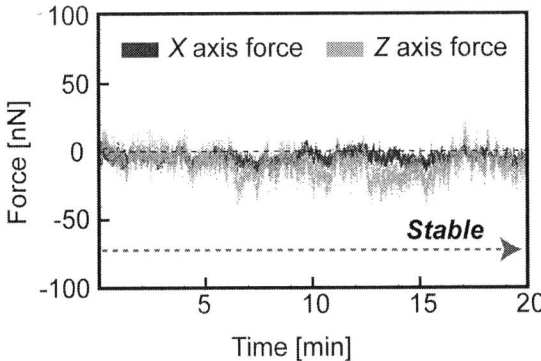

Figure 8: We confirmed the output signals of the force sensor in the 37 °C medium without a cell. The signals were stable.

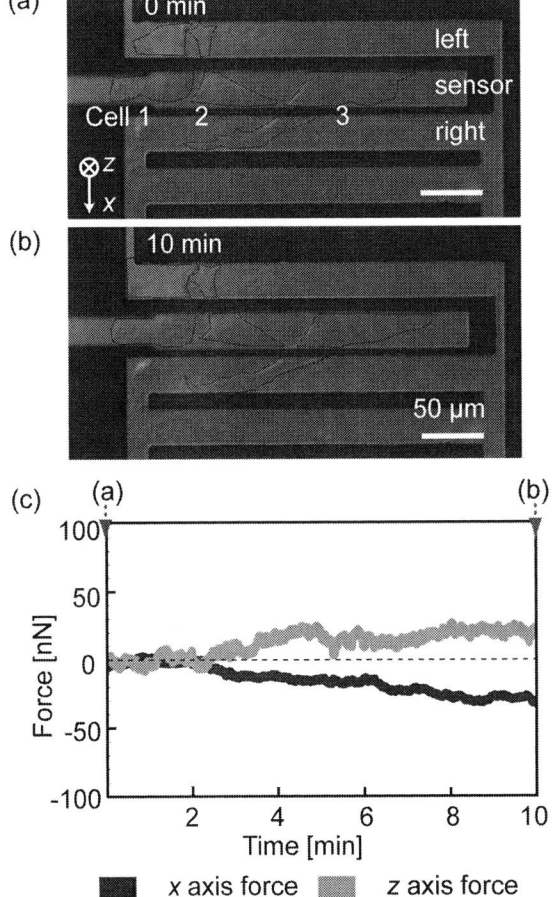

Figure 9 : Traction forces of the smooth muscle cells(BAOSMCs) were measured using the proposed sensor.

to the sensor pad and the surrounding frame. If cell 1 and cell 2 generated the forces, the sensor is dragged to the left frame. If cell 3 generated the forces, the sensor is dragged to the right frame. The traction force in x direction was increased by 30nN for 8 min. From the measurements, the sensor was

dragged to the left side.

In the case of z directional forces, the measured traction force was 20nN. The sensor was dragged to the downward. The sensor was located on the upper side from the surrounding frame of both sides due to the fabrication process. Therefore, we thought that the cell pulled the sensor pad to the x and z direction. Therefore, the total traction forces were generated from 20 to 50nN.

CONCLUSION

In conclusion, we measured traction forces of BAOSMCs during cell growth on a rigid substrate using the proposed piezoresistive force sensor. The size of the sensor pad is 125μm×15μm×5μm. We confirmed that three cells were adhered to the sensor pads. The measured traction force in x direction was 30nN for 8 min. In the case of z direction, the force was 20nN.

ACKNOWLEDGEMENTS

The photolithography masks were made using the University of Tokyo VLSI Design and Education Center (VDEC)'s 8 inch EB writer F5112 + VD01, which was donated by ADVANTEST Corporation. This work was conducted in Center for Nano Lithography & Analysis, The University of Tokyo, supported by the Ministry of Education, Culture, Sports, Science and Technology (MEXT), Japan. This work was supported by JSPS KAKENHI Grant Number 25000010.

REFERENCES

[1] W. Craelius, "The Bionic Man: Restoring Mobility," *SCIENCE*, vol, 295, no. 8, FEB 2002.

[2] M. Ghibaudo, "Traction forces and rigidity sensing regulate cell functions," *Soft Matter*, vol. 4, pp. 1836–1843, 2008

[3] S. Ghassemi, et al., "Cells test substrate rigidity by local contractions on submicrometer pillars," *PNAS*, vol. 109, no. 14, pp. 5328-5333, April 3 2012.

[4] L. Trichet, et al., "Evidence of a large-scale mechano sensing mechanism for cellular adaptation to substrate stiffness," *PNAS*, vol. 109, no. 18, pp. 6933-6938, May 1 2012.

[5] M. Prager-Khoutorsky, "Fibroblast polarization is a matrix-rigidity-dependent process controlled by focal adhesion mechanosensing," *Nature cell biology*, vol 13, no 12, pp.1457-65, 2011.

[6] U. G. Jung, et al., "A Piezoresistive Cellular Traction Force Sensor," *(MEMS '13)*, pp. 927-930, Taipei, Taiwan, 20-24 January, 2013.

[7] H. Takahashi, et al., "A triaxial tactile sensor without crosstalk using pairs of piezoresistive beams with sidewall doping," *Sensors and Actuators A: Physical*, vol. 199, pp. 43-48, 2013.

CONTACT

*Uijin G. Jung,
Mail:uj_jung@leopard.t.u-tokyo.ac.jp

A MICRO GAS CHROMATOGRAPH WITH INTEGRATED BI-DIRECTIONAL PUMP FOR QUANTITATIVE ANALYSES

Yutao Qin and Yogesh B. Gianchandani
Center for Wireless Integrated MicroSensing and Systems (WIMS[2])
University of Michigan, Ann Arbor, MI 48109, USA

ABSTRACT

This paper describes a micro gas chromatography (μGC) system that comprises a Knudsen pump with bi-directional capability (KP2), a two-stage preconcentrator-focuser (PCF2) and a separation column. In this valveless system, the bi-directionality of the pump allows flow reversal in the multi-stage preconcentrator. The KP2, PCF2, and separation column are arranged in a 4.3 cm^3 stack, and used with a commercial flame ionization detector. In preliminary experiments, the μGC system demonstrated quantitative separation of benzene, toluene, and xylene (BTX) in ambient room air.

INTRODUCTION

Typical μGC systems include several components: the preconcentrator, which accumulates vapor analytes and provides vapor injection; the column, which separates the vapor analytes; the gas detector, which quantifies the eluents (retention peaks) from the column; and the gas pump, which provides flow [1-3]. In the simplest architectures, the preconcentrator, column, detector, and pump are connected in series, and operated with unidirectional flow [4-5]. In more complex architectures, valves are used to create opposite flow directions (bi-directional flow) in the preconcentrator during sampling and separation; examples appear in [6-8].

While bi-directional flow is potentially beneficial for a single-stage preconcentrator, it is essential for a multi-stage (or multi-bed) one. A multi-stage preconcentrator has different sorbents packed in its stages that are connected in series (Fig. 1). Weaker stages – those packed with lower surface-area sorbents – are located upstream in the sampling flow and are intended to trap vapor analytes with lower volatility. Stronger stages – those packed with higher surface-area sorbents – are located downstream in the sampling flow, and are intended to trap vapor analytes with higher volatility that pass through weaker stages [9]. The vapor sample must flow from the weaker stages to the stronger stages to permit upstream capture of low volatility analytes during the sampling phase. In the analytical separation phase, the flow is reversed, allowing the low volatility sample to enter the separation column together with the high volatility sample that was captured deeper within the preconcentrator. In conventional systems, the flow reversal is accomplished by valves [6-8].

This effort explores a valveless μGC architecture, which is comprised of a bi-directional Knudsen pump (KP2), a two-stage preconcentrator (PCF2), and a separation column. During vapor sampling, vapor analytes enter the μGC system through the separation column and settle into the PCF2 (Fig. 1). During analytical separation, the flow is reversed by the KP2; the sampled vapor analytes are then thermally desorbed from the PCF2 and separated in the column. A commercial flame ionization detector (FID) is used as the detector for accurate quantification of the bi-directional operation. In the future, a micro gas detector can be integrated for realizing a complete, bi-directional μGC system.

DESIGN

Bi-Directional Knudsen Pump (KP2)

Knudsen pumps operate by thermal transpiration [10]. High reliability and simple configuration make these pumps attractive for integration with μGC systems [5, 11]. One example of a Knudsen pump implementation utilizes nanoporous mixed cellulose ester (MCE) membrane(s) (thickness ≈105 μm, pore diameter ≈25 nm, porosity ≈70%, Millipore, MA) [12]. In the presence of a temperature gradient, thermal transpiration flow is generated from the cold side to the hot side of the membranes. A previously reported bi-directional Knudsen pump used thermoelectric elements to provide reversible temperature gradients and gas flow [13]. In this effort, however, the bi-directional Knudsen pump is implemented simply by integrating resistive heaters on both sides of the MCE membranes. During operation, one of the sides is heated while the other is cooled (by a heat sink or natural convection), providing a temperature gradient.

The KP2 consists of four glass dice (Die 1a, 1b, 2a, and 2b, thickness =500 μm) sandwiching a stack of four MCE membranes (Fig. 2). Side-A of the KP2 consists of Die 1a

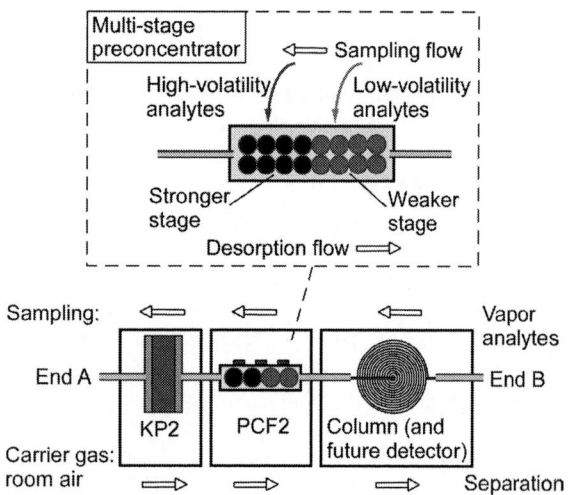

Fig. 1: Concept of multi-stage preconcentrator and the bi-directional operation of the μGC system in this effort.

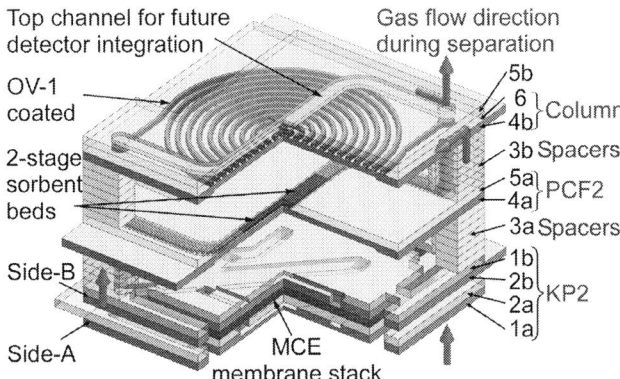

Fig. 2: Stack-integrated μGC architecture.

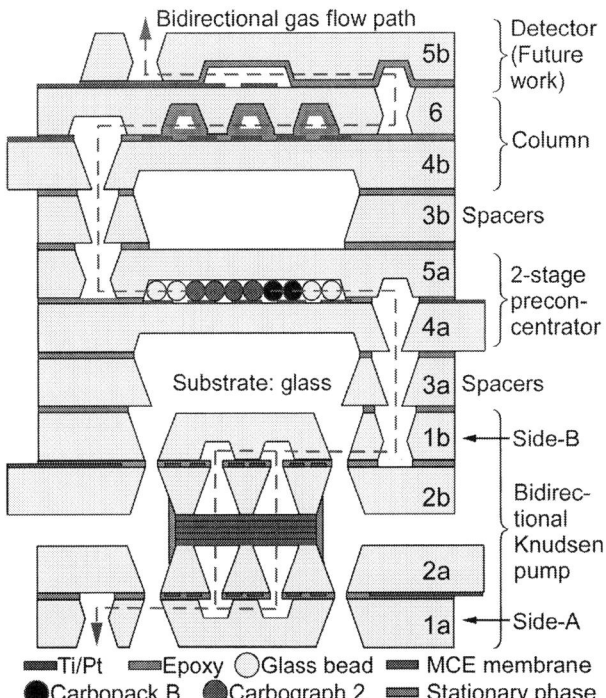

Fig. 3: Cross-section of the stack μGC architecture.

and Die 2a, whereas Side-B consists of Die 1b and Die 2b. Each side has on-chip heaters and thermistors for thermal control as well as multiple grooves and through-holes for gas flow. In the stacked μGC system (Fig. 2), Side-B is attached to the rest of the stack, whereas Side-A is attached to an external heat sink. During vapor sampling, Side-A (and the heat sink) is heated while Side-B is cooled by natural convection. During analytical separations, Side-B is heated while Side-A is cooled by the heat sink.

Two-Stage Preconcentrator (PCF2)

Prior reports of microfabricated multi-stage preconcentrators describe the use of multiple chambers, each containing one type of sorbent [9]. In this effort, however, the preconcentrator contains only one chamber designed in the shape of a channel; multiple stages are formed by packing multiple types of sorbents in sequence. Such a design mimics its macro-scale counterpart: multi-bed thermal desorption tubes. One advantage of this design is convenient scalability: as demanded by the application, the preconcentrator can easily be configured into more stages without redesign and refabrication.

In this effort, the PCF2 consists of a channel (width ≈1 mm, depth ≈200 μm) formed by Die 4a and Die 5a (Fig. 2). The channel is packed with four segments of particles in series: glass beads (with diameter 150-180 μm, Sigma Aldrich, WI); Carbograph 2 (with surface area ≈10 m^2/g, diameter 120-150 μm, Grace Davison Discovery Sciences, IL); Carbopack B (with surface area ≈100 m^2/g, diameter 112-140 μm, Sigma Aldrich, WI); glass beads (with diameter 150-180 μm, Sigma Aldrich, WI). The segment of Carbograph 2 is intended to trap vapors with lower volatilities, and the segment of Carbopack B is intended to trap vapors with higher volatilities. The two segments of glass beads, one at each end of the preconcentrator channel, are used to confine the sorbents in the central region of the preconcentrator, where the temperature is the highest during thermal desorption.

Separation Column and the μGC System

The separation column in this effort is a channel (length ≈25 cm, hydraulic diameter ≈230 μm) coated with a ≈0.2 μm thick non-polar polydimethylsiloxane (OV-1, Ohio Valley Specialty, OH) stationary phase, which separates the vapor analytes based on their boiling points. The column is formed by Die 4b and Die 6 (Fig. 2).

The KP2, PCF2, and column are integrated in the form of a stack (Fig. 2 and Fig. 3), together with 14 spacers (Dice 3a and Dice 3b). The dashed line in Fig. 3 illustrates the gas flow path of the μGC system.

FABRICATION

The fabrication process is similar to that described in our prior work [5]. All components of the μGC system (including the KP2, PCF2, column, and spacers) are

Fig. 4: Photographs of the fabricated components and assembled system.

fabricated on the same glass wafer using a three-mask process. First, a Ti/Pt thin film is deposited and patterned on the glass wafer to form heaters, thermistors, and discharge electrodes (Mask #1). Next, the glass wafer is micromachined on both sides by sandblasting (Bullen, Inc., OH) to create fluidic channels and thermal-isolating cutouts (Mask #2 and #3). After wafer dicing, all the components are assembled using epoxy adhesion layers. The separation column is coated with the stationary phase using the static coating method [14]. The PCF2 is packed by using a gentle vacuum to draw the sorbents [15] and glass beads into the channel in sequence. Photographs of the fabricated components and the assembled μGC system are shown in Fig. 4. The μGC system has a footprint of 1.8×2.0 cm^2 and a volume of 4.3 cm^3.

EXPERIMENTAL RESULTS

The operation of the μGC system was tested using benzene, toluene, and xylene (BTX), which are typical indoor pollutants. The testing vapors were prepared by evaporating 0.1-0.75 μL of liquid BTX in a 2 L dilution bottle filled with N$_2$, resulting in vapor concentration ranging from 43-328 mg/m^3.

During vapor sampling, the bottle was connected to End B of the system (Fig. 1) by capillary tubes. The KP2 provided a sampling flow of 0.05 sccm for 10 min, drawing the vapor through the separation column and into the PCF2. Then the system was allowed to cool down for 10 min by natural convection.

During analytical separation, End B of the system (Fig. 1) was connected to an FID in a commercial GC (Model # Agilent 6890); whereas End A was open to ambient room air, which served as the carrier gas. The KP2 provided a reverse room air flow 20 sec prior to the PCF2 desorption. This time period was used to allow sufficient gas flow (of 0.2 sccm) to be established. Component temperatures were measured by on-chip thermistors (Fig. 5).

In the chromatograms provided by the FID (Fig. 6), the BTX retention peaks are manifested in the chromatograms as being superposed on fluctuating baselines, which were caused by the BTX vapors remaining in the separation column during vapor sampling. To verify quantitative analysis, the FID peak areas (indicative of analyzed BTX quantities) in the resulting chromatograms were calculated with the subtraction of the baseline areas, and are plotted against their corresponding vapor concentration (Fig. 7). The data points are fitted by straight lines that are connected to the origin of the coordinate. The linear regression achieves R^2 values of 0.9987-0.9999, indicating good linearity of the system response for quantitative analysis.

DISCUSSION AND CONCLUSIONS

This paper describes a bi-directional μGC system for quantitative vapor analysis. The bi-directional operation required by the multi-stage preconcentrator is provided by an integrated bi-directional Knudsen pump, eliminating the need for valves. Quantitative analysis is demonstrated

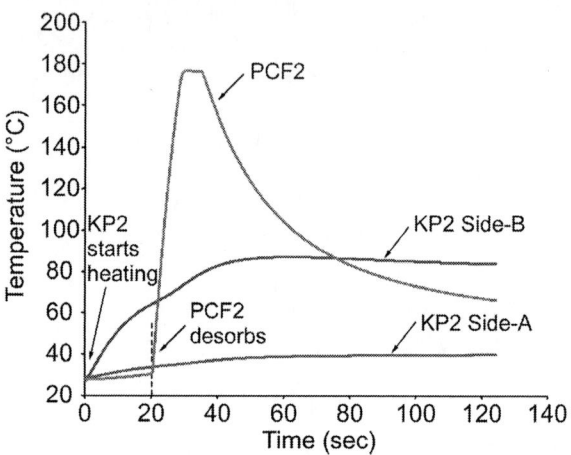

Fig. 5: Transient temperatures of the components measured by on-chip thermistors during analytical separation.

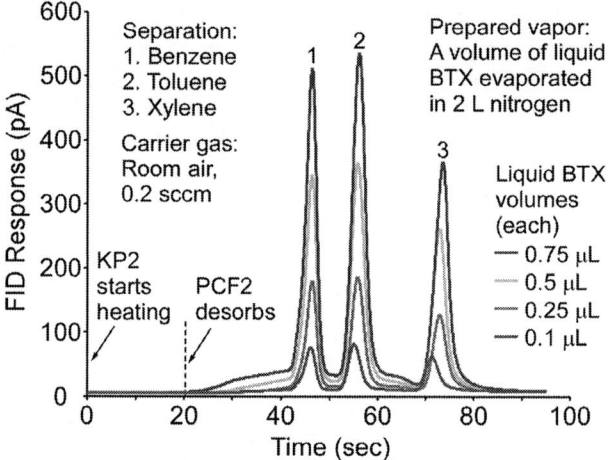

Fig. 6: Separation of BTX vapors (four concentration values) provided by the system.

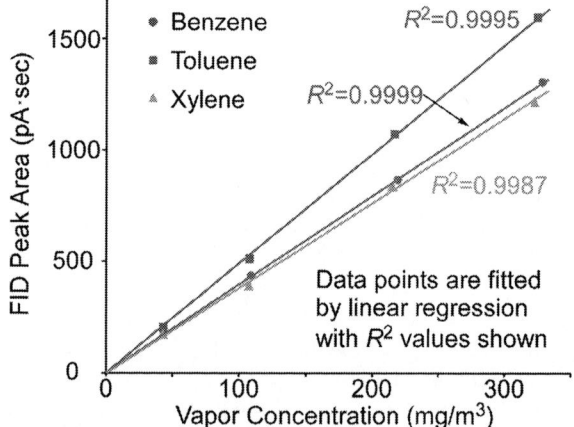

Fig. 7: The plot of each BTX peak area in Fig. 6 against corresponding vapor concentration.

experimentally, which is a significant functional improvement over our prior work [5]. It should be noted that the separation column present in the vapor sampling path only causes baseline fluctuations in the chromatogram but does not affect quantitative analysis.

The system can be easily reconfigured for further improvements. The PCF2 can be reconfigured into more stages for an expanded range of vapor analyte species, by simply packing more types of sorbents in sequence. An elevated temperature can accelerate vapor analytes passing through the column during sampling. In the future, an on-chip gas detector will be integrated for realizing a complete bi-directional μGC system.

ACKNOWLEDGEMENTS

The study was supported in part by the Microsystems Technology Office of the Defense Advanced Research Projects Agency High-Vacuum Program (DARPA Contract #W31P4Q-09-1-0011). Facilities used for this research included the Lurie Nanofabrication Facility (LNF) operated by the Solid-State Electronics Laboratory (SSEL) and the University of Michigan. The authors thank Prof. Ken Wise and Mr. Robert Gordenker for providing access to test facilities, and Dr. Seungdo An for wafer metallization.

REFERENCES

[1] S.C. Terry, J.H. Jerman, J.B. Angell, "A gas chromatographic air analyzer fabricated on a silicon wafer," *IEEE Trans. on Electron Devices,* 26, pp. 1880-1886, 1979.

[2] P. R. Lewis, R. P. Manginell, D. R. Adkins, R. J. Kottenstette, D. R. Wheeler, S. S. Sokolowski, D. E. Trudell, J. E. Byrnes, M. Okandan, J. M. Bauer, R. G. Manley, G. C. Frye-Mason, "Recent advancements in the gas-phase MicroChemLab," *IEEE Sensors J.,* 6, pp. 784-795, 2006.

[3] S. Zampolli, I. Elmi, F. Mancarella, P. Betti, E. Dalcanale, G. C. Cardinali, M. Severi, "Real-time monitoring of sub-ppb concentrations of aromatic volatiles with a MEMS-enabled miniaturized gas-chromatograph," *Sensors and Actuators: B Chemical,* 141, pp. 322-8, 2009.

[4] R.J.M. Gordenker, K.D. Wise, "A programmable palm-size gas analyzer for use in micro autonomous systems," *Proc. SPIE 8373, Micro- and Nanotechnology Sensors, Systems, and Applications IV,* May 2012, p. 83731O (6 pp.).

[5] Y. Qin, Y.B. Gianchandani, "A facile, standardized fabrication approach and scalable architecture for a micro gas chromatography system with integrated pump," *IEEE Int. Conf. on Solid-State Sensors, Actuators and Microsystems (Transducers),* Barcelona, Spain, June 2013, pp. 2755-2758.

[6] C.-J. Lu, W. H. Steinecker, W.-C. Tian, M. C. Oborny, J. M. Nichols, M. Agah, J. A. Potkay, H. K. L. Chan, J. Driscoll, R. D. Sacks, K. D. Wise, S. W. Pang, E. T. Zellers, "First-generation hybrid MEMS gas chromatograph," *Lab on a Chip,* 5, pp. 1123-31, 2005.

[7] E.T. Zellers, G. Serrano, H. Chang, L.K. Amos, "A micro gas chromatograph for high-speed determinations of explosive markers," *IEEE Int. Conf. on Solid-State Sensors, Actuators and Microsystems (Transducers),*

Beijing, China, June 2011, pp. 2082-2085.

[8] S.K. Kim, H. Chang, E.T. Zellers, "Microfabricated gas chromatograph for the selective determination of trichloroethylene vapor at sub-parts-per-billion concentrations in complex mixtures," *Analytical Chemistry,* 83, pp. 7198-7206, 2011.

[9] W.-C. Tian, H. K. L. Chan, C.-J. Lu, S. W. Pang, E. T. Zellers, "Multiple-stage microfabricated preconcentrator-focuser for micro gas chromatography system," *J. Microelectromechanical Systems,* 14, pp. 498-507, 2005.

[10] M. Knudsen, "Eine Revision der Gleichgewichtsbedingung der Gase. Thermische Molekularstromung," *Annalen der Physik, Leipzig,* 336, pp. 205-229, 1909. (in German).

[11] J. Liu, N.K. Gupta, K.D. Wise, Y.B. Gianchandani, X. Fan, "Demonstration of motionless Knudsen pump based micro-gas chromatography featuring micro-fabricated columns and on-column detectors," *Lab on a Chip,* 11, pp. 3487-3492, 2011.

[12] N.K. Gupta, Y.B. Gianchandani, "Thermal transpiration in mixed cellulose ester membranes: Enabling miniature, motionless gas pumps," *Microporous and Mesoporous Materials,* 142, pp. 535-541, 2011.

[13] K. Pharas, S. McNamara, "Knudsen pump driven by a thermoelectric material," *J. Micromechanics and Microengineering,* 20, paper 125032, pp. 1-7, 2010.

[14] S. Reidy, G. Lambertus, J. Reece, R. Sacks, "High-performance, static-coated silicon microfabricated columns for gas chromatography," *Analytical Chemistry,* 78, pp. 2623-2630, 2006.

[15] J.H. Seo, S.K. Kim, E.T. Zellers, K. Kurabayashi, "Microfabricated passive vapor preconcentrator/injector designed for microscale gas chromatography," *Lab on a Chip,* 12, pp. 717-724, 2012.

CONTACTS

Y. Qin; yutaoqin@umich.edu
Y.B. Gianchandani; yogesh@umich.edu

A MICRO TRACE HEAVY METAL SENSOR BASED ON DIRECT PROTOTYPING MESOPOROUS CARBON ELECTRODE

Fei Teng[1], Xiaohong Wang[1,2], and Caiwei Shen[1]

[1]Tsinghua National Laboratory for Information Science and Technology
Institute of Microelectronics, Tsinghua University, Beijing, P.R. China
[2]State Key Laboratories of Transducer Technology, Chinese Academy of Sciences, P.R. China

ABSTRACT

We present a micro trace heavy metal sensor based on bismuth-modified mesoporous carbon electrodes that are directly prototyped on silicon wafer. The novel device features a great electrochemical sensing platform for voltammetric analysis because the thicker mesoporous carbon electrode has high surface area, wide potential window, high electric conductivity, and can be integrated into microsystems. The sensor achieves excellent sensing performance, the detection limits are 0.03 μg/L for Cd(II), 0.05 μg/L for Pb(II) and Cu(II), which are about an order of magnitude lower than other reported recently. The peak current also exhibits well linear response over a concentration range from 0.5 to 50 μg/L.

INTRODUCTION

The increasing environmental pollution, in particular water pollution by heavy metal has dramatically attracted the public's attention due to its harm to human health. Exposure to heavy metal such as Cu(II), Pb(II) and Cd(II) can cause a variety of disorders in plants and human bodies, such as kidney injury and anemia [1-2]. Therefore, in situ detection of heavy metals in environment with accurate and efficient methods is of great importance.

The traditional methods mainly use atomic absorption spectrometry (AAS) and inductively coupled plasma mass spectroscopy (ICP-MS) to measure the heavy metal in environment with acceptable sensitivity and accuracy, but the large size and the long analysis time hinder the scaling-down use in microscale environment [3]. With the progress in electrochemical techniques, electrochemical stripping analysis has merged and been recognized as a powerful technique for detection of heavy metals. Square wave stripping voltammetry (SWSV) is a frequently used electrochemical stripping analysis technique because the background noise coming from the charging current is greatly reduced during the potential scan and generates high signal-to-background ratio. Mercury has been widely used as the working electrode in electrochemical stripping analysis with the unique ability to preconcentrate target metal by amalgam formation [4], which leads to high reproducibility and sensitivity. However, the extreme toxicity of mercury limits its application. Many researchers have attempted to introduce mercury-free electrode materials with satisfied performance, including gold [5] and iridium [6], but their performance rarely approached that of mercury.

The environmentally friendly bismuth material is proved to be the best alternative material for the electrochemical determination of heavy metals. Compared with mercury, it possesses similar electrochemical property owing to the analytical properties of fusing alloy formation with other metal [7]. The bismuth film is often electrodeposited on carbon substrates for their high electric conductivity and good chemical stability. Wang et al. successfully used a bismuth thin film coated glassy carbon electrode for simultaneous detection and determination of Cd(II) and Pb(II) [8]. Lezi et al. introduced the bismuth modified screen-printed carbon ink electrode for the rapid voltammetric screening of trace Pb(II) and Cd(II) [9]. Recently, mesoporous carbon material has become a research hotspot due to its good pore structure and large surface area. Lee et al. used the mesocellular carbon foam to immobilize enzymes and got a high sensitive and fast glucose biosensor [10]. Dai et al. used the screen printed mesoporous carbon for amperometric sensing of norepinephrine at picomolar concentrations [11]. Mesoporous carbon provides high adsorptive ability and large quantities of active sites, which potentiates heavy metal analysis compared to other carbon materials.

Therefore, in order to meet safe and high sensitive trace heavy metal detection in water environment monitoring, we develop a micro trace heavy metal sensor using direct prototyping mesoporous carbon as the electrode substrate. The electrode is plated with bismuth for the detection of lead, cadmium and copper. Additionally, simultaneous detection of lead and cadmium is performed by the proposed sensors.

PRINCIPLES AND MATERIALS

Carbon with high electric conductivity, surface area and wide potential window, is one of the preferred materials for trace heavy metal sensor electrodes. In terms of glassy carbon or screen-printed carbon ink, they only use the planar area as the active sites. We propose an effective strategy to improve the sensor performance is to make use of the vertical dimension. A thicker mesoporous carbon electrode is built up, on which more bismuth can be electrodeposited without increasing the planar area.

After the mesoporous carbon electrode is modified uniformly by electrodeposition bismuth (Figure 1(a)), the sample solution containing trace heavy metal flows through the electrode and accumulation potential will be applied to the working electrode (Figure 1(b)). Then the bismuth form alloys with the heavy metal (Figure 1(c)). Not only the outer surface, but also the inner wall of the mesoporous carbon is deposited with bismuth, which provides more active sites for the formation of multicomponent alloys.

Carbon ● **Heavy metal ion**

Buffer solution ○ **Bismuth ion**

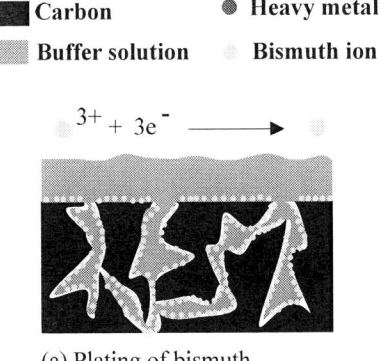

(a) Plating of bismuth

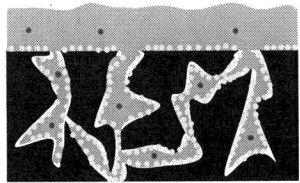

(b) Sample solution flowing through

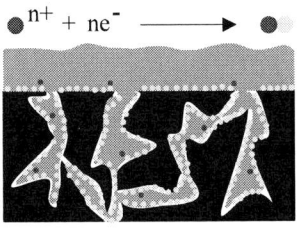

(c) Forming alloys with metal

Figure 1: Plating of bismuth film and the measurement mechanism of detecting trace heavy metal.

By combining nano templating method with microfabrication techniques, our group has successfully presented an approach on direct prototyping of carbon membranes with highly nanoporous structures inside [12]. Epoxy-based SU-8 is used as the photosensitive polymer precursor and nano silica spheres with a uniform diameter of 30nm are used as the template. To enable the sensor's performance, the mesoporous carbon membrane should build on an insulated, chemical and electrochemical stable substrate. Thus, silica is chosen as the material of this layer. The designed electrode is illustrated in Figure 2.

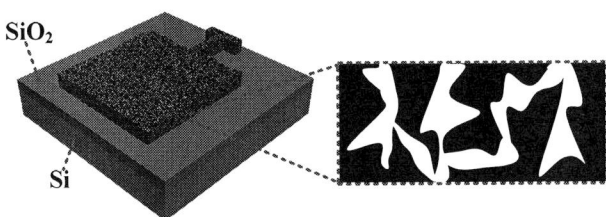

Figure 2: 3D schematic of the direct prototyping mesoporous carbon electrode.

FABRICATION

The fabrication process of the direct prototyping micro electrode is schematically shown in Figure 3. Silica layer was firstly deposited on 4-inch silicon wafer using plasma enhanced chemical vapor deposition (PECVD) method. Then the silica-templated SU-8 layer was spin coated on the SiO_2/Si substrate. Secondly, the layer was patterned by a single step of photolithography. After pyrolyzing at 900 centigrade in nitrogen atmosphere, the SU-8 molecules were turned into carbon. Finally, by removing of the silica spheres, the mesoporous carbon was obtained.

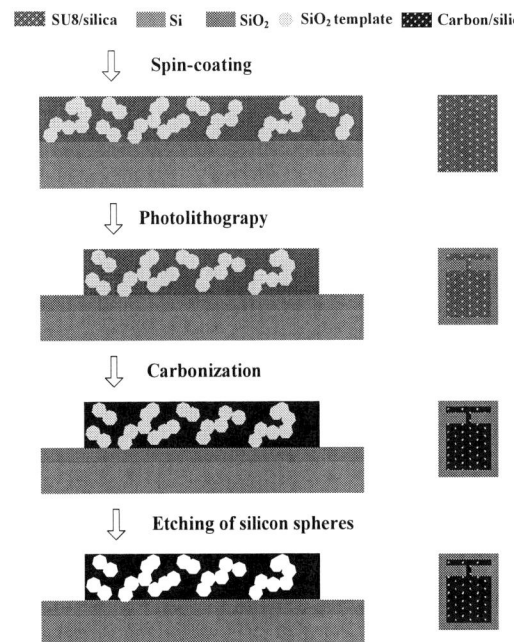

Figure 3: Fabrication process of mesoporous carbon electrode for micro heavy metal sensor.

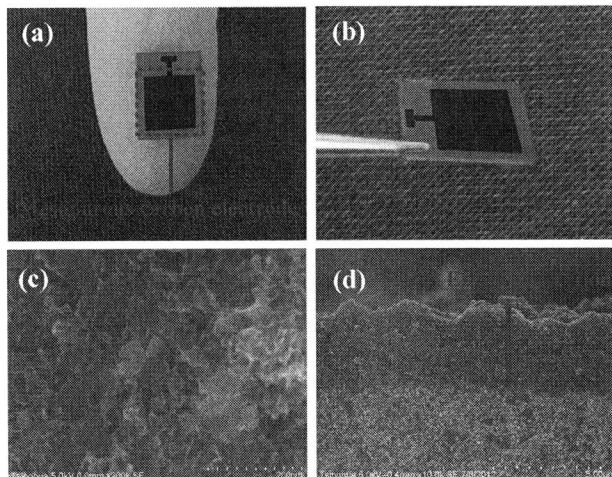

Figure 4: (a) Photos of a fabricated mesoporous carbon electrode surface; (b) Photos of the cross-section of a fabricated mesoporous carbon electrode; (c) The SEM image of mesoporous carbon surface; (d) The SEM image of the cross-section of a half-etched mesoporous carbon electrode.

The bismuth was plated onto the electrode in 500µg/L of Bi(III) solution with the -1.4V accumulation potential. Figure 4 shows a photo of the surface (a) and cross-section (b) of a fabricated mesoporous carbon electrode. The scanning electron microscopic (SEM) image gives more information about the structure of the electrode surface with pore size around 30nm as shown in Figure 4(c). Figure 4(d) shows the SEM image of the cross-section of a half-etched mesoporous carbon electrode and the membrane thickness is approximately 10 µm.

RESULTS AND DISCUSSIONS

Electrochemical tests were carried out by CHI860D electrochemical workstation. The platinum plate was used as counter electrode and Ag/AgCl was used as reference electrode. All measurements were carried out in 0.1M acetate buffer (pH=4.5). Square wave stripping voltammetry (SWSV), as the electroanalysis method, was applied under the following conditions: accumulation potential of -1.2 V, accumulation time of 300 s, frequency of 50 Hz, pulse size of 50 mV, step size of 5 mV and potential range from -1.0 V to 1 V. A 30s conditioning step at 0.7 V in 0.5 mol/L H_2SO_4 was used before each accumulation step to make the electrode surface clean.

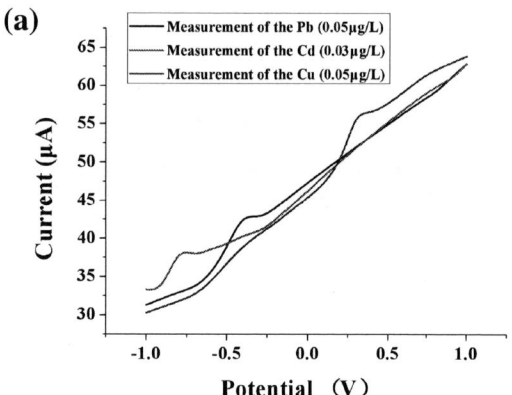

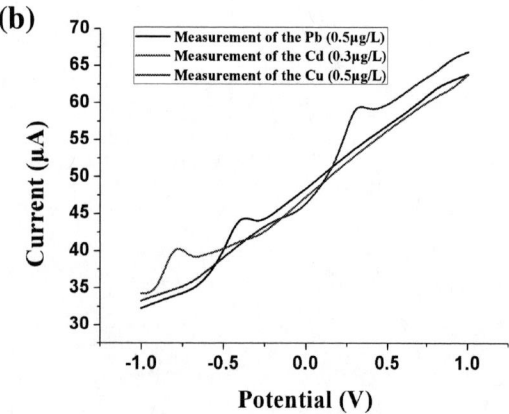

Figure 5: (a) SWSV scans at 0.05µg/L of Pb(II), 0.03µg/L of Cd(II) and 0.05µg/L of Cu(II) in the acetate buffer solution(pH=4.5); (b) SWSV scans at 0.5µg/L of Pb(II), 0.3µg/L of Cd(II) and 0.5µg/L of Cu(II) in the acetate buffer solution(pH=4.5).

Table 1. Comparison of the detection limits for different electrodes.

Electrode	Limit of detection (µg/L)	
	Cd(II)	Pb(II)
Mesoporous carbon	0.03	0.05
Glassy carbon [8]	0.12	3.4
Carbon ink [9]	1.1	0.9

Detection limit is an extremely important evaluation index for a trace heavy metal sensor. Three same bismuth-modified mesoporous carbon electrodes are used to detect the minimal concentration of Pb(II), Cd(II) and Cu(II) individually. In order to make sure the accuracy of the measure results, concentration of the target metal is increasing from the relatively low order of magnitude (1 ng/L). Figure 5(a) shows the minimal detectable concentration for trace heavy metal respectively, 0.03 µg/L for Cd(II), 0.05 µg/L for Pb(II) and 0.05 µg/L for Cu(II) with stripping peak at -0.8V, -0.4V and 0.3V respectively. Stripping peak at different potential reflects the existence of target metal and thus the detection limit is obtained. Figure 4(b) also shows the SWSV scans of 0.3 µg/L for Cd(II), 0.5 µg/L for Pb(II) and Cu(II). To evaluate the sensor performance, the detection limits of other carbon material sensors are listed in Table 1 for comparison. It is indicated that the aforementioned micro trace heavy metal sensor based on direct prototyping mesoporous carbon electrode can reach lower detection limit compared to other carbon material sensors reported.

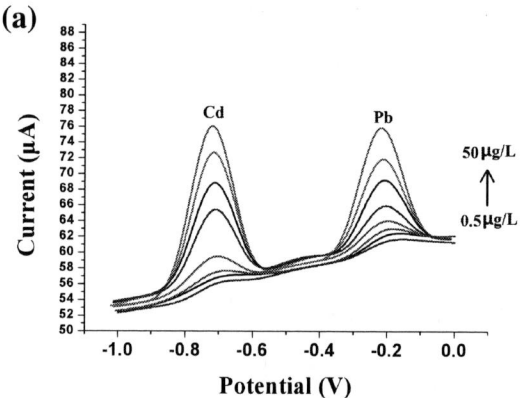

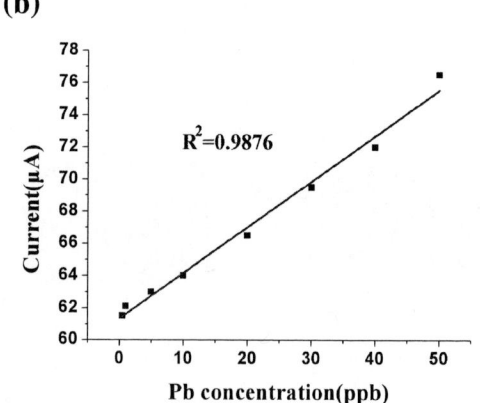

(c)

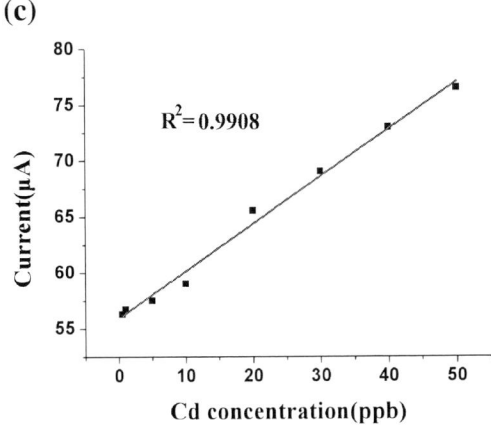

Figure 5: (a) SWSV sans for simultaneous determination of Pb(II) and Cd(II) with different concentrations in acetate buffer sulotion (pH=4.5), from bottom to top, the concentration are 0.5, 1, 5, 10, 20, 30, 40 and 50 µg/L; (b) The resulting calibration curve of Pb(II); (c) The resulting calibration curve of Cd(II).

For a micro trace heavy metal sensor, simultaneous detection of mutiple heavy metals with considerable current response is much more practical. Also, simultaneous detection over comparatively low concentration range contributes more to the practical application. Therefore, detection of heavy metals over different concentration range was tested. A series of stripping voltammograms curves against the concentrations of Pb and Cd from 0.5 to 50 µg/L are shown in Figure 6(a). Figure 6 (b) and (c) shows the corresponding calibration plot for lead and cadmium. The peak current increases linearly with the metal concentration with the correlation coefficient of 0.9876 and 09908 for lead and cadmium. The sensitivity is 0.28 µA/ppb for lead and 0.42 µA/ppb for cadmium.

CONCLUSIONS

A micro trace heavy metal sensor based on direct prototyping mesoporous carbon electrode is designed, fabricated and characterized for the detection of cadmium, lead and copper. Detection limits are 0.03 µg/L for Cd(II), 0.05 µg/L for Pb(II) and Cu(II), which are about an order of magnitude lower than the commonly-used glassy carbon and screen-printed carbon electrode sensor. Simultaneous detection of Cd(II) and Pb(II) is observed and exhibits well linear response over a concentration range from 0.5 to 50 µg/L. Furthermore, the direct prototyping method of the mesoporous carbon on silicon wafer allows scalable design, batch fabrication, and easy integration with microfluid chips and microsystems, and eventually applies to the in-situ water environment monitoring sensor network.

REFERENCES

[1] J. A. Rodrigues et al., "Increased Sensitivity of Anodic Stripping Voltammetry at the Hanging Mercury Drop Electrode by Ultracathodic Deposition", *Anal. Chim. Acta.*, vol. 701, pp. 152-156, 2011.

[2] M.F.M. Noh et al., "Development and Characterisation of Disposable Gold Electrodes, and Their Use for Lead(II) Analysis", *Anal. Bioanal. Chem.*, vol. 386, pp. 2095-2106, 2006.

[3] Z. Zou et al., "Environmentally Friendly Disposable Sensors with Microfabricated On-Chip Planar Bismuth Electrode for In Situ Heavy Metal Ions Measurement" *Sensors and Actuators B*, vol. 134, pp. 18-24, 2008.

[4] S. Legeai et al., "Economic Bismuth-Film Microsensor for Anodic Stripping Analysis of Trace Heavy Metals Using Differential Pulse Voltammetry", *Anal. Bioanal. Chem.*, vol. 383, pp. 839-847, 2005.

[5] J. Wang et al., "Mercury-Free Disposable Lead Sensors Based on Potentiometric Stripping Analysis at Gold-Coated Screen-Printed Electrodes", *Anal Chem.*, vol. 65, pp. 1529-1523, 1993.

[6] M. A. Nolan et al., "Microfabricated Array of Iridium Microdisks as a Substrate for Direct Determination of Cu^{2+} or Hg^{2+} Using Square-Wave Anodic Stripping Voltammetry", *Anal Chem.*, vol. 71, pp. 3567-3573, 1999.

[7] G. H. Hwang et al., "Determination of Trace Metals by Anodic Stripping Voltammetry Using a Bismuth-modified Carbon Nanotube Electrode", *Talanta*, vol. 76. pp. 301-308, 2008.

[8] Z. M. Wang et al., "Bismuth/Polyaniline/Glassy Carbon Electrodes Prepared with Different Protocols for Stripping Voltammetric Determination of Trace Cd and Pb in Solutions Having Surfactants", *Electroanalysis*, vol. 22, pp. 209-215, 2010.

[9] N. Lezi et al., "Disposable Screen-printed Sensors Modified with Bismuth Precursor Compounds for the Rapid Voltammetric Screening of Trace Pb(II) and Cd(II)", *Anal. Chim. Acta.*, vol. 728, pp. 1-8, 2012.

[10] D. Lee et al., "Simple Fabrication of a Highly Sensitive and Fast Glucose Biosensor Using Enzymes Immobilized in Mesocellular Carbon Foam", *Adv. Mater.*, vol. 17, pp. 2828-2833, 2005.

[11] M. Dai et al., "Amperometric Sensing of Norepinephrine at Picomolar Concentrations Using Screen Printed, High Surface Area Mesoporous Carbon", *Anal. Chim. Acta.*, vol. 788, pp. 32-38, 2013.

[12] C. Shen, et al., "Direct Prototyping of Patterned Nanoporous Carbon: A Route from Materials to On-chip Devices", *Sci. Rep.*, vol. 3, pp. 2294, 2013.

CONTACT

*X. Wang, tel: +86-10-62798432; wxh-ime@tsinghua.edu.cn

A NOVEL QUANTITATIVE DESIGN MODELING ON GAS SENSING PARAMETER OF NANO-MATERIALS BASED ON MICRO-GRAVIMETRIC THERMO-DYNAMIC EXPERIMENTS

Pengcheng Xu, and Xinxin Li*

State Key Lab of Transducer Technology, and, Science and Technology on Micro-system Lab,
Shanghai Institute of Microsystem and Information Technology, Chinese Academy of Sciences,
Shanghai 200050, CHINA

ABSTRACT

The study is aimed at building a novel quantitative adsorbing/sensing model for chemical gas sensing materials, with which various key sensing parameters can be comprehensively evaluated and optimally designed. Gravimetric resonant-cantilevers are used in experiment to real-time record sensing curves at different temperatures, which are further used to extract adsorbing/sensing performance of the specific materials for quantitative evaluation and optimal design of the sensor. The technique is well validated by choosing the best trimethylamine (TMA) sensing material among three similar mesoporous silica nanoparticles (MSNs).

INTRODUCTION

For on-site quick detection of chemical vapors such as volatile organic compounds (VOCs), micro/nano gas sensors have been intensively researched for a long period of time [1]. It is known that more physical-sensors have been successfully produced, as the physical-sensors can be with their sensing-parameters (like sensitivity) quantitatively designed before fabrication and the device testing data can agree well with the design. Comparatively, most gas-sensors cannot be so quantitatively designed before fabrication, i.e. the key performance (sensitivity, selectivity, etc.) can only be known after testing. Therefore, development of precise design model for gas sensors is highly demanded.

Being not as simple as physical micro-sensors, which only possess one transduction interface of pressure induced displacement to capacitance change, a chemical gas sensor normally contains two transduction interfaces. Besides the similar interface to that of the physical sensor for transduction from physical effect to electric signal, another molecule-recognition interface exists for translating specific molecules adsorption/absorption into a certain physical sensing phenomenon like heating, resistance change, mass addition or optical illumination, etc. To realize the molecule recognition, gas sensing material is anyway indispensable. In normal sense, the performance of a chemical sensor strongly depends on the sensing material used in the sensor. With the rapid development of nanoscience, nowadays many kinds of nanomaterials have been utilized for gas sensing [2-4]. Sensing nanomaterials normally feature high specific surface area that is beneficial for enhanced molecule adsorption/absorption and improved limit of detection (LOD) for sensing to trace-level gas. Now that the molecular interaction or specific adsorption between the targeted gas and the sensing nanomaterial determines the sensing performance at the molecule-recognition interface, the sensing nanomaterial needs to be designed comprehensively. For on-site repeatable detection, the material needs to exhibit well in terms of both the molecule adsorption (during sensing) and desorption (for recovery after detection). For optimal and comprehensive design of sensing materials to fulfill gas detection requirements, it is really crucial to build a relatively universal model where all the key criteria should be taken into account.

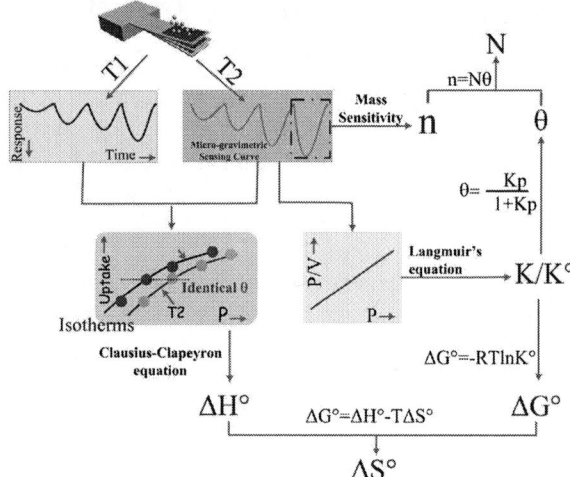

Figure 1: Nanomaterial sensing parameter obtaining sequence (top-down): resonant microcantilever gas adsorbing experiment getting isotherm data at different temperatures, (left) calculating enthalpy $\Delta H°$; (right) calculating adsorbing site number N, coverage θ and equilibrium constant K. Then Gibbs free energy $\Delta G°$ and entropy $\Delta S°$ are obtained.

Herein, we propose and develop a quantitative model based on micro-gravimetric experiment (schemed in Figure 1). The sensing-material to be evaluated is loaded on a resonant microcantilever to real-time record the adsorbing mass induced frequency-shift. The experiment is performed at two different temperatures by putting the microcantilever into a temperature-adjustable chamber. Two isotherms about sensing-response v.s. gas concentration (i.e. adsorbed molecule number v.s. gas partial pressure p) can be plotted together. Any horizontal line intersects with the two isotherms at two points that correspond identical adsorbate quantity (i.e. identical fractional-coverage θ) but different gas concentration. With Clausius–Clapeyron equation of $\Delta H^\circ = RT_1T_2(T_2-T_1)^{-1}\ln(p_1p_2^{-1})$, enthalpy ΔH° can be calculated that represents the adsorbing heat and can be used for evaluation of adsorption/desorption performance. Moderate $-\Delta H^\circ$ is preferred that indicates weak chemical-adsorption. Comparatively, lower $-\Delta H^\circ$ means pure physical-adsorption and poor specificity, and higher $-\Delta H^\circ$ of strong chemical-adsorption (or chemical-reaction) leads to poor recovery after sensing. The model has been successfully used to evaluate and select the best trimethylamine (TMA) sensing material among three similar mesoporous silica nanoparticles (MSNs): *i*) –COOH functional MSNs, *ii*) –SO₃H modified MSNs and *iii*) ≡Si-OH exposed unmodified MSNs, which look really similar. Each of the nano-materials features huge surface-area and some degree of specificity with TMA-molecule.

EXPERIMENTS

Synthesis of mesoporous silica nanoparticles (MSNs). Cetyltrimethylammonium bromide (CTAB) is purchased from Sigma-Aldrich. Carboxyethylsilane triolsodium (CES, 25wt% salt solution) and 2-(4-Chlorosulfonylphenyl) ethyltrimethoxy silane (CSPES, 50wt% solution in dichloromethane) are purchased from Gelest, Inc. Tetraethylorthosilicate (TEOS), methanol, concentrated hydrochloric acid (36 wt%) and NaOH are of analytical grade and purchased from Shanghai Chemical Reagent Corp.

The synthesis details of -COOH functionalized MSNs are summarized as follows: 1 g CTAB is dissolved into 480 mL deionized water under stirring at 80 °C. Then 3.5 mL of NaOH solution (2 mol/L) is added and allowed to react for 5 min. Thereafter 5 mL TEOS is slowly added into the solution for 10min. CES (200 μL) is then added into the solution. Subsequently, the resultant mixture is allowed to react for 2 hours at 80 °C. The product is filtered and washed with deionized water, then overnight dried at 80 °C to obtain -COOH functionalized MSNs precursor. The molar composition of the reaction mixture is varied in the range of 1TOES: (0~0.02) CES: 0.27NaOH: 0.1CTAB: 1000H₂O. Acid extraction of the CTAB surfactant is performed at 90 °C by placing the -COOH functionalized MSNs precursor in the mixture of methanol (150 mL) and concentrated hydrochloric acid (9.0 mL) for 24 hours. The -COOH functionalized MSNs is then filtered, washed with water/methanol and dried under vacuum for12 hours at 80 °C.

In control experiments, unmodified MSNs and –SO₃H functionalized MSNs are synthesized. For synthesis of the unmodified MSNs, CES is not added and the molar composition of the reaction mixture is 1TOES: 0.27NaOH: 0.1CTAB: 1000H₂O. For the –SO₃H functionalized MSNs, CSPES is employed to introduce –SO₃H groups and the molar composition of the reaction mixture is varied in the range of 1TOES: (0~0.02) CSPES: 0.27NaOH: 0.1CTAB: 1000H₂O.

Sensing Material Loaded on Cantilever. The fabrication procedure of the microcantilever can be found in our previous report [5]. The MSNs sensing material (about 0.01 g) is added into 1mL deionized water (under ultra-sonic) to form a crude suspension. About 0.1 μL suspension is loaded onto top-surface of the cantilever end region by using a commercial micro-manipulator (Eppendorf made, model: PatchMan NP2). The process control is aided by inspection under a microscopy. Then, the microcantilever together with the material is dried in an oven at 333 K (60 °C) for about 2 hours.

TMA Vapor Generator. The TMA vapor with desired concentration is prepared from a commercial available standard vapor generator (model: Molecular Analysis series 8000S, Taiwan), which is equipped with a temperature programmable oven. A TMA permeation tube (Kin-Tec made, La Marque, Texas), with the permeation rate calibrated by weight loss speed of 242 ng/min, is inserted into the oven. Highly pure N₂ is used as carrier, where the flow rate is set as 1000±1 sccm (standard cubic centimeter per minute). After stabilization at 303 K for 3 days, the N₂-diluted TMA vapor with constant concentration is generated. Using another mass flow controller, the TMA vapor can be further diluted by pure N₂ down to the desired concentration and, then, is introduced to a testing chamber where the cantilever is inside.

RESULTS & DISCUSSION

The three MSNs are all prepared using a modified "Stöber" protocol. N₂ sorption measurement shows that there are always two hysteresis-loops appeared in the N₂ sorption isotherms of the three MSNs (see Figure 2). Known as a type IV isotherm according to IUPAC classification, the first hysteresis-loop around $p/p_0=0.3$ is associated with capillary condensation and desorption in the open mesopores. Due to the interstitial pores, the second capillary condensation step is observed above $p/p_0=0.85$ of the adsorption branch, which indicates the ultra-small particle size of the MSN sample. The specific surface area is always measured as around 1050m²/g, and the average nano-pore size for the three mesoporous materials is always about 2.1 nm. As shown in Figure 3, the FT-IR spectra indicate that the organic groups are modified successfully onto the surface of the three materials.

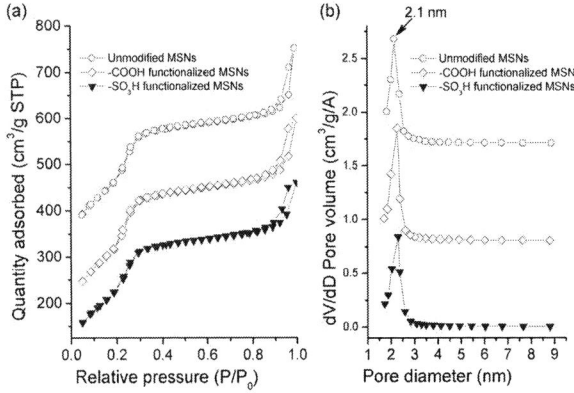

Figure 2: (a) Sorption isotherms of the three samples of unmodified MSNs, –COOH functionalized MSNs and –SO₃H functionalized MSNs, respectively. The specific surface area

is always calculated as around 1050m²/g. (b) The averaged nano-pore size for the three mesoporous materials is always calculated as about 2.1nm.

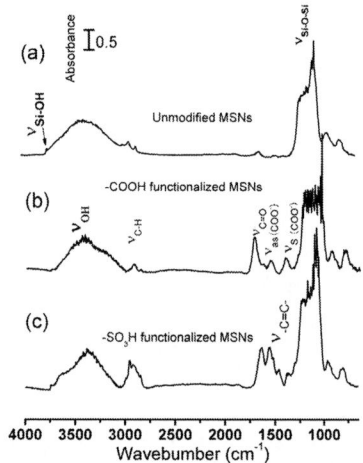

Figure 3: FT-IR spectra of the three sample. (a) unmodified MSNs; (b) –COOH functional MSNs; (c) –SO₃H functional MSNs.

After material preparing, the MSN sample is ultrasonically dispersed into de-ionized water to form cloud suspension. Then the suspension of accurate volume is loaded onto the microcantilever surface by using a micro-manipulator. Figure 4 shows the SEM image of the as-fabricated resonant microcantilever. After dried in an oven, the cantilever is put in a testing cell that is linked with a commercial TMA vapor generator. Stabilized with pure N_2 steam, the micro-gravimetric resonant cantilever is ready for experiment.

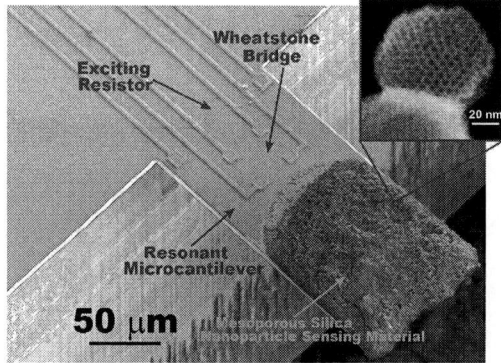

Figure 4: SEM image of the resonant microcantilever sensor. The inset shows the high-resolution SEM image of two nanoparticles with clearly visualized mesopores.

Under stable gas steam, the sensing signal of the resonant microcantilever is online recorded by a frequency counter. When the gas steam of the testing cell is switched to the TMA/N_2 mixture of desired concentration, the sensor outputs a frequency drop that implies the TMA adsorption on the MSNs. For the three MSN materials, Figure 5 shows the recorded sensing curves to TMA of various trace-level concentrations. For the –COOH MSNs, the sensing results obtained at 298 K and 318 K are shown in Figures 5(a1-a2),

respectively. With the TMA concentration increased from 90 ppb to 900 ppb, the sensing signals of the experiment at 298 K are in the range of 2.8~6.3 Hz and the signal noise-floor is lower than ±0.8 Hz. With so high the signal-to-noise ratio, the –COOH MSNs loaded sensor exhibits trace-level LOD of several tens *ppb*. To the best knowledge of the authors, this has been already the lowest LOD to amine VOCs among the reported results so far. However, only using the single parameter of LOD to evaluate the sensing material is still partial. For comprehensive evaluation, we would like to use the proposed model to select the optimal one among the three materials.

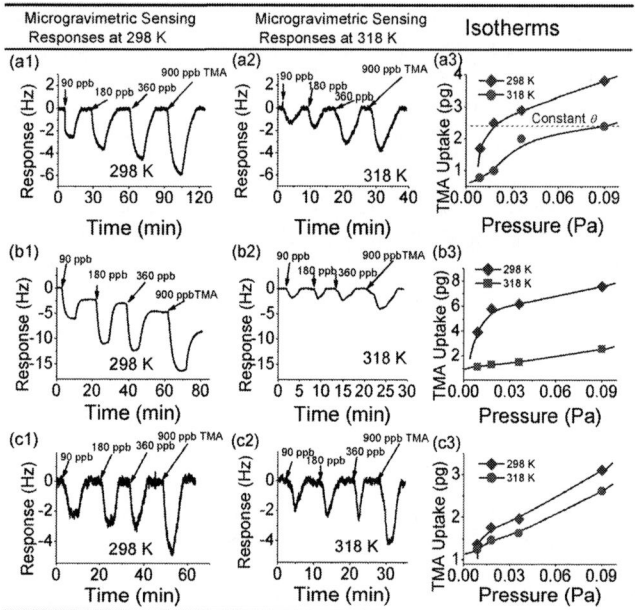

Figure 5: Adsorption induced gravimetric sensing responses and the isotherms of the MSNs loaded resonant-cantilever to TMA vapor of various trace-level concentrations are shown from left to right. The two experimental temperatures are 298 K and 318 K. The top raw of (a1)-(a3) is for the -COOH functionalized MSNs, the middle raw of (b1)-(b3) is for the –SO₃H functionalized MSNs, and the bottom raw of (c1)-(c3) is for the –OH covered unmodified MSNs.

Under certain gas concentration, the adsorbed mass of TMA molecules onto the –COOH MSNs is worked out by reading the frequency sensing signal and referring to the known mass sensitivity of the cantilever [5]. In addition, the partial pressure of TMA vapor is linearly associated with the concentration. By plotting the adsorbing mass in terms of partial pressure of TMA, the isotherm of the –COOH MSNs at 298 K can be obtained. Then the temperature for the same sensor is changed to 318 K and the sensing signal is recorded again and shown in Figure 5(a2). Accordingly, another isotherm of the material to TMA is obtained. The two isotherms are plotted together in Figure 5(a3). We randomly draw a horizontal line (see the dotted line in the figure). As long as the horizontal line intersects with the two isotherm curves that are obtained from the experiments under 298 K and 318 K, the two intersect points must feature the same quantity of adsorbed gas molecule, *i.e.*, feature identical fractional coverage θ. Based on Clausius-Clapeyron equation, the coordinate data of the two points can be used to calculate $\Delta H°$. The calculated $\Delta H°$ for TMA adsorption on the –COOH MSNs is derived as -62.5 kJ mol^{-1}. Similarly, the $\Delta H°$ values for TMA adsorption on the unmodified MSNs and the –SO$_3$H MSNs are -23 kJ mol^{-1} and -149.6 kJ mol^{-1}, respectively. Based on the $\Delta H°$ data, apparently TMA is physically adsorbed onto the unmodified MSNs, which is not preferred for specific adsorption. For the other two functionalized MSNs, chemo-adsorption occurs. However, too high the -$\Delta H°$ value for the –SO$_3$H functionalized MSNs indicates strong chemical adsorption that is not conducive to post-sensing rapid desorption for next-time detection. In comparison, the –COOH MSNs features a moderate -$\Delta H°$ value that indicates weak chemisorption. With the thermodynamic judgment, the –COOH MSNs may be preliminarily considered advantageous in both adsorption/sensing and desorption/recovery properties among the three materials.

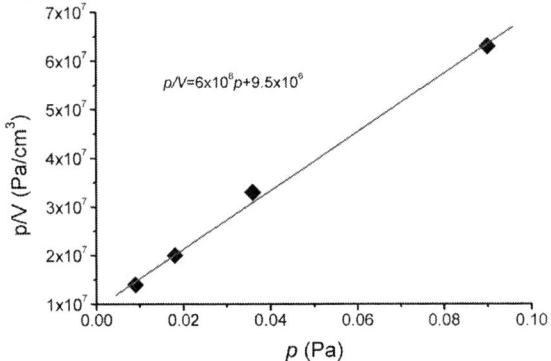

Figure 6: The plot of p/V versus p is converted from the isotherm (at 298K) in Figure 5(a3) and the straight red-line is linearly fitted as $p/V=6\times10^8 p+9.5\times10^6$.

With the $\Delta H°$ value obtained, more thermodynamic and kinetic parameters are further worked out. Containing nitrogen long pair electrons, the Lewis base of TMA may react with the proton of a material that possesses ≡Si–OH, –COOH or –SO$_3$H group to form a (CH$_3$)$_3$N···H type of hydrogen bond. Thus, low pressure TMA vapor tends to be mono-layer adsorbed onto the MSNs, which is in accord with the hypothesis of Langmuir adsorption theory.

Table 1: Thermodynamic and kinetic parameters of the three MSNs obtained from the micro-gravimetric experiment based modeling.

	unmodified MSNs	–COOH MSNs	–SO$_3$H MSNs
$\Delta H°$	-23.0 kJ mol^{-1}	-62.5kJ mol^{-1}	-149.6 kJ mol^{-1}
$\Delta G°$	-38kJ mol^{-1}	-38.8kJ mol^{-1}	-39.9 kJ mol^{-1}
$\Delta S°$	50.3 J K^{-1}	-79.5 J K^{-1}	-368 J K^{-1}
θ	0.29	0.36	0.47
$K°$	4.6×10^6	6.3×10^6	10^7
N	7.9×10^{-14}mol	8.1×10^{-14}mol	1.4×10^{-13}mol

Using a facile calculation, the isotherm (for sensing at 298 K) in Figure 5(a3) can be converted into a plot of p/V

versus p that is plotted in Figure 6. In the figure, the data can be linearly fitted as $p/V=9.5\times10^6+6\times10^8 p$. According to Langmuir equation, we have K=63 Pa^{-1}. Accordingly we get $K°=K\times p°=6.3\times10^6$ for the –COOH MSNs.

With the obtained K° for certain temperature of T, $\Delta G°$ can be calculated from the equation of $\Delta G°=-RT\ln K°$. Thereafter $\Delta S°$ can be calculated from $\Delta G°=\Delta H°-T\Delta S°$. From the data listed in Table 1, obviously the–COOH functionalized MSNs always take the moderate data.

CONCLUSION

A quantitative adsorbing/sensing modeling method is proposed and developed, with which key sensing parameters can be obtained for evaluation and optimization of gas sensing materials. Resonant microcantilever is loaded with the sensing material to be evaluated for micro-gravimetric gas-sensing experiment. With obtained sensing isotherm curves at different temperatures, thermodynamic and kinetic parameters of the sensing-materials are quantitatively modelled for comprehensive evaluation and optimal design. The model has been validated by identifying the best trimethylamine (TMA) sensing-material among three similar mesoporous silica nanoparticles (MSNs).

ACKNOWLEDGEMENTS

This research is supported by Chinese 973 Project (2011CB309503) and NSF of China (91023046, 61161120322, 61021064, 61102010) and the Project of National Science & Technology Pillar Plan (2012BAK08B05). Pengcheng Xu thanks to project support from the Key Laboratory for Micro/Nano Technology and System of Liaoning Province.

REFERENCES

[1] E. S. Snow, *et al.*, "Chemical detection with a single-walled carbon nanotube capacitor", *Science,* vol. 307, pp. 1942-1945, 2005.

[2] P. C. Xu, *et al.*, "Functionalized Mesoporous Silica for Microgravimetric Sensing of Trace Chemical Vapors", *Anal. Chem.,* vol. 83, pp. 3448-3454, 2011.

[3] R. A. Potyrailo, *et al.*, "Materials and transducers toward selective wireless gas sensing", *Chem. Rev.,* vol. 111, pp. 7315-7354, 2011.

[4] P. C. Xu, *et al.*, "*In situ* growth of noble metal nanoparticles on graphene oxide sheets and direct construction of functionalized porous-layered structure on gravimetric microsensors for chemical detection", *Chem. Commun.,* vol. 48, pp. 10784-10786, 2012.

[5] H. T. Yu, *et al.*, "Resonant-cantilever bio/chemical sensors with an integrated heater for both resonance exciting optimization and sensing repeatability enhancement", *J. Micromech. Microeng.,* vol. 19, p. 045023, 2009.

CONTACT

*X. X. Li, tel: +86-21-62131794
email : xxli@mail.sim.ac.cn

A POLYMER-BASED MEMS DIFFERENTIAL SCANNING CALORIMETER

Yuan Jia, Bin Wang, Jing Zhu, Qiao Lin
[1]Department of Mechanical Engineering, Columbia University, New York, NY, USA

ABSTRACT

We present a flexible, polymer based MEMS differential scanning calorimetric (DSC) device combining integrated microfluidic channels, highly sensitive thermoelectric sensing, and real-time temperature monitoring for thermodynamic characterization of biomolecular samples with minimized sample consumption. The device uses an inexpensive, commercially available polymer substrate and a novel fabrication approach to create a microstructure consisting of a pair of microchannels (containing the sample and reference buffer, respectively), which are integrated with resistive temperature sensors (for *in-situ* measurement of sample temperature) and an antimony-bismuth (Sb-Bi) thermopile (for measurement of the temperature difference between the sample and reference channels). We demonstrate the utility of this MEMS DSC device by measuring the unfolding of lysozyme in a small volume (1 μL), and at practically relevant protein concentrations (approaching 1 mg/mL). Thermodynamic properties including the total enthalpy change per mole of protein (ΔH) and melting temperature (T_m) at different protein concentrations during this conformational transition are determined and found to agree with published data.

INTRODUCTION

Differential scanning calorimetry (DSC) is a thermal analysis technique which has been used to measure the temperatures and heat flows associated with transitions in materials as a function of time and temperature [1]. It is particularly useful for studying the thermal behavior of materials as well as the thermodynamic characterization of biomolecular interactions by allowing universally applicable, direct, label-free measurements of thermodynamic properties of biomolecular interactions and conformational transitions [2]. Although widely used in both basic science and applications such as drug discovery, conventional DSC methods are constrained to a small number of very high-value measurements by their large sample consumption requirements and long measurement times [3]. MEMS-based DSC devices can potentially address these issues with improved thermal isolation as well as reduced thermal mass and sample volume, thereby significantly improving device sensitivity and minimizing sample consumption. Such devices, however, remain scarce, while the few existing MEMS DSC devices are limited by drawbacks such as an inability to process liquid samples with precise environmental control and inadequate sensitivity leading to impractically high sample concentrations [4].

We previously demonstrated a silicon-based MEMS DSC device that showed promise for addressing these drawbacks by integrating microfluidic calorimetric chambers and an antimony-bismuth (Sb-Bi) thermopile on a single chip [4], although its utility was limited by high material cost, expensive, and time consuming silicon-based fabrication processes. Furthermore, the device had rather limited poor robustness due to the use of fragile freestanding microstructures, which, at elevated DSC measurement temperatures, were prone to failure. In addition, the device was susceptible to cross contamination under repetitive use.

Here, we aim to address these limitations by presenting a new polymer-based MEMS DSC device. The device, integrates inexpensively fabricated polymer calorimetric microstructures, a layer of aligned temperature sensors, and highly sensitive Sb-Bi thermopile junctions all on a mechanically flexible polymer substrate that leads to significantly improved device robustness and fabrication yield. More importantly, by using polymeric materials, this device is well suited for potential disposable use, eliminating issues of cross contamination in the measurement of biomolecular energetics. We demonstrate this polymer MEMS DSC device with the characterization of unfolding of the protein lysozyme in a 1 μL volume at practically relevant concentrations (approaching 1mg/mL).

PRINCIPLE AND DESIGN

The MEMS DSC method monitors heat induced voltage differences between the hot and cold junctions of a thermopile during a biomolecular interaction as a function of temperature. The differential power (i.e., the difference between the power generated by the sample and reference) can be represented as $\Delta P = P_s - P_r$, where P_s and P_r are the thermal power generated in the sample and reference materials respectively. The differential heat capacity between the sample and reference, can be then calculated by [5]:

$$\Delta C_P = \frac{\Delta P}{\dot{T}} = \frac{\Delta U}{S\dot{T}} \qquad (1)$$

where ΔU is the measured thermopile output voltage, $\dot{T}$ the constant rate (scanning rate) at which the sample and reference temperatures are varied in a range of interest, and S the device responsivity determined via device calibration, i.e., the thermopile output voltage per unit differential power. Using ΔC_p from Eq. 1, the heat capacity of the sample, C_{sample} can be calculated by

$$C_{\text{sample}} = C_{\text{buffer}} \left(\frac{v_{\text{sample}}}{v_{\text{buffer}}} \right) + \left(\frac{\Delta C_p}{m_{\text{sample}}} \right) \qquad (2)$$

978-1-4799-3510-9/14 $31.00 © 2014 IEEE   306

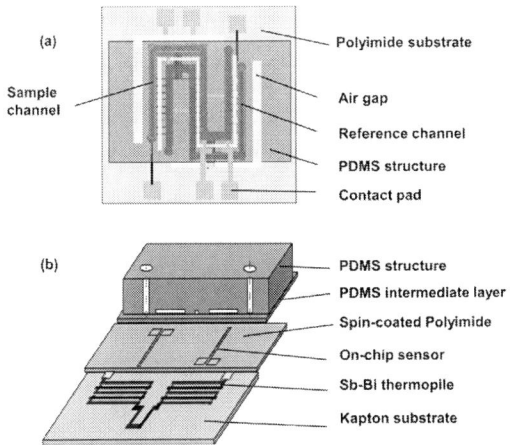

Figure 1: Schematic of the polymer MEMS DSC device: (a) top and (b) isometric view.

where v_{sample} and v_{buffer} are the partial specific volumes of the sample and the buffer respectively, m_{sample} is the mass of the biomolecule in the sample channel, and C_{buffer} is the partial specific heat capacity of the buffer. Integrating the sample partial heat capacity over a temperature range of interest allows the determination of the total enthalpy change per mole of biomolecules associated with the biomolecular interaction:

$$\Delta H(T) = \int_{T_0}^{T_1} C_{sample}(T)dT \qquad (3)$$

The polymer MEMS DSC device (Fig. 1) consists of identical serpentine microfluidic sample/reference channels separated with air-gaps to enhance thermal isolation. Antimony (Sb) and bismuth (Bi) were chosen (Seebeck coefficients: +43 and -79 μV/K, respectively) for their high thermoelectric transduction performance and ease of fabrication [4]. A suitable number (400) of thin-film antimony-bismuth (Sb-Bi) thermopiles hot and cold junctions, situated on the polymer substrate, are located under each channel. A layer of temperature sensors are integrated and placed underneath the center of the region containing the sample or reference channel. Polyimide is used as the substrate because of its excellent mechanical properties (Young's modulus > 2 GPa) and thermal stability (glass transition temperature >400 °C).

FABRICATION PROCESS

The polyimide film which was used as a substrate was purchased from DuPont (Kapton® 50HN, 12.5 μm thick). The fabrication began with the reversible binding of the substrate to a silicon wafer carrier by a spin-coated poly(dimethylsiloxane) (PDMS) layer (20 μm). After fully curing this PDMS adhesion layer, Sb and Bi (0.8 and 1 μm thick, respectively) were thermally evaporated and patterned on the substrate using a standard lift-off process to form a 400-junction thermopile, which was then passivated with a spin-coated polyimide thin layer (1.5 μm). Subsequently, a chromium/gold thin film (100 μm) was deposited and patterned to define the on-chip temperature sensor, and then

was passivated with another thin PDMS layer. The fabricated devices were mechanically released from the substrate, and the PDMS binding layer was peeled off so that the silicon carrier can be saved for reuse. In parallel, serpentine microfluidic channels (width: 200 μm, height: 200 μm, length: 25 mm; volume: 1 μL) were fabricated of PDMS via soft lithography. The released substrate was then bonded to the microfluidic structure via oxygen plasma. A packaged device is shown below in Figure 3(b).

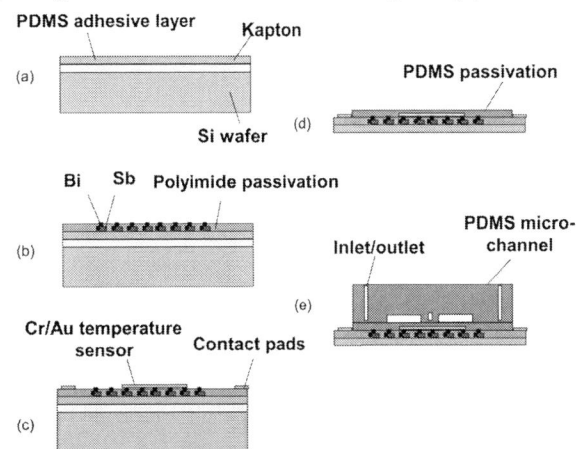

Figure 2: Fabrication process for the polymer MEMS DSC device.

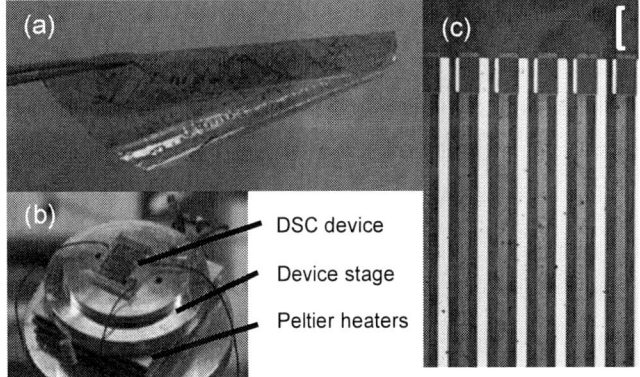

Figure 3: Images of the polymer MEMS DSC device and thermopile: (a) Wafer-level features on the Kapton substrate after separation from the silicon carrier, (b) packaged device, and (c) thermopile junctions (scale bar: 50 μm).

EXPERIMENTAL METHODS

Liquids with well-established heat capacities were chosen to calibrate the MEMS DSC device. A temperature controller (Lakeshore model 311) provides uniform temperature control for the calibration process while the temperature sensor, measured by a digital multimeter (Agilent 34410A), provides *in situ* temperature monitoring. The thermopile output voltage was measured by a nanovoltmeter (Agilent 34420A). During the calibration process water and glycerol were used for their relatively high boiling temperatures, and the device was placed inside of a thermal enclosure aimed to minimize environmental disturbances. The thermoelectric and resistive measurements were automated through a LabVIEW-based program. After calibration, the device was thoroughly washed with 0.1 M sodium hydroxide (NaOH) and deionized water, while the biological sample and buffer solutions were degassed with a vacuum pump, and introduced by a syringe.

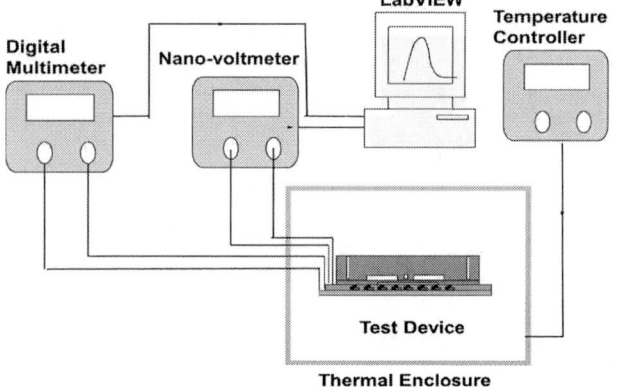

Figure 4: Polymer MEMS DSC device test set-up.

RESULTS AND DISCUSSION

The polymer DSC device was calibrated to determine the responsivity of the device output [6]. The device temperature was first scanned with both calorimetric channels filled with air. Then, water and glycerol were successively introduced into the sample channel, while the reference channel remained filled with air. The heat capacities of all materials were obtained from the literature [7, 8]. The device responsivity was determined from Eq. (1), and used later for determining the sample heat capacity. All calibration experiments were thermally scanned with a rate of 3 K/min to be consistent with reported measurements in the literature [7]. The device responsivity was then determined to be 4.78 mW/mV. As shown in Figure 5 below,

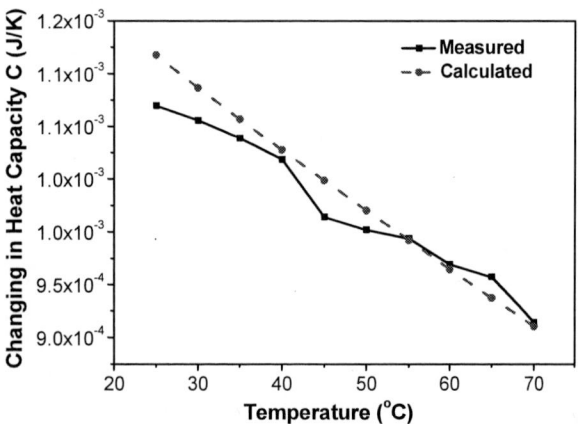

Figure 5: Comparison of measured (baseline-subtracted) and calculated differences in heat capacity between water and glycerol.

when the device responsivity, S, and scanning rate, $\dot{T}$, are substituted back into Eq. 1, the experimentally determined change in heat capacity between glycerol and water as a

function of temperature agree with the calculation using data reported in the literature. We also observed an experimental output voltage noise level of approximately 100 nV in the DSC detection system, which correlated to a differential power noise of 20 nW. The calibrated polymer MEMS DSC device was then exploited to characterize protein unfolding. Glycine-HCl buffer (0.1 M, pH 2.5) was filled in both sample and reference calorimetric channels while the device was scanned at a constant rate of 5 K/min. After the scan was completed, the device was allowed to cool to room temperature and a second experiment under identical conditions was performed to test the stability of the baseline. As shown in Figure 6, there was minimal fluctuation between the two baselines.

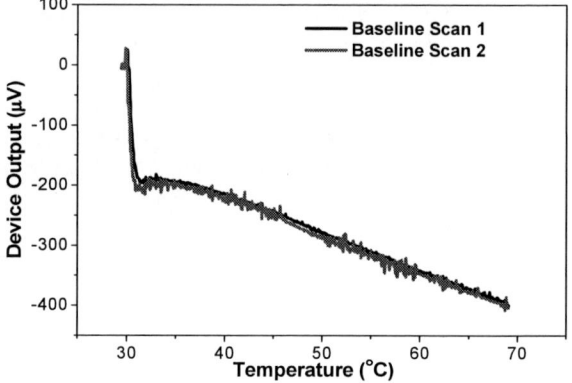

Figure 6: Baseline measurements of glycine-HCl buffer (pH 2.5) (constant temperature scan rate: 5 K/min).

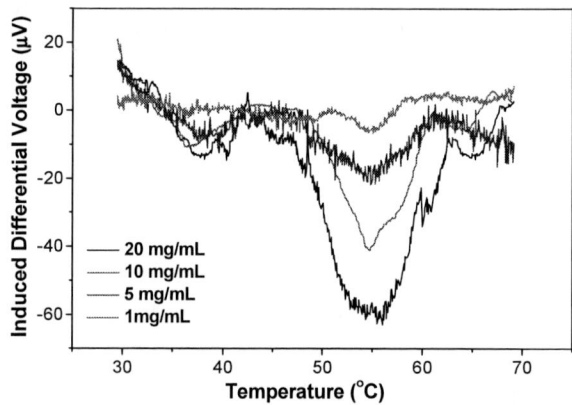

Figure 7: Baseline-subtracted differential voltage as a function of temperature for lysozyme unfolding at different concentrations.

Notably, a non-zero slope was apparent at elevated temperatures, possibly as a result of the volumetric mismatch between the reference and sample channels. Following the measurement of the baseline, lysozyme in 0.1 M Glycine-HCL buffer (pH 2.5) was introduced into the device sample channel while the reference channel remained filled only with buffer. The characterization of the unfolding of lysozyme was carried out with identical experimental conditions used in the baseline determination. The thermopile differential voltage as a function of temperature, corrected by baseline subtraction, was measured at varying concentrations ranging from 1 to 20 mg/mL (Fig. 7). The device output exhibited an endothermic thermodynamic profile within the 25-75 °C temperature range at all protein concentrations. Notably, the unfolding of lysozyme was detectable at 1 mg/mL. Furthermore, the differential heat capacity as a function of temperature and the calibrated device sensitivity can be calculated from the device voltage output by Eq. (1). The sample heat capacity as a function of temperature was determined using Eq. (2), and the interpretation of fundamental thermodynamic properties, such as the total enthalpy change per mole of lysozyme (ΔH) and melting temperature (T_m defined as the temperature at which the enthalpy change achieves 50% of ΔH) associated with this conformational transition can then be determined.

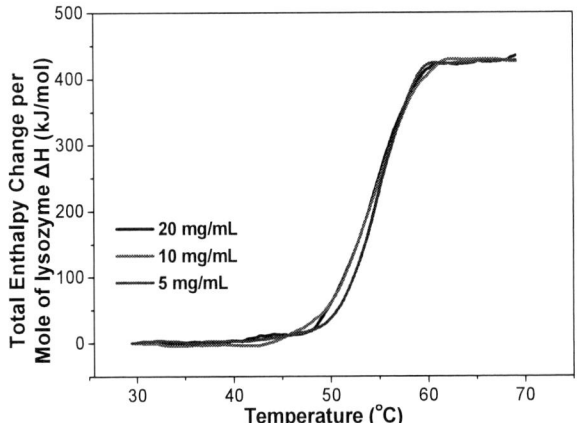

Figure 8: Total enthalpy change per mole of lysozyme during lysozyme unfolding as a function of temperature at different lysozyme concentrations.

The enthalpy change was determined using Eq. (3) fro all protein concentrations except 1 mg/mL, which was excluded from the calculations due to the high noise in the thermopile output. As shown in Figure 8, ΔH = 421 kJ/mol obtained consistently, with a corresponding melting temperature T_m = 54.71 °C. These results agree with the published data [9], and thus demonstrate the capability of the polymer MEMS DSC device for highly sensitive detection of biomolecular interaction.

CONCLUSION

This paper presents a polymer MEMS DSC device integrating microfluidic channels, highly sensitive thermoelectric sensing, and real-time temperature monitoring for thermodynamic characterization of liquid biomolecular samples with minimized sample consumption. The MEMS DSC device consists of a pair of microfluidic channels, integrates with resistive temperature sensors, and a Sb-Bi thermopile which are situated on flexible Kapton diaphragms. The polymer MEMS DSC device has been determined to have a device responsivity of 4.78 mV/mW, and a differential power noise of 20 nW. The device has been applied to the characterization of lysozyme unfolding in a small sample volume (1 μL) at detectable sample concentrations approaching 1 mg/mL. The total enthalpy

change per mole of lysozyme and melting temperature in the lysozyme unfolding process have been determined to be 421 kJ/mol and 54.71 °C respectively, which are in agreement with reported values of 377-439 kJ/mol and 55-58.9 °C [8].

ACKNOWLEDGEMENTS

The authors thank Timothy Olsen, Junyi Shang, and Zhixing Zhang for helpful discussions of the device design and testing.

REFERENCES

[1] P. S. Gill, S. R. Sauerbrunn, and M. Reading, "Modulated Differential Scanning Calorimetry," *Journal of Thermal Analysis*, vol. 40, pp. 931-939, 1993.

[2] S. Youssef, J. Podlecki, *et al.*, "MEMS Scanning Calorimeter with Serpentine-Shaped Platinum Resistors for Characterizations of Microsamples," *Journal of Microelectromechanical Systems*, vol. 18, pp. 414-423, 2009.

[3] F. E. Torres, P. Kuhnt, D. De Bruyker, A. G. Bell, *et al.*, "Enthalpy Arrays," *Proceedings of the National Academy of Sciences of the United States of America*, vol. 101, pp. 9517-9522, 2004.

[4] B. Wang and Q. Lin, "A MEMS Differential-Scanning-Calorimetric Sensor for Thermodynamic Characterization of Biomolecules," *Journal of Microelectromechanical Systems*, vol. 21, pp. 1165-1171, 2012.

[5] L. Wang, B. Wang, and Q. Lin, "Demonstration of Mems-Based Differential Scanning Calorimetry for Determining Thermodynamic Properties of Biomolecules," *Sensors and Actuators B-Chemical*, vol. 134, pp. 953-958, 2008.

[6] C. A. Cerdeirina, J. A. Miguez, *et al.*,"Highly Precise Determination of the Heat Capacity of Liquids by DSC: Calibration and Measurement," *Thermochimica Acta*, vol. 347, pp. 37-44, 2000.

[7] M. C. Righetti, G. Salvetti, and E. Tombari, "Heat Capacity of Glycerol from 298 to 383 K," *Thermochimica Acta*, vol. 316, pp. 193-195, 1998.

[8] D. C. Ginnings and G. T. Furukawa, "Heat Capacity Standards for the Range 14 to 1200 °K," *Journal of the American Chemical Society*, vol. 75, pp. 522-527, 1953.

[9] H. J. Hinz and F. P. Schwarz, "Measurement and Analysis of Results Obtained on Biological Substances with DSC.," *Journal of Chemical Thermodynamics*, vol. 33, pp. 1511-1525, 2001.

DESIGN AND MOTION CONTROL OF SELF-PROPELLED DROPLETS

Aya Suzuki[1], Shingo Maeda[2], Yusuke Hara[3], and Shuji Hashimoto[1]
[1]Waseda University, Tokyo, JAPAN
[2]Shibaura Institute of Technology, Tokyo, JAPAN
[3]National Institute of Advanced Industrial Science and Technology (AIST), Tsukuba, JAPAN

ABSTRACT

We reported a new oil droplet system that was autonomously driven by the energy of oil-water interactions, and its motion control. Two factors influenced droplet motion: the force from the ejection of products and the effect of convective flows following the chemical reaction of an anhydride in hexamethylenediamine solution. The micro- to milliscale oil droplets moved along designed flow channels by deforming in an amoeboid motion. Also, we specifically demonstrated their motions in microfluidic channels. Such droplets offer considerable potential for use as transportable actuators.

INTRODUCTION

Diverse approaches have been proposed for the development of micro- and milliscale autonomously propelled robots, especially with respect to underwater propulsion mechanisms and performance. Such self-propelled small scale robots are expected to play a role as transportable actuators in biomedical or industrial applications, ranging from targeted drug delivery [1, 2] to environmental remediation [3]. Although the bacterial biomotor is an example of a microrobot [4], micromotors based on biological materials are difficult to control and exhibit individual variability. In contrast to these biological systems, our goal is to develop a completely artificial and autonomous micro- or milliscale robot that do not rely on biological materials. In this study, we proposed a novel propulsion mechanism that gained its driving force chemically, through unconventional oil-water interactions.

Previous studies of self-propelled oil droplets that were driven by interface dynamics were related to the nonequilibricity of the systems, which was evidenced, for example, by the chemical Marangoni effect [5–8], a potential gradient due to differences in composition or density [9], a temperature gradient [10], or convective flows developed outside of the droplets [11]. These studies offered considerable discussion about theoretical models and the patterns of oil droplet-water interactions, but there were no attempts to perform physical work by conversion of the self-propulsion energy. The reason for the lack of applied development was likely due to the low energies involved and the short lifetimes of the self-running droplets. This paper describes new self-propelled droplets with the potential for use as transportable actuators in fluidic channels.

In a previous study, Hanczyc et al. showed that an oil droplet containing oleic anhydride in nitrobenzene moved autonomously in alkaline water containing oleic acid [12]. In this system, oleic anhydride was hydrolyzed to oleic acid as follows:

$$(C_{17}H_{33}CO)_2O + H_2O \rightarrow 2C_{17}H_{33}COO^- + 2H^+ \quad (1)$$

The conjugate base of oleic acid exhibited surfactant properties in the high pH, basic solution. When the oil was immersed in the basic aqueous medium, the oleate salt formed a surfactant coat on the oil droplet, facilitating surface contact with the medium. A pair of convective flows was generated inside the oil droplet through the process of hydrolysis.

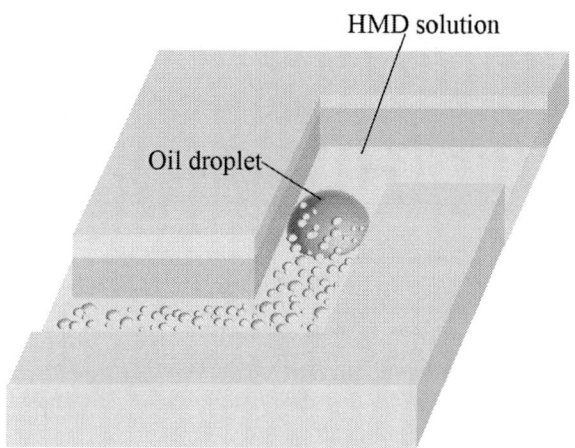

Figure 1: Illustration of a self-running oil droplet with ejected products in a fluidic channel filled with aqueous hexamethylenediamine (HMD) solution.

In this study, we employed an aqueous solution of oleic acid and hexamethylenediamine (HMD), and found that the oil droplets displayed a novel self-running motion. In particular, oleic acid and HMD were introduced to an oil droplet in the process of hydrolysis. The two oligomers were considered to be synthesized by dehydration condensation as follows,

$$2C_{17}H_{33}COOH + H_2N(CH_2)_6NH_2 \rightarrow$$
$$C_{17}H_{33}\text{-CO-NH-}(CH_2)_6\text{-NH-CO-}C_{17}H_{33} + 2H_2O \quad (2)$$

$$C_{17}H_{33}COOH + H_2N(CH_2)_6NH_2 \rightarrow$$
$$C_{17}H_{33}\text{-CO-NH-}(CH_2)_6\text{-NH}_2 + H_2O \quad (3)$$

Or, oleic anhydride and HMD formed above same oligomers as follows,

$$2(C_{17}H_{33}CO)_2O + H_2N(CH_2)_6NH_2 \rightarrow$$
$$C_{17}H_{33}\text{-CO-NH-}(CH_2)_6\text{-NH-CO-}C_{17}H_{33} + 2C_{17}H_{33}COO^- +$$
$$2H^+ \quad (4)$$

$$(C_{17}H_{33}CO)_2O + H_2N(CH_2)_6NH_2 \rightarrow$$
$$C_{17}H_{33}\text{-CO-NH-}(CH_2)_6\text{-NH}_2 + C_{17}H_{33}COO^- + H^+ \quad (5)$$

As a result, the oil droplet gained the driving force for movement by ejection of the reaction products as carried by convective flows. This process of product ejection resulted in an amoeboid motion.

In this paper, we developed a self-propelled oil droplet which moved spontaneously in micro- or milliscale fluidic channels filled with aqueous HMD solution (Figure 1). We also succeeded in the motion control of the oil droplets by manipulating a pH gradient in the aqueous HMD medium and by directing their movement through defined channels.

EXPERIMENTAL SETUP

The oil phase was prepared by adding oleic anhydride (TCI, Japan) to nitrobenzene (Kanto Chemical, Japan) at a concentration of 0.5 M. The aqueous medium was prepared by adding HMD (Kanto Chemical, Japan) to alkaline water (pH ~12) at a concentration of 0.2 M and the surfactant oleic acid (Wako, Japan) at a concentration of 5.0 mM. The milliscale fluidic channels were constructed in HMD solution by immersion of a mass of glass beads (No. 1, Toshinriko Co., Ltd., Japan) with radii of 1.0–1.5 mm. The motion of the self-propelled oil droplets was recorded using a CCD camera (WAT-231S2, Sugitoh Co., Ltd., Japan) controlled by a computer, and their velocities were analyzed using image processing software (Motion Analyzer, Keyence, Japan).

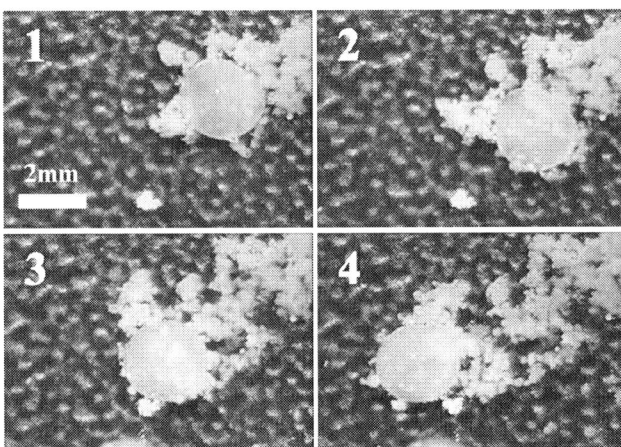

Figure 2: Self-propelled oil droplet in HMD solution: time-lapse images illustrating droplet propulsion at 2.5 s intervals.

RESULTS AND DISCUSSION

Self-propelled Droplets

The oleic anhydride-containing oil droplet (5.0 μL, 0.5 M) was added to aqueous HMD solution (0.2 M) via micropipette. The droplet moved spontaneously and randomly with product ejection, as shown in Figure 2. The velocities of the droplets ranged from 0.2 to 0.4 mm·s^{-1}, which corresponded to 0.1 body lengths·s^{-1}. As the hydrolysis reactions progressed, adducts were produced inside the droplets and ejected by internal flows within the droplet. Then, the ejected materials were left behind the oil

droplet, and the trajectory of the self-running droplet could be traced visually in the solution by the aggregates of trailing products.

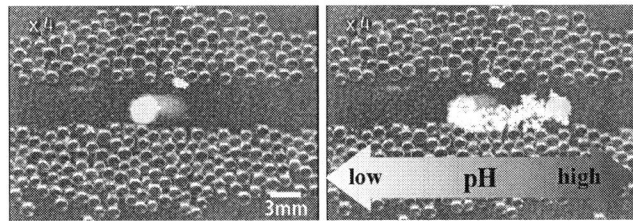

Figure 3: Directional control of an oil droplet utilizing a pH gradient. (Left): a droplet soon after introduction to the aqueous medium, and (Right): after 50 s.

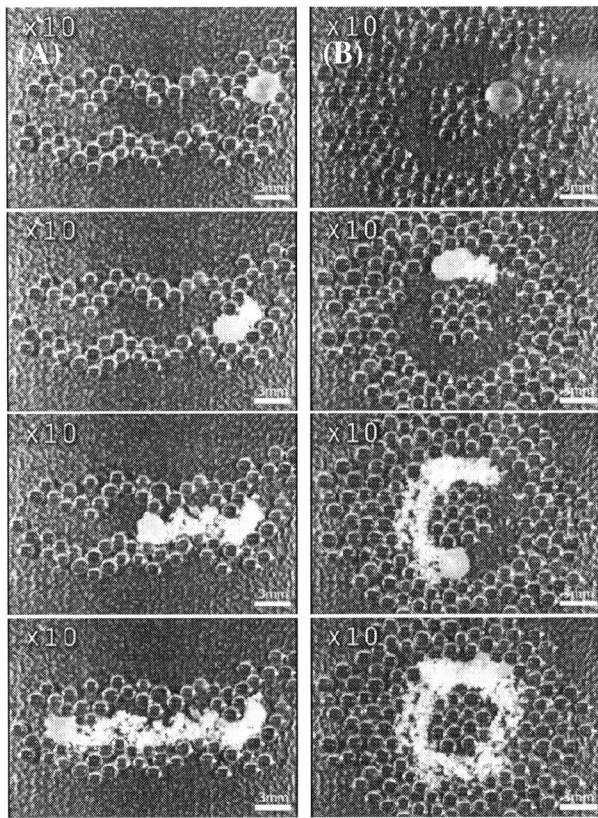

Figure 4: Time courses of milliscale oil droplet motion at 20–30 s intervals in (A) zigzag and (B) circular fluidic channels constructed from glass beads.

Motion Control

pH Gradient. To create a pH gradient in the aqueous HMD solution, a small block of agarose gel soaked in a solution of HMD (pH 13.0) was placed at one side of a Petri dish filled with HMD solution (pH 12.0). After allowing approximately 30 s to establish a pH gradient within the Petri dish, an oil droplet containing oleic anhydride (2.0 μL, 0.5 M) was placed at the low pH side, and the droplet moved toward the higher pH region with the ejection of product, as shown in Figure 3. Since we wanted to observe oil droplet motion in a single direction, a straight course was made by assembling glass beads in the HMD solution. In this experiment, we showed that the droplet motion direction could be controlled by a pH gradient in the HMD solution.

We believe that the directional property of the droplets can be attributed to the reaction of oleic anhydride. Since the self-propelled droplets showed directional motion, paddling through the aqueous phase via internal convective flows, the reactive precursor must have been localized toward the front of the direction of movement. The hydrolysis reactions (equations (1)) would proceed more efficiently in regions of higher pH. Therefore, the droplets were self-propelled toward the zone of higher pH from the zone of lower pH, promoting reaction of the oleic anhydride-containing oil droplet.

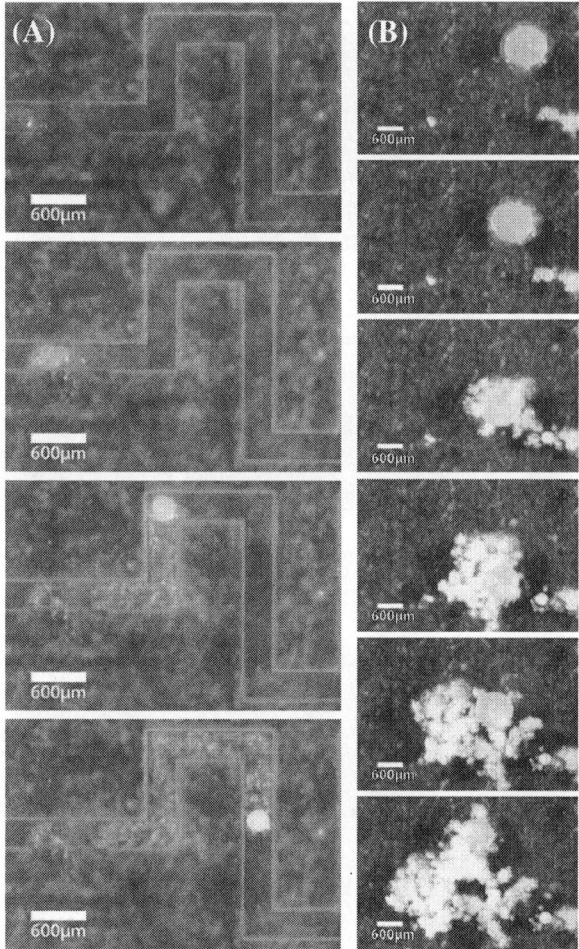

Figure 5: Time courses over approximately 40 s of microscale oil droplet motion: (A) in micro fluidic channel and (B) random.

Motion in fluidic channels

Channel experiments. We demonstrated the motion of self-propelled droplets in constrained environments. We constructed simplified milliscale fluidic channels using flexible assemblies of glass beads. As shown in Figure 4, two fluidic channels were manually constructed using glass beads in HMD solution; a zigzag structure and a circular structure.

After droplets containing oleic anhydride (2.0 μL, 0.5 M) were introduced into the glass beads channels, the droplets moved spontaneously along the lines of flow. In the circular course, after the droplet traversed one circuit, its motion ceased because of the lack of fresh HMD solution, which had been displaced by the products ejected inside the restricted channel.

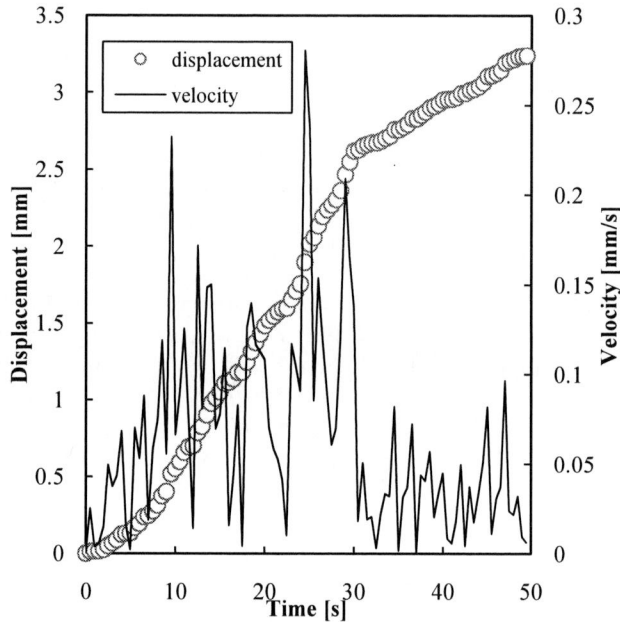

Figure 6: The displacement (red) and the velocity (gray) of a self-propelled droplet in a microfluidic channel (Figure 5(A)) over time.

In addition to the above milliscale droplet motion studies, we conducted the experiment in the commercialized microfluidic channel (I1-1, Fluidware Technologies Inc., Japan) with corners placed at short intervals and a width of 300 μm. When the micro fluidic channel was filled with HMD solution and an oil droplet containing oleic anhydride (7.0 nL, 0.5 M) was introduced, the droplet showed self-running motion with the ejection of products (Figure 5(A)), similarly to the cases of the milliscale fluidic channel experiments. For comparison, Figure 5(B) illustrates the random motion of a microscale (400 nL) droplet. Oil droplets moved autonomously along the lines of fluidic channels because they were self-propelled towards fresh HMD solution. The experiments showed that products were ejected from a droplet in the area opposite to its direction of motion, and there was fresh HMD solution in front of the droplet. Furthermore, we observed that a droplet never retraced its path.

Figure 6 charts the displacement $r(t)$ and velocity $v(t)$ of the self-propelled oil droplet in a microfluidic channel as a function of time (Figure 5(A)). The droplet remained in motion for approximately 50 s, and the maximum observed velocity was 300 μm·s^{-1}, or about 1.0 body length·s^{-1}. The droplet did not maintain a uniform velocity because its motion was not smooth, therefore, the velocity data varied over a wide range (Figure 6). The droplet velocity increased initially, and then decreased as the droplet approached its life span. Finally, the motion of the droplet stopped.

These experimental results indicated that self-propelled oil droplets can display autonomous motion in a single direction along the confines of various fluidic channels, ranging from milli- to microscale.

CONCLUSION

We demonstrated self-propelled oil droplets that eject the products for the first time. Our droplets were composed of completely artificial chemical substances and moved autonomously in fluidic channels ranging from milli- to microscale. In addition, we succeeded in controlling the droplet motion by utilizing a pH gradient in the HMD solution. Compared with previous self-propelled oil droplets, our droplets, with their ejected products, demonstrated advantages of energetically driving force, increased lifetimes, and improved visibility. We believe that their applications as micro- or milli-sized robots utilizing self-propelled droplets will be realized.

In future work, we will analyze the self-propelled droplet system to better understand the phenomenon and the droplet dynamics. We assume, for example, that parameters such as the concentration differential between the oil and the medium or other experimental conditions will exert significant influences on droplet movement. We also plan to conduct additional experiments involving cargo manipulation that may lead to the aforementioned applications. Our oil droplets can be utilized as transportable actuators in water on the milli- or microscale, such as in the propulsion of a target substance through a microfluidic channel.

ACKNOWLEDGEMENTS

This work was supported in part by: (1) The Global COE (Center of Excellence) Program, "Global Robot Academia," MEXT, Japan; (2) a Grant-in-Aid for Challenging Exploratory Research (24656177); (3) a Grant-in-Aid for Young Scientists (A) (23686043); (4) a Grant for Advanced Industrial Technology Development from the New Energy and Industrial Technology Development Organization (NEDO) and (5) and a Project 13L03 of RISE, Waseda University.

REFERENCES

[1] L. Zhang, T. Petit, K.E. Peyer, B.J. Nelson, "Targeted Cargo Delivery Using a Rotating Nickel Nanowire," *Nanomed. Nanotechnol. Biol. Med.* vol. 8, pp. 1074–1080, 2012

[2] W. Gao, D. Kagan, O.S. Pak, C. Clawson, S. Campuzano, E. Chuluun-Erdene, E. Shipton, E.E. Fullerton, L. Zhang, E. Lauga, J. Wang, "Cargo-Towing Fuel-Free Magnetic Nanoswimmers for Targeted Drug Delivery," *Small*, vol. 8, pp. 460–467, 2012.

[3] M. Guix, J. Orozco, M. Garcia, W. Gao, S. Sattayasamitsathit, A. Merkoçi, A. Escarpa, J. Wang,

"Superhydrophobic Alkanethiol-Coated Microsubmarines for Effective Removal of Oil," *ACS Nano*, vol. 6, pp. 4445–4451, 2012.

[4] B. Behkam, M. Sitti, "Bacterial Flagella-based Propulsion and On/off Motion Control of Microscale Objects," *Appl. Phys. Lett.,* vol. 90, pp. 1–3, 2007.

[5] Y. Sumino, N. Magome, T. Hamada, K. Yoshikawa, "Self-Running Droplet: Emergence of Regular Motion From Nonequilibrium Noise," *Phys. Rev. Lett.*, vol. 94, 068301, 2005.

[6] M. Dupeyrat, E. Nakache, "Direct Conversion of Chemical Energy Into Mechanical Energy at an Oil Water Interface," *Bioelectrochem. Bioenerg.*, vol. 5, pp. 134–141, 1978.

[7] N. Magome, K. Yoshikawa, "Nonlinear Oscillation and Ameba-like Motion in an Oil/Water System," *J. Phys. Chem.*, vol. 100, pp. 19102–19105, 1996.

[8] R.H. Farahi, A. Passian, T.L. Ferrell, T. Thundat, "Microfluidic Manipulation via Marangoni Forces," *Appl. Phys. Lett.*, vol. 85, p. 4237, 2004.

[9] Z. Jilin, H. Yanchun, "Shape-Gradient Composite Surfaces: Water Droplets Move Uphill," *Langmuir*, vol. 23, pp. 6136–6141, 2007.

[10] A.A. Nepomnyashchy, M.G. Velarde, P. Colinet, *Interfacial Phenomena and Convection*, Chapman & Hall/CRC, Boca Raton, 2002.

[11] I. Lagzi, S. Soh, P.J. Wesson, K.P. Browne, B.A. Grzybowski, "Maze Solving by Chemotactic Droplets," *J. Am. Chem. Soc.*, vol. 132, pp. 1198–1199, 2010.

[12] M.M. Hanczyc, T. Toyota, T. Ikegami, N. Packard, T. Sugawara, "Fatty Acid Chemistry at the Oil-Water Interface: Self-Propelled Oil Droplets," *J. Am. Chem. Soc.*, vol. 129, pp. 9386–9391, 2007.

CONTACT

*S. Hashimoto, e-mail: shuji@waseda.jp

DETECTION OF MUTATIONS IN THE BINDING DOMAIN OF TAU PROTEIN BY KINESIN–MICROTUBULE GLIDING ASSAY

S. P. Subramaniyan[1], M. C. Tarhan[2], S. L. Karsten[3], H. Fujita[2],
H. Shintaku[1], H. Kotera[1] and R. Yokokawa[1]

[1]Department of Micro Engineering, Kyoto University, Japan
[2]Institute of Indusrial Science, The University of Tokyo, Japan
[3]NeuroInDx Inc., USA

ABSTRACT

Tau protein is a biomarker for neurodegeneration. The microtubule (MT)–tau binding affinity varies according to the type of tau isoform and their degree of phosphorylation. We have utilized the difference in binding affinity of tau protein to MT to be evaluated by kinesin motor protein based MT gliding system, in order to detect and differentiate tau isoforms and their mutants. Evaluation parameters are landing rate and density of MTs on a kinesin-coated surface, and their strong correlation enables us to measure only landing rate to distinguish 2N3R and 2N4R (tau isoforms) and mutants. Secondly; we designed and fabricated a microstructure to detect tau-attached MTs, which is composed of a reservoir, parallel channels and collectors. Increase of fluorescent intensity by accumulation of MTs over time was successfully detected at the collector areas. Sensitive and rapid MT-kinesin based detection of tau isoforms (3R/4R) and mutated tau proteins on a microchip format will aid in differential diagnosis and early detection of neurodegenerative condition such as Alzheimer disease (AD).

INTRODUCTION

The tau proteins are bound to microtubules in the neurons and are found to play key roles in stabilization of microtubules and axonal transport. Abnormalities related to tau protein alone can lead to dissociation of MTs and formation of insoluble aggregates of tau proteins [1] that are found in neurofibrillary tangles (NFT) of AD brain [2]. Along with total tau protein elevation in cerebrospinal fluid (CSF), 4R:3R ratio is also found to increase in AD-CSF. Hence, tau is a potential biomarker for AD.

Tau protein is functionally divided into binding domain and projection domain (Figure 1). Binding domain is constituted by three or four pseudorepeat amino acid sequences encoded by exon (E) 9-12 at the Carboxyl-terminal (C) of the protein. The projection domain is encoded by exon 1-3, and is located at the amino-terminal (N) of the protein. Six tau isoforms are present in the human brain and they differ from each other by the presence or absence of exon 2 and/or 3 in N-terminal and by the presence of three or four repeats in the C-terminal [3].

In the adult human brain the ratio of 3R tau to 4R tau is ~1. Missense mutations in FTDP-17 (Frontotemporal dementia and Parkinsonism linked to chromosome 17) that change the splicing sites results in the alteration in the ratio of tau isoforms [4]. When 4R tau is increased *in vivo* it results in insoluble tau aggregates, because 4R is readily hyperphosporylated than 3R. Many mutations located in the binding domain results in dissociation of MTs, by either affecting the binding affinity of tau or by favoring hyperphosphorylation [5]. Hyperphosphorylated tau cannot bind to MTs efficiently which simultaneously leads to MT dissociation and tau aggregation, resulting in the elevated levels of tau protein in CSF (normal tau concentration in the CSF 135 − 230 pg/ml and during neurodegenerative conditions tau concentration is 804 − 244 pg/ml) [6].

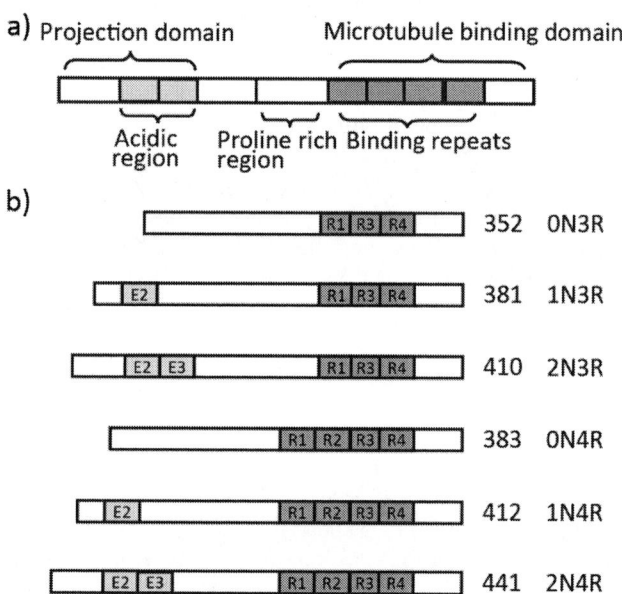

Figure 1. a) Schematic representation of functional domains of the longest tau isoform (2N4R): The amino terminal (N) - projection domain is negatively charged and relatively acidic, followed by the proline-rich region that interacts with the cytoskeletal elements. Carboxyl terminals (C) harbor the microtubule binding repeats-binding domain, and eventually play key roles in the polymerization and stabilization of microtubules. b) Schematic diagram showing the six tau isoforms. The number of amino acid in each isoform is indicated at the right. The isoforms are generated by alternate splicing of exon 2 and/or 3 (E2 and E3) in the projection domain and exon 10 (R2) in the binding domain of tau protein.

Previously, kinesin motor protein-based bead assay with tau-decorated MTs has been proposed as a diagnostic tool, in which bead velocity varies with respect to tau isoforms [7]. They have also studied modulation of MT gliding rate and trajectories on a kinesin-coated surface according to the tau isoforms, and found that tau-assembled, tau-stabilized and taxol-stabilized MTs showed different gliding behavior. Tau-assembled MTs displayed linear trajectories and taxol-stabilized MTs displayed curved trajectories. The gliding velocity was also altered with respect to the tau isoforms (3R, 4R), showing the tau isoforms have different influence on the kinesin-driven motility [8]. However, none of the previous methods enables to differentiate the tau mutants. Here, we have studied the kinesin motor protein based gliding assay to differentiate tau isoforms and some of its mutants located in the C terminal (V248L, G272V, P301L, V337M and R406W). Along with gliding velocities we have used gliding parameters such as the landing rates and the MT density of MTs on the kinesin-coated surface.

We have demonstrated the fluorescent intensity change with respect to the density of microtubules at the collector region of the micro fabricated device. This will aid in differentiation among tau isoforms and mutants.

MATERIALS AND METHODS

Buffers and solution: BRB80 solution containing: 80 mM PIPES, 1 mM EGTA and 1 mM $MgCl_2$, was used as buffer solition [9]. Motility solution included 216 µg/mL glucose oxidase, 36 µg/mL catalase and 1% (v/v) of 2-mercaptoethanol. Casesin solution (0.5 mg/ml) was used to passivate the glass surface to support kinesin motor proteins. Kinesin (amino sequence 1-573) was expressed, isolated and purified from *E. coli* [10]. Kinesin was diluted to 30 µg/mL in kinesin dilution buffer (casein 10 mg/ml, ATP 100 mM and $MgCl_2$ 100 mM). Tubulin (~55kD) and (Tetramethyl rhodamine-labeled) tubulin were polymerised and stablized by taxol [11].

Tau decoration on microtubule: Tau proteins were purchased from rPeptide, diluted and stored at -80°C. Five µM MTs solution was prepared with the motility solution comprising 1 mM ATP. MTs were sheared using syringe gauge and were then decorated with different tau proteins respectively (2N3R, 2N4R and mutant tau –V248L, G272V, P301L, V337M and R406W), by incubating, at 37°C for 30 min. They were further diluted by the motility solution, and were introduced into the flow cell.

Assay in flow cell: In brief, a glass slide and cover slip were sandwiched together using paraffin tapes. Firstly, 10 µL of casein was introduced into the flow cell and incubated for 3 min.

Secondly, 10 µL of kinesin was introduced into the flow cell and incubated for 5 min. Finally, 10 µL of tau-decorated MTs along with the motility solution were introduced into the flow cell and the flow cell was sealed immediately using grease and was incubated for 5 min prior to observation. MTs without tau were taken as control.

We have recorded 5 movies each for 60 sec for every flow cell, focusing different areas in the flow cell, in order to eliminate any error due to inhomogeneous distribution of kinesin molecules. The landing rate, MT density and gliding velocity were determined by analyzing ~100 to 300 MTs. The normalization was done with the control group.

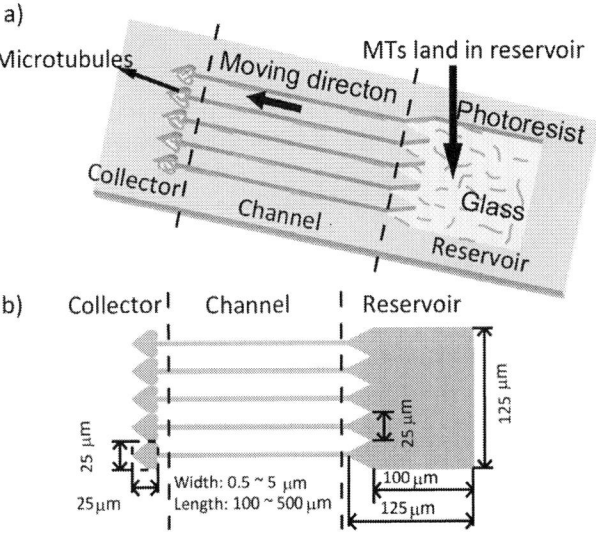

Figure 2. The tau detecting micro fluidic device. a) The device consists of a reservoir - where the MTs land, multiple channels and collector region. The device is coated with casein, kinesin and MTs decorated with tau is introduced into the device and fluorescent intensity is observed at the collector region. b) Schematic representation of the device. The channel width ranging from 0.5 to 5 µm and channel length varying from 100 to 500 µm.

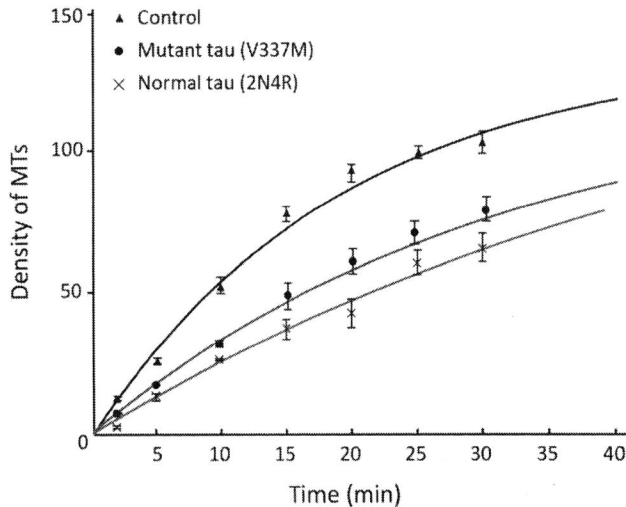

Figure 3. MT density over a time period of 0 – 30 min, right after MT are introduced into the flow cell. Blue line – normal tau (2N4R), red line – mutant tau (V337M), black line – control (no tau). The landing rate values were fitted by the equation $y = A(1 - e^{-\frac{t}{\tau}})$.

Preperation of micro device: A negetive photoresist, SU-8 was patterned on a KOH-cleaned glass surface [12] and The surface was made hydrophilic either by washing with a mixture of 28% NH₄OH, 30% H₂O₂ and DIW (1:1:5) at 120°C for 1 hour or by an exposure to plasma. Assay protocol was the identical to the previous section except that 80% ethanol and 2.0 mg/ml Pluronic were first infused into the device (Figure 2).

RESULTS

Landing rate of MTs decorated with tau proteins

As landing rate and MT density are the parameters to determine the type of tau proteins, we studied the interrelations between these two parameters. We find strong correlation between these two parameters. The landing rate was defined as the number of MTs landed per minute per unit area, which corresponds to the gradient of the graph in Figure 3. Therefore, the landing rate is decreasing in the time course, and reaches to zero at the saturated MT density.

MT density over a time period of 0–30 min, were determined and fitted by the equation (1).

$$y = A(1 - e^{-\frac{t}{\tau}}) \qquad (1)$$

Where, A-saturation co-efficient, t-time (min), y-MT density and τ-time constant. The value obtained by least square fitting are $A_c = 133.9$ and $\tau_c = 18.8$ for control, $A_{nt} = 157.3$ and $\tau_n = 55.0$ for normal tau (2N4R), and $A_{mt} = 123.5$ and $\tau_{mt} = 31.3$ for mutant tau (V337M), with $r > 0.98$ in all the cases.

We found that the difference in both these parameters with respect to the tau types were significant. And we could observe this difference at a very short interval of time that is right after incubation, thus we have recorded our assay between 5 to 10 min, when the MT density in the flow cell has not saturated and we can still observe significant difference in the landing rate with respect to each tau isoform and mutants, thereby making this assay more sensitive and rapid.

Landing rate and MT density of normal tau and mutant tau

Figure 4 shows that the landing rate (graph a) and the MT density (graph b) of normal tau decorated MTs were lowered when compared with the mutant tau -decorated MTs and the control (MTs without tau). Among the two normal tau isoform 2N4R (four binding repeats) had significantly lower landing rate than that of MTs decorated with 2N3R (three binding repeats) tau.

Landing rate of MTs decorated with mutated tau proteins

FTDP17 mutations that we have experimented are all located in the binding domain (carboxy terminal) of the tau protein; these mutants contribute in different ways for the loss of MT integrity in the neurons. It should be noted that these mutations have not caused any quantifiable change in their protein structure, hence making it more complicated to detect them [13]. The MTs decorated with mutant tau did not

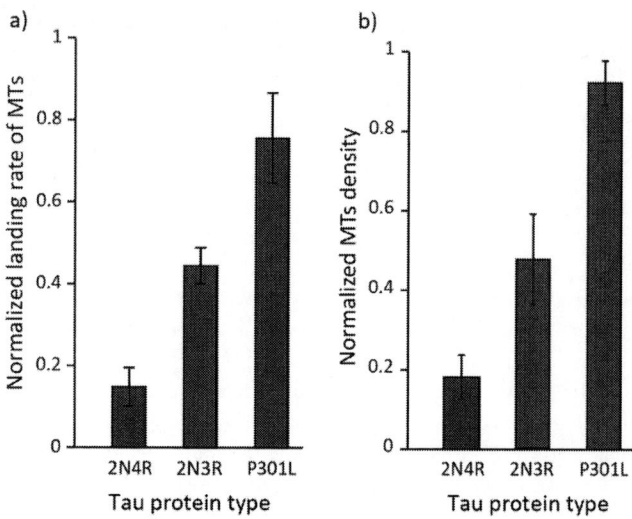

Figure 4. a) Landing rate of MTs decorated with 2N4R, 2N3R and mutated tau P301L. MT without tau decoration is taken as control. The landing rate of tau decorated MTs are significantly lower than the control and the mutated tau, and among the normal tau the landing rate of 2N4R is significantly lower than the 2N3R tau, (n = 10, p ≤ 0.01). b) The MT density of tau decorated MT after 5 min incubation showed significant difference among the control and the tau decorated MTs as well as among the tau isoforms (2N4R, 2N3R).

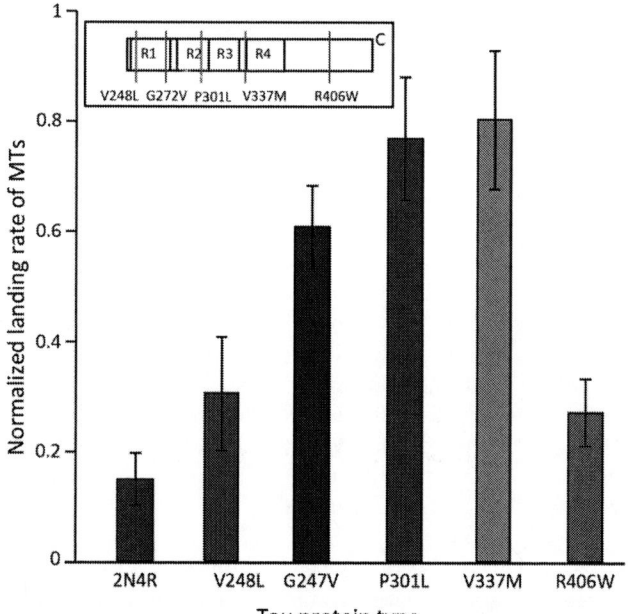

Figure 5. The landing rate of mutated tau decorated MTs. The inset shows the mutation sites in the 4R tau protein. The red line represents the mutation position. The landing rate of the MTs decorated with mutated tau was higher than the MTs decorated with normal tau (2N4R).

decrease the landing rate of the MTs like their non-mutant counterpart 2N4R tau (Figure 5). The G272V, P301L and V337M, had the least effect on the landing rate. The G272V mutation is located in R1 (exon 9), P301L is located in R2 (exon 10) and V337M mutation is located between R3 and R4 (Figure 5 inset) these three mutations affect the MT binding affinity in par with each other. The landing rate of R406W mutation located in the carboxy terminal away from the binding repeats (exon 13) and V248L located in R1 were found to significantly lower the landing rate compared to the other three mutants.

Velocity of tau decorated MTs

The gliding velocity of the normal tau (2N3R, 2N4R) decorated MTs were lower than that of the MTs with mutant tau decoration and the control group (Figure 6). But the differences were not statically significant among the two isoforms. We believe that the taxol stabilization could have masked some of the isoform specific features. Though the gliding velocity of some of the mutant tau type was different again they were of no statistical significance.

Microdecice for tau detection

The gliding assay was conducted in the microdevice with similar conditions and solutions. MTs landed in the reservoir region, glide through the channels and were collected at the collector regions over a period of time. Figure 7 shows the accumulation of fluorescent labeled MTs at the collector region at different time intervals, and the fluorescent intensity (FI) is plotted against time.

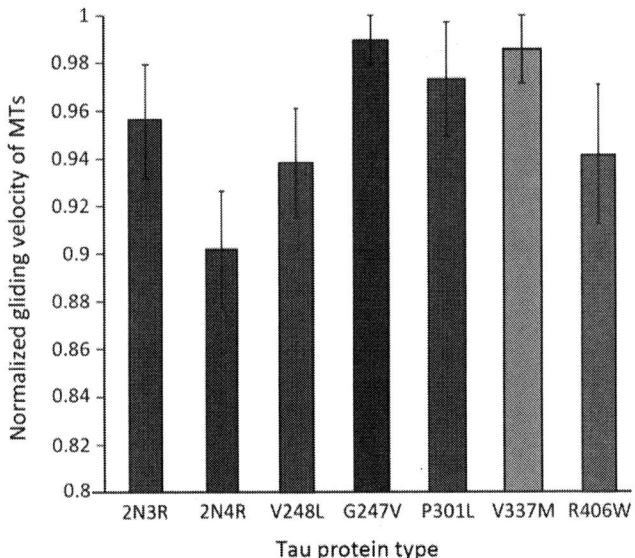

Figure 6: The gliding velocity of MTs decorated with normal tau (2N3R, 2N4R) and the mutant types (V248L, G272V, P301L, V337M and R406W). Gliding velocity of the normal tau decorated MTs were lower than the control (Without tau) and mutant, but the gliding velocity difference among the tau isoforms were not statically significant.(n = 3)

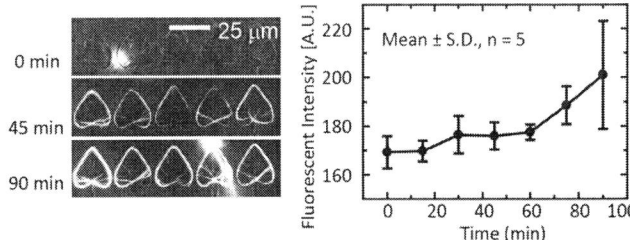

Figure 7: FI measurement. a) This image shows increase in the FI at the collector region, b) FI vs time; gradual increase of FI as the MTs density increases at the collector region.

CONCLUSION

The kinesin motor protein based gliding assay has been sensitive to detect the type of tau protein (isoforms & mutants) attached on the MTs by their landing rates. By able to differentiate between 3R & 4R and mutants make this method preferable for detecting tau protein in CSF of AD.

ACKNOWLEDGEMENTS

Authors acknowledge Mr. K. Fujimoto for kinesin purification, Mr. T. Nakahara for tubulin labeling and MEXT to S.P.S and Nakatani Foundation to R.Y for financial support.

REFERENCES

[1] H. W. Querfurth et al., New England Journal of Medicine, vol. 362, no. 4, pp. 329-344, 2010.

[2] Alvila, Jesus et al., Physiol rev 84, pp.361-384, 2004.

[3] N. Hirokawa et al., The journal of cell biology, vol. 107, pp. 1449-1459, 1988.

[4] Parimala Nacharaju et al., FEBS Letters, vol. 447, Pages 195-199, 1999.

[5] Pavan K. Krishnamurthy, et al., J. Biol. Chem. vol. 279, pp.7893-7900, 2003.

[6] M. Sjogren, et al., Journal of Neurology Neurosurgery and Psychiatry, vol. 70, pp. 624-630, 2001 .

[7] Mehmet. C, et al., Lab Chip, vol. 13, pp. 3217-322, 2013.

[8] Peck. A., et al., Cytoskeleton, vol. 68, pp. 44–55, 2011.

[9] H. Hagiwara, et al., J. Biol. Chem., vol. 269, pp. 3581-3589,1994.

[10] Olmsted, et al. Biochemistry. vol. 14, pp. 2996–3005, 1975.

[11] R. Yokokawa et al., Nanotechnology. vol. 19, pp. 125505 (7pp), 2008.

[12] S. G. Moorjani et al., Nano Lett., vol. 3, pp. 633-637.

[13] S. Barghorn et al., Biochemistry, vol. 39, pp 11714-11721, 2003.

CONTACT

* S.P.Subramaniyan, tel: +81-75-383-3687; subramaniyan.subhathirai.76x@st.kyoto-u.ac.jp

FROG EGG-ARRAY DEVICE INTEGRATED WITH FLUIDIC CHANNEL AND MICROELECTRODES FOR CHEMICAL SENSING

Mitsuyoshi Tomida[1], Yuji Murakami[1] and Nobuo Misawa[2]
[1]Department of Electrical and Electronic Information Engineering,
Toyohashi University of Technology, JAPAN
[2]Electronics-Inspired Interdisciplinary Research Institute, Toyohashi University of Technology, JAPAN

ABSTRACT

This paper describes a membrane protein-based chemical sensor that is consisted of fluidic channels and cells (*Xenopus laevis* oocytes) expressing chemical receptors. The fluidic device has Si-based microfabricated electrodes to measure the *Xenopus* oocyte's response to chemicals using two-electrode voltage clamping. The fluidic device was fabricated by combining fluidic channel made of acrylic resin and the electrode substrate. After the cell installing, the fluidic device can be separated to each fluidic channel that can measure an individual oocyte membrane potential change. We succeeded to array each oocyte in the device and to detect an individual *Xenopus* oocyte's responses to chemical stimulus.

INTRODUCTION

Recently chemical sensor based on an excellent biofunction of biomolecules and living things has been attracted a great deal of considerable attention. Advantage of chemical sensing using biomolecules is their high selectivity to a target molecule. Since applications of biofunction in various fields such as food analysis, medical care and care-worker support robot are expected, the research and development of various chemical sensors using biomaterials has been advanced by many scientists and engineers [1-6]. Response of a receptor in the cell membrane promotes the opening and closing of ion channels, and ion flow causes a rapid change in cell membrane potential. Some kind of receptor will be opened the ion channel by one molecule of a chemical. Since the number of ions across the cell membrane compared to the number of molecules to be received is large, the response of the receptor can be measured artificially as an amplified cell membrane potential change. Hence, we focused on the development of biochemical sensors with the aim of chemical detection by measuring membrane potential of cells expressing chemical receptors as conceptually shown in Figure 1.

Conventionally, in the field of pharmacology, *Xenopus* oocyte has been used as a tool for functional analysis of receptors [6]. *Xenopus* oocytes are frequently used as the host cell for expressing various chemical receptors [7]. They are easy to handle owing to their cell size (about 1 mm in diameter). *Xenopus* oocyte is also capable of high sensitivity measurements of weak ion current generated at the cells by electrophysiological techniques called two-electrode voltage clamping (TEVC) [8]. However, the conventional TEVC setup requires large footprint by using micromanipulators and a microscope. Recently, compact chemical sensors that combine the glass capillary electrodes for TEVC and the fluidic device made of acrylic resin or polydimethylsiloxane using *Xenopus* oocytes expressing olfactory receptors as transducers have been reported [9-11].

However, as shown in Figure 2(a), the previous device required a complicated manual operation for installing the glass tube electrodes, and the system was not suitable for using plural cells. To overcome the laborious process, we fabricated microelectrodes for TEVC integrated with a cell-array fluidic device made of acrylic resin that was capable of simple electrode positioning. As shown in Figure 2(b), the fluidic device can be separated to each fluidic channel that can measure an individual oocyte membrane potential change after electrode insertion to plural cells automatically by flowing. By the individual or combination measurement, identification of a variety of chemicals is facilitated even if cells expressing various chemical receptors were used. In this study, we fabricated a fluidic device that geometry was based on a result of the fluid simulation. Then we measured an oocyte membrane potential change caused by chemical irritation. Here, we demonstrate of oocyte–array and current traces of oocyte responses to chemical as the feasibility study of biochemical sensing.

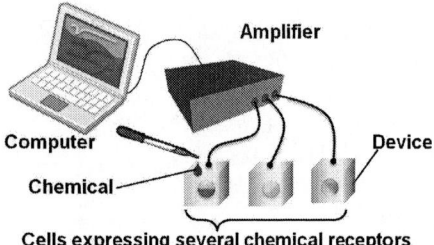

Figure 1: Concept image of chemical sensor using a number of living cells.

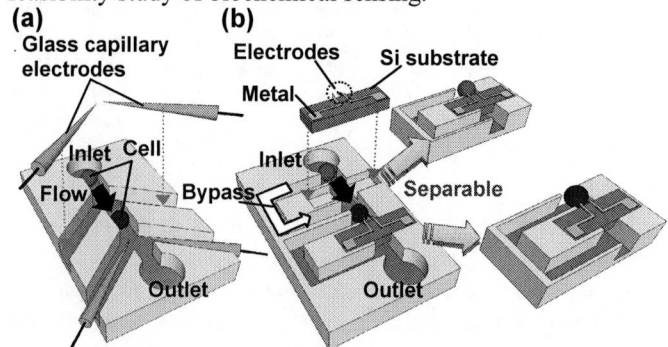

Figure 2: Schematic images of the configuration of fluidic device for chemical sensor using cells. (A): Previous device. (B): New separable device.

978-1-4799-3510-9/14 $31.00 © 2014 IEEE

DESIGN AND FABRICATION OF DEVICE

Fluid Simulation and Design of the Fluidic Channel

Bead-array technology and cell trapping fluidic channel which are capable of plural cells and particles trapping has been reported [12, 13]. With reference to them, we fabricated a bypass channel for arraying separately from the main channel having trapping regions of oocytes. Cell trapping regions were located 1.5 mm lower than the channel of the outlet and inlet to prevent leaking of solution. The width and depth of the fluidic channels were all 1.5 mm excepting cell trapping regions in reference to previous research [11]. Two electrodes in the shape of cantilever are located at the central portion of the flow path in trapping regions of the cell. Using a software for fluid analysis (Ansys® 14.0 Fluid Flow (CFX), Ansys Japan K.K.), we verified the streams in a part of the fluidic channel devised. We assumed that the body of fluid was water and a diameter of *Xenopus* oocyte was 1.3 mm. Figure 3 shows the simulation results of streamlines assuming at injection flow rate of 5 mL/min. In Figure 3(a), it was implied that the oocyte would be trapped at trapping regions before no objects at the trapping channel. Figure 3(b) shows the result assuming after an oocyte was trapped. In this case, since the flow of the bypassing channel direction is strengthened, subsequent oocytes would be bypassed and directed to next units downstream. According to the results of these simulations, it was suggested that *Xenopus* oocytes would be arrayed in order from upstream to downstream in the fluidic device.

We fabricated a fluidic channel made of acrylic resin by the cutting process with end milling based on the simulation results. As shown in Figure 4, one unit of device (enlarged illustration) is composed of a silicon substrate and two plastic plates that had open fluidic channels. We used four magnets to fix the two plastic plates in the one unit. The frame (green part) for fixing and alignment each unit were produced by milling process of acrylic resin, and the whole fluidic channel was configured by combining these parts.

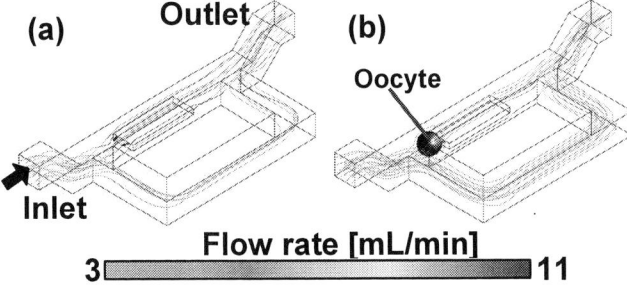

(a) Outlet
Inlet

(b) Oocyte

Flow rate [mL/min]
3 ▬▬▬▬▬▬▬▬▬▬ 11

Figure 3: Fluidic simulations of stream lines (a) before trapping and (b) after trapping of an oocyte.

Fabrication of Electrodes Substrate

Figure 5 shows the fabrication process of the electrodes substrate. The silicon substrate (Si(100), p-type, 350 μm-thick) was preliminarily surface oxidized of about 1 μm-thick. The substrate was double-sided etched by deep reactive ion etching (DRIE) to form two cantilever-like needle-shaped structures (Figure 5a). The metal layer Ti/Ag

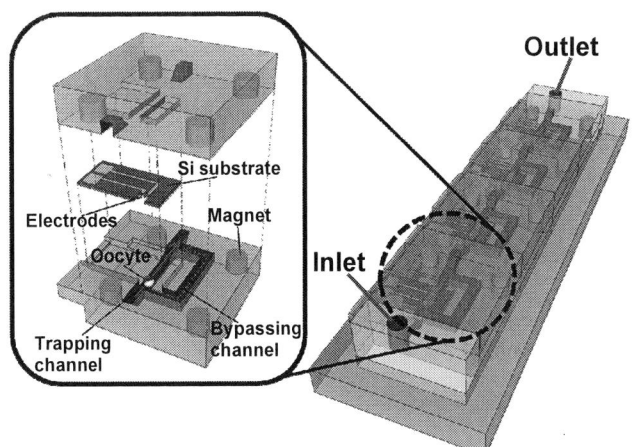

Figure 4: Schematic image of the separable cell-array fluidic channel for TEVC. (Left): One unit image of cell installing fluidic channel. (Right): Whole image of cell-array fluidic device.

was continuously deposited on SiO_2 surface of silicon substrate by sputtering (Figure 5b). Parylene C was deposited for insulation of entire substrate surface (Figure 5c), and areas of electrode's tips and contact pads were selectively exposed by O_2 ashing (Figure 5d). Tip width of these electrodes is 10 μm, and the thickness was about 50 μm. The distance between two electrodes tip was 570 μm and the angle between the electrodes was 30° to the left and right relative to the direction of flow using the previous study as a reference [8]. Figure 6 shows the picture of separated fluidic channel and scanning electron microscope (SEM) image of the fabricated electrode structure.

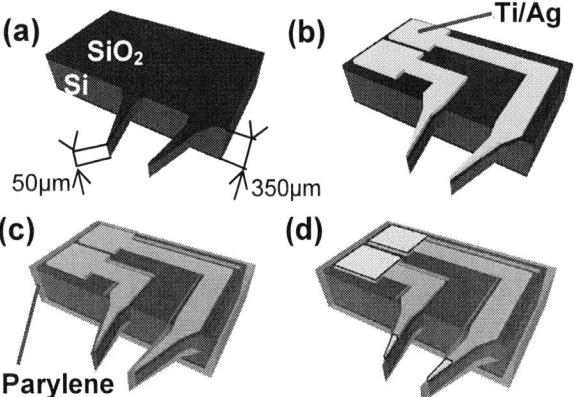

Figure 5: Fabrication process flow of the electrode substrate. (a) Surface oxidation and DRIE. (b) Selective metal deposition. (c) Parylene coating. (d) Removing of parylene.

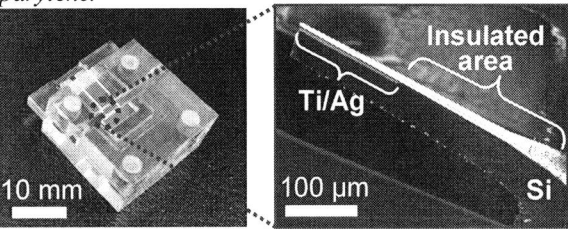

Figure 6: (Left): Picture of one unit of cell installing channel. (Right): SEM image of the electrode tip.

EVALUATION OF FABRICATED DEVICE
Demonstration of Cell-array

We verified an array of *Xenopus* oocytes using the fabricated device. Ringer's solution was manually injected with *Xenopus* oocyte by a disposable syringe. Figure 7 is the continuous images of *Xenopus* oocyte array that shows two trapping regions in the fluidic device. Ringer's solution was flowing from top to bottom of images. *Xenopus* oocytes were arrayed in order and successfully trapped in the units four of all linked. Figure 8 shows the continuous pictures of the moment a *Xenopus* oocyte is trapped and penetrated by two electrodes. These results match the simulation outcomes. Thus, we successfully demonstrated that our fabricated device worked as a cell-array fluidic channel with electrodes insertion.

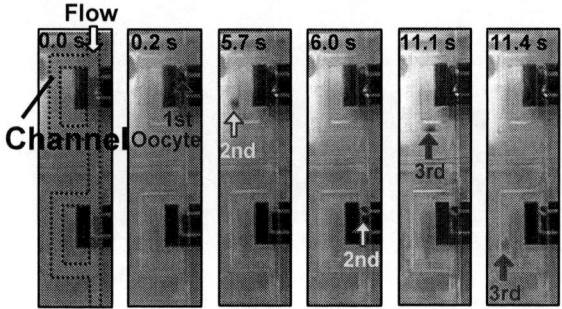

Figure 7: Continuous images of Xenopus oocytes array

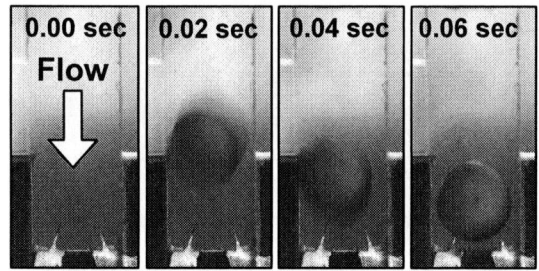

Figure 8: Enlarged view when the Xenopus oocyte is trapped and penetrated by electrodes.

Oocytes Response Measurement

As shown in Figure 9, the one unit could be separated from the fluidic device manually. *Xenopus* oocyte was fixed in trapping region of the one unit and maintained a state of soaking in the solution. The signal currents of the *Xenopus* oocytes were recorded using a custom-built amplifier (Pico 2, Tecella LLC) connected to the one unit. The currents were monitored every 100 μs at a hold voltage of -80 mV. The extracellular solution was grounded using Ag/AgCl wire independently. We used K^+ channel as a model of membrane protein which expressed endogenously in *Xenopus* oocyte. As the chemical stimulus, 20 μL of 3 M KCl solution was added to *Xenopus* oocytes inserting electrodes. Figure 10 shows that the current trace depending on the membrane potential change when the *Xenopus* oocyte responses to the KCl stimulus. Furthermore, we confirmed that the change in the current value was increased when the concentration of

KCl was higher (Figure 11). From these results, we found that the device we developed could perform cell-array device and electrophysiological measurement of individual *Xenopus* oocyte.

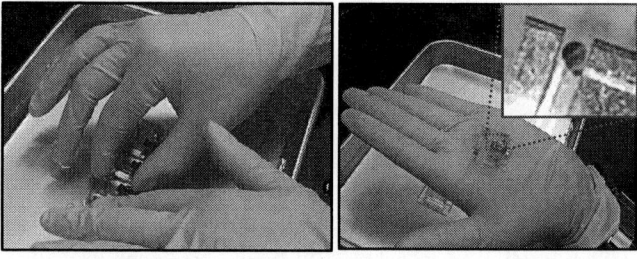

Figure 9: Picture of separating one unit from device installing an oocyte.

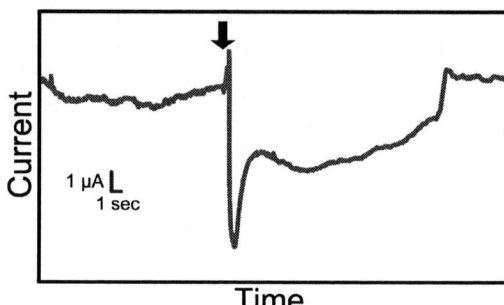

Figure 10: TEVC current trace of an oocyte response to KCl stimulus. KCl solution was added at the time indicated by the arrow.

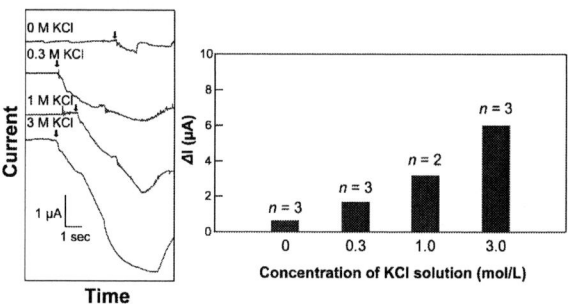

Figure 11: Dose-dependent increases in amplitude of K^+ -induced current.

CONCLUSIONS

Aiming at development of highly sensitive, selective and compact chemical sensor, we fabricated a fluidic device composed of acrylic resin and Si-based microfabricated electrodes for TEVC to measure the membrane potential changes of the *Xenopus* oocytes expressing chemical receptors.

The device can be separated to each fluidic channel with a trapped oocyte, and the fabricated electrodes could penetrate oocytes. We successfully placed the each oocyte into the device using our fabricated cell-array fluidic channel and detected the oocyte responses to chemical stimulus as a model of chemical sensing. Hence, we believe that we are capable of detection of various chemicals using *Xenopus* oocytes expressing several chemical receptors.

ACKNOWLEDGEMENTS

This work was financially supported by the Japan Science and Technology Agency, Core Research of Evolutional Science program partially. This study was also partly supported by the Ministry of Education, Cultures Sports, Science and Technology Japan with "Program to Foster Young Researchers in Cutting-Edge Interdisciplinary Research". And we really appreciate Prof. R. Kanzaki and Dr. H. Mitsuno of the University of Tokyo for kindly providing *Xenopus* oocytes.

REFERENCES

[1] J. Y. Lee, H. J. Ko, S. H. Lee, T. H. Park, "Cell-based measurement of odorant molecules using surface plasmon resonance". *Enzyme Microb Tech*, Vol.39, p.375, 2006.

[2] D. Kohl, L. Heinert, J. Bock, T. Hofmann, P. Schieberle, "Systematic studies on responses of metal-oxide sensor surfaces to straight chain alkanes, alcohols, ketones, acids and esters using the SOMMSA approach", *Sens Actuators B*, Vol.70, p.43, 2000.

[3] K. Yano, U. T. Bornscheuer, R. D. Schmid, H. Yoshitake, H. Ji, K. Ikebukuro, Y. Masuda, I. Karube, "Development of an odorant sensor using polymer-coated quartz crystals modified with unusual lipids", *Biosens Bioelectron*, Vol.13, p.397, 1998.

[4] R. Glatz, B. H. Kelly, "Mimicking nature's noses: From receptor deorphaning to olfactory biosensing", *Progress in Neurobiology*, Vol.93, p.270, 2011.

[5] N. Misawa, H. J. Lee, H. Mitsuno, R. Kanzaki, K. Sawada, "Cell array fluidic channel integrated with electrodes for cell-based multiple chemical sensing", *Transducers 2013*, p.2451, 2013.

[6] H. Kudo, X. Wang, Y. Suzuki, M. Ye, T. Yamashita, T. Gessei, K. Miyajima, T. Arakawa, K. Mitsubayashi, "Fiber-optic biochemical gas sensor (bio-sniffer) for sub-ppb monitoring of formaldehyde vapor", *Sens Actuators B*, Vol.161, p.486, 2012.

[7] T. Sakurai, T. Nakagawa, H. Mitsuno, H. Mori, Y. Endo, S. Tanoue, Y. Yasukochi, K. Touhara, T. Nishioka, "Identification and functional characterization of a sex pheromone receptor in the silkmoth *Bombyx mori*", *PANS*, Vol.101, p.16653, 2004.

[8] W. Stuhmer, "Electrophysiological recording from *Xenopus* oocytes", *Methods Enzymol*, Vol.207, p.319, 1992.

[9] N. Misawa, H. Mitsuno, R. Kanzaki, S. Takeuchi, "Microfluidic odorant sensor with frog eggs expressing olfactory receptors", *MEMS2009*, p.180, 2009.

[10] N. Misawa, H. Mitsuno, R. Kanzaki, S. Takeuchi, "Biological Noses for a Robot", *MicroTAS 2010*, p.1274, 2010.

[11] N. Misawa, H. Mitsuno, R. Kanzaki, S. Takeuchi, "Highly Sensitive and Selective Odorant Sensor using Living Cells Expressing Insect Olfactory Receptors", *PNAS*, Vol.107, p.15340, 2010.

[12] W. H. Tan, S. Takeuchi, "A trap-and-release integrated microfluidic system for dynamic microarray applications", *PNAS*, Vol.104, p.1146, 2007.

[13] T. Teshima, H. Ishihara, K. Iwai, A. Adachi, S. Takeuchi, "A dynamic microarray device for paired bead-based analysis", *Lab on a Chip*, Vol.10, p.2443, 2010.

CONTACT

*N. Misawa, tel: +81-532-81-5135; misawa@eiiris.tut.ac.jp

INTEGRATION OF DIAMOND MICROELECTRODES ON CMOS-BASED AMPEROMETRIC BIOSENSOR ARRAY BY FILM TRANSFER TECHNOLOGY

Takeshi Hayasaka[1], Shinya Yoshida[1], Kumi Y. Inoue[1], Masanori Nakano[1], Tomohiro Ishikawa[1], Tomokazu Matsue[1], Masayoshi Esashi[1], and Shuji Tanaka[1]
[1]Tohoku University, Sendai, Miyagi, JAPAN

ABSTRACT

This paper reports a complementary metal oxide semiconductor (CMOS)-based 20×20 amperometric biosensor array using boron-doped diamond (BDD) microelectrodes with excellent electrochemical properties. The BDD electrodes were once formed on a Si wafer at 800°C, and then transferred to a 0.18 μm CMOS wafer with a benzocyclobutene (BCB) bonding interlayer. As a result, the BDD microelectrodes were arrayed without damage in the CMOS LSI circuit. The fully-integrated device could detect histamine and dopamine owing to a wide potential window of the BDD electrode, and offered 2-dimensional real-time imaging of histamine diffusion in a solution.

INTRODUCTION

A CMOS-based amperometric sensor with a multiple electrodes array, which can be applied to electrochemical real time imaging and simultaneous multiple analysis of DNA, enzymes etc., is a prospective analysis platform for medical diagnostics, environmental measurements and basic biochemistry [1-5]. Thus far, noble metals such as Au or Pt have been utilized for the working electrode of CMOS-based amperometric sensors, considering process compatibility with CMOS circuits. However, their potential windows are relatively narrow compared with carbon electrodes, and thus detectable analyte is limited.

As a superior electrode material, boron-doped diamond (BDD) has been investigated recently. BDD has excellent electrochemical properties such as a wider potential window and smaller background current. Thus, it is suitable for detecting biomolecules with high oxidation potential such as histamine, glutathione etc. [6,7]. BDD also has long-term stability and superior biocompatibility. Therefore, the combination of BDD electrodes and the CMOS amperometric sensor is promising to open wider applications. However, high quality BDD is normally synthesized by chemical vapor deposition (CVD) at high temperature such as 800°C. Thus, the direct deposition of BDD is difficult on a CMOS LSI wafer due to its limited thermal budget.

To address the above problem, we propose to employ film transfer technology using an adhesive resin layer for monolithic integration of BDD microelectrodes on a CMOS LSI wafer. This paper reports the fabrication process and evaluation results of the CMOS-based amperometric biosensor array with BDD electrodes.

EXPERIMENTAL METHOD

Fabrication process

A 0.18 μm CMOS LSI was designed for the amperometric biosensor array and manufactured in a CMOS foundry. The CMOS-based biosensor has a 20 × 20 working electrode array with a 250 μm pitch [1]. The integration process is shown in Fig. 1 and described as follows.

BDD electrode fabrication on Si substrate

(A1) The electrophoretical seeding of diamond was performed on a Si substrate with alignment marks on the backside. For electrophoresis, a voltage of 75 V was applied to the Si substrate against a counter electrode in isopropyl alcohol containing diamond nanopowders under ultrasonic agitation.

(A2) A BDD layer with a thickness of ~1 μm was synthesized on the Si substrate by microwave-plasma-enhanced CVD (AX5200, CORNES TECHNOLOGIES Ltd.) under the condition shown in Table 1. Trimethylborate was used as a boron dopant.

(A3) An Al etching mask was formed on the BDD layer.

(A4) The BDD layer was dry-etched with O_2 plasma, and the Al mask was removed.

BDD electrode transfer to LSI wafer

(B1) A Cr/Pt/Au/Pt/Cr-stacked layer was patterned on the LSI wafer. The Cr layer on the top surface enhances adhesion to a resin used in the next process step.

(B2) AP3000 as an adhesion promoter and a benzocyclobutene (BCB) resin (CYCLOTENE3022-63, The Dow Chemical Company) were spin-deposited on the LSI wafer, and BCB was partially cured in N_2 atmosphere. The Si wafer with the BDD electrodes was also spin-coated with AP3000, and aligned to the LSI wafer.

(B3) The LSI and Si wafers were bonded in the condition shown in Table 2.

(B4) The Si wafer was removed by dry etching with SF_6 plasma. As a result, the BDD electrodes were transferred to the LSI wafer without damage to the circuit.

(B5) BCB was partially dry-etched with a mixture gas of O_2 and SF_6.

(B6) An Au/Cr metal layer was patterned to electrically connect between the BDD electrodes and the I/O pads of the LSI.

(B7) A SU-8 microwell array was fabricated to define sensing areas in the BDD electrodes.

(B8) The LSI wafer was diced, and the chip was mounted on a ceramic substrate with Ag paste.

(B9) The LSI chip and ceramic substrate were electrically connected by wire-bonding.

978-1-4799-3510-9/14 $31.00 © 2014 IEEE

(B10) Polydimethylsiloxane (PDMS) frames were mounted on the LSI and ceramic substrate.

(B11) Au wires were covered with PDMS. As a result, the sensing areas of the BDD electrodes were only exposed to an electrolyte in the PDMS pool, and the implementation of the CMOS sensor array were completed, as shown in Fig. 2.

BDD electrode fabrication on Si substrate

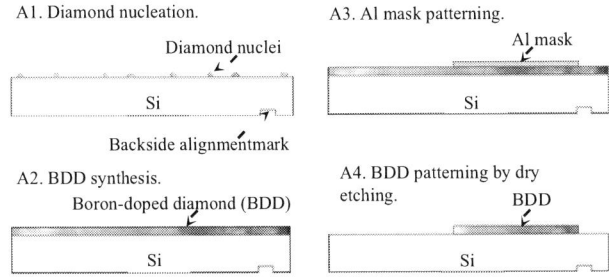

BDD electrode transferred on CMOS LSI wafer

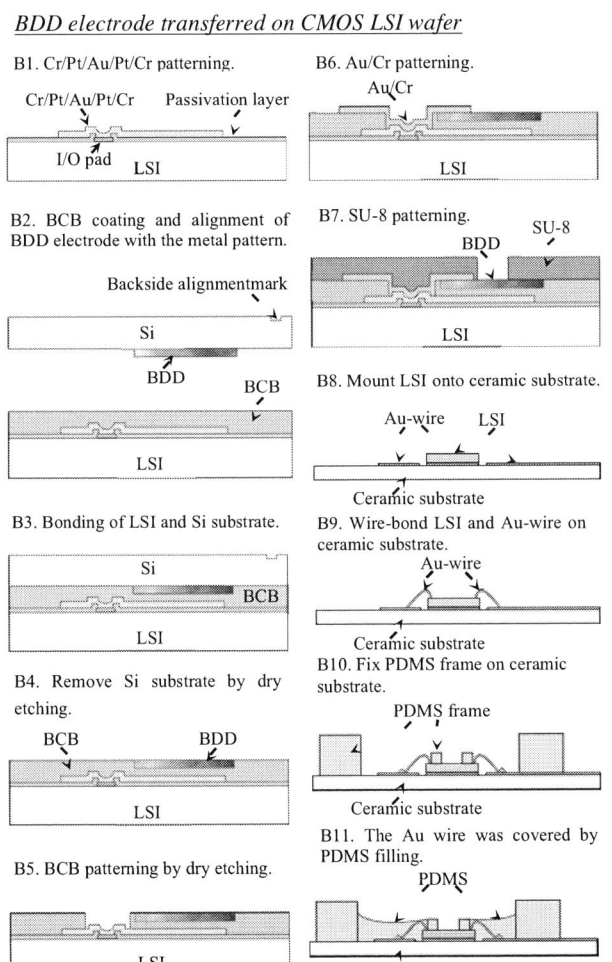

Figure 1: Fabrication process of a CMOS-based amperometric biosensor array with BDD electrodes.

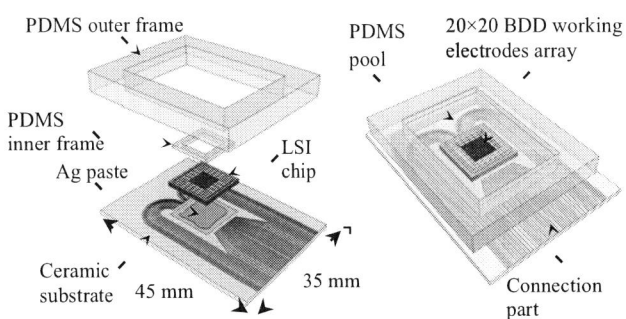

Figure 2: Mounting of the CMOS-based amperometric biosensor array on a ceramic substrate.

Table 1: Condition of BDD synthesis in CVD.

Power	Pressure	H_2 flow	CH_4 flow	$B(OCH_3)_3$ flow	Growth Temp.
1 kW	6.5 kPa	300 sccm	3 sccm	0.18 sccm	800 °C

Table 2: Condition of bonding process.

Atmosphere	Bonding temperature	Pressure to sample	Bonding time
N_2	270 °C	1.3 MPa	1 hour

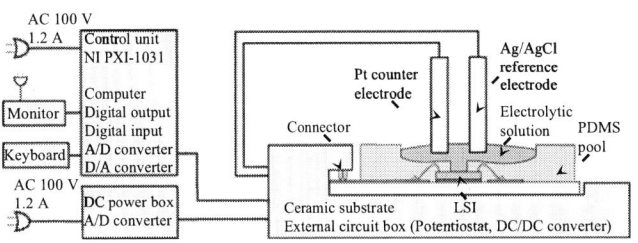

Figure 3: Schematic illustration of the setup for electrochemical measurement test.

Operation test

Figure 3 shows the operation test setup for the CMOS-based amperometric biosensor array with the BDD electrodes. The system consists of a potentiostat, a power supply, a control unit and so on. The biosensor array was treated with O_2 plasma for the hydrophilization of both SU-8 and PDMS in advance, and then connected to the external circuit. An electrolytic solution containing an analyte was impounded in the PDMS pool, and Ag/AgCl reference and Pt counter electrodes were dipped into the solution. In this system, 20×20 sensing areas of the BDD electrodes were operated as working electrodes.

RESULTS AND DISCUSSION

Fabrication result

Figure 4 (a) shows the scanning electron microscope (SEM) image of the BDD electrode surface on a Si wafer. The BDD film looks highly dense without pinholes. Figure 4 (b) shows the BDD electrodes transferred onto the LSI wafer using the BCB adhesion layer in success. The cure ratio of BCB before bonding was optimized to satisfy both

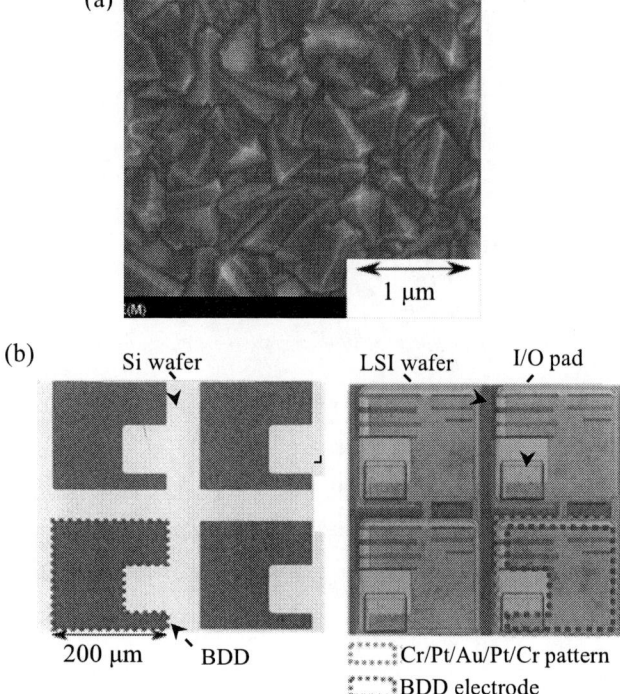

Figure 4: (a) SEM image of the BDD electrode surface. (b) Transfer of the BDD electrodes array from the Si wafer to the LSI wafer by use of a BCB adhesion layer.

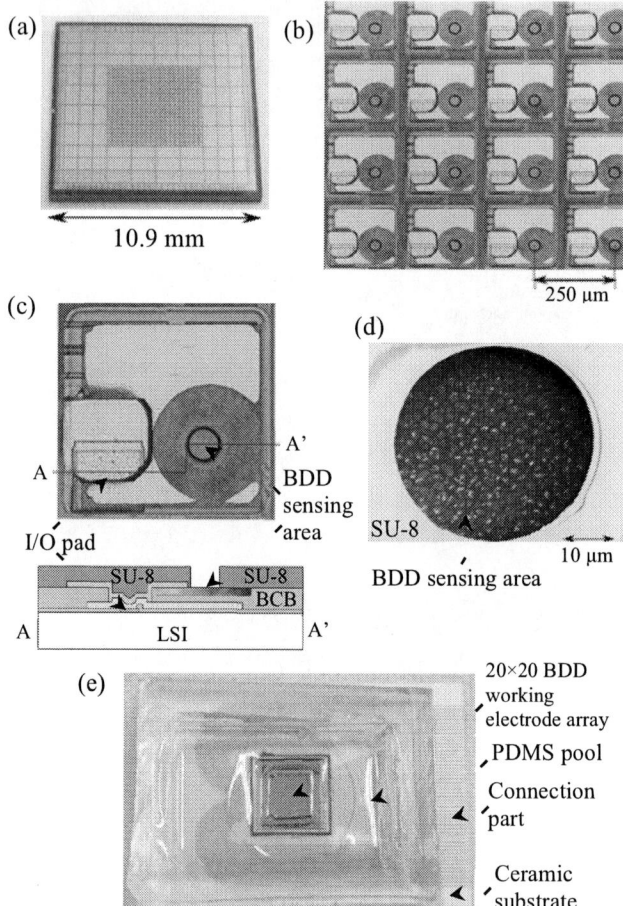

Figure 5: (a) LSI chip, (b) Unit cell array, (c) Optical microscope image and schematic cross sectional view of the single unit cell, (d) SEM image of the BDD electrode surface, (e) Optical microscope image of the completed device.

alignment accuracy and bonding strength. If the cure ratio was low, bonding was easy and uniform, but alignment error reached as large as several dozen micrometers. If the cure ratio was too high, the BDD electrodes did not well bonded to the BCC layer, and transfer yield was low. Finally, using 40%-cured BCB, 400 BDD electrodes were perfectly and reproducibly transferred with acceptable alignment error less than 10 μm.

Figure 5 (a) and (b) show the LSI chip with the 20×20 BDD working electrodes and a part of the electrode array, respectively. The top view image of the single unit cell and its cross-sectional structure are shown in Fig. 5 (c). The I/O pad of the LSI chip and the BDD electrode, which looked slightly darkish, were connected with the Au/Cr film. A 40-μm-diameter sensing area was defined with the SU-8 microwell. As found in Fig. 5 (d), any significant crack or destruction of the BDD electrode was not observed. The fully-implemented device is shown in Fig. 5 (e).

Operation test result

The electrochemical properties of the integrated BDD electrode were investigated. Figure 6 (a) the potential windows of the as-grown BDD film, transferred BDD electrode and Au electrode integrated onto the LSI. The transferred BDD electrode exhibited a narrower potential window of ~2.3 V compared with that of the as-grown BDD (~3.2 V). The reason is not clear, but one possibility is process damage to BDD owing to plasma exposure in the Si substrate removal etc. However, the potential window is still much larger than that of an Au electrode (~1.3 V). Furthermore, the transferred BDD electrode showed small background current, which was equivalent to that of the as-grown BDD film.

The cyclic voltammogram of ferrocenemethanol (Fig. 6 (b)) presents the ideal redox current profile specific to microelectrodes owing to smooth electron transfer, suggesting that the transferred BDD electrode is useful for electrochemical measurement. Next, the detection of histamine was demonstrated as shown in Fig. 6 (c). Histamine has a high oxidation potential of ~1.4 V, and is difficult to detect using a Au electrode because of large oxidation current of hydroxide ions in water. On the other hand, the BDD electrode with a large potential window clearly detected histamine oxidation current depending on the concentration. The biosensor also successfully detected dopamine and measured its concentration from oxidation current at ~1.1 V (Fig. 6 (d)).

Finally, 2-dimensional real time imaging of histamine diffusion was demonstrated. Figure 7 shows histamine diffusion process in a phosphate buffer (0.05 M KH_2PO_4,

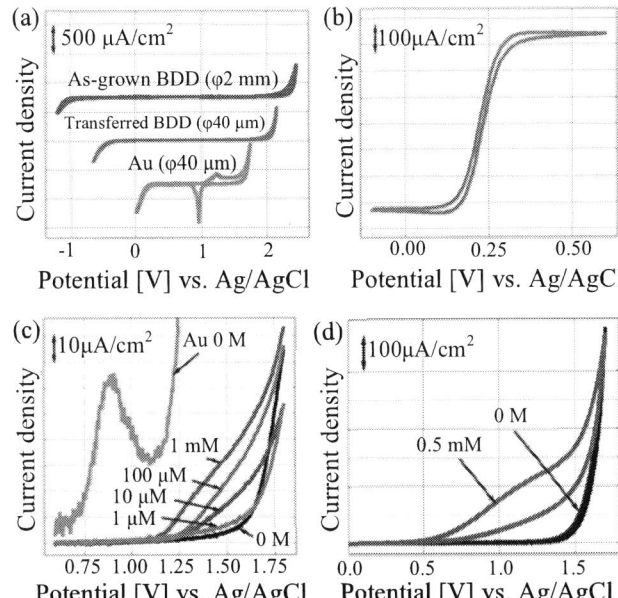

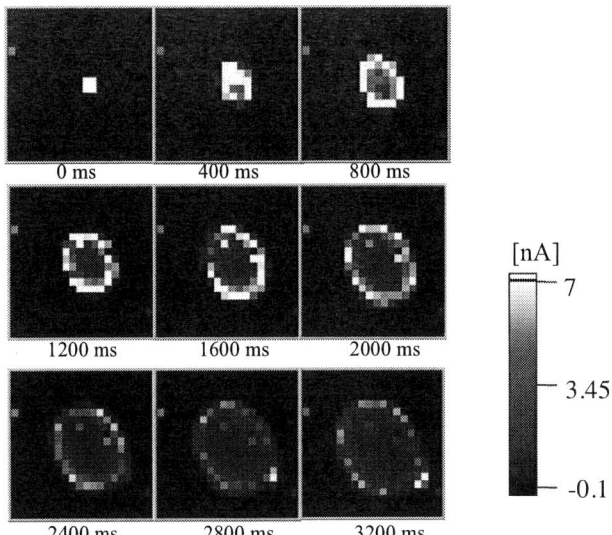

Figure 6: Typical cyclic voltammogram of (a) 0.5 M H_2SO_4, (b) 2 mM ferrocenemethanol in 0.1 M KCl, (c) 1 µM-1 mM histamine in phosphate buffer, (d) 0.5 mM dopamine in Dulbecco's phosphate buffer saline. Scan rates, 20 mVs^{-1}.

0.05 M K_2HPO_4), which was imaged by parallel measurement of oxidation current at 400 points in real time. For the experiment, histamine was dropped into the buffer solution near the center position of the BDD electrode array, to which 1.4 V was applied. The propagation of the large oxidation current area, i.e. histamine diffusion from the center area, was amperometrically monitored. A similar real time imaging is possible for other biomaterials such as living cells and sliced organs to investigate their mechanisms and functions.

Figure 7: 2D imaging of histamine diffusion dissolved in phosphate buffer (0.05 M KH_2PO_4, 0.05 M K_2HPO_4). Color maps correspond to the redox current intensities of 400 electrodes at 1.4 V.

ACKNOWLEDGEMENTS

This work was supported in part by World Premier International Research Center Initiative (WPI Initiative), MEXT, Japan, and the Formation of Innovation Center for Fusion of Advanced Technologies.

REFERENCES

[1] Kumi Y.Inoue, et al., "LSI-based amperometric sensor for bio-imaging and multi-point biosensing," *Lab Chip*, vol. 12, pp. 3481-3490, 2012

[2] Philipp Kruppa, et al., "A digital CMOS-based 24×16 sensor array platform for fully automatic electrochemical DNA detection," *Biosens. Bioelectron.*, vol. 26, pp. 1414-1419, 2010

[3] Meisam H.Nazari, Hamed Mazhab-Jafari, Lian Leng, Axel Guenther, and Roman Genov, "CMOS Neurotransmitter Microarray: 96-Channel Integrated Potentiostat With On-Die Microsensors," *IEEE Trans. Biomed. Circuits Syst.*, vol. 7, pp. 338–348, 2013

[4] Segyeong Joo and Richard B.Brown, "Chemical Sensors with Integrated Electronics," *Chem. Rev.*, vol. 108, pp. 638–651, 2008

[5] Brian N.Kim, Adam D.Herbst, Sung J.Kim, Bradley A.Minch, Manfred Lindau, "Parallel recording of neurotransmitters release from chromaffin cells using a 10×10 CMOS IC potentiostat array with on-chip working electrodes*Biosens. Bioelectron.*, vol. 41, pp. 736–744, 2013

[6] Akira Fujishima et al., *Diamond Electrochemistry*, BKC INC. and ELSEVIER B. V., 2005

[7] John H.T.Luong, Keith B.Male and Jeremy D.Glennon, "Boron-doped diamond electrode: synthesis, characterization, functionalization and analytical applications," *Analyst*, vol. 134, pp. 1965-1979, 2009

CONCLUSIONS

We developed the integration process of the BDD electrodes on the CMOS-based amperometric biosensor array. The BDD electrodes prepared by CVD at high temperature on a Si wafer was successfully transferred to the LSI wafer by film transfer technology with BCB as an adhesion layer. The transferred BDD electrode had a wider potential window compared with a Au electrode, and the background current was also small. Histamine, which has a high oxidation potential, was detected by the BDD electrodes. In addition, the 2-dimentional real time imaging of histamine diffusion process was successfully demonstrated by parallel measurement of oxidation current at 400 points. It is expected that the CMOS-based amperometric sensor array with BDD electrodes will be applied to the detection and/or monitoring of various important biomaterials, which cannot be detected by conventional sensors.

IONIC LIQUID-GATED GRAPHENE FET ARRAY
WITH ENHANCED SELECTIVITY FOR ELECTRONIC NOSE
Akira Inaba, Yusuke Takei, Kiyoshi Matsumoto, and Isao Shimoyama
The University of Tokyo, Tokyo, JAPAN

ABSTRACT

We report the selectivity enhancement of the graphene gas sensor using ionic liquids (ILs). Because the gas absorbability and selectivity of the IL depends on the types of its anion and cation, the graphene sensors covered with different ILs have different gas responses. We fabricated the IL-gated graphene field-effect transistors (ILGFETs) and assembled three ILGFETs with three kinds of ILs into an ILGFET array. Their responses to ammonia, hydrogen peroxide, and iodine were measured. The ILGFET with an IL ([EMIM][BF$_4$]) was more selective for ammonia and iodine and less selective for iodine than that with another IL ([EMIM][TFSI]). The response patterns obtained using the ILGFET array indicated the feasibility of the graphene electronic nose.

INTRODUCTION

Gas sensors based on graphene, atomically-thin hexagonal lattice of carbon, have high sensitivity and low detection limit [1]. The ultimate surface-to-volume ratio enables all the atoms to interact with the surrounding gas, and the highly-ordered crystal structure and high electron mobility reduce electrical noise. However, the attractive property that graphene strongly interacts with various gases diminishes the selectivity [1]. That is why graphene is difficult to apply for electronic nose, which detects gas species as well as concentration.

Several methods to serve selectivity to graphene were reported. For example, the Pt-coated graphene sensor was selective to hydrogen gas due to Pt's catalysis [2]. The reduced graphene oxide (rGO) gas sensor in which rGO flakes were randomly laminated had selectivity depends on the interlayer distance of the flakes caused by the fabrication error [3]. The former and latter have disadvantages in fabrication cost and reproducibility, respectively.

In this study, we focused on ionic liquid (IL) for selectivity enhancement of the graphene gas sensor. ILs are ionically-bonded salts melting at room temperature. The negligible vapor pressure and chemical stability of the IL are suitable for sensing material. Furthermore, IL absorbs various gases, and the gas absorbability selectivity depends on the type of IL [4][5]. Therefore, graphene gas sensor covered with an IL selectively interacts with the gas which is absorbed by the IL, as shown in Figure 1(a). In this method, the additional noble metal deposition is not required, and the selectivity can be controlled by an appropriate choice of the IL. Previously, we reported the fabrication and ammonia gas response of the IL-gated graphene field-effect transistor (ILGFET) at MEMS 2013, but the selectivity had not been studied [6].

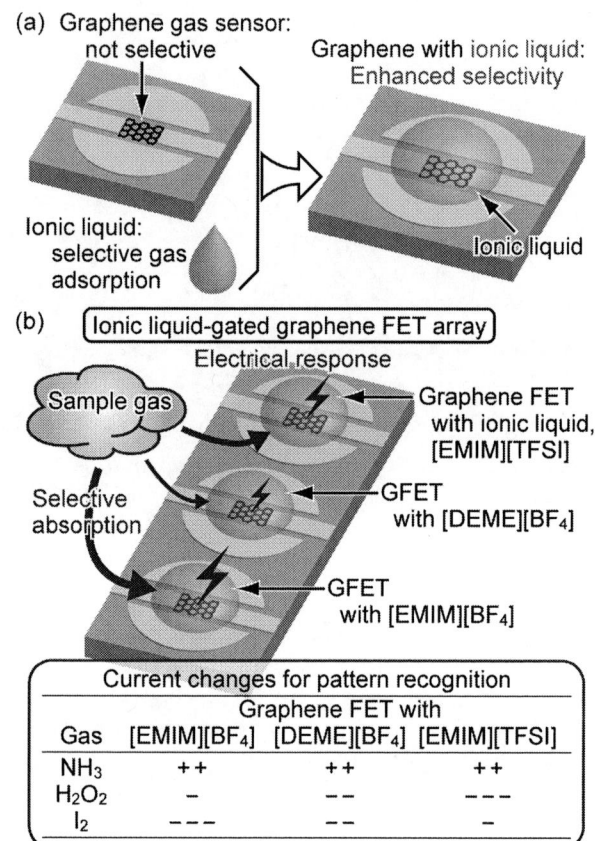

Figure 1: Illustrations of the proposed graphene gas sensor. (a) An ILGFET with enhanced selectivity and (b) an ILGFET array for electronic nose.

Gas	Graphene FET with		
	[EMIM][BF$_4$]	[DEME][BF$_4$]	[EMIM][TFSI]
NH$_3$	++	++	++
H$_2$O$_2$	–	– –	– – –
I$_2$	– – –	– –	–

This paper reports on the experimental investigation of the selectivity of the ILGFETs. The ILGFETs with three different ILs were fabricated and assembled into an ILGFET array. The current-voltage characteristic changes of the array in response to ammonia (NH$_3$), hydrogen peroxide (H$_2$O$_2$), and iodine (I$_2$) gases were measured. The measured response pattern verified the selectivity according to the IL's gas absorbability. We confirmed the feasibility of the graphene electronic nose using the ILGFET array, as shown in Figure 1(b).

FABRICATION

The fabrication process of the ILGFETs, which is shown in Figure 2, is based on that reported at MEMS 2013 but slightly different. First, bilayer graphene was formed on a Si wafer with a thermally oxidized SiO$_2$ layer (Figure 2(i)). We ordered Graphene Platform, Inc. (The Woodlands, Texas, USA) to transfer bilayer graphene onto the SiO2/Si wafer.

978-1-4799-3510-9/14 $31.00 © 2014 IEEE

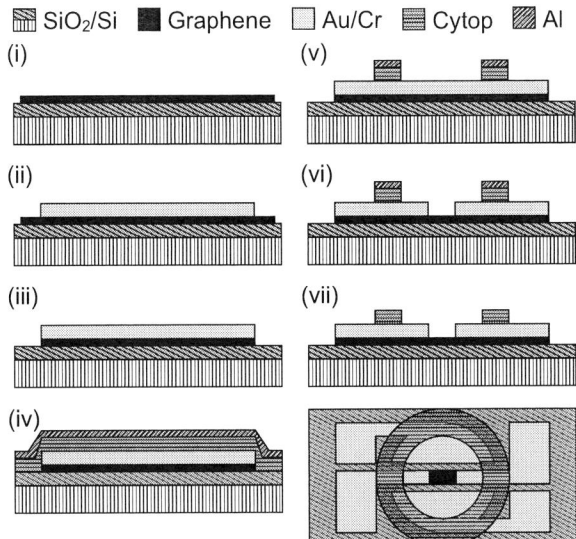

Figure 2: The fabrication process of the ILGFET.

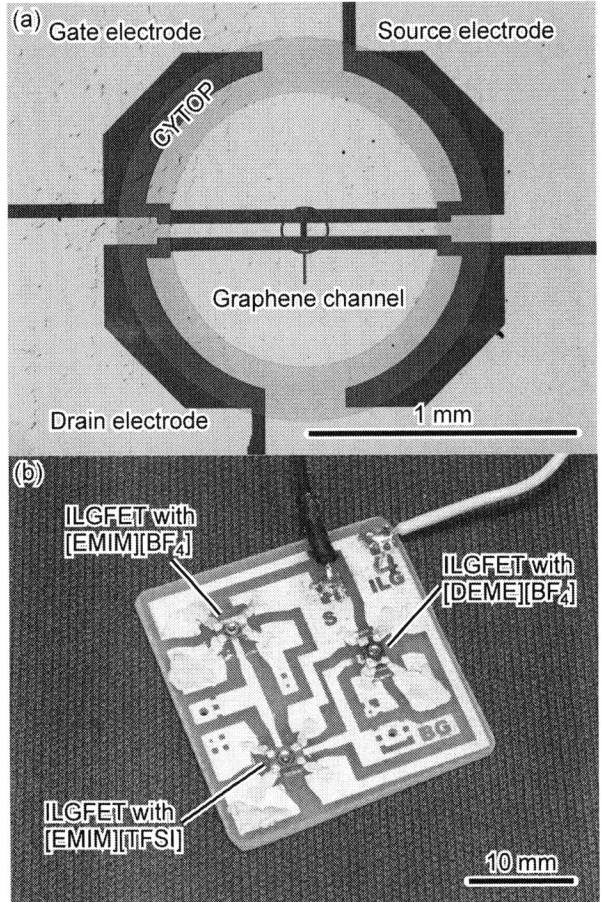

Figure 3: Photographs of (a)the fabricated chip, (b) the graphene channel, and (c) the ILGFET array.

Au/Cr was patterned for drain, source, and gate electrodes by lift-off (Figure 2(ii)). At that time, graphene channel was also covered with Au/Cr as a metal mask to prevent graphene from etching during the following O_2 plasma etching of the undesired graphene (Figure 2(iii)). CYTOP layer was formed

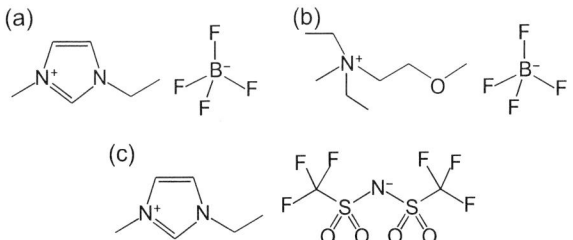

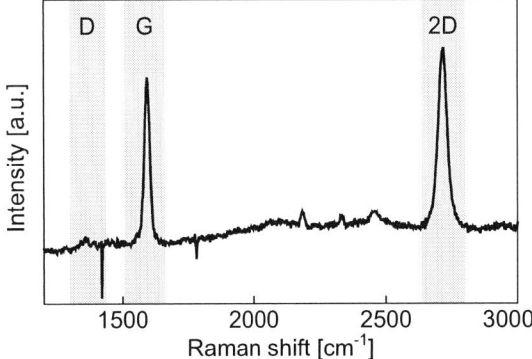

Figure 4: Chemical structures of three ILs used. (a) [EMIM][BF$_4$], (b) [DEME][BF$_4$], and (c)[EMIM][TFSI].

Figure 5: A Raman spectrum of the graphene channel.

by spincoating, and Al was deposited by vacuum evaporation because photoresist was not able to be coated on CYTOP (Figure 2(iv)). The CYTOP and Al layers were patterned by dry and wet etching, respectively (Figure 2(v)). After the Au/Cr mask on the graphene channel was removed by wet etching, Al remained on the CYTOP was etched (Figure 2 (vi,vii)). The doughnut-shaped CYTOP was prepared to hold IL inner the doughnut, because hydrophobic IL tends to expand on the Au/Cr electrodes. The photograph of the fabricated chip is shown in Figure 3(a). The channel was 20 μm in length and 50 μm in width. The average resistivity measured in ambient air was 180 Ω with a standard deviation of 29 Ω.

After the fabrication based on semiconductor process, we attached the processed chip on a printed-circuit board (PCB) using instant glue. Conductive Ag paste was used for the electrical contact between the electrodes and PCB. Finally, we covered the graphene channel and the electrodes with an IL droplet. Here we used three ILs, as shown in Figure 4: 1-ethyl-3-methylimidazolium tetrafluoroborate ([EMIM][BF$_4$]), N,N-diethyl-N-methyl-N-(2-methoxyethyl) -ammonium tetrafluoroborate ([DEME][BF$_4$]), and 1-ethyl-3-methylimidazolium bis(trifluoromethylsulfonyl) -imide ([EMIM][TFSI]). Three ILGFETs with the three ILs were assembled into an ILGFET array on a PCB, as shown in Figure 3(b).

The quality of the graphene was assessed by Raman spectroscopy. Figure 5 shows the Raman spectrum measured using a laser of 488 nm in wavelength and 40 mW in power before the IL placement. The intensity ratio of the 2D to the G band was 1.09, and the full width at half maximum (FWHM) was 39 cm^{-1}. The D band peak was enough small as

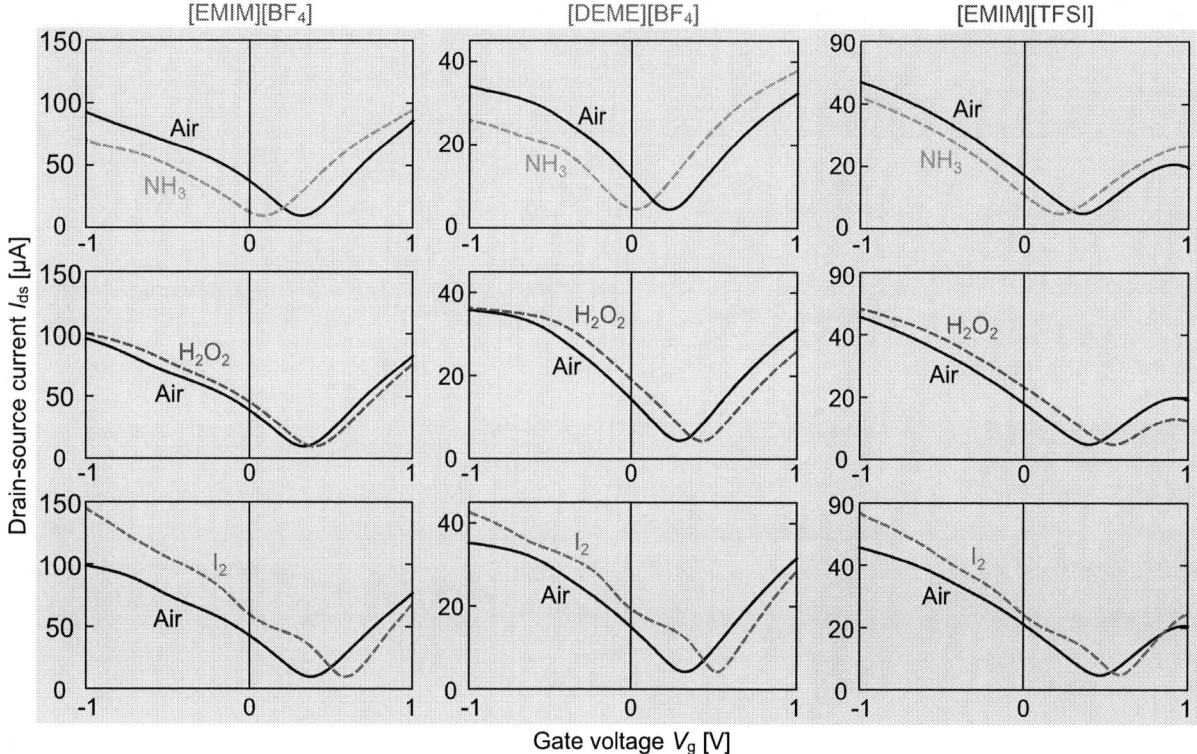

Figure 7: Measured current-voltage characteristics of the ILGFETs with three ILs in air, NH_3, H_2O_2, and I_2.

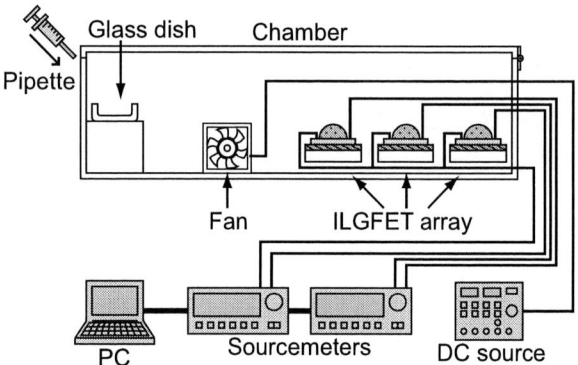

Figure 6: An illustration of the experimental setup.

compared to the G band peak, indicating that the crystal structure has few defects. The intensity ratio near 1 suggested the presence of bilayer graphene [7]. The measured FWHM was less than that of the reported value for bilayer graphene [8]. This result reflects the fact that the bilayer graphene used was fabricated by overlapping a monolayer graphene on another monolayer graphene, in which neither the direction nor the position was controlled. Therefore, the bilayer graphene used was doubly-laminated single layer graphenes rather than an ab-stacked bilayer graphene.

GAS RESPONSES
Experimental Setup

Figure 6 illustrates the diagram of the experimental setup. The fabricated ILGFET array was placed in an acrylic chamber with electrical connects. The electric properties of the ILGFETs were measured using two 2ch sourcemeters (2612A, Keithley Instruments, Inc., Cleveland, Ohio, USA). A drain-source voltage, V_{ds}, of 10 mV was induced to measure the drain-source current, I_{ds}, as the gate-source voltage, V_g, was scanned from 1 to -1 V.

After the I_{ds}-V_g characteristics were measured in ambient air, the sample gas, NH_3, H_2O_2, or I_2, was introduced in chamber by dropping the solution in which the gas dissolved. Then, the I_{ds}-V_g characteristics were measured again. 9.3 μL of 28% NH_3 solution, 13.9 μL of 30% H_2O_2 solution, and 27.4 μL of 0.5 mol/L I_2 solution were used for introducing NH_3, H_2O_2, and I_2, respectively. After dropping the solution into a glass dish, we circulated the atmosphere inner the chamber using an electric fan to quickly achieve a uniform equilibrium state. The gas concentrations were predicted from the solution concentration, volume, and the chamber volume. The calculated concentrations were 300, 300, and 30 ppm for NH_3, H_2O_2, and I_2, respectively.

Results and Discussion

The measured responses of the three ILGFETs for NH_3, H_2O_2, and I_2 were shown in Figure 7. All the I_{ds}-V_g characteristics were V-shaped, i.e. ambipolar, which was a typical feature of graphene field-effect transistors (GFETs). For NH_3, the I_{ds}-V_g characteristics were shifted toward the negative direction in V_g, as reported in the previous studies [6][9]. The negative charge transfer from NH_3 to graphene caused the negative shift. For H_2O_2 and I_2, the I_{ds}-V_g characteristics were shifted toward the positive direction. We considered that the positive charge transfer from the gas to the graphene occurred due to the oxidative property of H_2O_2 and I_2 gases.

978-1-4799-3510-9/14 $31.00 © 2014 IEEE   328

Table 1: ΔV_{dir} of the ILGFETs in response to three gases.

Gas	ILGFET with [EMIM][BF$_4$]	[DEME][BF$_4$]	[EMIM][TFSI]
NH$_3$	-0.24 V	-0.21 V	-0.13 V
H$_2$O$_2$	0.06 V	0.13 V	0.15 V
I$_2$	0.22 V	0.20 V	0.11 V

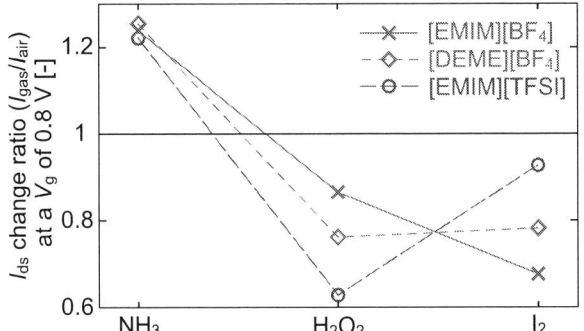

Figure 8: Current change patterns of the ILGFETs with three ILs in response to three gases.

In the case of [EMIM][BF$_4$], dirac voltages, V_{dir}, defined as the V_g for the minimum I_{ds}, were 0.32 and 0.08 V in air and NH$_3$, respectively. Therefore, the V_{dir} shift, ΔV_{dir}, was calculated to be -0.24 V. Such measured ΔV_{dir} were listed in Table 1. ΔV_{dir} in response to NH$_3$ for the ILGFET with [EMIM][BF$_4$] was larger than that for [EMIM][TFSI]. The previous studies about the NH$_3$ absorbabilities of both ILs showed that [EMIM][BF$_4$] can absorb larger amount of NH$_3$ than [EMIM][TFSI] [10][11]. The experimental result is consistent with this fact. In the case of H$_2$O$_2$ responses, ΔV_{dir} for [EMIM][TFSI] was the largest. In the case of I$_2$ responses, ΔV_{dir} for [EMIM][BF$_4$] was the largest. The gas responses of the ILGFET with [DEME][BF4] were the second largest for the three gases. We can concluded that the ILGFET with [EMIM][BF$_4$] is more selective for NH$_3$ and I$_2$ and less selective for H$_2$O$_2$ than that with [EMIM][TFSI].

In order to confirm the feasibility of the gas species detection, the change ratios from the I_{ds} in air, I_{air}, to that in the sample gas, I_{gas}, at a V_g of 0.8 V were plotted, as shown in Figure 8. The obtained response patterns are linearly independent. Therefore, we can estimate the gas species using the response pattern obtained by the ILGFET gas sensor array.

CONCLUSION

We fabricated the ILGFET array gated by three different ILs and measured the gas responses of the I_{ds}-V_g characteristics. ΔV_{dir} of each ILGFET for the same gas were different. The ILGFET with [EMIM][BF$_4$] exhibits selectivity for NH$_3$ and I$_2$ while that with [EMIM][TFSI] is selective for H$_2$O$_2$. We confirmed the feasibility of graphene electronic nose from the current change patterns obtained using the arrayed ILGFETs.

ACKNOWLEDGEMENTS

The photolithography masks were fabricated using the EB lithography apparatus of the VLSI Design and Education Center (VDEC) of the University of Tokyo. A part of this work was conducted in Center for Nano Lithography & Analysis, The University of Tokyo, supported by the Ministry of Education, Culture, Sports, Science and Technology (MEXT), Japan.

REFERENCES

[1] F. Yavari and N. Koratkar, "Graphene-based chemical sensors," *J. Phys. Chem. Lett.*, vo. 3, pp. 1746-1753, 2012.

[2] B. H. Chu, J. Nicolosi, C. F. Lo, W. Strupinski, S. J. Pearton, F. Ren, "Effect of coated platinum thickness on hydrogen detection sensitivity of graphene-based sensors," *Electrochem. Solid St.*, vol. 14, pp. K43-K45, 2011.

[3] A. Lipatov, A. Varezhnikov, P. Wilson, V. Sysoev, A. Kolmakov, A. Sinitskii, "Highly selective gas sensor arrays based on thermally reduced graphene oxide," *Nanoscale*, vol. 5, pp. 5426-5434, 2013.

[4] A. Finotello, J. E. Bara, D. Camper, R. D. Noble, "Room-temperature ionic liquids: temperature dependence of gas solubility selectivity," *Ind. Eng. Chem. Res.*, vol. 47, pp. 3453-3459, 2007.

[5] X. Jin, L. Yu, D. Garcia, R. X. Ren, X. Zeng, "Ionic liquid high-temperature gas sensor array," *Anal. Chem.*, vol. 78, pp. 6980-6989, 2006.

[6] A. Inaba, G. Yoo, Y. Takei, K. Matsumoto, I. Shimoyama, "A graphene FET gas sensor gated by ionic liquid," in *Proc. 26th IEEE Int. Conf. Micro Electro Mechanical Systems (MEMS 2013)*, Taipei, Taiwan, Jan. 20-24, 2013, pp. 969-972.

[7] Z. Ni, Y. Wang, T. Yu, Z. Shen, "Raman spectroscopy and imaging of graphene," *Nano Res.*, vol. 1, pp. 273-291, 2008.

[8] Y. Hao, Y. Wang, L. Wang, Z. Ni, Z. Wang, R. Wang, C. K. Koo, Z. Shen, J. T. L. Thong, "Probing layer number and stacking order of few-layer graphene by raman spectroscopy," *Small*, vol. 6, pp. 195-200, 2010.

[9] M. Gautam, A. H. Jayatissa, "Graphene based field effect transistor for the detection of ammonia," *J. Appl. Phys.*, vol. 112, 064304, 2012.

[10] G. Li, Q. Zhou, X. Zhang, LeiWang, S. Zhang, J. Li, "Solubilities of ammonia in basic imidazolium ionic liquids," *Fluid Phase Equilibr.*, vol. 297, pp. 34-39, 2010.

[11] A. Yokozeki, M. B. Shiflett, "Ammonia solubilities in room-temperature ionic liquids," *Ind. Eng. Chem. Res.*, vol. 46, pp. 1605-1610, 2007.

CONTACT

*A. Inaba, tel: +81-3-5841-6318,
mail: inaba@leopard.t.u-tokyo.ac.jp

LoC SENSOR ARRAY PLATFORM FOR REAL-TIME COAGULATION MEASUREMENTS

Onur Cakmak [1], Necmettin Kilinc [1,2], Erhan Ermek [1], Aref Mostafazadeh [1],Caglar Elbuken [1], Goksenin G. Yaralioglu [3] and Hakan Urey [1]*

[1]Koc University , Istanbul, TURKEY
[2]Gebze Institute of Technology, Kocaeli, TURKEY
[3]Ozyegin University, Istanbul, TURKEY

ABSTRACT

This paper reports a MEMS-based sensor array enabling **multiple clot-time tests in one disposable microfluidic cartridge** using **plasma**. The versatile LoC (Lab-on-Chip) platform technology is demonstrated here for real-time coagulation tests (activated Partial Thrompoblastin Time (aPTT) and Prothrombin Time (PT)). The system has a reader unit and a disposable cartridge. The reader has no electrical connections to the cartridge, which consists of multiple microfluidic channels and MEMS microcantilevers placed in each channel. Microcantilevers are made of electro-plated nickel and actuated remotely using an external electro-coil. The read-out is also conducted remotely by a laser and the phase of the MEMS oscillator is monitored real-time. The system is capable of monitoring coagulation time with a precision estimated at 0.1sec.

INTRODUCTION

Coagulation tests are performed at several scenarios. Clot time is measured before and during all cardiovascular surgeries. Patients with a risk of embolism, stroke or atrial fibrillation require their coagulation time to be measured and adjusted periodically by suitable medication. All patients on anticoagulant therapy should take frequent coagulation tests to keep their coagulation times within specified limits.

Although automated coagulation measurement systems exist in the market, these systems are for hospital or laboratory use. In general practice, patients need to visit a hospital or a central laboratory periodically for coagulation tests. At a clinical environment, a tube of blood is drawn from the patient. Sample is labeled and sent to clinical test laboratory. Upon measurement, results are interpreted by a doctor and the dose of anticoagulant is determined. Such a procedure puts a significant burden on the health-care provider and increases the return time and the cost of the test. To alleviate these problems, some portable clot time measurement devices have been developed [1, 2]. Although these systems are practical and require low sample volume, they measure clot formation indirectly with a potentiometric measurement. In addition, these devices require an electrical connection between the disposable cartridge and the analyzer unit which is prone to failure in long-term. On the

other side, the system developed in this study measures the clot-time directly by monitoring the mechanical properties of blood with a non-contact actuation and detection method.

There have been recent studies to measure clot-time with low sample volumes using MEMS technology. The researchers utilized microfluidic platforms [3], Quartz Crystal Microbalance (QCM) [4, 5] Film Bulk Acoustic Resonator (FBAR) [6] based techniques for clot-time measurements. But they are limited to single tests and not multiplexed.

In our recent works we measured blood plasma viscosity with high sensitivity using the proposed scheme [7, 8]. During blood coagulation, a sudden change in the blood plasma viscosity occurs due to fibrin generation. By monitoring the viscosity over a course of time, this increase can be detected as an indication of coagulation time for the given sample.

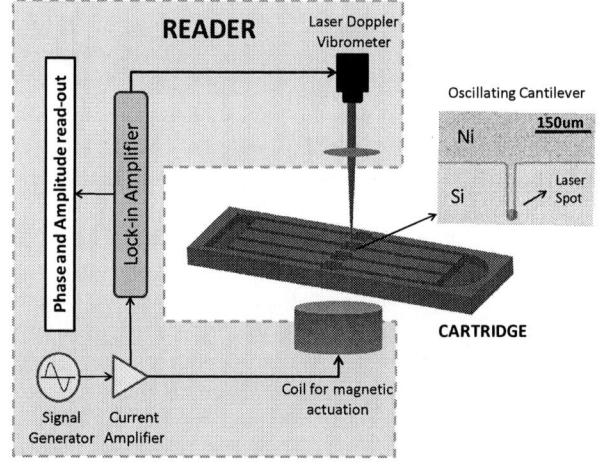

Figure 1: Schematic view of the measurement setup. No electrical connection between cartridge and reader.

The proposed system has a reader unit and a disposable cartridge as illustrated in Figure 1. The disposable cartridge is made of PMMA and the channel geometry is patterned with precision machining. The MEMS microcantilevers are placed in each channel. Microcantilevers are made of electro-plated nickel and actuated using an external electro-coil. The microcantilevers are driven close to their resonant frequencies. The vibration amplitudes are measured with an LDV, and the phase difference between the coil drive and LDV is monitored with a Lock-in Amplifier (SR830). The

---

* C. Elbuken is now with Bilkent University, UNAM, National Nanotechnology Research Center, Bilkent, 06800, Ankara, Turkey

aPTT and PT tests are conducted in functionalized, parallel channels of the same cartridge, one at a time. Simultaneous measurements can also be achieved using a photodetector array. The technology is also scalable to larger number of measurements and to other types of measurements.

MATERIALS AND METHODS
Sensor and Cartridge Fabrication

The nickel cantilevers are fabricated by a simple, one mask process utilizing sputter, photolithography, electrodeposition and wet etching steps, respectively as shown in Figure 2a [8]. RF sputtering is utilized to deposit 20 nm/100 nm Cr/Au layer on a 4″ <100> silicon wafer (step1). The Au layer serves as the seed layer for subsequent Ni electroplating. A positive photoresist (PR), AZ1514H®, layer is coated on the top of Au surface and patterned with UV lithography (step 2). Afterwards the electrodeposition of nickel layer is conducted (step 3). The remaining PR is stripped via AZ100 remover (step 4). To release cantilevers, wet etching of Cr and Au by commercial etchants and finally wet etching of Si in 35% KOH solution at 60°C are conducted (step 5). The etch depth is kept around 40 μm to avoid the squeeze film damping effect. The nickel cantilevers have 150 μm length, 15 μm width and 1.5μm thickness.

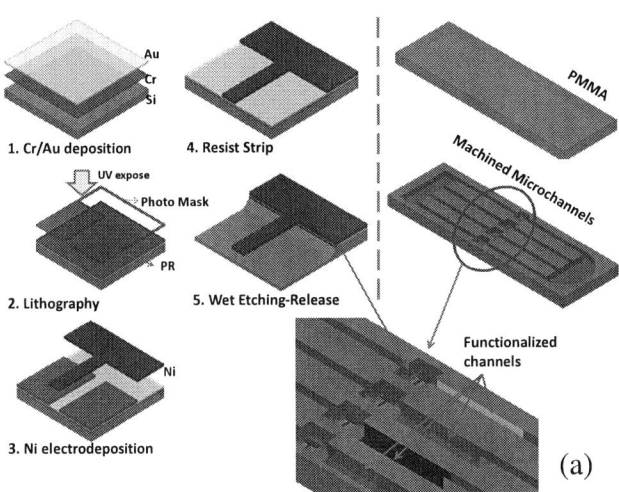

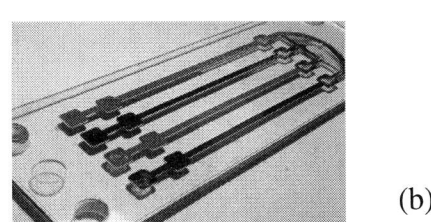

(a)

(b)

Figure 2: (a) Microfabrication of the MEMS and MEMS die assembly within the functionalized microchannels of the cartridge. (b) The cartridge channels filled with colored liquids

The microfluidic channels are patterned on PMMA by precision machining. Microchannels are functionalized by vacuum drying 10 μl of 250 mM $CaCl_2$ on channel sidewalls as shown in Figure 5. Since we used the citrated blood plasma, $CaCl_2$ is necessary to initialize the blood coagulation cascade. After channel functionalization, MEMS dies are epoxy bonded to the cells on the channels. A glass slide is glued on the PMMA cartridge by UV cured epoxy to provide sealing between the parallel channels. Sealing was tested by filling the parallel channels with different colored liquids (Figure 2b).

Experimental Procedure

When the cantilevers are operated in a viscous liquid media, the viscosity influences the resonant frequency and the quality factor of the system. In our previous works we measured the viscosity changes with a sensitivity of 0.01 cP, by tracking frequency [7] or by tracking phase changes at constant frequency [8]. During blood coagulation a sudden change in blood plasma viscosity occurs. When the coagulation is initialized, thrombin acts on fibrinogens which results in the fibrin generation, then fibrins aggregate and form insoluble clots. Changes in the viscosity of plasma were caused by the formation of the fibrin clot [5]. This significant change in the viscosity can be monitored by real-time tracking of the phase and amplitude.

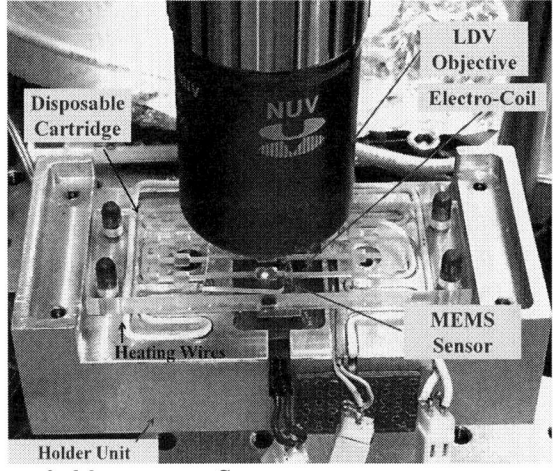

Figure 3: Measurement Setup

During the measurement, the microcantilevers are driven close to their resonant frequencies by using a coil. The resonance frequency of the cantilevers in blood plasma is around 20 kHz whereas the quality factors of the devices are varying between 5-7 for cantilevers on different chips. The vibration amplitudes are measured with a Laser Doppler Vibrometer (LDV). Optical measurement techniques based on grating interferometer, and knife-edge methods have also been successfully implemented in our group. Therefore, the LDV can be changed with other techniques to provide measurement parallelism. The phase difference between the coil drive and LDV is monitored with a Lock-in Amplifier (SR830) and recorded simultaneously using a Labview interface. All experiments are conducted at 37±0.1 °C. Temperature stability is an important aspect since it has a strong effect on viscosity.

978-1-4799-3510-9/14 $31.00 © 2014 IEEE   331

Hence a temperature controller unit with a 0.1 °C precision is built around the cartridge as shown in Figure 3.

Figure 4 shows the long-term phase stability of the system in blood plasma. The cantilever is driven at 21850 Hz. At this frequency the maximum amplitude is obtained. The phase is recorded for 30 minutes. The standard deviation of the phase during this measurement is 0.015° which can be defined as the noise floor of the system.

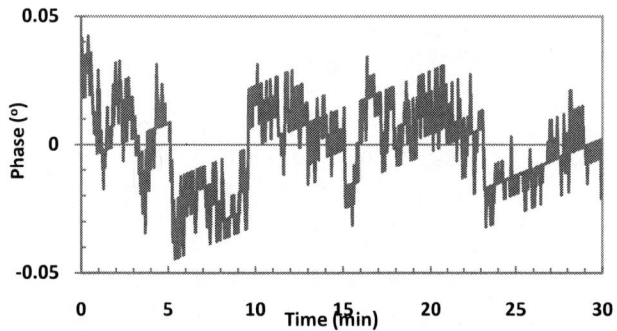

Figure 4: Phase stability result in blood plasma. f=21850 Hz, Standard Deviation=0.015°

During the blood plasma coagulation tests, different channels of the cartridge are treated for aPTT and PT tests as shown in Figure 5. Though the tests are conducted one at a time within the scope of this paper, tests can be performed simultaneously using multiple photodetectors for the readout. Since the electromagnetic actuation is at a constant frequency, using a detector array will enable simultaneous monitoring of the phases of the cantilevers in parallel channels.

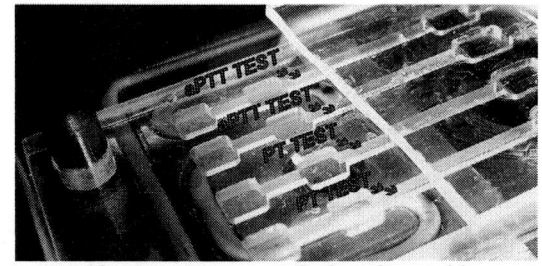

Figure 5 Close-up picture of the disposable cartridge during operation, aPTT and PT tests are conducted in different channels

RESULTS AND DISCUSSION

The coagulation tests are conducted with a sample taken from a healthy donor. The blood is drawn from the healthy donor in a citrated vacutainer tube. The sample is spun at 5000 rpm for 10 minutes in order to separate plasma. For the aPTT test; 70μl of plasma at 37 °C is taken and mixed with 35 μl of aPTT reagent (DIAGEN Micronised Silica Platelet Solution) which is also at 37 °C. This was the standard protocol proposed by the manufacturer. 10 μl of this mixture is pipetted into the first channel.

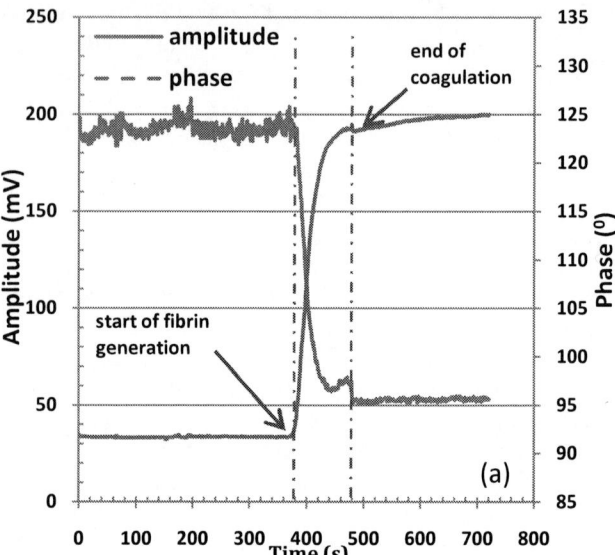

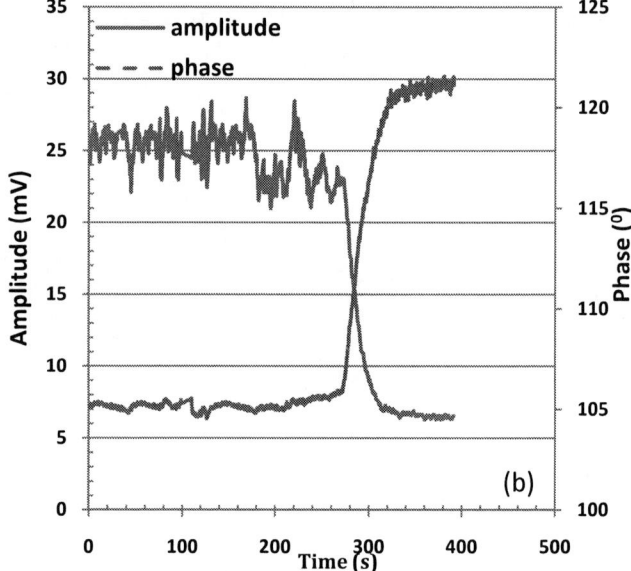

Figure 6 (a) aPTT Test. Phase gives accurate reading of coagulation time. (b)PT Test

Figure 6a shows the change of phase and amplitude. The start and the end of fibrin generation during coagulation can be clearly seen in real-time. The coagulation time, which is around 450 seconds, is significantly different from the standard clinic test reports which are generally 30-40 seconds for normal blood plasma. This disparity can be explained by the difference between the mixing conditions inside the microfluidic channel and manual mixing as well as $CaCL_2$ concentration. As described before, $CaCl_2$ was vacuum dried on the sidewalls of the channel which directly affects the reaction time. $CaCl_2$ concentration can be adjusted to get faster measurement times. For the future use of the proposed system as a clinical instrument, the coagulation time for normal blood plasma will be calibrated with various clinical tests. In the light of these clinical tests a normalized reference value will be obtained. Certain amount of deviation from this reference value will be reported as patient plasma clotting time.

Similar preparation is done for the PT test using PT reagent (DIAGEN Freeze Dried Rabbit Brain Thromboplastin), and 10 µl of PT mixture is pipetted into the second channel. In Figure 6b, the start of fibrin generation and end of the coagulation can be seen also for this test. In both of the tests the phase read-out gives clearer results than the amplitude read-out. Phase measurements are more immune to laser noise.

CONCLUSION

A LoC (Lab-on-Chip) sensor array platform enabling multiple blood test is introduced. The system has independent reader and cartridge units. The cartridge consists of microfluidic channels and MEMS microcantilevers placed inside them. The magnetic microcantilevers are made of electro-plated nickel. The reader unit contains actuation and read-out components. The actuation of the microcantilevers is conducted by an electro-coil in the reader unit and the read-out is done by an LDV. The phase between the input signal driving electro-coil and the output signal gathered from LDV is monitored simultaneously by a Lock-in Amplifier. During coagulation a severe change occurs in the viscosity of blood plasma. We real-time monitored this change by tracking the phase.

In different channels of the cartridge aPTT and PT tests are conducted with blood plasma taken from a healthy donor. Commercial reagents are used. In both tests, fibrin generation and end of coagulation process can be seen clearly. Both the phase and the amplitude can be used for monitoring the coagulation; however phase measurements are more immune to noise and give clearer signal. According to the noise floor, the coagulation time can be measured with a 0.1 s sensitivity which is sufficient for clinical observations. The sample volume used for each test is 10 µl which can be decreased with further design improvements.

Within the scope of this study, the tests are conducted one at a time. In the near future we are planning to handle simultaneous measurements by using a photodetector array. The ultimate goal of this project is to develop a system that is capable of making measurements from whole blood. Towards this goal, in this study we demonstrated the system using blood plasma. In the future, the tests will be conducted with whole blood samples.

ACKNOWLEDGEMENTS

This work is supported by TÜBİTAK 113S074 grant. The authors thank to Selma Bulut, Ibrahim Baris, İ. Halil Kavakli, Erdem Alaca, Umit Celik and Ahmet Oral for valuable discussions about this study. Necmettin Kilinc is supported by TUBITAK- BIDEB National Postdoctoral Research Fellowship Programme.

REFERENCES

[1] Coagucheck XS. (15.11.2013). *Roche, Germany.* Available:
http://www.coaguchek.com/coaguchek_patient/landing
[2] Alere-INRatio. (15.11.2013). *United Kingdom.* Available: http://www.alere.com/ww/en/product-details/inratio-pt-inr-monitoring-system-test-strips.html
[3] B. Ramaswamy, Y. T. Yeh, S. Zheng, "Microfluidic device and system for point-of-care blood coagulation measurement based on electrical impedance sensing," *Sensors and Actuators B: Chemical,* vol. 180, pp. 21-27, 2013.
[4] L. Muller*, et al.*, "Investigation of prothrombin time in human whole-blood samples with a quartz crystal biosensor," *Analytical Chemistry,* vol. 82, pp. 658-663, 2009.
[5] C. Y. Yao, L. Qu, W. L. Fu, "Detection of Fibrinogen and Coagulation Factor VIII in Plasma by a Quartz Crystal Microbalance Biosensor," *Sensors,* vol. 13, pp. 6946-6956, Jun 2013.
[6] W. Xu, J. Appel, J. Chae, "Real-Time Monitoring of Whole Blood Coagulation Using a Microfabricated Contour-Mode Film Bulk Acoustic Resonator," *Journal of Microelectromec. Sys.,* vol. 21, pp. 302-307, 2012.
[7] O. Cakmak*, et al.*, "Microcantilever based disposable viscosity sensor for serum and blood plasma measurements," *Methods,* vol. 63, pp. 225-232, 2013.
[8] O. Cakmak*, et al.*, "MEMS Based Blood Plasma Viscosity Sensor Without Electrical Connections," in *Sensors, 2013 IEEE*, Baltimore, Maryland, USA, 2013, pp. 853-857.

CONTACT

*O. Cakmak, tel: +90-212-338-1772;
ocakmak@ku.edu.tr

PARALLELIZATION OF FISSION AND FUSION+ OPERATIONS FOR HIGH THROUGHPUT GENERATION OF COMBINATORIAL DROPLETS

Helena C.Zec[1], Tushar D. Rane[1], Polly Ma[1], and Dr. Tza-Huei Wang[1,2,3]

[1]Department of Biomedical Engineering, [2]Department of Mechanical Engineering,
[3]Sidney Kimmel Comprehensive Cancer Center, Johns Hopkins University

ABSTRACT

We present a highly parallelized, programmable droplet-based device for high throughput generation of combinatorial droplets. The device features a multi-channel, parallel architecture. Throughput is increased by two orders of magnitude compared to conventional devices [1]. With 32 Hz droplet generation, the projected throughput of this parallelized design is nearly 3 million sample-probe droplets per day on a single device (with 4 replicates of 750 thousand different mixtures). This translates to 240 unique sample-probe mixtures with 4 replicates per minute. This device has the potential to meet the need for low-cost, high throughput screening applications.

INTRODUCTION

Recent research in digital microfluidics has burgeoned as droplets can function as miniaturized reactors in biological and chemical applications. Droplet microfluidic platforms boast the ability to generate many

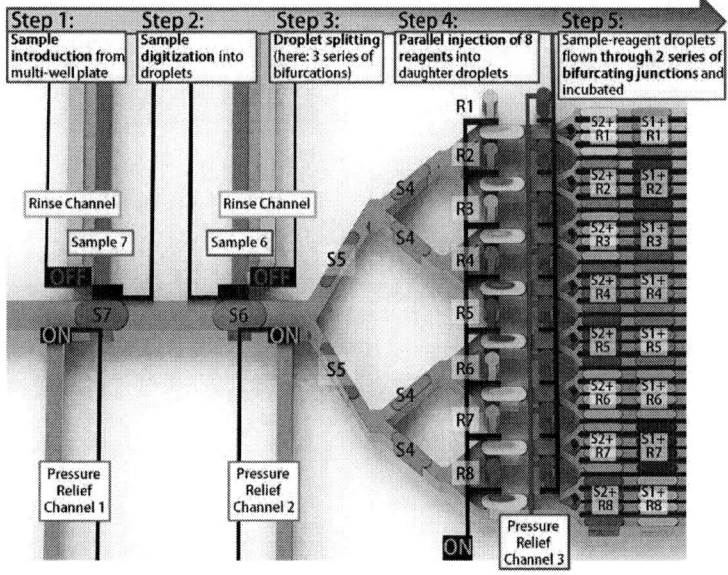

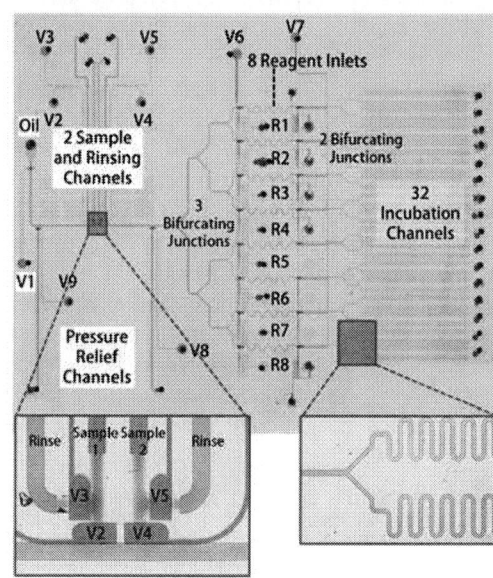

Figure 1: Parallelized droplet fission and fusion platform. Steps 1 – 5 executed in less than a second. Step 1: Samples are injected directly from a multi-well plate. After each sample is injected and processed, inlets are rinsed using rinsing channels running parallel to the sample inlets. An unlimited number of samples can subsequently be injected and processed. Step 2: On-demand digitization of incoming samples into droplets. Volume of sample droplets is controlled by valve opening time and back pressure on inlets. Pressure relief channels up- and down-stream contribute to droplet monodispersity by reducing downstream resistance. Step 3: Droplets undergo 3 bifurcating junctions, producing a total of 8 daughter droplets. A third downstream pressure relief channel ensures homogeneous droplet splitting. Step 4: Probes (R1-R8) are injected directly into the droplets. Probe volume is controlled by valve opening time and back pressure on inlets. Step 5: Sample-probe droplets undergo 2 additional bifurcating junctions to produce a total of 32 droplets of 8 different compositions. Detection can be performed using imaging or parallel confocal fluorescence spectroscopy systems[3].

Figure 2: Microfluidic Device Design Photograph of a prototype device. All fluidic channels are indicated with red. Valve layer (V1-V8) is indicated with green. The oil inlet is connected to the central channel with parallelfusion, fission and incubation regions. There are 2 sample inlets with corresponding rinsing channels. 2 pressure relief channels near the sample inlets ensure that the initial sample droplets are monodisperse by decoupling droplet size from flow resistance of the incubation channel. The third pressure relief channel after the fission regions decouples droplet splitting performance from flow resistance of the incubation channel.

reactions within short time periods. However, most droplet platforms digitize samples into discrete droplets and are limited to the analysis of single samples under homogeneous probe conditions [2]. Such platforms are incapable of addressing the needs of next generation applications which require large libraries of samples and probes. Examples include SNP analysis for crop improvement and genotyping required for identification of genes associated with common diseases. Here, we present a parallelized droplet-based platform for on-demand, combinatorial generation of nanoliter droplets.

By parallelizing fission and fusion modules, throughput is increased by two orders of magnitude. With 32 Hz droplet generation, the projected throughput of this parallelized design is nearly 3 million sample-probe droplets per day on a single device (with 4 replicates of 750 thousand different mixtures).This translates to 240 unique sample-probe mixtures with 4 replicates per minute.

DESIGN

We present a parallel microfluidic emulsification device (**Fig. 1**), which increases throughput while maintaining the ability to generate combinatorial mixtures. *Step 1:* The droplet platform is capable of

accepting an unlimited number of samples from a multi-well plate with a custom-build Serial Sample Loading (SSL) system [3]. *Step 2:* Sample droplets are digitized into smaller daughter droplets (~30 nL). Once a sample has been processed, the sample inlet is rinsed with buffer solution prior to injection of new samples to prevent cross-contamination. *Step 3:* Fission occurs as the daughter droplets flow through 3 serial bifurcating junctions and are split into 8 (2^3) droplets. Flow is halted once the daughter droplets reach the probe injection site by activating the oil valve. *Step 4:* A library of probes is then injected directly into the 8 sample daughter droplets simultaneously. *Step 5:* Post-injection, the 8 sample-probe drops are mixed in serpentine channels and flow through 2 additional serial bifurcating junctions, producing a total of 32 droplets of 8 unique compositions. This entire sequence of operations is carried out in less than a second. Furthermore, the sequence of droplets is maintained on the device. This permits spatial indexing for droplet identification. This precludes the need to include barcodes [4] in each droplet to identify its contents.

FABRICATION

Fig. 2 illustrates our implementation of the scheme

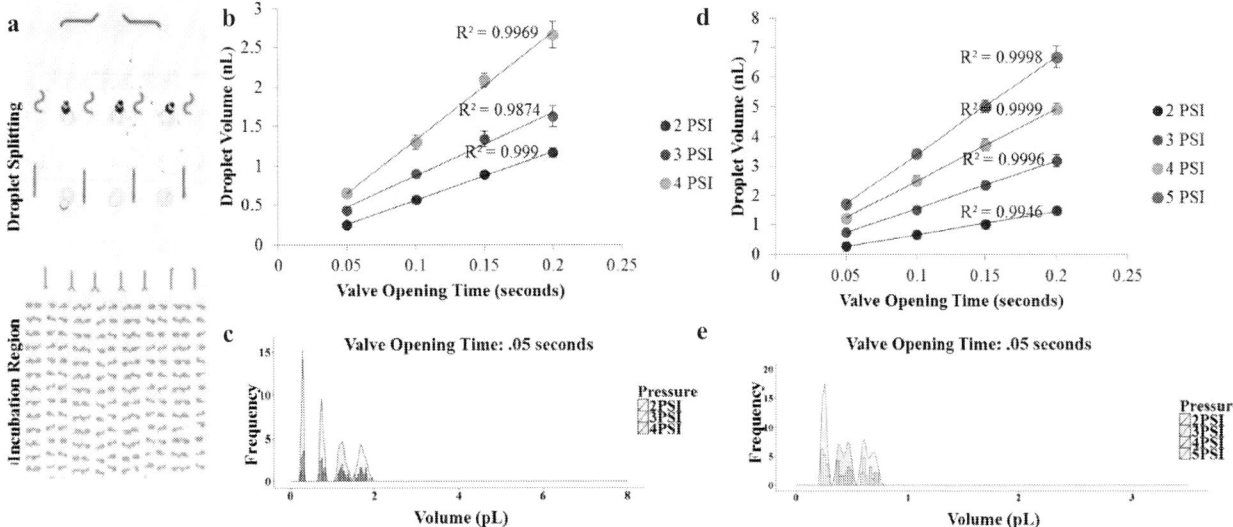

Figure 3: Droplet Fission Measurements a) Micrograph of a section of fission and incubation regions of the device shows sample droplets containing green food dye being split and incubated. b) Plot of sample droplet volume dependence on valve opening time and back pressure. Droplet volume was measured after droplet fission. Droplet volume varies linearly with the valve opening time. Small error bars indicate monodispersity. c) Example of histogram of sample droplet volumes (valve opening time .05 seconds). Histograms are overlayed with Kernel density plots. Three datasets are visible: droplet volumes for 2 PSI, 3 PSI and 4 PSI. All populations of droplets have a narrow distribution indicating monodispersity and are well-separated (no overlap in droplet volumes). d) Plot of probe droplet volume dependence on valve opening time and back pressure. Droplet volume was measured after droplet fission. Droplet volume varies linearly with the valve opening time. Small error bars indicate monodispersity. e) Example of histogram of probe droplet volumes (valve opening time .05 seconds). Histograms are overlayed with Kernel density plots. Four datasets are visible: droplet volumes for 2 PSI, 3 PSI, 4 PSI and 5 PSI. All populations of droplets have a narrow distribution indicating monodispersity and are well-separated (no overlap in droplet volumes).

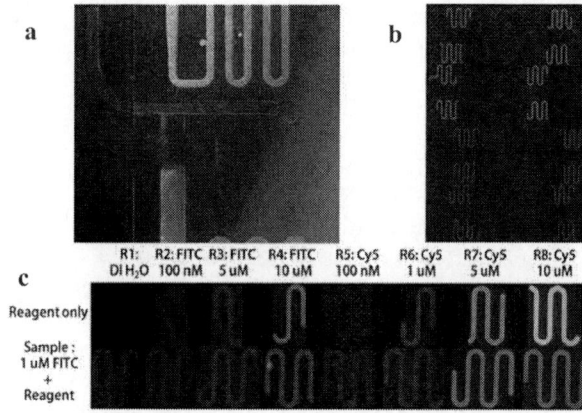

Figure 4: Fluorescent micrographs of the microfluidic device indicating the multiplexing capability of the device. a) Reagent injection: Fluorescent micrograph of sample droplet (green: FITC - 1 µM) at reagent (Cy5 - 5 µM) injection inlet. b) Merged sample-reagent droplets in incubation region. Top 4 rows of droplets were injected with Reagent 8 (Cy5 - 10 µM). Bottom 4 rows were injected with Reagent 7 (Cy5 - 5 µM). c) Fluorescent micrographs of combinatorial droplets. Top panel displays droplets containing only reagents (R1-R8). Bottom panel displays merged sample (1 µM FITC) and reagent (R1-R8) droplets.

proposed in Fig. 1. We designed a microfluidic chip capable of performing sample droplet generation, droplet splitting, droplet merging with probes and droplet detection on a single device.

Fabrication of Molds

Our device design involves regions with two different channel heights. Positive, shallow channels (25 µm) are incorporated near the sample introduction region and probe inlets to allow for valve actuation. The rest of the fluidic layer is 45 µm high. We used SPR220-7 (Rohm & Haas, 25 µm) and SU-8 (Microchem, 3000 series, 45 µm) photoresist as the structural material for fabricating the mold for our device.

Fabrication of PDMS Devices

The microfluidic devices were fabricated using multilayer soft lithography techniques. A modified three-layer fabrication process was developed. Soft lithography was used to make multiple devices from these molds. We used SYLGARD 184 Silicone Elastomer Kit for fabricating our chips. The elastomer and curing agent from the kit are mixed in 10:1 (PDMS supportive material), 15:1 (fluidic), 7:1 (valve) ratio by weight and degassed for approximately 30 minutes before pouring on the respective molds. Once the individual PDMS layers have been assembled, the entire assembly is baked at 80°C for 20 minutes. The solidified polymer is then peeled off and cut into individual chips. Fluidic access holes are then punched into individual chips and the chips are bonded with cover glass (No. 1) using O_2 plasma. All the devices were treated with Aquapel to render their surface hydrophobic. The carrier fluid used to maintain the separation between sample plugs consisted of a perfluorocarbon (FC-3283) and a non-ionic fluorous-soluble surfactant (1*H*,1*H*,2*H*,2*H*-perfluoro-1-octanol) mixed in a ratio of 4:1 by volume.

RESULTS

Device Control

All the inputs on the device were kept under constant pressure, with independent input pressures for 1) carrier fluid input, 2) both sample inlets and 3) all 8 probe inputs. The pressure applied to the sample inlets was directly controlled by the pressure controller used for the SSL system. All the valves on the device were controlled by an array of off-chip solenoid valves, as has been demonstrated earlier. We developed Matlab (Mathworks, Natick MA) software for computer control of the valve array. This software allowed us to execute a predetermined sequence of valve actuation with independent time control for each actuation. The opening of a valve corresponding to an input on the device led to the release of a sample droplet of fluid from that inlet into a central channel on the device. The volume of this droplet could be controlled through variation of the opening time of the valve as well as the back pressure.

Reagents

We estimated the volume of sample and probe droplets generated using the microfluidic device. This volume estimation was performed by processing the images of these droplets using the software ImageJ. For sample droplet volume estimation, we generated droplets made of blue food dye using one of the four reagent inlets on the microfluidic device, until the whole incubation region on the device was full of droplets. The whole device was then imaged using a DSLR camera. The image was imported in ImageJ and cropped to obtain an image of the incubation region on the device. This image was then converted to a binary image using colour thresholding to identify droplets over the background image. An estimate of the droplet area for each droplet in the image was then obtained using the 'Analyze Particles' function. This analysis was limited to particle areas larger than a lower threshold to exclude any particles and occasional satellite droplets from the analysis. The droplet areas thus estimated were then converted to droplet volume using the known depth of the incubation channel region (200 µm).

RESULTS

The device (**Fig. 2**) exhibits excellent sample droplet uniformity for identical droplet generation and fission conditions. We demonstrate the fine control of droplet size generated on the device from an individual sample inlet through variation of pressure and valve opening time in (**Fig. 3**). For these measurements, final droplet size after fission was measured. A unique feature of the device is 3 pressure relief channels. The pressure relief channels decouple both 1) the dependence of initial droplet size generated as well as 2) fission of droplets on the device from the flow resistance of the incubation channel. We demonstrate proof-of-concept generation of 8 combinatorial mixtures of sample plugs and probes on our device (**Fig. 4** using 1 sample and a library of 8 probes. We used different fluorophores with varying concentrations (FITC, Cy5, DI H_2O) to simulate different samples and probes.

CONCLUSIONS

We present an on-demand, parallelized nanoliter droplet-based platform that accepts an unlimited number of sample plugs from a multi-well plate, digitizes these plugs into smaller daughter droplets, performs droplet splitting and robust synchronization-free fusion with a library of probes in parallel. The sequence of sample-probe droplets on the device is maintained, permitting spatial indexing to identify droplet contents. This device combines the precision of valve-based devices while featuring increased throughput. This on-demand platform has the potential to meet the demand for flexible and cost-effective tools that can perform high throughput screening for next generation applications.

ACKNOWLEDGEMENTS

We would like to thank D.J. Shin for his help with writing the LabVIEW software for controlling the SSL system. We also thank the funding support from DARPA (Micro/Nano Fluidics Fundamentals Focus (MF3) Center) National Institutes of Health (R01CA155305, U54CA151838, R21CA173390) and National Science Foundation (0967375, 1159771) and Pioneer HiBred International, Inc.

REFERENCES

[1] Zec H, Rane TD, Wang TH, "A microfluidic droplet platform for multiplexed single nucleotide polymorphism analysis of an array plant genomic DNA samples", in IEEE MEMS, Taipei, Taiwan, January 20-24, 2013, pp. 263-266.

[2] Huebner A, Srisa-Art M, Holt D, et al. *Chem Commun (Camb).* 2007;(12)(12):1218-1220.

[3] Rane TD, Zec, H, Wang TH, *JALA* "A Serial Sample Loading System: Interfacing Multi-well plates with Microfluidic Devices" Journal of Laboratory Automation, 17(5), 370-377, 2012

[4] Brouzes E, Medkova M, Savenelli N, et al. *Proc Natl Acad Sci* U S A. 2009;106(34):14195-14200.

TOWARDS ON-CHIP CHEMICAL REACTION MONITORING BY EWOD IMPEDANCE MEASUREMENT

Xiaoxiao Ma[1,2], Supin Chen[3], Chang-Jin "CJ" Kim[3,4], R. Michael van Dam[1,2,3]*

[1]Department of Molecular and Medical Pharmacology, [2]Crump Institute for Molecular Imaging,
[3]Bioengineering Department, [4]Mechanical and Aerospace Engineering Department
University of California, Los Angeles (UCLA), Los Angeles, CA 90095, U.S.A.

ABSTRACT

This paper demonstrates the use of impedance sensing in an electrowetting-on-dielectric (EWOD) chip to monitor the progress of a chemical reaction. Sensing is achieved by measuring the current through a sample droplet with the aid of a lock-in amplifier. Monitoring of the reaction of HCl with NaOH (i.e., during a titration) was demonstrated. A sharp minimum in the impedance signal was observed at the titration equivalence point, enabling the endpoint to be determined electronically. Future work is aimed at extending reaction monitoring to additional acid-base systems and other classes of reactions.

INTRODUCTION

The EWOD platform has been shown to be well-suited to microscale chemical synthesis, especially when working with scarce reagents such as short-lived radiolabeled molecular imaging tracers and natural products[1, 2]. However, one limitation is the current lack of available on-chip analytical methods to measure reaction progress or outcome, and often chemical intermediates or products must be analyzed off-chip by conventional, macroscale instruments [3]. Doing so has many disadvantages. For example, when working with small volumes, it can be difficult to take only a small representative sample of the reaction mixture. In addition, transfer between microreactors and macroscale instruments can lead to substantial sample loss. Offline analysis also limits the efficiency of the entire synthesis workflow: the prolonged time for analysis or evaluation of each step makes the platform unsuitable for high-throughput optimization or multi-step fast syntheses.

Thus, there is an unmet need for on-chip sensors for *in situ* monitoring of reaction kinetics or progress. In addition to addressing the issues above, integrated analysis may enable real-time process control via feedback signals for optimal process performance.

The integration of electrical detection of liquid is a natural choice because the necessary electrodes can be fabricated by the same processes used to make electrodes for droplet actuation. In a two-plate EWOD device, by measuring the impedance between an actuation electrode in one substrate and the ground electrode in the other substrate, it is possible to determine the location of a droplet, to measure the droplet volume, or even to distinguish droplet compositions [4-7]. Such information has been used for feedback control of droplet actuation [4, 8].

Here we apply impedance sensing to *in situ* chemical

reaction monitoring. Using as an example reaction the neutralization of NaOH with HCl, we demonstrate a signal that varies with reaction progress and show that the equivalence point can be determined electronically.

SYSTEM AND EXPERIMENTAL

Impedance measurement system

Figure 1(a) shows a schematic of the EWOD impedance measurement system [7]. A single $1.2k\Omega$ resistor was placed in series with the actuation circuit of the EWOD device. Because the resistance is much smaller compared with the equivalent impedance of the EWOD chip ($>M\Omega$), the effect of this measurement resistor (R_m) on the EWOD actuation voltage is negligible. For impedance sensing, the voltage V_m across R_m is monitored while actuation potential is applied to the droplet to determine the total current in the circuit. A voltage, V_{ref}, was generated by digital acquisition interface (DAQ PCI-6115, National Instruments, Inc.) and amplified to a voltage V_{in} of 10V with a frequency of 50 kHz.

The signals V_{ref} and V_m were measured by the DAQ. (The DAQ has the capability for simultaneous sampling on multiple channels, avoiding the introduction of phase errors.) The amplitude and phase of V_m were measured by a software-implemented lock-in amplifier using V_{ref} as a reference signal.

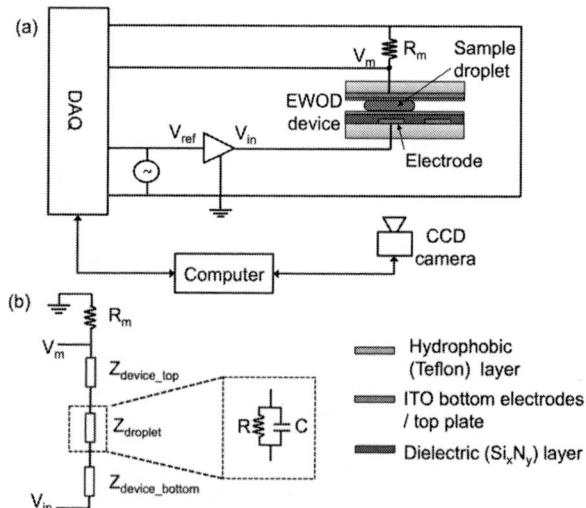

Figure 1: (a) Schematic of impedance measurement in EWOD system; (b) Electrical circuit model for a sample droplet within a two-plate EWOD device.

978-1-4799-3510-9/14 $31.00 © 2014 IEEE   338

EWOD chip

The EWOD chip was a typical two-plate construction, fabricated as described in Keng et al. [7], except that hydrophobic layers were omitted.

The size of each sensing electrode size was 2mm by 2mm. With a gap height, d, of 75 µm, the approximate sensing volume is 0.3 µL. To minimize the parasitic capacitance between the sensing site and other electrodes on the chip, an effect that is not yet well characterized, adjacent electrodes were spaced 5mm away. This spacing is too large for actuation of a sample from one electrode to another, and therefore reagents were manually introduced onto the chip via pipette. Omission of the hydrophobic layer facilitated this process because the samples could wet the exposed silicon nitride surfaces of the device.

Reagents

Sodium hydroxide standard solution (NaOH, 1.999M), hydrochloric acid standard solution (HCl, 1M), sodium chloride (NaCl, ≥99.999%), and phenolphthalein (ACS reagent grade) were purchased from Sigma-Aldrich.

Impedance measurements

NaOH and HCl solutions with concentrations of 0.1M, 0.08M, 0.05M, 0.03M, and 0.01M were prepared by dilution of stock solutions with deionized water. A 1M NaCl stock solution was also prepared and then diluted to the same concentrations.

First, impedance of individual samples was characterized. For each measurement, one droplet of solution (2µL) was loaded by pipette onto the sensing electrode nearest the edge of EWOD device. This volume ensured a full coverage of the entire sensing electrode. For volumes greater than the sensing electrode, the signal is expected to be independent of the droplet volume. After loading, voltage was applied to the droplet and the complex value of V_m was recorded. Each measurement was repeated in triplicate and the average and standard deviation calculated. This procedure was repeated for each concentration of each species (NaOH, HCl, NaCl).

For impedance measurements of the titration reaction, samples with different volume ratios of 0.1M HCl to 0.1M NaOH from 0.1 to 1.5 were mixed off-chip, allowed to equilibrate to room temperature and then loaded into the device for measurement.

To visually stage the progress of the titration reaction, 0.1M NaOH was premixed with 0.02% v/v pH indicator phenolphthalein (0.5% wt phenolphthalein in 50:50 ethanol/water). Phenolphthalein is pink when pH is in the range of 8.2~12 and turns to colorless when pH <8.2.

THEORETICAL
Electrical circuit model

Figure 1(b) shows an equivalent electrical circuit model for impedance measurement of a standard two-plate EWOD device containing a sample droplet fully covering a sensing electrode within the device.

Since the droplet lateral size is much larger than the droplet height (d), the droplet impedance can be approximated as a parallel plate capacitor (C) in parallel with the resistance (R) of the fluid. The equivalent capacitance and resistance can be expressed as:

$$C = \frac{\varepsilon_0 \varepsilon_r A}{d} \qquad R = \frac{\rho d}{A} \qquad (1)$$

where ε_0 is the vacuum permittivity, ε_r is the relative permittivity of the droplet, ρ is the droplet resistivity, and A is the area of the electrode. Thus the inverse impedance of the droplet is:

$$\frac{1}{Z} = \frac{1}{R} + j\omega C = \frac{A}{d}\left(\frac{1}{\rho} + j\omega \varepsilon_0 \varepsilon_r\right) \qquad (2)$$

where $\omega = 2\pi f$ and f is the frequency of the applied voltage V_{in}. At the typical kHz working frequency range, the magnitude of the inverse impedance can be approximated as:

$$\left|\frac{1}{Z}\right| = \frac{A}{d}\sqrt{\frac{1}{\rho^2} + \omega^2 \varepsilon_0^2 \varepsilon_r^2} \approx \frac{A}{d \cdot \rho} = \frac{A}{d} \cdot \kappa \qquad (3)$$

showing a linear dependence on the droplet electrical conductivity κ.

The total impedance in the measurement circuit is the sum of the impedances of the droplet, the two plates of EWOD device, and R_m. Thus the output voltage V_m is expected to be:

$$V_m = \frac{V_{in}R_m}{Z_{total}} = \frac{V_{in}R_m}{Z_{droplet} + Z_{device\_top} + Z_{device\_bottom} + R_m} \qquad (4)$$

If the droplet is non-conductive or has low conductivity (e.g., organic solvents, water, very dilute electrolytes), the droplet impedance can be many orders larger than the impedance of the EWOD device and the measurement resistor. Consequently, V_m is mainly determined by the droplet impedance, which is approximately linearly related to the droplet conductivity: $V_m \approx \dfrac{V_{in}R_m}{Z_{droplet}}$ (5)

If the droplet is highly conductive (e.g., strong electrolytes, some ionic liquids, liquid metals), the droplet impedance becomes negligible compared to other impedances and V_m is mainly determined by the device impedance, with approximation as:

$$V_m \approx \frac{V_{in}R_m}{Z_{device\_top} + Z_{device\_bottom}} \qquad (6)$$

Acid-base titration

Titration is a common method of quantitative chemical analysis that is used to determine the unknown concentration of an identified analyte. Acid-base titrations depend on the neutralization between an acid and a base in solution. Usually, in addition to the sample, an appropriate indicator is added to reflect the pH range of the equivalence point, i.e. the point when equivalent quantities of acid and base are mixed and the solution is neutralized. In this paper, the reaction of HCl with NaOH is demonstrated:

$$NaOH + HCl \rightarrow NaCl + H_2O \qquad (7)$$

NaOH is a strong base, meaning that in solution, the ions Na^+ and OH^- completely dissociate. Similarly, HCl is a strong acid, which dissociates to H^+ and Cl^- ions in solution. During the titration, HCl is added to a solution of NaOH until the concentrations of H^+ and OH^- are equal and the solution is a neutral (pH = 7) solution of NaCl. If insufficient acid is added, excess OH^- ions remain and pH > 7; and if excess acid is added, excess H^+ ions are in solution and pH < 7.

During the process of an titration, the pH value of the mixture is expected to change as in Figure 3(b) (blue curve) (pH = $-\log_{10}[H^+]$). Before the equivalence point, the solution contains excess NaOH and some NaCl and thus the pH value is similar to the original NaOH solution. Around the titration equivalence point, the H^+ and OH^- are almost neutralized, the concentration of NaOH is low, the solution primarily consists of NaCl. Just past the equivalence point, the solution is mostly NaCl plus a small amount of HCl. As more acid is added, the concentration of HCl increases and that of NaCl decreases slightly (due to increased total volume as HCl is added), and the pH approaches that of the HCl solution being added. The graph of theoretical pH exhibits the steepest slope at the equivalence point.

Knowing the initial volume of the NaOH sample, total volume of HCl titrant added to reach the equivalence point, and concentration of the titrant solution, the concentration of the original NaOH solution can be determined.

RESULTS AND DISCUSSION
Characteristic curves of pure electrolytes

The individual impedance signals for different concentrations of HCl, NaOH, and NaCl were measured (Figure 2) and observed to be increasing functions of concentration. Comparing the overall signals of these three species, the order of measured signal amplitude, $|V_m|$, from high to low is HCl, NaOH and NaCl, which also agrees with the sequence of theoretical molar conductivity values [9].

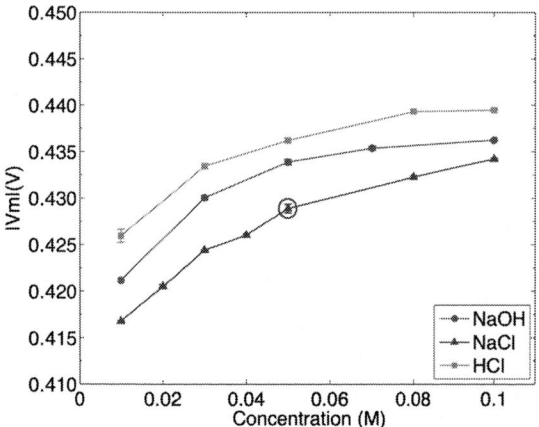

Figure 2: Impedance characterization of reactants (NaOH and HCl) and product (NaCl) for acid-base neutralization reaction. The data point circled is the expected impedance of the equivalence point.

It can be seen that the measurement sensitivity is lower at higher concentrations of acid, base, and salt. We suspect that this is mainly due to the coupling with impedance of EWOD device in the circuit. For low concentration electrolytes at typical EWOD working frequency range (kHz), the Equation 5 applies, and, combining with Equation 3, $|V_m|$ is approximately proportional to the droplet conductivity. In the concentration range used in our experiments, the conductivity of the three electrolytes is almost linear with concentration [9]. Thus, $|V_m|$ is expected to increase with increasing electrolyte concentrations. However, at higher concentrations, $|V_m|$ is mainly determined by other, fixed, impedances (Equation 6) and thus the curves "saturate".

Monitoring NaOH-HCl titration

We demonstrate the titration of a sample of NaOH with 0.1M HCl. In a typical titration, the concentration of NaOH would be unknown, but we choose a known concentration (0.1M) so the behavior of the impedance signal as a function of added volume can be compared to the known titration endpoint.

As HCl is added, $|V_m|$ is observed to decrease (Figure 3a), reach a sharp minimum, and then start to increase again. The minimum is reached when an equal volume of 0.1 HCl has been added to the original 0.1M NaOH solution, i.e. at the titration endpoint. The derivative of this data (Figure 3(b), red curve) exhibits a discontinuity at this minimum. We are currently developing a model to understand the relationship between the impedance signal of each individual component and the impedance signal of the mixtures of components (i.e. various points during the titration reaction). We are also performing additional experiments to determine whether the identification of equivalence point by an extremum or discontinuity can be extended to other acid-base systems.

At the endpoint, the impedance signal is expected to arise from the concentration of NaCl alone. The expected concentration of NaCl at this point is 0.05M, and indeed $|V_m|$ of the mixture matches well with the impedance signal for NaCl alone at this concentration.

Optical micrographs taken for each mixture (Figure 3a) ratio showed a pink color (indicating pH in the range of 8.2-12) when the amount of HCl added was less than the volume of NaOH. When the ratio of HCl to NaOH volumes was greater than or equal to 1, the droplet turned colorless (indicating pH < 8.2). The color change of the indicator very close to the equivalence point acts as further confirmation of the titration endpoint.

This demonstration shows that impedance measurements in an EWOD device can be used to monitor titration reactions in real-time and to determine the titration endpoint. Unlike the pH indicator used in a traditional titration, which provides a visual indication only when the equivalence point is reached, the impedance signal can be monitored during the entire titration process.

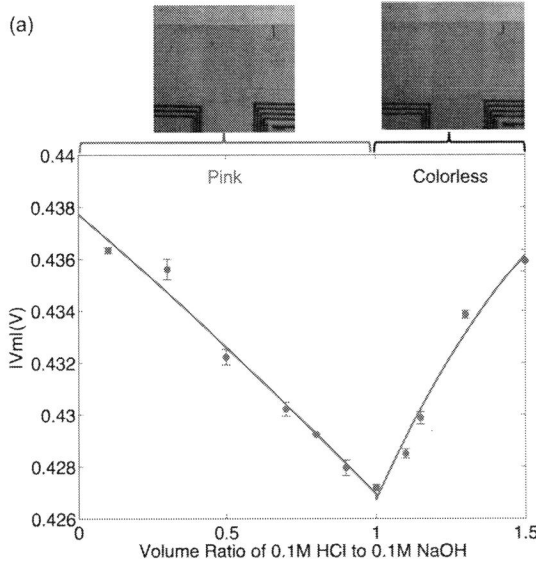

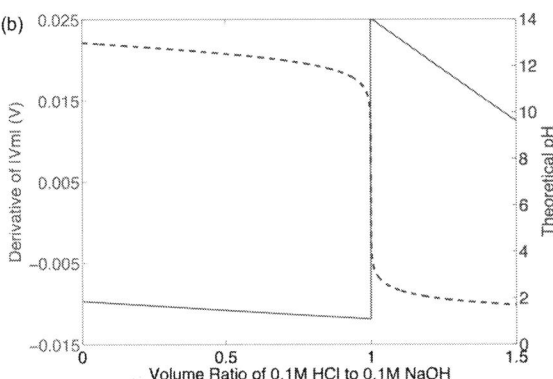

Figure 3: Demonstration of on-chip monitoring of NaOH-HCl neutralization by EWOD impedance sensing. (a) Measured signal amplitude (n=3) vs. the volume ratio of HCl to NaOH. Insets are example micrographs of droplets with the pH indicator phenolphthalein; (b) The discontinuity in the derivative of the measured signal amplitude (red) seems to be well correlated with the inflection point in the theoretical pH curve (blue).

CONCLUSIONS

Using an acid-base titration (HCl/NaOH) as an example, we demonstrated that the progress of a chemical reaction could be monitored by measuring the droplet impedance in an EWOD device, and that the equivalence point of this neutralization reaction could be identified. To the best of our knowledge, this is the first demonstration of on-chip measurements to monitor the progress of a chemical reaction on an EWOD chip.

Leveraging these results, as well as established methods of droplet actuation (dispensing, transporting, and mixing), we are developing a system for fully-automated on-chip titration assays.

ACKNOWLEDGMENTS

This work was funded by the Department of Energy, Office of Biological and Environmental Research (DE-SC0001249). We thank Dr. Saman Sadeghi, Huijiang Ding, Dr. Gaurav Shah and Dr. Pei Yuin Keng for helpful discussions, and Xiang Yin for the assistance in data processing.

REFERENCES

[1] P. Y. Keng, S. Chen, H. Ding, S. Sadeghi, G. J. Shah, A. Dooraghi, M. E. Phelps, N. Satyamurthy, A. F. Chatziioannou, C.-J. Kim, and R. M. van Dam, "Micro-chemical synthesis of molecular probes on an electronic microfluidic device," *Proc. Natl. Acad. Sci.*, vol. 109, no. 3, pp. 690–695, 2012.

[2] M. J. Jebrail, A. H. C. Ng, V. Rai, R. Hili, A. K. Yudin, and A. R. Wheeler, "Synchronized synthesis of peptide-based macrocycles by digital microfluidics," *Angew. Chem. Int. Ed.*, vol. 49, no. 46, pp. 8625–8629, 2010.

[3] A. E. Kirby and A. R. Wheeler, "Microfluidic origami: a new device format for in-line reaction monitoring by nanoelectrospray ionization mass spectrometry," *Lab. Chip*, vol. 13, no. 13, pp. 2533–2540, 2013.

[4] J. Gong and C.-J. Kim, "All-electronic droplet generation on-chip with real-time feedback control for EWOD digital microfluidics," *Lab. Chip*, vol. 8, no. 6, pp. 898–906, 2008.

[5] T. Lederer, S. Clara, B. Jakoby, and W. Hilber, "Integration of impedance spectroscopy sensors in a digital microfluidic platform," *Microsyst. Technol.*, vol. 18, no. 7–8, pp. 1163–1180, 2012.

[6] M. J. Schertzer, R. Ben-Mrad, and P. E. Sullivan, "Using capacitance measurements in EWOD devices to identify fluid composition and control droplet mixing," *Sens. Actuators B Chem.*, vol. 145, no. 1, pp. 340–347, 2010.

[7] S. Sadeghi, H. Ding, G. J. Shah, S. Chen, P. Y. Keng, C.-J. Kim, and R. M. van Dam, "On chip droplet characterization: a practical, high-sensitivity measurement of droplet impedance in digital microfluidics," *Anal Chem*, vol. 84, no. 4, pp. 1915–1923, 2012.

[8] S. C. C. Shih, R. Fobel, P. Kumar, and A. R. Wheeler, "A feedback control system for high-fidelity digital microfluidics," *Lab. Chip*, vol. 11, no. 3, p. 535, 2011.

[9] W. M. Haynes, D. R. Lide, and T. J. Bruno, *CRC Handbook of Chemistry and Physics 2012-2013*. CRC Press, 2012.

CONTACT

*Xiaoxiao Ma, Tel: +1-310-794-1731;
Email: xxma@mednet.ucla.edu

3D SOLID-STATE SUPERCAPACITORS OBTAINED BY ALD COATING OF HIGH-DENSITY CARBON NANOTUBES BUNDLES

Giuseppe Fiorentino[1], Sten Vollebregt, F.D. Tichelaar[2], Ryoichi Ishihara and Pasqualina M. Sarro
[1]Delft University of Technology, ECTM-DIMES, Delft, The Netherlands
[2]Delft University of Technology, TNW, Delft, The Netherlands

ABSTRACT

A three-dimensional solid-state miniaturized supercapacitor based on double conformal coating of Multiwalled Carbon Nanotubes (MWCNTs) bundles is presented. Atomic Layer Deposition (ALD) is used to deposit Al_2O_3 as dielectric layer and TiN as high aspect-ratio conformal counter-electrode on 2μm long MWCNTs bundles. The devices are realized using an IC wafer-scale manufacturing process and show a remarkable volumetric capacitance density value of 12mF/cm^3 with high reproducibility (≤0.3E-12F deviation). The small footprint (100μm^2 to 625μm^2), a thickness of only 2μm, the extremely high capacitance density and the novel and easy-to-integrate fabrication process make it possible to realize high performance energy storage micro-devices.

INTRODUCTION

There is a great demand for the manufacturing of high performance, light and small size electronic devices, such as mobile phones and wireless sensor networks. The energy requirement of such systems is increasing in parallel with their ability to perform a large variety of operations, driving the necessity of new generation of compact and efficient energy storage device. The overall properties of conventional electrical components can be effectively enhanced using the typical characteristic of nanomaterials, like the accurate dimensioning and the huge surface area to volume ratio. This feature is particularly relevant for the realization of thin and compact solid state supercapacitors [1,2]. These components are very attractive as power sources because the physical, rather than chemical energy storage avoids undesirable effects, such as hysteresis, short cycle-life and low rates of charge and discharge. In fact, compared to a conventional electrochemical supercapacitor that stores charges in electric double layers on the electrode area, solid state supercapacitors only use surface charge to accumulate the power. The electrochemical capacitors have then a larger energy density storage on the electrode surfaces, but power density is limited as mass transport of ions and/or redox reactions are less effective and slow [4]. Furthermore, the use of electrolyte capacitors, such as electric double-layer capacitors (EDLCs), is still a challenge in particular applications such as space and automotive[5]. Indeed, despite 60 years of studies, the different environmental conditions (e.g. high temperature) these applications are subjected to, make it difficult to find a limit the choice of suitable electrolyte [4]. The use of solid state supercapacitors with high aspect ratio nanostructured electrodes can then represent the solution to achieve both

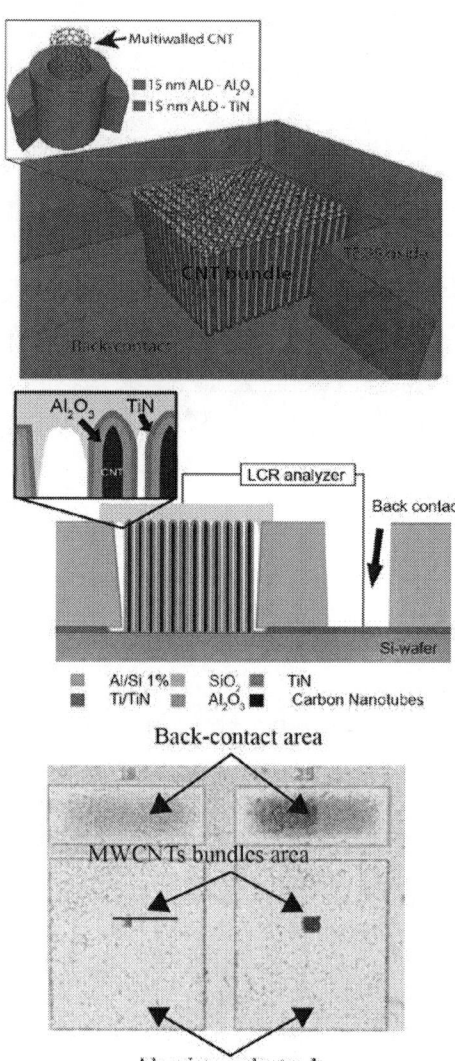

Figure 1 : Top) A 3D view of the supercapacitor. The coated CNTs bundle is buried in a TEOS oxide layer. The inset shows a detail of the double ALD coating of the CNTs. Cross-section view of the device. Bottom) Optical micrograph of the capacitors. Two different bundles sizes (10x10, and 25x25 μm^2) are shown. The red line indicates the cross section for the TEM shown in Fig.2.

high power and high energy density. Such high aspect ratio 3D scaffolding electrodes can indeed enhance the amount of stored charge, enabling at the same time a rapid motion of the electric charges, with speeds limited only by the external RCs circuit [2-4].

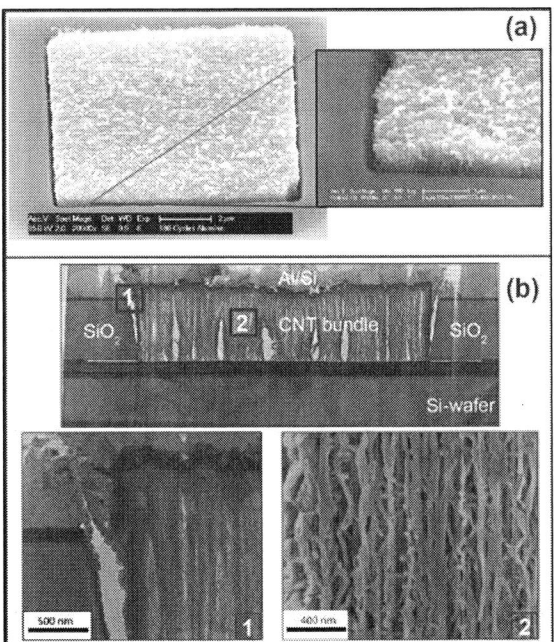

Figure 2 : a) Top view of the CNTs bundle after the double ALD coating; b) TEM cross-section of the supercapacitor. The close ups corresponding to the positions 1 and 2 clearly show the uniform ALD coating on the CNTs.

Furthermore, the ability to fabricate very thin (<5 μm) supercapacitors using a low temperature IC wafer-scale manufacturing process can represent a technological breakthrough in applications such as hybrid electrical engines for vehicles, pulse-power applications in the aerospace and defense industries [5].

Here we present three-dimensional solid state supercapacitors (Fig. 1) based on the use of multiwalled carbon nanotubes bundles (MWCNTs) as first electrode, Aluminium Oxide (Al_2O_3) as dielectric and Titanium Nitride (TiN), both deposited by Atomic Layer Deposition, as high aspect-ratio conformal counter-electrode. Impedance measurements are employed to estimate the performance of the devices, comparing the measured data with a circuital model and extracting the relevant parameters. Several devices with different geometries were measured, showing a very high capacitance density and a high reproducibility.

EXPERIMENTAL

The supercapacitors are realized starting from MWCNTs bundles that work as first electrode (Fig. 1). A100 mm p- type Si wafer substrate is coated with a 500 nm of Ti and 50 nm of TiN, sputtered using a Trikon (SPTS) Sigma sputter coater. A 2 μm layer of tetraethyl orthosilicate (TEOS) oxide is then deposited using a Novellus Concept One PECVD reactor. The wafer is then coated with 1.5 μm AZ nLOF2020 negative resist, exposed

and developed to define the cavities in which the CNT are later grown. The exposed oxide is etched using a Drytek etcher with fluorine chemistry. As catalyst, Fe deposited in a Solution CHA e-beam evaporator and patterned by lift-off process, is used. The CNTs are then grown at 500 °C in a BlackMagic Pro 4" inch CVD reactor. More details on the CNTs growth process can be found in Ref. [6]. The direct use of a dense (10^{11} CNT/cm^2) MWCNTs bundle as electrode is one of the key step to achieve a very high capacitance density device. The supercapacitor dielectric is a 15 nm thick layer of Al_2O_3 deposited by means of ALD technique. With the same technique, a conformal counter electrode is then realized depositing a 15 nm thick layer of TiN. ALD layers are deposited using an ASM F-120 reactor. The precursor used are Trimethylaluminium (TMA) and deionized water (H_2O) for the Al_2O_3 (15 nm) while for the TiN. ALD is becoming coatings on nanostructures because enable a precise control on the layer thickness and a high conformality on high aspect ratio structures[2-5]. To ensure the complete coverage of the dielectric and counter electrode on MWCNTs, fine tuning of the deposition parameters was performed..The optimum values are summarized in Table I and the main properties of the layers are given in Table II.

Table I : Deposition conditions of the ALD Al_2O_3 and the TiN layers.

	Precursor A	Purge A	Precursor B	Purge B
Al_2O_3	TMA		H2O	
Time (s)	2	6	2	8
TiN	TiCl$_4$		NH$_3$	8
Time (s)	4	12	4.8	8

Table II: The optical, mechanical and electrical properties of the Al_2O_3 and TiN layers.

	Al_2O_3	TiN
Temperature	300°C	400°C
Pressure	1 Torr	1 Torr
Growth-rate	0.9 nm/cycle	0.019 nm/cycle
Refractive index	1.61 @ 556 nm	1.92 @ 556 nm
Density	2.3 g/cm^3	4.8 g/cm^3
Stress	300 ± 60 MPa (tensile)	400 ± 60 MPa (tensile)
Resistivity	-	300 μΩ*cm
Dielectric constant	8.5	-

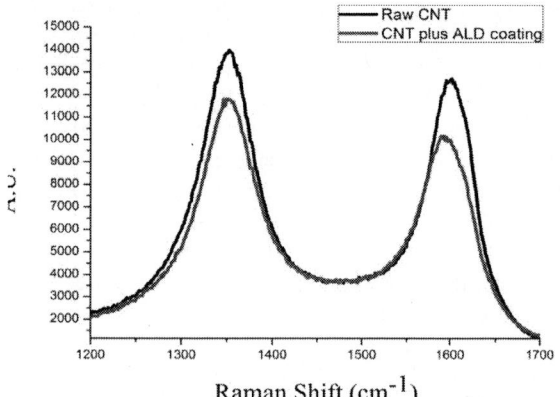

Figure 3 : Raman spectra performed on the CNT after the alumina coating. As can be seen from the D and G band, respectively at 1330 cm-1and 1580 cm-1, only minor changes on the spectra are visible after the ALD deposition.

This process results in MWCNTs bundles covered with dielectric layer and an electrical conductive layer that fully utilizes the huge surface area available [7]. High-resolution TEM and SEM pictures of the device cross-section confirm the coating of the carbon nanotubes (Fig.2). No damaging of the CNTs occurs during the ALD deposition as confirmed by Raman measurements (Fig.3). The TiN counter electrode is then covered by a 2 μm layer of Al/Si for a robust electrical contact between the nanostructure and the measurement probes, while the back electrode is contacted through an opening in the oxide, as indicated in Fig. 1.

RESULTS AND DISCUSSION

Impedance measurements using a conventional LCR meter are carried out. Two device sets with different area ($100 \mu m^2$ to $625 \mu m^2$) are here presented to study the scaling of device performance with size.

Prior to any measurement an open calibration step has been performed to cancel the contribution of any parasitic component of the measurement system. In Fig. 4, a typical measured impedance curve of the realized capacitor is presented. The device is purely capacitive over the entire frequency range under study (10^3-10^6 Hz). This behavior represents a significant improvement respect to previously presented work (see for example ref.4), where at high frequency no simple circuital model was able to explain the observed curves. The real and the imaginary part of our device impedance can be instead effectively modeled using a simple circuit scheme as represented in Fig. 5. According to this, the impedance can be written as :

$$Z_{real} = R_s + \frac{R_p}{1 + \omega^2 C^2 R_p^2}$$

$$Z_{imag} = -\frac{\omega R_p^2}{1 + \omega^2 C^2 R_p^2}$$

where C represents the capacitance component of the

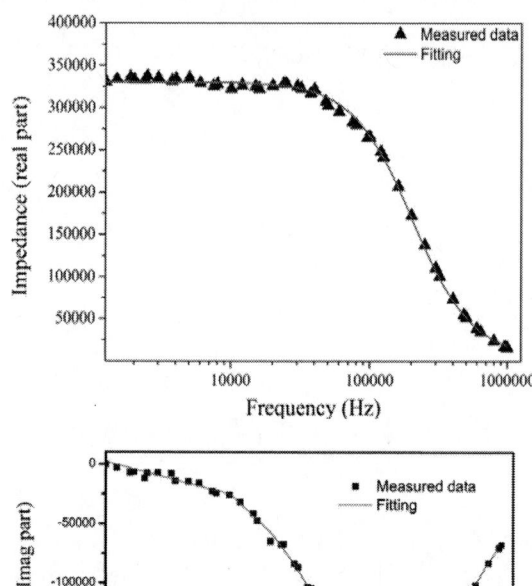

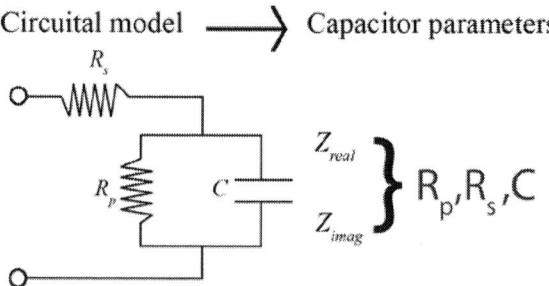

Figure 4 : Impedance of the supercapacitors. Both real (top) and imaginary part (bottom) are shown. The device presents the typical behaviour of a capacitor in the entire frequency range under study.

Figure 5: The parameters C (capacitance), Rs (series resistance) and R_p (parallel resistance) are extracted fitting the impedance data with a model that accounts for leakages due to the not ideal behaviour of the dielectric coating.

device, ω is 2π times the frequency, R_p is a resistor that take into account the leakage component eventually present, while R_s is the series resistance of the capacitor. In table III the best fit parameters of the capacitors impedance are reported. The relevant capacitors parameters are extracted from the presented model to take into account the non ideal behavior of the capacitors. Uncovered part of the CNTs by the Al_2O_3 or thinner layers can indeed generate an unwanted resistive behavior due to the electrical shorts between the capacitors electrodes, thus explaining the observed

Table III: Supercapacitors best fitting parameter. The very high parallel resistance of the 100 μm² indicates a very small number of pinholes in the aluminum oxide layer.

Area (μm²)	C (F)	Rs Ω	Rp kΩ	C-C$_{avg}$ (F)
100	2.5E-12	3500	382	± 0.3E-12
625	1.17E-11	110	11	± 0.47E-11

Table IV: Supercapacitors performances compared with a planar capacitor of the same footprint. The devices show a remarkable increment (up to 6.85 times) of the area and an extremely high volumetric capacitance density. A very small capacitance deviation (C-C$_{av}$) among measured samples (>16 for each size), is obtained.

Area (μm²)	Area increase factor	Density (F/cm³)
100	6.85	0.012
625	4.97	0.0094

impedance curves. The non uniform coating of the overall length of the CNTs is probably due/related to the relatively high base pressure inside the ALD reactor (1 Torr), which limits the precursors diffusions through the CNT bundles. The use of different reactors with a lower deposition pressure would reduce this effect, leading to a better performing capacitors.

From the data reported in Table III it is clear that the devices with the smaller area (100 μm²) present a very high parallel resistance (in the order 380 kΩ), meaning that the ALD coating of the CNTs has been quite effective. The series resistance of these devices is in the order of 3.5 kΩ. The second device set has a surface area of 625 μm², more than 6 times higher than the first device set. These supercapacitors present a very low parallel resistance of 11 kΩ, while the capacitance is almost 5 times higher. More than 16 devices for each size have been measured on the wafer area and only a small deviation of the capacitance and density values observed, indicating the good reproducibility of the fabrication process.

To verify the effective enhancement of the presented design, we compared the obtained values of capacitance with a planar capacitor of the same footprint and evaluated the capacitance density. The results are summarized in Table IV. Our measurements confirm that the ALD coating on the CNTs effectively increases the electrode area of a factor 6.85 and 4.97, respectively. The correspondent capacitance density is 0.012 F/cm³ for the 100 μm³ devices and 0.0094 F/cm³ for the 625 μm³.

CONCLUSIONS

A 3D solid state supercapacitor has been realized by means of atomic layer deposition coating of highly dense CNTs bundles. A 15 nm thick layer of aluminium oxide has been employed as dielectric layer, while a 10 nm thick layer of titanium nitride is deposited as conformal counter electrode on top of the oxide. Impedance measurements on these devices show that the high aspect ratio design of the capacitor electrodes effectively enhances the capacitor performances up to a factor 6.85 compared to a planar capacitor with the same footprint. The very high performance, the relatively simple and easy-to-integrate fabrication process, and the low thermal budget required (500 °C reached during the CNT growth) make these devices the first known example of large-scale manufacturable nanostructured supercapacitors.

ACKNOWLEDGEMENTS

The authors gratefully acknowledge the technical support and advice of the staff at the Dimes Technology Centre. Part of this work is financially supported by the Dutch Technology Foundation STW.

REFERENCES

[1] A.S. Arico et al.,"Nanostructured materials for advanced energy conversion and storage devices" *Nature Mater.* 4, pp.366–377, 2005.

[2] P. Banerjee et al. "Nanotubular metal–insulator–metal capacitor arrays for energy storage" *Nature Nanotechnology V.4*, pp.292-296, 2009.

[3] C. Shen et al. "Direct prototyping of 3D Micro supercapacitors based on in-situ fabricated nanoporous carbon electrodes"*Proc. MEMS 2013*, pp. 797 – 800.

[4] C. L. Pint et al. "Three dimensional solid-state supercapacitors from aligned single-walled carbon nanotube array templates"*Carbon V. 49*, pp. 4890 – 4897, 2011.

[5] S. Boukhalfa et al. "Atomic layer deposition of vanadium oxide on carbon nanotubes for high-power supercapacitor electrodes" *Energy Environ. Sci.*, 5, 6872, 2012.

[6] S.Vollebregt et al. "Integrating low temperature aligned carbon nanotubes as vertical interconnects in Si technology", *11th IEEE International Conference on Nanotechnology*, August 15-18, 2011.

[7] G. Fiorentino et al. *"12th IEEE International Conference on Nanotechnology"*, August 20-23, 2012.

CONTACT

*G. Fiorentino, tel:+31-1527 81871; g.fiorentino@tudelft.nl

A HIGH PERFORMANCE TRIBOELECTRIC GENERATOR FOR HARVESTING LOW FREQUENCY AMBIENT VIBRATION ENERGY

B. Meng[1], W. Tang[1], X. S. Zhang[1], M. D. Han[1], X. M. Sun[1], W. Liu[1] and H. X. Zhang[1]

[1]National Key Lab of Nano/Micro Fabrication Technology, Peking University, China

ABSTRACT

We present a novel triboelectric generator for vibration energy harvesting based on the mass production manufacture of flexible printed circuit (FPC). 10 pairs of friction surfaces were integrated on the zigzag-shaped FPC structure, which served as an elastic spring in the mass-spring system. This design makes the triboelectric generator simple and easy to be stimulated. Using industrial FPC manufacture makes the fabrication efficient, low-cost and with high yield. Load circuits can be integrated with the generator without additional assembly. The generator can be operated within a large frequency range from 1 Hz to 200 Hz. Low resonant frequency of 16 Hz and wide bandwidth of 37 Hz were achieved. The maximum effective output power of 77.3 µW was obtained with an optimal load of 10 MΩ at 16 Hz with an oscillation amplitude of 2 mm.

INTRODUCTION

Harvesting energy from the environment is widely considered an attractive approach of providing sustainable and green energy sources for applications like wireless sensor networks, implanted medical devices, wearable electronics and other low-power electronic systems. To date, numerous energy harvesting devices have been developed to convert mechanical energy into electrical energy [1]. Various transduction mechanisms have been adopted, including electrostatic induction, electromagnetic induction, piezoelectricity, and so on.

As a traditional phenomenon, triboelectricity was discovered by Thales with the experiments of amber charging against wool 25 centuries ago. Recently, it was exploited for harvesting kinetic energy from human motion and ambient. Novel energy harvesting devices termed triboelectric generator [2-8] have been developed and shown to achieve high power density and efficient. These generators works based on triboelectricity and electrostatic induction. As compared with the electret generators [9-11], triboelectricity provides a much simpler method for electrostatic charging and makes it feasible the low-cost mass production of electrostatic-induction-based generator.

Our previous work has presented a FPC-based triboelectric generator for harvesting kinetic energy of human motion [8]. Multiple friction pairs were integrated in a single generator. Employing well-developed industrial manufacturing of flexible printed circuit, the fabrication procedure is efficient and low-cost. Moreover, the use of FPC technology makes it simple to integrate load circuits with the generator.

In this work, we developed an optimized design of triboelectric generator to harvest low frequency ambient vibration energy. The zigzag-shaped FPC structure was employed to serve as an elastic spring in the mass-spring system. Polyimide film was laminated to compose the friction pair with gold, aiming to enhance the triboelectric charge density and stability on friction pairs. Low resonant frequency, wide bandwidth and high output power were achieved.

EXPERIMENTAL

The schematic of the triboelectric generator is outlined in Figure 1. The generator employed a simple mass-spring structure, consisting of a FPC spring and an aluminum mass. The FPC spring was designed as an elastic zigzag-shaped structure. 10 friction pairs are integrated in a FPC spring to improve the output power. Polyimide and gold were chosen to serve as the two kinds of friction surfaces for the large difference of abilities in attracting electrons between them. Micro-cubic arrays are patterned to facilitate triboelectricity. The induction electrodes have been designed as two counter combs corresponding to the two kinds of friction surfaces. The effective friction area of the friction pair is 1.5 cm × 3 cm. Owing to the excellent and adjustable elasticity, the zigzag structure can work well as a spring in the mass-spring system.

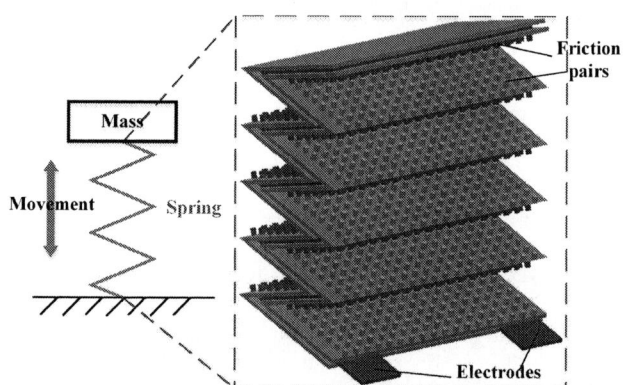

Figure 1: Schematic of the triboelectric generator incorporating an elastic zigzag-shaped FPC spring with 10 friction pairs integrated.

The zigzag-shaped elastic spring is fabricated using the industrial manufacture procedure of flexible printed circuit [12]. The fabrication process is illustrated in Figure 2. It starts from two sheets of flexible copper clad laminate (FCCL), one single-sided and anther double-sided. The FCCL is composed of a 25 µm thick flexible polyimide substrate and one or two 18 µm thick copper films with epoxy adhesive thin layers in between. The counter comb electrodes and copper micro-cubic arrays were patterned by

978-1-4799-3510-9/14 $31.00 © 2014 IEEE

photolithography and etched by FeCl$_3$ solution. Thin layers of electroless nickel and immersion gold were then plated on copper using an industrial electroless nickel immersion gold (ENIG) surface plating process. A polyimide thin film was then laminated on the gold-coated cubic arrays selectively with correspondence to one of the comb electrode. The two laminates were then aligned and bonded with the electrodes amid two polyimide substrates.

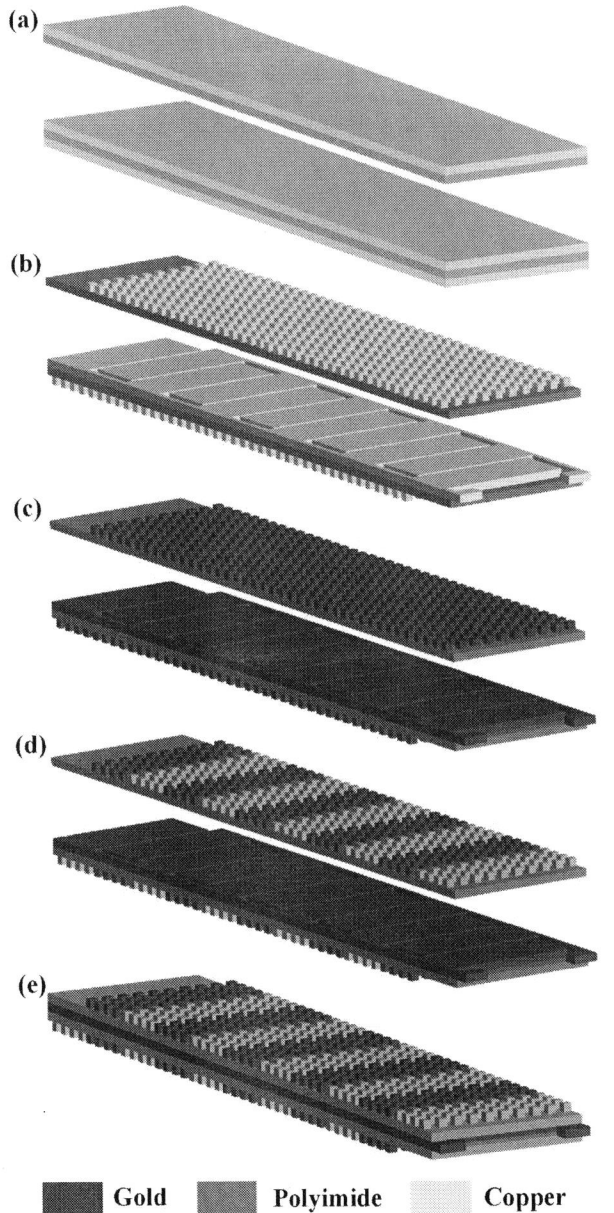

| Gold | Polyimide | Copper |

Figure 2: Fabrication process of the zigzag-shaped FPC spring. (a) two sheets of flexible copper-clad laminate, (b) FeCl$_3$ etching of copper film, (c) electroless nickel and immersion gold on copper electrodes and cubic patterns, (d) laminating PI film on gold-coated cubic patterns, (e) bonding of the FCCLs.

The bonded multiple laminate of FPC board were then folded into a zigzag-shape and compressed to 9 mm in height, as the zigzag-shaped FPC spring in natural state and the triboelectric generator with the compressed FPC spring shown in Figure 3a and Figure 3b. Figure 3c shows the top view SEM image of the gold-coated micro-cubic array. The micro-cubic pattern is 150 μm in side length with a period of 300 μm.

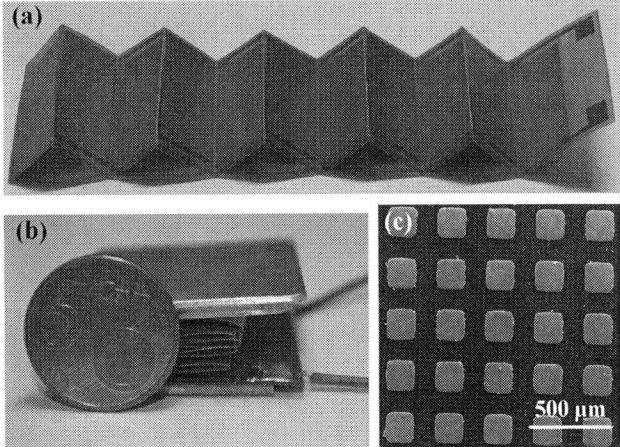

Figure 3: Photographs of (a) a fabricated FPC spring in natural state and (b) a triboelectric generator with a compressed FPC spring. (c) SEM image of the gold coated micro-cubic patterns.

RESULTS AND DISCUSSIONS

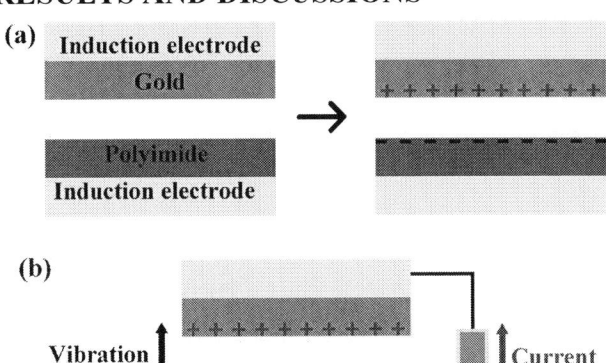

Figure 4: The two-stage operating principle of the triboelectric generator. (a) electrostatically charged by contact electrification, (b) out-of-plane gap closing.

The operating of triboelectric generator is actually a two-stage process, as shown in Figure 4. In the first stage, the friction pairs get electrostatically charged. Owning to the big difference in electron-attracting abilities between gold and polyimide, electrons will be transferred from gold surface to polyimide surface when gold and polyimide are brought into contact and separate. Under repeated cycles of contact and separation, the friction surfaces will get charged with negative charge on polyimide surface and positive charge on

978-1-4799-3510-9/14 $31.00 © 2014 IEEE   347

gold surface. During the contact, heat is produced, which will facilitate the contact electrification. In the second stage, after electrification charged, the generator operates in the very similar mode with an out-of-plane gap closing electrostatic energy harvester [13]. Thus, mechanical energy can be captured and converted into electric energy with the gap variation of each electrostatically charged friction pair under vibration.

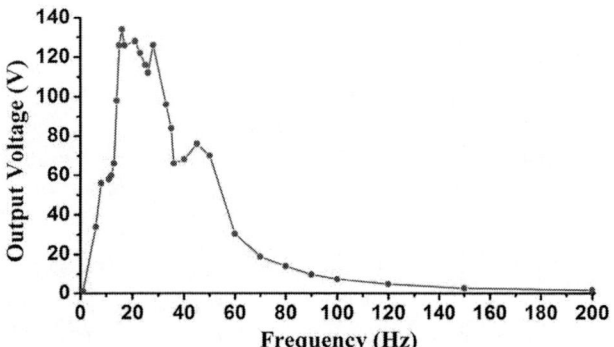

Figure 5: Output voltage versus frequency ranging from 1 Hz to 200 Hz.

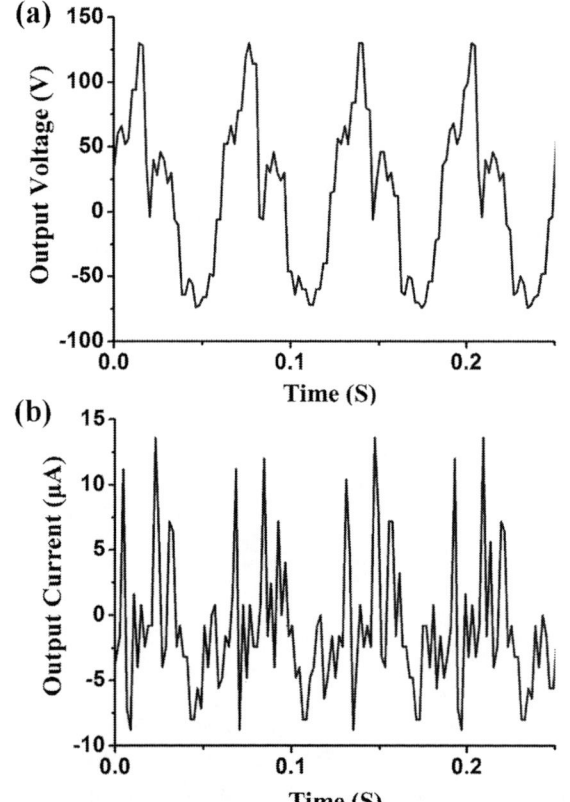

Figure 6: Time trace of (a) output voltage and (b) current at 16Hz showing a peak-peak voltage of over 200 V and a peak-peak current of 22 μA.

The performance of the generator were characterized via a vibration system. The vibration system incorporates a function generator (RIGOL DG1022), a power amplifier and a vibrator. The output voltage was measured using an oscilloscope (RIGOL DS1102E) with an input impedance of 100 MΩ. Figure 5 shows the frequency performance of the triboelectric generator. The device works within a large frequency range from 1Hz to 200 Hz. The maximum peak voltage reached about 134 V at the resonant frequency of 16 Hz. A wide working bandwidth was achieved. The FWMH is 37 Hz, ranging from 13 Hz to 50 Hz. At higher frequency of 100 Hz, a peak voltage of 7.5 V was still obtained.

Figure 6 shows the time trace of output voltage and current at 16 Hz with an oscillation amplitude of 2 mm. The current was measured by employing a 100 KΩ sampling resistor. It shows a peak-peak voltage of over 200 V (with positive peak of over 130 V and negative peak of over -70 V) and a peak-peak current of 22 μA (with positive peak of about 14 μA and negative peak of about -8 μA). Multiple peaks of output current were observed in the waveform of a single cycle.

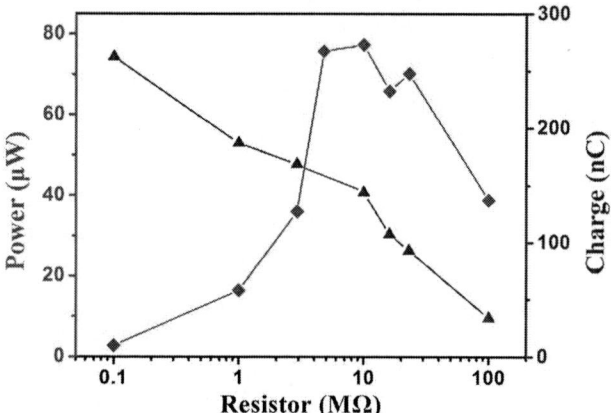

Figure 7: Output power and charge that generated in a single cycle versus external load at 16Hz with an oscillation amplitude of 2 mm.

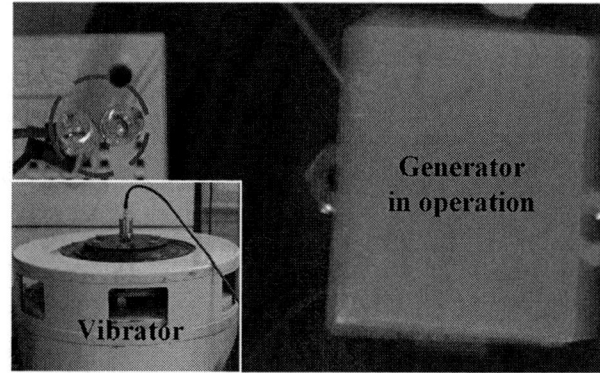

Figure 8: 2 LEDs directly illuminated by the generator when operating under vibration.

Figure 7 shows the output power and charge generated in a single cycle on the variation of load resistance at 16 Hz with an oscillation amplitude of 2 mm. The maximum effective output power of 77.3 μW is obtained with an optimal load of 10 MΩ. Taking the effective friction area (1.5×3 cm^2) of the friction pair into account, the maximum power density is $17.18\ \mu W/cm^2$. With a load of 100 kΩ, the quantity of charge generated in a single cycle is 262.2 nC. The quantity of charge reduces to 33.8 nC when the load resistance increases to 100 MΩ.

To demonstrate the considerable output power of this triboelectric generator, 2 LEDs were connected in series as the load. The LED bulbs were instantaneously illuminated by the generator when it was operating under vibration, as shown in Figure 8.

CONCLUSIONS

In summary, based on the industrial manufacture of flexible printed circuit, a novel triboelectric generator was developed for harvesting low frequency ambient vibration energy. A zigzag-shaped FPC spring and an aluminum mass were employed to compose a simple mass-spring system, making the generator easy to be stimulated. The fabrication of the generator is efficient, low-cost and with high yield. It's simple to integrate load circuits with the generator owning to the ease of mounting electronic components on the FPC board. The generator with 10 friction pairs integrated on the FPC spring has achieved a low resonant frequency of 16 Hz and a wide bandwidth of 37 Hz (from 13Hz to 50 Hz). The maximum effective output power reached 77.3 µW at 16 Hz with an optimal load of 10 MΩ. Using the triboelectric generator as the power source, 2 LEDs were instantaneously illuminated when the generator was operating under vibration.

ACKNOWLEDGEMENTS

This work is supported by the National Natural Science Foundation of China (Grand No. 91023045 and 61176103), 863 Project (Grand No. 2013AA041102), and Doctoral Program Fund (Grand No. 20110001110103).

REFERENCES

[1] P. D. Mitcheson, E. M. Yeatman, G. K. Rao, A. S. Holmes, T. C. Green, "Energy harvesting from human and machine motion for wireless electronic devices", *Proceedings of the IEEE*, vol. 96, pp. 1457-1486, 2008.

[2] Z. L. Wang, "Triboelectric Nanogenerators as New Energy Technology for Self-Powered Systems and as Active Mechanical and Chemical Sensors", *ACS Nano*, DOI: 10.1021/nn404614z, 2013.

[3] F. R. Fan, Z. Q. Tian, Z. L. Wang, "Flexible triboelectric generator" *Nano Energy*, vol. 1, pp. 328-334, 2012.

[4] X. S. Zhang, M. D. Han, R. X. Wang, F. Y. Zhu, Z. H. Li, W. Wang, H. X. Zhang, "Frequency-Multiplication High-Output Triboelectric Nanogenerator for Sustainably Powering Biomedical Microsystems", *Nano Lett.*, vol. 13, pp. 1168- 1172, 2013.

[5] B. Meng, W. Tang, Z. H. Too, X. S. Zhang, M. D. Han, W. Liu, H. X. Zhang, "A Transparent Single-Friction-Surface Triboelectric Generator and Self-Powered Touch Sensor", *Energy Environ. Sci.*, vol. 6, pp. 3235-3240, 2013.

[6] J. Chen, G. Zhu, W. Yang, Q. Jing, P. Bai, Y. Yang, T. C. Hou, Z. L. Wang, "Harmonic-Resonator-Based Triboelectric Nanogenerator as a Sustainable Power Source and a Self-Powered Active Vibration Sensor", *Adv. Mater.*, vol. 25, pp. 6094-6099, 2013.

[7] W. Tang, B. Meng, H. X. Zhang, "Investigation of Power Generation Based on Stacked Triboelectric Nanogenerator", *Nano Energy*, vol. 2, pp. 1164-1171, 2013,

[8] B. Meng, W. Tang, X. S. Zhang, M. D. Han, W Liu, H. X. Zhang, "Self-Powered Flexible Printed Circuit Board with Integrated Triboelectric Generator", *Nano Energy,* vol. 2, pp. 1101-1106, 2013.

[9] S. W. Liu, J. M. Miao, S. W. Lye, "High Q and Low Resonant Frequency Micro Electret Energy Harvester for Harvesting Low Amplitude Harmonic of Vibration", in *Digest Tech. Papers MEMS'12 Conference*, Taipei, January 20-24, 2012, pp. 837-840.

[10] Y. Chiu, Y. C. Lee, "Flat and Robust Out-of-Plane Vibrational Electret Energy Harvester", *J. Micromech. Microeng.*, vol. 23, pp.015012, 2013.

[11] Y. Suzuki, "Recent Progress in MEMS Electret Generator for Energy Harvesting", *IEEJ T Electr. Electr.*, vol. 6, pp. 101-111, 2011.

[12] R. L. Liang, *Flexible Printed Circuit*, Science Press, Beijing, China, 2008.

[13] S. Roundy, P. K. Wright, K. S. J. Pister, "Micro-electrostatic vibration-to-electricity converters", in *Digest Tech. Papers Proceedings of IMECE2002*, New Orleans, November 17-22, 2002, pp. 34309.

CONTACT

*H. X. Zhang, Tel: +86-10-62766570; zhang-alice @pku.edu.cn

A GAP-VARYING ELECTROSTATIC TRANSDUCER UTILIZING FERROFLUID-BASED ACTUATION FOR MOTION HARVESTING

T. Galchev, D. Barutçu, and O. Paul

Department of Microsystems Engineering (IMTEK), University of Freiburg, Germany

ABSTRACT

This paper provides the electrical characterization of the gap-varying ferrofluid-based electrostatic springless proximity inertial harvester (SPIH). The SPIH is a multi-axis motion harvesting structure capable of three-dimensional low-frequency operation from human or environmental application scenarios among others. The device structure consists of an array of electrostatic transducers that are interconnected and filled with a magnetic fluid. A spherical magnet serves as the proof mass. Mechanical energy is transferred to each transducer magnetically. A hydrostatic pressure in the magnetic fluid actuates the top plate of each variable capacitor. Each 2-mm-diameter transducer is capable of producing between 0.05-4.2 nJ of energy per actuation cycle at bias voltages of 10-100 V under controlled experiments. Harvesting multi-axial motion from random hand movements (including *x*- and *y*-axis translation and rotation) is demonstrated to produce peak power levels as high as 3 nW (with only one transducer from the array connected) and by using a 10 V initial bias.

INTRODUCTION

Energy harvesting power sources have been studied at length over the past decade as possible replacements and/or supplements to battery powered operation for wireless systems. Energy in the form of motion is among the most popular and important sources of renewable power. However, a great deal of the focus has been on harvesting linear vibrations from machinery, moving vehicles, and other man made sources. On the other hand, applications like body area networks (BAN) and environmental and structural health monitoring require a greater degree of harvester versatility with respect to bandwidth, input amplitude, and axis of motion. Multi-axis harvesters (MAH), important for human powered applications, have recently become a topic of increased investigation. Most can be grouped in two general categories including in-plane micromachined varieties [1-4] and devices that include a freely moving mass [5-8] that is not suspended to the casing.

A new MAH concept was recently presented [9] that uses a suspension-less spherical inertial mass to capture multi-axial human motion. The design features an array of transducer cells that can couple to the rolling magnetic proof mass, absorb mechanical energy, and convert it electrostatically. This arrangement has been shown to be less sensitive to the exact magnet position and orientation and thereby offers greater design flexibility and design versatility. The previously presented concept is further elaborated in this paper. Additionally, the electrostatic gap-varying transducer

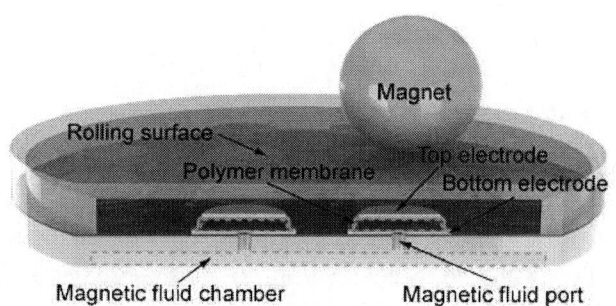

Figure 1. Illustration of the springless proximity inertial harvester (SPIH)

cells are electrically characterized and energy harvesting from simple hand movements is demonstrated.

DEVICE STRUCTURE AND FABRICATION

An illustration of the springless proximity inertial harvester (SPIH) can be seen in Figure 1. It has a spherical magnetic proof mass that is free to roll. The proof mass captures kinetic energy, after which an array of electrostatic transducer cells converts it into electricity. The magnet never comes in contact with the transducer cells, which is important for long-term reliability. Instead energy is transferred via magnetic coupling. The link between the transducers and the magnet is established via a ferrofluid. These are colloidal liquids made of nanometer-size magnetic particles. This liquid becomes magnetized in the presence of a magnetic field and produces a hydrostatic pressure against the movable top electrode of the transducer cell thereby varying the capacitance.

Fabrication of the transducers (Figure 2) was previously discussed in detail [9] and consists of a three-wafer process. The transducers are surface micromachined and use a 10 µm sacrificial photoresist to release a Parylene membrane, about 4 µm in thickness, after which fluidic interconnects are made using DRIE. Electrodes are made from 50/500 nm Cr/Au. The most critical process step is the bonding of the bottom recessed Pyrex cap that forms the ferrofluid chamber. In order to prevent air bubbles from forming inside the device, a process to bond the cap while inside the ferrofluid solution was developed. A UV curable epoxy is used to seal the fluid reservoir. The ferrofluid chamber has a combined depth of 400-500 µm taking into account the combined recess in the silicon and Pyrex wafers. A photograph of an assembled SPIH with 2-mm-diameter transducers is shown in Figure 3.

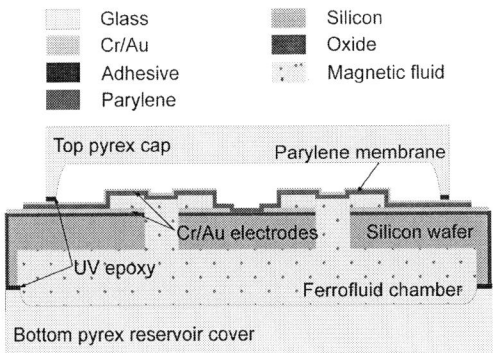

Figure 2. Process flow summary. The variable-gap transducers are surface micromachined on a silicon wafer and enclosed from both sides using Pyrex caps. '

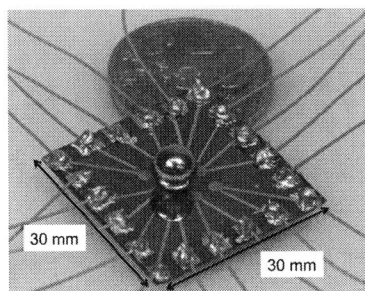

Figure 3. Photograph of the SPIH next to a 50 Euro cent coin.

FERROFLUID ACTUATION

The ferrofluid-based actuation process is illustrated in Figure 4. When a magnetic field is applied over the transducer cell a hydrostatic pressure, P_{mag}, is generated in the ferrofluid according to [10]

$$P_{mag} = \mu_o M_s \int_{H_b}^{H_t} dH, \qquad (1)$$

where μ_0 is the permeability, M_s is the saturation magnetization, and H_t and H_b refer to the magnitude of the magnetic field at the top and bottom of the fluid column respectively. The red (left) curve in Figure 5 shows the measured magnetic flux density generated by a cubic NdFeB magnet ($5\times5\times5$ mm^3) as a function of distance, d, from the top of the movable electrode. Measurements were made using a Bell 6010 Teslameter. In order to ease the calculation of hydrostatic pressure two simplifications are made: 1) The magnetic field is assumed to vary through the depth of the fluid column as a third-order polynomial function extracted from the measured data in Figure 5; 2) The magnetic field is assumed to be uniform across the plane of the wafer. The black (right) axis of Figure 5 is used to plot the hydrostatic pressure with respect to the fluid column height, h. The column height, or in other words, the depth of the magnetic fluid reservoir can be designed to be ≤ 1 mm in the present process. Therefore the expected hydrostatic pressure will be in the range of 4-6 kPa. A wealth of previous art exists that allows the designer to then convert this uniform pressure

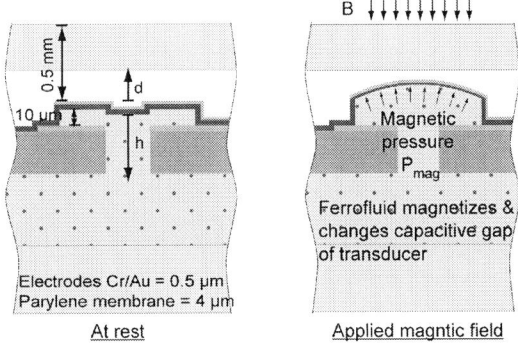

Figure 4. Schematic method of operation. An applied magnetic field magnetizes the ferrofluid and varies the capacitive gap in the transducer.

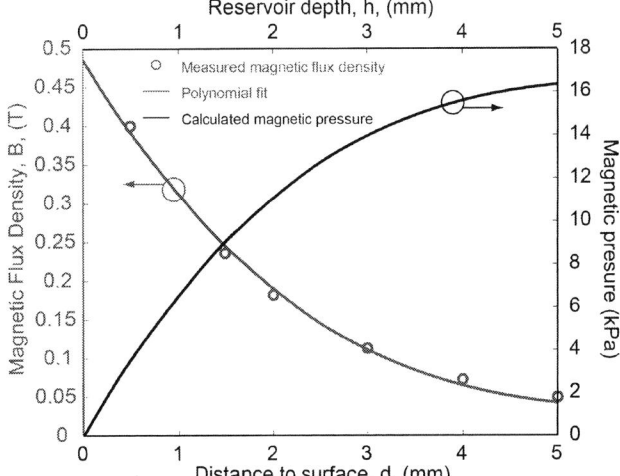

Figure 5. Red curve shows the measured magnetic flux density, B, as a function of distance from the surface of the transducer, while the black curve shows the simulated magnetic pressure as a function of ferrofluid reservoir depth at a fixed B.

applied on the thin membrane into a deflection, and subsequently into a capacitance change. Finally, the energy that can be generated per actuation cycle is given as

$$E_{cycle} = \frac{1}{2}\Delta CV^2, \qquad (2)$$

where V is the applied pre-bias voltage. Maximum capacitance change can be achieved through careful optimization of the Cr/Au layer thickness which defines the effective membrane stiffness, initial inter-electrode gap, and magnetic pressure.

TRANSDUCER CHARACTERIZATION

Harvested energy is experimentally measured. Sets of controlled experiments were performed where the SPIH was mounted on an automated xyz-stage while a cubic magnet ($5\times5\times5$ mm^3) was suspended above. A voltage-constrained cycle was carried out to measure the energy generation

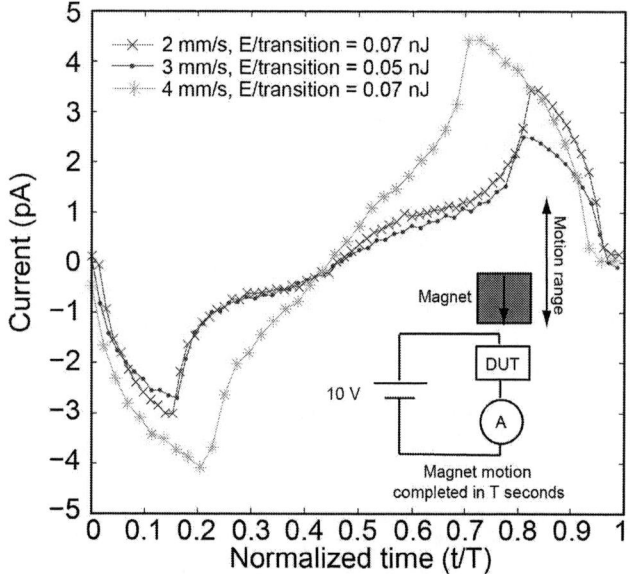

Figure 6. Current is measured while the magnet is translated perpendicular to the transducer surface (see inset) at different velocities. The capacitor transitions from ($C_{min} \rightarrow C_{max} \rightarrow C_{min}$).

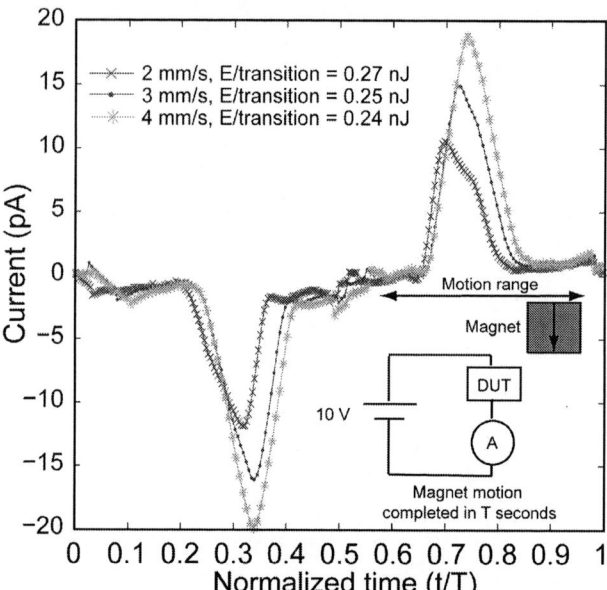

Figure 7. Current is measured while the magnet is scanned horizontally across the transducer surface (see inset) at different velocities. The capacitor transitions from ($C_{max} \rightarrow C_{min} \rightarrow C_{max}$).

capability. A Keithley 6487 picoammeter is used to measure current at a bias voltage of 10 V. Figure 6 shows the results of varying the capacitance by moving the magnet perpendicular to the transducer in the z-direction. The actuation cycle starts at the bottom where the membrane is deflected and consequently capacitance is minimum. One full cycle moves the magnet up and then back down again ($C_{min} \rightarrow C_{max} \rightarrow C_{min}$). This motion is repeated at different velocities. The bias voltage is kept fixed while the capacitance is varied. The plot is normalized so that currents generated at different speeds can be displayed on top of each other for comparison. Additionally one must note that since there is no diode to prevent current backflow there are two current spikes, one produced during energy generation and one produced during capacitor relaxation as charge is again brought back by the priming source. A comparable experiment is shown in Figure 7, however this time the motion of the magnet is parallel to the surface of the transducer. In this case the capacitance undergoes a slightly different cycle ($C_{max} \rightarrow C_{min} \rightarrow C_{max}$) since the magnet starts and ends away from the transducer. To compute harvested energy, half of the current (only the energy generation half) is integrated over one translation period and multiplied with the applied potential difference. Horizontal magnet motion is shown to produce a higher output energy, ≈0.25 nJ/transition, with magnet velocities of 2-4 mm/s, as compared with vertical magnet motion which generated ≈0.05 nJ/transition. In the case of horizontal motion, there is an observable reduction of energy with increasing magnet speed, which can be interpreted as being the result of damping in the dynamic response of the device. For vertical motion, harvested energy reduces significantly as compared with horizontal motion. The vertical motion experiments were performed mainly for

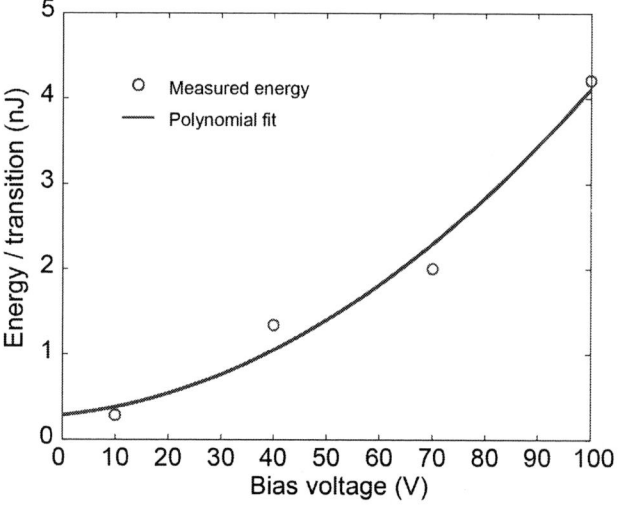

Figure 8. Energy produced per transition is measured as a function of bias voltage for the horizontal magnet translation. The velocity is 2 mm/s.

their academic value since the horizontal motion is more closely related to a spherical magnet rolling on the surface. The reduction in energy seen in the vertical translation may be from two possible sources. First, even when the magnet is at rest in the farthest position, there is still a non-zero external magnetic field resulting in a hydrostatic force at all times. Additionally, fluid motion is different in the two scenarios. In the case of horizontal translation the magnet can be thought of as dragging in the fluid and then helping to empty the thin and long capacitive-gap, whereas in the case of vertical motion the fluid may have a harder time moving out from within the electrodes resulting in lower capacitance

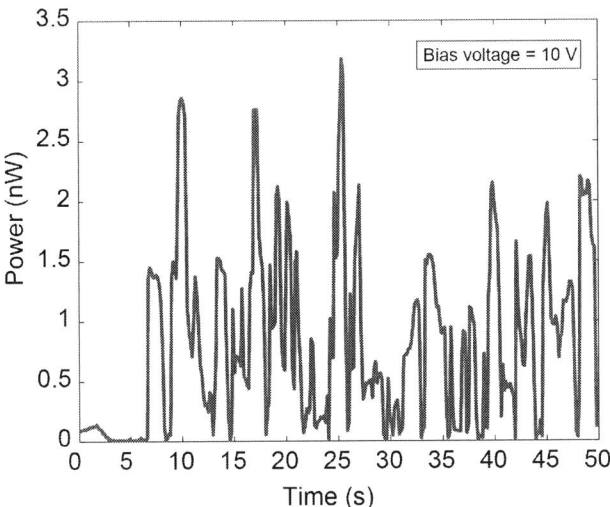

Figure 9. Power is measured while a 6 mm-diameter spherical magnet is placed on the SPIH and allowed to roll freely as the harvester is moved by hand.

change. Energy is plotted as a function of bias voltage in Figure 8. The measurements are made using horizontal translation at a velocity of 2 mm/s. There is a deviation from the square law relationship expected from (2). A higher increase in converted energy is expected. One possible explanation is that there is measurement error at higher bias voltages. There could be bias voltage instability or insufficient bandwidth to capture the larger current peaks. These issues have to be investigated further.

ENERGY HARVESTING TEST RESULTS

A preliminary experiment was performed to measure the energy producing capability of the SPIH under multi-axial motion. A 6-mm-diameter NdFeB spherical magnet was used. One transducer cell in the center of the SPIH was connected to the picoammeter in the same manner as the previous experiments. Current was measured continuously while the harvester was held in the operator's hand and subjected to lateral x- and y-axis translations as well as rotation around both axes in a random manner. While acceleration was not measured simultaneously, it is assumed to be quite low and well within the range expected from regular human motion. Tethering the SPIH to the sensitive current measurement equipment did not allow for large or rapid movements. The results of this experiment are shown in Figure 9, where a single transducer cell generated over 3 nW of instantaneous power.

CONCLUSION

This paper presented the electrical characterization of the gap-varying springless proximity inertial harvester. The multi-axis motion harvester uses a spherical magnetic proof mass to pass mechanical energy within a gap-varying electrostatic transducer by magnetizing an integrated magnetic fluid. In this manner the proof mass never comes in contact with the transducers. A prototype with an array of 2-mm-diameter transducers was shown to be capable of producing 0.05-4.2 nJ of energy per actuation cycle when biased in the range of 10-100 V. For the first time energy harvesting was demonstrated using the SPIH. Simple translational and rotational hand movements were able to generate power levels between 0.5-3 nW using only one connected transducer under a very weak bias of 10 V.

ACKNOWLEDGEMENTS

This work was supported by the Alexander von Humboldt Foundation.

REFERENCES

[1] U. Bartsch, J. Gaspar, and O. Paul, "A 2D electret-based resonant micro energy harvester," in *IEEE Int. Conf. on Microelectromechanical Systems (MEMS)*, USA, pp. 1043-6, 2009.

[2] U. Bartsch, J. Gaspar, and O. Paul, "Low-frequency two-dimensional resonators for vibrational micro energy harvesting," *J. of Micromech. and Microeng.*, vol. 20, p. 035016 (12 pp.), 2010.

[3] J. L. Fu, Y. Nakano, L. D. Sorenson, and F. Ayazi, "Multi-axis AlN-on-silicon vibration energy harvester with integrated frequency-upconverting transducers " in *IEEE Int. Conf. on Microelectromechanical Systems (MEMS)*, Paris, France, pp. 1269-1272, 2012.

[4] L. Huicong, S. Bo Woon, W. Nan, C. J. Tay, Q. Chenggen, and L. Chengkuo, "Feasibility study of a 3D vibration-driven electromagnetic MEMS energy harvester with multiple vibration modes," *J. of Micromech. and Microeng.*, vol. 22, p. 125020, 2012.

[5] M. E. Kiziroglou, C. He, and E. M. Yeatman, "Rolling rod electrostatic microgenerator," *IEEE Trans. on Industrial Electronics*, vol. 56, pp. 1101-8, 2009.

[6] P. Pillatsch, E. M. Yeatman, and A. S. Holmes, "A scalable piezoelectric impulse-excited generator for random low frequency excitation," in *IEEE Int. Conf. on Microelectromechanical Systems (MEMS)*, pp. 1205-8, 2012.

[7] B. J. Bowers and D. P. Arnold, "Spherical, rolling magnet generators for passive energy harvesting from human motion," *J. of Micromech. and Microeng.*, vol. 19, p. 094008 (7 pp.), 2009.

[8] T. Galchev, R. Raz, and O. Paul, "A new multi-dimensional low-frequency springless proximity inertial harvester for converting human and environmental motion," in *PowerMEMS*, Atlanta, USA, pp. 117-120, 2012.

[9] T. Galchev, R. Raz, and O. Paul, "An electrostatic springless inertial harvester for converting multi-dimensional low-frequency motion " in *IEEE Int. Conf. on Microelectromechanical Systems (MEMS)*, Taipei, Taiwan, pp. 102-105, 2013.

[10] R. E. Rosensweig, *Ferrohydrodynamics*. Cambridge University Press, 1985.

A HYBRID SUPERCAPACITOR USING VERTICALLY ALIGNED CNT-POLYPYRROLE NANOCOMPOSITE

Firas Sammoura[1,3], Kwok Siong Teh[2], Alina Kozinda[3], Xining Zang[3], and Liwei Lin[3]

[1]Department of Electrical Engineering and Computer Science, Masdar Institute of Science and Technology, Abu Dhabi, UAE
[2]School of Engineering, San Francisco State University, San Francisco, CA 94132 USA
[3]Department of Mechanical Engineering, University of California at Berkeley, Berkeley, CA 94720 USA

ABSTRACT

We have successfully demonstrated, for the first time, the fabrication of vertically aligned carbon nanotube (VACNT)-polypyrrole (PPY) nanocomposites as a "hybrid supercapacitor" material directly integrated on silicon-based electrodes. In contrast to previous works, three distinctive achievements have been accomplished: (1) a "hybrid supercapacitor" using VACNT forest with electroplated PPY and dodecylbenzenesulfonate (DBS) as a dopant in acetonitrile, (2) realizing 500% higher capacitance as compared to the capacitance of electrodes made of VACNT or DBS-doped PPY alone, and (3) highly reversible cycling between -1 V and +1 V with improved knee frequency at 797 Hz. As such this hybrid nanocomposite could become a new class of material for future supercapacitors.

INTRODUCTION

Power sources for portable 3D micro-systems including distributed sensor networks are critical for their operational life, performance, and reliability [1]. The two major power sources for portable devices are batteries and capacitors, which have historically lagged behind the development of micro-electronic circuits, in areas such as system integration density [2]. As supercapacitors store electrical energy by either fast ion absorption at the electrode/electrolyte interface as in Electric Double Layer Capacitor (EDCL) [3] or quick and reversible surface redox reactions as in Pseudo Supercapacitor (PSC) [4], they bridge the power density gap between dielectric capacitors and batteries [5]. Supercapacitors find myriad commercial successes in high power applications, such as backup power supplies, camera flash units, electric cars, or power load-leveling of heavy machinery [6].

It is desirable to enhance the energy density of supercapacitors by increasing either their specific capacitance or their operating voltage [7]. Fan *et al.* developed an asymmetric supercapacitor with different anode/cathode materials in order to maximize the operating voltage [8]. PSCs based on metal oxides such as RuO_2 and MnO_2 or conducting polymers including polypyrrole (PPY) and polyaniline (PANI) have been under extensive investigations over the past decades because their specific energy density is higher than that of their EDLC counterparts [9]. Organic electrodes based on electronically conducting polymers have many benefits in PSC formation, including environmental stability, low-cost, and high-energy density [10]. Specifically, PPY offers unique properties as a PSC material with the ease of process compatibility, redox-tunable conductivity as p-doped or undoped polymer, and biocompatibility, [11].

Previously, An *el al.* presented a high-capacitance supercapacitor based on a mixture of single-walled carbon nanotubes (SWNT) and PPY using polytetrafluoroethylene (PTFE) as the binding material and KOH as the electrolyte [12]. In order to overcome the large variations due to CNT/PPY composition, Fang *et al.* electrodeposited PPY on a pre-assembled CNT membrane using short potential pulses and tested the fabricated supercapacitor in the electrolyte solution of Na_2SO_4 [13]. Frackowiak *et al.* prepared the electroconducting PPY using a simple chemical polymerization technique on an entangled CNT mesh, where H_2SO_4 was later used as the electrolyte [14]. It is worth mentioning that the PPY/CNT supercapacitors demonstrated in the literature works were not optimized by mass or volume because of the requirement of a binding material during their construction and they suffered from a large contact resistance between CNT/PPY nanocomposite and the metal electrode. In addition, the operational voltage was limited to 0-0.8V range due to the selected negative doping ion and the nature of the aqueous electrolyte, which was either basic or acidic.

Recently, we have demonstrated nano-textured electrodes with enhanced specific energy density using vertically aligned carbon nanotubes (VACNT) as an EDCL [15], and VACNT with electroplated nickel nanoparticles as a PSC using vacuum infiltration technique [16]. In addition to the high specific surface area, VACNT offer other advantages including low contact resistance to the metal electrode, high conductivity, and binder-free integration. In order to overcome the drawbacks of the previous works, this paper presents a hybrid supercapacitor combining EDLC and PSC on the same electrode utilizing VACNT forests as the cores and electroplated conductive polymer PPY as the shells with dodecylbenzenesulfonate (DBS⁻) as a dopant ion in acetonitrile for high energy density and large voltage window of operation.

OPERATION PRINCIPLE

The operation principle of an EDLC is illustrated in Figure 1(a), where charge is stored electrostatically in the electrochemical double layer at the interface between the VACNT electrode and the electrolyte. When an electric

978-1-4799-3510-9/14 $31.00 © 2014 IEEE

potential is applied, the Na^+ and DBS^- ions within the electrolyte separate and are drawn to the respective electric double layers. The separator prevents the opposite electrodes from shorting and the capacitance is proportional to the electrode area. Figure 1(b) explains the operation principle of a PPY-based PSC, where the PPY transforms from the reduced (undoped) state to the oxidized (p-doped) state during charging and discharging, respectively. The electrical energy is stored electrochemically by the intercalation of the DBS^- dopant at the surface of the PPY coated CNTs with fast electrosorption/desorption rates. The operation principle of a VACNT-PPY supercapacitor as an EDLC/PSC hybrid supercapacitor is demonstrated in Figure 1(c). For positive input voltages, the supercapacitor operates as PSC, where the negative DBS^- ions are attracted towards the positively charged electrode such that the PPY is oxidized. When a negative input voltage is applied, on the other hand, the PPY is reduced and the positively charged Na^+ ions starts accumulating at the double layer that forms at the surface of the PPY-coated CNT, rendering an EDLC-mode of operation.

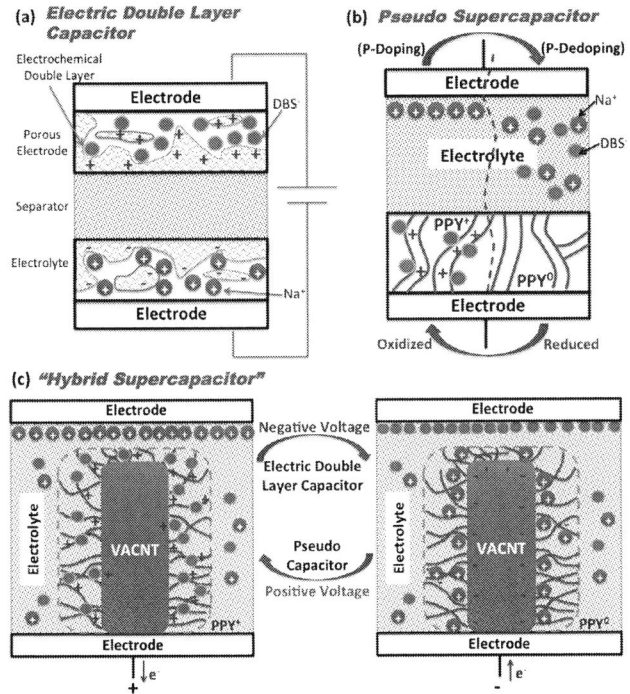

Figure 1: A schematic diagram illustrating the operation principles of: (a) Electric Double Layer Capacitor (EDLC), where the charge is stored in the electromechanical double layer; (b) Pseudo Supercapacitor (PSC), where PPY changes between oxidized (p-doped) and reduced (undoped) states during charging and discharging, respectively; and (c) Hybrid Supercapacitor, vertically aligned CNT (VACNT) coated with PPY is the nanocomposites material. For a positive input voltage, the negative charges are stored in the polymer mesh as dopants. The charge is stored in the electrochemical double layer under a negative input voltage.

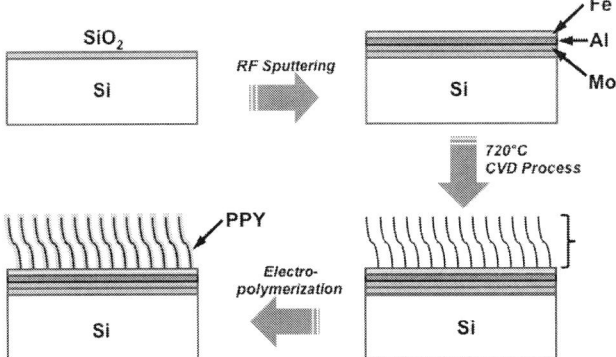

Figure 2: Fabrication process of a hybrid vertically aligned CNT-PPY supercapacitor electrode.

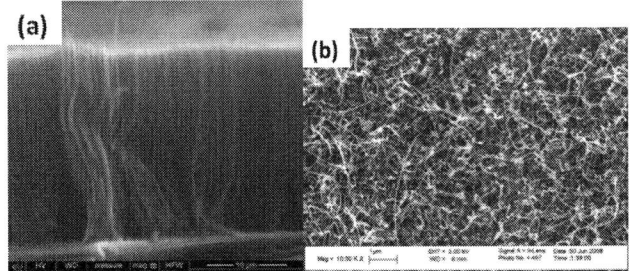

Figure 3: SEM of vertically aligned CNT forest (a) slide view and (b) top view.

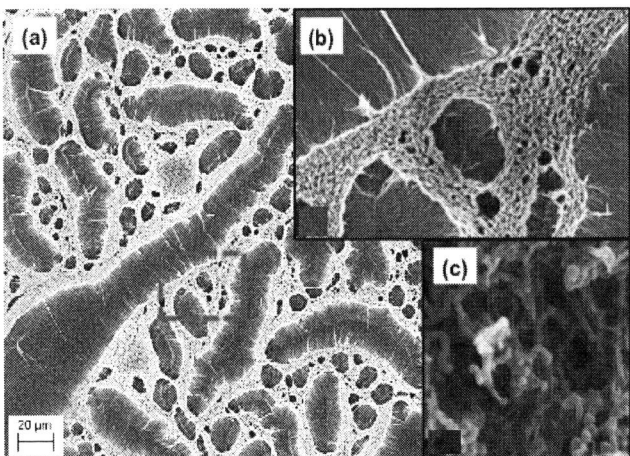

Figure 4: SEMs of (a) Vertically aligned CNT-PPY as-deposited on an $Si/SiO_2/Mo/Al/Fe$ electrode, (b) enlarged image of the boxed area in (a), and (c) further magnified image showing PPY-coated CNTs.

FABRICATION

We synthesized the prototype hybrid VACNT-PPY nanocomposite supercapacitor as outlined in Figure 2. The fabrication process begins by passivating a 1 cm by 0.5 cm silicon chip with 300nm-thick thermally grown oxide, followed by evaporating a catalytic layer of 50nm/10nm/5nm Mo/Al/Fe for VACNT growth in a thermal chemical vapor deposition (CVD) furnace (Thermo Electron Corp., model Lindberg/Blue M® three-zone tube). After purging the furnace of air, the temperature is gradually raised to 720°C while flowing hydrogen gas at 40 sccm. As soon as the temperature stabilizes around the set point, ethylene gas is pumped at a flow rate of 90 sccm and the hydrogen flow rate is raised to 611 sccm. VACNT is grown for 10 minutes, after which the sample is allowed to cool down to room temperature for 15 minutes and the furnace is purged with Argon. Figure 3 shows SEM micrographs of the as grown VACNTs with 20 μm in height. PPY is then electroplated at a constant current density of 1 mA/cm^2 using 0.1M pyrrole and 0.1M NaDBS solution in acetonitrile. Figure 4 shows the morphologies of post-drying VACNT forests coated with PPY, where capillary action-induced bundling is evident. VACNTs grew from 20nm to approximately 30nm in diameter (Figure 4(c)), indicating the PPY sheath thickness of ~5nm.

RESULTS AND DISCUSSION

Figure 5 shows the cyclic voltammograms (CV) of as-deposited PPY with three different thicknesses of PPY—50nm, 100nm, and 200nm—in 0.1M NaDBS solutions in acetonitrile. Comparing the CV graphs of each of the three films, the thickness of as-deposited PPY film appears to have negligible influence on the total capacitance of the pure PPY film. Based on this observation, we hypothesize that at these thicknesses (50nm – 200nm) the faradaic electron charge transfer in PPY—which contributes to its pseudocapacitance—is mostly a surface phenomenon rather than a volumetric phenomenon. Therefore, from a process point of view, the thickness-independent pseudocapacitance of PPY means that thinner PPY, with higher surface-area-to-volume ratio, would be more favorable given the shorter synthesis duration. Figure 6 compares the CVs of as-grown VACNT, pure PPY, and VACNT-PPY nanocomposites electrodes. In these measurements, the supercapacitor behaved as an EDLC between -1V to -0.2V, and a PSC between -0.2V to 1V. Figure 6 shows the measured specific capacitance of the VACNT-PPY is 5mF/cm^2, which is 500% higher than that of as-grown VACNT or pure PPY film. This phenomenon demonstrates that by combining the optimal operational voltage ranges of VACNT and PPY respectively, the useful voltage range of the hybrid supercapacitor is expanded and the increase in the currents between the maximum and minimum operating voltages can exceed the sum of VACNT and PPY operated individually. Figure 7 juxtaposes the CVs of VACNT-PPY with PPY deposited at 0.5, 1, 1.5, and 2 minutes—these graphs show that there is an optimal duration for PPY deposition that maximizes the total capacitance of a VACNT-PPY nanocomposite. Of the different durations, 1 minute is found to be the most optimal window of deposition. As both EDLC and pseudocapacitance are surface phenomena, the total capacitance of the hybrid material is maximized when both the available surfaces of VACNT and PPY are optimized and maximized. For the case of 1 minute of PPY deposition, the total surface area of PPY deposited maximizes the surface area and hence pseudocapacitance of PPY on VACNT without compromising the surface area of

VACNT that is available for electrochemical double layer capacitance. In fact, longer (more than 1 minute) deposition of PPY is likely to cause the bridging of the gaps between VACNT, which could reduce the hybrid material's electrochemical double layer capacitance with minor improvement on its pseudocapacitance. Finally, the VACNT vs. VACNT-PPY impedance is compared using impedance measurements. The measured Z-curves (Figure 8) show that the series resistance of VACNT-PPY is lower than that of comparable VACNT by 29%.

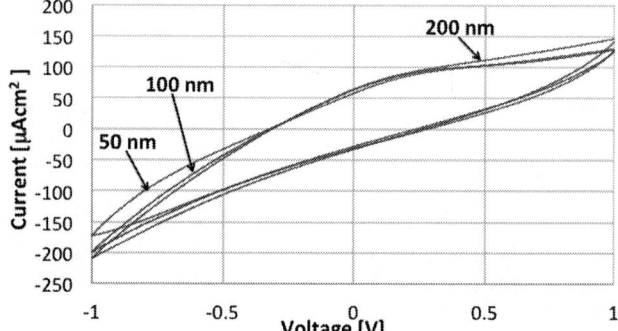

Figure 5: Cyclic voltammograms of a pure PPY film with a thickness of 50 nm, 100 nm, and 200 nm, respectively.

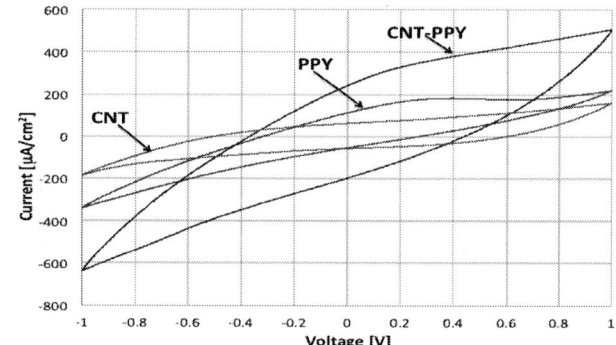

Figure 6: Cyclic voltammograms measured from: (1) vertically aligned CNT forest electrode, (2) DBS-doped PPY electrode, and (3) vertically aligned CNT coated with DBS-doped PPY electrode. Electrode made of CNT-PPY has achieved 500% higher capacitance than pure CNT or pure PPY.

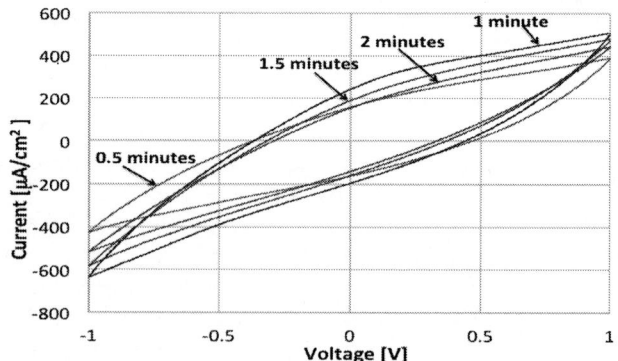

Figure 7: Cyclic voltammograms from vertically aligned CNT forest coated with PPY under 0.5, 1, 1.5, and 2 minutes of PPY electroplating process.

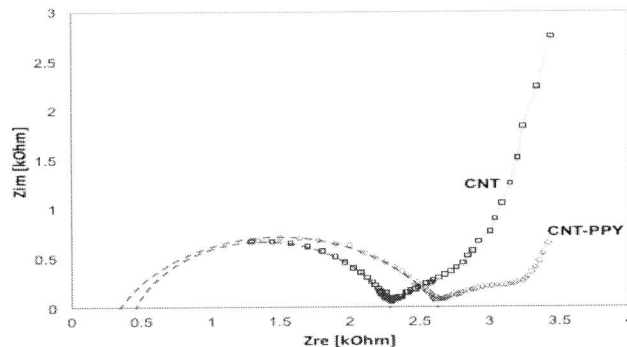

Figure 8: Impedance curves showing the reduction in series resistances of the vertically aligned CNT-PPY hybrid electrode as compared with the pure CNT electrode.

CONCLUSION

This paper detailed the development of a novel hybrid supercapacitor material that is made of a nanocomposite of VACNT and PPY. By combining VACNT and PPY to form a nanocomposite, total capacitance of the hybrid superconductor increases by five folds compared to its individual components due to its expanded operating voltage and current ranges. The enhanced energy storage in the VACNT-PPY nanocomposite is attributed primarily to fast surface redox reaction on PPY, a surface phenomenon that is unaffected by the thickness and hence volume of PPY deposited, as well as the use of a neutral acetonitrile solution. Optimal energy storage of the supercapacitor can be achieved with a PPY electrodeposition duration of 1 minute and at thickness as low as 50nm. This maximizes both the electrochemical double layer capacitance on VACNT and the pseudocapacitance on PPY, which yields an enhanced total capacitance. Finally, the series resistance of the VACNT-PPY nanocomposite decreases while its knee frequency increases.

ACKNOWLEDGEMENTS

This project is supported in part by the Advanced Technology Investment Company (ATIC, Abu Dhabi, United Arab Emirates) under the TwinLab project grant no. 12RAZB9, and Berkeley Sensor and Actuator under the general membership support. The devices were fabricated in the UC Berkeley Marvell Nanofabrication Laboratory. The authors would like to thank Dr. Jeff Clarkson from UC Berkeley for taking the SEM pictures and Dr. Yingqi Jiang from Analog Devices for his valuable discussions.

REFERENCES

[1] J. M. Rabaey *et al.*, "PicoRadios for Wireless Sensor Networks: The Next Challenge in Ultra-Low Power Design," *IEEE Int. Solid-State Circuits Conf.*, vol. 1, pp. 200-201, February 2002, San Francisco, CA, USA.

[2] D. Linden, *Handbook of Batteries and Fuel Cells*, McGraw-Hill Publishing Company, 1984.

[3] W. Cheng *et al.*, "Graphene and Carbon Nanotube Composite Electrodes for Supercapacitors with Ultra-

High Energy Density," *Phys. Chem. Chem. Phys.*, vol. 13, pp. 17615-17624, 2011.

[4] S. W. Lee *et al.*, "Carbon Nanotube/Manganese Oxide Ultrathin Film Electrodes for Electrochemical Capacitors," *ACS Nano*, vol. 4, no. 7, pp. 3889-3896, 2010.

[5] S. W. Lee *et al.*, "Nanostructured carbon-based electrodes: bridging the gap between thin-film lithium-ion batteries and electrochemical capacitors," *Energy Environ. Sci.*, vol. 4, no. 6, pp. 1972-1985, Jan. 2011.

[6] B. E. Conway, *Electromechanical Supercapacitors: Scientific Fundamentals and Technological Applications*, Kluwer Academic/Plenum Publishers, New York, 1999.

[7] S. W. Lee *et al.*, "Nanostructured Carbon-Based Electrodes: Bridging the Gap between Thin-Film Lithium-Ion Batteries and Electrochemical Capacitors," *Energy Environ. Sci.*, vol. 4, no. 6, pp. 1972-1985, Jan. 2011.

[8] Z. Fan *et al.*, "Asymmetric Supercapacitors Based on Graphene/MnO_2 and Activated Carbon Nanofiber Electrodes with High Power and Energy Density," *Adv. Functional Mater.*, vol. 21, no. 12, pp. 2366-2375, June 2011.

[9] P. Simon and Y. Gogotsi, "Materials for Electrochemical Capacitors," *Nature Materials*, vol. 7, pp. 845-854, 2008.

[10] A. Rudge *et al.*, "Conducting Polymers as Active Materials in Electrochemical Capacitors," *J. Power Sources*, vol. 47, pp. 89-107, 1994.

[11] K. S. Teh and L. Lin, "MEMS Sensor Material based on Polypyrrole Carbon-Nanotube Nanocomposite: Film Deposition and Characterization," *J. Micromechanics and Microengineering*, vol. 15, pp. 1777-1785, 2005.

[12] K. H. An *et al.*, "High-Capacitance Supercapacitor Using a Nanocomposite Electrode of Single-Walled Carbon Nanotube and Polypyrrole," *J. Electrochem. Soc.*, vol. 149, no. 8, pp. A1058-A1062, 2002.

[13] Y. Fang *et al.*, "Self-Suppoted Supercapacitor Membranes: Polypyrrole-Coated Carbon Nanotube Networks Enables by Pulsed Electrodeposition," *J. Power Sources*, vol. 195, pp. 674-679, 2010.

[14] E. Franckowiak *et al.*, "Supercapacitors Based on Conducting Polymers/Nanotubes Composites," *J. Power Sources*, vol. 153, pp. 413-418, 2006.

[15] Y. Jiang *et al.*, "A Two-Stage Self-Aligned Vertical Densification Process for As-Grown CNT Forest in Supercapacitor Applications", *Sensors and Actuators – A Physical*, vol. 188, pp. 261-267, 2012.

[16] Y. Jiang *et al.*, "Uniformly Embedded Metal Oxide Nanoparticles in Vertically Aligned Carbon Nanotube Forests as Pseudocapacitor Electrodes for Enhanced Energy Storage, *Nanoletters*, vol. 13, pp. 3524-3530, 2013.

CONTACT

*Firas Sammoura, Tel: (510) 529-9142; firas@berkeley.edu

A MEMS-ENABLED BIODEGRADABLE BATTERY FOR POWERING TRANSIENT IMPLANTABLE DEVICES

Melissa Tsang[1], Andac Armutlulu[2], Adam Martinez[3], Florian Herrault[3],
Sue Ann Bidstrup Allen[2], and Mark G. Allen[2,3,4]

[1]School of Biomedical Engineering, Georgia Institute of Technology, USA
[2]School of Chemical and Biomolecular Engineering, Georgia Institute of Technology, USA
[3]School of Electrical and Computer Engineering, Georgia Institute of Technology, USA
[4]Department of Electrical and Systems Engineering, University of Pennsylvania, USA

ABSTRACT

Active implantable medical devices (IMD) for the monitoring and treatment of transient disease states have garnered increasing interest in the medical research community. In order for these technologies to be fully viable, they require a similarly biodegradable energy source. This study presents a series of MEMS-enabled biodegradable batteries comprising Mg anodes and Fe cathodes in a 0.1 M $MgCl_2$ electrolyte. The anode was fabricated by electroplating Mg from a non-aqueous solution and passivated with either polycaprolactone or poly(glycerol-sebacate). Mg anodes coated with the biodegradable polymers hindered parasitic corrosion of the biodegradable anode and significantly enhanced the performance of the battery. The batteries demonstrated a capacity and power delivery capability of up to 0.7 mAh and 26 µW, respectively, which are sufficient for powering MEMS-based IMD systems.

INTRODUCTION

Recent developments in biodegradable electronics offer new challenges in the field of energy sources. Implantable medical devices (IMDs) for the monitoring and treatment of transient diseases, such as bone fracture or wound healing and drug delivery systems, have typically been limited to either passive designs or wireless powering [1]. The development of biodegradable batteries, however, would support this emerging technology by providing an alternative means to power active transient devices. The ideal power source for transient IMDs should be biocompatible, and comprise biodegradable chemistries that would satisfy device power requirements during the lifetime of the IMD and benignly degrade thereafter. For reference, the power consumption of currently available non-degradable IMDs ranges from 10-1000 µW [2].

Magnesium (Mg), iron (Fe) and zinc (Zn) are metals that may be considered for biodegradable battery chemistries [3]. These metals have already found applications in biodegradable implants, such as stents and bone screws, with demonstrated biocompatibility *in vivo* [4]. In particular, Mg offers desirable mechanical and electrochemical properties. Magnesium is a light, alkaline earth metal with a density of 1.74 g/cm^3, comparable to that of aluminum. Magnesium also features a standard electrode potential of -2.34 V *vs.* standard hydrogen electrode (SHE) and a theoretical capacity of 2.2 Ah/g. Magnesium films at thicknesses greater than 1 µm can be micropatterned by the subtractive etching of commercial Mg foil or electrodeposited from a non-aqueous solution [5]. Similarly, Fe films can be fabricated by physical vapor deposition and aqueous electrodeposition.

This study presents the design, fabrication and testing of a biodegradable Mg/Fe battery for the powering of transient IMDs. The battery features a Mg anode and Fe cathode in a 0.1 M magnesium chloride ($MgCl_2$) electrolyte solution. The governing chemistry of the battery is the cathodic protection of the Fe electrode through the oxidation of Mg and, thus, reduction of hydrogen on the Fe cathode surface (Figure 1). In addition to these discharge reactions, Mg naturally degrades in aqueous solutions. This effect, or parasitic corrosion of the Mg, would detract from the capacity and energy of the battery. To hinder this effect, the Mg surface was protected with either polycaprolactone (PCL) or poly(glycerol-sebacate) (PGS), both biodegradable and biocompatible polymers [6]. As the thickness and permeability of the polymeric coating may contribute to mass transfer resistance and impact discharge performance of the battery, PCL and PGS were compared at varying thicknesses to determine the relevant parameters for achieving a high coulombic efficiency and discharge potential. The timescale of PCL and PGS degradation (i.e., months) exceeds the lifetime of the battery, and the thickness of the polymer coating does not appreciably change during discharge of the battery. The electrolyte was selected as $MgCl_2$ because Mg^{2+} and Cl^- ions are constituents of physiological solutions [7].

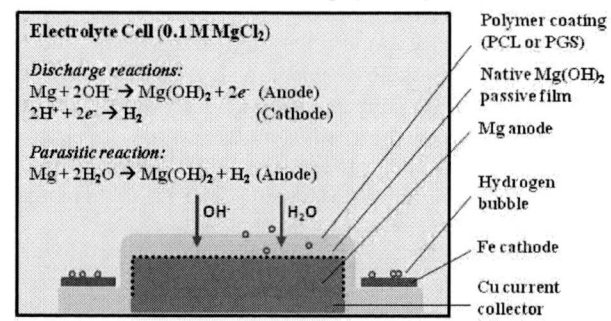

Figure 1: Schematic of the Mg/Fe biodegradable battery. Electrolyte diffuses across the polymer coating and native magnesium hydroxide (Mg(OH)$_2$) film to reach the Mg surface, where oxidation reactions that supply electrons for the battery and generate hydrogen occur. The Mg(OH)$_2$ formation is countered by mechanical disruption of this passive film from the discharge current. Hydrogen is reduced on the Fe cathode surface.

978-1-4799-3510-9/14 $31.00 © 2014 IEEE

In this initial study, a silicon battery substrate and acrylic electrolyte cell were used to isolate any effects of biodegradable packaging. The longer-term goals are to implement biodegradable packaging and substrate for these batteries as well.

EXPERIMENTAL
Fabrication of the Mg/Fe battery

Fabrication of the biodegradable battery began with through-mold electrodeposition of Mg to form the anode (Figure 2). Silicon dioxide (500 μm thickness) and metallic seed layers (50 nm Ti/ 500 nm Cu/ 50 nm Ti) were deposited onto a silicon wafer by plasma-enhanced chemical vapor deposition (PECVD) and sputter deposition, respectively. To achieve a mesh seed layer, the Ti and Cu were chemically etched through a photomask. Next, an electroplating mold was patterned using polyvinyl alcohol (PVA), a water-soluble polymer with low solubility in solvent solutions. The negative image of the electroplating mold was patterned with NR-21 photoresist (Futurrex), and PVA (33 w/v%) was spin-coated onto the wafer and cured. A brief oxygen plasma treatment removed any PVA coating the edges of the photoresist. The sample was sonicated in acetone to achieve a micropatterned water-soluble mold.

Magnesium was electroplated from a non-aqueous solution of methylmagnesium chloride and aluminum chloride at a 6:1 molar ratio in tetrahydrofuran (THF). The electrodeposition was performed in a moisture-free glove box under inert nitrogen atmosphere. Pulse plating was conducted with an average current density of 10 mA/cm^2 at 20% duty cycle to a thickness of 35 μm. The electroplating mold was removed by solubilizing in water. The Mg anode was coated with either PCL or PGS. Commercial 80 kDa PCL pellets were solubilized in trifluoroethanol at a concentration of 100 mg/mL. PGS was synthesized by a polycondensation reaction of sebacic acid and glycerol, as described in [6], and diluted in THF (33 w/v%). PCL or PGS

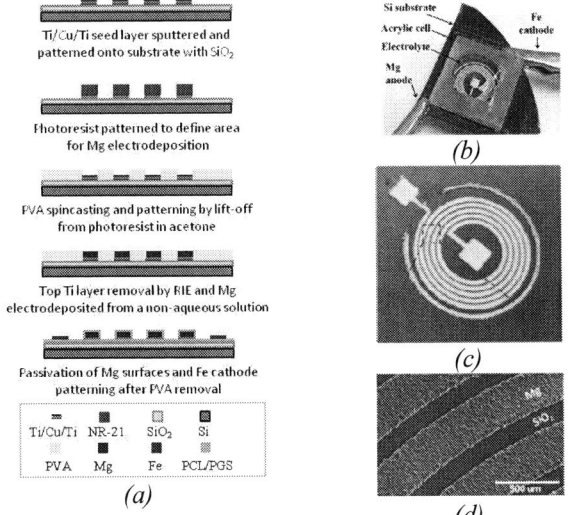

Figure 2: (a) Fabrication process and (b) testing setup of the Mg/Fe battery, and (c) optical and (d) SEM images of the electroplated Mg anode.

was spin-coated onto the Mg anode at varying thicknesses, and cured at room temperature and at 120°C under vacuum, respectively.

The Fe cathode was patterned by e-beam evaporation through a Kapton (DuPont) shadow mask to a thickness of 300 nm. Both the shadow mask and electrolyte cell were fabricated by laser micromachining with a CO$_2$ laser.

Characterization of battery components

The surface morphology and elemental composition of the Mg and Fe electrodes were characterized by SEM/EDX (Hitachi S-3700N). Potentiodynamic polarization tests were performed between -1.7 and -1.2 V at a scan rate of 1 mV/s to characterize the corrosion behavior of electroplated Mg and the effect of polymeric coatings. A three-electrode setup was implemented with Mg, Pt, and Ag/AgCl as the working, counter, and reference electrodes, respectively. Tests were conducted in 0.1 M aqueous MgCl$_2$ solution with a potentiostat (Wavedriver 10, Pine Instruments).

Water vapor permeability tests were performed to evaluate the barrier properties of the biodegradable polymers. Glass vials filled with calcium oxide, a hygroscopic desiccant, were fitted with polyethylene filter connectors containing either PCL or PGS membranes. The polymer membranes were 400 μm in thickness and cut with a CO$_2$ laser. The filters were filled with 1 mL of DI water to provide saturated conditions on one side of the polymer membrane and sink conditions within the vial. Silicone o-rings were used to provide a tight seal and the system was wrapped in parafilm. At periodic time points, the glass vials were removed and weighed to determine the rate of water permeation across the polymers.

Electrochemical testing of biodegradable batteries

A two-electrode-cell configuration, where Mg and Fe serve as anode and cathode, respectively, was utilized to test the performance of the batteries with a potentiostat (Model 263, EG&G Princeton Applied Research). The electrolyte was selected as 0.1 M aqueous MgCl$_2$ solution, which provided a conductivity of 0.9 mS/cm. Tests were performed under galvanostatic conditions at a discharge current of 25 μA.

RESULTS & DISCUSSION
Characterization of battery components

Figure 2 shows optical and SEM images of the microfabricated Mg anode. Surface elemental analysis confirmed the atomic composition of the electroplated metal as 95% Mg and 5% O (data not shown). While the theoretical standard electrode potential of Mg is -2.34 V, this value is not observed in practice due to a native Mg(OH)$_2$ film that forms on the Mg surface. Electrolyte must first diffuse across the passive film to react with the Mg. The film is further disrupted mechanically during current draw to expose more active material to the electrolyte [8]. Figure 3 shows the polarization curves of uncoated Mg and Mg coated with either PCL or PGS in the proposed electrolyte. Uncoated Mg demonstrated a corrosion potential of -1.32 V *vs.* SHE. This

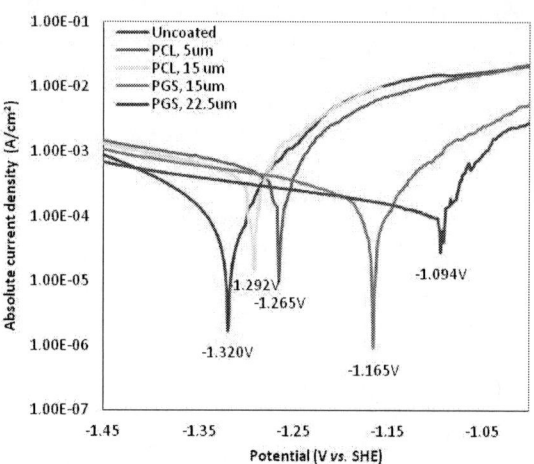

Figure 3: Polarization curves of the Mg anode coated with PCL and PGS at varying thicknesses, compared against uncoated Mg, in a 0.1M MgCl$_2$.

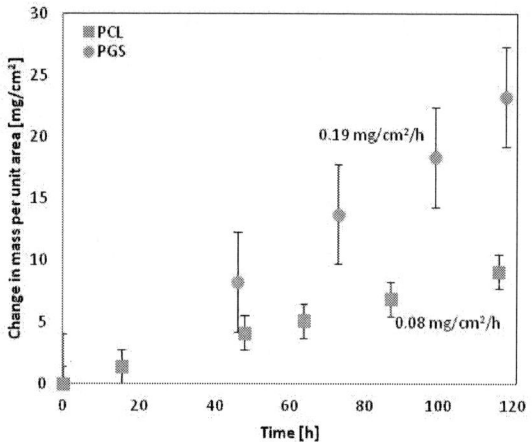

Figure 4: Water vapor permeability of PCL and PGS membranes (400 µm thickness). Data are mean $\pm$ SEM; n=3.

is in agreement with literature values for Mg in dilute chloride solutions [9]. In contrast, Mg coated with 15 µm of PCL or PGS exhibited corrosion potentials of -1.292 V and -1.165 V, respectively. The corrosion potential shifted towards the noble direction with PCL and PGS coatings, suggesting that the polymer coatings reduced the parasitic corrosion of Mg in the electrolyte solution [9-10]. The higher corrosion potentials observed with PGS coatings than PCL coatings further suggested that Mg coated with PGS may be more corrosion resistant. The fluctuations observed in the anodic curves were attributed to the pitting nature of Mg corrosion, with repeated breakdown and formation of the passivating film [9].

As a first step towards understanding ion transport across the biodegradable polymers, water permeability tests were conducted to evaluate the diffusion properties of the polymers (Figure 4). As shown in Figure 4, the rate of water transport across the PCL and PGS membranes were 0.08 mg/cm^2/h and 0.19 mg/cm^2/h, respectively. These findings confirmed that the polymer coatings would not eliminate water transport to the Mg anode surface.

Electrochemical testing of biodegradable batteries

Figure 5 demonstrates the effect of polymer coatings on the discharge profile and capacity of the Mg/Fe batteries. Uncoated batteries delivered the shortest service life and power, at 2.9 h and 15.5 µW, respectively. It was speculated that the hydrogen evolution, as well as the continuous breakdown and formation of the native passive film on the Mg surface, induced stress to the Mg film. The stressing may have been the cause of Mg delaminating from the substrate and consequently causing loss of electrical connection to the current collector. Hence, the polymer coating may provide a secondary purpose of mechanically stabilizing the Mg anode on the substrate.

Batteries coated with 5 and 10 µm of PCL showed higher discharge potentials than uncoated batteries. As expected, electrodes with thicker PCL coatings exhibited increased resistance to charge transfer and provided lower discharge potentials. With increasing polymer thickness, it becomes more difficult for OH$^-$ ions to travel to the Mg surface. This effect also governed the lifetime of the battery because with thicker polymer films, reaction products accumulated at the PCL-Mg interface with prolonged discharge and corresponded to a decline in discharge potential. This can be observed from batteries coated with 20-µm-thick PCL. Further, hydrogen formation at the Mg anode caused a buildup of internal pressure with thicker PCL membranes. This weakened the adhesion of the PCL film to the substrate and, in some instances, resulted in film delamination.

PGS-coated batteries delivered longer discharge lifetimes than the uncoated batteries. As PGS exhibited higher water permeability and PGS-coated Mg demonstrated lower corrosion potential, it was expected that batteries coated with PGS would also demonstrate better performance than those coated with PCL. This was observed with 10-µm-thick PGS, which showed the highest capacity and coulombic efficiency of 0.7 mA/h and 13.5%, respectively. This discharge lifetime was similar to that of PCL-coated batteries at 5 and 10 µm thickness. PGS also provided greater mechanical stability, as thicker PGS films did not delaminate during discharge, as with thicker PCL films. The lower average potential and less stable discharge profile obtained with PGS-coated batteries, especially with increasing polymer thickness, suggested an accumulation of reaction products at the PGS-Mg interface. This was confirmed with X-ray photoelectron spectroscopy (XPS) of PGS-coated batteries after discharge, which showed Mg(OH)$_2$ in regions where Mg was consumed (data not shown). As such, PGS-coated batteries provided slightly lower power at 25 µA discharge, and the highest power and energy were obtained from PCL-coated batteries. PCL-coated batteries showed a maximum of 26.2 µW of power and 2.6 J of energy. As these results fall within the range in power requirements for commercial IMDs, the presented battery chemistries show a promising direction towards providing viable energy sources for powering transient implantable electronics. The range in average discharge potential obtained with batteries coated

978-1-4799-3510-9/14 $31.00 © 2014 IEEE

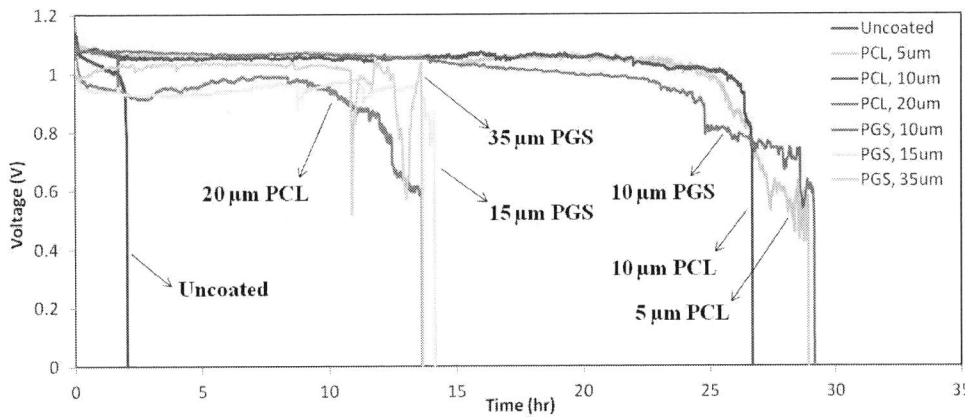

Figure 5: Galvanostatic discharge profiles at a current of 25 µA. The cutoff potential was 400 mV.

Table 1: Summary of battery performance at a discharge current of 25 µA.

Passivation	Uncoated	PCL			PGS		
		5um	10um	20um	10um	15um	35um
Discharge time (h)	2.9	28.9	26.7	13.6	29.1	14.1	13.6
Capacity (µAh)	73	722	667	341	729	353	341
Power (µW)	15.5	25.3	26.2	22.9	24.8	23.5	25
Energy (J)	0.2	2.6	2.5	1.1	2.6	1.2	1.2
Coulombic efficiency (%)	1.0	13.0	12.0	6.3	13.5	6.6	6.3

with PGS and PCL, which exceeded the average discharge potential of the uncoated battery, suggested that there exists an optimum coating and thickness for maximizing the power and performance of Mg/Fe batteries. The coating should hinder parasitic corrosion at the Mg anode without appreciably increasing resistance to mass transfer. Further, the coating should be mechanically robust to accommodate for hydrogen evolution at the polymer-Mg interface. Future work may further optimize the use of biodegradable polymer coatings or consider alternative means of enhancing corrosion resistance of the Mg anode. Biodegradable material sets should also be considered for the substrate and packaging. Further design constraints should be governed by the design requirements for a targeted transient disease state.

CONCLUSIONS

This study comprised the design, fabrication and characterization of a series of MEMS-enabled biodegradable batteries. The underlying principle of the battery operation is the cathodic protection of an evaporated Fe electrode through the oxidation of an electroplated Mg anode in an $MgCl_2$ electrolyte solution. Two different biodegradable polymer coatings, PGS and PCL, were utilized as passivation layers on the Mg anode to minimize the parasitic degradation of the Mg and, thus, to increase the coulombic efficiency of the system. It was demonstrated that the thickness, as well as the choice of the polymer coating, have a remarkable effect on the battery performance. Thinner coatings yielded higher energy densities while no significant change is observed in power densities. Power and energy values of more than 25 µW and 2.6 J, respectively, were obtained, which fall within range of the reported performance requirements for the commercial IMDs.

REFERENCES

[1] Hwang, S.W. et al., "A physically transient form of silicon electronics", Sci., vol. 337, pp. 1640-44, 2012.

[2] Wei, X. and J. Liu, "Power sources and electrical recharging strategies for implantable medical devices", Front. Energy Power Eng. China, vol. 2, pp. 1-13, 2008.

[3] Luo, M. et al., "A microfabricated wireless RF pressure sensor made completely of biodegradable materials", Dig. Solid-State Sensors, Actuators, and Microsystems Workshop, Hilton Head, 2012.

[4] Morajev, M. and D. Mantovanni, "Biodegradable metals for cardiovascular stent applications: Interests and new opportunities", Int. J. Mol. Sci., vol. 12, pp. 4250-4270, 2011.

[5] Tsang, M. et al., "Methods for the microfabrication of magnesium", MEMS 2013, Taipei, Taiwan, 2013.

[6] Sundback, C.A. et al., "Biocompatibility analysis of poly(glycerol sebacate) as a nerve guide material", Biomat., vol. 26, pp. 5454-64, 2005.

[7] Feyerabend, F. et al., "Ion release from magnesium materials in physiological solutions under different oxygen tensions", J. Mater Sci: Mater Med, v. 23, pp. 9-24, 2012.

[8] Ratnakumar, B.V., "Passive films on magnesium anodes in primary batteries", J. App. Electrochem., vol. 18, pp. 268-79, 1988.

[9] Gill, P. et al., "Corrosion and biocompatibility assessment of Mg alloys", J. Biomat. And Nanobiotech., v. 3, pp. 10-13, 2012.

[10] Walter, R. et al., "Influence of surface roughness on the corrosion behavior of magnesium alloy", Mat. And Des., v. 32, pp. 2350-54, 2011.

CONTACT

M. Tsang, tel: +1-404-894-5251; melissa_tsang@gatech.edu

A MICRO-SCALE MICROBIAL SUPERCAPACITOR

Hao Ren[1], He Tian[2], Tian-Ling Ren[2], and Junseok Chae[1]

[1]School of Electrical, Computer and Energy Engineering, Arizona State University, U.S.A.

[2]Institute of Microelectronics & Tsinghua National Laboratory for Information Science and Technology (TNList), Tsinghua University, Beijing, CHINA

ABSTRACT

We report a MEMS microbial supercapacitor, aiming for a carbon-neutral renewable miniaturized electrochemical power converter. Microbial electrochemical technologies have been studied for years, yet the current and power density of them are still significantly lower than those of existing energy conversion techniques, which limits their potential applications. This work presents a microbial supercapacitor with a graphene-inserted anode having current and power density of more than one order of magnitude enhancement over prior art, to meet high current and power demand. Current and power density of 450 A/m^2, and 202.5 W/m^2 are achieved, which is more than 15 and 29 folds of the previous records of microbial electrochemical techniques, delivering the micro-scale microbial supercapacitor as an attractive alternative to existing energy conversion and storage device.

INTRODUCTION

In the era of energy crisis, bioenergy becomes attractive due to the abundant biomass on earth, its carbon-neutrality, and renewability [1]. A microbial electrochemical technique is to convert biomass directly to electricity or fuel, harvesting electrons from specific bacteria species, exoelectrogen ("exo" means out of and "electrogen" means bacteria producing electrons), via their unique extracellular electron transport (EET). One of such techniques is a microbial fuel cell (MFC). Current and power densities of MFCs have improved by 10^4 folds, including our recent work demonstrating the record-high 30 A/m^2 and $6\text{-}7 \text{ W/m}^2$ [2, 3]. However in the past few years, performance improvement of MFCs has stalled [2, 4].

Various approaches have been implemented in improving current and power generation capability. For example, Inoue *et al.* and Mink *et al.* used CNT anode[5, 6], and Xie *et al.* and He *et al.*[7, 8] used graphene anode, respectively, to enhance current and power density of their MFCs. Despite the successful improvement, the current and power density of MFCs are still orders of magnitudes lower than those of other electrochemical methods such as hydrogen fuel cell and lithium ion battery.

Supercapacitors are known to have high current/power generation capability, meeting high current and power demand. Thus, microbial supercapacitors have the potential to bridge the low current/power of microbial electrochemical techniques to high current/power demand. Besides high current and power generation capability, the microbial supercapacitor is carbon neutral and renewable, in contrast to conventional supercapacitors based on non-renewable metal oxide, such as MnO_2 and RuO_2. Furthermore, the microbial supercapacitor may become a low cost energy converting and storage device, due to the abundance of biological entities. Capacitive behavior of exoelectrogen on electrode has been discovered recently and microbial supercapacitor has been presented [9-12]. However, the reported current generation capabilities are low, less than 2 A/m^2.

In this paper, we present a microbial supercapacitor with high current/power density, which can store and release the electrons produced in the metabolism of *Geobacter sp.* The microbial supercapacitor significantly enhances the current/power density by more than one order of magnitude, compared with conventional microbial electrochemical techniques. Stability of the microbial supercapacitor is characterized by high cycle stability.

OPERATION PRINCIPLE AND FABRICATION OF THE MICROBIAL SUPERCAPACITOR

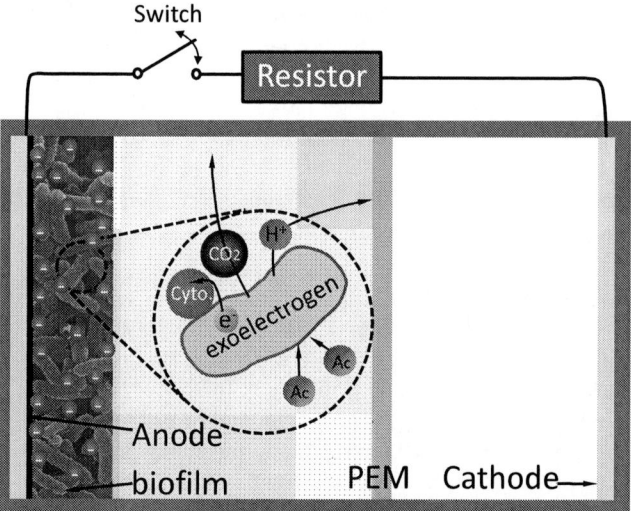

Figure 1. Schematic of the microbial supercapacitor: when a high-speed switch is off, electrons are stored in the biofilm on graphene anode, and when it is on, electrons stored inside biofilm are discharged very quickly, resulting an extremely high current and power density.

A schematic of the operation principle of the supercapacitor is illustrated in Figure 1. A layer of biofilm is grown on the anode, and the redox proteins in biofilm, such as cytochrome c, store and release electrons [9]. During charging process, the high speed switch is off, and

978-1-4799-3510-9/14 $31.00 © 2014 IEEE

exoelectrogen break down organic matter and store electrons in the redox cofactors in biofilm. During discharging process, the switch is on, and electrons stored inside biofilm are released at a high speed, resulting in a high current and power density.

MATERIALS AND METHODS
Electrode fabrication

Bare gold electrodes were deposited on a glass slide (micro slides, $4.6 \times 2.6 \times 0.1$ cm^3, VWR) with six through holes pre-drilled mechanically: one inlet, one outlet, and four for assembly. Afterwards, Cr/Au (20 nm/200 nm) films were sputtered on the glass slides. Afterwards, CVD graphene on copper foil was transferred with a PMMA protection film to Cr/Au on the slide. Then, PMMA was removed by acetone to leave the graphene on Cr/Au. An optical image of the slide is shown in Figure 2a, and Raman spectra were obtained from the graphene film on gold (Figure 2b), showing a typical monolayer feature with sharp G peak and a single 2D peak with higher intensity. The ratio of I_{2D} to I_G is ~2.0, which presents the good quality of single-layer graphene.

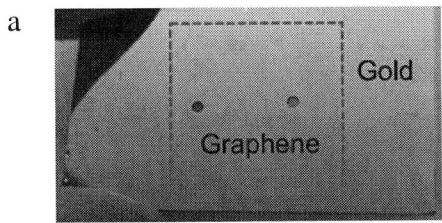

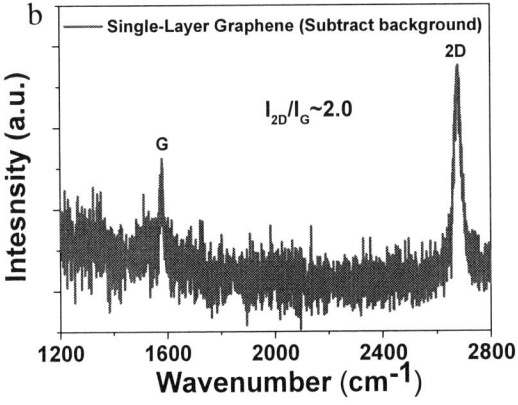

Figure 2 (a) Optical image of the graphene on Cr/Au on a glass slide (b) Raman spectra of graphene

Device assembly

The device has a proton exchange membrane (PEM) (Nafion 117, Sigma Aldrich), to permit cation transport and to avoid electrical short-circuiting and electrolyte cross mixing. Two silicone gaskets (250 μm thick, Fuel Cell Store Inc.) are sandwiched between anode and cathode. The anode and cathode are comprised of a graphene layer on Cr/Au and a Cr/Au film on glass slides, respectively. The volume of each chamber is 25 μL, and the size of the electrode is 1 cm^2. Two nanoports (10-32 Coned assembly, IDEX Health &

Science) were used to provide microfluidic pathways into and out of each chamber. The assembly process of the microbial supercapacitor started with the preparation of the anodes and cathodes, as described in the previous section. Then, the nanoports and fluidic tubing (PEEK polymer, IDEX Health & Science) were aligned and glued to the inlets/outlets to supply anolyte and catholyte. Next, silicone rubber gaskets were patterned to define the chamber. Finally, the MFC was assembled with four screw bolts and nuts to minimize oxygen/electrolyte leakage.

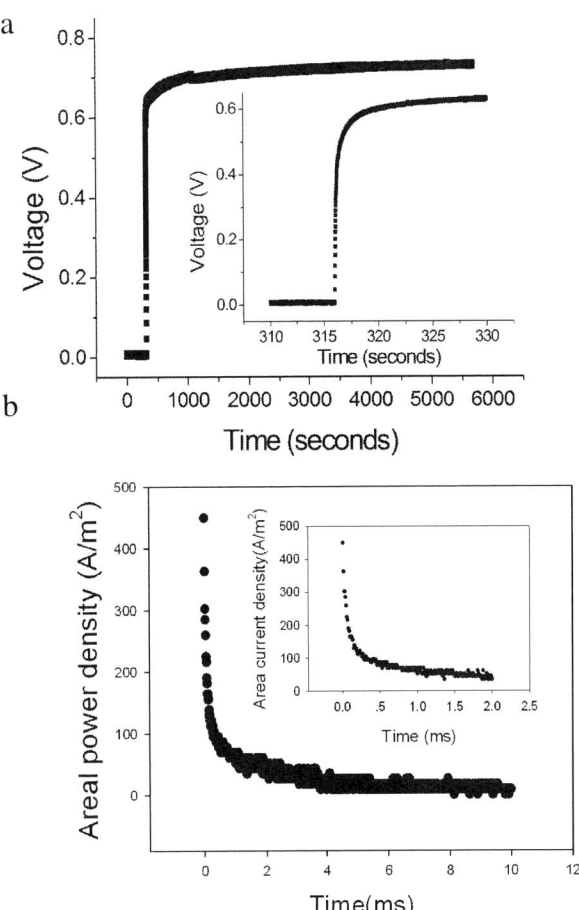

Figure 3 Charging and discharging characteristics of the microbial supercapacitor

Inoculum and electrolyte

The inoculum for the microbial supercapacitor was obtained from an acetate-fed microbial electrolysis cell (MEC) that had *Geobacter*-enriched bacterial community originally from anaerobic-digestion sludge. The anolyte was composed of a 25-mM sodium acetate medium with 1,680 mg KH_2PO_4, 12,400 mg Na_2HPO_4, 1,600 mg NaCl, 380 mg NH_4Cl, 5 mg EDTA, 30 mg $MgSO_4 \cdot 7H_2O$, 5 mg $MnSO_4 \cdot H_2O$, 10 mg NaCl, 1 mg $Co(NO_3)_2$, 1 mg $CaCl_2$, 0.001 mg $ZnSO_4 \cdot 7H_2O$, 0.001 mg $ZnSO_4 \cdot 7H_2O$, 0.1 mg $CuSO_4 \cdot 5H_2O$, 0.1 mg $AlK(SO_4)_2$, 0.1 mg H_3BO_3, 0.1 mg $Na_2MoO_4 \cdot 2H_2O$, 0.1 mg Na_2SeO_3, 0.1 mg $Na_2WO_4 \cdot 2H_2O$, 0.2 mg $NiCl_2 \cdot 6H_2O$, and 1 mg $FeSO_4 \cdot 7H_2O$ (per liter of deionized water) (pH 7.8 ± 0.2). For the start-up process, inoculum and anolyte were mixed at a volumetric ratio of 1:1. The catholyte was composed of 50-mM potassium ferricyanide in a 100-mM phosphate buffer solution (pH 7.4). The anolyte and catholyte were supplied into the microbial supercapacitor using a syringe pump.

RESULTS AND DISCUSSION

Charging and discharging characteristics

Charging and discharging characteristics of the supercapacitor are shown in Figure 3. The charging process takes 1000s of seconds. During this process, exoelectrogen break down organic matter and generate electrons, which are transferred to redox cofactors in the biofilm through EET. Thus redox cofactors are reduced, so that their potential decreases, and the voltage between anode and cathode increases. Discharging process takes place in a much shorter period. During this process, electrons stored inside biofilm are released at a high speed, resulting in a high current/power density, 450 A/m^2 and 202.5 W/m^2, which is more than one order of magnitude higher than those reported in prior literature [2, 3].

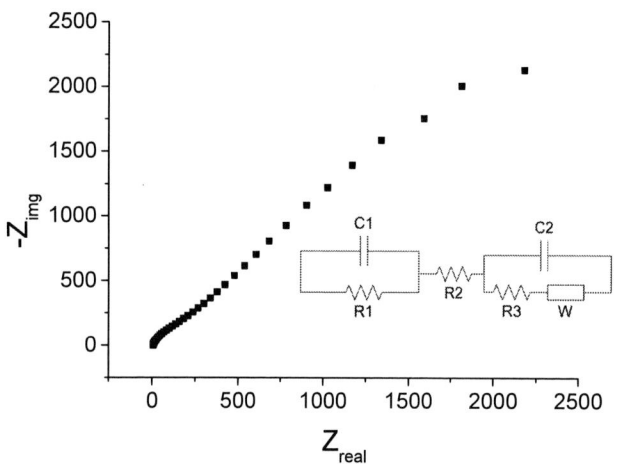

Figure 4: Nyquist plot of the microbial supercapacitor (inset: equivalent circuit model of the microbial supercapacitor)

Electrochemical impedance spectroscopy

Electrochemical impedance spectroscopy was used to study the impedance of the microbial supercapacitor using a potentiostat (Gamry Instruments). We fitted the impedance model to the measured Nyquist plot (Figure 4). The charge transfer resistances (R_1 and R_3 in the equivalent circuit model) of 13 Ω and 36 Ω are obtained for the anode and the cathode, respectively.

Cyclic voltammetry

Cyclic voltammetry of the microbial supercapacitor was also performed, as shown in Figure 5. The CV curve shows an obvious peak at approximately -0.3V versus Ag/AgCl reference electrode (in 3M NaCl solution), which suggests the psuedocapacitance of the microbial supercapacitor.

Cycle stability

In order to demonstrate the stability of the supercapacitor, cyclic charging/discharging were performed, as shown in Figure 6. The microbial supercapacitor demonstrates excellent cycle stability. The charges stored in each cycle remained almost constant up to 1,000,000 cycles with charging/discharging of 0.25 sec / 0.25 sec. The high cyclic stability is believed to be due to the excellent mechanical and electrochemical stability of graphene film and microbial biofilm.

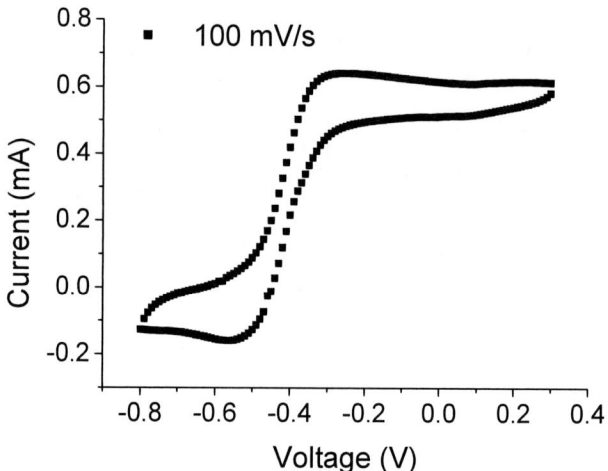

Figure 5 Cyclic voltammetry of the microbial supercapacitor at 100 mV/s.

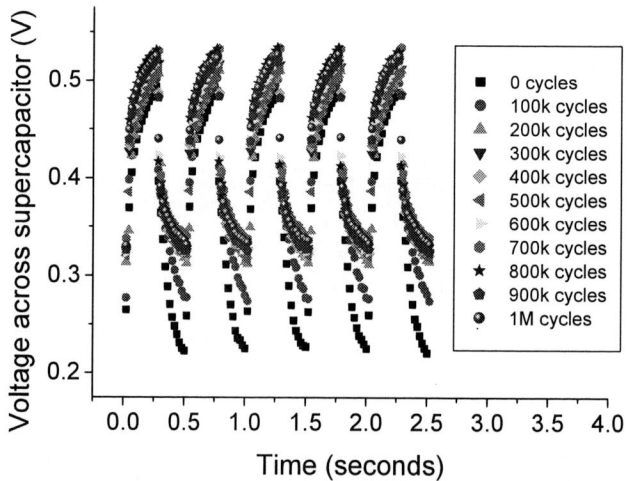

Figure 6: Charging/discharging of the microbial supercapacitor up to 1,000,000 cycles

Insight of microbial supercapacitor vs. MFC

According to Ohm's Law, the current of the microbial supercapacitor can be calculated by:

$$I(t) = \frac{V(t)}{R_i(t) + R_e} \qquad (1)$$

where $I(t)$, $V(t)$, $R_i(t)$, R_e are the current, voltage, internal resistance and external resistance of the microbial supercapacitor during the discharging process, and the first three terms are time dependent. Based on equation (1), we can calculate the change of $R_i(t)$ versus time based on the discharging characteristics, as shown in Figure 7.

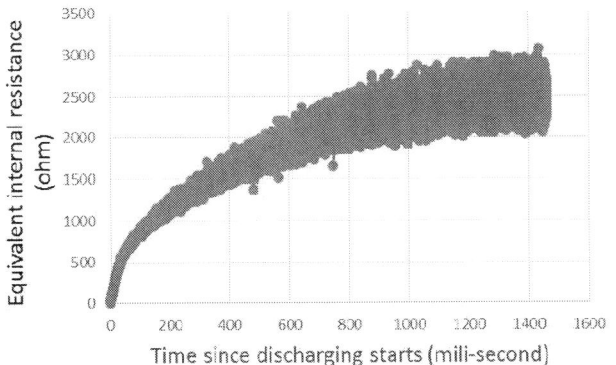

Figure 7 Equivalent internal resistance extracted from the microbial supercapacitor

At the very beginning of the discharging process, the equivalent internal resistance is very small, merely 6 Ω, while as the discharging progresses, equivalent internal resistance reaches 2-3 kΩ. In other words, as the microbial supercapacitors discharges, equivalent internal resistance increases and current generation capability reduces. The reduced redox cofactors inside the biofilm decrease as the discharge process proceeds. This may suggest the kinetic reactions of EET is studied by the charging and discharging microbial biofilm. Further work needs to be performed to study the EET, and consequently enhancing the steady state current/power generation of microbial electrochemical techniques.

CONCLUSION

A high current/power micro-scale microbial supercapacitor with a graphene-inserted anode is presented in this paper, aiming towards a carbon-neutral, renewable energy converter and storage device. Comprehensive characterization of the supercapacitor is performed, including charging/discharging, electrochemical impedance spectroscopy, cyclic voltammetry and cyclic stability. High current/power density, 450 A/m^2, and 202.5 W/m^2, which are more than one order of magnitude higher than those reported in prior literature, and a cycle lifetime of more than 1,000,000 is presented. The high performance and stability makes the microbial supercapacitor a strong candidate for future energy conversion and storage applications.

ACKNOWLEDGEMENTS

This work was supported by the National Natural Science Foundation of China (61025021, 60936002, 51072089, and 61020106006), the National Key Project of

Science and Technology (2011ZX02403-002) and the Special Fund for Agro-scientific Research in the Public Interest (201303107). He Tian is additionally supported by the Ministry of Education Scholarship of China.

REFERENCES

[1]   A. J. Ragauskas, C. K. Williams, B. H. Davison, G. Britovsek, J. Cairney, C. A. Eckert, *et al.*, *science*, vol. 311, pp. 484-489, 2006.

[2]   B. E. Logan and K. Rabaey, *Science,* vol. 337, pp. 686-690, 2012.

[3]   H. Ren, S. Rangaswami, H.-S. Lee, and J. Chae, MEMS 2013, pp. 869-872.

[4]   H. Ren, H.-S. Lee, and J. Chae, *Microfluidics and Nanofluidics,* vol. 13, pp. 353-381, 2012.

[5]   J. E. Mink, J. P. Rojas, B. E. Logan, and M. M. Hussain, *Nano Lett,* vol. 12, pp. 791-5, Feb 8 2012.

[6]   S. Inoue, E. A. Parra, A. Hinga and L. Lin *et al. Sensors and Actuators A-Physical,* vol. 177, pp. 30-36, 2012.

[7]   X. Xie, G. H. Yu, N. Liu, Z. N. Bao, C. S. Criddle, and Y. Cui, *Energy & Environmental Science,* vol. 5, pp. 6862-6866, May 2012.

[8]   Z. He, J. Liu, Y. Qiao, C. M. Li, and T. T. Y. Tan, *Nano letters,* vol. 12, pp. 4738-4741, 2012.

[9]   N. S. Malvankar, T. Mester, M. T. Tuominen, and D. R. Lovley, *ChemPhysChem,* vol. 13, pp. 463-468, 2012.

[10]   P. S. Bonanni, G. D. Schrott, L. Robuschi, and J. P. Busalmen, *Energy & Environmental Science,* vol. 5, pp. 6188-6195, 2012.

[11]   A. Deeke, T. H. Sleutels, H. V. Hamelers, and C. J. Buisman, *Environmental science & technology,* vol. 46, pp. 3554-3560, 2012.

[12]   N. Uría, X. Muñoz Berbel, O. Sánchez, F. X. Muñoz, and J. Mas, *Environmental science & technology,* vol. 45, pp. 10250-10256, 2011.

CONTACT

*Tian-Ling Ren, tel: +86-10-62798569; RenTL@tsinghua.edu.cn

* Junseok Chae, tel: +01-480-965-2082; Junseok.Chae@asu.edu

A THREE-DIMENSIONAL ELECTROSTATIC/ELECTRET MICRO POWER GENERATOR FOR LOW ACCELERATION AND LOW FREQUENCY VIBRATION ENERGY HARVESTING

K. Tao, S.W. Liu, J.M. Miao and S.W. Lye
School of Mechanical and Aerospace Engineering
Nanyang Technological University, Singapore

ABSTRACT

This paper presents the fabrication and characterization of a novel three-dimensional (3D) electret-based micro power generator, which is capable of converting low acceleration (<0.05g) and low frequency (<100Hz) ambient kinetic energy to electrical energy. A localized charging method integrating multiple needles is proposed. The experimental analysis shows that the proposed generator operates an out-of-plane direction at mode I of 66Hz and two in-plane directions at mode II of 75Hz and mode III of 78.5Hz with a phase difference of about 90°. It can be a potential candidate in the development of a 3D vibration energy harvester.

INTRODUCTION

In recent years, MEMS energy harvesters from ambient environment are gaining increasing research interest due to their broad applications and potential for integration with electronic circuitry [1, 2]. Vibration-based energy harvesters are normally based on electromagnetic, piezoelectric or electrostatic conversion mechanisms.

The development of MEMS-based vibration energy harvesters encounters several challenges. Firstly, current reported MEMS energy harvesters meet with difficulties in achieving low-resonant frequency structures (several tens of Hz level) in a comparatively small space, since the vibration frequency existing in the ambient environments is normally below 100 Hz [3]. To decrease the resonant frequency, silicon-on-insulator (SOI) wafer and high-aspect-ratio (HAR) silicon spring micromachining process are commonly utilized. The full wafer thickness is used for the seismic mass, while the suspension springs are only fabricated on the silicon layer above the buried oxide (BOX) layer [4-6]. However, the reported frequencies are still several hundreds of Hz. Two-sided DRIE and HF releasing would further increase the difficulty and complexity of the whole fabrication process. Moreover, the reported resonant energy harvesters are only single direction targeted, which comprise of a set of springs that are only flexible along the desired axis while keeps stiff in out-of-axis directions. They can only get the optimal power output while their oscillation is precisely aligned with the excitation vibration. However, the vibration directions in the ambient environment are unpredictable, even arbitrary or multidirectional, which poses a challenge to deal with alignment. To solve these problems, Bartsch et al. [6] proposed a two-dimensional resonator which has the potential to extract energy from ambient vibrations with arbitrary planar motion directions.

Based on the similar spring design, Liu et al. [4] developed another 3D electromagnetic MEMS energy harvester which can harvest energy from three-dimensional excitations by three sets of double-layer aluminum coils. However, the resonant frequencies reported are hundreds or thousands of Hz, which is still too high comparing with that of ambient vibrations.

In this work, a novel 3D electret-based micro power generator with spiral springs is proposed for scavenging energy from low-level ambient kinetic energy, which is readily available and ubiquitously exists in our daily life. Based on the parallel-spiral-spring design, the resonant frequency is much smaller than that of the 2D resonator previously proposed with similar dimensions. Thus, the low-frequency spring-mass system can be easily formed by one mask DRIE step without complicated process of SOI wafers.

OPERATING PRINCIPLES

In electrostatic/electret based energy harvesting systems, electrical power is generated from capacitance change between the two parallel plates. The particularity of electret-based energy harvester is utilizing an electret as a negative/positive permanent surface voltage source. Similar to the electrostatic energy conversion modes, the electret based power generator normally operates in two configurations: in-plane overlap varying (Figure 1(a)) and out-of-plane gap closing (Figure 1(b)). In this study, both in-plane and out-of-plane vibrations are investigated.

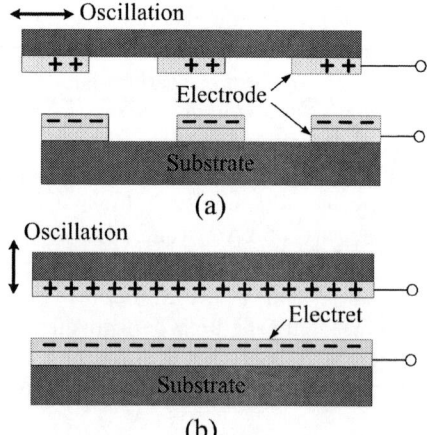

Figure 1: Two common operation modes of electret-based vibration energy harvester: (a) In-plane overlap varying and (a) Out-of-plane gap closing

978-1-4799-3510-9/14 $31.00 © 2014 IEEE

DESIGN AND SIMULATION

The schematic structure of the electret-based vibration energy harvester is shown in Figure 2. It consists of two parallel-silicon plates, stoppers, spacers and electret materials. The top plate composes of a movable spring-mass structure with gold electrode on it. The circular mass with a diameter of 6mm and a thickness of 300μm is suspended by a series of parallel spiral springs around. The spiral springs consist of three independent parallel beams with a width of 40μm and a height of 300μm at a spacing of 260μm. Liquid crystal polymer (LCP) is used as spacers to define the air gap thickness between electret and top electrode patterns. Ceramic spheres are dispersed on the surface of electret film as stoppers to avoid the stiction problems when out-of-plane vibration is excited. The electret material mounted on the bottom substrate can be charged as a negative/positive permanent surface voltage source. The electrode on bottom substrate serves as an anchored electrode. When the top plate is excited by an in-plane or out-of-plane oscillation, the capacitance between the top and bottom electrodes varies accordingly, which results an alternating current in the external circuit.

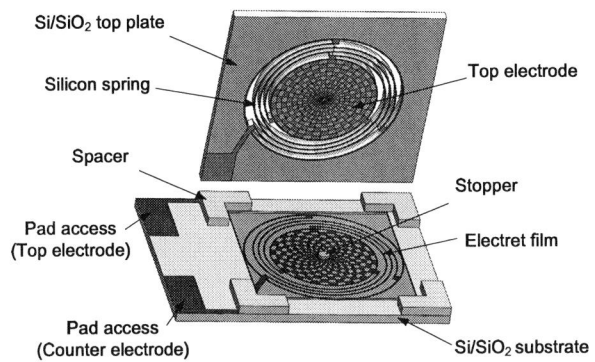

Figure 2: Schematic structure of the 3-D electret-based vibration energy harvester

Finite element method is employed to analyze the dynamic behavior of spring-mass system of the electret-based energy harvester. Modal analysis is performed by ANSYS simulation to find its first three basic vibration modes. Each end of three parallel spiral springs is considered as fixed terminal. Figure 3 shows the mode shapes of the first three vibration modes. Figure 3(a), 3(b) and 3(c) are corresponded to first three modes at the resonant frequencies of 66.9, 75.8 and 77.9Hz, respectively. It is seen that the circular-mass vibrates in out-of-plane direction at mode I, while oscillates in in-plane vibration directions at mode II and III. It's also observed that the oscillation at mode II occurs at the angle of 60°(240°), while mode III takes place at the angle of 150°(330°). It's about of 90° difference with each other. The resonant frequencies of two in-plane vibration modes are with little difference and similar vibration forms are observed, these may be due to the symmetrical architecture of the whole spring-mass structure and material properties in the simulation.

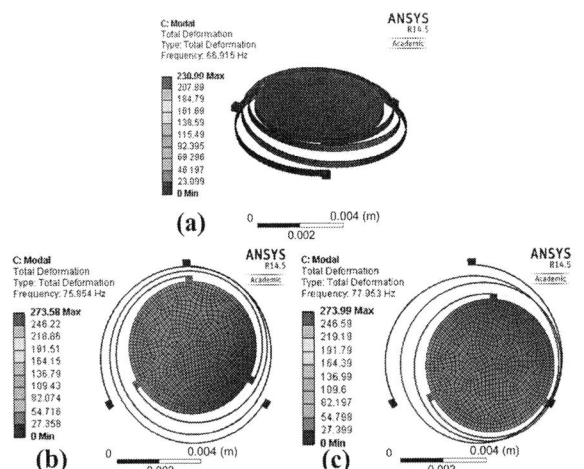

Figure 3: Modal simulation of Spring-mass structure: (a) mode I; (b) mode II; (c) mode III.

FABRICATION

Selective charging of electret films

A localized corona charging system integrating with a shadow mask and multiple discharge needles is developed to produce micro sized electret array in this study. Low-density polyethylene (LDPE) thin film is used as electret material in this prototype fabrication. It has a good dielectric strength up to 740kV/mm which is much higher than the commonly cited dielectric materials such as CYTOP (110 kV/mm) and Teflon AF (21kV/mm). The schematic structure of the charging system is shown in the figure 4.

Wafer-through DRIE process with a 150μm thick silicon wafer is used to get an array of micro-scale openings. Then 300nm gold is deposited by sputtering on one side of the wafer as conductive grid and 500nm silicon oxide is deposited by PECVD as insulation layer on the other side. Then the wafer is negatively biased which serves as a shadow mask to micro-patterning of the charge distribution.

An array of needles with high voltage, which functions as both discharge needles and metallic grid in conventional charging method, are utilized to enhance the uniformity of charge distribution. The electret thin film is then to be placed between the shadow mask on the top and the silicon substrate with gold electrode on the bottom.

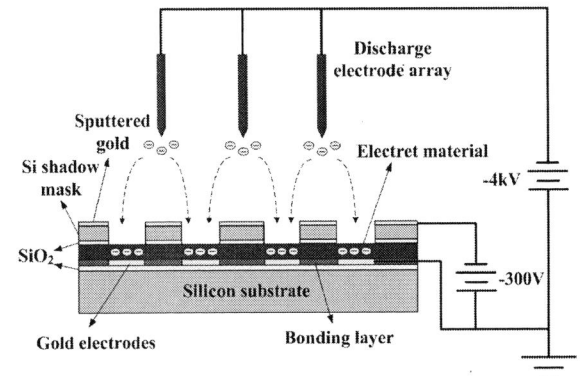

Figure 4: Corona localized charging by needle array and Si shadow mask

In this study, circular arrays of patterns are formed in both charging mask and the bottom electrode to facilitate charging. Figure 5 shows SEM image of micro-sized charge distribution on the LDPE thin films. The bright and dark areas correspond to negatively charged and uncharged areas, respectively. It is seen that good uniformity and high surface potential have been achieved through this method, which represents a significant improvement over the previously reported work [7, 8].

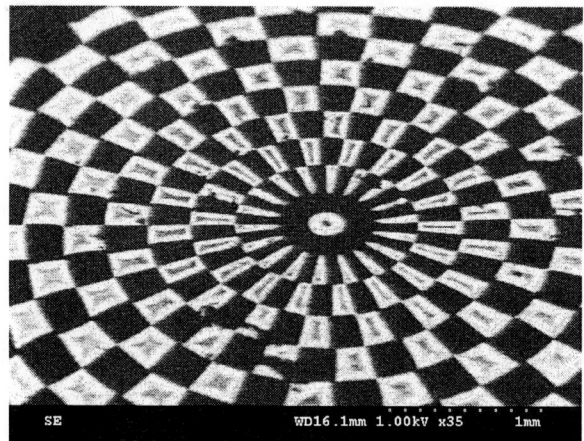

Figure 5: SEM image of charge distribution on the electrets films (negative charged)

Spring-mass structure fabrication

Both top and bottom substrates are fabricated on a (100) Si wafer with a 1μm oxide layer which is deposited by plasma-enhanced chemical vapor deposition (PECVD). Gold electrodes are patterned on oxide insulation layer by sputtering Ti (50nm)/Au (300nm) and lift-off process, and then followed by reactive-ion etching (RIE) and deep reactive-ion etching (DRIE) to define the in-plane geometry of the spring-mass structure.

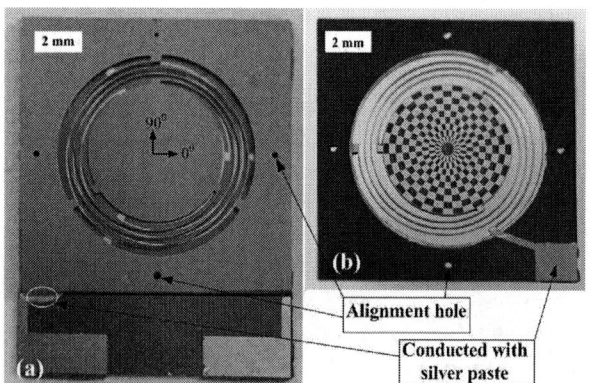

Figure 6: Photograph of (a) an assembled 3-D electret-based vibration energy harvester and (b) a fabricated spring-mass structure

After the top plate of spring-mass structure and the bottom electrode mounted with charged electret being prepared, the two plates are flip-chip packaged to form a two-parallel-layer structure. Optical photographs of an assembled 3-D electret-based vibration energy harvester and a fabricated spring-mass structure are shown in figure 6. The top electrode pad and one of the bottom electrode pads is connected by conductive silver paste, where the resistance of the connection is only several Ohms. Alignment holes on both the top and bottom plates with diameter of 300μm are etched to facilitate the plate alignment.

TESTING RESULTS

The fabricated energy harvester prototypes are tested by attaching to a linear vibrator, which is supported by a rotational frame. An L-shaped holder with a small rotational stage is fixed on the vibrator, so that both in-plane and out-of-plane vibration can be performed. An accelerometer is attached to the device holder to monitor the acceleration. A data acquisition (DAQ) system (NI USB-6289 M series) and a high speed camera system (PHOTRON FASTCAM-1024 PCI) are utilized to record the voltage output and mechanical motions of the prototype, respectively.

In-plane testing

For in-plane testing, excitation frequencies from 50 to 90Hz at a fixed acceleration of 0.05g are applied to the prototype for different excitation angles in increment of 15° at the load resistance of 10 MΩ. The resonant frequencies of in-plane oscillation at mode II and III are found to be at 75 Hz and 78.5 Hz, respectively. The peak-peak voltages as a function of excitation angles for mode II of 75Hz and mode III of 78.5 Hz are shown in figure 7(a) and 7(b), respectively. It is seen that the maximum output voltage of mode II reaches 116mV at the excitation angle of about 105°, while that of mode III obtains 153mV at the excitation angle of about 15°. With the help of high speed camera, it is also observed that the maximum amplitudes of the circular mass about 450μm at 105° of mode II and 560μm at 15° of mode III are obtained. Comparing with the simulation, the vibration directions at resonances shift about 45° for both mode II and mode III. These discrepancies are mainly due to the material properties and geometrical imperfections during the fabrication process. It is of a 90° phase difference between two in-plane vibration modes for both experimental results and modal analysis.

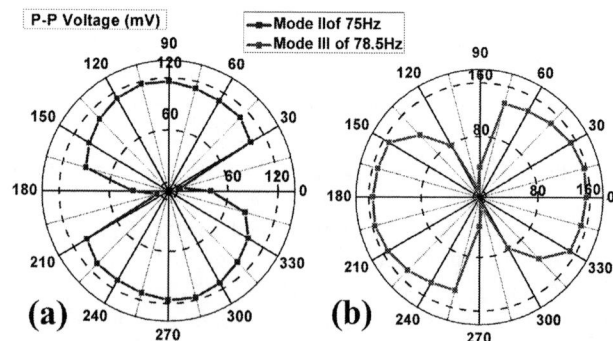

Figure 7: Peak-peak voltage output as a function of excitation angles for: (a) mode II of 75 Hz and (b) mode III of 78.5 Hz

Out-of-plane testing

For out-of-plane testing, a forced out-of-plane vibration with frequencies from 30 to 100Hz at acceleration of 0.05g with load of 20MΩ is applied on the prototype. The voltage responses versus excitation frequencies are shown in Figure 8(a). It indicates that the resonant frequency occurs at 66 Hz and maximum output of 330 mV can be obtained. Figure 8(b) shows voltage and power output against different resistances at 0.05g at the resonant frequency of 66 Hz. It indicates that the maximum power output of 4.8nW with an optimum load of 60MΩ can be obtained for out-of-plane vibration.

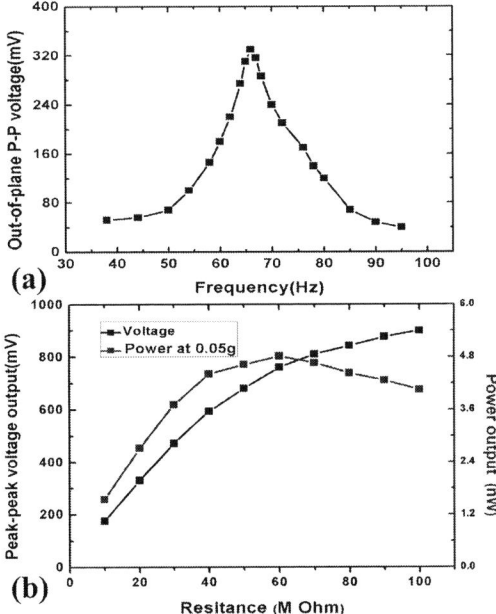

(a)

(b)

Figure 8: (a) Peak-peak Voltage output with a swept frequency 30~100Hz at 0.05g; (b) voltage and power output against different resistances at the frequency of 66Hz at 0.05g

To further study the performance of the device at mode I, the voltage responses at various accelerations ranging from 0.05g to 0.5g are investigated. The load resistance is 20MΩ and the excitation frequency is fixed at 66Hz. When the acceleration increases to 0.5g, the peak-peak output voltage is up to 2.93V. The voltage output as a function of time at an acceleration of 0.5g is shown in figure 9.

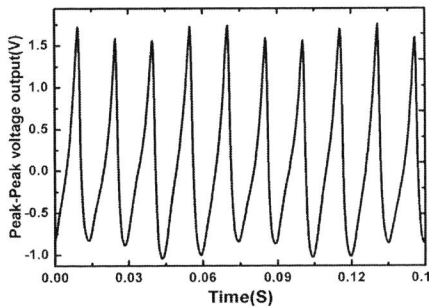

Figure 9: Peak-Peak voltage output as a function of time at an acceleration of 0.5g and a resistance of 20MΩ

CONCLUSION

A low-frequency three-dimensional resonant micro power generator using a novel spiral spring has been designed, fabricated and tested. The experimental analysis shows that the proposed generator operates an out-of-plane direction at mode I of 66Hz and two in-plane directions at mode II of 75Hz and mode III of 78.5Hz with a phase difference of about 90°. Although the performances of the devices here are limited currently, it provides an intriguing alternative for scavenging low-frequency ambient energy from three-dimensional vibration sources.

ACKNOWLEDGEMENTS

This research is supported by Singapore Energy Research Institute and Singapore–MIT Alliance for Research and Technology and conducted at Nanyang Technological University.

REFERENCES

[1] P. Mitcheson, E. Yeatman, G. Rao, A. Holmes and T. Green, "Energy Harvesting From Human and Machine Motion for Wireless Electronic Devices," *Proc. IEEE*, vol. 96, pp. 1457-1486, 2008.

[2] S. Beeby, M. Tudor and N. White, "Energy harvesting vibration sources for microsystems applications," *Meas. Sci. & Technol.*, vol. 17, pp. R175-R195, Dec 2006.

[3] Suzuki, Y., Miki, D., and Edamoto, M.: 'A MEMS electret generator with electrostatic levitation for vibration-driven energy-harvesting applications', *J. Micromech. Microeng.*, vol. 20, pp. 104002, 2010

[4] H. Liu, B. Soon, N. Wang, C. Tay, C. Quan, and C. Lee, "Feasibility study of a 3D vibration-driven electromagnetic MEMS energy harvester with multiple vibration modes," *J. Micromech. Microeng.*, vol. 22, pp.125020, 2012.

[5] B. Yang, C. Lee, R. Kotlanka, J. Xie, and S. Lim, "A MEMS rotary comb mechanism for harvesting the kinetic energy of planar vibrations," *J. Micromech. and Microeng.*, pp.065017, 2010.

[6] U. Bartsch, J. Gaspar and O. Paul, "Low-frequency two-dimensional resonators for vibrational micro energy harvesting," *J. Micromech. and Microeng.*, vol.20, pp. 035016, 2010.

[7] T. Fujita, T. Toyonaga, K. Nakade, K. Kanda, K. Higuchi and K. Maenaka, "Selective electret charging method for energy harvesters using biased electrode," Procedia Engineering, vol. 5, pp. 774-777, 2010.

[8] S. Liu, J. Miao and S. Lye, "High Q and low resonant frequency micro electret energy harvester for harvesting low amplitude harmonic of vibration," *in 26th IEEE Int. Conf. Micro Electro Mechanical Systems*, 2013, pp. 837-840.

CONTACT

*Kai Tao, tel: +65-84205461; taok0001@e.ntu.edu.sg

A WEARABLE SYSTEM OF MICROMACHINED PIEZOELECTRIC CANTILEVERS COUPLED TO A ROTATIONAL OSCILLATING MASS FOR ON-BODY ENERGY HARVESTING

Robert Lockhart, Pattanaphong Janphuang, Danick Briand and Nico F. de Rooij
École Polytechnique Fédérale de Lausanne (EPFL), Institute of Microengineering (IMT)
The Sensors, Actuators and Microsystems Laboratory (SAMLAB), Neuchâtel, Switzerland

ABSTRACT

In this paper, we present a compact, wearable piezoelectric on-body harvesting system that uses a small eccentric mass from a common watch movement to mechanically deflect a set of micromachined piezoelectric cantilevers when excited by the low frequency movements of the human body. The piezoelectric cantilevers are directly coupled to the rotating mass via a set of pins located near its rotational center. The energy produced by each pluck of a single cantilever is 545 nJ, corresponding to a maximum output power of 11 µW for continuous plucking; however, accounting for the periodic rest of typical human motion, the average output power over a full day cycle will be considerably less.

INTRODUCTION

Harvesting the mechanical energy of the human body through piezoelectric transduction could provide a means for powering portable and implantable systems. Human motion, however, is irregular and is limited to very low frequencies (a few Hz). Therefore, resonant type devices, where the natural frequency of the harvester is matched to the frequency of ambient vibrations, do not present a viable option. Impact-type harvesting in which the environmental motion is coupled to an inertial object which transfers its accumulated energy to the harvester through physical impacts provides a means of coupling low frequency irregular motion to high frequency piezoelectric oscillators.

Priya *et al.* first presented an impact harvester based on a windmill design in which a wind powered rotating wheel with notches plucks a series of harvesters extending radially outwards from its center [1]. Pozzi *et al.* inverted the windmill design to reduce the dimensions of the harvester in order to extract the energy from a bending knee [2]. Here, we miniaturize this concept using an eccentric mass that oscillates with the movement of the body. The movement of the mass is then transferred to the piezoelectric cantilevers through direct mechanical impact with a set of pins inserted near the rotational center of the eccentric mass.

The presented approach differs from recent work based on magnetic actuation [3] which still requires a relatively large mass, strong magnets and high rotational speeds to effectively actuate the piezoelectric cantilever - the absence of post-excitation oscillations in the case of low-frequency magnetic actuation reduces the electromechanical efficiency of the system. The concept presented here can be used with low frequency movements from a small rotational mass commonly used in a typical wristwatch.

We previously presented a novel efficiency characterization setup for rotational energy harvesters using a flywheel to store and quantify the mechanical input energy and compare energy losses with the electrical output [4]. With this characterization tool, we have demonstrated that the efficiency of mechanical plucking is greater than magnetic actuation for low to moderate frequencies. A potential drawback of a mechanical contact approach, however, is the reliability of the system. Nonetheless, we also demonstrate stable operation of the mechanically plucked cantilevers over long periods through accelerated lifetime tests [5].

Here, we present the design considerations related to developing a rotational micro-energy harvester for on-body applications, taking into consideration the forces available from an eccentric mass from a common wristwatch. We discuss the design and fabrication of the piezoelectric harvesters developed to satisfy the design constraints and we present the results of a demonstrator created to test this micro-energy harvesting concept.

SYSTEM DESIGN

The rotational micro-energy harvesting system is shown schematically in Figure 1. The piezoelectric cantilevers are fixed along the outer diameter of the system such that they extend inward towards the rotational center of the device. In the current configuration, pins, 200 µm in diameter, are inserted into the center ring of the oscillating mass, 3.5 mm from the rotational center.

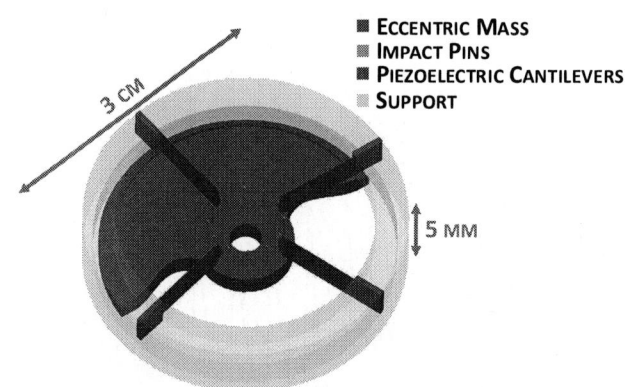

Figure 1: Schematic of the presented concept. Pins inserted into the center ring of an eccentric mass pluck the cantilevers as the mass rotates due to low frequency excitations caused by body movements.

978-1-4799-3510-9/14 $31.00 © 2014 IEEE

Designing the pin / cantilever interaction closer to the center of rotation increases the available torque applied by the eccentric mass, albeit, at the expense of a reduced circumference. The size of the circumference defines the maximum displacement and number of cantilevers that fit within a given system. This is because only one cantilever / pin interaction can occur at any given time due to the limited force available. Therefore, multiple cantilevers must be aperiodically staggered around the circumference of the face, as shown in Figure 1, to reduce the load applied to the mass at any given time. The torque generated by the mass is given by

$$\tau = mgr \sin \beta \qquad (1)$$

where $\beta = 180° - \theta$ refers to the angle extended between the gravitational vector and r, the radial line connecting the center of rotation to the center of mass; mg is the weight of the mass (Figure 2). Using a mass from the standard ETA 2824 watch movement, commonly used in many self-winding wristwatches, a maximum torque of $\tau = 0.0001$ Nm is generated when the mass is aligned along the horizontal such that $\theta = \beta = 90°$. At $\theta = 45°$, $\tau = 0.00007$ Nm.

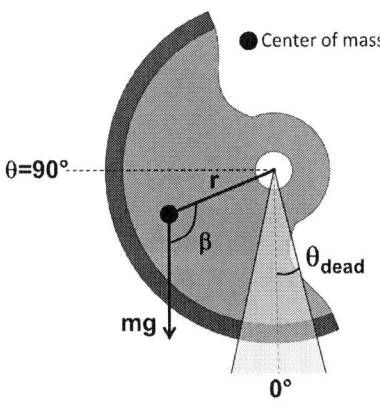

Figure 2: The torque produced by the eccentric mass depends on the angle of rotation. As a result, there is a zone between $\pm\theta_{dead}$ where the force of the mass is too weak to pluck the cantilever.

The required plucking force depends on the stiffness of the cantilever design and the amplitude of the deflection necessary to fully release the cantilever. The deflection amplitude depends on the insertion depth of the cantilevers as shown in Figure 3 and can be approximated by

$$\delta = \sqrt{R^2 - (x - R)^2} \qquad (2)$$

where R is the radius of the circle of pins and x is the insertion depth. In order to create a realistic design, an insertion depth of 50 μm was chosen, corresponding to a deflection amplitude of 500 μm. Therefore, taking the available force, the required deflection amplitude and the size of the system into consideration, a cantilever length of 9 mm was selected and the total thickness set to 50 μm: 20 μm of PZT and 30 μm of silicon, satisfying the optimum thickness ratio ($t_r \sim 0.6$) in terms of efficiency for an end-deflected piezoelectric cantilever [6].

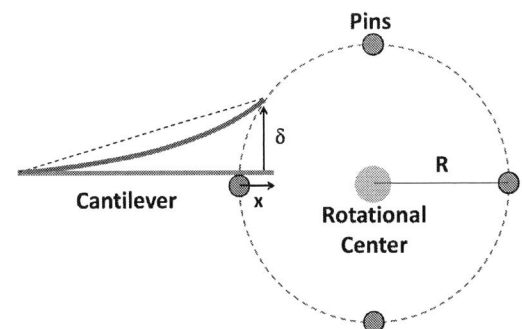

Figure 3: Schematic depicting the cantilever deflection required in order for it to be released by the pin.

As the angle of the mass approaches 0°, the torque produced by the mass vanishes. Therefore, there is a spread of angles around 0° where the mass does not provide an adequate force to fully pluck the cantilevers. The maximum angle in this spread is referred to as the dead angle and can be determined with the following equation

$$\theta_{dead} = \sin^{-1}(\frac{k\delta R}{mgr}) \pm 180° \qquad (3)$$

It is important that the dead angle be small in order to allow the mass to rotate with the motion of the body. Given the designed dimensions of the cantilever, the stiffness, k, is 13 N/m and the dead angle of the presented configuration is $\theta_{dead} = \pm13°$.

The mechanical plucking of the harvesters in this configuration is bi-directional, meaning rotation of the mass in either direction will excite the harvesters, deflecting them to a maximum amplitude and then releasing them, allowing them to freely oscillate as the stored mechanical energy is converted to electrical energy.

HARVESTER FABRICATION

The micromachined piezoelectric cantilevers are fabricated at the wafer-level by bonding a bulk, 130 μm thick, commercially available PZT-5A sheet onto a silicon substrate and thinning it down to a thickness of 20 μm using an automatic grinder (Disco DAG810) [7]. The cantilever tips are patterned with reactive ion etching and cavities are anisotropically etched into the silicon wafer in a potassium hydroxide (KOH) bath before dicing is used to release the cantilevers. Images of the front- and backsides of a processed wafer before the final dicing step are shown in Figure 4. The dimensions of the fabricated cantilevers are 9 x 3 x 0.05 mm³. A single cantilever fixed to a brass support is also visible in Figure 4. The support was used to easily insert and fix the cantilevers in the demonstrator presented in Figure 5. The brass support can be adjusted to vary the insertion depth of the cantilevers. The cantilever is fixed to the support with a conductive epoxy that connects to the bottom electrode of the PZT through a via etched in the Si.

978-1-4799-3510-9/14 $31.00 © 2014 IEEE   371

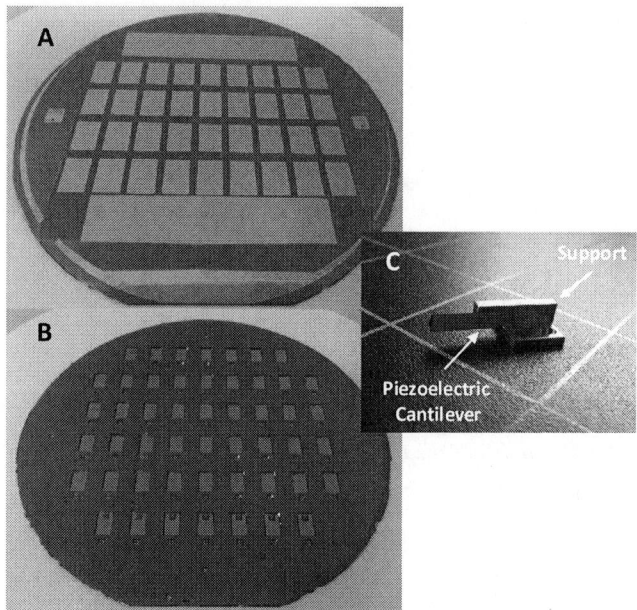

Figure 4: MEMS cantilevers fabricated at the wafer-level: A) a thinned PZT sheet on the frontside of the wafer; B) cavities on the backside of the wafer; and C) a mounted piezoelectric harvester.

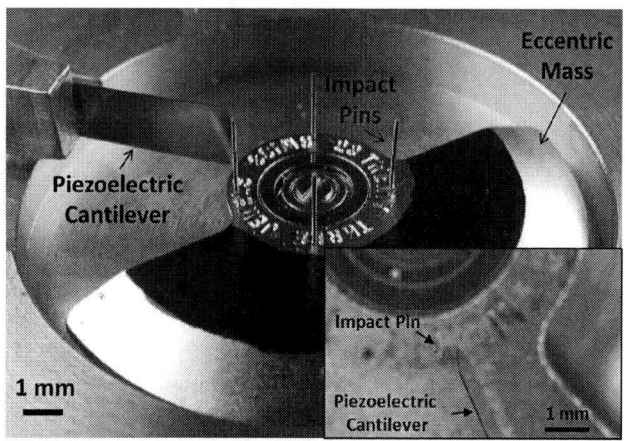

Figure 5: Optical image of an eccentric mass (ETA 2824) with 4 pins inserted in the central ring and a single piezoelectric cantilever mounted. (Inset: magnified view of an impact pin and the overlapping tip of the cantilever).

RESULTS

As the mass turns, the pins strike the tips of the piezoelectric cantilevers producing voltage pulses similar to the one presented in Figure 6; an initial deflection is followed by free oscillations of the cantilever. To measure the voltage pulse from a single pluck, a cantilever was aligned to an angle of 50° and the mass was rotated to 20° such that a single pin was positioned just above the cantilever. Upon dropping the mass, the pin plucked the cantilever once before the mass stabilized at 0°. 545 nJ of energy is generated by a single pluck across an optimum

load of 14.5 kΩ. The initial contact between the cantilever and the pin accounts for only 15 nJ while the majority of the electrical energy is generated by the free oscillations after the cantilever has been fully released by the pin as demonstrated in Figure 6.

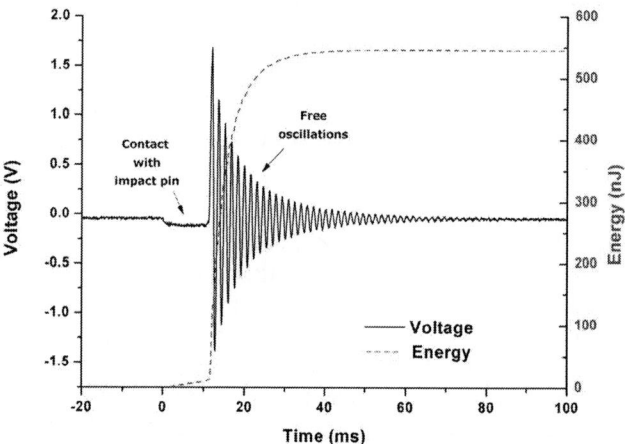

Figure 6: Voltage and energy output from a single pluck when connected to a load of 14.5 kΩ.

The transduction efficiency of the harvesters to convert mechanical energy into electrical energy can be determined by comparing the mechanical energy input to the electrical energy output. The mechanical energy is simply given by the work required to deflect the cantilever of a given stiffness, $k \sim 20$ N/m, by a certain displacement, $\delta \sim 500$ μm $(E_M = 0.5k\delta^2 = 2.5 \mu J)$. The electrical energy per pluck, as shown in Figure 6, was determined by integrating the instantaneous power dissipated by the resistive load over the duration of the pulse. Comparing these two values at the optimum load, the transduction efficiency of the harvester was determined to be $\sim 20\%$.

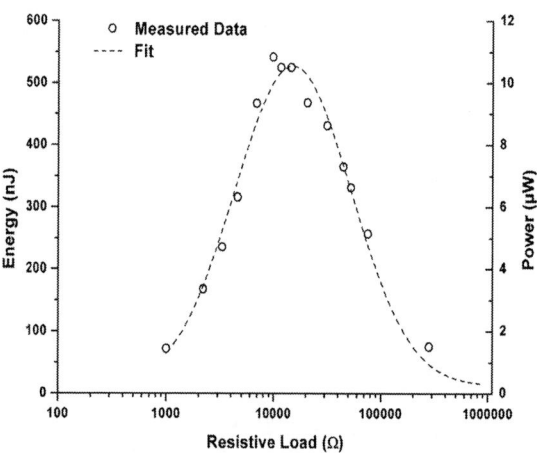

Figure 7: Energy per pluck from a single cantilever as a function of the connected load. The optimum load is 14.5 kΩ.

The optimum load was determined by varying the resistance connected across the electrodes of the harvester and measuring the energy produced for a single pluck. The energy produced per pluck as a function of the connected load is shown in Figure 7. The peak visible at 14.5 kΩ corresponds to the expected optimal load given by $1/\omega C$ where C = 20 pF is the capacitance of the piezoelectric layer and ω is the oscillation frequency, or the natural frequency of the cantilever in this case.

The ideal power for continuous, regular plucking shown on the right-hand axis of Figure 7 was determined using the energy produced and the duration of each pluck. This corresponds to an optimum case in terms of efficiency, but requires regular motion of the mass which is clearly not the case for human motion. An example of human motion is presented in Figure 8. The demonstrator with 4 pins inserted and a single harvester (Figure 5) was strapped to a human arm. The voltage signal produced by the harvester was recorded as the subject walked. Variation in the plucking frequency is clearly visible and the power produced is reduced by approximately half (6 µW) as a result of the gaps that are present between successive plucks. In reality, the human body is at rest for a significant portion of the day, and therefore, the average power produced over a 24 hour period will be considerably less than the power produced during a period of active motion.

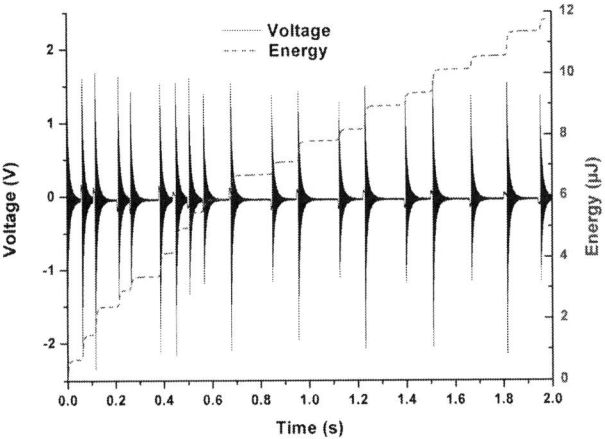

Figure 8: Voltage and energy output by the system while on the wrist of person walking. A 14.5 kΩ optimum load is connected to the harvester.

CONCLUSIONS

We have demonstrated a working concept for a compact, wearable energy harvesting system to convert the mechanical energy of the human body into useable electrical energy to provide a sustainable power source for on-body microelectronics and implantable devices. The system employs piezoelectric cantilevers actuated by an eccentric mass. The piezoelectric cantilever design has been optimized to work with a small mass commonly found in a standard automatic wristwatch and the concept is designed to work with the low actuation frequencies commonly encountered in human body motion.

We are presently working to optimize this concept by increasing the number of pins and the number of cantilevers while improving the compactness of the design. Through optimization, the presented concept could eventually be used to power on-body electronics.

AGKNOWLEGMENTS

We would like to acknowledge the Swiss National Science Foundation (SNF) for their support of this research. We would also like to thank the Swatch Group Research and Development, Division Asulab for their significant contribution to this work.

REFERENCES

[1] S. Priya, "Modeling of electric energy harvesting using piezoelectric windmill," Applied Physics Letters, vol. 87, pp. 184101, 2005.

[2] M. Pozzi and M. Zhu, "Plucked piezoelectric bimorphs for knee-joint energy harvesting: modelling and experimental validation," Smart Materials and Structures, vol. 20, no. 5, pp. 055007, 2011.

[3] P. Pillatsch, E.M. Yeatman, A.S. Holmes, "A wearable piezoelectric rotational energy harvester," Body Sensor Networks (BSN), 2013 IEEE International Conference on, pp.1-6, 6-9 May, 2013.

[4] P. Janphuang, R. Lockhart, S. Henein, D. Briand, and N.F. de Rooij, "On the experimental determination of the efficiency of piezoelectric impact-type energy harvesters using a rotational flywheel," Tech paper PowerMEMS Conference, London, UK, December 3-6, 2013.

[5] P. Janphuang, R. Lockhart, D. Briand and N.F. de Rooij, "On the optimization and performances of a compact piezoelectric impact MEMS energy harvester", Proc. IEEE 27th Int. Conf. MEMS, 2014.

[6] C. Mo, S. Kim, W. W. Clark, "Theoretical analysis of energy harvesting performance for unimorph piezoelectric benders with interdigitated electrodes," Smart Materials and Structures, vol. 18, no. 5, pp. 055017, 2009.

[7] P. Janphuang, R. Lockhart, D. Briand and N. F. de Rooij, "Wafer Level Fabrication of Vibrational Energy Harvesters using Bulk PZT Sheets", 26th European Conference on Solid-State Transducers (Eurosensors), Krakow, Poland, Procedia Engineering, 2012.

CONTACT

*R. Lockhart, tel: +41-21-6954572; robert.lockhart@epfl.ch

ALL-POLYMER PIEZOEELCTRET ENERGY HARVESTER WITH EMBEDDED PEDOT ELECTRODE

Yue Feng and Yuji Suzuki
The University of Tokyo, Tokyo, Japan

ABSTRACT

A novel all-polymer high-aspect-ratio piezoelectret energy harvester with embedded PEDOT electrode is proposed, and its performance for low-resonant-frequency in-plane vibration energy harvesting is demonstrated. Butterfly-shaped stop valves is devised to control the PEDOT capillary flow inside the parylene channels. With the present early prototype, 5.2 V open circuit voltage and 53 nW output power have been obtained at the resonant frequency of 205 Hz with 3.0 g acceleration.

INTRODUCTION

Recently, MEMS-based energy harvesting from environmental vibration attracts significant attention for replacing button batteries for battery-less wireless sensor nodes [1-2]. Energy conversion principles for such energy harvesting include piezoelectric, electromagnetic and electrostatic/electret [3-7]. In order to match the resonant frequency to the environmental frequency range (<100 Hz), a cantilever with heavy tip mass is often used, which vibrates in the out-of-plane direction. On the other hand, devices with the in-plane motion have the advantage of low profile and thus smaller volume figure of merits, but soft spring is required due to limited weight of the mass. In that sense, electrostatic/electret energy harvester is suitable to capture energy from in-plane vibration, because the spring design/material can be independent on the transduction mechanisms.

We previously proposed trench-filled high-aspect-ratio (HAR) cellular parylene spring for in-plane low-frequency piezoelectret transducers [8]. The sidewalls of the cellular parylene structure charged using soft X-ray irradiation [9], and high piezoelectric sensitivity up to 960 V/N at its resonant frequency of 149 Hz has been obtained. However, due to relatively-low bias voltage during the soft X-ray charging, the surface potential remains low.

The objectives of the present study are to develop piezoelectret structure with embedded PEDOT electrodes, and to further improve the sensitivity of the piezoelectret structure for low-resonant-frequency in-plane piezoelectret energy harvesters.

PIZELECTRET ENETRGY HARVESTER WITH EMBEEDDED PEDOT ELECTRODE

Figure 1 schematically shows the MEMS in-plane piezoelectret energy harvester with HAR parylene comb fingers with embedded electrodes. Pairs of the opposite PEDOT electrodes inside the HAR parylene beams provide large electric field during the soft X-ray charging [10], and thus higher surface potential. When the Si mass is vibrated

in the in-plane direction, air gap between parylene comb fingers are shrunk/extended, resulting in changes of the capacitance and thus change in the macroscopic dipoles. The induced charges are collected with electrodes embedded in the parylene comb fingers. The electrodes is made of PEDOT, which is introduced with the aid of the capillary effect through the void inside the parylene comb fingers.

Each pair of positive/negative HAR combs is connected with short beams, which serves as the mechanical support without electrical contact. For the electrical isolation through the mechanical supports, novel capillary stop valve structures are developed [10]. Geometrical angle in the butterfly-shaped stop valve is designed as 130° to provide enough barrier pressure. The capillary stop valves also serve as mechanical stoppers providing nonlinear spring behavior for broadband response; the spring constant of the comb finger is suddenly enhanced when the tip of the valve and the comb finger are in mechanical contact.

Spiral-shaped reservoirs are integrated into the frame and the Si mass as the inlet/outlet of the PEDOT capillary flow. Cross-shaped short insulation beams are devised between electrical pads on each side for decreasing the parasitic capacitance.

In order to realize a large variable capacitance and thus high power output, air gap between the comb fingers should be minimized. In the present study, the air gap t_g is designed as 60 μm. The thickness of polymer beam t_p is set as 20 μm for low resonant frequency. Capillary stop valves are devised along the polymer comb fingers. In order to compensate the weight loss of the Si mass, another 0.05 g Si mass (7x7 $(mm)^2$) is glued onto the original 0.08 g seismic mass (1x1 $(cm)^2$). Total dimension of device are 2.3 cm in length and 2.0 cm in width, respectively.

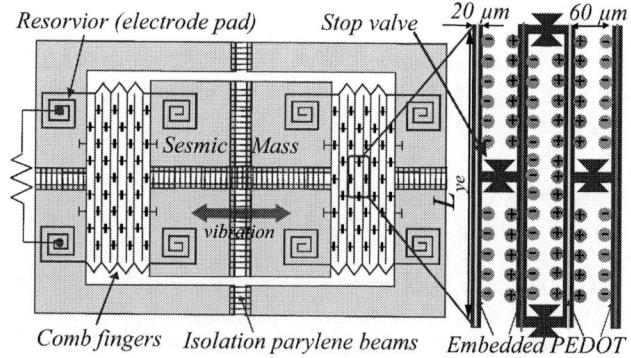

Figure 1: Schematic of all-polymer in-plane piezoelectret energy harvester with embedded PEDOT electrodes. Two sets of electrodes are mechanically connected with short beams, but electrically isolated by the capillary stop valves.

Table 1: The geometrical parameters of the piezoelectret energy harvester.

E (GPa)	m_1 (g)	m_2 (g)
3.2	0.05	0.08
h (μm)	t_p (μm)	t_g (μm)
350	20	60
L_{ye} (μm)	L_y (μm)	ε_p
1050	1100	3.1
ε_p	N_y	N_x
2.0	9	6

The geometrical parameters of device are shown in Table 1. The designed values of the spring constant and the resonant frequency of device estimated with

$$k_t = 16Eh(\frac{t_p}{L_{ye}})^3(\frac{N_y}{N_x}) \tag{1}$$

$$f_r = \frac{2\sqrt{Eh(t_p/L_{ye})^3 N_y/N_x(m_1+m_2)}}{\pi} \tag{2}$$

are 186 N/m and 190 Hz using following equations [7].

MICROFABRICATION PROCESS

Figure 2 shows the MEMS fabrication process. Firstly, 10 μm-thick AZ P4620 photoresist is patterned on a 4-inch silicon wafer (Fig. 2a) for deep reactive ion etching (DRIE) of 350 μm-deep and 25 μm-wide Si trenches (Fig. 2b). Then a 5 μm-thick parylene-C film is deposited into Si trenches using a chemical vapor deposition (CVD) method, forming 10 μm-wide parylene inherent voids inside the HAR Si trenches (Fig. 2c). A Cr/Au/Cr electrode is selectively sputtered near the deep parylene voids using a hard mask (Fig. 2d). Next, 10 μL DMSO-modified PEDOT:PSS droplet is injected to the inlet reservoirs (Fig. 2e). It takes several minutes for the liquid to flow through the whole parylene channel. After thermal curing at 100 °C for 0.5 hours in open environment, the second Cr/Au/Cr layer is

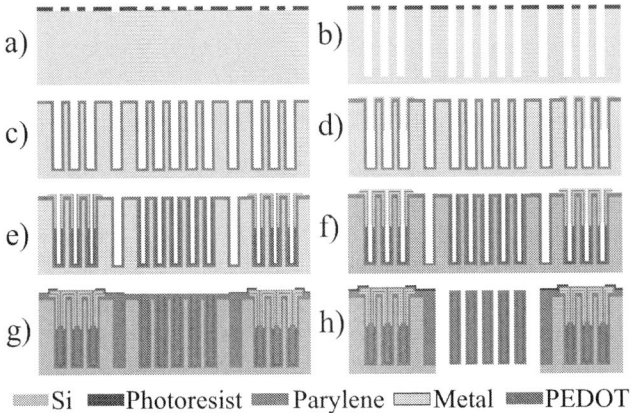

Figure 2: MEMS process for piezoelectret energy harvester with embedded PEDOT electrode.

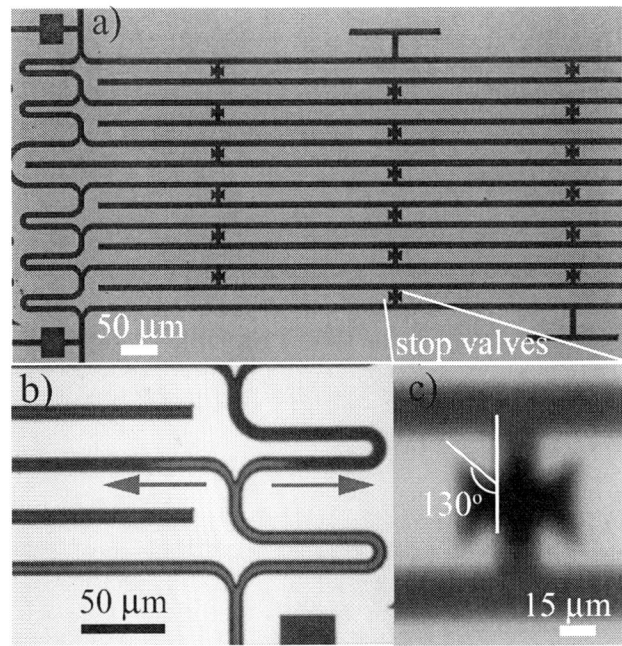

Figure 3: a) Top view of voids of parylene comb fingers with butterfly-shaped stop valves, where surface-tension driven PEDOT:PSS flow is stopped, b) Snap shot of flowing PEDOT:PSS inside the parylene channel, c) 130° capillary valve stopping liquid at its junction.

sputtered (Fig. 2f). This is followed by a 8 μm-thick parylene refilling to form a solid all-polymer spring (Fig. 2g). Then, the parylene layer on the surface is patterned with O_2 plasma to open etching windows. Finally, the device with free-standing all-polymer comb springs is released in XeF_2 (Fig. 2h).

Figure 3 shows the surface-tension-driven PEDOT:PSS flow through voids in the parylene comb fingers. More than 200 capillary-stop valves with angle of 130° effectively stop liquid flowing across the mechanical supports, and thus electrical isolation between positive/negative comb fingers has been realized. After thermal curing of PEDOT:PSS, the average resistance between contact pads on the inlet/outlet reservoirs through embedded PEDOT electrodes is around 300-500 kΩ.

POWER GENERATION EXPERIMENT

Soft X-ray charging process for artificial macroscopic dipoles formation is carried out at room temperature and in an air-filled enclosure [9]. The distance between X-ray tube and the device is set as 1.5 cm to effectively irradiate the whole device. To prevent electrical breakdown between a pair of PEDOT electrode, a 300 V bias voltage is applied across the opposite electrodes, which is 10 times higher voltage than our previous work [8]. The charging time is chosen set to one hour.

Power generation experiment is conducted for examination of electromechanical characteristics of the present device. Figure 4a shows the experimental setup, in which an in-plane sinusoidal oscillation is provided by an electromagnetic shaker (APS-113, APS Dynamics). The output voltage across a load resistance is measured through AC converter. The resonant frequency and the oscillation amplitude are measured with a high-speed CMOS camera (Phantom 5, Vision Research).

The resonant frequency is 205 Hz, which corresponds to the spring constant of 215 N/m. These results are in accordance with the designed values of 190 Hz and 186 N/m, respectively. Figure 4b shows snap shots of the deformation of the piezoelectret parylene spring, induced by the displacement of the Si seismic mass at the resonant frequency of 205 Hz.

Figure 5 shows the oscillation amplitude captured with the high-speed camera. Up to 130 μm amplitude is achieved at 3.0 g acceleration. Theoretical maximum amplitude is 720 μm when the entire air gap (12x60 μm) is collapsed. However, in the present study, only 18% strain in the parylene comb structure is obtained for 3.0 g acceleration.

Figure 6 shows the output power of the piezoelectret energy harvester versus the excitation frequency at different accelerations. Output power of 18 nW is obtained at its

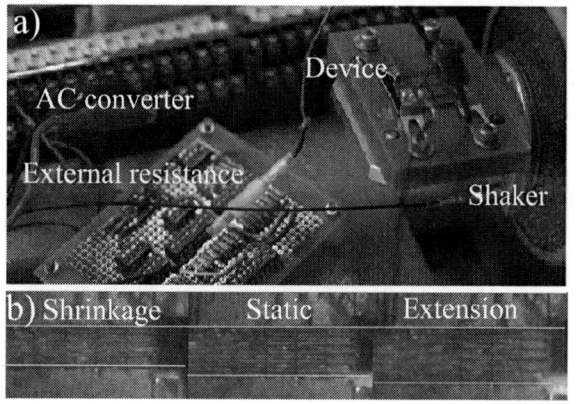

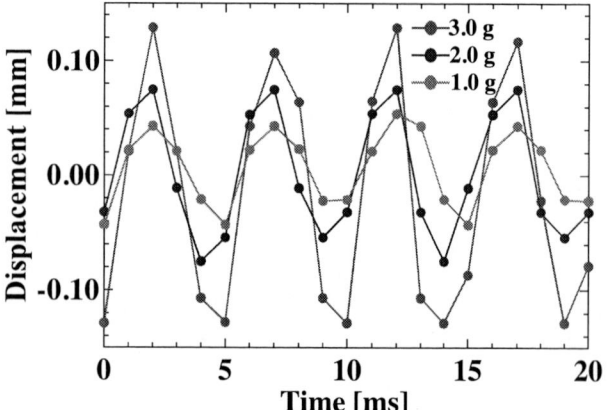

Figure 4: a) Experimental setup for testing the mechanical-electrical performance, b) Snap shots of comb-shaped HAR parylene comb drives at its resonant frequency of 205 Hz.

Figure 5: In-plane amplitude of the seismic mass at resonant frequency of 205 Hz.

resonant frequency of 205 Hz and 2.0 g acceleration. However, nonlinear behavior with a broadband response is

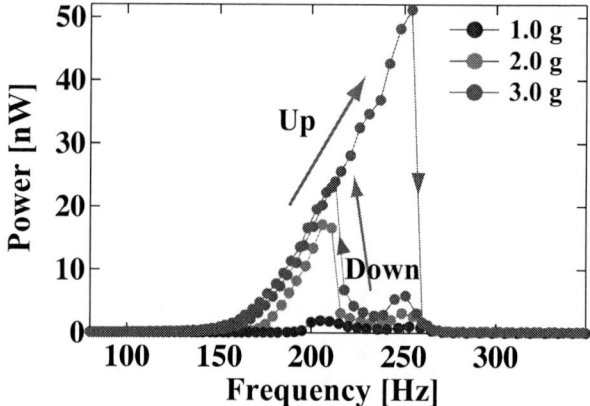

Figure 6: Output power versus in-plane oscillation frequency at 1.0, 2.0, 3.0 g accelerations.

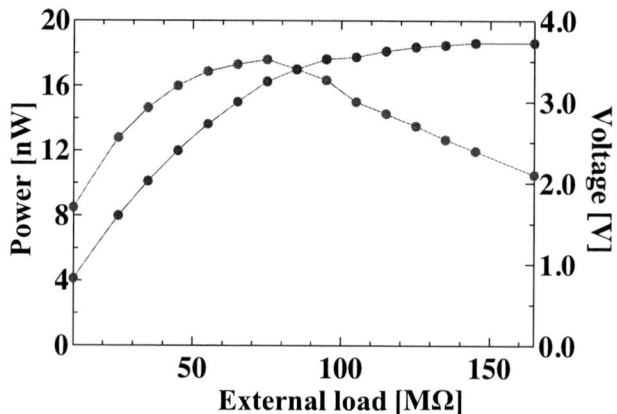

Figure 7: Output voltage and power versus external load at the resonant frequency of 205 Hz and 2.0 g acceleration.

observed when a relative large amplitude of 130 μm at acceleration of 3.0 g. Maximum output power of 53 nW is obtain when the frequency is swept up to 260 Hz. Obvious hysteresis of the power output is observed, showing the hardening spring characteristics. The nonlinearity of the spring is attributed to the butterfly-shaped valves, serving also as the mechanical stoppers. The damping ratio ζ and Q factor are 0.04 and 12, respectively.

Figure 7 shows the voltage amplitude and the output power versus external load at 2.0 g. Open-circuit voltage amplitude V_0 and the optimal load are 1.95 V (V_{p-p}=3.9 V) and 75 MΩ, respectively.

DISCUSSION

The target of the present research is to employ embedded PEDOT electrode to increase the effective bias voltage V_{bia} and thus higher surface potential V_s, higher surface charge density σ, and higher power output.

To evaluate performance of the novel comb finger piezoelectret, important physical parameters are examined for comparison using Eqs. (3) and (4):

Table 2: Comparison of piezoelectret with and without embedded PEDOT electrode.

Parameters	Previous work [8]	Present work
d_{33} (pC/N)	1522	1138
Bias voltage V_{bia} (V)	1000	300
Number of cells	72	12
Charge density σ (mC/m^2)	0.025	0.194
Surface potential V_s (V)	14	142

$$d_{33} = \frac{2V_0 \varsigma \varepsilon_{cell}}{ma} \qquad (3)$$

$$\sigma = \frac{4 d_{33} E (\varepsilon_g t_p^2 + \varepsilon_p t_p t_g)^2}{L_{ye}^3 L_y \varepsilon_g \varepsilon_p} \qquad (4)$$

Table 2 summarizes the comparison between the present device and our previous device. In previous work, 1 kV bias voltage is applied across 36 parylene honeycomb cells in series, which corresponds to the effective bias voltage of only 14 V per cell. On the other hand, with the 300 V bias voltage, up to 142 V surface potential (10 times higher) can be achieved for each comb finger, which is realized by embedded PEDOT electrode in each comb finger.

Since the power output is proportional to the piezoelectric constant d_{33} squared, which depends on both surface charge density σ and thickness of the air gap [8], it is believed that higher power output of the piezoelectret energy harvester can be realized by geometrical optimization with narrower air gap.

In the present device, the piezoelectric constant d_{33} of 1138 pC/N is obtained. In the present design, the air gap between parylene electret beams is 60 µm, which is much larger than 25 µm in our previous study. That is the main reason why d_{33} is lower than our previous one.

CONCLUSION

In conclusion, we have developed an MEMS-based all-polymer piezoelectret energy harvester based on HAR parylene comb fingers with embedded electrodes. Capillary stop valves are successfully implemented to mechanical supports for the comb fingers without electrical contact. Using the embedded PEDOT electrodes, surface potential and charge density of piezoelectret are significantly improved. Output power of 53 nW has been obtained at the low resonant frequency of 205 Hz with acceleration of 2.0 *g*. With the aid of butterfly-shaped stop valves, nonlinear behavior for a broadband response is realized at relative large displacement.

ACKNOWLEDGEMENTS

This work is supported through NEXT Program of JAPAN Society for the Promotion of Science (JSPS).

Photomasks are made using the University of Tokyo VLSI Design and Education Center (VDEC)'s 8-inch EB writer F5112+VD01 donated by ADVANTEST Corporation.

REFERENCES

[1] J. A. Paradiso, and T. Starner, "Energy scavenging for mobile and wireless electronics", *IEEE Pervasive Comp.*, Vol. 4, pp. 18-27, 2005.

[2] P. D. Mitcheson, et al., "Energy harvesting from human and machine motion for wireless electronic devices", *Proc. IEEE,* Vol. 96, pp. 1457–1486, 2009.

[3] M. Renaud, et al., "Fabrication, modeling and characterization of MEMS piezoelectric vibration harvesters", *Sensors Actuators A.,* Vol. 145–146, pp. 380–386, 2008.

[4] S. P. Beeby, et al., "A micro electromagnetic generator for vibration energy harvesting", *J. Micromech. Microeng.*, Vol. 17, pp. 1257-1265, 2007.

[5] D. Hoffmann, et al., "Fabrication, characterization and modelling of electrostatic micro-generators," *J. Micromech. Microeng.*, Vol. 19, 094001, 2009.

[6] Y. Suzuki, et al. "A MEMS electret generator with electrostatic levitation for vibration-driven energy harvesting applications," *J. Micromech. Microeng.*, Vol. 20, 104002, 2010.

[7] Y. Suzuki, "Recent progress in MEMS electret generator for energy harvesting", *IEEJ Trans. Electr. Electr. Eng.*, Vol. 6, pp. 101-111, 2011.

[8] Y. Feng, et al., "Trench-filled cellular parylene electret for piezoelectric transducer," *Appl. Phys. Lett.,* Vol. 100, 262901, 2012.

[9] K. Hagiwara, et al., "Electret charging method based on soft X-ray photoionization for MEMS applications", *Trans. IEEE, Dielectr. Electr. Insul.,* Vol. 19, pp. 1291-1298, 2012.

[10] Y. Feng, and Y. Suzuki, "All-polymer high-aspect-ratio spring with embedded electrode," *Proc. 17th Int. Conf. Solid-state Sensors, Actuators, and Microsystems (Transducers '13),* Barcelona, pp. 1569-1572, 2013.

CONTACT

*Yue Feng, tel: +81-3-5841-6419;
feng@mesl.t.u-tokyo.ac.jp

ELECTROMAGNETIC ENERGY HARVESTER WITH HIGH EFFICIENCY USING MICRO-MACHINING SI SPRINGS

T. Shirai, Y. Wakasa, T. Nakagawa, K. Nomura, and H. Yagyu
Panasonic Corporation, Osaka, Japan

ABSTRACT

The direct force input type of electromagnetic energy harvester with significantly high generating efficiency over 40% has been developed. This high efficiency was realized by using the 1mm-thick uniquely-structured Si springs in the harvester, which was fabricated by D-RIE (deep reactive ion etching) from both sides of a 1mm-thick Si wafer. The harvester showed that the peak voltage was 3.7 V and the output energy on the most adequate load resistance was over 1 mJ at an input energy of 2.7 mJ, resulting in a generating efficiency of 41%.

INTRODUCTION

Relating to the growing importance of sensor network systems, energy harvesting technology has been regarded as one of the solutions for the power supply of abundantly deployed sensors. The vibration energy harvesters are the devices extracting energy from oscillation in environments such as cars, bridges, plant machineries, human motion and so on. These devices are really attractive to power wireless sensor nodes and have been studied comprehensively over the recent years. Especially the direct force input type of energy harvesters from human motion like pushing, opening or pulling something are expected to be used in buildings and industrial automation and are edging closer to practical use [1]. In almost all of the direct input type of energy harvesters, the output power generated by one motion is used to actuate a sensor or to transmit a radio signal. Therefore, by realizing a high output the device can be applied to deal with various sensors and wireless communication standards. Additionally, a high generating efficiency can improve an operational feeling of pushing, opening or pulling something which an energy harvester is implemented due to the input energy decreasing. There are few reports about the direct force input type of energy harvester which can generate high output power with high efficiency and those characteristics are necessary to encourage for broad use.

This work demonstrates the electromagnetic vibration energy harvester with significantly high generating efficiency. This high efficiency was realized by using the 1mm-thick uniquely-structured Si springs in the harvester, which was created using micro-machining process.

THEORY OF ELECTROMAGNETIC VIBRATION ENERGY HARVESTER

The mechanisms about the electromagnetic vibration energy harvester fall into the general classification of a resonant type and a direct input type. The theory of the resonant type has been detailed earlier [2, 3]. Here, the theory of the direct input type is discussed. A schematic illustration

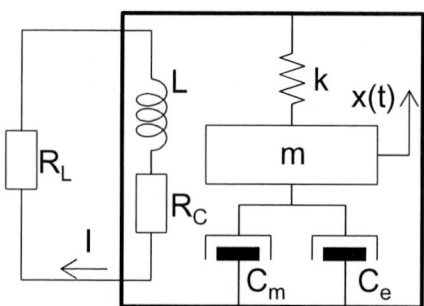

Figure 1: A schematic illustration of the vibration energy harvester system

of the vibration energy harvester system is shown in Fig. 1.

The voltage generated on the coil is modeled by a first order LR circuit. The dominant physical parameters are the motion and the mass of the magnet $x(t)$, m, respectively, spring constant k, mechanical damping coefficient C_m, electrical damping coefficient C_e, inductance of the coil L, resistance of the coil R_C, load resistance R_L and the electrical current I. In the case of the direct force input type of vibration energy harvester, only the mass of m is movable, which means the frame doesn't move, therefore the equation of motion can be described as:

$$m\frac{d^2 x(t)}{dt^2} + (C_m + C_e)\frac{dx(t)}{dt} + kx(t) = 0 \qquad (1)$$

Damping force caused by the electrical damping coefficient corresponds exactly to Lorentz force that the coil is subjected to by the magnetic field.

$$C_e \cdot \frac{dx(t)}{dt} = IBl \approx \frac{(Bl)^2}{R_L + R_C}\frac{dx(t)}{dt} \qquad (2)$$

Where B is the magnetic flux density, l is the active wire length of the coil and inductance of the coil L is omitted since the impedance ωL is commonly quite smaller than R_L or R_C in this system. Because equation (1) means a damped free vibration, the energy consumed at the load can be written as:

$$W_L = \int_0^\infty I^2 R_L dt \approx \frac{kx_0^2}{2}\frac{C_e}{C_m + C_e}\frac{R_L}{R_L + R_C} \qquad (3)$$

Where x_0 is the initial position of the mass and $kx_0^2/2$ means the initial energy input into the spring, the power conversion efficiency is:

$$\eta \approx \frac{C_e}{C_m + C_e}\frac{R_L}{R_L + R_C} = \frac{kx_0^2}{2}\frac{1}{\frac{C_m}{(Bl)^2}(R_L + R_C)+1}\frac{R_L}{R_L + R_C} \qquad (4)$$

Note that the equations above include some approximation because the magnetic flux density B is treated as a constant. However, it is enough to give the intrinsic physical phenomena of the harvester.

From equation (1), there are two approaches in order to

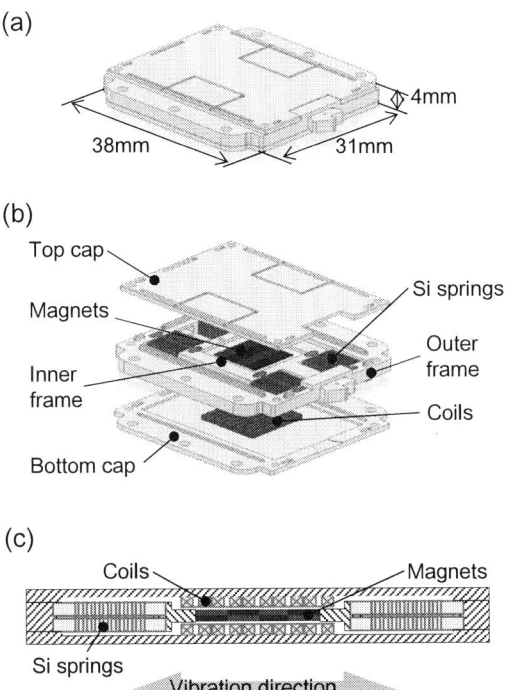

Figure 2: Schematic structure of the developed energy harvester. (a)External, (b)exploded, and (c)cross-sectional view.

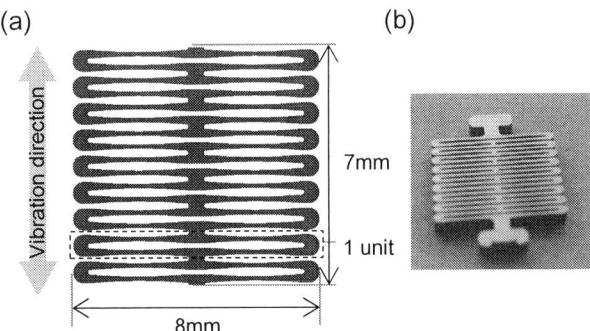

Figure 3: (a)Schematic top view and (b)photograph of the developed Si spring.

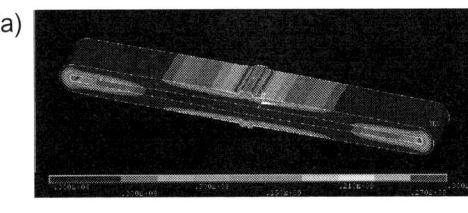

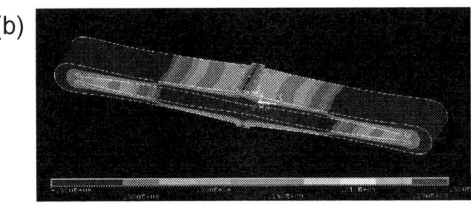

Figure 4: Stress distribution of two kinds of springs (FEM simulation). (a)The standard loop-shaped spring and (b)the improved spring in tensile direction force.

improve the efficiency; reducing C_m and improving Bl. In the system of such vibration energy harvesters, it is known that the mechanical damping coefficient C_m depends a great deal on the damping capacity of the material used for springs. The damping capacity of Si is considerably smaller than that of stainless steel, beryllium copper alloy or phosphor bronze, *etc*. Therefore, it is quite effective to use Si for the springs to improve the power conversion efficiency.

BASIC STRUCTURE AND FABRICATION PROCESSES

Basic Structure

Figure 2 shows the schematic of the developed energy harvester. Device size is approximately 38 X 31 X 4 mm. A magnet block, which consists of 4 pieces of 1mm-thick NdFeB magnets, is located at the center of the device. These magnets are arranged in alternating magnetic poles (N,S,N,S) with a pitch of 2.5 mm. This magnet block surrounded by an inner frame is connected to an outer frame via 8 springs made of Si and can oscillate freely by a direct force input in the direction indicated in Fig. 2(c). Two coil blocks, which consist of 5 air-core coils, are mounted on the top and the bottom caps which are on each side of the magnet block. These coils are arranged in line with a pitch of 2.5 mm along magnet arrangement and electrically connected in series, resulting in a high voltage output. The total resistance and inductance of the coils are approximately 90 Ω and 0.7 mH, respectively. The gap between the magnet block and coil blocks is 0.2 mm.

Design of Si Spring

The spring used for the direct force input type of vibration energy harvester needs to store a relatively large amount of mechanical energy (1-10 mJ). Although there some reports about developing Si springs [4, 5], these Si springs could not store such large energy. Our Si spring developed can store relatively large energy by designing to make stress distributed uniformly all over in Si spring. Figure 3 shows the schematic view of the developed Si spring.

Spring size is 7 X 8 X 1 mm and the direction of vibration is indicated in Fig. 3. This spring has a periodic structure consisted with 9 units of wide loops. Because the highest stress on applied is the inner surface of the edge, this bending region has wide width to prevent a concentration of stress. In addition, the middle region of the center and the edge has narrow width to reduce the second moment of area and be stressed more.

The 3-dimensional finite element method (3D-FEM) was used to model and analyze the stress of springs in ANSYS. Two models of spring were analyzed for comparison; one was the standard loop-shaped spring which had uniform line width and the other was improved spring. The analyses were carried out in one unit structure of each spring. The results of the stress analysis are shown in Fig. 4.

978-1-4799-3510-9/14 $31.00 © 2014 IEEE

Table 1: FEM simulation results of two types of the Si spring; the standard loop-shaped one and the improved one.

Spring characteristics		Improved spring	Standard spring
Spring constant (N/m)		5060	3388
In tensile direction	Displacement (mm)	0.32	0.63
	Stored Energy (mJ)	0.25	0.67
In compressive direction	Displacement (mm)	0.57	0.80
	Stored Energy (mJ)	0.82	1.1

The forces in tensile direction were adopted in two springs so that both of the springs stored the same energy. For the standard loop-shaped spring, there were high stress regions at the inner surface of the both sides of the edges. In contrast, the improved spring structure showed more uniform stress distribution. The characteristics of complete spring consisted with 9 units can be calculated by the results of 1 unit structure. Then, the displacement and stored energy were calculated on the condition that the highest stress in the spring became 250 MPa (sufficiently less than allowable stress of Si) in both tensile and compressive direction. The results are shown in Table 1. The stored energy particularly in tensile direction was highly-improved and the improved Si spring could store energy over twice as much as a standard one.

Furthermore, this improved spring structure showed higher spring constant in the thickness direction and the lateral direction than that of the meander structure spring which was commonly employed in MEMS devices. This characteristic of the spring can reduce influence of undesired vibration modes and the power conversion efficiency can be improved.

Fabrication processes and characteristics of Si Spring

Figure 5 shows the fabrication processes of the Si spring. The spring was fabricated in a 1mm-thick Si wafer. Initially, the top and the bottom surface of the Si wafer were thermally oxidized in a furnace (Fig. 5(a)). The silicon oxide layers on the both sides were patterned by photolithography and RIE (Reactive Ion Etching) processes, which served as a hard mask for the subsequent deep etching processes (Fig. 5(b)). The top side of the Si was removed to reach approximately 500 μm by ICP (Inductively Coupled Plasma) etching (Fig. 5(c)). The next step was backside processing. The top side of the Si was covered with a protection layer for penetration process of the Si wafer (Fig. 5(d)). Then the bottom side of the Si was removed up to penetrating Si wafer by ICP (Fig. 5(e)). Finally, the protection layer and the silicon oxide layers were removed (Fig. 5(f)).

The fabricated Si springs (N=25) were evaluated by a micro load meter, then the average of spring constant was 3198 N/m and the energy which the spring could store was over 0.6 mJ in both of tensile and compressive direction. The

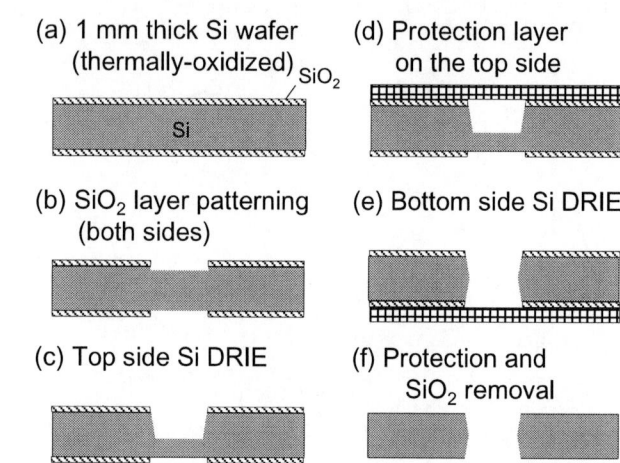

Figure 5: Schematic of the fabrication processes of the Si spring (cross-section).

developed harvester was designed to use 8 Si springs and total spring constant for the harvester became approximately 25600 N/m and the allowed input energy reached up to 4mJ.

CHARACTERIZATION OF HARVESTER

The fabricated energy harvesting devices were evaluated as follows. The harvester was fixed on a rigid plate and the tip of the inner frame was pulled by the load meter. The inner frame was released and the output voltage was measured when the force came to the point of the intended value. The input energy was estimated from the input force and the total spring constant of the harvester. For comparison, the same structured harvester with the springs made of stainless steel was prepared. The stainless steel spring had the 0.3 mm-thick meander structure and spring constant was approximately 16,000 N/m (measured value).

The measured output voltages across an open circuit for both harvesters for same input energy are shown in Fig.6. Both showed fine curves of damped oscillation, which meant that the undesired vibration modes were almost unexcited. The peak voltage for the harvester with the Si springs and that with the stainless steel springs were approximately equal (3.7 V and 3.9 V, respectively). However, the damping time of the harvester with the Si springs and the stainless steel springs were 0.25 sec and 0.04 sec, respectively, which meant the damping factor of the harvester with the Si springs was 6 times smaller than that of the stainless steel springs.

The output energy of the harvesters on the most adequate load resistance was evaluated. The results are shown in Fig. 7. The generated energy of the harvester with the Si springs increased in proportion to the input energy and the harvester generated over 1 mJ and the peak electrical power was 70 mW at an input energy of 2.7 mJ, resulting in a generating efficiency of 41%. The efficiency of harvester with the Si springs was about 3 times as large as that with stainless steel springs which was 15 %.

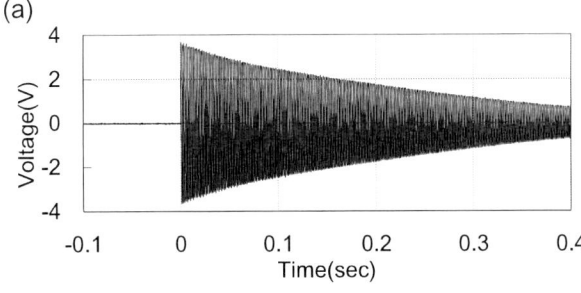

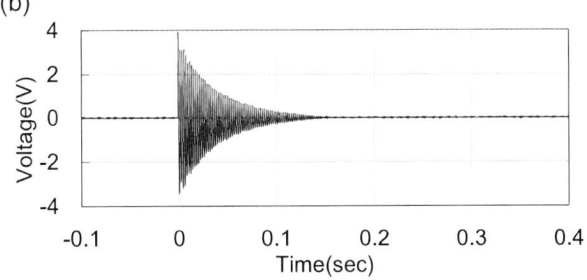

Figure 6: Output voltage across an open circuit. Results of the harvester with (a)Si springs, and (b)stainless steel springs.

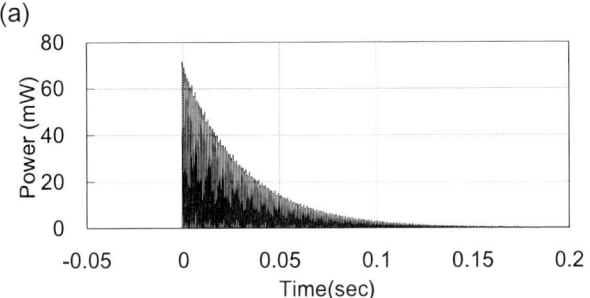

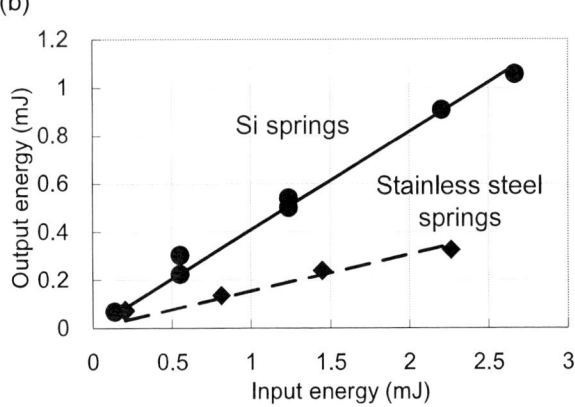

Figure 7: (a)Electrical power generated by the harvester with Si spring across the most adequate load resistance at an input energy of 2.7 mJ. (b)Output energy vs input energy. Filled circles and filled diamonds are results of Si springs and stainless steel, respectively.

CONCLUSION

The authors have presented the direct force input type of electromagnetic energy harvesters that showed significantly high generating efficiency. The harvester had the 1mm-thick uniquely-structured Si springs which was fabricated by D-RIE from both sides of a 1mm-thick Si wafer. The harvester showed that the peak voltage was 3.7 V and the output energy on the most adequate load resistance was over 1 mJ at an input energy of 2.7 mJ, resulting in a generating efficiency of 41%.

This work demonstrates substantial progress in development of the direct force input type of energy harvesters. Novel applications taking advantages of the high output energy and the high generating efficiency of the harvesters will likely be realized in the near future.

REFERENCES

[1] S. Roundy and E. Takahashi, "A Cost-Effective Planar Electromagnetic Energy Harvesting Transducer", in *Digest Tech. Papers PowerMEMS 2012* , pp. 10-13.

[2] N. N. H. Ching , G. M. H. Chan , W. J. Li , H. Y. Wong and P. H. W. Leong, "PCB integrated micro-generator for wireless systems", *Proc. Int. Symp. Smart Structures and Microsystems*, 2000

[3] L. Mateu and F. Moll, "Review of energy harvesting techniques and applications for microelectronics", *Proc. SPIE*, vol. 5837 (2005)

[4] S. Kulkarni, S. Roy and T. O'Donnell, "Vibration based electromagnetic micropower generator on silicon",*J. Appl. Phys*, vol. 99, iss. 8 (2006), pp. 511-513.

[5] M. Han, Q. Yuan, S. Zhang, X. Sun and H. Zhang, "An Electromagnetic Energy Harvester With Integrated Magnet Array", in *Digest Tech. Papers PowerMEMS 2012* , pp. 424-427.

CONTACT

*T.Shirai,
tel: +81-6-6900-0521;shirai.takeo@jp.panasonic.com

ELECTROMAGNETIC ENERGY HARVESTER WITH AN IN-PHASE VIBRATION BANDWIDTH BROADENING TECHNIQUE

Shih-Jui Chen, Jia-Yin Wu and Shu-Yu Liu

Department of Mechanical Engineering, National Central University, Taiwan

ABSTRACT

This paper reports a duo-mode vibration structure for increasing usable bandwidth in a micromachined electromagnetic energy harvester. The energy harvester is characterized by using finite element analysis on the vibration frequency and amplitude. Since a spiral diaphragm is much more flexible than a cantilever diaphragm for a given size, a spiral is inserted in to a U-shape cantilever, forming the proposed energy harvester. By designing the coupled structure, not only the resonance frequency is reduced, but also the usable bandwidth is broadened. Experimental results show that the device generates electromotive force of 1.00 and 1.03 mV, 7.66 and 7.43 nW (into 27 Ω load) at two resonance frequencies at 211 and 274 Hz, respectively.

INTRODUCTION

In recent years, renewable energy sources such as sunlight, heat and vibration have been explored as alternative power sources in portable sensor system. When a generator experiences vibration motion, three kinds of transduction mechanisms in MEMS technology are commonly used: vibration energy can be converted into electric power through electrostatic, piezoelectric, or electromagnetic operation. Though vibration energy can be a source of infinite power, it still suffers from limited bandwidth and output power.

Williams proposed the analysis method of a micro-electric generator [2]. For a 5 mm × 5 mm × 1mm micro generator, 0.1 mW power at frequency of 330 Hz can be generated. To further take advantage of the environmental vibration, Sari used frequency up-conversion method to convert frequencies for 95Hz to 2 KHz for a single cantilever [3]. Sari combined 35 different lengths of cantilevers to form an array of parylene-based micro generator. It generates power of 0.4 μW in frequency range between 4.2 – 5 kHz [4].

This paper reports a duo-mode electromagnetic energy harvester with electroplated thick copper coils. If there is relative movement of the coils to permanent magnets which provide the magnetic field perpendicular to the device, electromotive force (EMF) will be induced through the Faraday's law of induction. The proposed generators are designed to have large displacements at two low fundamental resonant frequency bands, increasing the usable bandwidth.

DESIGN

In the proposed design, copper coils are located on the U-shape cantilever and spiral diaphragm. When the device vibrates, the magnetic flux through the coils varies and induces electromotive force at the terminals. In order to achieve high voltage and power output at low frequency range, the rate of change of the magnetic flux should be high and the rigidity of the energy harvester should be low. From simulation, a spiral diaphragm is found to be much more flexible than a cantilever diaphragm for a given size, and offers a higher vibration displacement at the cost of more stringent control of the fabrication process [5]. By inserting a spiral structure inside a U-shape cantilever, the resonance frequency can be reduced and the usable bandwidth can be broadened.

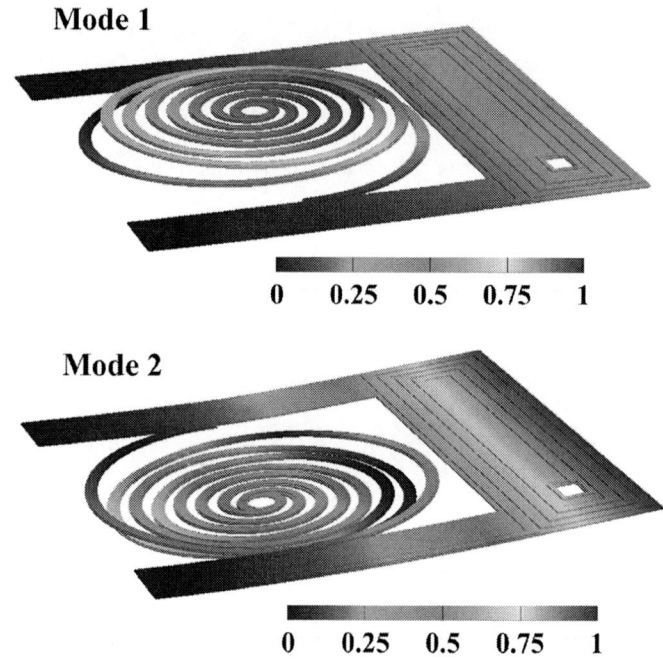

Figure 1: FEA simulation results of the proposed electromagnetic energy harvester: (upper) 1st mode shape; (lower) 2nd mode shape.

Finite element analysis is performed using CoventorWare™ to design the resonance frequency and mode shape of the energy harvester (Fig. 1). The results showed that there are two modes under excitations: the in-phase and out-of-phase vibration modes. In in-phase mode (mode 1), the vibration direction of the spiral part is the same as the U-shape cantilever part. In out-of phase mode (mode 2), the vibration direction of the spiral part is opposite to that of the U-shape cantilever part. Therefore, by carefully arranging the connections of the two coupled parts (Table 2), this harvester can be used as a single-band or duo-band energy harvester, which increases usable bandwidth of the device and potential applications.

978-1-4799-3510-9/14 $31.00 © 2014 IEEE

FABRICATION

The fabrication process of the electromagnetic energy harvester is schematically shown in Fig. 2. First, an 18-μm-thick silicon diagram was formed by KOH etching on the backside of the wafer. Then, Ti/Cu (20nm/150nm) was evaporated on the front side of the wafer as a seed layer. Thick photoresist of 18μm was spun and patterned to form a mold for electroplating of copper coils [6]. After 18μm-thick copper was electroplated, the photoresist and seed layer were removed, respectively. Finally, the diaphragm was released by RIE of the silicon from the front side.

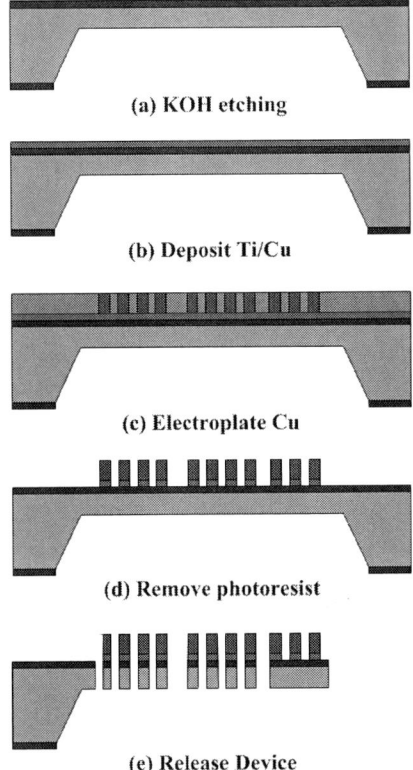

(a) KOH etching

(b) Deposit Ti/Cu

(c) Electroplate Cu

(d) Remove photoresist

(e) Release Device

Figure 2: Fabrication process of the electromagnetic energy harvester.

Table 1: Parameters of the fabricated energy harvester.

Device dimensions	$7.1 \times 7 \times 0.04$ mm^3
Resonance frequency (Mode 1)	211 Hz
Resonance frequency (Mode 2)	274 Hz
Total coil resistance	27 Ω
Magnet size	$2 \times 2 \times 1$ cm^3
Magnetic field strength	0.45 T

Photo of the fabricated energy harvester is shown in Fig. 3. The electroplated copper forms a pair of thick spiral coils, which eases the potential microstructural curling induced by inevitable residual stress. Table 1 shows detailed parameters of the fabricated device.

Table 2 shows the connection sequence of the fabricated device: for connection type 1, there will be two resonances at both 211 and 274 Hz.

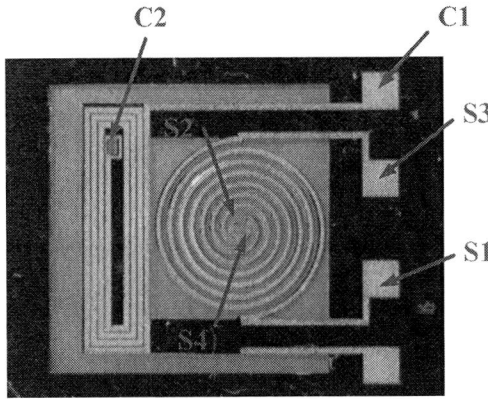

Figure 3: Photo of the fabricated energy harvester with definition of the connection ports.

Table 2: Connection definitions.

Type	Connection sequence	Mode 1	Mode 2
1	C2→C1→S4→S3 →S2→S1	out of phase	in phase
2	C1→C2→S4→S3 →S2→S1	in phase	out of phase

RESULTS

The fabricated energy harvester with a magnet was characterized with a testing system including a shaker, an oscilloscope and a preamplifier (Fig. 4). The energy harvested from the vibration of shaker was amplified and displayed on the oscilloscope. The unamplified open circuit voltage of the energy harvester as a function of vibration frequency was shown in Fig. 5. Since copper coils on spiral structures and U-shape cantilever are coupled together by connection type 1 or 2, the output voltage is summed up. In Fig.5, the measured open circuit voltage for device with connection type 1 shows two resonance peaks at 211 and 274 Hz. On the resonance frequency of 211 Hz, open circuit voltage of 1.00 mV is measured at acceleration of 1.42g. On the resonance frequency of 274 Hz, open circuit voltage of 1.03 mV is measured at acceleration of 1.23g.

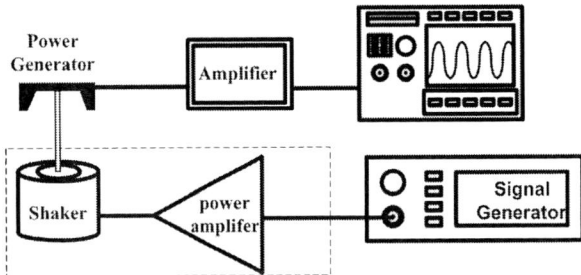

Figure 4: The experimental setup.

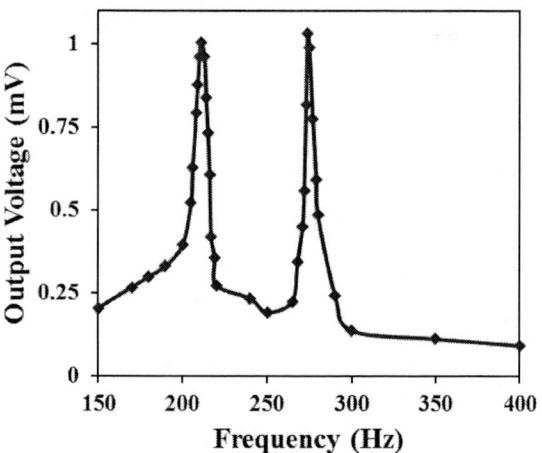

Figure 5: Measured open circuit voltage for device with connection type 1.

The output voltage and power of the energy harvester are measured as a function of the load resistance (Fig. 6). The output voltage increases as the load resistance increases. In case of the energy harvester with connection type 1, maximum power of 7.66 nW and 7.43 nW are delivered to a 27 Ω at resonance mode 1 and mode 2, respectively, when the load resistance is equal to the coil resistance.

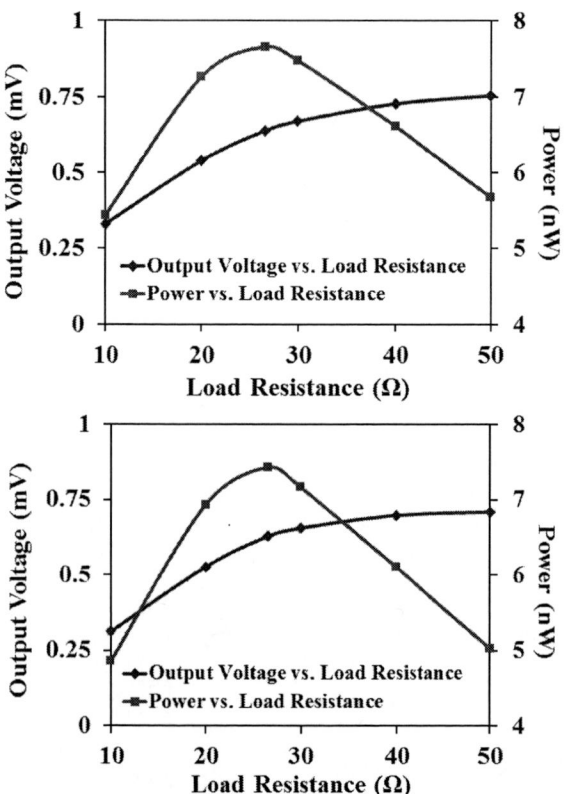

Figure 6: Measured output voltage and power across different load resistance for connection type 1 at (a) 211 Hz@1.42 g; (b) 274 Hz@1.23 g.

SUMMARY

A duo-mode electromagnetic energy harvester fabricated by MEMS fabrication process is presented in this work. The harvester generates an EMF of 1.00/1.03 mV and 7.66/7.43 nW power output (into 27 Ω load) at a resonance frequency of 211/274 Hz in response to a vibration acceleration of 1.42/1.23 g, respectively. Compared to a pure cantilever harvester, the proposed cantilever-spiral coupled energy harvester has lower resonance frequencies and larger bandwidth. The generated power can be further improved by adding the proof mass to enlarge the vibration amplitude. The bandwidth broadening technique showed a great prospect of potential applications in harvesting vibration energy in portable sensor system.

ACKNOWLEDGEMENTS

This material is based upon work supported by National Science Council, Taiwan, R.O.C. under Contract No. 101-2218-E-008-006 and 102-2221-E-008-110. We are grateful to the National Center for High-performance Computing, Taiwan, R.O.C. for providing software and facilities.

REFERENCES

[1] S.P. Beeby, M.J. Tudor, E. Koukharenko, N.M. White, T. O'Donnell, C. Saha, S. Kulkarni, S. Roy, "Design and performance of a microelectromagnetic vibration powered generator", *IEEE International Conference on Solid-State Sensors and Actuators*, Seoul, Korea, June 5-9, pp.780-783, 2005.

[2] C.B. Williams and R.B. Yates, "Analysis of a micro-electric generator for microsystems," *Sensors and Actuators A*, 52, pp. 8-11, 1996.

[3] I. Sari, T. Balkan and H. Kulah, "An electromagnetic micro power generator for low frequency environmental vibrations based on the frequency up-conversion technique," *IEEE International Micro Electro Mechanical Systems Conference*, Sorrento, Italy, January 25–29, pp. 1075 – 1078, 2009.

[4] I. Sari, T. Balkan and H. Kulah, "An electromagnetic micro power generator for wideband environmental vibrations," *Sensors and Actuators A: Physical*, 145–146(0), pp. 405-413, 2008.

[5] S.J. Chen, Y. Choe, L. Baumgartel, A. Lin, and E.S. Kim, "Edge-released, Piezoelectric MEMS Acoustic Transducers in Array Configuration," *Journal of Micromechanics and Microengineering*, 22, 025005, 2012.

[6] Q. Zhang and E.S. Kim, "Energy Harvesters with High Electromagnetic Conversion Efficiency through Magnet and Coil Arrays," *IEEE International Micro Electro Mechanical Systems Conference*, Taipei, Taiwan, January 20–24, pp. 110 – 113, 2013.

CONTACT

* S.J. Chen, tel: +886-3-426-7374; raychen@ncu.edu.tw

ELECTROSTATIC GENERATOR WITH FREE MICRO-BALL AND ELASTIC STOPPERS FOR LOW-FREQUENCY VIBRATION HARVESTING

F. Cottone[1,2], P. Basset[2], F. Marty[2], D. Galayko[3], L. Gammaitoni[1] and T. Bourouina[2]
[1]University of Perugia, NIPS Laboratory, Department of Physics, Italy
[2]Université Paris-Est / ESYCOM / ESIEE Paris, [3]UPMC-Sorbonne Universités / LIP6, France

ABSTRACT

We present a new MEMS electrostatic vibration harvester that exploits mechanical frequency amplification by multiple-mass impacts in combination with elastic stoppers. When the system is shaken at low frequency (10-60 Hz) a tungsten micro-ball impacts within the oscillating proof mass of the harvester transferring kinetic energy to the gap-closing comb transducer which in turns resonates at its natural frequency (92 Hz). In addition, elastic stoppers amplify the proof mass and ball velocity throughout collision with the fixed frame. Output power between 0.25 and 0.45 μW is achieved at 0.3-*g* amplitude and only 15 V bias in the range of 10-60 Hz with a -3db bandwidth of 50 Hz.

INTRODUCTION

Future wireless electronics must be independent from powering sources and have the capability to operate for very long time without maintenance in order to enable smart applications [1]. The possibility for wireless sensors to find power supply where they are deployed open vast sceneries of intelligent applications for healthcare, transportation, surveillance, military and industrial monitoring. In this regard, kinetic energy is the shape of vibrations is abundantly available in the real world such as industrial plants, transportations, infrastructures and biological beings. However, mechanical vibrations from natural and artificial sources are inconsistent and mostly located in between 2 and 100 Hz. Typical commercial vibration energy harvesters (VEHs) are based on linear mechanical oscillators that efficiently work within few hertz around the resonant frequency. Regardless of the type of conversion method, at MEMS dimensions, the harvester resonance quickly increases up to several kHz depending on the weight of the inertial mass and on the equivalent stiffness of the suspension springs. Therefore, the mismatch between real world vibrations frequency (2-100 Hz) make MEMS linear harvesters unpractical. New approaches have been recently investigated to widen the operating bandwidth, such as nonlinear piezoelectric oscillators [2], self-tuning resonators [3] and piezoelectric cantilevers arrays [4], although with centimeter-sized demonstrators. In order to address the problem of extracting energy from low-frequency vibrations, frequency−up conversion system have also been proposed by means of piezoelectric or electromagnetic generators [5]. However, these prototypes were quite bulky, whereas, electrostatic vibration energy harvesters (e-VEH) are more suitable for MEMS fabrication [6]. Recently, a multiple-mass electrostatic harvester exploiting frequency-up conversion

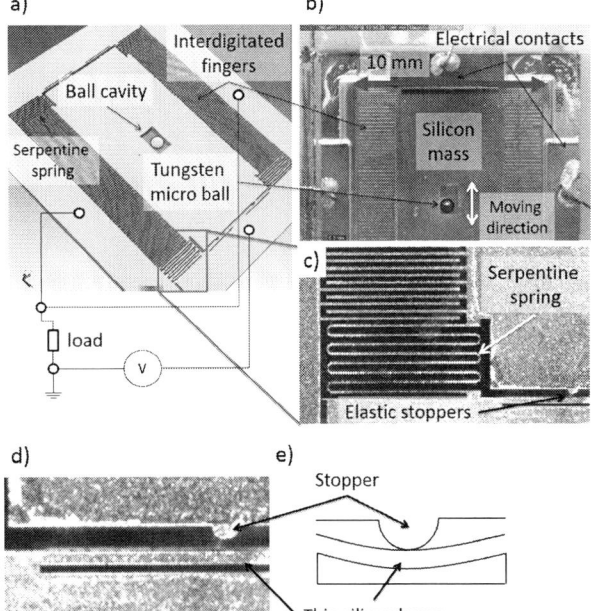

Figure 1: (a) Computer drawing of the gap-closing MEMS electrostatic VEH with electrical connection scheme of bias/output voltage. (b) Microscope photograph of the device (65x of magnification) with (c) magnified view (90x) of mechanical serpentine springs, (d) elastic stoppers with thin silicon beam and (e) sketch of stopper with beam.

have been proposed by Fu et al [7] to capture low frequency kinetic energy. However, this device shown few nano-watts of generated power.

The use of multiple impacting masses was shown to be advantageous for electromagnetic vibration harvesters [8, 9]. With reference to electrostatic generators, Lee and Halvorsen recently proposed a device exploiting internal mass impacts with end stops to widen the frequency response [10]. Galchev et al. [11] shown an original springless inertial harvester for converting low-frequency motion into electricity by means of magnetic ball. The authors of this work also presented a wideband spring-mass gap-closing MEMS e-VEH with nonlinear mechanical and electrical features.

The concept reported here improves the previous device by exploiting a mechanical frequency amplification through multiple-mass impacts of a heavy tungsten micro-ball in combination with elastic stopper bouncing effect. The design, fabrication and testing of a new prototype of silicon gap-closing MEMS electrostatic VEH assembled with a

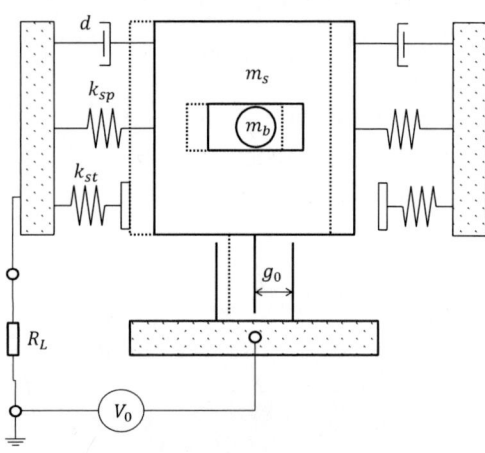

Figure 2: Model of the MEMS e-VEH.

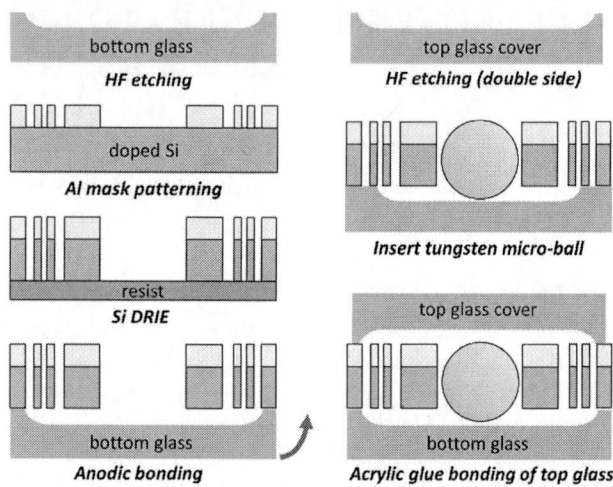

Figure 3: Batch fabrication process of the MEMS-eVEH with the tungsten micro-ball and top glass cover. The support glass was assembled through anodic bonding while the top cap was successively glued with acrylic glue.

tungsten micro-ball is here presented. The system takes advantage of the free movement of the micro-ball housed inside a rectangular cavity of the silicon proof mass that can capture low-frequency vibrations and transfer kinetic energy to the electrostatic transducer.

SYSTEM DESIGN AND PRINCIPLE

Figure 1(a) presents the 3D drawing of the MEMS-eVEH, while figure 1(b) and shows a microscope photograph of the generator with magnified (90x) particular of the serpentine support springs and elastic stoppers. Figure 1(d) shows an expanded view of elastic stoppers with a thin silicon beam of the fixed part and (e) schematics of the stopper when impacting on the silicon beam. The system model is illustrated in figure 2. The electrostatic transducer consists of a bulk silicon in-plane gap-closing interdigitated-combs with the proof mass suspended by serpentine springs of 20 μm of thickness. The comb capacitor is pre-charged by an initial bias voltage V_0 that span from 10 to 20 V. A tungsten carbide micro-ball of 0.8 mm of diameter is housed inside the proof mass within a rectangular cavity that measures 1 mm of width by 1.5 mm of length.

When the system vibrates at frequency below 60 Hz, the micro-ball freely moves within the cavity and impact on the upper and lower inner wall of the holed proof mass. Each collision provides a wideband mechanical impulse to the proof mass of the gap-closing comb transducer that in turns resonates at its natural frequency of 92 Hz. In this way, kinetic energy of the input vibration is transferred from 10-60 Hz to higher frequencies in the range of 92-120 Hz. The operation bandwidth of such a mechanical frequency-up conversion depends on the ball travelling distance and on the deflection height of the stopper beams. The cavity length of this first prototype is designed such that the maximum rate of impacts occurs around 15-20 Hz for a given acceleration of 0.3 g (where g=9.81 ms^{-1}). Based on calculations on the system model, a length of the ball cavity of 7 mm would

optimize the harvester for operation at 2 Hz. This configuration has a great potential for capturing energy from human movements. This is useful, for example, to recharge the battery of a pacemaker.

The fabrication process is shown in figure 3. The structure is obtained by DRIE through a 400-μm-thick silicon wafer, which is then sandwiched between two glass wafers: a support bottom glass and a cover glass that prevents the micro-ball to escape. Anodic bonding assembles the support glass, the micro-ball is then inserted and the top glass realized with double side mask process is subsequently attached with acrylic glue. Silicon beams (60-μm-thick, 2 mm of length) realized in correspondence of the stoppers allow quite good elastic impact of the mass with the rigid frame. The calculated equivalent spring stiffness of the stopper beams at midpoint results $k_{st} = 2.71 \times 10^4$ N/m, considering trapezoidal section like in [12] while the stiffness of the serpentine springs along the moving direction results $k_{st} = 2.10$ N/m. The overall die surface measures about 10 × 10 mm^2. The comb fingers have a length of 2 mm, width of 30 μm and an initial gap g_0 of 70 μm is in between. Table 1 shows the system parameters.

EXPERIMENTAL SETUP

The harvesting device prototype was tested onto an electromagnetic shaker that provided the vibrations (TMS, model K2007E01 with integrated power amplifier). The vibration input was generated and handled by a vibration controller (Brüel & Kjær) through the feedback of an accelerometer. The micro electrostatic VEH was pre-charged at constant voltage V_0 (5-20 V) and the output power is dissipated across a variable load resistance R_L (1 kΩ–0.5 MΩ) connected in series.

978-1-4799-3510-9/14 $31.00 © 2014 IEEE

Table 1: Model parameters for the VEH prototype.

Parameters	Value
Proof mass, m_s	57.2×10^{-6} Kg
Serpentine springs stiffness, k_{sp}	2.1 Nm^{-1}
Mechanical resonance, f_r	92 Hz
Tungsten micro-ball mass, m_b	4.19×10^{-6} Kg
Stopper beams stiffness, k_{st}	2.71×10^{4} Nm^{-1}
Active area, A_0	10×10 mm^2
Gap between fingers, g_0	70 μm
Optimal load resistance, R_{opt}	5.6 Mohm
Device thickness, t_d	400 μm
Fingers length, l_f	2 mm
Fingers width, w_f	30 μm

All the signals are then recorded through a data acquisition card (National Instruments, model USB-6211) handled with a PC with a LabView program.

The MEMS e-VEH prototype was initially characterized with both simulations and experimental testing for identifying the model parameters that are listed in Table 1. Subsequently, preliminary tests under sine sweeping for input acceleration $a_i = 0.3g$ rms and different bias voltage V_0 = 5-20 V were performed. All the measurements were carried out comparing the system behaviour both with and without the tungsten micro-ball.

RESULTS AND DISCUSSION

Frequency sweeping

Figure 4(a) and (b) display the output RMS voltage and the corresponding harvested power respectively across the optimal load resistance (5.6 MΩ) under frequency up- and down-sweep over 10-120 Hz. The comparison concerns the performance of the MEMS-eVEH prototype with and without the tungsten micro-ball in order to analyze the effect of the mechanical frequency multiplication. In this case, the used bias voltage was $V_0 = 15$ V. Since the support serpentine springs were very soft (2 N/m), above 20 V electrostatic pull-in was observed. Nevertheless, at this voltage level and at 0.3 g of acceleration the micro-ball collision were sufficiently strong to avoid permanent sticking of the parallel plates of the comb capacitor. It is important to notice that the system with micro-ball shows an extended response in frequency with respect to that without it. In particular, the -3db bandwidth increases up to 56% and in the intervals of 10-20 Hz and 40-60 Hz, a significant power enhancement (10 to 100 times) is shown. Most of all it shows to be capable of operating below 20 Hz producing a respectable output power around 0.2 μW with only 15 V of bias voltage. On the other hand, it is interesting to observe that the system without the micro-ball presents a multiple band response at both low (20-50 Hz) and high frequency ranges (90-100 Hz). This effect is due to the repeated bouncing of the proof mass when it collides to the elastic beams of the rigid frame, in correspondence to the stoppers, thanks to the very soft

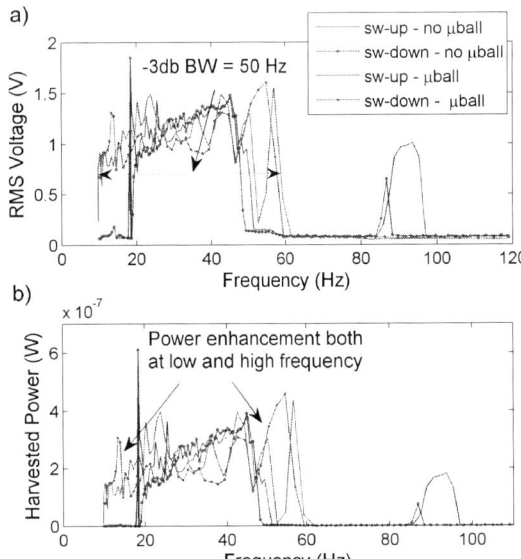

Figure 4: (a) RMS Voltage and (b) harvested power at acceleration amplitude of 0.3 g under sinusoidal frequency up- and down-sweep (10-120 Hz). The comparison regards the same structure with and without the tungsten micro-ball.

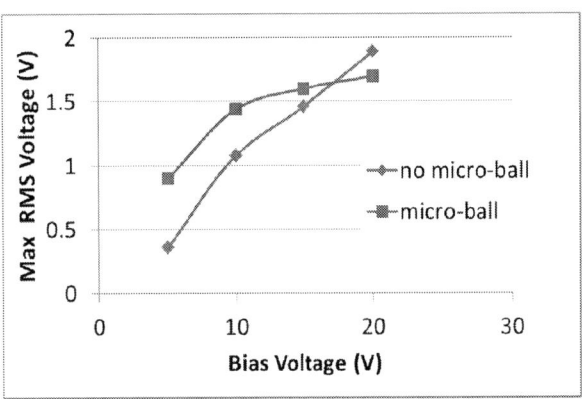

Figure 5: Maximum output voltage across the optimal load Rl=5.6Mohm on frequency sweeping for the device with and without micro-ball.

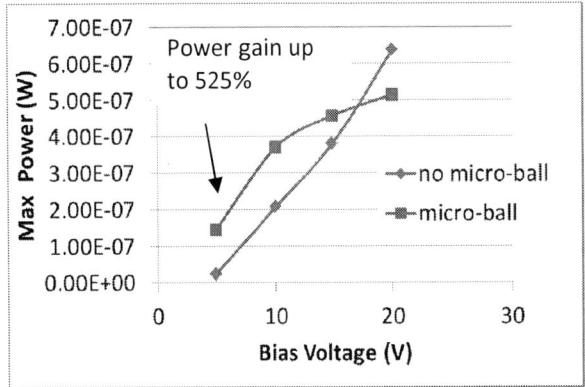

Figure 6: Maximum output harvested power over the optimal load Rl=5.6Mohm on frequency sweeping for the device with and without micro-ball.

serpentine springs.

Figures 5 and 6 respectively show the maximum output voltage and the harvested power across the optimal load resistance versus the given bias voltage respectively with and without the tungsten micro-ball inside the oscillating mass. It is relevant to observe that for bias voltage lower than 20 V the system with the micro-ball exhibits a power gain of 77.8% at V_0=10 V up to 525% at V_0 = 5 V. This concept is thus suitable for improving the power harvesting performance when using low voltage external sources to pre-charge the generator (i.e. with PVDF piezoelectric systems). In principle, this system could be designed to work at even higher bias voltage up to 100 V in vacuum by choosing stiffer support springs and heavier micro-ball to prevent a low pull-in voltage. Thus the generated power would be potentially of two orders of magnitudes higher (60-100 μW).

CONCLUSIONS

This work presented the design, fabrication and testing of a new MEMS electrostatic silicon gap-closing comb vibration energy harvester that exploits mechanical frequency amplification by means of micro-ball impacts with the walls of an holed proof mass in combination with elastic bouncing of the mass with the rigid frame. The main objective was to improve the power generation and extend the operating bandwidth of MEMS e-VEHs at very low frequency.

Under harmonic frequency up- and down-sweeping (10-120 Hz) at 0.3 g rms and with different bias voltages ranging in 2.5-20 V, the electrostatic generator shown a gain factor up to 525% at 5 V of bias voltage in harvesting energy at 10−40 Hz versus the one without the ball. The -3db bandwidth increased up to 56% and in the intervals of 10-20 Hz and 40-60 Hz. This result make the system suitable even for harvesting energy from human movements.

ACKNOWLEDGEMENTS

The authors gratefully acknowledge the support of FP7 Marie Curie Intra–European–Fellowship (IEF) funding scheme (NEHSTech, Grant No. 275437) at Université de Paris-Est, ESIEE Paris and the Galielo Galileo project No. 28150WA.

REFERENCES

[1] M. Kroener, "Energy harvesting technologies: Energy sources, generators and management for wireless autonomous applications," 2012, pp. 1-4.

[2] F. Cottone, H. Vocca, and L. Gammaitoni, "Nonlinear Energy Harvesting," *Physical Review Letters,* vol. 102, p. 080601, 2009.

[3] D. Zhu, S. Roberts, M. J. Tudor, and S. P. Beeby, "Design and experimental characterization of a tunable vibration-based electromagnetic micro-generator," *Sensors and Actuators A: Physical,* vol. 158, pp. 284-293, 2010.

[4] M. Ferrari, V. Ferrari, M. Guizzetti, B. And , S. Baglio, and C. Trigona, "Improved energy harvesting from wideband vibrations by nonlinear piezoelectric converters," *Sensors and Actuators A: Physical,* 2010.

[5] S. M. Jung and K. S. Yun, "Energy-harvesting device with mechanical frequency-up conversion mechanism for increased power efficiency and wideband operation," *Applied Physics Letters,* vol. 96, p. 111906, 2010.

[6] P. Basset, D. Galayko, A. M. Paracha, F. Marty, A. Dudka, and T. Bourouina, "A batch-fabricated and electret-free silicon electrostatic vibration energy harvester," *Journal of Micromechanics and Microengineering,* vol. 19, p. 115025, 2009.

[7] J. Fu, Y. Nakano, L. Sorenson, and F. Ayazi, "Multi-axis AlN-on-Silicon vibration energy harvester with integrated frequency-upconverting transducers," in *Micro Electro Mechanical Systems (MEMS), 2012 IEEE 25th International Conference on,* 2012, pp. 1269-1272.

[8] B. Ahmed Seddik, G. Despesse, E. Defay, and S. Boisseau, "Increased bandwidth of mechanical energy harvester," *J. Sensors & Transducers,* vol. 13, pp. 62-72, 2011.

[9] F. Cottone, R. Frizzell, S. Goyal, G. Kelly, and J. Punch, "Enhanced vibrational energy harvester based on velocity amplification," *Journal of Intelligent Material Systems and Structures,* 2013.

[10] C. P. Le, E. Halvorsen, O. Søråsen, and E. M. Yeatman, "Microscale electrostatic energy harvester using internal impacts," *Journal of Intelligent Material Systems and Structures,* vol. 23, pp. 1409-1421, 2012.

[11] T. Galchev, R. Raz, and O. Paul, "An electrostatic springless inertial harvester for converting multi-dimensional low-frequency motion," in *Micro Electro Mechanical Systems (MEMS), 2013 IEEE 26th International Conference on,* 2013, pp. 102-105.

[12] R. Guillemet, P. Basset, D. Galayko, F. Cottone, F. Marty, and T. Bourouina, "Wideband MEMS electrostatic vibration energy harvesters based on gap-closing interdigited combs with a trapezoidal section," in *IEEE 26th Int. Conf. on MEMS,* Taipei, 2013.

CONTACT

*F. Cottone; tel +39 075 585 2770,
Department of Physics, University of Perugia, Italy
Email: francesco.cottone@unipg.it

FLEXIBLE MICRO-SUPERCAPACITORS FROM PHOTORESIST-DERIVED CARBON ELECTRODES ON FLEXIBLE SUBSTRATES

Mun Sek Kim, Ben Hsia, Carlo Carraro, and Roya Maboudian

University of California, Berkeley, USA

ABSTRACT

This paper reports a simple and scalable technique for the fabrication of flexible micro-supercapacitors. The supercapacitor electrodes are synthesized via the pyrolysis of patterned photoresist on a SiO_2/Si substrate. The electrodes can then be moved to flexible substrates through a simple transfer process. The fabricated devices show excellent energy density of 0.3 mJ/cm², maximum power density of 5 mW/cm², and good performance under electrochemical cycling. The electrochemical performance is maintained after 300 mechanical bending cycles.

INTRODUCTION

Supercapacitors are energy storage devices that have high power density and more robust cycle lifetime relative to existing battery technologies [1, 2, 3]. These properties make microscale supercapacitors attractive for integration in applications such as mobile or autonomous microdevices. In particular, flexible micro-supercapacitors have received increasing research interest for applications including bioimplantable devices, flexible displays, and wearable electronics. In these applications, an integrated flexible energy storage device is necessary for maintaining device mobility, flexibility, and small form factors.

Supercapacitors are composed of two primary components, the two electrodes and the electrolyte. In order to store charge, the electrodes are polarized by an external voltage, and the mobile ions in the electrolyte preferentially migrate to form areas of high charge density at each electrode/electrolyte interface. Due to this interfacial electrostatic charge storage mechanism, supercapacitor electrodes must have extremely high surface area in order to store appreciable amounts of energy. Furthermore, electrode materials should have sufficient electrical conductivity in order to transport charge carriers without excessive resistance. In addition to these requirements of high surface area and good conductivity, microscale supercapacitor electrodes must also be easily deposited and patternable. For these reasons, photoresist-derived porous carbon is a promising material for micro-supercapacitors due to its porosity, good conductivity, and easy patternability [4, 5, 6].

While the electrochemical performance of this material has been successfully demonstrated on rigid substrates [4, 5], for incorporation into flexible devices, the porous carbon structures must be successfully transferred onto flexible substrates. This paper reports the fabrication and testing of devices fabricated from the transfer of patterned porous carbon structures.

METHODS

The fabrication of the micro-supercapacitor is illustrated schematically in Figure 1. First, the photoresist is patterned into a two-pad electrode configuration on a SiO_2/Si substrate. The sample is then pyrolyzed at 900 °C in a hot wall tube furnace in H_2 and Ar as described in Ref. [4]. A double-transfer technique is then used to transfer the patterned electrodes to a flexible substrate. First, a polycarbonate sheet is adhered to the sample upon heating to 300 °C. After cooling, the sheet is carefully peeled off from the substrate; the porous carbon preferentially adheres to the polycarbonate, transferring the patterned electrodes. As will be further discussed, this single transfer inverts the carbon electrodes, exposing the surface of the film that was in contact with the SiO_2 during pyrolysis, and yields a lower capacitance than the untransferred electrodes. Therefore, a second transfer is needed. This transfer is performed by adhering a polyethylene sheet to the polycarbonate at 200 °C and selectively etching the polycarbonate in methylene chloride.

The morphology of both surfaces of the film is probed via atomic force microscopy (AFM, Digital Instruments Nanoscope IIIa) in tapping mode. In order to test the electrochemical performance of the device, Cu tape is used to make electrical connection to the electrodes and a 0.5 M H_2SO_4 aqueous solution is used as the electrolyte. Cyclic voltammetry (CV), galvanostatic charge/discharge, and AC impedance (AC Instruments, 660D Model) are used to probe the electrochemical characteristics of the fabricated device.

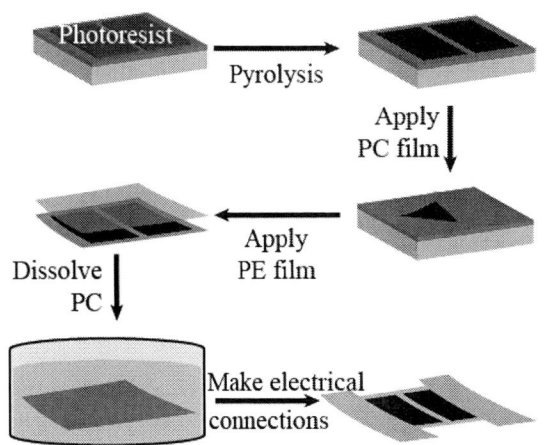

Figure 1: Schematic of flexible microsupercapacitor fabrication

RESULTS AND DISCUSSION

Comparison between Electrode Top and Bottom Surfaces

As previously mentioned, the morphological and electrochemical properties of the pyrolyzed photoresist surface are significantly different for the untransferred and singly transferred films. These differences arise during the pyrolysis process. The "top" surface of the photoresist is exposed to the H_2/Ar ambient during pyrolysis, while the "bottom" surface is in contact with the underlying SiO_2 layer on the substrate. The first transfer exposes the bottom surface to the electrolyte and thus shows different electrochemical behavior.

The AFM results (shown in Figure 2) show that the top surface is significantly rougher than the bottom surface. The RMS roughness is calculated to be about 1.3 and < 0.3 nm, respectively. This difference in surface morphology likely results from three factors. First, the bottom surface of the photoresist is in contact with the smooth thermal oxide surface and forms a matching smooth surface. Second, the pore formation mechanism during pyrolysis stems from the evaporation of volatile components from the film and this evaporation occurs through the top surface [7]. Third, pore formation and accessibility are assisted through a hydrogen etching process which occurs primarily at the top surface due to transport limitations [4, 7].

These morphological differences are manifested in the electrochemical testing of the films after one transfer and after a second transfer (which recovers the original top surface). CV tests show that devices fabricated after one transfer give areal capacitances of only 10 µF/cm² at a scan rate of 100 mV/s (all values normalized by the projected active area of both electrodes), while devices fabricated after the second transfer give capacitances about 75 times higher at the same scan rate, 0.75 mF/cm². Thus, a double transfer appears to be necessary to achieve high energy density devices.

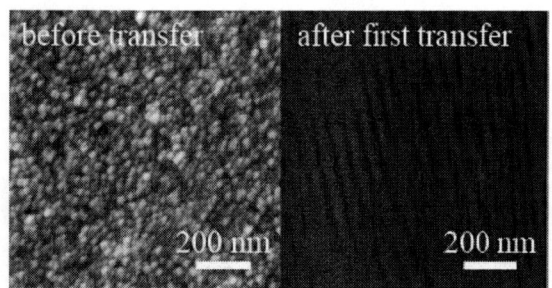

Figure 2: Atomic force micrographs of pyrolyzed photoresist before transfer and after the first transfer exposing the bottom surface. Z-scale is identical in both images (10 nm). The apparent ridges in the right-hand image are an experimental artifact.

Electrochemical Analysis

Cyclic voltammetry scans of the device show near-ideal electrochemical behavior at scan rates of up to 10 V/s, as evidenced by the rectangular CV plots shown in Figure 3.

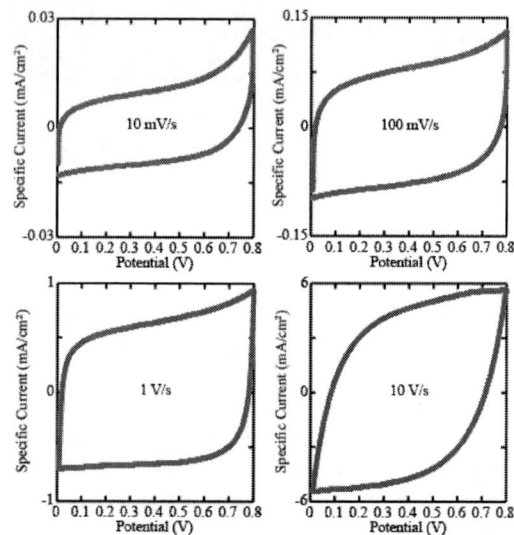

Figure 3: Cyclic voltammograms of fabricated device at a variety of scan rates. Current normalized by active area of the electrodes.

The specific capacitance, C, achieved for the device is calculated via Equation 1,

$$C = \frac{I_{avg}}{sA} \qquad (1)$$

where I_{avg} is the current averaged over both positive and negative sweeps, s is the scan rate, and A is the projected area of both electrodes which is in contact with electrolyte. C is calculated to be 0.75 mF/cm² at a scan rate of 100 mV/s, dropping to 0.62 and 0.36 mF/cm² at 1 and 10 V/s respectively. These values correspond to a material capacitance of between 1.4 and 3 mF/cm² which is calculated by multiplying the device specific capacitance by a factor of 2 to obtain the capacitance at one electrode and by another factor of 2 to normalize by active area of one electrode. These results show good agreement with Ref. [4] which shows a specific capacitance of 1.5 – 3.5 mF/cm² for three-electrode measurements on an unpatterned pyrolyzed film.

Galvanostatic charge/discharge tests also show near ideal capacitive performance, with linear charge and discharge curves at all probed scan rates. Three representative curves are shown in Figure 4a. A small IR drop is present in each discharge curve, indicating some ohmic resistance in the device, known as the equivalent series resistance (ESR). The ESR can be calculated via Equation 2,

$$ESR = \frac{V_{drop}}{2I} \qquad (2)$$

where V_{drop} is the magnitude of the IR drop and I is the applied current density. The ESR is about 8 $\Omega \cdot$cm² for the device. This ESR limits the maximum power and probably arises from the high resistance of the carbon film and the contact resistance to the Cu tape current collector.

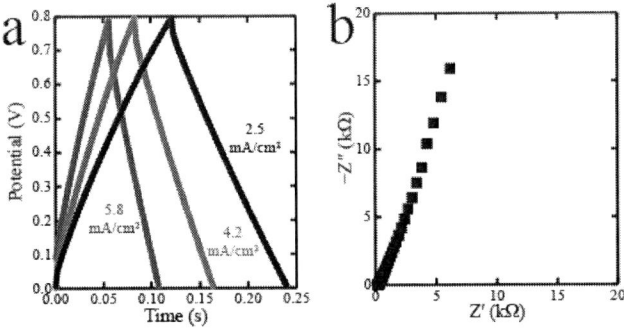

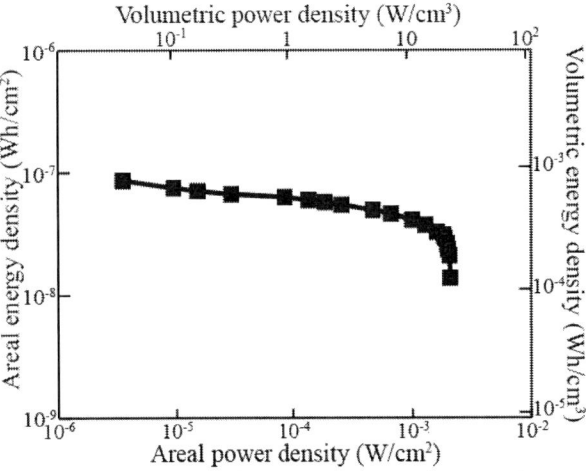

Figure 4: (a) Galvanostatic charge discharge curves of fabricated device at a variety of specific currents. (b) Nyquist plot showing results of AC impedance test.

AC impedance tests also show typical supercapacitor behavior, with a near-vertical line in the Nyquist plot [2, 8] (Figure 4b). The AC frequency at which the phase shift crosses –45° is ~122 Hz for the fabricated device. This value indicates the approximate highest frequency at which the device can yield capacitive behavior and compares favorably to other flexible micro-supercapacitors such as laser-scribed graphene based devices, which show a crossover frequency of about 50 Hz [9].

Cycle lifetime studies are performed through 10,000 repetitive CV cycles at a scan rate of 100 mV/s. The good capacitive performance of the device is maintained throughout the cycling, with a capacitance retention of 86%.

Energy and Power Density

The primary figures of merit for supercapacitors are the energy and power density. For microscale devices, the areal and volumetric densities are most frequently reported (rather than the gravimetric density). The energy density, E, and power density, P, are calculated from Equations 3 and 4,

$$E = \frac{1}{2}CV^2 \tag{3}$$
$$P = \frac{E}{\Delta t} \tag{4}$$

where C is the specific capacitance calculated from CV, V is the voltage window over which the device is cycled (0.8 V in this case), and Δt is the discharge time. Many CV scan rates are tested, yielding several points on the Ragone plot as plotted in Figure 5. Volumetric densities can easily be calculated by dividing the areal values by the film thickness (0.9 μm). The energy density achieved is extremely high, 10^{-8} to 10^{-7} Wh/cm^2 (0.05 to 0.3 mJ/cm^2) or 10^{-4} to 10^{-3} Wh/cm^3 (0.6 to 3.5 J/cm^3). For comparison, laser-scribed graphene based flexible micro-supercapacitors yield about 2×10^{-4} to 5×10^{-4} Wh/cm^3 energy density using an aqueous gel electrolyte [9]. These comparison values are obtained by normalizing by device area, rather than electrode area, as in this work, so the actual energy density of the electrode material is slightly underestimated for the laser-scribed graphene. Nonetheless, the energy densities achieved in this work are competitive with state-of-the-art flexible electrode materials.

Figure 5: Ragone plot showing energy and power density of fabricated micro-supercapacitor device.

Flexibility Testing

In order to probe the performance of the device under flexion, the 1×1 cm^2 sample is conformally wrapped around cylinders of various size and manually cycled from an unbent to a bent position. This cycle is repeated 100 times for a 17.5 mm radius of curvature. The device is then tested in an unbent configuration. The bending cycle is then repeated 100 more times for $r = 12.5$ mm, and the device is tested again. Finally, 100 more cycles are performed at $r = 5$ mm. Figure 6 shows the CV results before and after each 100 cycle test. The difference in initial capacitance and capacitance after mechanical cycling is $\sim 2\%$. This excellent retention of capacitance indicates that the pyrolyzed photoresist electrodes can withstand repeated flexing cycles and that this material is promising for integration in flexible energy storage applications.

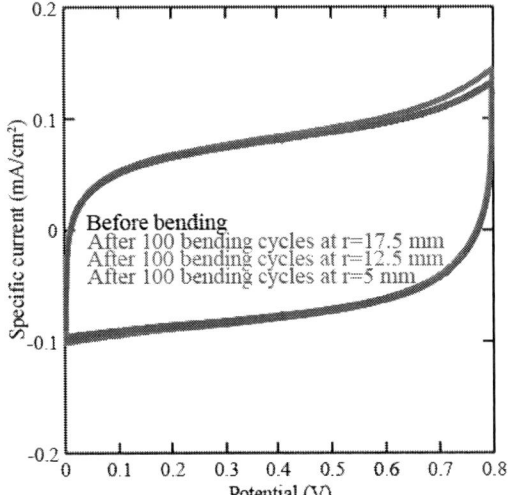

Figure 6: Cyclic voltammograms before and after bending cycles. Scan rate is 100 mV/s.

CONCLUSION

In summary, this work presents a scalable and simple technique to fabricate and transfer micro-supercapacitor electrodes for flexible microscale energy storage applications. The electrode material, pyrolyzed photoresist, is a promising material micro-supercapacitor material platform due to its high specific capacitance, good conductivity, and easy large-scale manufacturability. A double-transfer to a flexible substrate is necessary for achieving good capacitive performance and very high energy densities. Both the electrochemical and mechanical stability of the device is excellent. This fabrication technique is a promising step towards flexible microscale integrated energy storage devices.

ACKNOWLEDGEMENTS

We gratefully acknowledge the Siemens CKI program and National Science Foundation through Center of Integrated Nanomechanical Systems (COINS) and a Graduate Research Fellowship (BH). Portions of the device fabrication and characterization were performed in the UC Berkeley Marvell Nanofabrication Facility.

REFERENCES

[1] B. E. Conway, *Electrochemical Supercapacitors: Scientific Fundamentals and Technological Applications*, Kluwer Academic/Plenum Publishers, 1999.

[2] R. Kotz and M. Carlen, "Principles and applications of electrochemical capacitors", *Electrochim. Acta,* vol. 25, pp. 2483-2498, 2000.

[3] P. Simon and Y. Gogotsi, "Materials for Electrochemical Capacitors", *Nat. Mater.,* vol. 7, pp. 845-854, 2008.

[4] B. Hsia, M. S. Kim, M. Vincent, C. Carraro and R. Maboudian, "Photoresist-derived porous carbon for on-chip micro-supercapacitors", *Carbon,* vol. 57, pp. 395-400, 2013.

[5] B. Hsia, M. S. Kim, C. Carraro and R. Maboudian, "Cycling characteristics of high energy density, electrochemically activated porous-carbon supercapacitor electrodes in aqueous electrolytes", *J. Mater. Chem. A,* vol. 1, pp. 10518-10523, 2013.

[6] M. Beidaghi, W. Chen and C. Wang, "Electrochemically activated carbon micro-electrode arrays for electrochemical micro-supercapacitors", *J. Power Sources,* vol. 196, pp. 2403-2409, 2011.

[7] G. M. Jenkins and K. Kawamura, *Polymeric carbons - carbon fibre, glass and char*, Cambridge University Press, 1976.

[8] P. L. Taberna, P. Simon and J. F. Fauvarque, "Electrochemical characteristics and impedance spectroscopy studies of carbon-carbon supercapacitors", *J. Electrochem. Soc.,* vol. 150, pp. A292-A300, 2003.

[9] M. F. El-Kady and R. B. Kaner, "Scalable fabrication of high-power graphene micro-supercapacitors for flexible and on-chip energy storage", *Nat. Commun.,* vol. 4, p. 1475, 2013.

CONTACT

*R. Maboudian, tel: +1-510-6437957; maboudia@berkeley.edu

GRAPHENE ELECTRODES ENHANCE PERFORMANCE FOR MICRO-LITER SCALE MICROBIAL FUEL CELLS

Vishnu Jayaprakash*, Ryan D. Sochol, Roseanne Warren, Kosuke Iwai and Liwei Lin
Berkeley Sensor and Actuator Center, University of California, Berkeley, USA

ABSTRACT

Recently, microliter-scale microbial fuel cells (μMFCs) have garnered significant interest as effective energy harvesters for low power biological and electronic systems. Although researchers have attained high current densities and columbic efficiencies from such fuel cells, low power outputs and working potentials caused by the use of Au/Cr electrodes have limited the implementation of μMFCs in practical applications. To overcome these limitations, here we present a graphene-based μMFC (G-MFC) that utilizes laser synthesized graphene electrodes to generate open circuit potentials (OCPs) of 0.8 ± 0.05 V and power densities of 1820 ± 10 W/m^3. Furthermore, the G-MFC produces a maximum power output of 364 μW. The stackable and low cost design of our G-MFC allows for a wide range of applications and also serves as a platform for repeatable electrode and substrate based testing. These results suggest that our G-MFC methodology could offer an effective route to achieve viable energy harvesters for low power systems.

INTRODUCTION

Microbial fuel cells (MFCs) enable electricity harvesting directly from organic or inorganic matter through bacterial catalysis [1-11]. Benefits including relatively non-toxic components, flexibility of usable substrates, ambient operating temperatures, and mediator-less electron transfer from exoelectrogenic bacteria to solid electrode surfaces have all contributed to increased scientific interest in MFCs

as a renewable source of sustainable energy [2, 11, 12]. At present, the majority of MFCs are milliliter-scale devices, which allows for the easy use of carbon-based electrodes, air cathodes and simple operating techniques [4]; however, reducing MFCs to microliter-scales enables: *(i)* higher surface area to volume ratios, *(ii)* increased control over electrode and chamber environments, *(iii)* superior reaction kinetics, and *(iv)* greater mass transfer flux [13]. Additionally due to their smaller chamber volumes and low start-up times, μMFCs are ideal testing platforms for electrode, microbe, and substrate-based experiments [13]. As a result, μMFCs could potentially be used for a variety of electronic and biological applications, such as powering low power devices in remote areas and powering on-chip systems [14, 15]. Using microfabrication techniques, researchers have managed to obtain increased columbic efficiencies and current densities from μMFCs [13]. Unfortunately, current μMFCs produce orders of magnitude lower power outputs and much lower potentials than their milliliter-scale counterparts. This is caused predominantly by high internal resistances and inefficient biological systems in the anode. Gold has been the primary electrode material in most μMFC works thus far as it is easy to fabricate and control with microscale precision using existing microfabrication techniques; however, researchers have found that gold electrodes result in increased contact resistance at the bacteria-electrode interface [14, 15]. This increased contact resistance has been the primary cause

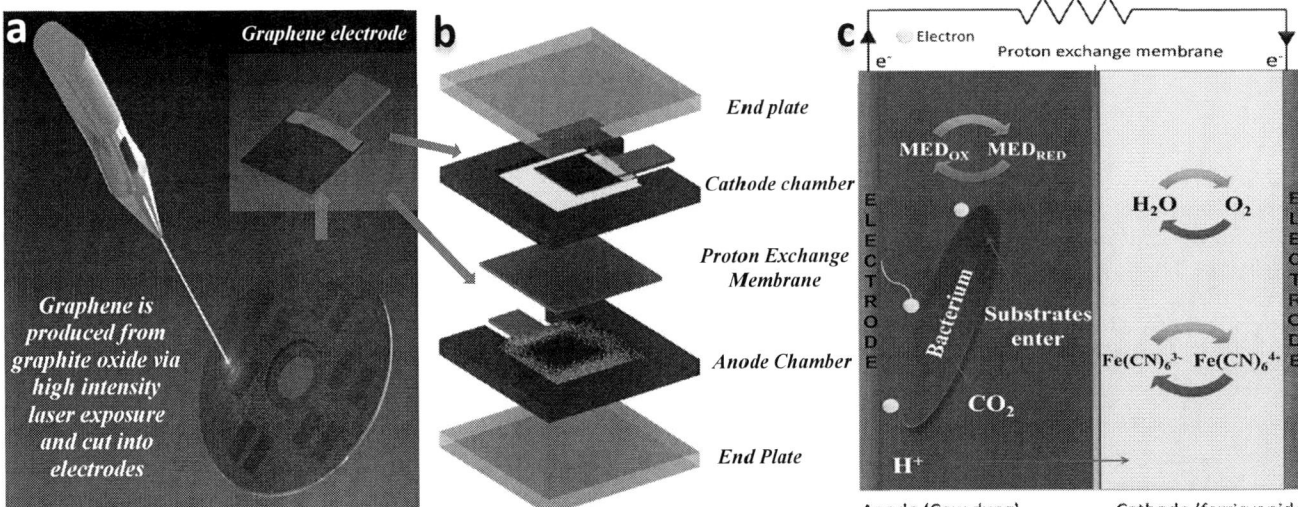

Figure 1: Conceptual Illustrations of the graphene-based microliter-scale microbial fuel cells (G-MFC). (a) Graphite oxide is converted into graphene using a programmable laser. Graphene sheets were attached to graphite leads to make electrodes. (b) Exploded schematic of the G-MFC. (c) Bacteria catalyze the decomposition of carbohydrates in the cow dung, releasing electrons and protons. The electrons are transferred to the graphene anode via nanowires/endogenous mediators/direct contact. The electrons reduce the catholyte (ferricyanide) while the protons traverse the proton exchange membrane (PEM) and re-oxidize it.

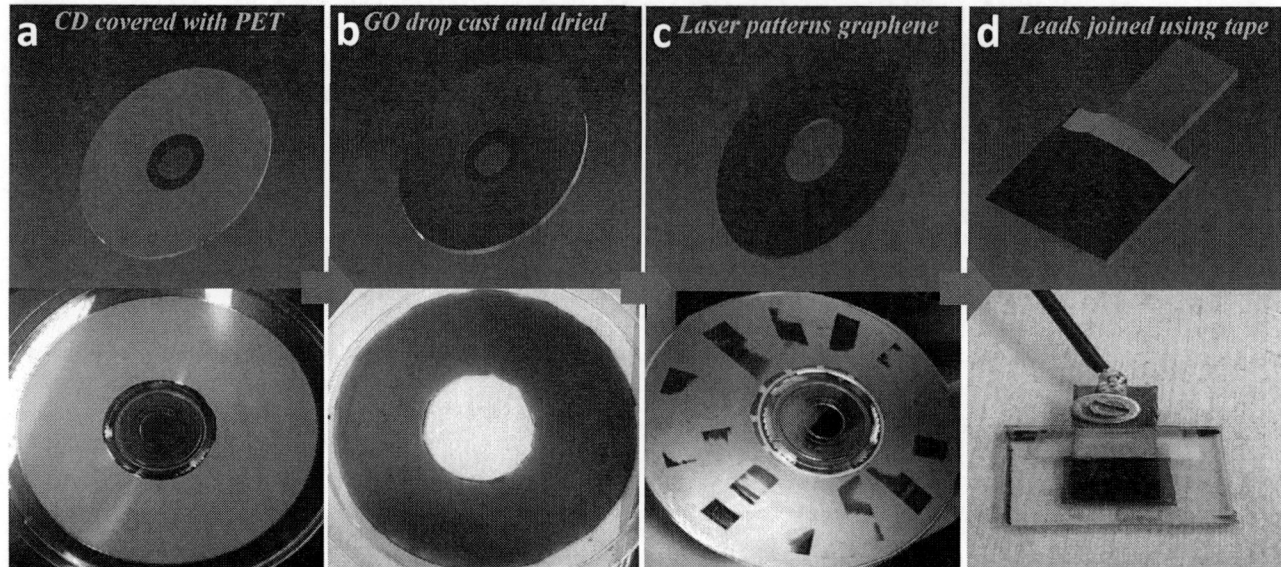

Figure 2: Graphene electrode fabrication process. (a) A standard LightScribe CD is covered with a PET sheet. (b) A 3 mg/ml graphite oxide (GO) solution in water is drop cast on to the PET substrate and dried. (c) After 6 runs in the LightScribe drive, graphene is obtained (d) Graphene is cut and made into an electrode.

of low power outputs in μMFCs, restricting their use to laboratory testing [4, 13]. To overcome these issues, here we present a G-MFC that exploits the advantages offered by μMFCs while utilizing graphene electrodes to reduce the internal resistance and increase power output.

CONCEPT AND MATERIALS

The basic concepts behind the G-MFC are included in figure 1. Figure 1a shows a conceptual illustration of the LightScribe method of converting graphite oxide into graphene under high intensity light [16]. Laser synthesized graphene offers a simple and controllable fabrication process combined with favorable electrical properties and high surface area, which make it an ideal electrode material for MFCs. Figure 1b shows an exploded schematic of the fuel cell. The end plates were made of acrylic and both the neoprene chambers had volumes of 200 μL. Neoprene was chosen as the chamber material to ensure that both chambers remained anaerobic during operation. In our previous work we demonstrated that cow dung was an ideal anodic substrate for microbial fuel cells as it: *(i)* contains a rich consortium of anaerobically respiring bacteria, *(ii)* has a diverse range of carbohydrates, *(iii)* is readily available in remote rural areas, *(iv)* is low cost, and *(v)* doesn't require complex inoculum preparation techniques [10]. Figure 1c shows the operating concept of G-MFC. A 1:1:1 (wt%) mixture of water, dried and anaerobically digested cow dung was used at the anode and a 100 mM potassium ferricyanide solution in water was used at the cathode. Anaerobically digested cow dung was obtained by placing fresh cow dung in air tight container for 48 hours as described previously [15]. The G-MFC utilized 1 cm^2 graphene electrodes and Ultrex CMI-7000 as the proton exchange membrane (PEM).

FABRICATION

Figure 2 shows the graphene electrode fabrication process and results. A standard LightScribe (LS) CD was cleaned and covered with a PET sheet (416-T PET sheet, MG Chemicals). A 3 mg/ml graphite oxide (GO) solution in water was then drop cast onto the PET substrate and left to dry overnight, resulting in a thin, golden brown film of GO. An ultrasonic bath was used to ensure the complete dispersion of the GO in the solution. The desired pattern was then programmed into the LS software and the assembly was exposed to the laser. The CD was exposed six times to obtain black graphene as shown in figure 3a. Finally, the PET substrate containing the graphene was cut out and the graphene was joined with graphite leads using insulating tape to make electrodes. Stainless steel screws and washers were then used to connect wires to the graphite leads to allow for easy operation and testing. Figure 3c shows an image of a fully assembled G-MFC prototype.

METHODS

Although several studies have shown that allowing time for biofilm formation results in better performance in MFCs [4, 5], here we conducted experiments immediately after fuel cell assembly and loading to examine the minimum outputs of the G-MFCs. OCPs were measured using a digital multimeter (Actron) and polarization curves were drawn with the help of a potentiostat (Ivium Technologies Inc). The potentiostat was operated in a two electrode setup, wherein both the reference and counter electrodes were connected to the anode while the working electrode was connected to the cathode. A scan rate of 1 mV/s was chosen as indicated in prior works [4, 5, 15]. Power density curves were subsequently plotted using the data. All current and power densities reported here were

978-1-4799-3510-9/14 $31.00 © 2014 IEEE

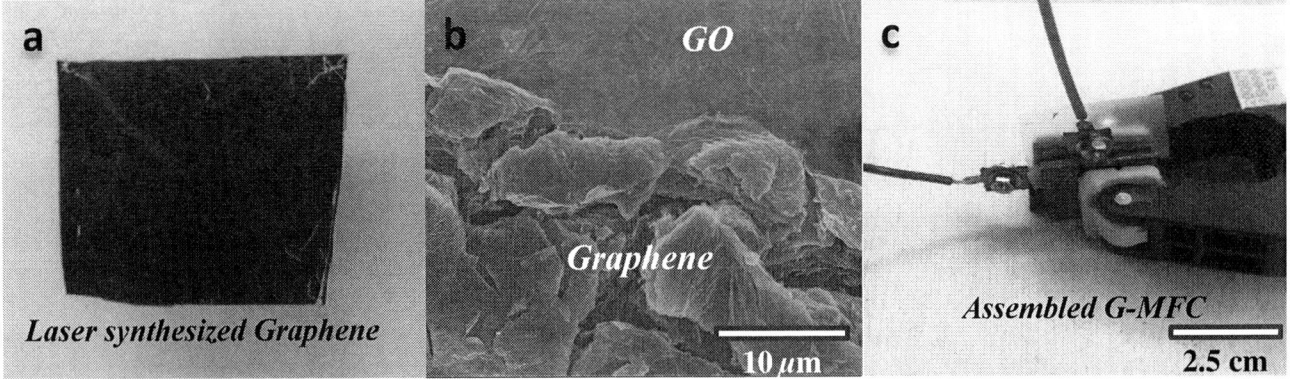

Figure 3: Photographs and SEMs. (a) Photograph of laser synthesized graphene. (b) SEM showing the GO-graphene boundary layer. (c) Photograph of assembled G-MFC.

calculated with respect to the volume of the G-MFC's chambers (200 μL). Each experiment was repeated five times to establish consistency of the results. SEM images were obtained using an FEI Nova NanoSEM scanning electron microscope. All experimental results are reported as mean ± s.e.m.

RESULTS AND DISCUSSION

Figure 4 shows the experimental results for the G-MFC. Figure 4a shows the OCPs of a single G-MFC and two G-MFCs in series plotted versus time. The G-MFC produced a maximum OCP of 0.8 ± 0.05 V and two cells in series produced a maximum OCP of 1.6 ± 0.05 V. The OCPs remained stable for approximately 2.5 hours before declining sharply. The G-MFC lasted approximately 108% longer than our previous work in which gold electrodes were used [15]. The drop in potential in the G-MFC was most likely caused by the inefficiency of the PEM as a relatively high quantity of ferrocyanide was found in the cathode chamber. When the ferricyanide was replaced, similar results were obtained. Where new gold cathodes were needed with every replacement of ferricyanide in our prior work, the graphene was surprisingly resistant to ware and flaking [15]. A buffer was

not required in the anode chamber to maintain the pH during operation. This suggests that the cow dung-water mixture contains naturally existing buffers; however, further investigation is required to identify the specific compounds responsible for this behavior [4, 5, 15]. Figure 4b shows the polarization and power density curves for a single G-MFC. The G-MFC produced a maximum power density of 1820 ± 10 W/m^3. This power density translates into a maximum power output of approximately 364 μW, which, to the authors' knowledge, represents the highest value reported for a μMFC in the literature at present. This performance can be attributed to the excellent electrical properties of graphene and to the high surface area of the synthesized electrode as evidenced by Figure 3b. Additionally, researchers have shown that the contact resistance associated with the bacteria-electrode interface is much lower in carbon-based electrodes than in gold-based electrodes [14]. The polarization curve's shape indicates that the voltage losses are caused predominantly by Ohmic resistance [4]. From the polarization and power density curves the internal resistance of the G-MFC was found to be approximately 336.5 Ω. This represents a 98% reduction from our previous work [15].

Although our G-MFC has demonstrated power densities

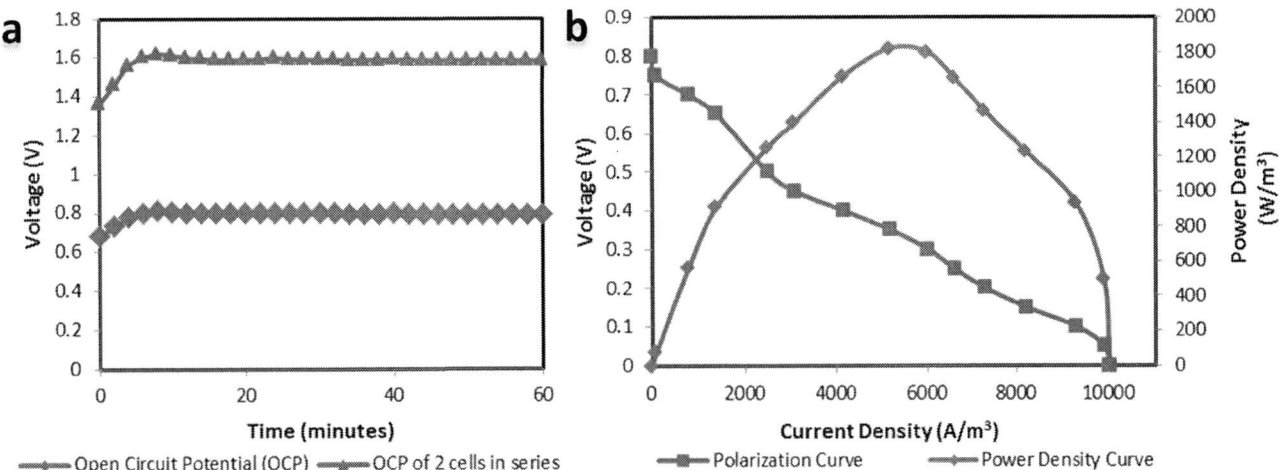

Figure 4: Experimental results. (a) Average Open circuit potentials of a single G-MFC and two G-MFCs in series plotted versus time. (b) Power density and polarization curves of the G-MFC.

comparable to macroscale microbial fuel cells, significant improvement of overall power output is still needed to enable use in practical applications. Established techniques to improve the power outputs of macroscale fuel cells could be employed to μMFCs as well. For instance, using air cathodes, moving away from PEMs, allowing for biofilm formation, and quantitatively analyzing different sources of internal resistance are all potential approaches that could lead to enhanced power outputs [4, 5, 13, 15]. Introducing additional mediators, functionalizing the anode, engineering more efficient membrane electrode assemblies and increasing electrode surface area are also methodologies that could improve performance.

CONCLUSION

Despite their vast potential, low power outputs and potentials have limited μMFCs from implementation in practical applications. In this work, we presented a G-MFC, the first stackable μMFC that utilizes graphene electrodes. The G-MFC utilized these high surface area electrodes and a complex natural anodic environment to produce high power outputs and potentials. Specifically, the G-MFC produced OCPs of 0.8 ± 0.05 V and power densities of 1820 ± 10 W/m^3, which represent some of the highest values reported [13-15]. Conventionally, biofilm formation is allowed when maximum power outputs are reported as power generally increases significantly with film thickness; however, the G-MFC produced 364 μW of power – one of the highest values recorded – even without complete film formation. These high power outputs were attained due to the low internal resistance of the G-MFC, which was approximately 336 Ω. The G-MFCs architecture also permits simple stacking, which allows for a wider range of applications and enables the G-MFC to potentially power devices with increased energy and power requirements. Established methods exist to further improve the scope of implementation of the G-MFC in terms of power outputs and longevity. Switching to air-cathodes or membrane-less architectures as well as employing techniques to reduce internal resistance could all substantially improve the applicability of the G-MFC. Consequently, the presented G-MFC represents an important step towards moving μMFCs out of the laboratory and into practical applications.

ACKNOWLEDGEMENTS

The authors thank Paul Lum, Luke P. Lee, the University of California, Berkeley, Marvell Nanofabrication Laboratory, the Biomolecular Nanotechnology Center (BNC), and members of the Liwei Lin Laboratory and the Micro Mechanical Methods for Biology (M3B) Laboratory Program for allowing us to use their facilities for fabrication and for their help, guidance, and support.

REFERENCES

[1] Berk, Richard S., and James H. Canfield. "Bioelectrochemical energy conversion." *Applied microbiology* 12.1 (1964): 10-12.

[2] Bond, Daniel R., et al. "Electrode-reducing microorganisms that harvest energy from marine sediments." *Science* 295.5554 (2002): 483-485.

[3] Lovley, Derek R. "Bug juice: harvesting electricity with microorganisms." *Nature reviews microbiology* 4.7 (2006): 497-508.

[4] Logan, Bruce E., et al. "Microbial fuel cells: Methodology and Technology." *Environmental science & technology* 40.17 (2006): 5181-5192.

[5] Logan, Bruce E. *Microbial fuel cells*. Wiley-interscience, 2008.

[6] Logan, Bruce E. "Exoelectrogenic bacteria that power microbial fuel cells." *Nature Reviews Microbiology* 7.5 (2009): 375-381.

[7] Logan, Bruce E. "Scaling up microbial fuel cells and other bioelectrochemical systems." *Applied microbiology and biotechnology* 85.6 (2010): 1665-1671.

[8] Chaudhuri, Swades K., and Derek R. Lovley. "Electricity generation by direct oxidation of glucose in mediatorless microbial fuel cells." *Nature biotechnology* 21.10 (2003): 1229-1232.

[9] Logan, Bruce, et al. "Graphite fiber brush anodes for increased power production in air-cathode microbial fuel cells." *Environmental science & technology* 41.9 (2007): 3341-3346.

[10] Liu, Hong, Shaoan Cheng, and Bruce E. Logan. "Power generation in fed-batch microbial fuel cells as a function of ionic strength, temperature, and reactor configuration." *Environmental science & technology* 39.14 (2005): 5488-5493.

[11] Fredrickson, James K., et al. "Towards environmental systems biology of Shewanella." *Nature reviews microbiology* 6.8 (2008): 592-603.

[12] Bretschger, Orianna, et al. "Current production and metal oxide reduction by Shewanella Oneidensis MR-1 wild type and mutants." *Applied and environmental microbiology* 73.21 (2007): 7003-7012.

[13] Wang, Hsiang-yu, et al. "Micro-sized microbial fuel cell: a mini-review." *Bioresource technology* 102.1 (2011): 235-243.

[14] Choi, S., and J. Chae. "A series array of microliter-sized microbial fuel cell." *Micro Electro Mechanical Systems (MEMS), 2011 IEEE 24th International Conference on*. IEEE, 2011.

[15] V.Jayaprakash, et al. "Stackable cow dung based microfabricated microbial fuel cells." *Micro Electro Mechanical Systems (MEMS), 2013 IEEE 26th International Conference on*. IEEE, 2013.

[16] El-Kady, Maher F., et al. "Laser scribing of high-performance and flexible graphene-based electrochemical capacitors." *Science* 335.6074 (2012): 1326-1330.

CONTACT

*Vishnu Jayaprakash, +14084253189; soorse@berkeley.edu

HIGH PERFORMANCE NONLINEAR MICRO ENERGY HARVESTER INTEGRATED WITH (K,Na)NbO₃/Si COMPOSITE QUAD-CANTILEVER

Le Van Minh, Motoaki Hara, and Hiroki Kuwano
Tohoku University, Sendai, JAPAN

ABSTRACT

We developed a lead-free (K,Na)NbO₃ nonlinear micro energy harvester. The harvester was densely integrated with a quatrefoil-shaped proof mass and quad-cantilevers using bulk micromachining. The KNN/Si composite cantilever was integrated with two-separated metal (Pt/Ti)/ KNN/ metal (Pt/Ti) structure to effectively collect charge. Clamped-clamped beam design was also adopted for wide band operation.

The harvester showed wide bandwidth of 253 Hz (fractional bandwidth: 12.9%) at acceleration of 6 m/s². The power density of 1623 µW/cm³ was achieved at the same acceleration. It is the highest value among the wide bandwidth piezoelectric energy harvesters.

INTRODUCTION

Energy source is a fundamental component in modern wireless sensor node. Harvesting electricity from natural resources, such as the solar, vibration, thermal gradient, or wind, reveals strong tendency for energy harvesting [1]. Vibration-based energy harvesters become promising to power smart sensor node, where long-time operation, miniaturization, environment-friendly materials take priority [1, 2]. At present, high-performance vibration energy harvesters focus on vibration collection and electromechanical transduction.

Nonlinearity has been introduced to energy harvester for harvesting effectively natural vibrations. Previous studies use magnetic or electrostatic forces to operate in nonlinear region [3, 4]. However, requirement of bulky magnet or high external voltage source causes technological issues. Mechanic theory shows that a simple clamped-clamped beam behaviors nonlinearly under large deflection [5]. Therefore, we designed nonlinear energy harvester with multiple clamped-clamped micro beams.

Piezoelectric transduction has been applied into energy harvester for highly efficient conversion and miniaturized possibility [6]. However, current research depends strongly on toxic lead-based piezoelectric materials [7, 8]. For developing the high-power and environment-friendly device generation, we applied lead-free (K,Na)NbO₃ (KNN) piezoelectric thin film. This material has high figure of merit (FOM = $e_{31}^2/\varepsilon_0\varepsilon_{33}$) [7], being compatible with MEMS process [8].

In this study, we developed a high-performance nonlinear micro-energy harvester having quad-cantilevers integrated with KNN thin film. A quatrefoil-shaped proof mass covered completely free space of the device. The mechanism, fabrication, and evaluation of the device were presented in this work.

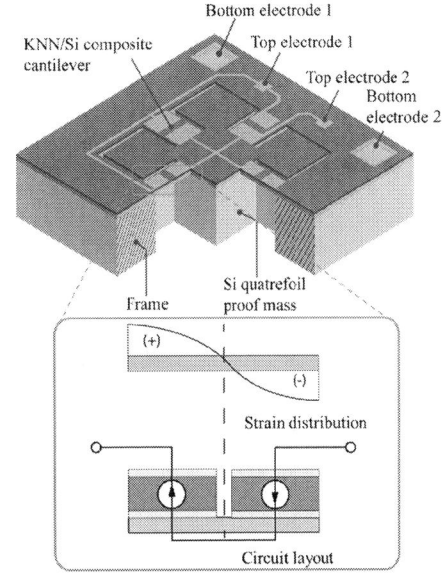

Figure 1: A proposed nonlinear micro energy harvester that consists of a quatrefoil-shaped proof mass and (K,Na)NbO₃/Si quad-cantilevers

Table 1: Geometric parameters of the proposed energy harvester

Length	1500 µm
Width	500 µm
KNN thickness	2 µm
Si thickness	10 µm
Total mass	90 µg
k_0	0.2146 µN/µm
k_1	0.5786×10^{-3} µN/µm³

STRUCTURE AND MECHANISM

Figure 1 illustrates the proposed nonlinear micro-energy harvester employing KNN thin film. The harvester consists of a big quatrefoil-shaped proof mass suspended by four KNN/Si composite beams. The Pt/KNN/Pt structures were isolated on the same strain polarity region of each beam. Such capacitors were wired in parallel by taking account of the current direction as shown in Fig. 1. The geometric parameters of the device are tabulated in Table 1.

The motion equation considering the nonlinearity [9] is expressed as

$$m\ddot{x} + C\dot{x} + k(x)x = F_0 \cos \omega t \qquad (1)$$

where *m, C, F₀,* and *k(x)* are mass, damping coefficient,

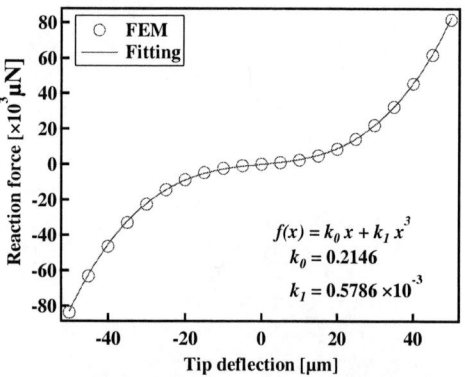

Figure 2: The nonlinear spring force of the proposed device simulated using ANSYS FEM

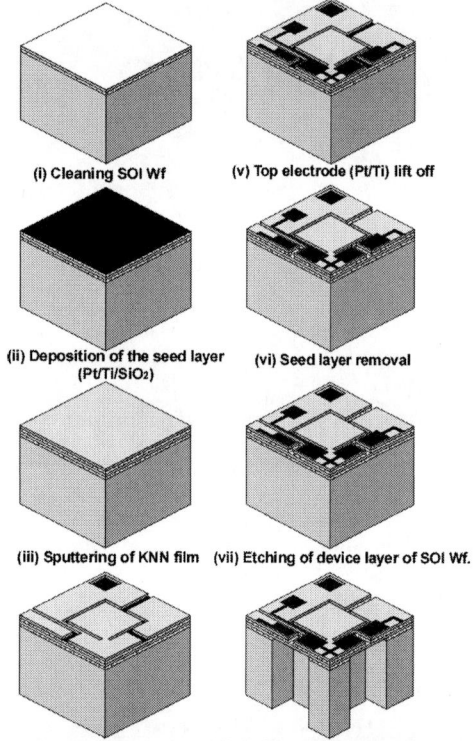

(i) Cleaning SOI Wf

(ii) Deposition of the seed layer (Pt/Ti/SiO₂)

(iii) Sputtering of KNN film

(iv) KNN patterning

(v) Top electrode (Pt/Ti) lift off

(vi) Seed layer removal

(vii) Etching of device layer of SOI Wf.

(viii) Etching of handle layer of SOI Wf.

Figure 3: Fabrication flow of the nonlinear micro energy harvester with integration of a quatrefoil-shaped Si proof mass and KNN/Si composite quad-cantilevers

amplitude of external harmonic force, and nonlinear spring constant of the device, respectively. $k(x)$ is defined as

$$k(x) = k_0 + k_1 x^2 \qquad (2)$$

where k_0 is linear spring constant and k_1 is nonlinear spring coefficient.

With the composite beam, k_0 and k_1 were determined simply from finite element method (FEM) using commercial software (ANSYS v.14.5). Figure 2 shows the nonlinear

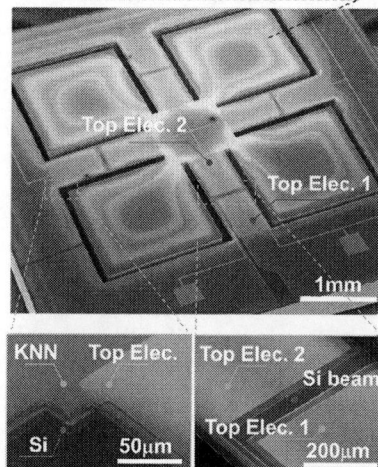

Figure 4: SEM images of the fabricated micro energy harvester. The sub-images show the layers of the KNN/Si composite beam and the etching trench to form two-separated KNN capacitors on each beam.

force as a function of the tip deflection. k_0 and k_1 were obtained from the curve fitting using Eq. (2).

The resonant frequency in the case of damping-less nonlinear system (see Eq. (1)) is expressed

$$\omega^2 = \frac{k_0}{m} + \frac{3}{4}\frac{k_1}{m}x_0^2 - \frac{F_0}{mx_0} \qquad (3)$$

where x_0 is the tip deflection.

FABRICATION

Figure 3 shows a manufacturing flow of the device. The device was fully microfabricated from KNN on silicon-on-insulator (SOI) wafer.

Similarity to our previous publications [8, 9], a stack of Pt/Ti/SiO₂ was deposited in sequence on silicon-on-insulator (SOI) wafer. Next, the piezoelectric KNN film of 2 μm was deposited using reactive sputtering technique (Fig.3 (iii)). The geometric piezoelectric capacitors and the contact windows were shaped using the two-step KNN etch [5]. The top electrodes Pt/Ti of the piezoelectric capacitors were fabricated using lift-off process (Fig.3 (v)). After annealing the sample at 500 °C in two hours, the residual Pt/Ti/SiO₂ areas were removed using reactive ion etching (RIE). Following the Si-front-side etching for beam forming, the Si-back-side etching was proceeded to make a quatrefoil-shaped proof mass and beam thickness (Fig.3 (vi-vii)). Finally, SiO₂ buried layer was completely removed by RIE CHF₃ for releasing the device structure (Fig.3 (viii)).

Figure 4 shows scanning electron microscopy (SEM) images of the fabricated nonlinear KNN-based micro energy harvester. Two sub-SEM images show the enlargement of the KNN/Si composite layer and the separated trench of two capacitors on each beam.

Table 2: Benchmarking of our device

Structure	a [g]	f_{jd} [Hz]	BW [Hz]	BW/f_{jd} [%]	PD [μW/cm^3]	NPD [μW/cm^3]
KNN d_{31} (This work)	0.6	1960	253	12.9	1623	209.4
AlN d_{31}[10]	2.0	2450	250	10.2	34	3.47
PZT d_{33}[11]	4.0	1350	250	18.5	582	107.7

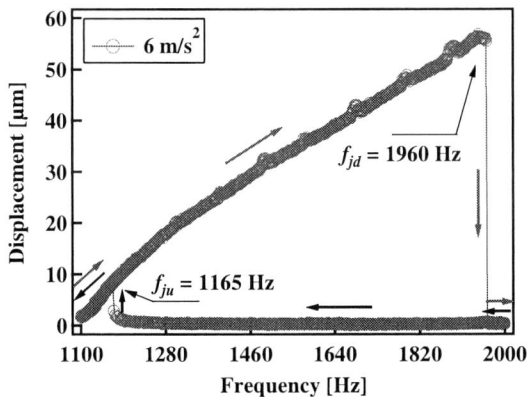

Figure 5: Displacement spectra of the energy harvester at the input acceleration amplitude of 6 m/s²

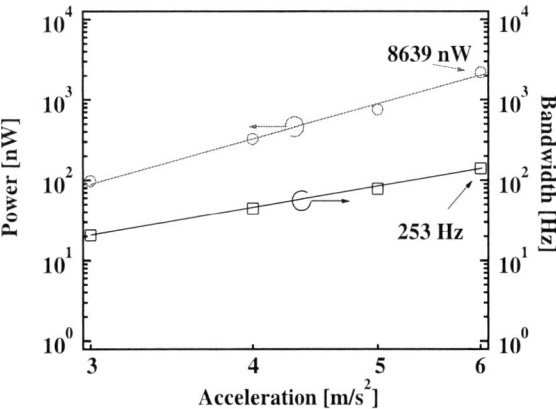

Figure 7: Variation of the maximum power and the bandwidth at 3 dB for various input acceleration amplitudes

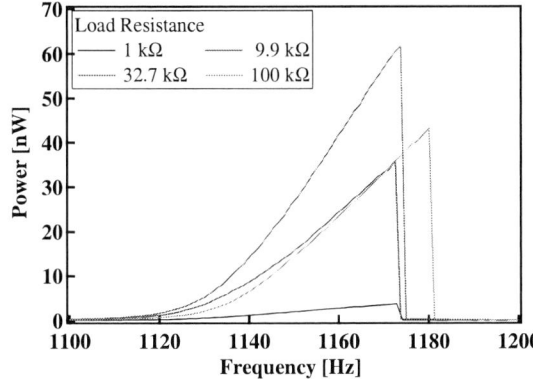

Figure 6: Power spectrum of the nonlinear harvester for different load resistances when keeping the input acceleration amplitude of 1 m/s²

RESULTS AND DISCUSSIONS

The dynamic operations in both mechanical and electrical features were conducted experiments. For the electrical device performance, a conventional resistive circuit was used in the experiments. Following the Ohm law, the output power spectra were determined.

Figure 5 shows the displacement operation of the device under the harmonic vibration. The displacement spectrum indicates the typical hardening spring effect expected from Eq. (3). It was confirmed the spectrum has a hysteresis. When sweeping up the frequency of the input vibration, the spectrum steeply jumped down at the jump-up frequency (f_{ju}) of 1165 Hz. Contrarily, the spectrum jumped up at the jump-up frequency (f_{ju}) of 1165 Hz, when sweeping down the input frequency.

Figure 6 shows the spectra of the output power for different load resistances at the input acceleration of 1 m/s². The optimal resistance was at 32.7 kΩ. Also, using the optimal resistance, the highest power of 60 nW.

Bandwidth (BW) was defined as the frequency range at a half of maximum output power. Figure 7 shows the maximum output power and bandwidth against the input vibration acceleration at the optimal resistance. The maximum power was 8.6 μW and bandwidth of 250 Hz at the low acceleration amplitude of 6 m/s².

Table 2 shows the performance of the state-of-the-art nonlinear piezoelectric micro energy harvesters. The power density (PD) was calculated by dividing the power to the effective volume. This volume is a sum of beams and proof mass volumes. The power density of our device was 1653 μW/cm³. This value was the highest in comparison with that of other piezoelectric nonlinear harvesters.

It is preferable that the frequency-normalized power density (NPD) was defined as the product of power density and the fractional bandwidth to compare the harvesters which have different frequency characteristics. The fractional bandwidth is the BW and the jump-down frequency ratio. In this study, the NPD was 209.4 μW/cm³. It was two times higher than that of the device disclosed in Ref. 11.

CONCLUSION

Wide band operation and high power density are requirements for high-performance vibration-based micro energy harvesters

In this study, we developed a high-performance, nonlinear KNN-based micro energy harvester. The harvester was densely integrated a quatrefoil-shaped proof mass and quad-cantilevers using bulk micromachining. The cantilever had a KNN/Si composite structure and separated electromechanical transducers on each beam to effectively collect the charge. Clamped-clamped beam design was also introduced to obtain the hard spring effect for wide band operation.

The experimental results show that the harvester had wide bandwidth of 253 Hz (fractional bandwidth: 12.9 %) at acceleration of 6 m/s^2. Power density of 1623 μW/cm3 was achieved at the same acceleration. It was the highest values among the wide bandwidth energy harvester using the nonlinearity

ACKNOLWEDGEMENT

We are grateful to Dr. Fumimasa Horikiri, and Dr. Kenji Shibata at Hitachi Cable, Ltd. for their excellent advice and support. A part of this work was performed at micro/nano-machining research and education center, Tohoku University, Japan. This work was conducted under the project "Research of a nano-energy system creation" (No.18GS0203), funded by the Ministry of Education, Culture, Sports, Science and Technology (MEXT).

REFERENCES

[1] S. Roundy, P. K. Wright, J. Rabaey, "A Study of Low Level Vibrations as a Power Source for Wireless Sensor Nodes", *Computer Commu.*, vol. 26, pp. 1131-1144, 2003.

[2] S. P. Beedy, M. J. Tudor, N. M. White, "Energy Harvesting Vibration Sources for Microsystems Applications", *Meas. Sci. Technol.*, vol. 17, pp. R175-R195, 2006.

[3] B. Ando, S. Baglio, C. Trigona, N. Dumas, L. Latorre, P. Noute, "Nonlinear Mechanism in MEMS Devices for Energy Harvesting Applications", *J. Micromech. Microeng.*, vol. 20, 125020-12, 2010

[4] D. S. Nguyen, E. Halvorsen, G. U. Jensen, A. Vogl, "Fabrication and Characterization of a Wideband MEMS Energy Harvester Utilizing Nonlinear Springs", *J. Micromech. Microeng.*, vol. 20, 125009-11, 2010

[5] D. S. Stephen, *Microsystem Design*, Springer, 2001.

[6] A. Marin, S. Bressers, S. Priya, "Multiple Cell Configuration Electromagnetic Vibration Energy Harvester", *J. Physics. D: Appl. Phys.*, vol. 44, 295501-11, 2011.

[7] K. Shibata, K. Suenaga, K. Watanabe, F. Horikiri, A. Nomoto, T. Mishima, "Improvement of Piezoelectric Properties of $(K,Na)NbO_3$ Films Deposited by Sputtering", *Jpn. J. Appl. Phys.*, vol. 50, 041503-7, 2011.

[8] L. V. Minh, M. Hara, F. Horikiri, K. Shibata, H. Kuwano, "Bulk Micromachined Energy Harvesters Employing $(K,Na)NbO_3$ Thin Film", *J. Micromach. Microeng.*, vol. 23, 035029-6, 2013.

[9] L. V. Minh, M. Hara, H. Kuwano, "Micro-Energy Harvesters Integrated with a Quatrefoil-Shaped Proof Mass Suspended by Multiple $(K,Na)NbO_3$ Beams", *Jpn. J. Appl. Phys.*, vol 52, 07HD08-4, 2013.

[10] M. Marzencki, M. Defosseux, S. Basour, "MEMS Vibration Energy Harvesting Devices with Passive Resonance Frequency", *J. Microelectromech. Syst.*, vol. 18, 6, pp.1444- 53, 2009.

[11] A. Hajati, S. G. Kim, "Ultra-wide Bandwidth Piezoelectric Energy Harvesting", *Appl. Phys. Lett.*, vol 99, 083105-3, 2011.

CONTACT

Le Van Minh,
Tel (Fax): +81-22-795-4771
E-mail: minhlv@nanosys.mech.tohoku.ac.jp

HIGH-ENERGY-DENSITY ON-CHIP LI-ION CAPACITORS

Siwei Li[1], and Xiaohong Wang[1,2]

[1]Tsinghua National Laboratory for Information Science and Technology
Institute of Microelectronics, Tsinghua University, Beijing, P.R. China
[2]State Key Laboratories of Transducer Technology, Chinese Academy of Sciences, P.R. China

ABSTRACT

This paper presents an on-chip Li-ion capacitor featured by higher energy density than that of supercapacitor under the same level of charge/discharge rate by building the hybrid cell containing a supercapacitor electrode and a Li-ion battery electrode, to take advantage of power of supercapacitor and capacity of Li-ion battery. Activated carbon (AC), a supercapacitor material, is used as the positive electrode, while graphite, an anode material of Li-ion battery, is used as negative electrode, and the electrolyte used in Li-ion battery serves as electrolyte in the device. The on-chip prototype with 100-μm-thick interdigital electrodes is fabricated and shows a capacity of $175\mu Ah/cm^2$ with an energy density of about $1550mJ/cm^2$ under a charge/discharge current of $0.5mA/cm^2$ and a cell voltage of 3.4V, which is higher than that of a symmetric AC-based supercapacitor with the same dimensions.

INTRODUCTION

On-chip energy storage devices, including supercapacitors and rechargeable batteries, are valuable with the trends of developing miniature energy storage devices as micro power suppliers for wireless sensor networks, and collectors of the energy generated the from harvesters, to form self-powered micro systems. The requirements to energy storage devices are adequate power, which matches the output of energy harvesters, or the power consumption of devices they powered, as well as high energy and long lifetime [1-2].

Both supercapacitors and batteries consist of two electrodes with a separator between them, immersed in electrolyte, and energy storage is conducted by charge storage. However, different electrochemical processes between the electrode and electrolyte make them different in properties. For batteries, bulk faradic reactions take place to store charge, providing stable output voltage, while for supercapacitors, electric double layer capacitance (EDLC) based on static charge accumulation and/or psuedocapacitance based on fast redox surface faradic reactions are principles to store charge, with nonconstant voltage dependent on the amount of charge during the processes. Batteries offer higher energy capacity than supercapacitors, but the power density is low and the cycle life is limited. For macro-scale use, supercapacitors often serve as complement to batteries, when high power is required. For on-chip use, the limited output power and capability to be frequently charge/discharged are disadvantages of batteries.

On the other hand, increasing the capacity of supercapacitor is of importance. The energy stored in a capacitor can be described by

$$E = \frac{1}{2}CV^2 \tag{1}$$

where C stands for the capacitance and V is the working voltage. Both C and V are key parameters to achieve high capacity. The capacitance mainly determined by the electrode material, and also affected by the electrolyte. The other factor, the working voltage range is limited by the electrochemical window of the electrolyte, and should also match the specific range where the electrode material shows good capacitive behavior.

EDLC is more widely used than pseudocapacitance due to better power performance and longer cycle life of non-faradic process, though the energy density is lower. Typical electrode materials are various carbon with high specific surface area and the electrolytes include aqueous solutions, organic electrolytes and ionic liquids (ILs). For aqueous solutions with high conductivity, the limitation in working voltage (generally lower than 1V) is an issue. For organic electrolytes, which can broad the working voltage range to about 2.7V, safety is a drawback. ILs, which are expected to further increase the voltage range, though have been widely explored in recent years, the practical use and commercialization are limited.

In aqueous systems, asymmetric supercapacitors with different active materials in two electrodes, generally an EDLC electrode and a pseudocapacitance electrode, are developed [3-5]. The type of devices can improve the performance compared with symmetric EDLC ones due to higher capacitance provided by pseudocapacitance electrodes, and extended voltage range beyond the thermodynamic limit by using different potential range of two kinds of electrode materials, while at the same time, power performance is also kept due to fast charge/discharge rate of EDLC electrode.

With similar consideration, the combination of a supercapacitor electrode with an electrode of battery, generally the widely used Li-ion battery, is a way to take advantages from each electrode [6-8]. The anode of a Li-ion battery provides higher charge capacity than a supercapacitor electrode, and a well-defined voltage plateau during charge/discharge process, while EDLC is feasible in organic Li-ion electrolyte and keeps reasonable power performance.

For on-chip supercapacitors, 3D interdigital electrodes with different materials have been explored [9-12], to increase the loading mass of advanced electrode material to satisfy the requirement of energy and power. The hybrid architecture is also a promising solution to improve the performance.

Here, we demonstrate the design, fabrication and

978-1-4799-3510-9/14 $31.00 © 2014 IEEE   401

characterization of an on-chip hybrid Li-ion capacitor. By the joint contribution of advanced activated carbon (AC)-graphite configuration in Li-ion electrolyte and designed 3D structure with SU-8 wall as supporter and separator, high-energy-density device is achieved with moderate power performance.

STRATEGIES AND EXPERIMENTS

Strategies

The mechanisms of EDLC supercapacitor and Li-ion battery are illustrated in Figure 1(a) and (b), respectively, and the principle of hybrid Li-ion capacitor is shown in Figure 1(c). The voltage of a cell is determined by the potential difference between positive electrode and negative electrode, so lower the potential of negative electrode is to increase the voltage of the cell. The anode of Li-ion battery works under a lower potential than that of an EDLC negative electrode, and shows a plateau during the process, as indicated in the figure, which is beneficial to increase the cell voltage. The charge/discharge curve of Li-ion capacitor contains two distinct stages, as shown in Figure 1(c). The first stage at high voltage range with an even slope, which is similar to that of a supercapacitor, appears when the potential of graphite negative electrode is on the plateau, and turns into the second stage at low voltage range with a steep slope when the potential of graphite electrode varies. In the first stage, the equivalent capacitance of the device, indicated by the slope of the curve, is mainly determined by AC positive electrode.

On the other hand, the specific capacity of a battery electrode is generally larger than a supercapacitor one, thus the charge capacity is also improved in the hybrid configuration compared with supercapacitors.

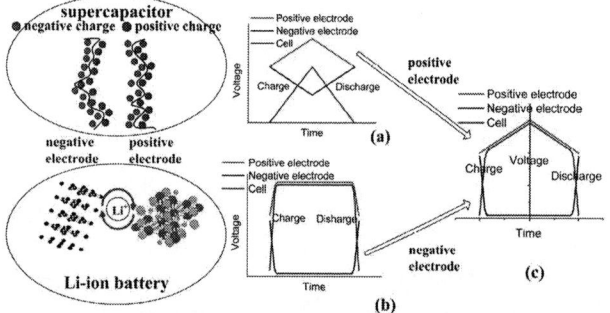

Figure 1: Schematic of the concept. Principles and the voltage profiles of the two electrodes and the whole cell during charge/discharge of (a) an EDLC supercapacitor and (b) a Li-ion battery. (c) Voltage profile of a Li-ion capacitor combining the positive electrode of supercapacitor and the negative electrode of Li-ion battery.

Structure and Materials

The overall diagram of the on-chip device is illustrated in Figure 2. A pair of 3D interdigital electrodes, supported and separated by a separator layer, is constructed. AC, a commonly explored and commercialized material used in supercapacitors, serves as positive electrode and graphite powder, the widely used anode material in Li-ion batteries,

with high capacity and a low and stable plateau during charge/discharge process, is adopted as negative electrode. The electrolyte is 1M $LiPF_6$ in ethylene carbonate (EC)/dimethyl carbonate (DMC) with a ratio of 1:1 (w:w).

An issue is the mass ratio between positive electrode and negative electrode. To maximum the capacity, approximate capacities of the two electrodes are expected. Although the theoretical capacity of graphite at low power density (372mAh/g) is much larger than that of AC (83mAh/g as an estimate based on a specific capacitance of 150F/g and a voltage range of 2V), when higher power is required, the capacity of graphite attenuated more than AC. As a result, a mass ratio of about 1:1 is a balance [7-8]. Since we cannot control the mass of each electrode precisely, 1:1 in volume ratio, i.e. the positive and negative electrodes have same dimensions, is carried out when demonstrate the concept.

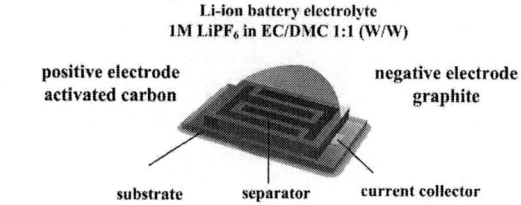

Figure 2: Schematic of the device.

Fabrication

A brief fabrication process is depicted in Figure 3. To begin with, an insulated glass substrate was cleaned. A 20nm/100nm Ti/Au layer was sputtered on the substrate and patterned to form the pair of interdigital current collectors and contact pads of the two electrodes, as shown in Figure 3(a). Then, 100-μm-thick SU-8 photoresist layer was coated and patterned to form the separator and the ring around the electrode area, which is illustrated in Figure 3(b). After the framework of the prototype was accomplished, different electrode materials were to fill in the channels constructed by SU-8, respectively. The active materials were mixed with conducting agent and binder followed by dispersal in water to form homogeneous slurries. The positive electrode material consisted of AC, acetylene black (AB), and carboxymethylcellulose sodium (CMC) with a mass ratio of 87:10:3, and the negative electrode material consisted of graphite, AB and CMC with the same ratio. The slurries containing different active materials were injected into corresponding channels and dried at 70 □. By the support of SU-8 and the guidance of the slurry in the channels, the materials could be injected apart. Repeated injection-drying operations were carried out to fill up the channels.

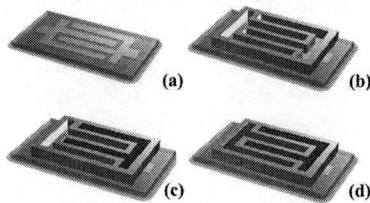

Figure 3: Fabrication process of the prototype. (a) Deposition and patterning of the current collectors. (b)

Formation of the channels for electrodes with SU-8 structure. (c) and (d) Injection of different material into different channels.

RESULTS AND DISCUSSIONS

The photographs of as-fabricated prototype before and after the injection process are presented in Figure 4. Serpentine SU-8 structure separating two electrodes can be seen in Figure 4(a) while in Figure 4(b), slight difference in colors of two electrodes can be seen, indicating different materials. The area of electrodes is about 0.1cm^2.

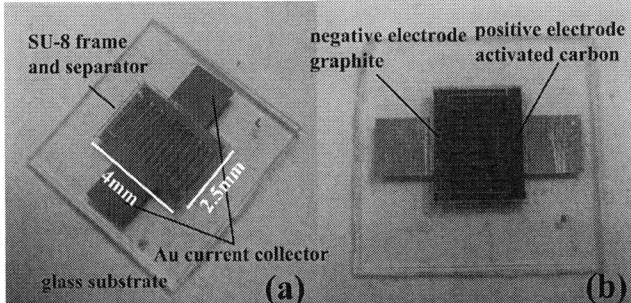

Figure 4: Photographs of the prototype (a) before and (b) after the injection of electrode materials.

The microscope image of the prototype is shown in Figure 5(a) and the difference between two electrodes can also be seen. Figure 5(b) and (c) show the scanning electron microscope (SEM) images of AC and graphite with microsized particles, respectively.

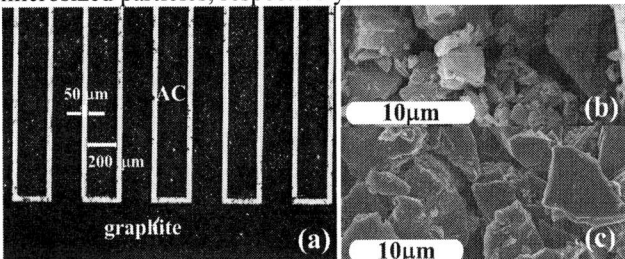

Figure 5: (a) Microscope image of the prototype. (b) and (c) SEM images of AC and graphite electrode materials.

Electrochemical performance was evaluated by galvanostatic charge/discharge technology in a two-electrode configuration. The prototype was charged to 3.2V and then discharged at different currents ranging from 0.02mA (0.2mA/cm^2) to 0.5mA (5mA/cm^2). All the discharge curves contain two stages correspond to the schematic shown in Figure 2(c). The capacitance of the prototype during the first stage can be approximately calculated by the slope of the curve by

$$C = \frac{It}{\Delta V} \qquad (2)$$

where I is the discharge current, t and ΔV are the time and the variation in voltage during a period of time. The results are listed in column 2 in Table 1. When the charge/discharge current increases, the decrease in capacitance is demonstrated.

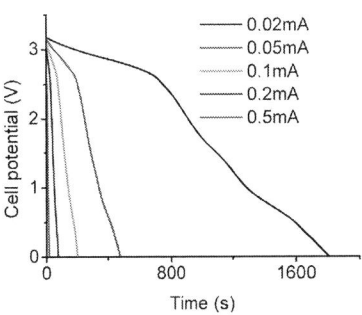

Figure 6: Galvanostatic discharge curves of the Li-ion capacitor prototype under different currents with the voltage range of 3.2V.

Then a symmetric supercapacitor prototype was also presented to give a comparison between two types of devices, and the test results are summarized in Figure 7. The prototypes were charge/discharged under the current of 0.05mA, 0.1mA and 0.2mA. For the supercapacitor prototype, the upper limit is 3.0V since obvious asymmetry appears during charge/discharge process when the cell potential goes higher. While for Li-ion capacitor prototype, the cell potential can reach 3.4V.

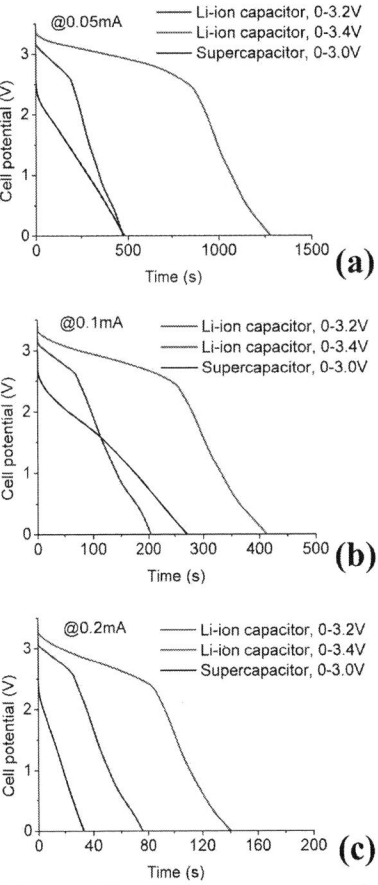

Figure 7: Galvanostatic discharge curves of the Li-ion capacitor prototype and the supercapacitor prototype, under different currents.

978-1-4799-3510-9/14 $31.00 © 2014 IEEE

The capacitances of supercapacitor under different current densities are also listed in Table 1, column 3. For a symmetric supercapacitor, the capacitance of the device is about half of each electrode due to series connection of two similar electrodes, and it should be about half of the equivalent capacitance of Li-ion capacitor in theory since the later one is almost the capacitance of the AC positive electrode. But the data deviate from the value. Possible reasons include the reduction in capacitance when current increases due to the rate of electrochemical processes and the contact of microsized particles inside unpressed electrodes, as well as the leakage current due to high specific area of electrode materials.

Table 1: Summary of the equivalent capacitance of the Li-ion capacitor prototype and the supercapacitor prototype.

Current Density (mA/cm^2)	Capacitance of Li-ion Capacitor (mF)	Capacitance of Supercapacitor (mF)
0.2	22.9	-
0.5	16.8	10.0
1.0	14.1	10.4
2.0	9.1	3.1
5.0	6.2	-

The charge and energy stored in the Li-ion capacitor during the whole discharge process, with the voltage range of 0-3.4V, are summarized in Table 2. Under the current density of 0.5mA/cm^2, the energy density is about 1550mJ/cm^2, while for the supercapacitor, the value is about 500mJ/cm^2. The different mainly due to higher equivalent capacitance and higher average voltage of the Li-ion capacitor.

Table 2: Capacity and energy density of the Li-ion capacitor with the upper limit voltage of 3.4V.

Current Density (mA/cm^2)	Capacity ($\mu Ah/cm^2$)	Energy Density (mJ/cm^2)
0.5	175	1550
1.0	114	950
2.0	78	630

The highest voltage presented here is about 3.4V, which is not as high as expected. The possible reason is the leakage current. The average leakage current, at the level of several μA, was estimated by open circuit voltage (OCV) test. The value is not negligible compared with the charge current, especially when the cell potential is high.

CONCLUSIONS

In this work, on-chip Li-ion capacitor is presented. The device combines the features of high power of supercapacitor and high energy of Li-ion battery by using an AC supercapacitor electrode as positive electrode and a graphite Li-ion battery electrode as negative electrode with LiPF$_6$ in EC/DMC as electrolyte to form a hybrid cell. Thick interdigital electrode structure with separator is carried out to ensure the loading mass and the detached filling of different

materials. The prototype is charge/discharged under the current density at the same level to that of a supercapacitor, while much higher energy density compared with symmetric supercapacitor is achieved.

ACKNOWLEDGEMENTS

The work is supported by the National Natural Science Foundation of China (No. 60936003), 973 Program (No. 2009CB320304), and 863 Program (No. 2009AA04Z319) of China.

REFERENCES

[1] J. W. Long, et al., "Three-Dimensional Battery Architectures", *Chem. Rev.*, vol. 104, pp. 4463-4492, 2004.

[2] M. Beidaghi, et al., "Recent Advances in Design and Fabrication of On-chip Micro-supercapacitors", in *Proc. of SPIE*, vol. 8377, pp. 837708-1-837708-10, 2012.

[3] M. S. Hong, et al., "Use of KCl Aqueous Electrolyte for 2V Manganese dioxide/Activated Carbon Hybrid Capacitor", *Electrochem. Solid-State Lett.*, vol. 5, pp. A227-230, 2002.

[4] V. Khomenko, et al., "Optimisation of an Asymmetric Manganese Oxide/Activated Carbon Capacitor Working at 2V in Aqueous Medium", *J. Power Sources*, vol. 153, pp. 183-190, 2006.

[5] C. Shen, et al., "A High-energy-density Micro Supercapacitor of Asymmetric MnO$_2$-Cabron Configuration by using Micro-fabrication Technologies", *J. Power Sources*, vol. 234, pp. 302-309, 2013.

[6] G. G. Amatucci, et al., "An Asymmetric Hybrid Nonaqueous Enenrgy Storage Cell", *J. Electrochem. Soc.*, vol. 148, pp. A930-A939, 2001.

[7] V. Khomenko, et al., "High-energy-density Graphite/AC Capacitor in Organic Electrolyte", *J. Power Sources*, vol. 177, pp. 643-651, 2008.

[8] M. D. Stoller, et al., "Activated Graphene as a Cathode Material for Li-ion Hybrid Supercapacitors", *Phys. Chem. Chem. Phys.*, vol. 14, pp. 3388-3391, 2012.

[9] D. Pech, et al., "Ultrahigh-power Micrometre-sezed Supercapacitors Based on Onion-link Carbon", *Nat. Nanotech.*, vol. 5, pp. 651-654, 2010.

[10] W. Gao, et al., "Direct Laser Writing of Micro-supercapacitors on Hydrated Graphite Oxide films", *Nat. Nanotech.*, vol. 6, pp. 496-500, 2011.

[11] C. Shen, et al., "A High-performance Three-dimensional Micro Supercapacitor Based on Self-supporting Composite Materials", *J. Power Sources*, vol. 196, pp. 10465-10471, 2011.

[12] C. Shen, et al., "Direct Prototyping of Patterned Nanoporous Carbon: A Route from Materials to On-chip Devices", *Sci. Rep.*, vol. 3, pp. 2294, 2013.

CONTACT

*X. Wang, tel: +86-10-62798432; wxh-ime@tsinghua.edu.cn

HIGH-ENERGY-DENSITY ON-CHIP SUPERCAPACITORS USING MANGANESE DIOXIDE-DECORATED DIRECT-PROTOTYPED POROUS CARBON ELECTRODES

Siwei Li[1], Xiaohong Wang[1,2], and Caiwei Shen[1]

[1]Tsinghua National Laboratory for Information Science and Technology
Institute of Microelectronics, Tsinghua University, Beijing, P.R. China
[2]State Key Laboratories of Transducer Technology, Chinese Academy of Sciences, P.R. China

ABSTRACT

High performance on-chip micro supercapacitors are presented, using manganese dioxide (MnO_2) decorated into direct prototyped porous carbon electrodes. By incorporating MnO_2 into carbon framework, both electric double layer capacitance (EDLC) and pseudocapacitance contribute to total capacity. The configuration is realized by combination of nano template into photolithography technology, carbonization, removing the template to form patterned porous carbon network, and designed electrochemical deposition process to grow MnO_2 in the carbon framework. The capacitance of the composite increases compared with pure porous carbon without obvious decay in cycle life. Then the on-chip prototypes using the composite as electrode material are designed and the volumetric capacitance of 2.9mF/(cm^2 μm) under the scan rate of 50mV/s is much higher than 0.8mF/(cm^2 μm) of carbon.

INTRODUCTION

Supercapacitors, also called electrochemical capacitors, are a kind of energy storage devices which play an important role in electric energy conversion, storage and application process. A supercapacitor generally consists of two electrodes immersed in electrolyte and a separator layer between them. The energy is kept by non-faradic charge accumulation forming EDLC and/or faradic fast reversible redox reaction leading to pseudocapacitance, happening on the interface of electrodes and electrolytes [1]. Supercapacitors are competent to output higher power and enable longer cycle life than rechargeable batteries, but offer inadequate energy density. Even though pseudocapacitance provides higher capacitance than EDLC, the capacity is still lower than that of batteries.

On-chip energy storage devices are desired with the development of various micro devices such as sensors, electric devices and harvesters. The goal of developing such devices on chips is to achieve adequate power performance to afford the output power of harvesters or the operating power of electric devices, and high capacity in limited on-chip space, with similar dimensions to other elements. Also, long cycle life, compatibility and safety are pluses. Intrinsic problems of batteries including low charge/discharge rate and poor cycle life are disadvantages in micro and integrated systems, limit the performance of micro batteries.

To achieve high-performance devices in limited space, both electrode structures and advanced materials are to be taken into consideration. Compared with 2D planner structure [2-3], 3D interdigital architecture [4-7] with various strategies to build electrodes with multiple materials is still the mainstream of current electrode structure, due to the extensibility in the vertical direction to increase the loading mass, and adjustability to design the patterns and dimensions of electrodes in the plane.

We have demonstrated in previous work the direct prototyped micro supercapacitors of patterned porous carbon which is a route from materials to on-chip devices, with moderate capacity [8]. However, the nature of EDLC of mesoporous carbon is not capable of high capacitance. To address this issue, the incorporation of pseudocapacitance material into the framework is promising.

Materials with pseudocapacitance typically include metal oxides and conducting polymers. Generally, they provide higher capacitance than EDLC but the charge/discharge rate of such materials is not satisfactory due to limited conductivity. The stability of such materials is also insufficient since the faradic redox process leads to change of microstructure.

MnO_2 is a kind of pseudocapacitance materials which has been extensively researched since the first report in 1999 [9]. The basic charge storage mechanism can be described as

$$MnO_2 + C^+ + e^- \longleftrightarrow MnOOC \qquad (1)$$

where C^+ stands for cations in the electrolyte, such as Li^+, Na^+, K^+, etc. MnO_2 has a high theoretical specific capacitance of 1370F/g, but the practical capacitances of bulk material reported in literature are generally much lower than the theoretical value due to low conductivity of 10^{-5}-10^{-6}S/m, and limited surface area. Designing the nanostructure of material and building composites are possible solutions to evaluate the performance of such material by breaking the bottleneck of the material and relieving the deformation during cycling [10].

Carbon/MnO_2 configuration is a trend in recent years to develop composite material based on high surface area and conductivity of carbon backbone and high pseudocapacitance of metal oxide active nanostructure. The method is promising to benefit from both materials. Carbon materials such as microbeads, carbon nanotubes (CNTs) and graphene are possible components in the composites [11-14].

Herein, we demonstrate the incorporation of MnO_2 into the monolithic disordered porous carbon framework to enhance the performance of on-chip prototypes. Electrochemical method is applied to control the deposition process and achieve reasonable loading mass. The MnO_2 thin layer is demonstrated to reside in the porous carbon and

978-1-4799-3510-9/14 $31.00 © 2014 IEEE

contribute pseudocapacitance to increase the total capacity, and the prototypes with increased capacitance are presented.

STRATEGIES AND EXPERIMENTS
Strategies

Designed composites as materials for supercapacitors take advantages from each component and are able to achieve good overall performance. The monolithic disordered porous carbon we developed previously can serve as conductive network and increase the specific surface area of the composite, and the pores of about 30nm in diameter is possible to accommodate active layer of several nanometers with a mesopore available for the electrolyte. It is likely to increase the capacitance compared with pure porous carbon and to develop MnO_2 efficiently.

A schematic of the configuration of carbon/MnO_2, and respective roles of two components are shown in Figure 1. After the formation of patterned of porous carbon electrodes demonstrated in Figure 1(a) and (b), MnO_2 layer is decorated on the inner face of pores in the bulk carbon, forming the structure shown in Figure 1(c). In the architecture, the specific surface area provided by porous carbon contributes EDLC while MnO_2 active layer contributes pseudocapacitance, as illustrated in Figure 1(d).

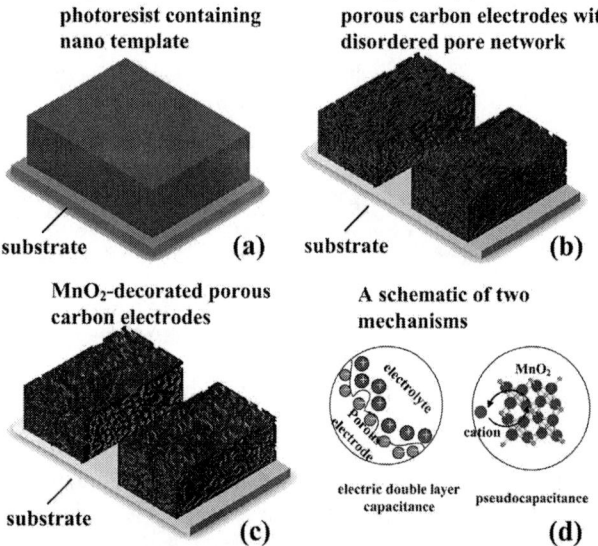

Figure 1: Schematic of the concept: MnO_2 decorated patterned porous carbon as the electrodes of supercapacitors. (a) Photoresist layer containing nano template. (b) Patterned porous carbon electrodes. (c) Incorporation of MnO_2 layer on the inner wall of the. (d) Principles of EDLC and pseudocapacitance. The composite combines the two mechanisms.

Porous Carbon

The patterned monolithic disordered porous carbon framework in Figure 1(a) and (b), as reported in the previous report [8], was fabricated with four basic steps, including (1) mixture of SU-8 photoresist and SiO_2 nano template, (2) spin-coating on the substrate and photolithography, (3) carbonization and (4) removal of the template.

Deposition of MnO_2

To deposit MnO_2 into carbon framework uniformly and efficiently in the disordered monolithic carbon, a key point is to control the deposition rate during the process, avoiding blocking the nanopores in the network. Chemical reaction is commonly used to grow MnO_2 layer on carbon surface due to the redox reaction between oxidizing Mn sources such as MnO_4^- and surface function groups on carbon, but its controllability is not satisfactory. On the other hand, electrochemical methods are also widely used to form thin films, with optional potential/current and operating time.

Basic principles of electrochemical deposition of MnO_2 are anodic process to oxide Mn sources with low valence, such as Mn^{2+}, or cathodic process to reduce Mn sources with high valence, typical MnO_4^-. However, cathodic process in Mn^{2+} is also a possible way to grow MnO_2 due to the induction process [13-14] and followed by oxidation.

The periods containing cathodic and anodic processes to deposit MnO_2 are shown in Figure 2(a) with the cathodic potential of -0.4V vs. saturated calomel electrode (SCE) for 10s and anodic potential of 0.3V vs. SCE for 20s, and 180s left between two adjacent periods for the diffusion of electrolyte. The reactions taking place during the two processes are illustrated in Figure 2(b). The cathodic process mainly evolves the nucleation and the anodic process mainly completes the oxidation of Mn.

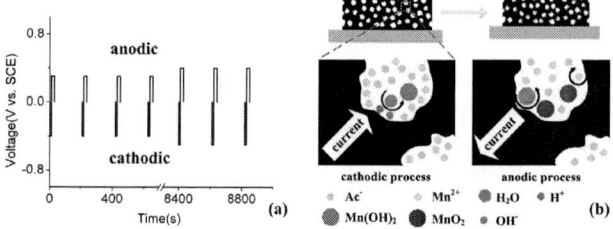

Figure 2: Illustration of basic deposition concept in this work. (a) Voltage curve containing both cathodic and anodic processes. (b) Reactions take place during cathodic process and anodic process.

The deposition was carried out in a three-electrode configuration, with the porous carbon material as working electrode, a Pt foil as counter electrode and SCE as the reference electrode. 0.1M $Mn(Ac)_2$ solution was adopted as the electrolyte, providing Mn^{2+} for the process.

Another issue during the deposition process is the resistance change of the composite. Since the conductivity of MnO_2 is low, when the film covers on the surface of carbon network, the electrochemical reactions are hindered. As a result, the MnO_2 will not keep growing after a certain periods and the capacitance will not keep increasing. However, when slightly increase the potential applied on the material, the deposition process continued. Thus, we increased the potential during the process gradually to increase the loading mass of MnO_2. When the capacitance stopped increasing at a specific potential, 0.1V increase in the magnitudes of both cathodic and anodic potential was applied.

Based on the considerations above, the comprehensive

potential profile is depicted in Figure 3. The potential up to -0.9V/0.8V vs. SCE with 10 periods is adopted. The MnO_2 loading mass and the capacitance will still increase if higher potential or more periods are applied. However, the power performance decreases palpably at the same time. As a result, the voltage profile here is a balance.

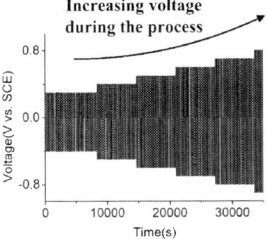

Figure 3: Overall voltage profile of during the deposition process. Stage 1 (-0.4V/0.3V) contains 40 cathodic/anodic periods, stage 2-5 (-0.5V/0.4V, -0/6V/0.5V, -0.7V/0.6V and -0.8V/0.7V) contain 30 periods and the last stage (-0.9V/0.8V) contains 10 periods.

The process is a successive procedure in solution without damage to the monolithic carbon network.

RESULTS AND DISCUSSIONS

To exam the effect of MnO_2 modification, film and on-chip prototypes with symmetric electrodes were presented and characterized.

An 80μm-thick single film, peeled off the substrate, was characterized as electrode material by scanning electron microscope (SEM), as shown in Figure 4. The low magnification image in Figure 4(a) gives an overall view of the film with energy dispersive spectrometer (EDS) results of different regions on the cross section. Mn is distributed in the whole framework, with an atom percentage of about 3%-4%. A high content of O atoms in the composite is possibly due to the crystal water, since the process forms $MnO_2 \cdot xH_2O$ rather than pure MnO_2. High magnification image in Figure 4(b) confirms the layer of about 5-7nm.

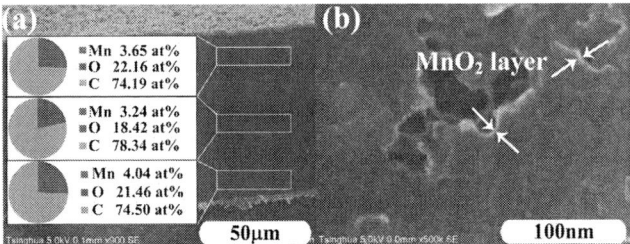

Figure 4: (a) SEM image of the cross section of a carbon/MnO_2 film and EDS analysis of different regions near the surface and in the middle on the cross section. (b) High magnification image of the composite.

Then the film before and after the incorporation of MnO_2 was tested to evaluate the electrochemical performance in a three-electrode configuration, with the film pasted on a Pt wire as working electrode, a Pt foil as counter electrode and SCE as reference electrode, in 0.2M K_2SO_4 aqueous solution. Cyclic voltammetry (CV) tests under different scan rates

were carried out under a uniform potential range of 0.0-0.6V vs. SCE, the typical range of pseudocapacitance behavior of MnO_2. Figure 5(a) shows the CV curves of carbon film and composite film at the scan rate of 10mV/s. The capacitance is evaluated from CV data by

$$C = \frac{Q}{V} = \frac{\int I(t)dt}{V} \qquad (2)$$

where Q stands for the charge during charge or discharge process, V stands for the working voltage range (0.6V in the testss) and I is the current varies with time. Calculated capacitance values under different scan rates are plotted in Figure 5(b). Under low scan rate of 10mV/s, the increase is about 6 times, while under higher scan rates, the capacitance of the composite decrease significantly mainly due to the low conductivity of MnO_2. However, at moderate scan rate of 50mV/s, the capacitance of carbon/MnO_2 is still about 4 times of pure carbon.

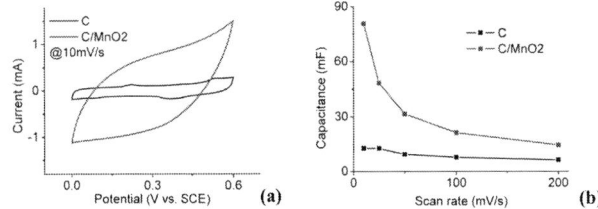

Figure 5: (a) CV curves under the scan rate of 10mV/s, for the film with/without MnO_2. (b) Summary of the capacitance of the film under different scan rates.

Cycle life test was also conducted. Capacitance change was recorded during continuous charge/discharge processes. The results are plotted in Figure 6. It is demonstrated that little change in the cycle life with and without MnO_2, and high retention ratios of above 90% in capacitance of both materials are kept after 5000 charge/discharge cycles. Two possible reasons are indicated: (1) the carbon framework and composite are both binder free, avoiding the loss of material due to invalid binder, and (2) the disordered porous network, which makes it hard to grow MnO_2 on the inner surface, can also serves as a holder keeping nanomaterials from escaping once the layer is formed. As a result, the stability of the composite is good and the incorporation of MnO_2 doesn't deteriorate the lifetime.

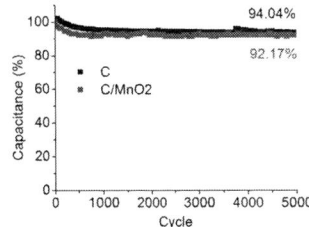

Figure 6: Capacitance retention ratio during repeated charge/discharge cycles of carbon and carbon/MnO_2.

Then the prototypes were presented, as shown in Figure 7, and the left one in the figure, was tested in a two-electrode configuration. The CV curves under scan rates of 50mV/s

978-1-4799-3510-9/14 $31.00 © 2014 IEEE

and 100mV/s are compared in Figure 8(a). The summary of the specific capacitance per volume, which stands for the capacitance value per footprint area per electrode thickness, is plotted in Figure 8(b). The trend is similar to that in Figure 5(b), indicating that the composite material keeps its electrochemical properties when served as electrodes. The specific capacitance per volume of about 2.9mF/(cm^2 μm) under the scan rate of 50mV is about 4 times of the pure carbon, about 0.8mF/(cm^2 μm), due to the contribution of pseudocapacitance of MnO_2. Under low scan rate of 2mV/s, where the pseudocapacitance of MnO_2 is more notable, the value is even higher. Compared with other on-chip supercapacitors, such as activated carbon based one in [4] (1.4mF/(cm^2 μm)), and polypyrrole based one in [5] (0.11mF/(cm^2 μm)), the specific capacitance per volume is competitive.

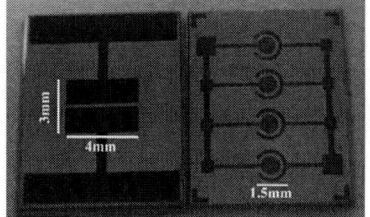

Figure 7: Photographs of two prototypes with different patterns.

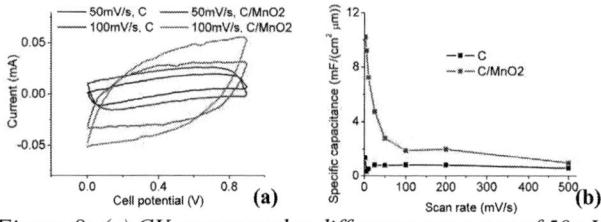

Figure 8: (a) CV curves under different scan rates of 50mV/s and 100mV/s for the left prototype shown in Figure 7 with/without MnO$_2$. (b) Summary of the specific capacitance per volume under different scan rates.

CONCLUSIONS

In this work, on-chip micro supercapacitors using MnO_2 decorated into direct prototyped porous carbon electrodes are presented. In the composite electrode material, both EDLC and pseudocapacitance contribute to the total capacity of the prototype. It is accomplished by MEMS fabrication process to form the monolithic porous carbon and a designed electrochemical deposition process to incorporate MnO_2. Characterizations demonstrate the existence of MnO_2 in the framework, and the increase in specific capacitance.

ACKNOWLEDGEMENTS

The work is supported by the National Natural Science Foundation of China (No. 60936003), 973 Program (No. 2009CB320304), and 863 Program (No. 2009AA04Z319) of China.

REFERENCES

[1] B.E. Conway, *Electrochemical Supercapacitors: Scientific Fundamentals and Technological Applications*, Kluwer Academic/Plenum Publishers, 1999.

[2] H. J. In, et al., "Origami Fabrication of Nanostructured, Three-dimensional Devices: Electrochemical Capacitors with Carbon Electrodes", *Appl. Phsy. Lett.*, vol. 88, 083104, 2006.

[3] H. X. Ji, et al., "Swiss Roll Nanomembranes with Controlled Proton Diffusion as Redox Micro-supercapacitors", *Chem. Commun.*, vol. 46, pp. 3881-3883, 2010.

[4] D. Pech, et al., "Elaboration of a Microstructured Inkjit-printed Carbon Electrochemical Capacitor", *J. Power Sources*, vol. 195, pp. 1266-1269, 2010.

[5] W. Sun, et al., "Symmetric Redox Supercapacitor Based on Micro-fabrication with Three-dimensional Polypyrrol Electrodes", *J. Power Sources*, vol. 195, pp. 7120-7125.

[6] Y. Jiang, et al., "3D Supercapacitor Using Nickel Electroplated Vertical Aligned Carbon Nanotube Array Electrode", in *Digest Tech. Papaers MEMS 2010 Conference*, Hong Kong, Jan 24-28, 2010, pp. 1171-1174.

[7] C. Shen, et al., "A High-performance Three-dimensional Micro Supercapacitor Based on Self-supporting Composite Materials", *J. Power Sources*, vol. 196, pp. 10465-10471, 2011.

[8] C. Shen, et al., "Direct Prototyping of Patterned Nanoporous Carbon: A Route from Materials to On-chip Devices", *Sci. Rep.*, vol. 3, pp. 2294, 2013.

[9] H. Y. Lee, et al., "Supercapacitor Behavior with KCL Electrolyte", *J. Solid State Chem.*, vol. 144, pp. 220-223, 1999.

[10] W. Wei, et al., "Manganese Oxide-based Materials as Electrochemical Supercapacitor Electrodes", *Chem. Soc. Rev.*, vol. 40, pp. 1697-1721, 2011.

[11] Z. Li, et al., "Manganese Dioxide-coated Activated Mesocarbon Microbeads for Supercapacitors in Organic Electrolyte", *Colloids and Surfaces A: Physicochem. Eng. Aspects*, vol. 366, pp. 104-109, 2010.

[12] D. Zhai, et al., "The Preparation of Graphene Decorated with Manganese Dioxide Nanoparticles by Electrostatic Adsorption for Use in Supercapacitors", *Carbon*, vol. 50, pp. 5034-5043, 2012.

[13] Z. Fan, et al., "High Dispersion of χ-MnO$_2$ on Well-aligned Carbon Nanobube Arrays and its Application in Supercapacitors", *Diamond & Related Materials*, vol. 17, pp. 1943-1948, 2008.

[14] J. Liu, et al., "Hybrid Supercapacitor Based on Coaxially Coated Manganese Oxide on Vertically Alligned Carbon Nanofiber Arrays", *Chem. Mater.*, vol. 22, pp. 5022-5030, 2010.

CONTACT

*X. Wang, tel: +86-10-62798432;
wxh-ime@tsinghua.edu.cn

MEMS VIBRATION ELECTRET ENERGY HARVESTER WITH COMBINED ELECTRODES

Qianyan Fu[1], and Yuji Suzuki[1]

[1]Department of Mechanical Engineering, The University of Tokyo, Tokyo, Japan

ABSTRACT

A novel in-plane MEMS electret energy harvester with combined electrodes of overlapping-area-change and gap-closing converters is proposed for large output power both at low and high vibration accelerations. An early prototype has been successfully micro-fabricated with the single layer silicon-on-insulator process. Soft X-ray charging is employed to establish uniform surface potential around 60 V on vertical electrets on the sidewall of the comb fingers. Up to 1.6 μW output power has been obtained, which corresponds to the effectiveness as high as 57%.

INTRODUCTION

Among various types of vibration energy harvesters, electrostatic/electret principle has been attracting much attention for its high output power in small dimensions [1, 2]. In our previous studies, we have employed a high-performance perfluorinated polymer electret CYTOP as the electret material, and developed an in-plane MEMS energy harvester, in which overlapping area between electret on the top moving mass and the counter electrodes on the bottom substrate is changed [3-4]. Up to 6 μW output power has been obtained at 40 Hz and 1.4 g acceleration [5].

Another configuration of in-plane electrostatic/electret energy harvesters is based on comb drives, which is micro-fabricated through one-mask SOI process without any assembling processes [6-8]. Nguyen et al. [7] developed a electrostatic energy harvester with overlapping-area-change comb drives and achieved output power up to 3.4 μW at 1-g acceleration and 150 V external bias. Guilllemet et al. [8] proposed an electrostatic vibration energy harvester with gap-closing interdigited combs and obtained up to 2.2 μW for an external acceleration of 1g at 150 Hz. However, external voltage source for the bias voltage is needed for those electrostatic generators.

The advantage of electret generator is that it has built-in bias voltage. When properly charged, electret films can retain electrical charges for a long period. Several methods are proposed for charging the electret films such as corona charging, thermal poling, electron-beam injection, and contact charging [9]. However, it is not a straightforward process to charge vertical electrets on the sidewall of comb drives because of the charge build-up near the gap opening. Suzuki et al. [10] employed SiO$_2$ highly doped with potassium as the electret film, and charged with a bias-temperature procedure. However, in their electret generator, the maximum output power is limited to 50 nW at 58 Hz, partially because of low surface potential.

Recently, we have developed a new charging method using soft X-ray irradiation [11], which can realize surface potential and thermal stability of electrets as good as those with corona discharge. We also prototyped an in-plane electret transducer, in which vertical electrets on the sidewall of comb drives are charged with the soft X-ray method [12].

In the present study, a novel in-plane MEMS electret energy harvester with combined electrodes with overlapping-area-change and gap-closing types is proposed for large dynamic range response, and its electromechanical performance is examined in detail in a series of experiments. Soft X-ray charging is employed to realize vertical electrets with high surface potential on the sidewall of comb drives.

DEVICE DESIGN

In electrostatic/electret energy harvesters, capacitance change per unit displacement should be maximized for large output power [2]. Capacitance of the gap-closing converters ((2) in inset of Fig. 1a) is inversely proportional to the displacement, and thus the gap-closing converter is suitable for large vibration acceleration. However, at small acceleration, the capacitance change diminishes. On the other hand, capacitance of the overlapping-area-change converters with a constant gap ((1) in inset of Fig 1a) is linearly changed with the displacement, and thus its change is higher than that of the gap-closing type at low acceleration. Therefore, we propose a novel electrode configuration combined with in-plane overlapping-area-change and gap-closing converters that provide relatively-high output power both at low and high vibration accelerations.

Fig 1a shows a schematic of the proposed structure. The mechanical resonator has a seismic mass of 4.25 mg, which is suspended by four suspensions at the corners. Based on the VDRG (velocity-damped resonance generator) model [13], the output power of vibration energy harvester is proportional to the seismic mass amplitude. Thus, the spring structure should be designed in such a way that the large travel range is obtained without lateral instability. For that purpose, we employ the tilted folded-beam suspensions [14]. As shown in Fig. 1b, the beam segments are slightly tilted. The tilted angle can be described as d/l for small d, where d is the initial shift in the oscillation direction and l is the beam length. The expressions for the spring constants of the proposed suspension, both in the stroke direction and perpendicular to it, can be described as:

$$K_x = 2Ehb^3/l^3 \tag{1}$$

$$K_y = \frac{50Ehb^3}{(3\delta_x - 5d)^2 l} \tag{2}$$

where E, b, and h are the Young's modulus, the beam width, and the beam thickness, respectively. The projected lengths of the beam along the driving direction (x) and the vertical

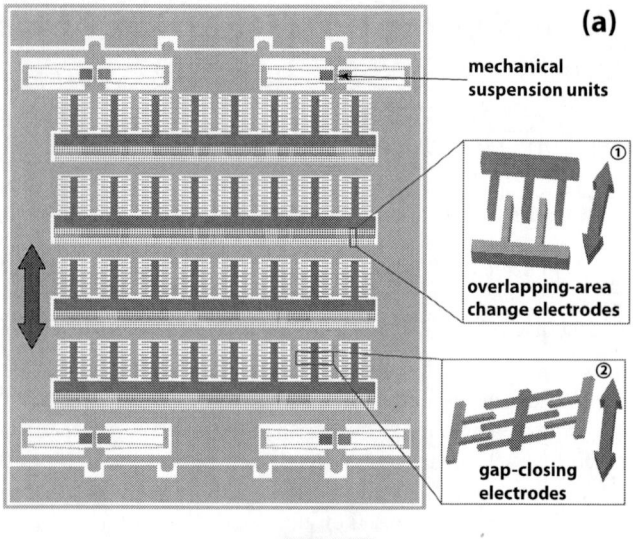

(a)

mechanical suspension units

① overlapping-area change electrodes

② gap-closing electrodes

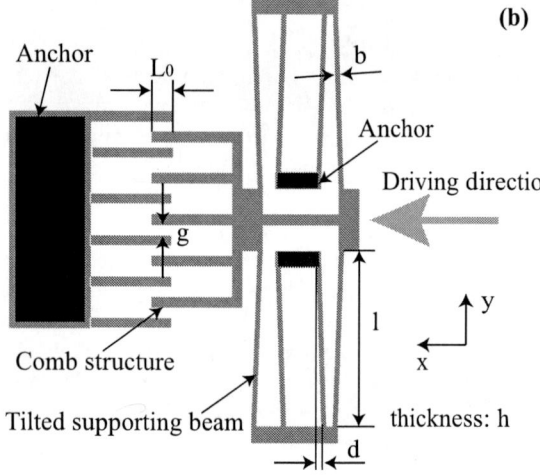

(b)

Anchor

L_0

b

Anchor

Driving direction

g

l

y

x

Comb structure

Tilted supporting beam

thickness: h

d

Figure 1: MEMS electret energy harvester: (a) Schematic view of the in-plane electret energy harvester with combined electrodes and (b) Tilted folded-beam suspension unit.

(y) directions are d and l, respectively. δ_x is the maximum stable travel range, which is considered in the present structural design.

The tilted folded-beam suspensions have a width of 10 µm and a length of 1835 µm with a tilt angle of 2 degree. The electrodes consist of two types of interdigitated comb drives: gap-closing and overlapping-area-change. For the overlapping-area-change type, fixed and movable comb fingers are designed with a gap (g) of 10 µm and an initial overlap L of 50 µm. For the gap-closing type, g and L are 39 µm and 75 µm, respectively. Thickness of the device layer is 70 µm. The maximum variation of capacitance ΔC is 14.7 pF in the harvester with 600 finger pairs for overlapping-area-change type and 452 finger pairs for the gap-closing one. The oscillation amplitude δ_x of the proof mass is limited to 36 µm by mechanical stoppers. The chip size is 1 cm by 1 cm.

MEMS FABRICATION AND EXPERIMENTAL RESULTS

The device was fabricated through the standard silicon-on-insulator (SOI) MEMS technology. The fabrication process is a full batch process, which requires only one lithography mask. It starts with a standard lithography on 4-inch SOI wafer using photoresist (Fig. 2a). 70 µm-thick device layer is etched with DRIE (Fig. 2b) to form springs, electrodes and etched holes. Then, the buried oxide layer is etched with vapor HF for releasing the structure through the etched holes on the seismic mass (Fig. 2c). This is followed by a 1.5-µm-thick parylene-C deposition as the electret material (Fig. 2d). Finally, soft X-ray charging using 9.5 keV acceleration voltages is applied with the bias voltage of 130V for 150 seconds (Fig. 2e).

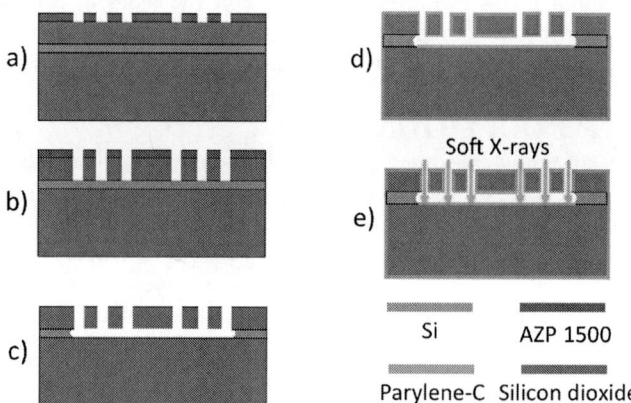

a)

b)

c)

d)

e) Soft X-rays

Si

AZP 1500

Parylene-C

Silicon dioxide

Figure 2: MEMS process flow.

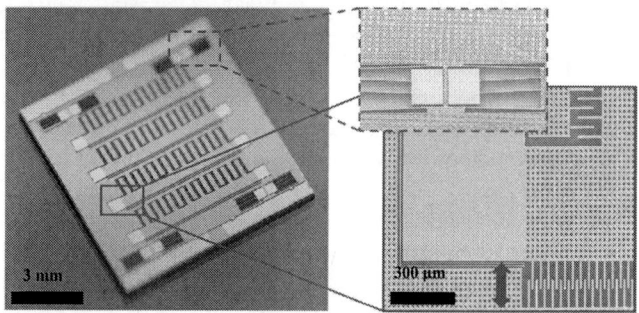

3 mm

300 µm

Figure 3: Overview of the present electret energy harvester prototype.

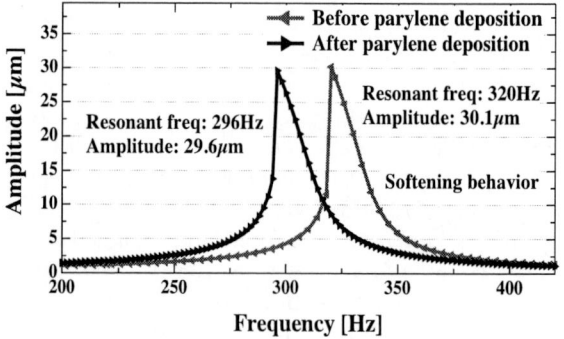

Figure 4: Frequency response before/after 1.5-µm-thick parylene deposition.

Table 1: Resonant frequency of the device before/after parylene deposition

Parylene deposition	Seismic mass [mg]	Designed resonant frequency [Hz]	Measured resonant frequency [Hz]
Before deposition	4.05	320	320
After deposition	4.77	295	296

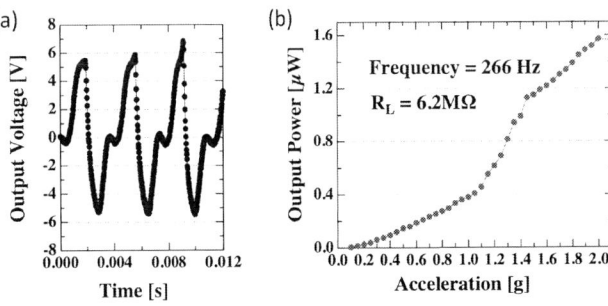

Figure 8: Power generation experiment results at 266 Hz. (a) Output voltage waveform at 2-g external acceleration and (b) Output power versus external acceleration.

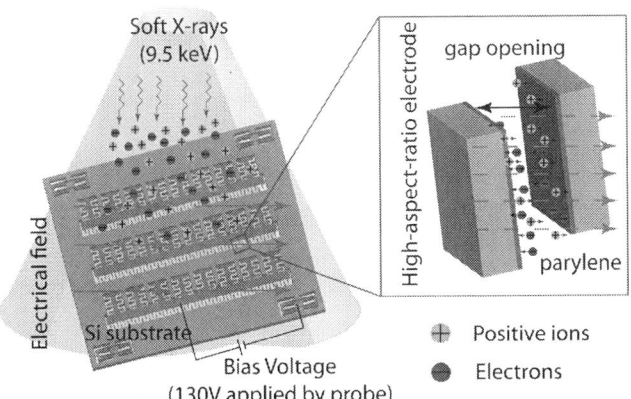

Figure 5: In-situ charging method for vertical electrets using soft X-ray irradiation. Bias voltage of 130 V is applied during the charging process.

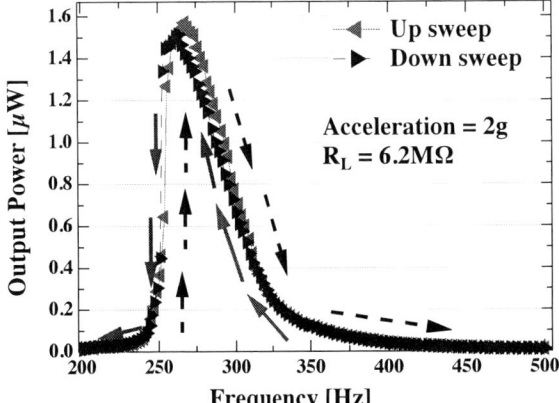

Figure 9: Output power versus the vibration frequency for both up- and down-sweeps at the external acceleration amplitude of 2 g.

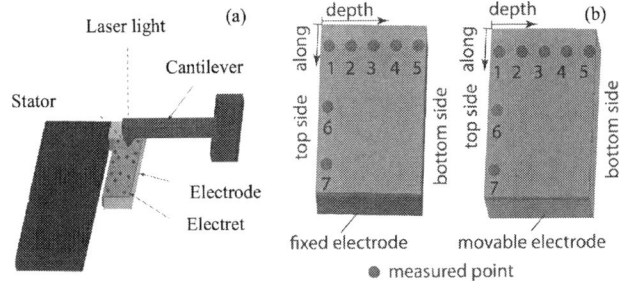

Figure 6: Kelvin force microscopy (KFM) measurement of surface potential for the vertical electrets. (a) Kelvin force microscopy with a nano-sized tip cantilever and (b) measurement points on the movable and fixed comb fingers that are intentionally broken after charging.

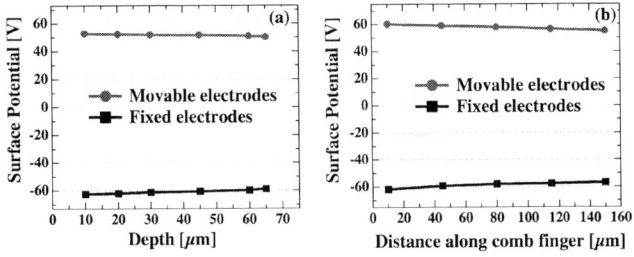

Figure 7: Surface potential distribution of the vertical electrets (a) in the depth direction, and (b) in the longitudinal direction along the comb finger.

Figure 3 shows the prototype device, which is then fixed onto a printed circuit board. After wire bonding for electrical connection, the PCB is screwed to a electromagnetic shaker. Figure 4 shows the mechanical response of the device, which indicates slight softening behavior. The resonant frequency is decreased from 320 Hz to 296 Hz after the 1.5-μm-thick parylene-C deposition, while the maximum amplitude is unchanged, which is as large as 30 μm. The quality factor is around 50. The estimated added mass of the parylene film is 0.72 mg, which results in the weight of the seismic mass of 4.77 mg after the parylene deposition. Thus, the estimate of the resonant frequency with the added mass is 295 Hz, which is in good agreement with the measure value (Table 1).

We employ soft X-ray irradiation for electret charging as shown in Fig. 5. When the soft X-ray is irradiated onto comb drives, positive and negative ions are equally generated in the air gap, which are dragged toward the electrets under an imposed electric field across the gap, transferring the charges to the electrets. The bias voltage and the charging time are 130 V and 150 seconds, respectively.

In order to evaluate the charging performance of vertical electrets, we employed a Kelvin force microscopy (KFM, SPM-9600, Shimazu) with a nano-sized tip cantilever (EFM-10, Nano world). After charging, the comb fingers are intentionally broken and one comb finger is fixed onto the KFM stage. Figure 6a schematically shows the KFM measurement setup for the surface potential of the vertical electrets. Figure 7 shows the surface potential distribution of the vertical electrets. High surface potential of around 60 V is uniformly distributed in the depth direction down to 65 μm. Along the longitudinal direction along the comb finger, uniform surface potential is confirmed, as shown in Fig. 7b.

Figure 8a shows the output voltage waveform at the resonant frequency of 266 Hz and 2 g acceleration. Peak-to-peak voltage of 12 V is obtained at the optimal load of 6.2 MΩ. Figure 8b shows the output power versus external acceleration at 266 Hz. Thanks to the gap closing electrodes, the output power is rapidly increased with the external acceleration, and over 1 μW is obtained at 1.5 g. Bi-stable behavior is found over 1.5 g acceleration, which is attributed to the mechanical response of the tilted folded-beam suspensions for large displacements, and the output power jumps from 1 μW to 1.17 μW. As shown in Fig. 9, the maximum output power of 1.57 μW is achieved at 258 Hz and 2-g, which corresponds to the effectiveness defined by the output power divided by the VDRG limit [13] as high as 57%. Through the combined effects of the electrical damping and the softening nonlinear spring of the tilted folded-beam suspensions, the bandwidth is increased to 50 Hz at 2 g.

CONCLUSION

A novel in-plane electret energy harvester with combined electrodes of overlapping-area-change and gap-closing converters has been proposed in order to realize large dynamic range of harvester. The proposed prototype is successfully micro-fabricated, and up to 1.57 μW output power has been obtained, which corresponds to the effectiveness as high as 57%.

ACKNOWLEDGEMENTS

This work is partially supported through NEXT Program of JAPAN Society for the Promotion of Science (JSPS). Photo-mask is made using the University of Tokyo VLSI Design and Education Center (VDEC)'s 8-inch EB writer F5112+VD01 donated by ADVANTEST Corporation.

REFERENCES

[1] P. D. Mitcheson, E. M. Yeatman, G. K. Rao, A. S. Holmes, T. C. Green, "Energy Harvesting from Human and Machine Motion for Wireless Electonic Devices", *Proc. IEEE,* Vol. 96, pp. 1457-1486, 2008.

[2] Y. Suzuki, "Recent Progress in MEMS Electret Generator for Energy Harvesting", *IEEJ Trans. Electr. Electr. Eng.,* Vol. 6, pp. 101-111, 2011.

[3] K. Kashiwagi, K. Okano, T. Miyajima, Y. Sera, N. Tanabe, Y. Morizawa, Y. Suzuki, "Nano-cluster-enhanced High-performance Perfluoro-polymer Electrets for Micro Power Generation," J. Micromech.

Microeng., Vol. 21, 125016, 2011.

[4] Y. Suzuki, D. Miki, M. Edamoto, M. Honzumi, "A MEMS Electret Generator with Electrostatic Levitation for Vibration-Driven Energy Harvesting Applications", *J. Micromech. Microeng.,* Vol. 20, 104002, 2010.

[5] K. Matsumoto, K. Saruwatari, Y. Suzuki. "Vibration-powered Battery-less Sensor Node Using Electret Generator", *PowerMEMS 2011*, Seoul, pp. 134-137, 2011.

[6] D. Hoffmann, B. Folkmer, Y. Manoli, "Fabrication, Characterization and Modeling of Electrostatic Micro-generators", *J. Micromech. Microeng.,* Vol. 19, 094001, 2009.

[7] S. D. Nguyen, E. Halvorsen, I. Paprotny, "Bistable Springs for Wideband Microelectromechanical Energy Harvesters", *Appl. Phys. Lett.,* Vol. 102, 023904, 2013.

[8] R. Guilllemet, P. Basset, D. Galayko, F. Cottone, F. Marty, and T. Bourouina, "Wideband MEMS Electrostatic Vibration Energy Harvesters Based on Gap-closing Interdigited Combs with a Trapezoidal Cross Section", *IEEE MEMS 2013*, Taipei, pp. 137-140, 2013.

[9] G. M. Sessler, *Electrets* (3rd ed.), Laplacian Press: California, 1998.

[10] M. Suzuki, H. Hayashi, A. Mori, T. Sugiyama, G. Hashiguchi, "Electrostatic Micro Power Generator Using Potassium Ion Electret Forming on a Comb-drive Actuator", *PowerMEMS 2012*, Atlanta, pp. 247-250, 2012.

[11] K. Hagiwara, M. Goto, Y. Iguchi, T. Tajima, Y. Yasuno, H. Kodama, K. Kidokoro, Y. Suzuki, "Electret Charging Method Based on Soft X-ray Photoionization for MEMS Applications", *Trans. IEEE, Dielectr. Electr. Insul.,* Vol. 19, pp. 1291-1298, 2012.

[12] M. Honzumi, A. Ueno, K. Hagiwara, Y. Suzuki, T. Tajima, and N. Kasagi, "Soft-X-Ray-Charged Vertical Electrets and Its Application to Electrostatic Transducers," *IEEE MEMS2010*, Hong Kong, pp. 635-638, 2010.

[13] P. D. Mitcheson, T. C. Green, E. M. Yeatman, A. S. Holmes, "Architectures for Vibration-Driven Micropower Generators", *J. Microelectromech. Syst.,* Vol. 13, pp. 429-440, 2004.

[14] G. Zhou, P. Dowd, "Tilted Folded-beam Suspension for Extending the Stable Travel Range of Comb Drive Actuators", *J. Micromech. Microeng.,* Vol. 13, pp. 178-183, 2003.

CONTACT

*Qianyan Fu, tel: +81-3-5841-6419;
qfu@mesl.t.u-tokyo.ac.jp

MICRO PATTERN OF CHARGE IN PTFE ELECTRET FOR ENERGY HARVESTERS

Wei Bian, Xiaoming Wu*, Xiaohong Wang

Tsinghua National Laboratory for Information Science and Technology
Institute of Microelectronics, Tsinghua University, Beijing 100084, China

ABSTRACT

This paper presents a novel fabrication process to pattern the charge in electret film for vibration energy harvester and sensor applications. Electret material is charged with pattern through a mask layer, and is not required to be etched, which simplifies the pattern process. The line width of the pattern is determined by photolithography process. Compared with previously reported techniques[1,2,3], PTFE electret material, which is inexpensive and has very high bulk resistivity, is used for the experiment. Experiment results show that the surface potential on the patterned charge zone of PTFE film is higher than -200V when the line width reaches 20μm. A demo electrostatic vibration energy harvester is built by using the pattern technique and tested.

INTRODUCTION

Stable and long lifetime power supply is a crucial challenge for autonomous wireless instruments, such as many implanted medical devices or wireless sensor nodes. Vibration energy harvester (VEH), as a very promising power solution for these instruments, attracts more and more attentions in recent years. There are three main mechanisms of VEH so far: piezoelectrical [4], electromagnetical [5], and electrostatical [6]. Many electrostatical VEHs have advantages of low working frequency, wide band, and micro fabrication process compatibility [7] [8]. To obtain higher output power and simpler device structure, comb shaped electret and electrode are widely used [9]. In this case, electret material is required to be patterned and charged in comb shape to form a periodic electrostatic field. As a kind of inexpensive material, PTFE material has high surface potential compared with other electret materials, such as PVDF, CYTOP and SiO_2/Si_3N_4. However, PTFE is hard to be patterned by conventional etching processes because of its inert chemical properties, which strongly limits its application in electrostatical VEHs.

In this paper, a novel fabrication process to pattern the charge in PTFE electret film without material etching is presented. This method can be applied on other electret materials.

PRINCIPLE AND FABRICAION

The concept of the charge pattern process is shown in figure 1. Firstly, electret material is covered by a mask which defines the charging areas. After charging (e.g. corona charging), charges are implanted in both mask material and opened electret areas. The covered electret area, however, is protected from charging by mask. Because of the very high bulk resistivity, electret material maintains the implanted charges for long time. But the charges in mask leak to ambient very soon. Then the charge pattern is formed on the electret material. To this end, mask material should be easy patterned, with relative small bulk resistivity, and avoid influencing the charging process. In corona charging case, the mask should be dielectric material which is almost not affect the distribution of charging electric field. Additionally, it is not necessary to remove the dielectric mask after charging because the thickness of the mask is very thin compared to the dimension of VHE. Therefore, there is no followed extra process may worsen the charging property of the electret. In this work, PECVD SiO_2 is used as mask material.

Figure 2 shows the process of the charge pattern in PTFE electret. Firstly, a PTFE tape (3M™ PTFE Film Tape 5480) is glued on a conductive substrate as electret film. After that, a 0.3μm thick PECVD SiO_2 layer is deposited and patterned on PTFE electret film by lift off process. As the surface energy of PTFE is quite low, photoresist cannot be coated on it directly. As figure 2b shows, PTFE film is pretreated by nitrogen plasma to improve the surface energy and ameliorate wettability before coated with photoresist. Then the sample is corona charged in a homogeneous electric field. The voltages on the needle and grid are -7000V and -700V respectively, which are generated by two DC voltage sources. The charging temperature is 110 °C.

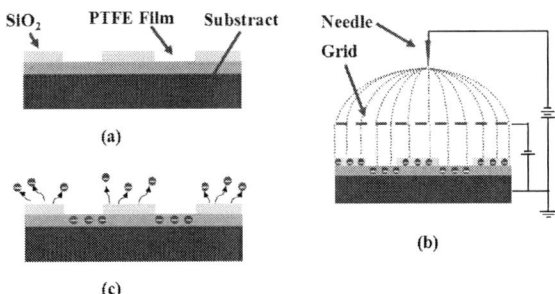

Figure 1: Concept of the electret charge pattern process.

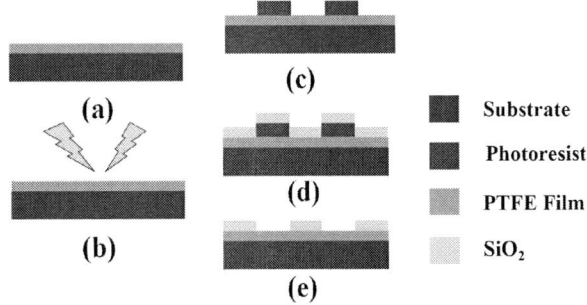

Figure 2: SiO_2 masked PTFE film fabrication process.

EXPERIMENTS AND DEMO DEVICE

Figure 3 compares the charge decay performance of the windowed PTFE electret and SiO_2 mask film on PTFE. The surface potential is measured by electrostatic voltmeter (Trek 541). When the thickness of PTFE electret and SiO_2 mask film are 100μm and 0.3μm respectively, and the charging grid voltage is -700 V, the as charged surface potential on PTFE and SiO_2 mask film are all -700V. However, the surface potential on SiO_2 decrease from -700V to lower than -10V in 80 minutes, while that of windowed PTFE does not change in more than 40 days.

Figure 4 presents a surface potential map of SiO_2 masked PTFE electret after corona charging process. The line width of the pattern "THU" is 3 mm. The surface potential where covered by SiO_2 mask film is lower than -10V, while the potential on the windowed PTFE is higher than -500V. It is observed that the highest surface potential on PTFE is a little bit lower than the test results shown in figure 3. The reason is that the line width of pattern "THU" is smaller than that on the sample of figure 3. This phenomenon implies that the surface potential of charge patterned PTFE electret is line width dependent.

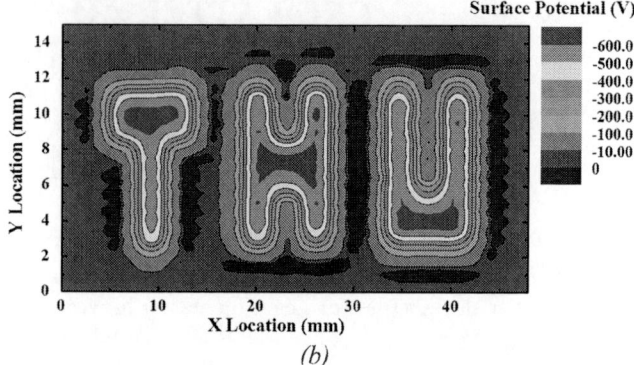

(b)

Figure 4: Schematic view and voltage potential distribution on sample with word "THU"

To find the relationship between surface potential and line width of the patterned charge, a test pattern is designed and the test results are shown in figure 5. The pattern consists of a series of fingers, which are with width varying from 3 mm to 20μm. After corona charging in the same condition, the surface potential distribution is measured along the charge fingers and shown in figure 5. Because of the limited spatial resolution of electrostatic voltmeter, only the average surface potential of charged PTFE fingers and SiO_2 mask are recorded when the charge finger width becomes small.

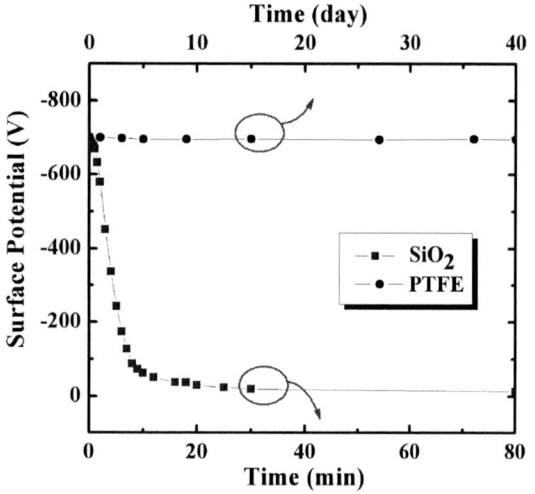

Figure 3: Time decay of surface potential of charged SiO_2 and PTFE films

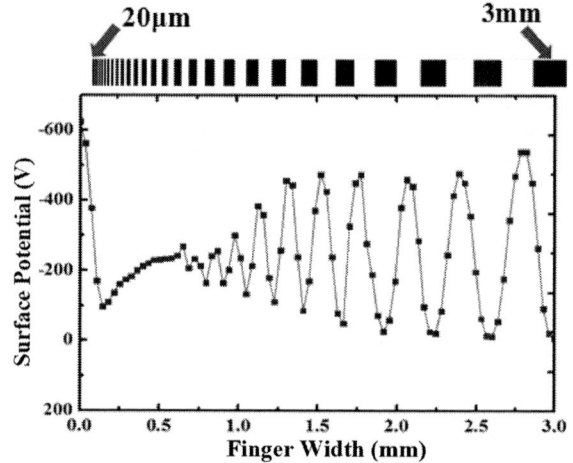

Figure 5: Potential distribution on test charge fingers

As the 50% duty ratio of the PTFE charge fingers, the surface potential of the thin fingers can be evaluated as high as doubled the measurement results. The evaluated values are shown in figure 6. When the finger is thinner, its surface potential becomes lower. The possible reason of this phenomenon is the charge diffusion and leakage from PTFE to SiO_2 through the boundary between them, which decreases the effective width of PTFE finger. As shown in figure 7, the effective duty ratio of PTFE charge fingers is less than 50%, and the average surface potential of the PTFE fingers decrease. When the PTFE finger is thinner, the effective duty ratio shrinkage is higher.

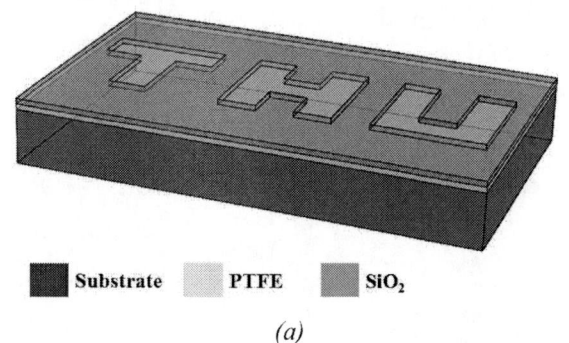

(a)

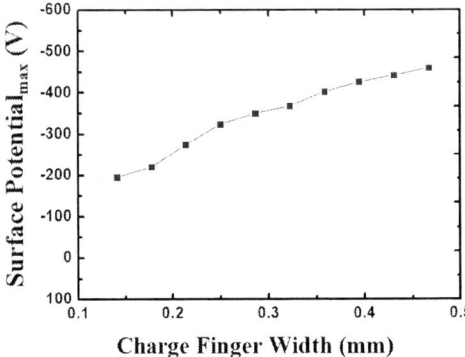

Figure 6: The evaluated surface potential of thin fingers

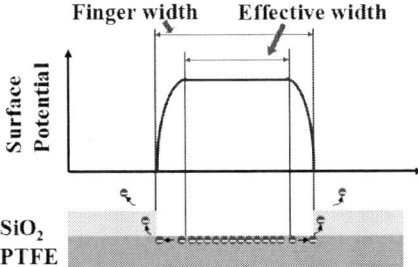

Figure 7: Effective width shrinkage caused by charge leakage of thin fingers

To verify the performance of the charge patterned PTFE electret film, a demo device for energy harvesting is fabricated and tested. As shown in figure 8, the device has two parts. One is a plate with the charge patterned PTFE electret, and another one is a plate with comb-shaped electrodes. The width of charge fingers and SiO_2 spaces on the electret plate is 0.5mm. The comb-shaped electrodes are with the same dimensions with that of electret charge pattern. The sizes of the two parts are all 25×25 mm². A load resistor is connected directly to the comb-shaped electrodes. Two channels of the oscilloscope are used to measure the voltages on the two sides of the load resistor. The oscilloscope displays the voltage difference between the two channels to suppress the common noises in the signals.

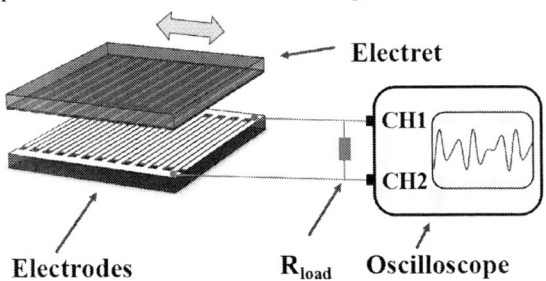

Figure 8: Schematic of the demo device and test circuit

Figure 9 shows the photography of the demo device and experiment setup. The electrode part is fixed on a XYZ stage, and the charge patterned electret part is fixed on an electromagnetic shaker. The two parts are placed face to face,

and the distance between them is adjusted to 0.1mm by the XYZ stage.

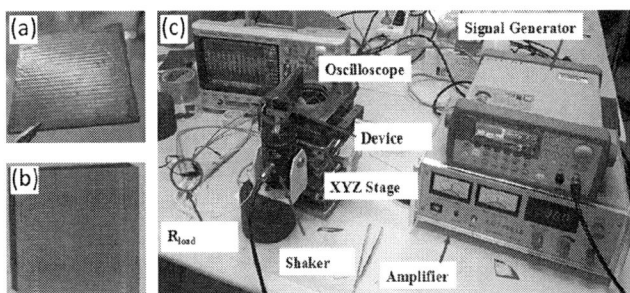

Figure 9: Photography of electret (a) and electrode (b) part of demo device, and test setup(c)

Figure 10 shows the output voltage traces of the demo device in different vibration conditions. The load resistance is 5MΩ, and the amplitude of vibration is 2.4 mm, 1.9 mm, 1.2 mm, and 0.2 mm, respectively. The vibration frequencies are all 30 Hz.

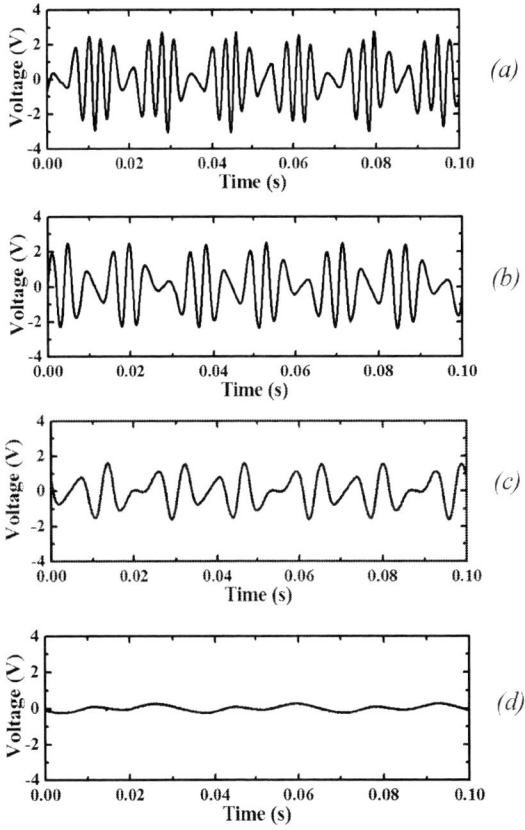

Figure 10: Output voltage traces of demo device, vibration frequency is 30 Hz.

When the vibration amplitude is smaller than the period of charge pattern, the output voltage is low and with the frequency of excitation vibration. When the vibration amplitude becomes larger than the period of charge pattern, the output electrodes pass by more than one charge fingers in one vibration period. Then a higher frequency signal with an envelop trace is observed. The frequency of the envelop signal is identical with that of excitation vibration. The frequency of the output voltage is determined by the number of charge fingers that the output electrodes pass by in one vibration period. This signal trace proves that the charges in PTFE film are patterned effectively.

The output voltage increases with the increase of vibration amplitude and frequency. Figure 11 shows the vibration condition dependence of the output voltage of the demo device when the load resistance is 5MΩ.

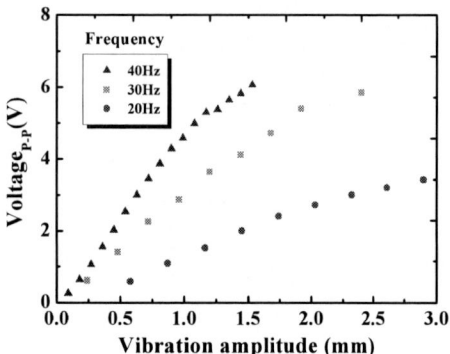

Figure 11: The output voltage vs. vibration amplitudes.

Figure 12 is the relationship between the output property of the demo device and load resistance. The optimized load of the demo device is 35MΩ. The maximum output power of 16μW is obtained in the vibration condition of 2.4mm @ 30Hz.

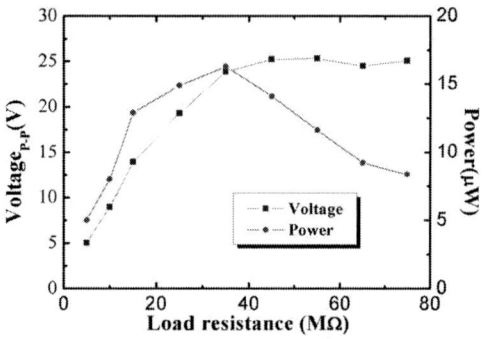

Figure 12: The load resistance dependence of the output voltage and power of demo device

CONCLUSION

In this paper, a novel fabrication process to pattern the charge in PTFE electret film for vibration energy harvester and sensor applications is presented. The electret material is charged with a patterned mask and no etching process is needed for electret material. The line width of the pattern is determined by photolithography process. Experiment results show that the surface potential on the patterned charge zone of PTFE film is higher than -200V when the line width is 20μm. The performance of the charge patterned PTFE electret material is verified by the output voltage trace of a demo VEH. This simple charge pattern method can be applied on other electret materials.

ACKNOWLEDGEMENTS

This work was supported by the State Key Development Program for Basic Research of China (Grant No. 2009CB320304), Key Program of Natural Science Foundation of China (Grant No.60936003), and Beijing Natural Science Foundation (Grant No.3122023)

REFERENCES

[1]   S. W. Liu, S.W. Lye, and J.M. Miao, "Sandwich structured electrostatic/electrets parallel-plate power generator for low acceleration and low frequency vibration energy harvesting", in *Proc. IEEE MEMS 2012,* pp. 1277-1280.

[2]   M. Suzuki, T. Wada, T. Takahashi, "Fabrication of Narrow Comb-Shaped Electret by Removing Charge Using Excimer Laser Beam from Charge-Implanted CYTOP Film for Avoiding Electrostatic Repulsion Problem", in *Proc. IEEE MEMS 2012,* pp. 1229-1232.

[3]   V. Leonov, R. van Schaijk, and C. Van Hoof, "Charge Retention in a Patterned SiO2/Si3N4 Electret", *IEEE Sensors Journal,* 13(9) 2013, pp3369-3376.

[4]   J. Kymissis, C. Kendall, J. Paradiso, N. Gershenfeld, "Parasitic Power Harvesting in Shoes", in *Proc. 4th Int. Conf. Mater.Eng. Resources,* 2001, pp.202-207.

[5]   C. R. Saha, T. O' Donnell, H. Loder, S. Beeby, and J.Tudor, "Optimization of an Electromagnetic Energy Harvesting Device", *IEEE Trans. Magnet.,* vol. 42, No. 10, Oct. 2006

[6]   Y. Naruse, N. Matsubara, K. Mabuchi, M. Izumi, K. Honma, "Electrostatic micro power generator from low frequency vibration such as human motion", in *Proceedings of PowerMEMS 2008+ microEMS 2008,* Sendai, Japan, November 9-12, (2008), pp 19-22.

[7]   M.Edamoto, Y.Suzuki, N.Kasagi, K.Kashiwagi, Y.Morizawa, T.Yokoyama, T.Seki, M.Oba "Low-Resonant- Frequency Micro Electret Generator for Energy Harvesting Application", in *Proc.IEEE MEMS 2009,* pp.1059–1062.

[8]   Y. Sakane, Y. Suzuki, and N. Kasagi, "The Development of High-performance Perfuluoriented Polymer Electret Film and Its Application to Micro Power Generation", *J.Micromech. Microeng.,* Vol. 18, 104011, 2008.

[9]   E. Halvorsen, E. R. Westby, S. Husa, A. Vogl, N. P. Ostbo, V. Leonov,T. Sterken, and T. Kvisteroy, "An Electrostatic Energy Harvester with Electret Bias" , in *IEEE Transducers 2009,* Jun. 21–25, 2009, pp. 1381–1384.

CONTACT

Xiaoming Wu: imewuxm@mail.tsinghua.edu.cn

NANO-POROUS SIO$_2$ ELECTRET WITH HIGH SURFACE POTENTIAL AND HIGH THERMAL RESISTANCE

Masato Suzuki[1], Takuya Wada[1], Tomokazu Takahashi[1],
Toshio Nishida[2], Yasuhiro Yoshikawa[2], and Seiji Aoyagi[1]
[1] 1 Kansai University, Suita, Osaka, Japan
[2] ROHM Co., Ltd., Kyoto, Japan

ABSTRACT

This paper proposes a new electret with high surface potential and high thermal resistance, which is made of nano-porous SiO$_2$. Electrical charge density in the nano-porous SiO$_2$ is higher than that in a normal (i.e., nonporous) SiO$_2$ because there are many voids in the nano-porous SiO$_2$ and the interface between void and SiO$_2$ traps electrical charges strongly. Therefore, decrease rate of the trapped charge density in the nano-porous SiO$_2$ is lower than that of normal SiO$_2$. Since SiO$_2$ is thermally stable, thermal stability of the electrical charge in nano-porous SiO$_2$ electret is better than that in a polymer electret. Output power generated by vibration energy harvesting using the nano-porous SiO$_2$ electret is also larger than that using the normal SiO$_2$ or polymer electret.

INTRODUCTION

Recently, vibrating energy harvester (VEH) which generates electrical power from environmental vibration was widely developed [1-11], because it is prospective power supply for low-power electronic devices such as wireless sensor networks [1, 2]. Since frequency of environmental vibration is usually on the order of a few to several tens of hertz, electrical power generated by electrostatic VEH using an electret are generally higher than that generated by a electromagnetic VEH using a magnet [3]. Electret is dielectric material that keeps electric charge semi-permanently. Previously, many type of electrostatic VEH using electret were developed. We also developed a novel VEH consists of a fixed electret on bade electrode and a high dielectric constant plate such as barium titanate (BaTiO$_3$) which is suspended by vertically vibrated spring with counter electrode with, as shown in Fig. 1 [4, 5].

Surface charge density of electret is most impotent characteristic for VEHs because generated power from a VEH is theoretically proportional to square of it. Conventionally, fluorocarbon polymer [1-8] and silicon dioxide (SiO$_2$) [9-11] were most frequently used as electret in VEHs because they keep more electrical charge than other materials. CYTOP is famous fluorocarbon polymer as an electret, of which surface charge density and its time stability are better than those of SiO$_2$ [10]. The charge density trapped in CYTOP can be enhanced by applying nano-clusters into CYTOP because charges are trapped at the interface between nano-clusters and CYTOP [8]. However, thermal resistance of CYTOP is lower than that of SiO$_2$ because CYTOP is an organic material.

Because of the above mentioned background, we propose a novel electret made of SiO$_2$ with nano-porous structure, in which many voids are expected to behave as same way as the nano-clusters in CYTOP. Therefore, it is expected the nano-porous SiO$_2$ electret satisfies both of high surface charge density and thermal stability. The concept of proposed nano-porous SiO$_2$ electret is shown in Fig. 2.

This paper presents the fabrication method and characteristics of nano-porous SiO$_2$ as an electret including thermal stability surface charge density. The output power of the VEH using nano-porous SiO$_2$ electret is also shown.

EXPERIMENT AND RESULTS

Fabrication of Nano-porous SiO$_2$ Electret

Process flow for fabricating nano-porous SiO$_2$ electret is shown in Fig. 3. In this study, the nano-porous SiO$_2$ was fabricated by thermal oxidation of nano-porous silicon. At first, a low resistivity silicon substrate (p-type, (100) orientation,

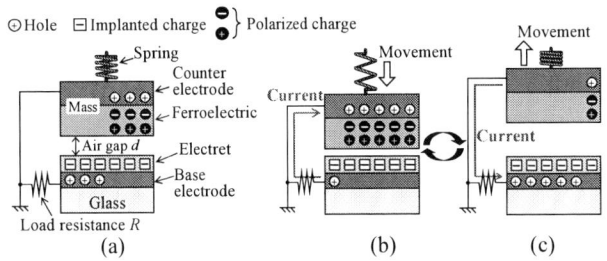

Figure 1: Schematic of our developed VEH which consists of electret and high-dielectric constant plate; (a) initial state, (b) mass is moving downward, and mass is moving upward.

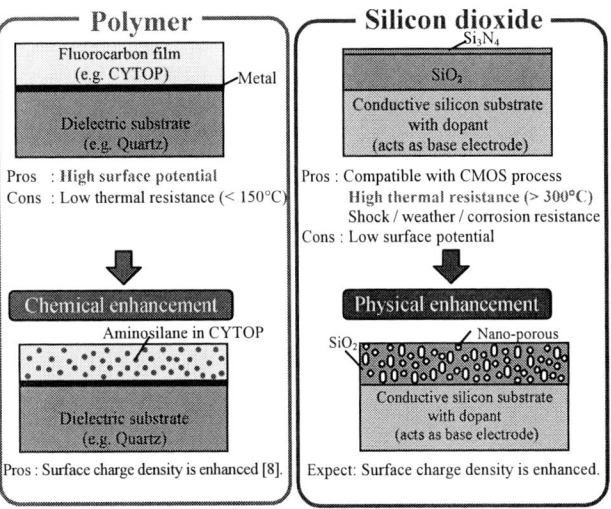

Figure 2: Concept of nano-porous SiO$_2$ electret with high surface charge density using porous structure.

<0.01 Ω·cm, 500 μm in thickness) was anodically-etched in hydrofluoric (HF) acid solution to form porous layer on the surface of substrate (Fig.3(a)) [12]. Then, the silicon substrate with porous layer was thermally oxidized in vapor of deionized (DI) water at 1,000°C (Figs. 3(b) and (c)). Since oxygen diffused rapidly in the porous silicon layer, nonporous silicon layer under the porous silicon layer was also oxidized concurrently with the oxidation of porous silicon layer, as shown in Fig. 3(d). In this paper, the normal (i.e. nonporous) SiO_2 layer under the porous SiO_2 is referred to as "base SiO_2". The thickness of nano-porous SiO_2 layer (t_{porous}) and that of base SiO_2 (t_{base}) were controlled by anodic etching time and oxidation time, respectively. In this study, two-types of samples were fabricated, in which the variable thickness of nano-porous SiO_2 is formed on thick (t_{base}=0.8 μm) or thin (t_{base}=0.2 μm) base SiO_2, as shown in Fig. 4.

Figure 5 shows field emission scanning electron microscopy (FE-SEM) images of fabricated nano-porous SiO_2 on base SiO_2. These figures show that the size of void in nano-porous SiO_2 is in the order of several tens nanometer. The negative charge (electrons) was implanted to each sample

by corona discharge method. Figure 6 shows the setup of corona discharge. Detail of the corona discharge was described in ref. [4]. The surface charge density of each sample was calculated from surface electric potential which was measured by an electrostatic voltmeter (Model 279, Monroe Electronics Inc.).

A normal SiO_2 and a CYTOP film were also fabricated as electrets for comparison. For nano-porous SiO_2 on base SiO_2 or normal SiO_2, low-resistivity silicon substrate acts as the base electrode. One the other hand, CYTOP film was formed on a quartz substrate with aluminum electrode.

Time Dependence and Thermal Stability of Implanted Charge in Nano-porous SiO_2 on Base SiO_2

Figure 7 shows the initial surface potential just after implanting charges versus thickness of nano-porous SiO_2. As shown in this graph, the initial surface potential, which is

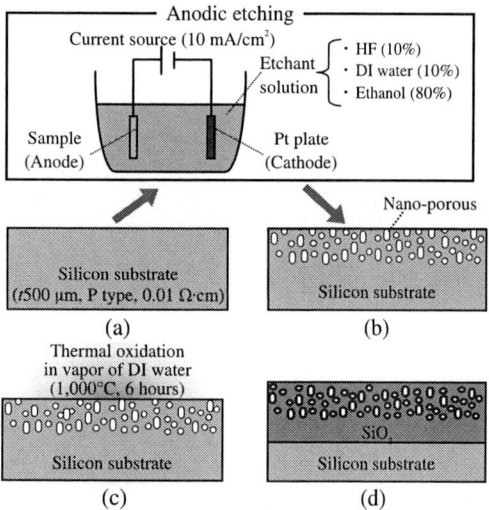

Figure 3: Process flow for fabricating porous SiO_2 electret. After (d), charge is implanted by corona discharge method.

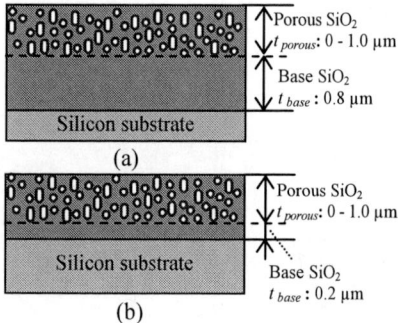

Figure 4: Cross sectional schematic of fabricated porous SiO_2 on base SiO_2 with (a) thick base SiO_2 of 0.8 μm, (b) thin base SiO_2 of 0.2 μm.

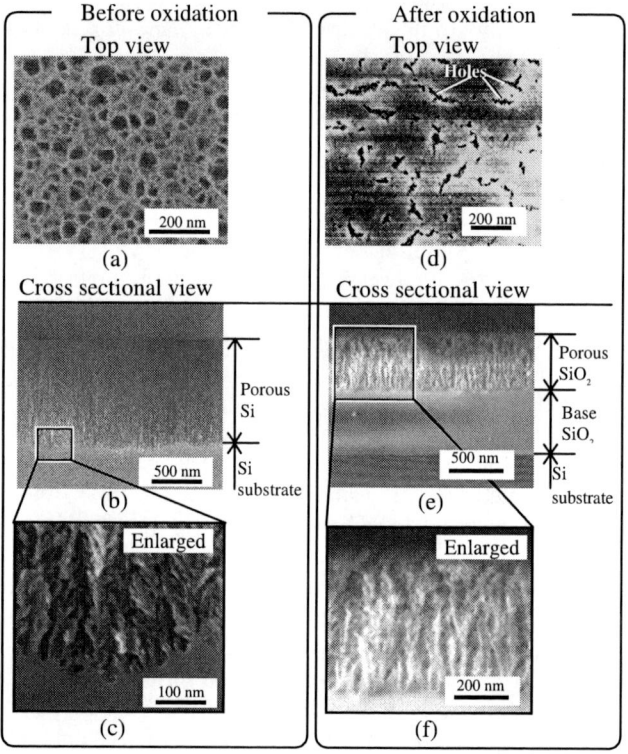

Figure 5: FE-SEM images of (a)-(c) porous silicon and (d)-(f) porous SiO_2 on base SiO_2.

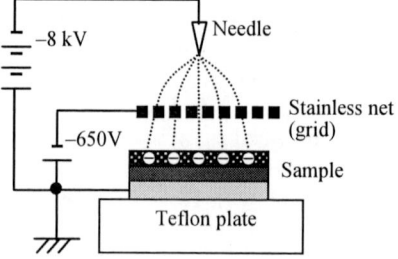

Figure 6: Setup of corona discharge for implanting electron into electret.

978-1-4799-3510-9/14 $31.00 © 2014 IEEE   418

proportional to the surface charge density, depends on both t_{base} and t_{porous}. It is noted that comparatively thick base SiO_2 is necessary to achieve high surface potential. Figure 8 shows time dependence of surface charge density of each sample. When the base SiO_2 is thick (t_{base}=0.8 μm), decreasing rate of charge density in nano-porous SiO_2 (t_{porous}=0.6 μm) was as low as that in CYTOP, which is much better than that of normal SiO_2. When the base SiO_2 is thin (t_{base}=0.2 μm), the charge density in porous SiO_2 rapidly decreased (Fig.6). Those results indicate that the base SiO_2 acts as a barrier layer of preventing charge dissipation from the porous SiO_2 to the underlying Si substrate.

Then degradations of surface charge density of nano-porous SiO_2 on base SiO_2, nonporous SiO_2, and CYTOP by heating samples were measured. The result of heating test is shown in Fig. 9. This result indicate that amount of charge in the nano-porous SiO_2 or normal SiO_2

is decreased gradually with increasing in the sample temperature when it is larger than approximately 130°C. On the other hand, the amount of charge in the CYTOP rapidly decreased when the sample temperature was around 150°C. It is proven that the thermal resistance of nano-porous SiO_2 is the best compared with nonporous SiO_2 and CYTOP.

Testing of Power Generation

Finally, output power of VEHs for free vibration were characterized, in which the nano-porous SiO_2 on base SiO_2, the normal SiO_2 or the CYTOP film was used as electret. Figure 10 shows the experimental setup for measuring the output power, of which the detailed operating principle was explained in ref. [4]. The mass was suspended to the case using a coil spring (spring constant; 0.34 N/mm). The steel stainless case was fixed on a vibration generator (PET-05, IMV Co., Ltd.). The thickness and area of dielectric (BaTiO_3) plate is 0.2 mm and 165 mm^2, respectively. The weight of aluminum mass is 70 gf. The overlapping area is same as the

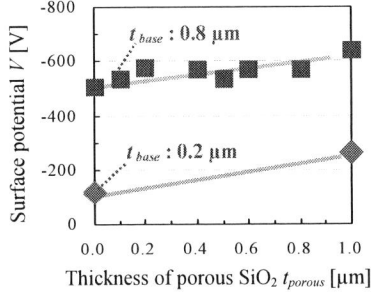

Figure 7: Surface potential of porous SiO_2 on base SiO_2 vs. thickness of porous SiO_2 (t_{porous}).

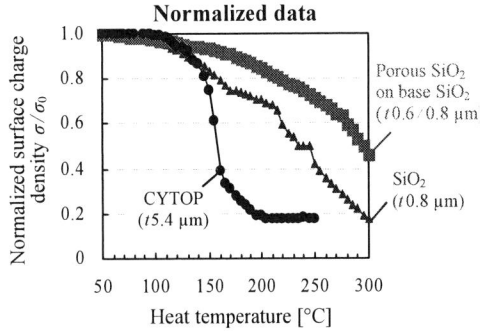

Figure 9: Degradation of normalized surface charge density of porous SiO_2, nonporous SiO_2, and CYTOP by heating samples.

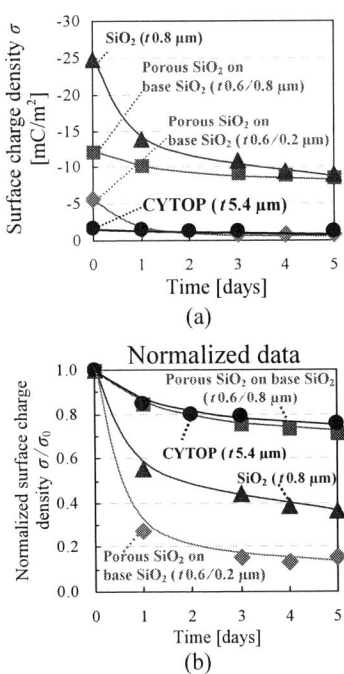

Figure 8: Time dependence of (a) surface potential and (b) normalized surface charge density of porous SiO_2 on base SiO_2.

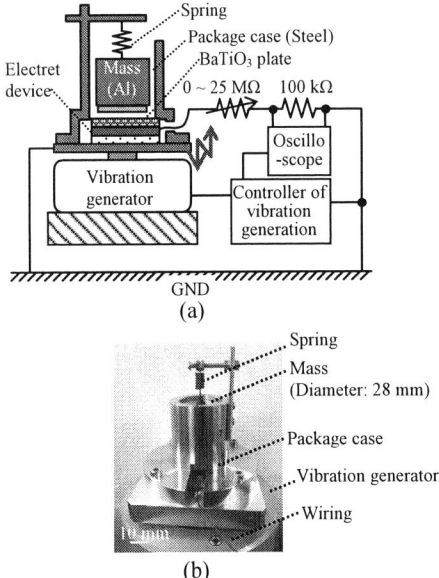

Figure 10: (a) Schematic and (b) photograph of Experimental setup for verifying power generation for free vibration.

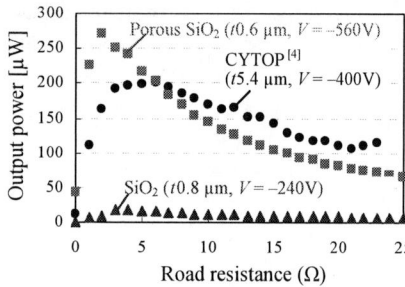

Fig. 11: Output power vs. load resistance at free oscillation at 20 Hz, 0.65 G.

Table 1: Performance comparison.

	This study	Y. Suzuki et al., Ref. [7]	T. Takahashi et al., Ref. [4] (Our previous study)
Frequency [Hz]	20	23	20
Acceleration [G]	0.65	0.4	0.65
Device area [mm^2]	165	118	165
Voltage of electret [V]	-560	-600	-400
Output [μW]	270	4	200

area of dielectric plate. A voltage across 100 kΩ was measured to estimate the output power. The oscillation amplitude and frequency were 0.4 mm and 20 Hz (acceleration: 0.65 G). Figure 11 shows the relationship between output power and load resistance. As indicated in the figure, the best performance of 270 μW was achieved when nano-porous SiO_2 (t_{porous}=0.6 μm) on thick base SiO_2 (t_{base}=0.8 μm) was used as the electret. The output power using nano-porous SiO_2 electret is not less than the reported ones [4, 7], as shown in Table 1.

CONCLUSION

The surface charge density of nano-porous SiO_2 electret and its thermal resistance was characterized in this study. The nano-porous SiO_2 was fabricated by anodic etching and thermal oxidation of a low resistivity silicon substrate. Size of void in the fabricated nano-porous SiO_2 is in the order of several tens nanometer. The thickness of base SiO_2 layer underneath the nano-porous SiO_2 layer influences time stability of surface charge density, because the base SiO_2 probably prevent implanted charge from dissipating through the substrate. When thickness of the base SiO_2 layer is 0.8 μm, time stability and thermal stability of the nano-porous SiO_2 on base SiO_2 as an electret is best compared with the nonporous SiO_2 and the CYTOP film. Finally, it is proven that output power generated by vibration energy harvesting using the nano-porous SiO_2 electret is also larger than that using the normal SiO_2 or polymer electret.

ACKNOWLEDGEMENTS

This work was supported in part by a grant of "Strategic Research Foundation Grant-aided Project for Private Universities": Matching Fund Subsidy MEXT (Ministry of Education, Culture, Sport, Science, and Technology, Japan),

2010-2014 (S1001048). This work was supported in part by JSPS (Japan Society for the Promotion of Science) KAKENHI (22310083).

REFERENCES

[1] S. P. Beeby, M. J. Tudor, and N. M. White, "Energy harvesting vibration sources for microsystems applications," *Meas. Sci. Technol.*, vol. 17, 175-195, 2006.

[2] K. Matsumoto, K. Saruwatari, Y. Suzuki, "Vibration-powered Battery-less Sensor Node Using MEMS Electret Generator," in *Digest Tech. PowerMEMS'11*, 134-137, 2011.

[3] T. Tsutsumino, Y. Suzuki, N. Kasagi, and Y. Sakane, "Seismic Power Generator Using High-Performance Polymer Electret," in *Digest Tech. IEEE Internat. Conf. MEMS'06*, Istanbul, 98-101, 2006.

[4] T. Takahashi, M. Suzuki, T. Nishida, Y. Yoshikawa, and S. Aoyagi, "Milliwatt Order Vertical Vibratory Energy Harvesting Using Electret and Ferroelectric –Discharge does not Occur with Small Gap and Only One Wiring is Required–", in *Digest Tech. MEMS'12*, 1265-1268, 2012.

[5] T. Takahashi, M. Suzuki, T. Nishida, Y. Yoshikawa, and S. Aoyagi, "Application of paraelectric to a miniature capacitive energy harvester realizing several tens micro watt –relationship between polarization hysteresis and output power–", in *Digest Tech. MEMS'13*, 877-880 2013.

[6] Y. Sakane, Y. Suzuki, and N. Kasagi, "The development of a high-performance perfluorinated polymer electret and its application to micro power generation", *J. Micromech. Microeng.* vol. 18, 104011 (6pp), 2008.

[7] Y. Suzuki, "Development of a MEMS Energy Harvester with High-Perfomance Polymer Electrets," in *Proc. of PowerMEMS'10*, Leuven, 47-52, 2011.

[8] K. Kashiwagi, K. Okano, N. Tanabe, Y. Sera, T. Miyajima, Y. Morizawa, Y. Sakane, Y. Hamatani, F. Nonaka, A. Asakawa, and Y. Suzuki, "Nano-cluster-enhanced High-performance Perfluoro-polymer Electrets," in *Proc. IEEE ISE14*, Montpellier, 71-72, 2011.

[9] T. Genda, S. Tanaka, M. Esashi, "Charging method of micropatterned electrets by contact", *Japanese Journal of Applied Physics*, vol. 44, no. 7A, 5062-5067, 2005.

[10] Y. Naruse, N. Matsubara, K. Mabuchi, M. Izumi, and S. Suzuki, "Electrostatic micro power generation from low-frequency vibration such as human motion", *J. Micromech. Microeng.* vol. 19, 094002 (5pp), 2009.

[11] T. Fujita, K. Fujii, T. Onishi, K. Kanda, K. Higuchi, K. Maenaka, "Evaluation of the Electret Based Energy Harvester by Using Multipurpose Data Logging Device", in *Digest Tech. PowerMEMS 2011*, 130-133, 2011.

[12] N. Koshida and H. Koyama, "Visible electroluminescence form porous silicon," *Appl. Phys. Lett.*, Vol. 60 347-349, 1992.

CONTACT

* M. Suzuki, tel: +81-6-6368-1115, fax: +81-6-6388-8785, e-mail: m.suzuki@kansai-u.ac.jp

NANOFLUIDIC REVERSE ELECTRODIALYSIS PLATFORM USING CONTROLLED ASSEMBLY OF NANOPARTICLES FOR HIGH POWER ENERGY GENERATION

Eunpyo Choi, Kilsung Kwon, Daejoong Kim, and Jungyul Park
Sogang University, Seoul, KOREA

ABSTRACT

This paper presents a novel microplatform for high power energy generation based on reverse electrodialysis. The effective cation-selective membrane for power generations is realized between two microfluidic channels using geometrically controlled *in situ* self-assembled nanoparticles with cost-effective and simple way. Nano-interstices between the assembled nanoparticles have a role as the collective three-dimensional nanochannel networks and they can increase the generated power, significantly. The proposed system can contributes to supply power sources to miniaturized devices and be also used to studies and investigate nanoscale electrokinetics by changing sizes, materials, and shape of the assembled nanoparticles, or geometry control of microchannel.

INTRODUCTION

Recently, a few types of ion-selective membrane in microsystems have been developed for power generation by reverse electrodialysis using silicon based nanochannel [1] and ion track-etched single nanopore [2]. However, these previously reported techniques need heavy fabrication process for nanochannels and have no material selectivity. The uniqueness of our proposed system is that the nanoporous ion selective membrane can be constructed with the desired position and shape in microchannel by simple and cost effective process using microdroplet control. Not only the benefit from efficient fabrication process, but also the advantage of the proposed system is that intrinsically high power generation can be achieved. Because nano-interstices between the assembled nanoparticles are the collective three-dimensional nanochannel networks compared to the typical nanochannel corresponding to thin one-dimensional pathway. The collective nanochannel networks maintain the transference number and the open circuit voltage and augment the ion current through the nanoporous membrane significantly, so that the harvesting power can increase. Another advantage in our system is that electrical performance such as maximum powers (P_{max}) and the energy conversion efficiency (η_{max}) can be adjusted quantitatively and improved by changing the geometry of microchannel and proper selection of size and materials of the assembled nanoparticles.

Figure 1 shows the proposed microplatform and working principle. The parallel straight microchannels are interconnected with the nanoporous membranes formed from self-assembled nanoparticles. Nano-interstices in the closed-packed nanoparticles with homogeneous sizes and negative surfae charges create cation-selective membranes. Since electric double layers are overlapped in these nano-interstices, cations are preferentially transported. Cations diffuse from high to low concentration and this asymmetric ion-transport induces the electrochemical redox reaction on electrodes surface to satisfy electro-neutrality. Finally, electrons are transferred along the external circuit and power is generated.

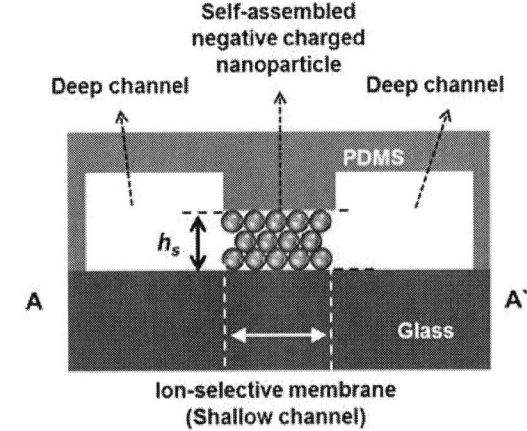

\<Cross sectional view\>

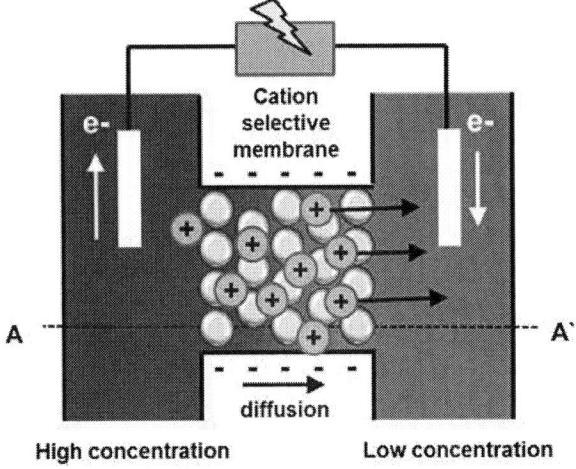

\<Working principle\>

Figure 1: Proposed microplatform and working principle

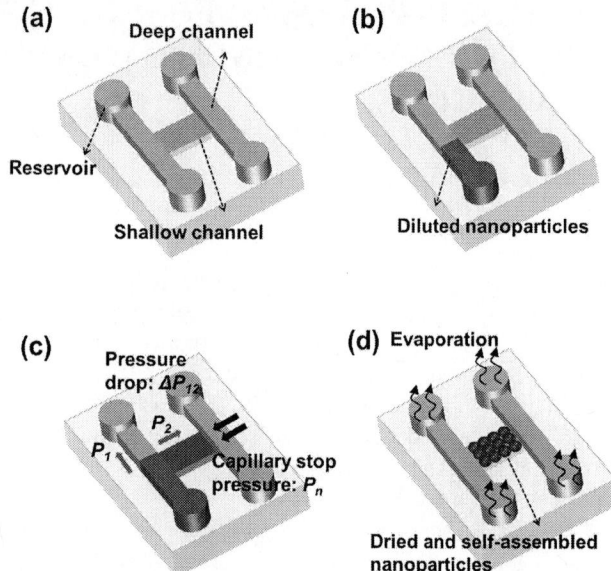

Figure 2: Fabrication process for the in situ formation of nanoporous membranes based on the self-assembly of nanoparticles within microchannel

MATERIALS AND METHODS

Materials and Experimental Setup

We used the diameter 100, 200, 300, and 700 nm of silica nanospheres and 100 nm of carboxylate polystyrene nanospheres. Before the initial use of nanoporous membranes, we fixed the concentration of KCl solution at 0.1 mM and gently filled this solution at the all channels. Subsequently, KCl solutions with different ionic strength (0.1, 5, 10, 50, and 100 mM) were sequentially introduced into the high concentration side of the deep channel. The picometer/voltage source (6487, Keithley Instrument) was used as the electronic load and connected with the device by Ag/AgCl electrodes.

Fabrication Process

Figure 2 describes the fabrication process: (a) The PDMS device with shallow and deep channels were molded from an SU-8 master with two different heights. (b) Diluted nanoparticles were introduced into the deep channels by capillary pressure. (c) The diluted nanoparticles try to drag the solution in the deep channel into the shallow-channel due to pressure differences [3]. The solution could not move over the interface during the crystallization because of effects in the capillary stop valve. (d) Nanoparticles were self-assembled within the shallow channel by evaporation.

RESULTS AND DISCUSSIONS

Power generation depending on nanoparticle size

Our experiments for the current-voltage (I-V) measurements were carried out using this fabricated platform. The open circuit voltage (E_{diff}) increased as the concentration

gradient increased at first (figure 3(a)). However, since the overlapped EDLs collapsed rapidly at higher concentration, the E_{diff} dropped eventually. Also, the membrane using the large size of nanoparticle created lower E_{diff} compared to using smaller size. The zero-voltage current (i_0) increased as the concentration difference increased due to the enhancement of ion diffusion (figure 3(b)). In addition, smaller size showed the higher i_0 because when the diameter decreased, the concentration of counter-ions in nanopore increased and then higher conductivity occurs. Figure 4(a) depicts P_{max} turned higher as the concentration gradient increased or the diameter decreased. η_{max} increased with the decrease of concentration or the smaller diameter (figure 4(b)).

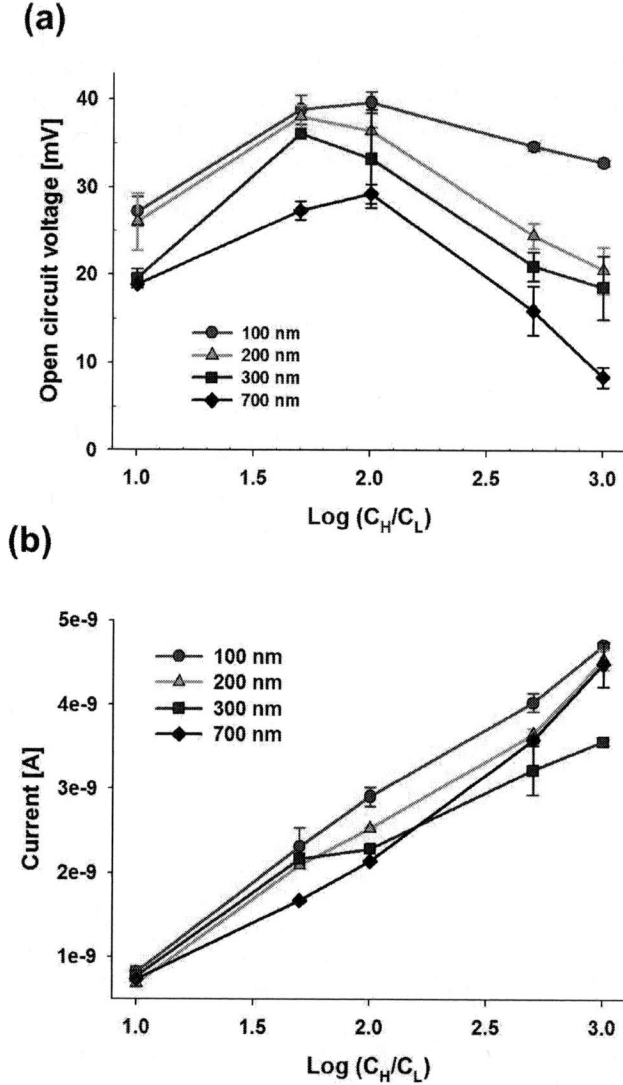

Figure 3: Dependence of the (a) open circuit voltage and (b) zero-voltage current on the concentration gradient and diameter of silica nanoparticles (S-NP:100, 200, 300, 700 nm)

(a)

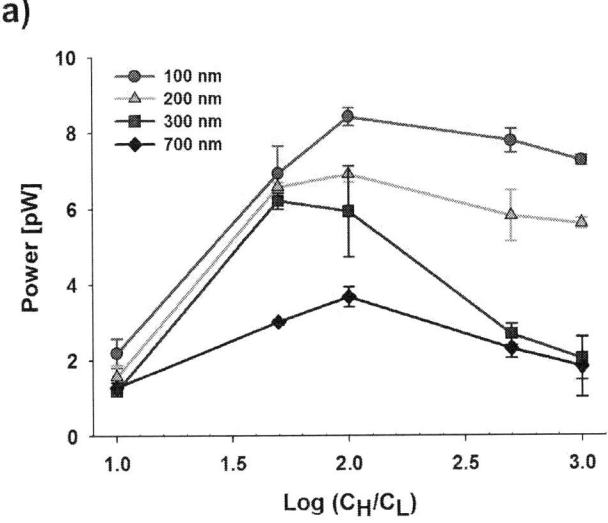

(b)

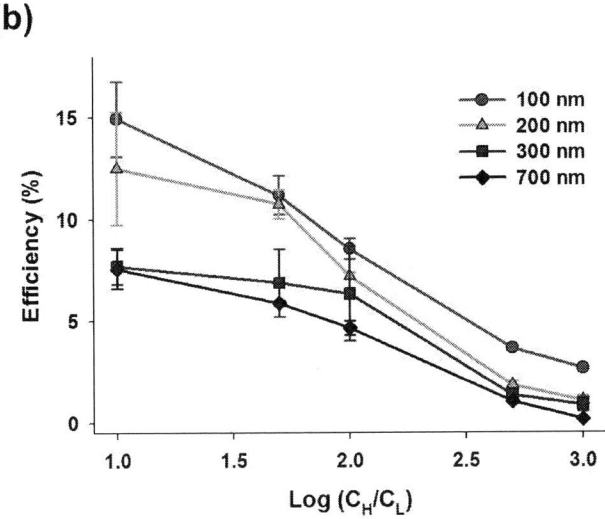

Figure 4: Dependence of the (a) maximum power, and (b) energy conversion efficiency on the concentration gradient and diameter of silica nanoparticles (S-NP:100, 200, 300, 700 nm)

Power generation depending on height of shallow channel

Power generation and efficiency can be further improved by changing height of shallow channel (h_s) and using the surface charge modification. As shown in figure 5(a), the values of E_{diff} were similar according to the variation of h_s, but i_0 became higher as h_s increased due to the achievement of high conductance (figure 5(b)). It can be easily expected, because higher h_s means that more nanochannel networks established and ion current increase more.

Naturally, P_{max} increased as h_s increased due to the growth of i_0, but η_{max} was obtained similarly as shown in figure 6(a) and (b), respectively.

Power generation depending on materials

We could observe the significant increase of P_{max} by changing the materials from silica nanoparticles (S-NP) to carboxylate polystyrene nanoparticle (C-PSNP). C-PSNP has higher E_{diff} because of the higher surface charge (figure 5(a)) [4]. However, i_o of C-PSNP tends to slightly lower than S-NP due to hydrophobic surface properties of C-PSNP (figure 5(b)).

Because of the slip length with water depletion layer [5], the lower efficient cross sectional area exists in the hydrophobic surface. Nevertheless, since the C-PSNP has very high E_{diff}, we could obtain higher P_{max} (42.38 pW) and η_{max} (27.03 %) as shown in figure 6(a) and (b), respectively.

(a)

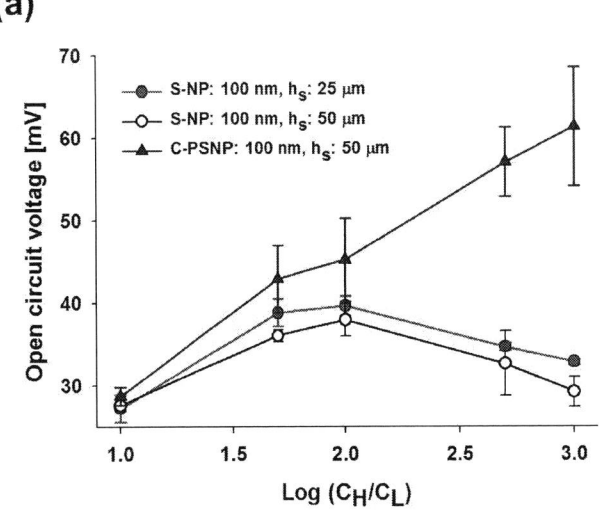

(b)

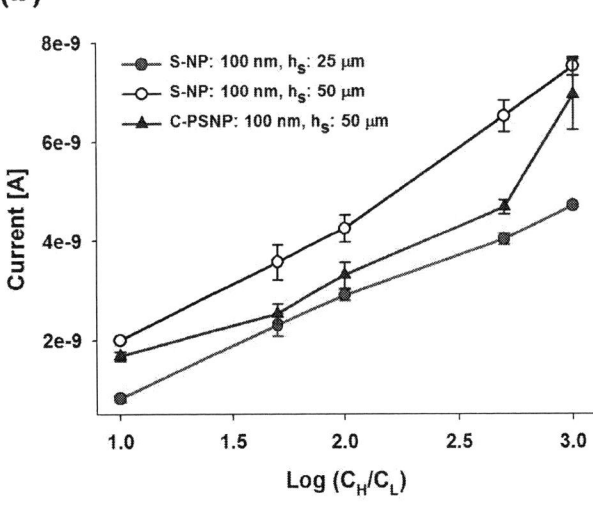

Figure 5: Dependence of the (a) open circuit voltage and (b) zero-voltage current on the concentration gradient, height of the nanoporous membrane (h_s: 25 μm (●) and 50 μm (○)), and different types of nanoparticles (carboxylate polystyrene nanoparticle, C-PSNP, ▲)

(a)

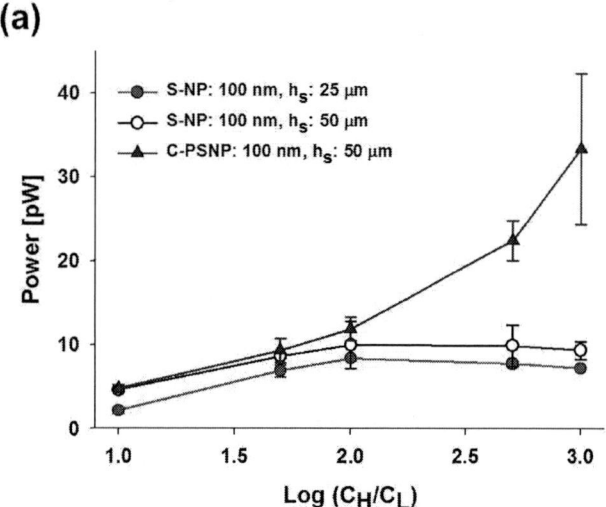

(b)

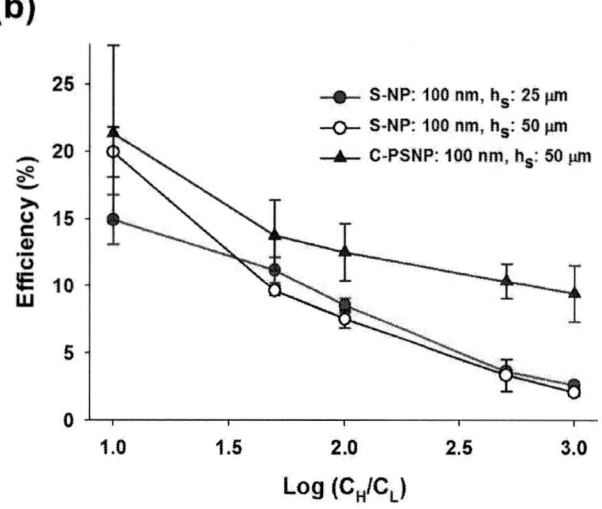

Figure 6: Dependence of the (a) maximum power and (b) energy conversion efficiency on the concentration gradient, height of the nanoporous membrane (h_s: 25 μm (●) and 50 μm (○)), and different types of nanoparticles (C-PSNP, ▲)

CONCLUSION

In this study, we demonstrated a nanofluidic reverse electrodialysis platform based on the geometrically controlled assembly of nanoparticles, which can efficiently convert the salinity-gradient energy into electricity with high power. Through the collective nanochannel networks among the nanoparticles, much higher power in microfluidics was generated (42.38 pW), compared to the silicon based nanochannel (~1 pW) [1] or ion track-etched nanopore (~2 6pW) [2]. In order to realize the efficient power generation system, we systemically investigated the dependence of power, open circuit voltage, current, and energy conversion efficiency on the size and materials of nanoparticles and microchannel dimension. The high power could be obtained when using small size and high surface charged nanoparticles that are close-packed in large area microchannel.

The proposed platform is promising for powering tiny microdevices, such as microelectromechanical systems, biomedical implant microdevice, portable personal electronics, and micro/nanorobots without the limitation of batteries or external power supplies. Moreover, it can provide a clear understanding of the ion transport behavior and the power generation mechanism, which is still lacking.

REFERENCES

[1] D.-K. Kim, C. Duan, Y.-F. Chen, and A. Majumdar, "Power generation from concentration gradient by reverse electrodialysis in ion-selective nanochannels," *Microfluidics and Nanofluidics*, vol. 9, pp. 1215-1224, 2010.

[2] W. Guo, L. Cao, J. Xia, F.-Q. Nie, W. Ma, J. Xue, Y. Song, D. Zhu, Y. Wang, and L. Jiang, "Energy Harvesting with Single-Ion-Selective Nanopores: A Concentration-Gradient-Driven Nanofluidic Power Source," *Advanced Functional Materials*, vol. 20, pp. 1339-1344, 2010

[3] S. Chung, H. Yun, and R. D. Kamm, "Nanointerstice-Driven Microflow," *Small*, vol. 5, 609-613, 2009

[4] R. Yongsunthon and S. K. Lower, "Force spectroscopy of bonds that form between a Staphylococcus bacterium and silica or polystyrene substrates," *Journal of Electron Spectroscopy and Related Phenomena*, vol. 150, pp. 228-234, 2006

[5] C. Sendner, D. Horinek, L. Bocquet, and R. R. Netz, "Interfacial Water at Hydrophobic and Hydrophilic Surfaces: Slip, Viscosity, and Diffusion," *Langmuir*, vol. 25, pp. 10768-10781, 2009

ACKNOWLEDGEMENT

This work was supported by Basic Science Research Program (2012R1A1A2003473) and Pioneer Research Center Program "Bacteriobot" (2012-0001032) through the National Research Foundation of Korea funded by the Ministry of Education, Science and Technology.

CONTACT

*Jungul Park, tel: +82-2-705-8642; sortpark@sogang.ac.kr or *Daejoong Kim, tel: +82-2-705-8644; daejoong@sogang.ac.kr

SPRINGLESS CUBIC HARVESTER FOR CONVERTING THREE DIMENSIONAL VIBRATION ENERGY

Mengdi Han[1], Wen Liu[1,2], Bo Meng[1], Xiao-Sheng Zhang[1], Xuming Sun[1], and Haixia Zhang[1]

[1]Institute of Microelectronics, Peking University, Beijing, China
[2]Peking University Shenzhen Graduate School, Shenzhen, China

ABSTRACT

This paper reports the design, fabrication and measurement of a springless cubic energy harvester. Based on the flexible printed circuits, coils are fabricated onto both sides of the polyimide substrate and then folded to a cubic box, meanwhile sealed a free magnet inside the box. As the coils are folded on three different directions, vibration in all dimensions can cause the change of magnetic flux and the energy is effectively harvested. Output performance of the device is both theoretically and experimentally investigated. Benefit from the springless structure, maximum output is achieved at low frequencies with a wide bandwidth. Moreover, the device can be placed on a backpack or wrist to harvest vibration energy from daily life.

INTRODUCTION

With the rapid development of microelectronic technology, devices such as portable electronics, wireless sensors have been applied in various fields. The power consumption of these devices has been lowered to the level of milli or micro watts thanks to the development of CMOS (complementary metal oxide semiconductor) low power consumption technique. Therefore, harvesting vibrational energy from the ambient environment is becoming a promising way to power microelectronic devices [1]. Based on different transduction mechanisms, various electromagnetic [2, 3], piezoelectric [4, 5], electrostatic [6, 7], and hybrid [8, 9] energy harvesters have been fabricated and investigated. Compared with other energy harvesters, electromagnetic energy harvester has been widely studied due to its low resistance and high output current. Moreover, the processing circuit of electromagnetic energy harvesters is relatively simple and efficient, which makes electromagnetic energy harvesters a promising substitute for traditional batteries.

However, previous electromagnetic harvesters have some drawbacks which limit the application field. Firstly, most of the electromagnetic harvesters are based on spring/mass configuration [10]. With this structure, only vibrational energy in a specific direction can be effectively harvested. Besides, fabrication of the spring/mass structure is complex and expensive. Etching process is usually required to fabricate such a structure. Finally, the substrates of previous electromagnetic energy harvesters are basically silicon or steel, which is not flexible and biocompatible.

Therefore, in this work, we fabricate an electromagnetic energy harvester on flexible polyimide substrate. Fabrication process of the device is based on the flexible printed circuits

(FPC). To effectively harvest vibrational energy in all direction, we specially designed a cubic structure with a freely located magnet. Theoretical calculation reveals that the total output power of this electromagnetic energy harvester remains unchanged in different vibration directions. Experimental results show that the device can operate at low frequencies with a wide bandwidth. Besides, three dimensional energy harvesting is obtained by this device and applications of harvesting energy in human body movement are demonstrated.

DESIGN OF THE DEVICE

For three-dimensional vibration energy harvesting, this electromagnetic energy harvester is designed to be cubic-shape. Figure 1(a) elucidates the structure of the cubic harvester, which mainly consists of the polyimide substrate, copper coils and a NdFeB permanent magnet. The substrate and copper coils are folded from a planner structure shown in Figure 1(b), thus forming a sealed cube. The permanent magnet is freely located inside the cube without the support of springs. This not only eliminates the sophisticated spring structures but also makes the device able to harvest vibrational energy from all direction, since the magnet is totally free inside the cubic box.

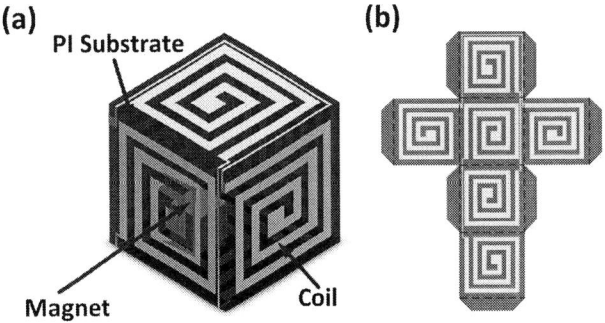

Figure 1: (a) 3D structure and (b) planar layout of the cubic electromagnetic energy harvester.

Compared with previous studies [11], this new device is improved in the following aspects. Firstly, the magnet is freely located rather than attached to a cantilever. Secondly, the planner coils are designed with proper spiral direction. In our design, the top and bottom copper coils have opposite helical direction and are in series connection. The same structure is applied to the left-right coils and front-back coils. In this way, number of the electrodes can be halved while turns of the coil will be doubled, thus enhancing the final output. Finally, copper coils are fabricated on both sides of

978-1-4799-3510-9/14 $31.00 © 2014 IEEE

the substrate, which is also beneficial to the output. Detailed geometry parameters of this electromagnetic energy harvester are listed in Table 1.

Table 1: Geometry parameters of the cubic energy harvester.

Components	Length	Width	Thickness/Height
PI substrate	10 mm	10 mm	0.025 mm
Copper coil	-	0.2 mm	0.018 mm ×2
NdFeB magnet	5 mm	5 mm	5 mm
Total device	**10 mm**	**10 mm**	**10 mm**

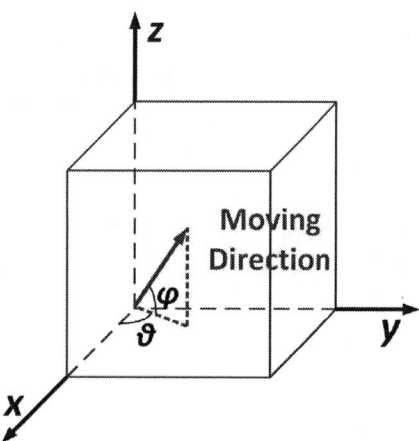

Figure 2: Theoretical calculation demonstrating that the total output power is proportional to the velocity.

To prove the ability of three-dimensional energy harvesting, theoretical calculation is conducted. First, we imagine the magnet moving in an arbitrary direction, with the velocity of v. The angle between the moving direction and XY plane is defined as φ. The angle between moving direction projection in XY plane and X-axis is defined as θ. Then, the velocity components along X-axis, Y-axis, and Z-axis can be expressed as:

$$v_x = v \cos \varphi \cos \theta, \qquad (1)$$
$$v_y = v \cos \varphi \sin \theta, \qquad (2)$$
$$v_z = v \sin \varphi. \qquad (3)$$

The induced voltage in each coil can be obtained from Faraday's law:

$$E_i(t) = -2 \times N \frac{d\phi}{dt} = -2 \times NA \frac{dB}{dz} v_i(t), \qquad (4)$$

where $E_i(t)$ is the induced voltage, N is the turns of each coil, $\frac{dB}{dz}$ is the changing rate of magnetic flux density, $v_i(t)$ ($i = x$, y, z) is the velocity along each axis. The factor 2 is caused by the series connection of two coils. By combining the output together, total output power generated by this cubic energy harvester can be expressed as:

$$P_{total}(t) = \sum_{i=x,y,z} \frac{E_i^2(t)}{R_i}, \qquad (5)$$

where R_i ($i = x$, y, z) is the inner resistance of the two series connected coils. Assuming that $\frac{dB}{dx} = \frac{dB}{dy} = \frac{dB}{dz}$ and $R_x = R_y = R_z$, we can obtain that $P_{total}(t) \propto v(t)^2$. The result shows that the total output power of this cubic energy harvester is only related to the magnet moving velocity, regardless of the moving direction.

FABRICATION PROCESS

Fabrication of this electromagnetic energy harvester is carried out using FPC method, which is robust, low-cost and mass-productive [12]. Detailed fabrication process is shown in Figure 2. First of all, a 25 μm polyimide substrate with copper foil on both sides is prepared (Figure 2a). Then, photoresist is patterned on the copper foil as a mask to etch the copper foil (Figure 2b). As the next step, polyimide substrate is etched to form a through-hole (Figure 2c). Later, catalyst activation is performed and copper is electroplated on both sides and the through-hole (Figure 2d). After that, photoresist is patterned on both sides and the electroplated copper is wet etched using FeCl₃ solution (Figure 2e). Finally, by removing the photoresist, the polyimide-based double-side copper coils if fabricated (Figure 2f). To make the electromagnetic energy harvester, the polyimide-based double-side copper coils is then folded to form a cube, in which the permanent NdFeB magnet is placed.

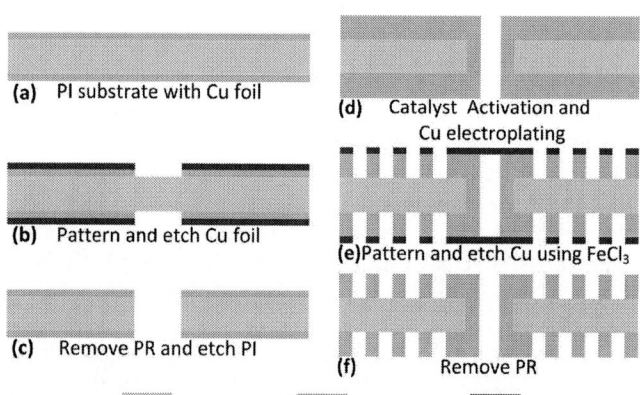

Figure 3: Fabrication process of the polyimide-based double-side copper coils.

From the above FPC-based fabrication process, we can conclude that there is no need to fabricate complex cantilever or spring structure. After fabricating the double-side planar coils, electromagnetic energy harvester can be obtained through folding and packaging. This greatly simplifies the fabrication process, thus improving the production efficiency. Moreover, this fabrication process is compatible with other

FPC fabrication process, which brings hope for further integration with the circuit.

MEASUREMENT AND DISCUSSION

Experimental measurement of the cubic energy harvester is conducted using a vibration system shown in Figure 4a. The dynamic signal analyzer (Agilent 35670A) is used to generate sinusoidal signal with tunable frequency. The signal is then amplified by a power amplifier to drive a vibrator. The cubic energy harvester is fixed on to the vibrator and the induced voltage is sent back to the dynamic signal analyzer.

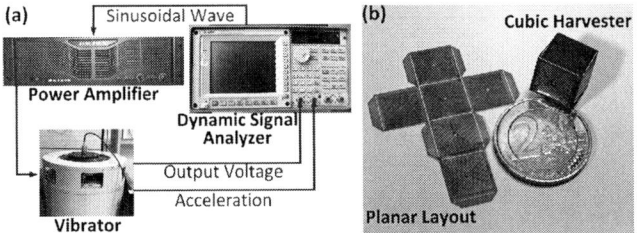

Figure 4: (a) Experimental setup. (b) Photos of the planar layout and the fabricated cubic harvester.

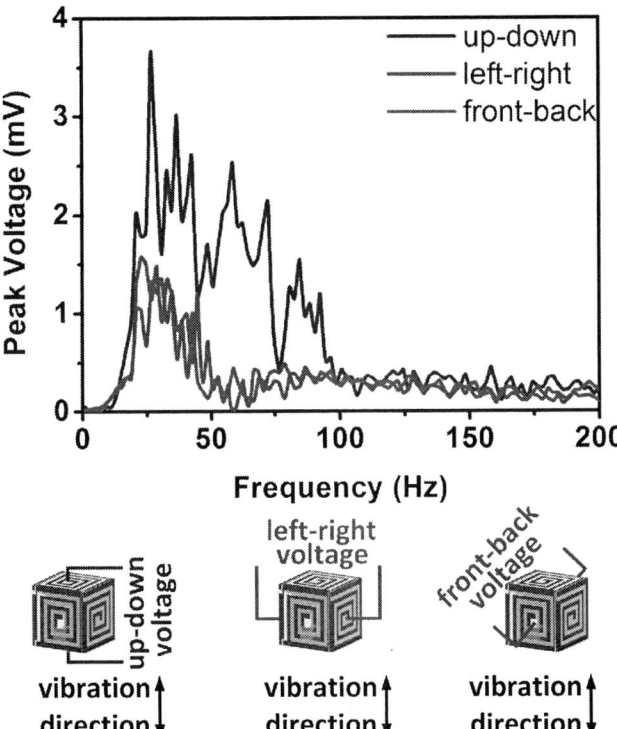

Figure 5: Output peak voltages versus vibration frequency at the acceleration of 0.5 g. Output voltages at three directions are illustrated.

Using the vibration system, output performance of this cubic energy harvester is investigated. First, a vibration is given in up-down direction, and the induced voltage in the up-down coils is recorded, as shown in the blue line in Figure 5. With the acceleration of 0.5 g, the maximum induced

voltage in up-down surface reaches 3.82 mV at 26.87 Hz and shows an ultra-wide working bandwidth between 20 and 100 Hz. This wide band behavior is caused by the springless vibration structure. Moreover, the vibration in up-down direction will inevitably make the magnet move in left-right direction and front-back direction, because the magnet is completely free inside the cube. The induced voltage and frequency response of the surfaces perpendicular to the vibration direction are also measured. As shown in the red and green lines in Figure 4, the maximum induced voltage is about 1.5 mV at resonance. Compared with the output voltage in up-down surface, the maximum voltage is lowered and the bandwidth is narrowed. By adjusting the vibration direction, the output voltage at up-down, left-right, and front-back surfaces will change according to the velocity components in each direction. Thus, three-dimensional vibration energy harvesting is achieved.

Output performance of this cubic energy harvester under different external load resistance is also investigated. As shown in Figure 6a, with the load resistance varying from 0 to 10 Ω, the induced voltage raises monotonously and tends to saturate. On the contrary, the current in the coil reduces as the load resistance increases. For the output power, maximum value of 0.75 µW is achieved with a 2.5 Ω optimal resistance, as shown in Figure 6b.

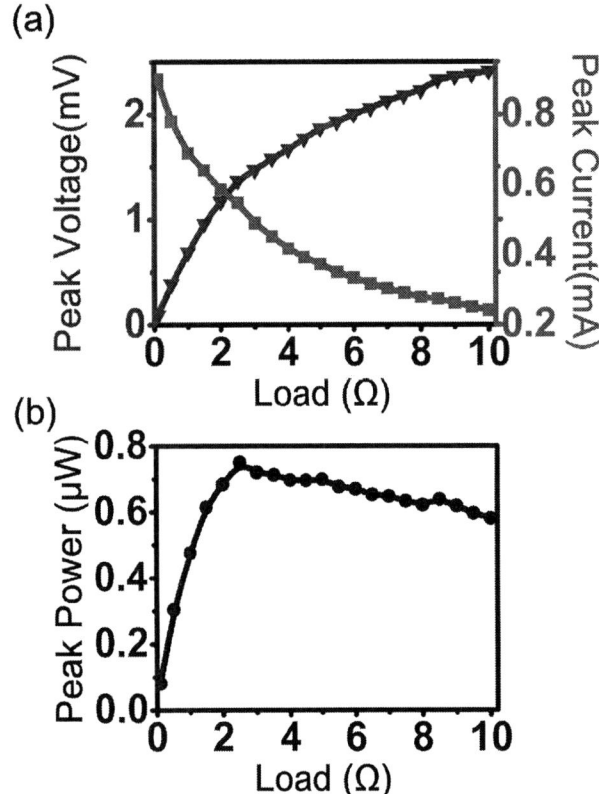

Figure 6: (a) Peak output voltage (left axis) and current (right axis) as a function of the external load resistance. (b) Peak output power as a function of the external load resistance.

Benefit from its cubic structure, this energy harvester offers great opportunity to harvest vibration energy from ambient environment. As an example, the device can be placed on a backpack to harvest the vibration energy from daily walking. As shown in Figure 7a, a gentle jump can generate voltage as high as 13.97 mV. Moreover, by fixing the device on the wrist, energy can be harvested in the writing process, as illustrated in Figure 7b.

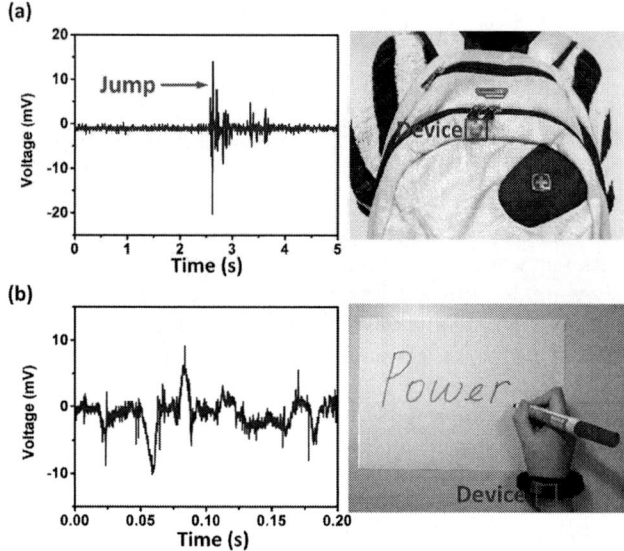

Figure 7: (a) Output voltage of the harvester placing at the backpack. (b) Output voltage of the harvester in the writing process.

CONCLUSION

In summary, we innovatively design a cubic energy harvester that can convert three-dimensional vibration energy into electricity. With a magnet freely located inside a cubic box, a springless structure is proposed. Trough, theoretical calculation, we obtain that the total output power of this cubic energy harvester is only related to the magnitude of the magnet's velocity, regardless of the moving direction. Fabrication of the device is based on FPC process, which is simple, reliable, and mass-productive. Experimental results indicate that this device can effective harvest vibration energy from all directions, with a relatively low resonant frequency and wide bandwidth. Additionally, by fixing this device onto a backpack and human wrist, vibrational energy harvesting from daily life is demonstrated.

ACKNOWLEDGEMENTS

This work is supported by the National Natural Science Foundation of China (Grant No. 91023045 and No. 61176103), Development Program ("863" Program) of China (Grant No. 2013AA041102), National Ph.D. Foundation Project (20110001110103).

REFERENCES

[1] S. Roundy, P. K. Wright, J. Rabaey, "A Study of Low Level Vibrations as a Power Source for Wireless Sensor Nodes", *Comput. Commun.*, vol. 26, pp. 1131–1144, 2003.

[2] P. Wang, K. Tanaka, S. Sugiyama, X. Dai, X. Zhao, J. Liu, "A Micro Electromagnetic Low Level Vibration Energy Harvester Based on MEMS Technology", *Microsyst. Technol.*, vol. 15, pp. 941–951, 2009.

[3] P. Glynne-Jones, M. J. Tudor, S. P. Beeby, N. M. White, "An Electromagnetic Vibration-Powered Generator for Intelligent Sensor Systems", *Sens. Actuators A, Phys.*, vol. 110, pp. 344–349, 2004.

[4] R. Elfrink, T.M. Kamel, M. Goedbloed, S. Matova, D. Hohlfeld, Y. van Andel, R. van Schaijk, "Vibration Energy Harvesting with Aluminum Nitride-Based Piezoelectric Devices", *J. Micromech. Microeng.*, vol. 19, pp. 094005-1–094005-8, 2009.

[5] Z. L. Wang, J. Song, "Piezoelectric Nanogenerators Based on Zinc Oxide Nanowire Arrays", *Science*, vol. 312, pp. 242–246, 2006.

[6] P. D. Mitcheson, P. Miao, B. H. Stark, E. M Yeatman, A. S. Holmes, T. C. Green, "MEMS Electrostatic Micropower Generator for Low Frequency Operation", *Sens. Actuators A, Phys.*, vol. 115, pp. 523–529, 2004.

[7] Y. Suzuki, D. Miki, M. Edamoto, M. Honzumi, "A MEMS electret generator with electrostatic levitation for vibration-driven energy-harvesting applications", *J. Micromech. Microeng.*, vol. 20, pp. 104002-1–104002 -8, 2010.

[8] M. D. Han, X. S. Zhang, W. Liu, X. M Sun, X. H. Peng, H. X. Zhang, "Low-Frequency Wide-Band Hybrid Energy Harvester Based on Piezoelectric and Triboelectric Mechanism", *Sci. China Tech. Sci.*, vol. 56, pp. 1835–1841, 2013.

[9] B. Yang, C. Lee, W. L. Kee, S. P. Lim, "A Hybrid Energy Harvester Based on Piezoelectric and Electromagnetic Mechanisms", *J. Micro/Nanolith. MEMS MOEMS*, vol. 9, pp. 023002-1−023002-10, 2010.

[10] Ö. Zorlu, S. Türkyılmaz, A. Muhtaroğlu, H. Külah, "An Electromagnetic Energy Harvester for Low Frequency and Low-g Vibrations with a Modified Frequency Up Conversion Method", in *Digest Tech. Papers MEMS 2013 Conference*, Taipei, January 20-24, 2003, pp. 805–808.

[11] S. H. Zhang, Q. Yuan, M. D. Han, H. X. Zhang, "Fabrication and Simulation of a Novel 3D Flexible Electromagnetic Energy Generator", in *Digest Tech. Papers PowerMEMS 2012 Workshop*, 2012, Atlanta, December 2-5, 2012, pp. 315–318.

[12] R. E. Taylor, C. M. Boyce, M. C. Boyce, B. L. Pruitt, "Planar patterned stretchable electrode arrays based on flexible printed circuits", *J. Micromech. Microeng.*, vol. 23, pp. 105004-1–105004-7, 2013.

CONTACT

*H.X. Zhang, Tel: +86-10-62766570; zhang-alice@pku.edu.cn

ON THE OPTIMIZATION AND PERFORMANCES OF A COMPACT PIEZOELECTRIC IMPACT MEMS ENERGY HARVESTER

Pattanaphong Janphuang, Robert Lockhart, Danick Briand, and Nico F. de Rooij

École Polytechnique Fédérale de Lausanne (EPFL), Institute of Microengineering (IMT)
The Sensors, Actuators and Microsystems Laboratory (SAMLAB), Neuchâtel, Switzerland

ABSTRACT

This paper presents the development of a compact energy harvesting configuration to convert low frequency, mechanical oscillations into usable electrical energy using AFM-like MEMS piezoelectric cantilevers coupled to a rotating gear. In this approach, one or several piezoelectric harvesters can be positioned above a rotating gear driven by an oscillating mass. In order to analyze the motion and the electrical power output from the harvester, analytical and finite element models have been developed. The harvester, with an active device volume of 3.5 mm^3 ($3{\times}5{\times}0.23$ mm^3), is able to produce an average output power of 12 µW measured across an optimal resistive load of 4.7 kΩ at a rotational speed of 19 rps, demonstrating the potential of the compact MEMS piezoelectric micro-power generator.

INTRODUCTION

Piezoelectric harvesters offer a promising solution for powering portable and implantable systems using ambient vibrations due to the ease of direct conversion from vibrational to electrical energy, achieving high power densities and straight-forward integration which is quite attractive for volume-limited applications. Piezoelectric harvesters are often implemented as inertial devices set to oscillate at their resonant frequencies; however, the typical vibration frequencies of large body movements are on the order of a few Hz (< 30 Hz) and can vary considerably over time. Thus, resonant type devices do not present a practical option. To operate resonant devices at such low frequencies, macro structures with bulky proof masses are required limiting their use for many applications.

A more promising solution for converting low frequency vibrations into usable electrical energy is impact-based energy harvesting, in which environmental motion is coupled to an inertial object through physical impacts. In recent years, impact based energy harvesting has received significant interest. Several configurations used to couple low frequency vibrations into piezoelectric transducers have been proposed and demonstrated [1]-[4]. Frequency up-conversion based on knocking or plucking has received considerable attention due to the improvement of the electromechanical coupling and efficiency of energy harvester at higher frequencies [2]. Using this mechanism, the piezoelectric harvester is struck or plucked by an inertial object exited at low frequency. The cantilever is then released to freely oscillate at its higher resonant frequency.

We previously presented a novel approach using an AFM-like MEMS piezoelectric cantilever to extract the energy from a rotating gear [4]. However, the presented configuration of the harvester assembly was bulky. Here, we present a truly compact design in which an AFM-like MEMS piezoelectric cantilever is placed directly above a rotating crown gear in order to keep the system as compact as possible. The tip of the cantilever is plucked by the vertical teeth as the gear rotates. Using the sloped vertical profile of the crown gear diminishes the amplitude of the force creating torsion in the beam. An analytical model has been developed to analyze the motion and the electrical power output from the harvester. The voltage generated is calculated by determining the stress on the deflected harvester in accordance with the motion of the gear. A more detailed description of the electromechanical behavior was studied through FEM simulations in ANSYS. Through modeling, simulation and experimental validation, we demonstrate that by modifying the gear tooth profile and the rotational speed of the gear, the performance of the harvester is significantly improved. In addition, the efficiency and longevity of the system are investigated.

CONCEPT AND MODELING

The configuration studied in this work is illustrated in Figure 1(a). A piezoelectric cantilever is placed directly above the rotating gear such that the tip at the end of the cantilever extends down between the vertically mounted gear teeth and is plucked as each tooth passes while the gear rotates in-plane.

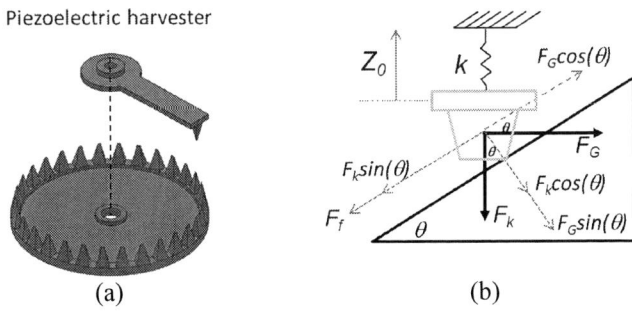

Figure 1: Schematic of (a) the proposed concept (b) the plucking action between the tip and the gear tooth.

Figure 1(b) illustrates the plucking mechanism. When a gear tooth comes in contact with the tip of the cantilever, the cantilever is deflected to a certain displacement (Z_0) following the angle (θ) of the sidewall of the gear tooth profile. The angles of the sidewalls of the cantilever tip and the gear tooth also direct the force in the vertical direction while reducing the wasted force directed along the horizontal direction which is lost or used to create torsional movement of the cantilever. The interactive force between the tip and the gear tooth is represented as F_G. According to its mechanical stiffness (k), a static force applied to the tip (F_k) is created. The frictional force (F_f) from the contact surfaces acts opposite to the direction of movement. The top of the gear tooth is the release point; the cantilever falls and is free to oscillate at its resonant frequency. The mechanical work done by the gear tooth is then given by

$$W_I = \int_0^{Z_0 \sin\theta} F_G x\, dx = \frac{kZ_0^2}{2\tan\theta}\left(\frac{\sin\theta + \mu_k \cos\theta}{\cos\theta - \mu_k \sin\theta}\right) \quad (1)$$

where μ_k is the coefficient of friction.

Analytical modeling

A unimorph piezoelectric cantilever in this work was structured by bonding a piezoelectric layer (PZT) to a support layer (Si) of the same length (L) and width (W). For simplification, the bonding layer is ignored. The bonding between the two layers is assumed to be ideal. The polarization direction of the piezoelectric layer is through the film thickness and perpendicular to the plane (d_{31}). The cross-section of the unimorph cantilever is shown in Figure 2. t_p and t_s are the thickness of the PZT and Si, respectively.

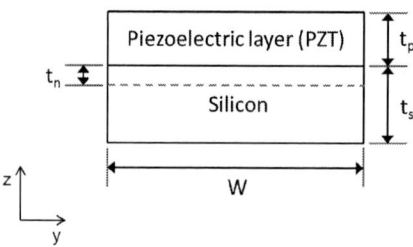

Figure 2: Cross-sectional view of the unimorph cantilever.

The lateral stress in the piezoelectric layer at a position (x, z) is given by [5]

$$\sigma = \frac{kZ_0 E_p}{(EI)_{composite}}(L-x)(z-t_n), \quad \text{for } 0 < x < L \quad (2)$$

where E_p is the Young's modulus of the piezoelectric, EI composite is the effective bending modulus of the composite beam, and the position of the neutral plane will be at distance t_n from the interface between PZT and silicon depending on the thickness and Young's modulus of each layer. The electric field E in the PZT layer at a position (x, z) is determined by

$$E(x,z) = \frac{d_{31}}{\varepsilon}\sigma = \frac{d_{31}kZ_0 E_p}{\varepsilon(EI)_{composite}}(L-x)(z-t_n) \quad (3)$$

The induced voltage generated from the PZT at x with respect to the thickness of PZT layer can be given as

$$V(x) = \int_0^{t_p} E(x,z)dz = \frac{d_{31}kZ_0 E_p}{\varepsilon(EI)_{composite}}(L-x)(\frac{t_p^2}{2}-t_n t_p) \quad (4)$$

where d_{31} is the piezoelectric strain coefficient and ε is dielectric constant of piezoelectric layer. The average voltage over the length L of the harvester can be define as

$$V_{ave} = \frac{1}{L}\int_0^L V(x)dx = \frac{d_{31}kZ_0 E_p L}{2\varepsilon(EI)_{composite}}(\frac{t_p^2}{2}-t_n t_p) \quad (5)$$

Deflection of the harvester as a function of the gear speed and gear tooth profile can be determined using the simple mass-spring system illustrated in Figure. 3 [6]. The force $f(t)$ from the gear tooth acts on a mass m (representing the cantilever) which is suspended by a spring constant with a stiffness k and a damping element c resulting in the displacement $z(t)$. The second order differential equation describing the motion of the cantilever can be written as

$$m\ddot{z} + c\dot{z} + kz = F \quad (6)$$

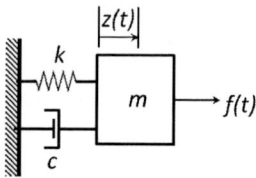

Figure 3: A generic model of impact (direct-force) generator.

In order to solve the differential equation for the dynamic displacement of the cantilever and therefore determine the output voltage from the harvester as a function of time, a fourth-order Runge-Kutta numerical model was developed in MATLAB.

FEM modeling

A transient 3D model was also developed in ANSYS 12.0 to investigate the electromechanical behavior of the designed harvester. The model predicts the voltage generated and the electrical energy produced by the harvester through a resistive load for a given input force. The analysis was based on SOLID5 (piezoelectric elements) and SOLID45 (structural elements) which include the adhesive and the silicon substrate. Using CIRCU94 elements, a resistive load (R) connects the top and bottom electrodes of the PZT layer. The boundary conditions consist of a displacement constraint on the clamped end of the cantilever. The bottom electrode of the PZT was set to ground while the top electrode was kept as a floating electrode.

The electromechanical behavior of the harvester and the plucking action between the tip and gear tooth was simulated in a transient analysis with the effects of viscous damping as illustrated in [2]. In the first step, a load is applied in which the cantilever is deflected to a certain displacement at a desired speed corresponding to the applied force. This was obtained using a ramped load solution in the model. The subsequent step is the release stage. This was simulated by removing the force from the tip of deflected cantilever allowing the cantilever to freely oscillate at its resonant frequency at the second stage of the transient analysis. The solution was simulated over several milliseconds in order to investigate the response of the harvester until the amplitude of vibration was diminished by damping. According to the simulation, the length and width of the harvester were first designed to be 5 mm and 3 mm allowing enough area to collect the generated charge while keeping a compact design. The thickness ratio of the PZT and silicon layers was designed to be close to 0.6 in order to maximize the output energy. Here, the thickness of PZT and silicon were 80 μm and 150 μm, respectively, as a compromise between flexibility and structural strength. Figure 4 gives an example of the simulated open circuit voltage (at 1 MΩ) from the designed harvester at different rotational speeds.

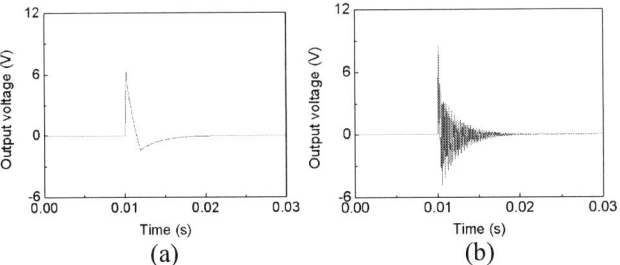

(a)   (b)

Figure 4: Simulated output voltage across resistive load of 1 MΩ as a function of time at the rotational speed of (a) 3 rps (b) 19 rps.

The average power and energy produced at the time t_N can be calculated using the following expression

$$P(t_N) = \frac{1}{t_N} \int_0^{t_N} \frac{V^2(t)}{R} dt \qquad (7)$$

$$E(t_N) = \sum_{i=0}^{N} P(t_i) \Delta t_i, \qquad \text{for } N > 0 \qquad (8)$$

Where t_i is the time at sub-step i.

RESULTS AND DISCUSSION

Piezoelectric harvesters were fabricated by bonding an 80 μm-thick PZT-5A layer onto a micromachined silicon AFM-like cantilever. More details about the fabrication process and experimental setup are illustrated in [4].

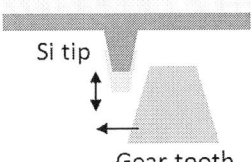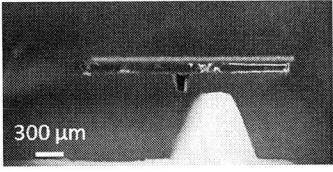

Figure 5: The harvester mounted on the X, Y and Z stage and positioned above the gear.

The power generation of the impact-type piezoelectric harvesters was observed by varying the tip insertion depth into a crown gear and adjusting the rotational speed of the

gear as shown in Figure 5. Figure 6 presents an example of the generated voltage and energy transferred from a harvester to an optimal load resistance. Post-excitation oscillations of the harvester were not observed at slow rotational speeds (<3 rps) - the tip of the cantilever simply follows the contour of the gear tooth as predicted by simulations (Figure 6(a)). Free oscillations of the harvester after plucking, however, are clearly observed for rotational speeds at 19 rps (Figure 6(b)).

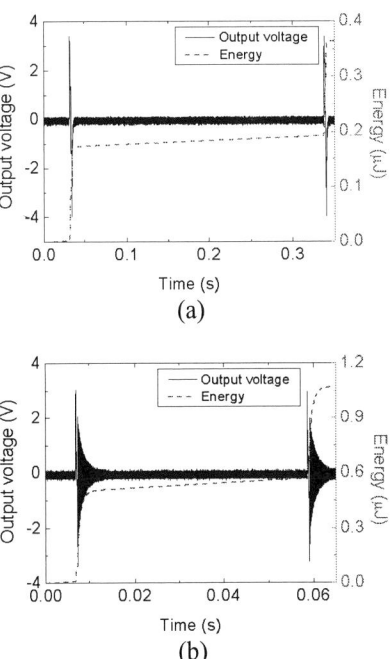

(a)

(b)

Figure 6: (a) Output voltage and energy dissipated in the optimal resistor at rotational speed of (a) 3 rps. (b) 19 rps.

At rotational speeds of 3 rps, most of the energy is produced during the deflection of the cantilever. For one cycle, the energy was found to reach a maximum of 0.2 μJ at an optimal load of 119 kΩ. At higher rotational speeds (19 rps) (Figure 5(b)), the harvester was capable to producing an energy of up to 0.6 μJ at 4.7 kΩ.

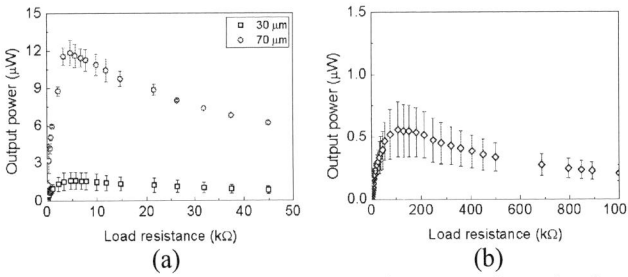

(a)   (b)

Figure 7: (a) Output power as a function of tip depth at rotational speed of 19 rps. (b) Output power at rotational speed of 3 rps and 70 μm tip depth.

The energy produced in the release step is significant. More than 80 % of the energy was produced during the free oscillations. The optimal load resistance (R_{opt}) can be approximated using $R_{opt} = 1/\omega C_p$ where ω is the natural frequency of the harvester and C_p is the capacitance of the piezoelectric layer. As the rotational speed dropped to 3 rps, the optimal load resistance increased to 119 kΩ due to a lack of oscillations after plucking. As a result, ω, which is a combination of the plucking frequency and the resonant frequency of the harvester [3], dropped causing the optimal load to increase.

Figure 7(a) presents the output power as a function of tip depth. At a rotational speed of 19 rps, the harvesters can generate an average output power of 12 μW (over one cycle) at a tip depth of 70 μm. As the rotational speed dropped to 3 rps, the output power also dropped to 0.6 μW due to limited oscillations of the harvester as describe above and to the fact that the time integral of the average power was increased as a result of the longer plucking period (Figure 7(b)). The results of the analytical calculation and the FEM simulations show good agreement with the experimental measurement as shown in Figure 8.

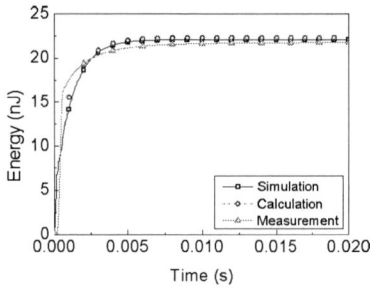

Figure 8: Comparison between simulation and experiment of the energy dissipated in the load resistance of 1 MΩ at rotational speed of 19 rps.

The overall efficiency of this approach was determined using a rotational flywheel as illustrated in [7]. The efficiency of the configuration was found to be 0.6 % at a tip depth of 70 μm for rotational speeds between 20 rps and 18 rps. The efficiency drops gradually as the speed is reduced.

Figure 9: Optical images of the silicon tip (left) and top view of a gear tooth (right) after the lifetime test.

The reason for low efficiency was mainly due to friction from the coupling between the two structures. Finally, the reliability of this configuration was investigated by monitoring the voltage generated as a function of the number of plucks. In the experiment, a crown gear with 36 teeth coupled to an electric motor rotating at 19 rps was used in the test.

The output voltage dropped approximately 33 % after 8 million plucks. The decrease in the output voltage was mainly caused by erosion of the acetal polymer gear by the silicon tip as shown in Figure 9. This demonstrates the significance of the materials used and the geometry of the tip/tooth profile for the reliability of this approach. Materials and wearless coatings are being investigated to reduce the wear problem and improve the efficiency and longevity of this harvesting system.

CONCLUSIONS

A compact configuration for harvesting energy from low frequency movement using the impact between a rotating gear and a piezoelectric MEMS harvester has been successfully demonstrated in this paper. This concept could be combined with an oscillating mass-gear mechanism to realize a compact autonomous micro-power generator from body movements as well as other rotating objects.

REFERENCES

[1] M. Renaud, P. Fiorini, R. van Schaijk, and C. van Hoof, "Harvesting energy from the motion of human limbs: the design and analysis of an impact-based piezoelectric generator", *Smart Mater. Struct.*, vol. 18, pp. 035001-035016, 2009.

[2] M. Pozzi and M. Zhu, "Plucked piezoelectric bimorphs for knee-joint energy harvesting: modeling and experimental validation", *Smart Mater. Struct.*, vol. 20, pp. 055007-055016, 2011.

[3] L.Gu, and C. Livermore, "Impact-driven, frequency up-converting coupled vibration energy harvesting device for low frequency operation", *Smart Mater. Struct.*, vol. 20, pp. 045004-045013, 2011.

[4] P. Janphuang, D. Isarakorn, D. Briand, and N.F. de Rooij, "Energy harvesting from a rotating gear using an impact type piezoelectric MEMS scavenger," in *Digest Tech. Papers Transducers'11 Conference*, Beijing - China, June 5-9, 2011, pp. 735-738.

[5] X. Gao, W.H. Shih, and W.Y. Shih, "Induced voltage of piezoelectric unimorph cantilevers of different nonpiezoelectric/piezoelectric length ratios", *Smart Mater. Struct.*, vol. 18, pp. 125018-125025, 20091.

[6] P.D. Mitcheson, E.M. Yeatman, G. Kondala Rao, A.S. Holmes, and T.C. Green, "Energy harvesting from human and machine motion for wireless electronic devices", *Proceedings of the IEEE*, vol. 96, pp. 1457-1486, 2008.

[7] P. Janphuang, R. Lockhart, S. Henein, D. Briand, and N.F. de Rooij, "On the experimental determination of the efficiency of piezoelectric impact-type energy harvesters using a rotational flywheel," *Tech paper PowerMEMS Conference*, London - UK, December 3-6, 2013.

CONTACT

*P. Janphuang, tel: +41-(0)2-6954435; pattanaphong.janphuang@epfl.ch

"ASSIST-FREE" ASSEMBLY TECHNIQUE OF STANDING OPTICAL DEVICES ON SOFT SPRING ACTUATOR STAGES

Yusaku Oka, Ryosuke Shinozaki, Kyohei Terao,
Takaaki Suzuki, Fusao Shimokawa, Fumikazu Oohira, Hidekuni Takao
Micro-Nano Structure Device Integrated Research Center, Kagawa University, JAPAN

ABSTRACT

In this study, a new assembly technique of separately fabricated MEMS optical devices on fragile MEMS actuator stages has been developed to realize novel functional optical-MEMS devices. This technique realizes the "assist-free" alignment and fixing of vertically mounted optical devices by combination of "micro spring slider" and "trapezoidal alignment slit". Various kinds of optical devices can be precisely aligned and stably fixed even on movable actuator stages supported by soft spring suspensions. In the experiments, micro mirrors were attached on electrostatic linear actuators and rotational actuators using this assembly technique, and a small average value of relative-angle error around 4/100 ° was successfully obtained.

INTRODUCTION

In recent years, many assembly techniques of 'vertically standing' micro components on a MEMS device have been reported. The previous assembly methods utilize springs [1, 2], thermal actuator [3], glue, and ditch [4] to fix the micro components. In addition, precise robot arm control is necessary to put the micro components on the proper position [5, 6]. In case the mounting base of the component is a rigid structure, additional assists of external forces in these methods can be applicable.

If micro optical components can be attached on fragile/movable MEMS actuators, high performance and complicated micro optical systems can be realized for various new applications [7]. However, existing assembly techniques of micro components are very difficult to apply to the fragile structures, since external force becomes a risk of fracture during the mounting process of component. In this study, "assist-free" assembly technique is newly proposed to solve the problem of fracture risk. It does not require alignment assist and external force to fix and adjust the component position on a movable stage. In other words, micro components can be precisely assembled on fragile/movable MEMS actuators by their self-alignment design structures. The assembled optical components are separately fabricated by perforating a silicon wafer. Using this technique, integrated micro optical systems with movable micro optical devices can be integrated on a silicon chip.

"ASSIST-FREE" ASSEMBLY TECHNIQUE

The assembly technique proposed in this study utilizes optical device with micro spring and trapezoidal alignment slit for precise mounting on movable stage. Fig. 1 explains the concept of our "assist-free" assembly technique to mount

an optional optical device (a mirror in this case) vertically on a movable stage suspended by tiny springs. The micro optical device is perforated from a silicon wafer using DRIE process, and assembled on various kinds of MEMS actuator stages. The micro optical device has two micro spring structures (spring slider) and a "reference plane" structure for vertical positioning. The trapezoidal slit works to fix the spring structures to proper positions on the micro stage without any alignment step. Using the silicon wafer surface in the perforated optical device, optical functions such as mirror, diffraction grating are realized. The vertically assembled structure is fixed only by frictional force between the spring sliders and the trapezoidal slit.

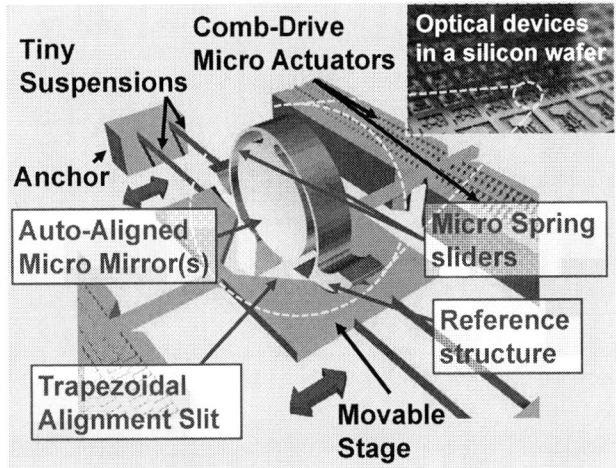

Figure 1: A conceptual diagram of the integrated micro actuator and vertically mounted optical device.

Vertical Positioning of the Optical Device

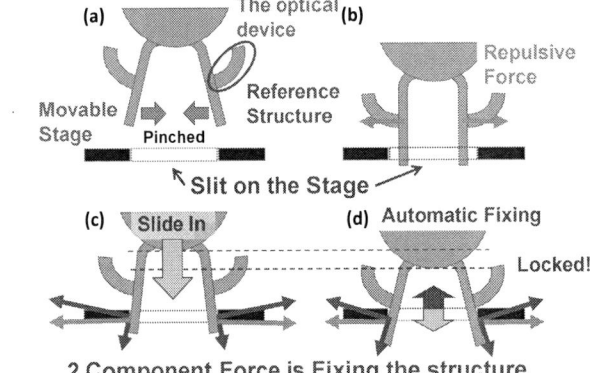

Figure 2: The principle of positioning and fixing of the "assist-free" assembly technique in vertical direction.

Fig. 2 shows the principle of positioning and fixing of the "assist-free" assembly technique in vertical direction. The initial distance of the two spring sliders on the optical device is wider than the width of the trapezoidal alignment slit. (a) First, the distance of the spring sliders is shortened by pinching. (b) The sliders are slotted into the slit on MEMS actuator stage. (c) The pinching force is released, and the spring sliders contact to the slit wall. The repulsive force generates two force components. One is normal force to generate frictional force and the other is parallel force to occur side slip along the spring slider surface. The optical device slides into the slit and moves downward in the figure. (d) Finally, the surface of the stage contacts to the reference structure, vertical position of the device is fixed precisely.

Horizontal Positioning of the Optical Device

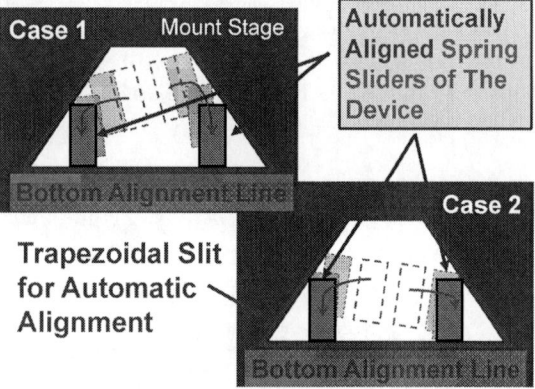

Figure 3: The principle of positioning and fixing of the "assist-free" assembly technique in horizontal direction

Horizontal positioning of the optical device is also performed by the repulsive force of the two spring sliders. In this case, trapezoidal slit on the movable stage has an important role for the assist-free alignment. Fig. 3 shows the principle of positioning and fixing of the "assist-free" assembly technique in horizontal direction. In the step of Fig. 2 (c), the released springs slide along the slit wall. As shown in the case 1 in Fig. 3, the spring slides along the trapezoidal slit, and finally contacts to the bottom alignment line. Even if the two springs are slotted into a different position (case 2), the final fixing positions of the springs are the same with the positions of the case 1. Therefore, the spring sliders are automatically aligned by the trapezoidal slit in horizontal direction, and the final position is insensitive to the initial position of the component.

DEVICE FABRICATION

The automatically aligned micro component is fabricated by perforating a silicon wafer with DRIE process. In this study, mirror is realized as the separated optical device. On the silicon wafer, Cr film is deposited and patterned as the hard mask of the DRIE process. Also, the mask layer is utilized as the optical surface of micro mirrors, since silicon

wafers have atomic-level flat surface. The optical device can be fabricated only by single lithography process and the following DRIE process. Fig. 4 shows the fabrication process flow of the optical device component. (a) First, Cr layer is deposited by spattering evaporation. (b) The Cr layer is patterned by photolithography and wet etching process. (c) The silicon bulk is perforated, and the optical devices are supported only by tiny silicon bars. Each optical device can be removed from the silicon wafer with a small force.

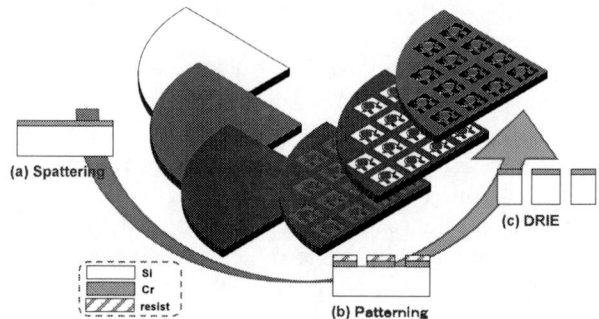

Figure 4: Fabrication process flow of the assist-free assembled optical device with single lithography process.

Fig. 5 shows a photograph of a fabricated silicon wafer including various kinds of the optical devices. On the surface of silicon wafer, Cr layer used as the hard mask is still remained, and the surface is utilized for the reflective surface of each micro mirror. The right-upper photograph shows a micro mirror device which is removed from the silicon wafer. The two spring sliders are pinched by a tweezers

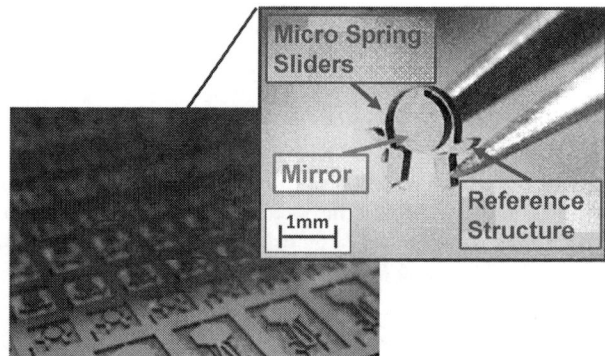

Figure 5: A photograph of a completed silicon wafer including various kinds of the optical devices.

EVALUATION
Mechanical Strength of Spring Sliders

The fabricated optical devices (micro mirror) with spring sliders were evaluated mechanically. First, deformable displacement of the spring sliders was evaluated by a manipulator. Fig. 6 shows a photograph of the optical device before and after the deformation. The tip of the slider can be

978-1-4799-3510-9/14 $31.00 © 2014 IEEE   434

deformed over 450µm. Since over 150µm displacement is typically required in the assembly step shown in Fig. 2 (b), 450µm is large enough to apply the structure to the assembly technique. After the compressive force is released, the shape of the structure completely corresponded with the initial shape thanks to the ideal silicon mechanical property.

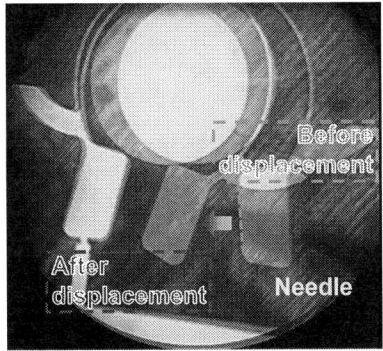

Figure 6: A photograph of the optical device before and after pinching.

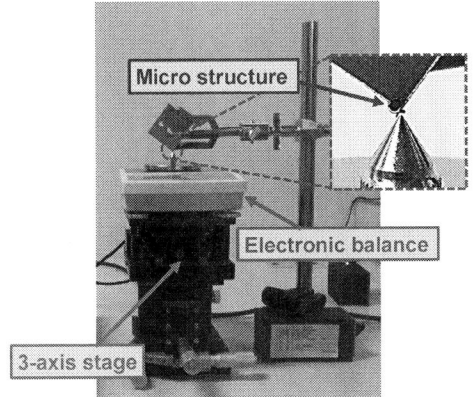

Figure 7: Experimental setup for the spring constant measurement.

The spring constant of the spring sliders was measured with micro force gauge and 3-axis optical stage. Fig. 7 shows the experimental setup for the spring constant measurement. Applying a small displacement to the micro structure, the repulsive force was measured by a micro-force gauge. As a result, the spring constant of the spring sliders in Fig. 5 was 34.6 N/m. Normal force which is generated under the assembled condition on the actuator stage can be calculated with the measured spring constant. Assuming the static frictional coefficient is 0.5, the frictional force between the stage wall and the spring slider can stand up to 85G acceleration.

Experiment of "Assist-Free" Assembly

Assist-free assembly of the micro components on MEMS

actuator stages was tried with the fabricated devices. Fig. 8 shows very simple experimental setup for the assist-free assembly. MEMS actuator device is put on a 3-axis stage, and the spring sliders of an optical device are pinched by the tweezers. Rough positioning and slot-in step to MEMS actuator stage were performed by moving the 3-axis stage. Releasing the pinched sliders after slotting in step, the optical device was automatically aligned and fixed by the frictional force between the slider and the trapezoidal slit wall.

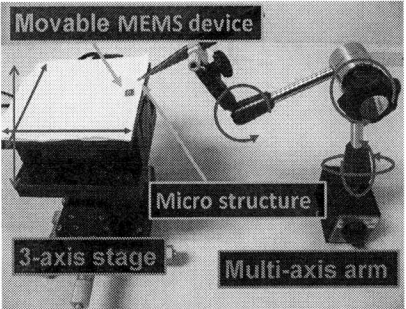

Figure 8: Very simple setup for the assist-free assembly.

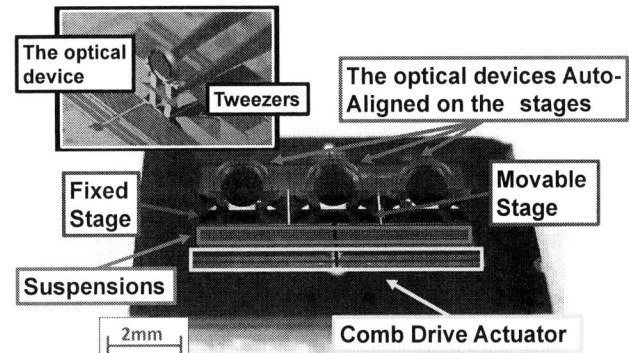

Figure 9: Fabricated light phase-shifter with three micro mirrors mounted on a MEMS actuator stage.

Fig. 9 shows a fabricated light phase-shifter with three micro mirrors mounted on a MEMS device. The center mirror is mounted on a fragile movable stage driven by linear electrostatic actuator. The other mirrors are mounted on the fixed stage as reference mirrors. Three mirrors are precisely assembled without any damages in the linear actuator.

Fig. 10 shows a vertical micro mirror successfully mounted on a MEMS rotational electrostatic actuator. The micro mirror is mounted at the center of the electrostatic rotary stage. Position accuracy of the assembled device was measured with the SEM image. Reference marking points are carved around the alignment slit in the CAD layout, and the typical error value was of the order of sub-micron using the "assist-free" technique.

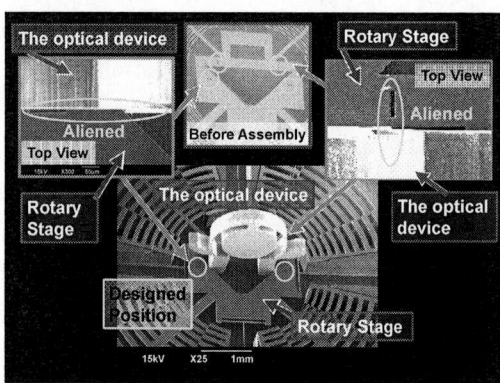

Figure 10: Successfully mounted vertical micro mirror on a MEMS rotational electrostatic actuator.

After the assembly experiment, the micro optical device was removed from the actuator stage. The trapezoidal alignment slit was observed by SEM to confirm the sliding motion of the spring slider. Fig. 11 shows an SEM image around the slide side of the alignment slit. As shown in the photograph, over-hanged oxide layer is removed along the side of the slit because the spring slider moved along it during the positioning and fixing of the "assist-free" assembly. Through the experiment, the principle of the assist-free assembly technique was successfully confirmed.

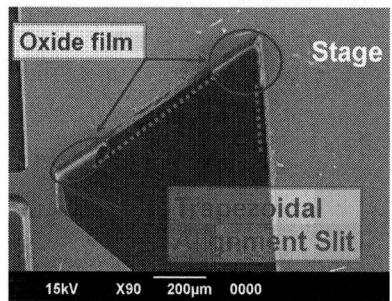

Figure 11: SEM image around the slide side of the trapezoidal alignment slit.

Evaluation of Relative Mounted Angle of the Structures

In order to evaluate the accuracy of the assist-free assembly technique, the perforated optical devices were assembled on trapezoidal alignment slits formed in a straight line. As shown in the photograph in Fig. 12, five micro mirrors were mounted on the same silicon device. Measuring the reflection angle of a laser light, relative angle variation of the optical surfaces was evaluated. The measured differences of angle among five micro mirrors are plotted in Fig. 12. The average variation of angle is suppressed below 4/100° without any additional adjustment. The measured maximum difference was 7/100°, and the minimum value was 1/100°.

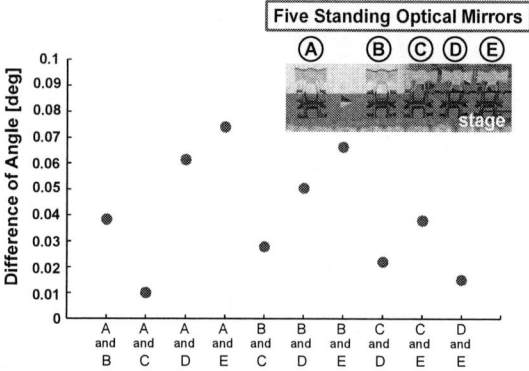

Figure 12: Measured differences of angle among five micro mirrors mounted in a straight line.

CONCLUSIONS

In this study, a new assembly technique of micro optical devices on fragile MEMS actuator stages has been developed. This technique realizes the "assist-free" alignment and fixing of vertically mounted optical devices by combination of "micro spring slider" and "trapezoidal alignment slit". By using this assist-free assembly technique, micro optical devices (mirrors) perforated from a silicon wafer were precisely aligned and stably fixed on electrostatic linear and rotary actuator stages. A small average value of relative-angle error around 4/100 ° was obtained.

REFERENCES
[1] J.Randall, et al., NSTI-Nanotech 2004, Vol.3, pp.499-502, 2004.
[2] S.Bargiel, et al., Journal of Micromechanic and microengineering20,4, p.20, 2010.
[3] S.Herwik, T.holzhammer, O.Paul and P.Ruther, Digest Tech. Papers Transducers'11 Beijing, china, June 5-9, 2011, pp.2323-2326
[4] Chee-Wei et al., IEEE Transactions on Advanced Packaging Vol.32, No.3, 2009.
[5] N.Dechev, Robotic Microassembly, The Institute of Electrical and Electronics Eng., Inc., pp.227-252, 2010.
[6] Dan. O. Popa et al., Robotic Microassembly, The Institute of Electrical and Electronics Engineers, Inc., pp.253-278, 2010.
[7] R.Shinozaki, K. Terao, T. Suzuki, F. Shimokawa, F. Oohira, H. Takao, Proc. of Optical MEMS and Nanophotonics, pp.107-108, 2012.

CONTACT
Prof. Hidekuni Takao,
Faculty of Engineering, Kagawa University,
2217-20, Hayashi, Takamatsu, Japan.
Tel/Fax: +81-87-864-2331;
E-mail: takao@eng.kagawa-u.ac.jp

3D NANOFABRICATION ON COMPLEX SEED SHAPES USING GLANCING ANGLE DEPOSITION

Hyeon-Ho Jeong[1], Andrew G. Mark[1], John G. Gibbs[1], Thomas Reindl[2], Ulrike Waizmann[2], Jürgen Weis[2], and Peer Fischer[1, 3,*]

[1]Max Planck Institute for Intelligent Systems, Stuttgart, Germany
[2]Max Planck Institute for Solid State Research, Stuttgart, Germany
[3]Institute for Physical Chemistry, University of Stuttgart, Stuttgart, Germany

ABSTRACT

Three-dimensional (3D) fabrication techniques promise new device architectures and enable the integration of more components, but fabricating 3D nanostructures for device applications remains challenging. Recently, we have performed glancing angle deposition (GLAD) upon a nanoscale hexagonal seed array to create a variety of 3D nanoscale objects including multicomponent rods, helices, and zigzags [1]. Here, in an effort to generalize our technique, we present a step-by-step approach to grow 3D nanostructures on more complex nanoseed shapes and configurations than before. This approach allows us to create 3D nanostructures on nanoseeds regardless of seed sizes and shapes.

INTRODUCTION

At the nanoscale, sizes, shapes and material compositions drive the performance of micro- and nano-devices. In particular, nanoelectronic devices and energy conversion devices including memristors [2], photocatalysts [3], and nanogenerators [4] are critically dependent on their structural dimensions which can directly influence their electrical properties, such as bandgap energies, electron transfer efficiencies, and signal-to-noise ratios (SNRs). For these reasons, fabricating 3D nanostructures for devices is highly desirable, but techniques for controlling 3D geometry at the nanoscale remain challenging.

In this study, we report a 3D fabrication scheme that combines two existing techniques, electron-beam lithography (EBL) and glancing angle deposition, to fabricate nanostructures with complex 3D shapes both parallel and perpendicular to the growth direction. GLAD is a physical vapor deposition (PVD) technique where evaporant is delivered to a substrate at a high angle of incidence [1, 5-8]. Local shadowing and azimuthal manipulation of the substrate allows a rich variety of complex 3D structures to be grown down to the nanoscale [1]. Using a substrate which has deliberately introduced seeds allows control over the local shadowing and hence the subsequent growth. However, previous work has been restricted to seeds with high symmetry circular or rectangular shapes, and with sizes in the micrometer regime [7, 9, 10]. Herein, we use EBL to write custom seed layers with complex shapes, and do so at the resolution limit for the GLAD technique.

We begin by optimizing the EBL writing conditions for fabricating periodic arrays of nanoseeds for use in the GLAD process at smaller scales than has been previously attempted

[5-8]. We use these nanoarrays to systematically characterize how different seed parameters (e.g. diameter, lattice spacing, and lattice symmetry) influence the properties of the GLAD structures that are grown upon them. Next, we demonstrate the growth of more complex structures including tilted rods, curved rods and helices. Finally, we combine the ability to write arbitrary seed geometries offered by EBL with 3D growth from GLAD to fabricate a variety of nanostructures.

DESIGN

For an effective local shadow effect between nanoseeds, the seed gap g_s must be less than the maximum seed gap g_{max} defined by the geometrical relation

$$g_s < h_s \cdot \tan (\alpha) = g_{max}, \qquad (1)$$

where g_{max} is the effective maximum seed gap, h_s is the seed height, and α is the incident flux angle [6]. When the seed gaps exceed the maximum seed gap g_{max}, the nanostructures on the nanoseeds grow drop-like shapes regardless of the azimuthal rotation of substrate φ due to the absence of the shadow effect between the seeded areas. Indeed, unseeded nanostructures grow between seeded areas because of the absence of the shadow effect in these regions. On the other hand, the columnar competition growth results in low uniformity of nanostructures when the nanoseeds are placed too closely to each other ($g_s \ll g_{max}$)[5, 11]. We experimentally demonstrate the different shadow effects by varying g_s and show which seed arrays are effective, when the seed arrays include complex shapes.

FABRICATION

As illustrated in Fig. 1, the fabrication scheme combines two processes, EBL and GLAD, to grow nanostructures on nanopatterned seeds. Nanoseeds were fabricated with hydrogen silsesquioxane (HSQ) negative tone electron beam resist by EBL (Jeol, JBX-6300FS) (See Fig. 1(a)). HSQ resist was spin coated onto Si substrate to produce h_s = 30 nm thick resist layer and baked at 90°C for 1 min. The designed seed patterns were written at an acceleration voltage of 100 kV and developed by immersion in a tetramethylammonium hydroxide (TMAH) based developer (Rohm and Haas, MF-322) for 80 sec.

Next, GLAD was used to grow nanostructures upon the nanopatterned substrate. Our GLAD system is based on electron-beam evaporation at room temperature with a base pressure of 1×10^{-7} mbar. During deposition, the flux angle α

978-1-4799-3510-9/14 $31.00 © 2014 IEEE 437

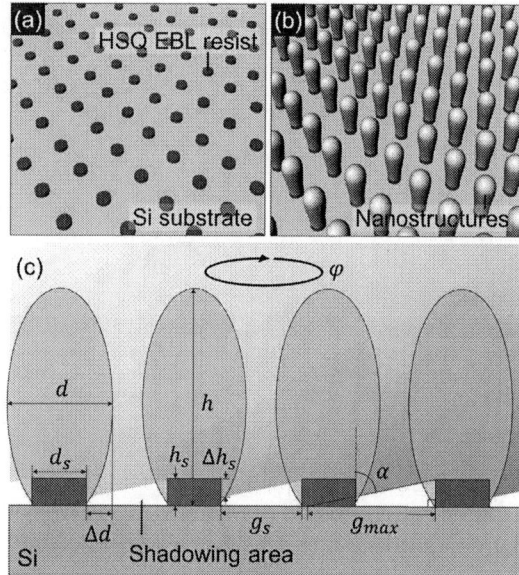

Figure 1: Illustrations of 3D nanofabrication process: (a) HSQ nanoseeds on Si substrate using EBL and (b) nanostructures array on the fabricated HSQ nanoseeds using GLAD. (c) Side view of nanostructures showing dimensions of the nanoseeds and the nanostructures which grow upon them.

and the azimuthal rotation rates per unit thickness $d\varphi/d\theta$ were constantly controlled by a computer running code developed in the lab. In order to examine the geometrical effect of the designed structures, in particular change of the seed gap, other deposition parameters including the temperature, the material, and the deposition rate were kept constant between different samples. A general growth process consisted of Ti deposition as an adhesion layer with a deposition rate ~ 0.1 nm/sec at $\alpha = 10°$, $d\varphi/d\theta = 210$ °/nm. After that, alumina (Al_2O_3) is deposited on such seeds to create 3D nanostructures by controlling the azimuthal rotation rate.

RESULTS & DISCUSSIONS

EBL excels at creating arbitrarily shaped objects at the nanoscale. We start by examining the role of exposure doses of e-beam in EBL, by writing seed layers with precise size control (Fig. 2). The exposure doses were varied from 1.5 mC/cm^2 to 6 mC/cm^2 with 0.5 mC/cm^2 intervals. Spherical seeds of 20 nm, 50 nm, 80 nm, and 110 nm in diameters are arranged in a hexagonal array with $g_s = 100$ nm. We measured seed diameters based on the top-view of SEM images using the Image-J software package [12]. The measured seed diameters were increased when the exposure doses were increased (Fig. 2(b)). Figure 2(a) shows the selected SEM images for the precisely written seeds. Each patterned area is 10 μm × 10 μm, but we only show 0.5 μm × 0.5 μm to avoid edge effects. In order to generalize our experimental conditions, we calculated the fill factor (FF), the surface density ratio between the surface area of single nanoseed S_s and S_a the Wigner-Seitz cell around it (Fig. 2(c)). This is given by,

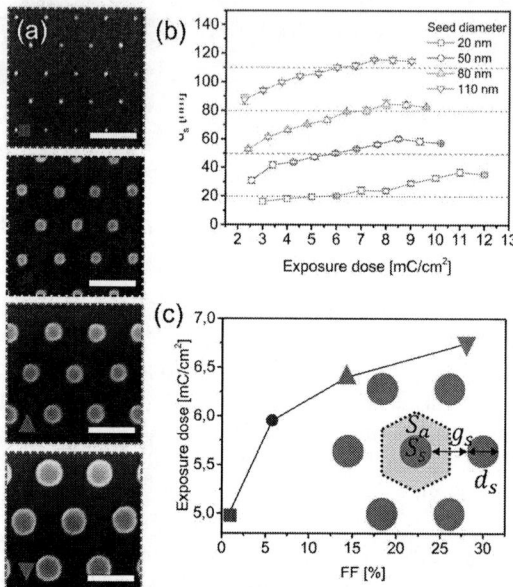

Figure 2: Optimization of e-beam writing condition by varying the e-bam exposure doses: (a) SEM images of the fabricated nanoseeds with 20 nm, 50 nm, 80 nm, and 110 nm diameters. Scale bars: 200 nm. (b) The measured diameters of the fabricated HSQ nanoseeds as a function of the exposure dose. (c) Exposure doses for the calculated fill factors at the precise pattern sizes.

$$FF\ [\%] = S_s/S_a \times 100$$
$$= \pi(d_s/2)^2/(3/2\sqrt{3})g_s^2 \times 100$$
$$= \sqrt{3}\pi d_s^2/6g_s^2 \times 100. \qquad (2)$$

From this equation and the experimental conditions, we can estimate the exposure dose required to write precise seed patterns by calculating the fill factor from the seed design parameters of d_s and g_s.

In order to confirm the ability of seed gaps to grow 3D nanostructures in the large extended array patterns for GLAD, we have grown nanorods on the nanoseed arrays for various seed gaps (Fig. 3). We deposited 500 nm of Al_2O_3 on the nanopatterned substrate at $\alpha = 85°$ and $d\varphi/d\theta = 18°/nm$. The maximum effective seed gap is expected to be $g_{max} = 342$ nm at $\alpha = 85°$ and $h_s = 30$ nm. Figure 3(a) shows the SEM images of the grown nanorods on 20 nm spherical seed arrays with various gaps. The nanorods are grown with good fidelity on smaller seeds what has previously been possible using GLAD [6]. At $g_s = 20$ nm and 50 nm, the high contrast variation in the SEM images between adjacent nanostructures indicates poor uniformity. This results from the columnar competition growth between seeds in this tightly packed arrangement. On the other hand, when g_s is wider than g_{max}, the unseeded nanostructures were grown between the seeded area due to the absence of the shadow effect in these areas. This experimentally demonstrates the effect of the geometrical relation of the seed arrangement for GLAD structures at the nanoscale.

Broadening $\Delta d\ (= (d - d_s)/2)$ is taken to be the difference between the nanostructure's radius, and the radius

978-1-4799-3510-9/14 $31.00 © 2014 IEEE 438

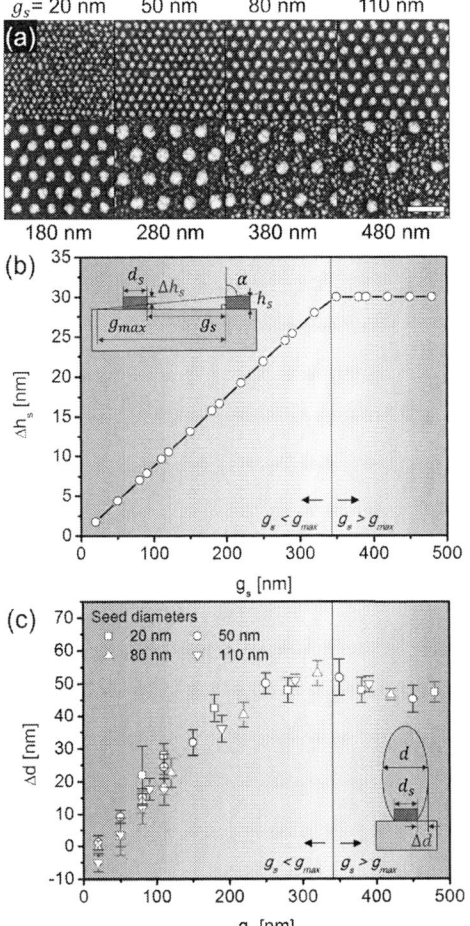

Figure 3: 3D nanofabrication process: (a) SEM images of nanorods grown by GLAD on 20 nm seed patterns for different gaps in the seed arrays. Scale bar: 400 nm. (b) The calculated heights of the nanoseeds exposed to the incoming flux for different seed gaps. (c) The measured broadening of the fabricated nanorod diameters as a function of the gaps of seed arrays.

of the seed it grows upon. We have confirmed that broadening is influenced by the gaps between the seeds, because this drives the exposed height of the nanoseed

$$\Delta h_s = g_s / \tan \alpha \quad (g_s < g_{max}) \qquad (3)$$
$$\Delta h_s = h_s \quad (g_s \geq g_{max}). \qquad (4)$$

These are given by the geometrical relation of the nanoseeds and plotted in Fig. 3(b). In the range of $g_s < g_{max}$, Δh_s increases linearly with increasing seed gap at the constant flux angle ($\alpha = 85°$). On the other hand, when the seed gaps exceed their maximum value, Δh_s is constant since the entire sidewall of the nanoseeds is exposed to the incident flux. Figures 3(b) and (c) show that the trend for broadening (Δd) clearly follows that of the exposed height (Δh_s). In particular g_{max}, where no portion of the seed sidewall is shadowed, leads to a distinct discontinuity in the broadening [5, 6]. Importantly however, the diameter of the seed seems to have

a little effect on the broadening.

Controlling 3D geometry is the biggest attraction of GLAD technique [1, 12-14]. Here, we present 3D shape control during perpendicular growth of the nanoscale objects on the nanoseed arrays by varying the azimuthal rotation rates $d\varphi/d\theta$ (Fig. 4). Here we show four different 3D structures, tilted rods (a), curved rods (b), 2.5-turn helices (c), and vertical rods (d), by controlling the rotation rates of $d\varphi/d\theta = 0$ °/nm, 0.18 °/nm, 1.8 °/nm, and 18 °/nm respectively. We deposited 500 nm thickness of Al_2O_3 on circular seed arrays in the range of $g_s < g_{max}$ and $\alpha = 85°$. The nanostructures produced well-defined 3D shapes expected from the growth program.

The ultimate goal of this study is perpendicular growth of 3D nanostructure on complex seed shapes. We deposited 500 nm Al_2O_3 upon 3 different complex shapes array including boomerangs, propellers, and crosses with $g_s = 250$ nm at $\alpha = 85°$ and $d\varphi/d\theta = 18$ °/nm (Fig. 5). Each

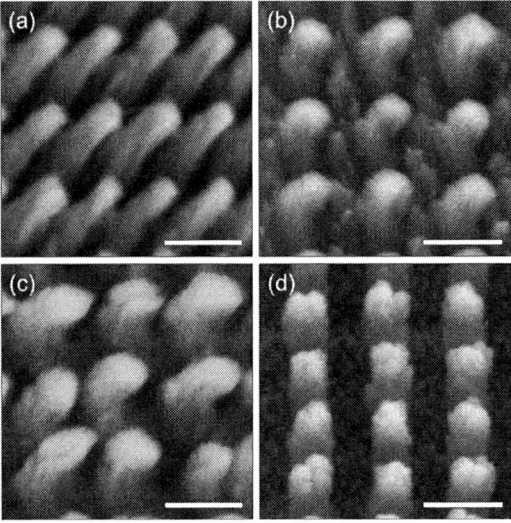

Figure 4: SEM images of 3D nanostructures grown by controlling the φ-angle rotation of the nanoseeds on Si substrate: (a) tilted rods, (b) curved rods, (c) helices, and (d) ver1tical rods. Scale bar: 200 nm.

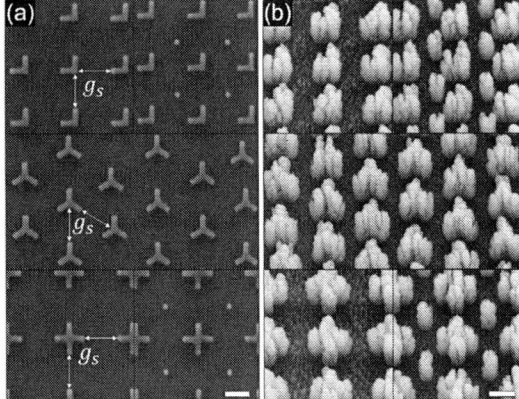

Figure 5: SEM images: (a) Complex seed arrays in the range of $g_s = 200$ nm, (b) 3D nanostructures grown on the complex seeds. Scale bar: 200 nm.

nanostructure mirrors the shapes of the nanoseeds. In the insets of Fig. 5(b), we have also shown the SEM images of the grown nanostructures on hybrid seed arrays which contain complex seeds (boomerangs or crosses) and circular seeds. Such structures were analogous to the grown nanostructures on each seed array (circular seed array in Fig. 3 and complex seed arrays in Fig. 5) because the circular seeds were placed between complex seed arrays with the gap in the range of $g_s < g_{max}$. Even though the graininess of the grown nanostructures remains challenging, this can be improved by further optimizing the seed designs and the growth processes such as the rotation speed, the incident angle, or the materials [15]. Future work will explore creating 3D-shaped nanoscale objects with smooth surfaces on more complex seed shapes which might be useful for electronic and/or photonic devices such as bit patterned memory systems [16], plasmonic antennas [17], and metamaterials [18].

CONCLUSIONS

We can accurately create 3D nanostructures on complex seed shapes with precise cross section and sweep trajectory. This can be extended to create 3D-shaped nanoscale objects on more complex seed shapes and configurations including randomly isolated seed patterns.

We anticipate that this technique will be of interest for realizing new nanoelectronic, photonic, and energy conversion devices, based on controlling 3D geometry and seed shapes and configurations.

ACKNOWLEDGEMENTS

The research was supported by the Max Planck Society and the European Research Council under the ERC Grant agreement Chiral MicroBots (278213).

REFERENCES

[1] A. G. Mark, J. G. Gibbs, T.-C. Lee, and P. Fischer, "Hybrid nanocolloids with programmed 3D-shape and material composition," *Nature Materials,* vol. 12, pp. 802-807, 2013.

[2] J. J. Yang, M. D. Pickett, X. Li, A. A. OhlbergDouglas, D. R. Stewart, and R. S. Williams, "Memristive switching mechanism for metal//oxide//metal nanodevices," *Nat Nano,* vol. 3, pp. 429-433, 07//print 2008.

[3] S. Mubeen, J. Lee, N. Singh, S. Kramer, G. D. Stucky, and M. Moskovits, "An autonomous photosynthetic device in which all charge carriers derive from surface plasmons," *Nat Nano,* vol. 8, pp. 247-251, 04//print 2013.

[4] Z. L. Wang and W. Wu, "Nanotechnology-Enabled Energy Harvesting for Self-Powered Micro-/Nanosystems," *Angewandte Chemie International Edition,* vol. 51, pp. 11700-11721, 2012.

[5] M. M. Hawkeye and M. J. Brett, "Glancing angle deposition: Fabrication, properties, and applications of micro- and nanostructured thin films," *J. Vac. Sci &*

Tech. A, vol. 25, pp. 1317-1335, 2007.

[6] M. O. Jensen and M. J. Brett, "Periodically structured glancing angle deposition thin films," *Nanotechnology, IEEE Transactions on,* vol. 4, pp. 269-277, 2005.

[7] D. X. Ye, T. Karabacak, R. C. Picu, G. C. Wang, and T. M. Lu, "Uniform Si nanostructures grown by oblique angle deposition with substrate swing rotation," *Nanotechnology,* vol. 16, p. 1717, 2005.

[8] J. Steele and M. Brett, "Nanostructure engineering in porous columnar thin films: recent advances," *Journal of Materials Science: Materials in Electronics,* vol. 18, pp. 367-379, 2007/04/01 2007.

[9] C. Patzig, T. Karabacak, B. Fuhrmann, and B. Rauschenbach, "Glancing angle sputter deposited nanostructures on rotating substrates: Experiments and simulations," *Journal of Applied Physics,* vol. 104, pp. -, 2008.

[10] C. Khare, B. Fuhrmann, H. S. Leipner, J. Bauer, and B. Rauschenbach, "Optimized growth of Ge nanorod arrays on Si patterns," *Journal of Vacuum Science & Technology A,* vol. 29, pp. -, 2011.

[11] B. Dick, M. J. Brett, and T. Smy, "Controlled growth of periodic pillars by glancing angle deposition," *Journal of Vacuum Science & Technology B: Microelectronics and Nanometer Structures,* vol. 21, pp. 23-28, 01/00/ 2003.

[12] D. Schamel, M. Pfeifer, J. G. Gibbs, B. Miksch, A. G. Mark, and P. Fischer, "Chiral Colloidal Molecules And Observation of The Propeller Effect," *Journal of the American Chemical Society,* vol. 135, pp. 12353-12359, 2013/08/21 2013.

[13] A. Ghosh and P. Fischer, "Controlled Propulsion of Artificial Magnetic Nanostructured Propellers," *Nano Letters,* vol. 9, pp. 2243-2245, 2009/06/10 2009.

[14] J. G. Gibbs, A. G. Mark, S. Eslami, and P. Fischer, "Plasmonic nanohelix metamaterials with tailorable giant circular dichroism," *Applied Physics Letters,* vol. 103, pp. -, 2013.

[15] M. A. Summers and M. J. Brett, "Optimization of periodic column growth in glancing angle deposition for photonic crystal fabrication," *Nanotechnology,* vol. 19, p. 415203, 2008.

[16] J. Moritz, G. Vinai, S. Auffret, and B. Dieny, "Two-bit-per-dot patterned media combining in-plane and perpendicular-to-plane magnetized thin films," *Journal of Applied Physics,* vol. 109, pp. 083902-4, 04/15/ 2011.

[17] M. W. Knight, H. Sobhani, P. Nordlander, and N. J. Halas, "Photodetection with Active Optical Antennas," *Science,* vol. 332, pp. 702-704, May 6, 2011 2011.

[18] C. M. Soukoulis and M. Wegener, "Past achievements and future challenges in the development of three-dimensional photonic metamaterials," *Nature Photonics,* vol. 5, pp. 523-530, 2011.

CONTACT

*P. Fischer, tel: +49-711-689- 3560; fischer@is.mpg.de

A FLEXIBLE TACTILE AND SHEAR SENSING ARRAY FABRICATED BY NOVEL BUCKYPAPER PATTERNING TECHNIQUE

Cheng-Wen Ma, Li-Sheng Hsu, Jui-Chang Kuo, and Yao-Joe Yang*
National Taiwan University, Taipei, TAIWAN

ABSTRACT

In this work, we present a flexible tactile and shear sensing array utilizing patterned buckypaper as the sensing elements. A novel fabrication process for patterning buckypaper with high aspect ratio was proposed. The fabricated sensing device possesses the advantages such as anisotropic sensing capability, flexibility, simple fabrication, and low cost. The measured resistance vs. applied shear force on a single sensing element shows that the element exhibits different sensitivities along different directions. This anisotropic sensing capability can be employed for better shear sensing. The sensing elements also give good sensitivity and repeatability.

INTRODUCTION

Carbone nanotubes (CNTs) have attracted significant attention because of excellent and unique electrical, mechanical [1], and thermal properties [2]. Building sensitive tactile sensors with CNT sensing elements has been proposed during recent years. Hu et al. reported a CNT-based flexible tactile sensor which is capable of detecting normal and shear forces [3]. Yimazoglu et al. proposed a simple method for the integration of flexible and vertically aligned multiwalled CNT arrays sandwiched between carbon layers. The electromechanical properties of the CNT arrays were also studied [4]. Lai et al. proposed a novel resistive sensing array which is capable of retaining and erasing tactile images. The sensing material was prepared by dispersing CNTs and silver nanoparticles through PDMS polymer with the assistance of the dielectrophoresis (DEP) technique [5].

CNT films with paper-like morphology, known as buckypapers [6], have been demonstrated for various applications, including energy storage, filtration, and so on. It has been reported that buckypapers is much lighter (about 10 times) than steel, but about 500 times stronger than CNT films. In addition, buckypaper exhibits excellent conductivity, which can be as high as 10^6 S/m [7].

A few fabrication techniques of buckypapers were proposed. Hennrich et al. reported a vacuum filtration method for fabricating large-area buckypaper with thickness of less than 200 nm [8]. Rigueur et al. proposed a buckypaper fabrication method by a two-step process using electrophoretic deposition (EPD) [9]. Wang et al. proposed an aligned CNT manipulation of arrays in the preparation of thick buckypapers with large areas [10].

The fabrication processes utilized in these aforementioned works are quite effective for fabricating a flat buckypaper with relatively large area, while the proposed processes are not capable of realizing buckypapers with desired patterns, which limits the applications of fabricated CNT films. Lim et al. proposed a filtration method which allows patterning buckypaper by using photoresist on an AAO filter. Extra step is required for transferring the patterned CNT film to an elastomer substrate for further applications [11].

In this work, we present a simple vacuum filtration method which creates patterned buckypapers on a flexible membrane filter. By using the method, sensing elements for a tactile and shear sensing array were realized. The sensing device possesses the advantages such as anisotropic sensing capability, flexibility, simple fabrication, and low cost.

SENSOR DESIGN

Figure 1 shows the schematic of the proposed tactile and shear sensing array. It consists of polymer substrate, a patterned buckypaper layer, a Cr/Au layer, and PDMS layer. The polymer substrate is a nylon membrane filter which is essential for patterning the buckypaper layer. In order to enhance the sensitivity, the PDMS bumps are bonded on the top of each sensing element for centralizing the applied force. A flexible printed circuit board (FPCB) is employed to provide electrical contacts for connecting to a scanning circuit. As shown in the figure, the meandering patterns are the sensing elements formed by patterning buckypapers using the proposed technique. Each shear sensing element consists of four resistance tactile sensing cells. This array is capable of sensing normal and shear forces.

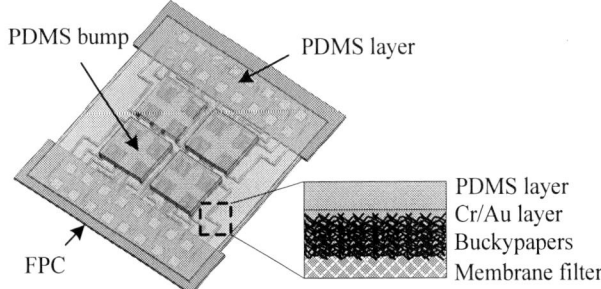

Figure 1: Schematic of the proposed tactile and shear sensing array.

FABRICATION

Figure 2 describes the fabrication process of the sensing array. Firstly, SU-8 photoresist is spin-coated on a handling wafer, as shown in Figure 2(a). Then, a nylon membrane filter (Nylon membrane filter, pore size: 0.8 μm, MS®) is placed on the top of the SU-8 film (Figure 2(b)). Again, SU-8 photoresist is spin-coated on the nylon membrane filter, as shown in Figure 2(c). After the wafer is soft baked, a standard lithography process is performed and the SU-8 photoresist is patterned (Figure 2(d) and 2(e)). Then, vacuum filtration with MWCNT-dispersed solution is performed (Figure 2(f) and 2(g)), and MWCNT molecules fill in the trenches between SU-8 walls.

After vacuum filtration process, the device was dried at room temperature for 2 hours. Then, as shown in Figure 2(h), chromium (200 Å) and gold (3000 Å) are deposited on the top of the multiwall CNT film with a shadow mask. In Figure 2(i), SU-8 was removed using remover PG (Microchem, USA) [12]. Note that the remover PG was heated to 50˚C for removing SU-8 effectively. Then, PDMS prepolymer and curing agent (Sylgard 184A and 184B, Dow Corning) were mixed at a 10:1 ratio. After stirring thoroughly and degassing in a vacuum chamber, the prepared PDMS was spin-coated on top of the device for strengthening the patterned buckypaper structures (Figure 2(j)). Finally, a PDMS layer with bumps, which was fabricated by using soft lithography technique, was bonded the top of the device for improving tactile and shear sensing, as shown in Figure 2(k).

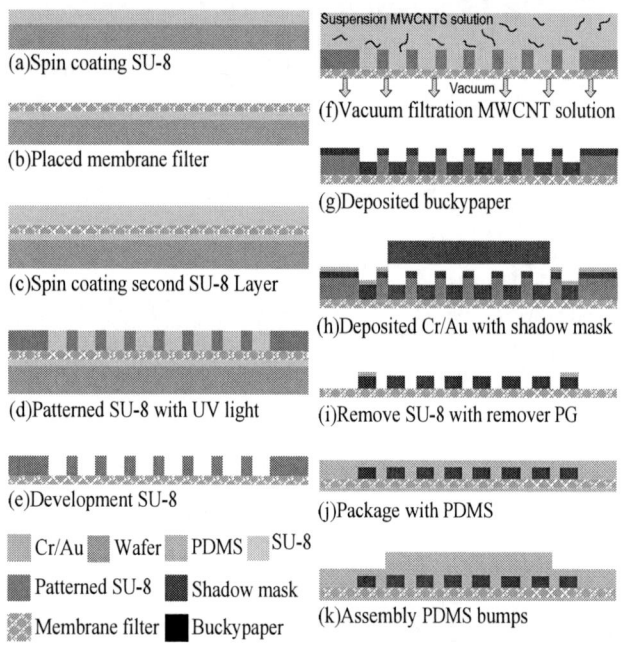

Figure 2: The fabrication process of the sensing array.

Figure 3 shows the pictures of the device at different fabrication process steps. Figure 3(a) shows the patterned SU-8 structure on the nylon membrane filter. Figure 3(b) is the patterned buckypaper after vacuum filtration. Figure 3(c) shows the device after Cr/Au deposition using a shadow mask. Each sensing cells were defined. Figure 3(d) shows fabricated device after removing SU-8 with remover PG.

Figure 4 presents the relationship of the buckypaper thickness vs. the concentration of MWCNT solution. The thickness was measured by a surface profiler. The total thickness increases with the volume of CNTs solution used during the filtration process. Note that the CNTs solution, whose concentration is of 0.01 wt%, is subjected to ultrasound for 2 h for reducing the CNT bundling tendency,

before the filtration process. Also, the multiwall CNTs used for this study were 10–20 nm in diameter and 1–2 μm in length (Golden Innovation Business, Taiwan).

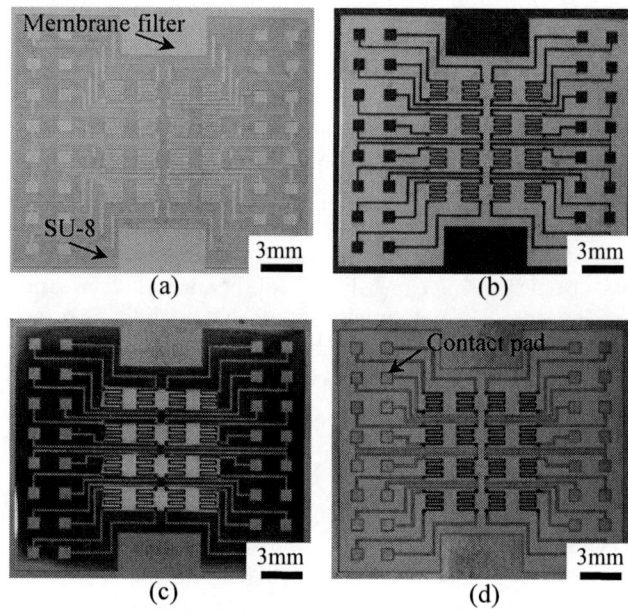

Figure 3: (a) Patterned SU-8. (b) CNT layer deposited. (c) Au/Cr layer deposited. (d) SU-8 layer removed.

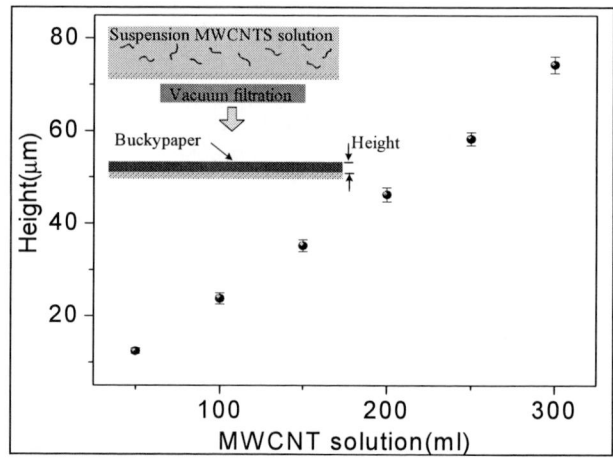

Figure 4: The buckypaper thickness vs. the concentration of MWCNT solution.

Figure 5(a) shows the SEM picture of the sensing element after removed SU-8. Numerous small pores on the nylon membrane filter structure were also observed. The thickness of buckypaper is about 50 μm. Figure 5(b) is the SEM image on the top of the patterned buckypaper, which shows the CNT intertwining network. The assembled sensing arrays are shown in Figure 6(a). Sliver paste was employed to glue a FPCB and the contact pads of the sensor array. Figure 6(b) shows the picture of the fabricated 2 × 2 flexible shear sensing array. The size of each shear sensing element is about 5 mm × 5mm. The size of each tactile sensing cell is about 1.8 mm × 2 mm.

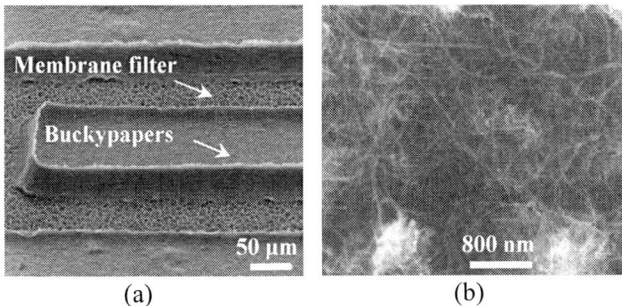

(a)　　　　　　(b)

Figure 5: SEM pictures: (a) Patterned buckypapers on membrane filter. (b) MWCNT networks.

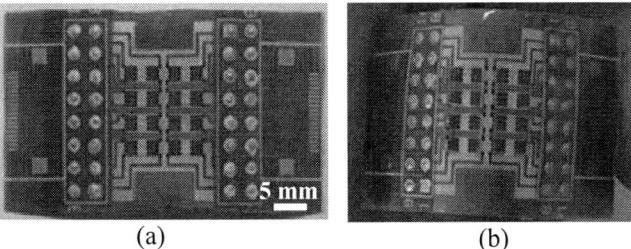

(a)　　　　　　(b)

Figure 6: (a) The assembled device. (b) The fabricated flexible 2 × 2 shear sensing array.

MEASUREMENT AND DISCUSSION

Figure 7 illustrates the experimental setups for testing the sensing arrays with applied normal and shear forces, respectively. A force gauge (HF-1, ALGOL Engineering Co.), whose maximum resolution is 1 mN, is used to measure the applied force. The force gauge is fixed at a vertical (z-axis) translational stage which has a displacement resolution of 1μm.

For normal force measurement, a PMMA rod is connected to the force gauge. As shown in Figure 7(a), the bump of the sensing element is pushed by the PMMA rod as the translational stage moves down. For shear force measurement, an L-shaped PMMA rod is connected to the force gauge. As shown in Figure 7(b), the bump of the sensing element is pushed by the L-shaped rod when the x-y stage moves. The resistance changes of the sensing elements can be detected by a multimeter or a scanning circuit.

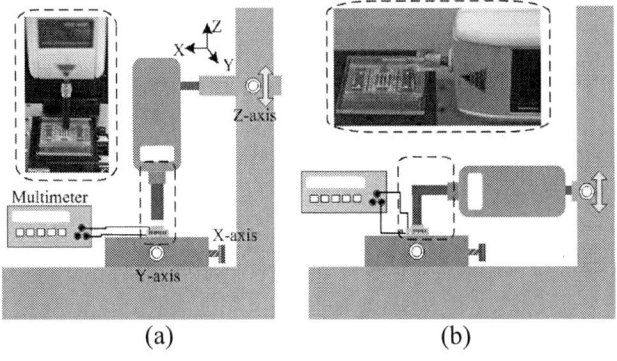

(a)　　　　　　(b)

Figure 7 Experimental setup for measuring (a) normal force, and (b) shear force.

Figure 8 shows the measured results of the device. The measured resistance vs. applied normal force is shown in Figure 8(a). The resistance value increases (Cells A, B, C, and D) almost linearly as the applied normal force increases. During the measurement for each data point, the displacement of the z-axis translational stage is slowly increased until the force applied on the sensing element reaches a steady value, and then the resistance value was recorded.

Figure 8(b) shows the measured resistance vs. applied shear force. When a shear force is applied in the x-direction, a torque is induced around the PDMS bump of the sensing element, the resistance values of Cell A and Cell B increases, while the resistances of the other two cells (Cell C and Cell D) were almost unchanged. By comparing the resistance change of these cells, the direction and magnitude of applied shear force can be determined.

Each data point in Figure 8 is the average value by measuring a sensing element 5 times. The error bars indicate the measured maximum and minimum values. The measured initial resistance of these meandering patterns without applying force is about 15.6 KΩ. These measured results indicate excellent repeatability and sensitivity.

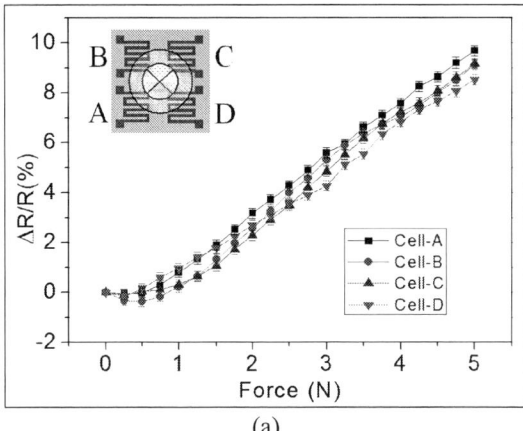

(a)

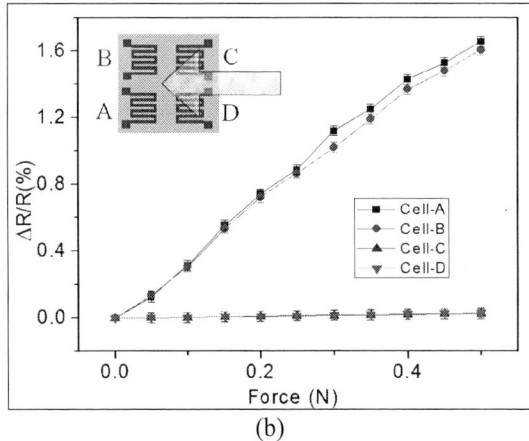

(b)

Figure 8: The measured results of device. (a) The measured resistance vs. applied normal force. (b) The measured resistance vs. applied shear force.

Figure 9 shows the measured resistance vs. applied shear forces along four different directions on a single sensing element. Each data point in the figure is the average result by measuring a sensing element 10 times. Different slopes along different directions were obtained (i.e., 0.03%/mN and 0.02%/mN). The results were also highly repeatable, and showed that the sensitivity of the sensing element is dependent on the direction of the meandering pattern because of the high-aspect ratio of CNT structures. This anisotropic characteristic can be further employed for better shear sensing capability.

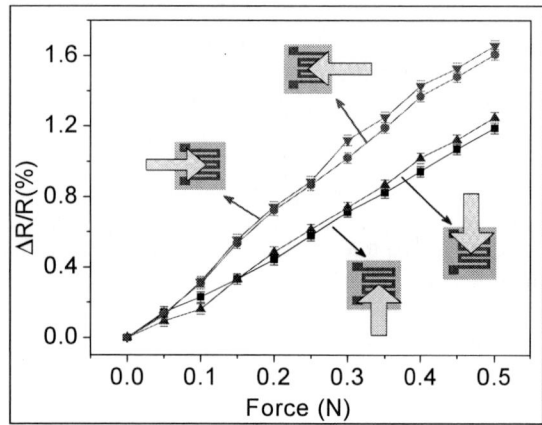

Figure 9: Measured resistance vs. applied shear force along four different directions on a single sensing element.

CONCLUSION

This work proposed a simple vacuum filtration method for patterning buckypapers on a flexible membrane filter. The proposed method was employed to realize the sensing elements for a tactile and shear sensing array. Due to relatively large aspect ratio of patterned buckypaper structure, the proposed sensing element possesses anisotropic sensing capability. Experimental setup for normal and shear force measurement was implemented. Measured results, including normal and shear forces vs. resistance changes, show excellent sensitivity and repeatability.

ACKNOWLEDGEMENTS

This work was supported in part by the National Science Council, Taiwan, R.O.C. (Contract No: NSC 100-2221-E-002-075-MY3).

REFERENCES

[1] T. Tong, Y. Zhao, L. Delzeit, A. Kashani, M. Meyyappan, and A. Majumdar, "Height independent compressive modulus of vertically aligned carbon nanotube arrays," *Nano Letters*, vol. 8, pp. 511-515, 2008.

[2] J. Gwinn and R. Webb, "Performance and testing of thermal interface materials," *Microelectronics Journal*, vol. 34, pp.215-222, 2003.

[3] C.-F. Hu, W.-S. Su, and W.-L. Fang, "Development of

patterned carbon nanotubes on a 3D polymer substrate for the flexible tactile sensor application," *Journal of Microelectromechanical Systems*, vol. 21, 115012, 2011.

[4] O. Yimazoglu, A. Popp, D. Pavildis, J. Schneider, D. Garth, F. Schuttler, and G. Battenberg, "Vertically aligned multiwalled carbon nanotubes for pressure, tactile and vibration sensing," *Nanotechnology*, vol. 23, 085501, 2003.

[5] Y.-T. Lai, Y.-M. Chen, and Y.-J. Yang, "A Novel CNT-PDMS-Based Tactile Sensing Array With Resistivity Retaining and Recovering by Using Dielectrophoresis Effect," *Journal of Microelectromechanical Systems*, vol. 21, pp. 217-213, 2012.

[6] M. Endo, H. Muramatsu, T. Hayashi, Y. Kim, M. Terrones, and M. Dresselhaus, "Buckypaper from coaxial nanotubes," *Nature*, vol. 433, pp. 476, 2005.

[7] H. Geng, K. Kim, K. So, Y. Lee, Y. Chang, and Y. Lee, "Effect of acid treatment on carbon nanotube-based flexible transparent conducting films," *Journal of the American Chemical Society*, vol. 129, pp. 7758-7759, 2007.

[8] F. Hennrich, S. Lebedkin, S. Malik, J. Tracy, M. Barczewski, H. Rosner, and M. Kappes, "Preparation, characterization and applications of free-standing single walled carbon nanotube thin films," *Physical Chemistry Chemical Physics*, vol. 4, pp. 2273-2277, 2002.

[9] J. Rigueur, S. Hasan, S. Mahajan, and J. Dickerson, "Buckypaper fabrication by liberation of electrophoretically deposited carbon nanotubes," *Carbon*, vol. 48, pp. 4090-4099, 2010.

[10] D. Wang, P. Song, C. Liu, W. Wu, and S. Fan, "Highly oriented carbon nanotube papers made of aligned carbon nanotubes," *Nanotechnology*, vol. 19, 075609, 2008.

[11] C. Lim, D. Min, and S. Lee, "Direct patterning of carbon nanotube network devices by selective vacuum filtration," *Applied Physics Letters*, vol. 91, 243117, 2007.

[12] M. Beidaghi, and C. Wang, "Micro-Supercapacitors based on interdigital electrodes of reduced graphene oxide and carbon nanotube composites with ultrahigh power handling performance," *Advanced Functional Materials*, vol. 22, pp. 4501-4510, 2012.

CONTACT

*Y.-J. Yang, tel: +886-2-33664941#16; yjy@ntu.edu.tw

A TECHNOLOGY FOR MONOLITHIC MEMS-CMOS INTEGRATION AND ITS APPLICATION TO THE REALIZATION OF AN ACTIVE-MATRIX TACTILE SENSOR

Fan ZENG and *Man WONG*

The Department of Electronic and Computer Engineering

The Hong Kong University of Science and Technology, Hong Kong, P. R. of CHINA

ABSTRACT

Presently described is an application of a technology based on the surface-migration of silicon for the monolithic integration of micro-mechanical devices and complementary metal-oxide-semiconductor (CMOS) electronic circuits. A cavity sealed with a cover-diaphragm is first formed without a sacrificial layer etch. The electronic devices are next fabricated. The issues of material- and process-incompatibility inherently present in many integration schemes are largely avoided. A 16×16 active-matrix tactile sensor integrating 256 force-sensing diaphragms, 512 pixel transistors and 512 piezoresistors was designed, realized and characterized. The spatial resolution of the sensor was ~145 "pixels per inch" and the pressure sensitivity was ~0.07 μV/V/Pa.

INTRODUCTION

Much effort has been devoted to the monolithic integration of micro-mechanical and micro-electronic devices for potential reduction in system cost and improvement in system performance and reliability. For an example of a "pre-CMOS" integration scheme [1], the mechanical structures are formed in a trench and buried in sacrificial oxide. The process is made complicated by the need of a planarization step to reduce surface topography. For an example of a "post-CMOS" integration scheme [2], the metals and the dielectrics of the multi-level interconnection system are utilized to build the mechanical structures. Low-temperature deposited poly-crystalline germanium [3] or silicon-germanium [4] has also been proposed for MEMS-CMOS integration, but they are not readily available in a conventional CMOS fabrication flow. In SCREAM [5], the mechanical structure is released by removing the underlying bulk silicon using an isotropic dry etch. Besides added process complexity, the lateral extension of the release-etch also limits device packing density. All of these schemes demand substantial etch selectivity among the structural, the sacrificial and other functional layers.

The application of a scheme, based on the surface-migration of silicon [6] [7] and using a conventional bulk silicon substrate, for the integration of micro-mechanical devices and CMOS electronic circuits is presently described. A buried cavity sealed with a silicon cover-diaphragm is first formed without a sacrificial layer etch. The electronic devices, both on and off the diaphragm, are next fabricated. With a sealed cavity, the issues of material- and process-incompatibility inherently present in earlier integration schemes are largely avoided. A 16×16 active-matrix tactile sensor integrating 256 force-sensing diaphragms, 512 transistors and 512 piezo-

resistors was designed, realized and characterized. The spatial resolution of the sensor was ~145 pixels per inch and the pressure sensitivity was ~0.07 μV/V/Pa.

SILICON MIGRATION TECHNOLOGY

The silicon-migration technology (SiMiT) starts with the formation of an array of "wells" (Fig. 1a) using a deep reactive-ion etcher (DRIE). During a subsequent heat-treatment, surface-diffusion of silicon leads to the widening of the void at the base and the closing of the opening at the top of a well (Fig. 1b). Upon further heat-treatment, the coalescence of the voids at the bases of adjacent wells and the closing of the space at the top of a well result in the formation of a continuous cavity buried under a cover-diaphragm (Fig. 1c). One distinction of SiMiT is that the time needed to complete the silicon-migration process is only weakly dependent on the lateral dimensions of the diaphragm and the cavity. The same is not true for a scheme involving the use of sacrificial-structural layers to form a suspended mechanical structure, since the time for the release-etch depends obviously on the lateral dimensions of the resulting structure.

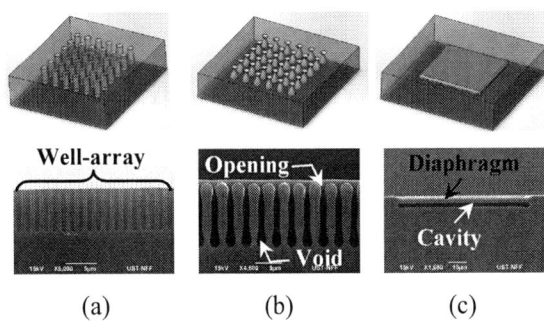

Figure 1. Schematic drawings (top row) and scanning-electron micrographs (bottom row) illustrating the surface-diffusion induced silicon-migration process: (a) well-array formed in a DRIE; (b) Respective widening of the void and closing of the opening at the bottom and the top of a well; (c) Formation of buried cavity and cover-diaphragm upon further heat-treatment.

Buried cavities with a multitude of topologies can be realized using SiMiT. These include (i) buried micro-channels obtained using either a narrow trench (Fig. 2a) or a linear array of wells (Fig. 2b); (ii) single (Fig. 3a) or stacked (Fig. 3b) buried cuboid cavities obtained using two-dimensional "rectangular" arrays of shallow and deeper wells, respectively. The resulting buried cavity and cover-diaphragm can be further processed or patterned to form a variety of micro-structures.

978-1-4799-3510-9/14 $31.00 © 2014 IEEE

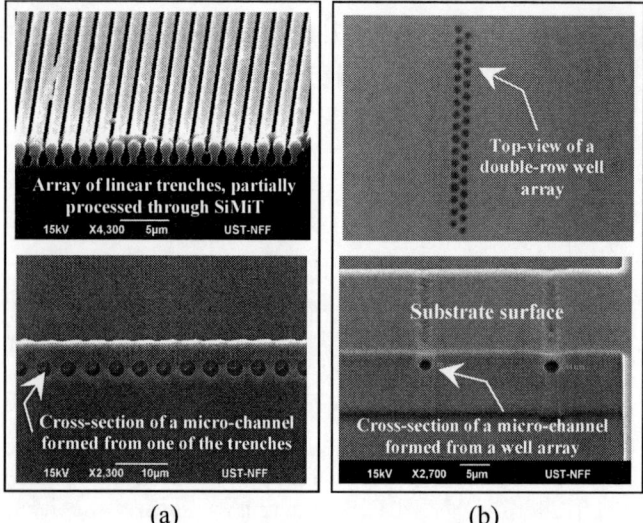

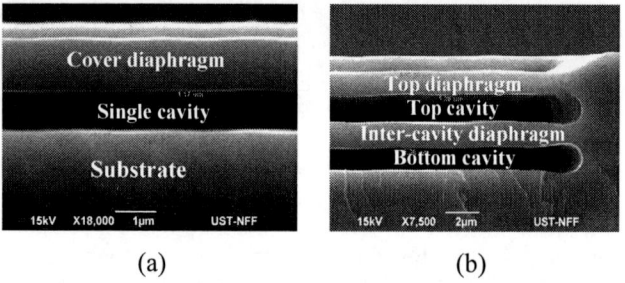

(a)   (b)

Figure 2. Buried silicon micro-channels realized using (a) linear trenches and (b) double-row of cylindrical wells.

(a)   (b)

Figure 3. Buried cavities realized using two-dimensional "rectangular" well arrays: (a) single cavity with shallow wells and (b) stacked cavities with deeper wells.

Oxide micro-channel [8] was formed by first exposing the ends of a buried silicon micro-channel. The wall of the silicon channel was subsequently thermally oxidized. Due to the volume expansion accompanying the oxidation, the diameter of the resulting oxide channel is smaller than that of the initial silicon channel and scalable down to tens of nanometers. Depending on the initial layout of the trench or well arrays, complex networks of oxide channel can be constructed. They were revealed by selectively removing the surrounding silicon substrate, as shown in Figure 4.

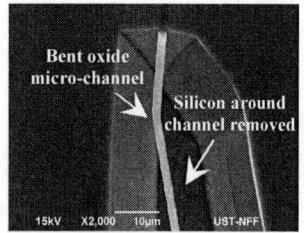 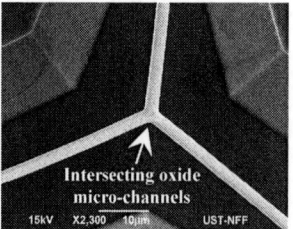

Figure 4. Bent (left) and intersecting (right) suspended oxide micro-channels realized using SiMiT and thermal oxidation.

The silicon cover-diaphragm can also be processed or patterned to form a variety of mechanical structures. All or part of the diaphragm can be thermally oxidized, resulting respectively in a pure oxide diaphragm (Fig. 5a) or, if only

the portion near the perimeter of the diaphragm was thermally oxidized, a suspended silicon plate (Fig. 5b) both electrically and thermally isolated from the substrate.

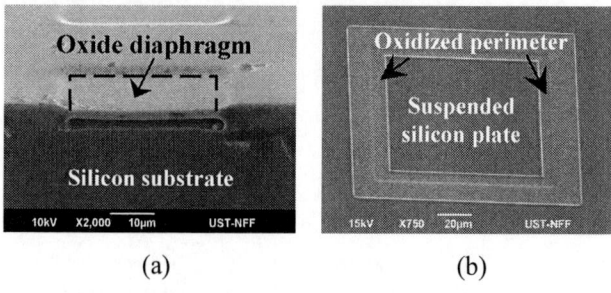

(a)   (b)

Figure 5. Micro-structures derived from the cover-diaphragm: (a) oxide diaphragm (bounded by the dashed line) and (b) oxide-isolated suspended silicon plate.

Detached silicon plates supported by cantilever beams (Fig. 6a) were formed by cutting through the cover-diaphragm in selected places. With the deposition of a thin layer of tensile silicon nitride prior to the cut-through etch, an intrinsic stress gradient was introduced on the connecting cantilever beams, thus leading to the formation of a tilted plate (Fig. 6b).

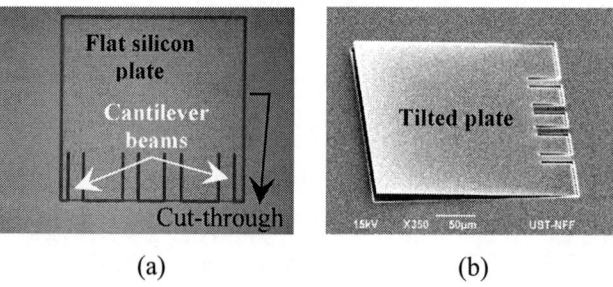

(a)   (b)

Figure 6. Supported micro-plates, (a) flat and supported by 4 cantilever beams and (b) tilted with the deposition of 300 nm silicon nitride with an intrinsic tensile stress.

Besides mechanical structures, electronic devices can be constructed on a SiMiT-processed wafer, both on and off the cover-diaphragm.

DESIGN AND FABRICATION

As a demonstration of the application of SiMiT to MEMS-CMOS integration, an active-matrix tactile sensor was designed, fabricated and characterized. The deformation of the cover-diaphragm resulting from the application of a force load was detected using two n-type piezoresistors, with respective length and width of 4 and 5 μm, and placed near the middle of two adjacent edges of a diaphragm. They were connected to form a voltage divider, with R_1 and R_2 (Fig. 7) configured respectively in the longitudinal and transverse modes. The respective piezoresistive coefficients, π_{11} and π_{12}, are -102.2 and 53.4 ($\times 10^{-11}$ Pa^{-1}). The longitudinal direction was oriented in the <100> direction, hence 45° w.r.t. the major flat of a conventional 100 mm p-type silicon wafer. A pair of n-channel transistors, each with respective channel length and width of 8 and 16 μm, was integrated per pixel. One (T_1) was used to control the application of the power supply V_{DD} to the voltage divider and the other (T_2) to control

978-1-4799-3510-9/14 $31.00 © 2014 IEEE

access to the "response" V_p of the voltage divider. When the 16 pixels of a row are "selected" by the corresponding row-scan, they are "powered" and their response can be read out in parallel.

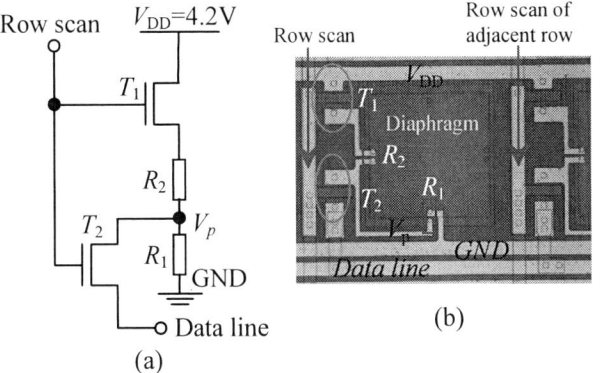

(a)

(b)

Figure 7. (a) A circuit schematic and (b) a photograph of a pixel.

The fabrication of the tactile sensor started with the definition of an 84×72 array of wells for each "pixel" of the sensor array (Fig. 8a). The wells were regularly arranged in a triangular lattice with an edge-to-edge separation of 0.5 μm between the nearest neighbors. The cross-sectional diameter and the depth of each well were respectively 0.9 and ~5 μm. The SiMiT heat-treatment was carried out at 1150 °C in argon for 6 minutes to form a 16×16 array of cuboid cavities, each with a 100×100 μm² base parallel to the wafer plane, a depth of ~1.2 μm and buried under a ~1.5 μm thick cover-diaphragm (Fig. 8b). A four-mask "NMOS" process was subsequently used to realize the pixel transistors: After the formation of the isolation structure (Fig. 8c) and the growth of 15 nm of gate oxide (Fig. 8d), polysilicon was deposited, heavily doped with phosphorus by diffusion and patterned (Fig. 8e). N-type piezoresistor arsenic implantation (Fig. 8f) was followed by heavy arsenic implantation of the source/drain and resistor contact regions (Fig. 8g). After the deposition of 500 nm of low-temperature oxide (LTO) insulation layer, the contact holes were opened (Fig. 8h) and aluminum-based interconnection was defined (Fig. 8i). For force sensing, no further processing of the diaphragm was necessary after the pads through the passivation layer were opened.

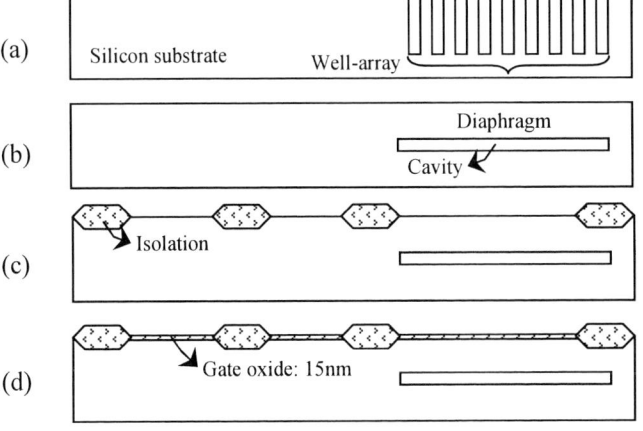

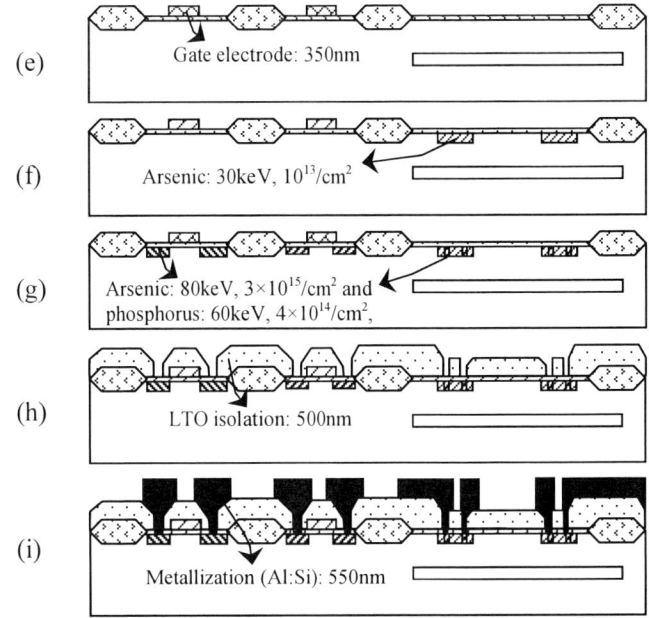

Figure 8. Schematic SiMiT-based MEMS-CMOS integration process: (a) Well-array definition; (b) SiMiT heat-treatment; (c) Isolation formation; (d) Gate oxidation; (e) Gate electrode doping and patterning; (f) Piezoresistor implantation; (g) Source/drain and contact implantation; (h) LTO deposition and contact hole opening; (i) Metallization.

CHARACTERIZATION

A photograph of a fabricated tactile sensor is shown in Figure 9. The measured resistance of R_1 and R_2 were ~5kΩ. The respective transfer and output characteristics of an n-channel transistor are displayed in Figures 10a and b, showing a threshold voltage of ~1.4V and a sub-threshold slope of ~81mV/decade. These characteristics are similar to those measured on a substrate not having been subjected to the SiMiT process, thus showing good preservation of the quality of the substrate.

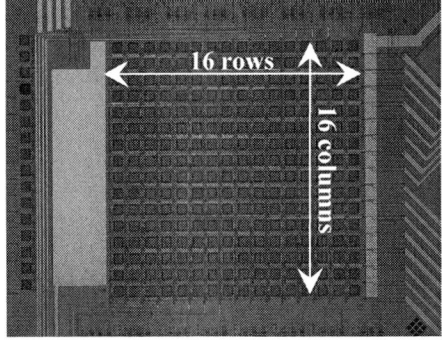

Figure 9. Photograph of a fabricated active-matrix tactile sensor.

The sensing diaphragm was characterized using a nano-indenter, with a concentrated point-force loaded at the center of the diaphragm. From the force-loading curve shown in

Figure 11a, an equivalent spring constant of ~1062 N/m was obtained for the diaphragm. The slope of the curve changes at a displacement of ~1.1 μm, when the diaphragm reaches the bottom of the cavity. The response of a diaphragm to uniform pressure load was measured using a custom-built pressure chamber. From the pressure-loading curve shown in Figure 11b, a sensitivity of ~0.07 μV/V/Pa was obtained.

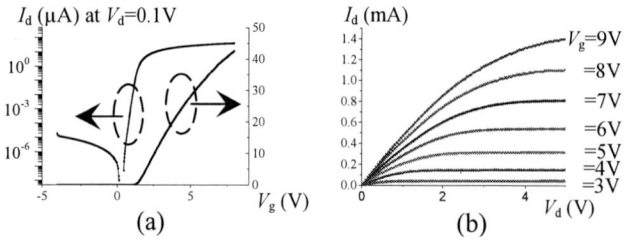

(a)   (b)

Figure 10. (a) Transfer and (b) Output characteristics of an NMOS.

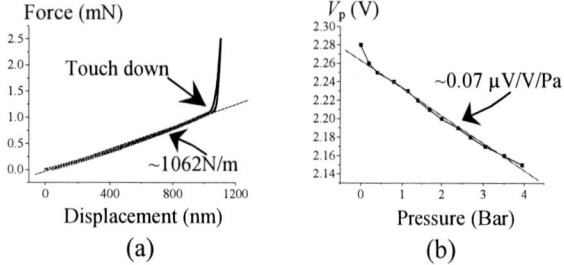

(a)   (b)

Figure 11. (a) Nano-indenter cyclic force-loading curve of a diaphragm. (b) Output voltage of a pixel subjected to uniform pressure-loading.

The functionality of the sensor-array was tested in combination with a printed-circuit board containing custom-designed periphery circuits for control, signal processing and read-out. The rows were sequentially scanned using an external shift register and the V_p values from the 16 columns of an activated row were digitized and sent to a computer for storage and display.

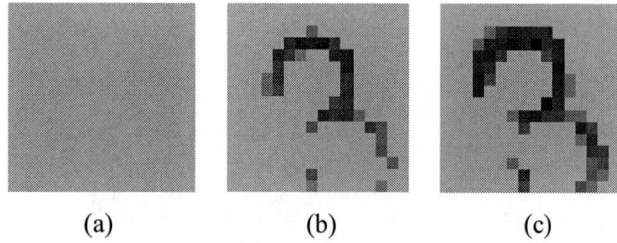

(a)   (b)   (c)

Figure 12. Images captured from the tactile sensor when pressed using a rubber stamp with a numeric character "3": (a) Before pressing; (b) After pressing; (c) After pressing with a stronger force.

Prior to the pressure-loading tests using a rubber stamp with a numeric character "3", several "frames" of the unloaded pixels were read, averaged and stored as the reference values to "zero" the sensor, resulting in the uniform image shown in Figure 12a. Frames of loaded pixels were then collected and

averaged. The differences between these and the reference values were computed and displayed, as shown in Figure 12b. When a heavier load was applied, a more striking image was obtained, as displayed in Figure 12c.

CONCLUSION

A general-purpose MEMS-CMOS integration technology based on the surface-diffusion of silicon is discussed. With the elimination of the conventional sacrificial layer etch, the issues of material- and process-incompatibility inherently present in many integration schemes are largely avoided. The application of this technology to the realization of a 16×16 active-matrix tactile sensor, integrating 256 force-sensing diaphragms, 512 pixel transistors and 512 piezo-resistors is also described. The spatial resolution of the sensor was ~145 "pixels per inch" and the pressure sensitivity was ~0.07 μV/V/Pa.

REFERENCES

[1] J. Smith, *et al.*, "Embedded micromechanical devices for the monolithic integration of MEMS with CMOS," in *International Electron Devices Meeting*, 1995, pp. 609-612.

[2] W.-C. Chen, *et al.*, "A generalized foundry CMOS platform for capacitively-transduced resonators monolithically integrated with amplifiers," in *IEEE 23rd International Conference on Micro Electro Mechanical Systems (MEMS)*, 2010, pp. 204-207.

[3] B. Li, *et al.*, "Germanium as a versatile material for low-temperature micromachining," *J. Microelectromech. Syst.*, vol. 8, pp. 366-372, 1999.

[4] A. E. Franke, *et al.*, "Polycrystalline silicon-germanium films for integrated microsystems," *J. Microelectromech. Syst.*, vol. 12, pp. 160-171, 2003.

[5] N. C. MacDonald, "SCREAM microelectromechanical systems," *Microelectronic Engineering*, vol. 32, pp. 49-73, 1996.

[6] T. Sato, *et al.*, "A new substrate engineering for the formation of empty space in silicon (ESS) induced by silicon surface migration," in *International Electron Devices Meeting*, 1999, pp. 517-520.

[7] R. Kant, *et al.*, "Experimental Investigation of Silicon Surface Migration in Low Pressure Nonreducing Gas Environments," *Electrochemical and Solid-State Letters*, vol. 12, p. H437, 2009.

[8] F. Zeng, *et al.*, "Self-formed cylindrical microcapillaries through surface migration of silicon and their application to single-cell analysis," *Journal of Micromechanics and Microengineering*, vol. 23, p. 055001, 2013.

ALD HONEYCOMB PLATES ENABLING ROBUST ULTRATHIN MEMS

Keivan Davami, Lin Zhao, and Igor Bargatin

University of Pennsylvania, Philadelphia, Pennsylvania, USA

ABSTRACT

This paper reports rigid MEMS structures made of ALD films, which had thicknesses as low as 10 nanometers and were patterned in the shape of a 3D honeycomb. Hexagonal honeycomb lattices offer a dramatically higher flexural stiffness compared to that of planar films, allowing us to fabricate large-area suspended devices without significant warping. Both alumina (Al_2O_3) and silica (SiO_2) structures were fabricated, each presenting a different set of fabrication challenges. The spring constants of the cantilever structures were measured and compared with the simulation results.

INTRODUCTION

Atomic layer deposition (ALD) has recently attracted much attention, largely because of the outstanding conformality of the resulting films and the large number of recipes for depositing an increasingly wide range of materials. While ALD films are often used as coatings for MEMS structures, there are still relatively few reported ultrathin devices that are made exclusively from ALD materials. Doubly-clamped suspended NEMS switches and tunneling devices were made from planar 30-nm-thick ALD tungsten layers [1]. An infrared bolometer was made using a structured ALD platinum film on top of an alumina seed layer with a total thickness of the order of 10 nm [2]. Finally, continuous membranes were fabricated out of nanometer-range-thick ALD alumina, whose Young's modulus was not much lower than in thicker films [3].

While MEMS devices made from ALD films have a number of advantages related to their low mass, flexibility, and transparency, there are also challenges associated with their extremely low thickness. In particular, planar ALD layers are subject to warping and damage caused by large stress gradients or other forces that may exist in ultrathin films during fabrication. However, these deformations may be mitigated by patterning the ALD films in three dimensions to include out-of-plane strengthening ribs. Honeycomb networks of such ribs are known from classical mechanics to offer a high isotropic bending stiffness per unit mass [4]. The performance of micro- and nano-scale honeycombs has been studied theoretically [5], but there is little or no data on the performance of nanoscale honeycomb MEMS structures.

MEMS made from ALD honeycomb plates are promising for a number of applications. For example, honeycombs made from refractive metals, such as ALD tungsten, can be used to create thermionic energy converters with a well-controlled gap [6]. Honeycomb cantilevers or beams can be used as resonant gas sensors, thanks to their low thickness and high stiffness, which lead to high resonance frequencies and high sensitivity to surface adsorbates. The transparency of the alumina ALD films may even enable their use as support films in electron microscopy.

FABRICATION

We designed and fabricated suspended honeycomb structures out of 10-nm-thick ALD alumina (Al_2O_3) and silica (SiO_2) in different geometries with lateral dimensions varying between 0.5 and 1 millimeter. Three clamping configurations have been used: cantilevers, doubly clamped beams, and rectangular plates clamped on all four sides (Figures 1(a) and 1(b)). A schematic of the fabrication procedure is outlined in Figure 1(c). The fabrication started with a double side polished Si wafer. SiN films with a thickness of 180 nm were deposited on both sides using PECVD. Honeycomb structures with a depth of 10 μm were patterned in silicon using photolithography and reactive ion etching (RIE). The back side was patterned via photolithography and the openings were obtained by RIE etching of SiN. The SiN mask was removed from the front side and the ALD layer was then deposited. For alumina deposition, trimethylaluminum (TMA) and water were used as precursors and two different temperatures, 150° C and 250° C, were used. The deposition rates were measured using an ellipsometer to be 0.91 Å/cycle at 150° C and 1.18 Å/cycle at 250° C. The deposition of SiO_2 was performed using 3-aminopropyltriethoxysilane (APTES), ozone, and water as precursors and the deposition rate was measured to be 0.72 Å/cycle at 150° C.

In order to pattern the ALD layer, a thick layer of SPR 220 resist was spin-coated on the structure. The thickness of the resist was measured to be 14 μm. After the spin coating and soft baking at 105° C the wafer was cooled down slowly to make sure the photoresist did not crack. After photolithography, inductively coupled plasma etching (ICP) with a BCl_3-based chemistry process was used to pattern the alumina ALD layer. In contrast, RIE was used to pattern the silica ALD layer.

Anisotropic KOH etching was the next step. Before placing the wafer in KOH, the top surface was covered with ProTek to prevent the ALD layer from being etched in the KOH solution. Silicon etching rate of 75 μm/hour was measured at 80° C in the 30% KOH solution. By accurately timing the KOH etching process, it was possible to stop the process ~ 20 μm from the top surface. The exact depth was measured using a Zygo profilometer. After that, the ProTek layer was removed and oxygen plasma was performed to make sure the surface of the wafer was completely clean and without any polymer residue. XeF_2 etching was used for the final release of the structure. Approximately 100 cycles (30 sec each) of XeF_2 etching with a ratio of 3.2:2 (XeF_2:N_2) was required for the complete release of the structures.

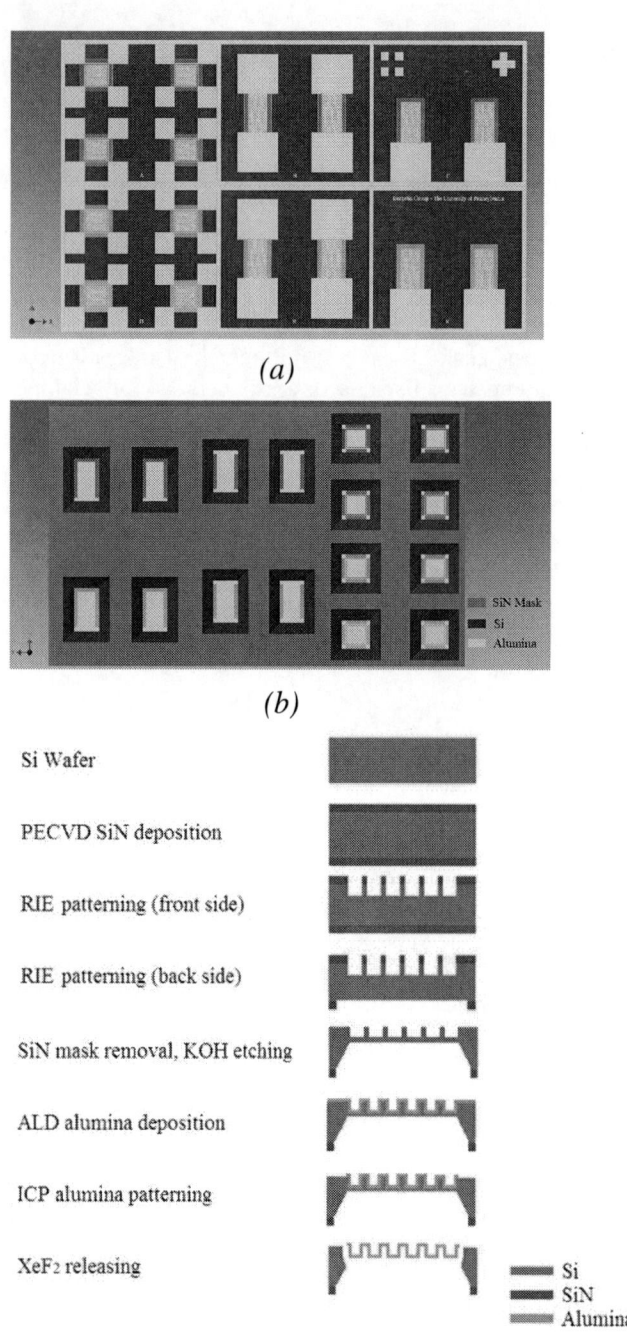

(a)

(b)

(c)

Figure 1: (a) Front view and (b) back view schematics of the final chip including three different types of devices, (c) a schematic of the fabrication steps.

XeF$_2$ etching is a fluorine-based isotropic dry etching method. It is a preferred method among other dry etching methods of Si due to its high selectivity, simplicity, and no requirement of expensive facilities. In general, SiO$_2$ and Al$_2$O$_3$ are resistant to XeF$_2$. While SiO$_2$ is usually used as a mask for XeF$_2$ etching, the etch selectivity is not infinite, which becomes particularly important during fabrication of ultrathin structures. Previously, it was reported that the etch selectivity of Si and SiO$_2$ is lower than 1000:1 [7]. In another

publication, an etch rate of ~ 2.5 Å/cycle was reported [8]. In our case, the etch rate of SiO$_2$ was measured to be ~ 6 Å/cycle. Since the thickness of the thinnest deposited ALD SiO$_2$ layer was 10 nm, the layer became completely etched before the structure can be released completely (Fig. 2).

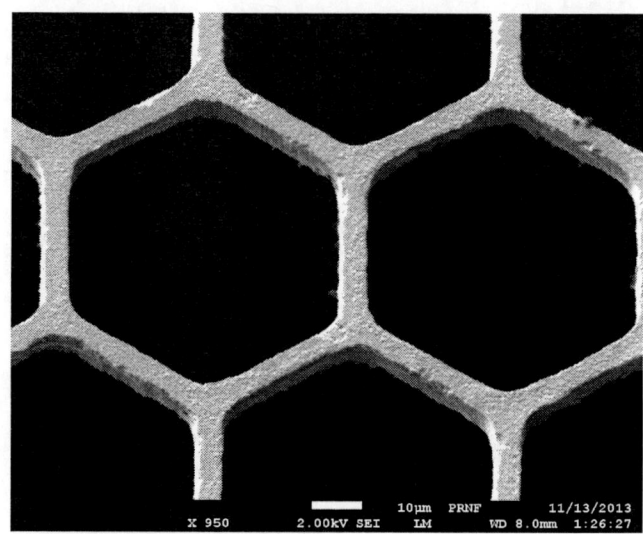

Figure 2: An SEM of the SiO$_2$ honeycomb structure. The SiO$_2$ ALD layer became etched before the complete release. The scale bar is 10 μm and the structure is tilted by 25□.

It was previously reported that for wafer-level XeF$_2$ etching of 3D structures, the average etch rate of SiO$_2$ typically increases with the number of etch cycles while the etch rate of Si decreases [7]. Thus, the etching selectivity between Si and SiO$_2$ deteriorates as the etching process proceeds. The high observed etch rate of SiO$_2$ was previously hypothesized to be caused by the formation of highly reactive XeF and atomic F species, which are capable of etching SiO$_2$ and are byproducts of the main reaction between XeF$_2$ and Si [9].

ALD alumina fared better than silica when exposed to XeF$_2$, but the selectivity was still not perfect. Alumina layers deposited at 150° C were etched in XeF$_2$ at a rate of ~ 4 Å/cycle and those deposited at 250° C at a rate of between 1 and 2 Å/cycle. Figure 3(a) shows a suspended beam structure after the release. Figure 3(b) shows the transparency of the suspended structure after the release in an ALD honeycomb with one hexagon broken out.

As shown in Figure 3(a), the released honeycomb lattice structures did not exhibit noticeable warping despite their large size. Honeycomb shells that formed individual units of the lattice were highly transparent both optically and in scanning electron micrographs, even for acceleration voltages as low as 5 kV (Figure 3(b)). However, they become opaque when the acceleration voltage in the SEM reduces to 1 kV (Figure 3(c)).

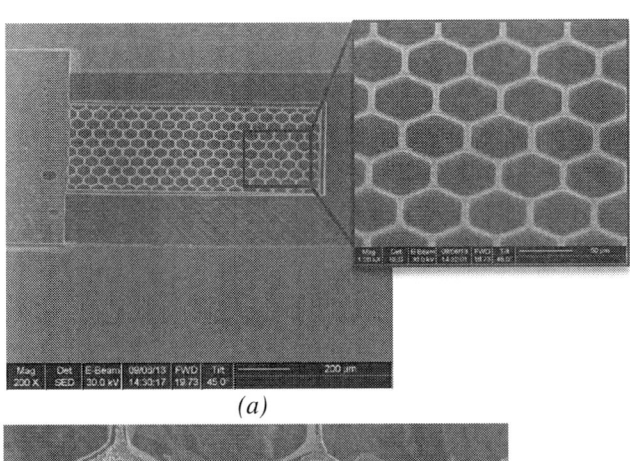

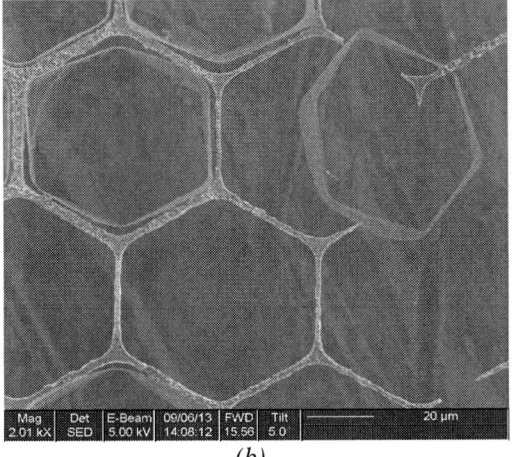

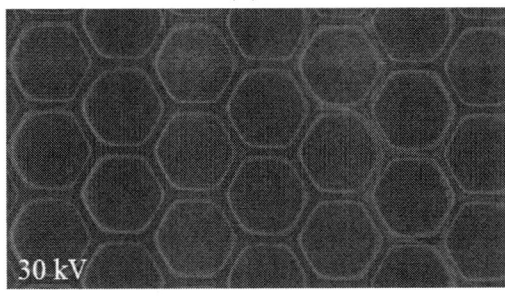

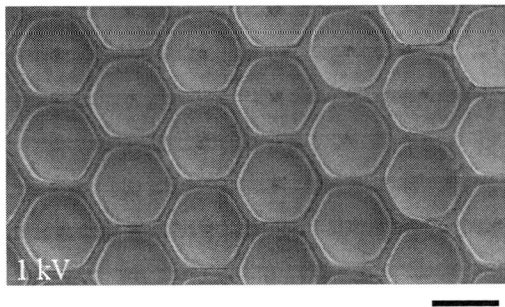

(c)

Figure 3: (a) SEM image of asuspended micro honeycomb cantilever (45° tilted). The inset zooms in on a section of the structure. The scale bar is 200 μm (b) ALD honeycomb with one hexagon broken out, illustrating the transparency of the ALD layer. The scale bar is 20 μm (c) A comparison of the transparency of the structures under different acceleration voltages of 30 kV and 1 kV. The scale bar is 50μm.

MECHANICAL CHARACTERIZATION

AFM measurements of force-displacement curves as well as finite element simulations of the fabricated honeycomb cantilevers were performed. In both experiments and simulations, one end of the cantilever was fixed and a force was applied on the other end (Figure 4). For honeycomb cantilevers with ALD film thickness of 100 nm, 50 nm, and 20 nm, the spring constant were predicted to be 0.09 N/m, 0.03 N/m, and 0.01 N/m, respectively, assuming Young's modulus of 400 GPa for the alumina layer.

For comparison, we also performed finite element simulations for planar cantilevers. When using the same film thickness, planar structures were approximately three orders of magnitude more compliant than honeycombs (Fig. 4). In other words, the planar structure needed to be much thicker than the ALD honeycomb to achieve the same spring constant. For example, a 50-nm-thick honeycomb cantilever had a spring constant of 0.036 N/m, similar to the spring constant of 0.032 N/m for a1-micron-thick planar cantilever.

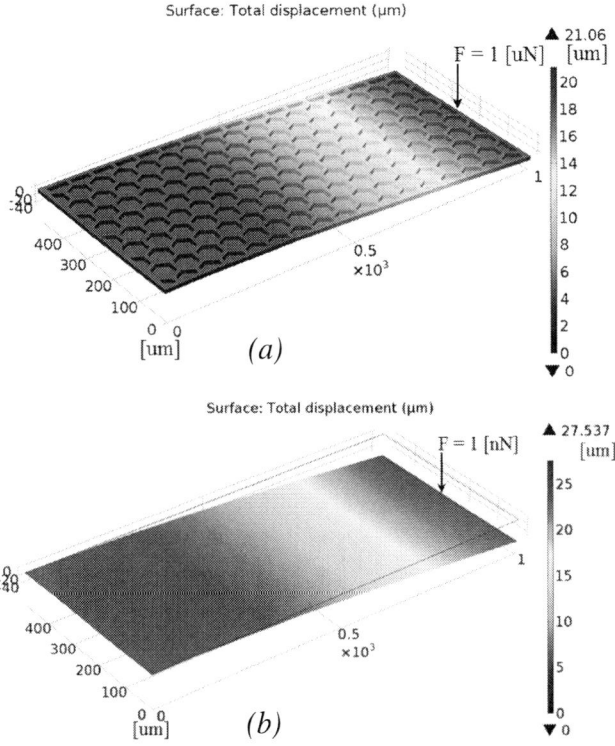

Figure 4: (a) A finite-element simulation of displacement for a honeycomb structure with a 1 micronewton normal load. (b) Same for a planar structure of the same thickness with a 1 nanonewton load (i.e., three orders of magnitude smaller).

The experimental results for the force-displacement behavior of two different fabricated cantilevers are shown in Figure 5 together with the predictions of finite-element simulations.The fabricated cantilevers had similar dimensions, but exhibited very different spring constants. This is attributed to the Si particles that remained attached to the ALD layer (Fig. 3(b)). For the same reason, the measured spring constants were larger than predicted by simulations. It is possible to remove the residual silicon particles using reactive ion plasma etching, as shown in Figure 3(c), but the resulting yield is low and we have not yet been able to measure such devices experimentally.

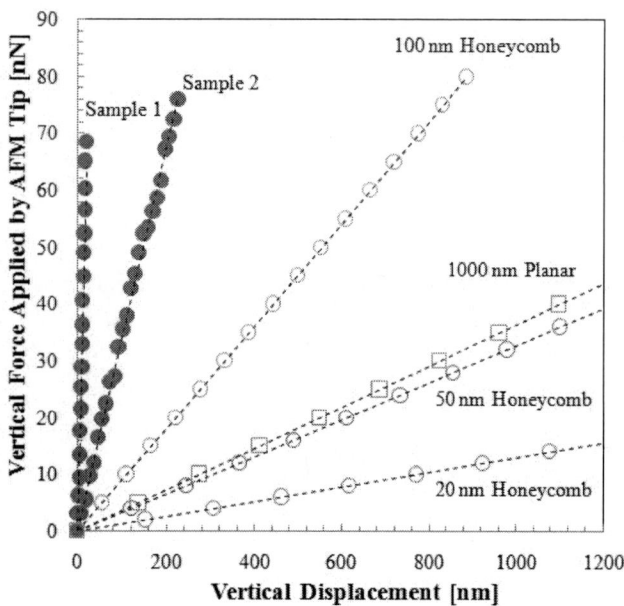

Figure 5: A comparison between the force-displacement curves measured for two 10-nm-thick microfabricated ALD honeycomb cantilevers (filled circles) and the predictions of finite element simulations for ALD honeycomb cantilevers with different thicknesses (open circles). Also shown for comparison is the predicted force-displacement curve of a 1000-nm-thick planar cantilever (open squares).

CONCLUSION

Freely suspended ultrathin ALD films are promising for a number of new applications because of their unique characteristics, such as low mass, flexibility, conformal deposition, and transparency. However, the warping and damage caused by stress gradients or other forces have largely limited the applications of ALD films to coatings that are deposited on stiffer structural materials. In this paper, we report freely suspended rigid MEMS structures made from ALD films that had a thickness of the order of 10 nanometers and were patterned in the shape of a 3D honeycomb, which successfully prevented warping and other types of damage.

The force-displacement curves of cantilevers with different thicknesses were calculated using finite element simulations. The spring constants of 3D honeycomb cantilevers were two to three orders of magnitude larger than those of planar cantilevers of the same thickness. To achieve a similar spring constant, planar cantilevers would need to be made from films about 20 times thicker than the 3D honeycomb cantilevers. The experimental spring constants of 3D honeycomb cantilevers—measured using an atomic force microscope—were found to be significantly larger than the predictions of finite element simulations due to the stiffening

effect of the residual silicon particles. Additional plasma etching steps have been shown to yield 3D honeycombs without any residual particles. Our future work will focus on energy and sensor applications of these versatile and robust nanoscale structures.

ACKNOWLEDGEMENTS

We thank Noah Clay, the director of the Quattrone Nanofabrication Facility at the University of Pennsylvania, for his helpful advice on nanofabrication.

REFERENCES

[1] B.D. Davidson, D. Seghete, S.M. George, V.M. Bright, "ALD Tungsten NEMS Switches and Tunneling Devices", *Sens. Actuators A: Phys.,* vol. 166, pp. 269-276, 2011.

[2] F. Purkl, T. English, G. Yama, J Provine, A. K. Samarao, A. Feyh, G. O'Brien, O. Ambacher, R.T. Howe, T. W. Kenny, "Sub-10 Nanometer Uncooled Platinum Bolometers via Plasma Enhanced Atomic Layer Deposition",*26th International IEEE Conference on Micro Electro Mechanical Systems (IEEE MEMS)*, 2013, pp. 185-188.

[3] L. Wang , J. J. Travis , A. S. Cavanagh , X. Liu , S. P. Koenig, P. Y. Huang , S. M. George,J. S. Bunch,"Ultrathin Oxide Films by Atomic Layer Deposition on Graphene", *Nano Lett*, vol. 12, pp. 3706-3710. 2012.

[4] L.J. Gibson and M.F. Ashby, *Cellular solids: Structure and Properties*, Cambridge University Press, 1997.

[5] H. X. Zhu, "Size-dependent Elastic Properties of Micro- and Nano-honeycombs", *J. Mechanics and Physics of Solids*, vol. 58, pp. 696-709, 2010.

[6] H. Lee, I. Bargatin, N.A. Melosh, R.T. Howe, "Optimal Emitter-Collector Gap for Thermionic Energy Converters", *Appl. Phys. Lett.,* vol. 100, pp. 173904-173908, 2012.

[7] D. Xu, B. Xiong, G. Wu, Y. Wang, X. Sun, Y. Wang, "Isotropic Silicon Etching With XeF_2 Gas for Wafer-Level Micromachining Applications",*J. Micro Electro Mechanical Systems*, vol. 21, pp. 1436-1444, 2012.

[8] K. Sugano, O. Tabata, "Etching Rate Control of Mask Material for XeF_2Etching Using UV Exposure", *Proc. SPIE* 4557, *Micromachining and Microfabrication Process Technology VII*, San Francisco, Sep 28, 2001.

[9] J.-F. Veyan,M. D. Halls, S. Rangan, D. Aureau, X.-M. Yan, Y. J. Chaball, "XeF_2-induced Removal of $SiO2$ Near Si Surfaces at 300 K: An Unexpected Proximity effect",*J. Applied Physics*, vol. 108, pp. 114914-114925, 2010.

CONTACT

*I. Bargatin, tel: +1-215-746-4887; bargatin@seas.upenn.edu

ARRAYS OF MICRO PENNING-MALMBERG TRAPS: AN APPROACH TO FABRICATE VERY HIGH ASPECT RATIOS

Alireza Narimannezhad, Joshah Jennings, Marc H. Weber, and Kelvin G. Lynn
Washington State University, USA

ABSTRACT

This paper reports on the progress of fabrication of very high aspect ratio (1000:1) micro-Penning-Malmberg trap arrays designed to store antimatter. The structure consists of thousands of 100μm diameter tubes etched by deep reactive ion etching through Si wafers. Cycles of thermal oxidation and wet etching in buffered oxide etch (BOE) minimized the sidewalls roughness and ensured a complete coating during gold sputtering. The wafers were then aligned and stacked in order to create the microtubes. Uniform plating with mean roughness of R_a=600nm was achieved by tuning the electroplating parameters.

INTRODUCTION

One of the authors (K. G. Lynn) has proposed a design, which consists of an array of microtraps [1] to increase positron storage by orders of magnitude as shown in Fig. 1. These traps have a large length to radius aspect ratio (1000:1) and a low confinement voltage (10V). The metallic microtubes screen the charge in each microtrap. Analytical calculations and simulation results have shown the trapped positron density is proportional to the inverse square of the trap radius [2]. Surko and Greaves [3] have proposed a multi-cell trap where each cell has a conventional aspect ratio of 10:1 with a diameter of one centimeter, with a confining voltage of a few kilovolts.

As the cylindrical symmetry is broken in a trap due to the fields and fabrication tolerances, the particles are lost over time. Studies have shown confinement times depend upon small trap asymmetries and subtle plasma effects [4]. With smaller radius microtraps, the space charge potential becomes negligible in the thermal equilibrium density distribution. However, as the electrodes become smaller in diameter, fabrication and alignment with respect to the magnetic field becomes more difficult.

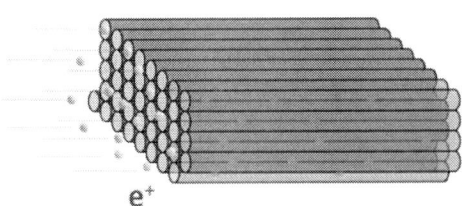

FIG. 1: Schematic configuration of an array of microtraps.

Other intrinsic asymmetries, such as patch effects, are also present. The patch effects encompass various phenomena, for instance, physically imperfect surfaces, chemical impurities, and random atomic lattice orientation. These all result in variation of the local surface work function [5] and induce local electric fields. The potential asymmetries affect the lifetime of a positron flying inside the microtrap [6]. Gold sputtering and electroplating of the wafers inside the vias helped reduce these effects.

DESIGN

Two hundred silicon dies of 530μm thickness and 38mm diameter provided by RTI International form the entirety of the trap as shown in Fig. 2 (each die contains thousands of 100μm diameter tubes etched by a deep reactive ion etching process). The dies were then aligned and stacked to create thousands of long tubes. More than one hundred million positrons can be trapped in one tube of this dimension when the applied electric field to the tube ends is only 10V [2]. Thus, the entire trap could hold more than 10^{12} positrons. The corresponding density, $1.6 \times 10^{11} cm^{-3}$, is comparable to the maximum reported density in conventional traps which uses kilovolts of end-electrode potentials [3].

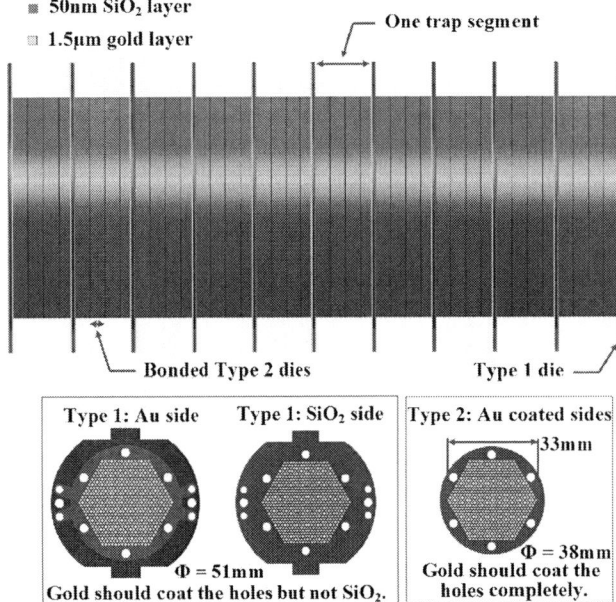

FIG. 2: The trap configuration showing axially stacked dies each including 20,419 number of 100μm vias.

Two types of dies were used to create a 10-section trap (10mm per section). Type 1 dies isolated each trap section from one another with a 50nm SiO_2 as an electrical barrier enabling application of different voltages on each section. They provide a constraint with tab features and an electrical connection on the Au side. Each trap segment between them was fabricated by bonding type 2 dies together.

978-1-4799-3510-9/14 $31.00 © 2014 IEEE   453

FABRICATION

Gold sputtering

To bond dies together and fabricate the segments shown in Fig. 2, we sputtered a thin gold film on the dies surfaces (1.5µm). The native oxide layer of silicon was stripped first by dipping into the BOE (Buffered oxide etch of NH_4-HF 10:1) solution for 10s. A 40nm thick TiW layer was sputtered to promote gold adhesion. The coatings were qualitatively checked by the typical tape test. It was crucial to have the gold coated on the micro-hole sidewalls to obtain a uniform potential through the vias. The solution was sputtering both die sides with an angle ($\tan^{-1}(D/L) \approx 11°$, where D is the vias' diameter and L is the length) from a horizontal state while it was rotating off-axis of the target as illustrated in Fig. 3. Flipping axis to the second side of the die was also important to get all the sidewalls coated.

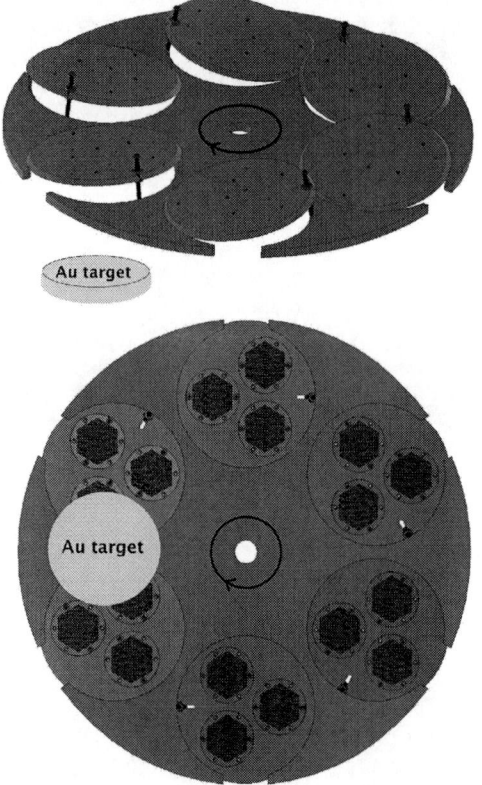

FIG. 3: 3D view and bottom view of the sputtering configuration illustrating the dies rotating off-axis of the gold target while it was kept with an angle from horizontal.

Electrical tests using an ohmmeter proved coating continuity after sputtering (0.4Ohm). Thus, the sample die was cleaved after sputtering both sides. By applying a 9.4VDC bias to the die and using a voltmeter with a very fine wire (48 AWG), we were able to probe the exposed vias. Each via exhibited the 9.4VDC bias. To further check the consistency, a 48 AWG wire was used to probe additional vias. A 2mm length was stripped at the wire end,

the wire was pushed all the way through a via, and the 2mm tip was then bent and slowly pulled back. This provided good contact with the sidewall while observing the voltage as the wire was pulled through a via. Several vias were checked in this manner across the die and all had a constant 9.4VDC.

The SEM images in Fig. 4 show sidewalls before and after gold sputtering. The scalloping size of the walls due to the Bosch process was measured as approximately 400nm. After gold sputtering the sample with the described technique, the gold grains were detected over the entire surface except in some "caves" of 500nm long at maximum. These caves were mostly found near the vias mid-section where the gold ions had difficulty reaching.

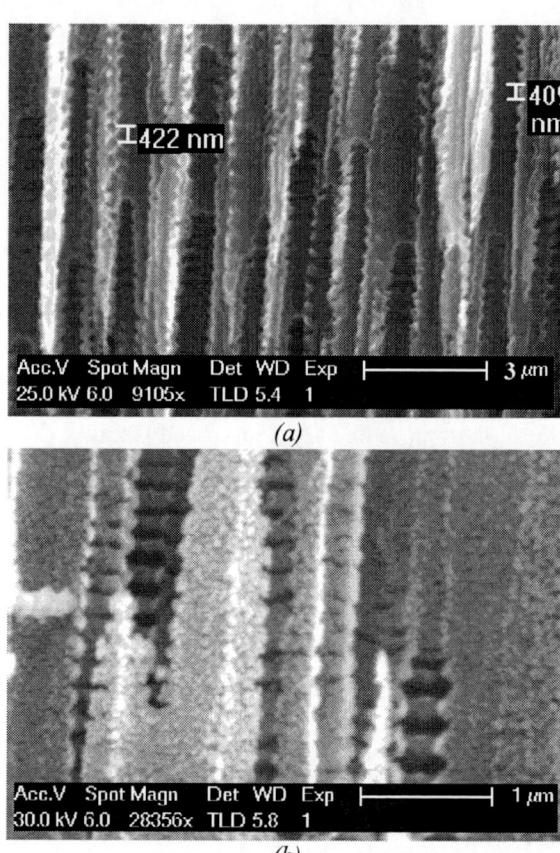

FIG. 4: (a) The sidewalls roughness in a bare silicon die due to the DRIE process with scalloping size of more than 400nm. (b) Gold grains are visible at the surface inside the tubes of gold-coated die except in some caves.

In order to minimize the patch effects on the lifetime of positrons, we circumvented the issue of uncoated caves by using two methods: improving the sidewalls roughness before sputtering, or electroplating the dies after bonding.

Roughness improvement

Prior to sputtering, thermal oxidizing and subsequent etching helped decrease the roughness of the sidewalls. We grew a 200nm oxide layer on the dies by high temperature

(1050°C) thermal, wet oxidation. The oxide layers were then etched by dipping the dies into the BOE solution for 10min. Repeating the oxidation and etching three times led to sidewalls with a reduced roughness, as shown in Fig. 5(a). We were then able to cover the entire surface with gold during the sputtering as shown in Fig. 5(b) as the caves' depths had been reduced dramatically.

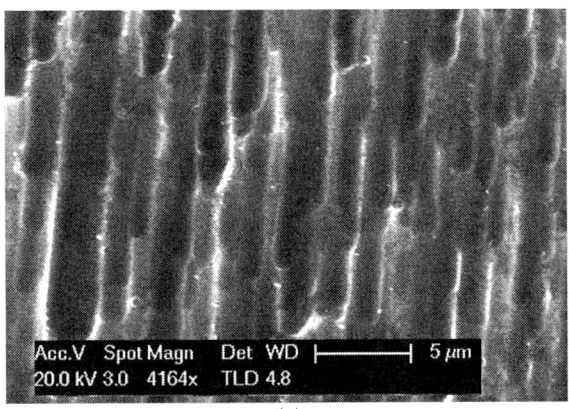

(a)

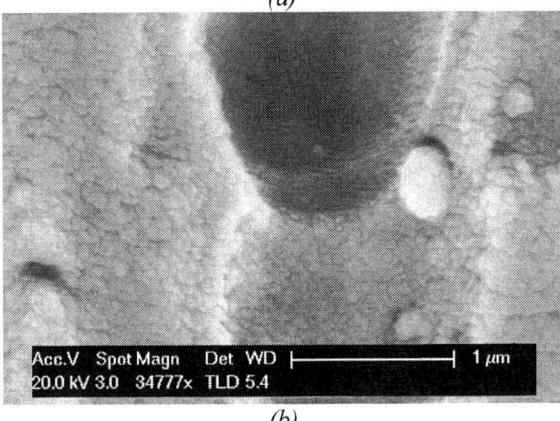

(b)

FIG. 5: (a) SEM of sidewalls that was subjected to three times of oxidization and etching cycle, (b) gold grains are visible even in deepest caves after sputtering.

Bonding

The dies were then bonded using a gold thermocompression bonding method. An alignment and bonding jig made of Invar36™ shown in Fig. 6 was precision-machined. Invar36 was chosen because it has a comparable thermal expansion to sapphire and silicon. The top plate (1) contained a screw that presses plate (2) against plate (3). The dies were slid onto the sapphire rods between plates (2) and (3). The sapphire rods used for bonding were 3.004 ± 0.001mm diameter and 50mm long with a straightness of 6µm over the entire length. There were two kinds of 3mm holes on the Si dies. Three out of six holes, tight-tolerance holes, had a diameter of 3.0096 ± 0.0005mm, and were utilized with the sapphire rods when bonding. The other three holes, regular holes, had a diameter of 3.0299 ± 0.0005mm, and were used for handling and coarse alignment of dies and bonded segments.

FIG. 6: The aligner-bonder jig. The dies were stacked through the sapphire rods in between plates (2) and (3).

The entire jig along with the dies was loaded into a furnace (Barnstead Thermolyne 1400™) that was preheated to the bonding temperature. The jig was baked for 60min to allow coming to equilibrium before adding pressure to the dies. Several bonding attempts were made at temperatures ranging from 200 to 300°C, pressures from 16 to 1230psi, and bonding times from 20 to 60min. Many authors report either much higher bonding temperatures (~400°C) or much higher pressures (~1,800psi) [7, 8]. Successful bonding in our jig occurred at a temperature of 250°C and a pressure of 1,230psi (9.7kN) for 60min and proved to be a reliable and reproducible process. A rudimentary drop test from a 5cm height was performed and demonstrated reliable bond strength in the bonded stack shown in Fig. 7(b). SEM images were taken from a single via. The misalignment of the dies was measured at ~4µm, as shown in Fig. 8.

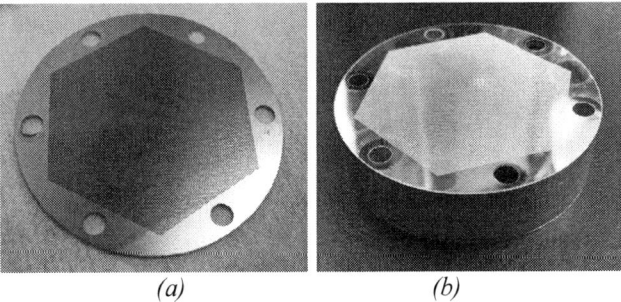

(a) *(b)*

FIG. 7: (a) One Si die etched in WSU. (b) A bonded stack of 18 sputtered dies with improved roughness.

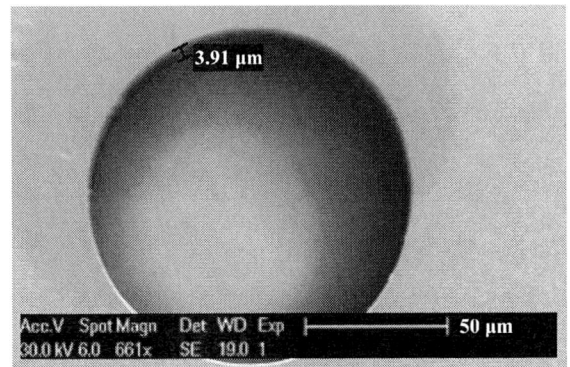

FIG. 8: SEM from a via in a stack of 13 dies focused on the middle where the maximum misalignment was detected.

Gold electroplating

The bonded stacks were then gold electroplated to fill the "micro-caves" which may have not been coated during gold sputtering (shown in Fig. 4(b)). The concern was that bare Si areas might have an effect on the charged particles in the microtrap. The "24k Pure Gold" gold plating solution (from Gold Plating Services company, Kaysville, Utah) designed to produce a gold electro-deposit with a higher purity than 99.9% was used in experiments. The surfaces were cleaned and activated with HCl acid for 5min prior to electroplating. Anodes were Platinized Titanium plates (also from Gold Plating Services). A series of experiments found optimum parameters that demonstrated good throwing power through the vias. The current density was an important parameter to get a uniform layer with lowest roughness, as shown in Fig. 9. The solution temperature was 55°C and the anode to cathode surface ratio was 3.5:1. A moderate agitation with a spin bar helped solution circulation inside the vias. The mean roughness after electroplating was $R_a = 600$nm when the current density was 1mA/in².

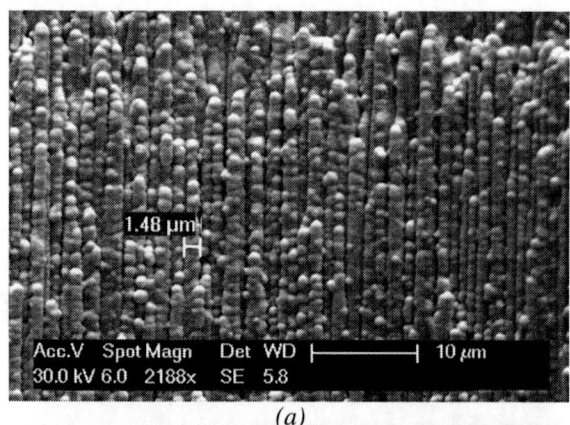

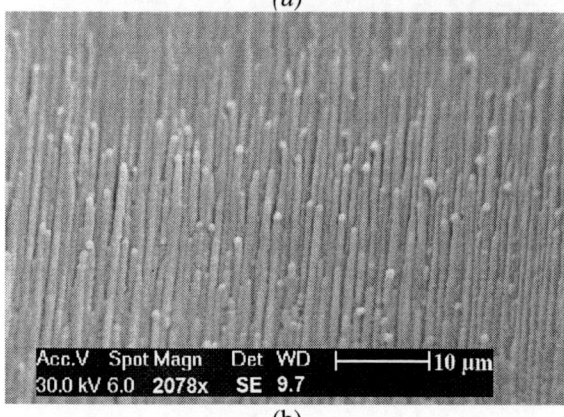

FIG. 9: SEM images of electroplated vias (a) using the current density of 20mA/in² exhibiting uniform but rough plating, (b) using the current density of 1mA/in² exhibiting a layer with a lower roughness.

Fig. 10 shows an electroplated stack of 13 dies. Due to the grain structure of the gold deposit, the reflective qualities of the surface are matte.

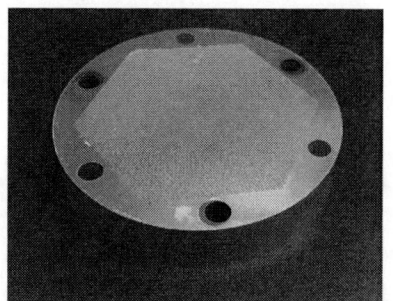

FIG. 10: The electroplated stack of 13 dies.

ACKNOWLEDGEMENTS

This work was supported by the Army Research Laboratory under contract W9113M-09-C-0075, and the Office of Naval Research under award #N00014-10-1-0543.

REFERENCES

[1] K. G. Lynn and R. G. Greaves, private communication, 2001.

[2] A. Narimannezhad, C. J. Baker, M. H. Weber, and K. G. Lynn, "Simulation studies of the behavior of positrons in a microtrap with long aspect ratio," *arXiv:1301.0030*, 2013.

[3] C. M. Surko and R. G. Greaves, "A multicell trap to confine large numbers of positrons," *Radiat. Phys. Chem.*, vol. 68, pp. 419-425, 2003.

[4] C. F. Driscoll, K. S. Fine, and J. H. Malmberg, "Reduction of radial losses in a pure electron plasma," *Phys. Fluids*, vol. 29, pp. 2015-2017, Mar. 1986.

[5] J. F. Jia, K. Inoue, Y. Hasegawa, W. S. Yang, and T. Sakurai, "Variation of the local work function at steps on metal surfaces studied with STM," *Phys. Rev. B*, vol. 58, pp. 1193-1196, 1998.

[6] A. Narimannezhad, J. Jennings, M. H. Weber, and K. G. Lynn, "Fabrication of High Aspect Ratio Micro-Penning-Malmberg Gold Plated Silicon Trap Arrays," *arXiv:1307.2335*, 2013.

[7] E. Jing, B. Xiong, and Y. Wang, "Low-temperature Au–Si wafer bonding," *J. Micromech. Microeng.*, vol. 20, pp. 095014-095019, 2010.

[8] Y. H. Wang, J. Lu, and T. Suga, "Low-temperature wafer bonding using gold layers", *Intl. Conf. EPT-HDP.* pp. 516-519, 2009.

CONTACT

*A. Narimannezhad, a.narimannezhad@wsu.edu
*K. G. Lynn, tel: +1-509-335-1131; kgl@wsu.edu

BATCH RELEASE OF MONODISPERSE LIPOSOMES TRIGGERED BY PULSED VOLTAGE STIMULATION

Toshihisa Osaki,[1,2] Koki Kamiya,[1] Ryuji Kawano,[1] and Shoji Takeuchi[1,2]
[1] Kanagawa Academy of Science and Technology, Kawasaki, JAPAN
[2] Institute of Industrial Science, The University of Tokyo, Tokyo, JAPAN

ABSTRACT

This paper describes a batch release technique for monodisperse liposomes immobilized on a substrate (Fig. 1). A single short-time pulsed voltage to the substrate induced detachment of the arrayed liposomes previously developed. Simultaneous release was observed shortly after the electrical stimulation. The release technique produced monodisperse and solvent-free liposomes freely suspended on the substrate, and allowed manipulation of the liposomes. The technique extends the application of our liposome array platform regarding collection/selection of the liposomes interested.

INTRODUCTION

Giant Liposome

Giant liposome or giant lipid vesicle is one of the self-assembled structures of lipid molecules in aqueous solutions, encapsulating the solution by a single or multiple lipid bilayer membranes with a diameter ranging between a few micrometers and a few hundred micrometers [1]. Such microcontainers find a number of biological and chemical applications in the fields of, for instance, drug delivery, cosmetics, and food products as well as the fundamental studies [2].

Selective Patterning for Giant Liposome Formation

Gentle hydration and electroformation are the most common methods for the giant liposome preparation [1], yet these methods have difficulty in controlling the shape and/or the size distribution of the formed liposomes [1,3,4].

Patterning of dried lipids prior to the hydration process will be one of the solutions of those problems; there were several techniques reported to obtain such lipid patterns [4–6]. In MEMS 2011, we also developed an alternative method that enables the precise control of the patterning process by the integration of an electrospray deposition (ESD) technique and a microfabrication process [7]; by the ESD, the spray of lipids was electrically led only to the bottom of microwells which was a conductive ITO-glass slide. With a simple hydration process of these dried patterns, we succeeded in the formation of giant liposomes on top of the microwells with a narrow range of the size distribution and obtained the desired sizes of liposomes by changing the microwell diameter, close to common cellular sizes.

Release of Immobilized Liposomes

The arrayed platform of formed liposomes was feasible for statistical and time-course observations by microscopy since the target liposomes were immobilized at the substrate surface. However, release and salvage of those liposomes have been hardly possible because the bilayer membrane easily deforms with mechanical stress; e.g. A glass capillary or a microfluidic channel may not be available due to a convection flow during the aspiration. Here we presented a technique that enables release of the arrayed liposomes from the substrate: A pulsed voltage on the substrate triggers to detach the liposomes at once as shown in Fig. 1.

EXPERIMENTAL

Substrate Fabrication & Lipid Patterning

A conductive/non-conductive patterned substrate was prepared as follows. First, a thin poly(chloro-*p*-xylylene) film was coated on an ITO-glass slide by a chemical vapor deposition method. Next, an array of microwells with 10 μm in diameter was fabricated on the film with a standard UV lithography process (Fig. 2a, left).

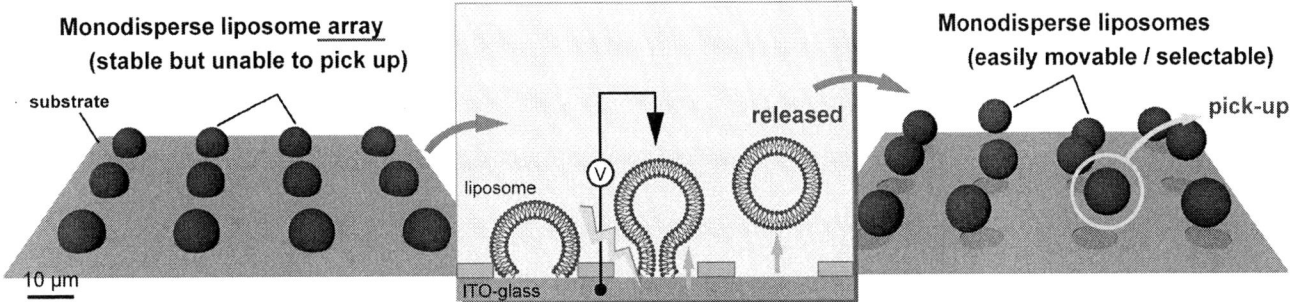

Figure 1: Conceptual diagram of monodisperse-liposome release triggered by a pulsed voltage. A monodisperse liposome array was reported previously (left, [7]); immobilized on the substrate, feasible for time-course observation, but unable to move or pick-up. In this work, we motivated batch release of the arrayed liposomes from the substrate by using a short-time pulsed voltage. The electrical stimulation enforces liposome detachment from the substrate (middle), and produces monodisperse liposomes floating on the substrate, applicable for salvage and move (right).

A selective deposition of lipids was performed on the substrate by using ESD [7]. 0.5 mg/mL DOPC/DOPS lipids (1:1) dissolved in a mixture of chloroform and methanol was used. A lissamine-rhodamine B labeled lipid was also mixed for fluorescent observations (Rhod- DPPE, 1 wt% to the total phospholipid weight). After the deposition, the samples were kept in vacuum (Fig. 2a, right).

Batch Release of Liposomes by Pulsed Voltage

Hydration of dried lipids was simply performed by infusion of an aqueous solution (20 mM sucrose) to the channels. After liposome array formation, a pulsed voltage was applied between the ITO electrode, i.e. the substrate, and Ag/AgCl electrode, which was set into the aqueous solution (Fig. 2b).

RESULTS

We confirmed that a pulsed voltage to the ITO electrode effectively released the liposomes from the substrate (Fig. 3). Examining the stimulation conditions, we found that constant DC voltages or a large pulse caused membrane rupture. The images taken by a high-speed confocal laser scanning microscope traced the simultaneous detachment of arrayed liposomes from the surface, although the detailed mechanism of the phenomena remained unclear. The uniformity of the liposome diameters was confirmed before and after the release. More detailed results and discussion will be presented on the poster presentation.

CONCLUSIONS

This work presented the release technique of the immobilized monodisperse liposome array using a pulsed voltage. We confirmed that the liposomes were instantly detached from the substrate by the electrical stimulation. The released liposomes kept their monodispersity. Patterning of the electrode on the substrate will implement selective release of the liposomes by the same technique.

ACKNOWLEDGEMENTS

We thank Ms. U. Nose for their kind technical support. This work was partly supported by JSPS (Grant-in-Aid for Young Scientists A; 25706015), Japan.

REFERENCES

[1] P. L. Luisi, P. Walde, *Giant Vesicles*, John Willy and Sons Inc., New York, 2000; N. Duzgunes, *Methods in Enzymology Volume 367 Liposomes Part A*, Academic Press, California, 2003.

[2] A. Karlsson, R. Karlsson, M. Karlsson, A-S. Cans, A. Strömberg, F. Ryttsén, O. Orwar, "Networks of Nanotubes and Containers", *Nature* **2001**, *409*, 150-152; I. A. Chen, K. Salehi-Ashtiani, J. W. Szostak, "RNA Catalysis in Model Protocell Vesicles", *J. Am. Chem. Soc.* **2005**, *127*, 13213-13219; G. Tresset, S. Takeuchi, "Utilization of Cell-sized Lipid Containers for Nanostructure and Macromolecule Handling in Microfabricated Devices", *Anal. Chem.* **2005**, *77*, 2795-

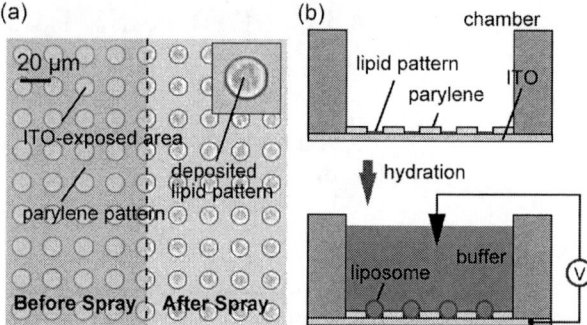

Figure 2: (a) Microscopic image of the substrate before and after the lipid patterning by using electrospray deposition. (b) Cross-sectional diagrams of a hydration chamber and a DC source for pulsed voltage stimulation.

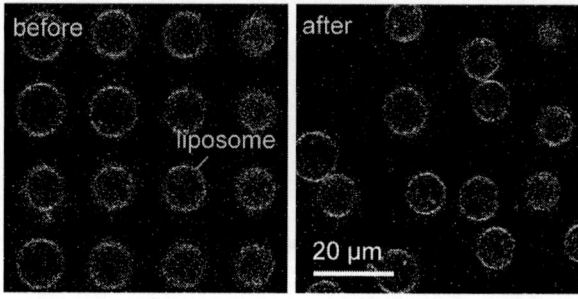

Figure 3: Confocal fluorescence images of liposomes before (left) and after (right) the application of the short pulsed voltage.

2801.

[3] K. Kuribayashi, G. Tresset, H. Fujita, S. Takeuchi, "Electroformation of Giant Liposomes in Microfluidic Channels", *Meas. Sci. Technol.* **2006**, *17*, 3121-3126.

[4] M. Le Berre, A. Yamada, L. Rech, Y. Chen, D. Baigl, "Electroformation of Giant Phospholipid Vesicles on a Silicon Substrate: Advantages of Controllable Surface Properties", *Langmuir* **2008**, *24*, 2643-2649.

[5] K. Kuribayashi, S. Takeuchi, "Electroformation of Solvent-Free Lipid Membranes over Microaperture Array", *Proc. IEEE MEMS* **2008**, Tucson, 296-299.

[6] P. Taylor, C. Xu, P. D. I. Fletcher, V. Paunov, "Fabrication of 2D Arrays of Giant Liposomes on Solid Substrates by Microcontact Printing", *Phys. Chem. Chem. Phys.* **2003**, *5*, 4918-1922.

[7] T. Osaki, K. S. Kuribayashi, R. Kawano, H. Sasaki, S. Takeuchi, "Uniformly-Sized Giant Liposome Formation with Gentle Hydration", *Proc. IEEE MEMS* **2011**, Cancun, 103-106; T. Osaki, K. Kamiya, R. Kawano, H. Sasaki, S. Takeuchi, "Towards Artificial Cell Array System: Encapsulation and Hydration Technologies Integrated in Liposome Array", *Proc. IEEE MEMS* **2012**, Paris, 333-3366.

CONTACT

*T. Osaki, Artificial Cell Membrane G, Kanagawa Academy of Science and Technology (Phone +81-44-819-2037, Email tosaki@iis.u-tokyo.ac.jp)

CLARITAS™ – A UNIQUE AND ROBUST ENDPOINT TECHNOLOGY FOR SILICON DRIE PROCESSES WITH OPEN AREA DOWN TO 0.05%

Oliver Ansell[1], Richard Barnett[1], Thomas Haase[2], Ling Xie[3], Steve Vargo[1] & Dave Thomas[1]

[1]SPTS Technologies Limited, UK
[2]Fraunhofer IPMS, Germany
[3]Harvard University, Centre for Nanoscale Systems, USA

ABSTRACT

Endpoint detection (EPD) is a critical control functionality for many etch processes, especially for deep silicon etches [1] that terminate on an underlayer. Where this device structure is employed, it is vital that the point at which the etch process reaches the underlayer is detected as promptly as possible. This allows for proper management of the overetch to clear all features to the underlayer without running for longer than necessary and introducing lateral notching to the base of the feature [2]. As device requirements have become more stringent, lower open areas and higher aspect ratios have necessitated development of more sensitive techniques to achieve successful endpoint detection and overetch control.

INTRODUCTION

Benefits of Endpoint Detection

There is little doubt that the ability to detect when an etch process reaches an intermediate, or stop, layer within a structure is an important capability to employ. The key benefits related to endpoint detection are well known. Primarily, EPD is a process control function that stops or adapts etch conditions when an aspect of the process changes, either as etch depth increases or as different layers in the wafer are reached. This can prevent damage to the feature sidewalls or to the underlayer itself, restricting the exposure of completed features to the plasma chemistry to no longer than is necessary to finish the etch across the wafer.

There are a number of other benefits that arise from the use of EPD. Through use of the EPD monitoring, users can review EPD data off-line and use software rather than wafers to optimize the treatment of endpoint signals. Fewer test wafers are used and process windows are generated much sooner. With a specific on-wafer behavior, or plasma chemistry, being used to detect the endpoint state, each wafer will be processed until the same pre-determined circumstances are reached. This means that every wafer will be processed in the same, repeatable manner.

Overetch conditions are tuned to the endpoint signal, either for the etch to terminate at that point, or for an additional time to be added to the process for completion. Applying this approach means that there is no extraneous time in the process, resulting in an optimized throughput..

Finally, the risk with timed etches is that the devices require re-work if underetching occurs or features receiving an excessive etch will be rendered unusable. With EPD control giving repeatable and expected on-wafer results, device yields are improved compared to the timed etch case, and levels of scrap significantly reduced.

CATEGORIES OF EPD

There are three main categories of endpoint detection. These are i) etch tool monitoring, ii) plasma, or gas, monitoring and iii) *in situ* wafer monitoring.

Etch Tool Monitoring

As most recipe parameters are datalogged, typically every second during every process, if any of these parameters change significantly at the point an endpoint occurs on the wafer, then an endpoint signal can be generated so as to trigger a change in recipe conditions.

The types of parameters that can be used include APC valve position, He backside flow, etc. Test wafers are required to correlate the datalogged parameter with the process endpoint.

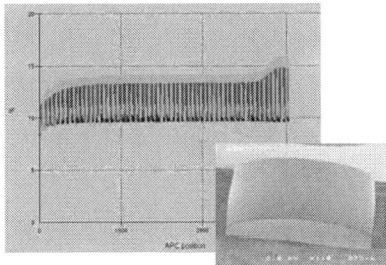

Figure 1: APC valve opening used as endpoint for a microphone cavity etch.

This is a low cost technique, reliant only on having appropriate software on the system. But, there are few applications where this technique can be applied, e.g through-wafer etches exposing the ESC or etches to large buried cavities.

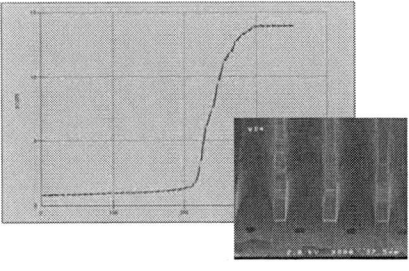

Figure 2: He backside cooling flow used as endpoint for etch through to ESC.

Plasma, or gas, monitoring

Here the plasma or gas is monitored to detect if any of the reactants or reaction products are affected when the endpoint condition is reached.

There are a number of techniques available to perform this particular function. None of the plasma monitoring methods are reliant on having additional target features present on the wafer. Plasma monitoring is only applicable to wafers where there are differing layer(s) within the etched feature. There is no step-change in signal generated if a bulk wafer is etched. The techniques are described, in Table 1, below;

Table 1: Comparison of plasma monitoring techniques.

Technique	Cost	Benefits (+)	Disadvantages (-)
LIF	$$$$	• Provides some detail of plasma chemistry	• Limited to species that fluoresce • Difficult to interpret data • Affected by plasma glow
Mass Spectroscopy	$$$	• Analyses all species	• Intrusive to plasma • Needs high vacuum (differential pumping) • Species are ionised/cracked before detection • Difficult to interpret data • Low filament lifetime in chemical plasmas
OES	$$	• Easy to use & interpret	• Limited to species that are emitting light
Langmuir Probe	$	• Easy to use	• Intrusive to plasma • Limited to ion & electron information • No chemistry information • Difficult to interpret

Optical emission spectroscopy (OES), is the most widespread of the plasma monitoring techniques in use. Whilst it is not the lowest cost method the OES approach is the easiest to implement and to interpret. The only drawback is that the OES equipment is limited to detecting those species that emit light. But, in an appropriately designed chamber [3] and chemistry this is not usually an issue.

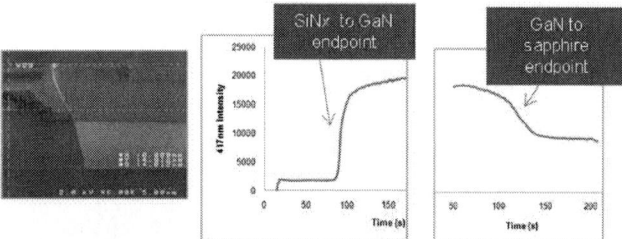

Figure 3: OES traces for SiN/GaN etches

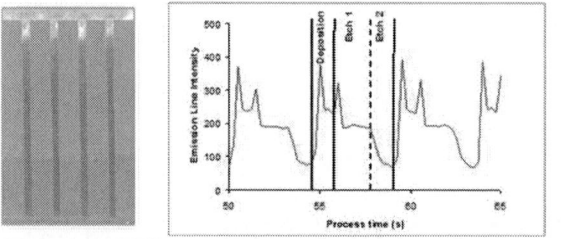

Figure 4: OES trace for Bosch Si etch, with repeating steps highlighted.

In-situ Wafer Monitoring

The third, and final, group of EPD methods relies upon being able to monitor events directly on the wafer. There are two main techniques; reflectometry and interferometry, and each one can be used with either a white-light or infra-red incident wavelength.

White-light reflectometry detects the changes in reflected light intensity as the light signals return from different layers. Since the illuminating light has a relatively large diameter, taking signals in from a large area, there is no need for a specific target feature to be included on the wafer pattern.

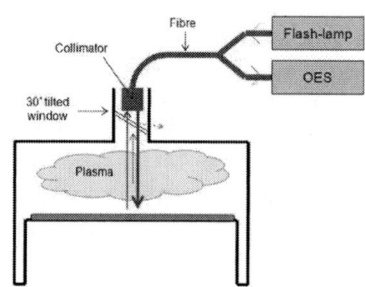

Figure 5: Schematic of plasma etch chamber with white-light inteferometry.

However, this method is only applicable to a sub-set of etches, including non-Bosch silicon etch. The Bosch process with a bright plasma due to the high pressures and relatively deep structures, does not provide a suitable environment for the white-light signals to be readily reflected back up to the detector.

White-light inteferometry uses the same hardware as the reflectomtery equivalent, but instead the detector analyses the interference fringes caused by the multiple reflections from the various layers on the wafer. E.g. the mask surface, the unetched material beneath the mask and the moving etch front. The various return signals can be de-convoluted and used to generate both in-situ etch rate data and mask selectivity information.

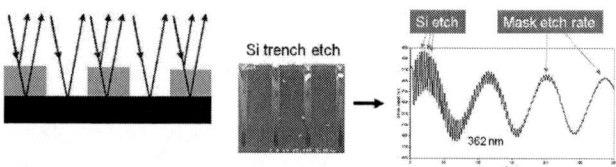

Figure 6: Schematic of white-light reflectometry.

For the same reasons as reflectometry, although ideal for endpoint detection of shallow, non-Bosch Si etch processes and without need of target features, the inteferometry approach cannot really be used for the Bosch processes.

Reflectometry and Inteferometry can be achieved with infra-red LASERs as well as white-light. By moving to the new incident light, there are some advantages, particularly in respect of the Bosch process. Table 2, below, compares the two LASER based EPD methods.

Table 2: Comparison of LASER EPD techniques.

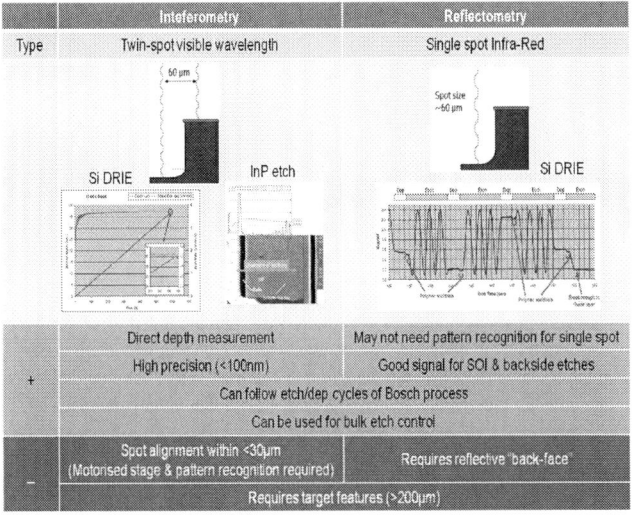

Neither of the LASER based EPD methods have successfully been implemented for deep Si etch. There remain concerns over the use of target features as well as the ability to accurately deliver and detect the signal through the highly illuminated Bosch plasma. Furthermore, the etch depths routinely used and the types of module geometries also seem to provide significant challenges for this concept.

OES FOR BOSCH ETCHING

Limitations of OES for deep Si etch

For many years, EPD has been explored and developed to produce the highest level of etch process control possible. It has already been described that bulk etch depth control for the Bosch process has not been successfully achieved. However, for etch processes using stop layers, OES has become pre-requisite for successful process control, but there remain some limitations to be overcome.

OES operates by detecting changes in emitting species,. There needs to be a ready source of material that can be energized to provide these species. Given the make up of the plasma within a DRIE reactor, it is more typical that the initial etch product, e.g. SiF* will get fluorinated quite quickly into something far more stable, e.g. SiF_3 or SiF_4 that does not emit.

However, the main source is the wafer pattern and there is a general trend for reducing the open area as MEMS designs, in particular, become more sophisticated. And so, as the open area reduces, any changes in the plasma that would be used for detecting an endpoint, are consequently also smaller and less obvious.

Also, the Bosch process tends to run at high process pressures, sometimes exceeding 100mT. This has the consequence of providing a background level against which it is difficult to detect any changes in the amount of emitting species. Overall, it can be said that by directly monitoring the plasma there is a restriction placed on the detection capability. These restrictions are based on open areas of <20% and at higher pressures, where the process is targeting high Si etch rates.

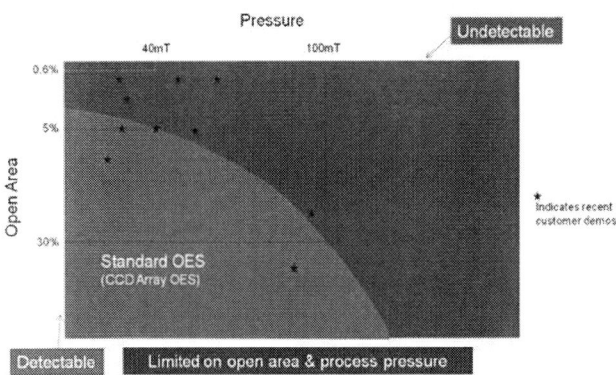

Figure 7: Operating window of standard OES

Claritas™ – Enhanced OES for Bosch Etching

SPTS developed the Claritas [4] enhanced OES to directly answer demands for clear, robust EPD control for etch processes of low open areas or high process pressures.

The use of Claritas decouples the signal detection from the actual etch plasma, eliminating the boundaries in place when monitoring the etch plasma directly. Compared to standard chamber mounted OES, the operating window is significantly enlarged.

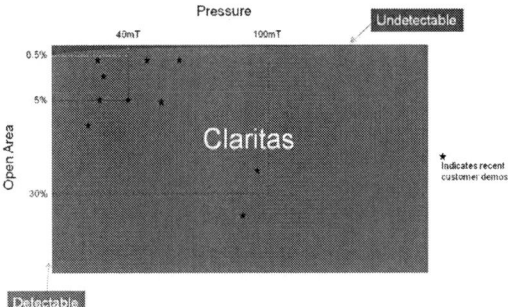

Figure 8: Extended OES detetction window for SPTS Claritas™

Claritas is an additional piece of hardware, positioned off the main plasma reactor. The plasma extinguishes en-route to the Claritas unit, where the gases are re-ignited using an RF supply independent of those running the main etch plasma.

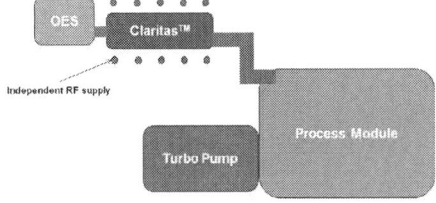

Figure 9: Schematic of Claritas installation

The Claritas RF supply can be tuned to enhance the concentration of the desired emitting species (e.g. SiF*). And since the OES detector head is attached to the Claritas unit, the monitoring is removed from the many events going on in the main plasma reactor.

Claritas significantly outperforms standard OES in all aspects. The chart below shows the comparison in signal strength when Claritas is used compared to standard hardware.

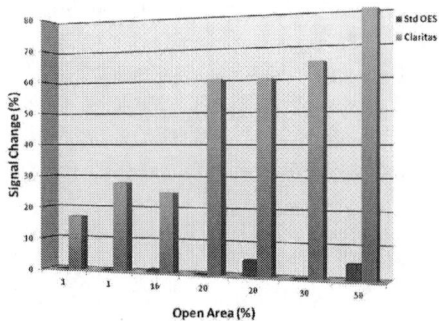

Figure 10: Chart showing Claritas detected signal change vs standard OES

The following examples show the types of signals generated by Claritas when used to monitor Si etching to buried layers, typically SiO2, and buried cavities.

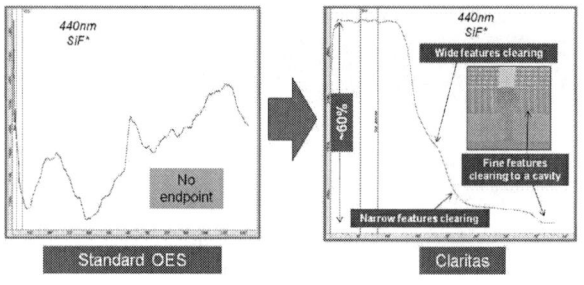

Figure 11: Comparison of standard OES and Claritas for detection of multiple features on a single pattern

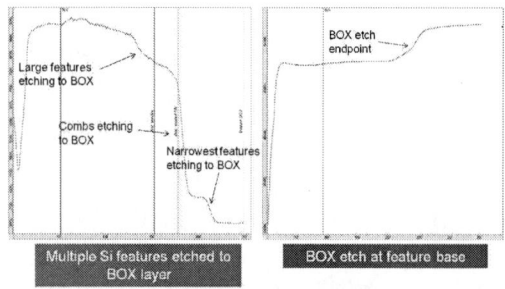

Figure 12: Use of Claritas to detect endpoint for Si etch to buried oxide layers and subsequent etch of the buried oxide, itself

There is also data showing the extreme end of the Claritas monitoring capability. When running wafers with open areas reducing from 3% to ~0.05% in an SPTS Rapier PM fitted with Claritas, it can be seen that detection of the endpoint is achieved at all levels of exposed Si.

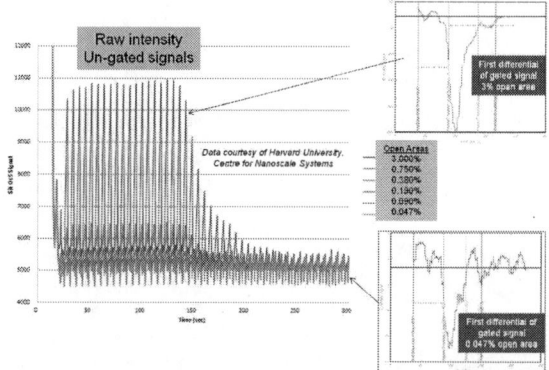

Figure 13: Claritas capability down to 0.05% open area

CONCLUSIONS

It can be seen that there is a wide range of endpoint techniques available for plasma etch processing. But, it is also evident that there are many limitations and compromises for these techniques when applied to the Bosch process. OES has been the most widely adopted endpoint method. However, the trend for low open area designs has reduced it's overall effectiveness.

SPTS developed an enhanced OES approach using the Claritas hardware which expanded the operating envelope to include very low open area process windows (~0.05%) as well as those at higher process pressures (>100mT). This capability has enabled MEMS device design to continue developing in order to meet the rapidly expanding demands from the consumer, in a robust and cost-effective manner.

ACKNOWLEDGEMENTS

I would like to thank Dr Oliver Ansell, Dr Steve Vargo, Dr Ling Xie and Dr Thomas Haase for the technical data provided for this paper.

REFERENCES

[1] Franz Lärmer, Andrea Schilp: Patents DE 4241045, US 5501893 and EP 625285
[2] Janet Hopkins, Ian Ronald Johnston, et al: Patent US 6,187,685
[3] "A New Plasma Source for Next Generation MEMS Deep Si Etching: Minimal Tilt, Improved Profile Uniformity and Higher Etch Rates" Proc. ECTC 2010, pp. 1056-1059
[4] Oliver Ansell, Anthony Barrass, et al: Patent Application US 2010/0230049

CONTACT

*Richard Barnett, tel: +44(0)1633474555;
richard.barnett@spts.com

CONCURRENT REACTIVE ION ETCHING EMPLOYING MICROMACHINED IONIC LIQUID ION SOURCE ARRAY

Ryo Yoshida[1], Motoaki Hara[1], Hiroyuki Oguchi[1], Tatsuya Suzuki[1], and Hiroki Kuwano[1]
[1]Tohoku University, Sendai, Miyagi, Japan

ABSTRACT

This paper describes concurrent reactive ion etching using micro ionic liquid ion source (ILIS) array. The system consists of micro needle emitters and a reservoir for the ionic liquid (IL) of 1-ethyl-3-methylimidazolium tetrafluoroborate ([EMIM]-[BF$_4$]). The ion beam etching of a (100) silicon substrate using the fabricated ILIS array was demonstrated. As a result of mass spectroscopy during the etching, the peaks of SiF$^+$, SiF$_2^+$, and SiF$_3^+$ were observed. The chemical reaction between the silicon and fluorine based ions from the IL was confirmed. Also, etching rate of the silicon using the ILIS array applying 5.1 kV ion-acceleration voltage was calculated from the etched dimple on the substrate and was 1.5 times larger than that of a conventional focused Ga$^+$ ion beam applying 30 kV ion-acceleration voltage.

INTRODUCTION

Today, a variety of MEMS devices such as accelerometers, gyroscopes, optical switches *et al.* are commercialized all over the world. These devices are manufactured massively employing photofabrication technologies for cost reduction as well as miniaturization. However, these MEMS devices are becoming complicated in order to improve the performance and create new functionalities. Since market is not as large as that of the LSI, it is not easy to suppress the manufacturing cost of the MEMS devices. Hence, the microfabrication technologies suitable to the high-mix low-volume production are strongly demanded for the MEMS devices.

From such demand, multiple focused ion beam (FIB) system as shown in Fig. 1 was proposed [1]. In the system, emitters, extractors and lenses are integrated in an array arrangement. Emitted ion beams are controlled individually by the extractors and lens plates. The emitter plate has multiple ion emitters immersed in the ionic liquid (IL) used as an ion source. In the multiple FIB system, various processing such as etching, deposition, surface modification, and image observation are achieved concurrently. Therefore, various multi-processing with high throughput in one vacuum chamber are possible.

In this study the IL of 1-ethyl-3-methylimidazolium tetrafluoroborate ([EMIM]-[BF$_4$]) was adopted as the source material of ions instead of the conventional liquid metals such as Ga or In. The IL is molten salt that is liquid over a wide range of temperature including room temperature. By using the IL, the FIB system can be simplified and miniaturized since the heating unit necessary for the liquid metals is no longer required [2]. Moreover, the IL is expected to increase the etching rate of silicon since the anion of the IL contains silicon reactive fluorine [3].

In this report, we investigated a method to increase the

ion emission from the ILIS array. We also investigated mechanism of the silicon etching using the *in situ* mass spectroscopy, energy dispersive X-ray spectrometry (EDX), and infrared spectroscopy.

FABRICATION OF THE ILIS ARRAY

The ILIS array was fabricated using the process flow reported in Ref. 1. In the process, micro needle emitter array and a reservoir for the IL were fabricated with single photomask based on silicon bulk micromachining. Then, the IL was poured into the reservoir. The wettability of the reservoir surface is important to uniformly spread the IL on the reservoir [4]. In this study, O$_2$ plasma was used to enhance the wettability of the surface of the reservoir. Figure 2 shows contact angle of water on the silicon surface before and after exposure to O$_2$ plasma. Decrease in contact angle indicated improvement of the wettability by the exposure.

Figure 3 shows scanning electron microscopy (SEM) images of the fabricated ILIS array. In this device, 25 (5×5) emitters were arrayed in the IL reservoir. These images confirmed that we succeeded to fabricate the sharp emitter tip on the nanometer scale.

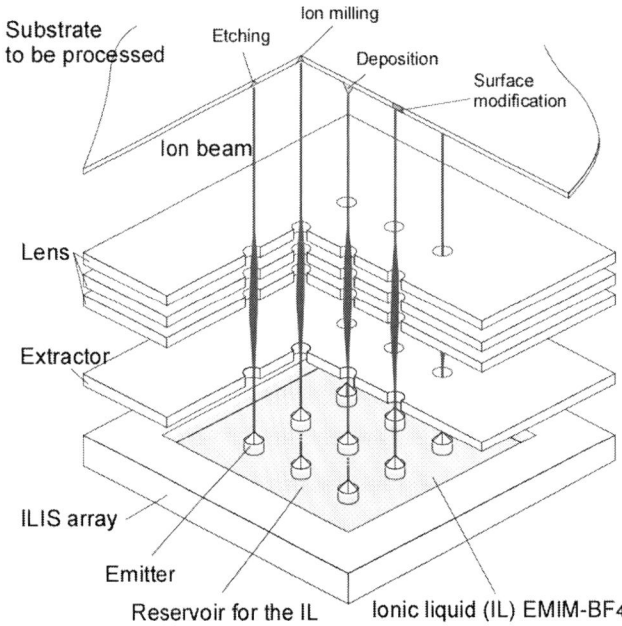

Figure 1: Schematic illustration of the multi focused ion beam system

978-1-4799-3510-9/14 $31.00 © 2014 IEEE 463

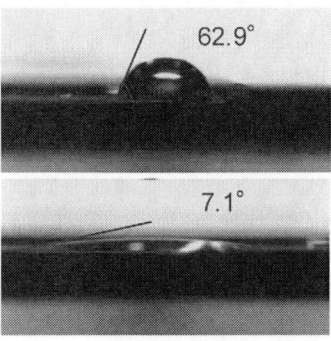

Figure 2: Contact angle of water on the device before O_2 treatment (upper) and after O_2 treatment (lower)

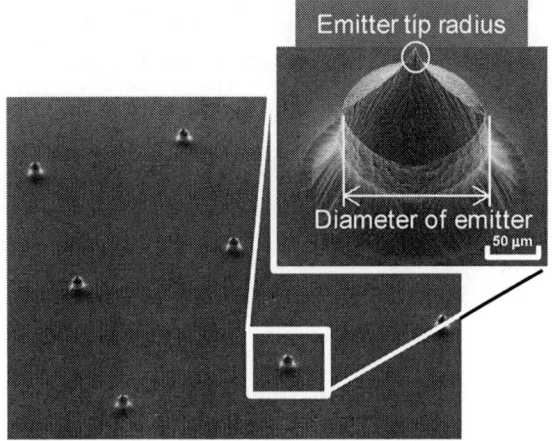

Figure 3: SEM images of the fabricated ILIS array

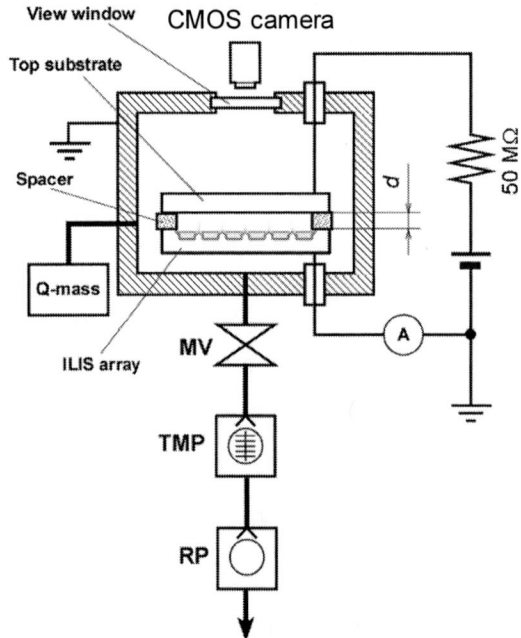

Figure 4: Experimental setup for the ILIS array

EXPERIMENTAL SETUP

The ion emission from the ILIS array was demonstrated using setup as shown in Fig. 4. The ILIS array was put inside a vacuum chamber with back pressure of 1×10^{-5} Pa. A Si (100) substrate used as an extractor and a target was fixed over the ILIS array. To emit ion beams from the ILIS array, high voltage was applied between the ILIS array and the top Si substrate. The emitted ions were measured as an ion current using an ammeter. Distance between the ILIS array and the silicon substrate was determined by the thickness of the spacer (500 µm or 800 µm).

RESULTS AND DISCUSSION

Ion emission characteristics

Figure 5 shows I-V characteristics measured using the setup shown in Fig. 4. In Fig. 5, with thickening the diameter of the emitter, ion current and threshold voltage V_{th} were increased.

Correlation between V_{th} and the emitter tip radius r_t can be expressed as [5]

$$V_{th} = \sqrt{\frac{\gamma r_t}{\varepsilon_0}} \ln \frac{2d}{r_t} \qquad (1)$$

where d, ε_0, and γ are the distance between the emitter tip and the extractor, the permittivity of free space, and the surface tension of the [EMIM]-[BF4] (0.053 N/m), respectively. This equation explains the fact that the V_{th} was decreased by sharpening the emitter tip. Since the etching time to fabricate the sharp tip depends on the diameter of the emitter, in thick emitter, etching time get long and r_t becomes larger. Hence, the V_{th} was increased due to increase of r_t.

Correlation between the ion current and the diameter of the emitter can be explained as follows. When voltage is applied between the ILIS array and the top substrate, the IL rises up on the emitter surface to the emitter tip, and then ions are emitted from the IL. However, when the diameter of the emitter was 500 µm, the voltage was not enough to raise the sufficient amount of the IL to the emitter tip. Therefore, measured ion current was small. To solve the problem, surface tension can be utilized. The IL rises up on the emitter surface due to the surface tension when the emitter immerses into the IL. Correlation between the rising height of the IL and the radius of the emitter can be expressed as following equation [6]

$$h = b \ln\left(\frac{2l}{b}\right) \qquad (2)$$

where h, b, and l are rising height of the IL, radius of the emitter, and capillary length, which is related to the surface tension. Therefore, in thick emitter, supply of the IL was facilitated, and the ion current was improved.

Reactive ion etching

Volatilization of the Si by the ion beam radiation was monitored using a quadrupole mass spectrometer (Pfeiffer: QMG220). Figure 6 shows mass spectra when changing the distance between the ILIS array and the extractor. In case that the distance was 500 μm, the peaks which indicate the silicon etching were weak and unclear. However, by changing the distance to 800 μm, the peaks of SiF^+, SiF_2^+, and SiF_3^+ were confirmed clearly. These peaks indicate that the silicon was etched by chemical reaction.

We monitored the inside of the device during the ion beam emission to clarify the difference of mass spectra as shown in Fig. 6. The top substrate was changed to the glass instead of the silicon to observe the inside of the device. The transparent electrode made of indium tin oxide (ITO) was deposited on the glass. Behavior of the IL during the emission was observed with a CMOS camera. As a result of monitoring, the transparent IL changed to brown in case that the distance was 500 μm by the emission. In contrast, this change was not observed when the distance was 800 μm. Figure 7 shows Fourier transform infrared (FT-IR) spectra of the IL after the emission test. From this figure, it was confirmed that the peaks which indicates B-F, N-H and O-H bond was suppressed for d=800 μm than for d=500 μm.

This result means that input energy was spent to generate the by-product due to the intense electric field when d was 500 μm.

Evaluation of etching characteristics

Figure 8 shows etching marks remained on a (100) Si substrate obtained for the ILIS array with the diameter $2b$=ϕ200 μm, the distance d=800 μm, and applied voltage of 5.1 kV. Using the optical profiler, cone-shaped dimples were observed as shown in Fig. 8 (a). The surface of the etched silicon was covered by the IL as shown in Fig. 8 (b). It is supposed from Fig. 8 (b) that the IL was emitted from the ILIS array as droplets or mixture of droplets and ions [7].

Figure 9 shows an SEM image and the element analysis of a vicinity of the etching mark obtained using an energy-dispersive X-ray spectroscopy (EDX). In Fig. 9, carbon (C), fluorine (F), nitrogen (N), oxygen (O), and silicon (Si) were detected around the etching mark.

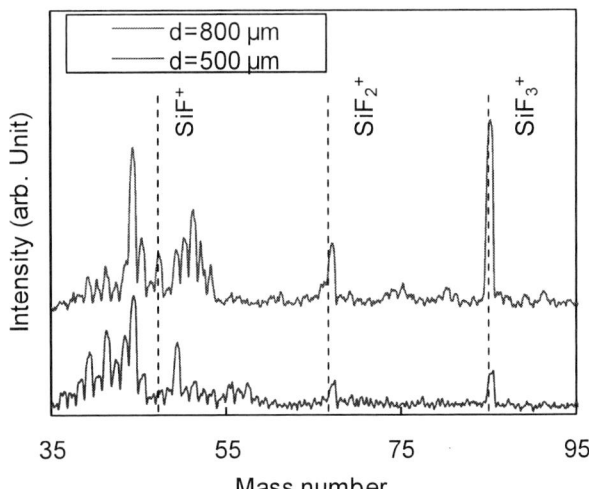

Figure 6: Mass spectra during the (100) silicon etching with the ILIS array

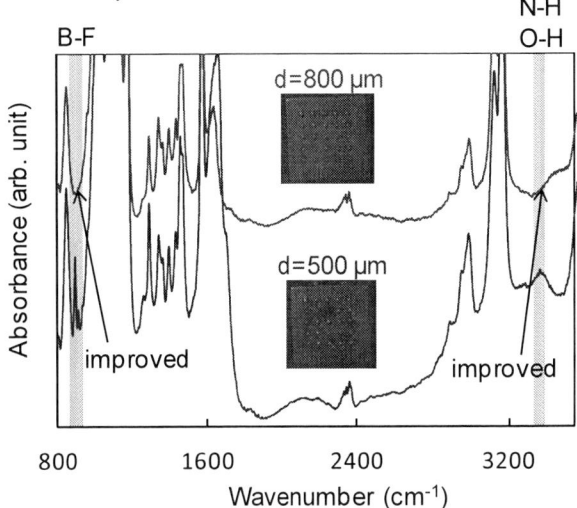

Figure 7: FTIR spectra of the IL after the emission test

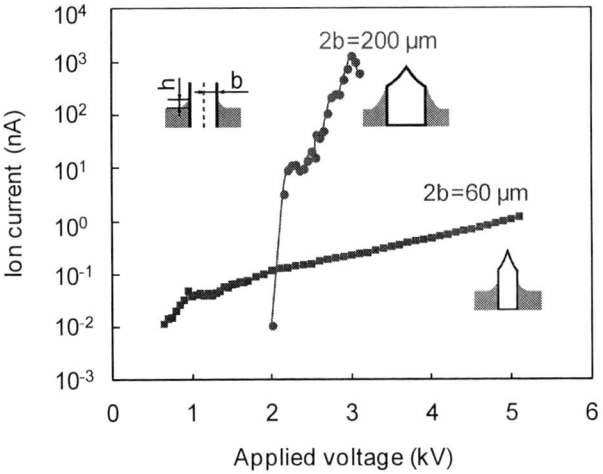

Figure 5: I-V characteristics of the ILIS array with various diameters

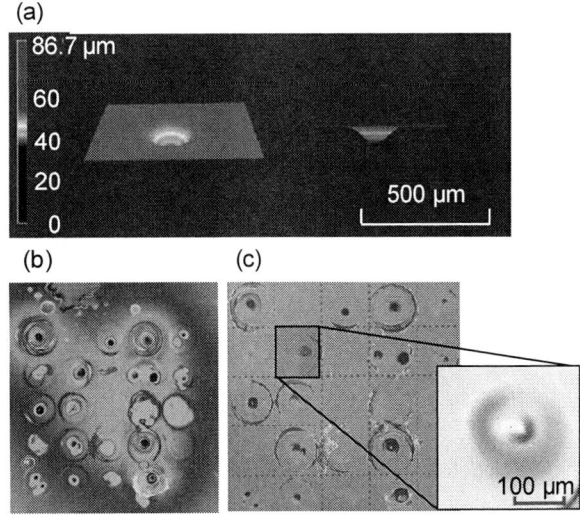

Figure 8: 3D images of a etching mark on silicon (a) and photos of etching marks on silicon before removing the IL (b) and after removing the IL (c)

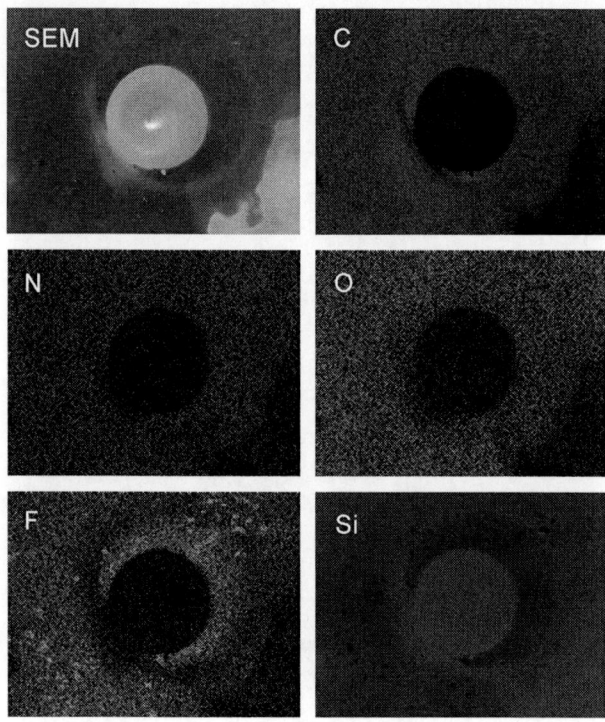

Figure 9: EDX elemental mapping of the etched Si substrate

However, only silicon was detected inside of the cone-shaped dimple. This result indicates that the Si can be etched even if the surface of the silicon was covered with the IL since the adhered IL was blown away due to the ion bombardment.

Etching rate of silicon per emitter calculated from the volume of the cone-shaped dimple was 24 μm^3/sec for the average ion current per emitter of 42 nA with ion-acceleration voltage of 5.1 kV. This value was 1.5 times larger than the reported etching rate of a conventional focused Ga$^+$ ion beam with ion-acceleration voltage of 30 kV [8].

CONCLUSION

Concurrent reactive ion etching of silicon using micro ionic liquid ion source (ILIS) array was demonstrated. In this device, the ionic liquid (IL) of 1-ethyl-3-methylimidazolium tetrafluoroborate ([EMIM]-[BF$_4$]) was employed as ion source material. Since [EMIM]-[BF$_4$] contains fluorine based ion, chemical etching is expected, and it could lead to higher etching rate of the silicon. The etching experiment of the silicon substrate was executed by using the fabricated ILIS array. Monitoring the ambient gas around the etching area during the emission using mass spectrometer, volatilization flux of SiF$^+$, SiF$_2^+$, and SiF$_3^+$ were detected. This result indicates that the silicon was chemically etched by the ions contained in the IL. After the emission, we confirmed the cone-shaped etching marks on the silicon substrate. By analyzing the etched substrate with energy dispersive X-ray spectroscopy (EDX), it was confirmed that the silicon substrate was bared only in the etching mark. This result indicates that the Si can be etched even if the surface of

the silicon substrate was covered with the IL since the passivated IL was blown away due to the ion bombardment. As an experimental result, etching rate of silicon by using the ILIS array with ion-acceleration voltage of 5.1 kV was 1.5 times larger than that of a conventional focused Ga$^+$ ion beam with ion-acceleration voltage of 30 kV.

ACKNOWLEDGEMENT

A part of this work was performed at micro/nano-maching research and education center, Tohoku University, Japan. This work was supported in part by the Challenging Exploratory Research from Ministry of Education, Culture, Sports, Science and Technology of Japan (MEXT) (Grant No. 12010289).

REFERENCES

[1] T. Suzuki, M. Hara, H. Oguchi, and H. Kuwano, "Ionic-liquid micro ion source array for flexible concurrent MEMS process", *In Proc. MEMS2013*, Taipei, January 20-24, 2013, pp. 315-318

[2] C. Perez-Martinez, S. Guilet, J. Gierak, and P. Lozano, "Ionic liquid ion sources as a unique and versatile option in FIB applications", *Microelectronic Engineering*, Vol. 88, Issue 8, pp. 2088-2091, 2011

[3] S. Guilet, C. Perez-Martinez, P. Jegou, P. Lozano, and J. Gierak, "Ionic liquid ion sources for silicon reactive machining", *Microelectronic Engineering*, Vol. 88, Issue 8, pp. 1968-1971, 2011

[4] L. F. Velásquez-García, A. I. Akinwande, and M. Martínez-Sánchez, "A planar array of micro-fabricated electrospray emitters for thruster applications", *J. Microelectromech. Syst.*, Vol. 15, No. 5, pp. 1272-1280, 2006

[5] P. D. Prewatt and G. L. R. Mair, *Focused ion beams from liquid metal ion sources*, Research Studies Press, 1991

[6] P. Gennes, F. Brochard-Wyart, and D. Quéré, *Capillarity and wetting phenomena drops, bubbles, pearls, waves*, Springer, 2003

[7] W. D. Luedtke, U. Landman, Y.-H. Chiu, D. J. Levandier, R. A. Dressler, S. Sok, and M. S. Gordon, "Nanojets, electrospray, and ion field evaporation: Molecular dynamics simulations and laboratory experiments", *J. Phys. Chem*, A, Vol. 112, Issue 40, pp. 9628-9649, 2008

[8] M. Lachab, M. Nozaki, J. Wang, Y. Ishikawa, Q. Fareed, T. Wang, T. Nishikawa, K. Nishino, and S. Sakai, "Selective fabrication of InGaN nanostructures by the focused ion beam/metalorganic chemical vapor deposition process", *J. Appl. Phys.*, Vol. 87, Issue 3, pp. 1374-1378, 2000

CONTACT

Ryo Yoshida,
Tel (Fax): +81-22-795-4771
E-mail: yoshida@nanosys.mech.tohoku.ac.jp

DEVELOPMENT OF MEMS PIERCE-TYPE NANOCRYSTALLINE SI ELECTRON-EMITTER ARRAY FOR MASSIVELY PARALLEL ELECTRON BEAM DIRECT WRITING

Hitoshi Nishino[1], Shinya Yoshida[1], Akira Kojima[2], Naokatsu Ikegami[3],
Nobuyoshi Koshida[3], Shuji Tanaka[1], Masayoshi Esashi[1]

[1]Tohoku University, Sendai, JAPAN
[2]CRESTEC CORPORATION, Tokyo, JAPAN
[3]Tokyo University of Agriculture and Technology, Tokyo, JAPAN

ABSTRACT

This paper mainly reports the process development of a Pierce-type nanocrystalline Si (nc-Si) electron emitter array for massively parallel electron beam (EB) lithography based on active-matrix operation using a large-scaled integrated circuit (LSI). The emitter array consists of 100×100 hemispherical emitters formed by isotropic wet etching of Si. EB resist patterning was demonstrated by 1:1 projection exposure using a discrete emitter array at CMOS-compatible operation voltages. To independently control each emitter using the LSI, isolation trenches filled with benzocyclobutene (BCB) were fabricated in the Si substrate. In addition, the integration process of the emitter array, the LSI and an extraction electrode plate was developed based on Au-In and polymer bonding technologies.

INTRODUCTION

Photo-mask cost for the latest large-scale integrated circuit (LSI) is enormous and increasing further with the scaling down of CMOS. Maskless electron beam (EB) lithography by massively parallel writing has a great potential for the cost reduction of small-to-medium volume LSIs. As one of high throughput EB lithography systems [1-3], we proposed a massively parallel EB lithography (MPEBL) system using a nanocrystalline Si (nc-Si) emitter array integrated with an active-matrix driver LSI [1]. The nc-Si ballistically emits EB with excellent straightness at CMOS-compatible voltage, and the active-matrix driven emitter array in a large scale (e.g. 100×100) provides higher throughput than the existing systems.

The conventional nc-Si emitter has a flat surface, and emits planar-shape EB. The EB must be focused by a stacked-type condenser lens array placed above the emitter array. However, it is difficult to well focus the planar-shape beam even using multiple lenses. The fabrication and integration of the multiple lenses are also difficult from process point of view. If the emitter is made smaller toward a point source, focusing becomes easier, but emission current decreases, leading to lower throughput.

For solving this problem, we decided to use hemispherical, i.e. Pierce-type nc-Si emitters, which had self-focusing function. In this study, the Pierce-type nc-Si emitter array was fabricated and tested without the driver LSI or an extraction lens. Then, the integration process of the emitter array, the LSI and the extraction electrode plate was developed.

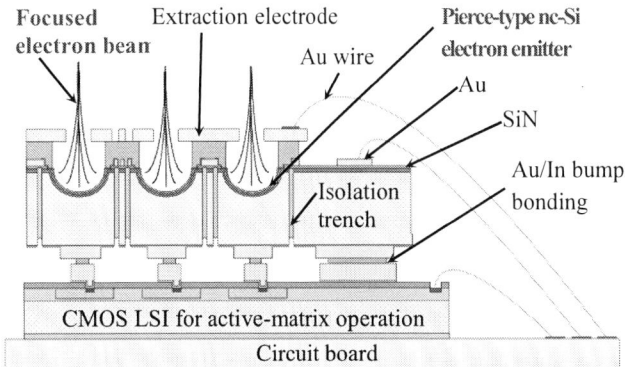

Figure 1. Schematic of MEMS Pierce-type nanocrystalline Si electron emitter array integrated with CMOS LSI.

FABRICATION PROCESS

Figure 2 shows the fabrication process of the Pierce-type nc-Si emitter array integrated with the LSI. (1) An n^+ silicon wafer is prepared, and (2) hemispherical shapes are formed by isotropic etching with a mixture of HNO_3, HF and CH_3COOH [4]. Then, nc-Si layer is formed on the hollows by anodization of a poly-Si film. A SiN layer is used as a mask. (3) After fabricating Au electrodes and bumps for aberration correction [5], the wafer is bonded to a handling wafer with a polyimide film. Next, Au bumps are formed on the backside by electroplating. (4) For electrical isolation of each electron emitter, isolation trenches are fabricated by deep reactive ion etching (RIE) and resin filling. Half cut lines are also formed.

(5) The wafer is then mounted on a LSI wafer by Au-In transient liquid phase (TLP) bonding [6]. (6) The supporting wafer and polyimide film are removed by etching, and a thin Au surface electrode is formed on the emitter. Then, the Si wafer was cut along with the cut lines. (7) An extraction electrode plate is fabricated in a silicon-on-insulator (SOI) layer, and the etched parts are filled with a photoresist. Then, the wafer is bonded on the nc-Si emitter wafer with a photo-sensitive adhesion polymer patterned on the SOI wafer. (8) The handling Si wafer and SiO_2 layer are dry-etched. (9) Al bonding pads are fabricated on the lens plate using a stencil mask, and the filling photoresist is selectively removed by acetone. As a result, the fully-integrated electron emitter is completed.

DEVELOPMENT OF THE FUNDAMENTAL

AND FABRICATION PROCESSES

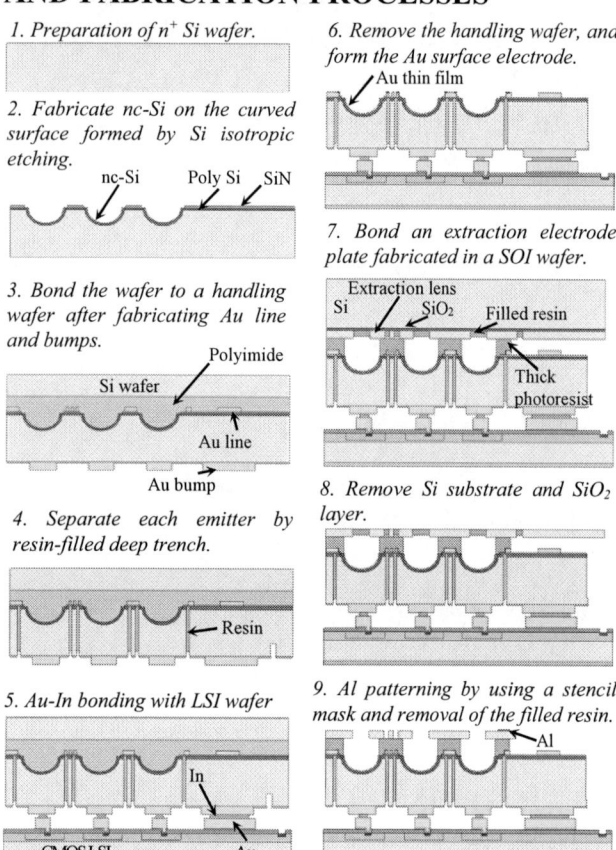

1. Preparation of n⁺ Si wafer.

2. Fabricate nc-Si on the curved surface formed by Si isotropic etching.

   nc-Si   Poly Si   SiN

3. Bond the wafer to a handling wafer after fabricating Au line and bumps.

               Polyimide
   Si wafer
                Au line
   Au bump

4. Separate each emitter by resin-filled deep trench.

               Resin

5. Au-In bonding with LSI wafer

               In
  CMOS LSI   Au

6. Remove the handling wafer, and form the Au surface electrode.

   Au thin film

7. Bond an extraction electrode plate fabricated in a SOI wafer.

  Extraction lens
 Si   SiO₂   Filled resin
                Thick photoresist

8. Remove Si substrate and SiO₂ layer.

9. Al patterning by using a stencil mask and removal of the filled resin.

                Al

Figure 2. Fabrication process of the Pierce-type nc-Si emitter array integrated on LSI.

Prototype and characterization of Pierce-type nc-Si emitter array

For the process development of the Pierce-type nc-Si emitter array and the characterization of its intrinsic electron emission property, a 100×100 emitter array with a pitch of 100 μm and an emitter diameter of 60 μm without the trench isolation or LSI integration was fabricated, as shown in Fig. 3. A 1.5-μm-thick poly-Si layer was deposited on a hemispherical Si surface and anodized by applying pulsing voltages ((2.5 mA/cm² × 1 sec + 25 mA/cm² × 2 sec) × 3 times) in a 1:1 mixture solution of HF and ethanol. A 300-nm-thick SiN film was used as a mask. Then, the electrochemical oxidation was performed by keeping a current density of 10 mA/cm² until the applied voltage reached to 30 V in a mixture of ethylene glycol, H₂O and KNO₃. As a result, a Pierce-type nc-Si emitter array was fabricated.

Figure 4 (a) shows the diode current density and emission current density vs. applied voltage. The electron emission started from about 8 V, and the current density reached 30 nA/cm² at 13 V, which exhibited a reasonable emission property as a nc-Si emitter.

EB resist patterning by 1:1 projection exposure was demonstrated for roughly evaluating the quality variation of the emitter array. Figure 4 (b) illustrates the experimental

setup. A 50-nm-thick EB resist (ZEP520, Nippon Zeon Co., Ltd.) on a Si wafer was exposed by EB from the emitter array by applying pulsing voltages of 15 V with a period of 1 s (duty ratio, 25 %). As seen in Fig. 4 (c), the arrayed pattern corresponding to the layout of the emitter array was transferred to the resist without a defect. This result suggest that the emitter array was fabricated with fair uniformity. On the other hand, each resist pattern had a ring shape, suggesting that the beam shape was a ring. One of the possible reasons is the site dependency of the quality of nc-Si in the hollow. For example, electrical field intensity in the anodization of the poly-Si film might be different between the bottom and the surface near the edge, and the thickness and/or porosity of the nc-Si layer were/was not uniform. Another reason might be a thickness variation of the Au surface electrode. The emission pattern would be improved by further optimization of the fabrication process.

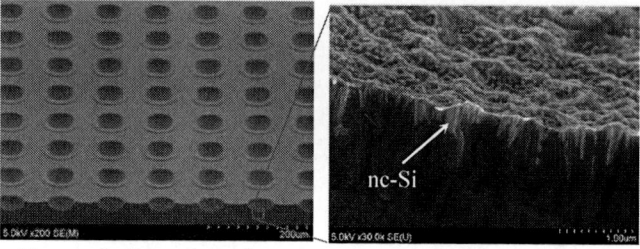

Figure 3. Scanning electron micrograph of the 100 ×100 Pierce-type nc-Si emitter. (Pitch 100 μm, Diameter ~40 μm)

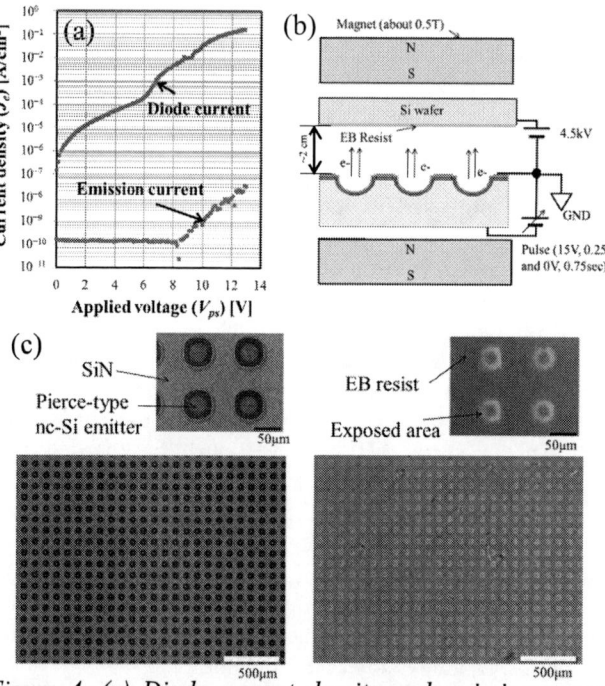

Figure 4. (a) Diode current density and emission current density vs. applied voltage. (b) Setup for exposure test. (c) [Left] Pierce-type nc-Si emitter array, [Right] EB resist after exposure test.

Process development of resin-filled isolation trench to separate each electron emitter

The fabrication process of the resin-filled isolation

978-1-4799-3510-9/14 $31.00 © 2014 IEEE

trench was developed. The resin must have chemical and thermal resistivity and good mechanical toughness, because it must stand the following process. In this study, polyimide (UR-3140, Toray Industries Inc.) and BCB (CYCLOTENE3022-63, The Dow Corning Corporation) were chosen as candidates of the resin, and the filling experiments were carried out. A 200-µm-thick Si wafer was bonded to a Si handling wafer with a polyimide adhesion layer. After deep trenches were fabricated in the Si wafer by deep RIE, the resin was infused and fully cured in N_2 atmosphere [7, 8]. Curing temperature for UR-3140 and BCB were 350 °C and 270 °C, respectively. As seen in Fig. 5 (a), UR-3140 generated many defects and did not fill the trenches completely, which was probably caused by low wettability and volume shrink in curing. Also, the isolated Si poles, where the emitters will be formed, surrounded by the trench was distorted (Fig. 5 (b)). On the other hand, BCB completely filled the trenches with an aspect ratio of 10, and the poles were not distorted (Fig. 5 (c) and (d)). Therefore, BCB is a suitable resin for this filling process, and the fabrication process of the isolation trench was established.

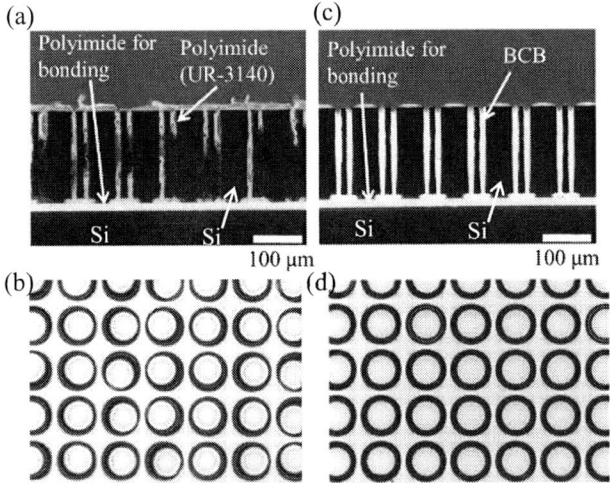

Figure 5. Cross sectional fluorescent micrographs of the trenches filled with (a) polyimide and (b) BCB trenches. Topside optical micrographs of the Si poles surrounded by the trenches with (a) polyimide and (b) BCB.

Development of Au-In TLP bonding for integration and electrical connection of the emitter array and LSI

Au-In TLP bonding was chosen for the integration of the emitter array and the LSI. It can be performed at low temperature and low bonding pressure. Thus, both nc-Si emitter and LSI may not to be damaged by the integration process. The mechanical strength and chemical resistivity enough to withstand the post-process are also expected. The electrical and mechanical reliabilities of the bonding were preliminarily investigated by fabricating daisy chain connections using the bonding process, as shown in Fig. 6. (1) A glass substrate was prepared, and (2) Al patterns were formed on it. (3) A SiO_2 layer deposited by plasma enhanced chemical vapor deposition was patterned. This substrate was used as a dummy LSI wafer. (4) Au pads and 5-µm-thick

electroplated Au bumps were fabricated. (5) Au bumps covered with an In layer were also fabricated on another glass substrate. (6) The TLP bonding was performed by applying a loading pressure of 20 MPa at 200°C under vacuum for 1 hour. As a result, an electrical path of the daisy chain was formed.

The electrical connections of the bumps were confirmed using probers, as illustrated in Fig. 7. In this experiment, 1468 of bumps forming the daisy chains were all electrically connected. In a shear test after the bonding, the glass substrate near the bumps broke, and the substrates were not separated at the bonding interface (Fig. 8), suggesting that the mechanical strength of the bonding was excellent.

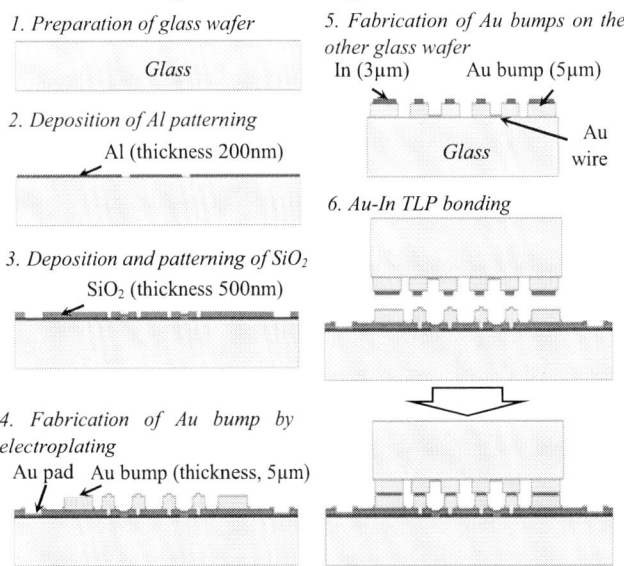

Figure 6. Development of Au-In TLP bonding for integration between the emitter array and LSI.

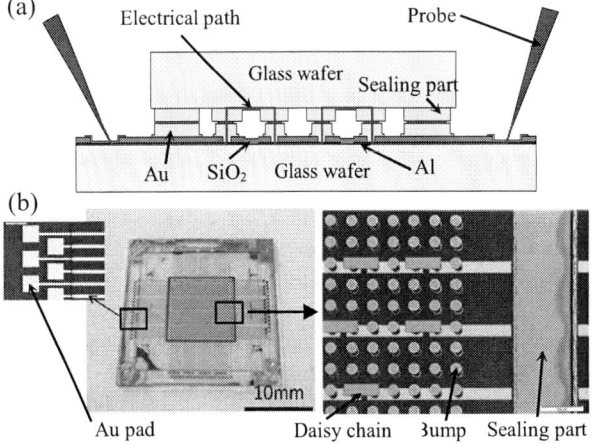

Figure 7. (a) Electrical connection check of the bumps by utilizing daisy chains. (b) Glass wafers bonded via Au-In TLP bonding, forming the daisy chains.

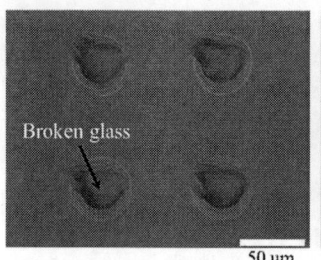 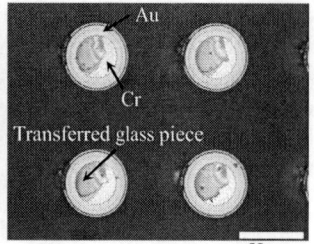

(Upper glass substrate) (Lower glass substrate)

Figure 8. Observation of the glass substrate surface after shear test.

Development of fabrication and integration processes of the extraction electrode plate

Finally, the fabrication and integration processes of the extraction electrode plate on the emitter array were developed through a preliminary experiment shown in Fig. 9. (1) A SOI wafer with a highly-doped n-type Si layer was prepared, and (2) the SOI layer was patterned into a lens shape by Deep RIE. (3) A photo resist filled the etched parts for increasing the mechanical rigidity. (4) A photo-sensitive adhesive polymer (TMMR N-A 1000, Tokyo Ohka Kogyo Co., Ltd.) with a thickness of 20 µm was patterned. (5) The wafer was bonded to a glass substrate, which is convenient to observe the bonding quality, by applying a pressure of 250 kPa at 120°C. (6) The handling layer and SiO_2 layer were dry-etched. (7) Finally, the reinforcing photoresist was selectively removed with acetone.

Figure 10 shows the result of the bonding experiment. The Si extraction electrode plate was successfully mounted on the glass substrate, and not distorted even after the bonding process due to the temporary-filling of the photoresist. The bonding polymer also successfully remained without the swelling and deformation after the removal process of the photoresist. This process will be applied to the actual Pierce-type nc-Si emitter array for the fully-integrated device..

CONCLUSION

The fabrication and integration processes for a Pierce-type nc-Si electron-emitter array on a active-matrix driver LSI were developed in this study. EB resist were patternd by 1:1 projection exposure utilizing electron emission from the fabricated 100 × 100 emitter array at the CMOS-compatible voltage. BCB was found to be a suitable resin to fabricate the resin-filled isolation trench. Au-In TLP bonding were tested for an integration method, and proved to be useful for our purpose. The extraction electrode plate was also successfully fabricated from a SOI wafer, and its integration process using a bonding polymer layer was developed. We believe that the development of these fundamental processes is a giant step toward the MPEBL system.

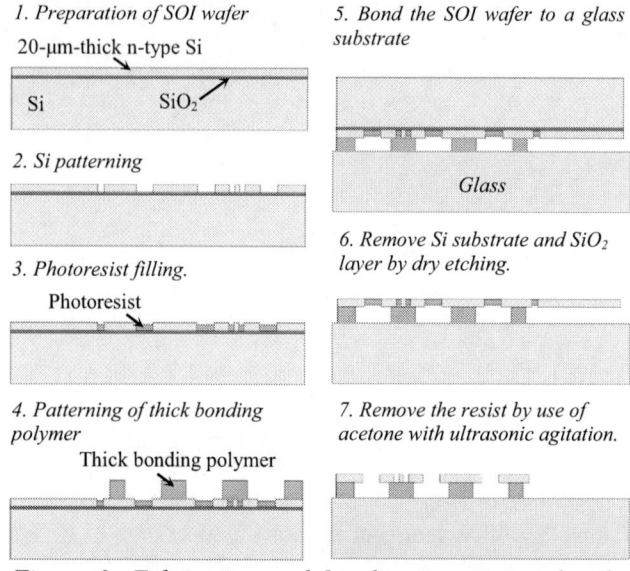

Figure 9. Fabrication and bonding experiment for the integration of the extraction electrode plate.

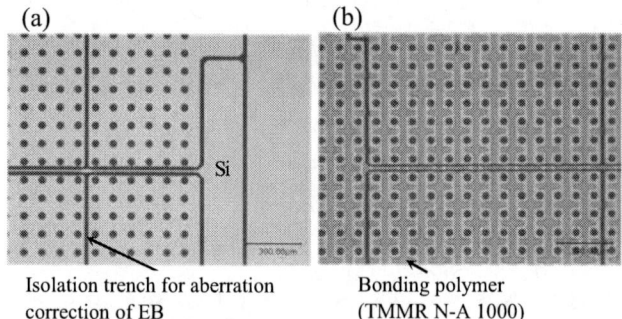

Figure 10. Optical micrograph of the extraction electrode plate bonded on a glass substrate (a) Top view, (b) Bottom view through the glass wafer.

REFERENCES

[1] A. Kojima *et al.*, Proc. SPIE 8680, Alternative Lithographic Technologies V, 86800I (March 26, 2013)

[2] D. Rio. *et al.*, Journal of Vacuum Science & Technology B, Vol.28, p. C6C14 – C6C20 (2010)

[3] Y. Tanaka *et al.*, Nanotechnology, Vol.24, 015203 (2013)

[4] S. Tanaka *et al.*, Journal of the Japan Society for Precision Engineering, Vol. 59, p. 72-76 (1993)

[5] K. Wataya *et al.*, Proc. Sensor and Micromachine Workshop IEEJ, 87-92 (2012)

[6] Welch, W.C. *et al*, Micro electro mechanical systems (MEMS 2008), Tucson, AZ, p. 806-809 (2008)

[7] T. Mtsumoto *et al.*, Jpn J. Appl. Phys, Vol. 37, p.1217-1221 (1998)

[8] Fabrice F.C. Duval *et al.*, IEEE Transactions on Packaging and Manufacturing Technology, Vol.1, No.6, p. 825-832 (2011)

Acknowledgments

This work was supported by the Japan Society for the Promotion of Science through the Funding Program for World-Leading Innovative Research and Development on Science and Technology (FIRST Program).

DIRECT LASER WRITING OF 3D PROTEIN STRUCTURES WITH NANOSCALE FEATURE SIZES

Daniela Serien [1], and Shoji Takeuchi[1,2]
[1] CIRMM-IIS, The University of Tokyo, Japan
[2] Takeuchi Biohybrid Innovation Project, ERATO, JST, Japan

ABSTRACT

We report preliminary results of three-dimensional (3D) protein structures with nanoscale feature sizes fabricated by two-photon direct laser writing (DLW). For this fabrication technology, we combine the established DLW technology with previously reported 3D protein structure fabrication by photosensitized crosslinking. We demonstrate the fabrication of 2D and 3D protein structures with nm-sized features. We report dependency of line height and independency of line width on writing parameters within the investigated range. Further, we demonstrate the biocompatibility of protein structures with nanoscale features to NIH/3T3 fibroblasts. This fabrication method enables the creation of biocompatible protein scaffolds down to the nm-range for cell studies and medical application.

INTRODUCTION

Direct laser writing is a mask-free lithography technique. By utilizing two-photon absorption polymerization of photoresists, arbitrary structures with nm-sized features can be fabricated [1]. For biological and medical applications such as 3D scaffolds [2] and magnetic microrobotic transporters [3], biocompatible materials are highly required. B. Kaehr et al. introduced photosensitized crosslinking of protein as a mechanism to utilize two-photon lithography technology with biomaterials [4 – 6]. Using dynamic mask-directed lithography, B. Kaehr et al. created submicron structures [5]. Here, we advance to sub-100nm feature size by using direct laser writing.

By 390nm UV light excitation or two-photon uptake of 780nm fs-pulsed photons, the photosensitizer flavin adenine dinucleotide (FAD) is excited to transit into its triplet state:

$$^1\text{FAD}+2h\nu \rightleftharpoons {}^3\text{FAD} \quad (1)$$

While usually transition into a triplet state is forbidden, the process is efficient with $\Phi=0.7$ [7, 8]. The excited photosensitizer then oxidizes specific amino acids: Tyrosine (Tyr), cystein, histidine, tryptophan and possibly others that are able to donate one electron [9]:

$$^3\text{FAD}+\text{Tyr-OH} \rightleftharpoons \text{FAD}\bullet^- +\text{Tyr-O}\bullet+\text{H}^+ \quad (2)$$

The neutral radicals of those amino acid residues crosslink via covalent bonds upon encounter, necessary hydrogen can be withdrawn from the immediate buffered environment:

$$\text{Tyr-O}\bullet + \text{Tyr-O}\bullet \rightleftharpoons \text{HO-Tyr-Tyr-OH} \quad (3a)$$
$$\text{Tyr-O}\bullet + \text{Tyr-O}\bullet \rightleftharpoons \text{Tyr-O-O-Tyr} \quad (3b)$$
$$\text{Tyr-O}\bullet + \text{Tyr-O}\bullet \rightleftharpoons \text{HO-Tyr-O-Tyr} \quad (3c)$$

The strong covalent C-C bond (3a) has been suggested by [9], but weaker covalent bonds should not be neglected.

As the most abundant protein in the circulatory system [10, 11], the small, globular protein serum albumin is a protein that is available abundantly and has been studied extensively. Serum albumin is found in blood plasma at typical concentrations of 5g/100mL, has a high affinity to various materials and it has been reported to transport solutes, regulate pH and osmotic pressure [11]. Due to those properties, it is a good candidate to introduce the method of protein crosslinking to the direct laser writing system for fabrication of 3D protein structures.

MATERIALS AND METHODS

Microfluidic components

Channels of about 100 μm width and 50 μm height were used to minimize sample usage and elongate writing times. Channels were structured with negative photoresist SU-8 (Micro Chem, USA). Polydimethylsiloxane and catalyst (PDMS, Silpot 184, Dow Corning Toray co., Ltd.) were mixed in the ratio 9:1 and hardened for two hours at 75 °C. Direct laser writing compatible cover glasses were cleaned with aceton and isopropanol. Glass and PDMS were not bonded, but instead simply adhered due to their smooth clean surfaces. Filling the channels via attached silicon tubes (Laboran) with protein solution was performed carefully.

Protein Crosslinking

The working solution contained 1, 4 or 8 mM FAD (F-6625, Sigma-Aldrich) with 4, 50, 200 or 400 mg/mL bovine serum albumin (BSA, A-7906, Sigma-Aldrich) in 0.01 mM HEPES (Sigma-Aldrich, H-3375) buffer at pH 7.4. We also used Phosphate Buffered Saline (PBS, diluted from 10x Dulbecco's PBS, D1408, Sigma-Aldrich) and Dulbecco's Modified Eagle's Medium (DMEM, D5796, Sigma-Aldrich) as buffer systems. The working solution was filled into the channels of the PDMS on glass directly prior to the experiment. Direct laser writing was performed with a Photonic Professional (Nanoscribe) that utilizes a fs-pulsed 780 nm laser. After the procedure, PDMS was removed and the sample was thoroughly washed with MilliQ water. Thereafter, we fixed and washed the samples stepwise for 15min each with 2.5% glutaraldehyde (111-30-8, Tokyo Chemical Industry co., LTD) in PBS, MilliQ water, 50:50 MilliQ water and methanol, 100% methanol. After carefully drying it, the sample was further dried in a vacuum chamber.

Cell Culture

NIH/3T3 cells were seeded at low cell concentration of approximately 210 cells/mm^2 and cultured for two days in DMEM medium at 37 °C with 5% CO$_2$. Cell staining was performed after fixation with 4% para-formaldehyde fixative (Muto Pure Chemicals co., ltd.): Consecutively, the sample was treated with 0.1% triton X-100 (A16046, Alfa Aesar) in PBS buffer for 5 min. Then, blocking with 1% BSA in PBS buffer was performed overnight. Cellular actin fibers were stained with Alexa 488 phalloidin (A12379, Invitrogen) overnight and nuclei were stained with Hoechst 33342 (Invitrogen) for 5 min.

RESULTS

Writing Conditions

To determine a good working solution, we wrote line structures and documented whether structures were fabricated. The preliminary results are summarized in Table 1. Additionally to prior reported HEPES buffer for DMD based fabrication [4-6], we were also able to fabricate BSA line structures in DMEM based working solutions. PBS buffer seems not suitable as working solution buffer. The ranges of concentration of FAD and BSA are not completely determined yet. It is interesting to note that an increase of FAD concentration above prior reported levels of 4 mM for DMD based fabrication [4] to 8 mM could not compensate the requirement of high minimal BSA concentration. Future work should focus on the optimization or enhancement of the photosensitive reaction mechanism because the current working conditions of 4 mM FAD and 400 mg/mL BSA are very costly.

Table 1: Working solution constituents

Buffer	FAD (mM)	BSA (mg/mL)	Structure
HEPES	4	400	Yes
HEPES	1	200	Yes
HEPES	1	50	No
HEPES	8	4	No
DMEM	1	200	Yes
PBS	1	200	No

Parallel to the screening of potential working solution conditions, we performed a screening of suitable writing parameters. Low laser power and high writing speed result in sub-threshold excitation, a condition that is not sufficient enough to create well-connected structures on the substrate. High laser powers were tested until 80mW, but above 20mW outbursts destroyed the original structure features due to extensive laser excitation. In comparison to prior reported power values of 30-60mW for the DMD based method [4], we observed that a laser power range around 15mW is suitable for fabrication (Fig. 1). Within the laser power range between 14 and 17 mW, outbursts occurred rarely. Those outbursts might have occurred rather due to air and microbubbles than due to extensive excitation.

Line Parameters vs. Writing Parameters

For fabrication of 3D structures, it is necessary to know the line properties and their dependencies on writing parameters. To study line width and line height, we wrote cross-like structures and measured their profile by 3D laser scanning microscopy (Fig. 2). Writing parameter of interest were laser power and writing speed. The dwell time was set

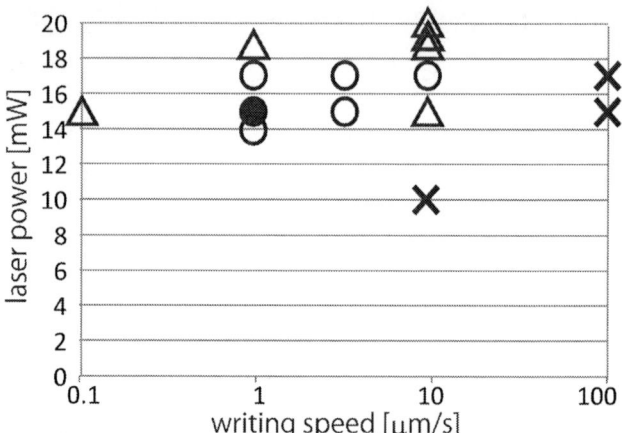

Figure 1: Screening of writing parameters. Diagonal lines over a 20×20 μm^2 area, dwell time 100 ms, 4 mM FAD 400 mg/mL BSA in HEPES buffer. No structures were achieved above 20 mW excitation. × no structure; results: Δ irregular ○ good ● best

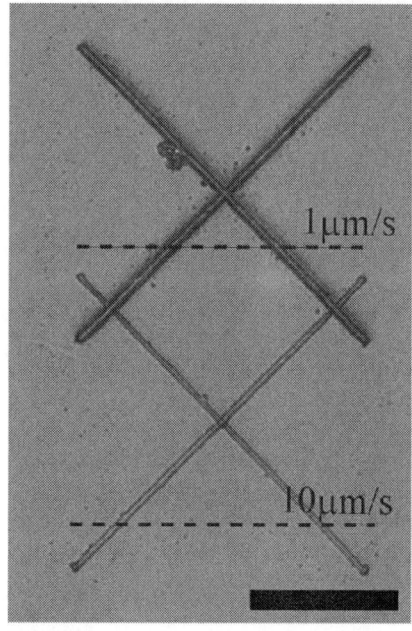

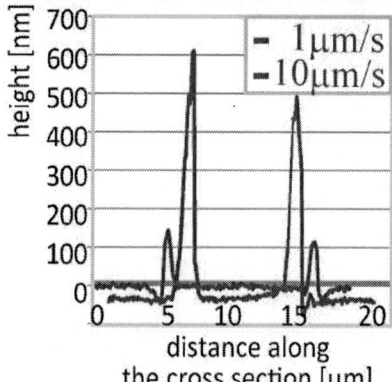

Figure 2: Two crosses, laser power 17 mW. Images and profiles along cross sections were acquired by 3D laser scanning microscope. Writing speed: 1 $\mu m/s$ (red), 10 $\mu m/s$ (green). Scale bar represents 10 μm.

to be constant at 100 ms after we observed no significant changes in a preliminary screening of different dwelling times.

Within the laser power range between 14 and 17 mW, preliminary results indicate that line height is proportional to laser power and inversely proportional to writing speed. With an excitation of 14 mW, we could not detect structures written with 5 and 10 µm/s, but those written with 1 µm/s were (218 ± 45) nm tall. The smallest line height observed in our preliminary study was (25 ± 12) nm written at 15 mW and 10 µm/s. This result indicates that by direct laser writing protein structures can be fabricated with sub-100nm feature sizes.

Line width, however, appears to be independent of laser power and writing speed within the currently investigated range. Because writing lines directly on the substrate requires anchoring the structures in the substrate by offsetting the laser focus, a possible dependency might have remained unobserved. More studies with free hanging horizontal strands are required, but the successful fabrication of such structures proved to be challenging due to structural fragility. In preliminary trials, the direction of writing along z-axis and the distance of the laser focus to the substrate were observed to influence line width. These observations remind of the extensively studied effects of focal region ("voxel") properties on photoresist DLW [1].

3D Protein Woodpile Structure

We fabricated an 8µm-tall block-woodpile structure and 2µm-wide, 8µm-tall free standing wall-woodpile structures with nm-sized lines. Due to constant writing conditions, the increase of line width with increasing distance to the substrate in z-axis was not corrected. Thus, the structural stability suffered from the imbalance between thin-lined base and thicker-lined top. The wall-woodpile structures collapsed during the washing and drying process, the block-woodpile structure withstood the entire process. The fabrication of woodpile structures with clearly distinguishable individual nm-sized lines demonstrates the feasibility of direct laser writing for the fabrication of 3D protein structures with nanoscale feature sizes.

Biocompatibility to the NIH/3T3 Cell Line

Finally, we report in addition to prior reports of biocompatibility [4, 6] that NIH/3T3 cells exhibited normal cell morphology and cell vitality when cultured on sparse BSA line structures and well-washed 3D BSA structures. Cells appeared to be unaffected, sometimes demonstrated enhanced adhesion to the BSA structures.

CONCLUSION

By combining direct laser writing with protein crosslinking, it is now possible to produce 3D protein structures mask-free and with sub-100nm feature sizes. Further characterization and optimization is required to broaden the applicability of this method.

ACKNOWLEDGEMENTS

We thank Yun Jung Heo for critical discussion, all lab members for supportive comments that contributed to the progress of this project.

REFERENCES

[1] Joachim Fischer, Martin Wegener, "Three-dimensional optical laser lithography beyond the diffraction limit", *Laser & Photon. Rev.* vol. 7, pp. 22–44, 2013

[2] Alexandra M. Greiner, Benjamin Richter, Martin Bastmeyer, "Micro-Engineered 3D Scaffolds for Cell Culture Studies", *Macromolecular Bioscience* 12, pp. 1301–1314, 2012

[3] Sangwon Kim, Famin Qiu, Samhwan Kim, Ali Ghanbari, Cheil Moon, Li Zhang, Bradley J. Nelson, and Hongsoo Choi, "Fabrication and Characterization of Magnetic Microrobots for Three-Dimensional Cell Culture and Targeted Transportation", *Advanced Materials*, pp. (1-6), 2013

[4] Bryan Kaehr, Richard Allen, David J. Javier, John Currie, and Jason B. Shear, "Guiding neuronal development with *in situ* microfabrication", *PNAS* vol. 101 (46), pp. 16104–16108, 2004

[5] Rex Nielson, Bryan Kaehr, and Jason B. Shear, "Microreplication and Design of Biological Architectures Using Dynamic-Mask Multiphoton Lithography", *small* vol. 5 (1), pp. 120–125, 2009

[6] Jason C. Harper, Susan M. Brozik, C. Jeffrey Brinker, and Bryan Kaehr "Biocompatible Microfabrication of 3D Isolation Chambers for Targeted Confinement of Individual Cells and Their Progeny", *Anal. Chem.* vol.84 (21), pp. 8985–8989 2012

[7] M.S. Grodowski, B. Veyret, and K. Weiss "Photochemistry of Flavins II. Photophysical Properties of Alloxazines and Isoalloaxines", *Photochemistry and Photobiology* vol. 26, pp. 341–352, 1977

[8] M. Sun, T.A. Moore, and P.S. Song "Molecular luminescence studies of flavins. I. The excited states of flavins", *Journal of American Chemical Society* vol. 94 (5), pp. 1730–1740, 1972

[9] J.D. Spikes, H.-R. Shen, P. Kopečková, and J. Kopeček "Photodynamic Crosslinking of Proteins. III. Kinetics of the FMN- and Rose Bengal-sensitized Photooxidation and Intermolecular Crosslinking of Model Tyrosine-containing N-(2-Hydroxypropyl)methacrylamide Co-polymers" *Photochemistry and Photobiology* vol. 70(2), pp.130–137, 1999

[10] J.R. Brown "Structure of Bovine serum albumin" *Federation Proceedings* vol. 34, pp. 591–591, 1975

[11] S. Sugio, A. Kashima, S. Mochizuki, M. Noda, and K. Kobayashi, "Crystal structure of human serum albumin at 2.5 Å resolution", *Protein Engineering* vol. 12 (6), pp. 439–446, 1999

CONTACT

*Daniela Serien, serien@iis.u-tokyo.ac.jp

FABRICATION OF ANISOTROPIC AND HIERARCHICAL UNDULATIONS BY BENCHTOP SURFACE WRINKLING

Kang Wei, and Yi Zhao

Department of Biomedical Engineering, the Ohio State University, Columbus, USA

ABSTRACT

This paper reports an electrical discharge assisted surface wrinkling process to produce hierarchical and anisotropic surface patterns on polymethylsiloxane (PDMS). Highly ordered sinusoidal wrinkles with tailored height (*h*) and wavelength (λ) are fabricated by surface modification of a uniaxially pre-strained PDMS using a hand-held corona discharger followed by strain relaxation. The resulting nanoscale wrinkles overlay on 1) microwrinkled substrate and 2) reflowed microstructures to create anisotropic and hierarchical surface topographies. This electrical discharge assisted wrinkling process for creating multi-scale roughness is accessible in general wet lab environments. Such structures have broad applications in optical gratings, cell mechanics studies, and micro/nanofluidics.

INTRODUCTION

Periodical structures with the characteristic feature size at microscale or nanoscale have attracted extensive research interests due to their broad applications including optical diffraction gratings [1, 2], stretchable electronics [3, 4], micro/nanofluidics [5, 6], cell adhesion and topographic cues [7], particle sorting [8], and self-cleaning surfaces [9]. Lithography-based fabrication approaches are most commonly employed to manufacture these intricate surface patterns because of the high fidelity and controllability. Microscale periodical structures with the characteristic feature size on the order of micrometers or submicron can be fabricated by conventional photolithography [10]; nanoscale periodical structures with the characteristic feature size below the optical diffraction limit can be patterned by nanolithography methods using electron beams, focused ion beams, or X-ray [11]. These periodical structures can also be transferred to soft materials by replica molding, pressure-driven hot embossing or alike methods. Besides, periodical structures can also be produced through non-lithographic and maskless means. In particular, mechanical buckling during the compression relief of a pre-strained bi-layered substrate comprised of a rigid thin film and a well adhered soft elastomeric foundation induces winkling structures on the substrate [12]. The buckling threshold is determined by the stiffness mismatch between the rigid thin film and the soft foundation. Such stiffness mismatch can be created by treating the surface with ultraviolet/ozone radiation, oxygen plasma, or focused ion beam, coating a polymeric layer on the base substrate or their combinations [13]. These maskless approaches outperform the lithography-based methods in terms of simplicity, low cost and versatility. However, cleanroom environment and special equipment including plasma asher/reactive ion etcher, e-beam/ion beam sources, or spin-coaters, are often required for generating the stiffness mismatch between the two layers to induce surface wrinkling. The limited accessibility of these facilities and the operation complexity compromise the manufacturing efficiency of these surface wrinkling processes.

In this work, we propose an electrical discharge assisted surface wrinkling process using a low-cost hand-held corona discharger. Periodically ordered micro/nanowrinkles and anisotropic binary wrinkling structures were created on polydimethylsiloxane (PDMS) surface within hours without the need of cleanroom environment and expensive equipment. This process can also generate surface wrinkles on top of microfabricated surfaces with continuous curvatures to produce hierarchically wavy patterns.

ONE LAYER WRINKLING PATTERN

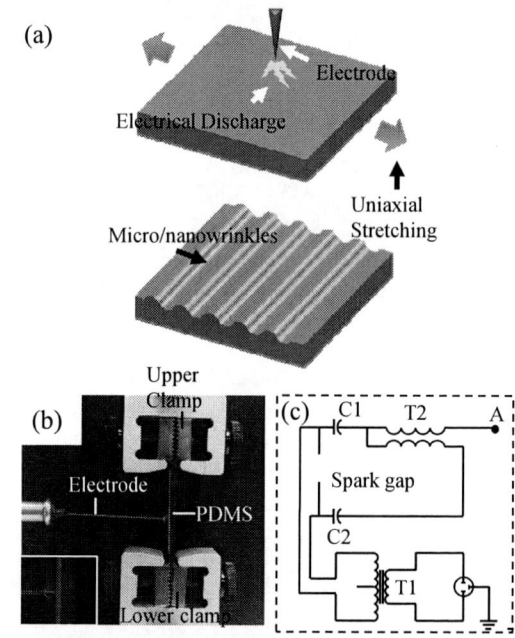

Figure 1: Electrical discharge assisted surface wrinkling process: (a) schematics, (b) experimental setup, and (c) circuit to generate electrical discharge.

The electrical discharge assisted surface wrinkling process is depicted in Figure 1. A PDMS membrane was first prepared by mixing the PDMS prepolymer and the curing agent at the weight ratio of 10:1. After cross-linking, the PDMS was cut into rectangular strips with 55mm in length, 10mm in width and 2mm in thickness. The strip was then clamped and stretched longitudinally to achieve a predetermined strain by a tensile test apparatus (100Q250-6, Testresources, Shakopee, MN). Atmospheric electrical

978-1-4799-3510-9/14 $31.00 © 2014 IEEE

discharge was generated at the room temperature by approaching the discharge tip connected to the output A of a hand-held high frequency corona discharger (BD-20AC Electro-technic Products, Chicago, IL) to the strip, while keeping another microelectrode B grounded. The power transformer T1 set up a high voltage which caused a spark gap to break down at the rate twice of the line frequency (100-120 Hz). The spark gap charged the capacitors C1 and C2 that were connected to the primary windings of the resonator coil T2 with an air core. Because of the inductance of primary windings of T2 and capacitors, an oscillating current of very high frequency was set up in the circuit. The spark gap was adjusted to reach the resonant frequency of the circuit about 3.8 MHz. High voltage was thus induced in the secondary windings of T2. During the electrical discharge treatment, the tip of the discharger was positioned 10 mm away from the PDMS strip and the operational power was set as 27W. The high electric field in the gap between the tip electrode and the PDMS strip ionized the surrounding air and created a plasma zone by generating oxygen free radicals and inducing free radical reactions. The ionized oxygen species within this zone attacked the bonds of molecules in the outmost PDMS surface, which generated water, CO_2, and other volatile carbonaceous compounds. The oxygen species formed covalent bonds with the silicon atoms, leaving a thin and stiff silica-like skin on the PDMS strip. After certain exposure time, the as-formed bi-layered composite was released to the original length at an unloading rate of 100 µm/s. Wrinkles formed spontaneously in the direction perpendicular to the pre-strain due to the stiffness mismatch between discharge-induced SiO_x thin film and the compliant PDMS strip. Both nanowrinkles and microwrinkles can be generated according to our previous trials [14]. The representative nanowrinkle pattern is shown in Figure 2. This was fabricated with an exposure time of 30s and a pre-strain of 40%. The nanowrinkle has a peak height of 135nm and the wavelength of 538nm.

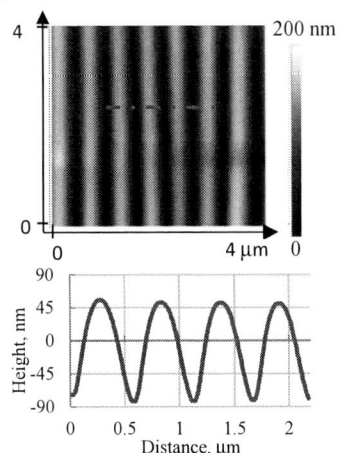

Figure 2: Nanowrinkling pattern.

BINARY ANISOTRPIC WRINKLES

To demonstrate the capability of this process in creating hierarchic and anisotropic wrinkling structures, binary wrinkling patterns were produced by repeatedly performing

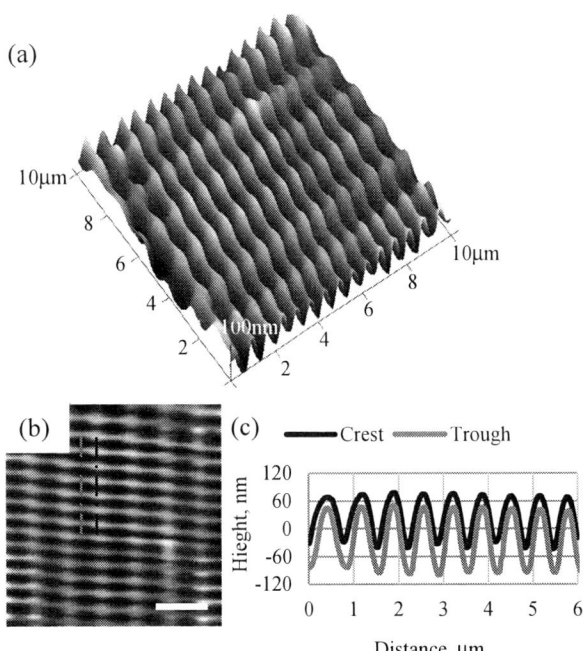

Figure 3: Binary wrinkling pattern: (a)&(b)3D and 2D topography, and (c)profile tracing. Black & green lines: nanowrinkles superimposed on the crest and trough of the microwrinkles. Scale bar: 2.5µm.

the aforementioned surface wrinkling process. In particular, linear wrinkles were first created on a PDMS substrate using the aforementioned electrical discharge assisted surface wrinkling process. These wrinkling features were transferred to another PDMS substrate through replica molding. Afterwards, the PDMS substrate with replicated wrinkles was subjected to a second wrinkling process, where it was stretched uniaxially along the direction of existing wrinkling pattern. Binary wrinkling structures were thus formed. Here, the wrinkling structures laid on the bottom were defined as the primary wrinkles, and those superimposed the primary wrinkles on the top were defined as the secondary wrinkles. Figure 3 shows the binary wrinkle pattern where the primary wrinkles and secondary wrinkles intersect at 90°. The exposure time of 240s and the pre-strain of 40% gave rise to the primary wrinkles with the average wavelength of 1.6 µm, while the exposure time of 60s and the pre-strain of 20% generated the secondary wrinkles with the average wavelength of 730nm. The secondary wrinkles superimposed on the crest and the rough of the primary wrinkles were in good phase. Elliptical protuberances formed at the crossing of two wrinkle groups. The surrounding wrinkling features formed elliptical depressions.

HIERARCHICALLY WAVY PATTERNS

The electrical discharge assisted surface wrinkling also allows easy integration with microstructures fabricated by standard photolithography. To demonstrate this, an array of semi-cylindrical microstructures was first fabricated on the PDMS substrate. This was implemented by reflowing the micropatterned rectangular AZ 9260 photoresist (AZ Electronic Materials, NJ) lines (Height: 10μm; width: 25 μm; pitch: 50 μm) on a silicon wafer. The reflow process was performed on a hot plate at 110 °C for 120s. The pattern was transferred to a PDMS substrate by replica molding. Figure 4 shows the micrograph of the reflowed microstructures on PDMS (SEM, Hitachi S-3000H). The reflowed structures were to ensure a continuous strain profile on the PDMS surface upon pre-straining so that the wrinkling structures were expected to be continuous.

Figure 4: SEM micrograph of the PDMS replica of the reflowed microstructures

The PDMS replica was stretched uniaxially with the strain (40%) applied at 90° to the longitudinal direction of the reflowed microstructures. It was then subjected to the electrical discharge assisted surface wrinkling process with 30s exposure time (Figure 5).

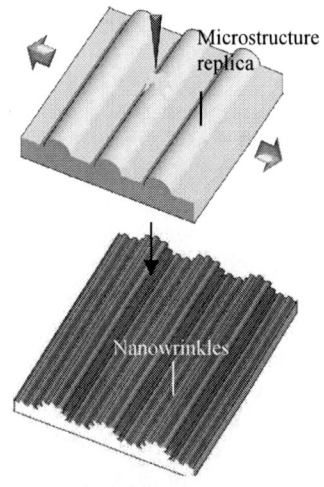

Figure 5: Fabrication schematics of hierarchically wavy pattern

The resulting hierarchically wavy pattern is shown in Figure 6. Nanowrinkling pattern formed both in the grooves and on the ridges of the reflowed microstructures with fairly good regularity. The nanowrinkles oriented parallel to the longitudinal axis of the reflowed microstructures, demonstrating an evident topographical hierarchy. Red profile traces revealed that the nanowrinkles propagated

along the curvilinear shape of the ridge of the microstructure, which contributed to a hierarchically wavy surface pattern. Blue traces show the course of the nanowrinkles on the flat surface of the groove of the microstructure. Such hierarchically wavy pattern mimics the surface patterns of many natural plants, e.g., daisy florets [15], and has the potential to become exhibit unique anisotropic wetting properties when properly treated.

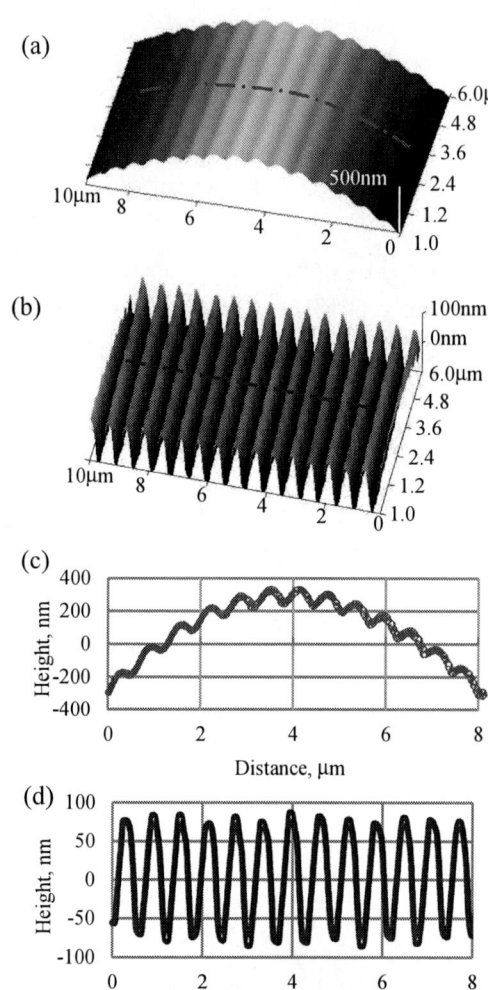

Figure 6: Hierarchically wavy pattern: (a) 3D profile of nanowrinkles on the ridge of reflowed microstructures, (b) 3D profile of nanowrinkles on the groove of reflowed microstructure, (c) profile tracing of the ridge, and (d) profile tracing of the groove.

CONCLUSIONS

In summary, an electrical discharge assisted surface wrinkling process using a hand-held corona discharger is reported. Linear micro/nanowrinkles ranging from a few hundred nm to several microns can be manufactured by electrical-discharge induced surface modification of a uniaxially pre-strained PDMS. This process can be accomplished in minutes without the need of clean room environment or the expensive facilities for deposition or coating. By sequentially performing surface wrinkling and replica molding processes, binary anisotropic wrinkling pattern was demonstrated. Increased surface hierarchy was demonstrated by superimposing continuous nanowrinkles with the characteristic feature size of a few hundred nm on PDMS reflowed microstructures with the characteristic feature size of a few tens microns. With the capacity of creating wrinkling features with different levels of hierarchy and anisotropy, and interfacing with microstructures for scaling up, this simple, versatile and inexpensive benchtop wrinkling process is expected to foster the applications of micro/nanowrinkles in a broad field.

ACKNOWLEDGEMENTS

This works is partially funded by a NSF CAREER Award under the award number DBI 0954013 and Institute for Materials Research at the Ohio State University. The authors also thank HHMI for Med into Grad program and Pelotonia graduate fellowship for the student supports. The authors also thank Dr. Jun Liu for her generous facilities support and Dr. David Lee for his technical assistance in experiments.

REFERENCES

[1] C. Harrison, C. M. Stafford, W. H. Zhang, and A. Karim, "Sinusoidal phase grating created by a tunably buckled surface," *Appl. Phys. Lett.,* vol. 85, no. 18, pp. 4016-4018, Nov, 2004.

[2] C. J. Yu, K. O'Brien, Y. H. Zhang, H. B. Yu, and H. Q. Jiang, "Tunable optical gratings based on buckled nanoscale thin films on transparent elastomeric substrates," *Appl. Phys. Lett.,* vol. 96, no. 4, Jan, 2010.

[3] D. H. Kim, J. H. Ahn, W. M. Choi, H. S. Kim, T. H. Kim, J. Song, Y. Y. Huang, Z. Liu, C. Lu, and J. A. Rogers, "Stretchable and foldable silicon integrated circuits," *Science,* vol. 320, no. 5875, pp. 507-11, Apr 25, 2008.

[4] S. P. Lacour, S. Wagner, Z. Y. Huang, and Z. Suo, "Stretchable gold conductors on elastomeric substrates," *Appl. Phys. Lett.,* vol. 82, no. 15, pp. 2404-2406, Apr, 2003.

[5] S. Chung, J. H. Lee, M.-W. Moon, J. Han, and R. D. Kamm, "Non-Lithographic Wrinkle Nanochannels for Protein Preconcentration," *Adv. Mater.,* vol. 20, no. 16, pp. 3011-3016, 2008.

[6] J. Guan, P. E. Boukany, O. Hemminger, N.-R. Chiou, W. Zha, M. Cavanaugh, and L. J. Lee, "Large Laterally Ordered Nanochannel Arrays from DNA Combing and Imprinting," *Adv. Mater.,* vol. 22, no. 36, pp. 3997-4001, 2010.

[7] Y. Zhao, H. Zeng, J. Nam, and S. Agarwal, "Fabrication of skeletal muscle constructs by topographic activation of cell alignment," *Biotechnol. Bioeng.,* vol. 102, no. 2, pp. 624-631, 2009.

[8] K. Efimenko, M. Rackaitis, E. Manias, A. Vaziri, L. Mahadevan, and J. Genzer, "Nested self-similar wrinkling patterns in skins," *Nature materials,* vol. 4, no. 4, pp. 293-7, Apr, 2005.

[9] B. Wang, Y. Zhang, L. Shi, J. Li, and Z. Guo, "Advances in the theory of superhydrophobic surfaces," *J. Mater. Chem.,* vol. 22, no. 38, pp. 20112-20127, 2012.

[10] M. J. Dalby, M. O. Riehle, S. J. Yarwood, C. D. Wilkinson, and A. S. Curtis, "Nucleus alignment and cell signaling in fibroblasts: response to a micro-grooved topography," *Exp. Cell Res.,* vol. 284, no. 2, pp. 274-82, Apr 1, 2003.

[11] M. J. Lopez-Bosque, E. Tejeda-Montes, M. Cazorla, J. Linacero, Y. Atienza, K. H. Smith, A. Llado, J. Colombelli, E. Engel, and A. Mata, "Fabrication of hierarchical micro-nanotopographies for cell attachment studies," *Nanotechnology,* vol. 24, no. 25, pp. 255305, Jun 28, 2013.

[12] S. Yang, K. Khare, and P.-C. Lin, "Harnessing Surface Wrinkle Patterns in Soft Matter," *Adv. Funct. Mater.,* vol. 20, no. 16, pp. 2550-2564, 2010.

[13] A. Schweikart, and A. Fery, "Controlled wrinkling as a novel method for the fabrication of patterned surfaces," *Microchimica Acta,* vol. 165, no. 3-4, pp. 249-263, 2009.

[14] K. Wei, and Y. Zhao, "Fast and Versatile Fabrication of PDMS Nanowrinkling Structures,", pp. 665-667, *Proceeding of 16th International Conference on Miniaturized Systems for Chemistry and Life Sciences*, 2012.

[15] K. Koch, M. Bennemann, H. F. Bohn, D. C. Albach, and W. Barthlott, "Surface microstructures of daisy florets (Asteraceae) and characterization of their anisotropic wetting," *Bioinspiration & biomimetics,* vol. 8, no. 3, pp. 036005, 2013.

CONTACT

*Y. Zhao, tel: +1-614-2477424; zhao.178@osu.edu

FABRICATION OF THIN STENCIL WITH BUFFER RESERVOIR UTILIZING THE COMBINATION OF AZ4620 AND SU-8 ELECTROPLATING MOLDS

Pi-Hsun Chen[1], Chun-Wei Huang[2], and Che-Hsin Lin[1]

[1] Department of Mechanical and Electromechanical Engineering, National Sun Yat-sen University, TAIWAN
[2] Energy and Agile System Department, Metal Industries Research and Development Center, Kaohsiung, TAIWAN

ABSTRACT

Screen printing is one of the major techniques for producing printing circuit board (PCB) in electronic industry. However, it is difficult to produce small patterns using conventional screen printing technique due to the limitation of the woven mesh. This work successfully develops a novel process for fabricating ultra-thin stencil with a buffer reservoir utilizing the combination of AZ4620 positive photoresist (PR) and SU-8 negative PR as the electroplating molds. The developed ultra-thin stencil is 2.5 μm in thickness, which is much thinner than the typical thickness of the conventional stencils. The stencil is produced with nickel plating process with the hardness and tensile strength of 250 HV and 70 kgf/mm², respectively. The printing result shows that the developed stencil capable to print high resolution and thin pattern. Good printing results will present in this paper. Silver paste line with the width of around 20 μm can be successfully printed on PET substrate. The method developed in this study provides a simple and low-cost way to produce high resolution metal stencil.

INTRODUCTION

Screen printing is a classic process for rapid and low-cost production of various documents such as newspapers, sticks and cloths. The advantages come with screen printing process is that this method is capable of printing different materials including metal, polymer, ceramic and dielectric materials. Nowadays, screen printing technology has become one of the most important methods for producing patterns. The substrate materials for screen printing are also flexible, depending on the glue used for printing. Screen printing is usually used in low-cost and mass production for printing circuit board (PCB) in modern electronic industry [1]. Recently, screen printing is also used in a varieties of manufacturing process such as solar cells [2], sensor electrodes [3], radio-frequency (RF) devices [4] and organic light-emitting diodes (OLED) [5].

There are a number of factors that would influence the quality for a screen printing process including the printing speed, blade pressure, printing direction, mesh count, ink composition and the substrate properties.[6, 7] The decision of these parameters is usually chosen by the rule of thumb learned from the experience if the pattern definition is not an important issue. On the other hand, the screen printing technology is tending to smaller line width, thinner printing layer and higher printing resolution. However, it is difficult of great challenge to produce fine line patterns using conventional screen printing technique due to some practical limitations of the woven mesh. First of all, the low

lithography resolution of the emulsion layer is an issue. Even a high resolution emulsion can be used, the diameter for the SS fabric to produce the woven mesh is usually greater than 40 μm such that the screen fabrics may shielding the injection of the paste then cause the pattern incomplete. Therefore, the critical dimension for typical screen printing process is limited between 50 μm to 100 μm [8]. Recently, smaller screen printed patters of around 40 μm have been achieved by the modification of the mesh surface[9]. However, surface modification could not deal with the pattern blocking by the physical dimension of the SS fabrics.

The trend for developing electronic devices with smaller size makes the urgent demands for the printing technologies capable of printing patterns with smaller critical dimension. Therefore, pattern printing with a single layer stencil has been developed to print the patterns of the dimension smaller than 50 μm [10]. Nevertheless, commercial available stencils are usually with the thickness of more than 150 μm due to the strength requirement. The thick metal film may limit the critical dimension of the patterns on the stencil and produces thicker patterns. Therefore, the printing quality and the printable critical dimension for conventional stencil printing are limited. In general, stencils are manufacture using chemical etching, laser-cutting or electroforming approaches[11]. Stencils fabricated with the isotropic etching of metal and high power laser cutting usually have the critical dimension bigger than around 50 μm. In addition, both of chemical etching method and laser-cutting method need electro-polishing process to improve the smooth of side walls and may increase the production cost. Therefore, there is a need to develop a new method to produce high quality stencils for fine pattern printing.

Figure 1 compares the stencil structures and the printed patterns for the conventional stencil and the proposed double-layer stencil. Due to the thicker substrate, the printed pattern exhibits a thicker paste such that the resolution of the printing pattern is limited. (Fig. 1A) The key to print fine patterns with thin paste is to reduce the thickness of the stencil. However, a thin stencil may not have enough strength to sustain the applied force during printing. Therefore, a double-layer structure composed of a thin layer with printing patterns and a thick metal layer to provide the strength for the stencil is proposed. The paste thickness of the printed patterns can be reduced due to the thinner pattern on the stencil. Moreover, the thick reservoir structure is also used as the buffer layer to prevent the pattern from damage during printing. This buffer reservoir design not only increase the printing life time but also improve the printing dimension because more uniform paste through rate and transfer pattern onto the substrate.

978-1-4799-3510-9/14 $31.00 © 2014 IEEE

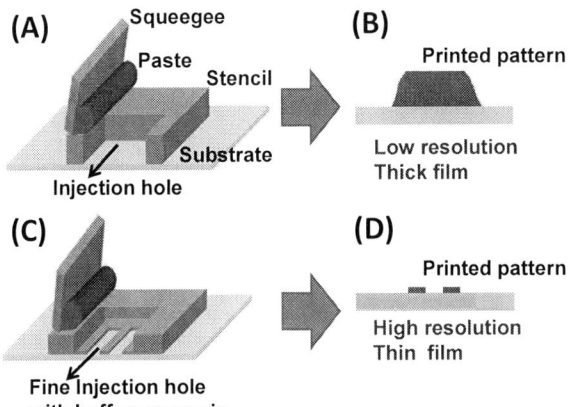

Figure 1: Comparison for the structure and the printing results for conventional stencil (A) and (B) the developed double-layer stencil.

In order to produce the thin stencil with buffer reservoir, this work develops a novel two-step electroplating process utilizing the combination of AZ4620 positive PR and SU-8 negative PR as the electroplating molds. The concept for developing the proposed double-layer stencil is presented in Fig. 2. A thin layer of AZ4620 PR of 5 µm in thickness was used to pattern the 1st metal layer then the other SU-8 layer of 50 µm in thickness was patterned as the second electroplating mold. The injection hole is defined by the AZ4620 PR since AZ4620 can well sustain the nickel plating bath in a short electroplating time. On the contrary, the SU-8 PR can sustain long electroplating time of the nickel plating bath then prevented the first metal layer damage in second electroplating process. The high transparency of SU-8 PR also makes it easy to align the two PR plating molds. A 30 µm thick nickel layer is then electroplated onto the first nickel layer. The second nickel layer is used to provide the necessary strength of the stencil and to define the buffer reservoir for the printing paste. The stencil will no longer exhibit the blocking problem by the screen mesh. Moreover, the developed stencil is capable of printing free-standing point patterns with good uniformity, which is difficult to be printed using conventional stencil printing method.

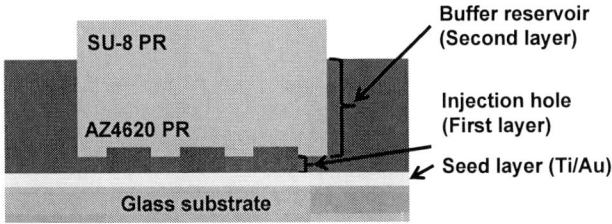

Figure 2: Design concept for producing the ultra-thin stencil using AZ4620 and SU-8 photoresist as the mold for nickel plating.

FABRICATION

Figure 3 presents the simplified fabrication process for producing the developed thin stencil for fine pattern printing.

Substrate

A low-cost microscope glass slide ($76 \times 26 \times 1.2$ mm^3) was used as the substrate for producing the stencil. In order to meet the requirement for metal electroplating and structure releasing, the Ti/Au layers of 50 nm in thickness were coated on the substrate by sputtering. (Fig. 3A) The titanium layer was used for both adhesion layer for gold seed layer and the sacrificial layer for structure release such that the thickness of titanium layer was thicker than typical metal adhesion layer of 3-5 nm in the present study.

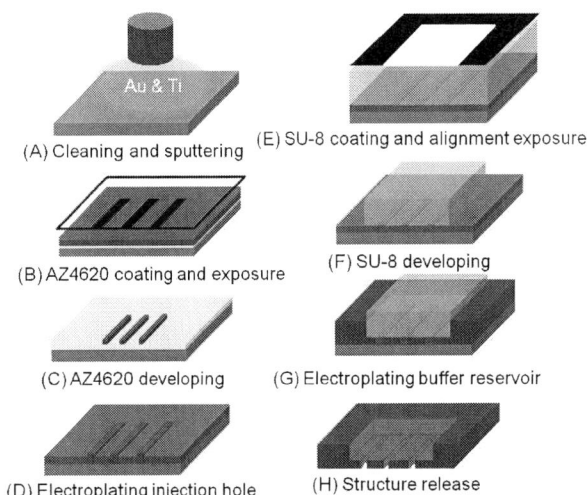

Figure 3: The simplified fabrication process for producing the developed thin stencil.

First metal layer formation

The first metal layer was precisely defined with a 5 µm thick of AZ4620 photoresist using a standard photolithography process. (Fig. 3B and 3C) A hard-bake on the patterned PR was carried out at 150°C for 10 min in order to further enhance the adhesion and the strength to sustain the nickel plating process. Prior to the nickel plating process, the patterned substrate was activated with O_2 plasma to enhance the surface wettability. A 2.5 µm thick nickel layer was then plated with a current density of 1.0 ASD in a nickel chloride bath of 50°C and pH 3.8. (Fig. 3D)

Second metal layer formation

Once the 1st metal layer was formed, a SU-8 layer of 50 µm in thickness was then spun on the electroplated substrate for the formation of the other thick nickel layer. During the soft baking process, the SU-8 was self-planarized since the low glass temperature of the unexposed SU-8. (Fig. 3E) The high transparency of SU-8 PR made it easy to align the two PR plating molds. The SU-8 was then aligned and patterned to form the buffer reservoir. The UV lithography was with an exposure dose of 1200 mJ cm^{-2} and the post exposure bake was carried out at a lower temperature of 60°C for 30 min. (Fig. 3F) The SU-8 patterned substrate was again treated with oxygen plasma and then plated with the second nickel layer of 30 µm in thickness. Note that the current density for the electroplating was ramped from 0.5 ASD to 3 ASD. (Fig. 3G)

Structure release and stencil mounting

After electroplating the 2nd metal layer, the plated substrate was immersed in a 75°C PG remover (Micro-chem Inc. USA) for 24 h to remove both the photoresists. The metal structure was then released from the glass substrate using a diluted (1.0 M) HF solution. The HF solution strongly reacted with titanium metal without rapid attacking the nickel such that the nickel plating structure could be released in a short period of time. The released structure was washed in DI water and then methanol to remove residual HF without stiction. The released structure was finally baked at 100°C for 3 min. Prior to the printing test, the produced stencil structure was mounted on a laser cut 1.5 mm thick PMMA substrate using UV glue. It was not necessary to stretch the stencil during the mounting process since the nickel plating layer exhibited the necessary tension force and was flat.

Stencil printing test

The stencil printing test was performed in a cleanroom to eliminate the particle contaminations. A commercial silver paste (TEC-PA-060, Inktec, Korea) was used to evaluate the printing performance of the stencil. The viscosity of the paste was 65000 cp and the standard printed thickness for the pattern was 3-4 μm using the emulsion of 10 μm in thickness. The patterns were printed on a 120 μm thick polyethylene terephthalate (PET) film. The stencil printing was operated with a speed of 150 mm/s and a normal force of 15 N using a home-built printing module.

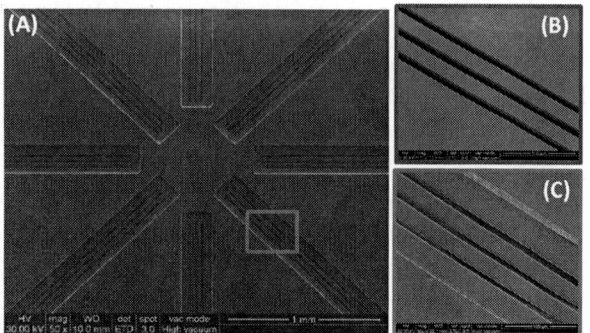

Figure 4: SEM images of the produced stencil (A), the close-up views of the injection hole (B) and the buffer reservoir structure (C).

RESULTS AND DISCUSSION

Figure 4 shows the SEM images of the produced stencil and the close-up views of the injection hole and the buffer reservoir structure. It is noted that the continuous length for the stencil pattern was 50 mm which was challenging for releasing the structure without stiction and distortion. The SEM image clear showed that the surface of the stencil was smooth and there was no significant bending or stretching caused by the induced residual stress during electroplating. (fig. 4A) Figure 4B shows the close-up view for the first metal layer after removing the photoresist. The line width for the injection hole was around 15 μm and the pitched metal structure was 25 μm. The second metal layer is used to

provide the necessary strength of the stencil and define the buffer reservoir structure. Figure 4C shows the close-up view for the buffer reservoir defined by the second metal layer. Results showed that the patterns are well defined with the developed process.

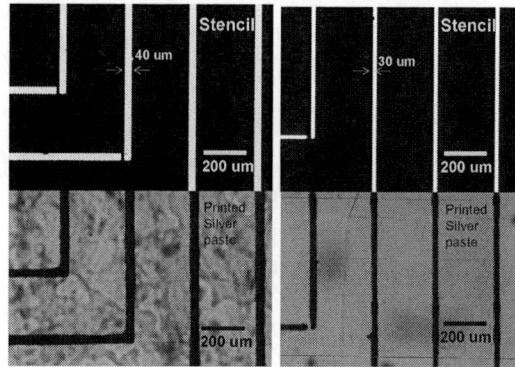

Figure 5: Comparison for the line width of the stencil and the printed silver paste patterns. (Left: 40 μm line width, right: 30 μm line width)

Figure 5 shows the optical images of the produced stencil and the printed silver paste lines. Good printing results confirmed that the stencil is capable of printing silver paste of small dimension. The silver paste was completely transferred onto the PET substrate while printing with the produced stencil. In addition, the printed paste patterns were uniform and the line edge was precisely confined. The printed patterns did not show the waveform signature that was usually observed for the patterns printed using screen printing due to the non-uniform paste extrusion via the screen mesh. Figure 6 presents the close-up SEM images for the printed pattern the vertical and parallel lines are printed with good line width control. Silver paste line with the width of around 20 μm was successfully printed on PET substrate. The right angle pattern was well printed, indicating the good orientation property for the developed thin stencil. High density and multiple directional patterns were capable to be printed using the developed stencil. The capability of the developed stencil will meet the requirement for the applications of modern electronic devices.

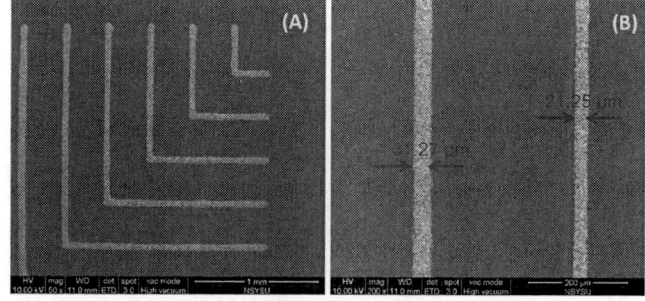

Figure 6: Close-up SEM images showing the printed silver paste on PET substrate.

In order to further evaluate the performance of the developed stencil, dot array with the dot diameter of 30 μm was printed with conventional screen printing and the stencil printing. Figure 7 compares the printing results of using screen mesh and the stencil for the printings. Figure 8(A) showed the patterned screen composed of 20 μm stainless steel wire and 20 μm thick emulsion. Note that the hole diameter was 50 μm. The screen mesh count was 400 which was the optima specification currently commercially available. It is clear that due to the shielding effect of the woven mesh, a number of dots were blocked and failed to be produced on the screen. (Fig. 7A) Therefore, several dots were missing from the printed pattern, resulting in an incomplete printed pattern. (Fig. 7B) On contrast, figure 8C shows the SEM image of a 30 μm dot array pattern on the stencil. The blocking issued was excluded for the stencil and the printed pattern. Although there was minor smear for the printed dots, this effect can be reduced by hydrophobic coating on the side contacting the printed substrate.

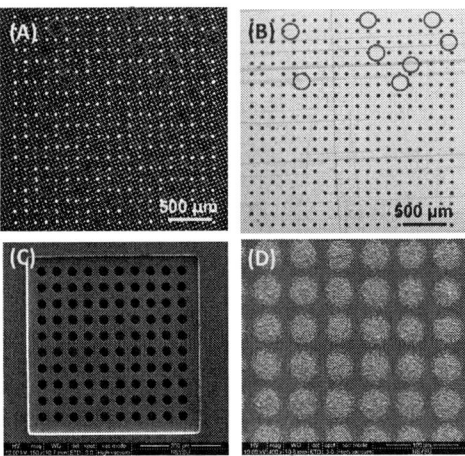

Figure 7: SEM image of a stencil with fine holes and the printing result.

CONCLUSION

A novel process for fabricating thin stencil with double-layer structure utilizing the combination of two different photoresist as the electroplating molds was reported in this paper. A thicker titanium adhesion layer was used to rapid releasing the metal structure in diluted HF solution after the electroplating process. The good pattern defined for the electroplated stencil significantly improved the printing resolution. Results showed that the developed stencil successfully printed silver paste with the pattern of around 20 μm in width and 3 μm in thickness. The complete right angle patterns confirmed that the developed stencil was capable for printing patterns with desired orientations. The method develop in the present study will give substantial impact on the modern printing technology.

ACKNOWLEDGEMENTS

The authors would like to thank the financial supports from National Science Council (NSC) and Metal Industries Research and Development Center (MIRDC) of Taiwan.

REFERENCE

[1] J. B. Pan, G. L. Tonkay, and A. Quintero, "Screen printing process design of experiments for fine line printing of thick film ceramic substrates," *Journal of Electronics Manufacturing,* vol. 9, pp. 203-213, Sep 1999.

[2] C. Park, T. Kwon, B. Kim, J. Lee, S. Ahn, M. Ju, *et al.,* "Front-side metal electrode optimization using fine line double screen printing and nickel plating for large area crystalline silicon solar cells," *Materials Research Bulletin,* vol. 47, pp. 3027-3031, Oct 2012.

[3] T. Schüler, T. Asmus, W. Fritzsche, and R. Möller, "Screen printing as cost-efficient fabrication method for DNA-chips with electrical readout for detection of viral DNA," *Biosensors and Bioelectronics,* vol. 24, pp. 2077-2084, 2009.

[4] S. Pranonsatit and S. Lucyszyn, "Micromachined screen printing (MaSPrint) technology for RF MEMS applications," in *High Frequency Postgraduate Student Colloquium, 2005,* 2005, pp. 3-6.

[5] G. E. Jabbour, R. Radspinner, and N. Peyghambarian, "Screen printing for the fabrication of organic light-emitting devices," *Ieee Journal of Selected Topics in Quantum Electronics,* vol. 7, pp. 769-773, Sep-Oct 2001.

[6] Y. T. Yen, T. H. Fang, and Y. C. Lin, "Optimization of screen-printing parameters of SN9000 ink for pinholes using Taguchi method in chip on film packaging," *Robotics and Computer-Integrated Manufacturing,* vol. 27, pp. 531-537, Jun 2011.

[7] D. Manessis, R. Patzelt, A. Ostmann, R. Aschenbrenner, and H. Reichl, "Technical challenges of stencil printing technology for ultra fine pitch flip chip bumping," *Microelectronics Reliability,* vol. 44, pp. 797-803, May 2004.

[8] D. Erath, A. Filipovic, M. Retzlaff, A. K. Goetz, F. Clement, D. Biro*, et al.,* "Advanced screen printing technique for high definition front side metallization of crystalline silicon solar cells," *Solar Energy Materials and Solar Cells,* vol. 94, pp. 57-61, Jan 2010.

[9] D. Schwanke, J. Pohlner, A. Wonisch, T. Kraft, and J. Geng, "Enhancement of Fine Line Print Resolution due to Coating of Screen Fabrics," *Journal of microelectronics and electronic packaging,* vol. 6, p. 13, 2009.

[10] R. W. Kay, S. Stoyanov, G. P. Glinski, C. Bailey, and M. P. Desmulliez, "Ultra-fine pitch stencil printing for a low cost and low temperature flip-chip assembly process," *Components and Packaging Technologies, IEEE Transactions on,* vol. 30, pp. 129-136, 2007.

[11] S. N. Bhat, G. A. Rao, N. S. Dinesh, and B. N. Baliga, "Photo-Defined Electrically Assisted Etching Method for Metal Stencil Fabrication," *Ieee Transactions on Components Packaging and Manufacturing Technology,* vol. 1, pp. 1116-1121, Jul 2011.

978-1-4799-3510-9/14 $31.00 © 2014 IEEE

FREE-STANDING SUBWAVELENGTH GRID INFRARED REJECTION FILTER OF 90 MM DIAMETER FOR LPP EUV LIGHT SOURCE

Yukio Suzuki[1,2], Kentaro Totsu[1], Masaaki Moriyama[1,2], Masayoshi Esashi[1], Shuji Tanaka[2]
[1]Micro System Integration Center, Tohoku University, JAPAN
[2]Graduate School of Engineering, Tohoku University, JAPAN

ABSTRACT

A subwavelength grid infrared (IR) filter as large as 90 mm in diameter was fabricated and tested on a 6 inch Si wafer for a laser-produced plasma (LPP) extreme ultraviolet (EUV) light source used in the next generation lithography tools. The IR filter is a grid type, which has higher thermal stability compared with a conventional multilayer metal membrane filter. The grid is a free-standing Mo-coated Si honeycomb structure with a thickness of 5 μm, a wire width of only 0.35 μm and a pitch of 4.5 μm. Such a large-size free-standing microstructure was successfully fabricated by carefully balancing film stress at each process step. The fabricated IR filter demonstrated 99.7 % rejection for 10.6 μm IR light.

INTRODUCTION

Extreme ultraviolet (EUV) lithography is expected as the ultimate on-wafer patterning technology for high volume production over many years. An EUV light source with a wavelength, λ, of 13.5 nm has been always a critical issue, and EUV power is still insufficient for practical throughput. At present, a laser-produced plasma (LPP) EUV light source is the most promising for the next generation lithography [1]. In the light source, Sn microdroplets are shot with CO_2 pulse laser (λ = 10.6 μm), and plasma is produced. EUV light is selectively extracted from the plasma using a spectral purity filter (SPF), as shown in Fig. 1. The SPF must have an IR light rejection of 99 % or higher. In addition, the transmittance of EUV light must be as high as possible.

At the intermediate focus (IF) point (see Fig. 1), IR to EUV light power ratio is 5 or higher. To obtain an EUV power of 350 W, which is expected for practical high volume production, the SPF must withstand at least 1750 W IR power input [1, 2]. To date, a $Mo/ZrSi_2$ multilayer membrane SPF [3, 4, 5] has been used in alpha and beta tools with much lower EUV light power, but it cannot withstand such high power IR light input. Actually, significant decrease in EUV light transmittance was observed due to metal interdiffusion under an IR light power density of 8 W/cm^2 [3].Compared to the multilayer membrane SPF, the subwavelength grid type is stronger against IR light input, because its working principle is based on diffraction not absorption [6, 7]. In addition, EUV light transmittance is sufficiently high, because EUV light is attenuated only by shadowing effect of submicron wide grid wires.

In the past research, a small-size metallized Si grid IR filter was fabricated for proof-of-concept, and 99.9% rejection of IR light was demonstrated even after 100 W/cm^2 laser irradiation [2]. However, scaling up the IR filter to a

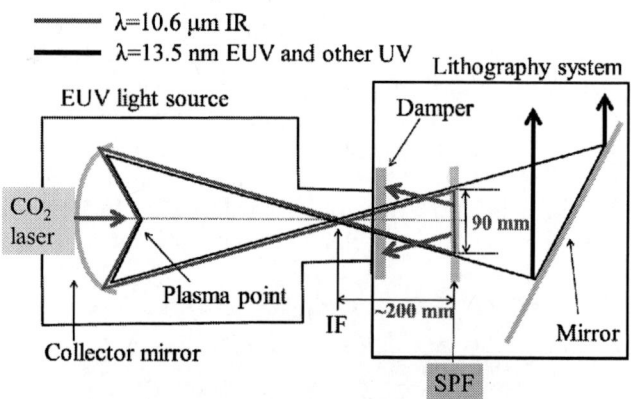

Figure 1: Schematic of LPP EUV light source with a spectral purity filter (SPF).

practical size is a difficult challenge [8], because the free-standing grid structure is mechanically fragile. In this study, we first fabricated the grid IR filter with a practical size (90 mm in diameter).

DESIGN

The structure of a honeycomb grid is shown in Fig. 2. This structure provides a close packing of identical apertures and is advantageous in mechanical strength because of small stress concentration at 120° intersection of each wall. The grid pitch, P, and wire width, W, were determined at 4.5 μm and 0.35 μm, respectively, based on IR light transmittance calculated by rigorous coupled wave analysis (RCWA). As found in Fig. 3, IR light transmittance rapidly increases, if the grid pitch is beyond 4.5 μm. A wire width, W, must be small enough to achieve an acceptable EUV light transmittance. On the other hand, the wire width must be large enough to keep mechanical strength. The EUV transmittance, T, can be calculated by considering shadowing effect as

$$T_0 = \left(1 - \frac{W}{P}\right)^2, \tag{1}$$

where T_0 is transmittance at an incidence angle, θ, of $0°$. T_0 at P of 4.5 μm and W of 0.35 μm is approximately 0.85 (85%).

In Eq. (1), the input of parallel light ($\theta = 0°$) is assumed, but actually incident light from the IF point enters the SPF obliquely. In addition, the grid has a height, H, which cannot be neglected compared with P and W. Therefore, the shadowing effect is underestimated by Eq. (1). EUV light transmittance is highest at the center of the filter, and

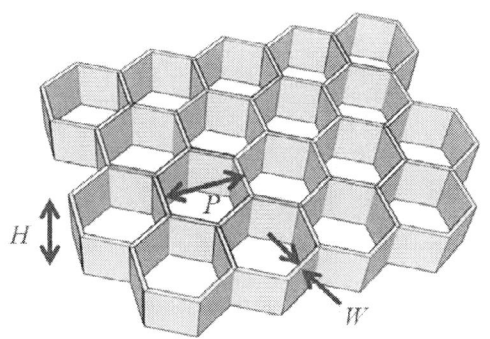

Figure 2: Structure of subwavelength grid IR filter (P: Pitch of aperture, W: Width of grid, H: Height of grid).

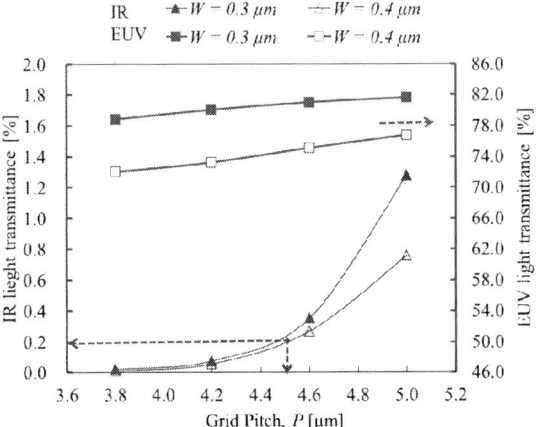

Figure 3: IR light transmittance calculated by RCWA and EUV light transmittance calculated by Eq. (3) (Grid height, H = 5 μm).

decreases toward the outside of the filter. Based on Fig. 4, EUV light transmittance at radius, r, is given by

$$T(r) = \frac{a}{c}T_0 = \frac{c-b}{c}T_0$$
$$= \left(1 - \frac{H\tan(\theta)}{c}\right)T_0 = \left(1 - \frac{Hr}{cD}\right)T_0 \qquad (2)$$

where a is the length of efficient open areas, b is the length of shadowing areas, c is the minimum length of apertures ($c = P - W$), and D is distance between the IF point and the filter, as shown in Fig. 4. For example, the IR filter of 90 mm in diameter is installed about 200 mm away from the IF point.

The total EUV light transmittance taking account of the shadowing effect by the grid height is obtained by integrating Eq. (2) over the filter area. For simplified calculation, the cross section of each honeycomb apertures is now considered as a circle. The total EUV light transmittance is given by,

$$T = \frac{\int_0^R 2\pi r\left(1 - \frac{H}{cD}r\right)dr}{\pi R^2}T_0, \qquad (3)$$

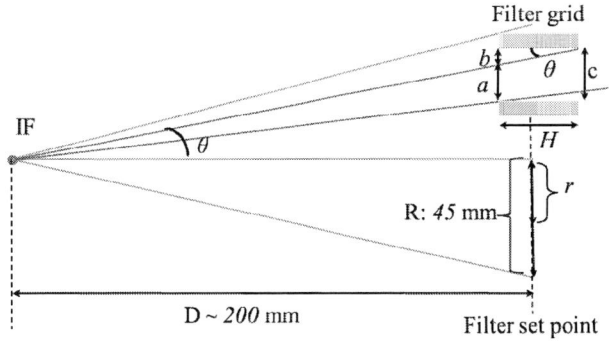

Figure 4: Schematic of optical path between IF point and the filter.

Figure 3 shows the EUV light transmittance calculated by Eq. (3), assuming that $H = 5$ μm. When $P = 4.5$ μm and W = 0.35 μm, an EUV light transmittance of about 78% is obtained. This value is higher than that of the existing Mo/ZrSi$_2$ multilayer membrane SPF.

FABRICATION

The starting material is a 6 inch SOI wafer with 5 μm thickness of device layer, 1 μm thickness of BOX layer and 400 μm thickness of handle layer. Since Si has a high IR transmittance, metal coating is needed to reflect IR light from both surface and sidewall of the grid wires. Molybdenum was chosen because of high reflectivity for IR light and high heat resistance.

The fabrication process is shown in Fig. 5. The grid pattern was defined on the SOI wafer by step-and-repeat g-line photolithography and deep reactive ion etching (DRIE) (a). In order to reduce so-called scallop patterns of sidewalls, which may cause scattered EUV light, we optimized DRIE condition and finally achieved a side wall roughness less than 30 nm peak to valley. At this step, the grid wire has a width of 0.5 μm, which is larger than the designed value ($W = 0.35$ μm) but close to the resolution of a g-line stepper. Therefore, the grid structure was once thermally oxidized (b) and then vapor HF etched to thin it down. 300 nm thick oxidation ate up 135 nm thick bulk Si, which corresponded to 44 % volume of SiO$_2$, and a 0.35 μm wide grid structure was finally obtained after etching SiO$_2$, as shown in Fig. 6.

The grid structure was released by Si backside wet etching with TMAH. Before that, 1.8 μm thick compressive SiO$_2$ was deposited on the front side by plasma CVD to balance the stress of the BOX layer across the grid structure (c). Otherwise, the grid structure bends or fractures by compressive stress of the BOX layer, as illustrated in (c').

As a result, the fragile thin grid structure did not fracture or bend after released, as illustrated in Fig. 7. The SiO$_2$ overcoat and the BOX layer were simultaneously etched using vapor HF with keeping stress balance across the membrane (d). Since the vapor HF etching rate of the plasma CVD SiO$_2$ is about 2 times as large as that of thermal SiO$_2$,

978-1-4799-3510-9/14 $31.00 © 2014 IEEE

both oxide films were almost simultaneously removed.

Finally, stress-controlled Mo was deposited on both sides of the free-standing grid structure (e). Molybdenum was deposited by radio frequency magnetron sputtering. The stress of sputter-deposited Mo was controlled to be nearly zero MPa by process pressure and wafer temperature. A wafer temperature of 300°C was finally chosen from stress control and film stability points of view. The IR filter was successfully fabricated with no significant defect, as shown in Fig. 8.

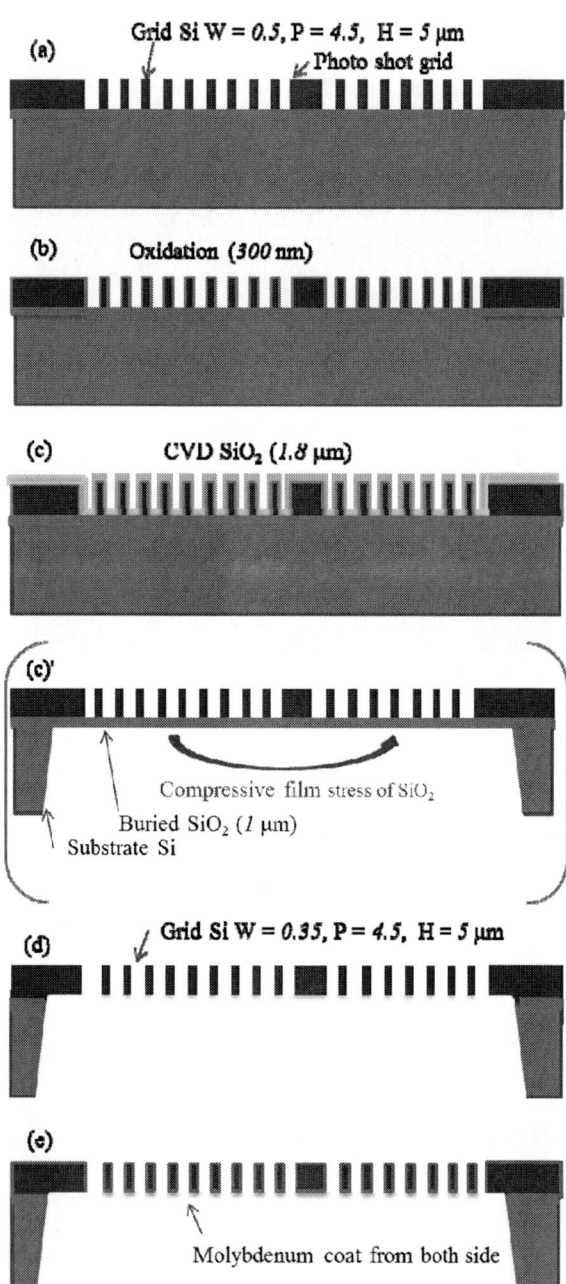

Figure 5: Fabrication process of the IR filter.

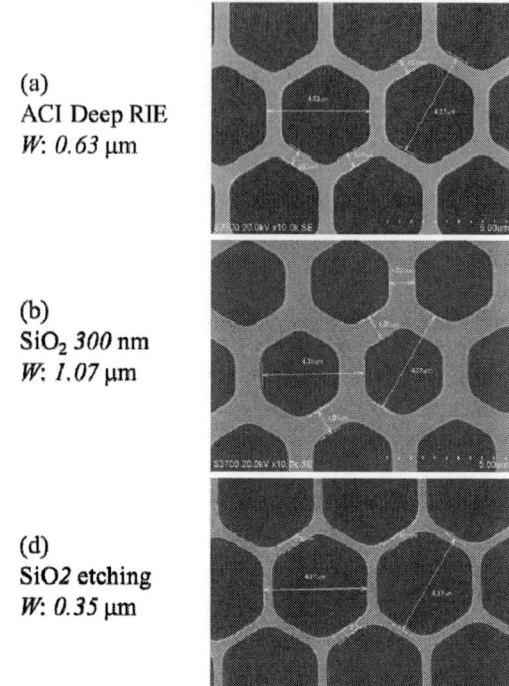

Figure 6: Scanning electron microscope (SEM) image of Si grid after (a) DRIE, (b) oxidation and (d) SiO₂ etching. .

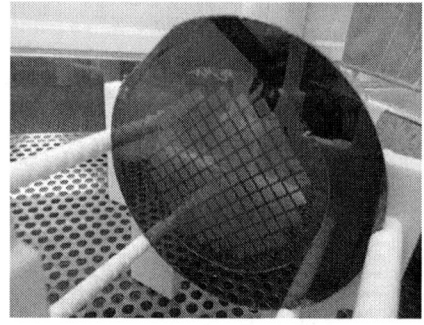

Figure 7: Filter wafer after backside wet etching, where film stress is balanced by both sides of SiO₂.

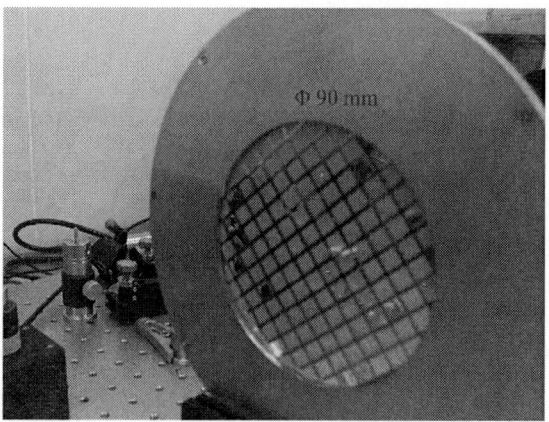

Figure 8: Subwavelength grid IR filter with a diameter of 90 mm installed in optical measuring

FILTER Evaluation

The IR filter was evaluated in terms of IR light transmittance using CO_2 laser. As shown in Fig. 9, the laser was split into halves by a beam splitter to monitor real-time input power with a power meter, PM (A). Transmitted light through the IR filter was detected by another power meter, PM (B). The IR filter was tilted at 5°, 10° and 15° to the normal. The measurement results were summarized in Table 1. Each data was an average of 3 point measurement. IR light transmittance was calculated from IR light powers measured by PM (A) and (B). Optical angles of incidence (AOI) of 5°, 10° and 15° were tested. IR light reflected at the IF filter was absorbed by a water-cooled dumper with a diameter of 2 inch. IR light transmittance of only $0.27 \pm 0.06\%$ (i.e. $99.73 \pm 0.06\%$ rejection) was experimentally confirmed, as shown in Table 1.

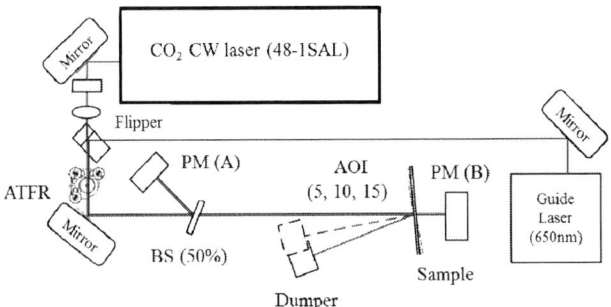

Figure 9: Schematic of optical measurement system. PM: Power meter (A, B), ATFR: Absorbing thin-film reflector, BS: Beam splitter (ZnSe), AOI: Angle of incidence

Table 1: IR light transmittance calculated from laser powers detected power meters.

AOI [deg.]	PM (B) [W]	PM (A) [W]	Transmittance [%]	
			Average	Error
5	0.00252	0.918	0.27 %	0.12 %
10	0.00254	0.921	0.28 %	0.11 %
15	0.00258	0.922	0.28 %	0.11 %

CONCLUSION

We designed and fabricated a subwavelength grid infrared filter as large as 90 mm in diameter for a LPP EUV light source used in the next generation lithography tools. The IR filter has a free-standing Mo-coated Si honeycomb grid structure with a thickness of 5 μm, a wire width of only 0.35 μm and a pitch of 4.5 μm. The grid wire was thinned down from 0.5 μm to 0.35 μm in width, which was smaller than the resolution of a g-line stepper, by thermal oxidation and SiO_2 etching. After releasing the grid structure, such a large-size free-standing microstructure was processed by carefully balancing film stress at each step. The fabricated IF filter showed 99.7% rejection for 10.6 μm IR light.

ACKNOWLEDGEMENTS

This study was granted by the Japan Society for the Promotion of Science (JSPS) through the "Funding Program for World-Leading Innovative R&D on Science and Technology" (FIRST Program) initiated by the Council for Science and Technology Policy (CSTP). All fabrication was performed in Hands-on Access Fabrication Facility a member of "Nanotechnology Platform of the Ministry of Education, Culture, Sports, Science and Technology (MEXT), Japan, at the Center for Integrated Nanotechnology Support, Tohoku University. Optical evaluation was supported by GIGAPHOTON INC.

REFERENCES

[1] T. Tomie, "Tin laser-produced plasma as the light source for extreme ultraviolet lithography high-volume manufacturing: history, ideal plasma, present status, and prospects", *J. Micro/Nanolith. MEMS MOEMS* vol. 11(2), 2012, 021109 1-9.

[2] W. A. Soer, M. J. Jak, A. M. Yakunin, M. M. van Herpen, V.Y. Banine, "Grid spectral purity filters for suppression of infrared radiation in laser-produced plasma EUV sources", *Proc. SPIE* vol. 7271, 2009, 72712Y 1-9.

[3] L. I. Chkhalo, M. N. Drozdov, E. B. Kluenkov, A. Y. Lopatin, V. I. Luchin, N, N, Salashchenko, N. N. Tsybin, "Free-standing spectral purity filters for extreme ultraviolet lithography", *J. Micro/Nanolith. MEMS MOEMS* vol. 11(2), 2012, 021115 1-7.

[4] F. R. Powell, T. A. Johnson, "Filter windows for EUV lithography", *Proc. SPIE* vol. 4343, 2001, 585.

[5] M. S. Bibishkin, N. I. Chkhalo, E. B. Kluenkov, A. Y. Lopatin, V. I. Juchin, A. E. Pestov, N. N. Salashchenko, L. A. Shmaenok, N. N. Tsybin, S. Y. Zuev, "Multilayer Zr/Si filters for EUV lithography and for radiation source metrology", *Proc. SPIE* vol. 7025, 2008, 702502.

[6] R. Ulrich, "FAR-INFRARED PROPERTIES OF METALLIC MESH AND ITS COMPLEMENTARY STRUCTURE", *Infrared Physics* vol. 7, 1967, 37-55.

[7] D. M. Byrne, A. J. Brouns, F. C. Case, R. C. Tiberio, B. L. Whitehead, E. D. Wolf, "Infrared mesh filters fabricated by electron-beam lithography", *J. Vac. Sci. Technol. B* 3(1), 1985, 268-271.

[8] A. Yakunin, E. Kluenkov, A. Lopatin, V. Luchin, N. Salashchenko, N. Tsybin. L. Sjmaenok, M. Markosov, R. Moors, V. Banine, "Spectral Purith Filter Development for EUV HVM", *Int. Symp. EUVL 2008*, Source II 59.

CONTACT

*Yukio Suzuki, tel: +81-22-229-4113; yukio.suzuki@mems.mech.tohoku.ac.jp

GRAPHENE SYNTHESIS VIA DROPLET CVD AND ITS PHOTONIC APPLICATIONS

Xining Zang and Liwei Lin

Berkeley Sensor and Actuator Center, University of California, Berkeley, USA

ABSTRACT

The process of "droplet CVD" for the synthesis of graphene sheets and its photonic applications have been demonstrated. Metal (Cu or Ni) droplets are naturally transformed from thin films under high temperature and used as the catalysts to grow graphene via the chemical vapor deposition (CVD) process. This work has achieved several advancements: (1) first demonstration of "droplet CVD" for graphene synthesis; (2) constructions of continuous graphene sheets with discontinuous metal droplets - readily available for graphene sensing applications; and (3) preliminary proof-of-concept demonstration of a photonic sensor based on the droplet CVD graphene. As such, this new class of fabrication process could open up various graphene-based device/system applications, including photonic sensors.

INTRODUCTION

Previously, CVD synthesis of graphene has been well-established on metal foils, plates or thin-films, typically using nickel or copper as the catalyst [1], and their synthesis mechanism has been extensively studied [2-3]. In general, the growth of graphene on Ni comes from carbon segregation and precipitation, while the growth on Cu is from the surface reaction. It is rather difficult to obtain single layer, uniform graphene on the nickel process, due to high carbon solubility in nickel with results in multiple layers of graphene growth. On the other hand, the surface separation mechanism makes Cu as a better catalyst to construct single layer graphene but the quality is affected by the grain boundaries and defects. In order to alleviate this problem, some researchers have tried to eliminate the grain boundary by employing liquid copper surface or copper vapor as the catalysts [4-6]. However, the surface tension variation on Cu-C interface can influence the domain and affect the nucleation of graphene crystal [4,7,8].

Here we studied the graphene synthesis via CVD on metal droplets of different size, surface energy and adhesion force. In contrast to prior works, the feasibility to grow graphene from copper or nickel droplets is demonstrated in this work. By adjusting the processing temperature and droplet sizes, electrically continuous graphene on top of discontinuous metal droplets has been constructed. We also show that under the activation of visible light, electron hole pairs can be generated to decrease the resistance of the graphene-metal contact as photonic sensors [9,10].

GRAPHENE VIA DROPLET CVD

Figure 1 illustrates the basic process for the synthesis of graphene via "droplet CVD." Thin metal films are deposited onto a silicon substrate with a layer of thermal oxide on top (Fig. 1a). Using CH_4 as the carbon source, H_2 as the

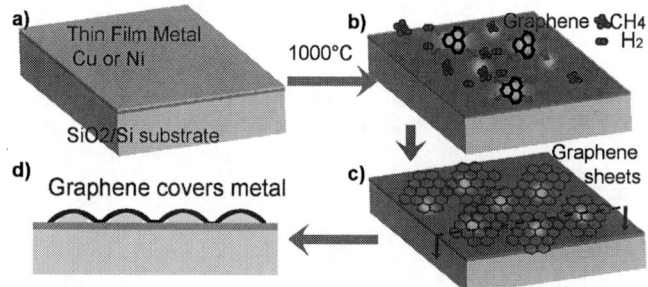

Figure 1: Conceptual illustration of graphene synthesized by the droplet CVD process. (a) Thin films of Cu or Ni deposited on Si substrate with SiO₂ on top. (b)Under high temperature, metal droplets are formed. (c)(d) Graphene sheets are synthesized with proper reaction gases and metal droplets are separated with each other during the cooling process.

reduction agent, and Ar as the carrier gas, graphene structures are grown under high temperature as thin films are melted to form droplets as the catalysts (Fig. 1b). Continuous graphene sheets can be constructed on top of discontinuous and shrunk metal droplets as temperature is cooled down (Fig. 1c & 1d).

The CVD process includes three stages, as shown in Figure 2. In the first stage, reduction gases of $Ar:H_2$ (300:55 sccm) flow through the furnace under a heating process at a speed of 20° C/min. When the temperature is stablized for pyrolysis reaction, 5 sccm pf CH_4 is added for 10 minutes for the synthesis of graphene. In the final stage, the furnace is cooled down at a speed of 20°C/min with the protection gases of H_2 and Ar flowing. Several different growth temperatures from 600 to 1000° C have been experimentally conducted and results are summarized in Table 1. It is found that the combination of right temperature, film thickness and metal type can produce successful growth of graphene, including continuous graphene films. In general, the pyrolysis reaction of CH_4 depends strongly on the processing temperature. When the temperature is under 750° C, there will be no reactions. If the temperature is higher than 950° C, pyrolysis happens and graphene sheets begin to grow from nucleus.

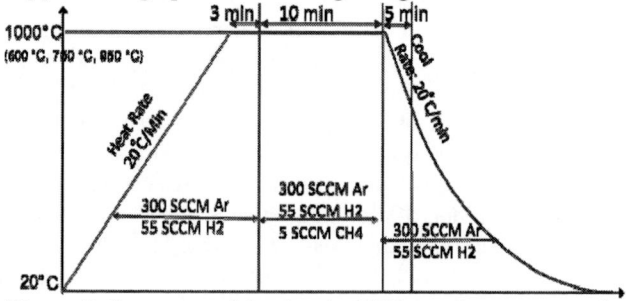

Figure 2: Sequence of the droplet CVD synthesis process for graphene, including gas flow, temperature and process time.

T (°C) Thickness	600		750		950		1000	
	Cu	Ni	Cu	Ni	Cu	Ni	Cu	Ni
20nm	×	×	×	×	×	×	☑	☑
30nm	×	×	×	×	×	×	☑	☑
40nm	×	×	×	×	☑	☑	☑	☑
50nm	×	×	×	×	☑	☑	☑	☑
75nm	×	×	×	×	▣	▣	▣	▣
105nm	×	×	×	×	▣	▣	▣	▣

Table I: *Summary of graphene synthesis results by droplet CVD.* ×-none, ☑-scattered, ▣-continuous graphene.

If the metal film thickness is smaller than 20nm, the graphene synthesis by droplet CVD is not successful. If the film thickness is higher than 120nm, droplet doesn't form the growth process is the same as the conventional graphene synthesis process. As shown in Table I that when the film thickness is between 20 and 105nm, the nickel and copper thin films will melt to become droplets at a temperature below 1000° C [8-10]. Figures 3 and 4 are the SEM images of graphene grown on different thicknesses of Ni and Cu thin films via droplet CVD at 1000° C. It is noted that the droplet size increases from about 100nm to 1 μm as the metal film thickness increases from 20 to 105nm while the distance between each droplet reduces.

Graphene has only one atomic layer but graphene sheets can be identified under SEM due to the contrast to substrate, especially if there are wrinkles. The red-color marks are examples of possible graphene structures in Figs. 3(a-c) and they are clearly larger than the bottom nickel droplets.

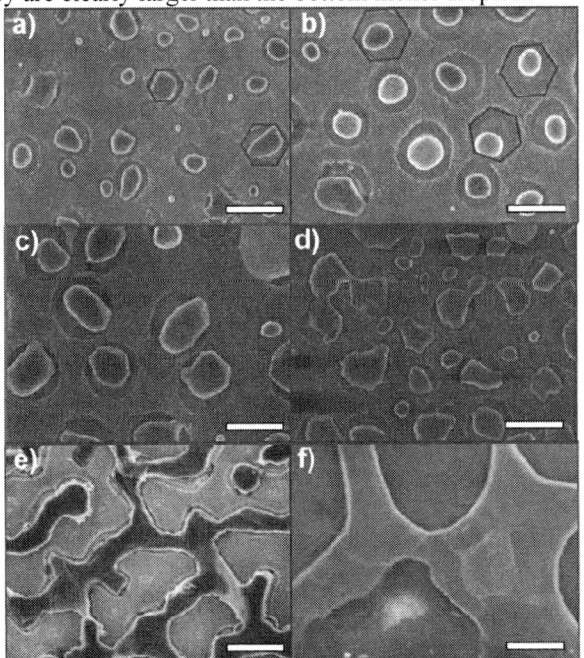

Figure 3: SEMs of graphene grown by Ni droplet CVD with different thickness of initial Ni film: (a) 20nm, (b) 30nm, (c) 40nm, (d) 50nm, (e) 75nm, and (f) 105nm. (Scale bar=1μm)

This phenomenon may be explained with a few reasons, including the condensation of metal droplets at low temperature, and higher diffusion speed and mobility of carbon atom at high temperature [11, 12]. These SEM images also suggest that droplets are discontinuous in Figs. 3 (a-d) and Figs. 4 (a-e). Since graphene sheets can grow larger than the size of the final metal droplets, there could be a "process window" in terms of the metal thickness that the graphene sheets are electrically connected while the bottom droplets are not. Resistance measurement results show that when the metal thin film initial thickness is 75 nm (either Ni or Cu), the CVD-droplet graphene has a sheet resistance $\sim 100\Omega/\square$, which is in the same order of magnitude as a single-layer graphene from conventional CVD by other groups [13]. Since samples made from the same process parameters without synthesizing the graphene sheets show resistance of infinity, it is concluded that droplet CVD can make graphene sheets electrically conductive while the bottom metal droplet are disconnected if the film thickness is about 75nm. It is also observed that Ni droplets adhere better to the substrate with lower contact angle as compared with the Cu droplets.

Figure 4: SEMs of graphene grown by Cu droplet CVD with different thickness of initial Cu film: (a) 20nm, (b) 30nm, (c) 40nm, (d) 50nm, (e) 75nm, and (f) 105nm. (Scale bar=1μm)

Raman Microscopy is broadly used in graphene study to determine the number of layers, defects, and distributions [14, 15]. Figures 5 and 6 are Raman analyses (Renishaw inVia Raman microscope) of graphene synthesized at 1000°C using different film thicknesses. It is found that graphene synthesized by the nickel droplet CVD has $I_G/I_{2D}>2$ and low I_D, implying single-layer graphene with low defects, respectively. It is observed in Fig. 5 that different metal thin films produce graphene sheets with the same wavelength shift and similar $I_G/I_{2D}>2$ ratios, which implies similar quality of graphene via the nickel droplet CVD process.

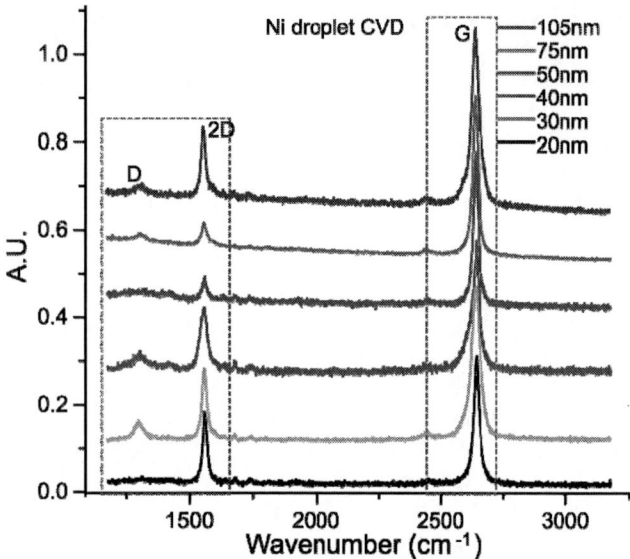

Figure 5: Raman results of graphene grown by Ni droplet CVD with different initial thickness, showing consistent single layer, few defects structures ($I_G/I_{2D} > 2$ and low I_D).

It is well known that Cu is more suitable to grow single layer graphene but single layer graphene is not observed in the Cu droplet CVD process. Instead, multilayer graphene structures with higher number of defects are synthesized in this work as shown in Figure 6. Our hypothesis is that cupper droplets have larger contact angles with respect to SiO_2 as compared to the nickel droplets as seen in the SEM photos. During the cooling and carbon segregation process, cupper droplets deform more severely than the nickel droplets and interrupt the nucleation of graphene and their domain growth. This results in increment of the D peak for more defects and small crystal size of graphene as compared to the nickel droplet CVD.

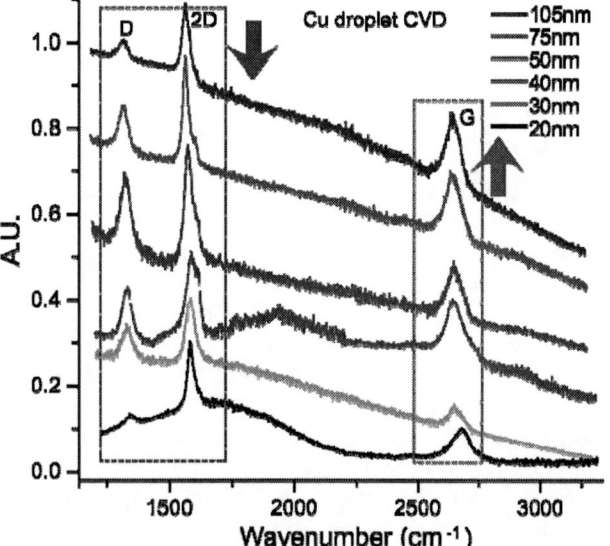

Figure 6: Raman results of graphene grown by Cu droplet CVD with different initial thickness, showing higher D peaks probably due to larger contact angles and deformations as compared with the Ni droplet CVD process.

GRAPHENE METAL DROPLET JUNCTION AS PHOTONIC SENSOR

Utilizing the property of continuous graphene sitting on top of the discontinuous metal droplets, graphene-based devices could be readily made and a graphene-based photonic sensor is demonstrated in this work. Figure 7 plots the electrical and band diagram of the droplet CVD graphene system.

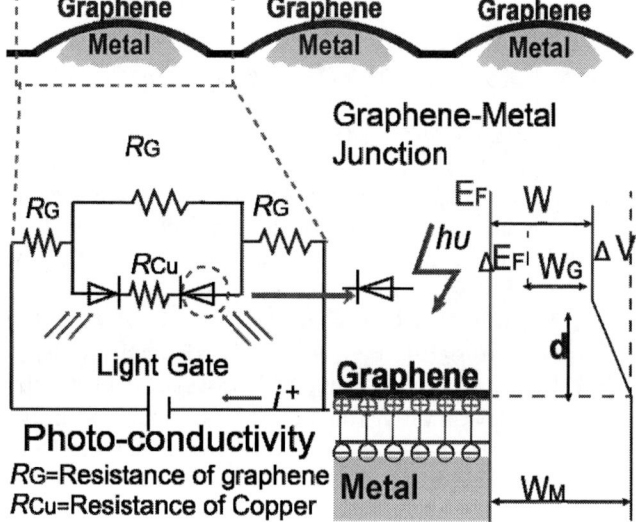

Figure 7: Light generate electron hole pairs increase the conductivity of graphene-metal diode.

Graphene has a work function $\chi\sim4.5eV$, and the work function of Cu is $\chi\sim4.7eV$, Ni $\chi\sim5.0eV$. As shown in Fig 7, graphene and metal form a junction which has characteristics of metal junctions Figure 8). Under incident light, the excited electron-hole pairs can increase the current outputs as the optical sensing mechanism. When two electrical probes are applied, the current is flowing though the graphene sheets as well as the metal droplets. As a result, metal-graphene junction and contacts are important parameters determining the overall resistance. The IV curve under dark and light for a resistor (by placing two probes 5mm apart on the fabricated structure) is shown in Figure 8 and a clear photo response is obtained. The visible light source we use here is a Motic Microscope 6v/30w halogen illumination (Motic AE2000). Light beam, which is focus on different spots on the sample.

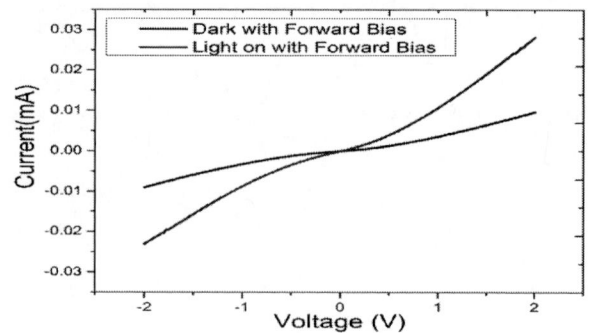

Figure 8: IV measurements with and without light illumination showing optical responses of a graphene-based resistor made from the Cu droplet CVD process.

978-1-4799-3510-9/14 $31.00 © 2014 IEEE

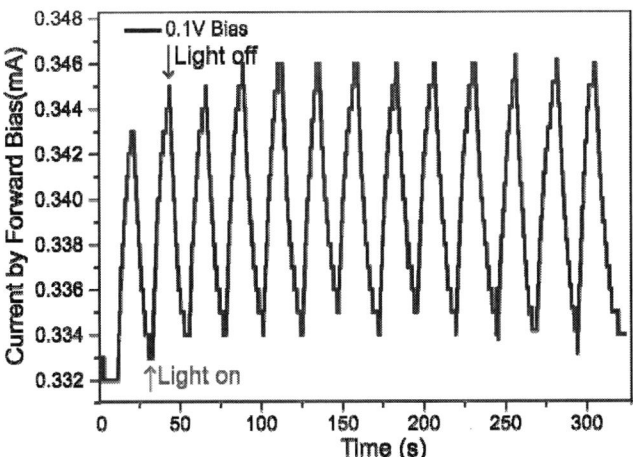

Figure 9: Preliminary optical sensing results from a photo-type graphene resistor made by the same process as figure 8 under a forward bias of 0.1V.

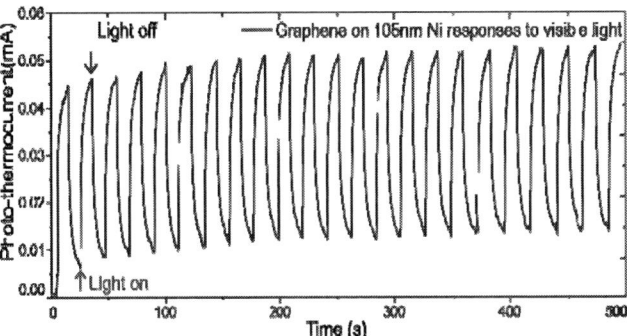

Figure 10: Preliminary optical sensing results from a graphene resistor made by the Ni droplet CVD process.

Figure 9 shows the experimental result of a graphene resistor based on the Cu droplet CVD process. It is observed that the overall resistance changes when the photocurrent is generated. The experiments are conducted under a 0.1V forward bias with on/off light illuminations. The response time of the sensor is 10~20 seconds, which is determined by the life time of the carriers generated on the interface of the graphene-metal junction. The dispassion of photocurrent is quite symmetrical to the generation of photocurrent. Similar results are also demonstrated in a graphene sensor made from the Ni droplet CVD process as shown in Fig. 10. It is found that the magnitude of resistance is larger than the one made from the Cu droplet CVD process in Fig. 9. The performance of resistance change is repeatable and stable, which further validates the feasibility of using the droplet CVD process to make photonic sensors.

CONCLUSION

In this paper, we present a new method to fabricate graphene via the metal droplet CVD process. A detailed characterization on the effects of temperature and initial metal film thickness has been performed to understand the growth mechanisms of graphene from liquid metal. It is demonstrated in the work that continuous graphene sheets can be synthesized on top of discontinuous metal droplets. Furthermore, the graphene sheets can be readily available for sensing applications, including the photonic sensors based on the graphene-metal droplet junction as shown in this work.

REFERENCES

[1] Y. Zhang, et al., "Review of Chemical Vapor Deposition of Graphene and Related Application", Acc. Chem. Res, vol. 46, pp. 2329-2339, 2013.

[2] M. Losurdo, et al., "Graphene CVD Growth on Copper and Nickel: Role of Hydroge in Kinetics and Structure", Phys. Chem. Chem. Phys., vol. 13, pp. 20836–20843, 2013.

[3] X. Li, et al., "Evolution of Graphene Growth on Ni and Cu by Carbon Isotope Labeling", vol. 9, pp. 4268-4272, 2009.

[4] D. Geng, et al., "Uniform Hexagonal Graphene Flakes and Films Grown on Liquid Copper Surface", *P. NATL. ACAD. SCI. USA.*, vol. 109, pp.7992-7996, 2013.

[5] G. Ding, et al., "Chemical Vapor Deposition of Graphene on Liquid Metal Catalysts", *Carbon*, vol. 53, pp. 321-326, 2013.

[6] H. Kim, et al., "Copper-Vapor-Assisted Chemical Vapor Deposition for High-Quality and Metal-Free Single-Layer Graphene on Amorphous SiO_2 Substrate", *ACS. Nano*, vol. 7, pp. 6575-6582, 2013.

[7] A. Ismach, et al., "Direct Chemical Vapor Deposition of Graphene on Dielectric Surfaces", *ACS. Nano.* vol. 10, pp. 1542-1548, 2013.

[8] K. Nanda, "Size-dependent Melting of Nanoparticles: Hundred Years of Thermodynamic Model", *Ind. Acd. Sci.*, vol. 72, pp. 617-628, 2013.

[9] F. Bonaccorso, "Graphene Photonics and Optoeketronics", *Nat. Photonics.*, vol. 4, pp. 611-622, 2013.

[10] K. Novoselov, et al., "A Roadmap for Graphene", *Nature*, vol. 490, pp. 192-200, 2012.

[11] A. Renia, et al., "Large Area, Few-Layer Graphene Films on Arbitrary Substrates by Chemical Vapor Deposition", *Nano Lett.*, vol. 9, pp. 30-35, 2009.

[12] D. Wang, et al., "Scalable and Direct Growth of Graphene Micro Ribbons on Dielectric Substrates", *Sci. Rep.*, vol.3, pp. 1-7, 2013.

[13] S. De, et al., "Are There Fundamental Limitations on the Street Resistance and Transmittance of Thin Graphene Film?", *ACS Nano*, vol. 4, pp. 271-2720, 2010.

[14] A. Ferrari, et al., "Raman Spetroscopy as a Versatile Tool for Studying the Properties of Graphene", *Nat. Nanotech.*, vol. 8, pp. 235-246, 2013.

[15] A. Ferrari, et al., "Supporting Information for: Raman Spetroscopy as a Versatile Tool for Studying the Properties of Graphene", *Nat. Nanotech.*, 2013.

CONTACT

*X.Zang, tel: +1-510-517-0356;
Xining.zang.me@berkeley.edu.

HIGH ASPECT RATIO, LARGE AREA SILICON-BASED GRATINGS FOR X-RAY PHASE CONTRAST IMAGING

J. Baborowski, V. Revol, C. Kottler, R. Kaufmann, P. Niedermann, F.Cardot, A. Dommann[1], A. Neels, M. Despont

CSEM SA, Neuchâtel, SWITZERLAND

ABSTRACT

This paper reports on the latest developments in the manufacturing of high aspect ratio silicon-based gratings used for X-ray phase contrast imaging (XPCI). Grating-based XPCI provides, in one measurement, unique information about the absorption coefficient, the index of refraction and the microscopic structure of a sample at hard X-ray frequencies. For this reason, XPCI can potentially overcome the limitations of classical absorption-based radiography, notably for weakly absorbing materials. New micro-fabrication processes were developed to manufacture full set of large area and high aspect ratio X-ray gratings with few defects. The complementarity of XPCI with conventional absorption-based radiography was experimentally demonstrated.

INTRODUCTION

X-ray imaging is an excellent and widely used tool to probe the internal structure of opaque objects. In fields as diverse as medicine, nondestructive testing, or security, it helps to diagnose, detect, or identify the sample without physical intrusion. Conventional radiographic systems, based on the absorption of x-rays in the sample, have however limited contrast for light materials such as plastics and biological tissues. XPCI, in the other hand, is able to reveal subtle changes in the microstructure of the samples, such as micro-cracks in composite materials [1] or micro-calcifications in breast tissues [2]. In fields as diverse as medicine, non-destructive testing or security, the gained information of this technique allows early diagnostic or detection of defects, tumors or explosives.

As shown in Figure 1, an interferometer is formed by assembling three gratings and allows the measurement of minute changes in the phase front after propagation in the sample [3–5]. Recently the method was embodied so that it is able to run at two different energy ranges [6], better sensitivity was achieved by using higher Talbot orders [7] or compact systems were achieved by bending the gratings in order to correct the shape of the spherical wave front [8].

Currently, industrial interest for XPCI is raising and the transition from laboratory instruments to industrial prototypes has started. Systems with a large field of view and no dead areas are now expected by the industry.

The achievable grating depth and hence the aspect ratio essentially defines the maximum X-ray energy that can be used for XPCI. As higher energy is necessary for larger and/or strongly absorbing samples, the range of opportunities offered by such technique depends, hence, highly on the achievable gratings parameters, such as periodicity, depth, duty cycle and aspect ratio. We have developed large gratings with a periodicity of down to 2 μm and a depth of up to 150 μm with extremely low defect density (<1 defect/cm^2). The fabricated gratings have been implemented on a XPCI set-up and used to demonstrate unprecedented imaging quality in material quality control.

EXPERIMENTAL

System requirements

X-ray phase contrast imaging (XPCI) has been documented in details in the literature and is only described shortly in the following paragraph. The setup (see Figure 1) consists of a conventional x-ray tube from COMET (MXR-160HP 20) with a focal spot size of 1×1 mm^2, a scintillation-based x-ray detector from Radicon (Shad-o-Box 2048 with a scintillator Min-R 2190) with 2048×1024 pixels of 48×48 μm^2 and three gratings G0, G1 and G2.

The phase sensitive part of the instrument is composed by G1 and G2, which form an array of collimated slits. Thus, any change in the beam propagation due to the object will induce a change in the interference pattern I_{fr} created by G1. Such changes in I_{fr}, like for example the lateral deviation of the fringes due to the refraction of the beam in the object, will finally reflect in the intensity $I(m,n)$ of the corresponding pixel. $I(m,n)$ thus depends, not only on the transmission of the sample, but also on the lateral displacement and amplitude of the interference fringes I_{fr}. As a consequence, a single image is not sufficient to separate these three contributions. This problem is solved by implementing a phase stepping method [9]. Finally, since the x-ray beam would not interfere without a sufficient spatial coherence, G0 forms a mask of individually coherent line sources (line width in the order of tens of microns), thus enabling the use of conventional x-ray source with a large focal spot.

---

[1] presently at EMPA, St.Gallen, SWITZERLAND

978-1-4799-3510-9/14 $31.00 © 2014 IEEE

Grating fabrication

For each of the three gratings, a different fabrication process was chosen, as the dimensions and X-ray optical requirements are quite different. The source grating G0 is fairly easy to make, as its period p_0 is typically on the order 10–150 μm, Moreover, the total area of G0 only needs to be large enough to cover the source size, i.e., a few square millimeters are sufficient. The main requirement of the grating structures is a sufficient height of the gold absorber. In our present setup, we use X-ray photon energies up to 30keV. Figure 2 presents the simplified process flow for the fabrication of this type of grating. Absorption gratings G0 are fabricated in highly doped Si wafers using a standard photo lithography process for the pattern definition. The trenches are defined by DRIE followed by the sidewall passivation. The oxide is etched from the bottom of the trenches followed by the Au electroplating to fill the Si molds. Absorption grating G0 with a periodicity from 10 to 100 μm, a duty cycle of 0.2 to 0.5 and depth 30 to 100 μm over an area up to $52x72mm^2$ were manufactured (Fig.3). The uniformity of the Au thickness is controlled either by polishing of the grating or by the precise partial filling of the Si mold. The requirements for the beam splitter grating G1 are more challenging. It should consist of low absorbing, phase shifting structures with a period $p_1 \approx 4\mu m$. The best performance is achieved when the structures have a duty cycle of 0.5, and the height is chosen such that they introduce a phase shift of π or $\pi/2$.. Furthermore, the grating area limits the field of view of the imaging device, so it should be many centimeters in size, depending on the application. The fabrication of the analyzer grating G2 is the most challenging part, as the grating period is typically only $p_2 = 2$ μm, and large areas have to be patterned with good uniformity. To fabricate the diffractive (G1) and analyzer (G2) gratings with a high aspect ratio (>50:1), pitch doubling technique and KOH etching on (110) of the silicon wafers have been used. At 80°C, we obtained an etch rate of 1.40 μm/min in the <110> direction. The etch rate along the <111> directions is about 80 times slower, resulting in nearly perpendicular side walls of the structures.

Diffractive grating G1 with periodicity 2 to 5 μm, with a duty cycle of 0.5 and depth 30 to 150 μm over an area up to $100x100mm^2$ have been achieved (Fig.5 and Fig.6). The sidewalls verticality is a challenging parameter and an angle below 0.2° for 150 μm deep trenches is observed. Fabrication process of G2 analyzer gratings is followed by subsequent Au electroplating of a meander-like structure. Analyzer grating G2 with periodicity 2 to 5 μm with a duty cycle of 0.5, depth 50 to 150 μm are currently manufactured with high reproducibility. A sensitive area of up to $100x100mm^2$ can be achieved. A large range of specifications is now available for industrial and academic customers. The final shape of the gratings die is obtained by the laser cutting.

XPCI EXPERIMENTS

The CSEM laboratory setup has been used to demonstrate the applicability of this technology for various applications. By enabling the measurement of small angle scattering of the X-ray beam the Phase Contrast imaging technology provides an innovative tool to probe the structure and detect efficiently the defects like: porosity, micro-cracks, foreign objects and inclusions, delaminations, resin rich / resin poor areas or cut and wavy fibers. In particular, the method is used for the non-destructive inspection and characterization of composite materials [10,11]. To illustrate the performance of XPCI, Figure 8 shows a comparison between the "conventional" X-ray radiography and the dark field image obtained by XPCI for the detection of micro-cracks induced by an particle impact onto a carbon fiber reinforced polymer (CFR) laminate as more and more used in the new, fueled efficient, generation of airplanes. The higher contrast of XPCI reveal details in the material transformation due to the impact that are simply not visible by conventional technique Besides aeronautics, the developed Phase Contrast imaging device will benefit a much wider range of applications such as the quality control of medical devices or automotive parts. Finally, Computerized Tomography (CT) techniques could be combined with XPCI to obtain three-dimensional information about the index of refraction and the linear diffusion coefficient. This imaging modality was illustrated [11] and helped to separate the different materials composing the composite object.

CONCLUSION AND OUTLOOK

We have developed and mature the microfabrication process of the various XPCI gratings for large number of interferometer applications. This technique can be used to reconstruct two new images in addition to the conventional absorption-based radiography. The differential phase contrast image is sensitive to the phase shift of the x-ray beam in the sample while the dark field image probes the small angle scattering power of the sample. The next step will be to extend the domain of usable x-ray energies even further. For that purpose, we are developing a new generation of gratings that will allow images to be taken with acceleration voltages up to 100kV while keeping a high visibility of the phase stepping curve. The new higher energy setup will thus be able to acquire images of objects which are thicker and/or made of dense materials, like metals, compared to what has been demonstrated in this paper. Such an instrument will enable XPCI applications that were previously out of reach.

REFERENCES

[1]   C. Kottler, V. Revol, R. Kaufmann, P. Niedermann, F. Cardot, A. Dommann, in:, International Conference on Industrial Computed Tomography, 2012, pp. 129–134.

[2]   M. Stampanoni, Z. Wang, T. Thüring, C. David, E. Roessl, M. Trippel, R. a Kubik-Huch, G. Singer, M.K. Hohl, N. Hauser, Investigative Radiology 46 (2011) 801.

[3]   F. Pfeiffer, T. Weitkamp, O. Bunk, C. David, Nature Physics 2 (2006) 258.

[4]   C. Kottler, F. Pfeiffer, O. Bunk, C. Grünzweig, J. Bruder, R. Kaufmann, L. Tlustos, H. Walt, I. Briod, T. Weitkamp, C. David, Physica Status Solidi (a) 204 (2007) 2728.

[5]   V. Revol, C. Kottler, R. Kaufmann, U. Straumann, C. Urban, Review of Scientific Instruments 81 (2010) 73709.

[6]   C. Kottler, V. Revol, R. Kaufmann, C. Urban, Journal of Applied Physics 108 (2010) 114906.

[7]   C. Kottler, V. Revol, R. Kaufmann, C. Maake, S. Stübinger, B. von Rechenberg, P.R. Kircher, C. Urban, 2011 IEEE Nuclear Science Symposium and Medical Imaging Conference, 2011, pp. 4437–4440.

[8]   V. Revol, C. Kottler, R. Kaufmann, I. Jerjen, T. Lüthi, F. Cardot, P. Niedermann, U. Straumann, U. Sennhauser, C. Urban, Nuclear Instruments and Methods in Physics Research 648 (2011) S302.

[9]   T. Weitkamp, A. Diaz, C. David, F. Pfeiffer, M. Stampanoni, P. Cloetens, and E. Ziegler, *Opt. Express* **13**, 6396 (2005).

[10]   V. Revol, C. Kottler, R. Kaufmann, A. Neels, A. Dommann, Journal of Applied Physics 112 (2012) 114903.

[11]   V. Revol, B. Plank, R. Kaufmann, J. Kastner, C. Kottler, A. Neels, NDT&E International 58 (2013) 64.

CONTACT

Jacek Baborowski, jba@csem.ch, +41327305279

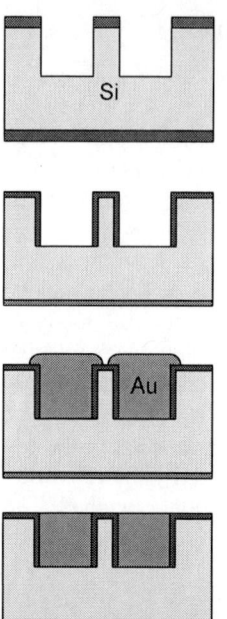

Figure 2: Simplified process flow for the fabrication of the absorption grating G0.

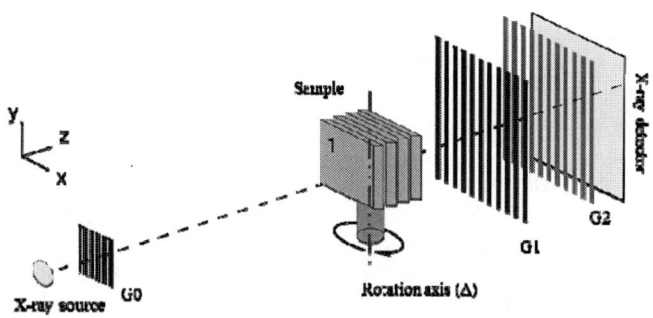

Figure 1: A classical schema of the XPCI set-up is composed of a X-ray source, a X-ray detector and three silicon-based gratings G0-G1-G2. The sample is typically placed between gratings G0 and G1 [11].

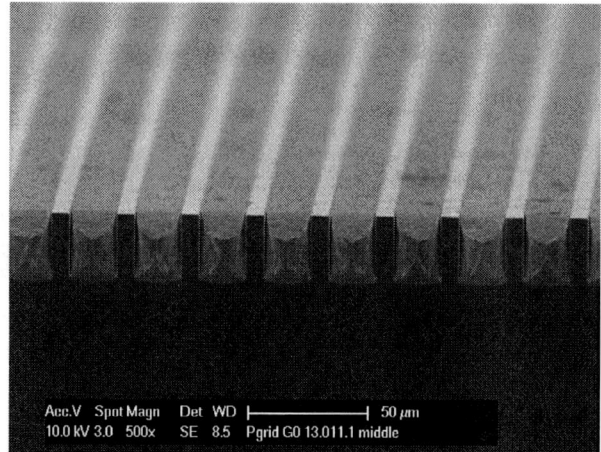

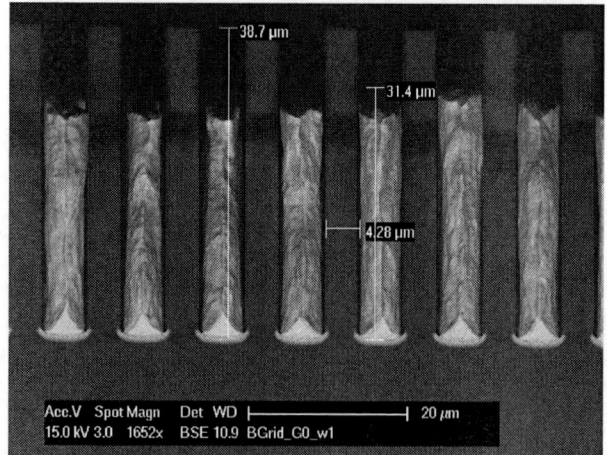

Figure 3: SEM Cross sections of an absorption grating G0 made by filling a silicon mold with gold for which the absorber thickness is defined by polishing (top) and by partial filling (bottom)

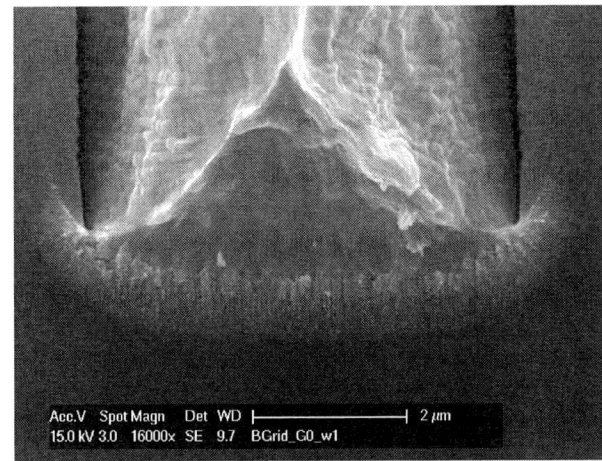

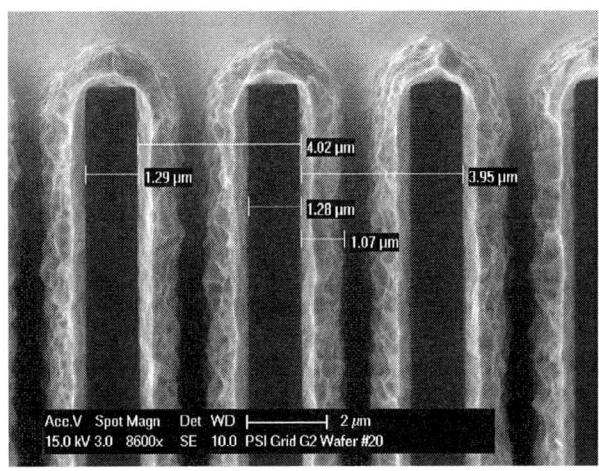

Figure 4: SEM close view of the bottom of G0 grating showing the sidewall passivation, seeding layer and the Au pillar.

Figure 7: SEM cross sections of G2 made by covering a silicon mold with a 1µm thin gold layer.

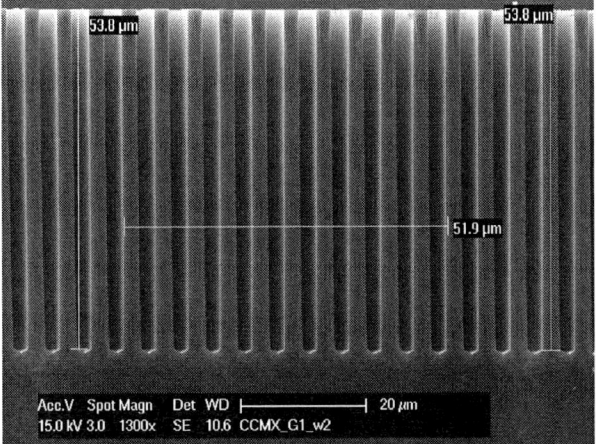

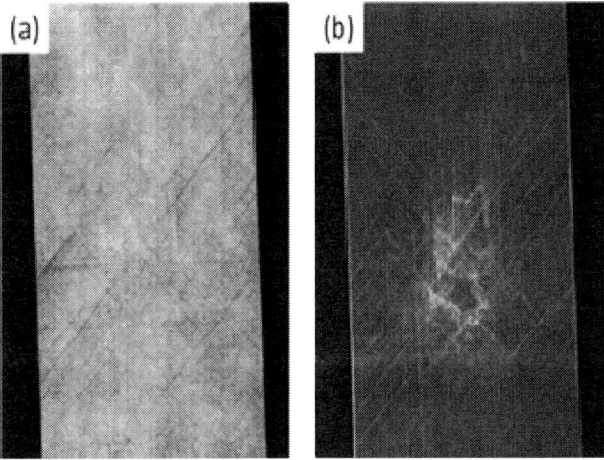

Figure 5: Cross sections of extremely high aspect ratio G1 made out of silicon by deep KOH etching.

Figure 6: Final large area 100x100mm² G1 grating after dicing.

Figure 8: (a) Conventional X-ray radiography based on the attenuation of the X-rays versus (b) X-ray scattering image obtained for a carbon fiber reinforced polymer (CFRP) laminate damaged by a particle impact.

HIGH RESOLUTION MICRO ULTRASONIC MACHINING (HR-μUSM) FOR POST-FABRICATION TRIMMING OF FUSED SILICA 3-D MICROSTRUCTURES

Anupam Viswanath, Tao Li, and Yogesh B. Gianchandani
Center for Wireless Integrated MicroSensing and Systems (WIMS²)
University of Michigan, Ann Arbor, USA

ABSTRACT

This paper presents the design and characterization of a high resolution micro ultrasonic machining (HR-μUSM) process suitable for post-fabrication trimming of 3-D microstructures made from fused silica and other materials. The process targets low machining rates, high resolution, and high surface quality. On flat fused silica substrates, the process achieves machining rates ≤10 nm/sec averaged over 1 minute. The average surface roughness (S_a) achieved is ≤30 nm. The process is successfully demonstrated for trimming hemispherical 3-D microstructures made from fused silica.

INTRODUCTION

Post-fabrication trimming is necessary for devices such as inertial sensors, timing references and mass-balance resonators to adjust stiffness, mass and potentially damping [1-2]. At the macro scale, trimming can be performed by grinding/polishing. For microstructures, alternatives include laser trimming, maskless plasma etching, and micromilling/drilling [3].

This paper presents a complementary technique based on micro ultrasonic machining (μUSM) for high resolution subtractive trimming of 3-D microstructures made from fused silica. Being transparent and brittle, fused silica is not well suited for trimming by laser or micromilling/drilling. The μUSM process is appropriate for micromachining both planar and 3-D structures of brittle materials without inducing stress or sub-surface cracks. It is maskless and involves no thermal process. Our previous work has demonstrated a μUSM process for die-scale pattern transfer in ceramics [4]. We have also presented a process that combines batch-mode μUSM, lapping, and micro electro-discharge machining (μEDM) for microfabrication of spherical structures [5]. In these μUSM processes, it is desired to rapidly remove material, with typical machining rates exceeding 200 nm/sec. The surface roughness is typically 200-400 nm [4-6].

The high resolution μUSM (HR-μUSM) process presented in this work aims for low machining rates, providing high resolution and high surface quality. Challenges/requirements in machining configurations and parameters to reach these goals are addressed. The process is characterized on flat fused silica substrates and demonstrated for trimming on 3-D, hemispherical microstructures made from fused silica. The following sections present the process description, experimental characterization, and the demonstration of post-fabrication trimming.

PROCESS DESCRIPTION

In μUSM, mechanical vibrations cause a tool to vibrate along its longitudinal axis (usually at a frequency of 20-40 kHz). An abrasive slurry is supplied around the cutting zone. The vibration of the tool causes the abrasive particles to impact the workpiece surface, causing material removal by microchipping. The dimensions of the machined features mimic the tool footprint. The material removal rate is determined mainly by the impact velocity of the abrasive particles. This velocity is significantly affected by the frequency and the amplitude of the tool vibration, as well as the separation between the tool and the workpiece [7]. The surface finish depends mainly on the particle size of the abrasive. In order to obtain low machining rates and high surface finish for HR-μUSM, several process conditions must be addressed, such as the vibration amplitude, tool miniaturization, tool-workpiece separation, and the size of abrasive particles.

The conceptual comparison of HR-μUSM and conventional μUSM is illustrated in Figure 1. For HR-μUSM, the tool tip is positioned at a predefined fixed distance (FD) from the workpiece, without tool feed toward the workpiece as in conventional μUSM. Low vibration amplitude and small abrasive particles are used to further reduce the machining rates and generate good surface finish. The details are provided below.

A small tool diameter is favorable for precision, but presents challenges in tool fabrication and handling. In this

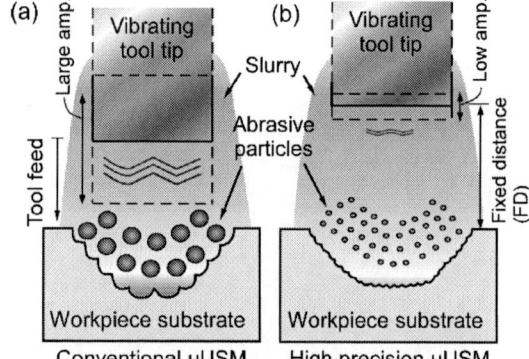

Figure 1: Conceptual comparison of μUSM used for conventional machining and for high resolution trimming. (a) Conventional μUSM provides rapid machining with rougher surfaces. (b) For high-resolution trimming, the tool tip is vibrated at a fixed distance (FD) from the workpiece. Smaller abrasive particles and lower tool vibration amplitude allow further decrease in machining rates and smoother profiles.

978-1-4799-3510-9/14 $31.00 © 2014 IEEE

work, stainless steel #304 (SS304) tools with tips of 50-μm diameter are fabricated by wire electro-discharge grinding (WEDG) of SS304 wires. The SS304 is selected for the tool as it provides high wear resistance, favorable elastic and fatigue strength properties, toughness, and hardness [4]. The length of the tools typically ranges between 2-5 mm. Tip diameters as small as ≈5 μm can be fabricated by this method. A customized tool holder holds the tool perpendicular to the workpiece. The shorter tools may be used for HR-μUSM of flat fused silica substrates, whereas longer tools are needed for machining hard-to-reach surfaces of complex 3-D workpieces.

Conventional μUSM uses abrasive particle sizes ranging between 1-10 μm. For this work, boron carbide and tungsten carbide abrasive powders with grain sizes as low as 100 nm are used. Commercially available diamond powder has grain sizes down to 10 nm and is used for further improvement of surface quality.

The amplitude of vibration of the tool depends on the ultrasound generator and the horn assembly that couples the vibration to the tool. Traditionally, the amplitude ranges from 10-50 μm. In this work, a commercial generator with a customized hardware assembly can provide adjustable vibration amplitude down to a few microns.

The HR-μUSM system custom-built for this work is shown in Figure 2. It provides ≈7 μm vibration amplitude and <1 μm repeatable alignment accuracy between the tool and the workpiece. A 20 kHz, stationary, benchtop USM machine (AP-1000™, Sonic-Mill®, Albuquerque, NM, USA) is used as the ultrasound generator. Two horizontal stages and one vertical stage (Horizontal: M-505.2DG, Vertical: M-501.1DG, from Physik Instrumente®, Auburn, MA, USA) are integrated to form the motorized XYZ stage system and resolves the workpiece movement to within 50 nm. An acoustic emission (AE) sensor monitors the trimming process

for feedback control [8]. A monoscope capable of 200x magnification is used to align the tool with the workpiece. A software program for process control and user interface has been developed for this purpose in Visual Basic 2012.

EXPERIMENTAL RESULTS

Process Characterization

The characterization of the HR-μUSM process was performed on flat fused silica substrates of 90 μm thickness. Tungsten carbide (WC) powder (Inframat Advanced Materials, Manchester, CT, USA) of 100 nm particle size or diamond powder (Sigma-Aldrich Co., MO, USA) of 10 nm particle size were used in the machining evaluations. The slurry concentrations were $WC:H_2O=1:1$ and diamond:$H_2O=1:5$ (by wt.).

The effect of proximity of the tool to the workpiece surface was experimentally evaluated and the results are presented in Table 1. This evaluation was performed using 100 nm WC powder for an FD varying from 25 μm to 40 μm in steps of 5 μm. The machining was performed for 1 minute in each case. The machined depth of features was measured using an interferometer (LEXT™, Olympus Corporation, PA, USA). The machining rates provided in Table 1 represents an average of 3 measurements clustered near the center of the machined feature. The machining rate increases with larger FDs. The increase in FD from 25 μm to 40 μm caused an 87% decrease in machining rate.

Table 1: Machining rate as a function of fixed distance.

Fixed distance (FD) (μm)	25	35	40
Machining rate (nm/sec)	86.5	75.2	10.5

Machining tests were also performed for an FD of 35 μm while varying the machining time from 1 minute to 10 minutes. These evaluations were performed using 100 nm WC powder or 10 nm diamond powder. The dependence of the machining rate as a function of machining time is shown in Figure 3. The machining rate reduces and tends to saturate

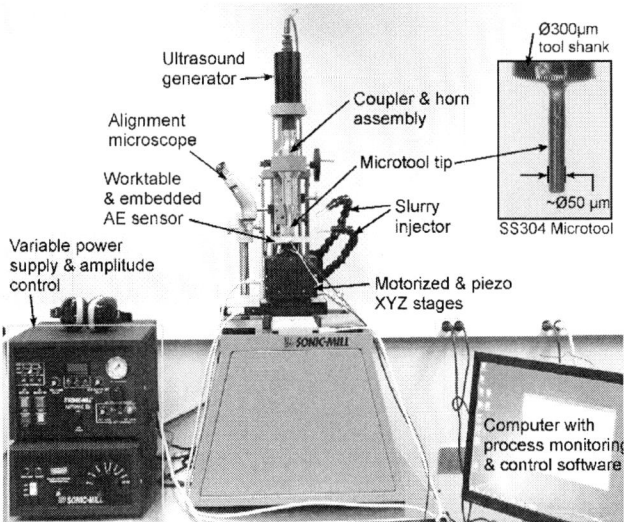

Figure 2: Photograph of a custom-built μUSM system for high resolution trimming, adapted from Sonic-Mill AP-1000 stationary USM machine. The inset shows a tool tip with ≈50 μm diameter as an example.

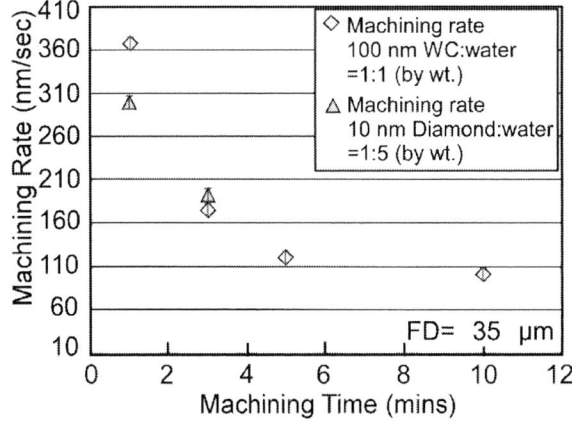

Figure 3: Measured machining rate on fused silica substrate as a function of machining time. Machining rates tend to saturate over time.

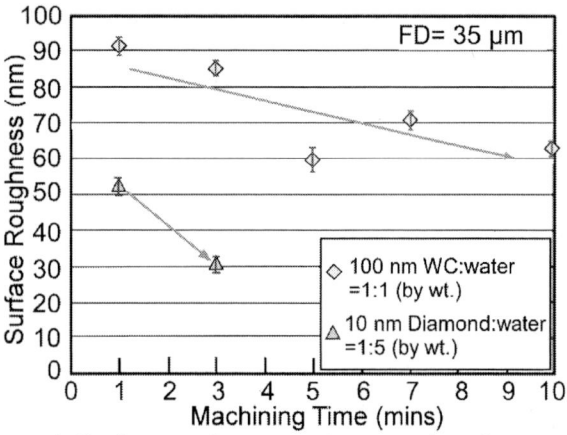

Figure 4: Surface roughness as a function of machining time. Minimum S_a is 30 nm with the use of 10 nm diamond powder and for machining time of 3 minutes.

over machining time. The average machining rate observed was ≈100 nm/sec.

The dependence of surface finish on the machining time and the slurry type was studied. Surface finish was evaluated by measuring the surface roughness (S_a) of different areas clustered near the center of the machined feature. The consistency of measurement was ensured by keeping the evaluation area for S_a constant during measurement. An average was calculated and set as the surface roughness of a machined feature. Measurements show that the surface roughness of features machined with 100 nm WC or 10 nm diamond particles decreased as machining time increased (Fig. 4). The finer diamond powder provided better surface finish (Fig. 5). The features that were machined for 3 minutes using 100 nm WC powder had a surface roughness of 85 nm. The corresponding surface roughness of features machined using 10 nm diamond slurry powder was ≈30 nm. In contrast, 1 µm WC powder, which is commonly used in traditional µUSM, provided a surface roughness of 245 nm (Fig. 5b). The surface roughness of the virgin flat fused silica substrate was ≈5 nm.

Table 2 summarizes the machining metrics achieved using the HR-µUSM process. At an FD of 40 µm, the HR-µUSM process achieved cutting rates as low as 10 nm/sec. The minimum S_a achieved was ≈30 nm using 10 nm diamond particles in the slurry.

Application to Trimming of 3-D Microstructures

The HR-µUSM process was demonstrated for trimming hemispherical 3-D microstructures made of fused silica. This work used bird-bath (BB) shells (Fig. 6), which are being

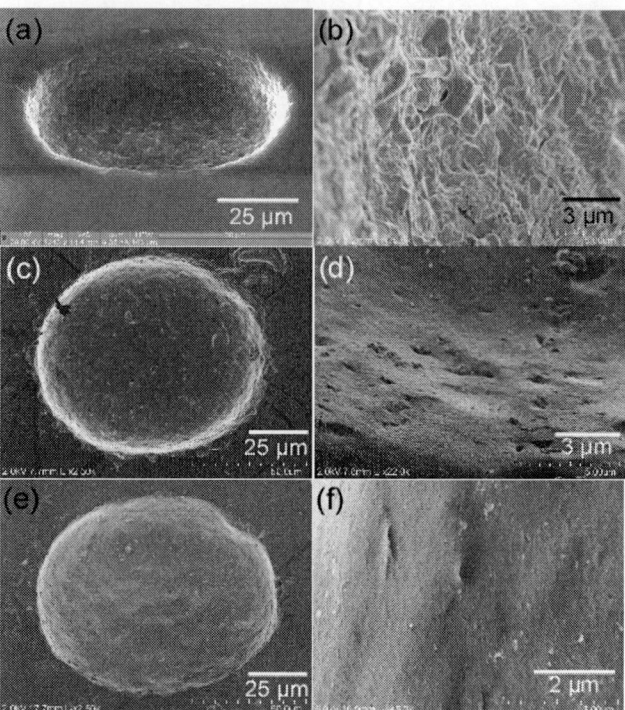

Figure 5: SEM images of machined features on fused silica substrates using: (a) Tungsten carbide (1 µm, $WC:H_2O=1:1$ by wt.). The machined feature was 73 µm in diameter. (b) S_a of (a): 245 nm. (c) Tungsten carbide (100 nm $WC:H_2O=1:1$ by wt.). The machined feature was 69 µm in diameter. (d) S_a of (c): 67 nm. (e) Diamond (10 nm $Diamond:H_2O=1:1$ by wt.) slurry. The machined feature was 63 µm in diameter. (f) S_a of (e): 30 nm. The closeup of the machined features shows better surface finish for diamond slurry when compared to WC slurry.

investigated for use in rate integrating gyroscopes [9]. These structures have a diameter of 5 mm, and a height of 1.55 mm, whereas the average thickness of the shell is only 70 µm. The trimming of these shells is needed for refining structural

Table 2: Machnining metrics for trimming by HR-USM.

Vibration amplitude (µm)	7±1.5*	
Vibration frequency (kHz)	20	
Abrasive: Avg. size (nm)	WC:100	Diamond: 10
Min. cutting rate (nm/sec)	10	≈10
Surface roughness (S_a) (nm)	>60	30 or better

*Accuracy limited by laser sensor. Further reduction possible.

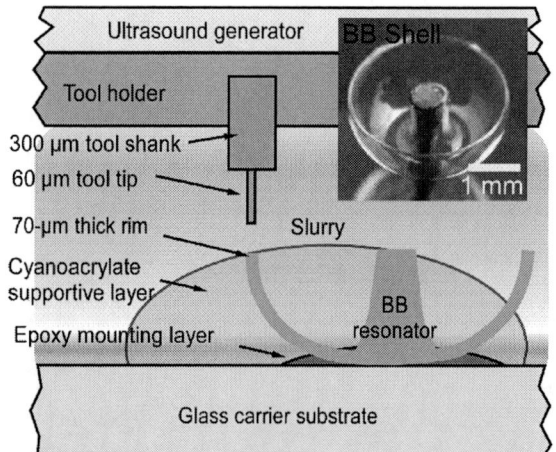

Figure 6: HR-µUSM on high-Q 3-D microshells. The bird-bath (BB) shell used here is fabricated by blow-torch molding from fused silica substrate, and has a rim thickness of ≈70 µm [9].

978-1-4799-3510-9/14 $31.00 © 2014 IEEE

isotropy and resonance characteristics. The BB shells were attached to a carrier substrate using standard 5-minute epoxy (5 Minute®, Devcon, MA, USA). The shells were potted in cyanoacrylate (Loctite®, Henkel Co., OH, USA), before immersing in slurry. This preliminary arrangement provided mechanical support for the 70-μm-thick shell walls, reducing the propensity for damage to the fragile shell. The mechanical support also reduced the topographical variation, allowing the slurry flow to be similar to that for a flat substrate.

Figure 7 shows a typical machined feature on the rim of the shell. Machining with 100 nm WC for 180 seconds provided an average depth of 18 μm, a diameter of 72 μm, and a surface roughness of 120-150 nm. The tool diameter and length were 60 μm and 2 mm, respectively. The average machining rate obtained during trimming of the rims of these shells was 80 nm/sec for an FD of 35 μm. This is consistent with the characterization of the HR-μUSM process.

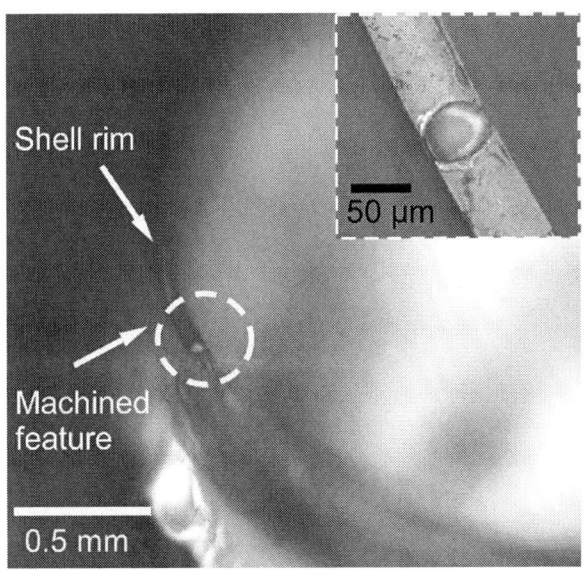

Figure 7: Results of trimming of BB shells at the rim using HR-μUSM. Machining was performed using 100 nm WC powder for 180 seconds. The average machining rate obtained was 80 nm/sec.

CONCLUSION

This work investigated a high resolution and high precision μUSM process intended for post fabrication trimming of complex 3-D microstructures made from fused silica. The typical machining rates of this process averaged up to ≈100 nm/sec, for an FD of 35 μm. The minimum machining rate was 10 nm/sec, for an FD of 40 μm. This allowed trimming with high resolution in the vertical (depth) direction. The HR-μUSM provides good surface finishes: an average surface roughness (S_a) of 30 nm was obtained using 10 nm diamond abrasive particles in the slurry. This is ≈7x smaller than that achieved using conventional USM. Further

decrease in vibration amplitude and abrasive particle sizes can facilitate lower machining rates and smoother profiles than those presented in this work. The process was demonstrated for trimming of hemispherical 3-D shells made of fused silica. Cavities were successfully formed on the thin shell rim with controlled depths and machining rates. The experimental results are very promising for further improvements, which will be pursued in future efforts.

ACKNOWLEDGEMENTS

The authors thank Prof. K. Najafi and Dr. J. Cho for providing the BB shells used in this work. This work was supported in part by DARPA MRIG award #W31P4Q-11-1-0002.

REFERENCES

[1] K V. Kempe, *Inertial MEMS Principles and Practice*, Cambridge University Press, 2011.

[2] J.T.M. VanBeek, R. Puers, "A Review of MEMS Oscillators for Frequency Reference and Timing Applications," *J. Micromech. Microeng.*, 22(2), 013001-1−013001-35, 2012.

[3] A. N. Samant, N. B. Dahotre, "Laser Machining of Structural Ceramics - A Review," *J. European Ceramic Society*, 29, pp. 969-993, 2009.

[4] T. Li, Y.B. Gianchandani, "A Micromachining Process for Die-Scale Pattern Transfer in Ceramics and its Application to Bulk Piezoelectric Actuators," *IEEE/ASME J. Micro Electro Mech. Sys.*, 15(3), pp. 605-612, 2006.

[5] T. Li, K. Visvanathan, Y.B. Gianchandani, "A Batch Mode Micromachining Process for Spherical Structures," *IOP J. Micromech. Microeng. (JMM)*, 2013 (accepted).

[6] X. Sun, T. Masuzawa, M. Fujino, "Micro Ultrasonic Machining and its Applications in MEMS," *Sensors Actuators A (Phys.)*, 57(2), pp. 159-164, 1996.

[7] D.C. Kennedy, R.J. Grieve, "Ultrasonic Machining- A Review," *The Prod. Engineer*, 54(9), pp. 481–486, 1975.

[8] T. Li, Y.B. Gianchandani, "A high-speed batch-mode ultrasonic machining technology for multi-level quartz crystal microstructures," *IEEE MEMS 2010 Conf.*, Hong Kong, 2010, pp. 348-351.

[9] J. Cho, J. Woo, J. Yan, R.L Peterson, K. Najafi, "A High-Q Birdbath Resonator Gyroscope (BRG)," *IEEE Intl. Conf. Solid-State Sensors, Actuators, Microsys. (Transducers) Barcelona, Spain*, pp.1847-1850, 2013.

CONTACT

A. Viswanath: anupamv@umich.edu
T. Li: litz@umich.edu
Y. Gianchandani: yogesh@umich.edu

FABRICATION OF CARBON NANOFIBROUS MICROELECTRODE ARRAY (CNF-MEA) USING NANOFIBER IMMERSION PHOTOLITHOGRAPHY

Pit Fee Jao[1], Eric Franca[1], Sheng-Po Fang[1], Junghae Yoon[1], Kun Cho[1,3,]
David E. Senior[2], Gloria Kim[1], Bruce Wheeler[1] and <u>Yong-Kyu Yoon</u>[1] *

[1]University of Florida, Florida, USA
[2]Universidad Tecnologica de Bolivar, Cartagena, Colombia
[3] Korea Basic Science Institute, Ochang, South Korea

ABSTRACT

Microelectrode arrays (MEAs) are widely used for stimulating and receiving electrical signals between human and machines and for *in vitro* neural study. This work demonstrates the fabrication process of nanofibrous 3D microelectrodes using immersion lithography. Oil immersion negates the diffraction effects intrinsic in the photopatterning of electrospun nanofibers to give increased aspect ratio microarchitectures. Nanofiber electrode resistivity is characterized and its performance compared to that of carbon thin film. *In vitro* testing of electrodes are performed using E18 cortical neurons and analyzed for cell density and cell viability

INTRODUCTION

Microelectrode arrays (MEAs) are used in various fields such as neural coding and decoding, recording field potentials for electrophysiology, multisite electroretinogram of explanted retinas, analysis of networks and cells in hippocampus brain slices and biorhythmic reading from cardiac myocytes [1]–[6]. The primary function for MEAs is stimulation of electric current and/or recording bioelectric signal from biological tissues or cells. Unlike microelectrode probes, microelectrode arrays are planar architectures with non-invasive electrodes that are used to measure the distributed electrical sensitivity of neural networks while still having the resolution of a single neuron. MEAs have to be both biocompatible to support tissue growth and also have low impedance (500 kΩ at 1 kHz) to measure the small signals (10~100 μV) of the neurons [1]. Earlier generation of MEAs were fabricated on glass substrates with patterned gold or indium tin oxide as conductors between the electrodes and read out probes, while electrodes were electroplated porous platinum (Pt black) followed by a biocompatible insulation layer. Porous conductive electrodes were used to enhance surface area for higher interface capacitance thereby reducing recording impedance [7]. High surface area electrodes not only decreased the electrical impedance but also enhanced interactions with cultured neurons [8]. But Pt black was not durable for long term culturing and was replaced by mechanically stronger and electrically superior material iridium oxide or titanium nitride [9], [10]. Other conductive polymers have also been electroplated such as polypyrrole (PPy), polythiopene (PT), polyaniline (PAni) and poly(3,4-ethylenedioxythiopene) (PEDOT) [11]. In addition to being a stable conductive polymer layer, they also functioned as a molecular counter-ion responsible for healthy tissue culture. To form porous PEDOT layers, polystyrene beads were used as template in the electrodeposition process on neural probes and then etched in toluene [8]. Another electrode material is chemically vapor deposited carbon nanotube (CNT) forests which have demonstrated high conductivity and large double layer capacitance inherent with large surface area [12], [13]. But the free standing CNT forests were fragile and had to be mechanically reinforced either with thermal oxide or electroplated Ppy [14]–[16]. Anodized aluminum oxide layers with their high aspect ratio pores have also been demonstrated to enhance the surface area of patterned electrodes and high frequency action potentials on cultured HL-1 cells have been demonstrated. [7].

In this work, the fabrication process of high aspect ratio carbon nanofiber microelectrode array is described using electrospun nanofibers, immersion photolithography, and carbonization. Figure 1 shows the schematic of the nanofiber based microelectrode array, where high aspect ratio carbon NF are the electrodes which interact with the test specimen and carbon thin film (CTF) are the trace electrodes which connect the nanofiber electrodes with the read out circuits. Oil immersion lithography has been exploited for high aspect ratio nanofiber pillars.

FABRICATION PROCESS

Figure 2 shows the fabrication process of the nanofiber microelectrodes. First, quartz substrates are sputtered with 120 nm thin chrome and patterned with positive resist Shipley S1818 to give a photolithographic mask of the electrodes. Second, 7μm thick SU-8 thin film is spin coated on the quartz mask and patterned with the trace patterns of the microelectrode array using a Karl Suss MA6 aligner. Third, SU-8 nanofibers are electrospun from a 60.87wt% solution with dimethylformamide with an electric field of 1.5 kV/cm and a flow rate of 1 ml/min directly on the patterned quartz substrate.

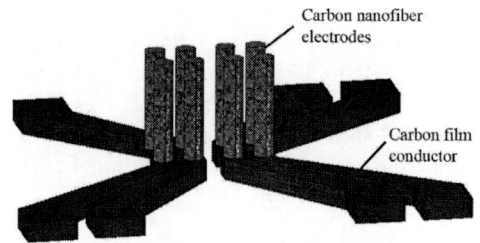

Figure 1: Schematic of nanofiber MEA architecture.

978-1-4799-3510-9/14 $31.00 © 2014 IEEE

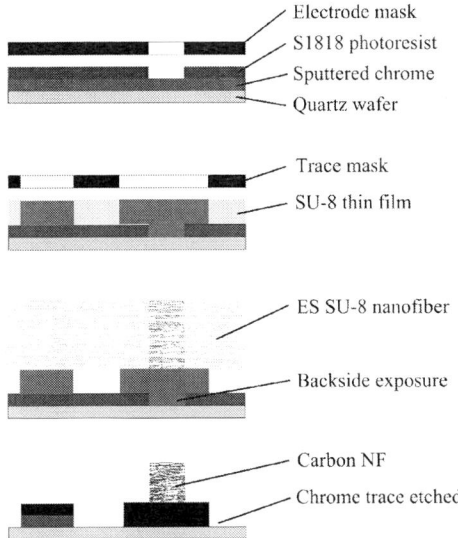

Figure 2: Fabrication process of nanofiber MEAs

Labels in Figure 2:
- Electrode mask
- S1818 photoresist
- Sputtered chrome
- Quartz wafer
- Trace mask
- SU-8 thin film
- ES SU-8 nanofiber
- Backside exposure
- Carbon NF
- Chrome trace etched

A reservoir was formed around the nanofibers, which are then immersed in an oil medium and exposed to UV light using the quartz wafer as the mask, followed by post exposure bake and developing steps. Fourth, the chrome layer is etched to electrically isolate between the patterned SU-8 nanofibers (SNF) and SU-8 thin film (STF) trace patterns. Then the MEAs are carbonized under forming gas atmosphere (4% hydrogen, BAL nitrogen) at a flow rate of 13 slm with the final carbonization temperature reaching 1000 ^{O}C with a ramp rate of 5~10 ^{O}C/min. Scanning electron microscope (SEM) (JEOL 5700) images were taken after sputtering a 20 nm Chrome layer on the samples. Nanofiber pillar height and diameters were measured from cross-sectional SEM images using ImageJ imaging software. While statistical distributions were plotted with 95 % confidence intervals using student's T distribution with N=5 and 2-tailed distribution. 7 day *In vitro* analysis on the fabricated MEAs was performed using E18 rat cortical neurons. The MEAs were treated with 0.1 % polyethylenimine (w/v) for supporting long term cell growth. Cell growth was analyzed via calcein and AM staining.

RESULT AND DISCUSSION

Optical simulations of wave propagation in the nanofiber matrix were performed using the COMSOL Multiphysiscs tool. Huygen Fresnel diffraction principles were applied to propagating wave functions to determine the accumulated UV light intensities. Figure 3 shows intensity plots of UV light in the nanofiber matrix in air and oil media. While air media demonstrates increased scalar diffraction due to the porous architecture of the nanofibers, the oil medium minimizes diffraction with near homogenous medium due to the matched refractive index of SU-8 (n_{SU-8}=1.67) and oil medium (n_{oil}=1.47). Figure 3c and 3d shows the patterned nanofibers in air and oil medium, respectively. The oil medium increases the height from 25.91 ± 4.13 to 55.50 ± 3.32 μm with a dose of 400 mJ/cm^2,

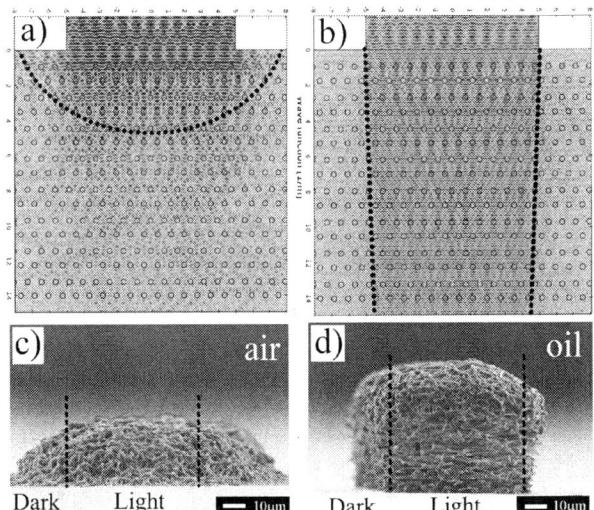

Figure 3: Oil immersion exposure. Optical simulation in (a) oil and (b) air medium. SEM crosssectional images of fabricated nanofibrous pillars in (c) air and, (d) oil medium.

while decreasing the diameter from 100.8 ± 24.08 to 73.28 ± 5.85 μm for a 60 μm pattern diameter giving an aspect ratio (height to width) of 0.76. The fabricated CNF based microelectrode arrays are demonstrated in Figure 4. In Figure 4a, the read electrodepads and nanofiber pillars in SU-8 are clearly delineated from the quartz wafer background. Two pillar diameters were used in the fabrication of the nanofiber electrodes, 60 μm and 30 μm on a 60 μm diameter trace electrode pad interfacing the nanofiber patterns with the readout electrodes.

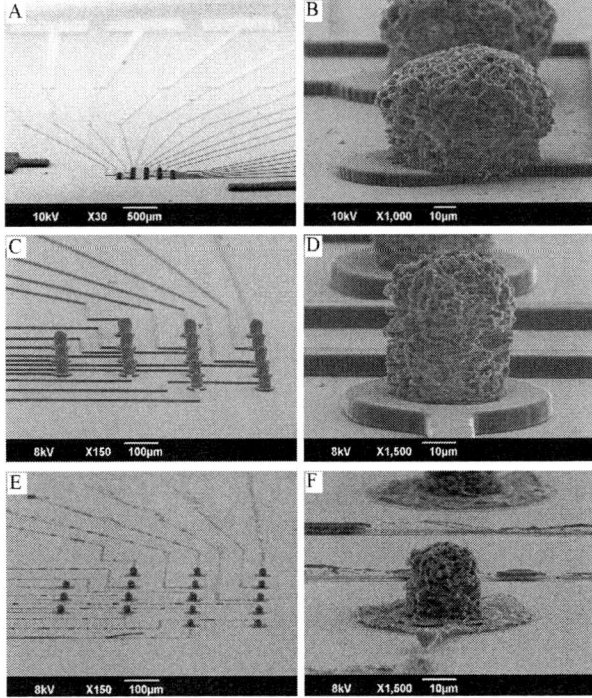

Figure 4: Fabricated nanofiber MEAs using A)-B) with 60 μm diameter pillars on SU-8 thin film, C)-D) with 30 μm diameter pillars and, E)-F) after carbonization of 30μm diameter pillars.

Figure 4b shows the patterned but slightly misaligned 60 μm nanofiber pillars in the SU-8 TF pad. Using 300 mJ/cm^2 exposure dosage gave a nanofiber average height and diameter of 43.5 ± 3.8 and 68.5 ± 9.5μm respectively. Figure 4c shows the 30 μm diameter nanofiber pillars on the trace electrodes of SU-8 demonstrating well aligned patterns. Figure 4d shows an angled cross-section of the 30 μm nanofiber patterns with near vertical cross-sections. Using an optimized dosage of 500 mJ/cm^2 gave a nanofiber pillar height and diameter of 37.3 ± 2.5 and 35.1 ± 2.3 μm, respectively. It is interesting to observe that a narrower mask pattern (d=30 μm) requires a higher exposure dosage when compared to the wider mask pattern (d=60 μm) for similar height of nanofibers. This is attributed to the increased reflectivity of smaller mask patterns which reduce dosage to the nanofibers. Figure 4e shows the nanofiber pillars and trace electrodes from Figure 4d after the carbonization process [17]. Immediately observable is the shrinkage in size of the structures in both the NFs and TF with the overall patterns still preserved. In the cross-section image in Figure 4f, we observe that the porous architecture of the nanofiber structure remains intact, while the carbon thin film forms a flaky but continuous structure. The decreased structural integrity of carbon thin film is attributed to the combined effect of under exposure of SU-8 thin film and edge delamination of SU-8 TF during the chrome etching phase. Figure 5 demonstrates the intermediate steps of the fabrication process. Figure 5a through 5c shows nanofiber pillars on a quartz substrate exposed with dosages 300, 400 and 500 mJ/cm^2, respectively. It is observed that with increasing exposure dosage not only does the nanofiber height increase but also the profile of the nanofiber pillar changes from tapered to a bulbous structure.

When an SU-8 TF is introduced between the quartz and the nanofibers and exposed with the same dosage as without the STF, we observe in Figure 5d through 5f that the height of the nanofibers decreases in comparison to that of nanofibers without the STF. The lower height of patterned

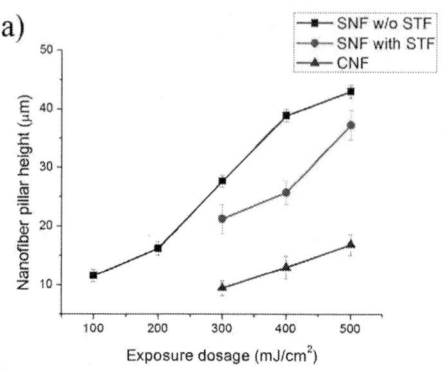

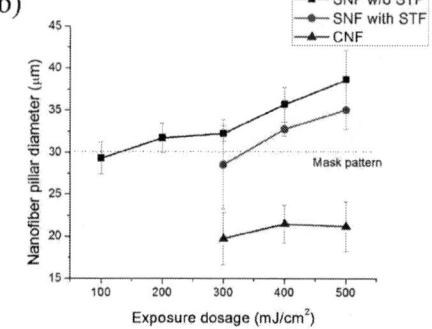

Figure 6: Effect of dosage on patterned dimension of a) height and b) diameter.

nanofiber in the case of STF is attributed to the 90~95% transmission loss of UV light in cross-linked SU-8 for 365 nm. Figure 5g through 5i show the cross-section of SNF-STF structures after carbonization to give CNF-CTF architecture. We observe that not only does the height of the nanofiber decrease, but also the diameter of the pillar decreases. This observation is different from that of only carbonized thin film on substrate, in which case only the height of the thin film decreases but not the width.

Quantitative analysis of height and diameter of nanofiber pillars in the intermediate stages at incremental exposure dosages is shown in Figure 6. As for the pillars directly on the substrate, it is observed that not only does the nanofiber height increase from 11.6 ± 1.1 μm to 43.03 ± 1.2 μm for exposure dosage from 100 mJ/cm^2 to 500mJ/cm^2, but also the diameter increases from 29.3 ± 1.9 μm to 38.6 ± 3.6 μm.

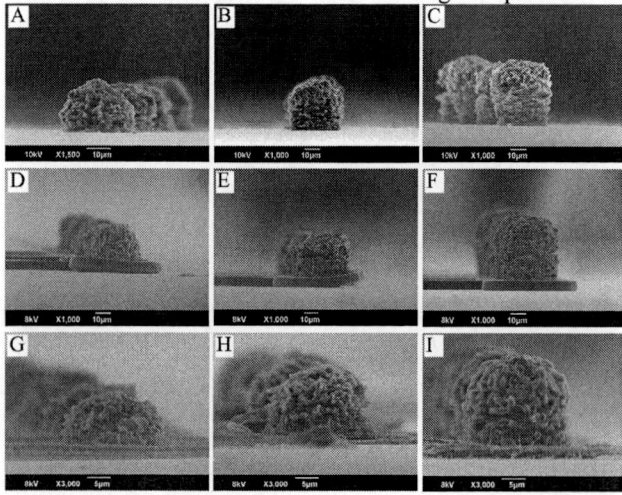

Figure 5: Patterned MEAs A)-C) without STF, D)-F) with STF and G)-I) after carbonization.

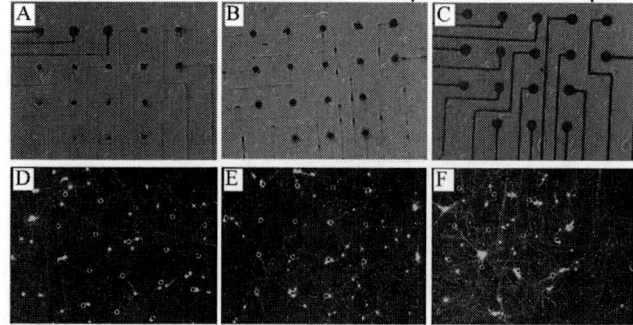

Figure 7: 7-day in-vitro E18 rat cortical neuronal culture on nanofiber MEAs; with height of MEAs at A) 10 μm, B) 12.5 μm, and C) 15 μm, with fluorescence imaging for nanofiber heights at D) 10 μm, E) 12.5 μm, and F) 15 μm

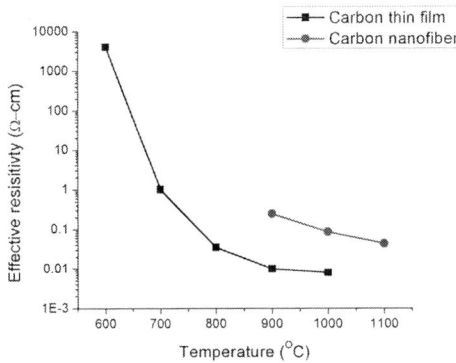

Figure 8: Effective resistivity of CNF and CTF with different carbonization temperatures.

In the case of STF between the nanofiber and mask, the height of the nanofibers decreases by an average height of 8.4 ± 4.1 μm and a diameter of 3.4 ± 0.5 μm when compared to that without the STF for the same exposure dosage. With the carbonization process the nanofiber pillar height and diameter further shrinks by 65.0 % and 46.7 %, respectively.

Figure 7 shows the images of culture E18 rat neurons on the fabricated nanofiber MEAs. Cells are observed to adhere well to the nanofiber electrodes with most of the cells preferring the periphery of the patterned electrodes. Multiple cell density clusters are observed on single electrode patterns with highest density observed in the case of 12.5 μm thick nanofiber pillars. Neural outgrowth is observed to prefer sharp edges of thin film electrodes as long strands of neurons are observed traversing the length of the line patterns. In Figure 7f, neuronal outgrowth is observed in long lines corresponding to the underlying trace electrodes formed in carbon thin film.

Figure 8 shows the measured effective resistivity of the CNF membranes at different final carbonization temperatures compared to that of carbon thin film. The carbon structures are observed to decrease the resistivity as increasing final process temperatures with the highest resistivity at 900 ^{O}C measured to be 0.247 Ω-cm and the lowest at 1100 ^{O}C to be 0.044 Ω-cm. The monotonous decrease in resistivity is similar to that of CTF [17]. But the CNFs are observed to have a higher resistivity than that of CTFs which is attributed to the increased porosity of CNFs.

CONCLUSION

Nanofiber microelectrode arrays have been fabricated using immersion lithography of electrospun nanofibers. High aspect ratio pillars are fabricated on thin film SU-8 and carbonized to give high aspect ratio carbon nanofiber pillars. Experimental data with different exposure dosages without and with thin film SU-8 are demonstrated. Cell culture on the fabricated MEAs demonstrates increased cell viability of neurons on carbon nanofibers and increased cell density on the periphery of the patterned nanofibers. Preferential growth of neurons has been observed along the length of patterned thin film.

REFERENCES

[1] S. M. Potter, *Advances in Neural Population Coding*, vol. 130, Chap. 4, M. A. L. N. B. T.-P. in B. Research, Ed. Elsevier, 2001, pp. 49–62.

[2] A. Stett, U. Egert, E. Guenther, F. Hofmann, T. Meyer, W. Nisch, and H. Haemmerle, *Anal. Bioanal. Chem.*, vol. 377, no. 3, pp. 486–495, 2003.

[3] B. C. Wheeler and J. L. Novak, *Biomedical Engineering, IEEE Transactions on*, vol. BME-33, no. 12. pp. 1204–1212, 1986.

[4] Y. Nam, J. Chang, D. Khatami, G. J. Brewer, and B. C. Wheeler, *IEE Proc. -- Nanobiotechnology*, vol. 151, no. 3, pp. 109–115, Jun. 2004.

[5] A. Blau, C. Ziegler, M. Heyer, F. Endres, G. Schwitzgebel, T. Matthies, T. Stieglitz, J.-U. Meyer, and W. Göpel, *Biosens. Bioelectron.*, vol. 12, no. 9–10, pp. 883–892, Nov. 1997.

[6] J. Pine, *J. Neurosci. Methods*, vol. 2, no. 1, pp. 19–31, Feb. 1980.

[7] M. Wesche, M. Hüske, A. Yakushenko, D. Brüggemann, D. Mayer, A. Offenhäusser, and B. Wolfrum, *Nanotechnology*, vol. 23, no. 49, p. 495303, 2012.

[8] J. Yang and D. C. Martin, *Sensors Actuators B Chem.*, vol. 101, no. 1–2, pp. 133–142, Jun. 2004.

[9] U. Egert, B. Schlosshauer, S. Fennrich, W. Nisch, M. Fejtl, T. Knott, T. Müller, and H. Hämmerle, *Brain Res. Protoc.*, vol. 2, no. 4, pp. 229–242, Jun. 1998.

[10] S. Gawad, M. Giugliano, M. Heuschkel, B. Wessling, H. Markram, U. Schnakenberg, P. Renaud, and H. Morgan, *Front. Neuroeng.*, vol. 2, p. 1, 2009.

[11] T. A. Skotheim, R. L. Elsenbaumer, and J. R. Reynolds, *Handbook of conducting polymers*. NY: Dekker, 1998.

[12] K. Wang, H. A. Fishman, H. Dai, and J. S. Harris, *Nano Lett.*, vol. 6, no. 9, pp. 2043–2048, Aug. 2006.

[13] V. Lovat, D. Pantarotto, L. Lagostena, B. Cacciari, M. Grandolfo, G. Spalluto, M. Prato, and L. Ballerini, *Nano Lett.*, vol. 5, no. 6, pp. 1107, 2005.

[14] J. E. Koehne, M. Marsh, A. Boakye, B. Douglas, I. Y. Kim, S.-Y. Chang, D.-P. Jang, K. E. Bennet, C. Kimble, R. Andrews, M. Meyyappan, and K. H. Lee, *Analyst*, vol. 136, no. 9, pp. 1802, 2011.

[15] P. U. Arumugam, H. Chen, S. Siddiqui, J. A. P. Weinrich, A. Jejelowo, J. Li, and M. Meyyappan, *Bios. Bioe.*, vol. 24, no. 9, pp. 2818–2824, May 2009.

[16] E. de Asis Jr., T. D. B. N-Vu, P. Arumugam, H. Chen, A. Cassell, R. Andrews, C. Yang, and J. Li, *Biom. Micr.*, vol. 11, no. 4, pp. 801, 2009.

[17] Pit Fee Jao, Kyoung Tae Kim, Gloria J. Kim, and Yong-Kyu Yoon, *Journal of Micromechanics and Microengineering*, no. 23, Oct. 2013, 114011

ACKNOWLEDGEMENTS

This work has been in part supported by the National Science Foundation grant, NSF ECCS 1132413.

CONTACT

*Yong-Kyu Yoon, Tel: +1- 352-392-5985; ykyoon@ecc.ufl.edu

INDUCTIVELY COUPLED PLASMA ETCHING OF BULK TUNGSTEN FOR MEMS APPLICATIONS

Lu Song, Nannan Li, Shibin Zhang, Jin Luo, Jia Hu, Yiming Zhang, Shuhui Chen and Jing Chen[*]
National Key Laboratory of Science and Technology on Micro/Nano Fabrication,
Institute of Microelectronics, Peking University, Beijing, China

ABSTRACT

Tungsten based MEMS devices have the potential to be used for many applications, such as tools for micro electrical discharge machining and ultrasonic machining, or mold for inject molding. For the first time, bulk tungsten inductively coupled plasma (ICP) etching was developed and characterized, which is capable of producing high aspect ratio (>13) structures with feature size below 3μm. Etching depth of 230μm has been achieved at an etch rate up to 2.2μm/min. This technology offers big opportunities for MEMS applications.

INTRODUCTION

Emerging metals are being considered as candidates for bulk MEMS in many applications. For example, high aspect ratio bulk titanium and molybdenum structures and devices have been realized by ICP deep etching, which are superior to silicon for certain applications [1-4]. This paper reports on the development of bulk tungsten ICP etching, which is capable of producing high aspect ratio (>13) structures with the feature size below 3μm. An etching depth of 230μm has been achieved at an etch rate up to 2.2μm/min.

Tungsten has the highest melting point (3,422 °C, 6,192 °F) and tensile strength of all metal, which is crucial in many high-temperature applications, such as emitter tips, micro-electrodes and rocket engine nozzles. It is also a good candidate for manufacturing micro-relays and micro-probes for the conductive properties and relative chemical inertia. Besides, tungsten is remarkable for its robustness - with hardness on the Mohs Hardness Scale of 9, it is wear-resistant and serves as an excellent mold material to produce plastic bio-MEMS chips in molding. Furthermore, tungsten is the main source for X-ray targets and also for shielding from high-energy radiations, which is necessary for anti-radiation devices. Since this element's thermal expansion is similar to borosilicate glass, it can also be used for making glass-to-metal seals.

Tungsten films have been identified as interconnect material for ULSI applications, and RIE etching of the film has been studied and reported by many researchers. However, the etching rates and aspect ratios are too low to satisfy the requirements of MEMS applications. For bulk tungsten MEMS process to become a competitive alternative to traditional micromachining methods, high etch rates, high aspect ratios, and high mask selectivity are essential. This paper reports on bulk tungsten ICP deep etching process, and the influence of the process parameters on etch rate, surface roughness and mask selectivity is discussed along with the optimization.

EXPERIMENTAL

The 4 inch double-side polished pure tungsten wafers of 400μm thick were used for the experiments. Aluminum of 1μm was deposited and patterned as the etching mask, and then the substrates were sectioned into 2×2 cm chips for process characterizations. The tungsten etching was carried out in Plasmalab System 100 ICP etcher from Oxford Instruments. SF_6 was used as the main feedstock of fluorine radicals. The samples were etched for 5 minutes with a specified parameter set. Only one parameter was varied for each run. Unless otherwise stated, all other parameters were held constant: 2000W ICP source power, 70W RF platen power, 100sccm SF_6 flow rate, a chamber pressure position at 90°, and a substrate temperature of 15°C, as the baseline recipe.

RESULTS AND DISCUSSION

According to the experiments, the tungsten etch rate and surface roughness data were plotted to determine the first-order trends for each etching parameter.

For the bulk Tungsten ICP etching, SF_6 is the main feedstock of fluorine radicals, which can react with the tungsten substrate forming WF_6, the volatile reaction product of tungsten and fluorine. This mechanism accounts for most trends.

RF platen power

The etch rate and roughness as a function of RF platen power were shown in Figure 1. The energetic ions generated by RF platen power may partially assist the removal of material, so etch rate increased with RF power from 10 to 130W, but then remained relatively constant.

Surface roughness increased significantly as RF power increased from 10 to 50W, and then decreased, which suggested that a stronger ion bombardment might help flatten the surface.

ICP source power

Etch rate increased appreciably with source power initially and then leveled off for powers above 2000W. This trend was shown in the plots of Figure 2. Below 2000W, the etching was ion and radical limited; while above 2000W, the etch rate was controlled by other process parameters such as gas flow.

The chemical etch was considered a leading factor for surface roughness [5]. With the increasing of ICP power, chemical effect was fostered, but the increasing ion flux reduced the roughness. The results were a balance of the two factors, which were also illustrated by Figure 2.

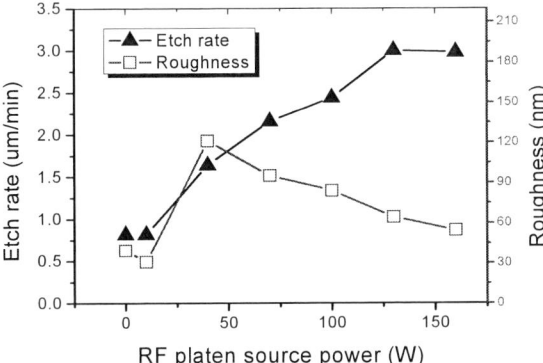

Figure 1: Etch rate & roughness as a function of RF platen power. The remaining process parameters were held constant at 2000W ICP source power, 8.4mTorr, and 100sccm SF_6. Samples were etched for 5 min with a chamber pressure position at 90° and a substrate temperature of 15°C.

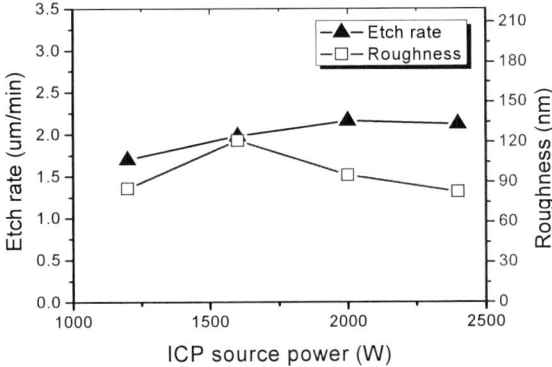

Figure 2: Etch rate & roughness as a function of ICP source power. The remaining process parameters were held constant at 70W RF platen power, 8.4mTorr, and 100sccm SF6. Samples were etched for 5 min with a chamber pressure position at 90°, and a substrate temperature of 15°C.

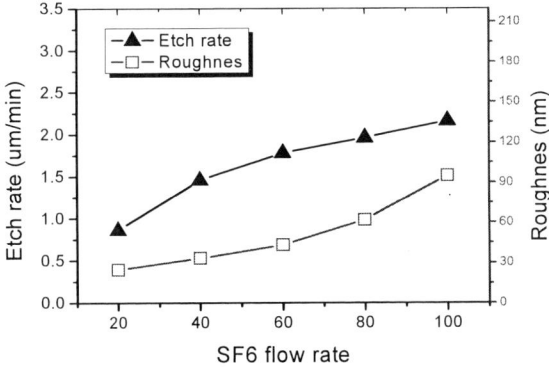

Figure 3: Etch rate & roughness as a function of SF_6 flow rate. The remaining process parameters were held constant at 2000W ICP source power, and 70W RF platen power. Samples were etched for 5 min with a chamber pressure position at 90° and a substrate temperature of 15°C.

Gas flow rate

The flow rate of SF_6 gas was the direct factor that determined the amount of gas feed in the chamber. Figure 3 showed the etch rate and roughness as a function of SF_6 flow rate. (SF_6 flow rate was limit to 100sccm on this equipment.) Etch rate raised up with the increasing of flow rate, accompanied by a higher surface roughness.

Optimized Recipes

According to the above results and a number of trials, variation of process parameters about the baseline recipe resulted in development of two optimized recipes listed in Table 1. Recipe 1 was optimized for etching structures with high etch depth, while recipe 2 was optimized for etching structures which require good surface roughness.

Table 1: Optimized etching recipes for higher etch depth (Recipe 1) and better surface roughness (Recipe 2)

Recipe	1	2
RF platen power	70W	130W
ICP source power	2000W	2000W
Chamber pressure (mTorr)	10	4.4
SF_6 flow rate(sccm)	100	50
Etch rate(μm/min)	**2.22**	2.06
Roughness(nm)	126	**30**
Aluminum mask selectivity	**82: 1**	25: 1
Aspect ratio	>10	>10

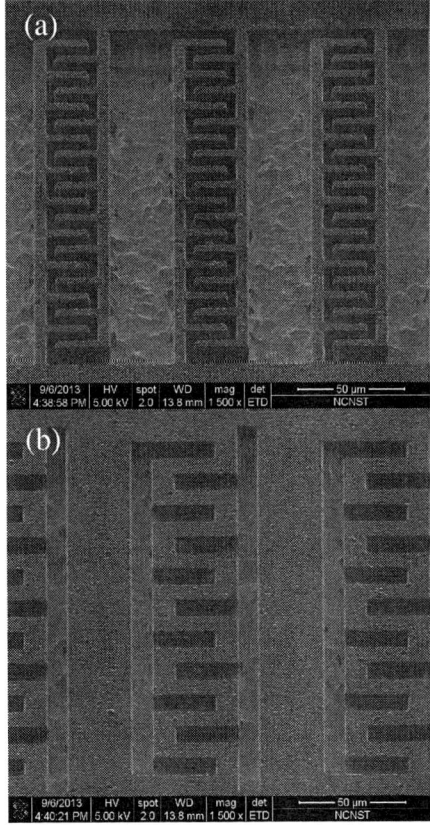

Figure 4: Typical MEMS structures etched by Recipe 1, with (a) comb drive structures and (b) micro mechanical springs. The scanning electron micrographs were taken at a 45° tilt.

(1) Recipe 1 optimized for higher etch depth:

Recipe 1 can be employed for the etching with large etch depths and high aspect ratios in MEMS. Figure 4 shows the scanning electron micrographs of the typical MEMS structures, including comb drive structures and micro mechanical springs, etched by Recipe 1.

Although the etch rate can be as high as 3μm/min with an aggressive platen power (Figure 1), more ion bombardments lead to lower mask selectivity [6]. The RF power was set 70W, with an etch rate still in excess of 2μm/min.

The minimum feature size achieved was less than 3μm, with an etching depth over 30μm. Figure 5 shows the densely patterned features with an aspect ratio of 13.7.

The process ability for fabricate deep structures is shown in Figure 6. Gear structures with an etching depth of 230μm were realized.

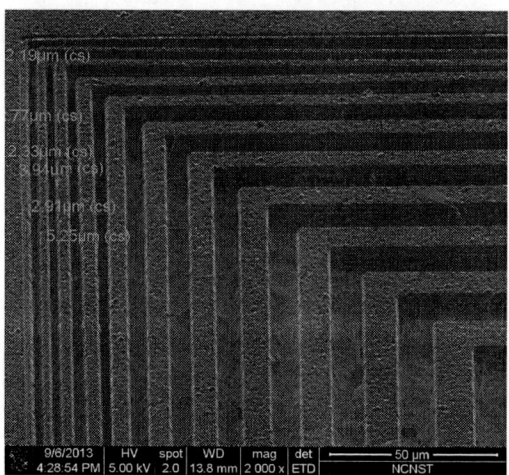

Figure 5: Densely patterned features. The minimum feature size achieved was 2.33μm, with an etching depth of 32μm. An aspect ratio of 13.7 was achieved. The scanning electron micrographs were taken at a 45° tilt.

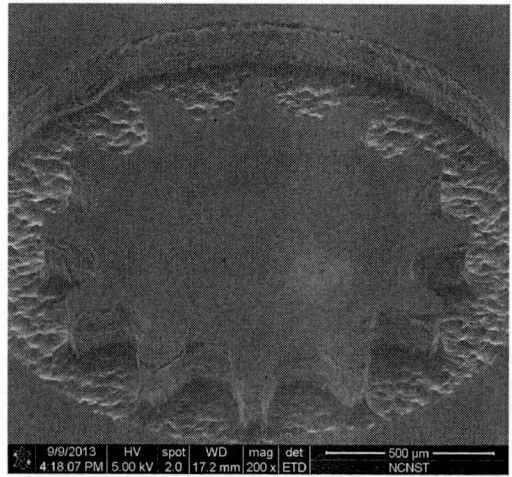

Figure 6: Gear structure etched by Recipe 1 with etching depth of 230μm. Aluminum mask was removed. The scanning electron micrographs were taken at a 45° tilt.

(2) Recipe 2 optimized for better surface roughness

Recipe 2 can be used to define a mold with the emphasis on good surface roughness. Figure 7 shows the structures as deep as 10μm etched by Recipe 2 with the bottom surface roughness of 30nm. According to the scanning electron micrographs, the side wall profiles are also fairly good.

Figure 7: Structures etched by Recipe 2 with etching depth of 10μm, and surface roughness 30nm. The scanning electron micrographs were taken at a 45° tilt.

The etch depth and roughness as a function of etching time were also plotted in the experiments. According to the trends shown in Figure 8, the etch rate uniformity with etch time was quite good, while the surface roughness was varied with etching time and depth, but stabilized for etchings deeper than 20μm.

CONCLUSION

In this paper, the characterization of bulk tungsten dry etching process using an ICP source is reported. Etching parameters such as RF platen power, ICP power and SF_6 flow rate were varied to determine their first-order effects on etching rate and surface roughness. After that, two optimized recipes were developed separately for etch depth and surface roughness. Etching depth of 220μm has been achieved at a rate up to 2.2μm/min. Feature structures in micrometers were realized with a high aspect ratio(>13) and good surface roughness(<30nm). The bulk tungsten deep etching offers numerous opportunities in MEMS applications.

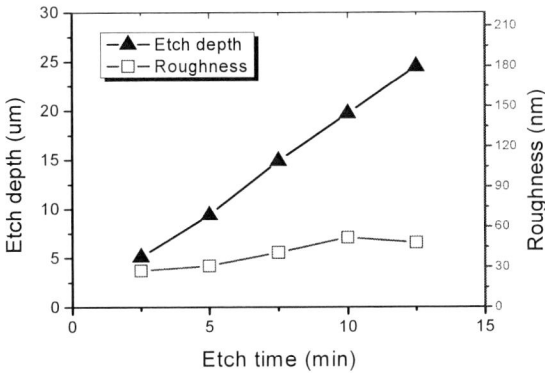

Figure 8: Etch depth& roughness as a function of etching time. The remaining process parameters were held constant at 2000W ICP source power, 70W RF platen power, 4.4mTorr, and 50sccm SF$_6$. Samples were etched with a chamber pressure position at 90° and a substrate temperature of 15°C.

ACKNOWLEDGEMENTS

The authors would thank the staff of Micro-nano Fabrication Platform, Institute of Microelectronics and Optoelectronics, Zhejiang University for the sample preparation.

REFERENCES

[1] Marco F. Aimi, Masa P. Rao, Noel C. MacDonald, Abu SamahZuruzi and David P. Bothman, "High-aspect-ratio bulk micromachining of titanium", *Nature Materials,* Vol. 3, 103 - 105 (2004)

[2] Jia Hu, Yiming Zhang, Shuhui Chen, Shuwei He, Nannan Li and Jing Chen, "Inductively coupled plasma etching of bulk molybdenum", *IEEE MEMS 2012,* pp. 267 - 270

[3] R. Löffler, M. Fleischer, D.P. Kern, "An anisotropic dry etch process with fluorine chemistry to create well-defined titanium surfaces for biomedical studies" *Microelectronic Engineering*, 97 (2012) pp. 361–364

[4] Yiming Zhang, Nannan Li, Bo Yan, Xiaoyang Feng, Jia Hu, Shuwei He,Yilong Hao and Jing Chen, "Fabrication of laterally driven bulk titanium devices on titanium-on-glass wafers" *J. Micromech. Microeng*, 23 (2013) 075026

[5] Byungwhan Kim and Byung-Teak Lee, "Relationships between etch rate and roughness of plasma etched surface", *IEEE Transactions on Plasma Science*, Vol. 30, No. 5, Oct. 2002

[6] E. R. Parker, B. J. Thibeault, M. F. Aimi,M. P. Rao,and N.C.MacDonald, "Inductively coupled plasma etching of bulk titanium for MEMS applications" *Journal of The Electrochemical Society*,152 (10)C675-C683 (2005)

CONTACT

*J. Chen, tel: +86-10-6275-2536; j.chen@pku.edu.cn

LARGE ARRAYS OF INKJET-PRINTED MEMS MICROBRIDGES ON FOIL

Francisco Molina-Lopez, Danick Briand, and Nico F. de Rooij
Ecole Polytechnique Fédérale de Lausanne (EPFL), Neuchâtel, SWITZERLAND

ABSTRACT

This works describes the fabrication process of an array of printed MEMS microbridges on polymeric foil, performed in only four easy steps. Each functional material was deposited exclusively by inkjet-printing technique, compatible with large-area fabrication. The array occupies an area of 2 mm x 2 mm and consists of 80 individual microbridges of 120 µm x 80 µm size each. When connecting all the bridges in parallel, the array displays a total capacitance value of 1.5 pF. The potential of the fabricated array has been demonstrated by employing it as a swelling-based capacitive humidity sensor where the polymeric substrate acts directly as the sensing layer.

INTRODUCTION

MEMS are gaining interest in the field of large-area systems due to the possibility to employ them in many new applications. For example, the fabrication of a "smart MEMS sheet" for the displacement of small objects on a flexible foil has been presented by Ataka *et al.* in 2013 [1]. MEMS are typically fabricated using clean room processes which limit their realization on a large-area and contribute to the increase of production costs. The application of printing fabrication techniques to MEMS production process would make it possible to combine the typical advantages of printing (such as low fabrication costs and compatibility with a broad range of materials and substrates) with the functionalities of MEMS. Among the different printing techniques, inkjet stands out for its simplicity and versatility due to its non-contact and digital character. A work combining inkjet-printed organic transistors with polymeric actuators on a large-area was reported in [2] for the fabrication of flexible ultrasonic systems. In that case, the actuators had sizes in the millimeter range and were not fabricated by printing methods. Nonetheless, inkjet-printed MEMS have been recently reported in [3], where the authors described the fabrication of cantilever-based switches through a relatively complex process involving also spin coating and the definition of via-holes.

In this work, we benefit from the advantages of inkjet-printing to develop arrays of fully printed MEMS microbridges on polymeric foil through four simple steps, all of them compatible with large-area fabrication. The microbridges have been characterized by electrostatic actuation and then, their operation as capacitive relative humidity sensors has been demonstrated. Fully inkjet-printed capacitive humidity sensors have already been fabricated on polymeric foil as reported in [4, 5]. Their operation was based on the changes in the permittivity of a sensing layer in the presence of moisture. Here, we propose a new working principle for gas sensing based on the swelling of a polymeric substrate which acts as a sensing layer. The application of the device as a capacitive sensor became possible owing to the high capacitance value provided by many microbridges connected in parallel. The microbridges array could also be adapted for detection of more specific gases by functionalizing the substrate, passivating it against permeation of moisture (or other gases), or inkjet-printing another custom-made swelling sensing layer onto the bridge. When compared to a membrane-like device providing an equivalent capacitance value, this device present advantages in terms of fabrication simplicity, mechanically robustness and little sensitivity to accelerations due to the small mass of every independent microbridge. Finally, the reported device could be turned into an array of micro-switches or micro-actuators by tailoring the geometry of the microbridges. This work foresees the potential of low-cost and large-area printed MEMS structures for sensing and actuating applications.

DESIGN

The presented device consisted of an array of 80 microbridges like the one sketched in Figure 1 that could function separately or altogether. Each bridge was designed to be formed by a 200 nm-thin, 65 µm-wide silver bottom electrode; and a thicker (~ 2 µm) 80 µm-wide suspended silver top electrode / microbridge. The microbridge was fabricated with the help of a sacrificial layer of photoresist placed in between the two electrodes. The thickness of the sacrificial layer was ~ 5 µm and corresponded with the height of the microbridge. The expected capacitance value of every single bridge was of ~ 20 fF.

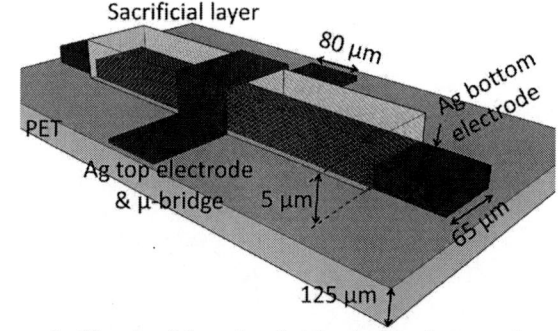

Figure 1: Sketch of the microbridge before being released.

In the array (see Figure 3), the bridges were arranged in 10 columns separated by a distance of 200 µm. Each column was composed by 8 rows of microbridges placed in line and separated by a distance of 240 µm. The bottom electrode was a continuous layer shared by the 10 microbridges in the same row. In order to reach a capacitance value easy to read with standard laboratory instruments, the 80 bridges in the device were connected in parallel. All the bottom electrodes were connected together by large pads placed at both sides of the array. Similarly, all the bridges were contacted by pads placed above and below the array. The size of the pads was 1 mm x 1 mm. The total expected capacitance was of 1.5 pF.

FABRICATION
Materials and Methods
All the chemicals employed during the fabrication of the MEMS array were purchased from *Sigma-Aldrich* and used as they were unless otherwise stated. The device was fabricated on a substrate made of 125 µm-thick polyethylene terephthalate (PET) *Melinex® ST506* from *Dupont Teijim Films*. Both bottom and top electrode (the bridge itself) were composed of silver-nanoparticles ink (*SuntTronic Jet EMD506* from *SunChemical*) with 20% solid content. The sacrificial layer was commercial photoresist *Microposit® 1813®* from *Shipley* diluted 90:10 (% weight) with propylene glycol monomethyl ether acetate (PGMEA) to make it inkjet-printable by the utilized printer *Dimatix Fujifilm DMP-2800™* with 10 pL drop volume cartridges.

Process Flow
Figure 2 shows the steps of the fabrication process, along with an optical picture of the device at the end of every step. Prior to the printing of the bottom electrode, the substrate was cleaned by successive immersion in acetone, isopropanol (both VLSI grade) and deionized water baths for 10 minutes. After drying with nitrogen, the substrate was subjected to a thermal treatment for 30 minutes at 140 °C for dehydration and thermal stabilization. Then, we treated the substrate with low frequency (13.56 MHz) oxygen plasma for 35 seconds at 50 W to increase the substrate surface wettability. An undesired side effect of the oxygen plasma treatment was a large spreading of the ink on the substrate, resulting in very wide printed lines. This effect was considerably reduced while maintaining good wettability by heating up again the substrate in the oven for 5 minutes at 120 °C. The bottom electrode was formed by inkjet-printing two consecutive layers of silver ink with a drop-to-drop separation of 40 µm (Figure 2 (a)). Following, the silver was sintered at 140°C for 30 minutes in a convection oven. In the second step, two layers of the diluted photoresist were also inkjet-printed, with a drop-to-drop spacing of 20 µm, on top of the bottom electrode to act as sacrificial layer (Figure 2(b)). Special care was taken during the alignment of the resist on the bottom electrode to ensure its total coverage and to avoid future short-circuits between the top and the bottom electrodes. After the resist was annealed on a hot plate at 115 °C for 10 minutes, the same silver ink was inkjet-printed on it to form the top electrode / microbridge. The annealing of the resist was needed to render it chemically resistant to the solvent of the silver ink. Nine consecutive layers with a drop-to-drop space of 20 µm were printed to form the microbridge, which was afterwards sintered on a hot plate at 115 °C for 30 minutes (Figure 2 (c)). Although this sintering temperature

was rather mild and led probably to bridges with poor mechanical properties, we could not increase it since we observed difficulties to dissolve the resist when it underwent temperatures higher than 115 °C. At this point, optional electrodeposition of Ni has been also demonstrated to thicken the bridge if necessary (Figure 2 (d)). Finally, the sacrificial layer was removed by dissolving it in acetone (Figure 2 (e)). Stiction problems were avoided by performing freeze-drying process: the acetone under the bridge was successively replaced by isopropanol and finally by cyclohexane (HPLC grade), which was frozen on a Peltier element (freezing point 6.47 °C) and sublimated under nitrogen atmosphere.

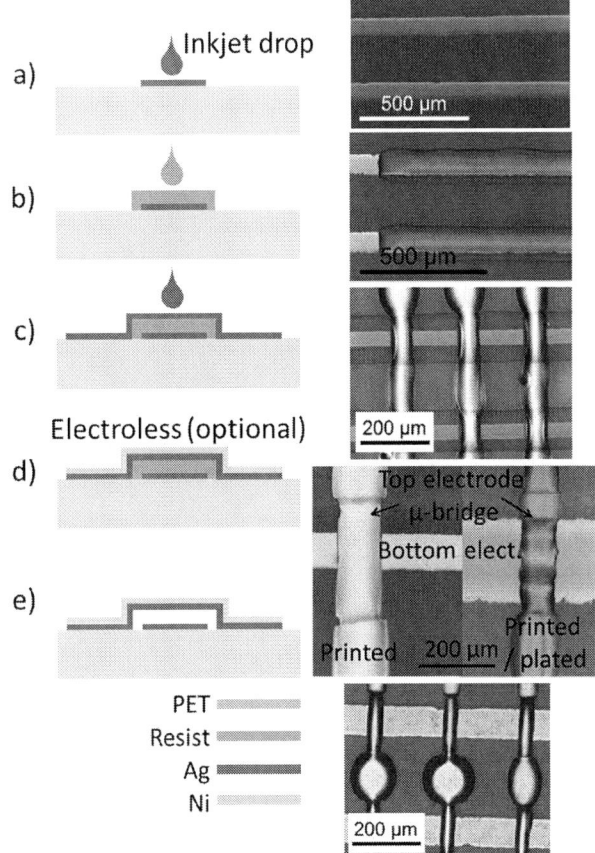

Figure 2: Sketch (left) and optical images (right) of every step of the microbridges fabrication process.

RESULTS
Fabrication of Arrays of MEMS Microbridges
Several arrays of MEMS microbridges were successfully inkjet-printed on PET foil. Figure 3 contains optical images of the arrays where the different parts have been labeled. The figure also includes a scanning electronic microscope (SEM) image of the array and a 3D image, obtained with a white light interferometer. The accumulation of ink between the bridges during the printing process formed pillars-like structures which were not found to be unfavorable for the proper operation of the device.

The use of inkjet-printing to fabricate MEMS makes it

easy to control the height of the bridge as well as its thickness by choosing the number of printed layers. Figure 4 depicts the relationship between the thickness and the number of printed layers for the sacrificial photoresist layer and the silver microbridge. The final selected values, ~ 5 μm thick photoresist and ~ 2 μm thick top electrodes, were chosen based on empirical observations, seeking to maximize the fabrication yield. Although Figure 4 (bottom) indicates that 6 printed layers were sufficient to print ~ 2 μm thick top electrode, the flow of the ink toward the pillars made necessary to print 9 layers as indicated in the process flow above. The bridges have a typical height of ~ 2.5 μm (smaller than the thickness of the sacrificial layer since the bridges tend to move down after being released), a length of 120 μm and a width ranging from 60 to 80 μm. The bottom electrode width resulted in 65 μm.

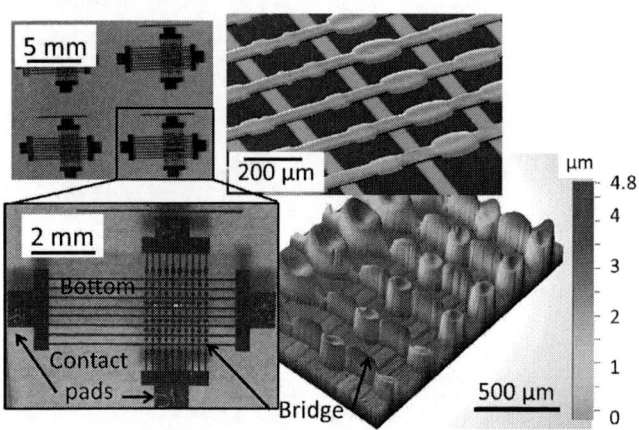

Figure 3: Optical image and magnification view of the array of microbridges (left). SEM image of the array (top right) and white light interferometer image of the same array (bottom right).

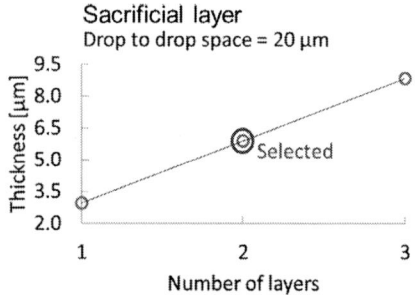

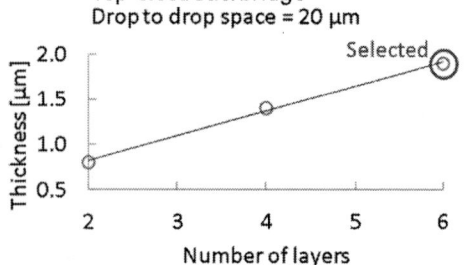

Figure 4: Thickness of the inkjet-printed sacrificial layer (top) and bridge (bottom) versus number of printed layers.

Electrostatic Actuation

In order to confirm the operation of the devices as suspended structures, they were electrostatically actuated by applying a voltage between their bottom and top electrode, while measuring at the same time the deflection of the top electrode. Figure 5 shows the deflection of a single device with applied voltage, measured with a white light interferometer. The distance between the top and the base of the released microbridge is 1.6 μm in the absence of any applied voltage.

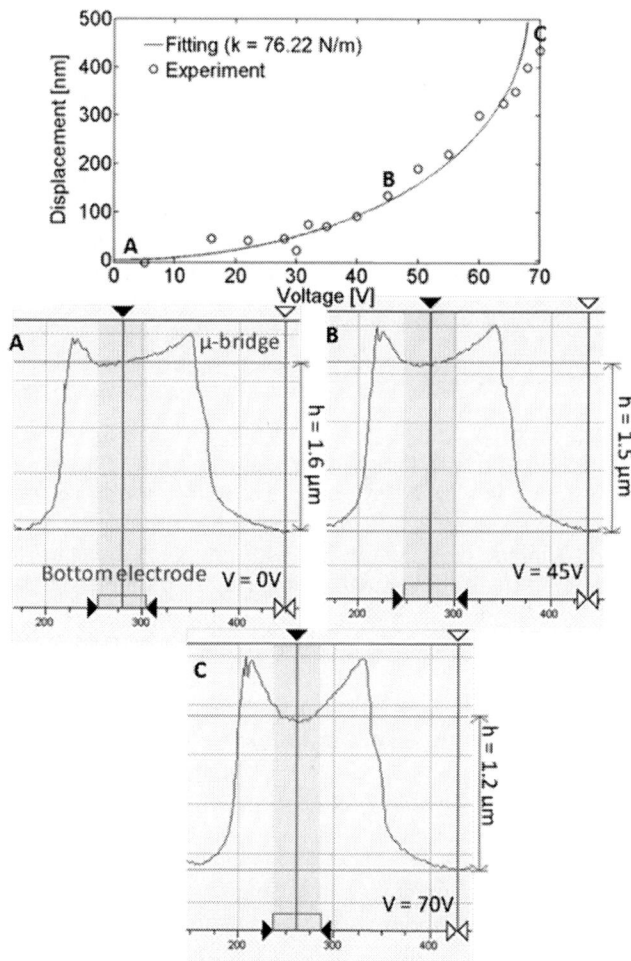

Figure 5: Microbridge displacement versus applied voltage: experimental points and fitting curve with a deduced effective spring constant (top). Microbridge profile measured at different voltages (bottom).

A mathematical fitting to the experimental values has been performed (Figure 5 (top)) using the parallel-plate model which states that in static mode, the mechanical force on the bridge F_M is equal to the electrostatic one F_E:

$$F_M = k\, d = F_E = \tfrac{1}{2}\, C\, V^2 / (g - d) \qquad (1)$$

where d is the deflection of the bridge, C the parallel plate capacitance formed between bottom and top plates, V the applied voltage and g the original gap in between the plates.

The effective spring constant for the shown bridge resulted in $k = 76 \pm 4$ N/m. This constant can be modified by changing the material of the bridge, its sintering conditions or the bridge geometry; making possible to adjust the stiffness of the bridge to the required application.

Application as Humidity Sensors

Since the substrate of the devices is a polymer, it swells upon gas absorption. We could then assume that the substrate bends and the bridges buckle up increasing the distance between the bottom and the top electrode. The capacitance value of the array should therefore decrease accordingly. In this way, the device can be used as a gas sensor. The results shown in Figure 6 (where C' < C) seem to confirm our assumptions. The tests were carried out for relative humidity for the sake of simplicity but the principle applies to every gas in general. The device was introduced in a climatic chamber and subjected to different steps of relative humidity (25%, 45% and 65% r.h.) while keeping the temperature fixed at 30°C. After letting the sensor to stabilize for 30 minutes, the capacitance value was registered at 100 kHz for every step using an LCR-meter. The device was tested for two downsweep and one upsweep cycles and it displayed a linear behavior with a coefficient $R^2 = 0.9757$ and a precision of 4.3% r.h. The sensitivity was -1.92 ± 0.14 fF / 1% r.h. As opposed to the most part of polymeric printed capacitive gas sensors already present in the literature (see as an example [4, 5]), the microbridge-based sensor does not rely on variations in electrical permittivity, but on mechanical swelling. Hence, they are expected to be more sensitive to large molecules, such as volatile organic compounds (VOCs), than to small water molecules, despite the high permittivity of the latter. These bridges-based sensors are thus a perfect complement to the ones already reported.

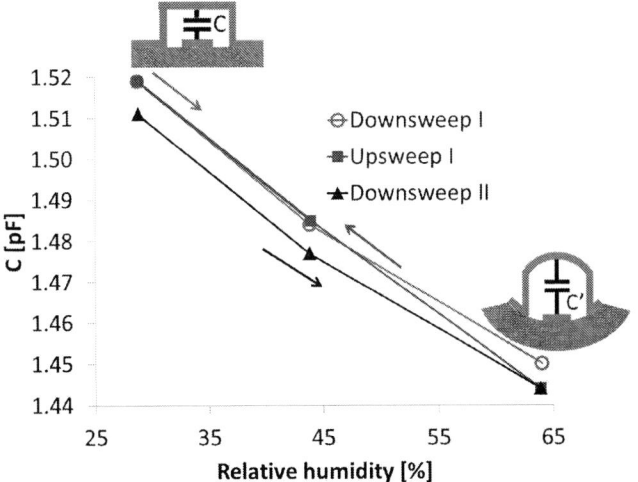

Figure 6: *Capacitance value versus relative humidity displayed by the MEMS array operating as a humidity sensor, along with a sketch of the expected bridge behavior.*

In order to expand the functionality of the sensor towards more specific analytes, it would be also possible to coat the bridge with a customized sensing layer. This would induce stress on the bridge itself too. Non-contact printing methods such as inkjet or laser induced forward transfer (LIFT) would be suitable methods for such purpose. Alternatively, the substrate could be functionalized or passivated at the same time to tailor the selectivity of the sensor.

CONCLUSIONS

Large arrays of fully inkjet-printed MEMS silver microbridges have been fabricated on PET foil through only four process steps. After confirming the electrostatic actuation of the bridges, we demonstrated the possibility of interconnecting the 80 bridges of the array in parallel to provide readable capacitance values of 1.5 pF. The potential of the device as a capacitive humidity sensor was then assessed by directly utilizing the polymeric substrate as a swelling sensing layer. By coating the bridges with a customized sensing layer or by functionalizing / passivating the substrate, the operation of the sensor could be expanded to detection of others analytes. For future improvements, a theoretical model should be elaborated to predict the sensor response to gases. Finally, by optimizing the pull-in voltage of every microbridge, the device could work as an array of MEMS switches.

ACKNOWLEDGEMENTS

EU FP7 Project FlexSmell, Marie-Curie ITN. Grant 238454. We also acknowledge the SNF R'Equip program.

REFERENCES

[1] M. Ataka, *et al.*, "Micro Actuator Array on a Flexible Sheet – Smart MEMS Sheet", in *Proc. MEMS 2013*, Taipei, January 20-24, 2013, pp. 536-539.

[2] Y. Kato *et al.*, "Large-area Flexible Ultrasonic Imaging System with an Organic Transistor Active Matrix", *IEEE Transaction on electron devices*, vol. 57 (5), pp. 995-1002, 2010.

[3] E.S. Park *et al*, *Nanoletters*, "A New Switching Device for Printed Electronics: Inkjet-printed Microelectromechanical Relay", vol. 13, pp. 5355-5360, 2013.

[4] F. Molina-Lopez *et al.*, "Large-area Compatible Fabrication and Encapsulation of Inkjet-printed Humidity Sensors on Flexible Foils with Integrated Thermal Compensation", *J. Micromech. Microeng.*, vol. 23, pp. 025012, 2013.

[5] F. Molina-Lopez *et al*, "Decreasing the Size of Printed Comb Electrodes by the Introduction of a Dielectric Interlayer for Capacitive Gas Sensors on Polymeric Foil: Modeling and Fabrication", *Sens. Act. B: Chem.*, vol. 189, pp. 89-96, 2013.

CONTACT

*F. Molina-Lopez, tel: +41-21-695-4434;
francisco.molinalopez@epfl.ch

LIGHTWEIGHT MICRO LATTICES WITH NANOSCALE FEATURES FABRICATED FROM PROJECTION MICROSTEREOLITHOGRAPHY

Xiaoyu Zheng[1], Joshua Deotte[1], John Vericella[1], Maxim. Shusteff[1], Todd Weisgraber[1], H. Lee[2], N. Fang[2], and Christopher.M. Spadaccini[1]*

[1]Lawrence Livermore National Laboratory, Livermore, CA, USA
[2]Department of Mechanical Engineering, Massachusetts Institute of Technology, MA, USA

ABSTRACT

Complex, three-dimensional lightweight cellular materials inspired by nature, such as honeycomb and foam-like structures are desirable for a broad array of applications such as structural components, catalysts supports and energy efficient materials. Additionally, when designed with interconnected porosity, the open volume in the architecture can be exploited for active cooling or energy storage, providing unique opportunities for multifunctionality. However, they are extremely difficult to fabricate with the current state–of-the-art fabrication techniques. This paper reports the fabrication of complex, three-dimensional cellular materials with nanoscale features using a novel additive manufacturing approach, namely Projection Microstereolithography (PµSL)

INTRODUCTION

Inspired by naturally occurring cellular structures such as honeycombs (wood, cork) and foam-like structures (trabecular bone, plant parenchyma, acorol, and sponge), light weight designed cellular foams made from a wide array of solid constituents are desirable for a wide range of applications including thermal insulation, catalyst supports, energy efficient materials, infiltration and biomaterials. Additionally, when designed with interconnected porosity, the open volume in the architecture can be exploited for active cooling or energy storage, providing unique opportunities for multifunctionality. However, the required outstanding properties have remained elusive on the bulk scale, constrained by the inherent coupling of material properties and the lack of suitable fabrication processes to generate these highly three-dimensional microarchitectures.

Only few efforts have been recently conducted on accessing ultra-light material (below 10mg/cm3, or with relative density equal or below 0.5%) at micron and sub micron-scales and investigate its mechanical properties, such as low density metallic lattice reported by T.A.Schaedler et.al,[1], nanoporous silica and carbon nanotube foams, reported by Kucheyev et al.[2], low density graphene monoliths by Qiu et al [3]. The critical challenge in these lightweight materials is the significant degradation of material property as the relative density reduces. As such, the mechanical properties of these materials degrade substantially with reduction of their weight via porosity. For example, the Young's modulus of low density silica aerogels with densities of <10mg cm-3 are reported to have degraded to 10kPa. In addition, the

functionality of these devices directly depends on their structural integrity and mechanical stability, driving the necessity to understand and to predict mechanical properties of materials at reduced dimensions in a quantitative manner. Unfortunately, the few existing experimental techniques for assessing mechanical properties at that scale are insufficient, not easily accessible, and are generally limited to thin films or rely on ingenious assembly techniques combined with high-temperature brazing.

In order to design reliable devices, a fundamental understanding of mechanical properties as a function of feature size is desperately needed. Based on the exploitation of unique phenomena arising from nano-scale structures into novel material architectures, it will be possible to define material design space with vastly superior properties than can currently be achieved. With the increasing use of novel fabrication technologies, a material's architectures can be manipulated over an ever increasing set of length scales. Manipulation of a material's properties can be achieved through prudent choice of design parameters. For example, the Poisson's ratio and the elastic modulus of a second order hierarchical honeycomb design can be altered through the choice of geometric structural parameters. Furthermore, these structures are found to have high strength relative to density when compared to more conventional designs.

THOERY

A stretch-dominated unit cell structure that consists of b struts and j frictionless joints satisfying Maxwell's criteria, $b - 3j + 6 > 0$, is significantly more mechanically efficient, with a higher stiffness-to-weight ratio (defined as E/□) than its bend-dominated counterpart. This is attributed to a portion of its struts carrying load under compression or tension rather than bending or buckling. A fundamental lattice building block of this type is the octet-truss unit cell, demonstrated on the macroscale [4]. Each unit cell consists of beams with an identical aspect ratio. Octet-truss structures in this study satisfy the deterministic equation for stretching dominated pin-jointed frame condition[5, 6]. The unit cell of the octet-truss stretch dominated lattice presented in this paper can be viewed as a regular octahedron core that is surrounded by eight regular tetrahedrons distributed on its eight faces. The cell has a Face Centered Cube (FCC) lattice structure with cubic symmetry generating a material with an isotropic behavior [3]; its nodes are similarly situated with 12 cell elements connectivity at each node. Octahedral cells can be stacked to

978-1-4799-3510-9/14 $31.00 © 2014 IEEE

synthesis the octet-truss structure, with each strut of an octahedral cell shared between two neighboring cells. The lattice materials are produced by the uniformly periodic replication of characteristic unit cell elements. These types of cubic-symmetric unit cells take advantage of the successive packing of each layer by sharing identical volume fractions with surrounding truss members. This packing system enables lattice materials to maximize stiffness and lightweight effect within the limited space.

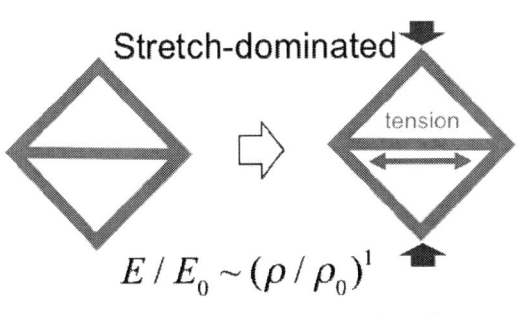

Stretch-dominated

$$E / E_0 \sim (\rho / \rho_0)^1$$

Maxwell's criteria, M = b − 2j + 6 < 0

Figure 1 Schematic illustration of stretch dominated architecture in 2-D

P□SL is a process that involves the chemical kinetics of photo-polymerization of UV curable resins consisting of monomer, photo-initiator and photo-absorber. First, under UV illumination, photo-initiators (PI) absorb incident photons and generate free radicals. The excited free radicals combine and react with the monomer molecules (M) to form larger reactive molecules. These reactive molecules continue to react with adjacent monomers to form longer polymeric molecules. The polymeric molecules keep growing until two of them combine and terminate the reaction. The solidified polymer structure eventually forms by the entanglement and cross-linking of these polymer chains.

Based on this photo-chemistry, P□SL is capable of rapidly fabricating complex three-dimensional microstructures in a bottom-up, layer-by-layer fashion. A computer-generated 3D CAD model is first sliced into a series of closely spaced horizontal planes. These two-dimensional slices are digitized as a bitmap image and transmitted to the LCoS chip which projects the image through a reduction lens into a bath of photosensitive polymer resin. Once the material in the exposed area is polymerized by the chemical reaction described above, the substrate on which the polymerized material rests is lowered to repeat the process with the next image slice.

Digital masking is the core technology that transforms a slow serial stereolithography process into an extremely rapid and effective parallel process. There are three types of digital mask technologies: 1) Liquid Crystal Display (LCD); 2) Digital Micromirror Device (DMD); 3) Liquid Crystal on Silicon (LCoS). LCD is the cheapest and earliest dynamic mask employed in stereolithography. The limitation of LCD as a dynamic mask for photocurable resins comes from the fact that LCD light absorption is significantly higher during the ON mode and, therefore, its use is limited to a small range of commercially available UV curable resins 17. The Digital Micromirror Device (DMD) offers improved contrast over the LCD and is better suited to handle UV light than LCD. However, DMD tends to mechanically fail when taken out of their sealed environment and is not well suited to a laboratory setup outside of a clean room. In this study, the LCoS chip was employed as the spatial light modulator due to its highly reflective surface coated with liquid crystals which provides high contrast ratio as well as high packing density, small pixel size, and adequate switching speed. In the P□SL system, the LCoS chip works both as a mask and as a shutter. Not only are the sliced bitmap images displayed on it to modulate the illumination light, but it also controls the polymerization time by providing the ability to switch completely off creating a non-reflective surface resulting in no light transmission to the resin bath.

A LED was chosen as the light source in the P□SL system. Among many others, the major advantage of LEDs over conventional light sources such as mercury arc lamps and lasers is their energy efficiency. LEDs consume much lower energy than lasers and mercury lamps for polymerization of the same volume of material. Moreover, they are tailored to a specific wavelength to match that of the UV curable resins. Other advantages include shorter starting time, longer lifespan, low cost, compact size, and low heat dissipation. Although the light output of an individual LED is modest, advanced arrays of high power LEDs delivers sufficient light power required for this study.

FABRICATION

PμSL is a layer-by-layer additive micromanufacturing process that is capable of fabricating arbitrary three-dimensional microstructures with high throughput using ultraviolet light (UV) and a digital mask (Figure 2). The process begins with a photosensitive polymer (1, 6-hexanediol diacrylate (HDDA)) resin bath. P□SL uses a spatial light modulator — a liquid crystal on silicon (LCoS) chip in this case — as a dynamically reconfigurable digital photomask. A three-dimensional CAD model is first sliced into a series of closely spaced horizontal planes. These two-dimensional image slices are sequentially transmitted to the reflective LCoS chip, which is illuminated with UV light at 405nm wavelength from a light emitting diode array. Each image is projected through a reduction lens onto the surface of the photosensitive resin. The exposed liquid cures, forming a layer in the shape of the two-dimensional image, and the substrate on which it rests is lowered, reflowing a thin film of liquid over the cured layer. The image projection is then repeated with the next image slice forming the subsequent layer. The penetration depth of the light is controlled by the light intensity, exposure time, and the concentration of photoabsorber and photoinitiator. No current technology is able to rapidly and sustainably produce structures with optimized three-dimensional architectures at the meso- and micro-scales while also being flexible enough to vary the geometry from part-to-part.

In order to fabricate nanoscale micro lattices, after successful fabrication of the polymer templates, the as-formed cellular structures can be coated with a thin layer of material of nanoscale thickness followed by removal of the polymer core. For example, the fabricated polymer templates through P□SL was subsequently coated with nickel-phosphorous film with 100nm-2□m thickness through eletroless plating, by controlling plating time. Alternatively, through Atomic Layer Deposition of Al2O3, polymer-ceramic hybrid cellular structures were generated with Al2O3 thickness ranging from 20-100nm, by controlling the deposition time. The base polymer templates was subsequently removed by thermal post-processing, leaving behind the hollow tube, ceramic microlattices shown in Figure 3.

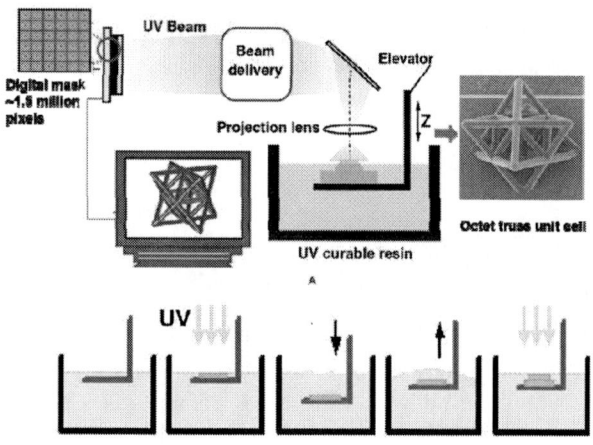

Figure 2 Schematic illustration of Projection microstereolithography system for fabrication of 3D microlattices

Nanoscale Al2O3 films were deposited by ALD using the well-established trimethyl-aluminum (AlMe3/ H2O) process in a warm wall reactor (wall and stage temperature of 125°C). Long pulse, pump, and purge times (90 seconds each) were used to ensure uniform coatings throughout the porous material. The film thickness was controlled by adjusting the number of ALD cycles, and the growth rate per cycle was calculated from the measured mass gain, the known surface area of lattice structures, and density of ALD Al2O3. This rate was found to be □0.25 nm per cycle. The bimaterial hybrid polymer-Al2O3 microlattices were then heat-treated in a box furnace in air at 500°C for 4 hours to remove the polymer template, leaving behind the Al2O3 hollow-tube octet-truss lattices with Es determined as 180 GPa from nanoindentation.

Figure 3 (a) Polymer (HDDA) octet-truss lattice generated from PmSL. (b) Hollow-tube octet-truss metallic octet-truss lattice with nanoscale tube thickeness (100nm). (c) hollow-tube ceramic octet-truss lattice with nanoscale tube thickness (50nm).

Figure 4 (a)-(d) illustrates the ability to fabricate arbitrary three-dimensional overhanging structures with defined complex curvature, multiple suspending structures, and cellular structures such as octet-truss architectures [2] and porous materials which were not achievable by previous 3D fabrication techniques. The octet truss material, shown in

Figure 4. Complex three-dimensional micro-structures manufactured with the PuSL system. (a) Micro-gripper structure with overhanging gripper and pre-defined curvature. (b) Micro-lattice unit cell structure. (c) Porous structure with Tetrakaidehedron architectures. (d)5X3X3mm porous structure consisting of multiple porous lattice from (c).

Figure 3,with its micro-struts carrying tension or compression when loaded, is expected to have superior mechanical performance under compression as compared to regular engineering foams and natural scholastic foams at

978-1-4799-3510-9/14 $31.00 © 2014 IEEE   512

the same relative density. Our proof-of-concept results shown in Figure 4 demonstrate superior compressive stiffness when compared to convention lightweight lattices on the material property chart.

CONCLUSION

We demonstrate the utility of the system by producing a variety of microstructures with complex geometries and explored the potential of using the system to build meso-

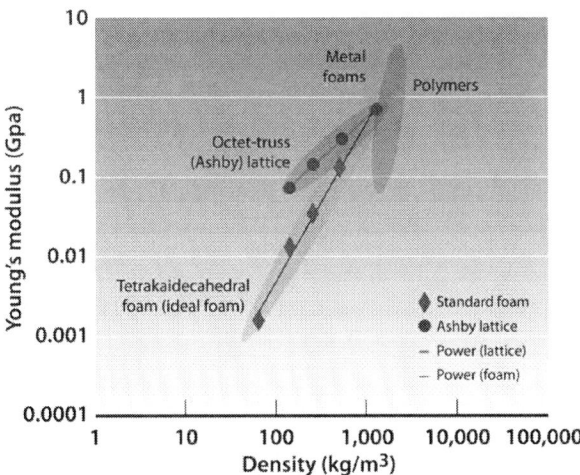

Figure 4 Compressive stiffness-weight density scaling comparison of light-weight octet truss material v.s. open cell foams.

scale structures with micro-scale architecture and nano-scale features. These achievements pave the way for large scale micro- and nano- manufacturing that extends the current state-of-the-art of three-dimensional fabrication technologies.

ACKNOWLEDGEMENTS

This work performed under the auspices of the U.S. Department of Energy by Lawrence Livermore National Laboratory under Contract DE-AC52-07NA27344. Funding support from LDRD Strategic Initiative 11-SI-005 is gratefully acknowledged. The authors would also like to acknowledge Christopher Harvey at Lawrence Livermore National Laboratory for his technical support (LLNL-JRNL-640334).

REFERENCES

[1] T.A. Schaedler et al Ultralight Metallic Microlattices. Science 18 November 2011: Vol. 334 no. 6058 pp. 962-965

[2] S. O. Kucheyev et al., Super-Compressibility of Ultralow-Density Nanoporous Silica

[3] M. A. Worseley et al., Mechanically robust and electrically conductive carbon nanotube foams, Appl. Phys. Lett. 94, 073115 (2009)

[4] Ashby, L.J.G.a.M.F., Cellular Solids: Structure and Properties. Book. 2001, Cambridge, UK: Cambridge University Press.

[5] Deshpande, V.S., M.F. Ashby, and N.A. Fleck, Foam topology: bending versus stretching dominated architectures. Acta Materialia, 2001. 49(6): p. 1035-1040.

[6] Renton, J.D., Elastic Beams and Frames. 2 ed. 2002, Chichester: Horwood Publishing Limited

[7] Solecki, R., Advanced Mechanics of Materials 2003, Oxford: Oxford University Press. 763.

[8] A., P.M., NIKE3D: A Nonlinear, Implicit, Three-Dimensional Finite Element Code for Solid and Structural Mechanics, 2012, Lawrence Livermore National Laboratory Livermore, California, USA.

CONTACT

*Xiaoyu (Rayne) Zheng, zheng3@llnl.gov

FABRICATION OF MICRO-HEATERS EMBEDDED IN PDMS USING A DRY PEEL-OFF PROCESS

Ikjoo Byun, Ryohei Ueno, and Beomjoon Kim
Institute of Industrial Science, The University of Tokyo, Tokyo, JAPAN

ABSTRACT

This paper describes a reliable fabrication method of micro-heaters embedded in polydimethylsiloxane (PDMS). Gold patterns are transferred and embedded to the PDMS from a silicon substrate, by peeling off. The surface adhesion among silicon substrate, Au patterns, and PDMS is modified with self-assembled monolayers. Therefore, micro-heaters embedded in PDMS can be fabricated by a simpler and easier way compared to conventional methods. The thermal characterization of micro-heaters and PDMS device is carried out by electrical and infrared thermo-microscopic measurements. The experimental results well agree with the numerical analysis performed by finite element method.

INTRODUCTION

The lab-on-a-chip (LoC) systems have been one of the dominant themes in analytical instrumentation for chemical and biomedical applications over the past decade [1]. Thermal control is one of the most critical factors for precise control of chemical reaction [2], protein synthesis [3], and polymerase chain reaction (PCR) [4], etc. Micro-heaters [3] enables more precise temperature control compared to heating blocks [4] or heating wires [5].

For LoC devices, polydimethylsiloxane (PDMS) is one of the most popular polymer because it is bio-compatible, chemically resistant, optically transparent, inexpensive and easy to be fabricated [2,3]. Usually, heaters for LoC devices are fabricated on a glass substrate, and assembled with PDMS micro-fluidic channels [3]. On the other hand, micro-heaters embedded in PDMS have the advantages such as flexibility, rapid prototyping, and greater compatibility with existing PDMS chips [6].

Some research groups have been trying to produce a conducting PDMS by mixing various fillers (e.g. carbon black powder or metallic powder) and PDMS prepolymers [6,7]. To pattern the conducting PDMS, gel-state conducting PDMS is filled into photoresist (PR) patterns, then the extra mixture is removed using a razor blade. However, this method suffers from two reasons: (1) a razor blade can damage PR patterns mechanically and (2) a conducting PDMS is sensitive to the change of ambient temperature due to a large coefficient of thermal expansion. Thus, it is difficult to control the temperature of micro-heater reliably even though it is calibrated. Recently, it has been demonstrated that (3-mercaptopropyl)trimethoxysilane (MPTMS) can considerably promote the adhesion between Au and PDMS using a liquid deposition method [8]. Also, Au micro-patterns can be embedded in PDMS using a dry peel-off process [9,10].

In this paper, we show a simple and reliable fabrication method of Au micro-heaters embedded in PDMS.

Evaluations including temperature with respect to applied voltages, spatial distribution and transient response were conducted. Finally, simulation results are compared to the experimental ones.

FABRICATION

The key point in fabricating the Au micro-heaters is the direct transfer of metal layer into PDMS carried out using a "dry peel-off" process, which involves modifying the surface properties of the substrate and metal patterns through self-assembled monolayer (SAM) treatment and peeling off the PDMS with embedded metal layers [10]. The fabrication process was based on the "dry peel-off" process, but improved compared to the previous research [10], as shown in Figure 1. A Si wafer was treated with a piranha solution for 10 min, followed by dehydration at 150 °C for 10 min. Then, a sparse MPTMS layer was formed on the Si substrate to make a moderate adhesion using a vapor deposition for 10 min. A thin Au layer (thickness: 100 nm) deposited by thermal evaporation onto the substrate was lithographically patterned. The substrate with both Au patterns and photoresist (PR) patterns were immersed in perfluorodecyl -trichlorosilane (FDTS) solution (5 mM in hexane) for 5 min for anti-adhesion between Si and PDMS. After PR removal, the substrate with Au patterns was treated with an ethanolic solution of 20 mM MPTMS for 2 h as a molecular adhesive between Au and PDMS. A 10:1 (by weight) mixture of PDMS base/curing agent was poured on the substrate, then heat-cured in an oven at 60 °C for 3 h, then maintained at room temperature for 12 h. The thickness of PDMS substrate was controlled to 3 mm. Finally, the PDMS with Au patterns was peeled off from the Si substrate.

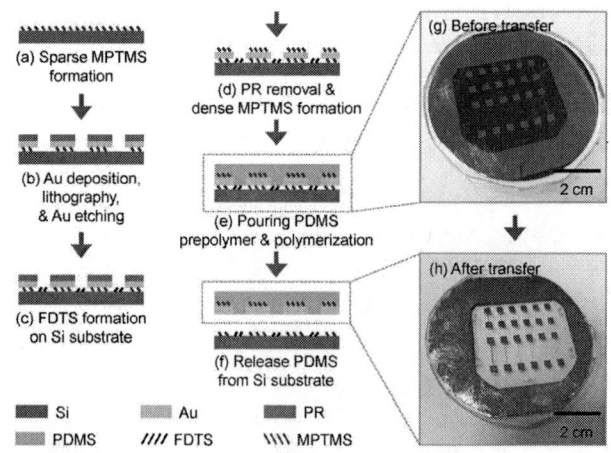

Figure 1: (a–f) Schematic illustration of the fabrication process, and optical images of Au micro-heaters (g) before and (h) after peeling off the PDMS from the Si substrate.

CHARACTERIZATION

The micro-heaters were designed with different geometric shapes (width: 40, 80 and 160 μm, length: 15 mm), as shown in Figure 2(a). Pads for electrical connection to power supply were designed for 3 × 3 mm. Joule heating was applied to the micro-heaters operations. To apply the voltage and measure the electrical resistance, DC voltage current source/monitor (6240A, ADCMT, Japan) was used. Copper wires and Au micro-heaters were electrically connected with silver paste (Electroconductives, D-362, Fujikura Kasei Dotite, Japan).

A simple setup was employed to characterize the micro-heaters, as shown in Figure 2(b). The temperature of micro-heaters and PDMS were observed by infrared (IR) thermo-microscopy (FSV-GX7700, Apiste). The emissivity for the IR imaging was determined to 0.86 for the PDMS [11]. It was difficult to measure the temperature both Au micro-heaters and circumferential PDMS, simultaneously because of a large difference of emissivity between Au (0.02) and PDMS (0.86). Thus, the highest temperature of the target area was measured and recorded (i.e. the temperature of circumferential PDMS adjacent to the Au micro-heaters). To investigate the transient response of heating and cooling, the power source was turned on for 100 s, then turned off with natural convective cooling. The temperature of Au micro-heaters was measured with 15 Hz of IR imaging.

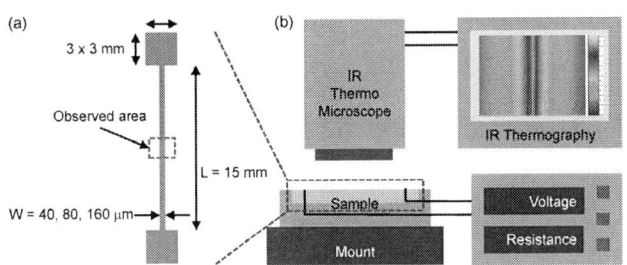

Figure 2: (a) Design of the Au micro-heaters and (b) schematic illustration of the experimental setup for measuring the temperature during Joule heating.

SIMULATION

Electro-thermo-mechanical models for steady-state as well as transient analysis were developed to investigate the temperature at certain applied voltages, spatial distribution, and transient response of temperature [12]. The simulation was conducted by finite element method using the commercial software FEMLAB® (COMSOL, version 4.3). The material properties applied to the simulation are shown in Table 1.

Table 1: Material properties for the simulation.

Materials	Thermal conductivity, k, W m^{-1}K^{-1}	Specific heat, C_p, J kg^{-1}K^{-1}	Density, ρ, kg m^{-3}
Au [13]	315	130	19320
PDMS [14]	0.18	1100	1030

RESULTS AND DISCUSSION

Temperature versus Applied Voltage

The temperature of micro-heaters was measured during the voltage applied-periods. Micro-heaters with wider width required lower voltage to be heated up compared to those with narrower width. The temperature of Au micro-heaters was proportional to the square of the applied voltage, as shown in Figure 3. The simulation results (dotted lines) well agreed with the experimental ones (solid lines).

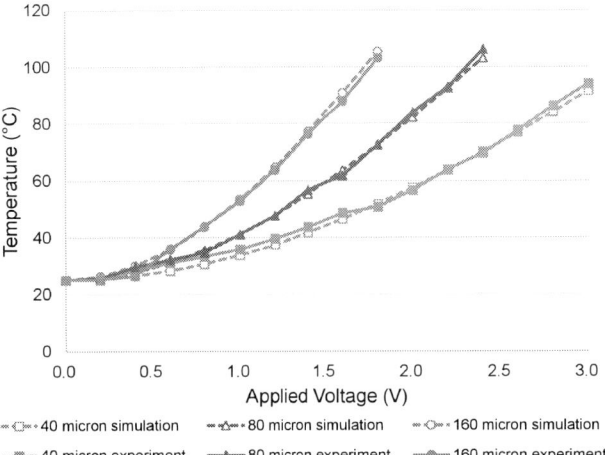

Figure 3: Experimental and simulation results of temperature of micro-heaters embedded in PDMS versus applied voltage.

Heating Efficiency

The temperature of micro-heaters was proportional to the power consumption. In order to simulate the temperature with respect to the power consumption, the power consumption was calculated based on the applied voltage and the theoretical resistance of Au micro-heaters. The resistivity (ρ) of Au micro-heaters was calculated by the equation (1) where ρ_0 is the resistivity of Au at 20 °C (23.5 nΩm), α is the temperature coefficient of resistance of Au (0.0034 K^{-1}) [15], T is the temperature of Au, and T_0 is 20 °C.

$$\rho = \rho_0(1 + \alpha(T - T_0)) \qquad (1)$$

The heating efficiency (H), the temperature heated up by the unit power, can be defined as the equation (2) where T is the temperature of micro-heaters, T_0 is the initial temperature, and P is the power consumption.

$$H = \frac{T - T_0}{P} \qquad (2)$$

The experimental and simulation results of the heating efficiency are shown in Table 2. The heating efficiency can be increased by decreasing the length of the micro-heaters. Approximately, a power of 1.3–1.5 mW was necessary to increase the 1 °C of temperature of micro-heaters. Interestingly, the heating efficiency was decreased at the micro-heaters with wider width. It can be assumed that the heat dissipation was increased as the surface area of the heater was increased. It was observed that the heating efficiencies estimated by simulation were higher than those measured by the experiment. The reason is assumed that the electrical properties of thin film Au are different with bulk state [15].

Table 2: Experimental and simulation results of heating efficiency in the temperature range from 25 °C to 100 °C when the length of micro-heaters is 15 mm.

Width (μm)	40	80	160
Experiment (K mW^{-1})	0.76	0.67	0.66
Simulation (K mW^{-1})	0.81	0.77	0.71

Resistivity and Temperature Coefficient of Resistance

From the experimental data of resistance with respect to the temperature, the empirical resistivity (ρ_{0E}) and temperature coefficient of resistance (α_E) of Au thin film (100 nm) embedded in PDMS can be calculated, as shown in Table 3. The average values are 22.9 nΩm for ρ_{0E} at 25 °C and 0.0024 K^{-1} for temperature coefficient of resistance. These values obtained from the thin film differ from the reported bulk value (23.5 nΩm for ρ_0 at 20 °C and 0.0034 K^{-1} for α), but similar to the property of the Au thin film reported previously (0.0017 K^{-1} [15] and 0.0022 K^{-1} [16] for α).

Table 3: The empirical electric properties of thin film Au (100 nm): resistivity at 25 °C and temperature coefficient of resistance.

Width (μm)	40	80	160
Resistivity (nΩm)	22.1	21.8	24.8
Temperature coefficient of resistance (K^{-1})	0.0029	0.0020	0.0023

Spatial Distribution

The spatial temperature distribution was investigated by measuring the surface temperature of Au micro-heaters and PDMS substrates, as shown in Figure 4. At the same applied voltage, Au micro-heaters with wider width consumed more electrical power, so that the temperature was increased more. It was observed that the temperature was decreased with increasing distance from the heaters. The temperature difference between the heater and the distant point about 1 mm from the heater was about 20 °C when the voltage of 1.2 V was applied to the micro-heater (width: 160 μm). At the position 0 mm, the temperature of experimental results was ~35 °C even though the temperature of peripheral PDMS was more than 60 °C that was caused by a mismatch of the emissivity between Au and PDMS. Also, the simulation of surface temperature distribution well agreed with the experimental results.

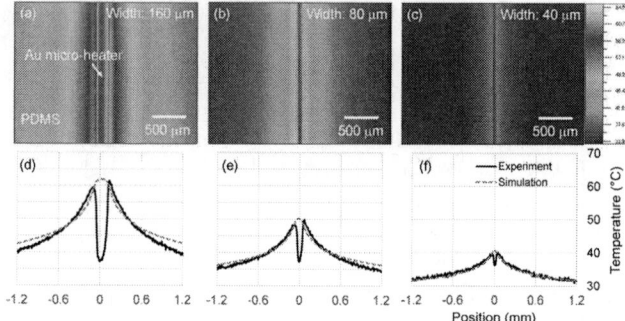

Figure 4: (a–c) IR thermo-microscopic images of Au micro-heaters embedded in PDMS during Joule heating by applying the voltage of 1.2 V, (d–f) experimental and simulation results of spatial temperature distribution.

Transient Response

The transient response of temperature rise and fall is an important characteristic to most thermal system. To investigate the transient response, a square signal of voltage was applied during the IR imaging of micro-heaters, as shown in Figure 5. When the signal (3 V) to the Au micro-heaters (width: 40 μm), the temperature was increased rapidly from the room temperature (25 °C) to 80 °C within 10 s, then slowly saturated to 90 °C. Finally, the temperature was saturated 60 s after the signal was applied. When the signal was turned off, the temperature was rapidly decreased to 35 °C within 10 s, then slowly decreased to the room temperature.

However, the simulation results showed 10 °C lower temperature compared to the experimental results (Simulation 1, applied voltage: 3 V). When the maximum temperature was set to 90 °C at the micro-heaters for the simulation (Simulation 2), the results are much similar to the experimental ones. Even though the maximum temperature was same at the 60 s after the signal on, the simulation results (Simulation 2) showed slower transient response compared to the experimental results. Furthermore, the reason of these mismatches will be investigated.

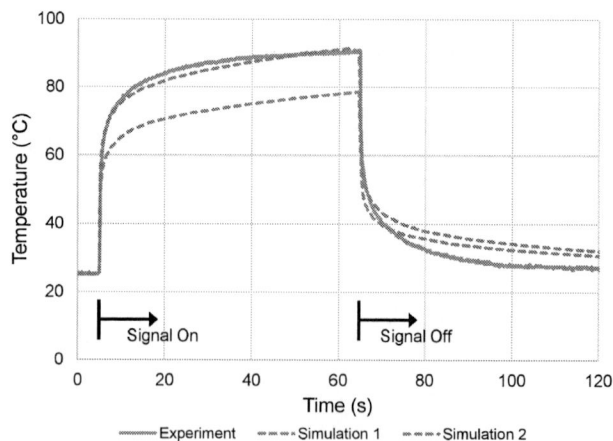

Figure 5: Experimental and simulation results of transient responses of Au micro-heaters embedded in PDMS. Signal duration was 60 s.

CONCLUSION

Micro-heaters patterned on a Si wafer were transferred and embedded in PDMS using a dry peel-off process. The fabrication method was simple including only conventional photolithography, metal deposition and etching, and surface modification through SAM. Moreover, the process did not cause any chemical swelling, contamination, or mechanical damage by a razor.

The temperature of micro-heaters depend on the power consumption, but does not depend on the width of micro-heaters. The temperature was saturated within 1 min, which is reasonable for many applications. The simulation results well agreed with the experimental results in terms of temperature at certain voltage or power consumption, surface distribution, and transient response.

As further works, peel-off transfer technique will be investigated to be applied to other polymers (e.g. polyimide or polyethylene terephthalate). This study would broaden not only the LoC, but also the fields of optical and electrical applications with low cost.

ACKNOWLEDGEMENTS

This work has been, partially, supported by the JSPS Core-to-Core Program A (Advanced Research Networks).

REFERENCES

[1] J. El-Ali, P. K. Sorger, K. F. Jensen, "Cells on Chips", *Nature*, vol. 442, pp. 403-411, 2006.

[2] J. W. Ha, A. Kundu, J. H. Hang, "Poly-dimethylsiloxane (PDMS) Based Micro-reactors for Steam Reforming of Methanol", *Fuel Process. Technol.*, vol. 91, pp. 1725-1730, 2010.

[3] T. Yamamoto, T. Nojima, T. Fujii, "PDMS-glass Hybrid Microreactor Array with Embedded Temperature Control Device. Application to Cell-free Protein Synthesis", *Lab Chip*, vol. 2, pp. 197-202, 2002.

[4] Y. S. Shin, et al., "PDMS-based Micro PCR Chip with Parylene Coating", *J. Micromech. Microeng.*, vol. 13, pp. 768-774, 2003.

[5] N. G. Wilson, T. McCreedy, "On-chip Catalysis using a Lithographically Fabricated Glass Microreactor—The Dehydration of Alcohols Using Sulfated Zirconia", *Chem. Commun.*, pp. 733-734, 2000.

[6] H. S. Chuang, S. Wereley, "Design, Fabrication and Characterization of a Conducting PDMS for Microheaters and Temperature Sensors", *J. Micromech. Microeng.*, vol. 19, pp. 045010, 2009.

[7] X. Niu, S. Peng, L. Liu, W. Wen, P. Sheng, "Characterizing and Patterning of PDMS-based Conducting Composites", *Adv. Mater.*, vol. 19, pp. 2682-2686, 2007.

[8] I. Byun, A. W. Coleman, B. Kim, "Transfer of Thin Au Films to Polydimethylsiloxane (PDMS) with Reliable Bonding Using (3-mercaptopropyl)trimethoxysilane (MPTMS) as a Molecular Adhesive", *J. Micromech. Microeng.*, vol. 23, pp. 085016, 2013.

[9] I. Byun, A. W. Coleman, B. Kim, "SAM Meets MEMS: Reliable Fabrication of Stable Au-patterns Embedded in PDMS Using Dry Peel-off Process", *Microsyst. Technol.*, 2013, DOI: 10.1007/s00542-013- 1923-8.

[10] I. Byun, A. W. Coleman, B. Kim, "Microcontact Printing Using a Flat Metal-embedded Stamp Fabricated Using a Dry Peel-off Process", *RSC Adv.*, vol. 3, pp. 24872-24876, 2013.

[11] J. L. Lin, M. H. Wu, C. Y. Kuo, K. D. Lee, Y. L. Shen, "Application of Indium Tin Oxide (ITO)-based Microheater Chip with Uniform Thermal Distribution for Perfusion Cell Culture Outside a Cell Incubator", *Biomed. Microdevices*, vol. 12, pp. 389-398, 2010.

[12] H. F. Arata, P. Low, K. Ishizuka, C. Bergaud, B. Kim, H. Noji, H. Fujita, "Temperature Distribution Measurement on Microfabricated Thermodevice for Single Biomolecular Observation using Fluorescent Dye", *Sens. Actuators B*, vol. 117, pp. 339-345, 2006.

[13] S. M. Lee, D. C. Dyer, J. W. Gardner, "Design and Optimisation of a High-temperature Silicon Micro-hotplate for Nanoporous Palladium Pellistors", *Microelectronic Journal*, vol. 34, pp. 115-126, 2003.

[14] D. Erickson, D. Sinton, D. Li, "Joule Heating and Heat Transfer in Poly(dimethylsiloxane) Microfluidic Systems", *Lab Chip*, vol. 3, pp. 141-149, 2003.

[15] F. Aviles, O. Ceh, A. I. Oliva, "Physical Properties of Au and Al Thin Films Measured by Resistive Heating", *Surf. Rev. Lett.*, vol. 12, pp. 101-106, 2005.

[16] R. B. Belser, W. H. Hicklin, "Temperature Coefficients of Resistance of Metallic Films in the Temperature Range 25° to 600°C", *J. Appl. Phys.*, vol. 30, pp. 313-322, 1959.

CONTACT

*Beomjoon Kim, tel: +81-3-5452-6224; bjoonkim@iis.u-tokyo.ac.jp

MICROPATTERNING OF BACTERIAL CELLULOSE AS DEGRADABLE SUBSTRATE FOR CELL CULTURE

Yuya Karita*, Kayoko Hirayama*, Hiroaki Onoe* and Shoji Takeuchi*
* Institute of Industrial Science, the University of Tokyo.

ABSTRACT

This paper describes microfabrication of bacterial cellulose membrane. Bacterial cellulose is a nanofibrous cellulosic material produced by the bacteria called *Acetobacter xylinum*. We micropatterned a bacterial cellulose membrane by utilizing MEMS process. This patterned bacterial cellulose worked as a scaffold for mouse embryonic fibroblast cells: The cells attached and grew on the patterned bacterial cellulose membrane. Moreover, formation of cell cluster was observed by the treatment of cellulose degrading enzyme. We believe that this micropatterned cellulose membrane would be useful as degradable microscaffolds for cell culture.

INTRODUCTION

Bacterial cellulose is a bacteria-derived material, known to work as scaffolds for various types of cells [1]. Cellulose has high stability in vivo because it is not degradable or biosynthesized by mammalian cells. Moreover, as the notable advantage, it is degradable by enzyme called cellulase and its degradation process does not change any culturing condition such as temperature or pH. Thus bacterial cellulose is promising as degradable scaffolds for cell.

The shape of bacterial cellulose membrane is usually regulated by harvesting the bacteria inside molds [2]. However, microscale regulation of the shape and size has not yet been accomplished. Therefore, bacterial cellulose has been used mostly in top-down tissue engineering, which uses millimeter-scale block of bacterial cellulose as a scaffold for constructing three-dimensional tissue. For applying to bottom-up process, microfabrication of bacterial cellulose has been necessary.

Here, we demonstrate micropatterning of 500-nm-thick bacterial cellulose membrane by utilizing photolithography process. The micropatterned bacterial cellulose finely works as scaffolds for mammalian cells.

METHOD

Preparation of bacterial cellulose membrane

Bacterial cellulose membrane was formed at air-liquid interface of culture medium (Hestrin-Schramm medium, pH6.0) by *Acetobacter xylinum subsp. sucrofermentas* BPR2001 provided by the RIKEN BRC through the National Bio-Resource Project of the MEXT, Japan. After 24 hours of cultivation, the membrane was sterilized and purified by autoclaving (120 °C, 1 bar, 20 min) and NaOH treatment (1 M, 75 °C, 3 hours). Then, the membrane was washed by deionized water and dried out (65 °C, 2 hour) to fix it on glass plates (Fig 1).

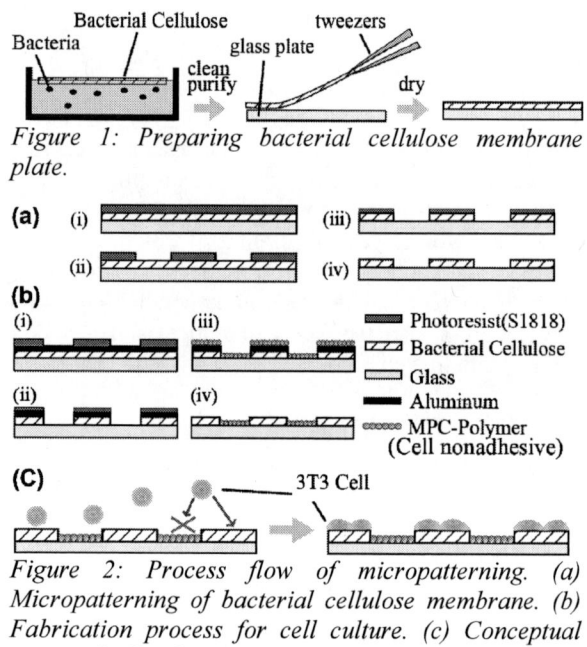

Figure 1: Preparing bacterial cellulose membrane plate.

Figure 2: Process flow of micropatterning. (a) Micropatterning of bacterial cellulose membrane. (b) Fabrication process for cell culture. (c) Conceptual image of cell culturing.

Microfabrication of bacterial cellulose membrane

After preparation of the bacterial cellulose membrane attached on the plate, photoresist (S1818, Shiply) was spin-coated and photolithographically patterned on the bacterial cellulose membrane by using a maskaligner and etched by oxygen (O_2) plasma (5 min, O_2 flow rate 10 mL/min, power 50 W, FA-1, Samco). The remained photoresist was removed by acetone treatment (Fig 2a).

The process was modified to fabricate bacterial cellulose micoplates for cell culturing (Fig 2b). In order to pattern cell non-adhesive polymer only on the glass surface, alminum was evaporated on a bacterial cellulose membrane, and S1818 photoresist was spin-coated and patterned on the aluminum layer. Then, the aluminum layer was etched by aluminum etchant through the patterned photoresist, and exposed bacterial cellulose membrane was etched by O_2 plasma (10 min, O_2 flow rate 20 mL/min, power 50 W). Cell non-adhesive polymer (MPC-polymer) was coated on the whole surface, and the aluminum layer was dissolved to liftoff the MPC-polymer over the cellulose membrane.

Evaluation of microfabrication

To evaluate etching rate, the height of bacterial cellulose membrane was measured by stylus profilemeter after O_2 plasma treatment (O_2 flow rate 10 mL/min, power 50 W) for 30 seconds, 1 minute, and 4 minutes, respectively.

The micropatterns of bacterial cellulose with various

width and gaps (5, 10, 25, 50, 75, 100 μm) were prepared to evaluate the patterning quality of our process. The width and gaps of the patterns were measured by 3D laser scanning microscopy, and compared with designed width.

Cell Culture

Mouse embryonic fibroblast cells (NIH3T3 cell) were seeded on micropatterned bacterial cellulose at the concentration of 5.0×10^5 cells/ml. NIH3T3 cells were cultured for 6 days under the condition of 37 °C and 5% CO_2 atmosphere. After 6 days of culturing, cellulose degrading enzyme was added to the culture medium at the concentration of 4 mg/ml to degrade bacterial cellulose micropatterns.

RESULTS

Evaluation of microfabrication

The data (Fig 3) showed that a 500 nm-thick bacterial cellulose membrane was completely etched within 4 minutes. Since the etching rate of S1818 was about 0.6 μm/min in our experimental condition (data not shown), 3-μm-thick S1818 etching mask was considered to be enough for protecting the cellulose membrane against O_2 plasma treatment.

Fig.4 shows the patterning property of our process. The difference between the actual width of pattern and designed width (5, 10, 25, 50, 75, 100 μm) were 0.78, 0.58, -0.70, 0.29, -0.19 μm, respectively (Fig.4a). The actual gap width of patterns differed from the designed gap width by -1.15, -0.75, -1.14, -1.68, -1.90 μm, respectively (Fig.4b). Thus the micropatterning of cellulose membrane was accomplished with a high degree of accuracy through our process.

Cell culture

NIH3T3 cells successfully attached to the micropatterned bacterial cellulose microplates and covered the surface of the cellulose within 6 days along the patterned shape (Fig 5ab). In an hour after the enzyme addition, the cells were partially detached from the microplates, and cell clusters were formed within 16 hours (Fig 5d).

CONCLUSION

The micropatterned bacterial cellulose membrane was successfully fabricated with a high degree of accuracy and suggested to work as degradable scaffold for cell. We believe that this bacterial cellulose microplate could be useful in tissue engineering.

ACKNOWLEDGEMENTS

This work was partly supported by Grant-in-Aid for Scientific Research on Innovative Areas "Bio Assembler" (23106008) from the Ministry of Education, Culture, Sports, Science and Technology of Japan. The authors gratefully acknowledge Prof. Kazuhiko Ishihara at the University of Tokyo for providing the MPC-polymer.

REFERENCES

[1] Helenius G, et al., *Biomed Mater Res A,* pp.431-438, 2006
[2] Putra A, et al., *Poymer Jounal,* pp.137-142, 2008

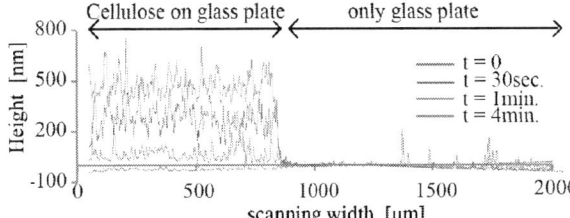

Figure 3: Etching rate of bacterial cellulose membrane by O_2 plasma.

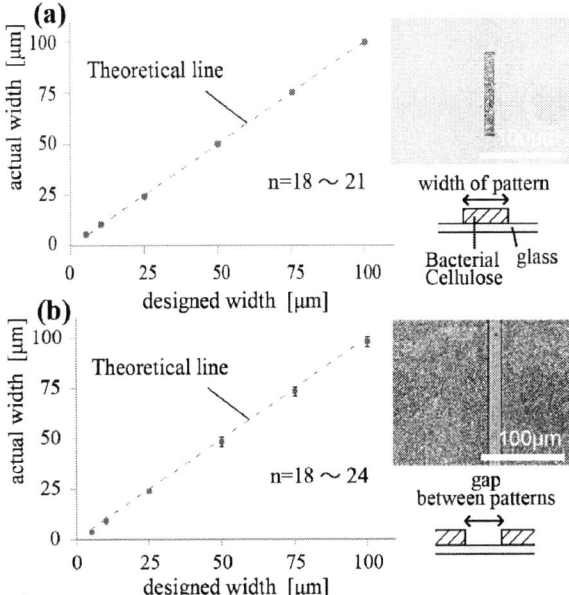

Figure 4: Patterning quality of bacterial cellulose membrane. The theoretical line (designed width = actual width) is shown as dashed line. At some points, an error bar is hidden within the marker. Images beside graph were taken by 3D laser scanning microscopy.

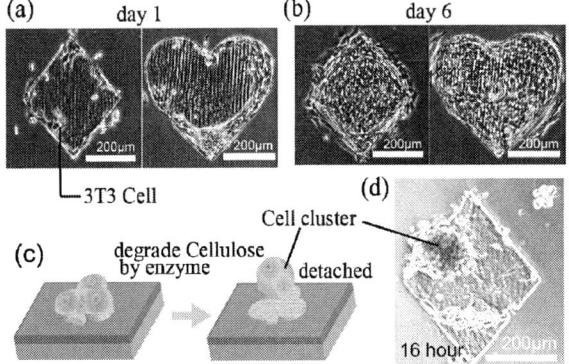

Figure 5: Cell culturing and degradation of bacterial cellulose. (a)(b)(d) Images taken by phase contrast microscope.

CONTACT

Prof. Shoji Takeuchi, takeuchi@iis.u-tokyo.ac.jp Tel: +81-3-5452-6650; Fax: +81-3-5452-6649.

MICROSCALE MAGNETIC PATTERNING OF HARD MAGNETIC FILMS USING MICROFABRICATED MAGNETIZING MASKS

A. Garraud[1], O.D. Oniku[1], W.C. Patterson[1], E. Shorman[1], D. Le Roy[2], N.M. Dempsey[2] and D.P. Arnold[1]

[1]University of Florida, Gainesville, Florida, USA
[2]Univ. Grenoble-Alpes/CNRS, Institut NEEL, Grenoble, FRANCE

ABSTRACT

We present a batch-fabrication process to imprint microscale magnetic pole patterns (perpendicular north/south poles) into hard magnetic films using field-shaping, soft magnetic "magnetizing masks". Using 7-μm-thick, electroplated Fe–Co magnetizing masks, magnetic stripes with widths down to 50 μm have been imprinted into both 15-μm-thick Co–Pt films and 5-μm-thick Nd–Fe–B films. These patterned films exhibit a sinusoidal stray magnetic field pattern with ~4 and ~7 mT_{pk-pk} variations and corresponding field gradients of 80 and 140 T/m, respectively. We also demonstrate the ability to transfer more complex patterns by showing magnetization of various geometric shapes.

INTRODUCTION

Magnetic microactuators/motors, vibrational energy harvesters, lab-on-a-chip devices, electromagnetic devices, and other microsystem applications have inspired growing interest in permanent magnet structures with size dimensions in the range of micrometers to hundreds of micrometers. With significant recent advancements in hard (permanent) magnetic thick film fabrication [1, 2], the next technological step is to develop methods to enable complex, multi-directional magnetic poles (magnetization patterns) in hard magnetic films. A common objective is to obtain large peak field intensities with fine periodic spacing, which also correlates with high spatial field gradients.

Previous studies have investigated the creation of multipole patterns in hard magnetic films [3-5] using electrical conductors or a sandwich of soft magnetizing heads to imprint sub-mm period magnetic north/south pole arrays. Moreover, thermomagnetic patterning has been used to realize patterns with lateral dimensions down to ~70 μm but only in the relative surface of the layer (1-μm deep) [6, 7]. Recently, a technique based on the use of a single laser-machined soft magnetic head to selectively reverse the magnetization direction in a hard magnetic layer was developed [8, 9]. Pole patterns down to 30 μm wide were successfully transferred in 15-μm-thick electroplated Co-rich Co–Pt films [9].

Building on these efforts, in this paper, we present the development of micro-machined soft magnetizing heads, enabling high-fidelity microscale magnetic features in contiguous hard magnetic films. Using 7-μm-thick electroplated Fe–Co microstructures as field-shaping magnetizing structures, we demonstrate the selective magnetization of north/south pole patterns on two different multi-micron-thick permanent magnet films: electroplated 15-μm-thick Co-rich Co–Pt ($Co_{80}Pt_{20}$) and 5-μm-thick sputtered Nd–Fe–B ($Nd_2Fe_{14}B$). The resulting magnetic pole patterns are investigated by both qualitative and quantitative measurements. Finite element simulations are performed to present the expected fields and compare with the experimental results.

CHARACTERIZATION METHODS

Magnetic Hysteresis Loops

The magnetic properties of each sample are measured using a vibrating sample magnetometer (VSM), ADE Technologies EV9, with a maximum applied field of ±1,800 kA/m. The two permanent magnet materials—Co-rich Co–Pt and Nd–Fe–B films—exhibit a preferred out-of-plane (perpendicular) magnetic hysteresis. In the case of the electroplated Fe–Co films, the in-plane magnetic properties are measured to reduce the influence of demagnetizing field.

Magnetic Array Characterization

Due to the small spatial size of the magnetic field patterns of the resulting magnetic arrays, characterization of the magnetic fields presents significant challenges. We employ two methods for magnetic measurements over the magnet array surface: a magneto-optical imaging technique and a scanning Hall probe microscope (SHPM).

Magneto-optical indicator films (MOIF) are used to provide images of the stray field pattern at the magnet array surface. A MOIF functions on the principle of the Faraday effect, a rotation of the polarization of light in proportion to the magnetic field [10]. The MOIFs used here are made of bismuth-substituted yttrium iron garnet (Bi:YIG) for its high Faraday rotation coefficient, and respond to the z-component of the stray field. Two types are used: "uniaxial MOIF," which provides binary indication of the pole direction, and "planar MOIF" which provides grayscale information for quantitative assessment of the magnetic field.

As a complement to the MOIF, the SHPM system raster scans a Hall sensor having a 1 μm x 1 μm active area over the patterned films to provide a detailed measurement of the magnetic stray field perpendicular to the surface (z-direction). Due to the magnetic period dimensions, the sensor must be in close proximity (< 20 μm) to the surface of the magnetic film, so it is mounted on a dithered piezo tuning fork, which is used to infer contact with the surface.

FABRICATION METHODS

Hard Magnetic Layers

The Co-rich Co-Pt layers are fabricated on silicon

substrates using the method reported in [11]. First, a (110) Si substrate is sputtered with a 100-nm-thick Cu(111) seed layer. Electrodeposition is carried out in a galvanostatic cell at 65 °C with a current density of 70 mA/cm^2 and a pH of 9.0. A 1-cm^2 sample is plated for one hour to obtain a thickness of 15 μm. Under the right deposition conditions, these films exhibit a strong magnetocrystalline anisotropy in the as-deposited state, without any post-deposition temperature treatment.

The Nd–Fe–B films are sputtered to a thickness of 5 μm at 400 °C onto silicon (100) substrates using high rate triode sputtering [12]. Films of 100-nm-thick tantalum are used as adhesion buffer and capping layers. A post-deposition annealing treatment at 750 °C for 10 min is needed to crystallize the hard magnetic phase, so as to achieve high values of coercivity.

The magnetic hysteresis curves of these two films—measured with the VSM—are shown in Figure 1 for a 15-μm-thick electroplated Co$_{80}$Pt$_{20}$ and a 5-μm-thick sputtered Nd–Fe–B film. They present a coercivity of 340 kA/m and 840 kA/m and a remanence of 0.55 T and 1.3 T, respectively.

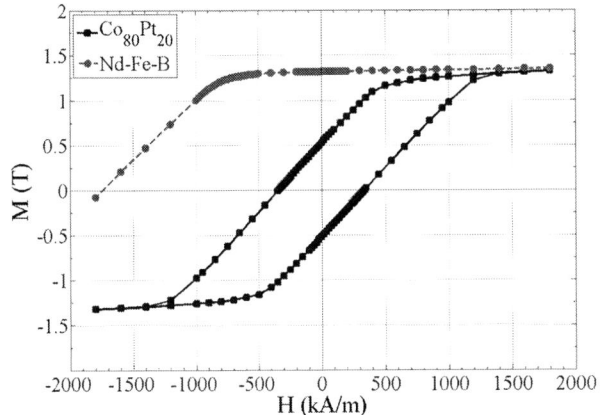

Figure 1: Magnetic hysteresis curves of the Co–Pt and Nd–Fe–B permanent magnetic films (maximum field applied = ±1,800 kA/m).

Soft Magnetizing Head

The fabrication process of the Fe–Co magnetizing mask is presented in Figure 2. A seed layer of Ti/Cu is sputtered onto a Si(100) substrate, followed by photolithography to define SU-8 plating molds using the methods in [13].

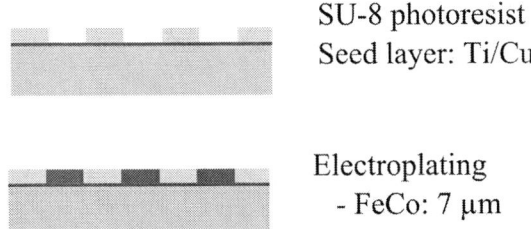

Figure 2: Fabrication of an electroplated Fe–Co magnetizing mask.

The electrodeposition is made for 45 min to obtain a 7-μm-thick Fe–Co layer. Figure 3 presents the magnetic hysteresis loop, showing the desirable soft magnetic properties of high magnetic permeability (μ$_r$ ~ 885) and saturation (~2.4 T). Figure 4 shows an example pattern of 60-μm-wide Fe–Co stripes with 40-μm-wide spaces (100-μm period). A variety of other stripe patterns and arbitrary geometric shapes are made concurrently.

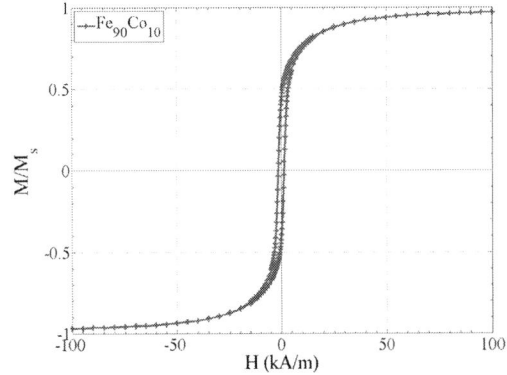

Figure 3: Magnetic hysteresis loop of the Fe–Co electrodeposited layer used for the magnetizing head.

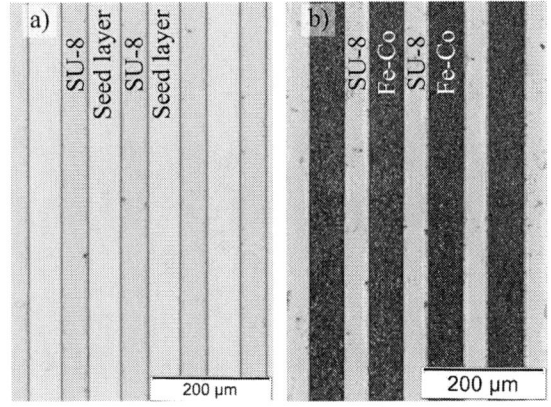

Figure 4: a) SU-8 mold for the electrodeposition; b) Resulting Fe-Co magnetizing mask.

Selective Magnetization Process

The selective magnetization is summarized in Figure 5. The initial step consists in the uniform magnetization of the hard magnetic layer (Co–Pt or Nd–Fe–B films, in this case) by applying a large pulsed magnetic field (3 T or 7 T, respectively) from a capacitive discharge pulse magnetizer. The microfabricated Fe–Co magnetizing mask substrate is then brought into contact with the hard magnetic layer, and a pulsed magnetic field is applied in the reverse direction (-0.5 T or -1.5 T, respectively). Owing to the difference in the relative permeability of Fe–Co and SU-8, the pulsed magnetic flux is concentrated in the regions with Fe–Co, allowing the magnetization to be reversed in the contacting areas. The magnetizing mask substrate is then removed, leaving the magnetic pattern in the permanent magnet film.

978-1-4799-3510-9/14 $31.00 © 2014 IEEE

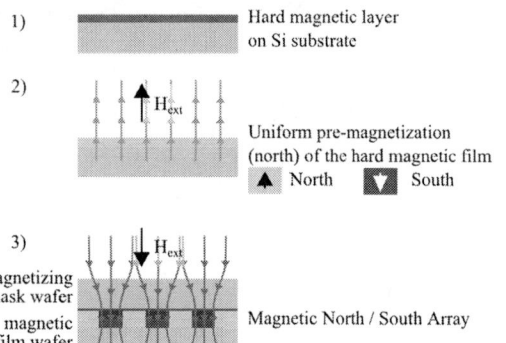

Figure 5: Microscale magnetic patterning process.

RESULTS

After selectively magnetizing the Co–Pt and Nd–Fe–B films with different patterns, the MOIF and SHPM methods are used to measure the resulting stray fields.

Magneto-Optical Characterization

The uniaxial MOIF is used to reveal the magnetic pole patterns (north or south) impressed in the films. Figure 6a shows the Nd–Fe–B film patterned with ellipses with a 500-µm magnetic period. Figure 6b and 6c show the stripe patterns on the Co-rich Co–Pt film and the Nd–Fe–B film, respectively. It is noted that the impressed patterns (50-µm-wide lines/spaces) have different dimensions than the (60/40-µm lines/spaces) magnetizing mask due to fringing field effects [9].

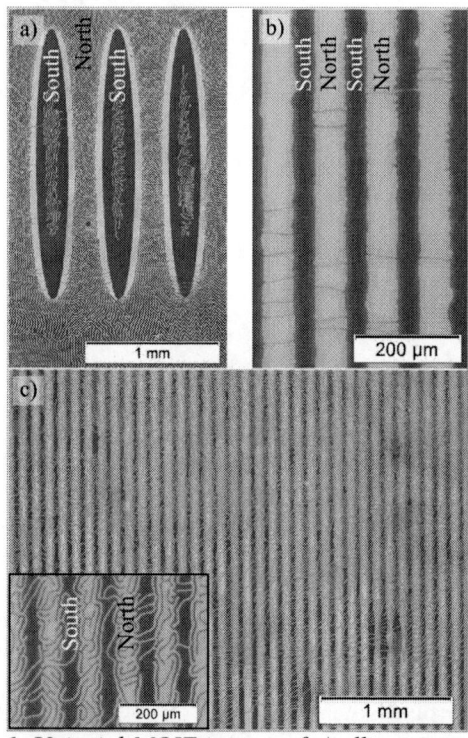

Figure 6: Uniaxial MOIF images of a) ellipses patterned in a 5-µm-thick Nd–Fe–B film; b) 50-µm-wide stripes in a 15-µm-thick Co–Pt film; and c) 50-µm-wide stripes in a 5-µm-thick Nd–Fe–B film. The squiggly "labyrinth pattern" is a MOIF artifact, and not from the imprinted magnetic pattern.

Measurements with the planar MOIF enable quantitative estimations of the magnetic field strength. Figure 7 presents the stray magnetic field over a 2-mm-long window for Co–Pt and Nd–Fe–B stripe patterns. The peak-to-peak variation at the surface of the patterned Nd–Fe–B film is close to 7 mT, while it is close to 4 mT for the Co–Pt film. The lower stray magnetic field is explained by the inferior magnetic properties of the Co–Pt compared to Nd–Fe–B.

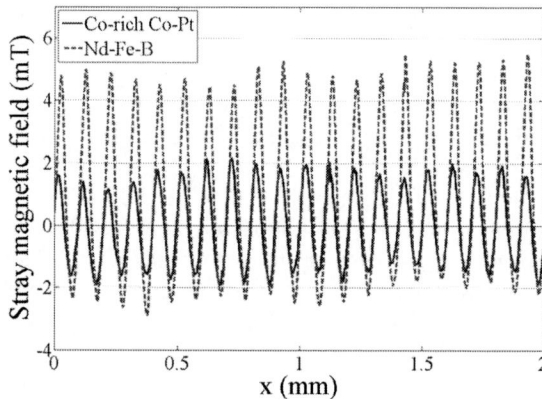

Figure 7: Experimental magnetic flux density (estimated from magneto-optical measurements) from the Co–Pt layer shown in Figure 6b and the Nd–Fe–B layer shown in Figure 6c.

Scanning Hall Probe Characterization

Figure 8 shows SHPM measurements at a height of 20 µm above the surface of the Nd–Fe–B film. The respective measurements on the Co–Pt film were not possible, due to the resolution of the SHPM. The peak-to-peak variation is close to 1.5 mT. Note that the DC offset value is also due to the SHPM setup. It shows that the stray field has the expected magnetic period (100 µm), in agreement with the MOIF measurements. The quick decay in the peak-to-peak variation is due to the thickness of the film (5 µm).

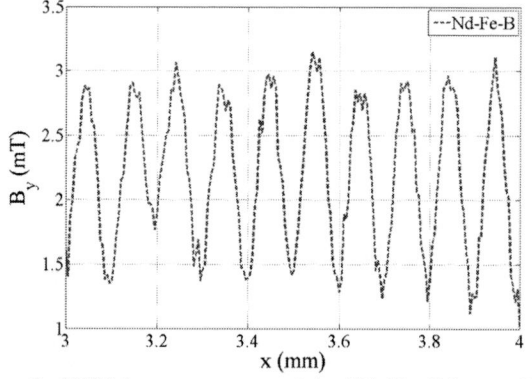

Figure 8: SHPM measurement of the Nd–Fe–B layer, 20 µm above the surface.

Numerical results

A finite-element simulation (COMSOL Multiphysics)

is made to predict the expected magnetic flux density from the permanent magnetic layer, when considering the actual magnetic properties (Figure 1). Figure 9 shows the stray magnetic field, 20 µm above the surface. A peak-to-peak variation of ~50 mT and ~105 mT is found for the Co–Pt and the Nd–Fe–B film, respectively. The two orders of magnitude difference between the experimental measurement and the numerical predictions could be linked to the fact that we do not fully saturate the layer in each magnetic direction.

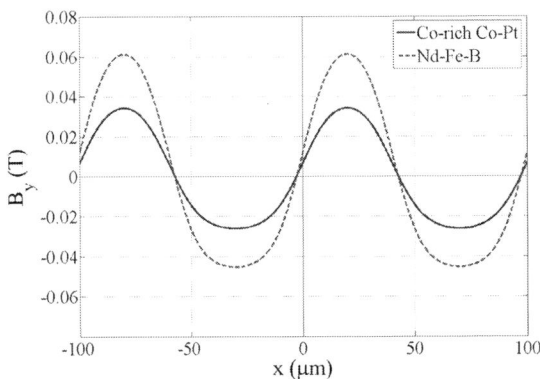

Figure 9: Simulation of the stray magnetic field from the Co–Pt and the Nd–Fe–B films, 20 µm above the surface.

CONCLUSIONS

A batch-fabrication method has been presented to imprint complex, fine-scale pole patterns (down to 50 µm) in microfabricated hard magnetic layers from an electroplated 7-µm-thick Fe–Co magnetizing mask. Depending on the coercivity of the permanent magnet film, the pulse reversal field has to be chosen carefully to imprint a magnetic north/south pole pattern. The stray magnetic field characterization shows a peak-to-peak variation of 7 mT and 4 mT at the surface of the Nd–Fe–B and the Co-rich Co–Pt respectively. The difference with the simulation results is likely linked to the inability to fully magnetize the films in each magnetic direction during the selective magnetization process.

In the future, smaller periods, down to 5 µm, will be micro-fabricated on the electroplated Fe–Co magnetizing mask to imprint magnetic periods in this range.

ACKNOWLEDGEMENTS

This work is supported by DARPA Grant #N66001-11-1-4198. The authors would also like to thank the technical staff of UF Nanoscale Research Facility for their assistance during the microfabrication process.

REFERENCES

[1] O. Gutfleisch, and N. M. Dempsey, "High Performance µ-Magnets for Microelectromechanical Systems (MEMS)." In *Magnetic Nanostructures in Modern Technology*. Springer, 2008, pp. 167-194.

[2] D. P. Arnold and N. Wang, "Permanent Magnets for MEMS," *Journal of Microelectromechanical Systems*, vol. 18, pp. 1255-1266, 2009.

[3] J. Topfer and V. Christoph, "Multi-pole Magnetization of NdFeB Sintered Magnets and Thick Films for Magnetic Micro-actuators," *Sensors and Actuators A-Physical*, vol. 113, pp. 257-263, 2004.

[4] I. Zana, F. Herrault, D. P. Arnold, and M. G. Allen, "Magnetic Patterning of Permanent-magnet Rotors for Microscale Motor/generators," *Proc. 5th Int. Workshop on Micro and Nanotechnology for PowerGeneration and Energy Conversion Apps* (PowerMEMS 2005), 2005, pp. 116-119.

[5] F. M. Rhen, E. Backen, and J. M. D. Coey, "Thick-film Permanent Magnets by Membrane Electrodeposition," *Journal of Applied Physics,* vol. 97, pp. 113908-113908-4, 2005.

[6] F. Dumas-Bouchiat et al., "Thermomagnetically patterned micromagnets," *Applied Physics Letters*, vol. 96, pp. 102511, 2010

[7] L. F. Zanini, O. Osman, M. Frenea-Robin, N. Haddour, N.M. Dempsey, G. Reyne, and F. Dumas-Bouchiat, "Micromagnet structures for magnetic positioning and alignment," *Journal of Applied Physics*, vol. 111, pp. 07B312, 2012

[8] O. D. Oniku, R. Regojo, Z. A. Kaufman, W. C. Patterson, and D. P. Arnold, "Batch Patterning of Submillimeter Features in Hard Magnetic Films Using Pulsed Magnetic Fields and Soft Magnetizing Heads," *Magnetics, IEEE Transactions on,* vol. 49, pp. 4116-4119, 2013.

[9] O. D. Oniku, P. V. Ryiz, A. Garraud, and D. P. Arnold, "Imprinting of Fine-scale Magnetic Patterns in Electroplated Hard Magnetic Films using Magnetic Foil Masks," *Journal of Applied Physics,* vol. In Press, 2014.

[10] R. Grechishkin, S. Chigirinsky, M. Gusev, O. Cugat, and N. M. Dempsey, "Magnetic imaging films," in *Magnetic Nanostructures in Modern Technology*, Springer, 2008, pp. 195-224.

[11] O. D. Oniku and D. P. Arnold, "Microfabrication of High-Performance Thick $Co_{80}Pt_{20}$ Permanent Magnets for Microsystems Applications," *ECS Transactions,* vol. 50, pp. 167-174, 2013.

[12] N. M. Dempsey, A. Walther, F. May, D. Givord, K. Khlopkov, and O. Gutfleisch, "High Performance Hard Magnetic NdFeB Thick Films for Integration into Micro-electro-mechanical Systems," *Applied Physics Letters,* vol. 90, pp. 092509 - 092509-3, 2007.

[13] W. C. Patterson and D. P. Arnold, "Evaluation of the Effects of Electroplating Conditions on the Material Properties of Iron Cobalt Thick Films Using Design of Experiments," *ECS Transactions,* vol. 50, pp. 133-139, 2013.

CONTACT

*A. Garraud, tel: +1-352-3289602; agarraud@ufl.edu

MONOLITHIC PIEZOELECTRIC IN-PLANE MOTION STAGE WITH LOW CROSS-AXIS-COUPLING

Sachin Nadig, Serhan Ardanuç, and Amit Lal

*Sonic*MEMS Laboratory, School of Electrical and Computer Engineering

Cornell University, Ithaca, NY, USA

ABSTRACT

We present a rotary dither stage that can provide rotation stimulus to objects mounted on it. The stage is in planar form-factor allowing compatibility with planar packages that can house both the stage and the inertial sensor. We used laser micromachining of bulk PZT-4 plates to form PZT beams to achieve monolithic integration of lateral actuators and flexures. This process enables high-aspect ratio PZT beams (500μm thick, 150μm wide) resulting in high out-of-plane stiffness, helping in reducing the out-of-plane motion to parts-per-thousand of the in-plane motion. The micro stage technology opens up a new design space of 100-micron scale lateral bimorphs for mm-scale stages with large motion. A dither stage was designed and fabricated that can achieve in plane dither of ~ 1.2 millidegree/V and dither rates up to 1800 degree/s. Low dither rates of 20 – 60 millidegree/s are demonstrated and measured.

INTRODUCTION

Precision motion stages are needed for many positioning applications [1] such as scanning probe microscopy, micro-optical parts, etc. Motion stages, if made small enough to be located inside an inertial sensor package, can also be used for calibrating inertial sensors by providing mechanical stimulus to stage-mounted inertial sensors. A rotational or dither stage to be used for micro-inertial sensor calibration in a compact manner should be planar, have high angular dynamic range, low power consumption, and minimal cross axis coupling as listed in Figure 1.

Previous attempts at implementing micro-motion stages with micrometer to millimeter length scales have used thermal or electrostatic actuators [2]. Conventional micro electrostatic actuators such as comb drives yield smaller forces. Additionally, electrostatic stages do not have sufficient out-of-plane rigidity, leading to considerable out-plane motion for in-plane actuation. For thermal actuators, the forces are considerably higher, but the high temperatures and power consumption can make systems impractical for portable applications. Piezoelectric actuators can generate high forces and support high rigidity structures for a given displacement. Since the actuator is capacitive in nature, one can recover the charge required to deform the structure in push-pull configurations. Amongst the many piezoelectric materials, PZT (Lead Zirconate Titanate) is known to have a very high electromechanical coupling (k_t) and piezoelectric strain coefficient (d_{31}). A common method of fabricating macro-scale PZT actuators such as bimorphs and unimorphs is to bond multiple layers of piezoelectric and elastic materials [4] in large sheets and then cut into millimeter to centimeter scale structures. However, this approach is not optimal for micro-scale PZT actuators as it suffers from roughness, de-lamination, non-uniformity, and issues such as stress mismatch between layers, which can lead to deformed structures post-fabrication [5]. In contrast, micro fabrication techniques for piezoelectric actuators enable actuators only several microns in thickness achieving comparatively smaller forces. Furthermore, micro-piezoelectric actuators often fail to match the piezoelectric properties and fatigue resistance of the bulk PZT materials [5]. Hence, the ability to realize bulk-PZT devices, but at the micron scale is desired. Between the bulk-PZT macro-scale actuators realized by bonding, and thin film actuators, the domain of mm-thick but 100-micron wide actuators has not been explored for actuation stages. In medical ultrasonic transducers, PZT plates are cut in two-dimensional pillars using high-speed diamond saws, but do not provide the design complexity necessary for micro-actuation and micro-motion stages. Compared to the existing techniques for processing bulk PZT, such as water-jet drilling, ultrasonic drilling, electrical discharge machining, and micro-CNC technologies [6] the laser micromachining process presented here has several advantages. In our process of micro machining bulk PZT, no additional deposition is necessary; it is a subtractive process and by using direct-write micro patterning technique, precision removal of PZT and/or electrode material, optionally with double-side (top-bottom) alignment capability can be achieved. Actuators formed monolithically by patterning top and bottom electrodes in addition to cutting through PZT can provide both in-plane and out-of-plane motion. By properly designing the electrodes and the through-cut patterns, one can achieve a wide variety of movable stages. This work focuses on having actuators with constrained in-plane motion. We have previously demonstrated that with dual side electrode patterning, the displacements are nearly doubled for the same applied voltages due to the increase in net applied electric

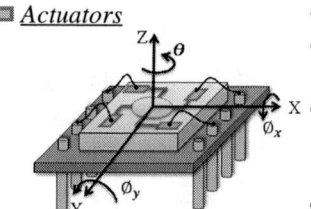

- Planar
- Low power consumption
- High angular speed dynamic range, ≈2000°/s
- Minimize cross axis coupling

$$minimize \left\{ \frac{u_z(V)}{\theta(V)}, \frac{u_x(V)}{\theta(V)}, \frac{u_y(V)}{\theta(V)}, \frac{\emptyset_x(V)}{\theta(V)}, \frac{\emptyset_y(V)}{\theta(V)} \right\}$$

Figure 1: Rotary stage design constraints. Here, u is the displacement vector and V is the applied voltage.

fields [7]. This may prove useful in reducing the operating voltage requirements of the system. PZT-4 plates with nickel as metal layer on both top and bottom side were used to fabricate the dither stages. PZT-4 was chosen due to its high strain coefficients and Curie temperature (325 °C) facilitating laser cutting without depolarization.

In order to achieve the stage constraints listed in Figure 1, one approach is to monolithically integrate planar flexures and in plane lateral actuators that have linear actuation as a function of applied voltages [7]. Several of these actuators can be configured to get a displacement amplifier, where the motion of the actuators is coupled to a central rotor carrying the payload (inertial sensor) via planar, hinge-like mechanisms. This yields in-plane dither, whose speed is a function of applied ramp voltage input and has multiple degrees of freedom owing to the several control electrodes.

DESIGN AND FABRICATION

The lateral bimorph [8] is formed by two-sets of electrodes on the top and bottom of the PZT beam. Voltages are applied such that the fields on the two sets result in opposing stresses creating a bimorph action in-plane of the PZT plate (Figure 2). In our designs, E_{air} is substantially less than E_{PZT} due to the much higher PZT dielectric constant.

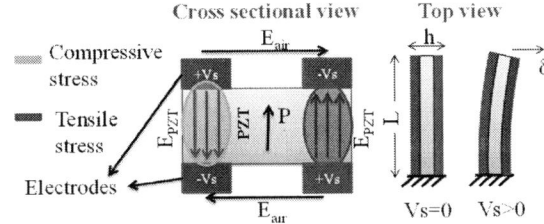

Figure 2: Bulk PZT lateral bimorph design for in-plane actuation. P and E represent polarization direction of PZT and the electric field, respectively.

In the laser micromachining process the laser cuts bulk PZT by repetitive scanning (hatching) of laser beam to define the structures as well as top/bottom electrodes using a commercial instrument with optical alignment capabilities (LPKF ProtoLaser-U). The variables optimized for PZT cutting include laser power, frequency of scan, hatch-density, and z-offset of the 3D stage of the instrument maneuvering the PZT sample. The process steps are detailed in [7].

The analytical model for tip displacement of the lateral multi-morph actuator with N layers is given by equations 1 and 2 below [8].

$$\kappa = \left[d_{31} E_{PZT} V_{in} \sum_{k=1}^{N} k + \frac{\beta - 1}{2} \right]$$

$$\times \left[h^2 \left\{ E_{PZT} \sum_{k=1}^{N} \left[\left(\frac{1-\beta}{12} \right)^3 + (1-\beta)\left(k + \frac{\beta-1}{2} \right)^2 \right] + E_{struc} \sum_{k=1}^{N} \left[\frac{\beta^3}{12} + \beta\left(k - 1 + \frac{\beta}{2} \right)^2 \right] \right\} \right]^{-1} \quad (1)$$

$$\delta = \frac{1}{2} \kappa L^2 \quad (2)$$

Here, κ is the curvature, E_{PZT} and E_{struc} are the Young's modulus of the PZT layer and the elastic layer, respectively, in the composite bimorph actuator, β is the ratio of the width of the structural layer to overall width, δ is the tip displacement, L is the actuator length, h is the overall width of the beam, and $N = 1$. In our case $E_{PZT} = E_{struc}$, as the middle layer in the actuator is PZT, β is 1/3, and h is 0.45mm. These parameters are illustrated in Figure 2. The dither of the central rotor is facilitated by spiral shaped displacement amplifiers with lateral in-plane bimorphs connected such that they improve the total actuation of the spiral (Figure 3). However, an out-of-plane motion is also generated due to the shear stress along the thickness of the beams. In order to minimize this undesired motion, S-shaped springs were designed to relieve the shear stress, while transmitting the displacement to the central rotor. These beams behave as planar hinges and facilitate in-plane rotation while maintaining the stiffness to cross-axis motion during the desired in-plane dither. One of the consequences of the S-Shaped springs is to reduce system rigidity and lower the resonance frequency of the stage.

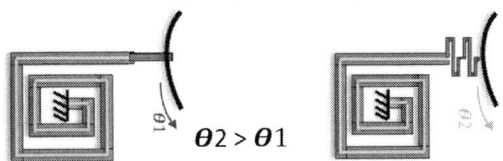

Figure 3: Comparison of two different coupling cases show that S-springs (right) enable better transmission.

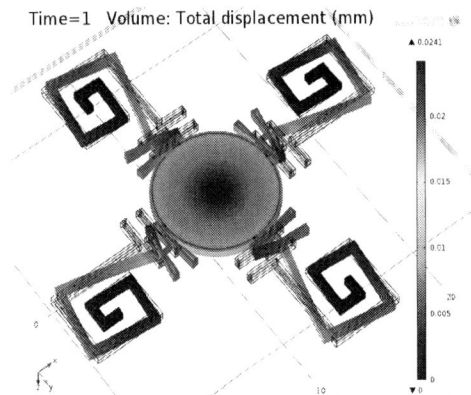

Figure 4: COMSOL simulation of the dual side electrode patterned PZT dither stage yielding ~16 μm of in plane rotation for 100 V_{dc}, which corresponds to ~0.24 degree in-plane dither.

A COMSOL model to simulate in-plane actuation is shown in Figure 4. Simulations indicate 16 μm of in-plane actuation of the rotor for 100V DC with dual side electrode patterning, and ~8 μm of in-plane actuation for single side electrode patterning. These displacements correspond to ~0.24 degrees and ~0.12 degrees, or ~2.4 millidegree/V and 1.2 millidegree/V for dual and single-side electrode designs of the rotor, respectively.

The fabricated device is shown in Figure 5 with SEM images in the insets focusing on side walls a certain regions. The stage is confined to a 25x25 mm PZT-4 plate. The

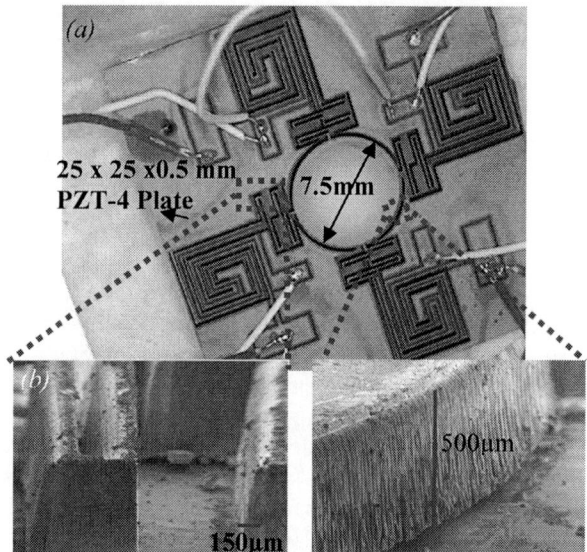

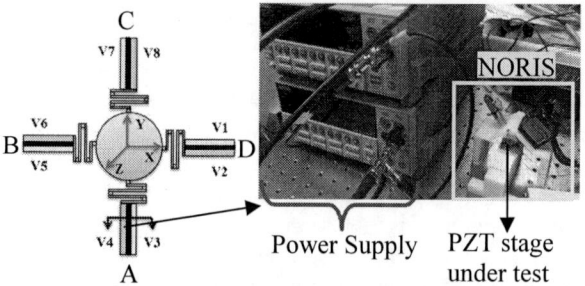

Motion	V1	V2	V3	V4	V5	V6	V7	V8
X	-V	-V	+V	-V	+V	+V	-V	+V
Y	+V	-V	+V	+V	-V	+V	-V	-V
Rotation about Z	+V	-V	+V	-V	+V	-V	+V	-V

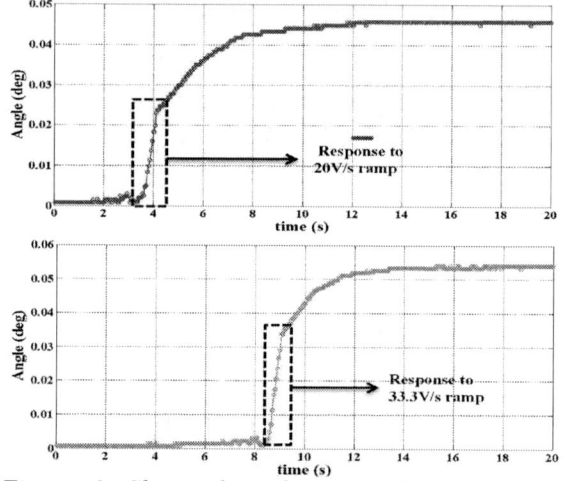

Figure 6: Shows the schematic of the stage and electrode drive pattern for different actuation modes. Plots show the angle vs. time for in-plane actuation for two different ramp inputs, measured using NORIS. These are the results for single-sided electrode case. Picture in the inset shows the experimental setup.

Figure 5: Fabricated PZT dither stage with single side electrode patterning: (a) Laser micro machined PZT dither stage (b) side walls of the S springs. (c) SEM image of the rotor's side wall

minimum feature that can be patterned through 0.5mm thick PZT is 150 μm with the developed recipe. The laser beam size and the currently used, non-optimal cutting parameters such as scan speed and hatch density limit through cuts to form finer features, but can be improved in future runs. The typical cut rates for PZT are ~16.6 μm/min. After the laser cutting, the device was cleaned in isopropyl alcohol and cleaned with cotton swabs to clear off remaining debris.

DITHER STAGE ACTUATION MEASUREMENT

Experiments were performed on a single side-electrode patterned version of the stage, which has eight electrodes on top surface allowing different degrees of freedom including X, Y as well as pure rotary motion. Schematic view of this device and experimental setup to measure the rotation of the stage are shown in Figure 6. Possible drive scenarios on the eight electrodes to realize different motion modalities are also tabulated in this figure. Labeling the four actuator legs of the stage A to D, X motion, for example, is realized by operating B and D as longitudinal actuators while A and C are operated as lateral actuators. Similarly, Y motion can be achieved by swapping the roles of B and D with A and C. Rotation about Z or in-plane dither can be achieved by operating all the actuators as in plane lateral bimorphs.

The actuation/rotation of the stage was measured using Nano Optical Ruler Imaging System (NORIS) discussed in [9], which involves measurement of in-plane displacement by means of diffractive optics. A diffraction grating was placed on the dithering stage and a 635nm laser was incident at a known angle to obtain diffracting patterns that accurately tracked the stage motion. Using a CCD imager mounted at an angle to image these patterns, the in-plane actuation was computed with image processing involving

sub-pixel interpolation and others which are described in [10]. Stage responses to DC ramp voltage applied on the eight electrodes measured using NORIS are shown in Figure 6 for two different DC ramp voltage inputs, which yields dither rates of ~0.02 °/s and ~0.06 °/s in average, for top and bottom plots respectively.

A key metric of stage performance is minimizing the cross-coupling between in-plane and out-plane motion. FEM simulation results for cross-axis actuation are shown in Table 1 and indicate less than 5 nm out-of plane displacement across the circular stage for 100 V_{DC}. The numbers presented are the ratio of undesirable actuation to desired actuation (μm) of the central rotary platform as a result of different possible actuation schemes mentioned in the table in Figure 6. The simulation data in Table 1 shows that the cross-axis coupling is on the order of parts-per-thousand between planar and transverse axes.

Experimentally, the out-of-plane motion during desired in-plane dither was measured using a ZYGO optical profilometer. The surface profile of the central region of the

978-1-4799-3510-9/14 $31.00 © 2014 IEEE

rotor was measured before and after excitation to measure the out-of plane actuations and the data is plotted in Figure 7. Since the optical profilometer relies on a reflective surface to measure the surface profile, a silicon piece in the shape of the rotor with groove on the top surface was adhesively bonded on top of the rotor. The out-of plane measurement precision of the optical profilometer was ~0.1nm as quoted by the manufacturer. The stage produces an RMS out-of-plane motion of only ~ 1 nm.

Since applications involving actuation for calibration require stable operation such that there are no spurious resonant modes excited in the device under test (say inertial sensors), the fundamental mode of the stage should not overlap with the resonant modes of the MEMS inertial sensors, and should be away from the signals of interest. Fundamental resonance frequency was measured using the HP4194A impedance analyzer and was found to be around 11.1 kHz as shown in Figure 8. This mode was simulated to be out-of-plane by FEM.

	OUTCOME			
Stage Sensitivity	**X**	**Y**	**Z**	**θ**
X	1	0.25	1.67×10^{-3}	NA
Y	0.25	1	1.67×10^{-3}	NA
θ	NA	NA	0.2×10^{-3}	1

Table 1: Stage actuation sensitivities derived from COMSOL simulation of the PZT dither stage

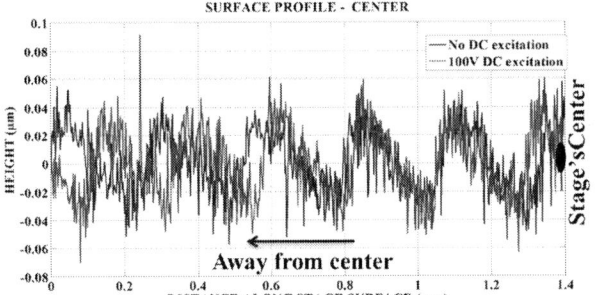

Figure 7: Surface profile measured off the reflections from the silicon attached on top of the central dither stage using ZYGO optical profilometer. Plot shows the fluctuations in height (the Z-axis motion). The RMS out-of-plane change was found to be ~1nm

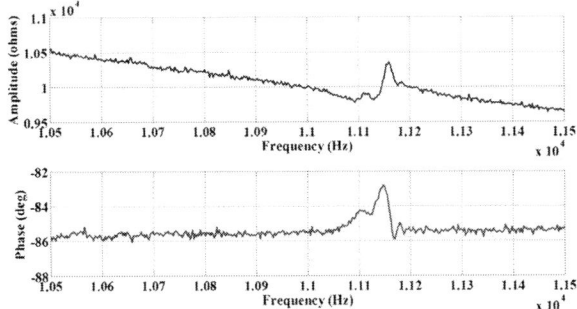

Figure 8: The 1ˢᵗ resonance of the PZT dither stage measured using HP 4194A impedance analyzer for 100mV excitation. The mode was found to be out-of-plane when simulated in COMSOL.

CONCLUSIONS AND FUTURE WORK

A PZT dither stage was fabricated using laser micro-machining of bulk PZT-4 plates and with a minimum feature size of ~150µm. The design includes displacement amplifiers and flexures to achieve in-plane dither actuation of a central rotor. Eight control electrodes defined on the top of the PZT actuator allows X, Y translation, and rotation about Z axis which is useful in MEMS Z-axis gyroscope calibration. The key metric of the stage is the low cross-axis actuation coupling, which is crucial for a calibration of inertial stages. The fabrication technique and the design space allow one to optimize to increase resonance frequency and reduce actuator size so that the stage is compatible with placing within a DIP package.

ACKNOWLEDGEMENTS

We would like to thank Defense Advanced Research Projects Agency (DARPA), PASCAL program, for funding this work and Cornell Nano Fabrication Facility (CNF) for testing and characterization equipment.

REFERENCES

[1]http://www.physikinstrumente.com/en/products/nanopositioning/nanopositioning_multi-axis_selection.php
[2] Judy, Jack W. "Microelectromechanical systems (MEMS): fabrication, design and applications." *Smart materials and Structures* 10.6 (2001)
[3] De-Yuan Zhang, Takahito Ono, Masayoshi Esashi, Piezoactuator-integrated monolithic microstage with six degrees of freedom, Sensors and Actuators A: Physical, Volume 122, Issue 2, 26 August 2005, Pages 301-306, ISSN 0924-4247
[4] Kenji Uchino, Sadayuki Takahashi, Multilayer ceramic actuators, Current Opinion in Solid State and Materials Science, Volume 1, Issue 5, October 1996, Pages 698-705, ISSN 1359-0286,
[5] Kommepalli, H. K., Mateti, K., Rahn, C. D., & Tadigadapa, S. A. (2011). Piezoelectric T-beam actuators. *Journal of Mechanical Design, 133*, 061003.
[6] Oldham, Kenn R., et al. "Thin-film PZT lateral actuators with extended stroke."*Microelectromechanical Systems, Journal of* 17.4 (2008): 890-899.M. Young, The Technical Writer's Handbook. Mill Valley, CA: University Science, 1989
[7] Sachin Nadig, Serhan Ardanuç, and Amit Lal "Planar Laser-Micro Machined Bulk PZT Bimorph For In-Plane Actuation" *2013 Joint Ultrasonics, Ferroelectrics and Frequency Control (UFFC), EFTF and PFM symposium, IEEE Transactions on* , July 2013
[8] Lee, Seung-Yop, Byeongsik Ko, and Woosung Yang. "Theoretical modeling, experiments and optimization of piezoelectric multimorph." *Smart materials and structures* 14.6 (2005).
[9] Sachin Nadig, Serhan Ardanuç, and Amit Lal "DOME-DISC: Diffractive Optics Metrology Enabled Dithering Inertial Sensor Calibration " *MEMS 2014* (2014).

POLYMER MICROMACHINING BASED ON CU ON POLYIMIDE SUBSTRATE AND ITS APPLICATION TO FLEXIBLE MEMS SENSOR

Yosuke Niimi, Shunji Shibata, and Mitsuhiro Shikida
Dept. of Micro-Nano Systems Engineering
Nagoya University, Aichi, JAPAN

ABSTRACT

As a way of making a variety of flexible MEMS sensors, we developed a process that uses a Cu On Polyimide (COP) substrate as a starting material and sacrificial Cu etching to produce a cavity and electrical feed-through structures on the substrate. The Cu etching characteristics of an iron (II) chloride solution were studied, and the results confirmed that etching in the depth and side directions strongly depended on the mask pattern and the etching time. A strip-shaped flexible thermal sensor was designed on the basis of the obtained etching results. One heater and two temperature sensors were designed for flow and acceleration sensors applications. The sensing metal on the polyimide thin membrane and the feed-through were successfully fabricated on the COP substrate.

INTRODUCTION

Micro Electro Mechanical Systems (MEMS) technologies enable us to miniaturize various physical and chemical sensors. Thermal convective phenomena have been widely used as sensing mechanisms in flow, shear stress, and acceleration sensors. The three different thermal principles, i.e., thermal anemometry, calorimetric flow sensing, and time-of-flight sensing, have been used in miniaturized thermal flow sensors, and they are summarized in refs. [1-6]. In one instance, a thermal sensor formed on a diaphragm was used to detect shear force of a fluid acting on a plane surface [1]. Moreover, thermal sensing was applied to accelerometers [2-4]. The sensor was composed of a heater and two temperature sensors placed on both sides of the heater. The applied acceleration was detected by the change in the temperature distribution pattern over the two sensors. This type of sensor has an advantage that it can detect accelerations without the need for a moving solid proof-mass. The sensing response was improved by changing the size of the cavity formed under the heater element as a thermal insulation [4].

Thermal sensors based on MEMS technologies have excellent space and time resolutions, and they have found many applications in the automobile industry. However, since they are fabricated on brittle Si or Silicon On Insulator (SOI) wafers, they have a difficulty being mounted on bendable surfaces. This has meant that Si-based MEMS sensors are difficult to be incorporated in human interfaces and wearable devices.

To overcome this problem, polymer materials, including polyimide, parylene, and silicone resin, have been used as substrate materials for flexible MEMS sensors. Various fabrication processes, including low-temperature film deposition, printing, and molding, have been developed. For example, a flexible pressure sensor called e-skin was developed on a polymer sheet based on an electrical switching device [7]. A fabric tactile sensor based on artificial hollow fibers was also developed as part of a wearable sensing system. The flexible sensor devices developed up to now are summarized in ref. [8].

In this paper, we propose to use a Cu On Polyimide (COP) substrate as a starting material, and sacrificial etching for producing a cavity and electrical feed-through structures on the substrate as a way of realizing various flexible MEMS sensors. We also describe the fabrication of a flexible thermal MEMS sensor on a COP substrate.

POLYMER MICROMACHINING BASED ON COP SUBSTRATE

A COP substrate and an example of its application to flexible MEMS sensors are shown in Figure 1. The substrate is composed of a base polyimide film and a Cu layer. A thin polyimide layer is also formed on the substrate, and the MEMS sensors are produced on its surface. A cavity and an electrical feed-through are produced by selectively etching part of the Cu layer in the same process (sacrificial etching). The advantages of polymer micromachining based on COP

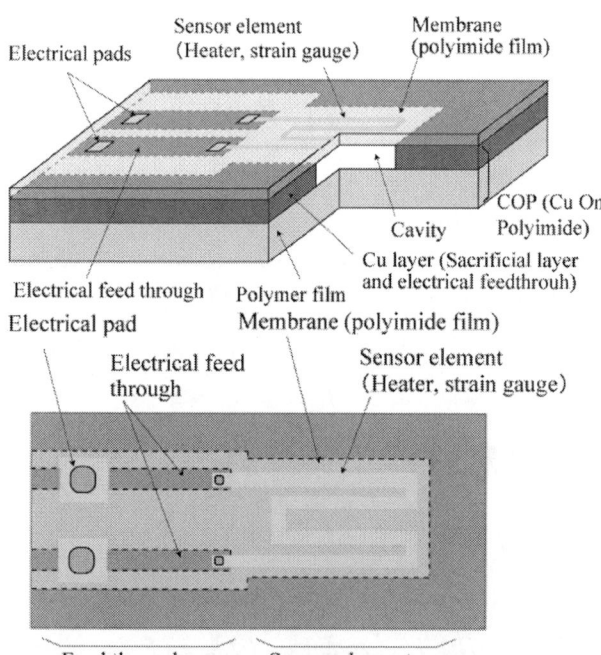

Figure 1: Polymer micromachining based on Cu On Polyimide (COP) substrate.

978-1-4799-3510-9/14 $31.00 © 2014 IEEE 528

substrate are as follows.

(1) It can produce flexible MEMS sensors.
(2) It can produce various physical sensors, such as flow, acceleration, and pressure sensors, because the cavity works as a thermal isolation in thermally operated flow and acceleration sensors and also works as a pressure standard room in pressure sensors.
(3) It can easily fabricate electrical feed-throughs having low electrical resistance, because part of the Cu layer works as the electrical wiring after the sacrificial etching.

FABRICATION PROCESS

Figure 2 illustrates a typical fabrication process for flexible MEMS sensors based on COP substrates. The COP substrate (Ube Exsymo Co., Ltd.) was used as the starting material. The thickness of copper and polyimide were 18 μm and 50 μm, respectively. First, the photosensitive polyimide solution (Photoneece, Toray Industries, Inc.) was coated to a thickness of 3 μm and patterned to define a structure in the Cu layer. This photosensitive polyimide works as an etching mask for forming the cavity and feed-throughs in the Cu layer in the final sacrificial Cu etching, and it becomes a membrane structure. Part of the pattern is used to form a connection area between the deposited metal film and the electrical feed-through in the Cu layer. A negative photoresist (ZPN1150-90, Zeon Corporation), which was specially developed for the lift-off process, was put on the

film surface and patterned with UV light to define the shapes of the Au/Cr film that would work as the heater and temperature sensors. The Au/Cr film was deposited by sputtering and patterned by selectively removing the

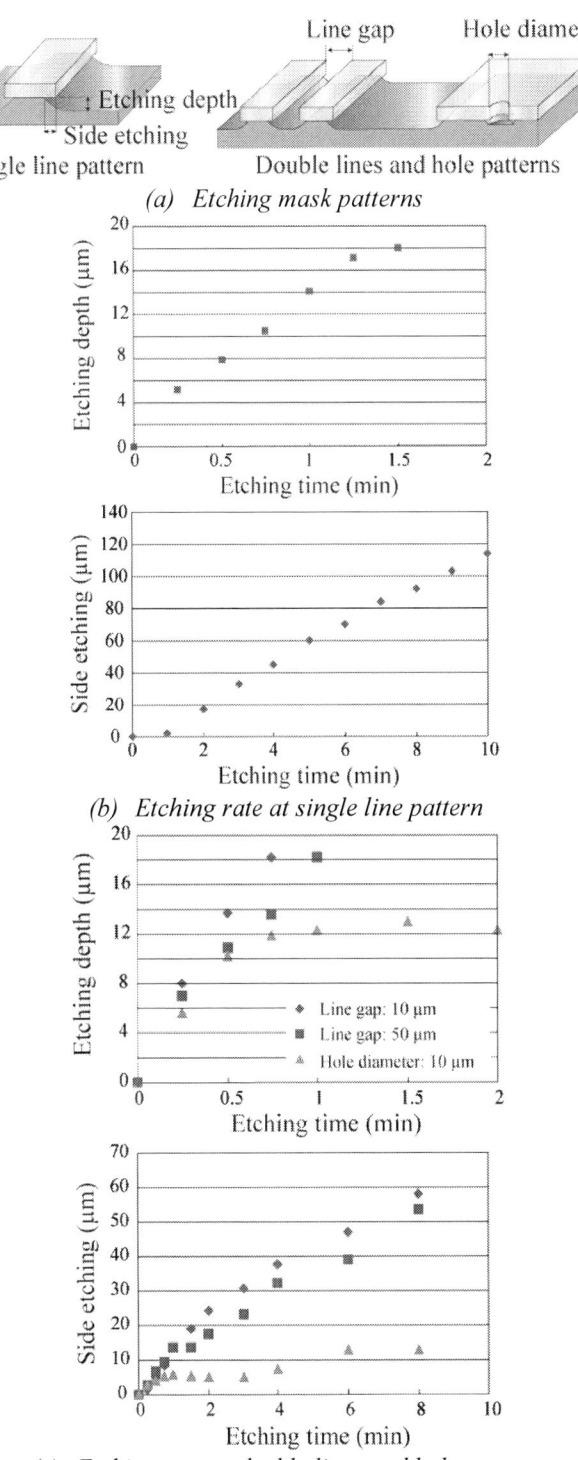

(a) Etching mask patterns

(b) Etching rate at single line pattern

(c) Etching rate at double lines and hole patterns

Figure 3: Etching characteristics of etching mask patterns having various shapes.

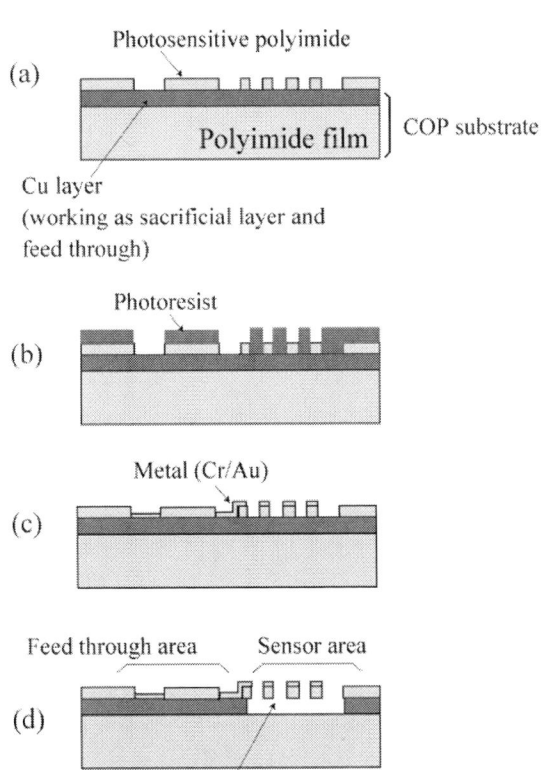

Figure 2: Fabrication process using COP substrate.

photoresist (lift-off process). The thickness of the Au and Cr were 250 nm and 10 nm, respectively. Finally, the cavity and electrical feed-throughs were formed by selectively etching part of Cu the layer based on the different etching rate depending on the polyimide mask pattern (sacrificial etching). An iron (II) chloride solution (Sunhayato Corp.) was used as the etchant. The etching temperature was set at 40°C. The etching resulted in metal patterns working as sensors being formed on the polyimide membrane and the electrical feed-throughs being formed on the COP substrate.

CHARACTERIZATION OF CU ETCHING

The sacrificial etching is the key technology in the polymer micromachining on COP substrate, because it defines the shapes of the cavity and the feed through. An iron (II) chloride solution was used as the Cu etching solution. Thus, the etching was dominated by the diffusion-limited reaction, and as a result, Cu layer was isotropically etched. The amount of this etching solution strongly depends on the mask pattern, because of the diffusion-limited reaction. Thus, the etching characteristics under the different shaped mask patterns were investigated in order to design the shapes of the cavity and feed-through into the Cu layer.

Three different mask patterns, i.e., single line, double line, and hole, were examined, as shown in Figure 3 (a). The gaps between the double lines were 10 μm and 50 μm, respectively. The hole diameter was 10 μm. The experimental results are shown in Figures 3 (b) and 3(c). The Cu etching characteristics were as follows.

(1) The etching amounts in the depth and side directions strongly depended on the mask pattern and etching time.
(2) In the case of a single line, the etching amounts in the depth and side directions linearly increased with the etching time.
(3) The etching rate depended on the width of the gap between the line, and it increased as the gap in creased in the case of the double line pattern.
(4) The etching rates in the depth and side directions in the case of the hole pattern were quite small compared with those of the line patterns.

STRIP-SHAPED FLEXIBLE THERMAL SENSOR

An etching mask pattern of photosensitive polyimide film was designed for a strip-shaped flexible thermal sensor. The pattern was based on the experimentally obtained characteristics of the Cu etching rate (Figure 4). The design included a heater and two temperature sensors for flow and accelerometer applications. Two different sized holes patterns were formed on the polyimide film. The large and small holes were 16 μm x 50 μm and 10 μm x 10 μm, respectively. Each hole was designed for forming the cavity and feed-through. Figure 5 shows a fabricated flexible thermal sensor. The sensing metal on the polyimide thin membrane and the feed-through were successfully fabricated on the COP substrate.

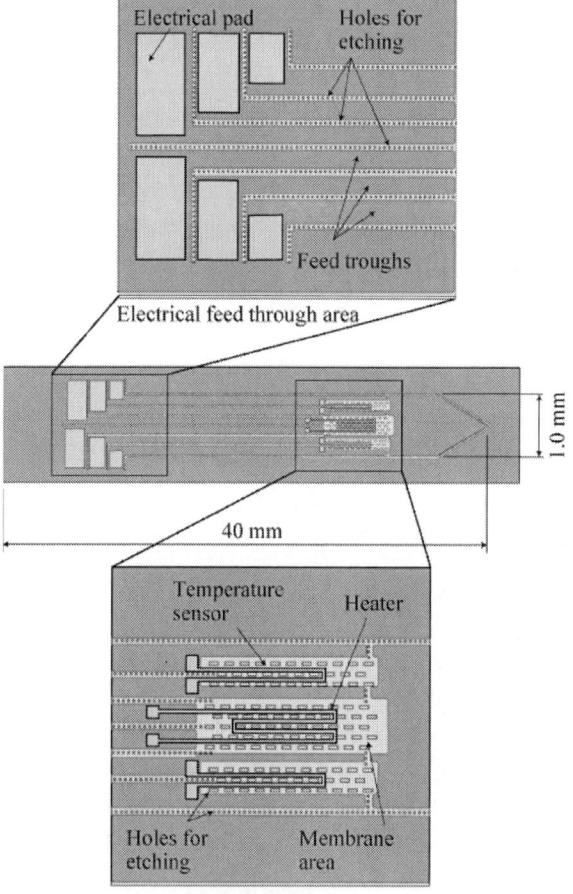

Figure 4: Strip-shaped flexible thermal sensor for sensing flow and acceleration.

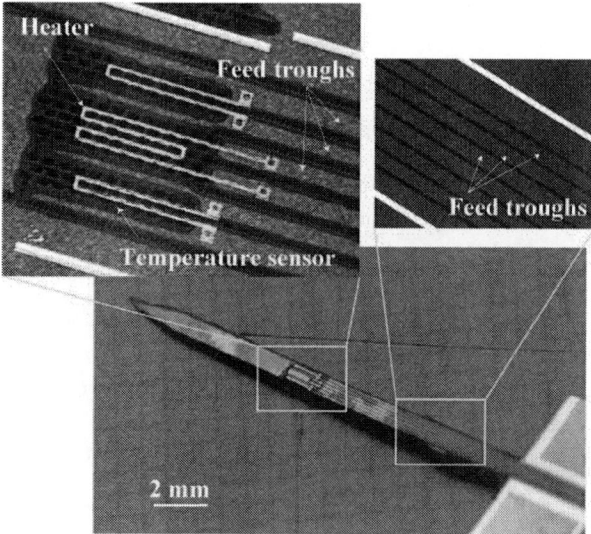

Figure 5: Fabricated strip-shaped flexible thermal sensor.

CONCLUSION

We used Copper On Polyimide (COP) substrate as a start material for producing flexible MEMS sensors and sacrificial etching for producing a cavity and electrical feed-through structures on the substrate. Cu etching in an iron (II) chloride solution was studied, and the following results were obtained.

(1) The etching amounts in the depth and side directions strongly depended on the mask pattern and etching time.
(2) The etching rate depended on the gap between the line, and it increased with the width of the gap in the case of the double line pattern.
(3) The etching rates in the depth and side directions in the case of the hole pattern were quite small compared with those obtained for the line patterns.

A strip-shaped flexible thermal sensor was designed on the basis of the results of the etching study. The design included a heater and two temperature sensors for flow and acceleration sensor applications. The sensing metal on the polyimide thin membrane and the feed-through were successfully fabricated on the COP substrate.

ACKNOWLEDGEMENTS

This research was supported by the Grant-in-Aid for Scientific Research (B) No. 23310091 from the Ministry of Education, Culture, Sports, Science and Technology (MEXT), Japan

REFERENCES

[1] J. B. Huang, Steve Tung, C. M. Ho, C. Liu, and Y. C. Tai "Improved Micro Thermal Shear-Stress Sensor," IEEE Transactions on Instrumentation and Measurement, vol. 45, no. 2, 1996.

[2] U. A. Dauderstädt, P.H.S. de Vries, R. Hiratsuka, and P. M. Sarro, "Silicon accelerometer based on thermopiles," Sensors and Actuators A, vol. 46, pp. 201-204, 1995.

[3] F. Mailly, A. Giani, A. Martinez, R. Bonnot, P. Temple-Boyer, and A. Boyer, "Micromachined thermal accelerometer," Sensors and Actuators A, vol. 103, pp. 359-363, 2003.

[4] J. Courteaud, N. Crespy, P. Combette, B. Sorli, and A. Giani, "Studies and optimization of the frequency response of a micromachined thermal accelerometer," Sensors and Actuators A, vol. 147, pp. 75-85, 2008.

[5] Y. Gianchandani, O. Tabata, and H. Zappe, "Comprehensive MEMS," 2, Flow sensor, pp. 209-272, Elsevier, 2008.

[6] M. Elwenspoek and R. Wiegerink, "Mechanical microsensors," Springer, 2001.

[7] T. Someya, T. Sekitani, S. Iba, Y. Kato, H. Kawaguchi, and T. Sakurai "A Large-Area, Flexible Pressure Sensor Matrix with Organic Field-Effect Transistors for Artificial Skin Applications," T. Proc. Natl. Acad. Sci. USA , vol. 101, pp. 9966-9970, 2004.

[8] C. Pang, "Recent Advances in Flexible Sensors for Wearable and Implantable Devices," Applied Polymer Science, vol. 130, Issue 3, 2013.

CONTACT

Y. Niimi, tel: +1-52-789-5224;
niimi.yousuke@g.mbox.nagoya-u.ac.jp

PRINTING AND ENCAPSULATION OF ELECTRICAL CONDUCTORS ON POLYLACTIC ACID (PLA) FOR SENSING APPLICATIONS

Andrés Vásquez Quintero, Nathalie Frolet, Daniel Märki, Alexis Marette, Giorgio Mattana, Danick Briand and Nico F. de Rooij

Ecole Polytechnique Fédérale de Lausanne (EPFL), Neuchâtel, SWITZERLAND

ABSTRACT

This paper presents the printing of resistive and interdigitated (IDE) capacitive devices for temperature and humidity sensing applications, respectively, on biodegradable polylactic acid (PLA) substrates. Inkjet and gravure printing were evaluated to transfer silver-based nanoparticles inks. Flash photonic ink sintering methodologies were employed to maintain the PLA mechanical integrity due to its low glass transition temperature (58 °C). Between the two printing techniques investigated, gravure-printed devices on 200 µm-thick PLA sheets were shown to have better resolution and higher sensitivities to temperature and humidity (1100 ppmK^{-1} and 5.6 fF/%RH). Additionally, we demonstrated the inkjet printing of IDE onto thin (25 µm) dissolved-PLA spin-coated substrates, to enhance the mechanical flexibility and to reduce the response time to humidity (from 238 s to 70 s). Finally, a low temperature encapsulation is proposed by embedding the printed structures within PLA sheets.

INTRODUCTION

Nowadays, there is a high interest to have electrically conductive materials on biodegradable substrates, mainly for biomedical and smart packaging applications. Such structures would allow the design of environmental transducers to track and monitor ambient or body parameters (e.g. temperature and humidity). Additionally, the disposal of such devices could be done at composting sites due to the biodegradable nature of the substrate and the low quantity of conductive material deposited. Polylactic acid (PLA) is preferred among other bio-plastics due to its inherent renewable source (maize starch) and production cycle.

Recently, global production capacity of PLA has increased specially in technical housing and packing due to its characteristics as bio-compatible, bio-degradable and bio-compostable. Some groups have already reported the sputtering and hot embossing of zinc [1], the lamination of copper [2] and the soldering at low temperatures [3] on PLA. However, the direct printing and sintering of electrical conductors on PLA has not been explored yet, due to its challenging low glass transition temperature (T_g: 58 °C).

In this study the sintering of printed silver-based inks on PLA by means of photonic techniques (near infrared–NIR wavelength: 500-1500 nm) is demonstrated. The latter are effective thanks to the low PLA absorption of NIR wavelengths which only heat the surface of the printed lines, maintaining the substrate integrity.

These printed and conductive lines were implemented and characterized as resistive temperature and capacitive humidity sensors. For the latter, the PLA substrate is also used as humidity sensing layer. Performances of the devices printed by both techniques on commercially available 200 µm-thick PLA sheets and 25 µm-thick spin-coated PLA films were compared and optimized. Finally, an encapsulation technique is proposed to fully embed the conductive structures within PLA sheets at low temperatures (< 60 °C). The printing, the sintering, as well as the encapsulation procedures are compatible with foil-level and large area fabrication methods.

FABRICATION

The conductive structures were fabricated using two different printing techniques, namely inkjet and gravure printing, onto two types of PLA substrates. The first one was the transparent 200 µm-thick film Bioclear® from the company Sodinor. The second one was a custom-made film obtained by spin-coating dissolved PLA beads (Ingeo 4032D), from the company NatureWorks. The dissolution was performed by mixing the PLA pellets with dioxane (from Sigma-Aldrich) at 40 °C with mechanical stirring. In order to decrease the dissolution time the pellets were added to the solvent after a vortex was created by the mechanical stirring. The solution (PLA/dioxane) was spin-coated onto 4'' silicon carrier wafers. Different concentrations and spin-coating speeds were used to obtain different film thicknesses. For this work a volume concentration of 15 % (7.5 g of PLA within 50 mL of dioxane) and a speed of 1000 rpm for 60 s were used as solution and spin-coating conditions. This combination was proven to give a homogenous film with a thickness of 24.5±0.1 µm.

After an oxygen plasma surface treatment (15 s), inkjet printing was performed on both types of PLA at room temperature. The non-contact technique was performed with a Dimatix® - DMP2800 printer using the silver-based nanoparticles ink SunTronic® JetEMD506 from SunChemical. The gravure printing technique was done onto the 200 µm-thick PLA with a TeslaColor-171® machine from Schläfli at 15 m/min using the silver-based nanoparticles ink TC-PR-020 from InkTec.

The sintering process was performed using near infrared (NIR) photonic tools, namely, NIR-120® from Adphos and PulseForge®-1200 from Novacentrix, referred as NIR and NIR-pulsed, respectively, in the document. The PLA films are mostly transparent to the NIR wavelength [4-5] but not the silver-based inks. Leading to an increment of temperature just at the surface causing ink sintering and protecting the film integrity (T_g: 58 °C). Figure 1 presents the PLA transmittance compared to a typical NIR energy source, showing a relatively high transmittance between 500 nm and 1500 nm. Inkjet-printed devices were sintered using 3 passes with an intensity of 55 % ($\approx$ 1650 W) at 0.12 m/s with the NIR-120 machine. Gravure-printed devices used 5 flash-pulses of 808 mJ/cm^2 with the PulseForge machine.

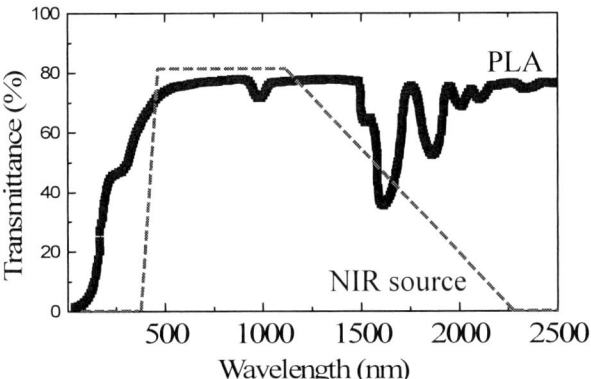

Figure 1: Transmittance profile of PLA (black-continuous line) compared to a NIR energy source (red-dotted line).

Following the procedure described above, two types of structures were designed and fabricated: resistors and IDE capacitors, used as temperature and humidity sensors, respectively. Figure 2 shows the optimal images of the inkjet- and gravure-printed devices on 200 µm-thick PLA film. The gap and pitch obtained were: 70µm/220µm and 30µm/160µm, respectively.

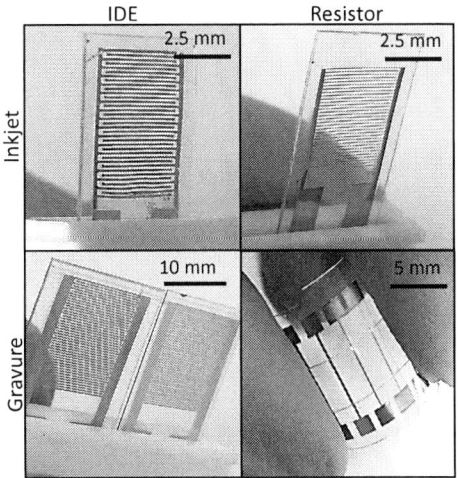

Figure 2: Optical images of the inkjet- and gravure-printed capacitive and resistive devices on PLA substrates (200 µm-thick).

CHARACTERIZATION
Photonic sintering process

The photonic sintering of inkjet-printed devices was performed with the NIR-120 (NIR) system (Figure 3a), which has as parameters the energy and speed under the lamp. Both variables need to be optimized depending on the type and thickness of the ink and substrate. It was observed that the PLA substrate was deformed plastically when energies too elevated were used. Additionally, the electrodes were damaged at the interface with the pads when the exposure took too long, as shown in Figure 3b. An optimal intensity and speed of 55 % (1650 W) and 0.12 m/s, respectively, were found optimal for this case.

The photonic sintering of gravure-printed devices was performed with the PulseForge (NIR-pulsed) system (Figure 3c), which has the energy dose as main parameter. It was observed that the PLA substrate started to deform and more interesting the ink started to crack above 830 mJ/cm^2, as shown in Figure 3d. For this reason several pulses at lower doses (808 mJ/cm^2) were found as optimal.

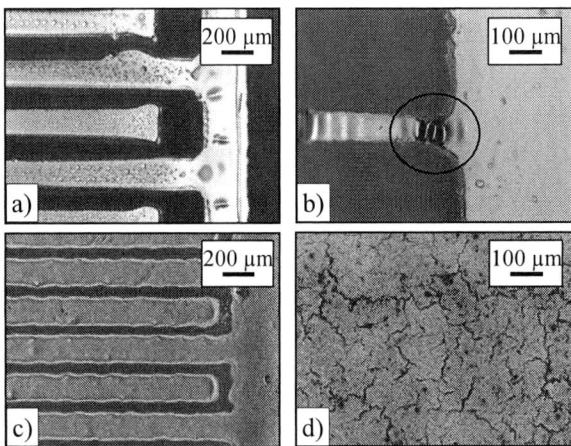

Figure 3: Optical images of inkjet-printed IDE sintered with: a) optimal speed, b) too low speed. Gravure-printed IDE sintered with: c) optimal dose, d) too high dose.

Electric properties

The resistivity of the sintered inks was calculated using their geometry parameters and measured resistances (4-points technique). Gravure + NIR-pulsed ink was shown to have 1/2 of the inkjet + NIR ink's resistivity, while the printed layers after curing had the same thickness (Table 1).

Table 1: Measured thicknesses (white light interferometer) and calculated resistivity for the inkjet- and gravure-printed structures.

Technique	Thickness (nm)	Resistivity (µΩ·cm)
Inkjet + NIR	700 ± 40	43.2 ± 2.1
Gravure + NIR-pulsed	700 ± 30	16.2 ± 0.3

The temperature coefficient of resistance (TCR) for the printed resistive structures was measured from -10 °C to 40 °C at 40 %RH. Figure 4 presents the up- and down-sweep curves showing a TCR value of 800 ppm·K^{-1} and 1130 ppm·K^{-1} for the inkjet and gravure silver-based inks, respectively.

978-1-4799-3510-9/14 $31.00 © 2014 IEEE   533

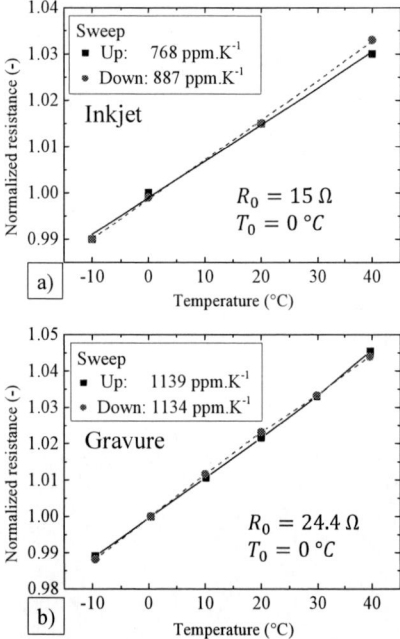

Figure 4: Temperature coefficient of resistance for a) inkjet + NIR ink and b) gravure + NIR-pulsed ink.

Humidity response

Using the demonstrated printing and sintering techniques, IDE capacitive structures were fabricated as humidity sensors with the PLA as sensing layer. Additionally, the sensor configuration (i.e. printing resolution and substrate thickness) was optimized

The devices were exposed to a humid environment in a climatic chamber with cycles from 30 %RH to 70 %RH of one hour at 25 °C. Figure 5a and 5b presents the curves for the inkjet- and gravure-printed devices on 200 μm-thick PLA, showing a sensitivity of 3.4 fF/%RH and 5.6 fF/%RH, respectively. Higher printing resolution of gravure IDE capacitors led to higher capacitance for the same area. The latter allowed higher absolute RH sensitivities (1.6 times better) and a better sensor performance. The dynamic behavior to relative humidity was measured in a climatic chamber using controlled RH steps from 10 % to 70 % at 25 °C. Figure 6a and 6b presents the curves for the inkjet and gravure devices on 200 μm-thick PLA, with response times of 238 s and 273 s (from 10 %RH to 30 %RH at τ = 63%.), respectively.

The same dynamic analysis was performed with the inkjet-printed devices on the spin-coated 25 μm-thick PLA films. The latter were printed and sintered using the Si wafer as a carrier substrate to maintain the planarity, and then cut and peeled-off, as shown in Figure 7. Since the rigidity is proportional to the cube of the thickness, it is interesting to point out that the 25 μm-thick film is about 500 times less rigid than its 200 μm-thick counterpart. This proportionates highly flexible and to some extend conformal films and printed structures.

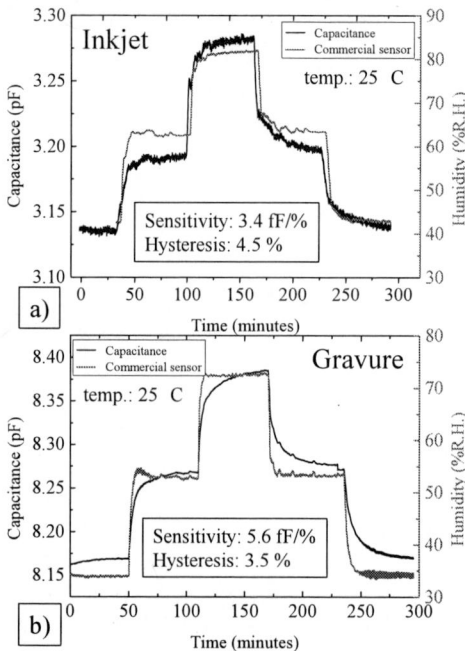

Figure 5: Response to humidity for a) inkjet- and b) gravure-printed IDE capacitors, showing the calculated-absolute sensitivity and hysteresis.

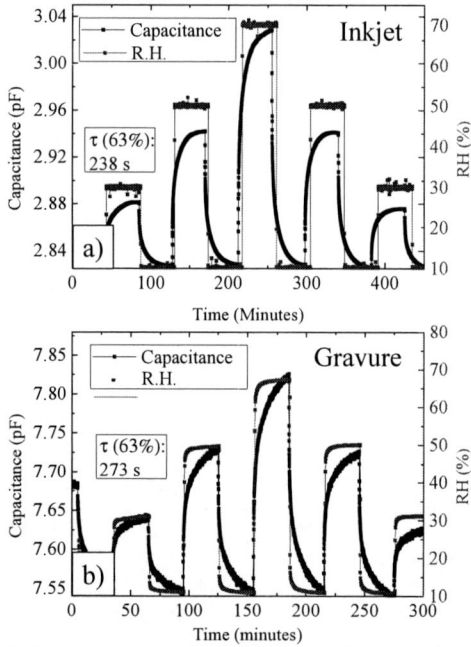

Figure 6: Response curves to RH steps for: a) inkjet- and b) gravure-printed IDE capacitors on 200 μm-thick PLA substrate, showing the response time (τ=63%).

Figure 8 shows the response to humidity of the 25 μm-thick PLA film at different RH levels (from 10 % to 70 % at 25 °C). It is noted that the device follows the input RH profile. However, its reversibility is hindered after a relatively high humidity exposure (70 %). The response time from 10 %RH to 30 %RH was found to be 70 s, which is 3.9 times faster than the devices on thick PLA.

The inset of Figure 8 shows the linear relationship between the capacitance and humidity, with a sensitivity of 0.2 fF/%RH, which is lower than the devices on thick PLA.

The improvement on the response time and the worsening of the sensitivity are due to the substantial reduction of the PLA thickness (8 times) keeping the same printing resolution (gap/pitch: 70 μm/220 μm). In order to maintain the fast response and increase the sensitivity, higher printing resolution and the deposition of an RH sensing layer (e.g. CAB, PLA, etc.) on top of the electrodes are required.

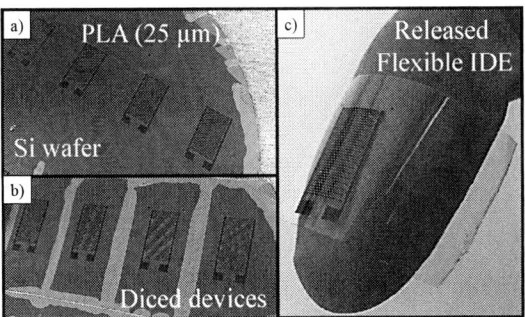

Figure 7: Photographs of: a) inkjet-printed devices on spin-coated PLA onto Si wafer, b) diced devices on Si wafer and c) released flexible device.

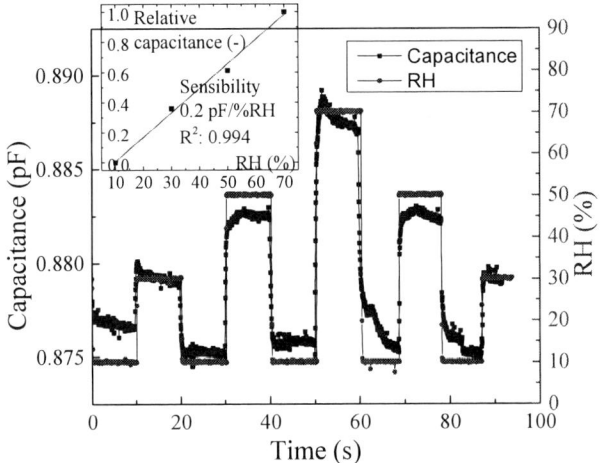

Figure 8: Response curve to RH steps for the inkjet-printed IDE capacitor on 25 μm-thick PLA substrate, showing the response time and the absolute sensitivity (inset).

ENCAPSULATION

A flexible encapsulation method is proposed to fully embed the printed structures onto PLA films at low temperatures (< 60 °C). After the printing of the conductive devices onto the thin PLA substrate, a second spin-coated film was pressed together to the first stack at 58 °C and 40 kPa for 2 minutes. Finally, the complete multilayer stack is cut and released from the Si carrier wafers, as shown in Figure 9. The encapsulation is used to keep the ink adhesion to PLA when in contact with the compost environment in biodegradation tests.

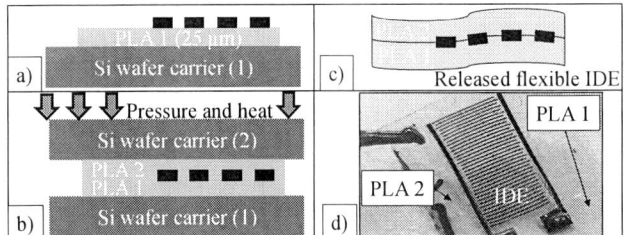

Figure 9: a-c) Schematics of PLA encapsulation process and d) photograph of an encapsulated IDE capacitor.

CONCLUSIONS

This paper presented the printing (inkjet and gravure) and photonic sintering of resistive and capacitive structures on polylactic acid (PLA). The procedures kept the integrity of the film which has a relatively low T_g (58 °C). Gravure-printed devices on 200 μm-thick PLA were shown to have higher sensitivity to temperature 1100 ppmK^{-1} and humidity 5.6 fF/%RH. Inkjet-printed devices on 25 μm-thick PLA were shown to have faster RH response time (70 s). Finally, the fully PLA encapsulation of printed devices was proposed at low temperatures. The latter are being tested for biodegradation in controlled compost environments.

ACKNOWLEDGEMENTS

Partly funded by the EU-FP7 project FlexSmell; a *Marie Curie* Initial Training Network (ITN), under the grant No. 238454.

REFERENCES

[1] M. Luo, C. Song, F. Herrault, M. Allen, "A Microfabricated Wireless RF Pressure Sensor Made Completely of Biodegradable Materials", *in Digest Tech. Papers Solid-State Sens., Act. and Microsys. Workshop*, Hilton Head Island, June 3-7, 2012, pp. 4.

[2] A. Géczy, M. Kovacs, I. Hajdu, "Conductive Layer Deposition and Peel Tests on Biodegradable Printed Circuit Boards, *in Digest Tech. Papers SIITME 2012*, Alba Iulia, October 25, 2012, pp. 139-142.

[3] A. Géczy, V. Léner, I. Hajdu, Z. Illyefalvi-Vitéz, "Low Temperature Soldering on Biopolymer (PLA) Printed Wiring Board Substrate", *in Digest Tech. Papers 34th ISSE 2011*, Tratanska, May 11, 2011, pp. 57-62.

[4] W. Mulbry, J.B. Reeves, P. Millner, "Use of mid- and near-infrared spectroscopy to track degradation of bio-based eating utensils during composting", *Bioresource Tech.*, vol. 109, pp. 93-97, 2012.

[5] C. Aulin, E. Karabulut, A. Tran, L. Waisgberg, T. Lindström, "Transparent nanocellulosic multilayer thin films on polylactic acid with tunable gas barrier properties", *ACS Appl. Mat. And Inter.*, vol. 5, issue 15, pp. 7352-7359, 2013.

CONTACT

*A. Vásquez Quintero, tel: +41 (0) 21 695 4428;
andres.vasquez@epfl.ch

RAPID PROTOTYPING OF RESISTIVE MEMS SENSING DEVICES ON PAPER SUBSTRATES

Thorsten Meiss[1], Roland Wertschützky[1] and Boris Stoeber[2]
[1]Technische Universität Darmstadt, GERMANY
[2]The University of British Columbia, Vancouver, CANADA

ABSTRACT

We have developed an inexpensive inkjet printing process to rapidly fabricate resistive sensor devices on paper substrates. We utilize the percolation based resistive change in inkjet-printed resistors based on carbon black (CB). We fabricated an ink with 2 wt% CB, 0.3 wt% binder and added 0.3 % sodium cholate as well as 0.5 wt% triton as surfactant to print strain sensitive resistors. To connect the resistors we inkjet-printed conductors with a resistivity of 1 Ω/sq. Cutting and folding the paper substrate yields a sensor with a 3-dimensional structure for use as a tactile input device as well as an inertial sensor with a sensitivity of 8 mV / V g.
Since we use a commercial inexpensive inkjet printer and the design can be modified and tested within minutes, the process is especially useful to easily develop and test MEMS sensor models. Additional applications encompass disposable medical sensors, sensors for paper packaging, as well as very low cost strain sensing.

INTRODUCTION

The prototyping of new MEMS-sensors involves well established and sophisticated design tools. These tools are merged into integrated design processes: Fast and user-friendly CAD programs interfaces with simulation software such as FEM-tools, and these tools then interface with sophisticated analysis software. Even though this design approach generates very predictable results of the subsequently fabricated structure and layout errors are efficiently reduced, it can take month to years until a developer has a first prototype on his desk.

Early access to a simple functional model can improve the understanding of the system to be designed and accelerate the system development. Therefore, a fast and simple method to generate functional models of sensors can be of great advantage. It is assumed that an easy, creative way of building and changing sensor structures within minutes will fosters the designer's creativity and will facilitate the testing of a great variety of different structures.

To address this need we consider inkjet printing of functional MEMS sensor models onto paper substrates. Inkjet printing has several advantages: It is a mask less process that allows for rapid change of the printed structures with conventional CAD-tools. Inkjet printers can be used in offices, whereby reducing the need for cleanroom space and shortening development cycles. Electrically conductive or nonconductive fluids, as well as emulsions and suspensions can be deposited through inkjet printing. Additionally, inkjet printers and paper substrates are very inexpensive in comparison to conventional MEMS fabrication equipment

and silicon substrates. To enable a fast on design processes, design changes can occur through reprinting of the sensor or through changes to the folding of an already manufactured sensor.

The resistive sensing principle permits the use of very small sensing elements that can be placed in nearly arbitrary locations to form physical models of a large variety of strain or deflection based sensors. The interconnection of several resistors permits simple but accurate voltage reading and first order error compensation.

The fabrication of paper-based resistive sensor models involves the four basic steps: (1) printing strain sensitive resistors, (2) printing electrical conductors, (3) cutting the sensor structure from paper, and (4) folding the sensor into a 3D-structure.

Inkjet-printed strain sensing

Different types of strain-sensitive resistors have recently been printed onto paper. Liu *et al.* deposited graphite ink through screen printing onto chromatography paper achieving gage factors around 4 and demonstrating cantilever-base force sensors [1, 2]. Recently, Khajeh *et al.* developed an inkjet deposition method for paper with high porosity using percolation-based strain-sensitive resistors based on suspensions containing conductive amorphous carbon (carbon black, CB), a binder (sodium carboxymethylcellulose, CMC) and a surfactant (sodium cholate, SC) [3]. Gage factors (GF) of up to 70 have been achieved with 3 wt% CB and 0.1 wt% binder in aqueous solution (Fig. 1). The high gage factor allows for signal acquisition with good quality while using simple electronics. The base resistance of these resistors can be four orders of magnitude higher than the resistance of printed conductors, so that the mechanical and thermal influence of the conductors on the measurement signal can be neglected.

Inkjet-printed conductors

Electric conductors on paper substrates have been part of intensive research. First electronic circuits where fabricated on phenolic resin bonded paper. Siegel *et al.* have built a paper plane with LEDs as position lights [4]; the conductors where deposited onto the paper by sputtering through a shadow mask, which led to the advantages of high conductivity and high tear resistance of the conductors. However, sputtering is not suitable for fabricating sensors within minutes in a designer's office and it requires high investment costs.

Several commercially available inks can be found for screen printing or painting highly conductive traces with a resistivity lower that 1 Ω/square on versatile paper substrates. These inks are loaded with a high content of conductive filler, e.g. metals such as silver [5] or nickel [6], or carbon [7]. The high concentration of the relatively large conductive particles with sizes in the micrometer range results in good conductivity even on rough paper surfaces, but these particles usually render these inks unsuited for inkjet printing. On the other hand, highly conductive inks are available especially made for inkjet printing. These include reactive inks [8], where silver is completely dissolved and the metal is formed on the printed surface. Another class is defined by silver nano inks with particle sizes around 50 nm to 150 nm [9, 10]. Unfortunately, most of the printable inks do not result in a conductive percolating network on standard paper surfaces. Shao *et al.* published the fabrication of low-cost RFID tags by inkjet printing highly conductive traces onto glossy photo paper with this kind of ink [11].

Cutting and folding the paper

3D-structures can be produced from plane paper sheets through folding or through folding after an initial cutting process. Architects commonly use cutting machines for paper, while laser cutters are another option for rapid prototyping. For a small number of devices knifes, scissors or scalpels are viable options. Many options are available to achieve the final 3D-structure, such as folding [12], folding and bonding, or stacking of several paper layers [13].

MATERIALS AND METHODS

To conduct the experiments we focused on low cost printer device for test purposes.

Preparation of the resistor ink

To print strain sensitive resistors we developed a percolation-based strain-sensitive ink based on suspensions containing conductive CB, the binder CMC and the surfactant SC as in [1]. To achieve maximum reproducibility of the sensor characteristics we selected 2 wt% CB and 0.3 wt% binder for the present work as suggested in Fig. 1. Increasing the amount of surfactant to 0.3 wt% SC and 0.5 wt% Triton makes the ink composition suitable for a commercial printer.

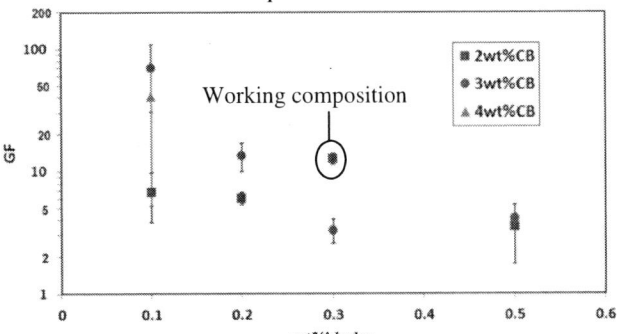

Figure 1: Gauge factor as a function of ink composition for the resistors; error bars correspond to +/- standard deviation; from [1].

All ink components are added together into a vial. They are mixed with a stir bar for 1 h, and they are sonicated for 30 minutes, before filling them into printer cartridges through a filter with an average pore diameter of 0.7 μm. This prevents any larger particles from entering the print head.

Preparation of the conductor ink

A commercial silver ink (Metalon 25P, Novacentrix, Austin) is chosen for the conductor traces. It allows for inkjet printing conductors onto photo paper (Premium Photo Paper Glossy, EPSON, Long Beach, CA, USA).

Printing system setup

To print different types of ink the piezojet printer principle was chosen. The experiments were conducted with a low cost Epson Workforce 30 printer with integrated paper position detector. This detector improves the repeatability of printing the same structure several times onto one piece of paper to better than 1 mm. By reinserting the paper and overprinting the resistors several times, the base resistance of the sensor can be adjusted by increasing the amount of ink deposited in the same location. The resistor and the conductor ink are filled in separate refillable ink cartridges for black ink and are printed successively by changing the cartridge.

Sensor design process and fabrication

For demonstration of the fabrication process we selected a 3-axial mechanical sensor as demonstration object. It can serve as a very cost efficient tactile input device, or, by applying a mass at the sensor platform, as a 3D-acceleration sensor.

For designing the sensor various 3-dimensional structures are folded from paper. Loads are applied in order to identify the regions of maximum strain. Then the paper is unfolded and the dimensions and the locations of maximum strain are transferred into a layout program for printed circuit boards (Eagle Layout, Cadsoft, Pleiskirchen, Germany). By assigning different layers and colors to the resistors, conductors, dimension labels, folding and cut lines the dimensional information and the functional elements are printed. To decrease the base resistance from 0.5 MΩ to 12.6 kΩ and to reduce the variance in resistor values, the resistors are overprinted six times. The conductors are formed by printing a single layer. Fig. 2 shows the printed structure.

After printing the photo paper is spray coated with a protective coating (ClearJet Type FA Low Gloss, Marabu, North Charleston, SC, U.S.A.) to increase the wear resistance of the inks on paper when exposed to touch.

First structures are cut out with a scalpel as shown in Fig. 3a and subsequently with a laser cutting machine (VLS3.50, Universal Laser Systems, Scottsdale, AZ, USA) for faster and precise cutouts.

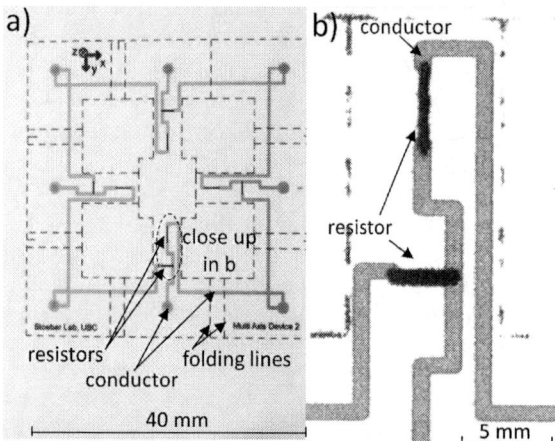

Figure 2: (a) The multi-axes device printed with resistor and conductor ink; b) close-up showing connected conductors and resistors.

To give structural stability the sensor device is bonded with instant adhesive onto a slightly smaller mounting frame. Thereby, the tactile element is raised out of the plane, and it is fixed flexibly in its position (Fig. 3b).

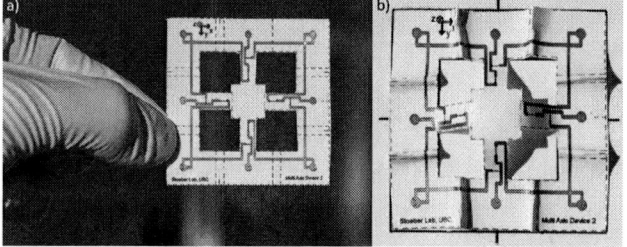

Figure 3: Multi axes sensor as input device; a) structure with cut outs, b) bonded onto a mounting frame.

Sensor system setup

The sensor connects to a mounting frame with electrical landing pads. If the sensor is folded with the resistors to the outside, then the electrical contacts are made with metal clamps. Alternatively, if the structure is folded with the printed side inside the sensor, then the contacts are made via the landing pads on the PCB; having the electrical elements inside protects the sensor better from its environment. For fast prototyping sensors are connected with wires and electrically conductive bonding material.

The sensor is evaluated with a bridge amplifier (5271, Kistler, Winterthur, Switzerland) with integrated voltage supply. This amplifier connects to a measurement board (KPCI-3108, Keithley, Cleveland, OH, USA). The output voltage of the strain sensing voltage divider (Fig. 2b) of one of the four sensor arms is evaluated and the information is transferred to a PC. An inertial measurement unit (IMU) (ADIS16835, Analog Devices, Norwood, MA, USA) is bonded onto the paper sensor device to serve as a reference sensor. The data of this three axis accelerometer is captured by a PC via USB with a separate software (ADiS 16385 Rev. 2, Analog Devices).

RESULTS

Fig. 3 shows a fabricated multi-axis sensing device that is sensitive to force and displacement in three orthogonal axes. The resistors have a base resistance of 12.6 kΩ. Their height is determined by profilometry (DEKTAK 8 Profiler, Plainview, NY) and is below 1 μm. Fig. 2b shows a close-up of the resistors. The deposited conductor ink forms a highly conductive percolation network at room temperatures with resistances of 0.2 Ω / square. Heating the substrate to 100°C for 10 minutes has no significant effect on the sheet resistance.

To demonstrate the function of the sensor as input device, the elevated platform is manipulated with a finger. The signal voltages of one arm are evaluated. The signals reflect the user's force directly with good resolution. To quantify this, the sensor is turned into an inertial device for detailed analysis. The sensor is mounted onto the PCB. Attaching a mass of 50 g to the platform adds the capability of resolving the orientation of the gravity vector (Fig. 4). The sensor signal achieved through rotating the sensor allows comparison with the reference sensor.

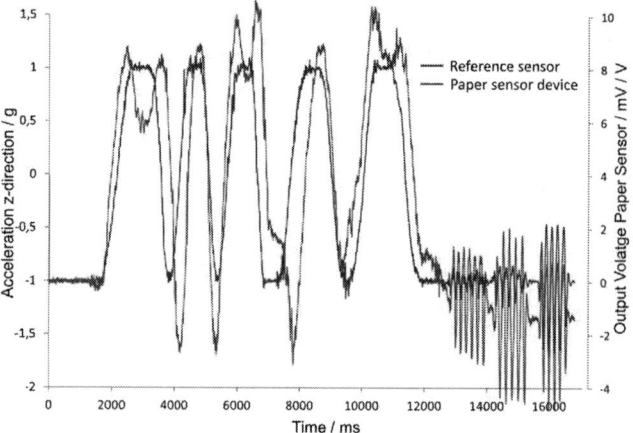

Figure 4: Measured z-acceleration of the multi axes device as inertial sensor with a mass of 50 g mounted to its platform; the device was rotated around the horizontal axis, and the test ends with shaking the device in three different axes.

DISCUSSION

In this experiment a sensitivity of about 8 mV / V g is achieved with a mass of 50 g. This amounts to a force sensitivity of 16.31 mV/ V N when used as a tactile input device or force sensor. Due to the design of the structure the sensitivity can be adjusted widely by using different beam widths and lengths and by changing the paper thickness.

It can be observed that the paper sensor reproduces the signal of the reference sensor. Nevertheless, the peak values of acceleration show a notch and do not resemble the signal curve properly. This can be due to the fact, that the sensors are not rotated properly around one single axis. This assumption is supported by the fact, that the signal shape differs particularly at the 180° turn points (e.g. at 3.000 ms) where the experimenter has to change his grip to rotate the sensor setup by another 180°. This is due to crosstalk from rotation about different axis because the paper sensor has not been compensated for cross-sensitivity by using all of its resistors.

Fig. 4 shows poor offset stability for the paper sensor over the course of the test. Two reasons are suggested: Since currently only one of the four bending arms is evaluated, no inherent temperature compensation of the resistors is implemented. Another reason can be that creeping is induced by the load applied to the sensor. To achieve a more stable signal the signals from all four bending arms should be evaluated and the signals should be paired. Forming the difference of signal voltages from opposite arms leads to higher signals for lateral forces with a first order compensation for disturbances; to derive the normal force component the output voltages can be summed.

CONCLUSIONS AND OUTLOOK

The described method shows the feasibility of printing resistive sensors onto paper substrates with low cost inkjet equipment. This allows for fast and simple design iterations for sensor models. This design flow can help the system designer to demonstrate and improve his ideas.

The technology additionally enables low-cost sensing devices. Even early sensor devices show a performance which could make them suitable for applications with reduced requirements on precision. This includes sensors for "smart" consumer goods in the future, such as interactive food containers, sensitive toothbrushes and the like. Especially sensors for medical application can benefit from a low cost and from a simple and sterile disposal method; the sensor tag can simply be burned.

This work demonstrates that high gage factors and low conductor resistance can be achieved with already available inks. To prove long-term stability and to investigate the influence of moisture and temperature, further experiments have to be performed. The gage factors of the current device will also need to be determined.

Future developments can focus on the replacement of conductive silver ink. Currently, conductive traces cannot be inkjet printed onto conventional document paper. Additionally, even though bending of the conductor traces is possible, folding these conductors likely disrupts the electrical path. For low cost sensing a mechanically flexible, highly conductive and environmentally friendly and cost effective material is favorable.

We have demonstrated a simple sensor fabrication process that supports the design process of complex resistive sensors by "creative hands on design" with functional physical models. Additionally, this technology enables low cost sensors for possible use in medicine or in interactive packages of consumer goods. Future research encompasses the increase of longtime stability by covering the resistors with hydrophobic or diffusion preventing coating.

ACKNOWLEDGEMENTS

The authors thank Ehsan Khajeh, William Lou and Hang Shi, who performed preliminary work on this project. They additionally appreciate the help of James Olson, Ata Sina and John Madden for sharing ideas and lending equipment. This work was supported by the Natural Science and Engineering Research Council (NSERC) of Canada through the Discovery Grant Accelerator Supplement program and by the German Research Foundation (DFG) associated with identification number WE 2308/3-3.

REFERENCES

[1] X.Y. Liu, M. O'Brien, M. Mwangi, X.J. Li, G.M. Whitesides "Paper-based piezoresistive MEMS force sensors" IEEE MEMS 2011, pp. 133-136.

[2] X.Y. Liu, M. Mwangi, X.J. Li, M. O'Brien, G.M. Whitesides, "Paper-based piezoresistive MEMS sensors." Lab on a Chip 11 (2011), pp. 2189-2196.

[3] E. Khajeh, W. Lou and B. Stoeber, "Paper-based strain sensing material" IEEE MEMS 2013, pp. 473-476.

[4] A. C. Siegel., S. T. Phillips, M. D. Dickey, N. Lu, Z. Suo, G. M. Whitesides, "Foldable printed circuit boards on paper substrates" Advanced Functional Materials, 20(1) (2010), 28-35.

[5] Product Description Silver Print 842, M.G. Chemicals Ltd., 9347 - 193 Street, Surrey, B.C., Canada

[6] Product Description Nickel Print 840, M.G. Chemicals Ltd., 9347 - 193 Street, Surrey, B.C., Canada

[7] Product Description Bare Conductive Paint, The Bare Conductive Studio, 98 Commercial St., London E1 6LZ, UK

[8] S. B. Walker, J. A. Lewis, "Reactive silver inks for patterning high-conductivity features at mild temperatures." Journal of the American Chemical Society 134.3 (2012): 1419-1421

[9] J. Perelaer, A. W de Laat, C. E Hendriks, U. S. Schubert, „Inkjet-printed silver tracks: low temperature curing and thermal stability investigation" Journal of Materials Chemistry, 18(27) (2008), 3209-3215.

[10] A. Kamyshny, J. Steinke, S. Magdassi, "Metal-based inkjet inks for printed electronics." Open Applied Physics Journal 4 (2011): 19-36.

[11] B. Shao, Q. Chen, Y. Amin, D. S. Mendoza, R. Liu, L. R. Zheng, "An ultra-low-cost RFID tag with 1.67 Gbps data rate by ink-jet printing on paper substrate" Solid State Circuits Conference (A-SSCC), 2010 IEEE Asian (pp. 1-4)

[13] S. Sally, "Paper Folding", Arcutus Publishing, New York

[14] Technical Documentation. How Paper-based 3D Printing Works. Mcor Technologies Dunleer, 2013

CONTACT

*T. Meiss, tel: +49-6151-16 3795; t.meiss@emk.tu-darmstadt.de

REAL-TIME DYNAMICALLY RECONFIGURABLE LIQUID METAL BASED PHOTOLITHOGRAPHY

Daeyoung Kim[1], Jun Hyeon Yoo[1], Wonjae Choi[2], Koangki Yoo[3] and Jeong-Bong (JB) Lee[1]
[1]Department of Electrical Engineering, The University of Texas at Dallas, TX, USA
[2]Department of Mechanical Engineering, The University of Texas at Dallas, TX, USA
[3]Department of Information and Communication Engineering, Hanbat National University, Daejeon, South Korea

ABSTRACT

We report real-time dynamically reconfigurable photomask by manipulating gallium-based liquid metal in microfluidic channel. As a demonstration of reconfigurable photomask, a polydimethylsiloxane (PDMS) based 7-segments microfluidic channel was designed and fabricated. With on-demand injection and withdrawal of gallium-based liquid metal in each segment channel, single digit numbers ('0' to '9') were dynamically reconfigured. For i-line and 400 nm wavelength UV lights, PDMS showed > 93% of light transmittance while PDMS + Galinstan® showed < 1 % of the light transmittance. In order to investigate achievable minimum feature size, various sizes of line shape, a horse shoe shape and Texas state map shape were demonstrated. The minimum feature size reliably and reproducibly created was 10 μm with the current approach.

INTRODUCTION

Driven by Moore's Law, numerous lithography techniques such as projection printing, immersion printing, absorbance modulation optical lithography, and nano imprint lithography, among others have been developed to achieve ever and ever smaller feature sizes. Although such advancement in lithography technique overall is absolutely phenomenal, it should be noted that the vast majority of lithography techniques only allow "passive" binary patterning of features. There have been a few efforts to realize 3D profiles in photoresist using a tilted and rotated mask [1], a moving mask [2] and light absorbing dyes in microfluidic channel [3]. However, none of these photolithography techniques was for real-time "reconfigurable" photolithography. A liquid crystal (LC) display was demonstrated as a real-time reconfigurable mask, but LC strongly absorbs UV lights and cannot be used for conventional photolithography [4]. MEMS is ideally positioned to provide unprecedented "active" options for truly unique lithography techniques.

In this paper, we report real-time dynamically reconfigurable photolithography technique using liquid metal Galinstan® as UV opaque material and PDMS as UV transparent material. To the best of our knowledge, this is the first demonstration of true real-time reconfigurable photolithography in UV wavelengths.

WORKING PRINCIPLES

Fig. 1a shows conceptual schematic of the liquid metal-based dynamically reconfigurable photolithography in 7-segments microfluidic channel. One of gallium-based liquid metal, Galinstan® (a ternary alloy of gallium, indium, and tin) [5] is selectively filled in each segment channel to create single digit number '2'. UV exposure process can be carried out with the liquid metal-filled 7-segments microfluidic channels as a photomask directly placed on positive photoresist on Si wafer. After developing, the number '2' is transferred to photoresist. It is known that gallium-based liquid metal have a challenging drawback that it gets easily oxidized and wets on almost any surfaces. However, once the liquid metal is exposed to HCl vapor, the oxide layer is removed so that liquid metal can recover non-wetting characteristic and thus it can be controllable [6, 7]. In this work, to reconfigure liquid metal pattern in the microfluidic channels, the 7-segments microfluidic channel is placed on top of a hydrochloric acid (HCl) reservoir, where HCl solution is stored, to have chemical reaction with HCl vapor through gas permeable PDMS and recover non-wetting characteristic of liquid metal. Based on the controllability by removing the surface oxide, another single digit number '3' is created by inserting/withdrawing liquid metal into/from a segment channel in the 7-segments microfluidic channel. After UV exposure and developing, the number '3' is transferred to photoresist.

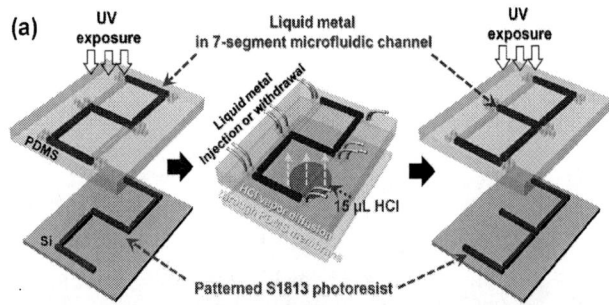

Figure 1: (a) Conceptual schematic of PDMS-based reconfigurable microfluidic photomask: liquid metal opaque pattern can be reconfigured on-demand by injection and withdrawal, and (b) optical images of reconfigured liquid metal Galinstan® in 7-segments microfluidic channel photomask for the demonstration of single digit numbers ('0' ~ '9').

Based on this working principle, single digit numbers ('0' ~ '9') reconfigured with liquid metal in 7-segments microfluidic channel were achieved as shown in Fig.1b.

FABRICATION

We fabricated the PDMS-based 7-segments microfluidic channel using a conventional SU-8 molding technique as shown in Fig. 2. The fabrication process was started with spin coating of SU-8 2025 photoresist (MicroChem, Corp.) on a thermally grown oxidized Si wafer to get approximately 25 μm thick photoresist. The SU-8 photoresist was soft-baked on a hot plate at 65°C for 2 min., 95°C for 3 min., and finally 65°C for 2 min. Then, an UV exposure dose of 140 mJ/cm^2 was applied by standard photolithography to pattern the 7-segments microfluidic channel shape, and finally a post bake was applied with the same process condition as those of the soft bake. Next, PDMS was casted over the SU-8 photoresist mold and cured at room temperature for 24 hours (Fig. 2b). PDMS was then peeled off from the SU-8 mold which has replicated inverse image of the channel shape (Fig. 2c). After through holes were made on the replicated PDMS, the replicated PDMS was bonded to 500 μm thick PDMS layer by applying oxygen plasma treatment. Polytetrafluoroethylene (PTFE) tubes were then connected to microfluidic channels (Fig. 2e). The fabricated 7-segments microfluidic channel (Fig. 2f) has seven straight 500 μm wide and 1 cm long segment channels. Each segment channel has an inlet (100 μm wide) and an outlet (20 μm wide) channel (Fig. 2g).

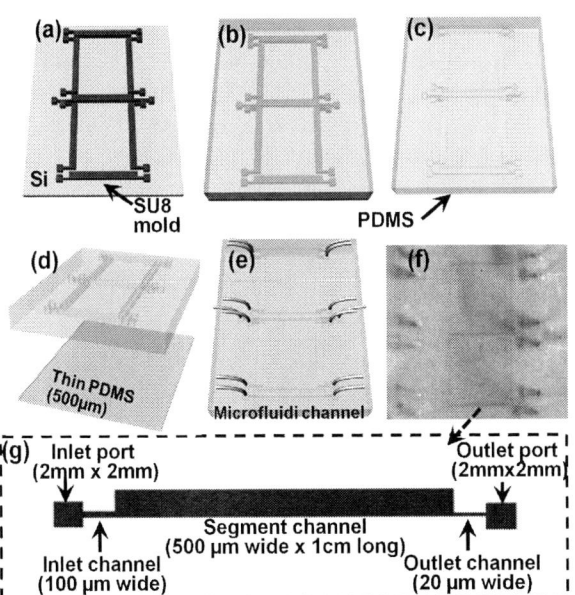

Figure 2: Fabrication sequence of PDMS-based 7-segments microfluidic channel: (a) SU-8 mold, (b) PDMS coating, (c) replicated PDMS, (d) PDMS-PDMS bonding, (e) tubing, (f) optical image of the fabricated microfluidic photomask and (g) top-view of the design of a segment channel.

We intentionally designed width of the inlet channel and the outlet channel to be different so that liquid metal can be easily inserted into the wider inlet channel and filled the straight segment channel. Within certain applied flow rate range, the liquid metal cannot escape from the segment channel to the narrower outlet channel due to its high surface tension. Both inlet and outlet channels are linked to 2mm x 2mm square shape ports where PTFE tubes were connected. Syringes were connected to PTFE tubes and a syringe pump was utilized to insert or withdraw liquid metal in the microfluidic channels on-demand.

RESULTS AND DISCUSSION

After fabrication of the PDMS-based 7-segments microfluidic channel, we studied the feasibility of real-time dynamically reconfigurable photomask. Since this PDMS-based 7-segment microfluidic photomask is not hermetically packaged, liquid metal Galinstan® is likely instantly oxidized as it fills the microfluidic channel. This native oxidation on the surface of the liquid metal Galinstan® may create huge problem of injection and withdrawal due to highly viscoelastic characteristic of the oxide skin of the liquid metal Galinstan®. To resolve this issue, the PDMS-based 7-segments microfluidic channel was directly placed on top of an HCl reservoir which contained 15 μL of 37wt% HCl solutions right before injection and withdrawal of the liquid metal. As reported in the earlier work [6, 7], HCl diffuses through 500 μm thick PDMS, reacts with oxidized liquid metal Galinstan® and recovers its non-wetting true liquid-like characteristic.

As a feasibility study, changing from number "7" to "9" to "4" of the liquid metal photomask was demonstrated as shown in Fig. 3a by injecting and withdrawing liquid metal Galinstan® into/from specific microfluidic channel segments. Liquid metal was injected in four specific segments of the PDMS-based 7-segments microfluidic channel to create number '7'. Next, by injecting additional liquid metal in the middle segment channel, number '9' was created. The average speed of injecting liquid metal into a microfluidic segment channel was 1.25 cm/sec. and the injection flow rate was ~10 μL/min. It should be noted that injection and withdrawal of liquid metal Galinstan® for 1 cm long segment channel takes less than 1 sec.

After the chemical reaction with HCl vapor, the liquid metal was withdrawn with an average speed of 3.34 cm/sec. and the flow rate of ~25 μL/min. We observed there was no residue after the liquid metal was withdrawn. Fig. 3b shows the optical images of the liquid metal photomask showing single digit numbers dynamically changed from '7' to '9' to '4'.

In order to transfer the numbers of liquid metal photomask, Shipley 1813 was spin coated on a Si wafer to have 1.3 μm thick photoresist. It was baked on a hot plate at 115°C for 1 min. Then, a UV exposure dose of 120 mJ/cm^2 was applied by standard photolithography using the liquid metal photomask. Next, it was developed in MF-319 (Microposit®, 2~3% tetramethylammonium hydroxide) for 1 min. under gentle stirring condition. Fig. 3c shows optical images of patterned S1813 corresponding to the reconfigured numbers from '7' to '9' to '4'of the liquid metal photomask.

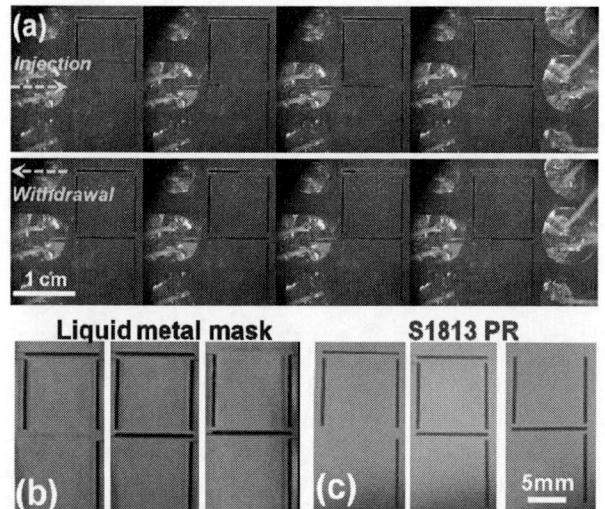

Figure 3: (a) A series of time-lapse images of injection and withdrawal of liquid metal Galinstan® to create number '7', '9', and '4'in a 7-segments microfluidic channel, (b) optical images of number '7', '9', and '4' in the 7-segment liquid metal reconfigurable photomask and (c) optical images of patterned S1813 PR using the photomask shown in (b).

In order to find out the minimum feature size of the patterned photoresist using liquid metal microfluidic photomask, the line shape channel with different width were designed and fabricated using a conventional SU-8 molding technique as explained in the previous section. Various line shape channel was designed to have different widths such as 1, 2, 5, 10, 25, 50, 100, 250, and 500 μm. Among them, we found that the liquid metal cannot be injected into microfluidic channels with widths smaller than 5 μm due to its high surface tension. Therefore, the minimum feature size reliably and reproducibly achieved in the microfluidic photomask was 10 μm.

Fig. 4 shows optical images of liquid metal-filled line shape channels with various widths and corresponding patterned S1813 photoresist.

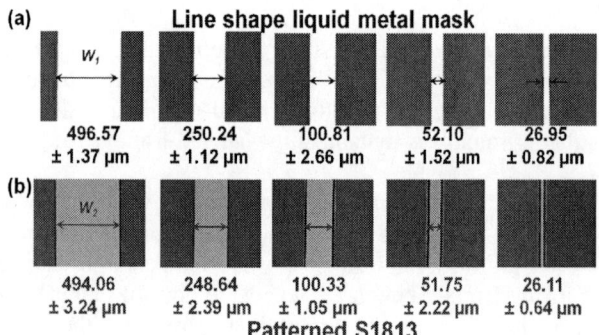

Figure 4: Optical images of (a) line shape liquid metal mask and (b) patterned S1813 PR.

We measured the width of the liquid metal mask and the patterned S1813 and compared each other as shown in Table 1. The width (W_1) of the liquid metal in the line shape channel was slightly larger than that of the target width (W) except 500 μm wide line channel. Specifically, as the target width was decreased, the deviation was increased. We believe that this is attributed to the expansion of the flexible PDMS microfluidic channel by applying higher flow rate to inject liquid metal into the narrower channels.

Table 1: Comparison of width of the filled-liquid metal and patterned S1813 PR.

Target Width (W) (μm)	Liquid metal Galinstan® width in channel (W_1)			Patterned S1813 width (W_2)			
	Average value (μm)	Error bar (±)	Deviation from 'W' (%)	Average value (μm)	Error bar (±)	Deviation from 'W' (%)	Deviation from 'W_1' (%)
10	11.63	0.92	14.01	9.98	2.65	0.20	14.19
25	26.95	0.82	7.24	26.11	0.64	4.44	3.12
50	52.10	1.52	4.03	51.75	2.22	3.50	0.67
100	100.81	2.66	0.80	100.33	1.05	0.33	0.48
250	250.24	1.12	0.10	248.64	2.39	0.54	0.64
500	496.57	1.37	0.70	494.06	3.24	1.19	0.51

The patterned width (W_2) of the photoresist is slightly smaller than that (W_1) of the liquid metal pattern in the photomask. The maximum deviation of the patterned feature size from the target width (W) and liquid metal width (W_1) was approximately 4.44 % and 14.19 %, respectively. It is believed that this photolithography is essentially a 'proximity printing' which is affected by 'Fresnel diffraction' [8]. Therefore, the patterned size of the photoresist turned out to be smaller than those of the liquid metal in the photomask channel. Fig. 5 shows the schematic of proximity printing applied to our study using liquid metal photomask. The 'g' indicates the gap between the liquid metal microfluidic photomask and the photoresist on a Si wafer. In this work, the gap was determined by the thickness of the bottom PDMS layer (500 μm).

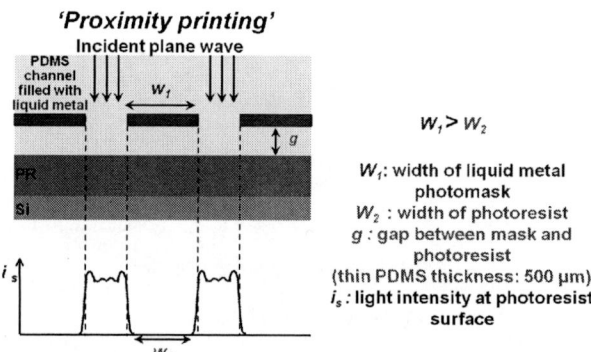

Figure 5: Schematic of working principles for proximity printing using liquid metal microfluidic photomask.

The 'W_1' and 'W_2' indicate the width of the liquid metal photomask and the patterned photoresist, respectively. The 'i_s' indicates the light intensity at the photoresist surface. When the incident plane wave passes through the mask aperture, light bends from the mask aperture edge due to light diffraction. Thus, it produces exposure outside of the mask aperture on the photoresist and it results in decrement of the width of the patterned positive photoresist.

In addition to straight line shape, a horse shoe shape and a Texas state map shape channel were fabricated to demonstrate feasibility of this reconfigurable photomask's patterning capability. The horse shoe shape channel was 25 μm wide and the Texas state map shape was 1.53 mm wide and 1.24 mm high. Fig. 6 shows optical images of the liquid metal-filled channels and corresponding patterned S1813 photoresist for those designs.

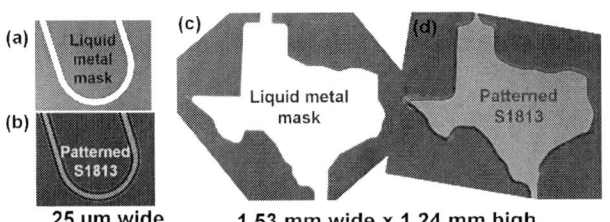

Figure 6: Optical images of (a) and (b) a 25 μm wide horse shoe shape, (c) and (d) Texas state map shape in liquid metal mask (a, c) and patterned S1813 PR (b, d).

To verify liquid metal's opacity at UV wavelengths, we compared light intensity of the i-line (365 nm) and 400 nm wavelength UV light after passing through 5 mm thick PDMS sheet, PDMS (5 mm thick) + liquid metal Galinstan® + PDMS (500 μm thick) and conventional chrome (Cr) opacity layer in soda-lime glass Cr mask as shown in Fig. 7. For i-line UV light, PDMS showed 93.6% of light transmittance while PDMS + Galinstan® + PDMS showed mere 0.62%. The trend was quite similar for the 400 nm UV light as PDMS showed 96.24% transmittance while PDMS + Galinstan® + PDMS showed only 0.22%. These are strong evidences that liquid metal Galinstan® is an effective opaque material in photomask application.

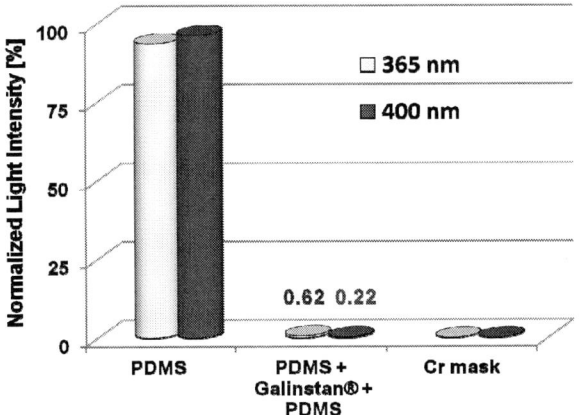

Figure 7: Normalized light intensity measured for i-line (365 nm) and 400 nm wavelengths of UV light. PDMS is 93~96 % transparent and liquid metal Galinstan® shows < 0.62 % transmittance which is very close to those of the Cr layer in conventional photomask (0 %).

CONCLUSION

In this paper, we demonstrated dynamically reconfigurable photomask in the PDMS-based 7-segments microfluidic channel. Liquid metal can act as an effective opaque material against i-line (365 nm) and 400 nm wavelengths of UV light. Based on the manipulation of liquid metal in the microfluidic channel, the 'active' reconfigurable single numbers ('0' to '9') were demonstrated. With this technique, we found that the attainable minimum feature size was 10 μm. We believe this novel real-time dynamically reconfigurable photolithography technique can enable myriads of unforeseen patterning applications and give far-reaching impact to a variety of disciplines.

ACKNOWLEDGEMENTS

The authors would like to thank Republic of Korea (ROK) Army for financial support. This research was also supported by MKE (The Ministry of Knowledge Economy), Korea, under the Brain Scouting Program (HB606-12-2001) supervised by the NIPA (National IT Promotion Agency).

REFERENCES

[1] M. Han, W. Lee, S.-K. Lee, S.S. Lee, "3D microfabrication with inclined/rotated UV lithography," *Sensors and Actuators A: Physical,* vol. 111, pp. 14-20, 2004.

[2] Y. Hirai, Y. Inamoto, K. Sugano, T. Tsuchiya, and O. Tabata, "Moving mask UV lithography for three-dimensional structuring," *Journal of Micromechanics and Microengineering,* vol. 17, p. 199, 2007.

[3] C. Chen, D. Hirdes, A. Folch, "Gray-scale photolithography using microfluidic photomasks," *Proceedings of the National Academy of Sciences,* vol. 100, pp. 1499-1504, February 18, 2003 2003.

[4] Q. Peng, C. Zhou, Z. Cui, S. Liu, Y. Guo, B. Chen, J. Du, Y. Zeng, "Real-time photolithographic technique for fabrication of arbitrarily shaped microstructures," *Optical Engineering,* vol. 42, pp. 477-481, 2003.

[5] Galinstan Safety Data Sheet [Online]. Available: http://www.rgmd.com/msds/msds.pdf

[6] D. Kim, P. Thissen, G. Viner, D.-W. Lee, W. Choi, Y.J. Chabal, J.-B. Lee, "Recovery of Nonwetting Characteristics by Surface Modification of Gallium-Based Liquid Metal Droplets Using Hydrochloric Acid Vapor," *ACS Applied Materials & Interfaces,* vol. 5, pp. 179-185, 2013.

[7] G. Li, M. Parmar, D. Kim, J.-B. Lee, D.-W. Lee, "PDMS based coplanar microfluidic channels for the surface reduction of oxidized Galinstan," *Lab on a Chip,* vol. 14, pp. 200-209, 2014.

[8] M. D. James D. Plummer, Peter B. Griffin, *Silicon VLSI Technology*: Prentice Hall, Inc., 2000.

CONTACT

*D. Kim, tel: +1-972-693-0988; daeyoung@utdallas.edu

RELEASE AND TRANSFER OF LARGE-AREA ULTRA-THIN PDMS

Jinsheng Gao[1], Dongzhi Guo[2], Suresh Santhanam[1], YingJu Yu[2],
Alan J.H. McGaughey[2], Shi-Chune Yao[2] and Gary K. Fedder[1]

[1] Department of Electrical and Computer Engineering, Carnegie Mellon University, Pittsburgh, PA USA
[2] Department of Mechanical Engineering, Carnegie Mellon University, Pittsburgh, PA USA

ABSTRACT

This paper reports on the fabrication of ultra-thin (~10 µm) polydimethylsiloxane (PDMS) films with embedded metal electrodes of 2 µm minimum feature size, as well as the release and transfer of large area films (>5 cm). The initial motivation for this work is the development of a miniature pump actuator for moving working fluid in an electrocaloric microcooler. PDMS diaphragms with electrodes are released and transferred onto contoured silicon chambers formed by gray-scale lithography and deep-reactive ion etching.

Keywords: PDMS transfer, diaphragm with electrode, electrocaloric cooling

INTRODUCTION

Polydimethylsiloxane (PDMS) has been widely used in flexible electronics, optics and biomedical research due to its excellent biocompatibility, physical properties and fabrication flexibility [1]. Its implementation includes a wide range of applications such as large-area conformable displays and stretchable integrated circuits [2]. In addition, due to their elastic properties, PDMS structures can function as an active mechanical component in MEMS devices [3].

However, due to its low Young's Modulus, poor microfabrication compatibility, and relatively high adhesion, micron electrode size (10 µm or smaller) and ultra-thin (25 µm or thinner) PDMS device fabrication is challenging. Traditional patterning techniques on PDMS include transfer printing and shadow masking. Those techniques limit the minimum patterned feature size to about 100 µm. Moreover, in literature reports on PDMS transferring to date, the thickness of PDMS is generally in the mm range.

In this paper, we report on the fabrication of 10 µm-thick PDMS films with embedded metal electrodes having minimum feature size of 2 µm, as well as the release and transfer of films with large area that can extend to greater than 5 cm diameter.

ELECTROCALORIC MICROCOOLER

The motivation for this work is the development of a miniature pump actuator for moving heat transfer fluid in an electrocaloric microcooler [4]. The conceptual electrocaloric microcooler cooling element, illustrated in Figure 1, is 1cm-long, 7 mm-wide, has a thickness of 300 µm, and includes two chambers made of cavities and diaphragms that are fabricated on a silicon wafer. A folded ferroelectric relaxor terpolymer layer of 10 µm thickness separates the two chambers thermally. The 2mm wide diaphragms are made of two PDMS layers embedded with electrodes and are driven electrostatically. Spacers made of SU-8 photodefinable epoxy are placed between the terpolymer layers to form channels to allow the passage of fluid. The working fluid is Galden HT-70 (Solvay Solexis, Inc), which is a heat transfer liquid and an electrical insulator.

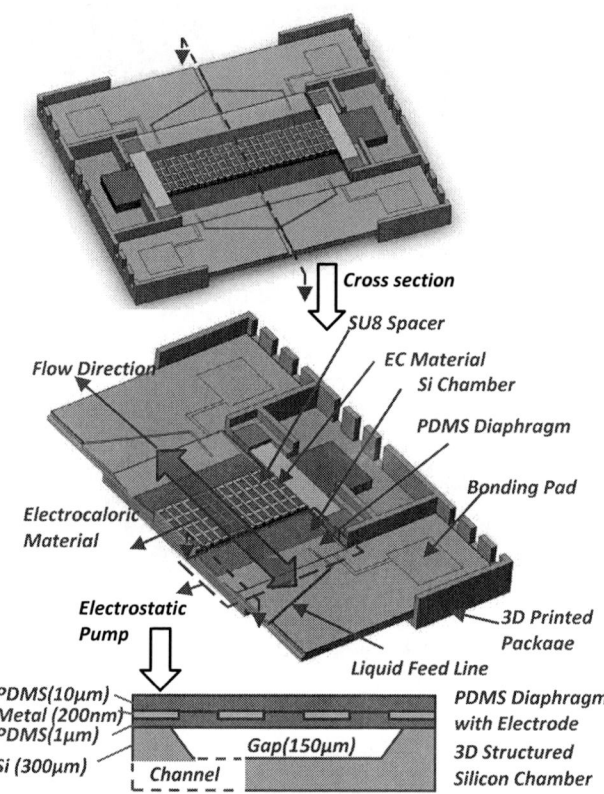

Figure 1: Solid-model view of the electrocaloric microcooler with dimensions 1 cm long, 7 mm wide 300 µm thick. The electrocaloric module is located between two electrostatic miniature pumps actuated electrostatically in the hot and cold chambers to drive the working fluid.

The designed refrigeration system is based on the electrocaloric effect in the terpolymer. The electrocaloric terpolymer provides reversible temperature and entropy change in the material due to polarization under the application and removal of an electric field. The diaphragms of the miniature pump on either side of electrocaloric terpolymer are actuated electrostatically in the hot and cold chambers to drive the working fluid, which transfers heat between the two chambers. Through synchronization of the electrical and mechanical cycling, the heat is extracted from the cold fluid chamber and released to the hot chamber to

978-1-4799-3510-9/14 $31.00 © 2014 IEEE

achieve cooling [5].

ULTRA-THIN PDMS DIAPHRAGMS
Fabrication with Embedded Electrodes

Inspired by the recent development of photolithography on the PDMS and encapsulation layers [6], we present a method to fabricate 10μm thick PDMS films having embedded metal electrodes with minimum feature size of 2 μm.

The PDMS film is made with three thin-film layers (sequentially: 1 μm PDMS, 0.2 μm Cr/Au/Cr, 10 μm PDMS) using the process flow in Figure 2a. A completed film with embedded metal electrodes, shown in Figure 2b, serves as the actuator diaphragm. In this release process, a thin polymer film, deposited on the substrate from a C_4F_8 precursor in a Surface Technology System (STS) inductively coupled plasma tool, acts as an anti-adhesion layer for eventual lift-off release of the completed PDMS diaphragm. The Cr/Au/Cr metal layers are then evaporated on the first PDMS layer, and are ion milled in a serpentine mesh pattern to prevent buckling and breakage that may be caused by intrinsic or actuation stresses [7]. Access openings to pads connected to the embedded wiring are created by aluminum masking and CF_4/O_2 plasma ashing of the PDMS. Low temperature treatments (<70°C) and slow temperature ramping rates are applied in the process steps to prevent potential metal cracking within PDMS films on the Si substrate due to different temperature expansion coefficients.

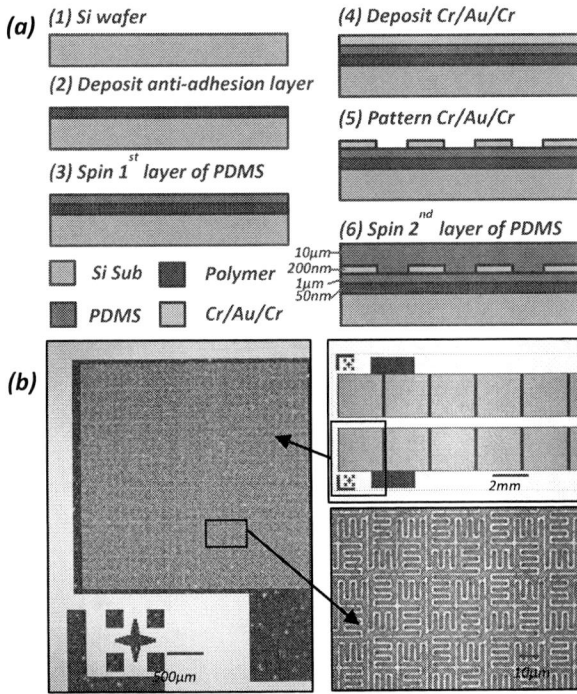

Figure 2: (a) The process flow for the fabrication of PDMS film with Cr/Au/Cr electrode embedded. (b) The optical microscope picture of fabricated PDMS film.

Initial Process for Transfer and Assembly

A silicon fluid chamber is formed by deep reactive-ion etching (DRIE) in the STS system and acts as the counter electrode for the diaphragm. The PDMS film is released and transferred onto the chamber by a lift-off process (Figure 3a). Heat release tape (Nitto Denko Corp.) is attached to the fabricated PDMS film which is then peeled from the Si substrate. Both the peeled-off film and fabricated chamber are treated in O_2 plasma for 30s to activate the PDMS to increase adhesion. The two parts are aligned and placed into contact using a device bonder. The heat release tape is then removed at 150°C. The final assembled miniature pump is shown is Figure 3b. The optical microscope picture shows the PDMS film with electrode embedded hanging over the Si chamber.

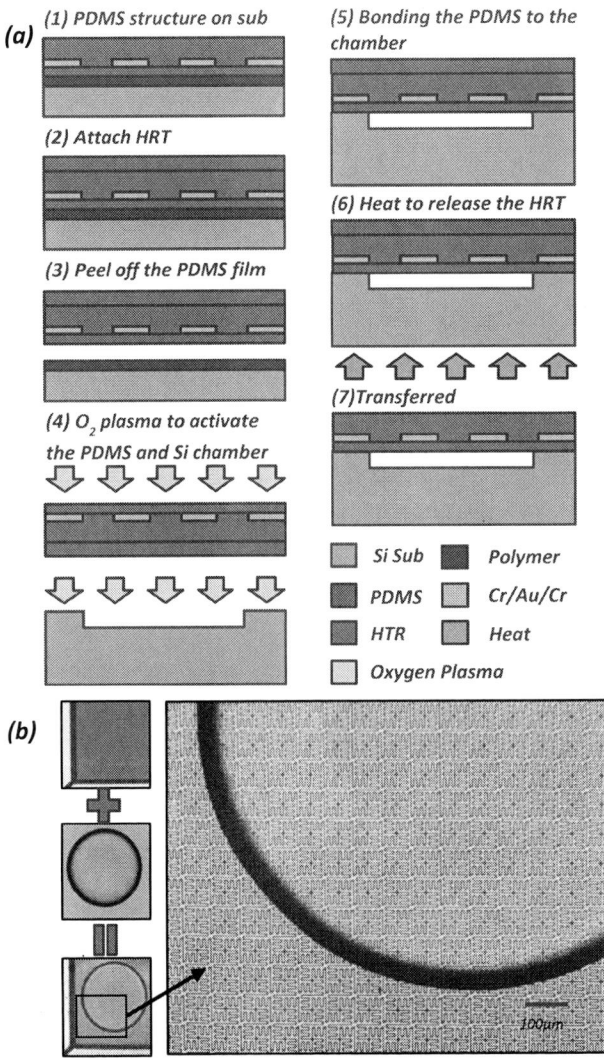

Figure 3: (a) The process flow for the transferring of ultra-thin PDMS film onto the fabricated Si chamber to create the electrostatic pump structure. (b) The optical microscope picture showing a PDMS film with electrode embedded hanging over a fabricated Si chamber.

Fabrication with Gelatin Sacrificial Layer

The above transfer procedure works reliably for devices around 1 mm in size, but the thermal expansion mismatch of the PDMS and heat release tape leads to buckled diaphragms and torn films for larger areas. An alternative way for releasing the PDMS film by using gelatin as a sacrificial layer enables PDMS transfer for much larger areas. The gelatin was chosen as a release layer due to its low processing temperature and high water solubility.

The gelatin powder (250 bloom, Modernist Pantry) is first dissolved in warm water (33 wt%). The solution is spun on the Si substrate with infrared heating applied simultaneously to prevent the gelatin from hardening, and then it is baked in a 40°C oven. The PDMS is spun on the gelatin surface. Then the diaphragm with embedded electrodes is fabricated by following the process flow described in the previous section and shown in Figure 2a. Due to the different thermal expansion coefficient values of the gelatin, PDMS and metals, stress is induced during evaporation, which leads to the ripples seen in the pad area in Figure 4b.

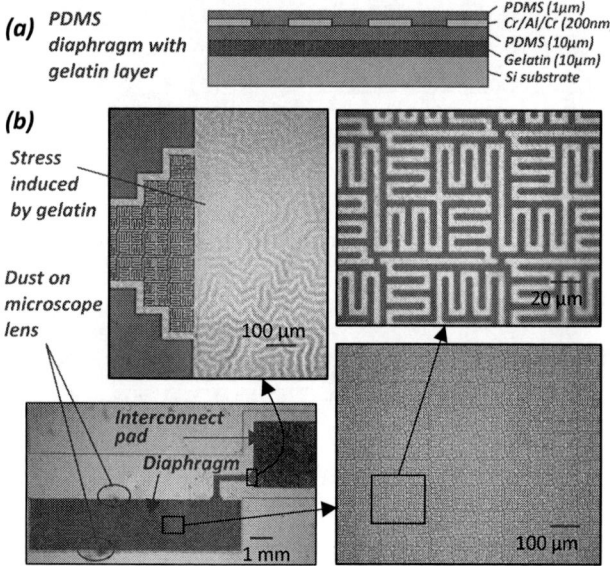

Figure 4: (a) Cross section of PDMS diaphragm with embedded electrode fabricated on the gelatin sacrificial layer. (b) Optical microscope photos of the fabricated diaphragm with bonding pad.

3D Structured Silicon Chamber

To optimize the performance of the electrostatic miniature pump, a "zipping" shaped substrate for the counter electrode is created by gray-scale lithography and is designed to reduce the pull-in voltage required to actuate the diaphragms. The silicon chamber also enables efficient heat transfer between the fluid and heat source/sink to improve the performance of the cooling element. Channels are opened at the outlet of chambers to allow the fluid to pass, as shown in Figure 5a.

Gray-scale technology has enabled the development of arbitrary 3D microstructures in various materials [8]. The process flow in Figure 5b shows 3D shaping of silicon performed in a single photolithography step with subsequent dry etching. AZ4210 photoresist is spun on the Si substrate at

2500 rpm to create a 2.7 μm layer followed by gray-scale lithography. The silicon chamber electrode shape is then formed by photoresist ashing and silicon DRIE to a depth of 150 μm. The channels to allow the fluid to pass are then created by using the back side DRIE. The resulting chamber is shown in Figure 5c.

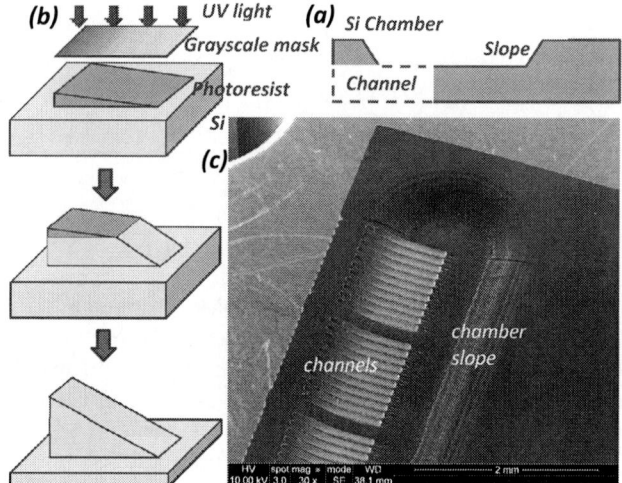

Figure 5: (a) Cross section of 3D structured Si chamber for the counter electrode. The zipping slope is designed to reduce the pull-in voltage requirement. (b) Process flow of the gray-scale lithography technology. (c) SEM picture of fabricated Si chamber with zipping slope and fluid channels.

Transfer to Pump Assembly

The process flow for assembly of the 10 μm PDMS diaphragm on the gray-scale Si chamber is shown in Figure 6a. After 30s O_2 plasma surface activation treatment for the PDMS diaphragm and the Si to increase adhesion, the two components are aligned and bonded using the device bonder. The assembled device is then dipped in 40°C warm water. The 10 μm gelatin layer dissolves after about 2 hours, releasing the miniature pump. The device is then dried in an oven at 40°C for 30 min. The bonding pads are created with an aluminum shadow mask in the reactive ion etching system by ashing the PDMS with CF_4/O_2 gas mixer. The bonding wires are then connected. The final fabricated miniature pump is shown is Figure 6b.

In the preliminary actuation test shown in the Figure 7, the diaphragm moves by electrostatic actuation using 200 V, which is below the breakdown voltage of the PDMS based on our tests. When the electrical field is on, the focus of the optical microscope blurs due to the movement of the PDMS film. In the actuation test, the stiffness of PDMS film is also found to be larger after evaporation and reactive ion etching due to surface oxidation [9].

TRANSFER OF LARGE AREA PDMS

We also tested the release and transfer ability for large area ultra-thin PDMS films. Shown in Figure 8, a large >7cm diameter, uniform 10 μm-thick PDMS film was successfully

978-1-4799-3510-9/14 $31.00 © 2014 IEEE

released from the substrate after attaching to an aluminum release ring coated with vacuum grease. Dissolution of the gelatin for such a large area takes about 20 hours.

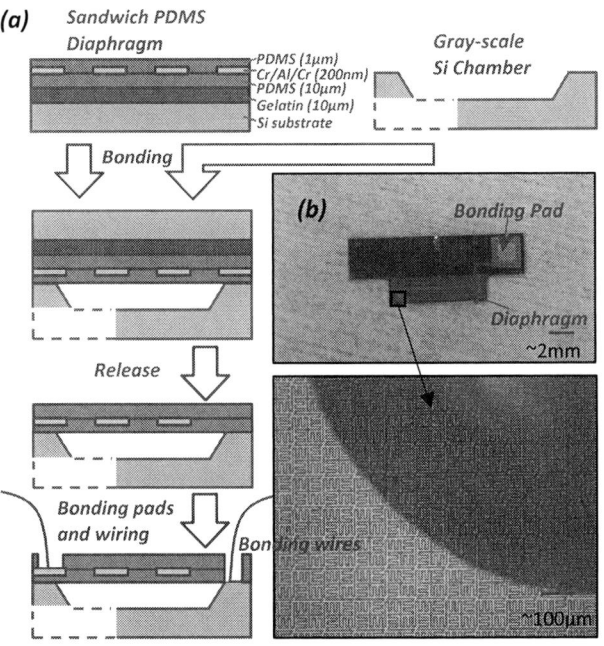

Figure 6: (a) Process flow for transfer and release of PDMS diaphragm with gelatin sacrificial layer, and assembly of the miniature pump. (b) Fabricated pump with bonding pads.

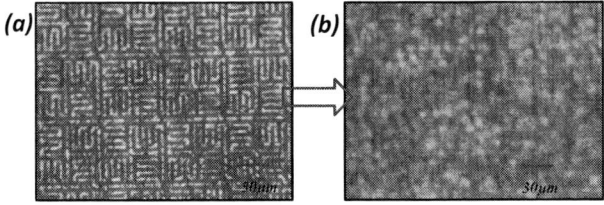

Figure 7: (a) Optical microscope picture of the PDMS film in focus when the electrical field is off. (b) When the electrical field is on, the focus of the optical microscope blurs due to the movement of the PDMS film.

CONCLUSION

The reported method of release and transfer of ultra-thin large area PDMS films, particularly with small feature size embedded electrodes, opens a promising route for fabricating new compliant MEMS and electronic devices.

ACKNOWLEDGEMENT

This work was supported by the Defense Advanced Research Projects Agency (DARPA) and the U.S. Army Aviation and Missile Research, Development, and Engineering Center (AMRDEC) under Grant No. W31P4Q-10-1-0015. The views and conclusions contained in this document are those of the authors and should not be interpreted as representing the official policies, either expressed or implied, of DARPA, the U.S. Army, or the U.S. Government.

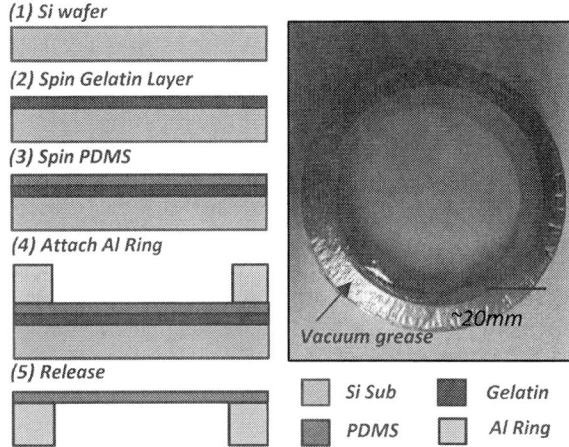

Figure 8: Release and transfer of large area (>7cm diameter) PDMS layer by using gelatin sacrificial layer and aluminum ring.

REFERENCES

[1] T. Sekitani, et al., "A Rubberlike Stretchable Active Matrix Using Elastic Conductors", *Science 321 1468*

[2] S. Lacour, et al., "Mechanisms of reversible stretchability of thin metal films on elastomeric substrates", *Appl. Phys. Lett. 88 204103 (2006)*

[3] S. Rosset, et al., "Metal ion implantation for the fabrication of stretchable electrodes on elastomers", *Adv. Funct. Mater. 19 470–8 (2009)*

[4] J. Gao, et al., "Stirling Microcooler Array with Elemental In-Plane Flow", *37th GOMACTech Conf., Las Vegas, 2012*

[5] D. Guo, et al., "Design of a Fluid-Based Micro-scale Electocaloric Refrigeration system", *ASME 2013 Summer Heat Transfer Conference, Minneapolis, MN, 2013*

[6] T. Adrega, et al., "Stretchable gold conductors embedded in PDMS and patterned by photolithography: fabrication and electromechanical characterization", *J. Micromech. Microeng. 20 (2010) 055025*

[7] J. Neumann, et al., "CMOS-MEMS membrane for audio-frequency acoustic actuation", *Sensors and Actuators A 95 (2002) 175-182*

[8] C. Waits, et al., "Microfabrication of 3D silicon MEMS structures using gray-scale lithography and deep reactive ion etching", *Sensors and Actuators A 119 (2005) 245–253*

[9] P. Gorrn, et al., "Topographies of plasma-hardened surfaces of poly(dimethylsiloxane)", *J. Appl. Phys. 108, 093522 (2010)*

CONTACT

*Jinsheng Gao, tel: +1-412-268-3059; jinsheng@andrew.cmu.edu

SOLID STATE MEMS DEVICES ON FLEXIBLE AND SEMI-TRANSPARENT SILICON (100) PLATFORM

Sally M Ahmed, Aftab M Hussain, Jhonathan P Rojas, and Muhammad Mustafa Hussain
Integrated Nanotechnology Lab, King Abdullah University of Science and Technology, Saudi Arabia

ABSTRACT

We report fabrication of MEMS thermal actuators on flexible and semi-transparent silicon fabric released from bulk silicon (100). We fabricated the devices first and then released the top portion of the silicon ($\approx$ 19 µm) which is flexible and semi-transparent. We also performed chemical mechanical polishing to reuse the remaining wafer. A tested thermal actuator with 3 µm wide 240 µm hot arm and 10 µm wide 185 µm long cold arm deflected by 1.7 µm at 1 V. The fabricated thermal actuators exhibit similar performance before and after bending. We believe the demonstrated process will expand the horizon of flexible electronics into MEMS world devices.

KEY WORDS

Flexible electronics, thermal flexure actuator, structural layer, sacrificial layer.

INTRODUCTION

Flexible and stretchable electronics, which can conform to non-planer surfaces, have gained increased attention recently. Many applications such as flexible displays [1-4], flexible energy generation [5, 6], energy storage devices [7-9], health monitoring devices [10] and artificial skin [11, 12] were illustrated on flexible substrates. Most of the work on flexible electronics is based on plastic substrates which are inherently flexible and transparent. However, plastic has low melting point and is incompatible with high performance CMOS fabrication processes. In order to integrate MEMS devices with high performance inorganic electronics, mono-crystalline (100) silicon is the best substrate option due to its low cost and well-established industry. However, one major challenge of using silicon in flexible electronics is its brittleness and inflexibility. Yet, if made thin enough, silicon wafers can be flexible. Different techniques were developed to use silicon in flexible electronics using silicon-on insulator (SOI) or silicon (111) wafers, which are expensive compared to traditional bulk silicon (100). In the recent past, our group has demonstrated a powerful technique for producing flexible semitransparent silicon sheets from mono-crystalline (100) silicon wafers using the standard micro-fabrication techniques [13-15]. Then chemical mechanical polishing is performed to reuse the remaining wafer.

In the MEMS field, few MEMS devices on flexible substrates have been demonstrated. F. Jiang *et al.* demonstrated a shear stress sensor on silicon islands formed by a combination of anisotropic wet and dry etching of the back side of a silicon wafer. The silicon islands were connected by polyimide strips [16]. However, the complexity of the fabrication process and resource wasting are the main drawbacks of this technique. MEMS temperature sensor was demonstrated by S. Xiao *et al.* on flexible polyimide substrates using Si wafer as a carrier substrate [17]. However, this technique is incompatible with CMOS processes because of the limited thermal budget of the polyimide.

In this paper, we customize our generic flexible electronics fabrication process using mono-crystalline (100) silicon to form free standing and movable MEMS structures on flexible, semi-transparent silicon sheets. Among the MEMS devices, thermal actuators were chosen to demonstrate the viability of the developed process.

MEMS thermal actuators are micro-electrical mechanical devices that generate mechanical movement due to thermal expansion. They can be classified into two categories based on the direction of the tip deflection: in plane [18] and out of plane [19]. We fabricated thermal flexure actuators that produce in plane tip deflection, as with flexing process it will be exposed to unfamiliar territory of non-planar surface oriented deflection.

DESIGN

A thermal flexure actuator consists of two arms: a hot thin arm and a cold wide arm connected at the tip and separated by a small air gap as shown in Figure 1. The hot arm has a smaller cross section area compared to the cold arm. Therefore, when current passes through the arms, it expands more than the cold arm due to its higher resistance and deflects towards the cold arm. The pads contain some etch holes required for silicon fabric release as described later on. In addition, a built-in scale was fabricated beside each actuator to measure the deflection of the tip.

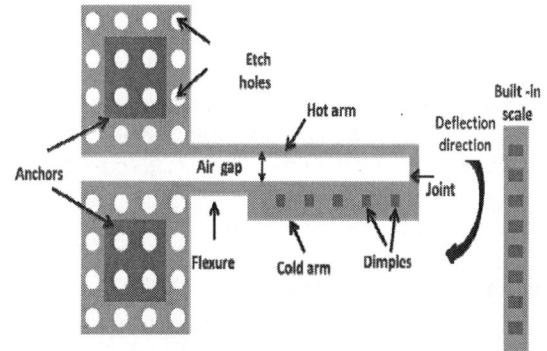

Figure 1: A thermal flexure actuator design.

The main reason for choosing thermal flexure actuator design over other types of thermal actuators is that it is anchored from one end only which reduces the chances of breaking the device while bending the substrate, unlike the chevron actuator which is anchored from both ends [19].

Design Limitations

There are some limitations on the thermal actuator design enforced by the etch holes width and separation. According to our previous study, the width of the etch holes and the separation between them affect the release time and the final thickness of the released silicon piece [13]. The etch holes are 10 μm wide separated by 20 μm. Therefore, the total width of the actuator arms and the air gap separation should be < 20 μm as shown in Figure 2. This restricted the hot arm width to 3 μm, cold arm width to 10 μm and the air gap to 3 μm. Increasing the separation between the holes will add flexibility in the design but will increase the release time to a couple of hours.

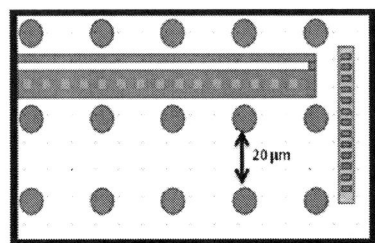

Figure 2: Thermal actuator arms location

MATERIAL SELECTION

Choosing the appropriate materials of the actuator that suits the silicon fabric release process was a challenging task. Silicon dioxide is used for insulation. The structural layer should be a material that is not etched by xenon difluoride (XeF_2). Therefore, polysilicon which is the most common material in MEMS actuators cannot be used and gold (Au) is chosen instead as a structural layer. On the other hand, the sacrificial layer should be a material that can be etched isotropically by a vapor phase etchant such as XeF_2 or vapor HF to avoid wet etching and critical point drying. However, HF cannot be used since it will etch SiO_2, the insulation layer. Therefore, amorphous silicon which is etched by XeF_2 is chosen as the sacrificial layer.

FABRICATION PROCESS

The fabrication process is illustrated in Figure 3. In all steps, conventional micro-fabrication and photolithography techniques are used. On mono-crystalline <100> silicon wafer, a 300 nm thick SiO_2 layer is thermally grown which acts as both the thermal and electric insulation layer followed by the deposition of 1 μm amorphous silicon sacrificial layer. Subsequently, the anchors are formed by etching through the amorphous silicon layer. The same photoresist used for anchor patterning is used to lift-off a bi-layer of Ti/ Au where Ti acts as an adhesion layer and the Au layer prevents Ti oxidation. After removing the photoresist with acetone, a thick layer of Au, which is the structural layer, is deposited by sputtering and patterned

using reactive ion etching (RIE). Next, 10 μm wide etch holes separated by 20 μm are formed by etching through all layers followed by deep RIE of silicon using the BOSCH process. To protect the side walls of the etch channels, a thin layer of aluminum oxide (Al_2O_3) is deposited by atomic layer deposition (ALD) and anisotropically etched to remove the oxide laterally and form vertical side walls around the inner side of the trenches. XeF_2 is then used to perform isotropic etch to release the thermal actuators by etching the amorphous silicon sacrificial layer and simultaneously release the ultra-thin silicon sheet. Next, the bottom portion of the bulk substrate was planarized using chemical mechanical polishing (CMP). It can be subsequently used for fabricating the next set of devices and to release again. This regenerative process continues until the whole wafer is consumed.

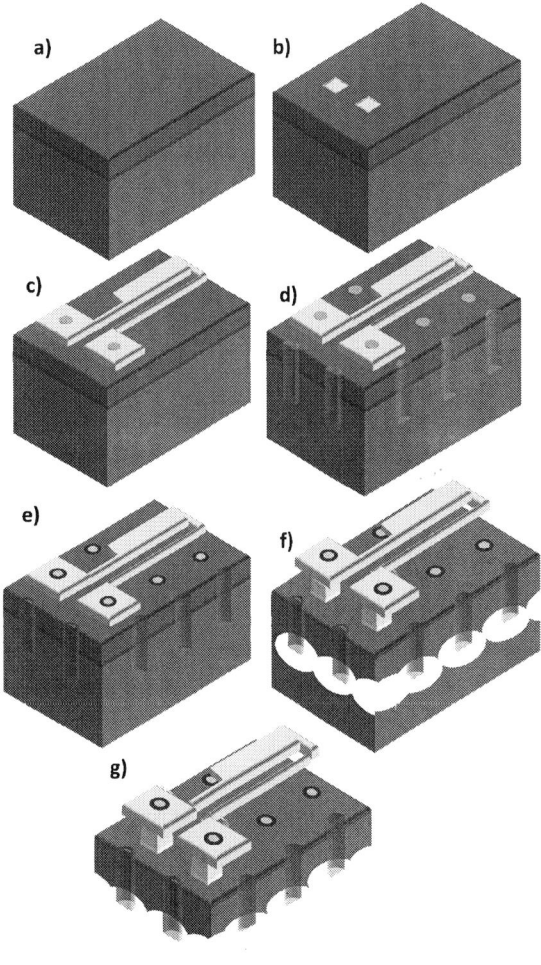

Figure 3: Fabrication Flow a) 300 nm thermal oxide growth followed by 1μm amorphous silicon deposition, b) anchor patterning and Ti/Au deposition, c) Au structural layer deposition and patterning, d) etch channel formation through RIE followed bye DRIE, e) Al_2O_3 deposition and etching for side wall protection, f) sacrificial layer etching and Si release and g) peeling off the released thin Si sheet.

RESULTS AND DISCUSSION

Figure 4.a shows a SEM image of a released thermal actuator on a flexible silicon piece and a cross section image of that piece (Figure 4.b). The released silicon sheet has a thickness of 19 μm or less, which makes it very flexible with a measured minimum bending radius of 6.14 mm. Figure 5 (a and b) show a 2.4 cm × 1.4 cm semi-transparent flexible silicon piece which was fully released in XeF$_2$ in approximately 3 hours. The semi-transparency is caused by the etch holes.

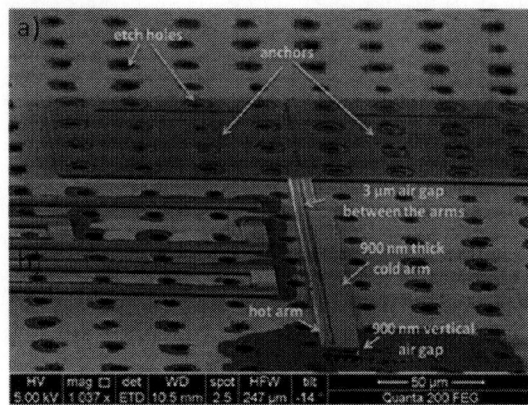

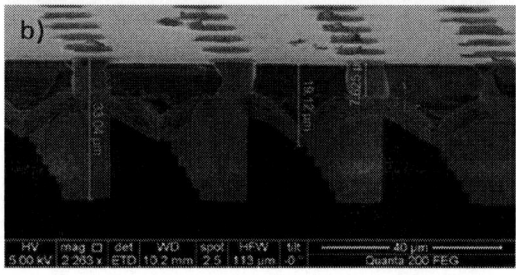

Figure 4: a) SEM image of released thermal actuator on flexible Si piece and b) SEM cross sectional image of the flexible Si piece.

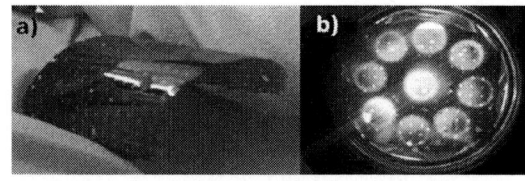

Figure 5: a) A 2.4 cm x 1.4 cm released Si piece and b) Semi-transparent Si piece.

A tested device with 3 μm wide 240 μm hot arm and 10 μm wide 185 μm long cold arm deflected by 1.7 μm at 43 mA and 1 V (Figure 6). As the current increases, the temperature difference between the arms increases causing larger tip deflections. After bending the flexible piece for more than 15 times, the same thermal actuator produced 1.5 μm deflection at the same current and slightly larger voltage (1.4 V) (Figure 7). The change in the performance of the device is due to a change in the resistance of the actuator after testing it several times and probably not due to substrate bending.

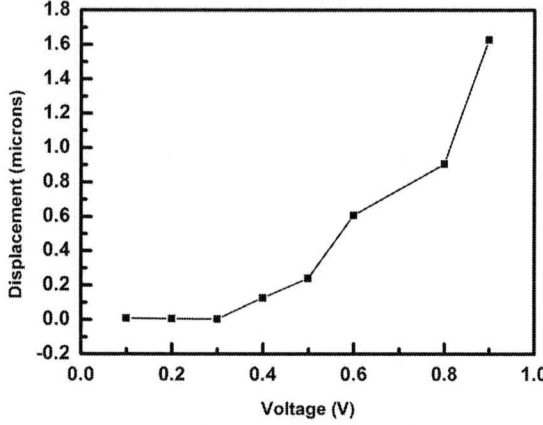

Figure 6: Displacement of fabricated thermal actuator at different applied voltages

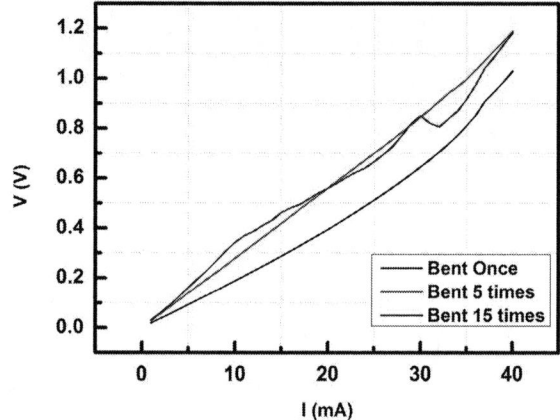

Figure 7: Voltage and current characteristics of a gold thermal actuator at different bending cycles of the ultra thin silicon sheet.

CONCLUSION AND FUTURE WORK

We have demonstrated a low-cost generic batch process to transform silicon based traditional MEMS into flexible and semi-transparent one. We fabricated MEMS thermal actuators which achieved a deflection of 1.7 um at 43 mA. Using similar architectures, other MEMS devices [20] and MEMS cantilevers for sensor applications [21] can also be fabricated using the same process. In addition, creating flexible MEMS based micro-robotics will enhance the field of microsurgery where they can be used for material delivery to the body and fighting cancer cells. We believe the demonstrated process will expand the horizon of flexible electronics into MEMS world.

ACKNOWLEDGEMENTS

This work has been financially supported by the KAUST OCRF CRG-1-2012-HUS-008 grant. We thank Dr Ian Foulds for allowing us to use characterization equipments in his lab. We also thank the support from KAUST Advanced Nanofabrication Cleanroom (KANF) staff.

REFERENCES

[1] G. H. Gelinck, H. E. A. Huitema, E. van Veenendaal, E. Cantatore, L. Schrijnemakers, J. B. van der Putten, et al., "Flexible active-matrix displays and shift registers based on solution-processed organic transistors," Nature materials, vol. 3, pp. 106-110, 2004.

[2] Y.-H. Kim, S.-K. Park, D.-G. Moon, W.-K. Kim, and J.-I. Han, "Active-matrix liquid crystal display using solution-based organic thin film transistors on plastic substrates," Displays, vol. 25, pp. 167-170, 2004.

[3] T. Sekitani, H. Nakajima, H. Maeda, T. Fukushima, T. Aida, K. Hata, et al., "Stretchable active-matrix organic light-emitting diode display using printable elastic conductors," Nature materials, vol. 8, pp. 494-499, 2009.

[4] C. Sheraw, L. Zhou, J. Huang, D. Gundlach, T. Jackson, M. Kane, et al., "Organic thin-film transistor-driven polymer-dispersed liquid crystal displays on flexible polymeric substrates," Applied Physics Letters, vol. 80, pp. 1088-1090, 2002.

[5] G. A. T. Sevilla, S. B. Inayat, J. P. Rojas, A. M. Hussain, and M. M. Hussain, "Flexible and Semi☐Transparent Thermoelectric Energy Harvesters from Low Cost Bulk Silicon (100)," Small, 2013.

[6] J. Yoon, A. J. Baca, S.-I. Park, P. Elvikis, J. B. Geddes, L. Li, et al., "Ultrathin silicon solar microcells for semitransparent, mechanically flexible and microconcentrator module designs," Nature materials, vol. 7, pp. 907-915, 2008.

[7] M. F. El-Kady, V. Strong, S. Dubin, and R. B. Kaner, "Laser scribing of high-performance and flexible graphene-based electrochemical capacitors," Science, vol. 335, pp. 1326-1330, 2012.

[8] L. Hu, H. Wu, F. La Mantia, Y. Yang, and Y. Cui, "Thin, flexible secondary Li-ion paper batteries," ACS nano, vol. 4, pp. 5843-5848, 2010.

[9] G. T. Sevilla, J. P. Rojas, S. Ahmed, A. Hussain, S. B. Inayat, and M. M. Hussain, "Silicon fabric for multi-functional applications," in Solid-State Sensors, Actuators and Microsystems (TRANSDUCERS & EUROSENSORS XXVII), 2013 Transducers & Eurosensors XXVII: The 17th International Conference on, 2013, pp. 2636-2639.

[10] J. Viventi, D.-H. Kim, L. Vigeland, E. S. Frechette, J. A. Blanco, Y.-S. Kim, et al., "Flexible, foldable, actively multiplexed, high-density electrode array for mapping brain activity in vivo," Nature neuroscience, vol. 14, pp. 1599-1605, 2011.

[11] T. Someya, T. Sekitani, S. Iba, Y. Kato, H. Kawaguchi, and T. Sakurai, "A large-area, flexible pressure sensor matrix with organic field-effect transistors for artificial skin applications," Proceedings of the National Academy of Sciences of the United States of America, vol. 101, pp. 9966-9970, 2004.

[12] K. Takei, T. Takahashi, J. C. Ho, H. Ko, A. G. Gillies, P. W. Leu, et al., "Nanowire active-matrix circuitry for low-voltage macroscale artificial skin," Nature materials, vol. 9, pp. 821-826, 2010.

[13] J. Rojas, A. Syed, and M. Hussain, "Mechanically flexible optically transparent porous mono-crystalline silicon substrate," in Micro Electro Mechanical Systems (MEMS), 2012 IEEE 25th International Conference on, 2012, pp. 281-284.

[14] J. P. Rojas, M. T. Ghoneim, C. D. Young, and M. M. Hussain, "Flexible High-κ/Metal Gate Metal/Insulator/Metal Capacitors on Silicon (100) Fabric," 2013.

[15] J. P. Rojas, G. Torres Sevilla, and M. M. Hussain, "Structural and electrical characteristics of high-k/metal gate metal oxide semiconductor capacitors fabricated on flexible, semi-transparent silicon (100) fabric," Applied Physics Letters, vol. 102, pp. 064102-064102-4, 2013.

[16] F. Jiang, Y.-C. Tai, K. Walsh, T. Tsao, G.-B. Lee, and C.-M. Ho, "A flexible MEMS technology and its first application to shear stress sensor skin," in Micro Electro Mechanical Systems, 1997. MEMS'97, Proceedings, IEEE., Tenth Annual International Workshop on, 1997, pp. 465-470.

[17] S. Xiao, L. Che, X. Li, and Y. Wang, "A cost-effective flexible MEMS technique for temperature sensing," Microelectronics journal, vol. 38, pp. 360-364, 2007.

[18] Q.-A. Huang and N. K. S. Lee, "Analysis and design of polysilicon thermal flexure actuator," Journal of Micromechanics and Microengineering, vol. 9, p. 64, 1999.

[19] L. Que, J.-S. Park, and Y. Gianchandani, "Bent-beam electro-thermal actuators for high force applications," in Micro Electro Mechanical Systems, 1999. MEMS'99. Twelfth IEEE International Conference on, 1999, pp. 31-36.

[20] D. COMSOL, C. B. Books, C. Consultants, S. Chart, C. S. Chart, C. Multiphysics, et al., "Design and Analysis of MEMS Micro Mirror using Electro Thermal Actuators."

[21] R. Raiteri, M. Grattarola, H.-J. Butt, and P. Skládal, "Micromechanical cantilever-based biosensors," Sensors and Actuators B: Chemical, vol. 79, pp. 115-126, 2001.

CONTACT

* sally.ahmed@kaust.edu.sa or, muhammadmustafa.hussain@kaust.edu.sa

WAFER SCALE FABRICATION OF HIGHLY INTEGRATED RUBIDIUM VAPOR CELLS

T. Overstolz, J. Haesler, G. Bergonzi, A. Pezous, P.-A. Clerc, S. Ischer, J. Kaufmann, and M. Despont
Centre Suisse d'Electronique et de Microtechnique (CSEM) SA, Neuchâtel, SWITZERLAND

ABSTRACT

This paper reports on wafer scale fabrication of MEMS rubidium (Rb) vapor cells used for spectroscopic measurements in miniaturized atomic clocks. The cell filling process is based on pipetting minute amounts of dissolved rubidium azide (RbN_3) into cavities etched in a silicon wafer, hermetic sealing of the cavities by anodic bonding of glass caps, and in situ UV decomposition of the RbN_3 into Rb and N_2. All relevant elements required for operation such as resistive heaters, temperature sensors, and coils are integrated onto the vapor cell using planar technology. Experiments showed short term frequency stability below 10^{-10} at 1 second integration time.

INTRODUCTION

Miniature atomic clocks (MACs) [1] providing high accuracy at small size (<1 cm^3) and low power consumption (<100 mW) have found an increased interest for applications in handheld communication and navigation devices requiring time-keeping at a superior level than achievable with widespread quartz oscillators. The unprecedented frequency stability of atomic clocks is achieved by a suitable interrogation of optically excited atoms which takes place in the so-called vapor cell.

Within the project Swiss–MAC, a current prototype has been developed (Figure 1) that integrates most of the desired functionalities. A close-up of the physics package (without vacuum encapsulation on the LTCC) is shown in Figure 2. The prototype is powered by a separate PCB equipped with a battery pack and different monitoring and debugging inputs/outputs. The main PCB has a dimension of 50x100 mm^2 as the focus of this first prototype was more on the proof of concept and not on its full miniaturization.

Figure 1: Picture of the full Swiss–MAC prototype (caps removed).

Figure 2: Picture of the current physics package with stacked PCB layers and the MEMS vapor cell on top. Internal magnetic shielding is not shown.

A main challenge of such technology is the fabrication of small and leak tight vapor cells and the incorporation of alkali metal, typically rubidium or cesium. Alkali metals are very volatile and readily oxidize when exposed to air. This has to be avoided since the alkali metal would then be no longer available for spectroscopic measurements.

Hence, to target this issue, we have developed and improved over the past years a new technology allowing batch fabrication of Rb vapor cells in a very stable fabrication process which is executed under normal atmospheric conditions.

VAPOR CELL FABRICATION
Cell filling

Rather than filling the vapor cells with metallic Rb [2], we have focused on another approach based on cell filling using RbN_3, which is stable in air at room temperature, and in situ decomposition under UV irradiation to produce Rb and N_2, a method known to yield very pure alkali metals [3]. We have developed a method similar to Woetzel et al [4] to deposit precise quantities of RbN_3 in a simple way by dissolving RbN_3 in water, and pipetting minute amounts of dissolved RbN_3 into cavities etched in a silicon wafer and closed at the bottom by a glass wafer. The solubility of RbN_3 in water is very good (114.1g/100g H_2O @ 17°C [5]) and hence a precise molar concentration can be achieved, i.e. the concentration is stable during the whole dispensing process and is not prone to precipitation. The pipetting is done using an automatic micro-dispensing system and takes place under normal atmospheric conditions which simplifies the handling enormously. The dispensing head used in our

Figure 3: Rubidium vapor cells of different external dimensions; 4x4x1.6 mm³ (left); 2x2x1.6 mm³ (middle); 1x1x1.6 mm³ (right).

setup ejects droplets of approximately 180 pl such that the total quantity of dispensed RbN_3 solution can easily be varied by adapting the total number of droplets per cavity. Typical quantities of dispensed liquid are in the range of some nanoliters corresponding to a few micrograms of RbN_3 per cavity. Due to the small volume of dispensed liquid, the solvent is evaporated quickly, leaving a uniform layer of recrystallized RbN_3 inside the cavities.

Although most of the work and all of the characterization has been done on cells of 4x4 mm² size and 1.6 mm height, we could recently successfully demonstrate the wafer scale fabrication of vapor cells having a size of 2x2 mm² and even 1x1 mm², both with 1.6 mm height (Figure 3). Hence more than 1000 respectively 3000 vapor cells are fabricated at once on a single 100 mm wafer.

Hermetic sealing

The cavities in the silicon wafer are sealed under Ar atmosphere with a second glass wafer by anodic bonding using standard wafer bonding equipment. As part of the bonding process, the bonding chamber is evacuated during several hours, while the filled cavity wafer and the top glass wafer are heated to 280 °C and maintained at this temperature during the pumping procedure. Upon reaching a pressure of $< 10^{-5}$ mbar inside the bonding chamber, the pumping procedure is stopped, and the bond chamber is backfilled with Ar to a level of some tenths of mbar. Successively the bonding procedure is initiated, i.e. a center pin brings the glass wafer in contact with the silicon wafer, the spacers at the border are taken out, a force is applied on the wafer stack, and the bonding is performed.

Decomposition

Although RbN_3 is reported to decompose either thermally at temperatures above 355 °C [3] or under UV irradiation, we have never been able to thermally decompose the encapsulated RbN_3 by heating the cells in a furnace, even when heating up to 650 °C. Rather, the RbN_3, initially clearly visible as a white deposit, disappeared and only reappeared as thin film deposited everywhere inside the cells once the temperature cooled down, a phenomenon also described by Ogden et al [6]. The only way to thermally decompose the RbN_3 was achieved by holding a vapor cell into the blue flame of a gas burner, i.e. upon shock heating, a process which is not fairly reproducible and

incompatible with mass production. We have therefore focused on UV photodecomposition [7-9] for which purpose a custom made irradiation chamber was built using low-pressure mercury vapor discharge tubular lamps emitting light at 254 nm wavelength. Up to 100 vapor cells can be irradiated in parallel. Upon irradiation, the RbN_3 deposits decompose into metallic Rb and N_2. The N_2, product of the decomposition, and the remaining Ar from the hermetic sealing process constitute the buffer gas inside the vapor cells whose purpose is to reduce sidewall collisions of the Rb atoms that causes degradation of the clock signal through atomic coherence loss. The ratio of Ar and N_2 partial pressure can be adjusted to allow for temperature drift compensation. We are currently conducting a study in order to reach the optimal gas partial pressure ratio for minimizing the temperature coefficient.

The decomposition under UV-lamp irradiation is a very slow process; typical irradiation times are on the order of 50 h or more. One reason for this long irradiation time is the fact that the borofloat glass which is required for anodic bonding has a very high absorption coefficient in the UV. Since the transmission depends on the glass thickness, wafers of 300 µm and more recently 200 µm thickness have been used. Further speeding up the decomposition process should be realized by using a deep UV-laser, which should also allow for a better control of the UV irradiance, thus to better control the amount of decomposed RbN_3 and hence to better control the amount of released N_2.

Functionalized glass lid

For spectroscopic measurements, and more specifically for integration in an atomic clock, the vapor cells have to be heated to roughly 100 °C in order to create the desired alkali vapor density, and a homogeneous magnetic field is required to achieve ground-state hyperfine splitting of the Rb atoms. To do so, resistive heaters and magnetic coils, along with a temperature sensor, have been integrated directly onto both glass lids using planar technology (Figure 4). Since the atomic clock frequency is very sensitive to undesired magnetic fields, the AlSi resistive heaters and the temperature sensors have been realized in two layers with identic shape sitting on top of each other, separated by a thin SiO_2 dielectric layer. This architecture allows reducing parasitic magnetic fields using opposite current flows in the two metal loops for both heater and temperature sensor. The diameter of the coils is chosen such that the coils on both sides of the cell fulfill the Helmholtz conditions required for the creation of a homogeneous magnetic field between two planar coils.

978-1-4799-3510-9/14 $31.00 © 2014 IEEE

Figure 4: Rubidium vapor cell of 4x4 mm² and 1.6 mm thickness including resistive heater, Helmholtz coils, and a temperature sensor realized in planar technology directly onto the two cell window.

Figure 5: 100 mm triple wafer stack (glass – silicon – glass) with 332 rubidium vapor cells of 4x4mm² equipped with heaters, temperature sensors, and magnetic coils integrated on both sides of the wafers.

RESULTS AND CONCLUSION

The atomic vapor cells were first integrated in a breadboard atomic clock. Typical absorption spectra measured in a miniature MEMS ^{87}Rb vapor cell are shown in Figure 6. The frequency stability of this atomic clock was measured to be below 10^{-10} at 1s integration time (Figure 7), showing that the developed atomic vapor cells meet the short term frequency stability specifications of miniature atomic clocks ($<6 \cdot 10^{-10}$ @ 1s).

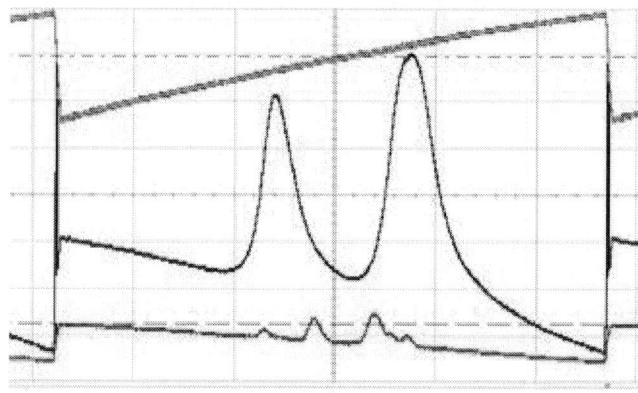

Figure 6: Blue: isotopic enriched ^{87}Rb absorption of a MEMS cell; Magenta: natural Rb absorption of a reference buffer gas free glass cell, with the two smaller peaks being related to the isotope ^{87}Rb (abundance 28%), and the two larger peaks being related to the isotope ^{85}Rb (abundance 72%).

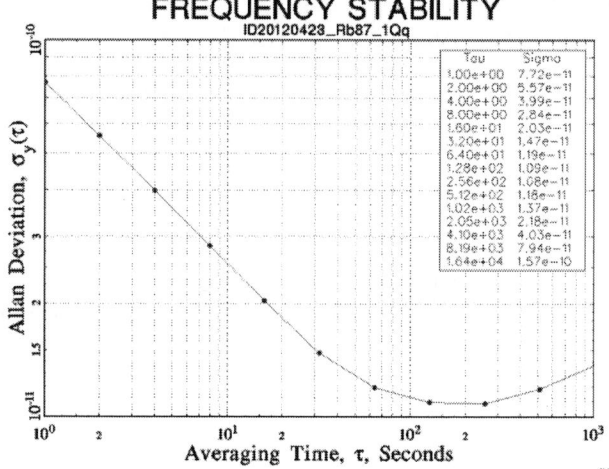

Figure 7: Frequency stability measured on a ^{87}Rb MEMS cell with a breadboard atomic clock.

Preliminary results show that the miniature vapor cells are very close to meet the challenging long term frequency stability specifications ($<1 \cdot 10^{-11}$ @ 1day), with a frequency drift measured to be close to $1 \cdot 10^{-10}$ per month. Work is ongoing to deeply characterize the miniature atomic vapor cells and to publish these very promising results.

The latest atomic vapor cells are currently being integrated in the Swiss-MAC prototype (Figure 1), which was already shown to sustain short term frequency stability as low as $6 \cdot 10^{-11}$ @ 1s [10]. Frequency stability measurements with the fully integrated Swiss-MAC prototype are expected for beginning of 2014.

ACKNOWLEDGEMENTS

This research activity is performed in the frame of a multidivisional research program and CSEM would like to thank the Swiss Confederation, the Canton of Neuchâtel, the Canton of Baselland, and the Cantons of Central Switzerland for their financial support.

REFERENCES

[1] J. Kitching, et al., "Microfabricated atomic clocks," IEEE International Conference on Micro Electro Mechanical Systems IEEE, pp. 1-7, Feb. 2005.

[2] M. Pellaton, Y. Pétremand, C. Affolderbach, G. Mileti, and N.F. de Rooij, Laser-pumped double-resonance clock using a micro-fabricated cell Proc. IEEE Int. Frequency control Symp. jointly with the European Frequency and Time Forum (IFCS-EFTF) (San Francisco, CA, 1–5 May 2011) pp 604–6.

[3] R. Suhrmann and K. Clusius, Über die Reindarstellung der Alkalimetalle, Z. Allgemeine Anorg. Chem., 151, pp 52-58 (1926)

[4] S. Woetzel, V. Schultze, R. IJsselsteijn, T. Schulz, S. Anders, R. Stolz, and H.-G. Meyer, Microfabricated atomic vapor cell arrays for magnetic field measurements, Rev. Sci. Instrum. **82**, 033111 (2011).

[5] Jiri Hála, H. Akaiwa, IUPAC-NIST Solubility Data Series, 79. Alkali and Alkaline Earth Metal Pseudohalides, J. Phys. Chem. Ref. Data, **33**(1), 17 (2004).

[6] J. S. Ogden, J. M. Dyke, W. Levason, F. Ferrante, L. Gagliardi, The Characterisation of Molecular Alkali-Metal Azides, Chem. Eur. J., **12**(13), 3580-3586 (2006).

[7] S. K. Deb, Ultraviolet Absorption Spectra of Alkali Metal Azides, J. Chem. Phys. **35**, 2122 (1961).

[8] S. K. Deb, Photoelectric Properties of the Alkali Metal Azides, Trans. Faraday Soc., **59**, 1414-1422 (1963).

[9] P. W. M. Jacobs, A. R. Tariq Kureishy, The Photochemical Decomposition of Rubidium and Caesium Azides, J. Chem. Soc., 4723-4730 (1964).

[10] J. Haesler, L. Balet, J.-A. Porchet, T. Overstolz, J. Pierer, R. J. James, S. Grossmann, D. Ruffieux, S. Lecomte, The Integrated Swiss Miniature Atomic Clock, Proc. IEEE Internatial Frequency Control Symposium jointly with the European Frequency and Time Forum (IFCS-EFTF) (Prague, 21–25 July 2013) pp 579–581.

CONTACT

*T. Overstolz, phone: +41-32-7205057; tov@csem.ch

WAFER-SCALE FLEXIBLE GRAPHENE LOUDSPEAKERS

He Tian[1,2], Ya-Long Cui[1,2], Yi Yang[1,2], Dan Xie[1,2] and Tian-Ling Ren[1,2]

[1]Institute of Microelectronics, Tsinghua University, Beijing 100084, CHINA

[2]Tsinghua National Laboratory for Information Science and Technology (TNList), Tsinghua University, Beijing 100084, CHINA

ABSTRACT

In this paper, wafer-scale flexible graphene loudspeakers are fabricated in one-step laser scribing technology. Current fabrication process for graphene devices is mainly based on chemical vapor deposition graphene, which needs several hours' graphene growth, transfer and patterning. By using this new laser scribing technology, wafer-scale graphene patterns can be obtained in 25 minutes with low cost. The flexible graphene loudspeaker is demonstrated to be high performance with wide-band sound generation from 1~50 kHz, which indicated that the laser scribed graphene could be widely used in integrating wafer-scale graphene-based acoustic devices.

INTRODUCTION

Graphene-based devices have great potential in complementing or replacing conventional silicon-based devices in varies applications due to its superior properties [1]. However, the graphene growth and patterning technology are still hostage to high cost, low efficiency and low quality [2]. As a result, complex process is needed in the fabrication of the graphene loudspeakers demonstrated by previous work [3, 4]. Here, wafer-scale loudspeakers are integrated by laser scribing technology. Compared with conventional graphene fabrication methods, the laser scribing technology has the advantages of direct growth graphene in designed shape at precise locations, low cost with large scale fabrication ability and simply process in 25 minutes. It is believe that the laser scribing technology could also have broad applications in the graphene-based electronics, sensors and actuators systems.

DEVICE FABRICATION

The DVD burner with laser scribe function has been used. It could produce 788 nm laser pulses with 5 mW maximum power to turn the stack single-layer graphene oxide (GO) film into graphene (Figure 1). In order to realize flexible device, a PET substrate is introduced. The fabrication process could be described as following. Firstly, cover a PET on DVD disc. Then, drop cast GO solution onto the PET. After the GO is dry, put into the DVD burner to write graphene patterns. Lastly, peel the PET from the DVD disc and graphene loudspeakers are done. In order to control the patterning position to form precise structure, a software is used to import a designed patterning. The software could control the DVD drive to reduce of GO into graphene selectively.

DVD Laser Scribing Technology

Figure 1: A schematic diagram showing the fabrication process of laser scribed graphene. It can make the wafer-scale of precise graphene patterns be obtained in 25 minutes.

Here shows two examples. The image of Tsinghua logo is imported into the software and 25 minutes later the laser scribed graphene reproduces the image precisely. Another image is graphene lattice with different color. The image could be reproduced by the laser scribed graphene with grey scale. The grey scale represents different reduction level of GO. Dark black patterns means high reduction level of GO under high laser power. And smaller laser power could form light dark patterns with lower reduction level of GO.

Original Image Laser Scribed Graphene

Original Image Laser Scribed Graphene

Figure 2: An original image and the same image reproduced by laser scribed graphene, which corresponds to a change in electrical properties. The resolution of this method could be 20 μm.

After import the electronic file into the computer software and ~25 minutes growth in DVD burner, the laser scribed graphene loudspeaker are directly patterned on a GO film (Figure 3). The device is 1 cm^2 with two triangle sides, which is suitable for electric single wire to lead out.

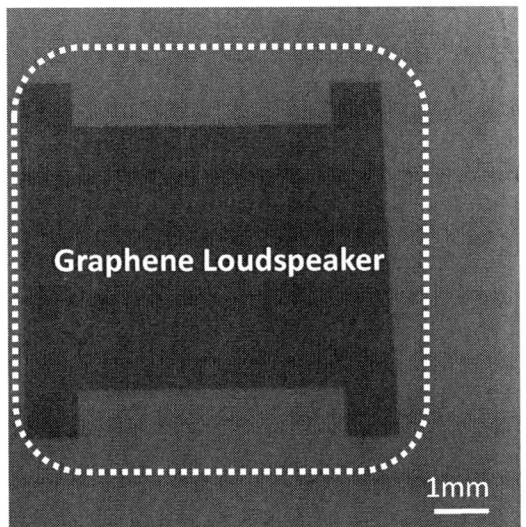

Figure 3: The laser scribed graphene loudspeaker. The device area is 1 cm^2. The thickness of the film could be 10 μm.

Wafer-scale different graphene patterns on PET substrate are also demonstrated. Figure 4 shows wafer-scale fabrication flexible graphene loudspeakers and other small graphene patters on PET substrate. Various kinds of flexible graphene devices could be fabricated easily by such method. It has the advantages of low cost, high efficiency and local growth graphene without transfer, which is much better than chemical vapor deposition method.

Figure 4: Wafer-scale fabrication flexible graphene loudspeaker after peel off the PET substrate from disc.

CHARACTERIZATION

To understand the structure information of before and after the laser scribing, the characterization of Raman are taken. As shown in Figure 5, the black line and red line show the Raman spectra of the GO and laser scribed graphene, respectively. GO exhibiting typical D, G, and amorphous 2D bands. After the laser scribing, the G band shift to smaller wavenumber. This is due to the reduction of oxygen functional group. The enhancement of the 2D band indicates the generation of few-layer graphene after the laser irradiation.

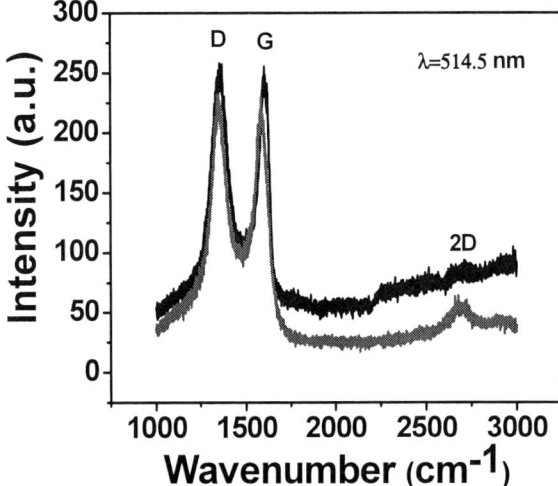

Figure 5: The Raman spectrum of the GO (black line) and laser scribed graphene (red line). Compared with GO, the increase in 2D band indicates that few-layers graphene generated after laser irradiation.

The GO film could be regard as insulator with the resistance up to 580 MΩ. The laser could reduce the GO and change it into electrical conducting materials. The I-V cure of graphene after laser scribing is tested by probe station to demonstrate this. As shown in Figure 6, after the reduction of GO, the resistance decreases to 8.2 kΩ significantly. This reveals that the laser scribing technology is a powerful method to realize tunable electrical properties.

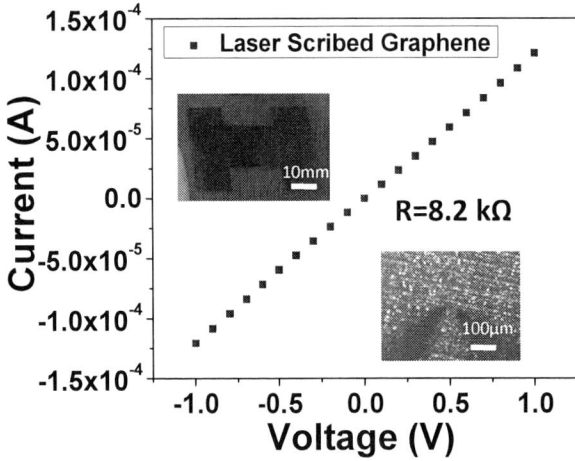

Figure 6: The I-V cure of graphene after laser scribing. The resistance of GO is 580 MΩ. After reduction, the resistance decreases to 8.2 kΩ significantly.

GRAPHENE LOUDSPEAKERS

Figure 7 shows a schematic diagram of experimental setup for a laser scribed graphene loudspeaker. 5 V sine signal sweep from 1 kHz to 50 kHz with 5 V bias is applied to graphene to generate the same frequency sound waves. A standard microphone is used to record the sound generated from graphene. A signal analyzer could make FFT to obtain the sound intensity vs. frequency.

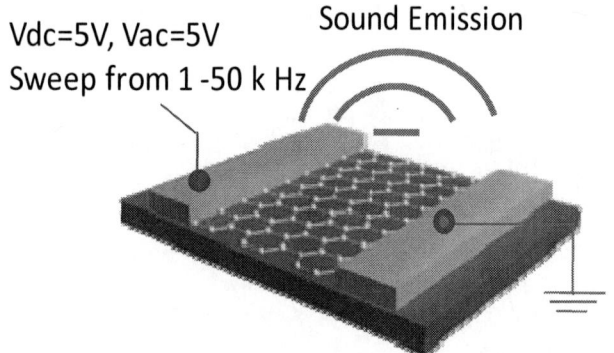

Figure 7: Schematic diagram showing the schematic for testing the graphene loudspeaker.

Figure 8 shows the wide-band sound generation from 1~50 kHz. From 1 kHz to 20 kHz, the sound pressure level (SPL) is increased with the frequency. It is noticed that the sound spectrum is quite flat under the 20 kHz to 50 kHz region.

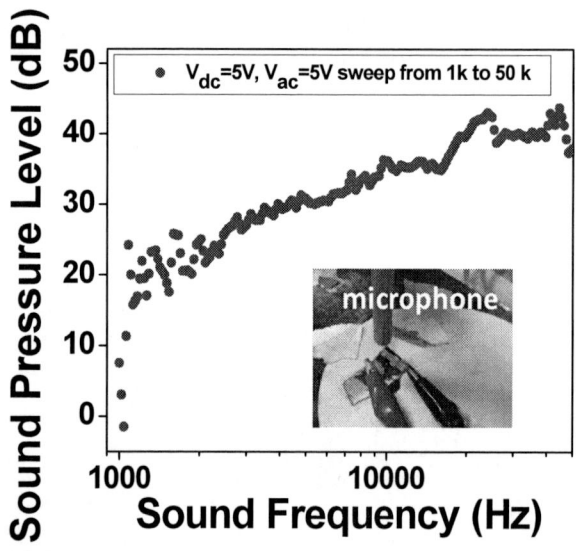

Figure 8: The sound frequency spectrum of the graphene loudspeaker. For 1 to 20 kHz region, the sound pressure levels (SPLs) increase linear with the frequency. For 20 kHz to 50 kHz region, the SPLs are quite flat. Inset showing the experimental setup.

The sound spectrum mapping by sweeping the input power and measure distance are shown in Figure 9 and 10 respectively. The warmer place represent higher sound intensity. Input power mapping in Figure 9 indicated that the lower left corner has smaller SPL and top right corner has higher SPL. So high input power under high frequency region has higher SPL than the others. The top left corner has smaller SPL and lower right corner has higher SPL, which is presented in the measure distance mapping in Figure 10. It is indicated that small measure distance under high frequency region has higher SPL than the others.

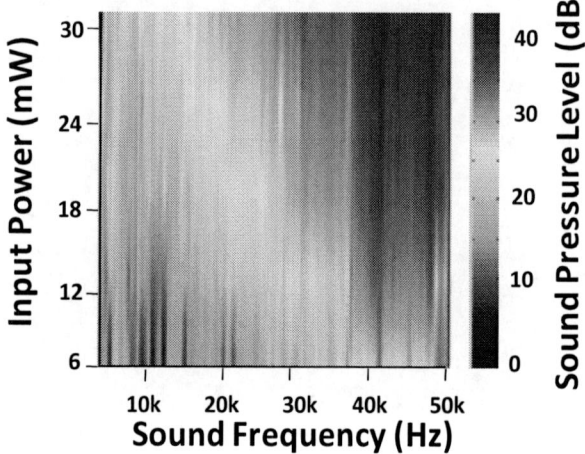

Figure 9: The sound spectrum mapping by sweeping the input power. The overall trend is that the SPLs increase with the input power.

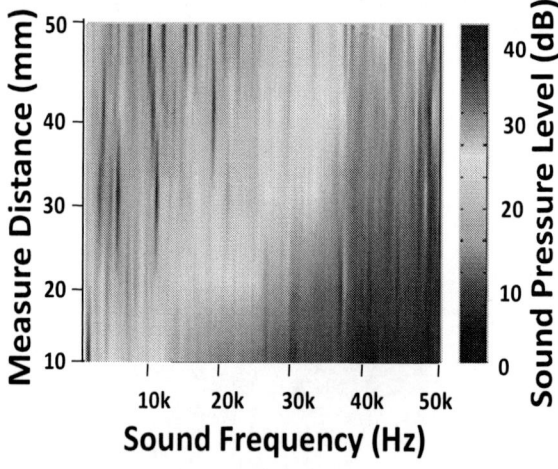

Figure 10: The sound spectrum mapping by sweeping the measure distance.

The output sound pressure increases linearly with the increasing of the input power is shown in Figure 11. The sound pressure has the linearly relation with the input power. After fitting, the relation of sound pressure vs. input power could be expressed as $Y= 0.031+0.059X$. The experimental results are fitting well with the linear line.

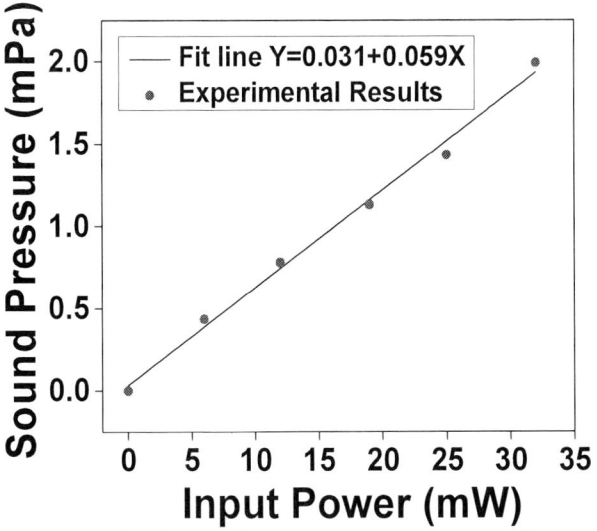

Figure 11: The plot of the sound pressure versus the input power at 20 kHz sound frequency. The line shows that the output sound pressure increases linearly with the increasing of the input power.

As shown in Figure 12, there is an inverse relationship between the output sound pressure and measure distance. The relation of sound pressure vs. measure distance could be expressed as $Y= 19.9/X$.

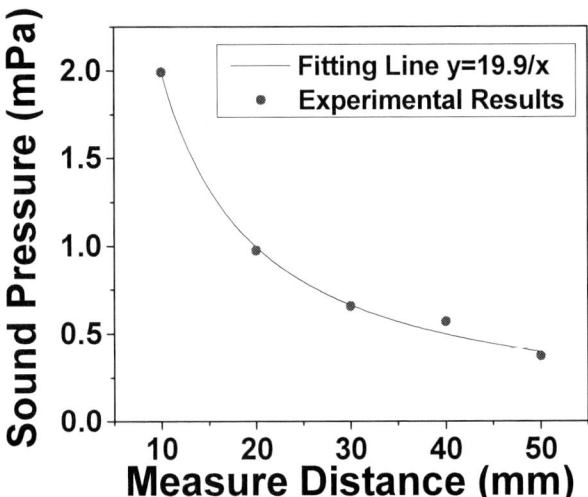

Figure 12: The plot of the sound pressure versus the measure distance. The line shows that there is an inverse relationship between the output sound pressure and measure distance.

The working principle of graphene loudspeaker is based on thermoacoustic effect, which is shown in Figure 13. When sine electric signal is applied on graphene. There generate sine Joule heating. The sine Joule heating conduct to gas molecules, which can make the air vibrate to produce sound. Resulting from the thermoacoustic effect, the graphene loudspeaker has the advantages of wide band sound generation.

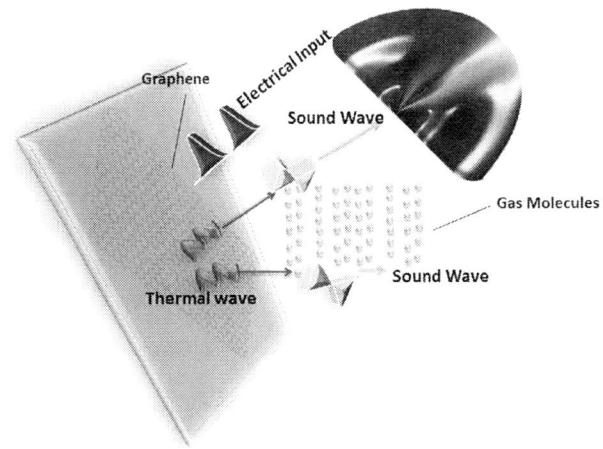

Figure 13: The working principle of graphene loudspeaker.

CONCLUSION

Wafer-scale graphene-based loudspeakers are realized by using a light-scribe DVD burner. Such laser scribing graphene method could produce graphene patterns on PET in 25 minutes without any transfer process. Sound generation of laser scribed graphene is well demonstrated ranging from 1~50 kHz sound frequency. The working principle is based on thermoacoustic effect, which is fundamental different from conventional acoustic devices. This work indicates that laser scribing technology could be a powerful and efficient method to make high-performance graphene acoustic devices.

ACKNOWLEDGEMENTS

This work was supported by the National Natural Science Foundation of China (61025021, 60936002, 51072089, and 61020106006), the National Key Project of Science and Technology (2011ZX02403-002) and the Special Fund for Agro-scientific Research in the Public Interest (201303107). He Tian is additionally supported by the Ministry of Education Scholarship of China.

REFERENCES

[1] A. K. Geim, "Graphene: Status and Prospects", Science, vol. 324, pp. 1530-1534, 2009.

[2] K. S. Kim et al., "Large-scale pattern growth of graphene films for stretchable transparent electrodes", Nature, vol. 457, pp. 706-710, 2009.

[3] H. Tian et al., "Graphene-on-Paper Sound Source Devices", ACS Nano, vol. 5, pp. 4878-4885, 2011.

[4] H. Tian et al., "Flexible and large-area sound-emitting device using reduced graphene oxide", Proc. 2013 IEEE 26th International Conference on Micro Electro Mechanical Systems (MEMS), Taipei, January 20-24, 2013, pp. 709-712.

CONTACT

*Tian-Ling Ren, tel: +86-10-62798569; RenTL@tsinghua.edu.cn

A POLYMER MEMS MIRROR FOR ON-DEMAND LIGHT DISTRIBUTION FABRICATED BY INJECTION MOLDING AND TRANSFER OF PRINTED LAYERS

Kazuma Kurihara[1], Osamu Nagumo[2], Hideki Takagi[1] and Ryutaro Maeda[1]
[1]National Institute of Advanced Industrial Science and Technology, Japan
[2]Designtech co. ltd., Japan

ABSTRACT

A high-throughput and low-cost fabrication process for an optical microscanner is proposed for lighting applications. The process utilizes mold replication and transfer of printed functional layers to form a film. We confirmed that a high-throughput fabrication lasting less than 30 s could be obtained by using a mold with a knife edge. A polymer microscanner with feature size of 0.6 × 2 mm2 was successfully fabricated using our process. The scanner was actuated by magnetic forces generated by an external coil, and the mirror tilt angle was successfully measured by a strain gauge formed by the functional layer transfer. By synchronizing the LED input signal and the polymer scanner tilt angle, the illuminance distribution could be controlled. For example, this distribution could be expanded from 10° to 50° or divided into two separated spots. We believe that our developed process can address new MEMS applications such as commercial lighting.

INTRODUCTION

Recently, the light-emitting diode (LED) has been widely employed in automobiles, houses, buildings, airplanes, and so on for the purpose of illumination and lighting. Achieving high luminance and illuminance distribution control are important for the effective use of LED lighting. In addition, several applications require active control of the illuminance distribution to include additional functions such as on-demand lighting. The microelectromechanical systems (MEMS) scanner is one of the possible candidates that can enable active illuminance control. However, the MEMS fabrication cost is still extremely high to be applicable to the commercial lighting industry because MEMS devices are basically fabricated using the semiconductor lithography process. In addition, commercial lighting applications require relatively large mirror scanners; this further increases the MEMS fabrication cost as the cost strongly depends on the size of the chip used.

Polymer-based MEMS devices have been considered as potential candidates to decrease the process cost, and such devices can consequently be used in several applications such as tactile sensors, optical mirrors, and energy-harvesting devices[1]. Liu et al. have reported a paper-based force sensor that can be fabricated using laser cutting and screen-printing processes[2]. Amaya et al. have reported a polymer-MEMS fabrication process utilizing a nanoimprinting and polishing technique[3]. The

nanoimprinting method enables the fabrication of a MEMS structure without involving the processes of lithography and etching. However, subsequent to nanoimprinting, a polymer skin layer inevitably remains in the fabricated device, and a polishing process is necessary to release the MEMS structure. Thus, in order to reduce the production cost of MEMs devices, a simpler and higher-throughput fabrication process is required. In this study, we propose a new MEMS fabrication process that utilizes an injection molding method combined with embossing replication using a mold with knife-edge structures and a layer transfer of printed functional layers onto a film[4,5]. We also demonstrate the working of a lighting-application optical scanner fabricated via the proposed process.

FABRICATION CONCEPT AND STRUCTURE DESIGN

Figure 1 shows the schematic of the polymer microscanner. It consists of a polymer microstructure along with stacked layers of mirror ink, magnetic ink, and electroconductive ink. The structure design includes a spring part and a mirror part. The conductive ink pattern on the spring part plays the role of a strain gauge to measure the bending deformation of the spring part, corresponding to the tilt angle of the mirror part. The polymer microscanner is actuated by a magnetic force between the magnetic ink layer and an external coil. The scanner can be actuated to a certain tilt angle at the mirror's resonance frequency by feedback control based on the strain-gauge signal. In our case, for the purpose of lighting application, the surface of the optical mirror scanner is required to be flat in order to reduce optical aberration even while the scanner is actuated. On the other hand, a large tilt angle obtained via a small magnetic force is also preferable. The bending deformation of a beam is estimated by

$$M = \frac{E}{R} I , \qquad (1)$$

$$I = ba^3/12 . \qquad (2)$$

Here, M denotes the mirror's bending moment, R denotes radius of curvature of the mirror, E denotes the Young's modulus, and I denotes the moment of inertia of the mirror area. For example, the value of I for a beam with a rectangular cross-section with thickness a and width b is obtained using Eq. (2). From Eq. (2), we note that the thickness of the structure largely influences the deformation as the cubic law. In order to allow a large tilt angle, the thickness of the spring part should be small. On the other hand, the mirror part must possess a thick structure to reduce optical aberration. For LED lighting applications, the resonance frequency is required to be above 50 Hz to minimize flicker. Based on these considerations, the thickness of the spring part and the mirror membrane were designed to be 50 and 500 μm, respectively.

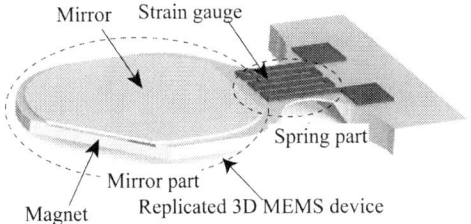

Figure 1. Schematic of polymer mirror-scanner for lighting application.

FABRICATION PROCESS OF POLYMER MICROSCANNER

Figure 2 shows the fabrication process of the polymer microscanner. The polymer microscanner was fabricated via an injection molding process using a film with printed functional ink layers. At first, a mold and a printed film were prepared. The printed film shown in Fig. 2(a) was prepared by coating a sacrificial layer and several screen-printed functional ink layers on a polyethylene terephthalate (PET) film. Because of the presence of the scarified layer, the PET film could be removed easily from the polymer microstructure after the mold replication process, and only the printed functional layers were transferred onto the polymer microstructure. The film thickness was 50 μm. Stacked layers of the mirror, magnetic, and electroconductive inks were screen-printed on the scarified layer. A mirror-ink layer consisting of MIR-51000 ink (Teikoku Ink) was printed to form a high-optical reflection layer. A ferrite-magnet-based ink layer was used as the magnetic layer. The silver ink FA-301CA (Toyo Ink) was employed for the electroconductive layer. Thus, we prepared the PET film with functional ink layers without using any vacuum process. On the other hand, the mold shown in Fig. 3(b) was fabricated via a computer numerical control (CNC) milling machine process. The MEMS microstructure was formed on the mold surface. Cr-steel alloy (Uddeholm, STAVAX) was used for the mold. Since the mold has a large hardness value of over 50 HRC, the mold can repeatedly be used in the injection molding process for over one million times. Next, the film was set to the mold as shown in Fig. 2(c). The positioning accuracy was evaluated to be about ±10 μm. After positioning the film, the melted polymer was injected into the mold, as shown in Fig. 2(d). In our micromirror device, the gap between the mold and the film was 50 μm for the spring section and 500 μm for the mirror membrane. In the injection molding process, a gap of 50 μm is insufficient to ensure successful polymer injection because the melted polymer is quickly refrigerated and solidified. Therefore, we combined an embossing process with injection molding. In the injection process, the gap between the mold and the film was kept wider than the designed thickness, and subsequently, we performed embossing by forcing the mold against the film, as shown in Fig. 2(e). In case of the conventional mold shown in Fig. 3(a), a thin polymer skin layer inevitably remains on the top of the mold surface. To resolve this problem, we employed a mold with a knife-edge structure, as shown in Fig. 3(b). This type of mold with a knife-edge structure can cut into the relatively soft PET film upon application of pressure, and subsequently, the injected polymer can be segmented as shown in Fig. 3(b). In our device fabrication process, polybutylene terephthalate (PBT) polymer was employed as the injected polymer material. The temperature of the injected polymer was estimated to be 245 °C. The mold temperature was set to 60 °C. Finally, the film and the replicated polymer microstructure were separated as shown in Fig. 2(f). The printed functional layers on the film are transferred onto the replicated polymer microstructure because the scarified layer is coated on the film. It is noteworthy that, using our proposed method, three-dimensional (3D) microscanner devices can be produced using only one replication process. This process facilitates reduction in cost and increased throughput in the fabrication of polymer microdevices.

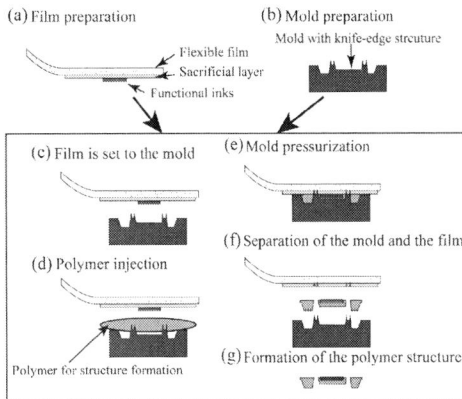

Figure 2. Concept of polymer scanner fabrication utilizing mold replication and pattern transfer from printed film.

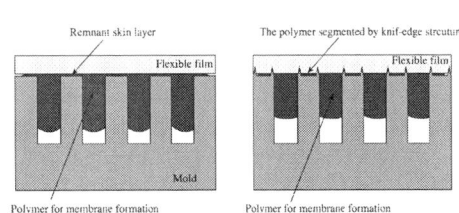

Figure 3. Remnant polymer skin layer. (a) Conventional mold structure and (b) knife-edge mold structure.

RESULTS AND DISCUSSION

Figure 4 shows the relation between the thickness of the

polymer mirror structures and the force applied to the mold during the microstructure formation. The projection of the knife-edge structure in the mold was evaluated to be 0.125 µm from the top surface of the mold. The angle of the knife-edge structure was designed to be 15°. As shown in Fig. 4 (a), the thickness of the mirror part linearly decreased as the applied force was increased. In contrast, the thickness of the remnant polymer layer at the knife-edge suddenly reduced to zero when the applied pressure exceeded 2000 kgf, as shown in Fig. 4(b). This result indicates that the injected polymer was segmented. We speculate that difference in pressure behavior between the mirror part and the knife-edge structure emerged from the elasticity properties of the flexible film. In addition, the knife-edge of the mold could cut into the relatively soft PET film.

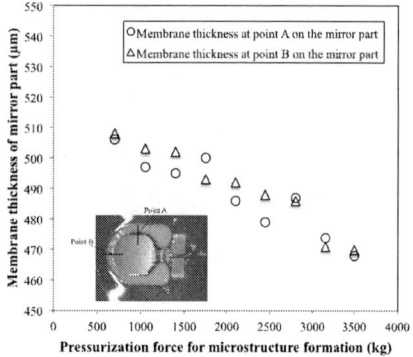

(a)

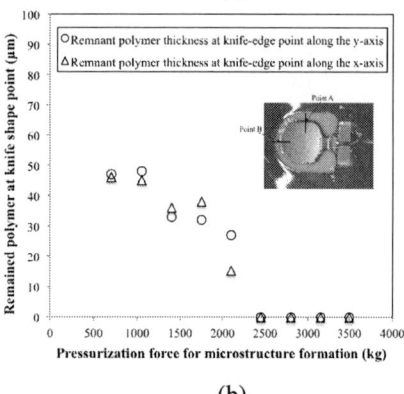

(b)

Figure 4. Relation between force applied during mold process and thickness of mirror membrane (a), and that for remnant polymer skin at knife-edge-shaped point.

Figure 5(a) shows the fabricated microscanner developed for lighting applications. The injection molding process enabled the formation of the polymer microstructure. We used the mold with knife-edge structure to fabricate the polymer microscanner by the replication process. The device fabrication time was only 30 s. The thickness of the fabricated polymer microstructure at the spring and the mirror was evaluated to be 70 µm and 480 µm, respectively. The reflectance of the mirror area was evaluated to be 70%, and the mirror diameter was about 10 mm. The thickness of the mirror ink and ferrite magnet ink was evaluated to be 55 µm and 405 µm, respectively. The width and thickness of

the silver-ink layer was evaluated to be 100 and 10 µm, respectively. The mirror resistance varied from 15 to 22 Ω, while the spring part of the polymer mirror deformed by 45°. This result indicates that the electroconductive ink pattern can function as a strain gauge. The resonance fre quency of the mirror was 65 Hz.

Subsequently, we used different types of mold structures to obtain a matrix pattern of polymer mirror devices, as shown in Fig. 5(c). The various types of polymer scanners were fabricated by using molds with different patterns. As regards addressing the challenge of fabricating a small feature size using the proposed fabrication process, we fabricated a microscanner with a width of 600 µm, as shown in Fig. 5(b). In our case, the feature size was constrained only by the smallest size in the machining process used for mold manufacture. Since the mold was fabricated using a conventional CNC milling machine, the feature size was limited to around 600 µm; the fabrication of sizes below this value led to damage of the milling tools. We believe that a feature size as small as 10 µm is possible if we employ a mold with this feature size.

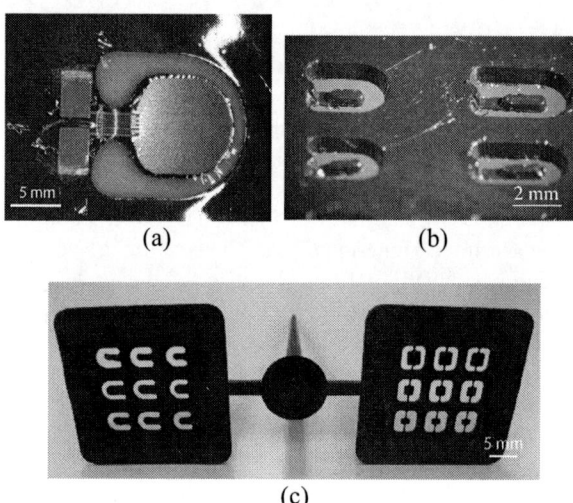

Figure 5. Photographs of fabricated polymer structures.

Figure 6 shows the schematic diagram of a light distribution control system for a lighting application. An LED with a light divergence angle of 15° (full width at half maximum) was employed as the light source. For control of the light distribution, the LED operation was synchronized with the scanner actuation. The scanner was actuated by a magnetic field. The actuation frequency was set equal to the resonance frequency of the mirror by detecting the signal from the strain gauge fabricated on the spring part. Figure 7 shows the results of our illuminance distribution control experiments. As shown in Fig. 7(a), illuminance distribution was focused to one spot when the polymer scanner was not actuated. In contrast, when the scanner was actuated, the illuminance distribution expanded over a larger region, as shown in Fig. 7(b). In addition, a divided illuminance distribution was enabled by synchronizing the LED input power and the scanner actuation, as shown in Fig. 7(c).

These results demonstrate that illuminance distribution can successfully be controlled using our developed polymer microscanner. In addition, the diameter of the reflected light spot was nearly identical under both the non-actuated and divided-illuminance conditions Figs. 7(a) and 7(c), respectively. This result indicates that the mirror part remained optically flat during the scanner actuation. The illuminance distribution expanded from 10 to 50°, and divided distribution was also enabled.

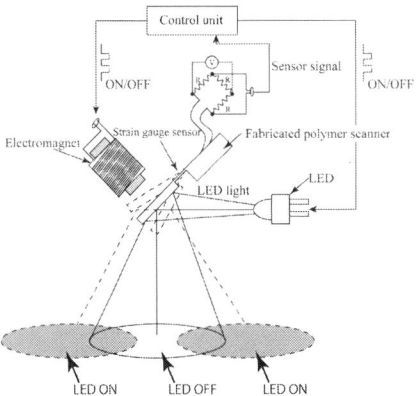

Figure 6. Schematic of light distribution control system for lighting application.

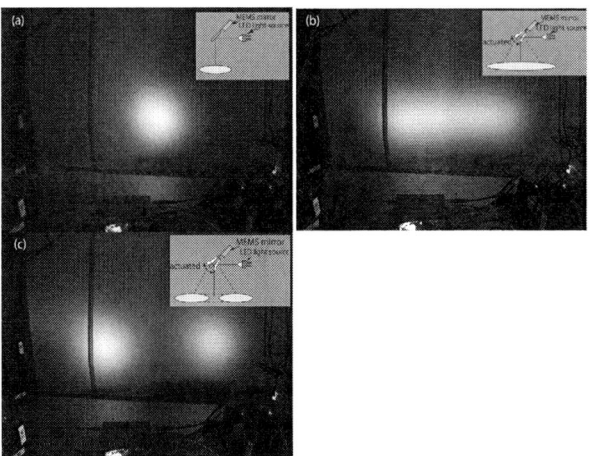

Figure 7. Photographs and measurement results of lighting intensity distribution control achieved using MEMS polymer. (a) Without mirror actuation, (b) with mirror actuation, and (c) LED output control synchronized with mirror actuation.

In summary, we fabricated a polymer mirror scanner using our proposed process via injection molding replication. The replication process provides the advantage of the easy formation of a 3D polymer structure. In addition, the reception process allows low-cost, high-throughput fabrication of relatively wide-area devices. We believe that our proposed process can be used in the realization of new MEMS applications and other industry requirements.

CONCLUSIONS

We fabricated a polymer microscanner using a replication process after transferring printed functional layers onto a polymer structure. The polymer microstructure was fabricated using only a one-step injection molding process. The replication process time was only 30 s. To realize the process, a mold with a knife-edge structure was employed. In addition, the knife-edge structure could cut into the relatively soft film containing the stacked functional layers. In order to demonstrate the possibility of fabricating small feature sizes using our approach, we successfully fabricated a 0.6 × 2 mm2 polymer microscanner. By changing the mold pattern, we were able to obtain various scanner patterns. In order to realize a lighting application with a scanner fabricated using our approach, we fabricated a polymer scanner with a 70-µm-thick spring part and a 480-µm-thick mirror part. The scanner comprised stacked layers of an electroconductive pattern, a permanent magnet, and an optical reflector. By synchronizing the LED light output and the actuation of the mirror scanner, the illuminance distribution was controlled to achieve various lighting patterns such as spot distribution, wide-area distribution, and divided-spot distribution. Our process provides the merits of high-throughput and low-cost device fabrication, and our approach is applicable to emerging fields such as large-area lighting and display.

ACKNOWLEDGMENTS

The authors would like to thank H. Hashimoto of AIST for her assistance. This research is supported by the Japan Society for the Promotion of Science (JSPS) through the "Funding Program for World-Leading Innovative R&D on Science and Technology (FIRST Program)," initiated by the Council for Science and Technology Policy (CSTP).

REFERENCES

[1] W Takayuki Fujita, Kazusuke Maenaka and Yoichiro Takayama, "Dual-axis MEMS mirror for large deflection-angle using SU-8 soft torsion beam", Sensors and Actuators A, 121 (2005) 16-21.

[2] C. Liu, "Recent Developments in Polymer MEMS", Adv. Mater, 19 (2007) 3783-3790.

[3] S. Amaya, D. V. Dao, and S. Sugiyama, "Development of polymerelectrostatic comb-drive actuatior using hot embossing and ultraprecision cutting technology" J. Micro/Nanolithogr. Mems and Moems 8 (2009) 43065.

[4] K. Kurihara, S. Takamatsu, T. Kobayashi, H. Takagi and R. Maeda "High Throughput Fabrication Process for Polymer MEMS using Molding and Printed Pattern Transfer" Conf. eurosensors 2011

[5] K. Kurihara, Osamu Nagumo, Seiichi Takamatsu, Hideki Takagi, and Ryutaro Maeda "High-Throughput and Low-Cost Fabrication of Polymer Microscanner for Lighting Application" Jpn. J. Appl. Phys. 52 (2013) 106701

CONTACT

*K. Kurihara, e-mail:k.kurihara@aist.go.jp

BI-CHAMBER ELECTROMAGNETIC FLUIDIC PUMP

Chris Gudeman[1], Paul Rubel[1], and John Foster[2]
[1]Innovative Micro Technology, Goleta, CA, USA
[2]Owl Biomedical Inc, Santa Barbara, CA, USA

ABSTRACT

This paper reports the design, fabrication and characterization of valve and piston motion for a novel bi-chamber fluidic micropump that is electromagnetically actuated. Unique properties of this pump include the ability to pump liquids over a very broad viscosity range (1 – 5000 mPa-s), extremely small size (0.3mm x 0.6mm x 0.6mm), the capability to be driven by a remote oscillating magnetic field source, fully integrated passive valving, large, straight fluidic paths for low clogging, integrated magnetic flux guides, and a completely monolithic wafer level process to enable low cost manufacturing.

INTRODUCTION

MEMS fluidic pump research and development has been described in several excellent review articles [1]. However, very little has been reported on MEMS devices that are capable of pumping very high viscosity fluids. Blanchard, et al [2] reports a large (~1cm diameter) device that is fabricated by conventional machining and is capable of pumping fluids 62 mPa-s in viscosity. Zordan and Amirouche [3] report a design for pumping blood (~30 mPa-s). Mao and Koser [4] report a very novel approach that uses ferrofluidics at a viscosity 5 mPa-s.

Ahn and Allen [5] have demonstrated an electromagnetic MEMS pump using a rotary actuator, achieving high fluid flow rates (24uL/min) with 1mP-s viscosity fluids. Although very elegant in its design, some assembly is required, thus increasing cost.

The device described here has demonstrated rates of 100nL/min at a viscosity of 5000 mPa-s [6] and is fabricated completely at wafer-level for improved yield and lowered cost.

First the pump design will be described, followed by a description of the Si wafer level process that is used to fabricate the micropump at low cost and high volume. The test set up that employs a stroboscopic video microscope and frame by frame analysis to measure the pumping piston motion relative to the valve motion will then be outlined.

DESIGN

A top down micrograph of the bi-chamber pump is shown in Figure 1. The central feature is the pumping

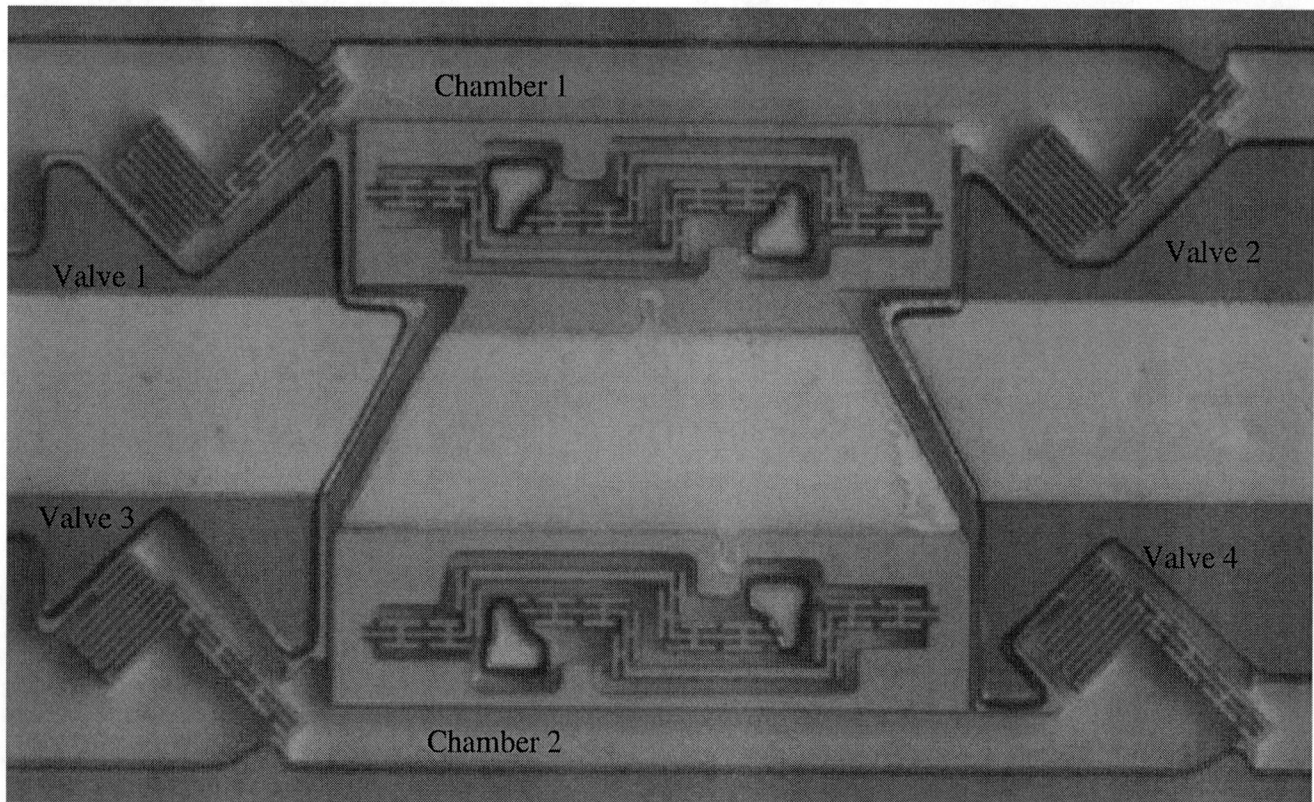

Figure 1: Micrograph of the Bi-Chamber Electromagnetic Pump. The bright slab that runs from left to right is the NiFe magnetic flux guide. The trapezoid is pulled upward when flux is injected into the fluxguide. Fluid flow is from left to right

piston, which consists of a Ni(45)Fe(55) (permalloy) trapezoid that is embedded in a single crystal Si frame and suspended on Si springs that are internal to the Si frame. Above and below this piston are fluidic channels that serve as the pumping chambers. During operation this permalloy block is oscillated with magnetic forces and spring forces. The permalloy is pulled into near contact with the permalloy chevron pole pieces, located at each end of the trapezoid when magnetic flux is injected into the pole pieces. This causes the pumping chamber on the top side to contract (chamber 1), while expanding the pumping chamber on the bottom (chamber 2).

As the chamber 1 contracts, fluid is forced out through the valve at the top right (valve 2), while the inlet valve at the top left (valve 1) is forced closed. These valves, which are formed in the same single crystal Si as the piston frame and springs, consist of flapper shafts on serpentine springs. Simultaneously chamber 2 expands, drawing fluid in through the inlet valve at the lower left (valve 3) and forcing closed the outlet valve at the lower right (valve 4).

At this point in time the magnetic flux is switched off, which allows the piston to return to its as-manufactured position under the force of the stored energy in the supporting springs. The functioning of the four valves described in the paragraph above is reversed.

Although a wide range of device variants have been fabricated, representative dimensions of the pump are as follows. The total pump length is 600 – 800 um and the width ranges from 300 – 400 um. The piston stroke is 7 um, and the valve stroke is 5 um. A longer valve stroke results in increased fluid displacement by the valve, which lessens the net volume displaced each cycle. The channels are 50 um deep by 29 um in width (mean). Device variants employing 40 N/m and 100 N/m piston springs have been fabricated for this study.

Performance scaling rules are generally as follows. The pumping volume per cycle is twice the volume swept by the piston minus the volume swept out by the valves. The pumping rate then scales as the frequency of the magnetic field, provided the piston and valves can move full-stroke in the cycle interval. As fluid viscosity is increased, the cycle internal must be increased. To first order optimum performance occurs when the force generated by the magnetic flux is twice that needed to deflect the piston springs to full-stroke in the absence of fluid. This then assures that the magnetic stroke and the return stroke drive the fluid equally. Note however that the magnetic and return strokes are not completely symmetrical since the return force of the spring falls to zero as the piston approaches its as-manufactured state, while the magnetic force reaches its maximum as the piston hits the crash stop at full-stroke. Asymmetry also stems from other factors including viscous effects, which will be addressed in a later publication.

FABRICATION PROCESS

Wafer level processing of MEMS devices enables several aspects of a manufacturing process that reduce cost,

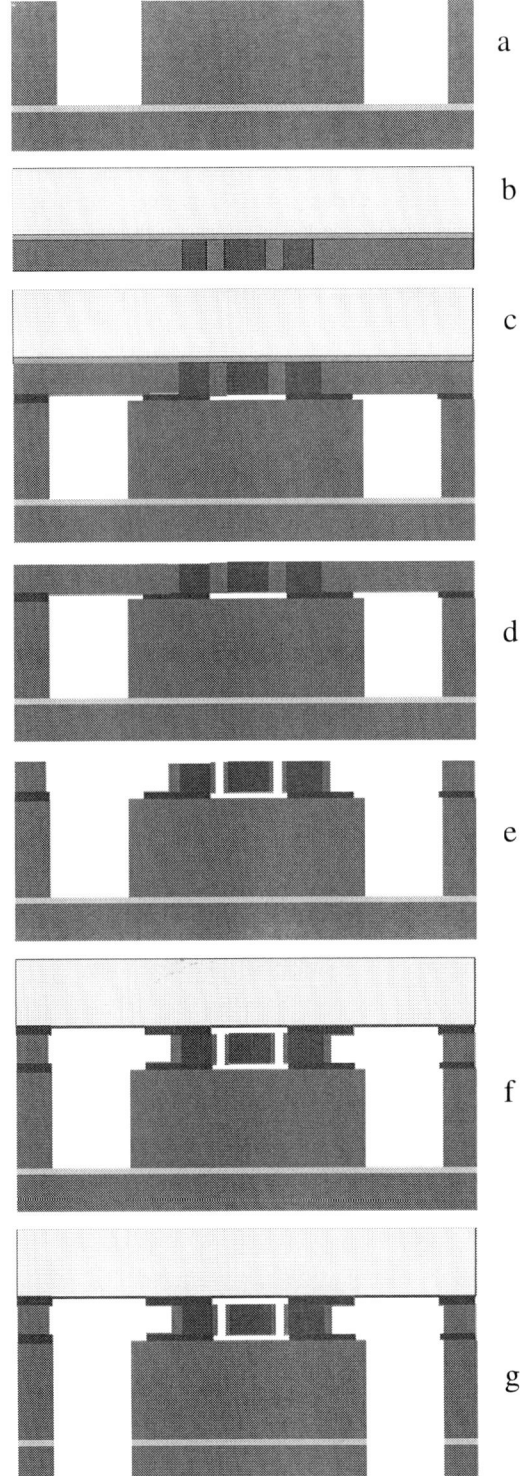

Figure 2: Si(blue), NiFe(violet), Glass(green), Adhesive(red). (a) Deep etch fluid connection wafer. (b) Temporary bond of Si to carrier, thin Si, deep etch trenches, electroplate NiFe, and planarize. (c) Apply photo-definable adhesive and bond to fluid connection wafer. (d) Remove carrier wafer. (e) Deep etch channels, piston, and valves. (f) Bond glass lid wafer. (g) Deep etch fluid ports and singulate.

increase yield, and increase volume and throughput. These include conventional dicing, reduced/eliminated assembly, reduced handling of fragile parts, and greatly improved alignment tolerance of components/features within the device. The bi-chamber pump reported here utilizes wafer level processing to the fullest extent. The process flow is based on a 6" wafer processing and uses conventional micro-fabrication tools that are commercially available. The flow diagram is shown in Figure 2. Starting material for this flow consists of a Silicon-on-Insulator (SOI) wafer, a standard grade Si wafer, and a Borofloat Glass wafer. .

Fluidics Wafer

The fluidics wafer uses the SOI starting material to provide a means to mount tubing to the completed device. The handle wafer is deep etched using the Bosch process to form holes (Figure 2a). This will later be bonded to the pump wafer.

Pump Wafer

The starting material is a standard grade Si wafer that is temporarily bonded to a carrier wafer. The Si is thinned to the desired channel depth and then deep etched to form cavities for the permalloy piston and poles, which are then filled with NiFe by electroplating and planarized with Chemical Mechanical Polishing (CMP). See Figure 2b. A photo-definable adhesive is patterned on the pump wafer and bonded to the fluidics wafer (Figure 2c), followed by the removal of the carrier wafer (Figure 2d). Deep Si etch is employed next to form the piston, channels, springs and valves (Figure 2e), followed by bonding of the Borofloat glass lid wafer, again using the photo-definable adhesive (Figure 2f). Lastly the fluidics ports are opened on the backside by deep Si etch (Figure 2g), and the wafer is singulated on a conventional dicing saw with a single pass to cut through the glass and Si layers simultaneously. The wafer stack is bonded to dicing tape before singulation to prevent coolant and debris incursion.

TEST METHODOLOGY

An optical microscope with stroboscopic flash lamp illumination is used as the test platform. The magnification is chosen so that a single valve and a portion of the piston are in the field of view, thus simplifying the extraction of valve motion with respect to piston motion from the series of micrographs. The field of view is shifted from valve to valve until videos for all four valves are recorded. The camera resolution is 1200 x 1600, yielding a pixel size of 0.28 um. For each video 100 frames are captured at a rate 10 frames/sec. The piston is driven at a rate of 500, 1000, and 1500 Hz with an external zero-based square wave magnetic field source. The strobe rate is adjusted about a nominal frequency of 500 Hz to create a beat note of 5 Hz.

Each individual frame of each video is stored as a bitmap file for further offline analysis, which is partially automated using LabView software, where the moving parts (valves and piston) are tracked. Non-moving parts are also

tracked so background motion due to heating and vibration can be subtracted out.

All tests are carried out at ambient temperature using degassed and deionized water. The pump is immersed in working fluid to minimize head pressure.

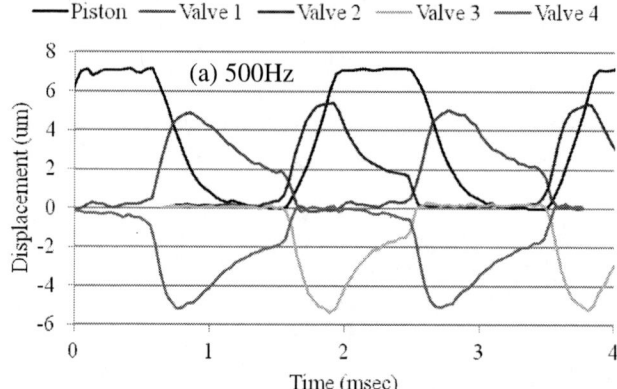

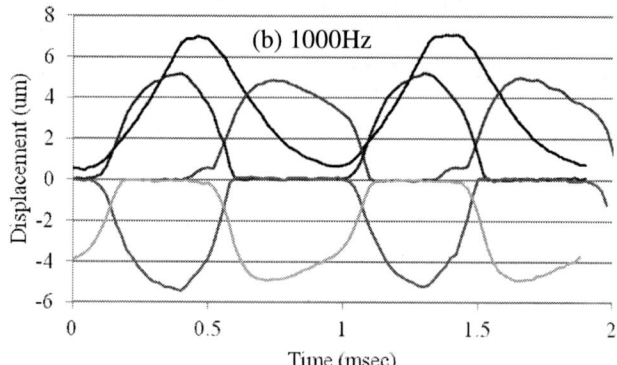

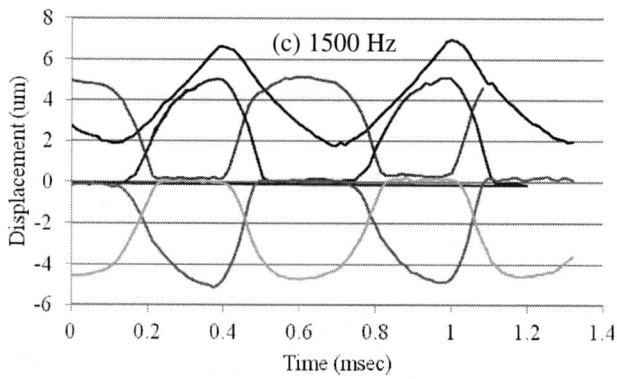

Figure 3: Displacement curves of the piston and four valves. The zero line of the y-axis represents the as manufactured position of the valves and piston. Negative and positive excursions are used in these plots simply to reduce congestion. Full range of motion for the valves is 5um and for the piston is 7um. (a) 500 Hz excitation, (b) 1000 Hz excitation, (c) 1500 Hz excitation. Note that the range of motion of the piston is degraded at 1500 Hz.

RESULTS

Analysis of 400 bitmap images (100 images for each valve) when the pumping piston is driven at 500 Hz yields the data displayed in Figure 3a, where the zero displacement line represents the unenergized position ("as-drawn"). In all cases negative and positive displacement of the valves represent displace in the open direction, but are shown this way to reduce congestion in the plots. At 500 Hz operation, it is evident that the range of motion of piston and valves are 7 um and 5 um, respectively (full stroke). It can also be inferred that the energized stroke of the piston slews at 85 um/msec, while the spring driven return stroke slews at 64 um/msec, indicating reasonable force balance.

Also evident is the shoulder in the displacement curves of the valves as they close. This suggests that fluid continues to flow even though the piston has reached its crash stop (in the energized stroke) or its as-drawn position (in the return stroke). This is not fully understood, but may be due to momentum effects of the working fluid, or simply slow closure of the valves resulting from their low spring constant.

Finally it is evident that for much of the cycle the pump is idle at 500 Hz, given that the piston is stationary and the valves are closed or nearly closed.

At 1 kHz the pump is operating near its peak in efficiency as can be seen in Figure 3b, where the range of motion of the valves spans the designed limits, and the piston motion hits the crash stop in the excited stroke, but falls slightly short (~1 um) of a complete return stroke. Also note that there is some asymmetry in the inlet and outlet valve duty cycle, which is likely due to the long traversal of the spring stroke relative to the excited stroke. Also, the shoulder in the valve displacement curves has vanished.

Further degradation of the piston range of motion is observed in Figure 3c, where displacement curves are plotted for 1.5 kHz operation. As seen for the 1 kHz case, the shortfall occurs in the spring return stroke, since theexcited stroke increases in force near its terminus, while the return stroke asymptotically approaches zero force near its end. Interestingly, the valves retain their full range of motion at 1.5 kHz, although the duty cycle asymmetry is increased over that observed at 1 kHz.

ACKNOWLEDGEMENTS

Thanks to Ruben Salvador for performing the test measurements and to Drs. Fardad Chamran and Yanting Zhang for fabrication process development.

REFERENCES

[1] F. Abhari, H. Jaafar and N. Yunus, *Int. J. Electrochem. Sci.,* 7 (2012), pp. 9765 – 9780 [2]

[2] D. Blanchard, P. Ligrani and B. Gale, J Fluids Eng, 128, (2006), pp.602-610.

[3] E. Zordan and F. Amirouche, *Proc Inst Mech Eng, Part H: J. Eng Med, 221 (2007), pp. 143-151.*

[4] L. Mao and H. Koser, *Nanotechnology,* 17 (2006), pp. 34-47.

[5] C. Ahn and M. Allen, Fluid Micropumps Based on Rotary Magnetic Actuators, in *Proc. 8th IEEE Int. Conf. on Micro Electro Mechanical Systems,* Amsterdam, the Netherlands, Jan20-Feb 2, 1995, pp.408-412.

[6] (to be published)

CONTACT

*C.S.Gudeman, tel: +1-805-717-2345; chris@imtmems.com

IMPROVED MECHANICAL RELIABILITY OF MEMS PIEZOELECTRIC VIBRATION ENERGY HARVESTERS FOR AUTOMOTIVE APPLICATIONS

M. Renaud, Z. Wang, M. Jambunathan, S. Matova, R. Elfrink, M. Rovers, M. Goedbloed, C. de Nooijer,
R. J. M. Vullers and R. van Schaijk
Holst Centre-Imec, The Netherlands

ABSTRACT

This paper addresses the issue of the mechanical reliability of MEMS piezoelectric vibration harvesters aimed at powering tire pressure monitoring systems. These harvesters generate sufficient power for the targeted application. However, for bringing them to the automotive market, their mechanical reliability has to be optimized, particularly in terms of shock resilience. Experimentally verified methods for improving the mechanical reliability of such devices are showcased in this article. These methods concern both the design of the harvesters (introduction of stoppers in the package) and their manufacturing process (release method of the MEMS structure).

INTRODUCTION

In many countries, tire pressure monitoring systems (TPMS) are or will soon be mandatory for newly build cars, with the aim of improving the safety of road users. Vibration energy harvesters, which act by converting the mechanical energy available in their environment into electrical power, are an alternative to electrochemical batteries for powering TPMS. Such harvesters eliminate the potential replacement costs inherent to batteries. Furthermore, the production costs of harvesters can be made low by using MEMS technologies that allow batch processing.

In a previous publication [1], our group demonstrated the functionality of the MEMS piezoelectric vibration energy harvester described in Figure 1. It consists of a silicon cantilever supporting an AlN based piezoelectric capacitor and attached to a proof mass. It is vacuum encapsulated in-between two glass wafers. It has a footprint of about 1 cm^2. When attached to the tire of a car, such a harvester combined with power managing electronics generates DC power in the range of several tens of μW. This is sufficient to power a wireless TPMS module.

The automotive industry has set stringent requirements on the reliability of electronic components. With regard to mechanical shocks, the standard JESD22-B104-B is usually adopted as a widely accepted reference [2]. According to the standard, samples are tested under a series of half-sine shocks, with maximum recommended test amplitude of 2900 g. This standard is used as general guideline for our reliability analysis.

Insuring the survivability of a MEMS device in a shock environment involves both design and manufacturing aspects [3-5]. In this article, an improving element of both categories is presented. The design improvement consists in the introduction of stoppers in the device. The manufacturing process improvement is related to the release method of the cantilevered structure.

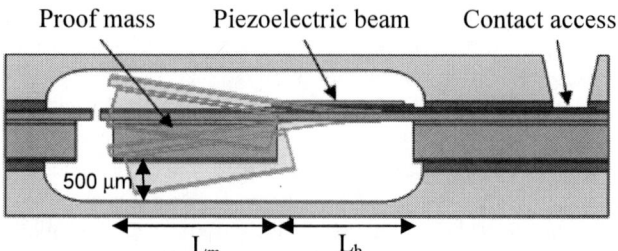

Figure 1: Schematic of the MEMS piezoelectric vibration energy harvesters investigated in this article. L_b and L_m are the length of the piezoelectric cantilever and proof mass, respectively. The width of the proof mass and cantilever are the same and is labeled as W.

IMPROVED SHOCK RESILIENCE WITH HARD STOPPERS IN PACKAGE
Concept

When the harvester is excited by a low amplitude shock, the structure deforms along its fundamental mode as illustrated in Figure 1. When the amplitude of the shock is large, the proof mass attached to the piezoelectric cantilever hits the encapsulating glass wafer. The first point to hit the encapsulating wafer is the free end of the seismic mass. Once the tip of the seismic mass is stopped, the mass center continues moving towards the glass package because of its inertia. In this process, the piezoelectric beam is deformed along a "S-shape" illustrated in Figure 2a. This leads to a large beam curvature at the junction between the cantilever beam and the seismic mass, resulting in a significant stress. We observed experimentally that, under the action of a large amplitude shock, most of our devices break at the junction between the piezoelectric beam and the proof mass.

To prevent the cantilever beam from being deformed along this S-shape, mechanical stoppers are introduced in the form of steps in the cavity, as schematically illustrated in Figure 2b. This concept has previously been discussed in [5]. Given well designed geometries of the stepped cavity, the impact between the seismic mass and the glass package takes place at two locations simultaneously, namely at the free-end of the mass and at the corner of the step. This allows reducing the "S" shape deformation of the cantilever beam after impact with the package. This limits in turn the stresses induced at the junction between the piezoelectric beam and the proof mass. In contrast to the devices without protective steps in the package, we observed experimentally that, under the effect of a large amplitude shock, most of these harvesters break at the anchoring point of the piezoelectric cantilever. A more detailed analysis of these elements is given in [6].

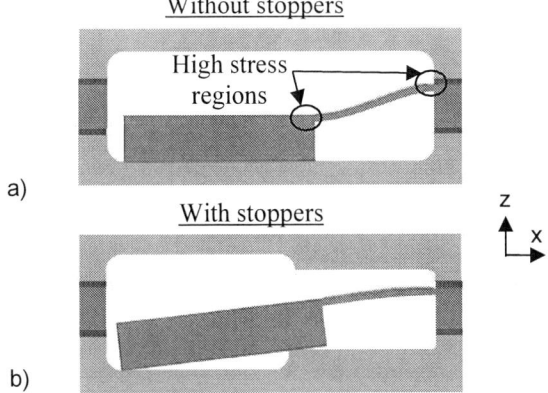

Figure 2: Deformation undergone by the silicon beam under the action of a large amplitude shock with and without protective steps in the package.

Experimental validation of the concept

In order to quantify the effect of the protective steps, a campaign of shock tests was carried out. The shocks were produced by a Lansmont M23 drop tester and have half-sine characteristics. The duration of the shocks was set to 0.7 ms which correspond to a worst case scenario with maximum mass displacement. This result is based on a shock response spectrum theoretical analysis described in [6]. The amplitude of the shock was varied from low to high till observing failure of the device. Failure of the device was determined by visual observations. The capacitance and the voltage output of the piezoelectric stack were also monitored.

Two sets of devices were characterized according to the procedure described above. The two sets were designed and manufactured in exactly the same way. One set was packaged with protective steps and the other set without. The dimensions of the tested devices were L_b=1400 μm, L_m=W=3000 μm. The results of the characterization are reported in Figure 3. While only 4% of the devices without the protective steps survived 1700 g shocks, 60% of the devices with the steps did, thus proving the effectiveness of the protective steps. In all the experiments, failure was due to the fracture of the silicon beam. It was checked by capacitance measurements that the piezoelectric capacitor was not damaged before fracture of the silicon.

IMPROVED FRACTURE RESILIENCE BY WAFER PROCESSING METHOD

Release of the cantilever

The fabrication process of the harvesters is based on bulk micromachining of SOI wafers. Details are given in [1]. After deposition and patterning of an AlN based piezoelectric capacitor on the top surface of the SOI wafer, the cantilever is shaped by front and backside DRIE of the silicon. The buried oxide (BOX) of the SOI wafer acts as an etch stop that allows to precisely control the thickness of the piezoelectric cantilever to 50 μm. The status of the structure at this level of the fabrication process is described in Figure 4.

The next fabrication step involves releasing the cantilever by etching the BOX layer from the trench. Two release methods, illustrated in Figure 4, were investigated. In the first method, the BOX is removed from the trench and from below the cantilever by plasma etching (Ar, CHF_3, CF_4). In the second method, the BOX is wet etched by a HF chemical solution. It is shown in the next sections that the wet BOX etching method leads to more fracture resilient devices than the dry BOX etch method. A possible explanation is that, compared to the wet method, the dry method leads to a rougher surface of the silicon located below the BOX. The fracture strength of micromachined Si devices is strongly dependent on the surface roughness [4, 7].

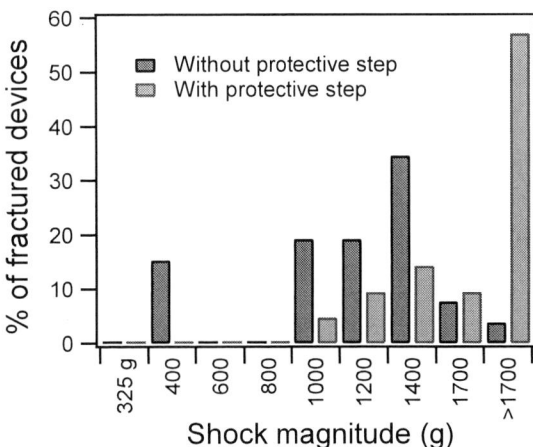

Figure 3: Shock resilience of a population of harvesters with (22 samples) and without (21 samples) protective steps for a half-sine shock with 0.7 ms duration.

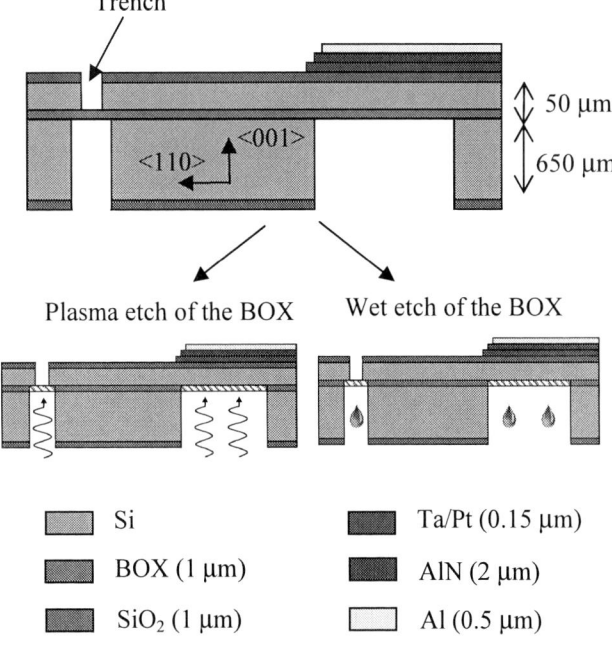

Figure 4: Status of the structure before release and the two tested release methods.

Experimental testing method

While we are ultimately interested in the shock resilience of the packaged cantilevers, the fracture properties of the samples were compared by a different method, illustrated in Figure 5a. It consists of a quasi-static bending test where a rigid needle bends the cantilever by applying a vertical displacement on the proof mass. The anchor of the cantilever is rigidly clamped to a holder. The reaction force of the structure is monitored by a force sensor. The cantilever is deflected till failure occurs. An example of measurement is given in Figure 5. This method is less demanding in terms of time and resources than shock testing of packaged devices and lead to information that can be correlated to the shock resilience.

The performed analysis is focused on the fracture properties of the silicon structure, as the drop tests described previously indicate that the silicon is damaged earlier than the piezoelectric capacitor. Therefore, the fracture experiments were performed on cantilevers where no piezoelectric capacitor was deposited. This again allows sparing resources and time while leading to meaningful results.

As illustrated in Figure 4b, the fracture deflection of the samples was measured by applying the displacement in the two opposite direction A and B. As shown later, the fracture deflection depends strongly on this direction.

From the fracture deflection to the fracture strength

In our manufacturing process, piezoelectric harvesters of different dimensions are fabricated on the same 150 mm wafer. The length L_b of the beam range from 1100 to 1700 µm. The length L_m and width W of the mass ranges from 3000 µm to 5000 µm. Details of the dimensions can be found in [6]. To obtain a statistically relevant population of samples and, in the same time, to limit the number of required tests, we choose to combine the fracture resilience measurements of all the samples on a same wafer. Because of the different dimensions, the fracture deflection cannot be used as a common variable quantifying the fracture resilience. Therefore, the maximum principal stress in the structure when fracture occurs is chosen as the variable to quantify the fracture resilience. In the following, we refer to this variable as the fracture strength σ.

To relate the fracture deflection to the fracture strength, we developed a simple FEM model reproducing the experimental testing conditions. In this model, the cantilever is represented in a 2D plain-stress representation. The layers other than silicon are not included and a simple geometry with ideal dimensions of the cantilever and proof mass is used. The orthotropic material properties of Si given in [7] are used. All motion is restrained at the anchor of the cantilever. The needle is represented by a perfectly rigid body with a rounded tip. In the initial situation, the tip of the needle just contacts the center point on the surface of the mass. The carried simulations are static and compute the final state of the system when the needle has been vertically displaced. As it moves vertically, the needle is allowed to slide freely along the surface of the proof mass it touches. This is also what happens during our experimental measurements and explains the nonlinear aspect of the force-deflection curves of Figure 6.

As expected from simple bending beam theory, the FEM model predicts that the maximum stress in the structure occurs on the clamped end of the cantilever. This is illustrated in Figure 5b. By combining the FEM model with the experimental measurements of the fracture deflection, it is now possible to compute the fracture strength for the samples of different dimensions present on the same wafer.

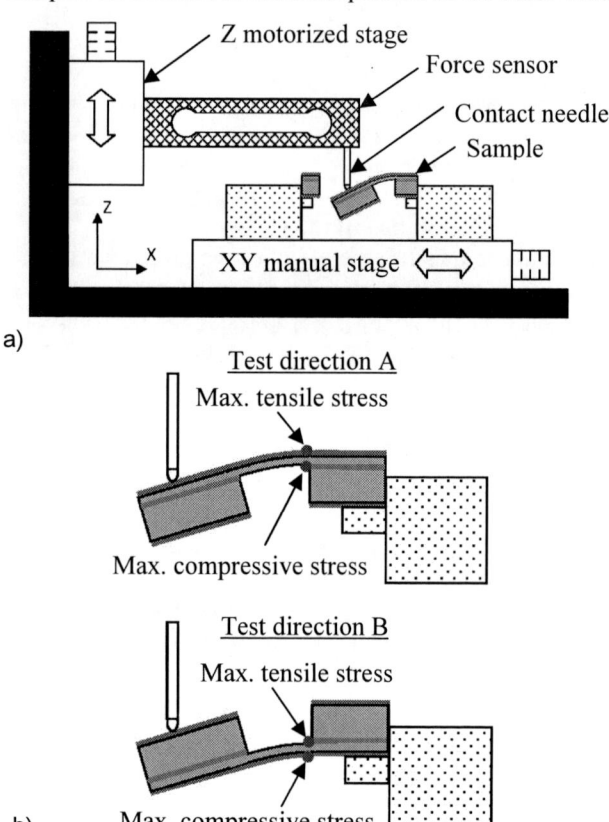

Figure 5: a) Bending test setup for fracture deflection measurement, b) directions of the applied displacement.

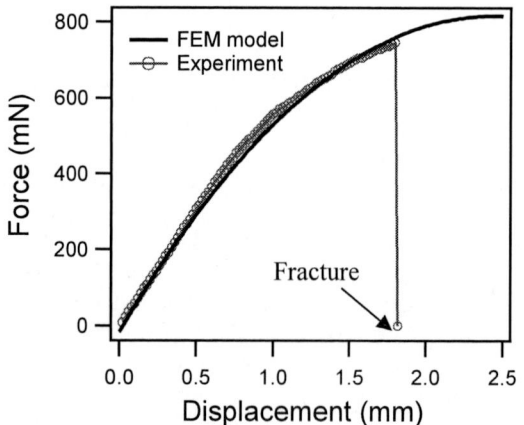

Figure 6: Example of a force-deflection measurement.

Release method vs. fracture strength

Two 150 mm wafers of silicon cantilevers were manufactured in precisely the same way, except for the release step where the BOX was removed by plasma etching in one case and by wet etching in the other.

First, we measured the fracture deflection by applying the displacement in direction A (see Figure 5). The corresponding fracture strength was then computed from the FEM model. 21 samples with plasma etched BOX and 17 samples with wet etched BOX were measured. The results are reported in Figure 7. The distribution of the fracture strength is very similar for both BOX etching method, with a median value $\bar{\sigma}$ around 2.1 GPa. Second, we determined the fracture strength with the displacement applied in direction B. 55 samples of each wafer were tested. From Figure 7, it is clear that, for both release methods, the fracture strength in direction B is much smaller than in direction A. Also, there is a strong difference between the samples with wet ($\bar{\sigma}$=1.05 GPa) and plasma etch of the BOX ($\bar{\sigma}$=0.59 GPa).

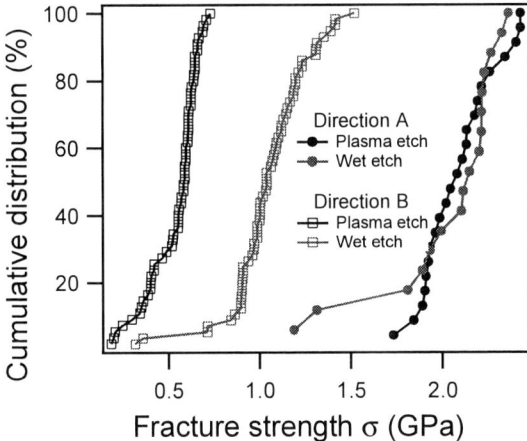

Figure 7: Cumulative distribution of the fracture strength.

To explain the obtained results, we reason in the following way: from Figure 7, it can be deduced that tensile stresses are more likely to damage the surface of the cantilever below the etched BOX than compressive stresses with the same magnitude. We hypothesize that this is also true for the unprocessed surface of the cantilever. This would explain why the fracture strengths measured in direction A are very similar independently of the BOX etch method: in this direction of the applied displacement, the part of the beam undergoing the maximum tensile stress belong to a surface that is unaffected by the BOX etch method. On the other hand, in direction B, the maximum tensile stress occurs on the surface located below the BOX. Microstructural damages are introduced on this surface during the BOX etching step, which make it weaker to tensile stress.

Our goal is however not to understand the details of these phenomena but rather to determine which release method of the cantilever will lead to the most shock resilient packaged harvesters. In a packaged harvester undergoing a shock, the surface of the silicon cantilever damaged by the BOX etching will inevitably undergo tensile stresses. The best etch method is then the one that leads to the highest

value of the fracture strength in test direction B. Clearly, wet etching of the BOX is the best option.

For the packaged harvesters whose shock survival rate is characterized in Figure 3, the cantilever was released with the plasma BOX etch method. Combining the protective step in the package with wet etching of the BOX, we expect to increase significantly the shock resilience of our harvesters, thus ensuring their survivability for in-tire applications.

CONCLUSIONS

While the functionality of MEMS piezoelectric vibration energy harvesters for TPMS applications has been demonstrated, the automotive industry has strong requirements in terms of mechanical shock reliability, which are not yet achieved by current devices. In this article, we describe two elements aimed at reaching this goal. The first element is related to the design of the structure and involves specific stoppers in the package of the device. This allows increasing the survival rate of devices tested with 1700 g shocks from 4% to 60%. The second reliability improvement is related to the manufacturing process of the harvesters. By using a wet etching method in place of a plasma etching method for the release of our MEMS device, the effective bending fracture strength of the samples was increased from 0.59 GPa to 1.05 GPa. This is expected to have an additional positive influence on the shock resilience of the harvesters.

REFERENCES

[1] R. Elfrink et al., "Shock Induced Energy Harvesting with MEMS Harvesters for Automotive Applications", in *Tech. Digest IEDM 2011,* Washington, December 5-7, 2011, pp. 677-680.

[2] JEDEC Standard Mechanical Shock JESD22-B104-B

[3] S. W. Yoon et al., "Shock Protection Improvement Using Integrated Novel Shock Protection Technologies", *IEEE J. of MEMS*, Vol. 20, pp. 1016-1031, 2011.

[4] E. Koukharenko et al., "Performance improvement of a vibration-powered electromagnetic generator by reduced silicon surface roughness", Materials Letters, Vol .62, pp.651-654, 2008

[5] C. Schröder et al., "Wafer Level Packaging of AlN Based Piezoelectric Micropower Generators", in *Tech. Digest Power MEMS 2012*, Atlanta, December 2-5, 2012, pp. 343-346.

[6] Z. Wang et al., "Shock Reliability of Vacuum Packaged Piezoelectric Vibration Harvester for Automotive Application", *accepted for publication in IEEE J. of MEMS.*

[7] K. Chen et al., "Controlling and Testing the Fracture Strength of Silicon on the Mesoscale", *J. of the Americ. Ceram. Soc.*, Vol. 83, pp. 1476-84, 2000.

CONTACT

M. Renaud, tel: +31-404020537;
Michael.Renaud@imec-nl.nl

LOW-COST MICROBOLOMETER WITH NANO-SCALED PLASMONIC ABSORBERS FOR FAR INFRARED THERMAL IMAGING APPLICATIONS

Fabian Utermöhlen[1], Daniel B. Etter[2], David Borowsky[1], Ingo Herrmann[1], Christoph Schelling[1], Franz X. Hutter[2], Shen Hue Sun[2], and Joachim N. Burghartz[2]

[1]Robert Bosch GmbH, Corporate Sector Research and Advance Engineering, Gerlingen, GERMANY
[2]Institut für Mikroelektronik Stuttgart (IMS CHIPS), Stuttgart, GERMANY

ABSTRACT

We present a scalable low-cost microbolometer technology platform which is based on separate fabrication of MEMS and read-out ASIC CMOS wafers. Mechanical, electrical and hermetical connection is achieved by Cu-based thermocompression bonding. The performance loss due to the resulting backside illumination of the sensor is compensated by an optimized microbolometer design including nano-scaled plasmonic absorbers, a dedicated pixel geometry and the use of highly temperature sensitive devices. The low-cost approach features CMOS compatible MEMS processes, wafer level packaging and uncooled operation of the sensor.

INTRODUCTION

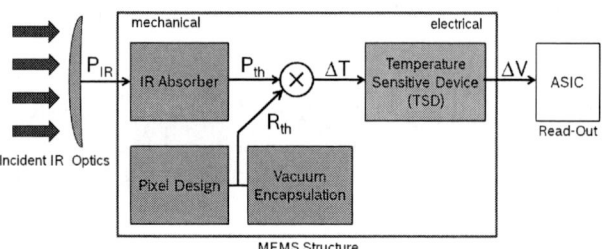

Figure 1: System model of our microbolometer including the signal path from the IR source to the read-out ASIC.

The basic principle of microbolometer operation is illustrated in Fig. 1: At first, incident IR radiation is focused by an optical system on a thermally insulated MEMS pixel containing a plasmonic absorber. The second step is a heating of the pixel ΔT according to $\Delta T = R_{th} P_{th}$ with the absorbed power P_{th} and the thermal resistance of the MEMS structure R_{th} which depends on the pixel design and the vacuum encapsulation of the sensor.

Afterwards, the temperature increase is converted into an electrical signal ΔV by a temperature sensitive device (TSD) which is directly placed in epitaxially grown Si on the pixel. In our low-cost concept, the TSD is represented by pn-junction diodes in constant-current forward mode operation due to the linear decrease of the anode-to-cathode voltage with rising temperatures, the temperature independent temperature sensitivity and the negligible current at reverse bias making additional switching elements for pixel selection in large arrays obsolete.

DESIGN, MODELING AND FABRICATION

Using the ASIC wafer as package for the MEMS wafer as demonstrated in Fig. 2, leads to a backside illumination of the sensor and a resulting loss of IR power in the bulk Si by approximately 50 % which has to be compensated. The system model (Fig. 1) indicates that there are three possibilities to increase the sensor performance, which are directly related to the MEMS structure: (i) Increase the absorbed IR power, (ii) increase the thermal resistance of the pixel and (iii) raise the temperature sensitivity of the pn-junction diode. Appropriate optimization approaches and an experimental validation are presented in this work.

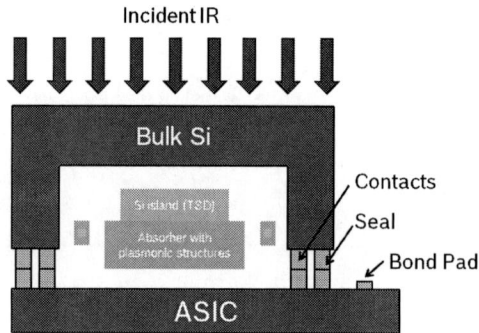

Figure 2: Cross-section of our sensor illustrating the ASIC as wafer cap. For reasons of clarity, only one pixel is illustrated.

Nano-scaled Plasmonic Absorbers

A first approach to increase the pixel performance is the use of dedicated absorption structures. It is demonstrated in [1] that plasmonic structures which are also known as metamaterial perfect absorbers (MPA) can be used for narrowband signal absorption. Therefore, the MPA absorber needs to be adapted to the MEMS fabrication process, the pixel design and the already existing layers (Ti and Al with intermediate SiO_2). Additionally, the absorber design has to be optimized in such a way that a broadband absorption in the FIR spectral range with wavelengths between 8 µm and 14 µm is maximized.

Therefore, frequency domain simulations with CST microwave studio were performed whereas both metals are modeled as Drude materials with parameters according to [2]. The Ti layer contains dots of varying sizes and distances and the Al layer acts as reflector.

During the simulation, the plasma frequency as well as the damping frequency are used to characterize the behavior

978-1-4799-3510-9/14 $31.00 © 2014 IEEE

of the metals due to an external electrical field. The optical characteristics of the material stack is derived from a S-parameter analysis. The sum of the absorption $A(\omega)$, the transmission $T(\omega)$ and the reflection $R(\omega)$ always equals one and allows to write $A(\omega)=1-T(\omega)-R(\omega)$.

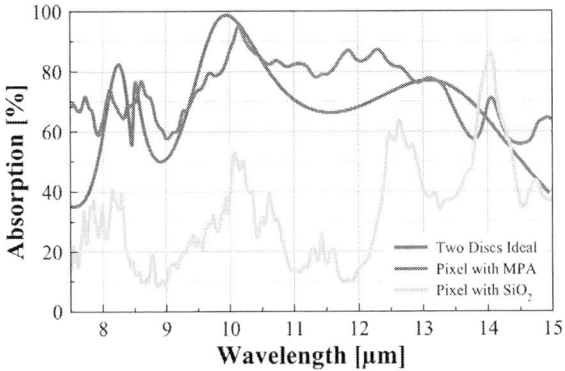

Figure 3: Results of the frequency domain simulation. This graph illustrates the significant absorption increase resulting from the ideal (red) and the pixel MPA (green), compared to a standard SiO2 absorber (yellow).

In order to find the best absorber design, an infinite self-repeating two disc structure was selected and optimized after Nelder and Mead [3]. The parameters to be optimized are the diameter of the first disc d_1, the diameter of the second disc d_2 and the distance between the centers of both discs g (repetition constant). The best broadband absorption in the FIR spectral range could be realized with $d_1/d_2=9/5$. Fig. 3 shows that a significant absorption increase can be realized by the MPA structure in contrast to the standard SiO_2 absorber.

In a second step, this design has been transferred to the pixel geometry with the corresponding process design rules as demonstrated in Fig. 4.

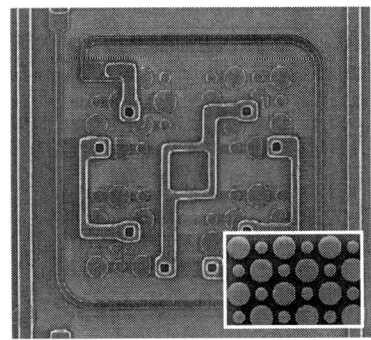

Figure 4: SEM image of one pixel after metallization. The dots and the interconnections are made of Ti. They are the first layer of the plasmonic absorber. The inset figure shows the ideal layout.

MEMS Pixel

The design of the pixels is always a trade-off between absorption area and thermal insulation. In order to find the design with the highest temperature increase at constant incident power density which does not exceed the predefined time-constant $\tau_{th}=R_{th}C_{th}$, we use an electrothermal MEMS pixel model [4] and a semiautomatic optimization tool.

The thermal insulation can be increased by using long and thin support arms holding the pixel and by a vacuum encapsulation of the complete sensor. Since the thermal conductivities of the used solid materials are strongly temperature dependent while the thermal transport through the residual gas environment depends on the pressure only, models for both heat transport mechanism have been developed. FE simulations have shown that radiation and convection can be ignored for typical microbolometer operating conditions and that a simple RC network is sufficient to model the dynamic thermal pixel characteristics. In order to validate the model, the temperature dependent thermal resistance can be extracted from the intersection points of isothermal and static electrical characteristics which can be determined from voltage response curves at varying ambient temperatures and current pulses. With a dedicated correction algorithm [4], the latter can also be used to extract the temperature independent thermal capacitance.

The basis of the featured microbolometer technology is a sintered porous Si (sPS) process followed by a standard MEMS processing including a self-aligned sub-micrometer hard-mask process and a final trench- and release-etch (Fig. 5), [5]. This allows for tuning of the thermal mass of the pixel and for a subdivision of the Si island to integrate more than one TSD per pixel.

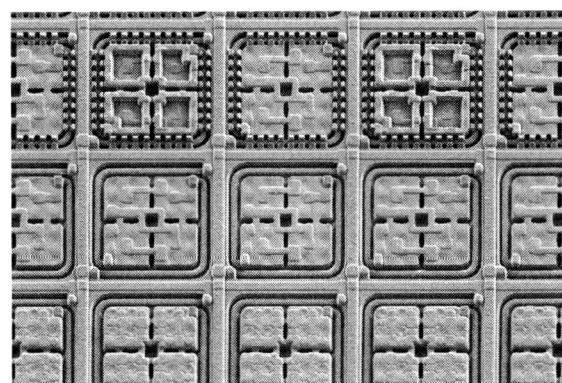

Figure 5: Top view on the pixel array, including reference pixels (top row), pixels without (middle row) and with complete plasmonic absorber (bottom row).

Bonding

The 3D integration of a MEMS and ASIC chip pair results in a significant foot print reduction. Therefore, the MEMS is encapsulated with the ASIC chip at wafer-level via Cu-based thermocompression bonding. Since wet cleaning with already released MEMS structures is not practical prior to bonding, the main challenge for Cu thermocompression bonding with MEMS structures is the protection of the Cu surface during intermediate lithography steps which are necessary for the MEMS release etch step.

Therefore, an additional thin Au layer is placed on top of the Cu to prevent from oxidation and delamination, as shown in Fig. 6.

The geometry of the bond interconnection consists of two parts which are fabricated simultaneously. For mechanical stability and hermetical encapsulation, the MEMS pixel array is surrounded by a rectangular bond frame. To connect the ASIC and MEMS dies electrically, bumps for every pixel row and column are placed outside the pixel array. The quality of the bond interconnection is assessed in two ways. First, shear strength measurements are conducted to evaluate mechanical stability of the bond. Secondly, for hermeticity evaluation, a thinner dummy wafer is used as a counterpart for bonding. After dicing, the deflection of the thinner chip due to an enclosed vacuum can be measured via white light interferometry.

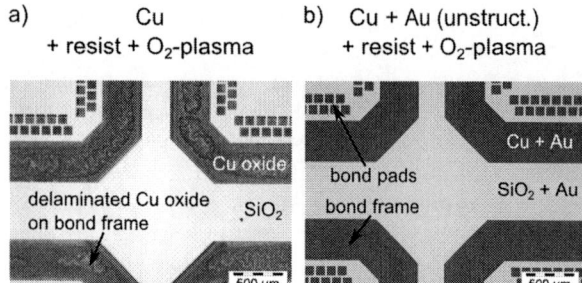

Figure 6: Effect of oxygen plasma strip. a) Exposed Cu oxidizes and delaminates. b) An additional Au layer successfully protects the Cu and prevents oxidation and delamination.

Temperature Sensitive Device

The model for the pn-junction diode is based on the well-known Shockley equation [6] $I=I_S(\exp(V/mV_T)-1)$ with the voltage V, the thermal voltage V_T, the saturation current I_S and the ideality factor m. A theoretical analysis [7] shows that the temperature sensitivity dV/dT at constant current linearly depends on both, the ideality factor and the natural logarithm of the inverse current density $\ln(1/J)$. Therefore, the target of the TSD design and processing is to boost the ideality factor significantly above one and to enforce low current density operation.

EXPERIMENTAL RESULTS

Nano-scaled Plasmonic Absorbers

In order to measure the impact of the MPA structures, a pixel array with and without nano-scaled plasmonic absorber in every second row has been fabricated (Fig. 5). Measurements have demonstrated that the signal-transfer function of a pixel with MPA is given by

$$\Delta V(\Delta T_{BB}) = 1.2\ \frac{mV}{\sqrt{K}} \cdot \sqrt{\Delta T_{BB}} \qquad (1)$$

with ΔV being the output signal difference between an active pixel and a reference pixel resulting from a 200 µs current pulse of 500 nA. ΔT_{BB} represents the temperature change of a blackbody source. Since the temperature coefficient of the TSD is already known, the temperature increase of the pixel resulting from the blackbody illumination can be calculated from the measured voltage difference ΔV. The absorption a_{Pixel} can afterwards be determined from the already described electrothermal pixel model

$$\Delta T = R_{th} \cdot P_{th} = R_{th} \cdot a_{Pixel} \cdot P_{IR} \qquad (2)$$

where P_{IR} marks the emitted IR power of the blackbody source (see Fig. 1).

Fig. 7 illustrates the relative absorption increase of a pixel with MPA compared to a pixel without MPA. In total, the broadband absorption could be increased by approximately 18 %.

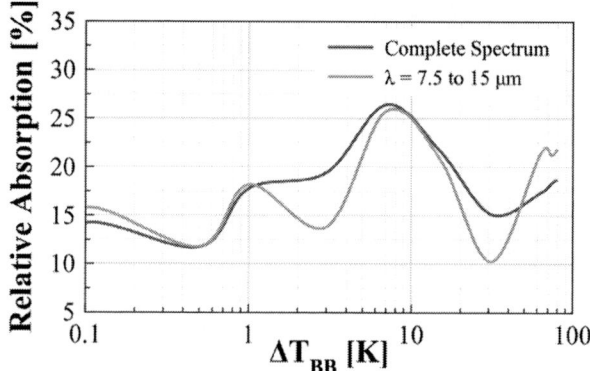

Figure 7: Measurement of the relative change in absorption of microbolometer pixels with MPA relating to a pixel without MPA.

Bonding

Introducing an intermediate lithography step including an oxygen plasma strip leads to a characteristic shear strength T_{63} of only 14 MPa for conventional Cu-Cu thermocompression bonding. An additional Au layer, deposited immediately after the Cu bump formation, increases the shear strength by 50 % to 21 MPa. According to Forsberg et al. [8], depending on application, this can already be sufficient. For hermeticity, the difference is even more significant since the delamination of the Cu surface oxide, as shown in Fig. 6a, completely prohibits the formation of a hermetical encapsulation. None of the measured chips showed a deflection. Adding the Au layer effectively prevents both, the oxidation as well as the delamination as shown in Fig. 6b and achieved a hermeticity yield increase to 75 %. To further increase the shear strength and the hermeticity yield, several annealing protocols were compared [9]. The best results were achieved by post-dice annealing for 1 min at 600 °C, resulting in an increase to 146 MPa, as shown in Fig. 8 and a hermeticity yield improvement to 100%.

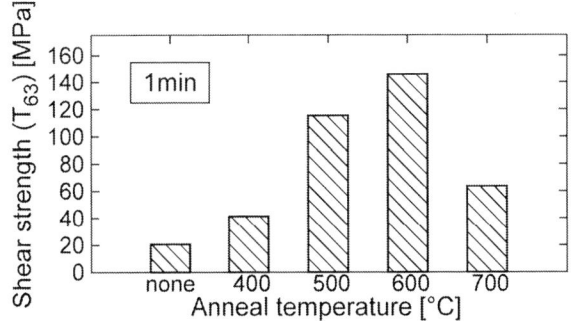

Figure 8: Shear strength T_{63} after 1 min post-dice anneal for samples with additional Au layer and intermediate lithography.

Temperature Sensitive Device

The ideality factors and the temperature sensitivities of n^+-p-p^+ diodes with a varying distance between the symmetry axis and the pn-junction r_p according to the inset of Fig. 9 have been extracted from DC measurements at varying ambient temperatures with self-heating suppression (only low bias currents and reduced thermal pixel insulation by passive air cooling). Fig. 9 contains the measurement data of two DUTs with r_p values varying by 20 % and illustrates that the ideality factor does not depend on the geometric design of the diode (dashed lines). It is clearly visible that the temperature sensitivities are significantly larger than the literature value (-1.7 mV/K). First, this is due to the high ideality factors which result from the onset of high injection effects at relatively low bias and second, the low current density operation (large r_p, black solid line) yields a further maximization.

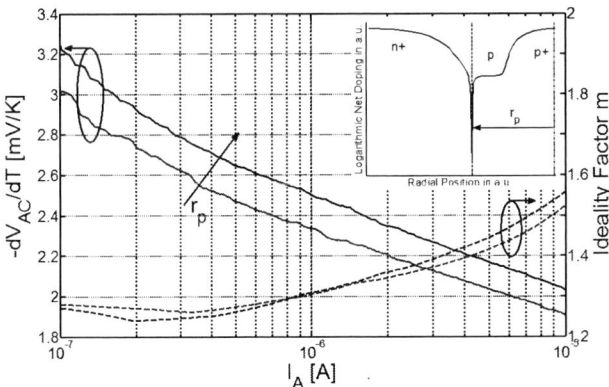

Figure 9: Temperature sensitivities and ideality factors of two diodes with small (blue curves) and large (black curves) r_p. The inset shows the rotation-symmetric doping profile.

With a low current density and a high ideality factor, we could develop a TSD with a better temperature sensitivity than other groups have published so far.

CONCLUSION

We have demonstrated that our microbolometer concept with backside illumination is feasible. The resulting IR radiation power loss can be well compensated by the plasmonic absorber, the optimized pixel design and the high temperature sensitive diodes. For that reason, the presented microbolometer is a novel approach to provide access to a broader market due to far lower system costs.

ACKNOWLEDGEMENTS

This work has been performed in context of the research project RTFIR within the German excellence cluster microTEC Südwest and with financial support by the German Federal Ministry of Education and Research (BMBF) under the grant number 16SV5129. The lab staff at IMS CHIPS, and in particular Dr. Martin Zimmermann, are acknowledged for fabricating the used DUTs in this paper.

REFERENCES

[1] Liu, Xianliang, et al., "Infrared spatial and frequency selective metamaterial with near-unity absorbance," Phys. Rev. Lett, 2010, 104. Jg., Nr. 20, S. 207403.

[2] M. Ordal et. al., "Optical properties of fourteen metals in the infrared and far infrared: Al, Co, Cu, Au, Fe, Pb, Mo, Ni, Pd, Pt, Ag, Ti, V, and W.," Appl. Opt. 24, 4493-4499, 1985.

[3] D. M. Olson and L. S. Nelson, "The Nelder-Mead simplex procedure for function minimization," Technometrics, vol. 17, no.1, pp. 45-51, 1975.

[4] F. Utermöhlen, I. Herrmann, "Model and measurement technique for temperature dependent electrothermal parameters of microbolometer structures," Design, Test, Integration and Packaging of MEMS/MOEMS (DTIP), 2013 Symposium on , 16-18 April 2013.

[5] D. B. Etter et al., "Low-cost CMOS compatible sintered porous silicon technique for microbolometer manufacturing," Micro Electro Mechanical Systems (MEMS), 2012 IEEE 25th International Conference on, pp. 273-276, Jan. 29 2012-Feb. 2 2012.

[6] R.-Y. Sah, R. N. Noyce, and W. Shockley, "Carrier generation and recombination in p-n junctions and p-n junction characteristics," Proceedings of the IRE, vol. 45, no. 9, pp. 1228-1243, 1957.

[7] F. Utermöhlen et al., "Temperature sensitivity modeling of pn-junction diodes for microbolometer-based thermal imaging applications", in International Semiconductor Conference Dresden Grenoble (ISCDG) 2013, 2013.

[8] F. Forsberg et al., "Very large scale heterogeneous integration (VLSHI) and wafer-level vacuum packaging for infrared bolometer focal plane arrays," in Infrared Physics & Technology 60, pp. 251-259, 2013.

[9] D. Borowsky et al., "Gold as an Enabler of Copper Thermocompression Bonding for MEMS", submitted to Microelectronic Engineering.

CONTACT

*F. Utermöhlen, tel: +49-711-811 48704; fabian.utermoehlen@de.bosch.com

2D PHOTONIC-CRYSTALS FOR HIGH SPECTRAL CONVERSION EFFICIENCY IN SOLAR THERMOPHOTOVOLTAICS

Andrej Lenert[1], Veronika Rinnerbauer[1], David M. Bierman[1], Youngsuk Nam[1,2], Ivan Celanović[1], Marin Soljačić[1] and Evelyn N. Wang[1]*

[1]Massachusetts Institute of Technology, USA and [2]Kyung Hee University, Korea

ABSTRACT

We present a novel solar thermophotovoltaic (STPV) device, which for the first time, incorporates a two-dimensional photonic-crystal (2D PhC) absorber-emitter to achieve spectral conversion efficiencies >10%. These results were achieved by tailoring the spectral properties of the absorber-emitter through surface nanostructuring of tantalum (Ta) and minimizing parasitic thermal losses through an innovative vacuum-enclosed experimental setup. By incorporating a sub-bandgap photon reflecting filter on the PV surface and optimizing the absorber-emitter ratio, we present how the demonstrated 2D Ta PhCs enable a realistic STPV configuration to exceed the Shockley-Queisser ultimate efficiency of a 0.55 eV cell.

INTRODUCTION

While PVs only utilize a portion of the solar spectrum efficiently, STPVs can overcome the Shockley-Queisser single-junction PV limit [1, 2]. In STPVs, the broad solar spectrum is converted into a narrow-band thermal emission spectrum tailored near the photovoltaic (PV) bandgap (Fig. 1a), *via* a hot absorber-emitter. Nevertheless, the high operating temperature (>1200 K) of the absorber-emitter poses significant challenges for spectral control and efficient collection of sunlight, limiting prior experimental demonstrations to low solar-to-electrical conversion efficiencies.

In this work, we experimentally demonstrated a novel STPV device, which for the first time, incorporates a two-dimensional photonic-crystal (2D PhC) absorber-emitter to achieve higher conversion efficiencies (Fig. 1b-c). We describe the design and fabrication of the 2D Ta PhCs absorber and emitter that enabled the high spectral performance. We also describe the experimental methods used in characterizing our 2D PhC absorber-emitter when coupled to a 0.55 eV InGaAsSb PV cell as a function of the solar irradiance. We further show how this absorber-emitter PhC surface can be used to exceed the ultimate efficiency of a 0.55 eV cell (as defined by Shockley-Queisser).

Ultimate and Real Spectral Conversion Efficiency

The performance of STPVs is highly dependent on the efficient conversion of sunlight to useful thermal emission, which the PV cell can then harness to excite electron-hole pairs and generate power. To identify the added value of placing an absorber-emitter in between the incident solar radiation and a (low-bandgap) PV cell, we consider the efficiency of the same cell when exposed to solar radiation (AM 1.5D), and when exposed to thermal emission from the solar-powered emitter. We demonstrate that unless the

absorber-emitter is properly designed, placing it in front of the PV cell may degrade the overall solar-to-electrical efficiency of the system.

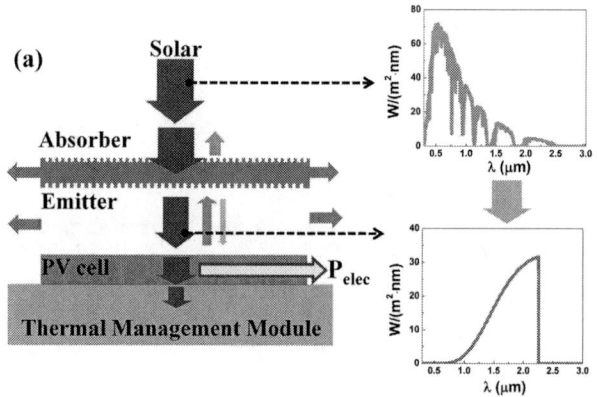
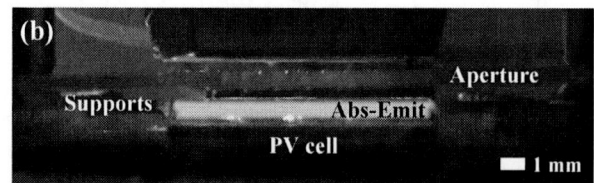
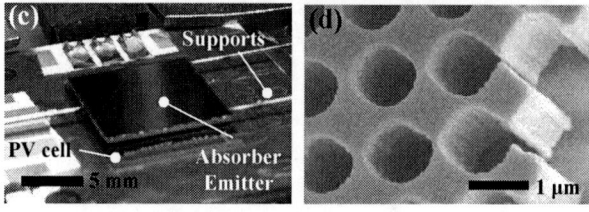

Figure 1: Operating principle of the STPV device: optically concentrated sunlight is converted into heat in the absorber; the absorber temperature rises; heat conducts to the emitter; the hot emitter thermally radiates towards the PV cell; radiation is ultimately converted into excited charge carriers in the cell and extracted. (a) Schematic of a 2D PhC absorber-emitter that converts solar radiation with a broad spectrum into a tailored spectrum matched to the spectral response of the PV cell. (b),(c) Optical images of our vacuum-enclosed devices composed of a 2D PhC absorber-emitter and a 0.55 eV bandgap PV cell (InGaAsSb). The visibly glowing absorber-emitter (b) indicates successful high temperature operation (~1300 K). The device was suspended by spring-loaded hollow metallic supports to minimize parasitic thermal losses. (d) SEM of the 2D PhC emitter showing an array of high-aspect ratio cavities etched in Ta.

To isolate the effect of the absorber-emitter mediated spectral-conversion process, we consider the limit in which the PV cell generates one electron-hole pair from each incident photon with sufficient energy ($>E_g$) and extracts it at the bandgap voltage (V_g). This limit is in accordance with the definition of ultimate efficiency, presented by Shockley-Queisser [2]. Here, we extend the definition of ultimate efficiency to account for the possibility of a converted incident spectrum using the ultimate spectral conversion efficiency (SCE_U):

$$SCE_U = \frac{E_g Q_s}{H_s} = \frac{\int_0^{\lambda_g} \frac{\lambda}{\lambda_g} H_{\lambda,pv} d\lambda}{H_s} \qquad (1)$$

where H_s is the total solar irradiance on the device, H_λ is the spectral irradiance on the PV cell, Q_s is the flux of above-bandgap quanta incident of the cell, and λ_g and E_g are the bandgap wavelength and energy of the PV, respectively. In the case where the spectral irradiance is equal to the solar irradiance, the SCE_U is equivalent to the ultimate efficiency. However, when the incident spectrum is modified through the *photo-thermal-photo* conversion process in the absorber-emitter, the ultimate efficiency can naturally be exceeded in the case that the thermal emission contains a greater fraction of useful near-bandgap radiation.

In real devices, the spectral conversion efficiency of the PV is proportional to the photocurrent density (J_{ph}) generated by the PV assuming a wavelength and angular independent external quantum efficiency. Hence, using easily measurable quantities, we will define the real spectral conversion efficiency as:

$$SCE_R = \frac{J_{ph} V_g}{H_s} \qquad (2)$$

where V_g is the PV bandgap voltage. In our experimental section, we will use SCE_R to compare the performance of the PV cell with and without the absorber-emitter.

PHOTONIC CRYSTAL DESIGN

The photonic crystal absorber-emitter consists of a 2D array of high-aspect ratio micro-cavities (Fig. 1d) etched into both sides of a Ta substrate. These structures were fabricated using interference lithography followed by deep reactive ion etching of Ta [3, 4]. Even at the high operating temperatures, the spectral properties of the Ta PhCs exhibit high-selectivity with a sharp cut-off enabled by the PhC surface structure (Fig. 2). For the absorber, the cut-off wavelength (λ_c) and the spectral properties were designed to optimize absorption (high ε_λ) of short-wavelength concentrated sunlight ($\lambda<\lambda_c$), and suppress re-emission (low ε_λ) of long-wavelength thermal radiation ($\lambda>\lambda_c$). For the emitter, the spectral properties were optimized to enhance thermal emission of useful radiation with energies above the PV bandgap ($\lambda<\lambda_g$) while suppressing sub-bandgap radiation ($\lambda>\lambda_g$).

The design of the PhCs is a square array of cylindrical holes with period (a), radius (r), and depth (h) created on a tantalum (Ta) substrate. Ta was selected due to its high

melting point (3290 K), low vapor pressure and low emissivity at long wavelengths. The emissivity of the substrate is selectively enhanced in the PhC by coupling to the cavity modes, and the cutoff wavelength is tunable by adjusting the fundamental cavity resonant frequency through changes in the dimensions of the cavities, while the maximum emittance of the first resonance peak below the cutoff is achieved via Q-matching [5].

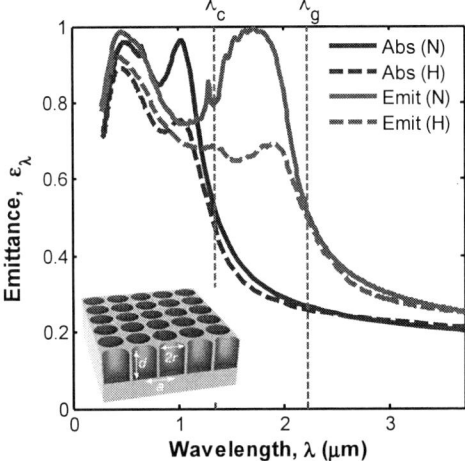

Figure 2: Simulated normal (N) and hemispherically-averaged (H) spectral emittance at 1300 K for the fabricated 2D Ta PhC absorber-emitter. For the absorber, cavity dimensions are a=0.65/r=0.25/h=4.6 µm. For the emitter, the cavity dimensions are a=1.3/r=0.55/h=8.0 µm.

Optimization

The optimization of the emitter was performed using the spectral efficiency η_{sp} as a figure of merit (FOM), which is defined as the number of photons emitted in the useful wavelength range relative to the total number of emitted photons. This is a measure of the spectral selectivity of the emitter at a given operating temperature T. The useful wavelength range is given by the bandgap and EQE of the PV cell; we used InGaAsSb PV cells with a bandgap of 0.55 eV and a useful wavelength range of about 1 µm-2.3 µm and we optimized the geometrical parameters of the emitter for a target operating temperature of 1300 K, using the hemispherical emissivity of the PhC. The hemispherical emissivity of the PhCs was computed using the Fourier Modal Method using a freely available software [6].

For the optimization of the absorber, the thermal transfer efficiency η_T was used as the FOM [7, 8]:

$$\eta_T = \overline{\alpha}_{abs} - \frac{\overline{\varepsilon}_{abs} \cdot \sigma T^4}{H_s} = \overline{\alpha}_{abs}\left(1 - \frac{\overline{\varepsilon}_{abs}}{\overline{\varepsilon}_{tot}}\right) \qquad (3)$$

where T is the operating temperature, $\overline{\alpha}$ is the absorptivity of the absorber averaged over the solar spectrum, $\overline{\varepsilon}$ is the emissivity of the absorber averaged over the blackbody radiation at T, σ is the Stefan-Boltzmann constant, H_s is the irradiance of the solar spectrum (AM 1.5D). In a STPV system, the irradiance needed to reach a certain operating T depends on the balance of the input and output power, and we can rewrite the equation using the sum of the averaged emissivities of all surfaces, including the absorber, emitter and sides. We approximate that all of the absorber-emitter losses are radiative.

Again, we optimize the design for a target operating of 1300 K. The spectral emittance of the 2D Ta PhC absorber and emitter is close to the black-body limit at short wavelengths and approaching that of polished metal at long wavelengths, with a sharp cutoff separating the two regimes. The spectral emittance of the fabricated PhCs, as determined from reflectance measurements at near normal incidence at room temperature, shows good agreement with the simulated spectra.

DEVICE CHARACTERIZATION

We designed the experimental system for characterizing high-temperature planar STPVs aiming to validate our previously-developed model [9, 10] and compare the performance to the PV cell without the absorber-emitter. The following experimental capabilities were developed to properly account for the coupled energy conversion steps and to measure the device characteristics: incident power of simulated solar radiation, absorber/emitter temperature (see *TPV Validation*), current-voltage characterization, and thermal load on the PV cell. The experimental layout (Fig.1b-c) minimizes parasitic heat losses while allowing for precise alignment and gap control between the absorber-emitter and the PV cell.

TPV Validation

An accurate measurement of the emitter temperature with minimal impact on the temperature distribution of the emitter was obtained using a thin gauge thermocouple bonded to the back of the emitter substrate (Fig.2b).

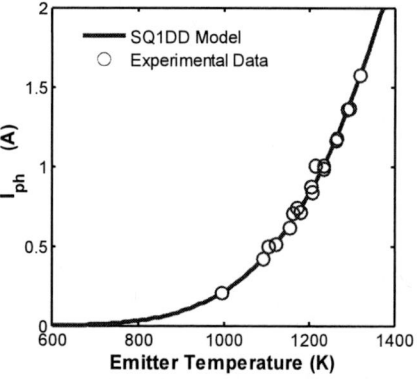

Figure 3: Photocurrent generated by two 0.55 eV InGaAsSb (0.5cm x1cm) cells connected in series as a function of the 2D Ta PhC (1x1 cm) emitter temperature. The cell and emitter were separated by a 300 μm vacuum gap. Line and symbols represent the results obtained from the model and experiments, respectively.

Using the experimental system, we investigated a solar-powered TPV system composed of the 2D Ta PhC as a high-temperature spectrally selective emitter paired with an InGaAsSb PV cell (0.55 eV). The IV characteristics of the TPV were experimentally characterized by controlling the solar power incident on the device, and accordingly varying the emitter temperature from room temperature to around

1300 K. The photocurrent generated by the PVs (at a slightly negative bias voltage) was measured as a function of the 2D PhC emitter temperature. As shown in Figure 3, predictions using a detailed numerical model [9, 10] match the results within experimental uncertainty, validating the numerical models used throughout the study.

STPV and PV Comparison

In a STPV configuration, we characterized the *SCE* of our 2D PhC absorber-emitter when coupled to the 0.55 eV InGaAsSb PV cell as a function of the solar irradiance (Fig. 4). The incident simulated sunlight was passed through an aperture with matching dimensions to the absorber, spaced 500 μm apart. As in the TPV experiment, the emitter was separated from the cell by a 300 μm vacuum gap. For comparison, we performed the same experiment but in a concentrated PV configuration where we measured the photocurrent generated by the PV cell without the absorber-emitter in front of it. To be consistent, the absorber and the PV cell were held in the same plane for both experiments (relative to the aperture). Furthermore, the temperature of the PV cell was consistent in both experiments.

In the STPV configuration, the *SCE* increased with increasing irradiance since the system efficiency was highly temperature-dependent. The highest overall *SCE* we measured for the STPV configuration was approximately 11% at 13 W/cm^2 of irradiance. At this point, the absorber-emitter was approximately at 1300 K, which is near the design temperature. On the other hand, the PV configuration achieved a *SCE* of 19-20%, suggesting that placing the absorber-emitter degrades the overall performance, primarily due to parasitic losses and sub-bandgap emission. Nevertheless, the experimental system represents a proof-of-concept STPV demonstration and integration of 2D Ta PhCs. When the system is scaled-up and optimized, the STPV performance can significantly exceed that of the PV configuration.

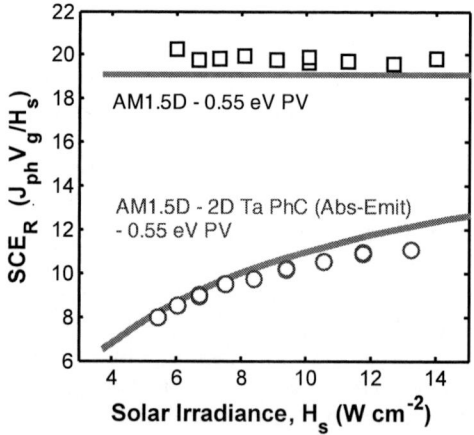

Figure 4: Spectral conversion efficiency comparing direct AM 1.5D illumination of the 0.55 eV cell (PV configuration), to the STPV configuration with a 2D Ta PhC absorber-emitter under the same conditions.

EXCEEDING THE S-Q LIMIT

The above experimental validation suggests that STPVs are rapidly approaching the performance of single junction PV cells. By incorporating a sub-bandgap photon reflecting filter on the PV surface [11] and scaling the emitter to ten times the area of the absorber (*i.e.*, area ratio of 10) [10], the ultimate spectral conversion efficiency (SCE_U) of the 2D PhC-based STPV will exceed ultimate efficiency (as defined by the Shockley-Queisser) for this 0.55 eV cell at optical concentrations as low as ~200 Suns (Fig. 5). At higher concentrations, the ultimate spectral conversion efficiency approaches 45% for this device. These promising results, along with the potential to incorporate thermal/chemical storage, suggest the viability of STPVs for next-generation, efficient, scalable and dispatchable solar power generation.

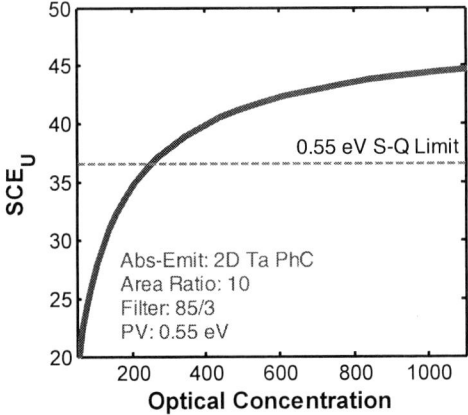

Figure 5: 2D Ta PhCs enable STPV to exceed the ultimate efficiency for a 0.55 eV cell (as defined by Shockley-Queisser), at relatively low optical concentrations when integrated in an optimized system design (i.e. by scaling up the emitter relative to the absorber and introducing a sub-bandgap reflecting filter).

CONCLUSION

We present the design and characterization of a planar STPV composed of 2D Ta PhC absorber-emitter and a 0.55 eV InGaAsSb PV cell. The experimental setup minimized parasitic heat losses and allowed precise alignment of the components. Through TPV and STPV experiments, we showed excellent agreement with our previously reported models. We also directly compared the performance with and without the absorber-emitter, showing that the absorber-emitter can degrade the overall performance if the system has substantial sub-bandgap emission and parasitic losses. By scaling-up the emitter relative to the absorber and incorporating a sub-bandgap filter into our model, we showed that a 2D Ta PhC enabled STPV can exceed the Shockley-Queisser ultimate efficiency at relatively low irradiances (~200 Suns). This study demonstrates the components and facilitates the design of such a high-efficiency system.

ACKNOWLEDGEMENTS

This work is supported as part of the Solid-State Solar Thermal Energy Conversion (S3TEC) Center, an Energy Frontier Research Center funded by the U.S. Department of Energy, Office of Science, Office of Basic Energy Sciences under DE-FG02-09ER46577. A.L. acknowledges the support of the Martin Family Society, the MIT Energy Initiative and the National Science Foundation GRF. Y.N. acknowledges the support from Basic Science Research Program through the National Research Foundation of Korea (NRF) funded by the Ministry of Science, ICT & Future Planning (No. 2012R1A1A1014845).

REFERENCE

[1] H. Nils-Peter and W. Peter,"Theoretical limits of thermophotovoltaic solar energy conversion", *Semicond. Sci. Technol.*, vol. 18, pp. 151-156, 2003.

[2] W. Shockley and H.J. Queisser,"Detailed balance limit of efficiency of p-n junction solar cells", *J. Appl. Phys.*, vol. 32, pp. 510-519, 1961.

[3] V. Rinnerbauer, et al.,"Large-area fabrication of high aspect ratio tantalum photonic crystals for high-temperature selective emitters", *J. Vac. Sci. Technol. B*, vol. 31, pp. 011802-7, 2013.

[4] V. Rinnerbauer, et al.,"High-temperature stability and selective thermal emission of polycrystalline tantalum photonic crystals", *Opt. Express*, vol. 21, pp. 11482-11491, 2013.

[5] M. Ghebrebrhan, et al.,"Tailoring thermal emission via Q matching of photonic crystal resonances", *Phys. Rev. A*, vol. 83, pp. 033810, 2011.

[6] V. Liu and S. Fan,"S4 : A free electromagnetic solver for layered periodic structures", *Comp. Phys. Comm.*, vol. 183, pp. 2233-2244, 2012.

[7] P. Bermel, et al.,"Design and global optimization of high-efficiency thermophotovoltaic systems", *Opt. Express*, vol. 18, pp. A314-A334, 2010.

[8] B.O. Seraphin, *Solar energy conversion: solid-state physics aspects*, Springer-Verlag, 1979.

[9] Y. Nam, et al. "Solar thermophotovoltaic energy conversion systems with tantalum photonic crystal absorbers and emitters", in *The 17th International Conference on Solid-State Sensors, Actuators and Microsystems (Transducers)*. Barcelona, Spain, June 16-20, 2013, pp. 1372-1375.

[10] A. Lenert, et al.,"A Nanophotonic Solar Thermophotovoltaic Device", *Nat. Nanotech.*, vol., 2013.

[11] M.W. Dashiell, et al.,"Quaternary InGaAsSb thermophotovoltaic diodes", *IEEE Trans. Electron* vol. 53, pp. 2879-2891, 2006.

CONTACT

A. Lenert , tel: +1-617-253-3355; alenert@mit.edu

AN OPTICAL IN-PLANE DISPLACEMENT MEASUREMENT TECHNIQUE WITH SUB-NANOMETER ACCURACY BASED ON CURVE-FITTING

Jaap Kokorian[1], Federico Buja[1], Urs Staufer[1] and W. Merlijn van Spengen[1,2]

[1]Delft University of Technology, The Netherlands

[2]Falco Systems B.V., The Netherlands

ABSTRACT

In this paper we present a new optical method for detecting in-plane displacements in microelectromechanical systems with deep sub-nanometer accuracy. The technique is based on curve fitting. We investigate the error sources that influence the measurement and show measurements to demonstrate that a position resolution of 130 pm can be obtained with a simple microscope and camera.

INTRODUCTION

When studying the mechanical behavior of microelectromechanical systems (MEMS), it is often necessary to measure in-plane displacements with high accuracy. For the measurement of out-of-plane motions, interferometry-based techniques such as laser-Doppler vibrometry are used routinely. Unfortunately, these techniques are often unpractical for in-plane measurements. In-plane motions can be easily observed with an optical microscope and a camera, but the resolution of any optical measurement of quantitative spatial information is fundamentally limited by the Rayleigh criterion. In practice, this limits the resolution of any 'simple' position tracking measurement to ∼ 0.5 μm.

Several methods exist to optically measure the in-plane motions of MEMS devices with higher resolutions than the Rayleigh criterion. Most widely used is digital image correlation, a method that has been investigated thoroughly by Davis & Freeman [1] and Petitgrand & Bosseboeuf [2]. Guo *et al.* [3] use a similar technique that they call the 'optical flow technique'. Powerful as these techniques may be, one drawback seriously limits the accuracy with which motion can be detected: Kleinemeier [4] and Davis *et al.* [5] have shown that measurements contain systematic errors when noise is present in the system. These errors are periodically varying over the displacement by less than a single pixel. Under extreme circumstances, this can result in systematic position determination errors up to a significant portion of a single image pixel.

In a technique developed by Yamahata *et al.* [6], the position of a feature is extracted from the phase of the discrete Fourier transform of an image. When analyzing a time-series of images, the movement of the feature can be tracked by observing how the phase-shift of the principal peak changes in the frequency spectrum. The technique works especially well with periodic features like multiple parallel beams. The resolution depends strongly on the number of feature periods and can become as precise as 0.2 nm (root-mean-square), which appears to be the lowest value currently reported in literature. Can we do even better?

Where digital image correlation techniques use a 'template' image for example, we will use a mathematical function that describes the shape of a displacing object and use curve-fitting to extract its position. We will show that even with a simple microscope and camera, it is possible to measure displacements as small as 0.13 nm.

THEORY AND METHOD

The curve fitting technique that we use to extract in-plane displacements is outlined in Figures 1, 2 and 3. We will explain our technique by analyzing the one dimensional motion of a single MEMS beam spring as shown in Figure 1. Each line of pixels has a similar 'intensity profile' that shows a small peak that corresponds to the position of the beam. When the beam moves along the x-direction of the image, the shape of the peak will stay roughly same, but its position will shift. The peak position can be found by fitting a mathematical function to it that depends on the horizontal shift x_0.

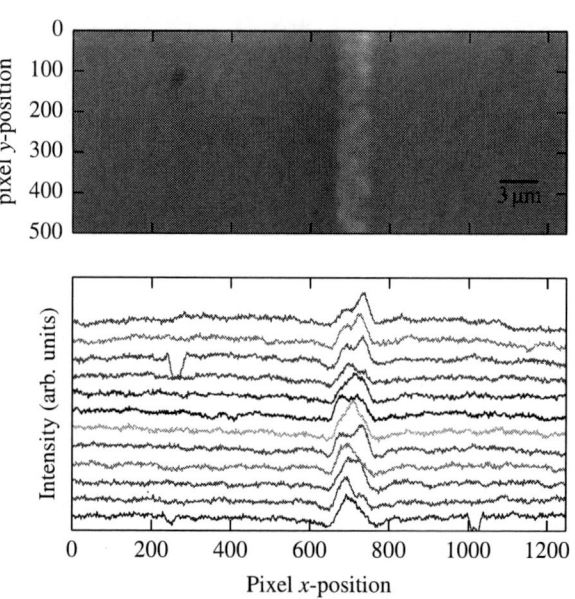

Figure 1: Single, free standing silicon beam of which the displacement in the x-direction is to be determined. Each line of pixels shows similar intensity profile with a peak that corresponds to the beam.

This fitting technique is commonly used to extract the peak positions (and widths) of e.g. Raman and XRD spectra, where the peak shapes can often be approximated by a Gaus-

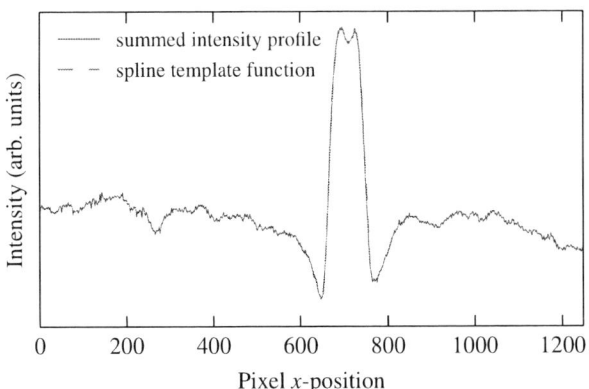

Figure 2: The intensity profiles of all images lines are summed to produce a single intensity profile with a higher signal to noise ratio than the individual profiles. This profile is used to create a 'template' spline function.

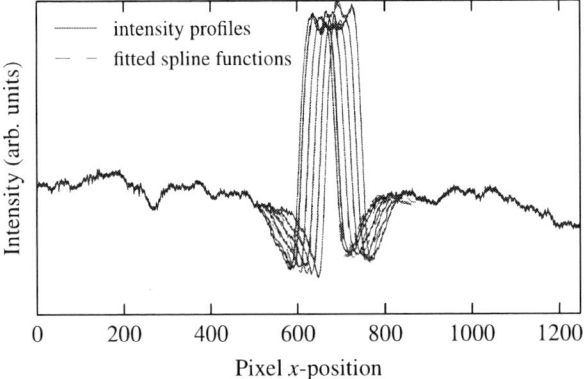

Figure 3: When the beam is displaced, the displacements are determined by fitting the spline function to the shifted intensity profiles.

sian or Lorentzian function [7]. Features of MEMS devices do not typically have such nicely defined shapes. Even though it is possible to manually 'design' a mathematical function to describe the peak shape, it is much easier to create a spline-based 'template function' $s(x)$ to approximate its shape. This can be done automatically by any scientific computing program. A position parameter x_0 can simply be inserted into the spline function's definition to be able to fit it to any shifted version of the template: $s'_{x_0}(x) = s(x - x_0)$. When fitting the shifted template function to the intensity profile to which it was modeled, the best fit will occur a $x_0 = 0$. All displacements of the intensity profile peaks of subsequent images are therefore relative to the position of peak in the template profile.

Noise

The resolution of the position measurement is determined by the amount of pixels the peak in the intensity profile is spread across and by the amount of light. W.M. van Spen-

gen [8] has shown that when a peak is spread across enough pixels to clearly show its shape, increasing the magnification has no effect on the position accuracy. This rule is only valid when the noise figure of each pixel is limited by photon shot noise.

Shot noise is caused by the fact that light consist of discrete photons. The amount of photons that is captured on a single pixel within the exposure time will vary between images. The total amount of photons that arrive at irregular intervals within a certain time is described by a Poisson distribution with mean value $= n_{\text{photons,px}}$ and standard deviation $\sigma = \sqrt{n_{\text{photons,px}}}$, where $n_{\text{photons,px}}$ is the amount of captured photons on a pixel. The signal to shot noise ratio for a single pixel is shown in (1),

$$\text{SNR}_{\text{shot}} = \sqrt{\frac{1}{n_{\text{photons,px}}}} \sim \sqrt{\frac{1}{I_{\text{px}}}} \qquad (1)$$

where I_{px} is the pixel intensity in arbitrary units. The amount of shot noise in the intensity profile can be lowered by making sure a pixel captures as much light as possible, by binning multiple pixels and by summing the pixel values of multiple images.

If the peak positions are equal across multiple lines of an image (as is the case in Figure 1), the intensity profiles can be summed to produce a new profile that has a factor $\sqrt{n_{\text{lines}}}$ less shot noise (Figure 2) than each of the originals.

Differential position measurement

When observing a single beam, in-plane movements of the camera and microscope with respect to the sample cannot be distinguished from movements of the beam with respect to the sample. This issue can be overcome by observing the displacement of two beams simultaneously, where one beam remains stationary with respect to the sample substrate. By fitting two separate functions to the corresponding intensity profile peaks and subtracting the resulting shift parameters, the differential displacement of the beams can be obtained. Any common-mode displacement caused by vibrations of the microscope or camera will be discriminated.

EXPERIMENTS

We have measured the movement of a 'nano-ram' MEMS adhesion sensor [9] using a Motic PSM-1000 bright field optical microscope fitted with a 3 megapixel uEye CMOS machine vision camera. The setup is housed inside an acoustic isolation booth on top of a vibration isolated table to damp external vibrations as much as possible.

The nano-ram device consist of a thick beam (the ram) that can be moved forwards by a comb drive actuator until it makes contact with a counter surface. When the ram moves backwards, it will momentarily adhere to the counter surface.

We have actuated the nano-ram by sweeping the forward comb drive actuator voltage from 0 V to 60 V and back in 1000 steps. At each voltage we have captured an image that shows a support beam connected to the ram and a support beam connected to an anchor. In order to clearly show the benefits of a differential position measurement of a moving beam with respect to a stationary one, we executed this measurement with the doors of the acoustic isolation booth fully opened and the pneumatic table dampers disabled.

To show that the position accuracy decreases with the amount of shot noise, we have repeated the measurement taking 16 exposures per voltage step instead of just one. We summed the pixel values of all 16 images, which effectively raises the pixel intensity I_{px} by a factor 16 and should lower the photon shot noise by a factor $\sqrt{16} = 4$ according to equation (1). For this measurement the acoustic isolation booth was closed and the pneumatic table supports enabled.

RESULTS

Figure 4b shows the voltage-displacement curve of the ram, using only the moving beam spring for the position analysis. The graph clearly shows the point where the ram touches the counter surface at an actuation voltage of 53 V. To determine the amount of nanometers per pixel we used the fact that the distance between the ram's rest position and the point where it touches the counter surface is exactly 2 μm. We determined the amount of nanometers per pixel to be 32.2 nm/px. This value remains approximately the same for all other measurements taken with the same camera and the same optical magnification. However, we have observed that because of small focus changes it can vary by a few tenths of a nanometer between measurements.

We have measured the position noise of the adhesion curve by subtracting a polynomial trend-line from the continuous quadratic part of the forward motion. The results are shown in the scatter plot and corresponding histogram in Figure 4. The standard deviation of the noise is $\sigma = 5.98$ nm.

Figure 5b shows a plot of the same measurement data that was used to produce the adhesion curve in Figure 4b, only this time relative to the position of a stationary beam that was captured simultaneously. The adhesion curve shows significantly more detail and this is confirmed numerically by the position noise analysis. The standard deviation of the noise has been lowered by more than a factor ten to $\sigma = 0.51$ nm. This dramatic reduction shows that the noise figure of a position analysis where only a single moving beam is considered, is limited by the mechanical vibrations of the microscope and camera with respect to the sample.

Figure 6b shows the adhesion curve for the measurement where the pixel intensities of 16 consecutively taken images where averaged for each applied actuator voltage. The position noise as low as 0.13 nm, a factor 3.9 lower than the noise level in a measurement with just a single exposure per applied voltage. This value corresponds well to the predicted value of 4 and confirms that the measurement is shot noise limited.

CONCLUSION

In-plane displacement measurements of silicon beams in MEMS can be measured with subnanometer accuracy using a simple microscope and CMOS camera. When the displacement of a single beam is measured relative to a stationary beam, the position noise level drops into the subnanometer regime. When photon shot noise is the dominant noise source, the resolution can be increased by summing the pixel values across multiple images and by summing the intensity profiles of multiple lines, which lowers the position noise by a factor $\sqrt{n_{px}}$ with n_{px} the number of summed pixels. Summing 489 lines and 16 consecutive images lowers the root-mean-square position noise to 0.13 nm, approximately the distance between two atoms.

ACKNOWLEDGEMENTS

This work has been financially sponsored by the Dutch NWO-STW foundation in the 'Vidi' program under ref no. 10771.

REFERENCES

[1] C. Q. Davis and D. M. Freeman, "Using a light microscope to measure motions with nanometer accuracy," *Opt Eng*, vol. 37, no. 4, pp. 1299–1304, 1998.

[2] S. Petitgrand and A. Bosseboeuf, "Simultaneous mapping of out-of-plane and in-plane vibrations of MEMS with (sub)nanometer resolution," *J Micromech Microengineering*, vol. 14, no. 9, pp. S97–S101, 2004.

[3] T. Guo, H. Chang, J. Chen, X. Fu, and X. Hu, "Micro-motion analyzer used for dynamic MEMS characterization," *Opt Lasers Eng*, vol. 47, no. 3-4, pp. 512–517, 2009.

[4] B. Kleinemeier, "Measurement of CCD interpolation functions in the subpixel precision range," in *Proc SPIE Int Soc Opt Eng*, vol. 1010, 1988, pp. 158–165.

[5] C. Q. Davis and D. M. Freeman, "Statistics of subpixel registration algorithms based on spatiotemporal gradients or block matching," *Opt Eng*, vol. 37, no. 4, pp. 1290–1298, 1998.

[6] C. Yamahata, E. Sarajlic, G. J. M. Krijnen, and M. A. M. Gijs, "Subnanometer translation of microelectromechanical systems measured by discrete fourier analysis of CCD images," *J Microelectromech Syst*, vol. 19, no. 5, pp. 1273–1275, 2010.

[7] W. M. van Spengen and J. B. Roca, "On the noise limit of stress and temperature measurements with micro-raman spectroscopy," *J. Raman Spectrosc.*, vol. 44, no. 7, pp. 1039–1044, 2013.

[8] W. M. van Spengen, "The accuracy of parameter estimation by curve fitting in the presence of noise," *J Appl Phys*, vol. 111, no. 5, pp. –, 2012.

[9] M. W. Van Spengen, E. Bakker, and J. W. M. Frenken, "A 'nano-battering ram' for measuring surface forces: Obtaining force-distance curves and sidewall stiction data with a MEMS device," *J Micromech Microengineering*, vol. 17, no. 7, pp. S91–S97, 2007.

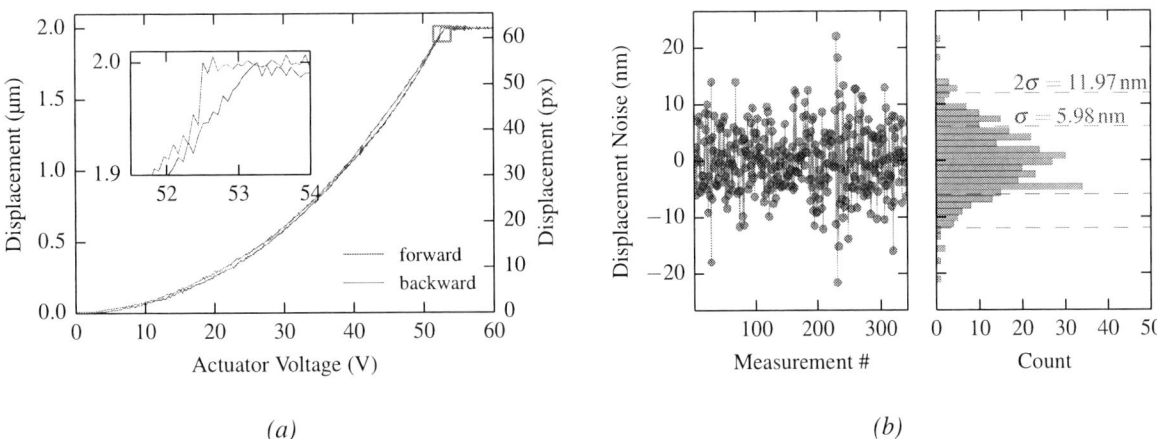

(a)   *(b)*

Figure 4: An adhesion curve of the nano-ram MEMS adhesion sensor measured by tracking the displacement of a single actuated beam.

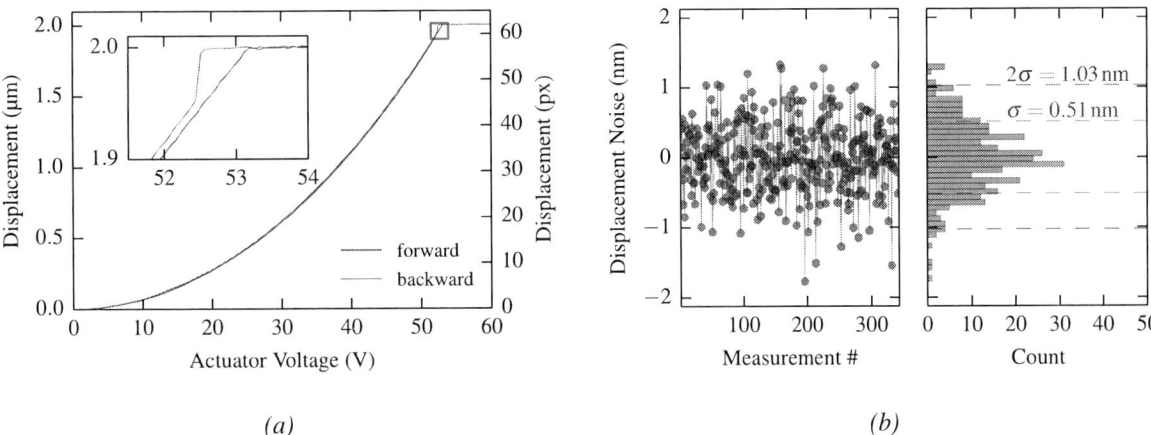

(a)   *(b)*

Figure 5: An adhesion curve of the nano-ram MEMS adhesion sensor measured by tracking the displacement of a an actuated beam relative to a stationary beam.

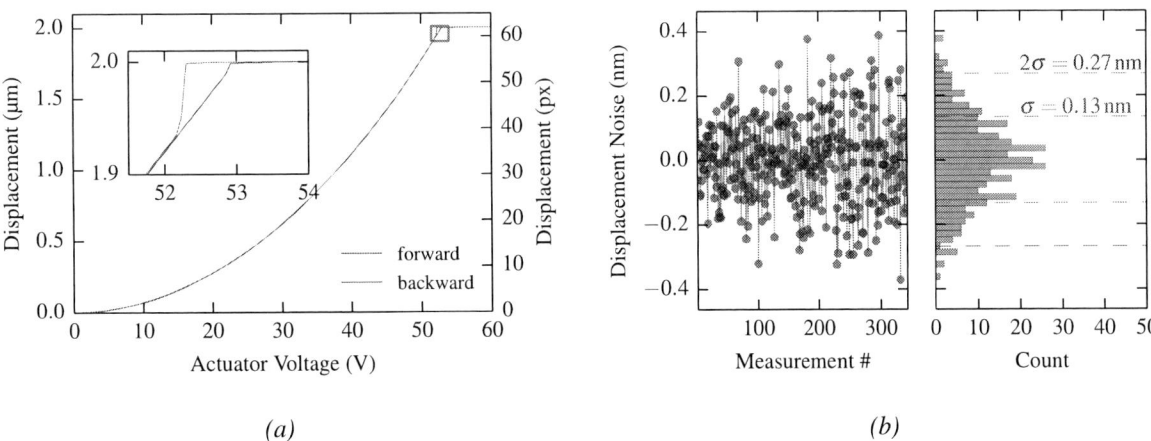

(a)   *(b)*

Figure 6: An adhesion curve of the nano-ram MEMS adhesion sensor measured by tracking the displacement of an actuated beam relative to a stationary beam using the sum of 16 consecutive exposures per applied voltage.

CHARACTERIZATION OF IMPROVED CAPACITIVE MICROMACHINED ULTRASONIC TRANSDUCERS (CMUTS) USING ALD HIGH-K DIELECTRIC ISOLATION

Toby Xu[1], Coskun Tekes[1] and F. Levent Degertekin[1,2]
[1]Georgia Institute of Technology, Woodruff School of Mechanical Engineering, USA
[2]Georgia Institute of Technology, School of Electrical and Computer Engineering, USA

ABSTRACT

Use of high-κ dielectric, atomic layer deposition (ALD) hafnium oxide (HfO_2) as an isolation layer material is demonstrated as an improvement over traditional plasma enhanced chemical vapor deposition (PECVD) silicon nitride (Si_3N_4) for Capacitive Micromachined Ultrasonic Transducers (CMUTs) fabricated by a low temperature, CMOS compatible, sacrificial release method. ALD HfO_2 dielectric properties are characterized to optimize CMUT design. Performance improvements are evaluated through parallel plate modeling which showed high gains especially for vacuum gaps of 50 nm and below. Experiments are performed on parallel fabricated test CMUTs with 50 nm gap and 16.5 MHz center frequency to measure and compare pressure output and receive sensitivity for both materials. Results show 6 dB improvement in receive sensitivity (Pa/V) with the collapse voltage reduced by one half, while in transmit mode, half the input voltage is needed to achieve the same maximum output pressure, both as predicted by the models.

INTRODUCTION

The driving principle behind Capacitive Micromachined Ultrasonic Transducers (CMUT) operation is that an electric field on the order of 10^8 V/m is achieved between the top and bottom electrodes to effectively move the membrane during transmit (Tx) and have high sensitivity during receive (Rx) mode [1]. To achieve this, most CMUTs are fabricated with vacuum gaps on the order of 100-300 nm [2].

Although it is convenient to use the same PECVD Si_3N_4 as the dielectric isolation material and membrane structure material in a low temperature process, it is not necessarily the optimum material in terms of generating large capacitances and electric fields for the CMUT. The goal is to design for a more sensitive CMUT device by choosing a high-κ dielectric material that can improve the efficiency of the CMUT in-terms of generating more electric field without increasing the applied voltage for Tx and Rx, and not reducing the maximum achievable pressure, i.e. vacuum gap thickness.

Many high-κ materials are cited in literature with dielectric constants ranging from 9-200 [3]. We choose HfO_2 as the high-κ dielectric for improvement due to its balance between a reasonably high-κ value and electrical breakdown strength (E_{BD}). Materials with higher dielectric constants were not chosen as they do not have the breakdown strength needed for an isolation thickness up to 200 nm.

COMPARISON OF ELECTROSTATIC FORCES FOR HFO_2 AND SI_3N_4 ISOLATION CMUTS

The maximum output pressure of a transmitting CMUT in non-collapse mode is determined by the ability to reach full gap swing when actuated. It is important to achieve the same full gap swing with a smaller amplitude voltage especially in the case of CMUT-on-CMOS integration applications when pulse amplitudes generated by IC are limited by the CMOS fabrication process [4]. To assess the improvement in electrostatic force generation by using HfO_2 instead of Si_3N_4 isolation, the ratio of electrostatic force (R) generated by each device to that of an ideal parallel plate model with no isolation layer (1) for the same input voltage is defined as a figure of merit. F_{el_CMUT} is the electrostatic force generated by a CMUT device given a particular thickness of dielectric, t_d, and relative permittivity, κ. F_{el_PP} is the dielectric force generated by a parallel plate without any isolation layer. The modified electrostatic force equation for a parallel plate with a layer of dielectric is shown in (2), where ε_0 is the permittivity of free space ($F \cdot m^{-1}$), g is the vacuum gap (m), and $V_{(DC+AC)}$ is the total input voltage (V). F_{el_PP} is achieved when $t_d = 0$.

$$R = \frac{F_{el\_CMUT}}{F_{el\_PP}} \quad (1)$$

$$F_{el} = \frac{\varepsilon_0 \cdot \kappa \cdot A \cdot V_{(DC+AC)}^2}{2 \cdot \left(g + \frac{t_d}{\kappa}\right)^2} \quad (2)$$

Figure 1 shows the variation R as a function of the vacuum gap for different HfO_2 isolation layer thickness values as well as 200 nm Si_3N_4 isolation layer. The breakdown field is not considered as a limitation in this graph, and the thickness of Si_3N_4 isolation layer is fixed to 200 nm due to limiting pin-hole effects during fabrication, and is further explained in the fabrication section. In the graph, it is observed that that thinner the isolation layer, the closer is its behavior to the ideal parallel-plate (R=1). Also, the importance of the isolation layer material and thickness is higher for smaller vacuum gaps. For gaps larger than 200 nm, all considered isolation layers achieve R > 0.75. As a relevant comparison, for a gap of 50 nm, the 2 types of

978-1-4799-3510-9/14 $31.00 © 2014 IEEE

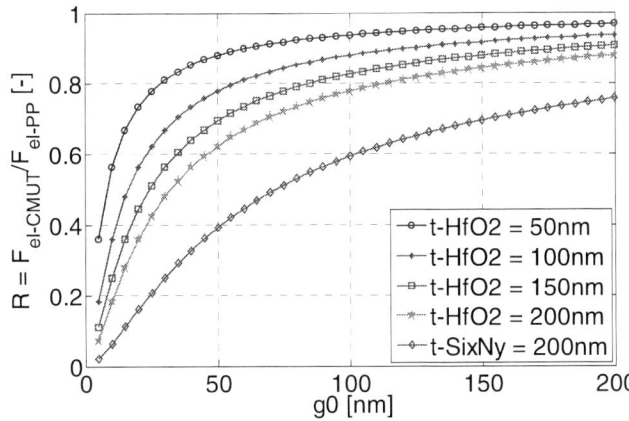

Figure 1: A plot of the gain of electrostatic force (R) for the same input pulse between HfO₂ Vs. Si₃N₄ isolation device compared to no isolation, varying the HfO₂ thickness and vacuum gap (g0). Si₃N₄ thickness is assumed constant at 200 nm. κ = 16 and 6.3 for HfO₂ and Si₃N₄ respectively used.

dielectric materials of t_d=200 nm, results in R = 0.64 and 0.3741 for HfO₂ and Si₃N₄ respectively. This means that a CMUT with HfO₂ will generate 71% more electrostatic force for the same input voltage as compared to Si₃N₄ device for a vacuum gap of 50 nm. This gap thickness is suitable for CMUTs operating around 20 MHz and above where pressure output can be large due to the high velocity of the membrane and does not require large vacuum gap swings to achieve reasonably high pressures.

HFO₂ DIELECTRIC MATERIAL PROPERTIES CHARACTERIZATION

For accurate prediction and simulation of CMUTs using HfO₂, the dielectric constant κ and breakdown strength (E_{BD}) needs to be characterized. A good prediction of E_{BD} is critical to the electrical reliability and the chosen design thickness of the dielectric layer, as the thinnest dielectric isolation that can withstand the maximum operational voltage applied for that CMUT is desired. In the case of non-collapsed CMUT operation, the maximum operational voltage to maximize receive sensitivity is a value close to $V_{COLLAPSE}$ (just before pull-in behavior is observed). To calculate κ and measure E_{BD} of the ALD HfO₂ thin film, we fabricated test capacitors with copper bottom and aluminum top electrodes. Since the nucleation rate of ALD HfO₂ may differ on various material surfaces, affecting its thickness per cycle of deposition [5], we choose to characterize the HfO₂ film on top of a copper bottom electrode to best replicate surface conditions used in the CMUT fabrication process. Both top and bottom electrodes are 1300Å and a 50 nm ALD HfO₂ dielectric layer was deposited using Cambridge Nanotech Fiji F202 system at 250°C. The thickness of the ALD film was measured with a Woollam M-2000 ellipsometer, with a recipe that was calibrated for the specific copper layer that the HfO₂ was

deposited on. Fluorine based Reactive Ion Etching (RIE) recipe (discussed in the fabrication section) is used to etch open the area over the bottom electrode copper bond pads used for external electrical characterization.

The capacitance of the test capacitors are measured using a standard probe station and Agilent B1500 Semiconductor Analyzer and shown in Table 1. The average capacitance value is obtained and is used to calculate the average dielectric constant of the ALD HfO₂ based on 50 nm thickness and capacitor electrode surface area. The variation in capacitance measured can be due to the non-uniformity of the deposited film.

Table 1: Test capacitor parameters and average κ value

Parameter	Value
Electrode Area	3.469e-7 m²
HfO₂ film thickness	50 nm
Measured Capacitance	0.86 - 1.1 nF
Dielectric constant (κ)	16 (+/-2)

Subsequently the dielectric breakdown of the ALD HfO₂ is measured and the corresponding I-V curve is shown in Figure 2. The measured E_{BD} can be calculated to be ~4MV/cm as shown in Figure 2 where V_{BD} is 20.5 V for a 50 nm film. This figure is close to reported values in the literature [3]. The retrace shows that the thin dielectric film has gone through breakdown phenomenon and behaves like a resistor instead of a capacitor post dielectric breakdown.

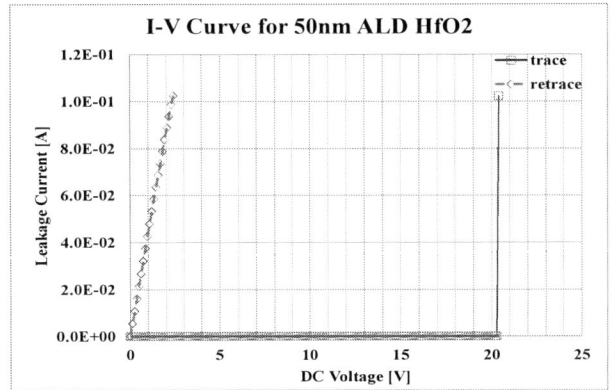

Figure 2: I-V curve for test capacitor with 50 nm of ALD HfO₂ as dielectric layer

FABRICATION OF CMUT WITH ALD HFO₂ ISOLATION LAYER

The CMUT fabrication method used in this investigation is similar to that of the low temperature silicon nitride process previously described in Knight *et al* in 2004 [6] and will be summarized in this paper along with the new steps. The novel steps are incorporating the ALD HfO₂ isolation layer and using a copper sacrificial layer as shown in Figure 3.

978-1-4799-3510-9/14 $31.00 © 2014 IEEE

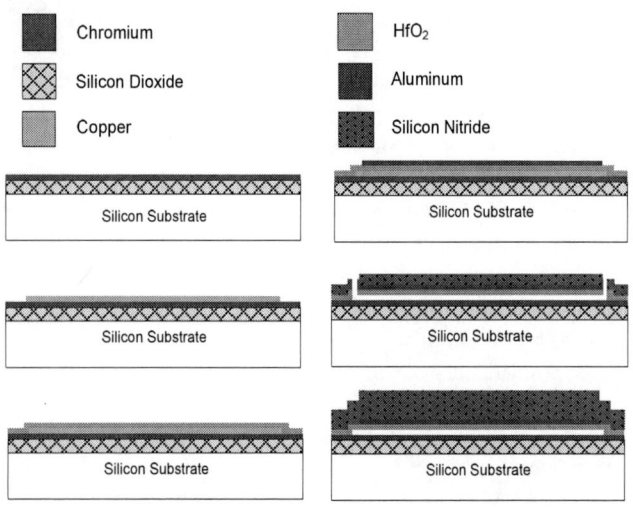

Legend:
- Chromium
- Silicon Dioxide
- Copper
- HfO$_2$
- Aluminum
- Silicon Nitride

Figure 3: CMUT fabrication process flow with copper sacrificial layer and ALD HfO$_2$ top isolation layer

Starting with a 4" <100> silicon wafer, a 3 μm thermal oxide is grown for passivation from the substrate. Next, a 120 nm of chromium is sputtered and patterned to form the bottom electrode of the CMUT. Then a 50 nm of copper is sputtered and patterned for the sacrificial or the later vacuum gap layer. Both electrodes are sputtered with the Unifilm DC Sputterer and have achieved high thickness conformity over the entire wafer. The gap thickness and its uniformity are critical because it affects the CMUT device performance, such as pull-in voltage and pressure output for the given input voltage. The chromium is wet etched via CR-7S Chromium etchant and the copper by Copper APS 100 diluted at 1:60 with dionized (DI) water. Both etchants etched in the range of 4-8Å/s at room temperature. Chromium was chosen as the bottom electrode because the copper etchant does not etch chromium, and thus is suitable for selectivity when patterning the sacrificial layer without the need to introduce an extra dielectric passivation layer in between, an improvement over the previous process. Next a 100 nm layer of thermal ALD HfO$_2$ is deposited uniformly across the surface of the substrate at 250°C to form the top dielectric isolation. The thickness of this isolation is designed to operate near V$_{COLLAPSE}$ without surpassing the threshold of E$_{BD}$. The precursor used in this study is tetrakis(dimethylamido)hafnium(IV) (TDMAHf) along with Cambridge Nanotech's Fiji F202 ALD system.

Next, a 300 nm of AlSi 1% is evaporated and patterned through lift-off as the top electrode. AlSi is chosen due to its good adhesion, relatively high electrical conductivity. After depositing 1 μm of Si$_3$N$_4$ for the membrane formation, small release holes are drilled via RIE to the copper sacrificial layer and soaked in copper etchant for a wet release. A separate SF$_6$ chemistry was used to etch the HfO$_2$ layer compared to the standard CHF$_3$ chemistry used to etch Si$_3$N$_4$. After the release, the wafer is placed back into the PECVD tool for sealing at 900 mTorr and further

thickening to a desired 2.2 μm total thickness. A final step of opening the electrical bond pad connections for testing is needed via RIE. The test array geometry parameters are summarized in Table 2.

Table 2: CMUT Array Geometry Parameters

Parameter	Value
Membrane Size	35x35 μm
Top Electrode Size	25x25 μm
Vacuum Gap	50 nm
Si$_3$N$_4$ Membrane Thickness	2.0-2.1 μm
HfO$_2$ Isolation Thickness	100 nm
Si$_3$N$_4$ Isolation Thickness	200 nm
No. of Membranes per element	4
No. of Elements per device	4
Total No. of Membrane per device	16

PERFORMANCE COMPARISON BETWEEN HFO$_2$ AND SI$_3$N$_4$ ISOLATION CMUTS IN TRANSMIT AND RECEIVE

Two wafers of CMUTs with geometries described in Table 2 were fabricated with one wafer using the 100 nm ALD HfO$_2$ as isolation and the other with 200 nm PECVD Si$_3$N$_4$ isolation. These two types of devices are packaged and wirebonded to the same PCB setup for conformity in measurement. Both devices were pulsed with the same voltage waveform with varying amplitudes and the hydrophone output was recorded and compared. Both hydrophone outputs produced similar temporal signal shapes inferring that both wafers are fabricated with good uniformity and have the same CMUT dynamics.

Transmit Performance Characterization and Comparison

To compare the transmit (Tx) performance between the HfO$_2$ and Si$_3$N$_4$ isolation CMUT array, each array is pulsed with the same 28 ns pulse from 0 V to the maximum pulse amplitude needed to achieve maximum output pressure respectively. Figure 4 shows that both arrays have the same maximum output pressure because both devices have 50 nm gaps and reached full gap swing with 32 V and 60 V pulse respectively. This result also shows the voltage squared dependence for electrostatic force and that more force per volt is generated for the 100 nm HfO$_2$ isolation device versus the 200 nm Si$_3$N$_4$ device for the same pulse amplitude. For example, at 30 V pulse amplitude, the HfO$_2$ isolation device generates ~3.5X larger output pressure compared to the Si$_3$N$_4$ isolation device. Since, maximum transmit pressure is the desired operating point to achieve the highest signal to noise ratio (SNR) during imaging mode, Figure 4 shows that there is approximately 2X reduction in voltage requirement to reach the same maximum output pressure. This is consistent with the predicted gain in electrostatic force from Figure 1 where at 50 nm gap thickness, R is approximate 2 times larger for 100 nm HfO$_2$ compared to 200 nm Si$_3$N$_4$.

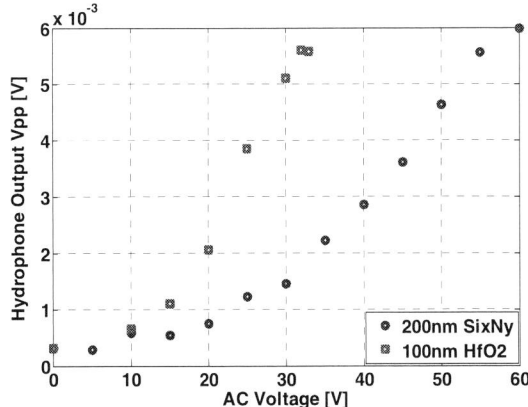

Figure 4: Experimental hydrophone data comparing the Tx efficiency of the 2 CMUTs. HfO$_2$ device can achieve the same maximum pressure output with half the input voltage compared to Si$_3$N$_4$.

Receive Performance Characterization and Comparison

To compare receive (Rx) sensitivity of the 2 different isolation layers, a 15 MHz piezoelectric transducer is used as a reference transmitter. The receive signals from the CMUT arrays are amplified by Panametrics Pulse/Receive (Model 5072) at +25 dB gain. The DC bias voltage is varied from 0V up to each CMUT array's collapse voltage. Figure 5 shows that the $V_{COLLAPSE}$ is 21 V and 42 V respectively for 100 nm HfO$_2$ and 200 nm Si$_3$N$_4$ isolation devices with the same 50 nm gap.

Two important conclusions can be made about the performances in receive for the 2 particular arrays. First, the 100 nm HfO$_2$ isolation device has half the collapse voltage of the 200 nm Si$_3$N$_4$ device. This is desirable since we want to operate the CMUT as close to collapse as possible in receive mode to get maximum receive sensitivity and increase overall SNR. Second, the maximum receive signal from the HfO$_2$ device is ~2.2X that of the Si$_3$N$_4$ device at collapse. This is due to the higher dielectric constant for HfO$_2$ compared to Si$_3$N$_4$ and the larger electric field as well as capacitance between the top and bottom electrodes.

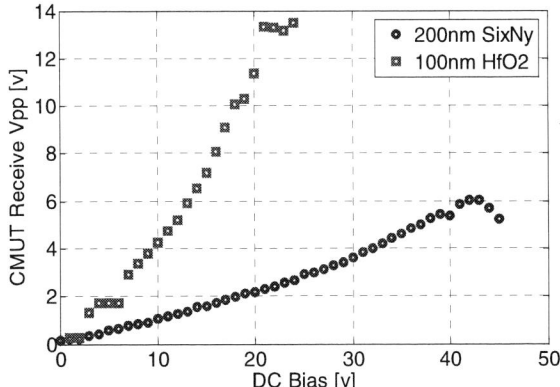

Figure 5: Experimental CMUT receive data comparing the Rx sensitivity of the 2 CMUTs.

CONCLUSION

The performance improvement by using HfO$_2$ isolation in CMUTs is significant compared to Si$_3$N$_4$ devices for vacuum gaps below 100 nm. Switching to ALD deposited HfO$_2$ can also allow low temperature fabrication of isolation layers with thickness below 100 nm without compromising the quality of the film as compared to PECVD Si$_3$N$_4$. Experiments on 15MHz CMUTs with 100 nm HfO$_2$ and 200 nm silicon nitride isolations layers show that maximum transmit voltage requirement is reduced by one half for the same 50 nm gap device The receive sensitivity is improved by 2 times while reducing collapse voltage requirements by one half.

REFERENCES

[1] O. Oralkan, A. S. Ergun, J. A. Johnson, M. Karaman, U. Demirci, K. Kaviani, T. H. Lee, and B. T. Khuri-Yakub, "Capacitive micromachined ultrasonic transducers: next-generation arrays for acoustic imaging?", *IEEE Trans. Ultrason., Ferroelec., Freq. Contr.*, vol. 49, no. 11, pp.1596–1610, 2002.

[2] Y. Huang, A. S. Ergun, E. Haeggstrom, M. H. Badi, and B. T. Khuri-Yakub, "Fabricating capacitive micromachined ultrasonic transducers with wafer-bonding technology," *J. Microelectromech. Syst.*, vol. 12, no. 2, pp. 128–137, 2003.

[3] J.W. McPherson, J. Kim, A. Shanware, H. Mogul, J. Rodriguez, "Trends in the ultimate breakdown strength of high dielectric-constant materials," *IEEE Transactions on Device Electronics*, vol.50, no.8, pp.1771,1778, Aug. 2003

[4] J. Zahorian, M. Hochman, T. Xu, S. Satir, G. Gurun, M. Karaman, F. Degertekin, "Monolithic CMUT-on-CMOS Integration for Intravascular Ultrasound Applications", *Ultrasonics, Ferroelectrics and Frequency Control, IEEE Transactions on*, vol.58, no.12, pp.2659, 2667, December 2011

[5] Q. Tao, G.M. Jursich, C. Takoudis, "Selective atomic layer deposition of HfO2 on copper patterned silicon substrates," Applied Physics Letters, vol.96, no.19, pp.192105,192105-3, May 2010

[6] J. Knight, J. McLean, F.L. Degertekin, "Low temperature fabrication of immersion capacitive micromachined ultrasonic transducers on silicon and dielectric substrates," *Ultrasonics, Ferroelectrics and Frequency Control, IEEE Transactions on*, vol.51, no.10, pp.1324-1333, Oct. 2004

CONTACT

*T. Xu, tel: +1-925-9631872; txu87@gatech.edu

CHARACTERIZATION OF STICTION FORCES IN ULTRA-CLEAN ENCAPSULATED MEMS DEVICES

David B. Heinz[1], Vu A. Hong[1], Eldwin J. Ng[1], Chae Hyuck Ahn[1], Yushi Yang[1], Thomas W. Kenny[1]*
[1]Department of Mechanical Engineering, Stanford University, Stanford, CA, USA

ABSTRACT

We show that contact between encapsulated MEMS devices and the bare silicon surrounding sidewalls generally results in a reversible adhesion with a consistent adhesion force. This force is small enough (25 mN) to be overcome by the restoring force of the springs in inertial sensors with resonant frequency above 4 kHz. Therefore, it should be possible to design and build stiction-free inertial sensors in this process – a significant advantage over approaches that rely on deposition, tuning and maintenance of chemical coatings for inertial sensors.

INTRODUCTION

Stiction in MEMS Devices

Stiction presents one of the largest outstanding problems to the development of reliable MEMS devices. Due to the high surface-area to volume ratio and small separation between surfaces that is intrinsic to MEMS devices, stiction can easily cause devices to fail. Stiction is largely caused by three fundamental forces: capillary action, Van der Waals forces, and 'chemical forces,' such as hydrogen bonding [1]. Traces of water cause capillary attraction between adjacent surfaces. Hydrogen bonding occurs between impurities containing hydroxyl groups (-OH), and is one of the strongest dipole-dipole interactions. Van der Waals forces are also caused by dipole-dipole interactions, but are much weaker than hydrogen bonds and occur between all surfaces.

Stiction forces are of particular importance for high displacement, low resonant frequency devices such as inertial sensors due to the higher likelihood of two surfaces coming into contact. Successful devices have relied upon a variety of chemical coatings, surface treatments and mechanical means to reduce the impact of this adhesion force. These methods have dramatically improved the usability of many MEMS devices [2, 3] but impose significant restrictions upon the design and operation of the devices.

Ultra-clean Encapsulated Devices

The epi-seal encapsulation process, developed by Robert Bosch GmBH and Stanford University, produces devices with many unique properties, especially as regards to stiction. Epitaxial polysilicon encapsulation is a wafer scale high-temperature, ultra-clean process; using standard CMOS foundry facilities, it is possible to package devices in a near-vacuum (<10 mTorr) inert environment free of oxygen, water and other contaminants [4].

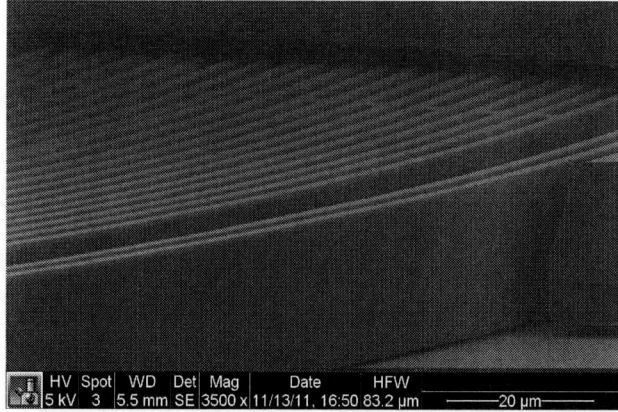

Figure 1 - Smooth and clean sidewall surfaces for encapsulated MEMS devices

The high-temperature seal takes place with H2 carrier gas, promoting a smoothing process that leaves clean, bare silicon sidewalls on all internal silicon surfaces [4].

These features have combined to enable ultra-stable resonators that are manufacturable at low cost, and which have been commercialized by SiTime. As integration of inertial sensors and timing references has become more essential, however, the inclusion of low resonant frequency devices with large displacements (such as accelerometers and gyroscopes) has necessitated considering the problem of stiction in this process.

The encapsulated environment free of water and hydroxyl groups immediately eliminates capillary attraction and hydrogen bonding – two of the three primary sources of stiction forces [4]. This suggests that the epi-seal process should exhibit a high natural resistance to stiction, but unfortunately, there exist some difficulties as well. The high temperature process precludes the use of anti-stiction coatings that have been found to greatly reduce stiction problems [2]. In addition, the exposed surfaces of the device have extremely low surface roughness (figure 1). Finally the common variant of the epi-seal process constrains gap sizes and sidewall profiles. These features imply that stiction may still be a problem despite the clean environment.

TEST METHODOLOGY

A series of test structures were designed and fabricated in order to further explore the adhesion forces in the epitaxially encapsulated MEMS devices. The test devices are electrostatically actuated pull-in MEMS switches, as shown in figure 2.

978-1-4799-3510-9/14 $31.00 © 2014 IEEE

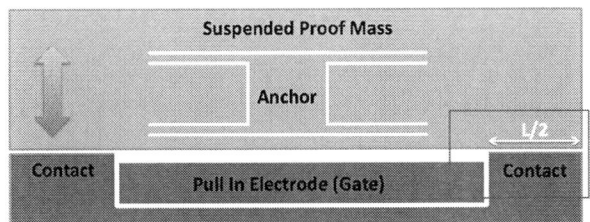

Figure 2 – Top-view schematic drawing of test device.

A series of these devices were fabricated on a silicon-on-insulator wafer, with a 40 μm thick <100> device layer. The proof mass has dimensions of 550 μm by 200 μm, with a 100 μm square anchor and resonant frequency of 10 kHz. The gaps were set such that electrostatic pull in would force the contact areas to impact one another, closing the switch. The contact area was varied from 0μm (tangential point contact) to 40μm. SEM images of the relevant contact area for these devices are shown in figure 3.

Figure 3 - SEM image of test device contact areas

All electrical measurements of these devices were conducted with a precision Semiconductor Parameter Analyzer (Agilent/HP 4156B). The HP 4156B has four high-resolution signal measurement units (HRSMU) that may be set to source or measure both current and voltage. In our configuration one HRSMU was set as the ground reference, and was connected to the proof mass. A second HRSMU was used to sweep the voltage on the electrostatic pull in electrode, and a third HRSMU applied a constant voltage to the contact. By measuring the current between the proof mass and the contact we were able to accurately detect pull in. A compliance current of 500pA minimized the effects of resistive heating at the contact point, which could otherwise reshape the device and damage the contact.

Both the pull-in and pull-out voltages for this simple configuration are well understood, and can be calculated from the following equations.

$$V_{pull-in} = \sqrt{\frac{8}{27}\frac{k \cdot g_0^3}{\varepsilon A}} \qquad (1, 2)$$

$$V_{pull-out} = \sqrt{2\frac{k \cdot a_0 (g_0 - a_0)^2}{\varepsilon A}}$$

Where k is the total spring constant, g_0 is the initial gap

between proof mass and electrode, a_0 is the height of the contact above the pull in electrode, A is the pull in electrode area and ε is the permittivity of free space.

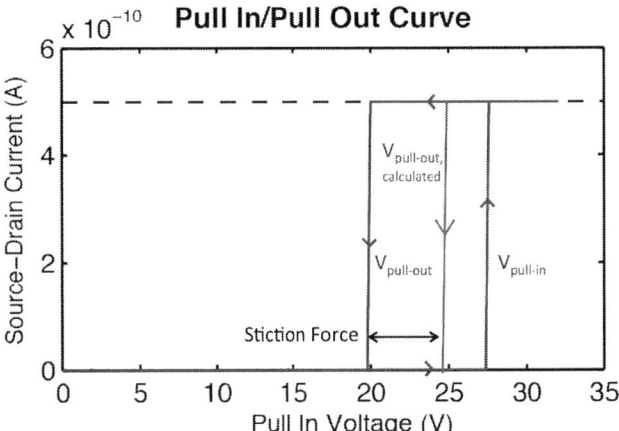

Figure 4 - Results of basic pull-in measurement

From equations 1 and 2, and the data shown in figure 4, it is possible to calculate the stiction force acting upon the device. First, the pull-in voltage (independent of stiction) is used to back-calculate the overetch. Once the parameters have been identified, the difference between the calculated and actual pull-out voltages can be used to determine the stiction force. This same process was repeated until the device failed for a number of test devices of different contact areas and from different wafers.

RESULTS

By repeating the basic measurement, we are able to gain some insight into the stiction behavior of encapsulated MEMS devices. Figure 5 shows the results of the repeated measurement over thousands of cycles for two different devices.

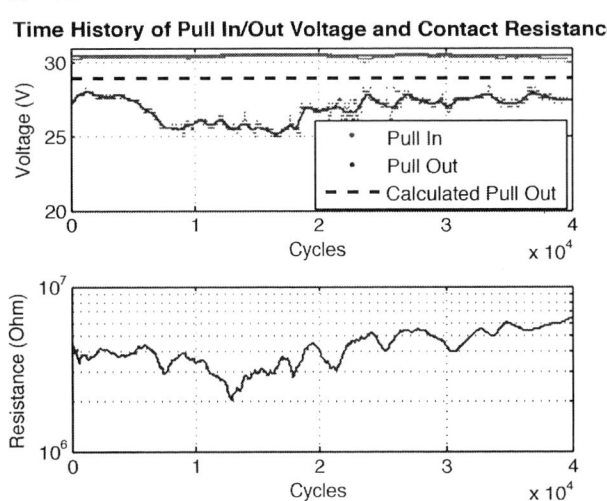

Figure 5a - Long-term evolution of stiction forces and contact resistance until switch failure (40 μm bump stop)

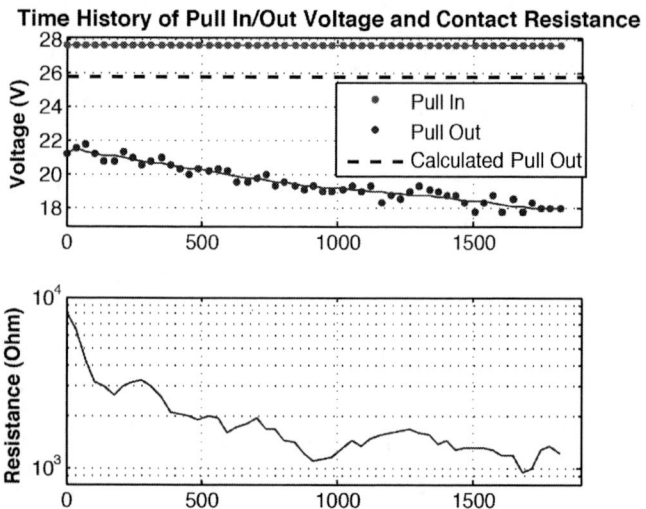

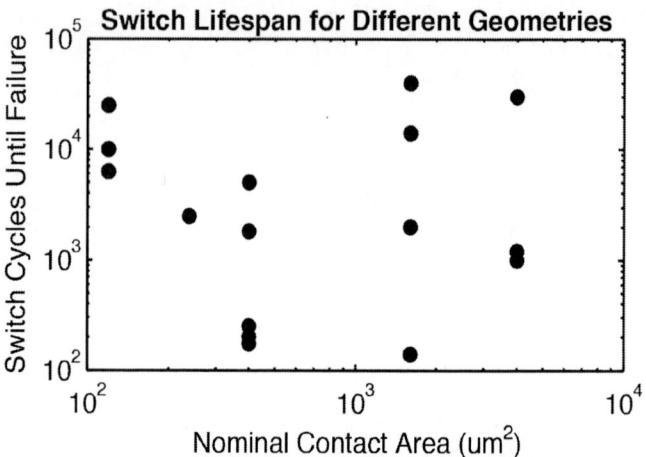

Figure 7 – Lack of correlation between switch lifespan and contact area over a range of devices

Figure 6b - Long-term evolution of stiction forces and contact resistance until switch failure for another device (10 μm bump stop)

From these trials we are able to measure the range of stiction forces in the test devices.

Table 1 – Measured stiction forces

Case	Stiction Force
Minimum	3.4 uN
Average	18.5 uN
Maximum	44.4 uN

Figure 6 compares the number of cycles the switch survived until it failed to exhibit standard switching behavior, with the nominal contact area. This plot defies initial expectations in that there is no obvious correlation. Stiction forces are typically directly proportional to area, so this is a surprising result, especially for devices with such smooth sidewall profiles, and correspondingly low atomic-scale roughness. In order to further explore this result, we look to the contact resistance measurement.

Both devices for which data is shown in figure 5 demonstrate a considerable change in resistance and stiction force over the duration of the test. This behavior is not consistent with a switch with constant contact area, and instead suggests that the real contact area may not be at all the same as the nominal or design contact area.

Another closer examination at an SEM image of a device cross-section helps to explain this result.

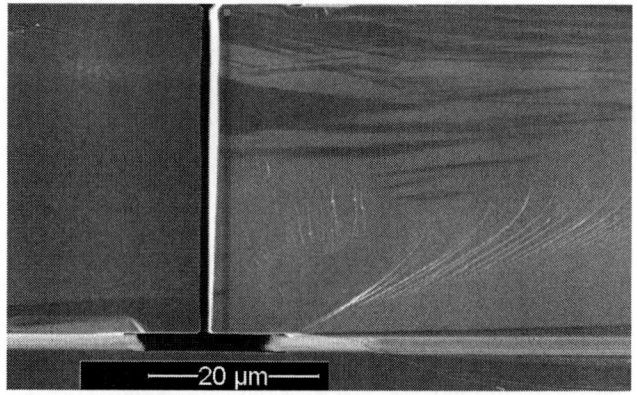

Figure 8 - SEM image of test device cross-section

Here we can see that though the device sidewalls are locally smooth, they are not perfectly flat, and have unpredictable protrusions and depressions. It is clear that a device like this will have asperity-dominated contact. This conclusion is completely consistent with the results discussed earlier. In order to once again confirm this result, the average contact resistance and stiction force over all repetitions for each device was recorded.

978-1-4799-3510-9/14 $31.00 © 2014 IEEE

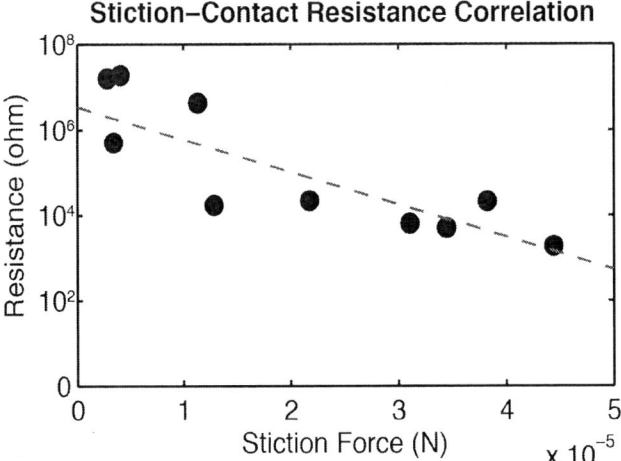

Figure 9 - Correlation between stiction force and contact resistance over all test devices

Figure 8 demonstrates a fairly strong trend; devices with high resistance tend to have the smallest stiction forces. This does serve to confirm our conclusion that the device contact is dominated by asperities, irrespective of the designed contact area.

CONCLUSIONS

The results provided by these test devices directly imply the following conclusions. Even though the devices were designed for worst-case stiction, with large, smooth, flat areas brought into contact under considerable force, the devices often proved to have only relatively modest stiction forces. This was largely due to the asperity-dominated contact area.

In order to give these results more relevance to the design of other inertial sensing devices, we may calculate some further results based upon the collected data. Typical inertial MEMS devices will include a spring that must be deflected from the neutral position to attain contact with a sidewall – if the restoring force of that spring can overcome the adhesive force, stiction will be overcome. Using the design parameters for the epi-seal process (maximum gap 1.5 μm, 40 μm device layer) we may extrapolate to the requirements for an accelerometer. Based on the average data from these experiments, we should require a spring constant of only 18 N/m, corresponding to an accelerometer with resonant frequency around 13kHz within a 1mm die. If, however, it proves to be possible to reliably replicate the best measured cases using specially designed bump stops to limit the contact area and assure a small asperity contact, a much softer spring will provide sufficient restoring force. In this case, an accelerometer with 4kHz resonant frequency could be successfully manufactured in this encapsulation process.

ACKNOWLEDGEMENTS

This work was supported by DARPA grant N66001-12-1-4260, "Precision Navigation and Timing program (PNT)," managed by Dr. Robert Lutwak, and DARPA grant FA8650-13-1-7301, "Mesodynamic Architectures (MESO)," managed by Dr. Jeff Rogers. Student support was provided to D.B. Heinz from the Stanford Graduate Fellowship. The authors would like to thank K.L Harrison, R.T. Howe, G.J. O'Brien and G. Yama for advice and guidance. This work was performed in part at the Stanford Nanofabrication Facility (SNF), which is supported by the National Science Foundation through the NNIN under Grant ECS-9731293

REFERENCES

[1] W. Merlijn van Spengen, "MEMS reliability from a failure mechanisms perspective", in *Microelectronics Reliability*, vol. 43, pp. 1049-1060, 2003.

[2] A. Partridge and M. Lutz, *U.S. Patent* 6 928 879, Aug. 6, 2005

[3] R. Maboudian and R. T. Howe, "Stiction reduction processes for surface micromachines," *Tribology Letters*, vol. 3, pp. 215-221 (1997)

[4] R. N. Candler et al, "Long-term and accelerated life testing of a novel single wafer vacuum encapsulation for MEMS resonators," *Journal of Microelectromechanical Systems*, 15-6 (2006) 1446-1456

CONTACT

*D.B. Heinz, tel: +1-650-736-00444;
e-mail: dheinz@stanford.edu

COMPOSITE OF THERMALLY RESPONSIVE SOLUTION AND LUBRICATING MICRO BEADS AS SEALING MATERIAL FOR PISTON-CYLINDER ACTUATOR

Takuya CHISHIRO, Shu HONDA and Satoshi KONISHI
Ritsumeikan University, Shiga, JAPAN

ABSTRACT

In this paper, we propose a novel sealing technique for a piston-cylinder actuator. Pneumatic actuators, especially piston-cylinder actuators are attractive in MEMS field because of its high force density. The sliding part of the piston actuator requires both high sealing and excellent lubrication. We reported parylene deposition of a piston rod to improve sealing. In contrast, this paper presents a smart sealing material composed of thermally responsive solution and lubricating micro beads. Micro beads contributed to provide lubricant between a sliding part and an inner wall when thermally responsive solution gels by heating. Heated composite of thermally responsive solution and micro beads gels around a piston rod while the composite works as pressuring liquid in the region without heating. We evaluate piston cylinder of restriction force, pressure resistance, and force. We observe inside of piston cylinder actuator to understand detail phenomenon. Piston actuator is integrated with a heater.

INTRODUCTION

Various robots using multiarticular manipulator were reported [1]. As minimally invasive surgery, endoscope is the most typical tool to inspect inside of the body [2]. Most of endoscope employs wire-driven mechanism. Wire-driven mechanism transmits the power to an end-effector. Wire-driven mechanism has compact structure by wire and uses external actuator. Wire-driven actuator, however, has drawbacks of wired stiff structure and large transmission loss of power. We previously proposed multiarticular manipulator integrated with distributed actuators to drive individual multiarticular elements [3]. The multiarticular manipulator adopted thin piston-cylinder actuator to drive individual multiarticular element. Piston cylinder actuator is pneumatically driven actuator.

The pneumatic driving principle has the characteristics of high force density and safety because of not using electrical or electromagnetic mechanisms. Various pneumatic driven actuators were reported such as contraction motion used pneumatic balloon actuator, Mckibben and piston cylinder actuator [4-5]. Among pneumatic actuators, contraction motion actuator and piston-cylinder are thin enough to be integrated into the surface of multiarticular manipulator: Contraction motion actuator requires motion space for swelling deformation. On the other hand, piston-cylinder can move linearly without crosswise deformation.

A sealing at opening around rod is important to increase output force of piston cylinder. Leakage at opening around a rod decreased output force of piston cylinder actuators. D. Volder et al. studied sealing technologies for piston cylinder such as contact seal, clearance seal and liquid seal. Contact seal was fabricated by a PDMS molding process [6]. Contact seal can seal at opening around rod. Clearance seal can keep very low leakage by small clearance at opening around a rod [7]. Piston-cylinder actuator can be sealed by clearance seal without sealing materials. Liquid seal is sealing technology by a gallium-based seal based on surface tension [8].

We have developed and reported thin piston-cylinder actuators for driving individual multiarticular elements. Thin piston-cylinder actuators were fabricated by Si-micromachining and high-speed machining center [9]. Leakage at opening around a rod, however, decreased output force of piston cylinder actuators. We previously proposed parylene deposition sealing at opening around a rod. Parylene deposition allows a good sealing of even microscopic clearance.

This paper will show a smart sealing material composing a thermally responsive solution and lubricating micro-beads as shown in Fig. 1. Our study employs Pluronic® F127 as thermally responsive solution whose viscosity can be controlled by temperature [10]. The micro-beads act as lubricant between the sliding part and inner wall when Pluronic® F127 is heated to form a gel (see Fig.1 (a)). Figure 1(b) shows the design of the piston actuator integrated with a heater. Heated composite of the thermally responsive solution and micro-beads gels around a piston rod while the composite works as pressuring liquid in the region without heating. This paper will show concept, design, and characterization of proposed sealing technique using a smart composite.

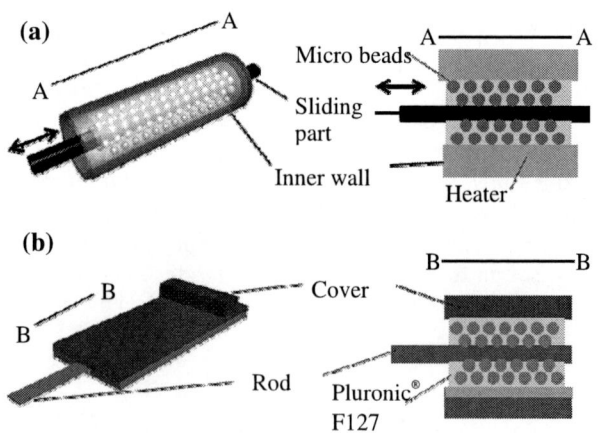

Figure 1: Conceptual image, (a) The sliding part and an inner wall, (b) Piston actuator integrated with a heater.

COMPOSITE OF THERMALLY RESPON-SIVE SOLUTION AND LUBRICATING MICRO-BEADS

We adopt thermally responsive solution (solution of Pluronic® F127) to seal at opening around rod. B. Stoeber et.al reported thermally responsive solution to control flow in a channel [10]. Pluronic® F127 solution has thermally responsive property that viscosity changes depending on a temperature. Pluronic® F127 solution changes gel state by heating while liquid state by cooling. The change in state is reversible process. The temperature of a change of state depends on mass percentage density of Pluronic® F127 solution (Fig. 2).

We use glass beads (Polysciences, Inc. Glass beads 3 - 10 μm) to provide lubricant between a sliding part and an inner wall when Pluronic® F127 gels by heating.

Figure 3 shows how to prepare a composite of 15 % Pluronic® F127 solution (Fig. 3(a)) and glass beads (φ 3 - 10 μm) (Fig. 3(b)). Figure 3(c) shows the composite of the above two materials when beads concentration was 0.3 %. Figure 3(c) shows a magnified view of the composite; distributed beads can be observed.

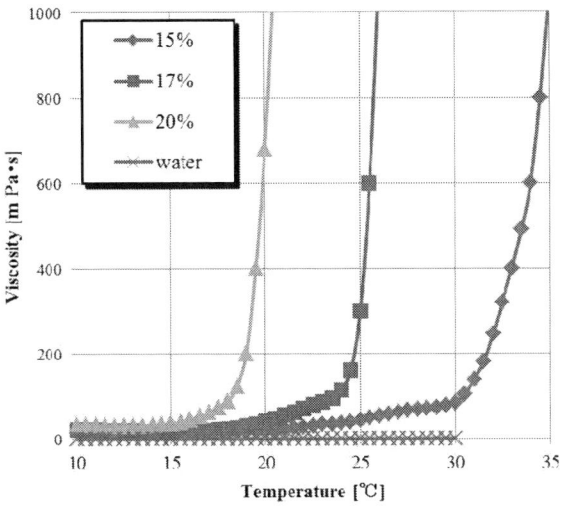

Figure 2: Viscosity-temperature characteristics of Pluronic® F127 solution.

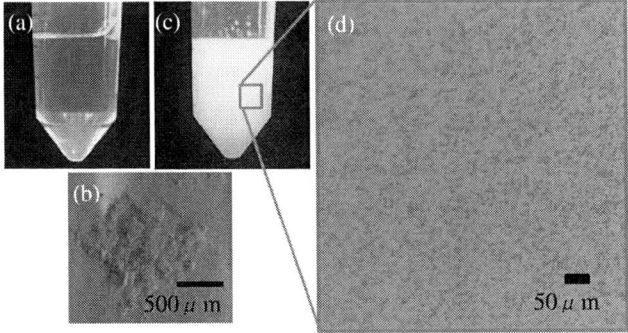

Figure 3: Composite of above two materials, (a) 15 % Pluronic® F127 solution, (b) Glass beads (φ 3 – 10 μm), (c) Beads 0.3 %, (d) Magnified view of beads 0.3 %.

PRELIMINARY EXPERIMENT OF PISTON-CYLINDER ACTUATOR

Process

Before integrating a piston-cylinder with heater, we confirm effect of Pluronic® F127 and glass beads by using only a piston cylinder. Figure 4 shows fabrication process of micropiston for preliminary experiment. Two layers were bonded together to form for a cylinder structure that accommodates the Si piston. The cover was made of a SUS by high-speed machining center (Fig. 4 (a1)). The piston was fabricated into a 250 μm thick Si substrate. The case was fabricated into a 800 μm thick Si substrate. Detailed process of the piston and case will be explained. Cr was deposited on a Si substrate as an etching mask of Deep RIE (Fig. 4 (b1), (c1)). Photoresist (OFPR800LB) was spin-coated and patterned by photolithography (Fig. 4 (b2), (c2)). Cr was patterned by wet etching and exposed. Si was etched by Deep RIE (Fig. 4 (b3), (c3)). Only Cr on the substrate for the case was removed by wet etching (Fig. 4 (c4)). The piston was aligned in the case. The cover was bonded on the case so as to complete process (Fig. 4 (d)). Si-micromachined micropiston (26 mm ×12 mm ×1.6 mm) was fabricated (Fig. 4). The piston has a cross-sectional area of 5 mm × 0.25 mm.

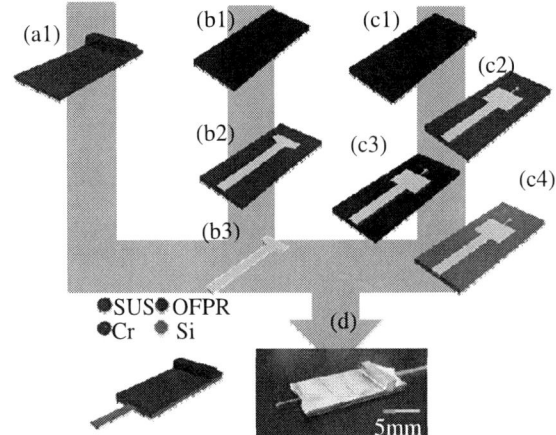

Figure 4: Fabrication process of piston-cylinder, (a1) SUS cover fabricated by high-speed machining center, (b1) Depositing Cr on Si, (b2) Piston patterning, (b3) DRIE, (c1) Depositing Cr on Si, (c2) Case patterning, (c3) DRIE, (c4) Cr etching, (d) Bonding.

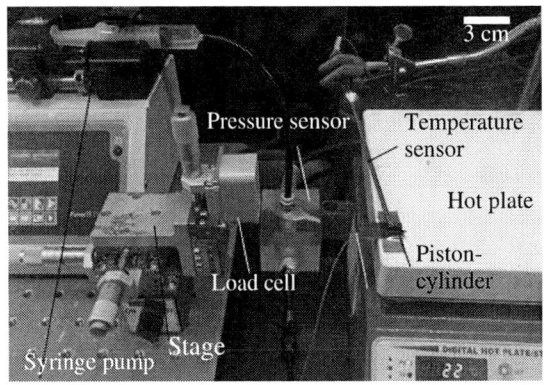

Figure 5: Experimental set-up.

978-1-4799-3510-9/14 $31.00 © 2014 IEEE

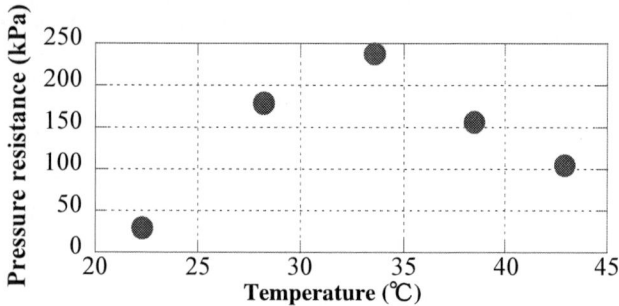

Figure 6: Pressure-resistant depending on the temperature of the opening end of piston-cylinder.

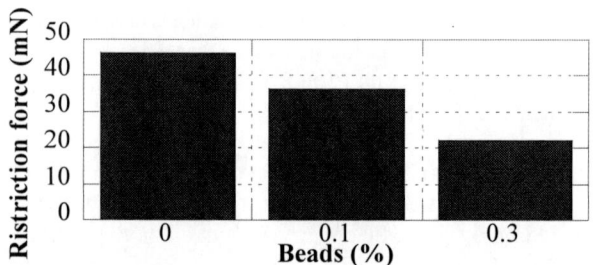

Figure 7: Evaluation of restriction force according to concentration of micro-beads at 34 ℃.

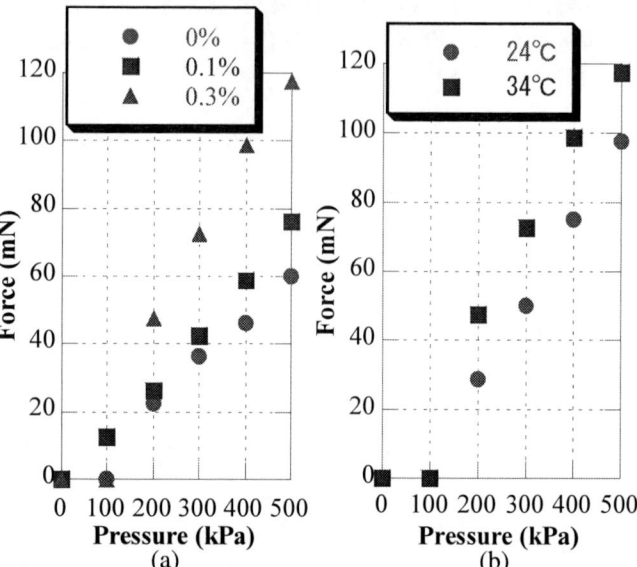

Figure 8: Output force according to pressure, (a) Output force according to concentration of micro beads at 34 ℃, (b) Output force according to temperature at 0.3 % concentration of micro-beads.

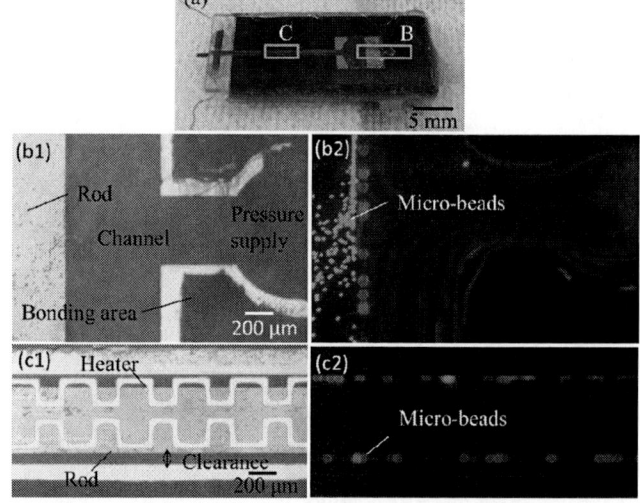

Figure 9: Observation of inside of piston cylinder, (a) Piston-cylinder fabricated by glass and Si, (b1) Magnified view of square B, (b2) Pressurizing composite of fluorescent beads and Pluronic® F127 in piston cylinder at 50 kPa, (c1) Magnified view of square C, (c2) Pressurizing composite of fluorescent beads and Pluronic® F127 in piston cylinder at 50 kPa.

Temperature dependence of pressure-resistance of thermally responsive solution

The pressure-resistance of the thermally responsive solution without micro-beads was evaluated using the experimental set-up in Fig. 5. The pressure applied by a syringe pump (HARVARD APPARATUS 70-4504) was monitored by a pressure sensor (KYOWA ELECTRONIC INSTRUMENTS PGM-5KH). The opening end of piston-cylinder actuator was placed on a hotplate (AS ONE, DEGITAL HOT PLATE DP-2S). We measured the temperature of top side of piston cylinder by temperature sensor. 15 % Pluronic® F127 solution was introduced into the piston-cylinder actuator as the working fluid. The pressure-resistance was evaluated depending on the temperature of the opening end as shown in Fig. 6. The withstanding pressure reached 238 kPa at 34 ℃. The pressure resistance increased with temperature because the viscosity of the Pluronic® F127 solution increased. The phenomena decreased pressure resistance over 34℃ is currently under study.

Characterization of composite material

A restriction force was generated by gelled solution. The sliding part of the piston actuator requires both high sealing and excellent lubrication. The composite of 15% Pluronic® F127 solution and glass beads (φ 3 - 10 μm) was estimated to satisfy both requirements. The restriction force against sliding piston rod was measured by changing the concentration of micro-beads at 34 ℃. A load cell (KYOWA ELECTRONIC INSTRUMENTS LVS-500GA) measured the restriction force in the experimental set up in Fig. 5. Figure 7 tells us that micro beads contributed to decrease restriction force against sliding rod.

Improvement of output force of piston-cylinder actuator

We applied the above results to a piston-cylinder actuator. Figure 8(a) shows the output force according to the concentration of micro-beads (0, 0.1, and 0.3 %) at 34 ℃. The force increased with the concentration of micro-beads. Figure 8(b) shows the output force according to the temperature (24 and 34 ℃) at 0.3 % concentration of micro-beads. The force increased with the temperature.

Observation of inside of piston cylinder

According to characterization of restriction force and output force, we confirmed that Pluronic® F127 provided sealing and glass beads had an effect on lubrication. We observed inside of piston cylinder by fluorescence microscope to understand detail phenomenon. Pluronic® F127 was mixed with fluorescent beads（Polysciences, Inc. Fluoresbrite® YG Microspheres 10 μm）in place of glass beads. Figure 9 shows the result of observing inside of piston cylinder through fluorescence microscope. Impressed pressure was 50 kPa. The case of piston cylinder was bonded on transparent glass plate so as to observe inside of piston cylinder. Figure 9(b) is magnified view of square B in Figure 9(a). Square B is a hole for pressurizing and pressure receiving area of rod (Fig.9 (b1)). We can see fluorescent beads between rod and inner wall (Fig.9 (b2)). Fluorescent beads increase a pressure receiving area of a rod. Figure 9(c) is magnified view of square C in Figure 9(a). There is a rod in square C (Fig.9 (c1)). We can also see fluorescent beads around rod (Fig.9 (b2)). Fluorescent beads decrease clearance around rod. As a result, it was found that beads acted as increased pressure receiving area and sealing at opening around rod.

PISTON-CYLINDER INTEGRATION OF HEATER

Figure 10 shows fabrication process of piston cylinder actuator integrated with heater. The heater was fabricated into a 1000 μm thick glass substrate. Cr/Au was deposited on a glass (Fig.10 (a1)). Cr/Au was patterned by wet etching to fabricate the heater on a glass (Fig.10 (a2-3)). Piston, case and cover were fabricated as Figure 4 (Fig.10 (b-d)). A hole in Si case was created by an ultrasonic drilling tool (Fig.10 (c)). Finally, the piston was aligned in the case. All parts were bonded so as to complete process (Fig.10 (e)). Piston-cylinder with heater (30 mm×12 mm×2.6 mm) was fabricated. Design of the heater was bellows for increasing resistance as shown in magnified view of the heater in Fig. 10(e). We evaluate force generation in future.

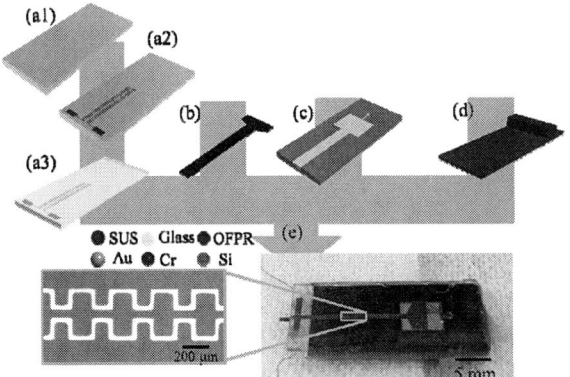

Figure 10: Fabrication process of piston-cylinder integration of heater, (a1) Depositing Cr/Au on Si, (a2) Heater patterning,(a3) Cr/Au etching, (b) Si piston, (c) Creating a hole in Si case by an ultrasonic drilling tool, (d) SUS cover, (e) Fabrication result.

CONCLUTION

We proposed a smart sealing material composed of thermally responsive solution and lubricating micro beads. Micro beads (φ 3 – 10 μm) contributed to provide lubricant between a sliding part and an inner wall of piston cylinder. We adopted Pluronic® F127 as thermally responsive solution to seal at opening around rod. The withstanding pressure of piston cylinder reached its greatest at 34 ℃. Piston cylinder heated at 34 ℃ generated almost 1.25 times as large force as actuator at 24 ℃. Pluronic® F127 had an effect on sealing at opening around rod. In addition, it was found that beads acted as increased pressure receiving area by observation of inside of piston cylinder as well as provided lubricant. Piston actuator integrated with a heater was fabricated.

REFERENCES

[1] K. Suzumori, T. Miyagawa, M. Kimura and Y. Hasegawa, "Micro Inspection Robot for 1-in Pipes", *IEEE/ASME TRANSACTION ON MECHATRONICS*, Vol. 4, No.3, pp.286-292, 1999.

[2] D. Azuma, J. Lee, K. Narumi, F. Arai, "Fabrication of Articulated Microarm for Endoscopy by Stacked Microassembly Process (STAMP)", *Journal of Robotics and Mechatronics*, vol. 21, no. 3, pp.396-401, 2009.

[3] T. Chishiro and S.Konishi, "Multiarticular Manipulator and its Multi Degree of Freedom Motion by Distributed Thin Piston-cylinder Actuators", *Proc. of IEEE Transducers 2013*, pp.1579-1582.

[4] M. D. Volder, A. J. M. Moers and D. Reynaerts, "Fabrication and control of miniature Mckibben actuator", *Sensors and actuator A166*, pp.111-116, 2011.

[5] T. Chishiro, T. Ono, and S.Konishi, "Pantograph Mechanism for Conversion from Swelling into Conversion Motion of Pneumatic Balloon Actuator", *Proc. of IEEE MEMS 2013*, pp.532-535.

[6] M. D. Volder, F. Ceyssens, D. Reynaerts and R. Puers,"A PDMS Lipseal for Hydraulic and Pneumatic Microactuators", *Journal of Micromechanics and microengineering*, Vol.17, NO.7, pp.1232-1237, 2007.

[7] M. D. Volder, F. Ceyssens, D. Reynaerts and R. Puers, "Microsized Piston-Cylinder Pneumatic and Hydraulic Actuators Fabricated by Lithography", *Journal of Microelectromechanical Systems*, Vol.18, NO.5, pp.1100-1104, 2009.

[8] M. D. Volder, J. Peirs, D. Reynaerts, J. Coosemans, R. Puers, O. Smal and B. Raucent,"Production and Characterization of a Hydraulic Microactuator", *Journal of Micromechanics and microengineering*, Vol.15, NO.7, pp.15-21, 2010.

[9] T. Obara and S.Konishi, "Multiarticular Actuator Composed of Serially Connected Micropistons For Wearable Actuator", *Proc. of IEEE MEMS 2012*, pp.76-79.

[10] Boris Stoeber, Zhihao Yang, Dorian Liepmann, and Susan J. Muller, "Flow Control in Microdevices Using Thermally Responsive Triblock Copolymers" *Journal of Microelectromechanical Systems*, Vol. 14, NO. 2, pp.207-213, 2005

CONTINUOUS DYNAMIC TIMING MEASUREMENTS TO MONITOR SPRING AND SURFACE FORCES IN MEMS SWITCH RELIABILITY

Cezary Kosla[1], Padraig Fitzgerald[2] and Martin Hill[1]
[1]Cork Institute of Technology, Cork, Ireland
[2]Analog Devices, Limerick, Ireland

ABSTRACT

In this work we demonstrate an automatic reliability detection/prediction system for industry manufactured ohmic MEMS switches based on dynamic time measurements. The developed method allows for constant device monitoring and highlights the influence of both restoring and surface forces evolution on switch reliability. Also for the first time it allows for an identification of an imminent device failure due to its continuous monitoring. We present the scalability of this approach by testing it on six different, industry-manufactured switch types and on a large number of samples.

INTRODUCTION

The growth in the market of personal electronics, as well as the trend of introducing "smart" systems to common everyday products, has driven the need for micro electromechanical systems (MEMS)[1], resulting in extensive research in this area. MEMS switches as one of the devices required for the development of modern mobile phones antennas[2] have seen especially high rise in their applications. But while majority of international study is focused on the development of said devices there is only a limited insight into their operation and cause of failure[3][4]. Working MEMS switches have to be kept under precisely controlled atmosphere and pressure to minimize the environmental effects on the structure thus there is usually no possibility to observe a working sample via optical methods. Because of that the most common means of observation are electrical methods such as capacitance, resistance measurements supplemented by SEM/AFM at certain steps of device lifecycle. Unfortunately measurements of this type are suited only for limited-small scale testing as they require the device to be stopped in its operation therefore do not allow for a real time health analysis and prediction of a device failure.

To overcome these issues our group has investigated a novel approach to health monitoring of MEMS ohmic switches. Based on the previous work of our group on the degradation effects in MEMS switches[5] we have identified that the change of health state in a working device can be determined by the observation of the evolution in closure time and opening time of the switch. The idea and proof of concept has been presented in our previous papers [6][7], while this work proves the feasibility of this method. Additionally to basic monitoring of the device health this approach provides for the first time an insight in the change of both restoring and surface forces of continuously working

MEMS switches simultaneously and allows to identify the main cause of impeding failure in a tested device.

This technique has been verified on a number of commercially available MEMS switches and is not tied to only one model. A large number of data has been obtained from multiple measurements and the results are consistent across all of the tests.

TESTING SETUP AND METHODOLOGY

The testing platform was based on the previously presented setup by Do et al.[8] for monitoring switch performance, The system consisted of following elements: tested samples, waveform generator, voltage amplifier, 500 mV signal source, oscilloscope and a PC to control the test parameters. A general schematic has been presented on Figure 1 below:

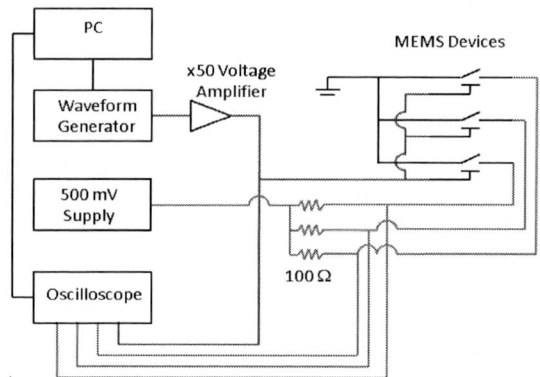

Figure 1: Test setup general schematic

The principle idea of this methodology is to detect the closure and opening events of the tested MEMS switches. To obtain this data a square waveform was supplied by the PC-controlled waveform generator to the actuation pads of the tested MEMS switches. This resulted in a controlled open/closure movement of the device. A closed state of the switch was dropping the delivered 500 mV signal, which was detected by the connected oscilloscope, while an open state was feeding the 500 mV signal directly to the oscilloscope.

Two general types of industry-manufactured MEMS switches have been used in this approach Type A – a single contact low-reliability switch and Type B – a multi contact high-reliability switch. For verification purposes the Type A switch has been designed in additional five different size variations with various expected lifetime. All of the tested devices have been fabricated in the same technological process using electroplated gold on high-resistivity silicon with platinum group metal on the contacting surfaces of

978-1-4799-3510-9/14 $31.00 © 2014 IEEE

beam and contact pads. The details of Type B switch and it's technology process was presented by Goggin et al.[9]. Figure 2 illustrates both of the general designs:

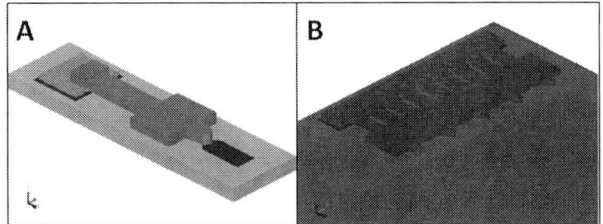

Figure 2: Two general types of tested sample. A – low reliability Type A switch, B – high reliability Type B switch.

The developed monitoring method uses continuous measurements of switch status in response to a step actuation voltage from which the switch closure time (Tc) and opening time (To) are extracted. An example of such a measurement can be seen on Figure 3 below:

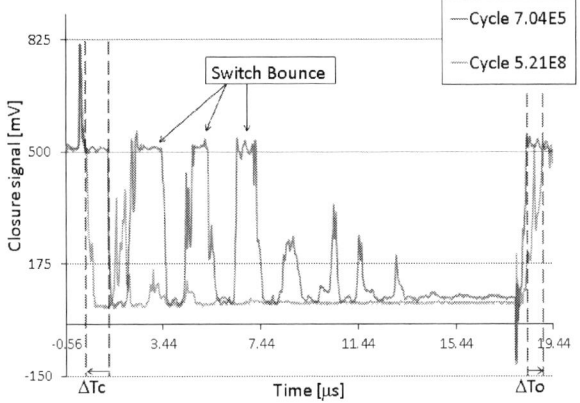

Figure 3: Example of a measurement of closure (Tc) and opening time (To) of a working switch and their change during the course of the experiment.

Due to the amount of data obtained from the measurements our group had to develop an automated system for its analysis. The developed algorithms consisted of a custom two-step filtering and categorizing method assisted with a mean-first derivative search to extract the sought reliability indicators. The work on the developed automatic process has been described in details in the work presented during the ISSC conference[7].

RESULTS

In the course of analyzing the timing data we have come to an observation that both Tc and To of the tested switches have represented a similar trend across multiple measurements, the Tc parameter was getting smaller the longer the switch has been cycled, while the To was increasing. Plotting both of those values on the same figure has yielded a result which was showing a degradation of the observed parameters as presented on Figure 4.

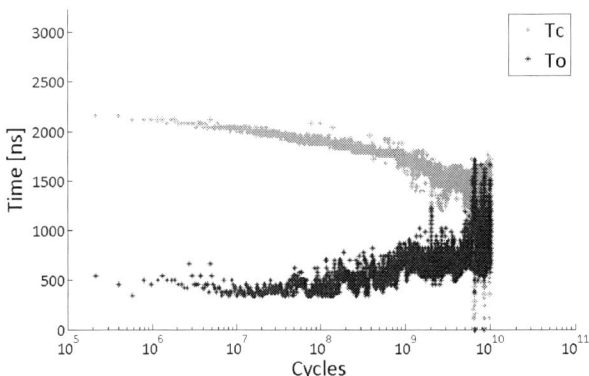

Figure 4: Observed Tc and To change in respect to device operating cycles – Type B switch

Further measurements on different samples of the same switch type have confirmed the existence of this effect with a similar trend, as shown on Figure 5:

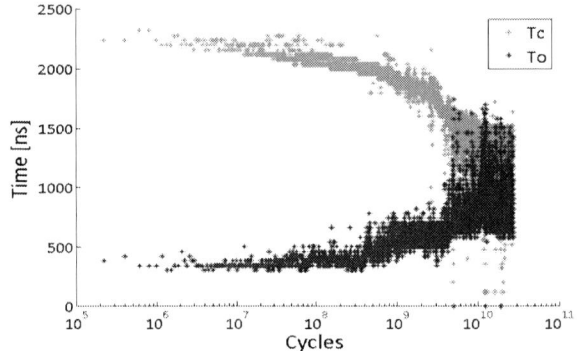

Figure 5: Observed Tc and To change in respect to device operating cycles – Type B switch, different sample

To verify the existence of this effect as independent of the used MEMS switch design, we have followed with a number of experiments on simpler Type A switches. The obtained data showed consistency with the previously observed effect of Tc/To deteriorating in Type B switches, though as expected from the switch structure the results were often highly variable. Figure 6 presents an example of the obtained results on a Type A switch.

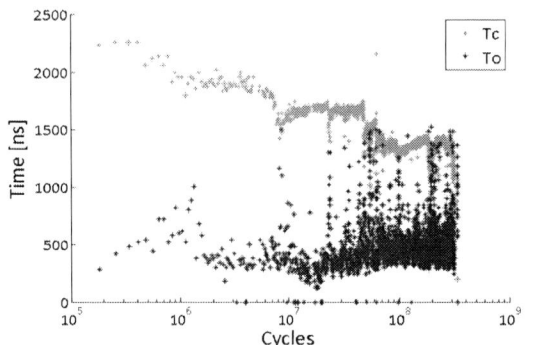

Figure 6: Example of a Type A (RF3) switch Tc/To change in respect to cycles

ANALYSIS

Evolution of spring restoring force

As it has been described in the literature[10][11] MEMS switches both ohmic as well as capacitive show an evolution of their pull-in voltage with respect to the number of cycles worked. Fruehling et al[12] has also presented the progression of bouncing behavior of MEMS switches as an effect depending on the cycles done. While Soma et al.[13] has proved that the evolution of structure damage due to cyclic fatigue has an influence on the pull-in voltage of the switches. In our previous work on explaining the Tc degradation[6] we have also presented a theory that the visible evolution of Tc can be linked to the weakening of spring restoring force as shown on Figure 7 below:

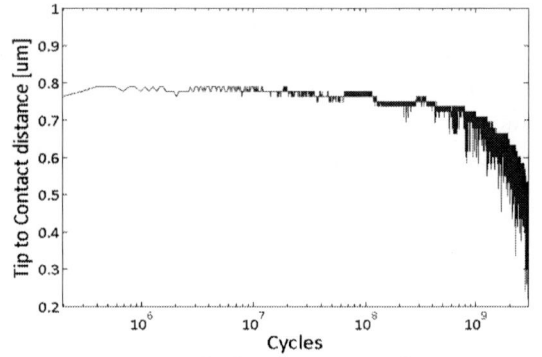

Figure 7: Calculated Tip_gap of MEMS switch vs Cycles, results of previous work presented in [6]

As all of the described effects are directly tied to restoring forces in MEMS switches and with respect to our results showing that the To data is consistent with our Tc analysis we can conclude that the observed symmetry in Tc/To of the switches is a representation of the weakening of switch restoring force. Therefore the difference between Tc and To can be employed as a simple indication of restoring force evolution as shown in Figure 8 and can be used in the prediction of MEMS switch failure in a continuous health monitoring system:

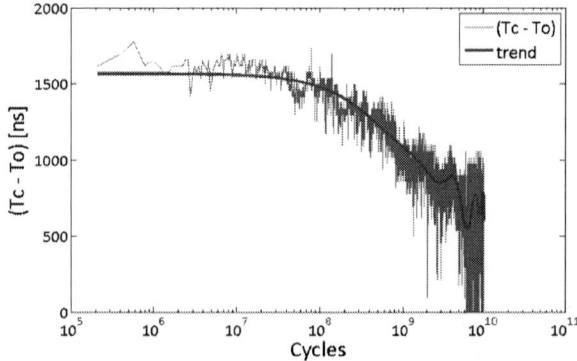

Figure 8: Observable effect of restoring spring force weakening through calculated Tc/To difference (Values of Tc - To below 0 omitted

Evolution of surface forces

In case of the presented Type A results the symmetry between the Tc and To is not as apparent as for the high reliability Type B switches. Results for this Type A switch have been chosen specifically to highlight a second failure mechanism in the tested devices which can be detected using this method.

As the opening time is not only depended on the spring restoring force but also on the surface adhesion forces between the contacts, a difference can be expected in the readings of To versus Tc. Surface adhesion in MEMS contacts is highly dependent on the topology of both contacting surfaces. Research in this area[14] has proved that those surfaces are rarely flat and most often consist of multiple asperities which in fact form the contact between the beam and the contacting pad of the device. As the area of asperities is very limited, they are subject to strong influences during the device cycling like shock damage and induced heating which in turn leads to constant modification of the contacting surface during each cycle. This has been investigated both experimentally[15][16] and theoretically[17][18] by many authors. Consensus of this research was that the number of asperities in contact increases with the number of the work cycles of the device, but until now it was impossible to continuously measure that change.

As it can be seen on the To data from the presented devices on Figure 46 all of the results show a faster increase in variability and larger spread of datapoints on the To then Tc measurements. This increase in variability is most evident in the low reliability Type A switch. Each of the tested devices have presented a similar trend in the variability of To with a usual spike near the end of the device lifecycle, an example of this can be seen on Figure 9:

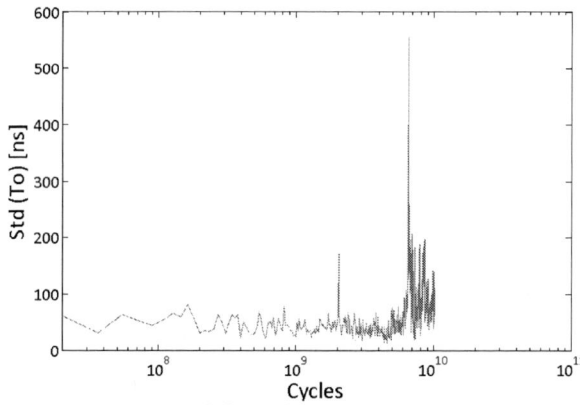

Figure 9: Variability of To measurement during a switch lifecycle – Type B switch

Because of the fact that the release movement of the switch is influenced by both restoring force and surface adhesion forces at the same time, the observed variability of the To measurement is a representation of the changing relation between those two forces. At the start of the lifecycle of the switch the restoring forces easily overcome their surface counterparts due to small number of asperities in contact as well to high potential energy stored in the spring structure. With increasing cycles the spring forces weaken, while the numbers of asperities in contact increases leading to a relation which produces higher variability of the To measurement and finishing when the adhesion forces overcome the spring restoring forces. This effect can be observed using the presented method while continuously monitoring the device in work.

CONCLUSSIONS

The developed method allows for the first time for a real time monitoring of health evolution in working, packaged MEMS switches. The system has been proved to be viable to automate and multiple measurements have been performed on various samples to confirm eligibility of this approach. The presented technique is scalable and obtained data proves that it is not design-specific. A novel insight into the failure mechanism was obtained through the To data. Analysis on the gathered measurements show that it is viable to create a fault prediction system based on combined Tc/To measurements. Two general types of failure modes in the tested switches have been identified based on the obtained results, previous research in this area and available literature. Closure time analysis of the device allows for the monitoring of spring restoring forces thus the structure change of the device, while open time analysis gives an insight in the change of contact surface and the relation between surface and spring forces.

REFERENCES

[1] ITRS, "International technology roadmap for Micro-Electro-Mechanical Systems 2011 edition," 2011.

[2] J. Bryzek, S. Roundy, and B. Bircumshaw, "Marvelous MEMS," *IEEE Circuits & Devices Magazine*, vol. 6, no. April, 2006.

[3] D. M. Tanner, "MEMS reliability: Where are we now?," *Microelectronics Reliability*, vol. 49, no. 9–11, pp. 937–940, Sep. 2009.

[4] W. Merlijn van Spengen, "MEMS reliability from a failure mechanisms perspective," *Microelectronics Reliability*, vol. 43, no. 7, pp. 1049–1060, Jul. 2003.

[5] C. Do, M. Lishchynska, K. Delaney, P. Fitzgerald, R. Goggin, and M. Hill, "Model-based analysis of switch degradation effects during lifetime testing," in *IEEE 25th International Conference on Micro Electro Mechanical Systems (MEMS)*, 2012, pp. 460–463.

[6] C. Kosla, C. Do, and M. Hill, "Monitoring MEMS switch closure time as a measure of reliability," in *Proceedings of the 23rd Micromechanics and Microsystems Europe Workshop*, 2012.

[7] C. Kosla and M. Hill, "Extraction of reliability indicators from large-scale , highly variable timing data of MEMS switches Tested Samples," in *Proceedings of ISSC 2012*, 2012.

[8] C. Do, M. Hill, M. Lishchynska, M. Cychowski, and K. Delaney, "Modeling, simulation and validation of the dynamic performance of a single-pole single-throw RF-MEMS contact switch," in *2011 12th Intl. Conf. on Thermal, Mechanical & Multi-Physics Simulation and Experiments in Microelectronics and Microsystems*, 2011, pp. 1/6–6/6.

[9] R. Goggin, P. Fitzgerald, J. Wong, B. Hecht, and M. Schirmer, "Fully integrated, high yielding, high reliability DC contact MEMS switch technology & control IC in standard plastic packages," in *2011 IEEE SENSORS Proceedings*, 2011, pp. 958–961.

[10] R. W. Herfst, P. G. Steeneken, and J. Schmitz, "Identifying degradation mechanisms in RF MEMS capacitive switches," in *2008 IEEE 21st International Conference on Micro Electro Mechanical Systems*, 2008, pp. 168–171.

[11] P. Fitzgerald and M. Hill, "Reliability modeling of MEMS cantilever switches under variable actuation stress level," in *Proceedings of the 21st Micromechanics and Microsystems Europe Workshop*, 2010, pp. 321–324.

[12] A. Fruehling, W. Yang, and D. Peroulis, "Cyclic evolution of bouncing for contacts in commercial RF MEMS switches," in *IEEE 25th International Conference on Micro Electro Mechanical Systems (MEMS)*, 2012, no. February, pp. 688–691.

[13] A. Soma and G. De Pasquale, "MEMS Mechanical Fatigue: Experimental Results on Gold Microbeams," *Journal of Microelectromechanical Systems*, vol. 18, no. 4, pp. 828–835, Aug. 2009.

[14] D. Hyman and M. Mehregany, "Contact physics of gold microcontacts for MEMS switches," *IEEE Transactions on Components and Packaging Technologies*, vol. 22, no. 3, pp. 357–364, 1999.

[15] A. Broue, J. Dhennin, F. Courtade, P.-L. Charvet, P. Pons, X. Lafontan, and R. Plana, "Thermal and topological characterization of Au, Ru and Au/Ru based MEMS contacts using nanoindenter," in *2010 IEEE 23rd International Conference on Micro Electro Mechanical Systems (MEMS)*, 2010, pp. 544–547.

[16] F. K. Chowdhury, H. Pourzand, and M. Tabib-Azar, "Investigation of contact resistance evolution of Ir, Pt, W, Ni, Cr, Ti, Cu and Al over repeated hot-contact switching for NEMS switches," in *2013 IEEE 26th International Conference on Micro Electro Mechanical Systems (MEMS)*, 2013, pp. 445–448.

[17] O. Rezvanian, M. a Zikry, C. Brown, and J. Krim, "Surface roughness, asperity contact and gold RF MEMS switch behavior," *Journal of Micromechanics and Microengineering*, vol. 17, no. 10, pp. 2006–2015, Oct. 2007.

[18] P. Shanthraj, O. Rezvanian, and M. A. M. A. Zikry, "Electrothermomechanical Finite-Element Modeling of Metal Microcontacts in MEMS," *Journal of Microelectromechanical Systems*, vol. 20, no. 2, p. 371, 2011.

978-1-4799-3510-9/14 $31.00 © 2014 IEEE

DIRECT MEASUREMENT OF SHEAR PIEZORESISTANCE COEFFICIENT ON SINGLE CRYSTAL SILICON NANOWIRE BY ASYMMETRICAL FOUR-POINT BENDING TEST

Taiki Kimura, Naoki Saito, Toshimitsu Takeshita, Koji Sugano and Yoshitada Isono[1]
[1]Graduate School of Engineering, KOBE University, JAPAN

ABSTRACT

This research evaluated the shear piezoresistance property of *p*-type single crystal silicon nanowire (SiNW) by the asymmetrical four-point bending (AFPB) technique proposed by the authors [1]. We fabricated the *p*-type SiNW on the AFPB test specimen with "V"-shaped notches (V-notches) made of single crystal silicon. Bending the specimen by the asymmetrical four point-supports, simple shear stress can be produced at the center of the specimen. Consequently, we have succeeded in evaluating the shear piezoresistance coefficient of SiNW directly, which was found to be π_{44}=203 × 10^{-11} Pa^{-1} at an impurity concentration of 7.3 × 10^{18} cm^{-3}. This value is 2.1 times larger than that of *p*-type piezoresistors used in conventional piezoresistance sensors on a micrometer scale. The proposed evaluation technique and obtained result will be effective for design application of high-sensitivity mechanical sensors integrating SiNW piezoresistance elements.

INTRODUCTION

One-dimensional nanowires have specific physical properties attributable to their extremely small size and low structural dimensions. Especially, SiNWs can be used as highly sensitive piezoresistance elements for mechanical force sensors [2], [3] because one-dimensional semiconducting nanowires show a drastic change in the electronic structural features such as a band gap and effective mass under enormous mechanical strain, which can be used to determine the electron mobility [4]. However, few data of piezoresistance properties for SiNWs have so far hindered the reliable designing and optimizing high-sensitivity mechanical sensors including SiNWs piezoresistance elements. Especially, the shear piezoresistance coefficient of SiNWs has not been evaluated sufficiently since it is difficult to apply simple shear load to SiNWs directly.

The authors have developed the shear loading tester that are able to apply simple shear stress to a micrometer scale specimen based on the asymmetrical four-point bending (AFPB) method, and have reported the shear fracture strength of single crystal silicon [1]. In this research, the piezoresistance effect of the *p*-type single crystal SiNW with four-terminals integrated on the micro silicon specimen was evaluated under shear stress using the AFPB method. The SiNW was oriented in the <100> direction on {001} plane. The shear and normal stresses of SiNW were accurately calculated by finite element (FE) method for estimation of shear piezoresistance coefficient of π_{44}.

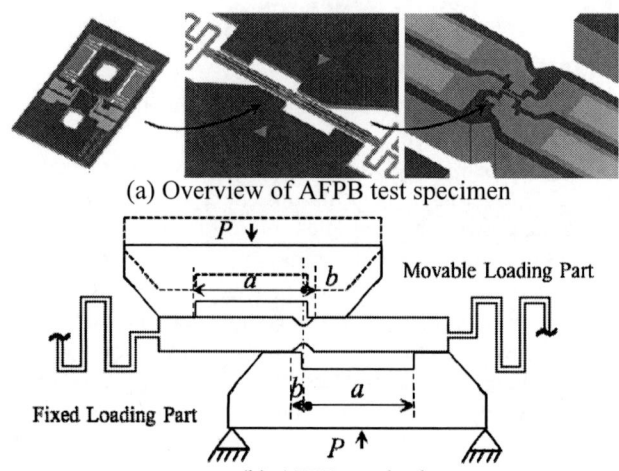

(a) Overview of AFPB test specimen

(b) AFPB method

Figs. 1 Schematics of AFPB test specimen including SiNW: (a) overview of AFPB test specimen and (b) AFPB method.

EXPERIMENTAL PROCEDURES
Asymmetrical Four-Point Bending (AFPB) Method

Figs. 1(a) and 1(b) show schematics of the AFPB test specimen including the four-terminal SiNW and the four-point bending method, respectively. In Fig. 1(a), the test specimen is composed of three parts; the movable loading part, the specimen part with V-notches, and the fixed loading part. The movable loading and the specimen parts are supported by suspension beams. In Fig. 1(b), the specimen part is sandwiched between the movable and fixed loading parts in order to apply a compressive load to the test specimen at four point-supports. The bending moment is completely canceled at the center of the specimen part between the bottoms of V-notches: thus shear stress is produced at the center area. The shear stress, τ_{12}, depends on the specimen dimensions, which is given by

$$\tau_{12} = \frac{P(b-a)}{wt(a+b)}, \qquad (1)$$

where, P is the total load, a and b are the distance from the center axis of the specimen to the loading points as shown in Fig. 1(b), w is the specimen width at the center of specimen, and t is the thickness of the specimen.

Fig. 2 shows photographs of the shear loading tester and the AFPB test specimen set on the tester. The shear force and displacement can be measured by the load-cell and the CCD camera system mounted on the optical microscope, respectively. The movable loading part of the test specimen is connected to the uniaxial PZT actuator installed in the tester for applying a compressive load to the specimen part.

978-1-4799-3510-9/14 $31.00 © 2014 IEEE

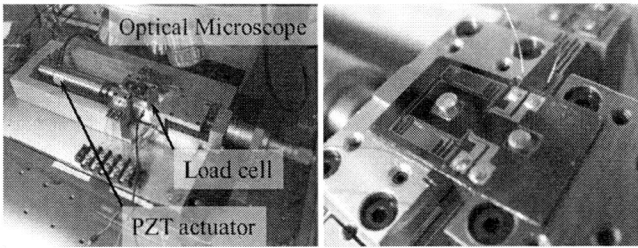

Fig. 2 Photographs of the shear loading tester and the AFPB test specimen set on the tester.

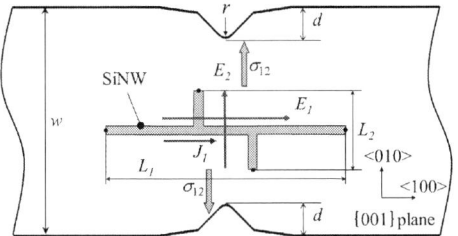

Fig. 3 Schematic of four-terminal SiNW. Crystallography orientations and the directions of the electrical field, current and shear stress are described.

Measurement of Shear Piezoresistance Property

Fig. 3 depicts a four-terminal SiNW integrated on the specimen part. When the voltage is applied to the SiNW in its longitudinal direction under a shear stress condition, the electric potential difference of SiNW in the transverse direction is induced in proportion to the stress and the current density. Therefore, we can obtain the shear piezoresistance coefficient, π_{44}, of SiNW. It is well known that the piezoresistive response of single crystal silicon is expressed as [5]

$$E_i = \rho J_i + \rho \pi_{ijkl} \sigma_{kl} J_j , \qquad (2)$$

where, E_i, J_i and σ_{kl} are components of the electric field, current density and stress acted on the material, respectively. ρ is the resistivity and π_{ijkl} are components of the piezoresistance coefficient. For the four-terminal SiNW, the above equation is transformed into

$$E_1 = \rho(1 + \pi_{11}\sigma_{11} + \pi_{12}\sigma_{22})J_1 , \qquad (3)$$

$$E_2 = \rho \pi_{44} \tau_{12} J_1 , \qquad (4)$$

$$E_3 = 0 . \qquad (5)$$

J_1 is assumed to be constant throughout the SiNW to make a first-order analysis. The potential difference between the current contacts of the SiNW in the longitudinal direction is expressed by the integral of E_1 as the following,

$$V_1 = \rho J_1 L_1 + \rho J_1 \int_0^{L_1} (\pi_{11}\sigma_{11} + \pi_{12}\sigma_{22})dx , \qquad (6)$$

where, L_1 is the length of the SiNW in the longitudinal direction. If the normal stresses of σ_{11} and σ_{22} are very small,

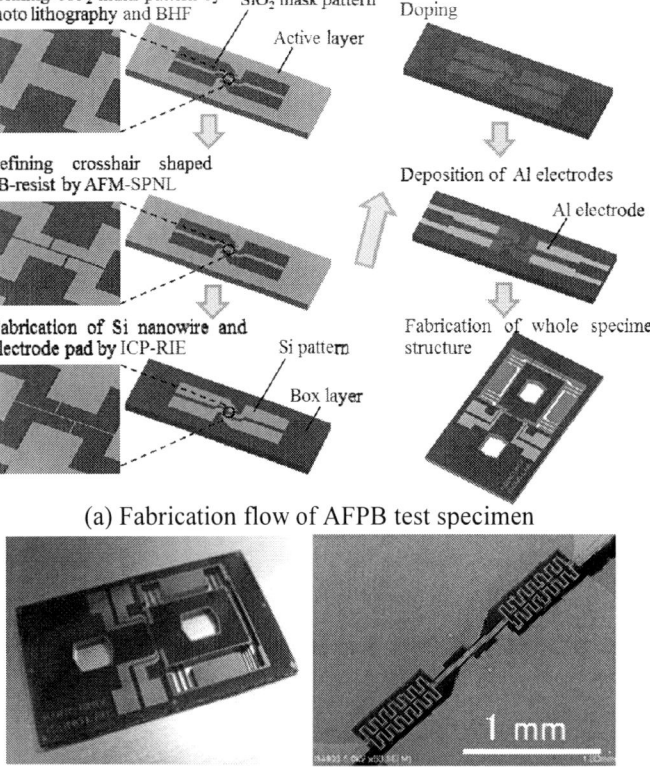

(a) Fabrication flow of AFPB test specimen

(b) Photograph of AFPB test specimen and SEM images of four-terminal SiNW on the specimen part.

Figs. 4 (a) Fabrication flow of AFPB test specimen and (b) SEM images of four-terminal SiNW on the specimen part.

the first term of equation dominates V_1. Thus, the above equation can be approximated as

$$V_1 \cong \rho J_1 L_1 . \qquad (7)$$

Using L_2 of the length between two transverse nanowires, the voltage of the SiNW in the transverse direction is given by

$$V_2 = \frac{L_2}{L_1} \pi_{44} \tau_{12} V_1 . \qquad (8)$$

In this research, applying a constant current to the SiNW in the longitudinal direction, voltages of V_1 and V_2 are measured under several shear stresses.

Fabrication of AFPB Test Specimen with SiNW

Fig. 4(a) shows the fabrication flow of the AFPB test specimen with the SiNW. The starting substrate is an SOI wafer with a thermal oxidized 100-nm-thick active layer on

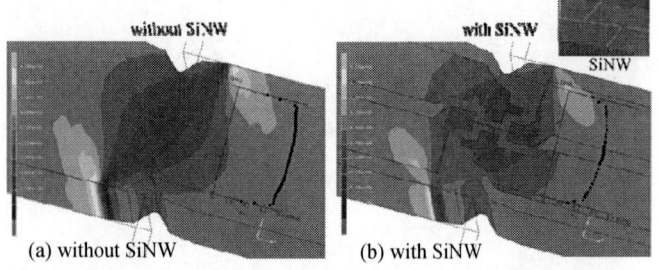

Figs. 5 Shear stress distributions at the center of AFPB specimen obtained from FEA: (a) without SiNW, and (b) with SiNW.

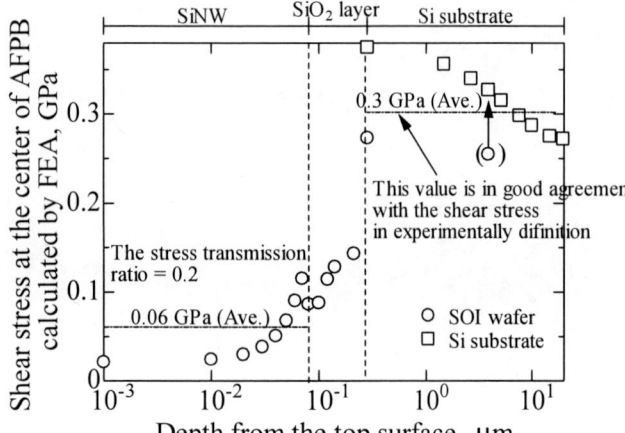

Fig. 6 Relationship between shear stress at the center of SiNW and the depth from the top surface.

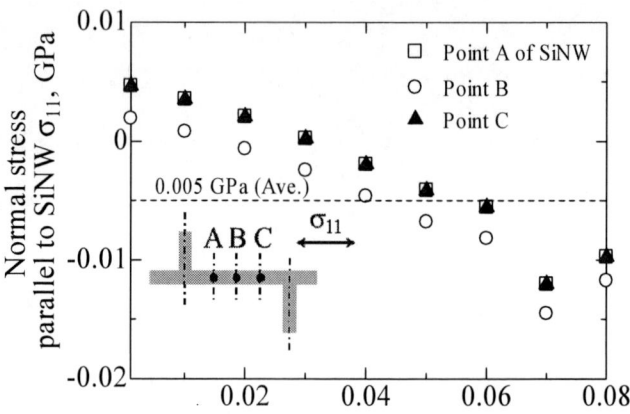

(a) σ_{11} at points of A, B and C on SiNW

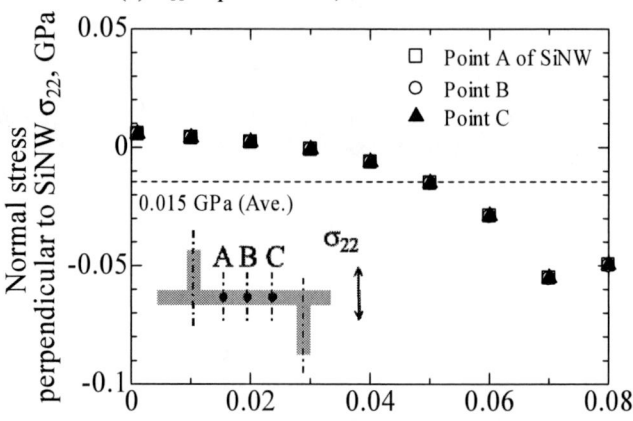

(b) σ_{22} at points of A, B and C on SiNW

Figs. 7 Relationship between normal stresses, σ_{11} and σ_{22}, at the center of SiNW and the depth from the top surface: (a) σ_{11}, and (b) σ_{22}.

200-nm-thick BOX layer and 400-nm-thick silicon substrate. The patterns for the wire and its transverse wires used as current contacts are formed by scanning probe nanolithography with the EB resist [6] after fabrication of four electrode pads on the top layer of SiO_2. The four-terminal SiNW is fabricated by ICP-RIE process. The conventional doping process is performed to make the p-type SiNW. Finally, the second ICP-RIE process forms the whole AFPB test specimen under the protection of the SiNW after making metal contact pads of Al films.

Fig. 4(b) shows SEM images of the four-terminal SiNW on the specimen part with a photograph of whole AFPB test specimen. The width and thickness of the specimen part are 29.8 μm and 39 μm, respectively. The depth of V-notch is 6.0 μm and the radius of the notch bottom is 2.33 μm on an average. The SiNW with a width of 85 nm and a height of 80 nm was successfully fabricated on the specimen part. The lengths of L_1 and L_2 in the SiNW are 8.66 μm and 2.07 μm, respectively.

RESULTS AND DISCUSSIONS
Finite Element Analyses

The shear stresses induced in the specimen part without and with the four-terminal SiNW were calculated by FEA, as shown in Figs. 5(a) and 5(b). Applying 5 μm displacement to the movable loading part from its initial position, the shear stress of about 0.36 GPa at the top of the silicon substrate without the SiNW is uniformly generated between the

V-notches in Fig. 5(a), whereas the stress decreases to 0.27 GPa at the top of substrate of the specimen with the SiNW in Fig. 5(b). Fig. 6 shows the variation of shear stress at the center of AFPB test specimen with the SiNW in the depth direction. Since the shear stress of the SiNW is generated only 0.06 GPa on an average, stress generated in the silicon substrate is not fully transmitted to the SiNW. Thus, we defined the stress transmission ratio as stress in the SiNW/stress in the silicon substrate for estimation of π_{44} after the test.

The normal stresses parallel and perpendicular to the SiNW, σ_{11} and σ_{22}, were also evaluated by FEA as shown in Figs. 7(a) and 7(b). The averaged σ_{11} and σ_{22} in the SiNW between two terminals for measurement of V_2 are 0.005 GPa and 0.015 GPa, respectively. σ_{11} is less than $\tau_{12} \times 10^{-1}$, while σ_{22} is a quarter of τ_{12}. However, the integral term in Eq. (6) was estimated to be of the order of $\rho J_1 L_1 \times 10^{-3}$ using π_{11} and π_{12} of bulk p-type silicon [7]: thus, we can ignore the second term in Eq. (6), and the approximation of Eq. (7) is appropriate.

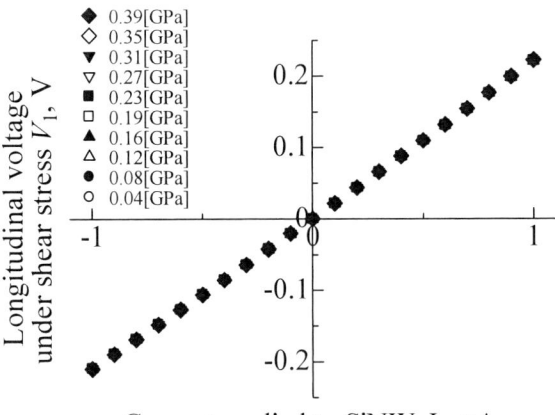

(a) V_1-I_1 characteristic

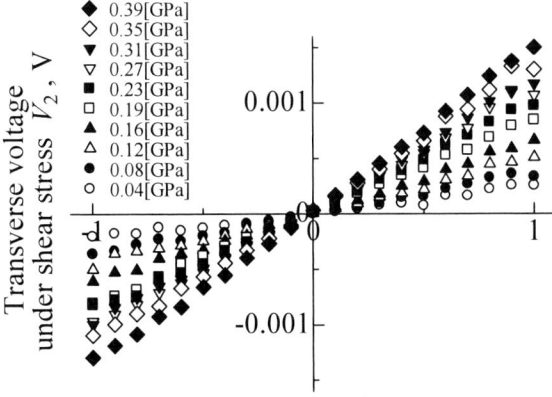

(b) V_2-I_1 characteristic

Figs. 8 *I-V* characteristics of SiNW under several shear stresses.

Estimation of Shear Piezoresistance Coefficient of SiNW

Figs. 8(a) and 8(b) show variations of the longitudinal and transverse voltages, V_1 and V_2, with increasing the current, I_1, under shear stresses, respectively. The ratio of V_1 to I_1 does not change under shear stresses applied in the test in Fig. 8(a), but the bigger stress grows, the larger increasing ratio of V_2 to I_1 becomes in Fig. 8(b). Then, considering the stress transmission ratio obtained from the FEA, we have directly determined π_{44} from the relationship between the relative change of $\Delta V_2/V_1$ and the shear stress, as shown in Fig. 9. The result is π_{44}=32.3 × 10^{-11} Pa^{-1} at an impurity concentration of 7.3 × 10^{18} cm^{-3} under a consideration of the stress transmission ratio. This is smaller than previously reported values [5]. Furthermore, we have considered the effect of the anisotropic electrical conductance on the

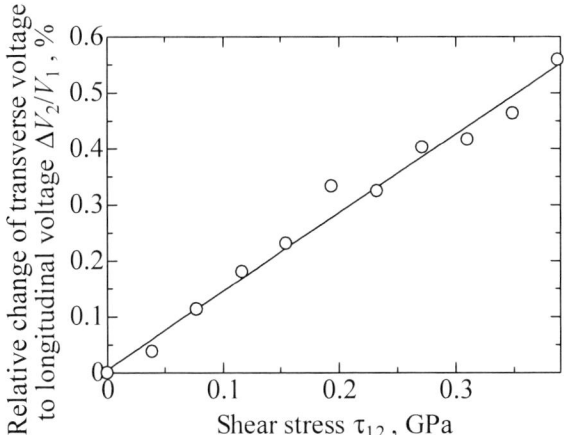

Fig. 9 Relative change of transverse voltage to longitudinal voltage with increasing shear stress.

transverse voltage [2], and then the huge coefficient, π_{44}=203 × 10^{-11} Pa^{-1} has been estimated. This is 2.1 times larger than that of *p*-type piezoresistors [5].

CONCLUSIONS

This research has quantitatively evaluated the shear piezoresistance coefficient, π_{44}, of the single crystal SiNW by the AFPB test. The shear piezoresistance coefficient of SiNW was found to be π_{44}=203 × 10^{-11} Pa^{-1} at an impurity concentration of 7.3 × 10^{18} cm^{-3}, which is 2.1 times larger than that of *p*-type piezoresistors used for piezoresistance sensors on a micrometer scale.

REFERENCES
[1] M. Ogawa, and Y. Isono, Proc. of MEMS 2007, (2007), pp. 259-262.
[2] T. Toriyama, Y. Tanimoto and S. Sugiyama, *J.MEMS*, Vol. 11, No. 5, (2002), pp. 605-611.
[3] A. Lugstein, M. Steinmair, A. Steiger, H. Kosina and E. Bertagnolli, *Nano letter*, Vol. 10 No. 8, (2010), pp. 3204-3208.
[4] K. Nakamura, Y. Isono, T. Toriyama, and S. Sugiyama, *Jpn. J. Appl. Phys.* Vol. 48, (2009), 06FG09 (5 pages).
[5] T. Toriyama, and S. Sugiyama, *J.MEMS*, Vol. 11, No. 5, (2002), pp. 598-604.
[6] L. Anggraini, B. Tanaka, N. Matsuzuka, and Y. Isono, *Jpn. J. Appl. Phys.* Vol. 52, (2013), 056501.
[7] C. S. Smith, *Phys. Rev.* Vol. 94, No. 1, (1954), pp. 42-49.

CONTACT
*Y. Isono, tel: +81-78-803-6145; isono@mech.kobe-u.ac.jp

DISSOLVABLE MATERIAL FOR HIGH-ASPECT-RATIO FLEXIBLE SILICON-MICROWIRE PENETRATIONS

S. Yagi[1], S. Yamagiwa[1], T. Imashioya[1], H. Oi[2], Y. Kubota[1], M. Ishida[1,2] and T. Kawano[1]

[1]Department of Electrical and Electronic Information Engineering, [2]Electronics-Inspired Interdisciplinary Research Institute (EIIRIS), Toyohashi University of Technology, Japan

ABSTRACT

High-aspect-ratio microwire array devices, which penetrates into a biological tissue, have widely been used in neuroscience, offering *in vivo/in vitro* electrophysiological stimulation and recording, drug delivery (e.g., DNA), and optogenetics applications. As low invasive and safe penetrations, these microwire devices are required to be further miniaturized and flexibility. However, tissue penetration with such high-aspect-ratio and flexible wires is problematic, because the wire buckles during the penetration. Here, we improve the penetration capability of high-aspect-ratio flexible wires by coating a dissolvable material of "silk fibroin" (Fig.1). The silk-fibroin is a material, which dissolves when the surface contacts with a wet biological tissue, resulting in that embedded wires are appeared and penetrated. We demonstrated the silk fibroin coating over high-aspect-ratio silicon-microwires (~720 μm in length), which was fabricated by vapor-liquid-solid (VLS) growth. The 420-μm-long silicon-wire with a ~200-μm-thick silk film exited the stiffness of 4.03 N/m, which is 72% improved value compared to that of the silicon-wires without silk (2.34 N/m). The effects of the silk support on the wire penetration were confirmed by demonstrating the gelatin penetrations. These results suggest that the numerous high-aspect-ratio flexible bioprobes can be penetrated by using the silk support.

INTRODUCTION

Silicon-microwire fabricated by MEMS technology offers electrophysiological recordings with a high spatial and temporal resolution. Such MEMS electrodes have been used in studies of numerous biological samples, including monkey- and human-brains [1-3]. Although these MEMS electrodes become powerful tools in electrophysiological studies, damage to neurons/tissue associated with the wire penetration is still problematic [4].

To reduce the damage, miniaturization of the diameter of the wire electrode is necessary. To address this technological challenge, we have proposed a micro-scale diameter wire arrays fabricated by selective vapor-liquid-solid (VLS) growth of silicon. The VLS growth offers an advantage of assembling high-aspect-ratio microwire arrays. Figure 2a shows VLS grown silicon-wires, each with the diameter of < 5 μm and the length of > 400 μm. Other advantages of the VLS growth based wire fabrication, compared to conventional MEMS technology are: (i) batch fabrication process for hundreds/thousands wires, (ii) fabricating small diameter wires ranging from nano-scale, (iii) making possible the construction of high-density wire arrays and (iv) on-chip complementary with CMOS electronics [4]. We have also demonstrated *in vivo* neuronal action potential recordings from a Long–Evans rat's brain using 200-μm-length wires [5, 6].

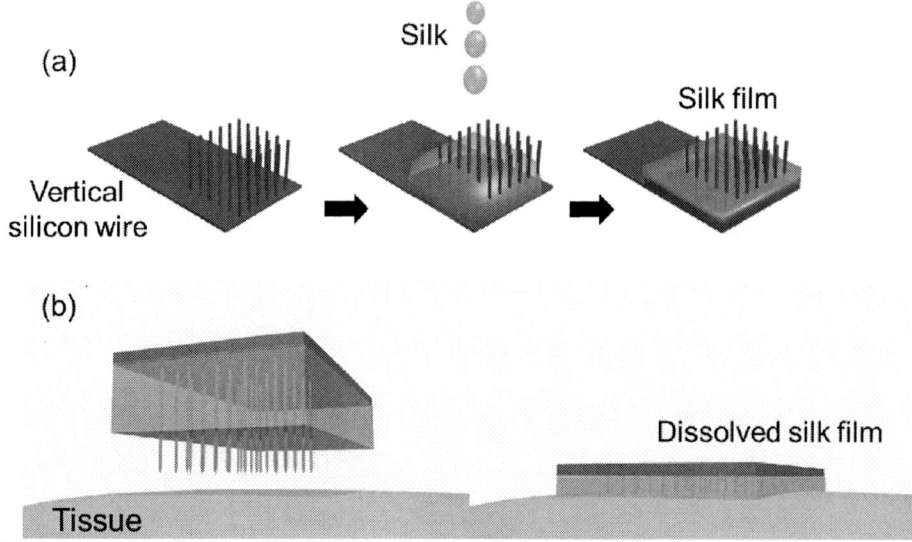

Fig.1 Schematics of dissolvable silk fibroin for the penetration of high-aspect-ratio silicon-microwires. (a) Coating and drying of silk fibroin. (b) Penetration of the silk supported silicon-wires into a tissue.

For investigating deep tissues, wire lengths more than 200 μm (~ 2 mm for human cortex) will be needed. However,

the silicon-microwires with the length of more than 400 μm cannot be penetrated because the wire buckles during the

978-1-4799-3510-9/14 $31.00 © 2014 IEEE

penetration [6, 7]. The buckling load of the silicon wire with a column shape can be depicted as

$$F_{Buckling} = \frac{2\pi EI}{l^2} \propto E\left(\frac{d^2}{l}\right)^2,$$

$$I = \frac{\pi d^4}{64},$$

where buckling load $F_{Buckling}$, Young's modulus E, wire length l, cross-sectional 2nd-order moment I and wire diameter d. As the formulas, the buckling load is proportional to $1/l^2$ and d^4, indicating that wires with higher aspect ratios show lower buckling forces than that of lower aspect ratio wires with a same diameter. Once the buckling force of a wire decreases less than the required force for the tissue penetration, the wire buckling occurs before the tissue punch. Thus the wire's stiffness is essential, particularly in tissue penetrations with high-aspect-ratio microwires.

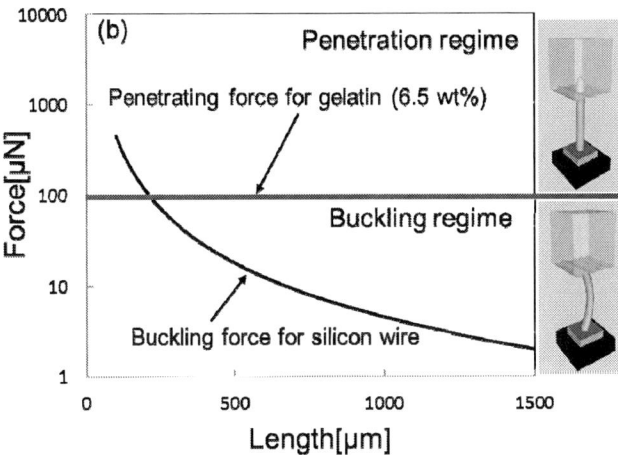

Fig.2 Buckled silicon microwire. (a) SEM observation of a buckled silicon-wire by FMS. (b) Wire buckling force–wire length curves measured by buckling a 2-μm-diameter silicon-wire with the FMS (a).

Figure 2b shows the buckling force–wire length curves measured by buckling a VLS grown 2-μm-diameter silicon-wire with a Force-Measurement-System (FMT-120, kleindiek). The graph also includes the experimentally measured penetration force (100 μN) for a gelatin (6.5 wt% in water), suggesting that silicon-microwires with the lengths of > 300 μm buckle before the penetration.

Here we propose a methodology to enhance the stiffness of the high-aspect-ratio silicon-microwire "temporary" by coating the wire with a dissolvable material (Fig. 1a). This dissolvable material coated over the wire dissolves when the surface contacts with a wet biological tissue, and the embedded silicon-wires are appeared and penetrated (Fig. 1b).

SILK FIBROIN COATING

We proposed silicon-microwire penetrations with a silk fibroin support. Silk fibroin is a candidate dissolvable material, because it is biocompatible, water soluble with programmable rates [8] and easy-to-use especially for coatings of such vertical silicon-microwires.

We fabricated silicon-microwire by Au-catalyzed VLS growth [9]. The wire length exceeds 1 mm, which value can be controlled with a growth time. For the wire coating with the silk fibroin, the silk solution (Silk protein extract, Matsuda farm) was dropped over the vertical silicon-wires, followed by the crystallization of the solution in air for 1 day (Fig. 1a). The thickness of the crystallized silk-film depends on the amount of the solution as well as the density of the silicon-wires. The thickness of the formed silk film exhibited ~ 90 μm for 120-μm-long silicon-wire, while the silk film for 420-μm-long wire showed thicker 200 μm film due to the dense wires (50 μm gap) compared to 120-μm-long wires (100 μm gap)(Fig. 3). Herein, a constant amount of the solution was 0.07 – 0.08 ml.

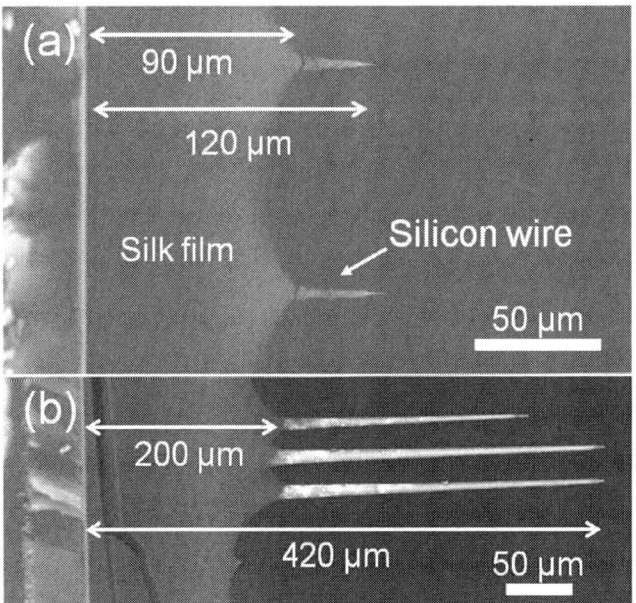

Fig.3 SEM images of silk-film embedded silicon-microwires. Silk formations for (a) 120-μm-long and (b)420-μm-long silicon-microwires. Gaps between the wires are 100 μm (a) and 50 μm (b). Note that constant amount of the silk solution was 0.07 - 0.08 ml.

BENDING TESTS

Improved stiffness of the silk embedded silicon-microwires was confirmed in the wire bending tests. Load–displacement characteristics of the wire were measured by Force-Measurement-System inside a SEM (Fig. 4a). Figure 4b shows the displacement–force curves of 400-μm-long silicon-wires with and without the silk-film support. The silicon-wire without forming the silk-film exhibited the stiffness of 2.34 N/m, while 72% improved stiffness of 4.03 N/m was observed with the silicon-wire embedded in the ~200-μm-thick silk film. The stiffness of the silk supported wire was consistent with that of a 200-μm-long silicon-wire with the same diameter. These results indicated that the wire section embedded in the silk-film does not bend and the displacement was mainly dominated by the wire section exposed from the silk support.

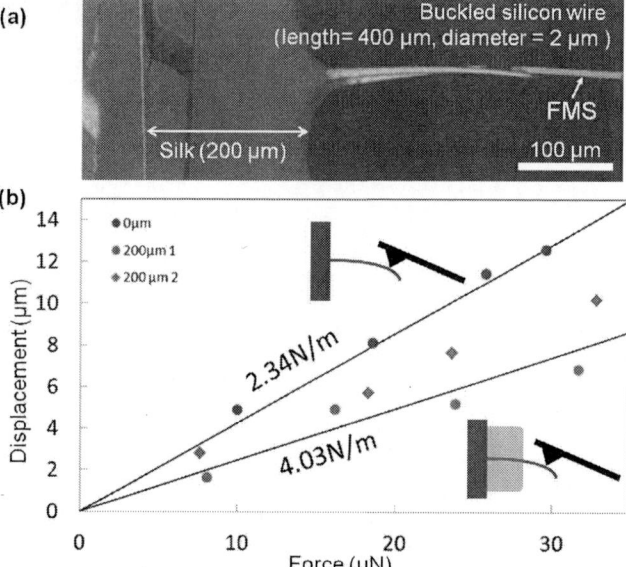

Fig.4 Bending tests on silk supported silicon wire. (a) SEM image showing a bending silicon-wire with FMS. The length and diameter of the measured wire are 400 μm, and 2 μm, respectively. The thickness of the silk film is ~200 μm. (b) Displacement–force curve taken from the silk supported silicon wire (a). The graph also includes the curve of un-supported silicon-wire for comparison.

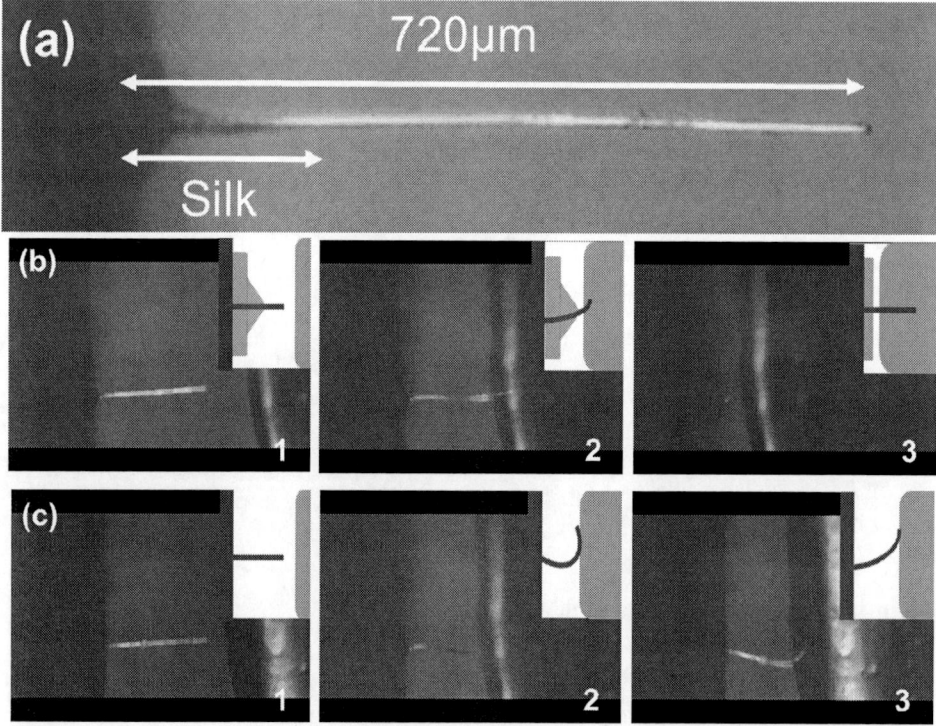

Fig.5 Penetration tests of silicon-microwires using a gelatin (6.5 wt% in water). (a) Photograph of a silk supported silicon microwire (2 μm in diameter and 720 μm in length) Thickness of the silk film is 150 μm. (b) Silk coated silicon-wire penetrated into the gelatin. (c) Silicon-wire penetration without silk support showing the un-penetrated silicon-wire.

PENETRATION TESTS

To confirm the effect of the silk support on the silicon-wire's stiffness, we demonstrated the penetration tests using a gelatin (6.5 wt% in water) (Fig. 5). The tested silicon-wire had the diameter of 2 μm and the length of 720 μm. The wire was supported by a 150-μm-thick silk-film at the wire base (Fig. 5a). Note that the hardness of the gelatin was similar to the cardiac muscle of a pig. The silicon-wire with the silk support was moved and contacted with the gelatin surface. During the further manipulation of the wire towards, the silicon-wire showed the buckling near the tip section (center panel in Fig. 5b), but the wire was eventually penetrated into the gelatin (right panel in Fig. 5b). On the other hand, the silicon-wire without the silk support showed the buckling of the entire wire section (center panel in Fig. 5c), resulting in the un-penetrated wire (right panel in Fig. 5c).

CONCLUSSION

We proposed a silk fibroin support for vertical silicon-micowires with a high aspect ratio, and demonstrated the wire penetrations using a gelatin. The bending and penetration tests of the silk supported wires indicated that this approach is a powerful methodology to penetrate high-aspect-ratio flexible silicon-wires. Although VLS grown silicon-microwires were used to demonstrate the penetration capability, this approach can be used for other numerous bioprobes, enabling further safe and easy tissue penetrations for the applications including *in vivo/in vitro* electrophysiological measurement (stimulation and recording), drug delivery, and optogenetics.

ACKNOWLEDGEMENTS

This work is supported by a Grant-in-Aid for Scientific Research (S), Young Scientists (B) and the PRESTO Program from JST.

REFERENCES

[1] L. R. Hochberg *et al.*, "Neuronal ensemble control of prosthetic devices by a human with tetraplegia", *Nature*, vol. 442, pp. 164-171, 2006.

[2] J. P. Donoghue, "Connecting cortex to machines: recent advances in brain interfaces", *Nat. Neurosci.*, vol. 5, pp. 1085-1088, 2005.

[3] A. Nicholas *et al.*, "Nanomaterials for neural interfaces", *Adv. Mater.*, vol. 21, pp. 3970-4004, 2009.

[4] D. H. Szarowski, M. D. Anderson, S. Retterer, A. J. Spence, M. Isaacson, H. G. Craighead, J. N. Turner and W. Shain, "Brain responses to micro-machined silicon devices", *Brain Res.*, vol. 983, pp. 23-25, 2003.

[5] A. Okugawa, K. Mayumi, A. Ikedo, M. Ishida and T. Kawano, "Heterogeneously integrated vapor-liquid-solid grown silicon probes/(111) and silicon MOSFETs/(100)", *IEEE Electron Device Lett.*, vol. 32, no. 5, pp. 683-685, 2011.

[6] A. Fujishiro, H. Kaneko, T. Kawashima, M. Ishida, T. Kawano, "A penetrating micro-scale diameter probe array for in-vivo neuron spike recordings", Proceeding of IEEE 24th International Conference onMicro Electro Mechanical Systems (MEMS), pp. 1011-1014, 2011.

[7] S. Morita, A. Fujishiro, A. Ikedo, M. Ishida and T. Kawano, "Fabrication of force sensitive penetrating electrical neuroprobe arrays", Proceeding of IEEE 25th International Conference onMicro Electro Mechanical Systems (MEMS), pp. 243-246, 2012.

[8] D. H. kim, *et al.*, "Dissolvable films of silk fibroin for ultrathin conformal bio-integrated electronics", *Nat. Mater.*, vol 9, pp. 511-517, 2010.

[9] A. Ikedo, T. Kawashima, T. Kawano and M. Ishida, "Vertically aligned silicon microwire arrays of various lengths by repeated selective vapor-liquid-solid growth of n-type silicon/n-type silicon", *Applied Physics Letters*, vol. 95, 033502, 2009.

CONTACT

*S. Yagi, yagi-s@int.ee.tut.ac.jp

DOME-DISC: DIFFRACTIVE OPTICS METROLOGY ENABLED DITHERING INERTIAL SENSOR CALIBRATION

Sachin Nadig[1], Serhan Ardanuç [1], Bill Clark[2] and Amit Lal[1]

[1] *Sonic*MEMS Laboratory, School of Electrical and Computer Engineering, Cornell University
[2] Analog Devices Inc.

ABSTRACT

We demonstrate 107-ppm accurate scale-factor and bias calibration of a commercial Coriolis force gyroscope, in which the typical un-calibrated scale factor variations are ~100,000-ppm. In this paper, we present a proof-of-concept result on calibration architecture - Diffractive Optics Metrology Enabled Dithering Inertial Sensor Calibration (DOME-DISC). DOME-DISC consists of a piezoelectric dither stage to provide built-in mechanical stimulus to the gyroscope attached to it. The motion of the dithering stage is measured by imaging an optical diffraction pattern created by an incident laser off diffraction gratings on the dither stage. In order to calibrate the gyroscope, the stage motion needs to be measured accurately and precisely. The motion of the stage is measured using the Nano Optical Ruler Imaging System (NORIS), with absolute accuracy of ~30nm over several millimeters, stable over several hours. NORIS provides parts-per-million stage motion measurement accuracy. In this paper, an angular dither motion of 0.1 to 0.5 degrees was optically measured with ~ 0.1 millidegree resolution. By measuring the scale factor and bias for a gyroscope on the dither stage mounted directly on a commercial rate table, and matching the gyroscope input-output curve to 100ppm, we demonstrate the capability to measure in-package gyroscope characteristics within the error limits of the commercial rate table.

INTRODUCTION

Micro inertial sensors are successful in rate sensing applications where power and size of the sensor have to be very small. However, long term precision navigation is not currently possible with existing sensors owing to the drift and aging of the scale factor and bias along with the inherent noise of the sensor. A miniature sensor that can provide long term self-calibrating capability to a MEMS sensor might provide high performance and low cost for applications such as personal navigation and GPS-denied navigation. There are several self-calibration approaches being investigated to measure the bias and scale factor during operation of the devices using rotary, dither, or electronic injection of rates [2]. In a calibration performed using a dither stage, the long term stability can be limited by the accuracy of the stage motion measurement. In this paper, we present a metrology method - Diffractive Optics Metrology Enabled Dithering Inertial Sensor Calibration (DOME-DISC) that can be integrated within an inertial sensor package. It consists of a dither stage to provide rotary mechanical excitation, to the gyroscope. The motion of the stage is measured by imaging the diffraction pattern

obtained from a grating placed on the stage along-side a commercial gyroscope, as shown in Figure 1. This architecture has the advantage that it is compatible with almost any gyroscope die that can be attached to the stage. Our group previously demonstrated the Nano Optical Ruler Imaging System (NORIS), which uses a rubidium-stabilized laser to implement long term stable metrology of <10 nm accuracy over 100 mm [3]. Here, NORIS is modified to measure translation and rotation of the dither stage via imaging the diffraction pattern on a CMOS imager. The position of the dither stage is measured with absolute accuracy of ~30 nm over several millimeters, over several hours [3], providing parts-per-million stage motion measurement accuracy.

In a long term stable and accurate calibration system, any reference used to measure length has to be stable over long time. The DOME-DISC architecture is compatible with the use of the recently available CSAC (Chip-Scale Atomic Clock [1]). CSAC provides a long terms stable wavelength in a package with a size compatible with 1-cc inertial sensor package. However, for demonstration of the calibration architecture, this work uses a standard off-the-shelf laser source.

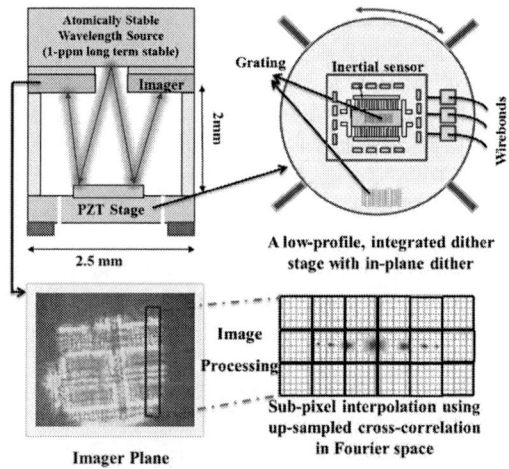

Figure1: Sketch of proposed calibration architecture.

IMAGER BASED OPTICAL METROLOGY OF GYROSCOPE RATE

Extraction of the input-output characteristics of a gyroscope is required for its calibration. Our approach to calibrate in-situ is by applying known excitations to the gyroscope and measuring the corresponding response that is used to compute the gyroscope instantaneous sensitivities and biases. The accuracy of our scale factor and bias measurements rely on the accuracy of the metrology system

978-1-4799-3510-9/14 $31.00 © 2014 IEEE

to measure the applied mechanical excitation. The displacement due to applied excitation is then computed from the diffraction pattern using an image processing algorithm. This approach also allows compensating for systematic misalignment of the laser source, grating, and the camera.

The algorithm for metrology [4] used here is implemented in MATLAB. It uses cross-correlation and FFT (Fast Fourier Transforms) to detect shift of the peaks (orders of diffraction pattern) of the diffraction patterns during translation and rotation. The cross correlation of two images f and g in x, y co-ordinate system can be calculated as

$$r_{fg}(x_o, y_o) = \sum_{u,v} F(u,v) G^*(u,v) \exp\left[i2\pi\left(\frac{u\,x_o + v\,y_o}{M}\right)\right] \quad (1)$$

where $F(u,v)$ and $G(u,v)$ are the FFT of two near in-time images (in u, v spatial frequency space), and each of dimension M and (*) denotes complex conjugate. With sub-pixel interpolation, shifts corresponding to $1/N^{th}$ of a pixel can be determined [4]. This accuracy is a result of cross-correlation algorithm applied to the highly correlated images during dithering. This algorithm uses all the information available in the image to compute initial estimate and each point in the up-sampled cross correlation grid, making it robust to noise originating from light scattering and vibration. It obtains an initial estimate of the cross-correlation peak by the FFT and then refines the shift estimation by up-sampling the DFT (discrete Fourier transforms) only in a small neighborhood of that estimate by means of a matrix-multiply DFT. This approach optimally uses all the image points to compute the up-sampled cross-correlation. By calculating the resultant peak shift in pixels (x_o, y_o) and multiplying the pixel shift with appropriate scale factor of the metrology system, the translation of images is computed. Since the grating is placed at a distance 'R' from the center, the corresponding rotation is given by equation (2)

$$\theta_{NORIS} = \cos^{-1}\left(1 - \frac{X^2 + Y^2}{2R^2}\right) \times \frac{180}{\pi} \quad (2)$$

where X and Y are displacements of the diffraction pattern peak (μm) calculated after multiplying the x, y pixel shifts with appropriate, experimentally obtained NORIS scale factors (pixel/μm). To test the accuracy and repeatability of the integrated position measurement and to acquire the imager's scaling factors along each direction of motion owing to the imperfect mounting angles, a commercial nano-scale precision motion stage (nPoint C300) was used. A 3D-printed mount was designed to house the imager and the laser. Figure 2 shows the experimental setup and a schematic explaining the arrangement of laser, imager and the precision reference stage used to obtain the NORIS scale factor (pixel shift/displacement). They are measured to be 0.1647 and 0.2151 pixel/μm for X and Y axis respectively. The up-sampling factor, N, used for sub-pixel registration here is 100. This means that for our setup, the smallest precisely detectable shift theoretically is ~50nm, which

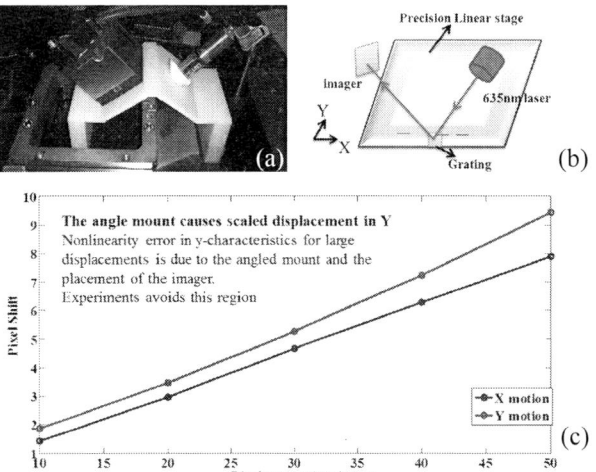

Figure 2: (a) Experimental setup with our metrology system , (b) corresponding schematic showing the laser (Thorlabs CPS180), imager (Thorlabs DCU223M) and grating arrangement, (c) data obtained using a commercial precision linear (X, Y) stage (nPoint C300) for verifying sensitivities of our diffraction optics metrology system.

corresponds to 4.65μm pixel size for the imager used.

OPTICAL MEASUREMENT OF STAGE ROTATION AND GYROSCOPE SIGNAL

To extract the input-output (voltage vs. rate) curve, the gyroscope is excited at different rotation rates while measuring the gyroscope output voltages. For this purpose,

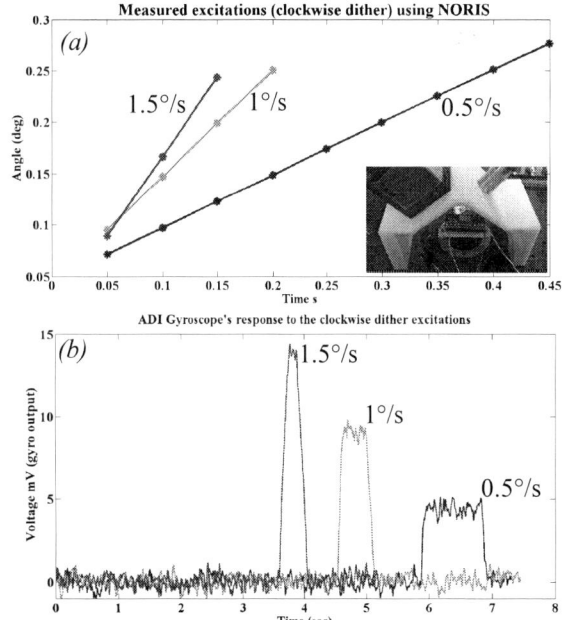

Figure 3(a) NORIS measured excitations (clockwise dither of different rates) (b) Plot for gyroscope voltage response for the corresponding applied clockwise dither excitation. Inset shows the experimental setup.

an off-the-shelf gyroscope (ADXRS646) was mounted on a commercial precision rotary stage (Newport URS50BPPV6) with speed and angle accuracy 0.2 millidegree, shown in the

978-1-4799-3510-9/14 $31.00 © 2014 IEEE   609

inset of the plot in Figure 3.

The motor was programmed to apply different clockwise dither rates around (0.5, 1 & 1.5°/s). For active calibration, the excitation applied to the gyroscope under test has to be known. A diffraction grating was placed on the moving rotor to compute the actual excitation using NORIS that was provided to the gyroscope. The plots for measured applied excitation using our metrology system (rounded to 1st decimal) and the corresponding gyroscope responses sampled at 10 kHz are shown in Figure 3. The noise on output was approximately 1 mV. As per the data sheet, the ARW of the gyroscope under test is 0.01°/√sec and bias instability of 12°/hr.

COMPARISON BETWEEN RATE TABLE AND DOME-DISC CALIBRATION

To compare DOME-DISC system to a known reference and verify the accuracy of calibration, our metrology system with commercial motor, DOME-DISC, was mounted on the center of the rate table as shown in the schematic and setup picture in Figure 4. The test bed was automated to run such that the rate table, which is our reference, is programmed to rotate for different angular rates. Each angular speed is maintained for 4 seconds, which is the time over which the output voltage of the gyro is averaged for the constant (1000ppm accurate based on rate table specs) rate applied. Immediately after the rate table run, DOME-DISC is activated to apply different clockwise dither rates. The corresponding gyroscope output voltages were acquired using NI-DAQ. Simultaneously, the moving diffraction patterns as a result of these dither excitations were recorded at 30 frames/sec. Using these frames, the excitation provided to the gyroscope was computed with image processing algorithms as described before. All the calculations were done for constant speed excitations. Each point on the input-output curve is obtained by averaging the gyro output voltage values over a stable/constant speed region for every excitation to alleviate the effect of the gyroscope noise.

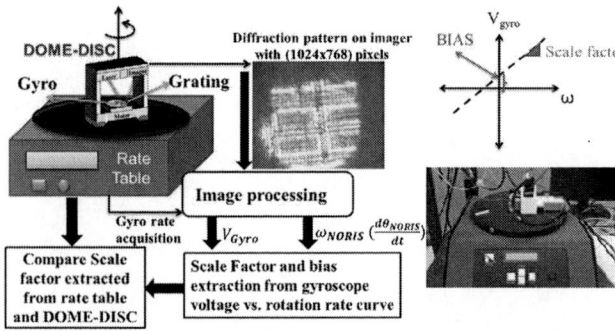

Figure 4: Schematic of the test bed used to calibrate the ADXRS646 gyroscope. Picture in the inset is the automated test bed.

Plots in Figure 5 are the comparison of the input-output curve extracted for this gyroscope. The commercial motor (URS50BPPV6) from Newport was programmed to dither

at different rates to generate the plot for the DOME-DISC system. With the help of the automated experimental setup described above, several runs were performed yielding many input-output curves used to arrive at statistical result presented in Figure 5. The plots shown in Figure 5 are the initial results obtained from 30 experimental runs of both the rate table and DOME-DISC with commercial motor. The error bars indicate the standard deviation of both the applied excitation and the measured gyroscope voltage responses in both of the systems. Upon averaging 30 runs and with linear fit to the obtained data points, 107 ppm match is achieved in scale factor between DOME-DISC estimates and the rate table for a gyroscope whose scale factor fluctuations are ~100,000-ppm due to aging and temperature variations. Furthermore, in our system, the bias variations during calibration can be cancelled, as the position and excitation of the stage on which the gyroscope is mounted is always known to parts-per-million accuracy. Hence, to the first order, any spurious output that result due to bias and many other unknown sources can be subtracted out.

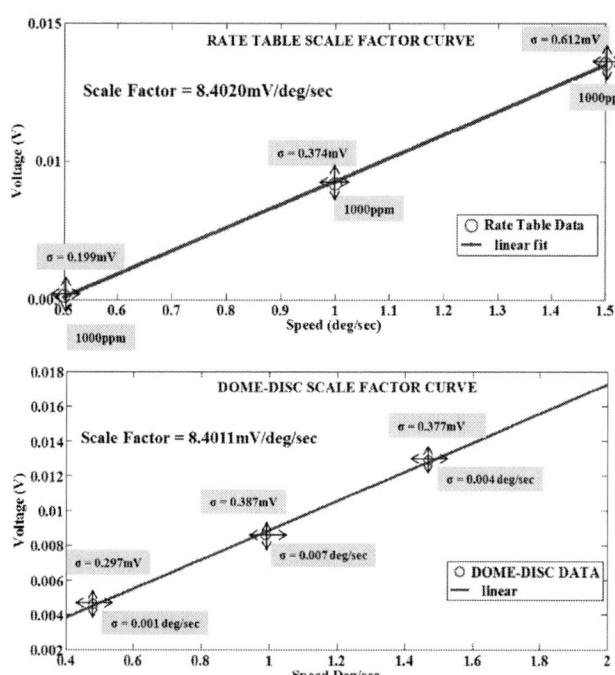

Figure 5: Plots comparing the scale factor of ADXRS646 gyroscope obtained from rate table and DOME-DISC system that has the gyroscope on a commercial motor. The data shown with error bars are the average of 30 experimental runs .The scale factor computed from our system matched within ~107ppm of the scale factor computed from the commercial rate table that is used as a reference.

PIEZOELECTRIC DITHER STAGE

Piezoelectric actuators are good candidates to realize precision calibration stages as they can generate high forces and support high rigidity structures for a given displacement

[5]. Figure 6 shows the fabricated piezoelectric dither stage used for this purpose. PZT (Lead Zirconate Titanate) was used in fabrication due to its higher electromechanical coupling (k_t) and strain coefficient (d_{31}) as compared to many other piezoelectric materials. The dither stage, which uses in-plane lateral actuators [6] as its elemental building block, was fabricated by laser micromachining bulk PZT 4 plates. An ADXRS646 gyroscope die from Analog Devices was mounted on the dither stage to demonstrate a proof-of-concept miniature calibration system. The stage design, which is described in [7], allows in-plane dither excitations necessary for a Z-axis gyroscope calibration and has sensitivity of ~ 1.2 millidegree/V. The fundamental resonant frequency of the dither stage is designed to be away from the resonant frequency of the ADI gyro mounted on it to

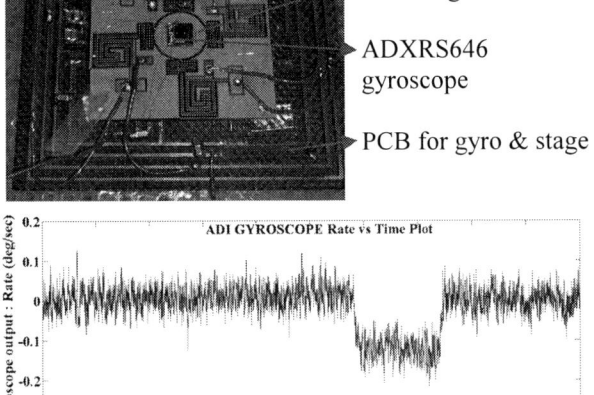

Figure 6: Fabricated PZT dither stage using laser micro machining bulk PZT4 plate. ADI gyroscope (ADXRS646) mounted on the stage. Plot below shows the gyro response for a clock-wise dither for 100V Dc

ensure stable operation during calibration.

Figure 6 illustrates the setup to test the operation of the gyroscope on the PZT stage, which is placed on a PCB with the gyroscope adhesively bonded at the stage's center. The gyroscope is mounted on the stage with the help of laser defined alignment marks for accurate axis alignment (~100μm), and essential gyroscope pins are wire bonded to external PCB circuitry. The bonding process of the gyro to the stage can introduce angular misalignment between the gyro and the stage due to uneven thickness of the adhesive resulting in additional off-axis sensitivity. This problem can be alleviated by automatic application of different off-axis stimulus to measure the sensitivities.

In order to show the basic operation, electrodes of the dither stage was driven with DC ramp like inputs approximating a constant angular rate mechanical stimulus to the gyroscope. The plot in Figure 6 shows the gyroscope response to the clockwise dither of around 0.15 °/sec.

CONCLUSION AND FUTURE WORK

A proof-of-concept result for a miniature inertial sensor calibration platform, which relies on a dithering scheme and a constant-of-nature optical system for metrology, is demonstrated. A MEMS commercial gyroscope from Analog Devices was calibrated in-situ, and the results obtained for the scale factors were compared with a known accurate reference, the rate table. Our result from DOME-DISC matched to within 107-ppm of the scale factor extracted from the rate table data. We are currently working towards further integration and miniaturization of the proposed architecture for a 1-cc implementation of the light source, imager, and the dither stage with gyroscope. An additional goal is to improve the grating design to confine the diffraction patterns within fewer pixels. This can facilitate accurate peak detection at a higher bandwidth and yield better accuracy (≈10ppm accuracy) in scale factor calibrations.

ACKNOWLEDGEMENTS

We would like to thank Defense Advanced Research Projects Agency (DARPA), PASCAL program, for funding this work and Cornell Nano Fabrication Facility (CNF) for testing and characterization equipment.

REFERENCES

[1] Lutwak, Robert. "The chip-scale atomic clock-recent developments." *Frequency Control Symposium, 2009 Joint with the 22nd European Frequency and Time forum. IEEE International*. IEEE, 2009.

[2] Casinovi, G.; Sung, W.K.; Dalal, M.; -Shirazi, A. N.; Ayazi, F., "Electrostatic self-calibration of vibratory gyroscopes," *Micro Electro Mechanical Systems (MEMS), 2012 IEEE 25th International Conference on* , vol., no., pp.559,562, Jan. 29 2012-Feb. 2 2012

[3] Yoshimizu, Norimasa, Amit Lal, and Clifford R. Pollock. "Nanometrology using a quasiperiodic pattern diffraction optical ruler." *Microelectromechanical Systems, Journal of* 19.4 (2010): 865-870.

[4] Guizar-Sicairos, Manuel, Samuel T. Thurman, and James R. Fienup. "Efficient subpixel image registration algorithms." *Optics letters* 33.2 (2008): 156-158.

[5] De-Yuan Zhang, Takahito Ono, Masayoshi Esashi, Piezoactuator-integrated monolithic microstage with six degrees of freedom, Sensors and Actuators A: Physical, Volume 122, Issue 2, 26 August 2005, Pages 301-306, ISSN 0924-4247

[6] Sachin Nadig, Serhan Ardanuç, and Amit Lal "Planar Laser-Micro Machined Bulk PZT Bimorph For In-Plane Actuation" *2013 Joint Ultrasonics, Ferroelectrics and Frequency Control (UFFC), EFTF and PFM symposium, IEEE Transactions on* , July 2013

[7]Sachin Nadig, Serhan Ardanuç, and Amit Lal "Monolithic Piezoelectric In-plane Motion Stage with Low-Cross-Axis-Coupling " *MEMS 2014* (2014).

ELECTRICAL CHARACTERIZATION OF ALD-COATED SILICON DIOXIDE MICRO-HEMISPHERICAL SHELL RESONATORS

Peng Shao, Vahid Tavassoli, Chang-Shun Liu, Logan Sorenson and Farrokh Ayazi
Georgia Institute of Technology, Atlanta, USA

ABSTRACT

This paper reports on electrical characterization of ALD-coated thermally-grown silicon dioxide micro-hemispherical shell resonators (µHSRs) with capacitive electrodes. A high aspect ratio silicon dioxide µHSR with a thickness of 2.6 µm and diameter of 910 µm, uniformly coated with 30 nm of platinum using ALD process, demonstrated Q of 19,100 at 19.17 kHz and 14,300 at 55.2 kHz for m=2 and m=3 wineglass modes, respectively. An optimized isotropic dry etching recipe was developed to create highly symmetric hemispherical molds in (111) silicon substrates, from which the oxide shells were thermally grown. This resulted in a significant improvement of frequency mismatch between m=2 degenerate modes, achieving 21 Hz split as fabricated for m=2 modes of an 8kHz SiO_2 µHSR that is 1240 µm in diameter and 2 µm in thickness. This creates a path for fabricating high Q and highly symmetric hemispherical shell resonators for microscale hemispherical resonator gyroscopes.

INTRODUCTION

Among high performance gyroscopes, hemispherical resonator gyroscope (HRG) is one of the most successful designs. After 14 years of production, the HRG boasts over 12 million operating gyro-hours with 100% mission success [1]. Applications of these gyroscopes include spacecraft stabilization, precision pointing, aircraft navigation, strategic accuracy systems, oil borehole exploration and planetary exploration. It consists of a precision-machined highly symmetric hemispherical shell made of an ultra-high-Q material such as fused quartz, with manually-assembled and fine-tuned electrodes. One reason for the superior performance of this gyroscope is that the intrinsic energy losses in the resonating shell are extremely low. Thermoelastic damping simulations that have been run on this structure yield a Q_{TED} value in excess of 10^{10}. However, the big drawback of this well-developed design is its cost, portability and the difficulty of manufacturing. With the latest micro-fabrication technology and the development of MEMS inertial sensors, there is a possibility that the conversional HRG can be miniaturized down to chip scale.

Implementation of micro-hemispherical shell resonator (µHSR) is the first step towards realization of micro-scale hemispherical resonating gyroscopes (µHRG). By miniaturizing HRG down to chip-scale, both mass and stiffness of the resonator are reduced by multiple orders of magnitude. When compared with other type of MEMS gyroscope such as tuning fork gyroscope [2], the µHSR presented in this paper has substantially smaller mass and stiffness (by ~10X), which makes them more susceptible to a variety of damping mechanisms.

Figure 1: 3D optical image of an ALD-coated oxide µHSR with assembled pillar electrodes.

The high-aspect-ratio silicon dioxide µHSRs are fabricated by growing thermal oxide in dry-etched silicon hemispherical mold. After releasing the structure by removing surround silicon, the µHSR is coated with ultra-thin ALD conductive layer, and followed by electrodes assembly process. The fabrication process of µHSRs was reported at Hilton Head 2012 [3], and an analysis on the effect of silicon dioxide growth on structure symmetry was reported at MEMS 2013 [4]. This paper shows a complete set of testing results on measured quality factors and frequency splits, and analysis on how the performance will change with fabrication and measurement parameters. The µHSR reported here differs from previously published works [5-7] on the molding and releasing technique as well as the structural material. It also shows 10X smaller form factor in volume compared to other shell resonators reported in [5] and [7]. In our approach, well-developed microfabrication technique with high throughput is used with the possibility of integration with co-fabricated electrodes [8].

FREQUENCY DOMAIN MEASUREMENTS

This section details the electrical testing of the resonances of fabricated µHSR that are assembled with excitation electrodes. Three batches of devices, with titanium nitride (TiN) coating (batch #1 & #3) and platinum (Pt) coating (batch #2), respectively, were tested in vacuum (µTorr range) as one-port resonators. A 3D optical image of

978-1-4799-3510-9/14 $31.00 © 2014 IEEE

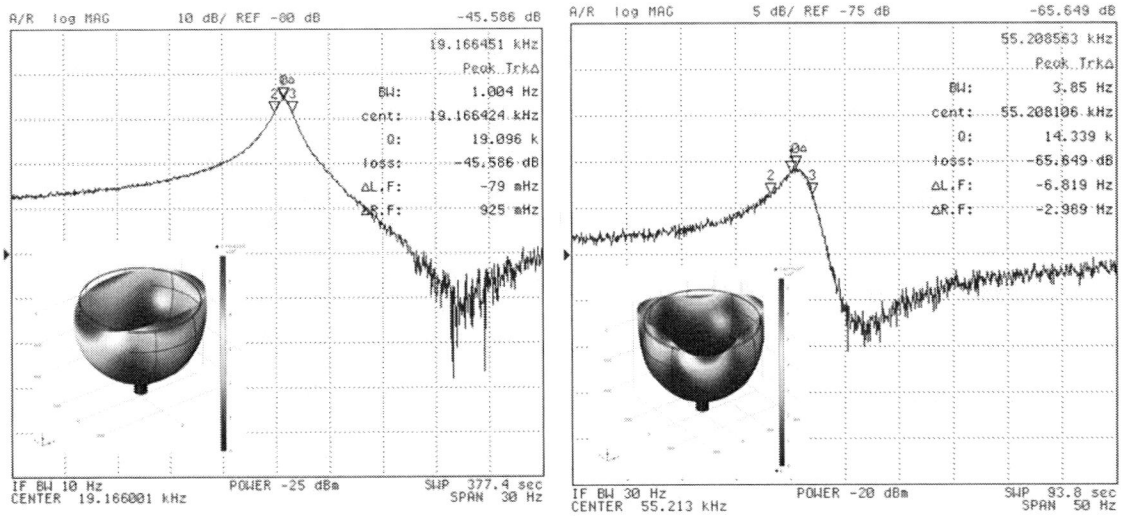

Figure 2: (left) m=2 resonance peak of Pt coated silicon dioxide hemispherical resonator, showing mechanical quality factor of 19.1 k at 19.17 kHz; (right) m=3 resonance peak of Pt coated silicon dioxide hemispherical resonator, showing quality factor of 14.3k at 55.2 kHz. Inset is the mode shape simulated by COMSOL FEA software.

Table 1: Geometric parameters of measured µHSR

Batch #	Coating Material	Shell Diameter (µm)	Shell Thickness (µm)	Support Diameter (µm)
1	TiN	740	1.6	77
2	Pt	910	2.6	82
3	TiN	1240	1.6	90

Table 2: Comparison of resonance frequency by (1), COMSOL FEA and measured results

Mode #	Equation (1) (kHz)	FEA (kHz)	Measured frequency (kHz)	Measured quality factor
Batch #1				
m=2	18.99	19.87	19.97	6,800
m=3	56.22	57.28	56.84	5,600
Batch #2				
m=2	18.84	18.38	19.17	19,100
m=3	55.76	53.22	55.20	14,300
Batch #3				
m=2	6.79	6.67	6.62	5,900
m=3	19.72	20.62	20.51	5,100

fabricated silicon dioxide µHSR is shown in Figure 1. The geometric parameters of these two devices are listed in Table 1. An Agilent 4395A network analyzer supplies AC drive voltage that is combined with DC polarization voltage to one electrode pillar, and a sense current is generated by the change in capacitance across the polarized gaps due to vibration of the shell. A trans-impedance amplifier (TIA) is used to amplify the signal from µHSR.

The resonance frequencies of a hemispherical shell can be estimated by [9]:

$$\omega = \frac{m(m^2-1)}{r^2}\sqrt{\frac{E \cdot I(m,h)}{3(1+v)\rho \cdot J(m,h)}}, \tag{1}$$

where

$$I(m,h) = h^3 \int_{\varphi_0}^{\varphi_F} \frac{\tan^{2m}(\varphi/2)}{\sin^3 \varphi} d\varphi,$$

$$J(m,h) = h \int_{\varphi_0}^{\varphi_F} \left(m^2 + 1 + \sin^2 \varphi + 2m \cos \varphi\right)\tan^{2m}(\varphi/2)d\varphi$$

E is Young's modulus, r is the radius of the shell, h is the shell thickness, φ_0 and φ_F are the boundary angles in spherical coordinates relative to the zenith axis, v is Poisson's ratio, ρ is the material density, and m is the mode number.

By substituting $I(m,h)$ and $J(m,h)$ into (1), we can clearly note that the resonance frequency scales as:

$$\omega \propto \frac{h}{r^2} \tag{2}$$

Eigenfrequency analysis by COMSOL Multiphysics FEA software was also run to predict the frequency of each resonance mode of the µHSRs. The support structure is also included in the FEA simulation to give more accurate prediction of resonance frequency. Table 2 illustrates the calculated results by (1), simulated results by COMSOL FEA, and measured frequency results along with the measured quality factor, showing good agreement between theory and experiment.

Batch #1 and #3 are coated with 30 nm ALD TiN as the conductive layer, while 30 nm ALD Pt is used for batch #2. Batch #2 with Pt coating shows the highest quality factor of 19,100 at 19.17 kHz for m=2 mode and 14,300 at 55.2 kHz for m=3 mode, illustrated in Figure 2. The insets are showing the mode shapes simulated by COMSOL eigenfrequency analysis. Compared with the measured results of batch #1 and batch #3, quality factor is showing a dependency on ALD coating material. The higher Young's modulus mismatch between TiN-SiO$_2$ interface compared to Pt-SiO$_2$ interface is potentially causing additional interfacial

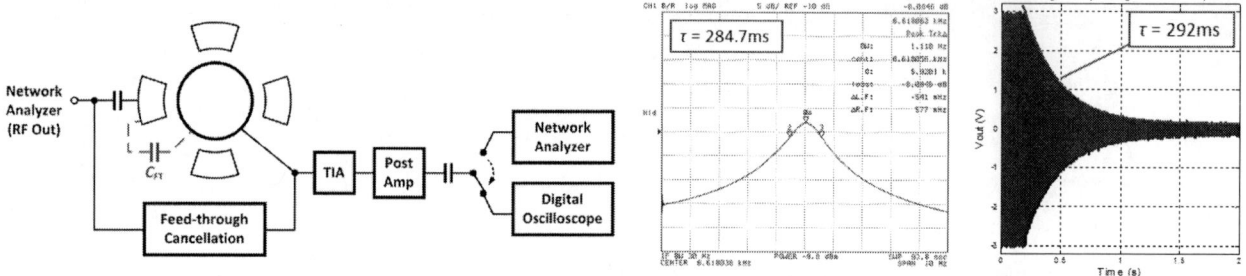

Figure 3: (left) Schematic view of ring down measurement setup; (middle) Resonance peak of a batch #3 device measured in frequency domain; (right) Ring down measurement of the same device, showing good agreement with theory

loss, and a reduced quality factor [10].

TIME DOMAIN MEASUREMENT

One of the operating modes of HRG is rate integrating mode (whole angle mode). In this operating mode, the resonator is excited by a delta function and allowed to precess freely without any further excitation. Due to the way that rate integrating mode works, long ringdown time of the resonator is preferred. Thus, a time domain ringdown test is also performed for µHSR.

Figure 3 illustrates the schematic view of ringdown test setup. A feed-through cancellation circuit is designed and connected in parallel with the input and output of the resonator. Trans-impedance amplifier (TIA) and post amplifier circuit enhance the signal for display on oscilloscope. The resonance peak is firstly measured by network analyzer with feed-through cancellation circuit. By setting frequency span to 0 Hz, the excitation is locked at the resonance frequency. Once the connection is switched from network analyzer to digital oscilloscope, the transient waveform can be recorded. Figure 3 also shows the measured resonance peak by network analyzer and corresponding ringdown measurement. For a µHSR from batch #3, a quality factor of 5,900 at 6.62 kHz is measured at frequency domain, while ring down time is measured to be 292 ms.

ENVIRONMENTAL BEHAVIOR

In order to study the loss mechanism on this device, quality factor is measured at various environment conditions. Figure 4 illustrates the measured quality factor of batch #2 device at different vacuum level. At sub-mTorr range, the quality factor does not show significant dependency on chamber pressure, while air damping will dominate when chamber pressure is above 1 mTorr. Therefore, the quality factor is not limited by air damping currently. For resonators that operated at Knudsen region, the air damping Q is inversely proportional to the air pressure. The proportional constant is extracted by the data at 30 mTorr (data point at highest pressure), and the air damping Q is plotted as the dashed line in Figure 4. This curve predicts that if other loss mechanism are all eliminated, vacuum level of 100 µTorr or better is needed to reach Qs in excess of one million.

The temperature behavior of the SiO_2 µHSR is measured to understand how resonance frequency and quality factor change over temperature. Vacuum chamber with µHSR inside is placed in a temperature oven with both heater and compressor. Measurement is done from 10°C to 90°C for batch #2. Four hours stabilization time was given between two temperature points. As shown in Figure 5, a linear and positive temperature coefficient of frequency (TCF) of 61.9 ppm/°C is extracted by linear regression of

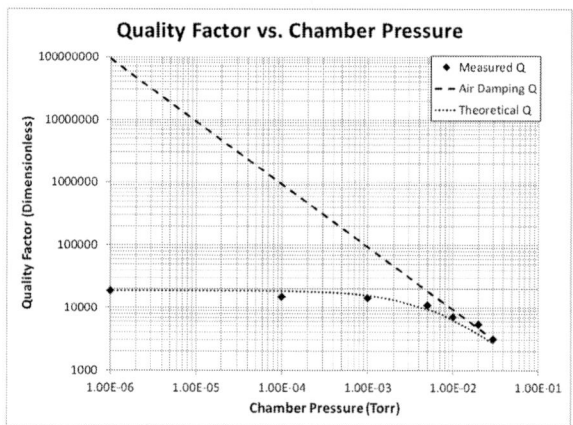

Figure 4: Measured quality factor of batch #2 at various air pressure, calculate air damping Q based on 1/p trend

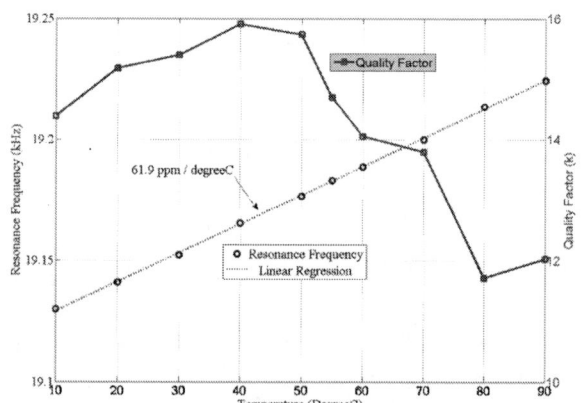

Figure 5: Temperature behavior measurement of batch #2. (blue left y-axis) Linear positive TCF of 61.9 ppm/°C is extraced; (green right y-axis) Quality factor at various tempertures

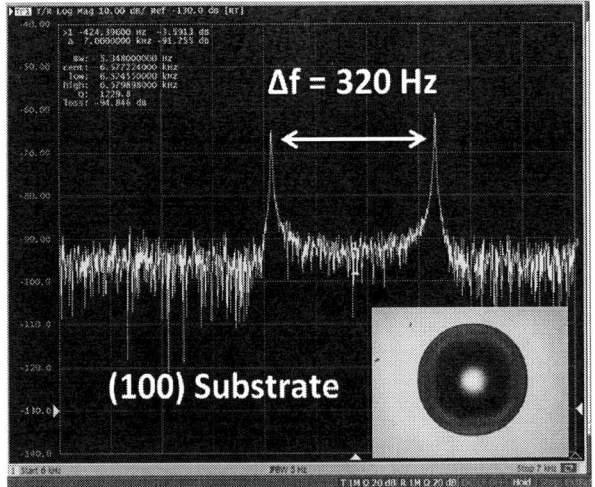

Figure 6: Frequency split between m=2 degenerate modes of μHSR fabricated from (100) silicon substrate, shows 320 Hz splits out of 6.5 kHz

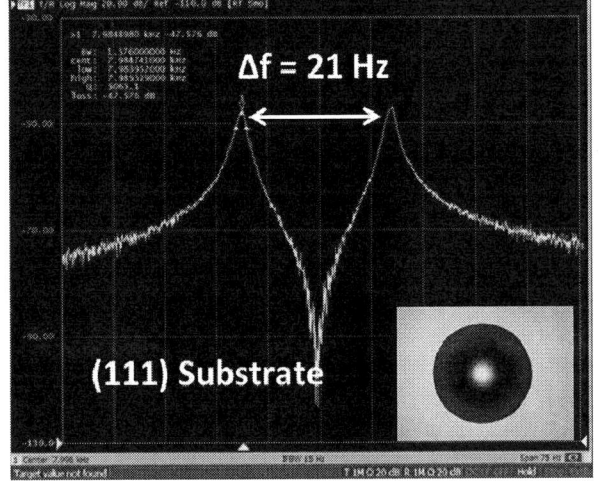

Figure 7: Frequency split between m=2 degenerate modes of μHSR fabricated from (111) silicon substrate, shows 21 Hz splits out of 8 kHz.

the resonance frequencies at different temperatures. The relatively smaller TCF value compared to a silicon dioxide resonator is due to the loading of negative TCF of Pt coating. Theoretical calculation shows a TCF of 59.8 ppm/°C, which matches the experimental result. Quality factor also shows an inverse trend at temperatures above 40°C.

FREQUENCY MISMATCH

For axis-symmetric gyroscopes, frequency mismatch between two degenerate fundamental modes is a critical performance. This mismatch reflects the level of symmetry of the structure as well as the difficulty in order for mode matching. Single crystal silicon wafer (100) and (111) are used as the starting substrates. Known as an anisotropic material with four fold symmetry, (100) silicon wafer shows more crystalline dependence in hemispherical molding process, while (111) silicon wafer shows an isotropic

pattern. Under optical microscope, the reflected light from hemispherical mold of (100) substrate shows a square-shape pattern, which significantly affects frequency split between m=2 degenerate modes. Figure 6 and Figure 7 show the measured frequency splits of μHSR fabricated from (100) substrate and (111) substrate respectively. Due to the square-shape pattern of (100) substrate, it is showing 320 Hz split out of 6.5 kHz (Δf/f = 4.9%). μHSR fabricated from (111) substrate demonstrates frequency split of 21 Hz out of 8 kHz (Δf/f = 0.26%), improving the frequency split by more than an order of magnitude.

CONCLUSION

Electrical characterization of ALD-coated silicon dioxide μHSR is reported, showing quality factors up to 19,100 and ring down time of 292 ms. Quality factor at various vacuum level and temperature is also studied. By optimizing the hemispherical molding process, frequency mismatch between m=2 is reduced to 21 Hz.

ACKNOWLEDGEMENTS

This work was supported by the DARPA Microsystems Technology Office, Microscale Rate Integrating Gyroscope (MRIG) program under contract #HR0011-00-C-0032 led by Northrop Grumman. The authors would like to thank the cleanroom staff at Georgia Tech's Institute for Electronics and Nanotechnology for processing assistance.

REFERENCES

[1] D.M. Rozelle, 19th AAS/AIAA Space Flight Mechanics Meeting, 2009, pp. 1157-1178.

[2] M.F. Zaman, Journal of Microelectromechanical Systems, 17 (2008) 1526-1536.

[3] P. Shao, Tech. Digest Solid-State Sensors, Actuators, and Microsystems Workshop, Hilton Head, SC, 2012, pp. 275-278.

[4] L.D. Sorenson, IEEE International Conference on Micro Electro Mechanical Systems (MEMS 2013), 2013, pp. 169-172.

[5] J. Cho, IEEE International Conference on Micro Electro Mechanical Systems (MEMS 2013), 2013, pp. 177-180.

[6] A. Heidari, IEEE International Conference on Solid-State Sensors, Actuators and Microsystems (TRANSDUCERS 2013), 2013, pp. 2415-2418.

[7] D. Senkal, IEEE International Conference on Micro Electro Mechanical Systems (MEMS 2013), 2013, pp. 469-472.

[8] L.D. Sorenson, IEEE International Conference on Micro Electro Mechanical Systems (MEMS 2012), 2012, pp. 168-171.

[9] S.-c. Fan, Applied Mathematics and Mechanics, 12 (1991) 1023-1030.

[10] A. Frangi, Sensors and Actuators A: Physical, (2012).

CONTACT

*Peng Shao, tel: +1-404-9885782; pengshao@gatech.edu

FABRICATION AND DEGRADATION CHARACTERISTIC OF SPUTTERED IRIDIUM OXIDE NEURAL MICROELECTRODES FOR FES APPLICATION

Xiao-Yang Kang[1,2,3], Jing-Quan Liu[1,2,3], Hong-Chang Tian[1,2,3], Jing-Cheng Du[1,2,3], Bin Yang[1,2,3], Hong-Ying Zhu[1,2,3], Yanna NuLi[4] and Chun-Sheng Yang[1,2,3]

[1]National Key Laboratory of Science and Technology on Micro/Nano Fabrication
[2]Key Laboratory for Thin Film and Microfabrication of the Ministry of Education
[3]Institute of Micro-Nano Science and Technology, Shanghai Jiao Tong University, Shanghai, PR China
[4]Department of Chemical Engineering, Shanghai Jiao Tong University, Shanghai, PR China

ABSTRACT

This paper shows the fabrication process of the reactively sputtered iridium oxide film (SIROF) microelectrodes under different oxygen flows and characters the electrochemical performances of the iridium oxide neural microelectrodes which are suffered from stimulus-evoked degradation. The SIROF microelectrodes prepared under 25 sccm oxygen flow shows the least degradation from continuous electrical stimulation (two million phases). That the charge storage capacity is only decreased by 9.6 % and the 1 kHz impedance is only increased by 4.23 %. Hence, the 25 sccm one can be an ideal microelectrode modification material for electrical stimulation with the least degradation.

INTRODUCTION

MEMS based microelectrodes offer higher spatial resolution for electrophysiological recordings and stimulation, compared to the conventional macro electrodes reported at MEMS 2013 [1]. Although iridium oxide modified microelectrodes have been widely used in neural recording and stimulating due to its good stability and large charge storage capacity (CSC) [2, 3], damage or degradation from continuous electrical stimulation is still problematic [4,

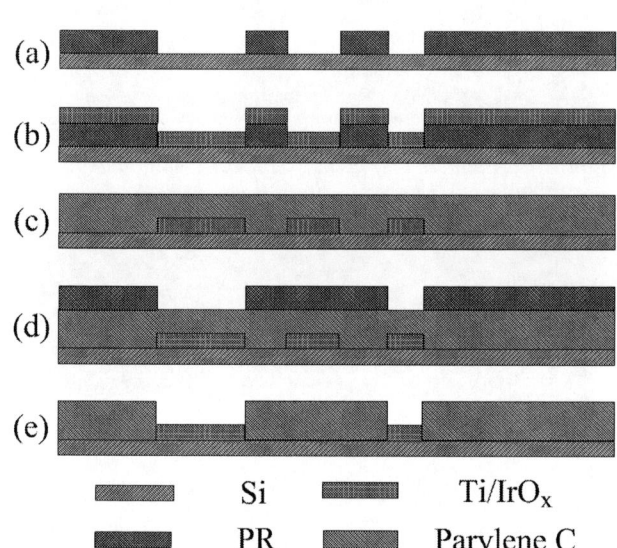

Figure 1: Fabrication process of the iridium oxide microelectrodes. (a) Photoresist spin coating and patterning, (b) Ti/ IrOₓ layer sputtering, (c) Lift-off and CVD Parylene C, (d) Photoresist spin coating and patterning, (e) RIE open microelectrode site.

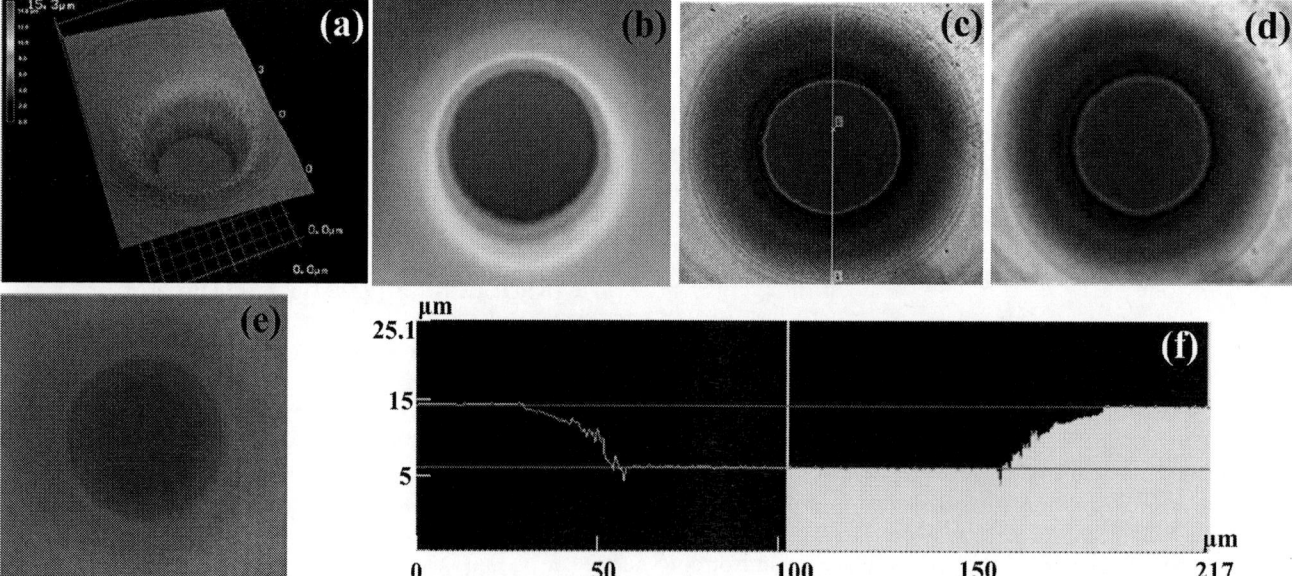

Figure 2: Laser scanning microscopy analysis of the fabricated microelectrode site. (a) 3D depth profile, (b) 2D depth profile, (c) Compound figure of the microelectrode site, (d) Laser figure of the microelectrode site, (e) Optical figure of the microelectrode site, (f) The depth profile of the scanning line in (c).

978-1-4799-3510-9/14 $31.00 © 2014 IEEE

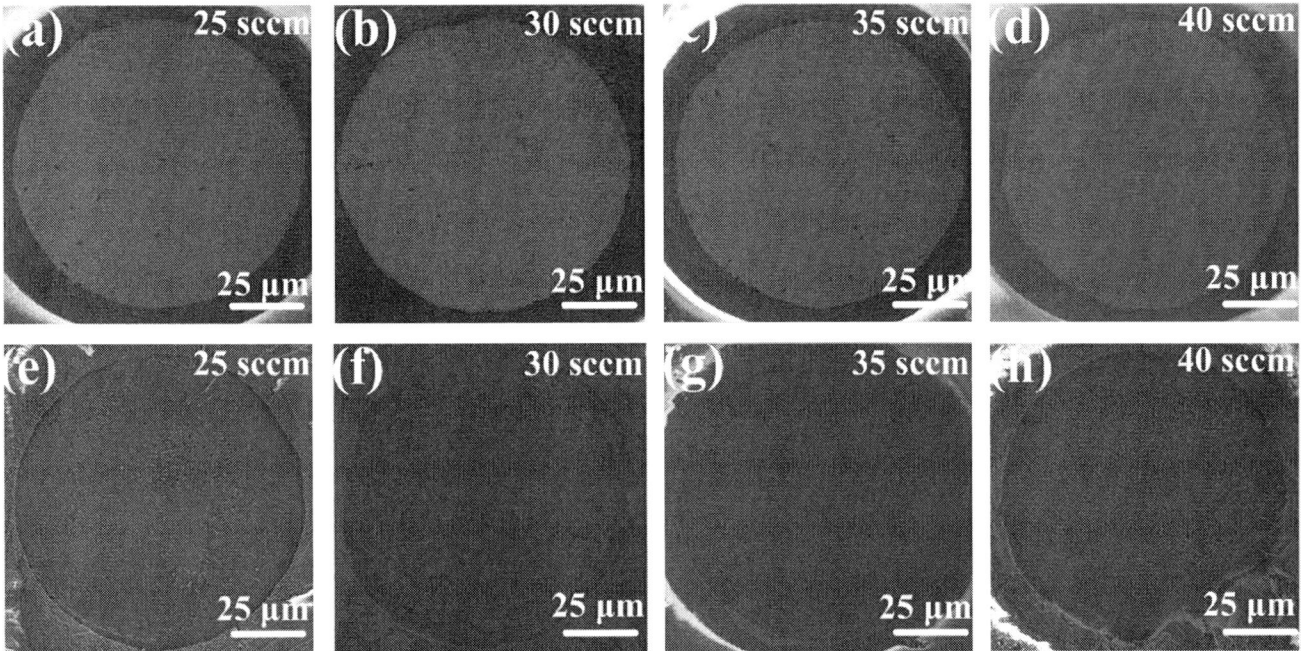

Figure 3: SEM micrographs showing the microelectrode sites of the SIROF stimulated before and after with 500 nC per phase having charge density of 6.369 mC/cm² per phase (two million phases). (a, e) 25 sccm, (b, f) 30 sccm, (c, g) 35 sccm, (d, h) 40 sccm.

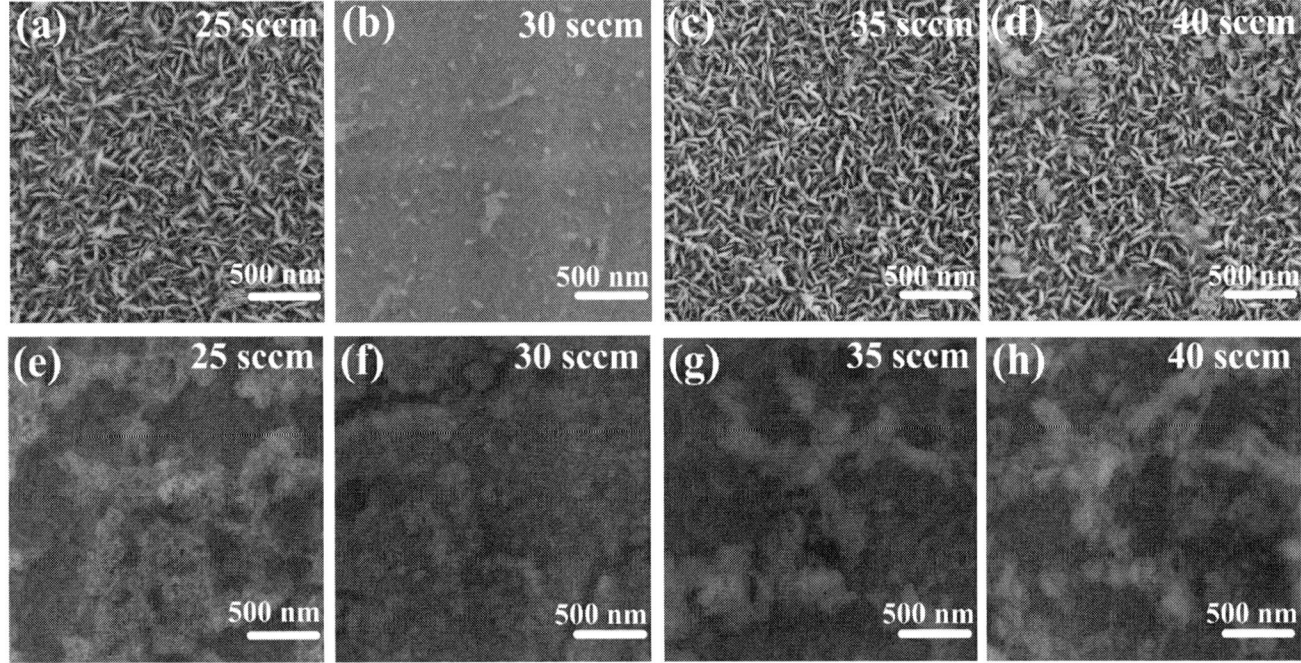

Figure 4: The high resolution SEM micrographs of the SIROF (prepared under different oxygen flow) stimulated before and after with 500 nC per phase having charge density of 6.369 mC/cm² per phase (two million phases). (a, e) 25 sccm, (b, f) 30 sccm, (c, g) 35 sccm, (d, h) 40 sccm.

5]. Seeking the least degradation of the reactively sputtered iridium oxide film (SIROF) is of key impact on the functional electrical stimulation (FES) [6]. Here, we have fabricated SIROF microelectrodes under different oxygen flows and the stimulus-evoked degradation properties are tested.

FABRICATION PROCESS OF THE

IRIDIUM OXIDE MICROELECTRODES

Figure 1 shows the fabrication process of the iridium oxide microelectrodes. Firstly, a 5 μm thick positive photoresist (PR) layer was spun on the silicon wafer (3 inches in diameter, 100 crystal orientation) and patterned by the UV exposure and developing (figure 1(a)). Then a Ti adhesive layer and iridium oxide (IrOx) thin film were sputtered (figure 1(b)). The Ti/IrOx layer are 50/300 nm thick. After that, they were patterned by a lift off process and a 10 μm thick Parylene-C film was deposited on the silicon wafer by a Parylene Deposition System (PDS-2010, Samco, USA) (figure 1(c)). A 10μm thick PR was spun, patterned and stored at 65 °C for 1 h in a vacuum baking oven (figure 1(d)). At last, the microelectrode site and bonding pad were opened by a reactive ion etching system (Nextral 100, Reactive Ion Etcher, Alcatel, France) (figure 1(e)).

LASER SCANNING MICROSCOPY ANALYSIS OF MICROELECTRODE SITE

Laser scanning microscopy analysis of the fabricated microelectrode site is shown in Figure 2. The 3D depth profile (figure 2(a)) and 2D depth profile (figure 2(b)) indicate the well-defined hole with 100 μm in diameter and 10 μm in depth. The iridium oxide microelectrodes have geometric surface area of 7850 μm². From the figure 2(c-e), we could see the SIROF is black or blue-black in color. The

delineation of the boundary between the SIROF and the Parylene C is clear. The depth profile (figure 2(f)) of the scanning line shows the sloped sidewall is about 45 degree. The planar MEMS based iridium oxide neural microelectrodes are successfully fabricated. Although the sidewall is not vertical, the electrochemical performances of the SIROF microelectrodes are not affected.

MICROSTRUCTURE AND DEGRADATION CHARACTERISTIC OF SIROF

The effect of the oxygen flow, ranging from 25 to 40 sccm using a mixture of Ar and O_2 (1:2.5 to 1:4) on the microstructure of the IrO_x films are characterized. The degradation from continuous electrical stimulation is conducted by applied two million phases' biphasic, symmetric square current pulses (500 μA pulse amplitude, 1 ms pulse width). Figure 3 shows the SEM micrographs of the SIROF microelectrode sites stimulated before and after with two million current pulses. On the micrometer scale, all the

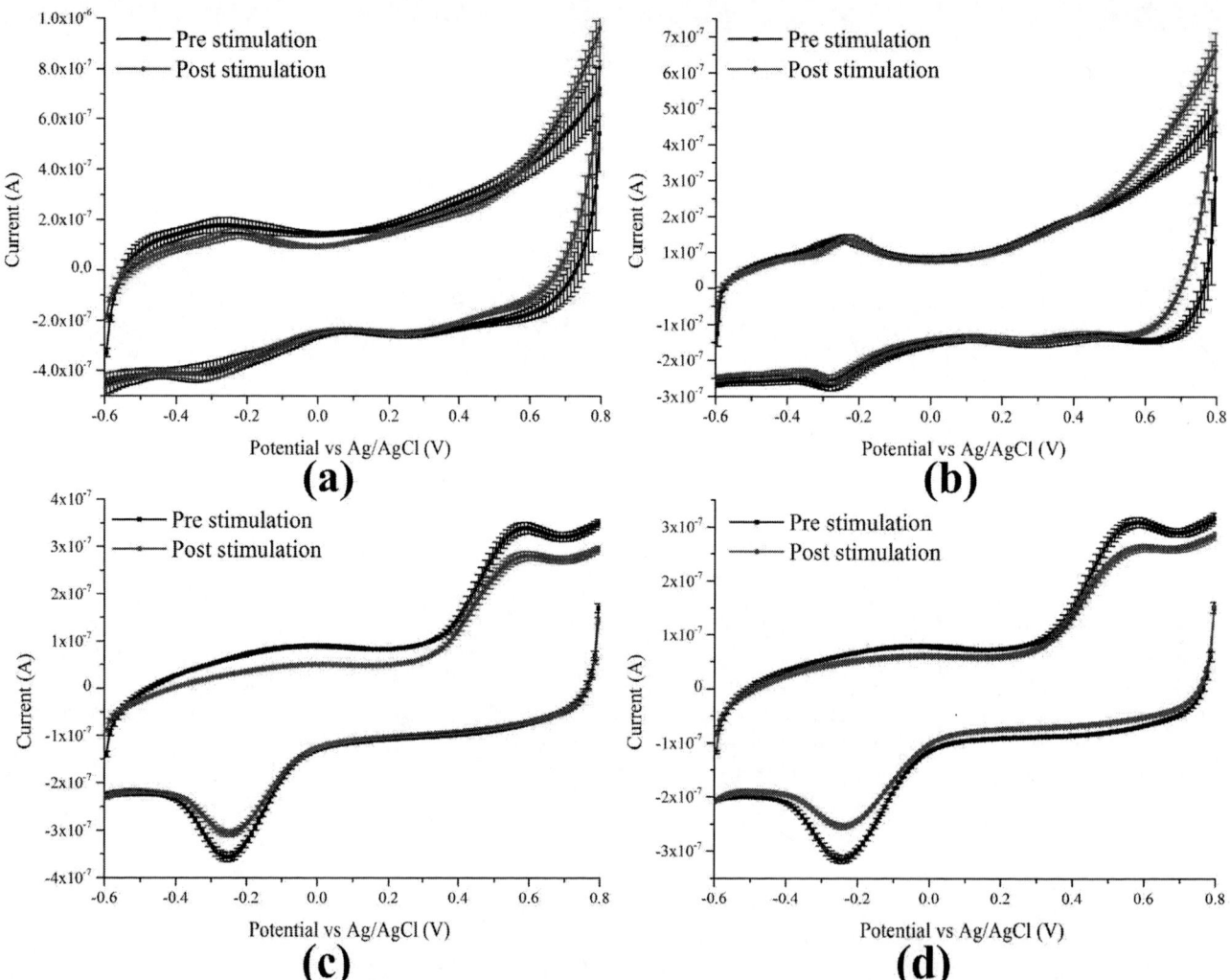

Figure 5: The cyclic voltammograms of the SIROF microelectrodes (n=5, prepared under different oxygen flow) stimulated before and after with 500 nC per phase having charge density of 6.369 mC/cm² per phase (two million phases). (a) 25 sccm, (b) 30 sccm, (c) 35 sccm, (d) 40 sccm.

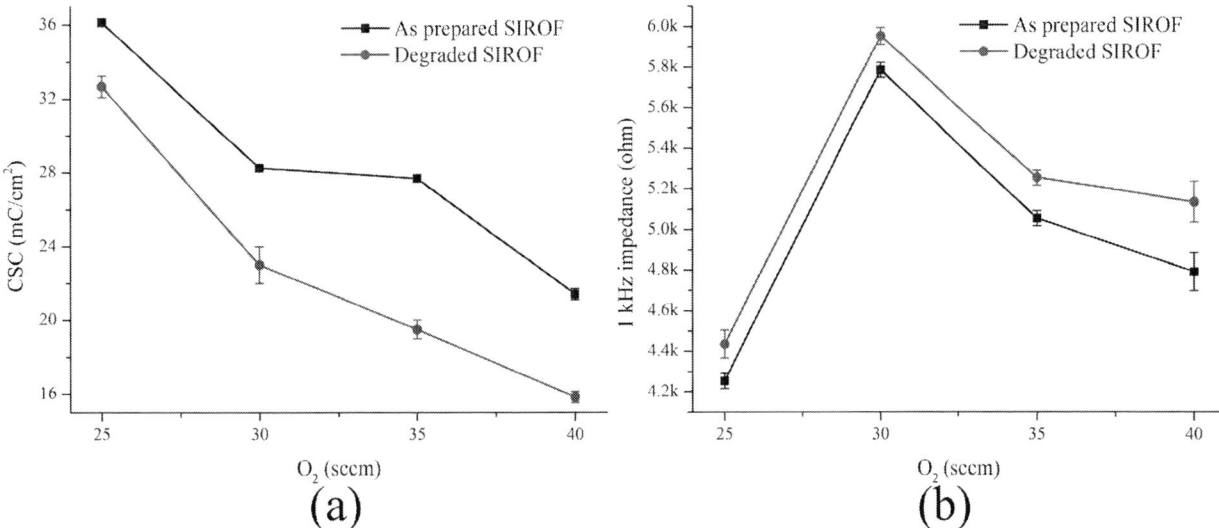

(a)   (b)

Figure 6: The electrochemical performances of the SIROF microelectrodes (n=5) stimulated before and after with 500 nC per phase having charge density of 6.369 mC/cm² per phase (two million phases). (a) The charge storage capacity (CSC), (b) The 1 kHz impedance (ohm).

tested samples remain the circular shape after degradation. The SIROF prepared under 25, 30 and 40 sccm oxygen flows exhibit clear changes in morphology. These are loose aggregates of a large number of primary particles of crystalline or grains. The 35 sccm one does not change in this scale. To investigate the surface microstructure of the SIROF stimulated before and after the degradation conditions, the high resolution SEM micrographs are shown in Figure 4. Figure 4(a) is characterized by a dendritic surface at 25 sccm oxygen flow. Although a dendritic surface microstructure is developed, the film obtains good adhesion property to the substrate. The 30 sccm one shows some big grains on the tiny crystalline. At 35 and 40 sccm oxygen flows, a dendritic surface with increasing aggregation is produced as shown in figure 4(c-d) which loses coherence and is mechanically unstable. After degradation from continuous electrical stimulation, the 25 sccm one shows the best structural porosity, which is more easily permeable for water and ionic species, resulting in a higher charge delivery.

The cyclic voltammograms of the SIROF microelectrodes stimulated before and after with 500 nC per phase (two million phases) are shown in Figure 5 (n=5). They all remain the similar shapes between the pre and post stimulation. Based on the electrochemical performances of the SIROF microelectrodes shown in Figure 6, the SIROF microelectrodes prepared under 25 sccm oxygen flow shows the least degradation from continuous electrical stimulation. That the CSC is only 9.6 % lost and the 1 kHz impedance is only 4.23% increase. Hence, the 25 sccm one can be an ideal microelectrode modification material for electrical stimulation with the least degradation.

CONCLUSIONS

Neural microelectrodes based on SIROF have been fabricated under different oxygen flows ranging from 25 to 40 sccm using a mixture of Ar and O_2 (1:2.5 to 1:4). Two

million phases' biphasic, symmetric square current pulses (500 μA pulse amplitude, 1 ms pulse width) are used for degradation. The SIROF microelectrodes prepared under 25 sccm oxygen flow shows the least degradation that the charge storage capacity is only decreased by 9.6 % and the 1 kHz impedance is only increased by 4.23 %. Hence, the 25 sccm one can be an ideal microelectrode modification material for electrical stimulation with the least degradation.

ACKNOWLEDGEMENTS

The authors thank to financial support from the National Natural Science Foundation of China (Nos. 51035005 and 61076104), 973 Program (2013CB329401), Shanghai Municipal Science and Technology Commission (Nos. 13511500200 and 11JC1405700), National Defense Pre-Research Foundation of China (Nos. 9140A 26060313JW3385), WUXI-SJTU project (2011JDZX017). The authors are also grateful to the collcagues for their essential contribution to this work.

REFERENCES

[1] T. Imashioya et al., *IEEE MEMS 2013*, pp. 365-368.
[2] S. F. Cogan, *Annu. Rev. Biomed. Eng.*, 10 (2008), pp. 275-309.
[3] X. Kang et al., *Sens. Actuators B: Chem.*, 190 (2014), pp.601-611.
[4] A. v. Ooyen et al., *J. Micromech. Microeng.*, 19 (2009), 074009.
[5] B. Wessling et al., *J. Electrochem. Soc.*, 154 (2007), pp. F83–F90
[6] A. A. Fomani et al., *J. Microelectromech. Syst.*, 20 (2011), pp.1109-1118.

CONTACT

*Jingquan Liu, tel: +86-021-34207209; jqliu@sjtu.edu.cn

FABRICATION AND TESTING OF PIEZOELECTRIC HYBRID PAPER FOR MEMS APPLICATIONS

Suresha K. Mahadeva, Konrad Walus, and Boris Stoeber
The University of British Columbia, Vancouver, BC V6T 1Z4, Canada

ABSTRACT

We have developed and demonstrated a new inexpensive and environmentally friendly functional paper based material that can be used as a piezoelectric substrate for sensing applications. The process involves embedding nanostructured barium titanate ($BaTiO_3$) into a stable matrix of wood fibers via fiber functionalization. This is achieved by employing a layer-by-layer approach, and results in the creation of a positively charged surface on the wood fibers. The treated wood fibers are then immersed in a $BaTiO_3$ suspension, leading to the electrostatic binding of the $BaTiO_3$ nanoparticles. We have investigated hybrid paper samples with five different $BaTiO_3$ concentrations (8 -48 wt%), and we have found the highest piezoelectric coefficient at 48 wt% $BaTiO_3$. Our study suggests that functionalizing wood cellulose fibers with nanostructured $BaTiO_3$ is a promising approach to enhancing the functionality and value of paper for developing low-cost microelectromechanical systems (MEMS).

INTRODUCTION

The MEMS industry has witnessed an enormous growth in recent years and many MEMS devices have been developed with applications in security, logistics, diagnostics, etc. [1]. However, most commercial MEMS devices are based on silicon substrates and are produced using conventional microfabrication techniques. Microfabrication processing requires cleanroom facilities and high-maintenance equipment, and it is time consuming and expensive. In addition, many applications require flexibility which limits the applications of silicon based MEMS devices. Although there have been efforts to utilize polymeric material [2], the material and process cost is still high due to high temperature processing and use of chemical solvents. Mainly driven by the potential lower cost and environmental benefits, significant research efforts have recently focused on using paper for sensing applications.

Several research articles have reported on piezoelectric zinc oxide (ZnO) paper that use paper merely as substrates. Most of the work relies on hydrothermal synthesis of ZnO nanostructures on these paper substrates. Gullapalli *et. al.* [3] have reported a technique for growing piezoelectric ZnO nanostructures on commercial printing paper by hydrothermal synthesis at low temperature and they have demonstrated its strain sensing capability. More recently, Costa *et al.* [4] have directly grown nanostructured ZnO on photographic paper and bacterial cellulose substrates without any surface modification. Ko and Yu [5] recently grew ZnO nanorod arrays on cellulose fibers via a hydrothermal method and demonstrated the material for an energy harvesting application. In these cases, ZnO nanostructures were hydrothermally grown only on the paper fibers on the surface of the sheet, which limits the piezoelectric properties of the resulting composite and the performance of the resulting devices. Furthermore, this process cannot be integrated into the existing paper making process to produce piezoelectric paper. The fabrication process can be expensive and time consuming as it requires several hours for ZnO growth. Our proposed process for fabricating piezoelectric hybrid paper is very simple, inexpensive and can be scaled up to industrial levels. The process involves anchoring nanostructured piezoelectric material to wood fibers via fiber functionalization, which is achieved with a layer-by-layer approach.

MATERIALS AND METHODS

Poly(diallyldimethylammonium chloride) (PDDA, 20 wt% in water, MW 100,000- 200,000) and poly(sodium 4-styrenesulfonate) (PSS, MW 70,000) were purchased from Sigma Aldrich, Sodium chloride (ACS certified) and tetragonal $BaTiO_3$ nanoparticles with particles size of 300 nm (99.9% purity) were purchased from Fisher and US Research Nanomaterials Inc., respectively. All the chemicals were used as received. Wood fibers from bleached softwood pulp (Confor Prince George Mills) were obtained after disintegration and washing with distilled water.

Hybrid Paper Preparation

Piezoelectric hybrid paper was prepared by fiber functionalization using a layer-by-layer (LbL) approach. In brief, the two aqueous solutions PDDA (1 % wt/v in 0.5 M NaCl) and PSS (1 % wt/v in 0.5 M NaCl) were prepared first. Then, wood fibers were immersed in the aqueous solution of PDDA (+) and then in PSS (-) and once again in PDDA (+), resulting in the creation of a positively charged surface on the wood fibers. After each immersion step (20 min), the wood fibers were rinsed with deionised water to remove excess and poorly adsorbed polyelectrolyte. The treated wood fibers were then immersed in a $BaTiO_3$ suspension, leading to the electrostatic binding of the negatively surface charged $BaTiO_3$ to the fibres. Finally, paper hand sheets were made according to the TAPPI method T-205 [6]. The resulting hybrid paper is then corona poled at 120 °C for 4 hrs with a needle voltage of 17 kV and a grid voltage of 5 kV to render it piezoelectric. A detailed description of the corona poling setup and procedure can be found in [7].

Hybrid Paper Characterization

Fourier transform infrared (FT-IR) spectra were recorded in the range of wavenumbers from 600 to 3800 cm^{-1} with a resolution of 4 cm^{-1}. The X-ray diffraction (XRD) patterns were recorded with an X-ray diffractometer

(Bruker D8 Advance) using a CuKα target at 40 kV and 50 mA, at a scanning rate of $0.015°$ /minute. The diffraction angle ranged from 10 to $70°$. Attachment of $BaTiO_3$ nanoparticles to the wood fiber is observed using FEI Technai G2 transmission electron microscopy (TEM). Zeta potential measurements were performed on the wood fibers and on the BaTiO3 nanoparticles under different conditions using a Zeta Sizer Nano Series (Malvern) instrument.

Sensitivity of piezoelectric paper to change in pressure and impact energy was measured using a charge meter (Type 5015 from Inter Technology Inc.), which was connected to a paper device.

Table 1: Zeta potential of wood fibers and $BaTiO_3$ Nanoparticles under different states.

Material	Zeta potential
Pristine wood fibers	-21.1 ± 3.99 mV
Pristine BaTiO3 nanoparticles	-23.5 ± 4.13 mV
Functionalized wood fibers	$+45.2 \pm 5.19$ mV
BaTiO3 NPs attached wood fibers	-21.8 ± 6.08 mV

RESULTS AND DISCUSSIONS
Hybrid Paper Fabrication and its Characterization

In our approach, hybrid piezoelectric paper was prepared by directly functionalizing nanostructured $BaTiO_3$ onto the fibers prior to forming the final paper sheet. The wood fibers are chemically modified since ionisable moieties such as carboxyl (1646 cm^{-1}) and hydroxyl (3328 cm^{-1}) groups [8] are expressed on the surface and evidenced by FT-IR (Figure 1). These expose negatively charged terminations in aqueous medium that are comparable to the intrinsic negative termination of the $BaTiO_3$. This is supported through measurements of the Zeta potential which was found for pristine cellulose fibers and $BaTiO_3$ to be -21.1 ± 3.99 mV and -23.5 ± 4.13 mV, respectively, as shown in Table 1.

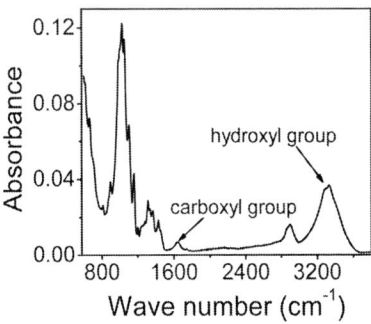

Figure 1: FT-IR scan of wood cellulose fibers.

Hence, the attachment of nanostructured $BaTiO_3$ onto wood fiber was facilitated through the creation of a positive polyelectrolyte interlayer, which results from the alternative deposition of PDDA (+), PSS (-), and PDDA (+) monolayers. This treatment leads to formation of positively charged surface on the wood fibers and ensures its

morphological homogeneity. Upon immersing the thus treated wood fibers in a $BaTiO_3$ suspension, $BaTiO_3$ nanoparticles strongly attached to the wood fibers due to the strong electrostatic interaction between the particles and fibers.

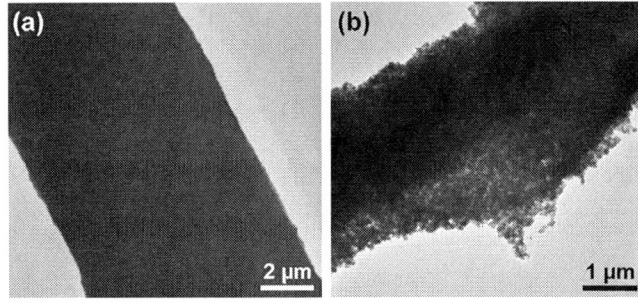

Figure 2: View of (a) pristine and (b) $BaTiO_3$ functionalized wood cellulose fiber under transmission electron microscopy.

Changes in the Zeta potential of functionalized fibers before and after activation in the suspension of $BaTiO_3$ (refer to Table 1) and TEM observations (Figure 2) suggest the attachment of $BaTiO_3$ onto wood fiber. Figure 3 shows the wide angle XRD pattern obtained from hybrid paper. The XRD pattern of the paper exhibits reflections at $2\theta = 15.6°$, $22.8°$ corresponding to (110) and (200) planes of cellulose I structure [9], and the remaining peaks at $2\theta = 22.12°$, $31.52°$, $38.76°$, $55.97°$ and $65.8°$ correspond to $BaTiO_3$ [10], while the two Bragg peaks located at $44.8°$ and $45.4°$ correspond to the (002) and (200) planes of the tetragonal phase of $BaTiO_3$ [11]. Comparison of the XRD patterns of wood fibers, before and after embedding nanoparticles (data not shown) revealed no change in intensity or position of the peaks corresponding to cellulose, suggesting that the crystal structure of the wood fibers remained intact.

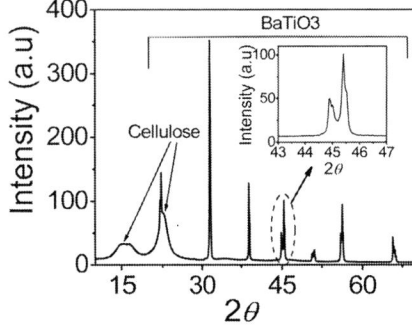

Figure 3: X-ray diffractogram of a hybrid paper hand sheet.

Piezoelectric Behavior of Hybrid Paper

The piezoelectric response of hybrid paper was assessed by measuring the charge induced on the electrodes due to the application of a load. A sample 6 mm × 30 mm in size was subjected to compressive loads of 2 N to 7 N in steps of 1 N, and the corresponding output charge was measured. The piezoelectric charge generated by the hybrid paper under such loading is shown in Figure 4. The hybrid paper produced a piezoelectric charge of 9.5 pC when loaded at 2 N and increased to 30 pC, when the force on the paper was increased to 7 N. Further, the piezoelectric coefficient d_{33} of the hybrid paper samples was determined. We investigated five different hybrid paper samples having different $BaTiO_3$ concentration (8, 10, 26, 42, and 48 wt.%). The amplitude of the charge generated from paper and its piezoelectric coefficient d_{33} tends to increase with nanoparticle concentration and at 48 wt.% $BaTiO_3$ loading, the hybrid paper showed the highest d_{33} of 4.8±0.4 pC/N. The $BaTiO_3$ nanoparticles generate a piezoelectric potential under the action of external stress. The hybrid paper exhibited a small piezoelectric coefficient compared to the large piezoelectric coefficient of bulk $BaTiO_3$ (d_{33} = 75- 190 pC/N) [12]. The smaller d_{33} of hybrid paper is due to different conformation of $BaTiO_3$ crystals in the wood fibers and these crystals may be stressed in numerous different modes when a piece of paper is loaded mechanically and also due to damping effect.

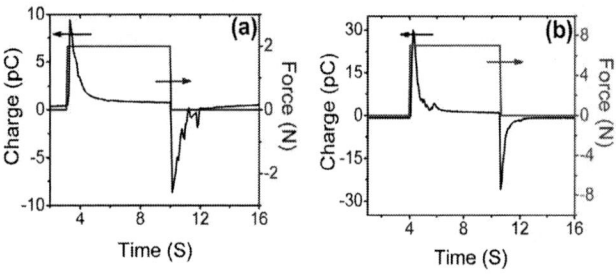

Figure 4: Typical charge response of hybrid smart paper in response to an applied force (a) 2 N and (b) 7 N.

In order to study the response of hybrid paper to a change in pressure and impact energy, devices were fabricated by depositing electrodes (conductive copper tape) on both sides of the hybrid paper (Figure 5). Pressure on the device was applied using Rheometer (Anton Paar), while the device is subjected to an impact by dropping an object having weight of 0.036 kg from defined height. The devices were connected to a charge meter and the piezoelectric charge induced due to application of pressure and impact energy was measured as a function of pressure and impact to characterize its sensitivity.

Figure 6 illustrates the pressure sensing characteristics of the hybrid paper as a function of $BaTiO_3$ concentration. From the pressure sensing curve, it can be seen that the piezoelectric charge induced by the hybrid paper increase

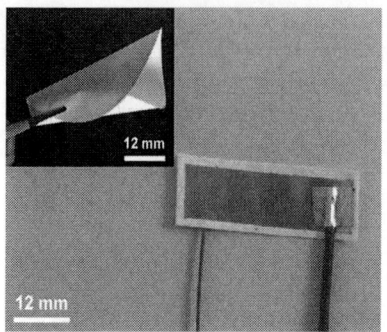

Figure 5: Photograph of a hybrid paper device. Inset: Photograph showing mechanical flexibility of hybrid paper.

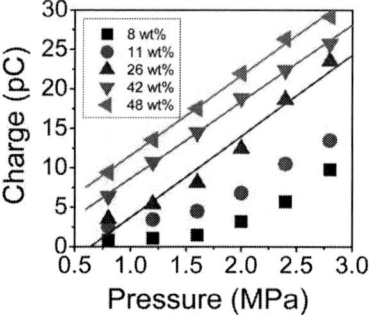

Figure 6: Effect of BaTiO₃ content on Hybrid paper sensor response to change in pressure.

with increasing magnitude of the applied pressure. For example, the piezoelectric charge generated by the hybrid paper with 42 wt% $BaTiO_3$ content under the application of 0.8 MPa is 6.4 pC and it increased to 22.4 pC, when the pressure on the paper increased to 2.4 MPa. A similar behavior was observed for all samples studied, while sensors made with large concentration of $BaTiO_3$ (26 wt% and above) showed a linear response to pressure while those with a low $BaTiO_3$ concentration (8 wt% and 11 wt%) exhibited a non-linear relationship. Similarly, when the hybrid paper was subjected to impact loading, the charge induced by the sensor increased linearly (R^2=0.97) with increasing the impact energy.

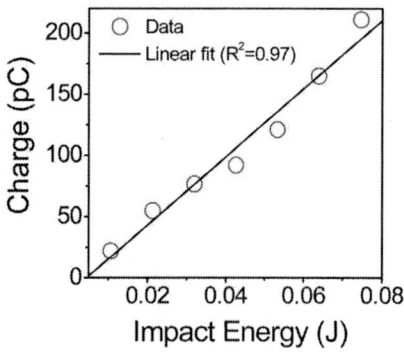

Figure 7: Hybrid smart paper (48 wt% BaTiO₃) sensor response to impact.

978-1-4799-3510-9/14 $31.00 © 2014 IEEE

SUMMARY AND CONCLUSIONS

A simple, cost-effective and new manufacturing method has been developed to fabricate functional piezoelectric hybrid paper, using wood cellulose fibers as base material and nanostructured BaTiO$_3$.

Zeta potential measurements on functionalized wood fibers before and after activation in nanoparticle suspension, microscopic observations and X-ray diffraction studies of hybrid paper suggest the successful attachment of BaTiO$_3$ onto wood fiber.

We investigated five different hybrid paper samples having different BaTiO$_3$ concentration (8, 10, 26, 42, and 48 wt%), among them hybrid paper with 48 wt% BaTiO$_3$ exhibited a large piezoelectric constant and also showed the highest sensitivity and a linear response to the change in pressure and impact loading

Our experimental results indicate functionalizing nanostructured BaTiO$_3$ onto wood cellulose fibers may be a promising approach to enhancing the functionality and value of paper in developing low cost and environment-friendly microelectromechanical systems. Also, the developed new functional paper may overcome the limitations of existing piezoelectric paper in terms of foldability, affordability and functionality.

ACKNOWLEDGEMENTS

This work was supported by BCFIRST Natural Resources and Applied Sciences (NRAS) endowment through the Research Team Program.

REFERENCES

[1] K. D. Wise, "Integrated sensors, MEMS, and microsystems: reflections on a fantastic voyage", *Sensor Actuat. A- Phys.*, Vol. 136, pp. 39- 50, 2007.

[2] C. Liu, "Recent developments in polymer MEMS", *Adv. Mater.*, Vol. 19, pp. 3783-3790, 2007.

[3] H. Gullapalli, V. S. M. Vemuru, A. Kumar *et. al.*, "Flexible piezoelectric ZnO-paper nanocomposite strain sensor", *Small.*, Vol. 6, pp. 1641- 1646, 2010.

[4] S. V. Costa, A. S. Goncalves, M. A. Zaguete *et. al.*, "ZnO nanostructures directly grown on paper and bacterial cellulose substrates without any surface modification layer", *Chem. Commun.*, Vol. 49, pp. 8096- 8098, 2013.

[5] Y. H. Ko, J. S. Yu, "Preparation of ZnO nanorods on cellulose fiber paper and their charge-generating application for waste paper recycling" *Phys. Status Solidi RRL.*, Vol. 7, pp. 985- 988, 2013.

[6] M. Agarwal, Y. Lvov, K. Varahramyan, "Conductive wood microfibers for smart paper through layer-by-layer nanocoating", *Nanotechnology.*, Vol. 17, pp. 5319- 5325, 2006.

[7] S. K. Mahadeva, J. Berring, K. Walus, B. Stoeber, "Effect of poling time and grid voltage on phase transition and piezoelectricity of poly(vinylidene fluoride) thin films using corona poling", *J. Phys. D:Appl. Phys.*, Vol. 46, pp. 285305, 2013.

[8] F. J. Kolpak, J. Blackwell, "Determination of the structure of cellulose II", *Micromolecules.*, Vol. 9, pp. 273- 278, 1976.

[9] S. K. Mahadeva, J. Kim, "Hybrid nanocomposite based on cellulose and tin oxide: growth, structure, tensile and electrical characteristics", *Sci. Technol. Adv. Mater.*, Vol. 12, pp. 055006, 2011.

[10] X. H. Zhu, J. M. Zhu, S. H. Zhou, *et. al.*, "Microstructural characterization of BaTiO3 ceramic nanoparticles synthesized by the hydrothermal technique", Solid State Phenom., Vol. 106, pp. 41- 46, 2005.

[11] J. F. Capsal, E. Dantras, L. Laffont, *et. al.*, "Nanotexture influence of BaTiO$_3$ particles on piezoelectric behavior of PA11/BaTiO$_3$ nanocomposite", *J. NonCryst. Solids.*, Vol. 356, pp. 629- 634, 2010.

[12] Y. Avrahami, "BaTiO3 based materials for piezoelectric and electro-optics applications", Ph.D thesis, Massachusetts Institute of Technology, pp. 78, 2003.

CONTACT

*Suresha K. Mahadeva, tel: +1-604- 827-4593; sure1977@mail.ubc.ca

GRAPHENE WOVEN FABRIC AS HIGH-RESOLUTION SENSING ELEMENT OF CONTACT-LENS TONOMETER

Yushi Zhang[1], Tianxing Man[1], Xiao Li[2], Hongwei Zhu[2, 3], and Zhihong Li[1, 3]

[1]National Key Laboratory of Science and Technology on Micro/Nano Fabrication,
Institute of Microelectronics, Peking University, Beijing, CHINA
[2]Key Laboratory for Advanced Manufacturing by Materials Processing Technology,
Department of Mechanical Engineering, Tsinghua University, Beijing, CHINA
[3]Center for Nano and Micro Mechanics (CNMM), Tsinghua University, Beijing, CHINA

ABSTRACT

In our work, the graphene woven fabrics (GWFs) are investigated as the sensing element of the contact-lens tonometer, which enables precisely monitoring IOP all the daytime. The current-voltage relationship of the device was tested under voltage sweep and the relationships between resistance change and deformation were calculated. Eight devices with GWF in different sizes and CVD conditions were fabricated and the relationship between the current changes of each device and effective IOP increasing, when keeping the voltage constant, was obtained. Combining the highly strain sensing sensitivity and transparency, the contact-lens tonometers with GWF as high-resolution sensing element have a promising prospective.

INTRODUCTION

Tonometers have great significance to detect intraocular pressure (IOP) for glaucoma patients. Various methods and instruments have been developed, including in hospital and by-self tonometry [1], but some serious patients need to be monitored all the daytime, even 24 hours. Many researches about contact lens sensing tonometers, which can measure the IOP continuously, have been reported and commercialized [2]. However, the sensing elements, such as resistance or capacitance, are not sensitive enough, and therefore the complex and costly circuits are required to build in the lens in order to amplify the signal and resist the noise.

In our previous work, the properties of graphene woven fabrics (GWFs), the combination of highly sensitive strain sensing and transparency, have been developed. The resistance-strain curves indicate that the resistance of GWF changes sensitively to the strain, and GWF on an elastic substrate can serve as a stretchable and highly sensitive strain sensor [3, 4].

In this paper, GWF is investigated as the sensing element of the contact-lens tonometer, which enables precisely monitoring IOP. The device was fabricated and tested with the mimic human eyeball and the syringe pump. The current-voltage relationship of the device was recorded and the relationships between resistance change and deformation were calculated. Different devices with GWF in different sizes and CVD conditions were fabricated and the relationship between the current changes of each device, and effective IOP increasing when keeping the voltage constant, was obtained. The results illustrate that using GWFs as the sensor on the contact-lens can make the device more sensitive to the deformation.

PRINCIPLE AND SIMULATION

The working principle was schematically shown in Fig. 1. A piece of GWF is mounted on the outside surface of contact-lens, which has been attached to the cornea tightly because of the homogeneous hydrophilicity. As the IOP increasing, the small deformation of the eyeball makes the contact-lens stretched, and therefore the piece of GWF stretches together. High-density cracks are appeared at weak points in the piece of GWF under small strains. With the increasing of the strain, the crack length and density gradually increase, which causes the resistance of the GWF increasing. In a similar way, the IOP recovery results in a decrease in resistance [4]. Thus, the IOP increasing and recovery can be monitored by testing the current change under the constant voltage.

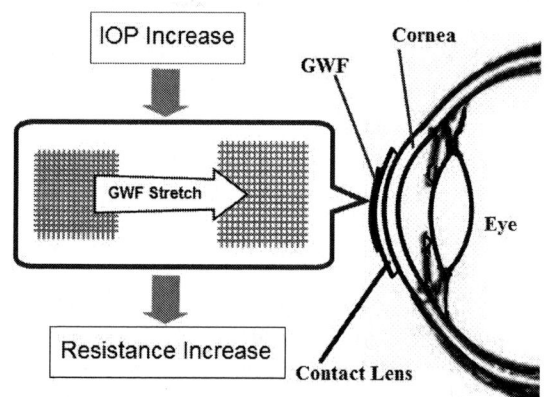

Figure 1: The working principle of the device

To understand the mechanical properties, the stress of the device was qualitatively simulated by the software of ABAQUS. In the model, a thin film of GWF was attached to the center square area on the outside surface of a hemispherical shell. As the elastic coefficient of GWF thin film is far smaller than that of the contact lens, it was assumed one order of magnitude smaller than that of the shell in the simulation. The simulated MISES stress is shown in Fig. 2(a), while the stresses in x, y and z directions are shown in Fig. 2(b)-(d), respectively. The equally distributed stresses indicate that the GWF contributes ignorable stress to the device in mechanical properties and can follow the deformation of the contact-lens very well.

978-1-4799-3510-9/14 $31.00 © 2014 IEEE

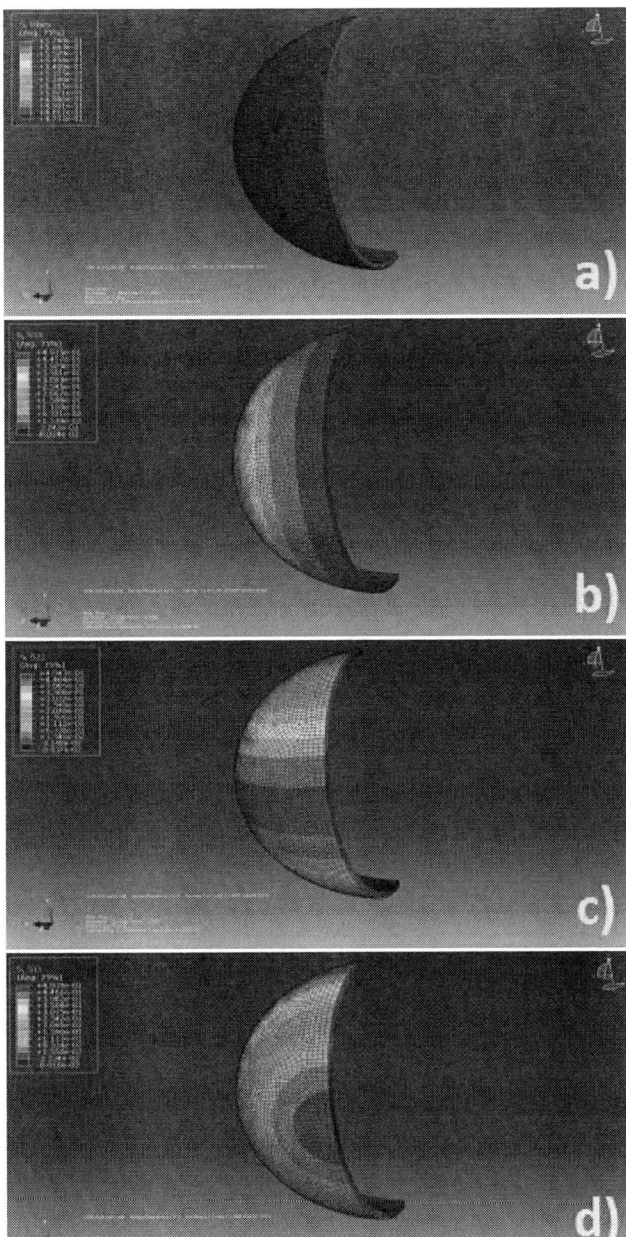

Figure 2: The mechanical analysis results of the device. MISES stress in a); stress in x, y, z directions respectively in b), c) and d)

FABRICATION AND EXPERIMENTS

The simplified process to fabricate the device is schematically illustrated in Fig. 3. A copper mesh with the size of 5×8 cm² was prepared as the substrate. After ultrasonically treated with hydrochloric acid and acetone to clean their surfaces and eliminate the thin surface oxide layer, the mesh was heated to 950°C in a CVD furnace under Ar (1000 mL/min) and H_2 (60 mL/min). Keeping 950°C, Ar was turned down to 200 mL/min and H_2 was turned off, and then 15 mL/min CH_4 was introduced into the chamber at ambient pressure to grow the multilayered graphene on the copper mesh for 20 min. After cooled down to room temperature rapidly, the mesh was cut into pieces in squares and put into $FeCl_3$/HCl mixture (1:1, mol/L) to dissolve the copper. About 2 hours later, the copper dissolved completely, and GWF was formed as planar mesh structure, as shown in Fig. 4(a). After transferred into DI water and floating for about an hour to clean, the GWF was collected by to a contact lens (Base Curve=8.6mm, with 24% water content) and bonded on the outer surface of the lens as the water evaporated. A copper foil tape was used to make two electrodes on the surface of GWF and the device was shown in Fig. 4(b).

The device was attached to a mimic human eyeball tightly because of the homogeneous hydrophilicity and could follow the deformation of the mimic eyeball well. After filled with water, the mimic human eyeball was connected with a syringe, which was controlled by a syringe pump to inject the water in, as shown in Fig. 4(c).

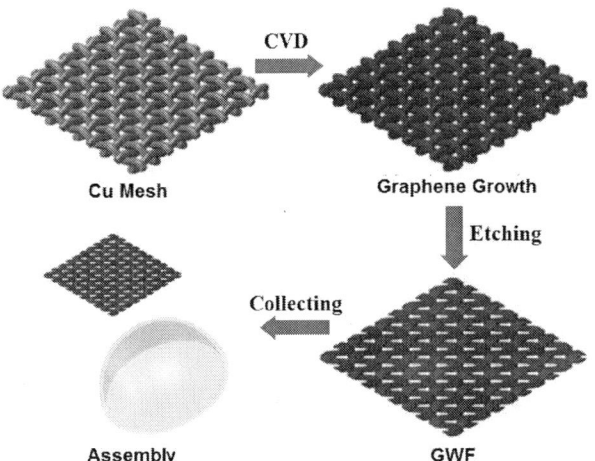

Figure 3: The process of device fabrication.

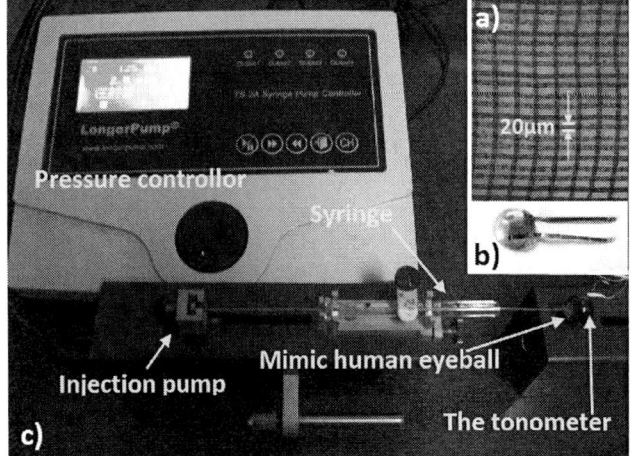

Figure 4: The photo of (a) GWF, (b) the device and (c) the testing setup

As the mimic human eyeball was dilated by the hydraulic pressure, the contact lens was stretched and the resistance of the attached GWF was increased. The

Table 1. The average current rate of change in different range of the IOP for the two group devices

The average current rate of change (%/mmHg)		The range of the effective IOP			
		0~5mmHg	5~10mmHg	10~15mmHg	15~20mmHg
Group 1 Device number (Side-length of GWF)	Device 1 (4mm)	14.7	3.89	1.98	0.69
	Device 2 (5mm)	23.6	3.68	2.28	1.02
	Device 3 (6mm)	23.6	3.83	2.51	0.98
	Device 4 (7mm)	16.7	4.20	1.87	1.04
Group 2 Device number (Side-length of GWF)	Device 5 (4mm)	1.84	2.68	2.88	3.02
	Device 6 (5mm)	3.16	2.42	2.79	2.86
	Device 7 (6mm)	2.92	3.14	2.91	3.06
	Device 8 (7mm)	3.34	2.50	2.65	3.33

current-voltage relationship of the device was tested under voltage sweep from 0V to 10V with the curvature radius change of the contact-lens.

Eight devices, numbered with Device 1 to Device 8, were fabricated to further characterize the GWF sensing properties. Among them, the sensing elements in Device 1~4 were cut from the same piece of GWF, which was fabricated in a condition tightly isolated from oxygen during the CVD; while the ones in Device 5~8 from another piece, which was in a condition with tiny oxygen flow. The GWFs in each group were cut into squares with different lengths of side in 4mm, 5mm, 6mm and 7mm, respectively shown in Table 1. Each device was also attached to the mimic human eyeball and the inner pressure of the mimic eyeball was increased by injecting water. The electrical properties of the device were tested by a semiconductor parameter analyzer (HP 4156B). Keeping the voltage constant in 10V, the varied current was recorded when increasing the pressure.

RESULTS AND DISCUSSION

The current-voltage relationship of the device tested under voltage sweep was shown in Fig. 5. The deformation in different injection volume can be calculated and the relationships between resistance change and deformation were obtained, taking the near linear region of 6~10V, as shown in Fig. 6. The results show that the device was more sensitive than the previously reported contact-lens tonometers.

Further, according to the current testing results of the eight devices in two groups, recorded by the semiconductor parameter analyzer, the relationships between the current changes and effective IOP increasing, shown in Fig. 7(a) and (b), can be obtained.

The average current rates of change in different ranges of effective IOP increasing for the two groups were calculated, as shown in Table 1. It shows that the first group devices were far more sensitive to IOP increasing when less than 5 mmHg, while the second group devices were less sensitive but with a high linearity. It indicates that the oxygen existence in the growing ambient may strongly affect the resistance-stress dependence of the GWF. The mechanism and quantitative relationship are under investigation. For tonometer application the measurement range of 5 mmHg in change is enough and the sensitivity is highly demanded, so the devices in the first group are preferred. In contrary, the devices in Group 2 are advantageous for the sensor that requires wide sensing range.

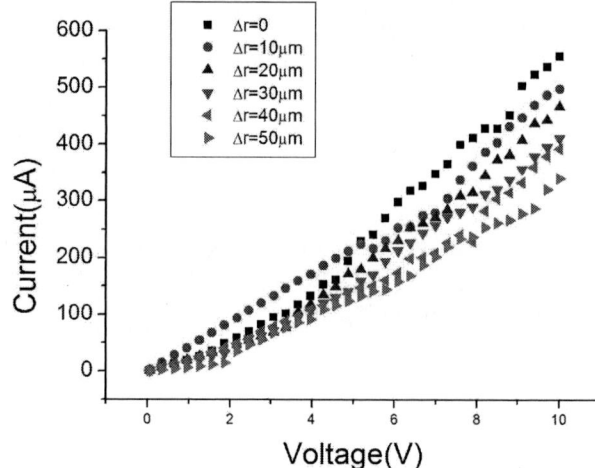

Figure 5: Current-voltage relationship of the device

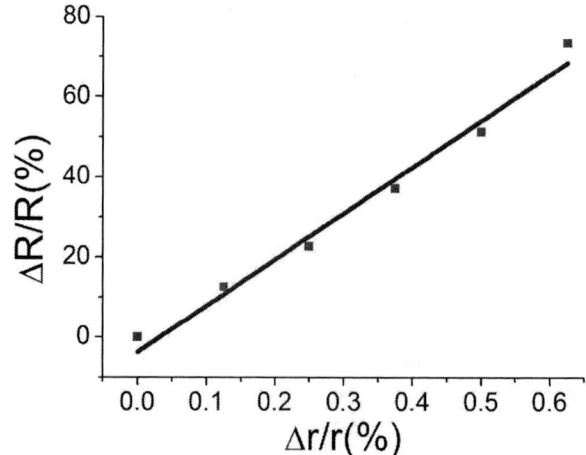

Figure 6: The device resistance change in percent ($\Delta R/R$) to the change in percent of sphere's curvature radius (Δr).

a)

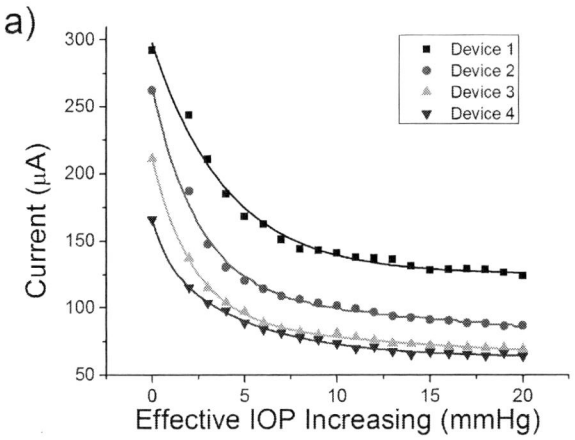

b)

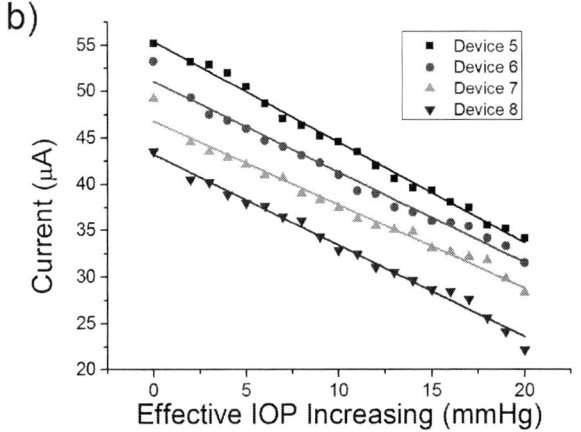

Figure 7: The relationship between the current and IOP increasing under 10V for each device

CONCLUSION

The resistance of the GWF on the contact-lens changes significantly with the deformation of the mimic human eye. Increase of the effective IOP can be monitored the current changes, with keeping the voltage constant.

Besides, the transparency of GWFs is more than 80%

and the transparency of contact lens substrate is about 95%. Combining the highly strain sensing sensitivity and transparency, contact-lens tonometers with GWF as high-resolution sensing element have a promising prospective.

REFERENCES

[1] S. Y.-W. Liang, Graham A. Lee, David Shields, "Self-tonometry in Glaucoma Management—Past, Present and Future", *Survey of Ophthalmology*, Vol. 54, No. 4, July-August 2009, pp. 450-462.

[2] M. Leonardi, P. Leuenberger, D. Bertrand, A. Bertsch, P. Renaud, "First Steps toward Noninvasive Intraocular Pressure Monitoring with a Sensing Contact Lens", *Investigative Ophthalmology & Visual Science*, September, 2004, Vol. 45, No. 9, pp. 3113-3117.

[3] X. Li, P. Sun, L. Fan, M. Zhu, K. Wang, M. Zhong, J. Wei, D. Wu, Y. Cheng, H. Zhu, "Multifunctional Graphene Woven Fabrics", *Scientific Reports*, 2: 395, DOI: 10.1038.

[4] X. Li, R. Zhang, W. Yu, K. Wang, J. Wei, D. Wu, A. Cao, Z. Li, Y. Cheng, Q. Zheng, R. S. Ruoff, H. Zhu, "Stretchable and highly sensitive graphene-on-polymer strain sensors". *Scientific Reports*, 2: 870, DOI: 10.1038, 1-6.

ACKNOWLEDGEMENTS

Thanks to Yuchao Ke and Xiaofei Zhang in School of Aerospace, Tsinghua University, for the help of the simulation. This work was supported by the National Basic Research Program of China (No. 2011CB707505).

CONTACT

* Zhihong Li, Peking University, Beijing, 100871, China, Phone: +86-10-62766581; Fax: 86-10-62751789; E-mail: zhhli@ime.pku.edu.cn

INCREASED THERMAL CONDUCTIVITY POLYCRYSTALLINE DIAMOND FOR LOW-DISSIPATION MICROMECHANICAL RESONATORS

H. Najar[1], A. Thron[1], C. Yang[2], S. Fung[1], K. van Benthem[1], L. Lin[2], and D.A. Horsley[1]
[1]University of California, Davis, USA
[2]University of California, Berkeley, USA

ABSTRACT

This paper reports an investigation of microcrystalline diamond (MCD) films deposited under different conditions to increase thermal conductivity and therefore mechanical quality factor (Q) in micromechanical resonators. Through a study of different deposition conditions, we demonstrate a three-fold increase in thermal conductivity and quality factor. Quality factor measurements were conducted on double ended tuning fork resonators, showing $Q = 241,047$ at $f_n = 246.86\ kHz$ after annealing, the highest Q reported for polycrystalline diamond resonators. We further present a study of the unique microstructure of hot filament chemical vapor deposition (HFCVD) diamond films and relate growth conditions to observed microstructural defects.

INTRODUCTION

From solid-state components in microelectronics to moveable mechanical structures in micro-electromechanical systems (MEMS), silicon is the main semiconductor material used for fabrication. However, compared to some other materials like diamond, silicon has its own limitations. Diamond is a promising material for MEMS due to its superior physical properties. In its single crystalline form, diamond is a material that has high thermal conductivity ($\kappa \sim 2200\ Wm^{-1}K^{-1}$), very low thermal expansion ($\alpha \sim 1.2\ ppm/K$), and extremely high elastic modulus ($E \sim 1050\ GPa$). These superior properties can result in high Q diamond resonators ($Q > 100,000$) due to diamond's low thermoelastic damping (TED). Relative to single crystalline diamond (SCD) resonators [1], polycrystalline diamond resonators have lower quality factor [2], but have the advantage of wafer-scale deposition processes. A variety of sources contribute to reduce the Q of polycrystalline diamond resonators, including reduced thermal conductivity [2, 3] caused by phonon scattering at grain boundaries [2], defect relaxation process [4], and surface and bulk defects [3-7]. Other works demonstrated that nanocrystalline diamond (NCD) resonators show lower Q in comparison to microcrystalline diamond (MCD) resonators [8] due an increase in the above mentioned dissipations. In previous work, we demonstrated that the measured thermal conductivity of MCD was $\kappa = 100\ Wm^{-1}K^{-1}$, an order of magnitude lower than that of SCD, and that the TED-limited quality factor (Q_{TED}) of diamond was directly related to this low thermal conductivity [9]. Thus, in order to increase Q, increasing thermal conductivity is crucial.

Here, through a study of different deposition conditions, we demonstrate a three-fold increase in κ and therefore QTED. We also present a study of the unique microstructure of hot filament chemical vapor deposition (HFCVD) diamond films and relate growth conditions to observed microstructural defects. The thermal conductivities of the deposited films were measured using time-domain thermoreflectance (TDTR) technique. Single-anchored double-ended tuning fork (SA-DETF) resonators were fabricated from these films to relate the measured thermal conductivities to measured Q-factors. The reason for this choice of resonator is to minimize, and ideally eliminate the anchor loss and reach the intrinsic damping of the material, Q_{TED}, and surface loss, $Q_{surface}$.

DIAMOND FILM CHARACTERIZATION

The diamond film growth process starts by seeding silicon wafers within a solution containing 4 nm nanocrystalline diamond powder. This pre-treatment step reduces the induction time for nucleation and increases the density of nucleation sites by (a) introducing scratches onto the surface to act as growth template and (b) embedding nano-metric diamond particles to act as seed crystals[10-13]. MCD films were grown in a large-area multi-wafer HFCVD system using H_2 and CH_4 as precursor gases and trimethylboron (TMB, $(CH_3)_3B$) to provide in-situ boron doping. All films grown here are $\sim 1\ \mu m$ thick and deposited with TMB:CH_4 concentration of 444 ppm and pressures near 25 Torr. Figure 1 shows a top view and a cross-section SEM image of the best quality diamond film in this study.

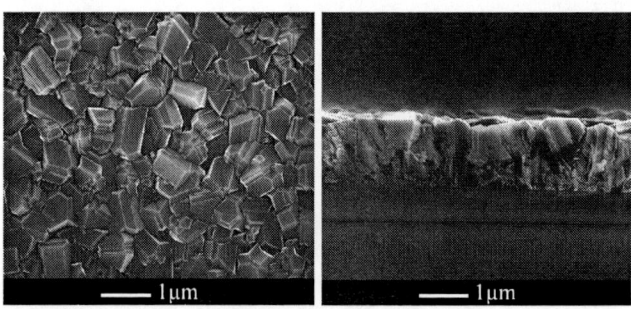

Figure 1: Top view (left) and cross-section SEM (right) of the MCD film indicating an average grain size of ~0.8μm and film thickness of ~1μm.

978-1-4799-3510-9/14 $31.00 © 2014 IEEE

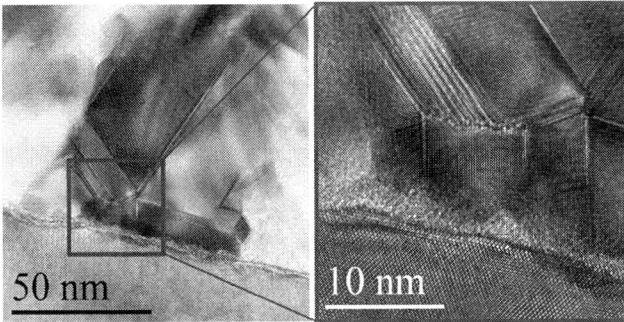

Figure 2: TEM image of the diamond-Si interface. The diamond film reveals grains with and without twin boundaries and stacking faults (left). The HRTEM image provides more details on the twin boundaries and stacking faults (right).

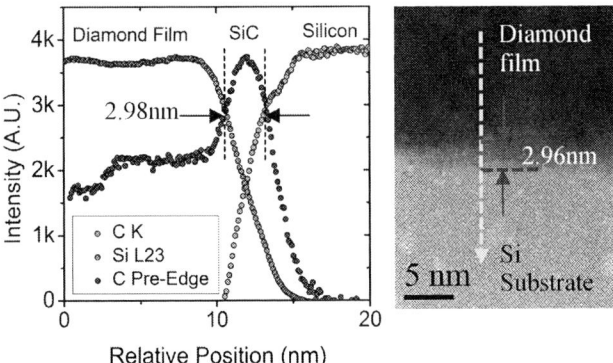

Figure 3: Integrated EELS intensity plot (left) acquired along the yellow line in the high resolution STEM (HRSTEM) image (right) reveals that C and Si intermix in a region that is the same width as the amorphous layer observed in the image.

High Resolution Transmission Electron Microscopy (HRTEM), Scanning TEM (STEM), and Electron Energy Loss Spectroscopy (EELS) were used to characterize the atomic structure and chemical composition of the diamond-Si interface. STEM images and EELS spectra were acquired in cross-section (with the interface plane parallel to the direction of the electron beam), at an accelerating voltage of 200kV. HRTEM images in Figure 2 reveal diamond grains initiating from the Si substrate contain a higher density of twin defects and stacking faults, compared with those initiating from the amorphous seed layer. EELS data acquired across the diamond-Si interface show an overlap between the Si $L_{2,3}$ edge signal and the C K edge signal. This suggests that intermixing between Si and C occurs at the diamond-Si interface. This ~3nm intermixed layer of Si and C is indicated in Figure 3. Furthermore, a pre-peak in the C K edge is observed at the interface. The presence of the pre-peak indicates unoccupied π^* states due to a transition from sp^3 bonding in diamond to a mixed sp^2-sp^3 bonding at interface [14]. We,

Table 1: Deposition conditions of diamond films.

	Temp (°C)	H$_2$:CH$_4$ (%)	κ (Wm^{-1} K^{-1})	Raman FWHM (cm^{-1})
V1	750	1%	130	13.83
V2	700	1%	120	12.28
V3	700	0.5%	300	8.38

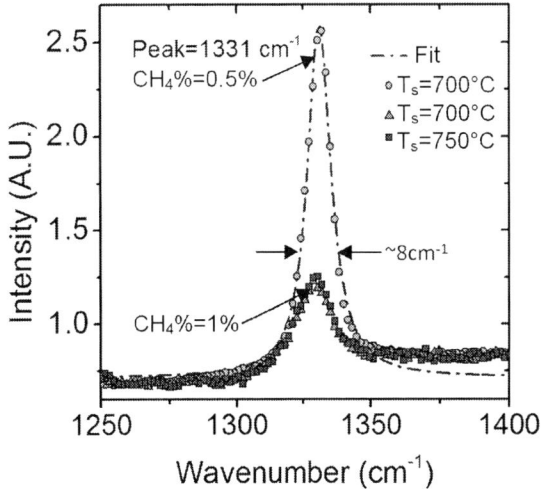

Figure 4: Raman spectra of three deposited diamond films. V3, the film deposited with 0.5% CH$_4$, has FWHM of 8.38 cm^{-1} at 1333 cm^{-1} indicating a higher fraction of sp^3 bonding and therefore higher quality diamond film.

therefore, consider the amorphous layer at the interface as a sub-stoichiometric SiC wetting layer.

Three different HFCVD growth conditions were studied, with variables being the substrate temperature (T_s) and methane concentration (c = CH$_4$:H$_2$%), in order to observe the change in thermal and mechanical properties of the films. Thermal conductivity of each film was measured using the time-domain thermoreflectance (TDTR) technique [15, 16]. Elastic moduli were further extracted by measuring the resonant frequency of cantilevers fabricated using the diamond films. Increasing temperature and decreasing CH$_4$ concentration were both observed to lead to increased elastic modulus, E.

Raman spectroscopy was used to evaluate the fraction of sp^3 diamond. As summarized in Table 1, lowering the methane concentration from 1% to 0.5%, drastically increased the thermal conductivity from ~120 Wm^{-1}K^{-1} to ~300 Wm^{-1}K^{-1}. This change is consistent with Raman spectra, Figure 4, in which the full width half maximum (FWHM) of the sp^3 diamond peak narrowed from ~14 cm^{-1} to ~8 cm^{-1}, indicating a greater fraction of sp^3 diamond and hence a higher quality diamond film. It should be also noted that the Raman peak for the film grown at 750°C and 0.5% methane concentration occurred at ~1332.78 cm^{-1},

978-1-4799-3510-9/14 $31.00 © 2014 IEEE

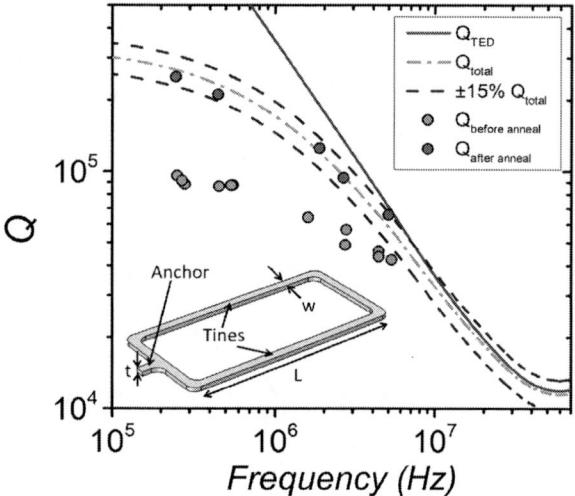

Figure 5: Q versus frequency of diamond DETF resonators. The green and magenta circles correspond to measurements before and after annealing in N_2 at 750°C for 1 hr, respectively. The solid red lines correspond to the theoretical Q_{TED} using $\kappa = 300\ Wm^{-1}K^{-1}$. The orange dashed-dotted line is $Q_{total} = Q_{surface} \| Q_{TED}$ and the dashed blue lines correspond to ±15% variation in Q_{total}. The lowest frequency DETF achieved $Q = 241,047$ at $f_n = 246.857$ kHz. The inset shows a schematic and dimensions of a DETF. Q measurements correspond to DETF's with tine length L that varied from 70 to 311 μm.

whereas the corresponding value for the films grown with 1% methane concentration was ~1331.19 cm^{-1} regardless of their deposition temperature. This wavelength shift indicates the presence of a greater strain in the film deposited at a lower CH_4 concentration than the other two diamond films. Although decreasing CH_4 flow rate had a dramatic effect on increasing the thermal conductivity, variation of temperature (i.e. from 700°C to 750°C) had almost no effect on thermal conductivity of the films.

DIAMOND MEMS RESONATOR

After the diamond films were characterized, DETF resonators were fabricated in the film with the highest thermal conductivity. DETF resonators were designed with 2.1 μm tine width and tine lengths from 70 μm to 311 μm such that their in-plane resonant frequencies spanned a range from 246 kHz to 5.2 MHz. To demonstrate the frequency-Q dependency, the measured Q's were plotted against the corresponding frequencies as shown in Figure 5. The solid line on Figure 5 corresponds to the theoretical TED limit based on the TDTR-measured $\kappa = 300$ Wm^{-1}K^{-1}, which follows the equation:

$$Q_{TED}(f) = \frac{Q_{min}}{2} \frac{1+(f/f_{min})^2}{(f/f_{min})} \qquad (1)$$

where Q_{min} is the minimum Q and f_{min} is the frequency at

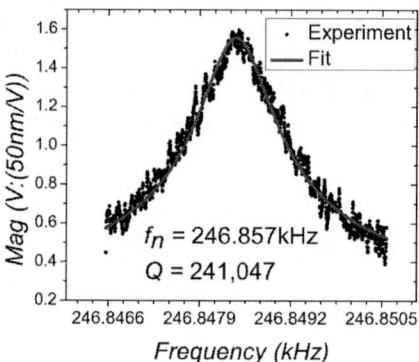

Figure 6: Frequency response of the double ended tuning fork at the anti-symmetric in-plane mode, with resonant frequency of 246.857 kHz and Q-factor of 241,047 at $P < 100$ μtorr vacuum.

which this occurs and are defined as

$$Q_{min} = \frac{2C\rho}{E\alpha^2 T_0} \qquad (2)$$

$$f_{min} = \frac{\kappa}{C\rho}\frac{\pi}{2w^2} \qquad (3).$$

In equations (2) and (3), E is the elastic modulus, α is the coefficient of thermal expansion, T_0 is the equilibrium temperature, C is the specific heat capacity, κ is the thermal conductivity, and w is the width along the direction of the resonator's tine movement. To minimize the effect of air damping, the resonators were placed in a vacuum probe station at a pressure $P < 100$ μTorr. The resonators were excited electrostatically and the responses were measured using a laser Doppler vibrometer (LDV).

The resonators were further annealed in a N_2 environment at 700°C for 1 hr. The heat treatment showed an improvement of more than 60% in Q. The experimental data show good agreement with the theoretical TED limit which was increased threefold relative to resonators made from $\kappa = 100$ Wm^{-1}K^{-1} films. The lowest frequency DETF achieved $Q = 241,047$ at $f = 246.857$ kHz, the highest Q observed among all the polycrystalline diamond resonators to date. Figure 6 shows the magnitude frequency response for the highest Q resonator. It should be noted that at lower frequencies, the Q's are limited by a frequency-independent damping mechanism. We attribute this to surface dissipation ($Q_{surface}$) and are actively investigating its origins.

CONCLUSIONS

Microcrystalline diamond films were deposited under various conditions to increase the thermal conductivity and thus increase the Q in diamond MEMS resonators. We demonstrated a three-fold increase of the thermal conductivity by reducing the methane flow rate from 1% to 0.5%. Raman measurements also confirmed a higher diamond film quality by comparing FWHM of the sp^3

diamond peaks. Increasing the temperature from 700°C to 750°C had negligible effect on both κ and sp^3 fraction.

This study further explored the unique microstructure of HFCVD diamond films through HRTEM and STEM, and showed an increase in twin boundary and stacking fault densities where the diamond grains meet the Si substrate. EELS images verified the presence of a thin ~3 nm of intermixed layer of Si and C. This layer enables a smoother transition from Si to diamond grains and therefore acts to relieve stress arising from lattice mismatch and substrate contraction.

Many SA-DETF resonators were fabricated with varying resonant frequencies and were tested in vacuum. The experimental data show good agreement with the theoretical TED-limit, which was increased threefold relative to resonators made from $\kappa = 100$ Wm^{-1}K^{-1} films. The lowest frequency DETF achieved $Q = 241{,}047$ at 246.857 kHz, the highest Q observed among all the polycrystalline diamond resonators to date.

ACKNOWLEDGEMENTS

This work was supported by DARPA under grant No.W31P4Q-11-1-0003 and NSF grant DMR-0955638. The authors thank Gerry Chandler and Nga Vu at SP3 Diamond Technologies Inc., for diamond film depositions, and David Cahill at Univ. of Illinois for TDTR measurements.

REFERENCES

[1] P. Ovartchaiyapong, L. M. A. Pascal, B. A. Myers, P. Lauria, and A. C. B. Jayich, "High quality factor single-crystal diamond mechanical resonators," *Appl. Phys. Lett.,* vol. 101, pp. 163505-4, 2012.

[2] N. Sepulveda, L. Jing, D. M. Aslam, and J. P. Sullivan, "High-Performance Polycrystalline Diamond Micro- and Nanoresonators,", *J. Microelectromec. Syst.,* vol. 17, pp. 473-482, 2008.

[3] V. P. Adiga, A. V. Sumant, S. Suresh, C. Gudeman, O. Auciello, J. A. Carlisle, *et al.,* "Mechanical stiffness and dissipation in ultrananocrystalline diamond microresonators," *Phys. Rev. B,* vol. 79, p. 245403, 2009.

[4] N. Sepulveda, D. Aslam, and J. P. Sullivan, "Polycrystalline diamond MEMS resonator technology for sensor applications," *Diam. Relat. Mater.,* vol. 15, pp. 398-403, 2006.

[5] O. Auciello and A. V. Sumant, "Status review of the science and technology of ultrananocrystalline

diamond (UNCD™) films and application to multifunctional devices," *Diam. Relat. Mater.,* vol. 19, pp. 699-718, 2010.

[6] A. Gaidarzhy, M. Imboden, P. Mohanty, J. Rankin, and B. W. Sheldon, "High quality factor gigahertz frequencies in nanomechanical diamond resonators," *Appl. Phys. Lett.,* vol. 91, 2007.

[7] J. W. Baldwin, M. K. Zalalutdinov, T. Feygelson, B. B. Pate, J. E. Butler, and B. H. Houston, "Nanocrystalline diamond resonator array for RF signal processing," *Diam. Relat. Mater.,* vol. 15, pp. 2061-2067, 2006.

[8] M. Akgul, R. Schneider, Z. Ren, G. Chandler, V. Yeh, and C. T. Nguyen, "Hot filament CVD conductive microcrystalline diamond for high Q, high acoustic velocity micromechanical resonators," IEEE IFCS, 2011, pp. 1-6.

[9] H. Najar, A. Heidari, M.-L. Chan, H.-A. Yang, L. Lin, D. G. Cahill, *et al.,* "Microcrystalline diamond micromechanical resonators with quality factor limited by thermoelastic damping," *Appl. Phys. Lett.,* vol. 102, p. 071901, 2013.

[10] R. J. Nemanich, "Growth and Characterization of Diamond Thin Films," *Annu. Rev. of Mater. Sci.,* vol. 21, pp. 535-558, 1991.

[11] K. Kobashi, "Chapter 10 - Diamond Nucleation," in *Diamond Films,* K. Kobashi, Ed., ed Oxford: Elsevier Science Ltd, 2005, pp. 121-153.

[12] X. Liu, J. F. Vignola, H. J. Simpson, B. R. Lemon, B. H. Houston, and D. M. Photiadis, "A loss mechanism study of a very high Q silicon micromechanical oscillator," *J. of Appl. Phys.,* vol. 97, pp. 023524-023524-6, 2005.

[13] P. W. May, "Diamond Thin Films: a 21st-Century Material," *Phil. Trans. Math. Phys. Eng. Sci.,* vol. 358, pp. 473-495, 2000.

[14] D. A. Muller, Y. Tzou, R. Raj, and J. Silcox, "Mapping sp2 and sp3 states of carbon at sub-nanometre spatial resolution," *Nature,* vol. 366, pp. 725-727, 1993.

[15] D. G. Cahill, "Analysis of heat flow in layered structures for time-domain thermoreflectance," *Rev. of Sci. Instrum.,* vol. 75, pp. 5119-5122, 2004.

[16] K. Kang, Y.-K. Koh, C. Chiritescu, X. Zheng, and D.-G. Cahill, "Two-tint pump-probe measurements using a femtosecond laser oscillator and sharp-edged optical filters," *Rev. of Sci. Instrum.,* vol. 79, p. 114901, 2008.

INTERFACE LOSSES IN MULTIMATERIAL RESONATORS

L.G. Villanueva[1,2,], B. Amato[1], T. Larsen[1] and S. Schmid[1]*

[1]DTU Nanotech, Technical University of Denmark (DTU), Lyngby, Denmark
[2]Advanced NEMS Group, École Polytechnique Fédérale de Lausanne (EPFL), Lausanne, Switzerland

ABSTRACT

We present an extensive study shedding light on the role of surface and bulk losses in micromechanical resonators. We fabricate thin silicon nitride membranes of different sizes and we coat them with different thicknesses of metal. We later characterize the 81 lowest out-of-plane flexural vibrational modes to achieve a total of more than 3000 experimental points that allow us to quantify the contribution of surface and volume intrinsic (material related) losses in MEMS resonators. We conclude that the losses in the interface between silicon nitride and aluminum is a very important contributor to the overall energy loss.

INTRODUCTION

With very high quality factors (Qs) values (up to 10^7) at room temperature and $Q \cdot f$ products (above 10^{13} Hz), stoichiometric Si_3N_4 membranes [1] and strings [2] have become a centerpiece of many research projects, particularly in opto-mechanics [3, 4]. Recently it has been shown that metallized membranes enable the design of exciting new opto-electro-mechanical systems that allow e.g. the optical detection of electrical signals with unprecedented sensitivity [5, 6]. For these applications and for MEMS resonators in general there has been a continuous effort to find materials and systems that provide as high Qs as possible. The thorough understanding of the underlying loss mechanisms is crucial to optimize Q.

Q can be defined as the ratio between the energy stored in a resonator over the energy loss every cycle. Due to their large intrinsic residual stress, resonating membranes and string are able to store more energy, thus increasing Q even though dissipated energy per cycle remains the same. Models based on this idea, considering only material losses, are able to reproduce the behavior of Q as a function of mode number (whenever neither of the indexes is lower than 3), and even suggest ways to control extra losses for multi-material resonators [7]. However, the data reported in the literature does not provide information on the relative importance of surface vs. bulk losses for these systems. In this work, we quantify both bulk and surface losses, evidencing the importance of proper surfaces, not only in the physical boundaries of the resonator, but also in the interface between different materials.

FABRICATION

We fabricate our membranes following the simple procedure outlined in Figure 1. We start by defining a set of Si_3N_4 square membranes ($L = 250, 500,$ and 1000 μm; $t_{Si_3N_4} = 50, 100,$ and 200 nm). Our initial substrates are double side polished P-doped silicon wafers (100mm in diameter) where the stoichiometric silicon nitride is deposited via Low Pressure Chemical Vapor Deposition (LPCVD) at 850°C. The backside is then patterned to define windows for membrane release, by standard photolithography (using flat alignment for a more precise definition of the membrane sizes) and dry etching, followed by KOH micromachining of Si wafers (KOH solution 40% in weight at 85°C), followed by a cleaning in a neutralization bath to remove KOH crystals residues. Aluminum is then deposited on top of some of the samples ($t_{Si_3N_4} = 50, 100,$ and 200 nm) by e-beam evaporation at 10^{-6} mbar at rates ranging between 0.5 and 1 nm/s. In the last step all samples are annealed at 400°C in order to help Al reflow, reduce intrinsic losses and make stresses uniform [7].

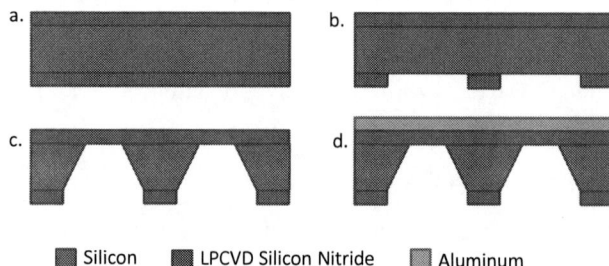

■ Silicon   ■ LPCVD Silicon Nitride   ■ Aluminum

Figure 1: Fabrication process flow for the membrane resonators. A layer of LPCVD stoichiometric Si_3N_4 is deposited on a Si wafer (a). This layer is patterned (b) on the backside to define the windows for the subsequent Si anisotropic etching using KOH (c). The front side is metallized with the desired thickness of aluminum using e-beam evaporation (d).

CHARACTERIZATION

Characterization is performed in vacuum ($P \leq 10^{-5}$ mbar), at room temperature. The silicon chips with membranes are located on top of a piezo-shaker disc for actuation purposes. To detect the motion of the membranes, a Polytec Laser-Doppler vibrometer with a bandwidth of 24 MHz is used. We study the 81 lower out-of-plane flexural vibrational modes measuring their resonance frequencies and quality factors. To determine the resonance frequencies, standard frequency sweeps are performed using a Zurich Instruments HF2LI lock-in amplifier. To determine the quality factor, we compare two different methods on the first membrane. We extract the quality factor from the frequency sweep scan performed with the lock-in amplifier and from ring-down experiments. We do this 5 times per mode for up to 20 modes and we observe that the variance of the

978-1-4799-3510-9/14 $31.00 © 2014 IEEE

experiment is larger than the difference between both methods. Therefore, for the rest of the modes we extract the quality factor from the same frequency sweep data we use to determine the frequency. This reduces considerably the experiment time.

Taking into account the different membrane geometries (both lateral and thickness) we obtain in total more than 3000 experimental points for both frequency and quality factor.

Reactive material properties

As expected for structures with such extreme aspect ratios, the frequency of the modes is very accurately described by standard thin plate theory (see Figure 2):

$$f_{n,m} = \frac{1}{2L}\sqrt{v_{eff}(n^2 + m^2)}\,; \qquad (1)$$

where v_{eff} is the effective speed of sound for each particular multimaterial stack and it is given by Eq. 2:

$$v_{eff} = \sqrt{\frac{\sigma_{Si_3N_4}t_{Si_3N_4} + \sigma_{Al}t_{Al}}{\rho_{Si_3N_4}t_{Si_3N_4} + \rho_{Al}t_{Al}}} \qquad (2)$$

Using equations (1) and (2), it is possible to use the measured frequency values to estimate the residual stress for both stoichiometric Si₃N₄ and Al, obtaining $\sigma_{Si_3N_4} \approx$ 1.125 GPa and $\sigma_{Al} \approx$ 100 MPa. The positive sign is kept to identify both intrinsic stresses being of tensile nature. In Figure 3 it is possible to see how these two values for stress allow us to recover all 12 v_{eff}.

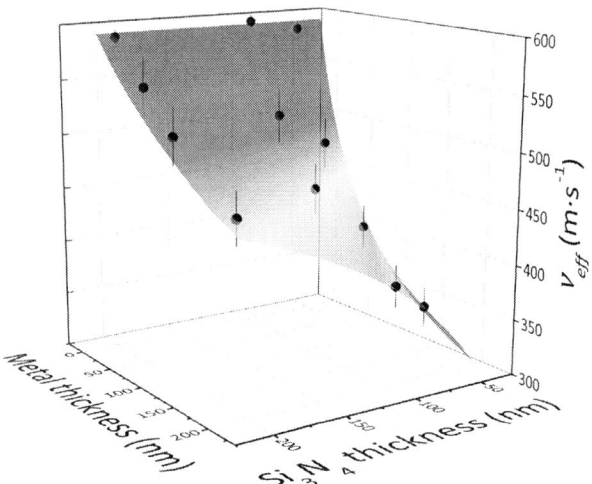

Figure 3: Using the 12 different material configurations for the fabricated membranes, it is possible to extract the residual stress and densities for both Si₃N₄ and Al. We do this by fitting the experimentally measured values for v_{eff}.

Dissipative material properties

In order to analyze the energy losses in the characterized resonators, we first separate the different modes into those limited by anchor losses (radiation of energy to the clamping substrate) and those limited by internal (material) losses. This is done following the same semi-empirical rule suggested in [7], which states that all the modes with any of the indexes smaller than 3 ($i \vee j < 3$). Figure 5 shows how the overall dispersion in the quality factors distribution is reduced considerably if we only consider modes limited by material losses (solid scattered data).

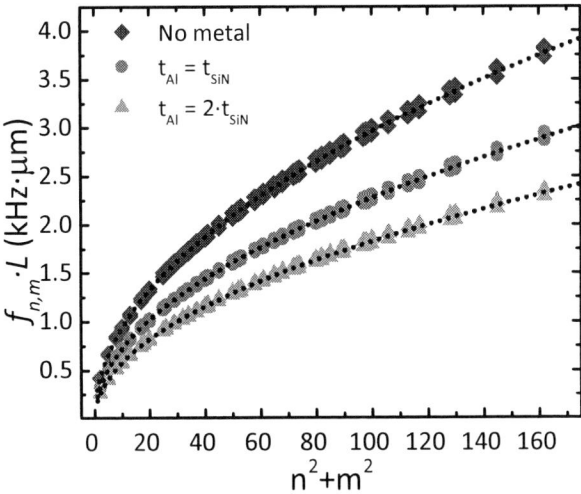

Figure 2: Experimentally obtained frequencies (scattered points, scaled by the length) for the 81 first flexural modes vs. mode number for 13 membranes with different dimensions. Only three theoretical curves (dotted black lines) are expected, depending on the different effective speed of sound (v_{eff}). A remarkable agreement to theory is obtained throughout the experimental data.

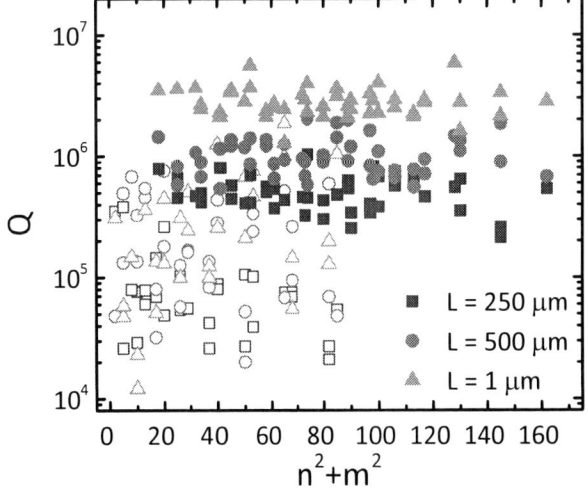

Figure 4: Experimentally obtained quality factors for the 81 first flexural modes vs. mode number for 3 membranes with the same $t_{Si_3N_4} = 50$ nm and different lengths. Following [7] we separate between the modes which are being limited by anchoring losses (outlined scattered symbols) and those that are limited by intrinsic material losses (solid scattered symbols).

Next, we use a model similar to those already established in the literature [2, 7] to account for material-related energy losses (in the subset of modes described above). This consists on the use of a Zener model, where the Young's modulus of the material has a real and an imaginary part, the former one being of reactive nature and keeping the phase of strain and stress fields the same; while the latter is dissipative because it creates a phase lag between the strain and stress fields.

We start by considering only bulk losses for both materials. This model is a modification of the one presented elsewhere [7], accounting for the fact that the metal thickness will cause the neutral axis to shift with respect to the monomaterial case. The results of our analysis using this model can be seen in Figure 5.

We find that the resonators purely made of Si_3N_4 can be represented by an imaginary Young's modulus of ≈ 0.2 GPa, which means that the behavior in this type of membranes that are characterized during our experiments can be purely explained using bulk losses. In Figure 5, top-left, it is possible to see how all the experimental data group together around the same value for the imaginary modulus.

However, when we put metal layers of different thicknesses, it is clearly visible that we need a more complex model. Figure 5 shows that for any of the nitride thicknesses the experimental counts show a mean value that depends strongly on the thickness of metal that the membrane has. This would mean that our deposited metal presents different intrinsic (bulk) losses depending on its thickness, which is very unrealistic.

Instead, our approach consists to include surface losses in addition to the already considered. In order to do this we follow the concept introduced in [8] and we consider the presence of extremely thin layer(s) at the interface between Al and Si_3N_4 and at the top of the Al layer. This layers have the same reactive elastic properties as Aluminum but they have different dissipative elastic modulus. We later scale out the thickness of this superficial layers, and that is why the surface loss modulus have units of N/m. The bottom surface of Si_3N_4 is not considered because, as it can be seen in the top-left of Figure 5, a model accounting only for bulk losses can explain the measured data. This point, however needs to be treated delicately as this conclusion contradicts previously reported data, where people showed that surface losses are also important for Si_3N_4 1-D structures [8-10]. Further measurements are being performed to check that the conclusion we reach is correct.

With the model we have described, we find that we are able to fit the loss parameters to:

$$
\begin{aligned}
E_{loss,Si_3N_4} &= 0.2 \pm 0.1 \text{ GPa}, \\
E_{loss,Al} &= 0.1 \pm 0.05 \text{ GPa}, \\
E^*_{Al-top} &= 2 \pm 0.5 \frac{N}{m}, \\
E^*_{interface} &= 20 \pm 5 \frac{N}{m},
\end{aligned}
\tag{3}
$$

with a confidence interval close to 75%. Figure 6 shows the difference between theory and experiment for all modes of metallized membranes limited by intrinsic losses. The estimations for the model are made using the central value of the parameters in Equation (3).

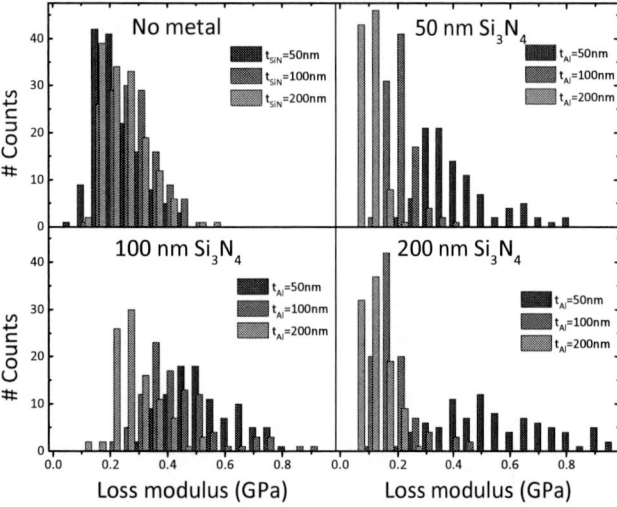

Figure 5: Histograms of the calculated imaginary components of the Young's modulus for Si_3N_4 (top-left) and Aluminum (rest of graphs) for different thicknesses. It is clear that a model considering bulk losses describes well our Si_3N_4 resonators (all counts are grouped together around the same value for the loss modulus), but not those that are multi-material (the event counts are dispersed, having mean values at different places depending on the metal thickness). This implies that an adjustment to the loss modulus needs to be done for different metal thicknesses.

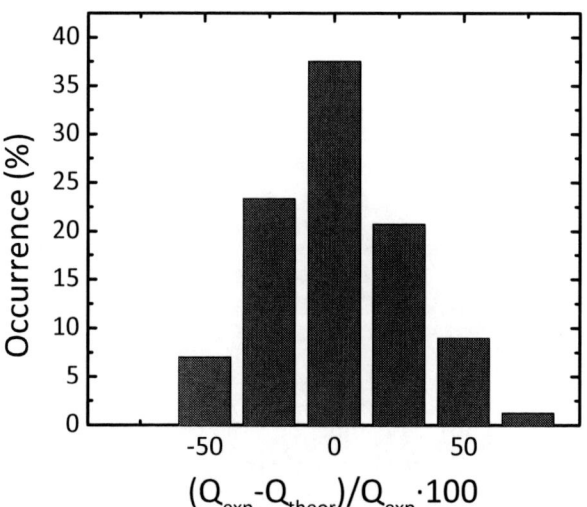

Figure 6: Relative difference between experimentally measured quality factors and theoretically estimated quality factors using our model and mean values for the fitted parameters. Close to 900 data points are considered in this plot, including values for the quality factors which differ in some orders of magnitude.

Figure 7 shows graphically how our model predicts the quality factor for a set of four membranes of 1 mm lateral dimensions and with a thickness of Si_3N_4 of 50 nm.

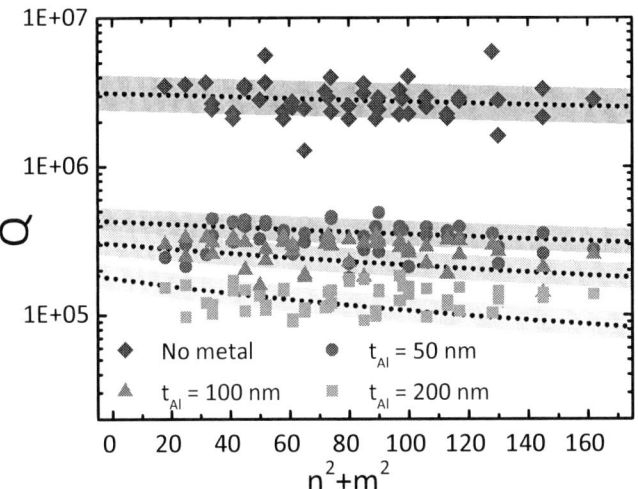

Figure 7: Experimentally obtained quality factors (scattered data) for 1mm long membranes with $t_{Si_3N_4} = 50$ nm and different metal thicknesses. Dotted black lines show the theoretical prediction using a model that accounts for surface losses. Shaded regions correspond to the (~ 75%) confidence intervals of the fit.

As a conclusion, we have quantified the importance of interface losses in multimaterial resonators which is of the utmost importance when piezoelectric [11, 12] and/or piezometallic [13] transduction are utilized or when a metal layer constitutes a functional part of the device, e.g. H_2 absorption into Pd [14]. In addition, are opening an important and interesting line or research to optimize the interfaces (by for example pre-deposition surface treatments) and the metallic material [15] in order to minimize dissipation.

ACKNOWLEDGEMENTS

The authors would like to acknowledge Prof. A. Boisen for her support and discussions, and the staff in DTU-Danchip for help in the fabrication of the membranes.

REFERENCES

[1] D.J. Wilson, C.A. Regal, S.B. Papp, et al., "Cavity Optomechanics with Stoichiometric SiN Films", *Phys Rev Lett*, vol. 103, pp. 207204, 2009.

[2] S. Schmid, K.D. Jensen, K.H. Nielsen, et al., "Damping mechanisms in high-Q micro and nanomechanical string resonators", *Phys Rev B*, vol. 84, pp. 165307, 2011.

[3] T.P. Purdy, R.W. Peterson, and C.A. Regal, "Observation of Radiation Pressure Shot Noise on a Macroscopic Object", *Science*, vol. 339, pp. 801-804, 2013.

[4] E. Gavartin, P. Verlot, and T.J. Kippenberg, "A hybrid on-chip optomechanical transducer for ultrasensitive force measurements", *Nat Nano*, vol. 7, pp. 509-514, 2012.

[5] T. Bagci, A. Simonsen, S. Schmid, et al., "Optical detection of radio waves through a nanomechanical transducer", *Arxiv*, vol. pp. 1307.3467, 2013.

[6] R.W. Andrews, R.W. Peterson, T.P. Purdy, et al., "Reversible and efficient conversion between microwave and optical light", *Arxiv*, vol. pp. 1310.5276, 2013.

[7] P.L. Yu, T.P. Purdy, and C.A. Regal, "Control of Material Damping in High-Q Membrane Microresonators", *Phys Rev Lett*, vol. 108, pp. 083603, 2012.

[8] K.Y. Yasumura, T.D. Stowe, E.M. Chow, et al., "Quality factors in micron- and submicron-thick cantilevers", *J Microelectromech S*, vol. 9, pp. 117-125, 2000.

[9] S.S. Verbridge, J.M. Parpia, R.B. Reichenbach, et al., "High quality factor resonance at room temperature with nanostrings under high tensile stress", *J Appl Phys*, vol. 99, pp. 2006.

[10] S.S. Verbridge, D.F. Shapiro, H.G. Craighead, et al., "Macroscopic tuning of nanomechanics: Substrate bending for reversible control of frequency and quality factor of nanostring resonators", *Nano Lett*, vol. 7, pp. 1728-1735, 2007.

[11] L.G. Villanueva, R.B. Karabalin, M.H. Matheny, et al., "A Nanoscale Parametric Feedback Oscillator", *Nano Lett*, vol. 11, pp. 5054-5059, 2011.

[12] L.G. Villanueva, E. Kenig, R.B. Karabalin, et al., "Surpassing Fundamental Limits of Oscillators Using Nonlinear Resonators", *Phys Rev Lett*, vol. 110, pp. 177208, 2013.

[13] L.G. Villanueva, R.B. Karabalin, M.H. Matheny, et al., "Nonlinearity in nanomechanical cantilevers", *Phys Rev B*, vol. 87, pp. 024304, 2013.

[14] J. Henriksson, L.G. Villanueva, and J. Brugger, "Ultra-low power hydrogen sensing based on a palladium-coated nanomechanical beam resonator", *Nanoscale*, vol. 4, pp. 5059-5064, 2012.

[15] S. Schmid, T. Bagci, E. Zeuthen, et al., "Graphene on silicon nitride for optoelectromechanical micromembrane resonators", *Arxiv*, vol. pp. 1305.5890, 2013.

CONTACT

*L.G. Villanueva, Tel.: +41 21 693 1187;
Guillermo.Villanueva@epfl.ch

INVESTIGATION OF DOMINANT FACTORS TO CONTROL C-AXIS TILT ANGLE OF ALN THIN FILMS FOR EFFICIENT ENERGY HARVESTING

Qi Wang, Hiroyuki Oguchi, Motoaki Hara, and Hiroki Kuwano

Graduate School of Engineering, Tohoku University, Sendai, JAPAN

ABSTRACT

We investigated the experimental conditions affecting degree of c-axis tilt angle, which is related to output power of piezoelectric energy harvesters, of the oriented aluminum nitride (AlN) thin films grown using electron cyclotron resonance (ECR) sputtering. The c-axis tilt angle of the AlN thin films deposited on silicon (100) substrates was measured by X-ray diffractometry with a 2-dimenaioanl detector (2D XRD). This study verified that 1) lowering the incident angle of the flux during thin film growth and 2) lower substrate temperature, and 3) increasing the buffer layer roughness are effective ways to increase the c-axis tilt angle of the AlN thin films.

INTRODUCTION

Vibration energy harvesters, as an attractive replacement of batteries, were widely studied in order to power sensor nodes used in wireless sensing systems [1]. Several piezoelectric films of such as lead zirconate titanate [$Pb(Zr_xTi_{1-x})$, PZT], zinc oxide (ZnO), aluminum nitride (AlN) were exploited as key materials determining performance of the vibration energy harvesters [2].

Among the piezoelectric thin films developed so far, the AlN thin film is expected to be one of the most promising thin films due to relatively high electromechanical coupling coefficient and no environmental pollution concerning by toxic Pb, problem of current best piezoelectric thin films PZT. However, output power of the energy harvesters with AlN thin films need to be increased to replace PZT. Considering that the output power of the energy harvesters is directly related to the electromechanical coupling factor k_{31} [3], that of the AlN thin film should be optimized. Our previous study found that one of the best ways to maximize k_{31} may be controlling orientation of the AlN thin films [4]. As shown in Figure 1(a), the AlN thin films usually have c-axis orientation normal to substrate surface. If we can tilt the c-axis from substrate normal as shown in Figure 1(b), we can obtain the AlN thin films with different k_{31} as indicated by our previous theoretical results shown in Figure 2: Defining the c-axis tilt angle to substrate normal as θ_c, the theoretical results shows clear dependence of k_{31} on θ_c. The maximum k_{31} is 0.244 at $\theta_c = 52.6°$ which is 206% as large as that of 0.118 at $\theta_c = 0°$.

The c-axis tilted AlN thin films have been grown by many researchers using methods such as RF-magnetron sputtering [5] or reactive DC magnetron sputtering [6]. These methods generally require high temperature to grow thin films with high crystallinity, which is essential to high piezoelectricity. Unfortunately, such high temperature process is generally not allowed in fabrication of MEMS integrated devices including the energy harvesters.

(a)   (b)

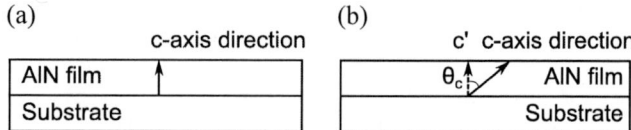

Figure 1. Schematic of (a) normal c-axis and (b) c-axis tilted AlN thin films.

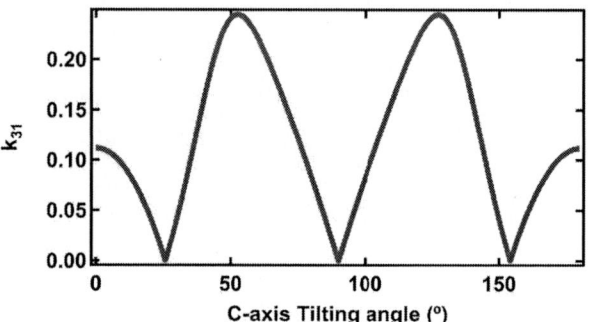

Figure 2. The theoretical relationship of k_{31} and c-axis tilt angle

We chose electron cyclotron resonance (ECR) sputtering preferable to the MEMS integrated devices by enabling lower temperature thin film growth than the method earlier mentioned without losing thin film crystallinity. ECR sputtering are also advantageous in the fact that it can generate very dense plasma at low operating pressure range which lowers contamination and improves surface smoothness of the thin films. However, there is no report of the growth of the c-axis tilted AlN thin films using the ECR sputtering.

Therefore, in this study, we aimed to investigate the conditions of the ECR sputtering affecting degree of c-axis tilt angle of the aluminum nitride (AlN) thin films.

EXPERIMENTAL,

In this study, c-axis tilted AlN thin films were deposited on Si (100) substrates using ECR sputtering by changing substrate tilt angles, substrate temperatures, and buffer layer roughness. The ECR sputtering system used in this study was shown in Figure 3. High density plasma was introduced by microwave of 2.45GHz. A cylindrical aluminum (Al) target with 99.99% purity was used as a metal source. The system also enabled sputtering at low pressure of (10^{-3}~10^{-1} Pa) which guarantees a low contamination and smooth surface of the deposited AlN thin films. The substrate can be heated from the backside by an infrared lamp located above the sample stage in the sputtering chamber. The chamber was pumped down to a base pressure of less than 1.6×10^{-5} Pa

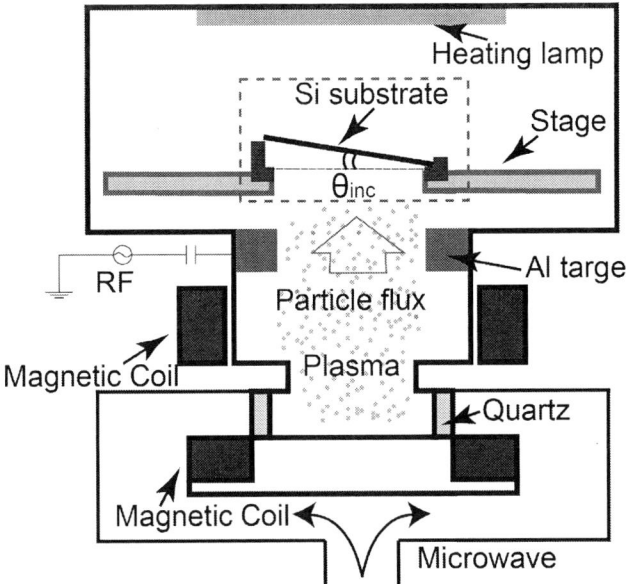

Figure 3. Schematics of ECR sputtering system with a special jig to change substrate tilt angle.

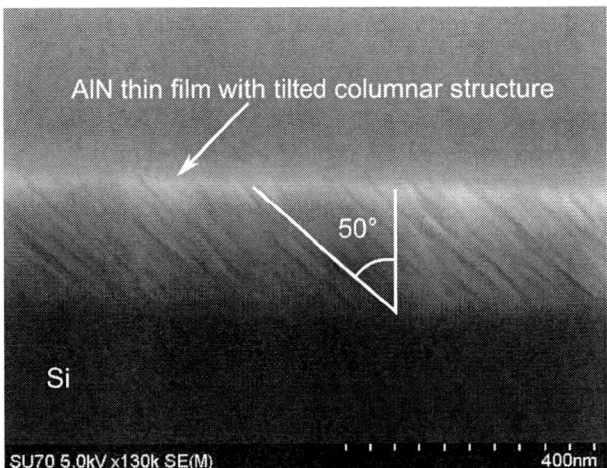

Figure 4. A Cross-sectional view of the c-axis tilted AlN thin film measured by SEM.

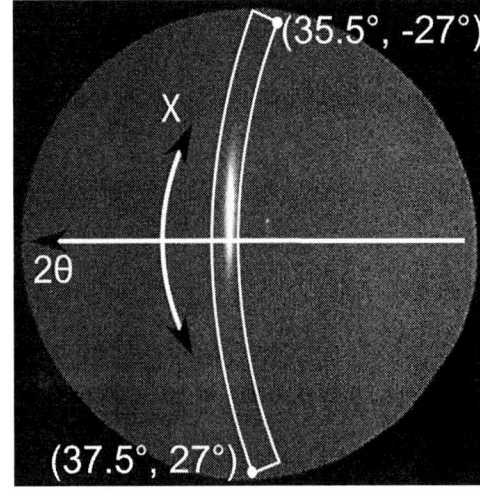

Figure 5. 2D XRD results

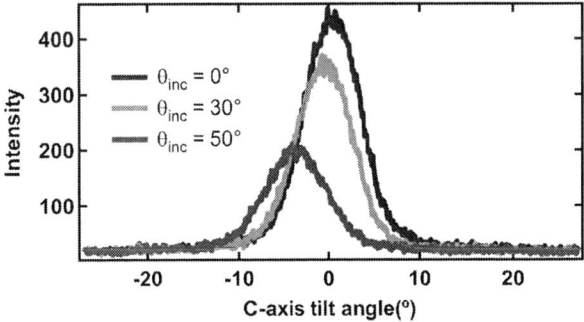

Figure 6. The c-axis tilting of AlN thin films under different surface roughness

before Ar and N_2 gases were introduced. The growth manner of the c-axis tilted AlN thin films were observed using scanning electron microscopy (SEM). The c-axis tilt angles of the AlN thin films were evaluated from diffraction images obtained using 2D XRD.

RESULTS AND DISCUSSION

Effects of substrate tilt angle

The Ordinary AlN thin films align their c-axis normal to the substrate surface which is the direction of thin film growth. If c-axis of the AlN has tendency to grow along the accumulation direction of the partible flux, the θ_{inc} should be controlled by the substrate tilt angle. Thus we deposited the AlN thin films at 300 ° C by changing substrate tilt angles using a tilting jig shown in Figure 3 holding a Si substrate with an tilt angle $\theta_{inc} = 0°, 30°,$ and $50°$. When the substrate is tilted, the particle flux impinges the substrate surface with an incident angle similar to θ_{inc}. The cross section of the AlN film deposited with $\theta_{inc} = 50°$ was observed by SEM as shown in Figure 4. The inclination angle of the columnar structure was about $50°$ from Si substrate normal. The inclination angle of the columnar structure indicates that the growth direction of the AlN thin film is similar to the jig tilt angle. As this morphological structure does not the same with θ_c, we evaluated θ_c by obtaining XRD diffraction images using the 2D-XRD. The result is shown in Figure 5. A bright spot located at $2\theta = 36.04°$ represented the AlN (002) orientation. As the spot was not located in the center along χ direction corresponding to axis fluctuation from the substrate surface normal, it proved that the c-axis of the AlN thin films was tilted. The specific tilt angle could be obtained by integrating intensity along χ direction and the results were shown in Figure 6. The absolute tilt angles were $0.31°, 0.62°$ and $3.74°$ for $\theta_{inc} = 0°, 30°$ and $50°$ respectively. It proved that

c-axis tilt angle of the AlN thin films could be increased by increasing substrate tilt angle during thin film growth using ECR sputtering.

Effects of substrate temperature

To further increase c-axis tilt angles, we investigated the

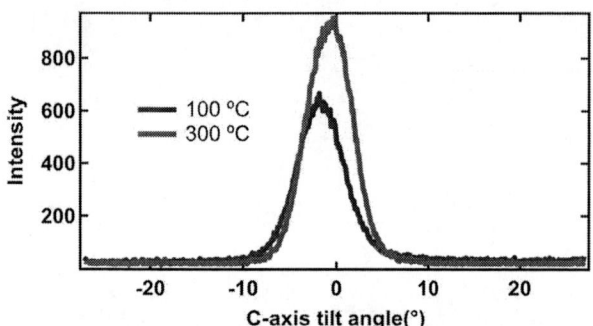

Figure 7. The c-axis tilting of AlN thin films under different substrate heating conditions

effect of substrate temperature. For this investigation, the substrate temperature was changed by turning on and off the heating lamp with fixing the substrate's tilt angle at 30°. When the heater was turned off, the substrate was heated up by the plasma during sputtering process and substrate temperature was stabilized at around 100°C.

As shown in Figure 7, the absolute c-axis tilt angle was 0.76° when the substrate temperature was 300°C. The absolute c-axis tilt angle increased to 1.72° when the substrate temperature was 100°C. The possible reason of the increase may be lower ad-atom mobility during thin film growth at lower temperature [5]. The results indicated that lower substrate temperature contributed to a higher c-axis tilt angle of the AlN thin films.

Effects of buffer layer roughness

We also investigated the effect of the AlN buffer layer roughness on the AlN c-axis tilt angles. As buffer layers, 250 nm thick c-axis oriented AlN thin films were deposited on Si(100) substrates. Then, NMD-3 developer (2.38% TMAH), Microposit developer (Na_2SiO_3 1-2.5%, Na_3PO_4 3-5%, H_2O), and Intervia developer additive (2.38% TMAH with surfactant) were used to etch the AlN buffer layers to obtain different surface roughness. As the active etching compound is tetramethylammonium hydroxide (TMAH), and NMD-3 developer had highest TMAH concentration, NMD-3 developer has the highest etching rate compared with other developers used in this experiment. Thus the buffer layers etched by NMD-3 developer had the roughest surfaces compared with other buffer layers. The surface roughness was measured by atomic force microscopy (AFM). The 3D images of the buffer layer surfaces were shown in Figure 8 and the arithmetic average roughness were 6.2nm, 4.8nm and 1.5nm for NMD-3 developer, Microposit developer and Intervia developer additive respectively. After the roughness evaluation, the c-axis tilted AlN thin films were deposited using ECR sputtering on the buffer layers fixing θ_{inc} = 30°. Then the AlN thin films orientations were evaluated by the 2D-XRD and the c-axis tilt angles were calculated by integrating the density along χ direction. The results were shown in Figure 9. It clearly indicated that when the average roughness of buffer layer increased from 1.5 nm to 6.2 nm,

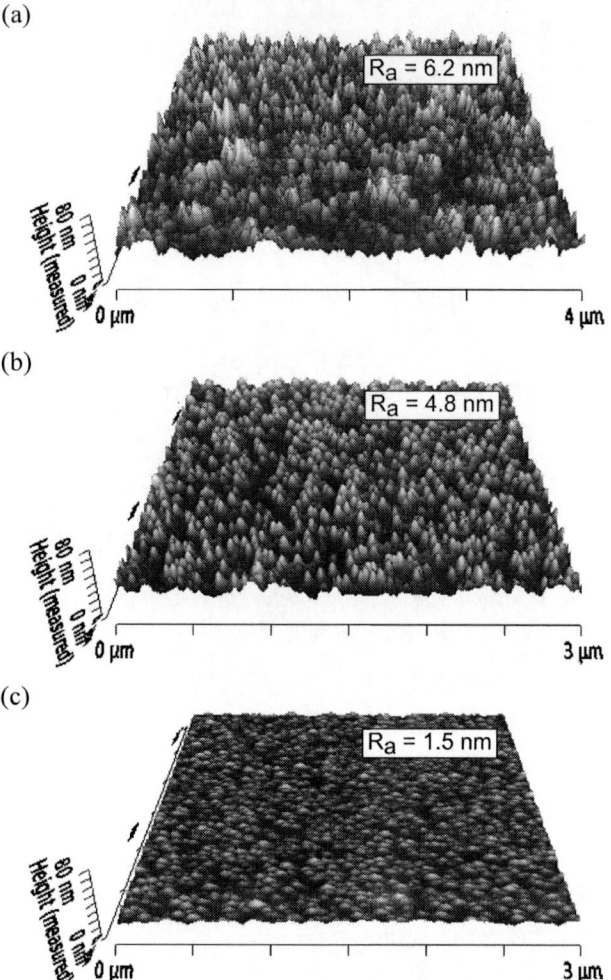

Figure 8. AFM figures for (a) AlN surface etched by NMD-3 developer, (b) AlN surface etched by Microposit developer, (c) AlN surface etched by Intervia developer additive.

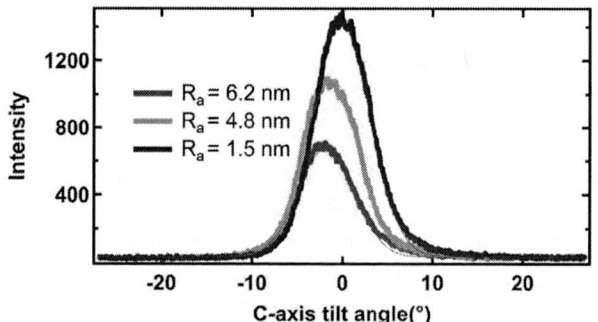

Figure 9. The c-axis tilting of AlN thin films under different surface roughness

the c-axis tilt angle of AlN c-axis increased from 0.05° to 1.71°. Thus roughening buffer layer surface can increase the c-axis tilt angle. The possible reason of increase in the c-axis tilt angle may be again lower ad-atom mobility during thin film growth on a rough surface of the buffer layers.

978-1-4799-3510-9/14 $31.00 © 2014 IEEE

CONLUSION

In order to increase the output efficiency of the vibration energy harvesters, this paper focused on investigating dominant factors to control c-axis tilt angle of the AlN thin films grown using ECR sputtering. The thin films were grown by changing substrate tilt angles, substrate temperatures and buffer layer roughness. The SEM image showed clear tilted columnar structure which proved an inclined impingement of particle fluxes during thin film growth. The AFM images showed buffer layers with different roughness. Finally, the c-axis tilt angles of the AlN thin films were calculated from the measurement results of 2D XRD. The results verified that increasing substrate tilt angle, lowering substrate temperature and increasing buffer layer roughness are benefit for higher c-axis tilt AlN thin films growing.

ACKNOWLEDGEMENTS

A part of this work was supported by NEDO (New Energy and Industrial Technology Development Organization, Japan) Project "Research and Development of Nanodevices for Practical Utilization of Nanotechnology": "Research and Development of Vibration-Based Micro Energy Harvester Using Lead-Free Piezoelectric AlN Thin Film (10003460-0, 10003461-0)."

REFERENCES

[1] S. Roundy, P. K. Wright, J. Rabaey, "Study of Low Level Vibration as Power Source for Wireless Sensor Nodes", *Computer Commun.,* vol. 26, pp. 1131-1144, 2003.

[2] S. T. McKinstry, P. Muralt, "Thin Film Piezoelectric for MEMS", in *J. Electroceram,* vol. 12, pp. 7-17, 2004.

[3] K. Najafi, T. Galchev, E. E. Akatakka, R. L. Peterson, J. McCullagh, "Microsystem for Energy Harvesting", in *Digest Tech. Papers Transducer 2011,* Beijing, June 5-9, 2011.

[4] Q. Wang, Z. P. Cao, H. Kuwano, "Vibration Energy Harvesters Based on C-axis Tilted AlN Thin Films", in *Digest Tech. Papers PowerMEMS 2012,* Atlanta, Dec. 2-5, 2012.

[5] A. F. Mameri, M. B. Assouar, O. Elmazria, C. Gatel, J. J. Fundenberger, B. Benyoucef, "C-axis Inclined AlN Film Grwoth in Planar System for Shear Wave Devices", in *Diam. Relat. Mater,* Vol 17, pp. 1770-1774, 2008

[6] J. Bjurstrom, G. Wingqvist, and I. Katardjiev, "Synthesis of Textured Thin Piezoelectric AlN Films With a Nonzero C-Axis Mean Tilt for the Fabrication of Shear Mode Resonators", in *IEEE Ultrason., Ferroelectr., Freq. Control,* Vol 53, pp. 2095-2100, 2006

CONTACT

Qi Wang

Tel (Fax): +81-22-795-4771

Email: wang@nanosys.mech.tohoku.ac.jp

INVESTIGATION OF THE FATIGUE ORIGIN AND PROPAGATION IN SUBMICROMETRIC SILICON PIEZORESISTIVE LAYERS

Giacomo Langfelder[1], Stefano Dellea[1], Patrice Rey[2], Audrey Berthelot[2], and Antonio Longoni[1]

[1]Politecnico di Milano, Milano, ITALY
[2]CEA-Leti, Grenoble, FRANCE

ABSTRACT

The work investigates fatigue damage accumulation in a 250-nm thick Silicon layer that can be integrated in surface micromachining processes to realize piezoresistive sensing elements, as an alternative to capacitive detection. The investigation is done through a suitably designed structure which combines a 20-μm-thick layer, used to apply cyclic stresses, to the nanometric layer, where stress accumulation is obtained. Fatigue results are compared to previous works on micrometric Silicon and put in the context of the current theories about the origin and propagation of fatigue in micro- and nano-machined Silicon.

INTRODUCTION

The combination of nanometric elements within a surface micromachining microelectromechanical system (MEMS) process has proved to be a potential way to boost the performance of inertial sensors. Silicon beams with a cross-section of about $(250 \text{ nm})^2$ can be used as highly-sensitive piezoresistive detecting elements, with potential advantages in terms of power dissipation and miniaturization with respect to commonly used capacitive readout. Demonstration of inertial sensors based on this concept can be found e.g. in [1-2]. As there are several classes of vibrating sensors that need to stand a long number of cycles during their lifetime, it is necessary, from a reliability point of view, to investigate the aging and fatigue performance of this Silicon layer. Besides, from a scientific point of view, the collection of fatigue behavior data on such a thin layer may provide the community with new evidences supporting the still debated theories behind the origin and propagation of fatigue in Silicon: indeed though it is a brittle material at the macro scale, with no known acting extrinsic toughening mechanism, clear evidences of fatigue behavior were found in Silicon at the micro scale [3-6].

In this work we present the experimental study performed on suitable structures designed and tested for the analysis of the long-term reliability and fatigue behavior of 250-nm thick crystalline Silicon. The process for the structures realization was already described in [1] and is exploited here for the design of 20-μm-thick suspended masses that apply mechanical stresses to 250-nm-thick notched specimens of various geometries. With a surface-to-volume ratio 10 to 100 times smaller than most of the previous researches in the literature, this work extends to the nanometric domain the still-debated data previously published about the origin and propagation of fatigue in Silicon and polysilicon at the micro scale. Among the theories proposed for the justification of this phenomenon, the ones which were for several aspects (even if not completely) validated by experiments are the so-called *environmental* model [4] and *mechanical* model [5]. The former ascribes the fatigue origin to the thin surface oxide layer (which natively covers the Silicon surface) through a mechanism known as stress-assisted oxidation and environmentally assisted crack growth within the Silicon oxide, up to a crack length which is critical for the whole structure [7, 8]. The latter ascribes fatigue to wear and debris (resulting from various process steps) that can accumulate in natively formed cracks inside the Silicon itself, and play a stress amplification role during the compressive portion of the cycles [9, 10].

The results obtained in this work, showing fatigue failure for stresses as low as the 47% of the nominal tensile strength in laboratory environment, will be put in the context of the mentioned theories and will also provide guidelines for a safe design of piezoresistive nanometric sensing elements within vibrating sensors.

TEST STRUCTURE DESCRIPTION

A schematic view of the designed test structure is shown in Fig. 1: a 20-μm-thick, 1.45-mm-long central frame, suspended through six springs, forms with the constrained parts a set of $N_F = 1408$ comb-finger driving electrodes (nominal gap $g_0 = 1$ μm, rest overlap $l_0 = 5$ μm) used to electrostatically apply a force to a lever system, designed at one end. The lever system is constrained by a 250-nm-thick specimen with a 700-nm curvature radius, such that the delivered force results in a stress concentration at its notch root. The stiffness in the motion direction given by the nanometric specimen is a relatively large fraction (~ 3%) of the overall structure stiffness. In the other in-plane direction and in the out-of-plane direction the structure stiffness is orders of magnitude larger and mostly set by the six suspending springs.

The structure displacement under applied forces can be monitored through another set of capacitive electrodes (same geometry as above), so that during fatigue, changes in the displacement under the same force can be related to fatigue damage accumulation in the nanometric constraint.

Differing from previous works, the electrostatic actuator is bidirectional. In this way the effects of different values of the load ratio R during fatigue cycles (defined as the ratio of the compressive to the tensile portion of the stress, compression taken as negative) on the specimen can be examined, which may be significantly relevant to assess the validity of the mechanical model.

978-1-4799-3510-9/14 $31.00 © 2014 IEEE

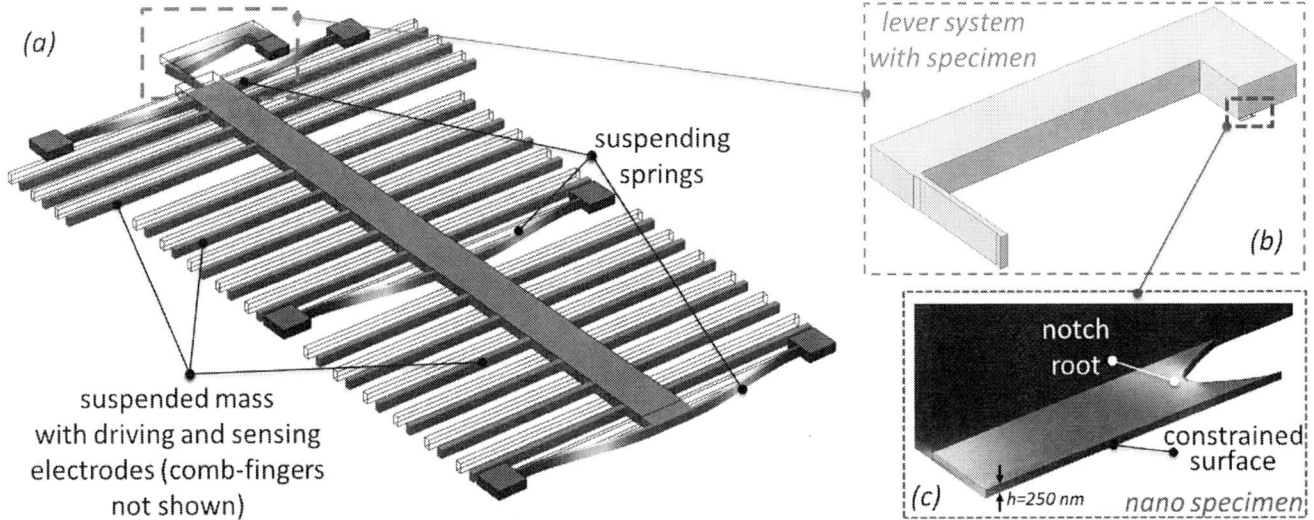

Figure 1: working principle of the test device: (a) the suspended mass (20 μm in thickness) applies – through a set of comb-finger actuators – a force to a lever system (b), which results in a stress concentrated at the notch root of a suitably designed specimen (c) with a thickness of 250 nm.

EXPERIMENTAL CAMPAIGN

Fig. 2 reports a top-view scanning electron microscope (SEM) picture of the realized device. The planned campaign consists in three steps which are 1) the statistical evaluation of the nominal tensile strength, 2) the fatigue campaign itself, at different load ratios and in different environmental conditions, and 3) an analysis and comparison of the failed specimens (and their crack paths) through SEM inspections. This paper reports the first step and part of the second one.

Evaluation of the nominal tensile strength

In a first part of the experiment, 12 samples are led to single-cycle failure in order to evaluate the nominal tensile strength. Fig. 3 reports typical measurements: the device is actuated with a quasi-stationary voltage ramp, thus a force proportional to the square of the applied voltage.

The measured displacement (seen as a capacitance variation at the sensing electrodes) follows a parabolic curve with a coefficient which depends on the overall structure stiffness. The evidence of the nanometric specimen failure is in the sudden change of the characteristic curve. Repeating the measurements after the failure confirms the change in the parabolic slope.

A further evidence of the specimen failure was performed by measuring, on a limited number of samples, the resonance frequency before and after the test through the step-response method described in [11]. A sample result is reported in Fig. 4, showing a relative change in the resonance frequency before and after the failure of ~ 1.5 %, in line with the predicted 3% ratio of the nanometric constraint to the overall stiffness ($\Delta f/f = \frac{1}{2} \Delta k/k$).

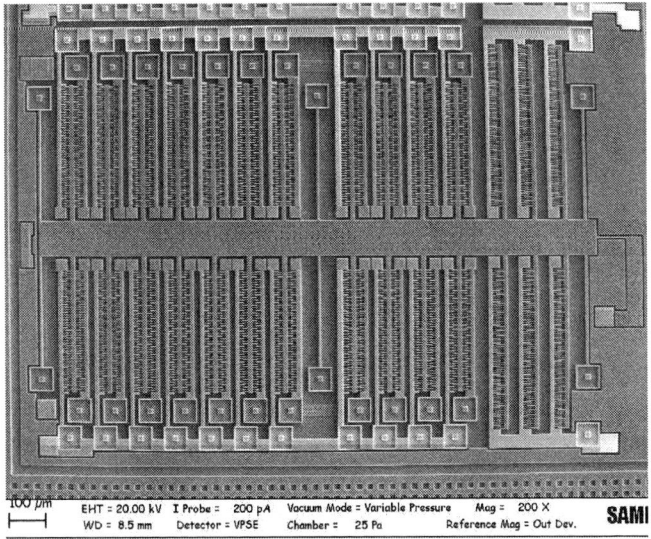

Figure 2: SEM image of the structure. The lever system is clearly visible at the right end.

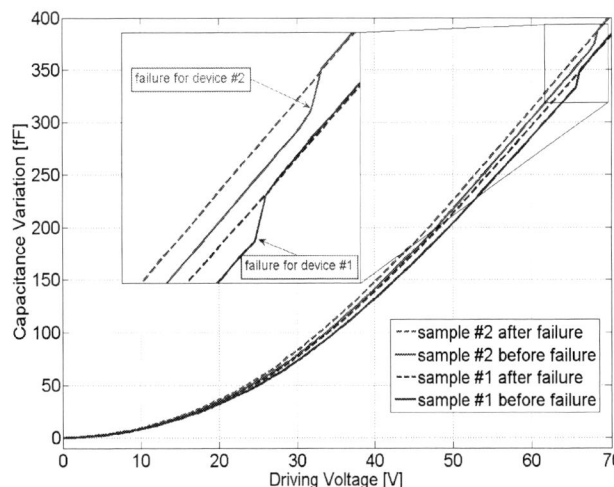

Figure 3: sample results for the nominal tensile strength evaluation procedure. Devices are driven up to single-cycle failure, which appears as a change in the CV curve when the nano specimen fails (see the inset).

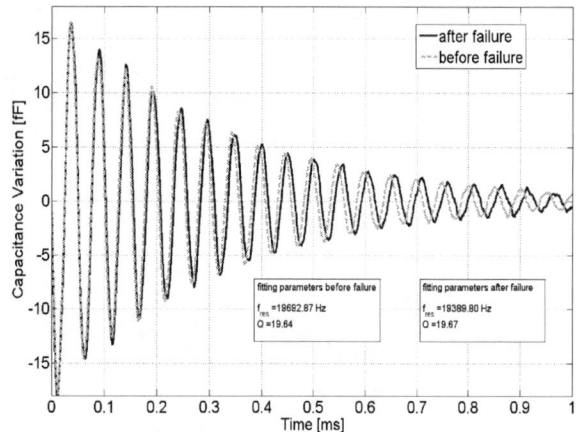

Figure 4: measured response to a step force for a sample structure, before and after the single-cycle failure. The change in the resonance frequency is symptomatic of the specimen failure.

The average failure voltage $V_{f,a}$ is then converted into a stress at the notch root through finite element simulations where a force corresponding to an applied voltage $V_{f,a}$ is applied to the device. The estimated nominal tensile strength is in the order of 8.5 GPa.

Fatigue campaign

In a second part of the experiment, 13 samples are actuated with voltage sine waves well below the nominal tensile strength. For each device, the actuation frequency is chosen slightly below the measured resonance frequency (typically ~ 400 Hz below) to facilitate damage accumulation monitoring (as in the testing principle described in [12]).

By choosing the distance from the resonance frequency and the voltage applied to the bidirectional actuators, one can set different values of the compressive and tensile displacements, in order to set a specific load ratio. First tests were performed at $R = -0.8$. As the maximum and minimum displacements of the structure during each cycle are monitored by the capacitive readout system, the actual load ratio for every test is known with very good precision. The transfer function between force and displacement can be thus real-time computed as the force from the comb fingers can be estimated from the geometry and the applied voltage.

Figure 5 reports a sample measurement of the relative change in the structure transfer function during a fatigue test and up to the device failure, for the specific case of a maximum stress during each cycle (normalized to the nominal strength) $\sigma_{max}/\sigma_N = 0.55$, a load ratio $R = -0.81$, and a total number of lifecycles $N = 9 \cdot 10^8$, at a temperature of ~ 22 °C in laboratory environment. The figure shows a clear damage accumulation, beginning to be evident roughly at $2 \cdot 10^6$ cycles, i.e. 104 s after the beginning of the experiment. The time evolution of this measurement (decreasing damage accumulation ratio with increasing lifetime) is in line with what found in previous works,

where thickened oxide was found at the end of the measurement, and therefore it is in line with the environmental model [7]: in this hypothesis some tens of seconds seem to be required for the oxide to significantly thicken.

Figure 6 collects the results for all the tested samples, plotting the number of cycles to fatigue failure N on the x-axis with respect to the maximum applied stress σ_{max}/σ_N on the y-axis (shown as a percentage of the found nominal tensile strength). Experimental points are fitted in terms of a Wöhler curve and compared to results on 22-µm and 15-µm-thick [6] samples of the same geometry, tested at similar load ratios, but realized in polysilicon.

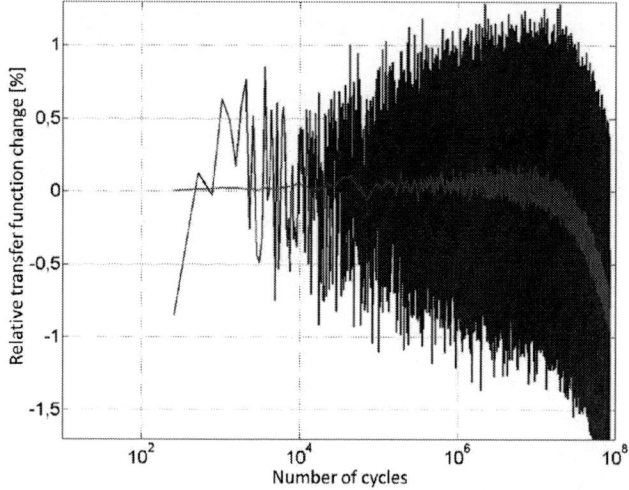

Figure 5: evolution of the transfer function change during a fatigue test. The clear decrease in the elastic stiffness indicates a damage accumulation at the notch root, that can be interpreted either in terms of crack growth or of oxidation at the notch root.

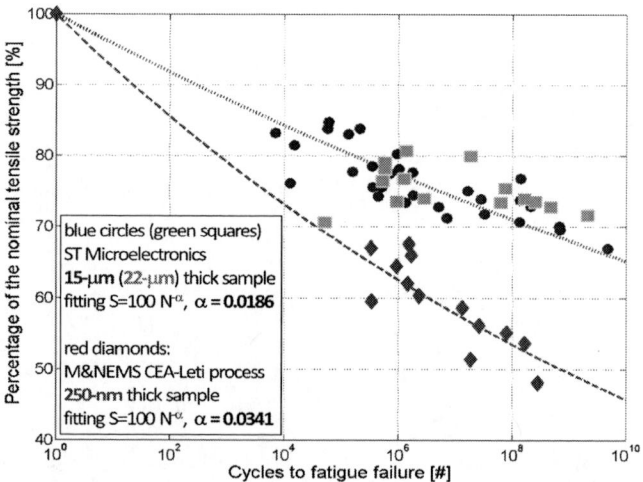

Figure 6: comparison of experimentally obtained data and fitting Wöhler curves, obtained on 15-µm [6], 22-µm, and 250-nm thick samples.

As a practical result, such a curve should be taken into account in the design phase of vibrating devices: given the vibration frequency and the expected device lifetime, the designers should be aware of the maximum stresses that the nano-gauge sensing elements can stand during each cycle, to avoid anticipated failures.

Effects of different load ratios: preliminary results

In order to extend the amount of collected data and to have therefore a deeper insight on the fatigue behavior, further measurements are planned at different load ratios and in different environmental conditions. In particular, load ratio values of $R = [0\ -0.8, -1.2, -1.6]$ will be investigated to put in evidence the possible role of the compressive portion of the stress. As a preliminary result, two samples tested at the 55% of the nominal strength and at $R = -1.2$ showed a lifetime of $5.9 \cdot 10^5$ and $2.2 \cdot 10^6$, which should be compared to the average value of $\sim 2 \cdot 10^7$ for the samples tested at $R = -0.8$. From this preliminary measurements the role of compression seems therefore to be quite relevant.

DISCUSSION AND CONCLUSION

As discussed in the literature, for a comprehensive analysis of fatigue, a lot of effects should be cross verified through different types of measurements in order to experimentally validate or exclude the proposed theories. Main dependencies to be verified are on *(i)* the environmental condition (presence of oxygen and/or water vapor), *(ii)* the amount of compression, *(iii)* the number of elapsed cycles rather than the elapsed time (static fatigue).

In this work the preliminary results of a new campaign performed on sub-micrometric Silicon layers have highlighted a marked difference in the behavior with respect to > 60 times thicker structures (e.g. $1.5 \cdot 10^6$ cycles vs $6 \cdot 10^9$ cycles when subject to only 68% of the nominal strength).

The observed trend in the damage accumulation seems in line with the environmental theory but does not exclude the mechanical model. On the contrary, the dependence on the load ratio (preliminary observed on a few samples only) and some observed very fast failures (e.g. < 1 s of total fatigue time) are in line with the mechanical model and cannot be wholly explained by the environmental theory as *(i)* it is hard to imagine such fast oxidation times and *(ii)* there is no definite explanation for the dependence on the load ratio for this theory. Measurements will continue with different load ratios and on packaged devices (absence of oxygen and water vapor), to increase the number of experimental evidences, to improve the statistics, and in turn to help the understanding of the origin of fatigue phenomena in micro- and nano-scale Silicon.

ACKNOWLEDGEMENTS

The authors thank Dr. N. Aresi and J. Della Bosca for helping with the experimental setup preparation and the measurement campaign. This work was supported by the European Union under the FP7-ICT grant agreement 288318 (NIRVANA project).

REFERENCES

[1] P. Robert, V. Nguyen, S. Hentz, L. Duraffourg, G. Jourdan, J. Arcamone, "M&NEMS: A new approach for ultra-low cost 3D inertial sensor," in *Proc. IEEE Sensors Conference*, Christchurch, New Zealand, Oct. 2009, pp. 963-966.

[2] A. Walther, M. Savoye, G. Jourdan, P. Renaux, F. Souchon, P. Robert, C. Le Blanc, N. Delorme, O. Gigan, C. Lejuste, "3-AXIS Gyroscope with Si Nanogage Piezo-resistive Detection," in *Proc. MEMS 2012*, Paris, France, Jan. 2012, pp. 480-483.

[3] W. W. Van Arsdell and S. B. Brown, "Subcritical crack growth in silicon MEMS," *J. Microelectromech. Syst.*, vol. 8, no. 3, pp. 319–327, Sep. 1999.

[4] D. H. Alsem, C. L. Muhlstein, E. A. Stach, and R. O. Ritchie, "Further considerations on the high-cycle fatigue of micron-scale polycrystalline silicon," *Scripta Materialia*, vol. 59, no. 9, pp. 931–935, Nov. 2008.

[5] H. Kahn, A. Avishai, R. Ballarini, and A. H. Heuer, "Surface oxide effects on failure of polysilicon MEMS after cyclic and monotonic loading," *Scripta Materialia*, vol. 59, no. 9, pp. 912–915, Nov. 2008.

[6] G. Langfelder, S. Dellea, F. Zaraga, D. Cucchi, M. Azpeitia Urquia, "The Dependence of Fatigue in Microelectromechanical Systems on the Environment and the Industrial Packaging," *IEEE Trans. on Industrial Electronics*, vol. 59, n. 12 (2012), pp. 4938-4948.

[7] C. L. Muhlstein, R. T. Howe, and R. O. Ritchie, "Fatigue of polycrystalline silicon for microelectromechanical system applications: Crack growth and stability under resonant loading conditions," *Mech. Mater.*, vol. 36, no. 1/2, pp. 13–33, Jan./Feb. 2004.

[8] M. Budnitzki and O. N. Pierron, "Highly localized surface oxide thickening on polycrystalline silicon thin films during cyclic loading in humid environment," *Acta Mater.*, vol. 57, no. 10, pp. 2944–2955, Jun. 2009.

[9] H. Kahn, R. Ballarini, J. J. Bellante, and A. H. Heuer, "Fatigue failure in polysilicon not due to simple stress corrosion cracking," *Science*, vol. 298, no. 5596, pp. 1215–1218, Nov. 2002.

[10] H. Kahn, R. Ballarini, and A. H. Heuer, "Dynamic fatigue of silicon," *Curr. Opin. Solid State Mater. Sci.*, vol. 8, no. 1, pp. 71–76, Jan. 2004.

[11] G. Langfelder, A. Tocchio, M. J. Thompson, G. Jaramillo, and D. A. Horsley, "Assessing micromechanical sensor characteristics via optical and electrical metrology," in *Proc. IEEE Sens. Conf.*, Waikaloa, HI, Nov. 2010, pp. 1765–1769.

[12] G. Langfelder, A. Longoni, and F. Zaraga, "Monitoring fatigue damage growth in polysilicon microstructures under different loading conditions," *Sens. Actuators A, Phys.*, vol. 159, no. 2, pp. 233–240, May 2010.

CONTACT

*G. Langfelder, tel: +39-02-2399-3425; giacomo.langfelder@polimi.it

MEMS 2014 KEYWORD INDEX

Scroll to the keyword and select a **Blue** link to open a paper. After viewing the paper, use the bookmarks to the left to return to the beginning of the Keyword Index.

3D 177, 366, 425, 1189
3D Fabrication 660
3D Ice Printing 52
3D Nanofabrication 437
3D Polydimethylsiloxane Foil 927
3D Silicon Microelectrodes 951

A

Accelerometer (s) 28, 761
Accelerometer Gyrometer Co-Integration 725
Acoustic Impedance Matching 845
Acoustic Sensor 849
Acoustics 897
Activated Carbon-Graphite Configuration 401
Active Scaffold 917
Actuator (s) 192, 909, 913
Adaptive Focusing 909
Additive Manufacturing 510
Adhesion Forces 596
Adjustable Sensing Ranges 713
Airflow Sensing 777
AlGaN/GaN HEMT-Based FET 250
Aligned 159
Alumina 449, 1011
Aluminum Nitride(AlN) 124, 636, 688, 729, 1265
Alzheimer Disease 314
Amorphous Carbon 143
Amplification 242
Anchor Loss 1257
Angular Acceleration Sensor 696
Annealing 825
Anodization 1011
Apparent Size Correlation 939
Aptamer 250
Arterial Diameter Monitoring 869

Arterial Dysfunction 869

Artificial Cell 17

Artificial Cilia 927

Artificial Finger 785

Assembly 433, 1015

Assist-Free 433

Asthma 935

Atomic Clock 552

Atomic Force Microscope 100, 128, 733, 789

Atomic Layer Deposition 167, 342, 449, 584

Atomization 1043

Autonomous 218

B

Back Action 725

Bacteria 246

Bacterial Cellulose 518

Band-Pass Filter (BPF) 1237

Battery 358

Bearings 200

Bent Cantilevers 177

Bi-Directional Gas Pump 294

Bioactuator 196

Biodegradable 358

Biofuel Cell 163

Bioimaging 322

Bioinspired 696, 741

Biomedical Application 901

BioMEMS 177, 238, 358

Biomimetics 741

Biomolecular Interactions 272

Biosensor (s) 1075, 1083

Birdbath Resonator Gyro 20

Bistable Micro Valve 1023

Blood Coagulation 330

Blood Storage 833

Boron-Doped Diamond 322

Bosch Process 459

Breath Diagnostics 935

BTX 294

Buckled Beam 692

Buckling 1099

Buckypaper 441
Bulk PZT 524
Bulk Tungsten 502
Bundles 48

C

C-Reactive Protein 250
Cancer Diagnosis Devices 889
Cantilever 769
Capacitance Measurement 222
Capacitive Gap Width 1245
Capacitive Micromachined Ultrasonic Transducers 584
Capacitive Sensing 680
Carbon Nanotubes 48, 167, 342
Cardiomyocyte 159
Cell 290
Cell Cell Interaction 943
Cell Characterization 901
Cell Culture 280, 518
Cell Culturing Device 181, 264
Cell Manipulation 1031
Cell Pairing 185
Cell Separation 951
Cell Surgery 885
Cell Targeting Aptamer 242
Cell Trapping 947
Cell Viability 865
Cellular Aggregate 925
Cellular Materials 510
Cellular Parylene 374
Cemented Carbide 652
Ceramic Micromachining 494
Chain-Like Structure 1197
Charge Bias 1277
Charge Pump 1273
Chemical Sensor 302, 318
Chemical Sensors & Systems 226
Chemical Synthesis 92
Chemiluminescence 1047
Chip Destruction 1123
Chip Scale Package 1213
Chip Security 1123

Circulatory	1007
Closed-Loop	761
Closest Packing	1163
Cluster	729
CMOS-Based Amperometric Biosensor Array	322
CMOS-IC	1139
CMOS-MEMS	676, 757, 1249
CMOS-MEMS Microphone	136
Co-Culture	943
Co-Integration	1071
Column	1007
Comb Drive	1229
Combinatorial	334
Combined Electrodes	409
Complementary	676
Complementary Metal Oxide Semiconductor	1071
Composite Membrane	1217
Confocal Hyperchromatic System	1167
Confocal Microscopy	881
Contact Lens Tonometer	624
Contact Material	143
Continuous Medium Exchange	234
Controllable Delivery	865
Convective Accumulation	1035
Cooling	1039
Copper On Polyimide	528
Coriolis Flow Sensor	975
Corner Lithography	1111
Coronary Artery Disease	893
Creep	660
Critical Bifurcation	749
Crosslinking	471
Cryopreservation	829
CTE	40
Cu Etching	528
Cubic	425
Cuff Electrode	9
Curve Fit	580
Curved Micromirror	1185
Curved Piezoelectric Micromachined Ultrasonic Transducers	124
CVD	486
Cylindrical Lens	1171

D

Deep Etching	502, 1185
Deformability	833
Degradable Scaffold	518
Degradation	616
Device Tuning	660
Diamond	1159
Diaphragm with Electrode	544
Dielectric	1011
Dielectric Liquid	713
Dielectrophoresis	268, 837, 951
Differential	1249
Differential Readout	1237
Differential Scanning Calorimeter	306
Diffraction Optics	608
Digital Microfluidics	272, 334, 829, 955
Direct Bonding	1119
Direct Laser Writing	471
Direct Prototyping	298
Disk Resonator Gyroscope	749
Displacement Sensor	1185
Dissolvable Material	604
Dither	608
DRIE	36, 459
Droplet (s)	92, 486, 664, 983
Droplet Merging	1003
Droplet Micro-Array	96
Drosophila Melanogaster	196
Drug Delivering	52
Dual Axis	797
Dual Axis Confocal Endoscope	805
Dual Mode	120
Duffing Nonlinearity	749
Dynamic Mask Lithography	733
Dynamics	1253

E

Eccentric Mass	370
Electret	366, 417, 704, 717, 1229
Electro Tactile Display	1183
Electrocaloric cooling	544

Electrocardiography 841

Electrochemical Etching 1183

Electrochemical Impedance 104

Electrochemiluminescence-Induced Fluorescence 108

Electrohydrodynamic 963

Electrolysis 809

Electromagnetic 378, 425, 564

Electromagnetic Actuation 1147

Electromechanic 218

Electron Beam Lithography 437

Electron Emitter 467

Electronic Article Surveillance 76

Electroosmotic Flow 1067

Electrophoresis 959

Electroplated Iron Cobalt 520

Electroplating 40, 64, 453, 1135

Electroporation 234, 268, 817

Electrospray 17, 100

Electrostatic 366

Electrostatic Actuation 805

Electrostatic Force 913

Electrostatic Generators 385

Electrothermal 789

Electrowetting 72, 222, 284, 1011, 1039

Electrowetting-On-Dielectric(EWOD) 338, 955

Electrowetting-On-Dielectric (EWOD), Recoverable 1047

Embryo Vitrification 829

Embryonic Body 181, 280

Emitter Array 463

Encapsulation 56, 588, 1281

Encased Cantilevers 128

Endoscopy 200

Endothelial Cells 238

Endpoint 459

Energy Harvest 159, 350, 366, 370, 374, 378, 382, 397, 413, 417, 421, 425, 429, 568

Energy Scavenging 897

Energy Storage Device 401

Enzyme 959

Epi-Seal 588, 1277

Escherichia Coli 175

Etch Hole 1257

Etching 656

Eutectic Bonding 64
EUV 482
Excitation 108
Extracellular Electron Transport (EET) 362
Extraction 995

F

Fabrication 478, 620
Fatigue 640
FENO 935
Ferrofluid (s) 200, 350
Fiber MEMS 60
Film Stress 482
Filter Termination 1249
Fine Line 478
FinFET 1063
Flapping Wing 648
Flexible 84, 306, 389, 556, 817
Flexible Device 737
Flexible Electronics 230, 548
Flexible Printed Circuit 346
Flexible Sensor 528
Flow Control 1031
Flow Meter 1139
Flow Regulator 935
Fluidics 56
Force Map 151
Force Sensor 680, 769
Frequency Divider 210
Frequency Split 612
Fresnel Lens 1175
Fuel Cell 393
Functional Electrical Stimulation 616
Fused Silica 20

G

G-Sensitivity 32
Galinstan 540, 967
Gas Analysis 171
Gas Chromatography 1007
Gas Sensing 230
Gas Sensor (s) 326, 506, 1179

Glancing Angle Deposition..437

Glass Reflow..1233

Gradient Force..721

Graphehe FET..230

Graphene..326, 362, 393, 486, 1055, 1075, 1159

Graphene Cantilever..1087

Graphene Loudspeaker..556

Graphene Woven Fabric..624

Gratings..490

Grayscale Mask Fabrication..1027

Grid Filter..482

Gyroscope..32, 612, 721

Gyroscope Quadrature..36

H

Hard Magnetic Layer..520

Heat Effects..825

Heat-Transfer Resistance..272

Heater Electrode..1047

Helicobacter Pylori..821

Hemispherical..672

Hemispherical Shell Resonator..40

Hemispherical Shells..494

HfO_2..1063

High Aspect Ratio..374

High Flow Knudsen Pump..112

High Frequency..955

High K Dielectric..584

High Quality Factor..24, 628

High Resolution..624

High Resolution Trimming..494

High Throughput Screening..334

HIV/AIDS Diagnostics..256

Honeycomb..449

Hot-Wire..777

Hotspot..1039

Human Detector..1213

Humidity Sensor..532

Hybrid..354

Hybrid Valves..260

Hydrodynamic Trapping..246

Hydrogel..733

HydrophilicityControl 925
Hydrophobic 979
Hyperspectral Imaging 1167
Hyperthermia Treatments 877
Hysteresis Window 1095

I

Ice Mould Microfluidics 52
Immersion 498
Immunosensor 809
Impedance Sensing 338
Implants 1127
In-Plane Capacitive Flow Sensors 971
In-Plane Gap Closing 136
Inclinometer 797
Induced Pluripotent Stem Cell 181
Inductively Coupled Plasma 502
Inertial Focusing 939, 1051
Inertial Measurement Units 28
Inertial MEMS 725
Infrared Absorptance 644
Infrared Focal Plane Array 1225
Injection Molding 560
Inkjet 963, 967
Inkjet Printing 506, 536
Insect 163
In-situ Gyroscope Calibration 608
Integrated 729
Integrated Heater 905
Integration 1189
Internet of Things 773
Intracellular Recording 155
Intrinsic Losses 632
Intrinsic Stress 177
Invar 40
Ion Assisted Breakdown 171
Ion Concentration Polarization 276
Ion Selective Membrane 421
Ion Source 463
Ionic Liquid 326, 463
iPS Cell 280
Iridium Oxide 616

J

Jet Flow 288

K

Kinesin 1107
Kinesin Motor Protein 314

L

Lab-on-a-Chip 272, 330
Lab-on-a-Disc 260
Label Free 1083
Laplace Trap 1003
Large Area Electronics 1135
Large Area Microfabrication 1217
Laser Micro Machining 524
Laser Scribing 556
Laterally Vibrating Resonators 1241
Lead Free Piezoelectric Thin Film 397
Lead ZirconateTitanate (PZT) 1155, 1237
Leukemia 837
Linear Acceleration Sensitivity 32
Lipid Bilayer 17, 457
Liposome 17, 288, 457
Liquid Based Transmitter 765
Liquid Bridge 769, 1115
Liquid Crystal Elastomer 905
Liquid Deposition 100
Liquid Filled Lens 1163
Liquid Lens 909
Liquid Medium 765
Liquid Metal 540, 664, 967
Liquid Sensor 132
Lithium Niobate 1241
Lithographic Microfabrication 112
Lithography 498
Live Bacteria Detection 5
Living Battery 163
Local Stimulation at Single Cell Level 264
Localized Nano-Electroporation 865
Lorentz Force 80, 700
Low Cost Sensor 536

Low Frequency 346, 429
Low Loss Radio Frequency Switch 1233
Low Phase Noise 1209
Low Power 717
Low Temperature Bonding 1131

M

Magnesium...358
Magnetic 564
Magnetic Bead 809
Magnetic Gradient 44
Magnetic Nanoparticle 917
Magnetic Resonance Force Microscopy 151
Magnetic Sensor..80
Magnetometer 80, 700
Magnetophoresis 256
Magnetorheological Effect..1197
Magnetostriction 1189
Manganese Dioxide 405
Mass Sensing 128
Mechanical Filter 1249
Mechanical Latch..1289
Mechanical Property 652
Mechanical Stress 917
Mechanomyogram 845
Membrane Protein...288, 457
Membranes..632
MEMS Aircraft 648
MEMS Force Sensor 979
MEMS Gyroscope 32, 749, 801
MEMS Infrared Devices 644
MEMS Oscillator..1285
MEMS Pump 564
MEMS Resonator...120, 1257
Meniscus Visualization 668
Mesoporous Carbon 298
Mesoporous Silica..995
Metamaterial (s) 664, 1221
Metamaterial Perfect Absorber 84
Metglas..76
Micro Beads 592
Micro Bearings..983

Micro Electrode Array **841**

Micro Electro Mechanical System (MEMS) **36, 132, 330, 362, 467, 506, 568, 632, 640,**
....**656, 741, 753, 777, 931, 971, 1055, 1241, 1289**

Micro Gas Chromatography **294**

Micro Heater(s) **514, 1179**

Micro Knife **885**

Micro Medical Robot **188**

Micro Plasma **171**

Micro Scale **393, 721**

Micro Shell **721**

Micro Supercapacitor **405**

Micro Thruster **999**

Micro Total Analysis System **5**

Micro Trace Heavy Metal Sensor **298**

Micro Variable Lens **72**

Micro Variable Prism **72**

Microactuator **905**

Microarray **254**

Microassembly **901**

Microbial Fuel Cell **393**

Microbial Supercapacitor **362**

Microbolometer **572**

Microbubble **104**

Microcantilever **330**

Microchamber **175**

Microchannels **1107**

Microelectrode **857**

Microelectrode Array **254**

Microfabricated Magnetizing Mask **520**

Microfabrication **56, 254, 490, 518**

Microfluidic (s) **5, 92, 185, 238, 246, 250, 256, 284, 809, 821, 939, 987, 995, 1027, 1051**

Microfluidic Channel **310, 947, 1139**

Microfluidic Chip **1035**

Microfluidic System **88**

Microhotplate **48**

Microlens **1015**

Microlens Array **1163**

Micromachined **753**

Micromachining Compatible **644**

Micromanipulator **901**

Microneedle **813, 817, 1183**

Microscale Magnetic Patterning **520**

Microscanner..805, 1185

Microsurgery...1147

Microtubule (s)...314, 1107

Microvalve..987

Millimeter-Waves..206

Millivolt Switching..1099

Minimal Invasive..885

Mirror...1229

Mirror Array...1155

Molecular Sensor..222

Monolithic...112

Motion Control..310

Motion Harvesting..350

Motor Unit EMG Recording...857

Movable Electrode...1245

MR Fluid Lens..1197

Multi-Band...1225

Multi-Channel Recording..857

Multi-Degree-of-Freedom Sensors...28

Multi-Parameter Measurement System..975

Multi-Physics Modeling...700

Multidrug Resistance Degree...837

Multilayer Magnetic anisotropy...44

N

Nano Channels...995

Nano Fountain Pen...100

Nano Opto Electro Mechanical...1261

Nano Opto Mechanical Systems..1091

Nano Pillar...147

Nano Technology..1083

Nanochannel...1019, 1059, 1067

Nanocomposite..354

Nanocrystalline Silicon...467

Nano Electro Mechanical (NEM)...1095

Nanoelectromechanical Relay(s)...143, 1099

Nanoelectromechanical Switch...143, 1103

Nano Electro Mechanical Systems (NEMS)..........................1055, 1071, 1079

Nanofiber...498, 917

Nanofiber Forests...644

Nanofluidic...1059

Nanofluidic Crystal Sensing..276

Nanofluidic Device 1019
Nanomanufacturing...474
Nanomechanical Resonators 1205
Nanoparticle 421, 1059
Nanoparticle Characterization 116
Nanoparticles Sorting...1015
Nanoporous Powder 1131
Nanoscale..471
Nanoscale Contact 1079
Nanostructure (s)..474, 648
Nanowire 155, 246, 652
Narrow Gaps 1281
Nd-Fe-B Magnet 151
NDT 490
NEMS Oscillator Stability 116
Neural Cell..254
Neural Interface 873
Neural Microelectrodes...616
Neural Probe 155, 853
Neural Recording 9
Neural Signal 853
Neural Stimulation 9
No Back-Plate 136
Non-Contact Pairing 1003
Non-Destructive Measurement 1151
Non-Equilibrium...310
Non-Linear Dynamics...1261
Nonlinear Harvester 397
Nonlinearity Tuning..801
Normal Physiological Condition...276

O

On-Chip Chemical Reaction Monitoring...338
On-Chip Control 1023
On-Chip Li-Ion Capacitor...401
On-Chip Mixing..829
Optical Coherence Tomography (OCT) 1147
Optical Control..1201
Optical Emission Spectroscopy 459
Optical Gradient Force..1091
Optical Measurement..580
Optical Memory 1091

Optical MEMS 433
Optical Resonators 1193
Optical Stimulation 853
Optical Switch 1155
Optical Tweezers 921
Optically Tunable 84
Optically-Induced Electroporation 234
Optics Array 72
Opto-Acoustic Oscillator 1209
Opto-Mechanical 1143
Opto-Mechanical System 869
Optofluidic (s) 797, 909, 1015
Optogenetics 196, 813
Optomechanics 1209
Organic 1217
Oscillating Droplet 897
Oscillator (s) 210, 214, 218, 1193, 1201, 1261
OSTE 96
OSTE+ 987

P

Package 1139
Packaging 1127
Paper 620
Paper Based Electronics 781
Paper Sensor 536
Parameter Optimization 700
Parametric Amplification 210
Parametric Resonance Excitation 805
Particle Tracking 939, 1051
Particle Transport 927
Particle Trapping 1019
Parylene 9, 132, 185, 841, 1127
Parylene C 104, 825
Pattern 413
Peel-Off 514
Periodic Heating Method 1151
Periprosthetic Joint Infection 5
pH Sensor 1063
Phase Change 1039
Phase Noise 210, 1193
Photo Detector 147, 1159

Photo Sensor	486
Photolithography	540
Photonic Crystal	576
Photonic Sintering	532
Photoresist	389
Photothermal Probing	1205
Pierce Typed Electron Emitter	467
Piezoelectret	374
Piezoelectric	159, 370, 429, 704, 745
Piezoelectric Actuator	1099, 1155, 1171
Piezoelectric Coefficient	620
Piezoelectric Effect	897
Piezoelectric Energy Harvester	397
Piezoelectric Material	620
Piezoelectric Resonator (s)	214, 688
Piezoelectronics	781
Piezoresistive	785, 789, 1201
Piezoresistive Cantilever	290
Piezoresistive Detection	640
Piezoresistor	1151
Pirani Gauge	676
Piston-Cylinder Actuator	592
Plasma FET	171
Plasmonic Nanostructures	1205
Plastic Deformation	660
PMUT	745, 753
Pneumatic Balloon Actuator	925
Pneumatic Force	913
Pneumatic Switch	1023
Point of Care	256
Poly-SiGe	1095
Polyacrylamide Gel Electrophoresis	1027
Polycide	1273
Polycrystalline Diamond	628
Polydimethylsiloxane(PDMS)	192, 514, 873, 987
Polydimethylsiloxane (PDMS) Transfer	544
Polylactic Acid	532
Polymer Dielectrics	230
Polymer MEMS	560
Polymer Micromachining	528
Polypyrrole	354
Porous Carbon	389, 405

Porous Silicon Dioxide 417

Positrons..453

Prandtl-Tube 999

Pre-Sealed Reagents..52

Precision..797

Pressure Sensor 104, 120, 717, 757, 845, 849, 893, 975

Print...560

Printed Electronics..506, 532

Printing...963, 967

Probe 861

Process Development...502

Protein Detection 1083

Protein Electrophoresis 1027

Protein Measurement 921

PTFE Electret..413, 684

Pull-In 1281

Pumping...1067

Q

Q-Enhancement..789

Quality Factor 612, 632, 725, 801, 1257

Quantitative Analyses 294

Quantum Tunneling 1103

Quinolones Resistance..821

R

Radiation Detection 793

Radiation Pressure 1209

Radio Frequency Microelectromechanical Systems..............................206, 1193

Radio Frequency Microelectromechanical Systems Switch................................1233

Radiotracer...284

Rarefied Plume..999

Rate Integrating Gyroscope 20, 24

Reactive Ion Etching..463

Reconfigurable 540

Reconfigurable Waveguide Iris 206

Red Blood Cell..1031

Reduction Mechanism 1031

Relay 1095

Reliability...568, 596

Replication 560

Resonant..132

Resonant Heating 877

Resonant Microcantilever 302

Resonant Pressure Sensor 120

Resonant Radiation Sensor 704

Resonant Switch...1273

Resonator **80, 612, 628, 672, 717, 1265, 1277**

Restenosis **877, 893**

Reverse Electrodialysis..421

RF Filters 1241

RF MEMS **1237, 1253**

Ring...951

Ring Gyroscope 24

RNA Purification 260

Rotary Motors 200

Rotating Gear 429

Rotational...684

Rubidium..552

Rubidium Azide 552

Rubrene Solution 108

Ruthenium Oxide 167

S

Sacrificial & Biodegradable..656

Sacrificial Layer...548

SAM...514

Sample Comparison Method..931

Scanner..1229

Scanning Probe Microscopy 128

Scanning Thermal Probe Microscopy..1111

Sciatic Nerve..873

Sealing Technique...592

Selection...242

Selective Formation 1131

SELEX...242

Self-Assembled...147

Self-Assembled Molecules 1103

Self-Assembly...56

Self-Oscillator...1071

Self-Propelled Droplets..310

Semicircular Channels 696

Sensing Material 302

Sensor (s) **729, 773, 1055**

Sensor Characterization 825

Sensor Device 536

SERS 1059

Shape Regulation 175

Shear Piezoresistance 600

Shear Stress 238

Shock Sensor 692

Sidewall 36

Sigma-Delta Modulator 761

Silica 449

Silicon 648

Silicon Carbide (SiC) 793, 1079

Silicon Microwire 604

Silicon Migration Technology 445

Silicon Nanowire (s) 600, 1175

Silicon Nanowire Probe 151

Silicon Optical Micro-Bench 881

Silicon Resonator 1245

Silicon Tweezer 13

Silk Fibroin 604

Silver Nanowire 785

Simulation 927

Single Cell 943

Single Cell Manipulation 1035

Single Cell Separation 1035

Single Layer Silicon-On-Insulator Process 409

Single Nucleotide Polymorphism Polymerase Chain Reaction 821

Sliding 979

Small Molecules 1075

Smart Cut 793

Soft X-Ray Charging 409

Solar 576

Solid Phase Extraction 284

Solid State Supercapacitor 342

Somatic Muscle 196

Sorting 92

Speed Sensor 684

SPF 482

Spheroplast 175

SPIH 350

Spiral 913

Spring Forces 596

Sputtering 453

Squeeze-Film Damping 28

Squegging 1273

Stability 1063

Stable...1277, 1281

Stamp Bonding..1131

Stencil Printing..478

Stent 877

Stiction 588

Stimulation 861

Stored Red Blood Cells..833

Strain Sensor 737, 785

Structural Layer 548

Subpixel Resolution 580

Supercapacitor..167, 354, 389

Superhydrophobic Surface 668

Surface Acoustic Wave 1043

Surface Energy Patterning 96

Surface Plasmon Polariton 1179

Surface Plasmon Resonance 147

Surface Ratchets..983

Surface Tension 222, 991

Suspended Nanochannel Resonators 116

Swallowing 849

Switch 1079, 1265, 1289

Switchable Cavity Resonator 206

Switched...218

T

Tactile Display 931

Tactile Sensor...441, 680, 709, 713, 1189

TASCL 280

Tau Protein...314

Temperature Compensation 214

Tensile Test...652

Terahertz 1221

Tetramethylammonium Hydroxide...857

Thermal Conductivity 628

Thermal Diffusivity 1151

Thermal Flexure Actuator..548

Thermal Flow Sensor...975

Thermal Imaging...572

Thermal Performance...48

Thermal Read-Out..696

Thermal Sensor...1213

Thermally Responsive Solution...592

Thermo-Acoustic Transmitter..765

Thermodynamic Parameters...302

Thermophotovoltaic..576

Thermopile Sensor..1213

Thermotropic..905

Thin-Film Magnesium (Mg)..656

Thiol-Ene Polymer..96

Thread..959

Through-Glass Vias (TGVs)..64

Through-Silicon Via...1115

THz..664

THz Polarization Rotation...88

Time Constant..833

Time-Domain...596

Timing References..214

Tissue Assay..13

Tongue Motion...849

Torque Sensor..680

Torsional Resonator..793

Touch Sensing..781

Traction Force..290

Transducer...745

Transfer Technology...322

Transience..1123

Transient Flow...1051

Transport..1107

Trehalose...163

Triboelectric Generator...346

Tunable...1163, 1221

Tunable Capacitor..1253

Tunable Laser..1143

Tunable Lens..881, 1171

Tunable Magnetic Anisotropy...44

Tunable Optical Liquid Lens...1147

Tunable Optical Microcavity...1217

Tunable Optical Modulation..1197

Tunable Spectral Filter...1167

Tuning..1201

Two-Phase...943
Two-Photon...471
Two-Photon Stereolithography...921
Two-Stage Solidification...44

U

Ultra-Thin Membranes...889
Ultrasonic...745, 769
Ultrasonic-Micromachining...494
Ultrasound...753, 885
Ultrasound Transducer...765
Ultraviolet...1159
Uncooled...1225
Uncooled Infrared Detector...688

V

Vacuum...761
Vacuum Filtration...441
Van der Waals Force...1119
Vapor Cell...552
Vapor-Liquid-Solid...604
Vapor-Liquid-Solid Growth...155
Variable Capacitor...1289
Velcro Principle...1119
Vertical Electrets...409
Vibration Energy Harvesting...385
Virtual Surface...931
Volume Precision...955

W

Wafer Bonding...453
Wafer Level Process...564
Wafer Level Sealing...24
Wafer-Scale...556
Water Droplet...979
Waveguide...813
Wavelength Selective Device...1167
Wavelength Selective IR Emission...1179
Wearable...370
Wetting Dynamics...668
Whole Angle Mode...20
Wineglass...672

Wireless Optrode 813

Wireless Sensor Node 773

Wrinkling 474

X

X-Ray 893

X-Ray Imaging 490

Z

Zebrafish ECG 841

Zinc Oxide Nanowires 781

MEMS 2014 Author Index

Scroll to the author and select a **Blue** link to open a paper. After viewing the paper, use the bookmarks to the left to return to the beginning of the Author Index.

A

Abdelgawad, M.	829
Abdolvand, R.	132, 1285
Abe, Y.	185
Abelmann, L.	100, 1111
Adachi, C.	108, 147
Adachi, J.	147
Ahamed, M.J.	24
Ahmadi, A.	939, 1051
Ahmed, S.M.	548
Ahn, C.H.	24, 80, 588, 749, 773, 1277, 1281
Aita, F.	1213
Ajiki, Y.	147
Akashi, T.	709
Akhbari, S.	124
Akin, T.	32
Akita, S.	737
Akiyama, Y.	163, 196
Allen, M.G.	1, 358
Alper, S.E.	32
Alveringh, D.	680
Amato, B.	632
Amjadi, M.	785
An, S.	112
Anand, S.V.	192
Ando, Y.	155
Ang, W.C.	688
Angelescu, D.E.	797
Ansell, O.	459
Aoki, R.	845
Aoyagi, S.	417
Aoyama, M.	1229
Appel, J.H.	889
Arai, F.	1031, 1035
Arcamone, J.	1071

Ardanuç, S. **524, 608**

Arie, T. **737**

Armutlulu, A. **358**

Arndt, G. **1071**

Arnold, D.P. **520**

Asadnia, M. **741**

Asano, T. **196**

Ashby, P.D. **128**

Ashizawa, H. **717**

Askari, S. **24**

Assadsangabi, B. **200**

Ataka, M. **717**

Averitt, R.D. **84**

Ayala, C.L. **143**

Ayazi, F. **28, 40, 120, 612**

Azgin, K. **32**

B

Baborowski, J.J. **490**

Bae, J. **72**

Baek, D. **1289**

Baghchehsaraei, Z. **206**

Bagherinia, M. **700**

Bahri, D. **128**

Banerjee, N. **1123**

Bargatin, I. **449**

Barnett, R. **459**

Barutçu, D. **350**

Basset, P. **385**

Bazigos, A. **1063**

Bedener, K. **1063**

Beh, S.P. **196**

Beld van den, W.T.E. **1067**

Bellamkonda, R.V. **873**

Benthem van, K. **628**

Berenschot, J.W. **100**

Berg van den, A. **1067**

Bergonzi, G. **552**

Berthelot, A. **640**

Beyazoglu, T. **1193**

Bhave, S.A. **1209**

Bhugra, H. **204**

Bian, W.	413, 684
Bidstrup, S.A.	358
Bienstman, J.	218
Bierman, D.M.	576
Bindiganavale, G.S.	1039
Binh-Khiem, N.	765, 769
Bleiker, S.J.	143
Boden, T.J.	721
Boer de, M.J.	696, 999
Boero, G.	177
Bohringer, K.F.	983
Boisen, A.	1205
Booth, R.	238
Borana, J.	865
Borowsky, D.	572
Boser, B.	753
Bourouina, T.	385, 725, 1185
Bouvet, D.	1063
Boyd, C.W.	20
Boyle, D.	260
Briand, D.	370, 429, 506, 532
Bright, V.M.	1201
Brookhuis, R.A.	680
Brücker, C.	927
Brugger, J.	56, 177
Buja, F.	580
Bulbul, A.	226
Bulovic, V.	1103, 1217
Bunch, J.S.	1201
Burtghartz, J.	572
Büttgenbach, S.	869
Byun, I.	514

C

Cai, H.	688, 692, 1091, 1143, 1261
Cakmak, O.	330
Campanella, H.	688
Cardot, F.	490
Carlborg, C.F.	987
Carraro, C.	389
Cassella, C.	1269
Celanovic, I.	576

Cermak, N.	116
Chae, J.	**362, 889**
Chae, J.B.	**1147**
Chang, C.-H.	**1189**
Chang, C.-I.	**136, 757**
Chang, C.-J.	**234**
Chang, C.-P.	**1189**
Chang, D.C.	**36**
Chang, D.T.	**721**
Chang, H.-C.	**757, 1189**
Chang, H.L.	**761**
Chang, J.	**230**
Chang, J.-H.	**1147**
Chang, J.H.-C.	**656, 1127**
Chang, K.W.	**250**
Chang, W.	**1217**
Chang, W.H.	**5, 250**
Chao, C.Y.	**821**
Chatziioannou, A.F.	**284**
Che, Y.J.	**821**
Chen, A.	**1023**
Chen, B.-A.	**861**
Chen, C.C.	**1135**
Chen, C.-H.	**1189**
Chen, C.-N.	**648**
Chen, C.-Y.	**1249**
Chen, C.Z.	**995**
Chen, D.P.	**644**
Chen, F.	**761**
Chen, J.	**502**
Chen, J.	**833**
Chen, P.-C.	**13**
Chen, P.H.	**478**
Chen, R.	**1197**
Chen, S.	**284, 338, 1011**
Chen, S.	**502**
Chen, S.J.	**382**
Chen, W.-C.	**44**
Chen, X.	**877**
Chen, Y.-C.	**943**
Chen, Y.J.	**644**
Chen, Y.-Y.	**861**

Chen, Z.J.	644
Cheng, C.-L.	676, 757
Cheng, Y.	943
Cheung, K.C.	939, 1051
Chia, E.M.	664
Chim, J.	1183
Chin, C.-H.	1249
Chin, L.K.	1015
Chiquet, M.	177
Chishiro, T.	592
Chmielewski, D.	128
Cho, I.-J.	853
Cho, J.W.	963
Cho, J.Y.	20
Cho, K.	498, 917
Cho, S.K.	188, 991
Choi, E.	421
Choi, K.	72
Choi, N.	853
Choi, S.	222
Choi, S.	72
Choi, W.	540, 967
Choi, Y.	873
Choi, Y.	72
Christensen, D.L.	1281
Chronis, N.	893
Chung, D.	256
Chung, S.	246
Chung, S.K.	897, 901, 1147
Chung, T.	1289
Chung, Y.	873
Clark, B.	608
Clerc, P.-A.	552
Colinet, E.	1071
Corigliano, A.	700
Cosemans, S.	1095
Coster De, J.	1095
Cottone, F.	385
Cu-Nguyen, P.-H.	1167
Cui, T.	833
Cui, Y.L.	556
Cunningham, S.	1253

D

Dahmardeh, M.	877
Dam van, R.M.	284, 338
Darvishi, M.	1237
Davami, K.	449
Davis, R.W.	1083
Decrop, D.	272
Degertekin, F.L.	584
Deimerly, Y.	725
Deléglise, S.	140
Dellea, S.	640
Demircan, Y.	837
Dempsey, N.M.	520
Deotare, P.B.	1103
Deotte, J.	510
DeReus, D.	1253
Despont, M.	143, 490, 552
Dickens, A.	36
Dijkstra, M.	999
Dimov, N.	260
Domicone, N.	909
Dommann, A.	490
Dong, B.	88, 692, 1091, 1261
Dong, L.	1225
Dooraghi, A.A.	284
Drechsler, U.	143
Droogendijk, H.	696
Du, J.-C.	616
Duan, X.	805
Ducrée, J.	256, 260
Duerig, U.T.	143
Duncombe, T.A.	1027

E

Edura, T.	108
Eijkel, J.C.T.	1067
Elbuken, C.	330
Elfrink, R.	568
Elwenspoek, M.	1237
Endrödy, C.	1167
English, A.W.	873

Erdem, M.	837
Ermek, E.	330
Ernst, T.	1071
Esashi, M.	322, 467, 482, 709
Esfandyarpour, R.	1083
Etter, D.	572
Eun, C.K.	64
Eun, Y.	1289
Everhart, C.L.	1277

F

Fan, K.	84
Fan, S.K.	1047
Fang, N.	510
Fang, S.P.	498, 917
Fang, W.	44, 136, 676, 757, 1135, 1189, 1197
Fatemi, H.	1285
Fedder, G.K.	544, 801
Feng, F.	1159
Feng, J.	188
Feng, P.X.-L.	793, 1079
Feng, Y.	374
Fernandes, J.	1175
Fernandez-Bolanos, M.	143
Fiorentino, G.	48, 342
Fischer, A.C.	1055
Fischer, P.	437
Fitzgerald, P.	596
Forsberg, F.	1055
Foster, J.S.	564
Franca, E.	498
Frolet, N.	532
Fu, C.-C.	1189
Fu, Q.Y.	409
Fujii, T.	652
Fujimoto, K.	1107
Fujimura, T.	1155
Fujita, H.	314, 713, 717, 1111
Fujiyoshi, M.	709
Fukuhara, T.	1035
Fukushi, H.	709
Funabashi, H.	709

Funahashi, A. 264
Fung, S. 628

G

Gadhari, N. 177
Galayko, D. 385
Galchev, T. 350
Gammaitoni, L. 385
Gao, J. 544
Garraud, A. 520
Gaughran, J. 260
Gavcar, H.D. 32
Geerlings, J. 100
Geisberger, A. 36
Ghannad-Rezaie, M. 893
Ghovanloo, M. 813
Gianchandani, Y.B. 64, 76, 112, 294, 494, 885
Gibbs, J.G. 437
Gibson, B.A. 1281
Glynn, M. 256
Goedbloed, M. 568
Goericke, F. 729
Gong, S. 1241
Gonzales, J.M. 132
Gore, R.K. 873
Goryu, A. 155
Green, S.R. 76
Grewe, A. 1167
Grine, A.J. 1193
Grinsven van, B. 272
Groenesteijn, J. 696, 975, 999
Grogg, D. 143
Grutter, K.E. 1193
Gu, J. 1115
Gu, Y.A. 688
Gu, Y.D. 692, 1091, 1143, 1261
Gudeman, C.S. 564
Guedes, A. 745, 753
Gulari, M.N. 893
Gullo, M.R. 56
Gündüz, U. 837
Guo, D. 544

H

Haase, T.	459
Haesler, J.	552
Hagleitner, C.	143
Hall, H.J.	1201
Hamada, A.	147
Hamano, H.	17
Han, C.H.	660
Han, M.D.	346, 425
Hans, H.	741
Hansson, J.	987
Hara, M	1043
Hara, M.	397, 463, 636
Hara, Y.	310
Harada, S.	737
Haraldsson, T.	96, 987
Harris, J.S.	1083
Harrison, K.L.	1277
Hasegawa, S.	931
Hashiguchi, G.	717, 1229
Hashimoto, S.	310
Hata, Y.	709
Hatakeyama, K.	1111
Hayakawa, T.	1035
Hayasaka, T.	322
Hayase, M.	60
Hayashi, M.	60
Hayashi, S.	280
He, T.	1079
Heidari, A.	745
Heinz, D.B.	588, 1277
Heo, Y.J.	175
Herr, A.E.	1027
Herrault, F.	358
Herrera, M.D.	869
Herrmann, I.	572
Hilbert, J.	1253
Hill, B.E.	773
Hill, M.	596
Hiraiwa, T.	264
Hirayama, K.	175, 518

Hiroi, N.	264
Hirooka, M.	196
Hoang-Giang, D.	765
Hoang-Phuong, P.	765
Honda, S.	592
Honda, W.	737
Hone, J.	1075
Hong, J.	947
Hong, S.J.	1147
Hong, V.A.	24, 80, 588, 749, 773, 1277, 1281
Hong, Y.	963
Hoof Van, C.	218
Hopcroft, M.A.	773
Horsley, D.A.	80, 124, 628, 700, 745, 749, 753
Hoshino, K.	196
Hoshino, T.	196
Hou, M.T.	648, 713
Howe, R.T.	1277
Hsia, B.	389
Hsieh, H.-C.	226, 1007
Hsieh, H.-S.	1189
Hsu, F.-M.	44, 1197
Hsu, L.-S.	441
Hsueh, H.-T.	809
Hu, J.	502
Huang, C.-C.	861
Huang, C.W.	478
Huang, J.G.	692, 1091, 1261
Huang, K.	1135
Huang, L.X.	721
Huang, P.-C.	861
Huang, R.F.	88
Hussain, A.M.	548
Hussain, M.M.	548
Hutter, F.	572
Hwang, J.	1233
Hwang, S.-H.	1233

I

Ichikawa, Y.	1151
Iimura, Y.	1163
Ikegami, N.	467

Ikeuchi, M. **181, 280, 921**
Ikuta, K. **181, 280, 921**
Imashioya, T. **604**
Imato, T. **108**
Inaba, A. **326, 1087**
Inoue, K.Y. **322**
Inoue, S. **652**
Ionescu, A.M. **1063**
Ischer, S. **552**
Ishida, M. **155, 604, 857**
Ishihara, R. **342**
Ishii, M. **1179**
Ishikawa, H. **931**
Ishikawa, T. **322**
Ishimatsu, R. **108**
Isono, Y. **600, 1059**
Isozaki, A. **1221**
Isozaki, A.I. **913**
Ito, K. **1035**
Itoh, T. **60**
Iwabuchi, K. **196**
Iwai, K. **393**
Iyer, S.S. **721**

J

Jacot-Descombes, L. **56**
Jalabert, L. **1111**
Jambunathan, M. **568**
Jamsaid, A. **92**
Janphuang, P. **370, 429**
Jao, P.F. **498, 917**
Javanmard, M. **1083**
Jayaprakash, V. **393**
Jennings, J. **453**
Jeon, E. **222**
Jeong, H. **885**
Jeong, H.-H. **437**
Jeong, Y. **28**
Jia, K. **36**
Jia, Y. **306**
Jiang, H. **1175**
Jiang, X. **1115**

Johansson, S.B. 935
Jourdan, G. 725
Joyce, R.J. 721
Juang, Y.Z. 1139
Jung, U.G. 290

K

Kamiya, K. 185, 288, 457
Kan, T. 147, 290, 1221
Kan, T.K. 913
Kanda, N. 1221
Kaneko, M. 1031
Kaneko, T. 845, 849
Kang, H.M. 1019
Kang, J.Y. 853
Kang, X.-Y. 616
Kang, Y.W. 250
Kao, C.Y. 821
Karita, Y. 518
Karlsson, J.M. 987
Karsten, S.L. 314
Kasahara, T. 108, 1131
Kashyap, K. 648
Katz, A.J. 917
Kaufman, R. 490
Kaufmann, J. 552
Kawai, K. 1229
Kawai, Y. 151
Kawano, R. 185, 288, 457
Kawano, T. 155, 604, 857
Keesara, V. 28
Kenny, T.W. 24, 80, 588, 749, 1277, 1281
Keshavarzi, S. 1119
Khalil, D. 1185
Khatri, B. 905
Kilinc, N. 330
Kim, B.J. 514
Kim, B.J. 825
Kim, C.J. 284, 338, 668, 1011
Kim, D. 421
Kim, D. 540, 967
Kim, D. 853

Kim, G.J.	498, 917
Kim, H.	226, 238, 672, 1007, 1123
Kim, H.	1003
Kim, J.	246
Kim, J.	1003
Kim, J.	1075
Kim, J.	1289
Kim, J.H.	1019
Kim, K.T.	917
Kim, M.-S.	389
Kim, M.O.	1289
Kim, S.	222
Kim, S.H.	963
Kim, T.G.	853
Kim, Y.	72
Kim, Y.K.	853
Kim, Y.-K.	1019, 1233
Kimura, H.	264
Kimura, T.	264, 600
Kinahan, D.J.	256, 260
Kippenberg, T.J.	140
Kirby, D.J.	721
Kitamura, N.	1183
Klaassen, A.	128
Koelmans, W.W.	143
Koh, Y.	1019
Kojima, A.	467
Kokorian, J.	580
Kometani, R.	1087
Konishi, K.	1221
Konishi, S.	592, 925
Koochak, Z.	1083
Kosemura, Y.	931
Koshida, N.	467
Kosla, C.	596
Kotera, H.	314, 1107
Kotler, C.	490
Kottapalli, A.G.P.	741
Kozinda, A.	167, 354
Kraft, M.	761
Krijnen, G.J.M.	680, 696
Kropelnicki, P.	688, 692, 1091, 1143, 1261

Kubota, Y. 155, 604
Külah, H. 837
Kumagai, S. 1179
Kumar, A. 648
Kumemura, M. 717
Kunisawa, E. 108
Kuo, J.-C. 441, 809
Kuo, P.-H. 809
Kuo, W.-C. 959
Kurabayashi, K. 805
Kurihara, K. 560
Kuroda, K. 652, 1031
Kuwano, H. 397, 463, 636, 1043
Kuwata-Gonokami, M. 1221
Kwon, D. 246
Kwon, D.S. 1289
Kwon, K. 421, 813
Kwon, Y. 222
Kwon, Y. 72

L

Ladhane, K. 793
Lai, C.-H. 1189
Lai, W.-C. 1189
Lai, W.-M. 44
Lai, W.C. 1135
Lake, J.J. 721
Lal, A. 13, 524, 608
Lammerink, T.S. 975
Lammertyn, J. 272
Lang, J. 1217
Lang, J.H. 1103
Langfelder, G. 640
Larsen, T. 632
Lasalandra, E. 700
Lazari, M. 284
Le Roy, D. 520
Lee, C.S. 963
Lee, G.B. 5, 250, 821
Lee, G.-B. 234
Lee, H. 510, 813, 853
Lee, I. 733

Lee, J.	116, 222, 733, 793, 947, 1079
Lee, J.B.	540, 967
Lee, J.E.-Y.	1257
Lee, J.	72
Lee, K.I.	963
Lee, M.S.	5
Lee, S.	204
Lee, S.	947
Lee, S.	1003
Lee, S.S.	704
Lee, Y.	967
Lee, Y.S.	1019
Lee, Y.-S.	1233
Lee, Y.R.	897, 901
Lei, C.	644
Lei, D.	204
Lemme, M.C.	1055
Lenert, A.	576
Leong, K.C.	688
Li, C.-S.	1249
Li, H.	52
Li, H.	805
Li, M.	80, 700
Li, M.-H.	1249
Li, N.	502
Li, S.	159
Li, S.	401, 405
Li, S.-S.	1249
Li, T.	494, 885
Li, W.	813
Li, W.	1273
Li, X.	624, 781
Li, X.X.	302, 995, 1115
Li, Z.H.	9, 52, 276, 624, 817
Liang, K.-C.	676
Liang, Z.	817
Liao, H.H.	1139
Liao, J.C.	889
Liao, K.W.	713
Liao, S.-C.	1189
Lim, B.	963
Lim, H.B.	777

Lin, C.F.	1139
Lin, C.H.	478, 959, 1139
Lin, C.L.	250
Lin, C.-T.	809
Lin, C.-W.	861
Lin, C.-Y.	861
Lin, L.	124, 159, 167, 230, 354, 393, 486, 628, 1159
Lin, N.	777
Lin, Q.	242, 306, 1075
Lin, Y.	1273
Lin, Y.C.	5
Liu, A.Q.	88, 664, 692, 1015, 1091, 1143, 1261
Liu, C.-S.	612
Liu, G.	1075
Liu, H.	1175
Liu, H.B.	777
Liu, J.	829
Liu, J.-Q.	616
Liu, L.	36
Liu, M.	1225
Liu, R.	210, 1273
Liu, S.W.	366, 777
Liu, S.Y.	382
Liu, W.	346, 425
Liu, X.	159, 1225
Liu, X.Y.	781
Liu, Y.	230
Liu, Y.	656, 1127
Llobera, A.	869
Lo, S.-C.	136
Lockhart, R.	370, 429
Longoni, A.F.	640
Lötters, J.C.	975, 999
Lu, M.-Y.	234
Lu, S.-S.	809
Lu, Y.	745
Lu, Y.-S.	1175
Luo, J.	502
Luo, X.	64
Luo, Y.	877
Lye, S.W.	366
Lynn, K.G.	453

M

Ma, C.-W.	**441, 809**
Ma, K.	**999**
Ma, P.	**334**
Ma, W.	**1225**
Ma, X.	**338**
Maboudian, R.	**389**
Maeda, R.	**60, 560**
Maeda, S.	**310**
Mahadeva, S.K.	**620**
Makihata, M.	**709**
Man, T.	**624**
Manalis, S.R.	**116**
Mansour, R.R.	**789**
Mansukhani, K.	**729**
Mao, H.Y.	**644**
Marelli, M.	**177**
Marentis, T.C.	**893**
Marette, A.	**532**
Mariani, S.	**700**
Mark, A.G.	**437**
Märki, D.	**532**
Martinez, A.	**358**
Marty, F.	**385, 797**
Mastrangeli, M.	**56**
Mastrangelo, C.H.	**672, 1123**
Masuno, K.	**1179**
Matchkov, V.	**869**
Matova, S.	**568**
Matsue, T.	**322**
Matsui, K.	**1087**
Matsumoto, K.M.	**147, 290, 326, 765, 769, 845, 849, 913, 979, 1087, 1151, 1221**
Matsumoto, Y.	**264**
Matsunami, S.	**108**
Mattana, G.	**532**
Maurer, K.	**1027**
McAuley, D.	**260**
McGaughey, A.J.H.	**544**
McNeil, A.	**36**
Mehanathan, N.	**40**
Meinert, T.	**881**

Meiss, T. 536
Meng, B. 346, 425
Meng, E. 104, 825
Mercier, B. 797
Mescheder, U. 1119
Metcalfe, G.D. 84
Meyer De, K. 1095
Miao, J.M. 366, 741, 777
Michelassi, F. 13
Miki, N. 185, 264, 931, 1183
Miller, N. 1269
Mimatsu, H. 1131
Ming, A.J. 644
Minh, L.V. 397
Minh-Dung, N. 769, 845
Misaki, T. 1155
Misawa, N. 318
Mitsuya, H. 717
Mizuno, J. 108, 1131
Molina-Lopez, F. 506
Moon, H. 955, 1039
Morana, B. 48
Morikaku, T. 652
Morishima, K. 163, 196
Moriyama, M. 482
Morris, A. 1253
Mostafazadeh, A. 330
Mugele, F. 128
Mukherjee, T. 801
Munoz-Berbel, X. 869
Murakami, Y. 318
Muramatsu, Y. 925
Murarka, A. 1217
Muroyama, M. 709

N

Nadig, S. 524, 608
Naghsh Nilchi, J. 210
Nagumo, O. 560
Naing, T.L. 210
Najafi, K. 20
Najar, H. 628

Nakagawa, T. 378
Nakamura, N. 163
Nakano, M. 322
Nakayama, T. 709
Nam, Y. 576
Namazu, T. 652
Narimannezhad, A. 453
Natarajan, S. 1253
Neels, A. 490
Nemoto, N. 1221
Ng, E.J. 24, 80, 588, 749, 773, 1277, 1281
Nguyen, C.T.-C. 210, 1193, 1265, 1273
Nguyen, H.D. 721
Nguyen, S.D. 971
Niedermann, P. 490
Nievergelt, A. 128
Niimi, Y. 528
Nikapitiya, N.Y.J.B. 955
Niklaus, F. 1055
Niroui, F. 1103
Nishida, T. 417
Nishijima, T. 181
Nishikawa, H. 1131
Nishino, H. 467
Nitzan, S.H. 749
Noda, K. 765, 849
Noh, S. 238
Nomura, K. 378
Nonomura, Y. 709
Nooijer de, C. 568
Norford, L.K. 777
Novelli, P. 893
NuLi, Y. 616
Numakunai, S. 92
Numano, R. 155

O

Oba, M. 1213
Oberhammer, J. 206
Oguchi, H. 463, 636, 1043
Oh, S.W. 963
Ohno, H. 163

Oi, H. 155, 604
Oka, K. 264
Oka, Y. 433
Olcum, S. 116
Oldham, K. 805
Ollier, E. 1071
Olsen, T. 242
Omura, Y. 709
Oniku, O.D. 520
Ono, T. 151, 1245
Onoe, H. 518, 1163
Oohira, F. 433
Osaki, T. 17, 185, 288, 457
Oshidari, Y. 1087
Osman, H. 1095
Osonoi, M. 280
Östling, M. 1055
Ota, H. 264
Ou, W. 644
Ouyang, W. 276
Overstolz, T. 552
Özgür, E. 837

P

Pai, M. 204
Pai, P. 171
Pan, S.S. 777
Pan, T. 1023
Pan, W. 204
Pandraud, G. 48
Paprotny, I. 971
Pardon, G. 96
Park, I. 246, 785
Park, I.S. 897, 901
Park, J. 421
Park, J. 656, 841
Park, J. 885
Park, S. 268
Park, S. 873
Patterson, W.C. 520
Paul, O. 350
Pavy, N. 797

Payer, K. 116
Pei, R. 242, 1075
Perahia, R. 721
Perez-Ruiz, E. 272
Petsch, S. 905
Pezous, A. 552
Phan, H.P. 769
Philippe, J. 1071
Piazza, G. 1099, 1241, 1269
Pichitpajongkit, A. 785
Pisano, A. 729
Plaza, J.A. 869
Pourkamali, S. 1201
Pruitt, B.L. 773
Przybyla, R. 753
Puers, R. 218
Purwar, P. 947
Pyne, D.G. 829

Q

Qalandar, K.R. 1281
Qin, Y. 112, 294
Qiu, Z. 805
Quang-Khang, P. 769

R

Rahman, M.M. 672, 1123
Rais-Zadeh, M. 214
Rajagopalan, J. 192
Ramezani, M. 1095
Randles, A.B. 688, 692, 1091, 1143, 1261
Rane, T.D. 334
Reindl, T. 437
Reinecke, H. 1119
Reith, P. 905
Ren, H. 362
Ren, M. 664, 1015, 1143
Ren, T.L. 362, 556
Renaud, M. 568
Revol, V. 490
Rey, P. 640, 725
Rhee, K. 1147

Rigante, S. 1063
Rindzevicius, T. 1205
Rinnerbauer, V. 576
Rix, R. 905
Robert, P. 725
Rocheleau, T.O. 210, 1193
Rochus, V. 1095
Rockenbach, A. 927
Rodríguez-Rodríguez, R. 869
Rojas, J.P. 548
Rooij de, N.F. 370, 429, 506, 532
Roozeboom, C.L. 773
Rottenberg, X. 218, 1095
Rovers, M. 568
Roxhed, N. 935
Rozen, O. 753
Rubel, P.J. 564
Ruh, D. 905
Ryu, S. 785

S

Saadany, B. 1185
Sabry, Y.M. 1185
Sader, J.E. 128
Saif, M.T.A. 192
Saito, M. 1131
Saito, N. 600
Sakuma, S. 1031
Sammoura, F. 124, 167, 354
Sanders, R.G.P. 100, 696
Santagata, F. 48
Santhanam, S. 544
Santra, T.S. 865
Sarajlic, E. 100, 1111
Sarkar, N. 789
Sarro, P.M. 48, 342
Sasaki, M. 1179
Sato, K. 254
Savoye, M. 1071
Sawada, T. 1179
Sawahata, H. 155, 857
Schaijk van, R. 568

Schatz, A.	1171
Schelling, C.	572
Schliesser, A.	140
Schmid, S.	632, 1205
Schnakenberg, U.	927
Schneider, R.A.	1265
Schröder, S.	1055
Schuhladen, S.	905
Sebastian, A.	143
Segovia-Fernandez, J.	1269
Seifert, A.	881, 1167
Seki, T.	1213
Sekiguchi, T.	92
Sekitani, T.	68
Senior, D.E.	498
Senkal, D.	24
Seo, H.W.	1147
Seo, Y.J.	151
Seren, H.R.	84
Serien, D.	471
Serrano, D.E.	28
Severi, S.	1095
Shahmohammadi, M.	1285
Shang, H.	917
Shao, P.	40, 612
Sheen, H.-J.	861
Shehata, N.	833
Shelton, S.	124, 745, 753
Shen, C.	405
Shen, C.W.	298
Shen, W.	116
Shi, Y.Z.	1015
Shibata, S.	528
Shih, W.-P.	861
Shikida, M.	528
Shimada, N.	921
Shimokawa, F.	433
Shimomura, S.	925
Shimoyama, I.S.	147, 290, 326, 765, 769, 845, 849, 913, 979, 1087, 1151, 1221
Shin, I.U.	1147
Shin, J.H.	897, 901
Shinozaki, R.	433

Shintaku, H.	314, 1107
Shiozaki, M.	1213
Shirai, T.	378
Shkel, A.M.	24
Shoji, K.	163
Shoji, S.	92, 108, 1131
Shorman, E.	520
Shusteff, M.	510
Siekman, M.H.	1111
Silvestri, C.	48
Simonsen, U.	869
Sin, L.Y.	889
Sinzinger, S.	1167
Sletten, E.M.	1103
Smith, A.D.	1055
Sochol, R.D.	393
Soljacic, M.	576
Someya, T.	68
Son, Y.	853
Song, L.	502
Song, Q.H.	88, 664
Song, S.H.	704
Sonmezoglu, S.	32
Sorenson, L.D.	40, 612, 721
Spadaccini, C.M.	510
Sparreboom, W.	975, 1067
Spasic, D.	272
Spengen van, W.M.	580
Staufer, U.	580
Steimle, R.	36
Stemme, G.	935
Sterner, M.	1055
Stoeber, B.	536, 620
Stojanovic, M.	242, 1075
Storey, M.J.	1209
Stuermer, M.	1171
Su, T.-H.	749
Subramaniyan, S.P.	314
Suekuni, K.	1059
Sugano, K.	600, 1059
Sugimoto, S.	1043
Sugiyama, T.	717, 1229

Sun, G. 668
Sun, S.H. 572
Sun, X.M. 346, 425
Sun, Y. 829, 833
Sun, Y.-C. 44, 136, 676
Sung, W.K. 28
Sung, W.-L. 1135, 1189
Suzuki, A. 310
Suzuki, M. 163, 417
Suzuki, T. 433
Suzuki, T. 463
Suzuki, T. 1151
Suzuki, Y. 374, 409
Suzuki, Y. 482
Swager, T.M. 1103

T

Tabata, Y. 925
Tabib-Azar, M. 171
Tabrizian, R. 120
Taguchi, Y. 264
Tai, J. 841
Tai, Y.-C. 656, 841, 1127
Tajima, S. 925
Takagi, H. 560
Takahashi, H.T. 290, 913, 979, 1087, 1221
Takahashi, T. 417
Takahata, K. 200, 877
Takahata, T. 765, 845, 1087, 1151
Takami, Y. 652
Takao, H. 433
Takei, K. 737
Takei, Y. 326, 849, 1087
Takenaka, Y. 264
Takeshita, T. 600, 1059
Takeuchi, S. 17, 175, 185, 254, 288, 457, 471, 518, 1163
Tallur, S. 1209
Tan, C.S. 688
Tanaka, J. 1213
Tanaka, S. 322, 467, 482, 709
Tanamoto, R. 264
Tang, J. 76

Tang, W.	346
Tao, G.W.	1047
Tao, K.	366
Tarhan, M.C.	314
Tas, N.R.	100, 1111, 1237
Tatar, E.	801
Tavassoli, V.	40, 612
Tee, M.H.	200
Teh, K.S.	354
Tekes, C.	584
Teng, F.	298
Teramachi, Y.	925
Terao, K.	433
Thakar, V.A.	214
Thanh-Vinh, N.	765, 979
Thomas, D.	459
Thron, A.	628
Tian, H.	362, 556
Tian, H.-C.	616
Tichelaar, F.D.	342
Tilmans, H.A.C.	218, 1095
Ting, S.K.	88
Toan, N.V.	1245
Toda, M.	151
Tomida, M.	318
Tonooka, T.	17
Torayashiki, O.	1155
Torri, G.B.	218
Totsu, K.	482
Triantafyllou, M.	741
Tsai, D.	861
Tsai, H.H.	1139
Tsang, M.	358
Tseng, F.-G.	865
Tsujimura, H.	196
Tsukagoshi, T.	290
Tsuwaki, M.	108
Tu, C.	1257
Tuan, Y.-T.	757
Turner, K.L.	1281

U

Uchino, R. 1155
Uematsu, S. 1179
Ueno, R. 514
Urey, H. 330
Utermöhlen, F. 572

V

Vandenryt, T. 272
Varel, C. 983
Vargo, S. 459
Vásquez Quintero, A. 532
Vaziri, S. 1055
Verhagen, E. 140
Vericella, J. 510
Verma, N. 947
Villanueva, L.G. 632
Viswanath, A. 494
Vollebregt, S. 48, 342
Vullers, R.J.M. 568

W

Wada, T. 417
Wagner, P. 272
Wahl, S. 797
Waizmann, U. 437
Wakasa, Y. 378
Wallrabe, U. 1171
Walus, K. 620
Wang, A.I. 1103, 1217
Wang, B. 306
Wang, C. 136
Wang, C. 833
Wang, C. 1075
Wang, C.H. 5, 821
Wang, E.N. 576
Wang, J.J. 1139
Wang, L. 1201
Wang, P.-C. 865
Wang, Q. 636
Wang, R. 817

Wang, S.	1225
Wang, T.D.	805
Wang, T.H.	334
Wang, W.	9, 52, 276
Wang, X.H.	159, 298, 401, 405, 413, 684
Wang, Y.-H.	781, 861
Wang, Y.L.	250
Wang, Z.	568
Wang, Z.	793, 1079
Warren, R.	167, 393
Washabaugh IV, E.P.	825
Wasserman, S.C.	116
Watanabe, J.	931
Weber, A.	813
Weber, M.H.	453
Weber, N.	881
Wei, K.	474, 909
Wei, Z.	817
Weis, J.	437
Weis, S.	140
Weisgraber, T.	510
Wen, J.-H.	1189
Wen, Y.	1225
Werthschützky, R.	536
Wheeler, B.	498
White, R.M.	971
Wiegerink, R.J.	680, 696, 975, 999
Wijngaart van der, W.	96, 987
Williams, B.J.	192
Winebarger, P.	36
Winer, M.H.	939, 1051
Wipf, M.	1063
Witters, D.	272
Wong, M.	445
Woo, J.-K.	20
Wouden van der, E.	975
Wraback, M.	84
Wright, P.K.	971
Wu, C.	13
Wu, J.J.	5, 821
Wu, J.Y.	382
Wu, K.	1205

Wu, M.C. 1193
Wu, M.X. 52
Wu, S. 200
Wu, W. 276
Wu, W.G. 644
Wu, X. 413, 684

X

Xie, D. 556
Xie, L. 459
Xie, Y. 672, 1123
Xing, X. 951
Xiong, S. 1015
Xiong, W.J. 9
Xu, M. 668
Xu, P.C. 302, 995
Xu, T. 584
Xue, F. 777

Y

Yabe, A. 1043
Yablonovitch, E. 1103
Yagi, S. 604
Yagubizade, H. 1237
Yagyu, H. 378
Yahiro, M. 147
Yamada, H. 709
Yamagiwa, S. 604, 857
Yamaguchi, U. 709
Yamamoto, K. 729
Yang, B. 616
Yang, C. 124, 628, 1159
Yang, C.-S. 616
Yang, H. 1115
Yang, R. 793
Yang, S.Y. 5
Yang, Y. 24, 80, 588, 749, 773, 1277, 1281
Yang, Y. 556
Yang, Y.-A. 959
Yang, Y.-J. 441, 809
Yao, K. 1159
Yao, S.-C. 544

Yaralioglu, G.G. ... 330
Yasukawa, A. ... 181
Yeh, J.A. ... 648, 713
Yobas, L. ... 951
Yokokawa, R. ... 314, 1107
Yoo, J.H. ... 540, 967
Yoo, K. ... 540, 967
Yoon, C.K. ... 704
Yoon, D.H. ... 92
Yoon, E. ... 943
Yoon, E.-S. ... 853
Yoon, J.-B. ... 660
Yoon, J.H. ... 498, 917
Yoon, Y.H. ... 660
Yoon, Y.K. ... 498, 917
Yoshida, R. ... 463
Yoshida, S. ... 254, 322, 467
Yoshikawa, Y. ... 417
Yossifon, G. ... 268
You, S.M. ... 955, 1039
Young, D. ... 793
Yu, H. ... 995
Yu, H.Q. ... 9, 52
Yu, J. ... 861
Yu, L. ... 104
Yu, X. ... 1225
Yu, Y.J. ... 544
Yuan, J. ... 991
Yuan, W.Z. ... 761

Z

Zaghloul, U. ... 1099
Zang, X.N. ... 354, 486, 1159
Zappe, H. ... 881, 905, 1167
Zec, H.C. ... 334
Zeimpekis, I. ... 761
Zeng, F. ... 445
Zeng, X.Y. ... 1047
Zentel, R. ... 905
Zhang, G.Q. ... 48
Zhang, H. ... 276
Zhang, H.X. ... 346, 425

Zhang, H.Z.	9, 52
Zhang, J.	84
Zhang, K.D.	1047
Zhang, S.	502
Zhang, W.	88, 664
Zhang, X.	84
Zhang, X.	841
Zhang, X.S.	346, 425
Zhang, Y.	60
Zhang, Y.	502
Zhang, Y.	624
Zhao, C.	781
Zhao, L.	449
Zhao, X.	84
Zhao, Y.	474, 909
Zhao, Y.	1225
Zheng, S.	817
Zheng, X.	510
Zheng, Y.	833
Zhou, J.	1047
Zhu, H.	624
Zhu, H.-Y.	616
Zhu, J.	242, 306, 1075
Zhu, W.M.	88, 664
Zhu, Y.	688
Ziaie, B.	704
Ziegler, D.	128

2014 IEEE 27th International Conference on Micro Electro Mechanical Systems

(MEMS 2014)

San Francisco, California, USA
26 – 30 January 2014

Pages 644-1292

IEEE Catalog Number: CFP14MEM-POD
ISBN: 978-1-4799-3510-9

Copyright © 2014 by the Institute of Electrical and Electronic Engineers, Inc
All Rights Reserved

Copyright and Reprint Permissions: Abstracting is permitted with credit to the source. Libraries are permitted to photocopy beyond the limit of U.S. copyright law for private use of patrons those articles in this volume that carry a code at the bottom of the first page, provided the per-copy fee indicated in the code is paid through Copyright Clearance Center, 222 Rosewood Drive, Danvers, MA 01923.

For other copying, reprint or republication permission, write to IEEE Copyrights Manager, IEEE Service Center, 445 Hoes Lane, Piscataway, NJ 08854. All rights reserved.

***This publication is a representation of what appears in the IEEE Digital Libraries. Some format issues inherent in the e-media version may also appear in this print version.*

IEEE Catalog Number: CFP14MEM-POD
ISBN 13:   978-1-4799-3510-9

Additional Copies of This Publication Are Available From:

Curran Associates, Inc
57 Morehouse Lane
Red Hook, NY 12571 USA
Phone: (845) 758-0400
Fax: (845) 758-2633
E-mail: curran@proceedings.com
Web: www.proceedings.com

MEMS 2014 PROGRAM SCHEDULE

Scroll to the title and select a Blue link to open a paper. After viewing the paper, use the bookmarks to the left to return to the beginning of the Table of Contents.

Sunday, January 26

17:00 - Registration and Wine & Cheese Welcome Reception
19:00

Monday, January 27

08:00 **Opening and Welcome Address**
Farrokh Ayazi, *Georgia Institute of Technology, USA*
Chang-Jin "CJ" Kim, *University of California, Los Angeles, USA*

Plenary Speaker I
Session Chairs:
F. Ayazi, *Georgia Institute of Technology, USA*
CJ Kim, *University of California, Los Angeles, USA*

08:20 MICROFABRICATED IMPLANTABLE WIRELESS MICROSYSTEMS:
PERMANENT AND BIODEGRADABLE IMPLEMENTATIONS 1
Mark G. Allen
University of Pennsylvania, USA

Session I – Biomedical Microdevices
Session Chairs:
K. Peterson, *Profusa, USA*
J. Lammertyn, *KU Leuven, BELGIUM*

09:00 AN INTEGRATED MICROFLUIDIC SYSTEM FOR RAPID ISOLATION AND DETECTION
OF LIVE BACTERIA IN PERIPROSTHETIC JOINT INFECTIONS 5
W.H. Chang[1], C.H. Wang[1], S.Y. Yang[2], Y.C. Lin[2], J.J. Wu[3], M.S. Lee[4], and G.B. Lee[1]
[1]*National Tsing Hua University (NTHU), TAIWAN,* [2]*Jabil Circuit Inc., Ltd, TAIWAN,* [3]*National Cheng Kung University, TAIWAN, and* [4]*Chia-Yi Chang Gung Memorial Hospital, TAIWAN*

An integrated microfluidic system was presented in this work, which could distinguish the existence of live bacteria within 1 hour. This is the first time that a microfluidic platform was reported to detect live bacteria in periprosthetic joint infection samples. The results demonstrated that the proposed system can detect live bacteria successfully in the micro-environment of clinical samples. The proposed system can be a promising tool for the clinicians with timely medical decisions.

09:15 A CABLE-TIE-TYPE PARYLENE CUFF ELECTRODE FOR
PERIPHERAL NERVE INTERFACES 9
H.Q. Yu, W.J. Xiong, H.Z. Zhang, W. Wang, and Z.H. Li
Peking University, CHINA

We design, fabricate and characterize a cable-tie-type parylene cuff electrode for peripheral nerve interfaces, whose diameter is adjustable to accommodate the nerve properly during implantation. Cuffs made of thin and flexible parylene minimize mechanical damage to surrounding tissues after implantation. Moreover, the integrated parylene cable and pads facilitate connection with external circuits through wired or wireless interfaces. The acute in vivo rat experiments were performed to verify the ability for the neural recording and selective stimulation of different nerve fascicles.

09:30 **A SILICON ELECTRO-MECHANO TISSUE ASSAY SURGICAL TWEEZER** 13
P.-C. Chen[1], C. Wu[2], F. Michelassi[1], and A. Lal[1]
[1]*Cornell University, USA and* [2]*University of Pennsylvania, USA*

This paper reports a first-ever silicon surgical tweezer for characterizing electromechanical properties of tissue. Unlike the other probe-like tissue stiffness tactile sensors, the tweezer structure provides a platform for the clinical use during surgery. We chose to pursue an all-silicon tweezer, instead of attaching sensors to existing tweezers, for repeatable tissue assessment across surgeries without external calibration.

09:45 **HIGHLY PACKED LIPOSOME ASSEMBLIES TOWARD SYNTHETIC TISSUE** 17
H. Hamano[1], T. Tonooka[1], T. Osaki[1,2], and S. Takeuchi[1,2]
[1]*University of Tokyo, JAPAN and* [2]*Kanagawa Academy of Science and Technology, JAPAN*

We present a highly-packed liposome assembly that implements lipid bilayer-lipid bilayer contact at the interfaces for mimicking cell-cell connection on living tissues. The closely packed liposomes, based on our previous technique producing a monodisperse liposome array, facilitate easy modification in size and components of the model structures as well as long-term observation of their interfaces. We believe that the assembly technique would help providing a synthetic tissue model.

10:00 **Break & Exhibit Inspection**

Session II – Gyros & Accelerometers
Session Chairs:
D. Horsley, *University of California at Davis, USA*
H. Külah, *Middle East Technical University, TURKEY*

10:45 **WHOLE-ANGLE-MODE MICROMACHINED FUSED-SILICA
BIRDBATH RESONATOR GYROSCOPE (WA-BRG)** 20
J.-K. Woo, J.Y. Cho, C.W. Boyd, and K. Najafi
University of Michigan, USA

We report the fused-silica birdbath resonator gyroscope (BRG) with a large angular gain, controlled in the whole angle (WA) mode. The BRG is fabricated using the micro-blow-torching process and exhibits good mechanical symmetry, which is ideal for WA mode operation. We adopted the control algorithm for the hemispherical resonator gyroscope (HRG). We report a large bandwidth and full-scale range of 700 deg/s with a large angular gain ($A_g = 0.27$).

11:00 **100K Q-FACTOR TOROIDAL RING GYROSCOPE IMPLEMENTED IN
WAFER-LEVEL EPITAXIAL SILICON ENCAPSULATION PROCESS** 24
D. Senkal[1], S. Askari[1], M.J. Ahamed[1], E.J. Ng[2], V. Hong[2], Y. Yang[2],
C.H. Ahn[2], T.W. Kenny[2], and A.M. Shkel[1]
[1]*University of California, Irvine, USA and* [2]*Stanford University, USA*

This paper reports a new type of degenerate mode gyroscope with measured Q-factor of > 100,000 on both modes at a compact size of 1760 μm diameter. The toroidal ring gyroscope consists of an outer anchor ring, concentric rings nested inside the anchor ring and an electrode assembly at the inner core. Devices were fabricated using high-temperature, ultra-clean epitaxial silicon encapsulation (Epi-Seal) process.

11:15 **SINGLE PROOF-MASS TRI-AXIAL PENDULUM
ACCELEROMETERS OPERATING IN VACUUM** 28
D.E.Serrano[1,2], Y. Jeong[1], V. Keesara[2], W.K. Sung[2], and F. Ayazi[1,2]
[1]*Georgia Institute of Technology, USA and* [2]*Qualtré, USA*

This paper reports on the design, fabrication and characterization of single proof-mass tri-axial capacitive accelerometers coexisting in a wafer-level packaged (WLP) low-pressure environment with high-frequency gyroscopes, for the implementation of monolithic 6-degree-of-freedom (6-DOF) inertial measurement units (IMUs). The accelerometers are designed to operate as quasi-static devices (i.e. non-resonant sensors) in high vacuum levels (1 – 10 Torr) by increasing squeeze-film air damping through the use of capacitive nano-gaps (< 300 nm).

11:30 **SIMULTANEOUS DETECTION OF LINEAR AND CORIOLIS ACCELERATIONS
ON A MODE-MATCHED MEMS GYROSCOPE** 32
S. Sonmezoglu, H.D. Gavcar, K. Azgin, S.E. Alper, and T. Akin
Middle East Technical University (METU), TURKEY

This paper presents a novel "in operation acceleration sensing and compensation method" for a single-mass mode-matched MEMS gyroscope. In this method, the amplitudes of the sustained residual quadrature signals on the differential sense-mode electrodes are compared to measure the linear acceleration acting on the sense-axis of the gyroscope. Measuring the acceleration along the sense-axis, the sensitivity of the gyroscope output to linear accelerations along this axis is eliminated without using a dedicated accelerometer.

11:45   **Lunch & Exhibit Inspection**

13:00   **Poster/Oral Session I**

Session III– Materials and Process Characterization
Session Chairs:
M. Despont, *CSEM, SA, SWITZERLAND*
H. Toshiyoshi, *University of Tokyo, JAPAN*

15:00   A STATIC CAPACITANCE PROBE STRUCTURE FOR RESOLVING THE SIDEWALL SKEW ANGLE OF SILICON DEEP REACTIVE-ION ETCHING 36
K. Jia, A. Geisberger, A. Dickens, R. Steimle, D.C. Chang, P. Winebarger, L. Liu, and A. McNeil
Freescale Semiconductor Inc., USA

This work presents a design that enables quantitative characterization of the silicon Deep Reactive-Ion Etching sidewall skew angle using static LCR prober at ambient pressure, which provides an easy, accurate and batch solution to the long existing challenge of resolving such process features in an industrial manufacturing environment.

15:15   INVAR-36 MICRO HEMISPHERICAL SHELL RESONATORS 40
N. Mehanathan, V. Tavassoli, P. Shao, L. Sorenson, and F. Ayazi
Georgia Institute of Technology, USA

We report the successful fabrication as well as operational characterization of electroplated Invar micro-hemispherical shell resonators. Additionally, the heat treatment of the samples and its effect on the quality factor of the resonators is studied. We show that thermal annealing shifts the coefficient of thermal expansion (CTE) of the alloy towards its minimum, as a result of which the Q increases at least 3 times and reaches ~7500. An annealed electroplated Invar µHSR shows Q of 7500, where unannealed electroplated Invar µHSRs have Qs in the range 2000-3000.

15:30   IMPLEMENTATION OF SINGLE/MULTI-LAYER MAGNETIC-ANISOTROPY MAGNETIC POLYMER COMPOSITES FOR MAGNETIC PROPERTY MODULATION 44
F.-M. Hsu, W.-C. Chen, W.-M. Lai, Y.-C. Sun, and W. Fang
National Tsing Hua University (NTHU), TAIWAN

This study extends the two-stage solidification technology to fabricate the isotropic/anisotropic magnetic polymer composites (MPC, polymer with magnetic particles). Multilayer magnetic-anisotropy/isotropic MPC film can also be implemented using the two-stage solidification process layer by layer. Merits of proposed technology: (1) material properties of magnetic-anisotropy MPC layer is realized using the two stage solidification and anisotropic-magnetization processes, and (2) film of various magnetic properties can also be implemented using the different combination of multilayer magnetic-anisotropy MPCs. In applications, the multilayer polymer-NdFeB magnetic composites are realized in silicon substrate and further integrate with MEMS structures. The 1-4 layers of different magnetic-anisotropy MPC are demonstrated. Performances enhancement of magnetic-anisotropy 30wt%-NdFeB MPC (vs isotropic MPC) are: coercivity force (3.4%), remanence (304%), and saturation magnetization (268%). Anisotropic magnetostatic shielding effect (reduce from 0.45Telsa to 0.3-0.35Telsa) is achieved. Moreover, change of magnetic field distributions after stacking of different magnetic-anisotropy MPC layers is also demonstrated.

15:45   CNT BUNDLES GROWTH ON MICROHOTPLATES FOR DIRECT MEASUREMENT OF THEIR THERMAL PROPERTIES 48
C. Silvestri, B. Morana, G. Fiorentino, S. Vollebregt, G. Pandraud, F.Santagata,
G.Q. Zhang, and P.M. Sarro
Delft University of Technology, THE NETHERLANDS

Vertically aligned Carbon Nanotubes (CNT) arrays were successfully grown on top of freestanding microheaters. This was made to investigate the thermal dissipation properties of CNTs bundles and their applicability as heat exchanger. The 70µm high bundles have a diameter of 20 and 200µm. A Platinum thin film microheater, integrated on a freestanding SiN membrane, is used as heat source and as temperature sensor. The power consumption of the micro-heaters with different CNTs patterns, is measured in air. At 300 °C a power increase up to 31% was recorded for the microheaters equipped with the CNTs.

16:00   Break & Exhibit Inspection

Session IV– Fabrication
Session Chairs:
J. Kim, *Pohang University of Science & Technology, SOUTH KOREA*
B. Pruitt, *Stanford University, USA*

16:30 **3D ICE PRINTING AS A FABRICATION TECHNOLOGY OF MICROFLUIDICS WITH PRE-SEALED REAGENTS** 52
H.Z. Zhang, H. Li, M.X. Wu, H.Q. Yu, W. Wang, and Z.H. Li
Peking University, CHINA

We propose a innovative and inexpensive method to fabricate 3D structure featured by ice printing.This "bottom-up" 3D fabrication method is achieved by printing water onto the cold substrate and turning into ice structure layer by layer.Through this method, fluid with reagents, such as drugs and nanoparticles, is sealed into microfluidics during fabrication which can be used for drug delivery or other medical care applications.Moreover, a complex microchannels is easily fabricated using 3D ice structure as soft lithography mould, which can be used for microfluid-mixer, three dimensional flow focusing ect.

16:45 **LIQUID-FILLED SEALED MEMS CAPSULES FABRICATED BY FLUIDIC SELF-ASSEMBLY** 56
M. Mastrangeli, L. Jacot-Descombes, M.R. Gullo, and J. Brugger
Ecole Polytechnique Federale de Lausanne (EPFL), SWITZERLAND

We present an innovative method based on fluidic self-assembly for the encapsulation of functional liquids into sealed picoliter MEMS capsules. Capsules self-assembly and liquid co-encapsulation are achieved through the interplay of global fluidic stirring and local capillary forces ensuing from a selectively-precipitated insoluble polymeric phase. Our encapsulation method is massively parallel, scalable and compatible with batch MEMS fabrication. It can address a large variety of applications, including distributed MEMS sensors, self-healing materials, fragrance release and drug delivery.

17:00 **3D MASK MODULES USING TWO-PHOTON DIRECT LASER WRITING TECHNOLOGY FOR CONTINUOUS LITHOGRAPHY PROCESS ON FIBERS** 60
M. Hayashi[1,2], Y. Zhang[1], M. Hayase[2], T. Itoh[1], and R. Maeda[1]
[1]*National Institute of Advanced Industrial Science and Technology (AIST), JAPAN and*
[2]*Tokyo University of Science, JAPAN*

In this paper, we report a new fabrication method of 3D mask modules using two-photon direct laser writing technology in 140 μm-diameter half-pipe structures on quartz substrates. For the first time, the two-photon direct laser writing technology is utilized for the high resolution patterning process on a curved surface. The minimum feature sizes of about 2 μm line and space are successfully fabricated across the whole 140 μm-diameter half-pipe structures. Using the new 3D mask modules, fine metal patterns are prepared on 125 μm-diameter fiber.

17:15 **FABRICATION OF A MONOLITHIC MICRODISCHARGE-BASED PRESSURE SENSOR FOR HARSH ENVIRONMENTS** 64
X. Luo, C.K. Eun, and Y.B. Gianchandani
University of Michigan, USA

We present a 6-mask monolithic fabrication process for a pressure sensor that uses a differential microdischarge signal to sense diaphragm deflection. Microdischarge-based transduction is advantageous for harsh environments because of its immunity to temperature and inherently large signals. This work reports the first monolithic fabrication process that successfully addresses a number of challenges for microdischarge-based pressure sensors. Compared to prior work, it results in a ≈30× smaller exterior volume (0.05mm3), a ≈30× wider pressure range (40MPa), and backside terminals for appropriate packages.

17:30 **Adjourn for the day**

Tuesday, January 28

08:00 ANNOUNCEMENTS

Plenary Speaker II
Session Chairs:
H. Toshiyoshi, *University of Tokyo, JAPAN*
X. Wang, Tsinghua *University, CHINA*

08:05 BIONIC SKINS USING FLEXIBLE ORGANIC DEVICES 68
Takeo Someya[1,2] and T. Sekitani[1,2]
[1]*University of Tokyo, JAPAN and* [2]*Japan Science and Technology Agency (JST), JAPAN*

We have fabricated ultrathin, ultra-lightweight, ultraflexible, organic devices, such as organic thin-film transistors (TFTs), organic photovoltaic (OPV) cells, and organic light-emitting diodes (OLEDs) on polymeric films with the thickness of only 1 μm. The ultrathin organic devices are utilized to fabricate human-machine interfaces such as a touch sensor and wearable electronic systems such as an electromyogram measurement sheet with a two-dimensional array of organic amplifiers. The transistor films exhibit extraordinarily tough mechanical robustness such as minimum bending radius of 5 μm for organic TFTs.

Session V – Optical & Magnetic Microdevices
Session Chairs:
N. Miki, *Keio University, JAPAN*
H. Zappe, *University of Freiburg, GERMANY*

08:45 DEVELOPMENT OF MICRO VARIABLE OPTICS ARRAY 72
Y. Kwon[1], Y. Choi[1], K. Choi[1], Y. Kim[1], S. Choi[2], J. Lee[2], and J. Bae[1],
[1]*Samsung Electronics Co., Ltd., SOUTH KOREA and* [2]*Seoul National University, SOUTH KOREA*

We develop a micro variable optics array which modulates the direction of a light beam. Each pixel of the device has an interface of two immiscible liquids at which the lights are deflected and the interface is actuated on the four separated wall electrodes of the pixel by electrowetting. The four separated electrodes enable the every single pixel to work independently with multiple degrees of freedom, e.g. various tilting angles in every direction for prism or a large number of curvatures for lens mode.

09:00 MINIATURIZED MAGNETOELASTIC TAGS USING FRAME-SUSPENDED HEXAGONAL RESONATORS 76
J. Tang, S.R. Green, and Y.B. Gianchandani
University of Michigan, USA

Magnetoelastic resonators are of considerable interest for passive wireless interrogation and detection. This paper presents miniaturized magnetoelastic tags using hexagonal resonators with an overall size of about ⌀1.3mm X 27μm, and a resonant frequency as high as 2.13MHz. The tags are 100X smaller than typical commercial tags. The frame-suspension results in ≈75X improvement in signal amplitude of hexagonal tags compared to that of non-suspended disc tags. This paper also demonstrates that the signal amplitude can be boosted by utilizing signal superposition of an ensemble of tags.

09:15 SINGLE-STRUCTURE 3-AXIS LORENTZ FORCE MAGNETOMETER WITH SUB-30 nT√Hz RESOLUTION 80
M. Li[1], E.J. Ng[2], V.A. Hong[2], C.H. Ahn[2], Y. Yang[2], T.W. Kenny[2], and D.A. Horsley[1]
[1]*University of California, Davis, USA and* [2]*Stanford University, USA*

This work demonstrates a 3-axis Lorentz force magnetometer for electronic compass purposes. The magnetometer measures magnetic flux in 3 axes using a single structure with sub-30 nT/√Hz resolution. Assuming 10 μT Earth's field, the magnetometer has an angular resolution of 0.17 deg/√Hz with 1 mW power consumption. Compared to the 3-axis Hall sensors currently used in smartphones, the 3-axis magnetometer shown here has the advantages of 10× lower noise floor and the ability to be co-fabricated with MEMS inertial sensors.

09:30 DESIGN, FABRICATION AND CHARACTERIZATION OF TUNABLE PERFECT ABSORBER ON FLEXIBLE SUBSTRATE 84
X. Zhao[1], K. Fan[1], J. Zhang[1], H.R Seren[1], G.D. Metcalfe[2], M. Wraback[2], R.D. Averitt[1], and X. Zhang[1]
[1]*Boston University, USA and* [2]*U.S. Army Research Laboratory, USA*

This paper reports our recent progress on a highly flexible actively tunable metamaterial (MM) perfect absorber at terahertz frequencies. The MM array on GaAs thin-film was patterned on 5μm polyimide substrate via transfer printing technique, and the backside of the substrate was coated with gold. THz time-domain-spectroscopy measurements show that the absorptivity at resonance frequency of 1.59THz can be tuned up to 60% by photo-excitation of free carriers in GaAs patches. Our flexible tunable MM perfect absorber has potential applications in energy harvesting, and imaging.

09:45   TUNABLE META-FLUIDIC-MATERIALS BASE ON
MULTILAYERED MICROFLUIDIC SYSTEM 88
W.M. Zhu[1], B. Dong[1], Q.H. Song[2], W. Zhang[1], R.F. Huang[3], S.K. Ting[3], and A.Q. Liu[1,2]
[1]Nanyang Technological University, SINGAPORE, [2]Xi'an Jiao Tong University, CHINA,
and [3]Temasek Laboratories, SINGAPORE

We demonstrate a multilayered microfluidic system with a flexible substrate, which has tunable optical chirality within THz spectrum range. The optical properties of the multilayered microfluidic system can be tuned by either changing the liquid pumped into each layer or stretching the flexible substrate. It is feasible for the multilayered microfluidic structure to be integrated to an optofluidic system, where strong or tunable optical chirality are needed, which not only can be used as traditional optic components such as THz polarizers and filters but also has potential applications on imaging and sensor of biomaterials.

10:00   **Break & Exhibit Inspection**

Session VI – Fluidic Microdevices
Session Chairs:
G.-B. Lee, *National Tsing Hua University (NTHU), TAIWAN*
A. Dietzel, *Technische Universität Braunschweig, GERMANY*

10:45   **MULTIPLE SIZE-ORIENTED PASSIVE DROPLET SORTING DEVICE AND BASIC
APPROACH FOR DIGITAL CHEMICAL SYNTHESIS** 92
S. Numakunai, A. Jamsaid, D.H. Yoon, T. Sekiguchi, and S. Shoji
Waseda University, JAPAN

This paper presents a multiple size-oriented passive droplet sorting utilizing a balance between surface free energy and flow force. We propose a multi-stage sorting structure and passive five different-sized droplets sorting of about 100 droplets/sec is achieved without any active elements. Also, we fabricated a prototype of the integrated micro fluidic system with droplet generation, merging and sorting for digital chemical synthesis.

11:00   **SURFACE ENERGY MICROPATTERN INHERITANCE FROM MOLD TO REPLICA** 96
G. Pardon, T. Haraldsson, and W. van der Wijngaart
KTH Royal Institute of Technology, SWEDEN

We report a novel surface energy patterning phenomenon, in which a novel polymer composition inherits the surface energy of the medium it is in contact with during polymerization. This process occurs via spontaneous alignment of hydrophilic and hydrophobic monomers contained in the prepolymer. This single-step method for simultaneous structuring and surface energy micropatterning of polymer structures is potentially more robust and lower cost than state-of-the-art. We further demonstrate the self-assembly of a liquid droplet array on the replicated polymer surfaces.

11:15   **ELECTROSPRAY DEPOSITION FROM AFM PROBES WITH NANOSCALE APERTURES** 100
J. Geerlings[1], E. Sarajlic[1,2], J.W. Berenschot[1], R.G.P. Sanders[1], L. Abelmann[1,3], and N.R. Tas[1]
[1]MESA+, University of Twente, THE NETHERLANDS, [2]SmartTip B.V., THE NETHERLANDS, and
[3]Korea Institute of Science and Technology (KIST) – Europe, GERMANY

In this contribution we present for the first time extraction of liquid from nano-sized apertures in fountain pen AFM probes by means of electrospray. This technique allows for contactless deposition and we show that droplets with radii in the order of one micrometer can be deposited. The required onset voltage for electrospray as function of gap spacing and applied pressure is studied and a simple model is presented which is in qualitative agreement with our measurements.

11:30   **A MICROBUBBLE PRESSURE TRANSDUCER WITH BUBBLE NUCLEATION CORE** 104
L. Yu and E. Meng
University of Southern California, USA

We present a microchannel-based microbubble (μB) pressure transducer (μBPT) with μB nucleation core for characterization of μB dynamics and pressure transduction in wet environments with low power consumption. The transducer leverages electrochemical impedance-based measurement to monitor the instantaneous response of μB size induced by hydrostatic pressure changes. We demonstrated on-demand μB nucleation and real-time pressure tracking (0-350 mmHg). Biocompatible construction and liquid-based operation of μBPTs are ideal for *in vivo* pressure monitoring.

11:45   MICROFLUIDIC ELECTROCHEMILUMINESCENCE (ECL) INTEGRATED
CELL FOR PORTABLE FLUORESCENCE DETECTION 108
M. Tsuwaki[1], J. Mizuno[1], T. Kasahara[1], T. Edura[2], E. Kunisawa[2], R. Ishimatsu[2],
S. Matsunami[2], T. Imato[2], C. Adachi[2], and S. Shoji[1]
[1]Waseda University, JAPAN and [2]Kyushu University, JAPAN

We propose a portable electrochemiluminescence (ECL)-induced fluorescence chip which consists of flow channels for fluorescence sample and multi-color emitting ECL excitation source. A prototype ECL-induced fluorescence chip was fabricated by conventional photolithography and bonding technique. Device performance was evaluated using ECL of rubrene as excitation source and resorufin as fluorescent dye. Fluorescence of 500 µM resorufin (600 nm) was successfully detected using 10 mM rubrene solution (560 nm) under the applied voltage of 4 V. The proposed principle is applicable for portable and on-demand multi fluorescence detection device using its freedom of choice for combination of the ECL light source.

12:00   **A MONOLITHIC KNUDSEN PUMP WITH 20 sccm FLOW RATE
USING THROUGH-WAFER ONO CHANNELS** 112
S. An, Y. Qin, and Y.B. Gianchandani
University of Michigan, USA

We report a lithographically microfabricated Knudsen pump for high gas flow. Knudsen pumps operate by thermal transpiration and require no moving parts. To achieve high gas flow, high-density arrays of microchannels are used in parallel (with over 4000 channels/mm^2). These vertically oriented microchannels have 2×120 µm^2 openings surrounded by 0.1 µm-thick silicon oxide-nitride-oxide (ONO) sidewalls. The thin ONO sidewalls provide thermal isolation between a heat sink formed within the Si substrate, and a Cr/Pt thin film heater formed above the microchannels that provides a temperature bias for thermal transpiration. The Knudsen pump is monolithically microfabricated on a single SOI wafer using a four-mask process. It has a total footprint of 8×10 mm^2. It produces a measured air flow of 20 sccm, with typical response times of 0.1-0.4 sec.

12:15   **MEMS 2015 Announcement**

12:30   **Lunch on Own**

14:00   **Poster/Oral Session II**

Session VII– Resonant Microdevices & Sensors
Session Chairs:
M. Rais-Zadeh, *University of Michigan, USA*
L. Buchaillot, *IEMN, FRANCE*

16:00   **SUSPENDED NANOCHANNEL RESONATORS AT ATTOGRAM PRECISION** 116
S. Olcum[1], N. Cermak[1], S.C. Wasserman[1], K. Payer[1], W. Shen[2], J. Lee[3], and S.R. Manalis[1]
*[1]Massachusetts Institute of Technology, USA, [2]Innovative Micro Technology, USA, and
[3]Sogang University, SOUTH KOREA*

We developed a nanomechanical resonator that can directly measure the mass of individual nanoparticles down to 10 nm in solution at room temperature with single-attogram precision, also enabling access to many of the engineered nanoparticles used in nanomedicine, most of the virions like HIV, HCV, and natural sub-cellular structures like exosomes. To achieve this, we demonstrate an oscillator system with frequency stability down to 4 ppb, approaching the fundamental limit imposed by intrinsic thermomechanical fluctuations of the resonator.

16:15   **DUAL-MODE VERTICAL MEMBRANE RESONANT PRESSURE SENSOR** 120
R. Tabrizian and F. Ayazi
Georgia Institute of Technology, USA

We present a novel dual-mode resonant pressure sensor operating based on mass loading of air molecules on transversely vibrating vertical silicon membranes. Identical piezoelectrically-transduced silicon bulk acoustic resonators are acoustically coupled through thin vertical membranes, resulting in two high-Q resonance modes with small frequency split, but large difference in pressure sensitivity. Being proportional to the flexural resonance frequency of the thin membranes, the small beat frequency (fb) extracted from subtraction of the two coupled modes shows amplified pressure sensitivity. A proof-of-concept device implemented on silicon substrate and transduced by aluminum nitride film shows an fb of 370 kHz with a linear pressure sensitivity of 280 ppm/kPa.

16:30 HIGHLY RESPONSIVE CURVED ALUMINUM NITRIDE PMUT 124

S. Akhbari[1], F. Sammoura[1,2], S. Shelton[3], C.Yang[1], D. Horsley[3], and L. Lin[1]
[1]*University of California, Berkeley, USA,* [2]*Masdar Institute of Science and Technology, UAE, and*
[3]*University of California, Davis, USA*

We have successfully demonstrated highly responsive, curved piezoelectric micromachined ultrasonic transducers (PMUT) based on a CMOS-compatible fabrication process using AlN as the transduction material for the first time. A prototype device using a 2μm-thick AlN layer on a curved diaphragm surface with a radius of curvature of 1065μm and physical size of 140μm in diameter has shown a measured resonant frequency at 2.19MHz. The DC response has been experimentally measured as 1.1nm/V, which is 50X higher than that of a planar device with same size and operation conditions. As such, this new class of curved PMUT could replace the state-of-art, planar PMUT to achieve high electromechanical coupling for various ultrasonic transduction applications, including gesture recognition and medical imaging.

16:45 ENCASED CANTILEVERS FOR LOW-NOISE FORCE AND MASS SENSING IN LIQUIDS 128

D. Ziegler[1], A. Klaassen[2], D. Bahri[1], D. Chmielewski[1], A. Nievergelt[1], F. Mugele[2],
J.E. Sader[3], and P.D. Ashby[1]
[1]*Lawrence Berkeley National Laboratory, USA,* [2]*MESA+, University of Twente, THE NETHERLANDS, and*
[3]*University of Melbourne, AUSTRALIA*

Viscous damping severely limits the performance of cantilever based sensing in liquids. Encased cantilevers achieve low damping in liquids by keeping the resonator dry. This is achieved by fabricating a hydrophobic encasement from which only few microns of the sensing tip protrude into the liquid. We achieve Q-factors and associated noise levels as if operating in air. We discuss fabrication of these devices and demonstrate successful application for low-noise mass sensing and gentle AFM imaging of soft matter in liquids.

**17:00 RESONANT MICRO-SENSOR PLATFORM FOR CONTACT-FREE
CHARACTERIZATION OF LIQUID PROPERTIES** 132

J.M. Gonzales and R. Abdolvand
Oklahoma State University, USA

This paper presents a novel resonant microsensor platform which maintains high quality factors(Q) when measuring ultrasonic properties of liquid samples such as blood. By avoiding the direct contact of the liquid with the resonator, significant losses due to liquid loading are mitigated and the physical properties of various fluids, including viscous samples, can be determined without adversely affecting the resonator performance. Devices have been fabricated and tested, achieving quality factors up to 6000 in air and the results show that the output signals measured from the device are sensitive to the properties of the liquid under test.

17:15 NOVEL IN-PLANE GAP CLOSING CMOS-MEMS MICROPHONE WITH NO BACK-PLATE 136

C.-I. Chang[1], S.-C. Lo[1], C. Wang[2], Y.-C. Sun[1], and W. Fang[1,2]
[1]*National Tsing Hua University (NTHU), TAIWAN and* [2]*MotionsTek Inc., TAIWAN*

The stacking of metal/tungsten layers as the sensing electrodes for CMOS-MEMS microphone without the back-plate has been proposed and demonstrated for the first time (Fig.1a). The acoustic pressure will deform the spring-diaphragm structure and further cause the in-plane gap-closing between sensing electrodes (Fig.1b). Thus, acoustic pressure and dynamic response of spring-suspension can be determined by the sensing capacitance changes. Such design has the following merits: (1) no back-plate is required, (2) bias voltage to pull diaphragm close to back-plate is not required, (3) in-use pull-in and process stiction between diaphragm and back-plate is also prevented, (4) easy integration with sensing circuits [1]. The design was implemented using the standard TSMC CMOS process. Typical microphone with 200μm-diameter diaphragm and 48-pairs sensing electrodes has been realized. Measurements show the sensitivity of microphone is -67.17dBV/Pa at 1KHz.

17:30 Adjourn for the day

Wednesday, January 29

08:00 **ANNOUNCEMENTS**

Plenary III
Session Chairs:
J. Brugger, *Ecole Polytechnique Federale de Lausanne (EPFL), SWITZERLAND*
W. van der Wijngaart, *KTH – Royal Institute of Technology, SWEDEN*

08:05 **CAVITY QUANTUM OPTOMECHANICS:**
COUPLING LIGHT AND MICROMECHANICAL OSCILLATORS 140
E. Verhagen, S. Deléglise, S. Weis, A. Schliesser, and **Tobias J. Kippenberg**
Ecole Polytechnique Federale de Lausanne (EPFL), SWITZERLAND

Session VIII – Nanodevices
Session Chairs:
L. Lin, *University of California, Berkeley, USA*
Y.-J. Yang, *National Taiwan University, TAIWAN*

08:45 **AMORPHOUS CARBON ACTIVE CONTACT LAYER FOR RELIABLE**
NANOELECTROMECHANICAL SWITCHES 143
D. Grogg[1], C.L. Ayala[1], U. Drechsler[1], A. Sebastian[1], W.W. Koelmans[1], S.J. Bleiker[2],
M. Fernandez-Bolanos[1], C. Hagleitner[1], M. Despont[1], and U.T. Duerig[1]
[1]IBM Research – Zurich, SWITZERLAND and [2]KTH Royal Institute of Technology, SWEDEN

This paper reports an amorphous carbon (a-C) contact coating for ultra-low-power curved nanoelectromechanical (NEM) switches. a-C addresses important problems in miniaturization and low-power operation of mechanical relays: i) the surface energy is lower than that of metals, ii) active formation of highly localized a-C conducting filaments offers a way to form nano-scale contacts, and iii) high reliability is achieved through the excellent wear properties of a-C, demonstrated in this paper with more than 100 million hot switching cycles.

09:00 **NEAR INFRARED PHOTO-DETECTOR USING SELF-ASSEMBLED FORMATION**
OF ORGANIC CRYSTALLINE NANOPILLAR ARRAYS 147
Y. Ajiki[1,4], T. Kan[2], M. Yahiro[3], A.Hamada[3], J. Adachi[3], C. Adachi[3], K. Matsumoto[2], and I. Shimoyama[2]
[1]Olympus Corporation, JAPAN, [2]University of Tokyo, JAPAN, [3]Kyushu University, JAPAN, and
[4]NMEMS Technology Research Organization, JAPAN

We proposed a near infrared photo-detector (NIR-PD) using self-assembled formation of organic crystalline arrays, which were formed on an n-type silicon (n-Si) substrate and covered with an Au film. These structures act as antennas for near infrared (NIR) light, resulted in an enhancement of the light absorption on the Au film. In this paper, the fabrication process of the NIR-PDs and the estimation results of photo-responsivity are described. The maximum value of the responsivity to NIR light (wavelength = 1.2 μm) was 1.79 mA/W without applying forward bias. This value is 10 times larger than that of a conventional Au/n-Si typed Schottky diode, which is fabricated as a reference.

09:15 **ULTRASENSITIVE SI NANOWIRE PROBE FOR MAGNETIC RESONANCE DETECTION** 151
Y.J. Seo, M. Toda, Y. Kawai, and T. Ono
Tohoku University, JAPAN

We have fabricated and evaluated an atto-newton-sensitive Si nanowire probe with a Nd-Fe-B magnet for magnetic resonance force microscopy. The width, thickness and length of the nanowire are 210 nm, 200 nm and 32 μm, respectively. The nanowire probe has a resonance frequency f0 of 11.256 kHz and a Q factor of order 12000. Then, we have demonstrated the measurement of force mapping based on electron spin resonance for three-dimensional imaging of radicals.

09:30 **A VERTICALLY INTEGRATED NANOSCALE TIPPED**
MICROPROBE INTRACELLULAR ELECTRODE ARRAY 155
Y. Kubota, H. Oi, H. Sawahata, A. Goryu, Y. Ando, R. Numano, M. Ishida, and T. Kawano
Toyohashi University of Technology, JAPAN

Here we report an integration of vertical 120-μm-long nanoscale tipped microprobe electrode (NTE) array and the intracellular recordings using a gastrocnemius muscle of a mouse. The tip diameter of the NTE was < 200 nm, with the height of 4 μm exposed from the parylene-shell. The impedance of the NTE exhibited 3.1 MΩ at 1 kHz in saline, with the output/input signal amplitude ratio of 50% for intracellular recordings. The penetrated NTE into the muscle of a mouse detected the residual potentials with the amplitude of ~ -200 mV, confirming the intracellular recording capability of the NTE.

09:45 **Break & Exhibit Inspection**

Session IX – Energy Harvesting & Power
Session Chairs:
Z. Li, *Peking University, CHINA*
J. Judy, *University of Florida, USA*

10:30 **ENERGY HARVESTING USING UNIAXIALLY ALIGNED CARDIOMYOCYTES** 159
X. Liu[1,2], X. Wang[1], S. Li[2] and L. Lin[2]
[1]Tsinghua University, CHINA and [2]University of California, Berkeley, USA

This study presents the concept of energy harvesting from uniaxially-aligned cardiomyocytes on a flexible substrate for the first time. Experimentally, synchronously contracting neonatal rat ventricular cardiomyocytes (NRVCs) at 0.5Hz have been found to cause the mechanical straining of a piezoelectric energy harvester to produce 87.5nA and 92.3mV of peak current and voltage, respectively. This work presents a successful step toward mechanical energy harvesting via living biological cells and tissues.

10:45 **DIFFUSION REFUELING BIOFUEL CELL MOUNTABLE ON INSECT** 163
K. Shoji[1], Y. Akiyama[1], M. Suzuki[2], N. Nakamura[2], H. Ohno[2], and K. Morishima[1]
[1]Osaka University, JAPAN and [2]Tokyo University of Agriculture and Technology, JAPAN

This paper reports an insect-mountable biofuel cell (imBFC) using trehalose, main sugar of insect hemolymph. The imBFC is refueled trehalose by diffusion from insect hemolymph automatically and generates electric power by oxidizing glucose which is obtained by hydrolyzing trehalose enzymatically. We fabricated the imBFC consisted of a connector, two chambers separated by a dialysis membrane and electrodes and succeeded in driving a light-emitting diode by the imBFC. The results have shown a potentially to be applied for a battery of novel ubiquitous robots such as insect cyborgs.

11:00 **ALD RUTHENIUM OXIDE-CARBON NANOTUBE ELECTRODES
FOR SUPERCAPACITOR APPLICATIONS** 167
R. Warren[1], F. Sammoura[1,2], A. Kozinda[1], and L. Lin[1]
[1]University of California, Berkeley, USA and [2]Masdar Institute of Science and Technology, UAE

This work presents the first demonstration of atomic layer deposition (ALD) ruthenium oxide (RuO_2) and its conformal coating onto vertically aligned carbon nanotube (CNT) forest as supercapacitor electrodes. The ALD method allows precise control over the RuO_2 layer thickness and composition without the use of binder molecules. The ALD RuO_2 coated CNTs achieve a specific capacitance of 100 mF/cm^2 and retain their high-performance over repeated cycling.

11:15 **SUB 3-MICRON GAP MICROPLASMA FET WITH 50 V TURN-ON VOLTAGE** 171
P. Pai and M. Tabib-Azar
University of Utah, USA

We report the smallest microplasma transistor reported till date that operates with a low turn-on voltage of 50V dc. The device achieves more than 3x reduction in the turn-on voltage and 100x reduction in size compared to devices reported by other groups in the past. Earlier work reported by our group used plasma from an external source to operate the transistor. Our recent work successfully generated direct current plasma within the device with a turn-on voltage of 180V. This paper reports gate field-effect characterization results performed under dc and rf excitation and draws a comparison.

11:30 **Lunch on Own**

13:00 **Poster/Oral Session III**

Session X– Microdevices for Cell Manipulation
Session Chairs:
I. Park, *Korea Advanced Institute of Science and Technology (KAIST), SOUTH KOREA*
Y. Sun, *University of Toronto, CANADA*

15:00 **FORMATION OF CROSS-SHAPED ESCHERICHIA COLI** 175
K. Hirayama, Y.J. Heo, and S. Takeuchi
University of Tokyo, JAPAN

We develop a method to regulate the shapes of bacteria by confining a single cell of bacteria into micro chamber. Escherichia coli has a cell wall structure which determines the rod-shape. We removed the cell wall of E. coli and then confined the spheroplasts into cross-shaped microchambers. If the bacteria re-synthesize its cell wall within the microchamber, we could obtain the cross-shaped bacteria. By analyzing the behavior of proteins and the cross-shaped bacteria, we believe that we could contribute to the comprehensive understanding of the shape regulation mechanism of E. coli.

15:15   SINGLE-CELL 3D BIO-MEMS ENVIRONMENT WITH ENGINEERED
GEOMETRY AND PHYSIOLOGICALLY RELEVANT STIFFNESSES 177
M. Marelli[1], N. Gadhari[2], G. Boero[1], M. Chiquet[2], and J. Brugger[1]
[1]Ecole Polytechnique Federale de Lausanne (EPFL), SWITZERLAND and
[2]University of Bern, SWITZERLAND

We present a 3D microenvionment for on-chip cell culture, made of stress-bent cantilevers and designed to mimic essential physical properties of the in vivo environment, at the single-cell scale and with a high degree of parallelization. In particular, we report on a combinatorial fabrication approach bringing to the realization of a palette of devices with constant sizes (fitting a single-cell), but with stiffnesses spanning two orders of magnitude and matching physiologically relevant values.

15:30   INTEGRATED MICRO CULTURE DEVICE FOR FULLY AUTOMATED
CLOSED CULTURE EXPERIMENT OF EMBRYONIC BODY 181
A. Yasukawa[1], T. Nishijima[2], M. Ikeuchi[1], and K. Ikuta[1]
[1]University of Tokyo, JAPAN and [2]Nagoya University, JAPAN

This paper reports formation, differentiation and analysis of embryonic bodies (EBs) from human iPS cells in a palm-size device "PASMA (Pressure Actuated Shapable Microwell Array)". By incorporating a transparent heat film and $CO2$ concentration adjusting system, PASMA realized miniaturization of the whole process of EB experiment in one chip. Moreover, the fully automated closed culture system can eliminate the risks of contamination due to manual operation. EBs from human iPS cells were successfully fabricated using this system.

15:45   MECHANICAL CELL PAIRING SYSTEM BY SLIDING PARYLENE RAILS 185
Y. Abe[1,2], K. Kamiya[1,3], T. Osaki[1,4], R. Kawano[1], N. Miki[1,2], and S. Takeuchi[1,4]
[1]Kanagawa Academy of Science and Technology, JAPAN, [2]Keio University, JAPAN,
[3]Japan Science and Technology Agency (JST), JAPAN and [4]University of Tokyo, JAPAN

This paper proposes a cell pairing system that is capable of defining the number and the position of trapped cells by mechanically sliding the parylene rail films (PRF). This system allows us to control the number as well as the order of lined-up cells. We successfully demonstrated lining up of three different cells in a designated order. The proposed system is readily applicable to study the cell-cell interactions using the single cell pairing.

16:00   **Break & Exhibit Inspection**

Session XI– Bio-Inspired Microactuators
Session Chairs:
Y. Yamanishi, *Shibaura Institute of Technology, JAPAN*
T.-H. Wang, *The Johns Hopkins University, USA*

16:30   TWO-DIMENSIONALLY STEERING MICROSWIMMER
PROPELLED BY OSCILLATING BUBBLES 188
J. Feng and S.K. Cho
University of Pittsburgh, USA

An acoustically excited and oscillated bubble can generate a propelling force in micro scale. We develop a simple and efficient method that allows a bubble-propelled micro swimmer to propel and steer two dimensionally as wirelessly and remotely commanded.

16:45   A MICROFABRICATED, BIOHYBRID, SOFT ROBOTICS FLAGELLUM 192
B.J. Williams[1], S.V. Anand[1], J. Rajagopalan[2], and M.T.A. Saif[1]
[1]University of Illinois, Urbana-Champaign, USA and [2]Arizona State University, USA

We develop a microfabricated soft robotics biohybrid swimmer utilizing the contractions of one to several cardiomyocytes to provide on-board actuation to a thin, deformable, polydimethylsiloxane (PDMS) filament. The actuated filament deforms passively in response to fluid drag, producing a time-irreversible deformation pattern and net propulsive force at low Reynolds number. We utilize an elastohydrodynamic model to determine appropriate filament parameters and realize a functional swimmer.

17:00 **EVALUATION AND OPTICAL CONTROL OF SOMATIC MUSCLE MICRO BIOACTUATOR OF CHANNELRHODOPSIN TRANSGENIC DROSOPHILA MELANOGASTER** 196
M. Hirooka[1], S.P. Beh[1], T. Asano[1], Y. Akiyama[1], T. Hoshino[2], K. Hoshino[3], H. Tsujimura[3], K. Iwabuchi[3], and K. Morishima[1]
[1]Osaka University, JAPAN, [2]University of Tokyo, JAPAN, and
[3]Tokyo University of Agriculture and Technology, JAPAN

In this research, we first developed light-activated somatic muscle of transgenic Drosophila melanogaster expressing a blue light sensitive cation channel, channelrhodopsin-2, and incorporated into a micro device. We successfully demonstrated that optogonetic stimulation using light pulses was able to control the contractile activity with a given temporal pattern. The contractile force was evaluated with varying light intensities and pulse widths. These results have shown that mechanical systems powered by muscles will be controlled more accurately under our strategy.

17:15 **FERROFLUID-ASSISTED MICRO ROTARY MOTOR FOR MINIMALLY INVASIVE ENDOSCOPY APPLICATIONS** 200
B. Assadsangabi, M.H. Tee, S. Wu, and K. Takahata
University of British Columbia, CANADA

Micro-scale rotary motors have a vast range of potential applications in broad areas. One promising area is medical applications. Minimally-invasive endoscopic catheters are an excellent example. The micromotors used in these catheters, however, generally have large axial lengths (e.g., >1 cm). This paper presents the first rotary micromotor enabled by ferrofluid bearing that drastically shortens the axial length of micromotors, suitable for medical catheter applications, owing to the extremely simple and reliable bearing mechanism.

17:30 **Adjourn for the day**

18:30 - **Conference Banquet**
21:30

Thursday, January 30

08:00 ANNOUNCEMENTS

Plenary IV
Session Chairs:
F. Ayazi, *Georgia Institute of Technology, USA*
CJ Kim, *University of California, Los Angeles, USA*

**08:05 COMMERCIALIZATION OF WORLD'S FIRST PIEZOMEMS
RESONATORS FOR HIGH PERFORMANCE TIMING APPLICATIONS** 204
Harmeet Bhugra, S. Lee, W. Pan, M. Pai, and D. Lei
Integrated Device Technology, Inc. (IDT), USA

Session XII – Resonators & RF MEMS
Session Chairs:
A. Seshia, *Cambridge University, UK*
K. Böhringer, *University of Washington, USA*

**08:45 MEMS-RECONFIGURABLE WAVEGUIDE IRIS FOR
SWITCHABLE V-BAND CAVITY RESONATORS** 206
Z. Baghchehsaraei and J. Oberhammer
KTH Royal Institute of Technology, SWEDEN

We present, for the first time, a reconfigurable waveguide iris based on a MEMS-reconfigurable surface integrated into a WR-12 rectangular waveguide (60–90 GHz, 3.099 mm × 1.549 mm). The reconfigurable surface is only 30 μm thick and incorporates 252 simultaneously switched contact points for activating or deactivating an inductive iris. The switchable irises can be utilized to implement components such as reconfigurable filters and cavity resonators. We also present for the first time a reconfigurable cavity resonator based on the novel MEMS-reconfigurable iris.

**09:00 A MICROMECHANICAL PARAMETRIC OSCILLATOR FOR FREQUENCY
DIVISION AND PHASE NOISE REDUCTION** 210
T.O. Rocheleau, R. Liu, J. Naghsh Nilchi, T.L. Naing, and C.T.-C. Nguyen
University of California, Berkeley, USA

A capacitive-gap RF MEMS resonator array demonstrates a first on-chip MEMS frequency divider with 61-MHz output generated from a 121-MHz electrical drive through use of a parametric oscillation effect. This provides not only the expected 6dB reduction in close-to-carrier phase noise, but also a remarkable 23dB reduction in far-from-carrier noise due to filtering by the high-Q mechanical resonator. In contrast with traditional frequency division, the parametric oscillator here requires no active devices, adds no noise to the signal and has essentially zero power consumption, limited in principle only by MEMS resonator loss.

**09:15 TEMPERATURE-COMPENSATED PIEZOELECTRICALLY
ACTUATED LAMÉ-MODE RESONATORS** 214
V.A. Thakar and M. Rais-Zadeh
University of Michigan, USA

In this work, we present a passive compensation strategy for Lamé-mode resonators using silicon dioxide refilled islands within the resonator body. With this technique we achieve compensation of the first-order TCF and further demonstrate that the turnover temperature (TOT) can be tuned across a wide range from -40°C to +120°C by optimizing the placement of oxide refilled islands.

09:30 A MEMS AUTONOMOUS SWITCHED OSCILLATOR 218
G.B. Torri[1,2], J. Bienstman[2], X. Rottenberg[1], H.A.C. Tilmans[2], C. Van Hoof[1,2], and R. Puers[1,2]
[1]*imec, BELGIUM and* [2]*KU Leuven, BELGIUM*

We design and measure an autonomous electrostatic MEMS oscillator that exhibits periodic and aperiodic behavior. The system consists of an electrostatic MEM, a dc voltage source, and a displacement dependent resistive circuit. The applied voltage above the pull-in and the position dependent circuit are responsible for sustaining the oscillations. Electronic and non-electronic information can be input or retrieved from the oscillator system. Applications for such device range from chaotic signals for communications, sensitive mass sensors, and signal processing.

10:00 Poster/Oral Session IV

Session XIII – Chemical Sensors & Systems
Session Chairs:
A. Llobera, *Centre Nacional de Microelectronica, SPAIN*
H. Moon, *University of Texas, Arlington, USA*

**12:00 SURFACE TENSION MONITORING IN A "SOFT" INTERFACE
USING ELECTROWETTING ON DIELECTRIC** 222
S. Choi[1], Y. Kwon[2], E. Jeon[1], S. Kim[1], and J. Lee[1]
[1]Seoul National University, SOUTH KOREA and [2]Samsung Electronics Co., Ltd., SOUTH KOREA

We report the first demonstration of interfacial tension monitoring across two immiscible liquids using electrowetting on dielectric (EWOD). Impedance measurement during EWOD reveals the variation of surfactant concentration at the liquid-liquid interface in real time. We also show such approach can be used for label-free monitoring of DNA hybridization. Our approach opens a new horizon of EWOD used as a molecular sensing mechanism across a "soft" interface between liquids.

12:15 MICROFLUIDIC BUBBLE-BASED GAS SENSOR 226
A. Bulbul, H.-C. Hsieh, and H. Kim
University of Utah, USA

We report a new class of a gas sensor that utilizes the variations in bubble sizes, when different gases are introduced into a liquid flow, to identify gas types and even quantify the amounts. To verify the feasibility of the concept, the sizes of discretely formed bubbles were optically monitored and analyzed through a custom-developed MATLAB software.

12:30 A FLEXIBLE GRAPHENE FET GAS SENSOR USING POLYMER GATE DIELECTRICS 230
Y. Liu, J. Chang, and L. Lin
University of California, Berkeley, USA

We have successfully demonstrated a Graphene FET (GFET) gas sensor on a flexible plastic substrate for the first time with a sensitivity of 0.00428ppm^{-1} ($\Delta R/R_0$) for ammonia detection. Compared with the state-of-art technologies, four distinctive advancements have been achieved: (1) first demonstration of a flexible graphene FET gas sensor; (2) a new fabrication process to achieve embedded-gate GFET on a flexible substrate; (3) proof of utilizing polymeric materials of parylene and polyethylenimine (PEI) as the gate dielectrics and channel dopant for graphene FET, respectively; and (4) validation of a real-time gas sensing mechanism utilizing n-type doping of graphene induced by ammonia exposure.

12:45 Awards Ceremony

13:00 Conference Adjourns

Poster/Oral Presentations

M – Monday (13:00 - 15:00)
T – Tuesday (14:00 - 16:00)

W – Wednesday (13:00 - 15:00)
Th – Thursday (09:45 - 11:45)

Bio MEMS (Bio)

M-001 A CONTINUOUS OPTICALLY-INDUCED CELL ELECTROPORATION DEVICE WITH ON-CHIP MEDIUM EXCHANGE MECHANISMS 234
C.-J. Chang, M.-Y. Lu, and G.-B. Lee
National Tsing Hua University (NTHU), TAIWAN

We present a novel design of continuously optically-induced electroporation (OIE) device capable of replacing culture medium and electroporation buffer in a seamless fashion. With this approach, the entire process for continuous cell electroporation could be performed automatically without human intervention. Furthermore, the survival rate of the cells could be greatly improved due to this fast, automatic procedure.

T-002 A HIGH-THROUGHPUT PERMEABILITY ASSAY PLATFORM FOR SHEAR STRESS CHARACTERIZATION OF ENDOTHELIAL CELLS 238
R. Booth, S. Noh, and H. Kim
University of Utah, USA

Here we present the first permeability assay platform to enable rapid characterization of shear stress effects fully spanning the physiologically relevant spectrum (1-60dyn/cm2) for endothelial cells in vitro. The structure comprised 4 parallel channels to enable independent permeability assays, and generated 15x shear stress range indicated by simulation and an integrated micro-flow sensor array. Endothelial cells exhibited decreased permeability of fluorescent tracers with shear stress, as well as increased elongation and cell alignment with increases in shear stress.

W-003 A MICROFLUIDIC DEVICE FOR ISOLATION OF CELL-TARGETING APTAMERS 242
J. Zhu[1], T. Olsen[1], R. Pei[2], M. Stojanovic[1], and Q. Lin[1]
[1]Columbia University, USA and [2]Chinese Academy of Sciences, CHINA

This paper presents a microfluidic device for synthetically isolating cell-targeting aptamers from a randomized single-strand DNA (ssDNA) library. The device integrates cell culturing with affinity selection of cell-binding ssDNA, which is then amplified by bead-based polymerase chain reaction (PCR). Coupling of the selection and amplification using pressure-driven flow realizes multi-round aptamer isolation on a single chip.

Th-004 A NANOWIRE-INTEGRATED MICROFLUIDIC DEVICE FOR HYDRODYNAMIC TRAPPING AND ANCHORING OF BACTERIAL CELLS 246
D. Kwon, J. Kim, S. Chung, and I. Park
Korea Advanced Institute of Science and Technology (KAIST), SOUTH KOREA

We develop nanowire-integrated microfluidic devices for hydrodynamic trapping and anchoring of bacterial cells and demonstrate that the mesh-like cages formed by integrating ZnO nanowires are effective in trapping and anchoring of Escherichia coli as a model bacterium. We present two anchoring modes, impaling and wedging, followed by irreversible damage or reversible deformation of the cell wall, respectively.

M-005 AN INTEGRATED MICROFLUIDIC SYSTEM USING FIELD-EFFECT TRANSISTORS FOR CRP DETECTION 250
C.L. Lin, Y.W. Kang, K.W. Chang, W.H. Chang, Y.L. Wang, and G.B. Lee
National Tsing Hua University (NTHU), TAIWAN

Cardiovascular diseases are responsible for 25-million deaths worldwide on a yearly basis. Timely diagnosis of the disease is therefore an important research area. Toward this end, C-reactive protein (CRP) is a general biomarker for inflammation and infection, and has become a good marker for evaluating risks of cardiovascular diseases. Previously, a CRP-specific aptamer (nucleic acid-based antibody, in a nutshell) with high sensitivity and specificity was used to detect CRP in microfluidic system, which was capable of performing the detection in an automated fashion while consuming tiny volumes of reagents and samples. In parallel, field-effect transistors (FET) have emerged as sensors to detect small molecules, proteins and even viruses, and have demonstrated rapid and highly sensitive detection in a compact system. In this work, an integrated device that combined the advantages of microfluidics, aptamers and FET-based sensors was developed to achieve rapid, sensitive and specific CRP detection. The developed integrated microfluidic system with FET sensor and CRP-specific aptamer can be promising for fast, sensitive and specific CRP detection.

T-006 ASSEMBLY OF NEURAL CELL-LADEN MICROPLATES ON A MICROFABRICATED BREADBOARD 254

S. Yoshida[1], K. Sato[1,2], and S. Takeuchi[1,2]

[1]University of Tokyo, JAPAN and [2]Japan Science and Technology Agency (JST), JAPAN

We developed a method to form an arbitrary network of neural cell-laden microplates using a breadboard-like microelectrode array (MEA). We cultured single neural cells on perforated microplates with control of their morphologies, then mounted the plates on the micro-sockets attached on the breadboard. Through a pore punched in the center of the microplate, the electrical activities of cells would be accessible with gold electrodes patterned on the MEA.

W-007 CD4 CELL ISOLATION FROM BLOOD USING FINGER-ACTUATED ON-CHIP MAGNETOPHORESIS FOR RAPID HIV/AIDS DIAGNOSTICS 256

M. Glynn, D.J. Kinahan, D. Chung, and J. Ducrée

Dublin City University, IRELAND

We have developed a microfluidic based disposable device, actuated solely by finger pressure from the operator, to repeatedly pass micro quantities of blood past a magnetic based cell capture structure to allow CD4 cell based HIV diagnostics in less than 1 minute following sample application. This is aimed for deployment in resource-poor regions where HIV is endemic but diagnostics is a challenge.

Th-008 CENTRIFUGALLY AUTOMATED SOLID-PHASE PURIFICATION OF RNA 260

N. Dimov, J. Gaughran, D. McAuley, D. Boyle, D.J. Kinahan, and J. Ducrée

Dublin City University, IRELAND

We report for the first time on a fully centrifugally automated solid-phase purification of RNA on an integrated microfluidic disc with sequential release of on-board reagents.

M-009 FABRICATION METHOD TO A HIGH RESOLUTION CONTROL IN THE SPACE OF CELL CULTURING ENVIRONMENT WITH MICROFLUIDIC SYSTEM 264

T. Hiraiwa[1], T. Kimura[1], Y. Takenaka[1], R. Tanamoto[1], H. Ota[3], H. Kimura[2], Y. Taguchi[1], N. Miki[1], Y. Matsumoto[1], K. Oka[1], A. Funahashi[1], and N. Hiroi[1]

[1]Keio University, JAPAN, [2]Tokai University, JAPAN, and [3]University of California, Berkeley, USA

We develop a reusable Cell Culturing Device, designed for single cells and cellular network analysis. This is the first success of combination of Microcontact Printing (μCP) and Vacuum Device. This device has following advantages: (1) cells stay within the micropatterns more than 12hrs, (2) cell culture environment is regulated precisely using laminar flow, (3) this device is reusable for further experiments.

T-010 INDIVIDUALLY ADDRESSABLE MULTI-CHAMBER ELECTROPORATION PLATFORM WITH DIELECTROPHORESIS-ASSISTED CELL POSITIONING 268

S. Park and G. Yossifon

Technion–Israel Institute of Technology, ISRAEL

We developed a novel, multifunctional platform with an array for high throughput screening which integrates DEP trapping/positioning and electroporation mechanisms in a multi-chamber array where each chamber may be individually activated. As a proof of concept, platforms with an array of 40 × 2 cell microchambers were fabricated and tested to demonstrate the feasibility of integrating the two functionalities: 1) the manipulation of the particles/cells using DEP, and 2) gradient electroporation using by producing a spatial gradient of electroporation parameters to optimize electroporation efficiency.

W-011 INTEGRATION OF HEAT-TRANSFER RESISTANCE MEASUREMENTS ONTO A DIGITAL MICROFLUIDIC PLATFORM TOWARDS THE MINIATURIZED AND AUTOMATED LABEL-FREE DETECTION OF BIOMOLECULAR INTERACTIONS 272

E. Perez-Ruiz[1], T. Vandenryt[2], D. Witters[1], D. Decrop[1], B. van Grinsven[2], D. Spasic[1], P. Wagner[2], and J. Lammertyn[1]

[1]KU Leuven, BELGIUM and [2]University of Hasselt, BELGIUM

In this work the successful integration of heat-transfer resistance measurements with a digital microfluidic chip is shown. The integrated miniaturized platform allows the automated label-free detection of biomolecular interactions. To immobilize biomolecules on the hydrophobic chip surface, hydrophilic gold sensing patches are created by means of a novel dry lift-off technique that leaves the chip surface unaffected. In order to validate the integrated device, DNA melting analysis was performed in the set-up.

Th-012 NANOFLUIDIC SENSING AT NORMAL PHYSIOLOGICAL CONDITION BY COUPLING ION CONCENTRATION POLARIZATION WITH A LIMIT OF DETECTION OF ONE FEMTOMOLE 276

W. Ouyang[1], W. Wang[1,2], H. Zhang[1], W. Wu[1,2], and Z. Li[1,2]
[1]Peking University, CHINA and
[2]National Key Laboratory of Science and Technology on Micro/Nano Fabrication, CHINA

Traditional nanofluidic sensing devices lose function at normal physiological condition (e.g. 0.1 M). This work enables nanofluidic sensing of Biotin at normal physiological condition by coupling ion concentration polarization (ICP) in a nanofluidic crystal (NFC) device. The enrichment effect of ICP was utilized for target molecule preconcentration for lower limit of detection, while the depletion effect of ICP was utilized for creating a nanofluidic regime in high ionic concentration buffer. A limit of detection of 1 fM was realized in our work.

M-013 MASS-PRODUCTION AND PROLONGED UNDIFFERENTIATED STATE OF EMBRYONIC BODIES BY USING A SEMI-PERMEABLE TAPERED MICROWELL ARRAY 280

M. Ikeuchi[1], S. Hayashi[2], M. Osonoi[2], and K. Ikuta[1]
[1]University of Tokyo, JAPAN and
[2]The Foundation for Promotion of State of the Art in Medicine and Health Care, JAPAN

We have demonstrated embryonic body (EB) formation and prolongation of undifferentiated state by using an improved tapered microwell array equipped with semipermeable bottom. The array had hydrophilic surface to prevent cell adhesion, and thus, the cells seeded onto the array aggregated into EBs in each microwell. The semipermeable bottom only permitted liquid to go through, and promoted exchange of culture medium in each microwell, resulting in prolonged term of undifferentiated state of EBs.

T-014 ON CHIP PRODUCT PURIFICATION FOR COMPLETE MICROFLUIDIC RADIOTRACER SYNTHESIS 284

S. Chen, A.A. Dooraghi, M. Lazari, R.M. van Dam, A.F. Chatziioannou, and C.J. Kim
University of California, Los Angeles, USA

We developed on-chip removal of excess radioactive fluoride to follow the radiolabeling of a neurotransmitter with an electrowetting-on-dielectric (EWOD) device. Solid phase extraction of fluoride was achieved by adding alumina particles in the radiolabeled droplet and filtering them out by passing the droplet through pillars in the device. Purification was analyzed both on chip with Cerenkov radiation imaging and off chip with radio-thin layer chromatography.

W-015 RECONSTITUTION OF FUNCTIONAL MEMBRANE PROTEINS INTO ASYMMETRIC GIANT LIPOSOMES BY USING A PULSED JET FLOW 288

K. Kamiya[1], R. Kawano[1], T. Osaki[1,2], and S. Takeuchi[1,2]
[1]Kanagawa Academy of Science and Technology, JAPAN and [2]University of Tokyo, JAPAN

We develop the reconstitution of functional membrane proteins (flippases) into asymmetric lipid giant liposomes that were prepared by deforming a planar asymmetric bilayer using a microfluidic pulsed jet flow. we observed the translocation of phosphatidylserine from the extracellular leaflet to the cytoplasmic leaflet which was catalyzed by flippases.

Th-016 TRACTION FORCE OF SMOOTH MUSCLE CELL DURING GROWTH ON A RIGID SUBSTRATE 290

U.G. Jung, T. Tsukagoshi, H. Takahashi, T. Kan, K. Matsumoto, and I. Shimoyama
University of Tokyo, JAPAN

This paper reports on the traction force of smooth muscle cell during cell growth on the rigid substrate, specially designed for measuring the x and z axis forces. The proposed piezoresistive force sensor is characterized by three points: 1) a rigid substrate, 2) high force sensitivity (10 nN force resolution), 3) small gaps between sensor pads. We measured the traction force of the smooth muscle cells during culture at 37°C.

Chemical Sensors and Systems (CSS)

M-017 A MICRO GAS CHROMATOGRAPH WITH INTEGRATED BI-DIRECTIONAL PUMP FOR QUANTITATIVE ANALYSES 294

Y. Qin and .B. Gianchandani
University of Michigan, USA

We report a micro gas chromatography (μGC) system that comprises a Knudsen pump with bi-directional flow capability (KP2), a two-stage preconcentrator-focuser (PCF2) and a separation column. In this valveless system, the bi-directionality of the pump allows flow reversal in the multi-stage preconcentrator. The KP2, PCF2, and separation column are arranged in a 4.3 cm³ stack, and used with a commercial flame ionization detector. In preliminary experiments, the μGC system demonstrated quantitative separations of benzene, toluene, and xylene (BTX) with concentrations of 43-328 mg/m³. The separations were completed in 80 sec using room air as the carrier gas.

T-018   A MICRO TRACE HEAVY METAL SENSOR BASED ON DIRECT PROTOTYPING MESOPOROUS CARBON ELECTRODE 298

F. Teng[1], X.H. Wang[1,2], and C.W. Shen[1]

[1]Tsinghua University, CHINA and [2]Chinese Academy of Sciences, CHINA

We present a micro trace heavy metal sensor based on bismuth-modified mesoporous carbon electrodes that are direct prototyped on silicon wafer. The proposed device features a great electrochemical sensing platform for voltammetric analysis because of the thicker mesoporous carbon electrode has high surface area, high electric conductivity, and can be integrated into microsystems. The novel sensor achieves excellent sensing performance, the limits of detection are an order of magnitude lower than other reported and the peak current also exhibits well linear response.

W-019   A NOVEL QUANTITATIVE DESIGN MODELING ON GAS SENSING PARAMETER OF NANO-MATERIALS BASED ON MICRO-GRAVIMETRIC THERMO-DYNAMIC EXPERIMENTS 302

P.C. Xu and X.X. Li

Chinese Academy of Sciences, CHINA

The study aims to build a novel quantitative adsorbing/sensing model for chemical gas sensing-materials, with which various key sensing-parameters can be comprehensively evaluated and optimally designed. Gravimetric resonant-cantilevers are used in experiment to real-time record sensing curves at different temperatures, which are further used to extract adsorbing/sensing performance of the specific materials for quantitative evaluation and optimal sensor design. The model is well validated by choosing the best trimethylamine (TMA) sensing-material among three similar mesoporous-silica nano-particles (MSNs).

Th-020   A POLYMER-BASED MEMS DIFFERENTIAL SCANNING CALORIMETER 306

Y. Jia, B. Wang, J. Zhu, and Q. Lin

Columbia University, USA

We present a new MEMS differential scanning calorimeter (DSC). The device, using polymer calorimetric microstructures inexpensively fabricated on a polymer substrate, is mechanically flexible and highly robust, and well suited to disposable use for measurement of biomolecular energetics. We demonstrate this polymer MEMS DSC device with the characterization of lysozyme unfolding in a 1 micro liter volume at low concentrations of practically useful levels.

M-021   DESIGN AND MOTION CONTROL OF SELF-PROPELLED DROPLETS 310

A. Suzuki[1], S. Maeda[2], Y. Hara[3], and S. Hashimoto[1]

[1]Waseda University, JAPAN, [2]Shibaura Institute of Technology, JAPAN, and
[3]National Institute of Advanced Industrial Science and Technology (AIST), JAPAN

We report a new oil droplet system that is autonomously driven by the energy of oil-water interactions and its motion control. This droplet moves by ejecting polymers in alkaline water and displays large driving force. The oil droplet can move along a flow channel by deforming in an amoeboid motion on the micro- to milli- scale, so we specifically demonstrated its motion in a micro fluidic channel. Such droplets offer considerable potential for the application as a transportable actuator.

T-022   DETECTION OF MUTATIONS IN THE BINDING DOMAIN OF TAU PROTEIN BY KINESIN– MICROTUBULE GLIDING ASSAY 314

S.P. Subramaniyan[1], M.C. Tarhan[2], S.L. Karsten[3], H. Fujita[2], H. Shintaku[1], H. Kotera[1], and R.Yokokawa[1]

[1]Kyoto University, JAPAN, [2]Tokyo University, JAPAN, and [3]NeuroInDx Inc., USA

We have studied the kinesin- microtubule (MTs) based gliding assay for its application as a diagnostic tool indetecting neuronal marker - tau protein. In this paper we report our findings; that the landing rate and density of MTs have depicted the type of tau protein decorated on them, we have discussed five major tau mutants located in thebinding domain of tau protein in addition to its isoforms, and we have also demonstrated a micro device to detect MTsby their landing rate and gliding density.

W-023   FROG EGG-ARRAY DEVICE INTEGRATED WITH FLUIDIC CHANNEL AND MICROELECTRODES FOR CHEMICAL SENSING 318

M. Tomida, Y. Murakami, and N. Misawa

Toyohashi University of Technology, JAPAN

This study describes a membrane-protein based chemical sensor device consisting of microfluidic channels and Xenopus laevis oocytes. The fluidic device has Si-based microfabricated electrodes to measure of the oocyte's response to each chemical by two-electrode voltage clamping method. After cell trapping, the fluidic device can be separated to each fluidic channel that can measure an individual oocyte membrane potential change. We successfully placed each oocyte into the device and detected individual Xenopus oocyte responses to chemical stimulus.

Th-024 INTEGRATION OF DIAMOND MICROELECTRODES ON CMOS-BASED
AMPEROMETRIC BIOSENSOR ARRAY BY FILM TRANSFER TECHNOLOGY 322
T. Hayasaka, S. Yoshida, K.Y. Inoue, M. Nakano, T. Ishikawa, T. Matsue, M. Esashi, and S. Tanaka
Tohoku University, JAPAN

This study reports on integration of boron-doped diamond (BDD) microelectrodes on CMOS-based 20×20 amperometric biosensor array. The BDD electrodes are once formed on a Si wafer at 800°C, and then transferred to a 0.18 µm CMOS wafer with a benzocyclobutene (BCB) bonding interlayer. The fully-integrated device successfully detects biomolecules such as histamine owing to a wide potential window of the BDD electrode, and offers 2-dimensional real-time imaging of histamine diffusion in a solution. This type biosensor promises sensing and imaging applications of various biological materials which cannot be detected by conventional sensors.

M-025 IONIC LIQUID-GATED GRAPHENE FET ARRAY WITH
ENHANCED SELECTIVITY FOR ELECTRONIC NOSE 326
A. Inaba, Y. Takei, K. Matsumoto, and I. Shimoyama
University of Tokyo, JAPAN

Graphene has high sensitivity to gases, but has poor selectivity. Therefore, graphene is hard to apply to electronic nose. Because the gas absorbability of ionic liquid (IL) depends on its type, graphene FETs (GFETs) gated by different ILs have different gas responses. The response pattern of the IL-gated GFETs (ILGFETs) enables gas species detection, i.e. ILs provide gas selectivity to graphene. We assembled three ILGFETs with three kinds of ILs into an ILGFET array. The response patterns to several gases were measured to demonstrate the feasibility of graphene electronic nose.

T-026 LoC SENSOR ARRAY PLATFORM FOR REAL-TIME COAGULATION MEASUREMENTS 330
O. Cakmak[1], N. Kilinc[1,2], E. Ermek[1], A. Mostafazadeh[1], C. Elbuken[1], G.G. Yaralioglu[3], and H. Urey[1]
[1]*Koc University, TURKEY,* [2]*Gebze Institute of Technology, TURKEY, and* [3]*Ozyegin University, TURKEY*

We report the first demonstration of MEMS-based sensor array enabling multiple tests in one disposable microfluidic cartridge using plasma. The LoC (Lab-on-Chip) platform technology is versatile and demonstrated here for real-time coagulation and clot-time tests (activated Partial Thrompoblastin Time (aPTT) and Prothrombin Time (PT)). The start and the end of fibrin generation during coagulation can be clearly seen in real-time for both of the tests. Magnetic actuation and optical read-out is used. Hence no electrical connection to the MEMS chip is required. This makes the system convenient for point-of-care tests.

W-027 PARALLELIZATION OF FISSION AND FUSION+OPERATIONS FOR
HIGH THROUGHPUT GENERATION OF COMBINATORIAL DROPLETS 334
H.C. Zec, T.D. Rane, P. Ma, and T.H. Wang
Johns Hopkins University, USA

In this paper, we present a parallelized droplet-based platform for on-demand, combinatorial generation of nanoliter droplets. By parallelizing fission and fusion modules, throughput is increased by two orders of magnitude. With 32 Hz droplet generation, the projected throughput of this parallelized design is nearly 3 million sample-probe droplets per day on a single device (with 4 replicates of 750 thousand different mixtures). This translates to 240 unique sample-probe mixtures with 4 replicates per minute each.

Th-028 TOWARDS ON-CHIP CHEMICAL REACTION MONITORING
BY EWOD IMPEDANCE MEASUREMENT 338
X. Ma, S. Chen, C.J. Kim, and R.M. van Dam
University of California, Los Angeles, USA

We develop an EWOD impedance measurement system for in situ chemical reaction monitoring to maximize the advantage of microscale chemical synthesis. As a demonstration, we measure the droplet impedance on EWOD at various stages of an acid-base titration, showing its capability of detecting the equivalence point of neutralization.

Energy Harvesting and Power MEMS (EHPM)

M-029 3D SOLID-STATE SUPERCAPACITORS OBTAINED BY ALD COATING
OF HIGH-DENSITY CARBON NANOTUBES BUNDLES 342
G. Fiorentino, S. Vollebregt, F.D. Tichelaar, R. Ishihara, and P.M. Sarro
Delft University of Technology, THE NETHERLANDS

A three-dimensional solid-state miniaturized supercapacitor based on double conformal coating of Multiwalled Carbon Nanotubes (MWCNTs) bundles is presented. Atomic Layer Deposition (ALD) is used to deposit Al_2O_3 as dielectric layer and TiN as high aspect-ratio conformal counter-electrode on 2µm long MWCNTs bundles. The devices are realized using an IC wafer-scale manufacturing process and show a remarkable volumetric capacitance density value of 12mF/cm^3 with high reproducibility (≤0.3E-12F deviation). The small footprint (100µm^2 to 625µm^2), a thickness of only 2µm, the extremely high capacitance density and the novel and easy-to-integrate fabrication process make it possible to realize high performance energy storage micro-devices.

T-030 A HIGH PERFORMANCE TRIBOELECTRIC GENERATOR FOR HARVESTING LOW FREQUENCY AMBIENT VIBRATION ENERGY 346
B. Meng, W. Tang, X.S. Zhang, M.D. Han, X.M. Sun, W. Liu, and H.X. Zhang
Peking University, CHINA

We present a novel triboelectric generator for vibration energy harvesting based on the mass production manufacture of flexible printed circuit (FPC). An elastic zigzag-shaped structure was employed as a natural spring, making the generator simple and easy to be stimulated. The use of FPC manufacture makes the fabrication efficient and low-cost. Low resonant frequency of 16 Hz and wide bandwidth of 37 Hz was achieved. The maximum effective output power of 77.3 µW was obtained at 16 Hz.

W-031 A GAP-VARYING ELECTROSTATIC TRANSDUCER UTILIZING FERROFLUID-BASED ACTUATION FOR MOTION HARVESTING 350
T. Galchev, D. Barutçu, and O. Paul
University of Freiburg – IMTEK, GERMANY

This paper provides electrical characterization of the gap-varying electrostatic springless proximity inertial harvester (SPIH). This is a new type of harvester for converting multi-dimensional motion from low-frequency sources such as humans or other environment. Each 2-mm-diameter transducer is capable of producing between 0.05-4.2 nJ of energy per actuation cycle at bias voltages of 10-100 V under controlled experiments.

Th-032 A HYBRID SUPERCAPACITOR USING VERTICALLY ALINGED CNT-POLYPYRROLE NANOCOMPOSITE 354
F. Sammoura[1,3], K.S. Teh[2], A. Kozinda[3], X. Zang[3], and L. Lin[3]
[1]*Masdar Institute of Science and Technology, UAE*, [2]*San Francisco State University, USA, and*
[3]*University of California, Berkeley, USA*

We have successfully demonstrated, for the first time, the fabrication of vertically aligned carbon nanotube (VACNT)-polypyrrole (PPY) nanocomposites as a "hybrid supercapacitor material directly integrated on silicon-based electrodes. In contrast to previous works, three distinctive achievements are accomplished: (1) a "hybrid supercapacitor" using VACNT forest with electroplated PPY and dodecylbenzenesulfonate (DBS) as a dopant in acetonitrile, (2) realizing 500% higher capacitance as compared to the capacitance of electrodes made of VACNT or DBS-doped PPY alone, and (3) highly reversible cycling between -1 V and +1 V with improved knee frequency at 797Hz. As such, this hybrid nanocomposite could become a new class of material for future supercapacitors.

M-033 A MEMS-ENABLED BIODEGRADABLE BATTERY FOR POWERING TRANSIENT IMPLANTABLE DEVICES 358
M. Tsang[1], A. Armutlulu[1], A. Martinez[1], F. Herrault[1], S.A. Bidstrup[1], and M.G. Allen[1,2]
[1]*Georgia Institute of Technology, USA and* [2]*University of Pennsylvania, USA*

A series of MEMS-enabled biodegradable batteries, composed of a Mg anode and Fe cathode in a 0.1M $MgCl_2$ electrolyte, were developed to power transient implantable medical devices (IMD). Biodegradable energy sources would enable active devices for the monitoring and treatment of transient disease states, such as bone fracture healing and drug delivery. The anode was fabricated by electroplating Mg from a non-aqueous solution and passivated with either polycaprolactone or polyglycerol sebacate. The batteries demonstrated a capacity and power of up to 1.1mAh and 22uW, respectively, which are sufficient for a typical IMD.

T-034 A MICRO-SCALE MICROBIAL SUPERCAPACITOR 362
H. Ren[1], H. Tian[2], T.L. Ren[2], and J. Chae[1]
[1]*Arizona State University, USA and* [2]*Tsinghua University, CHINA*

We report, for the first time, a micro-scale microbial supercapacitor to substantially enhance the current and power density, aiming for a carbon-neutral renewable miniaturized electrochemical power converter. Current and power density of 501.5A/m2, and 251.4W/m2 are achieved, which is more than 18 and 32 folds of the previous records, yielding the supercapacitor an attractive alternative to existing energy conversion and storage device.

W-035 A THREE-DIMENSIONAL ELECTROSTATIC/ELECTRET MICRO POWER GENERATOR FOR LOW ACCELERATION AND LOW FREQUENCY VIBRATION ENERGY HARVESTING 366
K. Tao, S.W. Liu, J.M. Miao, and S.W. Lye
Nanyang Technological University, SINGAPORE

In this work, a three-dimensional (3D) multimodal electret-based micro power generator is developed for scavenging energy from low acceleration (<0.05g) and low frequency (<100Hz) vibrations, which are ubiquitous existence and readily available in our daily life.

Th-036 A WEARABLE SYSTEM OF MICROMACHINED PIEZOELECTRIC
CANTILEVERS COUPLED TO A ROTATIONAL OSCILLATING
MASS FOR ON-BODY ENERGY HARVESTING 370
R. Lockhart, P. Janphuang, D. Briand, and N.F. de Rooij
École Polytechnique Fédérale de Lausanne (EPFL), SWITZERLAND

We present a compact, wearable piezoelectric on-body harvesting system that uses a small eccentric mass from a common watch movement to mechanically deflect a set of micromachined piezoelectric cantilevers when excited by the low frequency movements of the human body. The piezoelectric cantilevers are directly coupled to the rotating mass via a set of pins located near its rotational center. The energy produced by each pluck of a single cantilever is 545 nJ, corresponding to a maximum output power of 11 μW for continuous plucking. This concept could be used to power on-body electronics.

M-037 ALL-POLYMER PIEZOELECTRET ENERGY HARVESTER
WITH EMBEDDED PEDOT ELECTRODE 374
Y. Feng and Y. Suzuki
University of Tokyo, JAPAN

We develop a novel all-polymer high-aspect-ratio (HAR) piezoelectret energy harvester with embedded PEDOT electrode, and demonstrate its performance for low-resonant-frequency in-plane vibration energy harvesting. Butterfly-shaped stop valves is devised to control the PEDOT capillary flow inside the parylene channels. With the present early prototype, 5.2 V open circuit voltage and 53 nW output power have been obtained at the low resonant frequency of 205 Hz with 3 g acceleration.

T-038 ELECTROMAGNETIC ENERGY HARVESTER WITH
HIGH EFFICIENCY USING MICRO-MACHINING SI SPRINGS 378
T. Shirai, Y. Wakasa, T. Nakagawa, K. Nomura, and H. Yagyu
Panasonic Corporation, JAPAN

We developed electromagnetic vibration energy harvester with quite high generating efficiency over 40%. This high efficiency was realized by using the 1mm-thick uniquely-structured Si springs in the harvester, which was created using micromachining process.

W-039 ELECTROMAGNETIC ENERGY HARVESTER WITH AN IN-PHASE
VIBRATION BANDWIDTH BROADENING TECHNIQUE 382
S.J. Chen, J.Y. Wu, and S.Y. Liu
National Central University, TAIWAN

This paper develops a duo-mode vibration structure for increasing usable bandwidth in a micromachined electromagnetic energy harvester. Compared to a pure cantilever harvester, the proposed cantilever-spiral coupled energy harvester has lower resonant frequencies and larger bandwidth.

Th-040 ELECTROSTATIC GENERATOR WITH FREE MICRO-BALL AND
ELASTIC STOPPERS FOR LOW-FREQUENCY VIBRATION HARVESTING 385
F. Cottone[1], P. Basset[2], F. Marty[2], D. Galayko[3], L. Gammaitoni[1], and T. Bourouina[2]
*[1]University of Perugia, ITALY, [2]Université Paris-Est, FRANCE, and
[3]UPMC-Sorbonne Universités, FRANCE*

We present a novel MEMS electrostatic vibration harvester based on frequency amplification through multiple-mass impacts in combination with elastic stoppers. The harvester proof mass hosts a tungsten micro-ball free to travel along the vibration direction. At low frequencies (10-60 Hz) the micro-ball impacts with the oscillating mass of the generator transferring kinetic energy to the gap-closing comb transducer which in turn resonates at 92 Hz. In addition, the elastic stoppers amplify the proof mass and ball velocity by collision with the rigid frame. Output power between 0.25 and 0.45 μW is achieved at 0.3-g amplitude and only 15 V bias in the range of 10-60 Hz with a -3db bandwidth of 50 Hz

M-041 FLEXIBLE MICRO-SUPERCAPACITORS FROM PHOTORESIST-DERIVED
CARBON ELECTRODES ON FLEXIBLE SUBSTRATES 389
M.-S. Kim, B. Hsia, C. Carraro, and R. Maboudian
University of California, Berkeley, USA

We demonstrate a simple and scalable technique for the fabrication of flexible micro-supercapacitors. The flexible high surface area electrodes are fabricated via a photoresist pyrolysis process followed by transfer for the electrodes to a flexible substrate. An energy density of 1 mWh/cm^3 is measured and the mechanical stability of the device is demonstrated through mechanical cycling tests.

T-042 GRAPHENE ELECTRODES ENHANCE PERFORMANCE FOR MICROLITER-SCALE MICROBIAL FUEL CELLS 393
V. Jayaprakash, R.D. Sochol, R. Warren, K. Iwai, and L. Lin
University of California, Berkeley, USA

In this work, graphene electrodes are employed to increase the power output of a microliter-scale microbial fuel cell (μMFC) for the first time. Previously, researchers have predominantly used Au/Cr electrodes in μMFCs, and have operated these fuel cells under controlled anodic conditions to attain high current densities and columbic efficiencies. At present, relatively low power outputs and open circuit potentials have limited such fuel cells from implementation in practical applications. To improve such performance, here we introduce a graphene-based μMFC (G-MFC) that utilizes graphene electrodes.

W-043 HIGH PERFORMANCE NONLINEAR MICRO ENERGY HARVESTER INTEGRATED WITH (K,Na) NbO$_3$/Si COMPOSITE QUAD-CANTILEVER 397
L.V. Minh, M. Hara, and H. Kuwano
Tohoku University, JAPAN

We developed a lead-free (K,Na)NbO3 (KNN) nonlinear microenergy harvester. The harvester was densely integrated with a quatrefoil-shaped proof mass and quad cantilevers using bulk micromachining. The KNN/Si composite cantilever with two-separated KNN capacitors was to effectively collect charge. Clamped-clamped beam design was also adopted for wide band operation. The experimental results showed that the wide bandwidth of 253 Hz and the highest power density of 1623 uW/cm3 among the developed piezoelectric nonlinear MEMS harvesters were achieved at the low acceleration of 6 m/s2.

Th-044 HIGH-ENERGY-DENSITY ON-CHIP LI-ION CAPACITORS 401
S. Li[1] and X. Wang[1,2]
[1]Tsinghua University, CHINA and [2]Chinese Academy of Sciences, CHINA

This paper presents a new on-chip Li-ion capacitor featured by higher energy density than that of supercapacitor under the same level of charge/discharge rate. Activated carbon (AC), a supercapacitor material, is used as positive electrode, while graphite, an anode material of Li-ion battery, is used as negative electrode and the electrolyte used in Li-ion battery serves as electrolyte. The prototype with 100-μm-thick electrodes shows a capacity of 175μAh/cm^2 and an energy density of about 1550mJ/cm^2 under a charge/discharge current of 0.5mA/cm^2, and a cell voltage of 3.4V.

M-045 HIGH-ENERGY-DENSITY ON-CHIP SUPERCAPACITORS USING MANGANESE DIOXIDE-DECORATED DIRECT-PROTOTYPED POROUS CARBON ELECTRODES 405
S. Li[1], X. Wang[1,2], and C. Shen[1]
[1]Tsinghua University, CHINA and [2]Chinese Academy of Sciences, CHINA

This paper presents the high performance on-chip micro supercapacitors using manganese dioxide (MnO$_2$) decorated into direct-prototyped porous carbon electrodes. By a new method of incorporating MnO$_2$ into carbon framework, both electric double layer capacitance (EDLC) and pseudocapacitance contribute to total capacity. Therefore, about 4-time increase in volumetric capacitance (0.8mF/(cm^2 μm) vs. 2.9mF/(cm^2 μm) under the scan rate of 50mV/s) is achieved. The procedure makes such devices potentially to be integrated into multi-function microsystems.

T-046 MEMS VIBRATION ELECTRET ENERGY HARVESTER WITH COMBINED ELECTRODES 409
Q.Y. Fu and Y. Suzuki
University of Tokyo, JAPAN

We have developed a novel in-plane MEMS electret energy harvester with overlapping-area change and gap-closing electrodes that provides large output power both at low and high vibration accelerations. An early prototype has been successfully microfabricated with the single layer silicon-on-insulator process. Soft X-ray charging is employed to establish uniform surface potential around 60 V on vertical electrets on the sidewall of the comb fingers. Up to 1.6 μW output power has been obtained, which corresponds to the effective as high as 57%.

W-047 MICRO PATTERN OF CHARGE IN PTFE ELECTRET FOR ENERGY HARVESTERS 413
W. Bian, X. Wu, and X. Wang
Tsinghua University, CHINA

This paper presents a novel fabrication process to pattern the charge in electret film for vibration energy harvester (VEH) Applications. Compared with previously reported techniques, PTFE electret material, which is inexpensive and has highSurface potential, is used. The line width of the charge pattern is determined by photolithography process. Experiment results show that the surface potential on the patterned charge zone of PTFE is higher than -200V when line width reaches 20um. A demoed VEH is built by using the pattern technique and tested.

Th-048 NANO-POROUS SIO2 ELECTRET WITH HIGH SURFACE
POTENTIAL AND HIGH THERMAL RESISTANCE 417
M. Suzuki[1], T. Wada[1], T. Takahashi[1], T. Nishida[2], Y. Yoshikawa[2], and S. Aoyagi[1]
[1]Kansai University, JAPAN and [2]ROHM Co., Ltd., JAPAN

This study proposes a new electret with high surface potential and high thermal resistance, which is made of nano-porous SiO2. Electrical charge density in the nano-porous SiO2 is higher than that in a normal SiO2 because the interfaces between void and SiO2 trap electrical charges. Thermal stability of the nano-porous SiO2 electret is better than that of a polymer electret. Output power generated by vibration energy harvesting using the nano-porous SiO2 electret is also larger than that using the normal SiO2 or CYTOP electret.

M-049 NANOFLUIDIC REVERSE ELECTRODIALYSIS PLATFORM USING CONTROLLED
ASSEMBLY OF NANOPARTICLES FOR HIGH POWER ENERGY GENERATION 421
E. Choi, K. Kwon, D. Kim, and J. Park
Sogang University, SOUTH KOREA

This paper presents a novel microplatform for high power energy generation based on reverse electrodialysis. The ideal cation-selective membrane for power generations is realized using geometrically controlled in-situ self-assembled nanoparticles and it can be constructed with simple and cost effective process using microdroplet control containing nanoparticles in microchannel. Another advantage in our system is that maximum powers and the energy conversion efficiency can be improved by changing the geometry of microchannel and proper selection of size and materials in nanoparticles

T-050 SPRINGLESS CUBIC HARVESTER FOR CONVERTING
THREE DIMENSIONAL VIBRATION ENERGY 425
M.D. Han, W. Liu, B. Meng, X.S. Zhang, X.M. Sun, and H.X. Zhang
Peking University, CHINA

This paper reports the design, fabrication and measurement of a springless cubic energy harvester. Coils are fabricated onto polyimide substrate and folded to form a cubic box. Output performance of the device is theoretically and experimentally investigated. Vibration in all dimensions can be effectively harvested and the maximum output is achieved at low frequencies with a wide bandwidth. Moreover, the device can be placed on a backpack or wrist to harvest vibration energy from daily life.

W-051 ON THE OPTIMIZATION AND PERFORMANCES OF A COMPACT
PIEZOELECTRIC IMPACT MEMS ENERGY HARVESTER 429
P. Janphuang, R. Lockhart, D. Briand, and N.F. de Rooij
École Polytechnique Fédérale de Lausanne (EPFL), SWITZERLAND

This paper presents the development of a compact energy harvesting configuration to convert low frequency, mechanical oscillations into usable electrical energy using AFM-like MEMS piezoelectric cantilevers coupled to a rotating gear. The harvester, with an active device volume of 3.5 mm3 (3×5×0.23 mm3), is able to produce an average output power of 12 μW measured across an optimal resistive load of 4.7 kΩ at a rotational speed of 19 rps, demonstrating the potential of the compact MEMS piezoelectric micro-power generator.

Fabrication Technologies (FAB)

Th-052 "ASSIST-FREE" ASSEMBLY TECHNIQUE OF STANDING OPTICAL
DEVICES ON SOFT SPRING ACTUATOR STAGES 433
Y. Oka, R. Shinozaki, K. Terao, T. Suzuki, F. Shimokawa, F. Oohira, and H. Takao
Kagawa University, JAPAN

In this study, a new assembly technique of independently fabricated optical devices on fragile MEMS actuator stages has been developed to realize novel functional optical-MEMS devices. This technique realizes the "assist-free" alignment and fixing of vertically mounted optical devices by combination of "micro spring slider" and "trapezoidal alignment slit". In the experiments, micro mirrors were attached on electrostatic linear actuators and rotational actuators using this assembly technique, and a small average value of relative-angle error around 4/100° was successfully obtained."

M-053 3D NANOFABRICATION ON COMPLEX SEED
SHAPES USING GLANCING ANGLE DEPOSITION 437
H.-H. Jeong[1], A.G. Mark[1], J.G. Gibbs[1], T. Reindl[2], U. Waizmann[2], J. Weis[2], and P. Fischer[1,3]
[1]Max Planck Institute for Intelligent Systems, GERMANY, [2]Max Planck Institute for Solid State Research, GERMANY, and [3]University of Stuttgart, GERMANY

We report a 3D fabrication scheme that combines two existing techniques, electron beam lithography (EBL) and glancing angle deposition (GLAD), to fabricate nanostructures with complex 3D shapes both parallel and perpendicular to the growth direction. GLAD is a physical vapor deposition (PVD) technique where evaporant is delivered to a substrate at a high angle of incidence. Local shadowing and azimuthal manipulation of the substrate allows a rich variety of complex 3D structures to be grown down to the nanoscale. Herein, we use EBL to write custom seed layers with complex shapes, and do so at the resolution limit for the GLAD technique.

**T-054 A FLEXIBLE TACTILE AND SHEAR SENSING ARRAY FABRICATED
BY A NOVEL BUCKYPAPER PATTERNING TECHNIQUE** 441
C.-W. Ma, L.-S. Hsu, J.-C. Kuo, and Y.-J. Yang
National Taiwan University, TAIWAN

In this work, we present a flexible tactile and shear sensing array utilizing patterned buckypaper as the sensing elements. A novel fabrication process for realizing patterned buckypaper with high aspect ratio was proposed. The fabricated sensing device possesses the advantages such as anisotropic sensing capability, flexibility, simple fabrication process, and low cost. Measured results show excellent sensitivity and repeatability. In addition, the anisotropic sensing capability, which can be employed for better shear sensing, was also observed and discussed.

**W-055 A TECHNOLOGY FOR MONOLITHIC MEMS-CMOS INTEGRATION AND ITS
APPLICATION TO THE REALIZATION OF AN ACTIVE-MATRIX TACTILE SENSOR** 445
F. Zeng and M. Wong
Hong Kong University of Science and Technology, HONG KONG

A scheme of MEMS-CMOS integration based on the surface-migration of silicon is presently described. A cavity sealed with a cover-diaphragm is first formed, the electronic devices are next fabricated, and the suspended mechanical components are finally realized using the cover-diaphragm, without a sacrificial layer etch. With this scheme, the material and process incompatibility issues inherent in the existing integration techniques are largely eliminated. As a demonstration, a 16x16 active-matrix tactile sensor integrating a total of 256 force-sensing diaphragms, 512 transistors and 512 piezoresistors was designed, realized and characterized.

Th-056 ALD HONEYCOMB PLATES ENABLING ROBUST ULTRATHIN MEMS 449
K. Davami, L. Zhao, and I. Bargatin
University of Pennsylvania, USA

We report rigid MEMS structures made of ALD films, with a thickness of the order of 10 nanometers and patterned in the shape of a 3D honeycomb. Unlike planar ALD films, the 3D honeycomb plates do not warp due to fabrication stress gradients and are promising for a number of applications. For example, honeycombs made from refractive metals, such as ALD tungsten can be used to create thermionic energy converters with a well-controlled gap. Honeycomb cantilevers or beams can be used as resonant gas sensors, thanks to their low thickness and high stiffness, which lead to high resonance frequencies and high sensitivity to surface adsorbates. The transparency of the alumina ALD films may even enable their use as support films in electron microscopy.

**M-057 ARRAYS OF MICRO PENNING-MALMBERG TRAPS:
AN APPROACH TO FABRICATE VERY HIGH ASPECT RATIOS** 453
A. Narimannezhad, J. Jennings, M.H. Weber, and K.G. Lynn
Washington State University, USA

This paper reports on the progress of fabrication of very high aspect ratio (1000:1) micro-Penning-Malmberg trap arrays designed to store antimatter. The structure consists of thousands of 100μm diameter tubes etched by deep reactive ion etching through Si wafers. Cycles of thermal oxidation and wet etching in buffered oxide etch (BOE) minimized the sidewalls roughness and ensured a complete coating during gold sputtering. The wafers were then aligned and stacked in order to create the microtubes. Uniform plating with mean roughness of R_a=600nm was achieved by tuning the electroplating parameters.

**T-058 BATCH RELEASE OF MONODISPERSE LIPOSOMES TRIGGERED
BY PULSED VOLTAGE STIMULATION** 457
T. Osaki[1], K. Kamiya[1], R. Kawano[1], and S. Takeuchi[2]
[1]Kanagawa Academy of Science and Technology, JAPAN and [2]University of Tokyo, JAPAN

We present a batch release technique for monodisperse liposomes immobilized on a substrate. A single pulsed voltage to the substrate induced detachment of the arrayed liposomes previously developed. Simultaneous release was observed shortly after the electrical stimulation. The release technique produced monodisperse and solvent-free liposomes freely suspended on the substrate, and allowed manipulation of the liposomes.

**W-059 CLARITAS™ – A UNIQUE AND ROBUST ENDPOINT TECHNOLOGY FOR
SILICON DRIE PROCESSES WITH OPEN AREA DOWN TO 0.05%** 459
O. Ansell[1], R. Barnett[1], T. Haase[2], L. Xie[3], S. Vargo[1], and D. Thomas[1]
*[1]SPTS Technologies Limited, UK, [2]Fraunhofer Institute for Photonic Microsystems (IPMS,) GERMANY,
and [3]Harvard University, USA*

SPTS' Claritas is an enhanced method of OES endpoint detection for the Bosch process. It will be shown that Claritas has the capability to endpoint very low open area patterns (<1%), Bosch process recipes with high process pressures and show potential use of Claritas with other process solutions, including vapour phase etching.

Th-060 CONCURRENT REACTIVE ION ETCHING EMPLOYING
MICROMACHINED IONIC LIQUID ION SOURCE ARRAY 463
R. Yoshida, M. Hara, H. Oguchi, T. Suzuki, and H. Kuwano
Tohoku University, JAPAN

This paper describes concurrent reactive ion etching using micro ionic liquid ion source (ILIS) array. The ILIS array was fabricated using bulk micromachining and consists of micro needle emitters and a reservoir for the ionic liquid (IL) of 1-ethyl-3-methylimidazolium tetrafluoroborate. The ion beam etching of a (100) silicon substrate using the fabricated ILIS array was demonstrated. Monitoring mass spectra during the etching, the peaks of SiF+, SiF2+, and SiF3+ could be observed. These peaks indicate the chemical reaction between the silicon and fluorine based ions from the IL.

M-061 DEVELOPMENT OF MEMS PIERCE-TYPE NANOCRYSTALLINE SI ELECTRON-EMITTER
ARRAY FOR MASSIVELY PARALLEL ELECTRON BEAM DIRECT WRITING 467
H. Nishino[1], S. Yoshida[1], A. Kojima[2], N. Ikegami[3], N. Koshida[3], S. Tanaka[1], and M. Esashi[1]
[1]Tohoku University, JAPAN, [2]Crestec Corporation, JAPAN, and
[3]Tokyo University of Agriculture and Technology, JAPAN

This study reports on development of the fabrication process for 100×100 Pierce-type nanocrystalline Si electron-emitter array for massively-parallel electron-beam (EB) direct writing system based on active-matrix operation of large-scaled-integrated circuit (LSI). The 100-µm-pitch emitter array with each diameter of ~40 µm is prototyped and successfully demonstrates EB resist patterning by 1:1 projection exposure at CMOS-compatible operation voltages. This study also successfully establishes the integration process of the emitter array on a CMOS-LSI wafer. This achievement is a giant step for realizing the novel EB lithography system.

T-062 DIRECT LASER WRITING OF 3D PROTEIN STRUCTURES
WITH NANOSCALE FEATURE SIZES 471
D. Serien[1,2] and S. Takeuchi[1,3]
[1]University of Tokyo, JAPAN, [2]CIRMM-IIS, JAPAN, and [3]Japan Science and Technology Agency (JST), JAPAN

We report the fabrication of three-dimensional (3D) protein structures with nanoscale feature sizes by two-photon direct laser writing (DLW). For this fabrication technology, we combine the established DLW technology with previously reported 3D protein structure fabrication by photosensitized crosslinking. We demonstrate the fabrication of 2D and 3D protein structures with nm-sized features.

W-063 FABRICATION OF ANISOTROPIC AND HIERARCHICAL
UNDULATIONS BY BENCHTOP SURFACE WRINKLING 474
K. Wei and Y. Zhao
Ohio State University, USA

This paper describes a benchtop wrinkling process where highly ordered sinusoidal wrinkles with tailored wavelength and amplitude are created atop a PDMS foundation by atmospheric electric discharge. The method is used to fabricate hierarchical and anisotropic wrinkle-on-wrinkle and wrinkle-on-microstructure surface patterns. Its accessibility in general wet lab environments and simplicity to create multi-scale roughness are believed to facilitate applications in optical gratings, topography guidance for cell alignment, and micro/nanofluidics.

Th-064 FABRICATION OF THIN STENCIL WITH BUFFER RESERVOIR UTILIZING
THE COMBINATION OF AZ4620 AND SU-8 ELECTROPLATING MOLDS 478
P.H. Chen[1], C.W. Huang[2], and C.H. Lin[1]
[1]National Sun Yat-sen University, TAIWAN and [2]Metal Industries Research and Development Center, TAIWAN

This work develops a novel process for fabricating ultra-thin stencil with buffer reservoir utilizing the combination of AZ4620 positive photoresist (PR) and SU-8 negative PR as the electroplating molds. A 5 um thick AZ4620 layer is used to precisely define the printing patterns while a 3 um thick of nickel layer is electroplated. A SU-8 layer of the thickness 50 um is patterned as the second electroplating mold. The high transparency of SU-8 PR makes it easy to align the two PR plating molds. A 30 um thick nickel layer is then electroplated onto the first nickel layer. The developed stencil can be used to printing ultra fine line and thin film pattern.

M-065 FREE-STANDING SUBWAVELENGTH GRID INFRARED REJECTION
FILTER OF 90 MM DIAMETER FOR LPP EUV LIGHT SOURCE 482
Y. Suzuki, K. Totsu, M. Moriyama, M. Esashi, and S. Tanaka
Tohoku University, JAPAN

A subwavelength grid infrared filter as large as 90 mm in diameter was fabricated and tested on a 6 inch Si wafer for a laser-produced plasma extreme ultraviolet (EUV) light source used in the next generation lithography tools. The IR filter has a free-standing Mo-coated Si honeycomb grid structure with a thickness of 5 µm, a wire width of only 0.35 µm and a pitch of 4.5 µm, showing 99.7% rejection for 10.6 µm IR light. Such a large-size free-standing microstructure was successfully fabricated by carefully balancing film stress at each process step.

T-066 GRAPHENE SYNTHESIS VIA DROPLET CVD AND ITS PHOTONIC APPLICATIONS 486
X.N. Zang and L. Lin
University of California, Berkeley, USA

The process of "droplet CVD" for the synthesis of graphene sheets and its photonic applications have been demonstrated for the first time. Metal (Cu or Ni) droplets are naturally transformed from thin films in a high temperature furnace and utilized to grow graphene sheets via the chemical vapor deposition (CVD) process. As such, this new class of fabrication process could open up various graphene-based device/system applications, including photonic sensors."

W-067 HIGH ASPECT RATIO, LARGE AREA SILICON-BASED GRATINGS FOR X-RAY PHASE CONTRAST IMAGING 490
J.J. Baborowski, V. Revol, C. Kotler, R. Kaufman, P. Niedermann, F. Cardot,
A. Dommann, A. Neels, and M. Despont
CSEM SA, SWITZERLAND

The presented work reports on the latest developments in the manufacturing of high aspect ratio silicon-based gratings used for X-ray phase contrast imaging (XPCI. XPCI reveals subtle changes in the microstructure of the samples, such as micro-cracks in composite materials or micro-calcifications in breast tissues. In fields as diverse as medicine, non-destructive testing or security, the gained information of this technique allows early diagnostic or detection of defects, tumors or explosives. The range of opportunities offered by depends highly on the achievable gratings parameters, such as periodicity, depth, duty cycle and aspect ratio. We have developed large (100x100mm2) Au-Si-Au-Air gratings with a periodicity of down to 2 μm and a depth of up to 100 μm with extremely low defect density (<1 defect/cm2). The fabricated gratings have been implemented on a XPCI set-up and used to demonstrate unprecedented imaging quality in material quality control.

Th-068 HIGH RESOLUTION MICRO ULTRASONIC MACHINING (HR-μUSM) FOR POST-FABRICATION TRIMMING OF FUSED SILICA 3-D MICROSTRUCTURES 494
A. Viswanath, T. Li, and Y.B. Gianchandani
University of Michigan, USA

Post-fabrication trimming is interesting for devices such as inertial sensors, timing references, and mass-balance resonators to adjust stiffness, mass, and potentially damping. The trimming process should be capable of micro machining brittle materials, without inducing stress or subsurface cracks. We have developed and evaluated a subtractive trimming technique based on micro ultrasonic machining (μUSM), for high-resolution trimming of complex 3D microstructures made from fused silica. Machining rates as low as 10 nm/sec and surface roughness as low as 30 nm have been achieved.

M-069 FABRICATION OF CARBON NANOFIBROUS MICROELECTRODE ARRAY (CNF-MEA) USING NANOFIBER IMMERSION PHOTOLITHOGRAPHY 498
P.F. Jao[1], E. Franca[1], S.P. Fang[1], J. Yoon[1], K. Cho[1,3], D.E. Senior[2], G.J. Kim[1],
B. Wheeler[1], and Y.K. Yoon[1]
*[1]University of Florida, USA, [2]Universidad Tecnologica de Bolivar, COLOMBIA, and
[3]Korea Basic Science Institute, SOUTH KOREA*

Microelectrode arrays are used for stimulating and receiving neural electrical signals in vitro neural study. This work demonstrates the fabrication process of nanofibrous 3D microelectrodes using immersion lithography. Oil immersion negates the diffraction effects intrinsic in the photopatterning of electrospun nanofibers to give higher aspect ratio. Nanofiber electrode resistivity is characterized and its performance compared to that of carbon thin film. In vitro testing of electrodes are performed using E18 cortical neurons and cell density and cell viability analyzed.

T-070 INDUCTIVELY COUPLED PLASMA ETCHING OF BULK TUNGSTEN FOR MEMS APPLICATIONS 502
L. Song, N. Li, S. Zhang, J. Luo, J. Hu, Y. Zhang, S. Chen, and J. Chen
Peking University, CHINA

Tungsten based MEMS devices have the potential to be used for many applications, such as tools for micro electrical discharge machining and ultrasonic machining, or mold for inject molding. For the first time, bulk tungsten ICP etching was developed and characterized, which is capable of producing high aspect ratio (>13) structures with feature size below 3μm. Etching depth of 230μm has been achieved at an etch rate up-to 2.2μm/min. This technology offers big opportunities for MEMS applications.

W-071 LARGE ARRAYS OF INKJET-PRINTED MEMS MICROBRIDGES ON FOIL 506
F. Molina-Lopez, D. Briand, and N.F. de Rooij
Ecole Polytechnique Fédérale de Lausanne (EPFL), SWITZERLAND

This works describes the fabrication of an array of printed MEMS microbridges on polymeric foil in only four easy steps.Each functional material was deposited exclusively by inkjet-printing technique, compatible with large-area fabrication. Thearray consists of 60 to 80 individual microbridges, occupying an area of 2 mm x 2 mm, and displaying a total capacitancevalue of 1 - 1.7 pF when connected in parallel.

Th-072 LIGHTWEIGHT MICRO LATTICES WITH NANOSCALE FEATURES
FABRICATED FROM PROJECTION MICROSTEREOLITHOGRAPHY 510
X. Zheng[1], J. Deotte[1], J. Vericella[1], M. Shusteff[1], T. Weisgraber[1], H. Lee[2], N. Fang[2], and C.M. Spadaccini[1]
[1]Lawrence Livermore National Laboratory, USA and [2]Massachusetts Institute of Technology, USA

We demonstrate the utility of three-dimensional Projection Microstereolithography manufacturing system by producing a variety of microstructures with complex geometries and explored the potential of using the system to build meso-scale structures with micro-scale architecture and nano-scale features. These achievements pave the way for large scale micro- and nano- manufacturing that extends the current state-of-the-art of three-dimensional fabrication technologies.

M-073 FABRICATION OF MICRO-HEATERS EMBEDDED IN
PDMS USING A DRY PEEL-OFF PROCESS 514
I. Byun, R. Ueno, and B.J. Kim
University of Tokyo, JAPAN

We shows a reliable fabrication method of micro-heaters embedded in polydimethylsiloxane (PDMS), and shows characterization of the micro-heaters. Metallization of PDMS is achieved using a dry peel-off process which involves modifying the surface properties of the substrate and metal patterns through self-assembled monolayer and manually peeling off the PDMS with embedded metal layers. Thus, micro-heaters can be fabricated by a simpler and easier way compared to conventional methods.

T-074 MICROPATTERNING OF BACTERIAL CELLULOSE AS
DEGRADABLE SUBSTRATE FOR CELL CULTURE 518
Y. Karita, K. Hirayama, H. Onoe, and S. Takeuchi
University of Tokyo, JAPAN

This paper describes microfabrication of bacterial cellulose membrane, which is a nanofibrous cellulosic material produced by a bacterium, Acetobacter xylinum. We successfully micropatterned bacterial cellulose membrane by applying MEMS process and this patterned bacterial cellulose was confirmed to work as a scaffold for mouse embryonic fibroblast cells. Moreover, formation of cell cluster was observed by the treatment of cellulose degrading enzyme. We believe that this micropatterned cellulose plate would be useful in degradable microscaffolds for cell culture.

W-075 MICROSCALE MAGNETIC PATTERNING OF HARD MAGNETIC
FILMS USING MICROFABRICATED MAGNETIZING MASKS 520
A. Garraud[1], O.D. Oniku[1], W.C. Patterson[1], E. Shorman[1], D. Le Roy[2], N.M. Dempsey[2], and D.P. Arnold[1]
[1]University of Florida, USA and [2]University of Grenoble-Alpes, FRANCE

We present a batch-fabrication process to imprint microscale magnetic pole patterns (perpendicular north/south poles) into hard magnetic films using field-shaping, soft magnetic "magnetizing masks". Using 7-μm-thick, electroplated Fe–Co magnetizing masks, magnetic stripes with widths down to 50 μm have been imprinted into both 15-μm-thick Co–Pt films and 5-μm-thick Nd–Fe–B films. These patterned films exhibit a sinusoidal stray magnetic field pattern with ~4 and ~7 mT_{pk-pk} variations and corresponding field gradients of 80 and 140 T/m, respectively. We also demonstrate the ability to transfer more complex patterns by showing magnetization of various geometric shapes.

Th-076 MONOLITHIC PIEZOELECTRIC IN-PLANE MOTION
STAGE LOW-CROSS-AXIS-COUPLING 524
S. Nadig, S. Ardanuc, and A. Lal
Cornell University, USA

In this work we present a rotary dither stage which can provide rotation stimulus with high dynamic range of 1800-deg/s, and parts-per-thousand cross axis actuation, and is planar compatible with in-package inertial sensor calibration. We use bulk PZT-4 beams, laser cut out from plates, to achieve monolithic integration of lateral actuators and flexures. This process enables high-aspect ratio beams (500um thick x150um wide) resulting in parts-per-thousand in-plane to out-of-plane motion coupling

M-077 POLYMER MICROMACHINING BASED ON CU ON POLYIMIDE SUBSTRATE
AND ITS APPLICATION TO FLEXIBLE MEMS SENSOR 528
Y. Niimi, S. Shibata, and M. Shikida
Nagoya University, JAPAN

MEMS technologies have produced various types of MEMS sensors on a Si or Silicone On Insulator (SOI) wafers. To realize MEMS sensors in the flexible fashion, we newly proposed to apply a Cu On Polyimide (COP) substrate as a starting material, and introduced a sacrificial etching for producing a cavity and an electrical feed through structures on the COP substrate. Finally, a flexible thermal MEMS sensor was fabricated on COP substrate.

T-078 PRINTING AND ENCAPSULATION OF ELECTRICAL CONDUCTORS ON POLYLACTIC ACID (PLA) FOR SENSING APPLICATIONS 532
A. Vásquez Quintero, N. Frolet, D. Märki, G. Mattana, A. Marette, D. Briand, and N.F. de Rooij
Ecole Polytechnique Fédérale de Lausanne (EPFL), SWITZERLAND

This paper presents the printing of resistive and capacitive devices for temperature and humidity sensing applications, respectively, on biodegradable polylactic acid (PLA) substrates. Inkjet and gravure printing were assessed as direct silver-based nanoparticles inks transfer methods. Flash photonic ink sintering methodologies were optimized due to the low PLA glass transition temperature (58 °C) and maintain its mechanical integrity. An encapsulation method for electrical conductive structures is proposed by means of laminating PLA sheets at relatively low temperatures (< 60 °C). These fabricated structures are now exposed for long periods (months) in compost and high humidity environments to evaluate their degradation.

W-079 RAPID PROTOTYPING OF RESISTIVE MEMS SENSING DEVICES ON PAPER SUBSTRATES 536
T. Meiss[1], R. Werthschützky[1], and B. Stoeber[2]
[1]Technische Universität Darmstadt, GERMANY and [2]University of British Columbia, CANADA

We have developed an inexpensive inkjet printing process to rapidly fabricate resistive sensor devices on paper substrates. Since we use a commercial inexpensive inkjet printer and the design can be modified and tested within minutes, the process is especially useful to easily develop and test MEMS sensor models. Additional applications encompass disposable medical sensors, sensors for paper packaging, as well as very low cost strain sensing.

Th-080 REAL-TIME DYNAMICALLY RECONFIGURABLE LIQUID METAL BASED PHOTOLITHOGRAPHY 540
D. Kim[1], J.H. Yoo[1], W. Choi[1], K. Yoo[2], and J.B. Lee[1]
[1]University of Texas at Dallas, USA and [2]Hanbat National University, SOUTH KOREA

We report real-time dynamically reconfigurable photolithography technique using liquid metal Galinstan as UV opaque material and PDMS as UV transparent material. We demonstrated dynamically reconfigured on-demand patterning of single digit numbers in positive photoresists along with various patterns with minimum feature size of 10 μm. To the best of our knowledge, this is the first demonstration of true real-time reconfigurable photolithography in UV wavelengths.

M-081 RELEASE AND TRANSFER OF LARGE-AREA ULTRA-THIN PDMS 544
J. Gao, D. Guo, S. Santhanam, Y.J. Yu, A.J.H. McGaughey, S.C. Yao, and G.K. Fedder
Carnegie Mellon University, USA

This paper reports on a successful fabrication of ultra-thin 10μm PDMS films with embedded Au electrodes, as well as the releasing and transferring of large area films. The motivation for this work is in development of a miniature pump actuator for moving working fluid in an electrocaloric microcooler. It opens a promising route for fabricating ultra-thin and compliant MEMS and electronic devices.

T-082 SOLID STATE MEMS DEVICES ON FLEXIBLE AND SEMI-TRANSPARENT SILICON (100) PLATFORM 548
S.M. Ahmed, A.M. Hussain, J.P. Rojas, and M.M. Hussain
King Abdullah University of Science and Technology, SAUDI ARABIA

We report fabrication of MEMS thermal actuators on flexible and semi-transparent bulk silicon <100> substrate. The fabricated thermal actuators exhibit similar performance before and after bending. We fabricate the devices first and then release the top portion of the silicon ($\approx$ 19 μm) which is flexible and semi-transparent. Then we perform chemical mechanical polishing to reuse the remaining wafer. Prior demonstrations on flexible MEMS devices had limited thermal budget (<150 °C) compatibility and they did not have cost-saving wafer recycling process.

W-083 WAFER-SCALE FABRICATION OF HIGHLY INTEGRATED RUBIDIUM VAPOR CELLS 552
T. Overstolz, J. Haesler, G. Bergonzi. A. Pezous, P.-A. Clerc, S. Ischer, J. Kaufmann, and M. Despont
CSEM SA, SWITZERLAND

CSEM is developing a highly integrated chip scale atomic clock based on coherent population trapping (CPT) of 87-Rb atoms which are confined in a vapor cell. The vapor cells are batch fabricated, based on pipetting dissolved RbN3 into cell cavities etched into a silicon wafer, closing the cavities by anodic bonding, and UV decomposition of recrystallized RbN3 deposits into Rb and N2. The vapour cells are equipped with resistive heaters, temperature sensors, and Helmholtz coils integrated on both sides of the cell windows.

Th-084 WAFER-SCALE FLEXIBLE GRAPHENE LOUDSPEAKERS 556
H. Tian, Y.L. Cui, Y. Yang, D. Xie, and T.L. Ren
Tsinghua University, CHINA

Wafer-scale flexible graphene loudspeakers are fabricated in one-step laser scribing technology. Current fabrication process for graphene devices is mainly based on CVD graphene, which needs several hours' graphene growth, transfer and patterning. By using this new laser scribing technology, wafer-scale graphene patterns can be obtained in 25 minutes. The loudspeaker is demonstrated to be high performance with wide-band sound generation from 1~50 kHz. Our results show that the laser scribed graphene could be widely used in integrating wafer-scale graphene-based electroacoustic devices.

Industry

M-085 A POLYMER MEMS MIRROR FOR ON-DEMAND LIGHT DISTRIBUTION FABRICATED BY INJECTION MOLDING AND TRANSFER OF PRINTED LAYERS 560

K. Kurihara[1], O. Nagumo[2], H. Takagi[1], and R. Maeda[1]

[1]*National Institute of Advanced Industrial Science and Technology (AIST), JAPAN and* [2]*Designtech Co., Ltd, JAPAN*

A low-cost and wide mirror area polymer MEMS scanner for on-demand light distribution was fabricated by combined process of the injection molding and layer transfer of screen-printed patterns on a film. This fabrication process realizes low cost polymer MEMS. It is expected that the low-cost MEMS can be a killer application for new industry field.

T-086 BI-CHAMBER ELECTROMAGNETIC FLUIDIC PUMP 564

C.S. Gudeman[1], P.J. Rubel[1], and J.S. Foster[2]

[1]*Innovative Micro Technology, USA and* [2]*Owl Biomedical Inc., USA*

We describe the design, fabrication and performance of a MEMS pump that is actuated electromagnetically and is capable of pumping very high viscosity liquids. Valve motion relative to that of the pumping piston is described in detail.

W-087 IMPROVED MECHANICAL RELIABILITY OF MEMS PIEZOELECTRIC VIBRATION ENERGY HARVESTERS FOR AUTOMOTIVE APPLICATIONS 568

M. Renaud, Z. Wang, M. Jambunathan, S. Matova, R. Elfrink, M. Rovers, M. Goedbloed, C. de Nooijer, R.J.M. Vullers, and R. van Schaijk

Holst Centre-Imec, THE NETHERLANDS

We present a comprehensive approach to address the issue of the mechanical reliability of MEMS piezoelectric vibration harvesters. These harvesters generate sufficient electrical power for powering a tire pressure monitoring system. However, their reliability, particularly in terms of shock resilience, has to be optimized for in-tire applications. This paper showcases experimentally verified improvements of the mechanical reliability, which is achieved by optimizing both the package design and the wafer processing.

Th-088 LOW-COST MICROBOLOMETER WITH NANO-SCALED PLASMONIC ABSORBERS FOR FAR INFRARED THERMAL IMAGING APPLICATIONS 572

F. Utermöhlen[1], D. Etter[2], D. Borowsky[1], I. Herrmann[1], C. Schelling[1], F. Hutter[2], S.H. Sun[2], and J. Burtghartz[2]

[1]*Robert Bosch GmbH, GERMANY and* [2]*Institut für Mikroelektronik Stuttgart (IMS CHIPS), GERMANY*

We have developed a scalable low-cost microbolometer which can be used for automotive nightvision as well as consumer applications with a broad variety of requirements regarding image resolution and sensitivity. In contrast to state-of-the-art microbolometers which are based on a standard CMOS ASIC process with CMOS-compatible MEMS post-processing on the same wafer, we use a MEMS wafer with the microbolometer pixels and a standard CMOS wafer with the read-out ASIC which are mechanically, electrically and hermetically connected. This concept allows for significant cost reduction since the two dedicated technologies for the MEMS and the ASIC can be optimized and fabricated independently and because the ASIC chip can serve as a hermetic package for the MEMS.

Materials and Device Characterization (MDC)

M-089 2D PHOTONIC-CRYSTALS FOR HIGH SPECTRAL CONVERSION EFFICIENCY IN SOLAR THERMOPHOTOVOLTAICS 576

A. Lenert[1], V. Rinnerbauer[1], D.M. Bierman[1], Y. Nam[1,2], I. Celanovic[1], M. Soljacic[1], and E.N. Wang[1]

[1]*Massachusetts Institute of Technology, USA and* [2]*Kyung Hee University, SOUTH KOREA*

We present a high-efficiency 2D photonic crystal based solar thermophotovoltaic (STPV) device operating at high temperatures (~1300 K) under moderate solar concentration (~100 Suns). These results were only possible by tailoring the spectral properties of the absorber-emitter through surface nanostructuring of tantalum and minimizing parasitic thermal losses through an innovative vacuum-enclosed experimental setup.

T-090 AN OPTICAL IN-PLANE DISPLACEMENT MEASUREMENT TECHNIQUE WITH SUB-NANOMETER ACCURACY BASED ON CURVE FITTING 580

J. Kokorian[1], F. Buja[1], U. Staufer[1], and W.M. van Spengen[1,2]

[1]*Delft University of Technology, THE NETHERLANDS and* [2]*Falco Systems BV, THE NETHERLANDS*

We will show a technique, based on plain optical microscopy and curve fitting, for measuring in-plane displacements in MEMS applications. We modeled and experimentally verified how the measurement accuracy is influenced by quantization noise, photon shot noise, optical magnification, camera resolution and pixel binning. We found that when the noise figure was dominated by shot noise, the measurement error was lowered into the deep-subnanometer range.

W-091 CHARACTERIZATION OF IMPROVED CAPACITIVE MICROMACHINED ULTRASONIC TRANSDUCERS (CMUTS) USING ALD HIGH-K DIELECTRIC ISOLATION 584
T. Xu, C. Tekes, and F.L. Degertekin
Georgia Institute of Technology, USA

We show the advantages of high-k dielectric, ALD HfO2 over traditional PECVD silicon nitride isolation for Capacitive Micromachined Ultrasonic Transducers (CMUTs) fabricated by a low temperature, CMOS compatible, sacrificial release method. ALD HfO2 dielectric properties are characterized to optimize CMUT design in transmit and receive mode. Performances of the two different dielectric isolation devices are evaluated through parallel plate modeling and experimentally measured pressure outputs and receive sensitivities.

Th-092 CHARACTERIZATION OF STICTION FORCES IN ULTRA-CLEAN ENCAPSULATED MEMS DEVICES 588
D.B. Heinz, V.A. Hong, E.J. Ng, C.H. Ahn, Y. Yang, and T.W. Kenny
Stanford University, USA

We show that stiction in contact between encapsulated MEMS devices and the surrounding sidewalls generally results in a reversible adhesion with a consistent adhesion force. This force is small enough (25 µN) to be overcome by the restoring force of the springs in inertial sensors with resonant frequency above 4 kHz. Therefore, it should be possible to design and build stiction-free inertial sensors in this process – a significant advantage over approaches that rely on deposition, tuning and maintenance of chemical coatings for inertial sensors.

M-093 COMPOSITE OF THERMALLY RESPONSIVE SOLUTION AND LUBRICATING MICRO BEADS AS SEALING MATERIAL FOR PISTON-CYLINDER ACTUATOR 592
T. Chishiro, S. Honda, and S. Konishi
Ritsumeikan University, JAPAN

This paper proposes a novel sealing technique for miniaturized piston-cylinder actuator. The sliding part of a piston actuator requires both high sealing and excellent lubrication. This paper will show a smart sealing material composed of thermally responsive solution and lubricating micro beads. Micro beads are expected to contribute to provide lubricant between a sliding part and an inner wall when thermally responsive solution gels by heating. We will present the concept, design, and characterization of proposed sealing technique for a piston actuator.

T-094 CONTINUOUS DYNAMIC TIMING MEASUREMENTS TO MONITOR SPRING AND SURFACE FORCES IN MEMS SWITCH RELIABILITY 596
C. Kosla[1], P. Fitzgerald[2], and M. Hill[1]
[1]Cork Institute of Technology, IRELAND and [2]Analog Devices, IRELAND

We demonstrate an automatic reliability detection/prediction system for industry manufactured MEMS switches based on dynamic time measurements, allowing for non-invasive and continuous device monitoring. The developed method highlights the influence of both restoring and surface forces evolution on switch reliability and for the first time allows identification of an imminent device failure due to its continuous monitoring. Additionally we present the scalability of this approach by testing it on different switch types and on a large number of samples.

W-095 DIRECT MEASUREMENT OF SHEAR PIEZORESISTANCE COEFFICIENT ON SINGLE CRYSTAL SILICON NANOWIRE BY ASYMMETRICAL FOUR-POINT BENDING TEST 600
T. Kimura, N. Saito, T. Takeshita, K. Sugano, and Y. Isono
Kobe University, JAPAN

This research evaluated the shear piezoresistance property of p-type single crystal silicon nanowire (SiNW) by the asymmetrical four-point bending (AFPB) testing proposed by the authors. We fabricated the p-type SiNW on the AFPB specimen with "V" shaped notches made of single crystal silicon. Bending the specimen by the asymmetrical four point-supports, simple shear stress can be produced at the center of the specimen. Consequently, we have succeeded in evaluating the shear piezoresistance coefficient of SiNW directly."

Th-096 DISSOLVABLE MATERIAL FOR HIGH-ASPECT-RATIO FLEXIBLE SILICON-MICROWIRE PENETRATIONS 604
S. Yagi, S. Yamagiwa, T. Imashioya, H. Oi, Y. Kubota, M. Ishida, and T. Kawano
Toyohashi University of Technology, JAPAN

For realization of low invasive electrode penetrations into biological tissue, here we improved the penetration capability of high-aspect-ratio flexible silicon-micorwires by coating a dissolving material of silk fibroin. The silk fibroin was coated over vertically vapor-liquid-solid (VLS) grown silicon-microwires. The 420-µm-long silicon-wire with a ~200-µm-thick silk film exited the stiffness of 4.03 N/m, which is 72% improved value compared to that of the silicon-wires without silk (2.34 N/m). The effects of the silk support on the wire penetration were observed by demonstrating the gelatin penetrations. These results suggest that the numerous high-aspect-ratio flexible bioprobes can be penetrated by using the silk support.

M-097 DOME-DISC: DIFFRACTIVE OPTICS METROLOGY ENABLED DITHERING INERTIAL SENSOR CALIBRATION 608
S. Nadig[1], S. Ardanuç[1], B. Clark[2], and A. Lal[1]
[1]Cornell University, USA and [2]Analog Devices Inc., USA

We demonstrate ~100-ppm accurate scale-factor and bias calibration of a commercial gyroscope, in which the typical un-calibrated scale factor variations are 100,000-ppm. In this paper, we present a Diffractive Optical Metrology Enabled Dithering Inertial Sensor Calibration consisting of a novel piezoelectric dither stage, the motion of which is measured by imaging the diffraction pattern off the stage of a long-term stable wavelength laser. The architecture presented here illustrates how atomically stable lasers and CMOS imagers can be combined to form a miniature atomically stable self-calibrated inertial sensor platform.

T-098 ELECTRICAL CHARACTERIZATION OF ALD COATED SILICON DIOXIDE MICRO-HEMISPHERICAL SHELL RESONATORS 612
P. Shao, V. Tavassoli, C.-S. Liu, L.D. Sorenson, and F. Ayazi
Georgia Institute of Technology, USA

A micro-hemispherical shell resonator (µHSR) is the beating heart of a micro-scale hemispherical resonator gyroscope (µHRG). Small damping and high symmetry are two essential requirements for µHRGs. Damping can be quantified by mechanical quality factor (Q) of the resonance, and structural symmetry can be quantified by the frequency split between two degenerate resonance modes. This paper reports on important electrical characterizations of Q and frequency split of ALD coated thermally-grown silicon-dioxide µHSRs, and analysis on how the performance will change with fabrication and measurement parameters.

W-099 FABRICATION AND DEGRADATION CHARACTERISTIC OF SPUTTERED IRIDIUM OXIDE NEURAL MICROELECTRODES FOR FES APPLICATION 616
X.-Y. Kang, J.-Q. Liu, H.-C. Tian, J.-C. Du, B. Yang, H.-Y. Zhu, Y. NuLi, and C.-S. Yang
Shanghai Jiao Tong University, CHINA

We have fabricated reactively sputtered iridium oxide film (SIROF) microelectrodes under different oxygen flows and the stimulus-evoked degradation properties are also tested. The SIROF microelectrodes prepared under 25 sccm oxygen flow shows the least degradation from continuous electrical stimulation. That the charge storage capacity (CSC) is only 9.6 % lost and the 1 kHz impedance is only 4.23% increase. Hence, the 25 sccm one can be an ideal microelectrode modification material for electrical stimulation with the least degradation.

Th-100 FABRICATION AND TESTING OF PIEZOELECTRIC HYBRID PAPER FOR MEMS APPLICATIONS 620
S.K. Mahadeva, K. Walus, and B. Stoeber
University of British Columbia, CANADA

We have developed a new inexpensive functional paper based material that can be used as a piezoelectric substrate for sensing applications. In our simple method, nanostructured BaTiO3 is embedded onto the fibers prior to forming paper sheet, which involves immersion of wood fibers in aqueous solution of poly(diallyldimethylammonium chloride) PDDA and poly(sodium 4-styrenesulfonate) and once again in PDDA, and results in the creation of a positively charged surface on wood fiber. The treated wood fibers are then immersed in a BaTiO3 suspension, leading to the electrostatic binding of BaTiO3. The hybrid paper showed the highest d33 of 4.8±0.4 pC/N.

M-101 GRAPHENE WOVEN FABRIC AS HIGH-RESOLUTION SENSING ELEMENT OF CONTACT-LENS TONOMETER 624
Y. Zhang[1], T. Man[1], X. Li[2], H. Zhu[2], and Z. Li[1]
[1]Peking University, CHINA and [2]Tsinghua University, CHINA

In our work, the graphene woven fabrics (GWFs), the combination of highly sensitive strain sensing and transparency, is investigated as the sensing element of the contact-lens tonometer, which enables precisely monitor IOP. The relationship between the current changes when keeping the voltage constant and effective IOP increasing has been obtained.

T-102 INCREASED THERMAL CONDUCTIVITY POLYCRYSTALLINE DIAMOND FOR LOW-DISSIPATION MICROMECHANICAL RESONATORS 628
H. Najar[1], A. Thron[1], C. Yang[2], S. Fung[1], K. van Benthem[1], L. Lin[2], and D.A. Horsley[1]
[1]University of California, Davis, USA, and [2]University of California, Berkeley, USA

We report an investigation of microcrystalline diamond (MCD) films deposited under different conditions to increase thermal conductivity and therefore mechanical Q in MEMS resonators. Here, through a study of different deposition conditions, we demonstrate a three-fold increase in thermal conductivity (i.e. k = 100W/mK) and therefore Q-TED. We further present a study of the unique microstructure of hot filament CVD diamond films and relate growth conditions to observed microstructural defects.

W-103 INTERFACE LOSSES IN MULTIMATERIAL RESONATORS 632
L.G. Villanueva[1,3], B. Amato[1], T. Larsen[1], and S. Schmid[1]
[1]*Denmark Technical University, DENMARK and*
[2]*Ecole Polytechnique Federale de Lausanne (EPFL), SWITZERLAND*

We present an extensive study shedding light on the role of surface and bulk losses in micromechanical resonators. We fabricate a set of Si3N4 square membranes with different lateral dimensions, thickness and thickness of metal on top and characterize the 81 lowest flexural modes, obtaining more than 3000 experimental points to eventually quantify the importance of interface losses in multimaterial resonators.

**Th-104 INVESTIGATION OF DOMINANT FACTORS TO CONTROL C-AXIS
TILT ANGLE OF ALN THIN FILMS FOR EFFICIENT ENERGY HARVESTING** 636
Q. Wang, H. Oguchi, M. Hara, and H. Kuwano
Tohoku University, JAPAN

We investigated growth conditions to enhance the c-axis inclination of aluminum nitride (AlN) thin films grown on silicon substrates using the electron cyclotron sputtering. Higher substrate tilt angles, lower substrate temperature, and rougher buffer layer surface resulted in higher c-axis tilt angle, mainly due to decrease in ad-atom mobility on the surface. This study deepens the understanding of how to control c-axis inclination of AlN thin film to control the electro-mechanical coupling coefficient for larger output power of the AlN-based energy harvesters.

**M-105 INVESTIGATION OF THE FATIGUE ORIGIN AND PROPAGATION
IN SUBMICROMETRIC SILICON PIEZORESISTIVE LAYERS** 640
G. Langfelder[1], S. Dellea[1], P. Rey[2], A. Berthelot[2], and A.F. Longoni[1]
[1]*Politecnico di Milano, ITALY and* [2]*CEA - LETI – Minatec, FRANCE*

We present the study performed on structures designed and tested for the analysis of long-term reliability and fatigue of 250-nm-thick crystalline Silicon that can be used as piezoresistive sensing layer in low-power 10-axis inertial measurement units. With a specimen surface-to-volume ratio 100 times smaller than previous literature, this work extends to the nanometric domain the debate data previously published about the origin and propagation of fatigue in Silicon at the micro scale.

T-106 NANOFIBER FORESTS WITH HIGH INFRARED ABSORPTANCE 644
H.Y. Mao[1,3], C. Lei[1,4], Y.J. Chen[1,4], Z.J. Chen[2], W. Ou[1,3], W.G. Wu[2], A.J. Ming[1,3], and D.P. Chen[1,3]
[1]*Chinese Academy of Sciences, CHINA,* [2]*Peking University, CHINA,*
[3]*Jiangsu R&D Center for Internet of Things, CHINA, and* [4]*North University of China, CHINA*

Nanofiber forests with high infrared(IR) absorptance are reported in this work. In wavelength range from 1.5 to 5 μm, the absorptance of the nanofiber forests reaches a minimum of 96%, which is much higher than that of Si3N4-based IR absorbers and the polymer coatings from which the nanofibers are obtained. Such nanofiber forests are fabricated by using a plasma-stripping-of-polymer technique, which is fast, high-yield, and compatible with micro-fabrication. By introducing the nanofiber forests in MEMS IR devices, improved performance of the devices is expected to be acquired.

**W-107 NANOSTRUCTURED SILICON FLAPPING WING WITH HIGHER STRENGTH
AND LOW REFLECTIVITY FOR SOLAR POWERED MEMS AIRCRAFT** 648
K. Kashyap[1], A. Kumar[1], C.-N. Chen[1], M.T. Hou[2], and J.A. Yeh[1,3]
[1]*National Tsing Hua University (NTHU), TAIWAN,* [2]*National United University, TAIWAN, and*
[3]*National Applied Research Laboratories, TAIWAN*

We develop a novel way of higher strength silicon flapping wings design for MEMS aircraft achieved by silicon nanostructures, which breaks the limitation of silicon as a fragile material. Silicon flapping wings were designed for MEMS aircraft which increases the bending strength of wings by 6 times and reduces the reflectance to 2%. Both the benefits simultaneously were achieved from nanostructure surface texturing by low cost wet chemical etching.

Th-108 POSSIBILITY OF CEMENTED CARBIDE AS STRUCTURAL MATERIAL FOR MEMS 652
T. Morikaku[1], T. Fujii[1], K. Kuroda[2], Y. Takami[2], S. Inoue[1], and T. Namazu[1,3]
[1]*University of Hyogo, JAPAN,* [2]*Silveralloy Co., Ltd. JAPAN, and*
[3]*Japan Science and Technology Agency (JST), JAPAN*

We present the possibility of WC-Co cemented carbide as mechanical elements in MEMS. The cemented carbide is typically used as material for working tool because it has superior characteristics, such as very high Young's modulus, excellent rigidity, good chemical inertness, and good thermal stability. These are also very attractive as structural material in MEMS. We investigated the influences of specimen size and WC-Co composition ratio on mechanical properties of FIB-fabricated WC-Co cemented carbide nanowires by means of on-chip uniaxial tensile testing.

M-109 THIN-FILM MAGNESIUM AS A SACRIFICIAL AND BIODEGRADABLE MATERIAL 656
Y. Liu, J. Park, J.H.-C. Chang, and Y.-C. Tai
California Institute of Technology, USA

Magnesium (Mg) and magnesium alloys have drawn great attention as biodegradable materials. It means that magnesium could be an interesting dual "sacrificial and biodegradable MEMS material". This work then reports the first etching tests of the dual properties of ebeam-deposited thin-film Mg (i.e., 0.3 and 1.0 micron thick). Here we have tested etchants including diluted hydrochloric acid, saline, and culture medium. Data are fitted by "First-and-Second order" model. The initial results do show that thin-film Mg indeed is a promising dual sacrificial and biodegradable material."

T-110 THREE-DIMENSIONAL (3-D) RESHAPING TECHNIQUE IN MEMS DEVICES BY
SOLELY ELECTRICAL CONTROL WITH ULTRAFINE TUNING RESOLUTION 660
Y.H. Yoon, C.H. Han, and J.-B. Yoon
Korea Advanced Institute of Science and Technology (KAIST), SOUTH KOREA

We propose an innovative and simple three-dimensional (3-D) reshaping (plastic deformation) technique in MEMS devices by solely electrical control with ultrafine tuning resolution. While voltage input induces stress on the device, Joule heating is applied to make plastic deformation in the device, where the tuning resolution was demonstrated at a sub-100nm level. The proposed technique is expected to be favorably used in many integrated MEMS devices where reshaping feature is required avoiding any external instruments.

W-111 TUNABLE THZ FILTER BASED ON RANDOM ACCESS
METAMATERIAL WITH LIQUID METAL DROPLETS 664
Q.H. Song[1], W.M. Zhu[2], W. Zhang[2], E.M. Chia[2], M. Ren[2], and A.Q. Liu[1]
[1]Xi'an Jiao Tong University, CHINA and [2]Nanyang Technological University, SINGAPORE

We report a tunable THz filter based on random access metamaterial with liquid metal droplet, which is tuned by controlled electrowetting effects. The random access metamaterial consists of micro droplets formed by lotus effect. In experiment, it measures a near 0.01-THz frequency shift of the dipole resonance spectrum induced by changing of the droplets shape via electrowetting effect. The random access metamaterial is flexible in tuning and easy in fabrication, which has potential application on tunable filters, controllable beam steering and flat lens.

Th-112 WETTING DYNAMICS STUDY OF UNDERWATER SUPERHYDROPHOBIC
SURFACES THROUGH DIRECT MENISCUS VISUALIZATION 668
M. Xu, G. Sun, and C.J. Kim
University of California, Los Angeles, USA

We report the study of underwater wetting transition of superhydrophobic (SHPo) surfaces from dewetted (Cassie) to wetted (Wenzel) state through direct and continuous meniscus visualization. The result confirmed two meniscus states of pinning and wetting, the latter leading to the Wenzel state. Furthermore, the result revealed that the Cassie state can (or cannot) be indefinite if (or unless) the water is saturated with air and the hydrostatic pressure is low enough.

Mechanical Sensors and Systems (MECH)

M-113 3-D HEMISPHERICAL MICRO GLASS-SHELL RESONATOR WITH INTEGRATED
ELECTROSTATIC EXCITATION AND CAPACITIVE DETECTION TRANSDUCERS 672
M.M. Rahman, Y. Xie, C. Mastrangelo, and H. Kim
University of Utah, USA

This paper reports the development and performance of a 3D hemispherical micro glass-shell resonator with integrated electrostatic excitation and capacitive detection transducers. This paper presents the first performance results of the 3D shell resonator with integrated micro fabricated excitation and sensing units that produced the first vibration mode of resonance at 5.843 kHz with a quality factor of 730 at atmosphere with the time decay constant of 39.78ms.

T-114 A CMOS MEMS PIRANI VACUUM GAUGE WITH COMPLEMENTARY
BUMP HEAT SINK AND CAVITY HEATER 676
Y.-C. Sun[1], K.-C. Liang[1,2], C.-L. Cheng[1], and W. Fang[1]
[1]National Tsing Hua University (NTHU), TAIWAN and
[2]Taiwan Semiconductor Manufactury Company Ltd., TAIWAN

We design and manufacturing a new CMOS-MEMS Pirani vacuum gauge with complementary bump heat-sink and cavity heater. By using the bump heat-sink and cavity heater design, the active area of heat-sink and heater can be increased without changing device footprint size. In addition, the cavity in heater reduces the thermal mass for low power operation. The proposed design have larger dynamic range, higher sensitivity and lower power consumption as compare to the typical type.

W-115 A LARGE RANGE MULTI-AXIS CAPACITIVE FORCE/TORQUE SENSOR REALIZED IN A SINGLE SOI WAFER 680
D. Alveringh, R.A. Brookhuis, R.J. Wiegerink, and G.J.M. Krijnen
University of Twente, THE NETHERLANDS

A miniature silicon capacitive force/torque sensor is designed and realized to be used for biomechanical applications. The sensor is capable of measuring 5 degrees of freedom with a force range of 2 N in shear and normal direction and a torque range of 6 Nmm. The fabrication of the sensor requires only two masks, making the sensor cost-effective to fabricate. This is the first 5 degrees of freedom force/torque sensor in this force range made in a single SOI wafer.

Th-116 A NOVEL ELECTRET ROTATIONAL SPEED SENSOR 684
W. Bian, X. Wu, and X. Wang
Tsinghua University, CHINA

In this paper, a novel rotational speed sensor based on electrostatic variation is presented, which is fabricated by typical micro fabrication processes. Compared to the other rotational sensors, the merits of the presented sensor are its simple configuration, small size, and low cost.

M-117 ALN-BASED PIEZOELECTRIC RESONATOR FOR INFRARED SENSING APPLICATION 688
W.C. Ang[1,2], P. Kropelnicki[1], H. Campanella[1], Y. Zhu[1,2], A.B. Randles[1],
H. Cai[1], Y.A. Gu[1], K.C. Leong[3], and C.S. Tan[2]
[1]Agency for Science, Technology and Research (A*STAR), SINGAPORE, [2]Nanyang Technological University, SINGAPORE, and [3]GLOBALFOUNDRIES Singapore Pte Ltd, SINGAPORE

We develop a highly sensitive AlN-based resonant uncooled infrared (IR) detector utilizing the photoresponse and piezoelectric properties of polycrystalline AlN. The design, fabrication, and IR sensing characterization of the device are presented. Different from other reported works, photoresponse mechanism was proposed in this paper instead of thermal effect. Without the need of vacuum, AlN-based IR detector brings the great advantage in device packaging and thus further reduces the manufacturing and operation cost.

T-118 AN ALL OPTICAL SHOCK SENSOR BASED ON BUCKLED DOUBLY-CLAMPED SILICON BEAM 692
B. Dong[1], J.G. Huang[3], H. Cai[2], P. Kropelnicki[2], A.B. Randles[2], Y.D. Gu[2], and A.Q. Liu[1]
[1]Nanyang Technological University, SINGAPORE, [2]Agency for Science, Technology and Research (A*STAR), SINGAPORE, and [3]Xi'an Jiao Tong University, SINGAPORE

An all optical shock sensor is designed, fabricated and experimentally demonstrated. Fabricated with CMOS compatible process, this optical shock sensor can be easily integrated with other photonic devices. The opto-mechanical shock sensor can be potentially used at hash environment like in oil industry, or military usage in a complex electromagnetic environment. It also has potential applications such as inertial sensor, optical switch and other optomechanical devices.

W-119 AN ANGULAR ACCELERATION SENSOR INSPIRED BY THE VESTIBULAR SYSTEM WITH A FULLY CIRCULAR FLUID-CHANNEL AND THERMAL READ-OUT 696
J. Groenesteijn, H. Droogendijk, M.J. de Boer, R.G.P. Sanders, R.J. Wiegerink, and G.J.M. Krijnen
University of Twente, THE NETHERLANDS

We report on an angular accelerometer based on the semicircular channels of the vestibular system. The accelerometer consists of a water-filled circular tube, wherein the fluid flow velocity is measured thermally as a representative for the external angular acceleration. Measurements show a linear response for angular acceleration amplitudes up to $2 \times 10^5 \, °s^{-2}$.

Th-120 AN EFFICIENT EARTH MAGNETIC FIELD MEMS SENSOR: MODELLING AND EXPERIMENTAL RESULTS 700
M. Bagherinia[1], A. Corigliano[1], S. Mariani[1], D.A. Horsley[2], M. Li[2], and E. Lasalandra[3]
[1]Politecnico di Milano, ITALY, [2]University of California, Davis, USA, and [3]STMicroelectronics, ITALY

We present the experimental results and performance indexes of a new z-axis Lorentz force MEMS magnetometer with reduced dimensions and high efficiency and exploit an ad-hoc formulated multi-physics approach and its solutions to model the sensor dynamics. The obtained sensor has a good resolution for earth magnetic field detection and navigation, and is very efficient in terms of exciting current, surface area and bandwidth.

M-121 AN ELECTRET-BIASED RESONANT RADIATION SENSOR 704
S.S. Lee, C.K. Yoon, S.H. Song, and B. Ziaie
Purdue University, USA

In this work, an electret-biased resonant radiation sensor capable of measuring accumulated radiation dosage is presented. The sensor consists of a positive corona-charged Teflon electret placed underneath a ZnO piezoelectric cantilever. As ionizing radiation passes through the ambient air surrounding the electret, ions are generated in the air and drift toward the Teflon substrate. These ions neutralize the electret's surface charges and thus reduce the electrostatic force. The force reductions result in the cantilever's resonant frequency back to its natural frequency.

T-122 AN SOI TACTILE SENSOR WITH A QUAD SEESAW ELECTRODE FOR 3-AXIS COMPLETE DIFFERENTIAL DETECTION 709

Y. Hata[1], Y. Nonomura[1], H. Funabashi[1], T. Akashi[1], M. Fujiyoshi[1], Y. Omura[1], T. Nakayama[2],
U. Yamaguchi[2], H. Yamada[2], S. Tanaka[3], H. Fukushi[3], M. Muroyama[3], M. Makihata[3], and M. Esashi[3]
[1]Toyota Central R&D Labs., Inc., JAPAN, [2]Toyota Motor Corp., JAPAN, and [3]Tohoku University, JAPAN

This paper presents an SOI capacitive tactile sensor with a quad-seesaw electrode for 3-axis differential detection. For differentially detecting 3-axis forces, we propose a novel seesaw-electrode structure composed of four rotating plates individually suspended by torsion beams. We successfully fabricated the test device that integrates an SOI with seesaw electrodes and an LTCC with fixed electrodes. The test results demonstrated that the proposed sensor differentially detects 3-axis forces.

W-123 DIELECTRICAL LIQUID-BASED TACTILE SENSING ARRAY WITH ADJUSTABLE SENSING RANGES AND SENSITIVITY 713

K.W. Liao[1], M.T. Hou[2], H. Fujita[3], and J.A. Yeh[1]
[1]National Tsing Hua University (NTHU), TAIWAN, [2]National United University, TAIWAN, and [3]University of Tokyo, JAPAN

We present a novel tactile sensing array with adjustable sensing ranges. Each sensing element contains a low dielectric constant droplet covered with high liquid. We controlled the contact angle of the droplet by controlling the electric flux passing through the element. Then, the sensing ranges and sensitivity were also adjusted due to the variation of the droplet shape. The results show the sensor's the sensing range is easily adjusted from 0.04N ~ 0.60N to 0.33N ~ 1.05N. The sensitivity increases 1.9 times in at small force range from 1.47pF/N to 2.90.pF/N.

Th-124 ELECTRET-BASED LOW POWER RESONATOR FOR ROBUST PRESSURE SENSOR 717

H. Mitsuya[1], H. Ashizawa[1], T. Sugiyama[2], M. Kumemura[3], M. Ataka[3], H. Fujita[3], and G. Hashiguchi[2]
[1]Saginomiya Seisakusho, Inc., JAPAN, [2]Shizuoka University, JAPAN, and [3]University of Tokyo, JAPAN

We have developed a membrane-less pressure sensor based on squeeze-film damping in a 2-μm driving gap of a silicon ring-shape resonator. Its sensing range is from sub-atmospheric to over 1MPa; very wide-range pressure measurement is possible with one sensor element. An electret film having the 200-V-bias voltage was incorporated to the resonator; this allows the excitation of the resonator at very low AC voltage. This membrane-less pressure sensor has robust and low power consumption (nW-range) characteristics.

M-125 ELECTRIC GRADIENT FORCE DRIVE MECHANISM FOR NOVEL MICRO-SCALE ALL DIELECTRIC GYROSCOPE 721

R. Perahia, J.J. Lake, S.S. Iyer, D.J. Kirby, H.D. Nguyen, T.J. Boden, R.J. Joyce, L.X. Huang,
L.D. Sorenson, and D.T. Chang
HRL Laboratories, USA

This paper reports a novel drive mechanism used to excite a cylindrical, all-dielectric micro-shell gyroscope structure. The drive mechanism operates by generating a gradient electric-field force from a set of interdigitated electrodes placed adjacent to the gyroscope structure. This novel transduction mechanism enables mechanical actuation of a pristine dielectric structure without the need for direct metallization which could otherwise degrade mechanical performance. Design, fabrication, and experimental demonstration are presented.

T-126 ELECTROMECHANICAL DAMPING IN MEMS ACCELEROMETERS: A WAY TOWARDS SINGLE-CHIP GYROMETER-ACCELEROMETER CO INTEGRATION 725

Y. Deimerly[1], P. Rey[1], P. Robert[1], T. Bourouina[2], and G. Jourdan[1]
[1]CEA - LETI – Minatec, FRANCE and [2]Université Paris-Est, FRANCE

This work proposes a method for controlling mechanical damping in MEMS devices. By capacitively coupling a micro mechanical sensor to an electrical resistance, mechanical energy is dissipated by an additional damping sink. In this study, the damping rate of a MEMS accelerometer has been tuned under vacuum, in compliance with a simple electromechanical model that will be further detailed. Using this phenomenon, this presentation will discuss the possibility of co integrating accelerometers with gyrometers on a single chip inside a same cavity to form compact System In Package.

W-127 EXPERIMENTALLY VALIDATED ALUMINUM NITRIDE BASED PRESSURE, TEMPERATURE AND 3-AXIS ACCELERATION SENSORS INTEGRATED ON A SINGLE CHIP 729

F. Goericke[1], K. Mansukhani[1], K. Yamamoto[2], and A. Pisano[3]
[1]University of California, Berkeley, USA, [2]Murata Manufacturing Co., Ltd, JAPAN, and [3]University of California, San Diego, USA

This paper reports a unified fabrication process used to build multiple aluminum nitride (AlN) based micro-electromechanical system (MEMS) sensors on a single chip. A fully functional AlN-based sensor cluster has been demonstrated and is presented in this paper. This sensor cluster is a "five degree-of-freedom" cluster; it measures 3-axis acceleration, temperature and pressure fabricated on a 1 cm x 1 cm die. In addition to utilizing AlN as both the structural and active layer of the sensors, this work is novel because all sensors are fabricated in the same fabrication run.

Th-128 FABRICATION AND CHARACTERIZATION OF ALL HYDROGEL
CANTILEVERS FOR ATOMIC FORCE MICROSCOPY APPLICATIONS 733
I. Lee and J. Lee
Sogang University, SOUTH KOREA

We develop a novel method for fast and simple fabrication of hydrogel microcantilevers for atomic force microscopy applications. Fabricated hydrogel microcantilevers exhibit imaging performance comparable to that of commercial silicon microcantilevers in case of non-contact mode operation.

M-129 FULLY PRINTED, LARGE-SCALE, HIGH SENSITIVE STRAIN SENSOR
ARRAY FOR STRESS MONITORING OF INFRASTRUCTURES 737
S. Harada, W. Honda, T. Arie, S. Akita, and K. Takei
Osaka Prefecture University, JAPAN

We demonstrated a macroscale sensor sheet by fabricating the fully printed, large-scale, and high sensitive strain sensor array on flexible substrates, which can cover any surfaces, for the application of real-time secure infrastructure maintenance as a proof-of-concept. Printed strain sensor array exhibits that the impressively high gauge factor ~106 and successfully detects the small deformation <20μm distributions.

T-130 HARBOR SEAL INSPIRED MEMS ARTIFICIAL MICRO-WHISKER SENSOR 741
A.G.P. Kottapalli[1], M. Asadnia[1], H. Hans[1], J.M. Miao[1], and M. Triantafyllou[2]
[1]Nanyang Technological University, SINGAPORE and [2]Massachusetts Institute of Technology, USA

Harbor seal whiskers possess a unique geometry along the length of the whisker which is believed to perform vortex induced vibrations (VIV) in frontal flows. The geometry of the whisker appears to be well-tuned to offer maximum allowable sensitivity for sensing by minimizing the self-induced vibrations until an upstream stimulus is encountered. In this work we develop artificial MEMS versions of seal whiskers using stereolithography. These artificial sensors demonstrate a threshold velocity detection limit as low as 193μm/s which rivals the abilities of the Harbor seal's real whisker. Experiments conducted in water tunnel reveal VIV suppression by the whisker structure.

W-131 HIGH FREQUENCY PIEZOELECTRIC MICROMACHINED ULTRASONIC
TRANSDUCER ARRAY FOR INTRAVASCULAR ULTRASOUND IMAGING 745
Y. Lu, A. Heidari, S. Shelton, A. Guedes, and D.A. Horsley
University of California, Davis, USA

This paper presents a 1.2 mm diameter high fill-factor array of 1,261 piezoelectric micromachined ultrasonic transducers (PMUTs) operating at 18.6MHz for medical imaging applications. This process incorporates a sacrificial polysilicon release pit that precisely defines the PMUT diameter, thereby enabling 10× smaller device spacing and eliminating the need for through-wafer etching. Measurements show a large voltage response of 2.5nm/V and good frequency matching in air, a high center frequency 18.6MHz and wide bandwidth 4.9MHz when immersed in fluid, and phased array simulations based on measured PMUT parameters show high output pressure of the focused acoustic beam.

Th-132 IMPACT OF GYROSCOPE OPERATION ABOVE THE CRITICAL BIFURCATION
THRESHOLD ON SCALE FACTOR AND BIAS INSTABILITY 749
S.H. Nitzan[1], T.-H. Su[1], C.H. Ahn[2], E.J. Ng[2], V.A. Hong[2], Y. Yang[2], T.W. Kenny[2], and D.A. Horsley[1]
[1]University of California, Davis, USA and [2]Stanford University, USA

We investigate the impact of operating a vibratory rate gyro (VRG) at large oscillation amplitude where the VRG's driven becomes nonlinear. Nonlinearities arising at large amplitudes cause the resonator's amplitude-frequency response to become multi-valued above a level known as the critical bifurcation threshold, xc. Open-loop resonators operating at amplitudes above xc are subject to large amplitude instabilities. We demonstrate using closed-loop operation, that scale-factor and bias instability are not affected by operation above xc and angle random walk is reduced.

M-133 IMPROVED ACOUSTIC COUPLING OF AIR-COUPLED
MICROMACHINED ULTRASONIC TRANSDUCERS 753
S. Shelton[1], O. Rozen[1], A. Guedes[1], R. Przybyla[2], B. Boser[2], and D. Horsley[1]
[1]University of California, Davis, USA and [2]University of California, Berkeley, USA

A micromachined ultrasonic transducer (MUT) achieves maximum acoustic coupling when its radius approaches the acoustic wavelength. Previously, this fact posed a critical limitation on size for MUTs operating in air. We present a new approach to increase the acoustic coupling and bandwidth of MUTs using a resonant cavity etched beneath the MUT. The result is a 4x increase in sound pressure level for MUTs having radius equal to one-eighth the acoustic wavelength and an 8x improvement in the bandwidth, thereby enabling much smaller transducers.

T-134   MECHANICAL FORCE-DISPLACEMENT TRANSDUCTION STRUCTURE FOR PERFORMANCE ENHANCEMENT OF CMOS-MEMS PRESSURE SENSOR 757
C.-L. Cheng[1], H.-C. Chang[1], C.-I. Chang[1], Y.-T. Tuan[2], and W. Fang[1]
[1]National Tsing Hua University (NTHU), TAIWAN and [2]National Nano Device Laboratories, TAIWAN

This study implements a mechanical force-displacement transduction structure using the TSMC 0.18um 1P6M CMOS process to improve CMOS-MEMS capacitive pressure sensor. The membrane will be deformed by pressure and cause the sensing-gap change between undeformed movable-electrode and fixed-electrode. Feature of this study is CMOS-MEMS deformed membrane and undeformed movable-electrode to enable the parallel-plate gap-closing pressure detection. Thus, the performance of pressure sensor can be improved and stabilized.

W-135   LOW NOISE VACUUM MEMS CLOSED-LOOP ACCELEROMETER USING SIXTH-ORDER MULTI-FEEDBACK LOOPS AND LOCAL RESONATOR SIGMA DELTA MODULATOR 761
F. Chen[1], W.Z. Yuan[1], H.L. Chang[1], I. Zeimpekis[2], and M. Kraft[3]
[1]Northwestern Polytechnical University, CHINA, [2]University of Southampton, UK, and [3]University of Duisburg-Essen, GERMANY

We report a novel sixth-order sigma-delta modulator MEMS closed-loop accelerometer with extended bandwidth operating in a vacuum environment, which can coexist on a single die with other sensors requiring vacuum packaging. The sensing element was fabricated on a common SOI substrate, four electronic integrators with local resonators are cascaded with the sensing element form high-order noise shaping and notch to suppress the total in-band quantization noise. The feedback voltage signal was applied to the proof-mass to artificially damp the system, which guarantees stable operation in vacuum.

Th-136   MICRO LIQUID-BASED THERMO-ACOUSTIC TRANSMITTER FOR EMITTING ULTRASOUND IN LIQUID MEDIUM 765
D. Hoang-Giang[1], N. Thanh-Vinh[1], K. Noda[1], P. Hoang-Phuong[2], N. Binh-Khiem[1], T. Takahata[1], K. Matsumoto[1], and I. Shimoyama[1]
[1]University of Tokyo, JAPAN and [2]Griffith University, AUSTRALIA

We proposed a thermo-acoustic transmitter using a nanometer thickness metal layer encapsulated with a micrometer thickness liquid layer on thermal-insulator substrate for emitting ultrasound in liquid medium. To improve energy efficiency we take advantage of low specific heat capacity liquid which is encapsulated physically and thermally in small volume by a thin parylene film to fabricate the device. The experiment results demonstrated that by using silicone oil (HIVAC-F5) encapsulated on glass composite, we can obtain ultrasound with sound pressure 3Pa in water medium.

M-137   MULTI-AXIS FORCE SENSOR WITH DYNAMIC RANGE UP TO ULTRASONIC 769
P. Quang-Khang[1], N. Minh-Dung[1], N. Binh-Khiem[1], H.P. Phan[2], K. Matsumoto[1], and I. Shimoyama[1]
[1]University of Tokyo, JAPAN and [2]Griffith University, AUSTRALIA

We proposed a multi-axis force sensor that has a dynamic range up to ultrasonic. The sensor utilizes multilayer structure of elastomer/polymer/viscous liquid to conduct forces and acoustic vibrations to four piezoresistive cantilevers. Experiment results showed that the sensor was capable of measuring normal and lateral forces with high linearity in the range up to 40kPa. Moreover, the dynamic range of the sensor covers ultrasonic frequencies, with the first resonant frequency located at 170kHz.

T-138   MULTIFUNCTIONAL INTEGRATED SENSOR IN A 2X2 MM EPITAXIAL SEALED CHIP OPERATING IN A WIRELESS SENSOR NODE 773
C.L. Roozeboom[1], V.A. Hong[1], C.H. Ahn[1], E.J. Ng[1], Y. Yang[1], B.E. Hill[1], M.A. Hopcroft[2], and B.L. Pruitt[1]
[1]Stanford University, USA and [2]Hewlett-Packard Labs, USA

We present multifunctional integrated sensors that combine temperature, humidity, pressure, air speed, chemical gas, magnetic, and acceleration sensing on a single 2x2 mm chip. We fabricate the multi-sensor in a wafer scale encapsulation process to hermetically seal the sensor functions with moving parts at low vacuum, and then surface micromachine the environmental sensors on top of the sealed layer. We demonstrate the multi-sensor in a wireless sensor node that combines energy harvesting, power management, and low power electronics to transmit data using a cloud-based service.

W-139   OUT-OF-PLANE MICRO TRIPLE-HOT-WIRE ANEMOMETER BASED ON PYREX BUBBLE FOR AIRFLOW SENSING 777
S.W. Liu[1], S.S. Pan[1], F. Xue[1], N. Lin[1], H.B. Liu[1], J.M. Miao[1], L.K. Norford[2], and H.B. Lim[1]
[1]Nanyang Technological University, SINGAPORE and [2]Massachusetts Institute of Technology, USA

We report novel design and fabrication of out-of-pane micro airflow sensors based on the hot-wire sensing principle, i.e. gas cooling of electrically-heated hot-wires. With three micro Cr/Au/Cu hot-wire components fabricated on a Pyrex bubble, the anemometer has demonstrated the ability to detect velocity (<10m/s) and to determine flow direction with an error less than ±8° when the velocity is 10m/s.

Th-140 A PAPER-BASED PIEZOELECTRIC TOUCH PADS INTEGRATING ZINC OXIDE NANOWIRES 781
Y.H. Wang, X. Li, C. Zhao, and X.Y. Liu
McGill University, CANADA

We report a new type of paper-based piezoelectric touch pads integrating zinc-oxide nanowires (ZnO-NWs) as the sensing component. We directly grew ZnO-NWs on cellulose paper using a simple hydrothermal approach, and fabricated single-layer piezoelectric touch pads from ZnO-NW-coated paper. The presented piezoelectric touch pads are inexpensive, easy-to-fabricate, ultra-thin, lightweight and disposable, and will further enrich the tool set of paper electronics.

M-141 PIEZORESISTIVITY OF AG NWS-PDMS NANOCOMPOSITE 785
M. Amjadi, A. Pichitpajongkit, S. Ryu, and I. Park
Korea Advanced Institute of Science and Technology (KAIST), SOUTH KOREA

In this work, we developed a conductive silver nanowire (AgNW)-PDMS composite thin film for a flexible strain sensing application. The piezoresistivity of Ag NWs-PDMS nanocomposite thin film was experimentally investigated and analyzed by a computational model. Finally, a finger motion detection device was developed by using Ag NWs-PDMS nanocomposite thin film as a highly stretchable, flexible and sensitive strain sensor.

T-142 SINGLE-CHIP ATOMIC FORCE MICROSCOPE WITH INTEGRATED Q-ENHANCEMENT AND ISOTHERMAL SCANNING 789
N. Sarkar[1,2] and R.R. Mansour[1,2]
[1]University of Waterloo, CANADA and [2]ICSPI Corp., CANADA

We report on the design, fabrication, and imaging performance of a single-chip Atomic Force Microscope (AFM) that does not require any off-chip scanning or sensing hardware. The first AM-AFM images obtained with such a device reveal that 90nm vertical features (on an AFM calibration standard) can be resolved. The design comprises improved lateral and vertical actuators, an isothermal electrothermal scanner design that maintains constant tip-temperature while traversing a 50um x 10um area, and a Q-enhancement mechanism that improves the force resolution of the instrument.

W-143 SMART-CUT 6H-SILICON CARBIDE (SIC) MICRODISK TORSIONAL RESONATORS WITH SENSITIVE PHOTON RADIATION DETECTION 793
R. Yang[1], K. Ladhane[2], Z. Wang[1], J. Lee[1], D. Young[2], and P.X.-L. Feng[1]
[1]Case Western Reserve University, USA and [2]University of Utah, USA

We report on experimental demonstration of a new type of microdisk torsional resonators based on a smart-cut 6H-silicon carbide (6H-SiC) technology. We carefully calibrate these torsional mode resonances by employing highly sensitive multi-wavelength laser interferometric techniques. To utilize these first 6H-SiC torsional resonators, we further demonstrate sensitive detection of radiations from both blue and infrared (IR) photons. Toward force detection applications which are well suited for torsional resonators, our calibration measurements demonstrate impressive intrinsic force resolutions in these SiC torsional resonators.

Th-144 SUB-0.05° PRECISION OPTOFLUIDIC DUAL-AXIS INCLINOMETER 797
S. Wahl, F. Marty, N. Pavy, B. Mercier, and D.E. Angelescu
Université Paris-Est, FRANCE

We present a low-power bi-axial miniaturized inclinometer based on a mobile mass (spherical ball or fluidic droplet) positioned on a precision curved surface that is generated using a novel MEMS process. The detection of the mobile mass was implemented through an external optical system, using a quadrant photodetector. Nanotopography and chemical treatment of the curved surface have been implemented to increase accuracy when using a fluidic mobile mass, by tailoring wetting properties and minimizing contact angle hysteresis. Fluidic damping was also implemented to render the sensor less sensitive to vibrations.

M-145 TUNING OF NONLINEARITIES AND QUALITY FACTOR IN A MODE-MATCHED GYROSCOPE 801
E. Tatar, T. Mukherjee, and G.K. Fedder
Carnegie Mellon University, USA

This paper examines methods to electrically tune cubic nonlinearity and quality factor (Q) of a mode-matched MEMS gyroscope by changing the DC voltages across specially shaped combs. The gyroscope includes traditional combs for drive-sense and dedicated shaped combs for cubic nonlinearity and frequency tuning. In addition to nonlinearity, Q can be tuned by understanding the nature of the losses with the appropriate model. The electrical loss components are added to the electromechanical resonator model to account for the electrical losses which depend on the applied voltages.

Medical Microsystems (MEDM)

T-146 2D RESONANT MICROSCANNER FOR DUAL AXES CONFOCAL FLUORESCENCE ENDOMICROSCOPE 805

H. Li, Z. Qiu, X. Duan, K. Oldham, K. Kurabayashi, and T.D. Wang
University of Michigan, USA

We demonstrate a parametrically-excited 2D microscanner for a miniature dual axes confocal fluorescence endomicroscope. The scanner has a compact and robust gimbal structure which can perform resonant scanning with large tilting angle at high speed. A single-wafer based SOI process has been developed for improving the quality and the yield of the device. Ex vivo imaging on mouse colon is performed using the fabricated endomicroscope, and the near infrared fluorescence en-face image of dysplasia crypts over a large field-of-view of 800μm×400μm with subcellular resolution is obtained.

W-147 A CAPACITIVE IMMUNOSENSOR USING ON-CHIP ELECTROLYTIC PUMPING AND MAGNETIC WASHING TECHNIQUES FOR POINT-OF-CARE APPLICATIONS 809

J.-C. Kuo, P.-H. Kuo, H.-T. Hsueh, C.-W. Ma, C.-T. Lin, S.-S. Lu, and Y.-J. Yang
National Taiwan University, TAIWAN

This work presents a capacitive immunosensor using on-chip electrolytic pumping and magnetic washing techniques. The proposed device possesses the advantages such as simple operation, low power consumption, and portability. The proposed device was fabricated using typical micromachining process, and is suitable for mass-production. We also demonstrated the detection of N-Terminal pro-brain-Type natriuretic peptide (NT-proBNP) using the fabricated device integrated with a CMOS capacitance sensing chip. The proposed device potentially can be used as a portable system for point-of-care applications.

Th-148 A WIRELESS SLANTED OPTRODE ARRAY WITH INTEGRATED MICRO LEDS FOR OPTOGENETICS 813

K. Kwon[1], H. Lee[2], M. Ghovanloo[2], A. Weber[1], and W. Li[1]
[1]Michigan State University, USA and [2]Georgia Institute of Technology, USA

We develop a wireless-enabled, flexible optrode array with multichannel micro-LEDs for selective optical stimulation of cortical neurons and simultaneous recording of light-evoked neural activity. The array integrates wirelessly addressable micro-LED chips with slanted polymer waveguides for precise light delivery to multiple cortical layers simultaneously. A droplet backside exposure (DBE) method was developed to monolithically fabricate varying-length optrodes on a single polymer platform.

M-149 AN ELECTROPORATION CHIP BASED ON FLEXIBLE MICRONEEDLE ARRAY FOR IN VIVO NUCLEIC ACID DELIVERY 817

Z. Wei[1], R. Wang[2], S. Zheng[3], Z. Liang[3], and Z. Li[3]
[1]National Center for Nanoscience and Technology, CHINA, [2]North University of China, CHINA, and [3]Peking University, CHINA

We reports a flexible microneedle array electroporation chip for in vivo nucleic acid delivery. Silicon MNA is proposed to penetrate the high-resistant stratum corneum, while flexible parylene substrate is used to fit the natural shape of electroporated objects. Using the proposed chip, we successfully achieved plasmid DNA expression and siRNA delivery in living tissue with low voltage (30-40V), neither physical nor biological harm to skin was observed.

T-150 AN INTEGRATED MICROFLUIDIC SYSTEM FOR DIAGNOSIS OF QUINOLONES RESISTANCE OF *HELICOBACTER PYLORI* 821

C.Y. Chao[1], C.H. Wang[1], Y.J. Che[1], C.Y. Kao[2], J.J. Wu[2], and G.B. Lee[1]
[1]National Tsing Hua University (NTHU), TAIWAN and [2]National Cheng Kung University, TAIWAN

Helicobacter pylori play a crucial role in gastric diseases. The incidence rate of duodenal ulcer and gastric ulcer from H. pylori infected patients were found to be about 90-100% and 60-100%.Recently, some point mutations were found in gyrase genes against Quinolones. In this study a new method was therefore developed to perform molecular diagnostic techniques of SNP-PCR on an integrated microfluidic system to detect the Quinolones resistance of H. pylori.

W-151 ANNEALING EFFECTS ON FLEXIBLE MULTI-LAYERED PARYLENE-BASED SENSORS 825

B.J. Kim[1], E.P. Washabaugh IV[2], and E. Meng[1]
[1]University of Southern California, USA and [2]University of Michigan, USA

The mechanical and electrochemical properties and sensing performance of untreated and annealed Parylene-platinum electrochemical impedance-based force sensors were compared. Annealing reduced the height and increased the stiffness of the Parylene structure, and smoothed electrode surfaces, affecting sensor performance. Our results indicate that annealing effects cannot be ignored for Parylene-metal device systems and that mechanical and electrochemical properties and performance must be determined after heat treatment, such as annealing and sterilization.

Th-152 AUTOMATED VITRIFICATION OF MAMMALIAN EMBRYOS ON A DIGITAL MICROFLUIDIC DEVICE 829

D.G. Pyne[1], J. Liu[1], M. Abdelgawad[2], and Y. Sun[1]
[1]University of Toronto, CANADA and [2]Assiut University, EGYPT

We present, for the first time, the development of a digital microfluidic device to achieve automated vitrification of mammalian embryos for clinical in vitro fertilization (IVF) applications. Micro drops are used as vessels to move an embryo and subject it to a series of cyroprotectants of different concentrations, as required by the IVF vitrification protocols.

M-153 CHARACTERIZATION OF RED BLOOD CELL DEFORMABILITY CHANGE DURING BLOOD STORAGE 833

Y. Zheng[1], J. Chen[1], T. Cui[2], N. Shehata[3], C. Wang[3], and Y. Sun[1]
[1]University of Toronto, CANADA, [2]University of Minnesota, USA, and [3]Mount Sinai Hospital, CANADA

Deformability change of stored red blood cells over an 8 weeks' storage period was measured using a microfluidic device and high-speed imaging. Multiple parameters including deformation index (DI), time constant, and RBC circularity were quantified. Compared to previous RBC deformability studies, our results include a significantly higher number of cells (>1,000 cells/sample vs. a few to tens of cells/sample) and, for the first time, reveal deformation changes of stored RBCs when traveling through human-capillary-like microchannels.

T-154 DETERMINATION OF MULTIDRUG RESISTANCE LEVEL IN K562 LEUKEMIA CELLS BY 3D-ELECTRODE CONTACTLESS DIELECTROPHORESIS 837

Y. Demircan, M. Erdem, E. Özgür, U. Gündüz, and H. Külah
Middle East Technical University, TURKEY

We designed, fabricated and tested a MEMS based cell identification 3D-electrode contactless dielectrophoresis system. As an application for this system, the determination of multidrug resistance degree of K562 cells was presented in this study.

W-155 FLEXIBLE MEA FOR ADULT ZEBRAFISH ECG RECORDING COVERING BOTH VENTRICLE AND ATRIUM 841

X. Zhang[1], J. Tai[2], J. Park[1], and Y.C. Tai[1]
[1]California Institute of Technology, USA and [2]Tufts University, USA

We develop a parylene based MEA to monitor adult zebrafish ECG, for the first time, in both ventricle and atrium viewing angles, during its heart regeneration post injury. It is a novel tool to allow the discovery of fine bio-electrical activities in the entire heart.

Th-156 MEASUREMENT OF MECHANOMYOGRAM 845

T. Kaneko, N. Minh-Dung, R. Aoki, T. Takahata, K. Matsumoto, and I. Shimoyama
University of Tokyo, JAPAN

We proposed an approach for measuring mechanomyogram (MMG) by taking advantage of the acoustic impedance matching between liquid and human skin to convey the pressure signal of MMG to a piezo-resistive cantilever.In experiments, the sensor was placed on the skin surface above bicepcs brachii. The MMG signal, the frequency of which was 10-15 Hz, was able to be detected using silicone oil as the propagating medium, while it was not using air as the medium.Experiment results also indicated that the proposed sensor was able to detect the vascular oscillations.

M-157 MEASURING FLOW VELOCITY OF SWALLOWED LIQUID IN THE HUMAN PHARYNX BY TONGUE PRESSURE SENSOR AND SWALLOWING SOUND SENSOR 849

Y. Takei, T. Kaneko, K. Noda, K. Matsumoto, and I. Shimoyama
University of Tokyo, JAPAN

We measured flow velocity of swallowed liquid passing through pharynx. We put pressure sensor on palate and two acoustic sensors on the neck skin. From the output of these three sensors, we can know the timing of the liquid passing through each sensor points and can calculate the flow velocity of the swallowed liquid at the pharynx. In this paper, we compare the flow velocity between two swallowing positions, "sit straight position" and "look upward position." As a result, we found that the flow speed of the "look upward position" was 2.5 times faster than that of "sit straight position."

T-158 MEMS NEURAL PROBE ARRAY FOR MULTIPLE-SITE OPTICAL STIMULATION WITH LOW-LOSS OPTICAL WAVEGUIDE BY USING THICK GLASS CLADDING LAYER 853

Y. Son[1,2], H.J. Lee[1], D. Kim[1], Y.K. Kim[1], E.-S. Yoon[1], J.Y. Kang[1], N. Choi[1], T.G. Kim[2], and I.-J. Cho[1]
[1]Korea Institute of Korea Institute of Science and Technology (KIST), SOUTH KOREA and [2]Korea University, SOUTH KOREA

We present a MEMS neural probe array for multiple-site optical stimulation with low-loss SU-8 optical waveguides. The 20-μm-thick cladding layer was formed by glass reflow process and no additional thickness was required due to embedded structure. Furthermore, the low-loss optical waveguide enables multiple-site stimulation with the two-step optical splitter. We also demonstrate a successful in-vivo optical stimulation and recording of neural signals of a transgenic ChR2-YFP mouse. Recorded neural signals are synchronized with light pulses which confirm that neurons were successfully stimulated and recorded.

W-159 MICRO-ELECTRODE ARRAYS FOR MULTI-CHANNEL MOTOR UNIT EMG RECORDING 857

S. Yamagiwa, H. Sawahata, M. Ishida, and T. Kawano

Toyohashi University of Technology, JAPAN

We report an array of micro-electrodes, which can record motor unit (MU) electromyogram (EMG) signals. As a basic structure of the electrode, we prepared 200-μm-square Si-pyramids with the height of 200 μm by Tetramethylammonium hydroxide (TMAH), resulting in robust MU-EMG recordings without conductive gel. Platinum (Pt) was used as an electrode material and parylene-C was deposited as an insulator. Fabricated μEMG electrodes connected to a recording system clearly detected MU-EMG action potentials from a human forearm. In addition, different MU-EMG signals between μEMG electrodes were detected by crooking fingers. These results indicate that the μEMG array device becomes a powerful tool for medical applications including myoelectric prosthetic technologies.

Th-160 MICRO-WING AND PORE DESIGN IN AN IMPLANTABLE FPC-BASED NEURAL STIMULATION PROBE FOR MINIMALLY INVASIVE SURGERY 861

Y.-H. Wang[1], D. Tsai[1,2], B.-A. Chen[1], Y.-Y. Chen[1], C.-C. Huang[1], P.-C. Huang[1], C.-Y. Lin[1], J. Yu[1], W.-P. Shih[1], C.-W. Lin[1], and H.-J. Sheen[1]

[1]National Taiwan University, TAIWAN and [2]University of California, San Diego, CA

A bipolar porous probe for an implantable nerve stimulation treatment utilizing minimally invasive surgery is presented. The flexible printed circuit (FPC) probe features micro-wings, which can increase fixation after implantation, and contains porous structures for cell growth to promote permanence in the body. Two recording pairs detect whether or not cells grow into the pores, and one pair of stimulating pads stimulates the target nerve. This probe is composed of two SU-8 layers and one FPC layer, to form a 3-D porous structure.

M-161 NANOELECTROPORATION AND CONTROLLABLE INTRACELLULAR DELIVERY INTO LOCALIZED SINGLE CELL WITH HIGH TRANSFECTION AND CELL VIABILITY 865

T.S. Santra, J. Borana, P.-C. Wang and F.-G. Tseng

National Tsing Hua University (NTHU), TAIWAN

Here we demonstrate controllable nano-electroporation platform for HeLa and human Caucasian Gastric Adenocarcinoma (AGS) cell to achieve high efficient bimolecular delivery with high cell viability.

T-162 OPTO-MECHANICAL MICROBRIDLES FOR THE DETERMINATION OF STRUCTURAL AND FUNCTIONAL PROPERTIES OF SMALL RESISTANCE ARTERIES 869

R. Rodríguez-Rodríguez[1], J.A. Plaza[2], V. Matchkov[3], U. Simonsen[3], M.D. Herrera[1], S. Büttgenbach[4], A. Llobera[2], and X. Munoz-Berbel[2]

[1]University of Seville, SPAIN, [2]Centro Nacional de Microelectrónica (CNM), SPAIN, [3]University of Aarhus, DENMARK, and [4]Technische Universität Braunschweig, GERMANY

We develop and optimize an opto-mechanical system for monitoring diameter of arterial segments in vitro.

W-163 PDMS MICROCHANNEL SCAFFOLDS FOR NEURAL INTERFACES WITH THE PERIPHERAL NERVOUS SYSTEM 873

Y. Choi[1], S. Park[2], Y. Chung[3], R.K. Gore[4], A.W. English[4], and R.V. Bellamkonda[2]

[1]University of Texas – Pan American, USA, [2]Georgia Institute of Technology, USA, [3]University of Rhode Island, USA, and [4]Emory University, USA

Neural interfaces with the peripheral nervous system have been developed to provide a direct communication pathway between peripheral nerves and prosthetic limbs. This paper reports a regenerated peripheral nervous system which can control the reinnervated muscles and interpret neurological signals. The acquired bioelectrical signals can be used for the interpretation of mind which will be used to monitor prosthetic limbs. Transected nerves were regenerated through PDMS scaffolds and transferred signals through embedded microwires and acquisition systems.

Th-164 SELECTIVE RF HEATING OF RESONANT STENT TOWARD WIRELESS ENDOHYPERTHERMIA FOR RESTENOSIS INHIBITION 877

Y. Luo, M. Dahmardeh, X. Chen, and K. Takahata

University of British Columbia, CANADA

Stents have served as a critical device for minimally invasive treatment of cardiovascular disease, the leading cause of death in North America. Artery renarrowing (known as restenosis) often occurs after stent implantation due to excess growth of vessel tissue, blood clot(thrombus) formation, and/or other factors. This paper presents, for the first time, a novel active stent that serves as a resonant heater with high frequency selectivity controlled using external RF fields, offering a new therapeutic path to wireless endohyperthermia for in-stent restenosis.

M-165 TUNABLE MEMS FIBER SCANNER FOR CONFOCAL MICROSCOPY 881
N. Weber, T. Meinert, H. Zappe, and A. Seifert
University of Freiburg – IMTEK, GERMANY

We present an endoscopic probe with forward-looking piezoelectric fiber scanner for confocal optical imaging with a very short length of 13.1 mm. The system is based on Si bench technology with integrated fluidics for realizing tunable liquid-filled membrane-lenses. The tunability in focal length allows confocal depth scanning up to 100 µm without any movable optics or stages. The lateral and axial resolution were demonstrated to be 2 µm and 20 µm, respectively.

T-166 ULTRASOUND-ASSISTED MICRO-KNIFE FOR CELLULAR SCALE SURGERY 885
H. Jeong[1], T. Li[2], Y. Gianchandani[2], and J. Park[1]
[1]*Pohang University of Science and Technology (POSTECH), SOUTH KOREA and*
[2]*University of Michigan, USA*

We developed a microknife for cellular scalce surgery.The work includes modeling, fabrication, measurement and actual cell-cutting. The result showed that developed knife can cut cell monolayer successfully with 2 micro cut line width by utilizing the ultra-sharp edge and ultrasound assiat.

**W-167 WRINKLE CELLOMICS: SCREENING BLADDER CANCER
CELLS USING AN ULTRA-THIN SILICONE MEMBRANE** 889
J.H. Appel[1], L.Y. Sin[2], J.C. Liao[2], and J. Chae[1]
[1]*Arizona State University, USA and* [2]*Stanford University, USA*

We report a visualization platform comprised of an ultra-thin silicone membrane to differentiate between the biophysical properties of cancerous and healthy cells. Cancerous cells adhere to and spread on the membrane inducing deformation, termed 'membrane wrinkling', while healthy cells do not generate wrinkle patterns on the membrane. Quantitative measurement of wrinkling represents a powerful, non-invasive diagnostic tool for common cancers such as bladder cancer.

**Th-168 AN X-RAY DETECTABLE PRESSURE MICROSENSOR FOR
MONITORING CORONARY IN-STENT RESTENOSIS** 893
M.N. Gulari, M. Ghannad-Rezaie, P. Novelli, N. Chronis, and T.C. Marentis
University of Michigan, USA

We present a novel implantable X-ray-addressable MEMS Blood Pressure sensor, the X-BP, for the non-invasive and cost-effective surveillance of coronary in-stent restenosis. We successfully fabricated and tested the X-BP sensor and its pressure response curve. We placed the X-BP sensor in a coronary stent and prove adequate visibility in a clinically realistic scenario.

Micro-Actuators (ACT)

**M-169 A NOVEL ACTUATOR FOR ENERGY HARVESTING USING
AN ACOUSTICALLY OSCILLATING LIQUID DROPLET** 897
Y.R. Lee, J.H. Shin, I.S. Park, and S.K. Chung
Myongji University, SOUTH KOREA

This paper presents a novel actuator for energy harvesting from ambient acoustic noise using acoustically oscillating droplets. When a water droplet sitting on a piezocantilever is excited by an acoustic wave around its natural frequency, it oscillates and simultaneously bends the piezocantilever by the reaction of the droplet oscillation, resulting in electric power generation from the piezocantilever. The envisioned energy harvesting system can extract mechanical power from acoustic noise in a wide range of frequencies using liquid droplets in different sizes and natural frequencies and convert the mechanical power to electrical power for wireless electronic devices. This new type of actuation technique is a simple but useful tool not only for the energy harvesting system but also potential acoustic wave sensors and actuators.

**T-170 A NOVEL ON-CHIP MICROMANIPULATION METHOD USING A MICROBUBBLE
FOR SINGLE CELL MANIPULATION AND CHARACTERIZATION** 901
J.H. Shin, Y.R. Lee, I.S. Park, and S.K. Chung
Myongji University, SOUTH KOREA

This paper presents a novel on-chip micromanipulation method using a microbubble actuated by optical and acoustical excitation for single cell manipulation and characterization in a microfluidic chip, along with the experimental verification of bubble manipulation (generating and transporting operations) and micro-object manipulation (capturing, carrying, and releasing operations).

W-171 A THERMOTROPIC LIQUID CRYSTAL ELASTOMER MICRO-ACTUATOR WITH INTEGRATED DEFORMABLE MICRO-HEATER 905

S. Petsch[1], R. Rix[2], P. Reith[1], B. Khatri[1], S. Schuhladen[1], D. Ruh[1], R. Zentel[2], and H. Zappe[1]
[1]University of Freiburg – IMTEK, GERMANY and [2]University of Mainz, GERMANY

We present a large-stroke thermal actuator with an integrated, MEMS fabricated, deformable heater based on the phase transition in a thermotropic liquid crystal elastomer (LCE) material. The transition from nematic to isotropic phase in the LCE causes a contraction of 28% (1.15mm) when the integrated system is heated to 120°C. With the heater buried in the LCE, full contraction is reached after 19.7s at 320mW when heated from room temperature. Complete back actuation is achieved in 5.6s.

Th-172 A TUNABLE LIQUID LENS DRIVEN BY A CONCENTRIC ANNULAR ELECTROACTIVE ACTUATOR 909

K. Wei, N. Domicone, and Y. Zhao
Ohio State University, USA

We present a membrane-enveloped fluidic lens hydrostatically coupled to a concentric annular electroactive elastomer. Electrical activation deforms the annular elastomer, which induces fluid transmission between the lens part and the actuation part. The lens changes the shape and thereby the focal length from 12.5 mm to 105.2 mm within 1.0 kV. Compared to existing fluidic lenses driven by electroactive polymer, this lens implements a larger focusing range at a lower voltage. It finds applications in miniaturized optical components where adaptive focalization is at a premium.

M-173 LONG STROKE OUT-OF-PLANE ACTUATOR USING COMBINATION OF ELECTROSTATIC AND PNEUMATIC FORCES 913

T.K. Kan, A.I. Isozaki, H.T. Takahashi, K.M. Matsumoto, and I.S. Shimoyama
University of Tokyo, JAPAN

We propose an out-of-plane MEMS actuator with a large stroke length comparable to the size of the actuator itself. The proposed device is actuated by the combined forces of the electrostatic and the pneumatic forces. This combination of two independent forces enlarges a stable area during the actuation, resulting in a large stroke which is difficult to be achieved with only either force. The 3D profiling by the laser scanning microscopy confirmed the largest stroke of 103 μm was obtained with the 150-μm-diameter actuator area with the electric field of 5.8×10^5 V/m (330V between the electrodes) and the air pressure of 2.0 kPa.

T-174 MECHANO-ACTIVE TISSUE SCAFFOLD SYSTEM BASED ON A MAGNETIC NANOPARTICLE EMBEDDED NANOFIBROUS MEMBRANE 917

S.P. Fang[1], H. Shang[1], P.F. Jao[1], K.T. Kim[1], G.J. Kim[1], J.H. Yoon[1], K. Cho[1,2], A.J. Katz,[1] and Y.K. Yoon[1]
University of Florida, USA and Korea Basic Science Institute, SOUTH KOREA

A mechano-active nanofibrous scaffold system consisting of iron oxide nanoparticle embedded electrospun nanofibers, a membrane holder and an electro magnet, is designed and demonstrated. The scaffold provides mechanical stress on culturing cells by external AC magnetic fields. The mechanical properties of the nanoporous membrane including the density, the porosity, and the effective Young's modulus are characterized. Cell viability with and without magnetic nanoparticle embedded has been tested.

W-175 NANO-SCALE BIOMECHANICAL ANALYZER FOR STUDYING STIMULUS DEPENDENT SELF-ASSEMBLY OF ACTIN FILAMENT 921

N. Shimada[1], M. Ikeuchi[1,2], and K. Ikuta[1]
[1]University of Tokyo, JAPAN and [2]Japan Science and Technology Agency, JAPAN

We have developed nano-scale mechanical analysis system by using "optically driven nano-beam" to measure elasticity of self-assembled actin filament under dynamic mechanical stimulus. In this report, we worked on developing a new nano-beam to specifically capture actin on its surface. By using the new nano-beam, we have successfully measured elasticity of self-assembled actin filament in water. The nano-mechanical analysis system unravels cell life phenomenon which can't be dealt with through conventional methodologies."

Th-176 PINCHING AND RELEASING OF CELLULAR AGGREGATE BY MICROFINGERS USING PDMS PNEUMATIC BALLOON ACTUATORS 925

S. Shimomura[1], Y. Teramachi[1], Y. Muramatsu[1], S. Tajima[2], Y. Tabata[2], and S. Konishi[1]
[1]Ritsumeikan University, JAPAN and [2]Kyoto University, JAPAN

This paper proposes microfingers for manipulation of spherical cellular aggregates (φ200μm). The microfinger is driven by pneumatic balloon actuator "PBA" to pinch and release a spherical cellular aggregate directly. The paper presents the design, operation principle, fabrication, and characterization. The pinching force of developed fingers was estimated with the aim of evaluating the damage to the cellular aggregate. A series of operation of a real cellular aggregate by developed microfingers will be successfully demonstrated.

M-177 PNEUMATICALLY ACTUATED BIOMIMETIC PARTICLE TRANSPORTER 927

A. Rockenbach[1], C. Brücker[2], and U. Schnakenberg[1]

[1]*RWTH Aachen University, GERMANY and* [2]*Technical University Bergakademie Freiberg, GERMANY*

To prevent the adhesion of particles at surfaces by transporting them along the surface this paper reports on a pneumatically actuated new type of biomimetic particle transporter. Rows of flaps are positioned asymmetrically on movable membranes. Each flap row can be deflected separately by an induced pneumatic force. This membrane movement converts to a large deflection of the flaps in x-direction (lateral). Due to the high aspect ratio of the flaps the angle rotation results in a fluid movement parallel to the surface which prevents the particle deposition.

T-178 QUANTITATIVE ANALYSIS OF SURFACE TEXTURES CREATED BY MEMS TACTILE DISPLAY USING MICROFABRICATED TACTILE SAMPLES 931

Y. Kosemura[1], S. Hasegawa[1], H. Ishikawa[1], J. Watanabe[1], and N. Miki[1,2]

[1]*Keio University, JAPAN and* [2]*Japan Science and Technology Agency (JST), JAPAN*

This paper discusses characterization of the surface textures created by MEMS tactile display with a large displacement MEMS actuator array. The actuator consists of piezoelectric actuators and a hydraulic displacement amplification mechanism to achieve large enough displacement to stimulate human tactile receptors. In our prior work, we successfully displayed smooth and rough surfaces using MEMS tactile display by controlling the vibration frequency and the driving voltage of the actuators. In this paper, we propose "sample comparison method" to further characterize the virtually created surface textures, where microfabricated tactile samples are used. In this method, by requesting the subjects to select samples that they felt most similar to the displayed surfaces, the control parameters of the MEMS tactile display were successfully correlated with the surface properties of the samples.

Micro-Fluidic Components and Systems (μFLUIDIC)

W-179 A NOVEL CONSTANT FLOW REGULATION PRINCIPLE FOR COMPACT BREATH DIAGNOSTICS 935

S.B. Johansson, G. Stemme, and N. Roxhed

KTH Royal Institute of Technology, SWEDEN

Our work reports on a passive compact flow regulator designed to maintain a steady flow during breath diagnostics using a novel flow regulation principle. The fabricated prototype consists of two 3D-printed plastic parts with an integrated cantilever aligned in the direction of the flow, to control comparatively large air flows in the 50 ml/s regime suitable for asthma diagnostics.

Th-180 APPARENT SIZE CORRELATION: A SIMPLE METHOD TO DETERMINE VERTICAL POSITIONS OF PARTICLES USING CONVENTIONAL MICROSCOPY 939

M.H. Winer, A. Ahmadi, and K.C. Cheung

University of British Columbia, CANADA

We have developed and implemented a simple three-dimensional (3D) particle tracking method for use in particle focusing applications. Using conventional fluorescence microscopy and a multi-step image post-processing algorithm based on particle defocusing principles, this technique was experimentally verified with results comparable to theoretical predictions of (1) gravitational settling and (2) inertial focusing. Our technique determines particle positions to micron accuracy in microfluidic systems for Re < 100.

M-181 CELL-NICHE-ON-CHIP: PAIRED SINGLE CELL CO-CULTURE PLATFORMS USING IMMISCIBLE LIQUID ISOLATION AND SEMI-PERMEABLE MEMBRANES 943

Y.-C. Chen, Y. Cheng, and E. Yoon

University of Michigan, USA

The fundamental difficulty in single cell co-culture is to provide a controlled microenvironment. The cell culture chamber must be isolated for secreted cytokines to be accumulated inside the chamber over time. However, in an isolated environment nutrition factors will deplete. It is important to find a way to continuously supply nutrition factors while isolating cells sectional view of the proposed co-culture chip and its operation. We placed a semi-permeable membrane between the cell culture chamber and the media exchange channel. Nutrition can be supplied to the cells through the membrane, but the secreted cytokines are accumulated inside the chamber because their molecule sizes are too large to escape. The preliminary result demonstrated the capability of studying interaction between two cells and its potential to investigate modeling of more complicated cell niches.

T-182 DEVELOPMENT OF VACUUM ASSISTED MICROFLUIDIC CELL TRAPPING DEVICE FOR REPOSITIONING OF OOCYTE INTRACELLULAR CHROMOSOMES 947

J. Hong, P. Purwar, S. Lee, N. Verma, and J. Lee

Seoul National University, SOUTH KOREA

We report the design and fabrication of vacuum assisted microfluidic trapping device for the capture of single cell such as an oocyte. We also suggest an application of such device for an intracellular monitoring of a cell that has an interaction with external environments. Real time monitoring is enabled through the fabrication on a silicon-on-glass substrate, offering excellent optical imaging window. We demonstrate the single cell capture event and monitoring of chromosome activity. This result will provide a powerful tool for investigating the physiological and pathological cellular functions.

W-183 DIELECTROPHORETIC (DEP) SEPARATION OF LIVE/DEAD CELLS ON A GLASS SLIDE
FUNCTIONALIZED WITH INTERDIGITATED 3D SILICON RING MICROELECTRODES 951
X. Xing and L. Yobas
Hong Kong University of Science and Technology, HONG KONG

An elegant device with 3D interdigitated silicon ring electrodes is developed here for DEP-activated cell sorting. The integration of transparent glass substrate makes the device cost-effective and aids the coupling of DIC microscopy. The self-aligned lateral rings form multiple flow lines thus enhancing the throughput of the whole device. A capture efficiency of live mammalian cancer cells approaching 100% is achieved in separating them from a dead group with high flow rate.

Th-184 DROPLET DISPENSING AND SPLITTING BY ELECTROWETTING
ON DIELECTRIC DIGITAL MICROFLUIDICS 955
N.Y.J.B. Nikapitiya[1], S.M. You[2], and H. Moon[1]
[1]University of Texas at Arlington, USA and [2]University of Texas at Dallas, USA

This paper reports an experimental study of two essential capabilities of electrowetting-on-dielectric (EWOD) digital microfluidics (DMF) – 1) high precision and consistency in volume of unit nanodrop dispensed from a reservoir, and 2) reduction of time to dispense and split drops. These capabilities are sought in applications that need tiny but accurate volume of liquid delivery at high flow rate.

M-185 ENZYME-DOPED POLYESTER THREAD COATED WITH PVC MEMBRANE FOR ON-SITE
UREA AND GLUCOSE DETECTION ON A THREAD-BASED MICROFLUIDIC SYSTEM 959
Y.-A. Yang[1], W.-C. Kuo[2], and C.-H. Lin[1]
[1]National Sun Yat-sen University, TAIWAN and
[2]National Kaohsiung First University of Science and Technology, TAIWAN

This study presents a novel enzyme-doped thread with PVC membrane coating for on-site urea and glucose detection on a thread-based microfluidic device. The enzyme can be directly applied on the thread without delicate pretreatment or surface modification process. The passing biomolecules are digested by the enzymes and then electrochemically detected downstream. With this approach, CE-EC detection with on-site bio-reaction can be simply achieved. A thin layer of PVC membrane is coated on the enzyme-doped thread to further fix the applied enzyme and to prevent from the rapid evaporation of the running buffer due to the Joule heating effect. In addition, the PVC coated thread can be operated at a higher separation electric field of 500 V/cm due to the reducing buffer evaporation. Successfully on-site enzyme digestion, CE separation and EC detection of urea and glucose samples in a single test run is demonstrated with the enzyme-doped microfluidic system. Results also indicate that the developed system exhibits good linear dynamic range for detecting urea and glucose sample in concentrations from 0.1 mM – 10.0 mM (R2=0.9850) and 0.1 mM – 13.0 mM (R2=0.9668), which is suitable for adoption in detecting the BUN concentration in serum (1.78~7.12 mM) and the standard glucose fasting measuring range (3.89~6.11 mM).

T-186 FABRICATION OF HIGH ASPECT RATIO INSULATING NOZZLE ARRAY
USING GLASS REFLOW PROCESS AND ITS ELECTROHYDRODYNAMIC
PRINTING CHARACTERISTICS 963
K.I. Lee[1], B. Lim[1], S.W. Oh[1], S.H. Kim[1], C.S. Lee[1], J.W. Cho[1], and Y. Hong[2]
[1]Korea Electronics Technology Institute, SOUTH KOREA and
[2]Seoul National University, SOUTH KOREA

We develop micromachining process including glass filling to fabricate high aspect ratio glass nozzles which is more appropriate for electrohydrodynamic inkjet printing head. With this nozzle array, we print very narrow lines of various materials which is not obtained by conventional piezoelectric or thermal inkjet printing systems.

W-187 GALLIUM-BASED LIQUID METAL INKJET PRINTING 967
D. Kim[1], J.H. Yoo[1], Y. Lee[1], W. Choi[1], K. Yoo[2], and J.B. Lee[1]
[1]University of Texas at Dallas, USA and [2]Hanbat National University, SOUTH KOREA

We report clog-free and oxide-free metal inkjet printing using gallium-based liquid metal. Unlike typical metal nanoparticles or metal alloys, gallium-based liquid metal alloys are in liquid-phase at room temperature. Therefore, there is no need for heating or dispersing in solvent for inkjet printing. Another distinctive benefit is it maintains liquid-phase after printing if the substrate stays at around room temperatures. This is extremely useful to create 3D freeform rapid prototyping of metallic patterns that can conform to virtually any dynamic deformation of substrates.

Th-188 IN-PLANE CAPACITIVE MEMS FLOW SENSOR FOR LOW-COST
METERING OF FLOW VELOCITY IN NATURAL GAS PIPELINES 971
S.D. Nguyen[1], I. Paprotny[2], P.K. Wright[1], and R.M. White[1]
[1]University of California, Berkeley, USA and [2]University of Illinois, Chicago, USA

This paper presents the design, fabrication, and experimental results of an in-plane capacitive MEMS flow sensor that uses the displacement of a micro-fabricated paddle caused by dynamic pressure for measuring the velocity of the flow of surrounding gas. Simplicity of fabrication, combined with insensitivity to variations in ambient temperature makes this sensor ideal for widespread deployment in natural gas pipelines.

M-189 INTEGRATED MULTI-PARAMETER FLOW MEASUREMENT SYSTEM 975

J.C. Lotters[1], E. van der Wouden[1], J. Groenesteijn[2], W. Sparreboom[1], T.S. Lammerink[2], and R.J. Wiegerink[2]
[1]*Bronkhorst High-Tech BV, THE NETHERLANDS and*
[2]*MESA+ - University of Twente, THE NETHERLANDS*

We have designed and realised an integrated multi-parameter flow measurement system, consisting of an integrated Coriolis and thermal flow sensor and a pressure sensor. The integrated system enables on-chip measurement, analysis and determination of flow and several physical properties of both gases and liquids. With the system, we demonstrated the feasibility to measure the flow rate, density, viscosity and heat capacity of hydrogen, helium, nitrogen, air, argon and water.

**T-190 INTERACTION FORCES DURING THE SLIDING OF A
WATER DROPLET ON A TEXTURED SURFACE** 979

N. Thanh-Vinh, H. Takahashi, K. Matsumoto, and I. Shimoyama
University of Tokyo, JAPAN

Using a MEMS 2-axis force sensor array, we have directly measured the pressure and shear force during the sliding of a water droplet on a Su-10 micropillar array. The measurement results showed a fluctuation in the interaction forces when the micropillar was close to the trailing edge or leading edge of the droplet. Meanwhile, in the inner region of the contact line, both the normal and lateral interaction forces were relatively stable. These results indicate that the interaction forces at the edges of the droplet are important factors controlling the sliding motion of the droplet.

**W-191 LIQUID DROPLET MICRO-BEARINGS ON
DIRECTIONAL CIRCULAR SURFACE RATCHETS** 983

C. Varel and K.F. Bohringer
University of Washington, USA

This paper presents de-ionized water droplets used as torque-generating micro-bearings between a glass plate and a micromachined Si substrate. The pattern on the micromachined Si substrate includes circular tracks, which allow droplet motion in a single direction. When vertical vibration is applied to the system, a rotation in the transverse plane is triggered. The system can be tailored to respond to a specific vibration frequency, from 36.5 to 83 Hz by droplet volumes from 13 to 1 μL.

**Th-192 LOW GAS PERMEABLE AND NON-ABSORBENT RUBBERY
OSTE+ FOR PNEUMATIC MICROVALVES** 987

J. Hansson, J.M. Karlsson, C.F. Carlborg, W. van der Wijngaart, and T. Haraldsson
KTH Royal Institute of Technology, SWEDEN

We present an elastomeric, low gas permeable off-stoichiometric thiol–ene-epoxy (OSTE+) polymer fully compatible with standard micro-molding manufacturing and demonstrate its use in pneumatic pinch microvalves for lab-on-chip. The polymer is shown to have rubbery properties (similar to PDMS), low permeability to gases, low absorption of molecules from liquid samples, and the ability to bond layers in room temperature without the need for adhesives or plasma treatment.

**M-193 FREE SURFACE PROPULSION BY
ELECTROWETTING-ASSISTED 'CHEERIOS EFFECT'** 991

J. Yuan and S.K. Cho
University of Pittsburgh, USA

We combine electrowetting principle with Cheerios effect in order to manipulate floating objects in centimeter and millimeter scales. By turning electrowetting on/off, we attract or repel floating objects. Using an array of electrowetting electrodes, we generates translationally and rotationally continuous motions on floating objects.

**T-194 MESOPOROUS-SILICA NANO-CHANNELS INTEGRATED IN MICRO-FLUIDIC
CHIP FOR FAST LIQUID MICRO-EXTRACTION OF PESTICIDE RESIDUAL** 995

P.C. Xu, C.Z. Chen, H. Yu, and X.X. Li
Shanghai Institute of Microsystem and Information Technology (SIMIT), CHINA

The paper reports a micro-chip with nano-channels integrated as extraction-reservoir for quickly extracting analyt from aqueous solution to water-soluble organic solvent. Using this novel technology, trace-level residual of organophosphorus pesticide in water-solution can be micro-extracted to a common organic-solvent (e.g. ethanol) and, thereafter, quantitatively detected by GC-MS (Gas Chromatography-Mass Spectrometry) analysis.

W-195 MINIATURISED PRANDTL TUBE WITH INTEGRATED PRESSURE
SENSORS FOR MICRO-THRUSTER PLUME CHARACTERISATION 999
M. Dijkstram, K. Ma, M.J. de Boer, J. Groenesteijn, J.C. Lötters, and R.J. Wiegerink
MESA+ - University of Twente, THE NETHERLANDS

Micro chemical propulsion systems (µCPS) have been identified by ESA as emerging compact propulsion system. Within the PRECISE project a MEMS-based monopropellant propulsion system applying catalytic decomposition of hydrazine is being developed. Investigation of the micro-thruster rarefied plume flow as well as direct simulation Monte Carlo (DSMC) validation is of great importance for nozzle design and performance evaluation. A novel 6 mm long 40 µm diameter micro Pitot tube with integrated pressure sensors for characterisation of rarefied plume flow during hot-firing test has therefor been developed.

Th-196 MICROFLUIDIC-BASED DROPLET MERGING DEVICE WITH
A NON-CONTACT DROPLET PAIRING METHOD 1003
S. Lee, H. Kim, and J. Kim
Pohang University of Science and Technology (POSTECH), SOUTH KOREA

We developed a novel droplet merging method based on the deformability characteristic of a droplet in pressure-driven shear flow using only fluid flow control by a unique Laplace trap that performs a multi-step 'trapping–releasing–non-contact pairing–washing–and–merging' process. Using the unique Laplace trap array, parallel merging was successfully performed within a short time variation in non-contact pairing (SD ±4.3 s) compared with conventional contact pairing (SD ±136.4 s).

M-197 MINIATURE CIRCULATORY COLUMN SYSTEM FOR GAS CHROMATOGRAPHY 1007
H.-C. Hsieh and H. Kim
University of Utah, USA

We develop the first micro-scale circulatory column system for functioning gas chromatography and the resultant highest separation capacity demonstrated by any commercial and non-commercial GC column systems beyond the current state-of-art, by enabling the extension of the effective column length through the circulatory loop without increasing the device volume.

T-198 MIRRORED ANODIZED DIELECTRICS FOR RELIABLE ELECTROWETTING 1011
S. Chen and C. Kim
University of California, Los Angeles, USA

Anodized metal oxides are an attractive dielectric material for electrowetting-on-dielectric (EWOD) devices because of their ability to limit current leakage, high dielectric constants and low cost fabrication. However, the reliability is for only one actuation polarity because of their rectifying effect. To overcome this limitation, we developed parallel-plate EWOD devices using anodized aluminum on both plates so that one is always under the correct bias to limit the leakage current. Lifetime and current leakage were tested across a range of actuation biases.

W-199 NANOPARTICLES SORTING AND ASSEMBLY BASED ON
DOUBLE-AXICON IN AN OPTOFLUIDIC CHIP 1015
Y.Z. Shi[1,2], S. Xiong[2], L.K. Chin[2], M. Ren[2], and A.Q. Liu[1]
[1]Xi'an Jiao Tong University, CHINA and [2]Nanyang Technological University, SINGAPORE

We present a novel optofluidic system of sorting and assembly of nanoparticles by tunable interference patterns generated from injecting a Gaussian beam through a double-axicon. The tightly confined (several micrometers) Bessel beam is used to sort the 100-nm gold, 200-nm and 500-nm polystyrene nanoparticles massively and simultaneously by controlling the flow rate and the laser power (from 300 mW to 500 mW). In addition, the 500-nm polystyrene particles are assembled into a 2D array by the discrete interference pattern.

Th-200 NANOSLIT MEMBRANE INTEGRATED FLUIDIC CHIP FOR
MICRO/NANO PARTICLE TRAPPING AND SEPARATION 1019
Y. Koh[1], H.M. Kang[1], J.H. Kim[2], Y.S. Lee[1], and Y.K. Kim[1]
[1]Seoul National University, SOUTH KOREA and [2]Hanyang University, SOUTH KOREA

We propose a nanoslit fluidic chip that has a large number of nanoslit array membrane (Nanoslit-Chip) for trapping and concentrating particles of a desired size. The proposed Nanoslit-Chip has several benefits such as low flow resistance and little non-specific nanoparticle clogging for the separation of nanoparticles.

M-201 ON-CHIP CONTROL OF PNEUMATIC-BASED BISTABLE VALVE SWITCH 1023
A. Chen and T. Pan
University of California, Davis, USA

We present pneumatic-based, bistable valve (BSV) switches for immediate on-chip fluid-flow manipulation without the requirement of external microcontroller circuitries. The applicability of the on-chip controller is demonstrated in a 4-to-1 microfluidic multiplexor and its clinical relevance is further supported in a point-of-care ABO blood-typing diagnostic chip.

T-202 RAPID MICROFLUIDIC PROTOTYPING OF SOPHISTICATED PROTEIN
ANALYSIS PLATFORMS USING GRAYSCALE PHOTOPATTERNING 1027
T.A. Duncombe[1], K. Maurer[2], and A.E. Herr[1]
[1]University of California, Berkeley/University of California, San Francisco Joint Graduate Group in
Bioengineering, USA and [2]ETH Zurich, USA

We introduce, characterize, and demonstrate a novel grayscale fabrication technique for rapid prototyping of complex spatially varied hydrogels as lab-on-a-chip devices optimized to address important protein measurement questions. Our technique utilizes hydrogel photopatterning via grayscale masks to define non-uniform pore-size distributions from a single UV exposure and precursor solution. Using this method we realize two workhorse analytical electrophoresis platforms: (1) a 24-plex electrophoresis screening assay and (2) a 96-plex gradient gel-based protein sizing assay.

W-203 REALIZATION OF 240 NANOMETER RESOLUTION OF CELL
POSITIONING BY A VIRTUAL FLOW REDUCTION MECHANISM 1031
S. Sakuma[1], K. Kuroda[1], M. Kaneko[1], and F. Arai[2]
[1]Osaka University, JAPAN and [2]Nagoya University, JAPAN

For cell manipulation in a microchannel, it is often required to control the cell as accurate as possible. However, the issue for using a syringe pump is that the flow rate is geometrically amplified in microchannel. Therefore, we propose a virtual flow reduction mechanism in this paper. By using elastic feature of the PDMS chip, we designed and developed the total system for cell manipulation. Through experiments, we confirmed that the cell positioning resolution is 240 nm with the frequency up to 20 Hz.

Th-204 SINGLE CELL SEPARATION BY USING ACCESSIBLE MICROFLUIDIC CHIP 1035
T. Hayakawa[1], T. Fukuhara[2], K. Ito[1], and F. Arai[1]
[1]Nagoya University, JAPAN and [2]Tokyo University of Pharmacy and Life Science, JAPAN

In this paper, we proposed novel single cell separation method that used accessible microfluidic channel. Various single cell separation method by using microfluidic chip have been proposed. However, the biggest problem is that those microfluidic chip are closed and separated cells are tend to missed at interface of the chip to outer world. Therefore, we used cover opened microfluidic chip that can be accessible in order to collect the separated cell. And also, we proposed single cell pick-up tool to collect the separated single cell.

M-205 STUDY OF HOTSPOT COOLING USING ELECTROWETTING
ON DIELECTRIC DIGITAL MICROFLUIDIC SYSTEM 1039
G.S. Bindiganavale[1], S.M. You[2], and H. Moon[1]
[1]University of Texas at Arlington, USA and [2]University of Texas at Dallas, USA

This paper presents a novel digital microfluidic (DMF) cooling system using electrowetting on dielectric (EWOD) developed for demonstrating and studying hotspot cooling for applications in electronics thermal management. The merits of this cooling system lies in the fact that no mechanically moving parts such as valves, pumps and fans are required to achieve hotspot cooling, thus having smaller form factor than bulky heatpipes and other conventional cooling systems. This study reveals close profiles of temperature change during coolant drop motion over hotspot as well as importance of phase change in the proposed cooling system.

T-206 SURFACE-ACOUSTIC-WAVE DRIVEN POINT SOURCE ATOMIZER INTEGRATED
WITH PICOLITER MICRO PUMPS FOR POLYMERIC NANOPARTICLES SYNTHESIS 1043
S. Sugimoto, M. Hara, H. Oguchi, A. Yabe, and H. Kuwano
Tohoku University, JAPAN

We developed a surface acoustic wave (SAW) driven atomizer integrated with picoliter micro pumps for polymeric nanoparticles synthesis. The pumps consisted of the reservoir and a pair of interdigital transducer (IDT). The atomizer also consisted of arc-shaped IDT for focusing the SAW energy into the liquid. As an experimental result using water, when applying the burst signal to the IDT, discharge in which the rate was 0.3 pl a burst could be observed. Moreover, we succeeded in ejection of narrow mist spray from the atomizer.

W-207 TEFLON WETTING AND DEWETTING ON EWOD DEVICE
FOR CHEMILUMINESCENCE DETECTOR 1047
X.Y. Zeng[1], K.D. Zhang[1], G.W. Tao[1], S.K. Fan[2], and J. Zhou[1]
[1]Fudan University, CHINA, and [2]National Taiwan University, TAIWAN

We develop a hydrophobicity recoverable EWOD (electrowetting-on-dielectric) based chemiluminescence detector with an integrated signal and heater electrode. A series of experiments and X-ray photoelectron spectroscopic analysis are used to reveal the wetting and dewetting mechanism of Teflon in the EWOD device, and get the recovery relationships between the recovered contact angle, the recovery threshold time and heating temperature.

Th-208 TRANSIENT INERTIAL FLOWS: A NEW DEGREE OF FREEDOM
FOR PARTICLE FOCUSING IN MICROFLUIDIC CHANNELS 1051
M.H. Winer, A. Ahmadi, and K.C. Cheung
University of British Columbia, CANADA

We have investigated the unique effect of transient flow rate on inertial particle focusing in microfluidic systems. A comparative analysis was conducted using both constant and transient flow rates on polystyrene (PS) beads in various channel geometries. Results show that particle focusing equilibrium positions are affected by the use of a transient (changing) flow rate. Transient inertial flows provide a new degree of freedom for manipulation of particle positioning in microfluidic channels.

Nano-Electro-Mechanical Devices and Systems (NANO)

M-209 BIAXIAL STRAIN IN SUSPENDED GRAPHENE
MEMBRANES FOR PIEZORESISTIVE SENSING 1055
A.D. Smith[1], F. Niklaus[1], S. Vaziri[1], A.C. Fischer[1], M. Sterner[1],
F. Forsberg[1], S. Schröder[1], M. Östling[1], and M.C. Lemme[2]
[1]KTH Royal Institute of Technology, SWEDEN and [2]University of Siegen, GERMNAY

This work compares through both theory and experiment the effect of cavity shape and size on the sensitivity of piezoresistive pressure sensors based on suspended graphene membranes. Further, the paper analyzes the effect of both biaxial and uniaxial strain on the membranes.

T-210 FABRICATION OF GOLD NANOPARTICLE-EMBEDDED NANOCHANNELS
FOR SURFACE-ENHANCED RAMAN SPECTROSCOPY 1059
K. Suekuni, T. Takeshita, K. Sugano, and Y. Isono
Kobe University, JAPAN

A micro/nanofluidic device including linearly-arranged gold nanoparticles embedded into nanochannels was developed for highly-sensitive Surface-Enhanced Raman Spectroscopy (SERS) analysis. The nanochannels array was fabricated by a "photo" lithography-based process without costly and time-consuming process such as EB lithography. Then particles with diameters of 100 nm are arranged into the nanochannels by a nanotrench-guided self-assembly process. The device was successfully fabricated and it was active for SERS analysis with 4,4'-bypiridine as a target molecule."

W-211 FINFET WITH FULLY PH-RESPONSIVE HFO_2 AS
HIGHLY STABLE BIOCHEMICAL SENSOR 1063
S. Rigante[1], M. Wipf[2], A. Bazigos[1], K. Bedener[3], D. Bouvet[1], and A.M. Ionescu[1]
*[1]Ecole Polytechnique Federale de Lausanne (EPFL), SWITZERLAND, [2]University of Basel, SWITZERLAND, and
[3]Paul Scherrer Institute, SWITZERLAND*

We present a sensing platform based on high-stability low-power n-channel fully depleted FinFETs on Si-bulk. Efficient chemical and biological label-free sensing has been demonstrated, paving the way towards non-invasive simultaneous monitoring of human physiological signals such as pH and proteins. In contrast to other SiNW-based sensors, the use of scalable high-k dielectric FinFETs for both applications is in accordance with the material constraints which come along Moore's Law of scaling.

Th-212 FREQUENCY DEPENDENT AC ELECTROOSMOTIC FLOW IN NANOCHANNELS 1067
W.T.E. van den Beld, W. Sparreboom, A. van den Berg, and J.C.T. Eijkel
MESA+ - University of Twente, THE NETHERLANDS

We report frequency-dependent bidirectional AC electroosmotic flow (AC-EOF) in a nanochannel with double layer overlap. Simulations confirm the observed bidirectionality. By this frequency-dependent bidirectional pumping, nanochannel AC-EOF behaves in fundamentally different way than microchannel AC-EOF. The results are of importance for the understanding of ion and liquid transport in nanoconfinement.

M-213 FULLY MONOLITHIC AND ULTRA-COMPACT NEMS-CMOS SELF-OSCILLATOR BASED
ON SINGLE-CRYSTAL SILICON RESONATORS AND LOW-COST CMOS CIRCUITRY 1071
J. Philippe, G. Arndt, E. Colinet, M. Savoye, T. Ernst, E. Ollier, and J. Arcamone
CEA - LETI – Minatec, FRANCE

This work reports on the first experimental demonstration of a self-oscillator based on a single crystal silicon NEMS resonator monolithically co-integrated with a simple electronic circuitry manufactured with a very low-cost 0.35μm CMOS technology. This NEMS-CMOS self-oscillator pixel is as small as 50x70 μm² (pads excluded).

T-214 A GRAPHENE NANOSENSOR FOR DETECTION OF SMALL MOLECULES 1075
C. Wang[1,2], J. Kim[1], J. Zhu[1], R. Pei[3], G. Liu[2], J. Hone[1], M. Stojanovic[1], and Q. Lin[1]
[1]Columbia University, USA, [2]Nankai University, CHINA, and [3]Chinese Academy of Sciences, CHINA

We developed a graphene field effect transistor (GFET) biosensor for detection of an important small molecular hormone DHEA-S. In view of the low charged small biomolecules can't excite sufficient electrical response of GFET, we proposed a competitive dehybridization strategy based on the aptamer-target specific association. We experimentally demonstrated that on the graphene surface, aptamer dehybridization caused by DHEA-S specific association provides strong response of GFET. And the concentration of target DHEA-S can be quantitatively detected by observing the "half time period" of aptamer dehybridization kinetic process.

W-215 INTERROGATING CONTACT-MODE SILICON CARBIDE (SiC) NANOELECTROMECHANICAL SWITCHING DYNAMICS BY ULTRASENSITIVE LASER INTERFEROMETRY 1079
T. He, J. Lee, Z. Wang, and P.X.-L. Feng
Case Western Reserve University, USA

We report the experimental demonstration of probing the dynamics of nanoscale contacts in robust nanoelectromechanical switches based on silicon carbide (SiC) nanocantilevers. For the first time, we measure the dynamical behavior of contact-mode SiC nanoelectromechanical switches, in both frequency- and time-domain, by directly probing the tips of the SiC nanocantilevers, using ultrasensitive laser interferometric techniques.

Th-216 MATRIX INDEPENDENT LABEL-FREE NANOELECTRONIC BIOSENSOR 1083
R. Esfandyarpour[1,2], M. Javanmard[2], Z. Koochak[1], J.S. Harris[1], and R.W. Davis[2]
[1]Stanford University, USA and [2]Stanford Genome Technology Center, USA

We fabricated a novel, label free and real time electrical impedance biosensor, referred to as the nanoneedle biosensor. The nanoneedle is an ultrasensitive and localized device, which has the ability to directly measure biomolecular binding as a function of time (real-time). The utility of this sensor in affinity biosensing was demonstrated. As a practical example with clinical relevance, we demonstrated the detection of Vascular Endothelial Growth Factor (VEGF) for cancer diagnosis. Our demonstration of label-free and real-time detection of VEGF with this sensor can be envisioned to allow for one-step point-of-care cancer diagnosis. This work provides a strong starting point for a new class of electronic biosensing devices with the capability of rapid direct large-scale integration.

M-217 MECHANICAL PROPERTIES OF FEW LAYER GRAPHENE CANTILEVER 1087
K. Matsui, A. Inaba, Y. Oshidari, Y. Takei, H. Takahashi, T. Takahata, R. Kometani, K. Matsumoto, and I. Shimoyama
University of Tokyo, JAPAN

We report the spring constant measurement of few-layer (1-, 2-, and 3-layer) graphene (FLG) cantilevers by optical heterodyne interferometry. We fabricated FLG cantilever with a weight of diamond-like carbon using focused ion beam. The effective spring constants were obtained from the measured resonant frequency and the mass of the weight, and were calculated to be about 2.7×10^{-3} N/m. This result indicates FLG cantilever structure is more rigid than that predicted from the literature data.

T-218 NANO-OPTO-MECHANICAL MEMORY BASED ON OPTICAL GRADIENT FORCE INDUCED BISTABILITY 1091
B. Dong[1,2], J.G. Huang[3], H. Cai[2], P. Kropelnicki[2], A.B. Randles[2], Y.D. Gu[2], and A.Q. Liu[1]
[1]Nanyang Technological University, SINGAPORE, [2]Agency for Science, Technology and Research (A*STAR), SINGAPORE, and [3]Xi'an Jiao Tong University, CHINA

A bistable nano-opto-mechanical memory is designed, fabricated and experimentally demonstrated.Fabricated with CMOS compatible process, this optical memory can be easily packaged and integrated with otherphotonic devices. The nano-size of the memory enable for large scale integration, high speed operation and low powerconsumption. It has other potential applications such as optical switch, logic gate and actuator.

W-219 SUBMICRON THREE-TERMINAL SIGE-BASED ELECTROMECHANICAL OHMIC RELAY 1095
M. Ramezani[1,2], S. Cosemans[1], J. De Coster[1], X. Rottenberg[1], V. Rochus[1], H. Osman[1], H.A.C. Tilmans[1], S. Severi[1], and K. De Meyer[1,2]
[1]imec, BELGIUM and [2]KU Leuven, BELGIUM

We demonstrate functional cantilever switches based on a CMOS-compatible low-T(400°C) CVD SiGe process flow. Devices with dimensions in the micrometer range, thickness and gap smaller than 100nm were successfully fabricated and electrically characterized. Typical switches characteristics such as high I_{on}/I_{off} ratio, sharp sub-threshold slope and zero off-state leakage current were observed. A minimum of 1000 cycles device lifetime was demonstrated. The maximum current which can flow through the device without causing stiction due Joule-heating was investigated.

Th-220 PIEZOELECTRIC BUCKLING-BASED NEMS RELAYS
FOR MILLIVOLT MECHANICAL LOGIC 1099
U. Zaghloul[1,2] and G. Piazza[1]
[1]*Carnegie Mellon University, USA and* [2]*Electronics Research Institute, EGYPT*

We report on the design, fabrication, characterization, and scaling analysis of novel NEMS relays that use, for the first time, buckling piezoelectric actuators. The fabricated switches exhibit low actuation voltage (< 2 V) and reduced threshold voltage (~110 mV). Also, hysteresis in the switching process was observed and limits the minimum swing voltage to ~250 mV. A scaling analysis highlights the possibility of achieving milliVolt switching at aggressively scaled device footprints.

M-221 NANOELECTROMECHANICAL TUNNELING SWITCHES
BASED ON SELF-ASSEMBLED MOLECULAR LAYERS 1103
F. Niroui[1], P.B. Deotare[1], E.M. Sletten[1], A.I. Wang[1], E. Yablonovitch[2],
T.M. Swager[1], J.H. Lang[1], and V. Bulovic[1]
[1]*Massachusetts Institute of Technology, USA and* [2]*University of California, Berkeley, USA*

We propose and experimentally investigate nanoelectromechanical switches that operate via electromechanical modulation of tunneling current through compressible molecular films. This approach utilizes self-assembled molecular layers to define few nanometer-thick switching gaps, and has the potential to enable low-voltage operation while simultaneously mitigating device failure due to stiction.

T-222 TRANSITION OF Q-DOT DISTRIBUTION ON MICROTUBULE ARRAY
ENCLOSED BY PDMS SEALING FOR AXONAL TRANSPORT MODEL 1107
K. Fujimoto[1], H. Shintaku[1], H. Kotera[1], and R. Yokokawa[2,1]
[1]*Kyoto University, JAPAN and* [2]*Japan Science and Technology Agency (JST), JAPAN*

We developed an experimental system which enables kinesin driven transport on arrayed microtubules in enclosed micro channels. To avoid an expected difficulty of exchanging solution, surface fabricated micro tracks were encapsulated after reagents introduction using flow cells with deformable PDMS chip at its top. After enclosed micro channels were formed, directed transport and continuous accumulation of fluorescent labeled kinesin molecules were observed. These results indicate a possibility of application as in vitro model of intracellular transport as seen in axons.

W-223 WAFER-SCALE FABRICATION OF SCANNING THERMAL PROBES WITH INTEGRATED
METAL NANOWIRE RESISTIVE ELEMENTS FOR SENSING AND HEATING 1111
K. Hatakeyama[1], E. Sarajlic[2], M.H. Siekman[1], L. Jalabert[3], H. Fujita[3], N. Tas[1], and L. Abelmann[4]
[1]*MESA+ - University of Twente, THE NETHERLANDS,* [2]*SmartTip B.V., THE NETHERLANDS,*
[3]*University of Tokyo, JAPAN, and* [4]*Korea Institute of Science and Technology (KIST), GERMANY*

We present a novel scanning resistive probe aimed for thermal imaging and localized thermal analysis. The probe features an AFM cantilever with a sharp pyramidal tip. Metal nanowires are integrated at the inner edges of the pyramidal tip forming an electrical cross-junction at the apex. The cross-junction can be utilized both as a local temperature sensor and a heater.

Packaging Technologies (PCK)

Th-224 CAPILLARY EFFECT BASED TSV FILLING METHOD 1115
J. Gu, X. Jiang, H. Yang, and X. Li
Shanghai Institute of Microsystem and Information Technology (SIMIT), CHINA

We explore a capillary liquid solder through-hole filling method, which utilizes liquid bridge pinch-off effect. The filling is completed by first pushing solder into via holes from a solder pool through nozzle orifices, and then followed by cutting the solder pillars in the via holes off from the solder pool in the nozzle orifices. The whole TSV filling process can be completed by a cycle of pressure change. In addition, 'wafer sandwich' structure is utilized to neutralize pressure differential, which causes wafer breakage.

M-225 CHARACTERISATION AND SIMULATION OF LOW TEMPERATURE Si-Si-DIRECT
BONDING THROUGH VELCRO-LIKE SURFACES BASED ON POROUS SILICON 1119
S. Keshavarzi[1,2], U. Mescheder[1], and H. Reinecke[2]
[1]*Furtwangen University - IAF, GERMANY and* [2]*University of Freiburg – IMTEK, GERMANY*

We develop, characterize and model a new bonding technique based on Pours Silicon (PS) technology. PS allows strong permanent bonding between needle like surfaces as well as multiple bonding and un-bonding of chips similar to Velcro principle. This approach provides low temperature Si-Si direct bonding, a fully CMOS compatible approach suitable in system integration using the Si-motherboard concept.

T-226 FROM CHIPS TO DUST: THE MEMS SHATTER SECURE CHIP 1123
N. Banerjee, Y. Xie, M. Rahman, H. Kim, and C.H. Mastrangelo
University of Utah, USA

This paper presents the implementation of transience silicon microchips through post-processing microfabrication and micropackaging steps that transform almost any electronic, optical or MEMS substrate chips into transient ones. When transience is activated the chip mechanically shatters, and it is literally reduced to a heap of silicon dust. The massive cleavage action is achieved by the triggered release of mechanical energy stored within the silicon substrate in expandable microparticles.

W-227 LONG TERM GLASS-ENCAPSULATED PACKAGING FOR IMPLANT ELECTRONICS 1127
J.H. Chang, Y. Liu, and Y.C. Tai
California Institute of Technology, USA

This paper studies a new long-term packaging scheme for implant electronics using glass encapsulation featuring a controlled failure mode from fast diffusion to slow undercut. The experimental results show that this packagingscheme can easily survive for more than 10 years by accelerated "active" lifetime soaking test (i.e. with electric field applied) in 0.9 wt.% saline solution. This method provides advantages of easy employment, controllable long life time, and enhanced heat dissipation."

Th-228 LOW-TEMPERATURE GOLD-GOLD BONDING USING SELECTIVE FORMATION
OF NANOPOROUS POWDERS FOR BUMP INTERCONNECTS 1131
H. Mimatsu[1], J. Mizuno[1], T. Kasahara[1], M. Saito[1], S. Shoji[1], and H. Nishikawa[2]
[1]Waseda University, JAPAN and [2]Osaka University, JAPAN

We proposed low-temperature Au-Au bonding using nanoporous Au-Ag powders as an electrical connective adhesion between bump interconnects. The nano-porous powders were formed by de-alloying Au-Ag alloy sheet with Au:Ag. The influence of the annealing temperature on the porous structures was investigated. Selective formation of the powders on bumps was achieved by stamping. Bonding strength of about 2.4 MPa was achieved by using nanoporous Au-Ag powders at 150 °C. This result indicates that the proposed powder is a useful material for low-temperature Au-Au bonding.

M-229 MICRO DEVICES INTEGRATION WITH LARGE-AREA 2D CHIP-NETWORK
USING STRETCHABLE ELECTROPLATING COPPER SPRING 1135
W.L. Sung[1], W.C. Lai[1], C.C. Chen[2], K. Huang[2], and W. Fang[1]
[1]National Tsing Hua University (NTHU), TAIWAN and [2]imec Taiwan Inc., TAIWAN

This study presents a large-area multi-devices integration scheme using stretchable electroplating copper spring. Advantages of this approach: (1) using the existing process technologies and materials for semiconductor in large-area applications; (2) stretchable electroplating-copper spring with large maximum strain acts as mechanically and electrically connection; (3) Si-node acts as a hub for devices implementation and integration; and (4) the chip-network can apply to curved surfaces. The proposed expand network using stretchable spring integrated multi devices has been implemented and tested.

T-230 SOLID-STATE ISFET FLOW METER FABRICATED WITH A PLANAR PACKAGING
PROCESS FOR INTEGRATING MICROFLUIDIC CHANNEL WITH CMOS IC CHIP 1139
J.J. Wang[1], C.F. Lin[2], Y.Z. Juang[2], H.H. Tsai[2], H.H. Liao[2], and C.H. Lin[1]
[1]National Sun Yat-sen University, TAIWAN and [2]National Applied Research Laboratories, TAIWAN

We presents a solid-state ISFET flow meter fabricated with an innovative planar packaging process. The developed method provides a simple yet efficient method to integrate CMOS IC chip with microfluidic systems and the whole packaging process can be achieved in 40 min. The sealed ISFET chip is used for measuring the flow rate of non-ionic solutions including acetone, ethanol and glycerol of slow flow rate. And the flow rate measurement exhibited good reproducibility in the flow rate ranging from 66 to 1700 μm/s.

Physical MEMS (PHYS)

W-231 A TUNABLE LASER BASED ON NANO-OPTO-MECHANICAL SYSTEM 1143
M. Ren[1], H. Cai[2], Y.D. Gu[2], P. Kropelnicki[2], A.B. Randles[2], and A.Q. Liu[1]
[1]Nanyang Technological University, SINGAPORE and
*[2]Agency for Science, Technology and Research (A*STAR), SINGAPORE*

A tunable laser based on nano-opto-mechanical system is presented in this paper. A novel tuning approach is demonstrated which applies optical force to adjust the cavity mode via controlling the mechanical displacement of the silicon waveguide. In the experiments, a 24-nm wavelength tuning is realized due to a deflection of 14-nm. The optomechanical wavelength tuning coefficient is 214 GHz/nm. The demonstrated device has potential applications for optical communication system, pulse trapping/release, and chemical sensing, with easy on-chip integration on a silicon platform.

Th-232 A TUNABLE OPTICAL IRIS BASED ON ELECTROMAGNETIC ACTUATION FOR A HIGH-PERFORMANCE MINI/MICRO CAMERA 1147

H.W. Seo[1], J.B. Chae[1], S.J. Hong[1], I.U. Shin[1], K. Rhee[1], J.-H. Chang[2], and S.K. Chung[1]
[1]*Myongji University, SOUTH KOREA and*
[2]*Samsung Advanced Institute of Technology (SAIT), SOUTH KOREA*

This paper presents a tunable optical iris based on electromagnetic actuation for a high-performance mini/micro camera. In optics, an iris, an aperture stop, is placed in the light path of a lens or objective and regulates the amount of light that passes through the lens by controlling the size of the aperture, an opening at it center. The iris not only controls light flux, field of view, depth of field (DOF), but also blocks scattered light and improves image quality by limiting spherical aberration. Hence, the iris is an indispensable element in most optical systems. However, the conventional mechanical iris, consisting of movable sliding blades, requires a complicated sliding rotary mechanism that has to be operated by bulky motors and is therefore difficult to miniaturize. We develop a variable optical iris operated by electromagnetic actuation. According to electromagnetic induction, when an electrical current flows in an electric coil, a magnetic field is generated in its surroundings. In this work, the magnetic field is used to actuate or pull an optically opaque ferrofluid initially filled inside the sub-channel of the iris to the center of the main channel, resulting in controlling the diameter of an aperture.

M-233 CALORIMETRIC DEVICE FOR NON-DESTRUCTIVE MEASUREMENT OF THE THERMAL DIFFUSIVITY DEPENDENCY BY PHASE DELAY 1151

T. Suzuki, Y. Ichikawa, T. Takahata, K. Matsumoto, and I. Shimoyama
University of Tokyo, JAPAN

We developed a device for measuring thermal diffusivity dependency of the contacted surface layer non-destructively. The device was based on the principle that temperature phase delay between a heater and a resistance temperature detector (RTD) is affected by thermal diffusivity of the contacted surface layer. The device consisted of an Au wire, as oscillating heat source and a piezoresistance as a RTD. We exerted the simulation and the experiment for the device, and found that the phase delay decreased as thermal diffusivity increased.

T-234 CAPACITIVE FEEDBACK CONTROLLED PZT MICRO MIRROR ARRAYS FOR WAVELENGTH SELECTIVE SWITCH 1155

R. Uchino, T. Misaki, T. Fujimura, and O. Torayashiki
Sumitomo Precision Products Co., Ltd., JAPAN

We develop a single-axis mechanical micro mirror array used for gridless wavelength selective switch (WSS). The mirrors are driven by lead zirconate titanate (PZT) unimorph actuators, which is adequate for low-voltage actuation and low interference with adjacent mirrors in operation. In addition, the mirror tilt angle is feedback controlled using comb-shaped capacitance in order to realize high control resolution. We fabricated a prototype of the mirror array, and evaluated its basic performance.

W-235 CARBON SP^2-SP^3 TECHNOLOGY: GRAPHENE-ON-DIAMOND THIN FILM UV DETECTOR 1159

K. Yao[1], C. Yang[1], X. Zang[1], F. Feng[2], and L. Lin[1]
[1]*University of California, Berkeley, USA and* [2]*Chinese Academy of Sciences, CHINA*

We for the first time demonstrates the graphene-diamond-metal (GDM) vertical sandwich structure as a thin film UV detector. New scientific and engineering breakthroughs are: (1) first experimental investigation of the carbon-based sp2-sp3 junctions; (2) a peel-and-stick fabrication process to make flexible diamond films; and (3) first GDM vertical UV sensors. As such, the proposed detector/architecture can open up a new class of scheme to build diamond-based optoelectronic systems.

Th-236 CLOSE-PACKED LIQUID-FILLED TUNABLE MICROLENS ARRAY 1163

Y. Iimura[1], H. Onoe[1,2], and S.Takeuchi[1,2]
[1]*University of Tokyo, JAPAN and* [2]*ERATO Takeuchi Biohybrid Innovation Project,*

We develop close-packed liquid-filled tunable microlens arrays for optical devices such as integral imaging systems. These lenses are simply composed of poly(dimethylsiloxane)(PDMS) microchannels and applied pressure deforms the top membrane of microchannels to become convex lenses. These lenses have three advantages: (i)Uniform deformation by pressure-driven actuation, (ii)Adjustable optical characteristics without patterned electrode, (iii)High-density integration of tunable microlenses. We fabricated three types of lenses based on closed-packed structure and showed that the Spiderweb type packing is the most suitable for closer packing.

M-237 COMPACT TUNABLE HYPERSPECTRAL IMAGING SYSTEM 1167

P.-H. Cu-Nguyen[1], A. Grewe[2], C. Endrödy[2], S. Sinzinger[2], H. Zappe[1], and A. Seifert[1]
[1]*University of Freiburg – IMTEK, GERMANY and* [2]*Ilmenau University of Technology, GERMANY*

We demonstrate a compact tunable hyperchromatic lens system for imaging an object with highly resolved spectral information. This hybrid device is composed of a diffractive optical element, a tunable concave liquid-filled membrane lens, and an integrated magnetic actuator for hydraulically tuning the focal length of the refractive lens. The lens system can generate a hyperspectral datacube in the visible wavelength range, 400 – 730 nm, proved here with a spectral sampling interval of 2.4 nm.

T-238 CYLINDRICAL LENS WITH INTEGRATED PIEZO ACTUATION FOR FOCAL LENGTH TUNING AND LATERAL SCANNING 1171

M. Stuermer, A. Schatz, and U. Wallrabe

University of Freiburg – IMTEK, GERMANY

We present a cylindrical lens which features integrated piezo bending actuators for focal length tuning. The design is based on a PDMS membrane which encloses an optical liquid. We optimize the shape of the actuators for good cylindricity and show a process for prototype fabrication. The lens provides a large usable aperture of ca. 4 x 10 mm, a tuning range of more than 20 dpt, and the possibility to move the lens vertex along one axis. Therefore, it enables scanning of the line focus.

W-239 FRESNEL LENS BASED ON SILICON NANOWIRES 1175

Y.-S. Lu, J. Fernandes, H. Liu, and H. Jiang

University of Wisconsin-Madison, USA

We demonstrate silicon-based Fresnel lenses by photolithography techniques and metal assisted chemical etching, where the opaque zones are composed of 2 µm-tall silicon nanowires formed directly in silicon. The reflective Fresnel lens showed a high-contrast light intensity distribution between the bright and dark zones, leading to a focused spot with strong contrast above the lens. The lens has the potential to be integrated with dye-sensitive solar cells by reflecting and focusing light onto the photosensitive dye to improve their light absorption efficiency and photocurrent.

Th-240 ENHANCED WAVELENGTH SELECTIVE INFRARED EMISSION USING SURFACE PLASMON POLARITON AND THERMAL ENERGY CONFINED IN MICRO-HEATER 1179

T. Sawada[1], K. Masuno[2], S. Kumagai[1], M. Ishii[2], S. Uematsu[2], and M. Sasaki[1]

[1]Toyota Technological Institute, JAPAN and [2]Yazaki Corporation, JAPAN

A new surface plasmon polariton (SPP) based wavelength selective IR emitter is combined with microheater. IR emitted from the microheater is basically confined except SPP propagation on the metal grating carrying IR energy to the outside. The limited condition for SPP excitation realizes the narrow wavelength filtering. SPP related emission is obtained having the peak width similar order compared with the bandwidth of gas absorption. Since the microheater can minimize the thermal conduction loss, the high efficiently is expected at SPP related wavelength.

M-241 EFFECT OF NEEDLE SHAPE ON PERFORMANCE OF NEEDLE-TYPE ELECTRO TACTILE DISPLAY 1183

N. Kitamura[1], J. Chim[1], and N. Miki[1,2]

[1]Keio University, JAPAN and [2]Japan Science and Technology Agency (JST), JAPAN

In our prior work, we revealed that a needle-type electrotactile display that penetrates through a stratum corneum of a finger skin can display tactile information at 20 times as low voltage as that with flat electrodes. We discovered that the needle-tip shapes greatly affected the performance of the display. In this work, we experimentally deduced the optimum shape of the needle tip using titanium micro-needles patterned by electrochemical etching. The needles can be readily applicable to efficient electrotactile displays.

T-242 INCLINATION-INDEPENDENT TRANSFORMATION OF LIGHT BEAMS USING HIGH-THROUGHPUT UNIQUELY-CURVED MICROMIRRORS 1185

Y.M. Sabry[1,2], D. Khalil[2,3], B. Saadany[2], and T. Bourouina[1,2]

[1]Université Paris-Est, FRANCE, [2]Si-Ware Systems, EGYPT, and [3]Ain-Shams University, EGYPT

This paper reports a novel class of deeply-etched, specifically-designed curved micromirrors enabling phase-transformation of light beams independent of the inclination angle of the incident light with respect to the mirror surface. The micromirrors were fabricated on silicon by deep reactive ion etching technology. The profile of the specifically-designed mirrors' surfaces was controlled precisely, thanks to the photolithographic process. High optical throughput micromirrors exhibiting submillimeter focal lengths were fabricated with depth larger than 300 µm. Optical measurements show stable dimensions for the optical beam spot with less than ± 5% dependence on the inclination angle up to 60 degrees.

W-243 MAGNETOSTRICTIVE TYPE TACTILE SENSOR BASED ON METAL EMBEDDED POLYMER ARCHITECTURE 1189

H.-C. Chang[1], W.-L. Sung[1], H.-S. Hsieh[1], J.-H. Wen[1], C.-C. Fu[1], S.-C. Liao[1], C.-H. Lai[1], W.-C. Lai[2], C.-H. Chang[2], C.-P. Chang[2], C.-H. Chen[2], and W. Fang[1]

[1]National Tsing Hua University (NTHU), TAIWAN and [2]WinMEMS Technologies Co., Ltd., TAIWAN

This study presents new process scheme to fabricate polymer structure with embedded metal on silicon substrate. The primary merit of presented process scheme is: simple approach for the integration of 3D structures with different materials (e.g. metal, glass, polymer) on substrate. To demonstrate the feasibility, a tactile sensor design consisting of polymer structure with embedded 3D Ni inductor is demonstrated. As the polymer diaphragm deformed by tactile force, the magnetostriction effect of 3D Ni inductor will induce the permeability change. Thus, the permeability change as well as the tactile force can be detected by the inductance difference of embedded 3D Ni inductor.

Th-244 A MULTI-MATERIAL Q-BOOSTED LOW PHASE
NOISE OPTOMECHANICAL OSCILLATOR 1193
T. Beyazoglu, T.O. Rocheleau, K.E. Grutter, A.J. Grine, M.C. Wu, and C.T.-C. Nguyen
University of California, Berkeley, USA

We present a multi material Radiation Pressure driven Optomechanical Oscillator (RP-OMO) with simultaneously high mechanical Qm >22,000 and optical Qo >190,000 achieving best-to-date phase noise performance of -125 dBc/Hz at 5 kHz offset from its 52-MHz carrier, which is 12 dB better than the previous best RP-OMO constructed of silicon nitride alone. The device not only reduces phase noise, but does so with a lower input laser power of only 3.6 mW. The key to achieving this performance is the addition of polysilicon material as an inner ring that boosts the overall mechanical Qm of the total structure. The addition of polysilicon further provides a mechanism for voltage-controlled electrical stiffness tuning of the oscillation frequency.

M-245 NOVEL TUNABLE OPTICAL MODULATION LENS
USING MAGNETHORHEOLOGICAL EFFECT 1197
F.-M. Hsu, R. Chen, and W. Fang
National Tsing Hua University (NTHU), TAIWAN

This study extends the fluid dispensing and sealing technology to realize a novel MR-fluid lens (MR-fluid: liquid polymer with magnetic particles) for light intensity modulation. Merits of the device: Optical transmittance of lens is controlled by (1) weight fraction of magnetic powder, and (2) orientation of columnar particles controlled by magnetic field. In applications, the MR-fluid lens is realized on glass substrate and suspended MEMS structures. The light intensity modulation of MR-fluid lens (diameter: 2000μm) by magnetic field is demonstrated. Measurements show the NdFeB- liquid polymer (10wt%) has a 40% dark area change and 290% laser transmittance difference after applying magnetic field.

T-246 OPTICAL CONTROL AND TUNING OF THERMAL-PIEZORESISTIVE
SELF-SUSTAINED OSCILLATORS 1201
H.J. Hall[1], L. Wang[1], J.S. Bunch[1], S. Pourkamali[2], and V.M. Bright[1]
[1]University of Colorado, Boulder, USA and [2]University of Texas, Dallas, USA

We experimentally demonstrate the ability to frequency tune and provide on/off control of electrically driven thermal-piezoresistive self-sustained oscillators through the direct application of HeNe laser illumination at the device surface. These phenomena, which are unique to this class of oscillator, are explained by photoexcitation of charge carriers in the device's single crystal silicon structure inducing changes to the effective electrical resistivity and piezoresistivity.

W-247 PHOTOTHERMAL PROBING OF PLASMONIC HOTSPOTS
WITH NANOMECHANICAL RESONATOR 1205
S. Schmid, K. Wu, T. Rindzevicius, and A. Boisen
Technical University of Denmark, DENMARK

We present a novel technique to probe and image plasmonic structures with nanoscale resolution by measuring the photothermally induced frequency detuning of highly temperature sensitive nanomechanical resonators. We employ the high temperature sensitivity of a nanomechanical string resonator to directly probe the heating pattern produced by a gold nanoslit illuminated by a scanning laser beam. The experimental approach allows a sensitive heat mapping of single localized surface plasmons, thereby helping to shed light on the underlying thermal effects in hotspots.

Th-248 RADIATION-PRESSURE ENHANCED OPTO-ACOUSTIC OSCILLATOR 1209
M.J. Storey, S. Tallur, and S.A. Bhave
Cornell University, USA

We present a driving scheme for integrated chip-scale opto-acoustic oscillators (OAO) in silicon with improved phase noise performance. Through simultaneous incorporation of radiation-pressure (RP) and RF feedback oscillating mechanisms, we have demonstrated a silicon RP enhanced OAO with a 10dB close-to-carrier phase noise improvement and thereby 10dB improvement in the oscillator's figure of merit.

M-249 THERMOPILE INFRARED ARRAY SENSOR FOR HUMAN DETECTOR APPLICATION 1213
J. Tanaka, M. Shiozaki, F. Aita, T. Seki, and M. Oba
OMRON Corporation, JAPAN

This paper reports the design of thermopile infrared sensor for human detector application. Sensitivity and response time of thermopile infrared sensor element are important for human detector application. In order to fulfill the specification, we developed S-shaped structure for thermopile infrared sensor element and fabrication process of chip scale vacuum package for mass production of the thermopile infrared sensors. As the result, 140V/W sensitivity and 17msec response time of the thermopile infrared sensor element are achieved.

T-250 TRANSFER-PRINTED COMPOSITE MEMBRANES FOR ELECTRICALLY-TUNABLE ORGANIC OPTICAL MICROCAVITIES 1217

A. Wang, W. Chang, A. Murarka, J. Lang, and V. Bulovic
Massachusetts Institute of Technology, USA

We present a method for fabricating organic optical microcavities using a transfer-printed composite membrane which can be electrostatically actuated for dynamic tuning of the cavity emission spectra. Electrical actuation and optical characterization of a completed device show cavity mode tuning greater than 20 nm. The device structure and transfer technique is easily applicable to large area fabrication of electrostatically tunable organic lasers, and potentially allows single-point contactless-readout for large area pressure sensor arrays.

W-251 TUNABLE METAMATERIALS BY CONTROLLING SUB-MICRON GAP FOR THE THZ RANGE 1221

A. Isozaki[1], T. Kan[1], H. Takahashi[1], N. Kanda[2], N. Nemoto[1], K. Konishi[1], M. Kuwata-Gonokami[1], K. Matsumoto[1], and I. Shimoyama[1]
[1]University of Tokyo, JAPAN and [2]Institute of Physical and Chemical Research (RIKEN), JAPAN

We propose a tunable metamaterial actuated by pneumatic force. The tunable metamaterial has a double of sprit-ring-resonators (SRRs) whose gap between each other is controllable in sub-micron-order. Results of a terahertz (THz) spectroscopy confirmed that controlling gap in sub-micron-order or a few micron order was suitable for tuning resonant frequency of a metamaterial compared with that in 10 micron-order.

Th-252 UNCOOLED MULTI-BAND IR IMAGING USING BIMATERIAL CANTILEVER FPA 1225

W. Ma[1], S. Wang[1], Y. Wen[1], Y. Zhao[2], L. Dong[2], M. Liu[2], X. Liu[2], and X. Yu[1]
[1]Peking University, CHINA and [2]Institute of Beijing Technology, CHINA

A 256×256 bimaterial cantilever focal plane array (FPA), which is able to work in the three infrared atmospheric windows simultaneously, is fabricated and characterized. The FPA employs a silicon-framed structure by selectively etching away the substrate with Deep Reactive Ion Etching technique, and can be conveniently readout by an optical system. By combining the Chromium nano-films with silicon nitride as the multi-band IR absorber, the images of short wavelength, middle wavelength and long wavelength infrared are captured successfully with the same FPA.

M-253 VERY LOW POWER CONSUMPTION MEMS SCANNER WITH ALKALI ELECTRET COMB DRIVE 1229

T. Sugiyama, M. Aoyama, K. Kawai, and G. Hashiguchi
Shizuoka University, JAPAN

We developed the very low power consumption MEMS scanner that utilizes the electrostatic field generated by alkali-ion electret. The alkali-ion electret formed on comb electrodes of the scanner provides built-in potential for the electro-static actuator so that no bias voltage is necessary. The power consumption of prototype MEMS scanner was 0.57 μW (bias voltage: DC 0 V, driving voltage: AC 9 V_{pp}, deflection angle: 12°, resonance frequency: 1.4 kHz).

RF MEMS (RF)

T-254 A LOW-LOSS RF MEMS SILICON SWITCH USING REFLOWED GLASS STRUCTURE 1233

J. Hwang, S.-H. Hwang, Y.-S. Lee, and Y.-K. Kim
Seoul National University, SOUTH KOREA

This paper firstly reports on a low-loss RF MEMS switch that contains a reflowed glass structure beneath a contact metal. The reflowed glass structure is employed to reduce the electromagnetic wave loss brought about by the conductive silicon bulk underneath the contact metal. RF MEMS switch totally made of silicon is used as a reference model and the insertion loss is reduced as much as 0.26 dB for the proposed model in the frequency range of 5 to 30 GHz.

W-255 A UHF 4TH-ORDER BAND-PASS FILTER BASED ON CONTOUR-MODE PZT-ON-SILICON RESONATORS 1237

H. Yagubizade, M. Darvishi, M. Elwenspoek, and N.R. Tas
MESA+, University of Twente, THE NETHERLANDS

A novel RF-MEMS filter configuration around 700 MHz is proposed. It is based on a differential read-out of two in-phase actuated contour-mode resonators with slightly different resonance frequencies. The resonators are actuated independently in-phase and the outputs of the resonators are subtracted. This method is effective for improving the stopband rejection by canceling the feed-through signal. The BPF is presented using 50Ω termination with bandwidth of approximately 28.6 MHz and 35 dB stopband rejection. The ultimate rejection of the filter is improved by more than 20 dB compared to the individual resonators.

Th-256 AN 880MHZ LADDER FILTER FORMED BY ARRAYS OF LATERALLY VIBRATING THIN FILM LITHIUM NIOBATE RESONATORS 1241

S. Gong[1], and G. Piazza[2]

[1]University of Illinois, Urbana Champaign, USA and [2]Carnegie Mellon University, USA

This paper reports on the first implementation of a ladder filter using Lithium Niobate (LN) based laterally vibrating resonator (LVR) arrays. This demonstration is made possible by engineering the device orientation and using a distributed configuration of resonator arrays to simultaneously reduce spurious vibrations and insertion loss in a low impedance RF system. An almost spurious-free filter with < 3.5 dB of IL at 880 MHz was demonstrated by arraying properly sized devices into a ladder configuration. A total of 37 resonators were used for this demonstration. This work sets an important milestone in the development of a thin film LN technology platform for wide-band and frequency-agile RF filtering.

M-257 CAPACITIVE SILICON RESONATOR STRUCTURE WITH MOVABLE ELECTRODES TO REDUCE CAPACITIVE GAP WIDTHS BASED ON ELECTROSTATIC PARALLEL PLATE ACTUATION 1245

N.V. Toan and T. Ono

Tohoku University, JAPAN

This paper presents the design and fabrication of a capacitive silicon resonator with movable electrodes to obtain smaller capacitive gap widths, which results in smaller motional resistance and lower insertion loss. It also helps to increase tuning frequency range for compensation of temperature drift of silicon resonator.

T-258 COMBINED ELECTRICAL AND MECHANICAL COUPLING FOR MODE-RECONFIGURABLE CMOS-MEMS FILTERS 1249

C.-Y. Chen, M.-H. Li, C.-H. Chin, C.-S. Li, and S.-S.Li

National Tsing Hua University (NTHU), TAIWAN

This work presents a novel filter scheme which combines electrical and mechanical coupling implemented in a CMOS-MEMS filter to simultaneously attain narrow bandwidth and decent stopband rejection. As compared to the parallel-class filters and mechanically-coupled filters, the proposed filter structure features flexible electrical routing and non-conductive mechanical couplers, hence enabling single-ended to differential (SIDO) and differential to single-ended (DISO) reconfigurable modes in a single device. The proposed 8.6-MHz CMOS-MEMS filter was successfully demonstrated with narrow passband of 0.41% bandwidth and stopband rejection more than 20dB under proper termination.

W-259 DYNAMIC CHARACTERIZATION OF TUNABLE RF MEMS PRODUCTS 1253

D. DeReus, S. Cunningham, S. Natarajan, A. Morris, and J. Hilbert

WiSpry, Inc., USA

We present the dynamic characterization of tunable capacitors for RF MEMS products.The dynamic measurements have been made electrically by laser doppler vibrometry, and correlated with conventional finite element and high-order, parametric finite element models.

Th-260 ETCH-HOLE-ASSISTED ENERGY DISPERSION FOR ENHANCING QUALITY FACTOR IN SILICON BULK ACOUSTIC RESONATORS 1257

C. Tu and J.E.-Y. Lee

City University of Hong Kong, HONG KONG

This paper empirically demonstrates how the quality factor (Q) of a width-extensional mode silicon bulk-acoustic-resonator (SiBAR) can be enhanced by three times through strategic placement of holes on the structure. The holes serve to disperse the strain energy field concentrated around the nodal lines, ultimately re-distributing strain energy away from the anchors. This in turn reduces anchor loss and thus enhances Q. These results agree well with our finite-element (FE) simulations and we envisage the concepts herein to be transferable to other higher performance resonators like piezoelectric-AlN CMRs.

M-261 NANO-OPTO-ELECTRO-MECHANICAL (NOEM) OSCILLATOR WITH CONTROLLABLE NON-LINEAR DYNAMICS 1261

B. Dong[1,2], J.G. Huang[3], H. Cai[2], P. Kropelnicki[2], A.B. Randles[2], Y.D. Gu[2], and A.Q. Liu[1]

[1]Nanyang Technological University, SINGAPORE, [2]Agency for Science, Technology and Research (A*STAR), SINGAPORE, and [3]Xi'an Jiao Tong University, CHINA

An opto-mechanical oscillator with controllable non-linear dynamics is designed, fabricated and experimentally demonstrated. Fabricated with CMOS compatible process, this opto-mechanical oscillator can be easily packaged and integrated with other photonic devices. It has potential applications such as optical resonator type gyroscope, accelerometer and optical communication devices.

T-262 ON/OFF SWITCHABLE HIGH-Q CAPACITIVE-PIEZOELECTRIC ALN RESONATORS 1265
R.A. Schneider and C.T.-C. Nguyen
University of California, Berkeley, USA

AlN disk resonators having suspended (non-contacting) electrodes are demonstrated to have quality factors as high as 8,850 at 300 MHz, show no spurious modes, have single disk motional impedances of 3.0kOhm, and to possess an electrode collapse based off/on switching capability that operates via the application and subsequent removal of a strong bias voltage.

W-263 PARAMETRIC FILTERING SURPASSES RESONATOR NOISE
IN ALN CONTOUR-MODE OSCILLATORS 1269
C. Cassella, N. Miller, J. Segovia-Fernandez, and G.Piazza
Carnegie Mellon University, USA

We developed a new method for lowering the phase noise of oscillators where the intrinsic resonator frequency fluctuations represent the dominant noise source. We called this technique "parametric filtering".This method has been applied to a 227 MHz aluminum nitride contour-mode MEMS resonator that shows high level of intrinsic noise which limits the oscillator phase noise. By using this new approach we have obtained an improvement of more than 20 db and 26 db respectively at 1 khz and 10 khz offset. This resulted in the lowest phase noise level ever measured for any MEMS based oscillator.

Th-264 POLYCIDE CONTACT INTERFACE TO SUPPRESS SQUEGGING
IN MICROMECHANICAL RESOSWITCHES 1273
Y. Lin, R. Liu, W. Li, and C.T.C. Nguyen
University of California, Berkeley, USA

The use of a Pt-silicide-based contact interface has greatly reduced impact-induced energy loss in comb-driven resonant micromechanical switches (a.k.a., resoswitches) to the point where squegging phenomena (whereby impacts do not occur on every cycle) are eliminated, so no longer constrain the clock frequency of recently demonstrated mechanical charge pumps. This opens the application range of such charge pumps to higher power converters capable of delivering currents much higher than those of previously demonstrated version, which targeted low current-draw MEMS dc-biasing applications.

M-265 STABLE CHARGE-BIASED CAPACITIVE RESONATORS
WITH ENCAPSULATED SWITCHES 1277
E.J. Ng, K.L. Harrison, C.L. Everhart, V.A. Hong, Y. Yang, C.H. Ahn, D.B. Heinz, R.T. Howe, and T.W. Kenny
Stanford University, USA

We show that an electrically isolated silicon resonator within an epi-seal polysilicon encapsulation can retain a charge for prolonged periods of time with no noticeable leakage, even at elevated temperature. The charge is applied using a silicon contact switch that operates within the epi-seal cavity to isolate the resonator from the environment.

T-266 STABLE PULL-IN ELECTRODES FOR NARROW GAP ACTUATION 1281
E.J. Ng[1], Y. Yang[1], V.A. Hong[1], C.H. Ahn[1], D.L. Christensen[1], B.A. Gibson[2],
K.R. Qalandar[2], K.L. Turner[2], and T.W. Kenny[1]
[1]Stanford University, USA and [2]University of California, Santa Barbara, USA

We report on the use of pull-in electrodes for achieving narrower gaps than lithography/etch capabilities. Resonant devices with sub-ppm stability are demonstrated within the epi-seal polysilicon encapsulation process using pulled-in electrodes. The pull-in effect is reversible and can be made permanent by welding.

W-267 ULTRA-STABLE NONLINEAR THIN-FILM PIEZOELECTRIC-ON-SUBSTRATE
OSCILLATORS OPERATING AT BIFURCATION 1285
H. Fatemi, M. Shahmohammadi, and R. Abdolvand
Oklahoma State University, USA

Presented is a ~27MHz oscillator incorporating a thin-film piezoelectric-on-silicon (TPoS) resonator with a phase noise (PN) of -139 dBc/Hz at 1kHz and -157 dBc/Hz at 1MHz from the carrier. The close-to-carrier PN is equivalent to -148 dBc/Hz when normalized to 10MHz and is the lowest reported to date for MEMS oscillators. Additionally, it is experimentally proven that the PN significantly improves when the resonator is driven at or beyond the bifurcation point in the closed-loop oscillator circuit.

Th-268 VARIABLE CAPACITOR WITH SWITCHING MECHANISM FOR
WIDE TUNING RANGE AND LOW POWER CONSUMPTION 1289
D. Baek, Y. Eun, D.S. Kwon, M.O. Kim, T. Chung, and J. Kim
Yonsei University, SOUTH KOREA

We developed a variable capacitor with mechanical switching mechanism and reversible mechanical latching system to enhance tuning ratio and reduce power consumption. The switching mechanism could connect four sets of capacitors arranged in parallel sequentially by controlling the displacement of a microactuator for abrupt and coarse tuning of total capacitance. Continuous and fine tuning was also achieved by gap-closing mode of comb-finger type capacitors. The resultant maximum tuning ratio was 5.71 by combining coarse and fine tuning.

NANOFIBER FORESTS WITH HIGH INFRARED ABSORPTANCE

H.Y. Mao[1,3], C. Lei[1,4], Y. J. Chen[1,4], Z. J. Chen[2], W. Ou[1,3], W. G. Wu[2], A. J. Ming[1,3], and D. P. Chen[1,3]

[1]Key Laboratory of Microelectronics Devices & Integrated Technology, Institute of Microelectronics, Chinese Academy of Sciences, Beijing 100029, P. R. China; *Email: maohaiyang@ime.ac.cn
[2]National Key Laboratory of Science and Technology on Micro/Nano Fabrication, Institute of Microelectronics, Peking University, Beijing 100871, P. R. China
[3]Smart Sensor Engineering Center, Jiangsu R&D Center for Internet of Things, Wuxi 214135, P. R. China
[4]National Key Laboratory for Electronic Measurement Technology, North University of China, Taiyuan 030051, P. R. China

ABSTRACT

In this work, nanofiber forests with high infrared (IR) absorptance are reported. In wavelength range from 1.5 to 5 μm, the absorptance of the nanofiber forests reaches a minimum of 96%, which is much higher than that of Si_3N_4-based IR absorbers and the polymer coatings from which the nanofibers are obtained. Such nanofiber forests are fabricated by using a plasma-stripping-of-polymer technique, which is fast, high-yield and applicable to a wide range of polymers. Moreover, the technique is highly compatible with micro-fabrication. As a result, the nanofiber forests can be introduced into MEMS (Micro-Electro-Mechanical systems) IR devices, and it is expected that the performance of such devices can be improved.

INTRODUCTION

Infrared (IR) sensors can detect IR radiations from both living and non-living objects, they have wide applications in fields including gas analysis, temperature sensing, motion detection, imaging, and so on [1, 2]. Currently, MEMS (Micro-Electro-Mechanical systems) IR sensors based on various sensing principles have been presented. Though consist of different components, all these devices contain IR absorbers [3], and their performances are strongly dependent on the absorption capability of the absorbers. The usage of coating materials as IR absorbers, which include SiN_x or SiO_2-SiN_x-SiO_2 thin films, is limited by their relatively low absorption efficiency [4], and constrained by the interference effect of the coatings. In order to improve performance of the sensors, different types of highly efficient absorbers have been proposed, including quarter-wavelength structures [5], metal blacks such as gold-black, silver-black and platinum-black, as well as some others [6]. Because of the large porosity and the uneven distribution of the small structures, metal blacks exhibit relatively low heat capacities, high IR absorptance with negligible interference of the substrates. However, the fabrication compatibility of the metal blacks with micro-fabrication process remains an unsolved issue, which restricts their applications in high-yield MEMS IR sensors. In theory, quarter-wavelength structures can achieve extremely high IR absorption, but each of these structures can only detect few IR radiations with specific central wavelengths. Moreover, their absorptance depends largely on matching thickness of the dielectric mediums with quarter-wavelength of incident lights, and little difference between them may lead to a huge decay in the absorptance. As a result, precise control of parameters, both in structural design and fabrication process, is rigorously required.

As has been reported, nanostructure forests such as nanopillar forests can also offer high optical sensitivity in the IR spectral region [7]. Conventionally, the IR absorptive nanostructure forests were fabricated by irradiating silicon with femtosecond-laser pulses under atmosphere containing sulfur. The process is quite easy to implement, nevertheless, the semi-serial nature of this technology makes it both time- and labor-consuming. Several novel technologies aimed at convenient and cheap fabrication of similar nanostructure forests have been developed. Among them, the most commonly used technologies are vapor-liquid-solid (VLS) growth [8] and nanosphere lithography [9]. In VLS technology, however, nanoscale metal droplets are required, and they need to be scattered as catalysts, which makes the process complicated. Spin-coated or self-assembled monolayer nanospheres can be used as masks to generate nanostructure forests. Even so, forming a large-area of single nanosphere layer usually requires some specially designed complicated techniques.

In this work, highly IR absorptive nanofiber forests, a new type of nanostructure forest, are fabricated by using a novel plasma-stripping-of-polymer approach. The key technique of the approach is taking advantage of a combined action of plasma ashing and plasma polymerization. Such an approach is highly flexible, controllable and micro-fabrication compatible. In wavelength range from 1.5 to 5 μm, the absorptance of the nanofiber forests reaches a minimum of 96%, which is much higher than that of Si_3N_4-based IR absorbers and the polymer coatings from which the nanofibers are obtained.

FABRICATION

In conventional micromachining processes, plasma stripping is usually utilized to remove unwanted polymers. It has been observed that after a long period of bombardment, nano-materials could be generated [10-13]. Based on this phenomenon, nanofiber forests with different morphologies are successfully fabricated, the process is illustrated in Fig. 1. At the beginning, polymer is spin-coated and photo-patterned (Fig. 1(a)) on a substrate, and then plasma stripping is performed on the polymer patterns for a few minutes (Fig. 1(b)). In this way, the polymers are ashed away, and meanwhile, nanofiber forests are obtained in regions of the original polymer patterns (Fig. 1(c)). The nanofibers are assumed to form due to the combined action of plasma ashing and plasma polymerization. Monomer molecules generated in plasma ashing process are energized and dissociated into neutral particles and reactant fragments, which are further repolymerized in the following plasma stripping process. As stripping continues, highly branched and cross-linked networks are gradually generated and consequently appear in morphologies of nanofibers.

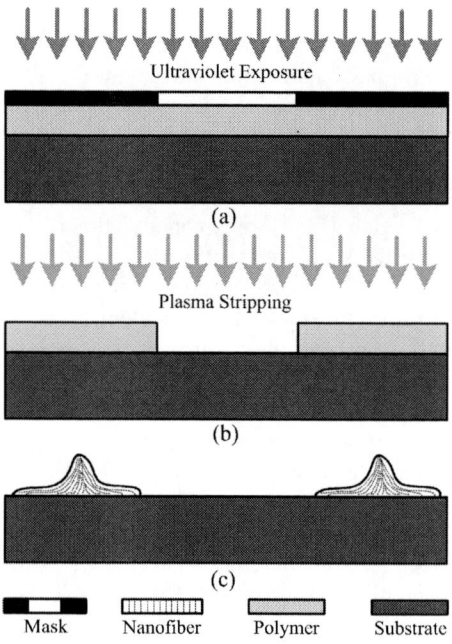

Figure 1. Fabrication of nanofiber forests by plasma-stripping-of-polymer technique.

In our experiments, oxygen and argon plasmas were employed to bombard different types of polymers, including SPR-220 photoresist (SPR), SU-8 3010 photoresist (SU-8), and polyimide (PI), thus to fabricate nanofiber forests of different kinds. Figure 2 illustrates the scanning electron microscopy (SEM) images of the obtained nanofiber forests. Figure 2 (a) shows nanofiber in morphologies of bunches, which were realized by oxygen-plasma-stripping-of-SPR. The fiberous bunches are 2-3 microns in dimension, while the original photoresist patterns were ~4.5 μm in thickness. Figure 2 (b) demonstrates fibers from oxygen-plasma-stripping-of-PI. The fibers present as individual nanowires with a length of ~3.3 μm, while the original PI layer was also ~4.5 μm in thickness. Figure 2 (c) and (d) show nanofiber in bushes generated from oxygen-plasma-stripping-of-SU-8, where the SU-8 patterns were ~8 μm in thickness. Results from argon plasma stripped SPR and SU-8 are shown in Fig. 2 (e) and (f), where the original thicknesses of the patterns were 3 and 6 μm, respectively. In our experiment, the RF power for oxygen plasma stripping process was 250 W, and the O_2 flow rate was ~30 sccm. For argon bombardment, the RF power was 230 W and flow rate of Ar was ~45 sccm.

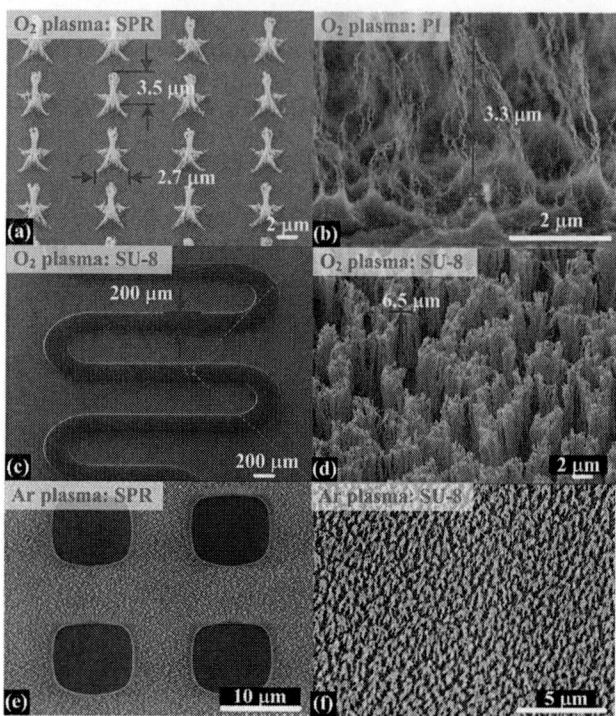

Figure 2. SEM images of the different nanofiber forests, and the nanofiber forests are obtained by (a) oxygen-plasma-stripping-of-SPR; (b) oxygen-plasma-stripping-of- PI; (c-d) oxygen-plasma-stripping-of-SU-8; (e) argon-plasma-stripping-of-SPR; (f) argon-plasma-stripping-of-SU -8.

Auger electron spectroscopy (AES, JEOL JAMP-9500F, Japan) was adopted to investigate the chemical changes from original polymers to the nanofibers. The results are demonstrated in Fig. 3. As shown in these curves, the chemical components are slightly changed. For the SPR-220 photoresist, the intensity of C decreases while that of O increases after oxygen and argon plasma treatment. Si and Al detected in the nanofibers from the SPR can be ascribed to the substrates on which the photoresist was coated and the atmosphere of the stripping machine. For the SU-8 3010 phtoresist, it is noted that a small amount of Sb content is detectable, which is assumed to be from the photoinitiator in the SU-8 photoresist.

CHARACTERIZATION

To measure the IR absorptance of the nanofiber forests, a large-area of nanofibers generated from oxygen-plasma-stripping-of-SU-8, as shown in Fig. 2 (d), were chosen as the analysis samples. IR transmittance and reflectance of the nanofiber forests were analyzed using Varian 660-IR (Varian Medical systems, USA), and then the data were analyzed by Varian Resolution Pro software package. Consequently, the IR absorptance was acquired according to the energy conservation principle. In the measurement, wavelength of the incident IR lights was within a range of 1.5 μm to 5 μm, which covers applications for gas sensing of different types [1, 2]. As shown in Fig. 4, the absorptance of the nanofiber forests reaches a minimum of 96%.

978-1-4799-3510-9/14 $31.00 © 2014 IEEE

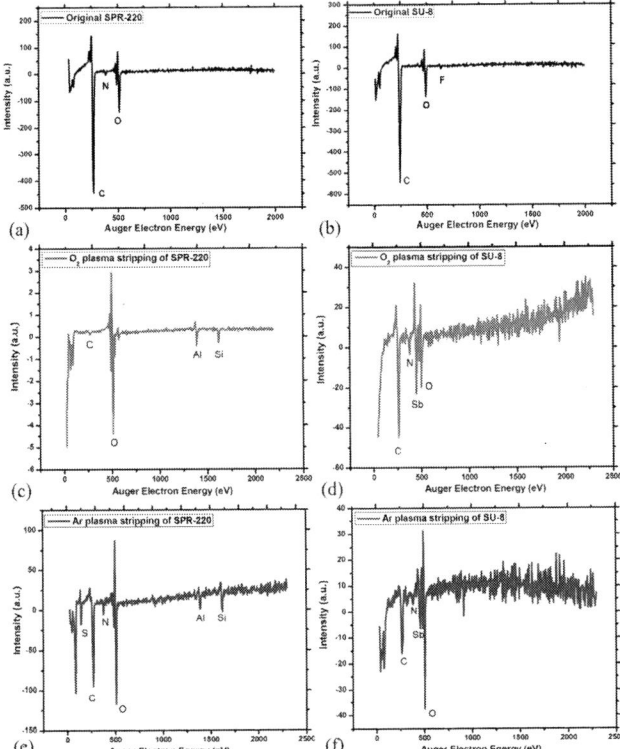

Figure 3. AES of SPR-220 and SU-8 3010 photoresist as well as the nanofiber forests obtained from the SPR and SU-8 by oxygen and argon plasma stripping.

To evaluate the advance of the nanofiber forests, IR absorptance of the original SU-8 layers (~8 μm in thickness) was also measured, shown as the red curve in Fig. 4. In the range of 1.5-5 μm, the absorption spectrum of the SU-8 layer resonates between 77 % and 95%. Besides, Si_3N_4 films with different thicknesses (400 nm, 530 nm and 720 nm) were prepared as well, and the IR absorptance analysis was conducted using the same Varian system. As illustrated in Fig. 4, the IR spectra of Si_3N_4 films resonate in a range of 47-80%, which is also much lower than that of the nanofiber forests. It should be noted that all these materials were deposited or fabricated on single crystal silicon substrates.

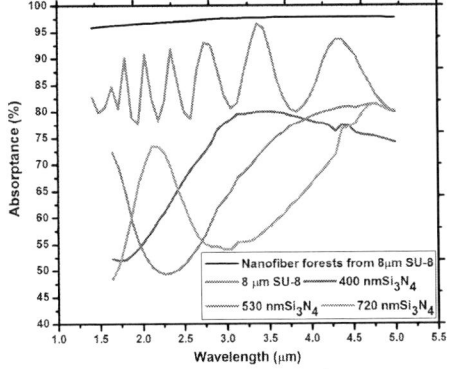

Figure 4. IR absorptance of nanofiber forests, SU-8 and Si_3N_4 coatings on Si substrates. In 1.5-5 μm, the absorptance of the nanofiber forests is larger than 96%, which is much higher and more stable than that of the other coatings.

The large IR absorptance of the nanofiber forests could be attributed to the optical properties of the forests for their randomly distribution and large density. The nanofibers standing on surface of a substrate may lead to multiple reflections of the incident IR lights. Namely, IR radiation incidenting onto the nanofiber forests is likely to undergo more than one reflections before leaving the surface, and IR absorption takes place at every position of the reflection. In this way, the optical energy absorbed by the wafer increases quickly, thus it leads to higher IR absorption of the wafer. Meanwhile, the spectral resonance of the original SU-8 layer and the Si_3N_4 films could be ascribed to interference between reflections at the interfaces of the coatings. While for the nanofiber forests, their surfaces are very rough and the thickness of the forest-layer varies at different positions, consequently, the interference effect is avoided.

APPLICATION

As the process for generating the nanofiber forests is highly CMOS compatible, the forests can further be introduced into MEMS IR devices as IR absorbers. Figure 5 demonstrates the process for integrating nanofiber forests into a thermopile-based IR sensor. After main components of the sensor, including the absorber, thermopiles, and electrodes, are fabricated, polymer (e.g. SU-8) is spin-coated and patterned on the Si_3N_4-based absorber, meanwhile, releasing holes are opened in both the polymer layer and the Si_3N_4 layer. Subsequently, the silicon substrate beneath the absorbing area is etched away by XeF_2 dry etching. After structural release, the device is further suffered to plasma stripping (e.g. oxygen plasma) for several minutes thus to form highly IR absorptive nanofiber forests. The fabricated thermopile-based IR sensor with nanofiber forests on its absorbing area is shown in Fig. 6.

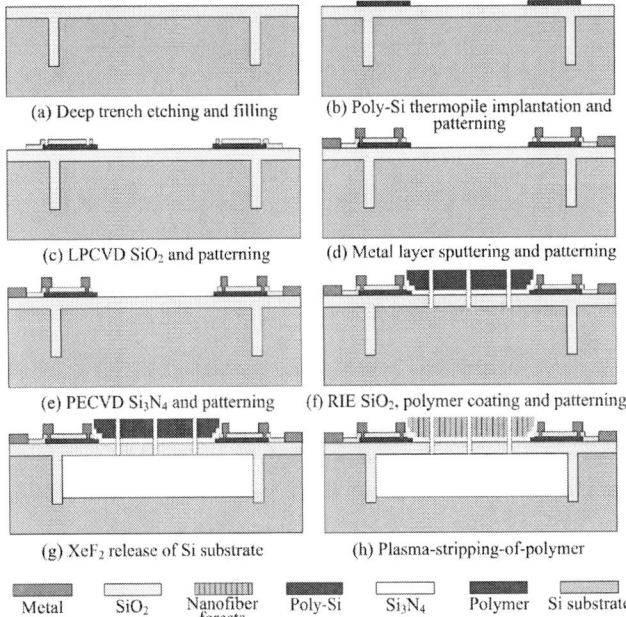

Figure 5. Fabrication process for integrating the nanofiber forests into thermopile-based MEMS IR sensors.

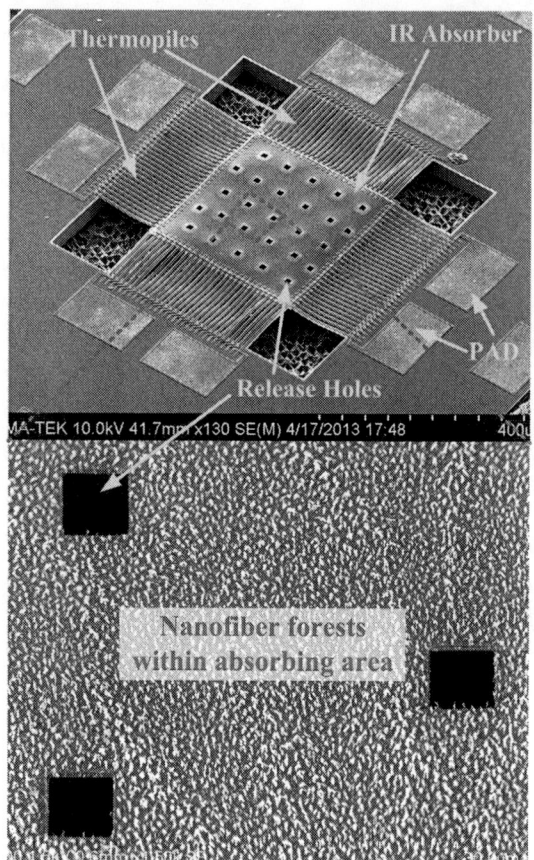

Figure 6. SEM images of a thermopile-based MEMS IR device, on the absorbing area of which, highly IR absorptive nanofiber forests are successfully introduced.

CONCLUSIONS

Nanofiber forests with different morphologies are fabricated by using a plasma-stripping-of-polymer approach, which is facile, flexible and applicable to a wide range of polymers. In wavelength range from 1.5 to 5 μm, the nanofiber forests can reach an IR absorptance of > 96%, which is much higher than that of conventional Si_3N_4 absorbers and the original polymer layers. In addition, such a method is highly compatible with micromachining process, therefore, the nanofiber forests can be introduced into MEMS IR sensors. By doing so, the IR sensors are expected to achieve higher performance, and thus they may meet more application requirements.

ACKNOWLEDGMENT

This work was supported in part by Jiangsu Natural Science Foundation (Grant No. BK20131098), the National Natural Science Foundation of China (Grant No. 61335008 & 61176114), and the National Basic Research Program of China (973 Program, Grant No. 2009CB320300).

REFERENCES

[1] S Udina, M Carmona, G Carles, J Santander, L Fonseca, and S Marco, "A micromachined thermoelectric sensor for natural gas analysis: Thermal model and experimental results," *Sensors Actuators* B, 134 (2008), pp. 551-558.

[2] J Frank and H Meixner, "Sensor system for indoor air monitoring using semiconducting metal oxides and IR-absorption," *Sensors Actuators* B, 78 (2001), pp. 298-302.

[3] H Mao, Y Chen, Y Ou, W Ou, J Xiong, C You, Q Tan, and D Chen, "Fabrication of Nanopillar Forests with High Infrared Absorptance Based on Rough Poly-Si and Spacer Technology," *J. Micromech. Microeng.*, 23 (2013) 095033.

[4] D Xu, B Xiong, Y Wang, M Liu, and T Li, "Integrated micromachined thermopile IR detectors with an XeF_2 dry-etching process," *J. Micromech. Microeng.*, 19 (2009), pp. 125003.

[5] S Sedky, P Fiorini, K Baert, L Hermans, and R Mertens, "Characterization and optimization of infrared poly SiGe bolometers," *IEEE T Electron Dev.*, 46 (1999), pp. 675-682.

[6] D J Advena, V T Bly, and J T Cox, "Deposition and characterization of far-infrared absorbing gold black films," *Appl. Opt.*, 32 (1993), pp. 1136-1144.

[7] M Stubenrauch, M Fischer, C Kremin, S Stoebenau, A Albrecht, and O Nagel, "Black silicon-new functionalities in microsystems," *J. Micromech. Microeng.*, 16 (2006), pp. S82-S87.

[8] H J Fan, P Werner, and M Zhcharias, "Semiconductor nanowires: from self-organization to patterned growth," *Small*, 2 (2006), pp. 700-717

[9] X Wang, C J Summers, and Z L Wang, "Large-scale hexagonal-patterned growth of aligned ZnO nanorods for nano-optoelectronics and nanosensor arrays," *Nano Lett.*, 4(2004), pp. 423-426.

[10] H Mao, D Wu, W Wu, J Xu, and Y Hao, "Fabrication of Diversiform Nanostructure Forests Based on Residue Nanomasks Synthesized by Oxygen Plasma Removal of Photoresist," *Nanotechnology*, 20 (2009), pp. 445304.

[11] H Mao, W Wu, D She, G Sun, P Lv, and J Xu, "Microfluidic Surface-Enhanced Raman Scattering Sensors Based on Nanopillar Forests Realized by Oxygen-Plasma- Stripping-of-Photoresist Technique," *Small*, 3013, DOI:10.1002/smll.201300036

[12] H Mao, D Wu, W Wu, J Xu, H Zhang, and Y Hao, "Fabrication of Nanopillars Based on Silicon Oxide Nanopatterns Synthesized in Oxygen Plasma Removal of Photoresist," *The 22nd IEEE International Conference on Micro Electro Mechanical System (IEEE MEMS 2009)*, Sorrento, Italy, January 25-29, 2009, pp. 677-680.

[13] H Mao, W Wu, Q Liu, Y Zhang, and Y Li, "Nanofiber-based Surface Microfluidic Structures for Cell and Nanoparticle Patterning," *The 14th International Conference on Miniaturized Systems for Chemistry and Life Sciences (MicroTAS 2010)*, Groningen, The Netherlands, October 4-7, 2010, pp. 500-502.

NANOSTRUCTURED SILICON FLAPPING WING WITH HIGHER STRENGTH AND LOW REFLECTIVITY FOR SOLAR POWERED MEMS AIRCRAFT

Kunal Kashyap[1], Amarendra Kumar[1], Chi-Nan Chen[1], Max T. Hou[2] and J. Andrew Yeh[1,3]

[1]Institute of Nanoengineering and Microsystems, National Tsing-Hua University, Hsinchu, Taiwan
[2]Department of Mechanical Engineering, National United University, Miaoli, Taiwan
[3]Instrument Technology Research Center, National Applied Research Laboratories, Hsinchu, Taiwan

ABSTRACT

This work presents silicon flapping wings design with higher bending strength and low reflectivity for MEMS aircraft. The design was bio-inspired by the cicada wings which are having numerous nanostructures upon it. The nanostructure on silicon increases the bending strength by ~6 times. The higher strength silicon samples can successfully demonstrate the 100 μm thick bendable silicon by nanostructure surface texturing used for flapping wings. Moreover, same nanostructures act as antireflective layer due to improved light trapping for solar cell. The reflectivity of the nanotextured surface reduces to ~2%.

INTRODUCTION

Flapping flight is very old concept inspired by many flying insects to design flapping wings for MEMS aircraft [1-3]. Nature makes it obvious that flapping flight can be very efficient and effective aerodynamically at low Reynolds number while still allowing for a high degree of control. Flapping wing devices fly by generating lift through oscillation or flapping of wings. Flapping wing needs dynamic movement in order to produce thrust and the lifting force [4]. MEMS aircraft also need a power module for driving the flapping wings. Solar cell or photovoltaic technology is a very promising self-powered technology when it comes for powering MEMS devices. The efficiency of the solar cell is important because it determines how much power a cell will produce based on amount of incident light being trapped. Solar cells as a photovoltaic power source limit the operation of the device during the daytime which requires additional power storage.

Most of materials used to construct the wings are flexible enough to fit the mechanical requirement. Using such wings (generally not photovoltaic) required an electronic module for power [5, 6]. Using photovoltaic materials instead to construct the wings may reduce or even eliminate the need of the electronic module. However, most of the photovoltaic materials, like silicon, are fragile [7]. Silicon is extensively used material for photovoltaic as well as for Micro-electro mechanical systems (MEMS) technology to create micrometer sized features and sub-millimeter sized devices. It is easily compatible with any kind of semiconductor process but not flexible enough to fit for flapping wings design.

This work presents an innovative method for attaining the higher strength silicon samples which results to envision even the 100 μm thick bendable silicon by nanostructure surface texturing used for flapping wings. Moreover, same nanostructures act as antireflective layer due to improved light trapping [8-10] for solar cell.

DESIGN AND FABRICATION

Recently, cicada (Cryptotympana takasagona Kato) is also considered as a bio-mimetic solution for designing the flapping wing for micro aerial vehicles (MAVs) [11]. The surface nanostructures distributed on the wing surface exhibit the superior physical properties for anti-reflective coverage [12]. However, the surface nanostructures can be viewed as parallel edge holes on the wing surface. The nanostructure morphology on the wing is shown in Figure 1. The SEM image was taken after cleaning the wings by acetone and DI water. In this study, the flapping wings design was inspired by the cicada wings.

Figure 1: Cicada wing having nanostructures on the surface

A method of simultaneously enhancing flexibility and light-harvesting efficiency for designing MEMS aircraft was achieved by fabricating nanostructures on the surface of silicon substrate (Figure 2). The nanostructures were fabricated on silicon substrates by metal-assisted wet chemical etching procedure. The ratio of $AgNO_3$, HF and H_2O in the etching solution dominates the formation mechanism of silicon nanostructures. The ratio of the etchant was varied to fabricate silicon nanostructures with very smooth profile at the interface of nanostructure and silicon bulk. The silver dentrites formed during the metal assisted etching were rinsed by HNO_3 and H_2O solution and further dehydrated by acetone and Isopropyl alcohol (IPA). Fabricated silicon nanostructures are shown in Figure 3, with a pitch of 200 ± 10 nm and width of 100 ± 10 nm respectively. The depth of nanostructures depending on etching time was varied from 1 μm to 8 μm for comparative study. The aluminum electrode fabricated after depositing the silicon nitride as a passivation layer upon silicon nanostructures. Low cost mass productive wet chemical etching process reduces the design complexity and cost.

978-1-4799-3510-9/14 $31.00 © 2014 IEEE

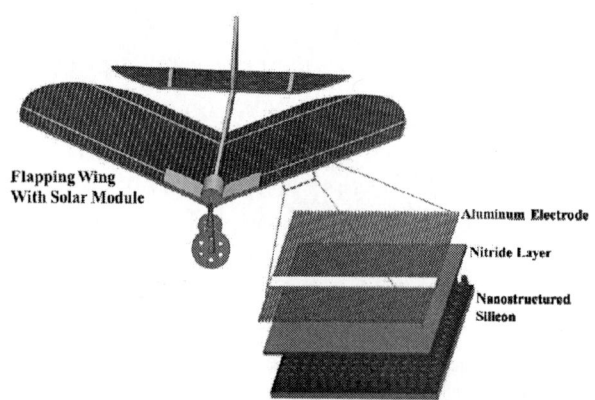

Figure 2: Schematic drawing of solar powered MEMS aircraft having nanostructure surface texturing on the wings.

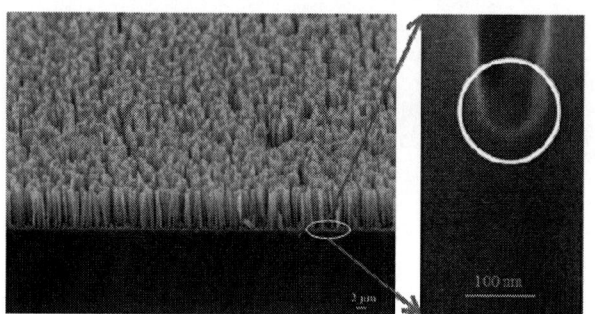

Figure 3: SEM image of nanostructured silicon flapping wing.

RESULTS AND DISCUSSION
Bending Strength Characterization

Silicon is brittle enough to fracture by initiating or propagating the crack under applied force. Most strength degrading defects of silicon substrates are surface-localized during manufacturing which leads to fracture due to stress concentration at defect under applied load. It is impossible to imagine a completely defect free silicon irrespective of their developing process. Silicon nanostructures on surface delocalize the stress concentrated at defects by redistributing the stress in nanostructure region resulting in enhancement of the bending strength. Three-point bending (3PB) test with ASTM E 855-90 standards were performed for measuring the bending strength by the material testing machine (HUNG TA HT-2102A) equipped with a 980-N load cell (HUNG TA 8336). The 3PB test was conducted at a displacement rate of 30 mm/min. The schematic of three point bending test is shown in Figure 4 (a). The bending strength of polished silicon and nanostructured silicon samples having length, width and thickness as 60 mm, 20 mm and 0.61 mm respectively were compared as shown in Figure 4 (b). The bending strength (σ_{br}) was calculated as equation 1[13].

$$\sigma_{br} = \frac{1.5 F_r L}{wt^2} \quad (1)$$

where F_r is load at rupture, L, w, t are respectively span length, width and thickness of specimen.

The polished silicon sample was having the bending strength

of 0.17 GPa. Bending strength increased as the depth of nanostructure increased and saturated at 6 μm depths of nanostructures showing the bending strength of 1.02 GPa which is about six times of polished silicon samples.

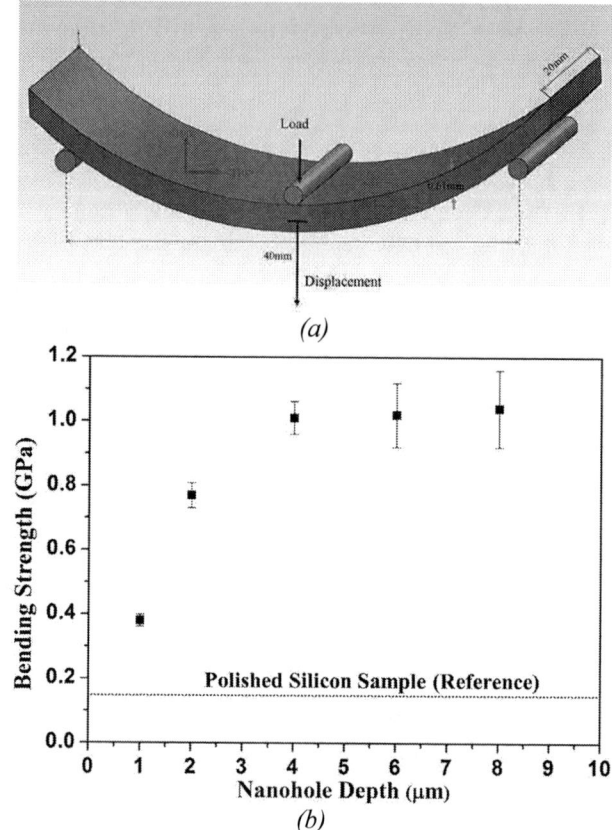

Figure 4: (a) Schematic of 3PB test (b) 3PB test showing 6 times bending strength enhancement of silicon sample.

This method can foresee the requirement of lighter flapping wing of MEMS aircraft by fabricating thinner silicon with larger strength and flexibility. Improved strength assured the better bendability of silicon samples. Figure 5(a) shows the bendability of 100 μm thick silicon wafer with 6 μm depth of nanostructures textured on the surface. Dynamic response is an issue essential to application of materials in vibration environment. The dynamic response for nanostructured silicon wings by applying cyclic vibration was bio-mimicked with the flapping behavior of cicadas. The nanostructure plate is attached to the transmission mechanism which successfully converts the rotary motion from the motor to the flapping motion of plate to experience the cyclic load. A series of animation for flapping behavior captured using high-speed camera (IDT –Y4) with frame rate of 10,000 frames per second under the illumination of 500-W halogen lamp. The nanostructure plate can sustain frequency of 60 Hz with 2 cm of maximum displacement without fracture as shown in Figure 5(b). Demonstrated nanostructured silicon wings for producing large displacement at reasonable frequency will overcome the limitations of silicon fragile nature in current silicon based

MEMS flapping wing aircraft technology.

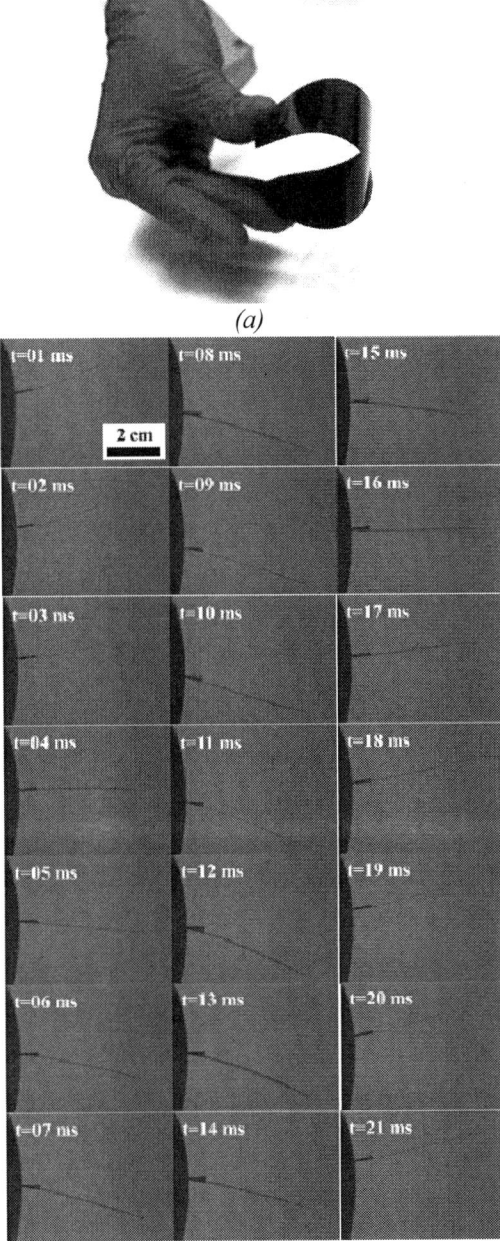

Figure 5: (a) Bendability of 100 μm thick silicon wafer (b) Dynamic response of flapping wing captured by high speed camera

Characterization of Nanostructure Antireflection

These silicon nanostructures are not only improving the flapping behavior although reducing the reflectivity of the samples which can be utilized for solar cells as power electronics for MEMS aircraft. Nanostructure surface textured wafer shown in Figure 6(a) is apparently noticeable as black wafer. Black silicon is extensively being researched for higher efficiency solar cells. Another advantage of this antireflection method is that the low reflectivity is insensitive

to the incident angle. The optical measurement for reflectivity was done by the spectrometer consisting of halogen light source with maximal power of 150 W and optical fibers able to transmit light within frequency from ultra-violet (UV) to near-infrared (NIR). The experiment set up for obtaining reflectivity is shown in Figure 6(b). The reflectivity spectrum from nanostructured silicon wing is compared with polished silicon sample as shown in Figure 6(c). The average reflectance of nanostructured silicon wings and polished silicon sample is about 2% and 40% respectively. The low reflectivity and simple fabrication has much potential to replace micro-texturing and another antireflection technique which is being used for solar cell as power option for MEMS aircraft.

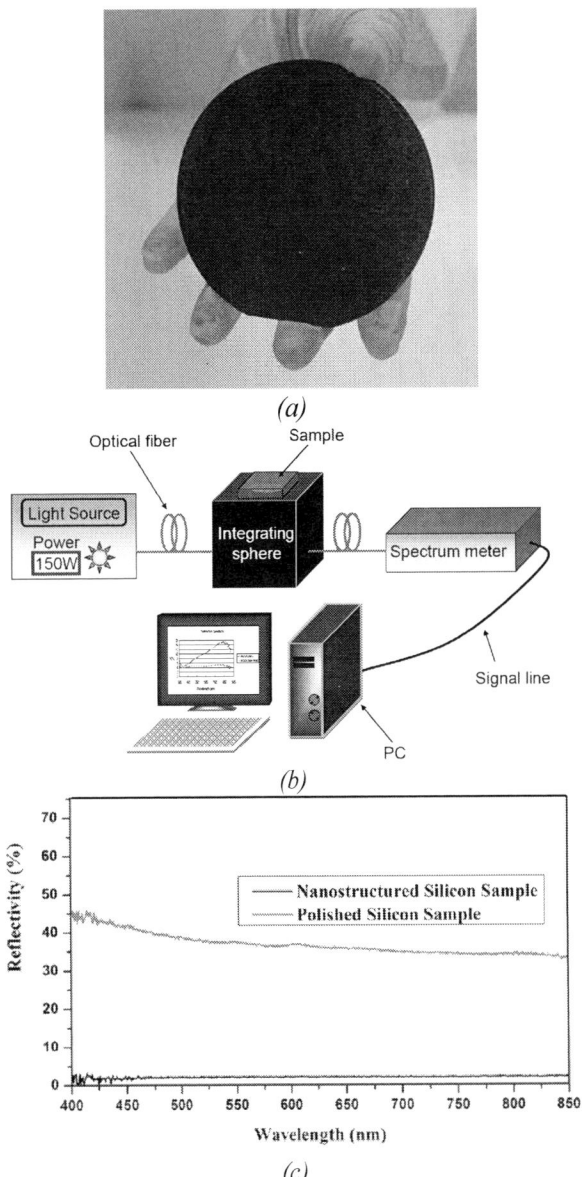

Figure 6: (a) Black wafer (b) Experiment set up for reflectivity measurement (c) Reflectivity comparison between nanostructured and polished silicon

978-1-4799-3510-9/14 $31.00 © 2014 IEEE

CONCLUSION

In summary, silicon flapping wings design for MEMS aircraft was proposed which increases the bending strength of wings by 6 times and reduces the reflectance to 2%. The wings thickness can be envisioned to be even less than 100 μm due to improved bending strength of silicon. This can effectively reduce the weight of the aircraft which will help to improve the MAV design criteria. Using silicon is very convenient for any kind of semiconductor and MEMS fabrication process. High strength and low reflectance simultaneously were achieved from nanostructure surface texturing by low cost wet chemical etching. The MEMS aircraft technology required flapping wing design with additional power generating source for long term usage. This technology achieves the requirement of flapping wing design and power generating source together with same module containing silicon nanostructures. Moreover, it can reduce the complexity and cost due to combining power generating source and wings together in a single module.

REFERENCES

[1] S. P. Sane and M. H. Dickinson, "The control of flight force by a flapping wing: Lift and drag production," *Journal of Experimental Biology,* vol. 204, pp. 2607-2626, Aug 2001.

[2] S. P. Sane and M. H. Dickinson, "The aerodynamic effects of wing rotation and a revised quasi-steady model of flapping flight," *Journal of Experimental Biology,* vol. 205, pp. 1087-1096, Apr 2002.

[3] J. Sirohi, "Microflyers: Inspiration from Nature," *Bioinspiration, Biomimetics, and Bioreplication 2013,* vol. 8686, 2013.

[4] X. Q. Bao, T. Dargent, S. Grondel, J. B. Paquet, and E. Cattan, "Improved micromachining of all SU-8 3D structures for a biologically-inspired flying robot," *Microelectronic Engineering,* vol. 88, pp. 2218-2224, Aug 2011.

[5] N. Pornsin-Sirirak, M. Liger, Y. C. Tai, S. Ho, and C. M. Ho, "Flexible parylene-valved skin for adaptive flow control," *Fifteenth Ieee International Conference on Micro Electro Mechanical Systems, Technical Digest,* pp. 101-104, 2002.

[6] H. Tanaka, K. Matsumoto, and I. Shimoyama, "A three-dimensional artificial insect wing by microcasting of thermosetting plastic using a thin elastic mold," *Proceedings of the Ieee Twentieth Annual International Conference on Micro Electro Mechanical Systems, Vols 1 and 2,* pp. 750-753, 2007.

[7] T. N. Pornsin-sirirak, Y. C. Tai, H. Nassef, and C. M. Ho, "Titanium-alloy MEMS wing technology for a micro aerial vehicle application," *Sensors and Actuators a-Physical,* vol. 89, pp. 95-103, Mar 20 2001.

[8] C. M. Hsieh, J. Y. Chyan, W. C. Hsu, and J. A. Yeh, "Fabrication of wafer-level antireflective structures in optoelectronic applications," *2007 Ieee/Leos International Conference on Optical Mems and Nanophotonics,* pp. 185-186, 2007.

[9] J. Oh, H. C. Yuan, and H. M. Branz, "An 18.2%-efficient black-silicon solar cell achieved through control of carrier recombination in nanostructures," *Nature Nanotechnology,* vol. 7, pp. 743-748, Nov 2012.

[10] E. Garnett and P. D. Yang, "Light Trapping in Silicon Nanowire Solar Cells," *Nano Letters,* vol. 10, pp. 1082-1087, Mar 2010.

[11] J. H. Park and K. J. Yoon, "Designing Cicada-Mimetic Flapping Wing with Composite Wing Structure and Application to Flapping MAV," *Intelligent Unmanned Systems: Theory and Applications,* vol. 192, pp. 119-133, 2009.

[12] F. Song, K. L. Lee, A. K. Soh, F. Zhu, and Y. L. Bai, "Experimental studies of the material properties of the forewing of cicada (Homoptera, Cicadidae)," *Journal of Experimental Biology,* vol. 207, pp. 3035-3042, Aug 2004.

[13] C. N. Chen, C. T. Huang, C. L. Chao, M. T. K. Hou, W. C. Hsu, and J. A. Yeh, "Strengthening for sc-Si Solar Cells by Surface Modification With Nanowires," *Journal of Microelectromechanical Systems,* vol. 20, pp. 549-551, Jun 2011.

CONTACT

*J.A. Yeh, +886-3-574-2912; jayeh@mx.nthu.edu.tw

POSSIBILITY OF CEMENTED CARBIDE AS STRUCTURAL MATERIAL FOR MEMS

Toshiyuki Morikaku[1], Tatsuya Fujii[1], Kazuki Kuroda[2], Yasuhiro Takami[2], Shozo Inoue[1], and Takahiro Namazu[1,3]*

[1]Department of Mechanical and Systems Engineering, University of Hyogo, Himeji, JAPAN
[2]Silveralloy Co., Ltd., Kasai, JAPAN
[3]JST PRESTO, Japan Science and Technology Agency, Kawaguchi, JAPAN

ABSTRACT

We discuss the possibility of WC-Co cemented carbide as structural material in MEMS. The cemented carbide is typically used as material for working tool because it has superior characteristics, such as very high Young's modulus, excellent rigidity, good chemical inertness, and good thermal stability. These are also attractive for mechanical elements in MEMS. We investigate the influences of specimen size and WC-Co composition ratio on mechanical properties of FIB-fabricated WC-Co cemented carbide nanowires by means of on-chip uniaxial tensile testing in FE-SEM. It is presented that Co binder size dependency on the Young's modulus and fracture strength has been obtained.

INTRODUCTION

As is well known, cemented carbides are a type of composite materials, which consist of hard carbide particles bonded together by a metallic binder. Since cemented carbides possess extremely excellent mechanical characteristics like very high Young's modulus, high hardness, high rigidity, and hard-wearing, in addition to good chemical inertness and good thermal stability, they are typically used for working tool. The representative cemented carbide is tungsten carbide (WC), the hard phase, together with cobalt (Co), the binder phase, as shown in Fig. 1, which can form the basic cemented carbide structure. When mechanical parts made of WC-Co are made at millimeter size or larger size, the structure can be regarded as continuum apparently. When the size is decreased to micron size or submicron size, however, the parts should be regarded as nonuniform because the cemented carbide includes submicron-sized WC particles. By downsizing the cemented carbide mechanical parts, the above-described superior mechanical characteristics might be reduced.

To judge whether micron- or submicron-sized cemented carbide structures can be used for MEMS or not, in specimen size ranging from nanometer to millimeter experimentally investigating the size effect on the mechanical properties is essential. We have several types of mechanical testing techniques for thin films and nanowires (NWs) [1]-[7]. By adopting them to evaluating size effect phenomena within nanometer to millimeter size, investigating the limit size that can show its excellent mechanical characteristics is significant, for realization of future MEMS structures made of the hard metal. In this paper, we report the results of uniaxial tensile tests using on-chip MEMS tension test device for evaluating FIB-fabricated WC-Co hard metal NWs, and

Figure 1: FE-SEM photograph of WC-Co hard metal surface. It is shown that WC particles are bonded with Co binder.

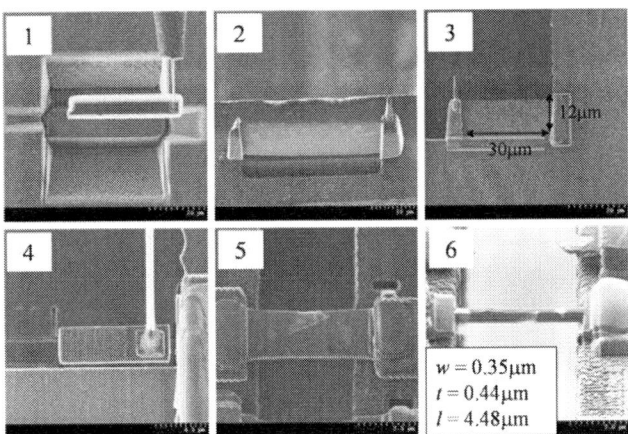

Figure 2: Snapshot of WC-Co hard metal NW fabrication using FIB.

discuss on the possibility of the use of the material as mechanical elements in MEMS.

EXPERIMENTAL PROCEDURE

WC-Co cemented carbide NWs were fabricated using FIB fabrication and microprobe sampling technologies that we have reported in MEMS 2013 [1]. As shown in Fig. 2, thin membranes made of WC-Co were fabricated using FIB from the hard metal bulk, and then these were attached to uniaxial tension MEMS device using FIB probe sampling and deposition functions. Just on the device, the membranes were formed using FIB to the NWs with various sizes (w = 247~508 nm; t = 269~1449 nm; l = 3.7~4.7 μm). In total 19

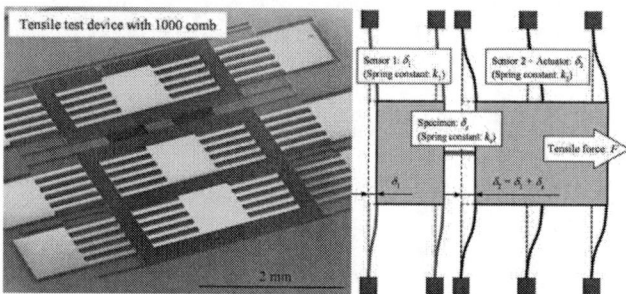

Figure 3: On-chip uniaxial tensile test device for NWs.

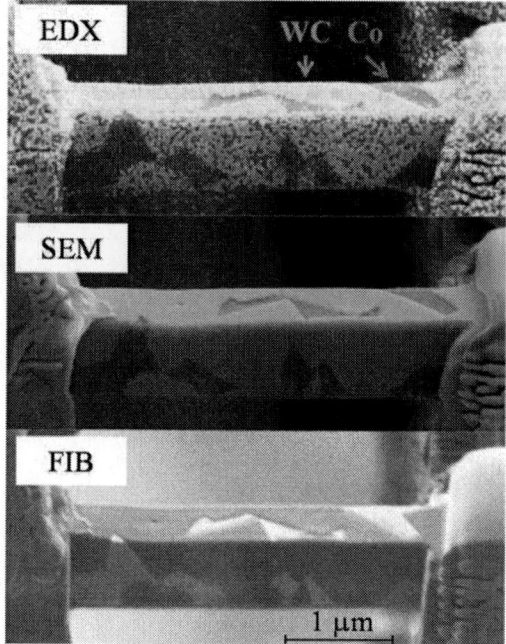

Figure 4: Comparison among EDX map, FE-SEM image, and SIM image.

NWs with different compositions were subjected to uniaxial tensile testing.

Fig. 3 shows developed MEMS tension test device made from SOI wafer. The device consists of many sets of comb structures used as electrostatic actuator for tensile force application and capacitive sensor for displacement and force measurements [2]. The elongation of the NWs is measured using both of the sensor and a digital image analysis system. All the tests were performed in a FE-SEM. Temperature was not controlled.

RESULTS AND DISCUSSIONS

Fig. 4 shows WC and Co mapping of a prepared WC-Co cemented carbide NW analyzed by EDX, along with FE-SEM and FIB images. In the EDX map, green and red colored portions are indicative of WC and Co, respectively. FE-SEM image shows that bright and dark portions indicate WC and Co, respectively, whereas FIB image shows that the relation between contrast and element is opposite.

Fig. 5 shows tensile stress-strain relations of all the WC-Co NWs tested, along with photographs of two representative NWs. All the NWs showed linear stress-strain

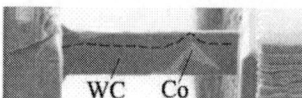

Type A: Non-segmentalized NW   Type B: Segmentalized NW

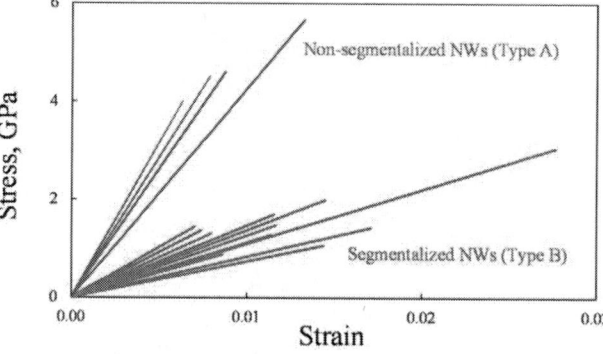

Figure 5: Tensile stress-strain relations of WC-Co NWs.

relations. The slopes were different in all the NWs because the ratio of WC to Co was different in every NW. This also indicates that the NWs' mechanical characteristics were not constant within the size range set in this work. The relations can be roughly categorized into two groups, which are indicated by blue and red colored lines. Blue (Type A) and red (Type B) colored data were obtained from the NWs where WC portions were visually non-segmentalized and segmentalized with Co, respectively, as exemplified in Fig. 5. This implies that the relation between WC grain size (or Co binder size) and specimen size strongly influenced the deformation of a whole specimen.

Fig. 6 (a) shows the relationship between WC area fraction and Young's modulus. The red and blue colored plots indicate WC-Co NWs with segmentalized and non-segmentalized WC, respectively. The black colored plot indicates WC-Co bulk values from the literature [8]. WC area fraction was derived from the mean value among area fraction values of the top and side surfaces of the NWs. When WC area fraction was 90 %, the Young's modulus was 587 GPa, which is approximately 12 % higher than the literature value for WC-Co cemented carbide [8]. The modulus value did not vary very much even though the area fraction increased to around 80%. The mean value for the segmentalized NWs (red colored plot) was 140 GPa, which is close to the Young's modulus for Co. At the area fraction higher than 80%, there were big differences in Young's modulus between the segmentalized and non-segmentalized NWs. The Young's modulus for the non-segmentalized NWs was 560 GPa on average, which was approximately 3 times higher than that for the segmentalized NWs. The Young's moduli for the non-segmentalized NWs were comparable to those for the literature value for WC-Co cemented carbide [8]. This indicates that at least these NWs possessed mechanical characteristics similar to the cemented carbide. On the other hand, in the NWs clearly showing Co segmentalization the deformation at Co portion was presumably dominant during tensile loading. The relation between WC area fraction and the fracture strength shown in Fig. 6 (b) showed a trend

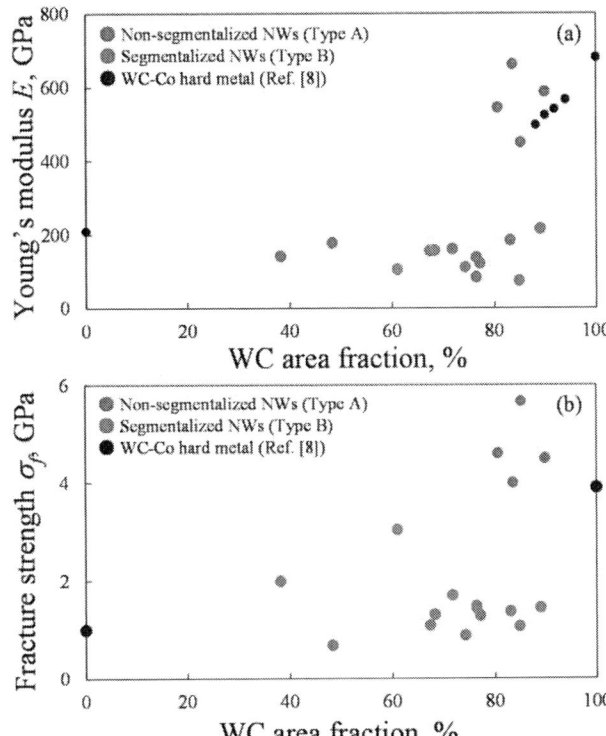

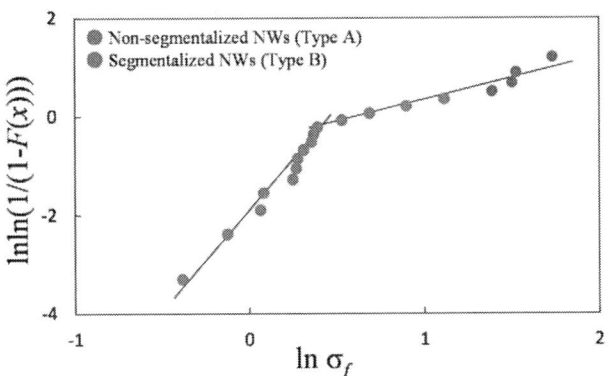

Figure 6: Relationships between WC area fraction, Young's modulus, and fracture strength of WC-Co NWs.

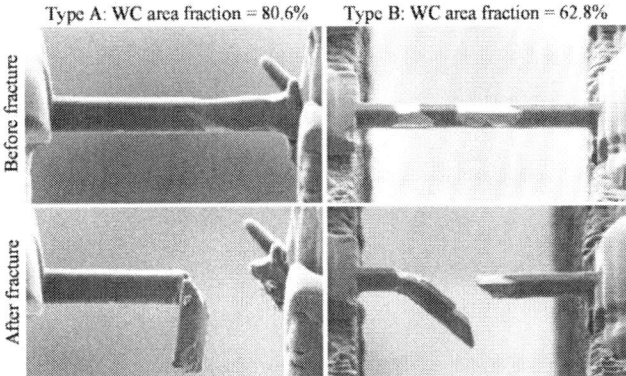

Figure 7: Weibull plot of tensile fracture strength.

similar to that between the fraction and the Young's modulus. When WC area fraction was higher than 80%, the fracture strength values scattered largely. The highest and lowest strength values were 5.66 GPa and 1.07 GPa, respectively, which are related to the degree of Co segmentalization. At the area fraction higher than 80%, therefore, we can make both of strong and weak structures with submicron sizes.

To discuss the NWs' fracture mechanism, the Weibull plot of the fracture strength was drawn, as shown in Fig. 7. The red and blue colored plots indicate WC-Co NWs with segmentalized and non-segmentalized WC, respectively. On the Weibull plot, all the strength data can be plotted on two straight lines. This implies that two different fracture mechanisms coexisted in the NWs tested.

For further discussion of fracture mechanism in more

Figure 8: FE-SEM photographs of WC-Co NWs before and after tensile test.

detail, fractured NWs were observed with FE-SEM. The representative results are shown in Fig. 8. Strong WC-Co NW with high WC area fraction is found to fracture at the WC and Co interface. It is also found that the fractured NW keeps its straight shape even after failure. Accumulated elastic strain energy during tensile loading would have been released very rapidly at the brittle fracture. The NW with low WC area fraction also fractured at the vicinity of WC and Co interface. The fractured NW was bent after fracture because deformable Co portion was relatively large. This phenomenon was found in many of the NWs with low WC area fraction. Just before fracture, a little plastic deformation might have happened in the NWs. All the fractured NWs tested in this work had an inclined fracture surface along the WC and Co interface. This indicates that, in all the NWs, a crack possibly propagated along the interface during shear deformation and then the NW fractured.

Based on the assumption that all the NWs follow the linear elastic theory, shear fracture strength, τ_f, can be roughly estimated using the equation, which is given by:

$$\tau_f = \frac{1}{2}\sigma_f \sin 2\phi \tag{1}$$

where σ_f is the tensile fracture strength and ϕ is the inclined angle of fracture surfaces.

Fig. 9 shows the relationship between inclined angle of fractured surface and estimated shear fracture strength. The red and blue colored plots indicate WC-Co NWs with segmentalized and non-segmentalized WC, respectively, as with Figs. 5-7. Shear fracture strengths for the segmentalized and non-segmentalized NWs were not dependent on the angle, although strength data scattering for respective NWs was observed. The mean values were 2.1 GPa and 0.63 GPa, which are close to estimated shear strength values, 2.0 GPa and 0.5 GPa, for WC and Co, respectively [9, 10]. Even in the segmentalized NWs, actually narrow Co portion between two WC particles is thought to probably exist because in WC-Co hard metal WC particles are bonded with Co binder. From the experimental facts obtained in this work, therefore,

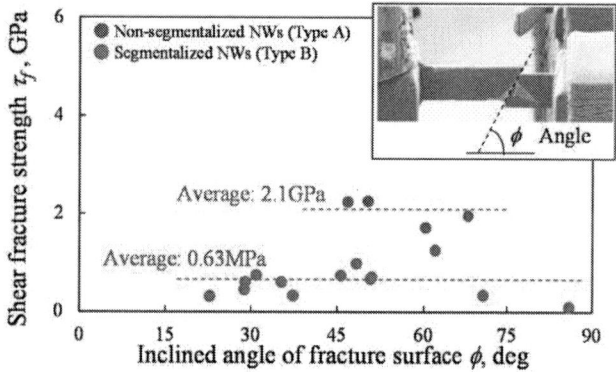

Figure 9: Relationship between inclined angle of fracture surface and shear fracture strength.

a threshold value of Co bonded size for showing excellent mechanical characteristics presumably exists. That is, only submicron-sized WC-Co hard metal structures, which have a Co bonded size smaller than the value, would be able to exert superior mechanical performances like typical hard metal materials.

CONCLUSIONS

In this work, we investigated mechanical properties of submicron-sized WC-Co cemented carbide NWs by means of on-chip uniaxial tensile testing, and discussed the possibility of the material as structural elements in MEMS. Since all the NWs showed linear stress-strain relation, they were found to fracture in a brittle manner. The Young's modulus and fracture strength depended on both WC area fraction and Co binder segmentalization. The Weibull plot of tensile fracture strength and FE-SEM observation suggested that two types of fracture mechanisms coexisted. All the NWs fractured at the vicinity of WC and Co interface, and these fracture surfaces were inclined along the interface angle to the longitudinal direction. Estimated shear fracture strength indicated that all the NWs fractured at shear deformation mode. The experimental results suggested that an optimum size of Co binder between two WC particles existed in order to make WC-Co cemented carbide MEMS components strong.

ACKNOWLEDGEMENTS

The authors express their gratitude to Dr. T. Takano, The New Industry Research Organization, Japan, for fruitful discussion on this study. Also we would like to thank Mr. C. Takami, President of Silveralloy Co., Ltd., Japan, for giving us helpful suggestions and physical supports in this collaborative work.

REFERENCES

[1] T. Fujii, K. Sudoh, S. Inoue, and T. Namazu, "Direct Tensile Testing of Sub-100nm-Size Silicon Nanowires Fabricated by FIB-Sampling of SON Membranes", *Proc. IEEE MEMS 2013, Taipei*, pp. 488-491, 2013.

[2] T. Fujii, K. Sudoh, S. Sakakihara, M. Naito, S. Inoue, and T. Namazu, "Nano-scale Tensile Testing and Sample Preparation Techniques for Silicon Nanowires", *Jpn. J. Appl. Phys.*, Vol. 52, No. 11, 110118 (9 pages), 2013.

[3] T. Fujii, T. Namazu, K. Sudoh, S. Sakakihara, and S. Inoue, "Focused Ion Beam Induced Surface Damage Effect on the Mechanical Properties of Silicon Nanowires", *Trans. the ASME, J. Eng. Mater. Technol.*, Vol. 135, Issue 4, 051002 (8 pages), 2013.

[4] M. Fujii, T. Namazu, H. Fujii, K. Masunishi, Y. Tomizawa, and S. Inoue, "Quasi-Static and Dynamic Mechanical Properties of Al-Si-Cu Structural Films in Uniaxial Tension", *J. Vac. Sci. Technol. B*, Vol. 30, No. 3, 031804 (9 pages), 2012.

[5] T. Namazu, H. Yamagiwa, and S. Inoue, "Tension-torsion Combined Loading Test Equipment for a Minute Beam Specimen", *Trans. the ASME, J. Eng. Mater. Technol.*, Vol. 135, Issue 1, 011004 (9 pages), 2013.

[6] T. Namazu, T. Fujii, M. Takahashi, M. Tanaka, and S. Inoue, "A Simple Experimental Technique for Measuring the Poisson's Ratio of Micro Structures", *J. Microelectromech. Syst.*, Vol. 22, Issue 3, pp. 625-636, 2013.

[7] T. Namazu, M. Fujii, H. Fujii, K. Masunishi, Y. Tomizawa, and S. Inoue, "Thermal Annealing Effect on Elastic-Plastic Behavior of Al-Si-Cu Structural Films under Uniaxial and Biaxial Tension", *J. Microelectromech. Syst.*, Vol. 22, Issue 6, pp. 1414-1427, 2013.

[8] T. Klünsner, S. Marsoner, R. Ebner, R. Pippan, J. Glätzle, and A. Püschel, "Effect of microstructure on fatigue properties of WC-Co hard metals", *Procedia Engineering*, Vol. 2, pp. 2001-2010, 2010.

[9] A.A. Karimpoor, U. Erb, K.T. Aust, G. Palumb, "High strength nanocrystalline cobalt with high tensile ductility", *Scripta Materialia*, Vol. 49, pp. 651-656, 2003.

[10] V. Livescu, B. Clausen, J.W. Paggett, A.D. Krawitz, E.F. Drake, M.A.M. Bourke, "Measurement and modeling of room temperature co-deformation in WC–10 wt.% Co", *Mater. Sci. Eng. A*, Vol. 339, pp. 134-140, 2005.

CONTACT

*T. Namazu: +81-79-267-4962; namazu@eng.u-hyogo.ac.jp

THIN-FILM MAGNESIUM AS A SACRIFICIAL AND BIODEGRADABLE MATERIAL

Yang Liu, Jungwook Park, Jay Han-Chieh Chang and Yu-Chong Tai
California Institute of Technology, USA

ABSTRACT

This work reports the study of ebeam-deposited thin-film magnesium (Mg) as a sacrificial and a biodegradable material. We have tested etchants including diluted hydrochloric acid (HCl), saline, and culture medium. Both vertical etching method and channel undercut method are used to characterize the Mg etching properties. The initial results confirm that thin-film Mg is a promising dual sacrificial and biodegradable material. In addition, an etching model, which fits accurately the etching length vs. time over a wide range of HCl concentrations (0.02-1M) is developed. This model is based on diffusion and a combined first-and-second order chemical reaction mechanism.

INTRODUCTION

Magnesium (Mg) and magnesium alloys have drawn a lot of attention as <u>biodegradable</u> materials because of their good mechanical properties as well as good biodegradability by Cl^--containing solution such as saline [1]. This means that magnesium could be an interesting dual "<u>sacrificial and biodegradable MEMS material</u>". It also can have a great potential to be applied to many parylene-based devices for implant applications, such as intraocular implantation [2-8]. To our knowledge, only Tsang et al. [9] has published a paper related to Mg in MEMS field, in which three ways of preparation of thick-film (>10μm) Mg were described including laser cutting, electrochemical etching (of commercial Mg foil), and electrodeposition (of Mg). Nevertheless, there's no reported work on thin-film Mg for MEMS applications. This work reports the first etching tests of the dual properties of ebeam-deposited "thin-film" Mg. Here we have tested etchants including diluted HCl, saline, and culture medium. Although preliminary, the initial results do show that thin-film Mg is a promising dual sacrificial and biodegradable material.

To study the etching of Mg, both vertical etching samples (Fig.1) and channel undercut [10] samples (Fig.2) are prepared. In vertical etching study, etching chemicals can be sufficiently supplied to the etching front so it's for the measurement of reaction-limited etching rates. Therefore, the etching time can be calculated just using the reaction-limited etching rate. However, if the etching length is long such as in channel etching, reactive chemicals at the etching front may be depleted. The etching rate then can be limited by diffusion mechanism. As a result, the model for channel etching has to include both diffusion and chemical reaction. In addition, a good etching model should also cover a wide range of etchant concentrations. A model which satisfies all these criteria will thereafter be universal and useful.

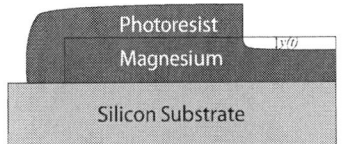

Figure 1: Schematic of magnesium vertical etching samples

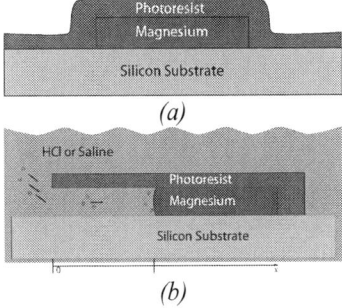

Figure 2: (a) Cross-section view of microchannel etching samples; (b) Schematic of etching sacrificial magnesium to form a microchannel.

FABRICATION

Like other sacrificial metal layers such as Al and Ni, Mg thin films can be prepared by ebeam evaporation and deposition because of its low melting temperature (~650°C). High-purity Mg pellets are also readily available. In this work, 0.06, 0.3 and 1.0-μm-thick Mg films are deposited on 4-inch silicon wafers.

The vertical etching samples (Fig.1) are then prepared with exposed etching windows in a 5-μm-thick photoresist. In this case, etching chemicals can be sufficiently supplied to the etched surface so it's for the measurement of reaction-limited etching rates.

We have also obtained the first diffusion-limited etching data of Mg using the channel undercut method. Sacrificial Mg strips are first formed to be 10μm wide. A 5-μm-thick photoresist is then spin-coated and patterned with an etching window (Figs.2a&b). In this case, the etchant from outside the window has to diffuse to the etching front inside the channel so diffusion can become an important factor that slows down the channel formation. All etching experiments are conducted at room temperature.

RESULTS AND MODELING

Vertical etching results

Fig.3a shows the data of the etched Mg thickness versus time with various HCl concentrations in the reaction-rate-limited condition. The dependence of magnesium etching rates on HCl concentrations are plotted in Fig.3b and a rate of 0.37μm/s·M is obtained.

978-1-4799-3510-9/14 $31.00 © 2014 IEEE

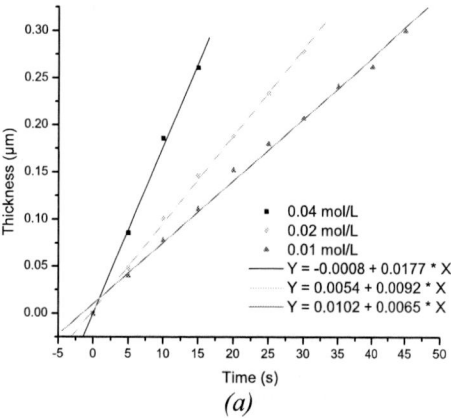

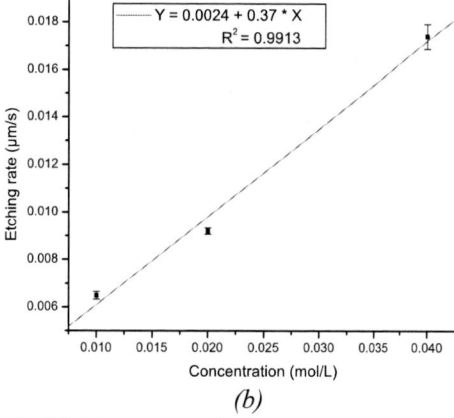

Figure 3: (a) Magnesium thickness versus time of etching in different hydrogen chloride concentrations; (b) Magnesium etching rate versus hydrogen chloride concentration.

Etching model of sacrificial Mg channel

Fig.2b shows the schematic of etching sacrificial magnesium to form microchannel. In order to study the model, we make the following assumptions: the diffusion coefficient is a constant; the heat generation during reaction is negligible to heat up the etching environment; convection inside the channel is negligible; and the etching process is one-dimensional. Tab.1 shows the symbols of all variables and coefficients in our model.

Table 1: Coefficients in the etching model.

Symbol	Expression
C	Concentration of reaction chemical
C_b	Bulk reaction chemical concentration
C_s	Etching front chemical concentration
u	Flow velocity
x	Etching length
D	Diffusion coefficient
J	Diffusive flux
t	Time
M	Molar mass
ρ	Density
k_1, k_2	Constants
a, b, e, f	Intermediate variables

Using the coordinates defined in Fig. 2b, the continuity equation is,

$$\frac{\partial C}{\partial t} + u\frac{\partial C}{\partial x} = D\frac{\partial^2 C}{\partial x^2} \tag{1}$$

Neglecting convection, (1) becomes,

$$\frac{\partial C}{\partial t} = D\frac{\partial^2 C}{\partial x^2} \tag{2}$$

According to Fick's first law,

$$J = -D\frac{\partial C}{\partial x} \tag{3}$$

Assume the chemical reaction is

$$Mg + 2HCl \rightleftharpoons MgCl_2 + H_2\uparrow$$

In a stoichiometric reaction,

$$J[Mg] = \frac{1}{2}J[HCl] \tag{4}$$

The etching rate is proportional to $J[Mg]$ at the etching front,

$$\frac{dx}{dt} = J[Mg]\frac{M[Mg]}{\rho[Mg]} \tag{5}$$

From (4) and (5),

$$\frac{dx}{dt} = \frac{1}{2}J[HCl]\frac{M[Mg]}{\rho[Mg]} \tag{6}$$

The other boundary conditions are,

$$C(0,t) = C_b \tag{7}$$

$$C(x(t),t) = C_s \tag{8}$$

In the combined First-and-Second model, the HCl concentration distribution is linear (9), and the reaction rate includes both a first and second order term (10),

$$J[HCl] = \frac{D(C_b - C_s)}{x} \tag{9}$$

$$J[HCl] = k_1 C_s + k_2 C_s^2 \tag{10}$$

Solving (9) and (10), we have,

$$J[HCl] = \frac{1}{2k_2}(\frac{D}{x})^2[1 - a + (k_1 + 2C_b k_2)(\frac{x}{D})] \tag{11}$$

In which,

$$a = \sqrt{1 + (\frac{k_1 x}{D})^2 + 2(k_1 + 2C_b k_2)(\frac{x}{D})} \tag{12}$$

Substitute Eq. (11) into (6), one has,

$$\frac{dx}{dt} = b(\frac{D}{x})^2[1 - a + (k_1 + 2C_b k_2)(\frac{x}{D})] \tag{13}$$

In which,

$$b = \frac{M[Mg]}{4k_2\rho[Mg]} \tag{14}$$

By integrate (13) from time 0 to t, we have $t=F(x)$,

$$t = \frac{D\{ae + 2k_1^2\frac{x}{D} + (k_1 + 2C_b k_2)[(\frac{k_1 x}{D})^2 - 1]\}}{8bC_b k_1^2 k_2 f} - \frac{D}{2bk_1^3}\log\frac{k_1 a + e}{2f} \tag{15}$$

Where,

$$e = k_1 + k_1^2\frac{x}{D} + 2k_2 C_b \tag{16}$$

$$f = k_1 + k_2 C_b \tag{17}$$

k_1, k_2 and D are determined by experimental data from the sacrificial Mg channel etching.

Sacrificial Mg channel etching results

Fig.4 shows the etched channel length versus time for different HCl concentrations. The initial etching rates derived from Fig.4 further verify the data from Fig.3. For instance, the etching rate for 0.04M HCl is 1.77E-2 µm/s (Fig.3b) while the initial etching rate from Fig.4 is 1.51E-2 µm/s. Coefficients k_1, k_2 and D can be determined experimentally. Nonlinear least squares fitting (NLSF) method, which is based on the Levenberg-Marquardt (LM) algorithm [11], is used to fit Eq. (15) to experimental data and the results are shown in Fig. 4. This model fits all data using $k_1 = 9.33$E-3 cm/sec, $k_2 = 275.71$E3 cm^4/mol·sec and $D = 3.86$E-6 cm^2/sec.

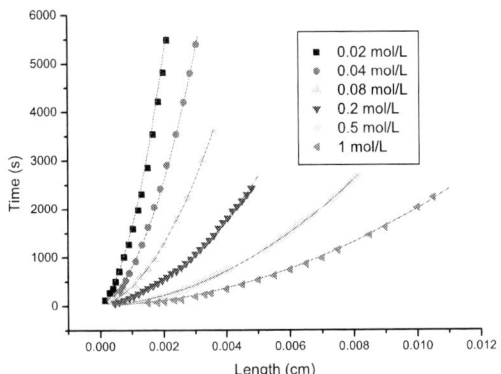

Figure 4: (1-µm-thick Mg) Etched microchannel length versus time for different hydrogen chloride concentrations.

Fig.5 shows the thickness effect with three different thicknesses (i.e., 0.06, 0.3 and 1.0 µm) of Mg in 0.04M HCl. The results show that the undercut rates for the thicker channels are bigger than the thinner ones, which is consistent with [10].

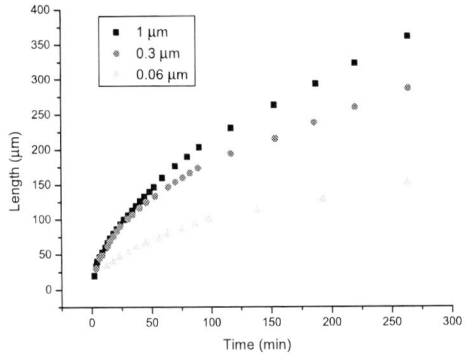

Figure 5: Microchannel etching length over time for three different thicknesses of magnesium in 0.04M HCl;

The hypothesis is that the etching chemical (i.e., Cl-) interacts with the inner photoresist surface so the overall effective diffusion constant is reduced. This remains to be proven but the current data shows that one should consider

the thickness effect if Mg is thinner than than 1µm. In comparison, Tab.2 lists the typical deposition rates, etching rates and etchants of Mg, Al and Ni. It shows that the ebeam deposition rates are similar among the popular sacrificial metals but Mg has a much higher etching rate.

Table 2: Comparison of preparation methods, etchants and etching rates of Mg, Al and Ni.

	Preparation methods	Etchants	Etching rates
Mg	Evaporation (30% power, 2.5 Å/sec)	hydrogen chloride	3700 Å/sec·M
		Saline	4.7 Å/sec
Al	Evaporation (47% power, 4.5 Å/sec)	Al etchant Type A (Transene©)	100 Å/sec at 50°C
Ni	Evaporation (45% power, 4.0 Å/sec)	Ni etchant TFB (Transene©)	30 Å/sec at 25°C

Fig.6 shows the snapshots of the etching of 1-µm-thick Mg microchannels using 0.04M hydrogen chloride at two different time intervals.

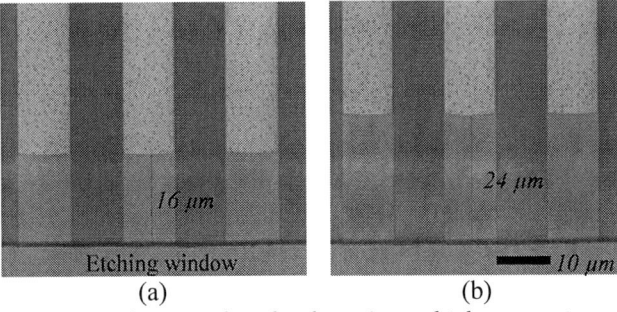

(a)   (b)

Figure 6: Photographs of etching 1 µm thick magnesium to form microchannels in 0.04M hydrogen chloride: (a) after 30 min; (b) after 60 min.

Biodegradable etching results

In addition to sacrificial etching, biodegradable etching in saline and culture medium is also demonstrated here. Fig.7 shows the etched length of microchannel versus time in saline.

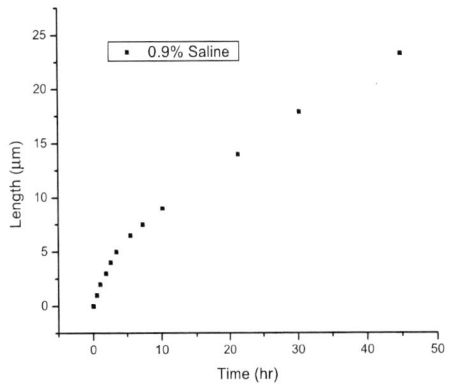

Figure 7: (a) Microchannel etching length over time in 0.9 wt% saline.

Fig.8 shows the open-surface Mg etching rates in saline and Dulbecco's modified Eagle medium (DMEM, a cell culture medium) mixed with FBS and PBS. The chloride ion concentrations are 0.15M in saline and 0.11M in DMEM, and the resulting etching rates are 1.7 and 1.27μm/hr respectively. The results show that the Mg biodegradation depends mainly on chloride ion concentration and has a converged rate of ~11.4μm/hr·M. Note that this rate is much faster than other widely-used biodegradable polymer materials such as PLA. For instance, from [12], the degradation time for PLA/PLGA (50:50) is only about 0.2μm/hr (i.e., 10 weeks of biodegradation time for microspheres of 100μm in diameter).

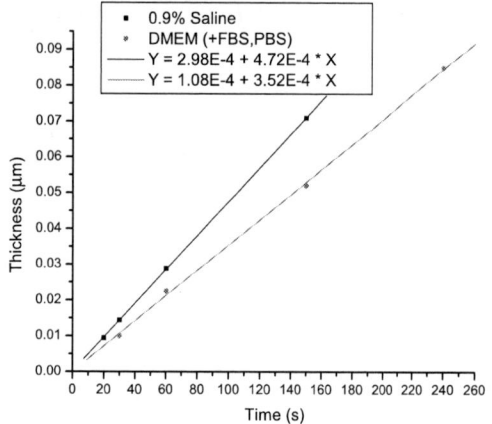

Figure 8: Thickness change over time in 0.9 wt% saline and DMEM.

CONCLUSIONS

This work reports the first etching tests of thin-film Mg in hydrochloric acid (HCl), saline, and culture medium. Both vertical etching and sacrificial Mg microchannel etching methods are applied to study the etching properties. Data of sacrificial Mg microchannel etching are fitted by the diffusion and a combined first-and-second order chemical reaction model. The initial results do show that thin-film Mg is a promising dual sacrificial and biodegradable material for MEMS applications.

ACKNOWLEDGEMENTS

The authors gratefully acknowledge the help of all the members in California Institute of Technology (Caltech) Micromachining Lab and experimental assistance of Mr. Trevor Roper. We also specially thank Mrs. Boyu Li and Mrs. Jionghui Li for their discussion.

REFERENCES

[1] X. Gu and Y. Zheng, "A review on magnesium alloys as biodegradable material", Front. Mater. Sci. China, vol. 4, pp. 111-115, 2010.

[2] J.H. Chang, R. Huang and Y.C. Tai, "High density IC chip integration with parylene pocket", in *Digest Tech. Papers NEMS'11 Conference*, Kaohsiung, February 20-23, 2011, pp. 1067-1070.

[3] J.H. Chang, R. Huang, and Y.C. Tai, "High density 256-channel chip integration with flexible parylene pocket", in *Digest Tech. Papers Transducers'11 Conference*, Peking, June 5-9, 2011, pp. 378-381.

[4] J.H. Chang, D. Kang and Y.C. Tai, "High yield packaging for high-density multi-channel chip integration on flexible parylene substrates", in *Digest Tech. Papers MEMS'12 Conference*, Paris, January 29-Febuary 2, 2012, pp. 353-356.

[5] J.H. Chang, Y. Liu, Y.C. Tai, "A low-temperature parylene-C-to-silicon bonding using photo-patternable adhesives and its application", in *Digest Tech. Papers Transducers'13 Conference*, Barcelona, June 16-20, 2013, pp. 2217-2220.

[6] J.H. Chang, Y. Liu, D. Kang, M. Monge, Y. Zhao, C.C. Yu, A. Emami, J. Weiland, M. Humayun, and Y.C. Tai, "Packaging study for a 512 channel intraocular epiretinal implant", in *Digest Tech. Papers MEMS'13 Conference*, Taipei, January 29-Feburary 2, 2012, pp. 353-356.

[7] Y. Liu, J. Park, R. J. Lang, A. Emami-Neyestanak, S. Pellegrino, M. S. Humayun, and Y.C. Tai, "Parylene origami structure for introcular implantation", in *Digest Tech. Papers Transducers'13 Conference*, Barcelona, June 16-20, 2013, pp. 1549-1552.

[8] J.H. Chang, Y. Liu, D. Kang, and Y.C. Tai, "Reliable packaging for parylene-based flexible retinal implant", in *Digest Tech. Papers Transducers'13 Conference*, Barcelona, June 16-20, 2013, pp. 2612-2615.

[9] M. Tsang, F. Herrault, R.H. Shafer, and M.G. Allen, "Methods for the microfabrication of magnesium", in *Digest Tech. Papers MEMS'13 Conference*, Taipei, January 29-February 2, 2012, pp. 347-350.

[10] J. Liu, Y.C. Tai, J. Lee, K.C. Pong. Y. Zohar, and C.M. Ho, "In situ monitoring and universal modelling of sacrificial PSG etching using hydrofluoric acid", in *Digest Tech. Papers MEMS'93 Conference*, Fort Lauderdale, February 7-10, 1993, pp. 71-76.

[11] W. H. Press, B. P. Fiannery, S. A. Teukoisky and W. T. Vetterling, *Numerical Recipes in Fortran 77: The Art of Scientific Computing (2 Edition)*, Cambridge University Press, 1992

[12] J. M. Anderson and M. S. Shive, *Adv Drug Deliv Rev*, "Biodegradation and biocompatibility of PLA and PLGA microspheres", vol. 28, 1997, pp. 5-24.

CONTACT

*Y. Liu, tel: +1-626-395-3885; ylliu@caltech.edu

THREE-DIMENSIONAL (3-D) RESHAPING TECHNIQUE IN MEMS DEVICES BY SOLELY ELECTRICAL CONTROL WITH ULTRAFINE TUNING RESOLUTION

Yong-Hoon Yoon, Chang-Hoon Han, and Jun-Bo Yoon
Department of Electrical Engineering, KAIST
373-1 Guseong-dong, Yuseong-gu, Daejeon 305-701, Republic of Korea

ABSTRACT

This paper reports an innovative and simple three-dimensional (3-D) reshaping (plastic deformation) technique in MEMS devices by solely electrical control with ultrafine tuning resolution. The proposed plastic deformation technique ado pted the creep phenomenon to achieve the desired plastic deformation. While voltage input induced stress on the device to start the creep phenomenon, Joule heating was applied to accelerate the creep phenomenon. Then, plastic deformation was successfully demonstrated with solely electrical control, where the tuning resolution was demonstrated at a sub-100nm level.

INTRODUCTION

Intentional plastic deformation has been considered in MEMS fields for versatile purposes. Out-of plane inductor and micro strip patch antenna have been successfully demonstrated for high Q-factor in the RF field [1-3]. Hot wire anemometer sensor and acoustic perception sensor have been also developed for making desired sensor topology [4-5]. Furthermore, plastic deformation has been used for angular comb drives that are hard to be made by conventional micromachining technology [6].

The common principle of the previous works is that excessive stress is applied to the MEMS devices to overcome the material's yield stress. To be specific, several hundred of MPa is generally required to cause the plastic deformation [7]. Therefore, it is insufficient to cause plastic deformation simply by electrostatic force generally ranging from several MPa to several tens of MPa.

For example, Jun Zou and et al. used external magnet to generate this high stresses [1]. The external magnet could cause enough stress to the MEMS devices by magnetic force. Other research group applies external mechanical apparatus with wafer bonding technique [6], where MEMS devices were fabricated at one wafer and the other wafer with pillars pushed the MEMS devices by wafer bonding technique. However, these special techniques to make high stress are not desired in integrated chip because this external instrument hinders the whole system to be integrated in a chip.

In this work, we paid our attention to a creep phenomenon to avoid this high stress problem. The creep phenomenon can generate plastic deformation even lower than the yield stress [7]. We found that by simply applying the creep phenomenon, we could achieve plastic deformation with general electrostatic force. Moreover, we advanced this concept through acceleration of the creep phenomenon by Joule heating.

CONCEPT

Fig. 1(a) represents the conventional plastic deformation technique induced by the stress. As depicted in Fig. 1(a), Initial spring has specific yield stress from material and structural characteristic. When stress is applied to the spring over the yield stress, the spring is permanently reshaped. In other words, high enough stress is required for plastic deformation. Contrarily, the proposed plastic deformation technique can make the plastic deformation using an elevated temperature in a controlled time, as shown in Fig. 1(b).

This difference of the conventional technique and the proposed technique come from the atom relocation mechanism. Conventional technique is based on the dislocation movement for the plastic deformation. In this model, the dislocation should overcome many obstacles such as grain boundary, contamination atoms, and other dislocation for movement. Therefore, high stress is inevitably needed to offer enough energy. Contrarily, the proposed plastic deformation technique is based on the atom diffusion mechanism (Creep). In this mechanism, the atom diffusion is controlled by time parameter. Then, it is remarkably accelerated by temperature. In other words, plastic deformation less depend on the stress parameter. It means that the proposed plastic deformation technique can use relatively weak electrostatic force.

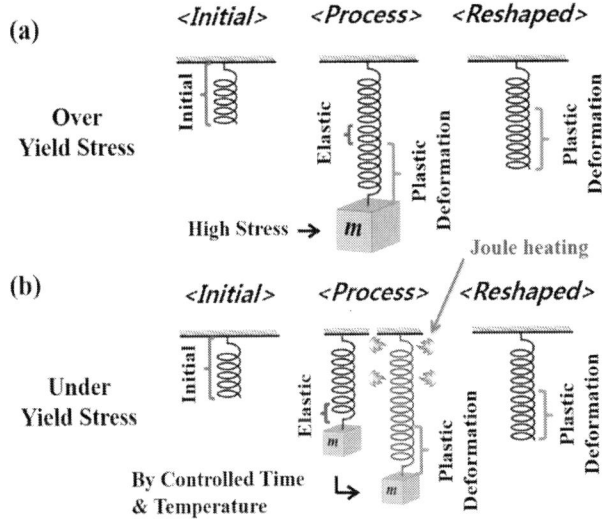

Fig.1 Conceptual schematics of the plastic deformation mechanism (a) conventional plastic deformation technique by high stress to overcome yield stress, and (b) proposed plastic deformation technique using elevated temperature

978-1-4799-3510-9/14 $31.00 © 2014 IEEE

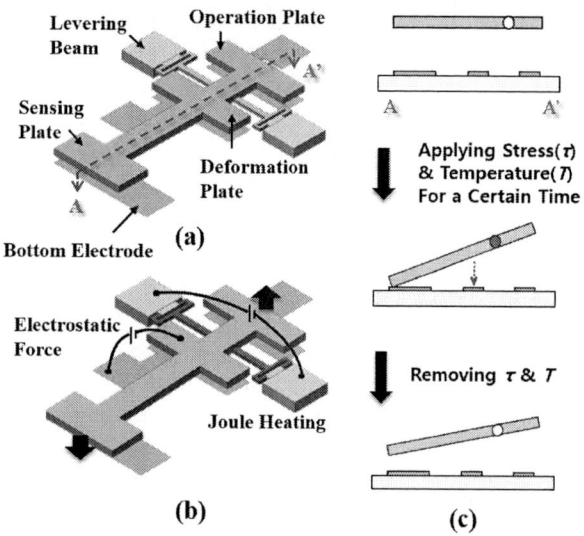

Fig.2 Schematic view of the (a) perspective overview of the test device and each part name, (b) electrical ports for applying stress and Joule heating for plastic deformation, and (c) plastic deformation sequence

Fig. 2(a) shows a schematic view of the test device to demonstrate the proposed concept. The test device adopts a levering structure that is highly sensitive to the creep phenomenon [8]. Also, it enables the sensing plate to be controlled in two directions upward and downward. The deformation plate is designed for starting plastic deformation by electrostatic force. The operation plate is intended to evaluate operation characteristic after plastic deformation without any interference. Finally, the sensing plate is designed to amplify test results by being positioned far away from the torsional axis through the capacitance measurement.

Fig. 2(b) represents electrical ports to induce stress and Joule heating, respectively. The sensing plate is lowered by the voltage input in the deformation plate to make stress for the creep phenomenon. Levering beam is Joule heated to accelerate the creep phenomenon. The Joule heating is locally restricted to the levering structure since the levering structure has much higher resistance than the other parts. Furthermore, "C" shape anchor is designed to prevent undesired thermal buckling from elevated temperature during the plastic deformation process.

Fig. 2(c) shows the proposed plastic deformation sequence. The test device is initially flat. Then, it experiences stress and elevated temperature by solely electrical controls for a precisely defined time. Finally, it will be deformed permanently (tilted) at a desired level.

FABRICATION

Fig. 3 represents the fabrication process of the proposed test device that uses electroplating Ni as a structural material and Cu as a sacrificial layer. A glass wafer was used for device isolation and accurate capacitance measurement. The

process starts with thermally evaporated Cr (200 Å)/Au (3000 Å) for the bottom electrodes. After a diffusion barrier Ti (2000 Å) is sputtered, the sacrificial layer was fabricated using thermally evaporated Cu (11000 Å). Then, the sacrificial layer was patterned to form support post. Next, thick photoresist film (PR) was coated and patterned as 1st PR mold for electroplating. The first mold was used to form the first electroplating Ni layer (10 µm). After removing the first PR mold, 2nd PR mold was coated subsequently. The second Ni layer (25 µm) was patterned by the second PR mold to form plate. In the last step, the second photoresist was removed and unnecessary sacrificial layer (Ti/Cu) was wet etched. Then, critical point dryer (CPD) is used for preventing undesired stiction.

The scanning electron microscopic (SEM) photograph of the fabricated test device is shown in Fig. 4. The perspective overview that shows the test device was successfully fabricated (Fig. 4(a)). The magnified view of sensing plate and the levering structure presents the suspended structure and "c" shape anchor are well fabricated as shown in Fig. 4(b) and Fig. 4(c).

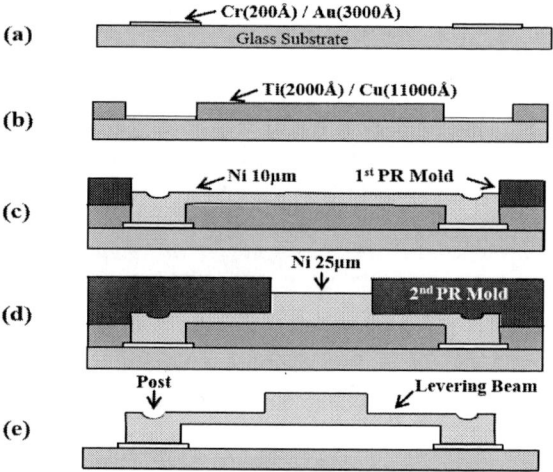

Fig.3 (a) Bottom electrode formation, (b) Ti/Cu Sacrificial layer deposition, (c) 1st Nickel electroplating, (d) 2nd Nickel electroplating, and (e) Removing the sacrificial layer.

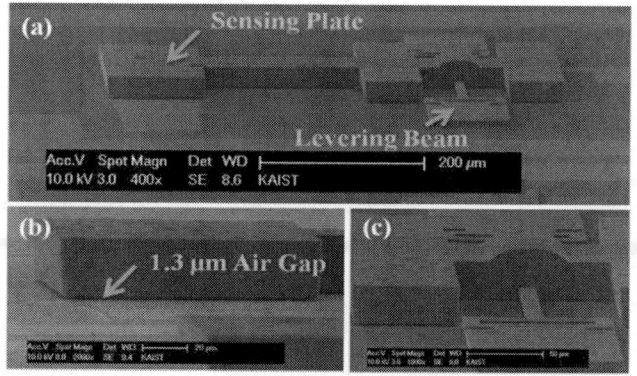

Fig.4 SEM photographs of the fabricated MEMS device (a) overview, magnified view of (b) sensing plate and (c) levering beam

978-1-4799-3510-9/14 $31.00 © 2014 IEEE

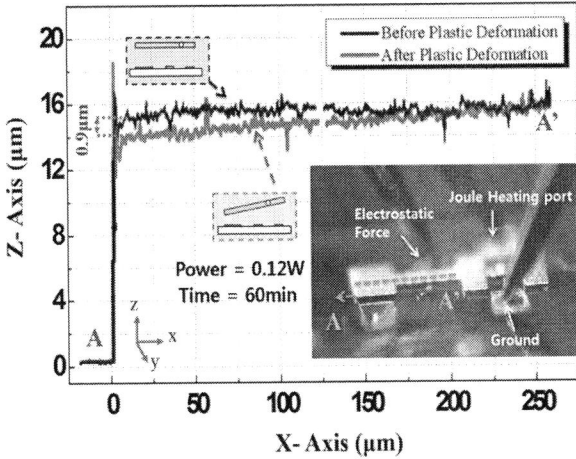

Fig.5 Surface profile of the fabricated device before and after the plastic deformation (permanent deformation happened with 0.9μm gap difference). Inset shows the plastic deformation experiment setup by electrical control.

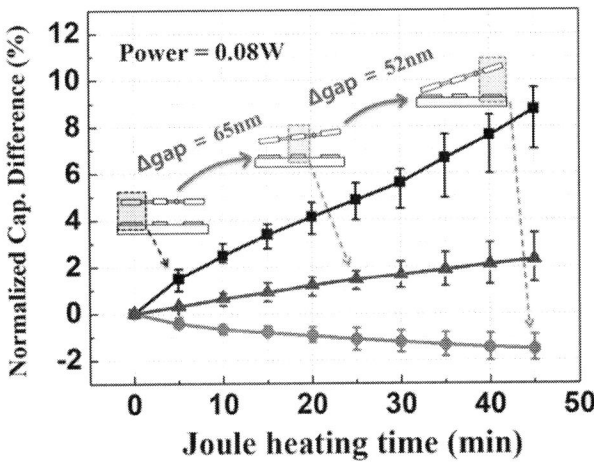

Fig.6 Normalized capacitance difference vs. Joule heating time when Joule heating power is 0.08W. Normalized capacitance difference means air gap difference by plastic deformation.

RESULTS AND DISCUSSIONS

We evaluated the plastic deformation characteristics of the fabricated MEMS device. Inset of Fig. 5 represents the plastic deformation setup with only electrical ports. First, voltage input is applied to the deformation plate. It makes sensing plate to more downward, inducing stress to the levering structure. Second, the levering beam is locally heated by electrical power applied between Joule heating port and ground.

Fig. 5 shows the measured surface profile change before and after the plastic deformation process. The beam end was lowered from 1.3μm of air gap to 0.4μm resulting in 0.9μm permanent deformation. This result demonstrates plastic deformation happened with solely electrical way. It is also confirmed that this plastic deformation is generated by the creep phenomenon since the maximum stress is lower than 20MPa in the MEMS device while the yield stress of nickel is over 140Mpa [9].

As shown in Fig. 5, near one micrometer gap difference was successfully observed by surface profiler. However, much fine gap difference measurement such as several hundred of nanometer is not easy to detect by the surface profile method. Therefore, we develop test setting based on capacitance measurement for more precise measurement.

Fig. 6 represents the measured capacitance change with respect to the Joule heating time. The normalized capacitance difference in sensing plate was increased as time goes by with constant power (square dot) while the one in operation plate was decreased (circle dot) as time goes by. It means that the air gap is changed as the MEMS device is tilted by the plastic deformation along the time.

Furthermore, the plastic deformation and time response can be used to achieve ultrafine tuning resolution through the precise time control. As shown in Fig. 6, the amount of the plastic deformation was precisely controlled in sub 100nm.

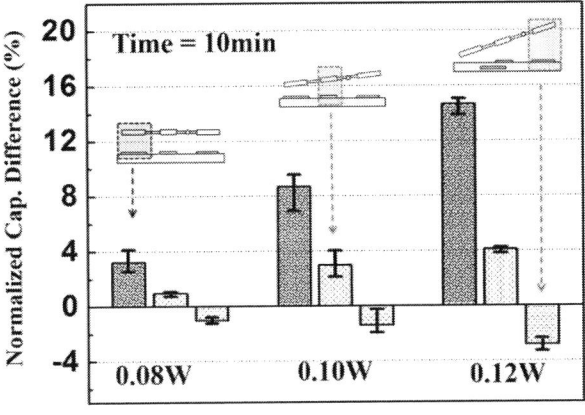

Fig.7 Normalized capacitance difference vs. Joule heating power during 10min. Temperature is increasing by increasing Joule heating power to accelerate plastic deformation process.

Fig. 7 shows that the plastic deformation can be accelerated by increasing the Joule heating power. The creep phenomenon happens very slowly in room temperature [10]. The test MEMS device also showed very small plastic deformation even over 10hrs at room temperature. However, the amount is dramatically changed with Joule heating. At 0.12W, plastic deformation amount is over 15% in just 10 minute, demonstrating that the creep phenomenon can be accelerated by elevated temperature.

Finally, the operation characteristic (C-V response) was evaluated. After the plastic deformation, the MEMS device works the same as the initial device does with only a capacitance offset generated by the plastic deformation technique, as shown in Fig. 8. It indicates that the proposed plastic deformation technique can be used independently while other device performances are unchanged.

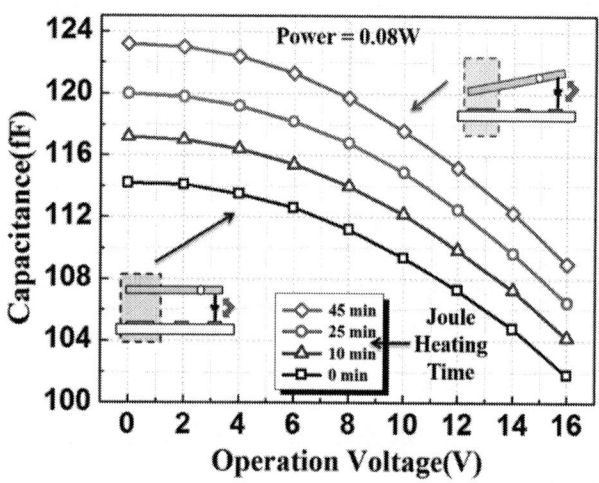

Fig.8 Capacitance vs. voltage response before and after the plastic deformation for checking up operation characteristic changes

CONCLUSION

A new 3-D reshaping technique (plastic deformation) using solely electrical control was firstly proposed and demonstrated with the experimental results. Furthermore, the plastic deformation is realized with ultrafine resolution of sub 100nm level. The proposed technique is expected to be favorably used in many integrated MEMS devices where reshaping feature is required without any external mechanisms or instruments.

ACKNOWLEDGEMENT

This work was supported by the National Research Foundation of Korea (NRF) grant funded by the Korea government (MEST) (No.2011-0028781)

REFERENCES

[1] J. Jou, J. Chen, C. Liu, and J. E. Schutt-Aine, "Plastic deformation magnetic assembly (PDMA) of out of plane microstructures: Technology and application", *IEEE J. Microelectromechanical Systems*, vol. 10, no. 2, pp. 302–309, June. 2001.

[2] J. Jou, C. Liu, D. R. Trainor, J. Chen, J. E. Schutt-Aine , and P. L. Chapman, "Development of three-dimensional Inductors using plastic deformation magnetic assembly (PDMA)," *IEEE Trans. Microwave Theory Tech.*, vol. 51, no. 4, pp 1067-1075, April 2003

[3] J. C. Langer, J. Zou, C. Liu, and J. T. Bernhard, "Micromachined reconfigurable out-of-plane microstrip patch antenna using plastic deformation magnetic actuation", *IEEE Microwave and Wireless Component Letters*, vol. 13, no. 3, pp. 120–122, March. 2003.

[4] J. Chen, and C. Liu, "Development and characterization of surface micromachined, out-of-plane hot-wire anemometer", *IEEE J. Microelectromechanical Systems*, vol. 12, no. 6, pp. 979–988, December. 2003.

[5] G. J M Krijnen, M. Dijkstra, J. J van Baar, S. S Shankar, W. J Kuipers, R. J H de Boer, D. Altpeter, T. S J Lammerink and R. Wiegerink, "MEMS based hair flow-sensors as model systems for acoustic perception studies", *Nanotechnology* 17, pp. s84–s89, January. 2006.

[6] J. B. Kim, and L. Lin, "Electrostatic scanning micromirrors using localized plastic deformation of silicon", *J. Micromech. Microeng.* 15, pp.1777-1785, July. 2005.

[7] M. F. Ashby, "A first report on deformation mechanism maps", *Acta Metallurgica*, vol. 20, pp. 887–897, July. 1972.

[8] A. B. Sontheimer, "Digital micromirror device hinge memory lifetime reliability modeling", *IEEE Internation Reliability Physics Symposium*, pp. 118–121, 2002.

[9] H. S. Cho, K. J. Hemker, K. Lian, J. Goettert, and G. Dirras, "Measured mechanical propertied of LIGA Ni structure", *Sensors and Actuators A: Physical 103*, pp. 59–63, 2003.

[10] H. Hsu, M. Koslowski, D. Peroulis, "An experimental and theoretical investigation of creep in ultrafine crystalline nickel RF-mems devices", *IEEE Trans. Microwave Theory Tech*, vol. 59, no. 10, pp. 2655–2664, October. 2011.

TUNABLE THZ FILTER BASED ON RANDOM ACCESS METAMATERIAL WITH LIQUID METAL DROPLETS

Q. H. Song[1,2], W. M. Zhu[2], W. Zhang[2], M. Ren[2], E. M. Chia[3] and A. Q. Liu[1,2†]

[1]School of Mechanical Engineering, Xi'an Jiaotong University, Xi'an 710049, China
[2]School of Electrical and Electronic Engineering, Nanyang Technological University
50 Nanyang Avenue, Singapore 639798
[3]School of Physical and Mathematical Sciences, Nanyang Technological University
50 Nanyang Avenue, Singapore 637371

ABSTRACT

Here we report a tunable THz filter based on random access metamaterial with liquid metal droplet, which is tuned through electrical bias controlled electrowetting effects. The random access metamaterial consists of 80 × 80 micro droplets, which are self-assembled in micro holes array due to lotus effect. The simulation results indicate resonant dip frequency shift of about 0.01THz induced by changing of the droplets shape via electrowetting effect and about 0.6 THz frequency shift when the droplets are connected in different forms. The random access metamaterial is realized through simple fabrication processes and can be tuned easily, which has potential application on tunable filters, tunable beam steering and flat lens.

INTRODUCTION

Metamaterials, rationally designed artificial materials with sub-wavelength scale metal elements, offers a new platform to control the electromagnetic (EM) field with designable and controllable functionalities. These sub-wavelength elements, typically with metal involved, response to the electric and magnetic field simultaneously, which result in many extraordinary physic phenomena, such as negative index [1-2], zero epsilon [3], giant chirality [4], or exotic and useful hyperbolic dispersion anisotropy [5]. Furthermore, metamaterials response to certain spectrum range depending on the size of the sub-wavelength elements, which can be engineered to function at certain frequency range, such as THz region [6, 7], where nature materials are out of choice for practical applications. Metamaterials are recently attracting wide research attentions due to enhanced nonlinear switching [8] and light emission [9, 10] performance of conventional active materials. For example, metamaterials are suitable candidate for waveform manipulation [11] and can be used for extraordinary applications such as cloaking [12, 13], wave guiding and localization of light. Driven by the promising technical prospects, tunable metamaterials are widely studied to control the EM wave using MEMS systems [14], phase change materials [15] and liquid crystals [16].

Of all the technics applied to tunable metamaterials, the tuning flexibilities, such as tuning range and the switching of the resonance modes, are highly depended on how the sub-wavelength elements are modulated during the tuning process. On the other hand, changing the geometry of the metal part of the sub-wavelength elements typically result in a dramatic EM properties change for the tunable metamaterials since the response of the metamaterials to the incident electric and magnetic fields are directly depended on the shape of the metal structures. Previous works on MEMS tunable metamaterials [17-18] target on the change of the geometry shape of the metal elements by changing the near field coupling of the metal parts anchored on the movable islands driven by micromachined actuators. However, it is difficult to reshape the metal structures once forged.

Liquid metal with sub-wavelength feature size are recently applied to tunable metamaterials due to their flexibility on reshaping the geometry [19]. This pioneer work involves complex microfluidic system for the tuning function. Although it offers an individual sub-wavelength element tuning without any metal contact, which can potential spoil the EM properties of the metamaterials and introduce extra losses, it still suffers many drawbacks due to the complexity of the system, which limit the tuning speed. Here an alternative technics electrowetting effects is applied for the tuning of the liquid metal structures as the metamaterials elements. In this paper, we demonstrated the experimental results on the shape tuning of the metamaterial elements both simultaneously and individually. Furthermore, the changes of the EM properties of the tunable liquid metal metamaterials are analyzed in the last section of this paper.

DESIGN OF THE METAMATERIALS

Figure 1 shows the schematic of the random access metamaterials, which consists of a square lattice array formed by mercury micro droplets with the period of 300 μm. The mercury droplets were confined in the holes, which are patterned on the 2.5-cm silicon substrate. The droplets array is formed by loading the mercury liquid on a silicon substrate with pre-etched cylindrical holes, which is then covered by a crystal quartz wafer on the top and make the mercury sandwiched in between. The mercury droplets array is thus formed and assembled by lotus effect. The electrowetting effects can be induced at the contact between the substrate and the mercury droplet through electrical bias. The electrowetting effect is used to control the radius of or the connection between the liquid droplets, which results in a reconfiguration of the droplet array. Therefore, the interaction between the incident THz wave and the droplet metamaterial can be manipulated in real time which tunes the resonance of the structure.

978-1-4799-3510-9/14 $31.00 © 2014 IEEE

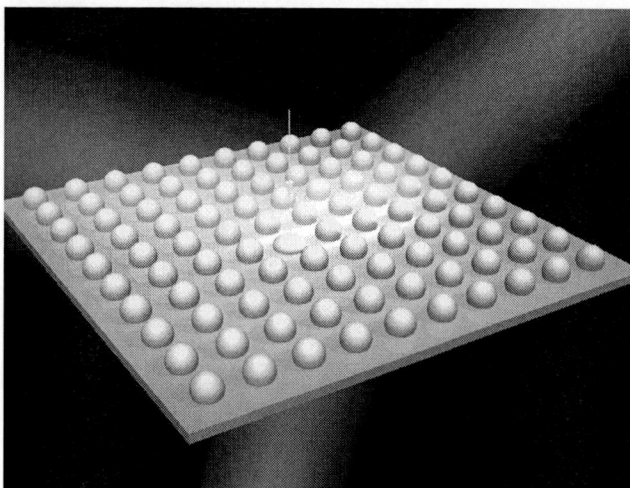

Figure 1: Schematic of random access tunable metamaterial for filter based on liquid metal micro droplets in terahertz regime.

The schematics and microscopic view of the mercury droplets manipulated by electrowetting effect are shown in Figure 2. Fig. 2(a) and Fig. 2(c) show the schematic and graphs of the droplet at initial state. The liquid metal droplets are formed by lotus effect and the silicon substrate is pre-etched with square-lattice cylinder holes array using Deep reactive-ion etching (DRIE) method. The holes are in the size of 240-μm in diameter and 50-μm in depth. When voltage is applied as shown in Fig. 2(b), the droplet is pulled down by electric field force created by the charged substrate, which is due to the electrowetting effect. This force tend to change the contact angle θ, which simultaneously enhance the contact area between the mercury droplet and the substrate by which means the radii of the droplet is changed as shown in Fig. 2(d).

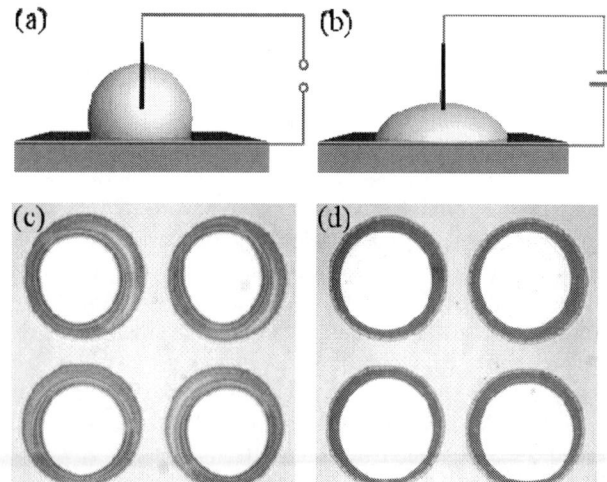

Figure 2: Schematics of the mercury droplets manipulated by electrowetting effect with (a) uncharged, (b) charged substrate, respectively. Corresponding graphs of the mercury droplets (c) at initial state, and (d) controlled by means of electrowetting effect when voltage applied, respectively.

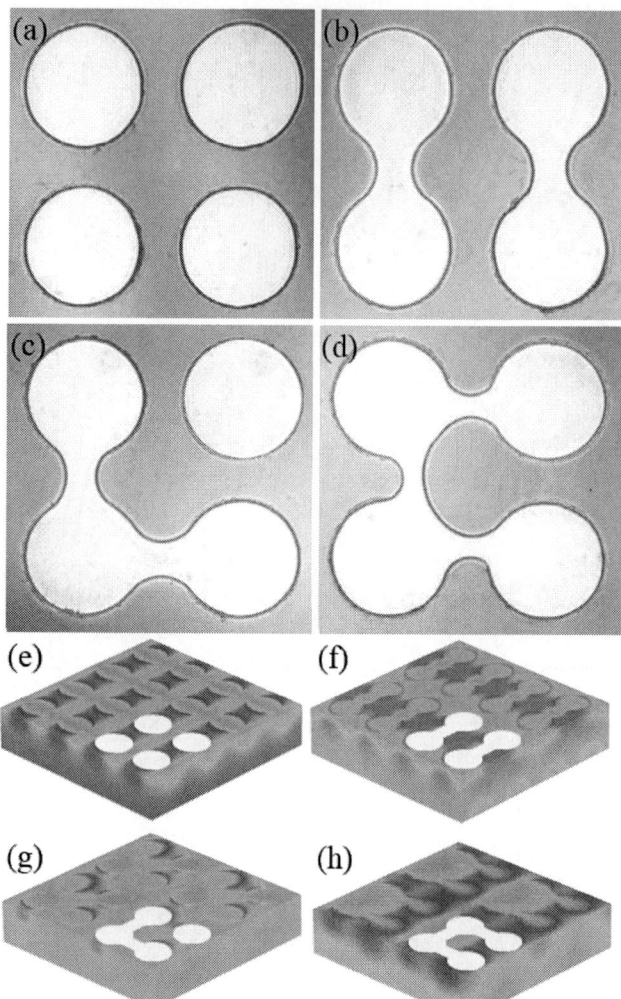

Figure 3: Top view of four mercury droplets connected with each other to form different type which was called (a) ":" type, (b) "‖" type, (c) "L" type, and (d) "C" type. Numerical results of electrical field distribution with different type of connection (e-h).

The phenomenon of Electrowetting can be interpreted by Young-Lippmann equation [20]:

$$\cos\theta = \cos\theta_0 + \frac{C}{2\gamma}V^2 \qquad (1)$$

where θ_0 is the initial contact angle, θ is the contact angle when voltage V applied, γ is the mercury surface tension, C is the areal capacitance of the substrate.

The resonant frequency can be effectively tuned through the droplet radius control, while the tuning range is limited. Therefore, we further explore an alternative approach which realizes a large frequency tuning through the droplet manipulation as shown in Fig. 3. The silicon holes are etched with the size of 80-μm in diameter and 20-μm in depth. In this case the droplets are shaped in cylinders and can be connected as ":" type (Fig. 3(a)), "‖" type (Fig. 3(b)), "L" type (Fig. 3(c)), and "C" type (Fig. 3(d)). The corresponding electrical field distribution are shown in Fig. 3 (e)-(h).

ANALYSIS OF THE EM RESPONSE

Figure 4(a) shows the numerical analysis of the transmission spectra at different radii of the mercury droplets. The resonant dip frequency is observed in the THz regime and shifts to the higher frequency region when the radii of the mercury droplets are increasing. The electrical field intensity of the structure is numerically investigated using CST microwave studio as shown in Fig. 4 (b-e). The droplet is modeled as a sphere for r = 80 μm and an ellipsoid for r = 90, 100, 110 and 120 μm with the same volume. For comparison, electrical field intensity at non-resonant frequency (Fig.4 (b) and (d)) and resonant frequency (Fig. 4(c) and (e)) are both plotted. Common dipole resonance is observed on the droplet at the non-resonant frequency, which is simply due to the incident linear electrical field. On the other hand, strong electrical field energy is confined in the space between the droplet and the substrate, which forms a resonant cavity and induces the absorption peak.

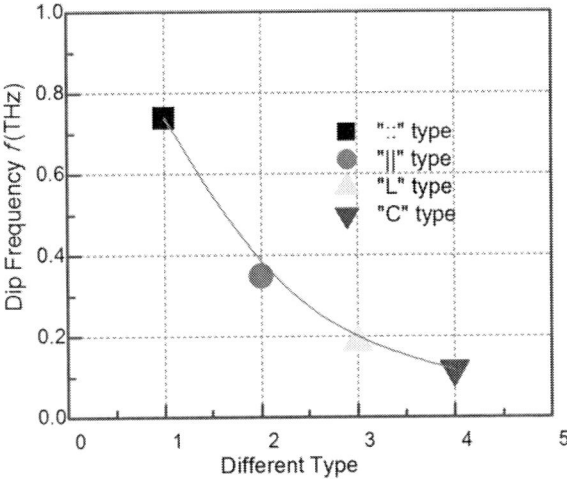

Figure 5: Numerical results of dip frequency at different connection type.

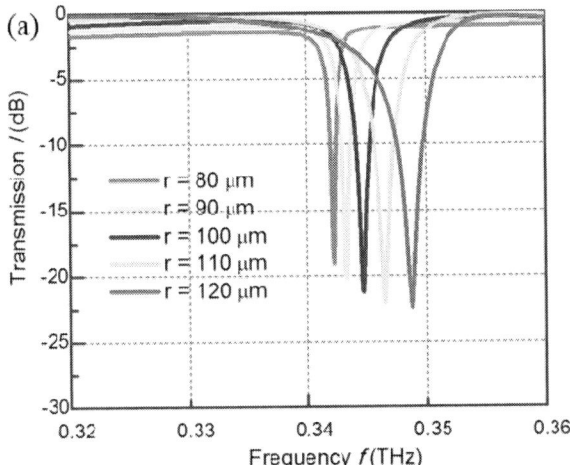

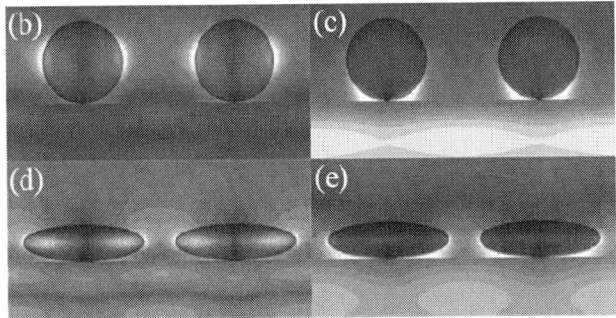

Figure 4: Numerical analysis of (a) the transmission spectra at different radii of unit cell and the electric field with different resonant mode at radii of 80 μm ((b), (c)) and 120 μm ((d), (e)).

The numerical analysis of the dip frequency at different connection type is shown in Fig. 5. The dip frequency is strongly decreased when the connection length increases. This tuning method achieves a 0.6 THz frequency shift which is much larger than the tuning method of radius control. This phenomenon can be interpreted by Fig. 6(a-d), which present the surface current of different connection types. In the "::" type (Fig. 6(a)), the surface current indicates a dipole

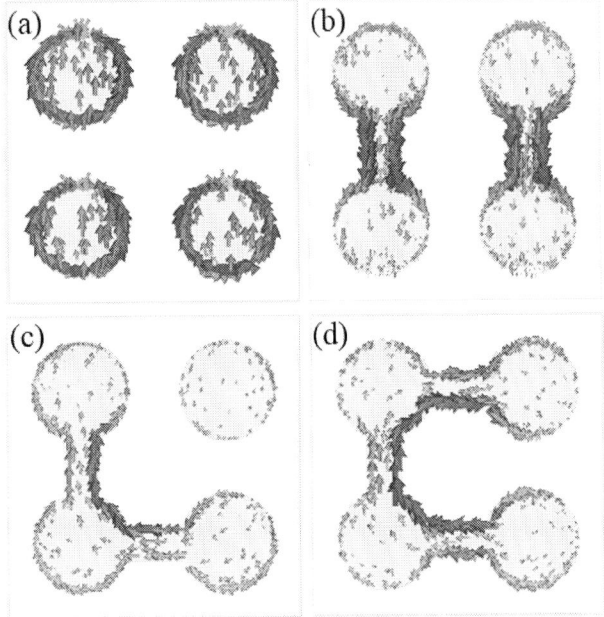

Figure 6: Numerical results of surface current at different connection type.

resonance on each isolated droplets. The small radius of the droplets results in a high resonant frequency at 0.735 THz. When two droplets are connected (Fig. 6(b)), the "::" type droplets is reshaped into a "‖" type structure. The surface current flows along the bridge between two droplets, where an electrical dipole is excited and a lower frequency at 0.35 THz is induced compared with the isolated droplets. On the other hand, the "::" type droplets can be connected into "L" type, as shown in Fig. 6(c), the surface current of which flows along the two connected bridges when interact with the linearly y-polarized incident light. The resonance frequency is then decreased to 0.185 THz. Furthermore, in the reconfigured "C" type metamaterial, the resonant frequency is decreased to 0.118 THz. Therefore, large frequency shifting is realized.

CONCLUSIONS

In conclusion, a THz random access metamaterial based on mercury droplets is designed, fabricated and experimentally demonstrated. In the experiment, the radii of each droplet are tuned from 80 μm to 120 μm, while the dip frequency is tuned from 0.342 THz to 0.349 THz. Furthermore, we also demonstrate a new tuning method by connecting the droplet, and the dip frequency is tuned from 0.75 THz to 0.118 THz, which has potential application on tunable filters, controllable beam steering and tunable flat lens.

ACKNOWLEDGEMENTS

The work is supported by the Environmental and Water Industry Development Council of Singapore (EWI), RPC programme (Grant No. 1102-IRIS-05-01 and 1102-IRIS-05-02)

REFERENCES

[1] D. R. Smith, J. B. Pendry and M. C. K. Wiltshire, "Metamaterials and negative refractive index", *Science*, vol. 305(5685), pp.788-792, 2004.

[2] T. Xu, A. Agrawal, M. Abashin, K. J. Chau, and H. J. Lezec, "All-angle negative refraction and active flat lensing of ultraviolet light", Nature, vol. 497(7450), pp. 470-474, 2013.

[3] M. Silveirinha and N. Engheta, "Tunneling of electromagnetic energy through subwavelength channels and bends using ε-near-zero materials", *Phys. Rev. Lett.*, vol. 97, pp. 157403, 2006.

[4] A. V. Rogacheva, V. A. Fedotov, A. S. Schwanecke and N. I. Zheludev, "Giant gyrotropy due to electromagnetic-field coupling in a bilayered chiral structure". *Phys. Rev. Lett.*, vol. 97, pp. 177401, 2006.

[5] J. Elser, R. Wangberg, V. A. Podolskiy, and E. E. Narimanov, "Nanowire metamaterials with extreme optical anisotropy", *Appl. Phys. Lett.*, vol. 89(26), pp. 261102-261102, 2006.

[6] W. Zhang, A. Q. Liu, W. M. Zhu, E. P. Li, H. Tanoto, Q. Y. Wu, J. H. Teng, X. H. Zhang, M. L. J. Tsai, G. Q. Lo and D. L. Kwong, "Micromachined switchable metamaterial with dual resonance". *Appl. Phys. Lett.*, vol. 101(15), pp. 151902-151902, 2012.

[7] W. Zhang, W. M. Zhu, H. Cai, M. L. J. Tsai, G. Q. Lo, D. P. Tsai, H. Tanoto, J. H. Teng, X. H. Zhang, D. L. Kwong and A. Q. Liu, "Resonance Switchable Metamaterials using MEMS Fabrications", *IEEE Journal of selected topics in quantum electronics*, vol. 19, pp. 4700306-4700306, 2013.

[8] N. I. Zheludev and Y. S. Kivshar, "From metamaterials to metadevices". *Nat. Mater.*, vol. 11, pp. 917, 2012.

[9] K. Tanaka, E. Plum, J. Y. Ou, T. Uchino and N. I. Zheludev. "Multi-fold enhancement of quantum dot luminescence in a plasmonic metamaterial", *Phys. Rev. Lett.*, vol. 105, pp. 227403, 2010.

[10] O. Hess, J. B. Pendry, S. A. Maier, R. F. Oulton, J. M. Hamm and K. L. Tsakmakidis, "Active nanoplasmonic metamaterials". *Nat. Mater.*, vol. 11, pp. 573-584, 2012.

[11] N. Yu, P. Genevet, M. A. Kats, F. Aieta, J. P. Tetienne, F. Capasso and Z. Gaburro, "Light propagation with phase discontinuities: generalized laws of reflection and refraction", *Science*, vol. 334(6054), pp. 333-337, 2011.

[12] D. Schurig, J. J. Mock, B. J. Justice, S. A. Cummer, J. B. Pendry, A. F. Starr and D. R. Smith, "Metamaterial electromagnetic cloak at microwave frequencies", *Science*, vol. 314, pp. 977-980, 2006.

[13] W. Cai, U. K. Chettiar, A. V. Kildishev and V. M. Shalaev, "Optical cloaking with metamaterials", *Nature photonics*, vol. 1(4), pp. 224-227, 2007.

[14] A. Q. Liu, W. M. Zhu, D. P. Tsai and N. I. Zheludev "Micromachined tunable metamaterials: a review", *Journal of Optics*, vol. 14(11), pp. 114009, 2012.

[15] Z. L. Samson, K. F. MacDonald, F. De Angelis, B. Gholipour, K. Knight, C. C. Huang, E. Di Fabrizio, D. W. Hewak and N. I. Zheludev, "Metamaterial electro-optic switch of nanoscale thickness", *Appl. Phys. Lett.*, vol. 96(14), pp. 143105-143105, 2010.

[16] I. C. Khoo, D. H. Werner, X. Liang, A. Diaz and B. Weiner, "Nanosphere dispersed liquid crystals for tunable negative-zero-positive index of refraction in the optical and terahertz regimes", *Optics letters*, vol. 31(17), pp. 2592-2594, 2006.

[17] W. M. Zhu, A. Q. Liu, T. Bourouina, D. P. Tsai, J. H. Teng, X. H. Zhang, G. Q. Lo, D. L .Kwong and N. I. Zheludev, "Microelectromechanical Maltese-cross metamaterial with tunable terahertz anisotropy", *Nat. commun.*, vol. 3, pp. 1274, 2012.

[18] W. M. Zhu, A. Q. Liu, X. M. Zhang, D. P. Tsai, T. Bourouina, J. H. Teng, X. H. Zhang, H. C. Guo, H. Tanoto, T. Mei, G. Q. Lo and D. L. Kwong. "Switchable magnetic metamaterials using micromachining processes" *Adv. Mat.*, vol. 23(15), pp. 1792-1796, 2011.

[19] T. S. Kasirga, Y. N. Ertas, M. Bayindir, "Microfluidics for reconfigurable electromagnetic metamaterials", *Appl. Phys. Lett.*, vol. 95(21), pp. 214102-214102-3, 2009.

[20] F. Mugele and J. C. Baret, "Electrowetting: from basics to applications", *Journal of Physics: Condensed Matter*, vol. 17(28), pp. R705, 2005.

CONTACT

*A. Q. Liu, tel: +65-67904336; eaqliu@ntu.edu.sg

WETTING DYNAMICS STUDY OF UNDERWATER SUPERHYDROPHOBIC SURFACES THROUGH DIRECT MENISCUS VISUALIZATION

Muchen Xu, Guangyi Sun and Chang-Jin "CJ" Kim
University of California, Los Angeles (UCLA), USA

ABSTRACT

We study wetting of an air-filled micro-cavity on hydrophobic surface submerged in water by developing an optically clear sample that makes the location of the liquid-air meniscus inside the cavity visible. The *plastron* state, i.e., the state of the trapped air under water, is a central issue for the superhydrophobic surface research today because of its importance for many important applications, such as drag reduction. By continuously observing the meniscus on and inside a single trench during the wetting process, we obtain deterministic dynamics of the meniscus for the first time, as opposed to the probabilistic data in the recent studies. Our results confirm that the meniscus is in one of two states – pinned at the mouth of the trench or sliding on the sidewall of the trench, the latter leading to the fully-wetting (i.e., Wenzel) state. Furthermore, the results reveal that the dewetted (i.e., Cassie-Baxter) state can (or cannot) be indefinite if (or unless) the water is saturated with air and the hydrostatic pressure is low enough.

INTRODUCTION

The extraordinary properties of superhydrophobic (SHPo) surface, including high water contact angle and low contact angle hysteresis, have fascinated researchers for decades [1]. The pioneer work of Wenzel [2] and Cassie and Baxter [3] showed two distinctively different wetting regimes on rough surfaces. In Wenzel regime, liquid penetrates into the surface roughness and is fully in contact with the solid. In Cassie-Baxter regime, air is trapped in the roughness, so that liquid and solid are in partial contact. In recent years, SHPo surfaces submerged in water has drawn a lot of attention due to its relevance to drag reduction [4] and anti-biofouling ability [5]. However, underwater Cassie-Baxter state is usually fragile [6], while its stability is essential for the SHPo surfaces to succeed in many applications. Unlike the wetting transition of a droplet on SHPo surfaces [7], the plastron on SHPo surfaces under water is isolated from the atmosphere. The air diffusion is believed to be the reason for the wetting transition from Cassie-Baxter state to Wenzel state for underwater SHPo surfaces [8].

Recently, the stability of underwater Cassie-Baxter state has been studied theoretically [9, 10] and experimentally [11, 12]: its longevity decreased with immersion depth [11] and increased with dissolved gas concentration [12]. However, all the experiments reported statistical information, such as the average wetting transition time, not teaching us the transition dynamics needed to design SHPo surfaces more robust against wetting. Moreover, even though long-term (>120 days) underwater Cassie-Baxter state was observed in the studies of underwater insects [13], stable underwater Cassie-Baxter state has not been observed on any engineered SHPo surfaces in recent long-term experiments [12, 14]. To understand why the insects retain the plastron better than the engineered surfaces, in this paper we studied the detailed process of underwater wetting transition from Cassie-Baxter state to Wenzel state by developing an optically clear SHPo surface. As one result of the study, we found that the underwater Cassie-Baxter state may be infinite on engineered surfaces as well.

EXPERIMENTAL METHODS

Sample Fabrication

SHPo surfaces made of a parallel array of micro-trenches are shown to have large drag reduction in both laminar and turbulent flows [4, 15, 16]. In this paper, we further chose to study a single trench because it would prevent the overlapping views of multiple menisci and allow clear images. Since all the micro-trenches are in parallel and isolated to each other, single trench represents the grating/trench/pillar-type SHPo surfaces when studying wetting transition dynamics. The scanning electron microscopy (SEM) image of the single trench sample is shown in Fig. 1.

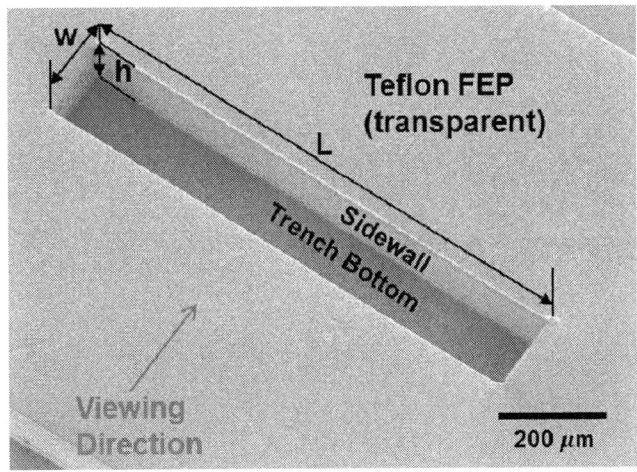

Figure 1: SEM image of a hydrophobic trench made of optically clear Teflon® FEP. The trench is w = 147 µm, h = 86 µm, L = 1 mm, in width, depth, and length, respectively.

The micro-trench structure was fabricated by hot-embossing Teflon® FEP into a silicon mold, as improved from [17]. The silicon mold was fabricated by deep reactive ion etching (DRIE), as shown in Fig. 2(a). The micro-feature was first defined with typical photolithography process. Then the photoresist was used as etching mask for the DRIE process to obtain the silicon micro-structures. The hot-embossing process is described in Fig. 2(b). The Teflon sheet was placed on the mold with a piece of glass slide on top. Then the mold was heated up to ~300°C and the Teflon sheet was pressed into the silicon mold. After the whole setup was cooled down to room temperature on another cold plate, the Teflon sheet was de-molded to obtain the testing sample. The glass slide was used to apply pressure during the molding process and prevent the Teflon sheet from bending due to stress during the cooling process. Teflon® FEP was chosen as the substrate material because it is intrinsically hydrophobic and optically clear.

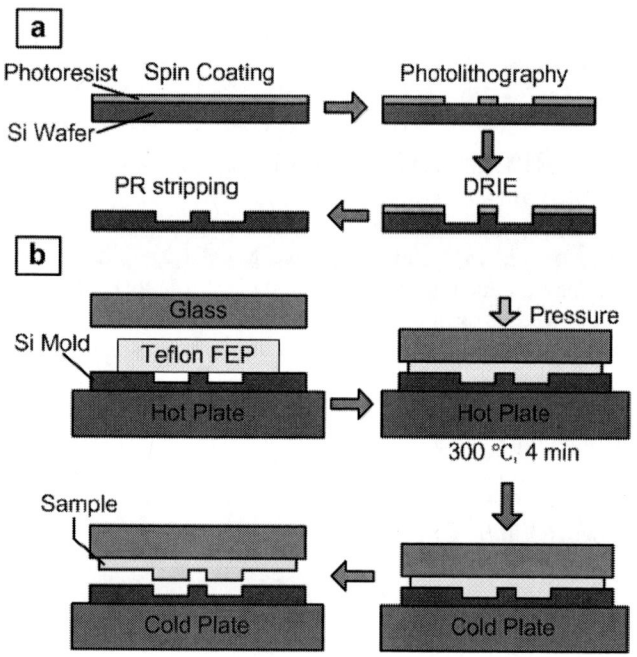

Figure 2: (a) DRIE process to fabricate silicon mold. (b) Hot embossing process to form an optically clear sample of hydrophobic micro-trench.

Visualization Technique

In recent investigations of plastron stability, different optical techniques were used to study the wetting transition, such as transmission diffraction technique [18] or light reflection density [11]. However, all these techniques could only visualize the surface from the top without the information of the meniscus vertical position as the wetting transition progresses. Confocal microscopy was used to provide the vertical position of the meniscus [11] but could not monitor the sample continuously over long time because of the laser heating.

To overcome the above problems, we developed an optically clear sample that allows a direct and continuous observation of the water-air interface from the side. The optically clear sample was placed at the bottom of an acrylic water tank filled with water to a certain height. The water tank was placed on an inverted microscope, illuminating the sample with a common LED light. The air-water interface could be directly visualized from the side, as schematically

illustrated in Fig. 3. The images of the air-water interface were captured via a CCD camera attached to the inverted microscope. Compared with the previous visualization methods, this technique was able to visualize the shape and position of air-water meniscus directly and continuously without local heating problem. The curved meniscus appears dark on the image, where the upper border is the triple contact line and the lower border indicates the lowest portion of the curved meniscus, as shown in Fig. 4(d).

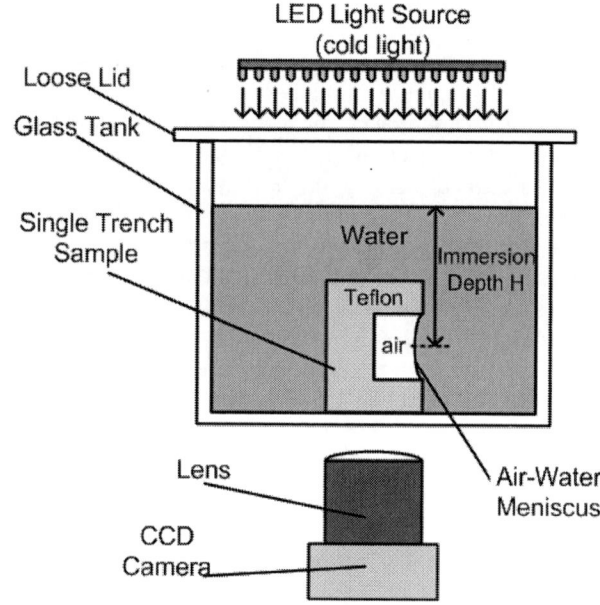

Figure 3: Schematic illustration of the experimental setup, drawn not to scale. Sample (~3 mm wide, ~3 mm high, ~5 mm long) is much smaller than tank (~6 cm wide, ~20 cm high, ~15 cm long), and trench is much smaller than sample body.

Environmental Condition

Because the concentration of dissolved air in water is very sensitive to temperature and other parameters, such as the agitation of water or biological activity [11], the experiment was carried out in stable lab environment with small temperature variations and no agitations. Deionized water was left in the lab environment for a minimum of 2 days before using so that the water is fully saturated with air [19]. The temperature (21.3±0.3°C) and relative humidity (53%±3%) of the environment were monitored throughout the experiments to make sure different tests were carried out under similar environmental condition. To reduce the decrease of water level caused by evaporation, the acrylic water tank was covered with a loose lid. It enabled air exchange with outside environment while minimizing the water loss due to evaporation. For long-term test, the tank was occasionally added with water to control the water depth variation within ~5 mm, using the water kept in the same testing environment.

EXPERIMENTAL RESULTS

Wetting Transition Dynamics

With the optically clear sample and observation setup described above, the evolution of air-water meniscus in the hydrophobic micro-trench under water was visualized over time. Figure 4 shows one exemplary test that eventually gets wetted. The sample with a 1 mm long, 147 µm wide, and 86 µm deep trench was immersed 100 mm below the water surface. In Fig. 4(a), the meniscus was initially flat and pinned at the mouth (i.e., top edge) of the trench since the compressed air in the trench resists the meniscus from bending into the cavity. Figure 4(b) shows that the meniscus started to bend down while still pinned at the top edge, as the trapped gas diffused out into the water. However, as more gas diffused out, the meniscus bent down more and finally the triple contact line started to depin from the left side, as shown in Fig. 4(c). The meniscus continued going down after depinning, as shown in Fig. 4(d). Once the meniscus touched the bottom, it suddenly broke into two, leaving two gas bubbles at the bottom corners of the micro-trench, as shown in Fig. 4(e). The gas bubble will disappear in a very short time, in correspondence with former study [11]. We define the lifetime of Cassie-Baxter state in this experiment as the time the meniscus takes to touch the bottom of the trench.

In order to fully characterize the wetting dynamics, the distance from the bottom of the micro-trench to the lowest point of the bent meniscus was measured from the images obtained through the inverted microscope. Figure 5 shows the distance change over time, presenting two distinctive cases. For low immersion depth (i.e., H = 55 mm), after the initial air diffusion the triple line stayed pinning at the top edge and the meniscus stabilized. For high immersion depth (i.e., H = 100 mm), the triple line depinned from the top edges at ~2.5 hour and started to slide down the sidewall. The speed of meniscus movement decreased before depinning and remained constant during the sliding. This observed trend is consistent with recent theoretical study [10]. The wetting transition dynamics greatly depends on the immersion depth. When the micro-trench was immersed at low depth, the air-water meniscus stabilizes without depinning, leading to "infinite" Cassie-Baxter state.

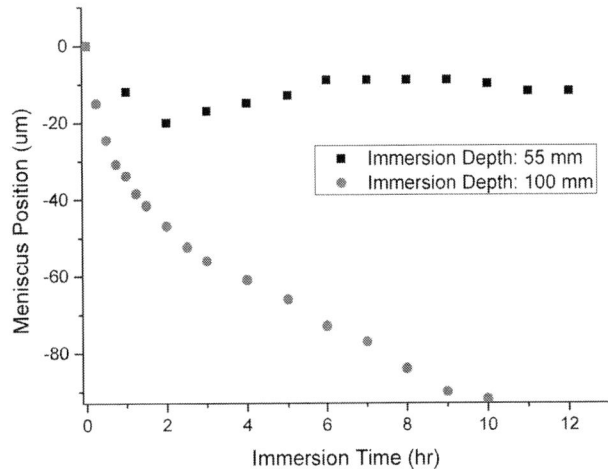

Figure 5: Plastron state on the sample of Fig. 1 (w = 147 µm, h = 86 µm, L = 1 mm) observed at two different immersion depths (H = 55 mm, H = 100 mm). The meniscus position is defined as the distance from the top edge of the trench, i.e., top surface of the sample.

Lifetimes of a micro-trench immersed at different water depths was collected and drawn in Fig. 6. As the immersion depth (i.e., hydrostatic pressure) decreased, the lifetime of underwater Cassie-Baxter state increased and approached "infinity" (measured >1000 hours) for immersion depth ~50 mm or smaller. This is the first experimental verification corroborating the recent theory that suggested a thermodynamically stable underwater Cassie-Baxter state [9].

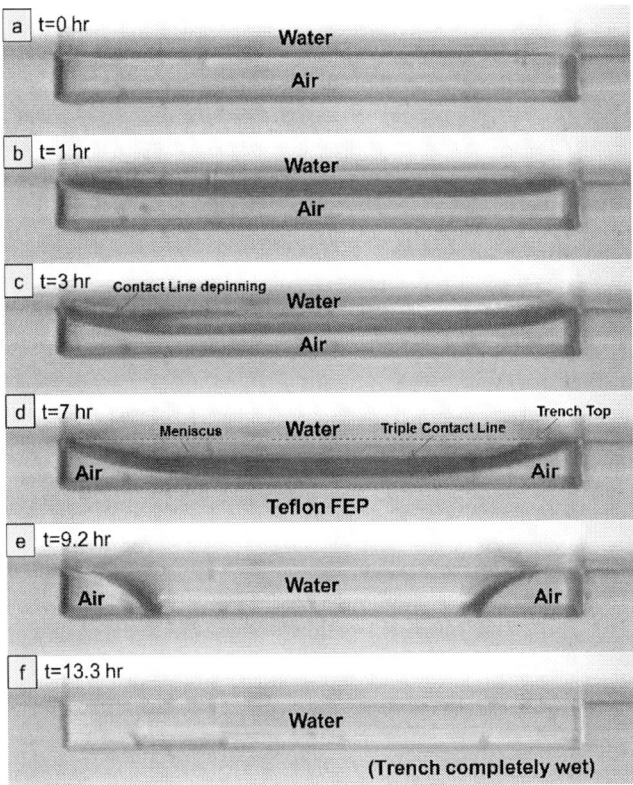

Figure 4: Snapshots of the meniscus taken over time from the side of the submerged trench reveal the wetting transition dynamics. The curved meniscus of air-water interface shows up as dark stripes. The wetting is caused by hydrostatic pressure and air dissolution.

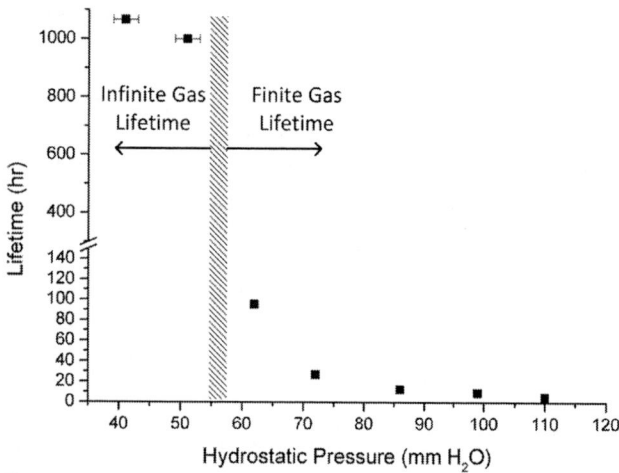

Figure 6: Longevity of underwater Cassie-Baxter state as a function of immersion depth. The maximum hydrostatic pressure allowable for infinite lifetime is determined mostly by trench width.

CONCLUSION

We have developed an optically clear sample of hydrophobic micro-trench and a technique to visualize the vertical positions of the air-water meniscus inside the cavity. Through directly and continuously monitoring the Cassie-Baxter to Wenzel wetting transition process under water, we resolved the details of the break-in dynamics. Furthermore, we verified the existence of stable "infinite" Cassie-Baxter state for underwater SHPo surfaces. The meniscus visualization method developed in this paper provides a simple, efficient way to monitor inside hydrophobic micro-cavities, especially to study underwater SHPo surfaces.

ACKNOWLEDGEMENT

This work has been supported in part by ONR (N000141110503) and NSF (CBET 1336966).

REFERENCES

[1] A. Lafuma and D. Quere, "Superhydrophobic states", *Nature Materials,* vol. 2, pp. 457-460, Jul 2003.

[2] R. N. Wenzel, "Resistance of solid surfaces to wetting by water", *Industrial and Engineering Chemistry,* vol. 28, pp. 988-994, 1936.

[3] A. B. D. Cassie and S. Baxter, "Large Contact Angles of Plant and Animal Surfaces", *Nature,* vol. 155, pp. 21-22, 1945.

[4] C. Lee, C.-H. Choi, and C.-J. Kim, "Structured surfaces for a giant liquid slip", *Physical Review Letters,* vol. 101, p. 64501, 2008.

[5] K. Koch and W. Barthlott, "Superhydrophobic and superhydrophilic plant surfaces: an inspiration for biomimetic materials", *Philosophical Transactions of the Royal Society a-Mathematical Physical and Engineering Sciences,* vol. 367, pp. 1487-1509, Apr 28 2009.

[6] L. Bocquet and E. Lauga, "A Smooth Future?", *Nature Materials,* vol. 10, pp. 334-337, 2011.

[7] G. Whyman and E. Bormashenko, "How to Make the Cassie Wetting State Stable?", *Langmuir,* vol. 27, pp. 8171-8176, Jul 5 2011.

[8] Y. H. Xue, S. G. Chu, P. Y. Lv, and H. L. Duan, "Importance of Hierarchical Structures in Wetting Stability on Submersed Superhydrophobic Surfaces", *Langmuir,* vol. 28, pp. 9440-9450, Jun 26 2012.

[9] M. Flynn and J. W. Bush, "Underwater breathing: the mechanics of plastron respiration", *Journal of Fluid Mechanics,* vol. 608, pp. 275-296, 2008.

[10] B. Emami, A. Hemeda, M. Amrei, A. Luzar, M. Gad-el-Hak, and H. V. Tafreshi, "Predicting longevity of submerged superhydrophobic surfaces with parallel grooves", *Physics of Fluids,* vol. 25, p. 062108, 2013.

[11] R. Poetes, K. Holtzmann, K. Franze, and U. Steiner, "Metastable Underwater Superhydrophobicity", *Physical Review Letters,* vol. 105, p. 166104, 2010.

[12] W.-Y. Sun and C.-J. Kim, "The role of dissolved gas in longevity of Cassie states for immersed superhydrophobic surfaces", in *Micro Electro Mechanical Systems (MEMS), 2013 IEEE 26th International Conference on,* 2013, pp. 397-400.

[13] A. Balmert, H. Florian Bohn, P. Ditsche - Kuru, and W. Barthlott, "Dry under water: Comparative morphology and functional aspects of air - retaining insect surfaces", *Journal of Morphology,* vol. 272, pp. 442-451, 2011.

[14] M. A. Samaha, H. V. Tafreshi, and M. Gad-el-Hak, "Sustainability of superhydrophobicity under pressure", *Physics of Fluids,* vol. 24, Nov 2012.

[15] R. J. Daniello, N. E. Waterhouse, and J. P. Rothstein, "Drag reduction in turbulent flows over superhydrophobic surfaces", *Physics of Fluids,* vol. 21, Aug 2009.

[16] E. Aljallis, M. A. Sarshar, R. Datla, V. Sikka, A. Jones, and C. H. Choi, "Experimental study of skin friction drag reduction on superhydrophobic flat plates in high Reynolds number boundary layer flow", *Physics of Fluids,* vol. 25, Feb 2013.

[17] K. N. Ren, W. Dai, J. H. Zhou, J. Su, and H. K. Wu, "Whole-Teflon microfluidic chips", *Proceedings of the National Academy of Sciences of the United States of America,* vol. 108, pp. 8162-8166, May 17 2011.

[18] L. Lei, H. Li, J. Shi, and Y. Chen, "Diffraction patterns of a water-submerged superhydrophobic grating under pressure", *Langmuir,* vol. 26, pp. 3666-3669, 2009.

[19] P. Forsberg, F. Nikolajeff, and M. Karlsson, "Cassie–Wenzel and Wenzel–Cassie transitions on immersed superhydrophobic surfaces under hydrostatic pressure", *Soft Matter,* vol. 7, pp. 104-109, 2011.

CONTACT

*M. Xu, tel: +1-424-6665326; morleyxjtu@ucla.edu

3-D HEMISPHERICAL MICRO GLASS-SHELL RESONATOR WITH INTEGRATED ELECTROSTATIC EXCITATION AND CAPACITIVE DETECTION TRANSDUCERS

Md Mahbubur Rahman, Yan Xie, Carlos Mastrangelo and Hanseup Kim
Electrical and Computer Engineering, University of Utah, Salt Lake City, UT.

ABSTRACT

This paper reports the development and performance of a 3-D hemispherical micro glass-shell resonator with integrated electrostatic excitation and capacitive detection transducers. A new fabrication method has been developed for a hemispherical micro glass-shell resonator with glass ball molding as well as a self-guided-alignment process to maintain the gap distance between the electrodes and the shell uniform. The fabricated micro glass-shell resonator produced the first vibration mode of resonance at 5.843KHz with a quality factor of 730 at atmosphere with the time decay constant of 39.78ms. The diameter of the fabricated glass-shell resonator was measured as 1mm with shell thickness of 1.2μm.

INTRODUCTION

Despite the inherent advantages of a macro-scale Hemispherical Resonator Gyroscopes (HRG) [1], its miniaturization has been challenged due to the difficulty in fabricating ultra-symmetric and 3-dimensional wineglass-like structures utilizing conventional planar microfabrication techniques. Conventional microfabrication techniques have been limited due to their thin-film based processes, such as thin film deposition, photolithography and wet etching. Recently, some innovative approaches have been reported for such purposes, including controlled deposition of polysilicon layer [2], 3-D SOULE fabrication of spherical and mushroom structures [3] and micro-glassblowing of borosilicate glass into spherical structures [4].
Recently we reported a glass-bead based molding technique to address the aforementioned challenges in a unique method, providing higher dimensional flexibility with variable diameters, ultra-smooth surface roughness ensuring better than 120ppm uniformity in shell thickness and less than ±0.125μm deviation from a perfect sphere, 3D symmetry and easy scale-up to cm ranges [5]. However, the fabrication technique did not fully construct sensing and actuation units for functioning resonating structures.

In this paper we present the simulation, fabrication and testing results on the integration of electrostatic excitation and capacitive detection units on top of the previously fabricated hemispheric glass-shell made of ULE (Ultra Low Expansion glass); and provide the first gauge of the performance as a potential gyroscope for the inertial navigation (Fig.1). Particularly, a new fabrication technique has been reported for the integrated electrodes of any size assembled around the center stem that eliminates the need of any simultaneous fabrication of the hemispherical shell with the electrodes as well as any size restrictions on the

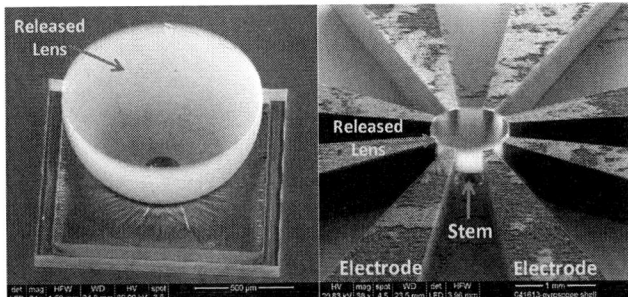

Figure 1: Integrated hemispherical shell (right) assembled with electrodes around (right) with uniform gap distance of < 5μm.

resonator structure. A self-guided-alignment process with a sacrificial layer also allowed uniform gap distance between the electrodes and the hemispherical shell. The simulation and its comparison to measurements are also discussed.

FABRICATION PROCESS

Figure 2 describes the complete fabrication process flow of the 3-D wineglass resonator with integrating Electrostatic transducers. Integrated electrostatic electrodes were first fabricated into a tall (>700μm) structure to provide an appropriate height for the 3D shell, by fusion-bonding a thin (>200μm) and a regular (500μm) silicon wafer. Specifically, a double side polished 200um-thick silicon wafer was patterned and etched by DRIE process (Fig.2-a). The deposition and patterning of a metal layer (Cr 150 nm) was followed in order to use it as a etch stop layer for later DRIE etching steps (Fig.2-b). This patterned, DRIE etched, metal embedded Si wafer was then bonded with a 500um-thick silicon wafer through a silicon fusion bonding process (Fig.2-c). For Si fusion bonding, pre-bonding cleaning was performed, such as piranha clean. Then surface activation with oxygen plasma (Oxford Plasmalab 80) at an oxygen flow rate of 6sccm, base pressure of 25mTorr for 30s and plasma power of 100W was performed to maximize the covalent bonding. Once a sufficient amount of force (1000N) was applied to the two wafers in contact for 5 minutes in the EVG 520 IS wafer bonder, a complete fusion bonding was achieved (Fig.2-c). The bonding was performed at room temperature, while annealing at an elevated temperature of 300°C was followed for 1 hour. Note that a metal layer was embedded within the two bonded silicon wafers.

To fabricate external metal leads of the electrodes for wire bonding, patterning of two metal layers of Ti (adhesion layer, 30nm) and Pt (Contact layer, 90nm) was followed on a separate glass wafer after their consecutive deposition (Fig.2-d). These metal patterns were sandwiched during the anodic bonding between the glass and the previously fusion-bonded stacked Si wafers (Fig.2-e). Before the anodic bonding, also pre-cleaning and baking steps were performed. During the anodic bonding, the substrates were heated up to 350°C and applied with electrostatic force (1000V) in the EVG 520 IS wafer bonder. Subsequently 500µm DRIE etching was performed by Oxford 100 ICP (Fig.2-f and g) with the embedded metal mask (Cr). For the DRIE etching, a gas combination of SF_6 (for etching) and C_4F_8 (for Polymerization the sidewall) was used with 9:5 ratios at 10°C.

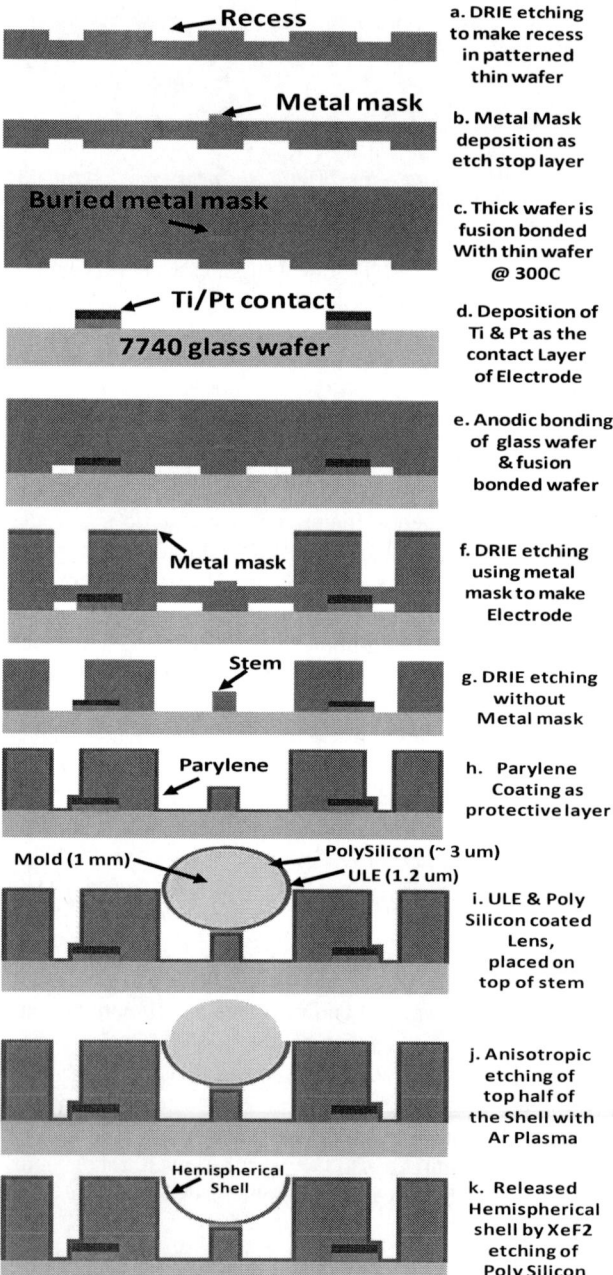

Figure 2: Fabrication process flow for the integrated electrodes around the hemispherical shell.

Figure 3 shows the fabricated 8 isolated electrodes as well as the center stem after finishing 700µm DRIE etching (Fig.2-g). All the electrodes were fabricated radially around the center stem (diameter 100µm) that was used as an anchor for the hemispherical shell. For a higher Quality factor and time decay constant, a minimum contact area between the resonating shell with the center stem was required. The stem radius was optimized by varying it from 10µm to 200µm, and eventually 100µm was selected for appropriate adhesive-based attachment.

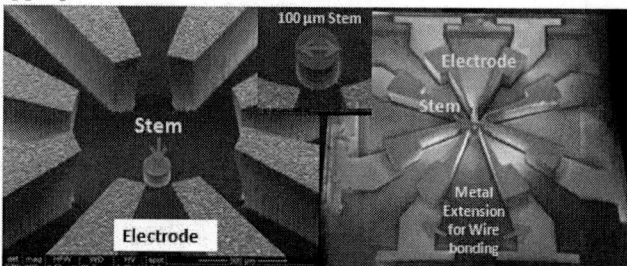

Figure 3: Fabricated electrodes & the center stem, before attached to the shell on top.

After parylene deposition as the protection layer (Fig.2-h), the ULE ball mold was dropped and attached to the stem utilizing adhesives (Fig.2-i). The high precision ball bearing mold was previously sputtered with uniform thin polysilicon (3µm) & ULE (1.2µm). After baking the adhesives for strong attachment, anisotropic Ar plasma etching was performed to remove the top half part of the coated mold including the ULE and polysilicon layer (Fig.2-j). As argon plasma was utilized as the anisotropic physical etching to remove the top half of the mold, the mold itself worked as an etch stop layer to protect the bottom half. Figure 4 shows the ball bearing attached to a stem (invisible) while surrounded by 8 electrodes, before (Fig.4-left) and after (Fig.4-right) the Ar plasma etching.

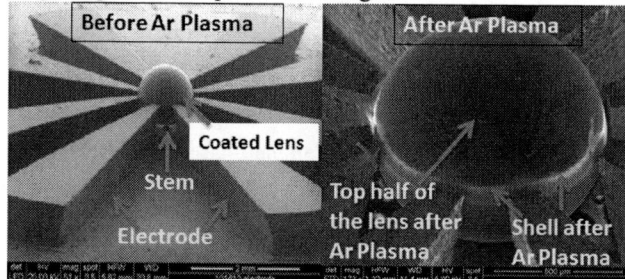

Figure 4: The ball bearing attached onto a stem before (a) & after (b) anisotropic Ar plasma etching.

Subsequently the inner polysilicon layer was etched by an isotropic XeF_2 etching, and the glass ball was released leaving the ULE glass shell attached to the stem forming the wineglass resonator structure (Fig.2-k). Figure 5 shows the released glass shell. For the timed XeF_2 etching, a gas combination of XeF_2 at 1torr for the polysilicon etching and N_2 at 10torr for residue cleaning was employed. XeF_2 etching was an isotropic chemical etching, thus being capable of under-etching all the inner polysilicon layer near the stem. The released glass bearing was carefully lifted by adhesives.

Figure 5: Fabricated resonator shell and eight electrodes after the ball bearing release.

The gap distance between the electrodes and the shell was maintained uniform through a self-guided-alignment process with Parylene coating (Fig.6). The parylene layer (10μm) was removed by oxygen plasma etching (Oxford Plasmalab 80), leaving a uniform and close gap distance around the shell. The tiny gap is necessary for electrostatic actuation and capacitive detection. In order to optimize the device performance, a few generations of devices were fabricated varying the gap distance between the shell and the electrodes. The minimum gap distance achieved was measured as 3.72μm around the shell with 120ppm uniformity for shell thickness and less than ±0.125μm deviation from a perfect sphere.

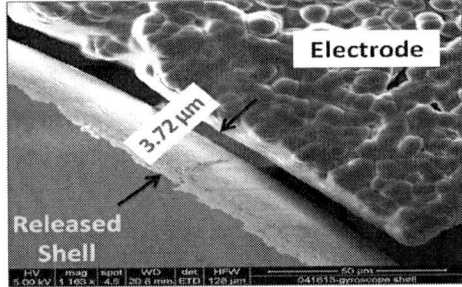

Figure 6: Gap distance between electrode and shell.

SIMULATION RESULTS

Finite Element Method analysis (COMSOL Multiphysics 4.3) of the hemispherical shell was performed to predict the natural frequency of wineglass modes. Eigen frequency analysis predicts natural frequencies of vibration in multiple modes of a given structure. For a hemispherical shell, vibration mode m=2 and m=3 are normally known as wineglass modes that have inherent rejection capacity of common mode accelerations and expected to have low anchor loss [3, 4]. For the hemispherical shell, the vibration modes m=2 and 3 consist of four and six antinodes (where the maximum displacement takes place) and four and six nodes of minimum displacement (Fig.7).

Table 1 lists all the simulated natural frequencies of the two wineglass modes for the shell being uncoupled with the stem. The shell was modeled with an ULE layer with diameter d=1.0mm, shell thickness h=1.2μm, Young's modulus E= 67.6e9Pa and Density ρ= 2210kg/m³. For Eigen frequency N=12, 12 natural frequencies were pursued to predict the resonance frequencies at each mode. Higher numbers of eigen frequencies (N) were utilized to confirm

the reliability of the data. At m=2 the simulated natural frequencies were 6023.63 and 6025.02Hz, showing less than 1.5Hz frequency split between the two elliptical modes. For m=3, the frequency split increased to 40Hz, and the resonance frequency was predicted at 16808.68Hz and 16848.41Hz.

Table 1: Simulated Resonance frequency for different vibration modes of the testing uHRG.

Number of Eigen frequency (N) used for COMSOL simulation	Wineglass mode Resonance Frequency (Hz)			
	Vibration mode m = 2		Vibration mode m = 3	
	For m=2, frequency split < 1.5Hz		For m=3, frequency split < 40Hz	
N=12	6023.634443	6025.027751	16808.688911	16848.411910
N=14	6023.633945	6025.027285	16808.688687	16848.411756
N=16	6023.633388	6025.027707	16808.688571	16848.411697
N=18	6023.633486	6025.027581	16808.688652	16848.412083

Four different conditions had been considered: couple and uncoupled of the shell to the stem for both the primary wineglass modes m=2 and m=3 (Fig.7). Simulation results (Fig.7) showed that the resonance frequency was predicted as 6.025KHz and 10.07KHz for the shell being uncoupled and coupled at vibration modes m=2, reasonably matching with experimental measurement of 5.843KHz. For m=3, the resonance frequency was predicted at 16.85KHz and 16.31KHz, with the experimental measurement of 17.532KHz. The simulation results also show that the resonance frequency increased with the increase in the shell thickness.

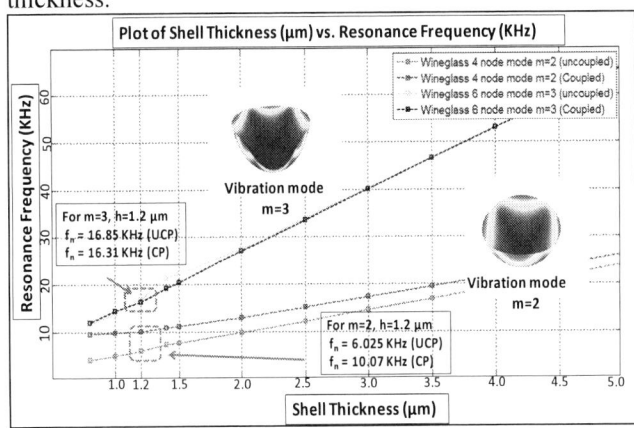

Figure 7: FEM Analysis of shell thickness vs. Resonance Frequency for two different wineglass modes of vibration.

EXPERIMENTAL

The fabricated device, placed into a ceramic DIP package (Fig.8), was electrostatically-actuated and capacitively sensed. After placing the device in the package with adhesives in the back of the supporting glass-bead, all the integrated electrodes were wire bonded between the metal extension of the electrodes and the contact pad of the package (Fig.8).

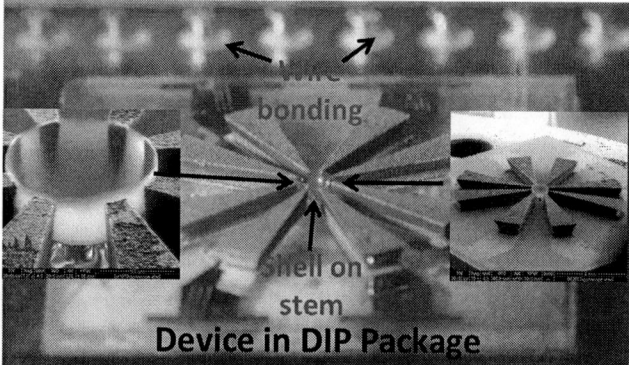

Figure 8: Ceramic DIP package with wire bonding of each electrode to resonate the shell.

To excite the device over a desired frequency range, the device was connected to a network analyzer through a transimpedance amplifier constructing the feedback loop for stable analysis. For the electrostatic excitation, a bias of 60V DC and a drive voltage of 100mV AC p-p were applied through an op-amp to a set of two driving electrodes. The resultant capacitance change was detected by a sense electrode that created the motional current. The motional current was converted into a voltage through the transimpedance amplifier with a gain equal to the feedback resistor value (Fig.9).

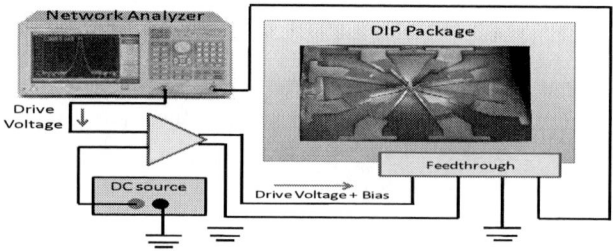

Figure 9: Schematics of the experimental setup for the electrostatic excitation of the hemispherical resonant shell.

The measurement results showed that for m=2 & m=3 wineglass modes, the resonance frequencies were measured respectively at 5.843KHz and 17.532KHz, which matched to the simulation results of 6.025KHz & 16.85KHz within ±4.0% (Fig.10). For measuring the quality factor, the -3dB frequencies were found to be around the resonance frequency f_o=5.843KHz for m=2. Then quality factor of the device was measured as Q = 730 at atmosphere, and the

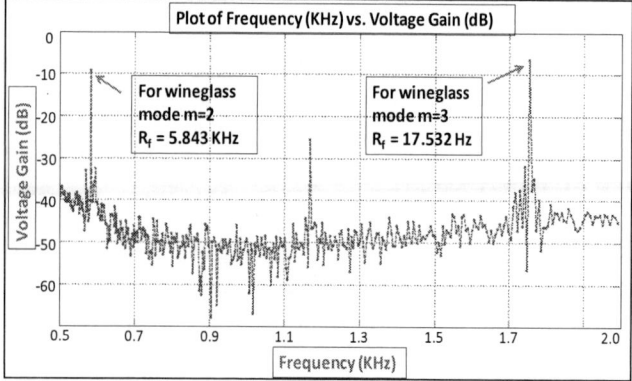

Figure 10: Frequency vs. voltage gain to show Resonant peaks for different wineglass modes m=2 & m=3 (in air).

time decay constant as τ = 39.78ms at 5.843KHz under DC bias and excitation (Fig.11). The quality factor was constrained primarily due to the air damping.

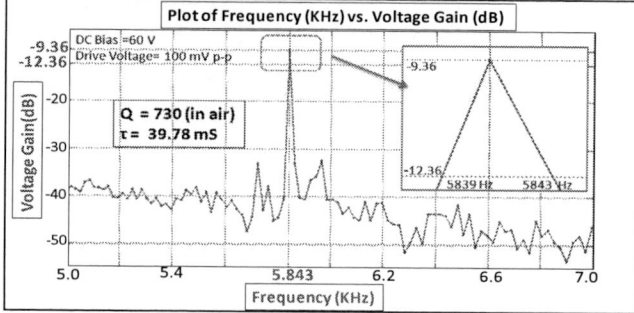

Figure 11: Frequency vs. voltage gain to show the Q factor & time decay constant (τ) for wineglass mode m=2 (in air).

CONCLUSION

This paper reports a 3-D hemispherical micro glass-shell resonator with integrated electrostatic excitation and capacitive detection transducers. New fabrication methods including glass ball molding and a self-guided-alignment process were reported. The fabricated micro glass-shell resonator produced the first vibration mode of resonance at 5.843KHz with a quality factor of 730 at atmosphere with the time decay constant of 39.78ms. The diameter of the fabricated glass-shell resonator was measured as 1mm with shell thickness of 1.2μm.

ACKNOWLEDGEMENT

This work was supported by the Defense Advanced Research Projects Agency (DARPA) Microscale Rate Integrating Gyroscope (MRIG) program. Microfabrication was performed at the Utah Nanofabrication Facility at the University of Utah.

REFERENCES

[1] D.M. Rozelle, "The hemispherical resonator gyro: From wineglass to the planets", AAS/AIAA Space Flight Mechanics Meeting, 2009, pp. 1157–1178.

[2] L.D. Sorenson, X. Gao, F. Ayazi, "3-D Micromachined Hemispherical Shell Resonators With Integrated Capacitive Transducers,"MEMS'12, pp.168-171.

[3] K. Visvanathan, Li Tao, Y. B. Gianchandani, "3-D soule: A fabrication process for large scale integration and micro-machining of spherical structures", MEMS'11, pp. 45-48.

[4] I.P Prikhodko, S.A. Zotov, A.A. Trusov A.M. Shkel, "Microscale Glass-Blown Three-Dimensional Spherical Shell Resonators, "IEEE J. Microelectromech. Syst.vol. 20, no. 3, 2011, pp. 691-701.

[5] Y. Xie, H. Hsieh, P. Pai, H. Kim, M. Tabib-Azar ,C. H. Mastrangelo," Precision Curved Micro Hemispherical Resonator Shells Fabricated by Poached-Egg Micro-molding, " IEEE Sensors 2012, pp. 28-31.

CONTACT

*M.M. Rahman, mahbubur.rahman@utah.edu.

A CMOS MEMS PIRANI VACUUM GAUGE WITH COMPLEMENTARY BUMP HEAT SINK AND CAVITY HEATER

Yi-Chiang Sun[1], Kai-Chih Liang[1,3], Chao-Lin Cheng[1], and Weileun Fang[1,2]

[1]Department of Power Mechanical Engineering, and [2]Institute of NanoEngineering and MicroSystems,
National Tsing Hua University, Hsinchu, Taiwan
[3]Taiwan Semiconductor Manufacturing Company Ltd., Hsinchu Science Park, Hsinchu, Taiwan

ABSTRACT

A novel CMOS-MEMS Pirani vacuum gauge with complementary bump heat-sink and cavity heater design has been proposed and demonstrated. This design using CMOS-MEMS process to offer the following advantages for Pirani gauge: (1) The bump heat-sink vertical integrates with cavity heater increases the dynamic range and sensitivity without changing device footprint size, (2) The cavity in heater reduces the thermal mass for low-power operation, and (3) Easy integration with packaged CMOS-MEMS devices for pressure monitoring [1]. The design is implemented using the standard TSMC 0.18μm 1P6M CMOS process. A 120μm×120μm die size with 0.53μm sensing gap is demonstrated. Measurement indicates the gauge has sensing range 0.3-100torr with sensitivity of 1.53×10^4(K/W)/torr. The power consumption is 67μW for 1% resistance change. In comparison, the gauge with typical heat-sink/heater design has sensing range 1-100torr with sensitivity of 0.99×10^4(K/W)/torr and power consumption of 119μW.

INTRODUCTION

Pirani vacuum gauge is a thermal-conductivity type vacuum sensor. Generally, the gauge consists of heater and heat-sink units. The heat generated by heater is transferred to the heat-sink through gas-molecules whose thermal conductivity depends on their pressure. Micromachined Pirani gauge have the advantage of smaller size, lower power consumption, lower operation temperature, faster thermal response, and a wider range of operating pressure when compare with conventional counterparts [2]. One application for micro Pirani gauge is the *in situ* measurement of the pressure inside the packaged devices such as resonator, inertial sensor, and some devices that require thermal isolation through vacuum environment [3,4]. Alternative methods for *in situ* pressure measurement are the helium leak test [5], and the monitoring of Q-factor variation for inertial sensors or resonators [6]. However, helium leak test require expensive equipment and are generally limited to a leak measurement resolution. As compare to resonator Q-factor approach, Pirani gauge is easier to calibrate and test and usually have higher pressure sensitivities [7].

Various approaches have been reported to improve the performance of Pirani gauge, such as increasing heater length to shift the measurement range toward lower pressure [7,8], process development for submicron/nano gap so that measurement range near atmospheric pressure can be obtained [9,10], and increasing active area by additional heat sinks to enhance the performance [11]. The standard CMOS process has been employed to realize Pirani gauge in [12]. In short, footprint of chip and available process for small air gap remain critical concerns while improving the Pirani gauge performances. Therefore, this study proposes a new structure design to increase the active area between heater and heat-sink to improve the performance of Pirani gauge.

The proposed design employs standard CMOS process and its feature of multilayer stacking and submicron gap to develop the Pirani vacuum gauge with complementary bump heat-sink and cavity heater. The bump heat-sink vertical integrated with cavity heater increases the active area between heater and heat-sink without changing device die size, yet the dynamic range and sensitivity can still be increased. The device is designed and implemented based on the standard TSMC 0.18μm 1P6M CMOS process and can easily integrate with other packaged CMOS-MEMS devices for pressure monitoring.

DESIGN CONCEPT

Fig.1a shows the sensors design, including the cavity heater and bump heat-sink. The typical design without bump heat-sink is also provided for comparison. As indicated in

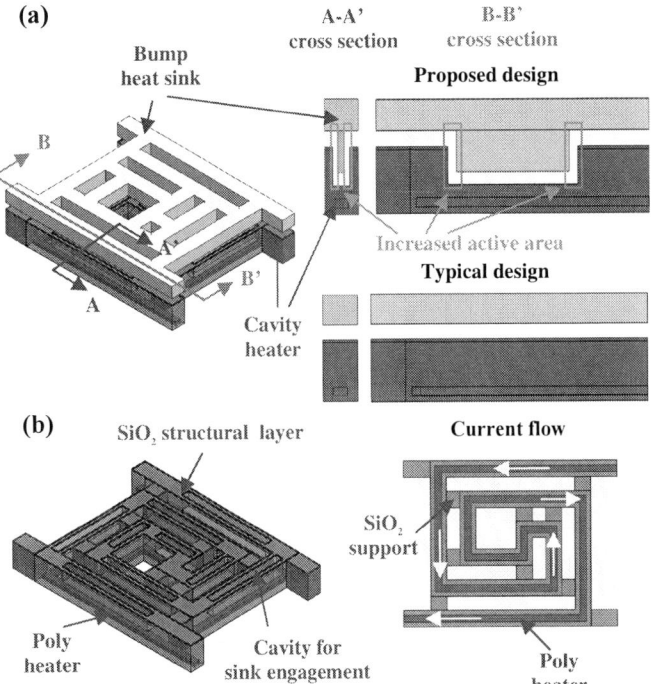

Figure 1: (a) Design concept of Pirani gauge and its cross-section view. (b) Heater structure and its current flow.

978-1-4799-3510-9/14 $31.00 © 2014 IEEE

the cross sections of structures, the proposed design has additional active area in side walls while the device die size remaining the same. By increasing the active area between heater and heat-sink, the sensitivity can be improved. As in Fig.1b, the polysilicon covered with SiO_2 are used for heater. Its shape and current flow are also indicated in figure. The heater is designed as a spiral shape in order to decrease the total device size. By using the spiral design, this study arranges a 620μm long heater inside a 120μm×120μm CMOS chip. Moreover, the cavity on heater is for the engagement of complementary bump on heat-sink, and such cavity will reduce the thermal mass of heater. The cavity heater design enables the reducing of thermal mass and further decreases the power operation. As indicated in Fig.1b, various auxiliary SiO_2 supports are employed to increase the stiffness of the suspended structure. Thus, the bending of heater caused by polysilicon/SiO_2 CTE mismatch is reduced, and the heat loss due to the contact between heater and heat-sink is prevented. In summary, the bump heat-sink design increases the active area for heat transfer and also decreases the thermal mass for heater. Thus, dynamic range and sensitivity of Pirani gauge is increased yet power consumption is decreased.

FABRICATION AND RESULTS

Fig.2 shows the fabrication process steps. Fig.2a shows the chip prepared by standard TSMC 0.18μm 1P6M CMOS process. In Fig.2b, H_2SO_4/H_2O_2 solution was used to remove metal and tungsten-via layers [1]. The sub-micron in-plane sensing gap between heater and heat-sink was defined by the sacrificial metal layers and tungsten vias. The metal films can be protected by the surrounding dielectric layers during wet etching. After that, the SiO_2 was removed by reactive ion etching to define bonding pads and the isotropic Si etching by XeF_2 was used to release MEMS structures, as in Fig.2c.

Fig.3 shows fabricated Pirani gauges. The total device size is 120μm×120μm and the total heater length is 620μm. The SEM micrographs in Fig.3a display the typical design. The heater and heat-sink unit is fully implemented. The sensing gap is 0.53μm defined by the thickness of metal film in standard CMOS process. The micrograph in Fig.3b shows the FIB sectioning of the structure marked with dash line. The stacking of CMOS layers and the vertical integration of the complementary bump heat-sink and cavity heater is clearly observed. The cavity heater only composes of polysilicon and SiO_2 in order to lower the influence of bending cause by CMOS materials CTE mismatch. The bump heat-sink contains metal and passivation layer. The metal film in heat-sink is used to enhance the heat conduction through heat-sink. The vertical sensing gap is also defined by the thickness of metal film (Metal 1 layer) in CMOS process.

MEASUREMENT RESULTS

Firstly, this study characterized the temperature coefficient of resistance (TCR) of the heater using the hotplate for temperature control and the source meter for resistance measurement. Measurements on resistance change

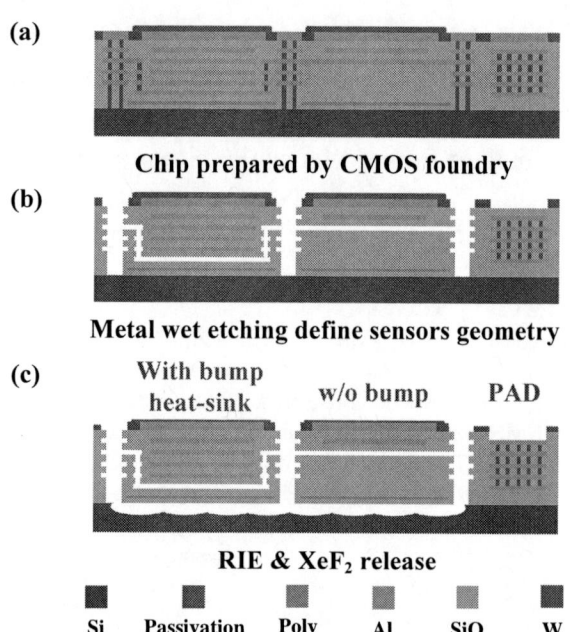

Figure 2: Fabrication process steps, (a) chip prepared by TSMC. (b-c) in-house post-CMOS metal wet etching and structure release.

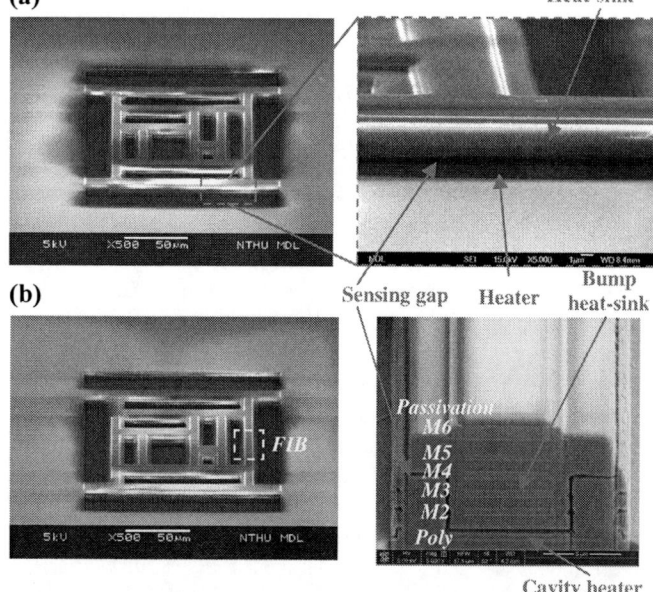

Figure 3: Micrographs of CMOS-MEMS pirani vacuum gauge (a) Typical design and its zoom-in configuration. (b) Bump heat sink, cavity heater design and its FIB cross-section view.

of devices at different temperature indicate the TCR of polyheater is 2.85×10^{-3} Ω/K. The characterization of Pirani gauge in this study uses the 4-point probe configuration [7]. Fig.4 shows the measurement setup used in this study. The pressure inside the chamber is specified by the controller and monitored by a commercial gauge. Two source meters are respectively employed to specify the input current and detect the voltage of the proposed Pirani gauge, and further determine the resistance change and applied power of the gauge. The temperature change is extracted by the relationship between resistance change and TCR. Thus, the temperature change versus applied power is shown in Fig.5. The slope of each curve indicates the thermal impedance of devices under different pressure. It clearly shows that the slope (thermal impedance) become larger as the pressure decrease. This phenomenon agrees with the operation principle of Pirani gauge.

The extracted thermal impedance versus ambient pressure is recorded in Fig.6. The results respectively indicate the measurements of the proposed and typical Pirani gauges. The linear region in the figure is the dynamic range, and the slope within dynamic range is the sensitivity of the device. As a result, the proposed Pirani gauge with bump heat-sinks design shows higher sensitivity and larger dynamic range due to increased active area. Furthermore, the dynamic range of proposed Pirani gauge extends to lower pressure range while the die size remaining the same. Fig.7 further shows the resistance change versus ambient pressure. In addition to the measurement results, the simulation results from numerical model in [13] are also available for comparison. The simulation model is based on the geometry parameter of typical design. The deviation between simulation and experiment could attribute to heat spreading through the additional SiO_2 support in the spiral heater structure [7]. Therefore, the measurement results shift toward lower pressure region. Moreover, the cavity heater design enables a higher resistance change under the same current heating due to reduced thermal mass. In other words, the power consumption is reduced by proposed Pirani gauge.

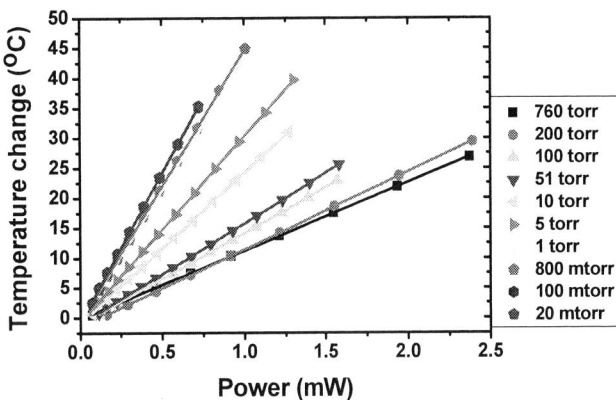

Figure 5: Temperature change versus applied power of the bump heat-sink Pirani gauge

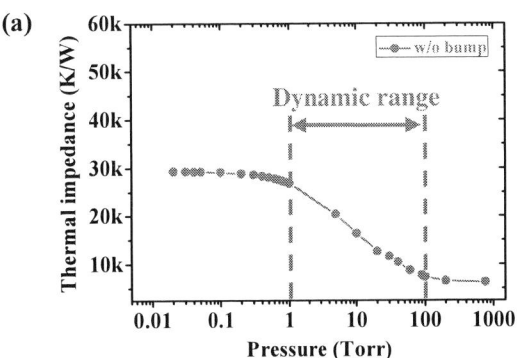

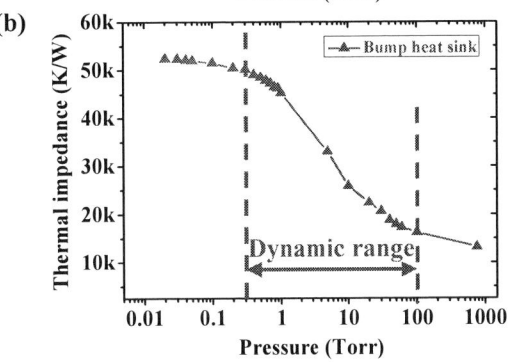

Figure 6: Thermal impedance of CMOS MEMS pirani vacuum gauge versus ambient pressure (a) Typical design. (b) Bump heat sink and cavity heater design.

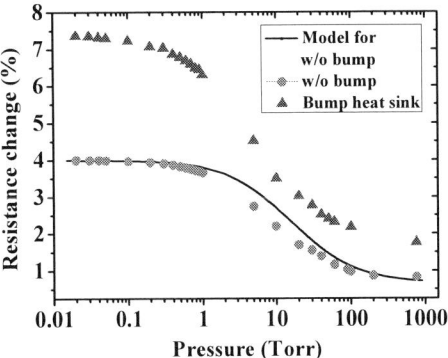

Figure 7: Fractional resistance change of CMOS MEMS pirani vacuum gauge versus ambient pressure under 1mA heating

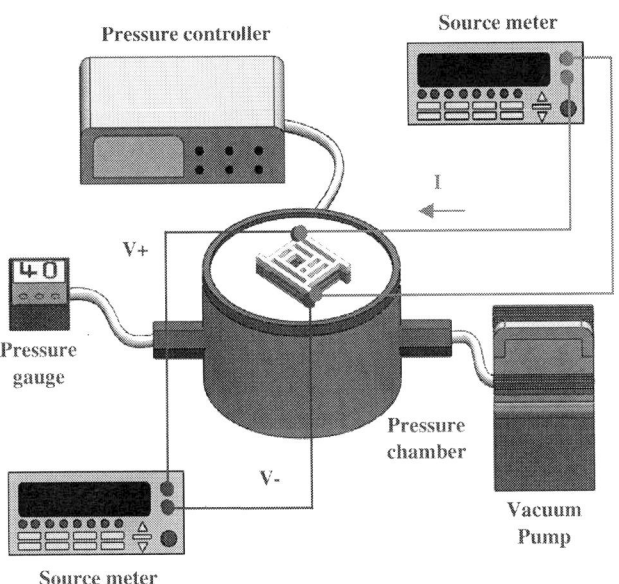

Figure 4: The schema of the 4-point probe measurement setup

978-1-4799-3510-9/14 $31.00 © 2014 IEEE

Table 1: Summary of this study

	Typical	*Proposed*
Die size	*120×120 μm²*	*120×120 μm²*
Total heater length	*620 μm*	*620 μm*
Power consumption (1% resistance change)	*119.75 μW*	*67.01 μW*
Dynamic range	*1~100 torr*	*0.3~100 torr*
Sensitivity (in a dynamic range)	*0.99×10⁴ (K/W)/torr*	*1.53×10⁴ (K/W)/torr*

Table 1 summarizes the performance of presented Pirani gauge. Note that a micro heater of 620μm long was employed in this study for concept proven of the proposed Pirani gauge. Measurements on lower pressure range can further be achieved using the presented Pirani gauge by extending its heater length. More applications for vacuum package monitoring become available.

CONCLUSION

In this study, the Pirani gauge with complementary bump heat-sink and cavity heater design is proposed and demonstrated based on standard TSMC 0.18μm 1P6M process. The bump heat-sink vertically integrates with cavity heater increases the active area without changing device footprint size. The cavity heater reduces thermal mass of heater for low power operation. Measurements indicate the proposed design have better performance than typical type. It has larger dynamic range, 45% power consumption decrease, and 53% sensitivity increase as compare with typical one. The heater length of presented devices is 620μm. The heater length can be further increased to detect lower pressure range for package monitoring. Because of the standard CMOS process, this device can easily integrate with other packaged CMOS-MEMS devices for *in situ* pressure monitoring.

ACKNOWLEDGEMENTS

This research was sponsored in part by the National Science Council of Taiwan under grant of NSC-102-2221-E-007-027-MY3, NSC-102-2622-E-007-014-MY3, NSC-102-2218-E-007-003-MY3, and NSC 101-2221-E-007-069-MY3. The authors wish to appreciate the TSMC and the National Chip Implementation Center (CIC), Taiwan, for the supporting of CMOS chip manufacturing. The authors would like to thank the National Center for High-Performance Computing for support of simulation tools. The authors would also like to appreciate the Center for Nanotechnology, Materials Science and Microsystems of National Tsing Hua University and the National Nano Device Lab. in providing fabrication facilities. The authors also wish to appreciate the Sensirion AG for the technical and financial support.

REFERENCE

[1] M.-H. Tsai, C.-M. Sun, Y.-C. Liu, C. Wang, and W. Fang, "Design and fabrication of a metal wet-etching post-process for the improvement of CMOS-MEMS capacitive sensors," *J. Micromech. Microeng.*, vol. 19, 105017, 2009.

[2] W. J. Alvesteffer, D. C. Jacobs, and D. H. Baker, "Miniaturized thin film thermal vacuum sensor," *J. Vac. Sci. Technol. A, Vac. Surf. Films*, vol. 13, no. 6, pp. 2980–2985, Nov. 1995.

[3] E. S. Topalli, K. Topalli, S. E. Alper, T. Serin, and T. Akin, "Pirani vacuum gauges using silicon-on-glass and dissolved-wafer processes for the characterization of MEMS vacuum packaging," *IEEE Sensors J.*, vol. 9, no. 3, pp. 263–270, Mar. 2009.

[4] B. H. Stark, Y. Mei, C. Zhang, and K. Najafi, "A doubly anchored surface micromachined Pirani gauge for vacuum package characterization," *in Proc. IEEE MEMS*, 2003, pp. 506–509.

[5] Department of Defense Test Method Standard Microcircuits, pp. 1–13, MIL-STD-883F, Jun. 18, 2004.

[6] Y.-T. Cheng, W.-T. Hsu, K. Najafi, C. T.-C. Nguyen, and Liwei Lin, "Vacuum packaging technology using localized aluminum/silicon-to-glass bonding," *J. Microelectromech. Syst.*, vol. 11, no. 5, pp. 556-565, Oct. 2002

[7] J. Mitchell, G. R. Lahiji, and K. Najafi, "An improved performance poly-Si Pirani vacuum gauge using heat-distributing structural supports," *J. Microelectromech. Syst.*, vol. 17, no. 1, pp. 93–102, Feb. 2008.

[8] F. Santagata, J. F. Creemer, E. Iervolino, L. Mele, A. W. van Herwaarden, and P. M. Sarro, "A tube-shaped buried Pirani gauge for low detection limit with small footprint," *J. Microelectromech. Syst.*, vol. 20, no. 3, pp. 676-684, June. 2011.

[9] M. Kubota, Y. Mita, T. Momose, A. Kondo, Y. Shimogaki, Y. Nakano, and M. Sugiyama, "A 50 nm-wide 5 μm-deep copper vertical gap formation method by a gap-narrowing post-process with Supercritical Fluid Deposition for Pirani gauge operating over atmospheric pressure," *in Proc. IEEE MEMS*, 2012, pp. 204-207.

[10] K. Khosraviani and A. M. Leung, "The nanogap Pirani—A pressure sensor with superior linearity in an atmospheric pressure range," *J. Micromech. Microeng.*, vol. 19, no. 4, p. 045007, Apr. 2009.

[11] J. Chae, B. H. Stark, and K. Najafi, "A micromachined Pirani gauge with dual heat sinks," *IEEE Trans. Adv. Package*, 2005, pp. 619-625.

[12] A. Haberli, O. Paul, P. Malcovati, M. Faccio, E Maloberti, and H. Baltes, "CMOS integration of a thermal pressure sensor system," *IEEE ISCAS 1996*, pp. 377–380.

[13] C. H. Mastrangelo and R. S. Muller, "Microfabricated thermal absolute pressure sensor with on-chip digital front-end processor," *IEEE J. Solid-State Circuits*, vol. 26, no. 12, pp. 1998–2007, Dec. 1991.

CONTACT

* W. Fang, Tel: +886-3-5742923; fang@pme.nthu.edu.tw

A LARGE RANGE MULTI-AXIS CAPACITIVE FORCE/TORQUE SENSOR REALIZED IN A SINGLE SOI WAFER

D. Alveringh, R.A. Brookhuis, R.J. Wiegerink, and G.J.M. Krijnen

MESA+ Institute for Nanotechnology, University of Twente, Enschede, THE NETHERLANDS

ABSTRACT

A silicon capacitive force/torque sensor is designed and realized to be used for biomechanical applications and robotics. The sensor is able to measure the forces in three directions and two torques using four parallel capacitor plates and four comb-structures. Novel spring and lever structures are designed to separate the different force components and minimize mechanical crosstalk. The fabrication process is based on deep reactive ion etching on both sides of a single silicon-on-insulator wafer and uses only two masks making it a straight-forward and robust process. The sensor has a force range of 2 N in shear and normal direction and a torque range of more than 6 N mm. It has a high sensitivity of 38 fF N^{-1} and 550 fF N^{-1} in shear and normal direction respectively.

INTRODUCTION

Many prostheses require safe and comfortable interaction with people who underwent an amputation. Bad fitting between the socket of the prosthesis and the residual limb may cause pain and injuries [1]. By measuring the force components between the socket and the residual limb, the prosthesis can adjust the shape of the socket, making the load distribution more comfortable. Besides biomechanical applications, force sensors are also very interesting for robotics.

For these applications, a few specific requirements are applicable:

- the sensor should measure multiple (preferably six) degrees of freedom;
- the sensor is small, preferably less than 1 cm^2 with a thickness of less than 1 mm;
- the sensor should be able to handle human forces, i.e. at least a few newtons.

Commercially available non-MEMS load cells support high force ranges, but are often too large to integrate in the applications mentioned above. There are MEMS-based force and torque sensors available in literature, e.g. [2, 3, 4, 5], but many lack the support for measuring torques, forces, high ranges or have a difficult fabrication process. We present a miniature easy to fabricate multi-axis capacitive force/torque sensor with a large range as a relatively low-cost alternative of the sensor we presented in [6] and [7].

PRINCIPLE OF OPERATION

The sensor consists of a suspended core which is fabricated in the handle layer of an SOI wafer. The core is supported by v-shaped silicon springs. An applied load to the suspended core will result in a displacement. In-plane displacement caused by a shear force is measured by comb-structures present in the device layer and results in a differential change in gap between the comb-fingers (figure 1).

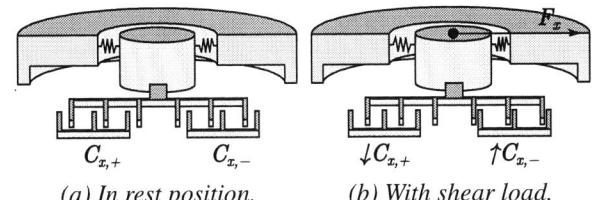

(a) In rest position.   *(b) With shear load.*

Figure 1: Principle of operation for shear forces, which results in a differential change of capacitance.

A normal force results in an out-of-plane displacement, which is measured by parallel plate electrodes (figure 2a). By differential measurement of two opposite electrodes (figure 2b), the applied torque is determined.

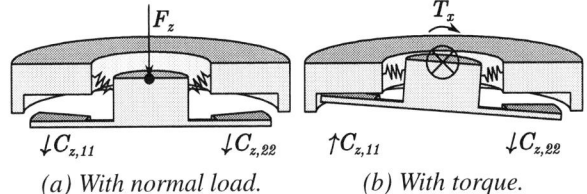

(a) With normal load.   *(b) With torque.*

Figure 2: Principle of operation for normal forces and torques, latter results in a differential change of capacitance.

MECHANICS OF THE SUSPENDED CORE

The proposed force/torque sensor uses the point symmetric v-shaped spring system shown in figure 3.

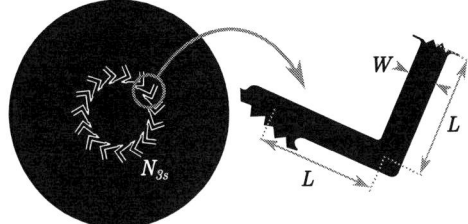

Figure 3: The point symmetric v-shaped spring system and the parameters of each spring realized in the handle layer with thickness T.

The stiffness in shear direction is equal [8] to:

$$k_x = \frac{45 N_{3s} E I_x}{2 L^3}, \quad \text{with} \quad I_x = \frac{T W^3}{12}, \quad (1)$$

with k_x the stiffness in x-direction, N_{3s} the number of

978-1-4799-3510-9/14 $31.00 © 2014 IEEE   680

spring triplets, E Young's modulus, I_x the second moment of area in x-direction, L the length of one spring part, T the thickness of the beam and W the width of the beam. The stiffness k_z in normal direction is derived from the guided beam theory from [9]:

$$k_z = \frac{12(3N_{3s})EI_z}{(2L)^3}, \qquad \text{with} \qquad I_z = \frac{WT^3}{12}, \quad (2)$$

The six degrees of freedom stage can be tuned for translations with parameters L, W and N_{3s}:

$$k_x \propto \frac{N_{3s}}{L^3}W^3, \qquad k_z \propto \frac{N_{3s}}{L^3}W. \quad (3)$$

The stiffness in x-direction compared to z-direction can be optimized by choosing the right value for the flexure width W, the stiffness in both directions can be tuned by the flexure part length L and the number of spring triplets N_{3s}.

CAPACITIVE SENSING STRUCTURES

Figure 4 shows where the capacitor structures are located. The parallel electrode capacitors for normal force and torque measurements and comb-structures for shear force measurements can both be modeled as gap closing parallel plate capacitors:

$$C = N_p\varepsilon\frac{A}{d_0 - \Delta d} = N_p\varepsilon\frac{A}{d_0 - \frac{F}{k}}. \quad (4)$$

With C the capacitance, N_p the number of parallel plates or finger pairs, ε the absolute permittivity, A the overlapping area of one plate or finger pair, d_0 the distance between the plates or fingers in rest, Δd the displacement, k the stiffness and F the force.

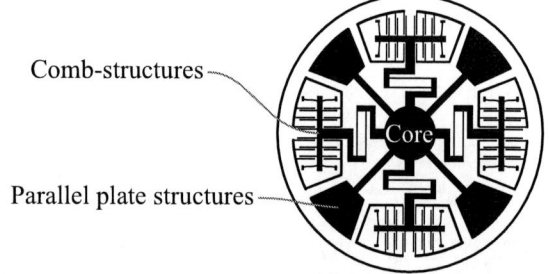

Figure 4: Schematic design of the device layer. The comb-structures are for shear force measurements and the parallel plate structures are for normal force and torque measurements.

The comb-structures (figure 5a) consist of combs mounted on a one degree of freedom shuttle which is supported by eight single flexures. Spring and lever structures are used to separate the different force components of the suspended core into comb-structure movements. The stator consists of two symmetric electrically isolated parts. The asymmetric positioning of the shuttle-fingers between the stator-fingers is optimized, for a smaller distance between the finger pairs allows more finger structures but increases the parasitic capacitance.

The parallel plate structures (figure 5b) consist of flat plates that form a capacitor with the handle layer. The plate is electrically connected to the bond pad with springs that are compliant in all directions.

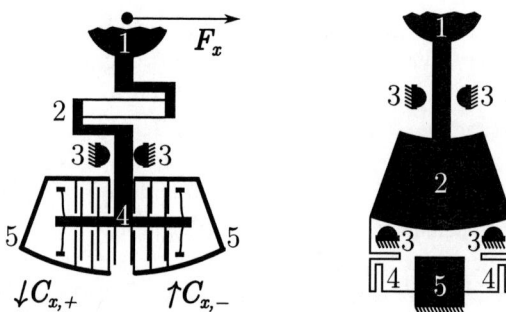

(a) Comb structures. Shuttle (4) is coupled for one direction via flexure (2) with core (1). Stators (5) can be measured differentially.

(b) Parallel plate structures. The plate (2) is directly coupled with the core (1). Flexures (4) electrically connect the plate to the bond pad (5).

Figure 5: Close up of the sensing structures. Bumps (3) protect the structures from snapping.

FABRICATION

A 100 mm p-type SOI wafer with a handle layer of 400 μm, a device layer of 50 μm and a box layer of 4 μm is used for the device. (figure 6a). The fabrication process needs two masks: the mask for etching the handle layer and the mask for etching the device layer.

The first step in the fabrication process is the oxidation (figure 6b) of the wafers for the formation of a hard mask. Patterning the oxide was done using photolithography and reactive ion etching (RIE). A standard Bosch process was used with a recipe based on an argon (Ar) and fluoroform (CHF_3) chemistry (figure 6c). Resist stripping was done in O_2-plasma and nitric acid (HNO_3).

The handle layer and the device layer were etched using DRIE (figures 6d and 6e). Sulfur hexafluoride (SF_6) was used as etchant and flurocarbon (C_4F_8) was used for the deposition of passivation layers. The fluorocarbon residues were removed using piranha cleaning and O_2-plasma.

The chips were pushed out of the wafer, eliminating the need for dicing and allowing a circular shaped chip. Because this resulted in particle contamination, all chips underwent ultrasonic cleaning. A wet etch with 50 % HF is performed and etched through the box layer of the SOI wafer (figure 6f). To prevent capillary forces making the structures snap to eachother, the final release etch was done using vapor HF (figure 6g).

A hole is drilled in a printed circuit board (PCB). The

978-1-4799-3510-9/14 $31.00 © 2014 IEEE

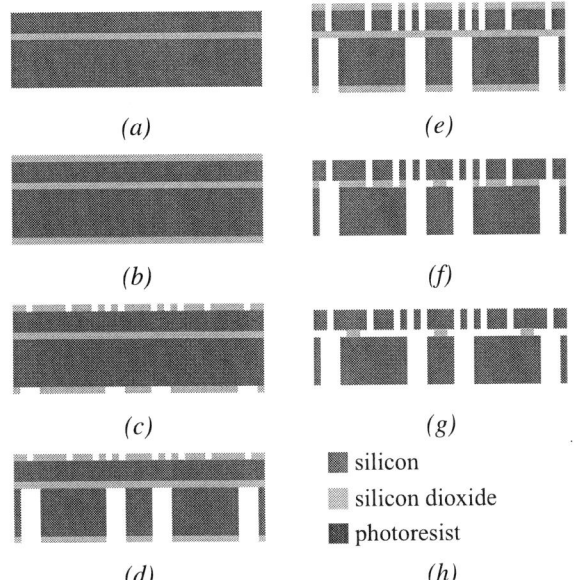

Figure 6: Fabrication process: (a) SOI wafer, (b) oxidation, (c) forming hard mask, (d) etching of handle layer, (e) etching of device layer, (f) oxide etch (wet), (g) release etch (vapor), (h) materials.

sensor is mounted with the handle layer on the PCB using adhesive bonding. The sensor is wire bonded and a stylus is mounted using epoxy glue on the top of the suspended core through the hole in the PCB (figure 8).

Figure 7: Impression of the device. The sensor has a diameter of 9.24 mm and a thickness of less than 0.5 mm.

CHARACTERIZATION

The force/torque sensor is characterized for all degrees of freedom, except for torques around normal axes (T_z). The mechanical measurement setups are shown in figure 9.

The sensor is characterized by applying loads in shear and normal direction. An extra stylus is mounted on the back

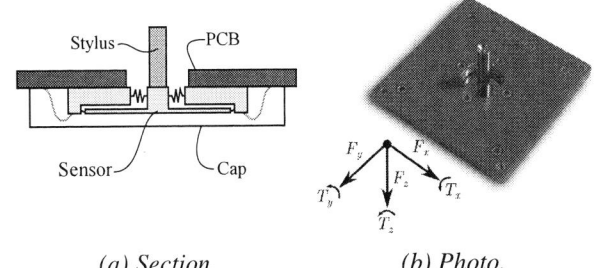

(a) Section.   *(b) Photo.*

Figure 8: The final assembly.

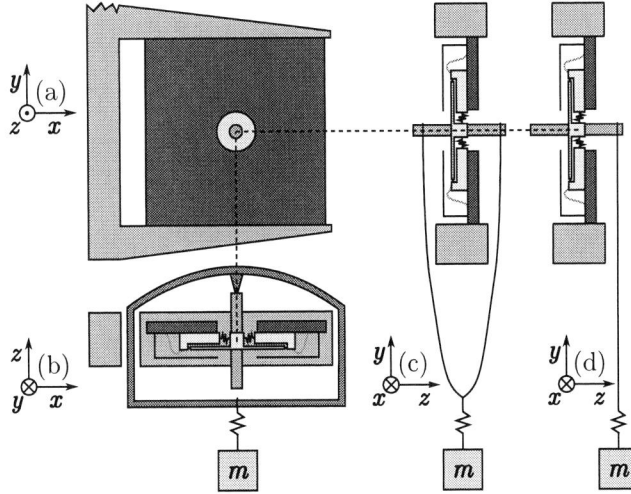

Figure 9: Measurement setups for applying loads to the sensor: (a) clamped assembled sensor, (b) measuring normal force, (c) measuring shear force and (d) measuring torque.

of the chip to make sure pure shear forces were applied. Torques around shear axes were measured by applying a load on the stylus at a defined distance from the sensor. The measurement electronics are schematically drawn in figure 10.

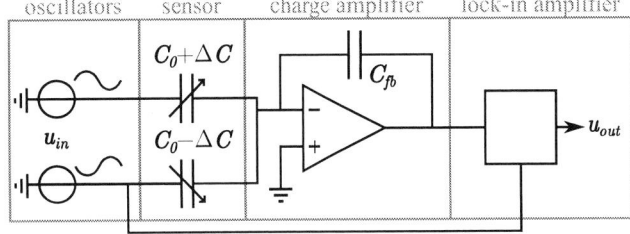

Figure 10: Electronic setup for differential measurements.

The output of the charge amplifier is as follows.

$$u_{out} = \frac{2\Delta C}{C_{fb}} u_{in}. \qquad (5)$$

With u_{out} the output voltage of the charge amplifier, u_{in} the input voltage, ΔC the change in capacitance of one side of the comb-structures or parallel plate structures and C_{fb} the feedback capacitance of the charge amplifier.

Shear force measurements (figure 11a) show a very linear ($>99\%$) differential change in capacitance with a sensitivity of $38\,\mathrm{fF\,N^{-1}}$. A slight crosstalk is observed when a shear force in orthogonal direction with respect to the measured direction is applied, mainly caused by misalignment in the measurement setup. The error bars represent misalignments from $-5°$ until $5°$. Normal force measurements (figure 11b) show a high sensitivity of $550\,\mathrm{fF\,N^{-1}}$ in the linear region. In figure 11c torque measurements around a shear axis are shown. All values are corrected for offset.

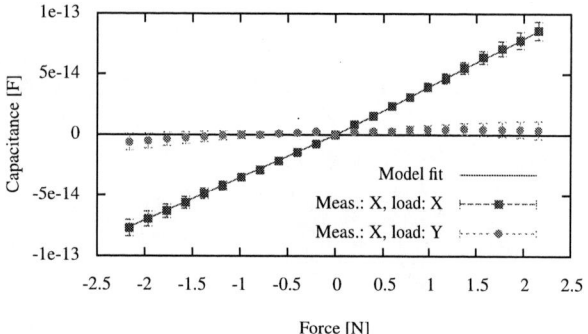

(a) Differential capacitance measurements of the comb-structures with varying shear forces.

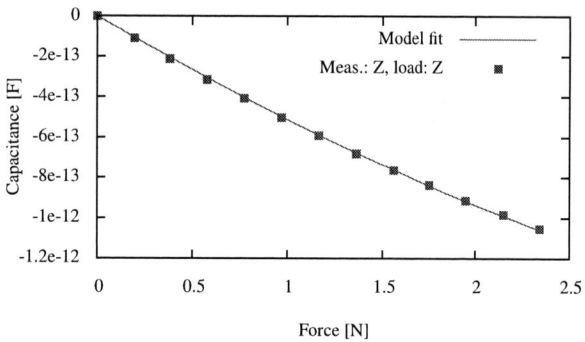

(b) Total capacitance measurements of the parallel plate structures with varying normal forces.

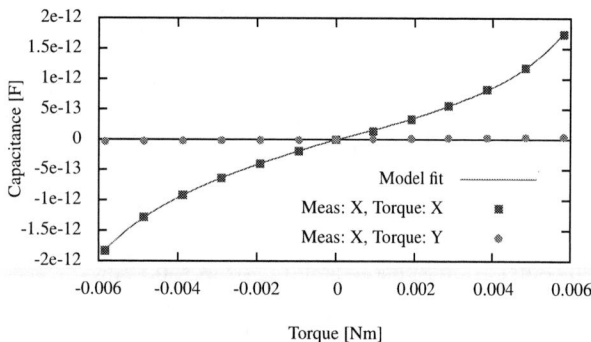

(c) Differential capacitance measurements of the parallel plate capacitors with varying torque.

Figure 11: Measurement results.

Mechanical robustness tests without stylus show that the sensor can be safely overloaded in normal direction with more than $15\,\mathrm{N}$ without causing damage to the sensor.

CONCLUSION

A miniature large range five degrees of freedom force/torque sensor is designed, realized and characterized. The first measurements were presented and are summarized in table 1. The sensor has a diameter of $9.24\,\mathrm{mm}$ and a thickness of less than $0.5\,\mathrm{mm}$. The proposed sensor is therefore suitable for biomechanical and robotic applications. The fabrication takes only two masks, making it a robust and relatively fast process. Future work will focus on further characterization of the sixth degree of freedom (rotation around normal axis) and increasing the range and sensitivity. For capacitive read-out, a relaxation oscillator will be used in the future [10].

Table 1: Summary of the sensor performance.

Quantity	Range	Sensitivity of linear region
F_x	$2.16\,\mathrm{N}$	$38\,\mathrm{fF\,N^{-1}}$
F_y	$2.16\,\mathrm{N}$	$38\,\mathrm{fF\,N^{-1}}$
F_z	$2.34\,\mathrm{N}$	$550\,\mathrm{fF\,N^{-1}}$
T_x	$5.84\,\mathrm{N\,mm}$	$23\,\mathrm{nF\,N^{-1}\,m^{-1}}$
T_y	$5.84\,\mathrm{N\,mm}$	$23\,\mathrm{nF\,N^{-1}\,m^{-1}}$

REFERENCES

[1] M. W. Legro *et al.*, "Issues of importance reported by persons with lower limb amputations and prostheses," *Journal of rehabilitation research and development*, vol. 36, no. 3, 1999.

[2] J. A. Dobrzynska *et al.*, "Capacitive flexible force sensor," *Procedia Engineering*, vol. 5, pp. 404–407, 2010.

[3] L. Beccai *et al.*, "Design and fabrication of a hybrid silicon three-axial force sensor for biomechanical applications," *Sensors and Actuators A: Physical*, vol. 120, no. 2, pp. 370–382, 2005.

[4] H. K. Chu *et al.*, "Design of a high sensitivity capacitive force sensor," in *7th IEEE Conference on Nanotechnology*, 2007, pp. 29–33.

[5] E. S. Hwang *et al.*, "A polymer-based flexible tactile sensor for both normal and shear load detections and its application for robotics," *Journal of Microelectromechanical Systems*, vol. 16, no. 3, pp. 556–563, 2007.

[6] R. A. Brookhuis *et al.*, "3D force sensor for biomechanical applications," *Sensors and Actuators A: Physical*, vol. 182, pp. 28–33, 2012.

[7] R. A. Brookhuis *et al.*, "Six-axis force-torque sensor with a large range for biomechanical applications," *Journal of Micromechanics and Microengineering*, accepted for publication.

[8] H. Soemers, *Design principles for precision mechanisms*, 2010.

[9] E. Oberg *et al.*, *Machinery's Handbook*, 27th ed. Industrial Press, 2004.

[10] R. A. Brookhuis *et al.*, "Three-axial force sensor with capacitive read-out using a differential relaxation oscillator," in *Proceedings IEEE Sensors 2013 Conference, Baltimore, MD*, 2013, pp. 1078–1081.

A NOVEL ELECTRET ROTATIONAL SPEED SENSOR
Wei Bian, Xiaoming Wu*, Xiaohong Wang
Tsinghua National Laboratory for Information Science and Technology
Institute of Microelectronics, Tsinghua University, Beijing 100084, China

ABSTRACT

In this paper, a novel rotational speed sensor based on electrostatic field variation is designed, fabricated and tested. Unlike traditional magnetic and optical rotational speed sensors, the presented sensor has the merit of simple configuration, small size, and low cost. The principle of the device is introduced through the equivalent circuit and the finite element model of electrostatic field. A demo sensor is fabricated by typical micro fabrication processes. Inexpensive PTFE film is corona charged as electret material to form the sensing electrostatic field. Experiment results show that the resolution of sensor is better than 6 rpm when the gear is with module of 3 and teeth number of 15. The rotation direction of gear is denoted by the sign of phase difference between the signals from two parallel output electrodes.

INTRODUCTION

Rotational speed sensors are widely used in automotive and industrial applications [1]-[3]. The mechanism of the conventional sensors includes magnetic field variation [4]-[6], and optical reflection [7]-[9].

The magnetic rotational sensor consists of a magnet to form a magnetic field. A magnetic sensor (e.g. Hall device) is used to measure the magnetic field variation induced by the movement of object to obtain rotation information. The material of the under measured mechanical part of these sensors is limited to ferromagnetic metals. The optical sensors need light source and photo detector which causes the configuration of the sensor complicated.

In this paper, a novel rotational speed sensor based on electrostatic filed variation is presented. The sensor consists of only passive elements. The merit of the novel sensor is its simple configuration, small size, and low cost. The demo sensor is fabricated by typical micro fabrication, and tested.

PRINCIPLE AND SIMULATION

Figure1 shows the schematic and operation principle of the rotational speed sensor. The sensor consists of an electret film which forms a stable electrostatic field. Beneath the film, several finger electrodes are distributed in parallel to collect the charges induced by the electret film.

When a mechanical part (e.g. gear tooth) passes by the sensor, the electrostatic field strength in the electrodes is varied, which causes the variation of the induced charges. Then the voltage change on the electrode can be read out to record the movement of gear tooth. In this case, the material of rotation object is not limited to ferromagnetic metals.

To suppress the ambient common noise in the output signal, the voltage difference between two electrodes, which are placed facing gear tooth and far from the gear tooth

respectively, is recorded as the output signal of sensor.

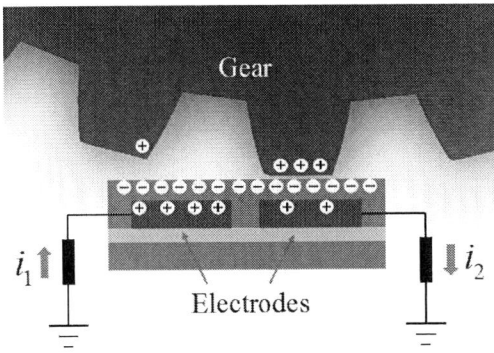

Figure 1: Operation principle of the sensor

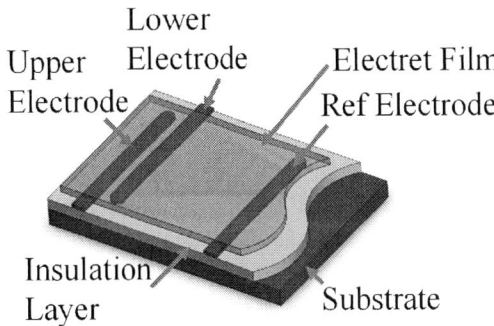

Figure 2: Schematic view of the sensor

Figure 3 is the equivalent circuit of the sensor[2]. The gap between gear and electrode is modeled as variable capacitance C_a, which is dependent on whether the gear tooth or space passes by the electrode. The electret film is modeled as capacitance C_e. C_a and C_e are connected in series. Q_O is the charge stored in the electret film and distributed to the capacitor C_a and C_e.

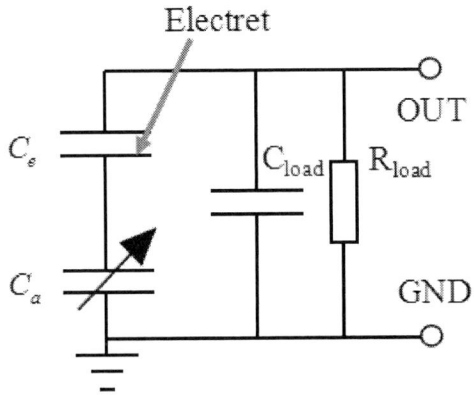

Figure 3: Equivalent circuit of the sensor

978-1-4799-3510-9/14 $31.00 © 2014 IEEE   684

Normally, the load of device consists of a capacitance C_{load} and a resistance R_{load} which are determined by the input electric property of the followed signal conditioning circuit. In this paper, the output electrodes of the sensor are connected directly to oscilloscope for signal recording.

As the whole charges in capacitance C_a and C_e are determined by which stored in electret material, one can obtain,

$$Q_O = Q_e + Q_a \qquad (1)$$

Where, Q_a and Q_e are the charges in capacitance C_a and C_e respectively. When the system is in static, there is no current flowing through load. Then the voltages cross the capacitance C_a and C_e are identical to each other.

$$\frac{Q_a}{C_a} = \frac{Q_e}{C_e} = U_O \qquad (2)$$

Based on equation 1 and equation 2, the charge in capacitance C_a and C_e can be deduced. Therefore, Q_e can be presented by equation 3, which is the charge induced on the electrode of sensor by the electret film.

$$Q_e = Q_O \times \frac{C_e}{C_a + C_e} \qquad (3)$$

When the capacitance C_a is changed followed the gear tooth movement, the charge Q_e is modified correspondingly. In this case, current flows through the load and the output voltage on the load is detected.

The relationship between Q_e/Q_O and C_a/C_e is presented in figure 4.

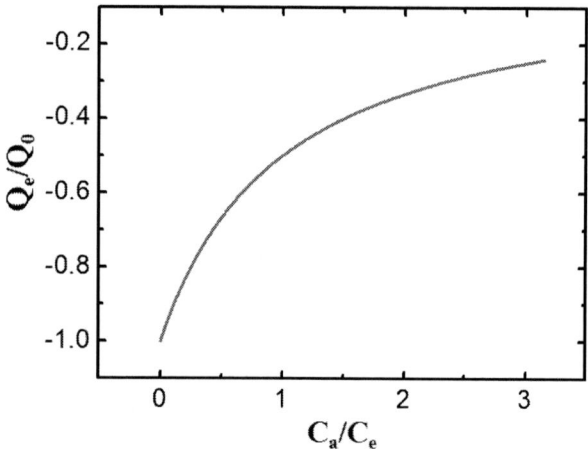

Figure 4: The relation between Q_e/Q_O and C_a/C_e

To understand the induced charge distribution in capacitance C_a and C_e, the finite element model of senor and gear teeth system is built and simulated using Ansys®. To simply the modeling process, a gear wheel is replaced by a toothed bar. Figure 5 is the simulation result of the electrostatic field distribution. The inset is the detailed structure of the sensor and gear tooth. Table 1 lists the parameters used in the FEM simulation.

As shown in figure 5, large field strength difference is observed between the conditions of electrodes facing gear teeth and teeth space respectively. Correspondingly, the induced charge on the electrodes in different conditions is different. Figure 6 shows the maximum charge variation on the electrodes. The variation of charge is strongly dependent on the width of electrode and the gap between sensor and gear tooth. When the sensor is closer to the gear tooth, the variation of charge on electrode is larger, and the sensitivity of the sensor is higher.

Table 1 Parameters used in FEM simulation

Parameter	Unit	Value
Electret surface potential	V	-700
Electret thickness	μm	100
Electret permittivity		2.55
Electrode width	mm	0.5
		1.0
		2.0
		2.5
Gear module	mm	3
Teeth thickness	mm	4.7

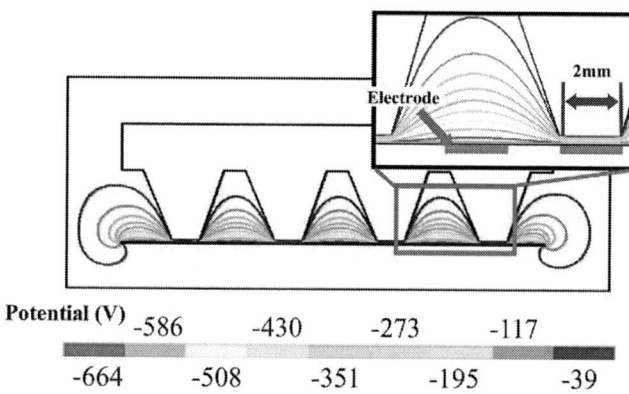

Figure 5: FEM simulation result of the electrostatic field distribution between rotational speed sensor and toothed bar.

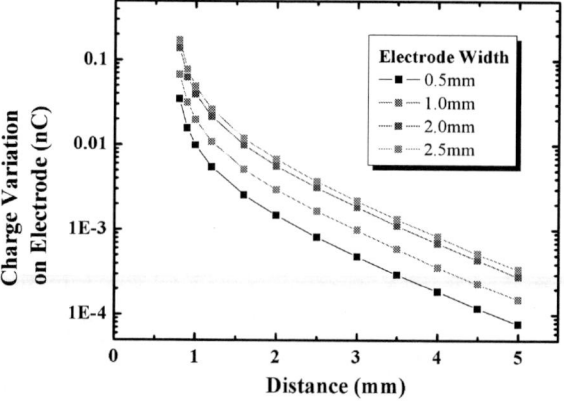

Figure 6: Variation of the induced charge on electrode vs. the gap width between the toothed bar teeth and sensor. The length of electrodes is 1 mm.

FABRICATION AND EXPERIMENT

Figure 7 shows the fabrication process of the device. The substrate is a 25mm×25mm×1mm Al plate. PTFE is used as electret material because it is inexpensive, and has high surface potential after charging. Firstly, 10μm thick PTFE is sprayed on the Al plate as insulation film (Figure 7.b). Then, an Al film (300nm thick) is sputtered on the insulation film (Figure 7.c), and patterned to form 2 mm wide and 20 mm long electrodes by lithography and etching (Figure 7.d). After that, a 100μm thick PTFE tape is glued partly on the electrodes (Figure 7.e). The end of the electrodes is uncovered to be used as pads for wire bonding. Finally, the PTFE tape is charged by corona method (Figure 4.f).

In corona charging process, the gap between grid and device is 1mm, and the distance between needle tip and grid is 150mm. The voltage on grid and needle is -700V and -7000V, respectively. The charging time and temperature is 15 minutes and 110°C. After charging, the surface potential of the PTFE electret tape reaches -700 V.

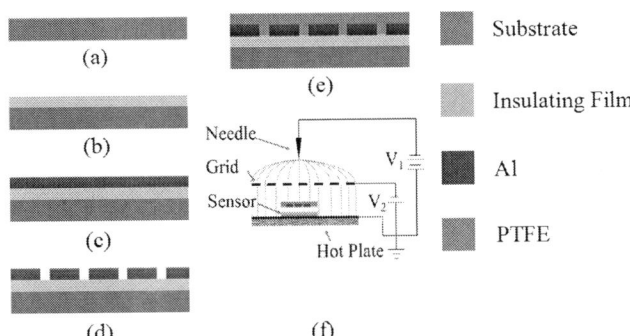

Figure 7: Fabrication process of the sensor.

Figure 8 is the schematic and optical image of the experiment setup. An aluminum gear is placed beside the sensor and driven by a DC motor as a rotational object. The module, number of teeth, and the thickness of the aluminum gear is 3mm, 15, and 4.7mm respectively. The upper electrode and the lower electrode have the same gap width to the gear teeth for the distinguishment of rotation direction. The reference electrode is located far from the two signal electrodes to reject the common mode noise. The rotational speed and direction of gear is adjusted by the voltage and sign of the power supply of the DC motor. The gap between the sensor and gear teeth is changed by a XYZ stage. The signals on the three electrodes are recorded simultaneously by an oscilloscope with input resistance of 10MΩ. The voltage difference between upper or lower electrode and reference electrode is traced to count the rotational speed of the gear. The phase difference of the two voltage traces indicates the gear rotation direction.

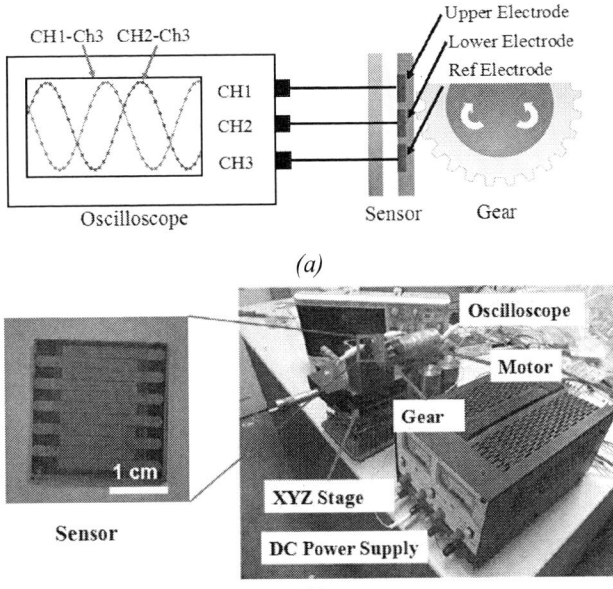

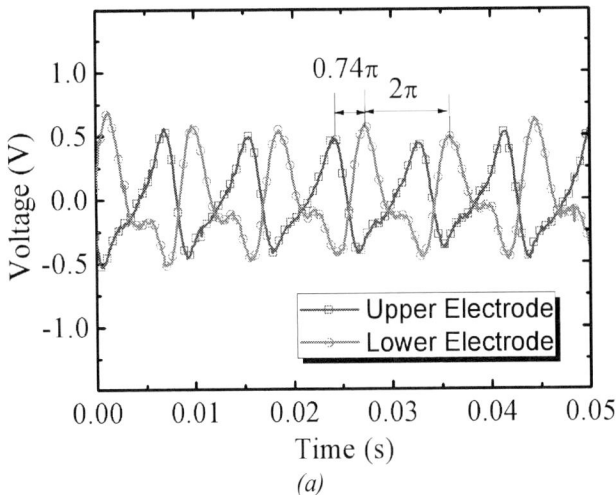

Figure 8 Schematic (a) and optical image (b) of the experiment setup.

Figure 9 shows the original output signal waveforms of the sensor when the gear rotates in 624 rpm and the gap between sensor and gear tooth is 0.5 mm. The frequency of the curves reflects the number of the times of gear tooth passing by the sensor. Therefore, the rotational speed of gear can be evaluated from the signal frequency and gear tooth number.

When the gear rotates clockwise, the voltage waveform of the lower electrode has 0.78π phase advance than that of upper electrode (figure 9.a). When the gear rotates anti-clockwise, the voltage waveform of lower electrode has 0.74π phase lag behind the upper electrode (figure 9.b). The small deviation between the absolute values of two phase difference comes from signal interference and noise.

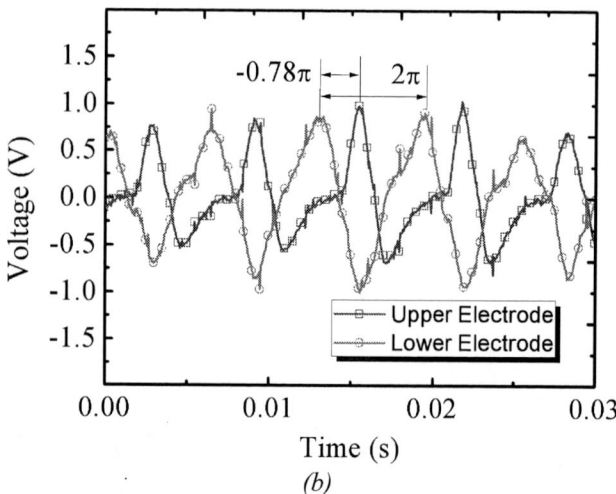

Figure 9: Output voltage waveforms of the sensor when the rotation speed of gear is 624 rpm.

Figure 10 shows the relationship between the output voltage of the sensor and the rotational speed of the gear. The gap width between the gear tooth and the sensor is 0.5mm, 1mm, 2.5mm, and 5mm, respectively. When the rotation speed is higher or the gap between sensor and gear tooth is smaller, higher output voltage of the sensor can be obtained. The output voltage reaches 1.7V when the gear tooth gap is 0.5mm and gear rotational speed is 624 rpm. In the case of 0.5mm gap width, the interference and noise level of the unshielded sensor is 26mV. When the rotational speed is 6 rpm, which is the lowest speed can be obtained by the DC motor, the output voltage of sensor is still 80mV. Therefore, the sensitivity of this demo sensor is better than 6 rpm. When the sensor is well packaged and shielded, higher resolution can be expected.

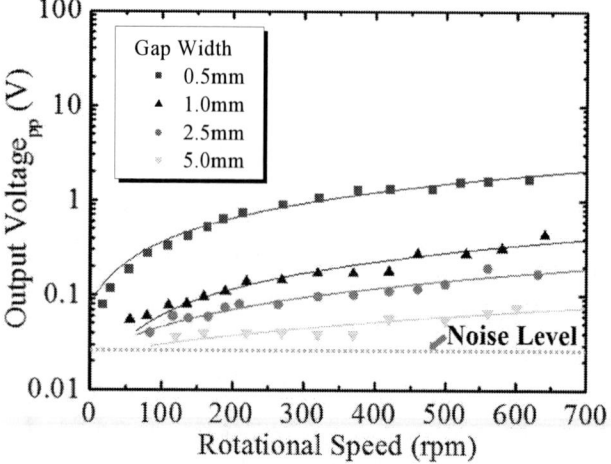

Figure 10: The relationship between the output voltage of sensor and the rotational speed of gear.

CONCLUSION

A novel electret rotational speed sensor, which is based on electrostatic field variation, is introduced. The operation principle of the sensor is verified by the simulation and experiment results. The resolution of the unshielded sensor is better than 6 rpm. Additionally, the size of the device can be further miniaturized by the micro fabrication process.

REFERENCES

[1] F. Gustafsson, "Rotational speed sensors: Limitations, Pre-Processing and Automotive Applications", *IEEE Instrum. Meas. Mag.*, vol. 13, no. 2, pp. 16-23, Apr. 2010.
[2] D.S. Smith and P.R. Jackman, "Optical Sensors for Automotive Applications", *IEE Colloq. Automotive Sensors*, Solihull, England, May 11, 1992, pp. 2/1–2/3.
[3] C. P. O. Treutler, "Magnetic Sensors for Automotive Applications", *Sensors and Actuators A*, vol. 91, no. 1-2, pp. 2-6, Jun. 2001.
[4] C.Giebeler, D.J.Adelerhof, A.E.T.Kuiper, J.B.A.van Zon, D. Oelgeschläger, G.Schulz, "Robust GMR Sensors for Angle Detection and Rotation Speed Sensing", *Sensors and Actuators A*,vol. 91, no. 1-2, pp. 2-6, Jun. 2001.
[5] Y. S. Didosyan, H. Hauser, H. Wolfmayr, J. Nicolics, P. Fulmeka, "Magneto-Optical Rotational Speed Sensor", *Sensors and Actuators A*,vol. 106, no. 1-3, pp. 168-171, 2001.
[6] G.Rieger, K.Ludwig, J.Hauch, W.Clemens, "GMR Sensors for Contactless Position Detection", *Sensors and Actuators A*, vol. 91, no. 1-2, pp. 7-11, 2001.
[7] K. Taguchi, K. Fukushima, A. Ishitani, and M. Ikeda, "Optical Inertial Rotation Sensor Using Semiconductor Ring Laser", *Electron. Lett.*, vol. 34, pp.1775 -1776,1998.
[8] P. S. Huang, S. Kiyono, and O. Kamada, "Angle Measurement Based on The Internal-Reflection Effect: A New Method", *Applied Optics*, vol. 31, issue 28, pp. 6047-6055, 1992.
[9] J. Rohlin, "An Interferometer for Precision Angle Measurements", *Applied Optics*, vol. 2, issue 7, pp. 762-763, 1963.

ACKNOWLEDGMENT

This work was supported by the State Key Development Program for Basic Research of China (Grant No. 2009CB320304), Key Program of Natural Science Foundation of China (Grant No.60936003), and Beijing Natural Science Foundation (Grant No.3122023)

CONTACT

Xiaoming Wu, Email: imewuxm@tsinghua.edu.cn

ALN-BASED PIEZOELECTRIC RESONATOR FOR INFRARED SENSING APPLICATION

Wan C. Ang[1,2], Piotr Kropelnicki[1], Humberto Campanella[1], Yao Zhu[1,2], Andrew B. Randles[1], Hong Cai[1], Yuandong A. Gu[1], Kam C. Leong[3], and Chuan S. Tan[2]

[1]Institute of Microelectronics, Agency of Science, Technology and Research (A*STAR), SINGAPORE
[2]Nanyang Technological University, SINGAPORE
[3]GLOBALFOUNDRIES Singapore Pte Ltd, SINGAPORE

ABSTRACT

This paper reports a highly sensitive aluminum nitride (AlN) based resonant uncooled infrared (IR) detector utilizing photo-sensitive and piezoelectric properties of polycrystalline AlN. The design, fabrication, and IR sensing characterization of the device are presented. Instead of resonant frequency shift, S_{21} magnitude shift was observed upon IR illumination under both vacuum and ambient measurements. Thus, photoresponse mechanism was proposed rather than thermal effect. An AlN resonator operating at 2.336 GHz with a quality factor (Q) of 830 exhibits an IR responsivity and detectivity of 166 kdB/W and 1.41×10^7 cm$\sqrt{Hz}$/W, respectively.

INTRODUCTION

With the discovery of exciting applications using miniaturized and portable IR detectors, micro-electro-mechanical-system (MEMS) based uncooled IR detectors are gaining increasing attention. As cooling system is not necessary, uncooled IR detectors have low weight, low power consumption, high reliability, low manufacturing and operation cost. With the broadband sensitivity, uncooled IR detectors are of particular interest in spectrometer applications. However, uncooled IR detectors have an overall lower performance and slower response compared with photon detectors.

Semiconductor-based resistive microbolometers are the most commercially successful uncooled IR detectors because they are relatively easy to fabricate compared with pyroelectric detectors [1] and have a better responsivity than thermoelectric detectors [2]. Since early 1990s, their performance has been improved with the advances in silicon micromachining technologies. However, they are tardily facing bottleneck in further enhancing their performance recently due to the inevasible high flicker (1/f) and Johnson noise. Hence, there is a necessity for novel technologies that could beat the sensing performance of photon detectors in addition to retain the advantages of uncooled detectors.

Resonant uncooled IR detector is the next promising candidate for IR detection due to the highest accuracy frequency readout. Furthermore, it draws much less power than a resistive microbolometer because no DC current is flowing through the sensing materials. Self-heating of a resonator can be easily controlled by minimizing the RF power level [3]. The device 1/f noise can be ignored owing to high frequency operation (from 100 MHz – 2.5 GHz). In 1985, IR sensing of quartz bulk acoustic wave (BAW)

resonators was first reported [4], followed by a detailed analysis in thermal imaging application in year 1994 [5]. With the advances in MEMS technologies, quartz film bulk acoustic resonator (FBAR) has been reported to exhibit a noise-equivalent-temperature-difference (NETD) of less than 5 mK in year 2011 [6]. Because of the non-scalability and non-CMOS compatibility, quartz BAW resonators are facing difficulty in mass production. Different piezoelectric materials have been investigated to replace quartz including GaN [7], ZnO [8], and AlN [9]. Promising IR sensing characteristic of these materials has been demonstrated.

In this work, AlN-based piezoelectric resonator was designed, fabricated and characterized. Post-CMOS integration is feasible because AlN piezoelectric thin films with high quality and uniformity can be deposited by sputtering process on silicon substrates at low temperature (~ 200 °C) [10]. It was found that AlN resonator exhibits photo-sensitive property in addition to thermal effects when exposed to IR source.

DESIGN AND FABRICATION

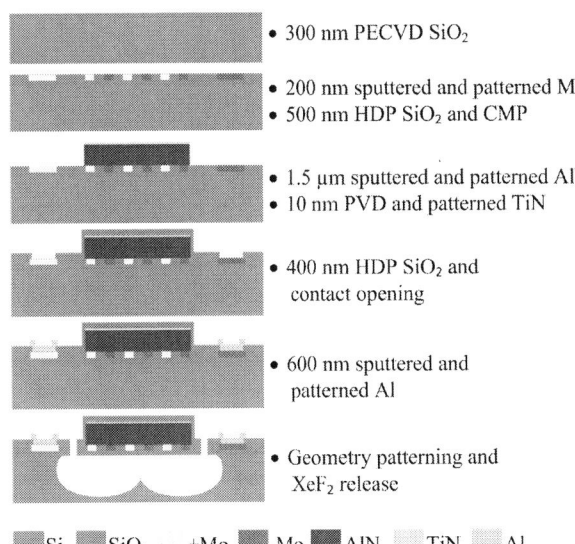

- 300 nm PECVD SiO$_2$

- 200 nm sputtered and patterned Mo
- 500 nm HDP SiO$_2$ and CMP

- 1.5 μm sputtered and patterned AlN
- 10 nm PVD and patterned TiN

- 400 nm HDP SiO$_2$ and contact opening

- 600 nm sputtered and patterned Al

- Geometry patterning and XeF$_2$ release

Si SiO$_2$ +Mo -Mo AlN TiN Al

Figure 1: The schematic process flow of the AlN resonant IR detector.

Figure 1 shows the schematic process flow of the AlN resonant IR detector. The device was fabricated on top of a PECVD SiO$_2$ support layer (300nm thick). After the sputtering and patterning of Mo bottom excitation electrodes

(200 nm thick with 2 μm pitch), a HDP SiO₂ layer (500 nm thick) was deposited and flatten until Mo electrodes by CMP. AlN piezoelectric thin film (1.25 μm thick) and TiN-absorber (10 nm thick) was then deposited by sputtering and patterned to define the IR detector pixel geometry (48 μm x 48 μm). The 10 nm TiN is designed to enhance the IR absorption by matching the atmospheric impedance (377 Ω/□) and also to act as floating electrode for the AlN resonator. Since the TiN layer is extremely thin, there is no tradeoff issue between absorptivity and Q-factor caused by the mechanical mass loading effect. Another layer of HDP SiO₂ (400 nm) was deposited to passivate the device, followed by Al contact pads (600 nm thick) formation and release hole opening. The device was finally released by XeF₂ to have a freestanding AlN resonator as illustrated by the SEM image in Figure 2.

Figure 2: The SEM image of the fabricated AlN resonant IR detector.

RESULTS AND DISCUSSION

The experiment setup in this work is shown in Figure 3. The device under test (DUT) was placed on a temperature chuck inside a vacuum chamber. Before the frequency response measurements of the fabricated device were taken, a Short-Open-Load-Thru (SOLT) impedance calibration of the GSG probes and network analyzer (Agilent E5071B) was carried out to ensure the results accuracy. The measured and modified-Butterworth van Dyke (MBVD) fitted admittance-frequency curves of the AlN resonator is plotted in Figure 4. The device has a resonant frequency of 2.336 GHz and Q-factor of ~ 830.

A blackbody source (BS) from ORIEL Instruments with different temperatures was used to emit a spectrum of IR radiation. An optical filter with transmitted wavelength of 0.5 μm to 20 μm was used to isolate ultraviolet (UV) light. To further eliminate visible light, a 725 μm thick Si wafer was placed above the optical filter. Figure 5 illustrates the change in S_{21} magnitude at resonant frequency with different BS temperatures. A smaller shift of S_{21} magnitude was observed with additional Si wafer filter which proves that the device is responsive to IR wavelength longer than 1.12 μm.

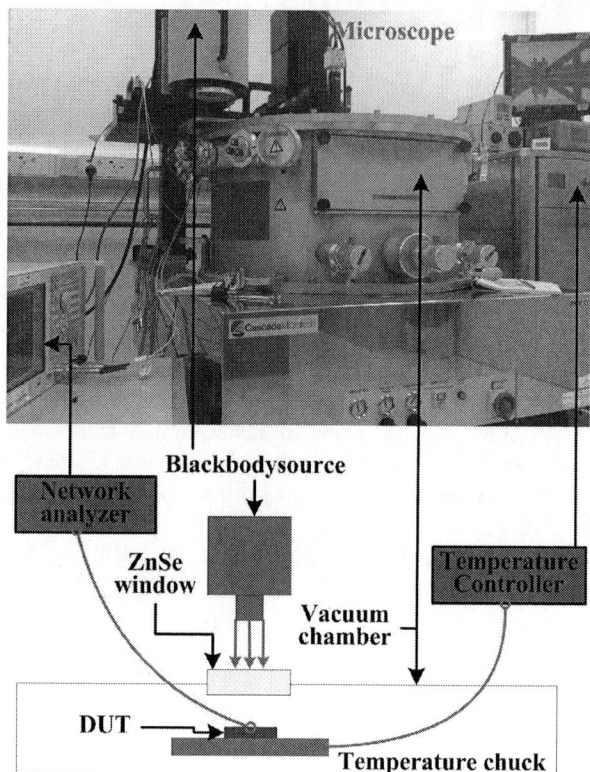

Figure 3: The experiment setup for frequency response measurement of the AlN resonant IR detector.

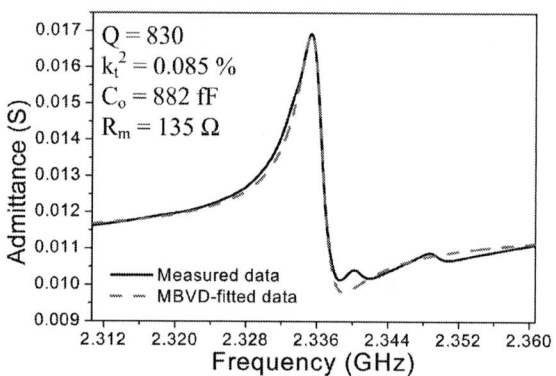

Figure 4: The measured and MVVD-fitted admittance of the AlN resonant IR detector.

It is worth noting that there is no significant shift in resonant frequency under different BS temperatures as observed from Figure 5. This observation is different from other works which reported on the shift of resonant frequency resulted from thermal effect [7-9]. To further investigate this phenomenon, the frequency response of the device was measured at different chuck temperatures and the effective temperature coefficient of frequency (TCF) was extracted as depicted in Figure 6. Due to the comparable thickness of SiO₂ with AlN, the positive TCF of SiO₂ dominates in the device effective TCF [11] which has a value of +19.69 ppm/K.

978-1-4799-3510-9/14 $31.00 © 2014 IEEE   689

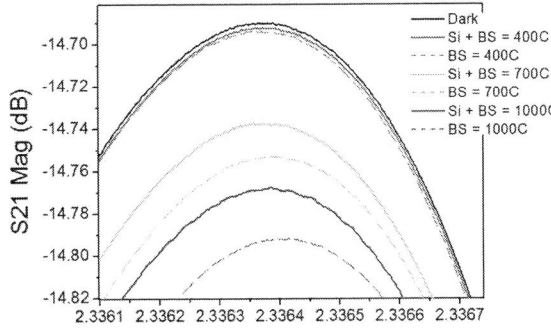

Figure 5: The measured S_{21} magnitude of the AlN resonant IR detector under different BS temperatures. (Si = Si wafer filter)

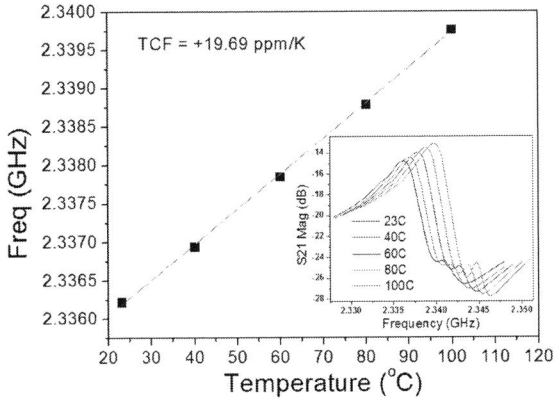

Figure 6: The temperature dependent resonant frequency of the AlN resonant IR detector.

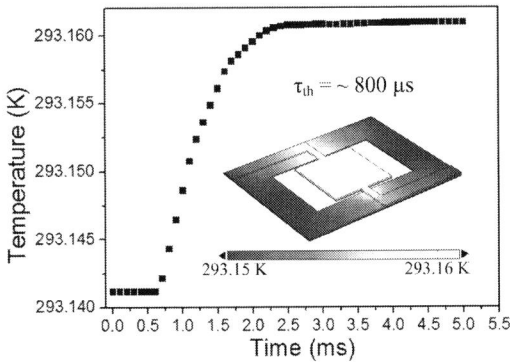

Figure 7: The simulated temperature profile and the thermal time constant of the AlN resonant detector at P_{IR} of 615 nW. .

COMSOL Multiphysics tool was employed to estimate the temperature profile of the device as shown in Figure 7. The material properties of AlN, SiO_2 and Mo used in this transient time simulation are summarized in Table 1. An IR power, P_{IR} of 615 nW (BS temperature of 1000 °C) rises the device temperature by less than 0.02 K which translated to a 920 Hz of upshift in resonant frequency assuming a full absorption. Detecting this small amount change of resonant frequency in a GHz-device is a great challenge due to temperature fluctuation, power supply noise and the network

analyzer resolution limit.

Table 1: Material properties of AlN, SiO_2 and Mo used in the thermal simulation.

Material properties	AlN	SiO2	Mo
Thermal conductivity [W/ (m·K)]	185	1.4	138
Mass density [kg/m³]	3260	2200	10200
Heat capacity [J/(kg·K)]	740	730	250

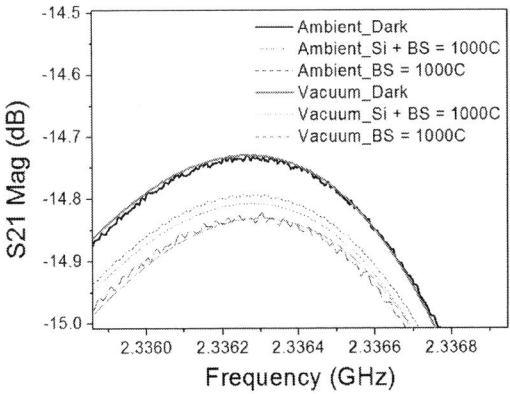

Figure 8: Comparison of IR sensing performance of the AlN resonant IR detector under ambient (black color curves) and vacuum (red color curves) condition.

The IR sensing performance of the device was then compared under ambient and vacuum conditions. The ambient measurement data is noisier due to the scratch of the device contact pads as it was done after the vacuum measurement. Figure 8 shows no significant difference in S_{21} magnitude shift between the ambient and vacuum measurement results. This observation strongly proves that there is no thermal effect involved in the IR responsivity. The reasons of the shift of S_{21} magnitude in response to the IR exposure is still unclear and has not been studied before. Similar observation has been reported in [12] for the wavelength range of visible light. It was explained that the defect-induced sub-bandgap absorptions is the main contribution of the photoresponse. However, it is believed that there are other mechanisms that could cause the IR sensitivity of AlN in this work including impurities-related energy states transitions [13] or polarization distortion caused by phonon absorption [14, 15], where both the occurrences can also alter the static capacitance of AlN and thus the admittance of the frequency response . In any event, a more in depth study and characterization are needed to explain this phenomenon.

At P_{IR} of 615 nW, a -0.102 dB shift of S_{21} magnitude was obtained corresponding to a IR responsivity, $\Re$ of 166 kdB/W. The root mean squared (RMS) noise, V_n of the measured S_{21} magnitude in 1 Hz bandwidth was obtained to be 56.9 µdB/√Hz. The device IR detectivity, D^* of 1.41×10^7 cm√Hz/W was calculated using the equation below.

$$D^* = \frac{\Re \cdot \sqrt{A_B}}{V_n} \qquad (1)$$

where A_B is the pixel area of the AlN resonant IR detector.

CONCLUSION

This paper demonstrates the capability of AlN-based IR detector to operate without the need of vacuum and cooling system. This brings the great advantage in device packaging and thus further reduces the manufacturing and operation cost compared with thermal detectors and photon detectors. Even though the estimated detectivity of the current design is about two orders of magnitude lower than the commercial resistive bolometers, it is believed that the performance can be boosted up by either device structure optimization or IR absorptivity enhancement. Thus, the detailed physics and mechanisms have to be first understood before the optimization steps.

ACKNOWLEDGEMENTS

The authors are grateful to the clean room staffs in Institute of Microelectronics, A*STAR, Singapore for their supports in fabrication process. GLOBALFOUNDRIES Singapore Pte Ltd and Singapore Economic Development Board are acknowledged for the scholarship of the PhD program offered to one of the authors at Nanyang Technological University (NTU). Funding is provided by A*STAR, Singapore (#1021650084). C.S. Tan is grateful for the support from the Silicon Technologies Center of Excellence (Si-COE) and Nanoelectronics Centre of Excellence (NOVITAS) at NTU.

REFERENCES

[1] X. Shao, J. Ding, X. Ma, Y. Yu, and J. Fang, "Design and thermal analysis of electrically calibrated pyroelectric detector," *Infrared Physics & Technology,* vol. 55, pp. 45-48, 2012.

[2] X. Dehui, X. Bin, W. Guoqiang, M. Yinglei, and W. Yuelin, "Uncooled Thermoelectric Infrared Sensor With Advanced Micromachining," *IEEE Sensors Journal,* vol. 12, pp. 2014-23, 2012.

[3] A. Tazzoli, M. Rinaldi, and G. Piazza, "Experimental Investigation of Thermally Induced Nonlinearities in Aluminum Nitride Contour-Mode MEMS Resonators," *IEEE Electron Device Letters,* vol. 33, pp. 724-6, 2012.

[4] J. E. Ralph, R. C. King, J. E. Curran, and J. S. Page, "Miniature quartz resonator thermal detector," in *IEEE 1985 Ultrasonics Symposium. Proceedings. (Cat. No.85CH2209-5), 16-18 Oct. 1985,* New York, NY, USA, 1985, pp. 362-4.

[5] M. R. Hamrour and S. Galliou, "Analysis of the infrared sensitivity of a quartz resonator application as a thermal sensor," in *Proceedings of IEEE Ultrasonics Symposium, 1-4 Nov. 1994,* New York, NY, USA, 1994, pp. 513-16.

[6] M. B. Pisani, K. Ren, P. Kao, and S. Tadigadapa, "Application of Micromachined y -cut-quartz bulk

acoustic wave resonator for infrared sensing," *Journal of Microelectromechanical Systems,* vol. 20, pp. 288-296, 2011.

[7] M. Rais-Zadeh, "Gallium nitride micromechanical resonators for IR detection," in *Micro- and Nanotechnology Sensors, Systems, and Applications IV, 23-27 April 2012,* USA, 2012, p. 83731M (6 pp.).

[8] Z. Wang, X. Qiu, S. J. Chen, W. Pang, H. Zhang, J. Shi, and H. Yu, "ZnO based film bulk acoustic resonator as infrared sensor," *Thin Solid Films,* vol. 519, pp. 6144-6147, 2011.

[9] Y. Hui and M. Rinaldi, "High performance NEMS resonant infrared detector based on an aluminum nitride nano-plate resonator," in *2013 Transducers & Eurosensors XXVII: 17th International Conference on Solid-State Sensors, Actuators and Microsystems (TRANSDUCERS & EUROSENSORS XXVII), 16-20 June 2013,* Piscataway, NJ, USA, 2013, pp. 968-71.

[10] A. Fardeheb-Mammeri, M. B. Assouar, O. Elmazria, J. J. Fundenberger, and B. Benyoucef, "Growth and characterization of c-axis inclined AlN films for shear wave devices," *Semiconductor Science and Technology,* vol. 23, p. 095013 (7 pp.), 2008.

[11] J. H. Kuypers, C.-M. Lin, G. Vigevani, and A. P. Pisano, "Intrinsic temperature compensation of aluminum nitride Lamb wave resonators for multiple-frequency references," in *2008 IEEE International Frequency Control Symposium, FCS, May 19, 2008 - May 21, 2008,* Honolulu, HI, United states, 2008, pp. 240-249.

[12] C. J. Zhou, Y. Yang, Y. Shu, H. L. Cai, T. L. Ren, M. Chan, J. Zhou, H. Jin, S. R. Dong, and C. Y. Yang, "Visible-light photoresponse of AlN-based film bulk acoustic wave resonator," *Applied Physics Letters,* vol. 102, p. 191914 (3 pp.), 2013.

[13] H. M. Huang, R. S. Chen, H. Y. Chen, T. W. Liu, C. C. Kuo, C. P. Chen, H. C. Hsu, L. C. Chen, K. H. Chen, and Y. J. Yang, "Photoconductivity in single AlN nanowires by subband gap excitation," *Applied Physics Letters,* vol. 96, p. 062104 (3 pp.), 2010.

[14] M. Kazan, B. Rufflé, C. Zgheib, and P. Masri, "Phonon dynamics in AlN lattice contaminated by oxygen," *Diamond and Related Materials,* vol. 15, pp. 1525-1534, 2006.

[15] C. Balasubramanian, S. Bellucci, G. Cinque, A. Marcelli, M. C. Guidi, M. Piccinini, A. Popov, A. Soldatov, and P. Onorato, "Characterization of aluminium nitride nanostructures by XANES and FTIR spectroscopies with synchrotron radiation," *Journal of Physics: Condensed Matter,* vol. 18, pp. 2095-104, 2006.

CONTACT

*W. C. Ang, Tel: +65-81235074; wcang1@e.ntu.edu.sg

AN ALL OPTICAL SHOCK SENSOR BASED ON BUCKLED DOUBLY-CLAMPED SILICON BEAM

B. Dong[1,2], J. G. Huang[3], H. Cai[2], P. Kropelnicki[2], A. B. Randles[2], Y. D. Gu[2] and A. Q. Liu[1†]

[1]School of Electrical & Electronic Engineering, Nanyang Technological University, Singapore 639798
[2]Institute of Microelectronics, A*STAR (Agency for Science, Technology and Research), Singapore 117685
[3]School of Mechanical Engineering, Xi'an Jiaotong University, Xi'an 710049, China

ABSTRACT

In this paper, an all optical shock sensor based on a buckled doubly-clamped silicon beam is demonstrated. A buckled silicon beam is in the middle of two ring resonator and it has two stable positions. The silicon beam encounters a snap-through process upon a shock force, which can be monitored by measuring the resonance wavelength of the ring resonators. During experiment, a 0.15 nm wavelength is observed for a > 50 g shock. It has merits such as fast response, low power consumption and immunity to electromagnetic interference. It can be applied to inertial navigation system and automotive industry.

INTRODUCTION

MEMS shock sensor has attracted numerous attentions due to its distinguished advantages such as low cost, small footprint, high sensitivity. However, most of the shock sensors have complicated structures and hardly compatible with standard ICs process [1-3]. With the advance of optical MEMS devices, an all optical shock sensor shows great potential for all optical devices integration and future all optical sensor and actuator systems. Recently, much progress has been made in the development of Nano-opto-mechanical Systems (NOMS), which provide a new approach for small footprint and low coast all optical devices [4-7]. In this paper, an all optical shock sensor based on a buckled doubly-clamped silicon beam is demonstrated, while the optical sensing is realized via high Q-factor ring resonators. Compared with traditional shock sensor, the proposed all optical shock sensors have merits such as simple structure, fast response, low power consumption and immunity to electromagnetic interference. It has potential applications in inertial navigation system and automotive industry.

DESIGN AND THEORY

The optical shock sensor consists of two ring resonators, a buckled doubly-clamped beam and a bus waveguide, which is shown in Fig. 1(a). Light is coupled into the two ring resonators through bus waveguide. The ring resonators both have a diameter of 50 μm. They are supported by the rib structure to reduce the propagation loss caused by the release window. There is a 40-μm long silicon beam in-between the two ring resonators, which is released from the substrate and anchored at both sides. The doubly clamped silicon beam is buckled and closer to one of the ring resonators due to the residual stress in the silicon structure layer. There are several arcs in the middle of the

silicon beam, which works as sensing element and inertial mass. There is a silicon arc structure close to each ring resonator, the gap between the act and ring resonator can

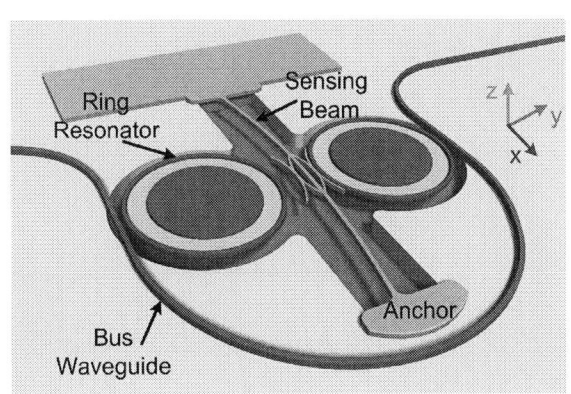

(a)

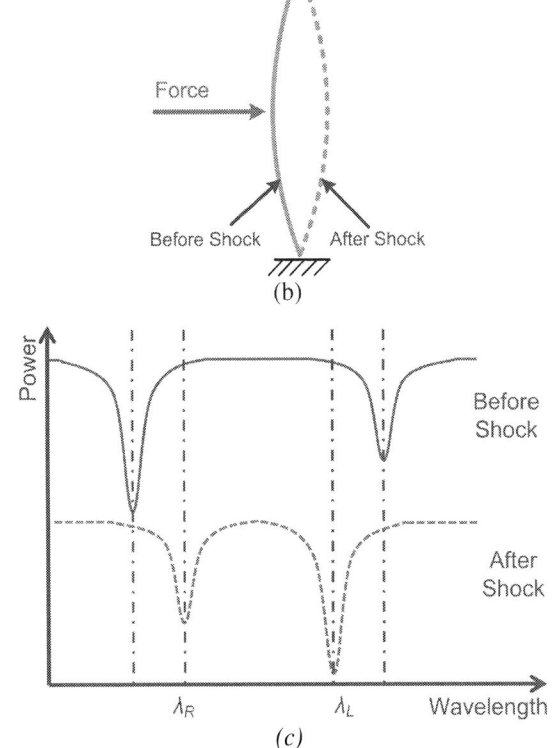

Figure 1: (a) Schematic of optical shock sensor, (b) illustration of the buckled doubly-clamped beam, and (c) spectrum demonstration of the shock sensor.

influence the resonance condition of the ring resonator. The buckled silicon beam has two stable positions, upon a shock occur, the force can switch the silicon beam from one side ("left") to the other side ("right") and change the resonance wavelengths of both ring resonators, as shown in Fig. 1(b). The resonance wavelength of the ring resonator shifts due to the gap change. For instance, when the beam bends toward the right ring resonator, it will cause red shift for right ring resonator and blue shift for left ring resonator. Therefore, the shock can be monitored by observing the resonance wavelength shift or the transmission power of a single wavelength laser, as shown in Fig. 1(c).

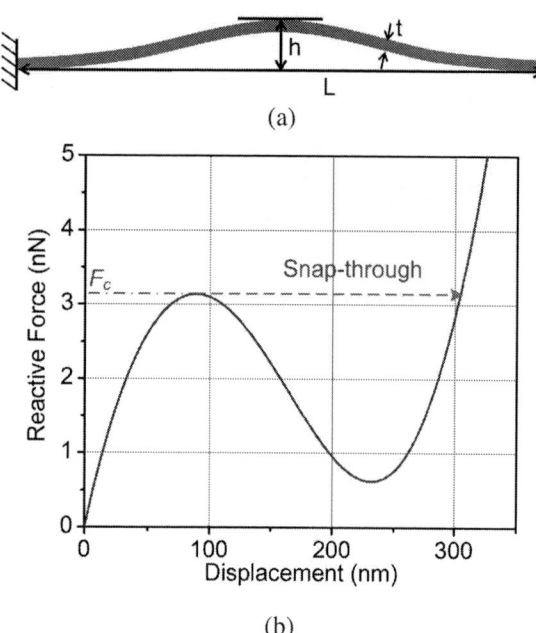

(a)

(b)

Figure 2: (a) Schematic of the buckled doubly-clamped silicon beam and (b) Snap-through process of the buckled doubly-clamped silicon beam.

For doubly-clamped silicon beam, buckling happens when the beam length is longer than the critical length, which is defined as [8]

$$l_c = \pi t \sqrt{\frac{E}{3\sigma_0}} \ , \qquad (1)$$

where t is the beam width, E is the young's module of silicon, and σ_0 is the compressive internal stress which is in the range of 10–100 MPa for normal SOI wafer. The critical length for an internal stress of 50 MPa is around 20 μm. Consequently, the 40-μm beam is buckled to one side. It is 200 nm in width, 340 nm in height and the central displacement h is around 100 nm as show in Fig. 2(a).

The buckled silicon beam has two stable positions, which are termed as "left" and "right". The force required to move the silicon beam form one stable position ("left") to another position ("right") can be estimated from the following equation

$$F_c \approx 8\pi^4 \frac{EIh}{L^3} \ . \qquad (2)$$

Once the force is large enough ($> F_c$), the silicon beam snaps through the middle position and reaches the other stable position. As a result, the shock forces that are larger than the threshold force can switch the silicon beam between the two stable positions. Figure 2(b) shows the simulated snap-through process of the buckled silicon beam using COMSOL and the critical force is around 3.2 pN.

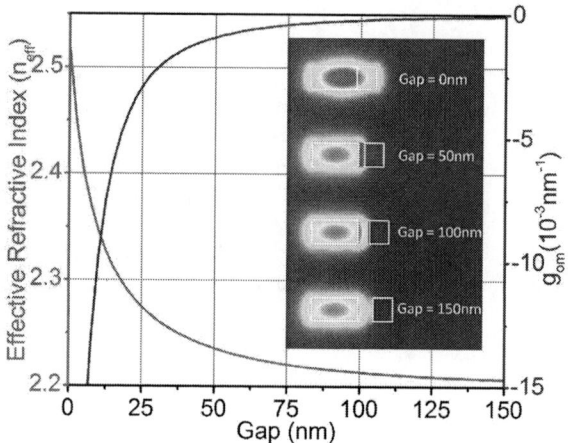

Figure 3: Effective refractive index and opt-mechanical coefficient as a function of gap between arc and ring resonator. The inset shows the electrical field distribution at various gaps.

Upon shock, the position shift of center doubly-clamped beam can affect the resonance condition of both ring resonators due to the evanescent wave perturbation. The effective refractive index of ring resonator as a function of the gap between the arc and ring resonator is shown in Fig. 3. When the gap is decreased, the effective refractive index increases, causing a red shift of the resonance wavelength. Conversely, the effective refractive index decreases with the increasing gap, causing a blue shift of resonance wavelength. Therefore, the resonance condition of each ring resonator changes with the gap, which can be used to monitor the position of the center doubly-clamped silicon beam.

FABRICATION

The shock sensor is fabricated on a 340-nm SOI wafer with CMOS compatible nanophotonic fabrication technology. Figure 4 shows the SEM images of the optical shock sensor. The optical shock sensor is fabricated by nano-photonic fabrication processes using standard silicon-on-insulator wafer with a 340 nm thick silicon structure layer and a 2 μm buried oxide layer. The waveguide structures have a width of 450 nm and a height of 340 nm for a single mode transmission. The silicon beam is designed to have a width of 200 nm. The waveguides and ring resonators are patterned by deep UV lithography, followed by plasma dry etching to transfer the photo resist pattern into the structure layer. After etching, a 2-μm SiO_2 layer is deposited on the structure layers to ensure a low optical loss. A 40-nm Al_2O_3 is deposited and patterned, which is used as the protection film to protect those fixed structures and leave the window area open for suspended structures. Finally, HF vapor selectively undercut the layer in the window area so as to release the movable structures. The silicon beam is bended towards one side due to the residual compress stress.

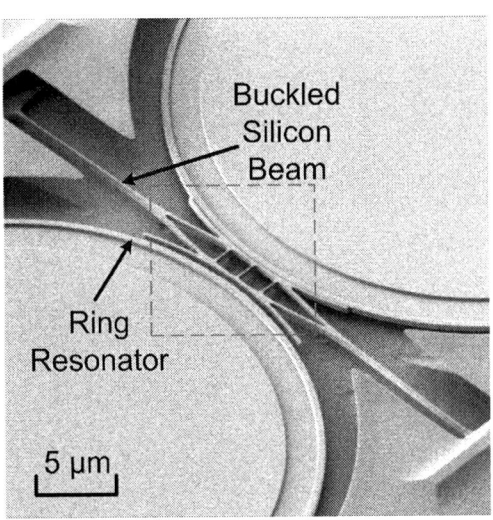

(a)

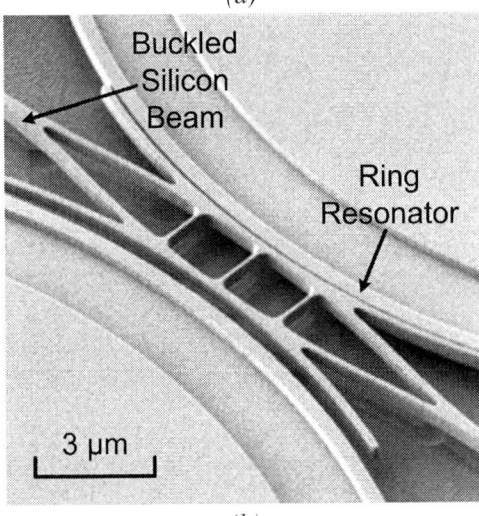

(b)

Figure 4: (a) SEM images of the doubly clamped beam and ring resonators, and (b) zoomed view of the sensing element.

EXPERIMENTS AND DISCUSSIONS

Figure 5 shows the transmission spectra of the shock sensor before and after the shock. The silicon beam is closer to ring 2 with resonance wavelength of 1593.25 nm before shock. After a 50-*g* shock in the *y* direction, the "snap-through" of buckled beam happens while the beam is switched to the other stable position. Due to the movement of the buckled beam, the resonance wavelength of ring 2 blue shift while resonance wavelength of ring 1 red shift. The resonance shifts for both rings are approximately the

same, which is 0.15 nm. Therefore, the shock status can be monitored by observing the wavelength shift or the power change at resonance wavelength.

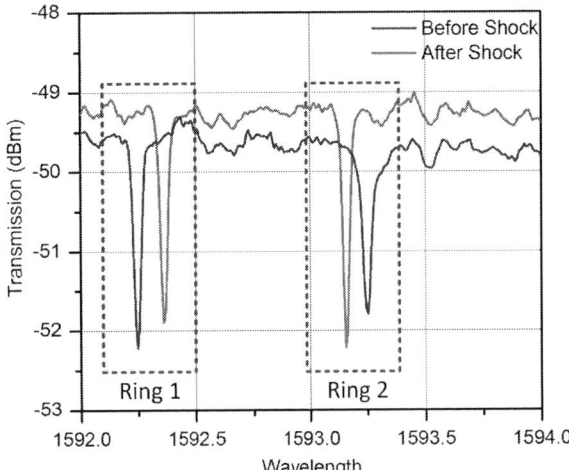

Figure 5: Transmission spectra of shock sensor before and after shock

The structure of silicon beam and middle structures can be modified to have different masses, so that the forces required for snap-through can be varied to meet different requirements for shock measurement.

CONCLUSIONS

In conclusion, an all optical shock sensor is designed, fabricated and experimentally demonstrated. Fabricated with CMOS compatible process, this optical shock sensor can be easily integrated with other photonic devices. A > 50 *g* shock can switch the buckled doubly-clamped beam from one stable position to another and induce a 15 nm wavelength shift. The opto-mechanical shock sensor can be potentially used at hash environment like in oil industry, or military usage in a complex electromagnetic environment. It also has potential applications such as inertial sensor, optical switch and other opto-mechanical devices.

ACKNOWLEDGMENTS

The authors would like to acknowledge the support from the Science and Engineering Research Council of A*STAR, Singapore, under SERC Grant 1021650084.

REFERENCES

[1] L. J. Currano, M. Yu, and B. Balachandran, "Latching in a MEMS shock sensor: Modeling and experiments," Sensors and Actuators A: Physical, vol. 159, pp. 41-50, 4, 2010.

[2] M. R. Whitley, M. S. Kranz, R. Kesmodel, and S. J. Burgett, "Latching shock sensors for health monitoring and quality control," pp. 185-194, 2005.

[3] B. Todd, M. Phillips, S. M. Schultz, A. R. Hawkins, and B. D. Jensen, "Low-Cost RFID Threshold Shock Sensors," Sensors Journal, IEEE, vol. 9, pp. 464-469,

2009.

[4] W. M. Zhu, T. Zhong, A. Q. Liu, X. M. Zhang and M. Yu, "Micromachined optical well structure for thermo-optic switching", Appl. Phys. Lett., vol. 91, 261106, 2007.

[5] A. Q. Jian, X. M. Zhang, W. M. Zhu and A. Q. Liu, "Liquid refractive index sensors using resonant optical tunneling effect for ultra-high sensitivity", Sensors and Actuators A, vol. 169, 347, 2011.

[6] B. Dong, H. Cai, G. I. Ng, P. Kropelnicki, J. M. Tsai, A. B. Randles, M. Tang, Y. D. Gu, Z. G. Suo and A. Q. Liu, "A nanoelectromechanical systems actuator driven and controlled by Q-factor attenuation of ring resonator," Applied Physics Letters, Vol 103, 181105, 2013.

[7] H. Cai, B. Dong, J. F. Tao, L. Ding, J. M. Tsai, G. Q. Lo, A. Q. Liu and D. L. Kwong, "A nanoelectromechanical systems optical switch driven by optical gradient force," Applied Physics Letters, Vol 102, 023103, 2013.

[8] I. Eiji, H. Pui-Chuen, W. David, W. R. Alejandro, G. J. Steven, C. Federico, et al., "Control of buckling in large micromembranes using engineered support structures," Journal of Micromechanics and Microengineering, vol. 22, p. 065028, 2012.

CONTACT

*A. Q. Liu, +65-67904336; eaqliu@ntu.edu.sg

AN ANGULAR ACCELERATION SENSOR INSPIRED BY THE VESTIBULAR SYSTEM WITH A FULLY CIRCULAR FLUID-CHANNEL AND THERMAL READ-OUT

J. Groenesteijn, H. Droogendijk, M. J. de Boer, R. G. P. Sanders, R. J. Wiegerink and G. J. M. Krijnen
MESA$^+$ Institute for Nanotechnology, University of Twente, Enschede, THE NETHERLANDS

ABSTRACT

We report on an angular accelerometer based on the semicircular channels of the vestibular system. The accelerometer consists of a water-filled circular tube, wherein the fluid flow velocity is measured thermally as a representation for the external angular acceleration. Measurements show a linear response for angular acceleration amplitudes up to $2 \times 10^5 \, ^\circ \, \mathrm{s}^{-2}$.

INTRODUCTION

In biology, the vestibular system is used to detect the head motion in space and results in stabilization of the visual axis, head and body posture [1]. Furthermore, the vestibular system helps with the sense for motion and change in orientation in space. The system consists of two parts: the two otolith organs (the saccule and utricle), which sense linear acceleration (gravity and translational movements), and the three semicircular canals (figure 1), which sense angular acceleration in three planes (pitch, roll, and yaw) [1, 2].

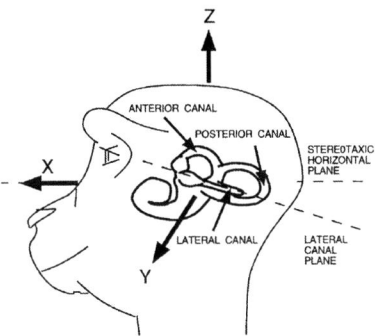

Figure 1: Position of the left labyrinth of a monkey, illustrating the three semicircular channels (image from [2]).

Each semicircular channel is filled with a fluid, where the ends of the channel are connected to a compartment with a sac, called ampulla, containing hair cells. These hair cells are composed of many cilia, which are embedded in a structure called the cupula. As the head rotates the channel moves, but the fluid within the channel lags behind due to its inertia. Consequently, the cupula is deflected and the cilia within as well. As a result, the bending of these cilia alters an electric signal that is transmitted to the brain and forms a measure for the angular acceleration.

Inspired by this vestibular system and the semicircular canals, a MEMS angular accelerometer has been proposed earlier by Arms and Townsend [3], in which they have a filled semicircular channel with a pressure transducer located at the

ends of the channel. Over the past years, also MEMS angular accelerometers have been developed based on other operation principles. Nasiri et al. [4] realized a MEMS angular accelerometer based on a conventional mass-spring system for measurements in three directions. A different approach is made by Li et al. [5], in which they propose to use a pendulum-based accelerometer.

Here, we describe a bio-inspired MEMS angular accelerometer, in which the channel is designed to be fully circular and thermal transduction principles are used for measuring the external angular acceleration. That is, by thermally measuring the flow velocity of the channel fluid, its subjection to angular accelerations can be determined. The concept of thermal read-out and a fully circular channel has been described earlier by Ploechinger [6], but has never been applied using MEMS technology.

THEORY AND MODELLING
Fluid dynamics

To describe the fluid dynamics inside the circular channel system, we start with the Navier-Stokes equations for incompressible flow [7]:

$$\rho \left(\frac{\partial \vec{v}}{\partial t} + \vec{v} \cdot \nabla \vec{v} \right) = -\nabla p + \mu \nabla^2 \vec{v} + \vec{f}, \qquad (1)$$

where the terms on the left-hand side comprise inertial terms and where the terms on the right-hand side are depending on pressure, viscosity and body forces represented by $\vec{f}$. Since the fluid flow is inside a cylindrical channel, we will use cylindrical coordinates to describe the fluid dynamics. We assume that there are no pressure gradients in the circular channel ($\nabla p = 0$) and the flow-velocity is only non-zero in the axial direction z with component v and the tube radius is much smaller than the radius of the vestibular circular system R_c. A harmonic angular acceleration will lead to a body force density f in the axial direction:

$$f_z = \rho R_c \alpha_{\mathrm{ext}} e^{i\omega t}, \qquad (2)$$

where α_{ext} is the amplitude of a harmonic angular acceleration with frequency ω. The Navier-Stokes equations in (1) can be simplified to

$$\mu \left[\frac{\partial^2 v}{\partial r^2} + \frac{1}{r} \frac{\partial v}{\partial r} \right] + f_z = 0, \qquad (3)$$

for Reynold's numbers in the fully laminar flow regime [8]:

$$\mathrm{Re} = \frac{\rho V_0 d}{\mu} < 200. \qquad (4)$$

978-1-4799-3510-9/14 $31.00 © 2014 IEEE

To find a solution for the fluid velocity-profile $v(r,t)$, we assume a harmonic flow with a parabolic profile [9, 10]:

$$v(r,t) = V_0 e^{i\omega t}\left(1 - \frac{4r^2}{d^2}\right). \qquad (5)$$

Substituting this expression into (3) gives

$$V_0 = \frac{\rho}{\mu}\frac{R_c d^2}{16}\alpha_{\text{ext}}, \qquad (6)$$

As a result, the flow velocity amplitude V_0 defined in (6) is valid when

$$\alpha_{\text{ext}} < \frac{3200\mu^2}{\rho^2 d^3 R_c}. \qquad (7)$$

Following (6), to achieve a sensitive angular acceleration sensor, we need a fluid with a high density ρ and low viscosity μ, to obtain a relatively large fluid flow velocity. Furthermore, the channel should have a large diameter d and the sensor benefits from a large system radius R_c.

Design

The design of the angular accelerometer is based on the fabrication process of the micro Coriolis mass flow sensor, developed earlier in our group [11]. The accelerometer (schematically shown in figure 2) consists of a tube with a diameter of 40 μm, whereas the diameter of the whole sensor is designed to be 5.5 mm. Further, two inlets are designed to fill the tube with a fluid, which is in our case water.

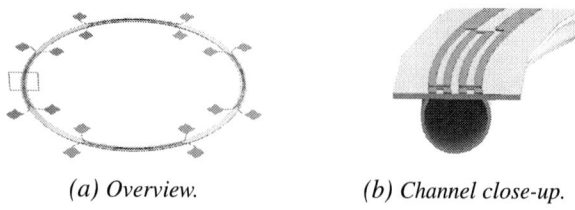

(a) Overview.   *(b) Channel close-up.*

Figure 2: Schematic view of the bio-inspired angular accelerometer with the heaters on top of the circular channel.

For measurement of the flow velocity, thermal read-out principles are used. Therefore, eight resistive elements are distributed along the circular channel. By applying a voltage over these resistors, heat is transferred to the fluid. In case of a fluid flow, the heat distribution will change and a change in resistance results [12].

The circular channel is supplied with eight equally distributed resistive elements for thermal measurement of the fluid flow velocity (figure 3). In case of a fluid flow, heat will be transferred from resistor R_1 to resistor R_2 or vice versa, depending on the direction of the fluid flow. Therefore, we can define the resistors R_1 and R_2 as:

$$R_1 = R_0 - \Delta R, \qquad R_2 = R_0 + \Delta R, \qquad (8)$$

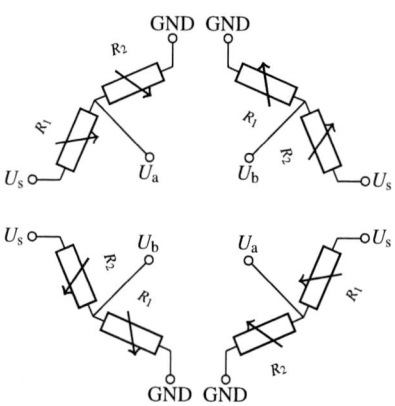

Figure 3: Schematical overview of the bridge configuration.

where R_0 is the intrinsic resistance in absence of fluid flow and resistive heating. Then, by implementing a Wheatstone-like bridge configuration for measurement of ΔR, the bridge voltage U_a and U_b are defined as:

$$U_{a,b} = U_s\left(\frac{R_{1,2}}{R_1 + R_2}\right) = U_s\left(\frac{1}{2} \pm \frac{\Delta R}{2R_0}\right). \qquad (9)$$

By using a differential amplifier, the change in resistance ΔR and thus the fluid flow velocity U_0 can be measured directly:

$$U_{\text{diff}} = U_a - U_b = U_s\left(\frac{\Delta R}{R_0}\right). \qquad (10)$$

Notice that in this configuration only two of the four available bridge output voltages are used.

FABRICATION

To fabricate the angular accelerometer, the fabrication process of the Coriolis flow sensor described by Haneveld et al. [11], schematically shown in figure 4, is used. First, a layer of low-stress LPCVD silicon-rich silicon nitride (SiRN) is deposited on a highly p-doped silicon wafer (a). Using deep reactive ion etching (DRIE), fluid inlet/outlet holes are etched from the backside using a photoresist (PR) mask, whereas the SiRN layer on top acts as a stop layer. Next, a 1 μm thick SiO_2 layer is deposited using TEOS and afterwards removed from the top side. A 50 nm layer of chromium is sputtered to create the centrelines of the channels. The pattern is then transferred into the nitride layer by reactive ion etching (RIE) and subsequently the channels are etched in the silicon using isotropic plasma etching by SF_6 (b).

The SiO_2 layer and chromium mask are then removed and another SiRN layer is grown with a thickness of 1.8 μm to form the channel walls and to seal the etch holes in the first nitride layer (c). A 10 nm layer of chromium and 200 nm layer of gold are sputtered (chromium serves as the adhesion layer for gold) and patterned to create the metal electrodes for thermal read-out (d). To thermally isolate the channels from the silicon bulk, release windows are created by reactive ion etching (RIE) of the SiRN layer (e). Then, the structure is released by isotropic etching of silicon using SF_6 (f). A

Si   SiRN   SiO$_2$ PR   Au/Cr

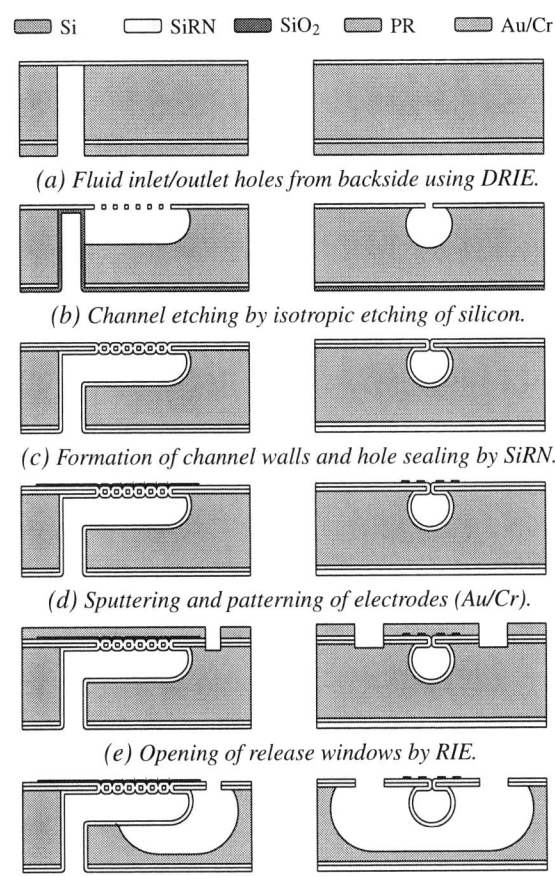

(a) Fluid inlet/outlet holes from backside using DRIE.

(b) Channel etching by isotropic etching of silicon.

(c) Formation of channel walls and hole sealing by SiRN.

(d) Sputtering and patterning of electrodes (Au/Cr).

(e) Opening of release windows by RIE.

(f) Release of device by isotropic etching of silicon.

Figure 4: Schematic view of the fabrication process. Left: through-wafer cross-section along the length of the tube. Right: through-wafer cross-section of the tube.

fabricated bio-inspired angular accelerometer is shown by the SEM image in figure 5.

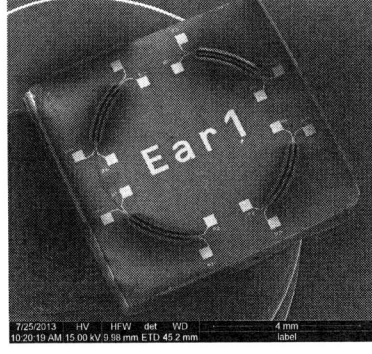

Figure 5: SEM image of the fabricated sensor.

EXPERIMENTAL

Setup

The accelerometer was tested using the rotational setup shown in figure 6. A small wheel is driven and connected

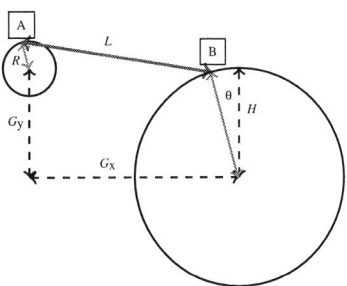

Figure 6: Schematical overview of the rotational setup.

by a lever with length L to a large wheel, on which the accelerometer is mounted. By driving the small wheel with a constant angular velocity the large wheel will show harmonic angular accelerations with constant amplitude and a frequency depending on the angular velocity ω of the small wheel. The latter is driven by a motor. For angles θ with amplitudes smaller than about 30°, the motion of the wheel can be considered sinusoidally. Consequently, the angular acceleration $\alpha(t)$ of the big wheel becomes

$$\alpha(t) \approx -\omega^2 \eta \sin(\omega t), \tag{11}$$

where η is a geometrical constant that depends on R, H, G_x, G_y and L.

Measurements

To demonstrate the sensing capability of our angular acceleration sensor, the setup shown in figure 7 was used and experiments were performed for rotational frequencies within a range of 7–14 Hz. The applied voltage U_s was generated sinusoidally using a Stanford SR 830 lock-in amplifier with its frequency set to 50 kHz and the amplitude to 0.7 V. The output of the bridge was measured differentially using the lock-in amplifier. Its output was demodulated by setting the amplifier time constant to 3 ms and the roll-off to 12 dB. The resulting envelope was then band-pass filtered using a Stanford SR 650 filter system with its band-pass set to 1–20 Hz, in order to improve the signal-to-noise ratio, and amplified with a gain of 20 dB. The filtered output was monitored using an oscilloscope (Agilent DSO1024A) and its RMS-value was measured using a multimeter (Keithley 2001). Using the geometrical properties of the rotational setup, the acceleration amplitude was calculated for every frequency. The obtained results for the measured RMS-voltage are shown in figure 8, together with a linear regression fit.

As we observe, the sensor's response is in good agreement with the expected linear response. The calculated full-scale error is found to be about 3.9 %, with the full-scale set at approximately 2×10^5 ° s^{-2}. To verify the sensor's response to angular accelerations, we changed the medium inside the tube from water to air. As a result, the output voltage dropped significantly and no clear sinusoidal waveform was observed when performing measurements identical to those shown in figure 8.

978-1-4799-3510-9/14 $31.00 © 2014 IEEE   698

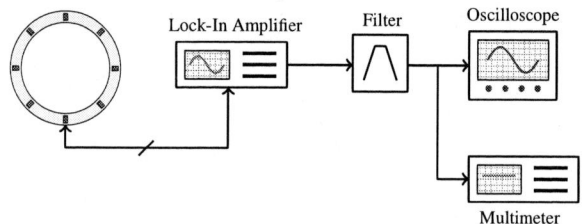

Figure 7: *Experimental setup for measurement of the output bridge voltage.*

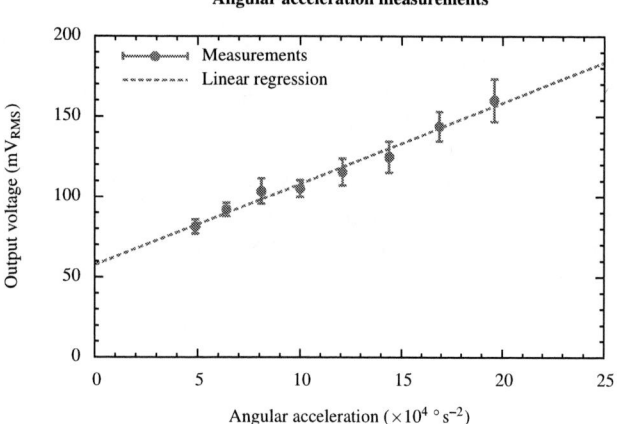

Figure 8: *Measured response versus angular acceleration amplitudes.*

DISCUSSION

The measured range of angular acceleration amplitudes is 5×10^4–$2\times10^5\,^\circ\,\mathrm{s}^{-2}$. These values are roughly comparable to commercially available angular accelerometers [13, 14]. However, by considering the lowest measured amplitude, our angular accelerometer turns out to be far less sensitive than the semicircular channel sensory system of both humans and monkeys. Groen and Jongkees [15] presented results indicating that humans can measure angular accelerations down to 0.5–$2\,^\circ\,\mathrm{s}^{-2}$, and Fernández and Goldberg [1] showed that the monkey's semicircular channel system responds to angular accelerations down to $5\,^\circ\,\mathrm{s}^{-2}$.

The aim of this work is to demonstrate the concept of an angular acceleration sensor with a fully circular fluid-channel. Future steps include optimization of the sensor. Especially the heater's geometrical design requires attention for this purpose, for which we expect that the sensor's responsivity towards angular accelerations can be significantly improved. Namely, the current heaters are designed based upon the heater design of channel-based flow sensors developed earlier in our group, which has proven its suitability for thermal read-out.

CONCLUSIONS

An angular accelerometer based on the semicircular channels of the vestibular system is developed. The accelerometer consists of a water-filled tube, wherein the fluid velocity is measured thermally as a representative for the angular acceleration. Measurements show a linear response for acceleration amplitudes up to $2\times10^5\,^\circ\,\mathrm{s}^{-2}$.

ACKNOWLEDGEMENTS

This work is carried out within the Coriolis-based SAS project of NanoNextNL and the STW/NWO funded BioEARS project.

REFERENCES

[1] C. Fernández and J. M. Goldberg, "Physiology of peripheral neurons innervating semicircular canals of the squirrel monkey. II. Response to sinusoidal stimulation and dynamics of peripheral vestibular system," *J. Neurophysiol.*, vol. 34, no. 4, pp. 661–675, Aug. 1971.

[2] S. B. Yakushin, T. Raphan, J.-I. Suzuki, Y. Arai, and B. Cohen, "Dynamics and kinematics of the angular vestibulo-ocular reflex in monkey: effects of canal plugging," *J. Neurophysiol.*, vol. 80, no. 6, pp. 8077–8099, 1998.

[3] S. W. Arms and C. P. Townsend, "MEMS based angular accelerometer," Patent, 2003, US 0047002.

[4] S. S. Nasiri, G. G. Yaralioglu, J. Seeger, and B. Taheri, "Vertically integrated 3-axis MEMS angular accelerometer with integrated electronics," Patent, 2011, US 7934423.

[5] J. Li, J. Fang, M. Du, and H. Dong, "Analysis and fabrication of a novel MEMS pendulum angular accelerometer with electrostatic actuator feedback," *Microsyst. Technol.*, vol. 19, pp. 9–16, 2013.

[6] H. Ploechinger, "Sensor for detecting a rotational movement or an angular acceleration," Patent, 2004, US 6722199.

[7] B. R. Munson, D. F. Young, and T. H. Okiishi, *Fundamentals of fluid mechanics*, 5th ed., J. Welter, T. Kulesa, and S. Dumas, Eds. USA: John Wiley & Sons, Inc., 2006.

[8] X. F. Peng, G. P. Peterson, and B. X. Wang, "Frictional flow characteristics of water flowing through rectangular microchannels," *Exp. Heat Transf.*, vol. 7, no. 4, pp. 249–264, Apr. 1994.

[9] E. R. Damiano and R. D. Rabbitt, "A singular perturbation model of fluid dynamics in the vestibular semicircular canal and ampulla," *J. Fluid Mech.*, vol. 307, pp. 333–372, 1996.

[10] R. Vega, V. V. Alexandrov, T. B. Alexandrova, and E. Soto, "Mathematical model of the cupula-endolymph system with morphological parameters for the axolotl (ambystoma tigrinum) semicircular canals," *Open Med. Inform. J.*, vol. 2, pp. 138–148, 2008.

[11] J. Haneveld, T. S. J. Lammerink, M. J. de Boer, R. G. P. Sanders, A. Mehendale, J. C. Lötters, M. Dijkstra, and R. J. Wiegerink, "Modeling, design, fabrication and characterization of a micro Coriolis mass flow sensor," *J. Micromech. Microeng.*, vol. 20, p. 125001, 2010.

[12] T. S. J. Lammerink, N. R. Tas, M. Elwenspoek, and J. H. J. Fluitman, "Micro-liquid flow sensor," *Sens. Act. A*, vol. 37–37, pp. 45–50, 1993.

[13] STMicroelectronics, "LIS1R02 (L6671)," 2002, Angular accelerometer.

[14] Endevco, "Model 7302BM4 – Piezoresistive angular accelerometer," 2009.

[15] J. J. Groen and L. B. W. Jongkees, "The threshold of angular acceleration perception," *J. Physiol.*, vol. 107, pp. 1–7, 1948.

AN EFFICIENT EARTH MAGNETIC FIELD MEMS SENSOR: MODELLING AND EXPERIMENTAL RESULTS

Mehrdad Bagherinia[1], Alberto Corigliano[1], Stefano Mariani[1]
David A. Horsley[2], Mo Li[2], Ernesto Lasalandra[3]

[1]Department of Civil and Environmental Engineering, Politecnico di Milano, ITALY
[2]BSAC, University of California, Davis, USA
[3]AMSGroup, STMicroelectronics, ITALY

ABSTRACT

The paper presents the experimental results and performance indexes of a new z-axis Lorentz force MEMS magnetometer with simple design, reduced dimensions and high efficiency. An ad-hoc formulated multi-physics approach is exploited to compute the sensor dynamics. Possible parasitic acceleration sensitivity is mechanically canceled in the proposed device. Optimality of the proposed configuration is discussed by means of an ad hoc formulated parameter optimization approach.

INTRODUCTION

Multi-axis MEMS sensors have been recently installed in various commercial products; they can reach nine and more axis sensing capabilities combining in a single packaged device three axis accelerometers, three axis gyroscopes, pressure and temperature sensors and magnetic field sensors. The present paper focuses on Lorentz force MEMS magnetometers [1-4], which can be particularly advantageous with respect to other devices for the measurement of magnetic fields in the direction orthogonal to the substrate (z-axis magnetometers). When compared to Hall effect sensors, Lorentz force devices have the advantage of lower power consumption and easier integration with standard, silicon-based, MEMS fabrication technologies.

The purpose of the present paper is to present a simple Lorentz force MEMS magnetometer fabricated with the industrial surface micromachining process THELMA[(C)] of STMicroelectronics. The device dynamics has been studied by means of an ad hoc formulated multi-physics model which takes into account, beside the Lorentz force contribution, the electrostatic loading due to capacitive sensing [5, 6]. Optimality of the device geometry is discussed making use of the multi-physics, multi-constrained optimization approach proposed in [6].

The paper is organized as follows: in Section 2 the multi-physics modelling is described; experimental results and the proposed device performances are discussed in Section 3. Optimization of the device is presented in Section 4; while closing remarks are given in Section 5.

MULTI-PHYSICS MODELLING

The resonating structure of the MEMS magnetometer is reported in Figure 1a. This sensor has been designed to sense a magnetic field aligned with the out-of-plane direction, as the two horizontal beams in the picture are excited by the Lorentz force resulting from the interaction between the current flowing along the beams and the magnetic field itself. Sensing is achieved through a couple of parallel plates connected to the mid-span cross-section of each beam, as shown in details in Figure 1b.

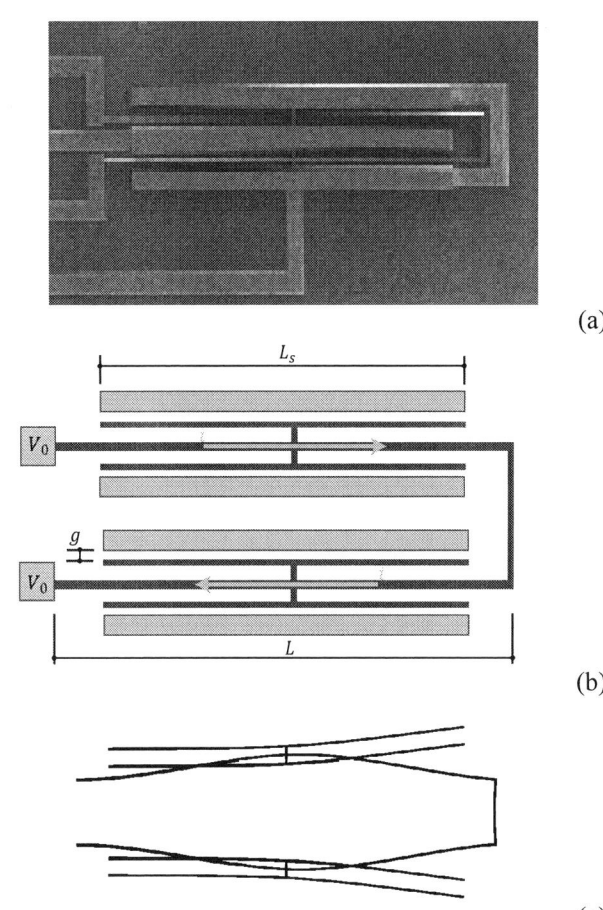

(a)

(b)

(c)

Figure 1: (a) SEM image of the resonating structure; (b) working principle and notation; (c) vibration mode at resonance (displacement magnified 50 times).

The two aforementioned beams are perfectly clamped at their left-ends, whereas at their right-ends they are free to axially deform (so as to avoid effects of the axial force or of residual stresses) and partially constrained to bend in-plane due to the vertical, short connecting beam. The resonance mode depicted in Figure 1c shows the effect of this connecting beam on the deformation of the horizontal ones;

for ease of analysis, in what follows we disregard the related compliance, and therefore assume the horizontal beams to deform as they featured clamped-clamped end cross-sections.

Due to symmetry at resonance, we focus now on a single beam featuring length L and constant cross-section area A. Its elastic response to external actions is modeled according to second-order theory, so as to account for lateral deflections affecting the equilibrium state. Dynamic equilibrium is enforced in weak form through (see also [7]):

$$\int_0^L \delta v'' \, EIv'' \mathrm{d}x + \int_0^L \delta v \, \eta \ddot{v} \mathrm{d}x - \int_0^L \delta v \, f \mathrm{d}x = 0, \quad (1)$$

where shear deformations have been disregarded due to the beam slenderness. In Eq. (1): $\delta\blacksquare$ stands for the variation of field $\blacksquare$; x is the coordinate along the longitudinal axis of the beam; $v(x)$ is the lateral displacement; $v'' = \partial^2 v/\partial x^2$ represents the curvature of the beam axis; $\ddot{v} = \partial^2 v/\partial t^2$, t being time, represents the lateral acceleration field; E is the effective Young's modulus of the beam material in the longitudinal direction and EI is the flexural stiffness of the beam; $\eta = \rho A$ is the mass per unit length of the beam, ρ being the mass density of the beam material; f is the magnitude of the external load per unit length.

As far as the external loading in Eq. (1) is concerned, load density f is the sum of two terms: the electrostatic one f_e, and the Lorentz one f_L. According to what shown in Figure 1b, where the beam and the attached sensing plates are held at potential V_0, possibly changing in time, and the sensing electrodes are instead held at $V = 0$, the term f_e is given by:

$$f_e = \frac{1}{2} V_0^2 \frac{\mathrm{d}C}{\mathrm{d}\mathcal{V}}, \quad (2)$$

where C is the sum of the capacitances per unit length between the beam and the two top and bottom electrodes, and $\mathcal{V}$ is the vertical displacement of the plates, which are assumed to undergo a pure translational motion without tilting. If V_0 is the sum of a constant, bias term V_{dc} and of a co-sinusoidally varying term $v = v_a \cos \omega t$, ω being the resonance circular frequency of the beam, we end up with:

$$V_0^2 = V_{dc}^2 + \frac{1}{2} v_a^2 + 2V_{dc} v_a \cos[\omega t] + \frac{1}{2} v_a^2 \cos[2\omega t]. \quad (3)$$

To compute $\mathrm{d}C/\mathrm{d}\mathcal{V}$ in Eq. (2), we now assume that the sensing plates do not deform at system resonance; symmetry thus allows modeling the sensing system as two parallel-plate capacitors. Accounting for variations of the gap g between the plates and the sensing electrodes up to third-order terms in the lateral displacement $\mathcal{V}$, we get:

$$\frac{\mathrm{d}C}{\mathrm{d}\mathcal{V}} \cong \varepsilon_0 b \left(4\frac{\mathcal{V}}{g^3} + 8\frac{\mathcal{V}^3}{g^5} \right), \quad (4)$$

where ε_0 is the permittivity of vacuum, and b is the out-of-plane thickness of beam and plates.

To clearly distinguish the response (and sensitivity) of the structure to the two sources of external actions, it is supposed that tests are carried out adopting a co-sinusoidally time-varying magnetic field with intensity B, whose circular frequency ω_B amounts to a few Hz only. This set-up specifically allows to drop the last two terms of Eq. (3) if the system response is monitored at $\omega \pm \omega_B$, instead of ω. Accordingly, the Lorentz force excitation f_L becomes:

$$f_L = \frac{iB_0}{2} \{ \cos[(\omega + \omega_B)t] + \cos[(\omega - \omega_B)t] \}, \quad (5)$$

where B_0 stands for the maximum amplitude of the out-of-plane magnetic field.

To build a reduced-order model of the structure, we now assume the beam to vibrate according to its (linearized) resonance flexural mode, which features a lateral displacement v varying along x in accordance with:

$$v(x,t) = \frac{1}{2}\left(1 - \cos\left(\frac{2\pi x}{L}\right)\right)\mathcal{V}(t), \quad (6)$$

where, because of the assumed symmetry in the boundary conditions, $\mathcal{V}(t)$ represents the time history of the maximum lateral displacement.

Besides the modeling of beam vibrations, we need to account for additional terms coming into play because of the sensing plates attached to the beam itself. Such plates provide a contribution to the mass of the system according to:

$$m_a = 2\eta^* L_s, \quad (7)$$

where coefficient 2 accounts for the two plates attached to each beam and $\eta^* = \rho A^*$, A^* being the cross-section area of each plate. They also provide a contribution to fluid damping, which is here assumed as the dominant one. By considering the interaction between the plates and the fixed sensing electrodes, a squeeze film damping source can be singled out. Due to the high L_s/b ratio featured by the sensing plates, the resultant viscous force acting on a single plate and resisting the beam motion is given by, see e.g. [8]:

$$F_d = \frac{\mu L_s b^3}{8g^3} \dot{g}, \quad (8)$$

where μ is the viscosity coefficient of the fluid and g is the gap between the two surfaces. In Eq. (8), $|\dot{g}| = |\dot{\mathcal{V}}|$ because of geometry and beam kinematics.

By exploiting the assumed deformation mode (6) in the variational formulation (1), the motion of the beam can be described by the following ordinary differential equation in the unknown function $\mathcal{V}$:

$$m\ddot{\mathcal{V}} + d\dot{\mathcal{V}} + K_1 \mathcal{V} + K_3 \mathcal{V}^3 = F(t), \quad (9)$$

where:

978-1-4799-3510-9/14 $31.00 © 2014 IEEE

$$m = \frac{3}{8}\eta L + 2\eta^* L_s$$

$$d = \frac{\mu L b^3}{8 g^3}$$

$$K_1 = \frac{2\pi^4}{L^3} EI - 2\frac{\varepsilon_0 b L_s}{g^3} V_0^2$$

$$K_3 = -4\frac{\varepsilon_0 b L_s}{g^5} V_0^2$$

$$F = \frac{L}{4} i B_0 \{\cos[(\omega + \omega_B)t] + \cos[(\omega - \omega_B)t]\} \tag{10}$$

respectively represent the effective mass, damping, linear and cubic stiffness, and external load terms.

The dynamics of the beam, while vibrating according to its working resonance mode, is therefore governed by relation (9), known as the Duffing equation. Accordingly, nonlinearities are a result of the coupled electro-magneto-mechanical physics of the problem.

SENSOR PERFORMANCE INDEXES AND EXPERIMENTAL RESULTS

The new designed device shown in Figure 1a is composed of a single polysilicon structural layer. The simplicity of the design and the reduced dimensions are the major relevant improvements with respect to other designs proposed in the literature. Another relevant feature of the proposed device is the fact that possible parasitic acceleration sensitivity is mechanically canceled; this is obtained thanks to the structural design and to the advantage of differential sensing, similar to the one presented in [1]. This important feature can be appreciated by observing, in Figure 1c, the sensor working vibration mode: the two beams, composing the half-double-ended tuning fork (DETF) sensor, are moving anti-phase to each other.

Frequency response, output spectrum and linear magnetic transfer characteristic of the device are shown in Figures 2a, 2b and 2c, respectively.

As the Brownian-noise-limited resolution of the sensor lies in the acceptable range for earth magnetic field detection and navigation, the proposed sensor can be judged very efficient in terms of dimensions, bandwidth and total mass with respect to Lorentz force magnetometers found in the recent literature [1-4], see Table 1. From the results of Table 1 it can be also remarked that the proposed sensor provides much larger bandwidth than other Lorentz force magnetometers with similar resolution. This is mainly due to the small modal mass.

During measurements, a very large feed-through signal was observed, probably due to the asymmetry in wire-bonding and in capacitance pick-off routing. This problem will be object of future design improvement.

Details on the experimental set-up and measurements will be given in a forthcoming paper.

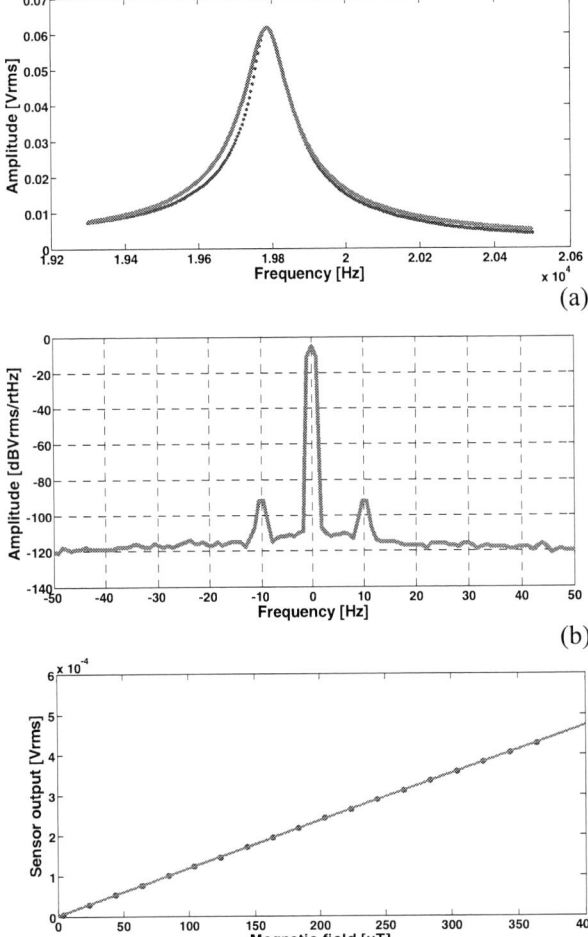

Figure 2: (a) frequency response of the sensor ($f_n = 19786$ Hz, $Q = 164$); (b) output spectrum for $\frac{\omega_B}{2\pi} = 10$ Hz; (c) measured magnetic transfer characteristic.

Table 1. Lorentz force magnetometers.

Publication	Axis	Resolution (nT mA/$\sqrt{\text{Hz}}$)	Dimensions (µm)	BW (Hz)	Total Mass (kg)
This work	Z	115	650×65	60	$5 \cdot 10^{-10}$
Li et al.[1]	X/Y	500	1080×800	69	$9.7 \cdot 10^{-9}$
Li et al.[2]	X/Y	88	200×1000	2.3	$8 \cdot 10^{-10}$
	Z	55	200×1000	7.4	
Thompson et al.[3]	Z	235	2000×1000	1	$6.8 \cdot 10^{-9}$
Kyynarainen* et al.[4]	Z	7	2000×400	2	$1.2 \cdot 10^{-9}$ **

* Consist of an extra metal layer
** Estimated data

STRUCTURAL OPTIMIZATION

The design parameters of the studied device were chosen by trial and error, considering the trade-offs between all performance indexes. From the data of Table 1, it appears that the fabricated device is potentially an optimal candidate as far as resolution and bandwidth are considered.

The optimality of the device was checked through a multi-physics approach to parameter optimization, initially formulated in [6] for another device.

The goal of the optimization process was to minimize the Brownian-noise limited resolution (objective function) while guaranteeing large bandwidth and sensitivity (lower bound constraints), a resonance frequency in a given range to avoid acoustic interference (lower and upper bound constraints) and geometrical optimization parameters in the technological process limits (lower and upper bound constraints).

The lower bounds on bandwidth and sensitivity are given by ($f_n = \omega/2\pi$):

$$\text{Bandwidth:} \qquad \frac{f_n}{2Q} > 60 \text{ Hz} \qquad (11a)$$

$$\text{Sensitivity:} \qquad 4\mathcal{V}\frac{dC}{d\mathcal{V}} > 0.246 \frac{\text{aF}}{\mu\text{T}} \qquad (11b)$$

and they were computed starting from the measured values for the tested device. The lower and upper bounds on the resonance frequency are instead providing:

$$20000 \ Hz < f_n < 30000 \ \text{Hz} \qquad (12)$$

The optimal values of resolution, frequency, bandwidth and sensitivity are given in Table 2.

Table 2. Optimal values of objective function and constraints.

Resolution	66 (nT $\frac{\text{mA}}{\sqrt{\text{Hz}}}$)
Frequency	$2.0245 \cdot 10^4$ (Hz)
Bandwidth	60 (Hz)
Sensitivity	$5.26 \cdot 10^{-19}$ ($\frac{\text{aF}}{\mu\text{T}}$)

The optimal values of the optimization parameters are given in Table 3.

Table 3. Optimal parameter values.

l_b (Beam length)	1000 (μm)
h_b (Beam width)	3.6 (μm)
l_s (Sensing plate length)	450 (μm)
h_s (Sensing plate width)	2 (μm)
V_0 (Voltage)	2 (V)
g (gap)	2 (μm)

Details concerning the formulation of parameter optimization will be given in a forthcoming paper.

CONCLUDING REMARKS

A high resolution, large bandwidth, Lorentz-force micro magnetic field sensor was designed, fabricated and tested. The device combines efficiency and reduced dimensions and could be used as sensor for out-of-plane hearth magnetic field detection in consumer market compasses. By means of multi-physics modelling and parameter optimization techniques, it was shown that the fabricated device is near to optimality as far as resolution is concerned.

ACKNOWLEDGEMENTS

The authors would like to thank A. Tocchio and H. Najar for their contribution to device design and measurements.

REFERENCES

[1] M. Li, V.T. Rouf, G. Jaramillo and D. A. Horsley, "MEMS Lorentz force magnetic sensor based on a balanced torsional resonator", *Proc. Transducers* 2013, pp. 66-69, Barcelona, 16-20 June, 2013.

[2] M. Li, V.T. Rouf, M.J. Thompson, D.A. Horsley, "Three-axis Lorentz-force magnetic sensor for electronic compass applications", *J.Microelectromech.Syst.*, vol. 21, 4, pp. 1002-1010, 2012.

[3] M.J. Thompson, D.A. Horsley, "Parametrically amplified Z-axis Lorentz force magnetometer", *J. Microelectromech. Syst.*, vol. 20, 3, pp. 702–710, 2011.

[4] J. Kyynarainen, J. Saarilahti, H. Kattelus, A. Karkkainen, T. Meinander, A. Oja, P. Pekko, H. Seppa, M. Suhonen, H. Kuisma, "A 3D micromechanical compass", *Sens. Actuators A, Phys.*, vol. 142, 2, pp. 561–568, 2008.

[5] L. Chen, W. Wang, Z. Li, W. Zhu, "Stationary response of Duffing oscillator with hardening stiffness and fractional derivative", *J. of Non-Linear Mechanics*, vol. 48, pp. 44-50, 2013.

[6] A. Corigliano, M. Bagherinia, M. Bruggi, S. Mariani, E. Lasalandra, "Optimal Design of a Resonating MEMS Magnetometer: a Multi-Physics Approach", *Proc. Eurosime 2013*, art. n° 6529931, Wroclaw, 15-17 April, 2013.

[7] X.Xi, Z. Yang, L. Meng, C. Zhu, "Primary resonance of the current-carrying beam in thermal-magneto-elasticity field", *Applied Mechanics and Materials*, vol. 29-32, pp. 16-21, 2010.

[8] M. Bao, H. Yang, "Squeeze film air damping in MEMS", *Sensors and Actuators* A vol. 136, pp. 3–27, 2007.

CONTACT

*A. Corigliano, Department of Civil and Environmental Engineering, Politecnico di Milano, Piazza Leonardo da Vinci 32, 20133 Milano, Italy; Tel: +390223994244; Fax: +390223994300; E-mail: alberto.corigliano@polimi.it

AN ELECTRET-BIASED RESONANT RADIATION SENSOR

Seung Seob Lee[1,2], Chang Keun Yoon[1,2], Seung Hyun Song[1,2] and Babak Ziaie[1,2]
[1]Birck Nanotechnology Center, West Lafayette, IN, USA
[2]Purdue University, West Lafayette, IN, USA

ABSTRACT

Resonance-based microcantilevers have been widely explored for various sensing applications. Subjecting the cantilever to an electrets-generated electrostatic field allows for self-resonant sensing of ionizing radiation. This paper reports the development of the resonant radiation sensor consisting of a ZnO microcantilever and a Teflon electret. The electrostatic force generated by the electric field shifts the self-resonant frequency of the cantilever. For a 125 (L), 55 (W), and 4 (T) μm (length) cantilever, the sensor displayed a sensitivity of 24.24Hz/Gy when exposed to 2Gy of gamma radiation.

INTRODUCTION

Ever since the discovery of radioactivity by Becquerel in 1896, health effects of ionizing radiation have been a constant source of concern for the public and healthcare workers [1,2]. Numerous adverse health effects of ionizing radiation and related diseases have been identified. The related diseases include cancers, leukemia, and radiation induced genomic instability which can be transmitted over many generations [3]. According to World Health Organization (WHO), people are exposed to natural source of ionizing radiation, such as soil, water, vegetation, and human-made sources, such as x-rays and medical devices. Such exposure, even at low dose levels, can increase the risk over a long term time window. For these health concerns, monitoring accumulated radiation dosage has been widely recommended (in particular for individuals dealing with higher levels on a daily basis).

Most commonly used radiation sensors on the market are thermo-luminescence dosimeters (TLD), optically stimulated luminescence dosimeters (OSLD), and electronic dosimeters such as avalanche diodes or RADFETs [4]. Passive dosimeters like TLD and OSLD do not provide a direct readout for real time monitoring. Electronic dosimeters, on the other hand, provide a direct readout capability; however, achieving the sensitivity of the TLD or OSLD still remains a challenge [4].

Resonance-based microcantilevers, due to their high sensitivity, simplicity, and low power consumption, have been widely used for various transduction applications such as chemical vapor [5] and biomolecules detection [6,7]. To date, however, a self-sensing cantilever has not been investigated for ionizing radiation dosimetry applications.

In this work, we present a novel electret-biased resonant radiation sensor composed of a Teflon electret and a ZnO cantilever capable of measuring an accumulated radiation dose. Teflon electrets, with their excellent electrical and mechanical stability under extreme environmental conditions, has been widely explored in MEMS community for energy

scavenging and microphone applications[8]. This is the first report of using microcantilevers in conjunction with Teflon electrets for real time measurement of radiation dose.

RESONANT RADIATION SENSOR DESIGN

ZnO Self-Sensing Microcantilever

The DMASP self-sensing and actuating ZnO piezoelectric microcantilever (Bruker Corp. CA. USA) is chosen for these experiments. These microcantilevers are primarily manufactured for tapping mode atomic force microscopy (AFM). The DMASP is fabricated with phosphorus doped silicon substrate and a piezoelectric ZnO film sandwiched between gold layers. The geometry of the cantilever is illustrated in Figure 1-a. The dimensions of the

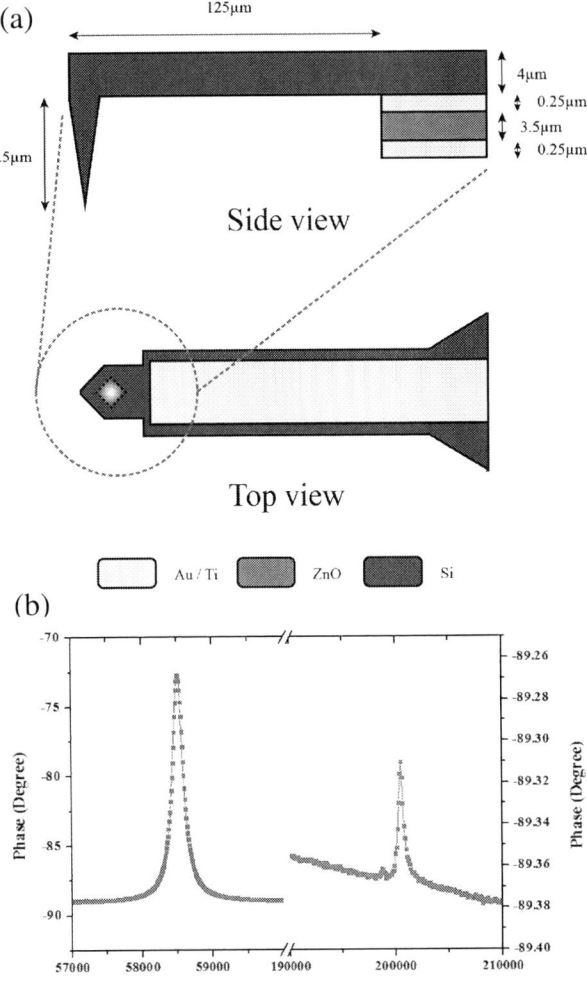

Figure 1: (a) Cross-section of the ZnO cantilever. (b) Phase vs. frequency spectrum showing the first two modes.

cantilever are 125, 55, and 4 μm (length, width and thickness). The fundamental and second harmonic resonance frequencies are measured to be 58 kHz and 200 kHz, respectively. The cantilever has a spring constant of 3 N/m.

The microcantilevers with integrated piezoelectric actuator allows for direct measurement of resonant modes through simple impedance measurements (bypassing the elaborate set-ups where external lasers and photo-detectors are normally required). Plots of impedance magnitude and phase versus the frequency of the AC signal exciting the piezoelectric cantilever are obtained using LCR meter (Agilent E4980A), Figure 1-b.

Teflon Electret

Understanding and optimizing the properties of Teflon electret are of primary importance for maximizing radiation sensor performance. The Teflon electret subjects the sensor to an electrostatic force which is directly related to the shift in resonant frequency. Electret's surface charge density, charge stability, and distance from the surface to the cantilever are the main design parameters related to the sensor's performance (dynamic range, stability, and sensitivity). For a maximal performance of the electret-biased resonant radiation sensor, an electret with a high surface charge density and a long lifetime is desired.

Teflon is the most widely used electrets material due to its excellent charge stability and high surface charge density. In literature, an extremely long charge storage lifetime (longer than 100,000 years [10]) and surface potential (higher than 3000 V [11]) have been reported. Using Monroe voltmeter, surface potential of corona charged Teflon electret samples were monitored over time. An initial surface potential higher than 3000 V was verified and a stabilization of surface potential after a rapid decay was observed. The stable surface potential measured after few days was about 1600V.

Electret-Biased Resonant Radiation Sensor: Operating Principles

The electret-biased resonant radiation sensor was fabricated by mounting the piezoelectric cantilever in close proximity and parallel to a positively charged Teflon substrate. The positively charged Teflon induces opposite charges at the metallic electrode of the cantilever and the phosphorus doped silicon tip, creating an attractive electrostatic force between the cantilever and the electret as illustrated in Figure 2. The oscillating cantilever adjacent to the electret experiences an electrostatic force gradient which reduces the effective spring constant of the cantilever. The reduced effective spring constant results in a reduced resonant frequency of the piezoelectric cantilever. As ionizing radiation passes through the ambient air surrounding the radiation sensor, direct interactions of photons with air molecules generate electrons which subsequently produce ion pairs within the air [12–14]. The ions produced by radiation drift toward the charged electret surface and neutralize the electret charges through annihilation of

opposite charge carriers. Therefore, the electric field strength of the electret steadily decreases resulting in an increase of resonant frequency of the cantilever. When the electret is completely discharged, the resonant radiation sensor recovers its natural resonant frequency.

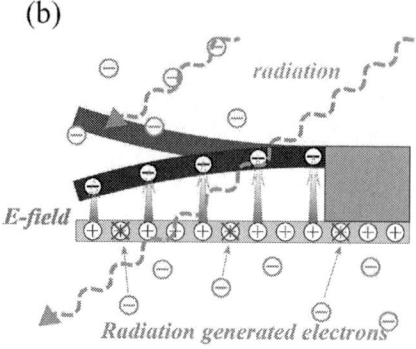

Figure 2: A schematic of the radiation sensor operation. (a) Illustration of oscillating ZnO cantilever experiencing the gradient of electrostatic force. (b) Illustration of ionizing radiation effects on the sensor: the surface charge neutralization of the electret and electrostatic force gradient reduction.

The physics of the reduced resonant frequency due to the electrostatic force gradient exerted on the cantilever can be understood by considering the second-order linear differential equation of the oscillating cantilever. In essence, any force proportional to the distance, z, can be considered as a part of spring constant. Ignoring the external dragging forces (due to friction or air resistance), the second-order linear differential equation of the freely oscillating cantilever can be written as

$$F = m\frac{d^2z}{dt^2} = -kz, \quad \frac{dF}{dz} = -k \qquad (1)$$

where k and m are the spring constant and mass of the cantilever. The electrostatic force the cantilever experiences at a separation z from the electret surface is a function of the biasing electret geometry and the cantilever tip. Given the distance between the cantilever tip and the PTFE electret is much greater than the tip size, the most dominant interaction the cantilever, not the tip, would experience most of the electrostatic force [15–17]. The total electrostatic force exerted on the cantilever can be written as

$$F(z) = \int_S dS \cdot \frac{\varepsilon_0}{2} \cdot E(x,y,z)^2 \qquad (2)$$

where S is the area of the cantilever, E is the electric field generated by the electret and ε_o is the permittivity of the vacuum. The electret generated electric field can be calculated by integrating over the electret, where σ is the charge density of the electret

$$E(x, y, z) = \int_A dE(x, y, z)$$

$$= \int\int_A (x, y, z) \cdot \frac{\sigma}{4\pi\varepsilon_0 (x^2 + y^2 + z^2)^{\frac{3}{2}}} \cdot dx \cdot dy \qquad (3)$$

Since the force in the z-direction is of the interest, electric field is written as

$$E(z) = z \int_{-\frac{L}{2}}^{\frac{L}{2}} \int_{-\frac{W}{2}}^{\frac{W}{2}} \frac{\sigma}{4\pi\varepsilon_0 (x^2 + y^2 + z^2)^{\frac{3}{2}}} \cdot dx \cdot dy \qquad (4)$$

where L, and W defines the dimension of the electret.

Since the distance, z, is much smaller than both L and W, the force can be approximated as

$$F(z) \sim \frac{\sigma}{2\varepsilon_0} \cdot s \cdot \left(1 + \frac{2z}{\sqrt{L \cdot W}}\right) \qquad (5)$$

Thus, the second-order linear differential equation of the system with electrostatic force can be written as

$$k' = \frac{dF}{dz} = -k + \frac{\sigma}{2\varepsilon_0} \cdot s \cdot \left(\frac{2}{\sqrt{L \cdot W}}\right) \qquad (6)$$

In (6), it is apparent that the electrostatic force gradient causes a change in spring constant of the cantilever. The reduced effective spring constant results in a reduced resonant frequency, a function of the electret charge density σ, which can be written as

$$f' = \frac{1}{2\pi} \sqrt{\frac{k'}{m}} \qquad (7)$$

FABRICATION

The fabrication process of the sensor is illustrated in Figure 3. A Teflon substrate (3.5 mm by 5 mm) was cut with a CO_2 laser (Universal Laser Systems) and cleaned with solvents to remove the organic residues, Figure 3a. The Teflon substrate was then corona-charged using a positive bias (12 kV for 20 minutes in ambient air, T = 21 °C) with the charging tips placed 2.5 cm above the substrate, Figure 3b. The existence and the magnitude of the surface charge density on the electret were verified with an electrostatic voltmeter (279L Isoprobe, Monroe Electronics Co.). The distance between the charged PTFE sheet and the cantilever was controlled by a tape spacer. Spacers were fabricated by stacking four layers of double sided tapes (75µm, 3M), Figure 3c. A piezoelectric bimorph microcantilever mounted and wire bonded on a specially designed board was assembled with spacers and corona charged electret, Figure 3d.

Figure 4 shows the completely assembled electret-biased resonant radiation sensor having final dimensions of 1.6 cm width, 1.9 cm length, and 5mm height. The head-on view of the tape spacer and the microcantilever is shown in Figure 4b. Figure 4c clearly shows the microcantilever's main body placed adjacent and biased by the Teflon electret.

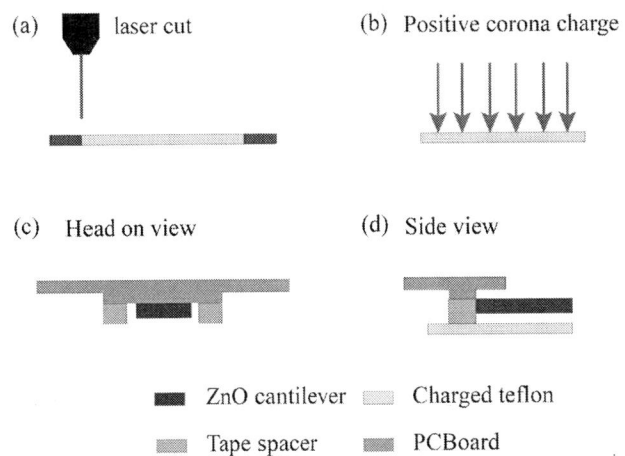

Figure3. Fabrication process of the radiation sensor. (a) CO_2 laser cut Teflon sheet. (b) Positive corona charging. (c) Spacers mounted besides the cantilever. (d) Charged Teflon electret placed on the spacers.

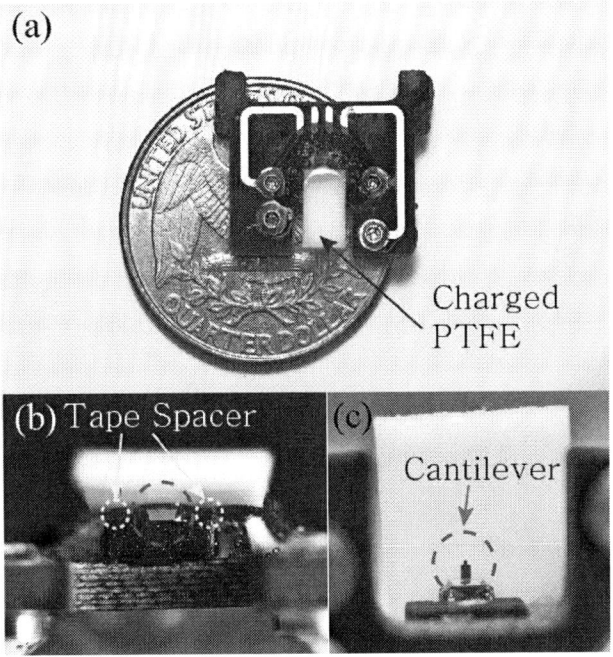

Figure 4: (a) A picture of the assembled radiation sensor. (b) Double sided tape spacers and the cantilever shown from the front view. (c) The cantilever and the Teflon electret placed on top.

EXPERIMENTS

To verify the sensing principle, the cantilever was first mounted on a micro-manipulator and placed parallel to a positive corona charged Teflon sheet of 3.5 mm by 5 mm. Initial distance between the Teflon sheet and the cantilever was 1mm. As the cantilever was lowered, the impedance magnitude and the phase over a frequency range were obtained using the LCR meter. It should be noted, however, due to the limitation of the manipulator resolution, ~10µm, the distance controls less than ~20μm were difficult to achieve. Subsequently, the assembled sensors were irradiated with Co[60] gamma ray source (1MeV) over dosage range of 0.1Gy to 2Gy with the resonant frequency being measured after each radiation fraction.

RESULTS AND DISCUSSIONS

A shifting of the resonant frequency of cantilever is observed as a function of distance between the PTFE sheet and the cantilever in Figure 5a. The resonant frequency is plotted verses the distance in Figure 5b. Three distinct regions are observed from the plot. For distance greater than 300μm, the resonant frequency stays relatively constant. For distance between 100μm and 300μm, the resonant frequency of the sensor shifts as the electric field gradient generated by the Teflon sheet becomes greater. As the distance between the two became comparable to the length of the cantilever (~100μm), the shift became much more profound, possibly due to the contribution of the tip geometry [9].

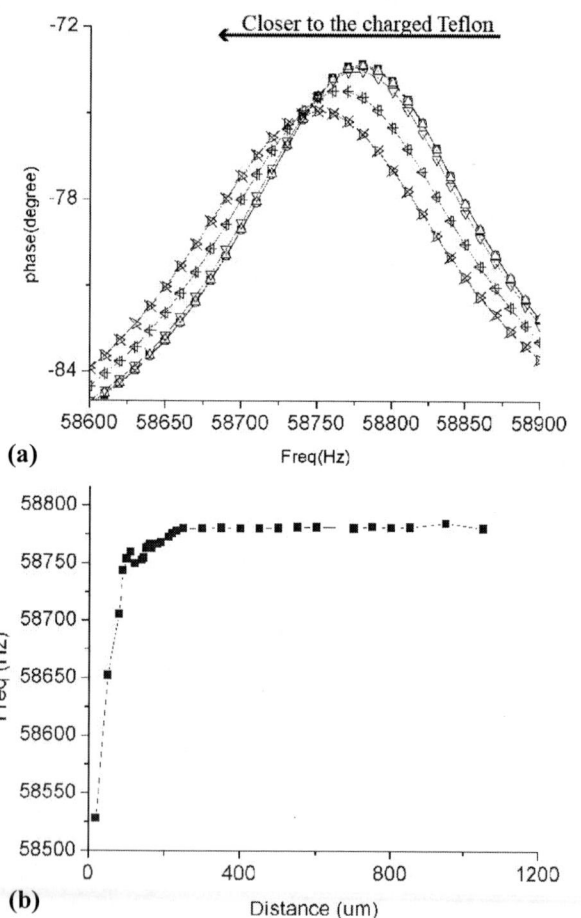

(a)

(b)

Figure 5: (a) Plot of Phase versus frequency illustrating resonant frequency shifts. (b) Plot of resonant frequency versus tip-electret distance.

The radiation response of the sensor is shown in Figure 6a. The resonant frequencies after exposure to different

ionizing radiation (Co[60] gamma source having 1 MeV photon energy) dosages were measured using an impedance analyzer. As expected, the resonant frequency increases in response to accumulated ionizing radiation dosage. Figure 6b shows the interpolated resonant frequencies versus the radiation dose. For the first 0.3 Gy of radiation, a linear response with a sensitivity of 24.24 Hz/Gy was obtained. A diminished sensitivity was observed for a higher dose of radiation and eventually saturated after 1Gy. The diminished sensitivity is anticipated and is the result of the dissipation of the electret charge of the Teflon electret.

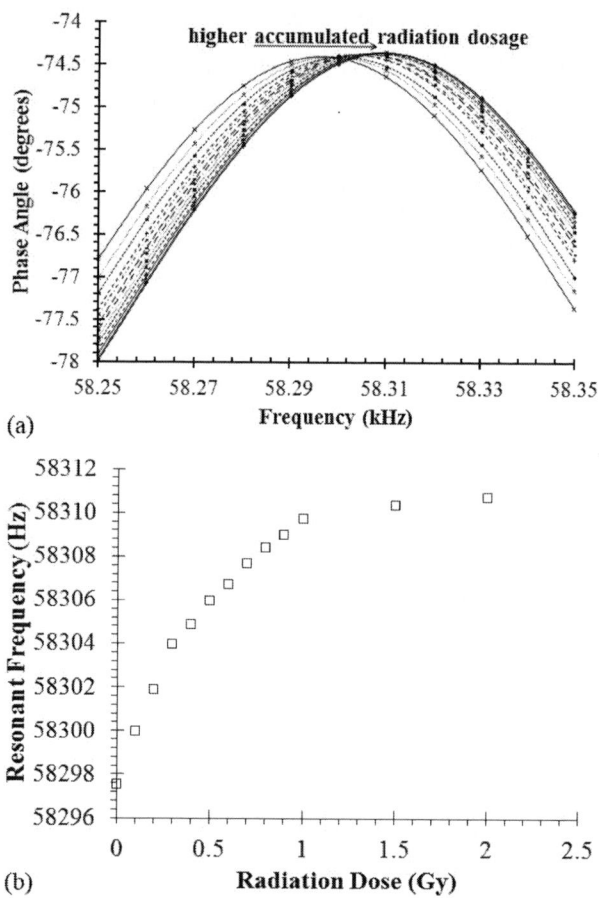

Figure 6: Radiation response of the electret-biased resonant radiation sensor.

CONCLUSION

A novel electret based resonant radiation sensor was designed, fabricated and characterized. The self-resonant frequency of the ZnO cantilever is reduced in the presence of a corona charged Teflon electret due to the electrostatic force gradient. The electret loses its initial surface charges under ionizing radiation, resulting in a reduction in the resonant frequency of the microcantilever. A prototype electret-biased resonant radiation sensor demonstrated a sensitivity of 24.24 Hz/Gy with a linear range of 0.3Gy when exposed to a Co[60] gamma ray source. The sensitivity of the resonant radiation sensor can be further improved through optimizations of geometry and electret properties.

ACKNOWLEDGEMENTS

Support for this work was provided by Landauer Corp. We would like to thank the staff of the Birck nanotechnology Center for their assistance.

REFERENCES

[1] J. Sutcliffe, "Radiation a new paradigm...Societal impacts.," *Mutat. Res.*, vol. 687, no. 1–2, pp. 67–72, May 2010.

[2] C. Grupen, "International Safety Standards for Radiation Protection," in *in Introduction to Radiation Protection*, Springer New York, 2010.

[3] K. Suzuki, M. Ojima, S. Kodama, and M. Watanabe, "Radiation-induced DNA damage and delayed induced genomic instability.," *Oncogene*, vol. 22, no. 45, pp. 6988–93, Oct. 2003.

[4] A. J. J. Bos, A. Rosenfeld, T. Kron, F. d'Errico, and M. Moscovitch, "Fundamentals of Radiation Dosimetry," vol. 5, pp. 5–23, 2011.

[5] B. Rogers, L. Manning, M. Jones, T. Sulchek, K. Murray, B. Beneschott, J. D. Adams, Z. Hu, T. Thundat, H. Cavazos, and S. C. Minne, "Mercury vapor detection with a self-sensing, resonating piezoelectric cantilever," *Rev. Sci. Instrum.*, vol. 74, no. 11, p. 4899, 2003.

[6] N. V. Lavrik, M. J. Sepaniak, and P. G. Datskos, "Cantilever transducers as a platform for chemical and biological sensors," *Rev. Sci. Instrum.*, vol. 75, no. 7, p. 2229, 2004.

[7] S. Faegh, N. Jalili, and S. Sridhar, "A self-sensing piezoelectric microcantilever biosensor for detection of ultrasmall adsorbed masses: theory and experiments.," *Sensors (Basel).*, vol. 13, no. 5, pp. 6089–108, Jan. 2013.

[8] G. Sessler and J. West, "Electret transducers: a review," *J. Acoust. Soc. Am.*, pp. 1589–1600, 1973.

[9] O. Cherniavskaya, L. Chen, V. Weng, L. Yuditsky, and L. E. Brus, "Quantitative Noncontact Electrostatic Force Imaging of Nanocrystal Polarizability," *J. Phys. Chem. B*, vol. 107, pp. 1525–1531, 2003.

[10] Z. Xia, A. Wedel, and R. Danz, "Charge storage and its dynamics in porous polytetrafluoroethylene (PTFE) film electrets," *Dielectr. Electr. Insul. ...*, vol. 10, no. 1, pp. 102–108, Feb. 2003.

[11] G.-J. Chen, H.-M. Xiao, and C.-F. Zhu, "Charge dynamic characteristics in corona-charged polytetrafluoroethylene film electrets.," *J. Zhejiang Univ. Sci.*, vol. 5, no. 8, pp. 923–7, Aug. 2004.

[12] M. A. Parada and A. de Almeida, "Teflon electret radiation dosimeter," *Nucl. Instruments Methods Phys. Res. Sect. B Beam Interact. with Mater. Atoms*, vol. 191, no. 1–4, pp. 820–824, May 2002.

[13] C. Son and B. Ziaie, "Electret Based Wireless Micro Ionizing Radiation Dosimeter," in *19th IEEE International Conference on Micro Electro Mechanical Systems*, pp. 610–613.

[14] B. Fallone and E. Podgorsak, "Production of foil electrets by ionizing radiation in air," *Phys. Rev. B*, vol. 27, no. 4, pp. 2615–2618, Feb. 1983.

[15] J. Colchero, a. Gil, and a. Baró, "Resolution enhancement and improved data interpretation in electrostatic force microscopy," *Phys. Rev. B*, vol. 64, no. 24, p. 245403, Nov. 2001.

[16] H. Jacobs, H. Knapp, S. Müller, and A. Stemmer, "Surface potential mapping: A qualitative material contrast in SPM," *Ultramicroscopy*, vol. 3991, no. 97, 1997.

[17] C. H. Lei, a Das, M. Elliott, and J. E. Macdonald, "Quantitative electrostatic force microscopy-phase measurements," *Nanotechnology*, vol. 15, no. 5, pp. 627–634, May 2004.

AN SOI TACTILE SENSOR WITH A QUAD SEESAW ELECTRODE FOR 3-AXIS COMPLETE DIFFERENTIAL DETECTION

Y. Hata[1], Y. Nonomura[1], H. Funabashi[1], T. Akashi[1], M. Fujiyoshi[1], Y. Omura[1], T. Nakayama[2], U. Yamaguchi[2], H. Yamada[2], S. Tanaka[3], H. Fukushi[3], M. Muroyama[3], M. Makihata[3], and M. Esashi[3]

[1]Toyota Central R&D Labs., Inc., Nagakute, Aichi, JAPAN
[2]Toyota Motor Corp., Toyota, Aichi, JAPAN
[3]Tohoku University, Sendai, Miyagi, JAPAN

ABSTRACT

This paper presents a novel SOI capacitive tactile sensor with a quad-seesaw electrode for 3-axis complete differential detection, which enables integration with a CMOS. For differentially detecting 3-axis forces, the tactile sensor is composed of four rotating plates individually suspended by torsion beams. In this study, to demonstrate the working principle, we fabricated a test device that integrates an SOI substrate with the quad-seesaw electrode and an anodically bondable LTCC substrate with fixed electrodes as an alternative to the CMOS. The experimental results of the test device successfully demonstrated the working principle as well as 3-axis differential detection with a matrix operation.

INTRODUCTION

Recently, tactile sensors have been studied for robot applications [1]. By embedding tactile sensors in a robot hand, the robot can perceive an object shape without visual information. Moreover, by embedding tactile sensors in the whole body of a nursing-care robot, the robot can make soft contact with a human. To realize these robot applications, we have been studying a tactile sensor system that is composed of capacitive tactile sensors integrated with CMOS and a serial sensor network, as shown in Figure 1 [2]. In a serial capacitive sensor network, many tactile sensors are set in a whole robot body with low power consumption and a small number of wires. In our previous work, the small sensor network was demonstrated by using single-axis capacitive tactile sensors [3]. In this paper, we present a new multi-axis capacitive tactile sensor device that is adaptable to our sensor system.

A key point for a tactile sensor is multi-axis force detection with high accuracy and reliability. Multi-axis detection enables a robot, including an industrial robot, to understand contact situations more precisely. As a capacitive multi-axis force sensor formed from reliable material, Brookhuis et al. reported a silicon multi-axis force sensor [4]. Although a complete differential detection is required for highly accurate detection, the force sensor did not have the differential structure for detection of the out-of-plane force F_z. This is because it is difficult to form the Z-directional differential structure due to the substrate configuration.

The present paper reports a novel tactile sensor, which is formed from a silicon on insulator (SOI) wafer, with a quad-seesaw electrode capable of 3-axis complete differential detection. Furthermore, the working principle is demonstrated using a test device that integrates an SOI

substrate with the quad-seesaw electrode and an anodically bondable LTCC substrate as an alternative to the CMOS.

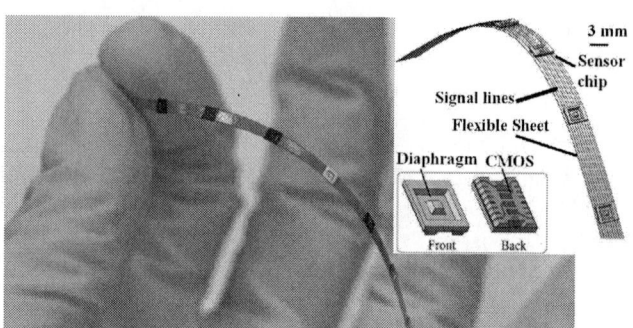

Figure 1: Concept of our tactile sensor and system.

DESIGN OF TACTILE SENSOR

Proposed Structure

Figure 2 shows the proposed sensor structure. The sensor is mainly composed of a quad-seesaw electrode in the upper silicon layer of an SOI substrate and a diaphragm with a boss in the lower silicon layer. The quad-seesaw electrode is composed of four rotating seesaw electrodes individually suspended by torsion beams. The torsion beams are connected to anchors fixed to the lower silicon by the buried oxide (BOX) layer. Each seesaw electrode rotates around the torsion beam axis and is mechanically connected to the boss by a connection beam and the center plate. By integrating with the CMOS, where fixed electrodes are formed, each seesaw electrode forms two parallel-plate capacitors. When the seesaw electrode rotates, one capacitance increases while the other capacitance decreases. Thus, the seesaw electrode enables complete differential detection.

Working Principle for 3-axis Detection

Here, the working principle of the proposed sensor for 3-axis detection is explained in Figure 3. In the left figures, cross-sectional deformations of the proposed sensor for the applied forces are depicted. The right figures show results of finite element method (FEM) analyses for the applied forces.

Figure 3(a) shows the working mode of the sensor for an out-of-plane force, $-F_z$. When $-F_z$ is applied to the boss, the boss and the center plate are translated in the Z-direction with deflection of the diaphragm. The translation motion of the center plate displaces all four connection beams with the deflections of the connection beams. These displacements induce rotations of all four seesaw electrodes around the torsion beam axes. Namely, the out-of-plane translation

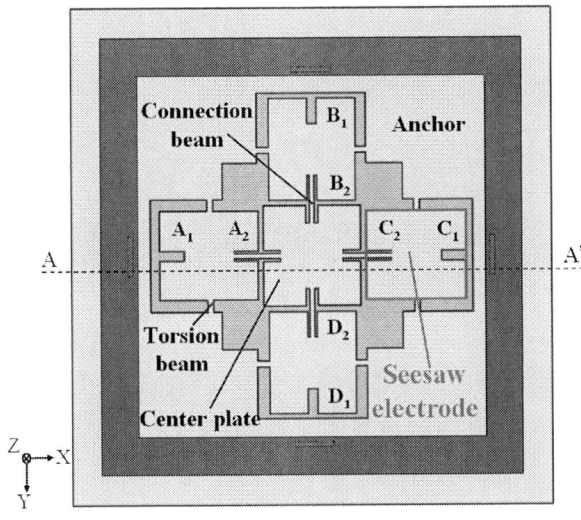

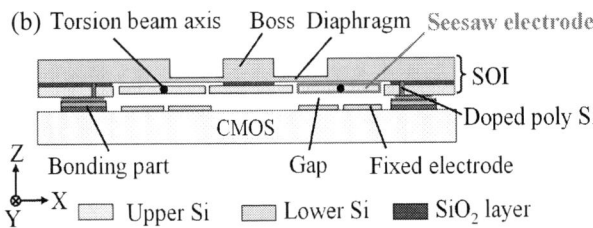

Figure 2: Proposed tactile sensor structure. (a) Bottom view (b) A-A' cross-sectional view. Upper and lower silicon layers are electrically connected by doped poly-Si through via.

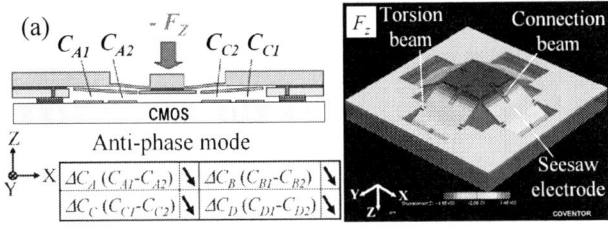

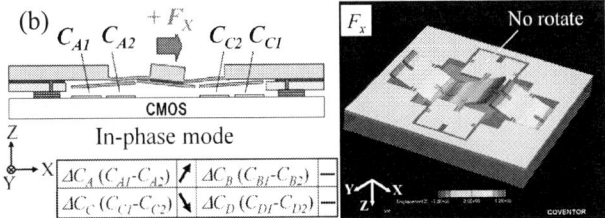

Figure 3: Working principle of the proposed tactile sensor. (a) Working modes induced by the out-of-plane force of $-Fz$. Left is cross-sectional deformation diagram, and right is FEM deformation image of the seesaw electrodes. (b) Working mode induced by in-plane shear force of $+Fx$.

motion of the boss is converted to the rotating motion of all four seesaw electrodes. In this case, the axisymmetrically located seesaw electrodes rotate in anti-phase mode, as shown on Figure 3(a) right. Therefore, all differential capacitances of the four seesaw electrodes, expressed as ΔC_A ($= C_{A1}-C_{A2}$), ΔC_B ($= C_{B1}-C_{B2}$), ΔCc ($= C_{C1}-C_{C2}$), and ΔC_D ($= C_{D1}-C_{D2}$), decrease as magnitude of $-F_z$ increases.

Figure 3(b) shows a working mode for the in-plane shear force of $+F_x$. When $+F_x$ is applied to the boss, the boss and the center plate are rotated around the Y-axis with deflection of the diaphragm. The rotation of the center plate displaces the connection beams located in the X-axis direction with deflections of the connection beams. On the other hand, the rotation twists the connection beams located in the Y-axis direction, but the displacements of the connection beams are negligibly small because the connection beams are located in the rotation axis of the center plate. Therefore, the pair of seesaw electrodes located in the X-axis direction rotates due to the displacement of the connection beams, while the pair of seesaw electrodes located in the Y-axis direction does not rotate. The two rotating seesaw electrodes rotate in the in-phase mode, as shown in Figure 3(b) right. Therefore, the one differential capacitance ΔC_A increases as magnitude of $+F_x$ increases, while the other capacitance ΔC_C decreases. The differential capacitances ΔC_B and ΔC_D of the two non-rotating seesaw electrodes do not change. The working mode for F_y is the same as that for F_x.

Based on the abovementioned working principle, 3-axis forces are identified by a matrix operation. When the differential capacitance ΔC is converted into differential voltage output ΔV with conversion circuits, the 3-axis force is expressed as

$$\begin{pmatrix} F_x \\ F_y \\ F_z \\ 0 \end{pmatrix} = \frac{1}{4} \cdot \begin{pmatrix} 2 & 0 & -2 & 0 \\ 0 & 2 & 0 & -2 \\ 1 & 1 & 1 & 1 \\ 0 & 0 & 0 & 0 \end{pmatrix} \cdot \begin{pmatrix} \Delta V_A \\ \Delta V_B \\ \Delta V_C \\ \Delta V_D \end{pmatrix}. \quad (1)$$

Coefficients in the matrix depend on the number of rotating seesaw electrodes for the direction of an applied force.

TEST DEVICE

Test Structure

A test device was fabricated to demonstrate the function of the proposed tactile sensor structure. To set up the fixed electrodes, an anodically bondable low-temperature co-fired ceramic (LTCC) substrate with Au vias and wet-etched cavities was used as an alternative to the CMOS substrate, as shown in Figure 4. With capacitance-voltage conversion circuits, the voltage outputs of the seesaw electrodes were obtained through an Au via. Moreover, the LTCC substrate reduced parasitic capacitances.

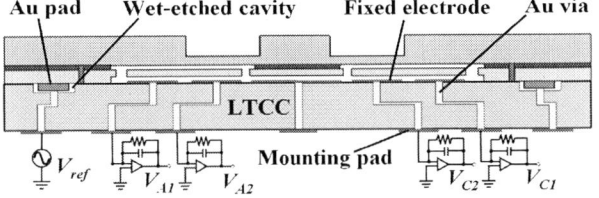

Figure 4: Cross-sectional diagram of the test device. Fixed electrodes were formed on the LTCC substrate.

Fabrication

Figure 5 illustrates the process flow for fabricating the test device. A starting SOI wafer was composed of a 50-μm-thick upper silicon layer, a 300-μm-thick lower silicon layer, and a 4-μm-thick BOX layer. The process steps are explained below.

In step (1), doped poly-Si through via was fabricated in the 50-μm-thick upper silicon layer of an SOI wafer to electrically connect the upper and lower silicon layers. Subsequently, to form gaps between the seesaw and fixed electrodes, the upper silicon layer was partially etched to a depth of 5 μm. Then, the bottom oxide layer was patterned as a mask pattern of the diaphragm with a boss. In step (2), Ti/Pt/Au contact pads were formed on the upper silicon layer. Subsequently, the quad-seesaw electrodes, connection beams, and the center plate were formed by deep-reactive-ion etching (DRIE) of the upper silicon layer. In step (3), the 50-μm-thick diaphragm with the boss was formed by DRIE of the lower silicon layer. In step (4), the BOX layer, the top un-doped silicate glass (USG) and the bottom oxide layers were etched by vapor hydrogen fluoride. The seesaw electrodes and the connection beams were released. In step (5), the upside-down flipped SOI wafer was anodically bonded to the LTCC substrate. On the top and bottom surfaces of the LTCC substrate, the fixed electrodes of Cr/Pt/Au and the mounting pads of Cr/Pt/Au were formed, respectively. Since the Au vias in the wet-etched cavities were porous, they were easily deformed to be electrically connected to the contact pads on the SOI wafer [5].

Based on these process steps, a 3.8-mm-square test device was successfully fabricated. Figure 6(a) shows a SEM image of the fabricated quad-seesaw structure. Figures 6 (b) and (c) show photographs of the fabricated test device.

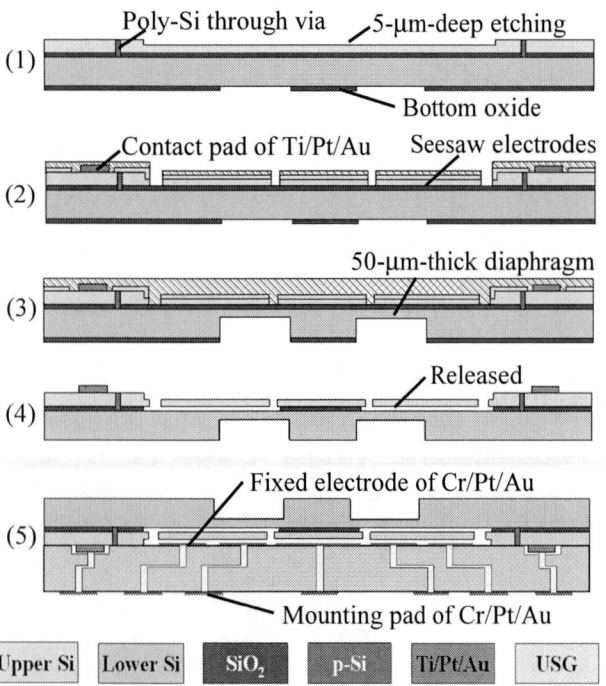

Figure 5: Process flow for fabricating the test device.

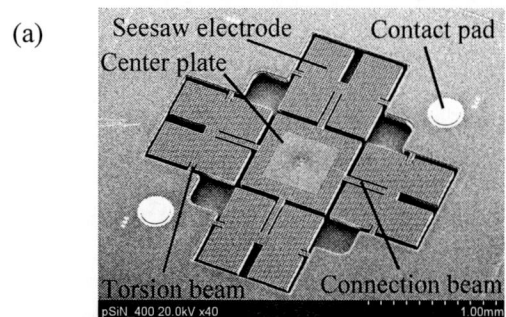

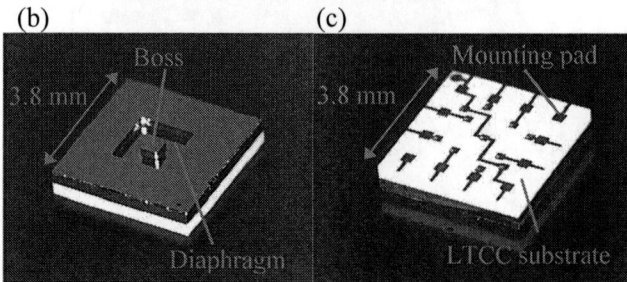

Figure 6: (a) SEM image of the fabricated quad-seesaw structure. (b) Top view of the 3.8-mm-square test device. (c) Bottom view.

MEASUREMENT
Measurement Setup

The test device was mounted to the capacitance-voltage conversion circuits in Figure 4 by flip-chip bonding. By applying reference AC voltage V_{ref} to the SOI structure, the outputs of the seesaw electrodes before the differential were measured. The V_{ref} was 300 mV$_{pp}$ with a frequency of 30 kHz, and the feedback capacitance C_f was 1 pF.

The four differential voltage outputs of the seesaw electrodes and the 3-axis force outputs were obtained by the differential operations of the measured outputs and the matrix operation of the differential voltage outputs, respectively, by using Labview. The relationship between differential voltage outputs ΔV and differential capacitance ΔC of each seesaw electrode is given as

$$\Delta V = \frac{\Delta C}{C_f} \cdot V_{ref}. \tag{2}$$

The 3-axis forces of F_z, F_x, and F_y were individually applied to the boss by jigs. A pin-shaped jig was used for the out-of-plane force of F_z, and a concave-shaped jig was used for the in-plane shear forces of F_x and F_y, as shown in Figure 7. The full scale of the applied force was set to 100 gf.

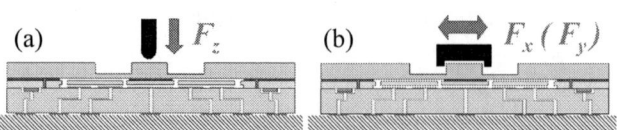

Figure 7: Force applied (a) with a pin-shaped jig for force Fz and (b) with a concave-shaped jig for forces Fx and Fy.

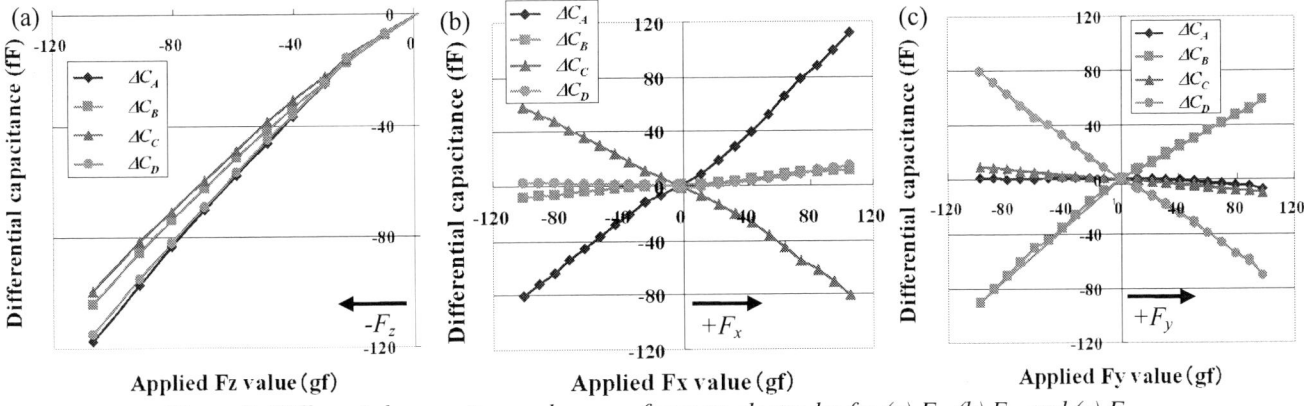

Figure 8: *Differential capacitance changes of seesaw electrodes for (a) F_z, (b) F_x, and (c) F_y.*

Experimental Results

Figures 8(a), (b), and (c) show the experimental differential capacitance changes of the four seesaw electrodes for the out-of-plane force F_z and the in-plane shear forces F_x and F_y, respectively. The differential capacitance values were calculated by Equation (2). For the applied force $-F_z$, all four differential capacitances decreased as magnitude of $-F_z$ increased. For the applied force F_x, the differential capacitances of the two seesaw electrodes A and C located in the X-axis direction changed. The differential capacitance ΔC_A of the seesaw electrode A increased as magnitude of $+F_x$ increased. On the other hand, the differential capacitance ΔC_C of the seesaw electrode C decreased. The differential capacitances of the two seesaw electrodes B and D located in the Y-axis direction did not change significantly. For the applied force F_y, the differential capacitances of the two seesaw electrodes B and D located in the Y-axis direction changed. The experimental results of the differential capacitance changes were in good agreement with the properties obtained by FEM.

Figure 9 shows the 3-axis force outputs after the matrix operation for the applied 3-axis forces. The 3-axis force outputs were successfully obtained. The nonlinear outputs were due to the imprecision force applied and large rotational displacements of the seesaw electrodes relative to the initial gap of 5 μm. The sensitivities for F_x, F_y and F_z were approximately 0.09, 0.08 and 0.11 mV/gf, respectively, on the full scale of 100 gf. The experimental results demonstrated 3-axis differential force detection and the working principle of the quad-seesaw electrode.

CONCLUSION

A tactile sensor structure with a quad-seesaw electrode was proposed to differentially detect the 3-axis force. To demonstrate the working principle of the proposed sensor, a test device was fabricated by using an LTCC substrate as an alternative to the CMOS. The experimental results demonstrated the working principle of the tactile sensor structure as well as the 3-axis force differential detection.

REFERENCES

[1] K. Noda, et al., "A Shear Stress Sensor for Tactile Sensing with the Piezoresistive Cantilever Standing in Elastic Material," *Sensors and Actuators A*, Vol. 127, 2006, pp. 295-301.

[2] M. Muroyama, et al., "LSI Design for Sensing Data Transmission by Interruption in Tactile Sensor," *IEEJ Trans. SM*, 129, 12, 2009, pp. 450-460.

[3] M. Makihata, et al., "A 1.7 mm^3 MEMS-on-CMOS Tactile Sensor Using Human-Inspired Autonomous Common Bus Communication," in *Digest Tech. Papers*, *Transducers'13 Conference*, Barcelona, Spain, June 16-20, 2013, pp. 2729-2732.

[4] R. A. Brookhuis, et al., "Scalable Six-Axis Force-Torque Sensor with a Large Range For Biomechanical Applications," in *Digest Tech. Papers*, *MEMS2012 Conference*, Paris, France, January 29-February 2, 2012, pp. 595-598.

[5] S. Tanaka, et al., "Versatile Wafer-Level Hermetic Packaging Technology Using Anodically-Bondable LTCC Wafer with Compliant Porous Gold Bumps Spontaneously Formed in Wet-Etched Cavities," in *Digest Tech. Papers*, *MEMS2012 Conference*, Paris, France, January 29-February 2, 2012, pp. 369-372.

CONTACT

*Y. Hata, tel: +81-561-71-7504;
yhata@mosk.tytlabs.co.jp

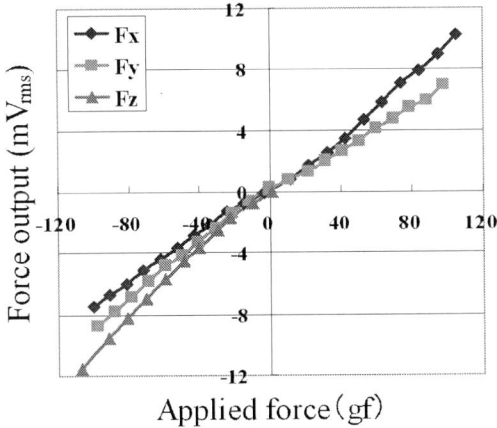

Figure 9: *3-axis force outputs after the matrix operation*

DIELECTRICAL LIQUID-BASED TACTILE SENSING ARRAY WITH ADJUSTABLE SENSING RANGES AND SENSITIVITY

Kai-Wei, Liao[1], Max T. Hou[2], Hiroyuki Fujita[3] and J. Andrew Yeh[1]*

[1]National Tsing Hua University, Hsinchu, TAIWAN
[2]National United University, Miaoli, TAIWAN
[3]The University of Tokyo, Tokyo, JAPAN

ABSTRACT

This paper presents a novel tactile sensing array with adjustable sensing ranges. Each sensing element contains a low dielectric constant (ε_r) droplet covered with high ε_r liquid. We controlled the contact angle of the droplet by controlling the electric flux passing through the element. Then, the sensing ranges and sensitivity were also adjusted due to the variation of the droplet shape. The results show the sensor's sensing range is easily adjusted from 0.33N ~ 1.05N to 0.04N ~ 0.60N. And the sensitivity is also adjusted from 1.47pF/N to 2.90.pF/N.

INTRODUCTION

Making physical contact with environment is important for robotics. Tactile sensors, which provide tactile information, are necessary for robotics and related applications [1-3]. The tactile information, such as weight and stiffness, would help robots to understand its surrounding environment. To perform different jobs requires different sensing range. For example, for grabbing different objects, such as eggs and rocks, robots require different sensing ranges to prevent damage.

In the field of tactile sensor, many researches study capacitive tactile sensors, which have highly linear and repeatable responses and low temperature drift. The capacitive tactile sensor not only able to increase the sensitivity by sandwiching the material with high dielectric coefficient between the electrodes [4-5], but also the sensing range is able to adjust. Fang et al. developed a capacitive CMOS-MEMS tactile sensor whose material can be changed during fabrication [6]. The sensing range can be tuned by ink-jetting deformable polymer with different stiffness. Takahashi et al. changed the percentage of the inflowing dielectric oil between two parallel electrodes to adjust the sensitivity [7]. Pan et al. developed the droplet-based pressure sensor [8]. The sensing range can be changed by using different volume of droplet. But these sensor's sensing range will be fixed after fabrication. Besides, a readout circuitry for adjusting the dynamic ranges is required. Yang et al. dispersed comprising carbon nanotubes in liquid crystals composites to form sensing material [9]. Applying different external field to change the conduction of sensing element can dynamically tune the sensing range without the readout circuitry. However, the sensor's sensitivity is not enough high for small force sensing.

In this paper, we proposed a novel tactile sensing array with adjustable sensing ranges as shown in Figure 1. Each sensing element contains a low dielectric constant droplet covered with high liquid. The sensing ranges and sensitivity were also adjusted due to the variation of the droplet shape. In the following sections, the sensing mechanism, design, results and conclusion will be explained.

DESIGN

To improve the above-mentioned drawbacks, a dielectric-liquid-based tactile sensing array with adjustable sensing range was developed. The volume of the each sensing element is 1.5π (mm^3). The 2×2 sensing array consists of the dielectric liquids as dielectric material, Polydimethylsiloxane (PDMS) structure, and two aluminum coating glass substrates as driving and sensing electrode. All liquid is sandwiched between top sensing electrode and bottom driving electrode. A dome shaped silicon oil droplet ($\varepsilon_r = 2$), which covered with mixed alcohol ($\varepsilon_r = 46$) is placed on the driving electrode of each element. The sensing and driving electrodes are in series of concentric rings. Fig. 2 shows the fabricated sensor.

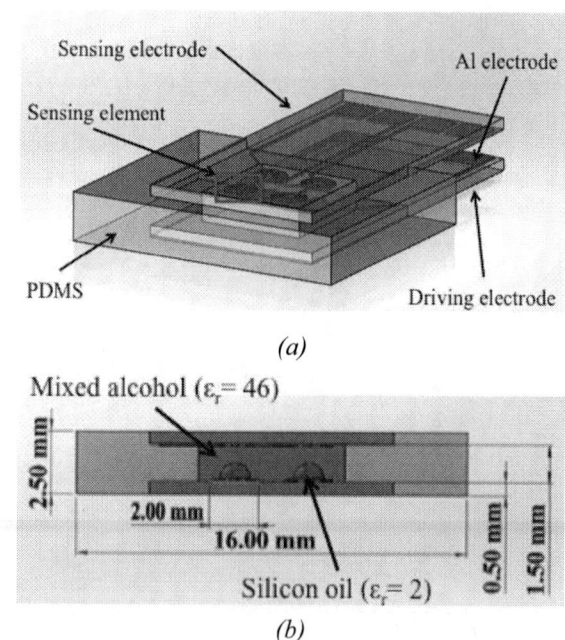

Figure 1: (a) The schematic of the sensing array, (b) The detailed dimensions of the sensing array.

Figure 2 is the schematic of the operation principle that the driving voltage adjusts the droplet shape. The contact angle of droplet can be adjusted under different driving voltage. Namely, the initial distance between droplet and sensing electrode is adjustable.

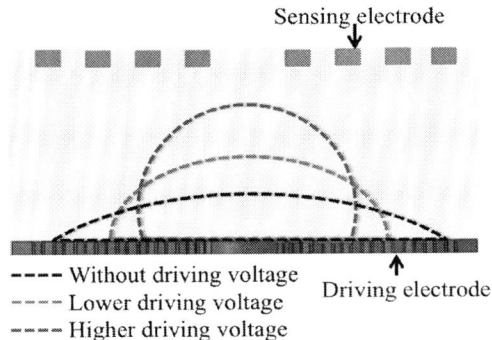

Figure 2: The operation principle of the designed tactile sensor.

The sensing mechanism is shown in Figure 3. When the force is applied, the droplet, which has lower dielectric constant, gets closer and then attaching the sensing electrode. The equivalent dielectric constant of the capacitance between sensing electrodes would be changed. Then, the capacitance would be decreased. The relation can be described as in (3)

$$\Delta C = \frac{\varepsilon_0 (\varepsilon_r - \Delta\varepsilon_r) A}{d} - \frac{\varepsilon_0 \varepsilon_r A}{d} \qquad (3)$$

where ε_0 is the permittivity of vacuum, ε_r, $\Delta\varepsilon_r$ are the initial equivalent dielectric constant and its variation , A is the area of the electrode and d is the initial gap between the sensing electrode.

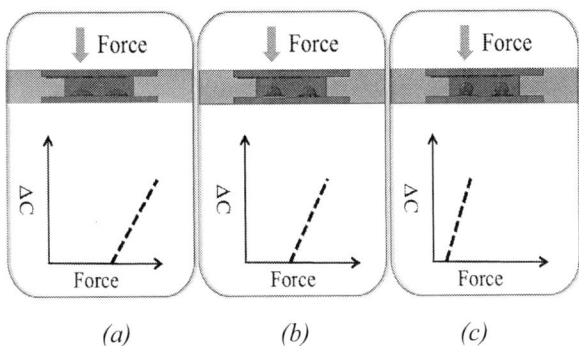

Figure 3: The operation principles of the sensing array. (a) Without driving voltage on driving electrode. (b) With lower driving voltage. (c) With higher driving voltage.

By applying different driving voltage, the initial distance between droplet and sensing electrode is adjusted due to the different contact angle of oil droplet. When applying higher driving voltage, the initial distance between

droplet and sensing electrode is smaller, as shown in Fig. 3(c). The required force which makes droplet closed enough to influence the equivalent dielectric constant would be less than that of the sensing element without driving voltage, as the case shown in in Fig 3(a). It means the sensing element is able to sense smaller force; corresponding sensing range and sensitivity can be tuned.

FABRICATION

The proposed sensor is fabricated base on microfabrication technology. Figure 4 shows the fabrication process of the sensor. Firstly, PDMS pre-polymer is mixed with curing agent at a ratio 10:1 (Sylgard 184A and 184B, Dow Corning). The PMMA master mold for PDMS spacer layer is fabricated using a CNC milling machine. Then, prepared PDMS mixture is poured onto PMMA master mold. After 24 hours at room temperature the cured PDMS spacer layer is prepared as shown in Figure 4(a). The 4(b) shows the fabrication process of the driving and sensing electrodes. The aluminum coating glasses are patterned by using photolithography and aluminum acid etch. Figure 4(c) shows the assembled process of proposed sensor. The bottom driving electrode and PDMS spacer are bonded together using oxygen plasma treatment. The prepared polyalcohol and oil are injected in the cell of spacer layer. Finally, the top sensing electrode is bonded with spacer layer. Figure 5 shows bottom electrodes and the fabricated sensor.

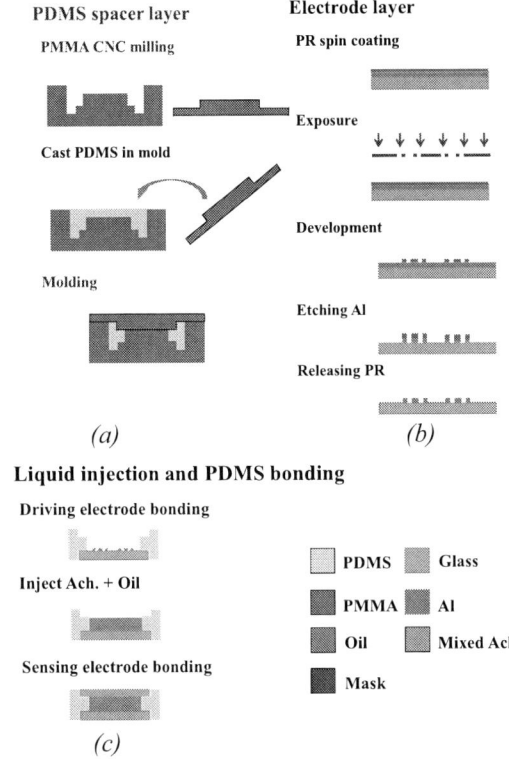

Figure 4: The fabrication process of the sensing array. (a) The electrode layer. (b) The PDMS layer. (c) The liquid injection and bonding process.

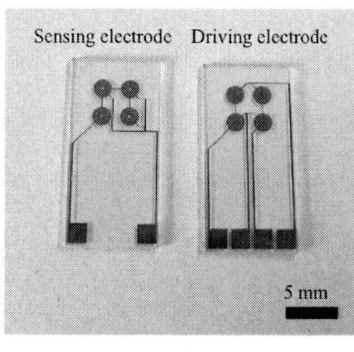

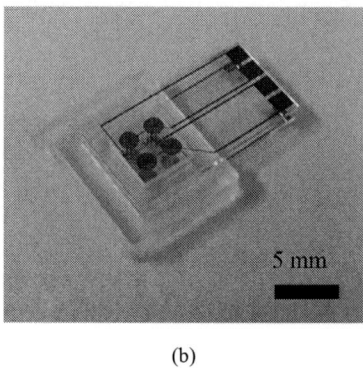

Figure 5: (a) The driving electrode and sensing electrode. (b) A fabricated 2 ×2 sensing array.

MEASUREMENT AND DISCUSSION

The experiment setup is as shown in Figure 6. By applying the different driving voltage on the driving electrode, the electrical field between the electrodes and the contact angle of droplet is adjusted. The function generator (33220A, Agilent Technologies) and a power amplifier were used to apply voltage. The applying force and the capacitances are measured by a force gauge (DTG-1 DigiTech Co.) and an impedance meter (E4980A, Agilent Technologies Co.), respectively.

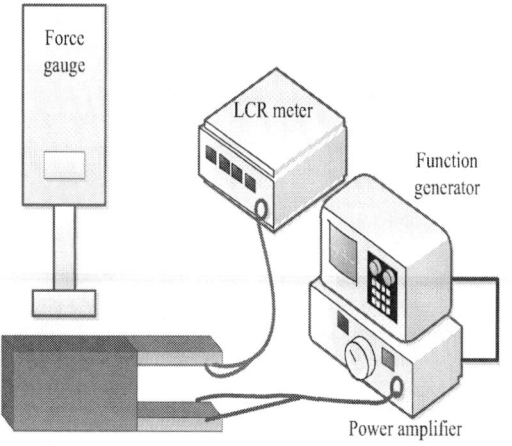

Figure 6: Experimental setup for measuring the sensing element.

The relationships between the applied forces and the absolute value of capacitance variation under different driving voltage are shown in figure 7. The sensing range is able to adjusted from large force range 0.33 ~ 1.05 N to small force range 0.04 ~ 0.60 N. And the sensitivity of the sensing array increases 1.9 times from 1.47 to 2.90 pF/N. The figure 8 shows the sensitivity, maximum and minimum force sensing ability under different driving voltage. The droplet shape changes by driving different voltage. Therefore maximum and minimum force sensing ability of sensing element is decreased compare with the sensing array with lower driving voltage. The sensitivity of sensing element is proportional to driving voltage.

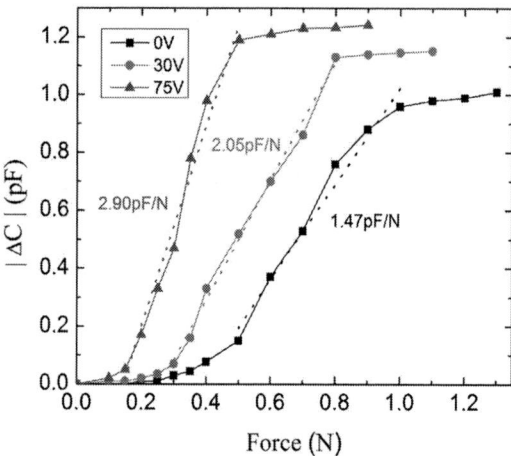

Figure 7: The relationships between the applied forces and the absolute value of capacitance variation.

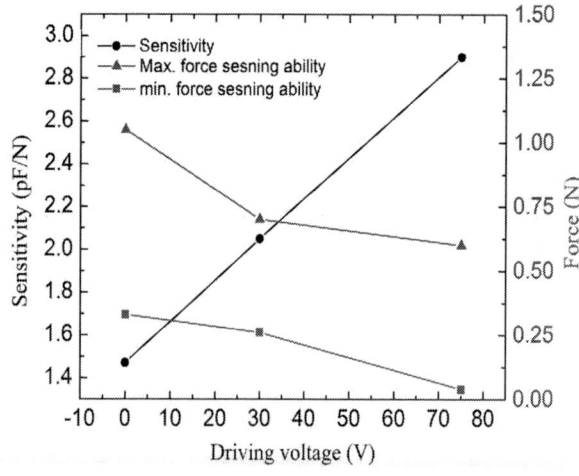

Figure 8: The relationship of sensitivity, maximum and minimum force sensing ability under different driving voltage.

CONCLUSION

An array of liquid-based tactile sensors with tunable force sensing range and sensitivity was developed. Each sensing element consists of a polyalcohol drop covered with two kinds of oil with different dielectric constants. The sensing range is able to adjusted form 0.33 ~ 1.05 N to 0.04 ~ 0.60 N. And the sensitivity of sensing element at 0.04~0.60 N force range is 1.9 times compare with 0.33 ~ 1.05 N force range. This design not only envisioned the adjustable force sensing range but also provide the high sensitivity in small force sensing range. The sensing range is adjustable after fabrication process. The proposed sensor array can be used on humanoid robots or artificial skin as a touch interfaces to satisfy different required sensing range.

ACKNOWLEDGEMENT

This work is financial supported by the National Science Council of Taiwan through grant numbers: NSC100-2628-E-007-013-MY3. Kai-Wei Liao is thankful for the support of "Summer Program, 2013" from the Interchange Association (Japan) and the National Science Council of Taiwan.

REFERENCES

[1] H.Z. Tan, M.A. Srinivasan, B. Eberman B. Cheng, "Human Factors for the Design of Force-Reflecting Haptic Interfaces," Dynamic Systems and Control, vol. 55, no. 1, pp. 353-359, 1994.

[2] T. Hoshi, H. Shinoda, "Robot Skin Based on Touch-Area-Sensitive Tactile Element", *Proc. of the 2006 IEEE International Conference on Robotics and Automation*, Orlando, May 15-19, 2006, pp. 3463-3468.

[3] S. C. B. Mannsfeld, B. C. K. Tee, R. M. Stoltenberg, C. V. H. H. Chen, S. Barman, B. V. O. Muir, A. N. Sokolov, C. Reese, Z. Bao, "Highly sensitive flexible pressure sensors with microstructured rubber dielectric layers", *Nature Materials*, vol. 9, no. 10, pp.859 -864 2010.

[4] Y. Hotta, Y. Zhang, N. Miki, "A Flexible Capacitive Sensor with Encapsulated Liquids as Dielectrics", *Micromachines*, vol. 3, pp. 137-149, 2012.

[5] K.W. Liao, Y.W. Huang, M.T. Hou, J.A. Yeh, "A dielectric liquid based capacitive tactile sensor for humanoid robots", *Networked Sensing Systems (INSS), 2012 Ninth International Conference on*, Antwerp, June 11-14 2012, pp.1-4.

[6] Y.C. Liu, C.M. Sun, L.Y. Lin, M.H. Tsai, W. Fang, "Development of a CMOS-Based Capacitive Tactile Sensor With Adjustable Sensing Range and Sensitivity Using Polymer Fill-In", *J. Microelectromech. Syst*, vol. 20, no.1, pp. 119-217, 2011.

[7] T. Takahashi, M. Suzuki, S. Iwamoto, S. Aoyagi, "Capacitive Tactile Sensor Based on Dielectric Oil Displacement put of a Parylene Dome into Surrounding Channels", *Micromachines*, vol. 3, pp. 270-278, 2012.

[8] B. Q. Nie, S. Y. Xing, J. D. Brandt, T. R. Pan, "Droplet-based interfacial capacitive sensing", *Lab on a Chip*, vol. 12, pp. 1110-1118, 2012.

[9] M.Y. Cheng, C.M. Tsao, Y.Z. Lai, Y.J. Yang, "A tactile sensing array with tunable sensing ranges using liquid crystal and carbon nanotubes composites", *Sensors and Actuators A: Physical*, vol. 177, pp. 48–53, 2012.

CONTACT

*J.A. Yeh, +886-3-574-2912;jayeh@mx.nthu.edu.tw

ELECTRET-BASED LOW POWER RESONATOR FOR ROBUST PRESSURE SENSOR

H. Mitsuya[1], H. Ashizawa[1], T. Sugiyama[2], M. Kumemura[3], M. Ataka[3], H. Fujita[3] and G. Hashiguchi[2]
[1]Saginomiya Seisakusho, Inc., Saitama, Japan, [2]Shizuoka University, Hamamatsu, Japan
[3]University of Tokyo, Tokyo, Japan

ABSTRACT

We have developed a membrane-less pressure sensor based on squeeze-film damping in a 2-μm driving/sensing gap of a silicon ring-shaped resonator. Its sensing range is from sub-atmospheric to over 1MPa; very wide-range pressure measurement is possible with one sensor element. An electret film having the 200-V-bias voltage was incorporated to the resonator; this allows the excitation of the resonator at very low AC voltage. This membrane-less pressure sensor has robust and low power consumption (nW-range) characteristics.

INTRODUCTION

Electret, a dielectric material storing electric charges semi-permanently, has been attracting more attention in recent years, particularly as electret power generators for energy harvesting. The electret power generators outputs higher power than electromagnetic ones in the frequency range of vibration existing in the environment (below 100 Hz) [1]. However, studies on transducing efficient energy from the mechanical to electrical domain for MEMS sensors and actuators are still missing even though electret is a useful material to generate the necessary electrostatic field. The main difficulty is to form strong electrets on the sidewalls of a narrow or a high-aspect-ratio gap. High-performance electrostatic MEMS devices tend to have a pair of opposing electrodes with narrow (less than 2-μm) gaps for high sensitivity. There are several methods to store electric charge to the electret films, e.g. corona discharge[2],

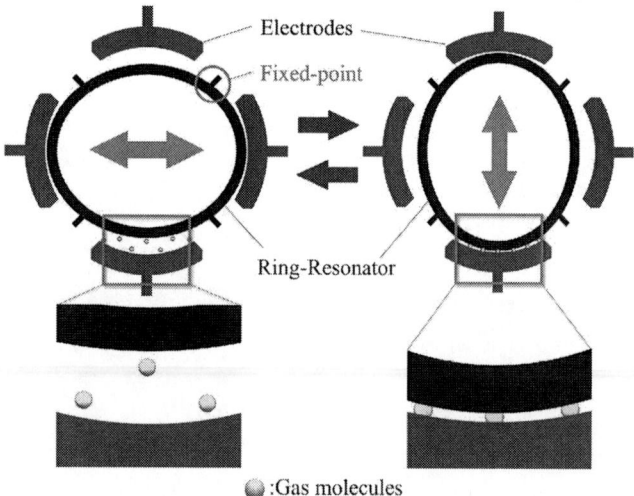

:Gas molecules

Figure 1: The ring-shaped resonator vibrates in ellipse-shape: squeeze-film damping effect in a narrow (2-μm) driving/sensing gap.

ion implantation [3], electron beam injection [4], molecule ionization by X-ray radiation [5], and potassium ion-doping during thermal oxidation [6]. However, only potassium ion-doping during thermal oxidation stores required electric charges to the electret films on the sidewalls of a narrow gap with a high-aspect-ratio.

As pressure sensors are used for various purposes including the demands of sensing operation in specific environment where external power cannot be available, e.g. TPMS (Tire Pressure Monitoring System), we propose a novel low power consumption and robust sensor using electret ascribed to potassium ions. This study provides a solution for suitable pressure sensors that can be applicable in harsh environments and can work in a wide pressure range with low power consumption.

PRESSURE SENSING METHOD

Most commonly used pressure sensors have a membrane deforming due to the pressure difference between the referential pressure (atmospheric pressure or vacuum pressure) and the pressure to be measured; the deformation is detected by either capacitive or piezo resistive elements. Due to the trade-off between sensitivity and robustness, these sensors operate in a limited pressure range. Membrane-based low-pressure sensor can be mechanically damaged when being used in a high-pressure environment, whereas high pressure sensor cannot detect low pressure.

To cover low to high pressure sensing capabilities, a ring-shaped resonator structure can be utilized as an alternative pressure sensing method. The ring is held by four beams located at the vibration nodes. Detection electrodes are placed close to the ring wall at antinodes. Applying AC actuation voltage with DC bias voltage, the ring-shaped resonator vibrates in an ellipse-shape shown in Figure 1. The pressure change alters the squeeze-film damping effect between the ring and the electrodes that is detected by the change of the amplitude at the resonance or the resonance frequency shift. The squeeze-film damping is observed when a surface moves perpendicular to the opposing surface (towards the fluid). Two components of this damping, i.e. viscous and elastic, vary with the viscosity of the air that depends on the pressure [7]. Consequently, the change in the damping results in a mechanical resonant frequency shift of the ring-shaped resonator.

FABRICATION AND SETUP

Conventional micromachining techniques were used to fabricate the device and then an electret film was formed between the ring and the electrodes. (a) The first step was

978-1-4799-3510-9/14 $31.00 © 2014 IEEE

photolithography and deep reactive-ion etching (DRIE) on a silicon-on-insulator (SOI) wafer with p-type silicon substrate for handle and device layers. The area where electrical pads were fabricated was covered with 100-nm-thick silicon-nitride (SiN) film by low-pressure chemical vapor deposition (LPCVD) for oxidation mask. (b) Then, the whole device was thermally oxidized at 930°C by bubbling a stream of KOH solution to form silicon oxide (SiO_2) film; this results in a uniform deposition of potassium ions on the surface. For electrical access to the electrodes, the SiN films were selectively removed by reactive-ion etching (RIE) using SF_6 gas after oxidation. (c) Finally, we conducted the bias-temperature (BT) procedure to store potassium ions in the SiO_2 film. Unlike the conventional BT procedure, our process utilized a hot plate for heating the device.

A schematic representation of the ring-shaped resonator is shown in Figure 3. The SiO_2 electret film by doping potassium ions generated extremely strong electric field (as high as 1×10^8 V/m). Figure 4 shows the relationship of the

resonance peak (Gain) and the bias voltage applied to the ring-shaped resonator. Two data in the figure were obtained before and after the BT procedure with the same device. Comparison of these two data reveals that the bias voltage (approximately 250V) was generated by the potassium ion-doped electret film.

Figure 5 shows the fabricated device having a 2-μm gap between the ring resonator and the electrodes. The ring had an outer diameter of 900 μm, a width of 50 μm and thickness of 30 μm. The electret generated 200-V-bias voltage in the 2-μm gap. Consequently we succeeded to excite the resonator at 4 V_{AC} without bias voltage from an external power supply.

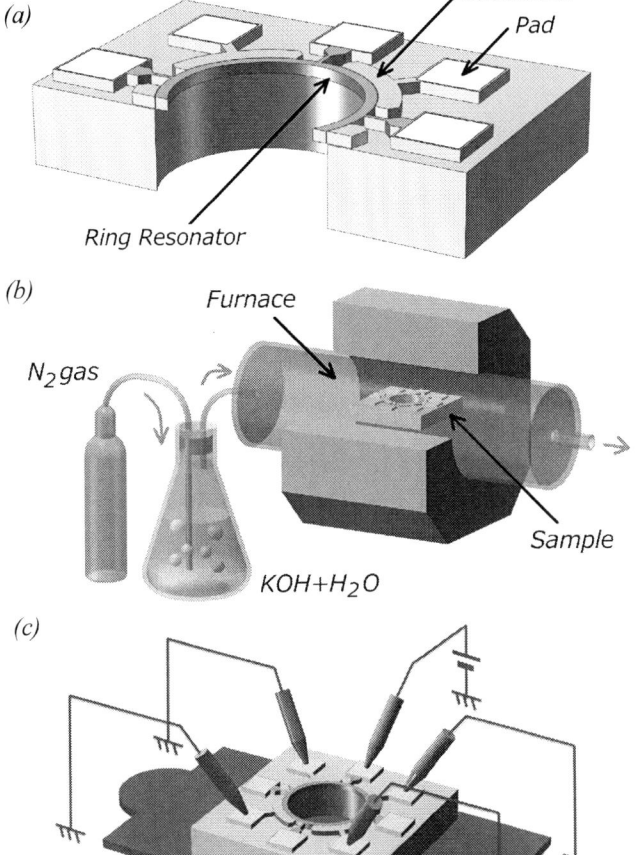

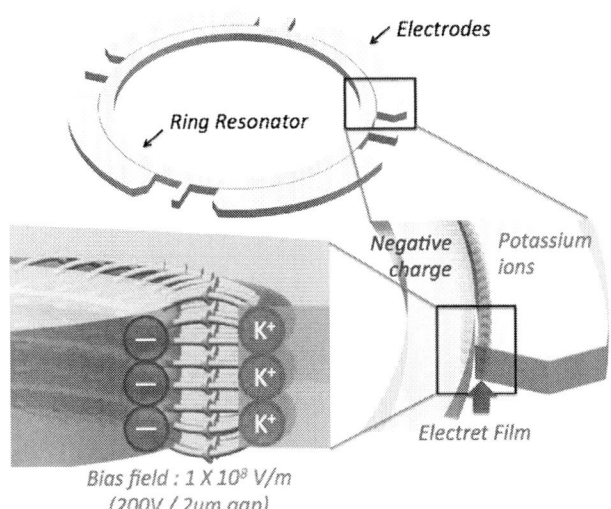

Figure 3: A schematic representation of the ring-shaped resonator: 200-V-bias voltage occurred between the ring resonator and electrodes.

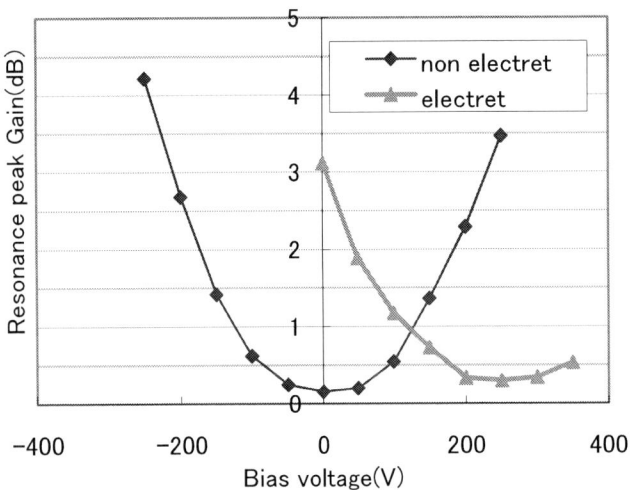

Figure 2: The process steps of the pressure sensor: (a) Conventional photolithography and DRIE. (b) Thermally oxidized by bubbling a stream of KOH solution. (c) Bias-temperature (BT) procedure.

Figure 4: The relationship of the resonance magnification (Gain) and the bias voltage applied to the ring-shaped resonator.

The experimental setup is shown in Figure 6. The device was mounted on a pressure sensor package where pressure could be modulated. The resonance frequency was monitored using a network-analyzer at certain pressure that was controlled by a pressure controller. The output signal of the network-analyzer was equally split between the input of the MEMS resonator and the reference (R-signal) by power splitter. Then, the signal output from the MEMS device (A-signal) was introduced back to the network analyzer to compare with the R-signal. The frequency sweeping allowed monitoring the resonance frequency and the gain (A/R in dB) as a function of the pressure ranging from 10kPa to 1MPa (limited by our pressure controller).

RESULTS AND DISCUSSION

Figure 7 shows the frequency characteristics at different pressures ranging from 10kPa to 1MPa. Increasing pressure could be detected as a decrease in the gain of the

and in the resonance frequency. We define the resonance magnification as the ratio between the maximum amplitude at the peak and the amplitude out of the resonance. Figure 8 plots the resonance magnification (in this study, compared with gain at 510kHz) at different pressures in the logarithmic scale. As seen, the sensor could perform effectively in a wide range.

Comparison of the resonance frequencies at different pressure levels using Figure 7 was not accurate due to noise. Therefore, we used the derivative of the gain with respect to frequency for better comparison as shown in Figure 9. The derivatives were well represented by the Gaussian fit. The frequencies at the minimum points of dG/df are shown with blue in Figure 10. Furthermore, to evaluate the sensor sensitivity, the shift of the minimum points is also depicted in Figure 10 with green.

We found that the sensor sensitivity was 1.8×10^{-3} Hz/Pa in the pressure range of 100kPa to 1MPa. We found that altering pressure changes resonance magnification and

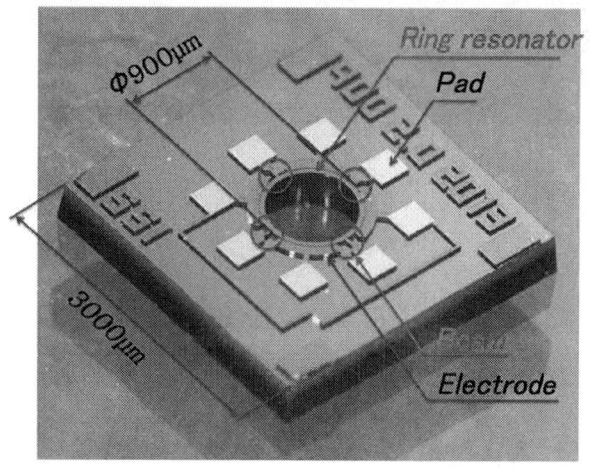

Figure 5: Optical microscope image of fabricated device: the whole device was oxidized by bubbling-KOH-solution.

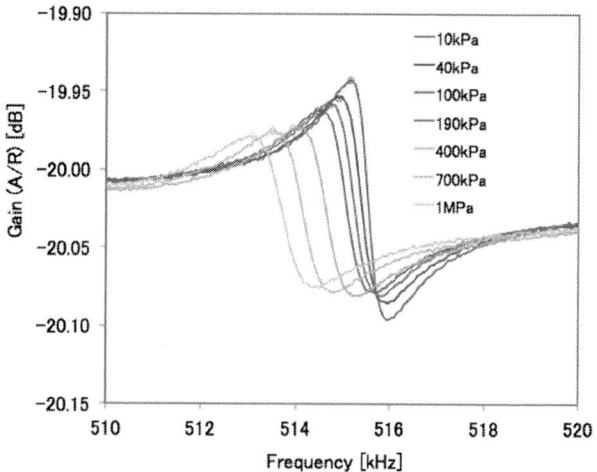

Figure 7: Monitored resonance frequency by network analyzer (10kPa to 1MPa in pressure).

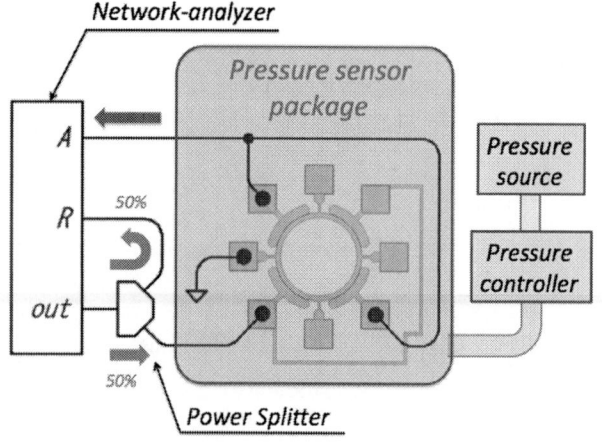

Figure 6: The experimental setup: pressure controller can be control from 10kPa to 1MPa by vacuum pump and N_2 gas cylinder.

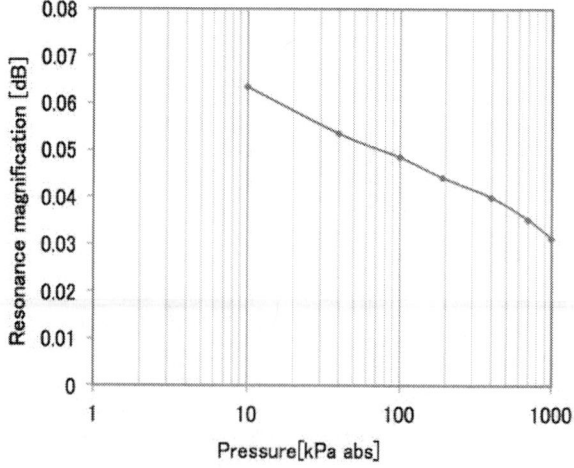

Figure 8: The resonance magnification: Relationship between gain of resonance and pressure.

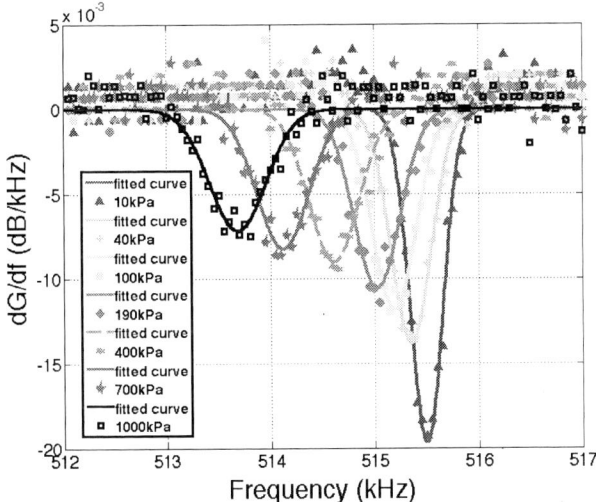

Figure 9: Derivatives of the curves in Figure 7: the relationship of dG/df and frequency.

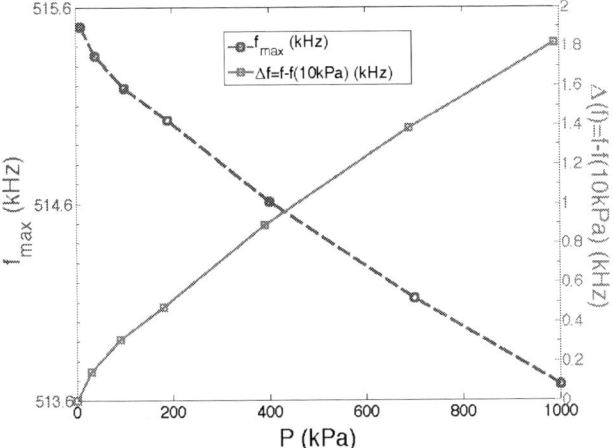

Figure 10: The frequency of minimum point of dG/df.

resonance frequency. The change of squeeze-film damping can explain the change in resonance magnification but not the change in resonance frequency. One possible explanation of the resonance frequency change is that altering pressure causes air density around the ring to change the effective mass of the ring. We are working on the explanation of this phenomenon.

CONCLUSIONS

In this study, we have developed a pressure sensor based on a silicon ring-shaped resonator. Its features are (1) wide-range and robust operation from 10kPa to 1MPa limited by our measurement equipment (but not by the sensor), and (2) small excitation voltage of 4V and low power consumption in nW-range owing to the superior performance of our electret film. The proposed technology allows us to develop portable equipment such as a wireless pressure sensor of TPMS (Tire Pressure Monitoring System) that requires robustness and low power consumption.

ACKNOWLEDGMENTS

We would like to thank VLSI Design and Education Center (VDEC, The University of Tokyo) for mask production and Dr. Laurent Jalabert for his help on data analysis.

REFERENCES

[1] Y. Naruse, N. Matsubara, K. Mabuchi, M. Izumi, and S. Suzuki, "Electrostatic micro power generator from low frequency vibration such as human motion" *J. Micromech. Microeng.,* Vol. 19, 094002, 5pp, 2009

[2] R. A. C. Altafim, J. A. Giacometti, J. M. Janiszewski, "A novel method for electret production using impulse voltages" *Proc. 7th International Symposium on Electrets (ISE 7),* Berlin, Sep.1991, pp. 267-271 (IEEE Catalog No. 91CH3029-6).

[3] U. Mescheder, P. Urbanovic, B. Muller, S. Baborie, "Charging of SiO2 electret film by ionimplantation for MEMS based energy harvesting systems" *Proceedings of PowerMEMS 2008+microEMS2008,* Sendai, Nov.2008, pp. 501-504.

[4] P. Günther, "SiO2 electrets for electric-field generation in sensors and actuators" *Sensors and Actuators A,* vol.32, pp.357-360, 1992.

[5] K. Hagiwara, M. Honzumi, M. Goto, T. Tajima, Y. Yasuno, H. Kodama, K. Kidokoro, K. Kashiwagi, Y. Suzuki, "Novel Through-substrate Charging Method for Electret Generator Using Soft X-ray Irradiation" *PowerMEMS 2009,* Washington DC, Dec.2009

[6] T. Sugiyama and G. Hashiguchi, "SiO2 Electret Generated by Potassium Ions on a Comb-Drive Actuator" *Applied Physics Express,* vol. 4, Issue 11, 114103(3pages), 2011

[7] J. J. Blech, "On Isothermal Squeeze Films", *J. Lubrication Tech.105,* 1983, 615-620

CONTACT

*H. Mitsuya, hiro-mitsuya@saginomiya.co.jp

ELECTRIC GRADIENT FORCE DRIVE MECHANISM FOR NOVEL MICRO-SCALE ALL-DIELECTRIC GYROSCOPE

Raviv Perahia, Jonathan J. Lake, Srikanth S. Iyer, Deborah J. Kirby, Hung D. Nguyen, Tracy J. Boden, Richard J. Joyce, Lian X. Huang, Logan D. Sorenson, David T. Chang

HRL Laboratories, USA

ABSTRACT

MEMS vibratory gyroscopes have recently shown great promise in the field of micro-scale position, navigation, and timing (μPNT), yet their performance often falls short of navigation grade due to losses in the vibratory structure. This paper reports a novel drive mechanism used to excite a cylindrical, all-dielectric micro-shell gyroscope structure. The drive mechanism operates by generating a gradient electric field force from a set of interdigitated electrodes placed adjacent to the gyroscope structure. This novel transduction mechanism enables mechanical actuation of a pristine dielectric structure without the need for direct metallization which could degrade the quality factor (Q) and mechanical performance. Mode spectroscopy in the range of 5-50 kHz is demonstrated with mode amplitudes as large as 0.3 μm for a 10 V drive signal. Quality factors of 12,000 have been measured. Design, fabrication, and experimental demonstration are presented.

INTRODUCTION

A tradeoff exists when considering optimal materials for micro-scale vibratory gyroscope structures. The use of a high Q, thermally-stable, dielectric material conflicts with the need for electrically-conductive materials or coatings necessary for use in traditional electrostatic drive/sense mechanisms [1-5]. A technique for driving a dielectric micro-shell without placing lossy conductive coatings directly on the structure will lead to higher performance by maintaining the high native mechanical Q of the dielectric structure.

It is well known that when a dielectric is partially inserted between two capacitor plates it will feel a force pulling it into the capacitor. The physical origin of the force is the fringing field at the edge of the capacitor plates. The field polarizes the dielectric which in turn feels a force pulling it into the capacitor [6]. Several experimental groups have recently used this gradient force to efficiently drive nano-mechanical cantilevers [7], control optomechanical cavities [8], and sense acceleration [9]. In this paper we report the first open loop demonstration of a gradient electric field drive mechanism for a micro-scale gyroscope.

ELECTRIC GRADIENT DRIVE

A test bed that allows for the demonstration of the electric gradient drive mechanism is shown in Fig. 1. The system is composed of an electrode chip positioned vertically next to a dielectric shell. The electrode chip comprises a set of alternating interdigitated gold electrodes connected to two bond pads via bus bars. The dielectric shell can vibrate freely as it sits on a pedestal 10 μm above the substrate. An applied AC voltage across the interdigitated electrodes generates a time-varying gradient force directly on the dielectric gyroscope structure, thereby exciting specific natural vibrational modes. The vibrational modes can be detected with a Laser Doppler Vibrometer (LDV) whose light is focused on the micro-shell roughly 90 degrees from the electrode chip. Simulation of the electric gradient force and experimental demonstration of mode spectroscopy will be discussed.

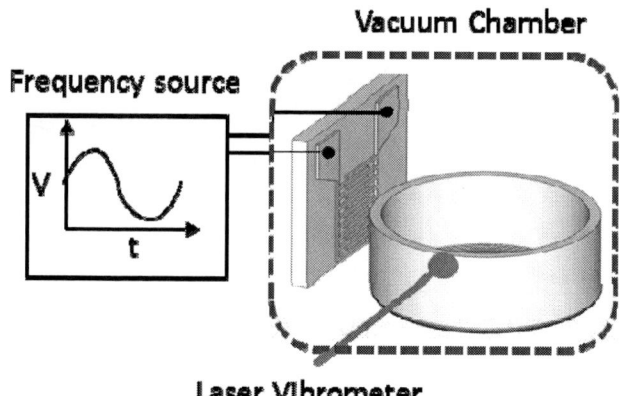

Figure 1: Schematic of electric gradient force drive mechanism in close proximity to a thin dielectric micro-shell. A sweeping AC source is used to excite the natural vibrational modes of the micro-shell and resulting vibration is observed using a Laser Doppler Vibrometer (LDV).

THEORY & SIMULATIONS

To estimate the force generated by the drive mechanism on a micro-shell, a representative two dimensional system is simulated using the finite element method (FEM). As shown in Fig. 2(a) the system consists of a radial cross section of the micro-shell and drive mechanism. The micro-shell is modeled as a rectangular SiO_2 slab with a thickness $T_S = 2$ μm. The drive mechanism is modeled as a periodic set of Au electrodes whose thickness $T_E = 60$ nm, width $W_E = 1$ μm, and spacing $S_E = 4$ μm. The drive electrodes are placed

978-1-4799-3510-9/14 $31.00 © 2014 IEEE

on a Si substrate, insulated from the substrate by 100 nm of thermal SiO$_2$. The electrode-micro shell gap d is variable. As the system is periodically repeated in the vertical dimension only a single set of electrodes needs to be simulated. A potential V = 10 V is placed between the two electrodes and the resulting potential profile for an electrode-shell gap d = 1 μm is shown in Fig. 2(b).

The force exerted on the dielectric due to the static electric field can be calculated using the Maxwell stress sensor T_{ij} [10]. In the case of an electric field gradient in the x-direction the dominant force contribution is from the T_{xx} component and is calculated from a line integral over the micro-shell boundary. To estimate the total gradient force of the 3D structure the unit force is scaled by the length of the electrodes L_E = 74 μm and by the number of electrode pairs N_{drive} = 60. The total force is plotted in Fig. 2(c) (blue circles) as a function of electrode-shell gap. The maximum force of more than 10 μN is achieved with zero gap. While the force decreases as the gap grows, even at a maximum simulated gap of d = 3 μm, a force of 44 nN is still exerted on the micro-shell.

When the gap is very small, the drive amplitude may be limited in the case that the restoring spring force is always less than the gradient force. The complementary behavior in capacitively actuated MEMS, known as snap down voltage, typically limits the range of motion to 1/3 of the gap. In this system the spring force is calculated for three different zero bias gaps of d = 1.0, 1.5 and 2 μm (red lines in Fig. 2(c)). The effective spring constant of k = 3 μN/μm was extracted from the eigenvalue simulation of a typical micro-shell wine-glass mode (Fig. 4(c)). For 10 V actuation the gradient force exceeds the spring force for the zero bias gap of d = 1 μm. For d = 1.5 and d = 2 μm the vibration amplitude before snap down is 1.2 and 1.8 μm respectively. While this actuation mechanism is limited by snap down, for gaps greater than 1.5 μm the range is increased above 80% of the gap. Furthermore, in the case of snap down, the dielectric shell will not short the electrodes, thus avoiding catastrophic damage to the device.

Vibrational modes, both low and high Q, can be observed using this drive mechanism as it takes only 1 nN to drive a low Q mode (Q = 1000) to 1 μm amplitude. The force generated at 10 V AC drive is more than sufficient to drive the system into resonance. This drive mechanism is therefore suitable for mode spectroscopy studies of uncoated micro-shell structures and will allow one to probe both wine-glass modes and other vibrational modes of the structure.

EXPERIMENTAL RESULTS

Both micro-shell and electrodes were fabricated using standard MEMS fabrication techniques. The SiO2 micro-shell shown in Fig 3(a) is 1.2 mm in diameter, 350 μm tall, and 2 μm in thickness. The micro-shell is elevated off the Si substrate by a 10 μm thick SiO$_2$ anchor 400 μm in diameter. The electrodes drive mechanism was fabricated on a different Si substrate and is shown in Fig. 3(b). Two sets of

interdigitated electrodes are fabricated from 100 nm of Au separated by 100 nm of insulating SiO2. The electrode spacing as fabricated is 3 μm and the overlap length between interdigitated electrodes is 150 μm. The electrodes are connected via two bus bars to large bond pads with thick Au coating suitable for wire bonding.

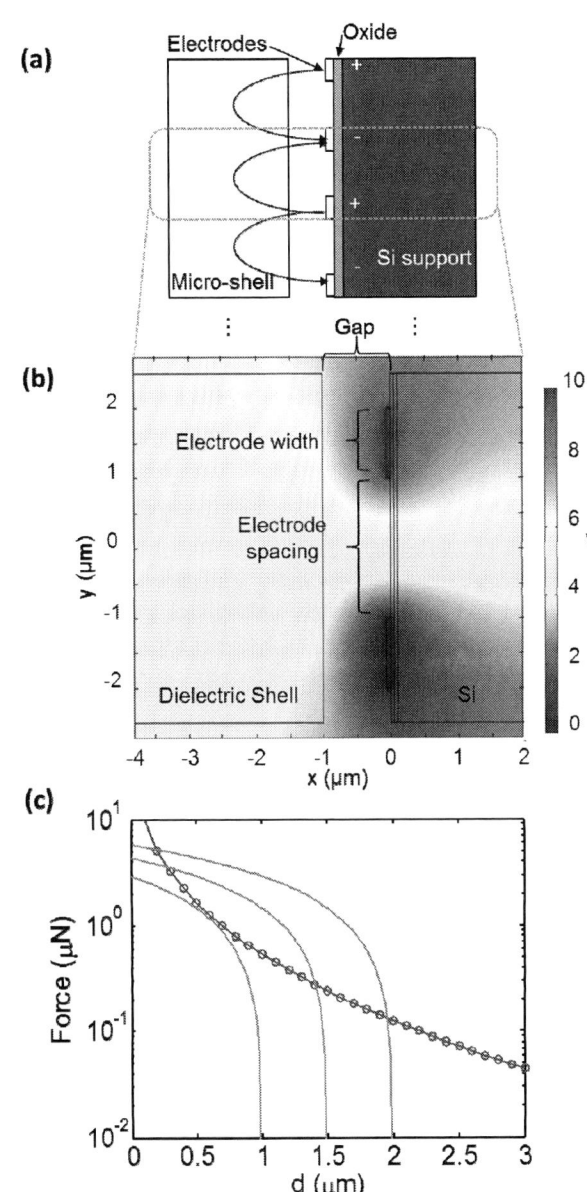

Figure 2: (a) 2D Schematic of a radial cross section of a dielectric shell in proximity to a drive mechanism. Drive mechanism is comprised of a periodic set of electrodes with alternating electric potential. (b) FEM simulated potential map of an electrode pair with V = 10 V applied across the two electrodes. (c) Force vs. electrode-shell gap shown in blue circles. Red overlays represent force-displacement curves for micro-shell motion with zero bias gap d = 1, 1.5, and 2 μm.

978-1-4799-3510-9/14 $31.00 © 2014 IEEE

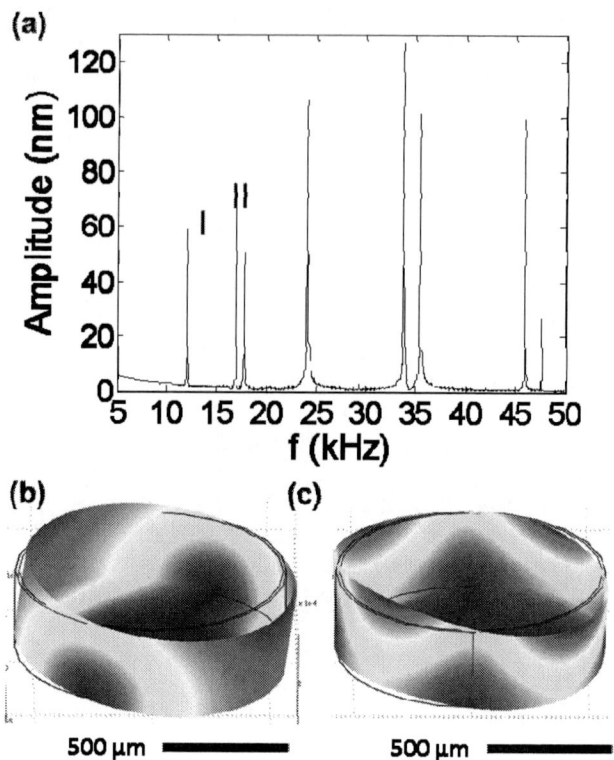

Figure 3: (a) Fabricated cylindrical micro-shell SiO$_2$ structure with flat base resting on SiO$_2$ pedestal (diameter = 1.2 mm, height = 350 µm, thickness = 2 µm, pedestal height = 10 µm) on a Si substrate. (b) Drive structure comprised of Au electrodes spaced 3 µm apart and overlapping length of 150 µm.

The experimental test configuration allowing for the placement of the micro-shell and drive electrode in close proximity is shown in Fig. 1(a). The micro-shell is kept fixed at the center of a vacuum chamber while the drive mechanism is placed tangentially to the shell. The vertical overlap between shell and drive mechanism is approximately 150 µm. The gap distance between micro-shell and drive mechanism can be controlled in real time. The LDV interrogation beam is fed into the vacuum chamber and focused on the normal surface of the micro-shell. For maximum detection the beam is focused onto the surface of the shell approximately 90 degrees from the drive chip. Electrical connection is made via a vacuum feedthrough to the drive mechanism through an intermediate circuit board. To avoid squeeze film and other air damping, experiments are carried out in vacuum P~1x10-4 Torr.

An experimental demonstration of the drive mechanism is shown in Fig. 4(a) for a micro-shell drive mechanism gap of d = 12.5 µm. A sinusoidal drive signal with V_{pp} = 10 V and V_{offset} = 5 V, is swept from 5 to 50 kHz. The micro-shell motion is measured with the LDV in velocity mode and the data is converted to displacement using the sensitivity scale factor and frequency. A very clear set of resonances arises, some with vibration magnitude of more than 100 nm. Initial simulations of the micro-shell indicate that the first mode (I) is a rocking mode of the structure and the 2nd pair (II) are the n = 2 order wineglass modes with frequency f = 17.3 kHz. FEM simulations of the rocking and wine-glass modes are shown in Fig. 4(b) and Fig. 4(c) respectively. The split nature of the wine-glass modes indicates asymmetry in the fabricated structure. The LDV beam is aligned normal the micro-shell surface and the drive mechanism is designed to excite only vibrations normal to the shell surface. Due to this configuration motion normal to the substrate is rejected both by the vibrometer and the drive mechanism and may not be observed.

Figure 4: (a) Frequency sweep from f = 5 kHz to f = 50 kHz (b) FEM simulated rocking mode (I) (c) FEM simulated n = 2 wine-glass mode (II)

The Q of individual modes can be probed by taking high resolution frequency sweeps. As an example, a high Q mode at f = 47.6 kHz is shown in Fig. 5(a). The amplitude of the mode is fit to a Lorentzian and a Q ~ 12,000 is achieved.

The power of this technique is shown by measuring the amplitude of vibration as a function of distance while the micro-shell is driven on resonance (f = 47.6 kHz) as shown in Fig. 2(c). Amplitude of 300 nm is achieved at 5 µm gap and amplitudes greater than 100 nm are achieved with gaps up to 12.5 µm for modest voltages < 10 V.

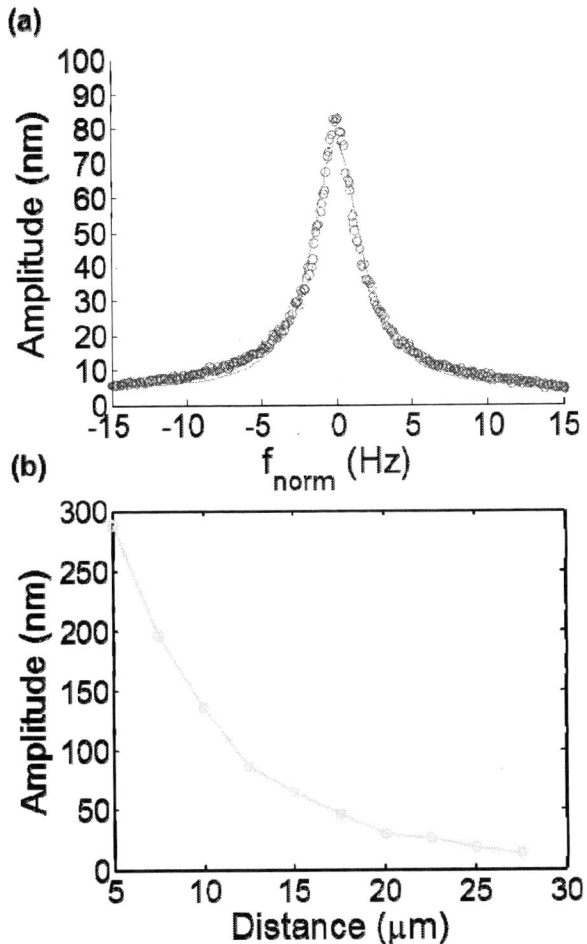

Figure 5: (a) Narrow frequency sweep (Δf = 30Hz) at f = 47.6 kHz (b)Vibration amplitude vs. micro-shell drive mechanism gap.

CONCLUSIONS

A novel mechanism for driving dielectric micro-shells has been simulated and experimentally demonstrated. The drive mechanism operates by generating a gradient electric field force from a set of interdigitated electrodes placed adjacent to the gyroscope structure. Mode spectroscopy and mechanical amplitudes as large as 0.3 μm have been demonstrated with the highest Q observed of 12,000. Motion can be driven at distances greater than 10 μm with low voltages. Using this drive mechanism, vibration amplitudes significantly larger than 1/3 of the gap are possible before snap down. In the event of snap down, the dielectric nature of the micro-shell will prevent catastrophic damage to the device. This new drive mechanism potentially clears the way for a new class of dielectric micro-scale gyroscopes that are not limited by conductive coatings.

ACKNOWLEDGEMENTS

The authors thank Dr. R. N. Candler at UCLA Electrical Engineering for assistance with measurements, B. Holden, M. Cline, and F. Stratton for the fabrication effort, and R. McElwain for assistance in testing.

REFERENCES

[1] N. Yazdi, F. Ayazi, K. Najafi, *Proc. IEEE*, Vol.86, pp.1640-1659,1998.

[2] S. A. Zotov, A. A. Trusov, A. M. Shkel, *J. MEMS*, Vol. 21, No. 3, pp. 509-510, 2012.

[3] L. D. Sorenson, P. Shao, F. Ayazi, *Proc. IEEE MEMS* 2013, pp. 169-172, 2013.

[4] J. Cho, J.-K. Woo, J. Yan, R. L. Peterson, K. Najafi, *IEEE Transducers 2013*, pp. 1847-1850, 2013.

[5] A. Heidari, M.-L. Chan, H.-A. Yang, G. Jaramillo, P. Taheri-Tehrani, P. Fond, H. Nanjar, K. Yamazaki, L. Lin, D. Horsley, *Proc. IEEE Transducers 2013*, pp. 2415–2418, 2013.

[6] J. D. Griffiths, *Introduction to Electrodynamics,* Saddle River: Prentice Hall, 1999.

[7] Q. P. Unterreithmeier, E. M. Weig, J. P. Kotthaus, *Nature*, Vol. 458, pp. 1001-1004, 2009.

[8] K. H. Lee, T. G. McRae, G. I. Harris, J. Knittel, W. P. Bowen, *PRL*, Vol 104, p. 123604, 2010

[9] S. Aoyagi, Y. C. Tai, *Proc. IEEE Transducers 2003*, pp. 1382-1386, 2003.

[10] J. D. Jackson, *Classical Electrodynamics*, John Wiley & Sons, 1999

CONTACT

*R. Perahia, tel: +1-310-317-5612; rperahia@hrl.com

ELECTROMECHANICAL DAMPING IN MEMS ACCELEROMETERS: A WAY TOWARDS SINGLE CHIP GYROMETER ACCELEROMETER CO-INTEGRATION

Yannick Deimerly[1,2], Patrice Rey[1], Philippe Robert[1], Tarik Bourouina[2], and Guillaume Jourdan[1]

[1]CEA, LETI, MINATEC Campus, 17 rue des Martyrs, 38054 Grenoble Cedex 9, France.
[2]Université Paris-Est, ESIEE Paris, 2 Bd Blaise Pascal, 93162 Noisy-le-Grand, France.

ABSTRACT

This paper reports a method for controlling mechanical damping in MEMS devices. It consists in coupling a micro resonator to an electrical resistance that provides an additional damping source. Quality factor of individual MEMS can then be individually controlled. Quality factor tuning offers an efficient solution to solve the co-integration issue of accelerometer with gyrometer inside a same MEMS cavity under low pressure.

INTRODUCTION

Motion tracking sensors have become a key function in many consumer electronic devices like smartphones, game consoles or tablets. In this context, Inertial Measurement Unit (IMU) development sets stringent constraints like size shrinking to improve sensor integration and to reduce MEMS cost, thus leading multi-axis sensors to be integrated today as System In Package (SIP). At present, IMU requires putting together 3 axis accelerometers, magnetometers with 3 axis gyrometers and pressure sensor on a same chip in order to decrease sensor size down to 3x3 mm² and to provide the consumer with cost-effective solutions. To address MEMS multi-sensors co-integration issue on a single chip along with footprint challenge, a new MEMS technological platform, called M&NEMS, has been proposed. It has been successfully applied to accelerometer [1], magnetometer [2] and gyrometer [3] as depicted in Figure 1. Besides, pressure sensor integration is under investigation [4]. To achieve the above-mentioned goals, the co-integration of quasi-static accelerometers with resonant gyrometers appears to be a simple and straightforward solution. However, as quasi-static accelerometers require quite large energy damping process, they are not compatible with vacuum packaging, and hence, their co-integration with a resonant gyrometer is not straightforward. This work aims at proposing an original approach to address quasi static and resonant sensors co-integration issue on a single chip for SIP. The paper describes a method that implements individual electromechanical damping control [5-6] to place both sensors inside a same cavity under low pressure

CO-INTEGRATION ISSUE

Gyrometer sensitivity is proportional to the driving amplitude of its sensing inertial masses. Gyrometer then requires operating under vacuum to decrease its mechanical energy losses and therefore to achieve high Quality Factor (QF) larger than 1000. High QF enables resonant sensors to generate large motion amplitudes, around 1 μm, at resonance with low actuation voltage around 1 V, which is usually available in ASIC chip. Unlike gyrometers, accelerometers and magnetometers need smaller QF below a few units, in order to keep parasitic motions at resonance as small as possible. Indeed, high QF can cause quasi-static sensors in operation to bump into its mechanical ends or to overload front end amplifier stage, because of its large sensitivity to parasitic signals at their resonance frequency. Operation under vacuum is not suitable, since it would require the overall sensor sensitivity to be decreased, to comply with the resonance phenomenon constraints, whereas the bandwidth of interest is located at low frequency. Thus, sensor signal to noise ratio would not meet usual requirements of consumer market. Another drawback is related to sampling frequency of analog digital converter, which needs to be larger than twice the resonance frequency to avoid aliasing and then to assure sensor the best signal to noise ratio. Oversampling can solve this issue, but it causes larger energy consumption, which is not suitable, for instance, for wireless sensors that transmit raw data directly to a remote base station [7]. Accelerometer reliability has been quoted as an additional motivation [8] to keep QF low, since it causes shock impacts on travel stops to be less damaging for the structural component than with high QF.

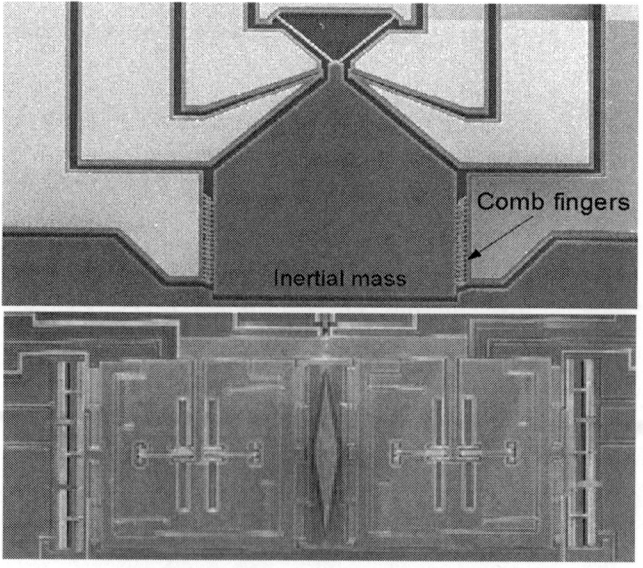

Figure 1: (Top) rotation motion M&NEMS accelerometer with comb fingers used as capacitive coupler. (Bottom) M&NEMS gyrometer.

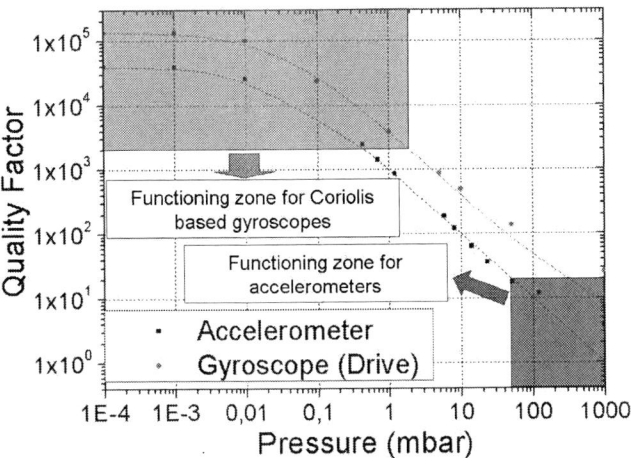

Figure 2: QF as a function of pressure for M&NEMS accelerometers and gyrometers. The green (blue) zone corresponds to the pressure environment that is compatible with gyrometers (accelerometers)

All the above-mentioned points lead us to the co-integration issue of quasi-static and resonant sensors on a same chip. As depicted in Figure 2, putting these two kinds of sensors in a same cavity cannot offer simultaneously proper operating points to each one: their pressure windows are incompatible. Accelerometer requires cavity pressure to be larger than 100 mbar, whereas gyrometer needs to operate below 1 mbar approximately. Multi pressure wafer level packaging (WLP) was proposed at MEMS 2010 [9] to separate sensors in different cavities under their proper pressure level. However it requires bond frame width in the range of 60-100 μm to ensure airtight property between cavities and a getter layer to ensure low pressure in gyrometer cavity: sealing is performed at higher pressure around 100 mbar to be compatible with accelerometer constraints. This work proposes a co-integration solution that allows to put together quasi-static and resonant sensors inside a same WLP cavity under low pressure (below 1 mbar). Thus, neither getter layer nor bond frame between cavities is required. QF control is performed on accelerometer or magnetometer in order to benefit from the same mechanical behavior under vacuum as under ambient pressure.

ELECTROMECHANICAL DAMPING

This section describes a controllable damping source that can decrease the QF of a mechanical resonator by several orders of magnitude with respect to an identical sensor in the same cavity without this damping source. This additional damping source is expected to provide enough damping to comply with the constraints on QF described in the previous section, to allow co-integration of quasi-static accelerometer and resonant gyrometer in the same cavity.

Electromechanical coupling

In order to selectively increase the mechanical damping of the quasi-static sensor, this work proposes to

electrostatically couple the sensor with an electric dissipative circuit. In figure 3, the variable capacitance C tied to the MEMS accelerometer couples the mechanical system to a resistive circuit. MEMS energy transferred to the resistance R is irreversibly dissipated through Joule effect. Physical phenomenon has already been observed and described in the field of Atomic Force Microscope [5] and NEMS [6]. Here the formalism is adapted to M&NEMS accelerometer [1], which exhibits rotational degree of freedom θ in operation. As a result, coupling strength is characterized by the first derivative of capacitance with respect to the rotation angle θ:

$$C' = \frac{\partial C}{\partial \theta}\bigg|_{r,z} \tag{1}$$

MEMS displacement is described by the following equation, where $M_{el} = 1/2\, C' V_C^2$ is an electrostatic torque, J is the moment of inertia, γ_0 is the damping rate and K the rotational stiffness of the MEMS resonator:

$$J\ddot{\theta} + J\gamma_0\dot{\theta} + K\theta = M_{el} \tag{2}$$

Charge Q on capacitance electrodes complies with:

$$RC\dot{Q} + Q = CV_0 \tag{3}$$

V_c and V_0 are the voltage drop across the capacitance and the voltage of DC source respectively. When considering small motion amplitudes $C_0 + \delta C$ at frequency $\omega/2\pi$, Equation (3) provides at first order $Q_0 + \delta Q$ with $Q_0 = C_0 V_0$ and

$$\delta Q = \frac{\delta C V_0}{1 + j\omega R C_0} \tag{4}$$

MEMS motion makes electric current $\dot{Q}$ flow through the resistive circuit. Mechanical energy is then turned into heat as a result of Joule effect. The energy dissipation is expected to reduce the QF of the quasi-static sensor.

Backaction and viscous mechanical damping

Capacitive torque M_{el} applied to the MEMS is affected by the voltage drop across the resistance. Since $V_C = V_0 - R\dot{Q}$, it comes $M_{el} = 1/2\, C'(V_0^2 - 2R\dot{Q} + R^2\dot{Q}^2)$.

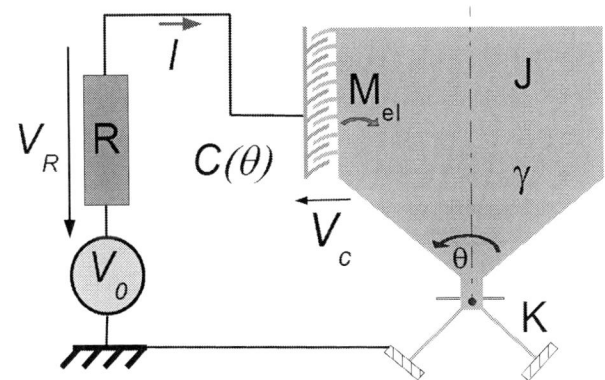

Figure 3: Electromechanical damping scheme. The MEMS is electrostatically coupled to a resistive circuit biased at V_0, thus increasing damping rate by γ_e.

978-1-4799-3510-9/14 $31.00 © 2014 IEEE

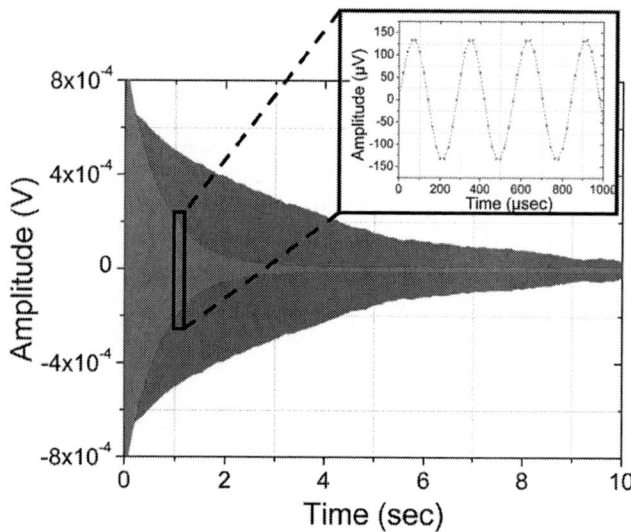

Figure 4: MEMS Response to a force step without (green) and with (red) electromechanical damping. Resonance frequency is 3300 Hz for MEMS sample 1.

At first order in $\delta C=C'\theta$, by considering Equation (4), the electromechanical torque generates an additional damping term in Equation (2). γ_0 is finally switched into:

$$\gamma = \gamma_0 + \frac{RC'^2 V_0^2}{J(1+\omega^2 R^2 C_0^2)} \qquad (5)$$

As energy is irreversibly dissipated in the resistance, damping rate can only increase and QF decrease. In Equation (5), the damping rate has to be evaluated at the resonance frequency $\omega_r/2\pi$, where the damping process regarding resonance phenomenon is the most relevant. It turns out that the resonance frequency of the MEMS needs to be smaller than the cutoff frequency of the RC circuit. At higher frequency, charge variations across the capacitance fall down. Thus no energy is transferred to the resistance and the damping rate is unchanged. Since mechanical behavior is dominated by inertia, whether under vacuum or under ambient pressure, the damping rate decrease has no strong impact on the co-integration solution, here discussed.

EXPERIMENTAL ANALYSIS

Equation (5) shows the larger the resistance, the coupling capacitance and the applied voltage, the lower the quality factor. Capacitance design of MEMS sample used for the study was not optimal, so it cannot benefit from very large damping rate. Nevertheless, resistance and bias voltage have been tuned to experimentally demonstrate electromechanical damping process on M&NEMS accelerometer under vacuum (10 µbar). Response to force steps [10] has shown method capability of decreasing the mechanical response time as depicted in Figure 4. Green curve provides the nominal damping rate of the MEMS set at 0.53 s^{-1} (QF=41800), whereas red curve exhibits a larger value at 2.3 s^{-1} with V_0=10 V and R=9.9MΩ. Damping rate depends on the square of V_0 in compliance with Equation (4) as shown in Figure 5. Accurate experiment to theory

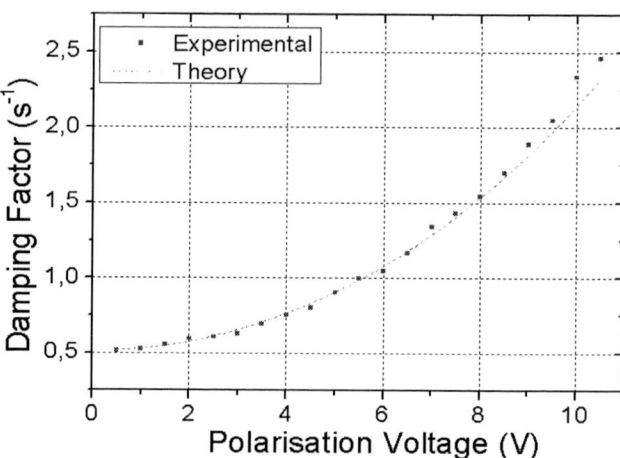

Figure 5: Effective damping rate at resonance $\gamma=\gamma_e+\gamma_0$ as a function of bias voltage V_0 (MEMS sample 1)

comparison can be performed on the basis of the Graph depicted in Figure 6, where both voltage V_0 and resistance R are tuned to change damping rate. Damping is maximum at R=2.25 MΩ due to the cutoff of the RC circuit. Fitting procedure based on the model described in Equation (5) provides a static capacitance C_0=20.4 pF when considering $\omega_r/2\pi$=3530 Hz and a coupling strength C'=1.12 pF/rad for an inertia moment J estimated at 1.07×10^{-16} kg.m^4. Agreement with the circular comb capacitance model of the variable capacitance, C_{th}'=1.14 pF/rad, is within the uncertainty range of MEMS thickness parameter t set at 0.5 µm (t=12 µm). The total capacitance C_0 is dominated by parasitic capacitances, around a few tens of pF, located between the MEMS and the off chip resistance: for the coupling capacitance, 50 fF are only expected. Figure 7 displays MEMS QF=ω_r/γ as a function of pressure under a bias voltage of V_0=10 V. The present experimental configuration clearly shows the method capability of controlling QF up to 0.1 mbar. Below this limit, the QF does not depend on pressure anymore.

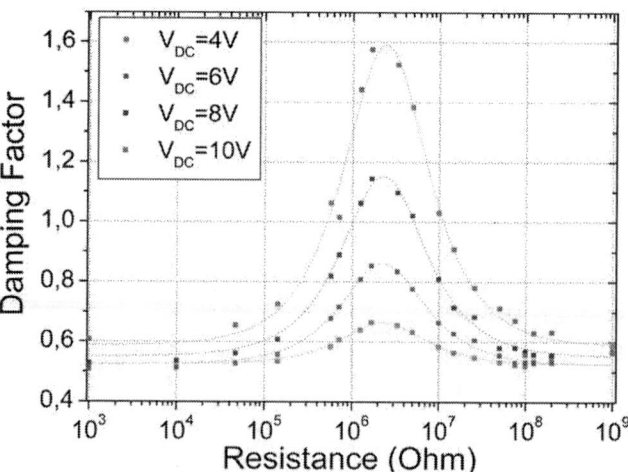

Figure 6: Damping rate $\gamma=\gamma_e+\gamma_0$ in s^{-1} as a function of resistance R and bias voltage $V_{DC}=V_0$ (MEMS sample 2).

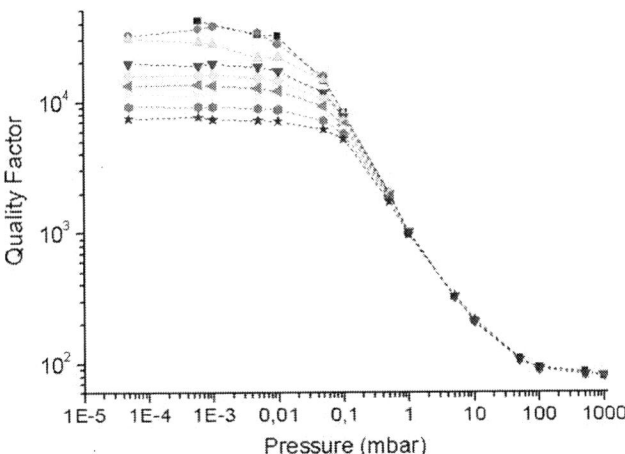

Figure 7: QF as a function of pressure and bias voltage. QF decreases when V_0 goes from 0 up to 10 V (sample 2)

CO-INTEGRATION PROSPECT

Based on the electromechanical damping model here described, a new MEMS accelerometer that can reach QF down to a few tens can be designed as shown in figure 8. A tradeoff between resistance, capacitance and bandwidth need to be addressed: to maximize the effect, R and C' have to be increased, while C_0 has to be kept as low as possible. To that purpose, the resistance R can be etched on chip close to the coupling capacitance, using low doping level silicon layer. One set of comb capacitance occupies approximately an area of 150 μm x 15 μm: it can be further slightly reduced. By implementing 10 comb capacitances on a dedicated MEMS design and by setting the on chip resistance at 200 MΩ, the damping rate can be improved by a factor 10000 with respect to the situation described in the present paper (as RC'^2). The coupling capacitance can be reduced to 20 fF for each capacitance set. As a result, the cutoff frequency of the RC circuit should remain around 3500 Hz so as not to reduce the electromechanical damping rate. QF below 1 can be expected with this method, under vacuum and a bias voltage of 10 V. QF in the range of 1-10 can be reached under 3 V and appears to be acceptable to address the co-integration issue of quasi-static sensor with resonant sensor inside a same MEMS cavity.

CONCLUSION

Mechanical response controlling greatly improves MEMS sensors operation: here, it makes it possible to co-integrate gyrometers and accelerometers inside a same cavity. Such control can be applied on accelerometer in order to selectively decrease its QF whatever environment pressure.

In contrast with force feedback approach [11], the electrical circuit does not require additional external electronics. As it can be shown with Equation (4) and (5), $< \dot{\gamma} \dot{x}^2 > = < RI^2 >$. On operation, energy consumption of electrical circuit corresponds to the energy drained from the mechanical system to the electric resistance through the

electromechanical damping effect. The system can then be considered as passive: it requires only an external bias voltage to load the capacitance. Comb capacitance geometry is suitable to reduce negative stiffness induced by the biasing voltage on the MEMS. Frequency shift was below 3% with V_0=10 V for the experiment presented in this paper. MEMS design then requires to keep the second derivative of the capacitance C'' as low as possible in order to ensure the mechanical stability of the system.

ACKNOWLEDGEMENTS

We thank A. Berthelot, F. Delaguillaumie, S. Verrun and H. Blanc for MEMS samples processing and preparation. This work was supported by the E.U. under grant number 288318 (FP7 NIRVANA project).

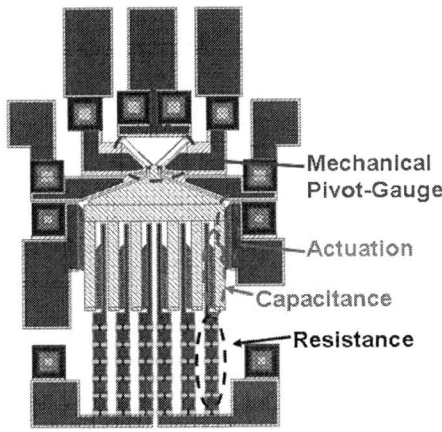

Figure 8: Proposed accelerometer design that integrates on chip resistances and comb capacitance to increase the coupling strength (MEMS dimensions about 300 x 400 μm²)

REFERENCES

[1] P. Robert et al, *Proceedings of IEEE Sensors 2009*, Christchurch, October 25-28, 2009, pp. 963-966.

[2] D. Ettelt et al, *Proceedings of IEEE Sensors 2011 Conference*, Limerick, October 28-31, 2011, pp. 2010-2013.

[3] A. Walther et al, *Proceedings of MEMS 2012 Conference*, Paris, January 29- February 2, 2012, pp. 480-483.

[4] Y. Deimerly et al, *Proc. Transducers 2013*, to be published

[5] G. Jourdan et al, *Nanotechnology*, 2007, Vol.18, Issue 47, (475502).

[6] T. Barois et al, *Physical Review B*, 2012, vol. 85, no. 7.

[7] O. Leman et al, *Proceedings of IEEE Sensors 2011*, pp 1429-1432.

[8] M. Dienel et al, *Vacuum*, 86(5), 2012, pp. 536-546.

[9] P. Merz, et al, *Proc. of MEMS 2010*, pp. 488.

[10] G. Langfelderet al, *Sensors and Actuators A* Vol.148 (2) pp.401–406, 2008.

[11] M. Yücetaş et al, *European Solid-State Circuits Conference 2011*, pp. 291.

978-1-4799-3510-9/14 $31.00 © 2014 IEEE

EXPERIMENTALLY VALIDATED ALUMINUM NITRIDE BASED PRESSURE, TEMPERATURE AND 3-AXIS ACCELERATION SENSORS INTEGRATED ON A SINGLE CHIP

Fabian Goericke[1], Kirti Mansukhani[1], Kansho Yamamoto[2] and Albert Pisano[3]

[1] University of California, Berkeley, USA
[2] Murata Manufacturing Co., Ltd., Kyoto, Japan
[3]University of California, San Diego, USA

ABSTRACT

This paper reports a unified fabrication process used to build multiple Aluminum Nitride (AlN) based micro-electromechanical system (MEMS) sensors on a single chip. A fully functional AlN-based sensor cluster has been demonstrated and is presented in this paper. This sensor cluster is a "five degree-of-freedom" cluster; it measures 3-axis acceleration, temperature and pressure fabricated on a 1 cm x 1 cm die. In addition to utilizing AlN as both the structural and active layer of the sensors, this work is novel because all sensors are fabricated in the same fabrication run.

INTRODUCTION

As MEMS sensor applications have grown from imaging, gaming and smart phones to smart appliances, fitness and health monitoring, there is a growing need for MEMS products to evolve from current SIP (System-in Package) devices to SoC (System-on-Chip) devices to add more functionality at smaller size and cost. Device designs and fabrication processes that support multi-sensor chips are an important step in this direction. The primary advantages of multi-sensor chips are lower cost, smaller footprint, lower power consumption, simpler packaging, and improved performance when compared with their discrete component counterparts. While some silicon based sensor clusters have been developed [1], similar efforts for AlN-based devices seem to be limited. Devices based on the inert, high melting point material AlN can withstand harsh environments including temperatures above 500 °C, high pressures, or reactive media.

SENSOR FUNCTIONAL PRINCIPLES

To integrate the different sensor components into one system, a technology based on resonating AlN sensing elements was selected. Utilizing the piezoelectric properties of AlN enables the excitation and measurement of the changing resonant frequencies.

The principle of operation of the accelerometer lies in the change in resonant frequency of a piezoelectric tuning fork with applied inertial force from a proof mass. The resonant frequency of the tuning forks is extremely big (hundreds of kHz) compared to traditional accelerometers resulting in an improved bandwidth of operation. Using symmetric devices in a differential configuration allows for cancellation of common mode signals, such as temperature and substrate stress. The differential elements are opposing

each other and their frequency changes due to acceleration are equal in magnitude and opposite in sign.

The resonant frequency of a membrane over a sealed cavity changes under the influence of a differential pressure across it. This phenomenon is used as the transduction scheme for the pressure sensors on this chip. In order to compensate for temperature effects on the resonant frequency, identical reference sensors are included on every chip that are not subjected to differential pressure. The measurements from the reference sensor can then be used to correct the pressure sensor measurements for temperature.

The temperature sensors built in this sensor cluster are resistive type and utilize the fact that resistance of a metal (in this case Molybdenum (Mo)) changes linearly with temperature. For this application, Mo was used because it provides good linearity over the required temperature range and for ease of fabrication at the Berkeley Nanofabrication laboratory.

FABRICATION PROCESS

An innovative fabrication process using AlN both as the structural and as the active layer of the sensors in the sensor cluster has been developed. The process includes a two-step backside etch of the wafer, which can create very large proof masses for improved acceleration sensitivity. A dual layer AlN process is used to create bimorph structures that can bend out of plane. The resulting triple-beam tuning forks [2] have advantages in electro-mechanical coupling and acceleration sensitivity compared to conventional double-ended tuning forks [3, 4]. The fabrication process is detailed in the following sub-sections.

Photo Mask Design

The fabrication process utilizes 5 photolithography steps. The first three steps are surface micromachining steps on the front side of the wafer and the last two steps are bulk micromachining steps from the backside of the wafer. Projection lithography is used for photoresist (PR) patterning for the three surface micromachining steps. The masks for these steps are shown in Figure 1. The features that are transferred into the AlN/Mo layer stack are shown in brown, the features that are transferred into the top electrode Mo are shown in green, and the features that are etched through the top AlN layer to open contact pads to the bottom Mo layer are shown in blue.

Contact lithography is used for creating the PR patterns on the backside of the wafer. The two photomasks for the

978-1-4799-3510-9/14 $31.00 © 2014 IEEE

bulk micromachining are shown in Figure 2. The photomask on the left side of the figure is used to create the silicon dioxide (SiO₂) hard mask on the wafer backside. These patterns remain for the whole duration of the deep silicon backside etching. The photomask on the right side of the figure shows the areas covered by the PR mask for the first portion of the deep silicon backside etching only. Areas that are open in both masks are etched all the way through to the bottom AlN layer. Areas that are covered by the right mask, but not by the left mask get partially etched and create proof masses for the accelerometers that are thinner than the silicon wafer thickness.

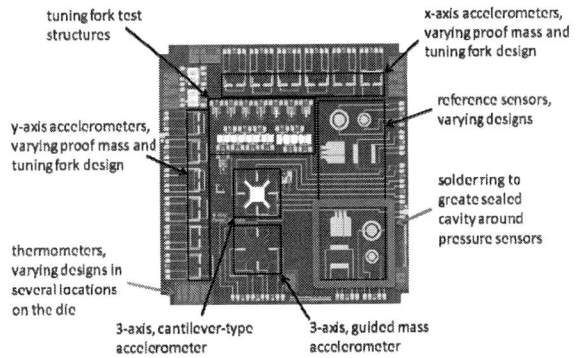

Figure 1: Die-level Mask layout showing Sensor Cluster.

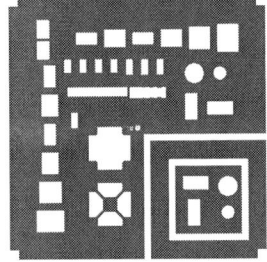

Figure 2: Photomask designs for backside bulk micromachining steps; (L): patterns transferred into the SiO₂ hardmask, (R): patterns transferred into the PR mask to create proof masses.

Process Flow

The fabrication process (depicted in Figure 3) begins with the deposition of the active thin film materials on double-sided polished silicon 15 cm-wafers. The following films are deposited in order: 1 μm AlN, 100 nm Mo, 1 μm AlN, 100 nm Mo, 7.5 nm titanium (Ti).

The two AlN layers form the bi-layer stack that enables out-of-plane motion of the devices when one of the layers is actuated independently from the other. In this case, the top AlN layer is the active layer that is actuated and sensed. The first Mo layer serves as a continuous bottom electrode to the active AlN layer. It is not patterned specifically, but gets etched in a self-aligned manner along with the AlN structure. The top Mo layer is a patterned metal layer that allows the formation of areas specifically for piezoelectric driving and sensing. The Ti film promotes the adhesion of the PR on the top layer.

Subsequent to the thin film deposition, PR is spun on the wafers, and patterned via projection lithography. The PR is then hardbaked (heat and UV). The PR patterns are

transferred into the top Mo layer using reactive ion etching (RIE) with good selectivity to the underlying AlN film. The PR is then removed and a short blanket etch is performed to remove the thin layer of Ti on the Mo electrodes.

PR is patterned as a continuous film with small square opening in selected places. It is hardbaked and used as a mask for wet etching the top AlN film. An ammonium hydroxide dip followed by a heated phosphoric acid bath is used to etch the AlN with good selectivity to the underlying Mo film. After removal of the PR, a 1.85 μm film of SiO₂ is deposited on the wafer. This film is used for two purposes. On the front side of the wafer, it is used as a hardmask for the patterning of the AlN devices. On the backside of the wafer it is used as a hardmask for etching the silicon to create trenches and proof masses.

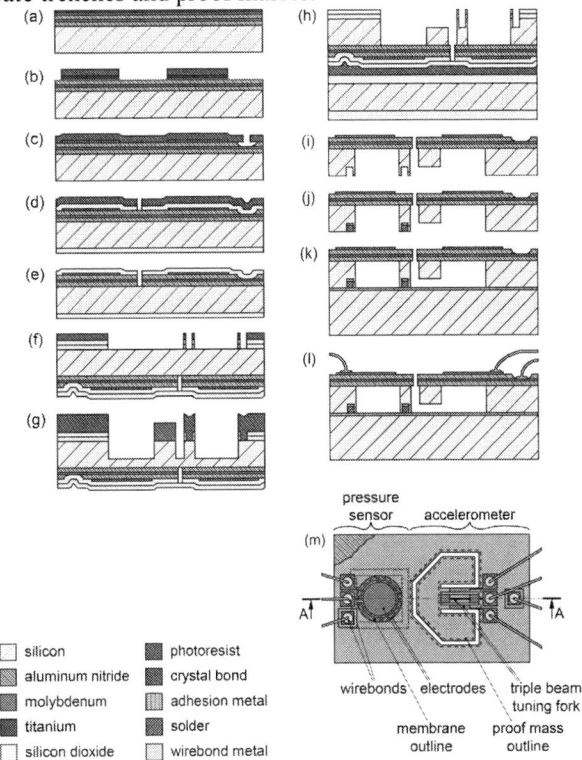

Figure 3: Abridged fabrication process for Sensor Cluster.

PR is patterned via projection lithography in the shape of the device outlines (tuning forks, etc.). The patterns are aligned to the top electrode features already present on the wafers. This is the most critical alignment step in the fabrication process as even small misalignments (greater than about 0.25 μm) can greatly affect the device performance. Areas that are especially critical are the small current traces on the out-of-plane tuning fork. However, the requirement is somewhat more relaxed than it would have been for in-plane tuning forks, as those devices would require much narrower beam widths. After PR patterning and hardbaking the pattern is transferred into the SiO₂ via RIE. Then the PR is removed in an oxygen plasma and the pattern is transferred from the SiO₂ hardmask into the AlN/Mo/AlN layer stack via another RIE process. This two-step etching process, which first creates a hardmask and then etches the device layers has been found to produce better feature reproduction and steeper sidewalls than using a single PR mask alone.

PR is patterned on the backside of the wafer via contact lithography. The features are aligned to the top electrode patterns on the frontside of the wafer via backside alignment. The PR is then hardbaked and the pattern is transferred into the SiO_2 layer via RIE. The PR is removed. The resulting SiO_2 patterns are the etch mask for the second deep silicon etch step on the backside of the wafer.

Thick (6 µm) PR is deposited on the backside and patterned via contact lithography. The alignment is performed with respect to the SiO_2 patterns already on the backside. The hardbake of this PR layer is performed in an oven for an 120 minutes at 120 °C (without UV light). Then deep silicon etching (Bosch process) is employed to transfer the PR pattern into the silicon wafer. After a period of time the PR is consumed and only the SiO_2 acts as a masking layer for the remainder of the etching. By using two masking materials together at first and only one mask in the end, two different etch depths are achieved. This process results in the through etch to the AlN film in some areas, while leaving silicon blocks in other areas that are used as proof masses for the accelerometers. When about 400 µm of etch depth is achieved the etching is interrupted to bond the wafer to a handle wafer.

A handle wafer is prepared by depositing about 4 µm of SiO_2 on it in a low temperature chemical vapor deposition (CVD) process. The handle wafer is heated and "crystal bond" is melted on its surface. The device wafer is then placed on the handle wafer and the wafer stack is cooled down slowly. The "crystal bond" forms a solid bond with good thermal conductivity once it hardens upon cooling.

Deep silicon etching is performed until silicon is removed in all desired areas. The device wafer is detached from the handle wafer by submersion in acetone. Since the deep silicon etching is also used to separate the individual sensor cluster chips from each other, the chips are now floating separately in the acetone. The individual chips are picked out of the acetone and transferred to a secondary acetone bath, then to an isopropanol (IPA) bath. The chips are removed from the IPA bath and put in a drying area. Once the IPA has evaporated, gaseous hydrofluoric acid (HF) is used to remove the SiO_2 layer from both front and back sides of the chips. The chips are then placed in "gelpak" chip carriers for temporary storage. All subsequent processing steps are on an individual chip basis.

A stack of tin and gold is deposited on the backside of the device chip and on a carrier chip. A solder preform is manually deposited on the device chip and the device chip is placed on the carrier chip. The chip stack is placed on a hotplate to melt the solder and create a permanent bond that creates the hermetic air cavity underneath the pressure sensors. The chip stack is then glued into a ceramic carrier and the desired devices are wire-bonded to the ceramic carrier's contacts. The ceramic chip carrier creates the interface to printed circuit boards.

TESTING

One device each for the pressure, temperature and inertial sensors have been tested so far and the results have been summarized in the following sections.

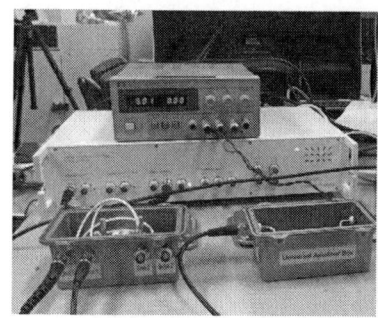

Figure 4: (L) Test Setup for actuation and sensing of devices. (R) Tilt setup for accelerometer testing

Each of the inertial and pressure sensors are operated as a 2-port piezoelectric resonator. Each channel is operated in a phase-locked loop (PLL) setup. The output of the PLL is connected to the drive port of the channel. The sense port of the channel is connected to a pre-amplifier that consists of a transimpedance amplifier to convert the output current into a voltage and a second stage that amplifies this voltage. The output of the pre-amplifier is phase-shifted and fed into the feedback port of the PLL. The PLL in conjunction with the phase shifter keeps the phase shift between the drive and sense ports of the sensor constant and thus lock to a specific pre-defined resonance frequency. The Zurich Instruments HF2LI measurement tool combines the functions of the phase shifter, PLL and frequency counter and is convenient to use for laboratory testing. The experimental setup is shown in Figure 4.

The accelerometer is operated at the resonant peak close to 460 kHz. The phase shifter for the subsequent measurements is set such that a phase relation of 0° is kept between sensor drive port and pre-amplifier output.

The same setup procedure was repeated for Channel B before the sensitivity characterization of the devices was performed. For the acceleration sensitivity testing, the sensor chip was mounted inside a rotatable enclosure with a reference tilt gauge (Figure 4). By turning the enclosure around a fixed axis of the sensor chip the acceleration on the sensor chip can be varied from +1G to -1G.

The operating frequency of the pressure sensor and reference sensor membranes is around 30 kHz. The sensor cluster chip was mounted in a vacuum chamber. The vacuum chamber was pumped down to about 0.6 atm and brought back to atmospheric pressure by stepwise filling with nitrogen. At each pressure level, the frequencies of the pressure and reference sensor were monitored.

For temperature testing, the sensor was mounted inside of a TPS Tenney environmental test chamber and connected to a Wheatstone bridge circuit.

RESULTS

Scanning electron microscope (SEM) images of each of the major sensors fabricated and tested as part of the sensor cluster are shown in Figure 5. The resonant accelerometers operate at about 460 kHz. Figure 6 shows that the y-axis acceleration sensitivity is 8708 Hz/G with good linearity in the measured range. The cross-axis rejection factor to the x-axis is about 100. The z-axis acceleration sensitivity is 7300 Hz/G (Figure 7). The pressure sensor has an operating frequency of about 30 kHz and a sensitivity of 250 Hz/psi. The reference sensor is significantly less sensitive to pressure changes than the pressure sensor (Figure 8), which enables it to be used for temperature compensation. The temperature sensor exhibits good linearity and a temperature sensitivity of 1.625 mV/°C (Figure 9).

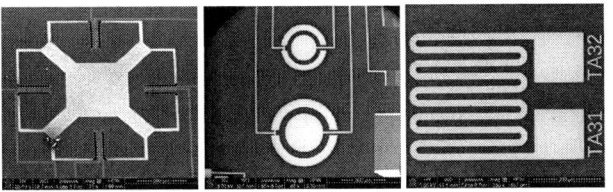

Figure 5: (L) SEM of 3-axis Accelerometer with 4 proof masses, each connected to the substrate by a tuning fork sensing element; (M) SEM of Bulk membrane pressure sensors; (R) SEM of Temperature sensor

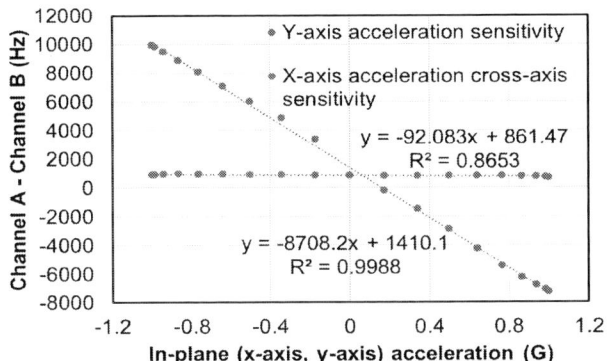

Figure 6: Measured Y-axis sensitivity and x-axis cross-axis sensitivity of 3-axis accelerometer.

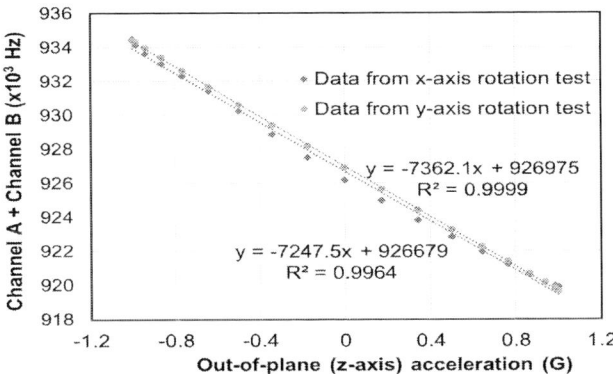

Figure 7: Measured Z-axis sensitivity 3-axis accelerometer.

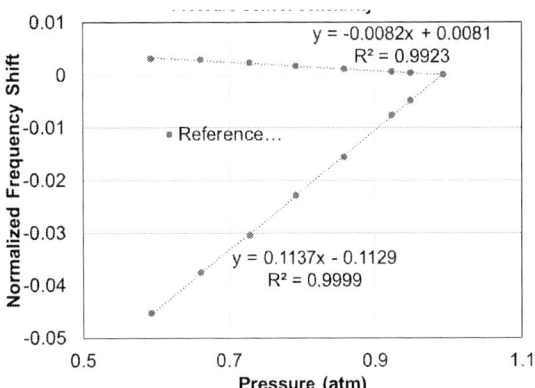

Figure 8: Measured Pressure Sensor Sensitivity.

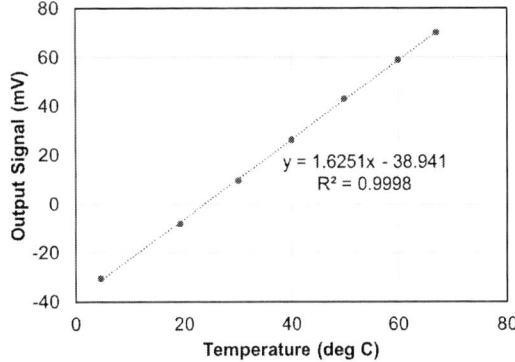

Figure 9: Measured Temperature Sensor Sensitivity.

CONCLUSIONS

A unified fabrication process used to build a fully functional "five degree-of-freedom" AlN based MEMS sensor cluster has been presented in this paper. The sensor cluster has been used to measure 3-axis acceleration, temperature and pressure and the preliminary experimental results have also been presented.

REFERENCES

[1] Xu Y., Chiub C-W., et al, "A MEMS multi-sensor chip for gas flow sensing", *Sensors and Actuators A* 2005, 121, pp. 253–261

[2] Goericke, F. T., G. Vigevani, et al., "Bent-beam sensing with triple-beam tuning forks", *Applied Physics Letters 2013*, 102(25): 253508, pp. 1-4.

[3] Olsson, R. H., K. E. Wojciechowski, et al., "Post-CMOS-Compatible Aluminum Nitride Resonant MEMS Accelerometers", *Journal of Microelectromech. Systems* 2009, 18(3): pp. 671-678.

[4] Vigevani, G., F. T. Goericke, et al., "Microleverage DETF Aluminum Nitride resonating accelerometer", *IEEE International Frequency Control Symposium (FCS)* 2012

CONTACT

K. R. Mansukhani, tel: +1-617-7805282; kirti@berkeley.edu

FABRICATION AND CHARACTERIZATION OF ALL HYDROGEL CANTILEVERS FOR ATOMIC FORCE MICROSCOPY APPLICATIONS

Il Lee and Jungchul Lee[*]

Department of Mechanical Engineering, Sogang University, Seoul, SOUTH KOREA

ABSTRACT

This paper reports a novel method for fabricating hydrogel based microcantilevers by using dynamic mask lithography. A hydrogel, polyethyleneglycol diacrylate (PEGDA), was introduced between two parallel polydimethylsiloxane (PDMS) guides then cured with ultra-violet (UV) exposure to intended shape and size defined by the dynamic mask; an image sent from a PC to a liquid crystal display projector. One PDMS guide has an embedded glass piece which serves as a handle for the microcantilever and the other guide is with or without an inverted pyramid tip mold to fabricate tip-integrated or tipless microcantilevers, respectively. After fabricated hydrogel microcantilevers were thoroughly characterized by using a stylus profilometer and an atomic force microscope (AFM), they were employed for both contact and non-contact mode AFM imaging. In case of non-contact mode, the imaging performance of hydrogel AFM cantilevers was comparable to that of commercial silicon AFM cantilevers.

INTRODUCTION

Microcantilever sensors have been typically fabricated with rigid materials such as silicon, silicon nitride or silicon dioxide. Since microcantilevers fabricated with those materials have high stiffness, their bending responses to small force or stress are generally limited. Moreover, such microcantilevers made from conventional microfabrication materials require surface coating or functionalization for specific biological or chemical sensing applications. Therefore, direct fabrication of microcantilevers with soft and responsive materials would be beneficial to specific applications. One of the most promising candidate materials is hydrogel considering manufacturability [1], stimulus-responsivity [2], and biocompatibility [3]. Hydrogels mixed with a photoinitiator are easily curable upon exposure to a UV source. Hydrogels sensitively respond to a variety of stimuli such as humidity, pH, UV, heavy metal ions, and temperature due to their swelling behaviors. Biocompatibility of hydrogels enables various biological applications including cell culture, drug delivery, and tissue engineering.

Recently, fabrication of all hydrogel structures has been reported for sensor and actuator applications. Photoresponsive hydrogel microcantilevers have been fabricated by two-photon initiated polymerization [4] and bioactuators have been demonstrated with hydrogel cantilevers fabricated by micro-streolithography [5]. In general, both two-photon polymerization and micro-streolithography are slow due to the scanning of focused lasers. Moreover, it is difficult to control the thickness and surface quality of fabricated structures due to the finite penetration depth and fluctuating power of the laser. Therefore, hydrogel cantilevers previously demonstrated were relatively thick and neither flat nor smooth. To extend applications of hydrogel microstructures, there is an urgent need to develop a fast and improved fabrication method.

This paper presents a simple and fast fabrication method of hydrogel microcantilevers which are thin, flat, and smooth by employing the dynamic mask lithography which has been mostly applied to particle synthesis in microfluidic platforms [6] and fabrication of three dimensional structures [7].

FABRICATION

Figure 1 shows our custom dynamic mask lithography setup. The experimental setup was constructed with a high power light emitting diode (405 nm and 10 W), a liquid crystal display projector, a 3-axis translation microstage, and various modular optical components including beam splitters, polarizers, a diffuser, and an objective lens. The optical system controls shapes and planar dimensions of hydrogel microcantilevers and the 3-axis microstage controls the thickness of hydrogel microcantilevers. The hydrogel used was PEGDA. The PEGDA was mixed with a photoinitiator, phenylbis (2, 4, 6-trimethylbenzoyl) phosphine oxide, at the weight ratio of 99:1, then, magnetically stirred for 24 hrs. All materials were purchased from Sigma-Aldrich and used as received.

Figure 2 shows fabrication processes for hydrogel

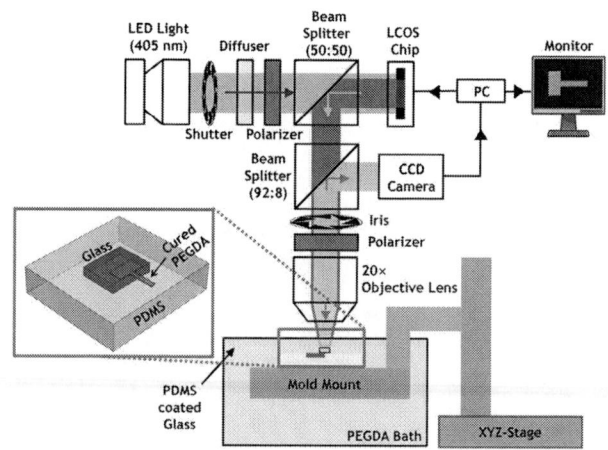

Figure 1: A schematic of the experimental setup which is constructed with a high power light emitting diode, a reflective type liquid crystal display (liquid crystal on silicon; LCOS) projector, a 3-axis translation microstage, and various modular optical components.

978-1-4799-3510-9/14 $31.00 © 2014 IEEE

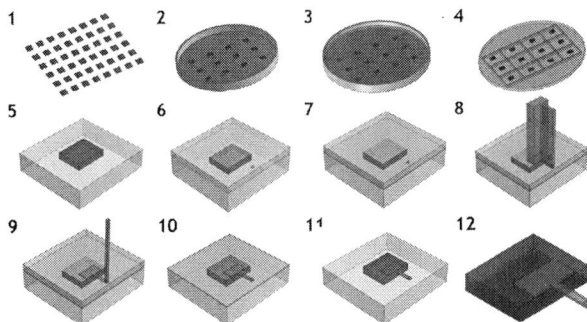

Figure 2: (1~4) Preparation steps for glass-embedded PDMS blocks and (5~12) fabrication processes for scanning tip integrated hydrogel microcantilevers.

microcantilevers. First, glass-embedded PDMS blocks was prepared by curing the PDMS with small glass pieces embedded (step 1~4). An individual glass-embedded PDMS block (step 5) was placed on the mount attached to the 3-axis microstage and a PDMS coated glass cover was placed on top of the glass-embedded PDMS block with a variable separation gap (approximately 20 ~ 70 μm). The PDMS layer on the glass cover was prepared with or without an inverted pyramid tip mold for tip-integrated or tipless microcantilevers, respectively (the PDMS layer with the tip mold is shown in step 6). The prepared PEGDA was then introduced into the gap between the glass-embedded PDMS and PDMS coated glass cover (step 7). The dynamic mask was a union of a large rectangle and a small cantilever part (either rectangle or V-shape). During exposures using the dynamic mask, the large rectangle and small cantilever part were placed over the glass and PDMS area in the glass-embedded PDMS block, respectively. In addition, the side of the large rectangle where the small cantilever part was connected was aligned with one side of the glass piece embedded in the PDMS. Upon the first exposure defining free-standing beam and anchoring structures on top of the glass piece (step 8), the large rectangle was securely bonded to the glass piece but the cantilever part was not attached to the PDMS due to the oxygen inhibition layer [8] exclusively near the PDMS. The second exposure confined at the tip mold (step 9) was only for tip-integrated microcantilevers thus skipped for tipless microcantilevers. The top cover was removed (step 10) and uncured PEGDA was washed away by gently applying an isopropyl alcohol (IPA) (step 11). Finally, the hydrogel microcantilever attached to the glass handle was taken out of the PDMS block (step 12).

RESULTS AND DISCUSSION

First, tipless microcantilevers were fabricated with PEGDA Mw 250 and Mw 575 and their mechanical properties were investigated by using a stylus profilometer [9] (Dektak XT, Bruker). Fabricated microcantilevers were 350 μm long, 50 μm wide, and 20 μm (PEGDA Mw 250) or 40 μm (PEGDA Mw 575) thick. Thicknesses of hydrogel

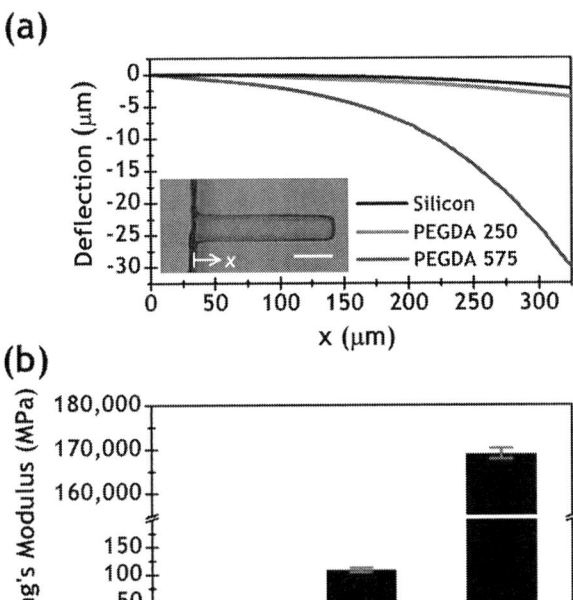

Figure 3: (a) Measured deflection of tipless rectangular microcantilevers made from silicon, PEGDA Mw 250, and PEGDA Mw 575. The constant loading forces are 29.4 and 9.8 μN for silicon and hydrogel, respectively. The inset in (a) shows a hydrogel microcantilever made from PEGDA Mw 250. The scale bar indicates 100 μm. (b) Elastic modulus of each material extracted from the bending experiment using the stylus profilometer.

microcantilevers were chosen to limit their deformation considering the minimum loading force of the stylus profilometer used (9.8 μN). A commercial silicon microcantilever which is 500 μm long, 50 μm wide, and 5 μm thick (Octo 500D, Micromotive) was also tested for validation. After the stylus made a contact near the clamping edge of each microcantilever with a constant loading force (9.8 and 29.4 μN for hydrogel and silicon microcantilevers, respectively), it scanned over the free standing region. As the stylus moved farther away from the clamping edge, the microcantilever bent more. Measured deflection was compared with the beam bending theory to extract the elastic modulus.

Figure 3(a) shows measured deflection of each microcantilever and Figure 3(b) shows the elastic modulus of each material obtained by comparing beam bending measurement and theory. Extracted elastic moduli of PEGDAs and silicon are E_{Mw250}~108.3±3.59, E_{Mw575}~9.8±0.34 MPa and E_{Si}~169±1.54 GPa, respectively. The average value for silicon is the exactly same as the known modulus of the silicon in the [100] crystalline direction [10]. This validates the measured elastic moduli of PEGDAs.

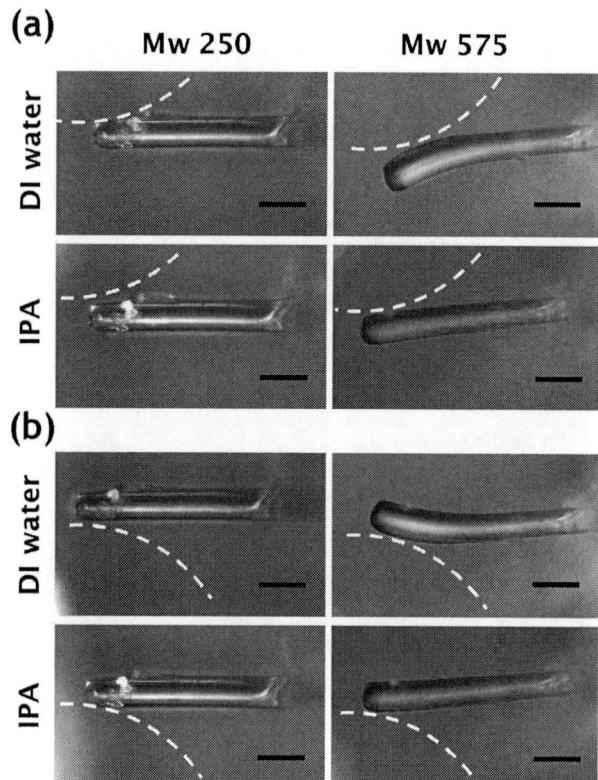

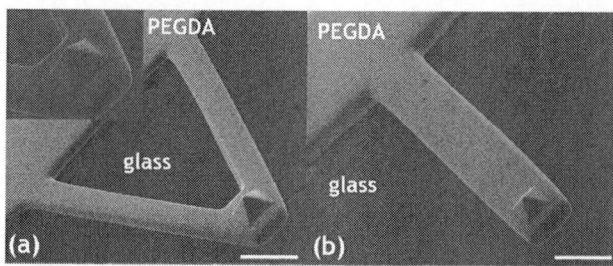

Figure 5: Scanning electron micrographs of scanning tip integrated (a) V-shaped and (b) rectangular hydrogel microcantilevers. The inset in (a) shows a zoom-in view near the scanning tip. All scale bars indicate 50 μm.

investigated. First, the force-displacement curve was obtained by deflecting a rectangular hydrogel microcantilever (350 μm long, 65 μm wide, and 50 μm thick) against a glass substrate in a commercial AFM (NX10, Park Systems) as shown in Figure 6(a). Then, the spring constant of the hydrogel microcantilever was measured to be 23.0 N/m by the thermal method [11]. Figure 6(b) shows the amplitude and phase spectra of the hydrogel microcantilever. The first and second flexural bending modes were found at 46.7 and 272.2 kHz, respectively. Besides these eigenmodes, there were several peaks in the amplitude spectrum which are

Figure 4: Swelling of rectangular hydrogel microcantilevers upon partial wetting with DI water or IPA droplets from (a) the top or (b) bottom surface of hydrogel microcantilevers. White dashed lines indicate the position of each droplet which is removed immediately after wetting. All scale bars indicate 75 μm.

Then, swelling behaviors of hydrogel microcantilevers were investigated by wetting surfaces of rectangular hydrogel microcantilevers (350 μm long, 50 μm wide, and 30 μm thick) with deionized (DI) water or IPA droplets. Figure 4 shows that swelling of hydrogel microcantilevers depends on the wetting direction, the hydrogel molecular weight, and the applied solvent. Images were taken immediately after the wetting droplet (~150 nL) was removed away from the microcantilever surface. Once the hydrogel wets, it swells thus makes the microcantilever bend toward the direction opposite to the wetting. PEGDA Mw 575 swells more than PEGDA Mw 250 does and DI water induces more swelling than IPA does. Therefore, PEGDA with lower molecular weight (e.g. Mw 250) is preferred when the cantilever structure should stay flat with minimum swelling in a high relative humidity or even in liquid. The swelling induced bending was recovered once hydrogel microcantilevers were dried. This took about 10 sec in ambient condition.

Next, tip-integrated hydrogel microcantilevers were fabricated with PEGDA Mw 250 for AFM applications. Figure 5 shows scanning electron micrographs of V-shaped and rectangular tip-integrated hydrogel microcantilevers.

Prior to AFM imaging, mechanical properties were

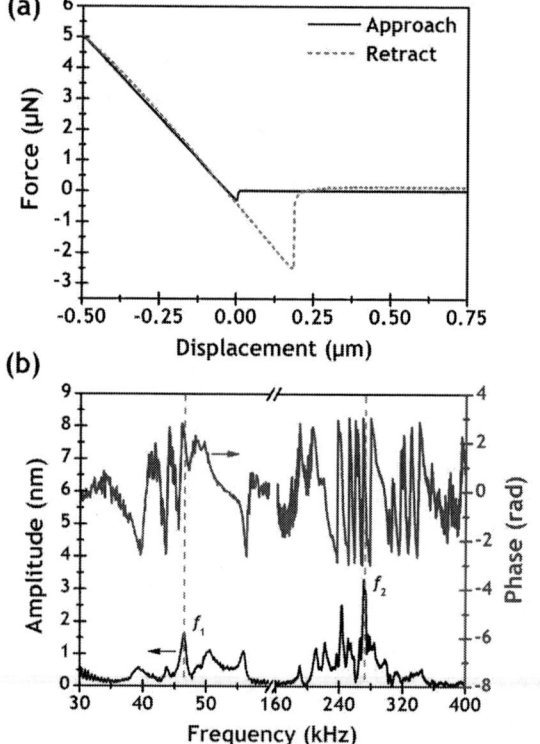

Figure 6: (a) Force-distance curves of the hydrogel microcantilever (b) Amplitude and phase sprectra of a hydrogel microcantilever. Amplitude and phase sprectra of the hydrogel microcantilever. Fundamental (f_1) and second (f_2) flexural bending modes are found at 46.7 and 272.2 kHz, respectively.

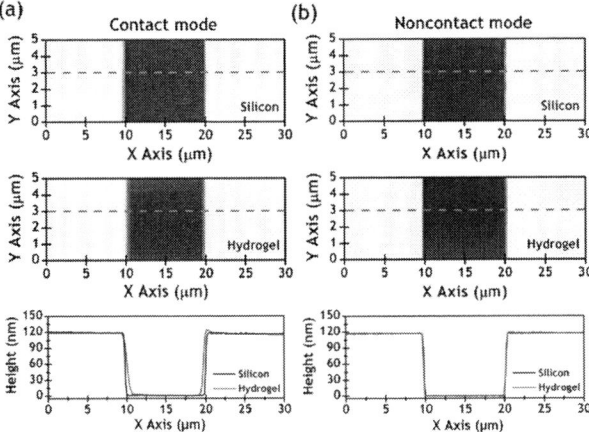

Figure 7: (a) Contact mode imaging for the calibration grating using commercial silicon contact mode and fabricated hydrogel microcantilevers (b) Noncontact mode imaging for the calibration grating using commercial silicon noncontact mode and fabricated hydrogel microcantilevers. Line scan data are for dashed red lines in topographic contours.

unique characteristics of hydrogel microcantilevers.

Then, the characterized microcantilver was employed to image a calibration grating of which average topographic height is 116 nm. Figure 7 shows contact and noncontact mode imaging results for the calibration grating obtained with fabricated hydrogel and commercial silicon microcantilevers. While one hydrogel microcantilever was used for both imaging modes, two silicon microcantilevers dedicated for each imaging mode were used. In contact mode, the imaging performance of the silicon microcantilever is superior to that of the hydrogel microcantilever. Encouragingly, however, the hydrogel microcantilever shows performance comparable to the commercial silicon microcantilever in case of noncontact mode imaging.

CONCLUSIONS

We developed a novel method for fabricating hydrogel based microcantilevers by using dynamic mask lithography and investigated mechanical properties of fabricated hydrogel microcantilevers for AFM applications. Thoroughly characterized hydrogel microcantilevers were employed for both contact and non-contact mode AFM imaging on a calibration grating. Encouragingly, the imaging performance of hydrogel microcantilevers was comparable to that of commercial silicon AFM cantilevers in case of the noncontact mode. Besides the topographic imaging, hydrogel microcantilevers may be useful for other AFM applications such as force spectroscopy on biological samples and swelling mediated delivery of biological and chemical matters from the tip to substrates.

ACKNOWLEDGEMENTS

This research was supported by Basic Science Research Program through the National Research Foundation of Korea

(NRF) funded by the Ministry of Science, ICT and Future Planning (NRF-2011-0012942) and by National Research Foundation Grant funded by the Korean Government (NRF-2011-220-D0014).

CONTACT

*Jungchul Lee, Dept. of Mechanical Engineering, Sogang University, Tel: +82-2-705-7973; Fax: +82-2-712-0799; E-mail: jayclee@sogang.ac.kr

REFERENCES

[1] K. W. Chun, J. B. Lee, S. H. Kim, and T. G. Park, "Controlled release of plasmid DNA from photo-cross-linked pluronic hydrogels," *Biomaterials*, vol. 26, pp. 3319-3326, 2005.

[2] Y. J. Lee and P. V. Braun, "Tunable inverse opal hydrogel pH sensors," *Adv. Mater.*, vol. 15, pp. 563-566, 2003.

[3] M. P. Lutolf and J. A. Hubbell, "Synthetic biomaterials as instructive extracellular microenvironments for morphogenesis in tissue engineering," *Nat. Biotechnol.*, vol. 23, pp. 47-55, 2005.

[4] T. Watanabe, M. Akiyama, K. Totani, S. M. Kuebler, F. Stellacci, W. Wenseleers, K. Braun, S. R. Marder, and J. W. Perry, "Photoresponsive hydrogel microstructure fabricated by two-photon initiated polymerization," *Adv. Funct. Mater.*, vol. 12, pp. 611-614, 2002.

[5] V. Chan, J. H. Jeong, P. Bajaj, M. Collens, T. Saif, H. Kong, and R. Bashir, "Multi-material bio-fabrication of hydrogel cantilevers and actuators with stereolithography," *Lab Chip*, vol. 12, pp. 88-98, 2012.

[6] S. E. Chung, W. Park, H. Park, K. Yu, N. Park, and S. Kwon, "Optofluidic maskless lithography system for real-time synthesis of photopolymerized microstructures in microfluidic channels," *Appl. Phys. Lett.*, vol. 91, pp. 041106, 2007.

[7] C. Sun, N. Fang, D. M. Wu, and X. Zhang, "Projection micro-stereolithography using digital micro-mirror dynamic mask," *Sens. Actuator A-Phys.*, vol. 121, pp. 113-120, 2005.

[8] D. Dendukuri, P. Panda, R. Haghgooie, J. M. Kim, T. A. Hatton, and P. S. Doyle, "Modeling of oxygen-inhibited free radical photopolymerization in a PDMS microfluidic device," *Macromolecules*, vol. 41, pp. 8547-8556, 2008.

[9] Y. C. Tai and R. S. Muller, "Measurement of Young's modulus on microfabricated structures using a surface profiler," in *Micro Electro Mechanical Systems, 1990. Proceedings, An Investigation of Micro Structures, Sensors, Actuators, Machines and Robots. IEEE*, 1990, pp. 147-152.

[10] M. A. Hopcroft, W. D. Nix, and T. W. Kenny, "What is the Young's modulus of silicon?" *J. Microelectromech. Syst.*, vol. 19, pp. 229-238, 2010.

[11] R. Levy and M. Maaloum, "Measuring the spring constant of atomic force microscope cantilevers: thermal fluctuations and other methods," *Nanotechnology*, vol. 13, pp. 33-37, 2002.

FULLY PRINTED, LARGE-SCALE, HIGH SENSITIVE STRAIN SENSOR ARRAY FOR STRESS MONITORING OF INFRASTRUCTURES

Shingo Harada, Wataru Honda, Takayuki Arie, Seiji Akita,and Kuniharu Takei

Osaka Prefecture University, Osaka, Japan

ABSTRACT

We demonstrate a macroscale sensor sheet by fabricating the fully printed, large-scale, and high sensitive strain sensor array on mechanically flexible substrates. This sensor sheet can conformally cover any surfaces for the application of real-time infrastructure stress monitoring as the first proof-of-concept. To realize this concept, a screen printing method is proposed to use by developing an ink for strain sensor. Printed strain sensor array exhibits impressively high sensitivity, and successfully detects two-dimensional strain distribution of small deformation <10µm.

INTRODUCTION

Monitoring and maintenance of infrastructures such as water, electrical, and gas pipelines, buildings, and vehicles are one of the most important things for convenient, comfortable, and secure human lives. However these infrastructures were usually installed in underground or manufactured without monitoring crack or stress, and gradually degraded in the decade of years after installations or fabrications. This would create critical problems due to defects or stress into the materials, resulting in that it causes shutdown of lifelines or accidents. In fact, there were several problems reported for airplanes, gas pipes, and some other infrastructures over the world. To prevent the problems, macro-scale cost-effective strain sensors covered over the structures are one of the most promising methods to monitor the stress and defects. Although there are many reports about the strain sensors using nanoparticles [1], nanotubes [2], piezoelectric materials [3], and transistor structures [4], none of them covers all above requirements to monitor the strain from the infrastructures.

To achieve all requirements, we here demonstrated the fully printed, large-scale, and high sensitive strain sensor array on flexible substrates as the sensor sheet, which can cover any surfaces, by developing the ink for the strain sensor and a printing method. By developing the nanomaterial-composite ink and a screen printing method, the device can be readily fabricated on macro-scale substrates with low cost, which is an important factor in the application of infrastructures.

FABRICATION PROCESS

Figure 1a shows the fabrication process of the strain sensor using a screen printing method, which allows us to fabricate macro-scale devices with low cost. First, two inks of carbon-nanotube (CNT) and Ag nanoparticle (AgNP) were mixed with the weight ratio of 5:4 as the strain sensing material. When the stress was applied into CNT-AgNP film,

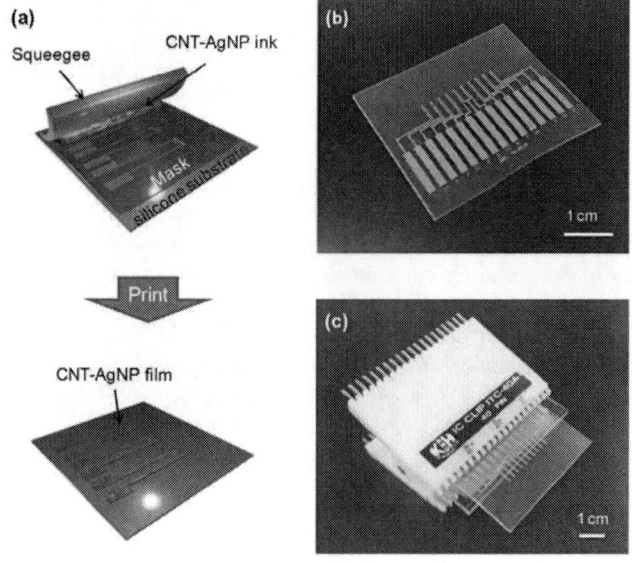

Figure 1: (a) Fabrication process of the strain sensor. (b) Picture of the fabricated strain sensor array. (c)Picture of the strain sensor array was packaged using a clip.

the resistance of this film was changed due to the change of AgNP distance. CNTs were entangled surrounding the AgNPs making that the electrical connect when high tensile stress was applied to the film. The strain in the film was extracted by measuring the resistance change. The fabrication process is simple. Briefly, the synthesized ink (CNT-AgNP) was printed onto the oxygen plasma-treated silicone rubber substrates via a screen print mask. The width of CNT-AgNP pattern is 200 µm with the thickness of the cured film ~1.5 µm in this study. After printing the ink, the film was cured at 70°C for 1hr. Figure 1b shows the strain sensor array printed on the silicone substrate, exhibiting that the ink was uniformly patterned. The strain sensor array was simply packaged using an electrical connector clip to measure the resistance of the strain sensor array (Fig.1c).

CHARACTERISTICS OF THE PRINTED STRAIN SENSORS

Sensitivity of strain sensor

First, to characterize the performance of the printed strain sensors, the resistance change was measured by bending the silicone substrates to apply tensile and compressive stress into the printed film. The design of strain sensor was 3 mm length of CNT-AgNP film on 8 mm-long silicone substrates as described in Fig. 2. Figure 2 reveals the resistance change (R/R_o) as a function of displacement (Ag ink only and CNT-Ag ink), where R_o and R are the resistance

978-1-4799-3510-9/14 $31.00 © 2014 IEEE

at the relax state and bending state of the film. Figure 2 depicts that the composite film shows large resistance change compared to only AgNP film. The resistance change from the relax state (i.e. displacement 0 mm) to the bending state ~2 mm was R/R_o ~80 for both tensile and compressive strains. The sensitivity was ~10 %/Pa that is one of the most sensitive strain sensors to date [5-6]. Based on the results, composition of CNT in AgNP film is very important to realize high sensitivity. In addition, the advantage of this film component is that the resistance was reversely changed by tensile and compressive strain, which is able to detect the bending direction.

Real-time measurements

The time dependence was also measured by bending the substrates to downward and upward with ~1 mm displacement as shown in Fig.3. The result exhibits that the response time is less than 0.4 sec that is high enough for the application of sensor sheet of infrastructure monitoring. This time delay were most likely due to 1) bending speed for this measurement and 2) elastic response of silicone rubber. If higher response speed is needed, the flexible substrate materials should be considered.

Silicone thickness dependence was also measured as shown in Fig. 4 when the sensor was bent to 0.5 mm displacement. The resistance change was clearly observed as a function of silicone thickness. This is because the strain in CNT-AgNP film was usually increased when the silicone substrate was thick, resulting that the resistance was increased (decreased) when thickness of silicone substrates was increased (decreased). These results were in good agreement with the fundamental mechanical theory.

TWO-DIMENSIONAL STRAIN MAPPING FOR PRACTICAL APPLICATIONS

As a proof-of-concept, two-dimensional strain distribution was demonstrated by integrating 5 strain sensor array. The experiment was carried out by bending 1 mm displacement on the edge of the PDMS film as shown in Fig. 5a. This experiment simulated the monitor of strain information on any surface such as a gas pipeline. To understand the bending displacement of each strain sensor, the finite element analysis (COMSOL Multiphysics 4.2a) was conducted, and it indicated that displacement of sensor #1 was ~82 μm and gradually decreased from #1 to #5. By measuring the resistance change experimentally, the displacement can be extracted. Figure 5c exhibited the displacement distribution of the silicone substrates. The experimental result indicates that the sensor #1 has ~56 μm displacement that is similar value to the simulation of Fig. 5b. #2-#5 strain sensors were also able to detect the displacement similar to the simulation results. It should be noted that the printed strain sensor reported here has the capability to detect <10 μm displacement.

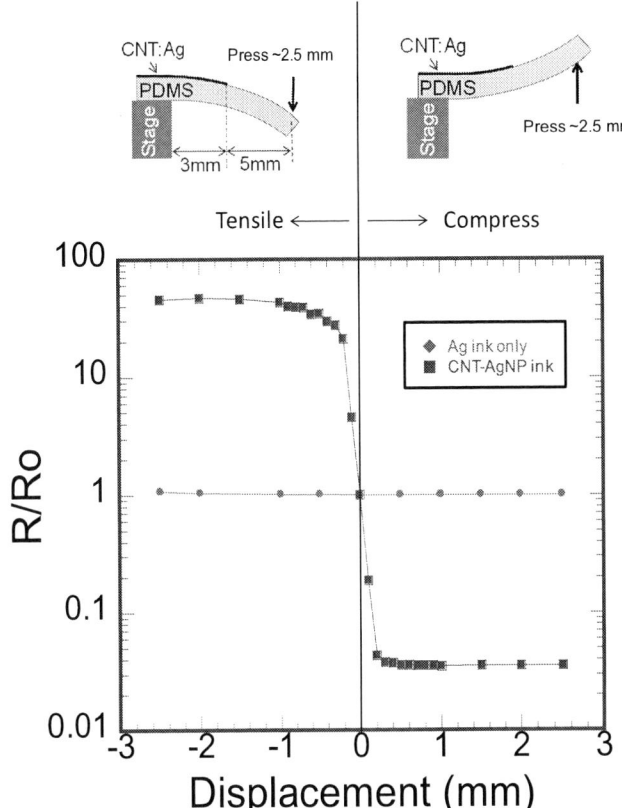

Figure 2: R/R_o as a function of displacement of CNT-AgNP film and only Ag film printed on silicone substrates. The tensile and compress stress were applied by bending the substrate to upward and downward.

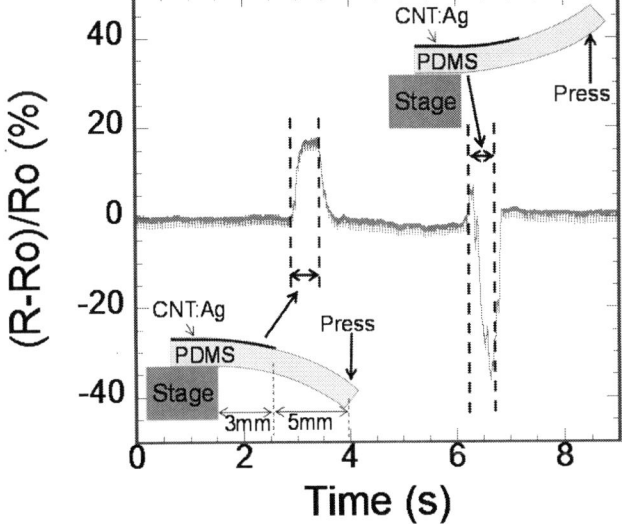

Figure 3: Real-time measurement of the strain sensor by bending upward and downward directions corresponding to compressive and tensile stress, respectively.

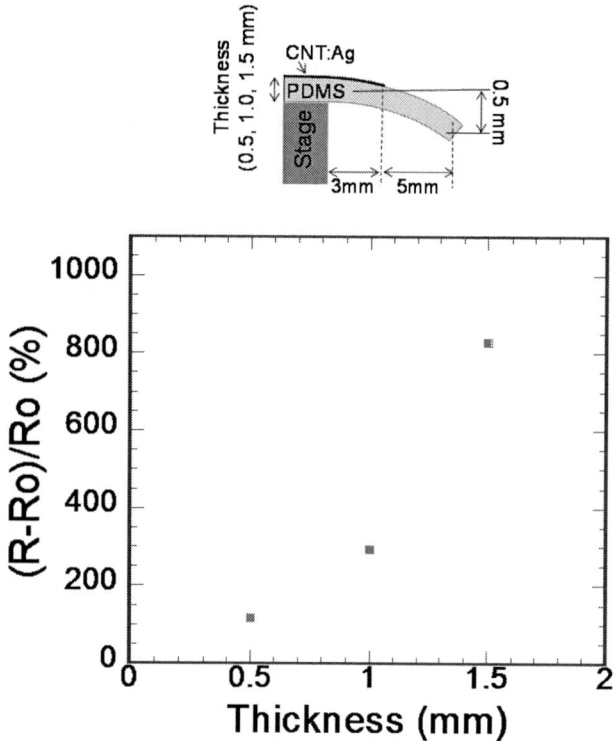

Figure 4: Normalized resistance change $(R-R_o)/R_o$ as a function of silicone thickness when the silicone substrates were bent to 0.5 mm to apply tensile stress in the CNT-AgNP films.

CONCLUSION

In this report, we have proposed and demonstrated a new type of fully printed high sensitive strain sensor array using nano-composite materials and a screen printing method for the application of the sensor sheet. The device exhibited high sensitivity ~10 %/Pa and successfully demonstrated the displacement/strain mapping of small displacement as a proof-of-concept for the real-time monitoring of infrastructures although the sensor was fabricated by the printing method. This method allows us to fabricate cost-effective, macro-scale, and high sensitive strain sensors to cover different kinds of infrastructures.

ACKNOWLEDGEMENTS

This work was partially supported by research grants from the Mazda Foundation, the Foundation Advanced Technology Institute, and JSPS KAKENHI (#25889048).

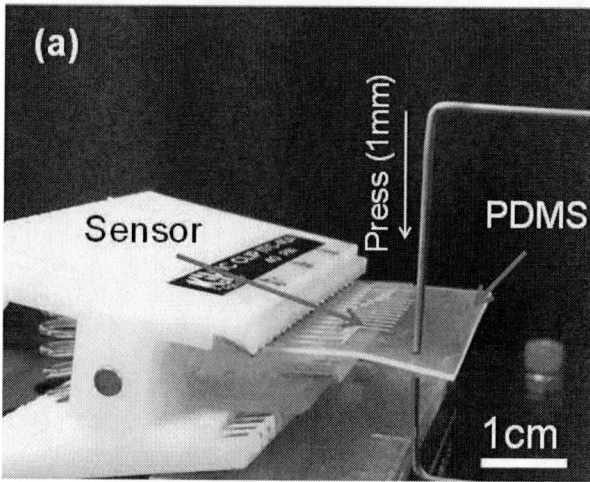

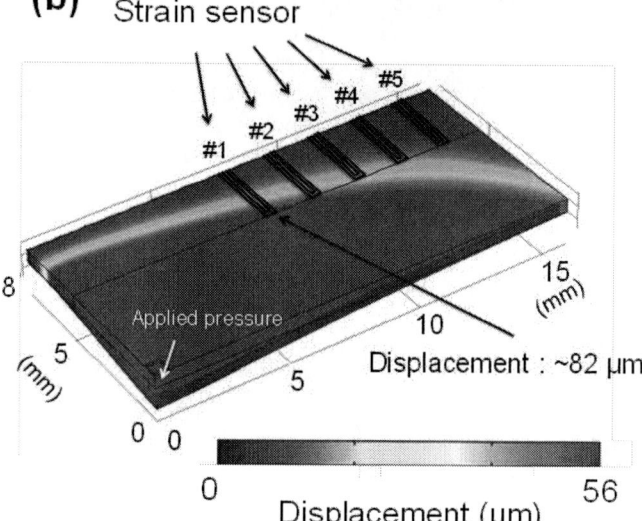

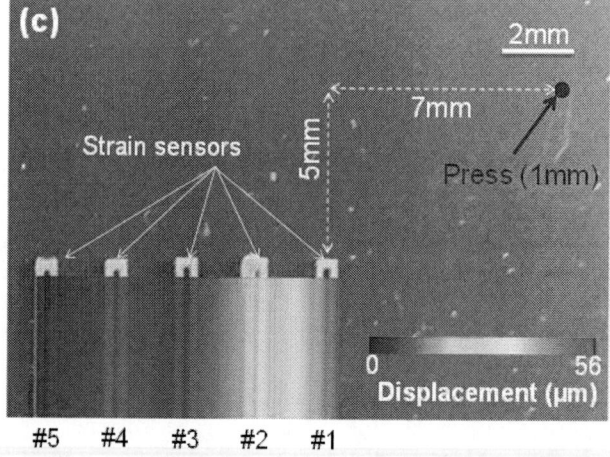

Figure 5: (a) Measurement setup of strain distribution. (b) Simulation result of the finite element method to estimate the displacement of sensors. (c) Displacement (strain) distribution when the area marked with a circle (applied pressure) was pressed ~1mm

REFERENCES

[1] J. Herrmann, K.-H. Muller, T. Reda, G. R. Baxter, B. Raguse, G. J. J. B. de Groot, R. Chai, M. Roberts, L.Wieczorek, "Nanoparticle films as sensitive strain gauge", *Applied Physics Letters*, vol. 91, p. 183105, 2007.

[2] T. Yamada, Y.Hayamizu, Y. Yamamoto, Y. Yomogida, A.Izadi-Najafabadi, D. N. Futaba, K. Hata, "A stretchable carbon nanotube strain sensor for human-motion detection", *Nature Nanotech.*, vol. 6, pp. 296-301, 2011.

[3] J. Zhou, Y. Gu, P. Fei, W. Mai, Y. Gao, R. Yang, G. Bao, Z. L. Wang, "Flexible piezotronic strain sensor", Nano Letters, vol. 8, pp. 3035-3040, 2008.

[4] T. Quang, N. T. Tien, D. Kim, M. Jang, O.J. Yoon, N.-E. Lee, "A flexible reduced graphene oxide field-effect transistor for ultrasensitive srain sensing", *Advanced Functional Mater.*, 2013, DOI: 10.1002/afm.201301845.

[5] G. Schwartz, B. C.-K. Tee, J. Mei, A. L. Appleton, D. H. Kim, H. Wang, Z. Bao, "Flexible polymer transistors with high pressure sensitivity for application in electronic skin and health monitoring", *Nature Comm.* 4:1859, 2013.

[6] [6] X. Liu, Y. Zhu, M. W. Nomani, X. Wen, T.-Y. Hsia, G. Koley, "A highly sensitive pressure sensor using a Au-patterned polydimenthylsiloxane membrane for biosesing applications", *J. Micromech. Microeng.*, Vol. 23, p. 025022, 2013.

CONTACT

*K. Takei, takei@pe.osakafu-u.ac.jp

HARBOR SEAL INSPIRED MEMS ARTIFICIAL MICRO-WHISKER SENSOR

A.G.P. Kottapalli[1], M. Asadnia[1], H. Hans[1], J.M. Miao[1], and M.S. Triantafyllou[2]*

[1] School of Mechanical and Aerospace Engineering, Nanyang Technological University, Singapore
[2] Department of Mechanical Engineering, Massachusetts Institute of Technology, Massachusetts, USA

ABSTRACT

Harbor seal whiskers possess a unique geometry along the length of the whisker which is believed to suppress vortex induced vibrations (VIV) in frontal flows. This paper presents the design, fabrication and experimental underwater characterization of a MEMS artificial whisker sensor. A bio-inspired artificial whisker fabricated by stereolithography is installed on a piezoelectric MEMS sensing membrane. Experimental results demonstrate that the whisker sensor is able to detect minute disturbances underwater with a velocity detection limit as low as 193μm/s.

INTRODUCTION

Unmanned underwater vehicles (UUVs) employ a plethora of sensors to achieve various purposes of control and feedback while maneuvering underwater. A major topic of research in the past few decades in this area is to develop sensors and sensory systems that could equip the UUV with an ability to intelligently assess its environment [1]. Most UUVs are space-constrained with limited inner volume and highly limited payload energy with utmost need to maintain neutral buoyancy. Therefore, there is a high demand for sensors that are light in weight, inexpensive, low-powered and small in size and could envision flows around the bodies of the vehicles and identify the underwater surroundings and thereby generate artificial vision. Sonar and optical methods have been used in the past to meet this purpose, but they suffer from serious disadvantages. They work fundamentally based on an active sensing principle and are large sized and heavy-weight systems, which often suffer poor resolution of detection in certain murky underwater conditions [2].

Some of the biological sensors found in nature portray the best designs with incomprehensible features. It has been found that even when visual and auditory senses are blocked, some pinnpeds are capable of tracking minute water movements such as those left in the wake of a fish [3]. Blind cave fishes demonstrate an exceptionally skilful ability of being able to swim at high speeds without colliding any surrounding underwater obstacles [4]. Relying on the lateral-line of pressure-gradient sensors, they are able to constantly monitor their surroundings and achieve energy-efficient maneuverability [5]. Some underwater animals like dolphins and tooth-whales use echolocation. Harbor seals neither have lateral-lines nor do they use echolocation, but they operate under similar conditions by depending on their highly sensitive whiskers (vibrissae) to track their surrounding objects [6]. While swimming, using their whiskers, harbor seals are able to detect water disturbances and perform hydrodynamic trail following [6]. More interestingly, the harbor seal whiskers possess a unique geometry along the length of the whisker (Figure 1)

which is believed to suppress vortex induced vibrations (VIV) in frontal flows [6]. Figure 1 shows the photographs of the real seal whisker and the artificial whisker fabricated in this work.

Figure 1: Optical photograph of the real seal whisker (center) and the artificial whisker (left and right) fabricated in this work.

Owing to their small size, light weight, low power consumption, bio-inspired arrays of MEMS flow sensors could be an ineluctable alternative to existing sensing technologies used on UUVs. In the past, a number of ultrasensitive flow sensors have been developed that are inspired by the superficial neuromast flow sensors present on the blind cave fish [7-10]. Most of these sensors feature a standing pillar structure (haircell) that extends into the surrounding flow [11]. The deflection in the haircell generated by an external disturbance is recorded as a voltage signal employing piezoresistors embedded at the base of the haircell. Such sensors employing cylindrical standing pillars that protrude into the flow experience noise generated by the VIV of the pillar. However, despite its significance, a MEMS flow sensor employing an artificial micro-whisker has never been developed before. In this work, we present the design, fabrication and characterization of a self-powered MEMS micro-whisker sensor that is inspired by the harbor seal whisker. Micro-whiskers that feature undulations similar to those present on the real harbor seal whiskers are developed.

BIO-INSPIRED MEMS WHISKER SENSOR DESIGN

The structure of the MEMS whisker sensor developed in this work consists of two major parts – the artificial micro-whisker fabricated by stereolithography (SLA) and the piezoelectric $Pb(Zr_{0.52}Ti_{0.48})O_3$ (PZT) MEMS sensing base fabricated by conventional microfabrication technologies. A schematic of the sensor structure is illustrated in figure 2. The

978-1-4799-3510-9/14 $31.00 © 2014 IEEE

artificial whisker mounted at the center of the PZT membrane extends into the external flow and responds to disturbances in water. Any flow disturbances or variations in the vicinity of the whisker cause the whisker to bend in response, which inturn causes the membrane at the base to buckle. Any displacement in the PZT membrane generates charges that are acquired by tapping voltage signal from the top and bottom electrodes of the PZT layer.

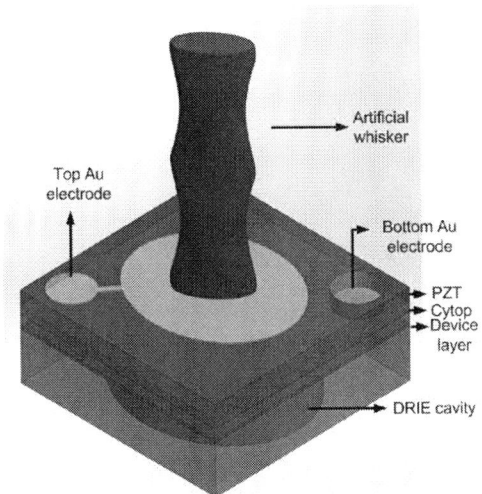

Figure 2: A schematic of the device structure of the artificial harbor seal whisker sensor developed in this work.

WHISKER SENSOR FABRICATION

The artificial micro-whisker and the PZT MEMS sensing base are fabricated separately and then assembled together. A polymer micro-whisker with dimensions similar to that of the real seal-whisker is fabricated by stereolithography (SLA) process. The whisker structure has a unique undulatory geometry which is almost impossible to fabricate using most conventional microfabrication technologies. It is extremely cumbersome to develop high-aspect SU-8 micro-structures that have continuously varying spatial features in the structural geometry. SLA is a very simple and promising technology to generate such structures with features varying along all the three dimensions. Three-dimensional models of a section of the seal whisker are created in Solid Works. The micro-pillar haircells are fabricated from Si60 polymer material on a built-in high resolution SLA VIPER machine. The designed 3D model is then sliced into a series of 2D layers of equal thicknesses which are executed to control an X-Y stage containing the UV curable Si60 solution. The high-aspect micro-pillars are formed by scanning a 355nm UV beam of spot size 0.01mm on the liquid monomer Si60 polycarbonate raisin, curing the raisin into a solid polymer structure layer by layer, and stacking all layers together.

The key fabrication steps for the piezoelectric sensing-base include a low-temperature bonding using Cytop, integration of the PZT layer on the silicon substrate, formation of the electrode interconnects and pads,

chemical-mechanical-polishing, wet etching of the PZT layer and releasing the diaphragm structure by using the DRIE technique. The PZT sensing-base is fabricated by bonding an SOI wafer (with 20µm thick device layer) to a 300µm thick $Pb(Zr_{0.52}Ti_{0.48})O_3$ (PZT) plate using a spin-on polymer (cytop) intermediate layer. After which, a 27µm thick PZT plate is formed by chemical mechanical polishing (CMP) of the 300µm thick PZT plate. The PZT plate is sandwiched between two Cr(10nm)/Au(300nm) metal electrodes that form the contact pads. A wet etching of PZT is performed alternatively in diluted HCl:HF and HNO_3 solutions with a photoresist mask to open a window to access the bottom electrode. A sensing membrane is defined by DRIE etching. The detailed fabrication of the PZT sensing base is presented in [12]. Figure 3 depicts a schematic that explains the various layers of the sensing membrane. Figure 4 shows the optical microscopic image of the fabricated device.

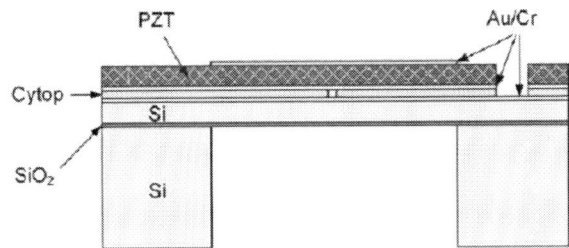

Figure 3: A schematic of the various layers in the sensing membrane formed during the device fabrication.

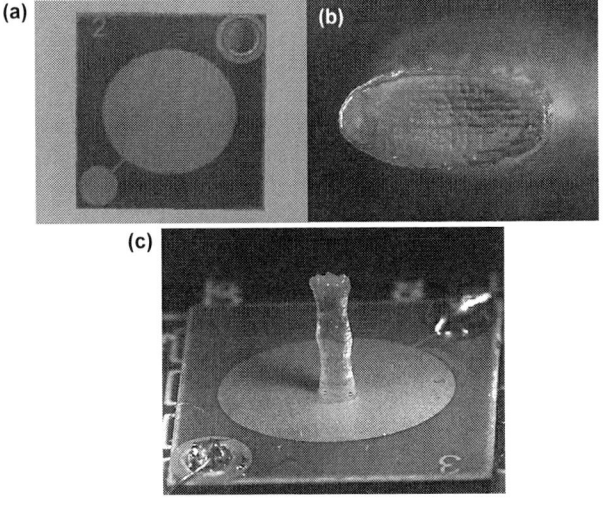

Figure 4: Optical images of (a)PZT membrane with two contact pads on top and bottom faces (b) flat end of the stereolithographically fabricated micro-whisker that is mounted at the center of the PZT membrane employing a precision XYZ control stage (c) PZT whisker sensor with contact pads wire-soldered for testing.

EXPERIMENTAL RESULTS

The goal of the experiments presented in this section is

978-1-4799-3510-9/14 $31.00 © 2014 IEEE

to evaluate the underwater sensing capabilities of the artificial micro-whisker sensor. The performance of the sensor in sensing minute underwater disturbances is evaluated using a vibrating sphere (dipole) stimulus [13].

A stainless steel sphere of 8mm diameter connected to a minishaker (model 4810, B & K, Norcross, GA) through a rod of diameter 2mm forms the dipole stimulus (source). The dipole can be driven from a function generator through a power amplifier and set to vibration of desired frequency and amplitude. The sensor, positioned underwater in the vicinity of the dipole at a distance of 30mm is observed to follow the dipole frequency and amplitude very well. The sensor's output is amplified by a gain of 500 using SRS560 low-noise preamplifier. Figure 5 shows a schematic of the experimental setup used. Figure 6 shows the output of the sensor for a dipole frequency of 10Hz, when the sensor is oriented at 0^0 and 90^0 with respect to the plane of vibration of the dipole.

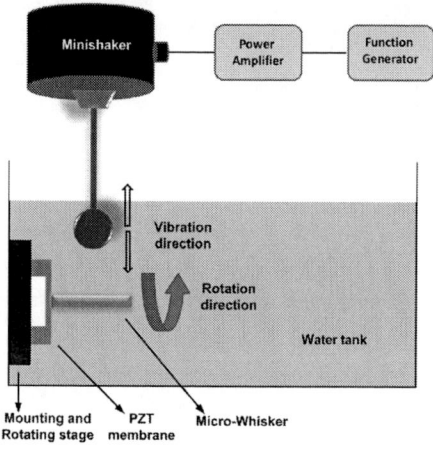

Figure 5: A schematic of the experimental setup used.

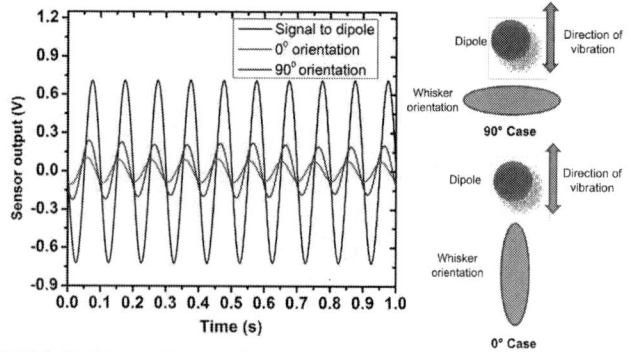

Figure 6 (a)Underwater response of the whisker sensor at 0^0 and 90^0 orientations to the dipole stimulus vibrating at 10 Hz (b)a schematic showing the top-view of the orientation of the whisker with respect to dipole.

In order to determine the resolution of directional sensing, we conducted an experiment where the stage on which the sensor is mounted is rotated in steps, which changes the orientation of the whisker with respect to the dipole. Since the whisker has an elliptical cross-section, at each angle the surface area it projects to the dipole changes and therefore the signal acquired from the PZT sensor changes. The results of this experiment are shown in figure 7.

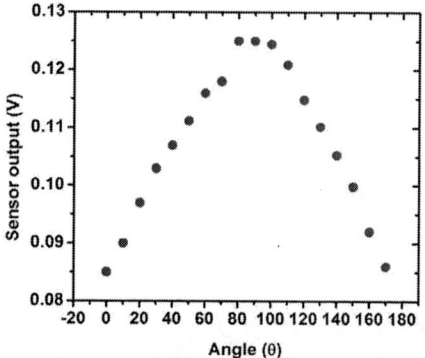

Figure 7: Experimental results showing the directional dependence of the output.

In order to determine the sensitivity and threshold velocity detection limit of the sensor, the velocity of the vibration of the dipole is varied from a very low velocity in the order of a few μm/s to a high velocity of 250mm/s by varying the amplitude of the sinusoidal signal supplied to the dipole. The sensors are mounted in such a way that the dipole moved perpendicular to the long axis of the whisker as described in the experimental setup shown in figure 5. The whisker sensors are tested for a dipole vibrating at constant frequency of 35Hz and varying sinusoidal source signal amplitudes. The signal from the dipole is collected at various velocities, and the peak-to-peak signal amplitude of the sensor output is plotted with respect to velocity in figure 8. The oscillatory flow velocity corresponding to a particular frequency and amplitude of vibration of the dipole is measured using a laser Doppler vibrometer (LDV).

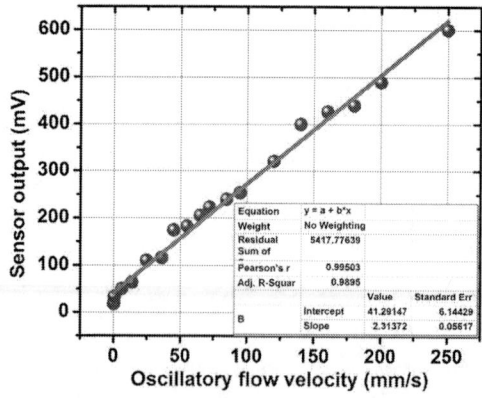

Figure 8: Oscillatory flow velocity sensing by the artificial MEMS whisker in water

The sensor demonstrates a linear response with a threshold velocity detection limit as low as 193μm/s (below which the sensor's output starts to become noisy) which rivals the abilities of the biological harbor seal whisker.

CONCLUSION

Most studies conducted in the literature, mainly focus on the theoretical fluid-dynamic analysis of the whisker geometry and fluid structure-interaction studies, but a functional seal-whisker inspired MEMS sensor has never been developed before. In this paper, the development of an artificial micro-whisker sensor that features a micro-whisker assembled on a MEMS PZT sensing membrane is presented. The three-dimensional simulation analysis conducted shows that the lift-coefficients are significantly higher (~43-times) in case of a cylinder as compared to a real seal-whisker. The underwater oscillatory flow sensing ability of the sensor is experimentally evaluated employing a dipole stimulus. More experiments are in progress to evaluate the VIV suppression abilities of a MEMS haircell sensor with whisker-like undulations on the haircell as compared to the conventional cylindrical haircell flow sensor.

ACKNOWLEDGEMENTS

This research was funded by the Singapore National Research Foundation (NRF) through the Singapore-MIT Alliance for Research and Technology (SMART) Centre, Center for Environmental Sensing and Modeling (CENSAM) IRG.

REFERENCES

[1] H. R. Beem and M. S. Triantafyllou, "Calibration and validation of a harbor seal whisker-inspired flow sensor," *Smart. Mater. Struct.,* vol. 22, p.014012, 2013.

[2] Y. C. Yang et.al., "Distant touch hydrodynamic imaging with an artificial lateral line," *Proc. Natl. Acad. Sci. USA,* vol.103, pp. 18891-18895, 2006.

[3] G. Dehnhardt, B. Mauck and H. Bleckmenn, "Seal whiskers detect water movements," *Nature,* vol. 394, pp. 235-236, 1998.

[4] J. C. Montgomery, S. Coombs, and M Halstead, "Biology of the mechanosensory lateral line in fishes," *Rev. Fish Biol. Fisher.,* vol. 5, pp. 399-416, 1995.

[5] J. C. Montgomery, S. Coombs, and C. F. Baker, "The mechanosensory lateral line system of the hypogean form of *Astyanax fasciatus*," *Evol. Biol. Fish.,* vol. 62, pp. 87-96, 2001.

[6] W. Hanke et.al., "Harbor seal vibrissa morphology suppresses vortex-induced vibrations," *J. Exp. Biol.,* vol. 213, pp. 2665-2672, 2010

[7] S. Peleshanko et.al., "Hydrogel-encapsulated microfabricated haircells mimicking fish cupula neuromast," *Adv. Mater.,* vol. 19, pp. 2903-2909, 2007.

[8] A. G. P. Kottapalli, M. Asadnia, J. M. Miao, G. Barbastathis and M. Triantafyllou, " A flexible liquid crystal polymer MEMS pressure sensor array for fish-like underwater sensing," *Smart. Mater. Struct.,* vol. 21, p. 115030, 2012.

[9] J. Dusek, A.G.P. Kottapalli, M. E. Woo, M. Asadnia, J Miao, J H Lang and M S Triantafyllou, "Development and testing of bio-inspired microelectromechanical pressure sensor arrays for increased situational awareness for marine vehicles," *Smart. Mater. Struct.,* vol. 22, p. 014002, 2013.

[10] M. Asadnia, A.G.P. Kottapalli, Z. Shen, J. M. Miao, and M. Triantafyllou, "Flexible, and surface-mountable piezoelectric sensor arrays for underwater sensing in marine vehicles," *IEEE Sensors J.* vol. 13, pp. 3918-3925, 2013.

[11] A. G. P. Kottapalli, M. Asadnia, J. M. Miao, and M. Triantafyllou, "Electrospun nanofibrils encapsulatedin hydrogel cupula for biomimetic MEMS flow sensor development," in *Proc. IEEE MEMS'13 Conference,* Taipei, January 20-24, 2013, pp. 25-28.

[12] C. W. Tan, A. G. P. Kottapalli, Z. H. Wang, X. Ji J. M. Miao, G. Barbastathis and M. Triantafyllou, "Damping characteristics of a micromachined piezoelectric diaphragm-based pressure sensor for underwater applications," in *Proc. Transducers'11 Conference,* Beijing, June 5-9, 2011, pp. 72-75.

[13] M. Asadnia, A.G.P. Kottapalli, Z. Shen, J. M. Miao, G. Barbastathis and M. Triantafyllou, "Flexible, zero-powered, piezoelectric MEMS pressure sensor arrays for underwater sensing in marine vehicles," in *Proc. IEEE MEMS'13 Conference,* Taipei, January 26-24, 2013, pp. 126-129.

CONTACT

*A.G.P Kottapalli, tel: +65-67904264; School of Mechanical and Aerospace Engineering, Nanyang Technological University, kott0002@e.ntu.edu.sg

HIGH FREQUENCY PIEZOELECTRIC MICROMACHINED ULTRASONIC TRANSDUCER ARRAY FOR INTRAVASCULAR ULTRASOUND IMAGING

Yipeng Lu, Amir Heidari, Stefon Shelton, Andre Guedes and David A. Horsley
University of California, Davis, USA

ABSTRACT

This paper presents a 1.2 mm diameter high fill-factor array of 1,261 piezoelectric micromachined ultrasonic transducers (PMUTs) operating at 18.6 MHz for intravascular ultrasound (IVUS) imaging and other medical imaging applications. At 1061 transducers/mm^2, the PMUT array has a 10-20× higher density than the best PMUT arrays realized to date. The PMUTs utilize a piezoelectric material, AlN, which is compatible with CMOS processes. Measurements show a large voltage response of 2.5 nm/V and good frequency matching in air, a high center frequency of 18.6 MHz and wide bandwidth of 4.9 MHz when immersed in fluid. Phased array simulations based on measured PMUT parameters show a tightly focused, high output pressure acoustic beam.

INTRODUCTION

Ultrasonic transducers are used in many applications, such as nondestructive testing, object detection, gesture recognition and real time medical imaging. Compared with conventional bulk piezoelectric ultrasonic transducers, micromachined ultrasonic transducers (MUTs) have a compliant membrane structure with low acoustic impedance, which is easy to match with the impedance of air and fluids and thereby generate good acoustic coupling. Compared with well-developed capacitive MUTs (CMUTs) [2-3], piezoelectric MUTs (PMUTs) do not require a polarization voltage (which can exceed 190V for CMUTs) to achieve the required transducer sensitivity. This is particularly important for catheter-based ultrasound applications as having a high voltage inside the body requires more complex packaging. Another advantage of PMUTs is that they have higher capacitance, which results in lower electrical impedance, allowing better matching to supporting electronic circuits and less sensitivity to parasitic capacitance. Most of the previous work on PMUTs has focused on lead zirconate titanate (PZT) films, because of its high piezoelectric coefficient. While PZT has higher piezoelectric constants than the aluminum nitride (AlN) film used in this work, the lower dielectric constant of AlN allows for comparable performance to be achieved, especially in terms of receiver sensitivity [1]. In addition, unlike PZT films that require high fabrication temperature (around 800 °C), AlN is deposited at a low temperature (<400 °C) and is a material with full compatibility with CMOS processes.

Micromachined transducers operating at 5 MHz, where the acoustic wavelength is about 300 μm in tissue, have little advantage over conventional saw-cut piezoelectric ultrasonic transducers. However, saw cutting cannot achieve an element pitch smaller than 100 μm, making micromachining an attractive option for high frequency (>10 MHz) arrays requiring half-wavelength (λ/2) element pitch. Previous PMUTs were fabricated by through-wafer etching [4-5], resulting in a low fill-factor, small element count, and therefore poor acoustic efficiency. Here we present micromachined transducers with 10-20× higher density (1061 transducers/mm^2) than the best PMUT arrays realized to date, 56 transducers/mm^2 [4] and 123 transducers/mm^2 [5]. This result was achieved using an AlN fabrication process originally developed for RF MEMS resonators, filters [6] and inertial sensors [7]. This process incorporates a sacrificial polysilicon release pit that precisely defines the PMUT diameter, thereby enabling 10× smaller device spacing and eliminating the need for through-wafer etching.

DESIGN

A PMUT cross-section with exaggerated deformation is shown in Figure 1. As a transmitter, the electric field between the top electrode (TE) and the bottom electrode (BE) creates a transverse stress in the AlN piezoelectric layer due to the inverse piezoelectric effect. The generated stress causes a bending moment which forces the membrane to deflect out of plane launching an acoustic pressure wave into the environment. In the receive mode, an incident pressure wave deflecting the plate creates transverse stress which results in a charge between the electrodes due to direct piezoelectric effect. Here, we use AlN as the active piezoelectric layer, which has a low dielectric constant and therefore higher receiving voltage sensitivity. SiO$_2$ is selected to be the passive layer and the layer stack (0.75 μm AlN/ 0.8 μm SiO$_2$) is chosen based on the analysis of neutral axis location and is optimized for high pressure sensitivity. Due to the thin layer stack, a small PMUT with 25 μm diameter has a center frequency of 25 MHz in air and 20 MHz in water. Because the PMUT radius is smaller than λ/2 (λ = 75 μm at 20 MHz in water), the design has the desirable advantage that there is no irregular near-field pressure pattern.

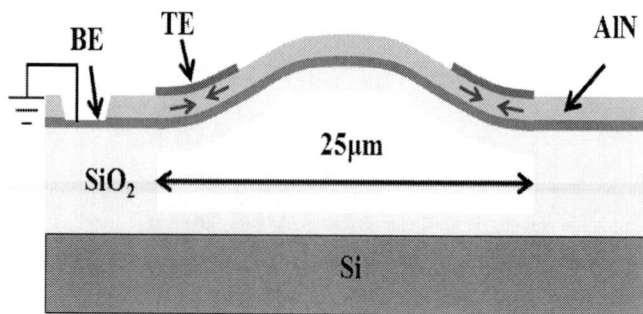

Figure 1: PMUT cross-section with exaggerated deformation. The PMUT membrane is composed of 750 nm piezoelectric AlN on top of a 800 nm SiO$_2$ passive layer.

978-1-4799-3510-9/14 $31.00 © 2014 IEEE

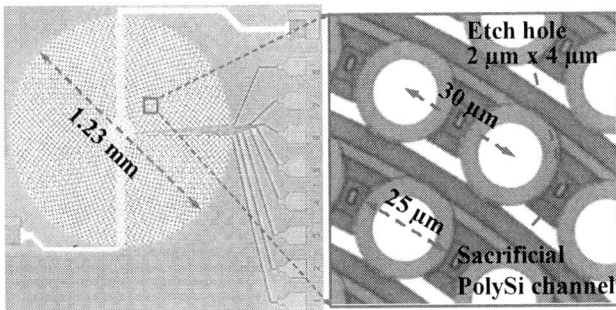

Figure 2: Optical image of the PMUT array and close-up picture of individual 25 μm diameter PMUTs. The sacrificial poly-Si is removed via 2 μm by 4 μm etch holes between each PMUT.

The optical image of the fabricated PMUT array and a close-up picture of individual PMUTs are shown in Figure 2. The PMUTs in the 1.2 mm diameter array are grouped into 8 annular rings to enable electronic control of the array's focus by phased array methods. Each group has approximately the same area and number of PMUTs to enable efficient focus depth control. The device is released using etch holes from the front side of the wafer, resulting in a 2 μm cavity beneath the PMUT membrane, similar to the structure used in a CMUT. However, to achieve the same electromechanical coupling (k_t^2) in a CMUT, it would require a 6 kV polarization voltage at this 2 μm gap or 50 V with a ~180 nm vacuum gap that is much more difficult to mass produce.

The small 5 μm spacing between PMUTs enables high fill factor. As a result, the array gain, which is the ratio of the array's output pressure to that of a single PMUT, is 20x higher than previously fabricated transducers with the same 25 μm diameter [4]. In addition, compared with conventional ultrasonic transducers using an acoustic lens to obtain a fixed focus, our PMUT array is capable of adjusting the focus depth electronically which improves the imaging resolution across a broad range of medical applications. The small array can be used for high resolution medical imaging, especially intravascular ultrasound (IVUS) imaging, which requires a

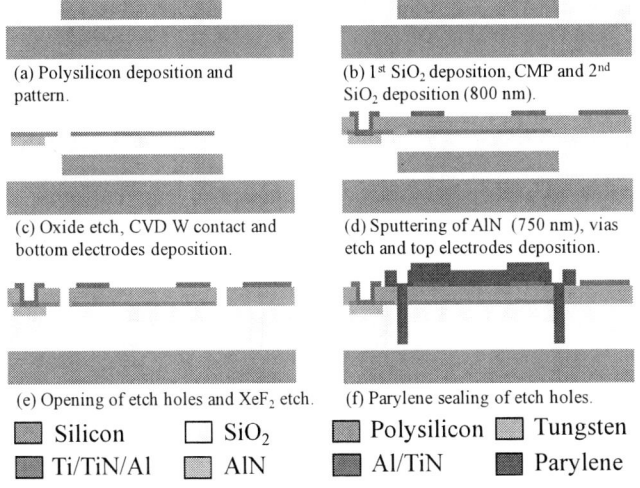

Figure 3: Fabrication process flow.

(a) Polysilicon deposition and pattern.

(b) 1st SiO₂ deposition, CMP and 2nd SiO₂ deposition (800 nm).

(c) Oxide etch, CVD W contact and bottom electrodes deposition.

(d) Sputtering of AlN (750 nm), vias etch and top electrodes deposition.

(e) Opening of etch holes and XeF₂ etch.

(f) Parylene sealing of etch holes.

■ Silicon   □ SiO₂   ▨ Polysilicon   ▨ Tungsten
▨ Ti/TiN/Al   ▨ AlN   ▨ Al/TiN   ▨ Parylene

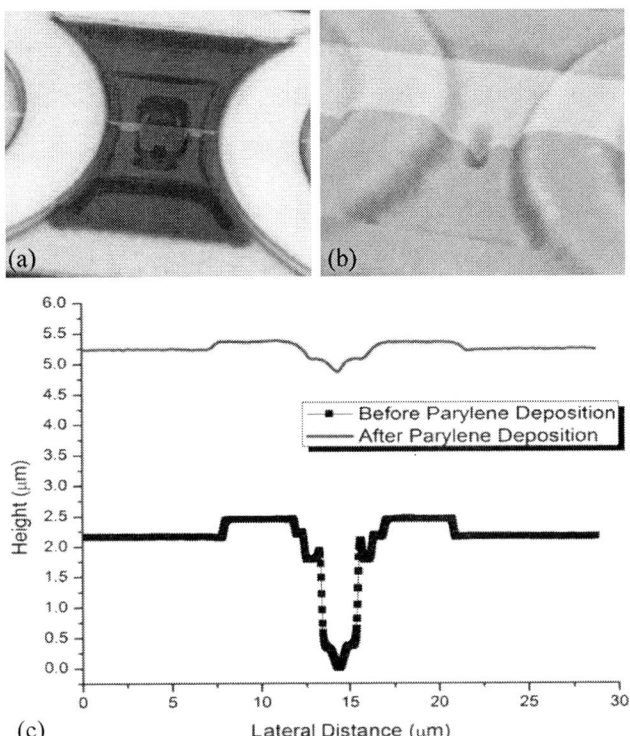

Figure 4: 3D confocal laser microscope measurements showing a 2 μm by 4 μm etch hole (a) before and (b) after Parylene sealing and (c) profile measurements.

small size (< 2 mm) of the transducer and high operating frequency (>10 MHz). In IVUS, close proximity to the target allows this PMUT array to pick up a relatively weak ultrasound signal to provide diagnostic information inaccessible from a noninvasive transducer.

FABRICATION

The process flow is shown in Figure 3, where steps (a-d) were performed in the Sandia National Labs AlN MEMS fabrication process and steps (e-f) were performed in the UC Berkeley Marvell NanoLab. This process incorporates a sacrificial polysilicon release pit that precisely defines the PMUT diameter, thereby enabling a small device size (25 μm and even smaller) with close spacing (5 μm) and eliminating the need for through-wafer etching. The sacrificial polysilicon is etched by vapor phase XeF₂, releasing the PMUT membranes as shown in step (e), after which the etch holes are sealed and the device is insulated via vapor-phase deposition of Parylene-C to enable fluid immersion, as shown in step (f). The etch holes must be small to enable them to be sealed without filling the cavity beneath the PMUT membrane. Dense arrays having etch holes as small as 2 μm x 4 μm were successfully released. Reducing the density to 155 transducers/mm² allowed successful release using smaller 1 μm diameter etch holes, suggesting that XeF₂ depletion occurs during the release of dense arrays. 3D confocal laser microscope images of an etch hole before and after Parylene sealing are shown in Figure 4, demonstrating that the 3 μm Parylene layer successfully seals the etch hole.

978-1-4799-3510-9/14 $31.00 © 2014 IEEE

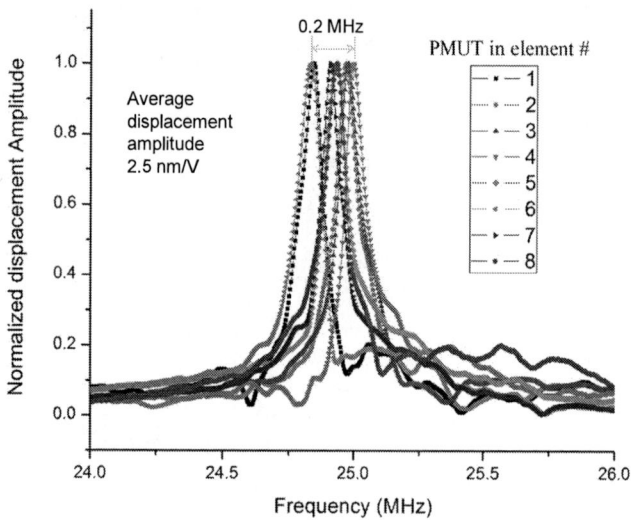

Figure 5: Frequency response measured in air showing 0.8% frequency mismatch across the PMUT array.

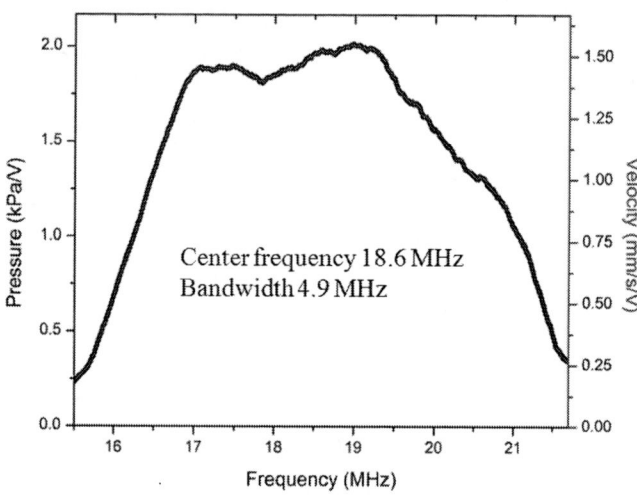

Figure 7: Measured frequency response in fluid (Fluorinert-70) with velocity converted to pressure on the surface of PMUTs.

EXPERIMENTS

A Laser Dropper Vibrometer (OFV 512 and OFV 2700, Polytech) is used in conjunction with a network analyzer (E5061B, Agilent Technologies) to measure the displacement frequency response in air, as shown in Figure 5. The peak displacement sensitivity is 2.5 nm/V at a center frequency of 25 MHz. Measurements of 8 PMUTs selected from each annular ring show a small center frequency mismatch of 0.2 MHz across the array, demonstrating good fabrication uniformity. Measured frequency responses of a single PMUT in air (before and after Parylene sealing) are shown in Figure 6. Although the measurements show that the Parylene layer reduces the dynamic displacement sensitivity from 2.5 nm/V to 0.36 nm/V, it is mostly caused by the reduction of the quality factor (Q) from 167 to 45 due to the additional Parylene layer. The static displacement sensitivity, which is related to the fluid-immersed performance, is only reduced by approximately 50%. Reducing the Parylene

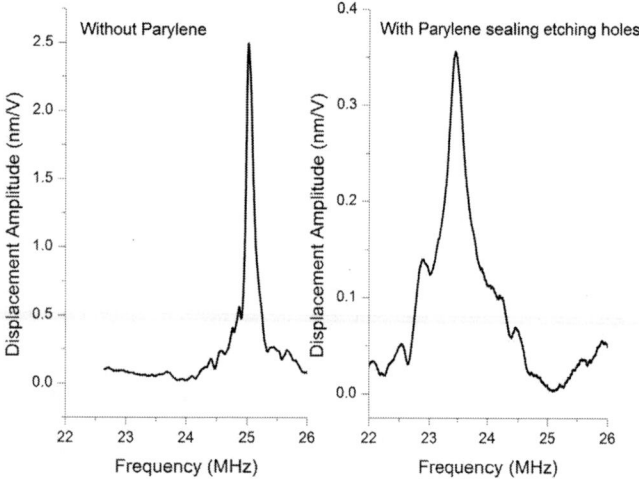

Figure 6: Frequency response of a single PMUT measured in air before and after Parylene sealing.

thickness is expected to improve performance while still sealing the etch holes.

The frequency response in fluid is shown Figure 7. Here PMUTs are immersed in Fluorinert-70, which has similar acoustic impedance with human tissue and a high electrical resistivity, eliminating the need for full insulation of all electrical connections to the device. The fluid-immersed transducer has a high 18.6 MHz center frequency and wide 4.9 MHz bandwidth. The center frequency is shifted from 25 MHz in air to 18.6 MHz in fluid because of the increased damping provided by the fluid. The wide bandwidth indicates that more energy will be transmitted into the acoustic domain and is indicative of good acoustic coupling. The 0.2 MHz variation of the center frequency across the array is a small fraction of the 4.9 MHz bandwidth and does not greatly affect the array performance. The peak PMUT membrane vibration velocity is 1.5 mm/s/V, which corresponds to a pressure of 2 kPa/V.

Phased array simulations based on the measured PMUT parameters are shown in Figure 8, where Figure 8 (a) is the pressure distribution without focus control and Figure 8 (b) and (c) are the pressure distributions with the focus set to 1 mm and 1.5 mm focal depth respectively. This simulation demonstrates the ability to vary the focal point by controlling the phase of the 8 annular rings. Furthermore it shows a small focus size of 100 - 150 μm, with acoustic pressures of 9 kPa/V and 6 kPa/V at the focus depths of 1 mm and 1.5 mm respectively. These focus points are ideal for the targeted IVUS imaging application.

CONCLUSION

A 1.2 mm diameter high fill-factor array of 1,261 CMOS compatible AlN PMUTs is fabricated and characterized. At 1061 transducers/mm², the PMUT array has a 10-20× higher density than the best PMUT arrays realized to date. The frequency response in air show a peak displacement sensitivity of 2.5 nm/V at the 25 MHz center frequency and

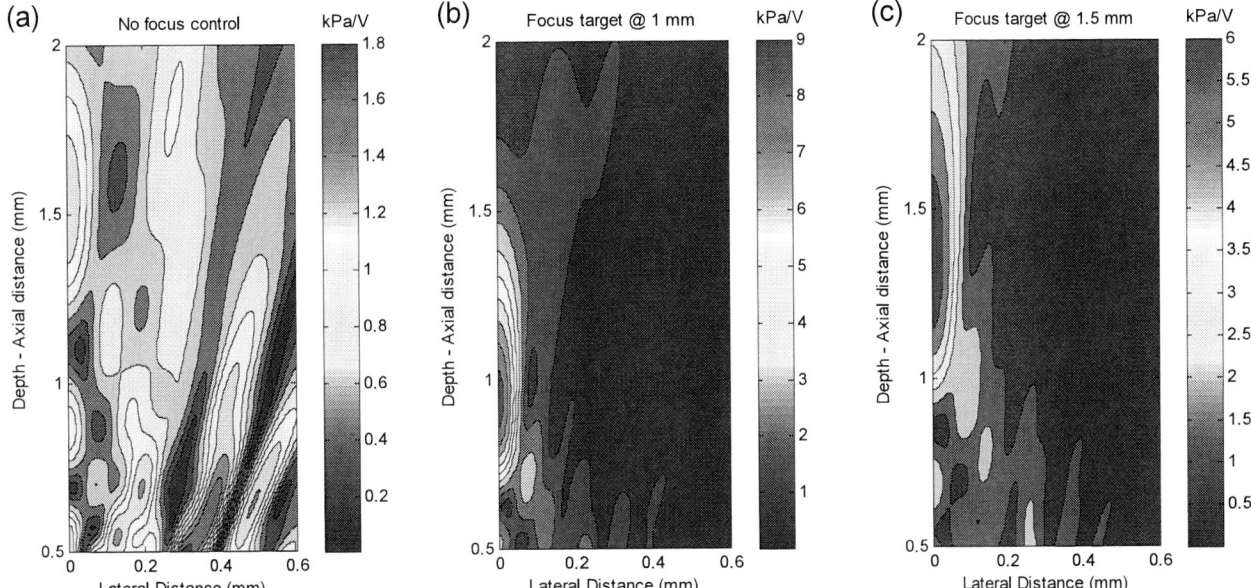

Figure 8: Simulated pressure distribution based on the measured PMUT parameters (a) without focus control, (b) focused at 1 mm, and (c) focused at 1.5 mm.

good frequency matching in air 0.2 MHz, demonstrating good fabrication uniformity. The measurements show that the Parylene layer reduces the dynamic displacement sensitivity from 2.5 nm/V to 0.36 nm/V, but it is mostly caused by the reduced quality factor (Q) from 167 to 45 due to the additional Parylene layer. However, the static displacement sensitivity, which is related to the fluid-immersed performance, is only reduced by approximately 50%. Reducing the Parylene thickness is expected to improve performance while still sealing the etch holes. Fluid (Fluorinert-70) immersed measurements reveal a high 18.6 MHz center frequency and wide 4.9 MHz bandwidth, indicating good acoustic coupling. The peak PMUT membrane vibration velocity in the fluid is 1.5 mm/s/V, which corresponds to 2 kPa/V. Phased array simulations based on measured PMUT parameters show high output pressure of the focused narrow acoustic beam, which demonstrates the feasibility of the array for use in IVUS application.

ACKNOWLEDGEMENTS

The authors thank Dr. Benjamin A. Griffin and Keith Ortiz at Sandia National Labs for AlN MEMS fabrication, the UC Berkeley Marvell Nanofabrication Laboratory for post-processing, and Berkeley Sensor and Actuator Center (BSAC) Industrial Members for financial support.

REFERENCES

[1] S. Shelton, M.L. Chan, H. Park, and D. Horsley, "CMOS-compatible AlN piezoelectric micromachined ultrasonic transducers", *Proc. IEEE International Ultrasonics Symposium,* Rome, September 20-23, 2009, pp. 402-405.

[2] F. L. Degertekin, R. O. Guldiken, and M. Karaman,

"Annular-ring CMUT arrays for forward-looking IVUS: transducer characterization and imaging", *IEEE Trans Ultrason Ferroelectr Freq Control*, vol. 53, pp. 474-482, 2006.

[3] D. T. Yeh, O. Oralkan, I. O. Wygant, M. O'Donnell, and B. T. Khuri-Yakub, "3-D ultrasound imaging using a forward-looking CMUT ring array for intravascular/intracardiac applications," *IEEE Trans. Ultrason., Ferroelect., Freq. Contr.*, vol.53, no.6, pp. 1202-1211, June 2006.

[4] W. Liao, W. Liu, J. E. Rogers, F. Usmani, Y. Tang, B. Wang, H. Jiang and H. Xie, "Piezoelectric Micromachined Ultrasound Transducer Array for Photoacoustic Imaging", *Proc. Transducers,* Barcelona, 16-20 June 2013, pp. 1831-1834.

[5] D.E. Dausch, K.H. Gilchrist, J.R. Carlson, J.B. Castellucci, D.R. Chou and O.T. von Ramm, "Improved pulse-echo imaging performance for flexure-mode pMUT arrays", *Proc IEEE International Ultrasonics Symposium*, San Diego, 11-14 Oct. 2010, pp. 451-454.

[6] R.H. Olsson, J.G. Fleming, K.E. Wojciechowski, M.S. Baker, and M.R. Tuck, "Post-CMOS Compatible Aluminum Nitride MEMS Filters and Resonant Sensors", *Proc. IEEE Frequency Control Symposium*, Geneva, May 29 2007-June 1 2007, pp. 412-419.

[7] R.H. Olsson, K.E. Wojciechowski, M.S. Baker, M.R. Tuck, and J.G. Fleming, "Post-CMOS-compatible aluminum nitride resonant MEMS accelerometers", *J. Microelectromech. Syst.*, vol. 18, no. 3, pp. 671-678, 2009.

CONTACT

*Yipeng Lu, tel: +1-530-752-5180; yplu@ucdavis.edu

IMPACT OF GYROSCOPE OPERATION ABOVE THE CRITICAL BIFURCATION THRESHOLD ON SCALE FACTOR AND BIAS INSTABILITY

S. Nitzan[1], T-H. Su[1], C. Ahn[2], E. Ng[2], V. Hong[2], Y. Yang[2], T. Kenny[2], and D.A. Horsley[1]

[1]University of California, Davis, Davis, USA
[2]Stanford University, Stanford, USA

ABSTRACT

This paper investigates the impact of operating a vibratory rate gyro (VRG) at large oscillation amplitude where the VRG's driven axis behaves like a nonlinear oscillator, described by the Duffing equation. Although open-loop resonators operating above a critical amplitude exhibit catastrophic jump instabilities, we demonstrate that through closed-loop operation, the drive axis can be stably operated at an amplitude above this threshold without impacting drive-axis stability or bias instability, resulting in decreased Angle Random Walk (ARW).

INTRODUCTION

Due to their small size, low power consumption, and ease of manufacture, MEMS vibratory rate gyros (VRGs) present a promising alternative to their macro-scale counterparts. In recent years, MEMS gyros have shown steadily increasing performance in terms of Angle Random Walk (ARW), however in many cases this performance comes at the cost of large die area. A VRG's thermomechanical noise-limited ARW, in (°/s)/√Hz, is given by

$$ARW = \frac{1}{A_\theta x}\sqrt{\frac{k_B T}{m \omega_n Q}}\frac{180}{\pi} \qquad (1)$$

where A_θ is the angular gain, x is oscillation amplitude of the driven axis, k_B is Boltzmann's constant, T is the temperature in K, m is the modal mass, ω_n is the resonant frequency (assuming mode-matched operation), and Q is the sense-axis quality factor.

Here, we investigate methods to achieve low ARW in a small gyro having a proof mass area less than 0.3 mm². Equation (1) shows that in order to achieve low ARW in a small (and therefore low mass) gyro, x, ω_n, and Q must be increased. As will be seen, however, increasing any of these quantities results in increased nonlinearity in the VRG's driven oscillator axis.

NONLINEAR OSCILLATORS

MEMS gyros, like other MEMS resonators, when driven to large drive amplitude exhibit cubic nonlinearities which can be modeled by the Duffing equation:

$$\ddot{u} + \frac{\omega_n}{Q}\dot{u} + \omega_n^2 + \alpha u^3 = F \qquad (2)$$

where u is the displacement, and α is the coefficient of cubic nonlinearity, which can be a positive or negative constant depending on whether the nonlinearity is due to geometric

spring stiffening or electrostatic spring softening, respectively. An approximate solution to this equation can be obtained through the method of multiple time scales [1], yielding frequency response functions such as those shown in Figure 1 for $\alpha < 0$. Here, as the oscillation amplitude increases, the resonant peak bends to the left. Above a critical amplitude threshold, x_c, where the phase-frequency relationship has infinite slope, and the amplitude-frequency relationship becomes multi-valued. Stable and unstable equilibria are indicated by solid and dashed lines respectively. Open-loop oscillation above x_c leads to dramatic jump instabilities, as well as hysteretic behavior.

As demonstrated by Kaajakari *et al.* [2] high-Q resonators reach x_c at a very small amplitude. This is illustrated in Figure 2, which shows the transfer functions of two resonators with $Q = 1000$ and $Q = 80,000$ juxtaposed. Furthermore, in mode-matched VRGs, increasing ω_n results in increased electrostatic nonlinearity because the frequency

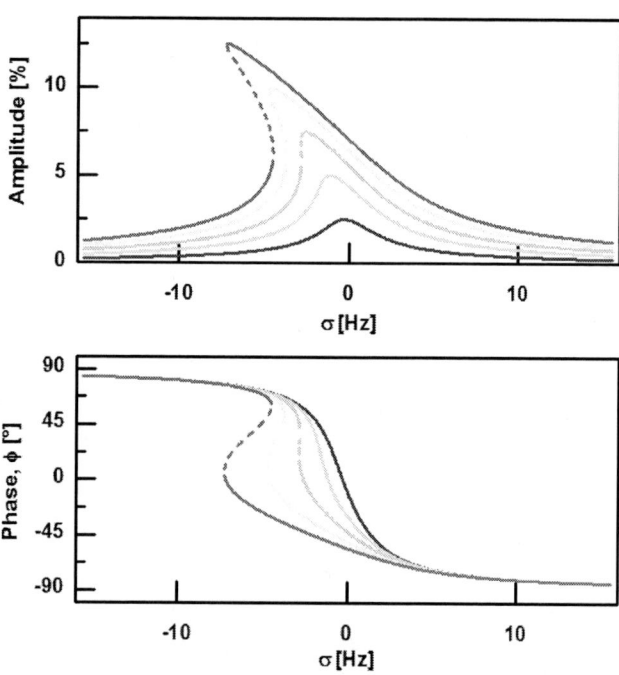

Figure 1: Theoretical frequency response of a resonator driven over a range of amplitudes. Here the amplitude is normalized to the nominal gap of the capacitive sensing electrodes and σ is the offset from nominal resonant frequency. Above x_c, the amplitude-frequency response becomes multi-valued, exhibiting two stable equilibria (solid lines) and one unstable equilibrium (dashed lines).

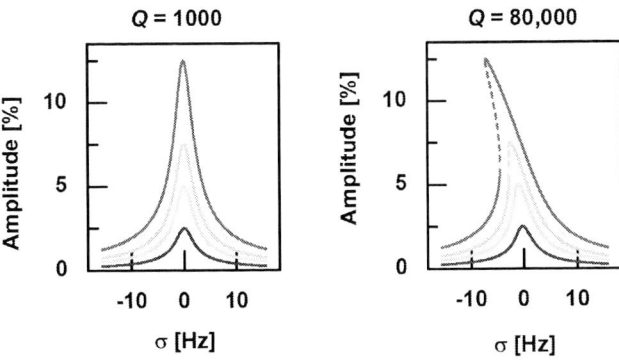

Figure 2: Amplitude-frequency response for a device with Q = 1000 (left) and Q = 80,000 (right) showing early onset of nonlinearity for devices with high Q.

difference between the drive and sense axes scales proportionally to ω_n, requiring either increased bias voltage or smaller gap to achieve mode-matched operation, both of which increase the nonlinearity, α. Thus, in order to achieve low ARW in a small gyro, nonlinearity must be addressed.

Because a VRG's scale factor depends on both the amplitude and frequency of oscillation, large amplitude is desirable. However, instability in the oscillator's amplitude or frequency will result in scale-factor and bias instability in the gyro's rate-sensing output. For this reason, and because it had been assumed until recently that it is impossible to operate a resonator stably in the highly nonlinear region above x_c, the conventional wisdom is that this limit defines the maximum amplitude for stable, low-noise gyro performance [2]. However, recent work by Lee *et al.*[3] showed that the phase-amplitude relationship, shown in Figure 3, is always single-valued, even at points above x_c, enabling stable closed-loop operation by maintaining a constant phase in the oscillation loop. Conveniently, the maximum amplitude always occurs at 0°, regardless of nonlinearity.

Operating at the point where the phase-frequency slope, $d\phi/df$, becomes infinite can result in reduced phase noise [4] [5], however the sensitivity of the oscillator's *amplitude* to phase variations, $dx/d\phi$, increases as displacement increases. This is illustrated in Figure 4, top. However, when this result is normalized to displacement amplitude, as

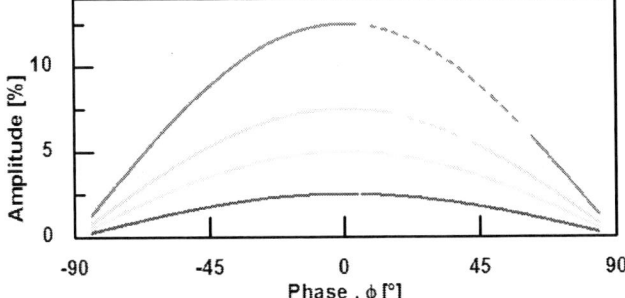

Figure 3: Amplitude vs. phase for increasing drive-axis displacement (x). The maximum amplitude always occurs at 0° phase shift regardless of displacement x.

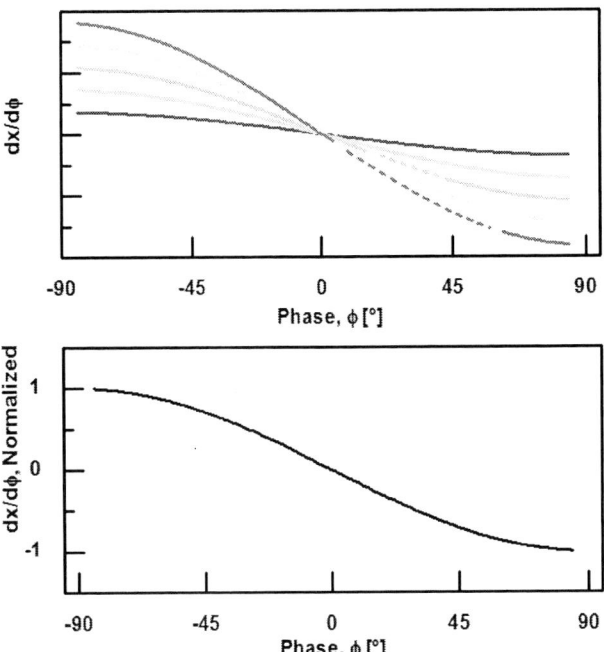

Figure 4: Top: Derivative of amplitude with respect to phase for increasing displacement. Amplitude sensitivity to phase variations increases as displacement increases, however, when normalized to peak displacement, (Bottom) the sensitivity is independent of displacement.

shown in Figure 4, bottom, it becomes clear that the fractional amplitude stability is unchanged when the oscillation amplitude is increased above x_c.

EXPERIMENTAL RESULTS

Experiments were conducted using the 0.6 mm single-crystal <100> frequency-compensated DRG described by Ahn *et al.* in [6], and shown in Figure 5. This device, with, a modal mass of only 4 μg, has roughly one hundredth of the mass of many MEMS gyros. High Q (80k), and resonant frequency (250 kHz), both help to compensate for this, but introduce large nonlinearities at small drive amplitudes. After performing electrostatic mode-match with a dedicated set of tuning electrodes, similar to those described in [7], the A (drive) axis is driven in a closed loop with a digital PLL as shown in Figure 5. By adjusting the phase of the drive-axis loop, the transfer function of the device can be experimentally measured, including regions that are unstable in the open loop. Figure 6 shows the device's drive-axis frequency response at varying amplitudes (x), demonstrating x_c is reached when the peak amplitude is 3% of the nominal capacitive gap. Despite changes in x, the maximum displacement occurs at 0°. With the PLL locked to 0° phase shift, the device is then operated as shown in Figure 5. The sense axis is IQ-demodulated, with the in-phase component used as a measure of rate, while the quadrature component is the input to a second quadrature-nulling feedback loop. Mode matching is performed by independently adjusting the frequency tuning voltage for

978-1-4799-3510-9/14 $31.00 © 2014 IEEE

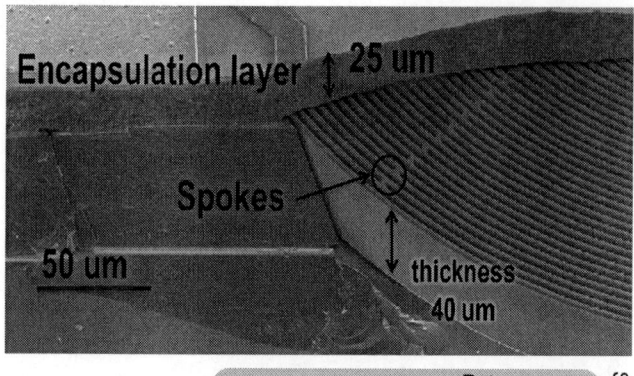

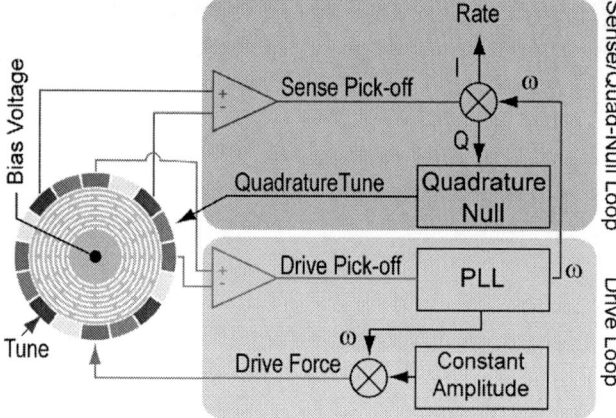

Figure 5: Top: SEM of device. Bottom: Diagram showing device operation. After electrostatic mode-matching, the Drive-axis is driven in a closed loop at constant input amplitude with a digital PLL, and the Sense-axis signal is IQ demodulated, with the in-phase component being used as a measure of rate, and the quadrature component being used to adjust the voltage on the quadrature-nulling electrodes. Both axes are differentially sensed.

each drive-amplitude x. The temperature-controlled, demodulated zero-rate output

(ZRO) is then collected over a 3-hour period for each value of x. In addition to recording the ZRO, the demodulated drive axis output is also recorded, so that the amplitude and frequency stability can be directly evaluated. These results are shown in Figure 7. Although 0.1 °C temperature variations result in small (<3 ppm) frequency variations, this effect is independent of driven amplitude. Similarly, variations in drive-axis amplitude due to the temperature-dependence of thermoelastic-damping limited Q are small and independent of driven amplitude.

Provided the axes are mode-matched for each drive amplitude, the scale factor will increase linearly with drive displacement ($SF \propto x$), Here we see a six-fold (linear) increase in scale factor as the drive displacement is increased from 0.6% to 3.8% of the nominal gap. Because thermo-mechanical and electronic noise are both independent of driven amplitude, the SNR increases with drive amplitude, resulting in decreased ARW. This result is

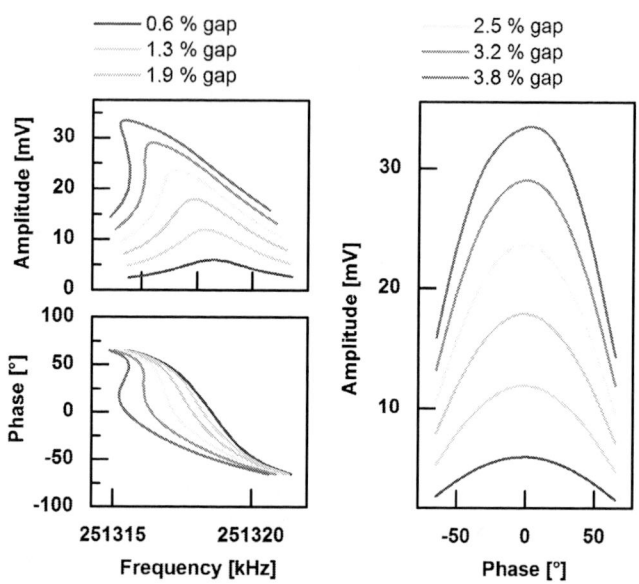

Figure 6: Left: Measured amplitude and phase vs. frequency at various drive-axis amplitudes (x). Increasing x results in increased nonlinearity. The critical bifurcation point, where the slope of the amplitude vs. frequency curve becomes infinite, first occurs at $x_c = 3\%$. Right: Amplitude vs. phase for varying drive-axis amplitudes. The maximum amplitude occurs at the same phase (0°) regardless of the amplitude x.

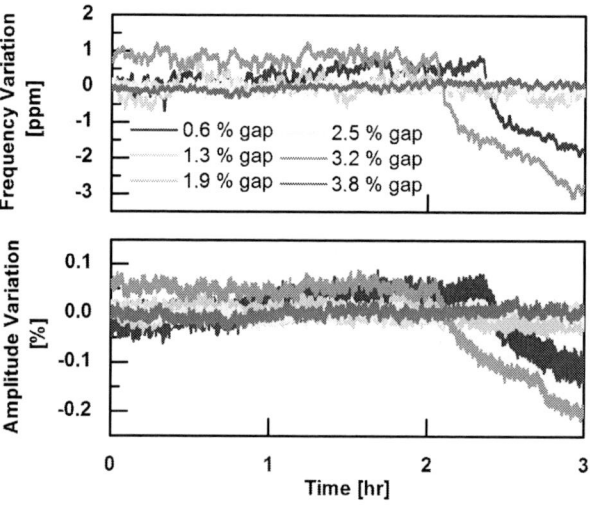

Figure 7: Frequency and amplitude variation vs. time for varying drive amplitude. Although 0.1 °C changes in room temperature result in small (<3 ppm) variations in frequency, this effect is independent of whether $x > xc$. Amplitude variations due to thermoelastically-limited Q result in ± 0.15% change in scale factor, and contribute to bias instability at long time frames.

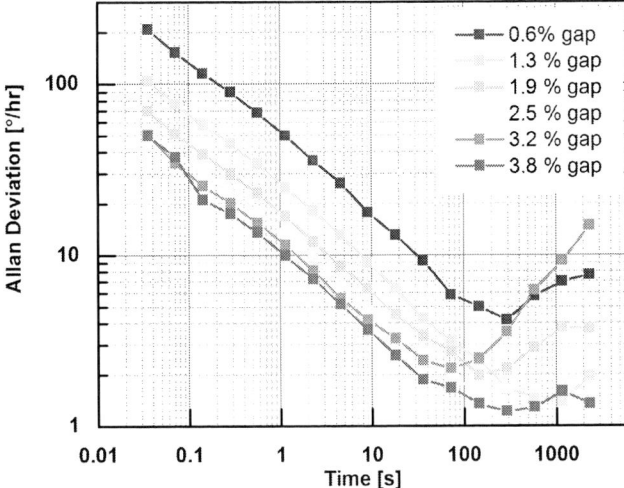

Figure 8: Allan deviation at various amplitudes. Performance improves for short time frames as amplitude increases, despite operation at $x > x_c$. Drive-axis amplitude variations, particularly pronounced for the 0.6% and 3.2% cases, as shown in Figure 7, contribute to bias instability at long time frames.

shown in Figure 8. The ARW, plotted separately in Figure 9, drops from 0.014 (°/s)/√Hz to 0.002 (°/s)/√Hz as the drive amplitude is increased, showing no negative effects due to operation above x_c. Bias instability, shown in Figure 9, right, is approximately 1.5 °/hr, and is independent of drive amplitude. This is because bias (β) in the gyro's output is due primarily to stiffness and damping coupling from the driven axis, and is therefore proportional to the drive-axis displacement ($\beta \propto x$). Increased scale factor, however, compensates for this effect, with the result that input-referred bias (β/SF) and bias instability, measured in °/s, remain unchanged.

CONCLUSIONS

In order to achieve low ARW with decreased sensor mass, x, ω_n and Q can be increased, however increasing any of these parameters leads to early onset of nonlinearity. Although operation above x_c leads to instability in the open-loop, closed-loop operation enables stable operation at and above x_c. Here we demonstrate that, counter to conventional wisdom and previous work by our group [7], operating above x_c can result in decreased ARW, without affecting bias instability. Drive-axis amplitude and frequency stability are also unchanged. Although these results have been demonstrated for a device experiencing electrostatic (spring-softening) nonlinearity, in theory they apply equally well to devices experiencing third-order spring-hardening nonlinearities.

ACKNOWLEDGEMENTS

The authors would like to thank Mo Li for discussion regarding nonlinearities. This project was funded by DARPA under contracts W31P4Q-12-1-0001 and N66001-12-1-4260. The authors would also like to thank

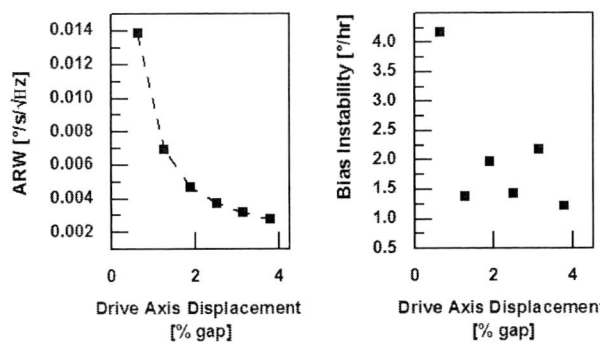

Figure 9: ARW and bias instability vs. drive axis displacement. Due to increased scale factor, ARW improves at large drive amplitudes, despite operation above $\dot{x}_c$. Bias instability is independent of increased drive amplitude.

Dr. Andrei Shkel and Dr. Robert Lutwak, MTO Program Managers responsible for the Micro Precision Navigation and Timing Program at DARPA. The device was fabricated at the Stanford Nanofabrication Facility.

REFERENCES

[1] M. Nayfeh and D. Mook, "Forced Oscillations of Systems Having a Single Degree of Freedom," in *Nonlinear Oscillations*, New York, John Wiley & Sons, 1979, ch. 4, pp. 161-196.

[2] V. Kaajakari et al, "Nonlinear Limits for Single-Crystal Silicon Microresonators," in *J. Microelectromech. Syst.*, vol. 13, no. 5, pp 715-724 2004.

[3] H.K. Lee, et. al, "Stable Operation of MEMS Oscillators Far Above the Critical Vibration Amplitude in the Nonlinear Regime", *J. Microelectromech. Syst.* vol. 20 no. 6 pp. 1228-1230, 2011.

[4] Pardo, M.; Sorenson, L.; Ayazi, F., "An Empirical Phase-Noise Model for MEMS Oscillators Operating in Nonlinear Regime," *Circuits and Systems I: Regular Papers, IEEE Transactions on* , vol.59, no.5, pp.979,988, May 2012.

[5] H.K. Lee et al., "Verification of the Phase-Noise Model for MEMS Oscillators Operating in the Nonlinear Regime," in *Intl. Conf. on Solid State Sensors, Actuators, and Microsystems (Transducers)*, Beijing, 2011, pp 510-513.

[6] C.H. Ahn et al., "Geometric Compensation of (100) Single Crystal Silicon Disk Resonating Gyroscope For Mode-Matching," in *Intl. Conf. on Solid State Sensors, Actuators, and Microsystems (Transducers)*, Barcelona, 2013, pp. 1723-1726.

[7] S. Nitzan et al., "Epitaxially-Encapsulated Polysilicon Disk Resonator Gyroscope," in *IEEE MEMS*, Taipei, 2013, pp. 625-628.

CONTACT

*D.A. Horsley, tel: 1-530-341-3236; dahorsley@ucdavis.edu

IMPROVED ACOUSTIC COUPLING OF AIR-COUPLED MICROMACHINED ULTRASONIC TRANSDUCERS

Stefon Shelton[1], Ofer Rozen[1], Andre Guedes[1], Richard Przybyla[2], Bernhard Boser[2], David A. Horsley[1]
[1] Berkeley Sensor & Actuator Center, University of California, Davis, CA, USA
[2] Berkeley Sensor & Actuator Center, University of California, Berkeley, CA, USA

ABSTRACT

Phased array imaging with micromachined ultrasound transducer (MUT) arrays is widely used in applications such as ranging, medical imaging, and gesture recognition. In a phased array, the maximum spacing between elements must be less than half of the wavelength to avoid large sidelobes. This places a limit on the maximum transducer size which is not attractive since the acoustic coupling drops rapidly for MUT diameters less than a wavelength.

Here, we present a new approach to increase the acoustic coupling of small radius MUTs using an impedance matching resonant tube etched beneath the MUT. Impedance, laser Doppler vibrometer (LDV), and acoustic burst measurements confirm a 350% increase in SPL and 8x higher bandwidth compared to transducers without the impedance matching tube, enabling compact arrays with high fill-factor and efficiency.

INTRODUCTION

Air-coupled micromachined ultrasound transducers (MUTs) achieve maximum output sound pressure level (SPL) and bandwidth when the radius, a, approaches the acoustic wavelength, λ. This requirement imposes a significant limitation on the minimum achievable transducer size for both piezoelectric micromachined transducers (PMUTs) and capacitive MUTs (CMUTs), particularly for transducers operating in array configurations. For transducers employing a compliant plate structure and operating less than 200 kHz this typically results in transducers with radius greater than 0.5 mm [1-5]. In this paper, we explore the use of a resonant tube, etched into the backside of a wafer, to enhance the acoustic coupling and bandwidth of small scale transducers ($a \sim \lambda/8$) enabling a significant size reduction and enhanced array fill factors, without loss in transducer output SPL or bandwidth.

THEORY AND MODELING

The PMUTs presented here are designed for air coupled operation in a 2D array configuration at a nominal frequency of 200 kHz [6]. The PMUT, shown in Fig. 1, is a unimorph design based on a SiO$_2$/AlN/Mo/AlN film stack. The thickness of the AlN layers is 1 um, the Mo and Al electrodes are 150 nm, and the SiO$_2$ is 100 nm. Detailed description of the design and fabrication of these devices can be found in our prior work [7, 8].

To illustrate the coupling performance we model the transducer, without the resonator tube, as a simple circular piston in an infinite baffle, the specific acoustic impedance (Rayl) of which is given by [9]:

$$Z_p(ka) = Z_0\left(1 - \frac{2J_1(2ka)}{2ka} + i\frac{2K_1(2ka)}{2ka}\right)$$

Where $Z_0 = 413$ Rayl is the characteristic impedance of air, a is the radius of the piston, $k = 2\pi/\lambda$ is the wave number, J_1 is the Bessel function of order 1, and K_1 is the Struve function of order 1. The normalized real and imaginary parts of Z_p are plotted in Fig. 2. The real part of the acoustic impedance, $Re(Z_p)$, determines the power radiated to the air. Small PMUTs ($ka < 2$) experience reduced acoustic impedance relative to larger transducers. Consequently, the stored mechanical energy is much larger than the radiated energy, resulting in narrow bandwidth and poor coupling. The PMUTs presented in this work have $ka \sim 0.7$ which results in $Re(Z_p)$ that is only 22% of Z_0.

Here, to enhance acoustic coupling and bandwidth, DRIE is used to create a tube resonator extending from the back of the wafer to the bottom of the PMUT membrane. Appropriate design of the tube dimensions results in an acoustic resonance that occurs at the same frequency as the mechanical resonance of the PMUT, thereby increasing the real acoustic impedance seen by the PMUT. The result is that the transducer's bandwidth is dramatically increased and the SPL emitted from the tube is several times greater than the SPL emitted from the front of the PMUT membrane.

Equivalent Circuit Model

An equivalent circuit model of the transducer and resonator tube acoustics is shown in Fig. 3. Building on our previous work with the electrical and mechanical portions of the model [10] the acoustic tube resonator is added as a transmission line element. The modal mass and stiffness used in the analytic model are extracted from impedance

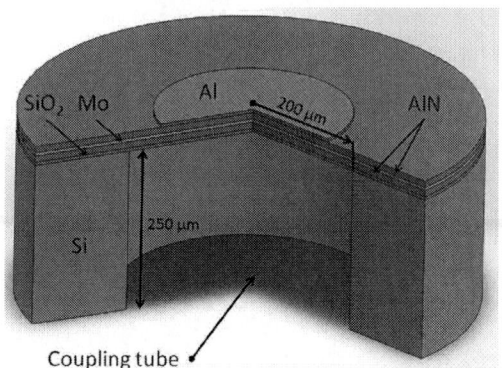

Figure 1: PMUT with coupling tube cross section and dimensions. The AlN layers are each 1 µm thick, the SiO$_2$ is 100 nm, and the Mo and Al layers are 150 nm.

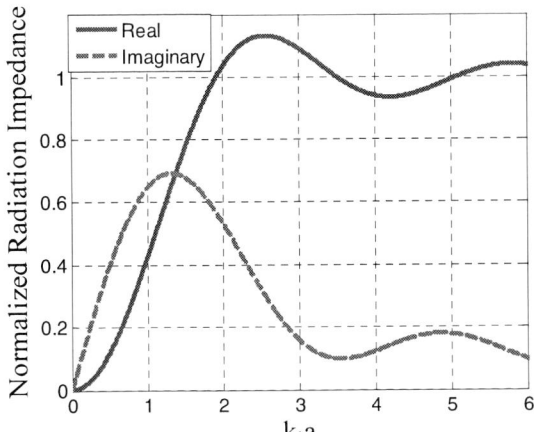

Figure 2: Real and imaginary parts of the radiation impedance normalized to the characteristic impedance of air and plotted versus ka = 2πa/λ, where a is the radius and λ is the acoustic wavelength.

and laser Doppler vibrometer (LDV) measurements [10] for transducers with resonant frequencies of ~230 kHz and are m_m=0.197 ng, and k_m=308 N/m respectively. In the acoustic domain we model the plate as a baffled piston. We assume a clamped plate model with a normalized plate deflection shape function of [11]:

$$\varphi(x) = (1 - x^2)^2$$

where x is the radial coordinate. Since we are modeling the acoustic source as a piston, we define an effective radius for this piston, a_{eff}, such that the piston, moving with the peak velocity of the plate, has the same volume velocity as that of the flexing plate. The acoustic impedance (Rayls/m^2) of the front side of the plate is also a baffled pison with effective radius, a_{eff}:

$$Z_{front} = Z_{baffle}(ka_{eff}) = \frac{Z_p(ka_{eff})}{\pi a_{eff}^2}$$

Although this impedance is a function of wavenumber k (or, equivalently, frequency f), in what follows we use compact notation and omit the explicit dependence on k. At the throat (open end) of the tube, moving air also encounters an acoustic impedance which is that of a baffled piston but the entire radius contributes:

$$Z_{th} = Z_{baffle}(ka) = \frac{Z_p(ka)}{\pi a^2}$$

The impedance of the acoustic transmission line is the acoustic impedance of the medium divided by the cross sectional area of the tube, $Z_{tube} = \frac{z_o}{\pi a^2}$. To calculate the

impedance seen by the backside of the plate we calculate the reflection and transmission coefficients for the throat of the tube as [9]:

$$R_{th} = \frac{Z_{th} - Z_{tube}}{Z_{th} + Z_{tube}}$$
$$T_{th} = \frac{2Z_{th}}{Z_{th} + Z_{tube}}$$

and the impedance of the backside of the plate:

$$Z_{back} = Z_{tube}\frac{e^{ikd} + R_{th} \cdot e^{-ikd}}{e^{ikd} - R_{th} \cdot e^{-ikd}}$$

The total acoustic impedance seen by the plate is then:

$$Z_{total} = Z_{front} + Z_{back}$$

and the total air damping is:

$$b_{air} = \left(\pi \cdot a_{eff}^2\right)^2 \cdot Re(Z_{total})$$

The acoustic damping is the dominant damping element and therefore the bandwidth is:

$$BW = \frac{b_{air}}{2\pi m_m}$$

As a result of the tube's acoustic resonance, the pressure at the tube's throat is greater than that on the front face of the PMUT, a phenomenon we describe as resonator gain. The impedance seen looking into the tube from Z_{th}, defined as Z_{fm}, is the sum of Z_{front} and the mechanical equivalent circuit impedances transformed into the acoustic domain using an ideal transformer:

$$Z_{fm} = Z_{front} + \frac{1}{\left(\pi a_{eff}^2\right)^2} \cdot \left(i2\pi f_n m_m + \frac{k_m}{i2\pi f_n}\right)$$

Similarly, the reflection and transmission coefficients at the plate end of the tube are given by:

$$R_p = \frac{Z_{fm} - Z_{tube}}{Z_{fm} + Z_{tube}}$$
$$T_p = \frac{2Z_{fm}}{Z_{fm} + Z_{tube}}$$

To calculate the tube gain we consider a Thévenin equivalent pressure source and impedance seen by the throat (load) impedance at the end of the tube:

$$P_{eq} = \frac{P_{in}}{\cos(kl)} \cdot \left(-\frac{iZ_{tube}\cot(kl)}{Z_{fm} - iZ_{tube}\cot(kl)}\right)$$

$$Z_{eq} = Z_{tube}\frac{e^{ikl} + R_p \cdot e^{-ikl}}{e^{ikl} - R_p \cdot e^{-ikl}}$$

Where $P_{in} = \frac{\eta}{\pi \cdot a_{eff}^2} \cdot V_{in}$ is the input pressure and l is the

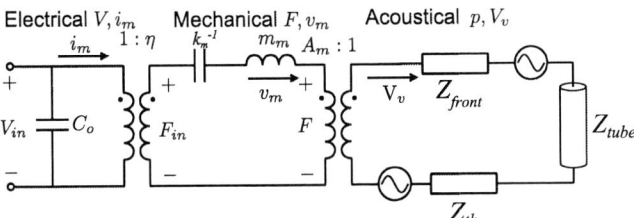

Figure 3: Equivalent circuit model of PMUT and acoustic resonator tube which is modeled as an acoustic transmission line with characteristic impedance, Z_{tube}.

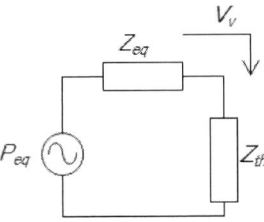

Figure 4: Simplified equivalent circuit model showing Thévenin equivalent circuit parameters.

length of the tube. The Thévenin equivalent circuit is shown in Fig 4. Using these parameters, we calculate the gain from the pressure on the tube-side transducer face to the pressure at the throat:

$$G_{tx} = \left|\frac{P_{eq}}{P_{in}}\right| = \left|\frac{-iZ_{tube}\cdot\cot(kl)}{[Z_{fm}-iZ_{tube}\cdot\cot(kl)]\cdot\cos(kl)}\right|$$

Finally, the resonator gain, defined as the ratio of the pressure magnitude at the throat of the tube to the pressure magnitude at the front side of the transducer, is calculated as:

$$G_{resonator} = \left|\frac{P_{th}}{P_{front}}\right| = \left|\frac{G_{tx}\left(Z_{back}+Z_{fm}\right)}{Z_{th}+Z_{eq}}\right|$$

Finite Element Model

The PMUT's dynamic response is modeled using a two dimensional axisymmetric piezo-acoustic finite element method (FEM) model implemented in COMSOL. The piezoelectric portion of the model is a fixed laminated circular plate that consists of two AlN layers, each of 1μm thick. The passive (bottom) layer is defined as a linear elastic material while the active (top) layer is defined as a piezoelectric material. An alternating voltage is applied across 70% of the radius of the active piezoelectric layer, corresponding to the radius of the PMUT's top electrode.

The simulation geometry and output pressure field are shown in Fig. 5. The tube boundary condition is defined as a perfectly reflecting wall. The air surrounding the transducer is modeled as two 5 cm radius half spheres (not fully shown in Fig. 5) bounded by a perfectly matched layer. The two spheres, as well as the interior of the tube are defined as air with a theoretical attenuation constant calculated at 200 kHz [9]. The SPL is probed at two points, one on each sphere's matched layer, located on the axis of symmetry.

Modeling results

The predicted resonator gain frequency response for different tube lengths is shown in Fig. 6 for both the FEM and equivalent circuit models. The maximum resonator gain

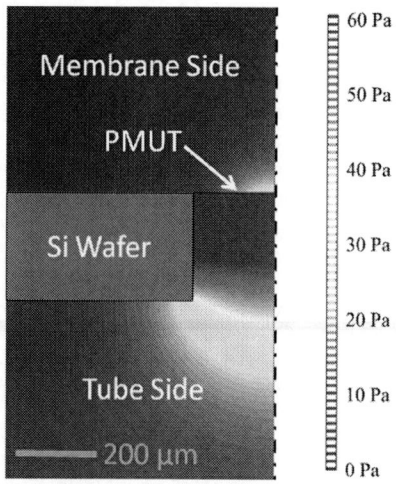

Figure 5: 2D axisymmetric FEM piezo-acoustic simulation geometry and output pressure field showing the gain from the acoustic resonator tube.

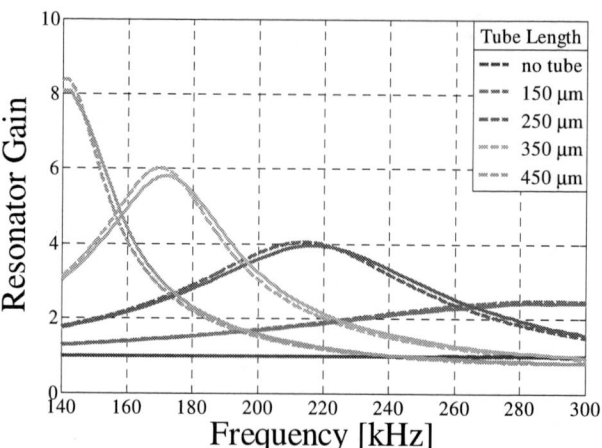

Figure 6: The analytic (dashed lines) and FEM (solid lines) resonator gain for different tube lengths. Without a tube the two sides are identical.

occurs at a frequency that varies with tube length. The analytic and FEM models are in good agreement and show that output pressure gains >8x are possible depending upon the frequency of operation and tube design. For our nominal design, we predict a resonator gain of 4 on both transmit and receive at a frequency of ~217kHz. This is equivalent to increasing the $Re(Z_{Baffle})$ to 88% of the high frequency value, a significant improvement.

EXPERIMENT

Acoustic testing was performed using a burst measurement technique. The transducer was excited at its resonant frequency with 50 cycles of a 220 kHz sine wave and the pulse was received by a microphone placed 5 cm away. The time domain signal was filtered, amplified, and captured with an oscilloscope. The transmit pulse and received signals from the membrane and tube side of the

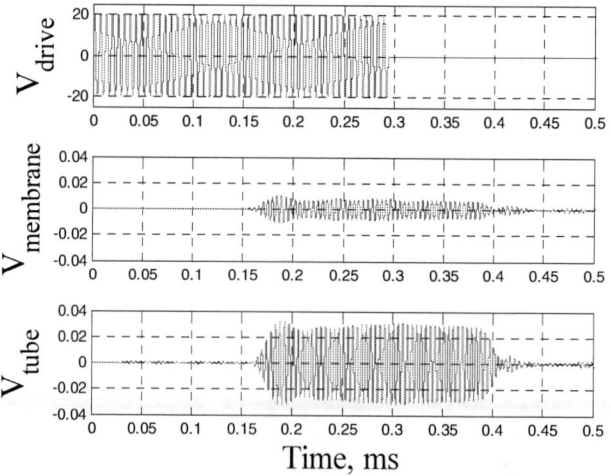

Figure 7: Burst measurement result showing the drive burst (V_{drive}) and microphone output voltage of the received acoustic signals from the membrane side ($V_{membrane}$) and tube resonator side (V_{tube}). The measured resonator-side output SPL is 3.5x higher than the membrane-side SPL.

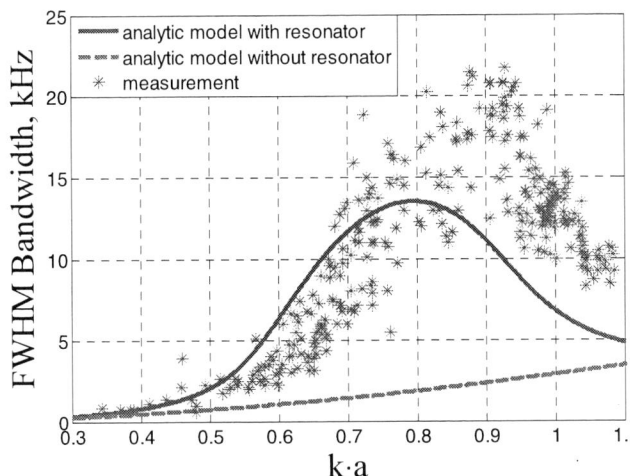

Figure 8: Measured and modeled bandwidth for PMUTs with and without an acoustic resonator. The acoustic resonator increases the bandwidth by a factor of 8.

transducer are shown in Fig 7. Comparing the received bursts we can see resonator gain of 3.5 which matches closely with the theoretical and FEM predicted gain of 4. In addition, the resonant tube increases the transducer bandwidth due to the increased acoustic coupling. In this way the tube can be thought of as an acoustic matching layer. To demonstrate the effect on the bandwidth we measured the frequency response, in air, of ~250 PMUTs with identical dimensions and varying acoustic wavelengths using a LDV. The acoustic wavelength variation in the transducers is due to residual stress variation in the thin film layers. Fig. 8 shows the measured bandwidth and the bandwidth predicted by the analytic model with and without the acoustic resonator. In the model, a constant mass m_m is assumed and the stiffness k_m is varied to yield varying resonant frequency. The analytic model results illustrate an increase in bandwidth of 8x over a transducer without the tube.

CONCLUSION

In this paper we have presented a method to improve the acoustic coupling and bandwidth of small radius micromachined ultrasound transducers. An equivalent circuit analytic model is presented and the results compared with a 2D-axisymmetric piezo-acoustic finite element simulation. Fully clamped circular PMUTs with a radius to wavelength ratio of 11% were fabricated and characterized in the electrical, mechanical, and acoustic domains. An acoustic gain of 350% is demonstrated using transducers with a resonant tube 250 µm in length. This is equivalent to increasing the $Re(Z_p)$ to 88% of the large diameter value, a significant improvement. In addition, measurements reveal peak bandwidths of >20 kHz (FWHM) for transducers with a resonator tube. Analytic modeling shows this is an 8x increase in bandwidth over transducers without the acoustic resonator tube. The increase in output pressure and bandwidth allows for the use of high fill factor arrays rather than a single large transducer, which opens up new applications for air coupled MUTs as well as increasing the performance of existing systems.

ACKNOWLEDGEMENTS

The authors would like to thank the Berkeley Marvell Nanolab fabrication facility, Texas Instruments, the UC Berkeley SWARM Lab, and the Berkeley Sensor and Actuator Center for their support of this research.

REFERENCES

[1] S. Horowitz, T. Nishida, L. Cattafesta, and M. Sheplak, "Development of a micromachined piezoelectric microphone for aeroacoustics applications," *J Acoust Soc Am,* vol. 122, pp. 3428-3436, 2007.

[2] M. D. Williams, B. A. Griffin, T. N. Reagan, J. R. Underbrink, and M. Sheplak, "An AlN MEMS piezoelectric microphone for aeroacoustic applications," *IEEE ASME J Microelectromech Syst,* vol. 21, pp. 270-283, 2012.

[3] B. A. Griffin, M. D. Williams, C. S. Coffman, and M. Sheplak, "Aluminum nitride ultrasonic air-coupled actuator," *IEEE ASME J Microelectromech Syst,* vol. 20, pp. 476-486, 2011.

[4] H. Lee, D. Kang, and W. Moon, "A micro-machined source transducer for a parametric array in aira)," *J Acoust Soc Am,* vol. 125, pp. 1879-1893, 2009.

[5] K. K. Park and B. T. Khuri-Yakub, "3-D airborne ultrasound synthetic aperture imaging based on capacitive micromachined ultrasonic transducers," *Ultrasonics,* vol. 53, pp. 1355-1362, 2013.

[6] S. E. Shelton, A. Guedes, R. J. Przybyla, R. Krigel, B. E. Boser, and D. A. Horsley, "Aluminum nitride piezoelectric micromachined ultrasound transducer arrays," *Hilton Head Solid-State Sensors, Actuators, and Microsystems Workshop*, Hilton Head, SC, 2012.

[7] A. Guedes, S. Shelton, R. Przybyla, I. Izyumin, B. Boser, and D. Horsley, "Aluminum nitride pMUT based on a flexurally-suspended membrane," *Transducers*, Bejing, China, 2011.

[8] S. Shelton, M.-L. Chan, H. Park, D. Horsley, B. Boser, I. Izyumin*, et al.*, "CMOS-Compatible AlN piezoelectric micromachined ultrasonic transducers," *IEEE Int. Ultrasonics Symp. (IUS)*, Rome, Italy, 2009.

[9] D. T. Blackstock, *Fundamentals of Physical Acoustics*: Wiley, 2000.

[10] R. J. Przybyla, S. E. Shelton, A. Guedes, I. I. Izyumin, M. H. Kline, D. A. Horsley*, et al.*, "In-air rangefinding with an AlN piezoelectric micromachined ultrasound transducer," *IEEE Sensors J.,* vol. 11, pp. 2690-2697, 2011.

[11] J. N. Reddy, *Theory and Analysis of Elastic Plates and Shells* vol. 1: Taylor and Francis, 2007.

CONTACT

*S.E. Shelton, tel: 1-707-799-0554; seshelton@ucdavis.edu

MECHANICAL FORCE-DISPLACEMENT TRANSDUCTION STRUCTURE FOR PERFORMANCE ENHANCEMENT OF CMOS-MEMS PRESSURE SENSOR

Chao-Lin Cheng[1], Heng-Chung Chang[1], Chun-I Chang[2], Yu-Tsung Tuan[3], and Weileun Fang[1,2]
[1]Power Mechanical Eng. Dept.,[2]NEMS Inst., National Tsing Hua Univ., Hsinchu, TAIWAN
[3]National Nano Device Laboratories, Hsinchu, TAIWAN

ABSTRACT

This study implements a mechanical force-displacement transduction structure in Fig.1a using the TSMC 0.18μm 1P6M CMOS process to improve CMOS-MEMS capacitive pressure sensor. The membrane will be deformed by pressure and cause the sensing-gap change between undeformed movable-electrode and fixed-electrode. Feature of this study is CMOS-MEMS deformed membrane and undeformed movable-electrode to enable the parallel-plate gap-closing pressure detection. Thus, the performance of pressure sensor can be improved. In comparison, the design with mechanical force-displacement transduction structure will increase sensitivity for 61% in pressure range (100kPa-60kPa). Moreover, sensitivity (non-linearity) of proposed design changes from 2.1fF/kPa (1.7%) to 1.9fF/kPa (2.5%) as pressure range changed from 100~60kPa to 100~20kPa. However, sensitivity (non-linearity) of existing design significantly drops from 1.3fF/kPa (1.5%) to 1.0fF/kPa (13.7%) as pressure range changed.

INTRODUCTION

Presently, MEMS pressure sensor has been extensively applied in consumer electronics, automotive systems, environmental monitoring, medical diagnostics, etc. Various fabrication approaches have been reported to meet the specifications of related applications [1-2]. Miniaturization of the die size is an important concern for cost. In this regard, to integrate the microstructures and sensing electronics in one chip is an attractive approach to implement the MEMS sensors. The standard CMOS process technology enables the monolithic integration of MEMS structures and sensing circuits, and thus is considered as a promising approach to implement micro sensors.

The CMOS-MEMS pressure sensor implemented using foundry available processes has been reported in [3-5]. These existing CMOS-MEMS pressure sensors mainly consist of sensing electrodes respectively in deformable membrane and fixed support (Fig.1b). As pressure applied, the membrane is deformed and the gap between sensing electrodes is changed. The pressure is detected by sensing capacitance change. However, the differences of existing designs are limited to membrane dimensions/shape and films stacking. Moreover, the capacitance change for large pressure is nonlinear. The concept to integrate the rigid-electrode with the flexible mechanical structure for capacitive micro sensors has been reported in [6-7]. The rigid-electrode and flexible-membrane are linked by a mechanical support to form the force-displacement transduction structure. Thus, the deformation of flexible membrane will be detected by the displacement of rigid capacitive sensing electrode.

This study exploits the concept in [6] to design the CMOS-MEMS capacitive pressure sensor with undeformed movable-electrodes on deformed membrane. The device are designed and implemented based on the standard TSMC 0.18μm 1P6M process. An in-house post-CMOS etching process is required to release the MEMS structure. The multilayer metal/dielectric films of CMOS process enables the design flexibility to implement mechanical structure, electrical routing, and sacrificial layer. Thus, performances of CMOS-MEMS pressure sensor are improved.

DESIGN CONCEPT

Fig.1 illustrates the pressure sensor with the mechanical force-displacement transduction structure. In comparison, the existing pressure sensor design with deformable membrane/electrode is also displayed. Fig.1a shows the presented design formed by the stacking of metal and dielectric layers of TSMC 1P6M process. The sensor consists

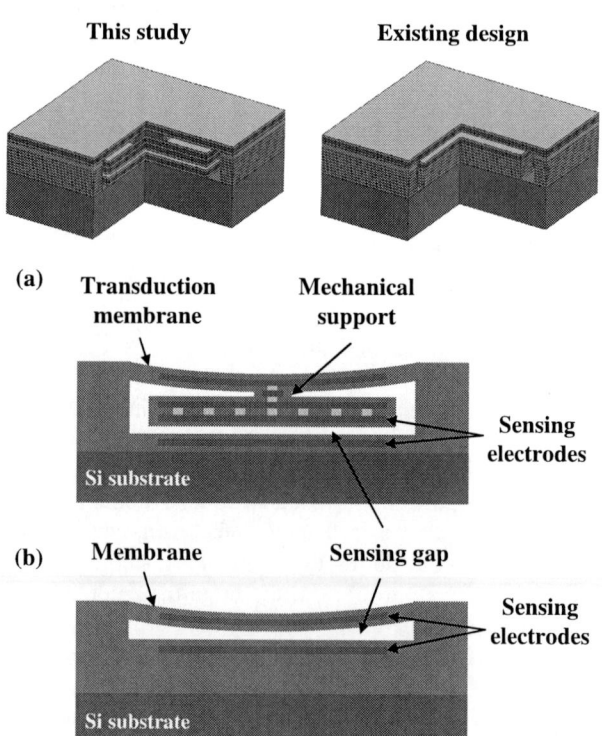

Figure 1: Design concept of pressure sensor, (a) proposed design (with transduction structure), and (b) existing design.

of membrane, mechanical support, movable structure with embedded sensing electrode (Metal-3), and fixed sensing electrode (Metal-1). Note the gap between sensing electrodes is defined by the sacrificial Metal-2 layer. The symmetric layer stacking in [8] is exploited to design the movable structure with embedded sensing electrode to reduce the deformations due residual stresses and CTE (coefficient of thermal expansion) mismatch of thin films. Thus, the movable structure with embedded sensing electrode consists of two metal layers (Metal-3 and Metal-4) and three dielectric layers (IMD2~4). The tungsten-via array is distributed between two metal layers for the electrical routing of movable sensing electrode.

For this absolute pressure design, the reference chamber is sealed in a vacuum environment. As indicated in Fig.1, a pre-deformation on membrane is caused by pressure difference between reference vacuum chamber and atmospheric pressure. The initial sensing capacitance is thus increased due to the reduction of initial gap between sensing electrodes defined by Metal-2. As the ambient pressure changed, the membrane deformation will be changed and cause the sensing gap variation between undeformed movable-electrode and the fixed-electrode. The pressure is then detected by the capacitance change between electrodes. Note the present sensor employs the undeformed movable-electrode design to increase the overlap area between sensing electrodes. Thus, the sensing capacitance is increased as compare with the pressure sensor with deformed sensing electrode in Fig.1b. Note that, these two designs have the same membrane dimension for fair comparison.

FABRICATION AND RESULTS

Fig.2 shows the process steps. Fig.2a shows the layer stacking and patterning after TSMC 0.18μm 1P6M process. Fig.2b shows the H_2SO_4/H_2O_2 solution was used to etch metal/tungsten-via [9] to define the sensing membrane and sub-micron out-of plane sensing gap of pressure sensor. The metal films for electrodes were protected by the dielectric films during the metal wet etching. Thus, the sensing membrane was suspended, and the sensing gap between electrodes is 0.53μm (metal-2 as the sacrificial layer). In Fig.2c, RIE used to open the pads for wire bonding. In Fig.2d, the chip was sealed by 2μm Parylene-C conformal coating for absolute pressure measurement. The vacuum chamber for Parylene-C conformal coating process was near 10Pa. The reference chamber of presented pressure sensor after sealed by Parylene-C was also near 10Pa. Note the chamber could also be sealed by other materials such as PECVD oxide, sputtering metal, etc. Finally the chip was wire-bonding on PCB for test.

The SEM micrographs in Fig.3a show the fabricated chip with proposed and existing designs. To make a fair comparison, the proposed and existing designs have the same membrane dimensions and sensing gap. The integration of 2×2 sensor array design is employed to increase the capacitance sensing signal. The dimension of each square membrane is 160μm×160μm. For fair comparison, the

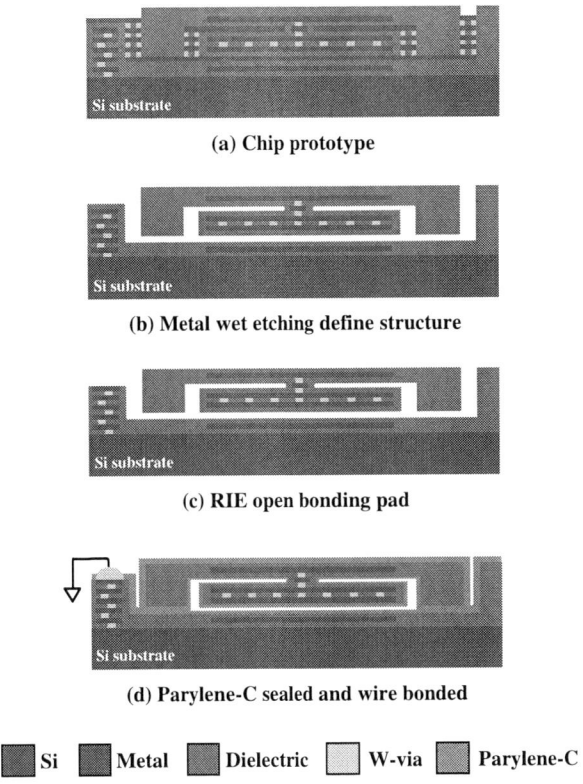

(a) Chip prototype

(b) Metal wet etching define structure

(c) RIE open bonding pad

(d) Parylene-C sealed and wire bonded

■ Si ■ Metal ■ Dielectric ☐ W-via ■ Parylene-C

Figure 2: Process steps, (a) chip prepared by TSMC, (b-d) post-CMOS process and vacuum sealing.

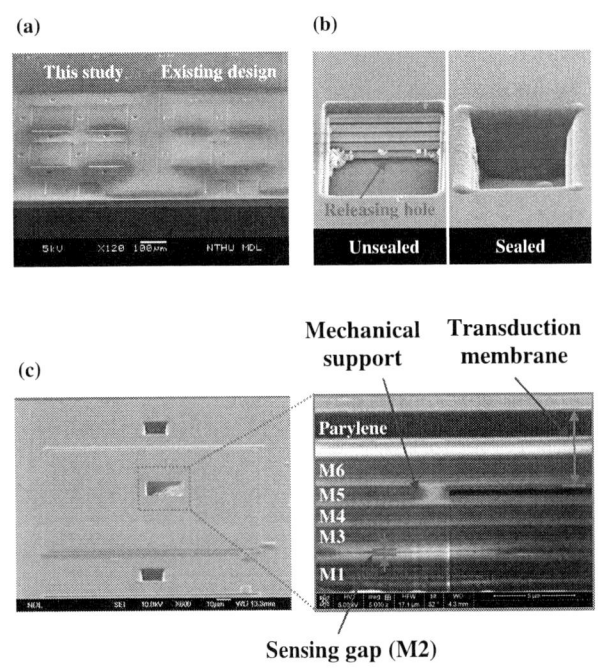

Figure 3: SEM micrographs of (a) two type pressure sensors, (b) Parylene-C deposit on releasing hole, (c) FIB cross-section of transduction membrane, mechanical support, and sensing electrodes.

membrane dimensions of these two designs are the same. Fig.3b shows the releasing hole was sealed by 2μm Parylene-C conformal coating. The focused-ion-beam (FIB) cross-section in Fig.3c displays the membrane, mechanical support, and sensing electrodes inside the chamber. The stacking of metal and dielectric layers in these structures is observed. The movable and fixed electrodes are covered by dielectric films to protect metal films during metal wet etching and also to prevent the electrical short.

MEASUREMENT RESULTS

The proposed and existing pressure sensors have been respectively characterized and measured for concept proven. Firstly, surface profile of pressure sensors (AA' cross-section shown in Fig.4a) were measured by optical interferometer. The measurements in Fig.4b show the membrane deformation of proposed and existing designs after vacuum sealing (~10Pa). Due to the thin film residual stresses and pressure load (~101kPa), the membranes (160μm×160μm) for both designs are slightly bent (central deformation of near 0.2μm). The results indicate the initial membrane deformations of proposed and existing designs are similar.

However, the membrane center of the proposed design has a slight concave deformation due to the influence of mechanical support.

Fig.5a illustrates the setup for pressure load testing. The test chip was wire bonding on PCB, and placed inside the vacuum chamber for performance characterization. During the test, the pressure controller was used to specify the pressure of vacuum chamber. The output capacitance change of pressure sensors were measured by the commercial LCR meter. Measurement in Fig.5b-c shows the results of output capacitance change versus the absolute pressure. The full-scale measurement range is 20kPa-100kPa for absolute pressure. Fig.5b show the signals characterized from sensor before/after Parylene-C sealing. The output capacitance did not change with the pressure loads for chip not sealed by Parylene-C (Fig.2c). For chip sealed by Parylene-C (Fig.2d), the output capacitance decreases as the ambient pressure decreases. In short, as the ambient pressure is reduced, the bending deformation of membrane is decreased and further lead to the increasing of sensing gap. The results also show the pressure sensor is successfully implemented after Parylene-C sealing. Measurements in Fig.5c depict the proposed pressure sensor design has the sensitivity of 2.1fF/kPa and non-linearity of 1.7% within the 100kPa-60kPa sensing range. In comparison, the existing design has the sensitivity of 1.3fF/kPa and non-linearity of 1.5% within the same sensing range. It indicates the sensitivity has been increased for 61% within the 100kPa-60kPa sensing range by the proposed design. Measurements in Fig.5c also indicate the sensitivity and non-linearity of proposed pressure sensor become 1.9fF/kPa

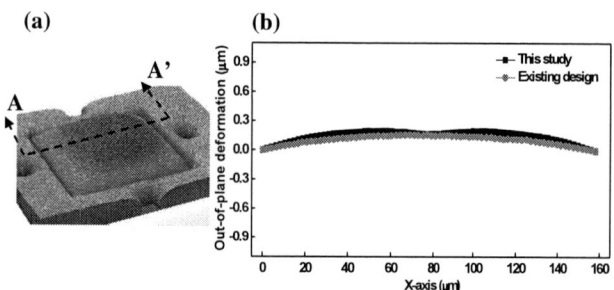

Figure 4: The membrane deformation of two designs after vacuum sealing.

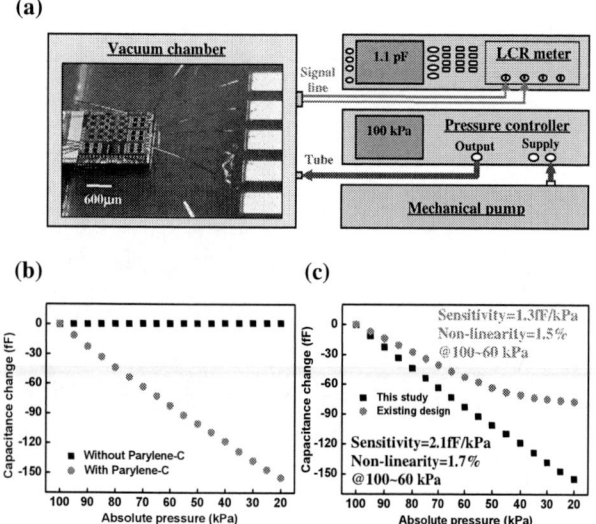

Figure 5: (a) The schematic of sensitivity measurement setup, (b-c) measured output capacitance change versus pressure change (5kPa).

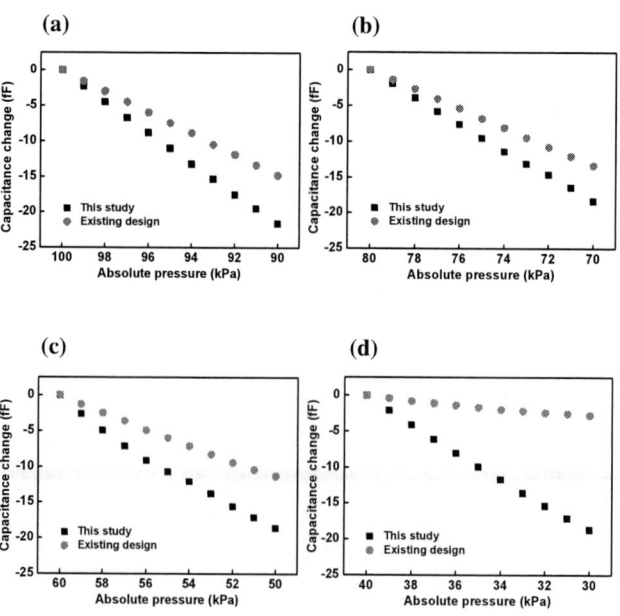

Figure 6: Measured output capacitance change versus pressure change in different pressure range.

and 2.5% within the 100kPa-20kPa sensing range. However, the sensitivity of existing design significantly drops to only 1.0fF/kPa and the non-linearity remarkably increases to 13.7% within the 100kPa-20kPa sensing range. As a result, the proposed pressure sensor design could increase both sensitivity and sensing range.

Measurements in Fig.6 further investigate the variation of sensitivity at different pressure range for proposed and existing designs. Figures 6a-d show the variation of sensing capacitance with a small pressure increment of 1kPa respectively at four different pressure ranges (100kPa-90kPa, 80kPa-70kPa, 60kPa-50kPa, and 40kPa-30kPa). The results show the sensitivity (non-linearity) of present design decreased from 2.2fF/kPa (0.7%) to 1.8fF/kPa (2.0%) as the sensing range of absolute pressure changed from 100kPa-90kPa to 40kPa-30kPa. However, the sensitivity (non-linearity) of existing design significantly decreases from 1.5fF/kPa (0.6%) to 0.3fF/kPa (7.2%) as the sensing range of absolute pressure changed from 100kPa-90kPa to 40kPa-30kPa. The improvements of sensitivity and linearity of the proposed sensor (as compare with the existing design) are higher than the prediction from simulation. Further investigations are required to find such differences.

CONCLUSIONS

In summary, this study successfully demonstrated the sensitivity enhancement of capacitive CMOS-MEMS pressure sensor with mechanical force-displacement transduction structure. The capacitance sensing area has been increased using a movable undeformed structure with embedded sensing electrode. The device has been designed and implemented using the TSMC 0.18μm 1P6M CMOS process. The proposed design has the sensitivity (non-linearity) of 2.1fF/kPa (1.7%) within the 100kPa-60kPa sensing range. In comparison, sensitivity (non-linearity) of existing design is 1.3fF/kPa (1.5%) within the same sensing range. Thus, the sensitivity of proposed design has been improved for 61% within 100kPa-60kPa pressure range. The sensitivity (non-linearity) of proposed and existing designs respectively become 1.9fF/kPa (2.5%) and 1.0fF/kPa (13.7%) as sensing range changed to 100~20kPa. It indicates the proposed sensor design offers a larger sensing range. Note the mechanical structure proposed in this design can be further employed to realize differential, fully-differential sensing mechanism.

ACKNOWLEDGEMENTS

This research was sponsored in part by the National Science Council of Taiwan under grant of NSC-102-2221-E-007-027-MY3, NSC-102-2622-E-007-014-MY3, and NSC-102-2218-E-007-003-MY3. The authors wish to appreciate the TSMC and the National

Chip Implementation Center (CIC), Taiwan, for the supporting of CMOS chip manufacturing. The authors would like to thank the Sensirion AG for the technical and financial support. The authors would also like to thank the National Center for High-Performance Computing for support of simulation tools. The authors also would like to appreciate the Center for Nanotechnology, Materials Science and Microsystems of National Tsing Hua University.

REFERENCES

[1] W. Eaton, and J. Smith, "Micromachined pressure sensors: review and recent developments," *Smart Material Structure*, vol. 6, pp. 530-539, 1997.

[2] T. Pedersen, G. Fragiacomo, O. Hansen, and E. Thomsen, "Highly sensitive micromachined capacitive pressure sensor with reduced hysteresis and low parasitic capacitance," *Sens. Actuator A-Phys.*, vol. 154, pp. 35-41, 2009.

[3] C.-M. Sun, C. Wang, M.-H. Tsai, H.-S. Hsieh, and W. Fang, "Monolithic integration of capacitive sensors using a double-side CMOS MEMS post process," *J. Micromech. Microeng.*, vol. 19, 015023, 2009.

[4] N. Narducci, Y.-C. Liu, W. Fang, and J. Tsai, "CMOS MEMS capacitive absolute pressure sensor," *J. Micromech. Microeng.*, vol. 23, 055007, 2013.

[5] T. Fujimori, H. Takano, S. Machida, and Y. Goto, "Tiny (0.72 mm^2) pressure sensor integrating MEMS and CMOS LSI with back-end-of-line MEMS platform," *Proc. Transducers'09 Conf.*, Denver, CO, June, 2009, pp. 1924-1927.

[6] Y. Zhang, R. Howver, B. Gogoi, and N. Yazdi, "A high-sensitive ultra-thin MEMS capacitive pressure sensor," *Proc. Transducers'11 Conf.*, Beijing, China, June, 2011, pp. 112-115.

[7] C.-K. Chan, W.-C. Lai, M. Wu, M.-Y. Wang, and W. Fang, "Design and implementation of a capacitive-type microphone with rigid diaphragm and flexible spring using the two poly silicon micromachining processes," *IEEE Sensors J.*, vol. 11, pp. 2365-2371, 2011.

[8] T.-H. Yen, M.-H. Tsai, C.-I. Chang, Y.-C. Liu, S.-S. Li, R. Chen, J.-C. Chiou, and W. Fang, "Improvement of CMOS-MEMS accelerometer using the symmetric layers stacking design," *Proc. IEEE Sensors'11 Conf.*, Limerick, Ireland, Oct., 2011, pp. 145-148.

[9] M.-H. Tsai, C.-M. Sun, Y.-C. Liu, C. Wang, and W. Fang, "Design and application of a metal wet-etching post-process for the improvement of CMOS-MEMS capacitive sensors," *J. Micromech. Microeng.*, vol. 19, 105017, 2009.

CONTACT

* W. Fang, Tel: +886-3-5742923; fang@pme.nthu.edu.tw

LOW NOISE VACUUM MEMS CLOSED-LOOP ACCELEROMETER USING SIXTH-ORDER MULTI-FEEDBACK LOOPS AND LOCAL RESONATOR SIGMA DELTA MODULATOR

Fang Chen[1], Weizheng Yuan[1], Honglong Chang[1], Ioannis Zeimpekis[2] and Michael Kraft[3]

[1]Northwestern Polytechnical University, Key Laboratory of Micro/Nano Systems for Aerospace, CHINA
[2]University of Southampton, School of Electronics and Computer Science, UK
[3]University of Duisburg-Essen, Faculty of Engineering Sciences, GERMANY

ABSTRACT

This paper reports on the design, implementation of a novel sixth-order sigma-delta modulator ($\Sigma\Delta M$) MEMS closed-loop accelerometer with extended bandwidth in a vacuum environment (~0.5Torr), which can coexist on a single die (or package) with other sensors requiring vacuum packaging. The fully differential accelerometer sensing element with a large proof mass ($4 \times 7mm^2$) was designed and fabricated on a Silicon-on-Insulator (SOI) wafer with 50μm-thick structural layer. Four electronic integrators were cascaded with the sensing element for high-order noise shaping ability. The local feedback paths created a local resonator producing a notch to further suppress the total in-band quantization noise. Measurement results show the overall noise floor achieved was -120dBg/√Hz, which is equivalent to a noise acceleration value of 1.2μg/√Hz in a 500Hz bandwidth; the scale factor was 950mV/g for input accelerations up to ±6g.

INTRODUCTION

MEMS accelerometers are being used in consumer, automotive, industrial and navigation applications, the market volume of accelerometers is constantly increasing. Integrating MEMS accelerometers with other types of micro-sensors including gyroscopes, resonators, barometers and magnetometers on a common substrate to form a multi-axis sensing microsystem cannot only reduce size and fabrication cost, but also broad the application field of these devices [1]; examples include ten-degrees-of-freedom (10-DOF) sensing microsystems and 6-DOF inertial measurement units (IMU). The 10-DOF sensing microsystem comprises accelerometers, gyroscopes, magnetometers and a barometer, which can be used to accurately map the motion and position of an object in space. The 6-DOF IMU consisting of accelerometers and gyroscopes are used in gaming, image stabilization in digital cameras, electronic stability control in automobiles and for robotics.

However, the vacuum operating requirements of high performance micro-sensors such as Coriolis gyroscopes and resonators are contradictory to the packaging requirements for accelerometers which should be critically damped. The high quality factor (*Q*) attained under vacuum provides lower mechanical (Brownian) noise and higher sensitivity for gyroscopes and resonators, but leads to unacceptably long settling times for accelerometers due to an under-damped response and thus ringing.

Open-loop accelerometers with narrow capacitive gaps can increase air damping to ensure stable operation at low-pressure levels (~1Torr), but overall system noise level, bandwidth and sensitivity are not suitable for high performance applications [2]. An alternative approach is to use a frequency modulated accelerometer which relies on tracking of the resonant frequency of MEMS resonators [3, 4]; the induced acceleration changes the resonant frequency of the device due to changes in the total effective stiffness [5]. In this configuration, gyroscopes and accelerometers can be integrated in the same vacuum environment. However, relatively large polarization voltages and complex compensation techniques are required.

In this paper, we propose a novel sixth-order $\Sigma\Delta M$ low noise MEMS closed-loop accelerometer, which can operate in vacuum. Four electronic integrators are cascaded with the accelerometer to provide sixth-order noise shaping ability. Furthermore, the local feedback paths create a local resonator which produces a notch to further suppress the total in-band quantization noise. A feedback voltage signal was applied to the proof-mass to artificially damp the system [6]. A similar high-order $\Sigma\Delta M$ control method can also be used for gyroscopes to improve its noise performance, linearity and bandwidth [7, 8]. The high-order loop filters consisting of electronic integrators and resonators can be implemented in a FPGA.

SYSTEM DESIGN
MEMS Accelerometer

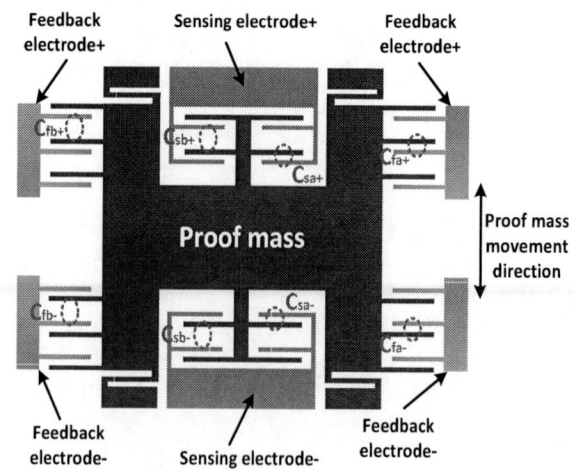

Figure 1: Schematic diagram of a single-axis fully differential accelerometer sensing element.

The proposed single-axis fully differential capacitive MEMS accelerometer was fabricated using a dicing free and HF vapour phase release process [9] on a SOI wafer with device layer thickness of 50μm. The schematic view of the accelerometer sensing element is shown in Figure 1. The proof mass is supported by four folded elastic beams, each side has two folded beams. It has two sets of sensing capacitors (C_{sa+} and C_{sa-}, C_{sb+} and C_{sb-}) and two sets of feedback capacitors (C_{fa+} and C_{fa-}, C_{fb+} and C_{fb-}) are connected in series, respectively. The capacitors C_{sa+}, C_{sa-}, C_{fa+}, C_{fa-} have a nominal gap of 7μm while C_{sb+}, C_{sb-} C_{fb+}, C_{fb-} have a gap of 30μm. When an acceleration input signal is applied, the proof mass displacement is converted into a capacitance change by the differential parallel capacitors C_{s_top} (C_{sa+}, C_{sb+}) and C_{s_bot} (C_{sa-}, C_{sb-}). The feedback excitation voltage acts on the differential feedback capacitor C_{f_top} (C_{fa+}, C_{fb+}) and C_{f_bot} (C_{fa-}, C_{fb-}) causing a feedback electrostatic force which acts on the proof mass to change the damping of the sensor, thus controlling the acceleration to displacement transfer function. The design parameters of the accelerometer sensing element are listed in Table 1.

Table 1: Design parameters of the MEMS accelerometer sensing element.

Parameter	Value
Mass of proof mass(m)	1.62μg
Damping coefficient (b)	1.2×10^{-3} Ns/m
Spring stiffness (k)	93.2N/m
Structural layer thickness	50μm
Number of sensing comb fingers	966
Number of feedback comb fingers	144
Length of sensing comb fingers (overlapping)	65μm
Length of feedback comb fingers (overlapping)	295μm

Closed-loop System

The Simulink model of the presented electromechanical ΣΔM interface is shown in Figure 2. The sensing element

embedded in a ΣΔM provides second-order integration, and is cascaded with four electronic integrators to achieve sixth-order noise shaping. A compensator $C_p(s)$ is required to provide phase lead to ensure loop stability. K_{po} is the pick-off gain. The scaling coefficients K_1, K_2, K_3 and K_4 are used for integrator output scaling, the feedback gains K_{f1}, K_{f2}, K_{f3} and K_{f4} determine the loop pole positions. The local feedback paths g_1, g_2 and g_3 create a local resonator that generates complex pairs of zeros resulting in an in-band notch.

For linear system analysis, the quantizer can be modeled by a variable gain K_q with additive white quantization noise [7, 8]. Using standard linear control system theory, the signal transfer function (*STF*), electronic noise transfer function (*ENTF*) and quantization noise transfer function (*QNTF*) of the closed-loop system can be derived as:

$$STF(s) = \frac{m \cdot K_m(s) \cdot \prod_{i=1}^{4} H_i(s) \cdot K_i}{L(s) \cdot K_{fb}} \quad (1)$$

$$ENTF(s) = \frac{K_m(s) \cdot \prod_{i=1}^{4} H_i(s) \cdot K_i}{L(s) \cdot K_{fb} \cdot M(s) \cdot K_{po}} \quad (2)$$

$$QNTF(s) = \frac{1 + \prod_{i=1}^{3} H_i(s) \cdot K_i \cdot H_{i+1}(s) \cdot g_i}{L(s)} \quad (3)$$

where

$$K_m(s) = K_{fb} \cdot M(s) \cdot K_{po} \cdot C_p(s) \cdot K_q$$

$$L(s) = 1 + K_m(s) \cdot \prod_{i=1}^{4} H_i(s) \cdot K_i$$

$$+ K_q \cdot \sum_{i=1}^{4} K_{fi} \prod_{j=i}^{4} H_j(s) \cdot K_j + \sum_{i=1}^{3} K_i \cdot g_i \prod_{j=i}^{i+1} H_j(s)$$

$$+ H_1(s) \cdot K_1 \cdot H_2(s) \cdot g_1 \cdot K_3 \cdot H_3(s) \cdot g_2 \cdot K_4 \cdot H_4(s) \cdot K_q$$

where $M(s)$ is the transfer function of the accelerometer sensing element and K_{fb} is the feedback transfer gain.

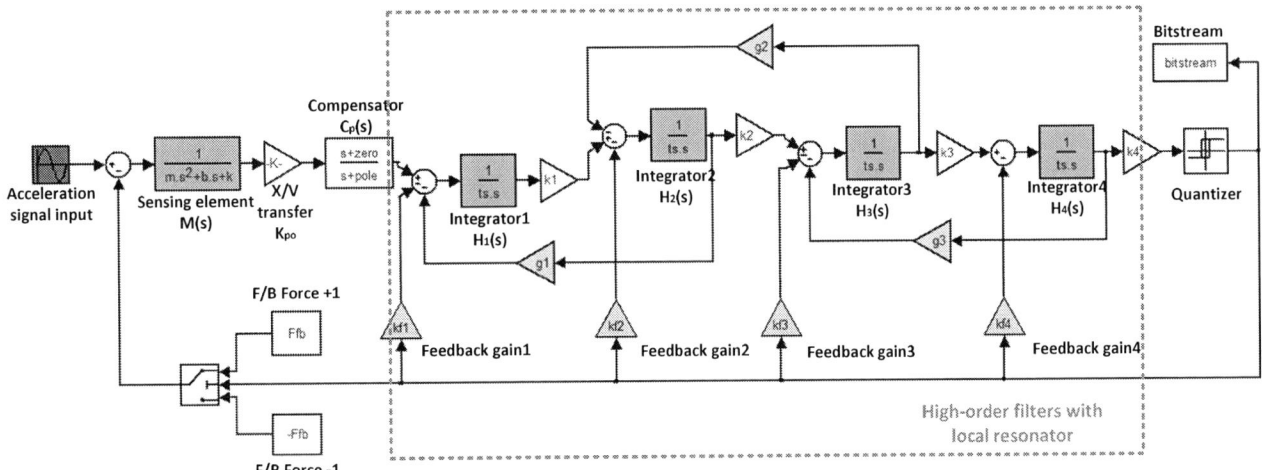

Figure 2: Simulink model of a MEMS accelerometer closed-loop system with the system noise shaped by a sixth-order ΣΔM control system.

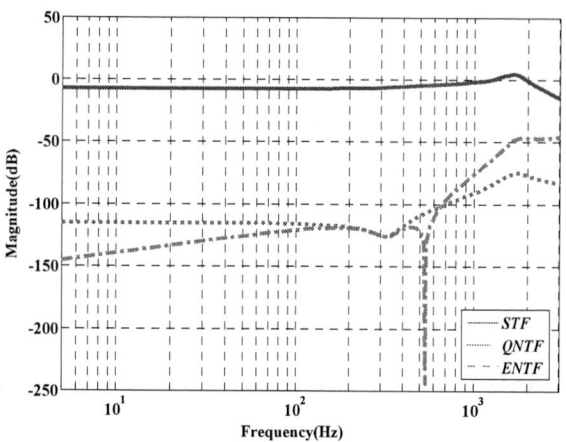

Figure 3: Magnitude plots of the transfer functions STF, QNTF and ENTF.

Figure 3 shows the bode diagram of *STF, QNTF* and *ENTF*. There are two notches in the QNTF, the first notch frequency is determined by the poles of the sensing element and the second notch is provided by the local resonator.

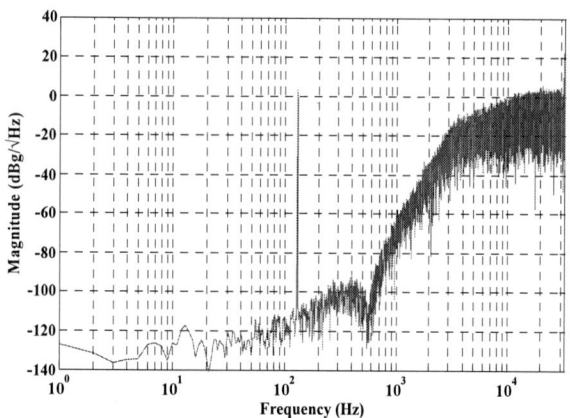

Figure 4: Simulated power spectral density of the simulated output bitstream with a 1g, 128Hz acceleration input signal.

The sampling frequency was chosen as 2^{17}Hz; assuming 2^{10}Hz signal bandwidth; this corresponds to an over-sampling ratio (OSR) of 64. For a sinusoidal acceleration input signal with 1g amplitude at 128Hz the simulated power spectral density (PSD) of the output bitstream of the sixth-order continuous time ΣΔM accelerometer is shown in Figure 4. An in-band notch can be clearly observed. The overall noise floor is about -125dBg/√Hz.

EXPERIMENTAL TESTS AND RESULTS

As shown in Figure 5, the accelerometer sensing element is mounted onto a carrier printed circuit board (PCB) and bonded using aluminium wires. A plastic and transparent cover is used to isolate the sensor chip from moisture and

particle contamination.

Figure 5: Picture of the wire bonded accelerometer sensing element without cover.

As shown in Figure 6 the system was implemented as a circuit with off-the-shelf electronic components on a PCB following the topology of the presented Simulink model. It used a power supply of ±12V. The output bitstream can be transmitted to a personal computer (PC) through a USB chip and interface. To reduce electro-magnetic interference on the PCB, analog and digital ground is connected together at a single point with a magnet bead.

A test was carried out in a vacuum chamber with a 1g, 128Hz acceleration excitation input. Figure 7 shows the shaker table setup, placed in a vacuum chamber in which the pressure can be adjusted. The input acceleration is applied by a commercial shaker. The output signal from the accelerometer is not only recorded by a spectrum analyzer (Agilent Inc., 35670A), but also transmitted to a PC and post-processed with Matlab.

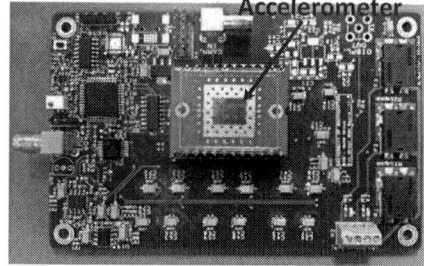

Figure 6: Photograph of the accelerometer with the sixth-order ΣΔM control circuit implemented on the PCB.

Figure 7: Experimental setup and testing circuit.

Figure 8 shows the output spectrum for the accelerometer at 0.5Torr vacuum for a 1g, and 128Hz acceleration excitation input. The measured overall noise floor achieved was -120dBg/√Hz, which is equivalent to a noise value in the 500Hz bandwidth of 1.2μg/√Hz, which is

978-1-4799-3510-9/14 $31.00 © 2014 IEEE

in excellent accordance with the simulation shown in Figure 4. Figure 9 shows the spectrum of the accelerometer with the local resonator switched on and off. It is obvious that the presented local resonators results in a wider baseband for the MEMS accelerometer. Figure 10 shows the static response for input accelerations up to ±6g; the measured scale factor is 950mV/g.

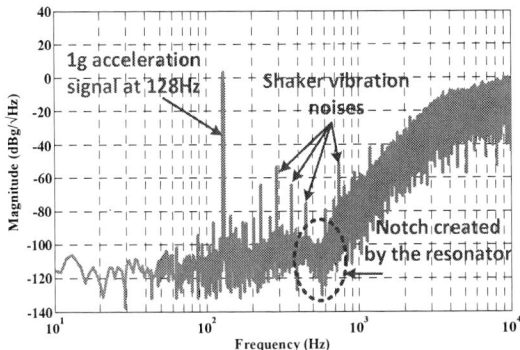

Figure 8: Noise spectrum of the measured closed-loop system at 0.5Torr vacuum with a 1g, 128Hz input signal, showing a noise floor level of 1.2μg/√Hz within 500Hz.

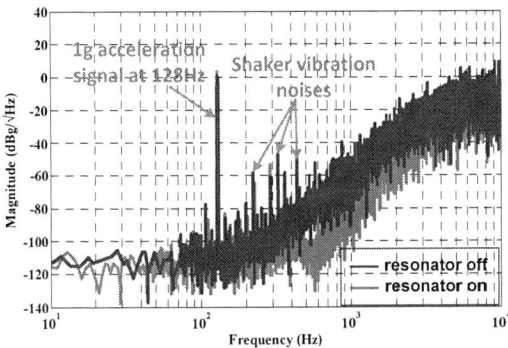

Figure 9: Measured spectrum of the presented closed-loop accelerometer at 0.5Torr vacuum with the resonator switched on and off.

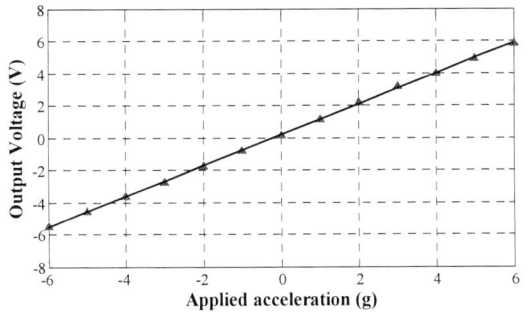

Figure 10: Static response for input accelerations up to ±6g. The measured scale factor is 950mV/g.

CONCLUSION

A fully differential single-axis capacitive accelerometer with sixth-order ΣΔM closed-loop control system was implemented. Results show that a high-order ΣΔM can stabilize the accelerometer system under low pressure levels, allowing co-existence on a single die with other

micro-sensors requiring vacuum packaging. This method provides a potential solution for high-performance single-chip multi-axis MEMS sensing systems.

ACKNOWLEDGEMENTS

This work was supported in part by the Chinese National Science Foundation under Grant 61273052, in part by the Chinese New Century Excellent Talents in University under Grant NCET-10-0077, and in part by the 111 project under Grant B13044.

REFERENCES

[1] F. Ayazi, "Multi-DOF inertial MEMS: From gaming to dead reckoning", *Transducers 2011 Conference*, Beijing, June 5-9, 2011, pp. 2805-2808.

[2] Y. Jeong, D. Serrano, V. Keesara, W. Sung, F. Ayazi, "Wafer-level vacuum-packaged tri-axial accelerometer with nano airgaps", *IEEE MEMS 2013 Conference*, Taipei, Jan 20-24, 2013, pp. 33-36.

[3] A. Seshia, S.A. Zotov, B.R. Simon, A.M. Shkel, "A vacuum packaged surface micromachined resonant accelerometer", *IEEE J. Microelectromech. Syst*, vol.11, pp. 784-793, 2002.

[4] A.A. Trusov, S.A. Zotov, B.R. Simon, A.M. Shkel, "Silicon accelerometer with differential frequency modulation and continuous self-calibration", *IEEE MEMS 2013 Conference*, Taipei, Jan 20-24, 2013, pp. 29-32.

[5] R. Hopkins, J. Miola, W.Sawyer, et al. "The silicon oscillating accelerometer: a high-performance MEMS accelerometer for precision navigation and strategic guidance applications", *in Proc. Nat.Tech.Meeting Inst.Navigation*, San Diego, Jan 24-26, 2005, pp. 970-979.

[6] M. Yucetas, M. Pulkkinen, A. Kalanti, et. al., "A high-resolution accelerometer with electrostatic damping and improved supply sensitivity", *IEEE J. Solid-State Circuits*, vol. 47, no.7, pp. 1721-1730, 2012.

[7] F. Chen, H. Chang, W. Yuan, R. Wilcock, M. Kraft, "Parameter optimization for a high-order band-pass continuous-time sigma-delta modulator MEMS gyroscope using a genetic algorithm approach", *J. Micromech. Microeng.*, vol. 22, no.10, 105006, 2012.

[8] F. Chen, W. Yuan, H. Chang, et al., "Design and Implementation of an optimized double closed-loop control system for MEMS vibratory gyroscope", *IEEE Sensors J.*, vol. 14, no.1, pp. 184-196, 2014.

[9] I. Sari, I. Zeimpekis, M. Kraft, "A dicing free SOI process for MEMS devices", *Microelectron. Eng.*, vol. 95, no.7, pp. 121-129, 2012.

CONTACT

*Honglong Chang, tel: +86-29-8849-2841; changhl@nwpu.edu.cn
*Michael Kraft, michael.kraft@uni-due.de

MICRO LIQUID-BASED THERMO-ACOUSTIC TRANSMITTER FOR EMITTING ULTRASOUND IN LIQUID MEDIUM

Dinh Hoang-Giang[1], Nguyen Thanh-Vinh[1], Kentaro Noda[1], Phan Hoang-Phuong[2], Nguyen Binh-Khiem[1], Tomoyuki Takahata[1], Kiyoshi Matsumoto[1], and Isao Shimoyama[1]

[1]The University of Tokyo, Tokyo, JAPAN
[2]Queensland Micro-Nanotechnology Centre, Griffith University, AUSTRALIA

ABSTRACT

We proposed a thermo-acoustic transmitter in liquid using a metal heater covered with silicone oil. Our experiments showed that the structure of silicone oil on metal heater emitted stronger ultrasound compared to the structure of water on metal heater. However, when used in other liquid, the silicone oil must be encapsulated to prevent the mixing of the two liquids. We therefore proposed the device structure of silicone oil on metal heater encapsulated by a thin film of Parylene (1 μm). The acoustic impedance gap between the silicone oil and liquid medium is small, resulting in the reduction of ultrasound reflection at the interface of the two liquids. Therefore, the thermal induction mechanism can emit short ultrasonic pulse without acoustic ringing. Our prototype device was confirmed to be able to emit short ultrasonic pulses (20 μs) over the range of 50-150 kHz. We also confirmed that, compared to the structure of metal heater put directly in water, the proposed structure enhanced the amplitude of the emitted ultrasound by 3 times.

INTRODUCTION

Nowadays, the generation of ultrasound in liquid medium attracts much attention due to its applications in high resolution underwater imaging, underwater acoustics communication and submarine flaw inspection [1-3]. In these applications, the ability to generate short ultrasonic pulses determines the measurement accuracy. Conventional methods to generate ultrasound in liquid medium use a piezoelectric material (typically PZT) with matching layers, which can only be optimized for a small number of operating mediums [3,4]. However, such sophisticatedly designed device cannot operate in different medium because the acoustic impedance mismatch between the piezoelectric material and the medium causes "acoustic ringing" [5].

We propose using thermal induction acoustic principle to generate ultrasound in liquid mediums. This method were reported to have many advantages, when used in air, such as resonance-free and flat frequency response [6-8]. In this method, a heater is immersed directly in the medium to heat up the medium material to cause mechanical vibration. Because of this mechanism, it can be assumed that in thermal induction of acoustic wave, the phenomenon of "acoustic ringing" will not occur. Our proposed liquid-based thermo-acoustic transmitter, shown in Fig. 1, consists of a metal heater covered with silicone oil. The silicone oil droplet is encapsulated by a thin Parylene film. This encapsulated liquid droplet is expanded and shrinked by the heat from the heater, causing ultrasound, which then

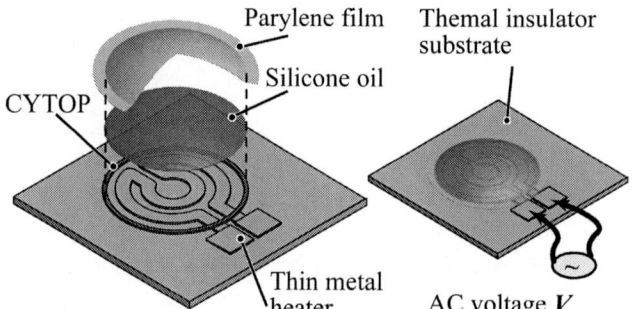

Figure 1: The structure of our proposed liquid-based thermo-acoustic device.

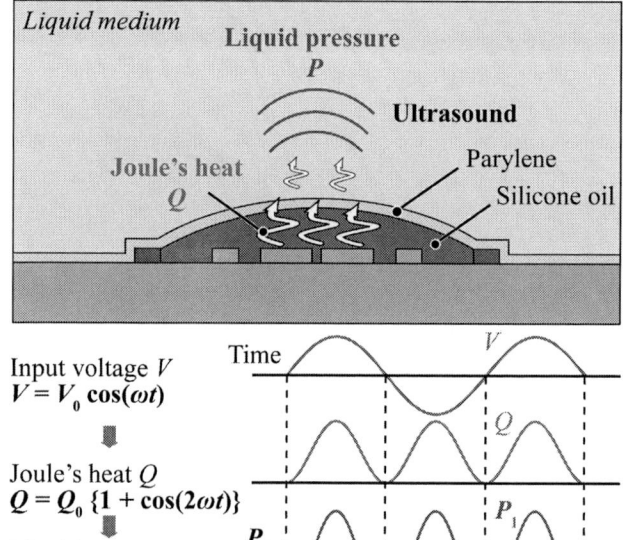

Figure 2: The working principle of our proposed device.

propagates to the outer liquid medium.

Our device is expected to have many advantages such as low cost, simple fabrication process, wide operating frequency range and possibility of fabricating transducer arrays for ultrasound focusing and scanning.

DESIGN AND PRINCIPLE

The structure of our device is depicted in Fig. 1. The device is composed of a thin metallic heater (100 nm order) formed on a thermal insulator substrate and a liquid cover. The liquid cover is encapsulated by a thin Parylene film (1 μm) which is deposited directly onto the liquid surface. We have employed two types of designs for the metal pattern: the square pattern adapted from the literature [6] and the circle pattern (Fig. 4).

The working principle of the device is shown in Fig. 2. By applying an alternating current voltage to the metal heater, the generated Joule's heat changes periodically with the doubled frequency of the applied voltage. Due to the small thermal capacity and the ultrathin thickness of the metal heater, its temperature also increases and decreases simultaneously with the Joule's heat, resulting in the expansion and shrinking of the liquid cover. This thermally induced oscillation of the liquid cover, in turn, generates a stress wave to the outer liquid medium through the thin Parylene film.

FABRICATION PROCESS

The fabrication process flow of our proposed device is shown in Fig. 3. First, a metal layer was deposited on a thermal insulator substrate by electron beam induced evaporation. Then, the metal layer was patterned and wet-etched to form the heater. In the next step, a ring of hydrophobic material (CYTOP, Asahi Glass Co.) around the heater was patterned to create a hydrophobic region border. After that, a drop of silicone oil was deposited on the heater. The spreading of the liquid droplet is restricted by the CYTOP ring, and thus, the droplet position is fixed on the heater. Finally, a 1μm-thick Parylene film was deposited directly on the surface of the droplet using chemical vapor deposition [9].

An image of the fabricated aluminum device is shown in Fig. 4 (a). The CYTOP ring around the heater was clearly observed. The liquid covered on the heater can be noticed by the light reflection on its surface. The shape of the heater is shown in Fig. 4 (b). The dimensions and electrical parameter of the devices were listed in Table 1.

EXPERIMENTS AND DISCUSSIONS

Experiment setup

We conducted experiments to characterize the contribution of the thermal effusivity of substrate and the type of the liquid to the efficiency of the ultrasound generation in term of wave amplitude. By this investigation, we can find the appropriate materials of the liquid, thermal insulator substrate which can enhance the ultrasound emitting efficiency of our device. The experiment setup for all these preliminary experiments is shown in Fig.5. Since the thermally induced ultrasound in liquid generated by the device with a metal heater on thermal insulator is relatively small, we have employed a long cylinder (diameter 16 mm) to focus the ultrasound. The sinusoidal pulse voltage with frequency 50 kHz was applied to the metal heater which then emitted ultrasound with frequency 100 kHz. The emitted ultrasound amplitude is measured by the hydrophone (type 8103, Brüel&Kjær) at distance 20 mm from the metal heater.

Effect of various types of liquids

The effect of various liquids in contribution to the

(1) Form metal layer on a substrate

(2) Etch metal layer to form heater pattern

(3) Pattern CYTOP to form hydrophobic ring

(4) Dispense silicone oil

(5) Encapsulate silicone oil by Parylene

■ Silicone oil ■ CYTOP
■ Metal □ Parylene
□ Glass composite

Figure 3: The process steps of proposed device.

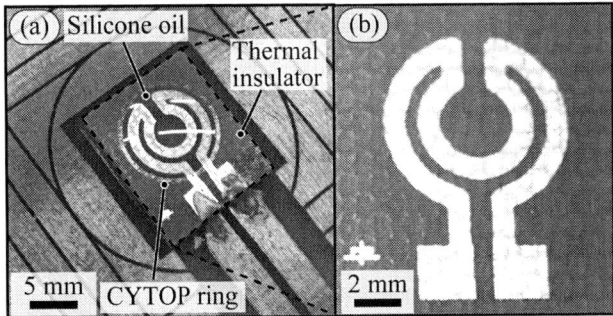

Figure 4: (a) The fabricated device with silicone oil encapsulated by the Parylene film. (b) The shape of the metal heater.

Table 1: The parameters of fabricated device.

Parylene film thickness	1 μm
Metal layer thickness	150 nm
Thermal insulator thickness (Glass composite)	1 mm
Heater resistance	15.2 Ω

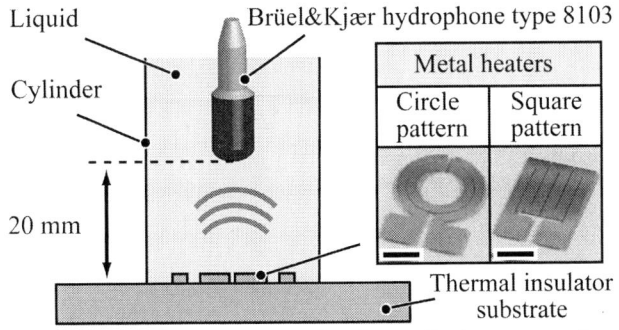

Figure 5: Experiment setup for the preliminary experiment (device without encapsulated liquid). Scale bars: 5mm.

thermally induced ultrasound emission were evaluated by using the experiment setup in Fig. 5. We have conducted the experiment with four types of liquids: HIVAC F5, HIVAC F4, Glycerol, and water. These liquids (except water) were chosen due to their capabilities to be encapsulated by the Parylene film [9] and their stable thermal properties. The experiment with water was performed to evaluate the thermal-induced ultrasound emitting efficiency of water in

978-1-4799-3510-9/14 $31.00 © 2014 IEEE

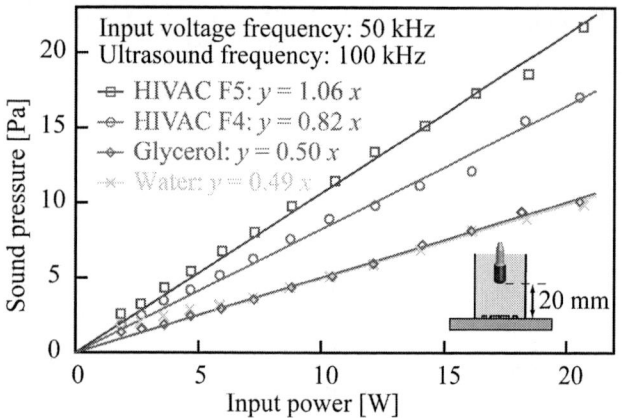

Figure 6: Effect of various liquids on the thermally induced ultrasound generation.

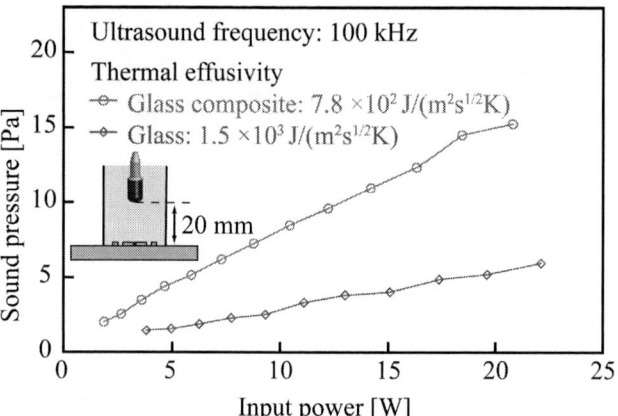

Figure 7: Effect of the thermal effusivity of substrate on the thermally induced ultrasound generation.

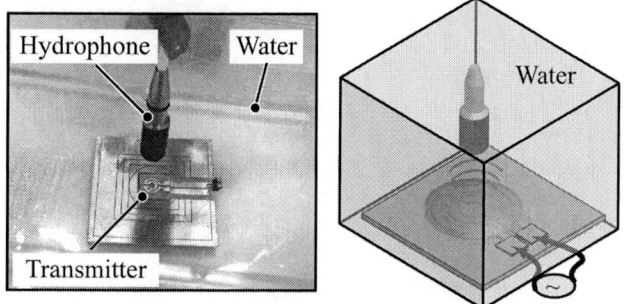

Figure 8: Experiment setup to measure the ultrasound generated by the fabricated device.

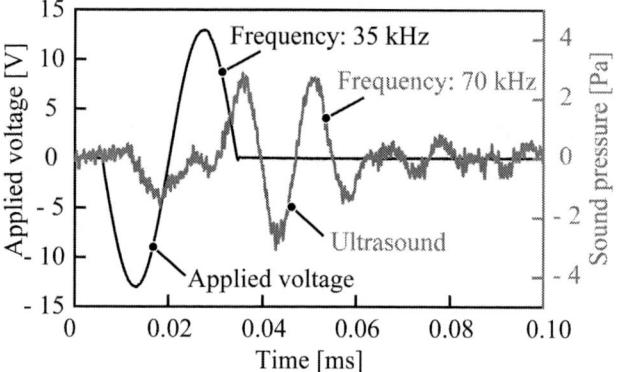

Figure 9: The thermal-induced ultrasound pulse of the fabricated device.

compared with other liquids. The experiment result is shown in Fig. 6. The result indicated that among the tested liquids, the silicone oil HIVAC F5 exposed the best enhancement on generating the thermal-induced ultrasound in liquid. Moreover, the generated ultrasound amplitude increased along with the increase of the input power. In our experiment, the correlation between the input power and the generated ultrasound amplitude is relatively linear, which was also observed for the thermal-induced ultrasound emission in air before by Shinoda *et al.* [6, 7].

Effect of the thermal insulator

Glass and glass composite were chosen as the substrate due to their stability to fabrication process and its high thermal insulation properties. We have fabricated the same pattern of the metal heater on both these two types of substrate to evaluate their efficiency in the enhancement of thermal-induced ultrasound emission. The experiment result as shown in Fig. 7, indicated that the device with the lower thermal effusivity substrate (glass composite) generated a stronger ultrasound. This can be explained that, due to the low thermal effusivity, the absorbed heat of the substrate reduced, and thus, most of the Joule's heat generated from the metal heater was converted to ultrasound. This result is

also in good agreement with the theory suggested by Shinoda *et al.* before [6, 7].

DEVICE EVALUATION

In order to evaluate the performance of our prototype device we have used the experiment setup shown in the Fig. 8. The fabricated device was immersed in a water tank (which is much larger than the cylinder used in the previous experiments) and the generated ultrasound was measured by the hydrophone (type 8103, Brüel&Kjær). The device was applied by an alternating current sinusoidal voltage pulse from a function generator which then, was amplified 10 times by an amplifier circuit. We have employed the power op-amp (PA 09A, Apex Microtechnology) in order to supply the power for the device.

Moreover, the effect of the encapsulation structure was also investigated. We have conducted the experiment with two devices which have the same metal heater pattern: one with silicone oil encapsulated, the other without encapsulated silicone oil. The alternating current sinusoidal pulse with frequency ranged from 25 kHz to 75 kHz was applied to these devices, which then emitted ultrasound ranged from 50 kHz to 150 kHz. Due to the limited range of the hydrophone used in this experiment, we couldn't perform the experiment with the higher frequency.

The real time data of the obtained ultrasound is shown in Fig. 9. The experiment result proved that our device emitted a short ultrasonic pulse with the doubled frequency of the input voltage's one as discussed in the working principle. It also

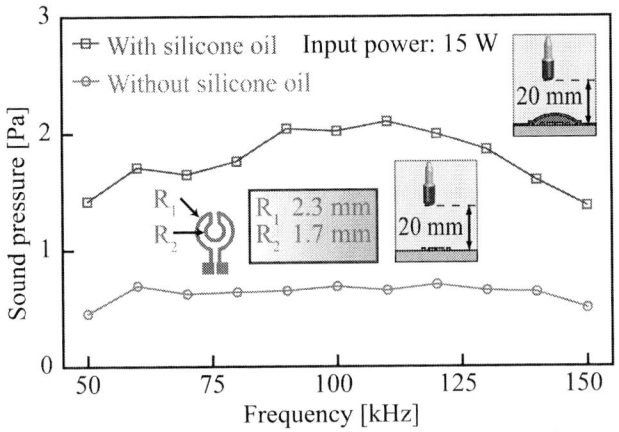

Figure 10: The operating frequency range of the fabricated device in comparison with the device of the same heater pattern without encapsulated silicone oil.

demonstrated the ability of our device to emit a clear short ultrasonic pulse without the unwanted tail (acoustic ringing phenomena) which is usually detected with the conventional solid ultrasound transducer [5, 8].

The operating frequency range of our proposed device is shown in Fig. 10. The result indicated that the emitted ultrasound from our prototype device was enhanced about 3 times than the device without encapsulated silicone oil. The experiment result also showed that our device was able to emit ultrasound at least from 50 kHz to 150 kHz. In compared with the broad-band underwater transducer using matching layer in [2], whose operating frequency range is reported of 47–159 kHz , our device exposed the advantage of working without the need of matching layer.

CONCLUSION

We proposed and characterized a thermo-acoustic transmitter based on encapsulated liquid on metal heater on thermal insulator substrate. Our prototype device, which emitted ultrasound 3 times stronger than the device without the encapsulated liquid, demonstrated the effectiveness of the encapsulation structure of silicone oil. The device was also confirmed to be able to emit a 20 μs short ultrasonic pulse without the need of matching layer. Moreover, the proposed device also has a simple fabrication, low cost, and a wide operating frequency range. Since the liquid in the device is encapsulated inside a film of Parylene, which prevents the silicone oil from leaking to the outer liquid medium, the fabricated device can be used in various types of liquid medium.

ACKNOWLEDGEMENTS

The photolithography masks were fabricated using an EB lithography apparatus (F5112+VD01) donated by Advantest Corporation at the VLSI Design and Education Center (VDEC) of the University of Tokyo.

REFERENCES

[1] Y. Li , "Position and time-delay calibration of transducer elements in a sparse array for underwater ultrasound imaging," *IEEE Trans. Ultrason., Ferroelect., Freq. Contr.*, pp.1458 - 1467, 2006.

[2] Kai Zhang, "The study of high-frequency broad-band underwater transducers," *Piezoelectricity, Acoustic Waves and Device Applications (SPAWDA)*, pp. 408 - 410, 2012.

[3] Q. Shufen, "Study of submarine pipeline corrosion based on ultrasonic detection and wavelet analysis," *Computer Application and System Modeling (ICCASM)*, pp. V12-440 - V12-444, 2010.

[4] M. I. Haller and B. T. Khuri-Yakub, "A surface micromachined electrostatic ultrasonic air transducer," *IEEE Trans. Ultrason., Ferroelect., Freq. Contr.*, vol. 43, pp.1 - 6, 1996.

[5] L. C. Lynnworth, Nguyen T.H., Smart C.D. , Khrakovsky and O.A, "Acoustically isolated paired air transducers for 50-, 100, 200-, or 500-kHz applications," *IEEE Trans. Ultrason., Ferroelect., Freq. Contr.*, vol. 44, pp.1087 - 1100, 1997.

[6] H. Shinoda, T. Nakajima, K. Ueno and N. Koshida, "Thermally induced ultrasonic emission from porous silicon," *Nature*, vol. 400, pp. 853 - 854, 1999.

[7] N. Koshida, D. Hippo, M. Mori, H. Yanazawa, H. Shinoda and T. Shimada, "Characteristics of thermally induced acoustic emission from nanoporous silicon device under full digital operation," *Appl. Phys. Lett.*, vol. 102, pp. 123504 - 123504 - 4, 2013.

[8] M. Daschewski, A. Harrer, J. Prager, M. Kreutzbruck, U. Beck, T. Lange and M. Weise, "Metallic nanofilm as resonance-free airborne ultrasound emitter," *Ultrasonics Symposium*, pp. 965 - 967, 2012.

[9] N. Binh-Khiem, E. Iwase, K. Matsumoto and I. Shimoyama, "Electrically Driven Varifocal Micro Lens Fabricated by Depositing Parylene Directly on Liquid," *Proc. MEMS*, pp. 305 - 308, 2007.

CONTACT

Dinh Hoang-Giang, Mechano-Informatics Department, Graduate School of Information Science and Technology, the University of Tokyo, Tokyo, Japan.
Address: 81B, Eng. Bldg. 2, 7-3-1 Hongo, Bunkyo-ku, Tokyo, JAPAN, 113-8656.
Tel: +81-3-5841-6318, Fax: +81-3-3818-0835
E-mail: giang@leopard.t.u-tokyo.ac.jp

978-1-4799-3510-9/14 $31.00 © 2014 IEEE

MULTI-AXIS FORCE SENSOR WITH DYNAMIC RANGE UP TO ULTRASONIC

Pham Quang-Khang[1], Nguyen Minh-Dung[1], Nguyen Binh-Khiem[1], Hoang-Phuong Phan[2]
Kyoshi Matsumoto[1], and Isao Shimoyama[1]
[1]The University of Tokyo, TOKYO, JAPAN
[2] Queensland Micro-Nanotechnology Centre, Griffith University, Australia

ABSTRACT

This paper reports on the design, implementation and characterization of a multi-axis force sensor that has a dynamic frequency range up to more than 1.2 MHz. The multilayer structure of elastomer/thin polymer/viscous liquid is utilized to conduct forces as well as acoustic vibrations to sensing piezoresistive cantilevers. Experimental results showed that this sensor is capable of measuring normal and lateral forces with high linearity in the range of 40 kPa. Moreover, by evaluating the dynamic response, this sensor is proved to have a dynamic range covered ultrasonic frequencies with the first resonant frequency at 170 kHz.

INTRODUCTION

The interest in the ability to monitor a structure and detect damage at the earliest possible stage is pervasive throughout the civil, mechanical, and aerospace engineering communities. Especially, research in vibration-based damage identification has been rapidly expanding over the last few years. In this method, changes in physical properties of the structure such as reductions in stiffness resulting from the onset of cracks will cause detectable changes in these modal properties, which are being used as indicators of damage by normal/lateral vibration-based damage detecting techniques. The sensor used in these real-time structural health monitoring is required to have the ability of, not only measuring the interactional normal and lateral forces applied on the surface of the structure, but also simultaneously detecting the minute acoustic waves emitted from the propagating cracks inside of the structural material. These acoustic waves are found to have frequencies up to ultrasonic range [1, 2, 3].

In conventional force sensors, the structure of piezoresistive sensing element embedded in elastomer, e.g. Polydimethylsiloxane (PDMS) has been widely used [4, 5]. In these sensors, the elastic cover enables the transmission of forces applied on the surface of objects to the sensing piezoresistive cantilevers inside. However, due to the stiffness of the rigid diaphragm of elastomer and sensing cantilever, these sensors are not able to detect such vibration at frequencies of ultrasonic range.

In this paper, we utilize a multilayer structure of elastomer/thin polymer/viscous liquid and fabricate a multi-axis force sensor that has a dynamic range up to ultrasonic frequencies. The concept of the proposed sensor is shown in Figure 1. The forces applied on the surface of the sensor are transmitted through the elastomer cover and make the packaged liquid deformed. In turn, the changes in shape

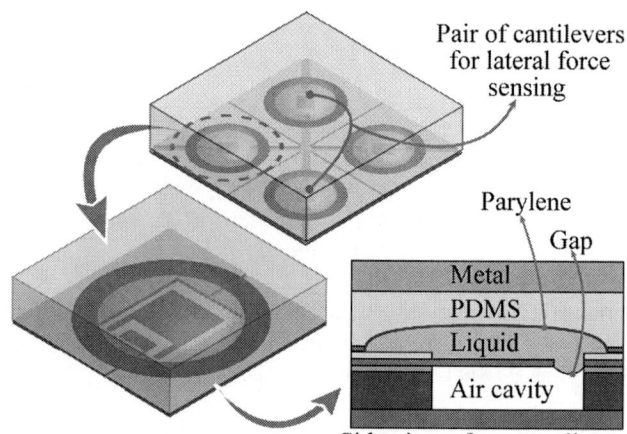

Figure 1: Schematic sketch of our multi-axis force sensor. Multilayer structure with small gap between cantilever and the wall was used to prevent liquid leaking and increase the sensitivity of the sensor.

of the liquid make the sensing cantilevers deformed. A thin layer of polymer was deposited on the surface of the liquid to package it before covered by elastomer. The air cavity provides more space for cantilevers to deform and since the width of gap between cantilever and the wall was down-scaled to 2 μm, the liquid can not leak through [6]. In addition, when vibration acts onto the elastomer cover, it is transmitted to the liquid and creates vibration at the boundary between liquid and air which in turn, vibrates the edge of cantilever. Therefore, the structure of cantilever placed on the boundary of liquid and air is able to detect vibration applied on the surface of sensor.

DESIGN AND SENSING PRINCIPLE

The concept of our device is shown in Figure 1. Two pairs of identical cantilevers are placed at four corners of the sensor chip. Each pair of cantilevers is responsible for measuring lateral force in one direction and together all four cantilevers simultaneously measure the normal force applied onto the sensor cover. The basic structure of each sensing unit is one piezoresistive cantilever placed on the interface between liquid above and an air cavity below for force measuring and vibration detecting. Liquid is constrained in a ring made of a hydrophobic material, in this research is CYTOP. In order to package liquid and prevent it from leaking, a thin layer of polymer (PARYLENE) was then deposited on to the surface of liquid. As cantilevers was designed to be ultrathin, with the thickness of 300 nm, we expected them to be ultra-flexible and therefore lead to

978-1-4799-3510-9/14 $31.00 © 2014 IEEE   769

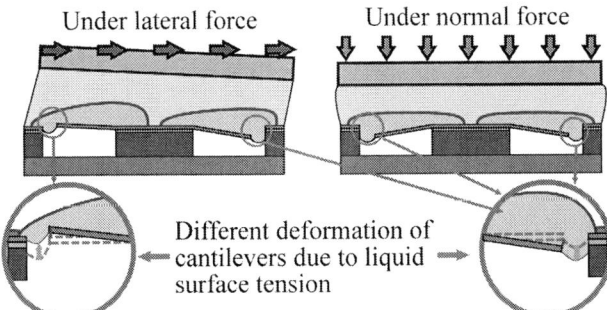

Figure 2: Normal and lateral forces sensing mechanism. Under normal force, cantilevers are deformed with the same proportion. Under lateral force, two cantilevers are deformed in opposite direction.

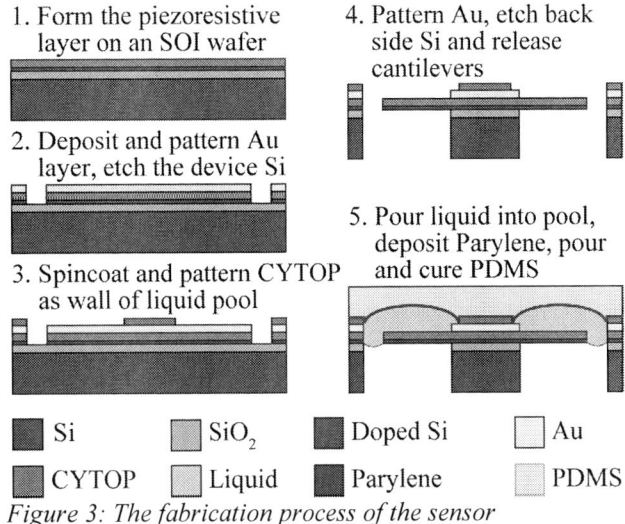

Figure 3: The fabrication process of the sensor

highly sensitive [7] to applied forces and vibrations.

The normal and lateral force sensing mechanism is shown in Figure 2.

When a normal force is applied on the cover of sensor, all four cantilevers deforms and their resistance increase at the same proportion. Hence, by measuring the fractional resistance changes of all cantilevers, we can calculate the normal force.

When the lateral force (in this case acting toward the right side) is applied on the sensor cover, as the bottom of the elastic cover was fixed to a rigid base, the external force and the counter torque appeared at the bottom of the cover will create a moment that causes the elastic cover to deformed as shown in Figure 2. The asymmetric deformation of elastic cover causes the packaged liquid to deform differently. The cantilever on the right side is forced to bend downward while the one on the left side is pulled upward due to conservation of the liquid volume and the liquid surface tension working at the small gap. In addition, due to the liquid surface tension the cantilevers are kept deforming until the lateral force is removed and the liquid return to its original shape. The difference in deformation between two cantilevers caused the difference in resistance change, from which we can calculate the lateral force in this direction.

FABRICATION PROCESS

The fabrication process steps of the piezoresistive cantilever and of the sensor are shown in Figure 3. In order to increase the sensitivity of the sensor, we used an SOI wafer with an ultrathin device Si layer of 0.3 μm/ 0.4 μm/ 300 μm. First, an N-type Si layer was formed on the surface of the wafer by rapid thermal diffusion [8]. Then a 30 nm thick Au layer was deposited and patterned. Next, the cantilever shape was formed by etching the surface Si layer. After that, a CYTOP layer was spin-coated onto the substrate and hydrophilic circular domains are patterned by O_2 plasma etching. The Au layer was then patterned again to form electrodes on the sensor chip surface. Next, the handle Si layer was deeply etched with ICP-RIE (Inductive Coupled Plasma-Reactive Ion Etching). The cantilevers were released by etching the oxide layer with hydrofluoric acid (HF) vapor.

After sensor chip was fabricated, small amount of liquid

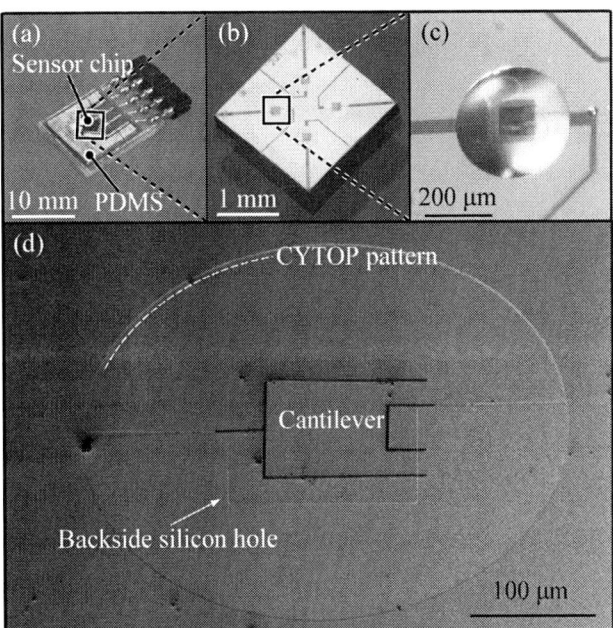

Figure 4: (a) Photograph of device. (b) Complete sensor chip. (c) Photograph of liquid droplet constrained in CYTOP ring. (d) SEM images of the fabricated structure before liquid deposit.

was deposited into the CYTOP rings. Due to the hydrophobic characteristic of CYTOP, liquid was kept inside of the ring and formed a spherical surface. In order to package the liquid inside of the elastic cover, a 1-μm-thin Parylene film was deposited directly onto the liquid surface [9]. Due to the vacuum condition during Parylene coating process and the effect of liquid surface tension on leaking problem [6], we chose HIVAC F-4 (Shin-Etsu Chemical Co. Ltd, Japan) as its characteristic of high surface tension and no vaporization in vacuum environment. Finally, a layer of 1 mm thick PDMS elastic cover was cure on top of the sensor chip.

Photographs of the sensor chip, the complete device and the SEM image of the fabricated structure are shown in Figure 4. The dimensions of the chip and the cantilever were

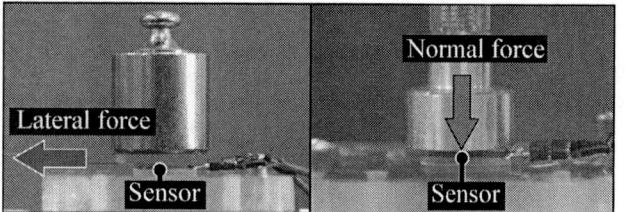

Figure 5: The experimental set up for lateral forces and normal force behavior test.

2 mm × 2mm × 0.3 mm and 125 μm × 100 μm × 0.3μm, respectively. The inner and outer radii of CYTOP rings are 150μm and 500μm, respectively. A cantilever structure with the inner circle of a CYTOP ring before liquid deposit was confirmed in the SEM image. Figure 4a shows the complete sensor chip with 10 mm × 10 mm × 1 mm PDMS cover. Figure 4b and 4c show the fabricated four cantilevers on chip before liquid deposit and the liquid droplet formed in a CYTOP ring after liquid deposit, respectively.

SENSOR EVALUATION

We conducted experiments to evaluate the behavior of the sensor due to static load and the dynamic frequency response to vibration.

Static load experiments

Figure 5 shows the experimental set up to evaluate the response of our sensor to static normal and lateral forces applied on the surface of the PDMS cover. The magnitude of the applied forces were measured by a digital force gauge attached to a linear stage. Applied forces were increased step by step in order to acquire a static state. Lateral forces were applied by pulling one acrylic plate attached onto the PDMS cover with a weight of 100 gram placed above.

In this study, we used three variables S_X, S_Y, S_Z to represent the behavior of the sensor to applied forces as shown in Figure 6a. The average resistance change of four cantilever S_Z represents the response of sensor to normal force (z-axis). The differences in resistance change of two cantilevers responsible for x-axis S_X and y-axis S_Y represent the response of sensor to lateral forces in x-axis and y-axis, respectively. We investigated the behavior of sensor under normal force up to 40 kPa and lateral forces up to 30 kPa.

The relationships between sensor outputs S_Z, S_Y, S_X and the forces applied in each axis are shown in Figure 6b, 6c and 6d, respectively. The fractional resistance change of our sensor response to lateral and normal forces are: $\Delta S_x/S_x = 3.4\times10^{-5}$ kPa^{-1}, $\Delta S_y/S_y = 3.0\times10^{-5}$ kPa^{-1}, $\Delta S_z/S_z = 83.4\times10^{-5}$ kPa^{-1}. Moreover, the experiment results demonstrated that when a force was applied in one direction, the sensor responses associated with forces in other directions were small. Furthermore, the good linear relationships between applied forces and sensor output were also observed. These results indicated that our sensor can measure normal and lateral forces independently.

The matrix of proportional coefficients between S_X, S_Y and S_Z and applied forces is shown in Equation 1. By using

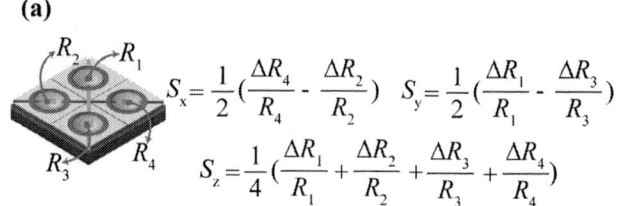

$$S_x = \frac{1}{2}\left(\frac{\Delta R_4}{R_4} - \frac{\Delta R_2}{R_2}\right) \quad S_y = \frac{1}{2}\left(\frac{\Delta R_1}{R_1} - \frac{\Delta R_3}{R_3}\right)$$

$$S_z = \frac{1}{4}\left(\frac{\Delta R_1}{R_1} + \frac{\Delta R_2}{R_2} + \frac{\Delta R_3}{R_3} + \frac{\Delta R_4}{R_4}\right)$$

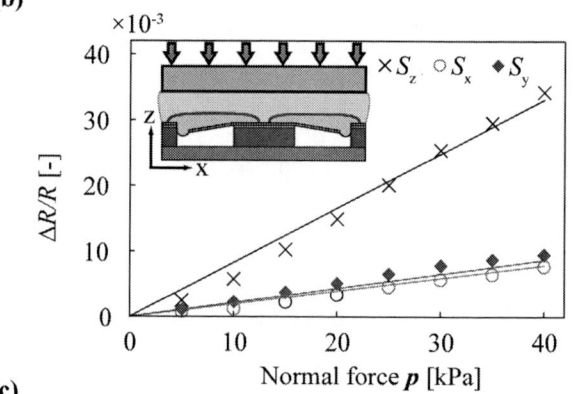

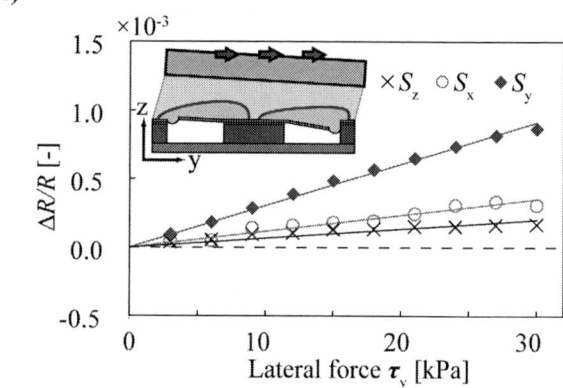

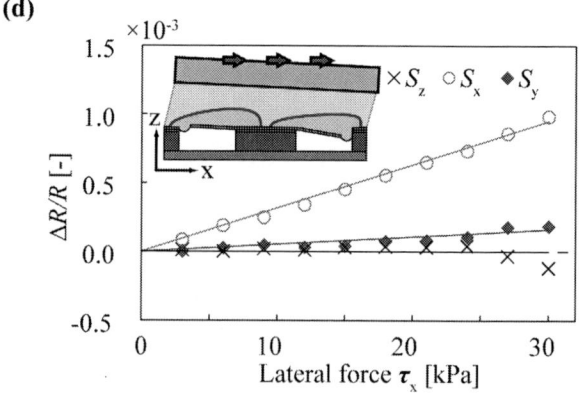

Figure 6: Response of sensor to applied normal and lateral forces.

this matrix, we can calculate three components of the force working on the surface of our sensor cover.

$$\begin{pmatrix} S_x \\ S_y \\ S_z \end{pmatrix} = 10^{-5} \times \begin{pmatrix} 3.4 & 0.6 & -0.3 \\ 1.2 & 3.0 & 0.7 \\ 19.5 & 22.7 & 83.4 \end{pmatrix} \times \begin{pmatrix} \tau_x \\ \tau_y \\ p \end{pmatrix} \quad (1)$$

(a)

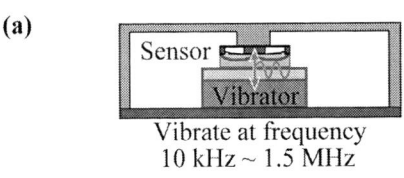

Vibrate at frequency
10 kHz ~ 1.5 MHz

(b)

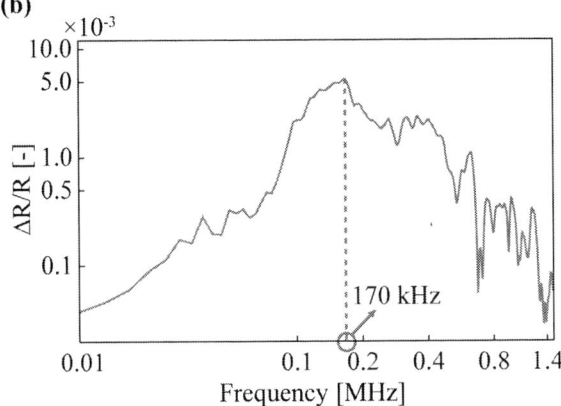

Figure 7: (a) Experimental setup of dynamic frequency response. (b) Response of sensor to vibration with frequencies from 10 kHz to 1.5 MHz.

Dynamic frequency response experiment

The experimental setup is shown in Figure 7a. Our sensor was fixed to a standing base. The PDMS surface of our sensor was fixed so as to contact to the surface of a piezoelectric vibrator that generated vibration with frequencies swept from 10 kHz to 1.5 MHz.

Figure 7b shows the relationship between the fractional resistance change of one out of four cantilevers in the sensor (cantilever number 1 in Figure 6a) and the frequencies applied to the vibrator. The results showed that the maximum response was at about 170 kHz. Therefore, our sensor had the first resonant frequency at about 170 kHz, much higher than the first resonant frequency 13.4 kHz of a cantilever with the same design placed in air [10]. In addition, the sensor also showed significant response at frequencies up to 1.2 MHz. This indicated that our sensor has a dynamic range up to ultrasonic frequencies.

CONCLUSIONS

A miniaturized multi-axis force sensors with dynamic range up to ultrasonic frequencies utilizing the multilayer structure of elastomer/ thin polymer/ viscous liquid on cantilevers was proposed and fabricated. We confirmed that our fabricated sensor is capable of measuring normal and lateral forces independently with good linearity in the range up to 40 kPa. Furthermore, we also experimentally proved that our sensor has a dynamic range covered a large range from 10 kHz to more than 1.2 MHz with the first resonant frequency at 170 kHz.

ACKNOWLEDGEMENTS

The photolithography masks were made using the University of Tokyo VLSI Design and Education Center (VDEC)'s 8 inches EB writer F5112 + VD01 donated by ADVANTEST Corporation.

This work was partially supported by JSPS KAKENHI Grant Number 23310089 and 24656162.

This work was partially supported by New Energy and Industrial Technology Development Organization (NEDO).

REFERENCES

[1] Gyuhae Park, Harley H. Cudney and Daniel J. Inman, "An integrated health monitoring technique using structural impedance sensors", *Journal of Intelligent Material Systems*, vol. 11, p448-455, 2000.

[2] Adrei N. Zagrai and Victor Giurgiutiu, "Electro-Mechanical impedance method for crack detection in thin plates", *Journal of Intelligent Material Systems*, vol. 12, p709-718, 2001.

[3] Victor Giurgiutiu, "Tuned Lamb wave excitation and detection with piezoelectric wafer active sensors for structual health monitoring", *Journal of Intelligent Material Systems*, vol. 16, p291-305, 2005.

[4] Kentaro Noda, Y. Hashimoto, Y. Tanaka, I. Shimoyama, "MEMS on robot applications", *Digest Tech. Papers Tranducers '09 Conference*, Denver, June 21-25, pp. 2176-2181.

[5] Nguyen Thanh-Vinh, N. Binh-Khiem, K. Matsumoto, I. Shimoyama, "High sensitive 3D tactile sensor with the structure of elastic pyramids on piezoresistive cantilevers", *Proceeding of 26th IEEE MEMS Conference (MEMS2013)*, Taipei, Jan. 20-24, pp. 41-44.

[6] Nguyen Minh-Dung, P. Hoang-Phuong, K. Matsumoto, I. Shimoyama, "A sensitive liquid-cantilever diaphragm for pressure sensor", *Proceeding of 26th IEEE MEMS Conference (MEMS2013)*, Taipei, Jan. 20-24, pp. 617-620.

[7] Nguyen Minh-Dung, Hideotoshi Takahashi, Takeshi Uchiyama, K. Matsumoto, I. Shimoyama "A barometric pressure sensor based on the air-gap scale effect in a cantilever", *Appl. Phys. Lett*, vol. 103, no. 143505, 2013.

[8] Murat Gel and Isao Shimoyama, "Force sensing sub micrometer thick cantilevers with ultra-thin piezoresistors by rapid thermal diffusion", *J. Micromech. Microeng.*, vol. 14, pp. 423-428, 2003.

[9] Nguyen Binh-Khiem, Eiji Iwase, K. Matsumoto, I. Shimoyama, "Electrically driven varifocal micro lens fabricated by depositing parylene directly on liquid", *Proceeding of 21st IEEE MEMS Conference (MEMS2007)*, Hyogo, Jan. 21-25, pp. 305-308.

[10] Hidetoshi Takahashi, N. Minh-Dung, K. Matsumoto, I. Shimoyama, "Differential pressure using a piezoresistive cantilever", *J. Micromech. Microeng.*, vol. 22, pp. 055015, 2012.

CONTACT

*Pham Quang-Khang, tel: +81-3-5841-6318; khang@leopard.t.u-tokyo.ac.jp

MULTIFUNCTIONAL INTEGRATED SENSOR IN A 2x2 mm EPITAXIAL SEALED CHIP OPERATING IN A WIRELESS SENSOR NODE

C.L. Roozeboom[1], V.A. Hong[1], C.H. Ahn[1], E.J. Ng[1], Y. Yang[1], B.E. Hill[1], M.A. Hopcroft[2], B.L. Pruitt[1]

[1]Stanford University, Stanford, California, USA
[2]Hewlett-Packard, Palo Alto, California, USA

ABSTRACT

We present multifunctional integrated sensors (abbreviated MFISES) that combine temperature, humidity, pressure, air speed, chemical gas, magnetic, and acceleration sensing on a single 2x2 mm chip. We fabricate MFISES in a wafer scale encapsulation process (called epi-seal) to hermetically seal the sensor functions with moving parts at low vacuum, and then surface micromachine the environmental sensors on top of the sealed layer. The encapsulation process provides very stable conditions for the pressure, magnetic, and acceleration sensors and enables the deployment of MFISES in dynamic environmental conditions without special packaging post-process. We demonstrate MFISES in a sensor node that combines energy harvesting, power management, and low power electronics to wirelessly transmit sensor data using a cloud-based service. The data can be processed and stored using server-side internet tools and then shared with any internet connected device. MFISES combined with the sensor node hardware demonstrates pivotal enabling technology in the emerging areas of wireless sensor networks and the Internet of Things.

INTRODUCTION

A recent *Wired* cover story highlighted multifunctional wireless sensor nodes in a variety of applications: radiation monitoring systems that measure gamma and beta radiation levels and weather conditions, parking space meters that detect the presence of a car using magnetic field and proximity sensors, and tsunami warning detectors that monitor wave height and sensor position [1]. DARPA has a program to develop unattended ground sensors that soldiers can deploy to measure acoustic, seismic, magnetic, and weather events [2]. The program goal is to develop a reusable hardware platform that is self-powered and compact. In industry, STMicroelectronics is developing HUB sensor platforms for environmental, motion, and acoustic sensing applications to reduce the size and cost of the sensor devices [3]. Current commercial integrated sensors, such as the InvenSense MPU-92150 and STMicro LSM9DS0, combine gyroscopes, accelerometers and compasses into 9-axis inertial measurement units. The previous multifunctional sensor examples, however, demonstrate that sensing capabilities beyond inertial measurements would benefit from chip scale integration.

We have previously reported on first generation (Gen. 1) MFISES devices that combined environmental and acceleration sensing in a low-cost integrated fabrication process [4]. The Gen. 2 MFISES devices incorporate design improvements from our previous investigations, are 25 times smaller, and have reduced cross-sensitivity compared to the Gen. 1 devices. MFISES

sensing capabilities are temperature, humidity, pressure, air speed, magnetic field, acceleration, and a chemical sensing platform that can be functionalized with various sensing films (Table 1 and Figure 1).

Table 1: MFISES image label and operation principles

Function	Fig.	Operation Principle
Chemical	a	Silicon heater, resistance measurement electrodes, sensing film
Anemometer	b	Heated bond wire resistor
Temperature	c	Aluminum RTD
	d	Resonant frequency shift of double ended tuning fork
Magnetic field	e	Vibrating Lorentz force
Acceleration- y	f	In-plane differential capacitive comb finger
Humidity	g	Polyimide sensing film between capacitive comb fingers
Acceleration- x	h	Out-of-plane teeter-totter differential capacitor
Pressure	i	Polysilicon membrane deflection, capacitive

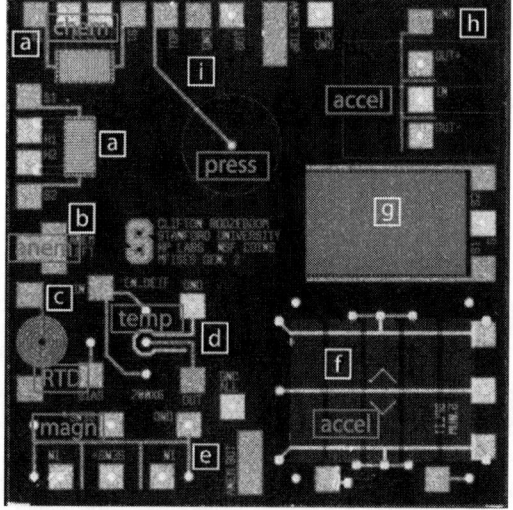

Figure 1: MFISES die is 2x2 mm and fabricated in the epi-seal process. Movable elements are vacuum encapsulated and environmental sensors are open to ambient conditions.

Integration of the sensor functions is enabled by the epi-seal wafer-scale encapsulation process [5]. Resonators sealed in the process have shown stability over many months of testing and temperature cycling and SiTime uses a similar process for high volume manufacturing of MEMS oscillators.

The MFISES devices target broad functionality and are useful for a variety of applications: climate monitoring in smart buildings, infrastructure monitoring in chemical plants and pipelines, and acoustic and seismic event monitoring for unattended ground sensors. We are using the concept of the "Swiss Army Knife" as the paradigm for the sensor. We are designing a device that is compact, low-cost, useful in many situations, and has built-in functionality for unexpected applications.

FABRICATION

MFISES devices were fabricated in the Stanford epi-seal process (Fig. 2). Fabrication begins with a silicon on insulator (SOI) wafer with a 40 μm-thick device layer. We use deep reactive ion etching (DRIE) to define the structures in the device layer silicon. The movable structures are sealed with an epitaxial polysilicon encapsulation layer and then released from the oxide layer with a vapor HF etch. An oxide passivation layer and metal layer are deposited and patterned on top of the encapsulation layer. In addition to in-plane electrodes, this work utilizes an extension to the base epi-seal process to include top out-of-plane electrodes in the encapsulation layer [6]. Nitride-filled isolation trenches allow for electrodes to be incorporated in the top encapsulation layer. The top electrodes are used for capacitive sensing for out-of-plane deflection of the z-axis accelerometer proof mass. The pressure sensor uses the epi-seal layer as the membrane and top electrode and the device layer as the stationary bottom electrode of a parallel plate capacitor. We also use the top electrode as a serpentine heater for the chemical sensor. In addition, isolation within the top electrode layer allows electrical routing of device layer electrodes through the encapsulation layer. This enables fabrication of a fully-differential y-axis accelerometer. The epi-seal process encapsulates the resonant temperature, magnetic field, pressure and acceleration sensors at approximately 1 Pa of pressure.

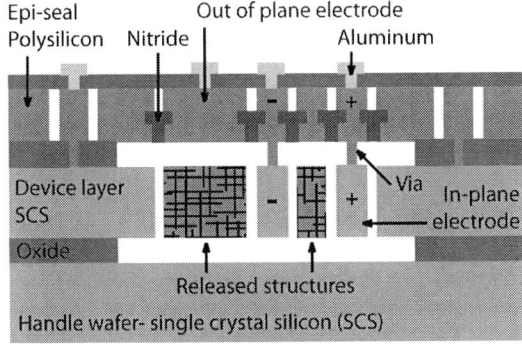

Figure 2: *Cross section of the process stack shows the movable sensor elements sealed by the epitaxial silicon encapsulation layer. The nitride plugs and air gaps enable the ability to create electrodes in the encapsulation layer.*

The electrodes for the humidity and chemical sensor and the resistor for the RTD are surface micromachined in the aluminum layer on top of the epi-seal and passivation layers. For the humidity sensor, we pipette 10 uL of PI-2555 polyimide (HD Microsystems, Parlin, NJ, USA) onto the comb finger electrodes and allow the solvent to evaporate. The anemometer is a looped aluminum wire bond suspended over the chip that we operate as a hot wire anemometer.

SENSOR DESIGN AND RESULTS
Chemical

Metal oxide semiconductor materials are commonly used for sensing carbon monoxide, ammonia, and hydrocarbon gases [7]. The sensing films adsorb gas molecules from the air, causing a reduction or oxidation reaction on the film surface, which leads to a change in conductance of the sensing film. The sensitivities of the metal oxide semiconductor films generally increase with temperature. Most gas sensors operate above 250 °C. In MFISES, we heat the chemical sensor with a serpentine polysilicon heater. We use an infrared microscope to measure the heater temperature and calibrate to applied power (Fig. 5a) and measure the time to a steady-state temperature (Fig. 5b). The chemical sensor has air gaps in the top electrode around the heater and is released from the device layer silicon to localize heating and increase thermal isolation. Aluminum comb fingers on top of the heater allow the deposition of metal oxide semiconductor films for sensing various chemicals.

Anemometer

We employ an aluminum wire bond as a hot wire anemometer to measure air flow over the sensor surface. Convective heat transfer from the heated bond wire to the passing air flow causes a decrease in wire temperature which we measure as a change in resistance or an increase in current for a set bias voltage (Fig. 5c).

Temperature

We measure temperature with two methods: an aluminum resistor operated as a resistant thermal detector (RTD) and a double ended tuning fork used as a resonant temperature sensor (Fig. 6d-e). The RTD sensitivity depends on the thermal coefficient of resistance which for aluminum is about $0.0037 \, C^{-1}$ [8]. The resonator temperature sensitivity is related to the temperature coefficient of frequency, TCF, which is a function of:

$$TCF = \frac{TCE + \alpha}{2} \quad (1)$$

where TCE is the material's temperature coefficient of Young's modulus (-64 ppm/°C for silicon depending on doping) and α is the coefficient of thermal expansion (2.6 ppm/°C) [9].

Magnetic Field

The magnetometer works on the principle that current flowing through the released beam in the presence of a magnetic field will cause a Lorentz force perpendicular to the magnetic field and current vectors (Fig. 3). The bias voltage is applied at the mechanical natural frequency of the beam, so that the displacement of the beam will be amplified by the mechanical quality factor. Parallel plate capacitors above and below the bar sense the change in displacement. The sense capacitors feed into transimpedance amplifiers and the signals are demodulated using a spectrum analyzer (Fig. 5f).

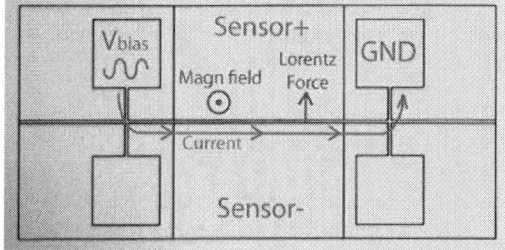

Figure 3: *The Lorentz force magnetometer consists of a released beam with bias current at the beam mechanical natural frequency. Capacitance sensing electrodes above and below the beam sense the mechanical vibrations.*

Acceleration

The *y*-axis accelerometer is a fully differential lateral comb drive design (Fig. 4a). The positive and negative sets of output combs are connected and routed through the top electrode layer. The *z*-axis accelerometer employs a teeter totter proof mass and out of plane sense electrodes in the epi-seal layer (Fig. 4b). For both accelerometers, the proof mass is biased with a sine wave signal above the mechanical natural frequency. The sensor outputs feed into transimpedance amplifiers, are demodulated using AD630 mixers, low pass filtered to remove the bias signal and then differentially amplified using an AD8226 instrumentation amplifier (Fig. 5g-h).

Humidity

A layer of polyimide absorbs and desorbs moisture from the ambient air causing a change in the dielectric constant of the polyimide. A set of interdigitated comb finger electrodes underneath the polyimide layer senses the change in dielectric constant as a change in capacitance. Polyimide is commonly used as the sensing layer in commercial sensors and shows linear response for relative humidity between 30 and 95% with little hysteresis over multiple humidity cycles (Fig. 5i).

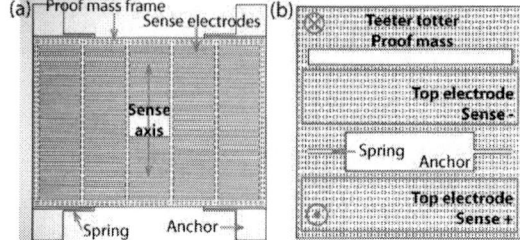

Figure 4: *Microscope images of (a) the y-axis and (b) the z-axis accelerometers show the proof masses, spring designs and placements of the sense electrodes.*

Pressure

The pressure sensor uses the epi-seal layer as the diaphragm and top electrode and the device layer as the bottom electrode of a parallel plate capacitor. The diaphragm is approximately 500 um in diameter and employs a reduced top electrode of 300 um to improve fractional capacitance change [10]. To measure the sensor capacitance, we apply an AC bias voltage at 100 kHz to the top electrode and connect the bottom electrode to a transimpedance amplifier and then demodulate the signal with an AD630 mixer and low pass filter (Fig. 5j).

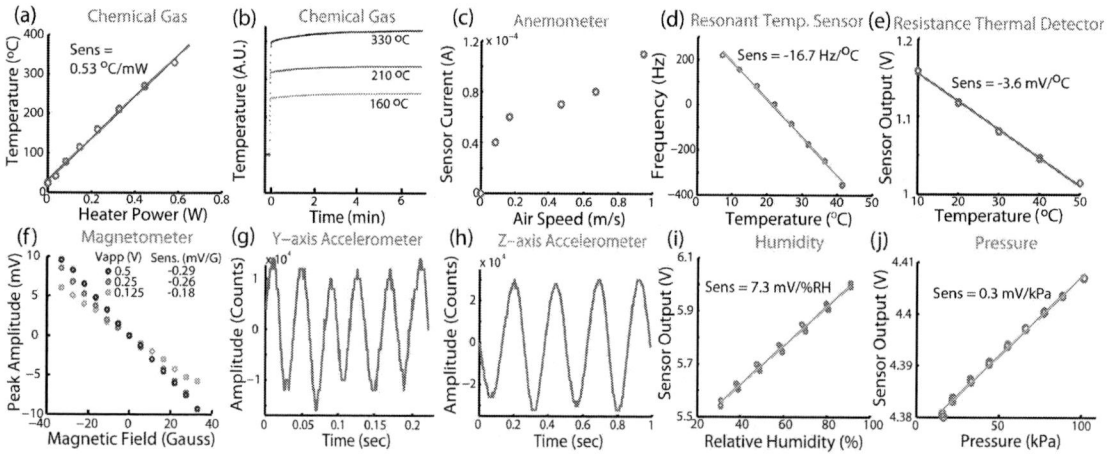

Figure 5: *Proof of concept data for each MFISES sensor function showcases the multifunctional capabilities of the integrated sensor. (a) Chemical sensor heater calibration shows the power required to heat sensor platform. (b) Time series data shows heating ramp for three temperatures. (c) Current flow through the hot wire anemometer increases as air speed and convective heat transfer to passing air flow increases. (d) Resonator and (e) resistance thermal detector calibrations demonstrate two transduction principles for temperature measurements. (f) Magnetometer calibration shows linear response to magnetic field strength and change in sensitivity with bias voltage. (g) Y-axis and (h) z-axis accelerometers demonstrate operation by measuring sinusoidal oscillations. (h) Humidity sensor uses polyimide sensing film and exhibits linear output from 30 to 95 %RH. (i) Pressure sensor responds to pressure change in vacuum chamber.*

WIRELESS SENSOR NODE DESIGN

Enabling wireless sensor nodes for emerging Internet of Things applications should include: (1) an energy harvesting device, (2) power management electronics, (3) energy storage, (4) multifunctional sensing capabilities, (5) low power microcontroller, and (6) low power wireless communications (Fig. 6a). The architecture of our sensor node implements these components with: (1) photovoltaic energy harvester (Sparkfun, PRT-07845), (2) LTC3105 DC/DC boost converter, (3) lithium ion battery, (4) MFISES multi-sensor, (5-6) Electric Imp card interfacing with an LTC2498 A/D converter and wirelessly transmitting data using 802.11 WiFi (Fig. 6b). The node can operate from just the solar cell power in direct sunlight and operates on the battery backup otherwise. The Electric Imp card is a microcontroller, wireless transmitter, and antenna packaged in an SD card form factor. The Electric Imp connects the node to the Imp Cloud for web based programming and data transmission. We are currently deploying these sensor nodes to monitor weather and other environmental conditions outdoors and for monitoring climate conditions and room occupancy indoors.

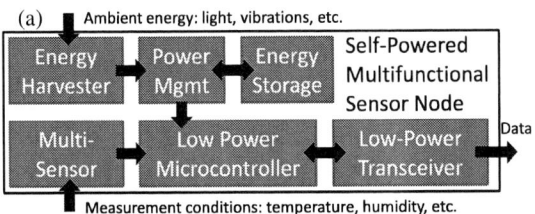

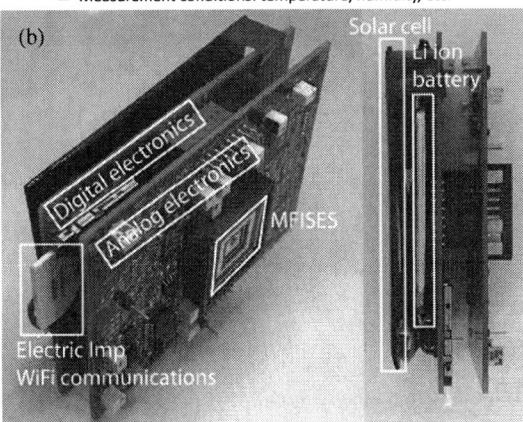

Figure 6: (a) Block diagram shows the components of a wireless sensor node for broad applications. (b) Our implementation of the sensor node using MFISES and a combination of custom and off-the-shelf components.

CONCLUSION

MFISES demonstrates the capability of integrating temperature, humidity, pressure, air speed, chemical gas, magnetic, and acceleration sensing in a 2x2 mm chip. The epi-seal fabrication process enables encapsulation of the pressure, magnetic, and acceleration sensors in a stable environment while allowing the environmental sensors to be exposed to ambient conditions. The operation of

MFISES in a complete sensor node showcases an enabling technology for emerging applications. Future work will be focused on demonstrating MFISES and the sensor node architecture in applications that need multifunctional sensing capabilities.

ACKNOWLEDGEMENTS

The authors would like to thank M. Tilghman for his assistance with the chemical sensing experiments. This work was supported by the Hewlett-Packard Labs Innovation Research Program Grant, the NSF Center of Integrated Nanosystems Grant No. ECCS-083281, the Sandia National Labs Excellence in Engineering Fellowship, and the Chevron-Stanford Energy and Environment Affiliates Program. Work was performed in part at the Stanford Nanofabrication Facility (a member of the National Nanotechnology Infrastructure Network), which is supported by NSF Grant No. ECS-9731293, its lab members, and the industrial members of the Stanford Center for Integrated Systems.

REFERENCES

[1] B. Gardiner, "How an army of sensors helps us track tsunamis and score parking spots," *Wired*, June 2013.

[2] K. McCaney. (2013, June). DARPA takes a smart-phone approach to Android ground sensors. [Online].

[3] B. Vigna, "MEMS & sensors for a new world," presented at the Shaping the Future of MEMS and Sensors Conference, Santa Clara, CA, 2013.

[4] C. Roozeboom, *et al.*, "Integrated Multifunctional Environmental Sensors", *J. of Microelectromech. Syst.*, vol. 22, pp. 779-793, 2013.

[5] B. Kim, *et al.*, "Frequency stability of wafer-scale film encapsulated silicon based MEMS resonators", *Sens. and Actuators, A: Phys.*, vol. 136, pp. 125-131, 2007.

[6] V.A. Hong *et al.*, "X-Y and Z-axis capacitive accelerometers packaged in an ultra-clean hermetic environment", *Transducers 2013*, pp. 618-621, 2013.

[7] V. Aroutiounian, "Metal oxide hydrogen, oxygen, and carbon monoxide sensors for hydrogen setups and cells", *Inter. J. of Hydrogen Energy*, vol. 32, pp. 1145-1158, 2007.

[8] N.M. McCurry, "Aluminum temperature coefficient of resistance vs. grain structure" MS thesis, Chem. Dept., Lehigh Univ., Bethlehem, PA, 1993.

[9] M. A. Hopcroft, *et al.*, "What is the Young's modulus of silicon?" *J. Microelectromech. Syst.*, vol. 19, no. 2, pp. 229–238, 2010.

[10] C.-F. Chiang, *et al.*, "A single process for building capacitive pressure sensors and timing references with precise control of released area using lateral etch stop", *MEMS 2012*, pp. 519-522, 2012

OUT-OF-PLANE MICRO TRIPLE-HOT-WIRE ANEMOMETER BASED ON PYREX BUBBLE FOR AIRFLOW SENSING

Shuwei Liu[1], Shanshan Pan[1], Fei Xue[1,2], Nay Lin[1], Haobin Liu[1], Jianmin Miao[1], Leslie K. Norford[3], H.B. Lim[2]

[1]School of Mechanical and Aerospace Engineering, Nanyang Technological University, Singapore
[2]Intelligent Systems Center, Nanyang Technological University, Singapore
[3]Massachusetts Institute of Technology, Massachusetts, USA

ABSTRACT

The paper reports novel design and fabrication of out-of-plane micro airflow sensors based on the hot-wire sensing principle, i.e. gas cooling of electrically-heated wires. Three micro Ti/Au/Cu hot-wire resistors have been fabricated on an out-of-plane Pyrex bubble with height of 300 μm. They are arranged 120 degrees apart on the sidewall of a bubble at height of 100 μm for airflow velocity and direction detection. Air velocity around the out-of-plane bubble structure has been investigated for square and circular package designs. The sensor with axisymmetric circular package has demonstrated the ability to detect velocity (<10 m/s) and to determine flow direction with an error less than ±8° when the velocity is 10m/s. The sensitivity could be further improved in a new design with increased bubble height (1 mm) and elevated hot-wire resistor position (500 μm), according to the modeling results.

INTRODUCTION

Flow detection is important in environmental monitoring. Airflow characteristics in urban areas affect pedestrian comfort, air quality, pollutant dispersion and energy performance of buildings [1]. MEMS sensors can reduce power consumption and have lower cost due to the nature of batch fabrication. Micro scale airflow sensors fabricated by MEMS technology are mainly based on two principles, namely mechanical and thermal. For mechanical principle, microcantilevers are employed as sensing elements. The passage of airflow causes the cantilever to deflect, leading to strain (stress) deformation detected by the piezoresistive/piezoelectric mechanism [2]. For sensors based on the thermal principle, the measurement relies on the detection of the convective heat transfer from an electrically heated resistive sensing element to flowing fluid [3]. Without introducing any movable structure, thermal principle is more commonly used in flow sensors due to its fast response, and robust and simple structure [4].

In order to detect airflow direction, the micro airflow sensor must have more than one sensing component to establish a flow-dependent thermal profile. In our previous work [5], we demonstrated that the configuration of three manually-assembled MEMS-based hot-wire resistors 120 degrees apart around a cylinder enables the sensor to detect airflow velocity and direction with high sensitivity. The hot-wire resistors function as heating and sensing element at the same time. The flow direction measurement is based on the relative output difference of the elements in response to temperature variation induced by flow. To improve assembly

effectiveness, we propose to fabricate three self-assembled micro hot-wire resistors on an out-of-plane Pyrex bubble that provides good thermal isolation. A schematic view of the sensor chip is shown in Figure 1. Three hot-wire resistors are arranged 120 degrees apart around a Pyrex bubble

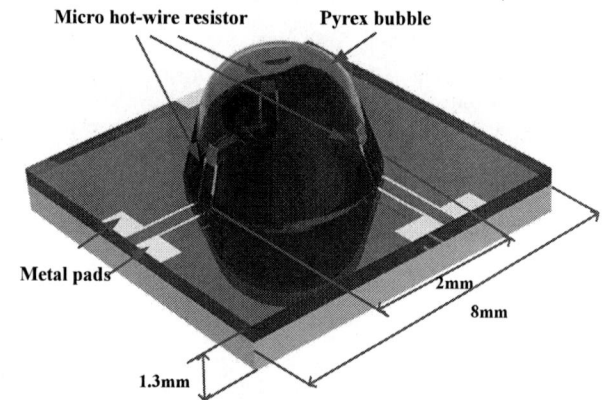

Figure 1: Schematic of airflow sensor on square chip

FABRICATION

Four-inch Silicon <100> P-type wafers with 1 mm thickness are cleaned in piranha solution ($H_2SO_4 + H_2O_2$) for 15 minutes to remove the organic residues and prepared for fabrication (Figure 2(a)). Circular patterns with diameter of 2 mm are defined on the wafer surface using 10 μm thick AZ 9260 photoresist in photolithography. Deep reactive ion etching (DRIE) process is then carried out to etch the silicon to obtain the cavities with depth of 500 μm (Figure 2(b)). Pyrex wafers of 300 μm thickness are anodically bonded to the etched silicon wafers at temperature of 400°C and atmospheric pressure of 1000 mbar (Figure 2(c)). Before anodic bonding, the wafers are cleaned in oxygen plasma for 10 minutes.

After the wafers are bonded, metal lines of hot-wire resistors are lithographically defined on five micrometer thick AZ9260 photoresist layers above cavities (Figure 2(d)). Ti/Cu/Au layers with the thicknesses of 20 nm/ 200 nm/ 200 nm respectively are deposited on the wafers. A titanium layer acts as adhesion layer between Pyrex and other metal layers. Copper is used as the structural material because it is easy to deform in the later bubble forming process due to the relatively low melting temperature of 1355 K, compared with commonly used Pt with melting temperature of 2041 K. Gold is used on the top to protect the metal from corrosion. After the deposition of metal layers, the lift-off process is used to obtain the hot-wire resistors on the Pyrex wafers (Figure 2(e)). Protective coating layer of 10 μm thick photoresist is spin-coated on the surface to protect the resistors before performing dicing to obtain the individual chips. After dicing, chips are then heated inside a furnace at a temperature (900 °C) above the softening point of the glass and for a 30 sec annealing period. The ramp rate for the annealing temperature is approximately 60 °C/min. Due to the expansion of the trapped gas in the silicon cavities during annealing, the Pyrex is blown into hemispherical bubbles with the hot-wire resistors on the sidewall (Figure 2(f)). Figure 3 shows the SEM image of the three micro hot-wire resistors 120-degree apart with approximate height of 100 μm around the perimeter of the Pyrex bubble with height of 300 μm.

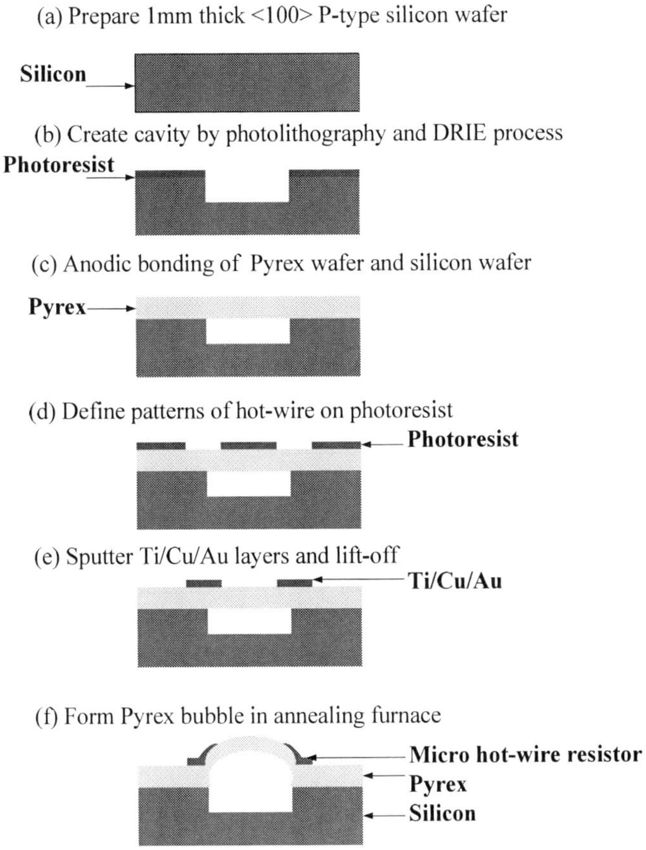

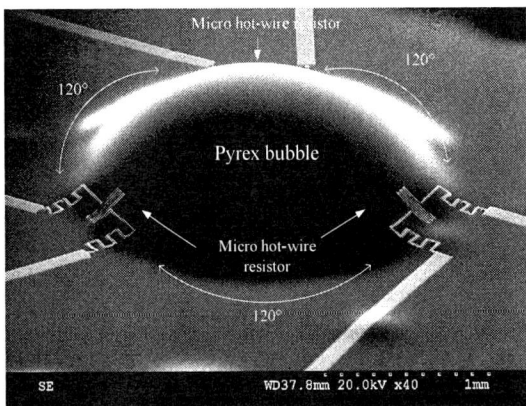

Figure 3: SEM of triple hot-wire resistors on Pyrex bubble

EXPERIMENTAL

Testing Interface

The complete testing rig, including sensing and data processing components, is placed in a wind tunnel, as shown in Figure 4. The sensor chip is placed on the top of a pillar with diameter of 6 mm which is mounted on a rotary station for flow direction characterization. The sensor is positioned in a wind tunnel such that the airflow is parallel to the sensor surface, as shown in Figure 5(a). The direction of airflow is always perpendicular to the pillar. The position at which hot-wire sensor 1 is facing the airflow direction is marked as 0°.

The measured resistances of the three hot-wire resistors are 66.9 Ω, 62 Ω and 51.3 Ω. Each resistor is connected to a constant temperature anemometry (CTA) circuit channel based on Wheatstone Bridge with the principle shown in Figure 5(b). Constant temperature means that the circuit actively keeps the resistance of each resistor constant by a closed-loop feedback. The voltage output from each channel can be displaced on LCD and stored in an SD card. Flow direction characteristics are analyzed at 10-degree intervals as the supporting pillar is rotated in the wind tunnel.

(a) Prepare 1mm thick <100> P-type silicon wafer

Silicon

(b) Create cavity by photolithography and DRIE process

Photoresist

(c) Anodic bonding of Pyrex wafer and silicon wafer

Pyrex

(d) Define patterns of hot-wire on photoresist

Photoresist

(e) Sputter Ti/Cu/Au layers and lift-off

Ti/Cu/Au

(f) Form Pyrex bubble in annealing furnace

Micro hot-wire resistor
Pyrex
Silicon

Figure 2: Fabrication process flow

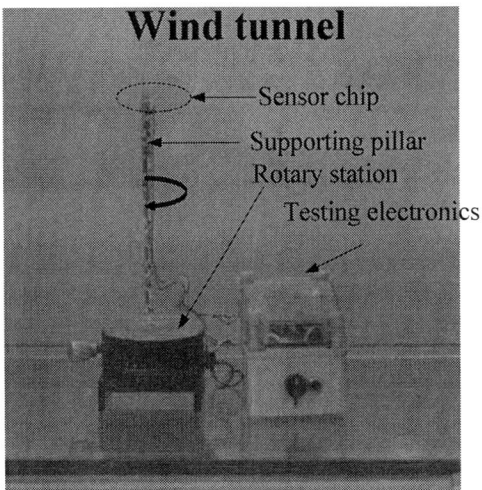

Figure 4: The testing rig in wind tunnel

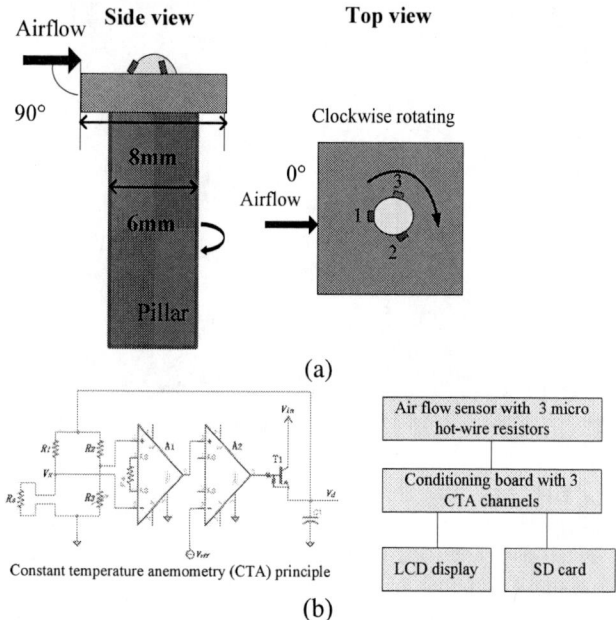

(a)

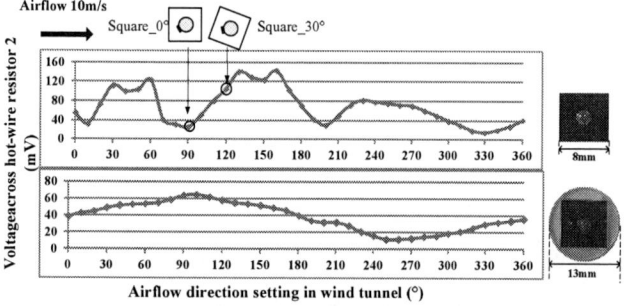

(b)

Figure 5: (a) Sensing and (b) Data processing components

Results and Discussion

The top plot in Figure 6 shows the voltage output from the channel of hot-wire resistor 2 when the square chip is rotated from 0° to 360°. Unexpected voltage valleys are observed as the chip rotates. However, the voltage valleys are removed and a smooth voltage profile is generated when an acrylic circular frame with thickness of approximately 1mm is added around the square plate, as illustrated in the bottom plot in Figure 6.

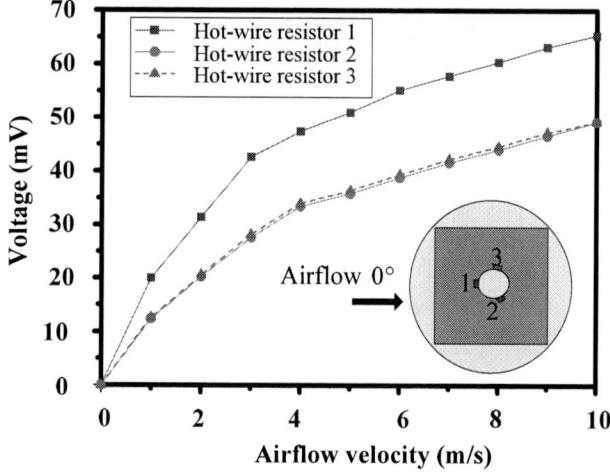

Figure 6: Comparison of measured voltage output when square plate and circular plate were used, respectively

This phenomenon can be explained by the CFD modeling in Figure 7. The flow around the sensor on a square plate was simulated with uniform input velocity of 10m/s at 0° and 30° incident angles. As the Pyrex bubble is fully immersed in the boundary layer formed on top of the sensor plate, the sensor system is very sensitive to the boundary layer properties. The non-axisymmetric square plate generates different boundary layers at different horizontal incident angles, resulting in different velocities at the same measurement point. For instance, the velocity in front of the bubble is 4.8 m/s at 0° but 6.8 m/s at 30°. This discrepancy results in difficulty in calibration. Adding an axisymmetric circular frame around the square plate eliminates this

variation. After adding a circular frame, the voltage magnitude is dropped from 130 mV (top plot in Figure 6) to 60 mV (bottom plot in Figure 6), due to the thicker boundary layer associated with the larger plate.

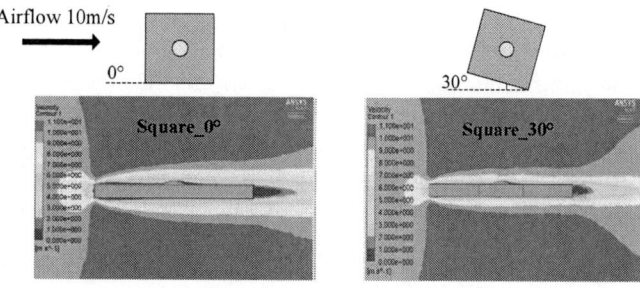

Figure 7: Velocity contours of the bubble on the square plate in CFD model at incidental angles of 0° and 30 °

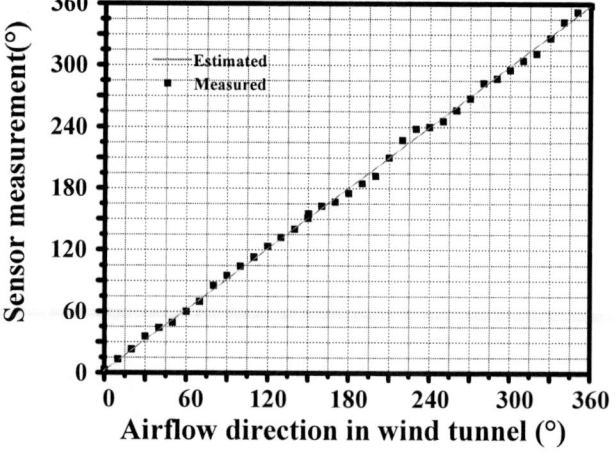

Figure 8: The measured voltage outputs of three hot-wire resistors at different airflow velocities (airflow direction is set at 0°)

Figure 9: Comparison of measured direction with reference airflow direction setting in wind tunnel

978-1-4799-3510-9/14 $31.00 © 2014 IEEE

The voltage outputs of the three hot-wire resistors on a circular plate at different airflow velocities are presented in Figure 8. The data are then processed with the algorithm developed in [4] for direction distinction. The direction measurement error is found to be less than ±8° at 10 m/s, as manifested in Figure 9.

CONCLUSION

We presented a novel out-of-plane micro airflow sensor with triple micro hot-wire sensors fabricated on the sidewall of a Pyrex bubble. We also studied the boundary layer effect on sensing accuracy caused by the shape of the sensor chip, concluding that the common square chip created by the standard wafer dicing process impairs the sensor's performance. The problem is solved by adding an acrylic circular frame to the square plate to create an axisymmetric circular package.

FUTURE WORK

The sensitivity and the sensing accuracy of the current device can be further improved if the boundary layer effect could be reduced. CFD modeling of a new design with increased bubble height (1 mm) and elevated hot-wire resistor position (500 μm) (Figure 10(a)) shows a larger velocity difference between two resistors e.g. at 90° and 210° respectively (Figure 10(b)). Larger velocity differences will result in higher accuracy in direction detection. Hence sensors with the new design will be fabricated in the future. More simulations and experiments on sensor structure optimization will also be performed.

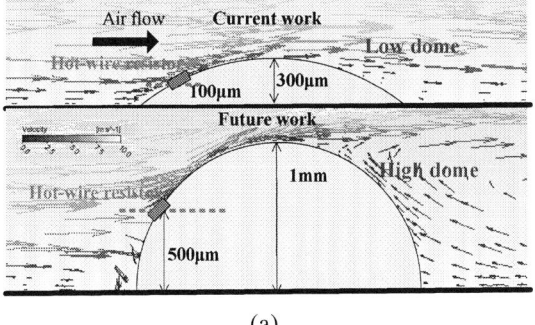

(a)

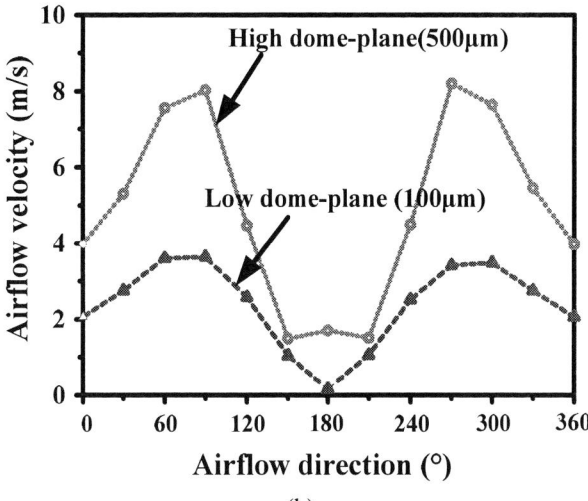

(b)

Figure 10: (a) Proposed high-dome design; (b) CFD modeling of airflow velocity variation (500 μm plane on the high dome and 100 μm plane on the low dome) with respect to airflow direction when airflow of 10 m/s is applied

ACKOWLEDGEMENT

The research was funded by the Singapore National Research Foundation (NRF) through the Singapore-MIT Alliance for Research and Technology (SMART) Center for Environmental Sensing and Modeling (CENSAM).

REFERENCES

[1] J. Chen, Z. Fan, J. Zou, J. Engel, C. Liu, "Two-dimensional micromachined flow sensor array for fluid mechanics studies", *J. Aerospace Eng.*, vol. 4, pp.85-97, 2004

[2] P. Zylka, P. Modrzynski, P. Janus, "Vortex anemometer using MEMS cantilever sensor", *J. J. Microelectromech. Syst.*, vol. 19, pp.1485-1489, 2010

[3] X.M. Jing, J.Y. Lu, J.M. Miao, H. Hans, H.A. Rahman, S.S. Pan, L. K. Norford, "An aerodynamically efficient sphere anemometer with integrated hot-film sensors for 2-D environmental airflow monitoring", in *Digest Tech. Papers Transducers '11 Conference*, Beijing, June 5-9, 2011, pp. 96-99, 2011

[4] H.B, Liu, N. Lin, S.S. Pan, J.M. Miao, and L.K. Norford, " High sensitivity, miniature, full 2-D anemometer based on MEMS hot-film sensors", *IEEE Sensors J.* vol.13, pp.1914-1920, 2013

[5] J.T.W. Kuo, L. Yu, E. Meng, " micromachined thermal flow sensors-A review", *Micromachines*, vol.3, pp.550-573, 2012

CONTACT

* J.M. Miao, Tel: 65-6790 6038; mjmmiao@ntu.edu.sg

A PAPER-BASED PIEZOELECTRIC TOUCH PAD INTEGRATING ZINC OXIDE NANOWIRES

Yu-Hsuan Wang[§], Xiao Li[§], Chen Zhao, and Xinyu Liu*
Department of Mechanical Engineering, McGill University

ABSTRACT

This paper describes a new type of paper-based piezoelectric touch pads integrating zinc oxide nanowires (ZnO-NWs), as user interfaces in paper-based electronics. The functionality of the touch pads is enabled by the piezoelectric property of ZnO-NWs grown on paper using a simple hydrothermal method. A piece of ZnO-NW paper with two screen-printed silver electrodes forms a touch button, and touch-induced electric charges from the button is converted into a voltage signal using a charge amplifier circuit. A touch pad consisting of an array of buttons can be readily integrated into paper-based electronic devices, allowing user input of information for various purposes such as programming, identification checking, and gaming. This novel design features ease of fabrication, low cost, and ultra-thin structure, and good compatibility with techniques in printed electronics, which further enriches the tool set available for developing paper-based electronic devices.

INTRODUCTION

Paper-based electronics has led to new types of inexpensive and flexible electronic products for applications such as flexible displays, disposable biosensors, smart packaging, and energy harvesting/storage devices [1]. Many paper-based electronic devices need to receive information from users for various purposes such as programing, identification checking, and gaming, and integration of touch-based user interfaces onto paper substrates is highly desired. While significant efforts have been spent on developing high-performance printable electronic materials and multifunctional device prototypes [2, 3], fewer advances are made on creating paper-based electronic user interfaces.

It is a natural choice to resort to micro-electro-mechanical systems (MEMS) based touch sensing mechanisms to conceive viable solutions for paper-based platforms. Piezoresistive, piezoelectric, and capacitive sensing mechanisms have been widely used in designs of flexible touch sensors; however, implementing them onto paper substrates in a cost-effective fashion is challenging. An attempt to develop paper-based touch pads were reported recently [4], which employed commercially available metallized paper to construct touch-sensitive capacitors. To date, there is no piezoelectric touch sensors made from paper. We report, for the first time, a new type of paper-based piezoelectric touch pads integrating zinc-oxide nanowires (ZnO-NWs) as the sensing component. ZnO, as a piezoelectric material, can be synthesized on various substrates and promises excellent sensing capabilities. We directly grew ZnO-NWs on cellulose paper (Figure 1A) using a simple hydrothermal approach [5], and utilized the ZnO-NW-coated paper to fabricate single-layer piezoelectric touch pads. The presented touch pads are inexpensive ($0.03/button), easy-to-fabricate, ultra-thin, lightweight, and disposable, and enriches the development tool set of paper electronics.

EXPERIMENTAL DESIGN

Design of Paper-Based Piezoelectric Touch Pads

The underlying mechanism of the proposed touch pad design is based on the piezoelectric response of ZnO-NWs on paper upon touching. Through a hydrothermal process, ZnO-NWs were grown on microfibers of cellulose paper. The ZnO-NWs, as shown in Fig. 1A, stand radially

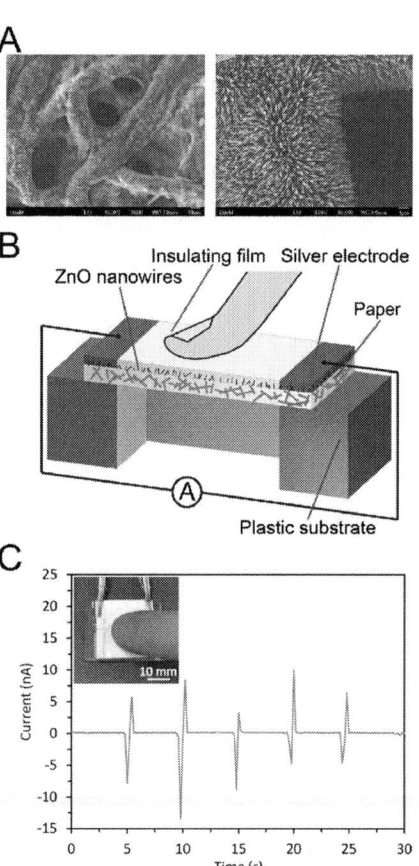

Figure 1: A paper-based piezoelectric touch button with hydrothermally grown ZnO-NWs. (A) SEM images of paper microfibers coated with ZnO-NWs. (B) Schematic view of the touch button. (C) A typical current response of the touch button upon repeated finger presses. Inset is a photo of the device.

[§] Both authors contributed equally to this work.

outwards on the cellulose microfibers at a high density. A piece of ZnO-NW paper suspended over a supporting substrate forms a touch button (Fig. 1B), and can produce electric charges in response to finger pressing. For device prototyping, we used an acrylic substrate with a square cavity as the support underneath the ZnO-NW paper, and the acrylic substrate can be replaced by thick cardboard to make the touch button completely paper-made. An adhesive tape covers the button top surface for electrical isolation.

During operations, figure pressing deforms the ZnO-NW paper, thus induces mechanical stresses and generates electric charges in the ZnO-NW layer on paper surface. Two silver electrodes screen-printed on top of the ZnO-NW paper collect the generated electric charges and feed them into a precision current meter or a charge amplifier circuit. Fig. 1C shows the typical current response of a touch button upon repeated finger presses, measured by a precision potentiostat (PGSTAT302N, Metrohm). A negative current peak at the nano-ampere level appears while pressing and then quickly dissipates through the closed circuit loop. Upon finger release, the deformed paper restores and generates a positive current peak.

Hydrothermal Growth of ZnO-NWs on Cellulose Paper

We grew ZnO-NWs on Whatman® 3MM chromatography paper (340 μm thick) through a hydrothermal process [6]. A seeding layer of ZnO NPs was first coated on oxygen- plasma-treated chromatography paper pieces (26 mm × 26 mm) in ethanol for subsequent growth of ZnO-NWs. The paper pieces were then dipped into 100 mL aqueous growth solution of 50 mM zinc nitrate hexahydrate (ZNH) and 25 mM hexamethylenetetramine (HMTA) at 86 °C for 1.5-15 hours. For better efficiency of the ZnO-NW growth, 5 mL of 30% ammonium hydroxide solution was added to the growth solution to suppress the homogeneous nucleation of ZnO-NWs [7]. After growth, the ZnO-NW paper was washed in deionized water and dried in a 90 °C oven.

Fabrication of the Paper-Based Touch Buttons

The 3D cross-section view of a touch button is illustrated in Fig. 1B. We selected Whatman® 3 MM chromatography paper to fabricate the touch buttons because: (i) its composition of pure cellulose makes the hydrothermal growth of ZnO-NWs more reproducible; and (ii) its relatively thick structure (340 μm) is more stable mechanically and can hence better withstand pressing-induced deformation. Nevertheless, ZnO-NWs can also be readily grown on other common paper substrates (packing paper [8] and plain printing paper [9]), and our touch button design, in principle, can be realized on many other types of paper as long as the paper substrate provides certain mechanical strength for finger pressing.

After the growth of ZnO-NWs, the paper pieces were screen printed with silver ink (E1660, Ercon) on their top surfaces to form 3 mm × 26 mm electrodes and dried at 80°C for 1 h. The silver electrodes form Ohmic contact with

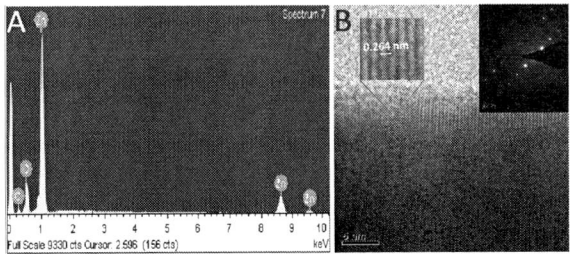

Figure 2: Characterization of ZnO-NWs grown on paper. (A) Chemical analysis of paper surface after ZnO-NW growth using energy dispersive spectroscopy (EDS). (B) Transmission electron microscopy (TEM) images of a ZnO-NW, inserted with an electron diffraction image showing the lattice orientation along [0001].

the ZnO-NW layer [10]. We covered a layer of insulating adhesive film on the top surface of the paper, and finally attached the paper to an acrylic piece (3 mm thick) with a central square cavity (20 mm × 20 mm) using double-sided tape. The cost of each touch button is $0.03.

RESULTS AND DISCUSSION
Quality of ZnO-NWs Grown on Paper

We analyzed the morphology and crystal structure of the ZnO-NWs grown on paper using scanning electron microscopy (SEM) and transmission electron microscopy (TEM). As shown in Fig. 1A, the cellulose microfibers were uniformly covered by the ZnO-NWs standing radially outwards. After 15-hour growth, the ZnO-NWs have an average length of 2.53 ± 0.19 μm and an average width of 68.24 ± 7.82 nm. Single ZnO-NWs have relatively uniform width along its length. Clear crystal lattice structures of ZnO-NWs were observed via high-resolution TEM imaging (Fig. 2A). It showed a 0.264 nm lattice spacing for the (0002) crystal planes. The electron diffraction image of a selected area in Fig. 2B indicated the lattice orientation along [0001], which is the c-axis of ZnO crystalline.

Characterization of Current Output

We measured the current output of the paper-based touch buttons using the potentiostat. The typical current response of a touch button to repeated finger presses are shown in Fig. 1C. Clear negative (upon finger pressing) and positive (upon finger releasing) peaks can be observed in the range of -15 nA to 10 nA, and the ground value of the output current without activation is 0.1~0.2 nA. The current peaks varied because of variations in finger pressing forces. We used the negative peaks as activation signal in the following experiments, and quantified it as the major output parameter of the touch buttons for device characterization.

Effect of ZnO-NW Growth Percentage on Device Current Response

It is a common observation that given extended growth time ZnO-NWs grow longer [7]. We defined the growth percentage of ZnO-NWs as the mass increase percentage of the paper pieces (vs. original mass) after growth, and

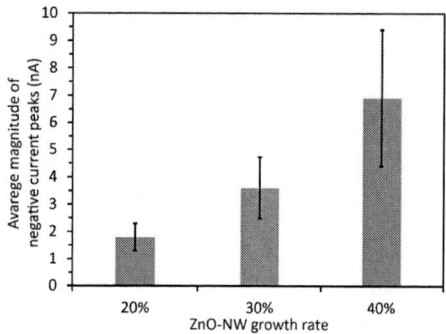

Figure 3: Experimental results of average negative current peaks vs. ZnO-NW mass growth percentage (n = 3 devices). ZnO-NWs were grown for 1.5 h, 3 h, and 15 h, yielding 20%, 30%, and 40% weight growth respectively.

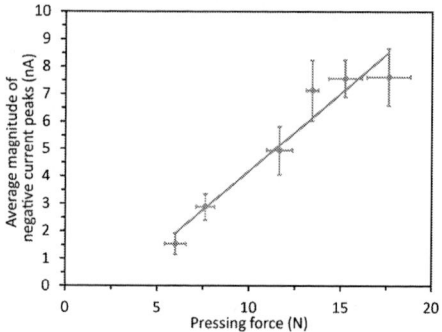

Figure 4: Experimental results of average negative current peaks vs. pressing force (n = 3 devices; 10 measurements per deformation). Equation of the linear regression: $y = 0.57x - 1.48$, ($R^2 = 0.937$).

investigated the effect of ZnO-NW growth percentage on the device current outputs. In our experiments, the paper pieces were weighed in dry form before and after growth. Similar with other reports [7, 11], we noted that the ZnO-NWs were grown quickly in the first three hours, and the growth slowed down after that and almost stopped after 15 hours. This growth profile was mainly due to the gradual depletion of chemicals in the growth solution. The growth percentages after 1.5 h, 3 h, and 15 h are 19.7 ± 0.5%, 30.3 ± 1.2%, and 40.3 ± 0.5%, respectively. Even higher growth percentages can be achieved by refreshing the growth solution during a longer period of growth time [12].

We measured current outputs of the touch buttons, with ZnO-NW growth percentages of 19.7% (1.5-hour growth), 30.3% (3-hour growth), and 40.3% (15-hour growth). We used a machine-shop-made metal stand with a finger-shaped tip to mimic finger pressing and apply consistent touch forces of 17.6 ± 1.2 N. As shown in Fig. 3, the average magnitude of negative current peaks shows an obvious increasing trend with the growth percentage. We speculate that the increased current outputs with higher growth percentages are due to the fact that longer ZnO-NWs deflect more under the same pressing force and thus generate more electric charges. Further research on nanomechanics modeling of the hierarchical structure of ZnO-NWs on cellulose microfiber network is required to better understand the pressing-induced stress/strain distribution and piezoelectronics in ZnO-NW paper.

Effect of Pressing Force on Device Current Response

We also investigated the current response of touch buttons at different pressing force levels. Hard and gentle presses deform the ZnO-NW paper to different extents, resulting in different current outputs. We adjusted the pressing force applied to the touch buttons and measured their current outputs. As shown in Fig. 4, the average magnitude of negative current peaks increases linearly with the pressing force, with a sensitivity of 0.57 nA/N. If a more sensitive response of the touch button is desired, one can choose a thinner and thus more flexible paper substrate and adopt a higher ZnO-NW growth rate in paper preparation.

Durability Testing

Performance degradation after repeated operations could be a concern if the paper-based touch buttons are designed for long-term uses. We tested the device durability through repeated pressing of a touch button made from paper with a ZnO-NW growth percentage of 30%. The button was continuously pressed for 2000 times using the metal stand at a high force level of 17.6 ± 1.2 N. After every 200 presses, the current output was measured for ten times to calculate the average. As shown in Fig. 5, the average magnitude of negative current peaks decreased gradually during the first 600 presses and started to stabilize after that.

We observed two causes associated with the output degradation. (i) Repeated presses resulted in unrecoverable (inelastic) deformation of the paper, which we started to observe after the first 100 presses. This irreversible deformation caused stiffening in the suspended paper structure, decreased the deformation/strain induced by subsequent presses, and thus lowered the current output. (ii) Repeated presses also permanently bent down the ZnO-NWs on paper, making them less stressed in the subsequent presses. This was revealed through SEM imaging of the ZnO-NWs after 600 presses (inset in Fig. 5). The current output stabilized after 600 presses, possibly because the suspended paper reached the limit of inelastic deformation and mainly underwent elastic deformation afterwards. After 2000 presses, the touch button still operated responsively and no mechanical damage were observed on the suspended paper structure. In application scenarios where extended uses are targeted, the paper touch buttons can be pre-loaded to reach stabilized performance.

Operations of a Ten-Key Touch Pad

After characterization of the touch button, we constructed a touch pad by arraying ten numbered buttons (Figs. 6C and 6D) on an acrylic frame. The touch pad also includes a 10-channel charge amplifier circuits (Fig. 6A; $1.4/channel) for converting electric charges from the buttons into voltage outputs, a microcontroller circuit (ATxmega32A4, ATMEL) for measuring the voltage outputs, and 11 LEDs (ten blues and one green) for touch-responsive display.

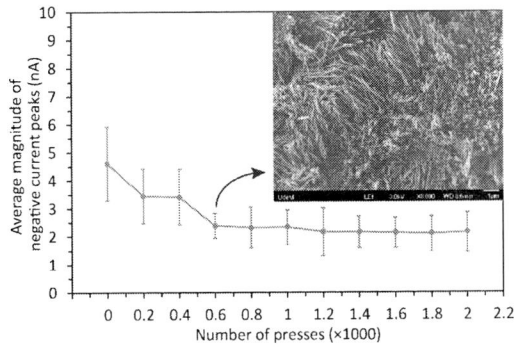

Figure 5: Experimental results of average negative current peak vs. number of presses (n = 10 measurements every 200 presses). The inset is a SEM image of a touch pad after 600 times of presses.

Fig. 6B shows the voltage outputs from the ten touch buttons when they were pressed sequentially by a human operator. The microcontroller was programmed to recognize finger pressing by detecting the negative voltage peaks from the touch buttons. Upon recognition of finger pressing on a specific button, a corresponding blue LED was lit up by the microcontroller (Fig. 6C). To highlight the potential use of our touch pads in electronic devices where input of information is needed, we demonstrated input of a six-digit numeric code on the touch pad. The microcontroller was programmed to compare the inputted code with the preset one and activate the green LED when there was a match (Fig. 6D).

CONCLUSIONS

This paper reported paper-based piezoelectric touch pads integrating ZnO-NWs, representing an inexpensive solution to developing new types of user interfaces for paper-based electronics. We successfully grew ZnO-NWs on cellulose paper and fabricated ZnO-NW paper into piezoelectric touch buttons. The ZnO-NW paper is responsive to finger pressing and generates electric charges which can be converted into a voltage signal using a charge amplifier circuit. We fully characterized the output performance of the touch button, and examined the device durability during 2000 cycles of pressing. We constructed an integrated touch pad with ten touch buttons and demonstrated its potential uses in user-device interaction. We believe that this technology has a significant potential for use in paper-based electronic platforms to improve user-device interactions and enable other sensing capabilities.

REFERENCES

[1] D. Tobjork and R. Osterbacka, "Paper electronics," *Advanced Materials,* vol. 23, pp. 1935-61, 2011.

[2] K. Kordás, T. Mustonen, G. Tóth *et al.*, "Inkjet printing of electrically conductive patterns of carbon nanotubes," *Small,* vol. 2, pp. 1021-1025, 2006.

[3] S. Chung, M. Jang, S.-B. Ji *et al.*, "Flexible high-performance all-inkjet-printed inverters: organo-

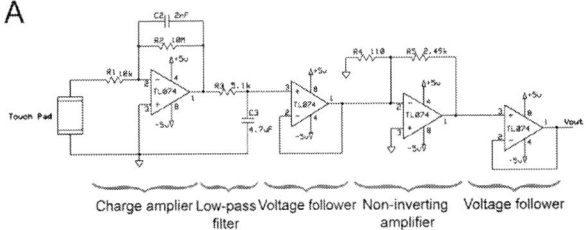

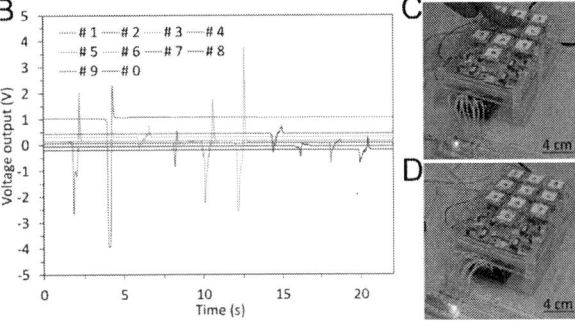

Figure 6: (A) Diagram of charge amplifier circuit for converting piezoelectric charges generated into measurable voltage. (B) Voltage outputs from 10 number keys while being pressed. (C)(D) Snapshots from demonstration videos showing: (C) the corresponding blue LED lights up when the number key is pressed; (D) the green LED lights up when a pre-programmed password is entered correctly.

compatible and stable interface engineering," *Advanced Materials,* vol. 25, pp. 4773-4777, 2013.

[4] A. D. Mazzeo, W. B. Kalb, L. Chan *et al.*, "Paper-based, capacitive touch pads," *Advanced Materials,* vol. 24, pp. 2850-2856, 2012.

[5] S. Baruah, M. Jaisai *et al.*, "Photocatalytic paper using zinc oxide nanorods," *Science and Technology of Advanced Materials,* vol. 11, p. 055002, 2010.

[6] L. Vayssieres, "Growth of arrayed nanorods and nanowires of ZnO from aqueous solutions," *Advanced Materials,* vol. 15, pp. 464-466, 2003.

[7] C. K. Xu, P. Shin, L. L. Cao *et al.*, "Preferential growth of long ZnO nanowire array and its application in dye-sensitized solar cells," *Journal of Physical Chemistry C,* vol. 114, pp. 125-129, 2010.

[8] Y. Qiu *et al.*, "Flexible piezoelectric nanogenerators based on ZnO nanorods grown on common paper substrates," *Nanoscale,* vol. 4, pp. 6568-73, 2012.

[9] H. Gullapalli, V. S. Vemuru, A. Kumar *et al.*, "Flexible piezoelectric ZnO-paper nanocomposite strain sensor," *Small,* vol. 6, pp. 1641-6, 2010.

[10] Z. L. Wang and J. H. Song, "Piezoelectric nanogenerators based on zinc oxide nanowire arrays," *Science,* vol. 312, pp. 242-246, 2006.

[11] J. J. Qiu, X. M. Li *et al.*, "Solution-derived 40 mu m vertically aligned ZnO nanowire arrays as photoelectrodes in dye-sensitized solar cells," *Nanotechnology,* vol. 21, 2010.

[12] M. Law, L. E. Greene, J. C. Johnson *et al.*, "Nanowire dye-sensitized solar cells," *Nature Materials,* vol. 4, pp. 455-459, 2005.

PIEZORESISTIVITY OF AG NWS-PDMS NANOCOMPOSITE

Morteza Amjadi[1,2], Aekachan Pichitpajongkit[1], Seunghwa Ryu[1], and Inkyu Park[1,2]
[1]Department of Mechanical Engineering, KAIST, South Korea
[2]Mobile Sensor and IT Convergence (MOSAIC) Center, KAIST Institutes (KI), KAIST, South Korea

ABSTRACT

In this work, we developed a conductive silver nanowire (AgNW)-PDMS composite thin film for a flexible strain sensing application. The piezoresistivity of AgNWs-PDMS nanocomposite thin film was experimentally investigated and analyzed by a computational model. The strain sensor shows a strong piezoresistivity with an average gauge factor in the ranges of 1.6 to 14 and a high stretchability up to 70 %. We found excellent agreement between our experiment and simulation results. Finally, a finger motion detection device was developed by using the AgNWs-PDMS nanocomposite thin film as a highly stretchable, flexible and sensitive strain sensor.

INTRODUCTION

Human motion recognition has stimulated much interest due to its potential applications such as rehabilitation and personal health monitoring [1-3], sport performance monitoring [4-5], and entertainment fields (e.g. motion capture for games and animation) [6]. Despite of numerous efforts for the human motion capturing by using vision based techniques (i.e. motion detection by set of cameras and image processing) [7, 8], vision based approaches are not widely applicable due to their space requirement, high cost and low resolution for a small movements (e.g. fingers' movement detection). Toward these objectives, demands are increasing for flexible, stretchable and wearable electronic devices due to their facile interactions with the human body, small size and higher accuracy [8-10]. For instance, flexible, wearable and stretchable strain sensors (i.e. devices that respond to the mechanical deformations by change in the resistance or capacitance) can be attached to the human body for the human motion detection. However, both high sensitivity (i.e. gauge factor (GF)) and high stretchability ($\varepsilon > 50\%$) are the minimum requirements of desired strain sensors for the human activity recognition; compare with that in the commercial strain gauges with stretchability (up to 5% maximum strain) and GFs~2 [10]. Several approaches have been reported to develop strain sensors with desirable properties (i.e. high stretchability coupled with sensitivity) by using nanomaterials and nanostructures. Among them carbon nanomaterial based sensors have shown outstanding performances due to the superior mechanical and electrical properties of carbon nanomaterials. Even though graphene based strain sensors pose a good sensitivity, they have very low stretchability due to brittleness of the graphene sheets [11, 12]. On the other hand, highly stretchable (up to 40%) strain sensors were developed by using single-walled carbon nanotubes (SWCNTs) on the elastomer substrate [10]; but, the strain sensors were suffering from approximately low GFs (~0.82). Although there have been numerous efforts to develop strain sensors with a high stretchability and sensitivity using carbon based nanomaterials, mostly reported strain sensors have high sensitivity coupled with approximately low stretchability and vice versa. Furthermore, strain sensors with desired characteristics particularly for the human motion detection are still remaining challenge.

Herein, we report a highly flexible and stretchable thin film based on the nanocomposite of PDMS with AgNWs as fillers. The thin film shows a strong piezoresistivity by external stimulus so that AgNW network with a sandwich structure (i.e. the AgNW network-PDMS nanocomposite layer laminated between two layers of PDMS (PDMS/AgNWs-PDMS nanocomposite thin film/PDMS)) was utilized as a highly stretchable and sensitive strain sensor. The characteristics of the sensors such as stretchability and response to both dynamic and static loads have been investigated. In addition, piezoresistivity of AgNW network was studied computationally using a 3D resistor network. We found an excellent agreement between our experiment and simulation. Finally, an artificial finger was developed to investigate the applicability of our strain sensor for the human fingers' gesture recognition. The resistance changes caused by the bending of strain sensors on the finger was measured and then positions of fingers were calculated based on the resistance changes, all in real-time.

MATERIAL AND SAMPLE FABRICATION

AgNWs were synthesized by polyol method according to Korte et al. [13] (see the reference for more details). AgNWs were stored in isopropyl alcohol (IPA) for further experiments. The average diameter and length of AgNWs were 150-200 nm and 10-20 μm, respectively.

The thin films were fabricated by drop-casting of the AgNW solution onto a glass slide which was patterned with a polyimide tape and cleaned (with acetone, ethanol, and DI water) beforehand. The uniformity of AgNW thin film is an important parameter for stable and predictable response of the strain sensors. Here, instead of using a hot plate or a vacuum oven, light heating has been employed as heating source. After drop-casting of the AgNW solution, the glass slide was exposed to a lamp light (OSRAM DR 51 50W 12V with Luminous intensity of 1450 cd) to dry AgNW solution and deposit the AgNWs onto the glass slide. Light heating provided uniform and gradual heating throughout deposited AgNW thin film and made it more uniform and homogenous. After drying the solution, polyimide tape was removed on the glass slide and patterned AgNW thin film was annealed at 200 °C for 20 min to increase the conductance of thin film. Furthermore, thermal annealing can reduce the resistance between AgNW junctions and improve the conductivity of film by removing the PVP surfactant and allowing fusion between AgNWs [14-15]. Sandwich structured samples were prepared by casting 0.5 mm layer of the liquid PDSM onto the patterned and annealed AgNW thin film. After partially curing the PDMS layer at 70 °C for 20 min, it was peeled-off and flipped. Then copper wires were attached to the ends of the AgNW thin film by silver paste. Finally, another layer of the liquid PDMS with a same thickness (~ 0.5 mm) was poured on the AgNW embedded PDMS film and cured at 70 °C for 2 hours. The fabrication process, photographs and SEM image of the sandwich structured sample are shown in Figure 1.

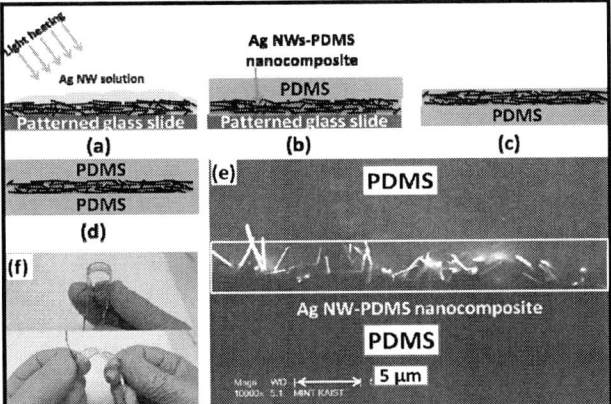

Figure 1: a-d) Fabrication processes of the sandwich type of AgNW thin films. a) Deposition of AgNWs on a glass slide under light heating and further annealing at 200 °C for 20 min. b) Casting the liquid PDMS onto the AgNW thin film and partially curing it at 70 °C for 30 min. c) Peeling-off and flipping the AgNWs embedded PDMS. d) Pouring another layer of PDMS and curing it at 70 °C for 2 hours. e) The cross-sectional SEM image on the sample; the AgNW nanocomposite layer is embedded between two layers of PDMS. f) Photographs of the fabricated sandwich structured sample when it is bended and twisted.

RESULTS AND DISCUSSION

Figure 1e shows the cross-sectional SEM image of the sandwich structured sample. As the figure shows, PDMS penetrated into the AgNW thin film network and filled the gaps between NWs, forming a composite of PDMS and AgNWs. The penetration of PDMS enhanced the contact of jointed NWs and made the total sparse network of AgNWs mechanically robust. The penetration enhanced the adhesion between the AgNW network and PDMS layers as well.

Figure 1f shows the photographs of sandwiched thin film which is bended and twisted without any damage to the AgNWs-PDMS nanocomposite thin film. The sample is very flexible and stretchable and it can be easily mounted onto the complex surfaces or directly attached to the human body.

To test the stretchability, samples were attached to a motorized moving stage. Strain/release cycles were applied to the conductive AgNW thin films while the current changes were measured with a potentiometer. Figure 2

shows the response of a sample to the cyclic loading (from $\square$=0 to $\square$=70 %). As figure illustrates the current of sample was recovered very well after releasing it from strain, indicating the applicability of the thin film as a highly stretchable strain sensor. Moreover, the position and orientation of NWs in the AgNW network-PDMS nanocomposite are changed by the applied strain. As strain increases, the numbers of disconnected NWs increase due to the PDMS elongation. Disconnection reduces the number of electrical passways, thereby causing the current to reduce and resistance to increase, accordingly. When the strain sensor is released from strain, all NWs slide back to their initial positions and the current of the strain sensor recovers to its original value.

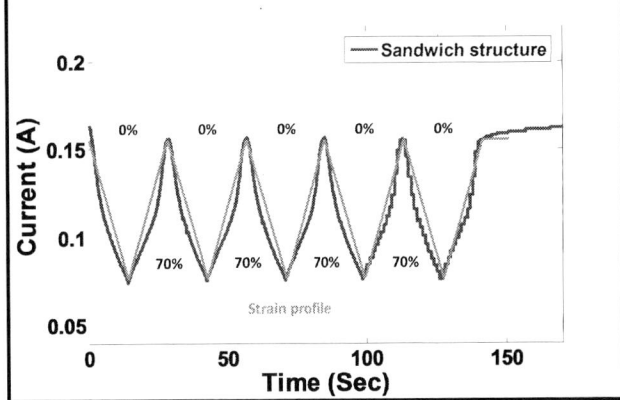

Figure 2: Response of the AgNWs-PDMS nanocomposite under 70% of stretch/release cycles.

The piezoresistivity of the AgNW network-PDMS nanocomposite is further investigated by a computational model. AgNWs with a constant diameter (D=150 nm) and length (L=20 μm) were randomly orientated in the PDMS matrix, as shown in Figure 3. The position and orientation of each NW can be determined by the coordinates of NW in the 3D space. Since the Young's modules of the AgNWs are much higher than that of the PDMS medium, the NWs can be assumed as rigid element [16, 17]. Therefore, the re-position and re-orientation of NWs by the applied strain can be calculated using the 3D equilibrium equation for the PDMS matrix.

We considered three electrical configurations for the two neighboring NWs including, (I) Complete connection without contact resistance; when the distance between centerlines of the adjacent NWs are less than the diameter of single NW, (II) Disconnected NWs; when the distance between the centerlines of NWs surpasses a cut-off distance (C), and (III) Tunneling junction; if the centerline distance of the two neighboring NWs is between diameter and cut-off distance so that electrons can tunnel through the PDMS matrix. We assign the cut-off distance as a distance between the two adjacent NWs whereby the resistance between NWs is 30 larger than the resistance of single NW (C~150.58 nm). The tunneling current can be estimated as [18]:

$$R_{Tunnel} = \frac{V}{AJ} = \frac{h^2 d}{Ae^2 \sqrt{2m\lambda}} \exp(\frac{4\pi d}{h} \sqrt{2m\lambda}) \qquad (1)$$

where J is tunneling current density, V is the electrical potential difference, e is the single electron charge, m is the mass of electron, h is Planck's constant, d is the distance between NWs, λ is the height of the energy barrier (1 eV for PDMS), and A is the cross-sectional area of the tunnel, which is assumed to be the same as the cross-sectional area of single NW.

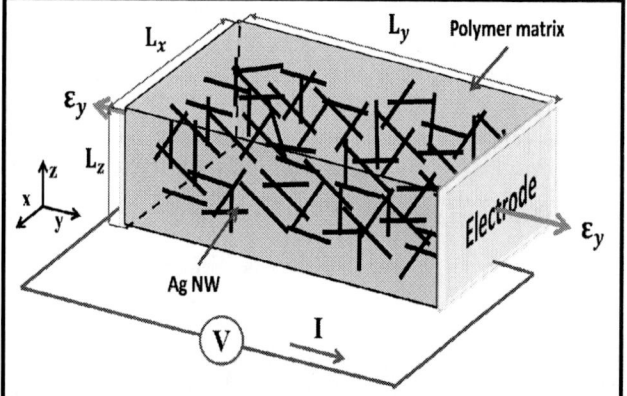

Figure 3: 3D random orientation of AgNWs in the PDMS medium.

A network resistor model was then constructed by the junction identification between all pairs of NWs in the network and the resistance of total network was calculated by using Kirchhoff's current law and Ohm's law. To obtain the total conductance change under strain, the positions and orientations of all NWs were re-calculated and the conductivity of thin film was analyzed again.

Figure 4 shows the relative change of resistance against the applied strain (up to 50%) for the nanocomposite, both experiment and simulation. As the figure demonstrates, there is an excellent agreement between our experiment and simulation results. Furthermore, the relative resistance of sensor gradually increases by the applied strain with almost linear manner.

Gauge factor (GF) showing the sensitivity of sensor to applied strain can be calculated by the following equation:

$$GF = (\Delta R / R_0) / \varepsilon \qquad (2)$$

where R_0 is the initial resistance at the zero strain, ΔR is the change of resistance caused by the applied strain, and ε is the applied strain. The GFs of all tested samples are in the ranges of 1.6 to 14; in comparison with conventional strain sensors (GF~2 with stretchability of 5%), these sensors have higher gauge factor and stretchability ($\Box$=70%).

The effects of the aspect ratio (L/D) of NWs on the sensitivity and stretchability of the strain sensor were investigated by our computational model. NWs with the aspect ratios of 115, 85.7, and 61.3 were randomly signed into the PDMS matrix. Size of the samples and volume fractions of NWs are assumed to be all the same. The relative change of resistance against the applied strain for the different aspect ratios is illustrated in Figure 5. As the figure shows, the sensitivity of sensor gains by the decrease of aspect ratio. On the other hand, both linearity and stretchability of the sensor enhance by the increase of aspect ratio. Furthermore, stretchability improvement by the higher

aspect ratios could be due to the better connection of long NWs in the network for larger strains. Since the NW-NW disconnection rates for longer NWs are slower than that of short NWs, reasonably the sensors possess lower sensitivity.

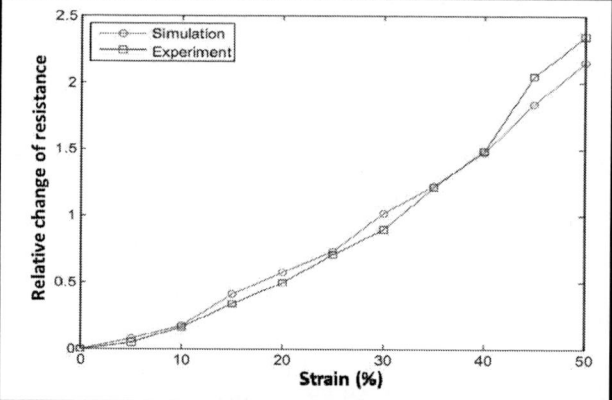

Figure 4: Relative change of resistance versus applied strain-both experiemnt and simulation.

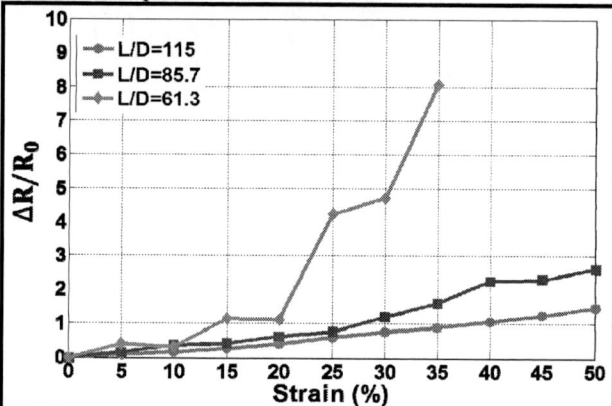

Figure 5: Relative change of the resistance versus strain for the different aspect ratios.

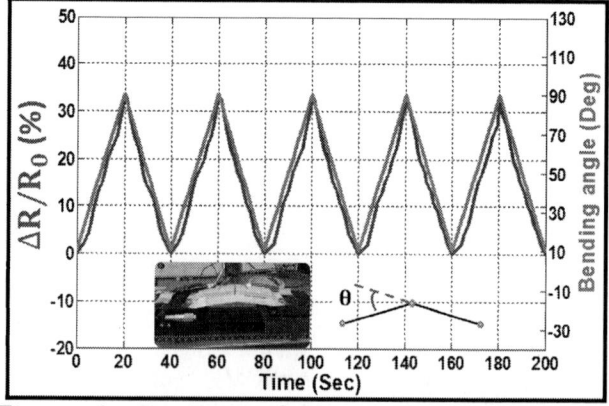

Figure 6: Response of the strain sensor to the bending/relaxation cycles; bottom inset, a photograph of the artificial finger device.

As an application of highly stretchable and sensitive Ag NWs-PDMS nanocomposite strain sensor, we constructed an artificial finger for the detection of human joints' angle, bottom inset of Figure 6. The strain sensor was attached on the artificial finger joint and then cyclic bending/relaxation loads (from 10° to 90°) were applied to the sensor while the current was measured. The more bending accommodates the more strain to the sensor, causing the resistance of sensor to increase accordingly. As Figure 6 shows, there is an excellent overlap between the response of sensor and loading profile. The strain sensor responds to cyclic loads very fast without considerable hysteresis and drift.

CONCLUSIONS

In this paper, a highly stretchable and sensitive strain sensor based on the Ag NW network-PDMS nanocomposite was fabricated. Stretchability and GFs of the sensors are 70% and in the ranges of 1.6 to 14, respectively. The Ag NWs-PDMS nanocomposite strain sensors exhibit a good bendability performance and they can be easily mounted on the body for the applications such as joint angle measurement and skin movement monitoring. Piezoresistivity of Ag NW network-PDMS nanocomposite was investigated by a computational model based on the network resistor. We found an excellent agreement between our experimental and computational results. Finally, an artificial finger was developed by assembling the stretchable strain sensor in the finger joint. The angle of joint can be calculated by the response of the strain sensor.

ACKNOWLEDGEMENTS

This work was supported by the Industrial Strategic technology development program (10041618, Development of Needle Insertion Type Image-based Interventional Robotic System for Biopsy and Treatment of 1-cm Abdominal and Thoracic Lesion with Reduced Radiation Exposure and Improved Accuracy) and by a grant from the Fundamental R&D Program for Core Technology of Materials funded by the Ministry of Knowledge Economy, Republic of Korea (N02120149).

REFERENCES

[1] C. X. Liu, J. W. Choi, "An Embedded PDMS Nanocomposite Strain Sensor toward Biomedical Application", *31st Annual International Conference of the IEEE EMBS*, 2009, pp. 6391-6394.

[2] T. Giorgino, P. Tormene, F. Lorussi, D. D. Rossi, S. Quaglini, "Sensor Evaluation for Wearable Strain Gauges in Neurological Rehabilitation", *IEEE Transactions on Neural Systems and Rehabilitation Engineering*, 2009, pp. 409-415.

[3] F. Lourussi, E. P. Scilingo, M. Tesconi, A. Tognetti, D. D. Rossi, "Strain Sensing Fabric for Hand Posture and Gesture Monitoring", *IEEE Transactions on Information Technology in Biomedicine*, 2005, pp. 372-381.

[4] R. J. N. Helmer, D. Farrow, K. Ball, E. Phillips, A. Farouil, I. Blanchonette, "A Pilot Evaluation of An Electronic Textile for Lower Limb Monitoring and Interactive Biofeedback", *Procedia Engineering*, pp.513-518, 2011.

[5] C. X. Liu, J. W. Choi, "Patterning Conductive PDMS Nanocomposite in An Elastomer using Microcontact

Printing", *J. Micromech. Microeng*, vol. 19, 085019, 2009.

[6] S. S. Rautaray, A. Agrawal, "Interaction with Virtual Game through Hand Gesture Recognition", *International Conference on Multimedia, Signal Processing and Communication Technologies*, 2011, pp. 244-247.

[7] R. Poppe, "A Survey on Vision-based Human Action Recognition", *Image and Vision Computing*, 28, pp. 976–990, 2010.

[8] D. Guan, T. Ma, W. Yuan, Y. lee, M. J. Sarkar, "Review of Sensor based Recognition Systems", *IETE technological review*, vol. 28, pp. 418-434, 2011.

[9] P. Mostafalu, S. Sonkusale, "Flexible and Transparent Gastric Battery: Energy Harvesting from Gastric Acid for Endoscopy application", *Biosensors and Bioelectronics*, dx.doi.org/10.1016/j.bios.2013.10.040

[10] T. Yamada, Y. Hayamizu, Y. Yamamoto, Y. Yomogida, A. Izadi-Najafabadi, D. N. Futaba, K. Hata, "A Stretchable Carbon Nanotube Strain Sensor for Human-Motion Detection", *Nature Nanotechnology*, vol. 6, pp. 296-301, 2011.

[11] X. W Fu, Z. M. Liao, J. X. Zhou, Y. B. Zhou, H. C. Wu, R. Zhang, G. Jing, J. Xu, X. Wu, W. Guo, D. Yu, "Strain Dependent Resistance in Chemical Vapor Deposition Grown Graphene", *Applied Physics Letters*, vol. 99, 213107, 2011.

[12] S. H. Bae, Y. Lee, B. K. Sharma, H. J. Lee, J. H. Kim, J. H. Ahn, "Graphene-based Transparent Strain Sensor", *Carbon*, vol. 51, pp. 23 6-242, 2013.

[13] K. E. Korte, S. E. Skrabalak, Y. Xia, "Rapid Synthesis of Silver Nanowires through a CuCl- or CuCl2-mediated Polyol Process", *J. Mater. Chem*, vol. 18, pp. 437-441, 2007.

[14] T. Kim, A. Canlier, G. H. Kim, J. Choi, M. Park, S. M. Han, "Electrostatic Spray Deposition of Highly Transparent Silver Nanowire Electrode on Flexible Substrate", *ACS Appl. Mater. Interfaces*, 5, pp. 788-794, 2013.

[15] F. Xu, Y. Zhu, "Highly Conductive and Stretchable Silver Nanowire Conductors", *Adv. Mater.*, 24, 5117-5122, 2012.

[16] X. Li, H. Gao, C. J. Murphy, K. K. Caswell, "Nanoindentation of Silver Nanowires", *Nano Lett.*, vol. 3, pp. 1495-1498, 2003.

[17] K. Keshoju, L. Sun, "Mechanical Characterization of Magnetic Nanowire–polydimethylsiloxane Composites, *J. Appl. Phys.*, 105, 023515-023519, 2009.

[18] S. Xu, O. Rezvanian, K. Peters, M. A. Zikry, "The Viability and Limitations of Percolation Theory in Modeling the Electrical Behavior of Carbon Nanotube–Polymer Composites", *Nanotechnology*, 24, 155706, 2013.

CONTACT

*I. Park, tel: +82-42-350-3240; inkyu@kaist.ac.kr

SINGLE-CHIP ATOMIC FORCE MICROSCOPE WITH INTEGRATED Q-ENHANCEMENT AND ISOTHERMAL SCANNING

Neil Sarkar,[1,2] and Raafat R. Mansour[1,2]
[1]University of Waterloo, Waterloo, CANADA
[2]ICSPI Corp, Waterloo, CANADA

ABSTRACT

We present the highest resolution imaging performance attained to date with a single-chip Atomic Force Microscope (AFM) that does not require off-chip scanning or sensing hardware. The marked improvement in sensitivity of the instrument stems in part from an internal quality (Q) factor enhancement mechanism that relies on the interplay between effects in the electrical, thermal and mechanical domains. In addition, careful matching of the strain sensor in an electrothermally actuated, piezoresistively detected resonant cantilever improves the dynamic range of the instrument. Furthermore, an integrated *isothermal electrothermal scanner* has been developed to scan a surface area of ~50μm x ~15μm while maintaining a constant temperature at the tip and sensor locations, thereby suppressing the deleterious thermal crosstalk effects that have plagued previously reported electrothermal scanner designs.

INTRODUCTION

Scanning Probe Microscopy (SPM) and MEMS

SPMs are the highest resolution imaging instruments available today and are among the most important tools in nanoscience. Micromachining technology has played a pivotal role in the progress of the AFM, in part because the instrument's performance is inexorably tied to the mechanics of a cantilever beam. The first micromachined cantilevers were built out of SiO2 and Si3N4 in 1990, and today's cantilevers are based on a design published in 1991 [1]. In 1993, atomic resolution AFM images were obtained using a cantilever with an integrated piezoresistive strain sensor [2]. MEMS cantilevers with integrated electrostatic actuation and optical gratings [3], electrostatic actuation and capacitive detection [4], and piezoelectric transduction [5] have also been employed in AFM studies.

Conventional SPMs suffer from several drawbacks owing to their bulky construction and to the use of piezoelectric materials that exhibit creep and hysteresis. The benefits of scaling such instruments have long been known. Microfabricated Scanning Tunneling Microscopes (STMs) were presented in [6] and [7]. An array of 1-D AFM cantilevers with integrated electronics was discussed in [8], and a 2-degree-of-freedom (DOF) probe with an integrated JFET was presented in [9]. A CMOS-MEMS probe was presented in [10]. We have reported the first imaging results with single-chip AFMs in the contact [11], intermittent contact [12], and frequency modulation (FM) [13] modes.

Motivation

Our work aims to improve the resolution and throughput of conventional AFM's by integrating all of the electromechanical components that are required to acquire an image onto a single chip fabricated with CMOS-MEMS technology (Figure 1). The result is an instrument that is volumetrically scaled by a factor of over 1 million. Integrated actuation obviates the need for piezoelectric materials. Position sensors that are collocated in the actuator beams are used to mitigate thermal drift and coupling effects. Piezoresistive cantilevers are used to measure tip-sample interaction forces, enabling the simultaneous operation of arrays of SPM's. An added benefit is the dramatically lowered cost of the instrument.

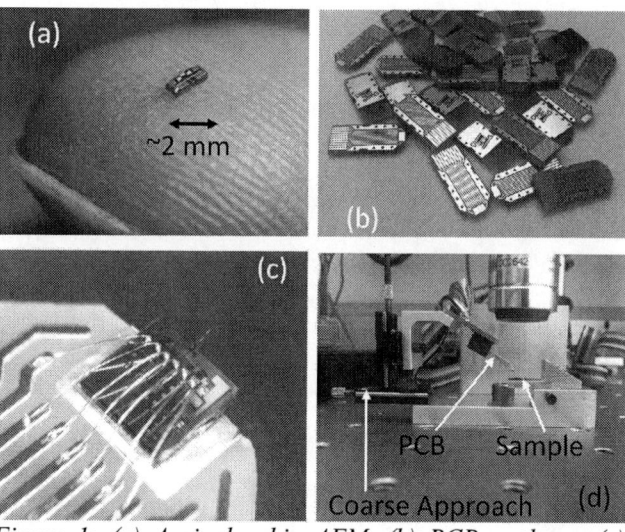

Figure 1: (a) A single chip AFM, (b) PCB package, (c) packaged instrument, and (d) manual coarse approach mechanism.

Several devices have been operated in the AM (Amplitude Modulation)-AFM mode that is the focus of the present work. The improved performance of these devices can be attributed to 1) the use of improved lateral actuators with larger work/unit volume, 2) the operation of scanners in an isothermal fashion, 3) matching of the piezoresistive strain sensors to reject thermal cross-talk, and 4) an internal Q-enhancement mechanism.

DEVICE DESIGN AND OPERATION

Single-chip AFMs consist of several components that must operate in a concerted fashion. Modeling effort has focused on the design of the following elements: self-heated resistors, lateral quasi-static actuators, flexural suspensions,

978-1-4799-3510-9/14 $31.00 © 2014 IEEE

thermal shunt paths and thermal isolation paths, vertical dynamic electrothermal actuators, resistive and thermocouple-based temperature sensors, flexural resonant cantilevers, and piezoresistive strain sensors. A resulting design is shown below in Figure 2.

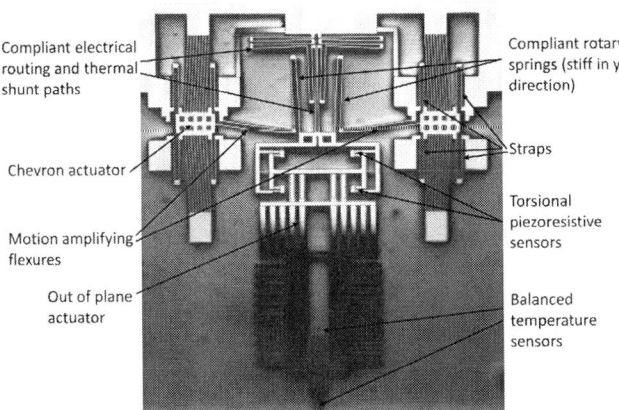

Figure 2: Optical microscope image of a single-chip SPM, highlighting several components in the design.

Chevron actuator with geometric advantage

Lateral scanning in the instrument is achieved with the use of electrothermal actuation. The folded bimorph design reported in [14] is well-suited to CMOS-MEMS processes that make use of the back-end-of-line materials, because it is designed to counter the parasitic effects of residual stress in the laminated structural layers. Chevron or bent-beam actuators [15], which are commonly used in silicon-on-insulator (SOI) and polysilicon MEMS processes, have not been reported in CMOS-MEMS, presumably because residual stress the causes out-of-plane buckling before any mechanical work can be performed by the actuator in the desired direction. This shortcoming is resolved with the use of folded springs to constrain the motion of the actuator in-plane. The springs also provide a thermal shunt path and electrical routing paths for interconnected devices. A flexure is used to provide sufficient geometric advantage to achieve the desired range of motion.

The design presented in the finite element analysis of Figure 3 attains 15μm of deflection and provides a blocked force of 40μN. In comparison, a lateral bimorph with the same footprint at the same temperature may achieve 7μm of deflection and exert a force of 4 μN. In addition, the out-of-plane stiffness of the strapped chevron design is an order of magnitude higher than the folded bimorph. A large vertical spring constant stands to improve the imaging performance of the AFM, since it lowers the likelihood of catastrophic tip crashes that are a byproduct of pull-in effects.

The augmented lateral stiffness of these actuators results in a high mechanical resonant frequency (>50kHz) and reduces the instrument's sensitivity to ambient vibrations. Strapped chevron actuators also have a short path to thermal ground, leading to an increased thermal -3dB frequency. The increased bandwidth comes at the expense

of higher power consumption; however, the higher voltages applied to the actuator improve the signal-to-noise ratio of position sensing with integrated self-heating resistors. This method of position sensing is discussed in the following section.

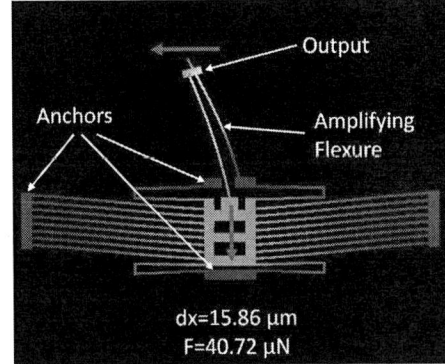

Figure 3: Strapped, mechanically amplified chevron actuator designed for a CMOS-MEMS process.

Scanner design

The scanner design depicted above in Figure 2 is used to position the tip over a lateral scan range of ~50μm x 15μm. The vertical actuator provides >30μm of deflection, easing the constraints on the coarse positioning stage that is used in imaging experiments.

With a single lateral actuator powered by a periodic signal, heat is coupled to the vertical actuator. An accurate measurement of this effect may be captured by monitoring the resistance of the self-heating resistor integrated in the z-actuator. A controller may be used to mitigate this effect, as shown in Figure 4 below.

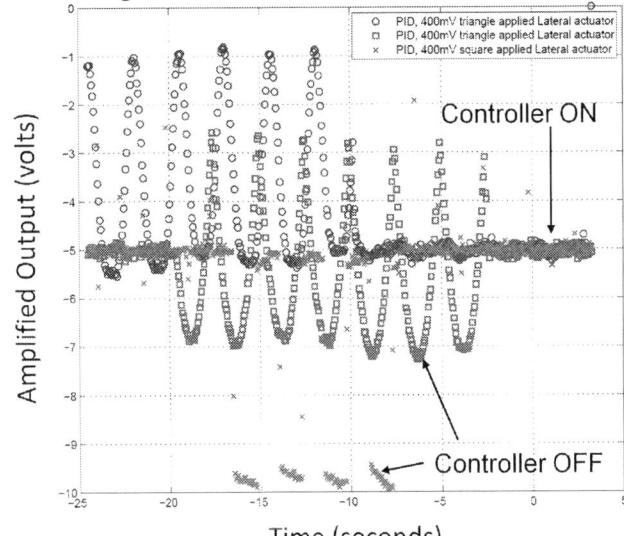

Figure 4: Position-control of a z-direction actuator experiencing thermal cross-talk from lateral actuation.

The scanner design presented above may be operated such that the temperature of the vertical actuator remains constant over the course of a line scan. The method consists of ensuring that the total power dissipated in both lateral actuators remains constant, and produces an *isothermal* line scan. Isothermal scanning obviates the need for temperature control in one lateral direction. However, in this design, it is still required in the opposing scan direction.

Balanced Piezoresistors

The thermal coupling from actuation results in a large parasitic signal from the piezoresistors, due to their appreciable temperature coefficient of resistivity. The z-direction thermal bimorph actuator is used to excite the piezoresistive cantilever into resonance. When operated in this fashion, an AC thermal component at the actuation frequency obscures the signal of interest. A *balance* piezoresistor is therefore added to capture only the thermal parasitic. Thus, the piezoresistive component is recovered using a differential measurement. Through symmetry, it is desirable to provide a thermal match while suppressing the mechanical resonance of the balance resistor.

When the sensors are placed in a Wheatstone bridge configuration, a potentiometer may be used to achieve good matching. In Figure 5 below, the *electromechanical Q-factor* is measured from the 3dB bandwidth of the frequency response of the circuit. Once the bridge is matched, the DC bias on the sensors may be increased to improve the match further. The effective electromechanical Q is increased from 50 to 50,000 in this fashion. This phenomenon improves the dynamic range of the system – a small shift in the natural frequency of the cantilever results in a large change in the output signal amplitude. However, the sensitivity of the instrument (the minimum detectable force, or MDF) is not improved, because measurement noise is also amplified by this mechanism. An empirical explanation may arise from the temperature dependence of the diffusivity of polysilicon, which modifies the phase of the thermal oscillation, thereby improving the bridge balance.

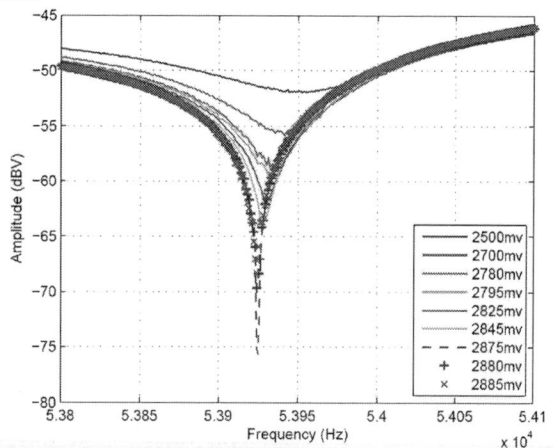

Figure 5: Electromechanical Q-factor enhancement. DC bias provides a means to compensate the phase of the thermal mismatch in the bridge. Effective Q varies from 50 (at 2500mV) to 50,000 (at 2875mV).

In order to improve the MDF of the sensors, the mechanical Q of the resonator must be increased. Q-control has been a topic of interest in the AFM community because higher Q values stand to improve the instruments'

resolution, while lower Q values stand to improve their bandwidth [16]. The present configuration allows for tuning of mechanical Q values by a factor of ~30%. Q-enhancement of piezoresistively transduced, electrothermally actuated devices has been reported in [17]. Figure 6 shows laser vibrometer data taken at various DC bridge bias voltages. The use of a laser vibrometer enables the measurement of the mechanical Q-factor exclusively. The envelope of the ring-down curves is fitted to an exponential decay curve in order to quantify the Q, as plotted in Figure 6.

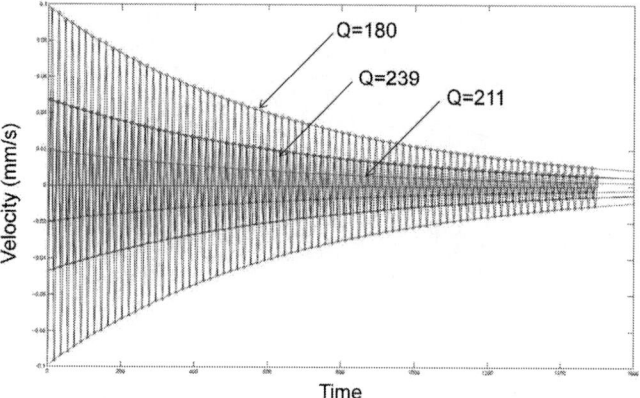

Figure 6: Laser vibrometer data from ring-down measurements with various DC bridge bias voltages. The envelope of the data is fit to an exponential to extract the mechanical Q-factor.

The nonlinearity of piezoresistively pumped resonant amplification is exemplified by the $f_0/8$ subharmonic excitation of resonance and the excitation of several superharmonics, as shown in Figure 7.

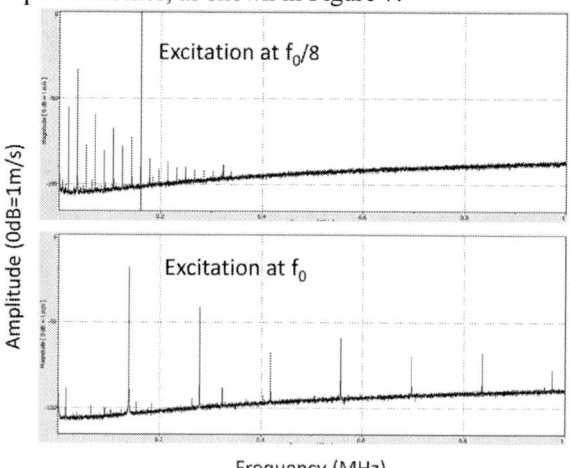

Figure 7: Subharmonic and superharmonic excitation by piezoresistive pumping as measured by laser vibrometry.

The presence of these harmonics suggests that parametric pumping may be exploited to further increase Q.

IMAGING OF CALIBRATION GRATINGS

A single-chip AFM was used in the AM mode to obtain images of a calibration standard with 90nm ridges on a 3μm

978-1-4799-3510-9/14 $31.00 © 2014 IEEE

pitch. The principle of detection of AM-AFM is as follows. The natural frequency of the cantilever increases when the tip experiences attractive tip-sample interaction forces in the *non-contact mode*. The opposite is true for repulsive interactions that occur in *intermittent contact or tapping mode*. The frequency of the forcing signal is fixed and the resulting amplitude variation is measured.

A manual coarse approach mechanism was used to lower the tip and to engage the sample, as discussed in [13]. Once the tip was in contact, a controller was used to maintain constant amplitude from the piezoresistive bridge circuit while the lateral actuators scanned over the surface of the sample. The resulting images are shown in Figure 8 (no post-processing).

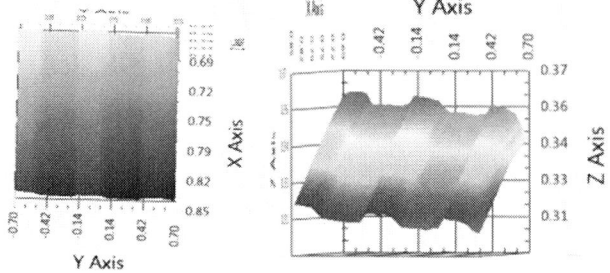

Figure 8: Images of an AFM calibration grating obtained with a single-chip instrument. Height of grating is 90nm, period is 3 microns.

CONCLUSION

AM-AFM is by far the most popular form of SPM because of its versatility and robustness. This work demonstrates, for the first time, that a complete AFM instrument that supports AM operation can be integrated onto a single chip.

The performance of such an instrument enjoys several benefits that are direct consequences of design improvements. Lateral actuators with higher spring constants produce a large scan range while increasing the bandwidth and robustness of the instrument. 1-D isothermal scanning operation reduces the effect of thermal crosstalk. Judicious matching of the piezoresistive sensors improves the dynamic range of the system. A built-in Q-factor enhancement mechanism enables the user to adjust the instrument for improved sensitivity or wider bandwidth.

It is believed that tiny, inexpensive, high throughput, and stable (with respect to drift and vibration) SPM arrays based on this work will benefit the nano-metrology community.

ACKNOWLEDGEMENTS

The authors would like to acknowledge the support of TowerJAZZ for their donation of fabrication area, and the Canadian Microelectronics Corporation.

CONTACT

*N. Sarkar, tel: +1-519-888-4567; nsarkar@uwaterloo.ca

[1]   O. Wolter et. al., "Micromachined silicon sensors for scanning force microscopy," *J. Vac. Sci. Technol. B Microelectron. Nanom. Struct.*, v. 9, no. 2, Mar. 1991.

[2]   M. Tortonese, R. C. Barrett, and C. F. Quate, "Atomic resolution with an atomic force microscope using piezoresistive detection," *APL*, vol. 62, no. 8, p. 834, 1993.

[3]   A. G. Onaran and et. al., "A new atomic force microscope probe with force sensing integrated readout and active tip," *Rev. Sci. Instrum.*, vol. 77, no. 2, 2006.

[4]   N. Blanc, J. Brugger, and N. F. De Rooij, "Scanning force microscopy in the dynamic mode using microfabricated capacitive sensors," v. 14, n. 2, 2007.

[5]   T. Itoh, C. Lee, and T. Suga, "Deflection detection and feedback actuation using a self-excited piezoelectric Pb(Zr,Ti)O3 microcantilever for dynamic scanning force microscopy," *Appl. Phys. Lett.*, v. 69, no. 14, p. 2036, 1996.

[6]   S. Akamine, T. R. Albrecht, M. J. Zdeblick, and C. F. Quate, "Microfabricated Scanning Tunneling Microscope," *IEEE Elec. Dev. Lett.*, v. 10, no. 11, 1989.

[7]   Y. Xu, N. C. MacDonald, and S. a. Miller, "Integrated micro-scanning tunneling microscope," *Appl. Phys. Lett.*, vol. 67, no. 16, p. 2305, 1995.

[8]   S. Hafizovic et.al. , "Single-chip mechatronic microsystem for surface imaging and force response studies.," *Proc. Natl. Acad. Sci. U. S. A.*, vol. 101, no. 49, pp. 17011–5, Dec. 2004.

[9]   K. Amponsah and A. Lal, "Multiple tip nano probe actuators with integrated JFETs," *Micro Electro Mech. Syst. (MEMS ...*, no. February, pp. 1356–1359, 2012.

[10]   J. Liu, M. Noman, J. a. Bain, T. E. Schlesinger, and G. K. Fedder, "CMOS-MEMS probes for reconfigurable IC's," *2008 IEEE 21st Int. Conf. Micro Electro Mech. Syst.*, pp. 515–518, Jan. 2008.

[11]   N. Sarkar and R. Mansour, "CMOS-MEMS atomic force microscope," *Transducers 2011*, pp. 2610–2613, Jun. 2011.

[12]   N. Sarkar, R. Mansour, and K. Trainor, "Forced oscillation and higher harmonic detection in an integrated cmos-mems scanning probe microscope," *Hilt. Head*, no. c, pp. 1–4, 2012.

[13]   N. Sarkar, G. Lee, and R. Mansour, "CMOS-MEMS dynamic FM atomic force microscope," *Solid-State Sensors, Actuators ...*, 2013.

[14]   P. Gilgunn and J. Liu, "CMOS–MEMS lateral electrothermal actuators," *MEMS, J.*, vol. 17, no. 1, pp. 103–114, 2008.

[15]   J.-S. Park and Y. B. Gianchandani, "Bent-Beam Electrothermal Actuators," *J. Microelectromechanical Syst.*, vol. Vol. 10, no. 2, pp. 247–254, 2001.

[16]   T. Sulchek et. al., "High-speed tapping mode imaging with active Q control for atomic force microscopy," *Appl. Phys. Lett.*, vol. 76, no. 11, p. 1473, 2000.

[17]   A. Rahafrooz and S. Pourkamali, "Active self-Q-enhancement in high frequency thermally actuated M/NEMS resonators," *JMEMS*, pp. 760–763, 2011.

SMART-CUT 6H-SILICON CARBIDE (SiC) MICRODISK TORSIONAL RESONATORS WITH SENSITIVE PHOTON RADIATION DETECTION

Rui Yang[1†], Kalyan Ladhane[2], Zenghui Wang[1], Jaesung Lee[1], Darrin J. Young[2], Philip X.-L. Feng[1†]
[1]Electrical Engineering, Case Western Reserve University, Cleveland, OH 44106, USA
[2]Electrical & Computer Engineering, University of Utah, Salt Lake City, UT 84112, USA
[†]Email: rui.yang@case.edu; philip.feng@case.edu

ABSTRACT

This digest paper reports experimental demonstration of a new type of microdisk torsional resonators based on a smart-cut 6H-silicon carbide (6H-SiC) technology. We carefully calibrate the thermomechanical vibrations of these torsional-mode resonators by employing highly sensitive laser interferometric readout techniques, with displacement sensitivities (*i.e.*, limits of detection) at the levels of $S_x^{1/2} \sim 0.1 \text{pm}/\sqrt{\text{Hz}}$ and smaller. To utilize these first 6H-SiC torsional resonators, we further demonstrate sensitive detection of radiations from both blue (405nm) and infrared (IR, 785nm) photons. Toward force detection applications which torsional resonators are suitable, our measurements demonstrate an intrinsic force resolution of $S_F^{1/2} \approx 5.7 \text{fN}/\sqrt{\text{Hz}}$, and a torque resolution of $S_T^{1/2} \approx 3.7 \times 10^{-20}$ (N·m)$/\sqrt{\text{Hz}}$, limited by the fundamental thermomechanical fluctuations.

INTRODUCTION

From torsional vehicle suspension bars to precise timing balances in mechanical watches, from Coulomb's torsion balance [1] to the apparatus in the Cavendish experiment [2], torsional mechanical devices have been employed for a variety of commercial and scientific applications. As devices continue to be miniaturized into micro- and nano-electromechanical systems (MEMS & NEMS), the resulting infinitesimal motional mass and ultrafine spring constant make MEMS/NEMS torsional resonators highly suitable for ultrasensitive force and torque sensing applications. MEMS/NEMS torsional resonators with charge sensitivity of $\sim 0.1 e/\sqrt{\text{Hz}}$ [3], torque sensitivity of $\sim 10^{-22}$ (N·m)$/\sqrt{\text{Hz}}$ [4], and mass sensitivity of 26fg [5] have been achieved, by employing various structural materials such as Si, SiN, Cr/Au, and carbon nanotubes (CNTs) [3-8]. However, MEMS torsional resonators based on advanced materials suitable for high-temperature and harsh-environment applications have not yet been demonstrated.

A leading material for harsh-environment electronics, 6H-SiC possesses an ensemble of outstanding physical properties, including very high elastic (Young's) modulus ($E_Y \sim 450$GPa), large refractive index (2.6 in visible range), wide bandgap (3.05eV), high-temperature durability, and chemical inertness, which are also highly attractive for making MEMS/NEMS devices, particularly high-frequency resonators and high-performance sensors that can survive and also respond to extreme stimuli [9,10]. Effort on realizing 6H-SiC resonant sensors, however, has been very limited, largely due to difficulties in micromachining bulk crystalline 6H-SiC [9-11], as compared to thin films in other conventional MEMS materials (*e.g.*, Si, SiN).

In this work, we demonstrate 6H-SiC circular microdisk torsional resonators by using a customized smart-cut 6H-SiC process, in combination with a surface nanomachining process involving high-resolution definition of device structures via focused ion beam (FIB). Undriven thermomechanical resonances are detected using sensitive laser interferometry techniques, showing excellent resonance performance in these devices. As a proof of concept, detections of device responses to radiations of 405nm and 785nm photon fluxes are demonstrated. This 6H-SiC torsional resonator platform holds strong promise for resonant sensing applications in harsh environments.

6H-SiC MICRODISK FABRICATION

We design fully suspended circular microdisks supported by thin tethers (Fig. 1), to operate in a torsional resonance mode, and to achieve high resonant sensitivities by engineering the trade-offs between low mass, high quality (*Q*) factor, and high frequency (speed). Figure 1a illustrates the device schematic, and Figure 1b shows the simulated mode shape of the torsional resonance mode, via finite element modeling (FEM, in COMSOL™). Figure 1c and 1d show scanning electron microscopy (SEM) images of a 15.9μm-diameter (*d*=15.9μm) 6H-SiC torsional resonator with a zoomed-in view of one of the tethers.

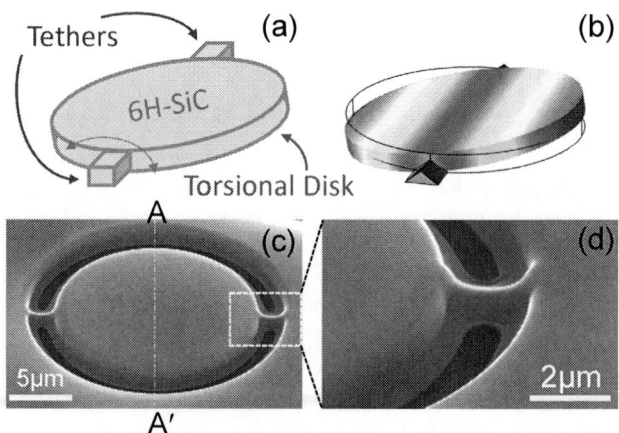

Figure 1: 6H-SiC microdisk torsional resonators. (a) Device schematic. (b) FEM simulation of the torsional mode. (c) & (d) SEM images of a device and zoom-in view of a thin tether.

The devices are fabricated from a 6H-SiC-on-SiO$_2$ wafer enabled by a 'smart-cut' technology. The process starts with H$^+$ ion implantation into a single-crystal 6H-SiC wafer (Fig.

2a). The ion implanted SiC wafer is then brought into contact with an oxidized Si wafer (with plasma activated surface) (Fig. 2b) in a wafer bonding machine (Fig. 2c). The bonded wafer is then split at 1000°C (Fig. 2d), yielding a 1.2µm-thick-SiC-on-SiO$_2$-on-Si structure (the SiC thickness can be tuned by varying H$^+$ ion energy). We then use focused ion beam (FIB) of high energy Ga$^+$ ion to define microdisks by milling SiC in patterned areas (Fig. 2e). Finally the SiO$_2$ underneath is removed by buffered oxide etch (BOE) through the patterned openings on the SiC layer (Fig. 2f).

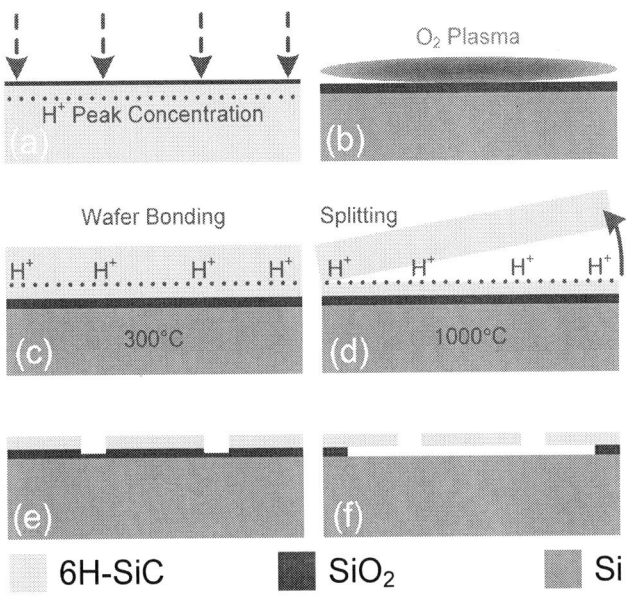

Figure 2: 'Smart-cut' 6H-SiC and microdisk resonators fabrication processes. (a) 200keV H$^+$ ion implantation into single-crystal 6H-SiC wafer with 120nm SiO$_2$ on top. (b) Surface activation with O$_2$ plasma on both SiC wafer and an oxidized Si wafer. (c) Wafer bonding of the SiC wafer to the oxidized Si wafer. (d) Annealing to split the 6H-SiC wafers, resulting in a 6H-SiC-on-SiO$_2$ wafer. (e) FIB pattering to define SiC devices. (f) Etching SiO$_2$ to release device. (e), (f) are cross section views at the AA' plane in Fig. 1c.

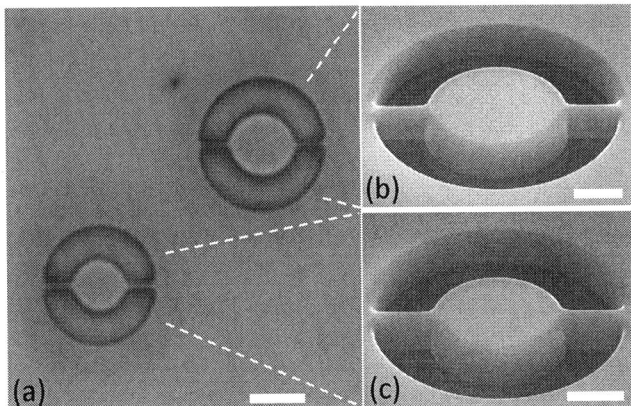

Figure 3: Devices possessing pairs of tethers with rectangular cross sections. (a) Optical image of two microdisks. Scale bar: 5µm. SEM images of the same devices are shown in (b): d≈6µm, and (c): d≈5µm. Scale bars: 2µm.

We fabricate microdisk resonators with different disk diameters and tether shapes/dimensions. Figure 3 shows two devices with similar tethers but different disk diameters. By tuning the parameters in FIB process, we are able to fabricate tethers with both rectangular (Fig. 3) and triangular (Fig. 4) cross sections. We can also adjust the tether dimensions. Figure 4 shows a microdisk device with the same disk diameter but different tether width and thickness than in Fig. 1(c)&(d). Adjusting device dimensions tunes resonance characteristics such as resonance frequency f_{res} and Q, allowing us to tailor devices specifications and performance.

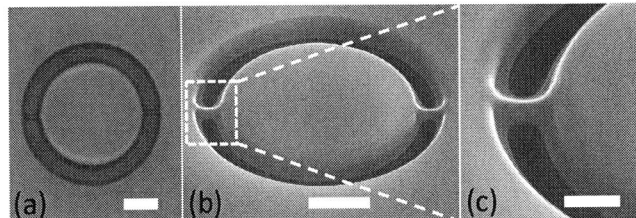

Figure 4: Devices with tethers of triangular cross section. (a) Optical image. Scale bar: 5µm. (b) SEM image. Scale bar: 5µm. (c) Zoom-in image of the tether. Scale bar: 2µm.

EXPERIMENTAL TECHNIQUES

We study these devices by using a specially engineered multi-laser interferometric detection system (Fig. 5). We finely focus a 633nm laser with laser power of 7mW onto one free edge (rim) of the disk, whose reflected light intensity is modulated by device motions and is thus transduced into electronic signal by a high-speed photodetector. Signal is analyzed and recorded by a radiofrequency/microwave spectrum analyzer. The device is mounted inside a chamber under vacuum (<10mTorr) at room temperature. With an X-Y translational stage, we map the mode shape of the torsional resonance by recording the resonance amplitude versus laser spot position. To study response to different photon radiations, we illuminate IR (785nm) or blue (405nm) laser on one edge of the device, while monitoring torsional resonance using the red (633nm) laser on the opposite edge.

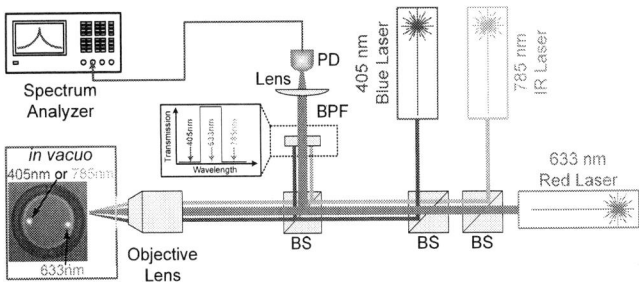

Figure 5: Multi-laser measurement system for testing SiC microdisk resonators and photon radiation detection. A 633nm laser is focused on the device near a free edge, and the resonant motions are detected with optical interferometry. Signal from thermomechanical motion of the device is read out with a spectrum analyzer. In radiation detection calibration, a 405nm or 785nm laser is focused onto the opposite edge (see lower left optical image).

978-1-4799-3510-9/14 $31.00 © 2014 IEEE   794

TORSIONAL RESONANCE RESULTS

Figure 6 and 7 demonstrate the measured undriven, thermomechanical resonance noise spectra from two devices with different tether widths (Fig. 6: $w_T \approx 1.5\mu m$; Fig. 7: $w_T \approx 2.1\mu m$. Device pictures are shown in Fig. 4b and 1c, respectively). Multimode thermomechanical resonances are observed within 1–10MHz. Comparing the measurement with the FEM simulations (insets of simulated mode shapes) allows us to discern from the observed resonances and associate the lowest-frequency one with the torsional mode.

The torsional spring constant of the tether is [12, 13]

$$k_T = \frac{I_T E_Y}{L_T (\nu + 1)}, \qquad (1)$$

where E_Y and ν are the Young's modulus and Poisson's ratio of 6H-SiC. L_T is the length of the tether and I_T is the torsional moment of inertia, which is

$$I_T = 0.0915 t_T^4 \left(w_T / t_T - 0.8592 \right), \qquad (2)$$

for triangular cross section when $\sqrt{3} < w_T / t_T < 2\sqrt{3}$. The wider-tether devices thus have higher torsional frequencies.

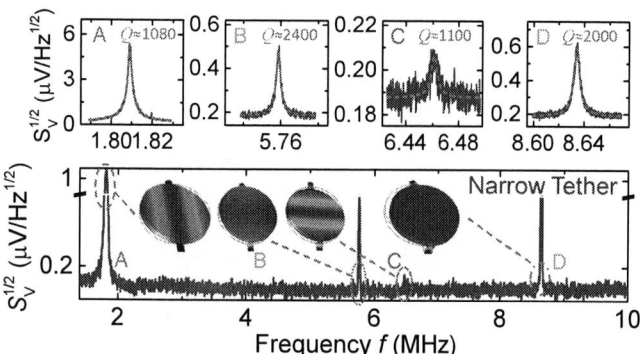

Figure 6: Multimode thermomechanical resonances measured from a narrow-tether device. Insets: COMSOL simulated mode shapes for the first 4 resonances. Zoom-in for each resonance with fitting of the Q is shown on the top.

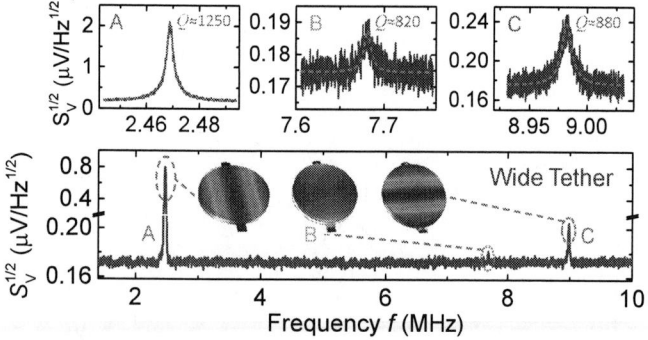

Figure 7: Measured thermomechanical resonance spectrum of a wide-tether device. Plot convention follows Fig. 6.

We perform direct spatial mapping of the mode shape (via scanning interferometric measurements) to confirm the torsional nature of the observed lowest resonance. Figure 8 illustrates the mapping scheme with a 3D plot of the actual measured data from the device in Fig. 6. As expected, both sides of the disk exhibit high motion amplitudes while the axial line connecting the tethers remains stationary, verifying the observation of the torsional mode.

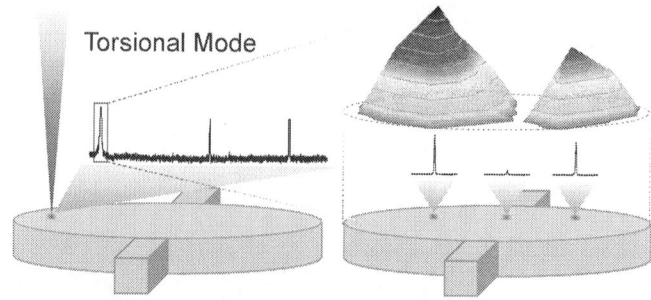

Figure 8: Spatial mapping of the torsional resonance mode shape. Schematic shows the mapping process. 3D data (top right) confirms the torsional mode resonance of the microdisk resonator shown in Fig. 4 (with data in Fig. 6).

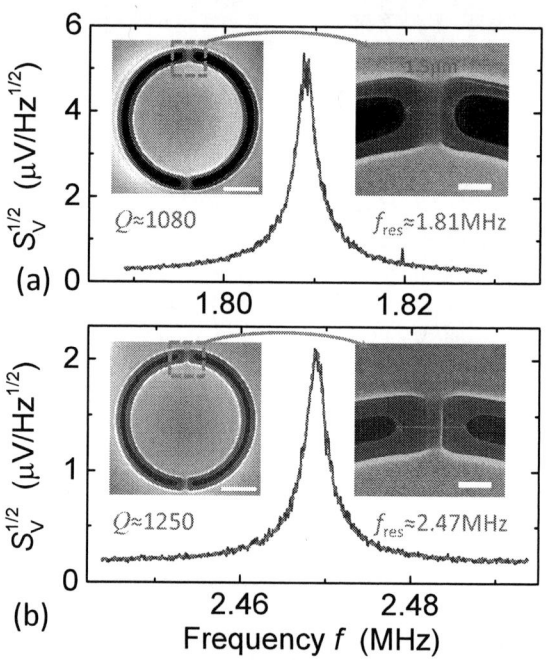

Figure 9: Effect of tether size on the torsional resonance. Measured torsional resonances are shown for devices with (a) narrower and (b) wider supporting tethers. Dashed line: fitting to a damped harmonic resonator model. Insets: SEM images of the microdisk resonators (scale bars: 5μm) and zoom-in view of their tethers (scale bars: 1μm).

We compare the thermomechanical torsional motion of the devices in Fig. 6 and 7. Precisely fitting the data (Fig. 9), we obtain the quality (Q) factors (both above 1000). SEM images (Fig. 9 insets) show the different tether widths of the two devices, defined as the base length of the triangular cross section. The device with wider tether exhibits both higher f_{res} and higher Q. By increasing the tether width from 1.5μm to 2.1μm, the torsional f_{res} shifts up from ~1.8MHz to ~2.5MHz due to an increase in torsional spring constant k_T; Q increases from 1080 to 1250, indicating less dissipation. From f_{res} and Q, we estimate an intrinsic force resolution $S_F^{1/2} \approx 5.7 \text{fN}/\sqrt{\text{Hz}}$ and a torque resolution $S_T^{1/2} \approx 3.7 \times 10^{-20}$ (N·m)/$\sqrt{\text{Hz}}$ for the device with narrower tether. For the device with wider tether; $S_F^{1/2} \approx 6.2 \text{fN}/\sqrt{\text{Hz}}$, and $S_T^{1/2} \approx 3.5 \times 10^{-20}$ (N·m)/$\sqrt{\text{Hz}}$.

DETECTION OF PHOTON RADIATIONS

We investigate the resonance responses to optical radiations to explore the potential of photon sensing in these 6H-SiC microdisk torsional resonators. Employing the measurement apparatus in Fig. 5, we monitor both f_{res} and Q with a series of incoming fluxes of photons of different energy (color), as shown in Fig. 10. For both infrared (IR) (785nm) and blue (405nm) laser radiations, f_{res} and Q both decrease with increasing photon fluxes, with the effects being more pronounced for the 405nm photon. From measurement we extract the f_{res}-to-single-photon-radiation *responsivity* to be $\Re_{ph,exp} \approx -2.38 \times 10^{-14}$ and -1.57×10^{-13} ppm/(photon·s^{-1}) for the 785nm and 405nm radiation, respectively.

Our FEM simulations show that the radiation pressure induced displacement of the resonator leads to extension of the supporting tether, which lowers the torsional spring constant and decreases f_{res}. If consider only radiation pressure, the estimated f_{res}-to-single-photon-radiation responsivity would be $\Re_{ph,pressure} \approx -0.67 \times 10^{-19}$ and -1.29×10^{-19} ppm/(photon·s^{-1}), respectively, for 785nm and 405nm radiation. The observed much higher responsivities suggest that the bolometric effect may be dominant. Even the 785nm photon energy is below the bandgap of 6H-SiC, lattice phonon and defects can effectively absorb the incoming photon energy and thus shift the resonance frequency. The demonstrated responses to photon fluxes, the estimated single-photon responsivities, and the clear dependence on wavelength, show that these microdisk resonators have potential for making sensitive photon radiation detectors.

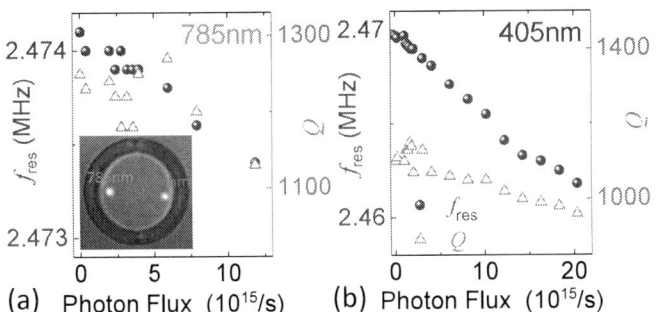

Figure 10: Demonstration and calibration of photon radiation detection by a 6H-SiC torsional resonator. f_{res} and Q are plotted against photon fluxes for both radiations of (a) 785nm (1.58eV) and (b) 405nm (3.07eV) photons.

CONCLUSIONS

In summary, 6H-SiC microdisk torsional resonators have been demonstrated with a smart-cut technique and FIB patterning. The thermomechanical resonance characteristics

of the torsional resonators are measured with sensitive scanning laser interferometry. Experimental data from different microdisk resonators show intuitive effects with varying tether dimensions. Detections of radiations of both 785nm and 405nm photon fluxes are demonstrated. By monitoring f_{res} and Q variations we determine different frequency responsivities for radiations with different photon energies and wavelengths. These devices show interesting potential for detecting radiations in harsh environments.

ACKNOWLEDGEMENTS

We thank Case School of Engineering, the Glennan Fellowship and the CSC Fellowship (No. 2011625071).

REFERENCES

[1] C. A. Coulomb, *Memoires de l'Academie Royale des Sciences*, Academie Royale des Sciences, 1784.

[2] H. Cavendish, "Experiments to Determine the Density of the Earth", *Phil. Trans. Royal Soc. Lond.*, vol. 88, pp. 469-526, 1798.

[3] A. N. Cleland and M. L. Roukes, "A Nanometre-Scale Mechanical Electrometer", *Nature*, vol. 392, pp. 160-162, 1998.

[4] X. C. Zhang, E. B. Myers, J. E. Sader, and M. L. Roukes, "Nanomechanical Torsional Resonators for Frequency-Shift Infrared Thermal Sensing", *Nano Lett.*, vol. 13, pp. 1528-1534, 2013.

[5] Y. Hwang, H. Sohn, R. N. Candler, *et al.*, "Dielectrophoresis-Assembled Zeolitic Imidazolate Framework Nanoparticle-Coupled Resonators for Highly Sensitive and Selective Gas Detection", *Nano Lett.*, vol. 13, pp. 5271-5276, 2013.

[6] S. J. Papadakis, A. R. Hall, S. Washburn, *et al.*, "Resonant Oscillators with Carbon-Nanotube Torsion Springs", *Phys. Rev. Lett.*, vol. 93, 146101, 2004.

[7] P. H. Kim, C. Doolin, J. P. Davis, *et al.*, "Nanoscale Torsional Optomechanics", *Appl. Phys. Lett.*, vol. 102, 053102, 2013.

[8] J. Losby, J. A. J. Burgess, M. R. Freeman, *et al.*, "Thermo-Mechanical Sensitivity Calibration of Nanotorsional Magnetometers", *J. Appl. Phys.*, vol. 111, 07D305, 2012.

[9] X. M. H. Huang, X. L. Feng, M. L. Roukes, *et al.*, "Fabrication of Suspended Nanomechanical Structures from Bulk 6H-SiC Substrates", *Mater. Sci. Forum*, vol. 457-460, pp. 1531-1536, 2004.

[10] P. Cong and D. J. Young, "Single Crystal 6H-SiC MEMS Fabrication based on Smart-Cut Technique", *J. Micromech. Microeng.*, vol. 15, pp. 2243-2248, 2005.

[11] T. K. Hossain, S. MacLaren, R. S. Okojie, *et al.*, "The Fabrication of Suspended Micromechanical Structures from Bulk 6H-SiC using an ICP-RIE System", *J. Micromech. Microeng.*, vol. 16, pp. 751-756, 2006.

[12] W. C. Young, R. G. Budynas, *Roark's Formulas for Stress and Strain*, McGraw-Hill, 2002.

[13] J. M. Gere, B. J. Goodno, *Mechanics of Materials*, 8th. Ed., Cengage Learning, 2013.

SUB-0.05° PRECISION OPTOFLUIDIC DUAL-AXIS INCLINOMETER

Sebastian Wahl, Frédéric Marty, Nicolas Pavy, Bruno Mercier, Dan E.Angelescu
Université Paris-Est, ESIEE Paris / ESYCOM Lab, Noisy-le-Grand, FRANCE

ABSTRACT

This paper details a low-power bi-axial miniaturized inclinometer based on a mobile mass (spherical ball or fluidic droplet) positioned on a precision curved surface that is generated using a novel MEMS process. The detection of the mobile mass was implemented through an external optical system, using a quadrant photodetector. Nanotopography and chemical treatment of the curved surface have been implemented to increase accuracy when using a fluidic mobile mass, by tailoring wetting properties and minimizing contact angle hysteresis. We achieve a range of ±1° with a true linear bi-axial measurement of precision better than 0.05°.

INTRODUCTION

Inclinometers have a large array of applications in multiple industrial fields including automotive, aerospace, mechanical and civil engineering. Certain applications, such as long-term continuous monitoring of major civil engineering works (large buildings, bridge pillars, dams) require autonomous sensors embedded within structures and responding to a combination of low measurement range (±2° typical), high accuracy (0.01°), low cost, low power consumption (<mW), small size, and reduced maintenance. Bi-axial measurements are often necessary, which often requires using two single-axis inclinometers, implying increase in power consumption, cost and logistics.

There have been numerous attempts at developing accurate inclinometers across different measurement ranges. Many miniaturized inclinometer concepts implement variations of the pendulum concept with various techniques of detecting changes in the pendulum state: resistive [1], capacitive [2], electrolytic [3] magnetic [4], optical [5], thermal [6] and photoelectric [7,8,9]. Commercial systems based on dual-axis MEMS accelerometers can deliver accuracies of order 0.1° over a full 180° range, albeit with relatively high power consumption (>30mW) [10]. Currently, the only systems that can achieve the satisfactory combination of the range and accuracy requirements for civil engineering monitoring are classical pendulum inclinometers (force balance tilt sensors), yet they do not satisfy the cost, size, and power consumption criteria.

We present a low-power bi-axial miniaturized inclinometer based on a mobile mass (spherical ball or fluidic droplet) positioned on a precision curved surface that is generated using a novel MEMS process and that does satisfy the above mentioned criteria of cost, size and power consumption. The detection of the mobile mass is implemented through an external optical system using a quadrant photodetector and a LED for shadow tracking of the mobile mass. Nanotopography and chemical treatment of the curved surface have been implemented to increase accuracy when using a fluidic mobile mass, by tailoring wetting properties and minimizing contact angle hysteresis.

We demonstrate the concept using a prototype based on a silicon membrane and a high-precision steel sphere as a mobile mass. We achieve a measurement range of ±1° with a true linear bi-axial measurement of precision better than 0.05°. The use of a nanostructured and chemically treated silicon membrane in combination with a mercury droplet resulted in even better precisions but also suffered from hysteresis effects.

OPERATION PRINCIPLE

Our inclinometer is composed of three main components: a sensitive element, a deflected silicon membrane of perfectly controlled geometry, and an optical detection system (Figure 1).

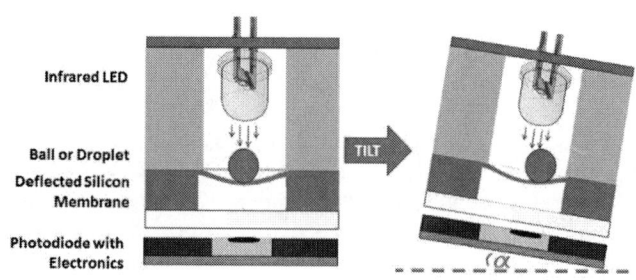

Figure 1: Schematic view of our inclinometer respectively in a non-tilted (left) and tilted state (right).

The sensitive element consists either of a metallic high-precision ball (with very low surface roughness and good sphericity) or alternatively a fluidic mercury droplet which eliminates imprecisions due to roughness but introduces other considerations such as contact angle hysteresis. We attempted to limit hysteresis by rendering the surface superhydrophobic using combined nano-structuring and chemical treatments.

The mobile mass is placed on the silicon membrane which is initially deflected due to the encapsulation of a vacuum cavity. The deflection of the membrane can be controlled by adjusting its geometry.

The shadow projected by the mobile mass is detected optically using a quadrant photodetector and a light source. The perfect control of the deflection and the very low membrane and mobile mass roughness enable accurate prediction of mobile mass displacement when the sensor is tilted and thus accurate tilt angle calculations.

FABRICATION

An aluminium thin film is photolithographically

978-1-4799-3510-9/14 $31.00 © 2014 IEEE

patterned on a 400μm silicon wafer to obtain circular openings that will be used as Deep Reactive Ion Etching (DRIE) etching masks for the silicon (Figure 2). The etch depth is chosen so as to preserve the desired membrane thickness. The geometric parameters of the membrane (diameter and thickness) directly influence its deflection profile and thus the overall measurement range of the inclinometer. After DRIE etching and aluminium removal, a second optional photolithography and cryogenic DRIE step can be performed on the opposite side of the wafer, in order to provide nano-scale topography. An additional 100nm thick fluoropolymer layer is deposited in a C_4F_8 plasma reactor, which results in a superhydrophobic surface as reported earlier (Figure 3) [11]. These additional steps are only performed when a fluidic droplet is used as a mobile mass in order to limit the contact between membrane and droplet. A last step involves the anodic bonding of a 500μm thick glass wafer to the silicon wafer in order to form a cavity in which vacuum is encapsulated. This pressure difference between the vacuumed cavity and the external environment causes the deflection of the silicon membrane in a perfectly controlled and reproducible way.

Figure 3: Left: Black-silicon nanostructuring of the membrane using a cryogenic RIE process. Right: Mercury droplet on Black Silicon membrane with fluoropolymer thin film.

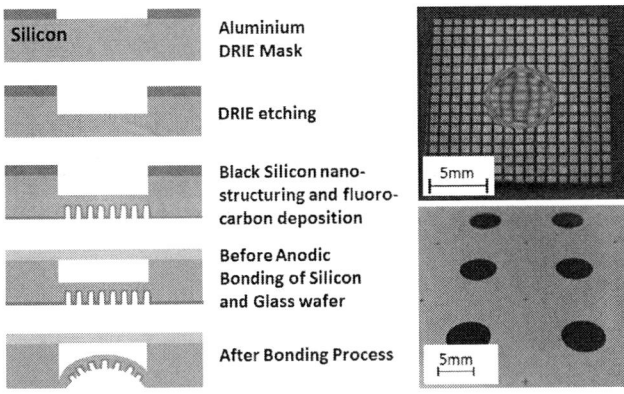

Figure 2: Left: Fabrication process of the nanostructured and chemically treated curved silicon membrane. Top Inset: Reflection of square grid seen in deflected unstructured membrane. Bottom Inset: Array of nanostructured membranes.

The mobile mass displacement is detected using an external optical system consisting of an infrared LED and a quadrant photodetector, which essentially enables the shadow tracking of the mobile mass displacement on the deflected silicon membrane. The wavelength of the infrared LED was chosen at 1070nm which ensures on one hand low silicon absorption and high spectral response of the quadrant photodetector to get an accurate signal of the shadow displacement.

We machined a mechanical aluminium packaging which integrated the quadrant photodetector with conditioning electronics (from First Sensor), the infrared LED (from Thorlabs), the deflected silicon membrane and the mobile mass.

DESIGN CONSIDERATIONS

Different parameters need to be taken into account for an optimal sensor design: silicon membrane geometry (diameter and thickness), characteristics of the mobile mass (size, surface roughness and sphericity) and size of the photodetector's sensitive area.

The deflection profile of the silicon membrane (for small scale deflections) can be described mathematically by the equation of a circular thin homogeneous isotropic clamped plate, subjected to a known uniform differential pressure as follows with axisymmetric boundary conditions[12]:

$$y(r) = -y_c + \frac{M_c r^2}{2D(1+v)} + LT_y \quad (1)$$

where r is the radial position coordinate, $y(r)$ is the deflection corresponding to r, $y_c = qa^4/64D$ is the maximal center deflection, q is the pressure difference, a is the membrane radius, $M_c = qa^2/[16(1+v)]$ is the moment at the membrane center, $D = E_y t^3/[12(1-v^2)]$ is the plate constant, E_y being the Young Modulus, v the Poisson's ratio, t is the membrane thickness, and $LT_y = -qr^4/64D$ is the loading term. This known deflection profile enables us to find the membrane inflection point r_{inf}, which is the maximal point of equilibrium for the mobile mass, and the associated maximal angle detectable by the sensor Θ_{max}:

$$r_{inf} = \sqrt{\frac{a^2}{3}} = a\frac{\sqrt{3}}{3} \quad (2)$$

$$\theta_{max} = \arctan(\frac{\sqrt{3}}{72}\frac{qa^3}{D}) = \arctan(1.54\frac{y_c}{a}) \quad (3)$$

When the ratio y_c/t is greater than 0.3, the small scale deflection equation doesn't apply anymore, and large scale deflection calculations have to be performed. However, the normalized profile remaining largely identical[13], equations 2-3 still hold but a larger effective thickness corresponding to the same maximal center deflection needs to be used.

The characteristics of the mobile mass in terms of size, surface roughness and sphericity are also an important point for the sensor performance. In order to be detectable, the mobile mass shadow must always be in contact with the four detector photoquadrants. The smaller the mobile mass, the sooner the shadow leaves the detector area, and the smaller the measurable tilt angle range. The optimal mobile mass radius needs to correspond to the distance from center to inflexion point, for the quadrant photodetector to be able to detect the position over the full measurement range. The mobile mass should be as spherical and as smooth as possible to ensure repeatable and predictable displacements. Surface roughness of the membrane is nearly atomic, corresponding to that of a high-quality polished silicon wafer.

The size of the photodetector's sensitive area must enable a measurement of mobile mass maximal displacement in order to be able to measure the full sensor tilt range. Thus its diameter must correspond to at least two times the distance from the center to the membrane inflection point.

A first set of circular membranes of 2.83mm radius and 20µm thickness have been produced, with a maximal center deflection of 89,8µm measured by interferometry (in good correlation with large scale deflection calculations done with the ANSYS software package predicting 90,4µm). The inflexion point of these membranes has been calculated to be 1.63mm (from equation 2), which implies the use of a mobile mass with a comparable radius and a maximal measurable tilt angle of 2.4° (computed from equation 3 with an effective membrane thickness of 48µm).

A second set of membranes with nanostructuring and fluorocarbon deposition has been produced with a residual membrane thickness of approximatively 16µm (Figure 3).

EXPERIMENTAL SETUP

The inclinometer has been tested for low tilt angles via a dedicated experimental setup platform described in Figure 4. It consists of a common plate on which the prototype and two uni-axial reference inclinometers (Schaevitz LSOP/LSOC-3 from Sherborne Sensors Limited [14]) are mounted. These reference inclinometers are force balance sensors, with a measurement range of ±3° and a precision of ±0.0006°.

The tilt of the entire platform can be uni-axially changed by adjusting a set of screws drilled through the common plate and in contact with the ground.

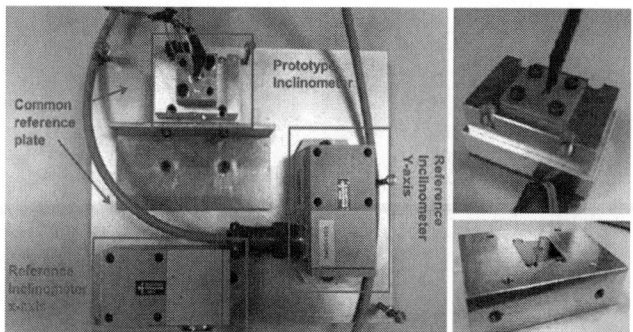

Figure 4: Left: Experimental set-up for inclinometer characterization. Top Inset: Picture of the complete prototype (5.5cm×4cm×3.5cm). Bottom Inset: Membrane holder.

The prototype inclinometer needs to be calibrated before beginning a measurement campaign in order to precisely align the prototype's zero tilt position with the zero position measured by the reference sensors. Any residual offset between the measurements of the reference sensors and the ones of the prototype is removed in software. A dedicated LabWindows software has been programmed that enables simultaneous visualization and acquisition of datapoints from the reference and prototype inclinometers through a National Instruments acquisition card (NI USB 62-59).

RESULTS

Different sets of experiments have been conducted with the two type of membranes previously described in the fabrication section.

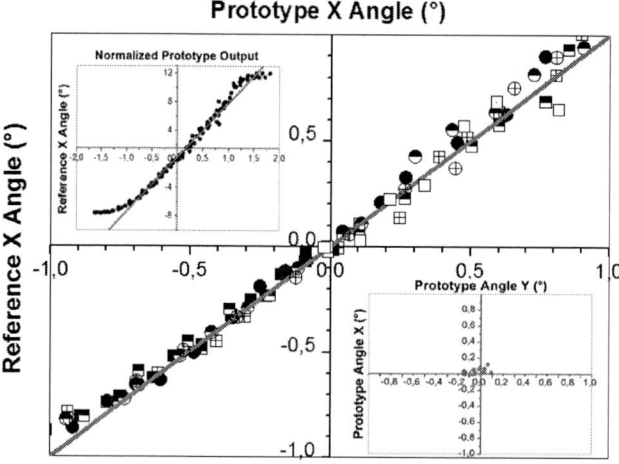

Figure 5: Experimental results of the unstructured membrane. Main graph: Plot showing good correlation between prototype and reference inclinometer measurements over the ±1° linear range. No sign of hysteresis (ascending sweeps with circles, descending with rectangles). Top left Inset: Normalized output of the prototype. Flattening outside the linear range is due to mobile mass shadow leaving center of quadrant detector. Linear fit to obtain calibration coefficient of 7,6/°. Bottom right Inset: Repeatability test following large angle tilt (>10°) showing standard deviation: y=0,033°, x=0,063°

A first measurement campaign (Figure 5) has been performed with an untreated silicon membrane and a 2.38mm diameter high precision steel ball (surface roughness variations of 0,025µm and sphericity error of 2.5µm). The prototype output data has been normalized by dividing the axis-sensitive voltage outputs of the photodiode (left minus right, top minus bottom) by its voltage sum. A linear fit has been performed on experimental data to obtain a slope calibration coefficient. The offset has been measured at 0° inclination and removed prior to plotting the data. The prototype output shows a ±1° linear range with good correlation to the reference inclinometer measurements. Imprecision is introduced through random surface roughness distribution and sphericity errors of the mobile mass. To estimate the precision of the sensor, repeatability tests have been performed. Repeatability tests have been done by placing the prototype in a no-tilt position and realizing repeated large angle tilts (>10°) and coming back to zero. Results show standard deviations of respectively 0.033° and 0.063° in y and x direction.

A second measurement campaign has been performed with a nanostructured and chemically treated silicon membrane and a mercury droplet of approximatively 2mm. The results are illustrated in Figure 6 and clearly show significant hysteresis effects between ascending and descending measurements (data points show a linear progression but are shifted by an offset dependent on the direction of angle change). Repeatability tests show improved standard deviation results of respectively 0.024°

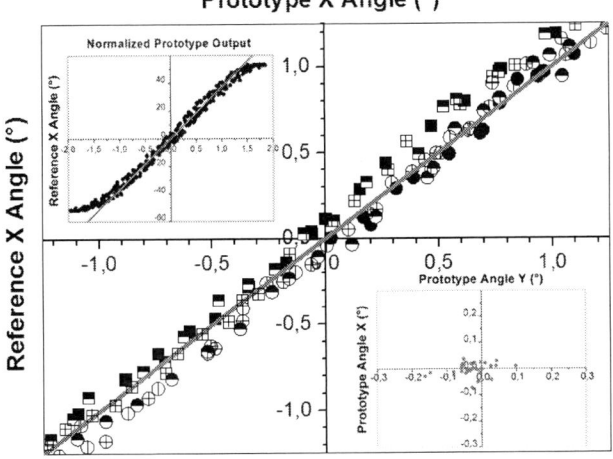

Figure 6: Experimental results of the structured membrane. Main graph: Plot showing significant hysteresis effects of prototype inclinometer measurements over the ±1° linear range (ascending sweeps with circles, descending with rectangles). Top left Inset: Full range normalized prototype output in x-direction. Linear fit to obtain calibration coefficient of 36,8/°. Bottom right Inset: Repeatability test following large angle tilt (>10°) showing standard deviation: y=0,024°, x =0,051°

CONCLUSION

This paper detailed a low-power bi-axial miniaturized inclinometer displaying linear behavior over a large part of the measurement range, which can be tuned by adapting the membrane diameter and thickness (by controlling etch time, or using a high-precision SOI substrate). The use of high precision balls showed sub-0.05° precision levels, the errors being due to mobile mass surface roughness and sphericity errors. Using a fluidic droplet as mobile mass resulted in even better precision, but introduced hysteresis in the system. Since the membrane deflection is influenced by pressure (equation 1) an external pressure sensor is needed for accurate operation. Future work will include a more in-depth analysis and understanding of underlying physics and better alignment of the optical system with the membrane

center through improved packaging in order to develop a sensor combining both high precision and low hysteresis.

ACKNOWLEDGEMENTS

We would like to thank Mr. Eric Laurent, Mr. Gilles Hovhanessian and Mr. Bernard Basile from Advitam for providing general technical assistance and civil engineering expertise, and for supplying reference inclinometers for this study. This work was supported by the Ile de France region through the SIPRIS project.

REFERENCES

[1] L.Tang et al., "MEMS Inclinometer based on a novel piezoresistor structure", *Microelectronics Journal*, vol. 40, pp.78– 82, 2009

[2] L.Shuangfeng, M.Tiehua, H.Wen, "Design and fabrication of a new miniaturized capacitive accelerometer", *Sensors and Actuators*, A 147, pp.70–74, 2008

[3] H.Jung, C.Kim, S.Kong, "An optimized MEMS-based electrolytic tilt sensor", *Sensors and Actuators*, vol. A139, pp.23-30, 2007

[4] B.Ando, S.Baglio, A.Beninato, "A ferrofluid inclinometer with a time domain readout strategy", Procedia Engineering vol. 47, pp.586 – 589, 2012

[5] J.S.Bajic et al., "A simple, low-cost, high-sensitivity fiber-optic tilt sensor", *Sensors and Actuators*, vol. A185, pp.33-38, 2012

[6] K.-M. Liao, R.Chen, B.Chou, "A novel thermal-bubble-based micromachined accelerometer", *Sensors and Actuators*, vol. A130–131, pp.282–289, 2006

[7] T.Constandinou, J.Georgiou, "Micro-Optoelectromechanical Tilt Sensor", *Journal of Sensors*, Article ID 782764, 2008

[8] H.Kato et al., "Photoelectric inclination sensor and its application to the measurement of the shapes of 3D-objects", *IEEE Transactions on Instrumentation and Measurement*, vol. 40, pp.1021-1026, 1991

[9] D.Welch, J.Georgiou, J.Christen, "Fully differential current-mode MEMS dual-axis optical inclination sensor", *Sensors and Actuators*, vol. A192, pp.133-139, 2013

[10] ADIS16209 High accuracy, dual axis digital inclinometer and accelerometer, Analog Devices datasheet, 2012

[11] N.Gaber et al., "On the free-space Gaussian beam coupling to droplet optical resonators", Lab Chip vol.13, pp.826-833, 2013

[12] W.C.Young, R.G.Budynas, *Roark's Formulas for Stress and Strain*, McGraw-Hill, 2002

[13] A.G.Striz, S.K..Jang, C.W.Bert, "Nonlinear Bending Analysis of Thin Circular Plates by Differential Quadrature", *Thin Walled Structures*, vol. 6, pp.51-62, 1988

[14] LSOP/LSOC DC-Operated, Gravity-Referenced Inclinometer, Sherborne Sensors datasheet, 2003

TUNING OF NONLINEARITIES AND QUALITY FACTOR IN A MODE-MATCHED GYROSCOPE

Erdinc Tatar[1], Tamal Mukherjee[1], and Gary K. Fedder[1, 2, 3]

[1]Department of Electrical and Computer Engineering
[2]Institute for Complex Engineered Systems, [3]The Robotics Institute
Carnegie Mellon University, Pittsburgh, Pennsylvania, USA

ABSTRACT

This paper examines methods to electrically tune quality factor (Q) and cubic softening nonlinearity of a mode-matched MEMS gyroscope. The gyroscope includes traditional drive-sense combs and dedicated shaped combs for cubic nonlinearity and frequency tuning. In addition to nonlinearity, Q can be tuned by understanding the nature of the losses. Interconnect resistances are added to the electromechanical resonator model to capture the electrical losses that depend on the applied DC proof mass potential.

INTRODUCTION

Quality factor (Q) is defined as the energy stored in a system divided by the energy dissipated per radian of vibration cycle [1], and is a measure of the dissipative losses for a system. Q is one of the important parameters for a MEMS gyroscope since it directly affects the thermal noise limit of the gyroscope. Matching the Q of the drive and sense modes eliminates one of the major drift sources for a whole angle gyroscope [2]. In this respect, methods to tune Q electrostatically will lead to a performance increase in whole angle gyroscopes. Having control over Q is also useful since it allows the designer to change the Q depending on the requirements. Q of a gyroscope can be tuned by electrical velocity feedback [3], at the cost of additional control circuits. However, understanding the nature of Q and losses of the gyroscope may lead to easier Q tuning. One focus of this paper is the effects of electrical losses on Q and their potential usage on Q tuning.

Force-displacement nonlinearity and its electrostatic cancellation is the other topic of this paper. As the displacement of a resonator increases, spring hardening or softening start to emerge [4]. The force displacement relation for a spring cannot be modeled as a pure linear relation ($F=kx$) anymore and a cubic spring constant is added to the relation ($F=k_1x+k_3x^3$) that results in hardening nonlinearity. Tip capacitances of a comb drive form parallel-plate capacitances and generate forces that result in softening nonlinearity. Hardening and softening effects have been modeled [4] and characterized for resonators [5-6] and in a gyroscope [7]. This study proposes a way to electrostatically tune the softening nonlinearity by introducing hardening nonlinearity through specially shaped combs. The goal behind the nonlinearity cancellation is mode matching the drive and sense modes at high displacement (10 μm) while keeping the gyroscope linear.

The 1st generation mode-matched SOI-MEMS gyroscope with large stroke shaped comb frequency tuning was described in [8]. This paper presents the 2nd generation gyroscope with traditional drive and sense combs, and a first attempt at dedicated shaped combs for cubic nonlinearity tuning.

ELECTRICAL MODEL

The gyroscope studied in this work is driven and sensed electrostatically. The gyroscope consists of two coupled resonators (modes) that are modeled as separate resonators. Fig. 1 shows the equivalent electrical model for one mode.

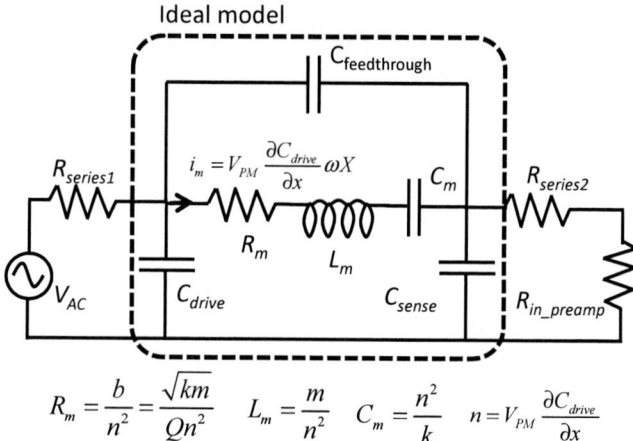

$$R_m = \frac{b}{n^2} = \frac{\sqrt{km}}{Qn^2} \quad L_m = \frac{m}{n^2} \quad C_m = \frac{n^2}{k} \quad n = V_{PM}\frac{\partial C_{drive}}{\partial x}$$

Figure 1: Equivalent electrical model for one mode of the gyroscope.

R_m, L_m, C_m, and i_m represent the motional resistance, inductance, capacitance, and current, respectively. $C_{feedthrough}$ represents the feedthrough capacitance from the drive to sense electrodes. C_{drive} and C_{sense} are the drive and sense electrodes, respectively. V_{PM} is the applied DC voltage for driving and sensing. $R_{series1, 2}$ are added resistors to the model to account for the electrical interconnect losses. The motional current output of the gyroscope is converted into voltage by a trans-impedance amplifier that has ideally zero input impedance. However, due to the finite gain of the op-amp, the circuit has a non-zero input impedance that is represented by R_{in_preamp}. m, k, and b represent the mass, spring constant, and damping respectively.

Q does not depend on the V_{PM} in the ideal circuit assuming damping (b) is constant. $R_{series1,2}$ and R_{in_preamp} are added to the ideal electrical equivalent model in Fig. 1 to observe the change of Q with changing V_{PM}. When V_{PM} is changed, R_m, L_m, and C_m changes accordingly but $R_{series1,2}$ and R_{in_preamp} stay constant. As a result, a change in the quality factor is expected since the equivalent motional resistance does not exactly scale with V_{PM}. As V_{PM} decreases R_m increases and the effect of $R_{series1,2}$ on Q decreases, so decreasing V_{PM} increases the Q.

978-1-4799-3510-9/14 $31.00 © 2014 IEEE

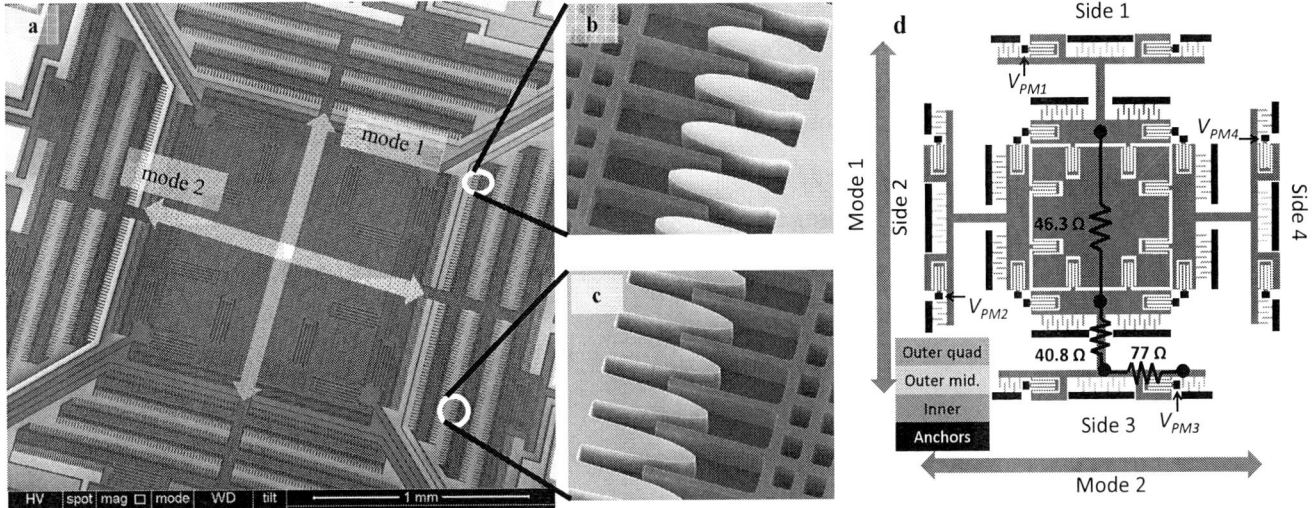

Figure 2.a-d: SEM image of the fabricated mode-matched SOI MEMS gyroscope tested in this study (a), along with the shaped combs for nonlinearity (b) and frequency tuning (c). Locations of the straight comb fingers are shown in (d).

The equivalent Q is

$$Q_{effective} \approx \sqrt{\frac{L_m}{C_m}} \frac{1}{R_m + R_{series1} + R_{series2} + R_{in\_preamp}} \qquad (1)$$

The same phenomena can also be explained physically; as V_{PM} increases so does the motional current, and the power losses on the $R_{series1,2}$ (I^2R) increases resulting in a decrease in Q. Table 1 shows the typical values of the equivalent circuit parameters for the gyroscope tested in this study.

Table 1: Typical values of the equivalent circuit parameters for the gyroscope tested in this study.

V_{PM}	10 V	b	2.7 e-8 Ns/m
dC/dx	1.45e-8 F/m	R_m	1.3 MΩ
m	3.15e-8 kg	L_m	1.5 MH
k	114.6 N/m	C_m	0.18 fF
Q	70000 (16 mTorr)	$C_{drive}=C_{sense}$	0.3 pF

TEST RESULTS ON Q

Fig. 2.a-c shows the SEM image of the mode-matched gyroscope tested in this study (a) along with the shaped combs for nonlinearity (b) and frequency tuning (c). The gyroscope was fabricated by a two mask in-house SOI-MEMS process [8]. The conventional straight finger combs are located on the inner and outer sides of the each side plate. Fig. 2.d shows the positions of the straight fingers (shaped combs are not shown) with the calculated side to side device resistances. There are three sets of straight fingers on each side that can be reached separately. The rotor is a one layer conducting silicon, and it can be reached from four different spring anchor locations (V_{PM1-4}).

Various resonance sweeps have been performed using different sets of fingers for different DC polarization voltages (V_{PM}) to extract Q. Fig. 3 shows the schematic of the tests. A DC potential is applied to the rotor, and the frequency of the AC signal that is applied to the stators of the fingers is swept, and the output voltage is recorded. Assuming the rotor is moving in the $+x$ direction, the motional current flows from stator to rotor on the drive side and from rotor to stator on the sense side. As a result, the motional current flows on the rotor and no current is drawn from the V_{PM} source. The electrical displacement current on the drive side can be ignored, assuming $V_{PM} \gg V_{AC}$.

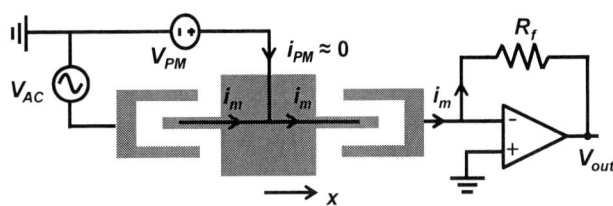

Figure 3: Resonance test-bench schematic: DC is applied to the rotor and V_{AC} is applied through the stator fingers.

Fig. 4 presents the Q of one mode of the gyroscope when tested from two different combs for different V_{PM}. Both modes exhibit similar Q-V_{PM} characteristics. The tests were done at a constant pressure of 16 mTorr. Only one set of the fingers is used in each set of experiments. For example, the gyro is driven from one side using the inner fingers and the gyroscope's current is sensed using the inner fingers on the other side. All the other stator fingers are connected to V_{PM} to assure they do not contribute any force. Q of the gyroscope changes by 14% when inner fingers are used and changes by 32% when outer quad fingers are used in a V_{PM} range of 10V to 40V. The motional resistance of the device decreases with increasing V_{PM} but it is still hundreds of kilohms for 40V, which is much larger than the device resistances shown in Fig. 2.d. Although the device resistances are not enough to cause such a Q change, the motional current flows through a smaller resistance for the inner fingers compared to outer fingers. V_{PM} connection is made through V_{PM2}.

The Q change with varying V_{PM} is repeatable and has been observed in many devices at constant pressure and at constant small (linear) amplitude. Fig. 5 shows the Q-V_{PM} curves for 3 different devices tested from the outer quad and outer mid. fingers. All of them exhibit similar Q-V_{PM} characteristics, but the 3rd device goes to a higher Q because of the lower pressure. In addition, the devices exhibit less Q − V_{PM} change if the inner fingers are used as demonstrated in Fig. 4. A similar Q-V_{PM} change is presented in [9].

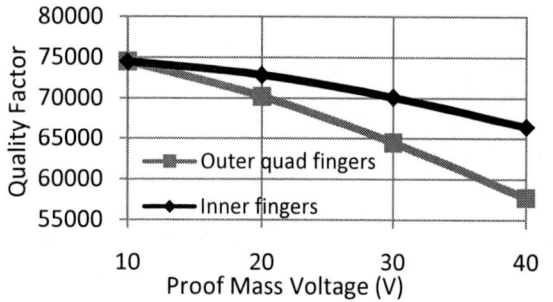

Figure 4: Q of one mode of the gyroscope when tested from different combs for different V_{PM}.

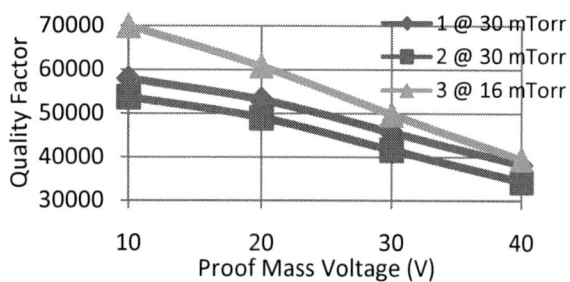

Figure 5: Q-V_{PM} curves for three different devices exhibiting similar behavior.

TEST RESULTS ON NONLINEARITY AND NONLINEARITY TUNING

High displacement (e.g., 10 μm) is desired for a MEMS gyroscope to increase the rate sensitivity. But nonlinearities start to occur at high displacements as shown in Fig. 6&7.

Fig. 6.a-b shows the high displacement up and down frequency sweeps for Mode 1 and 2 for a device mounted in a DIP-40 package and aligned with the square recess of the package. Mode 1 shows significant hardening behavior due to the concentrated packaging stress as explained in [8]. Fig. 7 shows the up and down frequency sweeps for a device aligned 45° with the package recess. The 45° mounting reduces the hardening nonlinearity and the two modes exhibit mainly softening nonlinearity. Shaped combs shown in Fig. 2.b are designed to tune the softening linearity by attempting to introduce an electrically controlled hardening nonlinearity.

Fig. 8 shows the fingers and gap function to generate a cubic force-displacement relation. Assuming the fingers are designed symmetrically the force-displacement relation is

$$F(x) = 2N\varepsilon h\left(-\frac{6x_{ov}^2 x}{g_0 x_{03}} - \frac{2x^3}{g_0 x_{03}}\right)V^2 \qquad (2)$$

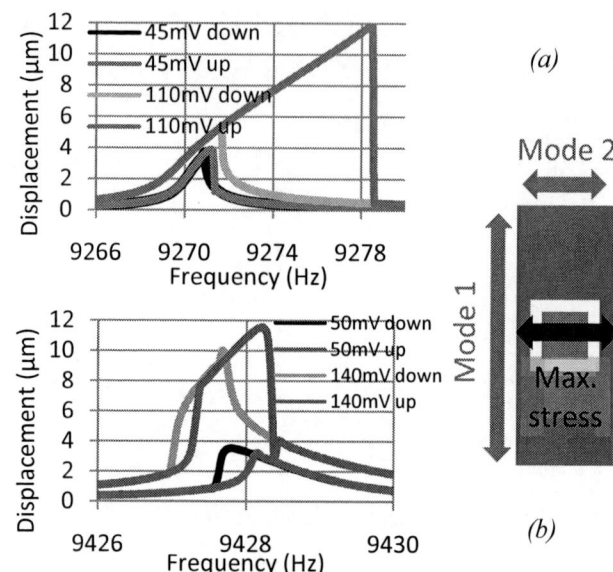

Figure 6.a-b: High displacement behavior of mode 1(a), and mode 2 (b) for a device mounted in a DIP-40 package and aligned with the package recess for different AC drive voltages with up and down sweeps for $V_{PM} = 40V$.

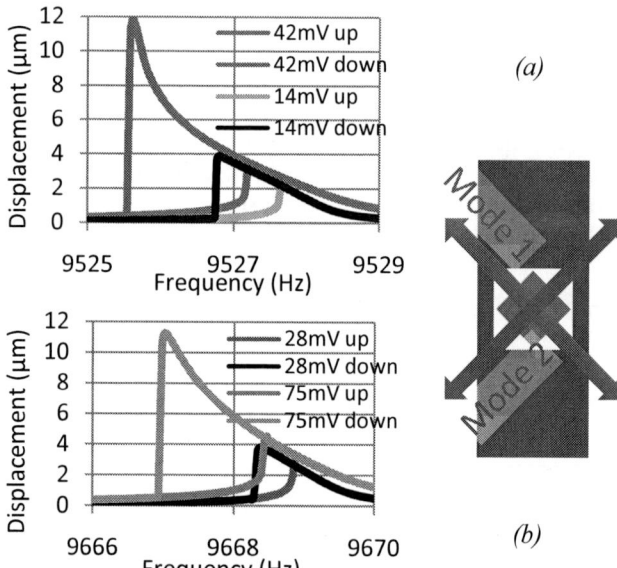

Figure 7.a-b: High displacement behavior of mode 1(a), and mode 2 (b) for a device aligned 45° with the package recess for different AC drive voltages with up and down sweeps for $V_{PM} = 40V$.

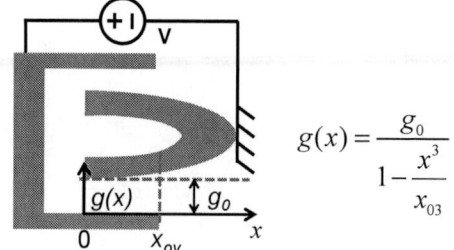

$$g(x) = \frac{g_0}{1 - \dfrac{x^3}{x_{03}}}$$

Figure 8: Fingers and shape function to generate a cubic nonlinearity.

In (2) N is the number of fingers, ε is the permittivity, h is the structural layer thickness, x_{ov} is the overlap between the fingers, g_0 is the gap at $x=0$, x_{03} is the gap coefficient, and V is the voltage between the rotor and stator. The force in (2) generates also a linear frequency hardening in addition to the cubic hardening which can be minimized by optimizing the design parameters. The initial oval-shaped comb design in Fig. 2.b has a gap of

$$g(x) = \frac{g_0}{1 + \frac{x}{x_{01}} - \frac{x^3}{x_{03}}} \qquad (3)$$

The intent for including the linear tuning term (x/x_{01}) in (3) is to obtain linear frequency tuning. Fig. 9 shows the response of one of the modes for different nonlinearity tuning potentials. The fingers are supposed to introduce hardening but in the measurements they do not contribute enough hardening. It has been found and verified by the finite element analysis (FEA) that (x/x_{01}) term in (3) causes a reduction in the cubic terms in the final force expression.

The updated design for the nonlinearity tuning fingers is as in Fig. 8, but with $x_{ov} = 0$; zero overlap allows decreasing the cubic gap coefficient (x_{03}) to have a larger cubic dependence. Zero overlap also minimizes the unwanted linear frequency hardening effect. Fig. 10 shows FEA results for the dC/dx vs. displacement relation for the updated fingers. The curve in Fig. 10 has both the linear frequency tuning and cubic term. The linear term is introduced because of the parallel plate capacitance formed by the finger tips.

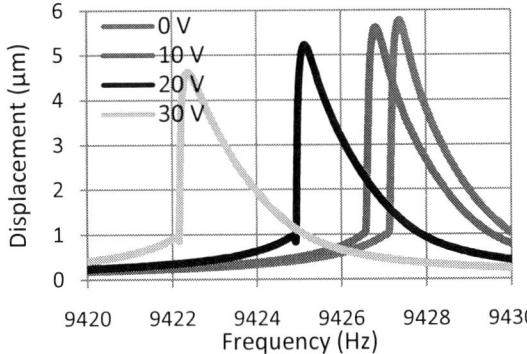

Figure 9: Frequency sweeps for various DC nonlinearity tuning potentials.

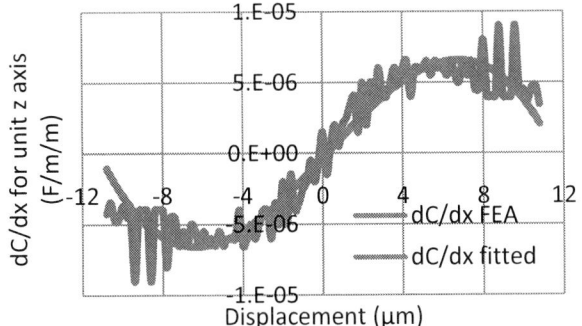

Figure 10: FEA result for the dC/dx vs. displacement relation for the updated finger design.

CONCLUSIONS

Q dependence on the applied DC potential has been experimentally demonstrated. However the calculated device resistances do not fully explain the measured changes in Q. The Q-V_{PM} change decreases as the resistance seen by the motional current decreases on the movable mass. We hypothesize that minimizing the motional current's resistive path on the rotor by placing the driving and sensing fingers side by side instead of placing them on opposite sides of the device would result in a more robust Q with respect to V_{PM} changes. Test results for the first generation shaped combs have shown that cubic capacitance has to be increased. A design update with no initial finger engagement increases the overall cubic capacitance.

ACKNOWLEDGEMENTS

This work was supported by Defense Advanced Research Projects Agency (DARPA) under agreement number FA8650-08-1-7824.

REFERENCES

[1] B. Kim *at al.*, "Temperature dependence of Quality factor in MEMS resonators," *J. Microelectromech. Syst.*, vol. 17, no. 3, pp. 755-766, 2008.

[2] I. P. Prikhodko *at al.* , "Foucault pendulum on a chip: angle measuring silicon MEMS gyroscope," *Proc. of MEMS 2011*, Cancun, Jan. 23-27, 2011, pp. 161-164.

[3] C. Jeong, S. Seok, B. Lee, H. Kim, and K. Chun "A study on resonant frequency and Q factor tunings for MEMS vibratory gyroscopes," *J. Micromech. and Microeng.*, vol. 14, pp. 1530-1536, 2004.

[4] A. M. Elshurafa *at al.*, "Nonlinear dynamics of spring softening and hardening in folded-MEMS comb drive resonators," *J. Microelectromech. Syst.*, vol. 20, no. 4, pp. 943-958, 2011.

[5] H. Jeong, and S. K. Ha, "Dynamic analysis of a resonant comb-drive micro-actuator in linear and nonlinear regions," *Sensors and Actuators A*, vol. 125, pp. 59-68, 2005.

[6] M. Agarwal *at al.*, "Nonlinear characterization of electrostatic MEMS resonators," *Proceedings of Int. Freq. Control Symp.*, Miami, June 2006, pp.209-212.

[7] M. Sharma, E. H. Sarraf, R. Baskaran, and E. Cretu, "Parametric Resonance: Amplification and damping in MEMS gyroscopes," *Sensors and Actuators A: Physical*, vol. 177, pp. 79-86, 2012.

[8] E. Tatar, C. Guo, T. Mukherjee, and G. K. Fedder, "Interaction effects of temperature and stress on matched-mode gyroscope frequencies," *Proc. of TRANSDUCERS 2013*, Barcelona, June 16-20, 2013, pp. 2527-2530.

[9] D. K. Agrawal, J. Woodhouse, and A. A. Seshia, "Modeling nonlinearities in MEMS oscillators," *IEEE Trans. on Ultrasonics, Ferroelectrics, and Freq. Control*, vol. 60, no.8, pp. 1646-1659, 2013.

CONTACT

E. Tatar, tel: +1-412-268-6606; etatar@andrew.cmu.edu

2D RESONANT MICROSCANNER FOR DUAL AXES CONFOCAL FLUORESCENCE ENDOMICROSCOPE

Haijun Li[1], Zhen Qiu[2], Xiyu Duan[2], Kenn R. Oldham[3], Katsuo Kurabayashi[3], and Thomas D. Wang[1,2,3]

[1]Department of Internal Medicine, School of Medicine, University of Michigan, Ann Arbor, USA
[2]Department of Biomedical Engineering, University of Michigan, Ann Arbor, USA
[3]Department of Mechanical Engineering, University of Michigan, Ann Arbor, USA

ABSTRACT

In this paper, we present a two-dimensional (2D) electrostatically actuated Micro-Electro-Mechanical System (MEMS) resonant microscanner for a miniature dual axes confocal fluorescence endomicroscope with an outer diameter (OD) of 5 mm, in which the scanner will be used to achieve a large field of view (FOV) in real-time for *en-face* imaging. The device has a compact and robust gimbal structure design, which enables a reflective mirror with high fill-in factor to perform high-speed 2D scanning with large tilting angles. Devices with a good optical quality mirror surfaces are produced using a three-step deep reactive-ion etching (DRIE)" silicon-on-isolator (SOI) micromachining process on a single wafer with three masks.

INTRODUCTION

Dual axes confocal endomicroscopy [1,2] is an emerging imaging technology that can achieve deep tissue penetration with sub-cellular resolution over a large field-of-view. The dual axes confocal geometry uses separate optical fibers and low-numerical aperture (NA) objectives for off-axis light illumination and collection to achieve a long working distance, high dynamic range and scalability to millimeter dimensions while preserving high axial resolution. A miniatured 2D scanner is critical for the endoscopic implementation of this optical configuration. MEMS technology is one of the most promising technologies for this application.

Presently, several types of MEMS-based scanners are being developed for endoscopic imaging systems that are based on different actuation mechanisms. However, a scanner that achieves large tilting angles with fast scan speeds at low driving voltages in a small footprint remains challenging. Electrothermally actuated MEMS scanners can achieve large tilting angles, but the scan speeds are limited by slow response times [3]. Electromagnetically actuated microscanners can achieve large tilting angles with fast scan speeds, but the dimensions are relatively large due to the external bulky magnet [4]. Thin-film based piezoelectric microscanners usually require relatively complicated fabrication processes [5,6]. Compared to others, the electrostatically actuated MEMS scanner [7,8] is preferred because of its ease of integration, high speeds and low power consumption. The high driving voltages needed for large tilting angles can be overcome using resonant modes.

A dual axes confocal fluorescence endomicroscope has been developed using a MEMS electrostatic scanner and low NA micro-optics [2,8]. 3D volumetric images can be acquired by stacking a set of *en-face* images using either a bulk piezo actuator or thin-film piezo z-axis actuator [6]. For the lateral scan, a staggered vertical combdrive (SVC) based 2D scanner [2,8] has been developed. Requirements for wafer bonding and fine alignment of the combdrive fingers can limit the yield of functional devices. Small tilting angles result in a small image FOV (362 μm×212 μm) [2].

In our previous work [1], a 1D resonant microscanner with a large optical scan angle of ±12° has been developed for real-time cross-section imaging with a large lateral FOV (800 μm) in a handheld near-infrared dual axes confocal fluorescence endomicroscope. Here, we present a novel 2D parametrically excited resonant microscanner with large optical scan angles (up to ±14° for inner axis, ±6° for outer axis) at low driving voltage (< 70V) in a small footprint (3.2 mm×2.9 mm) for a new dual axes confocal endomicroscope with an outer diameter of 5 mm.

SCANNER DESIGN AND FABRICATION
Scanner Design for Dual Axes Confocal Endomicroscope

The size and geometry of the 2D microscanner should accommodate the optical path and the instrument package of the OD 5 mm miniature dual axes confocal endomicroscope. Lateral xy-plane scanning using a ZEMAX (R12, Redmond, WA) simulation is shown in Fig. 1.

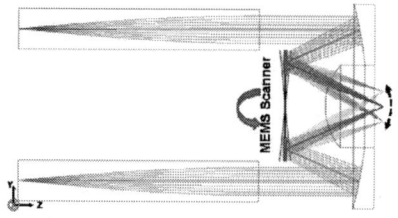

Figure 1: Schematic view of dual axes confocal endomicroscope with OD 5 mm package, in ZEMAX.

The design of the 2D resonant microscanner is shown in Fig. 2. A gimbal geometry is used to provide 2D scanning and two combdrives are used to drive the inner mirror and the gimbal frame along the inner and outer axes, respectively. The inner combdrives fully cover the two sides of inner mirror for large driving torque. On the outer axis, the combdrives are designed in two columns on each side of the outer torsional spring. Weight-release holes are located on the outer columns. Next to the corners of the gimbal frame, there are four isolated stoppers to avoid electrostatic collapse.

Unlike the SVC microscanners, the 2D resonant scanner

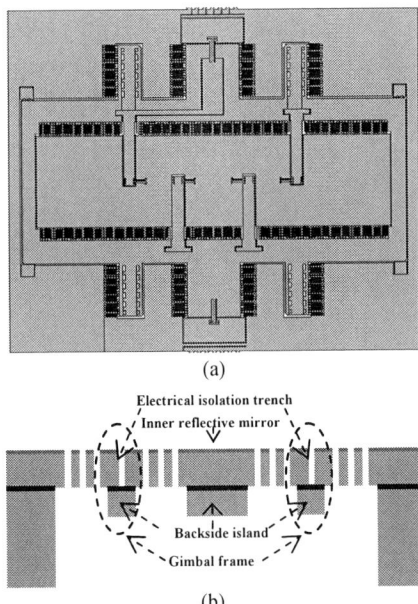

(a)

(b)

Figure 2: Schematic of the 2D resonant microscanner, (a) Top view (b)Cross-section view.

uses in-plane comb-drives to generate out-of-plane tilting by employing parametric resonance excitation[9], which significantly reduces the processing complexity. Two comb-drives with electrical isolation trenches can be formed on the device silicon layer of the SOI wafer by a single step DRIE etching.

As shown in Figure 1, a high fill-in factor mirror with a large reflective surface area can improve the collection efficiency of the illumination and collection beams. In this design, the inner springs are linked to the gimbal frame through 4 inner anchors, and their central location does not block the light beams. Additionally, the backside islands (Fig. 2b) provide mechanical support to reduce dynamic deformations that occur when the device is working at high speeds with large scan angles. This also enables a narrow (70 μm) gimbal frame design. Compared to previously reported micromirrors [8] developed for a OD 5 mm dual axes confocal fluorescence microendoscope, our scanner has a more compact structure with a footprint of 3.2 mm×2.9 mm and a 2.71 mm×0.8 mm inner mirror with two reflectors (675 μm×800 μm). The surface of the mirror is coated with aluminum to achieve a high reflectivity of >90% in the visible range and >70% in the near infra-red (NIR) regime).

Device Fabrication

A robust and high yield (>80%) three masks, three-step DRIE SOI process has been developed for the production of resonant scanners, as shown in Fig. 3. The process starts with an SOI wafer composed of 35 μm device silicon layer, 1 μm buried oxide layer and 500 μm handle silicon layer. First, a 1.5 μm oxide layer is deposited by plasma-enhanced chemical vapor deposition (PECVD) (GSI-PECVD) on the front-side blank surface of the SOI wafer. It is used not only as a hard mask for the front-side structure patterning, but

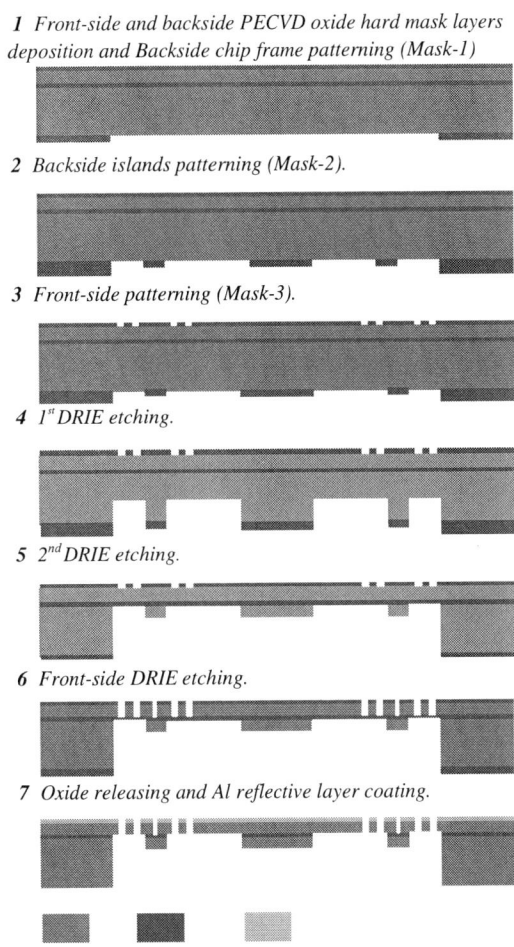

1 Front-side and backside PECVD oxide hard mask layers deposition and Backside chip frame patterning (Mask-1)

2 Backside islands patterning (Mask-2).

3 Front-side patterning (Mask-3).

4 1ˢᵗ DRIE etching.

5 2ⁿᵈ DRIE etching.

6 Front-side DRIE etching.

7 Oxide releasing and Al reflective layer coating.

Si   SiO₂   Al

Figure 3: Process flow.

also as a protection layer for reducing the contamination and scratching on the mirror surface in later processing steps, especially in backside processing steps. Then, 3 μm and 1.5 μm PECVD oxide layers are sequentially deposited and patterned on the backside by lithography and reactive-ion-etching (RIE) (LAM 9400), as shown in Fig. 3.1 and Fig. 3.2. Those layers serve as hard masks for silicon etching by DRIE (STS Pegasus 4). Next, the front-side oxide layer is patterned, in Fig. 3.3, by the double-sided lithography and RIE. During step 1-3, patterns of the three masks are all transferred to the backside and front-side oxide hard mask layers, defining the backside chip frame, backside island, and the front-side structures respectively. Then, the backside islands step depth is define by the 1st DRIE. The etching depth should be slightly larger than the designed thickness of the islands because of the loading effect for etching trenches with different widths. This effect is significant in our device fabrication because a large area of the SOI handle silicon layer should be etched through to release the gimbal frame, inner mirror, and comb-drives. Here, we chose an etching depth of ~200 μm at the narrowest (50 μm) backside trench as the endpoint ,as measured by optical surface profiler. Following a residual

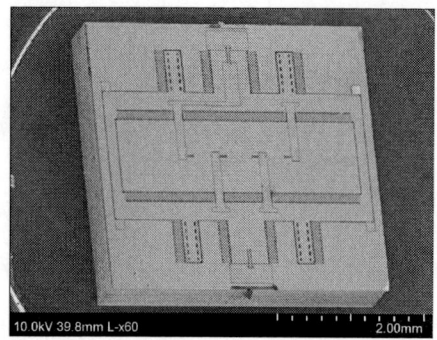

Figure 4: SEM of the fabricated 2D resonant scanner with dumb-bell shape mirror inside the gimbal frame.

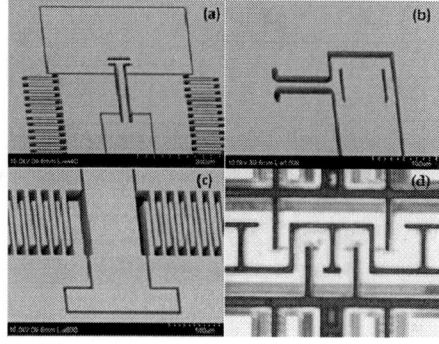

Figure 5: Photographs of structural features of the 2D resonant microscanner (a) Outer spring;(b) Inner spring; (c) Electrical isolation trench; (d) Backside island.

oxide on the island surface removal step by the RIE process, the silicon in trenches between the islands is etched away the 2nd step backside DRIE. By carefully timing the over-etching, nearly 50 μm thick backside islands are formed by the end of 2nd step backside DRIE. After removing front-side, backside and buried oxide layers with buffered hydrofluoric acid (BHF) wet etching, the movable structures are released by isopropyl alcohol (IPA) rinsing and drying. Finally, a 60 nm aluminum layer is sputtered onto the front-side blank silicon surface to achieve a high-reflectivity mirror with minimal defects.

Fabrication Results

Fig. 4 and Fig. 5 show photographs of the fabricated 2D mircoscanner and its detailed structural features. As shown in Fig.5a and 5b, fillets and stress releasing trenches are designed at the corners and ends of the springs which help ensure the high-speed scanning with large angular deflections. Fig. 5c shows electrical isolation trenches with a width of 5 μm. These narrow and deep isolation trenches, with an additional gap between the upper device silicon layer and the lower handle silicon layer formed during the buried oxide BHF etching, effectively avoid the short-current. Fig. 5d shows the stereomicroscope image of the detailed structural features of the flatness enhancement backside island (thickness ~50 ±10 μm) under the gimbal frame and the reflective mirror. The width reduction is about 5 μm caused by the undercut during a long-time

DRIE backside etching.

The curvature and the surface roughness of the inner scanning mirror's reflective surface are measured by a Zygo optical surface profiler (NewView 5000). The measurement shows that the mirror has a radius of curvature of ~1.7 m and a root mean square (RMS) roughness of ~2 nm, providing high optical quality flatness and surface.

DEVICE CHARACTERIZATION

Optical Scan Angle Characterization

The resonant scanner has been characterized by a custom-made optical scan angle test setup. For tilting amplitude test, position sensing detector (PSD) is used. He-Ne laser (633nm) (Model 1507 NovetteTM, JDSU, Milpitas, CA) beam is steered by the reflective surface of the tilting mirror and sensed by the PSD module (10 mm×10 mm window, PSM 2-10, On-Trak Photonics, Irvine, CA). The sensing voltage signal is amplified (OT-301, On-Trak Photonics, Irvine, CA) and acquired by data acquisition card (PCI-6115, National Instruments, Austin, TX). A custom-made LabVIEW® program is used to generate the driving signal and process the acquired signal. For testing the dynamic response, the frequency sweeping with adjustable voltage control is performed by this program. The swept-frequency square wave voltage from PCI-6115 is amplified using a TEGAM Model 2350 amplifier (gain 20×), and then the amplified signal is applied to the device.

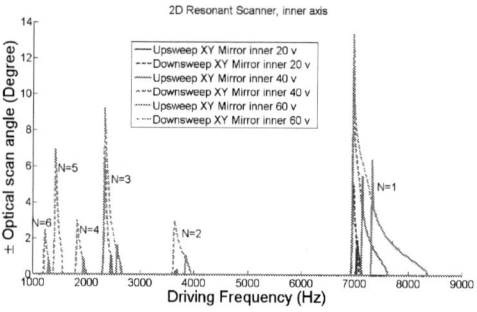

Figure 6: Optical scan angle versus driving frequency, inner axis, scan can be observed at N = 1 to 6, driving voltage is chosen at 20V, 40V, 60V.

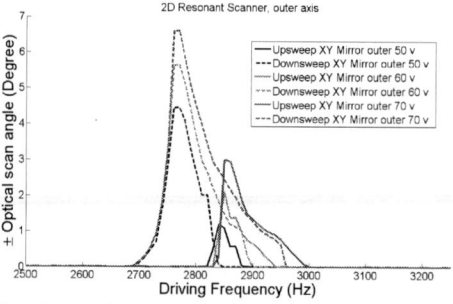

Figure 7: Optical scan angle versus driving frequency, outer axis, scan can be observed at N=1, driving voltage is chosen at 50V, 60V, 70V.

The optical scan angle versus driving frequency at different driving voltage for inner and outer axes are shown in Fig.7 and Fig.8, respectively. The device can achieve large scan angles (up to $\pm14°$ for inner axis, up to $\pm6°$ for outer axis) at a low driving voltage ($<70V$) with a tunable driving frequency bandwidth close to resonance (3.57kHz for inner axis, 1.36kHz for outer axis). It should be noted that the device is highly resistant to pull in. Its failure is caused by the fracture of springs rather than pull in when high voltage applied. The maximum optic scan angles are $\pm22°$ (180V) and $\pm27°$ (140V) for outer axes and inner axes, respectively, before failure of the springs.

Stability Characterization

The start and stop frequencies (f_{start}, f_{stop}) in the response curve determine the boundary of the unstable frequency range. The boundaries depend on the driving voltage. Fig. 8 shows the experimental results for stability region of the outer axis for observable parametric resonances at ambient pressures. The stability region of the inner axis is shown in Fig. 9. These regions will help us to quantify the mechanical characteristics of the resonant scanners. We will choose the scanner for use in the dual axes confocal endomicroscope based on best stability and tilting angle response.

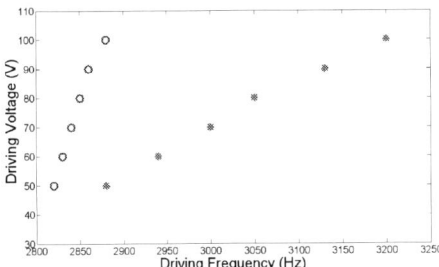

Figure 8: Stability region of outer axis, start and stop driving frequency versus driving voltage.

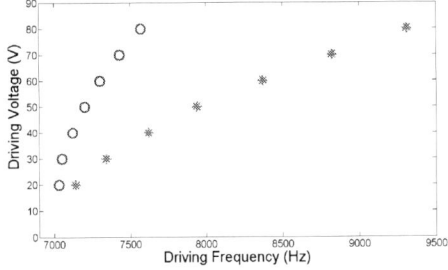

Figure 9: Stability region of inner axis, start and stop driving frequency versus driving voltage.

CONCLUSION

We have demonstrated a novel 2D resonant microscanner for a OD 5mm dual axes confocal fluorescence endomicroscope. The device has a compact and robust gimbal structure and is fabricated by a single wafer based SOI micromachining process which ensures a high optical quality. Using parametric resonance excitation, large tilting angles at high-speed is realized. This new 2D device will provide large field of view and sub-cellular resolution *in-vivo* imaging.

ACKNOWLEDGEMENS

We acknowledge support in funding from the National Institutes of Health R01 CA142750 and U54 CA136429 (T.D. Wang) and the Rackham international student fellowship (Z. Qiu and X. Duan). We would like to thank Dr. H. Schenk for helpful discussions on the MEMS resonant scanner design.

REFERENCES

[1] Z. Qiu, Z. Liu, X. Duan, S. Khondee, B. Joshi, M. J. Mandella, K. Oldham, K. Kurabayashi, T. D. Wang, "Targeted Vertical Cross-sectional Imaging with Handheld Near-infrared Dual Axes Confocal Fluorescence Endomicroscope", *Biomed Opt. Express*, 4(2), pp 322-330, 2013.

[2] W. Piyawattanametha, H. Ra, Z. Qiu, S. Friedland, J. T. C. Liu, K. Loewke, G. S. Kino, O. Solgaard, T. D. Wang, M. J. Mandella, C. H. Contag, "In vivo Near-infrared Dual-axis Confocal Microendoscopy in The Human Lower Gastrointestinal tract", *J. Biomed. Opt.*, 17(2), 021102 (2012).

[3] L. Wu, H. Xie, "A Large Vertical Displacement Electrothermal Bimorph Microactuator with Very Small Lateral Shift", *Sensors and Actuators A: Physical*, vol. 145-146, pp 371-379, 2008

[4] C. L. Arrasmith, D. L. Dickensheets, A. Mahadevan-Jansen, "MEMS-based Handheld Confocal Microscope for In-vivo Skin Imaging", *Opt. Express*, 18(4), 3805-3819, 2010.

[5] W. Liu, Y. Zhu, K.Jia, W. Liao, Y. Tang, B. Wang, H. Xie, "A Tip–tilt–piston Micromirror with a Double S-shaped Unimorph Piezoelectric Actuator", *Sensors and Actuators A: Physical*, vol. 193, pp 121-128, 2013.

[6] Z. Qiu, J. Pulskamp, X. Lin, C. H. Rhee, T. D. Wang, R. Polcawich, K. Oldham, "Large Displacement Vertical Translational Actuator based on Piezoelectric Thin Films", *J. Micromech. Microeng.*, 20(7), 075016, 2010.

[7] H. J. Shin, M. C. Pierce, D. Lee, H. Ra, O. Solgaard, Richards-Kortum R., "Fiber-optic Confocal Microscope using a MEMS Scanner and Miniature Objective Lens", *Opt. Express*, 15(15), 9113-9122, 2007.

[8] H. Ra, W. Piyawattanametha; Y. Taguchi,. S. Lee, M. J. Mandella, O. Solgaard, "Two-Dimensional MEMS Scanner for Dual-Axes Confocal Microscopy", *J. Microelectromech. Syst.*, vol. 16, pp. 969-976, 2007.

[9] H. Schenk, P. Durr, T. Haase, D. Kunze, U. Sobe, H. Lakner, H. Kuck, "Large Deflection Micromechanical Scanning Mirrors for Linear Scans and Pattern Generation", *J. Sel. Top. Quantum Electron.*, vol. 6, 715-22, 2000.

CONTACT

*H. Li, tel: +1-734-615-4834; haijunl@umich.edu

978-1-4799-3510-9/14 $31.00 © 2014 IEEE

A CAPACITIVE IMMUNOSENSOR USING ON-CHIP ELECTROLYTIC PUMPING AND MAGNETIC WASHING TECHNIQUES FOR POINT-OF-CARE APPLICATIONS

Jui-Chang Kuo, Po-Hung Kuo, Hsiao-Ting Hsueh, Cheng-Wen Ma,
Chih-Ting Lin, Shey-Shi Lu, and Yao-Joe Yang*
National Taiwan University, Taipei, TAIWAN

ABSTRACT

This work presents a capacitive immunosensor using on-chip electrolytic pumping and magnetic washing techniques. The proposed device possesses the advantages such as simple operation, low power consumption, and portability. The proposed device was fabricated using typical micromachining process, and is suitable for mass-production. We also demonstrated the detection of N-Terminal pro-brain-Type natriuretic peptide (NT-proBNP) using the fabricated device integrated with a CMOS capacitance sensing chip. The proposed device potentially can be used as a portable system for point-of-care applications.

INTRODUCTION

Biological assays are widely used in medical diagnostic testing for infectious diseases, cancer, heart attack, and so on [1-2]. Conventional biological assays, such as protein immunoblot, lateral flow immunochromatographic assay, and enzyme-linked immunosorbent assay (ELISA), generally require expensive and bulky equipment [3-4]. With the advances in microelectromechanical systems (MEMS) technologies, miniaturizing biological assays towards point-of-care (POC) diagnostics has become a mainstream research. Yoon et al. proposed a microcalorimetric sensor for Neisseria meningitides detection. High sensitivity and accuracy were achieved by integrating a split-flow microchannel and highly thermal resistivity layer with the microcalorimeter [5]. Park et al. proposed a hand-operated portable immunosensing device with latch mechanism. Multiple sample solutions could be driven and controlled without any external equipment. [6]. Skucha et al. proposed a design methodology for on-chip magnetic bead label detectors based on Hall-effect sensors. The proposed detection platforms are suitable for high accuracy point-of-care assays [7]. Chang et al. developed a capacitive sensor array with integrated microcoils for manipulation of magnetic beads. The device can be used for building large DNA arrays and immunoassays to provide multiple-sample detection [8]. In [9], Shim presented a disposable biosensor for biochemical clinical analysis by implemented an interdigitated electrode array on a polymer substrate. The device integrated with a microfluidic channel was demonstrated for the electrochemical detection of poly-aminophenol. In order to apply these biological assay devices for POC applications, it is extremely favorable that the devices are of low cost and of disposability. Therefore, the inexpensive fabrication process is essential for implementing a fully integrated device.

In the work, we present a capacitive immunosensor using on-chip electrolytic pumping and magnetic washing techniques. Compared with other microfluidic pumping mechanisms, an electrolysis bubble pumping mechanism requires a relatively simple structure with large pumping capability [10-11]. The proposed device, which is integrated with a CMOS capacitance sensing chip, is designed for detecting N-Terminal pro-brain-Type natriuretic peptide (NT-proBNP). The device possesses the advantages such as simple operation, low power consumption, and portability. In addition, the device is suitable for mass-production by using typical fabrication process.

DESIGN AND PRINCIPLE

Figure 1(a) shows the schematic of the proposed immunosensor. The device consists of an electrolytic reservoir and a sensing reservoir. The reservoirs were created by patterning a SU-8 layer on a silicon substrate. The solution dispersed with magnetic beads, which were coated with biotinylated antibodies, was filled in the electrolytic reservoir before the reservoir was bonded with an anodic aluminum oxide (AAO) membrane. The hydrophilicity of the AAO membrane causes wetting and spreading of sample solution into the reservoir [12]. In addition, the pores of the AAO membrane are sufficiently small to prevent blood cells from diffusing into the reservoir.

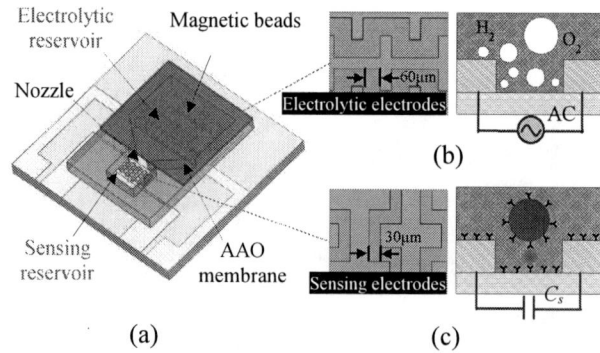

Figure 1: (a) Schematic of the proposed immunosensor. (b) Operational principle of the electrolytic fluid driving. (c) Schematic of detecting NT-proBNP by using magnetic beads

Figure 1(b) shows the operational principle of the electrolytic fluid pumping, in which electrolysis-induced oxygen/hydrogen gas is generated to push fluid. When the electrical signal is applied, electrolysis of water generates hydrogen gas at the cathode and oxygen gas at the anode

[10-11]. As a result, the solution of magnetic beads is pushed forward, and flows through the nozzle. Figure 1(c) shows the schematic of NT-proBNP detection. The surfaces of sensing electrodes are functionalized with capture antibodies that specifically bind to NT-proBNP. Magnetic beads bound with NT-proBNP will be bound to the sensing electrodes via the capture antibodies, which in turn changes the dielectric property around sensing electrodes, and can be capacitively detected.

The operation procedure of the proposed device is shown in Figure 2. First, a sample solution with NT-proBNP is dropped on the AAO membrane, and flows through the AAO membrane into the electrolytic reservoir (Figure 2(a)). Then, the magnetic beads in the reservoir will bind with the NT-proBNP in the sample solution. By applying ac voltage to the electrolytic electrodes, the liquid with the magnetic beads will be pumped to the sensing reservoir by the gas induced by electrolysis (Figure 2(b)). The magnetic beads can be specifically bound on the sensing electrodes via the capture antibody. Subsequently, the magnetic washing was performed to remove the unbound magnetic beads by using a magnetic coil (Figure 2(c)). Finally, the concentration of NT-proBNP can be detected by an integrated COMS capacitance sensing chip, as shown in Figure 2(d).

FABRICATION

Figure 3 illustrates the fabrication process for the proposed device. As shown in Figure 3(a) to Figure 3(c), the electrodes (electrolytic electrodes and sensing electrodes), which consist of a 200Å chromium (Cr) film and a 3000Å gold (Au) film, were fabricated on a silicon substrate by micromachining techniques. The chromium film served as an adhesion layer, and the gold film was the main material of the electrodes. Then, the SU-8 photoresist (SU-8 2050, MicroChem) was spin-coated and patterned for creating the structures of the electrolytic reservoir and the sensing reservoir of the device, as shown in Figure 3(d). In Figure 3(e), the capture antibodies were coated on the surface of the sensing electrode. The magnetic beads (M-PVA SAV1, Chemagen), which were coated with biotinylated antibodies, were filled in the electrolytic reservoir by using a micropipette, as shown in Figure 3(f). The typical diameters of the magnetic beads are about 0.5-1.0μm. Finally, an AAO membrane (Anodisc 13, Whatman) was bonded to the top of SU-8 photoresist to cover the electrolytic reservoir, using silicon-based adhesive (Figure 3(g)). Note that the AAO membrane was trimmed to fit the dimension of the chip by using a dicing saw machine. The typical diameter of the pores of the AAO membrane is about 0.2μm. The thickness of the AAO membrane is 60μm.

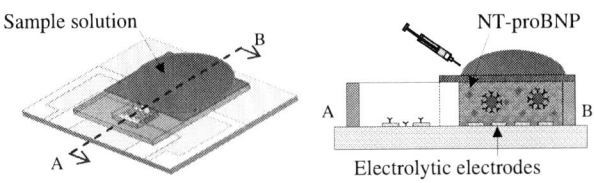

(a) The sample solution is dropped on AAO membrane

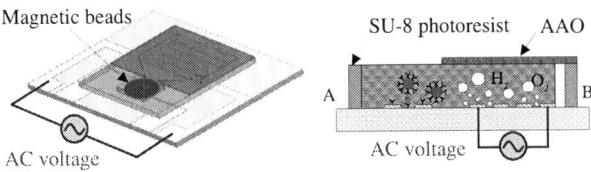

(b) Applying AC voltage to the electrolytic electrodes

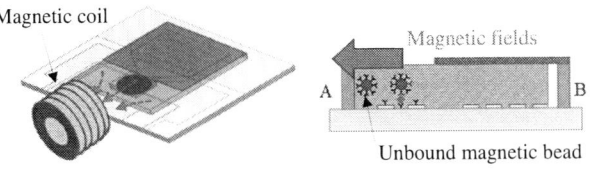

(c) Magnetic washing is performed by magnetic coil

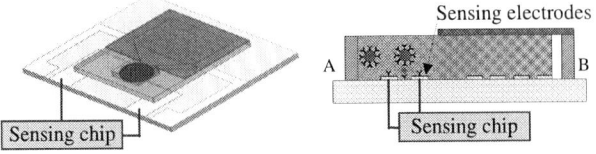

(d) Detection by a CMOS capacitance sensing chip

Figure 2: The operation procedure of the proposed immunosensor.

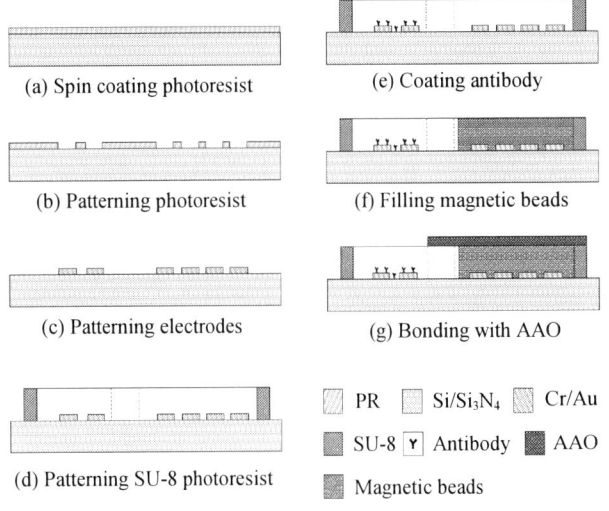

Figure 3: The fabrication processes of the proposed device.

Figure 4(a) and Figure 4(b) show the top and side views of the fabricated device. The size of the fabricated device is 15mm×14mm. The dimension of the sensing reservoir is 2mm×2mm. The width of the nozzle of the electrolytic reservoir is 700μm. The thickness of the fabricated SU-8 photoresist structure is 400μm. The inset in Figure 4(a) is the picture of the sensing electrodes. The gap between the electrolytic electrode pair is 60μm, and the gap between the sensing electrode pair is 30μm. The total thickness of the device is about 800μm.

978-1-4799-3510-9/14 $31.00 © 2014 IEEE

(a)   (b)

Figure 4: The pictures of (a) the top view and (b) the side view of the fabricated immunosensor.

MEASUREMENT AND DISCUSSION

Figure 5(a) shows the device integrated with the capacitance sensing chip fabricated by the TSMC (Taiwan Semiconductor Manufacturing Company) 0.35μm CMOS process. The immunosensor chip, which is designed to be disposable and low cost, can be connected to the CMOS sensing chip via connectors. The sensing chip was designed for low power consumption so that it can be powered by battery. The inset of Figure 5(a) shows the picture of the capacitance sensing CMOS chip. The size of the chip is 3mm×3mm. Figure 5(b) is the schematic of C-F (capacitance-frequency) converter of the capacitance sensing IC. The C-F converter, which consists of a sensing clock loop (SCL), a dynamic comparator and a delay stage, uses digital counting approach to reduce the power consumption [13].

The CCD images that present the sequences of pumping magnetic beads to the sensing reservoir are shown in Figure 6. Figure 6(a) shows the sensing reservoir before electrolytic bubble-activated pumping. The electrolytic reservoir was originally full of the sample solution. When an ac electrical signal of 3V was applied to the electrolytic electrodes, electrolysis bubbles were generated, as shown in Figure 6(b). Because the top side of the electrolytic reservoir was sealed, the bubble pressure was exerted on the sample solution, pushing the solution toward the nozzle into the sensing reservoir. Finally, the sensing reservoir was filled with the sample solution (Figure 6(c)), and the enzyme-linked immunosorbent assay was performed.

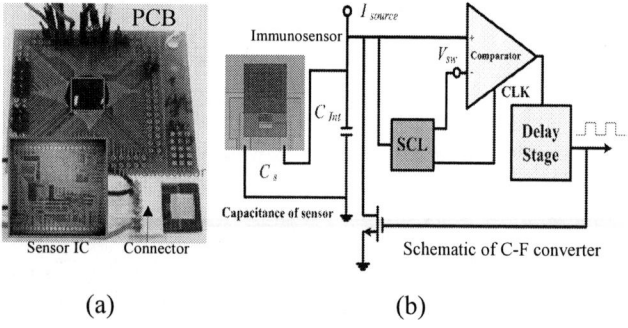

(a)   (b)

Figure 5: (a) The picture of the device integrated with the capacitance sensing chip. (b) The schematic of C-F converter of the sensor IC.

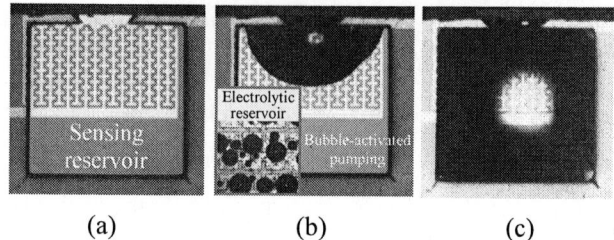

(a)   (b)   (c)

Figure 6: Sequences of pumping the magnetic beads to the sensing reservoir.

Figure 7 presents the pictures of the sensing electrodes during magnetic washing. A magnetic coil (200 turns) is placed beside the sensing reservoir at a distance of about 3mm. The coil was used to generate magnetic fields ($\vec{H}$) to attract the magnetic beads in the sample solution, as shown in Figure 7(b). The diameter of the magnetic coil is 5 mm, and it was wound using an insulated copper wire of 0.25 mm in diameter.

After the sample solution has been filled on the surface of the sensing electrode, magnetic bead labels in the solution bound to the surface via the capture antibodies. Then, the magnetic coil generated a magnetic field for removing all the nonspecifically bound magnetic beads away from the area of sensing electrodes.

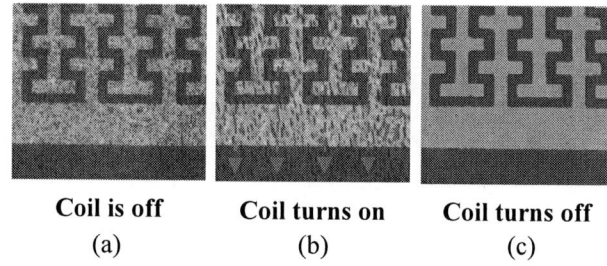

Coil is off   Coil turns on   Coil turns off
(a)   (b)   (c)

Figure 7: Pictures of the sensing electrodes during magnetic washing.

Figure 8 shows the measured capacitances vs. concentrations of magnetic beads. Each data point is the average value of 5-time measurements. The error bars indicate the maximum and minimum values. This figure shows the sensing range of the concentration of the magnetic beads is from 0.01 to 10 mg/ml. The results show that the capacitance of the device increases as the concentration of magnetic beads increases. The inset of Figure 8 shows the measured results of magnetic bead concentration from 0.01 to 0.1 mg/ml.

Figure 9 shows the measured frequencies vs. NT-proBNP concentrations. The frequencies were measured using the capacitance sensing chip with a C-F converter. As shown in the inset of Figure 9, the output frequency of the C-F converter is 706.4 Hz for the NT-proBNP concentration of 1000 pg/ml. The measured results indicate the detectable NT-proBNP concentration by using the proposed immunosensor is ranged from 10 pg/ml to 1000 pg/ml. The measured frequencies decrease as the concentrations of the NT-proBNP increase.

978-1-4799-3510-9/14 $31.00 © 2014 IEEE   811

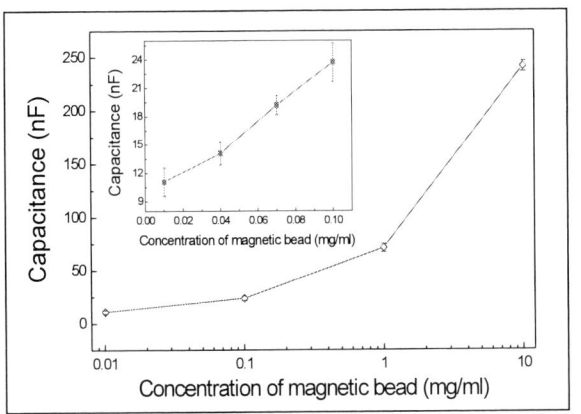

Figure 8: Measured capacitances vs. concentrations of magnetic beads.

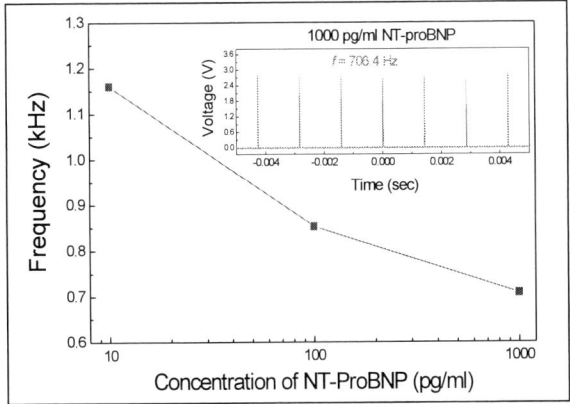

Figure 9: Measured frequencies vs. NT-proBNP concentrations.

CONCLUSION

A capacitive immunosensor using on-chip electrolytic pumping and magnetic washing techniques was presented. The advantages of the proposed device include simple operation, low power consumption, and portability. The proposed device can easily be fabricated by using micromachining process. The detection of N-Terminal pro-brain-Type natriuretic peptide (NT-proBNP) demonstrated by using the fabricated immunosensor integrated with a CMOS capacitance sensing chip. The measured results indicate the detectable NT-proBNP concentration by using the proposed immunosensor is ranged from 10 pg/ml to 1000 pg/ml. The proposed immunosensor potentially can be used for point-of-care applications.

ACKNOWLEDGEMENT

This project was supported in part by the National Science Council, Taiwan. (Contract number: NSC 100-2221-E-002 -075 -MY3).

REFERENCES

[1] J. Choi, K. Oh, J. Thomas, W. Heineman, H. Halsall, J. Nevin, A. Helmicki, H. Hendersona, and C. Ahn, "An integrated microfluidic biochemical detection system for protein analysis with magnetic bead-based sampling capabilities," *Lab Chip*, vol. 2, pp. 27-30, 2002.

[2] F. Myers and L. Lee, "Innovations in optical microfluidic technologies for point-of-care diagnostics," *Lab Chip*, vol. 8, pp. 2015-2031, 2008.

[3] S. Piletsky, E. Piletska, A. Bossi, K. Karim, P. Lowe, A. Turner, "Substitution of antibodies and receptors with molecularly imprinted polymers in enzyme-linked and fluorescent assays," *Biosens Bioelectron*, vol. 16, pp. 701-707, 2001.

[4] D. Lee and T. Cui, "An electric detection of immunoglobulin G in the enzyme-linked immunosorbent assay using an indium oxide nanoparticle ion-sensitive field-effect transistor," *J. Micromech. Microeng.*, vol. 22, pp. 1-9, 2012.

[5] S. Yoon, M.-H. Lim, S.-C. Park, J.-S. Shin, and Y.-J. Kim, "Neisseria meningitidis detection based on a microcalorimetric biosensor with a split-flow microchannel," *J. Microelectromech. Syst.*, vol. 17, pp. 590-598, 2008.

[6] S. Park, J. Lee, K. Kim, H. Yoon, and S. Yang, "An electrochemical immunosensing lab-on-a-chip integrated with latch mechanism for hand operation," *J. Micromech. Microeng.*, vol. 19, pp. 1-11, 2009.

[7] K. Skucha, S. Gambini, P. Liu, M. Megens, J. Kim, and B. Boser, "Design considerations for CMOS- integrated Hall-effect magnetic bead detectors for biosensor applications," *J. Microelectromech. Syst.*, 2259615, 2013.

[8] A.-Y. Chang and M.-C. Lu, "A CMOS magnetic microbead-based capacitive biosensor array with on-chip electromagnetic manipulation," *Biosens Bioelectron*, vol. 45, pp. 6-12, 2013.

[9] J. Shim, M. Rust, and C. Ahn, "A large area nano-gap interdigitated electrode array on a polymer substrate as a disposable nano-biosensor," *J. Micromech. Microeng.*, vol. 23, pp. 1-6, 2013.

[10] C.-M. Cheng and C.-H. Liu, "An Electrolysis-Bubble-Actuated Micropump Based on the Roughness Gradient Design of Hydrophobic Surface," *J. Microelectromech. Syst.*, vol. 16, pp. 1095-1105, 2007.

[11] S.-C. Chan, C.-R. Chen, and C.-H. Liu, "A bubble-activated micropump with high-frequency flow reversal," *Sens. Actuators A, Phys.*, vol. 163, pp. 501-509, 2010.

[12] M. Lei, A. Baldi, E. Nuxoll, R. Siegel, and B. Ziaie, "Hydrogel-based microsensors for wireless chemical monitoring," *Biomed. Microdevices*, vol. 11, no. 3, pp. 529-538, 2009.

[13] C.-H. Chen, R.-Z. Hwang, L.-S. Huang, S. Lin, H.-C. Chen, Y.-C. Yang, Y.-T. Lin, S.-A. Yu, Y.-H. Wang, N.-K. Chou, and S.-S. Lu, "A wireless bio-MEMS sensor for C-reactive protein detection based on nanomechanics," *IEEE Trans. Biomed. Eng.*, vol. 56, pp. 462–470, 2009.

CONTACT

*Y.-J. Yang, tel: +886-2-33664941#16; yjy@ntu.edu.tw

A WIRELESS SLANTED OPTRODE ARRAY WITH INTEGRATED MICRO LEDS FOR OPTOGENETICS

Ki Yong Kwon[1], Hyung-Min Lee[2], Maysam Ghovanloo[2], Arthur Weber[1], and Wen Li[1]
[1]Michigan State University, East Lansing, USA
[2]Georgia Institute of Technology, Atlanta, USA

ABSTRACT

This paper presents a wireless-enabled, flexible optrode array with multichannel micro light-emitting diodes (μ-LEDs) for bi-directional wireless neural interface. The array integrates wirelessly addressable μ-LED chips with a slanted polymer optrode array for precise light delivery and neural recording at multiple cortical layers simultaneously. A droplet backside exposure (DBE) method was developed to monolithically fabricate varying-length optrodes on a single polymer platform. *In vivo* tests in rat brains demonstrated that the μ-LEDs were inductively powered and controlled using a wireless switched-capacitor stimulator (SCS), and light-induced neural activity was recorded with the optrode array concurrently.

INTRODUCTION

Optogenetics has demonstrated the ability to optically control specific types of genetically modified neurons in the brain [1]. Optical neural stimulation based on optogenetics is considered to be more beneficial than electrical neural stimulation, because it permits activation or inhibition of specific types of neurons with sub-millisecond temporal precision, eliminates electrical artifacts, and potentially has a prolonged lifetime due to hermetic sealing of light sources [2]. Current approaches to optogenetics-based neural interfaces include laser- or light-emitting diode (LED)-coupled optical fiber, micro LED (μ-LEDs) arrays, and focused laser beam through a microscope [3-5]. However, for experiments with freely behaving subjects, only a few methods of light delivery are available, such as a laser-coupled optical fiber and a head-mountable single LED system. The spatial resolution of the existing systems is poor and the tethered optical fiber restricts natural behavior of the subjects.

To address these limitations, we developed a multichannel 3-D waveguide array that integrates μ-LED chips with microneedle waveguides to minimize light scattering in the tissue and achieve high spatial resolution [6]. Further improvement of the spatial resolution of optical stimulation in depth was achieved using varying-length waveguides that precisely deliver light to targeted cortical layers. The optical, electrical, and mechanical properties of these waveguide arrays were investigated to demonstrate the capability of the devices for optical neural modulation [7].

Building on our previous studies, in this paper, we present a wireless slanted optrode array with integrated μ-LEDs, which is capable of simultaneous light stimulation and electrical neural recording to achieve a truly untethered bi-directional neural interface.

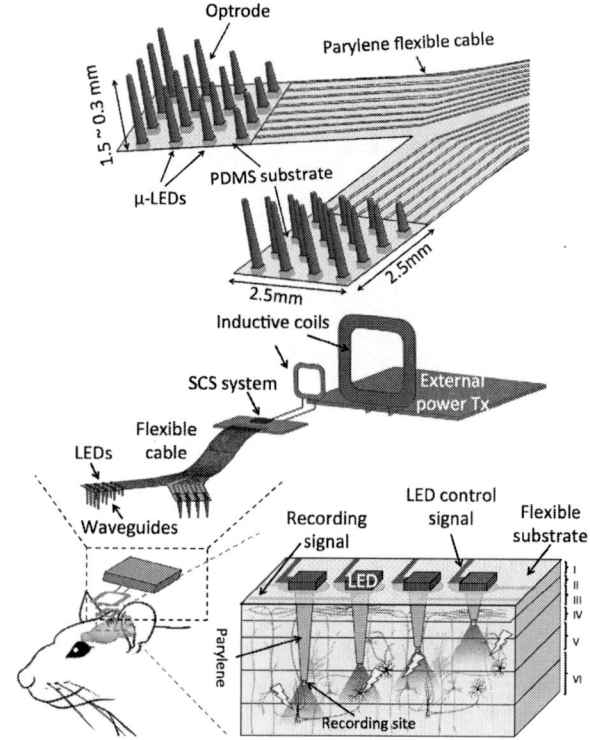

Figure 1: Concept diagram of wireless optrode array with integrated μ-LEDs.

Recording electrodes wrapped around the waveguide cores permit simultaneous measurement of neural responses from various depths upon optical excitation. Fig. 1 depicts the concept and design of the wirelessly powered-slanted optrode array, which contains 32 embedded μ-LED light sources on a 2.5×2.5mm² flexible substrate, with 4×4 channels per each hemisphere to cover both visual cortices in rats. Integrated LED light sources powered by the wireless switched-capacitor stimulator (SCS) enable a truly untethered system [8].

The slanted microneedle array was fabricated separately on a PDMS substrate and integrated with the multi-LED array using shape-matching assembly. A droplet backside exposure (DBE) method [9] was used to form a slanted microneedle structure, in which the lengths (300–1500μm) of individual microneedles were determined by the height variance along the droplet curvature. A multilayer metal/Parylene shell was fabricated on the waveguide core to prevent photoelectrical artifact and to improve biocompatibility.

In vivo acute experiments were conducted to demonstrate optical modulation of neural activity with the integrated LED array powered by the wireless SCS system.

978-1-4799-3510-9/14 $31.00 © 2014 IEEE

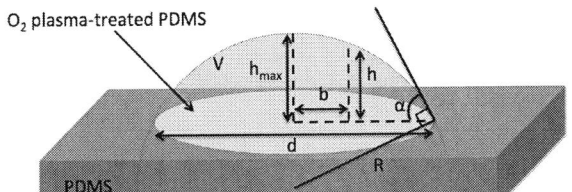

Figure 2: Geometry of ideal partial sphere droplet

METHODS

Droplet Backside Exposure (DBE)

In our previous work, a polymer based fabrication method (DBE) was developed, utilizing the height variance in a dome-shaped SU-8 structure to create out-of-plane microneedles with various lengths. This method can independently control the length, and tip/bottom diameters of individual microneedles without using a specialized machine or complex fabrication techniques [9]. Controlling these parameters is critical in waveguide application, because these parameters determine the coupling efficiency, irradiance and total flux of the needle-shaped waveguide [9].

In this method, the SU-8 droplet with a pre-defined base size was confined on O_2 plasma-treated hydrophilic PDMS surface with hydrophobic intact PDMS surroundings, owing to surface energy differences between two adjacent contact areas. Based on this principle, we designed an analytical model to determine the maximum volume of the SU-8 droplet sustained in a pre-defined pattern as well as to predict the length variation of the microneedles.

To simplify the model, the SU-8 droplet is assumed to be a perfect partial sphere as shown in Fig. 2. The volume of the droplet can be expressed as a function of α, the equilibrium contact angle of the droplet, and d, a diameter of the pre-defined O_2 plasma treated PDMS surface.

$$V = \frac{1}{12}\pi d^3 \frac{(2+\cos\alpha)(1-\cos\alpha)^2}{\sin^3\alpha} \quad (1)$$

Once, h_{max}, the maximum height of the droplet, is estimated from the applied volume, a parameter h, which is a height at the distance b from the center of the droplet, can be calculated using

$$h = \frac{\sqrt{\left(\frac{d}{2}\right)^4 + h_{Max}^2\left(2\left(\frac{d}{2}\right)^2 - 4b^2 + 1\right)} - \left(\frac{d}{2}\right)^2 + h_{Max}^2}{2h_{Max}} \quad (2)$$

With fixed V, three parameters, tip diameter, bottom diameter, and the length of the microneedle, can be controlled by the mask diameter and the distance from the center of the droplet of the mask aperture. Details of the DBE method are described in [9].

Thin Film Sandwich Structure

The optrode enables delivering light into the target area as well as recording neural activity simultaneously. The basic structure of the optrode consists of a SU-8 microneedle waveguide core covered by metallic cladding, which is used as a recording electrode and as a shielding layer to prevent light-leakage from the sidewall.

However, it was found that, if the metal layer is used for neural recording directly, light-induced artifacts will occur, which require post processing to recover the original neural data. This phenomenon is consistent with a classical photoelectrochemical finding, the Becquerel effect. To minimize the light-induced artifacts, four layers of oxide-polymer-metal-polymer sandwich structure were designed. A multilayer metal/Parylene shell was fabricated on the outer sidewall of the waveguide core, which contained an indium tin oxide (ITO) shielding layer to prevent photoelectrical artifact, an opaque metal layer to block light side-leakage and record neural activity, and Parylene as insulation between the ITO and metal, and as an encapsulation for improved biocompatibility. Opaque metal was removed from the tip of the waveguide to allow light delivery to the brain neurons. A profile of the sandwich structure is shown in Fig. 4 (b).

Switched-Capacitor Stimulation (SCS)

Fig. 3 shows the simplified block diagram of the wireless switched-capacitor stimulating (SCS) system for power-efficient optogenetics. Conventional inductively-powered current stimulators require a rectifier, a regulator, and an array of current sources to generate stimulation pulses from V_{COIL}, while power losses at each stage result in poor overall stimulator efficiency [10]. On the contrary, the proposed SCS system efficiently charges a bank of storage capacitors, C_{ST}, directly from V_{COIL} through the series capacitor, C_S, and the inductive charger without using any rectifiers and regulators, improving the capacitor charging efficiency [8]. Moreover, the charge stored in capacitors is delivered to the load through switches, creating stimulation pulses efficiently.

The SCS system is also capable of providing high instantaneous current, which is limited in conventional inductively-powered devices. Therefore, the wireless SCS system has been utilized for power-efficient optogenetics by periodically discharging the storage capacitors into the micro-LED arrays, which require high instantaneous power to emit sufficient light and evoke neural activity. The forward data telemetry also utilizes the same inductive link to wirelessly set the stimulation parameters in IMDs, so the proposed SCS system can be further used for freely behaving animal experiments as a chronic implant.

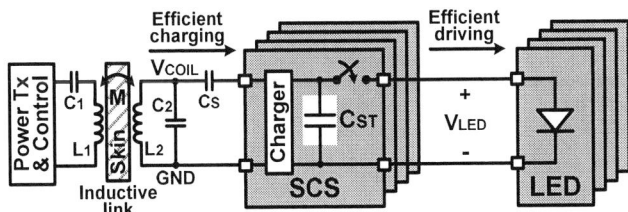

Figure 3: Simplified block diagram of the wireless switched-capacitor stimulating (SCS) system to efficiently drive the μ-LED array

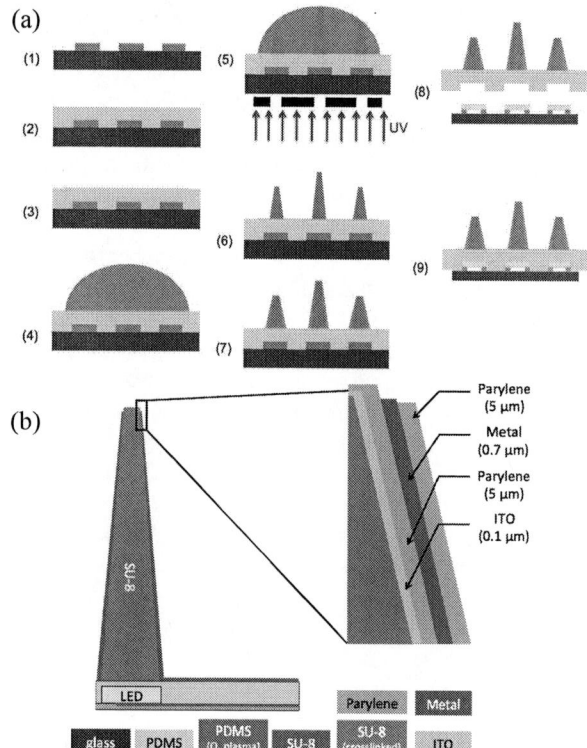

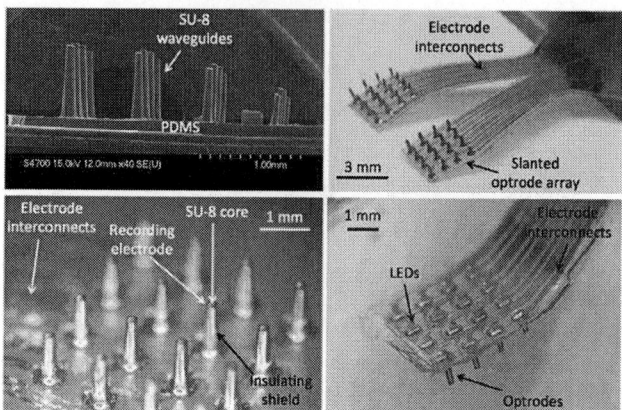

Figure 3: (a) Fabrication process flow for making the slanted optrode array with integrated μ-LEDs and (b) the profile of the optrode.

Figure 4: Fabricated prototypes of (a) SEM images of a slanted microneedle structure, (b) 32-channel optrode array, (c) microscopic image of individual optrode, and (d) optrode array coupled with μ-LEDs

FABRICATION

In order to reduce fabrication complexity, the multi-LED array and the slanted optrode array were fabricated and calibrated separately. Then the system was assembled by polymer bonding of individual components with SU-8.

Detailed fabrication was divided into two steps: flexible multi-LED array and slanted optrode array fabrication. Details of flexible multi-LED array fabrication process is described in [11], so, here we only focus on the fabrication process of the slanted optrode array (Fig. 4 (a)).

Slanted microneedle array fabrication and assembly

(1) A 3" glass wafer was cleaned and went through a dehydration bake, and an ~ 50μm SU-8 layer was spun onto the wafer and patterned as the mock LEDs. (2) A thin layer of PDMS was spun on the SU-8 master to make cavities that matched the shape of the LED. (3) After the PDMS was cured for 40min at 95°C, photoresist (PR) was patterned on the PDMS substrate to expose 7mm-diameter circles, followed by oxygen plasma treatment to convert the exposed hydrophobic areas to hydrophilic ones. (4) After removal of the PR mask, ~45μL SU-8 (SU-8 3005) was dispensed on top of the plasma treated PDMS surface using a micropipette and (5) patterned with the backside exposure to form the microneedles. (6) After SU-8 development, the array was polished by O_2 plasma etching. (7) DC sputtering of a 0.1μm thick ITO layer was performed in a Kurt Lesker Axiss PVD System, followed by deposition of 5μm Parylene-C in a

chemical vapor deposition (CVD) system (PDS2010 Parylene Coater, Specialty Coating System) and then 1μm thick Au layer in a thermal evaporator (Edward Auto306). Opaque metal was removed using wet etching from the tip of the waveguide to allow light delivery to the brain neurons and 5μm Parylene-C was deposited as a protection layer. (8) The Parylene-C film at the tip of the optrode was removed using reactive-ion etching (RIE), and the membrane was released from the glass wafer. (9) Finally, the microneedle array with the matched LED cavities was aligned onto the corresponding LED chips and bonded with polymer adhesive.

RESULTS

Fig. 5 shows images of the prototypes of slanted optrode arrays. The SEM image of a length-varying SU-8 microneedle structure (300-800μm) shown in Fig. 5(a) demonstrates the capability of the DBE method for making varying-length microneedles. The electrode-electrolyte interface impedances of the optrode array were measured at 1kHz using a built-in electrode-impedance-testing circuitry in an Intan evaluation board (RHD2132 and RHD2000-EVAL, Intan Tech. LLC). The impedance of the optrode array ranged from 10 to 500kΩ, which is suitable for local field potential recordings. In this study, the impedance of the optrode was controlled by size of the Parylene opening.

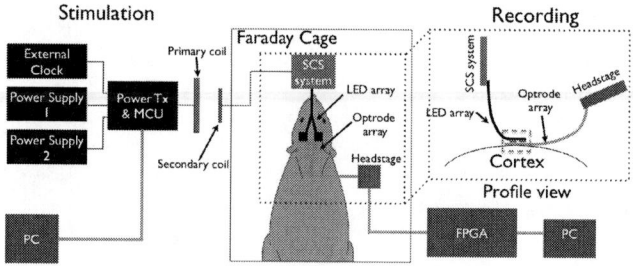

Figure 5: Schematic of in vivo animal experiment set-up with the wireless SCS system.

978-1-4799-3510-9/14 $31.00 © 2014 IEEE

In vivo animal experiments

In vivo acute animal experiments were performed to demonstrate the ability of the slanted optrode array to simultaneously modulate and record neural activities in the visual cortex (V1) of a rat. All procedures were approved by the Institutional Animal Care and Use Committee (IACUC) at Michigan State University. Prior to the acute experiments, rodent subjects (Sprague-Dawley rats: 250-400g) had neurons in their V1 transfected with channelrhodopsin-2 (ChR2) to enable light sensitivity.

As shown in Fig.6, the SCS system was used to power and control the 32-channels LED array and a 32-channels headstage (RHD2132, intan technologies) was used for simultaneous extracellular recording. A clear light-induced neural activity was observed in LFP (1-500Hz) in time domain driven by a 3.2V input (Fig.7 (a)). The 3.2V input voltage resulted in an average irradiance of 1.4mW/mm^2 at the tip of the optrode, which is above the minimal irradiance (1mW/mm^2) required for light-evoked neural response. We also tried lowering the input voltage to 2.7V (resulting the average irradiance of 0.35mW/mm^2), but no significant neural modulation was observed. To visualize neural oscillations induced by the optrode array clearly, we measured instantaneous phase of the two datasets (2.7V and 3.2V input voltages) of light-evoked LFP (1-25Hz) based on Hilbert Transform. The instantaneous phases of each trial (1s with 100ms LED on), labeled with different colors according to Fig. 7 (c), were aligned based on the stimulus ON time and stacked. With the 3.2V input voltage, the light-evoked neural activity showed clear phase synchronization, while no phase synchrony was observed in neural recordings following the 2.7V input voltage.

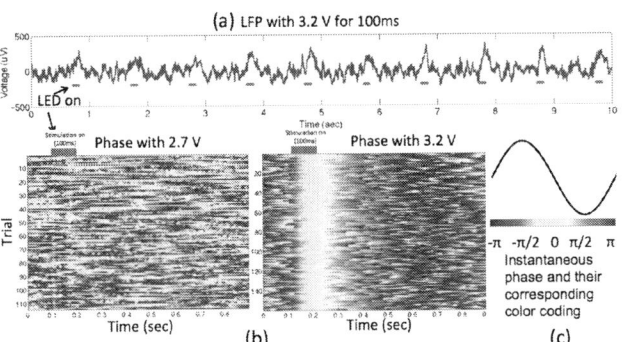

Figure 6: Light-induced LEP recording driven by 2.7V and 3.2V for 100ms (a) in time domain, (b) instantaneous phase in low frequency band (1-25Hz), and (c) their corresponding color coding.

CONCLUSION

We have presented the design and fabrication processes of a wireless slanted optrode array for a truly untethered bi-directional neural interface. The array is inductively powered and controlled by the wireless SCS system, which is designed to improve power efficiency. *In vivo* acute animal experiments successfully demonstrated optical neural modulation and simultaneous extracellular recording with the wireless SCS system. We are working on developing a chronic version for seamless, bi-directional, neural interface for freely behaving animals.

ACKNOWLEDGEMENTS

This work was supported in part by the Electrical, Communications and Cyber Systems and Chemical, Bioengineering, Environmental (ECCS), and Transport Systems (CBET) Division of the National Science Foundation under the Award Number ECCS-1055269 and CBET-1264772 respectively.

REFERENCES

[1] F. Zhang, A. Aravanis, A. Adamantidis, L. de Lecea, and K. Deisseroth, "Circuit-breakers: optical technologies for probing neural signals and systems," *Nature Reviews Neuroscience,* vol. 8, no. 8, pp. 577–581, 2007.

[2] V. Gilja, V. Gilja, C. A. Chestek, C. A. Chestek, I. Diester, I. Diester, J. M. Henderson, J. M. Henderson, K. Deisseroth, K. Deisseroth, K. V. Shenoy, and K. V. Shenoy, "Challenges and Opportunities for Next-Generation Intracortically Based Neural Prostheses," *IEEE Transactions on Biomedical engineering*, vol. 58, no. 7, pp. 1891–1899.

[3] J. Wang, F. Wagner, D. A. Borton, J. Zhang, I. Ozden, R. D. Burwell, A. V. Nurmikko, R. van Wagenen, I. Diester, and K. Deisseroth, "Integrated device for combined optical neuromodulation and electrical recording for chronic in vivoapplications," *Journal of Neural Engineering*, vol. 9, no. 1, p. 016001, Dec. 2011.

[4] A. Zorzos and E. Boyden, "Multiwaveguide implantable probe for light delivery to sets of distributed brain targets," *Optics letters*, vol. 35, no.24, pp. 4133-4135, 2010.

[5] B. McGovern, R. B. Palmini, N. Grossman, E. Drakakis, V. Poher, M. Neil, and P. Degenaar, "A new individually addressable micro-LED array for photogenetic neural stimulation," *Biomedical Circuits and Systems, IEEE Transactions on*, vol. 4, no. 6, pp. 469–476, 2010.

[6] K. Kwon and W. Li, "Integrated multi-LED array with three-dimensional polymer waveguide for optogenetics," *Micro Electro Mechanical Systems, IEEE 26th International Conference on*, Jan. 2013.

[7] K. Kwon and W. Li, "Integrated slanted microneedle-LED array for optogenetics," *Medicine and Biology Society, IEEE 35th Annual International Conference on*, Jun. 2013.

[8] H.-M. Lee and M. Ghovanloo, "A power-efficient wireless capacitor charging system through an inductive link," *IEEE Trans. Circuits Syst. II*, vol. 60, no. 10, pp. 707-711, Oct. 2013.

[9] K. Kwon, B. Xiaopeng, and W. Li, "Droplet Backside Exposure for Making Slanted SU-8 Microneedles," *Medicine and Biology Society, IEEE 35th Annual International Conference on*, Jun. 2013.

[10] H.-M. Lee, H. Park, and M. Ghovanloo, "A power-efficient wireless system with adaptive supply control for deep brain stimulation," *IEEE J. Solid-State Circuits*, vol. 48, no. 9, pp. 2203-2216, Sep. 2013..

[11] K. Kwon, B. Sirowatka, A. J. Weber, and W. Li, "Opto-µECoG Array: A Hybrid Neural Interface with Transparent µECoG Electrode Array and Integrated LEDs for Optogenetics," *Biomedical Circuits and Systems, IEEE Transactions on*, vol.7, no.5, pp.593-600, Oct. 2013.

AN ELECTROPORATION CHIP BASED ON FLEXIBLE MICRONEEDLE ARRAY FOR IN VIVO NUCLEIC ACID DELIVERY

Zewen Wei[1]*, Renxing Wang[2]*, Shuquan Zheng[3]*, Zicai Liang[3] and Zhihong Li[4]

[1] National Center for Nanoscience and Technology, Beijing, China

[2] Science and Technology on Electronic Test&Measurement Laboratory, Key Laboratory of Instrumentation Science and Dynamic Measurement, North University of China, Taiyuan, China

[3] Institute of Molecular Medicine, Peking University, Beijing, China

[4] National Key Laboratory of Science and Technology on Micro/Nano Fabrication, Institute of Microelectronics, Peking University, Beijing, China

* Contributed Equally

ABSTRACT

This paper reports a flexible microneedle array (MNA) electroporation chip for *in vivo* nucleic acid delivery, which is of great importance for gene therapy. Silicon MNA is proposed to penetrate the high-resistant stratum corneum, while a flexible parylene substrate is used to fit the natural shape of electroporated objects. The chip provides a sufficient electrical field beneath the skin for electroporation with low voltage, which is less likely to harm tissues. Using the proposed chip, we successfully achieved plasmid DNA expression and siRNA delivery in living tissue with low voltage (30-40V). Neither physical nor biological harm to skin was observed.

INTRODUCTION

Nucleic acid delivery has been proved to be a key issue for many frontier biological applications, such as gene therapy and DNA vaccines. Compared with chemical vector or viral based delivery, *in vivo* electroporation is a physical method, showing many advantages, such as less biological contamination and no selectivity of different tissues. As a common used laboratory technique, needle-based *in vivo* electroporation [1] is with some limits, such as invasive penetration and harmful high voltage, which prevent itself from practical human applications. Recently, microdevices were introduced to overcome above limits. The flexible planar microelectrode was reported to reduce electroporation voltage to 60V [2], but the skin should be removed before electroporation. MNA devices have been demonstrated for transdermal drug delivery [3] and *in vitro* gene delivery [4], but *in vivo* nucleic acid delivery by MNA has not been conclusively proved. The presented work provides a tangible evidence of *in vivo* nucleic acid delivery by MNA electroporation and offers a vast potential for applications of practical human applications, since the applied voltage is under the safety limit for human and delivery process is free of harmful invasive operation.

METHODS

Devices Fabrication

The prototype of the proposed MNA chip was shown in Figure 1a. A transparent parylene film was used as substrate for electrodes due to its good flexibility, which provides tight

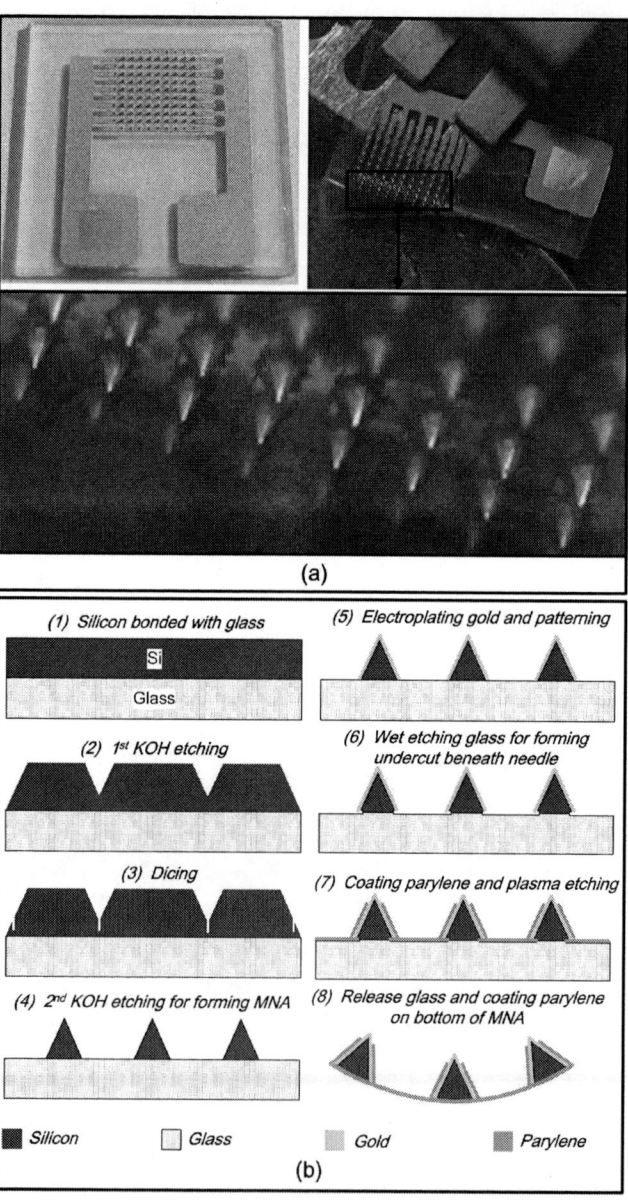

Figure 1: Flexible MNA electroporation chip, (a) photos of proposed chip and close-up view of micro-needle electrodes, (b) fabrication process.

attachment between proposed chip and living tissues with different profiles and sizes. The parylene layer also acts as insulator between electrodes. Silicon was employed as the material of the microneedle due to its good strength. Gold was chosen as the material of electrodes because of its good conductivity and bio-compatibility. Microneedle height was 190μm, with 340μm interval between needles.

The basic fabrication process was previously reported [5](Figure.1b). Briefly, a silicon wafer, bonded with glass, was wet etched twice to form MNA. Silicon dicing was introduced to reduce the interval of needles, therefore decreased the applied voltage for electroporation, and an Au layer was patterned as electrodes. To withstand high electroporation current, electroplating was employed to thickening Au layer to 6μm. A parylene layer (8μm) was then coated on the top side of MNA. After exposing the Au electrode by plasma etching, the flexible MNA was released from glass. Finally, another parylene layer (2μm) was coated on the bottom side of MNA for encapsulation.

Nucleic Acids and Animals

Cy-5 labeled siRNA used for fluorescent imaging and DNA oligonucleotides used for vector construction (pmRFP-C1) were from Invitrogen (Beijing, China). Before use, the siRNA and DNA plasmid were both diluted to 1μg/μl with DI water.

Male C57BL/6 mice (6-8 weeks, weight 18-22g) were form Vital River (Beijing, China). Mice were unhaired 1 day before assays.

Experimental Procedures

In the experiment to electroporate myocytes with DNA plasmid and siRNA, 40μl (1μg/μl) Sodium Hyaluronate was injected into muscle of the mouse thigh after unhairing. After reaction for 30minutes, 40μl RFP plasmid (or siRNA) was directly injected into the mouse thigh muscle. 10 minutes later, the skin of the mouse thigh was covered tightly by proposed MNA chip. The chip was pushed by hand 3 times to make sure that microneedles did penetrate stratum corneum. Then 5 electrical pulses were applied for tissue electroporation. The RFP fluorescence was *in vivo* imaged after 48 hours. The average fluorescence intensities were calculated by Kodak *In Vivo* Imaging System. Presented data are the average of three independent assays. SiRNA delivery shared the same protocols with DNA plasmid delivery.

RESULTS

Skin Penetration

First, we studied the skin penetration of MNA. The Cy-5 labeled siRNA, to which the skin is impermeable, was placed on unhaired leg skin of three individual mice. Then we pushed the MNA into the skin of two mice, while the third mouse was treated with flat glass. 10 minutes later, the MNA was removed and the skin was thoroughly washed. As shown in Figure.2, the fluorescent images illustrate MNA penetration makes the siRNA soak in the skin (Figure.2b), while repeatedly push of MNA makes the skin more permeable to siRNA (Figure.2c). The fluorescent area shows uniform fluorescent intensity, which indicates the flexible MNA chips adapt the tissue profile.

Then the strength of MNA was determined by SEM observation. By comparing the SEM photos of MNA before and after skin penetration and subsequent electroporation, the MNA maintained its profile and strength. For the entire MNA chip, no damage of the microneedle was observed. The gold layer also withstood the high current and kept intact on microneedles. Therefore, the potential biological hazard caused by MNA cracking in tissue could be avoided.

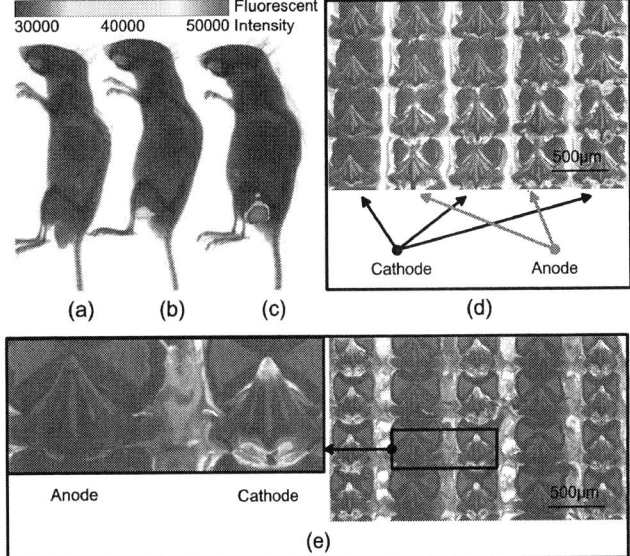

Figure 2: The skin penetration by MNA. After placing a drop of Cy-5 labeled siRNA on unhaired skin of mice leg, the MNA was pushed into skin once(b) or repeatedly(c), while the control mouse(a) was pushed by a flat glass, then both three mice were washed thoroughly. The fluorescent image (b) shows the siRNA sinks into the skin by MNA penetration, and repetitive MNA penetration increases the amounts of siRNA in the skin. The SEM photos of MNA before(d) and after(e) skin penetration and electroporation show no needle damage but small changes on surface appearance of cathode.

Nucleic Acid Delivery

According to MNA geometrical design and skin assays, the MNA should penetrate the skin and produce an electrical field beneath the stratum corneum. To determine the proper voltage to generate sufficient electrical field (300–800V/cm) for electroporation, the electrical field distribution of both MNA and previously reported planer electrodes were simulated by COMSOL ver.3.5a (Figure.3). The simulation indicates that the voltage, required for electroporation of tissue beneath the skin, could be remarkably reduced by penetrating high-resistance stratum corneum. Using MNA chip, 25V would produce a sufficient electrical field in a range of 100μm to 700μm beneath the skin.

Plasmid DNA (pmRFP-C1) delivery was processed to demonstrate the MNA's ability for *in vivo* electroporation. As shown in Figure.4a, left legs of both mice were injected with plasmid DNA. The right mouse, treated with 5 pulses (40V), showed obvious fluorescence, which indicated successful DNA expression in myocytes. In contrary, the left mouse, acting as the negative control, showed no fluorescence.

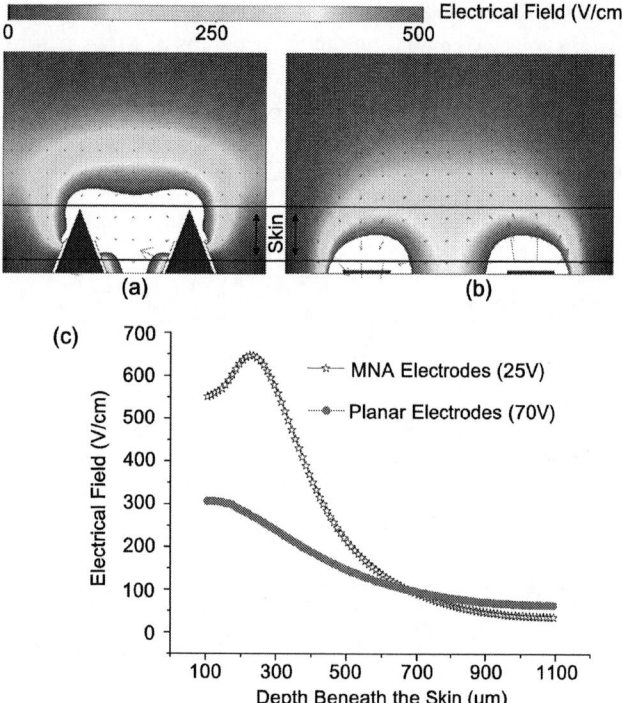

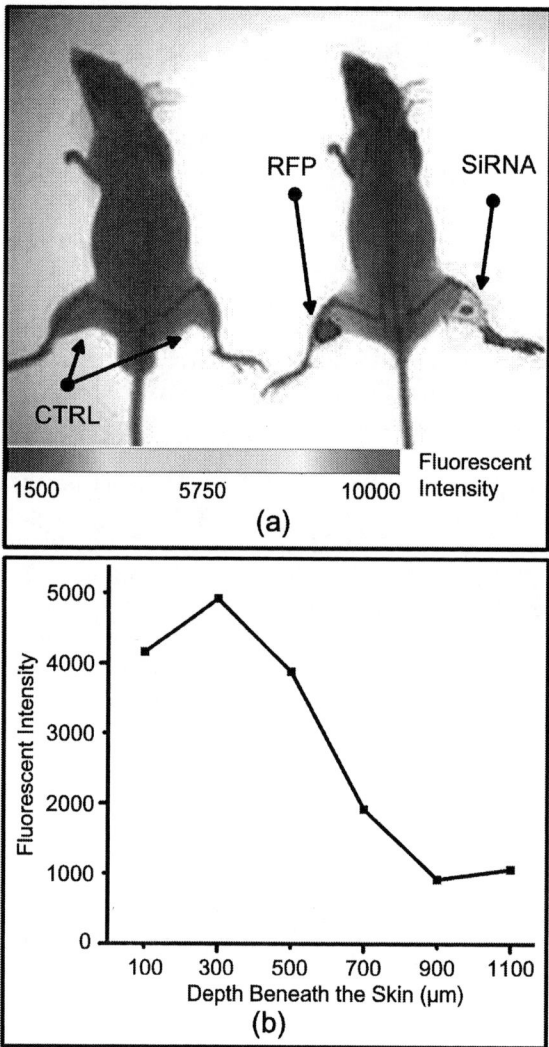

Figure 3: The simulation of electric field generated by MNA electrodes(a) and planar electrodes(b). Compared with planer electrodes, MNA electrodes could produce stronger electrical field beneath the skin with lower voltage. The relationship between depth and electric field (c) indicates the sufficient electrical field for electroporation by MNA could reach 500μm depth beneath the skin.

SiRNA delivery, which is more meaningful for gene therapy, was also demonstrated on the right legs with same protocol of DNA plasmid electroporation. Both mice were injected with same amount of Cy-5 labeled siRNA. For the right mouse, the MNA chip successfully delivered siRNA into myocytes, which prevented siRNA from being removed by blood circulation and metabolizing. Therefore, the right leg of right mouse maintained fluorescent for 48 hours. As contrast, the left mouse, without electroporation, metabolized all injected siRNA in only 2 hours.

For both DNA and siRNA delivery, benefited from non-invasive operation and low voltage, the skin showed no burning and the mouse muscle maintained physical health. It is worthwhile to note that the difference on fluorescent intensity of DNA and siRNA did not represent the difference of DNA and siRNA delivery efficiency. The reason is that Cy-5 dye produces stronger fluorescence than protein does. For further analysis of nucleic acid delivery in different tissue depth, the muscle tissue delivered with siRNA was further frozen sectioned and quantitatively analyzed by confocal fluorescent imaging. The result (Figure.4b) reveals delivery efficiency of different depth beneath the skin fits the simulated electrical field distribution. The effective depth is enough for DNA vaccine or treatment of near-skin tumor.

Figure 4: (a) The fluorescent image of plasmid DNA(RFP) expression and siRNA delivery by flexible electroporation chip. The left leg of right mouse shows a remarkable fluorescent signal, which means successful RFP expression. As a contrast, the control mouse (left one) shows no fluorescence at all. The in vivo electroporation also achieves successful siRNA delivery into the right leg of right mouse. The electrical pulse duration is 10ms and interval between pulses is 1s. Note that the skin of mouse leg was removed before imaging to make sure the fluorescence is from muscle tissue. The difference of fluorescent intensity of RFP and Cy-5 siRNA doesn't mean siRNA delivery is more efficient. The reason is Cy-5 dye produces stronger fluorescence than RFP. (b), the relationship between the depth beneath the skin and the delivery efficiency fits the simulation of electric field distribution.

Low Voltage Electroporation

The main obstacle preventing in vivo electroporation from human application is the unsafe voltage. To determine the optimized electroporation voltage of MNA chip, the relationship between applied voltage and RFP delivery efficiency was studied by quantitatively analyses of fluorescence intensity. As shown in Figure.5, expression efficiency increases with enhancement of applied voltage and reaches its peak value when voltage is 40V. A higher voltage brings no better RFP expression, but damage on skin. For all voltages, the fluorescence maintained uniform distribution.

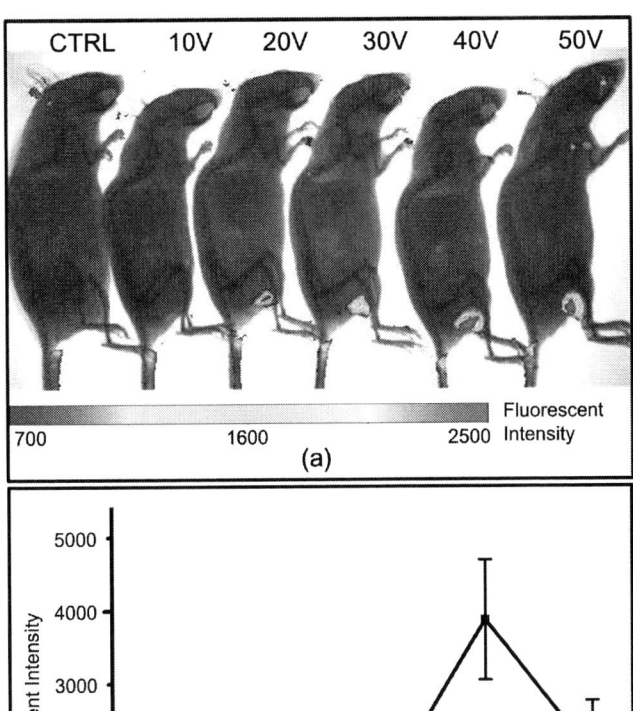

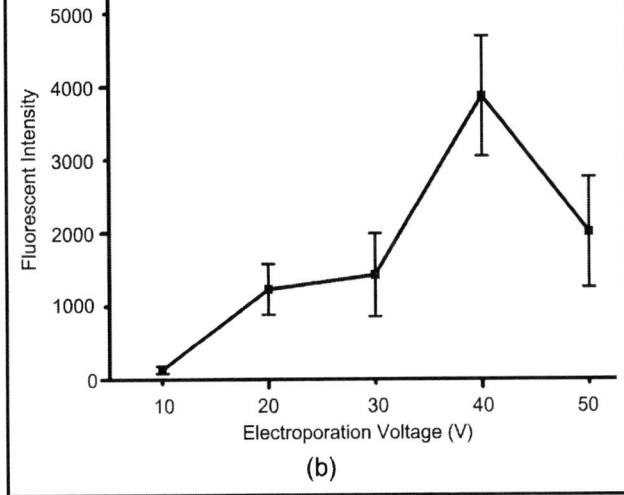

Figure 5: (a) The fluorescent image of RFP expression under different electroporation voltage. All mice were treated with same protocol, and the only difference is the voltage. The fluorescent signal represents RFP expression. Note that the skin of leg was removed before imaging to get clean and precise fluorescence signal for quantitative analyses. (b) The relationship between RFP expression and applied voltage was studied by quantitative fluorescence data from three individual assays. The RFP expression efficiency reaches its peak value when voltage is 40V, and higher voltage brings no enhancement of RFP expression, but dehydration of skin.

CONCLUSION

We have demonstrated a flexible electroporation chip based on MNA electrodes. The flexibility of the proposed chip provides good adaptability to different living tissues. With non-invasive operation and low voltage, both plasmid DNA and siRNA delivery to mouse muscle tissue have been efficiently achieved. The further quantitative analysis reveals the effective delivery depth could reach about 700μm beneath the skin, which is enough for many applications, such as DNA vaccines and treatment of near-skin tumors. More importantly, avoiding physical and biological hazards of *in vivo* nucleic acid delivery means broad prospects of practical human applications.

ACKNOWLEDGEMENTS

The project was financially supported by National Natural Science Foundation of China (Grant No. 61204118 and 61176111).

REFERENCES

[1] H. Aihara, J. Miyazaki, "Gene transfer into muscle by electroporation *in vivo*", *Nat. Biotechnol.*, vol.16, pp867-870, 1998.

[2] Z. Wei, Y. Huang, D. Zhao, Z. Liang and Z. Li, "A parylene-based flexible electroporation chip applicable for in vivo gene and siRNA delivery", in *Tech. Digest Transducers'11*, Beijing, China, June 5-9, 2011, pp. 1942-1945.

[3] M. J. Garland, E. Salvador, K. Migalska, A. D. Woolfson, R. F. Donnelly, "Dissolving polymeric microneedle arrays for electrically assisted transdermal drug delivery", *J. Control. Release*, vol. 159, pp. 52-59, 2012.

[4] S. O. Choi, Y. C. Kim, J. W. Lee, J. H. Park, M. R. Prausnitz, M. G. Allen, "Intracellular protein delivery and gene transfection by electroporation using a microneedle electrode array", *Small*, vol. 7, pp. 1081-1091, 2012.

[5] R. Wang, Z. Wei, W. Wang, Zh. Li, "Flexible microneedle electrode array based-on parylene substrate", *The 17th International Conference on Miniaturized Systems for Chemistry and Life Sciences, μTAS2012*, Okinawa, Japan, October 28 - November 1, 2012, pp. 1249-1251.

CONTACT

*Zh.H. Li, Tel: +86-10-62757163; zhhli@ime.pku.edu.cn.

AN INTEGRATED MICROFLUIDIC SYSTEM FOR DIAGNOSIS OF QUINOLONES RESISTANCE OF *HELICOBACTER PYLORI*

Chih-Yu Chao[1], Chih-Hung Wang[2], Yu-Jui Che[2], Cheng-Yen Kao[3], Jiunn-Jong Wu[4] and Gwo-Bin Lee[1,2*]

[1]Institute of Biomedical Engineering, National Tsing Hua University, Hsinchu, Taiwan

[2]Department of Power Mechanical Engineering, National Tsing Hua University, Hsinchu, Taiwan

[3]Institute of Basic Medical Sciences, College of Medicine, National Cheng Kung University, Tainan, Taiwan

[4]Department of Medical Laboratory Science and Biotechnology, National Cheng Kung University, Tainan, Taiwan

ABSTRACT

Helicobacter pylori (*H. pylori*) is a bacterium which can colonize the stomach mucosa and therefore play a crucial role in gastric diseases. Triple therapy treatment consisting of two antibiotics and a proton pump inhibitor has been routinely taken to eradicate *H. pylori*. Recently, some point mutations were found in gyrase genes against Quinolones. The highly frequent mutation sites were indicated as single amino acid substitution discovered in gyrase A subunit. The epsilometer test has been commonly used to confirm the antibiotic resistance after triple therapy treatment. However, the method is time-consuming and false negative results with trace amounts of *H. pylori* could be easily induced. Alternatively, conventional molecular diagnostic techniques such as polymer chain reaction (PCR) could be used to confirm the antibiotic resistance of *H. pylori*. However, this diagnostic process is relatively labor-intensive and requires expensive and bulky apparatus. In this study a new method was therefore developed to perform molecular diagnostic techniques of SNP-PCR on an integrated microfluidic system to detect the Quinolones resistance of *H. pylori*.

INTRODUCTION

H. pylori are spiral-shaped Gram-negative bacteria that colonize the stomach mucosa of human. The incidence rate of duodenal ulcer and gastric ulcer from *H. pylori* infected patients were found to be about 90-100% and 60-100%, respectively [1]. Symptom diagnosis and biopsy are commonly used to confirm *H. pylori* infections. However, *H. pylori* cannot be eradicated successfully because of the antibiotic resistance. For those who fail to antibiotic therapy, they have low desire to accept more biopsy treatment. For the reason, accurate detection of antibiotic resistant strain for *H. pylori* infections is important.

The quinolones inhibit bacteria growing by binding DNA gyrase which is necessary to DNA duplication. In recent reports, some point mutations found in *H. pylori* DNA could cause resistance to quinolones. The highly frequent mutation sites were indicated as single amino acid substituted at location of 87 (Asn to Lys, N→K) and 91 (Asp to Gly, D→N) in gyrase A subunit [2]. The epsilometer test is a routine method to confirm the antibiotic resistance. However, the test involves a time-consuming step for bacteria culturing. Moreover, if the bacteria number is too low, false negative might occur.

The polymerase chain reaction (PCR) is a molecular diagnostic technology for amplification of target DNA. After the PCR process, target DNA segments are sequenced to confirm the existence of the mutation sites of antibiotics resistance [3]. However, there are some disadvantages. For example, the process should be operated by well-trained personnel. Furthermore, some delicate apparatus are required and the labor-intensive process might cause contamination. A new detection method is therefore needed to improve those disadvantages mentioned above.

Single nucleotide polymorphism polymerase chain reaction (SNP-PCR) technology has been used to detect the single point mutations that express antibiotic resistance [4]. The technology involves a specific primer design. Because polymerase cannot distinguish the single mutation site from the normal gene segment, both two different sequences could be amplified by PCR. The primer is therefore designed to solve this problem by amplifying the mutation strain DNA segment without amplifying the normal strain DNA.

Recently, micro-electro-mechanical-systems (MEMS) technology has been successfully utilized for a variety of biomedical applications. For instance, with the technology, a microfluidic chip which integrated micro-scale reaction chambers and functional micro-components could be designed and fabricated to perform the entire process for molecular diagnosis [5]. In this work, SNP-PCR was performed on a single chip with less consumption of reagents and samples. Moreover, sample pre-treatment was also integrated on the microfluidic platform. The detection of amplified SNP-PCR products could be further measured on line by using a fluorescent scheme. In the study, *H. pylori* specific mutation sites of producing resistance to Quinolones can be detected by using the developed platform. This is the first time that the SNP-PCR was performed on an integrated microfluidic system for diagnosis of *H. pylori* antibiotic resistance.

MATERIALS AND METHODS
Experimental procedures

The specific single-point mutation may lead to expression of antibiotic resistance. The primers with an artificial mismatch of a single nucleotide at the third base closest to the 3'-end (SNP site) were designed [4]. The primers could be compliment to the mutation site but not to the normal site. If the template DNA is a normal strain, there are two mismatches between the template DNA and the primer. The operating principle is schematically shown in Figure 1. As a result, the DNA polymerase cannot stick on them and the elongation cannot start accordingly.

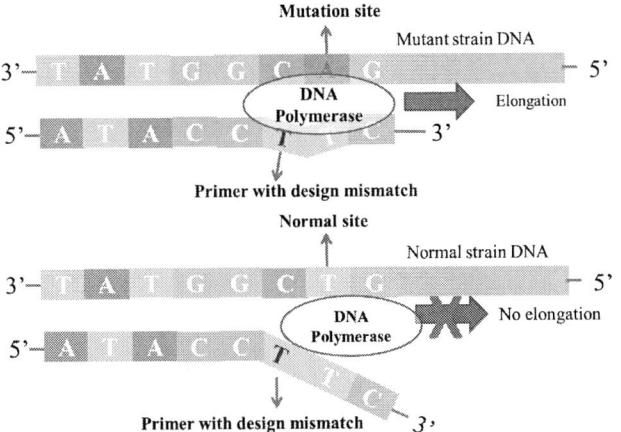

Figure 1: Schematic illustration of the operating principle of the specific primers for SNP-PCR to detect the single mutation point.

The specific primers were then added to a PCR kit (KAPA SYBR FAST ABI prism qPCR Kit, Kapa Biosystems, USA) for SNP-PCR. The PCR reagent contains SYBR green which is a fluorescent dye and can intercalate into double-stranded DNA and release fluorescent signal by laser excitation.

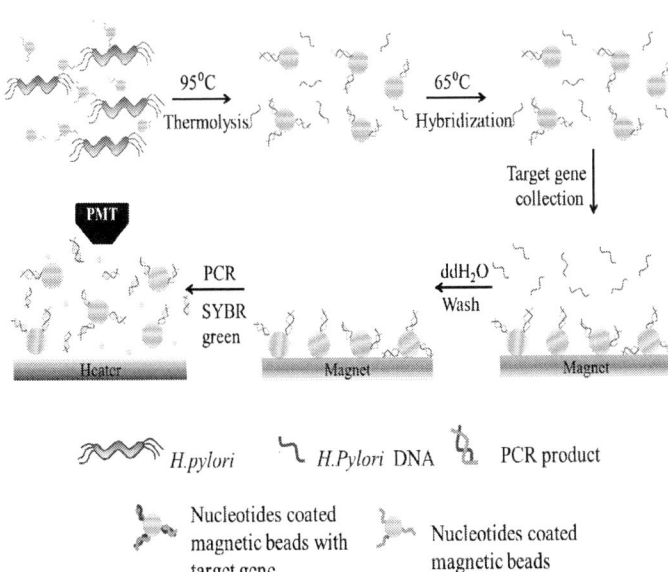

Figure 2: Schematic illustration of the experimental procedure for the detection of mutation sites related to Quinolones resistance of H. pylori.

The experimental procedure for diagnosis of Quinolones resistance of H. pylori in the developed microfluidic system was schematically shown in Figure 2. Bacteria sample was first loaded in the chamber with magnetic beads and then underwent thermolysis and DNA denaturation at 95°C. Then DNA was hybridized with the nucleotide probes on the magnetic beads at 65°C. Note that the H. pylori specific 16s rRNA nucleotide probe was conjugated onto the magnetic beads to capture the conserved gene of H. pylori. Unwanted substances and debris were then washed by ddH2O and removed from the waste channel. At the same time, target DNA captured by magnetic beads were collected by magnet placed underneath the chip. After washing, the PCR solution stored in PCR reagent chamber was transported to the reaction chamber. With the TE cooler which has a set temperature protocol underneath the chip, the PCR reaction was performed. Finally, the fluorescent signal was detected to diagnose the H. pylori resistant DNA.

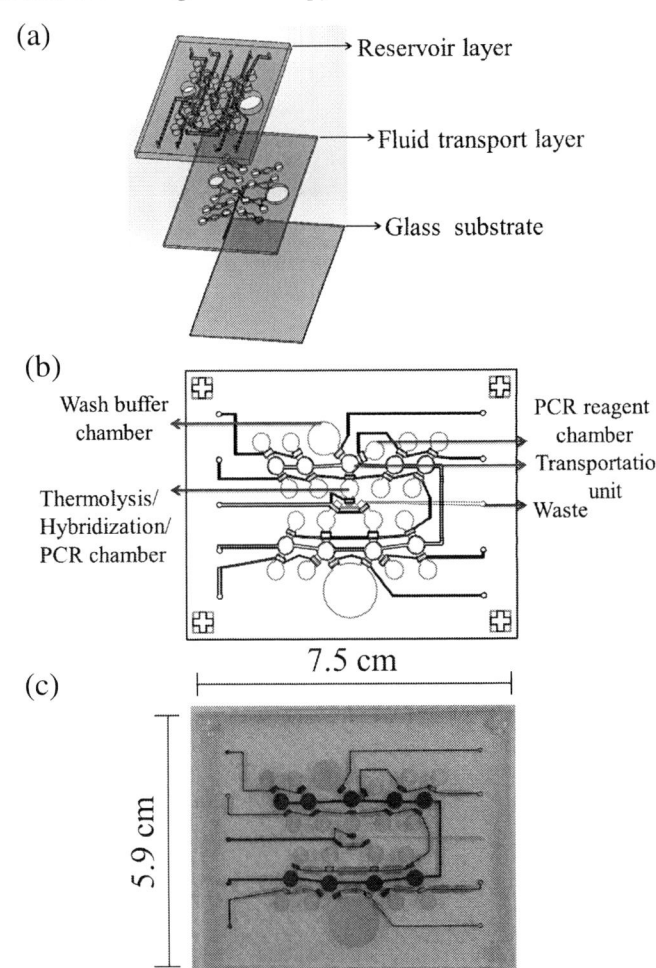

Figure. 3: (a) An exploded view, (b) a top view and (c) a photograph of the integrated microfluidic chip

An integrated microfluidic chip was designed and fabricated to perform the experimental process mentioned above. The schematic illustrations and a photograph of the microfluidic chip are shown in Figure 3. It was consisted of two PDMS layers and one glass substrate. The thick PDMS layer was composed of a fluid transport module (with blue ink) and the thin layer was consisted of reservoirs (with red ink), as shown in Figure 3(c), The dimensions of the developed chip were measured to be 7.5 cm (length) x5.9 cm (width). There were three identical chambers at the center of the chip with a magnet underneath. The three chambers could be used to load bacteria samples and magnetic beads. The PCR reagent could be mixed with different primers (16s rRNA, N87K, and D91N). The chambers next to the three chambers above-mentioned were designed to load negative control samples. Therefore, *H. pylori* and control samples could be confirmed and the mutation sites could be detected, respectively. Micropumps, microchannels, and microvalves were designed and programmed for transporting the samples and reagents between chambers [6]. Furthermore, reaction solution could be mixed properly by activating pneumatic components and electromagnetic valves (EMVs). Note that negative gauge pressure (suction force) was used to fill up the cavities to deform the PDMS membranes.

The fabrication of the microfluidic chip adopted a typical soft lithography process. First, two molds were designed and made by computer-numerical-control machining process on polymethylmethacrylate (PMMA). Then the PDMS were casted on the PMMA molds to form two PDMS layers. Finally, the oxygen plasma treatment was used to bond the two PDMS layers and a glass substrate together to form a microfluidic chip.

RESULTS AND DISCUSSION

There were five samples used to perform the experiments, including a normal strain *H. pylori* sample, two N87K mutation strain *H. pylori* samples, and two D91N mutation strain samples. The 16s rRNA specific primers were used as an internal control to confirm if the samples are *H. pylori*. The fluorescent intensity results indicated that the N87K and D91N mutations of *H. pylori* could be successfully detected in a microfluidic system after SNP-PCR, as shown in Figure 4. Note that the normal gene could not be amplified by specific single point mutation primers for SNP-PCR. All of the *H. pylori* samples were confirmed by 16s rRNA PCR product fluorescent signals. In addition, the electropherograms were used to confirm the result as shown in Figure 4.

(b)

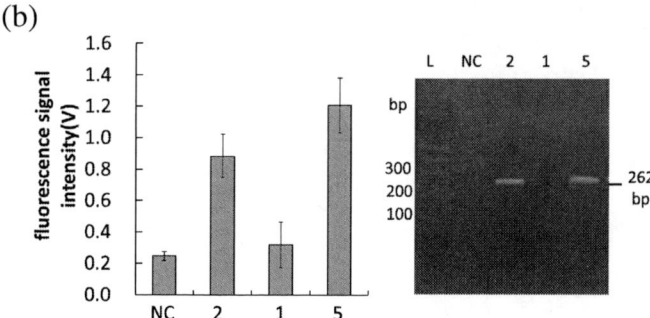

(c)

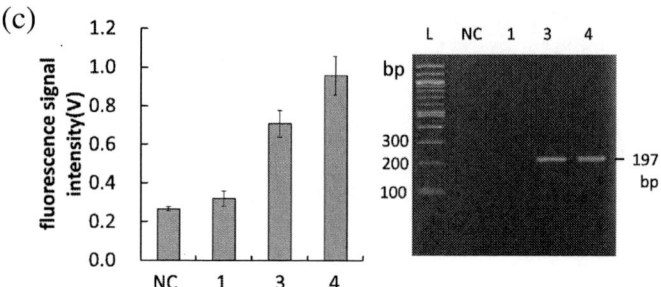

Figure. 4: The fluorescent signals from the PCR products; Sample1indicated H. pylori normal, samples 2 and 5 were N87K and samples 3 and 4 were D91N samples, respectively; (a) the PCR results by using 16s rRNA primers; (b) specific point mutation N87K primers used; (c) specific point mutation D91N primers used. And the electropherograms were used to confirm the results. The lane L indicated 100-bp DNA ladder; NC: negative control using ddH2O.

Detection limit of *H. pylori*

Moreover, the detection limit of *H. pylori* was further investigated. Serial dilution of purified DNA with target gene was used for the measurement of detection limit. Figure 5 shows the detection limit of each target gene fragments. 16s rRNA, N87K and D91N mutation DNA were measured to have a detection limit of 101, 102 and 103 copy numbers respectively. The PCR detect limit level is superior to the previous report [7]. Similarly, the electropherograms were used to confirm the result as Figure 5 shows.

(a)

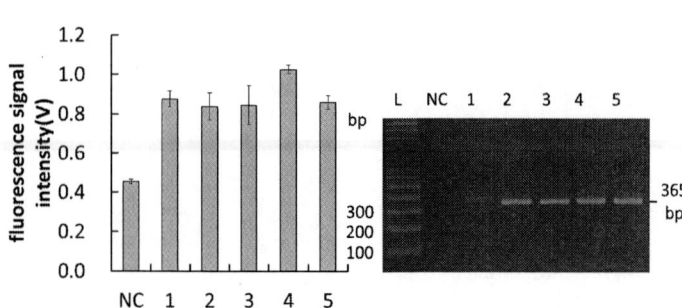

(a)

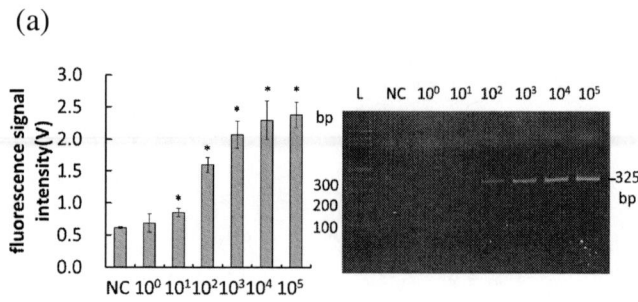

(b)

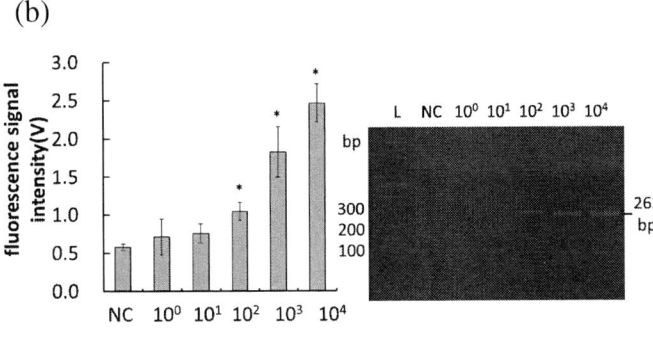

(c)

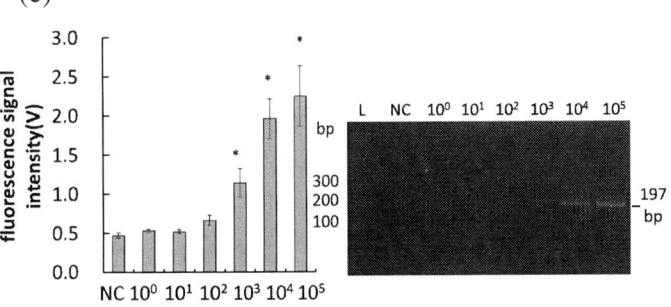

Figure. 5: The fluorescent signals from the PCR products of serial dilution sample. (a) the PCR results by using 16s rRNA primers; (b) specific point mutation N87K primers used; (c) specific point mutation D91N primers used. "*" Represents p value <0.05. The electropherograms are used to confirm the results. The lane L indicated 100-bp DNA ladder; NC: negative control using ddH2O.

CONCLUSION

We have successfully demonstrated that the specific primers could distinguish Quinolones resistant *H. pylori* and the amplified SNP-PCR products could be detected by fluorescent signals directly. The detection limit of using the developed method was further explored. Moreover, the entire experimental process could be automatically performed on an integrated microfluidic chip. Consequently, the integrated microfluidic system may provide a powerful platform for rapid, automatic diagnosis of antibiotic resistance of *H. pylori*.

ACKNOWLEDGEMENTS

The authors would like to thank the financial support from National Science Council in Taiwan.

REFERENCES

[1] E. J. Kuipers, J. C. Thijs, H. P. Festen, "The prevalence of Helicobacter pylori in peptic ulcer disease", *Alimentary Pharmacology & Ther.*, pp. 59-69, 1995.

[2] K. H. Hung, B. S. Sheu, W. L. Chang, H. M. Wu, C. C. Liu, J. J. Wu, "Prevalence of Primary Fluoroquinolone Resistance Among Clinical Isolates of *Helicobacter*

pylori at a University Hospital in Southern Taiwan", *Helicobacter.*, vol. 7, pp. 61-65, 2008.

[3] N. Akopyanz, N. O. Bukanov, T.U. Westblom1, S. Kresovich, D. E. Berg, "DNA diversity among clinical isolates of*Helicobacter pylori* detected by PCR-based RAPD fingerprinting" *Nucl. Acids Res.*, vol. 20, pp. 5137-5142, 1992

[4] T. Nishizawa, H. Suzuki, A. Umezawa, H. Muraoka, E. Iwasaki, "Rapid Detection of Point Mutations Conferring Resistance to Fluoroquinolone in *gyrA* of *Helicobacter pylori* by Allele-Specific PCR", *Clinical Microbiol.*, vol. 45, pp. 303–305, 2007.

[5] J.H. Wang, L. Cheng, C. H. Wang, W. S. Ling, S. W. Wang, G. B. Lee. "An integrated chip capable of performing sample pretreatment and nucleic acid amplification for HIV-1 detection.", *Biosens Bioelectron.*, vol. 41, pp.484–491, 2013.

[6] C. H. Wang, K.Y. Lien, J. J. Wu, G. B. Lee, "A magnetic bead-based assay for the rapid detection of methicillin-resistant Staphylococcus aureus by using a microfluidic system with integrated loop-mediated isothermal amplification", vol. , pp.1521–1531, 2011.

[7] H. Moyaert, F. Pasmans, R. Ducatelle, F. Haesebrouck, and M. Baele, "Evaluation of 16S rRNA Gene-Based PCR Assays for Genus-Level Identification of Helicobacter Species" , *Clinical Microbiology.*, vol. 46, pp.1867–1869, 2008

CONTACT

*Gwo-Bin "Vincent" Lee, Department of Power Mechanical Engineering, National Tsing Hua University, Hsinchu, Taiwan; Tel: +886-3-574-2834; Fax: +33-3-572-2840; E-mail: gwobin@pme.nthu.edu.tw

ANNEALING EFFECTS ON FLEXIBLE MULTI-LAYERED PARYLENE-BASED SENSORS

Brian J. Kim[1], E. Peter Washabaugh IV[2], and Ellis Meng[1]

[1]Department of Biomedical Engineering, University of Southern California, Los Angeles, CA, USA
[2]Department of Biomedical Engineering, University of Michigan, Ann Arbor, MI, USA

ABSTRACT

To mitigate long term, soaking-induced delamination failure of multi-layered Parylene C devices, a post-process annealing step can be employed to increase adhesion between the Parylene layers. However, it has been shown that annealing of Parylene thin films can alter the bulk properties of the polymer, and thus impact final device performance. To elucidate these effects, the mechanical and electrochemical properties, and sensing performance of untreated and annealed Parylene C-platinum electrochemical impedance-based force sensors were compared. Annealing reduced the height (~3%) and increased the stiffness of the Parylene C sensing channel structure (~1.6x), affecting the sensor's mechanical response. Furthermore, the electrode surface was smoothed as built-in residual stresses were removed during annealing, altering the sensor's electrochemical properties. Together, these phenomena resulted in a 24% reduction in sensor sensitivity. These results indicate that heat-based effects cannot be ignored for Parylene-metal device systems, including neural microelectrode implants, and that mechanical and electrochemical properties and performance must be determined after heat treatment, such as annealing and sterilization.

INTRODUCTION

Initially used as an encapsulation coating for implantable devices, Parylene C (from here on referred to as Parylene) is now frequently chosen as a structural material for implantable or other biomedical devices because of its biocompatibility, flexibility, and compatibility with microfabrication techniques [1]. The introduction of techniques to form three dimensional structures from Parylene thin films [2] has further expanded the potential for Parylene as a MEMS material; novel Parylene microdevices and microsystems have been developed that leverage the polymer's properties for biomedical applications [3].

However, it is well established that these multi-layered Parylene-MEMS encounter a failure mode under continuous soaking conditions (i.e. *in vivo*) via delamination between Parylene-Parylene and Parylene-metal interfaces that is attributed to liquid intrusion that weakens these layer interfaces. To mitigate this, many efforts to increase the adhesion between the Parylene layers have been explored [4]; one of note is a simple post-process annealing treatment at temperatures greater than the glass transition point of Parylene (~60-90°C [5]) to allow for increased entanglement between the chains of the Parylene layers [6]. However, heat treatment of Parylene thin films has been found to cause

significant chain reorganization and thus material changes to the bulk polymer [7], which may in turn impact device performance. These heat-induced modifications to Parylene can further extend beyond the annealing step; any process in which the film experiences temperatures greater than the glass transition point (e.g. autoclave sterilization) can alter the properties of the bulk polymer and thus the final device performance.

To explore the annealing effects on multi-layered Parylene device performance, we present an investigation of the post-annealing performance of a distributed Parylene-based force sensor array consisting of an electrolyte-filled Parylene microchannel sensing structure encasing a linear array of platinum (Pt) electrodes [8]. Deflection of the top surface of the microchannel by an applied force induces changes in the volumetric conduction path between the electrodes, altering the measured electrochemical impedance (Figure 1); thus device performance is closely associated with both the mechanical response of the Parylene microchannel and the electrochemical properties of the electrode surface. Annealing was found to induce changes in both aspects of the sensor, thereby affecting final sensor performance. Our results indicate that annealing effects cannot be ignored for mechanically or electrochemically sensitive Parylene-MEMS devices; device properties and performance must be characterized following any heat-based process.

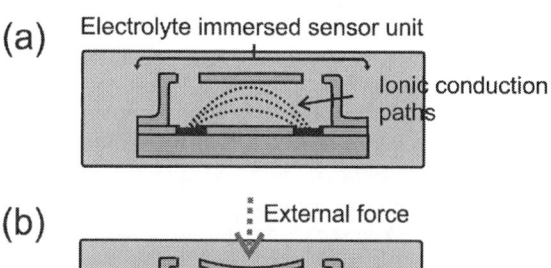

Figure 1: Illustration of sensor transduction mechanism (a) during steady state and (b) during an applied external force. Applied forces deflect the top surface of the microchannel to induce a change in the volumetric conduction path, and thus electrochemical impedance. This mechanism highlights the importance of the mechanical and electrochemical sensor properties on performance.

978-1-4799-3510-9/14 $31.00 © 2014 IEEE

EXPERIMENTAL METHODS

The Parylene-based force sensor array consists of a Parylene microchannel (100 μm x 4.15 mm x 24 μm) encasing a linear array of eight electrodes (each measuring 100 x 130 μm); electrode pairs constitute a single sensor (Figure 2). As mentioned previously, the sensors utilize an electrochemical transduction principle; force-induced disruptions of the ionic conduction path between the electrode pairs alters the measured electrochemical impedance. The impedance is acquired at a specific measurement frequency ($f_{measurement}$ = 1 kHz), selected such that it is sufficiently high enough to bypass the double layer capacitance (C_{dl}), isolate the solution resistance, and thus capture the disruptions in the ionic conduction path.

The sensor array was designed to measure the interfacial forces between an implanted rigid neural probe and surrounding tissue. To improve the sensor's *in vivo* lifetime, devices were annealed to prevent delamination resulting from chronic soaking in saline. More specifically, sensor arrays were placed within a vacuum oven (10 mTorr), which was ramped at 1.6°C/min to 200°C. Following a thermal soak time of 48 hours, the devices were cooled overnight (~15 hours) under vacuum, and then removed for post-process characterization. All untreated sensors were tested following removal off wafer.

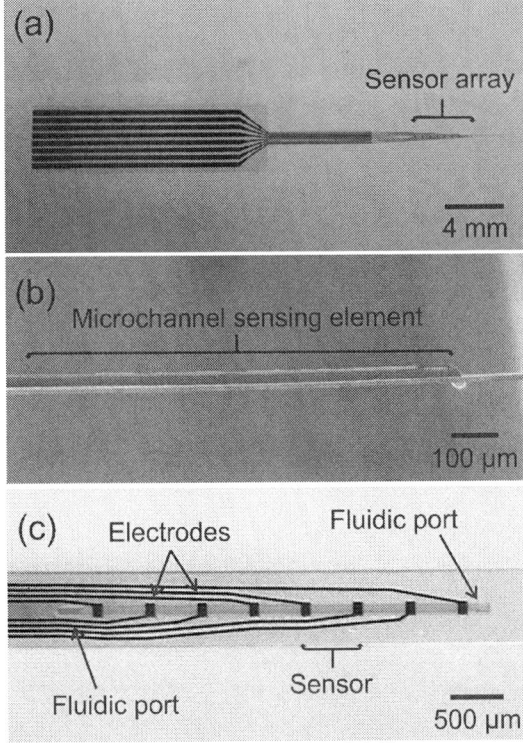

Figure 2: (a) Optical micrograph of full sensor array and integrated Parylene cable. Microchannel structure is highlighted in pink. (b) SEM image of a sideview of the 20 μm Parylene microchannel sensing structure. (c) Zoomed in top-down image of Parylene microchannel (pink) indicating the electrodes and fluidic ports on opposing ends of the structure.

The bulk material and mechanical properties of the Parylene-Pt sensors were assessed before and after annealing via scanning electron microscopy (SEM), profilometry, and load-deflection tests of the sensing structures immersed in an electrolyte solution (1x PBS). For load-deflection tests, a motorized z-axis stage and an inline load cell with a flat-tipped deflection probe was used to measure both channel deflection and load forces (Figure 3). A similar testing setup along with real-time electrochemical impedance measurement utilizing a precision LCR meter and LabVIEW graphical user interface was also used to assess device performance by analyzing sensor calibration sensitivities.

Two point electrochemical impedance spectroscopy (EIS; 1x PBS, 20 Hz–1 MHz), a commonly used electrochemical technique to assess electrode surface properties, was used between adjacent electrodes to determine changes to the electrochemical properties of the sensing electrodes.

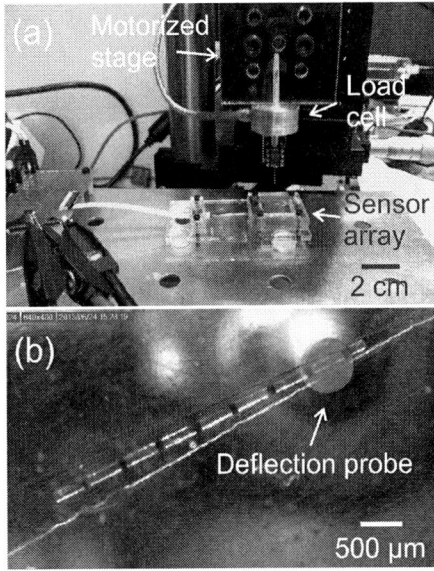

Figure 3: (a) Load-displacement setup for mechanical and calibration testing of Parylene force sensor array. (b) Optical micrograph taken using a camera underneath testing setup illustrating deflection probe (pink) displacing into sensor element.

RESULTS AND DISCUSSION

Mechanical Characterization

SEM analysis revealed no significant qualitative differences between the two samples (Figure 4a, b). However, profilometric measurements indicated a ~3% reduction in the height (~1 μm) of the microchannel structure following annealing (Figure 4c), analogous to bulk shrinkage effects previously observed in lateral thin film structures [9]. Analysis of the electrode spacing indicated no significant shrinkage of the pitch of the electrodes (data not shown).

Load-deflection tests revealed a ~1.6x increase in the stiffness of the structure following annealing (Figure 5), likely due to a combination of microchannel height shrinkage and an increase in the elastic modulus of annealed Parylene bulk polymer attributed to increased polymer crystallinity [10].

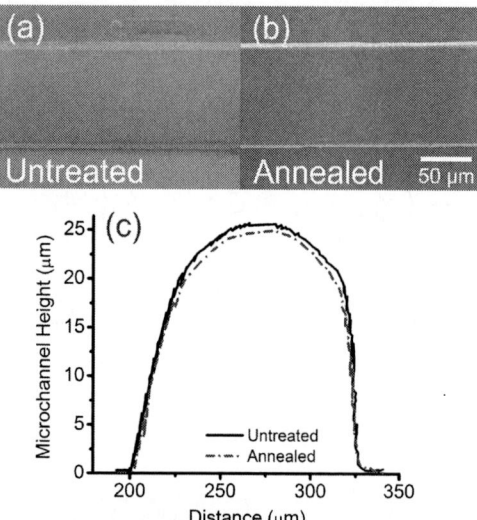

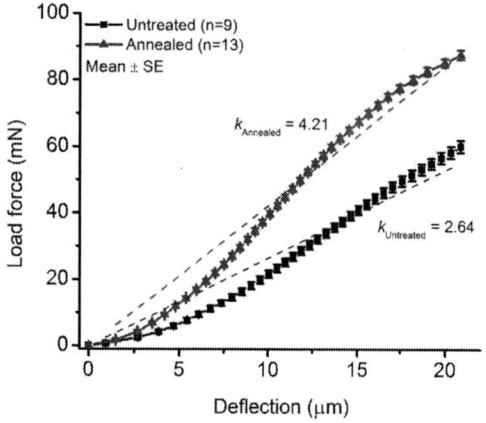

Figure 4: SEM images of top-down views of (a) untreated and (b) annealed sensing structures. (c) Representative profilometry measurement of untreated and annealed Parylene microchannel sensing structure illustrating ~3% shrinkage following annealing.

Figure 5: Results of load-deflection tests comparing untreated and annealed sensors illustrating an increased structural stiffness of ~1.6x after annealing.

Electrochemical Characterization

EIS results indicated a slight increase in the impedance magnitude and narrowing of the phase plot, attributed to a decrease in double layer capacitance, or C_{dl}, of the simplified Randles electrode-electrolyte system circuit model (Figure 6a, inset). C_{dl} reduction within the model is often seen as a consequence of the smoothing of the thin-film metal during annealing [11], which removes surface roughness but in turn also reduces electroactive surface area. Despite these slight differences, the measurement frequency (peak of phase plot) remained consistent following annealing (Figure 6b). The minimal effect on the electrochemical properties of the electrodes (relative to the mechanical changes) can be

hypothesized to be correlated with electrode size-dependent phenomena in altering electrode surface properties; annealing effects on electrochemical properties of Pt on Parylene electrodes have been demonstrated to be magnified in devices with smaller electrode sizes ($\varnothing = 40\ \mu m$) [12].

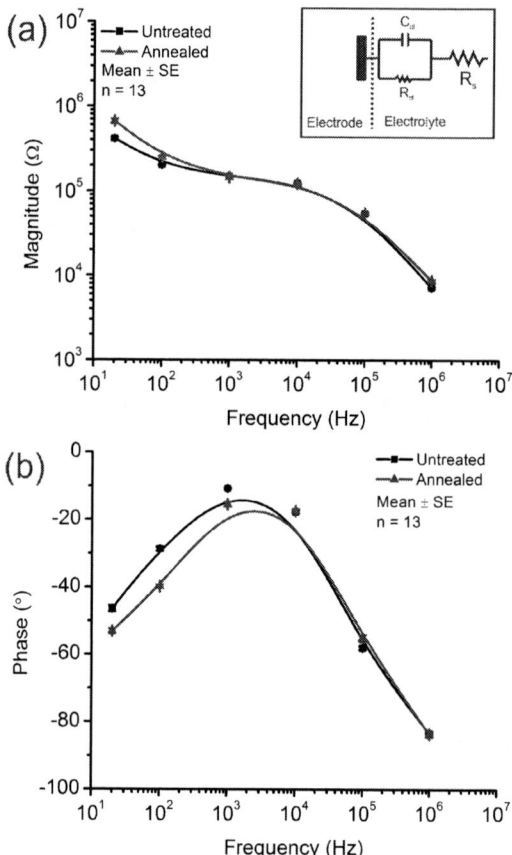

Figure 6: EIS plots of (a) impedance magnitude and (b) phase for electrodes of untreated (black) and annealed (blue) Parylene sensors. A slight increase in impedance magnitude and narrowing of the phase peak following annealing corresponds to a decrease in the C_{dl} within the simplified Randles circuit model (a, inset). Dashed line in (b) corresponds to the measurement frequency chosen as the peak of the phase plot.

Sensor Performance

A combination of the mechanical and electrochemical changes following annealing significantly altered sensor performance during calibration (Figure 7). Device sensitivity was decreased from $\alpha_{untreated} = 1.25 \times 10^{-3}$ (normalized impedance/mN) to $\alpha_{annealed} = 9.25 \times 10^{-4}$ (normalized impedance/mN). A reduction in the sensitivity of the sensor by 24% after the annealing process follows the increase in stiffness of the sensing structure. For a stiffer channel, applied forces induce smaller channel deflections, thus reducing sensor sensitivity. Finite element modeling studies utilizing COMSOL to model the observed changes following annealing elucidated that the increase in stiffness of the Parylene film contributed the most (rather than microchannel height reduction or decrease in electroactive surface area) to the differences of sensor performance for these devices (data that shown).

Additional work is underway to assess the benefits of annealing on sensor longevity in wet environments via soak testing. Also, previous work has demonstrated annealing temperature dependence on altering the properties of Parylene films [9]; experiments are underway to correlate the degree of intra-Parylene layer adhesion with annealing temperature to determine the optimal process parameters for improving chronic wet performance of multi-layered Parylene devices.

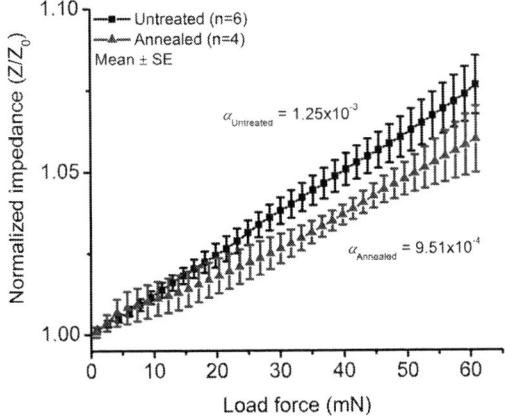

Figure 7: A marked difference in sensor calibration performance (reduction of ~24%) was observed due to the combination of mechanical and electrochemical effects of annealing on Parylene sensors. Sensitivity (α) units are in normalized impedance/mN.

CONCLUSION

Effects of annealing on a Parylene-based force sensor array utilizing the combination of a fluid-filled transduction microchannel and electrochemical sensing scheme was elucidated. Annealing was found to increase the stiffness of the Parylene microchannel, attributed to increased crystallinity of the polymer chains and the reduction of the height due to bulk shrinkage effects, and slightly decrease the electroactive surface area of the Pt electrodes, caused by the smoothing of the thin film metal surface. A combination of these effects was exhibited in a modified sensor calibration curve with a decreased sensitivity of 24% following annealing. The present study demonstrated that high-temperature processes or environments (relative to the glass transition point of the polymer) can modify device performance, which must be considered during sensor design and for proper sensor operation.

ACKNOWLEDGEMENTS

The authors would like to thank Mr. Louis Jug for aid in sensor calibration and Dr. Donghai Zhu and members of the USC Biomedical Microsystems Laboratory for their assistance. An OAI model 30 light source was used for processing of Parylene devices.

This work was funded in part by the Engineering Research Centers Program of the NSF under Award Number EEC-0310723, by the Defense Advanced Research Projects Agency (DARPA) under the auspices of Dr. Jack Judy through the Space and Naval Warfare Systems Center, Pacific Grant/Contract No. N66001-11-1-14207, and the University of Southern California Provost Ph.D. Fellowship (BK).

REFERENCES

[1] E. Meng, *et al.*, "Plasma removal of Parylene C," *Journal of Micromechanics and Microengineering,* vol. 18, p. 045004, 2008.

[2] K. Walsh, *et al.*, "Photoresist as a sacrificial layer by dissolution in acetone," in *MEMS 2001. The 14th IEEE International Conference on*, 2001, pp. 114-117.

[3] B. J. Kim, *et al.*, "3D Parylene sheath neural probe for chronic recordings," *Journal of Neural Engineering,* vol. 10, p. 045002, 2013.

[4] J. H. Chang, *et al.*, "Adhesion-enhancing surface treatments for parylene deposition," in *TRANSDUCERS 2011 16th International*, 2011, pp. 390-393.

[5] H.-S. Noh, *et al.*, "Parylene micromolding, a rapid and low-cost fabrication method for parylene microchannel," *Sensors and Actuators B,* vol. 102, pp. 78-85, 2004.

[6] D. C. Rodger, *et al.*, "Flexible parylene-based multielectrode array technology for high-density neural stimulation and recording," *Sensors and Actuators B: Chemical,* vol. 132, pp. 449-460, 2008.

[7] J.-M. Hsu, *et al.*, "Effect of Thermal and Deposition Processes on Surface Morphology, Crystallinity, and Adhesion of Parylene-C," *Sensors and Materials,* vol. 20, pp. 071-086, 2008.

[8] B. J. Kim, *et al.*, "Parylene-based electrochemical-MEMS force sensor array for assessing neural probe insertion mechanics," in *MEMS 2012 IEEE 25th International Conference on*, 2012, pp. 124-127.

[9] B. J. Kim, *et al.*, "Three dimensional transformation of Parylene thin film structures via thermoforming," in *MEMS 2013 IEEE 26th International Conference on*, 2013, pp. 339-342.

[10] R. Metzen and T. Stieglitz, "The effects of annealing on mechanical, chemical, and physical properties and structural stability of Parylene C," *Biomedical Microdevices,* pp. 1-9, 2013.

[11] M. Grosser and U. Schmid, "The impact of annealing temperature and time on the electrical performance of Ti/Pt thin films," *Applied Surface Science,* vol. 256, pp. 4564-4569, 2010.

[12] B. J. Kim, *et al.*, "Evaluation of Post-Fabrication Thermoforming Process for Intracortical Parylene Sheath Electrode," in *6th International IEEE EMBS Neural Engineering Conference*, 2013, pp. 379-382.

CONTACT

*E. Meng, tel: +1-213-8213949; ellis.meng@usc.edu

AUTOMATED VITRIFICATION OF MAMMALIAN EMBRYOS ON A DIGITAL MICROFLUIDIC DEVICE

Derek G. Pyne[1], Jun Liu[1*], Mohamed Abdelgawad[2,a], and Yu Sun[1,b]*
[1]University of Toronto, CANADA
[2]Assiut University, EGYPT

ABSTRACT

This paper presents the development of a digital microfluidic device to achieve automated sample preparation for the vitrification of mammalian embryos for clinical in vitro fertilization (IVF) applications. Individual micro droplets manipulated on a digital microfluidic device were used as micro-vessels to transport a single embryo through a complete vitrification procedure. The device showed cell survival and development rates of 77% and 90%, respectively, which are comparable to the control groups that were manually processed.

INTRODUCTION

The role of cryopreservation in clincal and research environments has played an increasingly important role in recent years. Stem cells, sperms, and embryos are now routinely frozen and preserved for use at a later time [1].

In the clinical setting, patients who undergo therapeutic procedures, such as chemotherapy which can place their fertility at risk, have the option of preserving their reproductive cells for future use through in vitro fertilization techniques (IVF). However, in these cases the number of viable cells for preservation can be low making the survivability rate and reproducibility of cryopreservation techniques critical. In particular, preservation of embryos or oocytes is challenging since these cells are highly sensitive, and the cell number is very small.

The two commonly used cryopreservation techniques for freezing embryos are the slow freezing method and the vitrification method. Conventionally, cells are frozen through the slow freezing method where cells are placed in a large freezer that can accurately control the freezing rate down to liquid nitrogen temperatures [2]. This procedure requires sophisticated equipment and produces a relatively poor survivability rate.

On the other hand, vitrification offers an alternate approach in which cells are frozen at extremely high rates, usually by directly plunging the sample into liquid nitrogen, after bathing them in a sequence of high concentration cryoprotectants [3]. Vitrification reduces intracellular ice formation, which is the primary cause of cell death, by freezing the sample in a glass-like state before the molecules have a chance to form crystal structures. This results in a higher cell survival rate after thawing compared to conventional slow freezing [4]. On the negative side, vitrification requires precise washing sequences and timings in each cryoprotectant medium (Fig. 1). Clinically, processing an embryo/oocyte in cryoprotectant medium costs a highly skilled embryologist ~15 minutes.

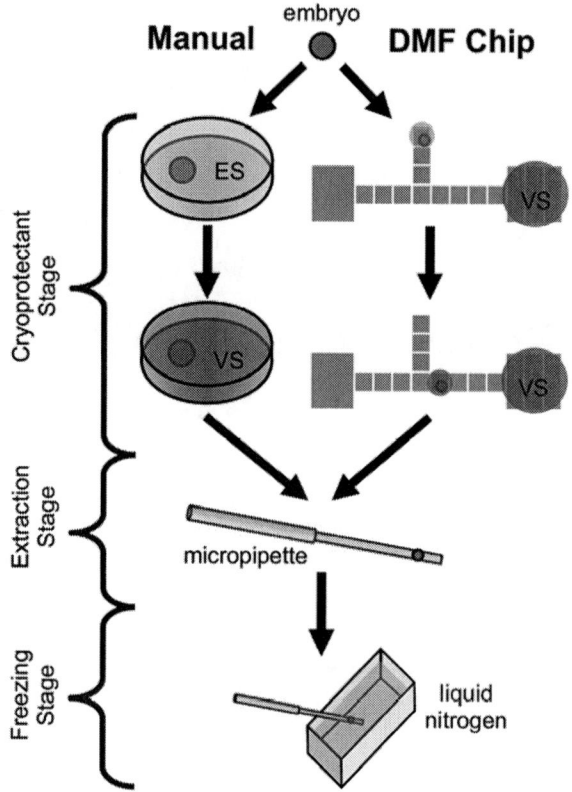

Figure 1: Schematic showing differences between manual vitrification approach, which requires manual pipetting between mediums in cryoprotectant stage, and the digital microfluidic (DMF) approach, which moves the embryo between mediums on chip. The chip automates the high skill portion of the procedure providing labor cost savings and opportunities for parallelization.

Digital microfluidics, which is powerful for sequential sample processing, is used in our work to automate embryo preparation for the vitrification procedure, lowering the high labor cost and possibly helping further spread the use of vitrification in IVF clinics. Here, we present the development of a digital microfluidic device to automate pre-processing of vitrification of mammalian embryos. Droplets on the microfluidic platform acted as micro-vessels to move an embryo and subject it to a series of cyroprotectants of different concentrations, as required by the IVF vitrification protocols. Compared to channel-based microfluidic vitrification [5], our device does not undesirably 'park' the embryo to a confined area and does not have intricacy of embryo introduction and retrieval onto and from the device.

978-1-4799-3510-9/14 $31.00 © 2014 IEEE

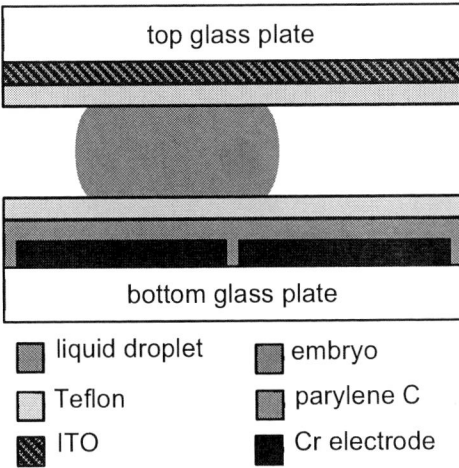

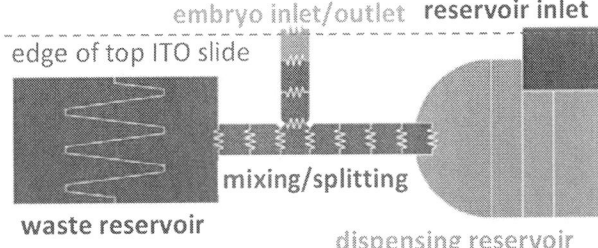

Figure 3: Chip design showing regions for vitrification medium dispensing and embryo inlet/outlet.

Figure 2: Device fabrication.

pipetted on the exposed half of the electrode and then actuated into the device through the covered half. For extraction, the embryo-carrying droplet was moved to this edge electrode where it bulges out of the device and is retrieved by a standard micropipette. This same mechanism was used to fill the dispensing reservoir before the embryo and reagents were loaded.

DEVICE DESIGN

Devices were fabricated in the cleanroom facilities of the Toronto Nanofabrication Centre (TNFC). As shown in Fig. 2, pre-coated chromium glass slides (Deposition Research Labs Inc., MO) were first primed with P-20 before spin coating Shipley S1811 photoresist (3000 rpm, 30 sec). Substrates were then baked to remove solvents (115°C, 2 min) and UV exposed through a transparency mask (10 sec). Substrates were next developed in MF-321 (2 min), hard baked (115°C, 1 min), etched with CR-4 (2 min) and photoresist removed with AZ-300T stripper (15 min in ultrasonic bath). A 2 µm thick dielectric layer of Parylene C was then deposited. Lastly, a hydrophobic coating of Teflon AF was spin coated on the device (2000 rpm, 1 min) and baked (160°C, 10 min). A second glass slide coated with un-patterned ITO (Deposition Research Labs Inc., MO) and Teflon AF was used as the ground electrode. Droplets were actuated in between the two glass slides separated by a spacer. Voltages applied to actuate droplets were 55-75 Vrms at 15 kHz. Cyroprotectant droplets were actuated inside silicone oil to reduce friction and evaporation.

As illustrated in Fig. 3, different regions of our device were designed to carry out the general digital microfluidic action of transport, mixing, dispensing, merging, and splitting. A large reservoir was used to hold and dispense the high concentration cryoprotectant medium. This reservoir was split up into many sections to handle variations in liquid volume in the reservoirs as droplets are dispensed during the vitrification protocol. The second reservoir was used as a waste reservoir and, thus, was less controllable than the dispensing reservoir. A central T-shaped array of electrodes was used for droplet transport, mixing, and splitting. Electrodes in this array were interdigitated to allow droplet overlap with adjacent electrode which increases electrodynamic forces applied on droplets. The top of this section was an input/output region where half of the edge electrode was exposed out of the ITO slide to enable embryo loading.

As shown in Fig. 4, the embryo-carrying droplet was

MATERIALS AND METHODS

Materials

Mouse embryos were gathered from the Canadian Mouse Mutant Repository (Toronto, ON). Embryos were produced by superovulating a female and were gathered 2.5 days past conception.

Vitrification solution usually contains antifreezing agents or cryoprotectant, such as dimethyl sulfoxide (DMSO), some small molecular size glycols (e.g., ethylene glycol), or sucrose [6]. A combination of DMSO and sucrose was used to follow the protocol used by our embryo suppliers. The vitrification solution (VS) was made by diluting DMSO in serum-free KSOM medium (EMD Millipore, Billerica, US) at 33% concentration, with 1.0 M sucrose. The equilibrium solution (ES) was at half concentration of VS (i.e., 16.5% DMSO + 0.5 M sucrose). VS was preloaded on the DMF chip before each experiment. The first mixing step, which mixes the VS with embryo culture medium (i.e., serum-free KSOM), generates the ES.

Embryo Loading and Retrieval

To input an embryo, it was selected and then aspirated into a micropipette with minimal volume. This small embryo-containing droplet was then pipetted onto the inlet column of electrodes and actuated into the device. This technique minimized exposure of the embryo to outside air and could be completed very quickly. Extraction of the embryo was completed in the opposite manner by transporting the embryo containing droplet into the micropipette (see Fig. 4). Once the embryo was extracted from the device it was either directly frozen in the micropipette.

Cryoprotectant Mixing

An embryo is input into the device in a small droplet of embryo culture medium, and 100% cryoprotectant is input into the device in larger volumes ('reservoir inlet' in Fig. 3). The cryoprotectant bathing procedure is then performed

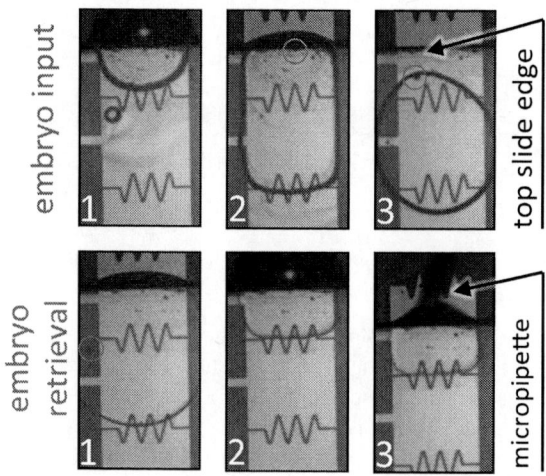

Figure 4: Embryo is input and extracted by actuating electrodes at edge of top glass slide.

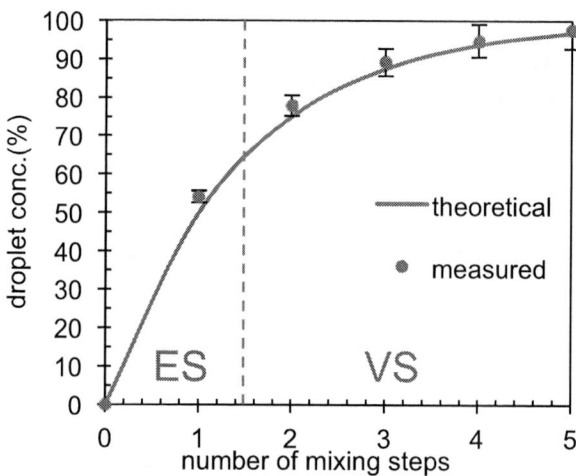

Figure 5: Mixing profile showing generation of ES medium and VS medium.

through a serial mixing/splitting process (Fig. 5 and Fig. 6). This is accomplished by mixing the embryo-containing droplet with a vitrification solution droplet (VS), thus increasing the concentration of cryoprotectant around the embryo. The resulting droplet is then split into two smaller droplets with the daughter droplet containing the embryo identified and kept, while the other droplet is moved to the waste reservoir. After the first mixing step the droplet reaches 50% cryoprotectant concentration (i.e., equilibrium solution or ES), the embryo is kept in the ES droplet for 10 minutes. Then the cryoprotectant concentration is increased again by droplet mixing and splitting. Contrary to ES medium, embryo volume sharply decreases in VS medium and does not recover (Fig. 7).

After complete transfer of the embryo into the VS medium, the droplet containing the embryo is moved toward the edge to be collected by a micropipette (Fig. 4), and then plunged into liquid nitrogen. To verify success of the vitrification process using digital microfluidics, embryos vitrified on device were thawed back and confirmed to have recovered in volume and have healthy morphology.

Contrary to conventional vitrification protocols and manual operation, which subject embryos to sudden changes in medium concentration, the digital microfluidic approach gradually increases the VS medium concentration, (Fig. 5), which is believed by IVF practitioners to induce lower osmotic stress on the embryo [7]. This gradual medium concentration increase is not feasible to achieve in manual operation.

RESULTS

We defined two measures to evaluate the performance of digital microfluidic vitrification, including survival rate and development rate. Survivability was measured by examining the morphology of the embryo before and after freezing. Dark colored cells, ruptured zona pellucida, and deformed or shrunken shaped cells were classified as dead embryos. The development rate was determined by

culturing survived embryos after freezing and thawing for an additional 24-48 hours. If the cell number within the embryo increased or developed to the blastocyst stage, it was counted as developed. A control sample of non vitrified embryos was also cultured to identify the base development rate of the mouse embryo population used.

Table 1: Summary of vitrification results

	Survival Rate	**Development Rate**
Control (Non vitrified)	100% (14/14)	93% (13/14)
Manual	73% (11/15)	91% (10/11)
DMF Chip	77% (10/13)	90% (9/10)

Table 1 summarizes the results, showing comparable survival and development rates between manual and automated trials. However, with a larger population size we hypothesize the digital microfluidic chip to produce a higher

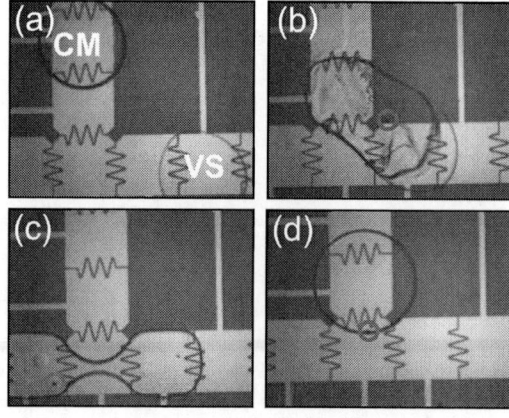

Figure 6: (a) Embryo (red circle) contained in culture medium (CM) droplet. (b) Embryo droplet mixed with VS droplet. (c) Droplet split into two droplets (left contains embryo). (d) Droplet containing embryo is kept and other droplet is sent to waste. Process is repeated to increase VS concentration.

978-1-4799-3510-9/14 $31.00 © 2014 IEEE   831

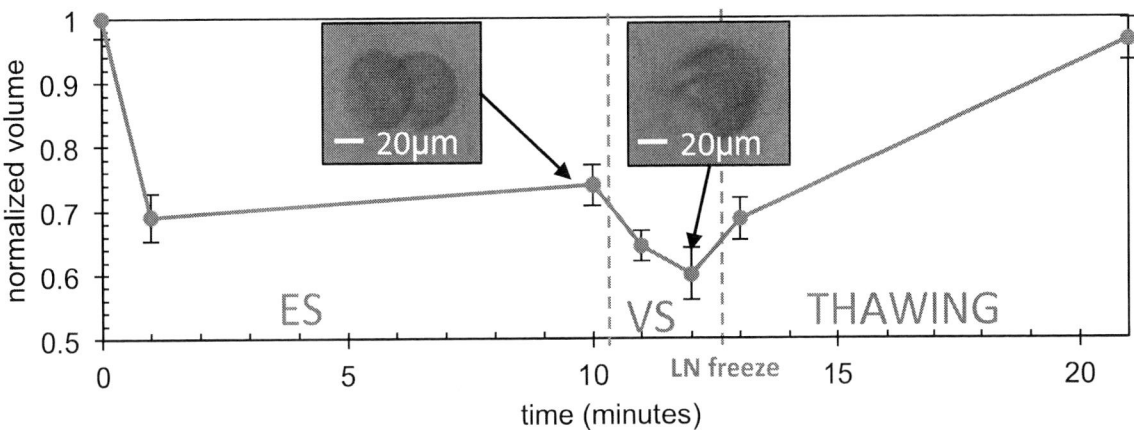

Figure 7: Embryo cell volume is monitored through vitrification procedure ensuring that embryo properly shrinks in vitrification medium and returns to original volume after liquid nitrogen (LN) freezing and thawing.

survival rate due to its gradient generation. Additionally, since the embryo is constantly imaged on video, its volume can be measured throughout the procedure and used to measure the quality of both the embryo and the protocol. Figure 7 shows one such assessment.

CONCLUSION

A digital microfluidic device was designed and tested to automate the liquid handling and timing tasks in embryo vitrification. Technical advantages of this approach, compared to manual operation and channel-based microfluidic vitrification, include automated operation, cryoprotectant concentration gradient generation, and feasibility of loading and retrieval of embryos. The device permits researchers to readily change/test protocols. Significant labor costs were also reduced by eliminating the need for highly skilled operators.

ACKNOWLEDGEMENTS

Y.S. acknowledges financial support from the Canada Research Chairs Program.

REFERENCES

[1] W. H. Tsang and K. L. Chow, "Cryopreservation of mammalian embryos: Advancement of putting life on hold.," *Birth defects research. Part C, Embryo today : reviews*, vol. 90, no. 3, pp. 163–75, Sep. 2010.

[2] D. G. Whittingham, "Survival of mouse embryos after freezing and thawing.," *Nature*, vol. 233, no. 5315, pp. 125–6, Sep. 1971.

[3] W. F. Rall and G. M. Fahy, "Ice-free cryopreservation of mouse embryos at -196 degrees C by vitrification.," *Nature*, vol. 313, no. 6003, pp. 573–5, 1985.

[4] G. Vajta and Z. P. Nagy, "Are programmable freezers still needed in the embryo laboratory? Review on vitrification," *Reprod. Biomed. Online*, vol. 12, no. 6, pp. 779–796, Jan. 2006.

[5] Y. S. Heo, H.-J. Lee, B. a Hassell, D. Irimia, T. L. Toth, H. Elmoazzen, and M. Toner, "Controlled loading of cryoprotectants (CPAs) to oocyte with linear and complex CPA profiles on a microfluidic platform.," *Lab Chip*, vol. 11, no. 20, pp. 3530–7, Oct. 2011.

[6] J. Ali and J. N. Shelton, "Design of vitrification solutions for the cryopreservation of embryos.," *J. Reprod. Fertil.*, vol. 99, no. 2, pp. 471–7, Nov. 1993.

[7] J. E. Swain, D. Lai, S. Takayama, and G. D. Smith, "Thinking big by thinking small: application of microfluidic technology to improve ART.," *Lab Chip*, vol. 13, no. 7, pp. 1213–24, Mar. 2013.

*These authors contributed equally to the work.

CONTACT

a) M. Abdelgawad; mohamed.abdelgawad1@eng.au.edu.eg
b) Y. Sun; sun@mie.utoronto.ca

CHARACTERIZATION OF RED BLOOD CELL DEFORMABILITY CHANGE DURING BLOOD STORAGE

Yi Zheng[1], Jun Chen[1], Tony Cui[2], Nadine Shehata[3], Chen Wang[3], and Yu Sun[1]
[1]University of Toronto, Canada, [2]University of Minnesota, USA, [3]Mount Sinai Hospital, Canada

ABSTRACT

Stored red blood cells (RBCs) show progressive deformability change during blood banking/storage. Their deformability change over an 8 weeks' storage period was measured in this work using a microfluidic device and high-speed imaging. Multiple parameters including deformation index (DI), time constant (shape recovery rate), and RBC circularity were quantified. Compared to previous RBC deformability studies reported in the literature, our results include a significantly higher number of cells (>1,000 cells/sample vs. a few to tens of cells/sample) and, for the first time, reveal deformation changes of stored RBCs when they travel through human-capillary-like microchannels. The correlation between deformability and morphology of stored RBCs is also presented.

INTRODUCTION

In transfusion medicine, red blood cells (RBCs) collected from blood donors are processed and stored in a blood bank. Every year in the U.S. and Canada, over 14 million units of RBCs are administered to more than 5 million patients. Present regulations specify 42 days as the shelf life for stored blood [1]. However, it has been suggested that during storage red blood cells undergo morphological, structural, and functional changes, which may induce clinical complications and adversely affect patient mortality.

The degradation of stored RBCs is known as the storage lesion. Although the clinical consequences of storage lesions remain controversial, intensive research has shown how parameters, which govern RBCs' metabolic ability and oxygen delivery capacity, such as 2,3-DPG, potassium, pH, HbO_2 saturation, RBC ATP, RBC NO, SNO-Hb, and haemolysis, change over the life span of stored RBCs [2]. In addition to these biochemical properties, the biconcave shape and high deformability of RBCs are also crucial for their physiological activities and functionality.

Conventional techniques for RBC deformability measurement (e.g., pipette aspiration and optical tweezers) are tedious and skill-dependent. More importantly, the slow measurement speed (minutes to tens of minutes for testing one cell) makes these techniques infeasible to obtain sufficient information of the highly heterogeneous blood cell population [3]. As stored RBCs age, the cells change their morphology progressively from biconcave to more spherical (spheroechinocytes). Even for the cells within the same stored blood sample, their morphology change varies significantly. Hence, testing only a few cells from a blood sample cannot objectively reveal deformability changes of the sample. Compared to micropipette and optical tweezers,

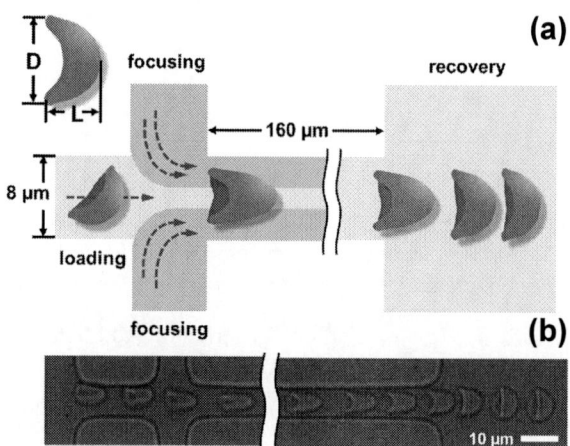

Fig.1 (a) Schematic of the microfluidic device for studying deformability changes of stored RBCs. (b) Experimental images showing the centering, orienting, folding, and shape recovering of an RBC.

ektacytometers are relatively easy-to-use. However, ektacytometry measurement is limited to approximately 50-60 RBCs per test. Importantly, the stretching deformation mode in ektacytometry testing is not physiologically relevant since in vivo RBCs are folded when flowing through human microcapillaries with diameters comparable to or smaller than RBCs. It is known that the mechanical properties of RBCs can differ significantly when deformed under different modes (e.g., extension or folding).

This paper describes a microfluidic device for studying the folding of stored RBCs, which is capable of characterizing over 1,000 RBCs within 3 minutes. Compared to existing stored RBC deformability studies, our results include a significantly higher number of cells (>1,000 cells/sample vs. a few to tens of cells/sample) and, for the first time, reveal deformation changes of stored RBCs when traveling through human-capillary-like microchannels. The correlation between deformability and morphology of stored RBCs is also reported.

Fig. 1(a) shows a schematic of the microfluidic device for quantifying RBC deformation behavior. The central channel is 160 μm long and has a cross-sectional area of 8 μm×8 μm. The recovery region has a cross-sectional area of 200 μm×8 μm, where the deformed RBCs gradually recover to their original shape. Two focusing channels are integrated to center and orient the RBCs. The loading and focusing channels are connected to a custom-developed precision pressure source using water tanks.

When RBCs flow through the central channel, RBC images are captured by a high-speed camera (5,000 frames/sec) through an inverted microscope. Fig. 1(b) shows the dynamic process of RBC deformation. When the cell entered the channel, its position was close to one of the microchannel walls. After the focusing unit, the cell was centered and adjusted to the 'standing' orientation. It then underwent symmetrical shear stress, resulting in a parachute-like shape near the exit. When the cell exited the microchannel, shear stress was released, and the RBC gradually recovered to its original shape.

In order to quantify the deformation behavior of RBCs, deformation index (DI) is defined as DI=L/D [see Fig. 1(a)] in the last frame of image right before the cell exits the microchannel and enters the recovery region. Recovery time constant (tc) is determined by exponentially fitting DI values with respect to time. Circularity, defined as circularity=$4\pi \times$Area/Perimeter^2 (circularity=1 means a perfect circle), is measured experimentally in the last frame of image right before the cell exits the region of interest to depict the morphology of each stored RBC.

METHODS AND MATERIALS

For each test, an RBC aliquot stored in a blood bank was removed and tested at desired time intervals. After an RBC sample was drawn from the aliquot, it was diluted with PBS with 1% w/v BSA (RBC sample:PBS=1:100) and incubated for 10 min at room temperature to prevent adhesion to channel walls. After the RBCs were loaded into the device inlet, a custom-developed water tank applied pressure (2.5 kPa) to drive the cells through the microchannels. Cell images were captured at a speed of 5,000 frames/sec via a high-speed camera, under a 60× objective of an inverted microscope. A custom-designed MATLAB image processing program was developed for automated RBC image analysis. Length, width, area, and perimeter are measured for each RBC to quantify its DI, time constant (tc), and circularity.

RESULTS AND DISCUSSION

Hydrodynamic focusing efficiency

The shape of RBCs flowing through small capillaries is a function of flow rate, initial position relative to the central line of the capillary, and the diameter of the capillary [4][5]. For consistency, RBCs need to be centered in the microchannel and deformed symmetrically. If cells are asymmetrically deformed, the definition of DI=L/D becomes less appropriate for describing their deformation behavior. Experiments demonstrate that RBCs close to the central line of the microchannel reveal a more symmetrically deformed shape, while cells near channel walls usually are asymmetrically deformed. Cells near the channel walls at the entrance typically stay near the channel wall until exiting the channel. When RBCs enter the channel, their positions are random. Hence, hydrodynamic focusing near the inlet (see Fig. 1) was used to better position RBCs close to the central line of the microchannel. Fig. 2 shows a histogram of central distance (i.e., the distance between the cell center and the microchannel's

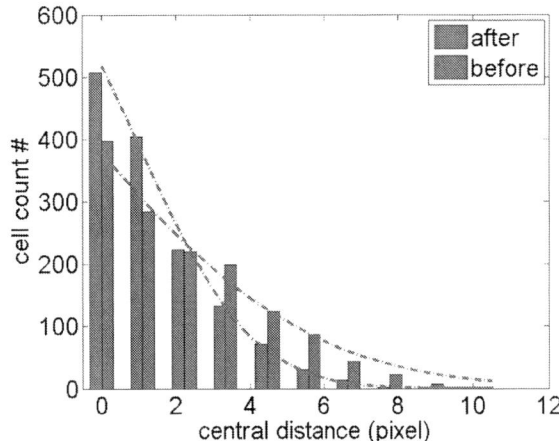

Fig.2 Efficiency of hydrodynamic focusing. Histogram of central distance (i.e., the distance between the center of the cell and microchannel's central line) of RBCs before (blue) and after (red) the focusing unit (n=1,500).

central line) of RBCs before (blue) and after (red) the hydrodynamic focusing unit (n=1,500). The percentage of RBCs near the microchannel center was increased significantly by hydrodynamic focusing, and the number of cells more than two pixels away from the channel center was significantly reduced. The data presented in this paper only include those RBCs that were no more than two pixels away from the central line of the microchannel.

Time constant

After the RBC enters the recovery region, the time constant (tc) of each RBC is determined by fitting its DI values over time to an exponential function, in order to characterize the recovery rate of the cell. The exponential model is adapted from the standard Kelvin–Voigt model for describing the recovery properties of RBCs. Fig. 3 shows two sets of data, one from a biconcave RBC (blue) and the

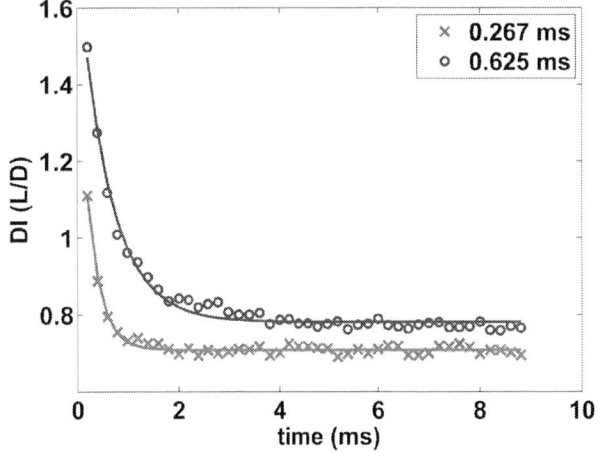

Fig. 3 Shape recovery of a biconcave RBC (blue) and spherical RBC (red) fitted to an exponential model. The spherical shape of the cell was due to morphological change during blood storage.

other from a spherical RBC (red) due to storage. The first data point on each curve was obtained from the image frame right after the cell exits the microchannel. For the RBCs shown in Fig. 3, time constant (tc) of the spherical RBC is 0.267 ms, and tc of the biconcave RBC is 0.625 ms.

It is worth noting that the time constant of RBCs reported in the literature was approximately 100 ms. In contrast, the time constant we quantified ranges from 0.1 ms to 1 ms. In previous studies, DI was typically measured starting from 100 ms after an RBC was released from deformation until the shape completely recovered (several seconds). In our experiments, enabled by high-speed imaging, the rapid recovery process immediately after an RBC is released was captured (from 0.2 ms to 10ms). During this period, RBCs recover much faster than in the later period (e.g., after 100 ms as in earlier studies). As a result, the captured fast dynamics of RBC deformation more authentically revealed the cell shape recovery behavior, and exponential fitting resulted in time constant values orders of magnitude lower than those from earlier results in the literature.

Deformability changes

Fig. 4 summarizes results from testing the same blood sample stored for 1 week (n=1,038), 3 weeks (n=1,027), and 6 weeks (n=1,001). The scatter plots show tc vs. circularity [Fig. 4(a)] and DI vs. circularity [Fig. 4(b)]. Within each sample, both tc and DI show a decreasing trend as circularity increases, indicating that morphological changes of RBCs are a factor that causes RBC deformability changes over blood storage. Fig. 4(c) shows the distribution of these three parameters. It can be seen that the average time constant becomes lower as the RBCs are stored longer. The lower time constant may be attributed to ATP loss. It has been proven that the depletion of ATP alters its binding to spectrin–actin, which modifies RBC's cytoskeleton structure, resulting in RBC stiffening and faster recovery. During storage, RBCs undergo a number of biochemical changes, including ATP depletion, which may be a cause of the lower time constants of older RBCs.

The average circularity of fresher and older RBCs is not significantly different; however, the distribution (i.e., standard deviation) becomes wider. Thus, the distribution width of circularity (circularity-DW) can possibly be used as an indicator of stored RBC quality or age. The alteration of circularity distribution over time is mainly contributed by RBC morphology changes. Based on experimental observation, as stored RBCs age, they first swell and then progressively change to a sphere-like shape. A portion of the aged RBCs became more spherical, making their circularity approach one. There were also RBCs appearing isotropically enlarged from storage, and these RBCs typically revealed a higher DI value. When they pass through the microchannel, their deformed shape resulted in low circularity. The more spherical cells and isotropically swelling cells increased circularity-DW.

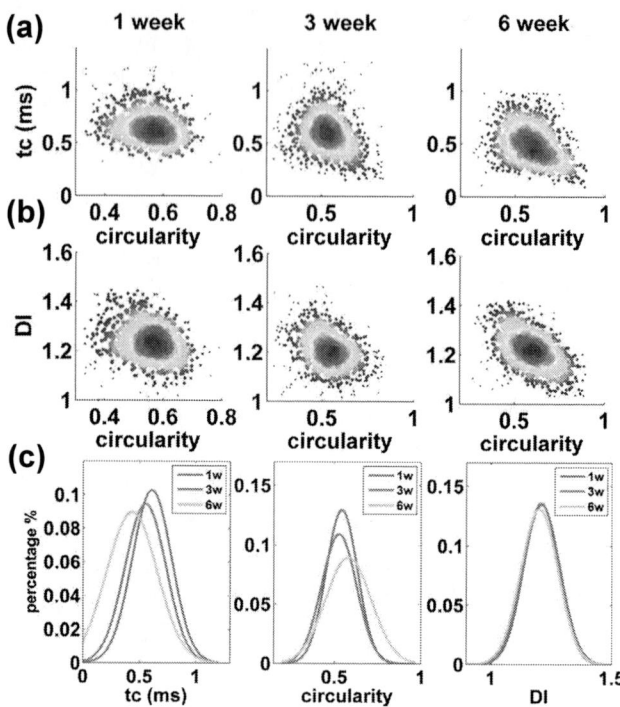

Fig. 4 (a)(b) Scatter plots of time constant vs. circularity; deformation index (DI) vs. circularity, for the same blood sample stored for 1 week (n=1,038), 3 weeks (n=1,027), and 6 weeks (n=1,001). (c) Distribution of time constant, circularity, and DI.

No significant difference in DI was found (DI: 1.227 for 1-week, 1.221 for 3-week, and 1.219 for 6-week) [Fig. 4(c)]. This implies that when stored RBCs are transfused into patients, they are able to flow through microcapillaries with a similar folding capability. However, their stretching capability might become poorer, according to the lower elongation index (EI) measured with ektacytometry [6][7]. The insignificant folding DI change of older RBCs (vs. fresher RBCs) as quantified in our work, and the significant

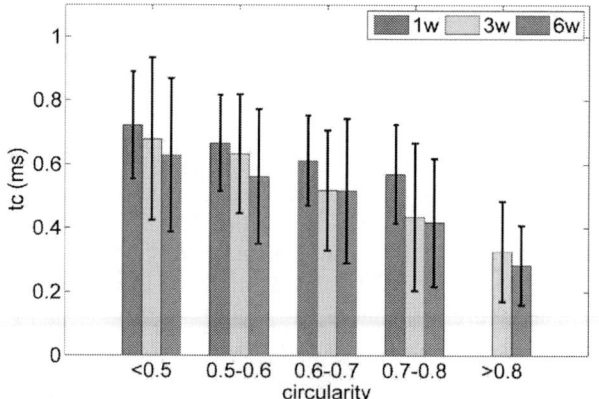

Fig. 5 Comparison of time constants for the same blood sample stored for 1 week (n=1,038), 3 weeks (n=1,027), and 6 weeks (n=1,001). Circularity is divided into five sub-ranges. Error bars represent standard deviation.

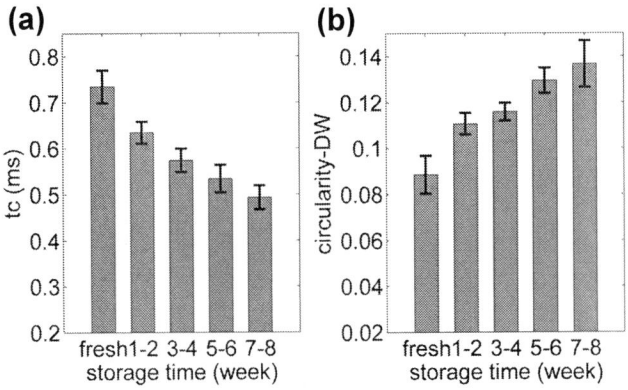

Fig.6 Time constant (a) and circularity-DW (b) alteration over time. Each data point was obtained from 5 blood samples and over 1,000 cells were tested within each sample. Significant difference exists between neighbor data points (p<0.05).

stretching EI change of older RBCs (vs. fresher RBCs) as reported in ektacytometry measurements can be due to the fundamentally different cell deformation modes. Although the depletion of ATP alters RBC cytoskeleton, the stretching EI change might not be a concern in transfusion medicine since the stretching mode is not in vivo like and can be physiologically irrelevant.

We further investigated the effect of RBC morphology change (circularity) on time constant. Fig. 5 shows the average and standard deviation of measured tc values within each circularity range (divided into five sub-ranges). Both fresher and older RBCs with higher circularity reveal lower time constant. Due to intrinsic property changes, within each circularity range, older RBCs show consistently lower time constants, compared to fresher RBCs.

Finally, five blood samples were tested from fresh to 8 weeks' storage, at time intervals of every two weeks (n>1,000 per time interval per sample). As shown in Fig. 6, time constant decreases and circularity-DW (distribution width) increases over blood storage. Significant differences exist between neighboring data points (p<0.05), demonstrating that time constant and circularity-DW can possible be used as indicators of RBC storage age or stored RBC quality. Standard deviations shown in Fig. 6 can be attributed to blood donor variations and variations in blood processing procedures.

CONCLUSION

The deformability changes of stored red blood cells (RBCs) were studied using a human-capillary-like microfluidic channel. High-speed imaging system and automated image processing were used to quantify multiple parameters, enabling a higher measurement speed. Fresh and stored blood samples (up to 8 weeks) were tested. Besides large sample sizes, our study, for the first time, revealed deformation behavior changes of stored RBCs when traveling through human-capillary-like microchannels. Although existing literature consistently reported stretching

deformability change of stored RBCs, our results show that no significant difference exists in their folding deformability. Furthermore, we report that significant changes in time constant (i.e., recovery rate) and circularity distribution width (i.e., heterogeneity of morphology) can be useful parameters for quantifying stored RBC quality or age.

REFERENCES

[1] E. Bennett-Guerrero et al., "Evolution of adverse changes in stored RBCs," Proceedings of the National Academy of Sciences of the United States of America, vol. 104, pp. 17063-17068, Oct 23 2007.

[2] J. Bonaventura, "Clinical implications of the loss of vasoactive nitric oxide during red blood cell storage," Proceedings of the National Academy of Sciences of the United States of America, vol. 104, pp. 19165-19166, Dec 4 2007.

[3] Y. Zheng et al. "Recent advances in microfluidic techniques for single-cell biophysical characterization," Lab on a Chip, vol. 13, pp. 2464-2483, 2013.

[4] G. Tomaiuolo et al.,"Red blood cell deformation in microconfined flow," Soft Matter, vol. 5, pp. 3736-3740, 2009.

[5] Y. Zheng et al., "Electrical measurement of red blood cell deformability on a microfluidic device," Lab on a Chip, vol. 13, pp. 3275-3283, 2013.

[6] B. Blasi et al., "Red blood cell storage and cell morphology," Transfusion Medicine, vol. 22, pp. 90-96, 2012.

[7] S. Henkelman et al., "Is red blood cell rheology preserved during routine blood bank storage?," Transfusion, vol. 50, pp. 941-948, 2010.

CONTACT

*Y. Sun, tel: +1-416-946-0549; sun@mie.utoronto.ca

DETERMINATION OF MULTIDRUG RESISTANCE LEVEL IN K562 LEUKEMIA CELLS BY 3D-ELECTRODE CONTACTLESS DIELECTROPHORESIS

Y. Demircan[1,2], M. Erdem[3], E. Özgür[2], U. Gündüz[3], H. Külah[1,2]

[1]Department of Electrical and Electronics Engineering, METU, Ankara, Turkey
[2]METU MEMS Center, Ankara, Turkey
[3]Department of Biology, METU, Ankara, Turkey

ABSTRACT

In this study, 3D-electrode contactless dielectrophoresis (DEP) system is utilized to drive a relationship between the drug resistance level and dielectrophoretic response of doxorubicin resistant K562 (K562/dox) cells. The number of trapped cells, stained fluorescently, is ascertained by measuring the light intensity with *Image J*. For accurate counting of cells through intensity analysis, an algorithm was developed, which exploits cell transparency. K562/dox cells resistant to 100, 300, 500 and 1000 nM doxorubicin are used for the analysis. A significant increase in trapped cell number is observed up to 300 nM. However, further increase in drug resistance level is not reflected in DEP response, considerably.

INTRODUCTION

Cancer cells may evolve resistance to different drugs/chemicals with distinct structures and functions, a phenomenon, named as multidrug resistance (MDR). To detect MDR in cancer patients, the levels of membrane-associated proteins (P-gp and MRP) are analyzed through three different detection methods: (1) *In vivo* imaging, (2) protein assays, and (3) flow cytometry [1]. *In vivo* imaging techniques, such as single-photon emission tomography and positron emission tomography, are exploited to provide the information about dynamic functions of P-gp and MRPs. In these techniques, substrates of P-gp and MRPs are tagged with different isotopes and injected to patient's body. Although labeling is made with safe dosages, repeated administration of such isotopes is not allowed due to side effects of radioactive isotopes. Protein assays give information about the existence of drug efflux pumps, not dynamic functions of them. Protein assays are *in vitro* studies, requiring isolation of cancer cells from the body. Therefore, they are invasive methods. In flow cytometry, isolated cancer cells are tagged with fluorescently labeled antibodies of membrane proteins to detect MDR. It cannot give information about the activity of the membrane proteins. Moreover, like *in vivo* imaging techniques, flow cytometry method is expensive and like protein assays, it is invasive. On the other hand, it provides quantitative information about the expression level of MDR proteins. The expression level of MDR proteins carries the information about the degree of resistance, critical in MDR for the success of therapy. There exists a commercial MDR detection kit based on flow cytometry (MultiDrugQuant (MDQ) kit), developed by Solvo Biotechnology [2].

As a more sensitive, cost-effective, easy-to-use and less invasive technique, DEP is a promising method for MDR detection since it was reported that the dielectric properties of MDR cancer cells are different from the sensitive ones. This difference is caused by the overexpression of P-gp, a membrane bound Cl⁻ channel, in resistant cells. For example, 100 nM doxorubicin resistant K562 (K562/dox100) cells have cytoplasmic conductivity that is 2.17 times greater than the sensitive ones [3].

In this work, K562/dox cells, resistant to 100, 300, 500 and 1000 nM doxorubicin, are trapped through 3D-electrode DEP at a flow rate of 5 μl/min and 9V$_{pp}$ voltage (*f*=48.64 MHz). Trapped cell number is determined by measuring the fluorescence intensity. Results show that DEP can also be used to determine the degree of MDR in K562 cells presenting a nonlinear relationship between the drug resistance level and DEP response of cells, for the first time in literature. The developed relationship can be used for label-free drug resistance degree determination in cancer cells.

DEP THEORY

Dielectrophoresis is defined as the relative movement of particles and suspending medium in nonuniform electric field. The time averaged DEP force for spherical particles is [4]:

$$\langle F_{DEP} \rangle = 2\pi\varepsilon_m r^3 Re\,(F_{CM})\nabla|E^2| \qquad (1)$$

where, ε_m is the medium permittivity, *r* is the particle radius, *Re(F$_{CM}$)* is the real part of Clausius Mossotti factor of the particle, and $\nabla|E|^2$ is the gradient of the external field magnitude square.

F_{CM} is defined for a spherical particle as [4]:

$$F_{CM} = \frac{\varepsilon_p^* - \varepsilon_m^*}{\varepsilon_p^* + 2\varepsilon_m^*} \qquad (2)$$

where, ε_p^* and ε_m^* are the complex permittivities of the particle and medium, respectively. Complex permittivity is a parameter, depending on conductivity (σ), permittivity (ε), and angular frequency (w).

Utilizing Eq. 2, the frequency at which the particle observes no net DEP force can be calculated. This frequency is named as crossover frequency (f$_{cross}$). Particles with similar sizes can be separated by DEP if they have different f$_{cross}$ values. To determine f$_{cross}$ values and *Re(F$_{CM}$)* characteristics of particles, modeling should be utilized as explained in the following part.

CELL MODELING

The real part of Clausius Mossotti factor of the particle depends on particle and medium dielectric properties. Medium properties can be adjusted but particle dielectric properties are out of control and should be measured. After the measurement of dielectric properties of the particles, they should be modeled to choose the proper structures for DEP operation [5].

Biological cell membrane and cytoplasm have different dielectric properties. These properties provide a cell-specific characteristics as cells vary in their molecular content. This can be utilized as a discriminatory feature in DEP cell identification, manipulation, or separation.

According to shell number around the cell, these modeling techniques are classified, such as single or double shell modeling. Effective complex permittivity (Eq. 3) is obtained with these modeling techniques and it is used instead of particle complex permittivity, ε_p^*, in the expression of F_{CM} (Eq. 2) [4].

$$\varepsilon_{eff,p}^* = \varepsilon_s^* \left\{ \frac{[(r+d)/r]^3 + 2[(\varepsilon_i^* - \varepsilon_s^*)/\varepsilon_i^* + 2\varepsilon_s^*]}{[(r+d)/r]^3 - [(\varepsilon_i^* - \varepsilon_s^*)/\varepsilon_i^* + 2\varepsilon_s^*]} \right\} \quad (3)$$

where, d is the thickness of the cell membrane, ε_i^* and ε_s^* are the cell interior and shell complex permittivities, respectively.

The $Re(F_{CM})$ characteristics of K562/dox and sensitive K562 cells are determined in MATLAB using double shell cell modeling (Fig. 1) and dielectric properties reported by Labeed et al. (Table 1) (considering r=7 μm, d= 10 nm) [3]. Based on the simulation results, 48.64 MHz was determined as the f_{cross} of K562 cells while K562/dox100 cells are pulled to 3D-electrodes at the same frequency. Therefore, this frequency is chosen as the working frequency for the trapping tests of K562/dox cells.

Table 1: Dielectric properties of sensitive K562 and K562/dox100 cells [3].

	σ_{cyto}(S/m)	ε_{cyto}	σ_{memb}(S/m)	ε_{memb}
K562	0.23	40	1.8x10^{-6}	8
K562/dox100	0.5	40	2x10^{-6}	9

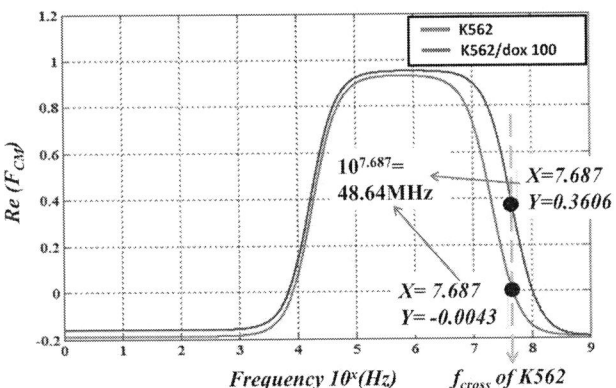

Figure 1: Re(f_{CM}) vs frequency graph of sensitive K562 and K562/dox100 cells in a medium with 2.5 mS/m conductivity.

DESIGN AND SIMULATION

A MEMS-based 3D-electrode contactless DEP device was developed for the detection of imatinib resistance in K562 cells, previously [6]. In this study, it is modified to determine the degree of MDR in K562/dox cells, quantitatively (Fig. 2). It has 27 sidewall patterned and sectional 3D-electrodes (with 30 μm thickness, 40 μm width). The gap between 3D-electrodes is kept as 15 μm, the minimum limit allowed by fabrication, to have a DEP force as high as possible. The total length of the DEP area is fixed to 1500 μm, containing 27 sectional electrodes in one side. Electrode height is chosen as 25-30 μm to obtain a uniform DEP force along the channel depth (30 μm). This provides a position-independent DEP effect for all cells passing through the channel. 300 μm channel width is utilized to obtain a high flow rate that is significant in analyzing higher sample volumes in a shorter period of time.

For hydrodynamic focusing of cells to DEP traps occurring around electrode arrays, 5 V-shaped parylene obstacles are used inside the channel. These V-shaped obstacles are placed at the center of the channel to obtain a better control on flow. The half-length of the channel width is 150 μm and cell diameter is 15 μm, on the average. Therefore, the gap between the tip of the arms of the 1st and 2nd obstacles and the channel wall is chosen as 60 μm, providing the passage of four cells without clogging. This is because the trapping starts at the first two obstacles where the cell concentration is high. After the first two obstacles, cell concentration decreases due to trapping. Therefore, the gap between arm tips and channel wall is decreased to 50 μm to increase hydrodynamic leading force on cells. The distance between obstacles is chosen according to the iterative simulations.

Electrodes are coated with very thin parylene layer (~0.3 μm, limited by current coating system) for insulating purposes. It prevents cell damaging, Joule heating, and electrolysis without a considerable loss in DEP force due to extremely low thickness of insulating parylene layer.

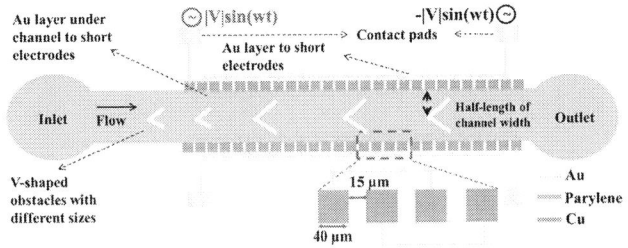

Figure 2: Device schematic.

Figure 3 presents the particle tracing simulations of K562/dox100 and sensitive K562 cells at 5 μl/min flow rate and 9 V_{pp} voltage utilizing finite element modeling. These simulations show that sensitive K562 and K562/dox100 cells can be separated at 48.64 MHz frequency.

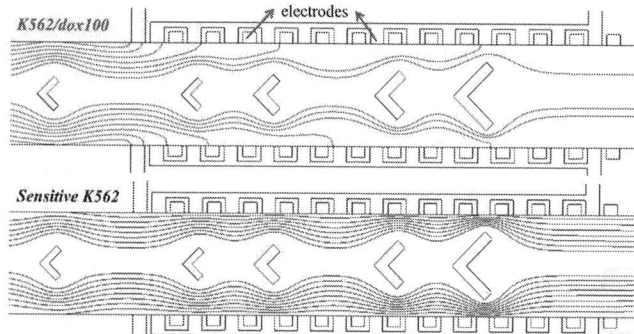

Figure 3: Particle tracing simulations for sensitive K562 and K562/dox100 cells (flow rate = 5 μl/min, voltage = 9 V_{pp}, f = 48.64 MHz).

FABRICATION

Figure 4 and 5 show the fabrication flow and fabricated devices, respectively. DEP devices are fabricated with parylene microchannel process on glass wafer. Au layer is sputtered and shaped with lithography techniques to form a seed layer for electroplating. A thin parylene layer (~5 μm) is deposited to prevent parasitic capacitances between extruded electrodes and Au layer. Holes are opened on this insulation parylene with RIE to determine the electroplated electrode area. Cu electroplating is performed and 3D-electrodes are obtained. Parylene microchannel and V-shaped obstacles are formed with lithography. Inlet and outlet openings are drilled with RIE, and release of photoresist is achieved with acetone. Finally, coating of very thin parylene layer (~0.3 μm) is performed to provide the isolation of 3D-electrodes.

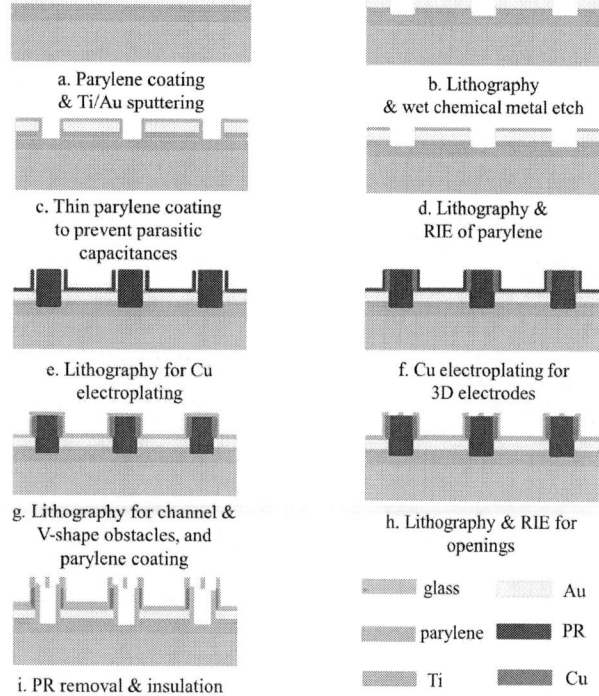

Figure 4: Fabrication flow.

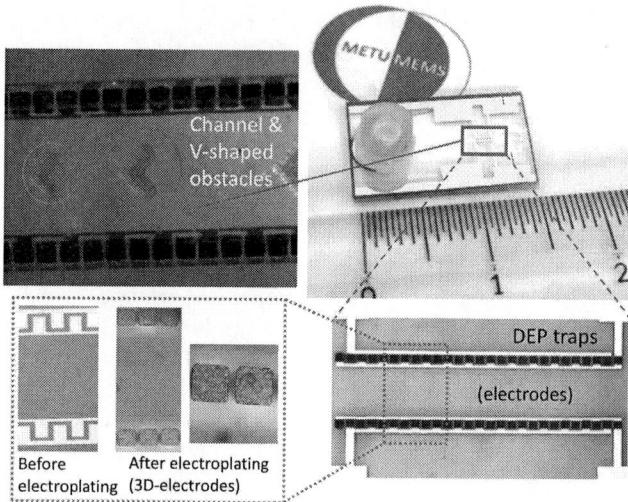

Figure 5: Fabricated device.

RESULTS AND DISCUSSION

100, 300, 500 and 1000 nM doxorubicin resistant K562 cells were centrifuged at 1000 rpm for 5 min. The pellets were washed twice and suspended in isotonic medium (8.5 % (w/v) sucrose and 0.3 % (w/v) dextrose) with a conductivity of 2.5 mS/m [7]. The final cell concentration was adjusted to 2.5×10^5 cells/ml. Cells were stained with fluorescein diacetate and monitored with Evolve 128 high speed camera. Figure 6 shows test setup.

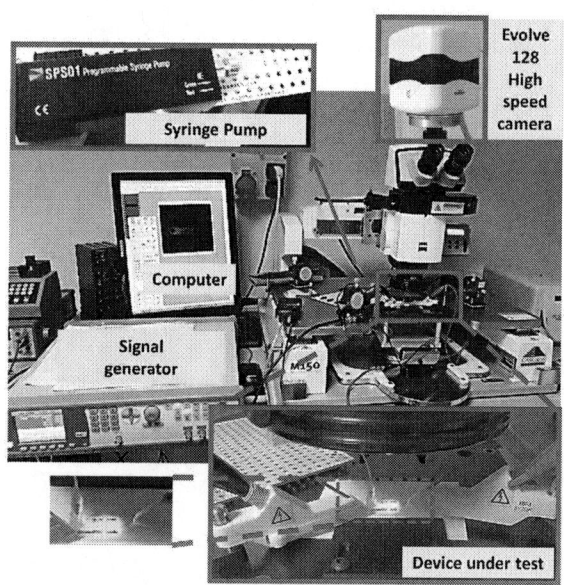

Figure 6: Test setup.

The flow rate was optimized experimentally as 5 μl/min, using the optimum cell concentration (2.5×10^5 cells/ml) and voltage (9 V_{pp}) determined for the separation of K562/IMA cells from sensitive ones [5].

Cell counting was performed by measuring the variation in light intensity, using *Image J*. The high speed camera (*Evolve 128*) has the ability to take screen shot images at every 10 ms. When a cell is trapped, light intensity changes. Utilizing differential analysis between images, cell counting was achieved. However, while the trapped cell number increases, discrimination of the difference between the peaks in light intensity analysis becomes challenging. The error ratio increases as the light intensity of a pixel under examination can be affected by the consecutive pixels. To decrease the error ratio, an algorithm was developed, taking care of the light intensity change in their consecutive pixels. In addition, cells trapped on top of each others can be discriminated with the developed algorithm taking advantage of cell transparency, leading to increased light intensity.

Cells were counted according to the above procedures and the average number of the trapped cells was calculated, for each cell type. A drug resistance coefficient (K_{MDR}), the ratio of the number of trapped cells to that of K562/dox100 cells, was determined to provide a better evaluation. Figure 7 shows the relationship between the drug resistance level and DEP response of K562-MDR cells, measured as K_{MDR} according to the trapped cell number. When the drug resistance level increases from 100 nM to 300 nM, trapped cell number rises almost two times (K_{MDR}~2). However, when 500 nM and 1000 nM resistant cells are compared, the rate of increase in trapped cell number does not correlate linearly with the increase in resistance level. K_{MDR} increases only by 2.3 and 2.5 times for 500 and 1000 nM cells, respectively. A nonlinear equation was derived based on these results, which can be used for the estimation of drug resistance level of cancer cells. As a result, DEP can be utilized for the quantitative analyses of MDR level besides the detection of MDR in cancer cells [6].

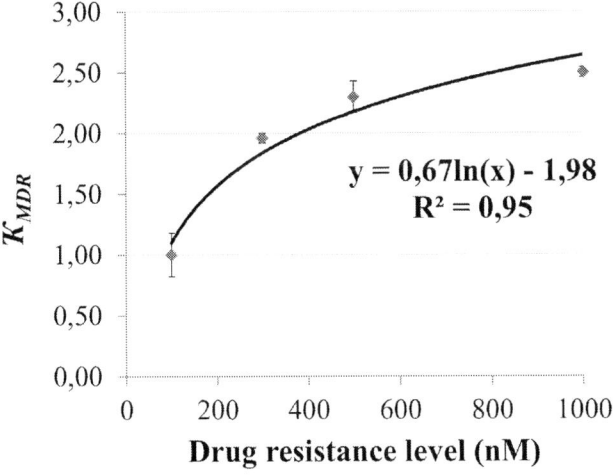

Figure 7: The relationship between the drug resistance level and DEP responses of K562 MDR cells. A nonlinear equation was derived based on these results, which can be used for the estimation of drug resistance level of cancer cells. (cell concentration = 2.5x10^5/ml)

CONCLUSION

To determine the drug resistance degree in K562/dox cells, a modified version of 3D-electrode contactless DEP device was utilized. It is experimentally verified that K562/dox cells can be trapped by applying 9 V_{pp} sinusoidal signal at 48.64 MHz and 5 µl/min flow rate. An algorithm was developed to determine the number of trapped cells exploiting the differential light intensity between consecutive images.

A drug resistance coefficient (K_{MDR}) was defined for the DEP response evaluation of K562/dox cells with different drug resistance level. A nonlinear relationship between K_{MDR} and drug resistance level was observed. This relationship can be further improved by increasing the number of cell types, i.e. DEP responses of K562/dox cells having different drug resistivity.

The observed relationship between drug resistance level and DEP response may imply that P-gp overexpression may not be responsible for MDR at high drug concentration and cells develop another mechanism for survival. Or, although P-gp level increases, cell conductivity cannot increase after a point since this may be destructive for the cell metabolism. To clarify the reason behind this, biological verification of P-gp expression should be performed.

ACKNOWLEDGEMENTS

The authors acknowledge The Scientific and Technological Research Council of Turkey (TUBITAK) for the financial support through project 111E194.

REFERENCES

[1] J. Zhou, Ed., *Multi-Drug Resistance in Cancer.* Humana Press, 2010.

[2] Solvo Biotech, from "http://www.solvobiotech.com" retrieved on Nov, 2013.

[3] F. H. Labeed, H. M. Coley, H. Thomas, and M. P. Hughes, "Assessment of multidrug resistance reversal using dielectrophoresis and flow cytometry," *Biophysical journal*, vol. 85, no. 3, pp. 2028–34, 2003.

[4] T. B. Jones, *Electromechanics of Particles.* Cambridge: Cambridge University Press, 1995.

[5] Y. Demircan, E. Özgür, and H. Külah, "Dielectrophoresis: applications and future outlook in point of care," *Electrophoresis*, vol. 34, no. 7, pp. 1008–27, 2013.

[6] Y. Demircan, A. Koyuncuoğlu, M. Erdem, E. Özgür, U. Gündüz, and H. Külah, "Detection of Imatinib Resistance in K562 Leukemia Cells by 3D-electrode Contactless Dielectrophoresis," *IEEE Transducers 2013*, pp. 2086-89, 2013.

[7] P. R. C. Gascoyne, J. Noshari, T. J. Anderson, and F. F. Becker, "Isolation of rare cells from cell mixtures by dielectrophoresis," *Electrophoresis*, vol. 30, no. 8, pp. 1388–98, 2009.

CONTACT

Yağmur Demircan, tel: +90 (312) 210 6081;
dyagmur@metu.edu.tr

FLEXIBLE MEA FOR ADULT ZEBRAFISH ECG RECORDING COVERING BOTH VENTRICLE AND ATRIUM

Xiaoxiao Zhang[1], Joyce Tai[2], Jungwook Park[1], and Yu-Chong Tai[1]
[1]California Institute of Technology, Pasadena, CA, USA
[2]Tufts University, Medford, MA, USA

ABSTRACT

This paper presents a flexible parylene micro-electrode-array (MEA) that records Electrocardiograms (ECG) from the Zebrafish heart *in-vivo*, covering both the ventricle and atrium area. ECG is a powerful tool for monitoring the heart activity. While ECG technology for human has been well established, this is not true for zebrafish. Our previous work demonstrated baseline ECG recording from zebrafish using MEMS MEAs [1, 2]. However, due to the body structure and small size of the zebrafish (e.g., the heart is roughly 1mm in size and its atrium is buried deep in the thoracic cavity, Fig.1b.), all zebrafish ECGs to date were only recorded from the ventricular side, making it easy to miss important electrophysiological signals from the atrium. To our knowledge, ECG from the atrial angles in Zebrafish has not yet been demonstrated. This work describes a flexible MEA implant (i.e., specially designed according to zebrafish heart anatomy) that records from both the ventricular and the atrial angles.

Furthermore, to demonstrate that this device is useful for heart regeneration monitoring, our work also includes ECG recording before and after laser damage on the ventricle (532nm green light, 32mJ/mm^2, 20mJ total). This chosen energy level of laser pulse is first calibrated using ablated heart histology by EthD-1 florescence staining. The post injury ECG data clearly show ST-wave depression, an indication of ventricular abnormal repolarization state. In addition, repeated missing T-wave is observed from the channels recorded from the atrial angles, which indicate abnormalities in atrial physiology. A hypothesis is that since absorption coefficient of 532nm light in body tissue is rather low, the laser beam penetrated deeply in the heart and created damage deep in the atrium as well as the ventricle. The MEA presented here shows potential for an effective tool to study long-term adult zebrafish heart development and regeneration.

INTRODUCTION

During the last decade, close resemblance between the zebrafish and human's heart physiology has been found [3] so zebrafish (Danio rerio) has become an emerging model for studying the heart. Recently the zebrafish was even proposed as an efficient platform for screening of drugs with potential electrophysiological effects on cardiomyocytes [4]. For instance, since the FDA required preclinical testing for QT prolongation for all new drugs in 2005 [5], zebrafish has been heavily used for drug screening of prolonged QT interval side-effect by identifying the QT interval length of its ECG data after administration of a certain drug. Furthermore, contrary to human's heart, the zebrafish heart has a remarkable ability to "regenerate" and completely recover after severe injury, making it a popular model for studies of regenerative medicine. Monitoring the ECG of the zebrafish during its recovery period has been a widely practiced methodology in heart studies.

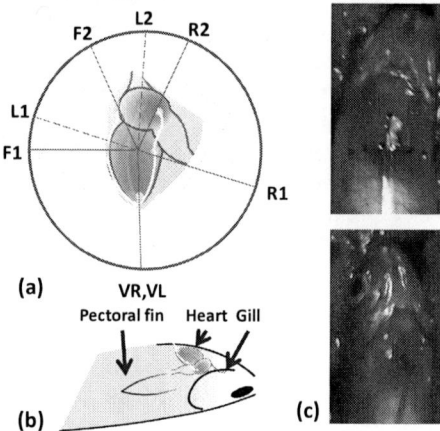

Figure 1: (a) Viewing angles from the 8-channel MEA. F1, L1 and R1 are viewed from mid ventricle. VR and VL are viewed from apical ventricle. F2, L2 and R2 are viewed from the atrium. ECG from each channel is recorded between the MEA electrode and the fish body. (b) Placement of device on fish heart illustration. The chest directly above the heart is opened and MEA is inserted into the chest, the 3 arms conform to the heart epicedial wall and slide into the previously described locations. (c) Photo of MEA surgical placement before and after MEA inserted into chest.

Currently for zebrafish ECG recording needle electrodes have been widely used [6]; one thin needle is inserted into the chest epidermis, another thin needle is inserted into the ventral epidermis in the abdomen as the reference electrode, and differential signal is recorded. The needle electrodes will suffice the basic need of reading the heart rate and the time intervals between the P-QRS-T waves, however the spatial resolution is low considering only one exploring needle can be placed around the heart. Furthermore, the viewing angle is restricted from the ventricle direction only. These limitations can lead to miss-diagnose of subtle abnormities in the atrium. In comparison, human ECG most often consists of 12 leads, 6 of which are chest leads placed in close proximity to the heart.

978-1-4799-3510-9/14 $31.00 © 2014 IEEE

Diagnostics are often made with information based on the multiple chest leads. In this paper, we produced and tested an 8 channel MEA tailored to the zebrafish heart anatomy, where both ventricular and atrial viewing angles are realized.

MEA CHEST LEADS DESIGN

A schematic of the viewing angles is shown in Fig.1a and the surgical placement of the implant is shown in Fig.1b. The flexible parylene MEA consists of 8 leads: 2 front leads (F1, F2), 4 symmetrical side leads (L1, L2, R1, R2), and two ventricle apical leads (VL, VR). F2, L2, and R2 are located at the end of the extended arms, resting on different perspectives of the atrium. F2, L2 and R2 are for the mid ventricle angles, and VL and VR for the apical ventricle angles. The reference electrode is 7mm from the center of the chest electrodes and will rest on the fish body away from the chest. The leads layout and dimensions are shown in Fig.2.

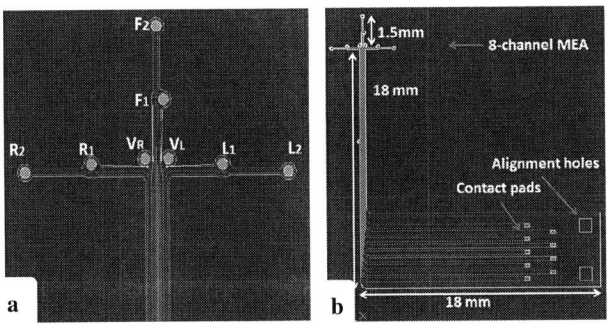

Figure 2: (a) Schematic of MEA and its 8 channels (L and R stand for "left" and "right" from operator's perspective during surgery. F means front. V stands for ventricle.) (b) Schematic of MEA device, the length of each arm is 1.5mm; electrode opening is 100µm in diameter.

MEA FABRICATION PROCESS

The device has a PA-C/Ti/Au/PA-C sandwich structure. The 4 layers in the sandwich structure are as follows: 5µm Parylene bottom passivation layer, 10nm Ti as adhesion layer, 300nm Au conductive layer, then 5µm Parylene top passivation layer. Finished device is treated with short oxygen plasma before released from Silicon wafer to create hydrophilic surface and hence can better conform to the heart tissue in order to reduce noise from motion artifacts. Fig. 3 shows the fabrication process. Parylene-C is deposited first on bare Si wafer, Ti and Au layers are then deposited with electron beam, and patterned by wet etching. Top parylene-C layer is deposited. Photoresist is then spin-coated and patterned as a mask for the opening of the electrode and contact pad openings. Reactive Ion Etching is used finally to etch the top parylene-C layer to expose the electrode, and to create the 2-step contact pad openings. The center of the 2-step contact pad is open to allow application of conductive epoxy through the opening onto the adaptor cable contact metal to be electrically connected [7, 8].

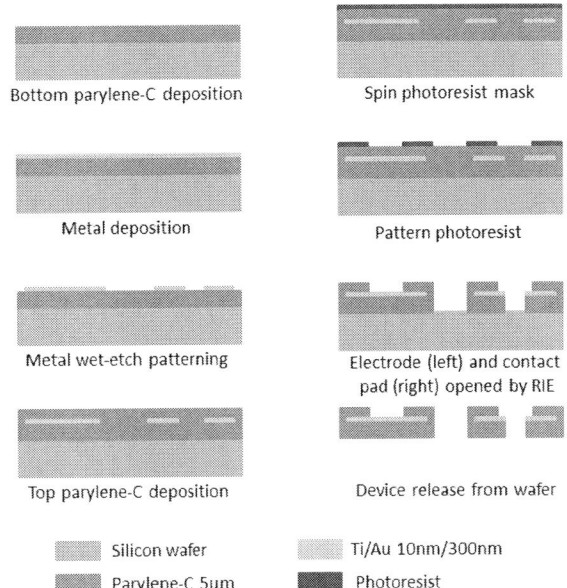

Figure 3: Fabrication process of the flexible MEA. Parylene-C is deposited first on bare Si wafer, Ti and Au layers are then deposited with electron beam, and patterned by wet etching. Top parylene-C layer is deposited. Photoresist is then spin-coated and patterned as a mask for the opening of the electrode and contact pad openings. Reactive Ion Etching is used finally to etch the top parylene-C layer to expose the electrode, and to create the 2-step contact pad openings.

Upon completion of fabrication, electrical connections are made between the MEA and a flexible flat cable (FFC) with conductive epoxy so that the MEA is compatible with commercial FFC connectors. For each channel, the MEA's contact pad is placed "face-down" on the FFC's metal contact. A small amount of conductive epoxy is then applied to the MEA contact pad opening. The conductive epoxy reflows during curing to make a connection between the FFC contact metal and the metal surrounding the MEA contact opening. The contact resistance is measured to be less than 2Ω consistently. An illustration of this method is shown in Fig. 4c. The photo after connection is shown in Fig. 4d. Fig. 4a and Fig. 4b show the MEA chest channel close-up and the assembled device respectively.

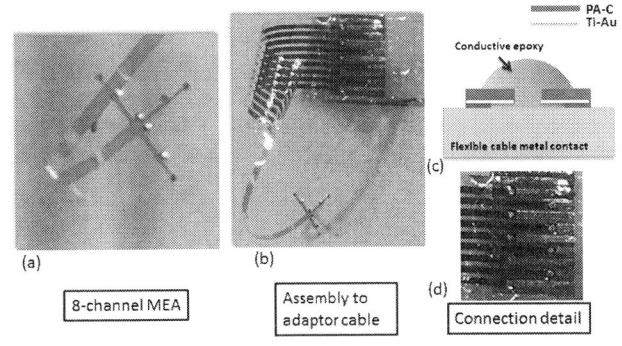

Figure 4: (a) MEA close-up, (b) Device assembled on flexible flat adaptor cable for electronics connection, (c)

978-1-4799-3510-9/14 $31.00 © 2014 IEEE

Flexible cable connection schematic, and (d) Close-up of the connected device on FFC.

RECORDING SYSTEM

Adult zebrafish is placed on its dorsal aspect along a wet sponge soaked with tank water after anesthesia. The zebrafish and the analog front end (signal filtering and amplification) circuit board are placed in a stainless steel shielding cage, shown in Fig. 5. The wet sponge and the fish body are connected to the shielding cage and to common ground along with the amplifier board, the digital PC interface board and the PC. The chest is opened by micro-scissors and fine tip tweezers so that the heart is visible, the 8 channel MEA is inserted carefully in place along the epicardium wall under a stereoscope. The reference electrode of the MEA is on the fish abdomen outside the fish body.

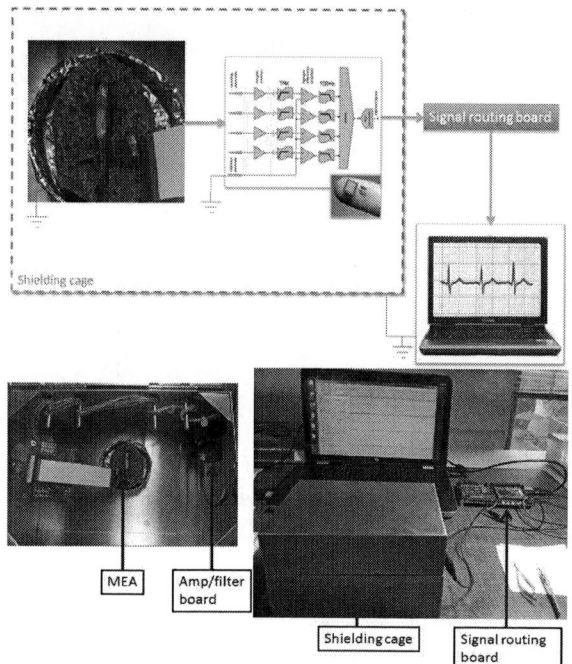

Figure 5: Up: Schematic of the recording system. Down: Photos of the experimental setup.

Intan RHD2000 series amplifier chip (Intan Technologies, Los Angeles, CA, USA) is used as the analog front end for amplification and filtering. In this experiment the band pass lower and upper corner frequencies are set to 0.1Hz and 100Hz respectively. The high pass filter is a simple one-pole filter, and the low pass filter is a three-pole Butterworth filter. The sampling frequency is 1 KHz. Two gain stages of 96 V/V and 2 V/V are used respectively on the Intan amplifier chip, with a total gain of 192 V/V. The A/D converter is 16 bit with step resolution of 0.195µV. Our software signal analysis utilizes a wavelet transform noise reduction algorithm by setting adaptive thresholds in different frequency bands from 0 – 100Hz. Detailed method is explained in our previous work [1]. The raw and filtered

recording is shown in Fig. 6. In the filtered waveform we can see distinct P-QRS-T waves which represent the depolarization of the atrium, depolarization of the ventricle and the repolarization of the ventricle respectively.

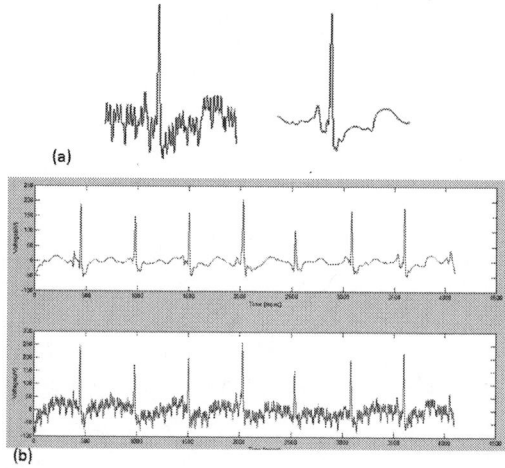

Figure 6: (a) a comparison of raw data and wavelet filter processed data; (b) recording of 4 seconds of raw (down) and processed (up) data.

LASER INJURY

To demonstrate that this MEA device is useful for heart regeneration monitoring and diagnoses, we induced laser injury on the ventricle of the live fish, and recorded ECG data before and after laser damage. The laser used is a class 3B green pulse laser with peak wavelength of 532nm, at energy density 32mJ/mm^2, 20mJ total. This chosen energy level of laser pulse is first calibrated with histology by EthD-1 florescence staining of the ablated heart. The damage is visible and the volume of ablation can be reproduced within reasonable tolerance. With the above energy density the ablated area is shown in Fig. 7.

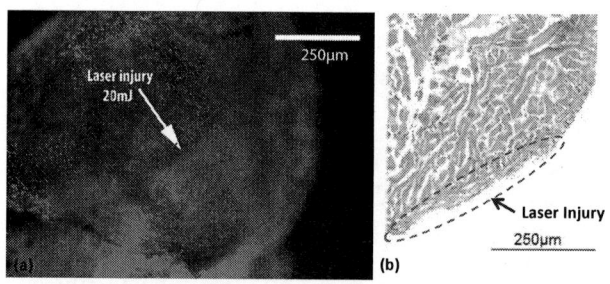

Figure 7: Apparent damage after pulsed laser shot (532nm green laser, 20mJ total energy at 32mJ/mm^2, 250µmx250µm focused spot). (a) EthD-1 staining, the bright red rectangle shows clear damage on the ventricle. (b) Histology showing the depth of the ablated tissue.

The before and post injury data are shown in Fig. 8. Post injury recording clearly shows ST-wave depression, a well-known indication of ventricular abnormal repolarization state. In addition, repeated missing T-wave is observed from the ECGs recorded from the atrial angles, which indicate abnormalities in atrial physiology. A hypothesis is that since absorption coefficient of 532nm light in body tissue is rather low, the laser beam penetrated deeply in the heart and created damage in the atrium as well as the ventricle. Nonetheless, the MEA captured both the ventricular and atrial ECG data, making it possible to confirm the abnormality in the atrium in situations like above whereas would be difficult or impossible to confirm based on a ventricle channel alone.

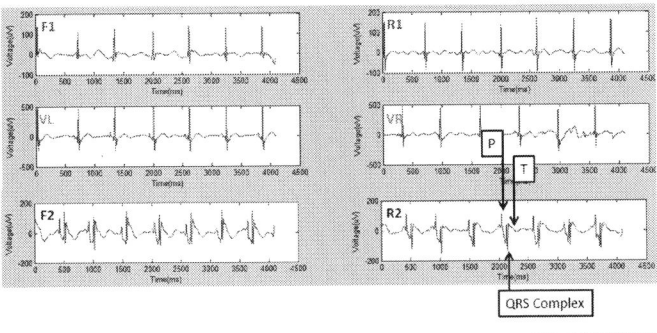

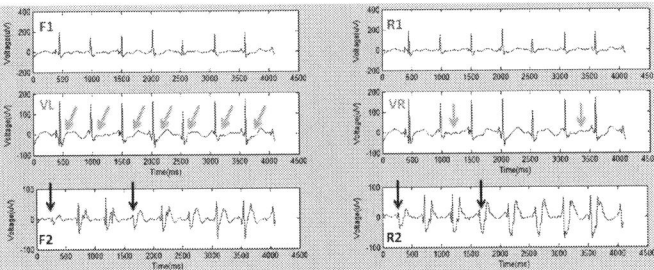

Figure 8: ECGs recorded pre-laser and post-laser. Up: Pre-laser normal ECG from channel F1, R1, VL, VR and F2, R2. Channel F2 and R2 shows a healthy heart activity from the atrial angles. Distinct P, QRS, T waves are pointed out in a period in channel R2, from which we can see a very pronounced P wave that is generated by the atrium depolarization. Down: Post-laser ECG from channel F1, R1, VL, VR and F2, R2. From channel VL and VR we can see a clear ST-wave depression, indicating abnormalities in the ventricle. Furthermore, from F2 and R2 we also see repeatedly missing P waves, which is a sign of atrium abnormalities.

CONCLUSIONS

We demonstrated a flexible parylene based MEA that covers both the ventricle and the atrium of the adult zebrafish heart. This MEA gives multiple viewing angles looking from both the ventricle and atrium into the heart. The wide viewing angles result in a more effective tool to monitor the activities of the zebrafish heart for both drug screening and regenerative studies. We also demonstrated the MEA's capability to capture subtle signals from the atrium to diagnose zebrafish heart abnormalities in the atrium region by comparing the 8 channel ECGs between before and after laser injury. The recorded data show pronounced difference in the S-T wave interval and the irregularities of the P wave in the post-injury case. To the

authors' knowledge, the MEA is the only zebrafish ECG device that is capable to capture both ventricle and atrium data.

ACKNOWLEDGEMENTS

This work is partially funded by the NIH, grant number R01 HL111437. The author would like to thank all Caltech MEMS group members for their fruitful discussions and suggestions.

REFERENCES

[1] F. Yu, Y Zhao, J. Gu, K Quigley, N. Chi, YC Tai and TK Hsai, "Flexible microelectrode arrays to interface epicardial electrical signals with intracardial calcium transients in zebrafish hearts," *in Biomedical Microdevices* 14:357-366, 2012.

[2] Y. Zhao, F. Yu, C. Hung, H. Chang, X. Zhang, T.K. Hsiai, YC. Tai, "A Wearable Percutaneous Implant for Long Term Zebrafish Epicardial ECG Recording", *in Solid-State Sensors, Actu. and Microsys, Transducers 2013 Conference*, pp. 756-759, 2013.

[3] I.U. Leong, J.R. Skinner, A.N. Shelling, and D.R. Love, "Zebrafish as a model for Long QT syndrome: the evidence, and the means of manipulating zebrafish gene expression", in *Acta Physiol.*, Oxf. 199, pp. 257–276, 2010.

[4] C-T. Tsai, C-K. Wu, F-T. Chiang, C-D. Tseng, J-K. Lee, et al. "In-vitro recording of adult zebrafish heart electrocardiogram - a platform for pharmacological testing", *In Clin Chim Acta* 412: 1963–1967, 2011.

[5] FDA (2005). ICH Guidance for Industry: S7B Nonclinical Evaluation of the Potential for Delayed Ventricular Repolarization (QT Interval Prolongation) by Human Pharmaceuticals: *U.S. Department of Health and Human Services.*

[6] DJ. Milan, IL. Jones, PT. Ellinor, CA. MacRae, "In vivo recording of adult zebrafish electrocardiogram and assessment of drug-induced QT prolongation", in *Am. J. Physiol. Heart Circ. Physiol.* 291(1):269-273, 2006.

[7] J.H. Chang, D. Kang and Y.C. Tai, "High yield packaging for high-density multi-channel chip integration on flexible parylene substrates", in *Digest Tech. Papers MEMS 2012 Conference*, pp. 353-356, 2012.

[8] J.H. Chang, Y. Liu, D. Kang and Y.C. Tai, "Reliable packaging for parylene-based flexible retinal implants", in *Digest Tech. Papers Transducers 2013 Conference*, pp. 2612-2615, 2013

CONTACT

*Xiaoxiao Zhang, +1-626-395-2254, xzzhang@caltech.edu

MEASUREMENT OF MECHANOMYOGRAM

Tomonori Kaneko, Nguyen Minh-Dung, Ryo Aoki, Tomoyuki Takahata,
Kiyoshi Matsumoto and Isao Shimoyama
The University of Tokyo, Tokyo, Japan

ABSTRACT

We have proposed an approach for measuring mechanomyogram (MMG) by taking advantage of the acoustic impedance matching between liquid and human skin. Using liquid, we could convey the pressure signal of MMG to a piezo-resistive cantilever efficiently. In experiments, the sensor was placed on the skin surface over biceps brachii (a large muscle that is located in the upperarm). The MMG signal, of which the frequency was in range of 10–15 Hz, was able to be detected using silicone oil as the propagating medium. On the other hand, there was no response in the case of using air as the medium. Experimental results also indicated that the proposed sensor was able to detect the vascular oscillations.

INTRODUCTION

Mechanomyogram (MMG) is a record of the acoustic waves generated by contracting muscle fibers. The waves propagate through human fat to the skin surface and can be detected as sound or vibration using a microphone [1] or an accelerometer [2]. In previous researches, electromyogram (EMG) was recorded simultaneously with MMG [3, 4]. These researches have confirmed that MMG does reflect muscle activity.

There are two explanations on the mechanism of acoustic waves generated by muscle fibers [5]. First, the lateral muscle fiber oscillations generate pressure waves that have one propagating direction [6]. Second, the changes of radii in muscle fibers generate pressure waves in all directions [4]. Though the mechanisms of MMG have not been explained completely, MMG signal reflects the mechanical characteristics of muscle fibers. MMG can also be used to evaluate the muscle properties.

In previous research measuring MMG with acoustic transducers, the propagating medium of acoustic waves was air [7]. Since the acoustic impedance (the product of a material's density and acoustic velocity) of air (4.0×10^2 kg/m^{2}s) and that of human skin (1.5×10^6 kg/m^{2}s) are largely different [8], the acoustic waves should decrease significantly at the boundary surface. In order to achieve measurement with high sensitivity, the measurement system should have the propagating medium that has good acoustic impedance matching with human skin.

In this paper, we propose an approach for measuring MMG with high sensitivity (**Figure 1**). We used a piezo-resistive cantilever as the pressure sensor and liquid as the propagating medium. Indeed, this air/cantilever/liquid structure has been proved to be able to measure acoustic waves with relatively high sensitivity, with measurement resolution of sub Pa [9].

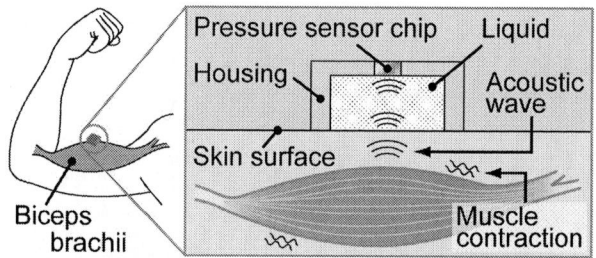

Figure 1: Conceptual sketch of mechanomyogram (MMG) measurement. The device's housing was filled with liquid as the propagating medium to convey MMG signal efficiently. Pressure sensor was composed of piezo-resistive cantilever, which had high sensitivity.

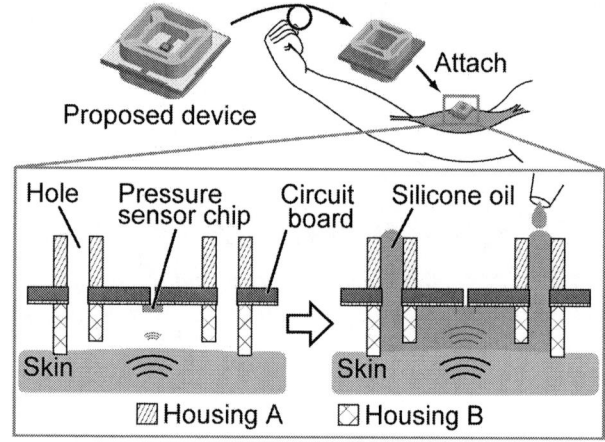

Figure 2: Acoustic impedance matching using silicone oil as the propagating medium. After the device was attached to the skin, silicone oil was injected into the hole of the housing A. Then, the space between the pressure sensor chip and the skin was filled with oil. Detailed description of the device assembly is shown in Figure 4.

PRINCIPLE AND DESIGN

Silicone oil for acoustic impedance matching

Figure 2 shows a conceptual sketch of acoustic impedance matching in this study. We used silicone oil (HIVAC F-4, Shin-Etsu Silicone) to fill the space between human skin and the pressure sensor chip. The density ρ and the acoustic velocity c of the silicone oil are 1.065×10^3 kg/m^3 and 981.6 m/s, respectively. The acoustic impedance of the silicone oil is calculated to be 1.0×10^6 kg/m^{2}s. Compared with the acoustic impedance of air, this value was much closer to that of human skin. Using this silicone oil, acoustic waves could be conveyed with little attenuation at the boundary surface between the skin and the propagating medium.

978-1-4799-3510-9/14 $31.00 © 2014 IEEE

Differential pressure sensor

The design of the differential pressure sensor, which was a piezo-resistive cantilever, is shown in **Figure 3**. The fabrication process of a piezo-resistive cantilever was described in the previous work [10]. In this study, the dimensions of the cantilever were designed as 150 μm×100 μm×0.3 μm.

There was a 5 μm gap between the cantilever and the surrounding walls. This gap size was previously reported as adequate distance for acoustic wave measurement with an air/cantilever/liquid structure [9].

Device design

Figure 4 shows the schematic design of our device. The device consisted of three layers. The top and bottom layers were housings. The middle layer was a circuit board on which the pressure sensor chip was attached.

The housing B was used to keep silicone oil between the skin and the pressure sensor chip. There were four same shaped holes through the two housings and the circuit board. Actually, when we placed and fixed the device onto the skin surface, the skin surface protruded up slightly. If we fill the housing B's cavity with oil without designing any escape space for the oil, relatively large static pressure caused by the skin protrusion will be applied on the cantilever. This pressure should cause oil leak through the gap of the cantilever, or damage the cantilever structure. The holes here were created to solve the above issues.

In experiments, silicone oil was injected through one hole of the housing A after the device was attached and fixed onto the skin. The housing A was used to prevent the oil from overflowing and covering the backside of the cantilever after the housing B's cavity was filled with oil.

Since the device had the housings with holes, acoustic waves could escape through them. The inner walls of housing B were aimed to create room to converge the acoustic waves to pressure sensor chip effectively.

The height difference in the walls of the housing B was 0.8 mm. This value was designed considering the skin protrusion which occurred when the device was attached to the skin surface over biceps brachii. We also designed the height difference so that silicone oil could go through the gap between the skin and the inner walls of the housing B.

Device fabrication

We used acrylic as the material of the housings. The acrylic housings were made by a 3D printing machine (ProJet HD3510, 3DSYSTEMS). The sensor chip was wire-bonded to the circuit board and the wires were covered and protected with UV curable resin (Norland optical adhesive-65, McCrone).

Photographs of the both sides of the complete device are shown in **Figure 5**. **Figure 5(a)** shows the view of housing A side. **Figure 5(b)** shows the view of housing B side. Three layers were attached using UV curable resin.

The size of the fabricated device was 15 mm×15 mm×7 mm (length×width×height).

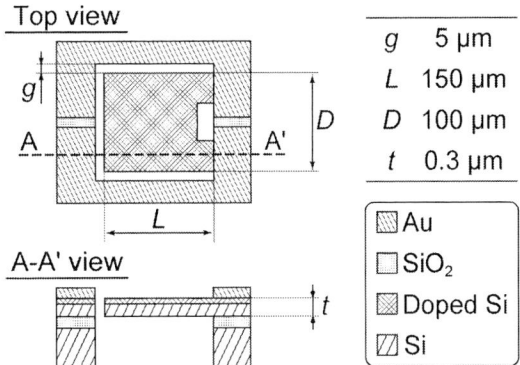

Figure 3: Design of the differential pressure sensor chip. The gap between the cantilever and surrounding walls was 5 μm, which is supposed to be efficient for acoustic wave measurement with liquid.

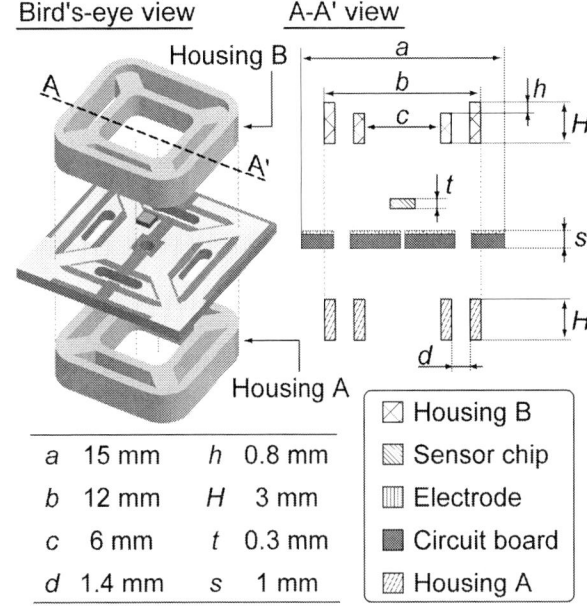

Figure 4: Schematic design of the proposed device. The device had a laminated structure with a pressure sensor chip, a circuit board, and two acrylic housings. The walls of the housing B had the difference in their height level. In experiments, the housing B was attached to the skin so that the space between the pressure sensor chip and the skin could be filled with liquid.

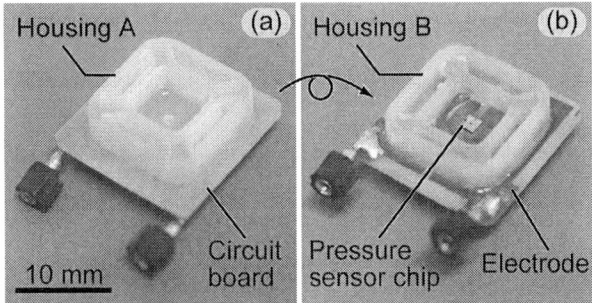

Figure 5: Photographs of the fabricated device. (a)View of housing A side. (b) View of housing B side. Three layers were attached using UV curable resin.

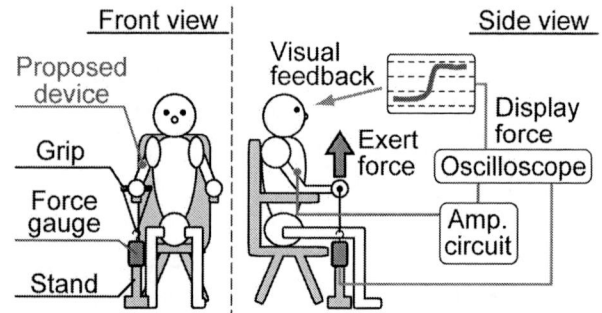

Figure 6: Schematic diagram of the experimental setup. The subject was seated in a chair and asked to fix the elbow angle at 120°. The force amplitude exerted by the subject was measured by a digital force gauge and showed to the subject as the visual feedback.

EXPERIMENTS AND RESULTS

Experimental setup

Figure 6 shows the schematic diagram of experimental setup for measuring MMG signals. The subject was a healthy adult male. The subject was seated in a chair with armrests, and asked to place the arms on the armrests and keep elbow angle at 120°. Note that the elbow angle was defined as 180° when an arm straightens completely.

The subject held a metal bar, which was connected with a digital force gauge (ZP-500N, IMADA) and was asked to perform force exertion upward in the vertical direction, keeping the elbow angle.

Our device was attached to the skin surface over the center of biceps brachii belly in the right arm of the subject and fixed using adhesive tape.

Experimental procedure and results

At the beginning of the experiments, the subject was asked to perform his maximum effort of isometric elbow flection lasting for 2 s for three times with 10 s intervals. The maximum voluntary contraction (MVC) was defined as the maximum force, which was 141 N.

Then, the subject was asked to perform two levels of muscle contractions, corresponding to 25% and 40% MVC for approximately 10 s. The force amplitude exerted by the subject was displayed on the oscilloscope monitor, so that the subject was able to adjust the force to the indicated value. We asked the subject not to shiver in the whole of the arm when the subject performed muscle contractions.

Figure 7(a) and **(b)** shows the results of MMG measurement. In each section, upper graph shows the exerted force and lower graph shows the response of the sensor. In graphs that are located in left or right column, the propagating medium of pressure waves was air or silicone oil, respectively. The vibrations could be measured at both 25% and 40% MVC contractions when the propagating medium was silicone oil, while almost no signal was detected using air as the propagating medium. Note that the vertical scales of **Figure 7(a)** and **(b)** are different.

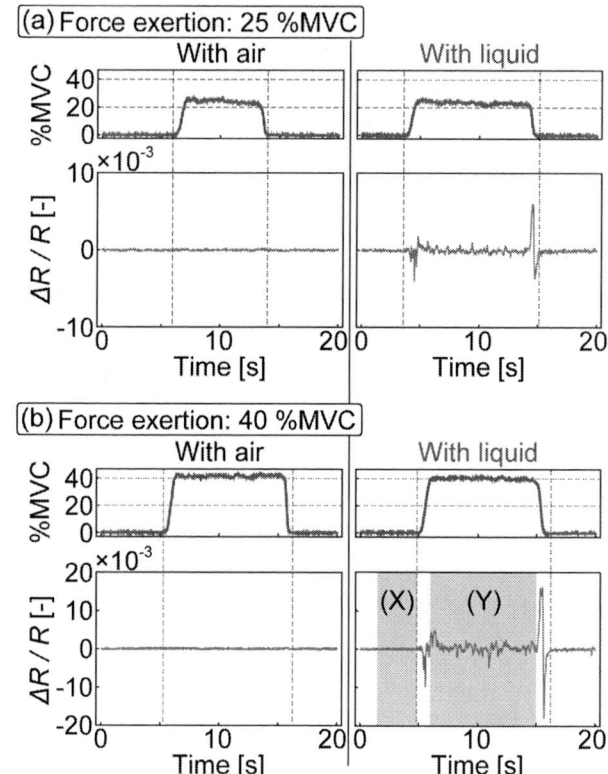

Figure 7: Responses of piezo-resistive cantilever during force exertion at (a) 25% MVC and (b) 40% MVC (Max Voluntary Contraction). In both graph areas, upper graphs show the exerted force and lower graphs show the response of the sensor. The propagating medium of pressure wave was air (graphs in left column) or silicone oil (graphs in right column). Pressure waves were able to be detected when the propagating medium was oil, while they were not with air. The sections (X) and (Y) will be referred in Figure 8.

When the target value for force contraction increased, the amplitude of the response of the sensor increased. These results indicated that the amplitude of MMG signal might increase in proportion to the force the subject performed.

Additionally, in the data obtained from the experiment where the subject performed 40% MVC and silicone oil was used as the propagating medium, a repetitive wave pattern was found approximately every 0.8 s in the section of 1.5–5 s (**Figure 8(a)**). In this 1.5–5 s section, the subject was not asked to perform muscle contraction. The subject measured vascular oscillations by hand on the carotid artery simultaneously and confirmed that the cycle of the vascular oscillations was the same as that of repetitive pattern. Therefore, the result indicated that our sensor could also measure vascular oscillations.

Figure 8(b) shows the FFT (Fast Fourier Transform) analytic result in the section of 6–15 s. In previous research, the main frequency component of MMG signal of biceps brachii was reported to be 10–15 Hz [11]. This FFT analysis result indicated that our sensor was able to measure MMG signal.

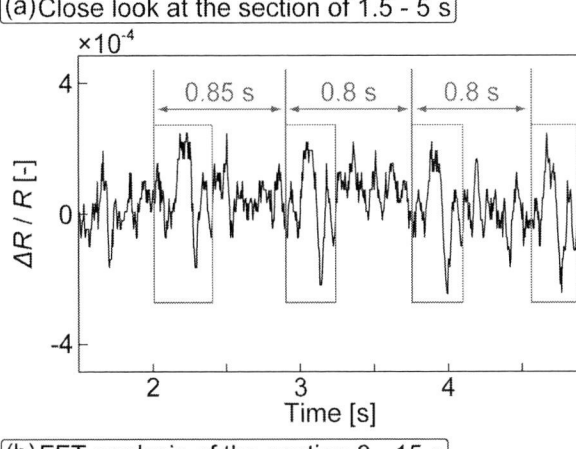

(a) Close look at the section of 1.5 - 5 s

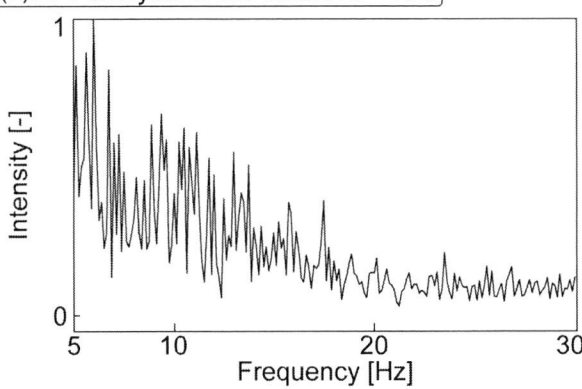

(b) FFT analysis of the section 6 - 15 s

Figure 8: Analysis of the data gained from the experiment where the subject performed 40% MVC and silicone oil was used as the propagating medium. (a) Close look at sensor response in the section of 1.5–5 s, where the subject did not perform muscle contraction (section "(X)" in Fig. 7). A repetitive pattern could be recognized every 0.8 s. (b) FFT analytic result in the section of 6–15 s (section "(Y)" in Fig. 7). The MMG signal, the frequency of which was reported to be 10-15 Hz, was detected.

CONCLUSION

We proposed an approach for measuring MMG by taking advantage of the acoustic impedance matching between silicone oil and human skin.

The fabricated device was placed on the skin surface over biceps brachii of the subject. While the subject was performing isometric muscle contractions, the MMG signal was able to be detected using silicone oil as the propagating medium. We conducted FFT analysis on obtained signal. Its frequency spectrum had 10–15 Hz component, which was reported to be the main frequency of MMG signal of biceps brachii.

Experimental results also indicated that the proposed device was able to detect the vascular oscillations while the subject was not performing muscle contraction. Those results show that our device will provide applications in medical measurement, such as measuring vibration or inaudible sound emitted from human body.

ACKNOWLEDGEMENT

The photolithography masks were made using the University of Tokyo VLSI Design and Education Center (VDEC)'s 8 inch EB writer F5112 + VD01 donated by ADVANTEST Corporation.

REFERENCES

[1] M. Petitjean, B. Maton, and J. C. Cnockaert, "Evaluation of human dynamic contraction by phonomyography," *Journal of Applied Physiology,* vol. 73, no. 6, pp. 2567-2573, 1992.

[2] K. Akataki, K. Mita, M. Watakabe, and K. Itoh, "Mechanomyogram and force relationship during voluntary isometric ramp contractions of the biceps brachii muscle," *European Journal of Applied Physiology,* vol. 84, no. 1-2, pp. 19-25, 2001.

[3] C. Orizio, R. Perini, and A. Veicsteinas, "Changes of muscular sound during sustained isometric contraction up to exhaustion," *Journal of Applied Physiology,* vol. 66, no. 4, pp. 1593-1598, 1989.

[4] C. Orizio, R. Perini, and A. Veicsteinas, "Muscular sound and force relationship during isometric contraction in man," *Journal of Applied Physiology,* vol. 58, no. 5, pp. 528-533, 1989.

[5] T. W. Beck, "The mechanisms underlying the surface mechanomyogram", *Applications of Mechanomyography for Examining Muscle Function,* Kerala, IN: Transworld Research Network, pp. 1-16, 2010.

[6] C. Cescon, P. Madeleine, and D. Farina, "Longitudinal and transverse propagation of surface mechanomyographic waves generated by single motor unit activity," *Medical & Biological Engineering & Computing,* vol. 46, no. 9, pp. 871-877, 2008.

[7] T-K. Kim, Y. Shimomura, K. Iwanaga, and T. Katsuura, "Comparison of an accelerometer and a condenser microphone for mechanomyographic signals during measurement of agonist and antagonist muscles in sustained isometric muscle contractions: The influence of the force tremor," *Journal of Physiological Anthropology,* vol. 27, no. 3, pp. 121-131, 2008.

[8] T. Iwamoto, M. Tatezono, and H. Shinoda, "Non-contact Method for Producing Tactile Sensation Using Airborne Ultrasound," *EuroHaptics '08 Proceedings of the 6th international conference on Haptics,* pp. 504-513, 2008.

[9] N. Minh-Dung, P. Hoang-Phuong, K. Matsumoto, and I. Shimoyama, "A hydrophone using liquid to bridge the gap of a piezo-resistive cantilever," *Proceedings of Transducers2013,* pp. 70-73.

[10] M. Gel and I. Shimoyama, "Force sensing submicrometer thick cantilevers with ultra-thin piezoresistors by rapid thermal diffusion," *Journal of Micromechanics and Microengineering,* vol. 14, no. 3, pp. 423-428, 2004.

[11] C. Orizio, R. Perini, B. Diemont, M. M. Figini, and A. Veicsteinas, "Spectral analysis of muscular sound during isometric contraction of biceps brachii," *Journal of Applied Physiology,* vol. 68, no. 2, pp. 508-512, 1990.

CONTACT

*Tomonori Kaneko, Mechano-Informatics Department, Graduate School of Information Science and Technology, The University of Tokyo, 7-3-1 Hongo, Bunkyo-ku, Tokyo, 113-8656, Japan.
E-mail: kaneko@leopard.t.u-tokyo.ac.jp
Tel: +81-3-5841-6318, Fax: +81-3-5841-6341

MEASURING FLOW VELOCITY OF SWALLOWED LIQUID IN THE HUMAN PHARYNX BY TONGUE PRESSURE SENSOR AND SWALLOWING SOUND SENSOR

Yusuke Takei, Tomonori Kaneko, Kentaro Noda, Kiyoshi Matsumoto, and Isao Shimoyama
The University of Tokyo, Tokyo, JAPAN

ABSTRACT

We measured flow velocity of swallowed liquid passing through pharynx. We put a pressure sensor on a palate to measure tongue pressure when swallowing the liquid. And we put two acoustic sensors on the neck skin in order to measure swallowing sound when liquid passes through the pharynx. From the output of these three sensors, we can know the timing of the liquid passing through each sensor points and can calculate the flow velocity of the swallowed liquid at the pharynx. In this paper, we compare the flow velocity between two swallowing positions, "look forward position" and "look upward position." As a result, we found that the flow velocity of the "look upward position" was 2.5 times faster than that of "look forward position."

INTRODUCTION

Japanese Ministry of Health, Labour and Welfare investigated cause of death in Japan at 2012. As the survey result, the third cause of the death was pneumonia. In the pneumonia death, 90% of the death was aged over 65. Especially, 70% of the pneumonia death in old aged people was caused by aspiration [1]. Aspiration is the phenomenon that food or drinks come into trachea when swallowing. These aspirations are mainly caused by the low tongue activities. To understand the mechanisms of the swallowing, we need to quantify lingual motions [2][3].

In MEMS2012 and 2013, we presented the triaxial force sensor which was applicable to human palate [4][5]. But when we think of the aspiration, we need to know not only the lingual motions but also the situations occurring in the pharynx. Because aspiration occurs at pharynx and the motion of the swallowed liquid are mainly determined by the lingual motions. Previously, swallowing sound was measured to understand the liquid flow around trachea [6]-[11]. And the liquid flow velocity around the pharynx is measured by ultrasonic pulse doppler method [12]. They measured only the flow velocity and cannot figure out the relationship between lingual motion and flow velocity around the pharynx.

In this paper, we measured the flow velocity of the swallowed liquid when the liquid passing through the pharynx. We fabricated two types of MEMS sensor. One is the tongue pressure sensor and the other is swallowing sound sensor. Tongue pressure sensor is attached to the palate to measure the tongue pressure when the tongue pushes the swallowed liquid into the pharynx. And two swallowing sound sensors are attached to the neck skin in order to measure swallowing sound when the swallowed liquid passing through the pharynx (Figure 1). From the output of these sensors, we can know the timing of the liquid passing through each sensor point and can calculate the flow velocity of the swallowed liquid at the pharynx. By measuring the tongue pressure and flow velocity at the same time, we can partly understand the aspiration mechanisms.

SENSOR DESIGN AND FABRICATION

Fig.2 shows our fabricated tongue pressure sensor. Tongue pressure sensor is composed of silicon beam buried in the silicon rubber (PDMS). Hinges of the beam are composed of piezo resistive silicon. When the stress was applied to the silicon rubber surface, the beam bends according to the deformation of the silicon rubber. And the strain of the piezo resistive silicon beam is translated into the resistance change. Thus, by measuring the resistance change of the silicon beam, we can calculate the strain applied on the tongue pressure sensor. The dimension of the pressure sensor is $7 \times 6 \times 0.8$ mm and the thickness and the width of the wiring are 80 μm, 4.5 mm respectively. The pressure sensor chip was fixed on the 80 μm thick wiring. To prevent the interruption of the inherent swallowing, we designed the wiring that comes out from the gap between upper-front-teeth and the lower-front-teeth. Since the

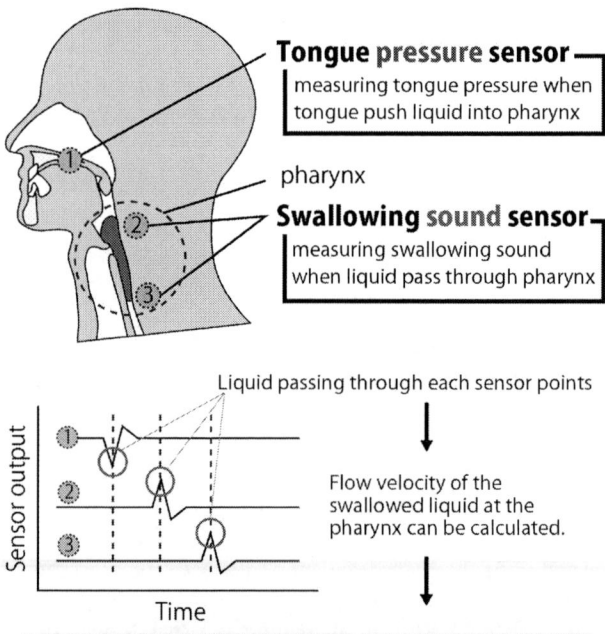

Fig.1 Concept of our research. We measure tongue pressure and flow velocity of swallowed liquid passing through pharynx by pressure sensor and sound sensors.

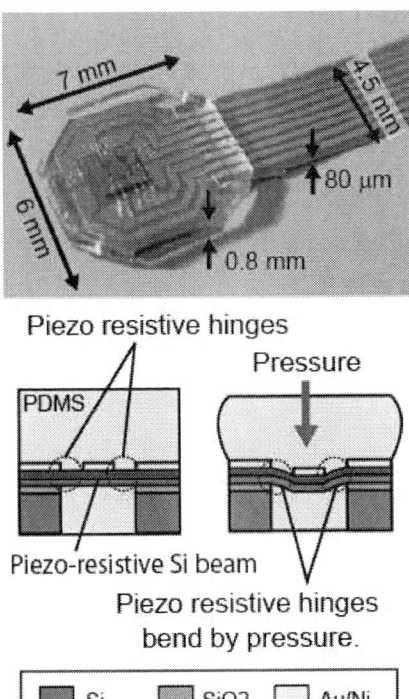

Fig.2 The tongue pressure sensor is composed of Si micro beam buried in PDMS. The beam has piezo resistive layers on its hinges.

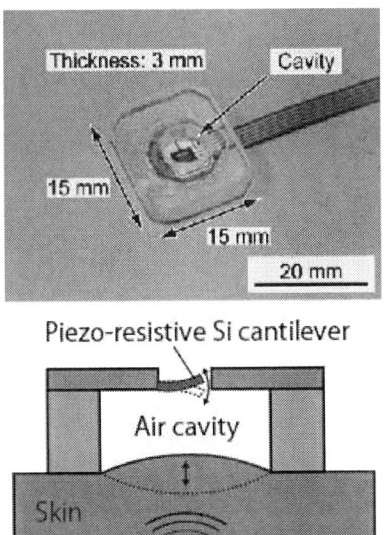

Fig.3 The swallowing sound sensor is composed of Si micro cantilever. When acoustic waves are applied, the hinges bend and the resistance would change.

thickness of the wiring substrate is 80 μm, which is much smaller than the gap between upper and the lower-front-teeth,

we can naturally swallow the sample liquid. Detail of the tongue pressure sensor fabrication is described in the reference [13]. After the sensor chip fabrication, we bonded the chip on the gold wiring patterned flexible polyimide substrate (thickness 80 μm) by epoxy glue. After the bonding, we connect the wiring between the sensor chip and the polyimide substrate by aluminum wire bonding. Then chip on the substrate was fixed at the bottom of the petri dish by double-faced tape and the PDMS（Polydimethylsiloxane）was poured into the dish. The mixture of the PDMS and its polymerization initiator was 10 : 1. After the PDMS curing (60 °C, 1 hour), we cut the cured PDMS into the shape. And finally, we conducted Parylene evaporation coating on whole surface of the PDMS coated sensor and wiring substrate. This coating prevents the peel-off of the PDMS from the substrate.

Swallowing sound sensor is composed of Si micro cantilever. At the hinges of the cantilever, impurity diffused piezo resistive layers were formed. Fig.3 shows the sensing mechanisms of the swallowing sound sensor. We put the cavity plate on the sensor so that skin vibration effectively propagates to the cantilever.

EXPERIMENTS AND RESULTS

To clarify the relation between the aspiration and the way of swallowing liquid, we conducted experiment as follows. We measured the tongue pressure and the flow velocity of the liquid passing through the pharynx in two swallowing position. First position is "look forward" position, which examinee sits on a chair and looks forward and swallows the liquid with his head and eyes locked. The other position is "look upward" position, which examinee sits on the chair and tips the head backward to look up and swallows the liquid. Generally, it is said that swallowing the liquid in "look upward" position has a high risk of aspiration. We think that this is because when looking upward, slope angle of inside the mouth is steeper than that of look forward position. And this steep slope accelerate the liquid in the mouth and cause the high flow velocity when the liquid passing the pharynx. And epiglottis cannot catch up the such high flow velocity and some of the liquid flew to the trachea.

Fig.4 is the photographs of our sensors fixed to the examinee. Tongue pressure sensor was fixed on the palate by medical glue. Two acoustic sensors are fixed on the neck by scotch tape (the distance of these two sensors are 7 cm). Examinee drink the normal viscosity (1mPa・s) water as the liquid sample. The volume of the swallowed water is determined by the examinee's optimum swallowing volume which is measured before the experiment (20 ml at this time).

Fig.5 shows the typical sensor outputs of the "look forward position" and "look upward position" swallowing respectively. From each data, we can assume the mouth movements as follows. Firstly, the water was placed on the tongue. The tongue pushed the water into throat by impressing the tongue to the palate and the impressing force was measured by the tongue pressure sensor. Secondary, the swallowed water passing the pharynx and the flowing sound is measured by the two swallowing sound sensors. Table 1 is

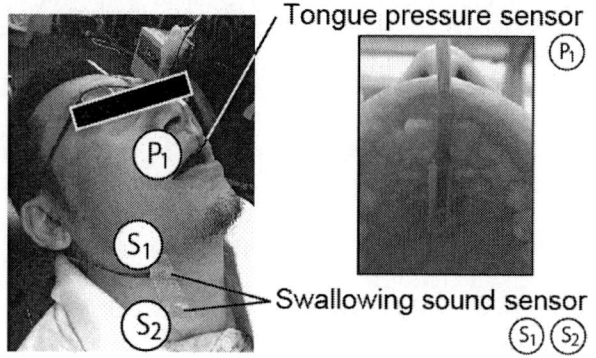

Fig.4 Photographs of our sensors fixed to the examinee. Tongue pressure sensor was fixed on the palate by medical glue. Two swallowing sound sensors are fixed on the neck by scotch tape (the distance of these two sensors are 7 cm).

showing the tongue pressure of swallowing the water and flow velocity calculated by the required time between two swallowing sound sensors.

From the tongue pressure data, "look forward position" swallowing shows high tongue pressure (15.9 kPa) than that of "look upward position (7.2 kPa)." This result indicates that when we swallow liquid in "look upward position," the liquid on the tongue is moving toward the pharynx mainly by the influence of the gravity, so that tongue does not need high pressure to feed them in to the pharynx.

From the flow velocity of the swallowed liquid, swallowing in "look upward position" showed 2.5 times faster flow velocity (0.92 m/sec) than that of "look forward position (0.37 m/sec)." This result indicates that when swallowing liquid in "look upward position," the liquid passing the pharynx much faster than the usual "look forward position". And this faster flow causes the high possibility of aspiration, which matches to the medical knowledge.

CONCLUSION

We fabricated the tongue pressure sensor and the swallowing sound sensor based on piezo resistive silicon cantilever and beam. By combining these sensors, we measured the tongue pressure and liquid flow velocity during the swallowing. We compared the tongue pressure and flow velocity between two swallowing positions, "look forward position" and "look upward position." As a result, we found that the flow velocity of the "look upward position" was 2.5 times faster than that of "look forward position." And the tongue pressure of the "look upward position" was half as large as that of "look forward position." These results indicate that when swallowing liquid in "look upward position," the liquid flows into the pharynx by its gravity so that tongue pressure is smaller and flow velocity is faster than "look forward position." These results are helpful in understanding the aspiration mechanisms which frequently occur in "look upward position" swallowing.

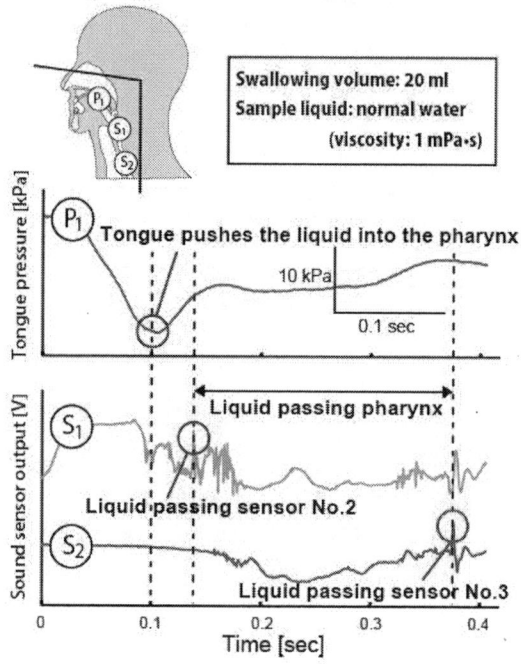

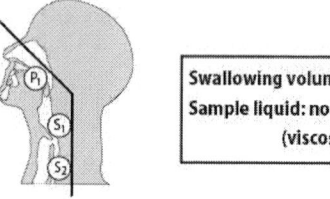

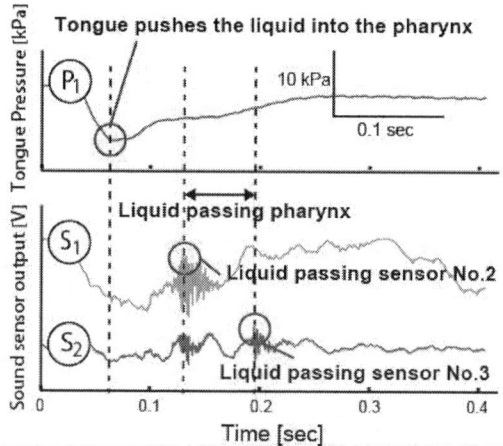

Fig.5 Sensor output of the "look forward" position and "look upward" position swallowing.

Table 1. Measured tongue pressure and flow velocity of each swallowing position.

swallowing position	tongue pressure (n=5)	flow velocity (n=5)
look forward	15.9 kPa (SD=12.7)	0.37 m/sec
look upward	7.2 kPa (SD=2.4)	0.92 m/sec

ACKNOWLEDGEMENTS

This research is supported by Grant-in-Aid for Young Scientists (B) (24700589), the Ministry of Education, Culture, Sports, Science and Technology (MEXT), Japan. The photolithography masks were made using the University of Tokyo VLSI Design and Education Center (VDEC)'s 8 inch EB writer F5112 + VD01 donated by ADVANTEST Corporation.

REFERENCES

[1] S. Teramoto, Y. Fukuchi, H. Sakai, K. Sato, K. Sekizawa, T. Matsuse, "High incidence of aspiration pneumonia in community-and hospital-acquired pneumonia in hospitalized patients: A multicenter, prospective study in Japan," *Journal of the American Geriatrics Society*, vol. 56, No. 3, pp.577-579, 2008.

[2] T. Tachimura, K. Nohara, Y. Fujita, T. Wada, "Effect of a speech prosthesis on electromyographic activity levels of the levator veli palatini muscle activity during syllable repetition," *Archives of Physical Medicine and Rehabilitation*, vol. 83, pp. 1450-1454, 2002.

[3] K. Satoh, T. Wada, T. Tachimura, S. Sakoda, R. Shiba, "A cephalometric study by multivariate analysis of growth of the bony nasopharynx in patients with clefts and non-cleft controls," *J. Cranio-maxillofacial surgery*, vol. 26, pp. 394-399, 1998.

[4] Y. Takei, K. Noda, T. Kawai, T. Tachimura, Y. Toyama, T. Ohmori, K. Matsumoto, and I. Shimoyama, "Triaxial force sensor for lingual motion sensing," *The 25th IEEE International Conference on Micro Electro Mechanical Systems (MEMS '12)*, pp. 128 - 131 , Paris, France, 29 January-2 February, 2012.

[5] Y. Takei, K. Noda, T. Kawai, T. Tachimura, Y. Toyama, M. Takai, K. Matsumoto, and I. Shimoyama, " Anterior And Posterior Tongue Activity Sensor Based On Triaxial Force Sensor," *The 26th IEEE International Conference on Micro Electro Mechanical Systems (MEMS '13)*, pp. 1093-1096, Taipei, Taiwan, 20-24 January, 2013.

[6] M. Taniwaki, Z. Gao, K. Nishinari, and K. Kohyama," Acoustic Analysis of the swallowing sounds of food with different physical properties using the cervical auscultation method," *Journal of Texture Studies*, vol. 44, pp. 169-175, 2013.

[7] Mohammad Aboofazeli, Zahra K. Moussavi, "Comparison of recurrence plot features of swallowing and breath sounds," *Chaos, Solitons and Fractals*, vol. 37, pp. 454–464, 2008.

[8] Samaneh Sarraf-Shirazi, Jonathan-F Baril, Zahra Moussavi, "Characteristics of the swallowing sounds recorded in the ear, nose and on trachea," *Med Biol Eng Comput*, vol. 50, pp. 885–890, 2012.

[9] Samaneh Sarraf Shirazi, Caitlin Buchel, Reesa Daun, Laura Lenton, and Zahra Moussavi, "Detection of swallows with silent aspiration using swallowing and breath sound analysis," *Med Biol Eng Comput*, vol. 50, pp. 1261–1268, 2012.

[10] K. Kohyama, H. Sawada, M. Nonaka, C. Kobori, F. Hayakawa, and T. Sasaki, "Textural Evaluation of Rice Cake by Chewing and Swallowing Measurements on Human Subjects," *Biosci. Biotechnol. Biochem.*, vol. 71, no. 2, pp. 358-365, 2007.

[11] Samaneh Sarraf Shirazi and Zahra M. K. Moussavi, "Acoustical Modeling of Swallowing Mechanism," *IEEE Transaction on Biomedical engineering*, vol. 58, no. 1, pp. 81-87, 2011.

[12] H. Moritaka and F. Nakazawa, "Flow velocity of a bolus in the pharynx and rheological properties of agar and gelatin," *Journal of Texture Studies*, vol. 41, issue 2, pp. 139-152, 2010.

[13] K. Noda, K. Hoshino, K. Matsumoto, I. Shimoyama, "A Shear Stress Sensor for Tactile Sensing with the Piezoresistive Cantilever Standing in Elastic Material," *Sensors and Actuators A*, vol. 127, no. 2, pp. 295-301, 2006.

CONTACT

Yusuke Takei, Department of Mechano-Informatics, Graduate School of Information Science and Technology, The University of Tokyo, 7-3-1 Hongo, Bunkyo-ku, Tokyo, Japan. Tel: (+81)-3-5841-0461, Fax: (+81)-3-3818-0835, e-mail: takei@leopard.t.u-tokyo.ac.jp

MEMS NEURAL PROBE ARRAY FOR MULTIPLE-SITE OPTICAL STIMULATION WITH LOW-LOSS OPTICAL WAVEGUIDE BY USING THICK GLASS CLADDING LAYER

Yoojin Son[1, 2], Hyunjoo Jenny Lee[1], Dohee Kim[1], Yun Kyung Kim[1], Eui-Sung Yoon[1], Ji Yoon Kang[1], Nakwon Choi[1], Tae Geun Kim[2] and Il-Joo Cho[1]
[1] KIST (Korea Institute of Science and Technology), Seoul, South Korea
[2] Korea University, Seoul, South Korea

ABSTRACT

We present a MEMS neural probe array for multiple-site optical stimulation with low-loss SU-8 optical waveguides. An embedded 20-μm-thick cladding layer was formed by glass reflow process; due to this embedded structure, no additional thickness was required. In addition, optical loss was reduced by using the thick cladding layer and integrating a thick SU-8 layer as a core layer. The low-loss optical waveguide enables multiple-site stimulation with two-step optical splitters. Using the presented probe array, we also demonstrate a successful *in-vivo* optical stimulation through recording of neural signals from the hippocampus of a transgenic ChR2-YFP mouse. Recorded neural signals were synchronized with light pulses, which confirms that neurons were successfully stimulated by the blue light and the integrated electrode array successfully recorded the neural signals from activated neurons.

INTRODUCTION

In recent years, there has been an active progress in human brain study to clarify the cause of neurological diseases such as Parkinson disease, epilepsy, and chronic pain. To investigate and treat these brain disorders, understanding numerous neural networks in brain is important. Because brain consists of 100 billions of neurons connected in complex structural and functional networks, it is still challenging to identify and characterize these networks [1]. Many neuroscientists have been utilizing implantable wire electrodes to simultaneously record neural signals at different brain regions to study functional connectivity among brain regions [2]. However, it is difficult to accurately control distance among the wires. Also, low electrode density inevitably requires a large number of wire electrode arrays and thus incurs large brain damage [3-5]. To overcome these limitations, MEMS neural probes have been introduced and have drawn a great attention as implantable devices because of various advantages, such as small size with high electrode density and capability to accurately position each electrode in an array structure, which is important for studying functional connectivity of brain. Recently, more functions have been integrated into MEMS neural probes including various stimulation capabilities such as optical and chemical stimulation [6, 7]. Specially, optical stimulation of neurons have attracted great attentions because genetically targeted neurons can be selectively excited or inhibited by light without stimulating neighboring neuron networks [7-9].

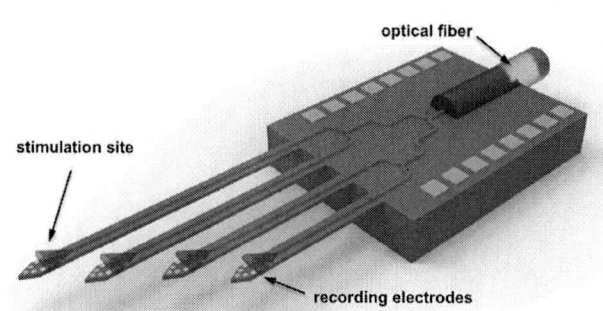

Figure 1: Conceptual diagram of the proposed multi-shank neural probe integrated with multiple optical stimulation sites and electrode arrays for simultaneous electrical recording.

Previously reported MEMS neural probe for optical stimulation suffered from few limitations, such as thick probe shank and inaccurate fiber positioning from direct manual attachment [10]. Our previous work achieved a thin neural probe with integration of waveguide but offered only two stimulation sites due to relatively high optical loss [11, 12].

In this paper, we present a MEMS neural probe array with new low-loss optical waveguides for simultaneous 4-sites stimulation based on a simpler fabrication process. The low-loss optical waveguide, the key advantage of this process, utilizes a thick glass as the cladding layer based on glass reflow process to reduce optical transmission loss. Use of the low-loss waveguides enables multiple-site optical stimulations with sufficient output power at each site where light from a single source is guided to multiple sites using two optical splitters (Fig. 1). Also, electrode arrays are integrated around stimulation sites to record neural signals generated by the stimulated neurons.

FABRICATION PROCESS

The fabrication process of the proposed neural probe is shown in Figure 2. First, a groove is patterned on a silicon-on-insulator (SOI) wafer and etched 20 μm deep using deep reactive-ion etching (DRIE) process to form cladding layer in the consequent steps. Then, the SOI wafer is anodically bonded with a 500-μm-thick borosilicate glass wafer in vacuum environment. Next, the glass is thinned down to obtain a 100-μm thickness by chemical mechanical polishing (CMP) process. The bonded wafers are then fully reflowed at 800°C for 2 hr in a furnace to form the cladding

978-1-4799-3510-9/14 $31.00 © 2014 IEEE   853

layer. This reflow process allows to freely control the cladding layer thickness limited by the desired final probe thickness. The unnecessary glass layer is then removed using CMP process. After depositing a 300-nm-thick SiO_2 insulation layer by plasma-enhanced chemical vapor deposition (PECVD), 300-nm-thick gold and 20-nm-thick chrome adhesion layers are deposited and patterned using lift-off process. The signal lines are protected through a 400-nm-thick SiO_2 layer, where only the microelectrode areas are opened for recording, followed by microelectrode patterning and 20-nm-thick Ti and 150-nm-thick Ir deposition. Next, a 15-μm-thick SU-8 layer is coated and patterned for the core of the waveguide. The refractive index of SU-8 is 1.58, which is larger than the refractive index of glass cladding layer. After patterning SU-8 waveguide on the cladding layer, a U-groove for fiber placement is patterned using DRIE. The depth of grooves is controlled by DRIE process for 125 μm diameter optical fibers to be aligned vertically. Finally, the probe is released from the backside using DRIE. Figure 3 shows scanning electron microscopy (SEM) images of the successfully fabricated neural probe array integrated with a two-step optical splitter, allowing for multi-site stimulation using a single light source. The cross-sectional SEM image illustrates the SU-8 core patterned on top of the thick glass cladding layer.

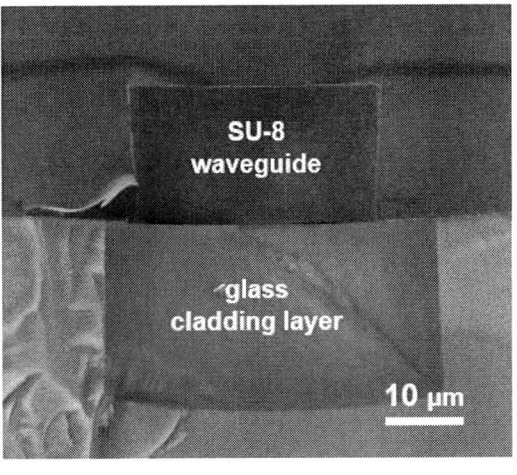

(a)

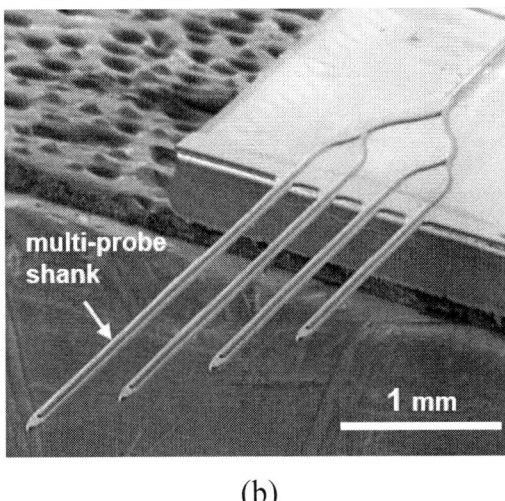

(b)

Figure 3: SEM images of the fabricated neural probes: (a) a cross-section of the probe showing the glass cladding layer and the SU-8 core layer and (b) probe array with multiple stimulation sites.

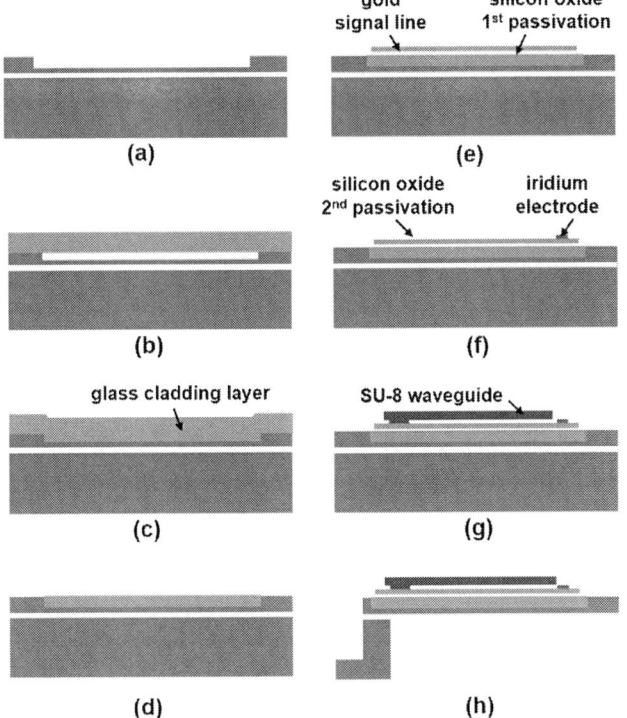

Figure 2: Fabrication process flow: (a) cavity formation, (b) anodic bonding and CMP of a glass wafer, (c) glass cladding layer formation using thermal reflow, (d) CMP, (e) 1st passivation layer deposition and signal line patterning, (f) 2nd passivation layer deposition and electrode patterning, (g) SU-8 core layer patterning, and (h) groove patterning and release.

EXPERIMENTAL RESULTS

Device performance characterization

The successfully fabricated 30-μm-thick neural probe consists of 4 shanks with 32 multiple recording electrodes (14 μm × 14 μm). 8 recording electrodes are located around each stimulation site to record the neural signals stimulated by delivered light. Also, two-step optical splitters split the light from a single source to 4-stimulations sites. Multi-mode optical fiber (GIF50, Thorlabs, Newton, NJ) was seated on the groove and self-aligned with the SU-8 waveguide to deliver the light from the light source (ADR-2301, RGBlase LLC, CA) to SU-8 waveguide. By using the low-loss optical waveguide, we successfully transmitted blue light (λ=473 nm) from the light source to 4-stimulation sites as shown in Figure 4. We also measured the maximum output

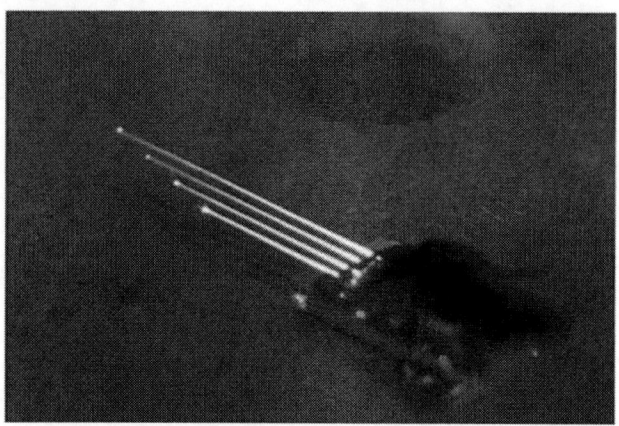

Figure 4: Optical picture of a multi-shank probe coupled with an optical fiber.

optical power of 1.2 mW, which was 3 times higher than that of the previous work [12].

In-vivo experiment

The fabricated MEMS neural probe was attached on a PCB and pads on the probe were electrically connected with pads on PCB through wirebonds. The impedance of 16 electrodes was measured using electrochemical impedance spectroscopy over a frequency range from 10 Hz to 100 kHz. An average impedance of iridium electrodes was ~0.5 MΩ at 1 kHz, which is low enough to measure neural spikes.

To verify the functionality of the probe, *in-vivo* recording of neural action potentials was performed using a transgenic ChR2-YFP mouse. We inserted the fabricated probe into the hippocampus region of an anesthetized mouse brain as shown in Figure 6. Fabricated neural probe was successfully inserted without bending or fracturing.

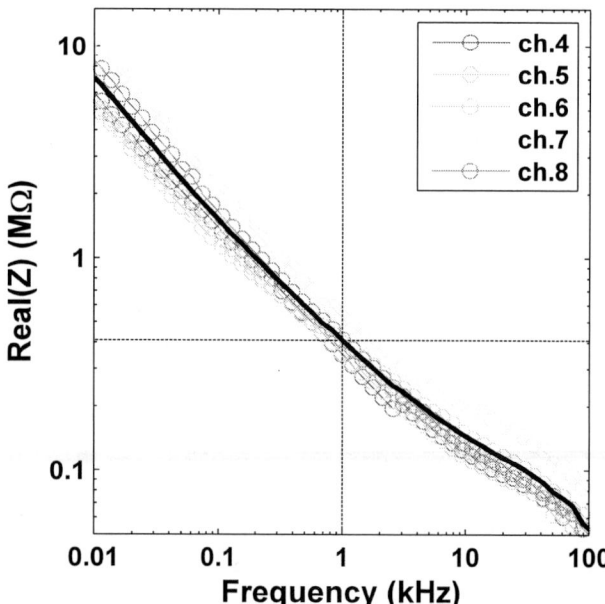

Figure 5: Impedance measurement of the 5 Ir electrodes in PBS solution.

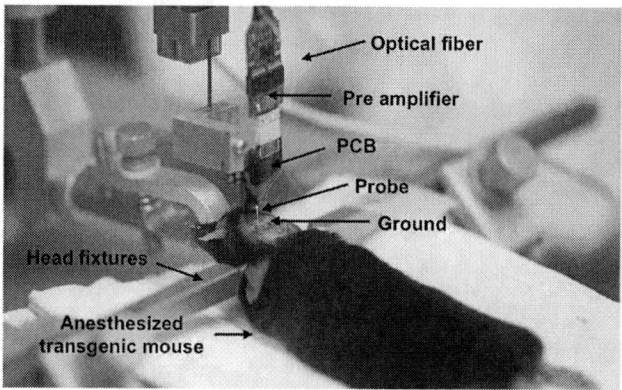

Figure 6: Optical picture of the in-vivo experiment setup.

For stimulating neurons, the light (λ=473 nm) was transmitted from the external light source (Model of Laser, Name of Company, Korea) to the stimulation site with an optical power of 0.2 mW, which is high enough to stimulate targeted neurons. The period of stimulation signal was 1-s and duty cycle was 20%. The signals from activated neurons on multiple electrodes were recorded using Neuralynx system. As shown in Fig. 7, recorded neural signals were synchronized with light pulses, which confirms that neurons were successfully stimulated in hippocampus by the blue light and signals were successfully recorded using the integrated electrode array.

Finally, the fluorescence image of the brain slice demonstrates the expression of ChR2-YFP confirming genetically modified mouse was used as shown in Figure 8.

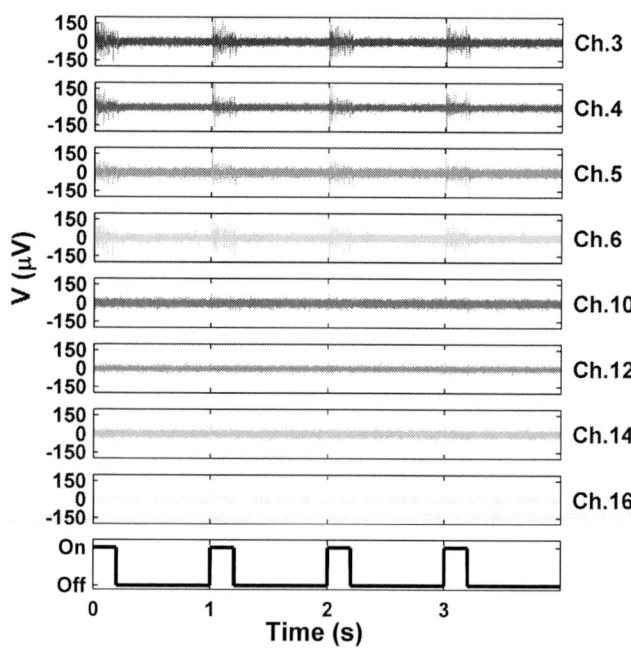

Figure 7: Transient plot of neural spikes from 8 representative electrodes, synchronized with light stimulation square pulses.

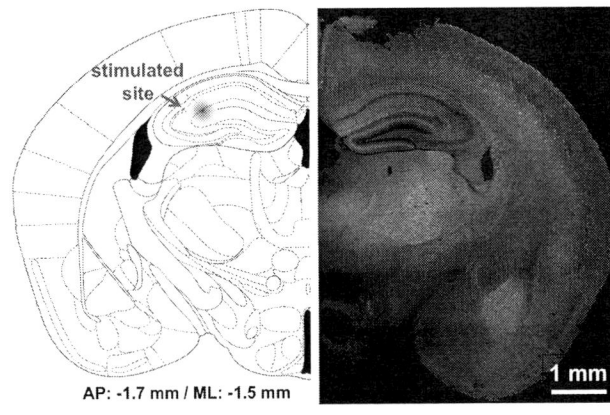

AP: -1.7 mm / ML: -1.5 mm

Figure 8: Florescence optical image of brain slice of the sacrificed transgenic mouse (B6.Cg–Tg(Thy1-COP4/EYF P)9Gfng/J), demonstrating the expression of ChR2-YFP.

CONCLUSION

In this paper, we present a MEMS neural probe array for multiple-site optical stimulation with low-loss optical waveguides. The 20-μm-thick cladding layer was easily formed by glass reflow process and no additional thickness was required due to embedded structure. Furthermore, the low-loss optical waveguide enables multiple-site stimulation with the two-step optical splitter. The presented neural probe array is a promising tool to study various brain circuits by simultaneously stimulating and monitoring neural signals in different regions.

REFERENCES

[1] J. G. Bernstein and E. S. Boyden, "Optogenetic Tools for Analyzing the Neural Circuits of Behavior", *Trends in Cogn. Sci.* Vol. 15, pp. 592-600, 2011.

[2] A.B. Schwartz, "Cortical Neural Prosthetics", *Ann. Rev. Neurosci.* Vol. 27, pp. 487-507, 2004.

[3] R. Bhandari, S. Negi, L. Rieth, R. A. Normann and F. Solzbacher, "A Novel Method of Fabricating Convoluted Shaped Electrode Arrays for Neural and Retinal Prostheses", *Sens. Actuators A-Phys.* Vol. 145, pp. 123-130, 2008.

[4] M. A. L. Nicolelis and M. A. Lebedev, "Principles of Neural Ensemble Physiology Underlying the Operation of Brain-machine Interfaces", *Nat. Rev. Neurosci.* Vol. 10, pp. 530-540, 2009.

[5] P. Bartho, H. Hirase, L. Monconduit, M. Zugaro, K. D. Harris and G. Buzsaki, "Characterization of Neocortical Principal Cells and Interneurons by Network Interactions and Extracellular Features", *J. Neurophysicol.* Vol. 92, pp. 600-607, 2004.

[6] K. D. Wise, A. M. Sodagar, Y. Yao, M. N. Gulari, G. E. Perlin and K. Najafi, "Microelectrodes, Microelectronics, and Implantable Neural Microsystems", *Proc. of the IEEE*, Vol. 96, No. 7, pp. 1184-1202, 2008.

[7] E.S. Boyden, F. Zhang, E. Bamberg, G. Nagel and K.

Deisseroth, "Millisecond-Timescale, Genetically Targeted Optical Control of Neural Activity", *Nature Neuroscience*, Vol. 10, pp. 663-668, 2005.

[8] D. P. O'Brien, T. R. Nichols and M. G. Allen, "Flexible Microelectrode Arrays with Integrated Insertion Devices", *Proc. of MEMS*, pp. 216–219, 2002.

[9] A. R. Adamantidis, F. Zhang, A. M. Aravanis, K. Deisseroth and L.d. Lecea, "Neural Substrates of Awakening Probed with Optogenetic Control of Hypocretin Neurons", *Nature*, Vol. 450, pp. 420-424, 2007.

[10] S. Royer, B. V. Zemelman, M. Barbic, A. Losonczy, G. Buzsaki and J. C. Magee, "Multi-array Silicon Probes with Integrated Optical Fibers: Light-assisted Perturbation and Recording of Local Neural Circuits in the Behaving Animal", *J. of Neurosci.*, Vol. 31, pp. 2279-2291, 2010.

[11] I. J. Cho, H. W. Baac and E. Yoon, "A 16-site Neural Probe Integrated with a Waveguide for Optical Stimulation", *Proc. 23th IEEE MEMS Conference, Hong Kong*, pp. 995-998, 2010.

[12] M. Im, I. J. Cho, F. Wu, K. D. Wise and E. Yoon, "Neural Probes Integrated with Optical Mixer/Splitter Waveguides and Multiple Stimulation Sites", *Proc. 24th IEEE MEMS Conference, Cancun*, pp. 1051-1054, 2011.

MICRO-ELECTRODE ARRAYS FOR MULTI-CHANNEL MOTOR UNIT EMG RECORDING

Shota Yamagiwa[1] Hirohito Sawahata[1] Makoto Ishida[1, 2] and Takeshi Kawano[1]

[1]Department of Electrical and Electronic Information Engineering, Toyohashi University of Technology, Japan
[2]Electronics-Inspired Interdisciplinary Research Institute (EIIRIS), Toyohashi University of Technology, Japan

ABSTRACT

We report an array of micro-electrodes, which can record multi-channel motor unit (MU) electromyogram (EMG) signals with a high spatial resolution. As a basic structure of the electrode, we fabricated an array of 200-μm-square Si-pyramids with the height of 200 μm by Tetramethylammonium hydroxide (TMAH), in order for robust MU-EMG recordings without conductive gel. Platinum (Pt) was used as the electrode material and parylene-C was used as the insulator for the Pt-electrode. The fabricated μEMG electrode array, which was connected to a recording system, clearly detected MU-EMG action potentials from a human forearm. In addition, different MU-EMG signals between the μEMG electrodes were detected by crooking the fingers. These results indicate that the μEMG array device becomes a powerful tool for medical applications including myoelectric prosthetic technologies.

INTRODUCTION

Conventional EMG electrodes, which is a large dish type electrode (diameter = 1 cm~), have been used for the diagnostic of muscle disorder[1] or other medical applications, such as power assisting suit[2], myoelectric prosthetic technology[3]. MU is known as a group of the muscle fibers, which are controlled by an axon of the motor nerve (Fig. 1). The shapes and firing rates of MU action potential provide the important information for medical applications such as neuromusclar disorder[1]. However, it is difficult to detect the single MU action potentials with conventional large EMG electrodes, since the centimeter-scale diameter electrode detects the EMG signals from many muscle fibers, and further analysis of the EMG signals are needed to clarify the MU action potentials.

On the other hand, there are two major methods to detect MU-EMG action potentials without subsequent signal analysis. One of the methods is the use of a wire electrode (diameter = ~350 μm), which penetrates a muscle. In this method, the MU-EMG action potentials can be recorded continuously from a same muscle fiber. However, since the examinee/patient moves his/her muscle while the electrode penetrates, this method results in damage to the muscle and has a risk of the infection diseases. Another method is the use of some tiny electrodes (diameter = ~1 mm) on the surface of a skin [4]. However, the size of these electrode devices is large to use for prosthetic technology. Here we propose a minimized noninvasive microelectrode array for multi-channel MU-EMG action potential recordings with a high spatial resolution.

FABRICATION PROCESS

Fig.2 shows the fabrication process of the proposed μEMG device. As a first step, a SiO_2 layer was prepared as a hard mask by thermal oxidation (~ 1 μm), and Si-pyramids were formed by TMAH (20% at 70°C) (Figs.2a-b). After the pyramid structure preparation, the SiO_2 hard mask was removed by buffered hydrogen fluoride (BHF). Next, SiO_2 was deposited on the wafer by plasma-enhanced chemical vapor deposition (PECVD) (Fig.2c) and the electrodes (Pt/Ti) were patterned by lift-off (Fig.2d). After parylene-C deposition (Fig.2e), the top of the pyramid electrode and the bonding pad were exposed by O_2 plasma (Fig. 2f). The thickness of SiO_2 insulator layer, patterned Pt/Ti, and deposited parylene-C were 1 μm, 200 nm, and 1 μm, respectively. The width and height of the pyramid electrode were set at 200 μm and 200 μm, respectively, and each electrode-site was spaced 1 mm apart for MU-EMG recordings with a high spatial resolution. Fig.3a shows the packaged μEMG device. The μEMG electrodes were connected to a flexible printed circuit (FPC) by wire bonding and the wires were covered with epoxy resin. Fig. 3a also includes a conventional EMG electrode (diameter = ~1 cm) and a one euro coin for size comparison. Figs. 3b and c show microscope images of a fabricated μEMG electrode array and the individual electrode site in the array.

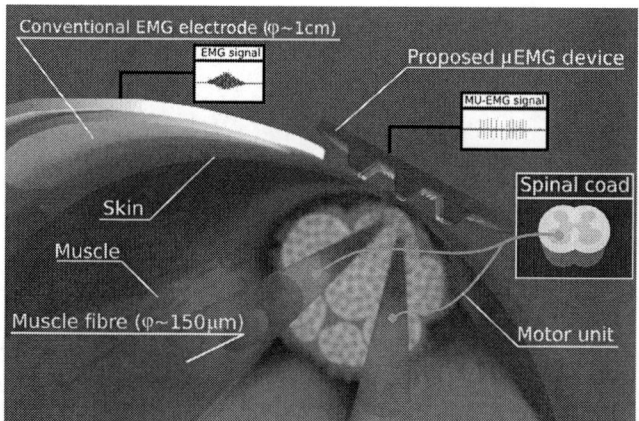

Figure 1: Schematic image of the device concept. High spatial resolution MU-EMG signals are recorded from the surface of a skin by a microelectrode array device.

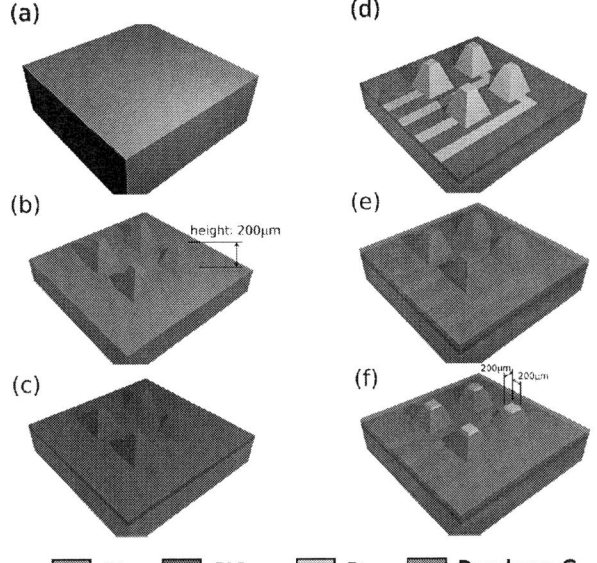

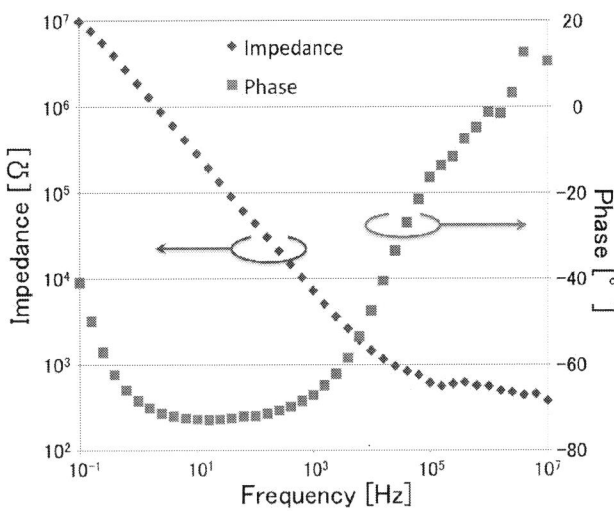

Figure 2: Fabrication process of µEMG array device: (a-b) Si-pyramid formed by TMAH, (c) SiO₂ deposition by CVD, (d) Pt patterning by lift-off, (e) parylene-C deposition, and (f) parylene C etching by O₂ plasma.

Figure 4: Magnitude and phase of impedance of a µEMG electrode measured in PBS. The width and height of the measured pyramid electrode are 200 µm and 200 µm, respectively.

RESULTS AND DISCUSSION

Electrode Impedance

After device fabrication, the impedance of the µEMG electrode was measured in phosphate buffered saline (PBS, room temperature) with an impedance analyzer (Model 1260A Impedance/Gain-Phase Analyzer, AMETIC. Inc.) from 0.1 Hz to 10 MHz (Fig.4). The measured impedances of the µEMG electrode were 280 kΩ - 1.5 kΩ at the frequency range of 10 Hz - 10 kHz. The measure impedance of a typical needle EMG electrode exhibited 1.4 MΩ- 6.6 kΩ at 10 Hz- 10 kHz, indicating that the impedances of the fabricated µEMG electrode are low enough to detect MU-EMG signals[5].

MU-EMG Signal Recording

To record the MU-EMG signals, the µEMG device, which was connected to the amplifier system (SH16, PZ2 and RZ2, Tucker Davice Technology inc.), was put on a human forearm (Fig.5). In this experiment, a set of differential amplifiers and band-pass filters (from 300 Hz to 8 kHz) was used to record the multi-channel EMG signals. Each channel of the µEMG electrode recorded MU-EMG signals while examinee was crooking the finger without conductive gel (Fig.6).

According to Henneman's size principal, when the load of muscles becomes larger, additional MU starts to the firing with the prior MUs which are working for weak load[6]. Our recording results were also consistent with the theory. The phenomenon was clearly observed from each µEMG channel individually (Fig.6a). Figs 6b-d show the recorded signals with short time scales (< 1 sec), indicating that recorded signals were action potentials of the MU (Channel No. 2, 4, and 6, in Figs. 6b-d).

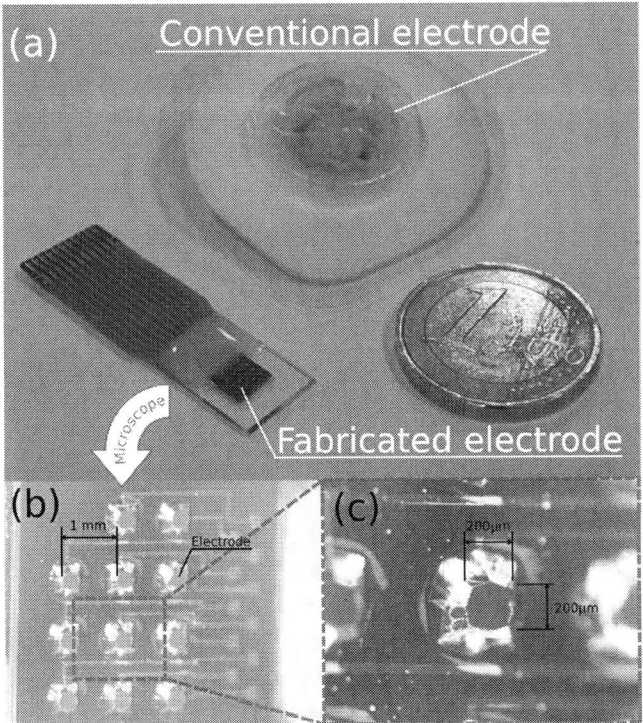

Figure 3: Fabricated device image. (a) Packaged device. Each µEMG electrode was connected to FPC by wire-bonding. Photograph also includes a conventional EMG electrode (diameter = ~1 cm) and a one euro coin for size comparison. (b, c) Microscope images of a fabricated µEMG electrode array and the individual electrode site in the array.

CONCLUSIONS

We proposed and fabricated μEMG electrode array device to detect multi-channel MU-EMG signals with a high spatial resolution. For the robust MU-EMG recording without conductive gel, we prepared an array of 200 μm square Si-pyramids with the height of 200 μm by TMAH. The measured impedance of the fabricated μEMG electrode was low enough to detect MU-EMG signals. The μEMG array device can detect multi-channel MU-EMG signals, as demonstrated in recordings using a human forearm.

These results indicate that the μEMG array device becomes a powerful tool for medical applications including myoelectric prosthetic technologies, power assisting suit and human-computer interface. Since proposed μEMG device is fabricated by using Si substrate, Si-CMOS (signal processor, channel selector, etc.) can be integrated on the same substrate, enhancing the performance of the multi-channel EMG recording. As a future work, the μEMG device will be able to be implanted on a bone to record MU-EMG signals from inner muscles.

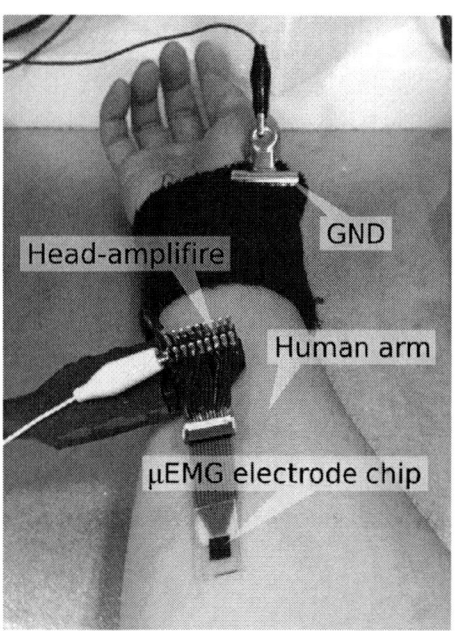

Figure 5: Motor-unit EMG recording using a human forearm. Fabricated EMG device, which was mounted on the surface of a human arm, was connected to head-amplifier of the recording system to record MU-EMG signals.

ACKNOWLEDGEMENTS

This work is supported by a Grant-in-Aid for Scientific Research (S), Young Scientists (B), and the PRESTO Program from JST.

REFERENCES

[1] M. B. I. Reaz, M. S. Hussain and F. Mohd-Yasin, "Techniques of EMG signal analysis: detection, processing, classification and applications", *Biological Procedures Online*, 2006, pp.11-35.

[2] T. Hayashi, H. Kawamoto and Y. Sankai, "Control Method of Robot Suit HAL working as Operator's Muscle using Biological and Dynamical Information", *Intelligent robots and Systems*, 2005, pp.3063-3068.

[3] T. Pistohi, C. Cipriani, A. Jackson and K. Nazarpour, "Abstract and Proportional Myoelectric Control for Multi-Fingered Hand Prostheses", *Annals of Biomedical Engineering*, 2013, pp.1-12.

[4] M. J. Zwarts and D. F. Stegeman, "Multichannel Surface EMG: Basic Aspects and Clinical Utility", *Muscle & nurve*, 2003, pp. 1-17.

[5] DO. Wiechers, JR. Blood and RW. Stow, "EMG Needle Electrode: Electrical Impedance", *Archives of Physical Medicine and Rehabilitation*, 1979, pp. 364-369.

[6] A. M. McPhedran, R. B. Wuerker and E. Henneman, "Properties of Motor Units in a Homogeneous Red Muscle (Soleus) of the Cat", *Journal of Neurophysiology*, 1965, pp. 71-84.

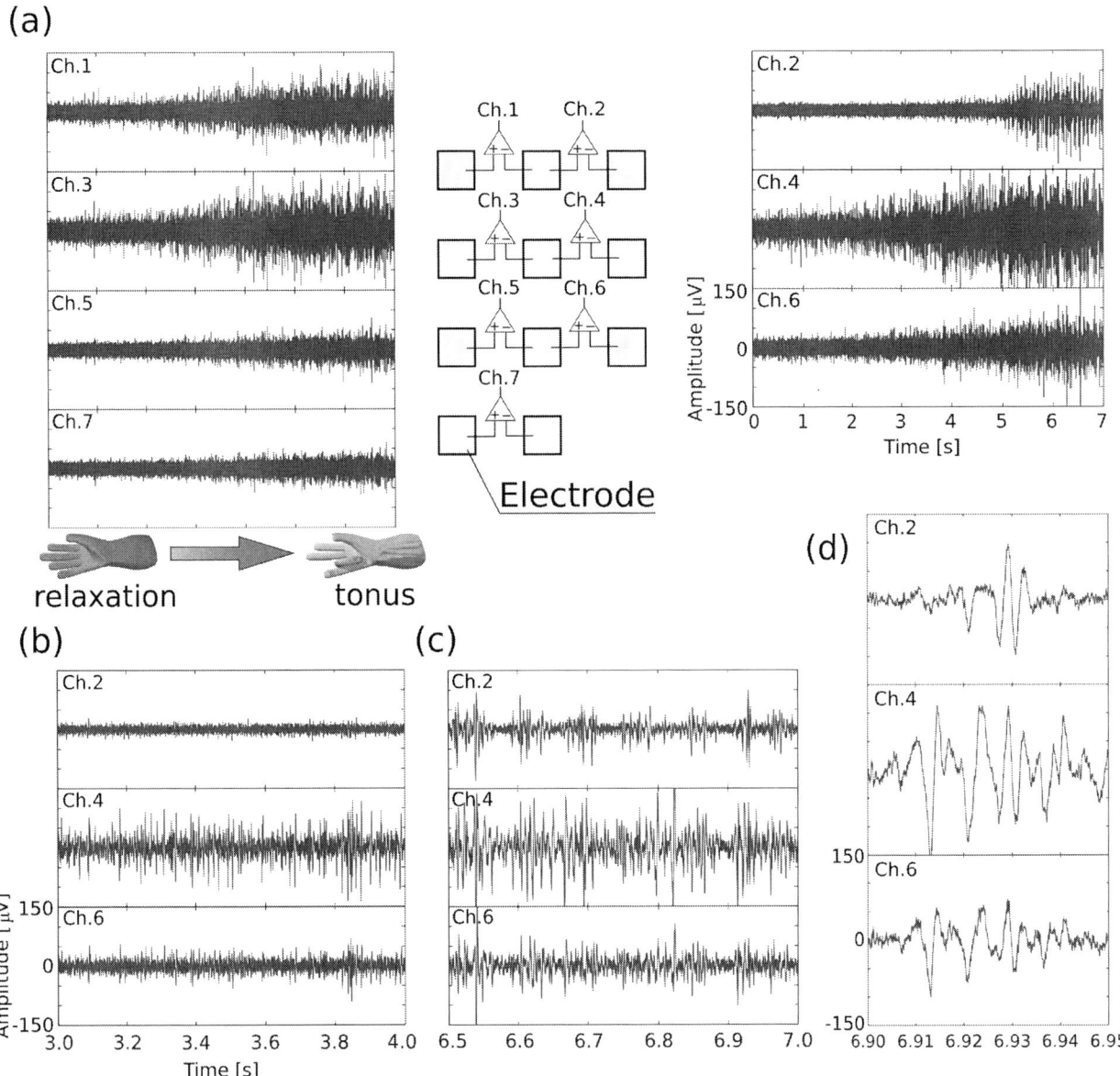

Figure 6: Motor-unit EMG action potential recordings. The examinee continued to crook his finger. (a) Recorded data of multi-channel (7chs) MU-EMG for relaxation (0 sec) and tonus (for 7s). (b) Data for crooked the finger. (c) Data for further crooked finger. (d) Recorded waveforms of the action potentials with a time scale of 0.05 sec.

MICRO-WING AND PORE DESIGN IN AN IMPLANTABLE FPC-BASED NEURAL STIMULATION PROBE FOR MINIMALLY INVASIVE SURGERY

Yu-Hsuan Wang[1], Daniel Tsai[1,2], Bo-An Chen[1], Yo-Yen Chen[1], Cheng-Chun Huang[1],
Pin-Chun Huang[1], Chih-Yeh Lin[1], Jiashing Yu[1], Wen-Pin Shih[1], Chii-Wann Lin[1], and Horn-Jiunn Sheen[1]

[1]National Taiwan University, Taipei, Taiwan
[2]University of California, San Diego, USA

ABSTRACT

This paper presents a bipolar porous probe for implantable nerve stimulation treatment utilizing minimally invasive surgery. The probe's design features micro-wings and pores for cell growth that promote long term fixation in the body. Two recording pairs detect whether cells grow into the pores, and one pair of stimulating pads stimulates the target nerve. The probe is composed of three layers: two SU-8 layers and one flexible printed circuit (FPC) layer. Results show that SU-8 films can increase the product of the area moment of inertia and Young's Modulus by 9.04% from 5.86×10^{-6} N·m^2 to 6.93×10^{-6} N·m^2 and that micro-wings can increase the force of fixation by 38.58% from 0.114 N to 0.158 N. From the impedance test, the impedance of the pores in gelatin is shown to be lower than the ones in air, demonstrating that the two recording pairs are promising for detecting cells growth.

INTRODUCTION

Neural probes have been widely used for recording and stimulating the specific sites of brains through an electrical signal. Studies showed that electrical stimulation on specific neural tissues can evoke different reactions [1]. Implantable stimulation systems have been developed for different long-term neural electrical stimulation treatments, such as for Parkinson's disease [2] and sciatica [3][4]. However, since implanting surgeries are often invasive, there is a shift toward performing minimally invasive surgeries by reducing the size of incisions [5]. In recent research, different types and geometries of probes have been developed for different functions to reduce tissue reaction [6][7].

Another issue with implantable stimulation systems is bodily movements that cause the probe to shift from its original place, making the therapy ineffective [8]. Thus, a probe's fixation in the body is an important issue. Scaffolds with different pore sizes and porosity are commonly investigated and applied to different cases of tissue engineering [9]. Cells can grow in suitable sized pores, so our probes were designed with such pores to allow for cell growth that can help anchor the probe and prevent it from shifting after implantation.

In this study, bovine aortic endothelial cells (BAEC) were used because of their similarities to human vascular endothelial cells. Vascular endothelial cells line the entire circulatory system from the heart to the smallest capillaries in humans. They are essential for maintaining blood fluidity and improving thromboresistance in cardiovascular implants such as stents, vascular grafts, and heart valves. Therefore, culturing BAECs is one of the most important aspects of this study.

Two current methods for Parkinson disease treatment are dopamine replacement therapy (DRT) and electrical deep-brain stimulation (DBS) [1], which both are effective but with their flaws. DRT is very useful for treating Parkinson's disease at early stages, but it is much less effective for long-term treatment. Electrical DBS is highly invasive and has crucial dependence on accurately targeting the right spot in small brain structures.

A current method for treating sciatica is with continuous radiofrequency (CRT) treatment [3]. The biggest drawback to CRT is that it can risk causing more nerve damage to the back when seeking to treat backache. This leads to the usage of a variant of the CRT, pulse radiofrequency (PRT) to alleviate pain without the consequences of tissue destruction and painful sequelae.

In this paper, a bipolar porous probe with micro-wings is presented. This probe's design seeks to be minimally invasive so that patients do not have to endure long recovery periods at hospital. The design also seeks to promote stable fixation to properly treat target sites. Its minimally invasiveness is evidenced by the size of the probe, which will fit into a guiding needle with a diameter of 1.6 mm. Its stable fixation is promoted initially by the micro-wings and then by cells when they grow into the pores. The experiments run in this paper will address the stable fixation of the probe and the detecting functions of the electrodes, but the demonstration of PRT with the stimulating pads will be done in the future.

CONCEPT AND FABRICATION
Materials

Since the probe is designed for minimally invasive surgery, the biocompatibility of its materials was largely considered. There are three main reasons for choosing FPC as the base material of the probe: FPC is modifiable so can achieve the acquired thickness for minimally invasive surgery, FPC is composed of polyimide and gold-plated copper which are preferred materials for implantation [10], and FPC is flexible so can resist damage from long-term body motion and can reduce tissue damage induced by shear stress [11]. In order to create 3-D interconnecting porous structure on the probe, two biocompatible SU-8 films [12][13] were attached on both sides of the FPC to form a 3-D porous structure.

978-1-4799-3510-9/14 $31.00 © 2014 IEEE

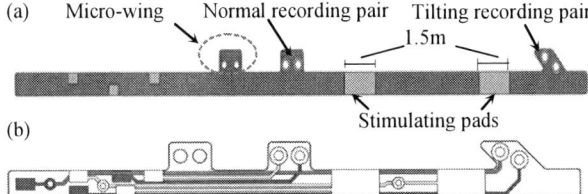

Figure 1: FPC electrode design. (a) FPC contains three conducting pairs: two recording pairs and one stimulating pair. Micro-wing design is used for further fixation. (b) Electric circuit design on the FPC.

Probe Design

The basic FPC probe features micro-wings, which can increase fixation after implanting surgery, and contains porous structures for cell growth, which can promote long term fixation in the body. Additionally, there are two recording pairs to detect whether cells grow into the pores and one pair of stimulating pads to stimulate the target nerve as illustrated in Figure 1. The space between two stimulating pads is designed for the pads to be positioned on both sides of target nerve in order to conduct an effective stimulation. This probe is composed of three layers: two SU-8 layers and one FPC layer to form a 3-D porous structure as shown in Figure 2. At the front-end of the probe, the wing tilts about 45 degrees to more easily insert the probe into the body.

Most animal cells are within the dimensions of 1 and 100 microns [13]. Thus, the size of the pores must be much larger than the cells for cells to grow through the pores. Several studies showed that the choice of materials, pore sizes, and porosity will have a large impact on cell growth in the scaffold [7]. Different types of cells require different suitable pore sizes, but there is no general consensus on the best pore size for cell growth. Lots of factors can allow cell growth. For example, pore size for bone formation can range from 150 to 710 μm. For smooth cells, a study showed that the pore size ranges from 38 to 150 μm [14]. In this paper, the dimensions of pores on the FPC and two SU-8 layers were picked based on the pore size for the smooth cells' scaffold, mentioned above.

The dimensions of the probe are as follows. The diameters of the pores are 250 μm on the FPC and 480 μm on cooper rings. The pore diameter on the SU-8 film is 60 μm with a center-to-center spacing of 100 μm. The length of both stimulating pads is 1.50 mm, and the thickness of the SU-8 film on the probe is 60 μm.

Fabrication of the SU-8 Film

The fabrication process and parameters of SU-8 film fabrication are listed in Table 1, and a diagram of the fabrication process of SU-8 film on one side of the probe is shown in Figure 3. As the manufactured FPC contains three pairs of holes, photoresist SPR-220 was applied on the glass wafer first so that SU-8 did not flow through the

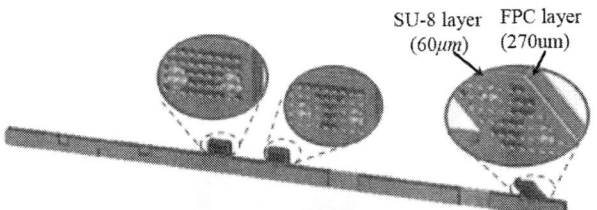

Figure 2: 3-D illustration of the implantable porous FPC-based neural stimulation probe. The front-end wing tilts about 45°, so as to easily insert the probe into the body.

Table 1: Process flow of SU-8 film

Process	Parameter
Coating (3 steps)	a. 500 rpm for 25 sec
	b. 1200 rpm for 25 sec
	c. 1500 rpm for 60 sec
Soft baking (2 steps)	a. 65 °C for 5 min
	b. 95 °C for 17 min
Exposing	3600 μW/cm^2 for 60 sec
Post exposure baking (2 steps)	a. 65 °C for 5 min
	b. 95 °C for 10 min
Developing	16 min
Hard baking	150 °C for 5 min

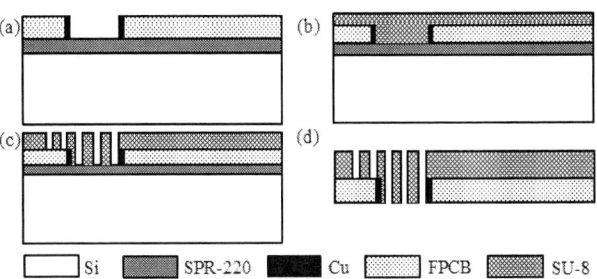

Figure 3: Fabrication process of stimulation probe. (a) Attaching FPC to the wafer. (b) Applying SU-8 on the wafer. (c) Patterning structure on SU-8. (d) Releasing FPC and SU-8 film from the wafer.

holes. Photolithography was then used to fabricate the structure of the SU-8 holes after SU-8 was applied on the FPC. Finally, the probe was released from the glass wafer by removing the SPR-220 film.

Completed Probe

As shown in Figure 4, the probe is flexible, which is ideal for long implantation in an active human body. A guiding needle is first inserted into the human body before the probe is implanted. The guiding needle then helps to direct a path for the probe and also helps prevent the probe from bending.

The procedure of implantation is shown in Figure 5. The probe is inserted through a guiding needle, which is close to the target site. As the guiding needle is retracted, the probe fixes at the target nerve. Once inside the body, the probe is now able to provide electrical stimulation to alleviate pain and record cell growth in the target nerve.

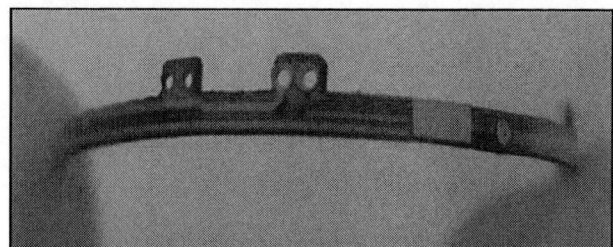

Figure 4: Fabricated flexible stimulation probe.

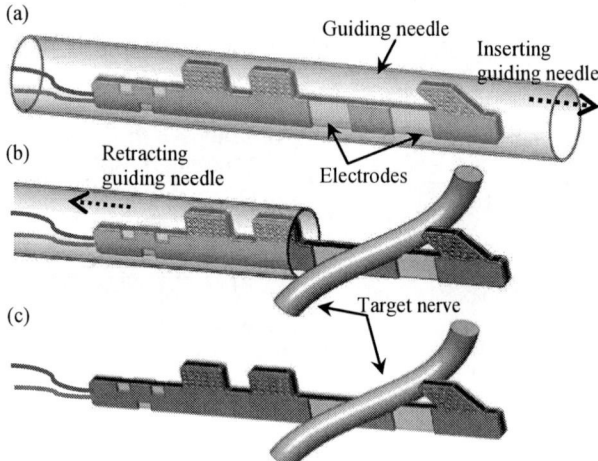

Figure 5: Steps of the implantation. (a) Inserting the probe through the guiding needle. (b) Removing the guiding needle from the body. (c) The probe fixes at the target nerve and provides electrical stimulation.

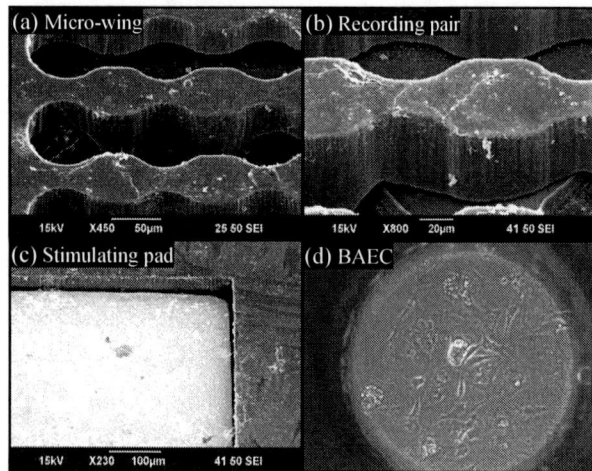

Figure 6: Structures on the probe: (a) 3-D pores on the micro-wing, (b) 3-D pores on recording pair, (c) The stimulating pad, and (d) the BAEC growing flat into the pore.

RESULTS AND DISCUSSION
BAEC Cell Simulation

Initially, BAECs were seeded in a tissue culture polystyrene (TCPS) using a 24-well plate dish (cell density: 5×10^5 cells/cm^2). A probe was then put into the well plate and left in the cell medium for two days. When

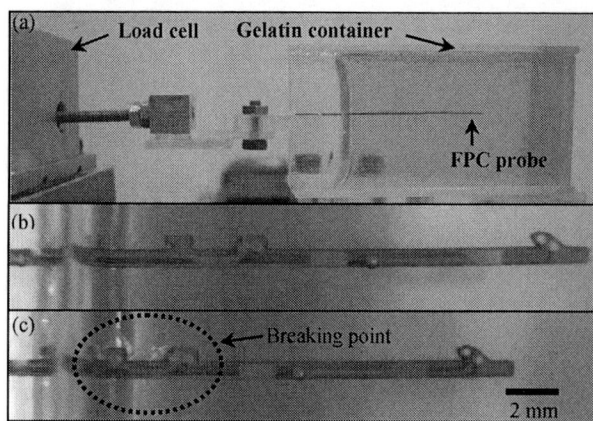

Figure 7. (a) Setup of the tensile test. (b) The probe held by gelatin at the beginning of the test. (c) The structures of the gelatin at the breaking point.

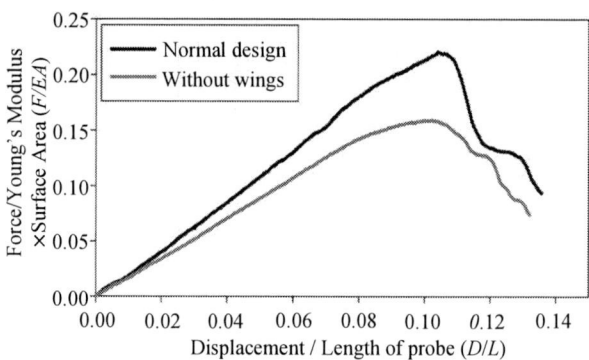

Figure 8: Tensile test of the different probes, where E is the Young's modulus of gelatin and A is the surface area of probe under the gelatin.

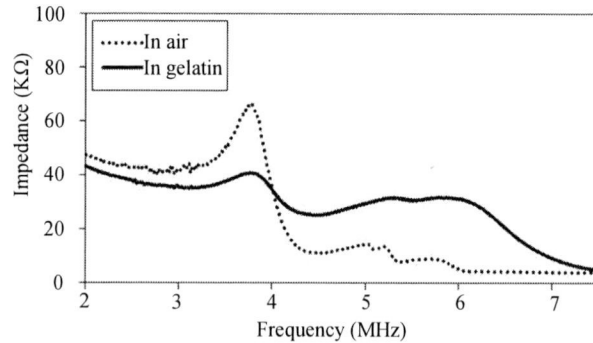

Figure 9: Impedance of recording pairs in air and gelatin.

the probe was taken out, BAECs were visibly present in the pores and observed by an optical microscope (Olympus CKX41) under 100x magnification (Figure 6).

Bending/ Tensile Test

A bending test was performed to determine the effect of SU-8 film on the probes. Treating the probes as cantilever beams, they were pushed by small forces to determine their Young's moduli (E). The product of the

978-1-4799-3510-9/14 $31.00 © 2014 IEEE   863

area moment of inertia and Young's modulus were calculated to be 6.93×10^{-6} N·m^2 for the probe with the SU-8 film and 5.86×10^{-6} N·m^2 without. The result shows greater difficulty in bending the probes with SU-8. From Euler's column formula, the critical buckling load is 0.22 N, so the probes are sufficiently flexible to accommodate body motion but are likely to buckle when inserted into a human spinal cord. Consequently, a guiding needle is used when implanting the probe.

The fixing ability of the probe was evaluated by conducting several tensile tests. In these tests, cells were assumed to fully grow into the SU-8 and FPC pores. To simulate the biological tissue, gelatin was used (Figure 7).

Comparing the breaking points, the probes with wings performed better than those without (Figure 8). The wings increased the fixation force by 38.58% from 0.114 N to 0.158 N, giving a better fixation for the probes at the target nerve. An impedance test was also performed, showing that the impedance of the pores in gelatin is lower than the ones in air since the gelatin provides conductive path of detecting signals (Figure 9).

CONCLUSIONS

We demonstrated and tested the merits of our implantable FPC-based neural stimulation probe for minimally invasive surgery. FPC and SU-8 were carefully chosen to construct the probe because they are biocompatible and therefore preferred materials for implantation and also because SU-8 films made the probe more durable, effectively increasing stiffness of the probe. Furthermore, tensile tests demonstrated that micro-wings indeed increased fixation, and the impedance tests showed that the electrodes can detect growth inside the pores. Overall, these tests show that this probe design has promise.

ACKNOWLEDGMENTS

This paper was funded by National Science Council with the contract number: 102-2320-B-002-040-MY2.

REFERENCES

[1] E. J. Tehovnik "Electrical stimulation of neural tissue to evoke behavioral responses," *Journal of Neuroscience Methods*, vol. 65, pp. 1-17, 1996.

[2] R. Fuentes, P. Petersson, W. B. Siesser, M. G. Caron, and M. A. L. Nicolelis, "Spinal cord stimulation restores locomotion in animal models of Parkinson's disease," *Science*, vol. 323, pp. 1578-1582, 2009.

[3] J. W. M. Geurts, L. Lou, C. A. Gauci, P. Newnham, and R. M. A. W. van Kijk, "Radiofrequency treatments in low back pain," *Pain Practice*, vol. 2, pp. 226-234, 2002.

[4] D. Byrd and S. Mackey, "Pulsed radiofrequency for chronic pain," *Current Pain Headache Reports*, vol. 12, pp. 37-41, 2008.

[5] F. Tendick, S. S. Sastry, R, S. Fearing, and M. Cohn, "Applications of micromechtronics in minimally invasive surgery," *IEEE/ASME Transactions On Mechtronics*, vol.3, pp. 34-42, 1998.

[6] F. Wu, M. Im, and E. Yoon, " A flexible fish-bone-shaped neural probe strengthened by biodegradable silk coating for enhanced biocompatibility," *Solid-State Sensors, Actuators and Microsystems Conference (TRANSDUCERS)*, pp. 966-969, 2011.

[7] F. Wu, L. Tien, F. Chen, D. Kaplan, J. Berke, and E. s Yoon, "A multi-shank silk-backed parylene neural probe for reliable chronic recording," *Solid-State Sensors, Actuators and Microsystems Conference (TRANSDUCERS)*, pp. 888-891, 2013.

[8] R. K. Shepherd, S. Hatsushika, and G. M. Clark, "Electrical stimulation of the auditory nerve: The effect of electrode position on neural excitation," *Hearing Research*, vol. 66, pp. 108-120, 1993.

[9] T. S. Karande, J. L. Ong, and C. M. Agrawal, "Diffusion in musculoskeletal tissue engineering scaffolds: Design issues related to porosity, permeability, architecture, and nutrient mixing," *Annals of Biomedical Engineering*, vol. 32, pp. 1728-1743, 2004.

[10] L. A. Geddes and R. Roeder, "Criteria for the selection of materials for implanted electrodes," *Annals of Biomedical Engineering*, vol. 31, pp. 879-890, 2003.

[11] Y. Zhong, X. Yu, R. Gilbert, and R.V. Bellamkonda, "Stabilizing electrode-host interfaces: a tissue engineering approach," *Journal of Rehabilitation Research and Development*, vol. 38, pp. 627-632, 2001.

[12] S. H. Cho, H. M. Lu, L. Cauller, M. I. Romero-Ortega, J. B. Lee, and G. A. Hughes, "Biocompatible SU-8-based microprobes for recording neural spike signals from regenerated peripheral nerve fibers," *IEEE Sensors Journal*, vol. 8, pp. 1830-1836, 2008.

[13] G. Voskerician, M. S. Shive, R. S. Shawgo, H. von Recum, J. M. Anderson, M. J. Cima, and R. Langer, "Biocompatibility and biofouling of MEMS drug delivery devices," *Biomaterials*, vol. 24, pp. 1959-1967, 2003.

[14] M. Lee, B. M. Wu, and J. C. Y. Dunn, "Effect of scaffold architecture and pore size on smooth muscle cell growth," *Journal of Biomedical Materials Research Part A,* vol. 87A, issue 4, pp. 1010-1016, 2008.

CONTACT

Wen-Pin Shih, Department of Mechanical Engineering, National Taiwan University, No.1, Sec.4, Roosevelt Road, Taipei, 106, Taiwan; Ph: +886-2-33664511; Fax: +886-2-23631755; E-mail: wpshih@ntu.edu.tw

NANOELECTROPORATION AND CONTROLLABLE INTRACELLULAR DELIVERY INTO LOCALIZED SINGLE CELL WITH HIGH TRANSFECTION AND CELL VIABILITY

Tuhin Subhra Santra[1], Jayant Borana[2], Pen-Cheng Wang[2] and Fan-Gang Tseng[1,2,3,4]

[1]Institute of Nano Engineering and Microsystems, National Tsing Hua University, Hsinchu, Taiwan

[2]Department of Engineering and System Science, National Tsing Hua University, Hsinchu, Taiwan

[3]Division of Mechanics, Research Center for Applied Sciences, Academia Sinica, Taipei, Taiwan

[4]Department of Medical Science, National Tsing Hua University, Hsinchu, Taiwan

ABSTRACT

Physical introduction of foreign biomolecules such as genes, proteins, DNA and RNA into living cells with high efficiency is a challenging task for biological and therapeutic research. Bulk electroporation technique, where high electric field pulses were applied to millions of cells together in-between two large electrodes, though widely employed, however is nonspecific resulting in variable efficiency with low cell viability. Here we demonstrate controllable nano-electroporation platform for HeLa cell and human Caucasian Gastric Adenocarcinoma (AGS) cell to achieve high efficient bimolecular delivery with high cell viability. Our system consists of 40nm triangular Indium Tin Oxide (ITO) metal tip with 60nm electrode gap to provide high intense electric filed into the local region of the single cell membrane. Therefore biomolecules can be delivered by much enhance electrophoresis and diffusion effects during pulsing process through a small specific nano-region of the single cell, where remaining other area of the cell membrane unaffected. This microfluidic device has great ability to offer spatial, temporal and qualitative dosage control as well as very high transfection efficiency and high cell viability.

INTRODUCTION

Introduction of foreign biomolecule into target living cells is an important phenomenon for biological cell studies and therapeutic research. For successful gene delivery, it is essential to maintain high transfection efficiency with high cell viability, low toxicity, lower cell damage with different variety of cells [1]. Viral mediated gene delivery has some limitations such as immune response, toxicity and high cost [2-3]. On the other hand physical method such as micro-injection [4], jet injection [5], sonoporation [6], and electroporation [7-11], are the potential substitute tools for gene transfer in compare with viral method [12]. Among all of these physical methods, electroporation has some advantages like easy operation, greater reproducibility due to determination of electrical parameters, avoidance of toxicity and it can control the size of the nanopores by adjusting pulse durations, number of pulses and time between two pulses [13]. For conventional electroporation, where two large electrodes (millimeter scale) positioned with suspension of millions of cells together. As results, very high electric field needed to transfect of each cell with

low cell viability. To understand cell to cell behavior with their cytosolic compounds and their biochemical effects, single cell analysis using electroporation technique has became a frontier issue in last couple of years, where bulk measurement with millions of cells together cannot provide proper cell to cell information. Thus, miniaturization of the device at single cell dimension is require with lower electrode surface area. Lower electrode surface can provide lower toxicity and higher transfection efficiency with very high cell viability [9,11,14]. Although larger electrode can provide higher transfection with high cell viability using electrophoresis driven ion transportation, where voltage requirement was higher [15-16]. Thus, to achieve high transfection with high cell viability, we have reduced the electrode size in submicron level in the past [8], to demonstrate localized single cell electroporation with membrane reversibility. In this paper, we are going to further demonstrate the power of introduction of substances into single cell within tens of nanometer regions on the cell membrane, as results of more reduction on cell toxicity, applied voltage and increase of cell viability.

MATERIALS AND MERHODS
Simulation

Figure 1 shows the schematic representation of localized single cell nano-electroporation (LSCNEP), where single cell seeded on top of the chip and due to intense high electric field in local area of the single cell, biomolecules can deliver from bottom of the channel to inside of the single cell. As intense electric field in local area of the single cell, membrane can deform only small area to deliver exogenous biomolecules from outside to inside of the cell with high cell viability, where remaining other area of the single cell membrane was unaffected. In our device, we fabricated triangular electrode shape with 60 nm electrode gap and 40nm tip diameter. As results electric field can intense only nano-electrode gap to transfect single cell. Figure 2(a) shows electric field with dielectric passivation layer (SiO_2) is approximately 4.5×10^7 V/m for 6V external applied voltage. This passivation layer can reduce hydroxyl and hydrogen ion generation during electroporation experiment resulting to enhance cell viability [10]. Due to application of external applied voltage outside of the cell membrane, the conductivity and permeability can enhance

978-1-4799-3510-9/14 $31.00 © 2014 IEEE

and it is completely different from outside to inside of the cell membrane. This potential difference is called transmembrane potential (TMP) across the cell membrane.

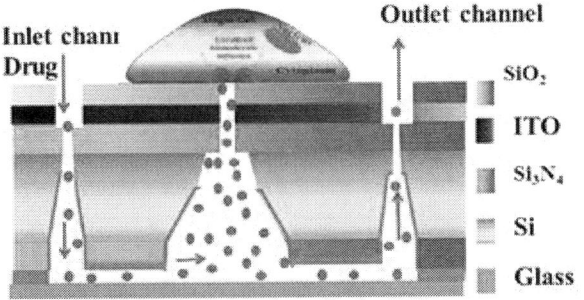

Single cell

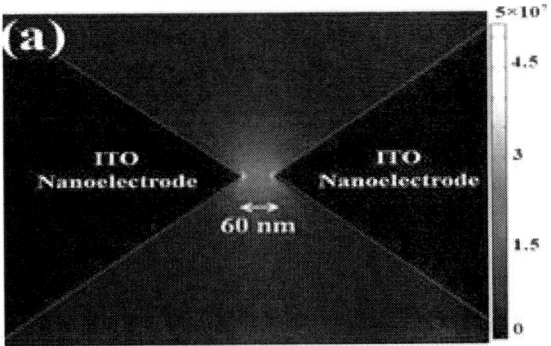

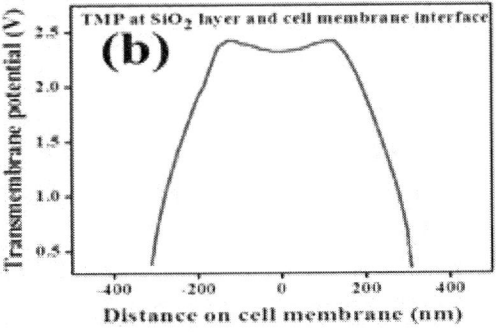

Figure1: Schematic representation of localized single cell nano-electroporation (LSCNEP) device including inlet/outlet channels

Figure 2: (a) Electric field distribution in-between nano-electrode gap with SiO$_2$ layer consideration (4.5×10^7 V/m) (b) localize TMP distribution in a single cell membrane.

For successful bimolecular delivery in single cell electroporation, TMP should be 0.2V ~1V. Figure 2(b) shows the TMP distribution across the SiO$_2$ layer and cell membrane interface (approximately 2.43V), where the area of the TMP distribution is in few nanometer range within a single cell membrane resulting very fast molecular delivery without affecting whole single cell. This novel idea can reduce the cell toxicity and enhance the cell viability.

Fabrication

To fabricate our LSCNEP device, initially we considered 225 μm silicon wafer with 100nm Si$_3$N$_4$ coating on both side of the wafer. Then we etched Si$_3$N$_4$ layer from bottom side by using standard lithography process and reactive ion etching (RIE) technique. The inlet/outlet channels were fabricated with KOH process and finally we open up the nitride membrane (from bottom to top side). After that, we deposited Indium Tin Oxide (ITO) film on top of the wafer using RF sputtering and patterned it by wet chemical etching to form as ITO lines (total 30 ITO lines were formed as an ITO array). Each ITO line width was 3μm with 80nm thickness. The resistivity of ITO film was 5 × 10^{-4} ohm/cm. Finally we deposited 50nm SiO$_2$ layer as dielectric passivation layer using Plasma Enhanced Chemical Vapor Deposition (PECVD) technique.

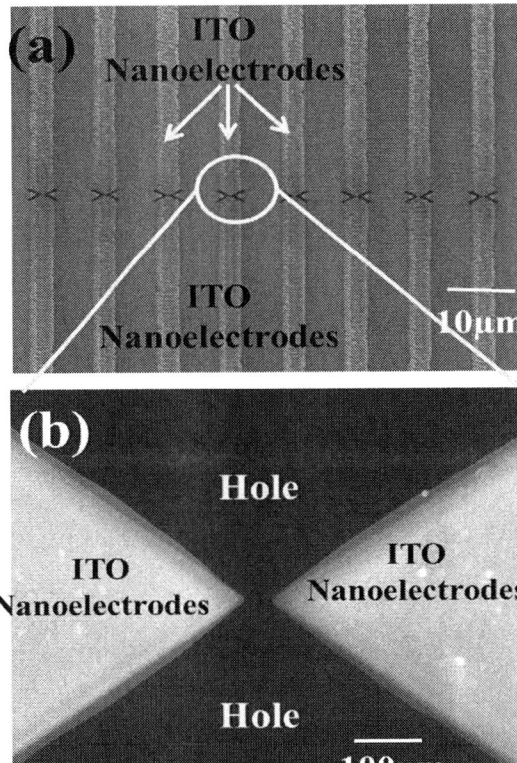

Figure 3: (a) Scanning electron microscopy (SEM) image of array of Indium Tin Oxide (ITO) based nano-electrode (b) nano-electrode tip with 60nm electrode gap and 40nm tip diameter

After passivation layer deposition, we fabricated ITO nano-electrode by using Focused Ion Beam (FIB) technique with 40 nm tip diameter and 60 nm gap between two ITO nano-electrodes. Figure 3(a) shows array of ITO nano-electrode and figure 3(b) shows ITO nano-electrode with 40nm tip diameter and 60nm gap between two nano-electrodes. After formation of nano-electrode, final chip was packaged with Printed Circuit Board (PCB).

Cell culture

To culture HeLa cell line and human caucasian gastric adenocarcinoma (AGS) cell line, initially 10 ml phosphate

978-1-4799-3510-9/14 $31.00 © 2014 IEEE   866

buffer saline (PBS) was added into the cells content disc surface to clean the cell surface properly. Then 1 ml trypsin (0.05% trypsin-EDTA, GIBCO) was added into the disc and immediately transferred it into the incubator (37 ^{0}C with 5% CO_2) for 5 minutes to detach the cells from the disc surface. Finally 9 ml DMEM (Dulbecco's Modified Eagle's Medium) medium was added on to the disc for suspension of cells. The final concentrations of the cells were 2.4 $\times 10^5$ cells/ml. After UV treatment of our chip surface, we introduce 500µl cells with DMEM medium into the chip and it incubated for 6-8 hour to achieve good adhesion between cells and chip surface prior to LSCNEP experiment.

RESULTS AND DISCUSSIONS

After successful cells adhesion onto the chip surface, cells were randomly distributed throughout the chip surface. But those cells were attached in between top of the nano-electrode gap, LSCNEP experiment was performed only on that single cell. Figure 4 shows randomly distributed AGS cell throughout LSCNEP chip surface.

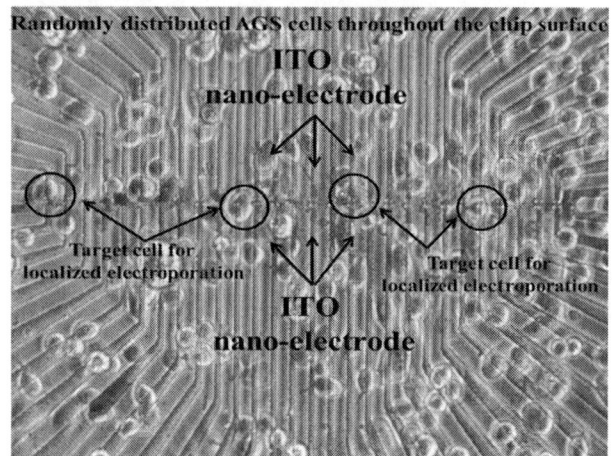

Figure 4: AGS cell distributed randomly throughout the chip surface. But some of them are in gap between two nano-electrodes, which was target cells for LSCNEP experiment.

For our experiment, we have used function generator (Tabor electronics, Israel) to apply square wave positive pulse. Before apply electrical pulse, we added phosphate buffered saline (PBS) mixed with propidium iodide (PI) dye into the chip. For live cell, PI day cannot enter inside the cell to give any fluorescence. But due to apply electrical pulse into the localized single cell, membrane can deform to create nanopores, which allow PI dye inside the single cell, resulting red fluorescence image. Figure 5 shows cell survival fluorescence image of PI dye uptake inside single HeLa cell due to 6Vpp 10ms single pulse. The dye uptake inside the single cell within 30seconds. As electric field intense only a small membrane region of the single cell, transmembrane potential can exceed easily to cell membrane threshold values resulting to deform cell membrane structure and deliver dye inside single cell, where

the reaming other area of the cell membrane was unaffected. Due to affected small area with high electric field, it might be possible to create larger nanopores size resulting fast delivery of dyes inside single cell. Thus LSCNEP technique provided 80% transfection efficiency with 95% cell viability (AGS cell)

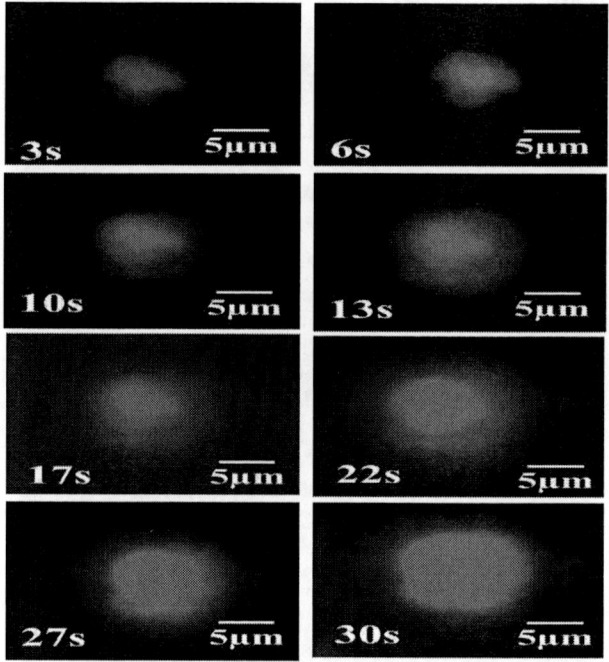

Figure 5: PI dye delivery inside single AGS cell (6Vpp, 10ms single pulse)

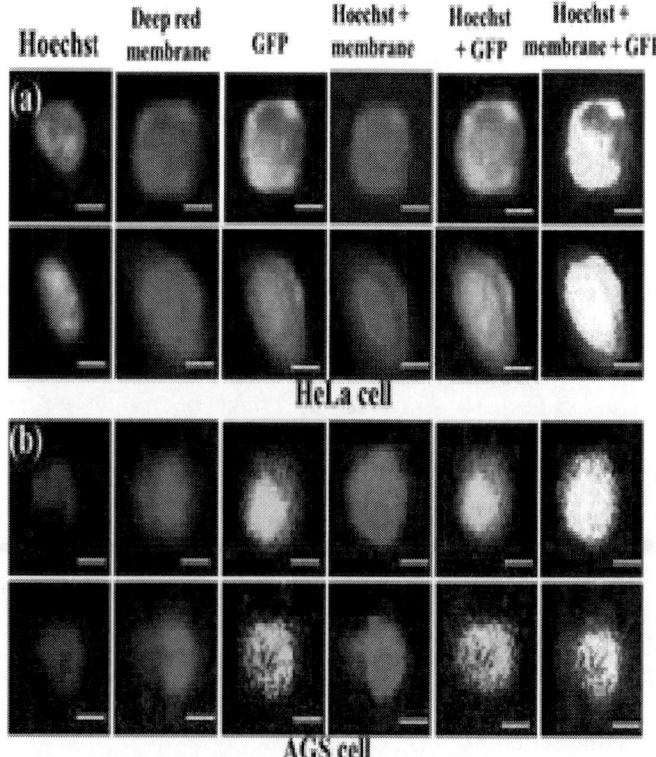

Figure 6: GFP delivery inside single HeLa cell and AGS cell after 12 hours.

Figure 6 shows the GFP (pMax-E2F1, Addgene, USA) delivery inside single HeLa and AGS cell with 5Vpp 20ms pulse (3 pulses) and 7Vpp 50ms pulse (3 pulses). After applications of electrical pulses, we washed out GFP from the chip and introduced DMEM medium. Then incubated this LSCNEP chip for 12 hours. After incubation, we added Hoechst 33342 for nucleus staining and deep red plasma membrane for membrane staining and put it again into the incubator for 15-20 minutes. Then we washed out all staining dyes again from the chip and finally captured fluorescence image. From figure 6, GFP successfully entered inside single HeLa cell and AGS cell line after 12 hours. The transfection efficiency for HeLa cell was 80% and for AGS cell was 70%. However the viability for HeLa cell and AGS cell were almost 90%.

CONCLUSIONS

We demonstrated well controlled localized single cell nano-electroporation with dielectric passivation layer, where electrical field intense only a small region of the single cell membrane. As a results biomolecules can delivery through a small specific membrane region of the single cell, where remaining other area of the cell was unaffected. This novel concept of nano-electroporation enhanced transfection efficiency as well as cell viability compared to other techniques, which potentially beneficial for medical diagnostics and therapeutic studies.

ACKNOWLEDGEMENTS

The authors greatly appreciate for financial support from National Science Council (NSC) of Taiwan ROC through Biomedical Engineering Program (NSC-101-2221-E-007-032-MY3), and National Nanotechnology and Nanoscience Program (NSC-101-2120-M-007-001).

REFERENCES

[1] D.C.Chang, P.Q.Gao, B.L.Maxwell, "High efficiency gene transfection by electroporation using a radio-frequency electric field", *Biochim. Biophys. Acta*, vol.1092, pp.153-160, 1991.

[2] R.G.Crystal, "Transfer of genes to humans: early lessons and obstacles to success", *Science,* vol. 270, pp.404-410, 1995.

[3] S.C.Hartman, D.M.Appledorn, A.Amalfitano, "Adenovirus vector induced innate immune responses: Impact upon efficacy and toxicity in gene therapy and vaccine applications", *Virus Res.,* vol.132(1-2), pp.1-14, 2008.

[4] J.A.O'Brien, S.C.R.Lummis, "Biolistic transfection of neuronal cultures using a hand-held gene gun" *Nature Protocol,* vol.1, pp.977-981, 2006.

[5] D.H.Fuller, P.Loudon, C.Schmaljohm, "Preclinical and clinical progress of particle-mediated DNA vaccines for infectious diseases", *Methods*, vol.40, pp.86-97, 2006.

[6] S.Ohta, K.Suzuki, Y.Ogino, S.Miyagawa, A. Murashima, D.Matsumaru, "Gene transduction by sonoporation", *Dev. Growth. Differ*, vol.50(6), pp.517-520, 2008.

[7] T.Y.Tsong, "Electroporation of cell membranes", Biophys. J, vol. 60(2), pp. 297-306, 1991.

[8] S-C.Chen, T.S.Santra, C-J.Chang, T-J.Chen, P-C.Wang, F-G.Tseng, " Delivery of molecules into cells using localized single cell electroporation on ITO micro-electrode based transparent chip", Biomed. Microdevices, vol. 14(5), pp.811-817, 2012.

[9] T.S.Santra, F.G.Tseng, "Recent trends on micro/nanofluidic single cell electroporation", Micromachines, vol. 4(3), pp.333-356, 2013.

[10] T.S.Santra, P-C.Wang, H-Y.Chang, F-G.Tseng, "Tuning of nano electric field to affect restrictive membrane area on localized single cell neon-electroporation", Applied Physics Letters, DOI: 10.1063/1.4833535.

[11] T.S.Santra, P-C.Wang, F-G.Tseng, Advances in micro/nano electromechanical systems and fabrication technologies, InTech, 2013.

[12] D.Luo, W.M.Saltzman," Synthetic DNA delivery systems", *Nature Biotechnol.*, vol.18, pp.33-37, 2000.

[13] G.L.Prasanna, T.Panda, "Electroporation:basic principles, practical considerations and applications in molecular biology", *Bioprocess Engineering*, vol.16, pp. 261-264, 1997.

[14] J.A.Kim, K.Cho, M.S.Shin, W.G.Lee, N.Jung, C.Chung, J.K.Chang, "A novel electroporation method using a capillary and wire-type electrode", Biosensors and Bioelectronics, vol.23, pp. 1353-1360, 2008.

[15] P.Y.Boukany, A.Morss, W-C.Liao, B.Hensiee, H.C.Jung, X.Zhang, B.Yu, X.Wang, Y.Wu, L.Li, K.Gao, X.Hu, X.Zhao, O.Hemminger, W.Lu, G.P.Lafyatis, L.J.Lee, "Nanochannel electroporation deliverys precise amounts of biomolecules into living cells", Nature Nanotechnology, vol.6, pp.747-754, 2011.

[16] X.Xie, A.M.Xu, S.L-Ortiz, Y.Cao, C.C.Garner, N.A.Melosh, "Nanostraw-electroporation system for highly efficient intracellular delivery and transfection", ACS Nano, vol.7, pp.4351-4358, 2013.

CONTACT

*Fan-Gang Tseng, Tel: +886-3-571-5131 ext. 34270. Email: fangang@ess.nthu.edu.tw

OPTO-MECHANICAL MICROBRIDLES FOR THE DETERMINATION OF STRUCTURAL AND FUNCTIONAL PROPERTIES OF SMALL RESISTANCE ARTERIES

Rosalia Rodriguez-Rodriguez[1], Jose Antonio Plaza[2], Vladimir Matchkov[3], Ulf Simonsen[3], Maria Dolores Herrera[1], Stephanus Büttgenbach[4], Andreu Llobera[2] and Xavier Muñoz-Berbel[2]

[1]University of Seville, SPAIN
[2]Institut de Microelectrònica de Barcelona (IMB-CNM, CSIC), SPAIN
[3]University of Aarhus, DENMARK
[4]Technische Universität Braunschweig, GERMANY

ABSTRACT

This work presents a polydimethylsiloxane (PDMS) system (named microbridle) monolithically integrating optical (microlenses, air mirrors etc.) and mechanical elements (cantilever) for the real time monitoring of dilatation and contraction events in small resistance arteries. Structural (response to intraluminal pressure changes) and functional properties (response to vasoactive substances) of mesenteric arteries were determined with the microbridle structure and compared to those obtained with conventional myograph systems. Both systems provide comparable data although microbridles were found advantageous in terms of precision, resolution (below the µm) and reliability for the poor contrast between sample and surrounding medium that impede reliable myography recordings.

INTRODUCTION

Cardiovascular diseases (CVDs) are the main cause of mortality and morbidity in developed countries nowadays [1]. One of the major problems for CVDs prevention is that current cardiovascular risk factors (e.g. age, hypertension) have been demonstrated poorly effective in the prediction of new cardiovascular events [2]. Endothelial dysfunction (impairment of the function of the vessel endothelial layer) has been positioned as one interesting alternative to conventional systemic risk factors [3]. In fact, it has been already proved to occur earlier than system risk factors, for example, in the course of atherosclerosis [4], and has been observed in patients with arterial hypertension [5], smokers [6], obese individuals [7] and other individuals under high cardiovascular risk. However, endothelial dysfunction is providing different information according to the role of each arterial bed [3]. That is, endothelial dysfunction in conduit arteries is more important in patients with existing atherosclerosis, whereas microvascular dysfunction (such as in small resistance arteries) may be an earlier indicator of cardiovascular risk [3, 8]. Therefore, the deep analysis of functional and structural properties of small resistance arteries may be extremely informative for early detection of cardiovascular risk.

Nowadays, most of structural and functional analyses of small resistance arteries are made in vitro with arterial segments using microvessels myography [9]. With this technique, the dilatation/contraction of the vessel in response to external stimuli (either vasoactive drugs, for functional properties, or mechanical forces, for structural properties) is monitored with videomicroscopes and suitable softwares. Although it has been recently applied to the determination of structural and functional properties of resistance arteries in vivo (in anesthetized animals) [10], this method present important drawbacks, mainly low resolution (> 5-10 µm) and lack of precision (videomicroscopes focus on one XY plane and small changes in the position of the vessel or contrast may alter the recording). Additionally, conjunctive tissue and fat surrounding the vessels can also interfere in the recordings obtained with myograph systems.

In an attempt to solve this open issue, a polydimethylsiloxane (PDMS) system (named microbridle) monolithically integrating optical (microlenses, air mirrors etc.) and mechanical elements (cantilever) for structural and functional evaluation of isolated segments of mesenteric arteries was fabricated and compared with conventional myograph systems.

EXPERIMENTAL

Microbridle design

The microbridle design is schematized in Figure 1.

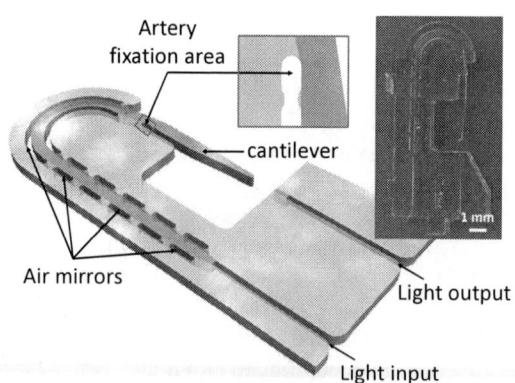

Figure 1: Scheme of the opto-mechanical microbridle. A detail of the artery fixation area is also included. In red, illustration of the main optical path of the confined light inside the microbridle. Inset an image of the microbridle.

Three basic elements are here monolithically integrated: (i) micro-optical elements, self-alignment elements (for the correct positioning and alignment of the optical fibers), microlenses (for numerical aperture correction) and air mirrors (for light confinement and guiding) (ii) the artery fixation area and (iii) a waveguide-cantilever.

Fabrication protocol

Microbridles were fabricated using conventional soft-lithography [11]. Briefly, the master was fabricated using the negative tone SU-8 polymer (MicroChem, Corp., Newton, MA, USA), developed by immersion in propylene glycol methyl ether acetate solution (PGMEA, MicroChem Corporation, Newton, MA, USA) and bake. PDMS (Sylgard 184 elastomer kit, Dow Corning, Midland, MI, USA) replicas of the master (prepared as indicated by the supplier) were cured for 20 min at 80^0C. The structured PDMS was irreversibly bonded to PDMS after treatment with oxygen plasma and temperature (20 min, 60^0C). An image of the final microbridle structure is shown in Figure 1, inset.

Polydimethylsiloxane was chosen as constituent material in this case for presenting high transmittance from the visible to the near infra-red [12] and low Young's modulus (300-800 kPa) [13], similar to biological systems (e.g. arteries = 100-700 kPa) [14].

Microbridle performance

The microbridle performance is illustrated in Figure 2.

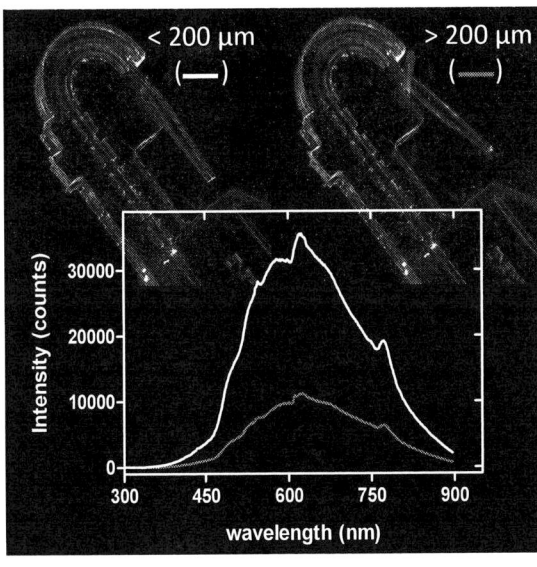

Figure 2: Representation of the microbridle performance. Below 200 μm, waveguide-cantilever and optical fiber optics are aligned (left image, white spectrum). Above 200 μm, there is misalignment between them (right image), which reduces the collected light intensity (red line spectrum).

In a few words, a fiber optics inserted in the input self-alignment element couples the light into the microbridle structure (after correction of the numerical aperture by the micro-lenses integrated on that). This light is confined with the help of the air mirrors towards the free end of the waveguide-cantilever, where it is focused by a microlens to the fiber optics fixed in the output self-alignment element. Light intensity at the output is modulated by the relative position between the waveguide-cantilever and the output fiber optics. Waveguide-cantilever and output fiber optics are perfectly aligned when the structure clamped in the fixation area is < 200 μm in diameter. Structures with higher diameters displace the cantilever, misaligning it and decreasing the collected light intensity, as shown in Figure 2. Thus, contraction-dilatation processes produce a displacement of the waveguide-cantilever, misaligning it with the output optical fiber.

Experimental set-up

Opto-mechanical measurements were performed in a conventional pressure myograph system (111P, Danish Myo Tech, Aarhus, Denmark) as described above. A 230 μm diameter multimode optical fiber (Thorlabs, Dachau, Germany) already connected to a broadband halogen lamp (HL-2000-FHSA, Ocean Optics, USA) was inserted into the input self-alignment element. On the other hand, an identical fiber was inserted in the output self-alignment element and connected to a microspectrometer (USB2000, Ocean Optics, USA) for light intensity recording. This microspectrometer has a spectral resolution of 3 nm and was operating at an integration time of 1 second.

The microbridle was fixed at an isolated rat mesenteric artery already mounted and pressurized in the myograph [15]. Artery isolation was performed following the protocol detailed in [15]. The mesentery was collected and placed in cold modified Krebs-Henseleit solution (KHS) (in mM: NaCl 118, KCl 4.75, NaHCO₃ 25, MgSO₄ 1.2, CaCl₂ 1.8, KH₂PO₄ 1.18 and glucose 11). Segments of isolated small mesenteric arteries were cannulated at each end with glass micropipettes, which were fixed in a pressure myograph. Intraluminal pressure was raised to 120 mmHg and the artery was unbuckled by adjusting the cannulas. The vessel was then set to 70 mmHg pressure and allowed to equilibrate for 30 min at 37°C in KHS [15]. The measurement set-up is illustrated in Figure 3.

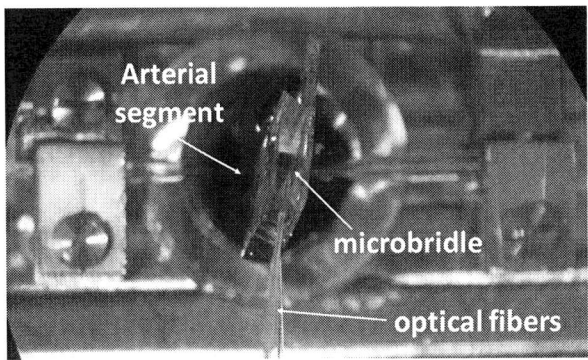

Figure 3: View of a microbridle fixed on a pressurized arterial segment mounted in a myograph system.

RESULTS

Structural and functional properties of resistance arteries were determined using the microbridle structure and compared with conventional myograph systems.

Structural properties

Structural properties of isolated segments of mesenteric arteries were evaluated by following the arterial response (external diameter changes) to intraluminal pressure increases. Intraluminal pressure was increased from 5 to 120 mmHg producing a decrease in the collected light intensity (at 600nm) as a consequence of the increase in the arterial diameter (Figure 4).

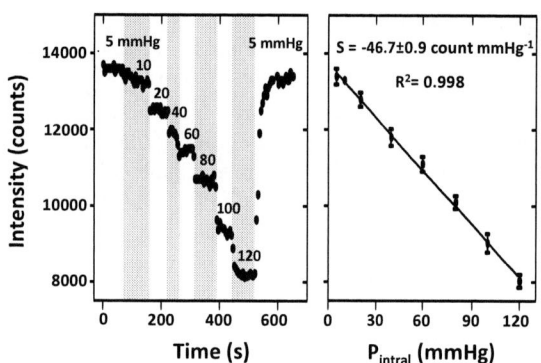

Figure 4: Measurement of light intensity in response to changes in intraluminal pressure of the vessel. Left: Intensity variation vs time. Right: Calibration curve considering intensity variation in response to intraluminal pressure increases (n=3). S corresponds to sensitivity.

These results coincided with those recorded using the conventional myography where the increase in the intraluminal pressure was shown to produce a change in the arterial diameter (external) from 240 to 320 μm.

Additionally, it should be noted that with microbridles, diameter changes were recorded at real time (around 1 s), with high repeatability (STDV below 2 %) and sensitivity, being capable to record the changes produced by an intraluminal pressure variation around 1 mmHg (~50 counts) which corresponds to changes around 1 μm in the artery diameter.

Functional properties

Functional integrity of the artery was determined by evaluating the vasodilatation and vasoconstriction processes produced by vasoactive substances with time. In this case, the response to phenylephrine (Phe) and acetylcholine (ACh) of arterial mesenteric segments under intraluminal pressure close to physiological conditions (70 mmHg) [15], was monitored with both the microbridle and the videomicroscope coupled to the myograph system. Initially, 5 μM Phe was added to produce a submaximal contraction of the artery. Subsequently, endothelium function was evaluated by adding increasing concentrations of ACh (from

10 nM to 3000 nm) and evaluating endothelial-dependent vasodilatation produced by this drug. The changes in the collected light intensity during this process are represented in Figure 5.

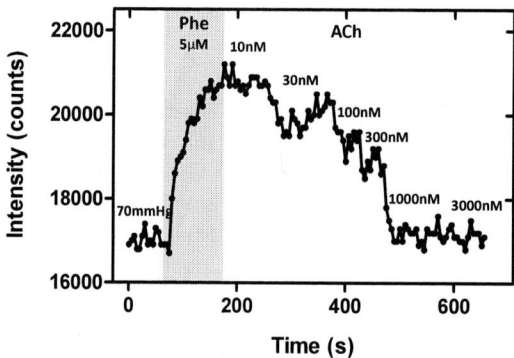

Figure 5: Representation of the endothelial-dependent vasodilatation response to acetylcholine (ACh) in phenylephrine (Phe)-induced contraction in a pressurized artery segment (70 mmHg).

As shown, microbridles allowed the real time monitoring of the vasocontraction/vasodilatation produced by the addition of these drugs. Intensity changes were again verified by myography records that confirmed the expected changes in the arterial diameter. Additionally, even in the presence of the microbridle, the artery responded to both agonists, confirming the integrity of the endothelial layer and the low invasive nature of the microbridle.

CONCLUSIONS

Opto-mechanical systems, called microbridles, here presented have been demonstrated a good alternative to convential myograph systems for the evaluation of functional and structural properties of small resistance arteries in vitro. Microbridles have been shown advantageous in terms of repeatability (STDV below 2%), sensitivity (being capable to determine intraluminal pressure changes below 1 mmHg, around 1 μm arterial diameter), simplicity and cost, when compared with conventional methods.

ACKNOWLEDGEMENTS

The research leading to these results has received funding from the GISCERV program (NGG-227) of the "Ministerio de Economia y Competitividad". Dr. Xavier Munoz-Berbel was supported by the "Ramón y Cajal" program from the Spanish Government. Some authors (U. S., A. Ll. and X. M.-B.) gratefully acknowledges the financial support of the European Commission's 7th Framework programme for funding via the consortium LIPHOS project (grant agreement no. 317916). (S. B.) acknowledges financial support of the Volkswagen Foundation and the German Research Foundation (DFG) in the framework of the Collaborative Research Group mikroPART FOR 856 (Microsystems for particulate life-science products). R. R-R and M. D. H. acknowledges funding from the "Ministerio de Ciencia e Innovación" (AGL2009-11559). V. M. acknowledges funding from the the Danish Heart Foundation (nr. 12-04-R90-A3969-22716).

REFERENCES

[1] World Health Organization. Cardiovascular diseases (CVDs), Fact sheet No. 317, 2011; http://www.Who.Int/cardiovascular diseases/en/

[2] P. Brindle, J. Emberson, F. Lampe, M. Walker, P. Whincup, T. Fahey, S. Ebrahim, "Predictive accuracy of the Framingham coronary risk score in British men: prospective cohort study", *BMJ*, vol. 327, pp. 1267, 2003.

[3] A.J. Flammer, T. Anderson, D.S. Celermajer, M.A. Creager, J. Deanfield, P. Ganz, N.M. Hamburg, T.F. Lüscher, M. Shechter, S. Taddei, J.A. Vita, A. Lerman, "The assessment of endothelial function: from research into clinical practice", *Circulation*, vol. 126, pp. 753-767, 2012.

[4] J.A. Vita, C.B. Treasure, E.G. Nabel, J.M. McLenachan, R.D. Fish, A.C. Yeung, V.I. Vekshtein, A.P. Selwyn, P. Ganz, "Coronary vasomotor response to acetylcholine relates to risk factors for coronary artery disease", *Circulation*, vol. 81, pp. 491-497, 1990.

[5] L. Linder, W. Kiowski, F.R. Bühler, T.F. Lüscher, "Indirect evidence for release of endothelium-derived relaxing factor in human forearm circulation in vivo: blunted response in essential hypertension", *Circulation*, vol. 81, pp. 1762-1767, 1990.

[6] D.S. Celermajer, K.E. Sorensen, D. Georgakopoulos, C. Bull, O. Thomas, J. Robinson, J.E. Deanfield, "Cigarette smoking is associated with doserelated and potentially reversible impairment of endothelium-dependent dilation in healthy young adults", *Circulation*, vol. 88, pp. 2149-2155, 1993.

[7] H.O. Steinberg, H. Chaker, R. Leaming, A. Johnson, G. Brechtel, A.D. Baron, "Obesity/insulin resistance is associated with endothelial dysfunction: implications for the syndrome of insulin resistance", *J. Clin. Invest.*, vol. 97, pp. 2601-2610, 1996.

[8] J.P. Halcox, A.E. Donald, E. Ellins, D.R. Witte, M.J. Shipley, E.J. Brunner, M.G. Marmot, J.E. Deanfield, "Endothelial function predicts progression of carotid intima-media thickness", *Circulation*, vol. 119, pp. 1005-1012, 2009.

[9] M.J. Mulvany, C. Aalkjaer, "Structure and function of small arteries," *Physiological Reviews*, vol. 70, pp. 921-961, 1990.

[10] V.S. Dam, D.M. Boedtkjer, J. Nyvad, C. Aalkjaer, V. Matchkov, "TMEM16A knockdown abrogates two different Ca2+-activated Cl - currents and contractility of smooth muscle in rat mesenteric small arteries", *Pflugers Arch.* 2013. DOI 10.1007/s00424-013-1382-1

[11] Y. Xia, G.M. Whitesides, "Soft lithography," *Angewandte Chemie - International Edition*, vol. 37, pp. 550, 1998.

[12] G.M. Whitesides, E. Ostuni, S. Takayama, X. Jiang, D.E. Ingber, "Soft lithography in biology and biochemistry," *Annual Review of Biomedical Engineering*, vol. 3, pp. 335-373, 2001.

[13] D. Fuard, T. Tzvetkova-Chevolleau, S. Decossas, P. Tracqui, P. Schiavone, "Optimization of polydimethylsiloxane (PDMS) substrates for studying cellular adhesion and motility," *Microelectronic Engineering*, vol. 85, pp. 1289-1293, 2008.

[14] M.F. O'Rourke, J.A. Staessenb, C. Vlachopoulosc, D. Duprezd, G.E. Plantee, "Clinical applications of arterial stiffness; definitions and reference values," *Am. J. Hypertension*, vol. 15, pp. 426-444, 2002.

[15] R. Rodríguez-Rodríguez, P. Yarova, P. Winter, K.A. Dora, "Desensitization of endothelial P2Y1 receptors by PKC-dependent mechanisms in pressurized rat small mesenteric arteries," *British J. Pharmacol.*, vol. 158, pp. 1609-1620, 2009.

CONTACT

*X.Munoz-Berbel, tel: +34 93 594 77 00; xavier.munoz@imb-cnm.csic.es

PDMS MICROCHANNEL SCAFFOLDS FOR NEURAL INTERFACES WITH THE PERIPHERAL NERVOUS SYSTEM

Yoonsu Choi[1], Sohee Park[2], Yoomin Chung[3], Russell K. Gore[4], Arthur W. English[4], and Ravi V. Bellamkonda[2]

[1]University of Texas – Pan American, Edinburg, TX, USA
[2]Georgia Institute of Technology, Atlanta, GA, USA
[3]University of Rhode Island, Kingston, RI, USA
[4]Emory University, Atlanta, GA, USA

ABSTRACT

Neural interfaces with the peripheral nervous system have been developed to provide a direct communication pathway between peripheral nerves and prosthetic limbs. This study reports a regenerated peripheral nervous system which can control the reinnervated muscles and interpret neurological signals. The acquired bioelectrical signals can be used for the interpretation of mind which will be used to monitor prosthetic limbs. Transected nerves were regenerated through PDMS scaffolds and transferred signals through embedded microwires and acquisition systems.

INTRODUCTION

A neural interface system is used for the communication between a neuroscience system in a human body and an outside acquisition apparatus to interpret biological neural signals to meaningful behavior patterns or a variety of sensing feedbacks. Although *in vitro* neural interface system have been developed intensively and have built a strong foundation of the neuroscience research [1, 2], here the study focuses on an *in vivo* neural interface system, specifically the peripheral neural interface system. The fundamental tool for the neural interface is an electrode, a conductive material which can acquire action potentials electrochemically from a vicinity of firing axons.

While targeted muscle reinnervation has been brought to the attention of the medical profession as a reliable clinical approach using bioelectrical signals to control prosthetic limbs, the limited number of reinnervated muscles makes it hard to match sophisticated upper and lower limb movements [3, 4]. Although it is still in early stages of research, advanced approaches have already been reported as a neural interface that communicates directly with the peripheral nervous system using microelectrode arrays and microfluidic channels [5, 6].

This study reports a regenerated peripheral nervous system which can control the reinnervated muscles and interpret neurological signals matching behavior patterns in awake animals. Transected nerves were regenerated through polydimethylsiloxane (PDMS) microchannel scaffolds and transferred bioelectrical signals through embedded microwires and acquisition systems. Figure 1 shows the schematic view of the animal study. One neural interface scaffold, two EMG electrodes, and two cuff electrodes are combined in the whole study.

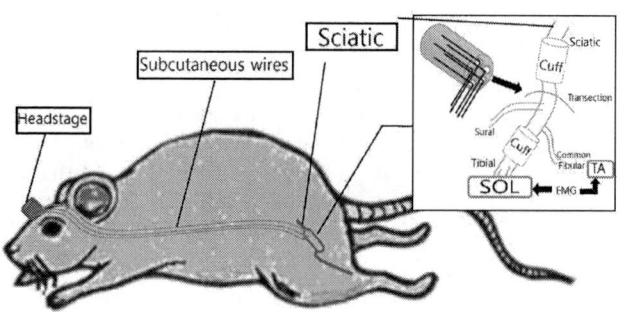

Figure 1: Schematic view of an animal study using a sciatic nerve model for the peripheral nerve interface. A headstage is required to protect the cables from animal bites and other damage. All EMG electrodes and microwires of the neural interface will be connected through the headstage to the outside data acquisition system.

MATERIALS AND METHODS

Animal Study

An animal surgery for neural interface implantation to bridge a transected sciatic nerve was performed on anesthetized Lewis rats (250~300 g), as described in a reference [7]. Methods for construction of the chronic nerve stimulating cuff, chronic muscle electrode implantation, EMG recordings in freely moving rats were similar to those previously described in detail [8-11]. Briefly, the rats were anesthetized with inhaled isoflurane gas, and the surgical site was shaved and sterilized. Marcaine (0.25% w/v, Hospira, Inc.) was next administered subcutaneously for post-surgical pain relief (0.3 ml/animal). A skin incision was then made along the femoral axis, and the underlying thigh muscles were delineated with a blunt probe to expose the sciatic nerve. After the nerves were freed from overlying connective tissue, microscissors were used to transect the sciatic nerve branch and the nerve stumps were pulled 1.5 mm into each end of a guidance channel and fixed into place with a single 9-0 nylon suture (Ethicon Inc.) to connect the neural interface and the peripheral nervous system. The muscles were then reapposed with 4-0 vicryl sutures (Ethicon Inc.) and the skin incision was clamped shut with wound clips (Braintree Scientific, Inc.). After the surgery, the rats were placed under a warm light, allowed to recover from anesthesia, and then housed separately with access to food and water in a colony room maintained at constant temperature (19~22 °C) and humidity (40~50%) on a 12:12 h light/dark cycle. When further action was required, treatment with a mixture of New Skin (MedTech) and Metronidazole (ICN Biomedical Research Products) as described in a reference [12] was administered and found to be effective. All procedures conformed to the Guide for the Care and Use of Laboratory Animals of the Institute of Laboratory Animal Resources, Commission on Life Sciences, National Research Council (National Academy Press, Washington, DC, 1996). Animals were maintained in facilities approved by the Institutional Animal Care and Use Committee (IACUC) at Emory University and in accordance with the current United States Department of Agriculture, Department of Health and Human Services, and National Institutes of Health regulations and standards.

Neural Interface Fabrication

The design of neural interface systems consisted of a microchannel scaffold with integrated microwire electrodes. A PDMS micro-molding technique was used to fabricate 5 mm long microchannel scaffolds. Polysulfone tubes (Koch Membrane Systems, 50,000 molecular weight cutoff – 19 mm long and 1.5 mm inner diameter) were used as PDMS casting molds into which commercially available microwires were inserted and used as microchannel molding materials. The 1.5mm diameter mold was used as an approximation of the diameter of the rat's sciatic nerve. Both 200 µm and 75 µm diameter microchannels were incorporated into the scaffold with 17 and 65 microchannels of each diameter, respectively. After mixing and degassing PDMS (Sylgard 184, Dow), a bundle of microwires were inserted into a polysulfone tube which was filled with PDMS. The microwires were then pulled out of the PDMS structure after the liquid PDMS was cross-linked in a convection oven at 90 °C. In order to integrate recording electrodes, 75 µm thick microwires (Stablohm 800A, California fine wire) were inserted into four of the 200 µm microchannels. Integrated microwires, inserted into the distal end of the scaffold, were curved to avoid obstructing axons regenerating out of the distal end of the scaffold. These wires were then directed along the outside of the scaffold and secured using medical grade UV glue (1187-M-SV01, DYMAX). A polysulfone suture guide was fitted around the neural interface prior to implantation to facilitate alignment and attachment of the transected proximal and distal nerve stumps. The neural interface system is composed of a scaffold for nerve regeneration and embedded electrodes for neural interface. Figure 2 shows the scaffold structure with 75 µm diameter microchannels which is not combined with interface function to optimize the nerve regeneration capability. There are two different size microchannels in the scaffold shown in Figure 3. A 75 µm microwire will be placed inside the bigger channel (200 µm in diameter) for the functionality of the neural interface.

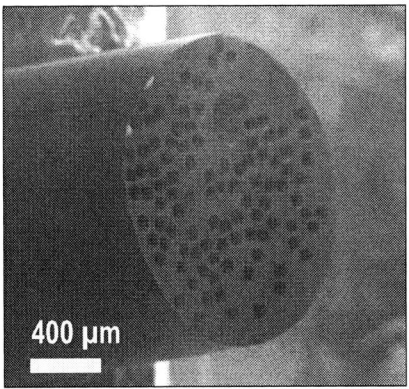

Figure 2: SEM image of a PDMS scaffold with 75 µm diameter microchannels. Although one bubble was observed in the cross section, microchannels were not merged and interfered with it. The scaffold has 100 microchannels and still has space to have more channels. Although more than 180 microchannels have been developed using the same technique, 100 microchannels are enough for the nerve regeneration study and the yield rate of fabrication is 100%.

Figure 3: Bright field image of the cross-section of the scaffold with two different microchannels (200 µm and 75 µm). Wave lines on PDMS came from manual razor blade section.

Immunohistochemistry Assay

After electrophysiology experiments, animals were euthanized with an overdose of sodium pentobarbital and perfused through the heart with saline followed by 4% paraformaldehyde. The integrity of the EMG electrodes, nerve cuffs, and neural interfaces were examined and the regenerated sciatic nerve and the intact neural interface were explanted and post-fixed for 24 hours in 4% paraformaldehyde (Sigma-Aldrich). The samples were rinsed again in PBS and transferred to 30% sucrose in PBS solution for cryoprotection. The samples were then embedded in O.C.T. gel (Tissue Tek) and stored at -40°C until the time of cryosectioning. A cryostat (CM30505, Leica) was used to collect 100 µm thick cross sections and 200 µm thick longitudinal sections.

Sections were later reacted for immunofluorescent demonstration of markers on axons (NF160, 1:500 dilution, Sigma–Aldrich) and Schwann cells (S100, 1:250, Dako). Nuclei were labeled with DAPI (Invitrogen) in PBS at a concentration of 10 mM. The following secondary antibodies were used: Goat anti-rabbit IgG Alexa 488/594, goat anti-mouse IgG1 Alexa 488/594. Immunohistochemistry techniques were conducted as previously described [13]. Briefly, sections were first incubated for one hour at room temperature in a blocking solution of 4% goat serum (Gibco) in PBS containing 0.5% Triton X-100 (Sigma). Sections were then incubated overnight at 4 °C in a mixture of primary antibody and blocking solution, then washed and incubated once more for 1 h at room temperature in a solution of secondary antibody, diluted 1:220 in 0.5% Triton in PBS. This incubation was followed by a 10-min incubation in DAPI solution. Finally, the sections were washed once more, dried, and cover-slipped for evaluation.

RESULTS

Peripheral nerve regeneration within microchannels was evaluated using an immunohistochemistry assay (longitudinal sections stained with NF160 for regenerated axons and S-100 for Schwann cells). Figure 4 showed that PDMS microchannel structures facilitated the nerve regeneration of transected sciatic nerves across 5 mm nerve gaps. The PDMS microchannel scaffolds including 75 µm microchannels were explanted 15 days after nerve transection.

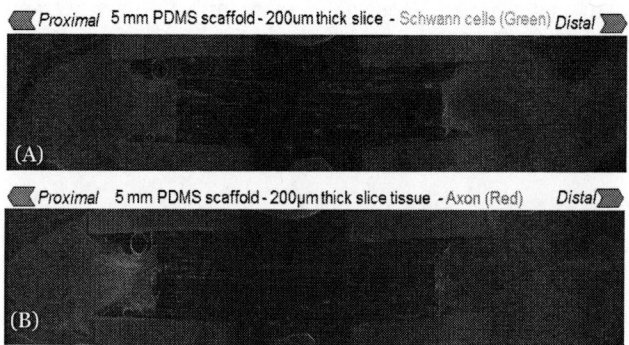

Figure 4: Immunohistochemical analysis of nerve regeneration through PDMS microchannels in vivo (longitudinal section). (A) Schwann cells infiltration through 5mm microchannels from both proximal and distal nerve stump (Green, S-100) (B) Representative axon regeneration through microchannels (Red, NF160).

Figure 4(A) showed that Schwann cells infiltrated into the microchannels from both proximal and distal nerve stumps. The transected NF160 positive axons entered into the proximal end of the microchannels, regenerated through entire length of the scaffold, and moved into the distal nerve stump. The brighter red tissues of the proximal side comparing with the distal side showed no degeneration of the axons in the proximal nerve stump (Figure 4(B)).

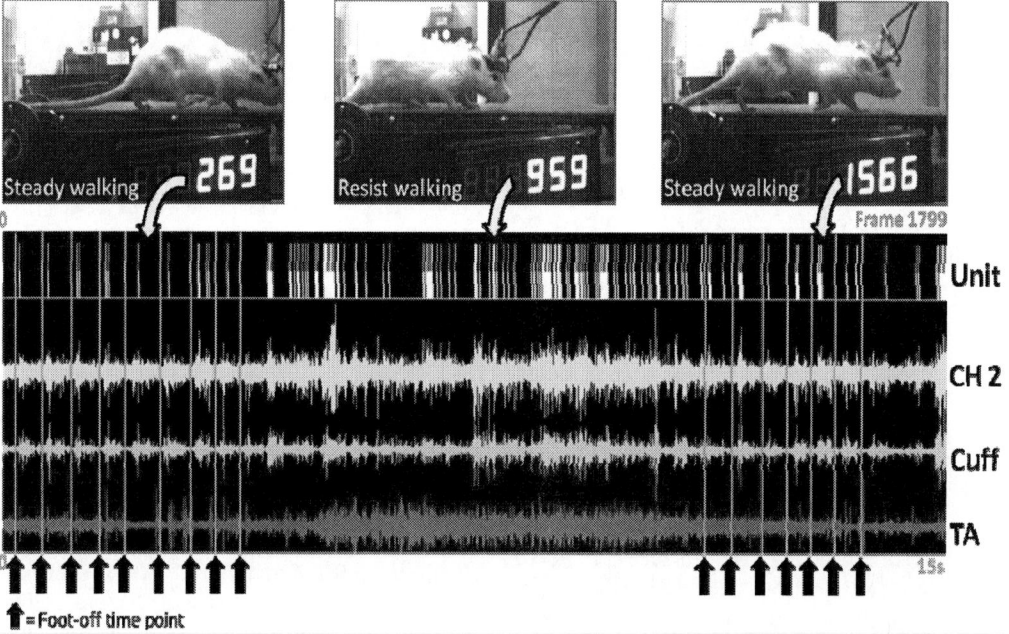

Figure 5: Video analysis of locomotion step cycles. This 15 second locomotion trial contains 14 complete step cycles, the animal resisted locomotion during the central portion of the trial making a unique trial with different patterns. The firing of this motor unit occurs at appropriate times during the step cycle suggesting tibialis anterior (TA) motor innervation. Resist working generates greater muscle and nerve reaction. Selected and repeated action potential compared with, microchannel, TA muscle, and cuff data clearly demonstrating the step cycles.

The spotted brighter red tissues on the distal side showed that two weeks were enough for the transacted nerves to regenerate through 5 mm microchannels and other original axons on the distal side degenerated after the nerve transaction. The light intensity of the fluorescent microscope for NF160 and S-100 was controlled at a little higher level above the saturated light intensity which could illuminate other nerve structures and polysulfone suture guides as well. These 5x images with little more background images helped to see regenerated nerve patterns inside microchannels and structural configuration of the neural interface together while only the brighter parts of the green and red color tissues were the specifically bound nerve structures.

PDMS microchannels are rubber-like structures and flexible even in a frozen OCT mold with tissues. To reserve the original regenerated nerve shape, PDMS microchannels and regenerated nerves were sectioned together. Although the study showed the best results with the 40 μm thickness, 100 μm thickness was chosen as the consistent and reliable sample thickness. Confocal microscope (Zeiss LSM 700) has been used efficiently to get the details of the tissue structures for 100 μm and 200 μm thick tissue samples. The morphological structures of the regenerated tissues were preserved in the molecular level and the target biomarkers were successfully labeled using antibodies inside 100 μm thick tissue samples. 75 μm diameter microchannels contain nerve tissues with significant immunoreactivity to neurofilament proteins and Schwann cells indicating that axons had regenerated into the microchannels and they had become myelinated.

As another analysis for the neural interface, the study incorporated microwires into the PDMS scaffolds to record the bioelectrical activity of the regenerated nerves through the neural interface and of the reinnervated muscles during walking. Action potentials from individual groups of regenerated nerves were recorded from microwires inserted inside a subset of microchannels (Figure 5).

CONCLUSION

PDMS microchannel scaffolds supported endogenous peripheral nerve repair mechanisms. Two week sciatic nerve regeneration made transacted axons infiltrated through 5 mm length microchannel scaffolds. After four weeks, sciatic nerves were regenerated through microchannel scaffolds and reinnervated to muscles showing the functional recovery by an electrophysiological analysis.

REFERENCES

[1] A. Novellino, P. D'Angelo, L. Cozzi, M. Chiappalone, V. Sanguineti, and S.Martinoia, "Connecting Neurons to aMobile Robot: An In Vitro Bidirectional Neural Interface," *Computational Intelligence and Neuroscience,* vol. 2007, pp. 1-13, 2007.

[2] J. N. Y. Aziz, K. Abdelhalim, R. Shulyzki, R. Genov, B. L. Bardakjian, M. Derchansky, *et al.*, "256-Channel Neural Recording and Delta Compression Microsystem With 3D Electrodes," *IEEE Journal of Solid-State Circuits,* vol. 44, pp. 995-1005, 2009.

[3] T. A. Kuiken, G. Li, B. A. Lock, R. D. Lipschutz, L. A. Miller, K. A. Stubblefield, *et al.*, "Targeted Muscle Reinnervation for Real-Time Myoelectric Control of Multifunction Artificial Arms," *JAMA,* vol. 301, pp. 619–628, 2009.

[4] S. P. Agnew, A. E. Schultz, G. A. Dumanian, and T. A. Kuiken, "Targeted Reinnervation in the Transfemoral Amputee: A Preliminary Study of Surgical Technique," *Plastic and Reconstructive Surgery,* vol. 129, pp. 187-194, 2012.

[5] J. J. FitzGerald, S. P. Lacour, S. B. McMahon, and J. W. Fawcett, "Microchannel Electrodes for Recording and Stimulation: In Vitro Evaluation," *Ieee Transactions on Biomedical Engineering,* vol. 56, pp. 1524-1534, May 2009.

[6] J. L. Seifert, V. Desai, R. C. Watson, T. Musa, Y.-t. Kim, E. W. Keefer, *et al.*, "Normal Molecular Repair Mechanisms in Regenerative Peripheral Nerve Interfaces Allow Recording of Early Spike Activity Despite Immature Myelination," *IEEE Transactions on Neural Systems and Rehabilitation Engineering,* vol. 20, pp. 220-227, 2012.

[7] Y. T. Kim, H. V. K., K. S., and R. V. Bellamkonda, "The role of aligned polymer fiber-based constructs in the bridging of long peripheral nerve gaps.," *Biomaterials,* 2008.

[8] M. J. Sabatier, N. T. Bao, J. Nicolini, and A. W. English, "Effect of Axon Misdirection on Recovery of Electromyographic Activity and Kinematics after Peripheral Nerve Injury," *Cells Tissues Organs,* vol. 193, pp. 298-309, 2011.

[9] M. J. Sabatier, N. Redmon, G. Schwartz, and A. W. English, "Treadmill training promotes axon regeneration in injured peripheral nerves," *Exp Neurol,* vol. 211, pp. 489-93, Jun 2008.

[10] X. Y. Chen and J. R. Wolpaw, "Operant conditioning of H-reflex in freely moving rats," *J Neurophysiol,* vol. 73, pp. 411-5, Jan 1995.

[11] A. W. English, "Enhancing axon regeneration in peripheral nerves also increases functionally inappropriate reinnervation of targets," *J Comp Neurol,* vol. 490, pp. 427-41, Oct 3 2005.

[12] Y. P. Zhang, S. M. Onifer, D. A. Burke, and C. B. Shields, "A topical mixture for preventing, abolishing, and treating autophagia and self-mutilation in laboratory rats," *Journal of the American Association for Laboratory Animal Science,* vol. 40, pp. 35-36, 2001.

[13] I. P. Clements, V. J. Mukhatyar, A. Srinivasan, J. T. Bentley, D. S. Andreasen, and R. V. Bellamkonda, "Regenerative scaffold electrodes for peripheral nerve interfacing," *IEEE Trans Neural Syst Rehabil Eng,* vol. 21, pp. 554-66, Jul 2013.

CONTACT

* Y. Choi, tel: +1-956-665-7822; choiy@utpa.edu

SELECTIVE RF HEATING OF RESONANT STENT TOWARD WIRELESS ENDOHYPERTHERMIA FOR RESTENOSIS INHIBITION

Yi Luo, Masoud Dahmardeh, Xing Chen, and Kenichi Takahata
University of British Columbia, Vancouver, CANADA

ABSTRACT

This paper reports a novel active stent targeted at the application to endohyperthermia treatment for in-stent restenosis problems. The stainless-steel stent designed to function as an electrical inductor is integrated with a flexible capacitor strip to form a resonant circuit, which serves as a frequency-selective wireless heater controlled using a tuned radio-frequency (RF) magnetic field applied externally. The fabricated stent device with the initial diameter of 2 mm is expanded up to 6 mm in diameter inside an artificial artery using a balloon catheter. The expanded device is revealed to show efficient heat generation with temperature rise of >30 °C when resonated using an RF power of 320 mW. Temporal and frequency characteristics are evaluated to demonstrate rapid heating ability with strong frequency sensitivity. These promising results validate the feasibility of wireless stent hyperthermia that potentially offers a novel therapeutic path to long-term inhibition and management of stent restenosis.

INTRODUCTION

Stents are mechanical implants that have been used with angioplasty and other catheter-based medical treatments for many years. Each year, three million stents are implanted worldwide, for both vascular (e.g., coronary, carotid, renal, and peripheral arteries) and non-vascular (e.g., urinary and biliary ducts) applications [1]. These implants are most widely used for cardiovascular disease, the number one cause of death in North America [2]. Most of commercialized stents have tubular bodies of stainless steel or shape-memory alloy, which are designed to be expanded radially in order to physically scaffold diseased sites of arteries narrowed by atherosclerosis, i.e., plaque deposition on the vessel walls [3]. Although stent implantation is effective in atherosclerosis treatment, a long-term complication called restenosis, re-narrowing of blood vessels, occurs in stented patients with substantial rates (e.g., up to 60% [4]), mainly due to scar-tissue proliferation within the stent. When severe, it may lead to complete blockage of blood flow [5]. Studies in hyperthermia treatments of restenosis have shown that moderate heating of stents (to temperatures of ~50 °C) is effective in limiting cell proliferation without thrombosis induction [6]. This promising result was, however, obtained using a special catheter to heat a stent, requiring an invasive procedure that not only limits the implementation of the method but also increases the treatment cost.

This study explores a novel method and technology to wirelessly control heating of implanted stents. The aim is to offer a reliable therapeutic means of stent-based wireless endohyperthermia toward long-term prevention and management of restenosis, an inherent issue involved in the implantation of present stents including drug-eluting stents [7]. Wireless MEMS devices have been reported to utilize inductor-capacitor (L-C) resonant circuits as wireless heaters controlled using external radio-frequency (RF) fields for the actuation of thermoresponsive microstructures [8, 9]. The resonant heaters have been demonstrated to provide efficient RF-to-thermal energy conversion and high frequency selectivity in heating, enabling low-power and precision wireless RF operation of the devices performed by modulating the frequency of the external field. This wireless heating mechanism is applied to develop the wireless active stent by constructing the stent to form an L-C tank circuit. When implanted, the device fully deploys the tank circuit to function as a frequency-sensitive wireless heater. By tuning an external field radiated to the implanted device through the skin, the device is precisely activated to locally apply heat stress to inhibit neointimal hyperplasia, the major cause of in-stent restenosis [6]. The efficient resonant heating principle is expected to enable precision control of wireless thermal treatment while suppressing RF radiation power exposed to the body.

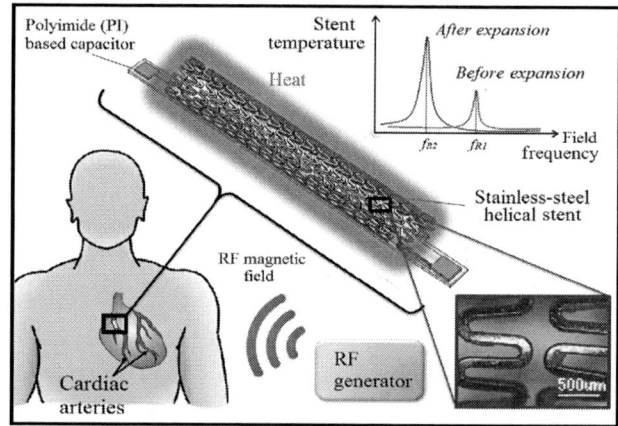

Figure 1: Conceptual illustration of the stent-based wireless endohyperthermia for in-stent restenosis treatment.

DEVICE PRINCIPLE AND DESIGN

The active stent is constructed by electromechanically coupling an inductive stent [10] with a planar capacitive element to establish an L-C circuit that behaves as the wireless heater. An AC current is generated in the circuit when exposed to an AC magnetic field due to the electromotive force induced by the field. The field power transferred to the circuit is maximized when the frequency of the current, or that of the external field matches the resonant frequency of the circuit (defined as $(2\pi\sqrt{LC})^{-1}$) and, thus, the circuit is operated in resonance. The power transfer at this specific condition can be expressed as v^2/R, where v is the induced electromotive force and R is the parasitic resistance

978-1-4799-3510-9/14 $31.00 © 2014 IEEE

of the circuit. As the power transfer is maximized by aligning the field frequency to the resonant frequency of the circuit, the field power is effectively converted to Joule heat generated due to R of the circuit, exhibiting a strong dependence of heating on the field frequency.

For the stent device developed in this study, the resonant frequency of the device changes as the device is expanded and implanted in a blood vessel (refer to Figure 1). This change of frequency can be caused by different factors, including a change in the inductance of the stent due to a radial expansion of its structure (will be described) and capacitive effects due to the change of device ambient (from air to blood). Therefore, it will be important to determine the final resonant frequency after the implantation. This determination can be performed through a well-established wireless technique, i.e., inductively coupling an external antenna coil with the inductive stent through the skin and detecting a dip in the impedance of the antenna using a spectrum analyzer, where the dip frequency represents the resonant frequency of the device [8]. Once this process is completed, the connection of the above antenna can be switched to a field generator to perform a tuned, resonant excitation of the device.

The inductive stent is fabricated by laser micromachining of medical-grade (type 316L) stainless-steel tubing (2.0-mm outer diameter, 100-μm wall thickness) followed by electropolishing, the fabrication approach commonly used in the stent industry. The design of the proof-of-concept device developed is shown in Figure 2. The stent is fabricated to have a solenoid-like geometry, with 17 turns of meandering wire, as its overall shape so that it works as an electrical inductor that can be radially expanded using an angioplasty balloon. The inductance level provided by the stent is 191 nH and 388 nH before and after the expansion (to 6 mm in diameter), respectively. The planar capacitor is formed to have two copper electrodes built on top and bottom of a thin, flexible polymer film in a form of narrow strip (Figure 2). This strip is designed to provide a capacitance of 2.9 pF. The capacitor is coupled with the inductive stent using

the bonding terminals created at both ends of the stent. The top capacitor electrode is connected to a bonding pad on the bottom of the strip through a via so that single-sided bonding of the strip establishes its electrical connection to the stent.

FABRICATION

Figure 3 shows the process flow developed for the prototype fabrication. The planar capacitor strip is fabricated using single-sided Cu-clad polyimide (PI) film with a thickness of 50 μm. The Cu-clad layer with 4-μm thickness on the PI film is photolithographically patterned by wet etching using a dry-film photoresist (SF-306, MacDermid Co. CO, USA) as a mask to create the top capacitor electrode (Figure 3, step (a)). The other Cu electrode (500-nm thick) is formed on the other side of the film by evaporating Cu through a lithographically patterned mask followed by a lift-off process (Figure 3, step (b)). Next, the via contact is formed in the PI film by wet etching of PI using a KOH-based solution (Figure 3, step (c)) and then refilling the hole with Cu electroplating (Figure 3, step (d)). Before performing the stent-capacitor bonding process, the stent is uniformly coated with Parylene-C (Specialty Coating Systems, IN, USA) for a thickness of 8 μm for electrical insulation except the end terminals. After temporarily fixing the stent onto the capacitor strip using spin-coated liquid photoresist (SPR 220-7, Rohm and Hass Co., PA, USA) and hard baking for 4 hours at 120 °C (Figure 3, step (e)), Cu electroplating is again performed for 2 hours to electromechanically bond the stent's end terminals to the strip's bonding pads; dissolving all photoresist layers completes the fabrication (Figure 3, step (f)). The fabricated device is shown in Figure 4(a).

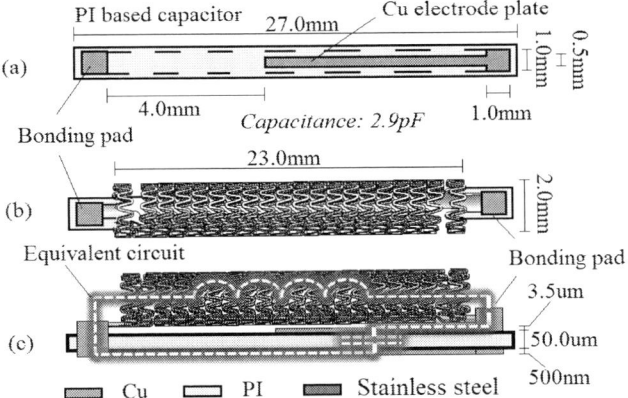

Figure 2: Design of the prototype device: (a) Top view of the capacitor strip; (b) and (c) respectively top and side views of the stent coupled with the capacitor strip forming an LC circuit as displayed in (c).

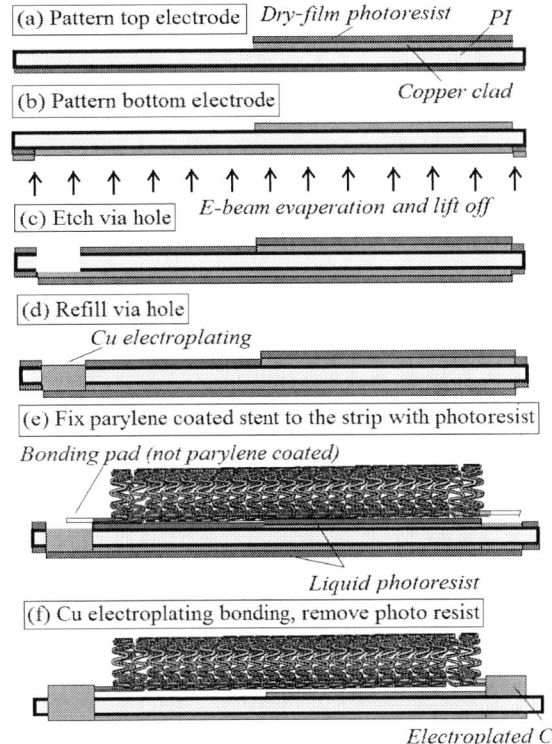

Figure 3: Fabrication process developed.

EXPERIMENTAL RESULTS

Device Expansion

The expansion test of the fabricate stent device was performed using commercial balloon catheters. The device was mounted on the balloon in its contracted form and then expanded up to 6 mm in diameter, by inflating the balloon with pressures up to 15 atm. The expansion process was observed to be completed across the entire structure without any failure both mechanically and electrically. The device deployment was also successfully performed inside a silicone-based artificial artery with 6 mm in diameter (Dynatek Labs, TX, USA) as displayed in Figure 4(b).

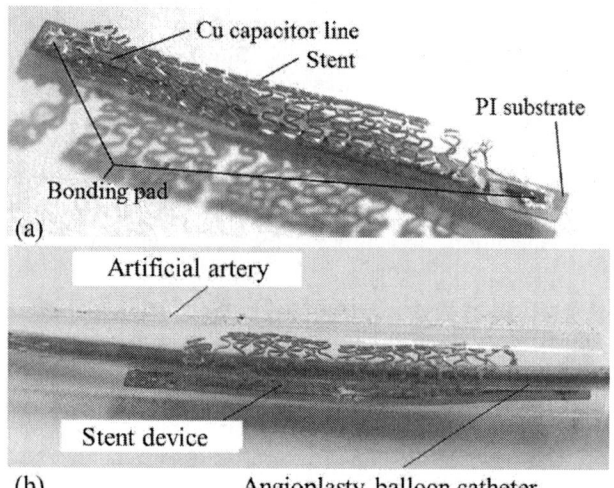

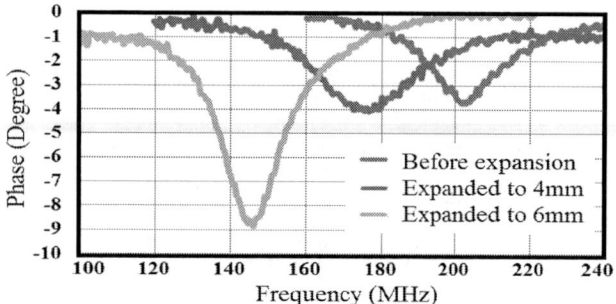

Figure 4: (a) Fabricated device before expansion; (b) the device expanded with the balloon up to 6 mm in diameter inside an artificial artery tube.

Electromechanical Characterization

As noted earlier, the resonant frequency of the device depends on the radial size to which the stent is expanded. This dependence was first characterized, by expanding the device in a manner described above while wirelessly monitoring the resonant frequency of the device through the inductive coupling technique described previously. Figure 5 shows the result measured using a spectrum-impedance analyzer (Agilent 4396B) to which a reader antenna was connected, showing the resonant dips observed in the antenna's impedance phase for three different conditions, i.e., before

Figure 5: Recorded shifts in the resonant frequency of the device led by its expansions to different diameters.

the expansion, after a partial expansion (to 4-mm diameter), and after the full expansion (to 6-mm diameter). As can be seen, the resonant frequency shifted to lower levels as the expansion progressed; this behavior is anticipated, as the larger the stent's diameter, the higher the inductance of the stent (note the inductance of a solenoid is proportional to its cross-sectional area), which lowers the resonant frequency. The measured resonant frequencies matched well with the theoretical values (e.g., theoretical 213.8 MHz and measured 205.3 MHz for the unexpanded state and theoretical 150.1 MHz and measured 145.6 MHz for the fully expanded case). As also can be seen in Figure 5, the increase in the inductance also enhanced the inductive coupling between the device and antenna, resulting in larger phase dips with expansion. This result suggests that the RF power transfer to the device will be maximized when the device is fully deployed upon implantation, an ideal condition for the targeted application.

RF Heating Tests

The experimental set-up used for heating tests is shown in Figure 6. RF magnetic fields were generated using a loop antenna (diameter 1-2 cm) aligned to the tubular axis of the stent and connected to an RF signal generator (HP 8657A) through an amplifier. The device temperature was monitored and analyzed using an infrared (IR) camera (Junoptik VarioCam HiRes 1.2 M, Germany) and its custom software.

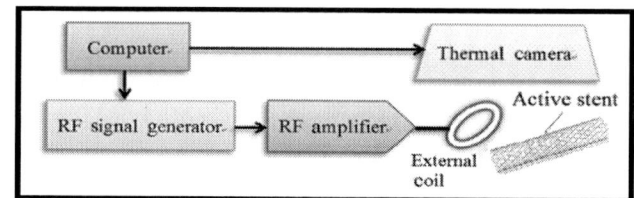

Figure 6: Experimental set-up used for wireless heating tests.

The testing was first conducted with the stent device at the pre-expansion state by radiating an RF field, with the frequency tuned to the resonant frequency (205.3 MHz) of the unexpanded device and an RF output power of 320 mW. A temperature increase was clearly observed in the stent (Figures 7(a) and 7(b)); the maximum temperature was seen around the central region of the stent, reaching up to 57.7 °C. The heating process was similarly observed in the device expanded inside the artificial artery, with the maximum temperature of 63.8 °C on the surface of the artery tube, again around the center region of the stent, using the same RF power (Figures 7(c) and 7(d)). The temperature rise observed in the expanded stent was substantially (22%) higher than that for the unexpanded case, likely due to the predicted effect of enhanced power transfer. The temporal response of heating in the expanded device was characterized by exciting it with the tuned field. As can be seen in the result (Figure 8(a)), temperature of the stent rapidly increased as soon as the RF field was turned on, providing the maximum temperature rise of 33 °C, in which 90% of it was completed in 10 seconds. The frequency dependence of heating was also evaluated for the expanded device by scanning the field frequency as it crossed the resonant frequency of the device. The result shown in Figure 8(b) indicates a strong peak of temperature at the field frequency corresponding to the resonant frequency of the device, validating the frequency selectivity in the operation of the active stent as expected.

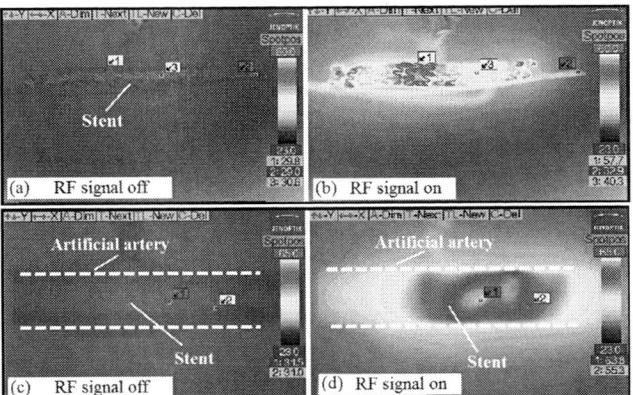

Figure 7: IR images of the stent device before expansion (a) without and (b) with RF excitation and the device expanded inside artificial artery (c) without and (d) with the excitation.

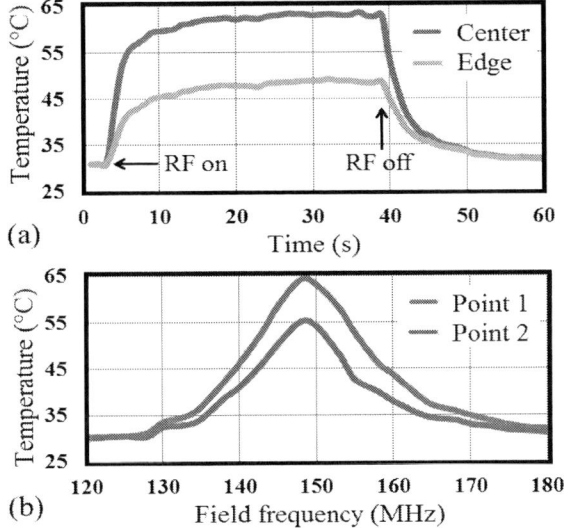

Figure 8: Measured thermal responses: (a) Temporal behavior of stent temperatures (at the center and edge regions); (b) temperature vs. field frequency for the device inside artificial artery (points 1 and 2 correspond to those indicated in Figures 7(c) and 7(d)).

CONCLUSION

A novel electrothermal stent has been studied and experimentally demonstrated. A prototype device was constructed by integrating a lithographically fabricated flexible capacitor strip with a stainless-steel inductive stent to form a resonant circuit. The stent device was shown to function as a frequency-dependent wireless heater controllable using external RF field through its frequency tuning. The device design was arranged to maintain the compatibility with standard stenting tools, demonstrating its electromechanical deployment inside an artificial artery using

a commercial angioplasty balloon. The device expanded in the artificial artery tube was revealed to increase temperature of the tube walls to 60 °C or more, well above the target temperature of ~50 °C, by radiating an RF field when its frequency was tuned to resonate the device. The obtained results verify the essential functionality of the device for endohyperthermia therapy applications.

ACKNOWLEGMENT

This work was partially supported by the Natural Sciences and Engineering Research Council of Canada, the Canada Foundation for Innovation, the British Columbia Knowledge Development Fund, and the Canadian Microelectronics Corporation (CMC). K. Takahata is supported by the Canada Research Chairs program.

REFERENCES

[1] D. Stoeckel, A. Pelton, T. Duerig, "Self-Expanding Nitinol Stents: Material and Design Considerations," *Eur. Radiol.*, vol. 14, pp. 292-301, 2004.

[2] A.G. Logan and D. Bradley, "Sleep Apnea and Cardiovascular Disease," *Curr. Hypertens. Rep.*, vol.12, pp.182-190, 2010.

[3] H.W. Roberts, S.W. Redding, "Coronary Artery Stents: Review and Patient-Management Recommendations," *J. Am. Dent. Assoc.*, vol. 131, pp. 797-801, 2000.

[4] M.R. Bennett, "In-Stent Stenosis: Pathology and Implications for the Development of Drug Eluting Stents," *Heart*, vol. 89, pp. 218-224, 2003.

[5] V. Rajagopal, S.G. Rockson, "Coronary Restenosis: A Review of Mechanisms and Management," *Am. J. Med.*, vol. 115, pp. 547-553, 2003.

[6] C. Brasselet, E. Durand, F. Addad, F. Vitry, G. Chatellier, C. Demerens, M. Lemitre, R. Garnotel, D. Urbain, P. Bruneval, A. Lafont, "Effect of Local Heating on Restenosis and In-Stent Neointimal Hyperplasia in the Atherosclerotic Rabbit Model: A Dose-Ranging Study," *Eur. Heart J.*, vol. 29, pp. 402-412, 2008.

[7] G.D. Dangas, B.E. Claessen, A. Caixeta, E.A. Sanidas, G.S. Mintz, R. Mehran, "In-Stent Restenosis in the Drug-Eluting Stent Era," *J. Am. Coll. Cardiol.*, vol. 56, pp. 1897-1907, 2010.

[8] M.S. Mohamed Ali, B. Bycraft, A. Bsoul, K. Takahata, "Radio-Controlled Microactuator Based on Shape-Memory-Alloy Spiral-Coil Inductor," *J. Microelectromech, Syst.*, vol. 22, pp. 331-338, 2013.

[9] S. Rahimi, E.H. Sarraf, G.K. Wong, K. Takahata, "Implantable Drug Delivery Device Using Frequency-Controlled Wireless Hydrogel Microvalves," *Biomed. Microdev.*, vol. 13, pp. 267-277, 2011.

[10] A.R. Mohammadi, M.S.Mohamed Ali, D. Lappin, C. Schlosser, K. Takahata, "Inductive Antenna Stent: Design, Fabrication, and Characterization," *J. Micromech. Microeng.*, vol. 23, 025015, 2013.

CONTACT

Yi Luo, tel: +1-604-3633468; luoyikey@ece.ubc.ca

TUNABLE MEMS FIBER SCANNER FOR CONFOCAL MICROSCOPY

Niklas Weber, Tobias Meinert, Hans Zappe, and Andreas Seifert
Department of Microsystems Engineering – IMTEK
University of Freiburg, GERMANY

ABSTRACT

We present an endoscopic probe with a forward-looking piezoelectric fiber scanner for confocal optical imaging. The probe, with an outer diameter of 2.5 mm and a length of 13 mm, features a tunable lens to allow confocal depth scanning up to 100 μm without any movable optics or stages. Imaging below 2 μm lateral, and 20 μm axial resolution was successfully demonstrated.

INTRODUCTION

To distinguish healthy from cancerous tissue, the standard medical test is a biopsy, for which small portions of tissue are removed from suspicious areas. This procedure is unpleasant for the patient and bears the risk of infection, injury and hemorrhage. Optical biopsy is a non-destructive and partly non-invasive diagnostic technique and can replace this standard approach [1]. Two prominent methods are optical coherence tomography (OCT) and confocal microscopy which both allow three-dimensional investigation of tissue by laterally scanning the volume of interest.

Bulky instruments with conventional optics are used for optical biopsy of suspicious skin lesions. However, to perform similar analysis within the interior of the body, an endoscopic confocal microscope is indispensable, despite the challenge of implementing high quality optics within a minimal volume. We developed and present here a forward-looking piezoelectric fiber scanner for confocal imaging. Compared to other approaches with implemented fixed-focus lenses [2, 3, 4], this system is based on silicon bench technology with integrated fluidics for realizing tunable liquid-filled membrane-lenses [5]. The tunability in focal length allows depth scanning without any movable parts.

SYSTEM DESIGN

The basic principle of fiber-based confocal imaging is illustrated in Figure 1. The tunable lens focuses light to a specific depth into the sample under test, and only from a small depth area the back reflected light can enter the facet of the single mode fiber to deliver the tissue-specific information.

Figure 2 shows the overall concept of the micro-probe based on an optofluidic Si micro-bench with integrated fiber; piezo tube; hydraulically tunable lens with fluidic connections; and GRIN lens. The fiber is actuated by a piezoelectric tube, thereby performing spiral scans for lateral scanning.

The concept of the optofluidic Si micro-bench allows a highly modular and flexible arrangement of different optical systems to be integrated in a probe with diameters below 2.7 mm [5]. Integrated fluidic channels connect the fluidic chamber of the tunable lens with an external actuation

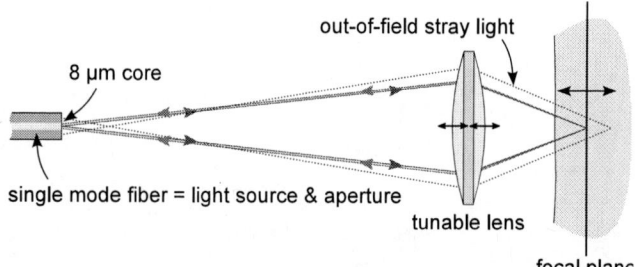

Figure 1: Principle of confocal scanning with a fiber-based optical micro-probe. Axial scanning is achieved by tuning the focal length of the lens.

scheme. The variable liquid-filled membrane lens [5], in combination with the high refractive power of the GRIN lens, is designed to provide a focal tuning over a depth range of about 100 μm at a working distance of 430 μm.

For the optical design, a GRIN lens with a length of 2.41 mm, a diameter of 1 mm and a numerical aperture of 0.5 is chosen; the tunable lens has a clear aperture of 1 mm and a distance of 0.3 mm to the GRIN lens in the non-actuated state. The membrane of the tunable lens is deflected by maximally 200 μm for achieving the shortest required focal length. The final arrangement of the lenses, as a result of the design optimization, is sketched in Figure 3.

Axial resolution in confocal microscopy is determined by the numerical aperture *NA* of the scanning fiber and the magnification of the lens system. For a single mode fiber for $\lambda = 633$ nm with $NA = 0.13$ and a system magnification of 0.72, the total *NA* is 0.18, yielding a resolution of [6]

$$\Delta z = 1.28 \frac{n\lambda}{NA^2} = 25\,\mu\text{m}. \qquad (1)$$

The lateral resolution in confocal microscopy is calculated by

$$\Delta x = 0.43 \frac{\lambda}{NA} = 1.5\,\mu\text{m}. \qquad (2)$$

The specified field of view has a diameter of 200 μm and is realized by a fiber deflection of ±150 μm at a mean magnification of the lens system of 0.69. The protruding length of the fiber is 3 mm and the calculated voltage for maximum deflection is 90 V for a piezo tube with 4 mm length, 0.9 mm inner diameter, and 1.5 mm outer diameter [7]. The inner cylindrical surface is completely covered with one electrode, the outer surface features four electrodes covering cylindrical segments of 45° over the entire length.

PROCESS TECHNOLOGY

Fabrication of the probe is divided into three different elements: the Si platform (the micro-bench); the Si chips, in

978-1-4799-3510-9/14 $31.00 © 2014 IEEE

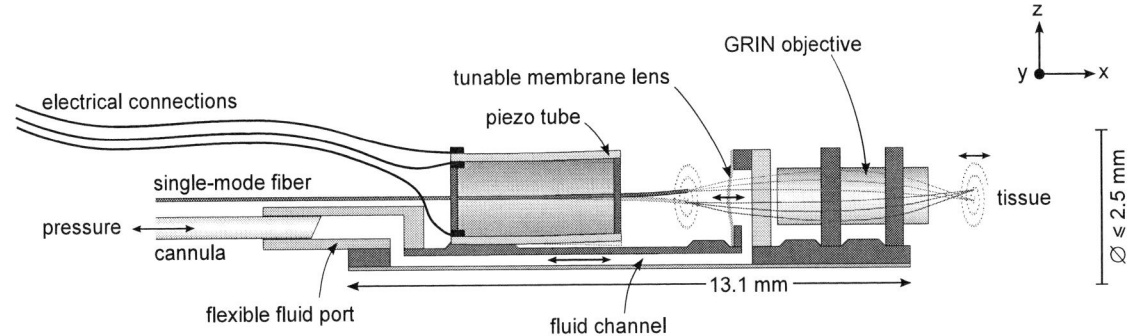

Figure 2: Schematic of the 3D confocal micro-scanner with a laterally scanning fiber driven by a piezo tube, a tunable liquid-filled lens and a GRIN lens, all mounted onto a Si micro-bench with fluidic structures for actuating the tunable lens.

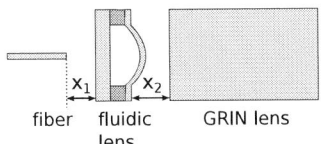

Figure 3: Lens arrangement and distances between fiber, tunable lens and GRIN lens: $x_1 = 0.7$ mm, $x_2 = 0.3$ mm.

particular holders and tunable lens; and the piezo tube with optical fiber. We discuss these separately.

Silicon optical bench

The process sequence for the Si micro-bench is described in detail in Figure 4. The basic steps are: surface passivation with SiO/Si_3N_4 by PECVD (plasma-enhanced chemical vapor deposition); photolithography with dry and wet etch processes on the front- and backside of the bench; and finally covering the fluidic channels on the backside by Ordyl, a dry film resist. To ensure a precise alignment of the Si chips, such as the holders and the tunable lens, a KOH etch step defines highly-precise alignment grooves.

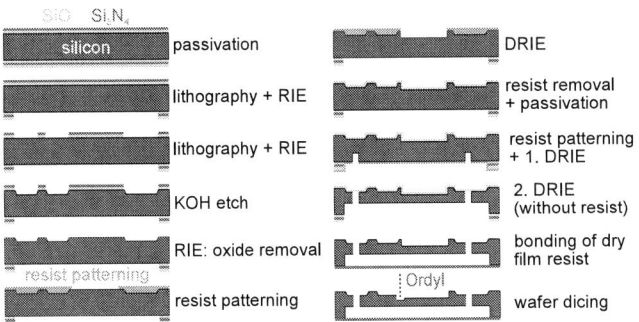

Figure 4: Process sequence for fabrication of the Si micro-bench; the holders and lenses are glued to this substrate.

Silicon components

All Si components are fabricated using standard Si micromachining as described in [5]. Figure 5 shows a selection of individual Si components before integration. Each component is integrated onto the Si micro-bench. A

PDMS (polydimethylsiloxane) membrane with a thickness of 22 µm is bonded on top of the Si lens chip to cover the optical chamber and define the refractive interface. The cavity of the lens is connected through fluidic vias, channels and ports in the bench to the external hydraulic actuation system to enable the deformation of the membrane and hence a change in refractive power. The optical chamber is filled with 3M™ Fluorinert™ Electronic Liquid FC-40 as an optical liquid with a refractive index of 1.29.

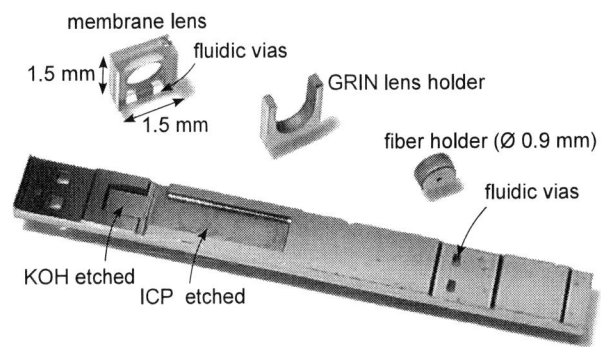

Figure 5: Photo of individual fabricated Si components including the optical bench and the tunable lens.

Piezo unit

Assembly and alignment of the individual components are done in a highly precise V-groove to guarantee centricity and perpendicular orientation. In the same way, the piezo tube is mounted, as sketched in Figure 6, with special care taken to define the orientation of the outer electrodes, which must be connected electrically. Finally, the optical fiber is inserted into the circular Si holder, which was glued into the piezo tube. The fiber is fixed by a drop of the epoxy adhesive Araldite 2020. The adjustment of the fiber is undertaken with the aid of a microscope to protect the fiber facet from damage.

Figure 7 shows a photo of the completely assembled probe with a total length of 13.1 mm, a width of 1.5 mm, and a maximum outer diameter of 2.5 mm.

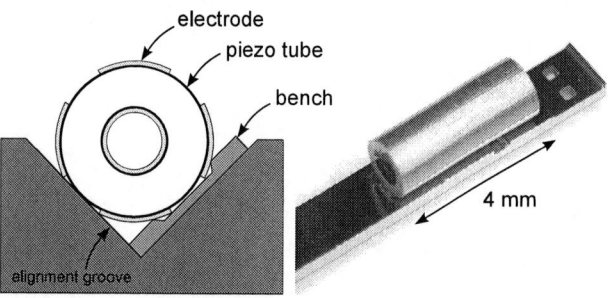

Figure 6: Integration of the piezo tube onto the Si optical bench. The photo on the right shows the integrated module.

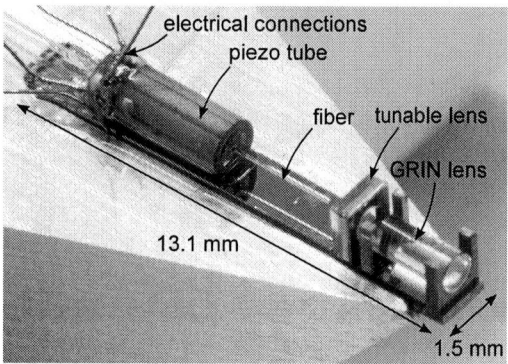

Figure 7: Completely assembled probe with all components on the optical micro-bench.

RESULTS

Mechanical characterization

The fiber scanner is designed to perform circular or spiral scans with a roughly 1 μm focal spot on the sample. The outer electrodes at the piezo tube are driven in pairs with a phase shift of $\pi/2$. The frequency response of the fiber was determined in a steady-state mode at different actuation voltages by imaging the focal spot behind the two-lens system with a magnification of 20 onto a CCD camera by a telescope. The radius of the circular oscillation is plotted in Figure 8 as a function of the driving frequency at different discrete voltage amplitudes. The resonance frequency is around 9.9 kHz with a shift to lower frequencies at higher actuation amplitudes.

In the non-stationary mode of a spiral scan, where the amplitude varies, a non-linear response is expected when the system is driven in resonance.

Optical characterization

The optical characterization comprises three different parts: determination of the sensitivity of the focal tuning for axial or depth scanning; the axial resolution; and finally the lateral resolution from a scanned confocal image. All measurements were carried out with a HeNe laser at a wavelength of 633 nm.

The optical design allows a focus shift of 100 μm. Within this axial scan range, the liquid volume in the tunable lens has to be controlled precisely. A stepper motor with a highly

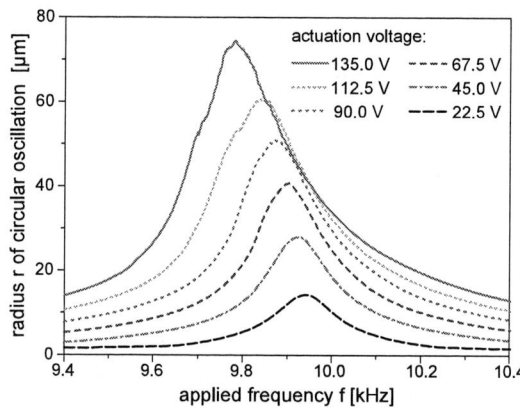

Figure 8: Mechanical characterization of the fiber scanner. The curves show the radius of the circular oscillation of the fiber as a function of oscillation frequency for different applied actuation voltages at the piezo electrodes.

precise gear ratio is used to displace the liquid in a syringe pump which is connected to the fluidic ports of the probe. The achieved incremental volume displacement is 0.65 nl. The total required volume for achieving the full range of the focal shift is about 250 nl. From the measurement in Figure 9, we see that the focal adjustment follows the actuation volume linearly with a sensitivity of ca. 0.6 μm/nl. The displacement of the first 80 nl leads to a well-defined prestressed state; the focal adjustment becomes useful only above this value.

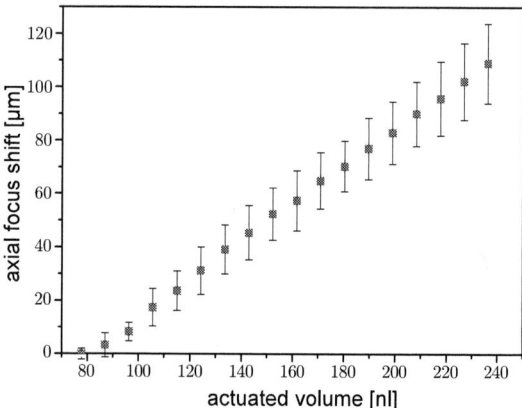

Figure 9: Axial focus shift in the confocal approach as a function of the displaced liquid volume in the tunable lens.

To determine the axial resolution of the system, a confocal 3D image of a reflective target was taken and the averaged intensity of each axial layer was calculated. For the axial scanning, the membrane lens was actuated as described above. At the beginning, the focus was placed behind the target and then shifted by steps of 3.5 μm towards the probe. The measured data were fit by a Gaussian, as illustrated in Figure 10. Since the target has almost no axial extent, the full width at half maximum (FWHM) of the fit corresponds to the axial resolution of the system. The scanner was designed for an axial resolution of 25 μm, the measured data yield a resolution of 19.4 ± 3.1 μm and fulfill the specification. The standard deviation results from the uncertainty in the actuation of the lens. Lateral resolution as well as imaging aberra-

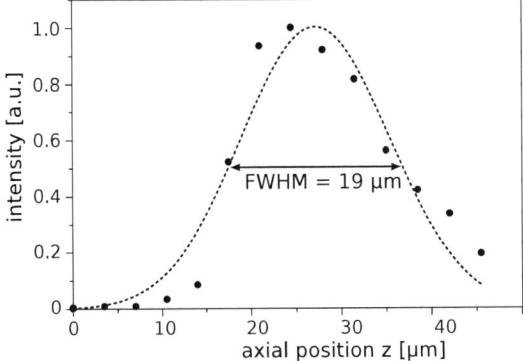

Figure 10: Axial light intensity when the focus is shifted through a target as a measure of the axial resolution.

tions were investigated by generating a two-dimensional confocal image of a grating structure. A rectangular grid with reflecting lines with a width of 1.7 μm at a period of 10 μm was scanned by spiral movement of the fiber. The field of view was reduced to about 70 μm because of the specular properties of the target lines. For high deflection angles of the fiber, the reflected light cannot enter the aperture of the fiber, in contrast to the case for scattering samples such as biological tissue or turbid media. Hence, the light intensity of the scanned specular grating lines decreases strongly with increasing distance from the center of the image, which is the optical axis, as clearly seen in Figure 11.

As indicated in the mechanical characterization, non-linear effects of the frequency response for a variable excitation amplitude are responsible for the distortion of the scanned image. Another reason for the distorted image is due to the fact that the resonance frequency might show an asymmetric behavior for the two superimposed perpendicular actuation directions due to imperfections in the assembled probe. For a known target, as for two-dimensional, right-angled grating in the present case, the measured data can be used for calibration or correction of the measurement.

The system was designed to resolve lateral structures of 1.5 μm. Figure 11 indicates that the design requirements could be successfully fulfilled.

CONCLUSION

A miniaturized confocal microscope based on Si optical bench technology has been presented. The forward-looking fiber scanner allows confocal depth scanning over 100 μm without the need of movable optics or stages. The designed axial and lateral resolution of the system could be confirmed by experiment. The piezo-driven fiber scanner provides fast spiral scans with about 10 kHz, thereby scanning a circular

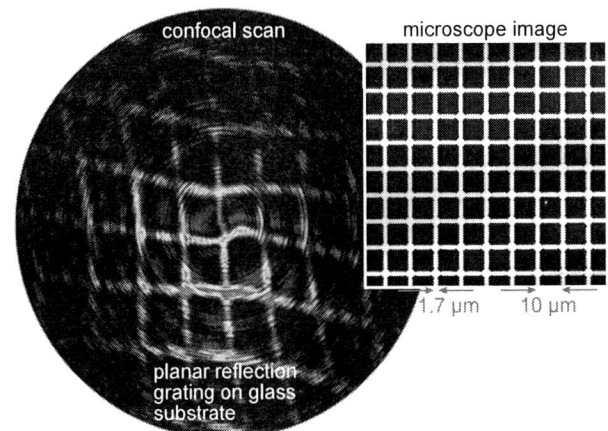

Figure 11: Confocal image of a reflective grating with line widths of 1.7 μm and 10 μm period. The inset shows a microscope image of the grating. The distorted image comes from non-linear and uneven deflection of the piezo tube in x and y direction.

200 μm field by 500 windings within 50 ms.

ACKNOWLEDGMENT

This work was supported by the German Research Foundation DFG in the project 'Endoscopic tomography of intracorporeal cellular microstructures' (Mikrotom, GZ: SE 2045/1-1).

REFERENCES

[1] T. Wang and J. Van Dam, "Optical biopsy: A new frontier in endoscopic detection and diagnosis," *Clinical Gastroenterology and Hepatology*, vol. 2, pp. 744–753, 2004.

[2] C. M. Lee, C. J. Engelbrecht, T. D. Soper, F. Helmchen, and E. J. Seibel, "Scanning fiber endoscopy with highly flexible, 1 mm catheterscopes for wide-field, full-color imaging," *Journal of Biophotonics*, vol. 3, no. 5-6, pp. 385–407, 2010.

[3] H.-C. Park, Y.-H. Seo, S.-B. Yang, M. Choi, S. Lee, W. Kim, and K.-H. Jeong, "Forward-viewing endoscopic OCT catheter using asymmetrically resonant fiber scanner," in *Proceedings of the 2013 IEEE International Conference on Optical MEMS and Nanophotonics*. IEEE, 2013, pp. 7–8.

[4] E. Seibel, R. Johnston, and C. Melville, "A full-color scanning fiber endoscope," in *Proc. SPIE*, vol. 6083. SPIE, 2006, pp. 608 303–608 303–8.

[5] N. Weber, H. Zappe, and A. Seifert, "A tunable optofluidic silicon optical bench," *Journal of Microelectromechanical Systems*, vol. 21, no. 6, pp. 1357–1364, 2012.

[6] J. Pawley, *Handbook of Biologic Confocal Microscopy*. Springer, 2006.

[7] C. Chen, "Electromechanical deflection of piezoelectric tubes with quartered electrodes," *Appl. Phys. Lett.*, vol. 60, pp. 132–134, 1992.

ULTRASOUND-ASSITED MICRO-KNIFE FOR CELLULAR SCALE SURGERY

Hwapyeong Jeong[1], Tao Li[2], Yogesh B. Gianchandani[2] and Jaesung Park[1]
[1]Pohang University of Science and Technology, Republic of Korea
[2]Center for Wireless Integrated MicroSensing and System (WIMS[2])
University of Michigan, Ann Arbor, MI 48109, USA

ABSTRACT

An ultrasound-assisted micro-knife with a 500nm-thick silicon nitride blade is described. The sharp blade can be operated to cut soft cells without lysing. The operation conditions are optimized by finite element analysis and experimental evaluation. For validation of the cutting precision Si_xN_y blades without and with ultrasonic actuation are compared to a commercial scalpel. The commercial scalpel causes lysing of hepatocytes in a mono-layer; the Si_xN_y blade without ultrasonic actuation cut these cells with a ragged cut line; the Si_xN_y blade with 1 V_{pp} and 70.1kHz ultrasonic actuation cuts these cleanly, as narrow as 2 μm. Due to the controlled ultrasonic mode shape, the high operating frequency, and low applied power (1V_{pp}), the micro-knife performs highly precise dissection at the cellular scale without the need for high compressive force on the target cell. The Si_xN_y blade with harmonic actuation has potential applications as a tool for minimally invasive surgery.

INTRODUCTION

Micromachined incision tools have been developed to minimize inadvertent damage to tissues. Microelectromechanical systems (MEMS) techniques have been used to develop neural probes that measure neural signals with minimal incision, exploiting their ultra-sharp incisive tip [1, 2]. Additional reports describe piezoelectrically actuated silicon nitride micromachined blades that are integrated with cauterizing electrodes or suction holes for hemostasis or tissue removal [3, 4]. The major advantages of silicon nitride blades are that these provide sharper blade edges and simultaneously higher mechanical strength [5]. For example, the edge of Si_xN_y blades can be as thin as a deposited thin film and this thickness determines the level of blade edge sharpness. However, the incision made by micromachined silicon knives is based more on pressure waves rather than cutting methods such as slicing, and the pressure waves cause undesired tissue damage due to the difficulty in focusing precisely on target tissue [3, 4]. Therefore, to fully exploit the advantage of sharp blade tips, a controlled blade movement is desired in the slicing direction. However, after conventional microfabrication, a blade width is typically much larger than its thickness, and out-of-plane displacement is dominant over other displacements at its fundamental natural frequency when it is harmonically oscillated [6]. Therefore, to fully exploit the advantages of the sharp cutting edge that are provided by micromachined silicon nitride blades, proper control over the blade movement is necessary.

This paper describes an unltrasonically actuated micro-knife with a 500nm-thick Si_xN_Y blade for minimally invasive surgery. To control blade movement, we analyze the resonance modes and displacements of the micro-knife to determine the optimal operating frequency at which transverse slicing movement is dominant. The precise incision of the micro-knife is demonstrated by cutting single cells with clean cut-lines. Because this blade is compatible with conventional micro fabrication processes, the micro-knife can be integrated with other electrical and mechanical sensors and may be used as an end-effector for surgical robots.

THEORY AND SIMULATION

The micro-knife consists of a Si substrate (500μm-thick, 1cm-wide and 4cm-long) with a Si_xN_y film deposited on it as a whole-body blade (Fig.1). A bulk PZT plate (1 mm-thick, 1cm-wide and 3cm-long) is bonded to the bottom of the blade, then the blade is fixed to a supporting jig that is mounted vertically on an xyz stage. The blade shape can be regarded as a cantilever with a fixed-end boundary condition, overhanging from the PZT plate on the top of the jig. For small deflections, the equation of motion of motion of the blade can be described as

$$EI \frac{\partial^4 v(x,t)}{\partial x^4} + \rho bh \frac{\partial^2 v(x,t)}{\partial t^2} = f(x)e^{j\omega t} \quad (1)$$

where v is the displacement, x is the distance from the blade edge. E is the modulus of elasticity, I is and moment of inertia, ρ is density, w is the width of the Si and t is its thickness, and ω is the harmonically oscillation frequency; this equation can be used to calculate the resonant frequencies and displacement mode of the blade. As an alternative method, the displacements is analyzed using a

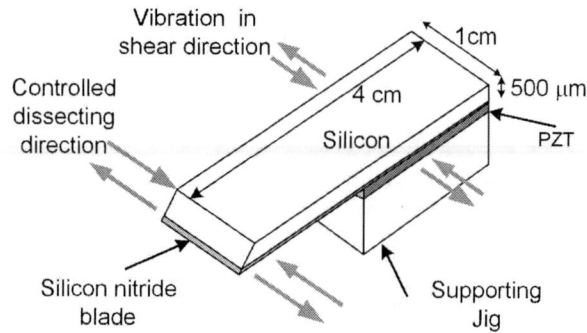

Figure 1: *Schematic diagram of micro-knife system.*

978-1-4799-3510-9/14 $31.00 © 2014 IEEE

finite element analysis tool (COMSOL™). The constitutive equation can be written as

$$\{S\} = [S^E]\{T\} + [d]\{E\}$$
(2)

where S is a strain tensor, S^E is a compliance matrix, d is the matrix for the direct piezoelectric effect, T is a stress tensor and E is the electric field. The superscript E indicates a constant electric field. The assembly of micro-knife and PZT is modeled and meshed with 4286 elements (Fig. 2a). In analysis, the relative displacements with respect to the static transverse displacement are compared.

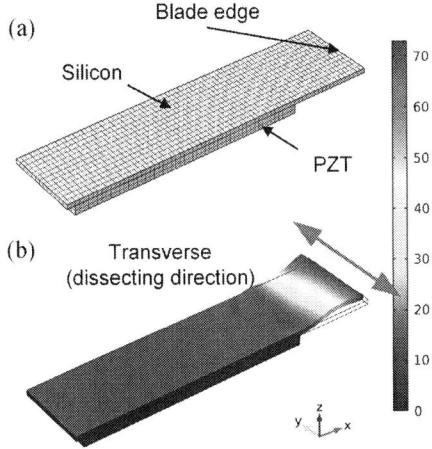

Figure 2: Modal analysis of micro-knife at 69.9 kHz, where transverse displacement is dominant over other directions. (a) Geometry and mesh of micro-knife with bulk PZT; (b) in transverse direction.

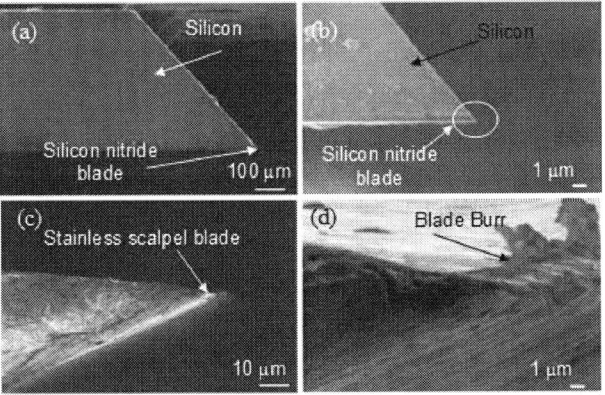

Figure 3: SEM image. (a),(b) cross-sectional views of the fabricated blade. (c),(d) cross-sectional views of the commercial scalpel.

Relative displacements of free-vibration were calculated at frequencies from 0 to 75 kHz in 50-Hz increments. Fundamental resonance occurs at 6.25 kHz, and a second resonant frequency occurs at 38.3 kHz. At these lower frequencies, the displacement of the blade edge in bending (out-of-plane) and longitudinal direction dominate over displacement in the transverse direction. However, at the third resonant frequency, 69.8 kHz, the displacement of the blade edge in the transverse direction becomes dominant over the other displacements.

Relative displacements of forced-vibration by the PZT attached to the blade were calculated using the same model. A displacement in the transverse direction induces coupled displacements of bending and longitudinal motion, although the PZT is oriented to induce maximal displacement in d_{15} (transverse direction). However, the resonant frequencies are found at similar to the free-vibration resonant frequencies. Additionally, at the two lower resonant frequencies, the displacements in bending and longitudinal directions dominate over the displacement in the transverse direction. In contrast, at 69.95 kHz, the displacement in the transverse direction is ≈1000x larger than the bending displacement, although the displacement in the bending direction also shows a local maximum. Therefore, it is desirable to operate the micro-knife at the third resonant frequency to increase the transverse displacement and reduce the bending. However, structural instability is generally induced by resonance, so the preferred operating frequency for the micro-knife should be close to 69.95 kHz, but slightly offset from this.

FABRICATION AND EXPERIMENTS

The micro-knife was fabricated using conventional micromachining techniques. A p-type silicon wafer was polished on both sides, and the final thickness was adjusted to 500μm with variation less than 10 μm. After cleaning and dehydration, a 200nm-thick silicon dioxide layer was thermally grown on the silicon wafer, and a 500nm-thick layer of low-stress silicon nitride was deposited on the wafer. The photoresist-patterned silicon nitride and oxide layers were dry-etched. After wet-etching through the wafer using KOH, the backside silicon nitride layer was aligned and patterned by using lithography and dry-etching. During wet etching, the backside Si_xN_y film was not protected. Due to the characteristics of anisotropic etching, the fabricated silicon nitride blade formed clean and sharp cutting edges (Fig.3a-b), compared to commercial stainless steel edges, which have burrs and irregular shapes (Fig.3c-d). The fabricated blades were sharper than a stainless-steel scalpel. Additionally, the thickness of the micro blade tip was less than 500nm, which is smaller than a single cell. Subsequently, bulk PZT (d_{15}, shear-dominant direction) was bonded to the silicon blade body with epoxy, and the assembly was clamped in an xyz-stage.

To confirm the cutting precision, hepatocytes are isolated from a 10-week-old female DS-RED mouse [7]. The viability of hepatocytes is higher than 95%. To protect both the dish and the silicon nitride blade, a 9:1 mixture PDMS elastomer base and curing agent is coated on the tissue culture dish. Approximately 2×10^5 hepatocytes are seeded on the dish with a concentration of 10^5/ml in Dulbecco's Modified Eagle's Medium (DMEM) (10% fetal bovine serum, Gibco), and incubated for 24 h in 10% CO_2 at 37°C. The culture dish containing hepatocytes is mounted below the blade. All cutting procedures are observed and recorded by an inverted fluorescence microscope (IX70, Olympus). A function generator (DS335, Stanford Research Systems) is used to apply power. The volume of liquid, DMEM, was maintained as 2mm-thickness.

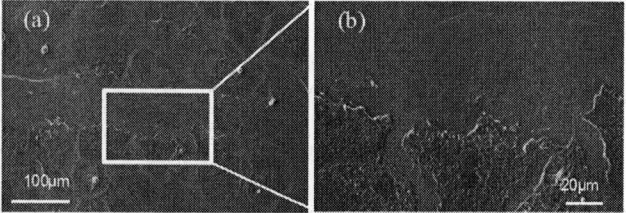

Figure 4: *SEM image of cut hepatocytes using commercial scalpel. After cutting, the width of cut line is about 100μm.*

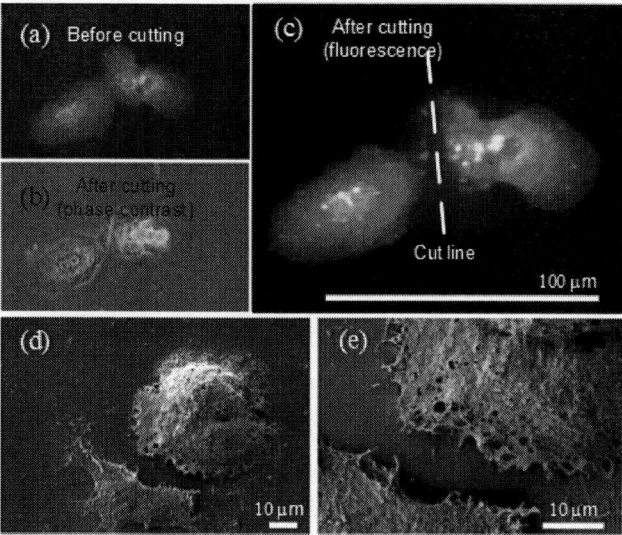

Figure 5: *Image of cut hepatocytes using micro-knife without actuation. (a) a fluorescence image of cell before cutting (b) a phase image of cell after cutting (c) a fluorescence image of cell after cutting (d),(e) SEM images of cells after cutting. The cut line is ≈9.5μm wide.*

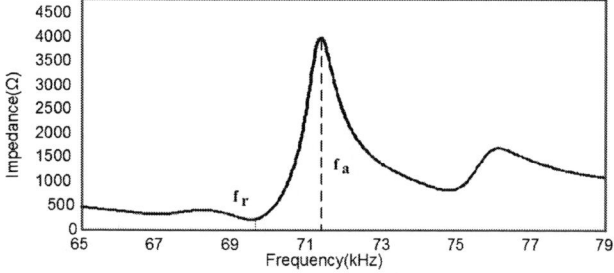

Figure 6: *Impedance analysis showing that the resonance frequency is ≈69.95 kHz.*

RESULTS

Hepatocytes are sensitive to mechanical stress [8]. Hepatocytes that are spread on polystyrene culture dishes are cut to compare the fabricated micro-knives to a commercial scalpel (Surgical blade No. 11, Feather). Both types of blades are clamped on a stage, and advanced toward the dish in the vertical direction. The cutting procedure is observed in real-time using a CCD camera through a fluorescence microscope. After the cells are cut, the hepatocytes are fixed using paraformaldehyde for 30 min, then dried by sequential immersion in immersion in 25%, 50%, 75%, 90%, and 100% ethanol:water (v:v) solutions.

The commercial scalpel and the silicon nitride blade without harmonic actuation are compared first. The commercial scalpel can successfully cut a few hepatocytes, but the cut lines are irregular and ragged. The cut hepatocytes are sparsely distributed because most cells lysed (data not shown) (Fig. 4a-b). After cutting, the width between cut segments is ≈100μm. This result suggests that the commercial scalpel cannot cut single hepatocytes because the cutting line is larger than a diameter of hepatocyte, which was about 20μm.

In contrast to the commercial scalpel, the cut lines of the silicon nitride blade without actuation are cleaner and narrower. When the silicon nitride blade starts cutting, the hepatocytes do not deform, unlike the commercial scalpel (Fig. 5a-c). The width of cut line is <10μm. However, the cell membrane still appears ragged (Fig. 5d-e), and the fluorescence signal in the cytoplasm around the cut line degrades after cutting because the cytoplasm is squeezed from the cut membrane during incision (Fig. 5c).

Finally, the ultrasonically actuated silicon nitride blade is evaluated. The resonance frequency of the transverse mode is found by measuring impedance (HP 4194A) at 69.8kHz (Fig. 6). According to the FEA result, the displacement in transverse direction is larger than in the other directions. To prevent mechanical instability at this resonant frequency, the operating frequency is chosen to be 0.3 kHz higher, i.e., 71.1kHz. At this off-resonant frequency, two advantages are expected: the displacement in bending direction is reduced; and if the resonant frequency shifts when the blade tip contacts target cells, this off-resonance frequency buffers the operation.

Actuation voltage is another important factor that deserves consideration. When power is applied, the blade generates acoustic energy via pressure waves; cells are lysed in a manner similar to high intensity-focused ultrasound. In this experiment, cells detach from the culture dish with $3V_{pp}$ at 70.1 kHz. Even the hepatocytes that remain attached to the dish are perturbed and then lysed upon contact with the blade edge. Additionally, most of the fluorescence protein in cytoplasm is lost due to the vibration energy of the blade (data not shown). However, when the silicon nitride blade is actuated at $1V_{pp}$ and 70.1 kHz, there is no damage observed. Consequently, the preferred operating voltage and frequency is established as $1V_{pp}$ at 70.1 kHz. Under these conditions, the cut line is clean, well-defined, and narrower than in any other case. The measured width of the cut line is <2μm and cut line is straight (Fig.7a-e).

To evaluate the cutting performance of the actuated silicon nitride blade, hepatocyte tissue in a mono-layer is cut. The blade is actuated at $1V_{pp}$ and 70.1 kHz. The cut line is clean and the blade does not cause any unnecessary damage other than slicing the target hepatocytes (Fig. 7f). This result successfully demonstrates that under the specified operating conditions, the fabricated blade can minimize damage during surgery at the cellular scale.

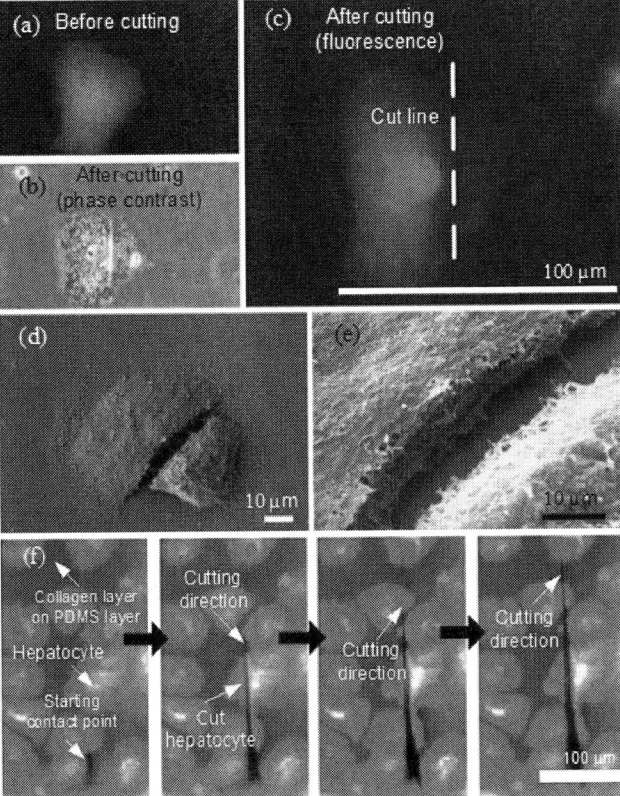

Figure 7: *Image of cut hepatocytes using micro-knife with actuation at 1V_{pp} and 70.1 kHz. (a) a fluorescence image of cell before cutting (b) a phase image of cell after cutting (c) a fluorescence image of cell after cutting (d),(e) SEM images of cells after cutting (f)Time-sequential fluorescence image of cutting mono-layer hepatocyte. The width of cutting line is about 2µm. The cutting line is clear and sharp.*

In this paper, the silicon nitride blade was analyzed assuming that it is operated in air. However, many biological applications, including the experiments in this paper, arise in liquid environments. When the blade is oscillating in the transverse direction in viscous liquid, a velocity boundary is formed from the surface of blade. Minimizing this boundary layer is desirable so as to not cause secondary intracellular damage. The boundary layer thickness δ around an oscillating blade is [6]

$$\delta = \sqrt{\frac{2\eta}{\rho_L \omega}} \tag{3}$$

where η is the kinetic viscosity of the fluid, ρ is its density and ω is oscillating frequency. According to this equation, higher oscillating frequency reduces the boundary thickness and localizes the effect of boundary layer from the surface of the blade. The micro-knife in this paper is operated near the third resonant frequency, 69.95 kHz. Compared to the fundamental and second resonant frequency, 6.25 and 38.30 kHz, the operating frequency is much higher and provides less boundary thickness in liquid. In addition to controlled displacement in direction and the sharp blade edge, the higher operating frequency is another important reason for the minimal cell damage observed in this work.

CONCLUSION

An ultrasonically actuated micro-knife with silicon nitride blade is described. It can be used to cut at the scale of a cell because the blade is thinner than the size of a cell. The performance micro-knife is demonstrated by cutting mono-layered hepatocyte tissue as well. The proposed blade can be miniaturized for attachment to the end-effector of a laparoscope, and integrated with other useful transducers such as force sensors and cauterizers, due to compatibility with conventional microfabrication processes. These advantages provide the blade with versatile functions as an end-effector for robotic surgery in future.

ACKNOWLEDGEMENTS

This work was supported by the National Research Foundation of Korea (NRF) grant funded by the Korea government (MSIP) (No. 2011-0030075, 2011-0028845)

REFERENCES

[1] A. C. Hoogerwerf and K. D. Wise, "A three-dimensional microelectrode array for chronic neural recording, " IEEE Transactions on Biomedical Engineering, vol. 41, pp. 1136-1146, 1994.

[2] R. J. Vetter, J. C. Williams, J. F. Hetke, E. A. Nunamaker, and D. R. Kipke, "Chronic neural recording using silicon-substrate microelectrode arrays implanted in cerebral cortex," IEEE Transactions on Biomedical Engineering, vol. 51, pp. 896-904, 2004.

[3] A. Lal and R. White, "Silicon micromachined ultrasonic micro-cutter," IEEE Proceedings in Ultrasonics Symposium, 1994, pp. 1907-1911.

[4] A. Lal *et al.*, "Silicon-based ultrasonic surgical actuators," IEEE International Conference of the Engineering in Medicine and Biology Society, 1998 pp. 2785-2790.

[5] K. E. Petersen, "Silicon as a mechanical material," Proceedings of the IEEE, vol. 70, pp. 420-457, 1982.

[6] R. Cox, F. Josse, S. M. Heinrich, O. Brand, and I. Dufour, "Characteristics of laterally vibrating resonant microcantilevers in viscous liquid media," Journal of Applied Physics, vol. 111, pp. 014907-014907-14, 2012.

[7] P. O. Seglen, "Preparation of isolated rat liver cells," Methods cell biol, vol. 13, pp. 29-83, 1976.

[8] J. Park, F. Berthiaume, M. Toner, M. L. Yarmush, and A. W. Tilles, "Microfabricated grooved substrates as platforms for bioartificial liver reactors," Biotechnology and bioengineering, vol. 90, pp. 632-644, 2005.

978-1-4799-3510-9/14 $31.00 © 2014 IEEE

WRINKLE CELLOMICS: SCREENING BLADDER CANCER CELLS USING AN ULTRA-THIN SILICONE MEMBRANE

Jennie Appel[1], Mandy L.Y. Sin[2], Joseph C. Liao[2], Junseok Chae[1]

[1]School of Electrical, Computer and Energy Engineering, Arizona State University, USA
[2]Department of Urology, Stanford University School of Medicine, Stanford, USA

ABSTRACT

We report a visualization platform, comprised of an ultra-thin silicone membrane, to differentiate between the biophysical properties of cancerous and non-cancerous cells from human patients. Cancerous cells adhere to, spread on, and induce deformation of this membrane to produce wrinkles while non-cancerous cells fail to generate wrinkles. Wrinkle patterns—number, length, and direction of wrinkles—can be visualized by a conventional microscopy. Quantitative measurement of these wrinkling patterns represent a powerful, non-invasive diagnostic tool for prevalent cancers, such as bladder cancer.

INTRODUCTION

In the United States, bladder cancer is the fifth most common cancer with an estimated 72,570 new cases in 2013 [1]. Current diagnostic strategies include cystoscopy, the visual inspection of bladder lumen with an endoscope and urine cytology, the microscopic inspection of cellular morphology derived from urine samples [1]. Cystoscopy is invasive and may not be able to differentiate between malignant and benign tumors [1]. Urine cytology, while non-invasive, has poor sensitivity and requires complex sample preparation and an experienced cytopathologist to interpret the results. Other non-invasive techniques include detection of cancer biomarkers in urine samples. These methods depend on the number of cancerous cells present in the urine sample [2].

Different cell types and cancerous variants, have been shown to have distinct biophysical properties. Numerous methods have been explored to quantify the biophysical properties of cells, permitting the analysis of biological systems as mechanical systems. For example, Lekka *et al.* measured the Young's modulus—a measure of the stiffness of a material—of a variety of cell types with scanning force microscopy. The Young's modulus of a bladder cancer cells were determined to be approximately an order of magnitude smaller compared to normal [3]. Cancer cells have a tendency to spread out more, have a high degree of focal adhesions in contact with a surface and exert a stronger force on the surface than normal cells. Traction and adhesion forces are commonly studied to determine the force a cell exerts when it is interacting with a surface. In 2003, Tan *et al.* reported a micro-pillar MEMS structure to study these forces. The apex of the micro-pillars would bend due to the applied cellular adhesion forces, and through pillar characterization the adhesion force could be determined [4]. This unique micro-pillar method, however, requires sophisticated fabrication techniques.

As an alternate approach, the wrinkling of a thin membrane has been applied recently to exploit the difference of the Young's modulus of a cell. Huang *et al.* developed an ultra-thin polystyrene membrane and induced a sinusoidal wrinkle pattern on the membrane using water droplets [5]. Harris *et al.* reported previously a silicone membrane that visualized the traction forces generated during cellular locomotion. They approximated these forces using a flexible weighted probe [6]. Further refinement of this approach has included embedding fluorescent beads within the membrane to aid in tracking cell locomotion [7]. Despite the considerable effort expended for measuring the Young's modulus and cellular forces of different cell types, these approaches have yet to be implemented for the design of a diagnostic device to differentiate between cancerous and healthy cells.

Here we report a device for rapid and accurate differentiation of cancer from non-cancer bladder cells. In contrast to urine cytology, our system requires minimal sample preparation and does not require significant expertise to interpret the results. Our system, is comprised of an ultra-thin silicone membrane and exploits the distinct biophysical properties of various cell types and cancerous variants. The silicone membrane is deformed by cellular forces applied by cancerous cells and not by the non-cancerous controls. Thus a wrinkle pattern is generated selectively by cancerous cells. Currently, no prior work has applied wrinkle patterns to screen for cancerous cells from complex bodily fluids, such as blood or urine. Our work here is the foundation for the development of a small-size and non-invasive device that avoids the complexity of sophisticated instrumentation to identify presence of cancerous cells in a clinical samples. The rapid screening of urine samples, as well as potentially various other sample types, will improve the early detection of cancer, permit less invasive therapies and enhance human health.

VISUALIZATION PLATFORM

Fabrication

Fabrication of the ultra-thin silicone membrane begins with filling a 15 mm diameter by 0.5 mm deep well in a glass slide with high viscosity (12,500 cP) liquid silicone, as shown in figure 1(a). Then we invert the slide over a low flame (Bunsen burner) effectively cross-linking the polymer chains in the liquid silicone forming a thin crust or membrane, figure 1(b). After sanitizing (20% ethanol) the culture chamber we apply a layer of vacuum grease to it and press it onto the glass slide creating a water tight seal in the area surrounding the membrane, figure 1(c).

978-1-4799-3510-9/14 $31.00 © 2014 IEEE

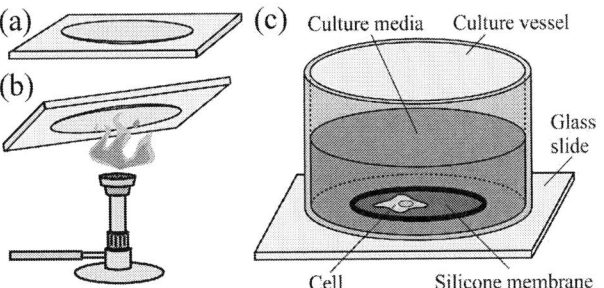

Figure 1: (a) Fill the 1.5 cm diameter, 0.5 mm depth well with 12,500 cps liquid silicone. (b) Flash heat the liquid silicone, effectively cross-linking the polymer to form an ultra-thin membrane. (c) Using vacuum grease to adhere cell culture chambers to glass slide and fill the chamber with culture media and cells.

Characterization

The physical properties of our ultra-thin silicone membrane were characterized with the addition of minute water droplets to its surface; the water droplets deformed the membrane producing a discrete wrinkle pattern [5]. The thickness of the membrane was determined to be approximately 28 nm from the observed number of wrinkles caused by deformation of the membrane. To obtain predictive estimation of the wavelength, amplitude and number of wrinkles, we consider the 1D out-of-plane displacement of an initially flat sheet of area (=WxL, W is width and L is length) as a function of spatial dimension (=$\xi(x,y)$) (Figure 2 (a)). The thickness of the sheet, t, is much smaller than W and L, where $0<y<W$ also, for

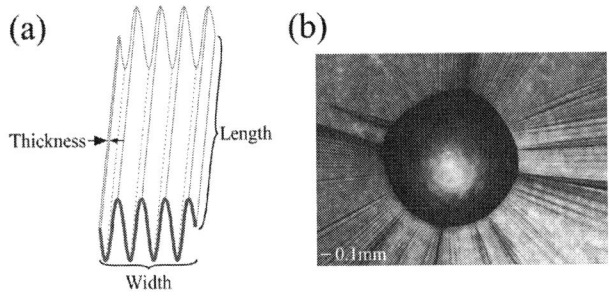

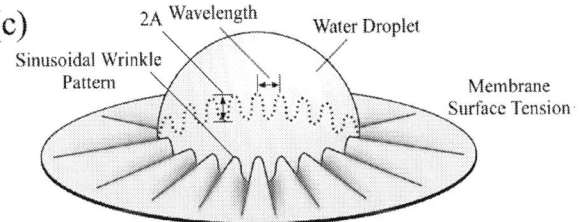

Figure 2: (a) 1D schematic of wrinkles in a flat soft sheet induced by a force in the x direction (b) Schematic demonstrating the sinusoidal wrinkle pattern, due to capillary forces acting upon an ultra-thin membrane (c) A ~ 0.2 mg, 1.1 mm diameter water droplet acting on ultra-thin membrane causing the formation of 148 wrinkles (approximately >0.4 mm long).

simplicity, $W \ll L$. When a stretching strain ε is applied in

the x direction, then the total energy of the system is:

$$U = U_B + U_S - L \tag{1}$$

Here U_B is the bending energy due to deformation in the y direction:

$$U_B = \tfrac{1}{2} \int B (\partial y^2 \zeta)^2 dA \tag{2}$$

where B is the bending stiffness. U_S is the stretching energy in presence of tension $T(x)$:

$$U_S = \tfrac{1}{2} \int T(x)(dx\zeta)^2 dA \tag{3}$$

Boundary conditions:

$$\int_0^L \left[\tfrac{1}{2} (\partial y^2 \zeta)^2 - \tfrac{\Delta(x)}{w} \right] dy = 0 \quad \zeta(0,y)=0 \quad \zeta(L,y)=0$$

After applying boundary conditions we determine:

$$\lambda = \frac{\sqrt{2\pi L t}}{\sqrt[4]{3(1-v^2)\varepsilon}} \tag{4}$$

$$A = \left(\sqrt{vLt} \right) \left(\sqrt[4]{\frac{16\varepsilon}{3\pi^2(1-v^2)}} \right) \tag{5}$$

where λ is the wavelength and A is the amplitude of the sinusoidal wrinkling shown in figure 2(c), L is the length of the membrane, t is the thickness of the membrane, v is the Poisson's ratio of the membrane, and ε is the strain of the membrane. We approximated the number of wrinkles by dividing the wavelength of the wrinkle pattern by the circumference of the water droplet ($\pi d/\lambda$), where d is the diameter of the droplet.

Table 1: Derived equations for λ, A and number of wrinkles for patterns generated by capillary action of a 1.1 mm diameter water droplet (~0.2 mg). The calculated number of wrinkles with approximated v and ε had an error of 24% compared with the measured value.

Property	Calculated	Measured
Wavelength	30.7 μm	~23.3 μm
Amplitude	7.1 μm	N/A
# of Wrinkles	113	148
Thickness	N/A	~28 nm

The wavelength, amplitude and number of wrinkles were calculated with an assumed thickness of 50 nm, a Poisson's ratio of 0.48 and a strain of 0.299. With a membrane length of 2.5 mm and a water droplet diameter of 1.1 mm, we calculated the wavelength, amplitude and number of wrinkles (Table 1). The measured wavelength, amplitude and number of wrinkles for a ~0.2 mg water droplet was determined by applying the water to our membrane and observing wrinkle formations under Nomarski imaging (figure 2(b)). Our calculated 113 wrinkles had a 24% error when compared to the measured 148 wrinkles.

MATERIALS AND METHODS
Cell Culture

RT4 transitional bladder cancer cells were grown in McCoy's 5A modified media (Life Technologies Corp. 16600-082) supplemented with 10% fetal bovine serum (FBS Life Technologies Corp. 10082-139) and 1X penicillin/ streptomycin (Sigma-Aldrich Co. P4333-20ML). Human embryonic kidney cells, HEK293f, were grown in eagle's minimum essential medium (ATCC 30-2003) supplemented with 10% FBS and 1X penicillin/streptomycin incubated at 37°C in humidity and 5% CO_2.

Sample Preparation

Cells at 70% confluency were trypsinized for 5 min at 37°C, inactivated with 10% FBS, pelted by centrifugation at 300 x gravity for 5 min, resuspended in culture media and $5x10^5$ cells were applied to each device. For buffy coat device testing, $5x10^5$ cells of buffy coat (Innovative Research) was first applied to the ultra-thin membrane prior to applying cancerous (RT4) or non-cancerous (HEK293f) cells.

RESULTS AND DISCUSSION

Following the characterization of our ultra-thin silicone membrane, we explored our device for discerning cancerous cells. We performed a time-lapse experiment for 0-hr, 18-hr, 20-hr, and 22-hr, to determine the onset and persistence of membrane wrinkling induced by RT4 cells (figure 3 (a)-(d)) Initially, at 18 hours incubation (b) we have 7 wrinkles with an average length of 180 μm, two hours later at 20 hours incubation (c) the same 7 wrinkles remain with an average length of 170 μm, and finally after an additional two hours at 22 hours incubation (d) 6 wrinkles remain with an average length of 170 μm, showing that the number and length of wrinkles produced by RT4 cells after 18 hours were consistent and did not change over the time interval.

We separately incubated $5x10^5$ bladder cancer cells RT4 and human embryonic kidney epithelial HEK293f cells on our membranes. One hour following adhesion to the membrane, RT4 cells induced wrinkle formation with two distinct wrinkles formed under these cancerous cells (figure 4(a)), while 14 hours following adhesion to the membrane, HEK293f cells failed to induce any wrinkles (figure 4 (b)). Cancerous cells exerted significantly stronger adhesion forces on the membrane when compared to non-cancerous cells.

Clinical urine samples often contain a high concentration of white blood cells when infection or other diseases such as cancer are present. These white blood cells have the potential to obscure or influence membrane wrinkle formation by cancerous cells. To more accurately replicate these patient sample conditions, we explored testing in the context of commercially available buffy coat, a combination of white blood cells and platelets purified from human blood. After 6 hours of incubation buffy coat cells fell out of suspension and appeared to have adhered to the membrane. After an additional 14 hours these cells failed to induce any wrinkles (figure 4 (c)).

To determine if a heterogeneous cell population would affect cancerous cell wrinkle formation on the ultra-thin membrane. We tested cancerous RT4 cells in combination with non-cancerous HEK293f and buffy coat cells. HEK293f cells are an epithelial cell line, thus they approximate the cells found in human urine samples. The presence of HEK293f and white blood cells did not

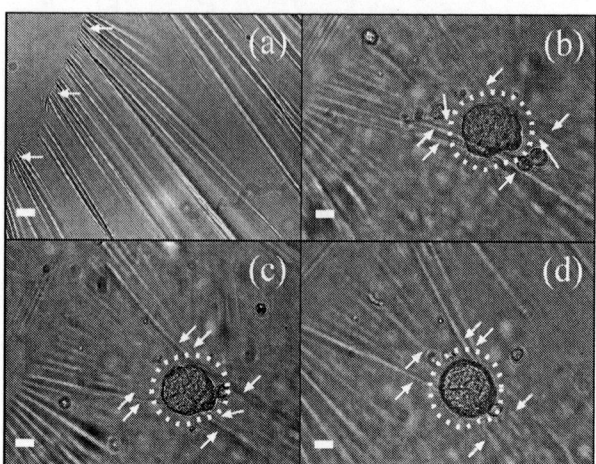

Figure 3: (a) Representative membrane after adding 1 mL of media. Membrane fabrication and adding media sometimes induces pre-existing wrinkles (white arrows). Cancerous (RT4) cells after (a) 18-hr, (b) 20-hr, and (c) 22-hr incubation in a 37°C 5% CO_2 : RT4 cell cluster distorts pre-existing wrinkles, deforming their natural shape (dotted enclosure indicates the cluster). Scale bar is 20 μm.

influence the membrane wrinkle formation, figure 4(d).

We next tested the ability of cancerous cells to induce membrane wrinkles in the presence of a high population of non-cancerous cells. After 16 hours incubation the cancer cells in each mix ratios produced a similar number and

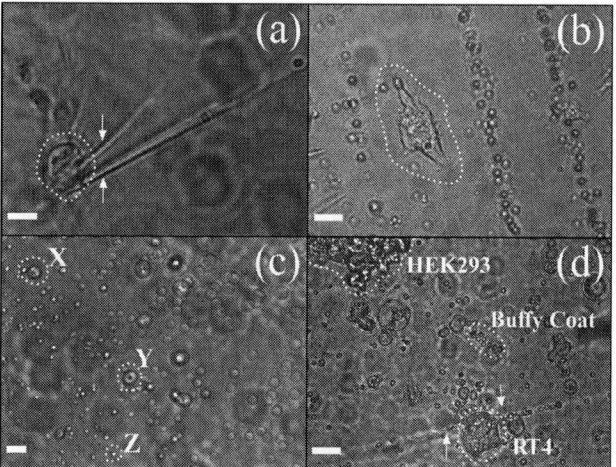

Figure 4: (a) Cancerous (RT4) cells, after 5 hours of incubation, form wrinkles at the outer edges of the RT4 cell cluster (dotted enclosure indicates the cluster and arrows indicate the wrinkle). (b) Non-cancerous (HEK293f) cells after 24 hrs of incubation: no membrane deformation was observed. (c) Buffy coat after 20 hours of incubation do not form wrinkles (X, Y, Z show a white blood cell, a red blood cell, and a platelet, respectively) (d) All cell types after 20 hours of incubation, only the cancerous cells, circled by a dotted enclosure, wrinkle the membrane. Scale bar is 20 μm.

length of membrane wrinkles (figure 5(a)). This suggests the wrinkle characteristics on the membrane are rather independent of mixture ratio.

We further tested the ability of cancerous cells to induce membrane wrinkles in the simultaneous presence of a high population of non-cancerous and white blood cells. Furthermore, these mixed samples of 1:1:1, 1:2:2, and 1:5:5, RT4 to HEK293f to buffy coat, are more realistic to patient urine samples. After 20 hours of incubation, the cancer cells in each of the mix ratios produced a similar number and length of membrane wrinkles (figure 5 (b)).

CONCLUSION

We present that an ultra-thin silicone membrane can serve as a visualization platform to differentiate between cancerous and non-cancerous cells based on membrane wrinkle patterns. Our system exploits the fundamental difference in cell biophysical properties, ensuring the accuracy and sensitivity for cancer detection. Further development will led to rapid cancer detection, early diagnosis, and improved human health.

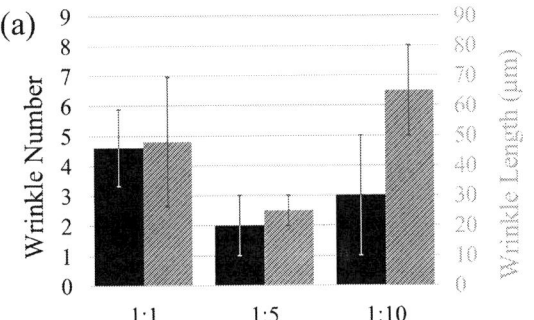

Cancerous RT4 : Non-cancerous HEK293

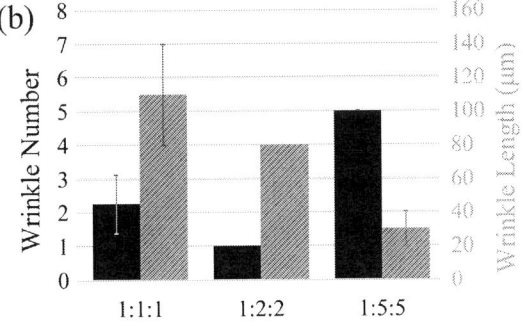

Cancerous RT4 : Non-cancerous HEK293 : Buffy Coat

Figure 5: (a) Characterizing wrinkles (number of wrinkles: solid bars, wrinkle length: dashed bars) induced by cancerous RT4 cells. Mixed samples of RT4 and HEK293f cells at ratios of 1:1, 1:5, and 1:10, respectively were incubated on ultra-thin silicone membrane for 16 hours at 37°C 5% CO₂. (b) Characterizing wrinkles induced by cancerous RT4 cells. Mixed samples of RT4, HEK293f and buffy coat cells at ratios of 1:1:1, 1:2:2, and 1:5:5, respectively were incubated on ultra-thin silicone membrane for 20 hours at 37°C 5% CO₂. Wrinkle lengths were measured by approximation based on a 20 μm fixed scale bar.

ACKNOWLEDGEMENTS

We would like to thank Dr. Julien Chen for HEK293f cell culture protocols and Joshua Podlevsky for his careful review and discussion of this manuscript. This material is based upon work supported by the NSF Graduate Research Fellowship under Grant No. DGE-1311230.

REFERENCES

[1] American Cancer Society. Cancer.org. 2013.

[2] G. Cheung, A. Sahai, M. Billia, P. Dasgupta, and M. S. Khan, "Recent advances in the diagnosis and treatment of bladder cancer.," *BMC medicine,* vol. 11, p. 13, 2013.

[3] M. Lekka, P. Laidler, D. Gil, J. Lekki, Z. Stachura, and a. Z. Hrynkiewicz, "Elasticity of normal and cancerous human bladder cells studied by scanning force microscopy.," *European biophysics journal : EBJ,* vol. 28, pp. 312-6, 1999.

[4] J. L. Tan, J. Tien, D. M. Pirone, D. S. Gray, K. Bhadriraju, and C. S. Chen, "Cells lying on a bed of microneedles: an approach to isolate mechanical force.," *Proceedings of the National Academy of Sciences of the United States of America,* vol. 100, pp. 1484-9, 2003.

[5] J. Huang, M. Juszkiewicz, W. H. de Jeu, E. Cerda, T. Emrick, N. Menon*, et al.*, "Capillary wrinkling of floating thin polymer films.," *Science (New York, N.Y.),* vol. 317, pp. 650-3, 2007.

[6] A. Harris, P. Wild, and D. Stopak, "Silicone Rubber Substrata : A New Wrinkle in the Study of Cell Locomotion Author (s): Albert K . Harris , Patricia Wild and David Stopak Reviewed work (s): Source : Science , New Series , Vol . 208 , No . 4440 (Apr . 11 , 1980), pp . 177-179 Publish," *Science (New York, N.Y.),* vol. 208, pp. 177-179, 1980.

[7] R. J. Pelham and Y. L. Wang, "High resolution detection of mechanical forces exerted by locomoting fibroblasts on the substrate.," *Molecular biology of the cell,* vol. 10, pp. 935-45, 1999.

AN X-RAY DETECTABLE PRESSURE MICROSENSOR FOR MONITORING CORONARY IN-STENT RESTENOSIS

Mayurachat Ning Gulari[1], Mostafa Ghannad-Rezaie[2], Paula Novelli[4],
Nikos Chronis [1,2,3] and Theodore Cosmo Marentis[4]
[1]Macromolecular Science and Engineering, [2]Biomedical Engineering,
[3]Mechanical Engineering, [4]Radiology
University of Michigan, Ann Arbor, USA

ABSTRACT

We present a novel implantable X-ray-addressable MEMS Blood Pressure sensor, the X-BP, for the non-invasive and cost-effective surveillance of coronary in-stent restenosis. We successfully fabricated and tested the X-BP sensor and its pressure response curve. We placed the X-BP sensor in a coronary stent and prove adequate visibility in a clinically realistic scenario.

INTRODUCTION

Coronary Artery Disease (CAD) is the leading cause of mortality in the United States [1]. CAD arises from narrowing in the coronary arteries commonly due to plaque formation. The narrowing reduces myocardial blood perfusion leading to infarction [2]. One of the two procedures commonly used to treat CAD is percutaneous coronary intervention (PCI). A stent is placed across the arterial stenosis to recanalize it. Common complication of PCI are plaque reaccumulation and neointimal hyperplasia that lead to in-stent restenosis [3] which is associated with significant morbidity and mortality.

Efforts to minimize restenosis have focused on modifying the material and/or the design of the stent. Bare metal stents (BMS) have a 5-year restenosis rates of 30% [2]. Finer stent struts have ~1.7 times lower restenosis rates [3]. Drug eluting stents (DES) have early restenosis rates lower than BMS, but meta-analyses have reported a fivefold greater risk of late stent thrombosis, while 5-year follow-up data revealed little difference in patient mortality or stent thrombosis [4][5].

The Fractional Flow Reserve (FFR) is defined as the ratio of the blood pressure distal (P_d) over the pressure proximal (P_p) to the stenosis of a coronary artery. The FFR is strongly correlated to the functional severity of coronary stenosis. A threshold of FFR= 0.75 is 88% sensitive and 100% specific for the identification of reversible ischemia [6]. Multiple studies have validated its use to triage a patient in medical management or intervention. Knowing the FFR is one of the most important pieces of information the clinician can have to treat intermediate grade stenosis.

Currently, the FFR is acquired through invasive angiography and there are no available technologies that can non-invasively determine it. Wireless MEMS sensing approaches have been developed in the past for measuring blood pressure but not FFR [7][8]. The large size is the biggest limitation of these microsensors when compared to the diameter of an artery (2-4 millimeters), which makes them incompatible for coronary implantation.

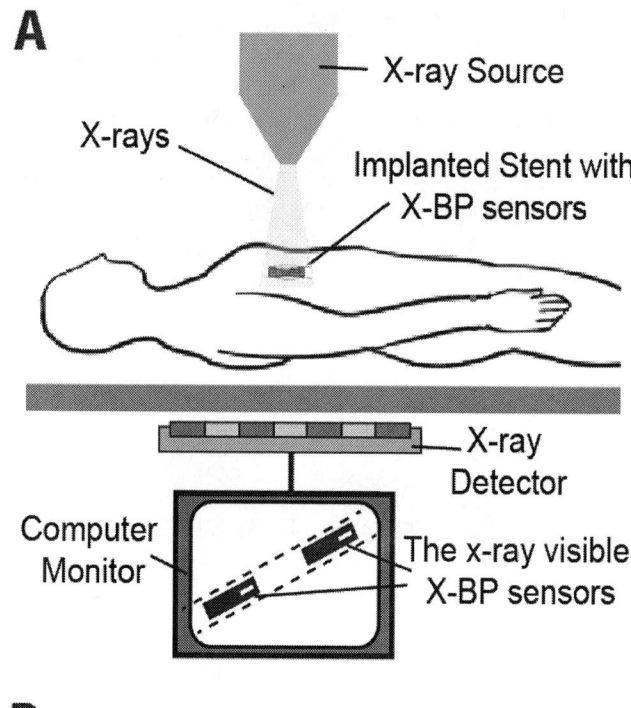

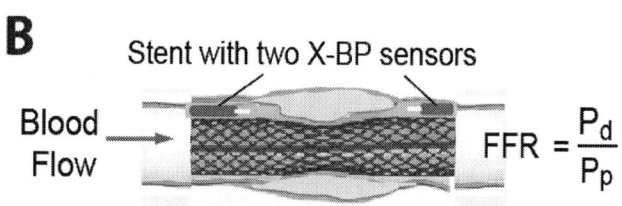

Figure 1. (A) X-ray based pressure sensing technology for monitoring restenosis non-invasively. (B) Two X-BP microsensors integrated in a stent provide a direct measure of the Fractional Flow Reserve (FFR).

Here, we describe the development of a novel implantable X-ray-detectable MEMS Blood Pressure sensor (the X-BP microsensor) for measuring the FFR non-invasively and therefore assessing coronary in-stent restenosis without a catheterization procedure. The X-BP changes its x-ray visible size with pressure. Blood pressure is obtained by simply measuring the x-ray footprint of the X-BP microsensor on a clinical chest radiograph. By integrating two X-BP microsensors into a stent, one on the proximal end and one on the distal end, one can deduce the FFR across the stent. The FFR is equivalent to the ratio of the two measured pressures. The integrated stent/X-BP microsensor unit has therefore a dual function: it increases the effective cross area of the artery (due to the presence of the stent) and measures FFR at any desired time point by simply visualizing the two X-BP microsensors in a radiograph. We believe that the X-BP microsensor due to its unique advantages, such as small size, absence of any electronic components and easy readout, can become an ideal solution to the long-standing need for periodically assessing restenosis non-invasively.

THE X-B MICROSENSOR

<u>X-BP Microsensor Design</u>. The X-BP microsensor contains a radiopaque liquid that is visible with x-rays. It is surrounded by a radiolucent bulk material (silicon/glass) that is not visible with x-rays (figure 2A). The X-BP microsensor works by changing the size of its active radiopaque liquid with pressure. The result is a small object in the radiograph with an easily- recognizable appearance that changes its size with blood pressure. The microsensor has a planar symmetry and it is designed with the clinician in mind: a single measurement along the long-axis of the sensor will give the pressure reading regardless of the sensor planar orientation in the body or to the x-ray beam.

Structurally the X-BP microsensor has three components: a blood pressure-exposed membrane, a micro-reservoir and a microfluidic gauge. The micro-reservoir below the membrane is filled with radiopaque fluid (isovue-370). The microfluidic gauge is initially filled with air, which is displaced as the gauge fills with the radiopaque fluid (figure 2B). By measuring the length of the radiopaque fluid inside the microfluidic gauge (indicated as 'gauge length' in figure 2, B(III)), one can estimate the blood pressure that is applied to the membrane. The micro-reservoir is 8.5 mm long by 600 µm wide and the microfluidic gauge is 4 mm long by 250 µm wide. Both the micro-reservoir and the microfluidic gauge are coated with a hydrophobic material and they are 250 µm deep. They are connected through a narrow, tapered microfluidic channel which is 100 µm wide and 250 µm deep. At atmospheric pressure, the membrane is not deflected and the radiopaque fluid stops at the tapered part of the microchannel (figure 2, B(I)).

<u>Microfabrication</u>. We used a three mask process to fabricate the X-BP microsensor (figure 3). We first performed a two-step deep reactive ion etching (DRIE) on a 400 µm thick silicon wafer to create the micro-reservoir, the microfluidic gauge and the tapered microchannel. We then deposited 0.2 µm of thermal oxide and 0.5 µm of low stress LPCVD silicon nitride. We anodically bonded a 200 µm thick Pyrex wafer to the DRIE-etched side of the silicon wafer to seal the X-BP sensor. Subsequently, we patterned the nitride/oxide films on the backside of the wafer using RIE. The bonded wafer was then diced to create side fluidic ports, followed by a backside DRIE step to release the low

stress silicon nitride membrane of the micro-reservoir. Subsequently, the sensor was soaked in a hydrophobic coating (Rain-X) for 5 minutes and rinsed with isopropyl alcohol. Finally, a radiopaque fluid (isovue-370) was injected using a glass micropipette and a syringe in the micro-reservoir and the fluidic ports were sealed with an FDA-approved adhesive (permabond ET5145 epoxy).

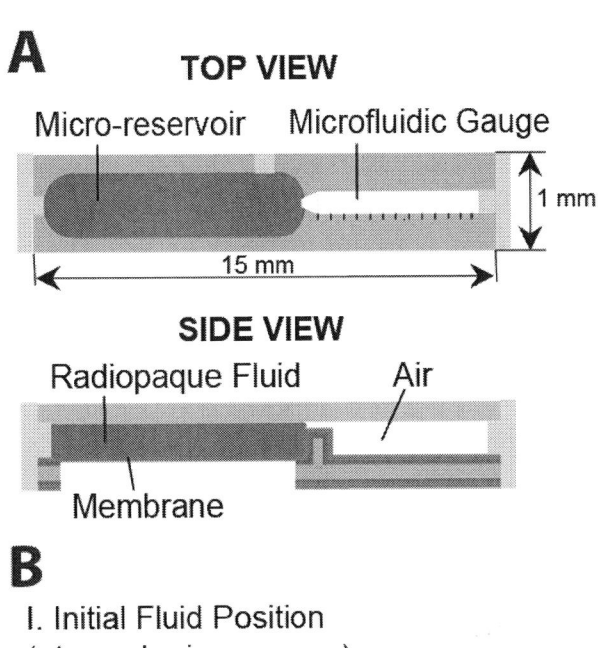

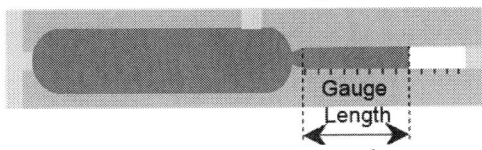

Figure 2. (A) Top and cross section diagrams of the X-BP sensor. The 3 fluidic ports are sealed with an epoxy. (B) The working principle of the X-BP sensor: the microfluidic gauge is filled with a radiopaque fluid when a pressure is applied to the membrane.

A

1. Two-step frontside DRIE

2. Oxide/Nitride Deposition

3. Pyrex Wafer Bonding

4. Backside RIE patterning

5. Dicing, DRIE and oxide removal

6. Isovue Filling and Sealing

■ SiO₂   ■ Si   ■ Radiopaque Fluid
▨ Glass   ■ SiN   ▨ Adhesive

B

Micro-reservoir   Microfluidic Gauge

Microfluidic Ports

Figure 3. (A) The 3-mask process of the X-BP microsensor. (B) the microfabricated device. Scale bar, 1.5 mm.

RESULTS AND DISCUSSION

<u>Pressure response.</u> In order to evaluate the pressure response of the X-BP microsensors, we placed them in a custom-made pressure chamber and applied pressures between 0 and 225 mmHg with an electronic pressure controller. We monitored the displacement of isovue-370 in the microfluidic gauge under a regular microscope. DRIE-etch marks located along the length of the microfluidic gauge (see figure 2) were used to accurately measure the length of the gauge that was filled with isovue-370 (indicated as 'gauge length' in figure 4). The pressure response shows a linear regime at low pressures (<100 mmHg), while at higher pressures there is a logarithmic trend. We anticipate that at even higher pressures (although high pressures are not clinically relevant), the response will saturate. We should also note that an air leak from the side port of the microfluidic gauge is possible as the FDA-approved adhesive might not be appropriate for sealing the silicon/glass ports. Therefore, the pressure response presented here, should be considered a rough estimate of the expected response if no leak is present.

At the maximum pressure tested, the maximum gauge length is ~ 1 mm, resulting in a resolution of 11 mmHg per 50 μm of gauge length. We report the resolution 'per 50 μm of gauge length change', as the detectable resolution of a typical x-ray imaging system is ~50 μm. At pressures below 15-20 mmHg, there is no gauge length change. That is probably due to the residual tensile stress of the silicon nitride membrane that needs to be overcome.

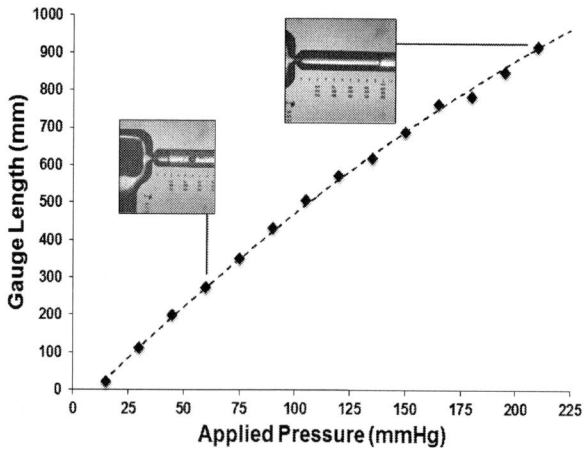

Figure 4. Applied pressure versus gauge length. The microfluidic gauge is 250 μm wide and 250 μm deep.

<u>X-ray imaging.</u> We imaged the X-BP microsensor in plain radiographs (Shimadzu Dart, 8MP Cannon DR detector). Isovue-370 was easily detectable inside the micro-reservoir (figure 5A). As expected, the microfluidic gauge is filled with air at atmospheric pressure and therefore it was not visible in the radiographs. The ratio of the x-ray signal intensity of the isovue-filled micro-reservoir over signal intensity of the air-filled gauge was ~1.5. In addition, we

imaged the X-BP microsensor in a custom-made pressure chamber at the 0-225 mmHg pressure range. The movement of isovue-370 inside the microfluidic gauge was clearly detectable (figure 5B).

Finally, we attached the X-BP microsensor to a coronary stent and image the integrated unit with a portable x-ray system over a chest phantom (figure 5C). The chest phantom represents a realistic scenario of the expected background noise in human tissue. Isovue-370 was clearly visible in the micro-reservoir of the X-BP microsensor. Future improvements in the radiopaque fluid and the geometry of the indicator channel can increase the signal in order to achieve an accurate measurement of the FFR.

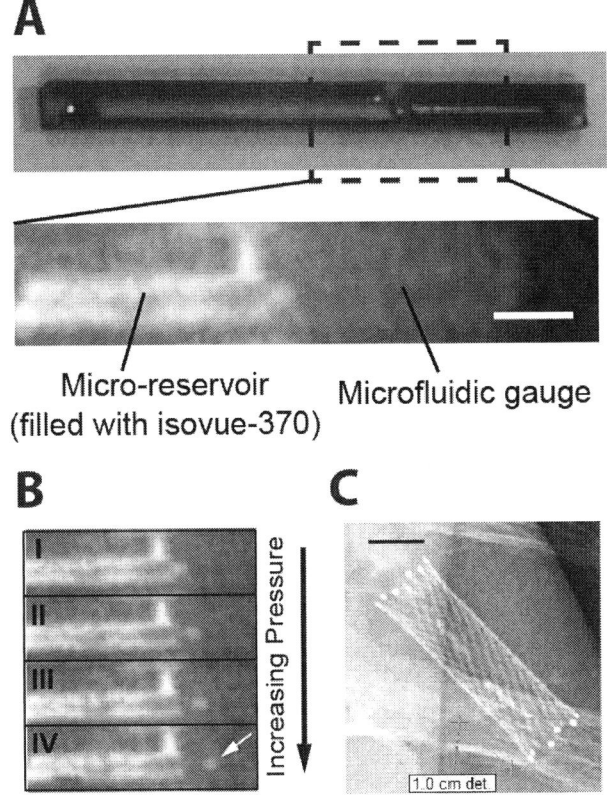

Figure 5. (A) An x-ray image of the X-BP microsensor. Scale bar, 1 mm. (B) X-ray snapshots of the micro-reservoir/gauge region at different pressures. (C) The X-BP microsensor integrated to a coronary stent, placed on top of a human phantom. Scale bar, 1 cm.

CONCLUSIONS

We successfully designed, fabricated and tested a novel, x-ray detectable blood pressure microsensor, the X-BP sensor for periodical surveillance of coronary in- stent restenosis with clinical chest radiographs. The results can be obtained in a few minutes in the ambulatory care setting. Short-term interval follow-up every 3 to 6 months is possible thanks to the low cost of radiographs and the negligible radiation exposure. With the availability of x-ray infrastructure in every hospital, we believe the X-BP sensor offers an elegant, cost-effective method to noninvasively assess coronary in-stent restenosis in a way not previously feasible. We anticipate this novel technology to solve other medical problems, such as peripheral arterial stent failure or ventricular shunt failure.

ACKNOWLEDGEMENTS

We are grateful to Onnop Srivannavit for helping in microfabricatioin, the Microfluidics in Biomedical Sciences Training Program of the National Institutes of Health, NIH T32 EB005582-05 for fellowship support, University of Michigan Radiology Department, the Radiological Society of North America R&E Foundation and the National Institute of Health for their encouragement, support and funding that has supported this research work.

REFERENCES

[1]   "What Is Coronary Heart Disease? - NHLBI, NIH." [Online]. Available: http://www.nhlbi.nih.gov/health/health-topics/topics/cad/. [Accessed: 02-Sep-2013].

[2]   M.-C. Morice et al, "Long-term clinical outcomes with sirolimus-eluting coronary stents: five-year results of the RAVEL trial.," *J. Am. Coll. Cardiol.*, vol. 50, no. 14, pp. 1299–304, Oct. 2007.

[3]   C. Briguori et al, "In-stent restenosis in small coronary arteries: impact of strut thickness.," *J. Am. Coll. Cardiol.*, vol. 40, no. 3, pp. 403–9, Aug. 2002.

[4]   A. Caixeta et al, "5-Year Clinical Outcomes After Sirolimus-Eluting Stent Implantation Insights From a Patient-Level Pooled Analysis of 4 Randomized Trials Comparing Sirolimus-Eluting Stents With Bare-Metal Stents.," *J. Am. Coll. Cardiol.*, vol. 54, no. 10, pp. 894–902, Sep. 2009.

[5]   S. G. Ellis et al, "Long-term safety and efficacy with paclitaxel-eluting stents: 5-year final results of the TAXUS IV clinical trial (TAXUS IV-SR: Treatment of De Novo Coronary Disease Using a Single Paclitaxel-Eluting Stent).," *JACC. Cardiovasc. Interv.*, vol. 2, no. 12, pp. 1248–59, Dec. 2009.

[6]   N. H. Pijls et al, "Fractional flow reserve. A useful index to evaluate the influence of an epicardial coronary stenosis on myocardial blood flow.," *Circulation*, vol. 92, no. 11, pp. 3183–93, Dec. 1995.

[7]   "Welcome to CardioMEMS." [Online]. Available: http://www.cardiomems.com/.

[8]   "Issys Home." [Online]. Available: http://www.mems-issys.com/.

A NOVEL ACTUATOR FOR ENERGY HARVESTING USING AN ACOUSTICALLY OSCILLATING LIQUID DROPLET

Young Rang Lee, Jae Hun Shin, Il Song Park and Sang Kug Chung
Myongji University, Yongin, Gyeonggido, South Korea

ABSTRACT

This paper presents a novel actuator for energy harvesting from ambient acoustic noise using acoustically oscillating droplets. When a liquid droplet sitting on a piezocantilever is excited by an acoustic wave around its natural frequency, it oscillates and simultaneously bends the piezocantilever by the reaction of the droplet oscillation, resulting in electric power generation owing to the piezoelectric effect. The oscillation amplitudes of three different sized water droplets actuated by acoustic waves in a wide range of frequencies generated from a cylindrical piezoactuator are firstly investigated using a high speed camera. The results show that the droplet oscillation amplitude is strongly dependent on the applied frequency and proportional to the droplet size. And the maximum droplet oscillation amplitude occurs at the droplet natural frequency. The energy harvesting based on an acoustically oscillating droplet is separately tested using a commercial piezocantilever. The bending displacement and generating voltage of the piezocantilever by an acoustically oscillating water droplet with three different sizes in different distances between the droplet and piezoactuator are measured by using a high speed camera and digital oscilloscope, respectively. The both bending displacement and generating voltage are also highly affected by the applied frequency and proportional to the droplet size but inversely proportional to the distance. The force generated from an acoustically oscillating droplet is also measured using a load cell. The maximum force generated from the acoustically oscillating droplet is about 120 μN at the maximum bending displacement (about 1.5 mm). Finally, as the concept proof, the output voltage and power generated from the piezocantilever actuated by an acoustically oscillating droplet (6 μl) are measured using a custom-made electric circuit mainly consisting of voltage rectifier and load. The maximum generated power for the load (470 kΩ) is measured about 72.2 μW.

INTRODUCTION

As interest in portable wireless devices grows, development of micro energy harvesting technology has become an important task[1-3]. Energy harvesting also called as power harvesting or energy scavenging is defined as capturing energy from one or more of the surrounding energy sources, accumulating them and storing them for later use. Although portable wireless devices offer several advantages such as flexibility and ability to facilitate the placement of sensors in previously inaccessible locations, the use of battery as power source limits their potentials due to short battery life[2-6].

Hence, various micro energy harvesting technologies powered by mainly four energy sources—light, radio-frequency (RF) electromagnetic radiation, thermal gradients, and mechanical motion—have been investigated and developed as the potential alternative for the battery[3, 7-10].

Among various types of energy harvesting technologies, piezoelectric energy harvesting based on mechanical vibration is most popular owing to its simple structure[11-17]. The piezoelectric effect converts mechanical strain into electric voltage and current. In this paper, a novel actuator for energy harvesting from ambient acoustic noise using acoustically oscillating droplets is presented. When a liquid droplet sitting on a piezocantilever is excited by an acoustic wave around its natural frequency, it oscillates and simultaneously bends the piezocantilever by the reaction of the droplet oscillation, resulting in electric power generation from the piezocantilever in Fig. 1[18-20].

The envisioned energy harvesting system can extract mechanical power from acoustic noise in a wide range of frequencies using liquid droplets in different sizes and natural frequencies and convert the mechanical power to electrical power for wireless electronic devices. This new type of actuation technique is a simple but useful tool not only for the energy harvesting system but also potential acoustic wave sensors and actuators.

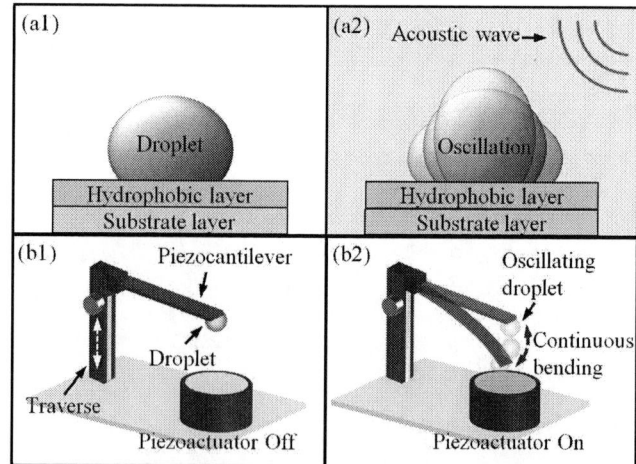

Figure 1: An acoustically oscillating droplet induced motion-powered energy harvester: (a) When a liquid droplet is acoustically excited by a piezoactuator around its natural frequency, it oscillates; (b) When an acoustically oscillating droplet is placed on the tip end of a piezocantilever, the oscillating motion of the droplet induces the continuous bending of the piezocantilever as the reaction, resulting in electric power generation from the piezocantilever.

EXPERIMENTAL RESULTS

Dynamic Behavior of an Acoustically Excited Water Droplet

First, dynamic behavior of an acoustically excited water droplet is investigated using a high speed camera (Phantom Miro eX4, Vision Research Inc.). When a water droplet hanging on a Teflon coated plate is acoustically excited by a cylindrical piezoactuator (PIC151, Physik Instrumente, Inc.) around its natural frequency, it continuously deforms (oscillates) with respect to the applied frequency, as shown in the inset of Fig. 2. The oscillation amplitudes of three different sized water droplets actuated by acoustic waves in a wide range of frequencies generated from the piezoactuator are measured using the high speed camera, and the results are plotted in Fig. 2. The droplet oscillation amplitude is strongly dependent on the applied frequency and proportional to the droplet size. And the maximum droplet oscillation amplitude occurs at the droplet natural frequency, which is a function of the droplet size[21-23].

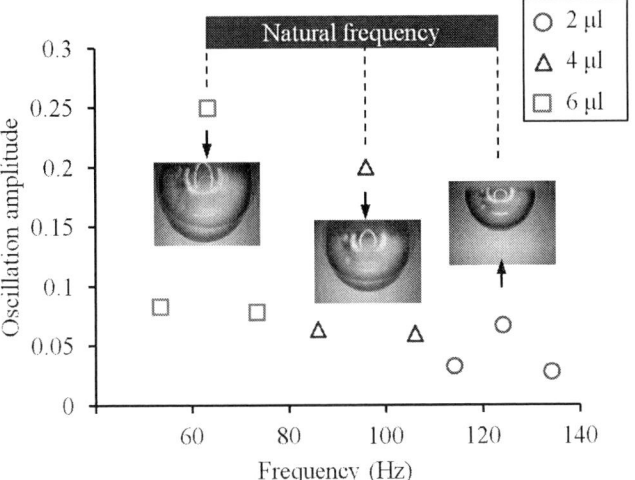

Figure 2: Measurement of the oscillation amplitude of an acoustically oscillating droplet in different frequencies at a fixed distance (4 mm) using a high speed camera.

Energy Harvesting Using a Piezocantilever Actuated by an Acoustically Oscillating Water Droplet

Energy harvesting induced by an acoustically oscillating water droplet is tested using a commercial piezocantilever (LDT2-028K/L w/rivets, Measurement Specialties, Inc.). A small water droplet (4 μl) is injected by a microsyringe (600 Series MICROLITER™ Syringes model 62, Hamilton Co.) and attached on the bottom tip end of the piezocantilever ($73(L) \times 1.5(W) \times 0.2(T)$ mm³). When the droplet is acoustically excited by a piezoactuator at 21 Hz, it oscillates and simultaneously makes the piezocantilever bend up and down, as shown in Fig. 3. And when the piezoactuator is turned off, the droplet and piezocantilever stop oscillating and bending motions.

The piezocantilever's bending displacement and generating voltage by an acoustically oscillating water

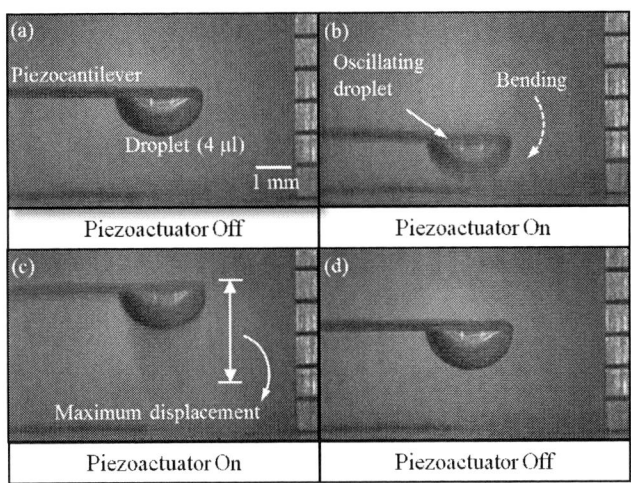

Figure 3: Snap shots of the continuous bending of a piezocantilever by the oscillation of a droplet attached on the tip end of the piezocantilever: (a) Initial; (b-c) When the drop is acoustically excited by a piezoactuator at its natural frequency (21 Hz), it oscillates and simultaneously bends the piezocantilever up and down; (d) When the piezoactuator is turned off, the motions of the droplet and piezocantilever stop.

droplet in three different volumes (2 μl, 4 μl, 6 μl) are measured by a high speed camera and digital oscilloscope (TDS3012, Tektronix, Inc.), respectively. The both bending displacement and generating voltage are also highly affected by the applied frequency and proportional to the droplet size in Fig. 4. And the maximum bending displacement and generating voltage occur at the natural frequency, which shows the close relationship with the oscillation amplitude of the droplet placed on a solid substrate in Fig. 2; however, the natural frequency is shifted due to the use of the piezocantilever for energy harvesting.

The effect of the distance between the droplet and piezoactuator is separately investigated, and the results are plotted in Fig. 5. The both bending displacement and generating voltage are inversely proportional to the distance,

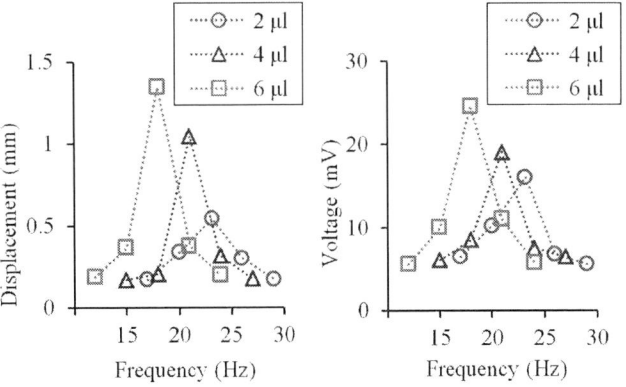

Figure 4: Measurement of the bending displacement and generating voltage of a piezocantilever actuated by an acoustically oscillating droplet in different frequencies: (a) Displacement vs. frequency; (b) Voltage vs. frequency.

which clearly shows that the envisioned actuator is powered by acoustic energy sources.

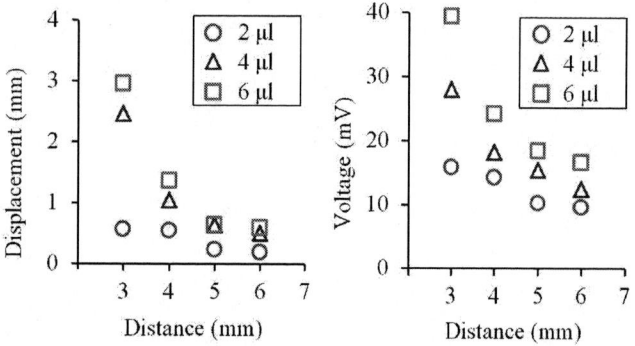

Figure 5: Measurement of the bending displacement and generating voltage of a piezocantilever actuated by an acoustically oscillating droplet in different distances from the piezoactuator: (a) Displacement vs. distance; (b) Voltage vs. distance.

Measurement of the Force Generated from an Acoustically Oscillating Droplet Using a Load Cell

The force generated from an acoustically oscillating droplet is measured using a load cell (GSO-100, Transducer Techniques, Inc.), as shown in Fig. 6. When the same piezocantilever used for the acoustically oscillating droplet induced energy harvesting is tested (pushed and bended) by the load cell attached by a precise traverse system, the bending force is measured in each different bending displacement and compared with the bending displacement induced by an oscillating droplet in Fig. 6. The maximum force generated from the acoustically oscillating droplet (4 μl) is about 120 μN at the maximum bending displacement (about 1.5 mm).

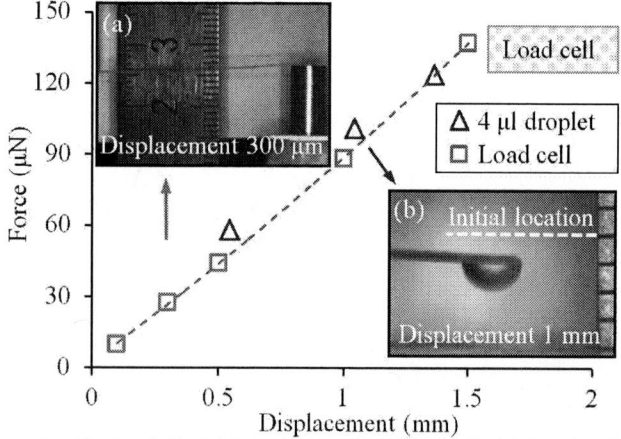

Figure 6: Measurement of the force induced by an acoustically oscillating droplet using a load cell.

Measurement of the Output Voltage and Power Using a Custom-made Electric Circuit

As the concept proof, the output voltage and power generated from a piezocantilever actuated by an acoustically oscillating droplet (6 μl) are measured using a custom-made

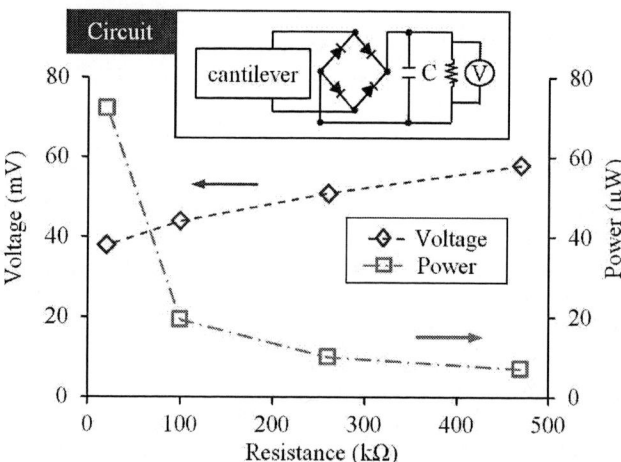

Figure 7: Output voltage and power generated from a piezocantilever actuated by an acoustically oscillating droplet using a custom-made electrical circuit mainly consisting of voltage rectifier and load.

electric circuit mainly consisting of voltage rectifier and load in Fig. 7. To rectify the alternating current (AC) signal generated from the piezocantilever, a full wave-bridge type rectifying circuit consisting of 4 diodes is used. Note that the inset in Fig. 7 shows the used circuit black diagram[24]. The results show that the output voltage increases as the electrical load increases, while the output power decreases as the load increases, and the maximum power for the load (470 kΩ) is measured about 72.2 μW.

CONCLUSION

A novel actuator for energy harvesting from ambient acoustic noise using acoustically oscillating droplets has been developed. First, the dynamic behavior of a water droplet actuated by acoustic excitation in different frequencies is investigated using a high speed camera. The droplet oscillation amplitude is strongly dependent on the applied frequency and proportional to the droplet size. And the maximum droplet oscillation amplitude occurs at the droplet natural frequency. Second, the energy harvesting using a commercial piezocantilever actuated by an acoustically oscillating droplet is separately tested. The piezocantilever's bending displacement and generating voltage by an acoustically oscillating water droplet in three different sizes are measured by a high speed camera and digital oscilloscope, respectively. The both bending displacement and generating voltage are also highly affected by the applied frequency and proportional to the droplet size. The effect of the distance between the droplet and piezoactuator is also investigated. The both bending displacement and generating voltage are inversely proportional to the distance, which clearly shows that the envisioned actuator is powered by acoustic energy sources. Third, the force generated from an acoustically oscillating droplet is measured using a load cell. The maximum force generated from the acoustically oscillating droplet is about

120 μN at the maximum bending displacement (about 1.5 mm). Finally, the output voltage and power generated from a piezocantilever actuated by an acoustically oscillating droplet (6 μl) are measured using a custom-made full wave-bridge type rectifying circuit consisting of 4 diodes. The maximum generated power for the load (470 kΩ) is measured about 72.2 μW. This new type of actuation technique is a simple but useful tool not only for the energy harvesting system but also potential acoustic wave sensors and actuators.

ACKNOWLEDGEMENTS

This work was supported by the Fundamental Research Supporting Program (2011-0012100) and Basic Science Research Program (2011-0025039) of National Research Foundation of Korea(NRF).

REFERENCES

[1] S. R. Anton and H. A. Sodano, "A review of power harvesting using piezoelectric materials (2003–2006)," *Smart Materials and Structures,* vol. 16, p. R1, 2007.

[2] H. S. Kim, J. H. Kim, and J. Kim, "A review of piezoelectric energy harvesting based on vibration," *International Journal of Precision Engineering and Manufacturing,* vol. 12, pp. 1129-1141, 2011.

[3] P. D. Mitcheson, E. M. Yeatman, G. K. Rao, A. S. Holmes, and T. C. Green, "Energy harvesting from human and machine motion for wireless electronic devices," *Proceedings of the IEEE,* vol. 96, pp. 1457-1486, 2008.

[4] Y. C. Shu and I. C. Lien, "Analysis of power output for piezoelectric energy harvesting systems," *Smart Materials and Structures,* vol. 15, p. 1499, 2006.

[5] M. Philipose, J. R. Smith, B. Jiang, A. Mamishev, S. Roy, and K. Sundara-Rajan, "Battery-free wireless identification and sensing," *IEEE Pervasive Computing,* vol. 4, pp. 37-45, 2005.

[6] K. A. Cook-Chennault, N. Thambi, and A. M. Sastry, "Powering MEMS portable devices—a review of non-regenerative and regenerative power supply systems with special emphasis on piezoelectric energy harvesting systems," *Smart Materials and Structures,* vol. 17, p. 043001, 2008.

[7] S. Saadon and O. Sidek, "A review of vibration-based MEMS piezoelectric energy harvesters," *Energy Conversion and Management,* vol. 52, pp. 500-504, 2011.

[8] H. A. Sodano, D. J. Inman, and G. Park, "A review of power harvesting from vibration using piezoelectric materials," *Shock and Vibration Digest,* vol. 36, pp. 197-206, 2004.

[9] H. A. Sodano, D. J. Inman, and G. Park, "Comparison of piezoelectric energy harvesting devices for recharging batteries," *Journal of Intelligent Material Systems and Structures,* vol. 16, pp. 799-807, 2005.

[10] H. A. Sodano, D. J. Inman, and G. Park, "Generation and storage of electricity from power harvesting devices," *Journal of Intelligent Material Systems and Structures,* vol. 16, pp. 67-75, 2005.

[11] S. P. Beeby, M. J. Tudor, and N. M. White, "Energy harvesting vibration sources for microsystems applications," *Measurement Science and Technology,* vol. 17, p. R175, 2006.

[12] H. U. Kim, W. H. Lee, H. V. Rasika Dias, and S. Priya, "Piezoelectric microgenerators-current status and challenges," *IEEE Transactions on Ultrasonics Ferroelectrics and Frequency Control,* vol. 56, pp. 1555-1568, 2009.

[13] C. Knight, J. Davidson, and S. Behrens, "Energy options for wireless sensor nodes," *Sensors,* vol. 8, pp. 8037-8066, 2008.

[14] S. Priya, "Advances in energy harvesting using low profile piezoelectric transducers," *Journal of Electroceramics,* vol. 19, pp. 167-184, 2007.

[15] S. Dash, N. Kumari, and S. V. Garimella, "Frequency-dependent transient response of an oscillating electrically actuated droplet," *Journal of Micromechanics and Microengineering,* vol. 22, p. 075004, 2012.

[16] R. Amirtharajah and A. P. Chandrakasan, "Self-powered signal processing using vibration-based power generation," *IEEE Journal of Solid-State Circuits,* vol. 33, pp. 687-695, 1998.

[17] S. Roundy, P. K. Wright, and J. Rabaey, "A study of low level vibrations as a power source for wireless sensor nodes," *Computer Communications,* vol. 26, pp. 1131-1144, 2003.

[18] H. Y. Kim, "Drop fall-off from the vibrating ceiling," *Physics of Fluids,* vol. 16, p. 474, 2004.

[19] J. H. Moon, B. H. Kang, and H. Y. Kim, "The lowest oscillation mode of a pendant drop," *Physics of Fluids,* vol. 18, p. 021702, 2006.

[20] X. Noblin, A. Buguin, and F. Brochard-Wyart, "Vibrations of sessile drops," *European Physical Journal-Special Topics,* vol. 166, pp. 7-10, 2009.

[21] E. D. Wilkes and O. A. Basaran, "Forced oscillations of pendant (sessile) drops," *Physics of Fluids,* vol. 9, p. 1512, 1997.

[22] O. A. Basaran and D. W. DePaoli, "Nonlinear oscillations of pendant drops," *Physics of Fluids,* vol. 6, p. 2923, 1994.

[23] J. M. Oh, D. Legendre, and F. Mugele, "Shaken not stirred—on internal flow patterns in oscillating sessile drops," *EPL,* vol. 98, p. 34003, 2012.

[24] G. K. Ottman, H. F. Hofmann, A. C. Bhatt, and G. A. Lesieutre, "Adaptive piezoelectric energy harvesting circuit for wireless remote power supply," *IEEE Transactions on Power Electronics,* vol. 17, pp. 669-676, 2002.

CONTACT

* S. K. Chung, tel: +82-31-330-6346; skchung@mju.ac.kr

A NOVEL ON-CHIP MICROMANIPULATION METHOD USING A MICROBUBBLE FOR SINGLE CELL MANIPULATION AND CHARACTERIZATION

Jae Hun Shin, Young Rang Lee, Il Song Park and Sang Kug Chung
Myongji University, Yongin, Gyeonggido, South Korea

ABSTRACT

A novel on-chip micromanipulation method has been developed where a gaseous microbubble is applied to physically manipulate micro-objects in an aqueous medium through optical and acoustical excitations. This micromanipulation concept is experimentally verified: bubble manipulation (generating and transporting operations) and micro-object manipulation (capturing, carrying, and releasing operations). Optically induced microbubble generation is firstly tested for different optical powers in a microfluidic chip with an amorphous silicon layer as an optically absorbent material. And microbubble transportation is also demonstrated using optically induced thermocapillary effects. Micro-object manipulation is separately demonstrated using a gaseous microbubble (300 µm dia.) with glass beads (100 µm dia.) in an aqueous medium. When the microbubble is acoustically excited by a piezoactuator attached on the bottom of a chip, it oscillates and simultaneously captures the neighboring glass beads owing to the oscillating bubble induced radiation forces. Finally, the manipulation of a single glass bead (100 µm dia.) using a microbubble (300 µm dia.) is successfully achieved with integration of optical and acoustical excitations. This novel micromanipulation technique may be an efficient tool for single cell manipulation and characterization in a microfluidic chip.

INTRODUCTION

Decipherment of intercellular signals is critical to understand cellular behaviors and functions and one of the most important tasks in cellular biology and tissue engineering[1-4]. Nevertheless, the complexity and heterogeneity of the intercellular signaling processes has defied investigation on intercellular communication[2, 5]. Hence, a single-cell level co-culture system is highly required to study single-cell pair interactions with reducing the complexity. To realize the reliable system, the development of an effective tool with the ability to grasp cells with high manipulating (transporting and positioning) accuracy to avoid any damage to the manipulated cells is highly demanded[6-9].

Various micromanipulation tools operated by different actuation schemes such as electrostatic actuators, piezoelectric actuators, shape-memory alloy actuators, pneumatic actuators, and thermal actuators have been developed[10-12]. However, most of these tools cannot be operated in an aqueous environment compatible with general culture media for cells. Particularly, these tools may cause unwanted physical contact with manipulating cells and damage them. To minimize contact damage, various non-contact manipulation techniques without physical structures have been developed, which include optical tweezers, optoelectronic manipulators, bubble tweezers, acoustic tweezers, and electrical-force-assisted manipulators such as electrophoresis and dielectrophoresis[13-19].

Recently, Ohta research group have developed optically controlled bubble microrobots where gaseous microbubbles are applied to physically manipulate micro-objects in an aqueous medium[20-22]. The bubble microrobots consist of gaseous microbubble instead of solid materials like most other microrobots and manipulate micro-objects by direct physical contact or by the bubble induced flow. Various micro-object manipulations such as drawing patterns with

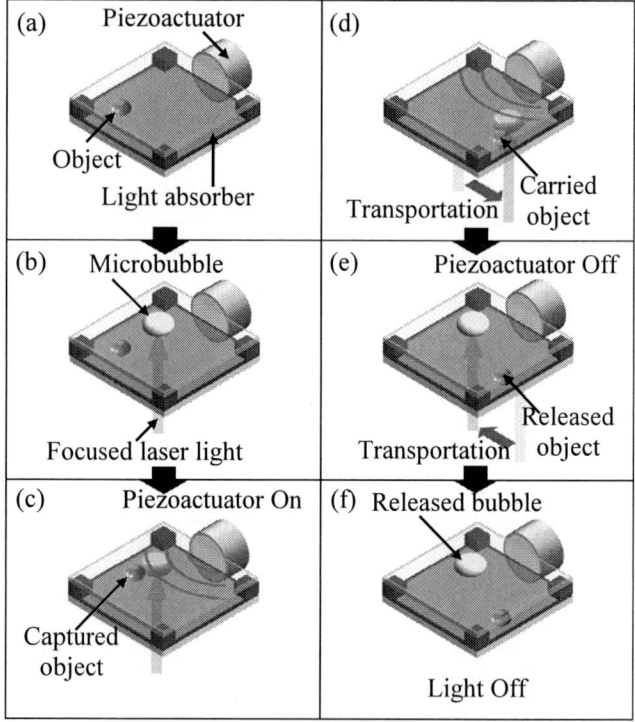

Figure 1: On-chip micromanipulation using a microbubble actuated by optical and acoustical excitations: (a) Initial; (b) When a laser light is concentrated on a spot, an optically induced microbubble is generated on the spot due to a light absorbing layer; (c) When the microbubble is acoustically excited, the acoustically oscillating microbubble captures a neighboring object due to the oscillating microbubble induced radiation forces; (d) The captured object is carried to a target location with the oscillating microbubble; (e-f) When the microbubble is transported without the acoustic excitation, the carried object is released from it.

microparticles on a surface and assembling glass beads into patterns have been demonstrated using the optically controlled bubble microrobot[22]. This micromanipulation provides the parallel and independent control of multiple microbubbles using the projection of multiple light patterns, increasing throughput.

However, micro-object manipulation relying on physical contact (pushing) with a microbubble is difficult to provide the precise position control of manipulating objects, resulting in low operation efficiency. Hence, adding a new function to grasp micro-objects like hands to the bubble microrobots is highly demanded. In this paper a novel on-chip micromanipulation method using a microbubble actuated by optical and acoustical excitations is presented for single cell manipulation and characterization in a microfluidic chip, along with the experimental verification of bubble manipulation (generating and transporting operations) and micro-object manipulation (capturing, carrying, and releasing operations).

Figure 1 describes the concept of the proposed micromanipulation. When a laser light is concentrated on a spot in a microfluidic chip, a microbubble is thermally generated and grown on the spot, because the bottom substrate is covered with an amorphous silicon layer as a light absorber in Fig. 1(b). When the microbubble is acoustically excited around its natural frequency, it oscillates and simultaneously captures a neighboring object owing to the oscillating microbubble induced radiation forces in Fig. 1(c). And the captured object can be carried to a target place with the oscillating microbubble by controlling the laser light using a traverse system, because the illuminated microbubble follows the light due to optically induced thermocapillary effects and the captured object also follows the oscillating microbubble due to the acoustically oscillating microbubble induced radiation forces in Fig. 1(d). When the object reaches the target place, the captured object can be released from the microbubble by turning the acoustical excitation off, and then the microbubble can be transported to any location or free itself by turning the light off in Fig. 1(e, f).

EXPERIMENT RESULTS

For optically induced microbubble generation and transportation tests, an amorphous silicon layer with 500 nm thickness is deposited on the bottom substrate as an optically absorbent material, UV treated for hydrophilic surface treatment, and integrated with a transparent cover glass using four spacers with 250 μm thickness, as shown in Fig. 2. For visual results, a charge coupled device (CCD, EO-1312C, Edmund Optics) with a zoom lens (VZMTM 450i eo, Edmund Optics) is used, and the image data is saved on a personal computer.

First, the microbubble generation is investigated using a microfabricated chip in Fig. 3. When a laser light (532 nm and 550 mW) is concentrated on the bottom subtract covered by an amorphous silicon layer with 500 nm thickness, a single microbubble is generated and continuously grown by the optically induced heat. The experiment is repeated for

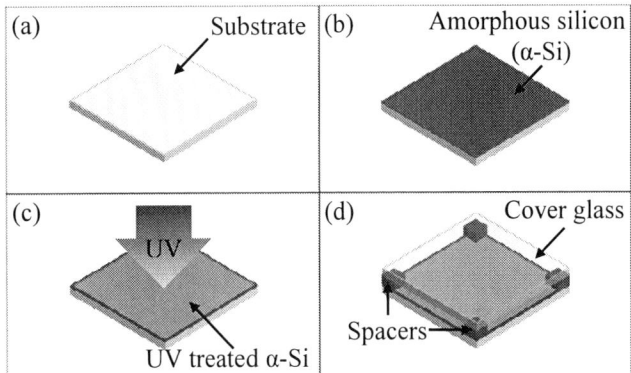

Figure 2: Microfabrication processes for testing chips.

three different optical powers, and the results are plotted in Fig. 3(b). The microbubble volume is proportional to the light illumination time and power. Note that the generated microbubble volume is in the range of about 10 nl to 60 nl.

Second, the transportation of a microbubble is demonstrated in Fig. 4. For the test, a microbubble (250 μm dia.) is initially generated inside a chip by the same process in Fig. 3. When a laser light (about 350 mW) illuminates on the spot of the bottom substrate where the microbubble is placed, the spot is heated up due to an optically absorbent material, generating temperature gradients around the microbubble.

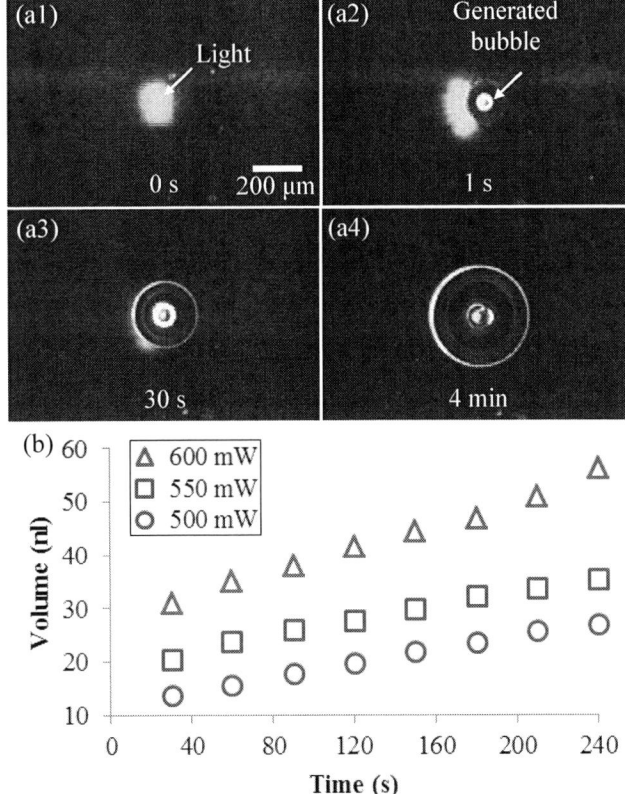

Figure 3: On-chip microbubble generation: (a) Sequential snap shots of a microbubble generated and grown by optically induced heating; (b) The volume of the microbubble vs. the light illumination time for three different optical powers.

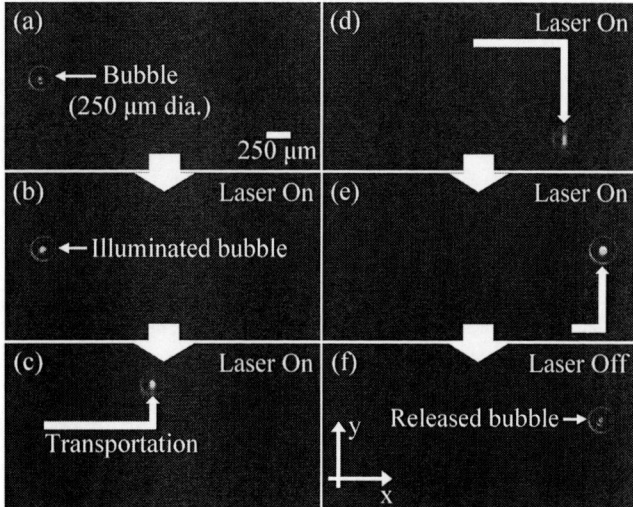

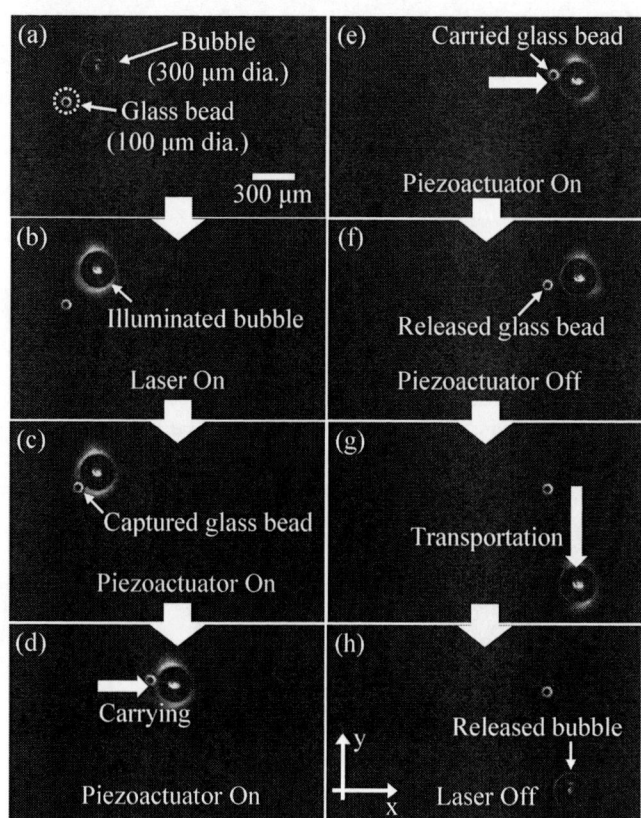

Figure 4: Sequential snap shots of microbubble transportation: (a-b) When a microbubble is illuminated by a laser light (532 nm and 350 mW), the microbubble is trapped by the light; (c-e) When the light moves, the microbubble follows the light path; (f) When the light is turned off, the microbubble is released from the light.

And the light induced heat also generates surface tension gradients and drives a fluid motion from the heated side of the microbubble to the unheated side, resulting in a net microbubble movement towards the heated side, which is thermocapillary effects.

Third, the micromanipulation of glass beads is conducted using an acoustically oscillating microbubble in Fig. 5. For the test, a microbubble (300 μm dia.) is also initially generated inside a chip by the same process in Fig. 3. When the microbubble is acoustically excited by a cylinder-type piezoactuator (PIC151, Physik Instrument Inc.) attached on the side of a microfluidic chip at the natural frequency (12 kHz), it oscillates and simultaneously captures neighboring glass beads (100 μm dia.) owing to the

Figure 6: Sequential snap shots of the micromanipulation of a single glass bead using a microbubble actuated by optical and acoustical excitations: (a) Initial; (b-f) Micro-object manipulation – capturing, carrying, releasing a glass bead (100 μm dia.); (g-h) Microbubble manipulation – transporting and releasing a microbubble (300 μm dia.).

acoustically excited microbubble induced radiation forces in Fig. 5(b, c). When the excitation is turned off, the captured glass beads are released from the microbubble and remain the microbubble rim in Fig. 5(d).

Finally, as the concept proof, the micromanipulation of a single glass bead depicting single cell micromanipulation is achieved using a microbubble actuated by optical and acoustical excitations in Fig. 6. When a microbubble (300 μm dia.) is acoustically excited at 12 kHz, it pulls and captures a single glass bead (100 μm dia.) in Fig. 6(c). And then the microbubble carries the glass bead to a target place with the assistance of the optically induced thermocapillary and acoustically induced radiation forces in Fig. 6(d-e). When the carried glass bead reaches the target place, it is released from the microbubble by turning the acoustical excitation off and transporting the microbubble down in Fig. 6(f-g); this demonstrates the high potential of the proposed micromanipulation method for future practical applications such as single cell manipulation and micro device assembly.

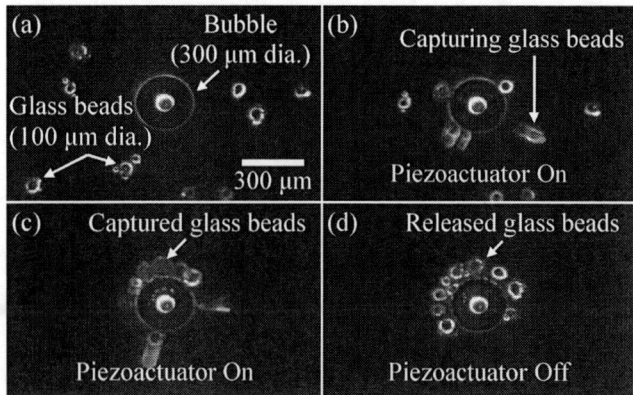

Figure 5: Micromanipulation of glass beads: (a) Initial; (b-c) When a microbubble is acoustically excited, it captures neighboring glass beads by radiation forces; (d) When the acoustic excitation is turned off, the captured glass beads are released from the microbubble.

CONCLUSIONS

This paper presents a novel on-chip micromanipulation method where a gaseous microbubble is applied to physically manipulate micro-objects in an aqueous medium

978-1-4799-3510-9/14 $31.00 © 2014 IEEE

through optical and acoustical excitations. Two excitation schemes—optical excitation for bubble manipulation (generating and transporting operations) and acoustical excitation for micro-object manipulation (capturing, carrying, and releasing operations)—are studied and applied. First, optically induced microbubble generation and the microbubble transportation using optically induced thermocapillary effects are tested. Second, micro-object manipulation is separately demonstrated using a gaseous microbubble (300 μm dia.) with glass beads (100 μm dia.) in an aqueous medium. Finally, the manipulation of a single glass bead (100 μm dia.) using a microbubble (300 μm dia.) depicting single cell micromanipulation is successfully achieved with integration of optical and acoustical excitations. This micromanipulation method can be applied to single cell manipulation and characterization in a microfluidic chip and micro device assembly.

ACKNOWLEDGEMENTS

This work was supported by the Fundamental Research Supporting Program (2011-0012100) and Basic Science Research Program (2011-0025039) of National Research Foundation of Korea(NRF).

REFERENCES

[1] B. Alberts, D. Bray, K. Hopkin, A. D. Johnson, J. Lewis, M. Raff, K. Roberts, and P. Walter, *Essential cell biology*. New York: Garland Science, 2009.

[2] S. Hong, Q. Pan, and L. P. Lee, "Single-cell level co-culture platform for intercellular communication," *Integrative Biology,* vol. 4, pp. 374-380, 2012.

[3] W. Hu, Q. Fan, and A. T. Ohta, "An opto-thermocapillary cell micromanipulator," *Lab on a Chip,* vol. 13, pp. 2285-2291, 2013.

[4] C. M. Waters and B. L. Bassler, "Quorum sensing: cell-to-cell communication in bacteria," *Annual Review of Cell and Developmental Biology,* vol. 21, pp. 319-346, 2005.

[5] D. Wang and S. Bodovitz, "Single cell analysis: the new frontier in 'omics'," *Trends in Biotechnology,* vol. 28, pp. 281-290, 2010.

[6] J. Cecil, D. Powell, and D. Vasquez, "Assembly and manipulation of micro devices - a state of the art survey," *Robotics and Computer-Integrated Manufacturing,* vol. 23, pp. 580-588, 2007.

[7] J. P. Desai, A. Pillarisetti, and A. D. Brooks, "Engineering approaches to biomanipulation," *Annual Review of Biomedical Engineering* vol. 9, pp. 35-53, 2007.

[8] E. W. H. Jager, O. Inganäs, and I. Lundström, "Microrobots for micrometer-size objects in aqueous media: potential tools for single-cell manipulation," *Science,* vol. 288, pp. 2335-2338, 2000.

[9] K. H. Lee, J. H. Lee, J. M. Won, K. Rhee, and S. K. Chung, "Micromanipulation using cavitational microstreaming generated by acoustically oscillating twin bubbles," *Sensors and Actuators A: Physical,* vol. 188, pp. 442-449, 2012.

[10] J. H. Lee, K. H. Lee, J. B. Chae, K. Rhee, and S. K. Chung, "On-chip micromanipulation by AC-EWOD driven twin bubbles," *Sensors and Actuators A: Physical,* vol. 195, pp. 167-174, 2013.

[11] J. Cecil, D. Vasquez, and D. Powell, "A review of gripping and manipulation techniques for micro-assembly applications," *International journal of production research,* vol. 43, pp. 819-828, 2005.

[12] A. Vasudev, A. Jagtiani, L. Du, and J. Zhe, "A low-voltage droplet microgripper for micro-object manipulation," *Journal of Micromechanics and Microengineering,* vol. 19, p. 075005, 2009.

[13] A. Ashkin, J. M. Dziedzic, and T. Yamane, "Optical trapping and manipulation of single cells using infrared laser beams," *Nature,* vol. 330, pp. 769-771, 1987.

[14] S. K. Chung and S. K. Cho, "On-chip manipulation of objects using mobile oscillating bubbles," *Journal of Micromechanics and Microengineering,* vol. 18, p. 125024, 2008.

[15] S. K. Chung, Y. Zhao, and S. K. Cho, "On-chip creation and elimination of microbubbles for a micro-object manipulator," *Journal of Micromechanics and Microengineering,* vol. 18, p. 095009, 2008.

[16] J. E. Curtis, B. A. Koss, and D. G. Grier, "Dynamic holographic optical tweezers," *Optics Communications,* vol. 207, pp. 169-175, 2002.

[17] J. Hu, J. Yang, and J. Xu, "Ultrasonic trapping of small particles by sharp edges vibrating in a flexural mode," *Applied Physics Letters,* vol. 85, p. 6042, 2004.

[18] J. O. Kwon, J. S. Yang, S. J. Lee, K. Rhee, and S. K. Chung, "Electromagnetically actuated micromanipulator using an acoustically oscillating bubble," *Journal of Micromechanics and Microengineering,* vol. 21, p. 115023, 2011.

[19] J. Voldman, "Electrical forces for microscale cell manipulation," *Annual Review of Biomedical Engineering,* vol. 8, pp. 425-454, 2006.

[20] W. Hu, K. S. Ishii, Q. Fan, and A. T. Ohta, "Hydrogel microrobots actuated by optically generated vapour bubbles," *Lab on a Chip,* vol. 12, pp. 3821-3826, 2012.

[21] W. Hu, K. S. Ishii, and A. T. Ohta, "Micro-assembly using optically controlled bubble microrobots," *Applied Physics Letters,* vol. 99, p. 094103, 2011.

[22] W. Hu, K. S. Ishii, and A. T. Ohta, "Micro-assembly using optically controlled bubble microrobots in saline solution," in *IEEE International Conference on Robotics and Automation,* 2012, pp. 733-738.

CONTACT

S. K. Chung, Tel: +82-31-330-6346; skchung@mju.ac.kr

A THERMOTROPIC LIQUID CRYSTAL ELASTOMER MICRO-ACTUATOR WITH INTEGRATED DEFORMABLE MICRO-HEATER

S. Petsch[1], R. Rix[2], Patrick Reith[1], Bilal Khatri[1], Stefan Schuhladen[1], Dominic Ruh[1], Rudolf Zentel[2], and H. Zappe[1]

[1]Department of Microsystems Engineering – IMTEK, University of Freiburg, GERMANY
[2]Institute of Organic Chemistry, University of Mainz, GERMANY

ABSTRACT

We present a liquid crystal elastomer (LCE) actuator with large stroke and fast reaction time. LCEs show a large macroscopic shape change when heated above the phase transition ($\approx 120\,°C$). Buried wafer-level fabricated micro-heaters offer optimal thermal reaction times and compact design of the actuators. A relative length change of $\lambda = 1.28$ was obtained with 320 mW power consumption. Heating the device from room temperature takes $\tau_{rise} = 19.7$ s, cooling below the phase transition temperature from the fully contracted state needs $\tau_{fall} = 5.6$ s. We verify that the displacement may be accurately controlled by varying electrical input power.

INTRODUCTION

Artificial muscle actuators, first envisaged by de Gennes in 1993, undergo large displacement, in a manner similar to mammalian muscles, when exposed to external stimuli [1, 2, 3]. Potential applications are found in biomimetics, robotics and endoscopic devices for minimal invasive surgery, i.e. all applications where large displacements are necessary and space is limited. Proposed MEMS applications may be found in a wide variety of research fields [4, 5, 6]. We intend to use artificial muscles as actuators for biomimetic solid-body elastomeric lenses [7].

Many different approaches for muscle-like actuators have been studied: thermal shape memory alloys (SMA) show high work densities but only moderate contraction, up to 8%. The continuous control of the displacement is challenging due to the narrow phase transition region. Moreover, integration of SMAs into MEMS systems remains challenging. Dielectric elastomer actuators, relying on electrostatic attraction, are able to provide very large displacements, up to 380%, and high work densities, but require extremely high electric fields. This limits the range of applications for such actuators.

LCEs combine the self-organizational properties of liquid crystals with the mechanical properties of elastomers. In an LCE, anisotropic mesogens (the basis of a liquid crystal) are covalently linked to cross-linked polymer chains, as shown in Figure 1. At low temperatures, the mesogens are in the nematic phase (Fig. 1a), an ordered phase in which the mesogens are aligned; the polymer chains to which they are bound are thus stretched. A transition to the isotropic phase takes place at higher temperatures: the mesogens are no longer aligned and the polymer chains contract into a disordered coiled state. It is the transition between these two

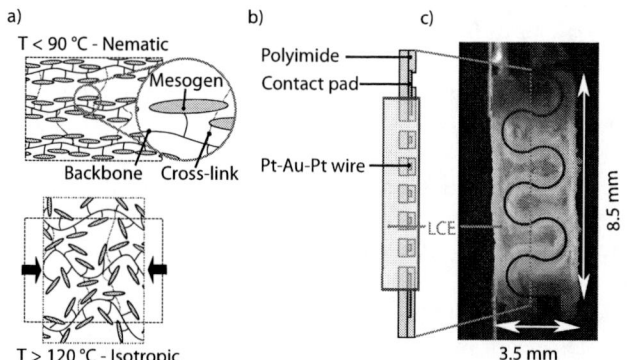

Figure 1: Illustration of the thermotropic actuator with integrated deformable wires. (a) Illustration of the liquid crystal elastomer actuation mechanism. At low temperatures the polymer chains are in the oriented state, imprinted by cross-linking. At the phase transition, the mesogens lose their order and the polymer chains coil, leading to a macroscopic shape change. (b) Cross-sectional view of the system: deformable heaters with robust polyimide cladding are buried in the LCE. (c) Photograph of the completed actuator. The deformable heater is adapted to the size of the LCE stripe for homogeneous heat distribution.

phases which forms the basis for the mechanical actuation employed here.

The preferred deformation direction of the LCE is orientated by applying a magnetic field to the material while cooling the liquid crystal monomer from the isotropic into the liquid crystalline, nematic phase. Well-oriented LCEs show contraction up to 35% and high work densities [8, 9, 10].

Despite the advantages of LCEs as artificial muscle actuators, reliable and homogeneous heating has not yet been achieved. Direct Joule heating of the material, using conductive PEDOT:PSS layers deposited on top of the LCE, results in considerable stiffening of the active material and has therefore only been shown to be useful for bending operation [11]. In contrast to the LCE thin film stacks with integrated NiCr heaters published before, we use here robust wafer-level fabricated heaters [12]. This new approach results in faster response times, despite thicker actuator films and a higher phase transition temperature. The heater design is easily adaptable to the requirements of a specific application, since structuring is done by standard MEMS processes.

The orientation of the LCE samples is defined by moderate magnetic fields. In this paper, we use rectangular shape actuator stripes and horseshoe shaped platinum-gold-

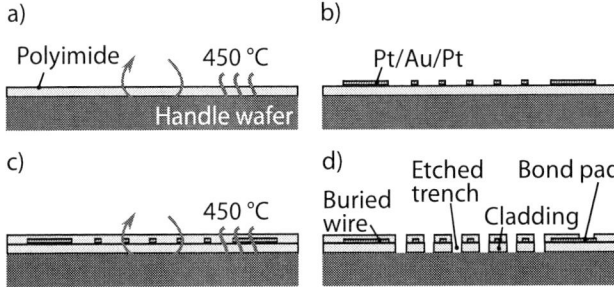

Figure 2: Schematic view of the fabrication of the deformable conductive heaters. (a) Spin-Coating of 5 μm PI and curing for 10 min at 450 °C. (b) The metal conducting lines are structured by evaporation and lift-off. (c) A second PI layer covers the metal structures. (d) The PI cladding is defined by reactive ion etching. The bond pads are opened in the same process step.

platinum deformable micro-heaters with electrical resistance $R = 30\,\Omega$, as depicted in Figure 1b and c. The overall fabrication process of the actuator is straightforward and, due to the very compact integration of the heater inside the active material, adaptable to a wide range of applications.

FABRICATION
Deformable heaters

We embed deformable conductive heater structures, based on a robust polyimide (PI) thin film technology published previously [13, 14], directly into the LCEs. This technology yields reproducible and robust deformable structures with high conductivity compared to other flexible wiring technologies. Lithographic structuring allows a large range of shapes to be fabricated and resistances to be realized.

The essential steps of the fabrication process are depicted in Figure 2. First step we deposit a 5 μm thick layer of polyimide (PI) (U-Varnish-S, UBE, Tokyo, Japan) by spin-coating on a silicon handle wafer (Fig. 2a). After curing for 10 min at 450 °C in a nitrogen atmosphere, the metal conducting lines are structured by evaporation and lift-off (Fig. 2b). Platinum (Pt) acts as an adhesion promoter to polyimide; gold (Au) may be used to increase the conductivity.

A second layer of PI, burying the metal structures, is spin-coated and cured on top (Fig. 2c). At the end, the PI cladding is structured by reactive ion etching (Fig. 2d). Since platinum acts as an etch stop layer against the reactive oxygen plasma, the contact pads are uncovered in the same process step. A photograph of the structures is shown in Figure 3. Due to low adhesion of the cured PI to the handle wafer, the structures are easily detachable.

Liquid crystal elastomer actuator

Subsequently, the liquid crystal elastomer (LCE) is cross-linked around these just-fabricated deformable polyimide conductive tracks to create optimal thermal contact between the two. First, the PI foil is

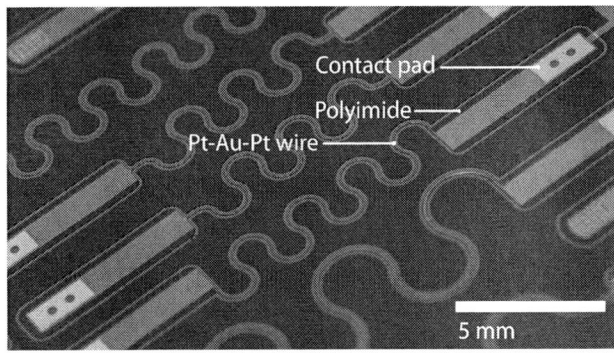

Figure 3: Photograph of the deformable thin film heaters embedded in polyimide on the handle wafer.

placed into a poly(dimethylsiloxane) (PDMS) mold (Fig. 4a). The LCE monomer (4'-acryloyloxybutyl 2,5-(4'-butyloxybenzoyloxy)benzoate is synthesized as described by Thomsen [9]. It is mixed with 10 mol % 1,6-hexanediol diacrylate (Alfa Aesar GmbH & Co KG, Karlsruhe, Germany) and 1 mol % Diphenyl(2,4,6-trimethylbenzoyl)phosphinoxid (Sigma-Aldrich Co. LLC., St. Louis, MO, USA). The mixture, which is solid at room temperature, is poured into the 95 °C preheated mold. When reaching the melting point, the LCE precursor wets the PDMS surface immediately and results in an even surface if the right amount of precursor is filled into the mold.

Letting the mold slowly cool to 60 °C under application of a magnetic field results in an orientation of the mesogens parallel to this applied field (Fig. 4b). Due to the phase transition to the nematic phase, this macroscopic orientation is maintained. Exposure to UV radiation (365 nm) then leads to polymerization and cross-linking of the oriented liquid crystal monomer (Fig. 4c). This leads to polymer chains oriented parallel to the magnetic field in the nematic low temperature phase of the LCE.

After demolding, the actuator is ready for use. When heated above phase transition temperature, the contraction of the actuator stripe takes place in the orientation direction (Fig. 4d).

EXPERIMENTAL SETUP

To quantify the movement of the actuator, we used the automatic measurement setup depicted in Figure 5. One end of the LCE actuator was held at a fixed position and a constant 2 g test load was applied to the other end. A digital camera (PL-A741, PixeLINK Ottawa (ON), Canada) equipped with a C-Mount objective (Xenon 25/0.95, Schneider Kreuznach, Bad Kreuznach, Germany) automatically recorded images of the dissension with high magnification at a rate of approximately 1 Hz.

The absolute displacement ΔL is calculated by

$$\Delta L = (x_{T_{Max}} - x_{RT}) \cdot M, \qquad (1)$$

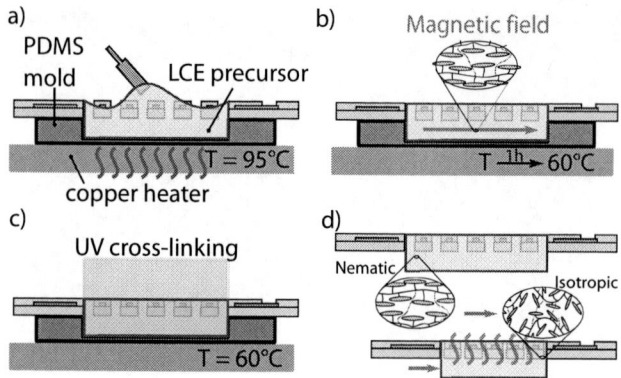

Figure 4: Schematic view of the fabrication of LCE actuators with buried heaters. (a) Pouring the LCE precursor in a preheated PDMS mold (95 °C). (b) Let the mold slowly cool down to 60 °C in presence of a magnetic field for orientation of the liquid crystaline monomer. (c) Photopolymerization with UV light at 365 nm. (d) Heating the LCE above the phase transition temperature, results in contraction along the orientation axis.

where x is the position of the edge in pixels and M the magnification in $\frac{mm}{px}$. The relative length λ of the actuator strip is then defined as

$$\lambda = \frac{L}{L_{Iso}} = 1 + \frac{\Delta L}{L_{Iso}}, \qquad (2)$$

where L denotes the current length of the actuator and L_{Iso} the length of the actuator when completely contracted.

In a second measurement, the force generated by the actuator was measured by attaching the second end of the actuator to a force sensor with a resolution of 2.5 mN (KD-78, ME-Messsysteme Gmbh, Hennigsdorf, Germany).

RESULTS

Applying 320 mW to the microheater results in complete contraction of the actuator. The time-dependent contraction and passive relaxation of this actuator is shown in Figure 6.

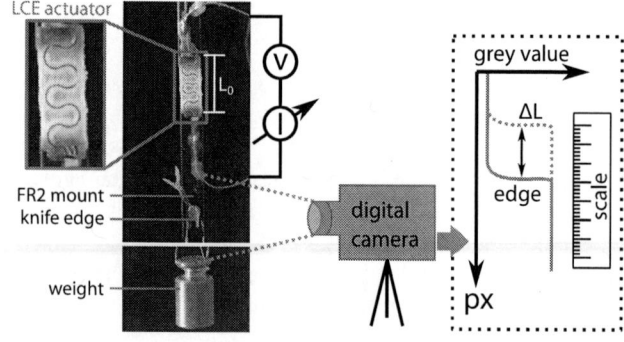

Figure 5: Illustration of the experimental setup. A digital camera records images of a knife edge suspended to the movable end of the actuator. The current position of the edge is calculated by edge filtering of the image.

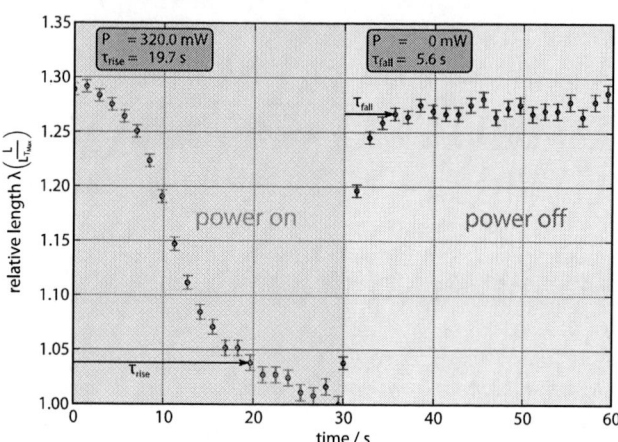

Figure 6: Contraction of the LCE actuator under application of 320 mW of electrical power to the micro-heater. 90% contraction is reached after 19.7 s. 90% relaxation is reached 5.6 s after switching off the power.

Starting from room temperature, the actuator shows continuous increase of displacement with time. When the thermal equilibrium is reached, the movement stops. The actuator reaches 90% contraction after $\tau_{rise} = 19.7$ s. Since the actuator starts cooling just above the phase change temperature, contraction happens much faster; 90% relaxation is reached $\tau_{fall} = 5.6$ s after switching off the power.

Figure 7 shows the LCE actuator at different times during the actuation cycle. The polyimide heater structure follows the movement of the surrounding. The transparency of the LCE in the contracted state indicates the transition to the isotropic phase.

The shape of the LCE depends on the Liquid Crystal order, which varies with temperature, whereby the steepest change happens around the phase transition temperature. In addition, the phase transition temperature varies, to some extent, over the actuator length due to differences in the cross-linking density. Both effects combine to lead to a change of

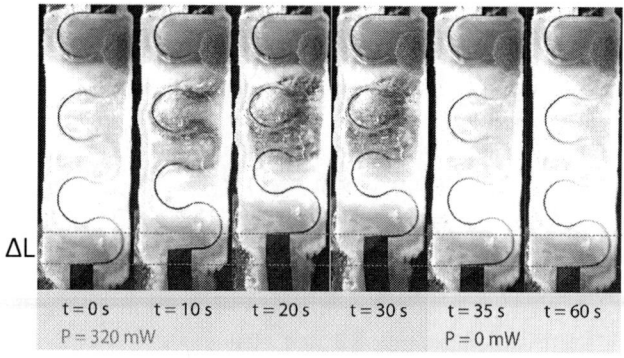

Figure 7: Photographs of the LCE micro-actuator at different times of the contraction and relaxation cycle for 320 mW. Transparent regions in the LCE material indicate the phase transition to the isotropic phase. The deformation of the integrated heater is clearly visible.

978-1-4799-3510-9/14 $31.00 © 2014 IEEE

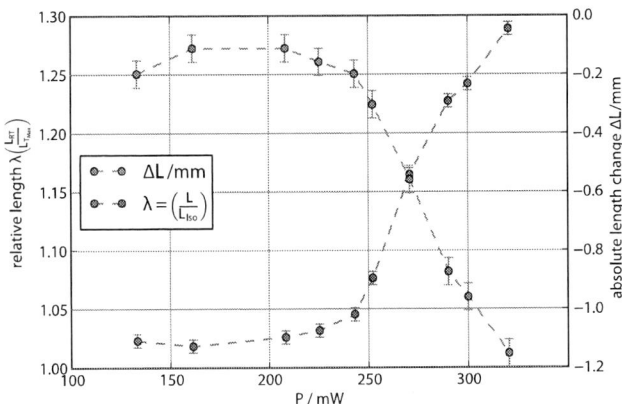

Figure 8: Maximum measured contraction of the LCE actuator for different applied heating powers. Maximum stroke is -1.15 mm, corresponding to $\lambda = 1.28$.

the actuator shape over a broad temperature range, allowing control of actuator displacement with applied heating power.

Figure 8 shows the maximum relative and absolute contraction as a function of applied electrical power. Up to 243 mW, only small movement is observed. At higher powers, contraction varies strongly with power. At 320 mW, a contraction of 1.15 mm was measured, which corresponds to $\lambda = 1.28$. The energy density,

$$\frac{W_{pot}}{V} = \frac{m \cdot g \cdot \Delta L}{V}, \tag{3}$$

with a calculated value of $1.52\,J/mm^3$, is almost as high as for SMAs, which are at present unmatched in work density [15].

A maximum generated force of 45 mN, corresponding to 25.7 kPa of stress, was determined for this actuator in the force sensor measurement.

CONCLUSIONS

A novel thermotropic actuator with large displacement and high work density has been presented. The combination of LCEs as active material and robust MEMS-fabricated deformable heaters, yields a compact integrated actuator, ideal for all applications where large displacements and high work densities are required in a confined space. The characteristics of this actuator are similar to those of mammalian muscles, which makes it suitable for biomimetic applications. Design variations of the flexible conductors may not only be undertaken to fit a specific actuator size, but also to implement sensing elements for smart actuators working in closed loop operation.

REFERENCES

[1] P. G. de Gennes, J. Prost, *The Physics of Liquid Crystals*, 2nd ed., ser. International Series of Monographs on Physics (83). Oxford: Clarendon Press, 1993.

[2] R. Zentel, "Liquid Crystalline Elastomers," *Angew. Chemie*, vol. 101, no. 10, pp. 1437–1445, Oct. 1989.

[3] H. R. Brand and H. Finkelmann, *Handbook of Liquid Crystals*, Demus, D. and Boodby, J. and Gray, G. W. and Spiess, H. W., Ed. Wiley-VCH, 1998, vol. 3, ch. 5.

[4] A. Sánchez-Ferrer *et al.*, "Photo-Crosslinked Side-Chain Liquid-Crystalline Elastomers for Microsystems," *Macromol. Chem. Phys.*, vol. 210, no. 20, pp. 1671–1677, Oct. 2009.

[5] C. J. Camargo *et al.*, "Batch fabrication of optical actuators using nanotube–elastomer composites towards refreshable Braille displays," *J. Micromechanics Microengineering*, vol. 22, no. 7, p. 075009, Jul. 2012.

[6] L. T. de Haan *et al.*, "Accordion-like Actuators of Multiple 3D Patterned Liquid Crystal Polymer Films," *Adv. Funct. Mater.*, pp. n/a–n/a, Oct. 2013.

[7] S. Schuhladen *et al.*, "Miniaturized tunable imaging system inspired by the human eye," *Opt. Lett.*, vol. 38, no. 20, p. 3991, Oct. 2013.

[8] M. Warner, E. M. Terentjev, *Liquid Crystal Elastomers*, ser. International Series of Monographs on Physics (120). Oxford: Clarendon Press, 2009.

[9] D. L. Thomsen *et al.*, "Liquid Crystal Elastomers with Mechanical Properties of a Muscle," *Macromolecules*, vol. 34, no. 17, pp. 5868–5875, Aug. 2001.

[10] E.-K. Fleischmann and R. Zentel, "Liquid-Crystalline Ordering as a Concept in Materials Science: From Semiconductors to Stimuli-Responsive Devices." *Angew. Chem. Int. Ed. Engl.*, pp. 8810–8827, Jul. 2013.

[11] F. Greco *et al.*, "Bending actuation of a composite liquid crystal elastomer via direct Joule heating," *2012 4th IEEE RAS EMBS Int. Conf. Biomed. Robot. Biomechatronics*, pp. 646–651, Jun. 2012.

[12] C. M. Spillmann *et al.*, "Stacking nematic elastomers for artificial muscle applications," *Sensors Actuators A Phys.*, vol. 133, no. 2, pp. 500–505, Feb. 2007.

[13] B. Rubehn *et al.*, "A MEMS-based flexible multichannel ECoG-electrode array," *J. Neural Eng.*, vol. 6, no. 3, p. 036003, Jun. 2009.

[14] R. Verplancke *et al.*, "Thin-film stretchable electronics technology based on meandering interconnections: fabrication and mechanical performance," *J. Micromechanics Microengineering*, vol. 22, no. 1, p. 015002, Jan. 2012.

[15] J. Madden, "Artificial muscle technology: physical principles and naval prospects," *IEEE J. Ocean. Eng.*, vol. 29, no. 3, pp. 706–728, Jul. 2004.

CONTACT

* S. Petsch, tel: 0049-761-203-7574; sebastian.petsch@imtek.uni-freiburg.de

A TUNABLE LIQUID LENS DRIVEN BY A CONCENTRIC ANNULAR ELECTROACTIVE ACTUATOR

Kang Wei, Nicholas W. Domicone, and Yi Zhao
Department of Biomedical Engineering, the Ohio State University, Columbus, USA

ABSTRACT

We present a membrane sealed fluidic lens that is hydrostatically connected to a concentric ring-shaped electroactive elastomer. Electrical activation deforms the ring-shaped elastomer, which induces fluid passage between the lens part and the actuation part of the membrane. This changes the lens's shape and therefore the focal length. The focal length ranges from 25.4 mm to 105.2 mm can be obtained with a voltage bias of 1.0 kV. Comparing to the existing fluidic lenses driven by electroactive polymers, this lens has a wider range of refractive power at a significantly lower actuation voltage. Our device finds applications in miniaturized optical components where adaptive focalization is paramount.

INTRODUCTION

Fast and accurate focal adjustment is essential for a camera-lens imaging apparatus (webcam, digital camera, microscope, endoscope, etc.) to produce a sharp image of a subject and to distinguish subjects positioned at varied distances. Mechanical focusing has existed for years as a conventional way of focalization, where an electromagnet or a piezoelectric motor is utilized to alter the relative distance between adjacent lens units and thus the effective focal length [1, 2]. The compound-lens construction integrated with the driving mechanism usually makes the optical system heavy, bulky and costly. This hinders the miniaturization and integration within portable products such as cellular phone cameras, mini projectors, eyeglasses and other lab-on-chip components [3]. Evolved from the human camera eye, the fluidic lens has achieved adaptive focusing without any moving components. This is done by simply tuning the surface curvature of the fluid using electrowetting [4-7], dielectrophoretic force [8], ferrofluidic plug [9, 10], reduction-oxidation reaction [11] or acoustic oscillation [12], or tuning the surface curvature of the elastomeric membrane enclosing the optical medium by applying mechanical [13, 14], magnetic [15] or electric stimuli [16]. Among various mechanisms of adaptive focusing, dielectric elastomer (DE) actuators have emerged in recent years as good candidates for driving membrane-sealed fluidic lenses because of their compactness, operation efficiency, fast response and low cost [17]. Researchers have demonstrated a fluid-filled elastomeric lens surrounded by a ciliary-muscle like DE actuator, which, upon electrical activation, radially stretches and presses against the central lens to tune its surface profile and therefore the focal length [18]. An array of membrane lenses with one made into a DE actuator was also reported [19]. Electrical activation forces the DE actuator to buckle upward, which causes the other membrane lenses to deflate. Other examples include the use of transparent DE actuators that are directly made into tunable lenses [20, 21]. However, these existing devices driven by DE actuators often require a high input of voltage (up to 5.0 kV) and a large extent of DE pre-straining, to achieve a certain range of focal length[20, 21]. When transparent actuators are used, the mechanical instability of the DE at high voltages (i.e. surface wrinkling or non-concentric bulging) degrades the imaging quality because the DE actuator is along the optical axis of the fluidic lens.

In this study, we report a tunable fluidic lens driven by a concentric ring-shaped DE actuator that can achieve a large focal length change at relatively lower operation voltages, without the need for pre-staining of the DE membrane. By reducing the mechanical instability of the DE membrane, no significant optical aberrations were observed. This fluidic lens with better performance and enhanced reliability thus promises a potential solution of adaptive focusing in various miniaturized optical components.

FLUDIC LENS DESIGN

Device Configuration and Working Principle

The perspective view of the fluidic lens is shown in Figure 1. It is constructed from two supporting frames and a thin membrane of DE that is sandwiched in between. For better illustration, the top frame is not shown. The bottom supporting frame sealed by the DE dividing the entire device into two functional modules: a circular elastomeric fluidic lens in the center and a concentric ring-shaped DE actuator. The DE actuator comprises of a ring-shaped reservoir sealed by the annular section of the DE membrane (hereafter referred as the actuation membrane), whose top and bottom surfaces are coated with compliant electrodes (shown in the inset). The fluidic lens consists of the circular elastomeric membrane (hereafter referred as the lens membrane) and the underlying optical medium. The lens membrane is uncoated. The circular reservoir and the ring-shaped reservoir are connected by rectangular channels so that fluid can transport between the two reservoirs.

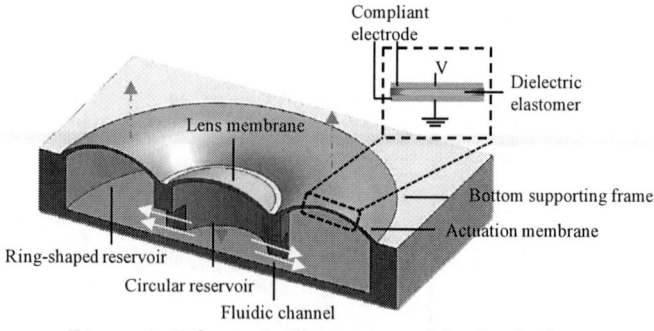

Figure 1: Schematic illustration of the fluidic lens.

It is known that a planar DE in combination with compliant electrodes expands laterally and contracts in thickness in response to an electrical signal. Because the actuation membrane here is constrained by the supporting frames at the borders and is subjected to a positive distributive load, the lateral expansion of the membrane leads to an upward deflection upon electrical activation. Consequently, the fluid is drained from the circular reservoir into the annular reservoir via the channels. This deflates the lens membrane and thereby reduces the lens sagitta (sag) of the fluidic lens (which is defined as the distance between the vertex of the lens membrane and the center of its base). The focal length changes as a function of the lens sag (Figure 2).

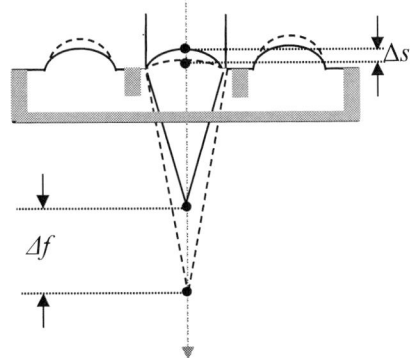

Figure 2: Principle of focal length adjustment

Theoretical Analysis of the Focal Length Adjustment

Since the out-of-plane deflection of the lens membrane is relatively small in comparison with its aperture, its surface profile can be simplified as a spherical cap. The focal length (*f*) of the fluidic lens in air is expressed as a function of the lens sag (*s*):

$$f = \frac{r^2}{2(n_1 - 1)s} \qquad (1)$$

where n_1 is the refractive index (RI), and r is the radius of the lens.

Because the lens membrane deflates upon activation, the initial sag of the fluidic lens determines its shortest focal length. The range of the tunable focal length is determined by the extent to which the lens sag decreases (Δs). It is also worth noting that for the same sag change, lenses of smaller initial sags have larger focal ranges than those of larger initial sags.

DEVICE FABRICATION AND ASSEMBLY

Silicon rubber (TC-5005 A/B-C, BJB Enterprises, CA) was selected as the material to construct the actuation membrane and the lens membrane. It possesses a lower Young's modulus and higher permittivity than other commercially available silicone-based DEs. It was prepared by mixing the resin ('A' component) with the curing agent ('B' component) at a 10:1 ratio by weight, and was placed in a vacuum for 20 minutes to remove air bubbles. The mixture was casted onto a glass wafer coated with a thin layer of photoresist (Shipley 1813, Microchem, MA) and spin-coated

at 2000 rpm/s. The photoresist reduced the adhesion between the prepolymer and the glass wafer. After curing at room temperature for 24 hours, the membrane has a thickness of 76 µm. To prepare the compliant electrode, carbon grease (846, MG Chemicals, Canada) was dispensed in isopropanol at the weight ratio of 1:3 and stirred on the hotplate for 1 hour. The supporting frames (top: 25 mm×25 mm×1.5mm; bottom: 25mm×25mm×5mm) were printed by a 3D printer (Objet 3D 24, Stratasys) and the bottom frame was bonded to the membrane on the glass wafer. The compliant electrode solution was deposited by an airbrush system onto the fluidic side of the ring-shaped region (*inner radius:* 4 mm; *outer radius:* 9 mm) of the membrane, which was defined by the frames. The uncoated region that served as the lens membrane had a diameter of 5 mm. Then the bottom frame together with the DE membrane was peeled off of the glass wafer, followed by electrode deposition on the air side of the ring-shaped region and the addition of the top frame. Copper foil tape was introduced to connect the carbon electrode. For good electrical connection, the contact point between the copper tape and the carbon electrode was manually wetted with the carbon grease. Finally, the bottom frame was sealed by a 1.2mm thick microscope slide. Figure 3 shows the assembled 3D-printed electroactive fluidic lens.

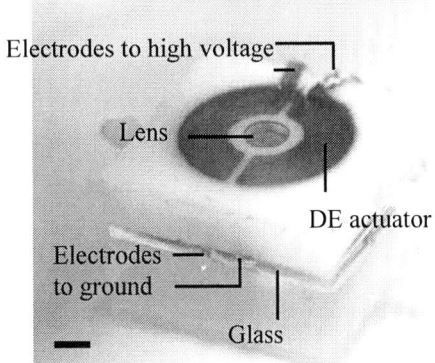

Figure 3: The assembled fluidic lens (scale bar: 5mm).

RESULTS AND DISCUSSION

Lens Sag Measurement

The reservoirs under the lens membrane and the actuation membrane were filled with $\geq$ 99wt% glycerol (G5516, Sigma Aldrich, MO) by a syringe pump (NE-1000, New Era Pump Systems Inc., NY). The glycerol has a refractive RI of 1.47, which is slightly larger than that of the DE (RI = 1.4). Its immiscibility with carbon electrodes and insignificant dielectric properties (ε_r = 42.5) also ensure the safety of electrical activation. After glycerol replaced all of the air in the reservoirs, the fluid outlet port was sealed and an extra volume of 96 µl of fluid was supplied through the inlet port to give the lens membrane an initial sag of 273 µm. During fluid addition, the lens sag was examined using an upright optical microscope with a 20X objective by focusing on the top surface of the lens membrane. Afterwards, the inlet port was also sealed so that the encapsulated fluid could only

978-1-4799-3510-9/14 $31.00 © 2014 IEEE 910

redistribute between the lens reservoir and the actuation reservoir via the channels.

The lens sag reduction was examined while the applied electrical voltage was changed from 0 kV to 1.0 kV at a 0.25 kV increment (Figure 4). The lens sag exhibited a nonlinear change upon electric activation. Voltages below 0.25kV changed the lens sag only to a modest extent. However, the slope increases with increasing voltage. The average total sag reduction at 1.0 kV was measured to be 210μm.

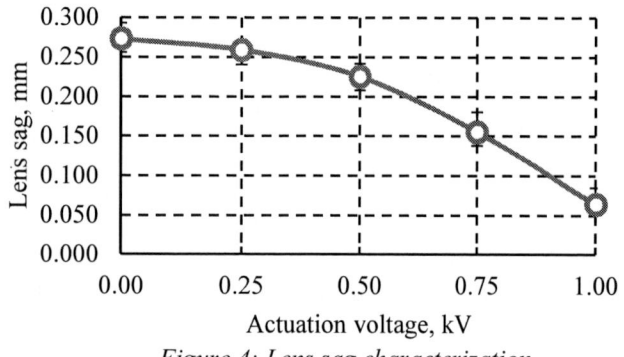

Figure 4: Lens sag characterization

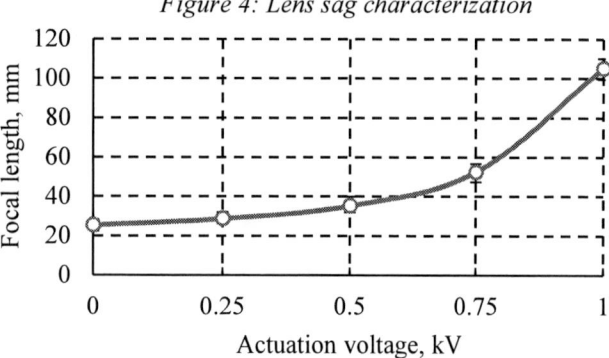

Figure 5: Focal length characterization.

Focal Length Measurement

The focal length change of the fluidic lens upon electrical activation was examined using a custom-built testing rig that consisted of a laser diode, a spatial filter, a plano-convex lens, and a detection screen. A green laser beam (wavelength: 532 nm) generated by the diode was converted into a spherical wave by the spatial filter. A plane wave was generated by a plano-convex lens. The detection screen was mounted on a linear translation stage. The focal length was approximated as the distance between the lens and the detection screen that yielded the minimum spot size of the laser. As expected, the focal length increased with the applied actuation voltage (Figure 5). The results also validated that the initial lens sag determined the maximum focusing power of the lens. Specifically, given the maximal allowable actuation voltage, the lens with an initial sag of about 273 μm exhibited a focal length change from about 25.4 mm to 105.2mm. This corresponded to more than 300% focal length variation. The result suggested a great performance improvement over prior studies where the largest focal length change reported is from about 36 mm to 75 mm at larger than 4.9 kV electrical activation, yet with image quality degradation [20].

Focusing/Defocusing Test

The focal length change was experimentally validated by observing characters printed on a transparency film (CG3300, 3M, MN) through the fluidic lens (Figure 6a). The transparency film was positioned in front of the lens to bring the characters "OSU" on the film into focus when the lens was set at an initial sag of 325 μm. As an actuation voltage of 1.0 kV was applied, the characters on the film were significantly blurred, which suggested the change of focal length (Figure 6b).

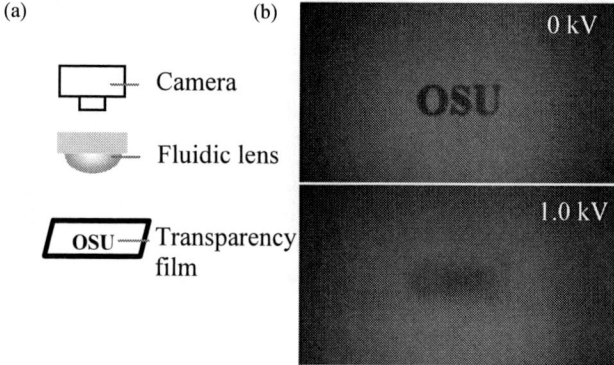

Figure 6: Single object focusing/defocusing test: (a) the optical testing rig and (b) the acquired images showing focusing and defocusing at varied voltages.

The ability of the fluidic lens to change its focal length to distinguish objects positioned at different depths was also demonstrated. Two transparency films printed with the letters 'IO' and 'OH' were used as the subjects and positioned at a distance of 101mm and 90.5mm from the lens, respectively. The fluidic lens was actuated to focus on the letters at different focal planes, and a CCD camera helped acquire the images captured by the lens. The lens, at rest, was first used to focus on the letters 'OH' while the distant letters 'IO' appeared blurred (Figure 7a). At around 0.5 kV of electrical activation, the image of the letters 'IO' became acceptably clear, while the image of the letters 'OH' was significantly blurred (Figure 7b). The fluidic lens successfully distinguished the letters positioned at two different focal planes by changing the actuation voltage. Since the DE membrane in this study was not pre-strained, no surface wrinkling was observed on the lens membrane. As a result, no significant optical aberration was observed during the electrical activation.

CONCLUSIONS

In summary, an adaptive fluidic lens driven by a concentric ring-shaped DE actuator was reported in this paper. The results showed that such design is able to implement a large focal range at a relatively lower actuation voltage compared to existing DE-driven fluidic lenses. Since no-prestraining of the DE membrane is needed, the mechanical instability of the lens is greatly reduced. The device's simple and compact construct, low cost and actuation efficiency promises wide industrial, medical and consumer applications.

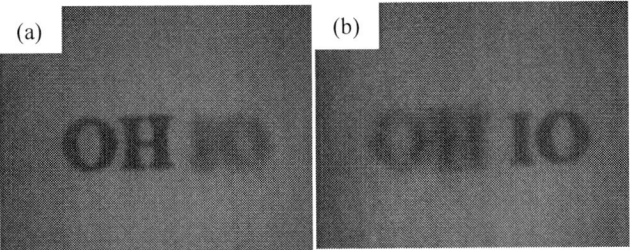

Figure 7: Two objects focusing/defocusing: (a) 0 kV, and (b) ~0.5kV.

ACKNOWLEDGEMENTS

This work was partially funded by a NSF grant under the award number 1138236. The authors also thank the HHMI Med into Grad program and Pelotonia program for the student fellowship supports. The authors also thank Matthew Steven Rudy for his technical support in experiments.

REFERENCES

[1] C.-S. Liu, S.-S. Ko, and P.-D. Lin, "Experimental characterization of high-performance miniature auto-focusing VCM actuator," *Magnetics, IEEE Transactions on,* vol. 47, no. 4, pp. 738-745, 2011.

[2] H.-C. Yu, T.-Y. Lee, S.-J. Wang, M.-L. Lai, J.-J. Ju, D.-R. Huang, and S.-K. Lin, "Design of a voice coil motor used in the focusing system of a digital video camera," *Magnetics, IEEE Transactions on,* vol. 41, no. 10, pp. 3979-3981, 2005.

[3] H. Ren, and S.-T. Wu, *Introduction to adaptive lenses,* Hoboken, N.J.: Wiley, 2012.

[4] B. Berge, and J. Peseux, "Variable focal lens controlled by an external voltage: An application of electrowetting," *Eur. Phys. J. E,* vol. 3, no. 2, pp. 159-163, 2000/10/01, 2000.

[5] T. Krupenkin, S. Yang, and P. Mach, "Tunable liquid microlens," *Appl. Phys. Lett.,* vol. 82, no. 3, pp. 316-318, 2003.

[6] S. Kuiper, and B. H. W. Hendriks, "Variable-focus liquid lens for miniature cameras," *Appl. Phys. Lett.,* vol. 85, no. 7, pp. 1128-1130, 2004.

[7] R. Shamai, D. Andelman, B. Berge, and R. Hayes, "Water, electricity, and between... On electrowetting and its applications," *Soft Matter,* vol. 4, no. 1, pp. 38-45, 2008.

[8] H. Ren, H. Xianyu, S. Xu, and S.-T. Wu, "Adaptive dielectric liquid lens," *Opt. Express,* vol. 16, no. 19, pp. 14954-14960, 2008.

[9] B. A. Malouin Jr, M. J. Vogel, J. D. Olles, L. Cheng, and A. H. Hirsa, "Electromagnetic liquid pistons for capillarity-based pumping," *Lab Chip,* vol. 11, no. 3, pp. 393-397, 2011.

[10] W. Xiao, and S. Hardt, "An adaptive liquid microlens driven by a ferrofluidic transducer," *J. Micromech. Microeng.,* vol. 20, no. 5, pp. 055032, 2010.

[11] C. A. López, C.-C. Lee, and A. H. Hirsa, "Electrochemically activated adaptive liquid lens," *Appl. Phys. Lett.,* vol. 87, no. 13, pp. 134102-134102-3, 2005.

[12] C. A. López, and A. H. Hirsa, "Fast focusing using a pinned-contact oscillating liquid lens," *Nature Photon.,* vol. 2, no. 10, pp. 610-613, 2008.

[13] G. Beadie, M. Sandrock, M. Wiggins, R. Lepkowicz, J. Shirk, M. Ponting, Y. Yang, T. Kazmierczak, A. Hiltner, and E. Baer, *Tunable polymer lens,* DTIC Document, 2008.

[14] K. Wei, and Y. Zhao, "A three-dimensional deformable liquid lens array for directional and wide angle laparoscopic imaging," *IEEE, Micro Electro Mechanical Systems (MEMS),* pp. 133-136, 2013.

[15] S. W. Lee, and S. S. Lee, "Focal tunable liquid lens integrated with an electromagnetic actuator," *Appl. Phys. Lett.,* vol. 90, no. 12, pp. 121129-121129-3, 2007.

[16] S. T. Choi, J. Y. Lee, J. O. Kwon, S. Lee, and W. Kim, "Varifocal liquid-filled microlens operated by an electroactive polymer actuator," *Opt. Lett.,* vol. 36, no. 10, pp. 1920-1922, 2011.

[17] P. Brochu, and Q. Pei, "Advances in dielectric elastomers for actuators and artificial muscles," *Macromol. Rapid Commun.,* vol. 31, no. 1, pp. 10-36, 2010.

[18] F. Carpi, G. Frediani, S. Turco, and D. De Rossi, "Bioinspired Tunable Lens with Muscle-Like Electroactive Elastomers," *Adv. Funct. Mater.,* vol. 21, no. 21, pp. 4152-4158, 2011.

[19] M. Niklaus, S. Rosset, and H. Shea, "Array of lenses with individually tunable focal-length based on transparent ion-implanted EAPs," *SPIE Smart Structures and Materials,* pp. 76422K-76422K-12, 2010.

[20] S. Shian, R. M. Diebold, and D. R. Clarke, "Tunable lenses using transparent dielectric elastomer actuators," *Opt. Express,* vol. 21, no. 7, pp. 8669-8676, 2013.

[21] S.-i. Son, D. Pugal, T. Hwang, H. R. Choi, J. C. Koo, Y. Lee, K. Kim, and J.-D. Nam, "Electromechanically driven variable-focus lens based on transparent dielectric elastomer," *Appl. Opt.,* vol. 51, no. 15, pp. 2987-2996, 2012.

CONTACT

*Y. Zhao, tel: +1-614-2477424; zhao.178@osu.edu

LONG STROKE OUT-OF-PLANE ACTUATOR USING COMBINATION OF ELECTROSTATIC AND PNEUMATIC FORCES

Tetsuo Kan[1], Akihiro Isozaki[1], Hidetoshi Takahashi[1], Kiyoshi Matsumoto[2], and Isao Shimoyama[1, 2]

[1]Department of Mechano-Informatics, Graduate School of Information Science and Technology,
The University of Tokyo, Tokyo, Japan
[2]IRT Research Initiative, The University of Tokyo, Tokyo, Japan

ABSTRACT

We propose an out-of-plane MEMS actuator with a large stroke length comparable to the size of the actuator itself. The proposed device is actuated by the combined forces of the electrostatic and the pneumatic forces. This combination of two independent forces enlarges a stable area during the actuation, resulting in a large stroke which is difficult to be achieved with only either force. The 3D profiling by the laser scanning microscopy confirmed the largest stroke of 103 μm was obtained with the 150-μm-diameter actuator area with the electric field of 5.8×10^5 V/m (330V between the electrodes) and the air pressure of 2.0 kPa.

INTRODUCTION

Enlargement of the stroke length of the actuator is one of fundamental properties pursued in the MEMS field. There have been proposed many out-of-plane MEMS actuators with a long stroke [1-8]. Most of the out-of-plane actuators have employed the electrostatic force to move the structures. Although the electrostatic actuation can offer many fascinating advantages such as low power consumption and fast actuation speed, it has a major drawback concerning the pull-in effect, limiting the obtainable stroke[9]. Due to this effect, the position of the actuator becomes unstable when the working voltage between the actuator and the electrode is near the pull-in voltage.

In this paper, we propose an out-of-plane MEMS actuation with not only the electrostatic force but also the pneumatic force. Because these two forces are based on different physics, the limit of the controllable area of each force is relatively independent. Thus, it is possible to combine a largest stroke with each different force to increase the attainable stroke further. In order to verify this concept, we prepared a planar spiral structures as a test type actuator which can be actuated in the out-of-plane direction with both electrostatic and pneumatic forces. We applied two forces simultaneously, and checked whether the attainable stroke increased within a stable condition by the combination of two forces.

ACTUATION PROCEDURE

A structure of a proposed actuator is a planar archimedian spiral formed on a thin suspended Si membrane (fig. 1 left). A counter planar electrode is placed above the spiral structure with the gap of several hundreds of micrometer. The spiral structure and the counter electrode are electrically connected so that the tunable voltage can be applied between them. When there is a voltage difference between the actuator and the counter electrode, the electrostatic force is generated so that the planar spiral is elevated toward the counter electrode.

The pneumatic force, one of the two actuation forces, is applied from the backside of the spirals by supplying the air pressure to the backside of the spiral. The pressure difference between the upper and bottom sides of the spiral actuate the structure[10][11]. Simultaneous application of the electrostatic and pneumatic force makes an out-of-plane actuation by these two forces.

The large stroke actuation often suffers from instability. For example, the pull-in effect is a well-known limit of the

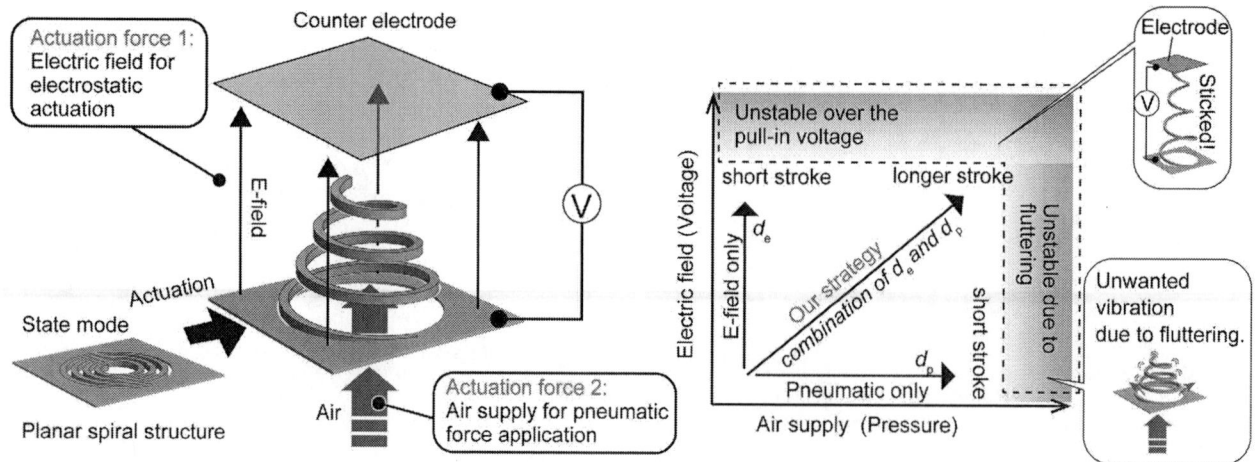

Figure 1. Configuration of the proposed out-of-plane actuator, which is deformed by the application of two forces; electrostatic force and pressure.

978-1-4799-3510-9/14 $31.00 © 2014 IEEE

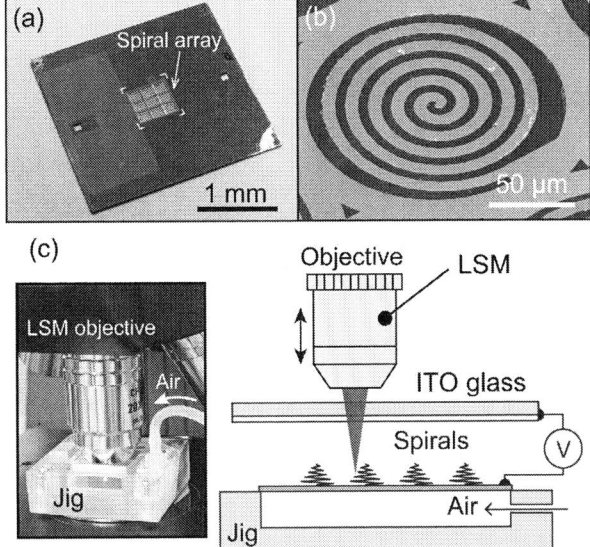

Figure 2. (a) An SEM image of the fabricated spiral structure, (b), (c) A photograph and a schematic of the experimental setup.

electrostatic actuation, where the actuator is stuck to the counter electrode above the pull-in voltage (see fig. 1 upper right). In addition, fluttering of the structure is seen when the excessive pressure is applied (see fig. 1 lower right). In the fluttering pressure zone, the structure vibrates so that the amplitude of the actuation stroke saturates there. The fluttering is thought to be attributed to the turbulence caused by the leak air from gaps between the actuator beams above the pressure limit.

If we define d_e and d_p as the displacements obtainable just before these limits, our strategy can provide the longest stroke of $d_e + d_p$, indicated by an diagonal arrow in a graph of fig. 1 right. This longer stroke cannot be obtained for the actuation with either one force alone.

FABRCATION AND SETUP

The fabricated structure was a planar 5-turn spiral on a 300-nm-thick suspended Si membrane. The diameter and the beam width of the spiral were 150 μm and 8μm, respectively. A similar structure was reported in [12][13]. A starting material was an SOI wafer (top 300 nm/ BOX 400 nm/ handle 200 μm). The spiral structures were first formed by RIE on a top Si membrane. Then, the handling Si and BOX layers beneath the spiral structures were removed by RIE and HF vapor etching, respectively, so that the structures became free-standing on a suspended top Si membrane. The actuators were formed on a 1 × 1 inch square chip. The device area is seen around the center of the chip in fig. 2(a). An SEM image of the spiral structure is shown in fig. 2(b).

The fabricated device was mounted on an experimental jig, which had an air supply channel and an air chamber as shown in fig. 2(c). The counter electrode was placed above the device chip with a spacing of 569 μm. An ITO coated glass was employed as a material of the electrode such that

the laser light (λ = 488 nm) went through it while the laser scanning microscopy (LSM). Air pressure was supplied from a N_2 cylinder, and the pressure was regulated to be around several kPa, monitored by a digital pressure sensor. Both the air pressure and the electric voltage were supplied in a DC manner.

EXPERIMENTAL RESULTS

Displacement measurement

The displacement of the center of the spiral was measured as an actuator stroke by a LSM. The obtained LSM images are shown in fig. 3(a). The actuator was initially flat without the pressure and voltage (fig. 3(a)-(α)). When the

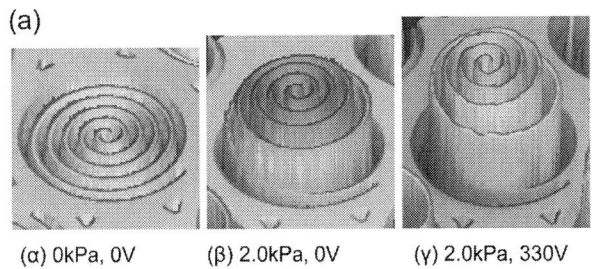

(α) 0kPa, 0V   (β) 2.0kPa, 0V   (γ) 2.0kPa, 330V

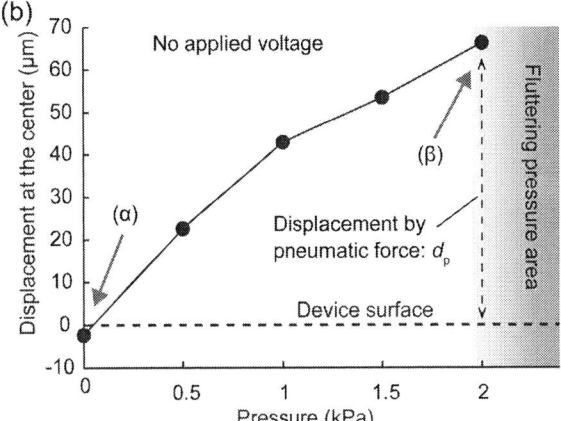

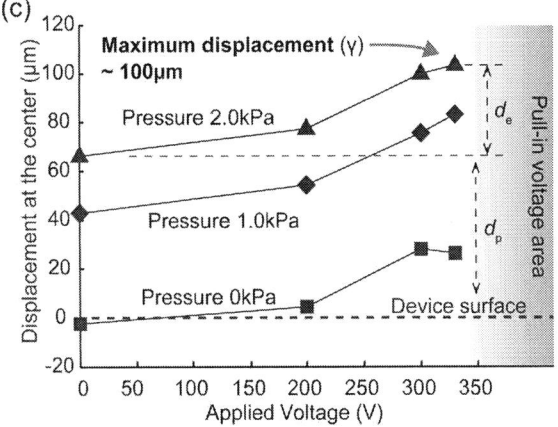

Figure 3. (a) Laser profile images of the spiral actuators with three different conditions, (b) The experimentally obtained deformation property with respect to the applied pressure, (c) The experimentally obtained deformation properties with respect to the applied voltage while keeping the applied pressure constant.

978-1-4799-3510-9/14 $31.00 © 2014 IEEE

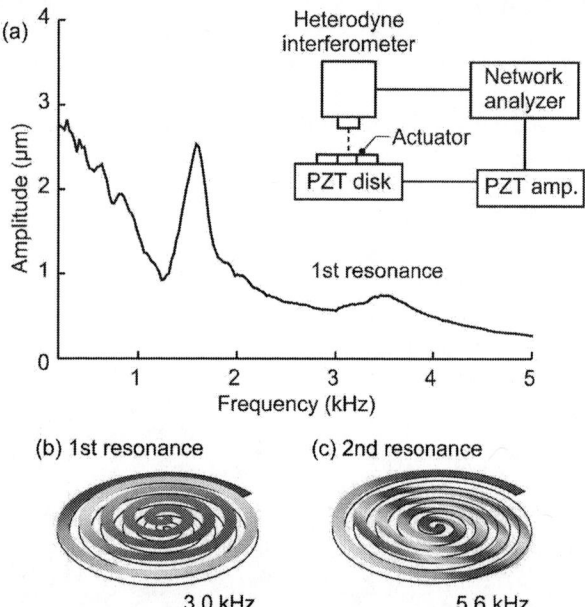

Figure 4. (a) Mechanical resonance point measurement by a heterodyne interferometer, (b), (c) COMSOL Multiphysics FEM simulations for the mode analysis for 1st and 2nd resonance, respectively.

pressure increased to be 2.0 kPa, the structure went up to form a three dimensional spiral (fig. 3(a)-(β)). The actuation property with respect to the applied pressure was plotted on fig. 3(b). The displacement at the center of the spiral was measured as an index of the actuator stroke. In this case, no voltage was applied during the measurement. The Si membrane surface was regarded as an origin of the displacement, and was indicated as a horizontal dotted line in the graph. When the pressure was applied, the displacement monotonically increased. The spiral started fluttering at the pressure over 2.0 kPa. The displacement d_p reached 66 μm at the air pressure of 2.0 kPa (fig. 3(a)-(β)).

In order to verify whether the stroke can be enlarged with the other force application, the electrostatic force was then applied in addition to the pneumatic force. In this time, the pressure was kept constant, and only the voltage was changed while the displacement measurement. Three air pressure conditions were adopted; 0, 1.0, and 2.0 kPa (fig. 3(c)). In all pressure conditions, the displacement almost monotonically increased with an increase in the voltage, and the pull-in voltage was about 330 V, which corresponded to 6.2×10^5 V/m corresponding to the fig. 3(a)-(γ). It was found that the displacement by the electrostatic force d_e was almost constant to be around 37 μm regardless of the air pressure. Trend of the electrostatic actuation was thus concluded to be not influenced by the initial out-of-plane displacement around the measured range. The combined maximum displacement, which is the combination of the stroke $d_p + d_e$, was 103 μm with 2.0 kPa and 330V applied. We therefore could obtain a large out-of-plane stroke which could not be attainable with only either forces, the pneumatic or the electrostatic forces.

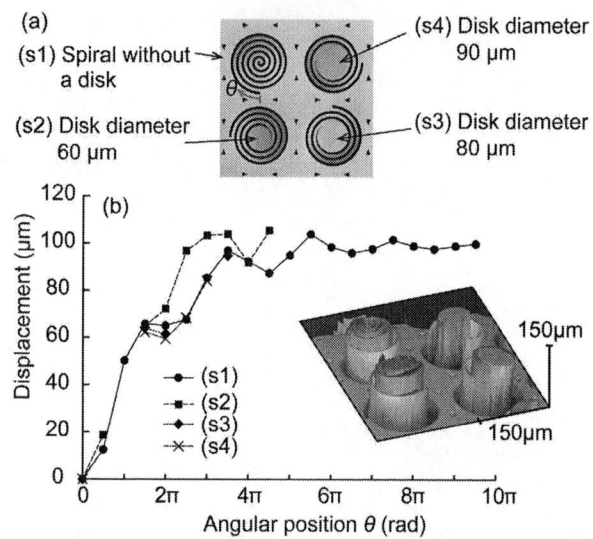

Figure 5. (a) Four different shapes of spiral actuators, (b) Angular height profile of the four spirals.

Mechanical resonance

The mechanical resonance was measured with a heterodyne interferometer to investigate the dynamic response of the actuator, shown in fig. 4(a). The frequencies of the first and second resonances were seemingly found around 1.6 kHz and 3.6 kHz. In order to verify that these resonant points are actually reflecting the vibration, finite element method calculations were conducted using COMSOL Multiphysics ver. 4.3. The prepared physical model was constructed as being composed of an isotropic media for simplification, having Young's modulus of 140×10^9 Pa. The simulation results exhibited the first resonance point at 3.0 kHz, indicating that the resonance point found in the heterodyne interferometer measurement at around 1.6 kHz seemed to be an artifact. The first resonant point of the spiral structure was thus found at around 3.5 kHz.

Dependency on the spiral dimensions

We prepared four different shapes of spiral actuators to investigate the effect of the spiral shapes on the deformability. Parameter on the shape was the diameter of the disk located on center of the spiral structure; 0 (a normal spiral), 60, 80, 90 μm, corresponding to S1 through S4, respectively (fig. 5(a)). The displacement along the spiral beam angular position θ was measured (fig. 5(b)). The angular position θ is 0 rad at the root of the outermost beam, and increases as the spiral goes into the inner direction. The measurement was performed in a condition of 2.0 kPa and 330 V applications. There were found no significant difference in maximum displacement with respect to the actuator design. The deformation of the spiral structures almost saturated around 100 μm. It was suggested that the area of the spiral did not play a determinant role in increasing the stroke. It was observed that the almost all displacement was attributed the outer two spiral beams, i.e. first and second turn beams from the outer side.

978-1-4799-3510-9/14 $31.00 © 2014 IEEE

CONCLUSION

We proposed a MEMS actuation method which combines two different forces to increase the stroke of the actuator in the out-of-plane direction. In order to verify the effectiveness of this concept, we fabricated spiral actuators and actuated them with both the pneumatic and electrostatic forces. The proposed actuator exhibited large out-of-plane stroke as large as 103 μm with the air pressure of 2.0 kPa and the voltage of 330 V. This stroke is significantly large and comparable to the actuator size itself, 150 μm in diameter. This technology will contribute to the actuation of optical components, such as a mirror and a polarization modulator [12][13].

ACKNOWLEGEMENT

This research was partially supported by The Murata Science Foundation. The photolithography masks were made using the University of Tokyo VLSI Design and Education Center (VDEC)'s 8 inch EB writer F5112 + VD01 donated by ADVANTEST Corporation.

REFERENCES

[1] T. Fukushige and S. Hata, "A MEMS Conical Spring Actuator Array," *J. MEMS*, Vol. 14, No. 2, pp. 243-253, 2005.

[2] S. He and R. B. Mrad, "Large-Stroke Microelectrostatic Actuators for Vertical Translation of Micromirrors Used in Adoptive Optics," *J. Micromech. Microeng.*, vol. 52, No. 4, pp 974-983, 2005.

[3] S. He, R. B. Mrad, and J. Chong, "Repulsive-force out-of-plane large stroke translation micro electrostatic actuator," *J. Micromech. Microeng.*, Vol. 21, 075002, 2011.

[4] M. S. M. Ali, B. Bycraft, C. Schlosser, B. Assadsangabi, K. Takahata, "Out-of-plane spiral-coil inductor self-assembled by locally controlled bimorph actuation," *Micro & Nano Letters*, Vol. 6, Iss. 12, pp. 1016-1018, 2011.

[5] M. A. Rosa, D. D. Bruyker, A. R. Völkel, E. Peeters, and J. Dunec, "A novel external electrode configuration for the electrostatic actuation of MEMS based devices," *J. Micromech. Microeng.*, Vol. 14, pp. 446-451, 2004.

[6] Y-S. Kim, N. G. Dagalakis, and S. K. Gupta, "Creating large out-of-plane displacement electrothermal motion stage by incorporating beams with step features," *J. Micromech. Microeng.*, Vol. 23, 055008, 2013.

[7] F. Hu, W. Wang, and J. Yao, "An electrostatic MEMS spring actuator with large stroke and out-of-planeactuation," *J. Micromech. Microeng.*, Vol. 21, 115029, 2011.

[8] F. Hu, Z. Li, Y. Qian, J. Yao, X. Xiong, J. Niu, and Z. Peng, "A multi-electrode and pre-deformed bilayer spring structure electrostatic attractive MEMS actuator with large stroke at low actuation voltage," *J. Micromech. Microeng.*, Vol. 22, 095023, 2012.

[9] S. Chowdhury, M. Ahmadi, and W. C. Miller, "A closed-form model for the pull-in voltage of electrostatically actuated cantilever beams," *J. Micromech. Microeng.*, Vol. 15, pp. 756-763, 2005.

[10] Hidetoshi Takahashi, Nguyen Minh Dung, Kiyoshi Matsumoto, and Isao Shimoyama, "Differential pressure sensor using a piezoresistive cantilever," *J. Micromech. Microeng*, Vol.22, 055015, 2012.

[11] Hidetoshi Takahashi, Kiyoshi Matsumoto, and Isao Shimoyama, "Differential pressure distribution measurement for the development of insect-sized wings," *Measurement Science and Technology*, Vol. 24, 055304, 2013.

[12] T. Kan, A. Isozaki, N. Kanda, N. Nemoto, K. Konishi, M. Kuwata-Gonokami, K. Matsumoto, and I. Shimoyama, "Spiral Metamaterial For Tunable Circular Dichroism," *The 26th IEEE International Conference on Micro Electro Mechanical Systems (MEMS '13)*, pp. 701-704, Taipei, Taiwan, 20-24 January, 2013.

[13] T. Kan, A. Isozaki, N. Kanda, N. Nemoto, K. Konishi, M. Kuwata-Gonokami, K. Matsumoto, and I. Shimoyama, "Spiral metamaterial for active tuning of optical activity," *Appl. Phys. Lett.*, Vol. 102, 221906, 2013.

CONTACT

*T. Kan, tel: +81-3-5841-0461; kan@leopard.t.u-tokyo.ac.jp

MECHANO-ACTIVE TISSUE SCAFFOLD SYSTEM BASED ON A MAGNETIC NANOPARTICLE EMBEDDED NANOFIBROUS MEMBRANE

Sheng-Po Fang[1], Hulan Shang[1], Pit Fee Jao[1], Kyoung-Tae Kim[1], Gloria J. Kim[1], Jung H. Yoon[1], Kun Cho[1,2], Adam J. Katz[1], and Yong-Kyu "YK" Yoon[1]

[1]University of Florida, Gainesville, Florida, USA
[2]Korea Basic Science Institute, Ochang, South Korea

ABSTRACT

A mechano-active nanofibrous scaffold system for *in vitro* active cell culture is fabricated and demonstrated using electrospun nanofibers with magnetic nanoparticles embedded, and an electromagnet. The electrospun nanofiber consists of polycaprolactone and iron oxide nanoparticles. The magnetic nanofibrous membrane is held by micromachined printing circuit board (PCB) O-rings and remotely actuated by an electromagnet, which generates alternating current (AC) magnetic fields. The scaffold provides mechanical stress and strain on culturing cells in response to external AC magnetic fields. The mechanical properties of the magnetic nanoporous membrane including the density, porosity, and effective Young's modulus are characterized. Cell viabilities on the nanofibrous membrane with and without magnetic nanoparticles embedded have been tested.

INTRODUCTION

Electrospun nanofiber has gained popularity in many scientific disciplines and engineering applications primarily due to its nanoporous morphology, i.e. its unique physical and chemical properties to differentiate it from bulk materials. Nanofiber provides the large surface area and allows nanoparticles to transfer through. Tissue engineers have found great cell viability and density enhancement in nanofiber due to its nanoporous architecture. Nanofiber serves as a mechanical support for the tissue and enables cells to be attached and communicate each other. It has the potential to regulate the division, migration, and shape of the cell. Many recent reports have confirmed significant improvement of cell culturing on nanofibrous scaffolds [1]-[4]. Several studies have demonstrated that a nanofiber scaffold is suitable for *in vivo* cell culture [5], especially for cells that require a highly porous and large surface area scaffold such as the bone cell [6]. A fabricated biocompatible nanofiber scaffold that has the nanoporous morphology is highly demanded in biomaterials and tissue engineering fields.

The primary goal of a nanofiber scaffold is producing an *in vitro* tissue that can be ultimately implanted *in vivo*. Unlike a conventional passive scaffold, which only provides a function of mechanical support to cells, an active tissue scaffold with various external stimulation modalities such as stress, strain, electrical, acid/base and optical field becomes increasingly important for advanced tissue engineering researches [7]-[9]. Especially, such an active tissue scaffold can be used for studying differentiation conditions of stem cells. Understanding the influence of different active stimulations on cell viability and differentiation is critical for stem cell research.

Alignment of nanofibers in electrospinning has been demonstrated using directional guided electrical fields. Cultured cells show directional growth, along with the aligned nanofibers [10]-[11]. Alternating external electric fields at low frequency has also been studied to enhance neural cell viability in electrokinetically driven flow. It is known that the electrokinetical driven flow frequency can either enhance or suppress neural cell differentiation [12]. Recently, it is reported that the efficiency of stem cell harvesting is maximized with 91.6-95% cell viability when cultured in a condition with a vibrating frequency of approximately 5 Hz [13].

In this work, a mechano-active nanofibrous scaffold system, which can provide mechanical stress and strain on culturing cells by external AC magnetic fields, is demonstrated. The system consists of an iron oxide nanoparticles embedded electrospun polycaprolactone nanofiber membrane, a membrane holder, and an external electromagnet producing AC magnetic fields. The system is designed to operate at 4.5 Hz and a nominal displacement in an order of 100 nm.

DESIGN PRINCIPLE

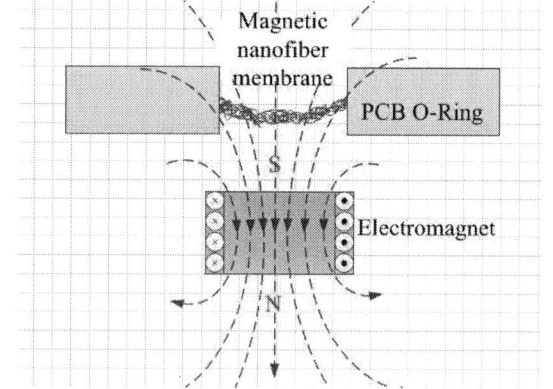

Figure 1: Schematic of a magnetic nanofibrous scaffold placed over the magnetic field generated by an electromagnet.

Figure 1 shows the schematic of the active magnetic nanofibrous scaffold system. The magnetic nanofibrous membrane is placed 2 cm above an electromagnet which generates magnetic fields for membrane actuation. The movement of the magnetic nanofibrous membrane can be controlled by the AC magnetic field of the electromagnet.

In order to study the effect of pure mechanical stimulation on cells, the external magnetic field should be small. The magnetic flux is set to be less than 10 gauss which is a typical EM value that human body may encounter in daily life [14]. Cell viability and density are known to be improved with mechanical oscillation from 50 nm to 240 nm during culturing [15]-[16]. Therefore, the targeted displacement is 100 nm, and the designed frequency is 4 ~ 5 Hz. Typical human cells are in the range of 10 μm in diameter, which is 100 times larger than the designed displacement [17]. Oscillating the cells within 1% of their volume size should prevent damaging cells during culturing. With the described design parameters, the dimension of the magnetic nanofibrous membrane is calculated to be a diameter of 2 cm and a thickness of 50 μm. The fine tuning of the displacement and resonant frequency is obtained via numerical analysis using COMSOL multiphysics.

FABRICATION PROCESS

Polycaprolactone (PCL) (Sigma Aldrich Inc.), a biodegradable polymer, is dissolved in dimethylformamide (Sigma Aldrich Inc.) and dichloromethane (Sigma Aldrich Inc.) for 24 hr to give a solution concentration of 16 wt/vol%. Then, 5 wt% of iron oxide nanoparticles (Sigma Aldrich Inc.) are mixed with the PCL solution. Iron oxide nanoparticles have an average diameter of less than 50 nm, which is also favorable to cell adhesion. The magnetic nanoparticle embedded polymer solution is prepared by mechanically stirring the contents by a DC motor at 500 rpm in an air tight bottle.

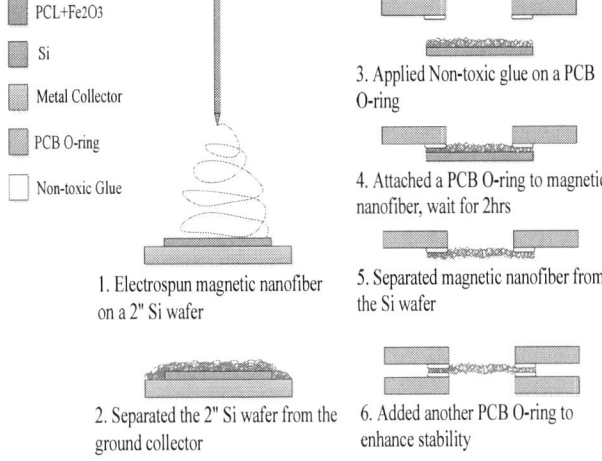

Figure 2: Fabrication process of the active magnetic nanofibrous scaffold.

Figure 2 shows the fabrication process of the active magnetic nanofibrous scaffold. First, magnetic nanofibers are electrospun on a 2 inch Si substrate for 240 second to collect a 50 μm thick nanofibrous membrane with an electric field of 1.16 kV/cm and a flow rate of the solution dispense of 1 ml/min (1, 2). Second, an O-ring shape membrane holder is machined out of printing circuit board (PCB) with an inner diameter of 2 cm and an outer diameter of 5 cm. The O-ring holder is glued on the magnetic nanofibrous membrane with

non-toxic water soluble glue (3). Third, after drying the glue, the magnetic nanofibrous membrane glued to the membrane holder is separated from the Si substrate (4, 5). Then the magnetic nanofibrous membrane is sandwiched with another membrane holder to increase stability (6). The fabrication process is scalable, manufacturable, and cost effective. The magnetic nanofibrous scaffold is placed in vacuum for 24 hr to remove any solvents that remain in the nanofiber. A 5 day *in vitro* cell culture analysis on the magnetic nanofibrous scaffold is performed using mouse cells and cultured with Dulbecco's Modified Eagle Medium with Nutrient Mixture F-12 (DMEM/F12) and 10% phosphate buffered saline (PBS) solution.

RESULT AND DISCUSSION

The images of the fabricated scaffold and the scanning electron microscope (SEM) (JEOL 5700) images of the magnified magnetic nanofiber are shown in Figure 3. The SEM image shows iron oxide nanoparticles embedded in polycaprolatone nanofiber. Electrospinning has been performed in different electric field conditions, leading to various nanofiber diameters from 550 ± 36 nm to 750 ± 42 nm with the electric field from 0.7 kV/cm to 1.0 kV/cm.

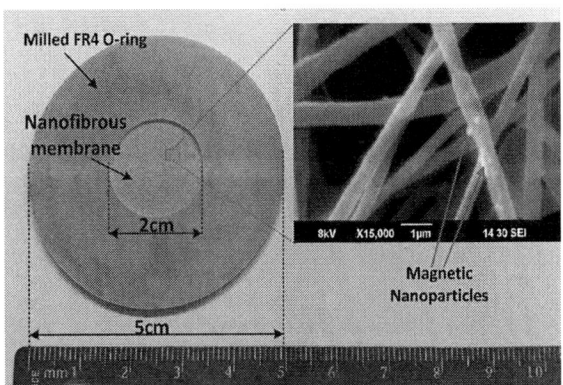

Figure 3: Photography of the fabricated magnetic nanofibrous scaffold. The insert shows an SEM image of the electrospun nanofibers with iron oxide nanoparticles embedded.

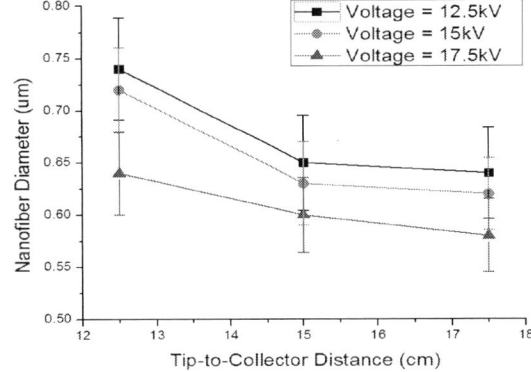

Figure 4: Nanofiber diameter variation at different electric field strengths

The hysteresis loops of the magnetic nanofibrous membranes produced with different electrical fields in electrospinning are measured by Vibrating Sample Magnetometer (ADE technologies) with the magnetic field feeding perpendicular to the membrane i.e. out of plane feeding. The magnetic moment and magnetic field relationship curve is shown in Figure 5.

The effective relative permeability of magnetic nanofibrous membrane can be calculated with the hysteresis loop. The magnetization is defined by the magnetic moment divided by the sample volume.

$$B = \mu_o(H + M) = \mu_o(H + XH) = \mu_o(1 + X)H \qquad (1)$$

where the relative permeability is defined by:

$$\mu_r = 1 + X = 1 + \frac{M}{H} \qquad (2)$$

The calculated effective relative permeability varies from 1.05 to 1.07 for the fibers collected in an electric field of 1.16 kV/cm to 0.83 kV/cm, respectively. As increasing the electric field strength of electrospinning, the resultant nanofiber diameter decreases and the porosity of the membrane increases. Therefore, less amount of iron oxide nanoparticles are embedded in the membrane.

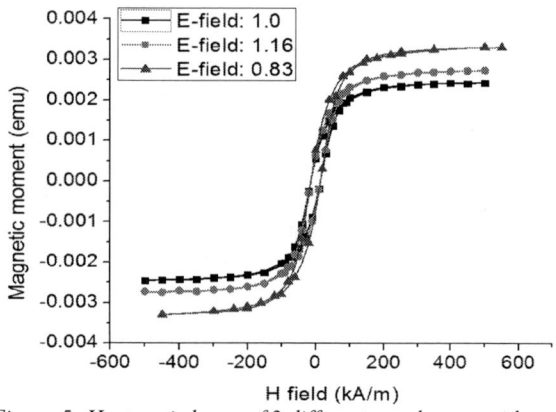

Figure 5: Hysteresis loops of 3 different membranes with nanofibers collected by electrospinning with different electric field strengths: sample 1: 1.16 kV/cm, sample 2: 1 kV/cm, sample 3: 0.83kV/cm.

The maximum displacement of the membrane at the center responding to the DC magnetic field is simulated and measured as shown in Figure 6. The maximum center deflection of the membrane shows almost linear relationship in the small deflection case. For example, the deflection changes from 1 μm to 4 μm as the magnetic flux density varies from 15 mT to 35 mT. The measurement result shows a similar trend and reasonably well matches with that of simulation. Figure 7 shows the AC response of the fabricated magnetic nanofibrous membrane with a self-resonant frequency of 4.43 Hz, a maximum deflection of 91 nm, and a Q-factor of 14. Laser vibrometer (Polytec Inc.) has been used for the displacement measurement. By fitting the simulated first self-resonant frequency to the measured self-resonant frequency, the effective Young's modulus of the nanofibrous magnetic membrane is determined as 0.127 MPa.

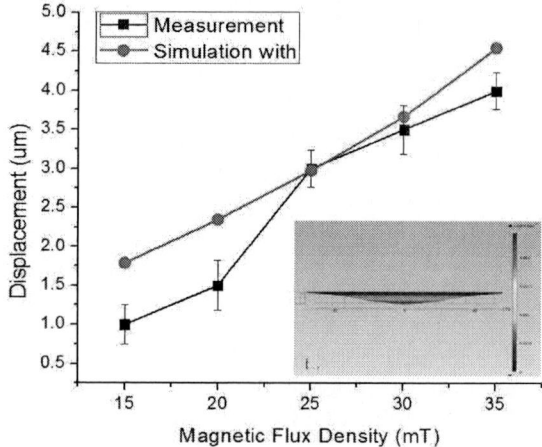

Figure 6: Simulated and measured maximum deflection of the magnetic nanofibrous membrane responding to the DC magnetic field. Inset shows a simulation plot.

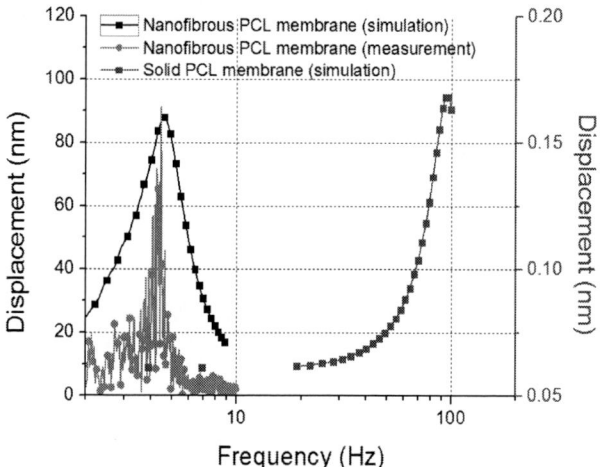

Figure 7: Simulated and measured resonant frequencies of the nanofibrous membranes (black and red) compared to the simulated resonant frequency of a solid PCL membrane (blue).

A cell viability test has been performed with mouse cells cultured for 5 days using the polycaprolatone nanofibrous membranes with and without magnetic nanoparticles embedded. The external magnetic field has not been applied on neither cases. Mouse cells are pre-cultured in Dulbecco's Modified Eagle Medium with Nutrient Mixture F-12 (DMEM/F12) and 10% PBS solution for 3 days to stabilize the condition and ready to be seeded. Then, cells are labeled with DiI and seeded on both nanofibrous membranes. After 5 day culture in DMEM/F12 and 10% PBS solution, cells are fixed and fluorescent stained with 4',6-diamidino-2-phenylindole (DAPI) and 488 fluorochrome for cytoplasm. Cells attached on both the pure polycaprolactone nanofibrous membrane and the one with magnetic nanoparticles embedded. The magnetic nanofibrous membrane shows much higher cell attachment and cell viability results as shown in Figure 8.

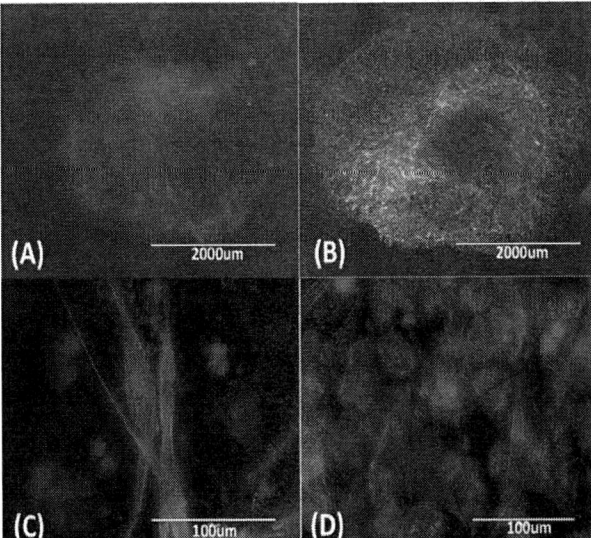

Figure 8: Images of the mouse cell viability test for 5 days: (A) on polycaprolactone nanofibers (2x) (B) on magnetic nanoparticle embedded nanofibers (2x) (C) on polycaprolactone nanofibers (40x) (D) on magnetic nanoparticle embedded nanofibers (40x). Magnetic nanoparticle embedded nanofibers show superior cell viability to polycaprolactone nanofibers. Blue: nuclei, Green: cytoplasm.

The magnetic nanoparticles embedded nanofibrous membrane has a lower density, a lower self-resonant frequency, and a lower effective Young's modulus due to its nanoporous morphology compared with its solid counterpart. The mechanical properties of both nanofibrous and solid membranes are summarized in Table 1.

Table 1. Mechanical characteristics of the magnetic nanofibrous and solid membranes.

	Nanofiber membrane	Solid PCL membrane
Density	572.85 kg/m^3	1145 kg/m^3
Porosity	72%	0%
Relative Permeability	1.05-1.07	1
Self-Resonant Frequency	4.5 Hz	80.5 Hz
Effective Young's Modulus	0.127 MPa	90 MPa

CONCLUSION

Magnetic nanofibrous membranes have been fabricated by electrospinning the solution consisting of iron oxide nanoparticles and polycaprolatone. The resonant frequency and mechanical properties of nanofibrous membranes have been studied and compared with those of a solid polycaprolatone membrane. The active scaffold dimension is designed for the effective proliferation of stem cells. Cell culture on the fabricated magnetic nanofibrous scaffold shows increased cell viability compared to the pure polycaprolatone nanofibrous scaffold. Cell culture study with active membrane operation is on-going.

REFERENCES

[1] N. Bhattarai, D. Edmondson, O. Veiseh, F. A. Matsen, and M. Zhang, *Biomaterials,* vol. 26, no. 31, pp. 6176–6184, 2005.

[2] C. P. Barnes, S. A. Sell, E. D. Boland, D. G. Simpson, and G. L. Bowlin, *Advanced Drug Delivery Reviews*, vol. 59, no. 14, pp. 1413–1433, 2007.

[3] O. Castaño, M. Eltohamy, and H.-W. Kim, *Methods in Molecular Biology Nanotechnology in Regenerative Medicine Methods and Protocols*, vol. 811, pp. 127–140, 2011.

[4] S. Kim and C. B. Park, *Advanced Functional Materials*, vol. 23, no. 1, pp. 10-25, Jan. 2013.

[5] V. I. Sikavitsas, G. N. Bancroft, and A. G. Mikos, *Journal of Biomedical Materials Research*, vol. 62, no. 1, pp. 136–148, 2002

[6] Y. Hu, D. W. Grainger, S. R. Winn, and J. O. Hollinger, *Journal of Biomedical Materials Research*, vol. 59, no. 3, pp. 563–572, 2002.

[7] J. M. Razal, M. Kita, A. F. Quigley, E. Kennedy, S. E. Moulton, R. M. I. Kapsa, G. M. Clark, and G. G. Wallace, *Advanced Functional Materials*, vol. 19, no. 21, pp. 3381–3388, Nov. 2009.

[8] J. T. Kannarkat, J. Battogtokh, J. Philip, O. C. Wilson, and P. M. Mehl, *Journal of Applied Physics*, vol. 107, no. 9, pp. 09B307–3, May 2010.

[9] Y.-S. Lee and T. Livingston Arinzeh, "Electrospun Nanofibrous Materials for Neural Tissue Engineering," *Polymers*, vol. 3, no. 4, pp. 413–426, Feb. 2011.

[10] C. Dang, N. Bhattarai, D. Edmondson, A. Cooper, and M. Zhang, *Journal of Undergraduate Research in Bioengineering*, pp. 29–32, 2007.

[11] J. M. Razal, M. Kita, A. F. Quigley, E. Kennedy, S. E. Moulton, R. M. I. Kapsa, G. M. Clark, and G. G. Wallace, *Advanced Functional Materials*, vol. 19, no. 21, pp. 3381–3388, Nov. 2009.

[12] M. A. Matos and M. T. Cicerone, *Biotechnol Prog* 2010; 26 (3): 664 - 670 .

[13] H. Kasuto, N. Drori-Carmi, and B. Zohar, *EU Patent* WO2012140519 Mar. 2013.

[14] J. L. Pipkin, W. G. Hinson, J. F. Young, K. L. Rowland, J. G. Shaddock, W. H. Tolleson, P. H. Duffy, and D. A. Casciano, *Bioelectromagnetics*, pp. 347-357, September, 1999.

[15] P. CLARK, P. CONNOLLY, A. S. G. CURTIS, J. A. T. DOW, and C. D. W. WILKINSON, *Journal of Cell Science 99*, pp. 73-77, 1991

[16] Y. Yamada, G. Umegaki, T. Kawashima, M. Nagai, T. Shibata, T. Masuzawa, T. Kimura, and A. Kishida, *J. Phys.: Conf. Ser.*, Vol. 352, Conf. 1, 2012

[17] E. P. Widmaier, H. Raff, and K. T. Strang, *Vander, Sherman, Luciano's Human Physiology: The Mechanisms of Body Function*, 7th edition.

ACKNOWLEDGEMENTS

This work has been in part supported by the National Science Foundation grant, NSF ECCS 1132413. Arnold's group at UF is acknowledged for the VSM characterization.

CONTACT

Sheng-Po Fang, Tel: +1-352-278-6884; max308@ufl.edu
Yong-Kyu Yoon, Tel: +1-352-392-5985;
ykyoon@ece.ufl.edu

NANO-SCALE BIOMECHANICAL ANALYZER FOR STUDYING STIMULUS DEPENDENT SELF-ASSEMBLY OF ACTIN FILAMENT

Naoya Shimada[1], Masashi Ikeuchi[1,2], and Koji Ikuta[1]
[1]The University of Tokyo, Tokyo, JAPAN
[2]PRESTO, Japan Science and Technology Agency, Saitama, JAPAN

ABSTRACT

We have developed nano-scale mechanical analysis system by using "optically driven nano-beam" to measure elasticity of self-assembled actin filament under dynamic mechanical stimulus. In this report, we worked on developing a new nano-beam to specifically capture actin on its surface. By using the new nano-beam, we have successfully measured elasticity of self-assembled actin filament in water. The nano-mechanical analysis system unravels cell life phenomenon which can't be dealt with through conventional methodologies.

INTRODUCTION

Cell changes reactions such as differentiation and migration depending on geometry to which external force is applied. For instance, Benjamin et al. showed that cultured stem-cell on PDMS surface differed from cultured stem-cell on nano-size grating in differentiation [1]. Additionally, Brendan et al. indicated that micro-construction of three dimensional scaffolds influences cellular migration ability via junction interactions [2].

Although such phenomena have been researched at a cell level, study at a protein level has not been extensively studied. For examining the mechanical properties of cellular proteins depending on the contact geometry, the shape and the area of the manipulator in contact with proteins need to be precisely defined. Conventional methods, however, cannot perform such experiments [3-4]. For example, optical tweezers using a micro-bead [4-6] cannot define the shape and the size of the contact area because of the micro-bead's spherical configuration (Fig.1).

To solve this problem, we have developed "optically driven nano-beam" as a quasi-static elasticity analyzer for

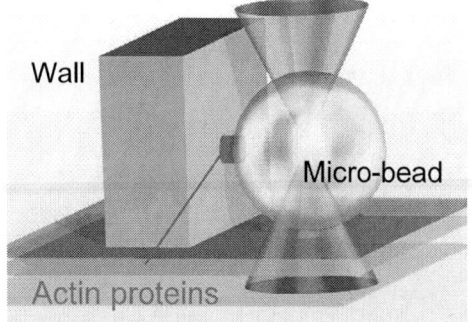

Figure 1: Conventional optical tweezers using micro-bead method cannot change force applied area between micro-bead and cellular proteins.

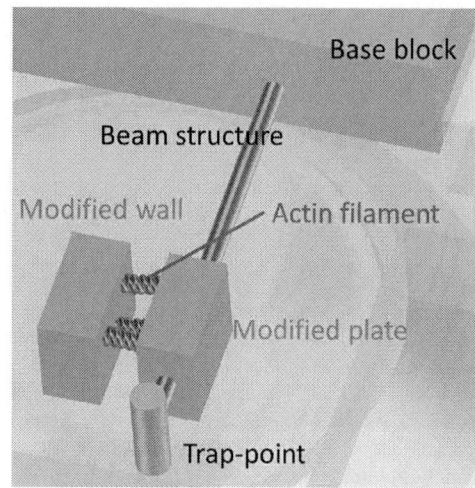

Figure 2: Concept image of new optically driven nano-beam. The new nano-beam can capture target proteins only to surfaces of its wall and plate.

cellular proteins and measured elasticity of cytoskeleton proteins using the nano-beam [7-8] (Fig.2). However, in order to bond its surface and target proteins on the measurement, adhesive proteins such as myosin is indispensable. Thus, the nano-beam is not able to measure the interaction of the networks only by actin or two kinds of proteins.

In this reports, we developed a new nano-beam to specifically capture target proteins without adhesive proteins and performed quasi-static measurement of self-assembled actin filaments with the new nano-beam.

DEVELOPMENT OF A NEW NANO-BEAM
Optically driven nano-beam

Optically driven nano-beam has been developed as a novel elasticity analyzer for small size cellular proteins. It is made of photo curable resin SCR701 (D-MEC) and fabricated by two-photon micro-stereolithgraphy with submicron resolution [9 - 10]. The nano-beam has a fixed wall and a movable plate whose shapes and sizes can be designed arbitrary since the fabrication method can make arbitrary structure. The plate was attached to end of a beam structure extended from a base block. In addition, it has a cylindrical structure called "Trap-point". The trap-point is trapped by optical tweezers to maneuver the plate. The trap-point, which was 3 μm height and 1μm diameter, was designed specifically to maximize the optical trapping force.

978-1-4799-3510-9/14 $31.00 © 2014 IEEE

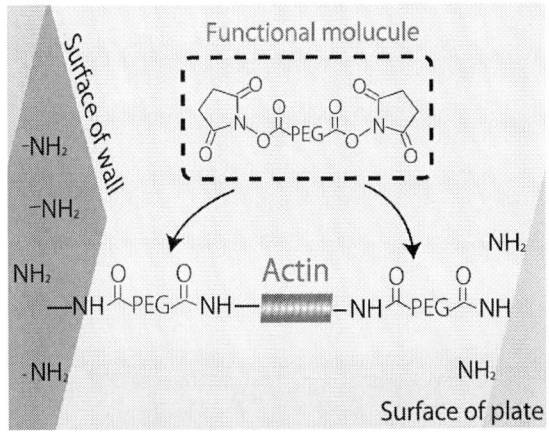

Figure 3: Schematic of how to modify NHS ester group to surface of solidified resin. By modification of NHS group, actin can be bonded to the surface.

Figure 4: Verification of the proteins capture ability of the NHS modified resin and normal resin. The modified resin was only illuminate by labeled actin.

Concept of a new nano-beam with protein capturing ability

When proteins are measured by the nano-beam, it was coated with hydrophilic molecules to prevent non-specific adhesion. Hence it is difficult to capture cellular proteins suspended in liquid between the wall and the plate. To overcome this problem, the fixed wall and the movable plate should have functional group such as N-hydroxysuccinimide ester (NHS-ester) group reactive to amine group abundant in many proteins (Fig.3). To modify the surfaces with NHS-ester group, the wall and the plate are made of another resin containing amino group (amino-resin) [11]. The amino-resin is made by mixing photo curable resin SCR701 (D-MEC) with silane coupling agent KBE-9103 (Sinetsu-silicone).For fabricating the new nano-beam, silane coupling agent was thoroughly mixed with photo curable resin at 10 weight percent. By treating the surface with a functional polymer which has NHS-ester group at either end, one NHS-ester group can be obtained on the surface of the solidified resin.

Substantiative experiment

To test the protein capture ability, micro-size structures were fabricated with NHS-ester group on their surfaces. In this experiment, rhodamine labeled actin (cytoskeleton.inc) was used as target protein. The labeled actin was suspended in buffer which the micro-size structures are placed in. After 30 minutes, the fluorescence of the labeled actin was successfully observed on the modified structure, while the non-modified structure was not fluoresced (Fig.4).

Fabrication method of the new nano-beam

By combining the amine-functionalized resin and the normal resin, a new nano-beam is fabricated using hybrid two-photon micro-stereolithgraphy [11] (Fig.5). In this process, firstly, the plate and the wall are fabricated with amino-resin on a glass substrate. After rinsing, normal photo curable resin is poured on the substrate, and other parts of the new nano-beam is fabricated with the normal resin. By rinsing the normal resin at least, the new nano-beam having the proteins capture ability at its wall and plate can be obtained.

Moreover, an array of nano-beam having different sized wall and plate was fabricated (Fig.6A). By using the array, we can measure elastic modulus under the same conditions of target protein.

MEASUREMENT OF SELF-ASSEMBLED ACTIN FILAMENTS

System setup

Figure 7 show the entire system setup. The system can control laser focus position under micro-scope by a pair of galvano scanner mirror which are maneuvered by computer. Additionally, microscopic image can be acquired from a high speed CCD camera fixed to micro-scope.

The optical trapping force was measured in real-time by visual processing, where distance between the laser focus and the trap-point position was monitored and calculated. The relationship between the laser focus and the trap-point position was monitored and calculated. The relationship

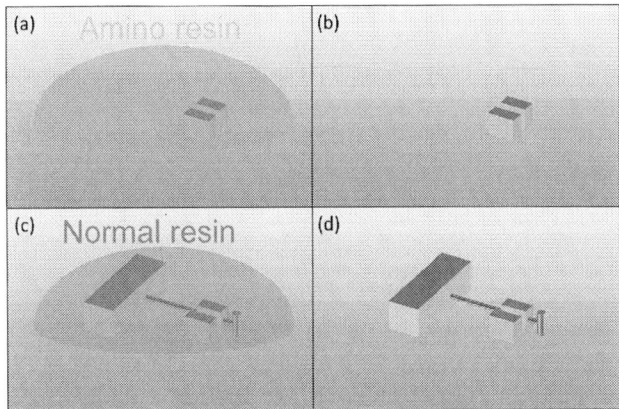

Fig.5 Hybrid 2-photon micro stereolithography (a) Fabricate amino parts with an amino photo curable resin (b)Rinse the resin (c)Fabricate another part by a normal photo curable resin (d) Rinse the resin and obtain the final structure

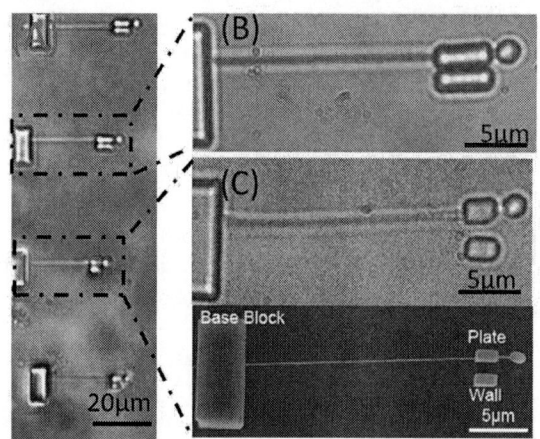

Figure 6: (A) Optical microscope image of array of the nano-beams. The wall and plate sizes gradually increase upper side on image. (B) Magnification of large nano-beam. (C)Upper: Microscope image of the nano-beam having small plate. Lower: SEM image.

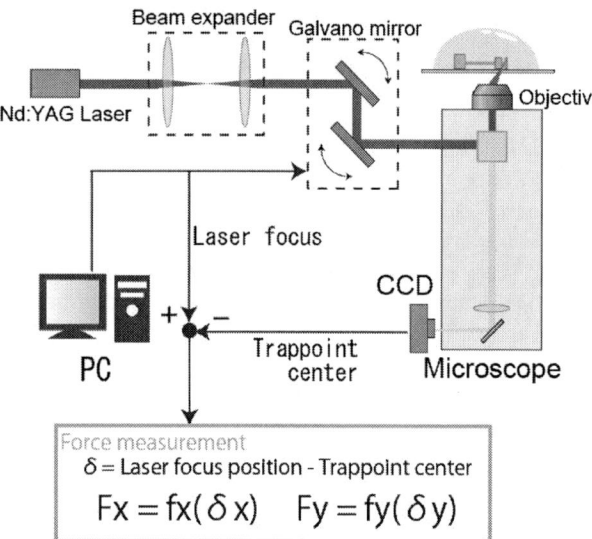

Figure 7: Experimental setup of mechanical analysis system. Force measurement is actualized by relationship between trapping force and deviation of laser focus calculated by computer and trap-point position sensed by image-processing.

between the distance and the optical trapping force has been calibrated in our previous work [12].

Target Protein Selections and preparation

In this measurement as a demonstration, we choose actin protein as target protein. Globular actin (G-actin) is among the most abundant proteins in the eukaryotic cell cytoskeleton and the famous protein found in muscle cells. G-actin polymerize in the presence of ion. Polymerized G-actin is called filamentous actin (F-actin). In vivo, F-actin rarely exists as isolated single filaments but instead associate into bundles or networks. F-actin networks are one of the most important factors to influence cell shape, motility, and adhesion. Therefore, mechanical properties of f-actin networks are traditionally measured to uncover cellular life phenomena by researchers of mechano-biology.

For this measurement, G-actin (Cytoskeleton.inc) which has been purified from rabbit skeletal muscle is used. To prepare f-actin for measurement, polymerization is performed at a G-actin concentration of 10μM. The lyophilized G-actin which is greater than 99 % pure was firstly re-suspended in a the following buffer: 2 mM HEPES-KOH pH 7.5, 0.2 mM MgCl$_2$, 0.5 mM ATP, 0.1 mM DTT. After re-suspension, 100 mM KCl is added to G-actin suspended buffer. Then, G-actin is polymerized over 8 hours at 4 degree and become filamentous actin (F-actin). After that, we use F-actin on this condition for experiment.

Preparing the New Nano-beam

In this measurement, we used the nano-beam whose rectangle shaped plate and wall were 3μm high, 2μm long and 1μm deep. Additionally beam structure of the nano-beam is 20 μm long, 147nm wide and 647 nm height in average. After fabricating the new nano-beam, its surface needed to be hydrophilized by a surfactant (Pluronic F-68, 10 wt%) over 8hours because its surface is originally hydrophobic and cannot be driven in liquid. Then, the new nano-beam is placed in 25mM aforementioned cross-linker suspended liquid for 30 min. After washing by follow polymerized buffer: 20 mM HEPES-KOH pH 7.5, 100 mM KCl, 2 mM MgCl^2, 1 mM ATP, 1 mM DTT, F-actin is suspended at F-actin concentration of 1 μM. After that, The nano-beam is maneuvered by optical tweezers until F-actin netweorks are self-assembled by external force.

Quasi-static Tensile Test

After above processes, the plate of the nano-beam slowly pulled the actin placed between the plate and the wall. It means that we carried out quasi-static tensile test of polymerized actin. On this system, elastic modulus of the nano-beam's beam structure needs to be calibrated on ahead. Hence we have already calibrated its elastic modulus [7] and subtract its elastic modulus from result data. As a result, the elastic modulus of the actin filament was measured to be

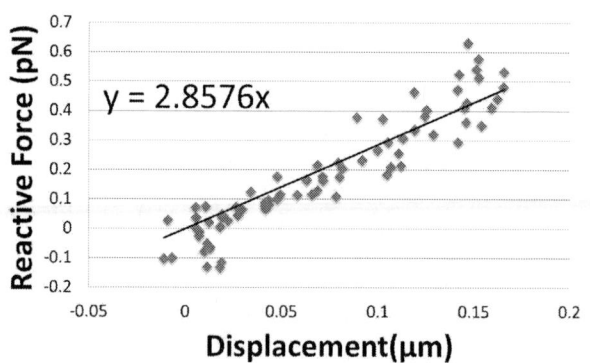

Figure 8: Elastic modulus of actin proteins measured by the nano-beam. Slope of black line is 2.8576[pN/μm].

about 2.85 [pN/μm] (Fig.8).

According to this data, we successfully developed a new type of optically driven nano-beam for measuring elastic modulus of actin proteins.

CONCLUSION

In this paper, we proposed and developed the new optically driven nano-beam for studying stimulus development self-assembly actin filament which have proteins capturing ability at its surfaces.

To realize proteins capturing ability, we used amino-resin, which is mixed photo curable resin and silane coupling agent, and cross-linker which has NHS-ester groups at both ends of polyethyleneglycol. By using both materials, we successfully developed a new optically driven nano-beam to specifically capture actin proteins on its surface.

Additionally, we carried out quasi-static measurement of actin filament network by developed nano-beam. From this result, we have verified the usefulness of the new nano-beam in biomechanical analysis of proteins.

Owing to develop the new biomechanical analyzer with the new nano-beam, the measurement which cannot conventionally be measured can be realized.

ACKNOWLEDGEMENTS

This study was supported by JST CREST and JSPS Frant-in Aid for Scientific Research (S) 22220008 for which Koji.Ikuta was the principal investigator.

REFERENCES

[1] Benjamin Kim Kiat Teo, Sum Thai Wong, Choon Kiat Lim, Terrence Y. S. Kung, Chong Hao Yap, Yamini Ramagopal, Lewis H. Romer, and Evelyn K. F. Yim, Nanotopography Modulates Mechanotransduction of Stem Cells and Induces Differentiation through Focal Adhesion Kinase, *ACS Nano 2013* 7 (6), 4785-4798

[2] Brendan A.C. Harley, Hyung-Do Kim, Muhammad H. Zaman, Ioannis V. Yannas, Douglas A. Lauffenburger, Lorna J. Gibson, Microarchitecture of Three-Dimensional Scaffolds Influences Cell Migration Behavior via Junction Interactions, *Biophysical Journal*, Volume 95, Issue 8, 15 October 2008, Pages 4013-4024

[3] Y. Tseng and D. Wirtz, "Mechanics and multiple-particle tracking microheterogeneity of alpha-actinin-cross-linked actin filament networks.," Biophysical journal, vol. 81, no. 3, pp. 1643–56, Sep. 2001.

[4] D. Mizuno, C. Tardin, C. F. Schmidt, and F. C. Mackintosh, "Nonequilibrium mechanics of active cytoskeletal networks.," Science (New York, N.Y.), vol. 315, no. 5810, pp. 370–3, Jan. 2007.

[5] A. Ashkin, "Acceleration and Trapping of Particles by Radiation Pressure",Phisical Review Letters, vol.24, pp.156-159, 1970.

[6] A. Ashkin, "Optical trapping and manipulation of neutral particles using lasers", Proc. Natl. Acad. Sci., Vol.94, pp. 4853-4860, 1997

[7] N. Shimada, M. Ikeuchi, K Ikuta, "Nanoscale Dynamic viscoelasticity analyzer by using optically driven nanobeams for study of biomolecules", Proc. IEEE MEMS'12, 2012.

[8] N. Shimada, M. Ikeuchi, K. Ikuta, "Mechanical properties analyzer for nano-size protein with optically driven nano-beam," The 17th international Conference on Solid-State Sensors, Actuators and Microsystems (TRANSDUCERS & EUROSENSORS XXVII), 2013 Transducers & Eurosensors XXVII , pp.550,553, 16-20 June 2013

[9] S. Maruo, O. Nakamura, and S. Kawata, "Three-dimensional microfabrication with two-photon-absorbed photopolymerization.," Optics letters, vol. 22, no. 2, pp. 132–4, Jan. 1997.

[10] S. Maruo, K. Ikuta, and H. Korogi, "Submicron manipulation tools driven by light in a liquid," Applied Physics Letters, vol. 82, no. 1, p. 133, 2003.

[11] M. Ikeuchi, K. Isozaki, K. Kyue, H. Sunabe, N. Shimada, H. Sasago and K.Ikuta, "Multifunctional Optically Driven Micro-Robot for realtime 3D bio-manipulation and Imaging", Proc. IEEE MEMS'11, January 23-27, 2011

[12] K. Ikuta; F. Sato, K. Kadoguchi, S. Itoh, "Optical driven master-slave controllable nano-manipulator with real-time force sensing," Micro Electro Mechanical Systems, IEEE 21st International Conference on MEMS 2008 , 13-17 Jan. 2008

CONTACT

*Koji Ikuta, Research Center for Advanced Science and Technology, The University of Tokyo, 4-6-1 Komaba, Meguro-ku, Tokyo 153-8904, JAPAN, Tel: +81-3-5452-5160, Fax: +81-3-5452-5161; E-mail:Koji_Ikuta@rcast.u-tokyo.ac.jp

PINCHING AND RELEASING OF CELLULAR AGGREGATE BY MICROFINGERS USING PDMS PNEUMATIC BALLOON ACTUATORS

S. SHIMOMURA[1], Y. TERAMACHI[1], Y. MURAMATSU[1],
S. TAJIMA[2], Y. TABATA[2] and S. KONISHI[1]
[1]Ritsumeikan University, Shiga, JAPAN
[2]Institute for Frontier Medical Sciences, Kyoto University, Kyoto, JAPAN

ABSTRACT

Micro manipulator for three dimensional cell cultures such as a cellular aggregate will be presented. We recently reported on a catcher device to manipulate and transport a cellular aggregate in a microwell by using flow control. This paper proposes microfingers to pinch and release a spherical cellular aggregate (φ200 μm). We have studied pneumatic balloon actuator which is featured by its small, soft and safe characteristics suitable for biomedical applications. This paper will present microfingers and their operation, followed by design, fabrication, and characterization. This paper will also show the significance of surface treatment of a fingertip. A series of operations demonstrated on a real cellular aggregate by using newly developed microfingers will be successfully demonstrated.

INTRODUCTION

Tissue engineering is expected as a new technology in regenerative medicine [1-6]. It would become possible to provide personalized medicine using cultured cells in future. Cultured cells in a dish or chip however are prepared in planar environment, while real tissues in the human body consist of three dimensional cellular structures. Scaffold techniques allow for the laboratory production of three dimensional cellular structures [3]. Three dimensional cell cultures such as a cellular aggregate are studied in the aim of drug screening as well as regenerative medicine [4-6].

Devices for culturing and transporting a cellular aggregate have been reported in the field of μTAS, for example [7]. We recently reported cellular aggregate catcher device using fluidic manipulation [8]. In literature, we can find other studies for cell manipulation [9, 10].

We have developed various devices using Pneumatic Balloon Actuator (hereafter PBA). PBA have small, soft, and safe (S3) characteristics because of PDMS (polydimethylsiloxane) as its structural material and pneumatic pressure as its safe driving principle [11]. All PDMS PBA is composed of two different PDMS layers to form cavities for balloons and pneumatic supply channels. PDMS allows for a soft and flexible structure. PDMS is suitable for devices to manipulate living tissue such as a cell aggregate. Compared with direct manipulation devices using rigid structures, this paper proposes flexible microfingers to pinch and release a cell aggregate.

MICROFINGERS

Concept and Design

A schematic view of microfingers is shown in Fig. 1(a). The bending motions of two opposing PBAs facilitate opening/closing motion of fingertips. Fingertips close normally while open by pressurization. Figure 1(b) explains a series of operation: (1) Initial state; (2) Introducing microfingers into a microwell and opening fingertips by actuators; (3) Pinching of an object by stopping actuation; (4) Moving to a desired microwell; (5) Positioning in a microwell: (6) Releasing of an object by opening fingertips. Corresponding operation in the sequence using developed device will be shown in Fig. 3.

Fabrication of Microfingers

Fabrication process is explained in Fig. 2. First, PDMS is spin-coated on SU-8 mold and thermally cured (Fig. 2(a)). The PDMS film is peeled off and placed on the PDMS film (Fig.2 (b, c)). Second, The PDMS film is pre-cured at this step. Third, two PDMS films are thermally cured completely so as to be bonded together (Fig.2 (c)). After masking, parylene was deposited on the surface of fingertip to prevent two opposing fingertips from sticking (Fig.2 (d)). In addition, the parylene surface is modified by VUV to control hydrophilicity (Fig.2 (e)). Two microfingers were positioned face-to-face and bonded together (Fig.2 (f)). PBA at the fingertip is 160 μm ×600 μm and a single finger is 560 μm ×900 μm.

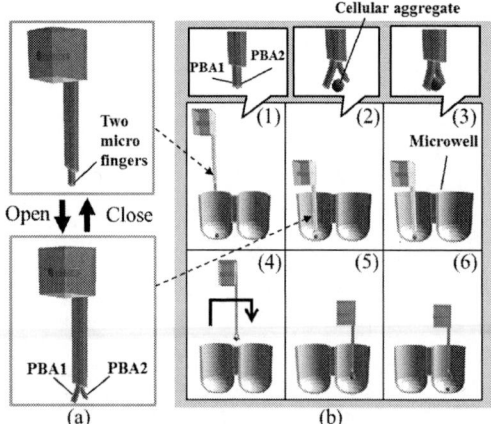

Figure 1: (a)A schematic view of microfingers. (b)(1) Initial state, (2) Introducing microfingers into a microwell and opening fingertips, (3) Pinching of an object, (4) Moving to a desired microwell, (5) Positioning in a microwell (6) Releasing of an object.

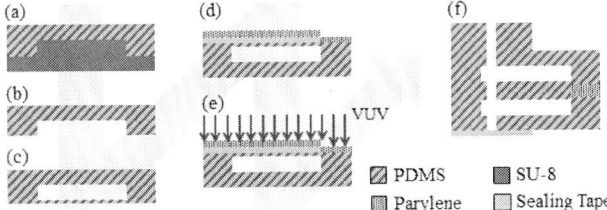

Figure 2: Fabrication process. (a) PDMS on SU-8 mold, (b) Peeling off, (c) Bonding to PDMS film, (d) Deposition of Parylene to fingertips, (e) Surface hydrophilic treatment by VUV, (f) Two microfingers are faced and bonded together.

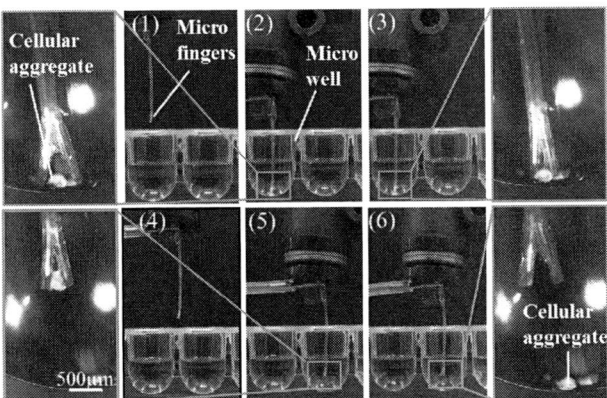

Figure 3: Experiment of Pinching cellular aggregate (200 μm). Figure 2correspond to Fig.1

Characterization

Fabricated microfingers were driven in the air as well as in phosphate buffered saline (PBS). PBA was actuated to open fingertips. An object such as a cellular aggregate is pinched by restoring force of a PDMS structure. We measured restoring force of the PDMS structure. The pinching force against a φ200 μm cellular aggregate was estimated about 30 μN. Damage evaluation of pushing force to a cellular aggregate also showed that obvious damage could not found up to 1 mN. Releasing of a cellular aggregate is affected by hydrophilic property of the contact surface. Low hydrophilicity restrains the finger from releasing an object. On the other hand, too high hydrophilicity of the surface causes difficulty in holding of the object. We examined various surface conditions and employed VUV treatment of parylene-coated PDMS providing 40 degree as the contact angle.

Operation

Figure 3 corresponds to Fig. 1(b) and shows a series of operation using optimized microfingers. Microfingers were observed both from the side and bottom by two microscopes and positioned by a XYZ stage. The magnified photographs in the upper left/right-hand corners show a φ200 μm cellular aggregate together with opening/closing microfingers (see Fig. 3). The gap between opening fingertips was about 450 μm. We could successfully execute pinching and releasing operation. The cellular aggregate could also be moved to another microwell as designed.

CONCLUSION

This paper proposed microfingers to pinch and release a cell aggregate using PBA. Proposed microfingers could pinch a cellular aggregate (φ200 μm) in a micro well (6.94 mm in diameter, 10.2 mm in depth). We could carry and release the cellular aggregate in another well. The microfingers allowed opening and closing of fingertips by two PBAs. It was found that surface characteristics of fingertips had critical influence in holding and releasing a cell aggregate. We could succeed in holding and releasing of a cellular aggregate by surface treatment using VUV.

ACKNOWLEDGEMENT

This work was partially supported by JSPS KAKENHI Grant Number 24240075.

REFERENCES

[1] L. A. Solchaga, J. E. Dennis, V. M. Goldberg, A. I. Caplan, "Hyaluronic Acid-Based Polymers as Cell Carriers for Tissue-Engineered Repair of Bone and Cartilage", J Orthop Res,17, pp.205-213, 1999

[2] J. D. Hartgerink, E. Beniash, S. I. Stupp, "Self-Assembly and Mineralization of Peptide-Amphiphile Nanofibers", Science 294, pp.1684-1688, 2001

[3] H. Hosseinkhani, T. Azzam, H.Kobayashi, Y. Hiraokae, H. Shimokawa, A. J. Dombb, Y. Tabata, " Combination of 3D tissue engineered scaffold and non-viral gene carrier enhance in vitro DNA expression of mesenchymal stem cells ", Biomaterials 27, pp.4269-4278, 2006

[4] N. Mohan, P. D. Nair and Y. Tabata, " A 3D biodegradable protein based matrix for cartilage tissue engineering and stem cell differentiation to cartilage ", J Master Sci,2009

[5] C. P. Vepari, D. L. Kaplan, "Covalently Immobilized Enzyme Gradients Within Three-Dimensional Porous Scaffolds", Biotechnol, Bioeng, pp.1130-1137, 2006

[6] H. Kurosawa, "Methods for Inducing Embryoid Body Formation: In Vitro Differentiation System of Embryonic Stem Cells", The Society for Biotechnology, pp.389-398, 2007

[7] A. Yasukawa, M. Ikeuchi, and K. Ikuta, " Combinatorial Differentiation Induction of Embrionic Bodies in "PASCL (Pneumatically Actuated Spheroids Culture Lab-on-chip)", MEMS 2013, pp.931-934, 2013

[8] Y. Teramachi , S. Shimomura, W. Tonomura, S. Tajima, Y. Tabata and S. Konishi, "Cellular Aggregate Catcher using Fluidic Manipulation in High Compatibility with Wide Spread Micro-Well-Plate" Proc. Transducers 2013, pp.2209-2212.

[9] J. Ok, Y.-W. Lu and C.-J. Kim "Pneumatically Driven Microcage for Microbe Manipulation in a Biological Liquid Environment" Journal of Microelectromechanical Systems , Vol.15, No.6, pp. 1499-1505, 2006.

[10] T. Tanikawa and T. Arai, "Development of a micro-manipulation system having a two-fingered micro-hand", IEEE Transactions on Robotics and Automation, Vol.15, No.1, pp.152–162, 1999.

[11] O. C. Jeong and S. Konishi, "All PDMS Pneumatic Microfinger With Bidirectional Motion and Its Application", Journal of Microelectromechanical Systems , Vol.15, No.4, pp. 896-903, 2006.

PNEUMATICALLY ACTUATED BIOMIMETIC PARTICLE TRANSPORTER

A. Rockenbach[1], Chr. Brücker[2], and U. Schnakenberg[1]

[1]Insitute of Materials in Electrical Engineering 1, RWTH Aachen University, Germany
[2]Institute for Mechanics and Fluid Dynamics, TU Bergakademie Freiberg, Germany

ABSTRACT

To prevent the adhesion of particles at surfaces by transporting them along the surface, a new type of pneumatically actuated particle transporter is introduced. The biomimetic approach is based on the transportation principle of particles by cilia arrays due to the generation of metachronal waves. Rows of flaps, which mimic the cilia, are asymmetrically positioned on flexible membranes. The membranes are individually deflected by applying a well-defined pressure profile to achieve a metachronal wave.

Detailed simulations of the membrane and flap deflections as well as a description of the proof-of-concept by applying metachronal waves to the flap arrays are presented.

INTRODUCTION

In microfluidic devices, a couple of methods to transport and separate particles in closed channels exist, like electrophoresis [1], electromagnetism [2], or peristalsis [3]. In some cases, however, particle transport in open channels is required (absence of a second wall). Here the traditional transport mechanisms are not applicable due to the absence of a second border.

The transportation of the fluid along the surface can be achieved by using cilia like structures to push the fluid along the surface by generation of metachronal waves. In nature many ciliated surfaces are known, such as the respiratory tract [4] or the fallopian tube [5]. In addition, the use of cilia as propelling mechanism for bacteria [6] or Ctenophore are known and well-studied in numerical analysis [7,8].

First studies to mimic cilia-type actuation using magnetic actuation principle were published in [9,10], whereas an oscillating force principle at the free end of the artificial cilia was investigated in [11].

The objective is the development of particle transportation near a surface by a biomimetic adaptation of the cilia movement principle. Here, the actuation of the artificial cilia array is carried out by using a micro pneumatic-based excitation.

DESIGN CONCEPT

The basic concept of the particle transporter chip is shown in Figure 1. Rows of cilia-like flaps are positioned asymmetrically on bendable membranes. Each flap row can be deflected separately by an induced pneumatic force which bends the supporting membrane. The membrane movement converts to large deflections of the flaps in x-direction (lateral) and in smaller movements in z-direction

(vertical). Due to the high aspect ratio of the flaps the angle rotation results in a fluid movement parallel to the surface which prevents the particle deposition.

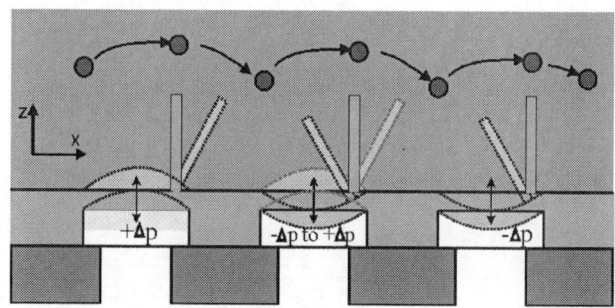

Figure 1: Concept of the pneumatically-actuated particle transporter (cross-section, not in scale): rows of flaps are asymmetrically located on flexible membranes supports. The pressures p to deflect the membranes individually are applied to the underlying channels.

Membrane bending is achieved by changing the pressure in the underlying channel from a negative differential pressure to a positive one. The bending cycle must be exactly coordinated to ensure a metachronal wave for an accurate particle transport. The phase difference of pressure in adjacent channels is one of the most important parameters for the transport mechanism.

SIMULATIONS

Simulations were carried out using ANSYS 12 Structural program in two-way coupling with CFX program. Both models were connected to ANSYS internal Fluid-Solid-Interaction (FSI) module. The simulations were based on a 2D model with great deformation. The material data for water was taken from the ANSYS Material Database. PDMS (PolyDiMethylSiloxane, here Sylgard 184), which serves as suitable material for the membrane and flaps, was defined as an isotropic material model. The material parameters were generated by tensile testing [12]. Optimum membrane bending conditions, shown in Figure 2, were extracted for a 600 µm wide and 100 µm thick PDMS membrane. The supporting structure between two membranes was optimized to 400 µm. The optimum flap position on the membrane was determined between 75 µm and 100 µm adjacent from the supporting structure.

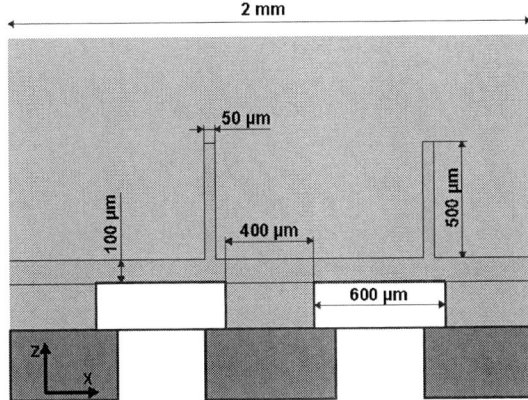

Figure 2: Design details of the particle transporter: the pressure channel under the 100 µm thick membrane is 600 µm wide. In combination with the 400 µm wide supporting structure between two channels a pitch of 1000 µm is defined. The flaps are 500 µm high, 50 µm wide, and 1000 µm long.

The van Mises stress distribution of the PDMS membrane is shown in Figure 3, when the membrane is deflected under a negative pressure of 30 kPa. With respect to the soft membrane material, the bending deforms the membrane shape and increases the membrane area. These deformations appear in both bending directions of the membrane and are responsible for a shift of the tip deflection.

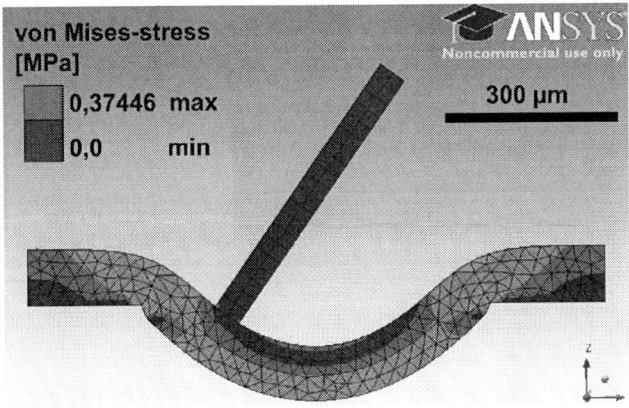

Figure 3: ANSYS 12 FEM simulation of van Mises stress distribution of a 100 µm thick PDMS membrane deflected by a negative pressure of 30 kPa.

To enhance the pumping effect, the upward bending pressure must have a higher pressure difference from ambient than the downward bending pressure. This leads to a faster flap movement during the upward bending of the membranes.

The asymmetric flap position on the membrane results in an asymmetrical movement of the flap. Figure 5 shows the simulated movement of the flap by setting a minimum

and maximum pressure difference of 40 kPa and -30 kPa compared to ambient pressure. The flap moves 480 µm in x-direction and 75 µm in z-direction. This behavior generates a lateral movement in the fluid. A similar characteristic movement was published by Khaderi et al. [13].

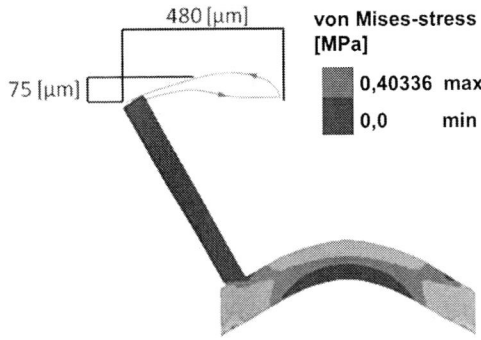

Figure 4: FEM simulation of the flap movement during a forward and backward displacement of the deflected membrane. The blue line represents the movement of the flaps tip during a whole membrane bending cycle, where the applied pressure under the membrane was set to a difference of   -30 kPa and +40 kPa compared to ambient pressure. The horizontal displacement was determined to 480 µm, the lateral displacement to 75 µm, respectively. In addition, the van Mises stress distribution in the membrane is shown.

The simulated tip deflection was confirmed by measuring the tip displacement experimentally, as shown in Figure 5.

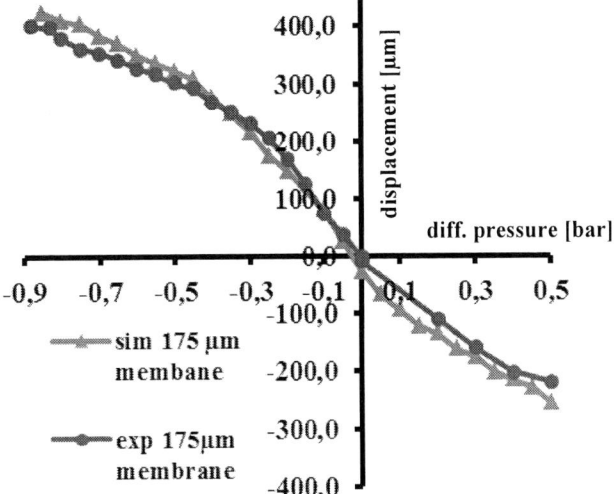

Figure 5: Simulated (green) and experimentally determined tip deflection (blue) of a 175 µm thick PDMS membrane with a 500 µm long flap on top. The values of the displacement are with respect to the flaps at rest.

Figure 6 shows two different, but characteristic streamline profiles. Particles in the fluid follow the circular streamlines whereas particles in vicinity to the flaps are transported by a net flow. 1000 μm above the membrane surface the flow stops almost completely. For larger distances to the surface no particle transport occurs.

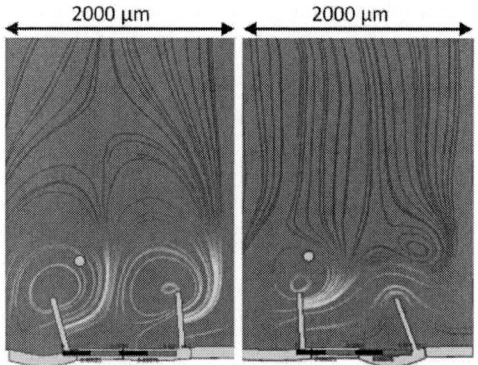

Figure 6: Two different, but characteristic streamline profiles resulting from the FSI-simulations. The membranes are shown at the bottom of the figure. Two bending arrangements are depicted. Above the flaps a particle is shown. The colored lines are streamlines of the fluid. The flow slows down rapidly above the flap. Left: Two rotating flows at the flaps tip. Right: One rotating flow and one laminar streamline profile.

Depending on the dynamic streamline characteristics of the fluid particles follow a net movement, which is shown in Figure 7 where two particle positions are compared. The left green particle corresponds to the start position and the right red particle to the position of the same particle after two membrane bending cycles, corresponding to a net movement of the particle of 40 μm. This value is about 6% of the amplitude of the tip movement of the flaps.

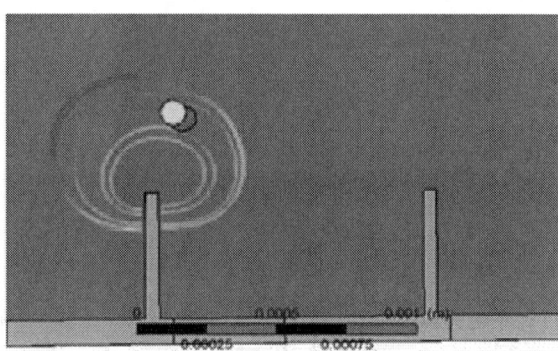

Figure 7: Displacement of a particle (100 μm in diameter) after two membrane bending cycles. The left green particle corresponds to the start position and the right red particle to the end position after two complete load cycles. The distance between the two positions is about 40 μm. The colored lines are streamlines of a specific time step.

Based on these promising results FEM simulations were carried out to generate a metachronal wave by combination of the dynamic bending behavior of 8 membranes. The membranes are individually deflected by

applying a well-defined pressure profile (from -30 kPa in 50 ms to a pressure of +40 kPa for duration of 50 ms, then again down to -30 kPa in 100 ms) in combination with a defined phase-shift 50 ms between adjacent channels to achieve a metachronal wave, which is necessary for a controlled particle transport [9]. One characteristic simulation result of membrane bending and flap deflection is shown in Figure 8. The stream lines are omitted, because of clarity.

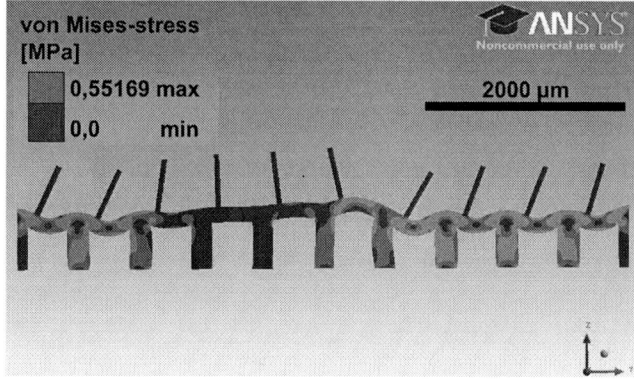

Figure 8: FEM-results of membrane and corresponding flap deflections for establishing a metachronal wave.

FABRICATION

The biomimetic particle transporter chips were fabricated in PDMS (Sylgard 184, Dow Corning) using standard UV lithography and soft-lithography techniques in combination with SU-8 micro molds.

Figure 9 shows the realized flap array. Transparent membranes are arranged in 20 rows. On top of each membrane 20 violet colored transparent flaps with a width of 900 μm are arranged in a line. Supporting structures between the channels are shown in light grey.

Figure 9: Photograph of the particle transporter chip. 20 transparent membranes are arranged in rows. Looking through the membrane the vertical walls of the supporting structures (dark gray) and the bottom of the chip (black) are shown. On each membrane 20 flaps (violet) are realized. The image magnification in the right bottom corner shows three PDMS-based transparent flaps on one

membrane over one channel. The light gray areas show the top area of the supporting structures for the membrane. Scale bar: One flap is 1000 μm wide. The overall dimension of the chip is 3200 μm x 3200 μm.

A detail of the realized flap array is shown in Figure 10. Looking at the top of the chip, two membranes arranged vertically is shown. Rows of flaps are arranged asymmetrically on the membrane.

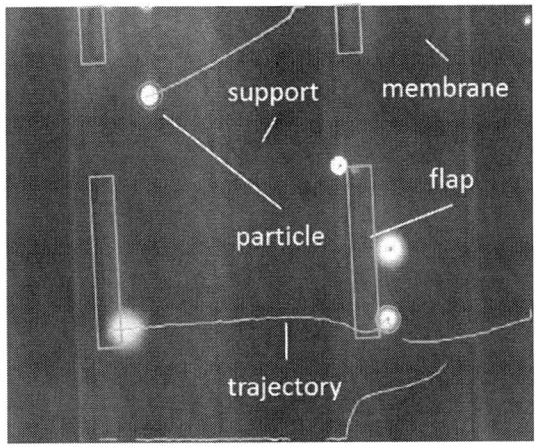

Figure 10: Photograph of a detail of membrane and flap arrangement. On the vertically arranged membranes flaps are arranged vertically. In addition, fluorescence-marked 60 μm polystyrene particles and their corresponding tracked path are shown. Scale bar: The flap is 50 μm wide and 900 μm long.

RESULTS

The particle transporter chip was connected to the custom-made pneumatic triggering module and integrated into a microfluidic chamber. A glycol-water mixture of 3:10 was used as fluid to ensure levitating particles. Fluorescence-marked 60 μm polystyrene particles were used for experiments. Figure 10 shows a snap-shot of the proof-of-concept (stroboscopic picture for not-deflected flap). Several particles are shown in the flap array. The different diameters correspond to the different vertical positions the particles have, with respect to the focus depth of the objective lens (focus is on top of the flap). The trajectories of some particles are shown in green. The particle transportation direction for the given pressure profile to realize a metachronal wave was from the left to the right and the net transport velocity was determined to 30 μm/s depending on relative position between particle and flap.

CONCLUSIONS

A new concept for active particle transport in single interface fluids is introduced. The active movement of cilia-like flaps induce a fluid flow. The flaps are actuated by an interface membrane, bent through pneumatic pressure from the backside. Simulations are carried out to determine an optimized design: Best membrane bending conditions are found for a PDMS membrane with width of 600 μm and a wall width of 400 μm by setting the membrane thickness to 100 μm. An optimum flap position on the membrane is determined between 75 μm and 100 μm from the attachment. The asymmetric position of the flaps on the membrane results in an asymmetric movement of the flaps. The whole horizontal distance of the flap movement is about 480 μm and the vertical movement about 75 μm, using 500 μm high and 50 μm wide flaps, respectively. The simulations show that dispersed particles in the fluid will be transported by the flow. They will move almost circular, but the net particle transport will be unidirectional.

ACKNOWLEDGEMENTS

Funding is given by the German Federal Ministry of Education and Research under grant #16SV5341 (PaTra). We gratefully acknowledge the cooperation with our project partners P. Uhlmann, A. Rollberg, and M. Kunder from Leibniz Institute of Polymer Research Dresden e.V. - Dept. Physical Chemistry und Polymer Physics, Dresden, Germany.

REFERENCES

[1] X. Xuan et al., Electrophoresis, (2005), 26, pp. 3558-3560

[2] A. Yang et al., Nano Letters, Vol. 9, No. 3, (2009), pp. 1182-1188.

[3] C.-H. Wang et al., J. Micromech. a. Microengineering, 16, (2006), pp. 341-348

[4] M.A. Sleigh et al., *Am. Rev. of Respir. Dis*, 137 (3) (1988), 726.

[5] R.A. Lyons et al., *Human Reproduction Update*, 12 (4) (2006), 363.

[6] B. Behkam et al., *Appl. Phys. Lett*, 90 (2) (2007), 023902.

[7] A. Dauptain et al., *J Fluids Structures*, 24 (8) (2008), 1156.

[8] A. Vilfan et al., *Phys. Rev. Lett*, 96 (5) (2006), 058102.

[9] S.N. Khaderi et al., *Lab on A Chip*, 11 (12) (2011), 2002.

[10] F. Fahrni et al., *Lab on A Chip*, 9 (23) (2009), 3413.

[11] A. Alexeev et al., *Langmuir*, 24 (21) (2008), 12102.

[12] I. Klammer et al., *Tech. Dig. IEEE MEMS 2008*, (2008), 626.

[13] J. Elgeti et al., PNAS 110 (12) (2013), 4470.

QUANTITATIVE ANALYSIS OF SURFACE TEXTURES CREATED BY MEMS TACTILE DISPLAY USING MICROFABRICATED TACTILE SAMPLES

Yumi Kosemura[1], Shoichi Hasegawa[1], Hiroaki Ishikawa[1], Junpei Watanabe[1] and Norihisa Miki[1,2]
[1]Keio University, Yokohama, JAPAN
[2]JST PRESTO, Tokyo, JAPAN

ABSTRACT

This paper reports quantification of virtual surface textures produced by the MEMS_based tactile display. Tactile display consists of large displacement MEMS actuator array which is composed of piezoelectric actuators and hydraulic displacement amplification mechanism (HDAM) for large displacement to stimulate human tactile receptors. Tactile display can provide various surface textures such as "rough" "soft" "elastic" etc. to fingertip by controlling the voltages, the vibration frequencies and the time of actuator's driving time. In previous work, we proved that tactile display could create smooth and rough textures by controlling the actuator's parameters. In this study, to quantify tactile feelings virtually produced by the MEMS tactile display, we newly propose "Sample comparison method". In this method, subjects require to compare between the virtual surface texture and tactile samples and select the sample most similar to displayed tactile feeling. In this paper, we firstly conducted comparing experiment using unprocessed samples to investigate what kind of surface characteristics can be controlled on the tactile display. Then, we experimented using microfabricated tactile samples to quantify the tactile feelings. At last,we successfully show the specific characteristics of surfaces virtually created by the MEMS tactile display.

INTRODUCTION

Tactile information exchanges have gained recognition recently because it's possible to be a new efficient way to communicate information comprehensibly. And now, the tactile display which can provide diverse tactile feeling to people are studied to be available as important human interfaces [1-4]. However, the achievement of displaying real tactile feeling is difficult. Because tactile display requires the actuator which can drive with high frequency, large displacement and high density due to the characteristics of human tactile receptors [5-8]. Then, we developed large displacement MEMS actuator which can implement these required conditions [9-10].

In prior work, we fabricated the MEMS tactile display which consists of large displacement MEMS actuator array and conducted a perception experiment. As a result, we found that tactile display could create rough and smooth surfaces and these differences of textures depend on vibration frequencies. But we could not understand how rough the created surfaces are and how other tactile factors, such as hardness and wettability, are displayed.

When the relationship between the tactile characteristics and any parameters is investigated, rating methods, such as semantic differential method and paired comparison method, have been widely used. However, these methods do not have objective standards and quite subjective. Therefore, they cannot quantify the surface properties created by the tactile display. Therefore, we propose a sample comparison method to quantify the tactile sensations. In this method, examinees requested to compare between created surface textures and the samples and to select the best similar sample to the created textures. The surface properties of the samples, in this work average roughness Ra and Young's moduls, were measured by a laser microscope and a compression testing machine in advance. We presume the created surface textures have similar surface properties of the selected sample. In this study, first we conducted sample comparison experiment using real material samples such a wood, sandpaper, net, etc, which have been typically used in tactile sensation experiments and defined what kind of characteristics could be easly controled by tactile display's parameters which are the driving voltage, vibration frequency, and actuation patterns of the actuators arrayed in the tactile display. After that, we quantified the tactile feelings by comparing with microfabricated tactile samples whose characteristics can be controlled by material and fabrication process. The unprocessed samples may contain various properties to affect the tactile sensation, such as surface topography including roughness and patterns, hardness, surface energy, thermal conductivity, etc. We can prepare samples that have identical surface properties but a single parameter we want to investigate by microfabricatino technology. In this paper, we successfully quantify the virtually created surfaces with respect to the roughness.

DESIGN AND FABRICATION
Design of HDAM

The large displacement MEMS actuator is composed of piezoelectric actuator and HDAM as shown Figure 1. The HDAM has a chamber in 2.2 mm diameter and encapsulates incompressible fluid with largely deformable PDMS membrane (DC 3145 Clear, Dow Corning) and a latex rubber membrane (PCR-518, Regitex Co., Ltd). A titanium cover which has small holes 0.48 mm in diameter are bond to the top of the HDAM to determine the cross-sectional area. The ratio of the cross-sectional areas determine amplification actuator's displacement [10-13].

978-1-4799-3510-9/14 $31.00 © 2014 IEEE

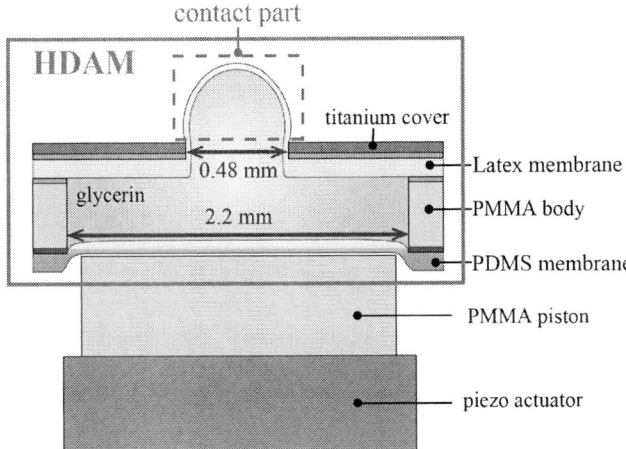

Figure 1: Schematic view of the large displacement MEMS actuators consisting of HDAM and piezoelectric actuators.

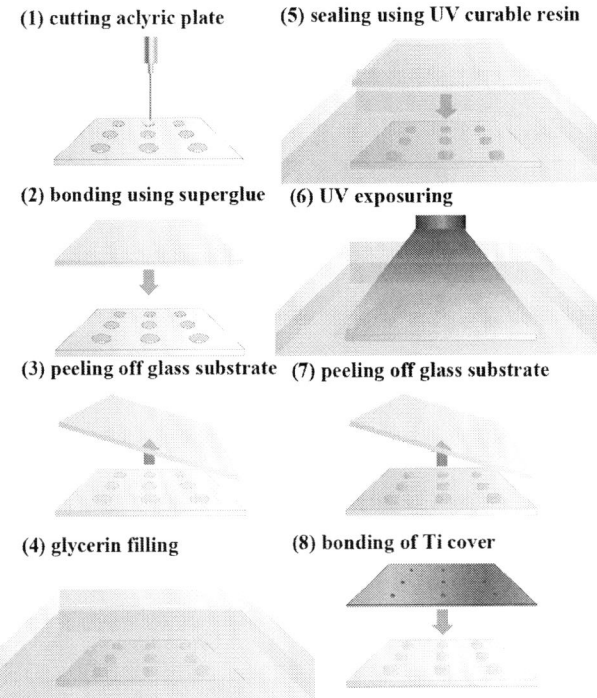

Figure 2: Fabrication process of HDAM

Fabrication

Figure 2 shows the fabrication process of HDAM. First we cut a PMMA plate by laser cutting machine. Next we bonded the latex rubber membrane to the surface of PMMA plate. Latex membrane was formed on a glass substrate by spin-coating. After the peeling grass substrate, we encapsulated glycerin using Largely deformable PDMS membrane by Bonding-in-Liquid Technique (BiLT) [14]. This method was developed by our laboratory and prevent bubbles from interfuing into a chamber. And then we bonded Ti cover to the latex rubber membrane. Ti cover was fabricated by a NC milling machine and etched by BHF for 10 hours for deburr.

Table 1: The list of samples and grouping according to Ra and fc.

No.	Sample	Ra [μm]	fc [N/mm^2]
1	Wood	13.6	45.0
2	Polystyrene foam A	9.11	5.00
3	Polystyrene foam B	49.3	14.0
4	Urethane	46.0	15.0
5	Sponge	45.2	0.65
6	Lumpy rubber A	45.1	76.4
7	Lumpy rubber B	20.5	60.2
8	Flat rubber A	12.8	38.7
9	Flat rubber B	21.9	17.6
10	Lumpy acrylic plate A	36.9	126.5
11	Lumpy acrylic plate B	49.0	126.5
12	Net A	46.25	57.1
13	Net B	27.8	76.5
14	Techno wipe	36.1	0.43
15	Kim towel	6.26	0.58
16	Denim fabric	44.4	9.43
17	Sandpaper A	38.9	245.0
18	Sandpaper B	45.5	255.1

EXPERIMENT AND RESULTS

First, we conducted comparing experiments using 18 real material samples as shown in Table 1. All the samples were investigated with respect to the roughness and Young's modulus. We tested 48 patterns, which were combinations of the driving voltages of 90, 120, and 150 V, the vibration frequencies from 10 to 100 Hz, and the actuation patterns of no overlap for 0.1 and 0.3 s switching time and with overlap of 0.05 and 0.15 s for 0.1 and 0.3 s switching time. Subjects were requested to select the sample which was most similar to displayed surface texture. 9 subjects joined the experiments. In order to find the relationships between the created tactile feelings and actuator's parameter easily, we categorized 18 samples into three groups with respect to roughness and softness respectively. And we assigned the RGB color to these groups as shown Figure 3 to grasp the trend visually. Each experimental condition is colored according to the subjects' selection. For example, half of the subjects selected rough and the other half slightly rough, the condition is colored purple (50% B and 50% R and 0% G). The results with respect to the roughness and Young's modulus are shown in Figure 4. In case of hardness, we can see the colors become more blue from green from the left figure to the right. It indicates that higher driving voltages led

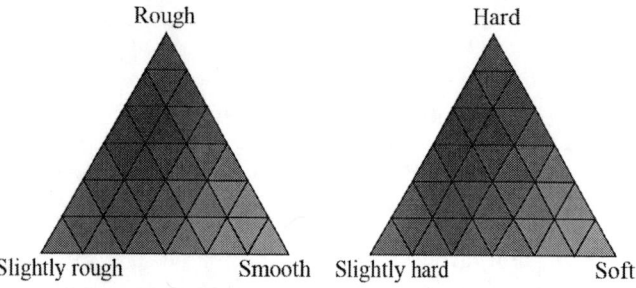

Figure 3: color-coding of "roughness" and "softness"

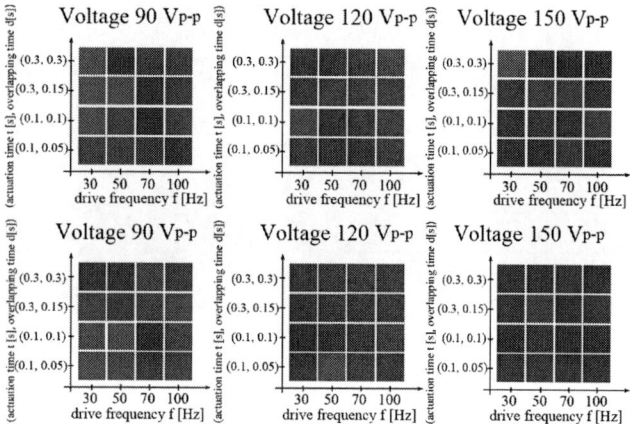

Figure 4: Results of perception experiment. Texture result that is color-coded by by "softness" (upper) and "roughness". The vertical axis is for actuation time and overlapping time, and the horizontal for the frequency.

to harder surfaces. Also we could see that low vibration frequencies, such as 30 and 50, present softer surfaces.

Interestingly, when the voltage was 150 V and the switching time of 0.3 s with no overlap showed green, meaning soft surface. This may relate to the characteristics of tactile receptors, such as phantom sensation. The roughness showed much clearer trend. As the voltage increases, colors in the figure changed from red toblue, indicating that the driving voltage can control the roughness and the rougher the surface can be displayed with larger voltage. The vibration frequencies also affected the roughness. Lower frequencies led to smoother surfaces.

Using the proposed sample comparison method and analysis using RGB-color coding successfully correlated the trend between the control parameters and the created surfaces. The surfaces can be quantified using the measured surface properties shown in Table 1. However, as we mentioned above, the unprocessed samples may have multiplecharacteristic surface properties to affect the tactile sensation. Therefore, next, we quantified the roughness of created surface texture using microfabricated tactile samples. Figure 5 shows the microfabricated tactile samples we used in the experiments. They are made of SU-8 and have stripe patterns with a depths of 30, 60 and 90 μm and definite pitch width of900 μm as shown in Figure 5. The tactile samples

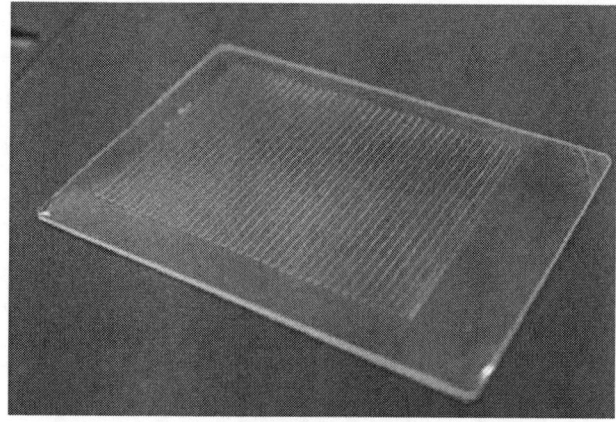

Figure 5: Fabricated tactile samples made of SU-8.

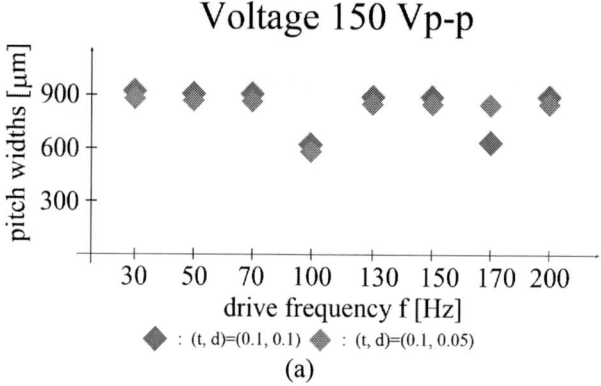

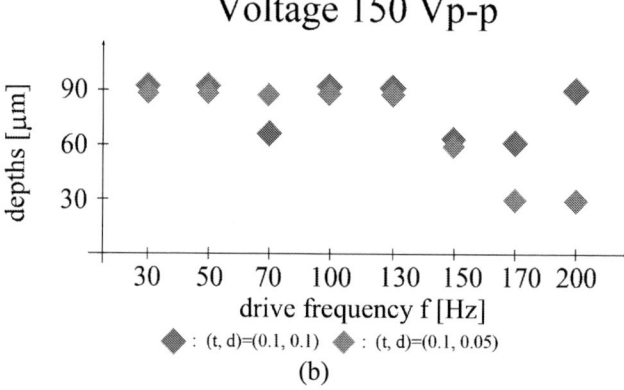

Figure 6: Results of quantification with respect to (a) pitch width and (b) depths.

were microfabricated by conventional photolithography.

While we set the driving voltage to be 150 V and actuation patterns of 0.1 s actuation time and with (0.05 s) or without overlapping, we changed actuators vibration frequencies from 30 to 200 Hz. The subjects were requested to select the most similar tactile samples. The results are Shown in Figure 6. According to Figure 6 (a), when the vibration frequency was 100 Hz, the pitch width of the virtually created surfaces were smaller than those at other frequencies in both cases with or without overlapping. This was not found in the previous experiments. When the actuator's actuation time $t = 0.1$ s and overlapping time $t = 0.05$ s, apparent depths started to decrease from the frequency $f = 130$ Hz and the texture had a depth of 30 μm at 200 Hz, which was not seen for the actuation pattern without overlapping.

In addition, at some frequencies, the created surfaces showed different features, such as 70 and 100 Hz. This may related to the characteristics of finger receptors and need to be further investigated.

CONCLUSION

We successfully showed correlation between the surface textures created by the developed MEMS tactile display and the control parameters of the display including the driving voltage, vibration frequencies, and actuation patterns of the arrayed large-displacement MEMS actuators by the newly proposed sample comparison method. The experiments revealed that the driving voltage or the displacement of the actuators had strong correlation with the roughness. In addition, when the adjacent actuators were driven with overlapping, the apparent depth of the created surface decreased when the vibration frequency was greater than 130 Hz. The greatest advantage of the sample comparison method is that the physical property of the created surfaces can be estimated by the compared samples. The quantification using the microfabricated tactile samples which was introduced by MEMS research group will be a breakthrough to make the practical use of MEMS tactile displays.

ACKNOWLEDGEMENTS

This work was supported in part by JST PRESTO (Information Environment and Humans), Ministry of Education, Culture, Sports, Science and Technology (MEXT), Grant-in-Aid for Young Scientists (B) (21760202), Ministry of Internal Affairs and Communications (MIC), Strategic Information and Communications R&D Promotion Programme (SCOPE) (092103005) and Miki Lab. Members.

REFERENCES

[1] Dimitrios A. Kontarinis and Robert D. Howe, "Tactile Display of Vibratory Information in Teleoperation and Virtual Environments", Presence 4 (4), pp.387-402, 1995.

[2] C. R. Wagner, S. J. Lederman and R. D. Howe, "A tactile shape display using RC servomotors", Haptic Interfaces for Virtual Environment and Teleoperator Systems, 2002, pp. 354-355, Mar. 24-25, 2002.

[3] G. Moy, C. Wagner and R. S. Fearing, "A compliant tactile display for teletaction", Robotics and Automation, 2000. Proceedings. ICRA '00. IEEE International Conference on, pp. 3409-3415, Apr. 24-28, 2000

[4] Morris A. Zlotnik, Crew System Design, "Applying electro-tactile display technology to fighter aircraft-flying with feeling again", Aerospace and Electronics Conference, 1988. NAECON 1988., Proceedings of the IEEE 1988 National,1988.

[5] A. V. Citrina, D. E. Stem Jr., E. R. Spangenbergb and M. J. Clark, "Consumer Need for Tactile Input: An Internet Retailing Challenge", *Journal of Business Research*, Vol. 56, Issue 11, pp. 915-922, Nov. 2003.

[6] M. Konyo, S. Tadokoro, A. Yoshida and N. Saiwaki, "A Tactile Synthesis Method Using Multiple Frequency Vibrations for Representing Virtual Touch", *Intelligent Robots and Systems, IROS*, pp. 3965-3971, Aug. 2-6,

[7] H. J. Kwon, S. W. Lee, S. S. Lee, "Braille Dot Display Module with A PDMS Membrane Driven by A Thermopneumatic Actuator", *Sensors and Actuators A*, Vol. 154, No. 2, pp. 238-246, 24 Sep. 2009.

[8] J. Streque, A. Talbi, P. Pernod and V. Preobrazhensky, "Electromagnetic Actuation Based on MEMS Technology for Tactile Display", EuroHaptics 2008, LNCS 5024, pp. 437-446, 2008.

[9] T. Ninomiya, Y. Okayama, Y. Matsumoto, X. Arouette, K. Osawa and N. Miki, "MEMS-based Hydraulic Displacement Amplification Mechanism with Completely Encapsulated Liquid" *Sensors and Actuators A: Physical 2011*, pp. 277-282, 2011.

[10] T. Ninomiya, Y. Okayama, Y. Matsumoto, K. Osawa and N. Miki, "MEMS Tactile Display with Hydraulic Displacement Amplification Mechanism" MEMS 2009, pp. 467-470, Jan. 2009.

[11] X. Wu, S. H. Kim, C. H. Ji, M. G. Allen, "A Piezoelectrically-Driven High Flow Rate Axial Polymer Microvalve with Solid Hydraulic Amplification", *MEMS 2008*, pp. 523-526, Jan. 2008.

[12] X. Arouette, Y. Matsumoto, T. Ninomiya, Y. Okayama and N. Miki, "Dynamic characteristics of a hydraulic amplification mechanism for large displacement actuators systems," Sensors, vol. 10, pp. 2946-2956, 2010.

[13] J. Watanabe, H. Ishikawa, X. Arouette, Y. Matsumoto, and N. Miki, "Demonstration of vibrational Braille code display using large displacement micro-electro-mechanical system actuators" Japanese Journal of Applied Physics, 51, 06FL11, 2012.

[14] Y. Okayama, K. Nakahara, A. Xavier, T. Ninomiya, Y. Matsumoto, Y. Orimo, A. Hotta, M. Omiya, and N. Miki, "Characterization of a Bonding-in-Liquid Technique for Liquid Encapsulation into MEMS Devices," Journal of Micromechanics and Microengineering, vol.20, no.9, 095018 (6pp), 2010.

CONTACT

* Y.Kosemura, tel: +81-45-563-1141; yumikosemura@gmail.com

A NOVEL CONSTANT FLOW REGULATION PRINCIPLE FOR COMPACT BREATH DIAGNOSTICS

Staffan B. Johansson, Göran Stemme, and Niclas Roxhed
Micro and Nanosystems, KTH Royal Institute of Technology, Sweden

ABSTRACT

This work reports on a passive compact flow regulator designed to maintain a steady flow during breath diagnostics using a flow regulation principle where a cantilever is directed towards the direction of the flow. A theoretical model has been developed describing the flow behavior and a prototype has been fabricated for proof of concept. The prototype uses a single integrated 300 µm thick 3D-printed plastic cantilever to control comparatively large air flows in the 50 ml/s regime suitable for asthma diagnostics.

INTRODUCTION

There is a trend towards quantifiable diagnostics exemplified by current breath diagnostic tools such as a handheld tuberculosis diagnostic tool (Otago Innovation), a carbon dioxide meter (MD Diagnostics Ltd.) for diagnosing health status of smokers and a lung cancer detection tool (Metabolomx). Quantifying diseases by point-of-care breath monitoring devices is convenient and painless for the patient and gives immediate results, potentially improving therapy and reducing health care costs. To ensure accurate and reproducible measurements, a number of physical parameters need to be considered such as temperature, flow rate and pressure, and these parameters should conform to certain protocols under which the test should be made [1]. Specifically for asthma monitoring using the fractional exhaled NO (FE_{NO}) detection method, measurements should be made at an exhaled flow rate of 50±5 ml/s according to regulatory guidelines [2]. Current e-nose technologies are unable to target some applications due to their size and power consumption while personal breath analyzers concurrently would be of considerable value to patients with chronic health problems like asthma [3]. Hence, there is a need for a miniaturized, cost-effective and preferably passive flow handling system that can operate in the relatively high flow rate of exhaled breath.

MEMS-based passive flow regulators have been presented for regulating small liquid flows in drug delivery systems and microfluidic applications [4-5] and gas flows in fuel cell and cooling applications [6-7]. However, the large flow rate associated with breath diagnostics requires a straight flow channel to avoid suppression of the flow regulation by static losses occurring when changing direction of high velocity flows. A miniaturized flow regulator with a straight flow channel and parallel flow restricting membranes have been suggested for regulating liquid flows, but the demonstrated flow rate was three orders of magnitude lower than the requirement for asthma monitoring and additionally the pressure range was at least one order of magnitude higher, making it unsuitable for

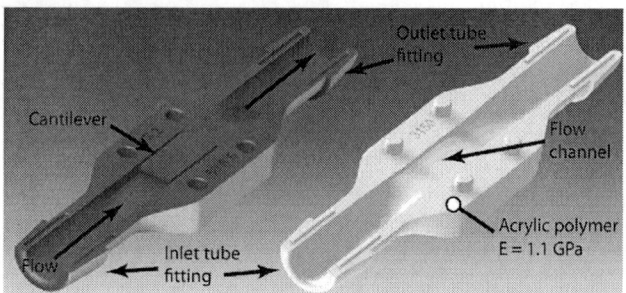

Figure 1: Illustration of the two flow regulator parts showing the cantilever, flow channel and fixating posts and holes.

breath diagnostics [8].

Based on the requirements for portable breath diagnostics we recently introduced the first straight channel MEMS-based passive flow regulator suitable for portable breath diagnostics in general and specifically for asthma monitoring [9-10]. However, while the flow rate of the demonstrated device was controlled in the target regime for asthma monitoring, it showed shortcomings in the working-pressure range and required fairly complex fabrication. Here we present a simple flow regulating principle that improves the possibility to maintain a constant flow over a wider pressure range. In addition the all-plastic design enables significantly simplified and potentially more cost effective fabrication.

PRINCIPLE OF OPERATION

The prototype consists of a cantilever that constitutes one of the walls in a flow channel, as can be seen in figure 1. When a pressure is applied at the inlet, the inlet pressure will act along the full length of the top of the cantilever. A flow in the narrow channel beneath the cantilever will result in a pressure drop resulting in a net force distribution along the length of the cantilever, causing it to bend down and restrict the flow in the channel, as can be seen in figures 2-3. The spring force of the cantilever effectively balances against the flow induced downward bending forces acting on the cantilever, resulting in a predictable deflection at any inlet pressure. In addition to the primary cantilever-controlled flow, a leak flow occurs which can be utilized to avoid too much flow restriction. A simplified analytical model for the non-linear flow regulation principle was developed and used to calculate the approximate dimensions of the prototype to target the flow regime of interest.

The total pressure drop in the device can be approximated and divided into three regions: a sudden contraction region where the flow meets the tip of the

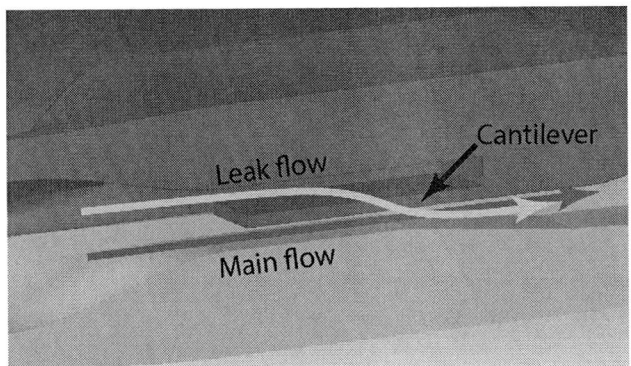

Figure 2: Illustration of the flow regulator cross section along the length of the cantilever showing the main flow and the leak flow that occurs at the edges of the cantilever.

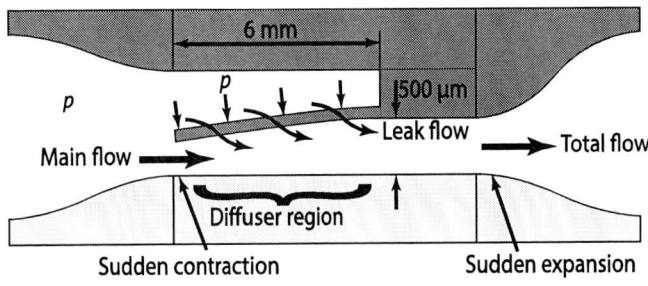

Figure 3: Illustration showing the cross section along the length of the cantilever and flow channel.

cantilever, a diffuser region along the length of the cantilever and a sudden expansion region at the exit, as shown in figure 3. The total pressure drop caused by the cantilever restriction and the flow channel along the bottom of the cantilever can then be described by the following expression:

$$p_c = \Delta p_i + \Delta p_d + \Delta p_o \tag{1}$$

where Δp_i is the inlet pressure drop caused by the sudden contraction of the flow channel, Δp_d is the diffuser region pressure drop along the length of the cantilever and Δp_o is the outlet pressure drop caused by sudden expansion of the flow channel. The pressure drop at these regions can be approximated by the following expressions [11]:

$$\Delta p_i = \xi_i \frac{\rho v_i}{2} \quad ; \xi_i \approx 0.4 \tag{2}$$

$$\Delta p_d = \xi_d \frac{\rho v_i}{2} \quad ; \xi_d = 1 - \left(\frac{A_i}{A_0}\right)^2 - C_p \tag{3}$$

$$\Delta p_o = \xi_o \frac{\rho v_i}{2} \quad ; \xi_o = \left(\frac{A_i}{A_0}\right)^2 \tag{4}$$

where ρ is the density of the fluid, v_i is the mean flow velocity at the contracted inlet, ξ_i, ξ_d and ξ_o are pressure loss coefficients, A_i and A_0 are cross sectional areas of the contracted inlet and the non-contracted channel respectively and C_p is the pressure recovery coefficient. Using Bernoulli's equation it can then be shown that the main flow can be approximated by the following expression:

$$\phi_c = \frac{\sqrt{p}}{B}\left(1 - \frac{C}{B^2}p\right) \tag{5}$$

where constants B and C are:

$$B = \sqrt{\frac{(1.4 - C_p)\rho}{2A_o{}^2}} \tag{6}$$

$$C = \frac{L^4 \rho (21 - 4C_p)}{20Eb^2t^3h_0{}^3} \tag{7}$$

where L is the length of the cantilever, E is the elastic modulus, b and t are the width and thickness of the cantilever, respectively, and h_0 is the channel height at zero cantilever deflection. The constants B and C can be optimized for a specific flow rate at two pressure levels using the following expressions:

$$B = \frac{\frac{P_2}{P_1} - 1}{\frac{\phi_1}{\phi_2}\left(\frac{P_2}{P_1}\right)^{3/2} - 1} \cdot \frac{\sqrt{P_2}}{\phi_2} \tag{8}$$

$$C = B^2\left(1 - \frac{B\phi_1}{\sqrt{P_1}}\right)\frac{1}{P_1} \tag{9}$$

where P_1 and ϕ_1 are the pressure and flow rate, respectively, for the first target point and P_2 and ϕ_2 are the pressure and flow rate for the second target point. While equation 5 describes an approximation of the main flow, the leak flow will add to the total flow through the device. Assuming an ideal pressure source and no losses in the tubings connecting the device the total flow through the device can be approximated by:

$$\phi = \phi_c + \phi_l \quad ; \phi_l \approx D\sqrt{p} \tag{10}$$

where ϕ_l is the leak flow and D is a constant defined by the geometry of the leak gaps. However possible disturbances of the main flow when the leak flow enters the flow channel could potentially reduce the total flow. Additionally, if pressure losses occur along the top of the cantilever due to the leak flow, the main flow regulating effect will be reduced, resulting in an increased total flow and reduced control. A FEM simulation was performed to evaluate the pressure distribution along the top of the cantilever. However, in this work where a first prototype has been developed for proof of concept, the coupled effect of the main flow, the leak flow and pressure losses on the cantilever is not explicitly investigated.

EXPERIMENTAL

For proof of concept, a flow regulating device was fabricated from acrylic polymer using multi jet modeling (MJM) 3D-printing technique. The prototype was printed in two separate parts featuring fixating posts and holes, respectively, as can be seen in figure 4. Although the precision of MJM enables simple press fit of the regulator parts, they were additionally fixated together by tape to eliminate any possible leak between the parts. The integrated cantilever is 6 mm long, 3.1 mm wide and 300 μm thick. The flow channel is 500 μm deep and 3.2 mm wide resulting in approximately 50 μm wide leak flow gaps on the sides of the cantilever. For characterization of the flow rate at different pressures, a standard piston air compressor with pressure regulator was used. The flow regulator was connected after a flow meter (Honeywell AWM5101VN) and a pressure sensor (Motorola MPX5050DP) connected to digital voltmeters. To improve the performance of the non-ideal pressure source, an open ended T-junction was connected between the pressure regulator and the sensors, shunting air out in the open and reducing the impedance of the circuit considerably. Pressure was then swept in discrete steps and the pressure and flow rate was read directly off the voltmeters. For reference, the flow rate was also characterized with the cantilever fixated in non-deflected and in fully deflected positions, respectively. For the non-deflected flow measurements, a small hole was drilled in the bottom flow regulator part, and a needle was inserted to prevent the cantilever from bending. For the fully deflected flow measurement, a small hole was drilled in the top regulator part and a needle was

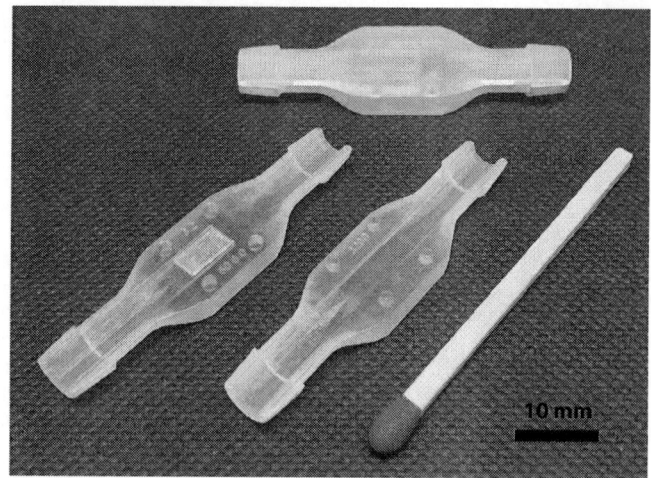

Figure 4: Photograph of the two prototype parts and an assembled device in comparison with a standard match.

inserted and used to force the tip of the cantilever against the bottom of the flow channel.

RESULTS

A simulation of the pressure distribution on top of the cantilever showed that a small pressure drop of approximately 0-10% occurs on top of the cantilever from the tip to the center while 10-15% pressure drop occurs from the center to the base, as can be seen in figure 5. Measurements show that the cantilever in the prototype is deflecting and regulating the flow close to the target for asthma monitoring as shown in figure 6. It can be seen that the cantilever starts deflecting at approximately 800 Pa and then increasingly restricts the flow as pressure increases until approximately 2500 Pa where the cantilever touches

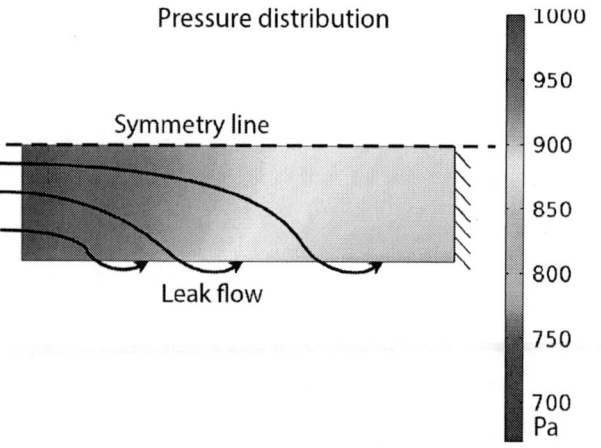

Figure 5: Simulated pressure distribution on top of the cantilever using a leak gap width of 50 μm and 200 μm headroom on top of the cantilever. It can be seen that at an applied pressure of 1000 Pa, the pressure on the right half of the cantilever is reduced by approximately 10-15%.

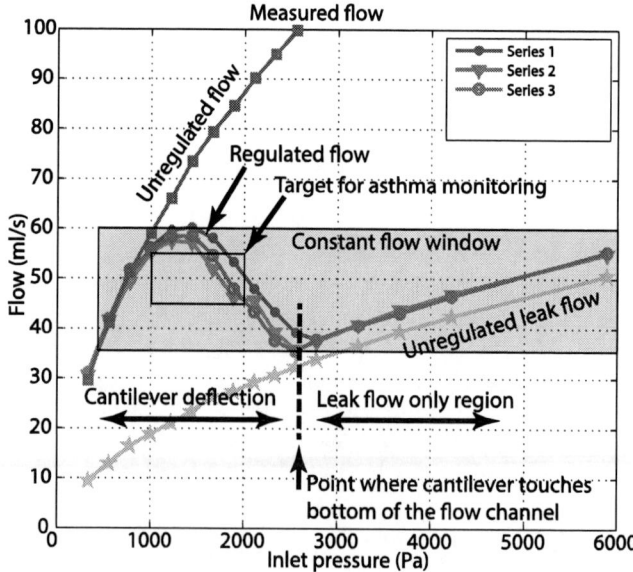

Figure 6: Measured flow rate with the cantilever fixated in undeflected position, fully deflected position and with a freely moving cantilever. It can be seen that the device is actively regulating the flow between approximately 800-2500 Pa.

978-1-4799-3510-9/14 $31.00 © 2014 IEEE   937

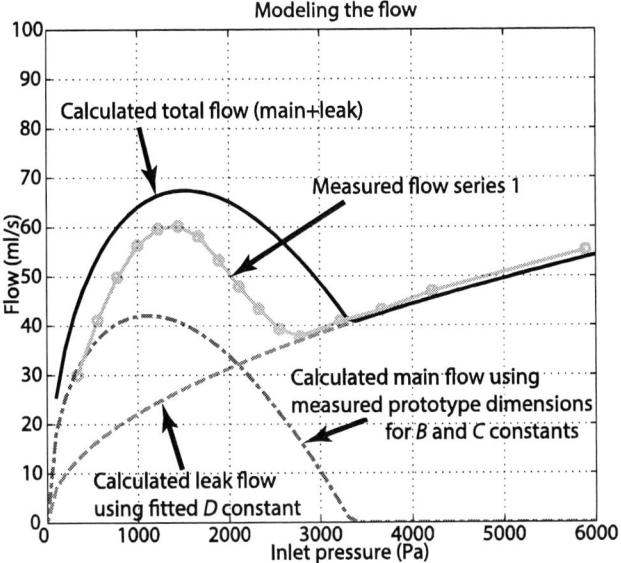

Figure 7: Measured flow rate compared with analytically calculated flow. The calculated main flow was superpositioned on the leak flow to get the total flow. It can be seen that the measured flow correlates with the calculated flow.

the bottom of the channel. It can also be seen that when the cantilever is fully deflected by inlet pressure, the flow is higher than when the cantilever is fully deflected by external force. A possible cause for this effect is that the contact surfaces between the cantilever and the flow channel is not perfectly flat and an external force is required to deform the plastic for full contact between the surfaces. The measured flow was also compared against the theoretical model. For this comparison, the constants B and C were calculated from measured actual dimensions and constant D was fitted against measured data in the leak flow region. The flow regulation behavior of the prototype is in reasonable agreement with the theoretical model, both in the regulating regime and the leak flow regime, as can be seen in figure 7. A possible cause for the lower measured slope at low pressures is that the model is not taking into account the additional losses in the channel before the tip or after the base of the cantilever. In the pressure range between 1500 and 2500 Pa the measured slope is more negative than the theoretical value indicating that the plastic could potentially have lower elastic modulus than specified.

CONCLUSION

This paper demonstrates a flow regulation principle suited for controlling air flows in portable breath diagnostic tools. A prototype was fabricated using a 3D-printing technique and measurements show that the flow rate is in the regime of interest and that the behavior is according to the theoretical model. This first prototype proves the function of the flow regulation principle and the feasibility of cost-effective constant flow controlling devices for handheld breath monitoring. The measured controlled flow is in the target regime, but can be further optimized.

ACKNOWLEDGEMENTS

This work was supported by Aerocrine AB, producer of handheld diagnostic tools for breath monitoring.

CONTACT

S.B. Johansson, tel: +46-8-7909059; stajoh@kth.se

REFERENCES

[1] Y. Zrodnikov, K. Zamuruyev, J. D. Pedersen, A. G. Fung, D. J. Peirano, M. J. Schirle, A. Panigrahy, A. Pasamontes, W. H. K. Cheung, A. A. Aksenov, M. Schivo, N. J. Kenyon, J.-P. Delplanque, and C. E. Davis, "Design criteria for portable point-of-care breath analysis systems," *proc. Transducers* 2013, pp. 1629–1632.

[2] American Thoracic Society and European Respiratory Society, "ATS/ERS recommendations for standardized procedures for the online and offline measurement of exhaled lower respiratory nitric oxide and nasal nitric oxide, 2005," *Am. J. Respir. Crit. Care Med.*, vol. 171, no. 8, pp. 912–930, Apr. 2005.

[3] Int.Techn. Roadmap for Semiconductors 2011 Ed.-Micro-Electro-Mechanical-Systems (MEMS), www.itrs.net

[4] V. Pasquier, N. M. White, P. Renaud, Y.-S. Leung Ki, 107144, 109479, C. Madore, C. Amacker, and M. Haller, "Passive micro-flow regulator for drug delivery system," *EUROSENSORS XII, VOLS 1 AND 2*, pp. 591–594, 1998.

[5] B. Yang, J. W. Levis, and Q. Lin, "A PDMS-based constant-flowrate microfluidic control device," *proc. MEMS*, 2004, pp. 379–382.

[6] A. Debray, T. Nakakubo, K. Ueda, S. Mogi, M. Shibata, and H. Fujita, "A passive micro gas regulator for hydrogen flow control," *J. Micromech. Microeng.*, vol. 15, no. 9, p. S202, Sep. 2005.

[7] M. McCarthy, N. Tiliakos, V. Modi, and L. G. Frechette, "Temperature-Regulated Nonlinear Microvalves for Self-Adaptive MEMS Cooling," *J. Microelectromechanical Systems*, vol. 17, no. 4, pp. 998–1009, 2008.

[8] I. Doh and Y.-H. Cho, "Passive flow-rate regulators using pressure-dependent autonomous deflection of parallel membrane valves," *Lab Chip*, vol. 9, no. 14, pp. 2070–2075, Jul. 2009.

[9] S. B. Johansson, G. Stemme, and N. Roxhed, "A compact passive air flow regulator for portable breath diagnostics," *proc. MEMS*, 2013, pp. 157–160.

[10] S. B. Johansson, G. Stemme, and N. Roxhed, "A MEMS-based passive air flow regulator for handheld breath diagnostics," *Sensors and Actuators A: Physical*, Sep. 2013.

[11] A. Olsson, "Valve-less diffuser micropumps," dissertation, KTH, 1998.

APPARENT SIZE CORRELATION: A SIMPLE METHOD TO DETERMINE VERTICAL POSITIONS OF PARTICLES USING CONVENTIONAL MICROSCOPY

Michael H. Winer[1], Ali Ahmadi[2], and Karen C. Cheung[1]
[1]The University of British Columbia, Vancouver, CANADA
[2]The University of British Columbia Okanagan, Kelowna, CANADA

ABSTRACT

This paper reports a simple and widely applicable particle tracking technique using conventional fluorescence microscopy to determine vertical positions of particles suspended in a fluid. Previous experimental work in this area has been restricted to two-dimensional intensity profiles or confocal and stereoscopic setups for three-dimensional tracking. Using a calibration-based defocusing method, our technique has been experimentally verified and compared with previously observed trends in inertial flow focusing. The technique has proven to be a simple and valuable tool for determining locations of micron-sized particles in microscale flows.

INTRODUCTION
Particle Tracking

Particle tracking has become an essential tool for characterizing microfluidic systems. Recently, several two- and three-dimensional methods have been developed to obtain flow and particle position information in lab-on-a-chip microfluidic systems. Two-dimensional fluorescence particle tracking has made significant advancements, particularly with applications related to micro-particle image velocimetry (μ-PIV) [1]. Although 2D techniques are simpler to implement, 3D particle tracking gives a more complete picture of particle motion that is advantageous in many applications including micro-mixers or particle focusing devices. To address the limitations of 2D tracking, several 3D techniques have been introduced. These include single camera (i.e. confocal scanning microscopy, deconvolution microscopy) and multi-camera (i.e. stereoscopic imaging, tomographic imaging) approaches [2-6]. Several of these methods have successfully tracked the three-dimensional path of particles as small as 100 nm with high accuracy [7-9]. However, these techniques require expensive and intricate optical setups including numerous lenses, aperture configurations, cameras and light sources.

Defocusing Techniques

Particle defocusing, in which particles are imaged at various vertical positions relative to a stationary focal plane, lends itself to a simple experimental setup. In this method, hereafter described as apparent size correlation (ASC), the apparent size of the particles is correlated to their vertical position in space relative to the focal plane chosen. Previous examples of this method show that a resolution of several hundred nanometers is possible using a three-hole aperture design [10]. Some disadvantages of this method include the

prerequisite for a sophisticated understanding of optics and a custom aperture mask designed over the objective lens. Other calibration-based defocusing methods exist without the need for aperture alterations; however these methods typically require high magnification or high-NA objective lens setups as they are reliant upon fitting the intensity profile of the particles to a point-spread function (PSF) of a single fluorescing particle [8, 11]. Diffraction ring methods such as PSF are relatively commonplace, but little work has been published for cell-sized particles (1-50 microns in size).

Inertial Flow Focusing

Our primary application for this work is inertial flow focusing. This fluid mechanics phenomenon has garnered significant interest due to its relatively simple implementation compared to hydrodynamic or other focusing methods. Experimental analysis of inertial flows is typically restricted to two dimensions using methods such as microparticle streak velocimetry (μ-PSV). Work in 3D particle tracking related to inertial focusing has been limited to confocal microscopy, with work in other focusing methods also using stereoscopic techniques [12, 13].

This paper demonstrates a simple and efficient 3D particle tracking technique using ASC to find the position of cell-sized particles in microfluidic channels for focusing applications. The method is calibration-based, and has been experimentally validated in finite Reynolds number conditions (Re $\leq$ 75) with high accuracy. Although the validation of the ASC method was completed using inertial focusing applications with micron-sized particles, we believe that this technique could be extrapolated to other applications in which particles must be tracked or located within a flow.

EXPERIMENTAL DESIGN
Particle Solutions

All experiments described in this paper used 15.5 μm (± 1.52 μm) green (emission wavelength = 502 nm) fluorescent polystyrene (PS) beads (Bangs Laboratories, CC FS07F, USA). This size was chosen to mimic the size of many cells typically analyzed in microfluidic assays including white blood cells or circulating tumor cells. Solutions were made to 0.1 wt% beads using deionized (DI) water. 1 vol% of Tween-20 surfactant (Sigma-Aldrich, P1379, Canada) was also added to reduce particle aggregation.

978-1-4799-3510-9/14 $31.00 © 2014 IEEE

Microfluidic Device Fabrication

Devices were fabricated using soft lithography [14]. Lithography masters were fabricated using conventional techniques on 4-inch silicon wafers (University Wafer, USA) with SU-8 3050 (MicroChem, USA) spin-coated to create a 100 x 85 μm straight rectangular channel design (width x height respectively). Devices were then fabricated using the masters with poly(dimethylsiloxane) (PDMS) elastomer (Sylgard 184, Dow Corning, USA). The final PDMS chip was plasma-bonded to a 75 x 50 x 1 mm No. 1 glass slide (VWR Scientific, PA, USA). Consider Reynolds number,

$$Re = Q \frac{\rho_f}{\mu_f} \frac{D_H}{HW} \quad (1)$$

where Q is the applied flow rate, ρ_f is the fluid density, μ_f is the fluid viscosity, D_H is the hydraulic diameter of the channel and H and W are the height and width of the channel respectively. The cross-sectional area of the device was chosen to approximate a 1:5 ratio of the particle diameter to the hydraulic diameter of the channel, which from previous work has been shown to produce a wide variety of inertial focusing positions [12, 15].

Calibration Curve for Apparent Size Correlation

To correlate the vertical (z-axis) position of a particle in space to its apparent size, a calibration curve experimentally validating the relationship was found. An agarose gel suspension was made using the PS beads and injected via syringe into the microfluidic device to randomly orient the beads within the channel. The suspension was made using DI water, 3 wt% agarose powder (Sigma-Aldrich, Canada) and 0.1 wt% PS beads. To image the beads, a Nikon Eclipse TE2000-U microscope was used, with a standard 10 X objective lens ($NA = 0.3$, $WD = 16$ mm). A sCMOS camera (LAVision, USA) with an exposure time of 100 μs was used to capture images. Data was acquired in three steps. First, the bottom of the channel was located and set as $z = 0$. The z-position of each suspended particle was determined in relation to the bottom of the channel. Finally, each particle was imaged with the optical focal plane set at $z = 0$, thereby gathering data on the apparent size of a particle relative to its known z-position. These steps were repeated for over 100 suspended particles to derive an experimental calibration curve. In order to theoretically validate this curve, the following single-lens approximation equation was used [16]:

$$d_e = M \left[d_p^2 + 1.49\lambda^2 \left(\frac{n_0^2}{NA^2} - 1 \right) + 4z^2 \left(\frac{n_0^2}{NA^2} - 1 \right)^{-1} \right]^{\frac{1}{2}} \quad (2)$$

where d_e is the expected/observed diameter of the particle, M is the magnification of the objective lens, λ is the wavelength of the fluorescence source, NA is the numerical aperture, z is the z-position of the particle with $z = 0$ as the origin, and n_o is the refractive index of the materials between the objective and the particle. The calibration curve determined experimentally matches well with this theoretical approximation, as can be seen in Figure 1.

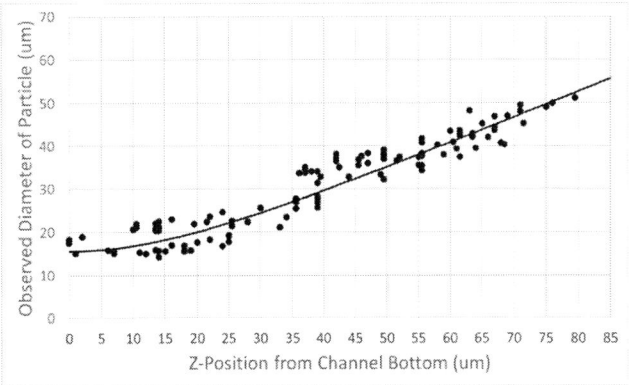

Figure 1: Calibration experimental data and theoretical curve relating z-position of a particle in the channel to its observed diameter. As predicted from equation (2), the observed diameter of the particles (mean observed diameter = 30.82 μm) steadily increases with increasing distance from the focal plane (at z = 0). $R^2 = 0.9090$.

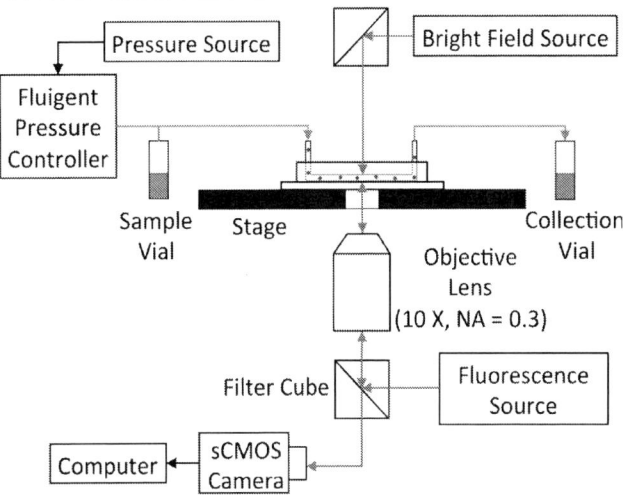

Figure 2: Experimental schematic. The microfluidic chip is placed on the stage, where both bright field (for finding the channel edges) and fluorescence sources (for imaging PS beads) are able to capture information. Data is collected using DaVis 7.2.2 software (LAVision, Göttingen, Germany). Schematic is not to scale.

Image Acquisition

Data for the inertial focusing study was collected using the same optical setup as the calibration curve. As shown in Figure 2, the bead solution described earlier was fed into the inlet of the microfluidic device using the Fluigent pressure control system to specify the pressure and thereby the Reynolds number via correlation with flow rate. Beads then travel through the channel and across the sCMOS frame of view, where the x- and y-axis positions and apparent size are collected over time. This frame of view was placed 1.5 cm from the inlet of the channel to ensure inertial focusing positions had been reached [17]. Data was collected over a wide range of flow rates (1 – 330 μL/min or $0.246 \leq Re \leq 75.356$) to determine the inertial effects on particle focusing positions as a form of validating the ASC method. The sCMOS camera captured sets of data consisting of 300 images over 10 second intervals, collecting several thousand (5000-10000) individual particle trajectories for each Reynolds number (7 in total).

Image Post-processing Algorithm

Once data sets were acquired, each set was analyzed using the free image software ImageJ and MATLAB. Specifically, a bright field image was taken for each set to determine the edges of the channel walls, as well as a background fluorescence image to subtract noise from all images in the set. A spatial bandpass filter, including a convolution of Gaussian and Boxcar functions, was used to remove background and smooth the intensity profiles of the particles [18]. The filter was then supplemented with dilation and erosion functions to more accurately resolve the edges of each particle. Finally, the centroid and diameter of each particle were determined using regionprops MATLAB functions. A final overlay image of the centroid and diameter positions onto the original image was used to visually confirm the accuracy of the algorithm. Figure 3 illustrates the entire post-processing exercise. A particle tracking algorithm was also implemented to remove duplicates of the same particle in frames taken within a single data set.

RESULTS

Inertial Focusing of Particles as Validation of ASC

2D plots of the cross-section of the channel were developed using MATLAB to map particle positions for all Reynolds numbers. Figure 4 shows the results for the 7 Re chosen for this study. From the results, the particles tend to migrate away from the center of the channel towards the walls along the y-axis, also spreading towards the top and bottom walls at higher flow rates. This generally follows trends indicated from previous work; however, particles in our results tend to stay along the central vertical plane rather than migrating towards the top and bottom walls [15, 17, 19]. Comparing our results to the more typical 2D methods such

as μ-PSV, our method gives added information in the z-axis with comparable results in the x- and y-axes [15, 17]. Furthermore, 2D methods are difficult to normalize, as they are reliant on average intensity of many hundreds of particles viewed in a channel. The simple 3D method discussed in this paper allows for individual particle tracking to create an overall cross-section of particle positioning in the channel as seen in Figure 4.

CONCLUSIONS

The ASC method described in this paper can be used for three-dimensional particle tracking in a microscale flow volume with relative accuracy and a simpler setup than previously recorded. The developed technique provides a convenient means of determining particle locations, specifically along the vertical axis. The method can be easily translated to a variety of experimental setups without custom lenses, apertures, added cameras or optical pieces such as prisms or beam separators. This method was designed and tested specifically for micron-sized particles such as cells and other particles (i.e. bacteria); however it could also be used in μPIV applications to better understand specific microflows such as microreactors or mixers.

Future work will explore use of this technique with biological samples, and as a method for determining unique Reynolds number regimes for controlling particle flows for separation by density, size, morphology, or biochemical characteristics.

ACKNOWLEDGEMENTS

This work was funded in part by the Natural Sciences and Engineering Research Council of Canada (NSERC), CMC Microsystems, and the Canada Foundation for Innovation (CFI). Special thanks to Benjamin Mustin, Eric Cheng, Samantha Grist, Fatehjit Singh, Garnet Martens, and Kevin Hodgson for expertise in MATLAB, and general experimental advice.

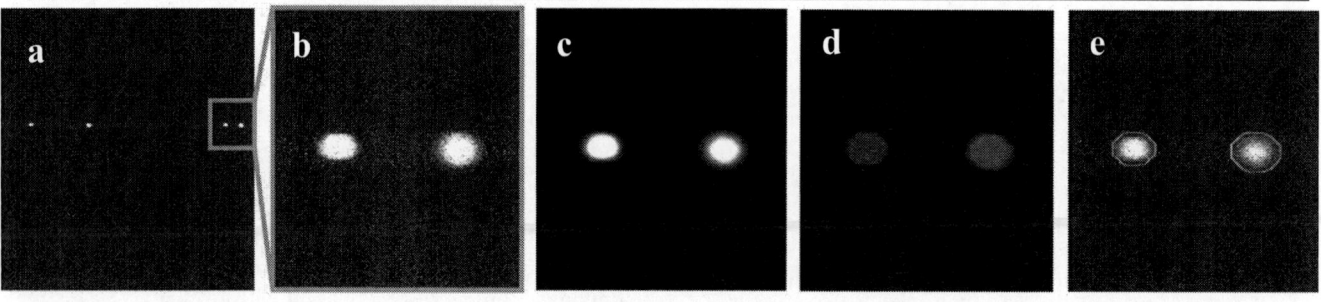

Figure 3: Visualization of image post-processing algorithm. (a) An original image from the sCMOS camera is imported into MATLAB. (b) A zoomed-in portion of the total image, no post-processing conducted at this step. (c) A bandpass filter is applied. (d) Dilation and erosion functions applied (an octahedral geometry is used rather than a higher-sided polygon to reduce computational time). (e) The diameter and centroid are overlaid on the original image to verify accuracy of the algorithm.

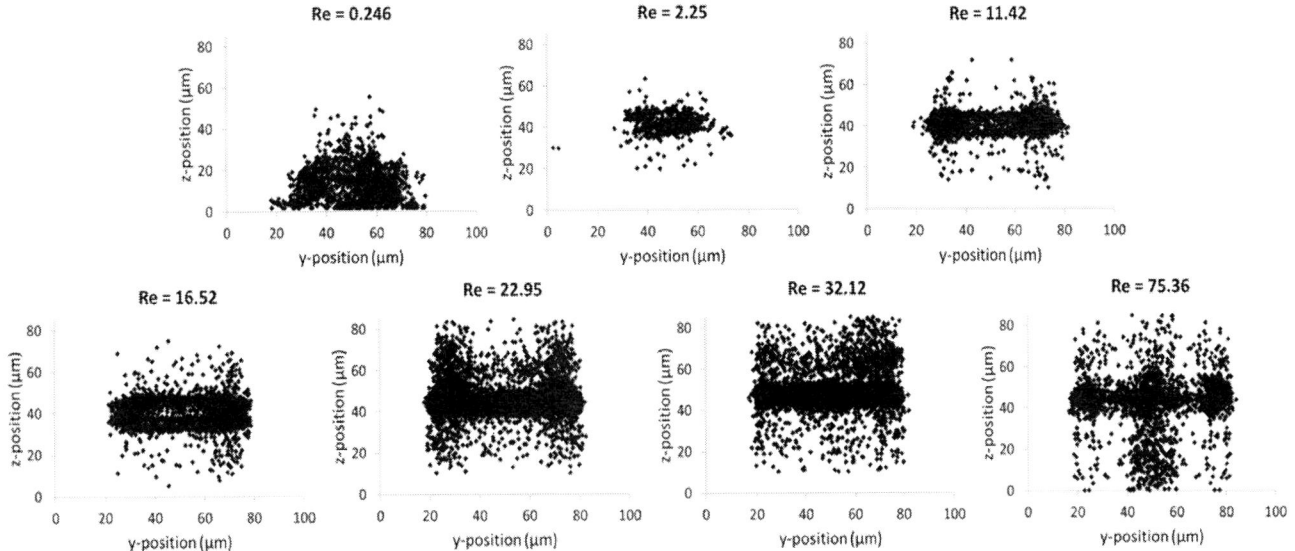

Figure 4: Plots of experimental data for inertial focusing using the ASC method.

REFERENCES

[1] S. T. Wereley and C. D. Meinhart, "Recent Advances in Micro-Particle Image Velocimetry," *Annual Review of Fluid Mechanics,* vol. 42, pp. 557-576, 2010.

[2] F. Scarano, "Tomographic PIV: principles and practice," *Measurement Science and Technology,* vol. 24, p. 012001, 2013.

[3] S. A. Klein, *et al.*, "Three-dimensional three-component particle velocimetry for microscale flows using volumetric scanning," *Measurement Science and Technology,* vol. 23, p. 085304, 2012.

[4] J.-B. Sibarita, "Deconvolution Microscopy," in *Microscopy Techniques.* vol. 95, J. Rietdorf, Ed., ed: Springer Berlin Heidelberg, 2005, pp. 201-243.

[5] S. Lee and S. Kim, "Advanced particle-based velocimetry techniques for microscale flows," *Microfluidics and Nanofluidics,* vol. 6, pp. 577-588, 2009.

[6] S. Williams, *et al.*, "Advances and applications on microfluidic velocimetry techniques," *Microfluidics and Nanofluidics,* vol. 8, pp. 709-726, 2010.

[7] F. Pereira, *et al.*, "Microscale 3D flow mapping with μDDPIV," *Experiments in Fluids,* vol. 42, pp. 589-599, 2007.

[8] J. S. Park and K. D. Kihm, "Three-dimensional micro-PTV using deconvolution microscopy," *Experiments in Fluids,* vol. 40, pp. 491-499, 2006.

[9] E. Toprak, *et al.*, "Three-Dimensional Particle Tracking via Bifocal Imaging," *Nano Letters,* vol. 7, pp. 2043-2045, 2007.

[10] S. Y. Yoon and K. C. Kim, "3D particle position and 3D velocity field measurement in a microvolume via the defocusing concept," *Measurement Science and Technology,* vol. 17, p. 2897, 2006.

[11] M. Speidel, *et al.*, "Three-dimensional tracking of fluorescent nanoparticles with subnanometer precision by use of off-focus imaging," *Optics Letters,* vol. 28, pp. 69-71, 2003.

[12] D. Di Carlo, *et al.*, "Continuous inertial focusing, ordering, and separation of particles in microchannels," *Proceedings of the National Academy of Sciences,* vol. 104, pp. 18892-18897, November 27, 2007.

[13] D. Di Carlo, *et al.*, "Particle Segregation and Dynamics in Confined Flows," *Physical Review Letters,* vol. 102, p. 094503, 2009.

[14] D. C. Duffy, *et al.*, "Rapid Prototyping of Microfluidic Systems in Poly(dimethylsiloxane)," *Analytical Chemistry,* vol. 70, pp. 4974-4984, 1998.

[15] A. Bhagat, *et al.*, "Inertial microfluidics for continuous particle filtration and extraction," *Microfluidics and Nanofluidics,* vol. 7, pp. 217-226, 2009.

[16] M. Rossi, *et al.*, "On the effect of particle image intensity and image preprocessing on the depth of correlation in micro-PIV," *Experiments in Fluids,* vol. 52, pp. 1063-1075, 2012.

[17] J. Zhou and I. Papautsky, "Fundamentals of inertial focusing in microchannels," *Lab on a Chip,* vol. 13, pp. 1121-1132, 2013.

[18] J. C. Crocker and D. G. Grier, "Methods of Digital Video Microscopy for Colloidal Studies," *Journal of Colloid and Interface Science,* vol. 179, pp. 298-310, 1996.

[19] B. Chun and A. J. C. Ladd, "Inertial migration of neutrally buoyant particles in a square duct: An investigation of multiple equilibrium positions," *Physics of Fluids,* vol. 18, p. 031704, 2006.

CELL-NICHE-ON-CHIP: PAIRED SINGLE CELL CO-CULTURE PLATFORMS USING IMMISCIBLE LIQUID ISOLATION AND SEMI-PERMEABLE MEMBRANES

Yu-Chih Chen[1], Yu-Heng Cheng[1] and Euisik Yoon[1,2]

[1]Center for Wireless Integrated Microsensing and Systems, Dept. of EECS, University of Michigan, Ann Arbor, MI, USA

[2]Dept. of Biomedical Engineering, University of Michigan, Ann Arbor, MI, USA

ABSTRACT

We present a novel design of single cell-cell interaction platform for studying tumor-stromal interaction. In cell-cell interaction assays, monitoring single cell behavior is critical to study highly heterogeneous population (e.g., cancer). For example, metastatic cancer cells are typically transported as a single circulating tumor cell; therefore, tumorigenesis should be induced from a single cell because it might be quite different from co-culturing many cells. In this work, we report a microfluidic chip that allows for isolating a pair of cells by using immiscible liquid isolation. We present the method of using a semi-permeable membrane to retain secreted proteins for interaction, while maintaining the exchange of nutrition for long-term cell-cell interaction studies. We verified proliferation enhancement of tumor cells when co-cultured with stromal cells, demonstrating the feasibility of our device as a generic platform of cell-cell interaction studies.

INTRODUCTION

The cancer cell niche is a complex microenvironment where cancer cells, endothelial cells (EC), macrophages and mesenchymal stem cells (MSC) are coexistent [1], and tumor-stromal interactions can determine the development of tumor [2]. It is believed that tumor cells can exploit nearby normal cells to enhance growth, metastasis and drug resistance. Conventionally, cell interactions can be studied by co-culturing two different cell types in the same petri dish. However, this dish based co-culture model lacks in several key aspects to understand comprehensive cancer development. First, metastatic cancer cells are typically transported as a single circulating tumor cell (CTC); therefore, tumorigenesis should be induced from a single cell because it might be quite different from co-culturing many cells [3]. As cancer metastases account for more than 90% of cancer-related mortality [2, 4], modelling tumorigenesis process in an appropriate microenvironment is essential in metastatic analyses. Conventional dish culture cannot provide an accurate model of tumorigenesis processes because cell behavior can be affected by multiple neighboring cells in an uncontrollable manner. [5]. The second issue is imprecise spatial control. In conventional interaction assays, two cell populations are simply mixed in a dish, so the spatial distribution of two cell types is not uniform, resulting in significant variation from one place to another. Some cells may be surrounded by many of the other type of cells in one region, while others may aggregate with the same type of cells in another region. It is difficult to achieve precise ratio controlled co-culture in the conventional culture platforms. Third, dish-based culture lacks in the ability to use small samples (<1000 cells). This important because CTCs and primary samples are more often available in a small sample. Finally, for highly heterogeneous cancer populations, dish-based co-culture can only monitor the average behavior rather than tracking individual cell behavior. This can be an issue because some sub-populations in tumors have different interaction pathways.

There are a number of previous works reporting on microfluidic platforms for cell-to-cell interaction studies, but most of them still load hundreds or thousands of cells in the device. As a result, these platforms suffer from the same issues as conventional co-culture in Petri dishes [6-9]. Droplet based technology can provide high-throughput combinatorial co-culture at single cell level [10-13]. However, there are two fundamental limitations in the droplet based cell culture. First, it has low cell viability. Most mammalian cells are adherent cells; therefore, suspension in a droplet can lead to anoikis, resulting in cell apoptosis. Second, it is difficult to continuously provide fresh media. Thus, the nutrition in the media depletes over time. This limited capability of long-term culture, droplet-based cell culture is confined to suspension cell culture and short-term (typically less than one day) assays.

In the past, our group reported two different cell-to-cell interaction chips [14-15]. The first one used pneumatic valve for microwell isolation. As secretion based cell-to-cell interactions need to accumulate cytokines from cell secretion, continual media perfusion is blocked by pneumatic valve operation using the deformation of a thin PDMS membrane in the microfluidic channel [14]. In the second chip, we realized valve-less isolation by using electrolytic bubble generation [15]. This work utilized a novel mechanism of isolating the microwells without using an external pump; however, it still requires electrical control to generate bubbles and would not be optimal for long-term cell-cell interaction studies.

In this work, we propose to use a semi-permeable membrane for long-term cell-cell interaction studies without using any external control of microwell isolation. The semi-permeable membrane under each microwell can provide stable supply of nutrition for cells, while retain the secreted signaling proteins for interaction. We incorporate the immiscible oil isolation to achieve stable channel isolation in a simple and robust way. As a result, the device can operate without any external components such as micropumps or electrical control signals. For proof of feasibility, we demonstrated the interaction between UMSCC1 (head and neck cancer) cells and Endothelial cells

(EC) by secreted growth factors. The secretion of ECs can boost the growth of UMSCC1 cells as compared with the control experiment of culturing UMSCC1 cells alone.

MATERIALS AND METHODS

Microfluidic Device Operation

The presented platform composes of cell-culture chambers sandwiched by two parallel oil channels, media exchange channels for nutrition supply, and semi-permeable membranes (2k Daltons Cut-off molecular weight) between the cell culture chamber and the media exchange channel (Fig. 1). Nutrition can be supplied to the cells through the semi-permeable membrane, but the secreted cytokines are accumulated inside the chamber because their molecule sizes (typically tens of kDa) are too large to escape. First, cells are loaded and captured in each chamber. During this process different cells are paired. After that, the oil is introduced from left to right in the upper layer to isolate culture chambers by immiscible oil. Secreted cytokines are accumulated inside the chamber for cell-cell interaction, while the nutrition can be steadily supplied through a semi-permeable membrane.

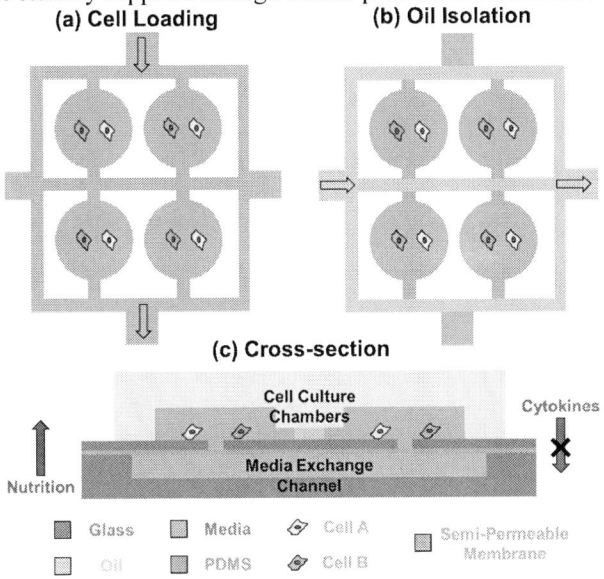

Figure 1: Schematics of the proposed cell-niche-on-chip: (a) cell loading, (b) oil isolation and (c) cross-sectional view of the device.

Device Fabrication

Three masks were used to fabricate a SU8 master mold: the first mask for a shallow (15 μm) interaction bridge, the second mask for microfluidic channels and cell culture chambers (40 μm), and the third mask for oil isolation channels (100 μm). The PDMS (PDMS, Sylgard 184, Dow Corning) layer was fabricated by the standard soft lithogrpahy processes. For vertical connection layers, we used a 100 μm-thick fused silica wafer. A 50μm SU-8 layer was patterned as the mask for DRIE. The fused silica was etched through by DRIE (Pegasus glass etcher), and the residual SU-8 was removed by PG Remover (Microchem). The media exchange channel was formed by HF etching of a glass substrate. The media exchange layer has many pillars to support the semi-permeable membrane on top. The PDMS

channel layer and the vertical connection layer were treated by oxygen plasma and then aligned and bonded together. Finally, the bonded PDMS-fused silica, semi-permeable membrane and media exchange layer were all assembled and sealed by UV cured Epoxy (OG147, Epoxy technology).

Cell Culture Experiment

USMCC1 cells were cultured with DMEM (Gibco 11965), 10% FBS (Gibco 10082), and 1% Pen/Strep (Gibco 15140), and endothelial cells were cultured with MEGM (Lonza CC-3150). In the device preparation, the substrate was coated with Collagen (BD 354236) overnight before the cell loading to enhance the cell adhesion. Before cell loading, trypsin/EDTA (Gibco 25200) was used to detach cells from their polystyrene culture dishes, and the detached cells were re-suspended to 10^5 cells/mL in culture media, and 100 μL of cell solution was pipetted into the inlet of the microfluidic platform. We mixed two cell types by a ratio of 1:1 to achieve cell pairing when loaded into the chambers. The culture media was pipetted into the inlet, and the loaded cells were cultured in an incubator. After cell adhesion (one day), the media was replaced by serum free MEGM, and the oleic acid (immiscible oil) was injected from left to right by negative pressure (1psi). We changed the media in the media exchange layer every day to supply nutrition to the cells in each chamber through the semi-permeable membrane.

RESULTS AND DISCUSSION

Cell Capture

In order to capture specific number of cells in each culture chamber, cellular valving mechanism are used to load the cells hydrodynamically at each capture site [16-17]. To capture one cell in each capture site, two paths are created in the design: a central path and a serpentine path. The hydraulic resistance of each path is inversely proportional to its flow rate. The long serpentine structure is designed to increase hydrodynamic resistance, so that the flow rate in these paths is lower than that of the central path. Thus, the cells are likely to be guided to the central path. Since the opening of the central path is slightly smaller (Height: 10 μm, Width: 10 μm) than the size of typical mammalian cells, the cells are sterically captured and plugs the gap, blocking the flow through the central path. With proper geometry design, a capture rate of ~90% can be achieved [17].

We have two capture sides in each chamber. 90% of capture sites can capture exactly one cell, therefore we can achieve a high cell-pairing rate in each chamber. Once two capture sites capture sites are filled with cells, the flow resistance through the central path is higher than that of serpentine paths, so the next coming cells will flow through the serpentine paths to the downstream. The same mechanism can work for higher number of capture pairing in the next chamber. As the size of most mammalian cells is similar, there is no selectivity on cell type. Thus, the ratio of captured cells will be similar to the composition of cells in cell solution. For co-culture of two cell types, we loaded the mixed cell solution of 1:1 ratio to maximize the probability of cell-pairing. Fig. 2a shows a pair of one UMSCC1 cell and one endothelial cell for interaction. Using two capture sites in each chamber, 70% chambers capture either one or two cells (Fig. 2b), so we can generate different combinations of cell pairing simultaneously in a single assay. The cell behavior of different combinations can be compared side by side to obviate device-to-device variation. As more cells (up to five) can be loaded in the same manner, the platform grants the potential of modeling cancer cell niches by co-culturing up to five different cell types or different ratios of cell types in a chamber (Fig. 2b).

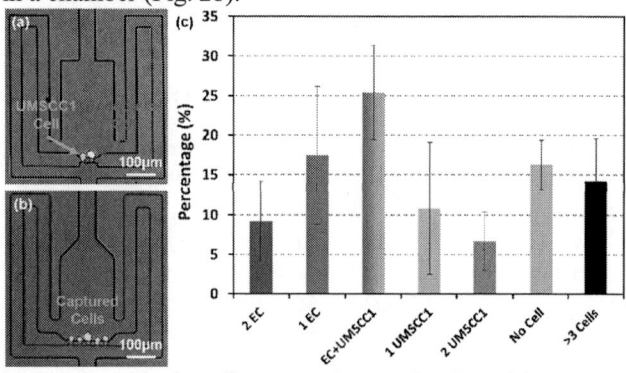

Figure 2: Multiple cells captured in a chamber: (a) a pair of UMSCC1 and endothelial cells (EC), (b) five UMSCC1 cells, and (c) the capture rate of different cell combinations.

Oil Isolation

The immiscibility between oil and water can be as an ideal way of chamber isolation (oil-water two-phase isolation). Previous works demonstrated isolation of water droplets in oil by optimizing channel geometry and hydrophobicity. Pico-liter water droplets can be generated in oil [18-20]. Each single droplet can be used as a nano-lab for assay [21]. Mammalian cells were cultured in droplets, but the cell anoikis and media depletion in the droplet confine it to short term (less than one day) culture [10-13]. On the contrary, the conventional media perfusion platform can allow cell culture longer than two weeks without affecting cell viability; however, isolating microenvironments needs bulky external components such as pneumatic pumps or function generators. In this work, we combine the advantages of two platforms, integrating immiscible isolation for adherent cell culture [17, 18] by incorporating a semi-permeable membrane under each chamber for continuous media perfusion.

In order to provide high cell viability in the long-term culture in our application, we optimize channel geometry to control the oil flow. We designed a higher and wider channel (100 μm by 100 μm) for oil isolation paths and a narrower design (30 μm by 40 μm) for cell loading channels. In this channel configuration, although oil has poorer affinity to the protein-coated hydrophilic channels, the oil flow driven by the negative pressure can still fill the wider channel and thus completely isolate the cell culture chambers. It is difficult for oil to invade the cell culture chamber because the collagen coated PDMS is hydrophilic. As a result, the channel geometry can guarantee good oil isolation while protecting cells inside the culture chamber. Fig. 3 shows the immiscible

oil isolation process in the channel. A pair of cells were loaded in the chamber as shown in the Fig. 3a. When negative pressure was applied from the left, the oil filled the horizontal channels to isolate the culture chamber. As we balanced the pressure difference between all horizontal channels, isolation process did not affect the cells loaded in the device.

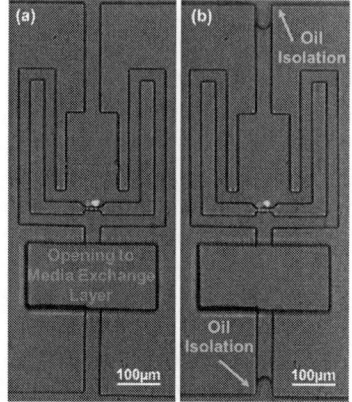

Figure 3: Oil isolation: (a) before and (b) after oil introduction. The culture chamber forms an isolated microenvironment.

Continuous Media Supply and Cytokine Accumulation

The fundamental dilemma in cell-cell interaction is to achieve the balance between accumulating secreted signals and supplying enough nutrition. It is difficult to decide the proper isolation time, especially for cancer cells, a highly heterogeneous population. Low metabolism rate cells, which are likely to be quiescent and drug resistant, may need a long interaction time, while high metabolism rate cells, which contribute to rapid growth, may need a short interaction time. After culturing several days, the difference in proliferation (some chambers with more cells and others with less cells) will make the situation even more complicated. Instead of struggling with taking a delicate balance, we present a platform that is capable of retaining the secreted signal proteins while providing stable nutrition.

The selectivity of semi-permeable membrane, which allows the small molecular weight nutrition to pass while retains the large molecular-weight signaling proteins, gives a viable solution. The membrane is sandwiched between the media exchange layer and the cell-culture chamber, so it can regulate the media exchange. As molecular weights of secreted proteins are typically larger than 2k Daltons, they will accumulated inside the culture chamber, inducing cell-cell interaction. Only the small molecules such as glucose and amino acids can pass the membrane, allowing continuous perfusion of nutrition from the media.

Cell-Cell Interaction Experiments

As a proof of concept, we demonstrated cell interaction between UMSCC1 cells and endothelial cells (EC). Endothelial cells are known to secrete a number of growth factors to enhance the growth of tumors [22]. Two conditions: co-culture of "one UMSCC1 and one EC" and single cell culture of "one UMSCC1" were compared in the experiment. After cell loading, the chambers were isolated by immiscible oil scheme for three days. The EC was labelled with orange fluorescent (Invitrogen, C2927), while the UMSCC1 was labelled by green fluorescent dye (Invitrogen, C2925).

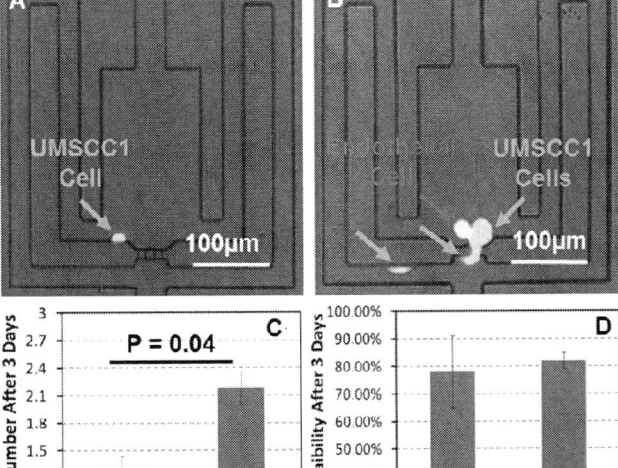

Figure 4: Cell Interaction between UMSCC1 and EC for three days: (a) a single UMSCC1 cell after 3-day culture, (b) a pair of one UMSCC1 and one EC - UMSCC1 after 3-day culture, (c) The proliferation rate, and (d) viability of UMSCC1 in the chamber after 3days.

With the EC in the chamber, the secreted cytokines were accumulated and boosted the growth of UMSCC1. Fig. 4 shows the proliferation results after three days. The isolated single cell (UMSCC1) barely proliferated, while the UMSCC1 cell co-cultured with one EC proliferated to three cells (Fig. 4b). The proliferation rate of co-cultured UMSCC1 cells is by two times higher than that of the isolated UMSCC1 cells (Fig. 4c). Both isolated and co-cultured UMSCC1 cells showed good cell viability, implying stable nutrition supply through a semi-permeable membrane (Fig. 4d) during the course of experiments. These preliminary results successfully demonstrated the capability of our device to retain the secretion factors for interaction while provide stable media perfusion through semi-permeable membrane to maintain good cell viability.

CONCLUSIONS

We have successfully implemented a cell-cell interaction platform that can co-culture arbitrary number of cell combination of 1-5 cells in one chamber. High cell capture rate over 90% and reliable oil isolation have been achieved. A semi-permeable membrane has been integrated to between the cell culture chambers and media exchange channels to retain the secreted proteins inside the chamber for cell-cell interaction, while allowing continuous nutrition supply from the media channels. The preliminary experiments have confirmed the increase in proliferation of cancer (UMSCC1) cells when co-cultures with endothelial cells (EC), demonstrating the feasibility of the proposed microfluidic platform for studying tumor-stromal interaction by controlling microenvironments in cell niches.

ACKNOWLEDGEMENTS

This work was supported in part by the Department of Defense (W81XWH-12-1-0325) and in part by the National Institute of Health (1R21CA17585701). The cells are provided by Prof. Nor's lab at the Dental School, the University of Michigan.

REFERENCES

[1] R. Peerani, et al, J. Clin. Invest, 120, 60–70, 2010.
[2] D. Hanahan, R. A. Weinberg. Cell, 144(5),646-74, 2011.
[3] M. Cristofanilli, et al, N Engl J Med, 351(8), 781-91, 2004.
[4] J.B. Mina, et al, Nature Medicine, 17, 320-329, 2011.
[5] P. Mignatti, T. Morimoto, D.B. Rifkin, Proc Natl Acad Sci U S A, 15, 88(24), 11007-11, 1991.
[6] A.Y. Hsiao, et al, Biomaterials, 30(16), 3020-7, 2009.
[7] M. Bauer, G. Su, D. J. Beebe, A. Friedl, Integr Biol., 2(7-8), 371-8, 2010.
[8] Y. Gao, et al, Biomed Microdevices, 13(3), 539-48, 2011.
[9] D. Majumdar, et al, J Neurosci Methods, 196(1), 38-44, 2011.
[10] E. Tumarkin, et al, Integr. Biol., 3, 653-662, 2011.
[11] J. Pan, et al, Integr. Biol., 3, 1043–1051, 2011.
[12] T. P. Lagus, et al, RSC Adv, 3, 20512, 2013.
[13] A. Huebner, et al, Lab Chip, 9, 692–698, 2009.
[14] P. Ingram, Y. J. Kim, T. Bersano-Begey, X. Lou, A. Asakura, and E. Yoon, Proceeding of MicroTAS, 277-279, 2010.
[15] Y.-C. Chen, X. Lou, P. Ingram, and E. Yoon, Proceeding of MEMS, 792-795, 2012.
[16] J. Chung, Y. J. Kim and E. Yoon, Appl. Phys. Lett, 12, 3701-3703, 2011.
[17] Y.-C. Chen, P. Ingram, X. Lou, and E. Yoon, Proceeding of MicroTAS, 1241-1244, 2012.
[18] S.-Y. The, et al, Lab Chip, 8, 198–220, 2008.
[19] W.-A. C. Bauer, et al, Lab Chip, 10, 1814–1819, 2010.
[20] W.-H. Tan, et al, Lab Chip, 6, 757–763, 2006.
[21] T. G. Mira, et al, Lab Chip, 12, 2146–2155, 2012.
[22] G. Kathleen, et al, Neoplasia, 11(6), 583-593, 2009.

DEVELOPMENT OF VACUUM ASSISTED MICROFLUIDIC CELL TRAPPING DEVICE FOR REPOSITIONING OF OOCYTE INTRACELLULAR CHROMOSOMES

Juhee Hong, Prashant Purwar, Sungjoon Lee, Neha Verma, and Junghoon Lee
Seoul National University, Seoul, Republic of Korea

ABSTRACT

We report the design and micro-fabrication of vacuum-assisted microfluidic trapping device for the rapid capture of single cell such as an oocyte. We also suggest an application of such device for monitoring an intracellular chromosome inside cell that has an interaction with external environments. Real time monitoring is enabled through the fabrication on a silicon-on-glass substrate, offering excellent optical imaging window. We demonstrate the single cell capture event and monitoring of chromosome activity. This result will provide a powerful tool for investigating the physiological and pathological cellular functions.

INTRODUCTION

Trapping of a cell plays a key role in biological applications such as molecules/drug screening and investigation of biological information at cellular level [1, 2]. High through put devices are required to understand the biophysical and biochemical phenomena in general with statistical significance. Thus, various types of cell trapping techniques have been reported, such as dielectrophoresis, laser, acoustic, and magnetic techniques [3-6]. However, there are some challenges to overcome limitations in most of these manipulative methods because they require complicated preparation and expensive equipment, and can damage live biological samples. Although the dielectrophoresis based cell tapping devices have high through put but there is a need of AC electric field which can lead to cell damage. Laser based optical tweezers could be a choice for highly controlled movement and trapping of the cell but high intense laser beam may also damage the cell. Moreover equipment is quite complex, expensive and lack of handling multiple cells simultaneously.

Recently, the pneumatic technique has attracted attention because of capabilities in high throughput and selective capture. Liu *et. al* developed a similar vacuum-assisted cell trapping device for microinjection in mouse zygote [7]. However, it was hard to control the roughness of glass surface and etching profile induced by isotropic wet etching [7]. This would result in the distortion of image when a microscopy is used to investigate the state of cells. Thus, we used dry etching process to etch glass with additional steps for high yield fabrication of cell trapping device, and demonstrated the cell trapping and chromosome relocation using this device. This result would provide a powerful tool for investigating the physiological and pathological cellular functions associated with the spatiotemporal distribution of cellular components without interfering with the normal functions of other cytosolic components. Here we have shown intracellular chromosome relocation in live oocyte using cell trapping device.

DEVELOPMENT OF VACUUM ASSISTED MICROFLUIDIC CELL TRAPPING DEVICE

Considering the size of mouse oocytes (ø: 70-80 μm), holes of various sizes (ø = 20~40 μm) were designed to ensure the optimal capturing condition for mouse oocyte. Figure 1(a) and (b) show schematics of side and top views of the device, respectively. Device has two chambers such as cell capturing chamber and suction chamber. Suction chamber is used to control the pressure of cell capturing chamber by application of negative pressure through it. The device is designed to handle nine oocytes simultaneously. Pressure of the chamber was controlled by a micromanipulator attached to an inverted microscope through suction part. Dimensions of pattern are shown in Figure 1(b).

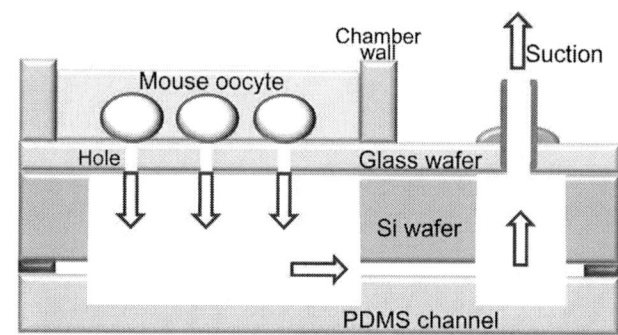

(a)

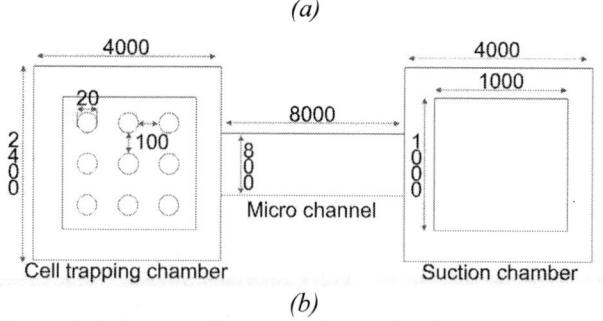

(b)

Figure 1: Schematic diagram of microfluidic oocyte trapping device. (a) side view of the device, (b) top view diagram of microfluidic cell trapping device. Dimensions are in micrometer and not in scale.

978-1-4799-3510-9/14 $31.00 © 2014 IEEE

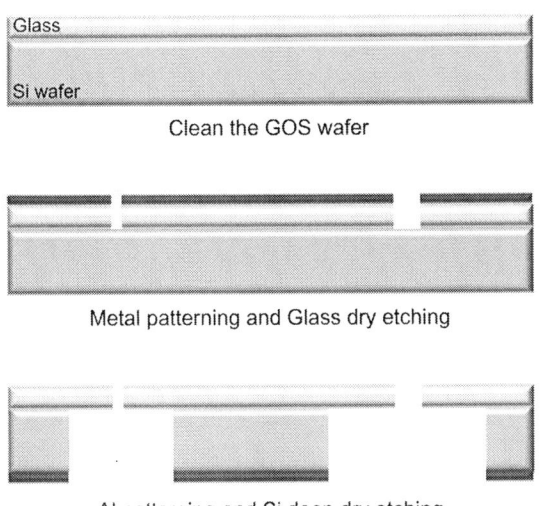

Clean the GOS wafer

Metal patterning and Glass dry etching

Al patterning and Si deep dry etching

Micro-trapping part

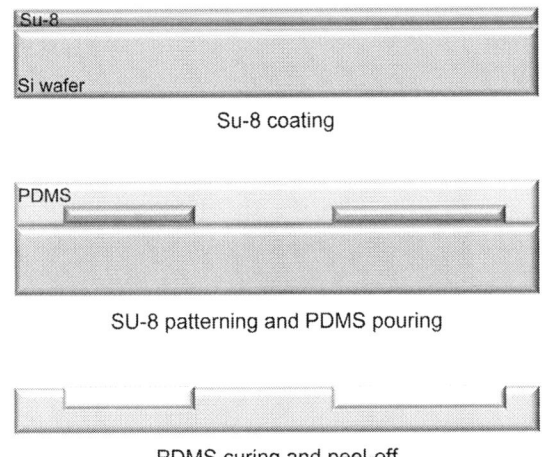

Su-8 coating

SU-8 patterning and PDMS pouring

PDMS curing and peel-off

Microfluidic-channel part

(a)

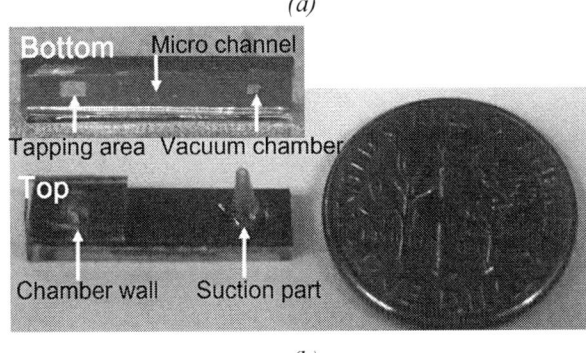

(b)

Figure 2: Microfluidic cell trapping device (a) Process flow for fabrication, (b) Fabricated microfluidic trapping device.

Figure 2(a) shows the fabrication process flow. The device is fabricated in two parts and integrated together. First part consists of the fabrication of cell trapping and suction through-in holes in a glass on silicon (GOS) wafer. Aluminum layer (2 and 0.3 μm, respectively) was used as a patterning mask to etch the glass and silicon from top and bottom sides, respectively. Bottom substrate was fabricated by transparent poly dimethyl-siloxane (PDMS) replica molding process on SU-8 mold. A mold was created with SU-8, spin-coated and patterned. The PDMS mixture (curing agent: PDMS = 1: 10) was placed in a vacuum chamber for 1 hr to remove the trapped gases, and then poured onto the mold. After PDMS curing, the separate parts were assembled into the final microfluidic trapping device using a biocompatible UV epoxy.

EXPERIMENTAL SETUP

Before applying the microfluidic cell trapping device to oocyte, we tested it by using micro poly bead (~70 μm) similar size to mouse oocytes. Pressure of the chamber was controlled by a micromanipulator (Narishige, Japan) attached to inverted microscope (TE 2000-E, Nikon, Japan) as shown in Figure 3(a).

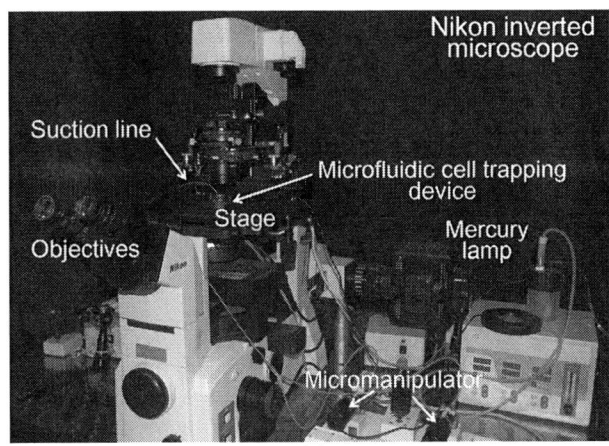

(a)

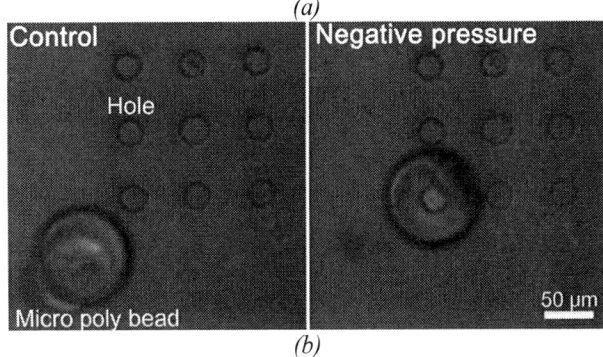

(b)

Figure 3: Micro poly bead trapping in hole. (a) Experiment setup for oocyte trapping on inverted microscope, (b) Micro poly bead trapping in the microfluidic trapping cell device. When negative pressure was applied, micro bead was trapped on hole till the negative pressure was continued.

Figure 3(b) shows the micro poly bead and the holes in trapping area captured by the inverted microscope. It is found that the micro poly bead image on trapping area is clear without the distortion of image. When the negative pressure was applied to the device, the micro poly bead was rapidly attracted towards the holes.

REAL TIME CHROMOSOME REPOSITIONING

After testing cell trapping device, chromosome manipulation in live oocyte was performed in real time. To end this, we used bacterial magnetic particles (BMPs) functionalized with Fluorescein isothiocyanate (FITC) fluorescence-labeled H1 histone antibodies (H1-BMPs), specifically targeting the chromosomal histone H1 in mouse oocytes. Produced by magnetic bacterium, the BMPs (~50 nm) are biocompatible and well dispersed in aqueous solutions because they are enclosed by lipid bi-layer membranes [8]. Functionalization can be easily carried out by using various proteins embedded in the enclosing lipid bilayer [9]. Furthermore the BMPs have much higher magnetic moments than most commercially available paramagnetic particles due to the presence of ferromagnetism with a strong magnetic susceptibility [10, 11]. Histone H1 protein in chromosome is the key candidate to target the chromosome by magnetic particles since it is in abundance at chromosomal surface. FITC was used to track the BMP position inside oocyte. H1-BMPs were produced through labeling and conjugation processes. The fluorescence labeling of the antibodies was carried out prior to the BMP-antibody conjugation. The FITC dye was linked to the H1 histone antibody (IgG, Chemicon, USA) with a protein labeling kit (Pierce, USA), followed by a dialysis process to remove excess FITC. For a specific targeting of the chromosomes inside a cell, the BMPs were conjugated with the fluorescence-labelled H1 histone antibodies by using glutaraldehyde as cross-linker since the terminal group of BMP surface is amine protein. The H1-BMPs created this way were then concentrated with a magnetic bar to increase the purity.

BMPs are 50 nm in diameter so they are unable to enter inside the cell nucleus since the pore size of nucleus membrane is smaller than 50 nm [12]. Thus, chromosome targeting efficiency can increase with the absence of nuclear membrane. To remove nuclear membrane, oocyte were first treated with nocodazole chemical which depolymerizes the microtubule and hence arrest the cell at prometaphase. At this phase, there is no nucleus membrane and hence it is easy to target the chromosome by H1-BMPs.

To monitor the chromosome movement inside live cell, portable incubator was used for long cell culture as shown in Figure 5. To target chromosome, H1-BMPs were delivered into live oocyte for 12 hrs.

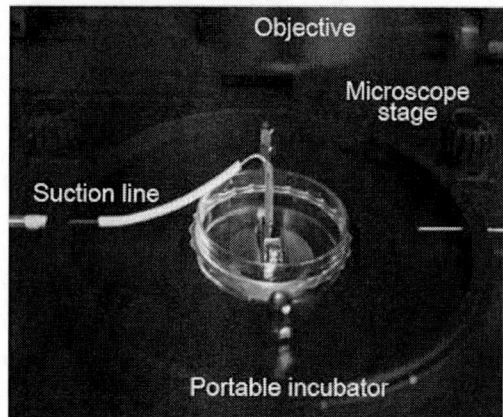

Figure 4: Experimental setup of oocyte trapping in the microfluidic trapping device

After targeting chromosome, whole extracellular H1-BMPs were removed with pipetting method to prevent the unexpected chromosome movement by any extra stimuli. And then, oocytes were immobilized on cell trapping device after applying negative pressure through the control of pressure. The mineral oil was used to prevent the evaporation of culture medium. After all conditions were stable, the experiment was carried out.

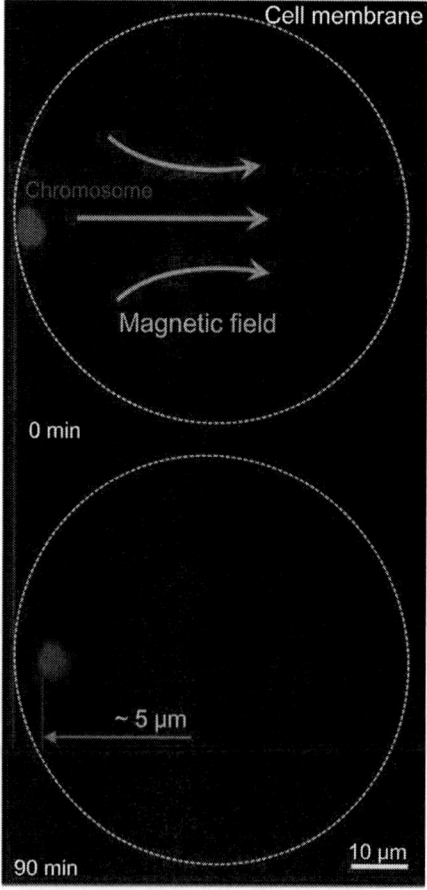

Figure 5: Real time H1-BMP bound chromosome movement (Hoechst labeled; blue in color) under an external lateral direction magnetic field gradient

For real time observation of chromosome movement, the chromosome was stained by Hoechst (blue color dye staining DNA specifically). Strong non-uniform static magnetic field was applied through neodymium magnetic at a distance of ~5 mm. It was calculated that the pico-newton range magnetic force was exerted on single BMP. Chromosome relocation was observed with the fluorescence microscopy for 90 min at interval of 5 min. When attracted by the magnetic field gradient, the chromosomes slowly moved in the direction of the magnetic field as shown in Figure 6. During 90 min, chromosomes moved up to ~5 μm despite the disturbance of cytosol viscosity and entanglement resistance of cytoskeleton filaments

RESULT AND DISCUSSION

Good working high throughput vacuum assisted microfluidic device is fabricated. Device can hold 9 oocytes at a time. Moreover device is capable to hold oocyte tightly compare to other existing cell trapping techniques to prevent trans-rotational motion at the time of real time chromosome relocation. In chromosome relocation, H1-BMP conjugate was successful synthesized and targeted the chromosome with high specificity. Since BMPs are good in magnetic property around 5 μm relocation of chromosome was observed in 90 min interval under external magnetic field as shown in Figure 6.

CONLCUSION

We demonstrated design and fabrication of vacuum assisted microfluidic cell tapping device, targeted delivery of magnetic nanoparticles, the cell trapping and real time chromosome relocation using microfluidic cell trapping device. This result would provide a powerful tool for investigating the physiological and pathological cellular functions associated with the spatiotemporal distribution of cellular components without interfering with the normal functions of other cytosolic components.

REFERENCES

[1] D. Castel, A. Pitabal, M.A. Debily, X. Gidrol, "Cell microarrays in drug discovery", *Drug Discov.,* Today, vol. 11, No. 13/14, pp. 616-622, 2006.

[2] D.S. Chen, M.M. Davis, "Molecular and functional analysis using live cell microarrays", *Curr. Opin. Chem. Biol.* vol. 10, Issue 1, pp. 28-34, 2006.

[3] A. Rosenthal, B.M. Taff, J. Voldman, "Quantitative modeling of dielectrophoretic traps", *Lab Chip,* Issue 6, pp. 508–515, 2006.

[4] A. Ashkin, "Optical trapping and manipulation of neutral particles using□lasers", *Proc. Natl. Acad. Sci.,* vol. 94, No. 10, pp. 4853–4860, 1997.

[5] N. R. Harris, M. Hill, R. Townsend, N. M. White, S. P. Beeby, "Performance of a micro-engineered ultrasonic particle manipulator", *Senosr and Acuators*, vol. 111-112, pp. 481-486, 2005.

[6] H. Lee, A. M. Purdon, R. M. Westervelt, "Micromanipulation of biological systems with microelectromagnets", *IEEE Trans. Magn.* vol. 40, No.4, pp. 2991–2993, 2004.

[7] X. Liu, Y. Sun, "Microfabricated glass devices for rapid single cell immobilization in mouse zygote microinjection", *Biomed Microdevices,* vol. 11, pp.1169-1174, 2009.

[8] Y. A. Gorby, T. J. Beveridge, R. P. Blakemore, "Characterization of the bacterial magnetosome membrane", J. Bacteriol., vol. 170, pp. 834-841, 1998.

[9] J. X. Kai Chen, X. Chen, "Production, modification and bio-applications of magnetic nanopraticles gestated by magnetotactic bacteria", *Nano Res.* vol. 2(4), pp. 261-278, 2009.

[10] R. P. Blackmore, "Magnetotactic bacteria", *Ann. Rev Microbiol.* vol. 36, pp. 217-238, 1982.

[11] Q. A. Pankhurst, J. Connolly, S. K. Jones, J. Dobson, "Applications of magnetic nanoparticles in biomedicine", *J. Phys. D: Appl. Phys.* vol. 36, pp. R167-R181, 2003.

[12] R. H. Kirschner, M. Rusli, T. E. Martin, "Characterization of the nuclear envelope, pore complexes, and dense lamina of mouse liver nuclei by high resolution scanning electron microscopy", *The Journal of Cell Biology*, vol. 72, pp. 118-132, 1977.

CONTACT

*J. Lee, tel: +82-2-880-9101; jleenano@snu.ac.kr

DIELECTROPHORETIC (DEP) SEPARATION OF LIVE/DEAD CELLS ON A GLASS SLIDE FUNCTIONALIZED WITH INTERDIGITATED 3D SILICON RING MICROELECTRODES

Xiaoxing Xing, and Levent Yobas

The Hong Kong University of Science and Technology, Hong Kong, P. R. of CHINA

ABSTRACT

This paper describes a microfluidic device which offers a dielectrophoretic solution to cell sorting via interdigitated 3D silicon (Si) electrodes on a transparent glass substrate. The Si electrodes with a self-aligned array of lateral rings were fabricated through a single mask process on an anodically bonded Si-glass wafer. The device has been characterized for separation of live/dead human colorectal carcinoma cells at select flow rates and voltages. The live cell capture and dead cell removal efficiencies both reached 90% at an AC activation of $5V_p$ 450kHz applied against a flow rate up to 0.25ml/hr. The device disclosed here, with successful demonstration of continuous live/dead cell separation, offers great potential as a bioparticle sorting front-end module for a lab-on-a chip system.

INTRODUCTION

Isolation of cellular subpopulation from heterogeneous cell mixture is essential prerequisite for various biological and biomedical assays. Food security, for instance, is an important public health concern that has promulgated the development of fast and sensitive methods for isolating and detecting pathogenic organisms. More recently, isolation of rare cells such as stem cancer cells has drawn increasing attention. Microfluidic or lab-on-a-chip (LOC) systems that enable miniaturization, integration and automation of such assay protocols offer versatile solutions to cell separation. One ubiquitous LOC technique is based on immunocapture and works through conjugation between the target cells and their specific cell ligands [1]. This approach achieves high purity but requires extra labeling procedure and faces difficulties with cell retrieval. Moreover, some cell types express low or no surface ligands. To address the issue, label free techniques using mechanical filters or inertial forces are proposed and yet applicable to only those cells with distinguishable size and stiffness [1, 2]. Besides, viability of the cells is compromised given the large shear forces that they are susceptible to in the fluidic channels [1].

Among various label-free approaches, dielectrophoresis (DEP) is a powerful and widely accepted technique that separates cells according to their dielectric fingerprints. DEP exerts a net electric field force on polarizable neutral particles in a non-uniform electric field. This force, based on the dielectric contrast between the particles and the suspension medium, either attracts the particles to the field maxima under positive DEP (pDEP) or repels them from such regions under negative DEP (nDEP). The fact that DEP force is strongly dependent on the cell viability makes it inherently suitable for the separation of live/dead cells. The cytoplasmic membrane is highly insulating at low frequencies but, once damaged, permeates ions and exhibits a conductivity rise by several orders of magnitude. Thus, the ion leakage through the permeable membrane of a non-viable cell alters the cytoplasmic conductivity [2, 3]. Such alteration in cell dielectric properties leads to a distinct DEP response and the separation of live/dead cells.

In DEP, the microelectrodes are of central importance as they define the non-uniform electric field distribution and the subsequent force field map. Conventional DEP devices for cell sorting exploit thin-film surface electrodes directly laid on the channel floor [3]. Such surface electrodes, although they are simple to fabricate, generate an electric field that exponentially dies off away from the surface. Placing the electrodes on both top and bottom surfaces addresses the challenge by generating a 3D more effective electric field [4], however, also requires the electrodes to be precisely aligned. As an alternative, insulator-based DEP (iDEP) has been proposed with a simplified fabrication process. It has been widely utilized for live/dead bacteria sorting, including Escherichia coli (E.coli) [2]. Recently, contactless DEP (cDEP), with the liquid electrodes physically isolated from the sample liquid, has been successfully demonstrated for live/dead separation of human leukemia cells [5]. The iDEP and cDEP, though fabricated in a simplified process opting out the electrode integration, need considerably high voltage supply which not only requires complex electronics but also imposes potential electric hazard.

Previously we have demonstrated an innovative DEP device with interdigitated 3D silicon (Si) ring electrodes and proved its functionality on polystyrene beads [6]. We fabricated the device on silicon-on-insulator (SOI) substrate through a single mask process. Here, we take a further step to integrate these 3D Si ring electrodes on a transparent glass slide to avoid the use of costly SOI wafers. Moreover the transparent substrate is compatible with the differential interference contrast (DIC) microscopy and thus more suitable for observing the cells. As shown in Figure 1, the interdigitated 3D Si microelectrodes present comb fingers perforated with an array of well-defined lateral rings on a glass substrate. The self-aligned ring array allows for many parallel streams, enabling a high throughput cell separation. We investigate the device capability for selective isolation of a cell subpopulation from a heterogeneous mixture by trapping viable human colorectal carcinoma cells (HCT116) while removing non-viable HCT116 cells. With a low activation peak voltage of $5V_p$ applied against a flow rate as high as 0.25ml/hr, nearly 90% efficiency has been achieved for both trapping live cells and removing dead cells.

978-1-4799-3510-9/14 $31.00 © 2014 IEEE

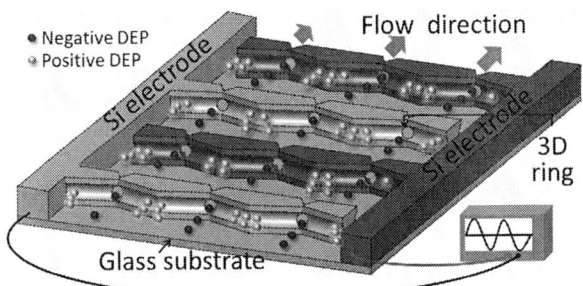

Figure 1: Device schematic illustrating interdigitated 3D Si microelectrodes with lateral rings on a glass substrate. Particles showing opposite DEP behaviors are arranged themselves into separate patterns.

THEORY AND SIMULATION

Dielectrophoresis (DEP)

DEP force is exerted on the induced dipole moment of a neutral body in a non-uniform electric field. The time averaged DEP force acting on a spherical particle (radius R) suspended in a medium with electric permittivity ε_m can be expressed as

$$\vec{F}_{DEP} = 2\pi\varepsilon_m R^3 Re[K^*(\omega)]\nabla\vec{E}_{rms}^2 \qquad (1)$$

where $\nabla\vec{E}_{rms}^2$ is the gradient of the squared electric field intensity. $Re[K^*(\omega)]$ is the real part of complex Clausius-Mossotti (CM) factor $K^*(\omega)$ which is described as

$$K^*(\omega) = \frac{\varepsilon_p^* - \varepsilon_m^*}{\varepsilon_p^* + 2\varepsilon_m^*} \qquad (2)$$

where ε_p^* and ε_m^* are the complex permittivity of the particle and the suspension medium, respectively. The complex permittivity is defined as $\varepsilon^* = \varepsilon - j\sigma/\omega$, where $j^2 = -1$, σ is the electric conductivity and ε is the dielectric permittivity. The complex permittivity of mammalian cells can be stated using single-shell model [7]:

$$\varepsilon_{cell}^* = C_m^* R[\frac{j\omega\tau_c + 1}{j\omega(\tau_c^* + \tau_c) + 1}] \qquad (3)$$

Here, $C_m^* = C_m + g_m/j\omega$ is complex capacitance with C_m the membrane capacitance per unit area and g_m the transmembrane conductance. g_m is negligible ($<10S/m^2$) for live cells yet not so for dead cells. The subscripts m and c refer to suspension medium and cytoplasm respectively. τ_c and τ_c^* are time constants defined as $\tau_c = \varepsilon_c/\sigma_c$ and $\tau_c^* = C_m R/\sigma_c$. The CM factor of cells can be calculated as

$$K^*(\omega) = -\frac{\omega^2(\tau_m\tau_c^* - \tau_c\tau_m^*) - j\omega(\tau_m g_c - \tau_c g_1 - \tau_m^* +}{\omega^2(2\tau_m\tau_c^* + \tau_c\tau_m^*) - j\omega(2\tau_m g_c + \tau_c g_1 + \tau_m^* +} \quad (4\text{-}1)$$

$$\frac{\tau_c^* + \tau_m) + g_1 - g_c - 1}{\tau_c^* + 2\tau_m) - g_1 - g_c - 2} \qquad (4\text{-}2)$$

where, $\tau_m = \varepsilon_m/\sigma_m$ and $\tau_m^* = C_m R/\sigma_m$ are time constants. g_1 and g_c are factors given by $g_1 = g_m R/\sigma_m$ and $g_c = g_m R/\sigma_c$. For a given cell suspension, the sign of $Re[K^*(\omega)]$ varies with frequency. A positive value suggests that cells experience pDEP and will be trapped near the regions of field maxima while a negative value implies nDEP with cells being repelled to the field minima.

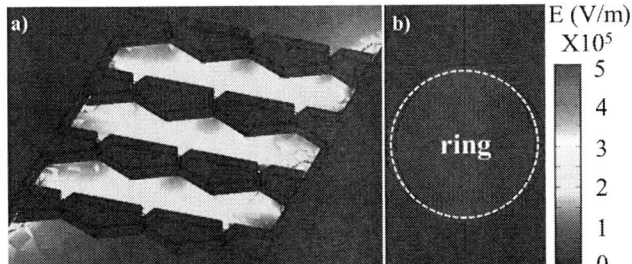

Figure 2: Numerical modeling for 3D distribution of electric field (a) between the electrodes and (b) inside the ring.

Numerical modeling

Figure 2 shows the 3D electric field within the device simulated through the finite element method (FEM) software (COMSOL Multiphysics 3.5) where the governing equation $\nabla\cdot((\sigma+j\omega\varepsilon)\nabla V) = 0$ was solved with the boundary condition set at an AC activation of $20V_p$ 450kHz. Electric parameters for solid/liquid domain are assigned with the values according to the experiments. As shown in figure 2a, the field gets intensified near the electrode fingers around the convex corners while being diminished near the rings reaching minimum at the ring centers (Figure 2b).

MATERIAL AND METHODS

Microfabrication

The fabrication process briefly described here has been adopted from [6]. First, a 200μm thick Si wafer with (100) orientation and 1Ω-cm resistivity was anodically bonded onto a 500μm thick glass wafer (800V 420°C and 600mbar). Second, a SiO_2 hard mask was patterned on the Si layer to designate the 3D electrodes and then the exposed Si area was etched in a deep reactive ion etching (DRIE) tool. Third, the entire topography was deposited with a thin layer of tetraethylorthosilicate (TEOS) upon which the TEOS was removed from the trench bottom through reactive ion etching (RIE), leaving the sidewalls passivated with TEOS. The bulk Si was subsequently isotropic dry etched to form the lateral rings, followed by the removal of the sidewall TEOS passivation in wet etching. Finally, the Si layer was further etched through a second DRIE step down to the glass layer, isolating the electrodes and defining the reservoirs.

Cells and reagents

HCT116 cells (ATCC) were cultured in a 5% CO_2 incubator at 37°C. Dulbecco's modified Eagle medium (DMEM) was used as a culture medium supplied with 10% fetal bovine serum (FBS) and 0.05% Pluronic F-68 (GE Healthcare Lifescience,Inc) to enhance the membrane stability. Before the experiment, cells were detached with trypsin-EDTA treatment and then suspended in the fresh culture medium. The harvested cells were divided into two batches for preparing live/dead cell mixture. The dead cells were prepared by heating one batch in 65°C water bath for 15min. Then the live and dead cells were respectively stained with Calcein AM (Life Technologies, Inc) at 2μg/ml and propidium iodide (PI, Sigma-Aldrich) at 50μg/ml in darkness

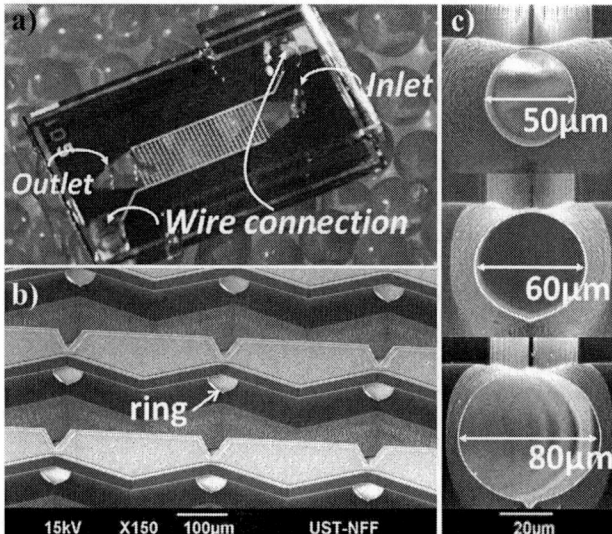

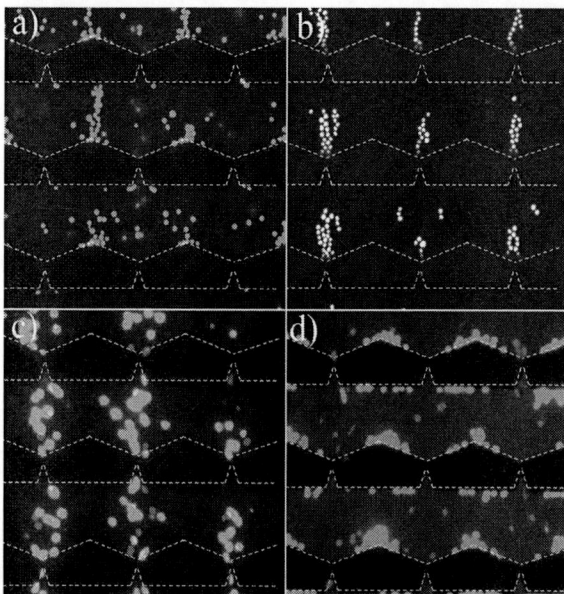

Figure 3: (a) Photograph of the overall device capped with PDMS. (b-c) SEM images of (b) the Si electrodes with rings partially covered by the DRIE profile on top and (c) the lateral rings fabricated with various diameters.

Figure 4: (a-b) Fluorescent micrographs showing live cells under (a) pDEP trapping at $10V_p$, 450kHz and (b) nDEP focusing at $10V_p$, 1kHz. (c-d) Superimposed images showing live (green) and dead (red) cell mixture at 0.15ml/hr (c) before and (d) during DEP activation at $5V_p$, 450kHz.

for 20min. DEP buffer (300mM D-Mannitol with a conductivity tuned to 100μS/cm using phosphate-buffered saline) was used to wash the cells twice at 100g for 5min, and to resuspend the live/dead cells at equal concentrations around 10^6 cells/ml. Prior to experiment, the chip was coated with 5% bovine serum albumin (BSA) in DEP buffer for 30min to minimize cell adhesion to the channel surface, followed by a priming step with DEP buffer. Live-cell capture efficiency was described as the percentage of captured live cells to total live cells injected. Similarly, dead-cell removal efficiency was defined as the percentage of removed dead cells to the total injected dead cells.

Experimental

The device was interfaced through a pair of inlet/outlet fluidic ports and of electrical vias in polydimethylsiloxane (PDMS) cover as depicted in Figure 3a. The sample injection rate was controlled by a syringe pump (Harvard Apparatus). The activation AC signal was delivered through a transformer connected to a power amplifier (AL-50HFA, Amp-Line Corp., NY) supplied by a function generator (Tektronix CFG250). The signal was monitored through an oscilloscope (Tektronix TDS 2012C). The image sequences were acquired by an epi-fluorescence microscope (Nikon ECLIPSE FN1) equipped with a colored CCD camera (SPOT, RT3 Mono).

RESULTS AND DISSCUSSION
Fabrication result

Figure 3b shows the interdigitated Si electrodes on glass from an oblique view where the rings are seen in part beneath the DRIE profile. The rings emerge laterally through the Si comb fingers upon the isotropic etch-fronts pinching underneath the necked segments from both sides. The ring could be tuned by adjusting the design parameters and

process time. As shown in Figure 3c, the electrode rings can be fabricated at various diameters within the same etching process, each corresponding to a different width of the necked segments (30μm, 20μm and 10μm).

Separation between live/dead HCT116 cells

Transition in DEP response has been studied using a homogeneous suspension of a single cell type. Crossover frequencies of live cells and dead cells have been separately assessed to determine a proper frequency band for their separation. As shown in Figure 4a-b, the live cells exhibit distinct DEP patterns at select frequencies. At intermediate frequencies (0.1 to 10MHz), the cytoplasmic membrane appears transparent to the field. Given the cytoplasm at a higher conductivity than the DEP buffer, this renders cells more polarizable. The cells polarized undergo pDEP and get trapped near the electrode edges where the field is maximum. At lower frequencies (<10kHz), the cytoplasmic membrane screens the field and leaves the cells less polarizable than the DEP medium. Therefore, the cells experience nDEP and remain focused at the ring centers where the field is minimum. As for the dead cells, they project a less clear transitional DEP pattern due to the fact that their leaky membrane leads to reduced cytoplasm conductivity and poor polarization contrast with the medium. At fairly high frequencies (>100MHz), the dielectric polarization governs the cell behavior and, with the cytoplasm at a lower dielectric permittivity than the DEP medium, cells show nDEP regardless of their viability.

Separation of live cells from dead cells is carried out at $5V_p$ and 450kHz. Live cells under this frequency show a pDEP response strong enough to overcome the fluid drag

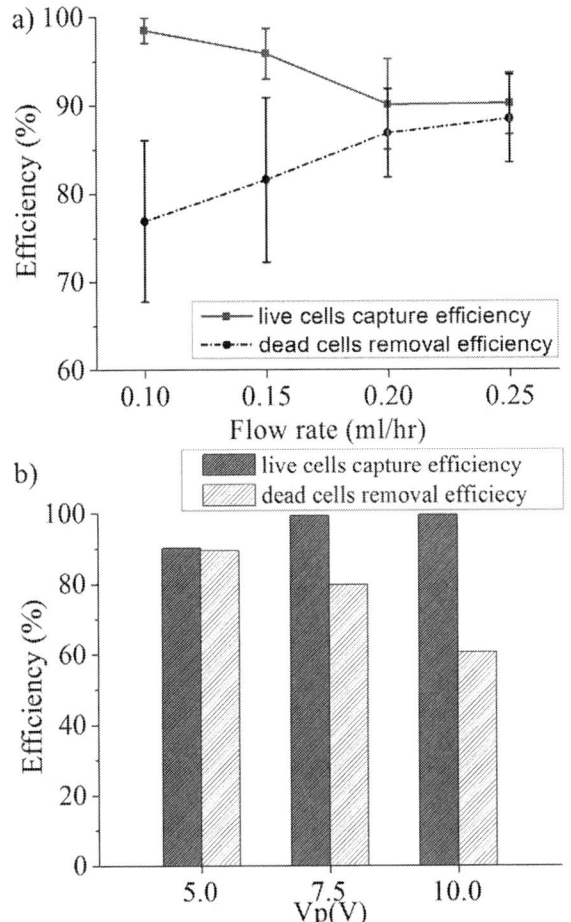

Figure 5: Device efficiency in trapping live cells and removing dead cells as a function of (a) flow rate at a fixed activation AC signal of $5V_p$ 450kHz and (b) peak voltage at a fixed flow rate of 0.25ml/hr and frequency of 450kHz.

force and to immobilize them around the electrode edges. On the contrary, the dead cells experience a weak pDEP that keeps them in the fluid streams. Superimposed fluorescent images in Figure 4c-d demonstrate these cell patterns.

The device has been further tested for the capacity to capture live cells and remove dead cells at select flow rates and peak voltages with a frequency at 450kHz. The device activated at $5V_p$ against 0.1ml/hr can achieve a capture efficiency of up to 98.50% (±0.0142) as shown in Figure 5a. With the increased flow rate up to 0.25ml/hr, the fluid drag force becomes more comparable to DEP force and hence the capture efficiency drops but remains over 90%. Sedimentation of the dead cells in the flow chamber accompanied by the overcrowding of the enriched live cells compromise their removal efficiency to a moderate value of 76.92% (±0.0916) measured against 0.1ml/hr. Increasing the flow rate up to 0.25ml/hr improves this value to 88.48% (±0.0498). Meanwhile, raising the activation voltage to $7.5V_p$ or $10V_p$ with the flow rate kept at 0.25ml/hr brings the capture efficiency up to nearly 100% while lowering the removal efficiency down to 61% due to stronger pDEP (Figure 5b).

CONCLUSION

This work presents a microfluidic chip with an elegant design involving Si ring electrodes on a transparent glass substrate. The 3D structure is fabricated through a single mask process, which allows for electrode rings of various diameters that could address a broad range of cell types such as bacteria, yeast, protoplasm and chloroplasts. A highly effective live/dead cell separation has been demonstrated with efficiency values considerably larger than those previously reported against a flow rate at an order of magnitude larger [5]. Given such high performance, the device shows great potential towards the isolation of rare cells such as circulating tumor cells (CTCs) and cancer stem cells (CSCs).

ACKNOWLEDGEMENTS

This work was supported by the Research Grant Council of Hong Kong under Grant 621711.

REFERENCES

[1]   D. R. Gossett, W. M. Weaver, A. J. Mach, S. C. Hur, H. T. K. Tse, W. Lee, H. Amini, and D. Di Carlo, "Label-free cell separation and sorting in microfluidic systems," *Analytical and Bioanalytical Chemistry,* vol. 397, pp. 3249-3267, Aug 2010.

[2]   B. H. Lapizco-Encinas, B. A. Simmons, E. B. Cummings, and Y. Fintschenko, "Dielectrophoretic concentration and separation of live and dead bacteria in an array of insulators," *Analytical Chemistry,* vol. 76, pp. 1571-1579, Mar 15 2004.

[3]   K. Khoshmanesh, J. Akagi, S. Nahavandi, J. Skommer, S. Baratchi, J. M. Cooper, K. Kalantar-Zadeh, D. E. Williams, and D. Wlodkowic, "Dynamic Analysis of Drug-Induced Cytotoxicity Using Chip-Based Dielectrophoretic Cell Immobilization Technology," *Analytical Chemistry,* vol. 83, pp. 2133-2144, Mar 15 2011.

[4]   S. H. Ling, Y. C. Lam, and K. S. Chian, "Continuous Cell Separation Using Dielectrophoresis through Asymmetric and Periodic Microelectrode Array," *Analytical Chemistry,* vol. 84, pp. 6463-6470, Aug 7 2012.

[5]   H. Shafiee, M. B. Sano, E. A. Henslee, J. L. Caldwell, and R. V. Davalos, "Selective isolation of live/dead cells using contactless dielectrophoresis (cDEP)," *Lab on a Chip,* vol. 10, pp. 438-445, 2010.

[6]   X. X. Xing, M. Y. Zhang, and L. Yobas, "Interdigitated 3-D Silicon Ring Microelectrodes for DEP-Based Particle Manipulation," *Journal of Microelectromechanical Systems,* vol. 22, pp. 363-371, Apr 2013.

[7]   T. B. Jones, *Electromechanics of Particles*: CAMBRIDGE University Press, 1995.

CONTACT

*L.Yobas, tel: +852-2358-7068; eelyobas@ust.hk

DROPLET DISPENSING AND SPLITTING BY ELECTROWETTING ON DIELECTRIC DIGITAL MICROFLUIDICS

N. Y. Jagath B. Nikapitiya[1], Seung M. You[1], and Hyejin Moon[2]*
[1]The University of Texas at Dallas, Richardson, USA
[2]The University of Texas at Arlington, Arlington, USA

ABSTRACT

The present study investigates two essential capabilities of electrowetting on dielectric digital microfluidics – 1) high precision and consistency in volume of a unit nanodrop dispensed from a reservoir, and 2) reduction of time to dispense and split drops. These capabilities are sought in applications that need tiny but accurate volume of liquid delivery at high flow rate. For this purpose, we mainly focused on geometry of electrodes and developed novel shapes and arrangements of electrodes in electrowetting on dielectric microfluidic device. To dispense liquid droplets in higher volume precision and consistency, a novel reservoir design named TCC reservoir is introduced. To reduce time to dispense a droplet, L-junction is developed. Y-junction is designed to split a droplet into two at the same speed as droplet dispensing speed by the L-junction, and hence increase the number of droplets arrive to the target area (i.e. at high frequency delivery).

INTRODUCTION

In an electrowetting on dielectric (EWOD) digital microfluidics (DMF), discretization of picoliter to microliter volume droplets and their motions are individually controlled by an applying external electric field to the designated electrodes within a device. Since compartmentalization (e.g. creating discretized droplet) and their motion on the surface (thus pumping of liquids) are solely done by electric field application, intricate systems to drive and regulate flow such as pumps and valves are not required in EWOD DMF. In addition, each created droplet is controlled individually, therefore positioning and multiplexing many droplets on 2-dimensional surface is achievable with ease. Due to these advantages of EWOD DMF, it has been developed to various applications including chemical [1] and enzymatic reactions [2], immunoassays [3], PCR [4], clinical diagnostics, proteomics [5], and electronics thermal management [6].

In most of these applications, flow rate, concentration (i.e. mass flow rate), and energy rate are totally dependent on the volume of a unit nanodroplet and their arrival frequency to a designated area. Therefore, precision and consistency of a unit nanodroplet and its dispensing frequency as well as motion speed are extremely important. Volume precision is defined by the difference between dispensed volume and volume subtended by the drop dispensing electrode. Precise drop dispensing is required to dispense predefined volumes and control of the volumes. Volume consistency is defined by the standard deviation associated with dispensed volumes. Consistent drop dispensing is important to maintain the regular drop size in repeating drop dispensing.

Several researchers have reported methods to enhance the volume precision and consistency of EWOD DMF [7-9] but further enhancement is still highly desirable. Moreover, speed of droplet dispensing has not been properly studied yet. Therefore, in this paper, we report these two essential capabilities of EWOD DMF. Although there are multiple parameters, we focused on geometry of electrodes and development of novel shapes and arrangements of electrodes in EWOD DMF to achieve the goal.

For precise and consistent droplet dispensing, a new reservoir electrode design, named "TCC" reservoir is developed. This design reduces the cutting length and the tail and to ensure the regular pinching-off location. To reduce the time to dispense a droplet, "L" shape electrode junction is presented. In addition, "Y" shape electrode junction is used to demonstrate fast droplet splitting into two, therefore the frequency of droplet supply to the designated area in the chip is achieved. All the operations are tested in parallel plate EWOD device in air environment. EWOD DMF chip fabrication can be found elsewhere [10]. Deionized water is used as the model liquid.

JUSTIFICATION OF DESIGNS
Precise and Consistent Droplet Dispensing Using TCC Electrodes

Controlled droplet dispensing on EWOD DMF chip is conventionally performed by extruding a liquid finger from the reservoir through activation of adjacent serial electrodes

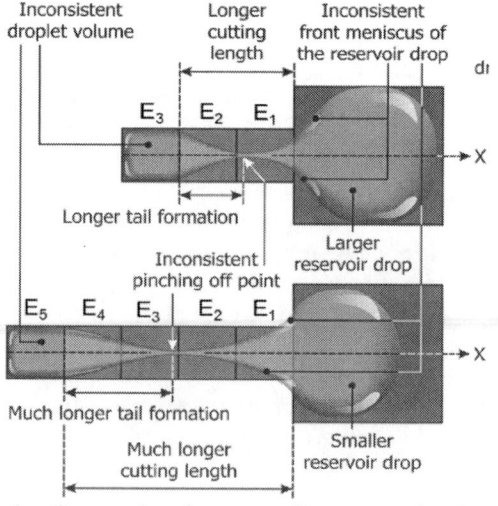

Figure 1: Conventional way to dispense a droplet in an EWOD DMF device.

[7 - 9] as shown in Fig. 1. The location of pinch-off depends on various factors such as the volume of the reservoir drop, applied voltage and surface conditions etc. After pinch-off, the irregular liquid tail adds extra amount of liquid to the already formed liquid droplet; hence the volume of unit droplet is poorly controlled. We speculate that irregular cutting length and pinch-off locations are due to the uncontrolled internal pressure difference between the liquid in the reservoir and the front of the liquid finger.

A novel design of reservoir electrodes is introduced to control the pressure difference between liquids in the reservoir and the extruded finger. We name this design TCC reservoir since it consists with one T-shape electrode and two C-shape electrodes (C_1 and C_2). One half of the square electrode is placed inside the T to reduce the cutting length and the tail formation. All the electrodes are arranged to be symmetric over the x-axis. Figure 2 illustrates the pinching-off of liquid drop in the TCC reservoir. The key of TCC reservoir design is to form fixed menisci (lines y_{1B} and y_{2B}) of liquid in the reservoir (on C_1) that are parallel to another fixed meniscus of extruded finger (line d). Such design ensures that curvatures (in topview) of both reservoir liquid and finger liquid are infinite so that the pressure difference between reservoir liquid and finger liquid are controlled the same. Therefore, dewetting menisci on T-electrode symmetrically move toward each other, and pinching-off always occurs on the fixed position (i.e. consistent in volume). In addition, the short cutting length results in higher precision in volume. Further elimination of cutting length and tail formation is achieved by introducing circular shape of electrode instead of square shape (square abcd) for liquid finger extrusion formation.

liquid and finger liquid forms passively relying on natural dewetting and unstable liquid column due to hydrophobicity of the deactivated surface, L-junction design actively forms liquid neck by using angular electrode. Moreover, L-junction design does not need opposite direction pulling force of liquid column but only forward force at the front leading meniscus is enough to form a neck and pinch-off. As shown in Fig. 3, while front meniscus of liquid moves through rectangular electrodes (E_{15}-E_{20}), de-wetting meniscus follows through angular electrodes (E_{11}-E_{14}) and forms small radius of curvature due to the forced acute angle of angular electrodes. Small radius of curvature creates higher airside pressure, which facilitates quick pinching-off.

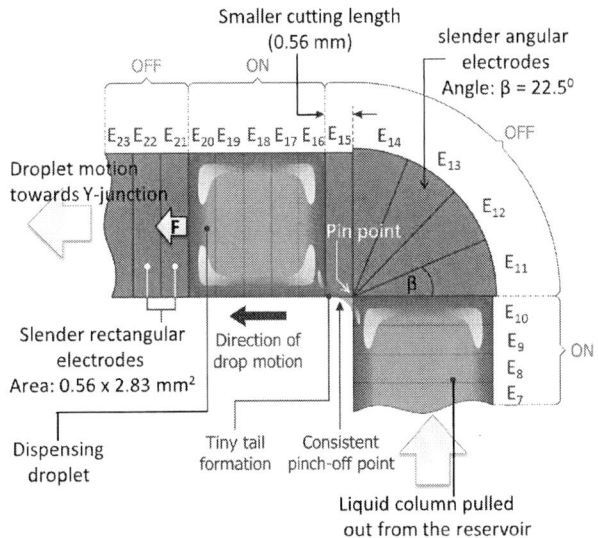

Figure 3: Dispensing a droplet from the L-junction electrode design. No backward force (downward direction) was required for pinching-off but only forward force at the front leading meniscus (block arrow with F) is needed.

Splitting a Droplet at Y-junction

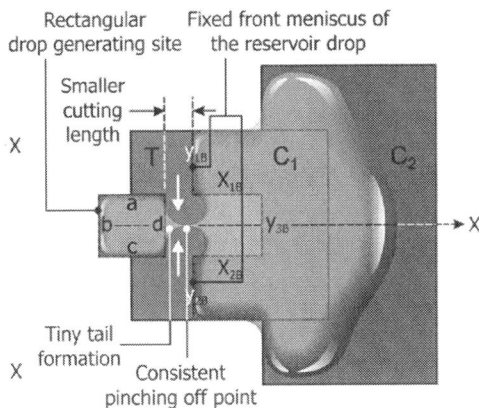

Figure 2: Droplet dispensing sequence using TCC reservoir electrodes design.

Fast Dispensing Using L-Junction

L-junction design is used to reduce the time to dispense a droplet while maintaining higher volume precision and consistency (Figure 3). L-junction consists of slender rectangular and angular electrodes. All the electrodes have the same area. Unlike the dispensing in both conventional and TCC reservoir where liquid neck between reservoir

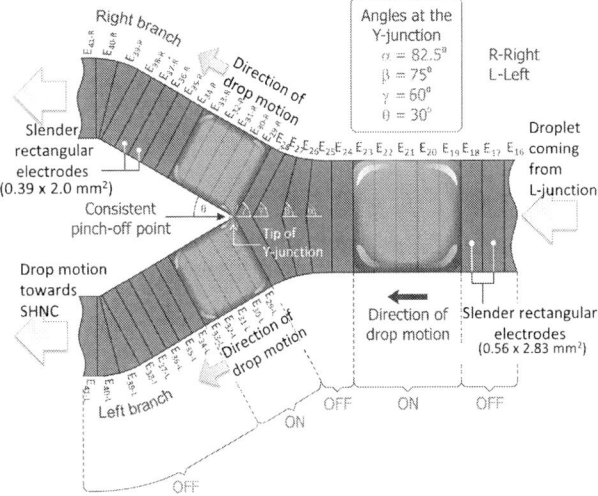

Figure 4: Splitting a droplet into two at the Y-junction

978-1-4799-3510-9/14 $31.00 © 2014 IEEE

Y-junction is designed to split a droplet into two at the same speed as it is dispensed by the L-junction. Entire Y-junction consists of slender rectangular electrodes. (Fig. 4) Corresponding electrodes at the left and right branches are connected together. In Y-junction splitting, the rear boundary of the droplet is always wetting with obtuse angles while the front meniscus is forced to make an acute angle by the inner edges of the Y-junction. This acute angle meniscus creates a large pressure and de-wetting force resulting in quick splitting of a droplet. Instead of generating a drop in small volume and delivering it to the designated area one by one, by using splitting a droplet into two by Y-junction, the droplet arrival frequency can be increased while decreasing the droplet volume.

RESULTS AND DISCUSSION

The experimental results are demonstrated in Figures 5-8. All micrographs are captured from high-speed video (frame rate: 1000 fps). Figure 5 demonstrates enhanced volume precision by using TCC reservoir design. As the reservoir drop is pulled by C_2, the front meniscus is fixed over the wetting edges a,b and c of square electrode and the edges y_{1B} and y_{2B} of the C_1. EWOD force supplied by C_2 is higher enough to push the de-wetting menisci on T towards each other until they meet and finally pinch off.

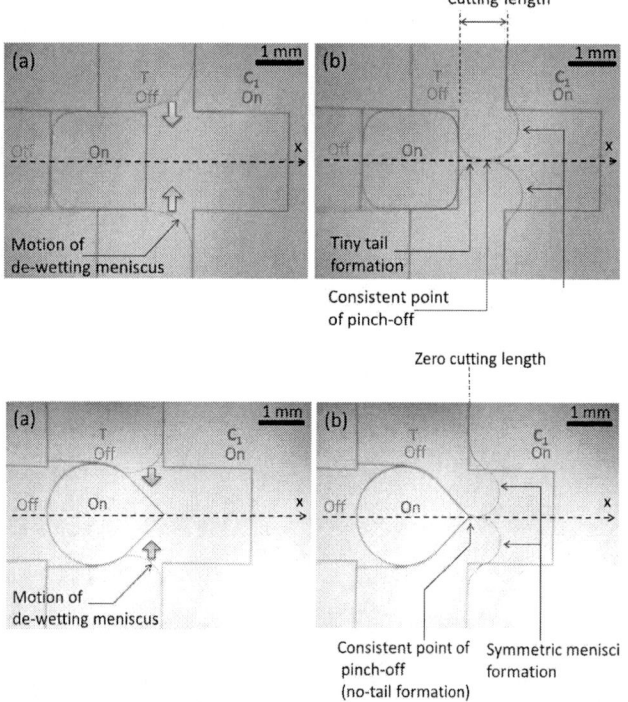

Figure 5: Dispensing a droplet from the TCC reservoir with – a square generating electrode (top row), and a circular generating electrode (bottom row). For both rows: (a) menisci motion towards each other, and (b) right before the droplet pinching-off.

The bottom row of Fig. 5 demonstrates further enhancement of volume precision by using circular drop

generating electrode. There is no area mismatch between the footprint of the dispensed droplet and the drop generating electrode. The cutting length and the liquid tail have been completely eliminated.

Figure 6 shows the average volume error of 50 droplets from each reservoir design. While a conventional design has 34 % error, TCC reservoir design with circular electrode shows less than 1 % error.

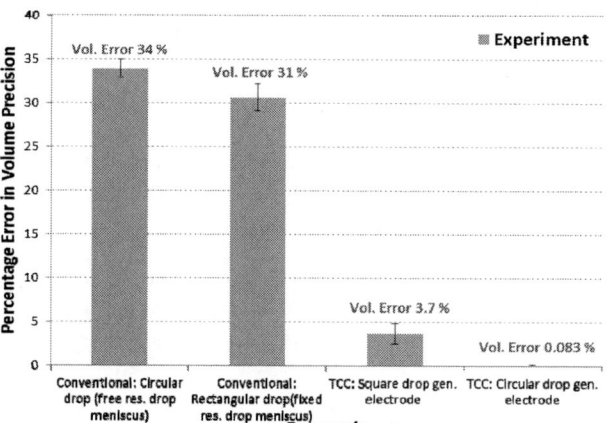

Figure 6: Average volume error (%) of 50 droplets dispensed from each of 4 different reservoirs – (from left) conventional/free reservoir drop meniscus, conventional/fixed reservoir drop meniscus, TCC/square drop generation electrode, and TCC/circular drop generation electrode.

Figure 7 shows dispensing a droplet from L-junction. While TCC reservoir creates 3 droplets per second (= 300 ms/drop), L-junction device creates 90 droplets per second (= 11 ms/drop) with 0.81 % error in volume.

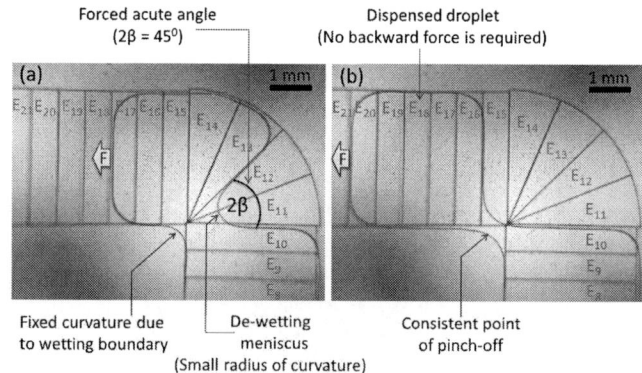

Figure 7: Dispensing a droplet from the L-junction electrodes design. (a) A forced acute angle actively forms a neck with small curvature; (b) Quick pinch-off at the pin-point only with forward EWOD force.

Figure 8 shows splitting the droplet into two by Y-junction. Conventional splitting by two square electrodes in a typical EWOD DMF takes 125 ms but splitting at Y-junction takes only 5 ms.

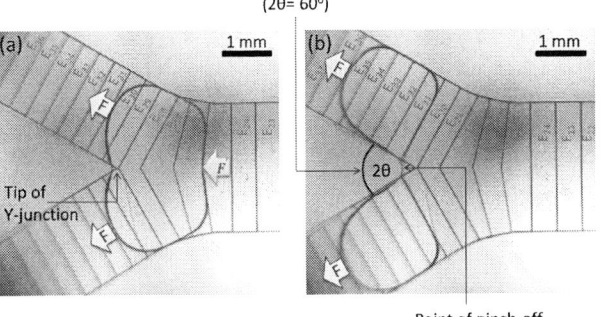

Forced acute angle
(2θ= 60°)

Point of pinch-off

Figure 8: Splitting a droplet into two at the Y-junction. (a) A droplet is pushed towards the tip of the Y-junction by slender angular electrodes, while droplet is pulled by slender rectangular electrodes into top and bottom branches; (b) Right before the droplet pinching-off.

CONCLUSIONS

In this paper, we report completion of two essential requirements of the EWOD DMF: 1) high precision and consistency in volume of unit droplet dispensed from a reservoir, and 2) reduction of time to dispense and split drops. To enhance the volume precision and consistency of the droplets, a novel reservoir named TCC reservoir is designed. The cutting length and the tail formation are minimized and a regular location of pinch-off was achieved. L-junction is designed to reduce time to dispense a droplet while maintaining higher volume precision and consistency. Angular shape electrodes array about the pivot point forces the formation of a neck (thus quick pinch-off) and regulates the pinch-off location (thus high volume precision). Y-junction is designed to split a droplet into two; and hence, increasing the droplet arrival frequency to the designated section while decreasing the droplet volume.

ACKNOWLEDGEMENTS

This study was supported by the Defense Advanced Research Projects Agency/Microsystems Technology Office (DARPA/MTO) under the supervision of the program manager, Dr. Avram Bar-Cohen (grant no. W31P4Q-11-1-0012).

REFERENCES

[1] E. M. Miller, A. R. Wheeler, "A Digital Microfluidic Approach to Homogeneous Enzyme Assays", *Anal. Chem.,* Vol. 80 (5), pp. 1614-1619, 2008.
[2] R. Sista, Z. Hua, P. Thwar, A. Sudarsan, V. Srinivasan, A. Eckhardt, M. Pollack and V. Pamula, "Development of a digital microfluidic platform for point of care testing", *Lab Chip*, Vol. 8 (12), pp. 2091–2104, 2008.
[3] Y. H. Chang, G. B. Lee, F. C. Huang, Y. Y. Chen and J. L. Lin, "Integrated polymerase chain reaction chips utilizing digital microfluidics", *Biomed Microdevices*, vol. 8, pp.215–225, 2006.
[4]. V. Srinivasan, V. K. Pamula, R. B. Fair, "An integrated

digital microfluidic lab-on-a-chip for clinical diagnostics on human physiological fluids", *Lab Chip*, vol. 4, pp. 310–315, 2004.
[5] V. N. Luk, A. R. Wheeler, "A Digital Microfluidic Approach to Proteomic Sample Processing", *Anal. Chem.,* vol. 81(11), pp.4524–4530, 2009.
[6]. P. Y. Paik, V. K. Pamula, K. Chakrabarty, "Adaptive Cooling of Integrated Circuits Using Digital Microfluidics", *IEEE Transactions on Very Large Scale Integration (VLSI) Systems*, vol. 16 (4), pp. 432-443, 2008
[7] H. Ren, R. B. Fair, "MicroNano Liter Droplet Formation and Dispensing by Capacitance Metering and Electrowetting Actuation", *IEEE-Nano 2002*, pp. 369-372, 2002
[8] S. K. Cho, H. Moon, C.-J. Kim, "Crating, Transporting, Cutting, and Merging Liquid Droplets by Electrowetting-Based Actuation for Digital Microfluidics Circuits", *Journal of Microelctromechanical Systems,* vol. 12 (1), pp. 70-80, 2003
[9]. J. Gong, C.-J. Kim, "Two dimensional Digital Microfluidic System by Multi-layer Printed Circuit Board", *Proceedings of 18th IEEE International Conference on Microelectromechanical Systems*, Miami, pp. 726-729, (2005)
[10] P. Wijethunga, Y. Nanayakkara, P. Kunchala, D. Armstrong, H. Moon, *Anal. Chem*, vol. 83, pp. 1658-1664, 2011

CONTACT

*Hyejin Moon, +1-817-272-2017; hyejin.moon@uta.edu

ENZYME-DOPED POLYESTER THREAD COATED WITH PVC MEMBRANE FOR ON-SITE UREA AND GLUCOSE DETECTION ON A THREAD-BASED MICROFLUIDIC SYSTEM

*Yu-An Yang[1], Wen-Cheng Kuo[2], Che-Hsin Lin[1, *]*

[1]National Sun Yat-sen University, Kaohsiung, 804, TAIWAN
[2]National Kaohsiung First University of Science and Technology, Kaohsiung, 824, TAIWAN

ABSTRACT

This study presents a novel method to produce enzyme-doped thread with polyvinylchloride (PVC) membrane coating for on-site urea and glucose detection. The enzyme can be directly applied on the thread without delicate pretreatment or surface modification process. The passing biomolecules are digested by the enzymes and then electrochemically detected downstream. With this approach, liquid transportation, biocatalytic reaction and CE-EC detection of biosamples can be simply achieved on a low-cost thread device. The thin PVC membrane on the enzyme-doped thread can further prevent from the rapid evaporation of the running buffer caused by Joule heat during CE separation. Results show that the PVC coated thread can be operated at a higher separation electric field of 500 V/cm. Successful on-site enzyme digestion, CE separation and EC detection of urea and glucose samples in a single test run is demonstrated. The measured results also indicate that the developed system exhibits good linear dynamic range for detecting urea and glucose samples with the concentration of 0.1 mM~10.0 mM (R^2=0.9850) and 0.1 mM~13.0 mM (R^2= 0.9668), which is suitable for detecting blood urea nitrogen (1.78~7.12 mM) and glucose in serum (3.89~6.11 mM).

INTRODUCTION

For the past decade, the applications of Bio-MEMS grow exponentially in various fields including medical detection, food detection, agriculture and other industrials. Over the developed bio-chip devices, capillary electrophoresis (CE) chip is one of the most promising schemes since it is capable of separating and detecting the target bio-samples in the microfluidic system. Recently, several researchers have developed the so-called Labchip systems for in-column enzymatic reactions and detections of multiple samples. Microfluidic devices integrated with the ion-selective membrane or modified nano-electrodes for high-performance analyzing the concentration of urea and glucose in human serum were also reported [1, 2]. However, the fabrication for the sensing electrodes was comparatively delicate. However, most biochip devices aim to perform one biocatalytic reaction and detection in one test. Therefore, it is beneficial to develop a microfluidic device which is capable of performing simultaneous multi-reactions and detections for medical diagnosis applications.

In general, silicon wafer, glass, polydimethylsiloxane (PDMS), polymethylmethacrylate (PMMA), Teflon are used as the substrate materials for bioanalytical detections [3, 4]. For example, PMMA-based capillary electrophoresis chip

embedded with the gold nano-electrode (GNEE) was reported for CE-EC of ammonia concentration of urine [5]. Some other CE-chips also have also been used for simultaneously measuring hydrogen peroxide, ascorbic acid and uric acid in glass-based microfluidic device [6]. The sensing performance of the in-column enzymatic reactions has also been experimentally investigated [7]. Recently, the use of the polyester thread as the microfluidic channel has shown a great potential for rapid and low-cost detection of biosamples [8]. Biomolecules and ions could be transported and electrochemically detected on a thread such that time-consuming fabrication processes for producing sealed microfluidic channel can be excluded.

Monitoring the blood urea nitrogen (BUN) and blood glucose are important for patients with diabetic or kidney diseases. A microfluidic system integrated with a solid-state sensor incorporated with enzyme-carrying alginate microbeads for on-site analyze of urea, glucose and creatinine in the human serums [9]. Nevertheless, the operation procedure of this device was comparatively delicate. In general, urea and glucose are stable molecules such that they are difficult to detect with the typical electrochemical procedure without the assist of enzymatic reaction. In general, urea can be rapidly converted into ammonium ion with in the presence of urease. These systems were integrated with electrochemical detection for in situ analyzing of the urea concentration [10]. In practice, ammonium ion is a cation which has a reduction potential of around -0.2 V ~ -0.3 V and can be easily detected using a negative potential state. Alternatively, glucose can be converted into hydrogen peroxide (H_2O_2) using the glucose oxidase (GOD). However, the H_2O_2 is an oxidation regent and has a standard reduction potential of around 1.776 V. Therefore, it is difficult to simultaneously detect these two products with a single set of electrochemical detecting electrodes. It is essential to apply the secondary enzyme of horseradish peroxidase (HRP) to further react with hydrogen peroxide to form the reduction products at a low negative voltage [11]. With this approach, the reaction products of these two biomolecules can be simultaneously detected with the same applied potential of -0.28 V.

EXPERIMENTAL DESIGN

Working principle

This study proposes a novel PVC coated membrane of the thread-based microfluidic system, with a variable volume injection capability and 3-dimensional (3D) electrodes for the CE-EC detection of urea and glucose. The developed thread-based CE-EC system is capable for on-site detection of urea and glucose on a single thread. Figure 1 presents the schematic illustration of the working principle of the developed CE-EC microdevice. Since enzymes can be attached on the polyester thread simply by direct applying a drop of enzyme solution on the thread, the thread can be doped with various enzymes for different sample detections. The passing samples in the thread will be digested with the immobilized enzymes on the thread and electrochemically detected. A thin layer of PVC solution is then applied on the wetted thread. Prior to the application of PVC solution, the enzyme-doped thread is immersed in buffer solution to fully soak the buffer solution on the thread to prevent form the infusion of PVC solution into the fiber bundle of the thread. Due to the immiscible property between the PVC solution and the running buffer, a thin PVC layer on the thread can be formed after drying the organic solvent. The formed PVC membrane not only prevents the buffer solution from rapid evaporation during CE test but also confines the fiber bundle of the polyester thread. Therefore, the applied electric field can be increased for sample separation, which would enhance the quality for the electrochemical signals of the samples.

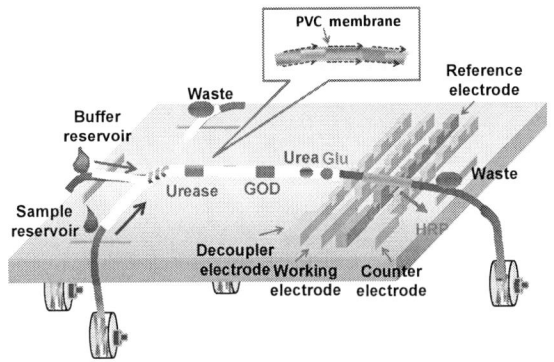

Figure 1: Schematic of the enzyme-doped thread coated with PVC membrane for high-performance CE-EC detection.

Chip design and fabrication

Figure 2 shows the simplified fabrication process for producing the proposed thread-based microfluidic system. The details for the fabrication process can be found in the previous report [12]. A concave master mold was first fabricated on a 2-mm thick aluminum plate using a micro CNC machine. The convex electrode structures were then produced on a PMMA substrate by hot embossing (Fig. 2A). The working and counter electrodes for electrochemical detection were then produced by sputtered Cr/Au layers and the reference electrode was with Cr/Pt layers (Fig. 2B). Polyester threads of around 200-μm in diameter were fixed on the PMMA substrate as the liquid routes for CE operation (Fig. 2C). Specific enzymes including urease, glucose oxidase (GOD) and horseradish peroxidase (HRP) were then directly applied on different sites of the thread with a 2-μL pipette (Fig. 2D). A thin PVC solution was sprayed on the desired region with the assistance of a plastic mask (Fig. 2E). The PVC solution would form a thin membrane covering the thread after evaporating the solvent. Enzyme-doped

polyester thread coated with PVC membrane was finally produced for simultaneously EC detection of urea and glucose. (Fig.2 (F))

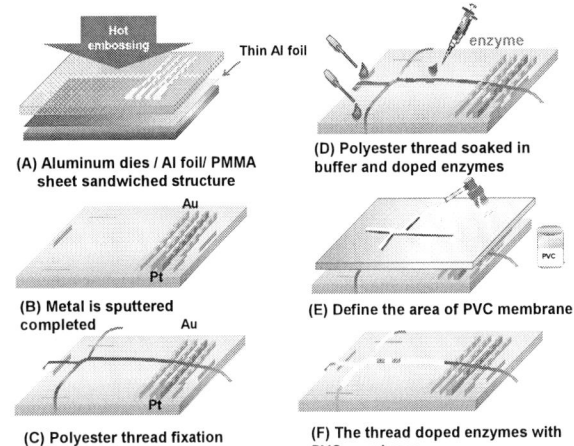

Figure 2: Schematic for the simplified fabrication process for the thread-based microfluidic device

Figure 3 presents the SEM images of the threads with and without PVC coating. It is clear that a PVC layer was successfully formed on the thread surface (Fig. 3A). From the cross-section view, the fiber bundle was well constrained by the PVC membrane of around 50 μm in thickness (Fig. 3B). Alternatively, a rougher surface can be observed for the thread without PVC coating (Fig. 3C). The fiber bundle was also loose (Fig. 3D) such that the CE separation performance was hindered in compare with the thread with PVC coating.

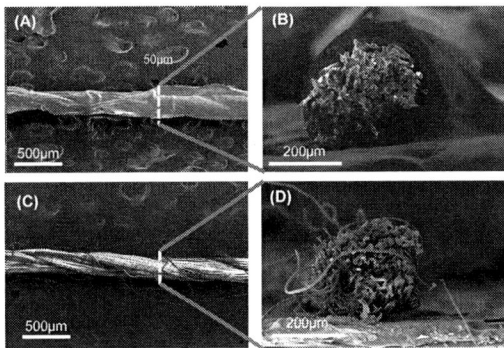

Figure 3: SEM images showing the polyester threads with (A)(B) and without PVC coating (C)(D). Note that (A)(C) are eagle view and (B)(D) are cross section view.

Figure 4 presents the experimental setup for the developed CE-EC detection system using the PVC coated polyester threads as the microfluidic channel. A high voltage power supply (MP-3500-250P, Major Science, Taiwan) was used for applying the electric potential for the CE injection and separation processes. A DC power supply (DP-3630S, HILA, Taiwan) was used to control the motors and provided the necessary tension force on the polyester threads. The reaction products for urea and glucose were detected using a commercial electrochemical analyzer (Model CHI611C, CH instruments, U.S.A). During operation, a 3-μL sample solution was applied on the sample loading site and the other sites were applied with 3-μL running buffer (0.1 mM MES).

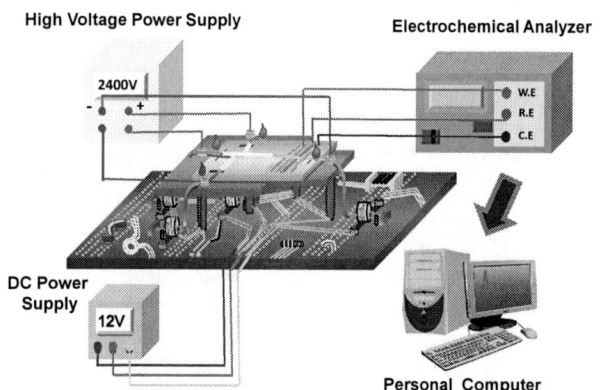

Figure 4: Experimental setup for the developed thread-based CE-EC detection system.

RESULTS

Electroosmotic flow (EOF) mobility is a critical concern for evaluating the capability of an electrokinetically driving system such as the capillary electrophoresis. The CE system with a higher EOF mobility is able to provide a faster sample transportation rate. In this regards, this study compared the electroosmotic mobility for the conventional glass-based microfluidic channel and the polyester thread coated with the PVC membranes of different thicknesses. The PVC solutions for coating the threads were diluted with different amounts of tetrahydrofuran (THF). Figure 5 presents the measured EOF mobility of the threads with and without PVC coating and the conventional glass channel. Results show that the EOF mobility of the thread-based microfluidic system is much higher than that of conventional glass-based microfluidic system. It is noted that the thread with PVC coating might have lower EOF mobility since the fibers in the bundle would be slightly coated with hydrophobic PVC due to the infusion of the PVC solution.

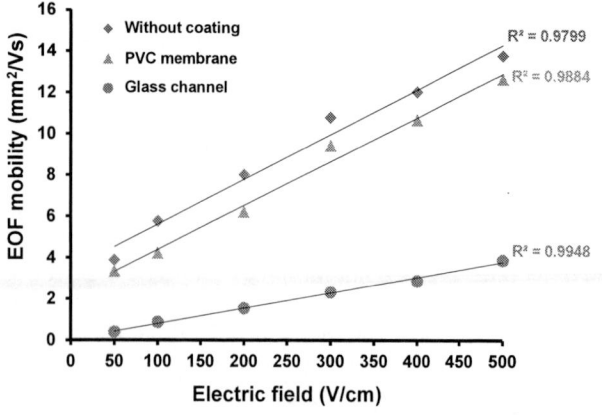

Figure 5: Measured EOF mobility of the threads with and without PVC coating and conventional glass channel.

Figure 6 shows the measured current responses for measuring urea and glucose samples using the thread modified with different concentrations of urease and GOD. Note that the secondary enzyme of horseradish peroxidase (HRP) was with the concentration 5 mg/ml and 0.5 mM catechol (mediator). The optimized enzyme concentrations on the thread were determined while the current responses reaching saturation. Results indicated that the saturated concentrations for urease and GOD were 20 mg/ml and 10 mg/ml respectively.

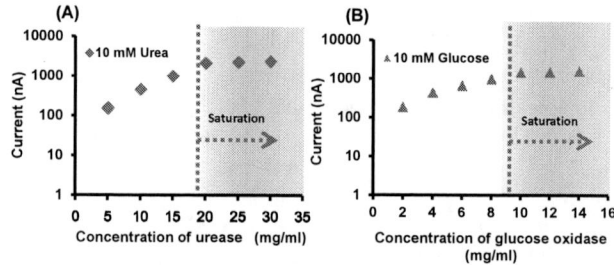

Figure 6: Measured current responses for detecting urea and glucose with the threads coated with different concentrations of urease and GOD.

Figure 7 shows the electropherograms of measuring a mixed sample composed of 1 mM urea and 1 mM glucose using the threads without (Fig. 7A) and with (Fig. 7B) PVC coating. The applied potential for this potential state measurement was set at -0.28 V for detecting the reduction signals of the reaction products. The applied voltages for sample injection and separation were 300 V/cm for the thread without PVC coating since higher operation voltage might burn the thread due to Joule heat. Alternatively, the PVC coated thread could sustain a higher operation electric field up to 500 V/cm, resulting a faster separation and higher signal response for detecting the same sample. In addition, due to the well constrain, the coated thread appeared a lower and stable background noise during electrochemical detection in compare with the uncoated thread.

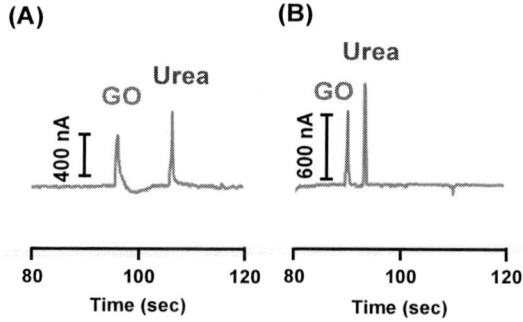

Figure 7: Electropherograms for on-site digesting and EC detecting 1mM of glucose and urea using (A) thread without coating and (B) thread coated PVC membrane.

In general, the normal concentrations for blood urea

nitrogen (BUN) and blood glucose (GLU) in serum are 1.78~7.12 mM and 3.89~6.11 mM, respectively. Note that 1 mg/dL of detected urea in-vitro is equivalent to 2.14 mg/dL of BUN in serum. In order to evaluate the sensing performance of the system, mixed samples composed of urea and glucose with thee concentrations of 0.1 mM~10.0 mM and 0.1 mM~13.0 mM, respectively, were measured using the developed thread-based microfluidic system. Note that the enzyme concentrations for urease and GOD applied on the thread were 20 mg/mL and 10 mg/mL, respectively. As described above, 5 mg/mL of HRP was applied on the region above the EC electrodes for further reducing the detection potential of hydrogen peroxide. Figure 8 presents the measured current responses for EC detecting different concentrations of urea and glucose using the thread-based microfluidic system. Results show that the developed system exhibited nice sensing performance for simultaneously detecting these two biomolecules. The reaction products of urea and glucose were successfully electrochemically detected with the developed system.

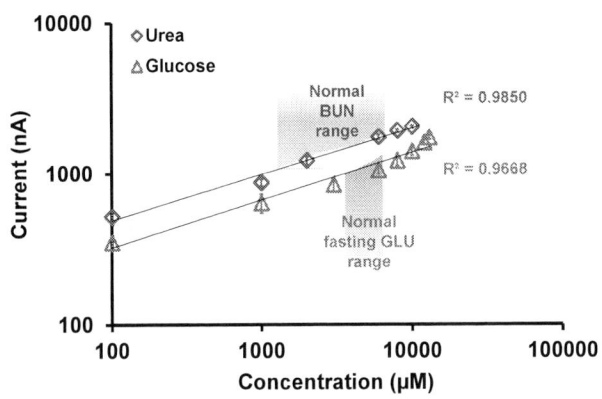

Figure 8: Measured current responses for detecting various concentrations of urea and glucose solutions.

CONCLUSIONS

This study developed a novel technique to form a PVC coated thread doped with various enzymes of urease, glucose oxidase and horseradish peroxidase for on-site bio-sample separation, bio-catalytic reaction and electrochemical detection. Enzymes modified polyester thread coated with a thin layer of PVC was used as the liquid route and the reactor for converting urea and glucose into ammonium ions and hydrogen peroxide. The products were then separated anddetected using the CE-EC detection scheme. With this approach, a sealed microfluidic channel embedded with various enzymes could be easily produced. Results showed that the thread with PVC coating exhibited higher electroosmotic mobility in compare with that of the conventional glass microfluidic channel. In addition, the PVC coated thread could sustain a higher operation voltage than the uncoated one, resulting in a better detection performance. The novel device developed in the present study provided a simple yet high performance way for the rapid detection of bio-samples.

REFERENCE

[1] L. Zhu, R. Yang, J. Zhai and C. Tian, "Bienzymatic glucose biosensor based on co-immobilization of peroxidase and glucose oxidase on a carbon nanotubes electrode", *Biosensors and Bioelectronics*, 23, 528-35, 2007.

[2] T.-Y. Chiang and C.-H. Lin, "Enzyme-doped ion selective membrane (ed-ism) formed with surface force and microstrucures for high perforlance urea detection," presented at Micro Electro Mechanical Systems (MEMS), 2013 IEEE 26th International Conference .

[3] C.-H. Lin, G.-B. Lee, Y.-H. Lin and G.-L. Chang, "A fast prototyping process for fabrication of microfluidic systems on soda-lime glass", *Journal of Micromechanics and Microengineering*, 11, 726, 2001.

[4] P. D. Voegel and R. P. Baldwin, "Electrochemical detection in capillary electrophoresis", *Electrophoresis*, 18, 2267-78, 1997.

[5] C.-M. Chen, G.-L. Chang and C.-H. Lin, "Performance evaluation of a capillary electrophoresis electrochemical chip integrated with gold nanoelectrode ensemble working and decoupler electrodes", *Journal of Chromatography A*, 1194, 231-36, 2008.

[6] J. Wang, M. P. Chatrathi, B. Tian and R. Polsky, "Microfabricated electrophoresis chips for simultaneous bioassays of glucose, uric acid, ascorbic acid, and acetaminophen", *Analytical chemistry*, 72, 2514-18, 2000.

[7] R. Wilke and S. Büttgenbach, "A micromachined capillary electrophoresis chip with fully integrated electrodes for separation and electrochemical detection", *Biosensors and Bioelectronics*, 19, 149-53, 2003.

[8] D. R. Ballerini, X. Li and W. Shen, "Flow control concepts for thread-based microfluidic devices", *Biomicrofluidics*, 5, 014105, 2011.

[9] Y.-H. Lin, S.-H. Wang, M.-H. Wu, T.-M. Pan, C.-S. Lai, J.-D. Luo and C.-C. Chiou, "Integrating solid-state sensor and microfluidic devices for glucose, urea and creatinine detection based on enzyme-carrying alginate microbeads", *Biosensors and Bioelectronics*, 43, 328-35, 2013.

[10] J. Wang, M. P. Chatrathi and B. Tian, "Microseparation chips for performing multienzymatic dehydrogenase/oxidase assays: Simultaneous electrochemical measurement of ethanol and glucose", *Analytical chemistry*, 73, 1296-300, 2001.

[11] T. Ferri, S. Maida, A. Poscia and R. Santucci, "A glucose biosensor based on electro- enzyme catalyzed oxidation of glucose using a hrp - god layered assembly", *Electroanalysis*, 13, 1198-202, 2001.

[12] Y.-C. Wei, L.-M. Fu and C.-H. Lin, Electrophoresis separation and electrochemical detection on a novel thread-based microfluidic device", *Microfluidics and nanofluidics*, 14, 583-90, 2013.

FABRICATION OF HIGH ASPECT RATIO INSULATING NOZZLE USING GLASS REFLOW PROCESS AND ITS ELECTROHYDRODYNAMIC PRINTING CHARACTERISTICS

Kyoung Il Lee[1,2], Byungjik Lim[1], Se Wook Oh[1], Seong Hyun Kim[1], Churl Seung Lee[1], Jin Woo Cho[1], Yongtaek Hong[2]

[1]Korea Electronics Technology Institute, Seongnam, KOREA
[2]Seoul National University, Seoul, KOREA

ABSTRACT

We apply a glass reflow process to fabricate tiny glass nozzles with inner diameter of 20 microns and with a height of 150 microns which is impossible to achieve with a conventional etching process. Due to the enhanced aspect ratio of the nozzle, we observe better printing performances such as narrow line width below 10 microns and anti-wetting of ink around a nozzle. High frequency drop jetting up to 10 kHz is observed by applying high voltage pulse signals on ink inside the nozzle. Silver particle ink is also printed well and continuous line patterns are fabricated.

INTRODUCTION

As electronics industry becomes mature, need for low cost fabrication technology get stronger. One of the candidates for reducing the fabrication cost is to substitute traditional photolithography by direct printing. Due to the recent development of inkjet printing technology, there are several applications using piezoelectric inkjet printing process in a mass production of liquid crystal display (LCD) such as color filter or polyimide layer. But the conventional piezoelectric inkjet printing technology has several problems such as nozzle clogging, limited viscosity range of ink, and difficulty for printing of continuous lines.

Electrohydrodynamic (EHD) printing technology gathered much attention for higher resolution comparing to traditional piezoelectric inkjet technology and various materials have been fabricated by EHD printing technology [1]. Glass capillaries or metallic tubes were used in most studies [1-2] because the sharp shape of capillaries is required for the strong field concentration at the end of the capillaries. But the throughput of EHD printing process with a single capillary is not enough for commercialization. Previous studies utilized glass capillaries and multi nozzle printing modules composed of capillaries were reported [2] but it is not practical to assemble large number of capillaries with high position accuracy to make multi nozzle arrays required for commercial applications. The cost of capillaries also matters. So an integrated multi nozzle array by batch process is indispensable for the commercialization of EHD printing technology. Some groups have shown silicon based micromachined nozzle for EHD printing since silicon is easy to make a nozzle-shape structure using deep silicon etching process. But the jetting stability of silicon nozzle is not good due to electrical conductivity of silicon [3], which reduces the concentration of the electric field at the meniscus. Also, the bias voltage required to eject droplets from the nozzle is

higher than several kV. So an insulating substrate like glass wafer is a good candidate for the fabrication of nozzles for EHD printing. Unfortunately, the etching process of insulating substrate with a high aspect ratio is not available unlike silicon. We reported a single and multi-nozzle printing head with batch processed tapered glass nozzle [4]. The tapered nozzle has a ink wetting problem which reduces an electric field concentration and shows poor jetting stability. Meanwhile, a glass reflow process for copper filled through glass via was reported also at MEMS 2013 [5]. In this paper we show the enhancement of the EHD printing with a high aspect ratio nozzle using this reflow process.

DESIGN

The principle of EHD printing is show in Fig. 1. Large electric field should be concentrated at the sharp end of nozzle tip to eject inks before electrical breakdown happens. So the fabrication of high aspect ratio nozzle array is key factor of EHD printing. Some groups show a deep etching of quartz wafer but the aspect ratio is very limited. The sandblasting process used in the previous study[4] has a limit to get a higher aspect ratio. Also the surface of the nozzle is not smooth due to the large sands used for the process, which results in the non-uniformity of the printing process. So we choose a bottom up approach instead of etching process.

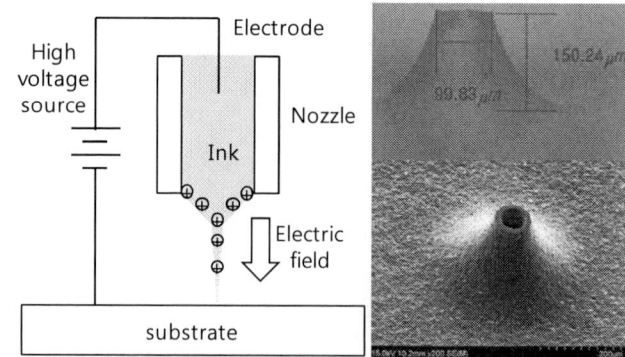

Figure 1: Principle of EHD printing and the nozzle in the previous study[4].

The structure of the nozzle is shown in Fig. 2. Tiny nozzle is connected to a large tapered hole in glass wafer. The outer diameter of the nozzle is 40 microns (type A) or 100 microns (type B) while the height is above 150 microns. The nozzles are separated with a pitch of 1 mm or 3 mm.

978-1-4799-3510-9/14 $31.00 © 2014 IEEE

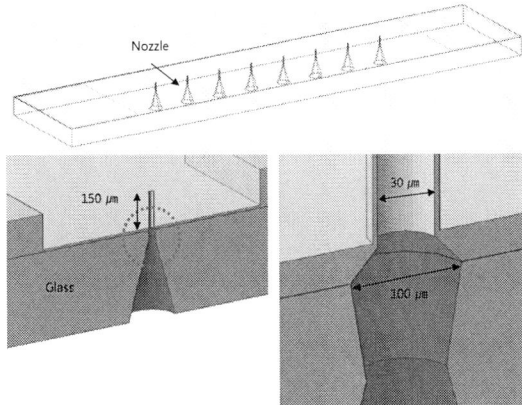

Figure 2 : Schematic of glass nozzle array.

FABRICATION

See Figure 3 for the fabrication process of glass nozzle die. We use a 500 micron thick 4" silicon wafer as a starting material. At first, 500 nm thick SiO2 film is deposited by PECVD and patterned by conventional photolithography. A photoresist (PR, red) is patterned on it again and the wafer is etched by Bosch process down to 200 microns. Additional silicon etching is done only for nozzle type A after the removal of PR layer with SiO2 pattern as a etching masking layer. Additional short isotropic silicon etching process is done in HNA solution to reduce the silicon grass and to make smooth sidewalls with small scalloping. After cleaning and stripping of SiO2 film, the wafer is bonded to a bare borosilicate-33 glass. With an annealing of it at 1000 °C for 5 hour, the melting glass fills the holes in the silicon wafer. Then the wafer is grinded until the thickness of the glass layer becomes 50 microns. The holes are fabricated in the glass layer by sandblasting with dry film resist (DFR) as a masking layer until the silicon poles are exposed.

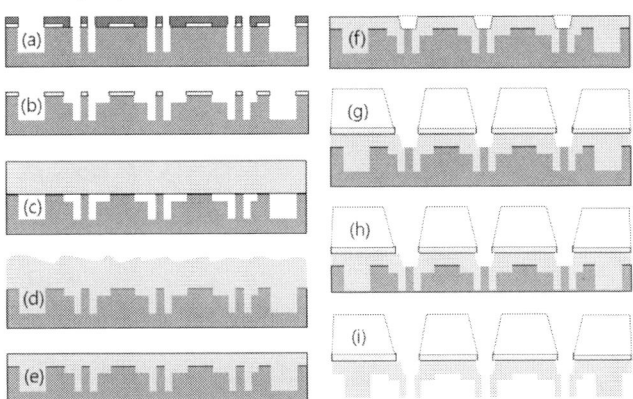

Figure 3: Fabrication process (a) photolithography on the pre-patterned SiO2 film and silicon deep etching (b) removal of the photoresist and additional etching (c) anodic bonding to a borosilicate glass (d) glass melting and filling into the hole (e) CMP (f) sandblasting with dry film resist (g) glass frit bonding to a glass wafer with holes (h) CMP (i) Si etching in TMAH solution.

A handling glass wafer has holes corresponding to the holes in the nozzle wafer, fabricated by sandblasting process also. They are bonded to each other with glass frit at 550 °C. The frit (Asahi glass, Japan) layer is formed on the handling glass wafer by screen printing before the bonding process. The thickness of the frit is around 10 microns after firing. Then, the thickness of the silicon layer is reduced by mechanical lapping until the hole patterns appear. The wafer is half-diced into each nozzle or nozzle array at this moment.

The remaining silicon is etched in hot TMAH solution. The silicon outside the nozzle is etched easily but the removal of silicon inside the nozzle takes several hours. So the sidewall of glass frit exposed to TMAH solution is etched several microns during the process. But strong bonding is maintained firmly since the frit pattern is larger than 200 microns in lateral. To reduce the outer diameter of the nozzles, additional diluted HF etching is also done. It also helps the removal of any wandering glass particles inside the nozzle.

Now the wafer is cleaned again and Teflon layer is spray-coated on the front side with Teflon AF 1601 (DuPont, USA) solution to make the side wall of the nozzles hydrophobic and smooth. The thickness of the layer is around 50 nm. The contact angle becomes 110° after Teflon coating while it was 30° on a bare glass wafer for deionized water. After separating to dies, they are bonded to a plastic (PEEK) adaptor by inert adhesive, and the adaptor is bonded to a metal tube connected to ink reservoir to complete nozzle head. The tube is wired to a high voltage source to apply high voltage on the ink inside the nozzle.

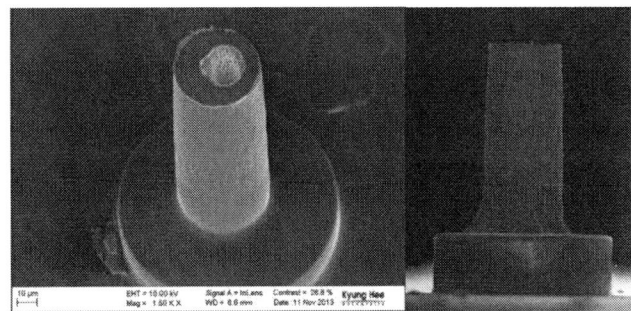

Figure 4:SEM image of fabricated nozzle :type A (up) and type B (down)

Fig. 4 shows the individual nozzle or nozzle array fabricated by the above process. Tiny high aspect ratio nozzle structure with an outer diameter of 40 microns or 100 microns and with a length of 150 microns has been fabricated. The aspect ratio is about 4 while it was below 1 in the previous study [4]. The vertical wall looks very smooth after

978-1-4799-3510-9/14 $31.00 © 2014 IEEE

Teflon coating process. Some holes in the nozzle don't look circular and it can be due to the non-uniform silicon deep etching process which should be enhanced. The uniformity of outer diameter is very important for multi nozzle array since the jetting depends on it strongly. The variation of the outer diameter from the above process was below 1 micron while it was more than 3 microns in the previous study [4].

PRINTING TEST

We tested our printing module with a system illustrated in Figure 5.

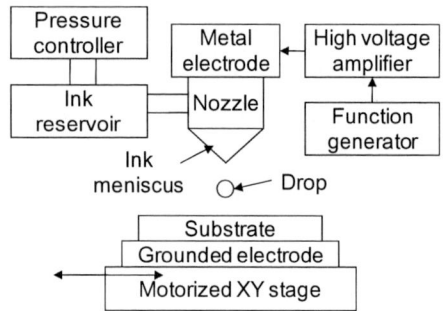

Figure 5: Schematic diagram of EHD printing system.

The metal electrode is connected to a high voltage amplifier (Trek, 10/40A) controlled by a function generator (NF 100, NF). We can control the meniscus by controlling air pressure in the ink reservoir with a pressure controller (PPM, Unijet). We tested our nozzle in a constant pressure mode instead of constant flow rate mode because the constant pressure mode is more appropriate for drop on demand printing process. In a constant flow rate mode, unwanted drop dripping could occur after a long time of rest. The jetting behavior was monitored by high speed video camera (Phantom v12.1, Vision Research) with an optical fiber illumination or LED stroboscope type drop watcher equipped in a lab-scale inkjet printing system (UJ-200, Unijet). The substrates are placed on a vacuum chuck grounded electrically and PC-controlled motorized XY stage equipped in the inkjet printer are used for a substrate scanning.

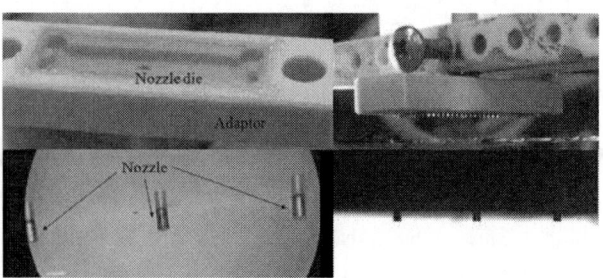

Figure 6:Fabrication results (a) nozzle assembly (b) mounting of 16 nozzle array (c) nozzles : side view (d) image from a high speed camera.

Fig. 6 shows the printing module after assembly. We tested a common mode operation which eject at every nozzle simultaneously. So there is no individual electrode for each nozzle.

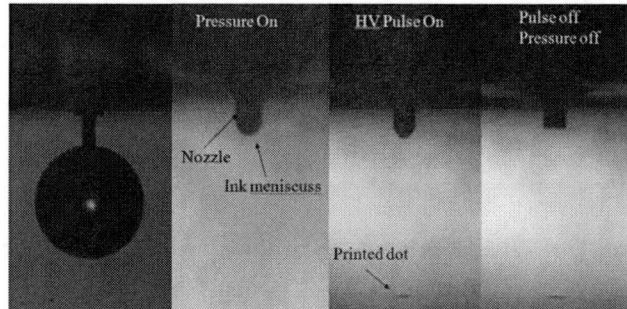

Figure 7:Anti-wetting characteristics of the nozzle(left) control of meniscus by pressure or voltage(right)

Since the sidewall of the nozzle is hydrophobic, most inks don't expand on the surface of the nozzle wall (Fig. 7). We use a carbon black ink (Infochem) developed for printing black matrix in color filter array of LCD since it has a suitable properties such as electrical conductivity. We can control the meniscus by applying moderate pressure on the ink in the reservoir. And very tiny drop can be jetted by applying high voltage pulse signal, which is observed using the high speed camera. The maximum jetting frequency we could observe is 10 kHz (Fig. 8) for both type A and type B. Above 10 kHz, the signal generated from the HV amplifier is no more sharp pulse shape.

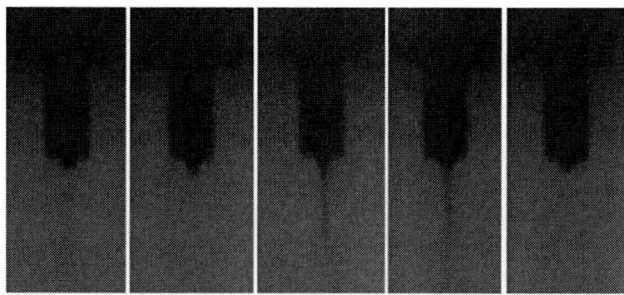

Figure 8:Jetting with high voltage pulse at 10 kHz (type A).

Continuous lines with a width of 10 ~ 140 microns on a silicon wafer were printed with a DC bias voltage of 2.0 ~ 2.8 kV at various nozzle-stage gap (Fig. 9) in the inkjet printing system with a nozzle type B. We could get a 10 micron width narrow carbon black line at 2.8 kV bias and at the stage speed of 500 mm/s without nozzle wetting. So we can get much finer patterns with our glass nozzles than that with a previous tapered nozzle.

With a nozzle type A, we could print a 5 micron width line at the stage speed of 500 mm/s as shown in Fig. 10. The line pitch is set to 35 microns but we could observe a slight discrepancy which we don't understand fully right now.

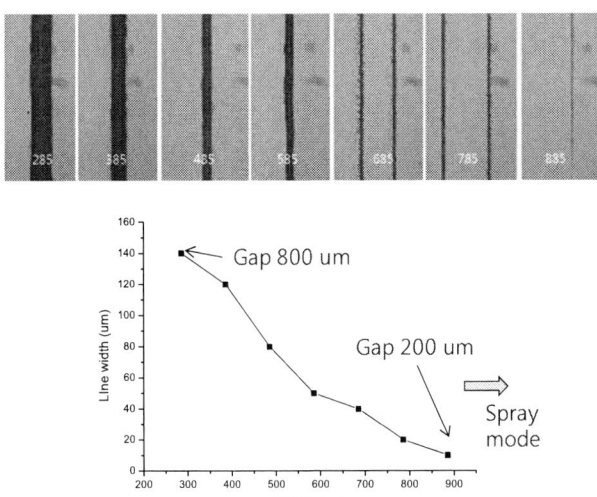

Figure 9: Printed line patterns vs. substrate height with DC bias with a nozzle type B.

The relationship between applied bias and the line width is proportional only in a limited bias range . The printed line width depends on the applied bias voltage as you can see in the below.

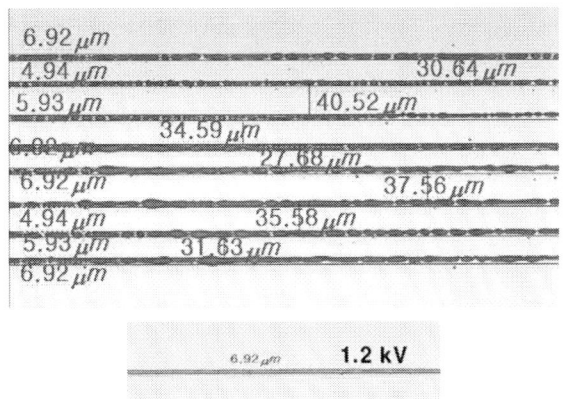

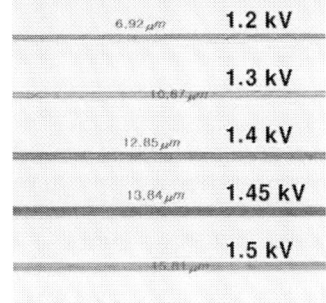

Figure 10: Printed line patterns with a pitch of 35 microns(up) and line width dependence on the DC bias.

One of the promising application of this device the high speed fabrication metal line on insulating substrates such as glass or PET film used as a transparent conducting film for larger display devices. Using silver nano-particle ink (Amogreentech, B-13, Korea), we could get a fine Ag line at the stage speed of 1 m/s. The line width varies from 10 to 20 microns depending on the printing conditions (Fig. 11)

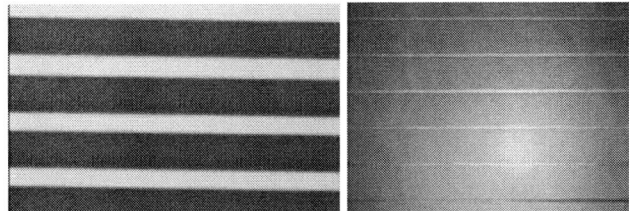

Figure 11: Printed silver line patterns with a line width of 20 micron(left) and 10 micron(right) with DC bias.

CONCLUSIONS

In summary, we apply a glass filling process to fabricate a high aspect ratio glass nozzle array for EHD printing modules. The module shows good printing capability with a lot of advantages such as fine line width below 10 microns at the high stage speed of 1 m/s, high speed drop jetting up to 10 kHz, comparing with the conventional piezoelectric inkjet printing technology and with the previous studies also.

REFERENCES

[1] J.-U.Park, M. Hardy, S. J. Kang, K. Barton, K. Adair, D. K. Mukhopadhyay, C. Y. Lee, M. S. Strano, A. G. Alleyne, J. G. Georgiadis, P. M. Ferreira, J. A. Rogers, "High-resolution electrohydrodynamic jet printing", Nat. Mater, vol. 6 pp. 782-789, 2007.

[2] A. Khan, K. Rahman, M.-T. Hyun, D.-S. Kim, K._H. Choi, "Multi-nozzle electrohydrodynamic inkjet printing of silver colloidal solution for the fabrication of electrically functional microstructures", Appl. Phys. A, vol. 104, no. 4, pp.1113-1120, 2011.

[3] J.-S. Lee,S.-Y. Kim, Y.-J. Kim, J. Park,Y. Kim,J. Hwang, Y.-J. Kim, "Design and evaluation of a silicon based multi-nozzle for addressable jetting using a controlled flow rate in electrohydrodynamic jet printing", Appl. Phys. Lett., vol. 93, pp. 243114 (1-3), 2008.

[4] K. I. Lee, B. Lim, H. Lee, S. H. Kim, C. S. Lee, J. W. Cho, Y. Hong , "Multi nozzle electrohydrodynamic inkjet printing head by batch fabrication", Proc. MEMS 2013, pp. 1165-1168, 2013.

[5] J.-Y. Lee, S.-W. Lee, S.-K. Lee, J.-H. Park , "Wafer level packaging for RF MEMS devices using void free copper filled through glass via", Proc. MEMS 2013, pp. 773-776, 2013.

CONTACT

*K. I. Lee, tel: +82-31-789-7455; leeki@keti.re.kr

GALLIUM-BASED LIQUID METAL INKJET PRINTING

Daeyoung Kim[1], Jun Hyeon Yoo[1], Yunho Lee[2], Wonjae Choi[2], Koangki Yoo[3] and Jeong-Bong (JB) Lee[1]

[1]Department of Electrical Engineering, The University of Texas at Dallas, TX, USA
[2]Department of Mechanical Engineering, The University of Texas at Dallas, TX, USA
[3]Department of Information and Communication Engineering, Hanbat National University, Daejeon, South Korea

ABSTRACT

We report clog-free and oxide-free metal inkjet printing applicable to flexible electronics using gallium-based liquid metal alloy. Inkjet printing has been developed and expanded to make a pattern of either non-conductive or conductive materials. In order to print typical conductive material, it utilizes metal nanoparticle dispersed in solvent or melts the metal. However, those methods often encounters clogging and oxidation problem. We fabricated a simple polydimethylsiloxane (PDMS) based inkjet printer incorporated with hydrochloric acid (HCl)-impregnated paper as orifice material. A constant stream of gallium-based liquid metal alloy droplet was demonstrated using the inkjet printer. Depending on the applied flow rate, pinch off and Rayleigh instability phenomena were observed. We printed beads-on-string shape gallium-based liquid metal alloy line on various flexible substrates such as Si wafer, PDMS, and a paper. Finally, it was demonstrated that the inkjet-printed gallium-based liquid metal can maintain its line shape without disconnection even with the significant deformation of a flexible paper.

INTRODUCTION

Inkjet printing is omnipresent powerful computer printing technique nowadays. Recently, inkjet printing went far beyond the computer printing and expanded to various applications such as thin film transistor [1], radio frequency identification [2], sensor [3], and even biomedical device [4]. In these applications, inkjet printing was contributed to make a pattern of conductive materials. Different types of conductive materials applicable to inkjet printing such as solder [5], metallic nanoparticle (NP) [6], and conductive polymer [7] have been studied. Among them, inkjet printing solder for chip-scale packages [5] and silver/gold NP for 3D MEMS devices [6] utilize molten solder or metallic colloid NPs dispersed in solvent. Once they are printed, solder is cooled/solidified while metallic colloid NPs are cured at elevated temperatures (> 100 °C) to remove solvent. Some of the critical issues in these metal inkjet printing are clogging of inkjet nozzle and oxidation of metals [6].

Unlike typical molten metal by applying high temperature or metal NPs in solvents, gallium-based liquid metal alloys are in liquid-phase at room temperature [8]. Therefore, there is no need for heating or dispersing in solvent for inkjet printing. Another distinctive benefit is that it maintains liquid-phase after printing if the substrate stays at around room temperatures. This is extremely useful to create 3D freeform rapid prototyping of metallic patterns that can conform to virtually any dynamic deformation of substrates.

Since it is not based on the colloidal NPs, gallium-based liquid metal inkjet does not have clogging issue. However,

gallium-based liquid metal alloy is readily oxidized in air and it behaves like gel rather than true liquid. It was reported that Galinstan® (ternary alloy of gallium) behaves like true liquid in < 1 ppm oxygen environment [8]. Establishing such an environment, however, is not an option as inkjet printing should work in open atmosphere.

In this paper, we report a method of the gallium-based liquid metal alloy printing without clogging and oxidation problem. We also show feasibility of the metallic line formation on flexible substrates using the proposed technique.

WORKING PRINCIPLES

The formation of liquid droplet is critical process for inkjet printing. Inkjet printing operates either in continuous mode or drop-on-demand (DOD) mode. Currently, the majority of inkjet printers are based on DOD inkjet printing mode where droplets are generated only when required. DOD inkjet works based on either piezoelectric or thermal droplet generation. For piezoelectric droplet generation, a voltage is applied to piezoelectric electrodes and it sequentially causes mechanical contraction. And thus, it squeezes ink chamber and generates droplets out of the orifice. For thermal droplet generation, current is passed through a resistive heater located in the ink chamber. It results in expanding vapor bubble which changes the volume of the ink chamber and consequently droplets are generated.

Unlike current inkjet printing, in our study, by manipulating flow rate of liquid metal in microfluidic channel, surface tension driven droplet generation is demonstrated. Fig. 1 shows conceptual schematics of liquid metal droplet generation principles based on pinch off and Rayleigh instability. At low flow rate (< 0.5 mL/min.), the liquid metal was ejected from the orifice as discrete droplets as shown in Fig. 1a. When oxidized liquid metal was inserted through microfluidic channel with an applied low flow rate, the liquid metal can have enough time to chemically react with either HCl solution or vapor from the HCl-impregnated paper before it was ejected. Thus, the viscoelastic oxide layer on liquid metal was removed resulting in the recovery of high surface tension (523.8 mN/m) [9]. After it recovers high surface tension, at the outside of orifice, the volume of liquid metal droplet increases until it reaches certain size with the applied flow rate while it keeps its spherical shape. Then, the liquid metal droplet can be pinched off from the confined orifice due to its surface tension.

At high flow rate (≥ 0.5 mL/min.), however, the liquid is ejected from the orifice as a jet stream as shown in Fig. 1b. The laminar jet subsequently breaks up into small droplets due to the Rayleigh instability [10] where small perturbation of wavelength (λ_d) can grow the disturbance exponentially with time and sinusoidally with space. With a fixed jetting velocity, the jet break-up time and jet break-up length (l_j) can be determined by the density and surface tension of liquid, and size of orifice. In addition, the jet diameter (d_j) and droplet diameter (d_d) are primarily determined by the size of orifice.

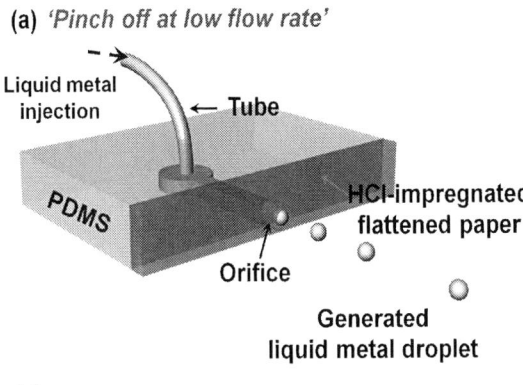

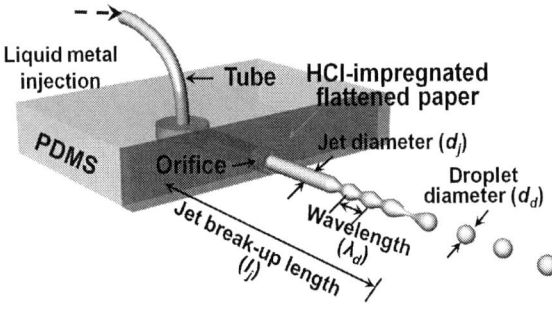

Figure 1: Conceptual schematics of liquid metal droplet generation based on (a) 'pinch-off' and (b) 'Rayleigh instability'.

FABRICATION

A proof-of-concept prototype inkjet printing device fabrication was started with formation of a PDMS microfluidic channel using SU-8 molding technique. SU-8 2025 photoresist (MicroChem Corp.) was spin coated on a thermally grown oxidized Si wafer to get approximately 25 µm thick photoresist. The SU-8 photoresist was soft-baked on a hot plate at 65°C for 2 min., 95°C for 3 min., and finally 65°C for 2 min. Then, an UV exposure dose of 140 mJ/cm^2 was applied to pattern the microfluidic channel shape, and finally a post bake was applied with the same conditions as those of the soft bake. Finally, it was developed in SU-8 developer (propylene glycol methyl ether acetate) for 3 min. under gentle stirring condition (Fig. 2a). After developing, the PDMS was casted over the PR mold and it was cured in a convection oven at 95°C for 2 hours (Fig. 2b). After PDMS was peeled off from the SU-8 mold, the replicated PDMS was cut to have the channel exposed to outside (Fig. 2c). The channel was designed to have either 200 or 500 µm wide channel with 3 mm channel length. It has a 3 mm diameter inlet port where a polytetrafluoroethylene (PTFE) tube can

be connected. It was bonded to a glass slide by applying oxygen plasma treatment. This bonded PDMS with a glass slide was placed in a 115°C convection oven for 10 min. to enhance adhesion strength (Fig. 2d).

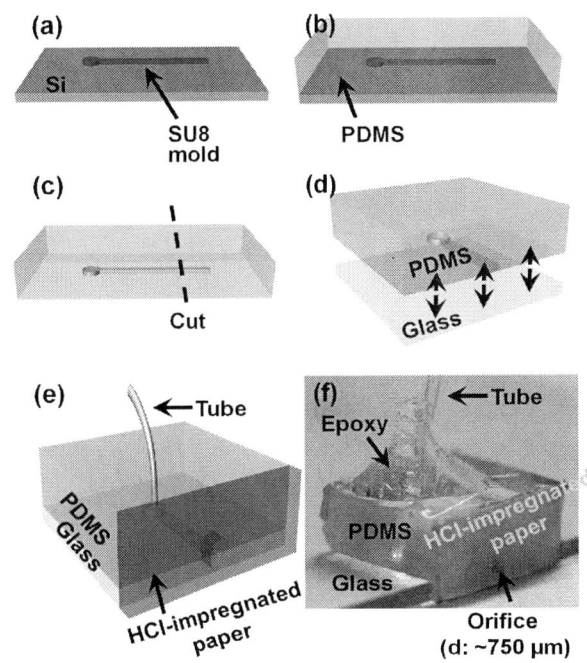

Figure 2: Fabrication sequence of prototype liquid metal inkjet printer: (a) SU-8 mold, (b) PDMS coating, (c) replicated PDMS and cutting, (d) PDMS-glass bonding, (e) HCl-impregnated paper bonding to sidewall of PDMS and glass, (f) optical image of the fabricated prototype of liquid metal inkjet printer.

HCl-impregnated paper with an orifice (~700 µm diameter) was bonded to the PDMS surface sideways using epoxy. The orifice on the paper was made by mechanical drilling. Then, PTFE tube was inserted into inlet port formed in the PDMS/glass (Fig. 2e). The interconnected area was tightly sealed with epoxy to avoid any leaking (Fig. 2f). Then, a syringe (NORM-JECT, 1mL) was connected to the PTFE tube with a hypodermic needle.

RESULTS AND DISCUSSION

With a fabricated PDMS-based microfluidic channel integrated with an HCl-impregnated paper orifice plate, in order to characterize the phenomena of generating droplets out of the orifice, we utilized a high-speed camera (Photron SA4) with different number of frames for a second (from 1000 frames/sec. to 6000 frames/sec.) as shown in Fig. 3. The high-speed camera was connected to the optical microscopy to have close-up top view. A syringe pump was used to inject liquid metal Galinstan® on-demand with specific flow rates. The image of the generating droplet was captured by a high-speed camera and the captured movie was analyzed frame by frame.

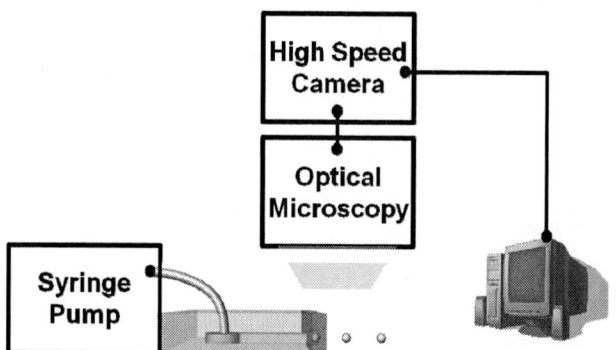

Figure 3: A schematic of droplet generation characterization set-up.

At flow rates < 0.5 mL/min., Fig. 4a shows that PDMS-based inkjet printer without the HCl impregnated-paper orifice plate cannot eject liquid metal droplet. Instead, it just continuously increased the volume of the liquid metal at the orifice. This is due to viscoelastic oxide layer causing it wets on the PDMS surface. In contrast, HCl-impregnated paper integrated PDMS-based inkjet printer can reproducibly pinch off true liquid metal droplets as the oxide skin of Galinstan® is removed at the instant moment of ejection from the orifice by either HCl solution or vapor (Fig. 4b). The diameter of the pinched off droplet was ~ 500 μm and its volume was ~ 50 nL. The shape of the pinched off droplet in the still image was not spherical though. We believe that this was due to the vibrating motion of the ejected liquid metal droplet. This vibrating motion of the droplet is direct result of the chemical reaction of the liquid metal Galinstan® with HCl.

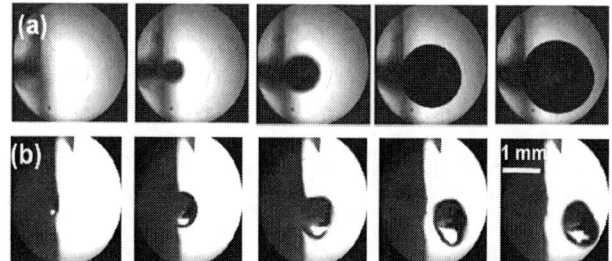

Figure 4: (a) No reproducible ejection of Galinstan® droplet from the PDMS only inkjet, (b) reliable/reproducible ejection of true liquid Galinstan® droplet from the PDMS + HC-impregnated paper orifice plate inkjet due to instantaneous oxide skin removal.

At flow rates ≥ 0.5 mL/min., Rayleigh instability based jetting (Fig. 5a) and subsequent droplet break-up (Fig. 5b, see the inside of blue-colored rectangular dot line) were observed. The break-up droplet size varies as there is combination of break-up droplets (Fig. 5c) before hitting the substrate. Note that there is no clogging and oxidation issue when generating liquid metal droplets based on either pinch off or Rayleigh instability in this approach.

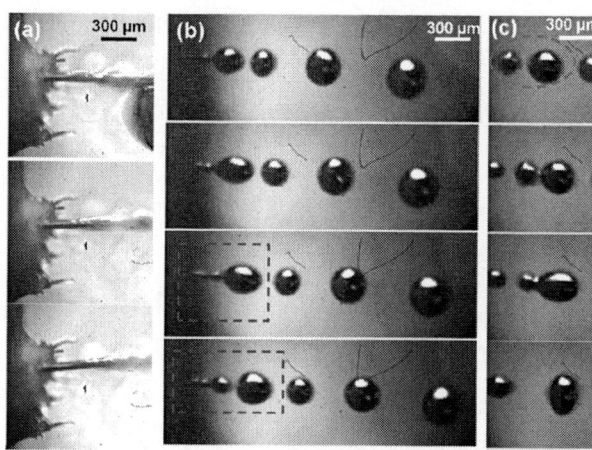

Figure 5: Rayleigh instability based (a) liquid metal jetting, (b) droplet generation, and (c) combination of two droplets.

By varying flow rates higher than 0.5 mL/min., we could readily create various diameter sizes of liquid metal Galinstan® droplets in the rage of 1 ~ 200 μm whose volume corresponds from ~ 50 attoliter (10^{-18} L) to ~ 4 picoliter (10^{-12} L) as shown in Fig. 6. With higher flow rates, smaller droplets were created. We expect that higher flow rate induces faster jetting velocity which can reduce the wavelength (λ_d) as shown in Fig. 1b. With smaller wavelength, the probability of creating smaller droplets increases.

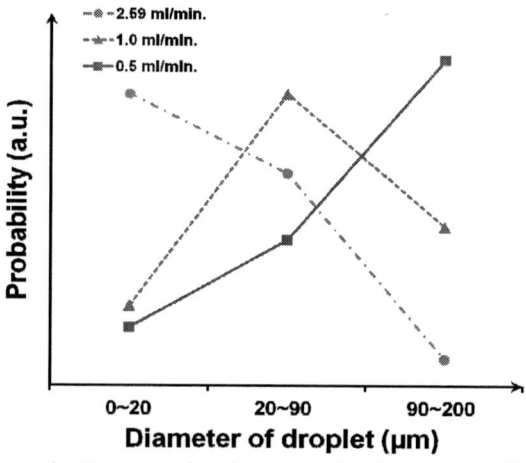

Figure 6: Varying droplet size distribution profile with different flow rates based on Rayleigh instability.

As a demonstration of the practicality of this liquid metal inkjet printing, we printed a liquid metal Galinstan® line (~ 45 μm wide and 5.4 cm long) on a Si wafer, a piece of paper and a PDMS sheet (Fig. 7). Note that while printed Galinstan® maintains liquid-phase, it does not freely flow due to the balance between adhesion and surface tension.

We further demonstrated inkjet printed liquid metal line on a paper as a direct writing of metallic interconnect on a flexible printing paper as shown in Fig. 8. The printed liquid metal line on a piece of paper was beads-on-strings shape which has ~ 90 μm wide (maximum width) beads and ~ 45 μm wide strings. The paper was bent and attached on a curved surface to show its flexible nature. Using the printed liquid metal line as metallic interconnect, we demonstrated powering a light emitting diode (LED) (Fig. 8a). This demonstrates that the inkjet-printed gallium-based liquid metal can maintain its metallic interconnect line without disconnection even with the deformation of the flexible substrate.

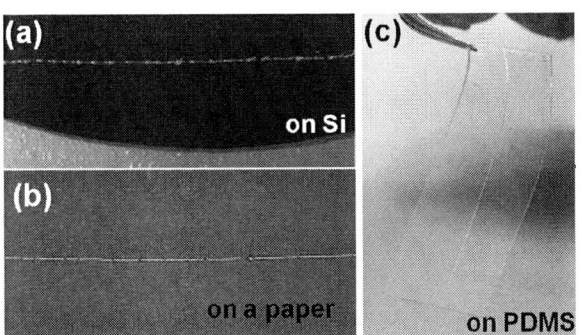

Figure 7: Generated liquid metal line on (a) Si, (b) paper, and (c) PDMS.

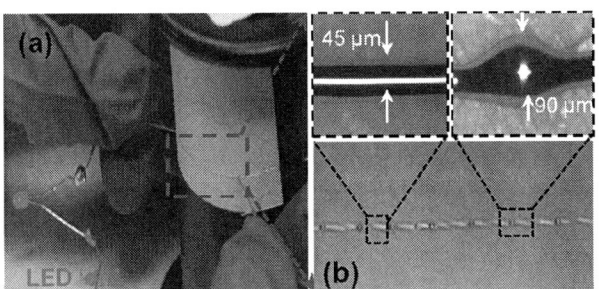

Figure 8: (a) Printed liquid metal line as a metallic interconnect on a paper and demonstration of powering a LED (battery hidden), (b) close-up images of the printed liquid metal line (inset images show close up image of a string and a bead).

CONCLUSION

In this paper, we demonstrated gallium-based liquid metal inkjet printing without any clogging and oxidation problem by incorporating with HCl-impregnated paper orifice plate. By manipulating flow rate, the liquid metal droplet was generated based on either pinch off or Rayleigh instability. Various sizes of droplets were created by varying flow rates. In addition, we demonstrated inkjet-printed liquid metal line on a flexible paper can be used as a metallic interconnect even with significant deformation of the flexible substrate. We believe this demonstration can apply to the patterning of various shapes of liquid metal and has a great potential to be utilized in flexible electronics.

ACKNOWLEDGEMENTS

The authors would like to thank Republic of Korea (ROK) Army for financial support. This research was also supported by MKE (The Ministry of Knowledge Economy), Korea, under the Brain Scouting Program (HB606-12-2001) supervised by the NIPA (National IT Promotion Agency).

REFERENCES

[1]   C. M. Hong and S. Wagner, "Inkjet printed copper source/drain metallization for amorphous silicon thin-film transistors," *Electron Device Letters, IEEE,* vol. 21, pp. 384-386, 2000.

[2]   M.L. Allen, K. Jaakkola, K. Nummila, H. Seppa, "Applicability of Metallic Nanoparticle Inks in RFID Applications," *Components and Packaging Technologies, IEEE Transactions on,* vol. 32, pp. 325-332, 2009.

[3]   R. Beccherelli, E. Zampetti, S. Pantalei, M. Bernabei, and K.C. Persaud, "Design of a very large chemical sensor system for mimicking biological olfaction," *Sensors and Actuators B: Chemical,* vol. 146, pp. 446-452, 2010.

[4]   B. Derby, "Bioprinting: inkjet printing proteins and hybrid cell-containing materials and structures," *Journal of Materials Chemistry,* vol. 18, pp. 5717-5721, 2008.

[5]   D.J. Hayes, M.E. Grove, and W.R. Cox, "Development and application by ink-jet printing of advanced packaging materials," in *Advanced Packaging Materials: Processes, Properties and Interfaces, 1999. Proceedings. International Symposium on,* 1999, pp. 88-93.

[6]   S.B. Fuller, E.J. Wilhelm, and J.M. Jacobson, "Ink-jet printed nanoparticle microelectromechanical systems," *Microelectromechanical Systems, Journal of,* vol. 11, pp. 54-60, 2002.

[7]   M. Singh, H.M. Haverinen, P. Dhagat, and G.E. Jabbour, "Inkjet Printing—Process and Its Applications," *Advanced Materials,* vol. 22, pp. 673-685, 2010.

[8]   T. Liu, P. Sen, and C.-J. Kim, "Characterization of Nontoxic Liquid-Metal Alloy Galinstan for Applications in Microdevices," *Microelectromechanical Systems, Journal of,* vol. 21, pp. 443-450, 2012.

[9]   D. Kim, P. Thissen, G. Viner, D.-W. Lee, W. Choi, Y.J. Chabal, and J.-B. Lee, "Recovery of Nonwetting Characteristics by Surface Modification of Gallium-Based Liquid Metal Droplets Using Hydrochloric Acid Vapor," *ACS Applied Materials & Interfaces,* vol. 5, pp. 179-185, 2013.

[10] L. Rayleigh, "On The Instability Of Jets," *Proceedings of the London Mathematical Society,* vol. s1-10, pp. 4-13, November 1, 1878 1878.

CONTACT

*D. Kim, tel: +1-972-693-0988; daeyoung@utdallas.edu

IN-PLANE CAPACITIVE MEMS FLOW SENSOR FOR LOW-COST METERING OF FLOW VELOCITY IN NATURAL GAS PIPELINES

Son D. Nguyen[1], Igor Paprotny[2], Paul K. Wright[1], and Richard M. White[1]
[1]University of California, Berkeley, USA
[2]University of Illinois, Chicago, USA

ABSTRACT

This paper presents the design, fabrication, and experimental results of an in-plane capacitive MEMS flow sensor that uses the displacement of a micro-fabricated paddle caused by dynamic pressure for measuring the velocity of the flow of surrounding gas. The fabrication process is simple; the device is fabricated on Silicon-On-Insulator wafers using only three photolithographic masks. A comb-drive capacitance is used as the transducer for the flow sensors. This capacitive mechanism is virtually insensitive to changes in ambient temperature. Simplicity of fabrication, combined with insensitivity to variations in ambient temperature makes this sensor ideal for widespread deployment in natural gas pipelines.

INTRODUCTION

Cost is currently the prohibitive factor preventing widespread deployment of natural gas flow meters, and thus the development of inexpensive solutions is of importance. MEMS flow sensors are usually designed to utilize the principles of asymmetric thermal conduction [1, 2, 3, 4]. However, placement of a heated filament in a potentially combustible gas mixture is not possible due to intrinsic safety requirements of natural gas pipelines. Dynamic pressure based MEMS-based flow sensors have been designed to measure force induced on an out-of-plane structures immersed in the surrounding gas flow [5]. However, the fabrication of such structures often requires complex out-of-plane fabrication process. In-plane dynamic pressure-based flow sensors have been previously fabricated [6, 7]. However, they used piezoresistive transduction which is highly sensitive to changes in ambient temperature and thus not well suited for outdoor applications. In this work, an in-plane MEMS flow sensor based on a dynamic pressure on a paddle and comb-drive readout mechanisms is presented. The fabrication is simple, resulting in low overall cost of the device, and the capacitive pickup mechanism offers superior temperature characteristics compared with piezoresistive devices.

DESIGN

Figure 1 shows a schematic drawing of the capacitive MEMS flow sensor. A paddle supported by two cantilevers will deflect out-of-plane under a dynamic pressure generated by the surrounding gas flow. Variable parallel-plate capacitors with movable electrodes, attached to the paddle, will change capacitance as the structure is deflected. By measuring the change of the capacitance, the velocity of the gas flow can be calculated. To increase the sensitivity of

the device, but still keep the critical dimension large (10 μm), three rows of digital capacitors are designed.

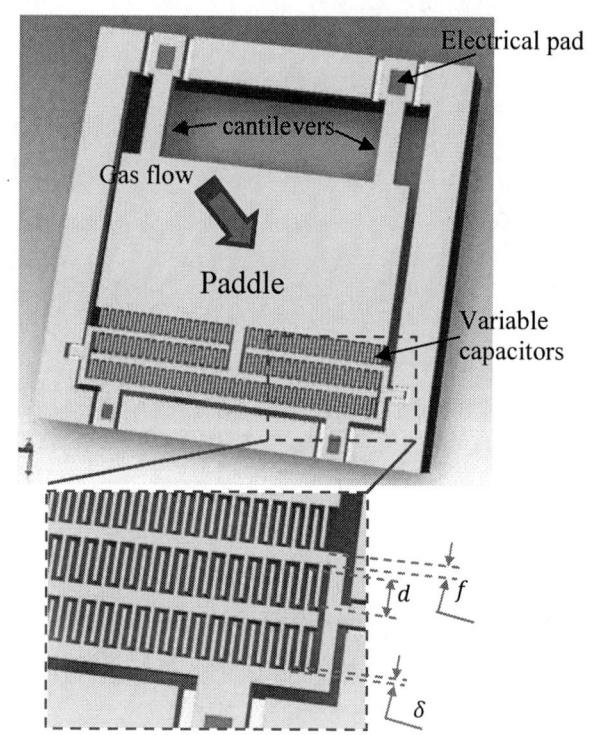

Figure 1: Simple schematic drawing of capacitive MEMS flow sensor. The drawing is not to scale. The paddle is perpendicular to the gas flow direction.

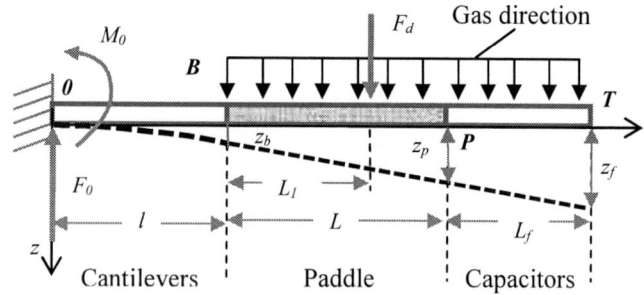

Figure 2: Free-body diagram of the flow sensor cross-section.

In turbulent conditions, with a large Reynolds number, the drag-force on the paddle can be simplified as

$$F_d = \frac{1}{2} C_D \rho A \, v^2 \qquad (1)$$

where C_D is the local drag coefficient calculated by empirical formulae depending on the structure of material, A

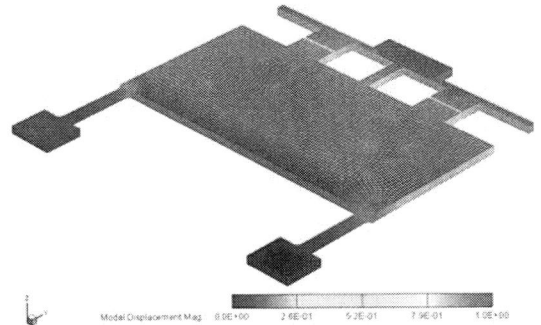

Figure 3: Deflection of the paddle and the capacitor fingers, calculated by FEM. The movable electrodes are replaced by a rectangular plate with equal surface areas for the purpose of this analysis

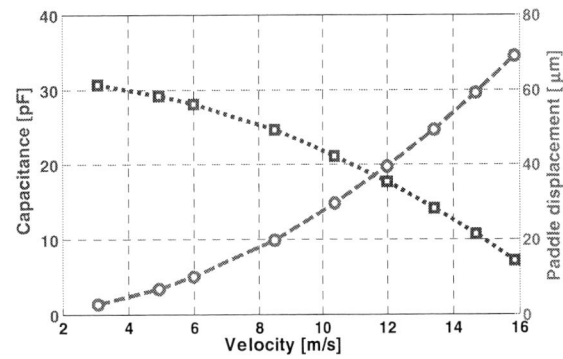

Figure 4: FEM results show capacitance vs. gas flow velocity at 1000 psi pressure and $25^0 C$. The deflection is calculated by FEM.

is the area of the paddle, capacitor fingers and the cantilevers in the direction facing the gas flow, ρ is the density of the gas and v is the mean velocity of the gas. For a wide range of Reynolds number, the drag coefficient of a flat plate is approximately $C_D = 1.28$.

Figure 2 shows the free-body diagram of the cross-section of the flow sensor with force distribution and deflection. The displacement due to drag force at the end of the beams (cantilevers), the paddle, and the capacitor fingers are respectively denoted z_b, z_p, and z_f. By ignoring the drag force on the beams, since the beam area is about 0.25% of the paddle area, the total drag force can be replaced by a concentrated force F_d at a distance L_1 from the free end of the cantilevers. It is assumed that the paddle and capacitor fingers have the same drag coefficients, L_1 can be calculated as

$$L_1 = \left(\frac{1}{2} + c\right) L \qquad (2)$$

where $c = \left(\frac{1}{2} + \frac{9d+6f+19\delta}{6L}\right) \frac{A_f}{A_f + A_p}$ (3),

with δ is gap between capacitor fingers and their frames, d is overlap length of the capacitors, and f is the width of capacitor frames, A_p is the paddle area, and A_f is the total area of movable capacitor fingers, given by $A_f = Nw_f(d + \delta)$, N is the amount of capacitors, w_f is finger width. Because A_f is very small compare to A_p, c is small, ≈ 0.077.

The slope and the deflection at the end of the cantilevers due to the drag force F_d is given by

$$\theta_b = \frac{F_d l^2}{2EI}\left(1 + \frac{2L_1}{l}\right) \qquad (4)$$

$$z_b = \frac{F_d l^3}{3EI}\left(1 + \frac{3L_1}{2l}\right) \qquad (5)$$

where l, E, and I are the length, the Young's modulus, and the area moment of inertia of the cantilevers, respectively. Since the paddle and the capacitor fingers do not bend, the deflection slope of them is equal to the slope of deflection at the end of the cantilevers. The deflection at the end of the paddle are given by

$$z_p = z_b + \theta_b L \qquad (6)$$

or $$z_p = \frac{2F_d l^3}{Ewt^3}\beta \qquad (7)$$

with $\beta = 1 + 3\frac{L_1}{l} + 3\frac{L_1^2}{l^2}$ (8)

The variable capacitance for a deflection z_p is calculated by

$$C = C_0\left[1 - \frac{z_p}{t}(1 + D)\right] + C_1\left[1 - \frac{z_p}{t}(1 + D')\right] + C_p \quad (9)$$

where t is the device thickness, C_0 and C_1 are the initial capacitance (no deflection), C_p is parasitic capacitance, and D and D' are the adjust coefficients due to the different overlap area of capacitor fingers for different rows.

$$C_0 = 2N\varepsilon_r\varepsilon_0 \frac{t*d}{g} \qquad (10)$$

$$C_1 = N\varepsilon_r\varepsilon_0 \frac{t*w_f}{\delta} \qquad (11)$$

$$C_p = \varepsilon_d\varepsilon_0 \frac{A_e}{t_{do}} \qquad (12)$$

$$D = \frac{1+2L_1/l}{\beta}\left(\frac{9d+6f+18\delta}{4l}\right) \qquad (13)$$

$$D' = \frac{1+2L_1/l}{\beta}\left(\frac{6d+3f+9\delta}{2l}\right) \qquad (14)$$

ε_r, ε_0, and ε_d are permittivity of gas, vacuum and dioxide, respectively, g is the gap between capacitor fingers, A_e is total area of electrical pads (approximately 0.415 mm^2), and $t_{do} = 2$ μm is the thickness of burry oxide layer.

The dimensions of the MEMS flow sensor are given in Table 1.

Table 1: Dimensions of the MEMS flow sensor

Description	Symbol	Layout dimension
Chip size	A_0	30.25 mm^2
Paddle length	L	2 mm
Paddle width	W	4 mm
Device thickness	t	100 μm
Cantilever length	l	1 mm
Cantilever width	w_b	95 μm
Capacitor finger width	w_f	10 μm
Capacitor finger gap	g	10 μm
Gap between fingers and frames	δ	10 μm
Capacitor overlap length	d	280 μm
With of capacitor frames	f	100 μm
Number of capacitor fingers	N	287

Figure 3 shows the deflection of the paddle and movable electrodes calculated using finite element method (FEM). Based on the deflection of the paddle, the variable capacitance versus the velocity of gas flow can be calculated

978-1-4799-3510-9/14 $31.00 © 2014 IEEE

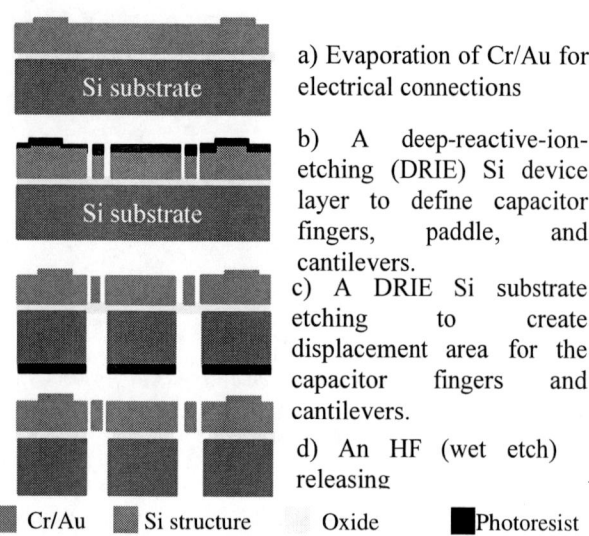

a) Evaporation of Cr/Au for electrical connections

b) A deep-reactive-ion-etching (DRIE) Si device layer to define capacitor fingers, paddle, and cantilevers.

c) A DRIE Si substrate etching to create displacement area for the capacitor fingers and cantilevers.

d) An HF (wet etch) releasing

■ Cr/Au   ■ Si structure   Oxide   ■ Photoresist

Figure 5: Fabrication process of MEMS flow sensor on Silicon-on-Insulator (SOI) wafers.

using Eq. (9), assuming the drag coefficient of the rectangular paddle is about 1.28 (Figure 4). Analysis suggests the device will respond with a sensitivity of about 1.84 pF/(m/s) for methane gas and a full-scale nonlinearity of about 11% without nonlinear compensation.

FABRICATION

Figure 5 shows the fabrication process of the sensors using Silicon-On-Insulator (SOI) wafers. The fabrication process is simple and requires only three photolithographic masks. First, 200-nm Cr/Au is evaporated using high-vacuum e-beam evaporation and patterned to form electrical bond pads (Figure 5a). Next, a Deep-Reactive-Ion-Etching (DRIE) is used on both sides of SOI wafers to define cantilevers, digital capacitors, and paddle (Figures 5b and 5c). A thick ultra-violet-baked layer of photoresist SPR-220 is used as mask during the DRIE steps. Last, the movable parts are released by wet etching buried oxide using buffered hydrofluoric acid (Figure 5d).

Figure 6 shows an optical micrograph (center) of the device after fabrication. Scanning-Electron Microscopy (SEM) images of specific components of the sensor are displayed in the insets. The capacitor fingers and their frame are shown in Figures 6a and 6b. To protect the sidewall of the cantilevers in DRIE, dummy structures are added along the cantilevers (Figure 6c). Mechanical end-stops to avoid unexpected motions of the paddle in lateral direction are shown in the figure inset 6d. The gap between the paddle and the end-stops was designed smaller than the gap between capacitor fingers to avoid the capacitor fingers touching together in the lateral direction.

EXPERIMENTS

The sensors were characterized using an air flow in a 150-mm diameter duct at standard atmospheric pressure. The Reynolds number is given by

$$R = \frac{Vd}{v} \tag{11}$$

where V is mean air velocity, d is duct diameter, and v is kinematic viscosity ($v \approx 1.568 \cdot 10^5 m^2/s$ at room temperature and atmospheric pressure). The Reynolds number is larger than 2300 when V is larger than 0.24 m/s, indicating turbulent air flow. The experimental setup for the flow measurement is shown in Figure 7. The air flow velocity was measured using a hot-wire anemometer TPI 575C1. The capacitance meter BK Precision 890C was used to measure the capacitance between two electrodes of the flow sensor. An air fan with adjustable speed was used to generate the air flow.

Figure 8 shows the output capacitance of the flow sensor versus the airflow velocity at room temperature. The data were best fit by a quadratic curve including the first order and the second order of velocity.

It is important to note that ***all*** flow sensors are sensitive to the variations in the density of the gas, which in turn is a function of its temperature. It would be undesirable if the transduction mechanism was also sensitive to changes in the ambient temperature. The benefit of the capacitive

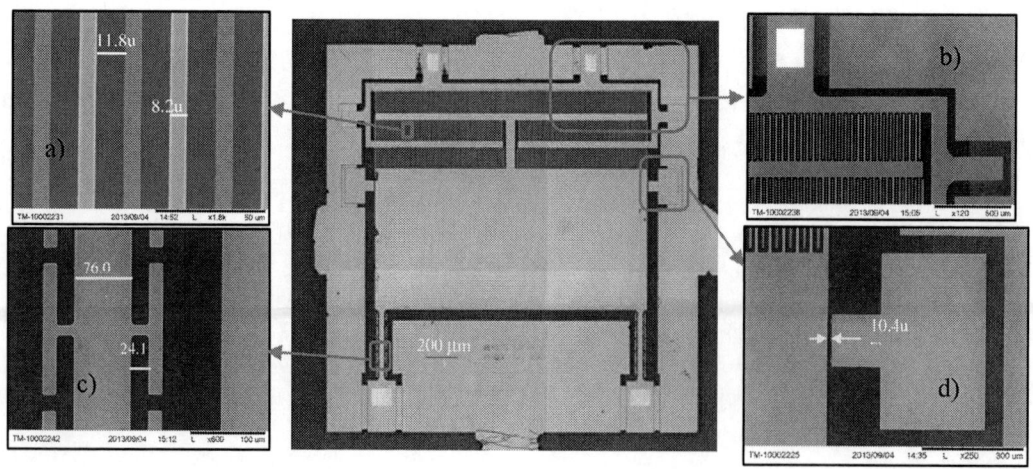

Figure 6: Micrograph of the MEMS-capacitive flow sensor after fabrication. The inset figures: SEM images of (a) the capacitor fingers, (b) a corner of the sensor, (c) cantilever and its DRIE protections, (d) mechanical end-stops of the paddle in lateral

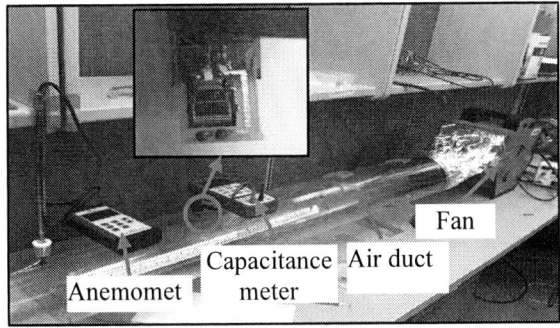

Figure 7: Experimental set up for testing the MEMS flow sensor using an air flow. Inset: the flow sensor on a circuit board.

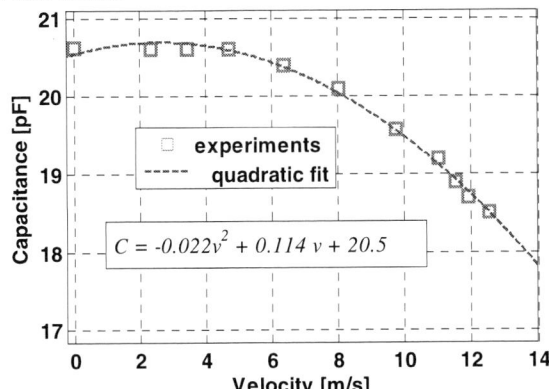

Figure 8: Measurement of variable capacitance vs. air velocity at the atmospheric pressure (14.7 psi) and the temperature of 25^0C

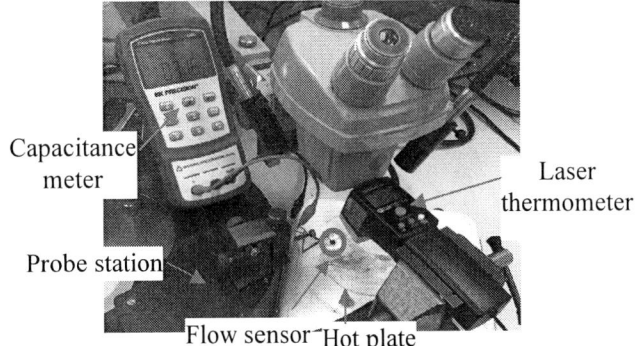

Figure 9: Experimental set up for thermal testing.

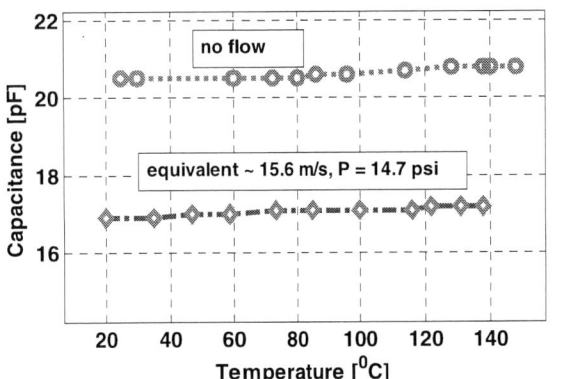

Figure 10: Response of the flow sensor under changing temperatures.

(compared to e.g., piezoresistive) transduction is the insensitivity to the temperature fluctuations. Figure 9 shows an experimental set-up to test the response of the flow sensor under changing temperatures. Figure 10 shows the capacitance versus the temperature on the device for different of paddle deflections, corresponding to different dynamic pressure. For no deflection, equivalent to no flow, the output capacitance remains unchanged for a wide range of temperatures. For a deflection that is equivalent to airflow of 15.6 m/s, the capacitance varied about 2% when temperature from 20^0C to 140^0C. Note that this 2% variability can be further reduced significantly by measuring, and calibrating for, the ambient temperature.

CONCLUSION

In-plane capacitive MEMS flow sensors have been designed, modeled, and fabricated using a bulk micromachining technology with only three photolithographic masks. Experimental results with an airflow showed that output capacitance of the device is a quadratic function of air velocity. Experimentation also showed that the capacitance readout mechanism is insensitive to temperature changing. A shield to protect the device from particles in the gas stream, as well as a corrosion coating, will be considered as in future work.

ACKNOWLEDGEMENTS

This project is sponsored by the California Energy Commission, contract number 500-10-044. The devices were fabricated in Marvell Nano-fabrication Lab, University of California, Berkeley. The author would like to thank Christine E. Gregg for valuable suggestions on language issues.

REFERENCES

[1] Y.H. Wang, C.P. Chen, C.M. Chang, C.P. Lin, C.H. Lin, L.M. Fu, C.Y. Lee, "MEMS-based gas flow sensors", *J. Microfluid Nanofluid*, Volume 6, pp 333-346, 2009.

[2] Christian G.J. Schabmueller, "Flow sensors", Chapter 9, *Artech House*, pp. 213 – 248, 2004.

[3] N. T. Nguyen, "Micromachined flow sensors—a review", *Flow Meas. Instrum.*, Vol. 8, pp. 7–16, 1997.

[4] S. Wu, Q. Lin, Y. Yuen, and Y.C. Tai, "MEMS flow sensors for nano-fluidic applications", *Proc. IEEE MEMS 2000*, Miyazaki, Japan, Jan 23-27, 2000, pp.745-750.

[5] Y. Yang, N. Chen, C. Tucker, J. Engel, S. Pandya, and C. Liu, "From Artificial Hair Cell Sensor to Artificial Lateral Line System: Development and Application", *Proc. IEEE MEMS 2007*, Kobe, Japan, Jan 21-25, 2007, pp. 577-580.

[6] V. Gass, B. H. van der School and N. F. de Rooij, "Nanofluid Handling by Micro-Flow-Sensor Based on Drag Force Measurements", *Proc. IEEE MEMS 1993*, Florida, USA, Feb 7-10, 1993, pp.167-172.

[7] L. Du, Z. Zhao, C. Pang, Z. Fang, 'Drag force micro solid state silicon plate wind velocity sensor', *Sens. Actuators A, Phys.*, vol. 151, pp. 35-44, 2009.

978-1-4799-3510-9/14 $31.00 © 2014 IEEE

INTEGRATED MULTI-PARAMETER FLOW MEASUREMENT SYSTEM

J.C. Lötters[1,2], E. van der Wouden[1], J. Groenesteijn[2]
W. Sparreboom[1], T.S.J. Lammerink[2] and R.J. Wiegerink[2]
[1]Bronkhorst High-Tech BV, The Netherlands
[2]MESA+ Institute for Nanotechnology, University of Twente, The Netherlands

ABSTRACT

We have designed and realized an integrated multi-parameter flow measurement system, consisting of an integrated Coriolis and thermal flow sensor, and a pressure sensor. The integrated system enables on-chip measurement, analysis and determination of flow and several physical properties of both gases and liquids. With the system, we demonstrated the feasibility to measure the flow rate, density, viscosity, specific heat capacity and thermal conductivity of hydrogen, helium, nitrogen, air, argon, water and IPA.

INTRODUCTION

Knowledge of both the flow rate and its composition is essential in medical infusion pumps, especially in neonatology, where a newborn baby should receive both the right and the right amount of medicine and/or nutrient. Since the flow rates involved are extremely small, typically in the range of 0.1 - 1 ml/h, it is important to have a very compact single chip integrated system rather than a system composed of separate devices, to maximally reduce the internal volume of the system. Other examples of applications are flow chemistry for the production of specialty drugs, production of the right mix of gases for medical purposes, and measurement of the composition of fuel gas to determine its energy content.

Van Baar [1] demonstrated the feasibility to measure several physical properties with different thermal sensors. Enoksson [2] showed density measurements with Coriolis flow sensors. Jakoby [3] demonstrated viscosity measurements with a resonant structure.

Up to now, no systems have been reported in which different types of sensor principles are integrated to measure the flow rate and several physical properties of both gases and liquids.

SYSTEM STRUCTURE

The basic structure of the integrated multi-parameter flow measurement system is shown in figure 1. The system consists of an integrated Coriolis and thermal flow sensor, and an additional differential pressure sensor. Fluid flow enters the system at the inlet, passes through the Coriolis and the thermal flow sensor, and leaves the system at the outlet. The differential pressure between the inlet and outlet is measured by the pressure sensor.

OPERATING PRINCIPLE

The output signal of the thermal flow sensor is a measure for the flow rate, the pressure is measured by the pressure sensor. The output signal of the Coriolis mass flow sensor provides both the mass flow and information about the density of the medium [4, 5].

As shown in figure 2, the other parameters can be obtained from the output signals via a calculation model. By comparing the output signals of the Coriolis flow sensor and the pressure sensor, and taking the density into account, the viscosity of the medium can be calculated. By comparing the output signals of the thermal and the Coriolis flow sensors at low flows, the heat capacity of the medium can be calculated. The thermal conductivity can be determined by comparing the output signals of the thermal and Coriolis flow sensors at higher flows.

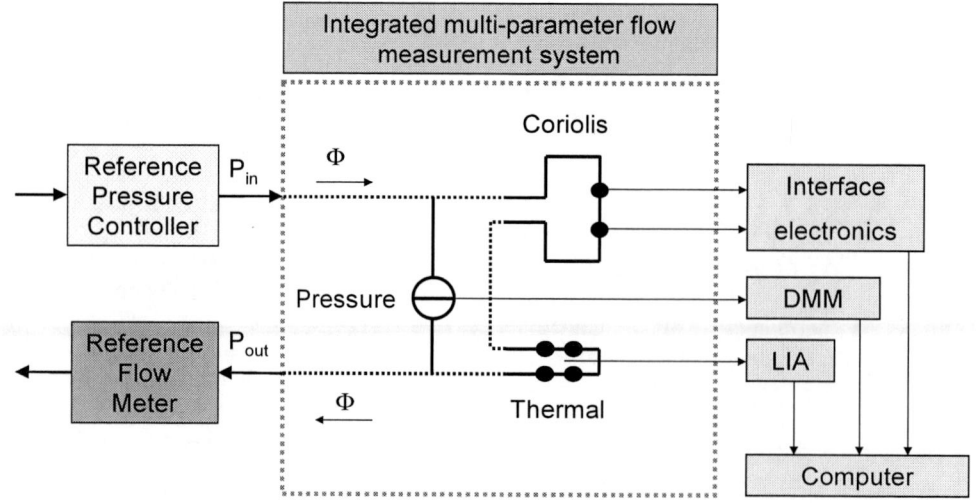

Figure 1: Basic structure of the integrated multi-parameter flow measurement system

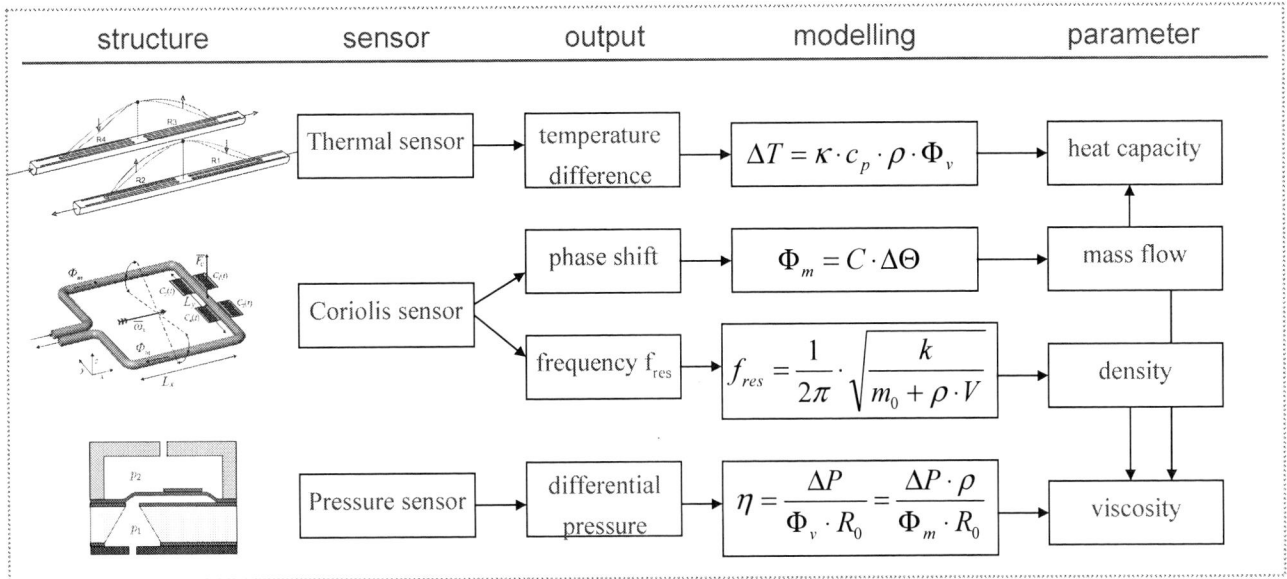

Figure 2: Operating principle of the integrated multi-parameter flow measurement system

MODELLING

In this section, it is explained how the different physical parameters are derived from the sensor signals.

Density

The resonance frequency of the Coriolis flow tube is dependent on and therefore a measure for the density of the medium that is inside the tube [4, 5]. The equation is shown in figure 2.

Specific Heat Capacity

When the output signal of the thermal flow sensor is plotted against the output signal of the Coriolis flow sensor, the slope of the curve at low flows is a measure for the heat capacity of the medium that is inside both sensors.

The resulting curve can be described by a third order polynomial function:

$$S = C_1 y^3 + C_2 y^2 + C_3 y + C_4 \qquad (1)$$

where the constants C_{1-4} are medium independent sensor parameters, their value is determined with air as reference medium, y is the output signal of the Coriolis flow sensor, and S the output signal of the thermal flow sensor.

When another medium is present in the system, we can solve equation (1) for a certain value of S. The ratio of the solved value for y and the measured value for y, multiplied with the value of the specific heat capacity of the reference medium air, provides the value of the specific heat capacity of the actual medium.

Viscosity

When the output of the Coriolis flow sensor is plotted against the output of the pressure sensor, the slope of the curve is a measure for the viscosity of the medium that is inside both sensors.

The viscosity can be calculated with the following equation [6]:

$$\eta = \frac{\Delta P}{\Phi_v * R_0} = \frac{\Delta P * \rho}{\Phi_m * R_0} \qquad (2)$$

The medium independent hydraulic resistance R_0 is determined by using air as reference medium. The mass flow Φ_m and the density ρ are measured by the Coriolis sensor, the pressure drop ΔP is measured by the pressure sensor. For gases, we have to correct the measured density for the compressibility of the gas.

Thermal Conductivity

When the output signal of the thermal flow sensor is plotted against the output signal of the Coriolis flow sensor, the slope of the curve at higher flows is a measure for the thermal conductivity of the medium that is inside both sensors.

The resulting curve can be described by [7]

$$V = S_o c_p \Phi_m \left(1 - \frac{S_1}{\lambda} c_p \Phi_m - S_2 c_p \Phi_m \right) \qquad (3)$$

where V is the output signal of the thermal flow sensor, S_{0-2} are medium independent sensor constants, c_p is the specific heat capacity, Φ_m the mass flow and λ the thermal conductivity. It should be noted that λ is determined from equation (3) by curve fitting, so, no direct analytical relation can be provided, and therefore equation (3) is not shown in figure 2.

FABRICATION

A detailed description of the fabrication process can be found in [8]. A picture of the fabricated system is shown in figure 3.

Figure 3: Fabricated system, showing the thermal (left) and Coriolis (right) flow sensors

MEASUREMENT SET-UP

For the gas flows, a pressurised vessel was used to generate air, hydrogen, helium, argon and nitrogen flows in the range of 1 up to 20 ml_n/min. For the liquid flows, a syringe pump system was used to generate water and IPA flows in the range of 1 up to 35 mg/h through the system. Pressures in the range of 1 through 7 bar were provided to the system. A photograph of the measurement set-up is shown in figure 4.

Figure 4: Measurement set-up

During the measurements the output signals of the pressure, thermal flow and Coriolis flow sensor were recorded simultaneously, together with the output signals of the reference instruments.

MEASUREMENT RESULTS

In figure 5, the relation between the output of the thermal flow sensor and the Coriolis flow sensor is shown, which is a measure for the heat capacity of the medium. All via equation (1) derived heat capacities were within 5% of their value as found in literature.

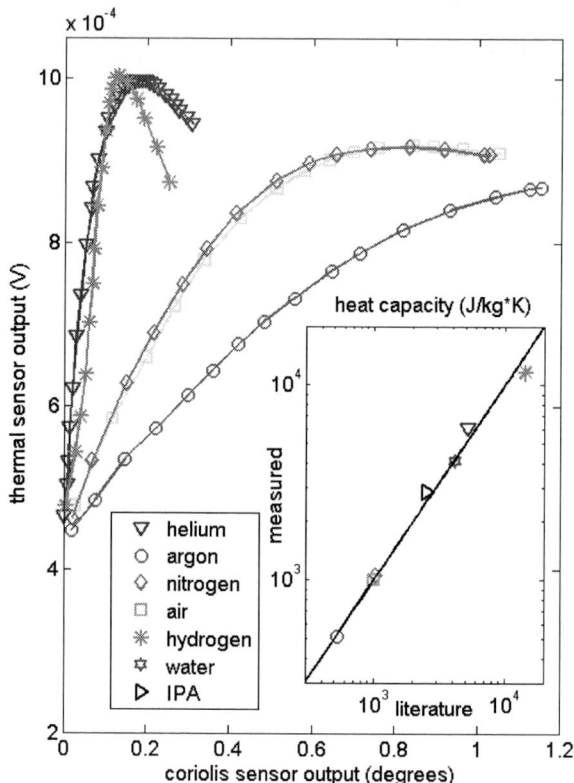

Figure 5: Relation between the output of the thermal flow sensor and the output of the Coriolis flow sensor for several gases (liquids not shown). According to the modelling, this represents the heat capacity of the medium, as shown in the smaller graph.

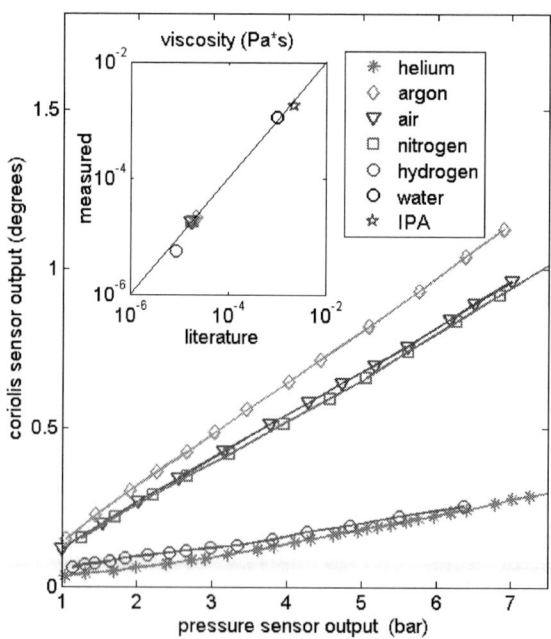

Figure 6: Relation between the output of the Coriolis flow sensor and the output of the pressure sensor for several gases (liquids not shown). According to the modelling, this represents the viscosity of the medium, as shown in the smaller graph

978-1-4799-3510-9/14 $31.00 © 2014 IEEE

In figure 6, the relation between the output of the Coriolis flow sensor and the pressure sensor is shown, which is a measure for the viscosity of the medium. All via equation (2) derived viscosities were within 10% of their value as found in literature. The biggest deviations occur for hydrogen and helium, as it is difficult to fill the system with these gases, and a mixture between hydrogen or helium and air is likely to occur.

In figure 7, the values for the thermal conductivity of the measured gases are given. All via curve fitting of equation (3) found values are within 10% of their value as found in literature, except for helium, which is within 20% of the literature value.

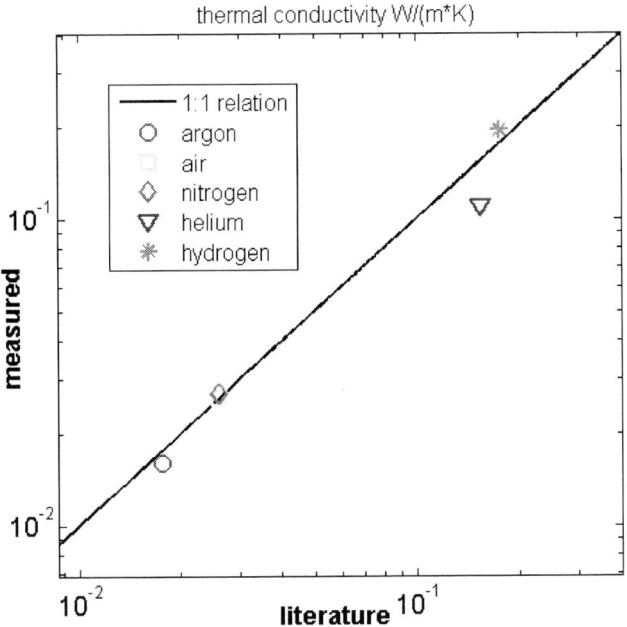

Figure 7: values for the thermal conductivity as a function of the literature value, black line indicates the 1:1 relation.

Currently, we are working on measuring the thermal conductivity of water and IPA.

CONCLUSIONS

We have designed and realised an integrated multi-parameter flow measurement system, consisting of an integrated Coriolis and thermal flow sensor, and an additional pressure sensor. The integrated system enables on-chip measurement, analysis and determination of flow and several physical properties of both gases and liquids. With the system, we demonstrated the feasibility to measure the flow rate, density, viscosity, specific heat capacity and thermal conductivity of hydrogen, helium, nitrogen, air, argon, water and IPA. Future research will focus on improving the accuracy of the measured parameters and integration of further functionalities in the system.

ACKNOWLEDGMENTS

This research was partly financed by the Dutch NanoNextNL program. The authors would like to thank the industrial partners in this project for their in-kind contributions and many fruitful discussions.

REFERENCES

[1] J.J. van Baar, et al., Micromachined structures for thermal measurements of fluid and flow parameters, *J. Micromech. Microeng.*, 11 (2001), pp. 311-318

[2] P. Enoksson et al., A silicon resonant sensor structure for Coriolis mass-flow measurements, *Journal of microelectromechanical systems*, Vol. 6, No. 2 (1997)

[3] B. Jakoby et al., Miniaturized sensors for the viscosity and density of liquids – performance and issues, *IEEE Transactions on Ultrasonics, Ferroelectrics and Frequency Control*, 57 (2010), No. 1, pp. 111 - 120

[4] J. Haneveld et al., Modelling, design and characterization of a micro Coriolis mass flow sensor, *J. Micromech. Microeng.*, 20 (2010), 125001, pp. 1-10

[5] W. Sparreboom et al., Compact mass flow meter based on a micro Coriolis flow sensor, *Micromachines*, 4 (2013), pp. 22 - 33

[6] R.W. Fox et al., Introduction to fluid mechanics, *John Wiley & Sons*, New York, Third edition (1985)

[7] W. Jouwsma, Marketing and design in flow sensing, *Sensors & Actuators A*, 37-38 (1993), pp. 274 - 279

[8] T.S.J. Lammerink et al., Single chip flow sensing system with a dynamic flow range of more than 5 decades, *Technical Digest of Transducers '11*, Beijing, China, June 5 - 9, 2011, pp. 890 - 893

CONTACT

*J.C. Lötters, tel: +31-53-4894655; j.c.lotters@utwente.nl

INTERACTION FORCES DURING THE SLIDING OF A WATER DROPLET ON A TEXTURED SURFACE

Nguyen Thanh-Vinh, Hidetoshi Takahashi, Kiyoshi Matsumoto, and Isao Shimoyama
The University of Tokyo, Tokyo, JAPAN

ABSTRACT

We have directly measured the interaction forces (in normal and lateral directions) during the sliding of a water droplet on a surface decorated with a micropillar array. The measurement was carried out using a MEMS-based two-axis force sensor array. The advantages of our sensor array were the high sensitivity and miniaturized size of the ultrathin cross-shaped piezo-resistive silicon structure fabricated under a micropillar to detect both the normal and lateral forces acting on the micropillar. A demonstrating measurement using a water droplet with volume of 18 µL sliding on the micropillar array was carried out. The measurement results showed a fluctuation in the interaction forces when the micropillar was close to the trailing edge or leading edge of the droplet. Meanwhile, in the inner region of the contact line, both the normal and lateral interaction forces were relatively stable. These results indicate that the interaction forces at the edges of the droplet are important factors controlling the sliding motion of the droplet.

INTRODUCTION

A smooth surface can only achieve a hydrophobicity with the contact angle less than 130°. However, it is known as lotus effect that a textured hydrophobic surface (a surface with microstructures coated with hydrophobic material) can exhibit the superhydrophobicity with the contact angle up to 175° [1-5]. This superior water-repellent property due to the surface roughness can be found in many biological surfaces, as well as artificial surfaces achieved by microfabrication techniques [1-4]. Many promising applications of these superhydrophobic surfaces include self-cleaning, anti-frog and friction drag reduction in microfluidic system. To design an optimized hydrophobic surface for these applications, it is important to understand the sliding mechanism of a water droplet on such surfaces with microstructures. In the case of the droplet sliding on a textured surface, the interaction between the droplet and each microstructure is the major factor controlling the sliding behavior of the droplet. Therefore, understanding this liquid-solid interaction becomes important to reveal the sliding mechanism of the droplet.

In previous researches, the sliding mechanism of droplets on a textured surface was studied using optical observations [6-9], or theoretical or numerical analysis [8-12]. By optical observations, contact behavior at the leading and trailing edges of the droplet and internal flow of a droplet can be studied. However, the interaction of the droplet and the textured surface inside the contact area, which significantly controls the sliding motion of the droplet, is not observable by this method. On the other hand, it is generally difficult to provide a detailed theoretical model

Measure the interaction force betwen the droplet and a pillar during the sliding

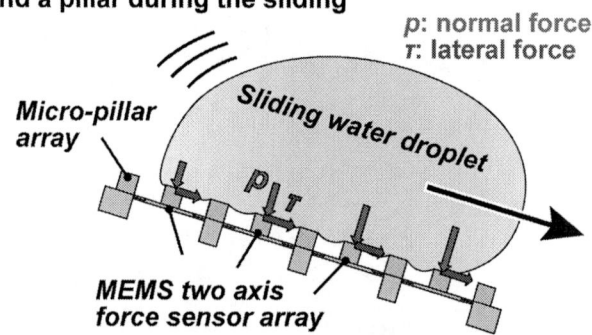

Figure 1: Conceptual sketch of this study. The interaction forces in normal and lateral directions between a water droplet and a micropillar during sliding were directly measured by a MEMS two-axis force sensor array.

(a) Design of a two axis force sensor

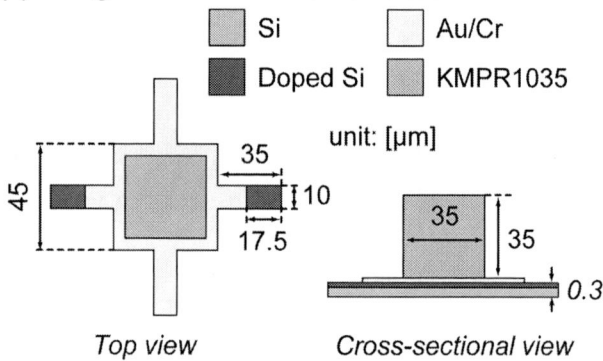

(b) Two axis force sensing principle

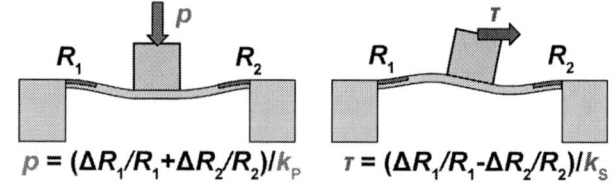

$$p = (\Delta R_1/R_1 + \Delta R_2/R_2)/k_p \qquad \tau = (\Delta R_1/R_1 - \Delta R_2/R_2)/k_s$$

Figure 2: Design and sensing principle of a sensor. The normal and lateral force are detected from the resistance changes of two piezoresistors.

which can describe exactly the forces acting on each microstructure during sliding. One of the reason is that both slipping and rolling, which have different corresponding mechanical models, can coexist in the sliding of the droplet.

In this paper, using a MEMS-based two axis force sensor array, we have carried out the direct measurement of the interaction forces between a droplet and a micropillar array during the sliding. The normal and lateral forces acting on a

978-1-4799-3510-9/14 $31.00 © 2014 IEEE

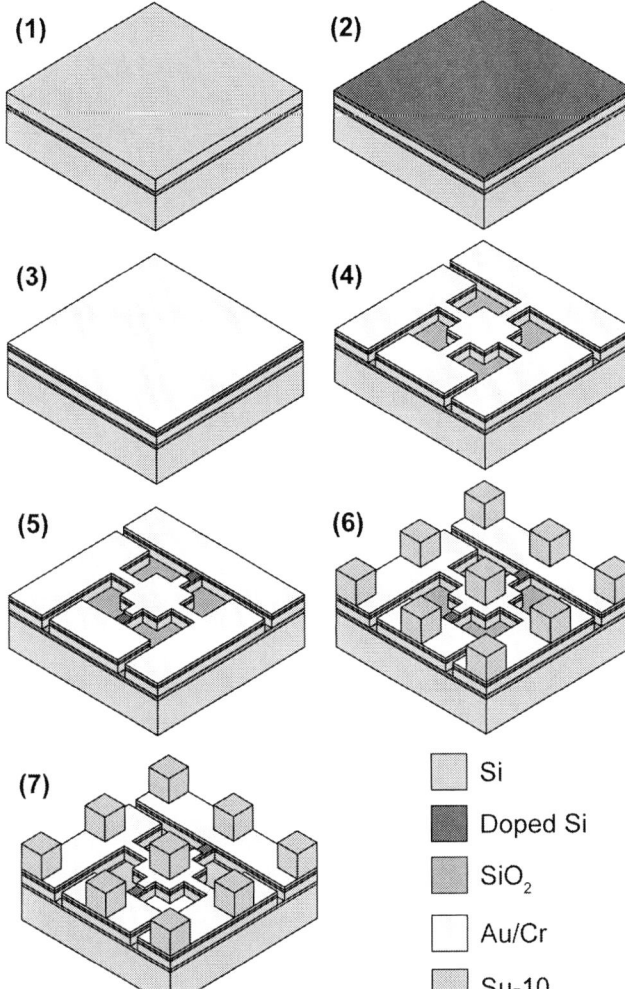

Figure 3: Fabrication process of the sensor array.

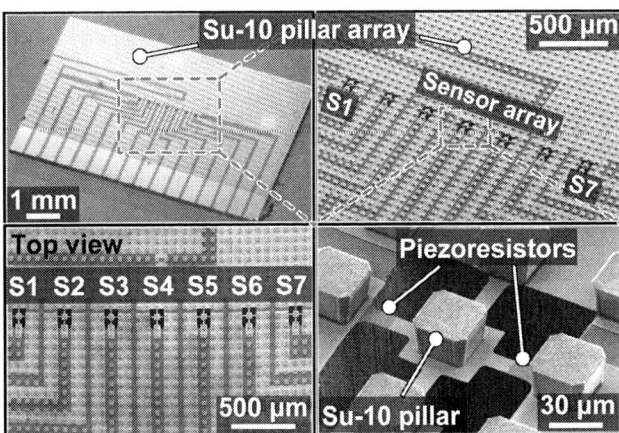

Figure 4: Photographs of the fabricated device.

single micropillar are measured by an ultrathin piezoresistive silicon structure fabricated underneath (Figure 1). The design, fabrication and calibration of our sensor array are first introduced. Then a demonstration experiment on the sliding of a water droplet (volume: 18μL) is described.

DESIGN AND FABRICATION

The design of a sensor is shown in Figure 2. The sensor consists of a 300nm-thick silicon cross-shaped structure and a KMPR1035 micropillar at the center. The normal and lateral force acting on the micropillar are detected by two piezo-resistors formed the roots of two beams in the structure. In our design, seven sensors were arrayed on a line with a pitch of 375μm. The size and pitch of micropillar array were 35μm × 35μm × 35μm and 75μm, respectively.

In the fabrication process is shown in Figure 3. First, the device layer of an SOI (Silicon On Insulator, 0.3/0.4/300μm) wafer was doped using ion implantation (As, dose: 10^{15}/cm^2, 10 keV for 585 sec and 40 keV for 303 sec). Then an Au/Cr layer (30nm/3nm) was deposited via evaporation in vacuum. In the next step, the Au/Cr was patterned and the top layer was etched by Inductive Coupled Plasma-RIE (ICP-RIE) to form the cross-shape structure. After that, the Au/Cr layer was patterned again to reveal the piezoresistors. To fabricate the micropillar array, a 35μm-thick KMPR1035 photoresist was spincoated (3000 rpm for 60s) and patterned on the device layer of the wafer. Next, the handle Si layer was patterned and etched. Finally, the glass layer underneath the cross-shaped structure was removed by HF vapor etching. Moreover, a thin layer of C_4F_8 was deposited on the micropillar array as a hydrophobic treatment.

The fabricated device are shown in Figure 4. The total sizes of the device and micropillar array were 10mm × 6mm and 10mm × 5mm, respectively. The area which is not covered by micropillars was used as electrodes to connect the sensor with an outer printed board using wire-bonding.

SENSOR CALIBRATION

Since our sensor has a high flexibility due to the ultrathin silicon structure, it is difficult to calibrate the sensor using commercial load cells. To obtain the sensitivities of the fabricated sensor regarding to normal and lateral force, both simulation and experimental results were used. We defined the sensitivities of our sensor in response to normal and lateral force k_P and k_S as shown in Figure 2. First, a simulation on the deformation of the cross-shaped structure under normal and lateral forces (both 100 Pa) was conducted (Figure 5 (a)). The material properties of silicon and KMPR1035 used in the simulation are shown in Table 1. The tetrahedral type elements were used for meshing and the number of elements was 299871. We define the normal and lateral forces in this study as the values on a unit area of the micropillar surface. From the simulation results, the spring constant of the sensor in response to normal force applied on the micropillar was obtained to be 8.0 (N/m). Besides, the relationship between resistance changes of two piezoresistors of the sensor and the vertical displacement of the middle micropillar was obtained from the experiment shown in Figure 5 (b). Based on these results, the sensitivity of the sensor in response to a normal force k_P can be calculated to be 2.9×10^{-6} (Pa^{-1}).

Moreover, also based on the simulation results, the ratio k_S/k_P was obtained to be 0.8. Therefore, the sensitivity in

Table 1: Properties of materials used in the simulation.

Material properties	Silicon (Cross-shaped structure)	KMPR1035 (Micropillar)
Density	2329 (kg/m^3)	1200 (kg/m^3)
Young's modulus	170 (GPa)	2 (GPa)
Poisson ratio	0.28	0.22

(a) Simulation results

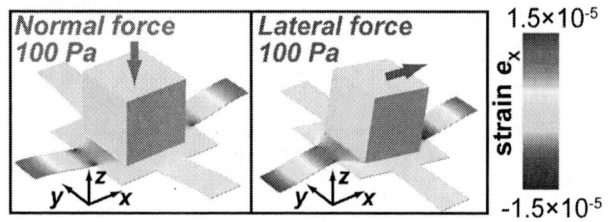

(b) Response to the normal displacement

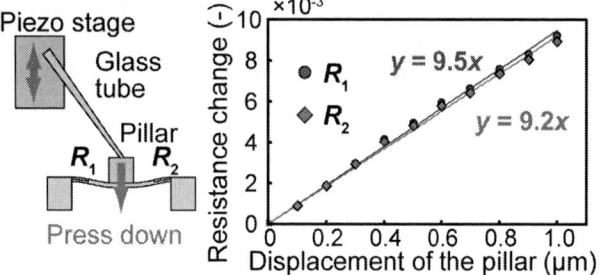

Figure 5: (a) Simulation results showing the deformation and strain of the sensor under normal and lateral load. (b) The relationship between vertical displacement of the pillar and the resistance changes of two piezoresistors.

response to a lateral force was calculated to be 2.3×10^{-6} (Pa^{-1}).

EXPERIMENT AND RESULTS

Experimental setup

The experimental setup to measure the interaction force of a sliding droplet on the fabricated micropillar array is shown in Figure 6. The sensor array was attached to a rotational stage, by which the angle of the sensor array to direction of gravity can be controlled. At the beginning, a water droplet with the volume of approximately 18 µL was placed on the micropillar array while it is adjusted to horizontal angle. Then the stage was rotated manually to let the droplet slide through the sensor array. The sliding motion of the droplet was captured using a high speed camera (PHOTRON, FASTCAM 1024 PCI) at 1000 fps while the outputs of sensor array were measured by a multi-channel oscilloscope (ScopeCorder DL850, Yokogawa Inc.) after amplified by 1000 times. The images of the high speed camera and outputs of the sensor array was synchronized by an LED used as an indicator.

Experimental results

As the micropillar array was rotated, the water droplet started to slide down. The angle of the micropillar array to

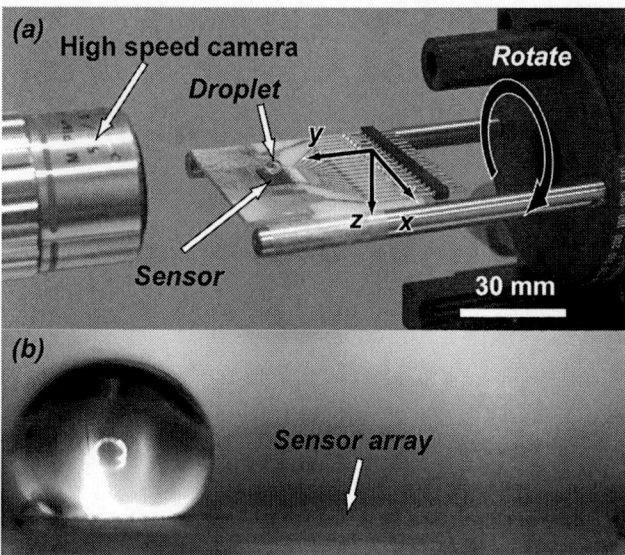

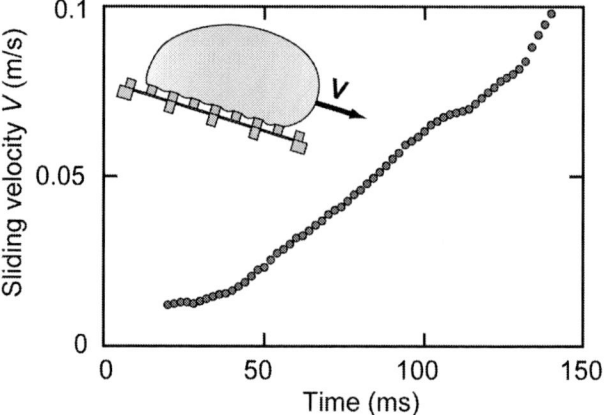

Figure 6: (a) Experimental setup. (b) An image of the high speed camera showing the initial position of the water droplet placed near the left edge of the micropillar array.

Figure 7: Sliding velocity of the droplet calculated from the images of the high speed camera showing that the droplet sliding with a constant acceleration.

horizontal direction when the droplet slid through the sensor array was approximately 18°. Figure 7 shows the sliding velocity of the droplet calculated from the images of the high speed camera. The sliding velocity increased linearly with time, which means that the droplet was sliding with a constant acceleration.

The image sequence of droplet sliding motion and the interaction forces between the droplet and micropillars of the sensor array are shown in Figure 8 (a). The positive direction of normal and lateral forces were defined to be the same with that of z- and x-axis, respectively. The plot-markers in each graph indicate the moment when the center of contact line crossed each sensor. The results suggested that the contact mechanism of a micropillar and the droplet can be described in three steps as followed (Figure 8(b)). First, when the droplet was approaching a micropillar, a capillary bridge between that micropillar and the previous micropillar was formed and the pillar was pulled backward, resulting in the

(a) Interaction forces at position of each sensor

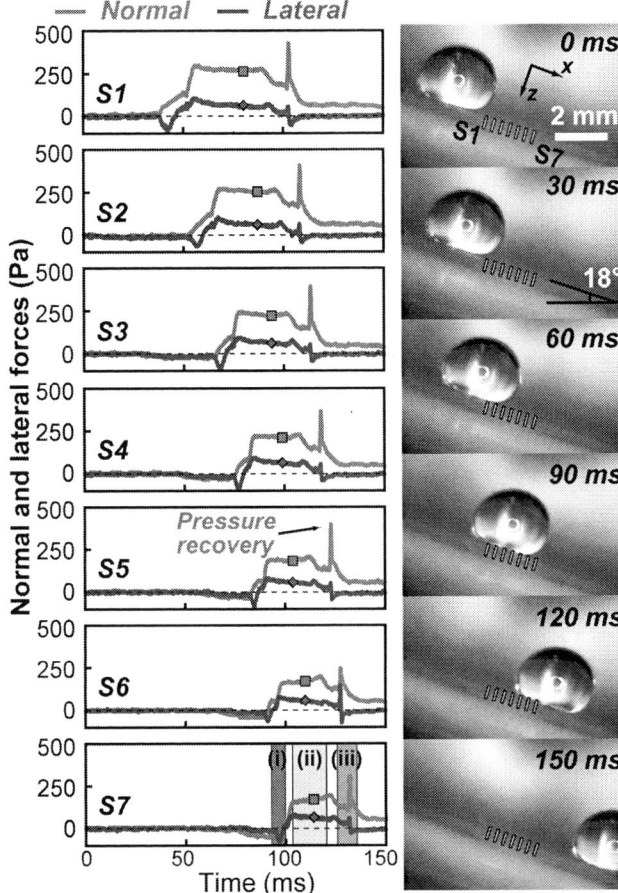

(b) Contact mechanism

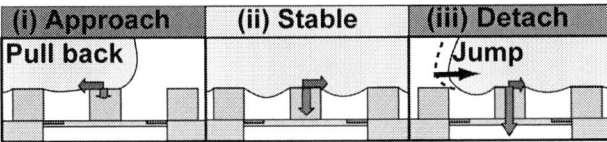

Figure 8: (a) Snapshots of the high speed camera showing the sliding of the droplet and the interaction forces between the droplet and the micropillar of each sensor. (b) The contact mechanism between the droplet and the micropillar array during sliding.

negative lateral force. In the second step, as the droplet continued sliding forward, the normal and lateral forces became stable. In the detachment step, first, both normal and lateral forces decreased. However, interestingly, our results showed that right before the detachment of the droplet from a micropillar, the normal force at that pillar increased. The reason is thought to be the rupture of the capillary bridge at the rear of the micropillar and the mass of that capillary bridge jumped forward causing more normal force on the micropillar.

CONCLUSIONS

In conclusions, we have directly measured the interaction forces in normal and lateral direction between a water droplet and the array of micropillar during sliding. The results showed that there were fluctuations in the interaction forces when a pillar contacted with the leading edge or trailing edge of the droplet. When the droplet started contacting that pillar, surface tension could cause the pillar to be pulled backward. On the other hand, before detachment of the droplet's trailing edge from a pillar, we observed a recovery of the normal force which is thought to be caused by the jumping of the capillary bridge from the previous pillar. Our results indicated that the interaction forces at the edges of the droplet are important factors which control the sliding motion of the droplet.

ACKNOWLEDGEMENT

The photolithography masks were made using the University of Tokyo VLSI Design and Education Center (VDEC)'s 8 inch EB writer F5112 + VD01 donated by ADVANTEST Corporation. This work was supported by JSPS KAKENHI Grant Numbers 25000010, 23310089, 24656162 and NSK Foundation for Advancement of Mechatronics.

REFERENCES

[1] B. Bhushan, Y. C. Jung, "Natural and biomimetic artificial surfaces for superhydrophobicity, self-cleaning, low adhesion, and drag reduction", *Prog. Mater Sci.*, vol. 56, pp. 1-108, 2011.

[2] M. Ma, R. M. Hill, "Superhydrophobic surfaces", *Curr. Opin. Colloid Interface Sci.*, vol. 11, pp. 193-202, 2006.

[3] Z. Guo, W. Liu, B-L. Su, "Superhydrophobic surfaces: From natural to biomimetic to functional", *J. Colloid Interface Sci.*, vol. 353, pp.335-355, 2011.

[4] R. Blossey, "Self-cleaning surfaces – virtual realities", *Nat. Mater.*, vol. 2, pp.301-306, 2003.

[5] D. Quere, "Wetting and roughness", *Annu. Rev. Mater. Res.*, vol. 38, pp.71-99, 2008.

[6] C. Lv, C. Yang, P. Hao, F. He, Q. Zheng, "Sliding of Water Droplets on Microstructured Hydrophobic Surfaces", *Langmuir*, vol. 26, pp.8704-8708, 2010.

[7] P. Hao, C. Lv, Z. Yao, F. He, "Sliding behavior of water droplet on superhydrophobic surface", *EPL*, vol. 90, pp. 66003-6, 2010.

[8] P. Aussillous, D. Quere, "Shapes of rolling liquid drops", *J. Fluid Mech.*, vol. 512, pp.133-51, 2004.

[9] P. Olin, S. B. Lindström, T. Pettersson, and L. Wågberg, "Water Drop Friction on Superhydrophobic Surfaces", *Langmuir*, vol. 29, pp. 9079-9089, 2013.

[10] B. M. Mognetti, H. Kusumaatmaja, J. M. Yeomans, "Drop dynamics on hydrophobic and superdydrophobic surface", *Faraday Discuss.*, vol. 146, pp. 153-166, 2010.

[11] M. Reyssat, D. Richard, C. Clanet, D. Quere, "Dynamical superhydrophobic", *Faraday Discuss.*, vol. 146, pp. 19-33, 2010.

[12] L. Mahadevan, Y. Pomeau, "Rolling droplets", *Phys. Fluids*, vol. 11, pp. 2499-2454, 1999.

CONTACT

*Nguyen Thanh-Vinh,
Tel: +81-3-5841-6318; vinh@leopard.t.u-tokyo.ac.jp

LIQUID DROPLET MICRO-BEARINGS ON DIRECTIONAL CIRCULAR SURFACE RATCHETS

Çağdaş Varel and Karl F Böhringer
University of Washington, Seattle, USA

ABSTRACT

This paper presents de-ionized water droplets used as torque-generating micro-bearings between a glass plate and a micromachined Si substrate. The pattern on the Si substrate includes circular tracks, which allow droplet motion in a single direction. When vertical vibration is applied to the system, a rotation in the transverse plane is triggered. The system can be tailored to respond to a specific vibration frequency, from 36.5 to 83 Hz for droplet volumes from 13 to 1 μL. The system is tested for its frequency response at different droplet sizes and droplet counts. A figure of merit is determined to quantify the responsiveness of the system. The largest angular speed is recorded as 0.302 rad/s for a vibration at 61 Hz.

INTRODUCTION

Micromachined surface patterns have been used to repel liquid away from specific regions by changing the apparent contact angle of the liquid, in analogy to the "lotus effect" in nature [1]. A special case of such micro-patterns are surface ratchets – asymmetric texture, which is employed for liquid transportation in droplet-based microfluidics by using a low frequency (≤150 Hz) vibration [2]. Surface ratchets translate the input vibration into a droplet motion in a single direction, which is determined by this asymmetry. Their response is frequency-sensitive and can be tailored to match the ambient frequency. They are standalone structures that do not require any electrical connection as long as a vibration source is present. In an environment that is rich of low-frequency vibration, they can be employed for translating this vibration into directed linear or rotational motion, which then can be used for other purposes such as energy harvesting and unpowered vibration sensors.

The design of surface ratchets for the purpose of controlling droplet transportation has been presented by our group before [2]. This paper shows a new application of droplets on surface ratchets acting as directional micro-bearings, and it provides design criteria for good system performance. Using micro-balls as bearings has attracted attention since the demonstration of MEMS micro-motors [3]. Liquid bearings are very promising for MEMS motors because of their low friction [4-6]. Furthermore, a micro-motor which uses electrowetting, a method of changing liquid contact angle by applying voltage, for generating motion has also been demonstrated [7]. All of these systems focus on generating motion from electrical energy while using liquid bearings. In contrast, the system presented here aims at translating the ambient vibration into a controllable and usable rotary motion. Asymmetry in the surface ratchets provides control over the direction of the rotation while their frequency-sensitive characteristics facilitate selectively controlling their response by specific input signals.

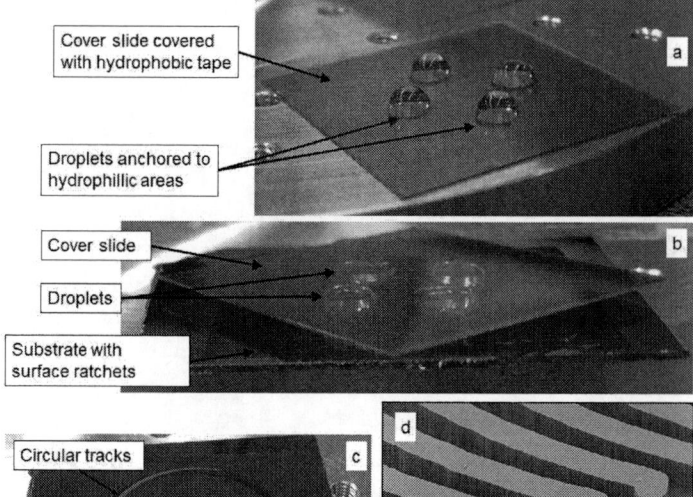

Figure 1: Overview of the droplet micro-bearing setup: a) de-ionized water droplets on a patterned microscope cover slide, b) the cover slide on the circular tracks suspended by the liquid micro-bearings, c) two concentric Si surface ratchets, d) scanning electron microscope image of the ratchet shows the rungs of the circular track and pillars forming a more hydrophobic region around the track.

DESIGN AND FABRICATION

The micro-bearing setup is shown in Figure 1 and includes two pieces; the rotary glass plate, which is covered by a hydrophobic tape sheet (Figure 1a) and the Si substrate with the micromachined surface ratchets (Figure 1b). Droplets are employed as micro-bearings between them.

The design and working principle of surface ratchets are explained in [1]. A circular ratchet track is formed of periodic semicircular rungs (Figure 2). Round pillars are placed in the surrounding area in order to restrict droplet motion. A droplet sitting on the ratchet experiences a difference between its advancing and receding contact angles due to the asymmetry in the track when vibration is applied in the vertical direction. That forces the droplet to move in a specific direction defined by the surface ratchet. Pillars around the tracks render a more hydrophobic surface in order to prevent the droplet from going off the track. That provides a directional droplet motion self-aligned to the track, which is also translated to the rotary plate motion.

The surface ratchets are fabricated using a combination of photolithography and deep reactive ion etching (DRIE) with an etch depth of 30 μm. Rungs have a length (L) of ~1.3 mm, width (W) of 30 μm and an edge-to-edge distance (E) of 20-45 μm as shown in Figure 2. Pillars have a diameter (D) of 30 μm and they are separated (S) by 40 μm. After DRIE, they are cleaned with piranha solution and coated with a layer of hexamethyldisilazane (HMDS) forming a hydrophobic surface.

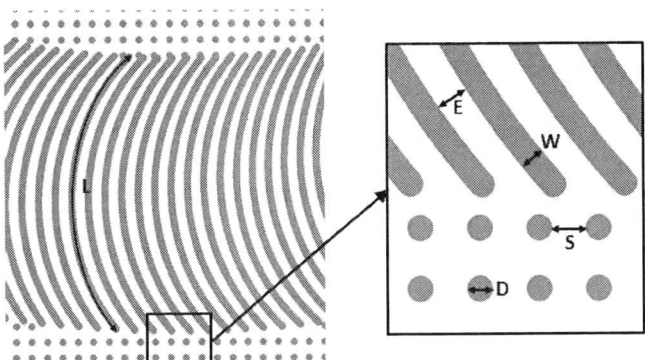

Figure 2: Schematic for rungs forming the track surrounded by pillars.

The rotary glass plate is a light-weight substrate with evenly spaced droplet anchor locations. A 22 mm × 22 mm microscope cover slip covered with a hydrophobic tape is used for this purpose. The hydrophobic tape sheet is patterned with disks using CO_2 laser cutting for exposing the underlying glass, which forms hydrophilic locations for anchoring droplets on the rotary plate. Anchored droplets, 13 μL each in Figure 1, are aligned to the surface ratchets and they can suspend the glass plate when it is placed upside down above the substrate.

Table 1: Design parameters for the micro-bearing system

n	D (mm)	v (μL)	Rotation
4	3	2	No
4	3	4	Yes
4	3	5	Yes
4	3	6	Yes
4	3	7	Yes
4	3	8	Yes
4	3	13	Yes
8	2	1	Yes
8	2	2	Yes
8	2	4	Yes
8	2	6	No
8	2	8	No

The system is vibrated in the orthogonal direction using a linear electromagnetic shaker (Brüel & Kjær Type-4809) and its rotation is captured with a high-speed camera. A square wave of 0-350 mV_{p-p} is supplied to the electromagnetic shaker to initiate and control the vibration.

The vibration on the substrate surface is also measured using a laser vibrometer. Captured videos are processed using the MATLAB™ image processing toolbox to measure the angular speed. The vibration frequency generating the fastest rotation is determined using a frequency sweep and named as "natural frequency" of the system. Number of droplets (n), diameter of the anchoring area (D) and droplet volume (v) are employed as the design parameters (Table 1).

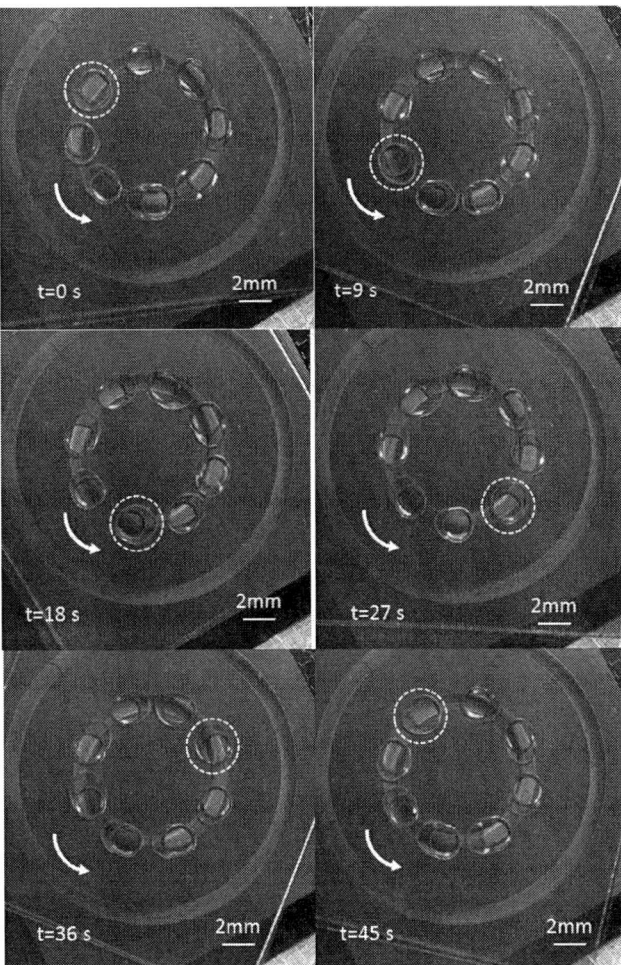

Figure 3: Time-lapsed image sequence of a micro-bearing system employing 8 2μL-droplets. It is rotating in the counterclockwise direction due to a vertical vibration at 72 Hz. Yellow disks in the image are droplet anchoring locations.

RESULTS

Space limitation and weight of the rotary plate are two factors that determine the rotation of a given system. Droplet volumes are restricted to an interval that is known to work successfully for this type of ratchet [1]. As seen in Table 1, space is the limiting factor for 8-droplet systems with large droplet volumes as they tend to join and pin the rotary plate to the substrate. The weight of the rotary plate becomes the limiting factor when the droplet volume is very small and it cannot carry the rotary plate. This is observed for the 4-droplet system with a droplet size of 2 μL.

However, the same droplet size can be used for an 8-droplet system since the load per droplet is halved for the 8-droplet system.

Time-lapsed images of a micro-bearing system employing 8 2 µL-droplets are presented in Figure 3. The natural frequency for this particular system is found as 72 Hz. The rotary plate is moving in the counterclockwise direction as determined by the surface ratchet. Its averaged angular speed is measured as 0.138 rad/s for a vibration speed of 101.6 mm/s.

The change in natural frequency with droplet volume/number is studied in Figure 4. 4 and 8 droplet systems with droplet volumes of 4-13 µL and 1-4 µL are employed. A frequency range of 36.5-83 Hz is covered by changing the droplet volume from 13 to 1 µL. It is observed that the natural frequency is decreasing with increasing droplet size for both systems. This is anticipated since the increased droplet volume results in a less stiff body. When systems are compared for droplet numbers, it is seen that the 8-droplet system is excited at a lower natural frequency for the same droplet volume. The change in natural frequency for the 8-droplet system (14.1 Hz/µL) is almost twice as fast as the 4-droplet system (7.9 Hz/µL for 4-8 µL range).

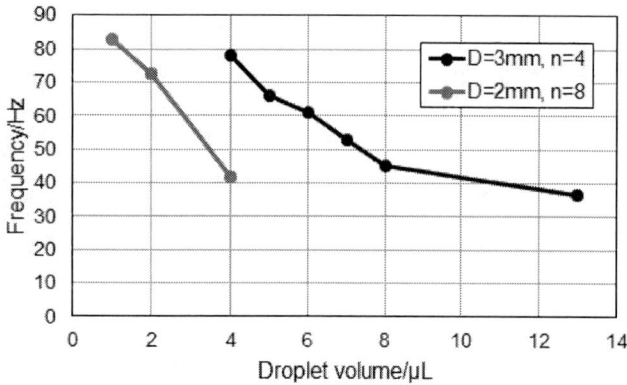

Figure 4: The change in the system's natural frequency with changing droplet volume.

The system can be modeled as a platform (rotary glass plate) sustained by springs (droplets) connected in parallel. Then, its natural frequency (ω_s) can be written as:

$$\omega_s = \sqrt{\frac{nk_d}{m}}$$

(1)

In this equation, m is the mass of the system, n is the number of droplets and k_d is the spring constant of a single droplet at a given volume. The mass can be thought as constant in this equation since the mass of the rotary plate dominates. Then, the change in system frequency should be proportional to $\sqrt{n}$ for a unit increase in droplet number. When the changes in natural frequency for both systems are scaled by $\sqrt{n}$, 4.98 Hz/µL and 3.95 Hz/µL are obtained for 8-droplet and 4-droplet systems, respectively. The significant difference in those values can be explained by a

nonlinear relationship between a droplet volume, its spring constant and applied load.

Transient behavior of the rotary glass plate is studied in the next step. Angular speed of the system is measured using image processing tools on recorded videos. It is measured whenever a droplet travels 1/4 of the track in a 4-droplet system and 1/8 of the track in an 8-droplet system. Those are plotted versus time in Figure 5.

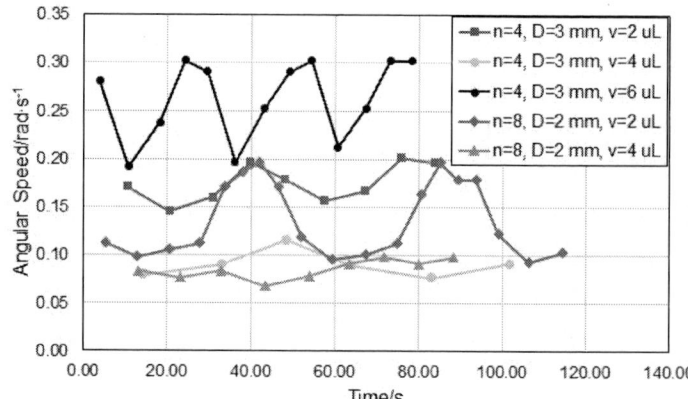

Figure 5: Angular speed variation over time observed on 5 devices.

It is seen that angular speed stays within ±19% of the average speed for three combinations while the others vary significantly. Those variations are somewhat periodic. For the third system (n=4, D=3 mm, v=6 µL), low speeds are recorded at the 2nd, 6th and 10th measurements. Those correspond to the same measurement location for a particular droplet, when it has traveled half of the track from its initial position, suggesting that the low speed can be explained by an imperfection in the track at that location. However, it is also surprising that this imperfection is only experienced by that particular droplet. A similar trend is observed with the fourth system (n=8, D=2 mm, v=2 µL).

Lastly, a figure of merit is defined to quantify the system response for a given vibration input. A scaled version of system efficiency is used for that purpose. Efficiency of this system can be defined as rotational energy obtained per unit energy input. Then, it can be scaled to Z since mass of the rotating object, dominated by the mass of the rotary plate, is kept constant among different combinations. It is calculated using the following equation:

$$Z = \omega_d^2 / E_{in}$$

(2)

Here, ω_d is the angular speed and E_{in} is the vibrational energy input to the system. The vibrational energy input is measured using a laser vibrometer aligned with the stage. The change in Z with system parameters is plotted in Figure 5. The most responsive, highest Z, system is built with 6 µL droplets with n=4 and D=3 mm. This is also the combination which has resulted in the largest angular speed of 0.302 rad/s. The most favorable system for n=8 can be built with 2 µL droplets.

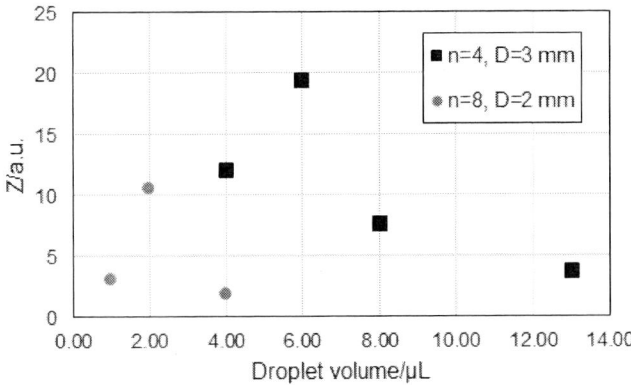

Figure 6: Figure of merit (Z) for various system parameters. 6 μL droplets for n=4, D=3 mm and 2 μL droplets for n=8, D=2 mm provide the most favorable results.

DISCUSSION

A novel application of surface ratchets which employs liquid droplets as micro-bearings is presented. The method makes use of circular micro-machined surface ratchets for directional motion of droplets when a vertical vibration is present. This directional motion is translated to a rotation of a glass plate which is anchored to those droplets. Thus, a liquid droplet micro-bearing system which translates a vertical vibration into a planar rotation is built.

The frequency-selective response of the system enables frequency sensing applications and tailoring system properties to match an ambient vibration frequency. Various systems are built in order to study the effect of droplet volume, anchor location size and number of droplets on system behavior. Spatial limitations dominate for large droplet volumes while weight of the rotary plate becomes the limiting factor for small droplet volumes. Frequency response of the system is studied for different droplet volumes. A decrease in "natural frequency" of the system is observed as the droplet volume increases. However, that decrease occurs at different rates for different numbers of droplets. A mass-spring model is proposed to explain this behavior. Further experiments are desired in order to achieve a complete understanding of the droplet dynamics. That will also be useful for modeling droplet systems for future applications.

The transient analysis of the rotation shows an acceptable variation in the angular speed for most of the systems. Large variations in the angular speed are thought to be related to imperfections in the ratchet track. Finally, a figure of merit is described in order to quantify how well a chosen system has responded to input vibration. Favorable combinations for both 4-droplet and 8-droplet systems are determined using that figure of merit.

The presented system can be employed for applications where ambient vibration is harvested and translated into a rotary motion. We believe that energy harvesting and unpowered vibration sensors are potential applications.

ACKNOWLEDGEMENTS

This work was supported in part by the National Science Foundation under grant ECCS-1308025 and by the John M. Fluke professorship to Karl Böhringer. The authors would like to thank Dr. Ji Hao Hoo for his help with image processing and Dr. Rajashree Baskaran for guidance and helpful discussions. Part of this work was conducted at the University of Washington Nanofabrication / Nanotechnology User Facility, a member of the NSF National Nanotechnology Infrastructure Network.

REFERENCES

[1] A. Murmur, "The Lotus effect: superhydrophobicity and metastability," *Langmuir*, vol. 20 (9), pp. 3517-9, 2004.

[2] T. A. Duncombe, E. Y. Erdem, A. Shastry, R. Baskaran and K. F. Böhringer, "Controlling Liquid Drops with Texture Ratchets," *Advanced Materials* vol. 24, pp. 1545-1550, 2012.

[3] N. Ghalichechian, A. Modafe, M. I. Beyaz, and R. Ghodssi, "Design, Fabrication, and Characterization of a Rotary Micromotor Supported on Microball Bearings," *Journal of Microelectromechanical Systems*, vol. 17, pp. 632-642, 2008.

[4] B. E. Yoxall, C. Mei-Lin, R. S. Harake, P. Tingrui, and D. A. Horsley, "Rotary Liquid Droplet Microbearing," *Journal of Microelectromechanical Systems,* vol. 21, pp. 721-729, 2012.

[5] M. L. Chan et al., "Design and characterization of MEMS micromotor supported on low friction liquid bearing," *Sensors and Actuators A: Physical*, vol. 177, pp. 1-9, 2012.

[6] T. L. Liu, S. Guangyi, K. Jong Jin, C. K. K. Yang, and C. J. Kim, "Electrostatic bottom-driven rotary stage on multiple conductive liquid-ring bearings," in *IEEE 26th International Conference on Micro Electro Mechanical Systems (MEMS)*, pp. 86-89, 2013.

[7] A. Takei, K. Matsumoto, and I. Shomoyama, "Capillary motor driven by electrowetting," *Lab on a Chip*, vol. 10, pp. 1781-1786, 2010.

CONTACT

* K. F. Böhringer, University of Washington, Department of Electrical Engineering, 185 Stevens Way, Paul Allen Center - Room AE100R, Seattle, WA 98195-2500, USA; Tel: +1-206-221-5177; Fax: +1-206 543-3842; E-mail: karlb@u.washington.edu.

LOW GAS PERMEABLE AND NON-ABSORBENT RUBBERY OSTE+ FOR PNEUMATIC MICROVALVES

Jonas Hansson, J. Mikael Karlsson, Carl Fredrik Carlborg,
Wouter van der Wijngaart and Tommy Haraldsson
KTH Royal Institute of Technology, Stockholm, SWEDEN

ABSTRACT

In this paper we introduce a new polymer for use in microfluidic applications, based on the off-stoichiometric thiol–ene-epoxy (OSTE+) polymer system, but with rubbery properties. We characterize and benchmark the new polymer against PDMS. We demonstrate that Rubbery OSTE+: has more than 90% lower permeability to gases compared to PDMS, has little to no absorption of dissolved molecules, can be layer bonded in room temperature without the need for adhesives or plasma treatment, can be structured by standard micro-molding manufacturing, and shows similar performance as PDMS for pneumatic microvalves, albeit allowing handling of larger pressure.

BACKGROUND

OSTE+ is a polymer system developed specifically for lab-on-a-chip applications. It has the potential to bridge the gap between research prototyping and commercial manufacturing [1-3] due to a number of attractive characteristics, including a fast turn around fabrication process, Young's modulus similar to common thermoplastics, tunable surface properties, and adhesive-free low temperature bonding to a large range of substrates [4-5]. Moreover, OSTE+ has demonstrated preferable material properties for nucleic acid amplification detection compared to PDMS and PMMA [6]. However, large-scale integration of microfluidic devices often uses on-chip integration of actuators such as valves and pumps, most often accomplished by integrating a rubbery membrane material as actuator; and macro to micro fluid connectors, often sealed by a rubbery material. In academia, PDMS is commonly used as the device material [7] because of its rubbery properties and ease of fabrication. However, PDMS is severely limiting in many applications, due to its high gas permeability, sample absorption and the need for plasma treatment prior to bonding to glass or itself. Similar limitations were seen in the PDMS-based off-stoichiometry thiol-ene (OSTE) rubbery material previously integrated in OSTE devices [1].

In this paper we introduce a new polymer, based on the OSTE+ polymer system, with a modified monomer composition that allows for obtaining rubbery properties while avoiding some of the limitations of PDMS. We will refer to previously demonstrated OSTE+ polymers with high Young's modulus as "stiff OSTE+" and to the new polymer as "rubbery OSTE+".

DESIGN

Rubbery OSTE+ formulation

To create the rubbery properties, compared to previous OSTE+ mixtures [2, 3], we lowered the crosslink density of the new polymer by reducing the amount of functional groups per monomer and by substituting the epoxy monomer with a rubbery copolymer variant, as shown in table 1: comparing the monomers between rubbery and stiff OSTE+, the thiol monomers are tetra- instead of tri-functional; the allyl monomers are tri- instead of di-functional; and the epoxy monomer is Albipox® 1000, which is a copolymer consisting of long rubbery segments and short epoxy functional group segments that bind into the polymer matrix upon cure. The mechanical properties were further tuned by modifying the stoichiometric ratio and with the use of a stabilizing tri-azine ring on the thiol instead of on the allyl as used in the stiff OSTE+.

Table 1: Monomer composition of stiff OSTE+ and rubbery OSTE+

Valve design

Pinch valves [7] were designed as a pneumatic control channel and a liquid flow channel, each in a different microfluidic layer, separated by an elastic membrane at the position where they cross each other. The pressure in the pneumatic channel controls the valve state: a pressurized control channel deflects the membrane and closes the flow channel; a non-pressurized control channel results in an open valve. As a test device, (see figure 2a) a device with several flow channels closed by the same control channel was designed.

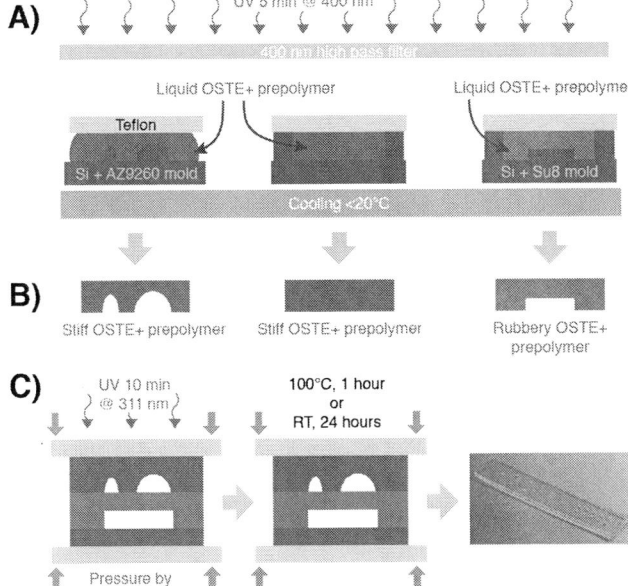

Figure 1: Fabrication process and final device. A) The 1st cure, which mold the prepolymers. B) Layers demolded after 1st cure. C) The 2nd cure for bonding of assembled layers (left and middle) and photography of final device (right).

FABRICATION

Molds were fabricated using photoresist on silicon. Square pneumatic control channels were defined in 50 µm thick SU-8 and the 60 µm membrane thickness was obtained using spacer structures. The liquid flow channels were defined in AZ-9260 photoresist that was reflowed at 125°C for 2 min to define channel molds with circular cross-section [8]. The height of the flow channels was measured to be between 20 and 30 µm, depending on the channel width. The molds were coated with a thin (<100 nm) layer of Teflon AF 1600 (DuPont™, USA) to facilitate demolding.

The stiff OSTE+ prepolymer was prepared using Triallyl-1,3,5-triazine-2,4,6-trione (Mercene Labs, Sweden), Tetrakis(3-mercapto-propionate) (Sigma-Aldrich), and D.E.N. 431 (Dow, Germany) at a stoichiometric ratio of 1,5:1:0,5. The rubbery OSTE+ was prepared using Trimethylolpropane diallyl ether (Sigma-Aldrich), Tris[2-(3-mercapto-propionyloxy)ethyl] isocyanurate (Sigma-Aldrich), and Albipox 1000 (Evonik) at a stoichiometric ratio of 1,2:1:0,2 (table 1). Both stiff and rubbery OSTE+ polymers used TPO-L (BASF) and photolatent DBN as photoinitiators, both at 1% (w/w).

OSTE+ chips containing an array of pneumatic test valves with different footprint size were fabricated by sandwiching a rubbery OSTE+ layer between two stiff OSTE+ layers (figure 1) using the OSTE+-typical dual cure mechanism [4]: 1) molding of the liquid prepolymer and first curing by UV illumination (collimated 12 mW cm-2 @ 365 nm near-UV short arc mercury lamp from OAI, Milpitas, USA) for 5 min above 400 nm wavelength (glass filter: GG400 Longpass, Schott), and cooled to below 20°C; 2) manually aligning and contacting of the three layers; 3) bonding during second cure by 10 min UV illumination (without filter) under clamping followed by 1 h at 100°C (temperature accelerated bonding). An additional valve was manufactured using the same method as above, but substituting the 1h 100° to 24h in room temperature (room temperature bonding). For comparison, microvalves in two PDMS layers on top of a glass microscopy slide were fabricated with standard PDMS casting, curing and plasma bonding, using the same molds.

1 mm and 0.5 mm thick unstructured rubbery OSTE+ membranes were fabricated for mechanical and gas permeability characterization, respectively, using the same dual cure mechanism as described above.

PDMS membranes for characterization (1 and 0.5 mm) and devices were all fabricated using a ratio 1:10 of curing agent:base (Sylgard 184, Dow Corning, USA), degassing, and curing at 70°C for 3 h. For PDMS valve structure manufacturing, the PDMS layers were bonded using a treatment with oxygen plasma (Femto, Diener, Germany) prior to layer contacting, and 10 min on a 70° C hotplate after contacting.

EXPERIMENTS
Material characterization

Storage moduli and loss factors in PDMS and Rubbery OSTE+ were measured at room temperature using a Dynamic Mechanical Analysis Q800 instrument from TA instruments on fully cured samples measuring 15 mm x 5 mm x 1 mm using the vertical tension film mode DMA instrument configuration, 15 µm oscillation amplitude, and 1 Hz frequency.

Water vapor permeation was measured by weight loss of water enclosed in a container, created by clamping membranes (500 µm thick PDMS and Rubbery OSTE+, and 1 mm thick glass for reference) with O-rings of area 1100 mm^2 area, for 72 hours at 40°C (figure 2).

Permeability to small molecules was tested by fluorescence microscopy of PDMS and OSTE+ channels, bonded to glass and stiff OSTE+ respectively, and filled with Rhodamine B (dissolved in ethanol) for 30 min.

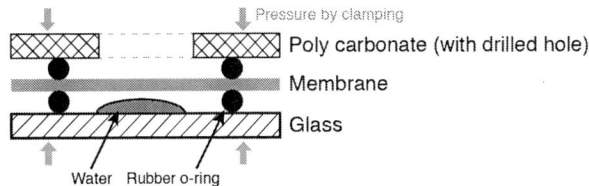

Figure 2: Gas permeability measurement setup. Membranes tested were rubbery OSTE+, PDMS, and glass for reference, all at 40°C for 72 hours.

Valve testing

The PDMS and OSTE+ valve devices were tested using relative air actuation pressures between 0 and 300 kPa (limited by failure of the pneumatic port connectors at higher pressures) and monitored using optical microscopy. The flow channels were filled with red dye for improved visibility of the threshold actuation pressure, and were left at atmospheric pressure. The valves were characterized as closed when a discontinuation of red fluid in the flow channel direction could be observed.

RESULTS AND DISCUSSION
Material characteristics

Dynamic mechanical analysis (table 2) revealed similar storage modulus and loss factor for rubbery OSTE+ and PDMS: 2.8 MPa vs 2.0 MPa, and 0.35 vs 0.18, respectively. I.e., the materials are very similar, with rubbery OSTE+ being slightly less elastic and with slightly more viscous losses.

Water vapor permeation (table 2) was measured to be 5.2 mg/h for PDMS but only 0.4 mg/h for rubbery OSTE, i.e. a reduction in excess of 90%. The reduced permeability to gases allows better control in devices with long incubation times and need for a controlled environment.

In the images of Rhodamine B incubated channels (table 2) an increased fluorescence outside the channel can be observed for PDMS while no increase in fluorescence can be observed for rubbery OSTE+. This confirms that PDMS is very permeable to small molecules while the rubbery OSTE+ has a significantly lower permeability to small molecules, possibly none. This property can avoid detrimental events such as analyte depletion, increased sample dispersion and contaminations between sequential liquid samples.

Table 2: Material characterization results, showing an order of magnitude reduction of gas permeability for rubbery OSTE+ compared to PDMS and no absorption of small molecules for OSTE+.

	PDMS	Rubbery OSTE+
Mechanical properties		
Storage modulus	2.0 MPa	2.8 MPa
Loss factor	0.18	0.35
Gas permeability		
H2O diffusion rate	5.2 mg/h	0.4 mg/h
Absorption		
Empty channel		
1 min incubation of Rhodamine B		
30 min incubation of Rhodamine B		

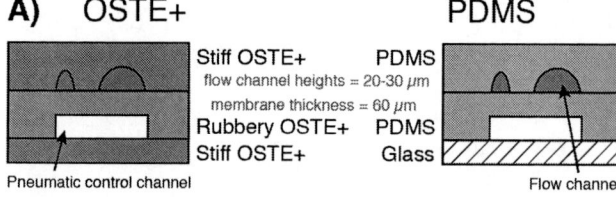

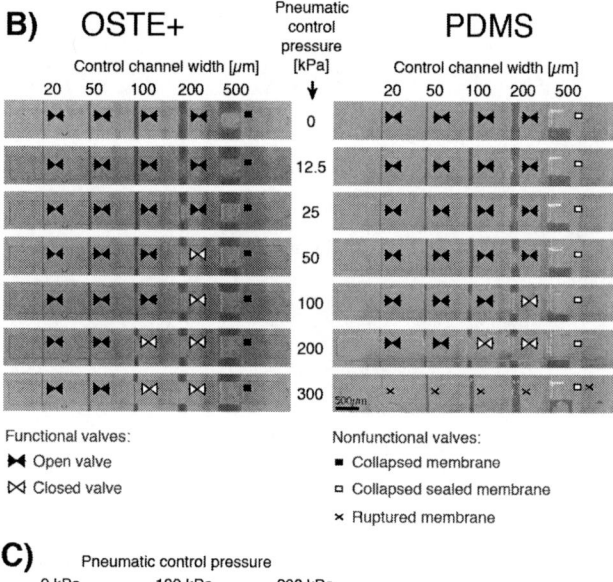

Figure 3: Pinch valve testing of OSTE+ (left) and PDMS (right) microvalves, flow channels filled with red dye. A) Cross-sectional schematic of valve devices. B) Photographs of valves under different control pressures. Valve footprint areas are 500 x 500, 200 x 500, 100 x 500, 50 x 500, and 20x500 μm². The OSTE+ device was fabricated with temperature accelerated bonding and the PDMS device was fabricated using standard micromolding and plasma bonding. C) Photographs of a 200 x 200 μm² pinch valve manufactured using room temperature bonding, actuated at different control pressures.

Valve performance

Valve structures were intact after manufacturing, except for rubbery OSTE+ and PDMS valves with 500 x 500 μm² membranes, which spontaneously collapsed. For PDMS, the membrane collapse occurred already during plasma bonding, hence no fluid could pass the valve in the 500 μm PDMS channel. The valve pressure test (see figure 2b) shows that valves made of PDMS and OSTE+ close at very similar control pressures. Several PDMS membranes ruptured at 300 kPa, whereas all OSTE+ membranes remained intact, indicating higher durability for high-pressure applications. After use, all valves returned to the normally open state when removing the pneumatic pressure.

The 200 x 200 μm² room temperature bonded OSTE+ valve in figure 2c closes with an actuation pressure between

100 and 200 kPa and withstands at least 300 kPa pressure (300 kPa not shown).

Discussion

While long term use and response time remains to be tested, the performed characterization indicates that rubbery OSTE+ has the potential of being used for valving and pumping in similar applications as PDMS components, but with the benefit of withstanding higher pressures. The lower gas permeability of Rubbery OSTE+ would prevent air bubble injection during operation of thinner membranes as described in [9], i.e. allowing pressurized gas as control medium in pneumatic membrane valves for thinner membranes and at higher pressures. The possibility to combine stiff and rubbery materials makes it easier to design valves that are completely leak tight and ensures that a complete microfluidic device withstands handling without deformation. The ability to bond layers of OSTE+ using room temperature bonding is especially useful in bio applications with sensitive functionalization inside of micro channels.

CONCLUSIONS

We have presented a rubbery, low gas permeability off-stoichiometric thiol–ene-epoxy (OSTE+) polymer fully compatible with standard micro-molding manufacturing. We demonstrated its use in pneumatic pinch microvalves for microfluidic applications. This is the first OSTE+ material with rubbery properties (similar to PDMS). The material shows low permeability to gases (magnitude orders less than PDMS) and low or no absorption of small molecules from the sample gases (magnitude orders less than PDMS). It can be bonded in room temperature without the need for adhesives or plasma treatment (unlike PDMS).

Thus, rubbery OSTE+ forms a viable alternative to PDMS, especially in applications where low temperature bonding or low permeability to gas and small molecules in solution are desired.

ACKNOWLEDGEMENTS

This work has been sponsored in part by the European Commission through the FP7 project ROUTINE.

REFERENCES

[1] C. F. Carlborg, T. Haraldsson, K. Öberg, M. Malkoch, and W. van der Wijngaart, "Beyond PDMS: off-stoichiometry thiol–ene (OSTE) based soft lithography for rapid prototyping of microfluidic devices," *Lab Chip*, vol. 11, p. 3136, 2011.

[2] X. Zhou, C. F. Calborg, N. Sandstrom, A. Haleem, A. Vastesson, F. Saharil, W. van der Wijngaart, and T. Haraldsson, "Rapid Fabrication of Of Oste+ Microfluidic Devices with Lithographically Defined Hydrophobic/Hydrophilic Patterns and Biocompatible Chip Sealing," in *MicroTAS'13*, Freiburg, October 27-31, 2013, pp. 134–136.

[3] A. Vastesson, X. Zhou, N. Sandstrom, F. Saharil, O. Supekar, G. Stemme, W. van der Wijngaart, and T. Haraldsson, "Robust microdevice manufacturing by direct lithography and adhesive-free bonding of off-stoichiometry thiol-ene-epoxy (OSTE+) polymer," in *Transducers'13*, Barcelona, June 16-20, 2013, pp. 408–411.

[4] F. Saharil, L. El Fissi, Y. Liu, F. Calborg, D. Vandormael, L. A. Francis, W. van der Wijngaart, and T. Haraldsson, "Superior Dry Bonding of Off-Stoichiometry Thiol-Ene Epoxy (OSTE (+)) Polymers for Heterogeneous Material Labs-on-Chip," in *MicroTAS'12*, Okinawa, October 28 - November 1, 2012, pp. 1831–1833, 2012.

[5] F. Saharil, F. Forsberg, Y. Liu, P. Bettotti, N. Kumar, F. Niklaus, T. Haraldsson, W. van der Wijngaart, and K. B. Gylfason, "Dry adhesive bonding of nanoporous inorganic membranes to microfluidic devices using the OSTE(+) dual-cure polymer," *J. Micromech. Microeng.*, vol. 23, p. 025021, 2013.

[6] F. Saharil, A. Ahlford, M. Kuhnemund, M. Skolimowski, A. Conde, M. Dufva, M. Nilsson, M. Brivio, W. van der Wijngaart, and T. Haraldsson, "Ligation-based mutation detection and RCA in surface un-modified OSTE+ polymer microfluidic chambers," in *Transducers'13*, Barcelona, June 16-20, 2013, pp. 357–360.

[7] M. A. Unger, "Monolithic Microfabricated Valves and Pumps by Multilayer Soft Lithography," *Science*, vol. 288, pp. 113–116, 2000.

[8] "Product Data Sheet: AZ 9200 Photoresist,", AZ Electronic Materials, 2001.

[9] M. Johnson, G. Liddiard, M. Eddings, and B. Gale, "Bubble inclusion and removal using PDMS membrane-based gas permeation for applications in pumping, valving and mixing in microfluidic devices," *J. Micromech. Microeng.*, vol. 19, p. 095011, 2009.

CONTACT

*W. van der Wijngaart, tel: +46 8 790 6613 ; wouter@kth.se

FREE SURFACE PROPULSION
BY ELECTROWETTING-ASSISTED 'CHEERIOS EFFECT'

Junqi Yuan and Sung Kwon Cho
University of Pittsburgh, Pittsburgh, USA

ABSTRACT

This paper presents how "Cheerios effect" is combined with EWOD (electrowetting on dielectric) in order to propel floating objects. Cheerios effect is a common phenomenon in which small floating objects are either attracted or repelled by the sidewall. This behavior is known to be highly dependent on the hydrophobicity of the walls. By electrically controlling the hydrophobicity (contact angle) of the sidewalls in the canal and thus distorting the adjacent interfaces, cheerios effect can be electrically controlled to manipulate floating objects. This work first verifies EWOD-assisted cheerios effect on propulsion. Using this effect, a linear propulsion of millimeter-sized floating objects in a canal is achieved. By sequentially applying a voltage to the arrayed EWOD electrodes on the sidewalls and by carefully controlling the time interval between electrode switching, floating objects can be propelled continuously along the mini canals. In addition, the similar propulsion is achieved in an even smaller canal whose dimension is comparable to the capillary length. Finally, circular rotation of floating objects is realized by arranging EWOD electrodes along a circular path and applying electrical signal sequentially.

INTRODUCTION

Some water-walkers are unable to climb menisci using their traditional propulsion methods.[1] Normally these small animals climb meniscus by capillary force that is generated by deformed water surface with their certain body postures, such as arcing the back, as shown in Fig. 1(a).[2] This phenomenon, which is also known as 'Cheerios Effect' (Fig.

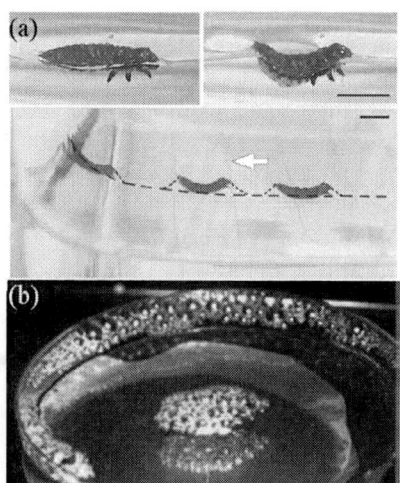

Figure 1: Cheerios Effect. (a) Meniscus climbing behavior of the water lily leaf beetle[2]; (b) Bubbles migrate to the wall[3].

1(b)), is a common phenomenon named after observations that breakfast cereals floating in milk tend to stick to the sidewall of the bowl.[3, 4] In addition to cereals, any small floating objects have tendency to move toward or away from the sidewall depending on the surface wettability and other properties.

A common misunderstanding of Cheerios effect is that the attraction or repulsion at the water-air interface solely depends on the surface wetting properties (hydrophobicity).[5-7] This is true if objects are large enough or have vertical walls in contact with water. In general, if objects have similar hydrophobicity to that of the sidewalls, the objects and wall attract each other; otherwise, they repel away. But for small particles, because of the complexity of the geometry and the effect of the particle density, this simple theory is not valid any more.[3, 8] For small spherical particles, relatively light particles have the similar behavior as large objects with hydrophilic vertical sidewalls do while heavy particles behave similar to large hydrophobic objects. So for hydrophilic walls, light spherical particles are attracted but heavy particles are repelled.

Recently, some attempts have been made to fabricate and investigate meniscus climbing devices. Hu et al. [9] and Yu et al. [10] used similar structures, a bent plastic or metal sheet, to mimic the meniscus climbing behavior of the natural insect.

Here, we present a novel mini/micro propulsion method based on the Cheerios effect principle. Combining electrowetting-on-dielectric (EWOD) with the Cheerios effect, we continuously propel mini floating objects along linear and circular paths in a controlled manner. Detailed fabrication, testings and results are as follows.

EWOD-ASSISTED PROPULSION

From the above discussion, we know that by switching wettability of vertical wall surface, attractive or repulsive force can be generated on the floating object such that it can be manipulated (translated).

Electrowetting-on-dielectric (EWOD) electrode is an excellent candidate to switch the wettability. EWOD electrodes are typically arranged horizontally to manipulate droplets or bubbles.[11, 12] But when the EWOD electrode is vertically inserted and partially immersed in water, surface wettability can also be adjusted by applying electrical potential to the electrodes. As shown in Fig. 2, The EWOD electrode is attached on the vertical wall. The wall surface is originally hydrophobic, so with EWOD off the floating hydrophilic object placed near the wall would be repelled away from the wall (Fig. 2(a)) while the hydrophobic object would be attracted toward the wall (Fig. 2(c)). With EWOD on, however, the wall is switched to a hydrophilic surface,

978-1-4799-3510-9/14 $31.00 © 2014 IEEE   991

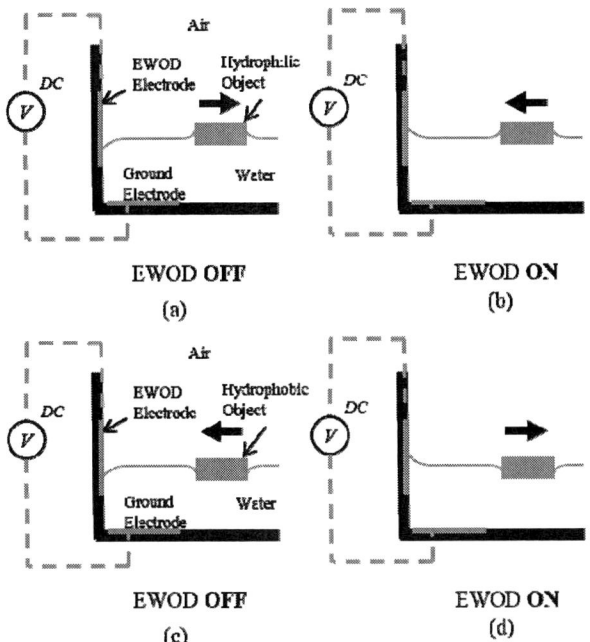

Figure 2: Sketch of EWOD-assisted cheerios effect propulsion. (a) & (b) Hydrophilic boat. (a) EWOD is OFF; (b) EWOD is ON. (c) & (d) Hydrophobic boat. (c) EWOD is OFF. (d) EWOD is ON.

which would attract the hydrophilic object (Fig. 2(b)) or repel the hydrophobic object (Fig. 2(d)).

The above prediction is experimentally confirmed as shown in Figs. 3(a) and 3(b). The hydrophilic object is

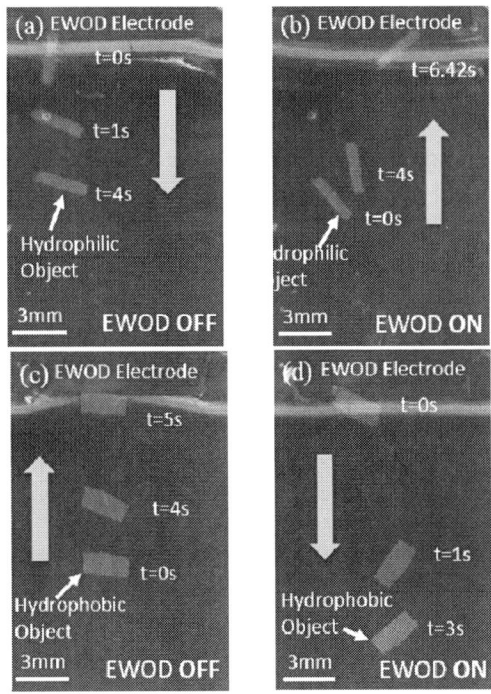

Figure 3: Experiment results for EWOD-assisted cheerios effect propulsion. Hydrophilic object ((a), (b)) and hydrophobic object ((c), (d)) propel when EWOD was ON or OFF.

pushed when EWOD is off and is pulled when EWOD is on. On the contrary, this attraction and repelling behaviors are reversed when a hydrophobic object is placed near the EWOD wall (Figs. 3(c) and 3(d)). The ground electrode is attached to the bottom of water container (not shown in Fig. 3). EWOD electrodes are prepared as follows: A Dupont Pyralux® flexible Cu product with 18 μm of Cu layer is coated with a 2.5 μm parylene layer. To make the surface hydrophobic, a thin layer (~2000 Å) of Teflon AF® is dip-coated on the top.

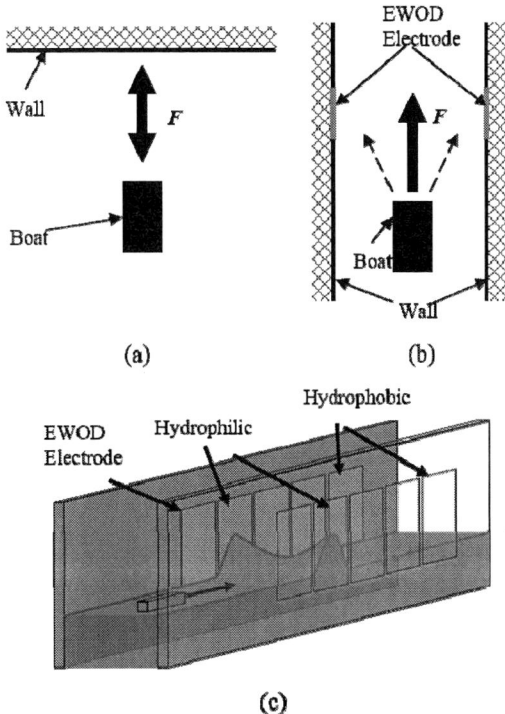

Figure 4: Comparison of two Cheerios effect propulsion methods. (a) Boat is attracted toward or repelled against the wall; (b) Boat propulsion in small canal; (c) 3-D sketch of EWOD-assisted cheerios effect continuous propulsion. Boat surface in (b) and (c) is hydrophilic.

CONTINUOUS LINEAR MOTION

However, by simply turning EWOD on or off with a single EWOD electrode, the motion of objects is limited to the space between the wall and objects (one-time movement toward or far away from the sidewall) as shown in Fig. 4(a). In order to realize continuous motions, Fig. 4(b) shows a modified propulsion method in a small canal. When a pair of EWOD electrodes (one in the left wall and the other in the right wall) are activated, the surfaces of electrodes change from hydrophobic to hydrophilic. As a result, the free surface near the electrodes is distorted and elevated. The distorted interface simulates the wall effect in Fig. 2 and generates a pulling force on the hydrophilic object. Due to symmetry, the lateral forces on the object are cancelled out. Only a net force on the object is toward the elevated interface, so the object is propelled along the canal path. Fig. 4(c) shows a

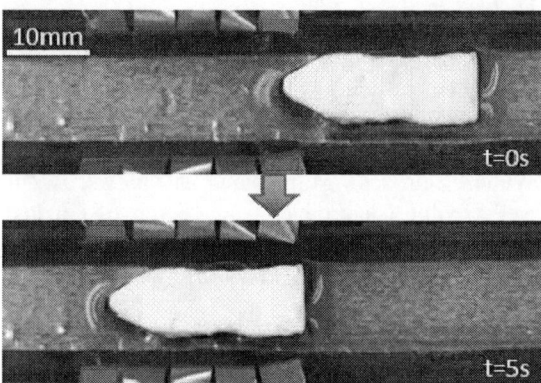

Figure 5: Hydrophilic boat propelling in 15mm wide canal. Pairs of electrodes were activated sequentially from right to left with duration of 1s. Red bars indicate the activated EWOD electrodes.

3-D construction of Cheerios effect in a small canal. When an array of EWOD electrodes is attached to each of two vertical sidewalls that are partially immersed in water, by shifting the EWOD activation from one pair to the next pair, a continuous movement can be realized.

It's worth mentioning that the boat surface in Fig. 4(b) and 4(c) needs to be hydrophilic. At the initial state, the surface of EWOD electrodes is hydrophobic, repulsive forces from both sides of the canal act on the boat and cancelled out. So the boat can align itself along the canal direction and stay in the center of the canal. When EWOD electrodes are activated, no lateral net force is generated, so the boat still maintains the position in the center. However, if the boat surface changes to hydrophilic, forces from side walls change to be attractive. There exists an equilibrium point in the center of the canal but it is not stable. So the boat is easily attracted toward either of the sidewalls.

Fig. 5 shows the hydrophilic object is propelled in the 15-mm wide canal with two arrays of EWOD electrodes (5 mm width). All the canal sidewalls were coated with Teflon to be initially hydrophobic, so the hydrophilic object automatically aligns and stays in the middle of the canal. An Al foil as the ground electrode is attached to the bottom of the canal. Upon shifting activation of EWOD electrode pairs to the left (160 VDC EWOD voltage) with duration of 1 sec, the

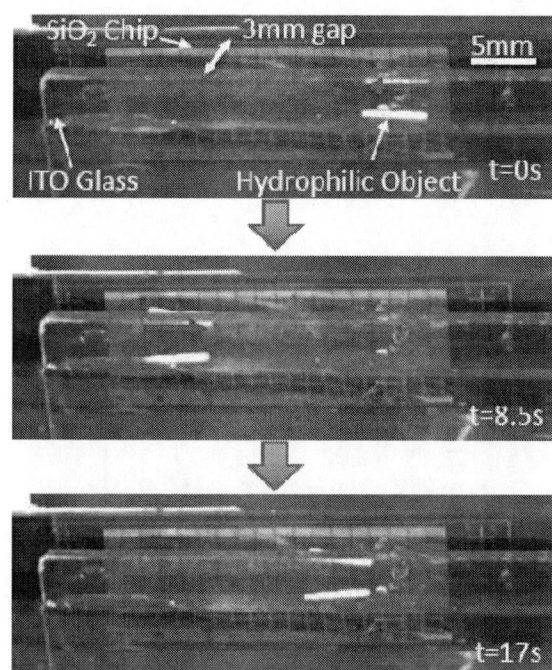

Figure 7: Propulsion in two directions with microchips. Duration of activation was 0.5s for each pair of electrodes.

object is step-by-step propelled to the left. A sequential signal is provided by a microcontroller (ATMEL ATtiny24A). Since the output of the microcontroller is in several volts range, which is much lower than the EWOD required voltage (> 50 V), relays are used for transmitting a high EWOD high voltage from amplifier to the EWOD electrodes.

The canal width in Fig. 5 is much larger than the capillary length for the air-water interface (2.7 mm). In order to confirm this principle in scales comparable to or smaller than the capillary length, two arrays with smaller EWOD electrodes (1.5 mm wide), one on a SiO_2 substrate and the other of ITO glass, are micro-fabricated (Fig. 6) and partially immersed in water vertically to form the small canal. The transparent ITO electrodes allow us to visualize motions of objects from the sides. The canal width can be adjusted by a linear traversing stage. For the SiO_2 wafer, a thin layer of Al is sputter-coated on the SiO_2 substrate. Photoresist (AZ®P4210) is spin-coated on the top. After exposure by UV light, unwanted photoresist and Al are removed by alkaline based developer simultaneously. As a result, the traditional acid wet etching process is not needed. For the ITO glass, 18% HCl is used to etch conductive ITO layer after photolithography. A 2.5 μm thick Parylene layer is coated for the EWOD dielectric layer after the photoresist is removed in both chips. Tapes are used to cover and protect the soldering area during the parylene coating process. Because some portion of the soldering area is needed immersed into water, silicone is used to cover the soldering area as insulating layer to avoid any electrical short after soldering process.

In Fig. 7, the spacing between the two arrays is reduced

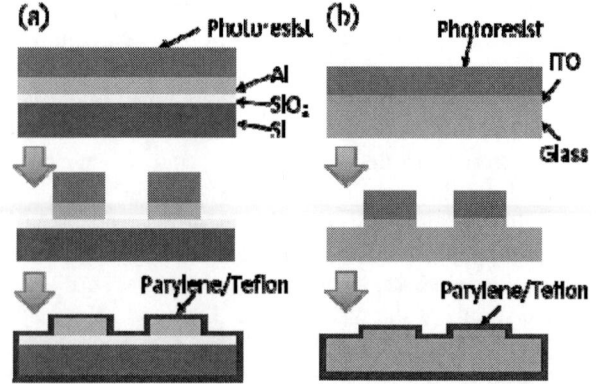

Figure 6: Fabrication flow diagram for (a) SiO_2 chip; (b) ITO glass chip.

978-1-4799-3510-9/14 $31.00 © 2014 IEEE   993

to 3 mm. The floating object is measured at 5.4 × 0.8 × 0.8 mm^3. Note that the shape of menisci at the walls, which is critical to the generated propulsion force, is related to the canal width and the contact angle that is determined by the applied voltage. Sequential images (perspective views in Fig. 7) show that the object is moved left first and then right with synchronization with the electrode activation shifts (80 VDC EWOD voltage with duration of 0.5 s).

ROTATION

Finally, a rotational motion is achieved by arranging 10 mm wide EWOD electrodes along the circular path (Fig. 8, 24 mm in radius). Two hydrophilic objects are connected by a rod and held in the center. As activations of diagonal electrode pairs are clockwisely shifted (duration of 1.5 sec), the net tangential force is generated and then pulls and rotates the floating objects. This can be applied to floating micro

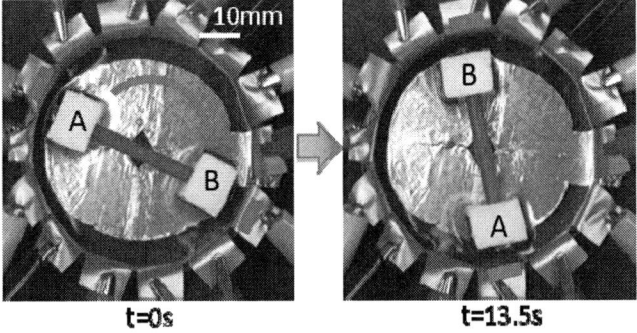

Figure 8: Rotation in 24mm radius circular container. Pairs of diagonal electrodes were activated clockwise with duration of 1.5s. Red bars indicate the activated EWOD electrodes.

motors.

CONCLUSION

This work presents a novel propulsion method combining Cheerios effect and EWOD (electrowetting on dielectric). By switching the wall surface between hydrophobic and hydrophilic states, a 1-D motion for the floating object toward or away from the wall is achieved. These results agree well with theory. Then, a continuous propulsion method in a canal is proposed and realized in mini scales. By sequentially activating microfabricated EWOD electrode arrays, linear translations of floating objects in the small scale canal is accomplished. Here, the canal width is comparable to the capillary length. Finally, a continuous rotational motion of the floating rod is achieved in a circular container on whose side walls EWOD electrodes are placed.

ACKNOWLEDGEMENTS

This research is supported by NSF Grant No. ECCS-1029318.

CONTACT

*Sung Kwon Cho, tel: +1-412-6249798; skcho@pitt.edu

REFERENCES

1. Baudoin, R., *La physico-chimie des surfaces dans la vie des Arthropodes aeriens des miroirs d'eau, des rivages marins et lacustres et de la zone intercotidale.* Bull. Biol. France Belg., 1955. **89**: p. 16-164.
2. Hu, D.L. and J.W. Bush, *Meniscus-climbing insects.* Nature, 2005. **437**: p. 733-736.
3. Vella, D. and L. Mahadevan, *The "Cheerios effect".* Am. J. Phys. , 2005. **73**(9): p. 817-825.
4. Walker, J., *The Flying Circus of Physics, 2nd ed.,* 2007, Wiley: New York.
5. Batchelor, G.K., *An Introduction to Fluid Dynamics.* 1967, Cambridge U.K.: Cambridge University Press.
6. Campbell, D.J., et al., *Spontaneous Assembly of Soda Straws.* Journal of Chemical Education, 2002. **79**(2): p. 201-202.
7. Berg, J.C., *An Introduction to Interfaces & Colloids: The Bridge to Nanoscience.* 2009, Singapore World Scientific.
8. Kralchevsky, P.A., et al., *Capillary Image Forces I. Theory.* Journal of Colloid and Interface Science, 1994. **167**(1): p. 47-65.
9. Hu, D.L., et al., *Water-walking devices.* Exp Fluids, 2007. **43**: p. 769-778.
10. Yu, Y., et al., *Meniscus-climbing behavior and its minimum free-energy mechanism.* Langmuir, 2007. **23**(21): p. 10546-50.
11. Cho, S.K., H. Moon, and C.-J. Kim, *Creating, transporting, cutting, and merging liquid droplets by electrowetting-based actuation for digital microfluidic circuits.* JOURNAL OF MICROELECTROMECHANICAL SYSTEMS, 2003. **12**(1): p. 70-80.
12. Chung, S.K., et al., *Micro Bubble Fluidics by EWOD and Ultrasonic Excitation for Micro Bubble Tweezers*, in *MEMS IEEE 20th International Conference*2007. p. 31-34.

MESOPOROUS-SILICA NANO-CHANNELS INTEGRATED IN MICRO-FLUIDIC CHIP FOR FAST LIQUID MICRO-EXTRACTION OF PESTICIDE RESIDUAL

Pengcheng Xu, Chuanzhao Chen, Haitao Yu, and Xinxin Li*

State Key Lab of Transducer Technology, and, Science and Technology on Micro-system Lab,
Shanghai Institute of Microsystem and Information Technology, Chinese Academy of Sciences,
Shanghai 200050, CHINA

ABSTRACT

The paper reports a micro-chip with nano-pore channels integrated in a micro-channel as reservoir for quickly extracting analyt from aqueous solution to water-soluble organic solvents. Using this novel technology, trace-level residual of organophosphorus pesticide in water-solution can be micro-extracted to a common organic-solvent (e.g., ethanol) for subsequent analysis by GC-MS (Gas Chromatography-Mass Spectrometry).

INTRODUCTION

Food safety not only becomes a severe issue that is associated with health of human-being, but also turns into a social issue in many countries [1-3]. Therefore, efforts from both research institutions and governments in the world are made on developing various effective technologies for food safety guarantee. It is known that many food-safety accidents in farm products are caused by micro-organism, pesticide residues and illegal use of veterinary drugs. Among the diverse kinds of bio/chemical contaminates, ultra-low concentration organophosphorus pesticides already exhibit quite high toxicity, thereby becoming important analytes to be detected.

Nowadays, GC-MS (Gas Chromatography-Mass Spectrometry) has been widely used to detect organophosphorus pesticides remained in agriculture products [4]. It needs to be pointed out that, GC-MS operation always requires the analyte pre-dissolved in an organic solvent (e.g., ethanol) instead of aqueous solution because water may cause: *i*) stationary phase on GC-column inner-wall is washed away; *ii*) vacuum system/electron-impact ionization is damaged; and *iii*) the MS filament is burn out. Since the organophosphorus pesticide residual is collected from plants or soil, the sample is normally dissolved in aqueous solution. Therefore, prior to GC-MS analysis, the sample has to be extracted into an organic solvent. Unfortunately, liquid-liquid extraction cannot be used herein, as the usable organic solvents like ethanol is normally intersoluble with water. Besides, nowadays available solid-phase extraction technologies are generally time-consuming, expensive and unfriendly to environment. Thus, new technologies for low-cost and high-efficiency micro-extraction from water to ethanol are highly demanded for development.

With the rapid development of lab-on-chip and micro-total-analysis system (μ-TAS), micro-fluidic chip is becoming an enabling tool for pre-treatment of ultra-low

concentration analyte. In high-efficiency micro-extraction for trace-level substance, the volume-aspect-ratio of the channel plays an important role. Nanostructures like nanoparticles, nanowires and nanopores have been recently employed to enhance the volume-aspect-ratio [5, 6]. The mesoporous silica thin-film with ultra-high specific surface area was ever employed to build micro/nano channel combined lab-chips [7]. However, this micro/nano combined technique was focused on utilizing the nanoscale size-effect of the nano-channels and the research purpose was for proton transport instead of any chemical analysis. Considering that the inner walls of the nano-pores are densely covered with silanol ($\equiv$SiOH) groups, which have specific interaction with the P=O groups of organophorous compounds, the nano-pore channels show the potential for capturing the organophosphorus pesticide. If an usable organic solvent can elute the pre-adsorbed organophosphorus pesticide molecules from the walls of the nano-channels, the a new extractor can be formed for the sample treatment prior to GC-MS.

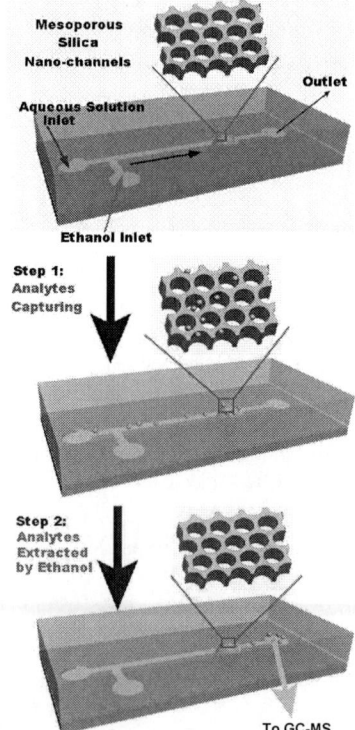

Figure 1: Schematic of the micro-extraction process of pesticide residue from water solution to an organic solvent (like ethanol) that is intersoluble with water.

This study develops a liquid-to-liquid extraction chip, where nano-channels are integrated in the extraction-reservoir of a micro-channel for quickly extracting analyt from aqueous solution to water-soluble organic solvent. Using this nano-technology combined fluidic chip, trace-level residual of organophosphorus pesticides (such as dichlorvos, paraoxon and chlorpyrifos) in water-solution is firstly trapped by $\equiv$SiOH groups of the nano-channels. Then, the trapped pesticide molecules will be micro-extracted to a commonly used organic solvent of ethanol for further analysis by GC-MS. The scheme is shown in Figure 1.

EXPERIMENTS

Chemicals. Pesticide samples of dichlorvos, paraoxon and chlorpyrifos (all with analytical standard) are obtained from Supelco (Bellefonte, PA, USA). APTES (3-aminopropyltriethoxysilane) and FAS-17 (1H,1H,2H,2H-Perfluorodecyltrichlorosilane) are purchased from Gelest, Inc. (Morrisville, PA, USA). Fuchsin dye, tetraethylorthosilicate (TEOS), ethanol, 37wt% HCl, 30%H_2O_2, 98wt%H_2SO_4 are of analytical grade and purchased from Shanghai Chemical Corp. Polydimethylsioxane (PDMS Sylgard 184) is obtained from Dow Corning (Wiesbaden, Germany).

Micro-channel fabrication. The micro-channel fluidic chip is fabricated with a standard MEMS (micro electro mechanical systems) technology. The micro-channels together with the extraction-reservoir and the outlet/inlet are wet etched in silicon wafer. The geometry of the extraction-reservoir is 3mm in diameter and 50μm in depth.

Preparation of mesoporous-silica precursor. The process for preparing precursor of the mesoporous silica is given as follows. Firstly, 50 mL TEOS is dissolved in 50 mL ethanol. Then, 4.14 mL H_2O and 1μL HCl (12 M) is added in an oil bath with vigorous stirring (60 °C, 90 min). The solution is taken off from the oil bath while stirring is kept. After the solution cools down to room temperature (25 °C), 16.5mL H_2O and 76 μL HCl (12 M) are drop-wise added under stirring for 15min. Thereafter, the solution is put into the oil bath again and stirred at 50 °C for 15 min. With the stirring kept at room temperature, the solution is taken off from the oil bath. Then, 250 mL ethanol is added to dilute the solution. 8.38 g CTAB are sequentially added into the solution. Finally, clear sol solution is obtained after the CTAB is absolutely dissolved. The molar composition of this sol solution is 1 TEOS: 0.1 CTAB: 23 ethanol: 5 H_2O: 0.004 HCl. This stock solution is prepared as the precursor to form nano-channels in the following procedure.

Nano-channels integrated in the micro-channel. The fabrication is schematically in Figure 2. (a): The micro-channel is wet etched in silicon-wafer, with the extraction-reservoir at center. (b): The micro-channel surface is self-assembled with hydro(oleo)phobic molecule-layer of FAS-17. (c) and (d): Through a pre-patterned hard-mask, ultra-violet (UV) light region-selectively oxidizes and removes the FAS-17 at the surface of the reservoir region [8]. (e): Another –NH₂ terminated molecule-layer of APTES is

re-grown at the region. (f): The precursor solution is dropped into the APTES-modified reservoir and, by evaporation induced self-assembly, mesoporous-silica nano-channels is locally constructed on top of APTES. After bonding with a PDMS cover plate, the nano-channel integrated micro-fluidic chip is formed.

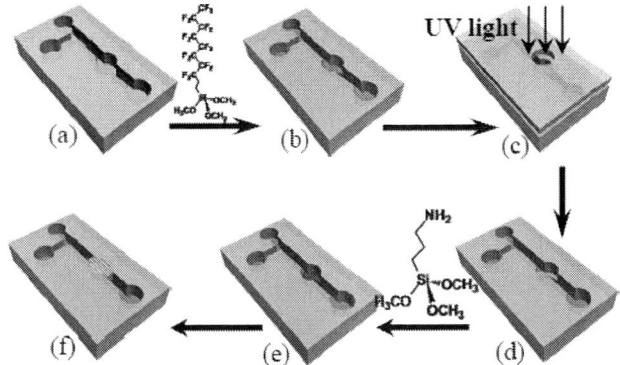

Figure 2: Fabrication processes of the nano-channel integrated micro-chip.

Preparation of pesticide standard solution. Dichlorvos, paraoxon and chlorpyrifos are selected as target pesticides in this research. For preparing the standard pesticide solution, 10 mg of pesticide is firstly added into an amber-colored volumetric flask (1000 mL). Then, about 100 mL of distilled water is added into the volumetric flask to dissolve the pesticide. Thereafter, distilled water is carefully added and made up to the desired volume-mark of the volumetric flask. The prepared standard pesticide stock solution can be used for the following extraction experiment.

Pesticide extraction experiment. The pesticide solution is introduced to the extraction chip by using a KDS-220 syringe pump (KD-scientific Inc., Holliston, MA, USA). Firstly, 1mL of pesticide solution is filled in a plastic syringe. Then, the syringe is placed in the syringe pump and a micro-line tube is connected to the inlet of the chip. With the flow rate is set at 100μL/min, the start button of the syringe pump is switched on. Thereafter, the pesticide solution is introduced to the micro channel with a constant flow rate, and the waster solution is collected by using a small vial. After the solution entirely flows through the chip, clear air is injected into the chip to blowing dry the chip. Then, 0.5mL absolute ethanol is introduced to the chip to elute the pesticide molecules that have been captured previously by the silica nano-channels in the reservoir. For the subsequently implemented GC-MS identification, the ethanol eluent is collected in a screw-cap vial (2mL volume, Agilent Technologies Inc., Palo Alto, CA, USA).

GC-MS identification for the extracted pesticide. The composition of the ethanol eluent is identified by using a capillary GC-MS apparatus (Agilent 7890A-5975C). The model of the commercially available column is DB-17MS (30 m × 0.25 mm × 0.25 μm). The used temperature program is 3 min at 40 °C, a ramp of 15 °C/min to 280 °C and hold-time of 5 min. 1μL ethanol eluent is injected into GC with a spilt ratio of 50:1. The flow rate of the He carrier gas is

1.1 ml/min. MS is operated in electron impact mode with the source temperature as 250 °C, quadrupole temperature as 150°C and ionising voltage as 70 eV. The scan is performed in the range of m/z 50 to 500. The composition of the eluent is identified using a NIST08 mass spectral library.

RESULTS & DISCUSSION

To observe the structure of the mesoporous silica nano-channels, a piece of the sample is grinded into thin pieces for TEM characterization. The TEM image in Figure 3 shows ordered meso-channels of the sample, with the pore diameter as about 2 nm. According to standard N_2 sorption measurements, the specific surface of the as-prepared mesoporous-silica-monolith is 339 m^2/g, the pore volume is 0.2 cm^3/g and the average pore size is 2.1 nm. The characterization results are shown in Figure 4.

Figure 3: TEM image showing the nano-porous structure of the as-prepared mesoporous silica.

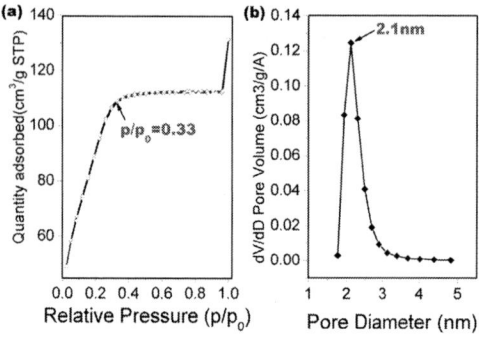

Figure 4: (a) Nitrogen sorption isotherm and (b) pore size distribution of the mesoporous-silica sample that is calculated according to BJH theory.

Figure 5 shows the fabricated micro-extraction chip that is made of silicon wafer by standard micromachining techniques. A PDMS plate is bonded with the silicon chip to close the channels and lead out the inlet/outlet pipes.

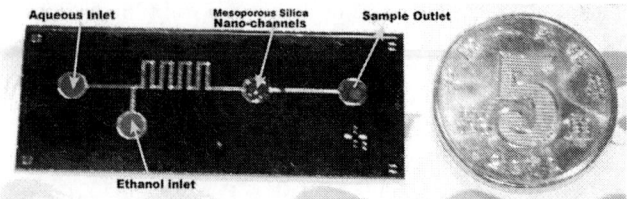

Figure 5: Optical photograph of the as-prepared micro-extraction chip, in which the mesoporous silica is clearly denoted for observation.

The experiment is performed to validate the extraction function of the nano-channels. Shown in Figure 6, aqueous acid fuchsin (showing pink-color in water solution) can smoothly flow through the nano-channels. With increase of the flowing time, the pink color in the reservoir becomes darker and darker that means effective capturing of the nana-channels to the pink-colored dye molecules. To visualize the extraction process, an experiment is conducted and shown in Figure 7. After the pink dye molecules captured by the nano-channels, ethanol is flowed through the nano-channels. The color at the nano-channels becomes to fade and the color of the ethanol solution at the outlet turns to be deepened. The results well confirm that the pre-concentrating/extracting process is successful.

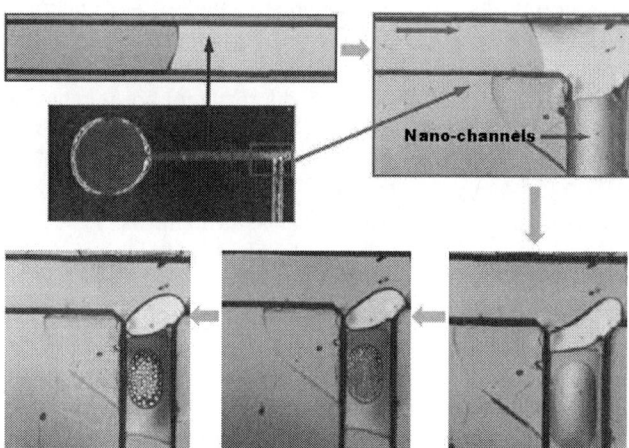

Figure 6: Experimental observation of fuchsin pre-concentrating process. The results verify that the pink molecules are effectively captured by the mesoporous-silica nano-channels.

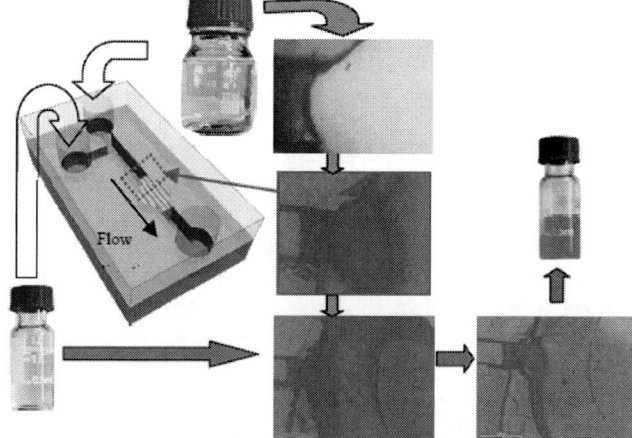

Figure 7: Experiment of the acid-fuchsin pink dye extracted from water to ethanol, by using the nano-channel integrated micro-chip.

Then the pesticide extraction experiment is implemented. Water solution containing three kinds of organophosphorus pesticides of dichlorvos, paraoxon and chlorpyrifos (all with µg/mL or ppb level concentration) flows through the nano-channels to introduce the pesticides for capturing and pre-concentrating. Since the organophosphorus pesticides have much higher solubility in ethanol than in water, the captured pesticides can be effectively extracted when ethanol flows through the nano-channels. Shown in Figure 8, the GC-MS results for the extracted ethanol solution clearly indicate that all the three pesticides have been well detected. According to the experimentally pre-calibrated standard-curve (see Figure 9), the detection limit to the pesticide of dichlorvos is as low as 200 µg/mL. Therefore, the nano-channel micro-extraction function is validated and the nano-channel integrated extraction chip is promising in food-safety applications.

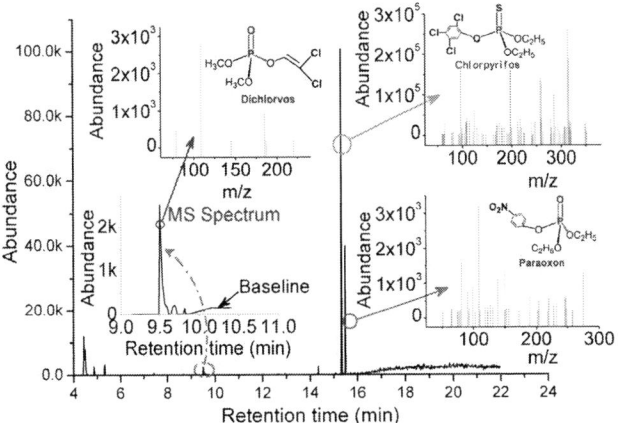

Figure 8: GC-MS qualitative analysis for the extracted ethanol solution, which contains the three kinds of pesticides.

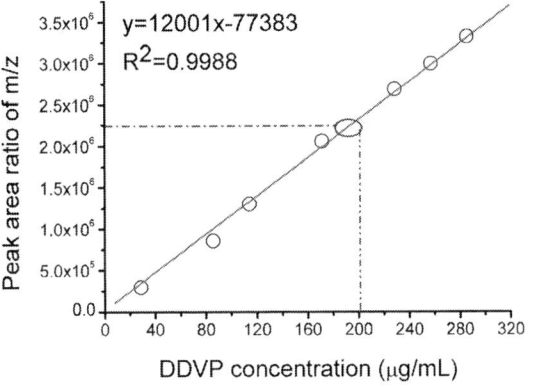

Figure 9: GC-MS quantitative analysis for the extracted dichlorvos (DDVP) in ethanol solution that was obtained by using the extraction micro-chip. The obtained linear relationship between the analyzed signal and the dichlorvos concentration indicates the high-efficiency of the extraction technology.

CONCLUSION

This paper reports a micro-chip with nano-channels integrated as extraction-reservoir for quickly extracting pesticeds from aqueous solution to ethanol solvent. Using this novel technology, µg/mL or *ppb* level residual of organophosphorus pesticides (such as dichlorvos, paraoxon and chlorpyrifos) in water-solution can be high-efficiency captured by ≡SiOH groups in the nano-channels. The trapped

pesticide molecules can be eluted by ethanol. The GC-MS quantitative analysis results for the extracted pesticide confirm that, by using such a new extraction technique for pre-treatment, the detection limit to pesticide of dichlorvos can be as low as 200µg/mL.

ACKNOWLEDGEMENTS

This research is supported by Chinese 973 Project (2011CB309503) and NSF of China (91023046, 61161120322, 61021064, 61102010) and the Project of National Science & Technology Pillar Plan (2012BAK08B05). Pengcheng Xu thanks to project support from the Key Laboratory for Micro/Nano Technology and System of Liaoning Province.

REFERENCES

[1] M. L. Xu, et al., "Determination and Control of Pesticide Residues in Beverages: A Review of Extraction Techniques, Chromatography and Rapid Detection Methods", *Appl. Spectrosc. Rev.*, vol. 49, pp. 97-120, 2014.

[2] Y. Wu, *et al.*, "Food safety in China", *J. Epidemiol. Commun. H.*, vol. 67, pp. 478-479, 2013.

[3] J.-Y. Yoon and B. Kim, "Lab-on-a-Chip Pathogen Sensors for Food Safety", *Sensors,* vol. 12, pp. 10713-10741, 2012.

[4] L. Alder, *et al.*, "Residue analysis of 500 high priority pesticides: Better by GC–MS or LC–MS/MS?", *Mass Spectrom. Rev.,* vol. 25, pp. 838-865, 2006.

[5] J. S. Shim and C. H. Ahn, "An on-chip whole blood/plasma separator using hetero-packed beads at the inlet of a microchannel", *Lab Chip,* vol. 12, pp. 863-866, 2012.

[6] J. Kim, *et al.*, "Direct synthesis and integration of functional nanostructures in microfluidic devices", *Lab Chip,* vol. 11, pp. 1946-1951, 2011.

[7] R. Fan, *et al.*, "Gated proton transport in aligned mesoporous silica films", *Nat. Mater.,* vol. 7, pp. 303-307, 2008.

[8] C. Chen, *et al.*, "Top-down batch-fabricating bottom-up self-assembles for region-selective multi-functionalization of microfluidic chips", *Transducers 2013*, pp. 2688-2691, 2013.

CONTACT

*Xinxin Li, tel: +86-21-62131794
email: xxli@mail.sim.ac.cn

MINIATURISED PRANDTL TUBE WITH INTEGRATED PRESSURE SENSORS FOR MICRO-THRUSTER PLUME CHARACTERISATION

M. Dijkstra[1,], K. Ma[1,2], M.J. de Boer[1], J. Groenesteijn[1], J.C. Lötters[1,3] and R.J. Wiegerink[1]*

[1] MESA+ Institute for Nanotechnology, University of Twente, THE NETHERLANDS
[2] MicroCreate B.V., THE NETHERLANDS
[3] Bronkhorst High-Tech B.V., THE NETHERLANDS

ABSTRACT

A miniaturised Prandtl-tube sensor incorporating a 6 mm long 40 μm diameter microchannel with integrated pressure sensors has been realised. The sensor has been designed for the characterisation of rarefied plume flow from a MEMS-based monopropellant propulsion system for high-accuracy attitude control of satellites. The 4.5 × 2.5 mm sensor chip incorporates 400 mbar full-scale capacitive pressure sensors. The capacitive pressure transducer is created by merging several microchannels into a rectangular pressure membrane, with outward-facing comb-fingers hinging on the microchannel sidewall. Additionally, thermistors can measure the temperature profile on the Prandtl tube and an integrated Pirani sensor can optionally measure vacuum pressures below 10 mbar. An electronics board and stainless steel probe, from which only the Prandtl tube protrudes has been realised. Initial plume measurements of dynamic pressure on the Prandtl tube have been obtained by flowing air from a metal tube with 0.7 mm diameter, demonstrating the feasibility of the miniaturised Prandtl-tube sensor.

INTRODUCTION

Micro chemical propulsion systems (μCPS) have been identified by ESA as emerging compact propulsion systems, targeting thrust levels in the order of 1-10 mN, having low power requirements and low system weight. The PRECISE project [1, 2] focuses on the research and development of a MEMS-based monopropellant μCPS, applying catalytic decomposition of hydrazine, for high-accuracy attitude control of satellites. The compressible Navier-Stokes equations are being evaluated with DLR TAU code for the design of transfer lines, catalytic reactor and nozzle, where the numerical calculation transitions to the Direct Simulation Monte Carlo (DSMC) method for the rarefied plume flow. With more micro-thrusters being developed [3-6] the investigation of the compact rarefied plume [7, 8] is becoming more important. A novel miniaturised Prandtl-tube sensor with high spatial resolution is proposed for the DSMC validation and performance evaluation of the μCPS nozzle design.

SENSOR DESIGN

High spatial resolution is obtained by a 40 μm diameter low-stress silicon nitride (SiRN) microchannel. Specifications require at least 4 mm length for the Prandtl tube not to interfere significantly with the μCPS rarefied plume flow. Expected pressures are 400 mbar at the nozzle and 1 mbar in

the plume down to 0.01 mbar in the μCPS backflow. The gas reaches 700 °C at the nozzle exit, expanding rapidly in the rarefied plume.

A 6 mm long Prandtl tube has been designed, connected to a 4.5 × 2.5 mm sensor chip containing at minimum two capacitive pressure sensors (Fig. 1). A dynamic-pressure sensor measures the total pressure inside the Prandtl tube in reference to the static pressure above the membrane being deflected [9]. A second reference-pressure sensor is mostly

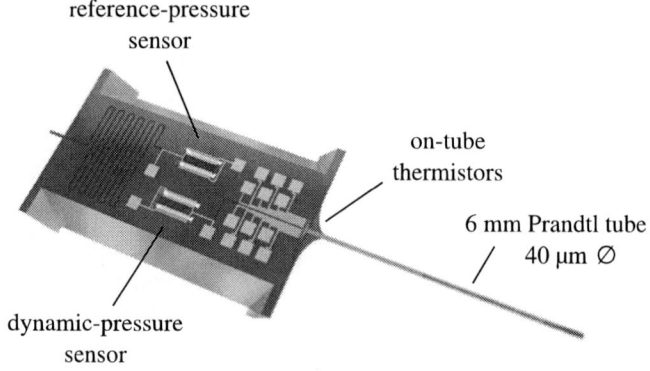

Figure 1: Miniaturised Prandtl-tube sensor chip measuring 4.5 × 2.5 mm, with integrated capacitive pressure sensors. The 6 mm long 40 μm diameter Prandtl tube contains three thermistors, allowing measurement of the thermal profile along the tube.

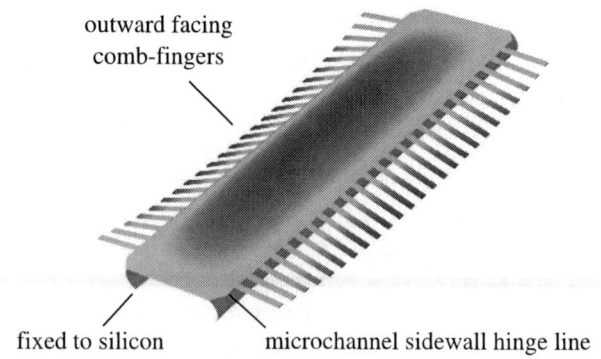

Figure 2: Numerical COMSOL result (deflection not to scale) of the microchannel pressure sensor, applying 400 mbar, revealing the comb-fingers effectively hinging maximally 0.6 μm downwards.

needed to compensate for long-term capacitance drift. Additionally, the connection to a long meandering microchannel opening to the backside of the chip allows transients in ambient static pressure to be captured. Thermistors are integrated on the Prandtl tube to possibly better understand the Boltzmann statistics of the rarefied flow.

A 400 mbar full-scale capacitive pressure sensor has been designed, which connects to the on-chip microchannel Prandtl tube (Fig. 2). A rectangular pressure membrane is created by merging several microchannels into a single cavity. The microchannel sidewalls of the pressure cavity are sufficiently flexible for outward facing comb-fingers to hinge downward during membrane deflection. The comb-fingers connected to the pressure membrane are lowered to provide linear sensitivity around the initial position, as well as increased sensitivity throughout. Numerical COMSOL simulation shows effective hinging around the microchannel sidewalls, provided that the bottom of the microchannel cavity is fixed to the silicon substrate (Fig. 2).

The thermal conduction to silicon across the 40 μm deep cavity is sensitive to pressures below 10 mbar, allowing integration of a Pirani pressure sensor [10] on the same membrane for measuring low vacuum pressures in the μCPS backflow.

SENSOR FABRICATION

A schematic overview of the process scheme for the miniaturised Prandtl tube with integrate capacitive pressure sensors is given in figure 3, where lowered comb-fingers are integrated with the existing surface microchannel technology [11]. Processing commences with the deposition of a low-stress 500 nm silicon-rich silicon-nitride (SiRN) layer and 100 nm Cr mask layer. Trenches for the lowered comb-fingers are created 3 μm deep by accurately timed 3 min KOH etching, making sure the native oxide is just previously removed by HF etching (Fig. 3a). The KOH-etched trenches allow for proper step-coverage of thin-film Cr/Au across the (111) slopes.

The KOH-etched trenches are covered with 5 μm SU-8 2005, while the 2 μm wide etch holes are exposed for the isotropic dry etching of the surface microchannels using high-density SF_6 plasma with zero self-bias (Fig. 3b). The inner surfaces of the microchannels and KOH-etched trenches are conformally coated by a low-stress 1.8 μm SiRN layer, sealing the microchannel etch holes. A 10 nm Cr adhesion layer and 200 nm Au metallisation layer are deposited and patterned using SU-8 2005, not yet defining the comb-fingers (Fig. 3c). The SiRN is patterned on the backside and a new 5 μm SU-8 2005 layer on the frontside is used to etch both the Cr/Au thin-film and SiRN layers simultaneously, thus minimising misalignment with the KOH-etched trenches (Fig. 3e-f). The Prandtl tube entrance is created by wafer dicing before final release by KOH etching (Fig. 3g).

Figure 4 shows a separately designed calibration chip, integrating two microchannel pressure sensors measuring 600 × 250 μm and 600 × 150 μm, with twenty-five comb-fingers on each side of the membrane. Additionally, thermistors for the Pirani sensor are integrated on the smallest membrane applying Kelvin contacts.

The lowered outward-facing comb-fingers connected to the microchannel sidewall are enlarged in figure 5. A small cut-out just outside the SiRN pressure membrane is made for the underetch of the (111)-plane during KOH etching,

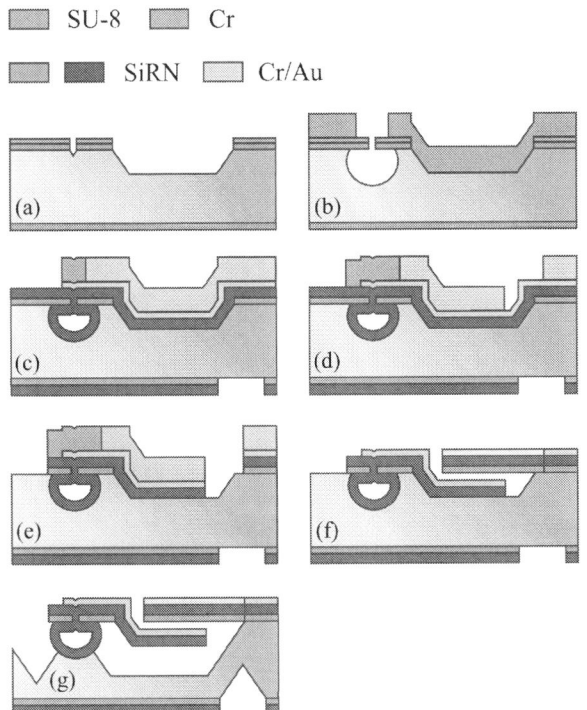

Figure 3: Process scheme for the fabrication of the miniaturised Prandtl tube sensor.

Figure 4: SEM micrographs of microchannel capacitive pressure sensors, with optional Pirani pressure sensor for measurement of low vacuum pressures.

releasing sufficient microchannel sidewall, such that a proper hinge is obtained. It can be observed that stress in the Cr/Au thin-film leads to some tearing on the KOH-etched (111)-sidewalls, but never as much as to result in the lowered comb-fingers to become open-circuited.

Figure 6a shows the apex of the Prandtl tube created after wafer dicing. Several sensor chips were fabricated, where 4 mm thin (111)-inclined silicon beams provide addi-

Figure 5: SEM micrograph of microchannel sidewall hinge line, with lowered outward-facing comb-fingers.

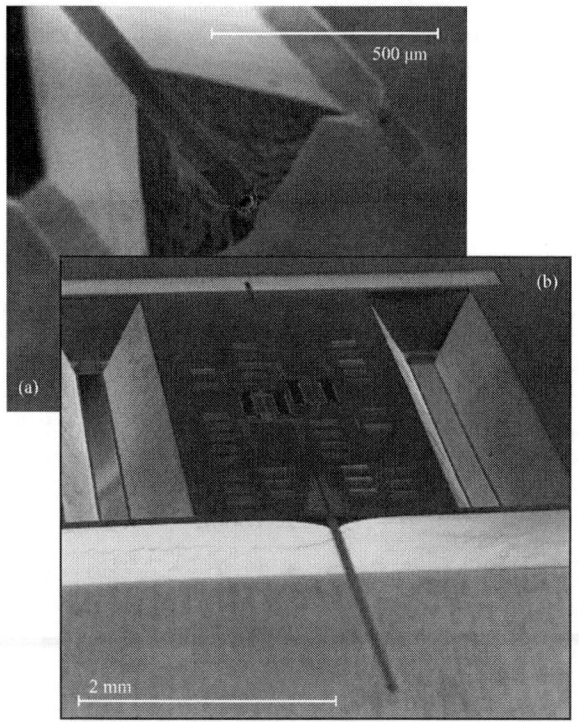

Figure 6: SEM micrographs of (a) Prandtl tube opening create by wafer dicing, with the microchannel partially supported by thin (111)-inclined silicon beams and (b) suspended Prandtl-tube sensor chip before breaking.

tional mechanical support for the microchannel. Figure 6b shows a suspended Prandtl-tube sensor chip, which can easily be released from the wafer. The sensor in figure 6b combines both the capacitive outward-facing comb-fingers with thermistors for the Pirani sensor on the same pressure membrane.

EXPERIMENTAL RESULTS

A stainless steel probe was realised (Fig. 7) from which only the 6 mm Prandtl tube protrudes. The probe can be mounted using a Swagelok® fitting. Two metal rings are slotted in the electronics board and a semicircular brass piece is glued to the electronics board facilitating accurate threading of the Prandtl tube through the probe opening. The electronics board includes four LM334M current sources, providing 3.6 mA current to the on-chip thermistors, where the voltage drop is measured by an LTC2493 analog-to-digital convertor (ADC). An AD7746 capacitance-to-digital convertor (CDC) is used to measure capacitance down to about 40 aF precision, requiring long-term output drift compensation through additional measurement of the reference-pressure sensor. A BeagleBone Black was used to communicate with the measurement ICs.

The 600×250 µm capacitive pressure sensor was calibrated using the calibration chip (Fig. 4), where external pressure could be applied to an on-chip microchannel. Figure 8 shows two subsequent calibration measurements using a Sensor Technics reference pressure sensor, measuring 3.3 aF· mbar^{-1} sensitivity up to 400 mbar.

Figure 9 shows plume measurements measuring dynamic pressure on the miniaturised Prandtl tube, due to airflow from a 0.7 mm metal tube mounted on a turntable setup to adjust for angle and distance to the sensor. The Prandtl-tube apex was positioned at the centre of rotation. Plume measurements for several distances were obtained, where the dynamic pressure shows a clear falloff for distances large than 3 cm, with a narrow angle of sensitivity of only 30° to one side, which might well be attributed to the highly directed airflow from the 0.7 mm metal tube.

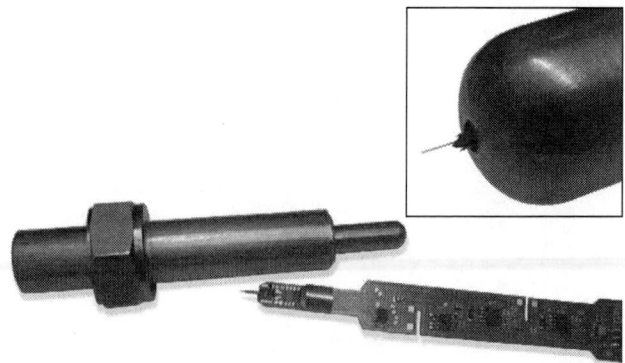

Figure 7: Miniaturised Prandtl-tube sensor chip mounted on electronics board next to stainless steel probe. The inset shows the Prandtl tube, partially supported by thin silicon beams, protruding from the probe tip.

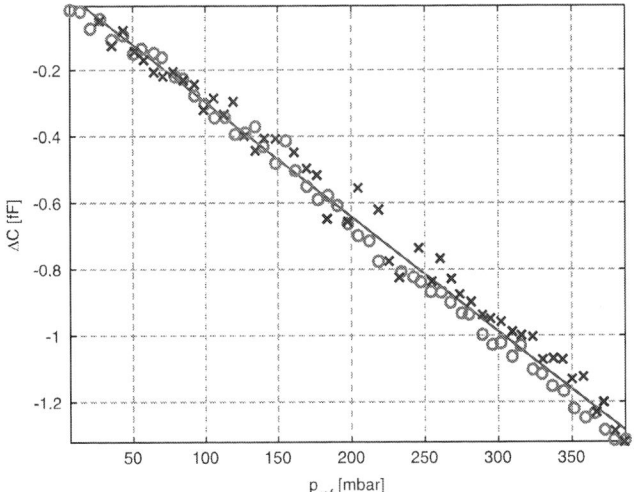

Figure 8: Two subsequent calibration measurements on the microchannel capacitive pressure sensor, measuring 3.3 aF· mbar^{-1} sensitivity. The calibration chip allows connection of external pressure to the capacitive pressure sensor.

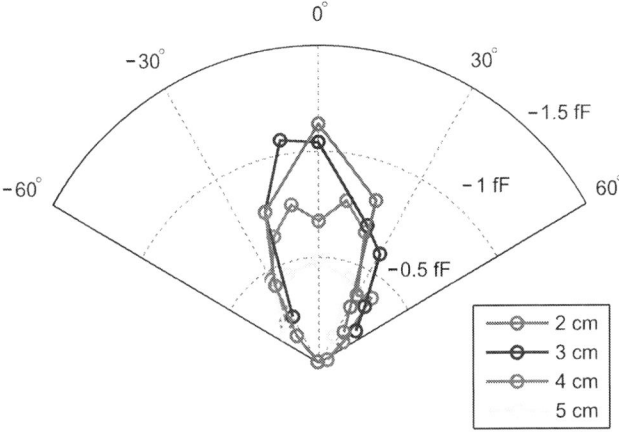

Figure 9: Plume measurement of dynamic pressure on the miniaturised Prandtl tube. A constant flow of air from a metal tube with 0.7 mm diameter was directed at different angles and distances towards the Prandtl-tube sensor.

CONCLUSIONS

Initial plume measurements flowing air from a metal tube have demonstrated the correct working of the miniaturised Prandtl-tube sensor. Without the supporting 4 mm silicon beams (Fig. 6a) the Prandtl tube shows tendency to flutter at highest dynamic pressures. In rarefied flow this might not even happen, however the additional silicon beams seem appropriate to withstand shocks when firing the µCPS. The Pirani sensor has to be calibrated in vacuum and further measurements in front of a cold-gas thruster are considered, before confronting the monopropellant µCPS.

ACKNOWLEDGEMENTS

This research is part of the PRECISE project, which is funded from the European Community's Seventh Framework Programme (FP7/2007-2013) under grant agreement n° 282948.

REFERENCES

[1] PRECISE – Chemical-µPropulsion for an Efficient and Accurate Control of Satellites for Space Exploration – http://www.mcps-precise.com/

[2] M. Gauer, D. Telitschkin, U. Gotzig, Y. Batonneau, H. Johansson, M. Ivanov, P. Palmer, R. Wiegerink, "First Results of PRECISE – Development of a MEMS-Based Monopropellant Micro Chemical Propulsion System", *Acta Astronautica,* vol. 93, pp. 77-83, 2014.

[3] J. Lee, K. Kim, S. Kwon, "Design, Fabrication, and Testing of MEMS Solid Propellant Thruster Array Chip on Glass Wafer", *Sensor. Actuat. A-Phys.,* vol. 157, pp. 126-134, 2010.

[4] M. Louwerse, H. Jansen, M. Elwenspoek, "A Modular Assembly Method of a Feed and Thruster System for Cubesats", *J. Micromech. Microeng.,* vol. 20, 115011, 2010.

[5] T. Takahashi, Y. Takao, K. Eriguchi, K. Ono, "Microwave-Excited Microplasma Thruster: a Numerical and Experimental Study of the Plasma Generation and Micronozzle Flow", *J. Phys. D: Appl. Phys.,* vol. 41, 194005, 2008.

[6] C-P. Chen, Y-C. Chao, C-Y. Wu, J-C. Lee, "Development of a Catalytic Hydrogen Micro-Propulsion System", *Combust. Sci. Technol.,* vol. 178, pp. 2039-2060, 2006.

[7] V. Lekholm, K. Palmer, G. Thornell, "Schlieren Imaging of Microthruster Exhaust for Qualitative and Quantitative Analysis", *Meas. Sci. Technol,* vol. 23, 085403, 2012.

[8] T. Zhuang, A. Shashurin, M. Keidar, "Microcathode Thruster (µCT) Plume Characterization", *IEEE T. Plasma Sci.,* vol. 39, no. 11, pp. 2936-2937, 2011.

[9] O. Berberig, K. Nottmeyer, J. Mizuno, Y. Kanai, T. Kobayashi, "The Prandtl Micro Flow Sensor (PMFS): A Novel Silicon Diaphragm Capacitive Sensor for Flow-Velocity Measurement", *Sensor. Actuat. A-Phys.,* vol. 66, pp. 93-98, 1998.

[10] F. Santagata, J.F. Creemer, E. Iervolino, L. Mele, A.W. van Herwaarden, P.M. Sarro, "A Tube-Shaped Buried Pirani Gauge for Low Detection Limit With Small Footprint", *J. Microelectromech. S.,* vol. 20, no. 3, pp. 676-684, 2011.

[11] M. Dijkstra, M.J. de Boer, J.W. Berenschot, T.S.J. Lammerink, R.J. Wiegerink, "A Versatile Surface Channel Concept for Microfluidic Applications", *J. Micromech. Microeng.,* vol. 17, pp. 1971-1977, 2007.

CONTACT

* Marcel Dijkstra, phone: +31-53489-2805, e-mail: m.a.dijkstra@utwente.nl

MICROFLUIDIC-BASED DROPLET MERGING DEVICE WITH A NON-CONTACT DROPLET PAIRING METHOD

Sanghyun Lee, Hojin Kim, and Joonwon Kim
Pohang University of Science and Technology (POSTECH), Republic of Korea

ABSTRACT

In this study, we present a simple, novel microfluidic device for merging two different droplets in parallel using only fluid flow control. By utilizing the deformability characteristic of a droplet in a pressure-driven shear flow, a unique Laplace trap with a multi-step "trapping–releasing–non-contact pairing–destabilization–contact" droplet merging mechanism was developed. Using a device comprising an array of these Laplace traps, parallel merging of two droplets was successfully performed within a short time variation ($n = 55$, mean = 5.7 s, SD ±4.3 s).

INTRODUCTION

Water-in-oil droplets, formed in a two-phase flow of immiscible fluids, are capable of compartmentalizing various substances into discrete volumes. With the advance of microfluidic droplet manipulation techniques such as generation, mixing, sorting, and merging, droplet-based microfluidics can be used in a diverse range of applications (e.g., biomolecule synthesis or antibody screening) [1]. In particular, merging different droplets is crucial for initiating reactions in droplet-based microfluidic platforms. Accordingly, various techniques for droplet merging have been developed [2–7].

However, in most droplet-based applications, surfactants are involved in the oil phase to stabilize the droplets, which prevents uncontrolled coalescence [8]. For this reason, conventional merging techniques normally require an additional (e.g., electrostatic or laser-induced) force apart from the driving (i.e., hydrodynamic) force for fluid flow in most microfluidic systems.

In the present paper, we propose a simple droplet merging method, which is enabled by a unique design of Laplace trap using only fluid flow control. Droplet merging was performed via a multi-step "trapping–releasing–non-contact pairing–destabilization–and–contact–merging" process. This process can be carried out by utilizing the deformability characteristic of a droplet in a shear flow with the Laplace trap structure.

Using a merging device composed of an array of these Laplace traps, the capability of the device for parallel merging was demonstrated. Parallel merging of two droplets was performed almost concurrently via the multi-step droplet merging process.

DESIGN AND THEORY

Figure 1a shows schematic views of the merging device with a close-up view of the Laplace trap array. The merging device was fabricated by standard soft lithography [9], with a channel height of 80 μm. Each Laplace trap has 2 necks and a storing chamber (Fig. 1b). Widths of the entrance neck and exit neck are 40 μm and 20 μm, respectively. The storing chamber was designed to accommodate up to two droplets (~60 μm in diameter).

Figure 1c shows the entire droplet merging process, which is performed via a multi-step "trapping–releasing–non-contact pairing–washing–and–merging" process. A key feature is that this process functions by utilizing the deformability characteristic of a droplet in a pressure-driven shear flow within a confined geometry.

As the droplets proceed towards the Laplace traps, they can be trapped in front of the entrance neck of each Laplace trap. In this situation (boxed section in Fig. 1c), a resistance

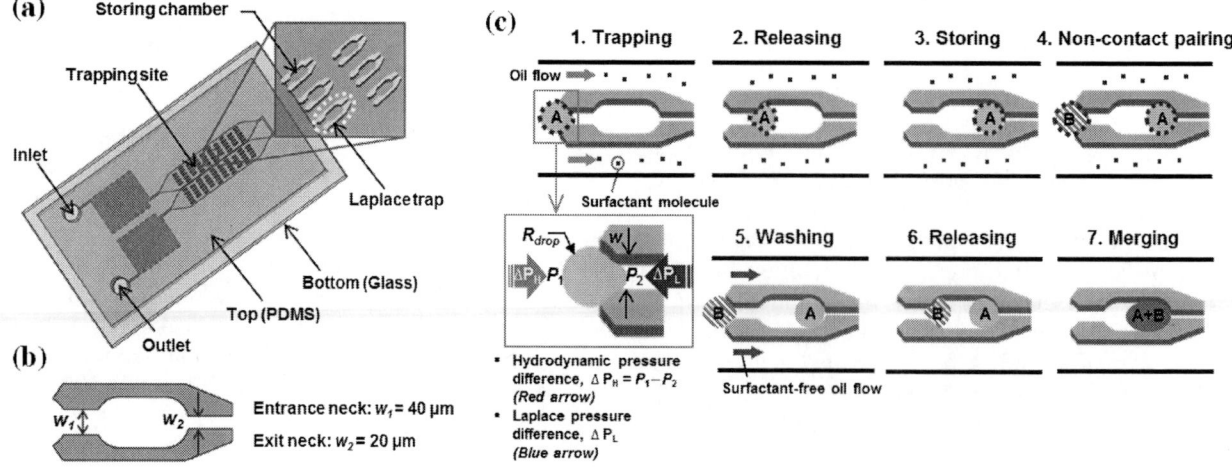

Figure 1: Schematic views of the droplet merging device and the entire droplet merging process.
(a) Overview of merging device, (b) dimensions of Laplace trap, and (c) entire droplet merging process.

pressure (i.e., a Laplace pressure difference, ΔP_L), which is generated by the difference in the radius of curvature between the front and back of the droplet, suppresses the droplet's forward movement against the hydrodynamic pressure difference across the droplet (ΔP_H).

The Laplace pressure difference, the maximum pressure that a droplet can resist without passing through the neck, can be expressed by using the difference between the Laplace pressure at the restriction side (channel neck) and that at the opposite side [10,11], as follows:

$$\Delta P_L = \gamma \left(\frac{2}{w} - \frac{1}{R_{drop}} \right). \quad (1)$$

where γ is the interfacial tension, R_{drop} and $w/2$ represent the radii of curvature of the droplet and of the part of droplet extended into the entrance neck of the Laplace trap, respectively.

Considering the trapped droplet again, it can either remain in front of the neck or pass through the neck depending on the pressure balance between hydrodynamic pressure difference across the droplet and Laplace pressure difference. If the hydrodynamic pressure difference is lower than the Laplace pressure difference, a droplet will remain in front of the neck (trapping). Otherwise, the droplet will pass through the neck (releasing). Therefore, the pressure conditions that determine the state of the droplet can be expressed as follows:

$$\Delta P_H < \Delta P_L \text{ for trapping}, \quad \Delta P_H > \Delta P_L \text{ for releasing.} \quad (2)$$

After the droplets are trapped, they can be released by applying sufficient hydrodynamic pressure. Subsequently, the released droplets can be re-trapped in front of the exit neck, where it is narrower than the entrance neck, because the Laplace pressure difference generated at the exit neck is larger than at the entrance neck (see equation (1)). Thereafter, a second set of droplet is infused and trapped, resulting in non-contact pairing of two different droplets. In the non-contact pairing state, the surfactant-added oil (carrier fluid) is replaced with the surfactant-free oil to desorb the surfactant molecules from the surface of the droplet. This can destabilize the droplet-oil interface. Then the two droplets can be merged by releasing the second droplet and bringing them into contact by increasing the hydrodynamic pressure.

EXPERIMENTS

Mono-dispersed spherical droplets (~60 μm in diameter) were generated using a typical T-junction device [12]. For generation of water-in-oil droplets, mineral oil containing surfactant (Span 80, 2% w/w) and deionized (DI) water containing dye (red or green color) were used for the continuous phase and discrete phase, respectively. The flow rates were 30 μℓ/h for water and 90 μℓ/h for oil, supplied by syringe pumps. The generated droplets are collected at the outlet of the T-junction device and then re-injected into the merging device.

The fluid flow in the merging device is controlled by a pneumatic pressure control system. The pressure control system comprises a diaphragm pump, a regulator to adjust the input pressure, and a solenoid valve to electrically switch the pressure to either positive or ambient. During the merging process, we used the pneumatic source to deliver droplets to the merging device, not using a syringe pump. There are some advantages to using a pneumatic source over a syringe pump, 1) the response of the pneumatically controlled flow is much faster than the flow driven by a syringe pump and 2) applying a discrete step-pressure (profile shown in Fig. 2a) is possible.

RESULTS AND DISCUSSION

We demonstrated the merging of two different droplets using only fluid flow control. The pressure difference profile between the inlet and outlet during the entire merging process is shown in Fig. 2a. The dashed line indicates the upper limit to ensure trapping. If the applied pressure for trapping droplets exceeds this limit, undesired releasing can occur. Dash–dot and dash–dot–dot lines indicate threshold values

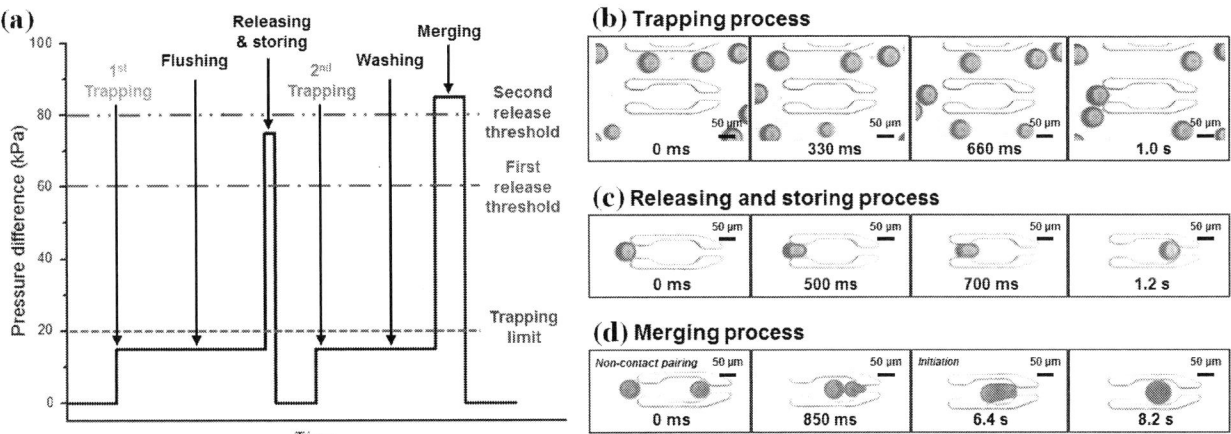

Figure 2: (a) Profile of applied pressure during the entire merging process, sequential images of the (b) trapping process, (c) releasing and storing process, and (d) merging process.

for releasing the first and second droplets, respectively. Those imply that pressure below these threshold values cannot ensure releasing of trapped droplets.

Figures 2b, c, and d show sequential images of trapping, releasing and storing, and merging, respectively. Under an applied pressure of 15 kPa, infused droplets were trapped along the streamline flowing into the storing chamber and by pushing of neighboring droplets (Fig. 2b). When a droplet is trapped, the streamline into the storing chamber is diverted, so the subsequent droplets bypass the Laplace trap. After completing the trapping step at all Laplace traps, the redundant droplets were flushed. When a step-pressure of 60 kPa was applied, the trapped droplet elongated and entered the storing chamber and remained there (Fig. 2c). When releasing the trapped droplets, increasing the pressure in discontinuous step (i.e., step-pressure) ensures the simultaneous release of the droplets for all the Laplace traps; this cannot be achieved by applying continuously increasing pressure.

Subsequently, the second batch of droplets (red color) were trapped in the same manner as the first batch (green color), resulting in a non-contact pairing state. In this state, two droplets are kept at two necks (entrance and exit neck) of the Laplace trap but isolated from each other. Then, the surfactant-added mineral oil was replaced with the surfactant-free mineral oil to rinse the surfactant molecules that are adsorbed on the surface of the droplets. The surfactant-free oil was infused into the device for 10 min to rinse surfactant molecules away sufficiently under a pressure of 15 kPa. By applying a step-pressure of 80 kPa, the trapped second droplet was released, and the two droplets were brought into contact and then merged (Fig. 2d).

Next, to demonstrate the capability of the device for parallel droplet merging, initiation times of merging events were measured in an array under non-contact and contact pairing states (Fig. 3). Because the Laplace trap structure can lead the pairing state of two droplets as either non-contact or contact, the comparison of both states was possible.

Parallel merging occurred within a short time variation in non-contact pairing (n = 55, mean = 5.7 s, SD ±4.3 s) compared with contact pairing (n = 55, mean = 191.2 s, SD ±136.4 s) (Fig. 4). The initiation time of merging was measured by the elapsed time from when the first merging event occurred in the array.

The merging process by means of the non-contact pairing state has the sequence of "non-contact pairing–washing–releasing–and–merging." In the non-contact pairing state, sufficient desorption of the surfactant molecules (i.e., complete depletion of surfactants on a droplet surface in an overall array) can be conducted before two droplets contact to achieve merging. Therefore, the merging events were carried out within a short time after the two droplets came in contact after the washing step.

On the contrary, the merging process by means of the contact pairing state has the sequence of "contact pairing–washing–and–merging." In the contact pairing state, the desorption of the surfactant molecules is not uniform for all

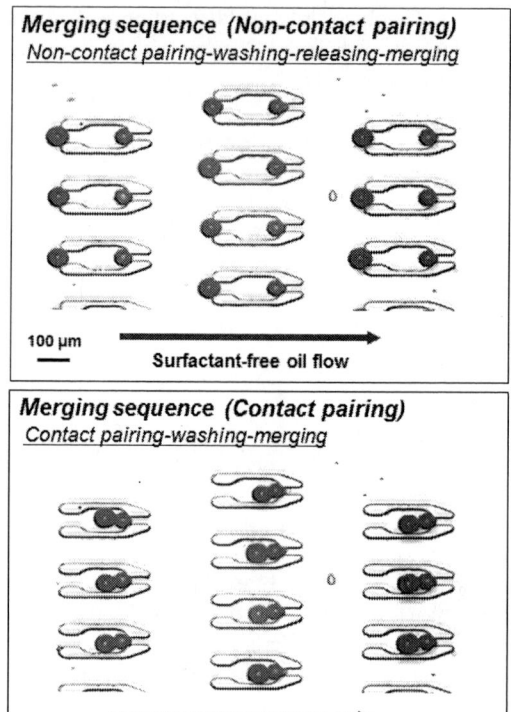

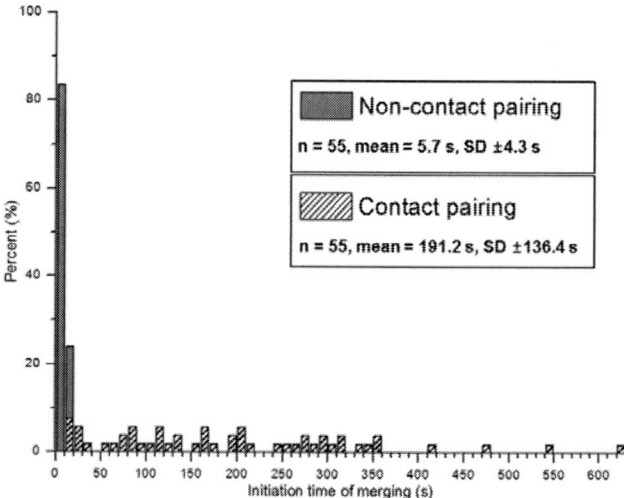

Figure 3: Two different pairing states.

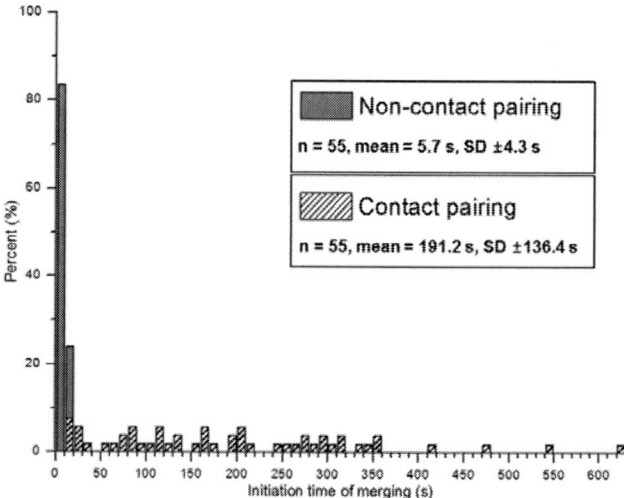

Figure 4: Frequency of the droplet merging event in an array format under the two pairing states.

the Laplace traps (i.e., partially desorbed surfactants). For this reason, the merging events in the array occurred randomly over a long period of time in the contact pairing state.

CONCLUSION

In this study, we have introduced and demonstrated a simple droplet merging method using the unique structure of the Laplace trap with a fluid flow control. The merging of two different droplets was performed through the multi-step "trapping–releasing–non-contact pairing–washing–and–merging" process. By exploiting the pressure balance between the hydrodynamic pressure difference and the Laplace pressure difference, the trap-and-release of a droplet could be controlled. The non-contact pairing state of two droplets facilitates the parallel merging of two droplets within a short time variation ($n = 55$, mean = 5.7 s, SD ±4.3 s) compared to the conventional contact pairing method ($n = 55$, mean = 191.2 s, SD ±136.4 s). We believe that the presented merging device can be used as a simple yet effective tool for quantitative biochemical analysis.

ACKNOWLEDGEMENTS

This work was supported by the National Research Foundation of Korea (NRF) Grant funded by the Korea government (MSIP) (No. 2011-0030075) and Basic Science Research Program through the National Research Foundation of Korea (NRF) funded by the Ministry of Education, Science and Technology (2012R1A1A2006305).

REFERENCES

[1] S. Y. Teh, R. Lin, L. H. Hung, A. P. Lee, "Droplet microfluidics", *Lab Chip*, vol. 8, pp. 198–220, 2008.

[2] C. Priest, S. Herminghaus, R. Seemann, "Controlled electrocoalescence in microfluidics: targeting a single lamella", *Appl. Phys. Lett.*, vol. 89, 134101, 2006.

[3] X. Niu, S. Gulati, J. B. Edel, A. J. deMello, "Pillar-induced droplet merging in microfluidic circuits", *Lab Chip*, vol. 8, pp. 1837–1841, 2008.

[4] B. C. Lin, Y. C. Su, "On-demand liquid-in-liquid droplet metering and fusion utilizing pneumatically actuated membrane valves", *J. Micromech. Microeng.*, vol. 18, 115005, 2008.

[5] L. M. Fidalgo, C. Abell, W. T. S. Huck, "Surface-induced droplet fusion in microfluidic devices", *Lab Chip*, vol. 7, pp. 984–986, 2007.

[6] A. M. Huebner, C. Abell, W. T. S. Huck, C. N. Baroud, F. Hollfelder, "Monitoring a reaction at submillisecond resolution in picoliter volumes", *Anal. Chem.*, vol. 83, pp. 1462–1468, 2011.

[7] E. Fradet, C. McDougall, P. Abbyad, R. Dangla, D. McGloin, and C. N. Baroud, "Combining rails and anchors with laser forcing for selective manipulation within 2D droplet arrays", *Lab Chip*, vol. 11, pp. 4228–4234, 2011.

[8] J.C. Baret, "Surfactants in droplet-based microfluidics", *Lab Chip*, vol. 12, pp. 422–433, 2012.

[9] Y. Xia, G. M. Whitesides, "Soft lithography", *Annu. Rev. Mater. Sci.*, vol. 28, pp. 153–184, 1998.

[10] B. Lee, J. Y. Yoo, "Droplet bistability and its application to droplet control", *Microfluid Nanofluid*, vol.11, pp. 685–693, 2011.

[11] W. Wang, C. Yang, C. M. Li, "On-demand microfluidic droplet trapping and fusion for on-chip static droplet assays", *Lab Chip*, vol. 9, pp. 1504–1506, 2009.

[12] T. Thorsen, R. W. Roberts, F. H. Arnold, S. R. Quake, "Dynamic pattern formation in a vesicle-generating microfluidic device", *Phys. Rev. Lett.*, vol. 86, pp. 4163–4166, 2001.

CONTACT

* J. Kim, tel: +82-542792185; joonwon@postech.ac kr

MINIATURE CIRCULATORY COLUMN SYSTEM FOR GAS CHROMATOGRAPHY

Hao-Chieh Hsieh and Hanseup Kim
University of Utah, Salt Lake City, Utah, USA

ABSTRACT

This paper presents the first micro-scale circulatory column system that enables the extension of the effective column length through the circulatory loop without increasing the device volume, thus ultimately achieving ultra-high separation capacity. The circulatory column system consists of two 25-cm micro columns and six mini valves to produce circulatory chromatographic separation. The fabricated column system (1) demonstrated the highest theoretical plate number per length-pressure of 206plates/m-kPa ever reported, enhancement in the detection capacity by 66% compared to the literature; (2) accomplished the longest effective micro column length of 5 meters ever reported; and (3) demonstrated the first micro-scale circulatory separation of target gas molecules up to 10 turns through a set of two 25-cm micro columns.

INTRODUCTION

Increasing attentions have been paid to the air quality monitoring because the impacts of air pollution to public health have critically contributed to the risk factors of diseases, such as cancers, malnutrition in children and various cardiopulmonary diseases. The Environmental Protection Agency (EPA) reported that the total death of 200,000 was presumably caused by the air pollution in 2010 in US only, and in worldwide 6.8 million deaths were estimated as attributable to the air pollution [1].

Among various air quality monitoring methods, the gas chromatograph (GC) has been regarded to be one of the most effective methods while its miniaturization has been rigorously pursued recently [2,3]. The main advantage of gas chromatograph (GC) is its capability to identify multiple gaseous compounds per analysis by incorporating the 'pre-separation' or 'chromatographic separation' step before sensing, ultimately achieving a wide range of detection capacity. Such a pre-separation step requires a separation column where different types of analysts are forced to race along a path until their becoming separated from each other. The miniaturization of a separation column has particularly been rigorously researched up to date [3,5], achieving comparable separation efficiency, represented by the plate number, to those of macro scale columns despite a tiny size and low power consumption.

However, the miniaturization of the 'high-capacity' gas chromatography system with all necessary components, such as a column, a sensor, and a pump, has not been achieved yet mainly due to the inevitable trade-offs between the detection capacity and the pumping limitations in the micro domain. To maximize the detection capacity, or the total number of detectable gas compounds, the length of a column needs to be maximized providing farther isolation among targets. This, however, would require higher pressure from a pump to flow gaseous compounds into the tiny micro column at the optimal flow rate, which is currently beyond the capacity of the best micropumps [4] reported yet. Thus, with current micropump technology, the miniaturized micro GC systems still remain as a low-capacity option [5,6].

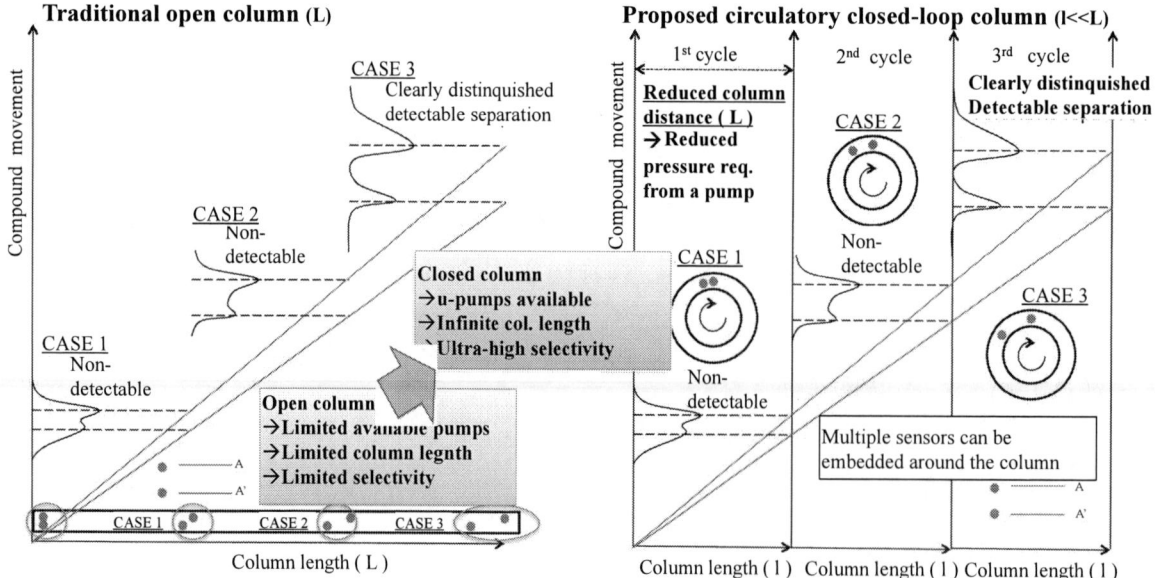

Figure 1. Working Principle of the proposed system and advantages compared to the conventional GC column.

978-1-4799-3510-9/14 $31.00 © 2014 IEEE

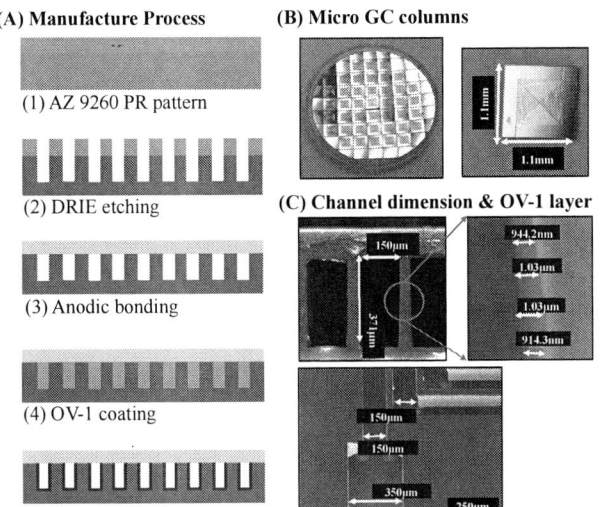

(A) Manufacture Process

(1) AZ 9260 PR pattern

(2) DRIE etching

(3) Anodic bonding

(4) OV-1 coating

(5) Functionalization

(B) Micro GC columns

(C) Channel dimension & OV-1 layer

Figure 2. (A) Manufacture process. (B) Appearances of micro GC columns. (C) Images of micro channel cross-section with OV-1 layer and outlet dimension.

To address such issues of the current state-of-art micro GC systems, one needs to achieve the enhanced length of the separation column operating within the limitations of the micropump technology. A circulatory column can extend an effective length of a column through circulation while requiring minimal fluidic pressure.

This paper reports the first micro-scale circulatory GC column system: concept, implementation and testing results. Particularly, it discusses the operation principle, column characterization, circulatory flow control results and the circulatory chromatographic separations results.

OPERATION PRINCIPLE

Figure 1 illustrates the operation principle of the developed circulatory column system. A conventional linear column achieves the separation of gas molecules (red and blue in Fig.1-(left)) when they travel a total distance of "L". The circulatory column consists of a circular loop with a short length of e.g. "L/3". It provides the same length of "L" by continuously circulating the gas molecules three times achieving the full separation (Fig.1-(right)). Thus, the

circulatory column system dramatically reduces the pressure required because of the shorter column length, clearly indicating the possibility of incorporating micro-scale pumps for a complete high-capacity micro GC system.

MICRO GC COLUMN DESIGN
Fabrication

The micro GC columns were fabricated on a 4-inch (100mm diameter) silicon wafer (Fig.2A&2B). The silicon wafer was first coated with hexamethyldisilazane (HMDS) to increase the adhesion between the wafer and the 14μm thick AZ 9260 photoresist. To achieve a high-aspect ratio channel structure, the patterned wafers were processed with a deep reactive ion etching (DRIE) process using an Oxford 100 ICP machine (Oxford Instruments, UK). After the DRIE process, the wafer was cleaned with oxygen plasma to remove the remaining polymer, then piranha solution (1:3 mixture of 30% H_2O_2 and 98% H_2SO_4). To seal the open channel, the wafer and Pyrex 7740 glass wafer were bonded by anodic bonding process at 350oC, 1000V condition in a 520 IS bonding machine (EVG Group, Australia). The combined wafer was diced into individual micro GC columns with DAD641 dicing machine (DISCO, Japan) and cleaned with Acetone/IPA to remove the debris produced during dicing. Each GC column was examined under a microscope to confirm that there is no defect in the channel walls, and that the bonding is completed between the glass and the silicon wafer, before the procedure of the OV-1 coating. The total footprint of each diced micro GC column was measured as $110 \times 110 mm^2$, containing a 25cm long spiral-shaped micro channel with a cross section of $150 \times 370 \mu m^2$ (Fig.2B). The inlet and outlet ports were 350μm wide, 370μm deep and 2mm long; and connected to fused silica capillary tubing with an OD of 360μm and ID of 250μm (Fig.2C)

Stationary Phase Coating

Nonpolar OV-1 polymer (OHIO VALLEY, USA) was chosen for the stationary phase in the micro GC column because it has a wide capability to separate various targets, for example from hydrocarbons to amine compounds. The OV-1 polymer was first dissolved in pentane and diluted.

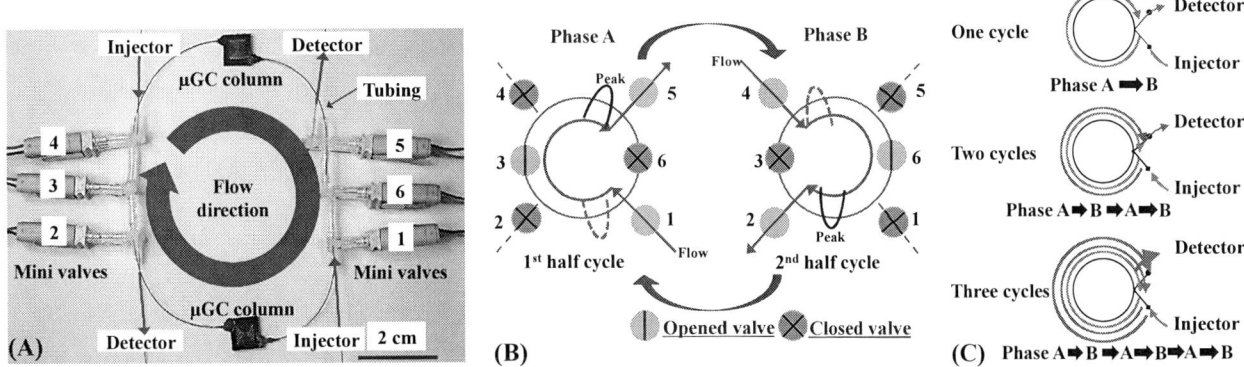

Figure 3. (A) Schematic view of the circulatory column system. (B) Operation timing for flow circulation, consisting of two half-perimeter flow circulation per timing. (C) Flow circulation paths for different cycles.

Table 1. Optimization of the measured separation efficiency of the fabricated micro GC columns at different flow rates and OV-1 thicknesses.

Flow rate (sccm)	OV-1 Stationary Phase Thickness (μm)					
	0.1	**0.2**	**0.4**	**0.8**	**1.1**	**1.5**
0.1	2315	526	3740	12720	1273	5048
0.2	1897	1510	4833	7911	1302	2393
0.4	630	639	2041	4436	704	1572
0.8	395	531	1285	2276	344	666
1.0	405	299	1115	1626	373	624

Then the solution was immediately filled into the columns by a syringe to avoid concentration changes, since pentane easily evaporates. The columns were dried overnight being placed in a vacuum oven at 60oC. After the OV-1 coating, the columns were installed in the Thermo FOCUS GC and heated with a ramp temperature program from 40oC to 150oC (ramping rate 5oC/min), and maintained at a fixed temperature of 150oC for 2 hours. The nitrogen flow rate of 1mL/min (sccm) was applied to completely remove the remaining solvent.

The OV-1 layer thickness varied at different points on the sidewall. It is assumed that such differences may have been caused by the varying evaporation rates at each location inside the micro GC chip. Since the inlet port was sealed with paraffin, the only exit for the solvent vapors was the outlet port. Thus, each location experienced different lengths of an evaporation path, resulting in variable thicknesses in relation to the distance from the outlet port.

Characterization

To evaluate the DRIE process and OV-1 coating uniformity, FEI Quanta 600 FEG scanning electronic microscope (FEI, USA) was applied. Micro GC columns were cut into half with diamond blade and cleaned with nitrogen gas for the observation of the DRIE etching angle and OV-1 thickness from the cross-section.

To optimize the separation efficiency of the fabricated micro column in a linear form, the relationship between the OV-1 thickness and nitrogen flow rates were investigated. As the standard test to calculate the theoretical plate number, 0.1μL of alkane mixture of decane ($C_{10}H_{22}$), dodecane ($C_{12}H_{26}$), tetradecane ($C_{14}H_{30}$) and hexadecane ($C_{16}H_{34}$) in a hexane solvent, was injected into the column and separated with a ramping temperature from 40°C to 150°C at a ramping rate 40°C/min. The peak of hexadecane was selected to calculate the plate number.

The measurement results showed that the highest column efficiency was measured as 12,720plates/m at the OV-1 thickness of 0.8μm and the flow rate of 0.1sccm (Table1). The measurement also showed that the column efficiency reached 28053plates/m for two columns connected in series. These plate numbers were sufficiently high for the testing of GC functions.

FLOW SYSTEM DESIGN
Flow Direction Control

To enable the circulatory gas flows, the flow control system employed (1) a set of the two fabricated micro columns of 25cm length and (2) six miniature solenoid control valves (S070 SMC, USA), as illustrated in Figure 3(A). The two micro GC columns were connected through capillary tubes forming a closed-loop circulatory flow path. The circulatory path was connected and disconnected at six locations by the six valves. Two valves alternatively opened and closed at the two inlet flow paths from the injector of the commercial Thermo FOCUS GC system; another two valves at the two outlet flow paths to the FID detector; and the last two valves at the two locations of the circular path. These six valves were actuated in two groups and two phases, creating a circulatory gas flow in a, for example, clockwise direction (Fig.3B). The timing control was precisely determined to create as many cycles as needed for chromatographic separation (Fig.3C). To efficiently manipulate the six valves, both the 8-channel relay module (SainSmart, USA) and the Arduino MEGA microcontroller

(Arduino, Italy) were connected to the valves. The valve

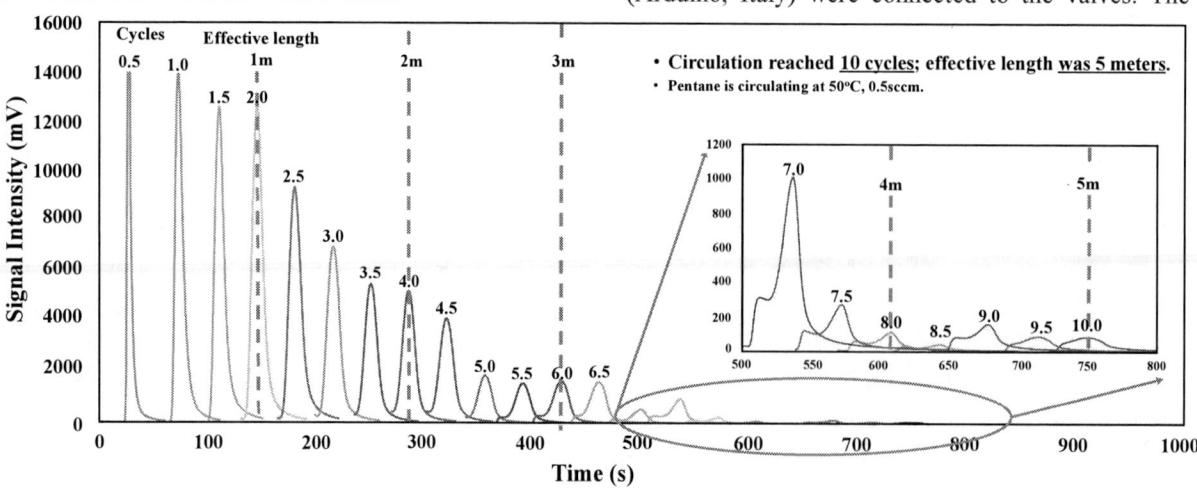

Figure 4. Demonstration of multiple circulations from 0.5 to 10.0 cycles at every 0.5 cycle.

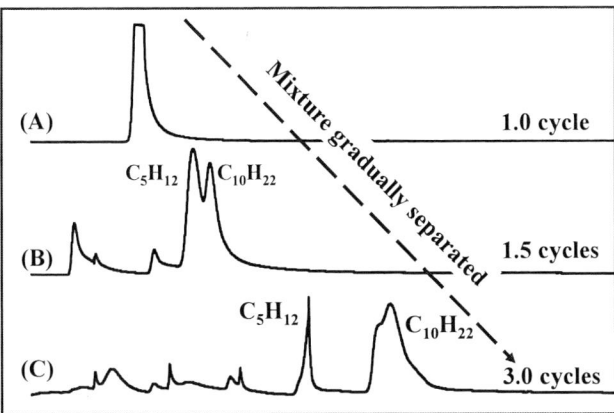

Figure 5. Circulatory separation of pentane and decane mixture after (A) 1.0 cycle, (B) 1.5 cycles, and (C) 3.0 cycles with nitrogen flow rate at 0.5mL/min with $50^{\circ}C$ isothermal condition.

timing was programmed with Arduino software and uploaded to the Arduino Mega microcontroller. The valve control timing was empirically determined by estimating the peak positions inside the column.

TESTING AND RESULTS

To demonstrate the circulation of the target molecules, the peak of pentane was monitored through the FID as they were circulated every 0.5 cycle up to 10 cycles. The volume of injected pentane was 0.1μL, and the operation was performed at a 50oC isothermal condition. For circulation, the six valves were alternately turned on and off every 33.0±2.0 seconds.

To demonstrate the circulatory separation, the width between the two peaks, respectively, of pentane and decane were monitored while the mixture was circulated every 0.5 cycles. The carrier gas was nitrogen at a flow rate of 0.5mL/min. The resultant chromatograms were analyzed to calculate the self-defined Figure of Merit (FOM), plate number/meter/pressure, for comparison.

Maximum Achievable Circulation

The measurement results demonstrated that the circulatory cycles reached up to 10 cycles, which is equal to the 5-meter effective column length, a record from any chip-scale micro columns (Fig.4). The measurement results also demonstrated the highest theoretical plate number per length-pressure of 206plates/m-kPa ever reported, which is an enhancement in the detection capacity by 66% compared to the reported numbers. The peak intensity decreased as the peak travelled through more cycles. The signal intensities were measured as 14000, 6000 and 81mV at the effective column lengths of 1, 2 and 5meters, which respectively corresponds to 2, 4 and 10 cycles.

Gas Compound Separation during Circulation

Figure 5 shows the successful circulatory separation of two closely located gaseous compounds of pentane (C5H12) and decane (C10H22). The measurements showed that the peaks of the pentane and decane mixture were not separated after 1 cycle, still appearing as a single peak. These peaks

started being split at 1.5 cycles; then further completely separated after 3.0 cycles with the separation time span of about 20seconds. This separation indicates the 'magnifying' ability of the circulatory GC system for higher gas selectivity. Note that this is the first micro-scale circulatory separation of target gas molecules up to 10 turns through a set of two 25-cm micro columns. The measured data also resulted in FOM was calculated as 206plates/meter-kPa, which is the highest among any reported GC column systems (both commercial and non-commercial).

DISCUSSIONS AND CONCLUSIONS

This paper reports the first micro-scale circulatory column system that enables the extension of the effective column length through the circulatory loop without increasing the device volume. The implemented circulatory column system consists of two 25-cm micro columns and six mini valves to produce circulatory chromatographic separation system. The fabricated column system (1) demonstrated the highest theoretical plate number per length-pressure of 206plates/m-kPa ever reported; (2) accomplished the longest effective micro column length of 5meters ever reported; and (3) demonstrated the first micro-scale circulatory separation of target gas molecules up to 10 turns through a set of two 25-cm micro columns.

ACKNOWLEDGEMENTS

This project is partially supported under the National Science Foundation (NSF) CAREER Award. The GC equipment became available through the Defense Advanced Research Project Agency (DARPA) Young Faculty Award (YFA) Program (N66001-11-1-4149).

REFERENCES

[1] S.S. Lim, et al., "A comparative risk assessment of burden of disease and injury attributable to 67 risk factors and risk factor clusters in 21 regions, 1990-2010: a systematic analysis for the Global Burden of Disease Study 2010", Lancet, v380, pp.2224-60, 2012.

[2] S.C. Terry, et al., "A Gas Chromatographic Air Analyzer Fabricated on a Silicon Wafer", IEEE Trans. on Electron Devices, v26, pp.1880-86, 1979.

[3] M. Agah, et al., "High-Performance Temperature-Programmed Microfabricated Gas Chromatography Columns", JMEMS, v14, pp.1039-50, 2005.

[4] H. Kim, et al., "A fully integrated high-efficiency peristaltic 18-stage gas micropump with active microvalves", IEEE MEMS 2007, Japan, pp. 131-4.

[5] J. Liu, et al., "Smart Multi-Channel Two-Dimensional Micro-Gas Chromatography for Rapid Workplace Hazardous Volatile Organic Compounds Measurement", LOC, v13, pp.818-25, 2013.

[6] H. Kim, et al., "A Micropump-Driven High-Speed Gas Chromatography system", Transducers '07, Lyon, France, Jun. 10-14, 2007, pp.1505-8.

CONTACT

haochieh.hsieh@utah.edu, hanseup@ece.utah.edu

MIRRORED ANODIZED DIELECTRIC FOR RELIABLE ELECTROWETTING

Supin Chen[1] and Chang-Jin "CJ" Kim[2]

[1]Bioengineering Department, [2]Mechanical and Aerospace Engineering Department,
University of California, Los Angeles (UCLA), USA

ABSTRACT

Anodized dielectrics in a mirrored arrangement are proposed and evaluated for the parallel-plate electrowetting-on-dielectric (EWOD) configuration. As valve metal oxides with current rectifying effects, anodized dielectrics previously could only be used for EWOD under a restricted range of voltages. However, in a mirrored configuration, one side of the anodized dielectric pair is expected to always be under the correct bias to restrict current. Both the mirrored and typical configurations were tested with current leakage measurements on anodized alumina samples under cycles of negative and positive bias and a range of electric fields. The mirrored configuration was effective in limiting current leakage over all voltages applied.

INTRODUCTION

Electrowetting-on-dielectric (EWOD)

Electrowetting decreases the observed contact angle when a voltage is applied between a substrate electrode and a liquid on it [1,2]. An insulating layer was added on the electrode to prevent electrochemical reactions between the liquid and electrode surface when voltages were applied [3], thus named electrowetting-on-dielectric (EWOD) [4]. The dielectric allows a higher electric field (and stronger electrostatic force) before electrical leakage or breakdown can occur. By choosing a dielectric material of a large dielectric constant and decreasing the thickness of the dielectric layer, the required voltage for a given electrowetting effect could be reduced [5]. When electrodes are arranged in a side-by-side pattern at the surface and voltages are sequentially applied, droplets can be moved over the electrodes. Because liquids can be manipulated as droplets (mixing, splitting, merging, etc.) simply by electric potentials [6], EWOD has been a useful technique for many biochemical lab-on-a-chip [7,8] and optical [9,10] applications as well as electronic [11] and mechanical [12].

An ideal dielectric for EWOD should have a high dielectric constant and high breakdown voltage and be easily deposited as a robust pinhole-free layer. Many dielectric materials have been used for EWOD including: photoresists, polydimethylsiloxane (PDMS), Parylene, silicon dioxide, silicon nitride, barium strontium titanite (BST), and bismuth zinc niobate (BZT). But none has been fully satisfactory, because there has generally been a tradeoff between ease of fabrication (e.g., low temperature deposition) and electrical performance (e.g., high dielectric constant, high breakdown voltage, pinhole free).

Anodized dielectrics for EWOD

Recently, progress has been made for CMOS compatible deposition of dielectrics with high dielectric constants by anodization [13]. Anodized metal oxides have several advantages as electrowetting dielectrics [14]. Because oxidation is faster along weaker insulation paths, the resulting dielectric layer is highly uniform with few pinholes. Also, metal oxides can be formed at room temperature with low cost, and their relatively high dielectric constants (8-110) lower EWOD actuation voltages, as the following electrowetting equation indicates:

$$cos\theta_v - cos\theta_0 = \frac{\varepsilon_0 \varepsilon_r}{2\gamma t} V^2 \qquad (1)$$

where θ_v and θ_0 are the contact angles with and without voltage, respectively, ε_0 is the vacuum permittivity, ε_r is the dielectric constant, γ is the surface tension of the liquid-gas interface, t is the dielectric thickness, and V is the applied voltage.

However, anodized dielectrics are also known as valve metal oxides (i.e., oxides grown from Al, Ta, Bi, Sb) because of their current rectifying effects, imposing a restricted range of EWOD voltages [14]. For anodized dielectrics, although the cathodic current is almost zero, anodic currents can be significant and increase with applied voltage [15].

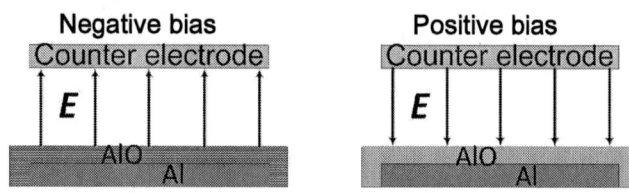

(a) Existing configuration

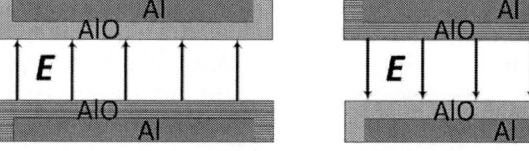

(b) Mirrored configuration

Figure 1: (a) Valve metal behavior for aluminum oxide (AlO) anodized from aluminum (Al). Under negative bias, the aluminum oxide (colored blue) blocks current. However, under positive bias, the aluminum oxide (colored red) passes current. (b) A mirrored configuration is proposed to assure one of the aluminum oxide layers is always under the correct bias to restrict current.

During EWOD operation, anodized dielectrics perform well under their formation bias; but under the reverse bias, the same anodized dielectrics pass current (Fig. 1(a)). EWOD devices with anodized Ta_2O_5 dielectric were found useful with negatively biased DC (negative potential applied to the droplet while the Ta electrode below the dielectric was grounded) or low frequency (<100 Hz) voltages, but deteriorated severely for positive DC or high frequency AC voltages [16].

Progress has been made to reliably use anodized dielectrics in EWOD devices. By sputtering Ta_2O_5 onto Ta before it was anodized, Huang et al. demonstrated a pinhole free dielectric that was insensitive to polarities [16]. Other work used valve metals for "self-healing" dielectrics [13], in which any exposed portion of the electrode was anodized during the EWOD operation instead of undergoing electrolysis. However, droplet composition and voltage restrictions were involved in the self-healing devices, which require electrolyte droplet solutions within a limited pH range and application of a voltage polarity suitable for anodization.

Mirrored configuration

In this report, we do not try to remove the rectifying effect of valve metal oxides, but instead include anodized dielectric on both plates of a parallel-plate EWOD device, so that one of the pair will always be under the correct (i.e., negative) bias to restrict current. This configuration is expected to enable the advantages of metal oxides as a dielectric for electrowetting without limiting the actuation voltages that can be applied for droplet movement.

The mirrored configuration (Fig. 1(b)) was tested by comparing its leakage current with that of the typical existing configuration (Fig. 1(a)), which utilizes dielectric on only one of the two parallel plates. Current leakages were measured while testing the breakdown voltage and lifetime of the two configurations.

EXPERIMENT
Alumina anodization

Aluminum (500 nm) was deposited by evaporation onto clean glass slides (0.7 mm). Alumina was grown from the aluminum by anodization in a room temperature 0.1 M solution of ammonium pentaborate in ethylene glycol using a constant current step of 0.5 mA and a constant voltage step of 50 V (for 68 nm thick alumina with 455 nm aluminum remaining) or 100 V for 1 hour (for 135 nm thick alumina with 410 nm aluminum remaining) [17].

Reference plates for the typical EWOD architecture were made by evaporating chrome/gold (20 nm/200 nm) onto glass slides. A hydrophobic layer of Cytop® (50 nm) was spin-coated onto all of the plates and patterned by lift-off to expose bare metal for electrical contact.

Current leakage measurement

Tests were performed in the parallel-plate EWOD configuration. The typical configuration refers to anodized alumina on one side and a reference plate on the other. In the mirrored configuration, both plates had anodized alumina.

Droplets were sandwiched between the plates with a gap of 80 μm during testing. The droplets consisted of a 1:1 mixture of glycerin (to prevent evaporation) and KCl in water standard solution (to ensure conductivity and exacerbate dielectric failure with its chloride) [18].

Both EWOD voltage actuation and leakage current measurements were performed by a sourcemeter (Keithley 2425). The Keithley was connected to a computer with a GPIB controller (GPIB-USB-HS, National Instruments) for control and data recording. For cycling tests, electrical connections to the plates were switched using photoMOS relays (AQW610EH PhotoMOS relay, Panasonic) controlled through a DAQ (NI USB-6255, National Instruments).

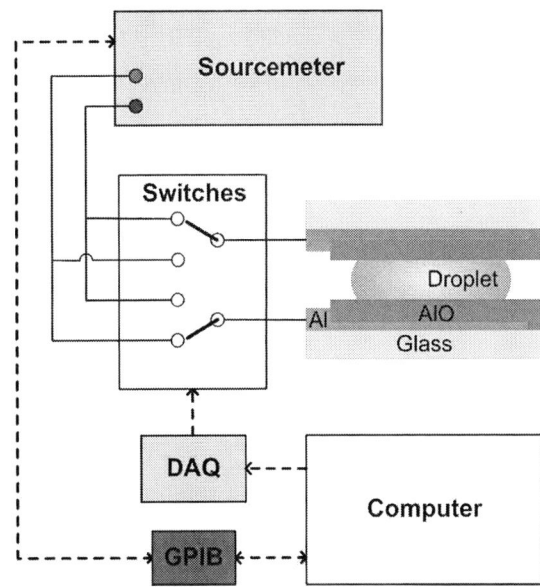

Figure 2: Setup scheme for current leakage measurement. The sourcemeter provided both voltage sourcing and low current measurement functions. It was programmed to stop supplying voltage if the measured current surpassed a limit of 50 mA. An open circuit current of 0.9 μA was measured in the setup before conducting tests.

RESULTS
Lifetime tests

In lifetime tests, 1 second actuation of 25 V was alternated with 1 second of 0 V for 1000 cycles. Current leakage was compared for three types of biases: positive bias, negative bias, and alternating bias (0.5 Hz).

Under positive bias, the leakage current climbed over repeated cycles, but the leakage current increase was not steady as electrolysis bubbles moved the droplet (Fig. 3). For negative bias, the leakage current was expected to decrease with repeated voltage cycling. However, the leakage current was too low to measure a noticeable decrease. For the 0.5 Hz alternating bias, no significant current change was measured.

978-1-4799-3510-9/14 $31.00 © 2014 IEEE

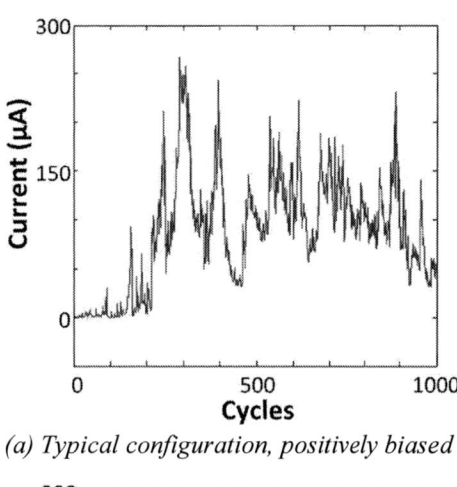

(a) Typical configuration, positively biased

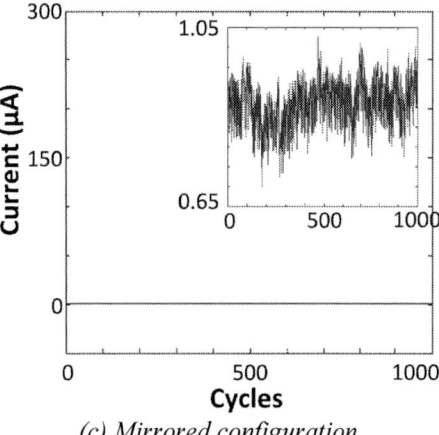

(b) Typical configuration, negatively biased

(c) Mirrored configuration

Figure 3: *(a,b) Typical configuration assessed, using a droplet on a single plate with anodic alumina. The leakage current was negligible for negative bias (b) but significant for positive bias (a). (c) For the mirrored configuration, the leakage current was negligible for both biases including an alternating bias (0.5 Hz).*

Breakdown tests

Currents were measured with respect to applied electric field. For electric fields less than 2000 V/cm (corresponding to half the anodization electric field), all of the measured currents were low. Above 2000 V/cm, the current

exponentially increased with electric field for positively biased devices. One side of each mirrored pair of dielectrics was also positively biased, but the current rectifying effect of its opposing plate restricted current from passing. However, because the leakage current is always low, the mirrored configuration does not exhibit the self-healing effect [13], as seen by the initially negative slope for the negatively biased data in Fig. 4.

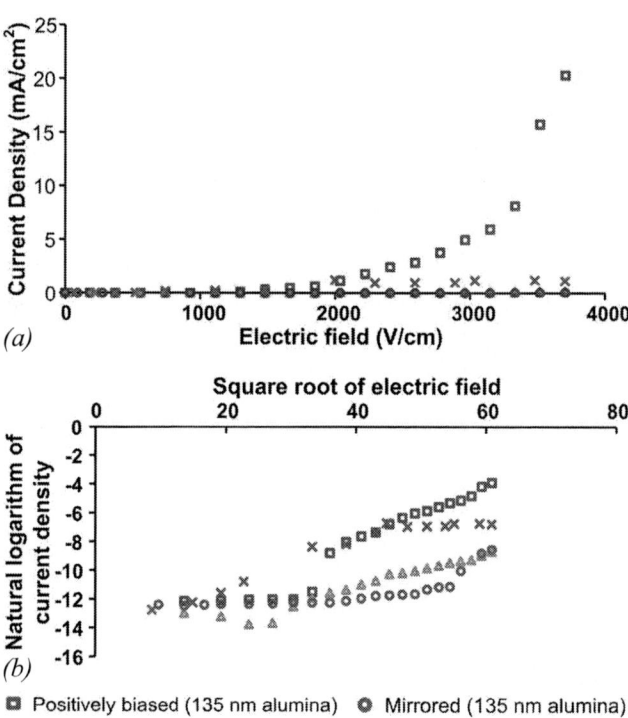

▣ Positively biased (135 nm alumina)   ◎ Mirrored (135 nm alumina)
▲ Negatively biased (135 nm alumina)   ✖ Mirrored (68 nm alumina)

Figure 4: *Current leakage measured as a function of electric field applied across typical EWOD and mirrored EWOD devices. (a) The mirrored devices show little leakage while the typical device shows an exponentially increasing leakage if positively biased. (b) A logarithmic graph of the same data to separate behavior of the mirrored devices from the negatively biased devices.*

CONCLUSION

By using anodized dielectrics in a mirrored arrangement, a parallel plate EWOD device that limits the current leakage has been proposed. Fabrication of the mirrored configuration device is simpler than adding sputtering step to deposit a valve metal oxide [16], although its use of an equal dielectric thickness on both plates is somewhat different from typical EWOD devices. In most parallel plate EWOD devices, the main dielectric of a necessary thickness is deposited entirely on the plate with patterned EWOD electrodes for simpler fabrication. The dielectric on the opposite plate is generally much thinner, and in some cases only consists of a hydrophobic top-coating.

Typically, the top plate serves as a reference electrode

978-1-4799-3510-9/14 $31.00 © 2014 IEEE   1013

for applying electric fields. It is common for the top plate electrode to be a transparent layer of conductive indium tin oxide (ITO) for visualization. Although it was not used in these tests, a transparent top plate can also be incorporated into the mirrored configuration if the valve metal thickness is sufficiently reduced during anodization. In a case in which an initial 105 nm thick aluminum was anodized at 100 V, the remaining aluminum after anodization was 15 nm thick and transparent.

Further work will investigate the performance of opposing valve metal anodized dielectrics in a coplanar EWOD configuration, where both the actuation and reference electrodes are fabricated on the same plate [19]. A coplanar EWOD device would only require anodization on one plate. Because the valve metal oxide dielectric would still cover both the actuation and ground electrodes, its rectifying current behavior should also oppose itself. With the assumption that there is no lateral current leakage within the dielectric, it is expected that there would be minimal current leakage during EWOD operation.

ACKNOWLEDGEMENTS

The authors would like to thank Lian-Xin Huang for helpful discussions on anodized dielectrics and EWOD and also Hoc Ngo and the UCLA Nanoelectronics Research Facility for the metal deposition. This work was supported in part by the Department of Energy [DE-SC0005056].

REFERENCES

[1] S. Hackwood and G. Beni, "Electrowetting Displays", *Applied Physics Letters*, vol. 38, pp. 207-209, 1981.

[2] W.C. Nelson and C.-J. Kim, "Droplet actuation by electrowetting-on-dielectric (EWOD): a review", *Journal of Adhesion Science and Technology*, vol. 26, pp. 1747-1771, 2012.

[3] B. Berge, "Électrocapillarité et mouillage de films isolants par l'eau (Electrocapillarity and wetting of insulator films by water)", *Comptes rendus de l'Académie des sciences, Série II*, vol. 317, pp. 157-163, 1993.

[4] J. Lee, H. Moon, J. Fowler, T. Schoellhammer, and C.-J. Kim, "Electrowetting and electrowetting-on-dielectric for microscale liquid handling", *Sensors and Actuators*, Vol. A95, pp. 259-268, 2002.

[5] H. Moon, S.K. Cho, R.L. Garrell, and C.-J. Kim, "Low voltage electrowetting-on-dielectric", *Journal of Applied Physics*, vol. 92, pp. 4080-4087, 2002.

[6] S.K. Cho, H. Moon, and C.-J. Kim, "Creating, transporting, cutting, and merging liquid droplets by electrowetting-based actuation for digital microfluidic circuits," *Journal of Microelectromechanical Systems*, vol. 12, pp. 70-80, 2003.

[7] M.J. Jebrail, M.S. Bartsch, and K.D. Patel, "Digital microfluidics: a versatile tool for applications in chemistry, biology, and medicine", *Lab on a Chip*, vol. 12, pp. 2452-2463, 2012.

[8] T.H. Lin and D.J. Yao, "Application of EWOD systems for DNA reaction and analysis", *Journal of Adhesion Science and Technology*, vol. 26, pp. 1789-1804, 2012.

[9] J. Heikenfeld, N. Smith, M. Dhindsa, K. Zhou, M. Kilaru, L.Hou, J. Zhang, E. Kreit, and B. Raj, "Recent progress in arrayed electrowetting optics", *Optics & Photonics News*, vol. 20, pp. 20-26, 2009.

[10] B.H.W. Hendriks, S. Kuiper, M.A.J. van As, C.A. Renders, and T.W. Tukker, "Electrowetting-based variable-focus lens for miniature systems", *Optical Review*, vol. 12, pp. 255-259, 2005.

[11] P. Sen and C.-J. Kim, "A Liquid-Solid Direct Contact Low-Loss RF Micro Switch", *Journal of Microelectromechanical Systems*, vol. 18, pp. 990-997, 2009.

[12] W. Nelson, H. P. Kavehpour, and C.-J. Kim, "A miniature capillary breakup extensional rheometer by electrostatically assisted generation of liquid filaments," *Lab on a Chip*, vol. 11, pp. 2424-2431, 2011.

[13] M. Dhindsa, J. Heikenfeld, W. Weekamp, and S. Kuiper, "Electrowetting without electrolysis on self-healing dielectrics", *Langmuir*, vol. 26m pp, 5665-5670, 2011.

[14] M. Mibus, C. Jensen, X. Hu, C. Knospe, M.L. Reed, and G. Zanragi, "Dielectric breakdown and failure of anodic aluminum oxide films for electrowetting systems", Journal of Applied Physics, vol. 22, pp. 253-255, 2013.

[15] M.M. Lohrengel, "Thin anodic oxide layers on aluminum and other valve metals: high field regime", *Materials Science and Engineering*, vol. R11, pp. 243-294, 1993.

[16] L.X. Huang, B. Koo, and C.-J. Kim, "Sputtered-anodized Ta_2O_5 as the dielectric layer for electrowetting-on-dielectric", *Journal of Micro-electromechanical systems*, vol. 22, pp. 253-255, 2013.

[17] C. Crevecoeur and H.J. de Wit, "The growth of anodic aluminum oxide layers after a heat-treatment", *Journal of the Electrochemical Society*, vol. 121, pp. 1465-1474, 1974.

[18] B. Raj, M. Dhindsa, N.R. Smith, R. Laughlin, and J. Heikenfeld, "Ion and liquid dependent dielectric failure in electrowetting systems", *Langmuir*, vol. 25, pp. 12387-12392, 2009.

[19] U.-C. Yi and C.-J. Kim, "Characterization of electrowetting actuation on addressable single-side coplanar electrodes", *Journal of Micromechanics and Microengineering*, vol. 16, pp. 2053-2059, 2006.

CONTACT

S. Chen, tel: +1-310-825-3977; supinchen@ucla.edu

NANOPARTICLES SORTING AND ASSEMBLY BASED ON DOUBLE-AXICON IN AN OPTOFLUIDIC CHIP

Y. Z. Shi[1,2], S. Xiong[2], L. K. Chin[2], M. Ren[2] and A. Q. Liu[1,2]

[1]State Key Laboratory for Manufacturing Systems Engineering, Xi'an Jiao Tong University, Xian 710049, CHINA

[2]School of Electrical and Electronic Engineering, Nanyang Technological University, SINGAPORE 639798

ABSTRACT

This paper presents a novel optofluidic system for nanoparticle sorting by using interference patterns generated through a double-axicon. The tightly confined Bessel beam is used to sort the 200-nm and 500-nm polystyrene nanoparticles massively and simultaneously by adjusting the flow rate and the laser power. Additionally, 2-μm polystyrene particles are assembled into a 2D array by utilizing the discrete interference pattern. This system first utilizes the interference patterns based on the on-chip double-axicon, and integrates the sorting and assembly abilities into a single chip. It has a great potential in bacterial and DNA sorting and cell assembly.

INTRODUCTION

Sorting and precise manipulating on micro/nanoparticles and biosamples such as cells and molecules in flowing streams is critical in the realm of biological and chemical analysis [1-4]. Optical tweezers, known as versatile and noninvasive tools, are able to manipulating particles from hundreds of micrometers to several nanometers efficiently [5, 6]. Conventional single beam optical tweezers are well known for sorting cells efficiently. They are also demonstrated to be capable of trapping and assembling nano-spheres and wires into 2D patterns one by one [5]. However, this assembly requires precise alignment and time-consuming manual works. Holographic optical tweezers have shown particularly success in sorting cells and assembling them into a designed array [7, 8]. Yet the sorting is only applicable to the microparticles, because of the small distinct gradient forces and relatively large hydrodynamic drag force on the nanoparticles. Besides, the assembly is only conducted in the still environment, and the samples are stick to the substrate, which causes samples contamination and measurement error. Recently, a novel optical switch is proposed to sort out two different cells with a high purity and yields [9]. The florescence of the samples are first detected and analyzed in the detecting zone, those with the chosen florescence signals will be pushed to a different outlet by the optical forces. This optical switch is widely used in the micrometer-sized biological sample separation. Meanwhile, many novel interference patterns in the optofluidic chips are used to sorting and assembly micro and nanoparticles. For example, Bessel beams and Laguerre-Gaussian beams are used to achieve 1D nanoparticle array in still water and flow stream, respectively [10, 11]. Nanoparticles and λ-DNA are trapped and transported on the "slot waveguide" by using evanescent wave [12].

DESIGN AND WORKING PRINCIPLE

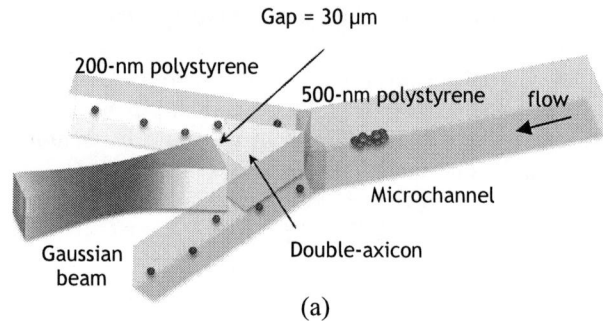

(a)

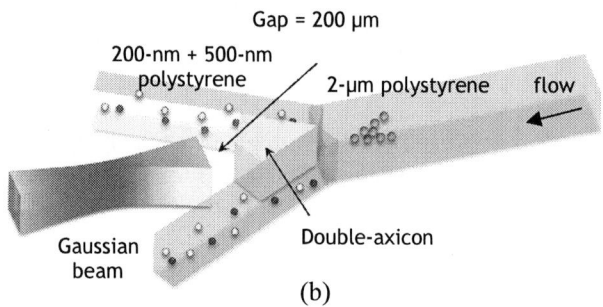

(b)

Figure 1: Illustration of optofluidic platform for nanoparticles sorting and assembly. (a) When gap is 30 μm, Qausi-Bessel beam is generated to retain large nanoparticles in the microchannel, whereas small nanoparticles are washed away by the flow. (b) When gap is 200 μm, the discrete interference pattern is formed to assembly large particles into a 2D array, whereas smaller particles are washed away.

The working principle of the optofluidic platform for particles sorting and assembly is illustrated in Fig. 1. The optofluidic chip consists of one inlet and two outlets. The aqueous suspension of particles is injected into the microchannel from the inlet. Light (532 nm) from a fiber (NA = 0.12) are coupled into the microchannel though a double-axicon optical elements. By increasing the gap between the fiber and double-axicon, the light beam can be switched from Bessel profile to different discrete interference patterns. When the gap is 30 μm, the interference pattern from the double-axicon is the quasi-Bessel beam. 500-nm polystyrene nanoparticles are trapped in the microchannel in

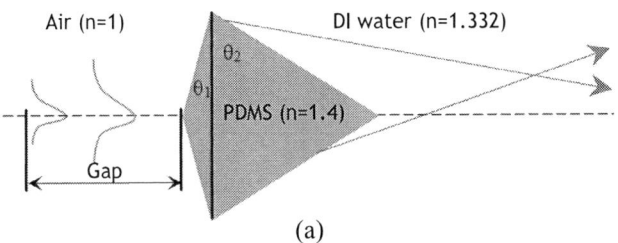

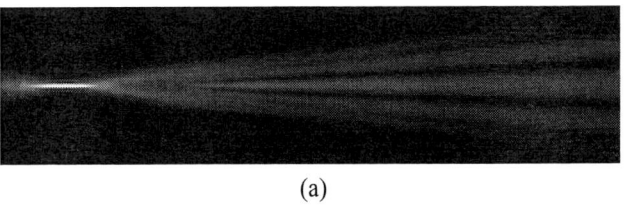

(a)

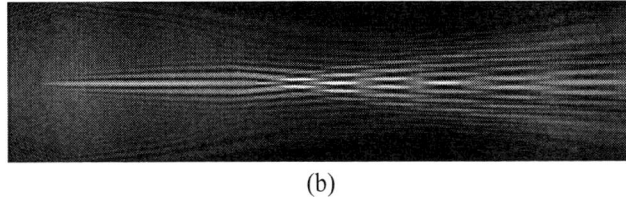

(b)

Figure 3: Simulated optical fields. (a) When the gap is 30 μm, a quasi-Bessel beam is generated. (b) When the gap is 200 μm, a discrete interference pattern is generated.

[13]. It is noticed that the length of the single light spot becomes larger as the light propagates.

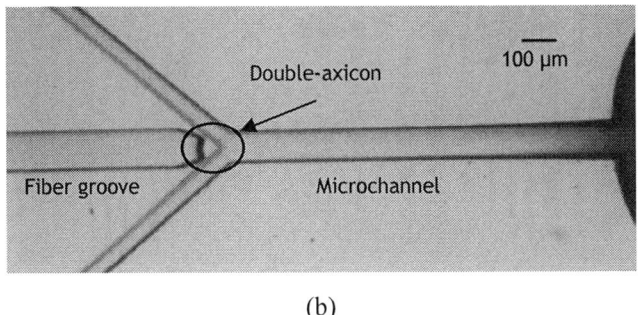

(b)

Figure 2: (a) Geometry of the double-axicon illuminated by a Gaussian beam. (b) Photography of the optofluidic chip.

the balance of the optical extinction forces and fluidic drag forces. Meanwhile, the 200-nm polystyrene nanoparticles are washed away by the flow, because they are seldom influenced by the optical field. When the gap is increased to 200 μm, the interference pattern switches to the discrete interference pattern. Large particles (e.g. 2-μm particles) will be assembled into a 2D array in the synergy of the optical extinction forces, optical gradient forces and fluidic drag forces. Nevertheless, the smaller polystyrene nanoparticles (e.g. 200-nm and 500-nm particles) are flushed out.

Figure 2(a) shows the geometry of a double-axicon illuminated by a Gaussian beam. The open angles (θ₁ and θ₂) of the two axicons are 8° and 55°, respectively. The lens length (from the left edge to the right edge) is 52 μm. The Gaussian beam is generated from the laser with a fiber coupled into the double-axicon. Figure 2(b) shows the fabricated optofluidic chip which is fabricated by polydimethylsiloxane (PDMS) utilizing the standard photolithography procedure.

DISTINCT INTERFERENCE FIELDS

The optical field in the microchannel is simulated based on the geometry in Fig. 2. Distinct interference patterns are generated by changing the gap between the optical beam waist and the double-axicon. Figure 3(a) shows that a Bessel profile beam is formed when the gap is 30 μm. Unlike the non-diffraction property of the Bessel beam, this beam diffracts slowly. The divergence of the beam is much smaller than that from the original low-NA (0.12) fiber. It is termed as quasi-Bessel beam. When the gap is increased to 200 μm, a discrete interference pattern emerges. This kind of discrete interference pattern is similar to the pattern in the optofluidic waveguide induced by the diffusion between two liquids

OPTICAL FORCES AND HYDRONAMIC DRAG FORCE ANALYSIS

Optical forces mainly consist of the optical scattering force and the optical gradient force. The optical scattering force transfers the momentum of light to the particles, pushing particles as light propagates. The optical gradient force tends to draw particles into the maximum of the electrical field.

For Rayleigh particles, the optical forces can be expressed as [2]

$$\mathbf{F}_{grad}(\mathbf{r}) = 2\pi n_2^2 \varepsilon a^3 \left(\frac{m^2 - 1}{m^2 + 1} \right) \nabla \left| \mathbf{E}^2(\mathbf{r}) \right|, \quad (1)$$

$$\mathbf{F}_{scat}(\mathbf{r}) = \frac{n_2}{c} \frac{128\pi^5 a^6}{3\lambda^4} \left(\frac{m^2 - 1}{m^2 + 2} \right)^2 I(\mathbf{r}), \quad (2)$$

where $\mathbf{r}$ is the position vector, ε is the dielectric constant in the vacuum, c is the speed of light, m is the ratio between the refractive index of the particle and that of the surrounding media n_2, and a is the radius of the particle. For particles in the Mie regime (e.g. 2-μm particles), the generalized Lorenz–Mie theory (GLMT) is invoked. The optical force can be expressed as

$$\mathbf{F}(\mathbf{r}) = \frac{n_2}{c} I_0 \left[\hat{x} C_{pr,x}(\mathbf{r}) + \hat{y} C_{pr,y}(\mathbf{r}) + \hat{z} C_{pr,z}(\mathbf{r}) \right], \quad (3)$$

Figure 4 shows the calculated results of the optical forces induced velocity on 200 and 500-nm polystyrene nanoparticles. 200-nm polystyrene particles experience much smaller optical scattering force. Thus they will be washed away by controlling a proper flow rate. At the same time, 500-nm polystyrene particles, which are more susceptible to the optical field, will retain in a position where optical forces and fluidic drag force balance. The light intensity distribution in the direction that perpendicular to the

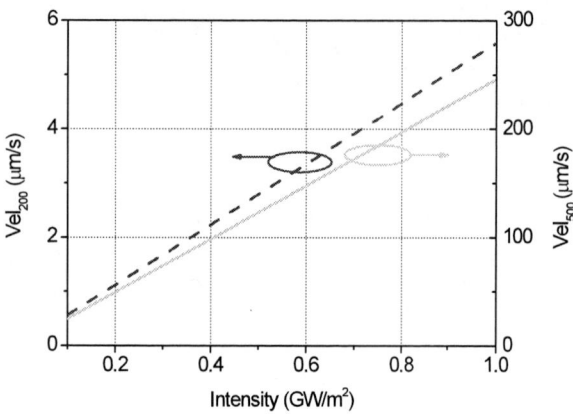

Figure 4: Optical forces on 200-nm polystyrene (a) and 500-nm polystyrene (b) nanoparticles induced velocity.

light propagating direction can be fitted well with Gaussian profile, forming an optical potential well. The form factor of a single Gaussian function is expressed as

$$f(\vec{r}) = \alpha a \exp\left(-\frac{\vec{r}^2}{2a^2}\right), \qquad (4)$$

where a is the particle radius and $\alpha = 2\pi \frac{n_2}{c}\left(\frac{n_2^2 - n_1^2}{n_1^2 + 2n_2^2}\right)$.

Thus the effective potential is expressed as

$$V(\vec{r}) = -I(\vec{r}) \otimes f(\vec{r}) = -\int f(\vec{x} - \vec{r}) I(\vec{x}) d^2 x \ . \qquad (5)$$

Hence, the gradient force in the light propagating direction is

$$F_{grad} = \vec{\nabla}\left(V(\vec{r})\right) = \vec{\nabla}\left[-\left(I(\vec{r}) \otimes f(\vec{x})\right)\right]$$

$$= 2\pi\alpha I \frac{a^3 \omega}{\sigma^3(a)} x \exp\left(-\frac{(x - x_1)^2}{2\sigma^2(a)}\right) \qquad (6)$$

The hydrodynamic drag force on a particle is caused by the flow velocity difference between the flow and the particle. It can be expressed as

$$F_{drag} = 6\pi\eta v a \ , \qquad (7)$$

where η is the viscosity of the liquid, v is the velocity difference between the flow and the particles, and a is the radius of the particle.

EXPERIMENTAL RESULTS

The aqueous suspension which carries particles was injected into the microchannel using a syringe pump (Genie, Kent Scientific Corporation, CT, USA). The flow velocity is set to 200 μm/s. The simulation results show that the optical scattering force on the 200-nm polystyrene particles is much smaller than the hydrodynamic drag force, which is almost impervious to the optical fields. However, the 500-nm polystyrene particles experience much larger optical scattering forces, which can be balanced against the hydrodynamic drag force at an equilibrium position.

Figure 5 shows that the 200-nm and 500-nm polystyrene nanoparticles are under bidirectional sorting in the optofluidic chip with a gap of 30 μm. The 200-nm

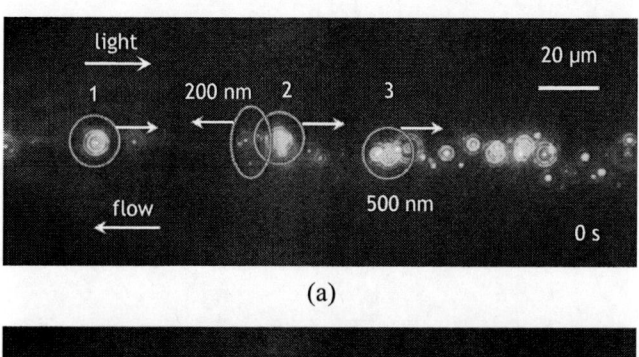

(a)

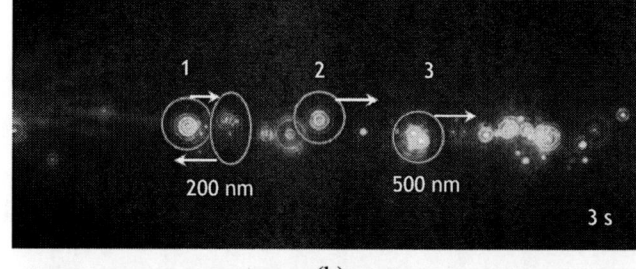

(b)

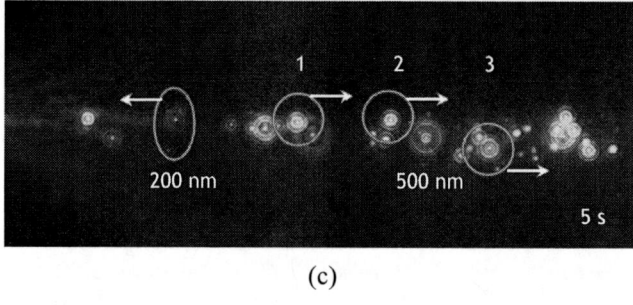

(c)

Figure 5: 200-nm and 500-nm polystyrene nanoparticles under bidirectional sorting.

polystyrene nanoparticles which experience much smaller optical forces flow to the left driven by the flow. At the same time, the 500-nm polystyrene nanoparticles are pushed back by the optical forces, and eventually retain in an equilibrium position. The optical gradient force is responsible for the confinement of particles in the middle of the light spot.

When the gap is increased to 200 μm, the vertical confinement of light in the vertical direction is weaker, resulting in the smaller optical gradient force. In order to assemble the particles in the optical pattern, the laser power is increased to 500 mW, which generates strong enough optical gradient force to conquer the Brownian motion.

Figure 6 shows a 2D array formed by 2-μm polystyrene particles in the flow. The patterned particles are almost trapped at the same position when the 500-nm nanoparticles are flushed away. This is the result of the equilibrium of the optical scattering force and the fluidic drag force in the horizontal direction and the confinement due to the optical gradient force in the perpendicular and vertical directions. The equilibrium positions are corresponding to the hot spot in the discrete interference pattern as shown in Fig. 3(b). If the solute is dilute enough, there will be only one particle occupy one hot spot in the optical pattern. This phenomenon has a great potential on single cell analysis.

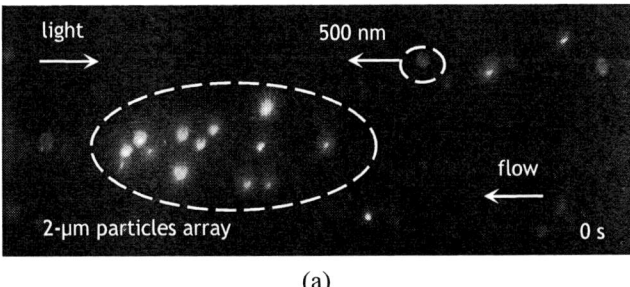

(a)

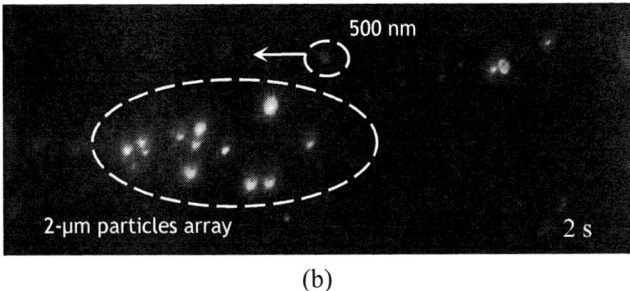

(b)

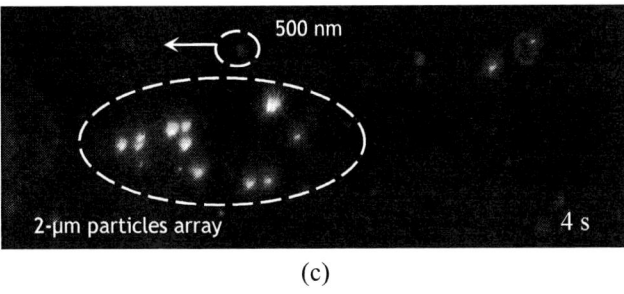

(c)

Figure 6: Assembly of 2-μm polystyrene particles into a 2D array in a flow stream.

CONCLUSIONS

In conclusion, an optofluidic chip with double-axicon is developed for sorting and assembling nanoparticles. The 200-nm and 500-nm polystyrene nanoparticles are massively and simultaneously sorted in the quasi-Bessel beam by controlling the flow rate and the laser power. The 2-μm polystyrene particles are successfully assembled into a 2D array in the discrete interference pattern in the flow stream. This optofluidic system can be applied to the single molecules sorting and single cells analysis and patterning.

ACKNOWLEGEMENT

This work is supported by the Environmental and Water Industry Development Council of Singapore (Research

project Grant No. 1102-IRIS-05-02).

REFERENCES

[1] C. Bustamante, Z. Bryant, S. B. Smith, "Ten years of tension: single-molecule DNA mechanics", *Nature,* vol. 421, pp. 423-427, 2003.

[2] Ashkin, J. M. Dziedzic, J. E. Bjorkholm, S. Chu, "Observation of a single-beam gradient force optical trap for dielectric particles". *Opt. Lett.,* vol. 11, pp. 288–290, 1986.

[3] Ashkin, J. M. Dziedzic, T. Yamane, "Optical trapping and manipulation of single cells using infrared laser beams", *Nature,* vol. 330, pp. 24-31, 1987.

[4] M. L. Juan, M. Righini, R. Quidant, "Plasmon nano-optical tweezers", *Nature Photon.,* vol. 5, pp. 349-356, 2011.

[5] K. C. Neuman, S. M. Block, "Optical trapping", *Rev. Sci. Instrum.,* vol. 75, pp. 2787-2809, 2004.

[6] K. C. Neuman, A. Nagy, "Single-molecule force spectroscopy: optical tweezers, magnetic tweezers and atomic force microscopy", *Nature Meth.,* vol. 5, pp. 491-505, 2008.

[7] M. P. MacDonald, G. C. Spalding, K. Dholakia, "Microfluidic sorting in an optical lattice", *Nature,* vol. 426, pp. 421–424, 2003.

[8] J. E. Curtis, B. A. Koss, D. G. Grier, "Dynamic holographic optical tweezers", *Opt. Commun.,* vol. 207, pp. 169-175, 2002.

[9] M. M. Wang et al., "Microfluidic sorting of mammalian cells by optical force switching", *Nature Biotechnol.,* vol. 23, pp. 83–87, 2005.

[10] T. Čižmár, M. Šiler, P. Zemánek, "An optical nanotrap array movable over a milimetre range", *Appl. Phys. B.,* vol. 84, pp. 197-203, 2006.

[11] T. Cizmar, V. Kollarova, Z. Bouchal, P. Zemanek, "Sub-micron particle organization by self-imaging of non-diffracting beams", *New J. Phys.,* Vol. 8, 43, 2006.

[12] H. J. Yang et al., "Optical manipulation of nanoparticles and biomolecules in sub-wavelength slot waveguides", *Nature,* vol. 457, pp. 71–75, 2009.

[13] Y. Yang et al., "Optofluidic waveguide as a transformation optics device for lightwave bending and manipulation", *Nature Commun.,* vol. 3, pp. 651, 2012.

CONTACT

*A. Q. Liu, Tel: +65-6790 4336; Email: eaqliu@ntu.edu.sg

NANOSLIT MEMBRANE INTEGRATED FLUIDIC CHIP FOR MICRO / NANO PARTICLE TRAPPING AND SEPARATION

Yul Koh[1], Homan Kang[2], Jong-Ho Kim[3], Yoon-Sik Lee[2], and Yong-Kweon Kim[1]

[1]School of Electrical Engineering & Computer Science, Seoul National University, Korea
[2]School of Chemical & Biological Engineering, Seoul National University, Korea
[3]Department of Chemical Engineering, Hanyang University, Korea

ABSTRACT

This paper describes a size selective micro- / nano-particle trapping in a nanoslit membrane integrated fluidic chip (Nanoslit-Chip) for trapping and concentrating target particles in a specific region of the chip for nanoparticle based assay. For size selective particle trapping, a uniform and large scale nanoslit fluid channel array was fabricated on a silicon dioxide membrane and integrated to the PDMS microchannel. A small amount of target fluorescent particles in a large volume of sample can be concentrated on the nanoslit membrane with continuous flow and successfully detected by a fluorescent microscopy. The size selective particle trapping are demonstrated by separating two different sized particles (440 nm / 1.8 μm) in the Nanoslit-Chip. The Nanoslit-Chip is applied to the sensitive detection of proteins by trapping protein induced nanoparticle agglomerates. The Nanoslit-Chip can be effectively used for trapping and concentrating a small amount of target nanoparticles in a large sample volume.

INTRODUCTION

Nanofluidic devices including planar nanochannels and nanopores have been interested for handling biological samples and nanoparticles. Specially, nanoparticle separation, trapping, and concentration in a nanofluidic device showed its potential for the sensitive bio-molecule detection and analysis [1, 2]. Therefore, various nanofluidic devices have been developed using silicon, glass, quartz, and polymers [3, 4]. As a structural point of view, the nanochannels are classified by a horizontal channel and a vertical channel. The horizontal nanochannels usually have planar structure. These nanochannels can be reliably fabricated by high resolution lithography technology such as electron beam lithography, focused ion milling, and nano-imprint lithography [5]. Although the fabrication of the horizontal nanochannels is reliable, they are not suitable for a nanoparticle trapping and concentration due to the high flow resistance. Because the orientation of the horizontal nanochannel is parallel to the substrate surface, the total cross section area of the nanochannel is limited even if the nanochannels are fabricated by 1-dimensional array [6]. Therefore, the volume flow rate of the horizontal nanochannel is very slow and nanoparticles are easily clogged at the entrance of the nanochannel by keystone effect [7, 8]. Compare to the horizontal structure, the vertical nanochannels have an advantage for the fluid transport because they could have low hydraulic resistance. The vertical nanochannels allow fabricating large scale 2-dimensional fluid channel array which could decrease the hydraulic resistance by increasing the total cross section area of the fluid channel. Therefore, vertical nanochannel array is desirable for nanoparticle trapping and concentration.

In this paper, we propose a nanoslit membrane integrated fluidic chip that has a vertical nanoslit fluid channel array (Nanoslit-Chip) for trapping and concentrating particles. The proposed device consists of top PDMS fluid channel, nanoslit patterned SiO_2 membrane, and bottom silicon fluid channel. Vertical nanoslit channel array are easily fabricated on the SiO_2 membrane by using a photolithography and a tetraethylorthosilicate (TEOS) chemical vapor deposition without high resolution lithography technology. Large numbers of vertical channels decrease the hydraulic resistance of the Nanoslit-Chip. Two different particles (440 nm / 1.8 μm) are selectively trapped on the nanoslit membrane and detected by the fluorescent microscopy. Finally, The Nanoslit-Chip is applied to the sensitive detection of biological molecules based on the protein-induced aggregation of particles.

DESIGN OF NANOSLIT-CHIP

The schematics of the Nanoslit-Chip and particle selective trapping process are shown in Figure 1. The proposed device is comprised of a PDMS top fluid channel, a silicon dioxide nanoslit array membrane, and a silicon bottom fluid channel. The sample fluid is injected to the inlet of the top PDMS microchannel and guided to the nanoslit membrane and bottom microchannel. Particles which are larger than the slit width are trapped on the nanoslit membrane. The nanoslit membrane was designed to have 4300 nanoslits for low hydraulic resistance. The bottom microchannels are connected to the outlet of PDMS microchannel.

The process for size-dependent particle trapping, concentration and detection is illustrated in Figure 1b. First, mixture of different sized fluorescent particles was injected into the Nanoslit-Chip. As shown in process (i) of Figure 1b, particles larger than the nanoslit width are trapped on the nanoslit while the smaller ones pass through to the bottom fluid channel. And then, the Nanoslit-Chip was washed by DI water for 2 min and dried in an oven. Finally, the PDMS top fluid substrate was removed and the trapped particles were detected using a fluorescent microscopy as shown in process (ii) of Figure 1b. Through the proposed process, particles of desired size could be separated from the smaller particles and concentrated on the nanoslit membrane. Because the low flow resistance of the Nanoslit-Chip allows filtering a large

978-1-4799-3510-9/14 $31.00 © 2014 IEEE

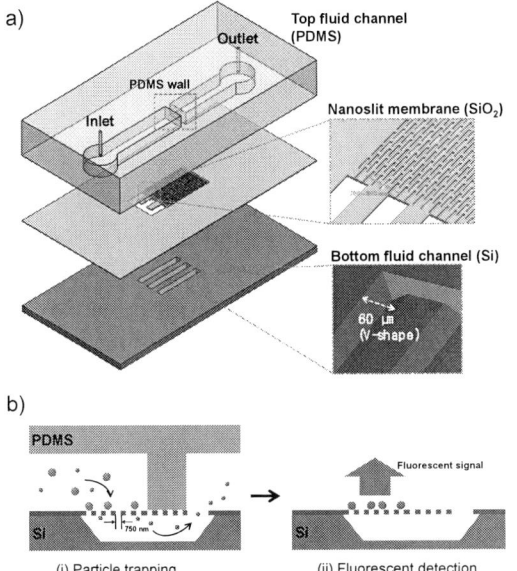

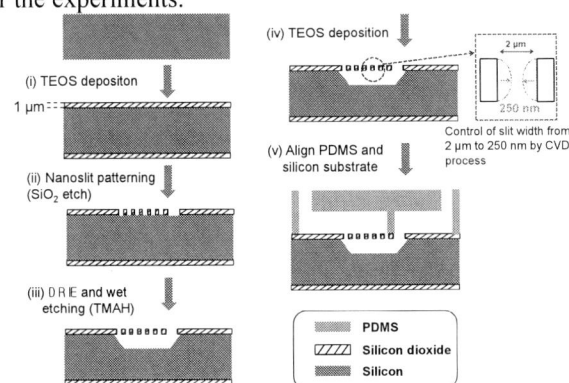

Figure 1: a) Schematics of the Nanoslit-Chip. b) Process for the particle separation and fluorescent detection

volume of the sample fluid to the nanoslit membrane, the proposed process is suitable for concentrating low population of target particles in a large sample volume.

FABRICATION PROCESS AND RESULTS

Figure 2 shows the fabrication process of the Nanoslit-Chip. First, 1μm of TEOS was deposited on the silicon wafer to form a SiO_2 layer for the nanoslit membrane. And then, the microslits (2 μm × 18 μm) were patterned on the SiO_2 layer by photolithography and dry etch process. After photoresist removal, the silicon layer below the nanoslit pattern was etched by a deep reactive ion etching (DRIE) and tetramethylammonium hydroxide (TMAH) wet etching process to form the bottom fluid channel. Finally, TEOS deposition (Applied Material, Precision 5000) reduced the microslit width from 2 μm to 700 nm for the selective trapping of 1.8 μm particles from a mixture of 1.8 μm and 440 nm particles, as shown in process (iv) of Figure 2. PDMS top substrate was fabricated using soft lithography. SU-8 2025 was used to fabricate mold for PDMS top fluid channel. The PDMS substrate was aligned to the nanoslit membrane for the experiments.

Figure 2: Fabrication process of the Nanoslit-Chip

The fabrication results were shown in Figure 3. The overall images and the cross section images of the nanoslit membrane were shown in Figure 3a. The 250 nm nanoslits were successfully fabricated from the microslits by the TEOS deposition process. However, the TEOS deposition process induces a nanoslit membrane deformation due to the compressive stress is increased by the TEOS deposition process. The cross section of the nanoslit membrane was measured after cutting the Nanoslit-Chip. The width of the nanoslit was gradually reduced from the bottom of the membrane by the conformal TEOS deposition process. Figure 3b is SEM images for the nanoslit during the TEOS deposition process. As shown in Figure 3c, the slit widths were linearly decreased by TEOS deposition process (the square of the correlation coefficient (R^2) is 0.9958). The proposed nanoslit fabrication process are simple and cost effective because it only uses the standard MEMS fabrication process to fabricated nanostructure on the SiO_2 membrane.

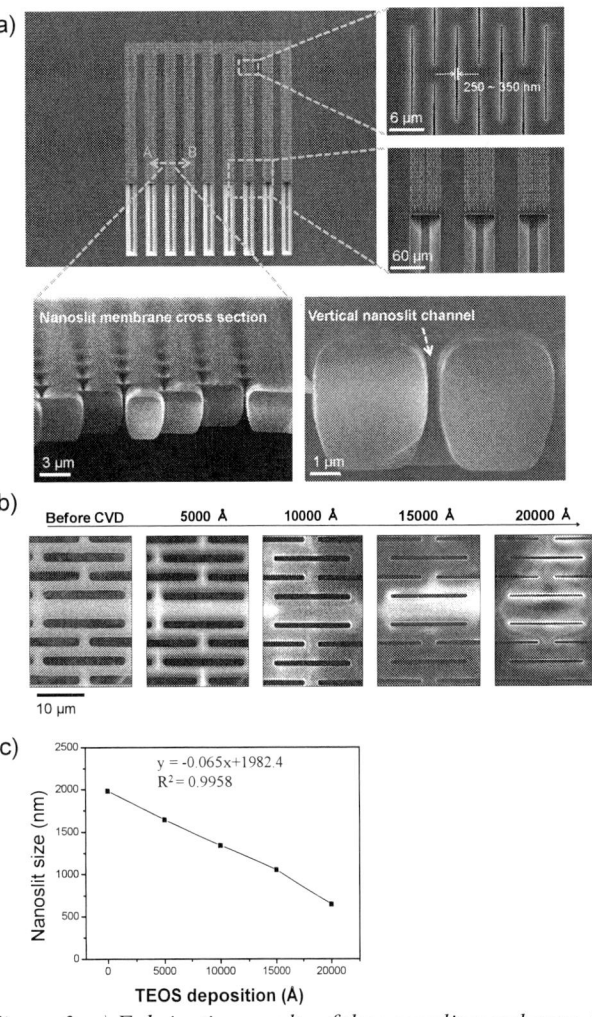

Figure 3: a) Fabrication results of the nanoslit membrane. b) SEM image and c) measurement data during the TEOS deposition process to control the nanoslit width.

MICRO- / NANO- PARTICLE TRAPPING AT THE NANOSLIT MEMBRANE

First, the trapping experiment for microparticles which are larger than the nanoslit width (700 nm) was performed in the Nanoslit-Chip. A 1.8 μm size particles that had a red fluorescent dye were injected into the Nanoslit-Chip at a flow rate of 40 μl/min for 150 s. Fluorescent images of the nanoslit membrane were acquired during the sample injection, as shown in Fig. 4a. The red fluorescence signals were increased according to the injection time, indicating that the particles were continuously injected into the nanoslits and successfully trapped on the nanoslit membrane. After the particle trapping experiment, the nanoslit membrane was measured by SEM to evaluate the fluorescence based particle detection. As shown in Figure 4b, the fluorescent detection is well matched to the SEM measurement. Therefore, the trapped particles can be reliably detected using the fluorescent microscopy.

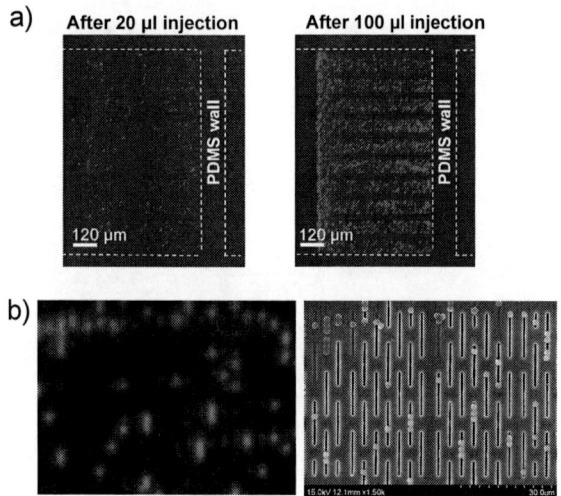

Figure 4: a) Fluorescent images of the 700 nm nanoslit membrane during the trapping of 1.8 μm red fluorescent particles. The fluorescent images were acquired when the 20 μl and 100 μl of the sample solution was injected b) Fluorescent image and the corresponded SEM image of the nanoslit membrane after 1.8 μm particle trapping.

Next, the trapping experiment for nanoparticles was performed in the Nanoslit-Chip. 450 nm size green fluorescent nanoparticles were injected into the 300 nm nanoslit membrane at a flow rate of 40 μl/min for 60 s. The nanoparticles were successfully trapped on the nanoslit membrane. The fluorescent images before and after the trapping experiment for 450 nm particles are shown in Figure 5.

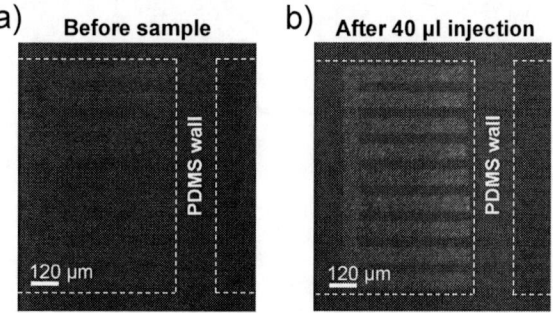

Figure 5: Fluorescent images of the 300 nm nanoslit membrane a) before and b) after the trapping experiment for 450 nm green fluorescent particles.

SIZE SELECTIVE PARTICLE TRAPPING AT THE NANOSLIT MEMBRANE

We demonstrate size selective trapping of microparticles (1.8 μm red fluorescent) and nanoparticles (440 nm, blue fluorescent). Mixtures of the 1.8 μm size red fluorescent and 440 nm blue fluorescent particles were prepared for size selective particle trapping. The concentrations of 1.8 μm-particles varied from 0 to 10^3 /ml, 10^4 /ml, and 10^5 /ml, while the concentration of the 450 nm-particles was fixed at 10^5 /ml. The particle mixtures were injected to the Nanoslit-Chip at a flow rate of 40 μl/min for 150 s. Figure 6 showed the experiment results for the selective particle trapping. The red fluorescent signal for 1.8 μm particles proportionally increased with their concentration. However, the blue fluorescent signal for 440 nm-particles showed a little increase because almost of the 440 nm particles were passed through the nanoslit membrane. Therefore, the Nanoslit-Chip can specifically trap and concentrate the target particles from the smaller particles.

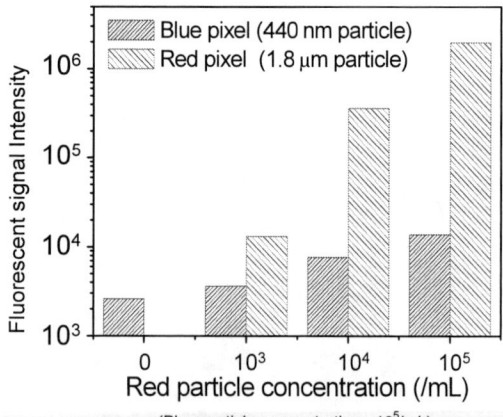

Figure 6: Experiment results for the size selective particle trapping (mixture of 1.8 μm and 440 nm particles).

PROTEIN DETECTION BY AGGREGATED PARTICLES TRAPPING

The Nanoslit-Chip was applied to the detection of a biotinylated BSA (b-BSA) by trapping the protein-induced nanoparticle aggregates. As illustrated in Figure 7a, the streptavidin coated particles were aggregated by the biotinylated BSA and the size and the number of agglomerates were increased depend on the amount of the target protein. For the agglutination assay, 100 µl of the biotinylated BSA solution and the streptavidin coated nanoparticles were incubated for 5 h at 25 °C. And then, the solution was diluted and injected to the Nanoslit-Chip for the selective trapping of the aggregated nanoparticles. The nanoparticle aggregates (larger than 800 nm) were successfully separated from the unbound ones (440 nm) and detected by the fluorescent microscopy. Figure 7b showed the detection results of the nanoparticle agglutination assay.

a)

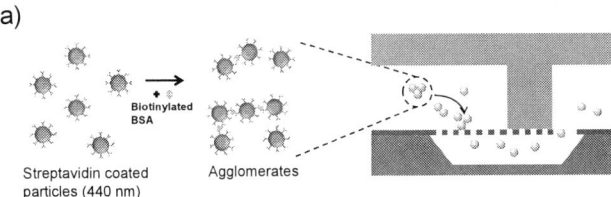

b)

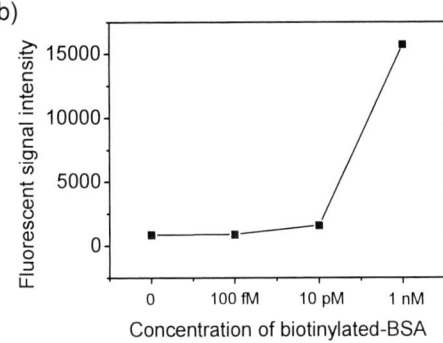

Figure 7: a) Protein-induced particle aggregation and detection process in the Nanoslit-Chip. b) Trapping results of the aggregated nanoparticles in the Nanoslit-Chip as the target protein (b-BSA) concentration.

CONCLUSION

In this paper, we demonstrated a nanoslit fluidic device (Nanoslit-Chip) for nanoparticle trapping and concentration. The Nanoslit-Chip has a large number of nanoslits as a vertical nanofluid channel array to trap target particles with low flow resistance. The vertical nanofluid channel array was simply fabricated on the SiO_2 membrane by using dry etching and TEOS deposition process. Mixtures of 1.8 µm size particles (red fluorescent) and 440 nm particles (blue fluorescent) were successfully separated in the Nanoslit-Chip and the 1.8 µm particles were concentrated on the nanoslit membrane. Finally, we successfully applied the Nanoslit-Chip to a sensitive detection of b-BSA by trapping of the protein-induced aggregated nanoparticles.

ACKNOWLEDGEMENTS

This research was supported by the Pioneer Research Center Program through the National Research Foundation of Korea funded by the Ministry of Science, ICT & Future Planning (NRF-2013-006163).

REFERENCES

[1] D. Huh, K. L. Mills, X. Y. Zhu, M. A. Burns, M. D. Thouless, and S. Takayama, "Tuneable elastomeric nanochannels for nanofluidic manipulation," *Nature Materials,* vol. 6, pp. 424-428, Jun 2007.

[2] J. L. Fraikin, T. Teesalu, C. M. McKenney, E. Ruoslahti, and A. N. Cleland, "A high-throughput label-free nanoparticle analyser," *Nature Nanotechnology,* vol. 6, pp. 308-313, May 2011.

[3] R. Chantiwas, S. Park, S. A. Soper, B. C. Kim, S. Takayama, V. Sunkara, H. Hwang, and Y. K. Cho, "Flexible fabrication and applications of polymer nanochannels and nanoslits," *Chemical Society Reviews,* vol. 40, pp. 3677-3702, 2011.

[4] C. H. Duan, W. Wang, and Q. Xie, "Review article: Fabrication of nanofluidic devices," *Biomicrofluidics,* vol. 7, Mar 2013.

[5] L. D. Menard and J. M. Ramsey, "Fabrication of Sub-5 nm Nanochannels in Insulating Substrates Using Focused Ion Beam Milling," *Nano Letters,* vol. 11, pp. 512-517, Feb 2011.

[6] H. Bruus, *Theoretical microfluidics.* Oxford ; New York: Oxford University Press, 2008.

[7] M. Wang, N. Jing, I. H. Chou, G. L. Cote, and J. Kameoka, "An optofluidic device for surface enhanced Raman spectroscopy," *Lab on a Chip,* vol. 7, pp. 630-632, 2007.

[8] E. Tamaki, A. Hibara, H. B. Kim, M. Tokeshi, and T. Kitamori, "Pressure-driven flow control system for nanofluidic chemical process," *Journal of Chromatography A,* vol. 1137, pp. 256-262, Dec 29 2006.

CONTACT

*Yul Koh, tel: +82-2-888-5017; tesadale@snu.ac.kr

ON-CHIP CONTROL OF PNEUMATIC-BASED BISTABLE VALVE SWITCH
Arnold Chen and Tingrui Pan
University of California, Davis, USA

ABSTRACT

Bistable valves are of particular interest due to its capability of remaining in open or closed states without energy consumption. This aspect is appealing for microfluidics transferring from benchside to bedside as input access and controls are limited. In this paper, we present pneumatic-based, bistable valve (BSV) switches for immediate on-chip fluid-flow manipulation without the requirement of external microcontroller circuitries. The applicability of the on-chip controller is demonstrated in a 4-to-1 microfluidic multiplexor. Furthermore, clinical relevance of on-chip BSV switches is displayed in point-of-care ABO blood-typing diagnostic chips.

INTRODUCTION

The introduction of lab-on-a-chip devices has revolutionized the way science is conducted by providing investigative methodologies that minimize target sample size, high resolution detection and sensitivity, rapid analytic processing, and most importantly, microscale control and manipulation. Driving the continual development of microfluidic chips are the translational applications it enables in the laboratory, clinic, and field. The fundamental building elements of a microfluidic system consist of micropumps for powering fluid flow and microvalves for manipulating flow directions. Microvalves can further be categorized into active or passive and mechanical or non-mechanical with subcategories determined by its principle operating mechanics such as magnetism, thermodynamics, electrochemistry, pneumatic, capillary effects, etc. [1].

Bistable microvalves are a type of mechanical valves of particular interest due to their unique property of remaining stable (no energy consumption) in either open or closed states and requiring power input only to transition between the two states. Within the valve, a deflectable membrane undergoes mechanical stress to permit or block liquid flow. Switching between the open/close bistable states in bistable microvalves are often achieved through principles involving electrostatics, electromagnetics, thermoelectrics, and/or pneumatics properties. One of the earlier bistable microvalves uses two pneumatically coupled membranes sharing the same air cavity to alternate deflections as electrodes generating electrostatic forces power on [2]. Electromagnetism has been used by several groups as the driving force to switch and maintain valves between on/off states [3]. Thermopneumatics is another widely implemented actuation mechanism as electric current flowing through microheaters provides the necessary electro-thermal expansion to displace microvalve membrane [4]. Liquid-solid phase changing material has been reported to be viable microvalves as materials such as paraffin [5] or Field's metal [6] were melted to liquid state and followed by

pneumatic manipulations to valve flow access.

Recently, the emerging trend has been focused on technology transfer, the process of taking science and engineering from research laboratory settings to less equipped environments such as patient's homes, rural areas, even third world countries. For new technologies to be more readily accessible to the general public, the device needs to conform to the concept of simplicity. Whitesides' group pioneered the platform of paper microfluidics, transferring technologies into applications requiring minimal prerequisites for broader impact [7]. However, certain processes prefer traditional laminar flow microfluidic channels. As such, one approach to simplifying microfluidic analysis systems is to reduce the number of external inputs/connections by integrating parts of the chip controls onto the device itself. A liquid-handling chip with functionalities of metering, mixing, incubation, and wash procedure was reported requiring only four state-selection vacuum inputs and one constant vacuum power source [8]. The chip pumping mechanism was actuated by a series of on-chip vacuum-powered oscillators. Another group reported an on-chip controlled "squeeze-chip" pump engineered by cascading one-way check valves [9]. Their biochemical assays exclude additional pumping instruments.

In this paper, we present pneumatic-based, bistable valve (BSV) switches for immediate on-chip fluid-flow manipulation without the need of external microcontroller circuitries. The design and structure of the BSV switch makes it simple to be incorporated into any existing microfluidic systems with a membrane component. The integration of BSV switch does not add substantial complexity to the fabrication process allowing it to be widely applicable. Metal coating or pattering steps are not present as electrodes for thermo or magnetic actuations are not necessary. On-chip nature of BSV switch promotes miniaturization of the entire device platform, which in term permits portability. The BSV switch requires only a single vacuum source, in this case a portable micro vacuum pump, to make a network of pneumatic switches functional. Vacuum is used to maintain the stability of bistable states and the transition between states is driven by the user's energy input through direct contact on one of the switch control pads. This gives the user absolute and instantaneous control over the operations of the device without involving any programming. Ultimately, the BSV switch permits the supply of negative vacuum suction to a pneumatic microfluidic valve thereby enabling on-chip control of liquid flow. The applicability of this on-chip control switch is demonstrated in a 4-to-1 microfluidic multiplexor where the supply of fluids from 4 different inlets can be determined simply by operating the BSV switches. Clinical relevance of BSV switch is further exemplified in a point-

978-1-4799-3510-9/14 $31.00 © 2014 IEEE

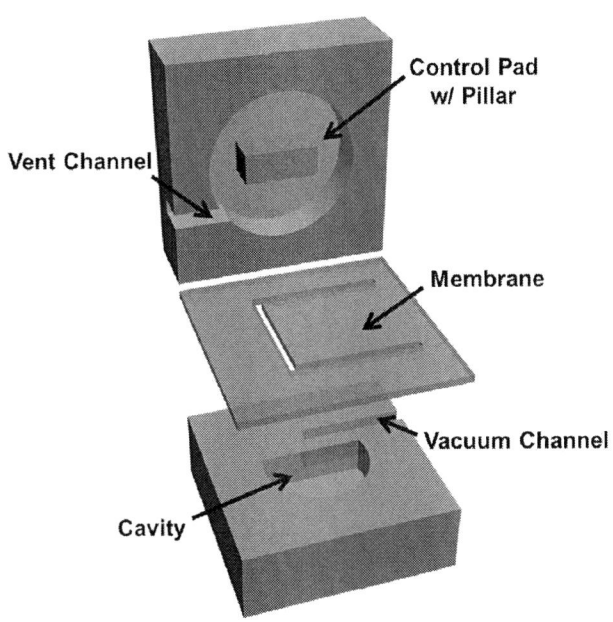

Figure 1: Structure of an on-chip control pad in BSV switch.

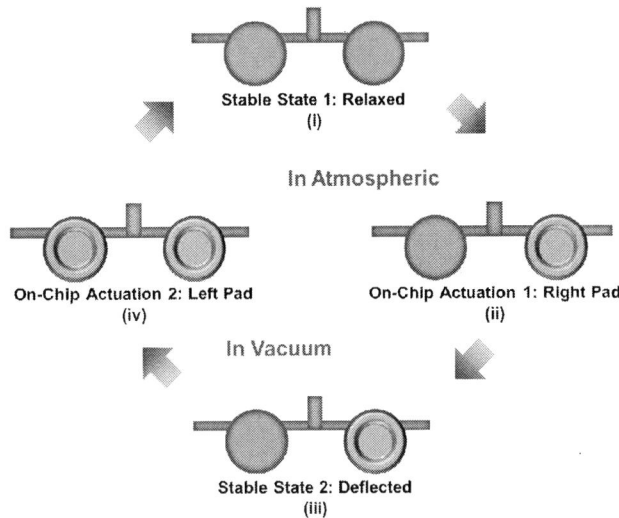

Figure 2: Mechanisms of BSV switch in bistable states.

of-care ABO blood-typing diagnostic chip where the flow direction of blood samples and antibody serums is controlled by two BSV switches each actuating a set of microvalves.

FABRICATION

The devices were fabricated using 3 layers of polydimethylsiloxane (PDMS) (Sylgard 184, Dow Corning) where the middle membrane (~200 μm) was oxygen plasma bonded between two thicker PDMS slabs (1.0-1.5 mm) that had microstructures etched into them via a CO_2 laser machine (VersaLaser, Universal Laser System) [10]. See Figure 1 for the structure of an on-chip control pad and how structures in each layer are aligned. Laser fabrication was implemented for rapid prototyping purposes; however, conventional template molding Softlithography could also be substituted. Fabrication of the membrane layer, however, is most convenient through laser technique as the laser can directly etch through the membrane to create structures such as a cantilever within this bistable switch. The top layer (thickness ~1.0 mm), has a 2.0 mm wide and <0.9 mm deep hole laser-etched leaving the center with a rectangular pillar. All PDMS layers were aligned and bonded under a stereomicroscope by hand. Lastly, the normally-closed microvalves on the device and the BSV switches themselves became functional when a micro vacuum pump (KPV-20A6V, Clark Solutions) was connected to supply the necessary negative suction source.

DESIGN

As previously mentioned, Figure 1 shows the devices are fabricated using 3 layers of PDMS. In the PDMS, there

is a circular control pad that alternates between two bistable states as the membrane with attached pillar is deformable by vacuum. A vent channel connects this chamber to a second pad or ambient environment, allowing the release of vacuum when switching between bistable states. The middle laser-etched-through membrane layer is a latchable cantilever-valve that covers over the vacuum supply channel at the bottom layer. Next to the vacuum channel on the bottom layer is a semicircular cavity that allows the top layer pillar to occupy the space during the deformed state.

An on-chip bistable microvalve switch is formed when two control pads are placed next to each other in series (where the vent channel of the first pad connects to the vacuum channel of the second pad as shown in Figure 2). Additionally, a channel branches off between the junctions at which the two control pads meet. This channel is under vacuum state according to the BSV switch; therefore when the channel is routed to a normally closed valve, the presence of vacuum can regulate fluid flow. Figure 2 further depicts the mechanism of two bistable states that are achieved through a user's direct contact with the switch control pads. In State 1, both control pads are in relaxed state under atmospheric pressure, only the first vacuum channel is under vacuum. When user presses the right control pad, the center pillar deforms the middle membrane layer thus uncovering vacuum channel and bringing the first control pad under vacuum pressure to State 2. When the user presses the left control pad, the membrane releases the vacuum state of right control pad back to atmospheric State 1. In another word, the user provides the necessary energy input required to alternate the BSV switch between the vacuum/atmospheric bistable states.

RESULTS AND DISCUSSION
On-chip controlled microfluidic multiplexor

The BSV switch is applicable in providing on-demand vacuum, therefore a demonstration device of a 4-to-1 microfluidic multiplexor was shown functional in

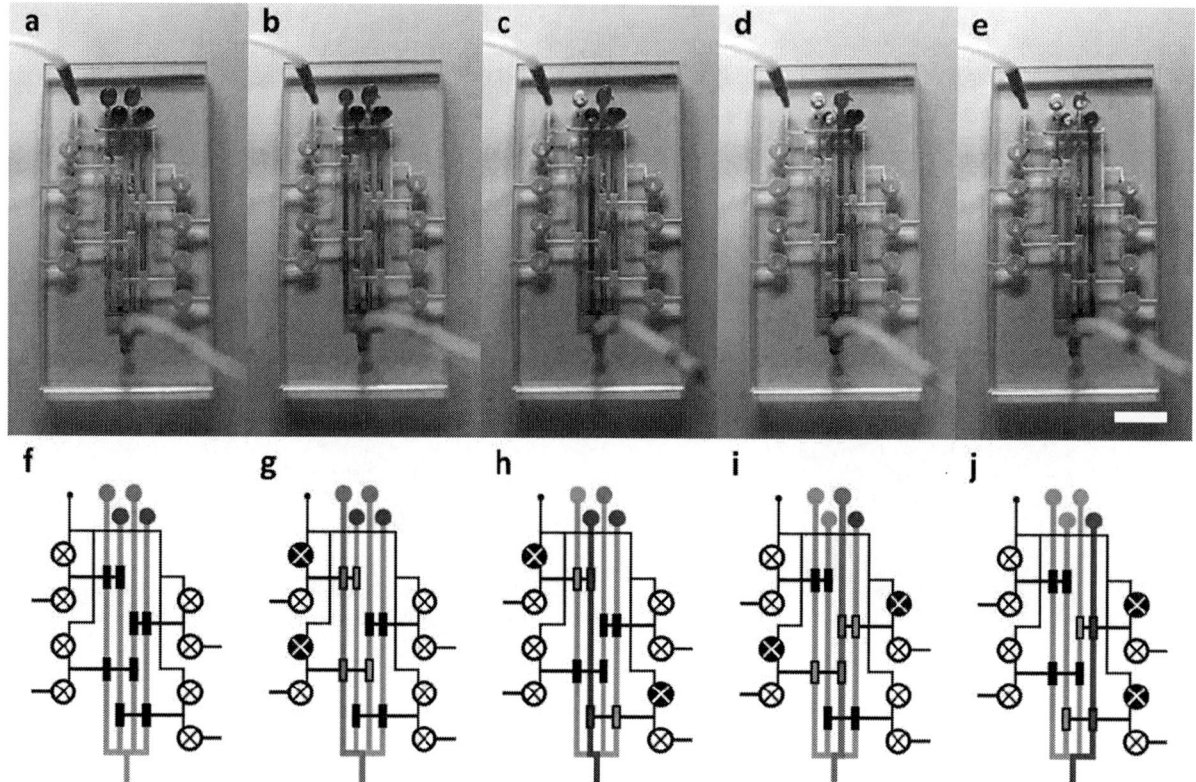

Figure 3: (a-e) Photographs of a 4-to-1 multiplexor chip as flow of red/blue dyed fluids switches from each microfluidic channel. (f-j) Cartoon illustrations depicting pairs of BSV switches controlling each channel. (scale bar: 5mm)

controlling vacuum actuated, normally closed microvalves. The device is shown in Figure 3. Using fabrication method described above, 4 parallel converging microfluidic channels and 4 BSV switches were incorporated onto a single device. Red and blue color dyed droplets were loaded to each channel inlet reservoir in an alternating fashion. The fluid flow in each channel is controlled by a set of 2 BSV switches operating similarly as a 2-bit digital address channel multiplexor. This means that 4 different combination pairs were possible with 4 BSV switches (i.e. 11, 10, 01, and 00). More importantly, this method of switch/microvalve layout was adapted because of its potential at larger system as the number of switches required to control 2^n channels is only $2*n$. A 2-bit system has $2^2=4$ channels and $2*2=4$ switches; 3-bit system has $2^3=8$ channels and $2*3=6$ switches; 4-bit system has $2^4=16$ channels and $2*4=8$ switches; etc. In our device, liquid were drawn from 4 individual reservoirs from left to right as shown in Figure 3a-e through activation of paired switches illustrated in Figure 3f-j. When the channel was opened by the switches, 5 μl of red or blue dye solutions flow through the multiplexing device to the outlet.

On-chip controlled blood typing chip

Lastly, 2 BSV switches were implemented into an ABO blood typing point-of-care diagnostic device to direct the on-chip fluidic operation processes as shown in Figure 4a. Device fabrication followed as described in earlier section.

The device design consist of a horizontal channel for blood flow/loading and 2 vertical crossing channels for addition of monoclonal anti-Blood Group A and anti-Blood Group B (SAB4700677 and SAB4700676, Sigma Aldrich) into assay detection chambers for antibody-antigen agglutination. There are a total of 7 normally-closed microvalves for directing liquid flow through the device, 3 along the horizontal channel and 2 along each vertical channel. One BSV switch actuates all 3 of the valves in the horizontal channel while another BSV switch actuates the rest of the 4 vertical channel valves. Essentially, switching the right BSV switch "on" opens the 3 horizontal valves thereby allowing blood flow and loading into the device as seen in Figure 4b. Afterwards, switching right BSV switch "off" and left BSV switch "on" opens the 2 vertical antibody channels each preloaded with 3 μl of antibody serum. The resulting antibody/blood mixture agglutinates in the downstream detection chamber and can be easily distinguished by presence of cell clumping indicating the corresponding ABO blood type (Figure 4c).

CONCLUSIONS

In this paper, a pneumatic-based, bistable valve switch has been presented to provide on-chip liquid control and manipulation. The requirement of external microcontrollers has been eliminated and the number of inputs has been greatly minimized to a single line of vacuum source, which

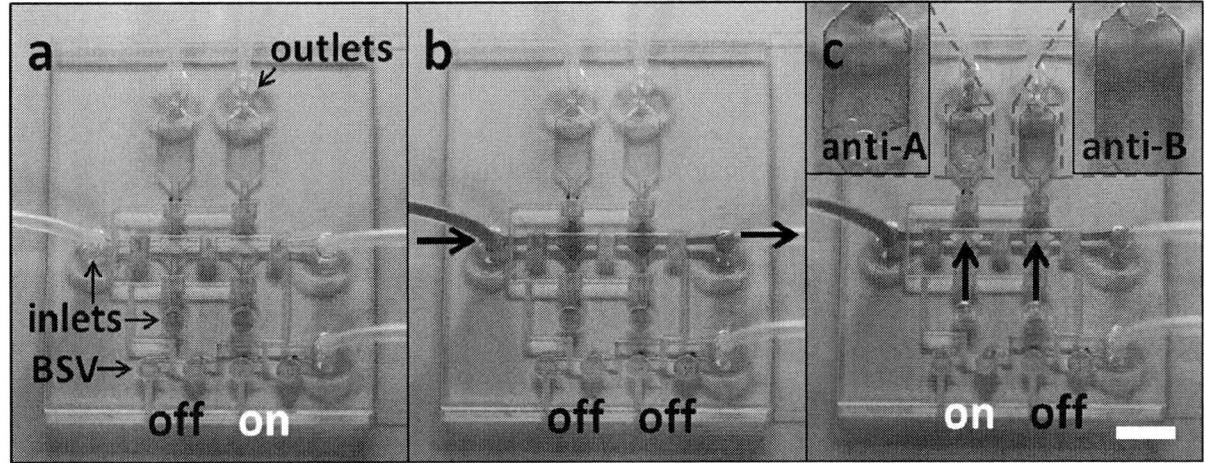

Figure 4: An ABO blood typing chip containing two BSV switches. (a) Right switch opens middle row valves allowing blood flow. (b) Both switches 'off' after loading. (c) Left switch 'on' to allow addition of antibodies. (scale bar: 5mm)

is generated by a portable micro vacuum pump and itself can drive a network of bistable, pneumatic switches. Switch operation works on-demand through the user's direct interactions, permitting a more immediate control of the microfluidic system. BSV switches have been demonstrated in microfluidic digital logic applications in terms of a 4-to-1 multiplexor using a 2-bit configuration. The on-chip control aspect of BSV switch permits portability necessary for point-of-care applications. BSV switch prove to exhibit clinical relevance when it is incorporated into an ABO blood typing diagnostic chip. In summary, the BSV switch is a bistable valve that offers on-chip controls while minimizing external device inputs, which is ideal for use in normal settings lacking precision instruments and by broad audience instead of specialists. It shows potentials in all types of point-of-care diagnostics and also in geographic locations including but not limited to developing countries, rural areas, remote extremities, and patient's home.

ACKNOWLEDGEMENTS

This work is in part supported by the National Science Foundation CAREER Program (ECCS-0846502).

REFERENCES

[1] K.W. Oh and C.H. Ahn, "A Review of Microvalves," *J. Micromech. Microeng.*, vol. 16, pp. R13-39, 2006.

[2] B. Wagner, H.J. Quenzer, S. Hoerschelmann, T. Lisec, and M. Juerss, "Bistable Microvalve with Pneumatically Coupled Membranes," *9th Ann. IEEE Int. MEMS Workshop*, 1996, pp. 384-8.

[3] S. Bohm, G.J. Burger, M.T. Korthorst, and F. Roseboom, "A Micromachined Silicon Valve Driven by a Miniature Bi-stable Electro-magnetic Actuator," *Sensors Actuators*, vol. 80, pp. 77-83, 2000.

[4] B. Yang, B. Wang, and W.K. Schomburg, "A Thermopneumatically Actuated Bistable Microvalve," *J. Micromech. Microeng.*, vol. 20, pp. 095024, 2010.

[5] R. Pal, M. Yang, B.N. Johnson, D.T. Burke, and M.A. Burns, "Phase Change Microvalve for Integrated Devices," *Anal. Chem.*, vol. 76, pp. 3740-8, 2004.

[6] K.A. Shaikh, S. Li, and C. Liu, "Development of a Latchable Microvalve Employing a Low-melting-temperature Metal Alloy," *J. Microelectromech. Syst.*, vol. 17, pp. 1195-203, 2008.

[7] A.W. Martinez, S.T. Phillips, and G.M. Whitesides, "Three-dimensional Microfluidic Devices Fabricated in Layered Paper and Tape," *Proc. Natl. Acad. Sci. USA*, vol. 105, pp. 19606-11, 2008.

[8] T.V. Nguyen, P.N. Duncan, S. Ahrar, and E.E. Hui, "Semi-autonomous Liquid Handling via On-chip Pneumatic Digital Logic," *Lab Chip*, vol. 12, pp. 3991-4, 2012.

[9] W. Li, T. Chen, Z. Chen, P. Fei, Z. Yu, Y. Pang, and Y. Huang, "Squeeze-chip: A Finger-controlled Microfluidic Flow Network Device and its Application to Biochemical Assays," *Lab Chip*, vol. 12, pp. 2012.

[10] A. Chen and T. Pan, "Three-dimensional Fit-to-flow Microfluidic Assembly," *Biomicrofluidics*, vol. 5, pp. 046505, 2011.

RAPID MICROFLUIDIC PROTOTYPING OF SOPHISTICATED PROTEIN ANALYSIS PLATFORMS USING GRAYSCALE PHOTOPATTERNING

Todd A. Duncombe[1], Kevin Maurer[2], and Amy E. Herr[1]

[1]University of California, Berkeley / UC San Francisco Joint Graduate Group in Bioengineering, USA
[2]Institute for Chemical and Bioengineering, ETH Zurich, CH-8093 Zurich, SWITZERLAND

ABSTRACT

We present a technique that utilizes hydrogel photopatterning via laser-printed grayscale masks to controllably define non-uniform pore-size distributions from a single <60s UV exposure and acrylamide precursor solution. This method enables facile and rapid fabrication of spatially complex chromatography tools that are impossible to realize using traditional methods. To highlight its utility we present two workhorse analytical electrophoresis platforms: (1) a 24-plex electrophoresis screening assay in adhesive microfluidics and (2) a 96-plex gradient gel-based protein sizing assay in a *free-standing* polyacrylamide gel platform.

INTRODUCTION

To achieve diverse functions, lab-on-a-chip devices often include hydrogels with spatially varying properties. Integration of these customized hydrogels in microfluidic channel networks has radically advanced lab-on-a-chip performance. Specifically, non-uniform hydrogels provide an important avenue for integration of both preparative and analytical functions (e.g., sample enrichment [1], high-resolution biomolecular separations [2], sample immobilization/capture [3]). Current fabrication protocols for spatially varying hydrogels remain inflexible – requiring either precursor exchange steps [2] or long diffusion times to establish monomer gradients [1]. Protocols that require diffusion are inherently slow, and, those that require precursor exchange are constrained to a limited subset of low-to-moderate complexity geometries (figure 1A). An alternative approach is to spatially vary the degree of polymerization of the precursor solution such that the final hydrogel product has a controlled variation in density – as has been previously applied to developing hydrogels to study cells [4] and fabrication of 3D structures [5]. In this

work, we demonstrate a grayscale mask photopatterning technique for rapid prototyping complex lab-on-a-chip devices optimized to address important protein measurement questions. This technique utilizes hydrogel photopatterning via laser-printed grayscale masks to controllably define non-uniform pore-size distributions from a single <60s UV exposure and acrylamide precursor solution. Additionally, we demonstrate grayscale photopatterning in two platforms that are extremely well suited rapid prototyping – in each the entire fabrication cycle is completed in just 15 minutes: (1) a 24-plex electrophoresis screening assay – from a single injection, sample can be assayed under 24 discrete gel conditions in an adhesive based lab-on-a-chip device and (2) a 96-plex gradient gel-based protein sizing assay in a *free-standing* polyacrylamide gel platform [6].

MATERIALS AND METHODS

Materials

Benzophenone and solutions of 40% (29:1) acrylamide/bis-acrylamide, isopropanol and glycerol were purchased from Sigma Aldrich (St. Louis, MO). Photoinitiator 2,2-azobis[2-methyl-N-(2-hydroxyethyl) propionamide] (VA-086) was purchased from Wako Chemical (Richmond, VA). Alexa Fluor 488 conjugated Trypsin Inhibitor (TI), Ovalubmin (OVA), Bovine Serum Albumin (BSA), Parvalbumin (PARV), and BenchMark™ Fluorescent Protein Standard were purchased from Life Technologies (Grand Island, NY). A 0.1% SDS Tris-glycine (pH 8.4) and a 0.1% SDS 500 mM Tris-HCL (pH 7.2) electrophoresis buffers (10x) were purchased from Bio-Rad Laboratories (Hercules, CA). Molecular Grade Water (DNase-, RNase-, and Protoase-free) was purchased from Corning (Corning, NY). The double sided adhesive (TESA 68575) was provided as a sample directly from TESA

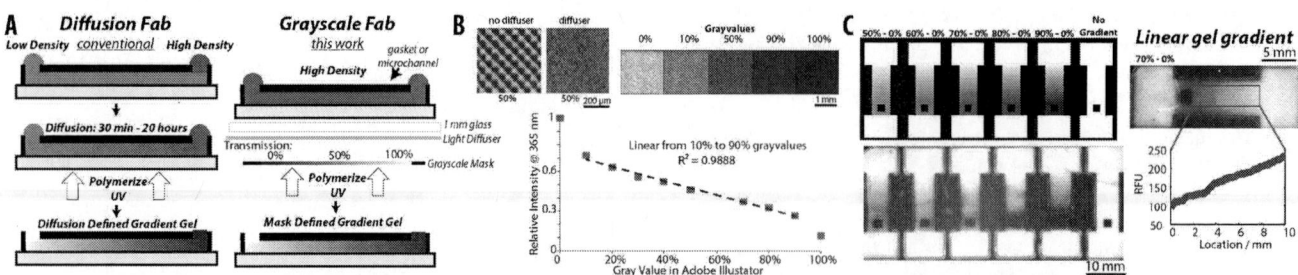

Figure 1: Direct gradient gel photo-patterning with a grayscale mask is a high throughput and design flexible alternative to diffusion based gradient gel fabrication (A). (B) By inserting an off-the-shelf photography light diffuser between the grayscale mask and the precursor solution – we mitigate negative effects from low resolution dots from laser printed grayscale masks. (C) Gradient gel formation was monitored for several gradient slopes by adding a fluorescent rhodamine methacrylate monomer to access relative monomer incorporation at different positions along the grayscale mask.

978-1-4799-3510-9/14 $31.00 © 2014 IEEE

(Charlotte, NC). GelBond® was purchased from Lonza (Basel, Switzerland).

Laser Printing Grayscale- and Photo-masks

Grayscale and photomasks were designed using Adobe Illustrator (2400 dpi, CMYK) and printed directly with a Brother MFC-9320 laser printer on a transparencies (PP2500, 3M, St. Paul, MN). To overcome resolution limits presented by printed grayscale dot patterns on the mask itself (i.e., 30-80μm dots, figure 1B) we incorporate a light diffuser (rosco PN# 3010) in the UV light path, thus allowing for the fabrication of linear density gel-gradients from ~5%-40% (figure 1C).

Gel Precursor Preparation

Gel precursor consisted of a 40% (w/v) acrylamide, with a bis-acrylamide crosslinker ratio of 3.33% (w/w), and 1% (w/v) VA-086 in Molecular Grade Water. In the adhesive microfluidics the solution also contained 0.1% SDS and 500 mM Tris-HCl. After mixing, the solution was degassed under a house vacuum in a sonicator for approximately 5 minutes. Immediately after degasing the gel precursor was used for UV photo-polymerization.

Adhesive Microfluidics Fabrication

We introduce a method for rapidly prototyping polyacrylamide gel electrophoresis (PAGE) microfluidic devices with adhesive defined microchannels (figure 2A). While adhesive microfluidics are extensively used for both rapid prototyping microdevices and cheap mass production, they have, to our knowledge, never been utilized for PAGE applications. This is likely due to the required surface

functionalization [7] to ensure covalently PAG incorporation on the surface. Treatments generally take an hour to complete and can require highly acid or basic environments that can dissolve the adhesive layer, effectively ruining the device. Here we utilized commercially available pre-functionalized substrate GelBond® (polyester based), such that we can photopolymerize almost immediately after device assembly.

The adhesive and GelBond® sheets are directly cut using a 60W CO_2 PLS6MW laser cutter (HDPFO optics, 80% power, 60% speed) from designs drawn in Adobe Illustrator. Following cutting, the adhesive features are cleaned of debris using an in-house vacuum. After the protective backing layers are removed from the adhesive, the GelBond®, adhesive, GelBond® sandwich is aligned and a manual pressure is applied to the ensure alignment is maintained. The assembled device is then twice laminated at 212°F with an Apache AL9 laminator. Once assembled and laminated, a solution of 6% benzophenone in isopropanol is wicked through the channel network and allowed to incubate for 30 seconds. The solution is then removed and the channels are washed twice with water. Immediately after the water wicks through all channels, it is vacuumed out and the channels are allowed dry. The device is then loaded with gel precursor and UV polymerization is commenced. The entire fabrication process, laser cutting, takes just 15 minutes to complete.

Free-Standing Polyacrylamide Gel Fabrication

The fabrication of uniform *free-standing* polyacrylamide gels has previously been published [6]. Briefly, a reservoir of gel precursor solution is established

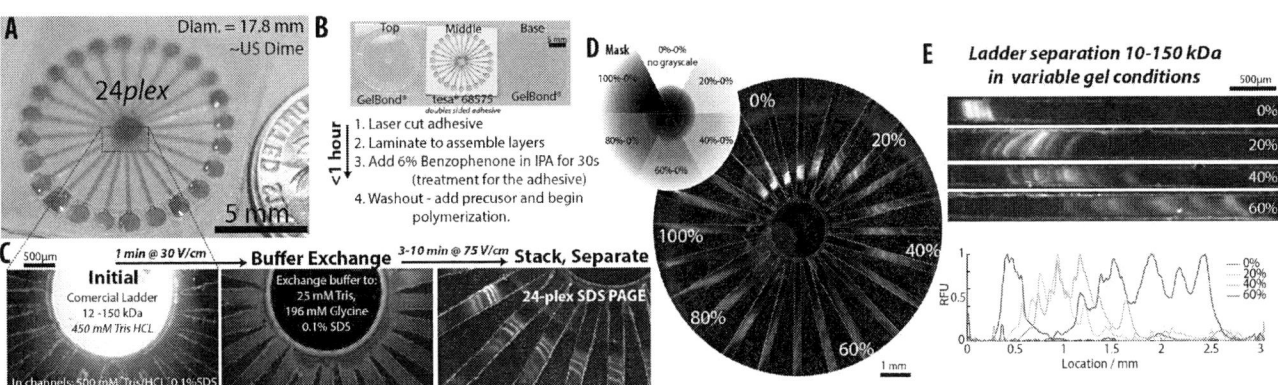

Figure 2: A 24-plex circular adhesive microfluidic device in the area of a US dime (A). (B) After adhesive assembly – only a 30s soak in 6% Benzophenone is needed before PAG polymerization. Importantly, GelBond® - a commercial product that covalently attaches to PAG during polymerization – is used as the top and bottom layer making the device PAG polymerization ready without channel surface modifications, supporting the rapid prototyping nature of the approach. (C) The 24 channel injection proceeds as follows: sample is added to central reservoir and an electric potential is applied uniformly towards the outer reservoirs causing sample to migrate into the channels. After loading, the remaining sample in the central reservoir is replaced with Tris/Glycine. Due to the conductivity difference between Tris/Glycine and the samples Tris/HCl buffer transient isotachophoresis occurs resulting in sample stacking initially and then a separation in each channel [3]. (D) An angularly varied grayscale mask is used to differentially pattern channels within the circular chip. A commercial fluorescent SDS PAGE ladder is separated in all channels to determine the optimal gel conditions. (E) The 60% to 0% gradient demonstrated the optimal separation – with incomplete gel polymerization observed for the 80% and 100% gel gradients.

above a piece of GelBond® using a glass gasket to define feature height. Photo-polymerization is performed through a photomask and/or grayscale mask to directly polymerize the desired gel microstructure as well as the gel density pattern. The gasket is then removed and the resulting PAG feature on GelBond® is soaked in water to remove excess unincorporated acrylamide monomer (a neurotoxin). The entire fabrication process takes less than 15 minutes. The gel can then soaked in the buffer of interest – in this work we used 20% glycerol, 0.1% SDS, 25 mM tris, and 192 mM glycine - and used immediately.

Image Acquisition and Processing

Trans-flourescence imaging of rhodamine methacrylate or AlexaFluor 488 labeled proteins in figure 1C and figure 3 was performed with a BioRad Chemidox XRS+ using an XciteBlue conversion screen and a 548-630 nm emission filter. Inverted epi-flourescence imaging of AlexaFluor 488 labeled protein in figure 2 was performed with a Peltier-cooled charge-coupled device (CCD) camera (Cool-SNAP HQ2, Roper Scientific, Trenton, NJ) and a 2×objective (PLN, N.A. = 0.08, Olympus, Center Valley, PA) on an Olympus IX-70 microscope. Camera exposure times ranged between 50 and 150 ms with 4 × 4 pixel

binning resulting in an acquisition resolution of 3.3 μm per data point. Light from an X-Cite® exacte mercury lamp (Lumen Dynamics, Mississauga, Canada) was filtered through a XF100-3 filter (Omega Optical, Battleboro, VT) for illumination. Image processing was performed using ImageJ.

RESULTS
24-Plex Circular Adhesive Microfluidics Device

Leveraging the rapid prototyping speed of adhesive microfluidics we developed a 24-plex electrophoresis screening assay in a small circular chip (diameter of US dime). The device contains 24 radial channels with a shared central reservoir. For operation, a sample is injected from the central reservoir as a discrete zone into all 24 radial channels using transient isotachophoresis – this had been previously demonstrated for single-channel injection [3] and is adapted to 24-parellel injections (figure 2C). Briefly, the device initially contains homogenous 500 mM Tris-HCl. Buffer in the central reservoir is removed using a kimwipe then replaced with protein sample containing 50 mM Tris-HCl, and an electric field is applied radially outward at 30 V/cm for one minute. Buffer in the central reservoir is then exchanged with a buffer 25 mM Tris 192 mM Glycine,

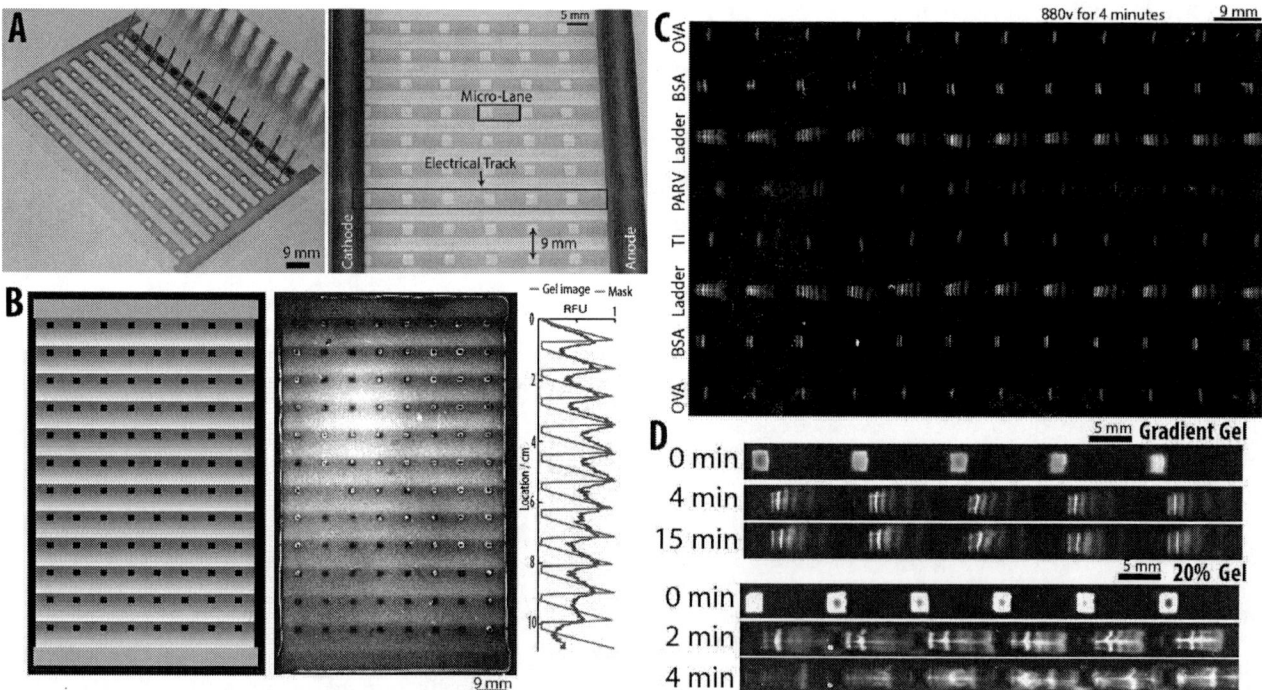

Figure 3: A 96-plex SDS PAGE gradient gel separation is performed for high resolution protein sizing within a large array. (A) We fabricate 96 separation lanes within the area of a 96 well plate. Each sample reservoir is spaced 9 mm enabling facile sample delivery using a multi-channel pipette. (B) A CAD image of the grayscale mask is compared to a fabricated gradient gel containing a rhodamine methacrylate monomer to visualize the spatial variation across the array. The apparent nonuniformity is caused by non-uniformity in the imaging system. (C) A 96-plex SDS PAGE gradient gel separation is performed in 4 minutes using four fluorescently labeled species: BSA (65 kDa, rows 2,7), OVA (45 kDa, rows 1,8), TI (21 kDa, row 5), PARV (12 kDa, row 4), and a ladder containing all four – in rows 3 and 6. (D) A key advantage of a gradient gel over a uniform gel array is the prevention of contamination from one lane to the next – we show five contiguous ladder separations in the gradient gel shown in (C) compared to in a uniform 20% gel. In the gradient gel samples are uncontaminated for a 15 minute separation – while in the uniform gel lane contamination can be seen as early as 4 minutes.

which has a significantly lower conductivity, and an electric field is applied at 75 V/cm. The resulting isotachophoresis stack acts to concentrate the protein species at the gel solution interface prior to the zone electrophoresis separation. To screen for optimal gel conditions, we utilized a grayscale mask to vary gel density angularly in six distinct regions (Figure 2D). As a proof of concept, we benchmarked a wide-size range protein ladder (BenchMark™ Fluorescent Protein Standard) and identified a 60%-0% gradient gel as the optimal separation condition.

96-Plex Gradient *Free-Standing* PAGE

In our previous work, we demonstrated 96-plex native protein separations in a *free-standing* uniform PAGE platform [6]. Here, in contrast, we demonstrate a 96-plex gradient SDS PAGE for protein sizing (figure 3). Adaption of direct gel density photo-patterning enables numerous assay improvements including the high resolution protein sizing over a large molecular weight range, the reduction of injection dispersion, and the mitigation of contamination between samples.

As discussed in previously published work [6], electroosmotic flow and sample adsorption in the injection reservoir can result in significant dispersion during electrophoretic injection that can reduce separation quality. Through the continuous stacking of a gradient gel we largely eliminate injection dispersion to achieve sharp protein bands (as seen in figure 3c) despite not using an electroosmotic flow suppressor, as had been done previously [6]. An additional advantage of a 96-plex gradient PAGE is the ability to place a high density, small pore size 'halting' region between consecutive lanes such that proteins cannot pass through the region and contaminate the subsequent lane. Thus, allowing a dense arrayed separation formats with contiguous, but distinct separation lanes on one device. In contrast, contamination can easily occur in a uniform gel array if the assay is run too long (figure 3D).

The 96-plex protein sizing assay using free-standing gradient gel separations produces clearly resolved protein peaks in just 4 minutes of electrophoresis - as compared to hours for conventional tools. With the gradient gel we see no contamination from neighboring lanes, even after even a lengthy 15 minutes separation.

DISCUSSION

Our methods for rapidly prototyping non-uniform hydrogel density, closed-channel adhesive and free-standing hydrogel microfluidic geometries combine to offer a powerful and highly versatile toolbox for the development and optimization of disruptive electrophoresis-based analytical tools.

ACKNOWLEDGEMENTS

The authors gratefully acknowledge members of The Herr Laboratory at UC Berkeley for assistance and helpful discussions. The authors also thank the generous TESA representative Kevin Olecki for the numerous donated adhesive samples. The authors thank Professor Albert Pisano and Dr. Jim Cheng for access and training to the CO_2 PLS6MW laser cutter. The authors acknowledge financial support from the US National Science Foundation (NSF) for a Graduate Research Fellowships (TAD). The authors thank the UC Berkeley Bakar Fellows Program (AEH) for additional financial support. A.E.H. is an Alfred P. Sloan research fellow in chemistry.

REFERENCES

[1] A.J. Hughes, A.E. Herr, "Quantitative enzyme activity determination with zeptomole sensitivity by microfluidic gradient-gel zymography", *Anal. Chem.*, 2010, 82, pp. 3803-11.

[2] A.V. Hatch, A.E. Herr, D.J. Throckmorton, J.S. Brennan, A.K. Singh, "Integrated preconcentration SDS-PAGE of proteins in microchips using photopatterned cross-linked polyacrylamide gels", *Anal. Chem.*, 2006, 78, pp. 4976-83.

[3] A.J. Hughes, A.E. Herr, "Microfluidic Western Blotting", *PNAS*, 2012, 109, pp. 21450-5.

[4] J.Y. Wong, A. Velasco, P. Rajagopalan, Q. Pham, "Directed Movement of Vascular Smooth Muscle Cells on Gradient-Compliant Hydrogels", *Langmuir*, 2003, 19, pp. 1908-13.

[5] S. Nicolas, E/ Dufour-Gegam, A. Bosseboeuf, T. Bourouina, J.P. Gilles, J.P. Grandchamp, "Fabrication of a gray-tone mask and pattern transfer in thick photoresists", *JMEMS*, 1998, 8, pp. 95-8.

[6] T.A. Duncombe, A.E. Herr, "Photopatterned free-standing polyacrylamide gels for microfluidic protein electrophoresis", *Lab Chip*, 2013, 13, pp. 2115-23.

[7] B.J. Kriby, A.R. Wheeler, R.N. Zare, J.A. Fruetel, T.J. Shepodd, "Programmable modification of cell adhesion and zeta potential in silica microchips", *Lab Chip*, 2003, 3, pp. 5-10.

CONTACT

*A.E. Herr, tel: +1-510-6663396; aeh@berkeley.edu

REALIZATION OF 240 NANOMETER RESOLUTION OF CELL POSITIONING BY A VIRTUAL FLOW REDUCTION MECHANISM

Shinya Sakuma[1], Keisuke Kuroda[1], Makoto Kaneko[1] and Fumihito Arai[2]
[1]Osaka University, Osaka, JAPAN
[2]Nagoya University, Nagoya, JAPAN

ABSTRACT

This paper presents the real-time precise positioning of a single cell with extremely high resolution. The positioning system is based on the visual feedback control of the syringe pump. The issue for using a syringe is that the flow rate is geometrically amplified in microchannel due to the ratio of cross sectional areas of the syringe and the microchannel. In order to overcome this issue, we introduce the virtual flow reduction mechanism. This mechanism utilizes the elasticity of poly-dimethylpolysiloxane (PDMS) microfluidic chip where the pressure peak is limited but the pressure response decreases with a sufficiently large time constant compared with the sampling time. By using this characteristic, we design and develop the system together with an online vision system. Through experiments, we could confirm that the cell positioning resolution is 240 nm corresponding to 1 pixel of the vision.

INTRODUCTION

Single cell manipulation has become popular along the progresses of micro/nano technologies in recent years [1-5]. A big advantage of using microfluidic devices is to prevent cell from contamination during the cell manipulation due to the closed space of the microchannel. The most popular approach for manipulating a cell in microchannel is to control fluid flow by an appropriate pump. By using a microchannel whose diameter is close to that of the cell, 3D manipulation of the cell eventually results in 1D manipulation, which makes the control much simpler and easier [6]. Conventional approaches can be classified into two groups where one is external actuation where a macroactuator [7-12], for example, a syringe pump is connected to microchip [12], and the second one is internal actuation where a microactuator [13-18], for example, a stirrer is installed inside of microchip [17]. We utilize the syringe pump because of its powerfulness and simple set up. Suppose that the cell manipulation is performed by the syringe pump, as shown in Fig.1. The syringe with the cross sectional area of A_1 with the diameter of 10 mm and connect to microchannel with the cross sectional area of A_2 with the diameter of 10 μm. Let δ be the resolution of actuator itself. Also, let $R = A_1/A_2$ be the reduction ratio. The velocity of output of the system is less than that of the actuator under $R>1$ and the resolution increases from δ to δ/R. Now suppose, $R = 10^6$ ($R \ll 1$), which results in the geometrically amplified mechanism. Also, suppose that we utilize an actuator with the order of 10 nm resolution where, $\delta = 10^{-8}$, δ/R results in the order of 1 mm, as shown in Fig.1. This δ/R is far from to ensure a high resolution in the order of 100 nm. Therefore, an appropriate

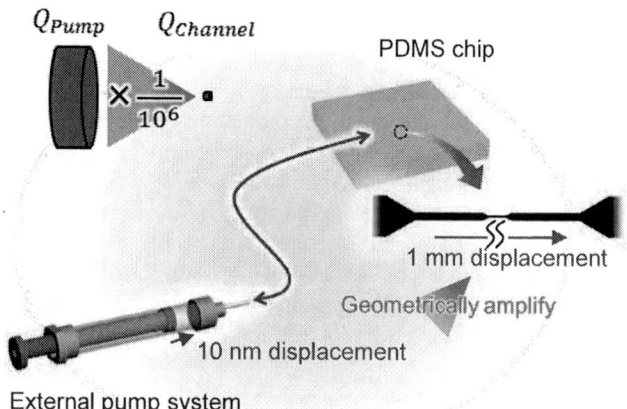

Figure1: Difficulty in precise position control due to geometrically amplification of flow rate.

flow reduction mechanism is highly required for the high resolution of cell positioning.

Through experiments, we found an interesting characteristic of the system where for a step input, the pressure increases quickly but keeps this pressure until the next input is given. From this preliminary experiment, we can expect the function of the flow reduction embedded into the microfluidic chip itself. We evaluate the performance of the developed cell positioning system, and the results shows that the positioning accuracy is the extremely high resolution of less than 240 nm corresponding of the limitation of a pixel of the vision sensor.

This paper is organized as follows. In second section, we briefly explain the basic idea of the flow reduction mechanism. We describe the experimental system and sample preparation of cells. Then, the results of the cell positioning are shown. Finally we conclude this work.

FLOW REDUCTION MECHANISM

Figure 2 explains the flow reduction mechanism, where a slightly displacement of the piston of the syringe pump makes a pressure increase in a PDMS microchannel, and it brings a small flow rate in the microchannel depending upon the flow resistance. Figure 3 shows possible pressure responses for a step position input to the piston. If there is no elasticity in the PDMS, the pressure should increase quickly and decrease also quickly. However, if there is an appropriate elasticity in PDMS, we can expect two different time constants T_1 and T_2 for both increasing and reducing speed of pressure, respectively. The key is that $T_1 \ll T_2$ under

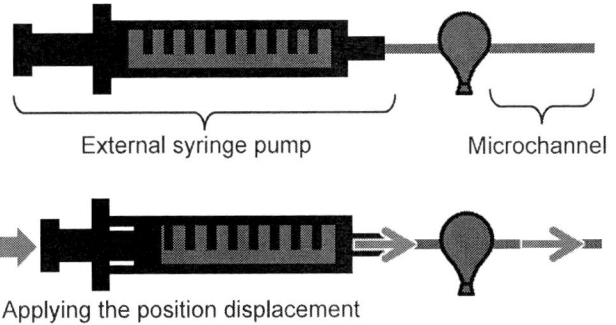

(a) Without flow reduction mechanism (under no elasticity)

Figure2: Outline of flow reduction mechanism.

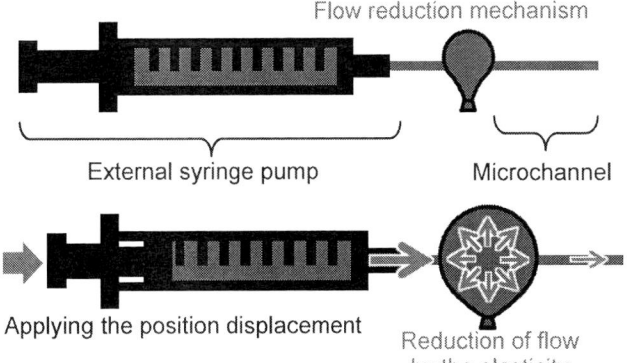

(b) With flow reduction mechanism (under elasticity)

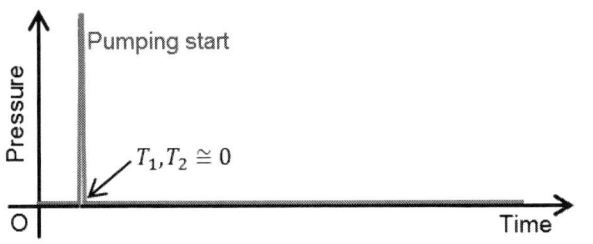

(a) Without flow reduction mechanism (under no elasticity)

Figure3: The key for the flow reduction mechanism.

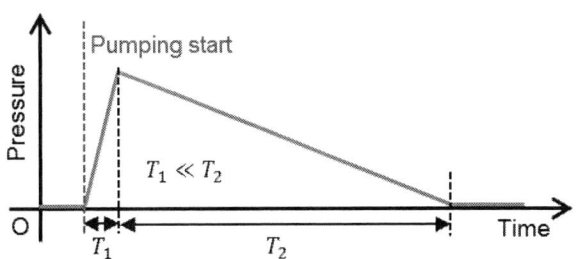

(b) With flow reduction mechanism (under elasticity)

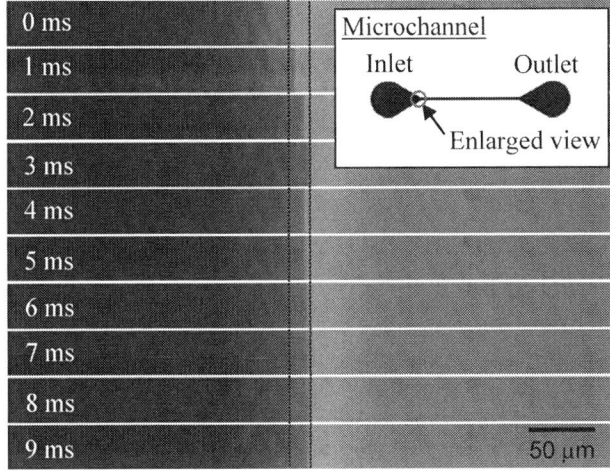

Figure4(a) : Deformation of inlet of PDMS chip.

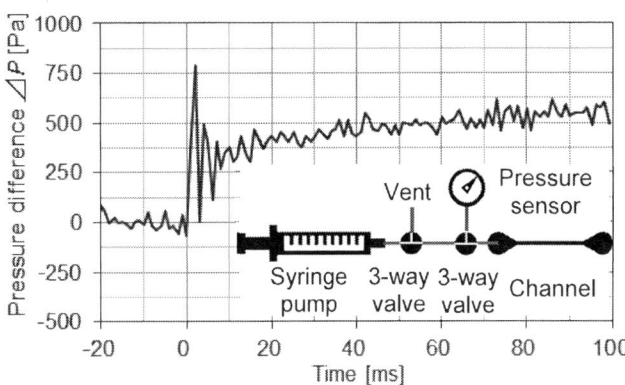

Figure4(b): Measured response of pressure in microchannel. The pressure is measured by the commercial pressure sensor (Copal electronics, PA-830-102A-10) through an additional 3-way valve (sub figure).

appropriate elasticity in PDMS can be achieved, while both T_1 and T_2 become nearly 0 under no elasticity. Figure 4(a) shows the deformation of the internal surface of microchannel at entrance part when a step input is given to the syringe pump. The surface is expanded with the deformation of roughly 20 μm and the time constant of roughly $T_1 = 4$ ms. A series of these photos provide us with an evidence showing that a step increase of pressure produces a volumetric change of the microfluidic system. Figure 4(b) shows the pressure in microfluidic system with respect to time after a step input is given. From Fig.4 (b), the pressure increases quickly and gradually decreases with the time constant $T_2 \gg 100$ ms. With these T_1 and T_2, we can expect

that a flow reduction mechanism is embedded into the microfluidic system itself, without any additional reduction mechanism. We call this interesting characteristics "virtual flow reduction mechanism", and the virtual flow reduction mechanism plays an essential role for the accuracy of the cell positioning system.

EXPERIMENT
Experimental System

Figure 5 shows the experimental system where it is composed of a syringe pump utilizing by the piezoelectric actuator, an online high speed vision and a microfluidic chip. Two tubes are connected to the inlet and outlet of the

Figure5: Experimental system.

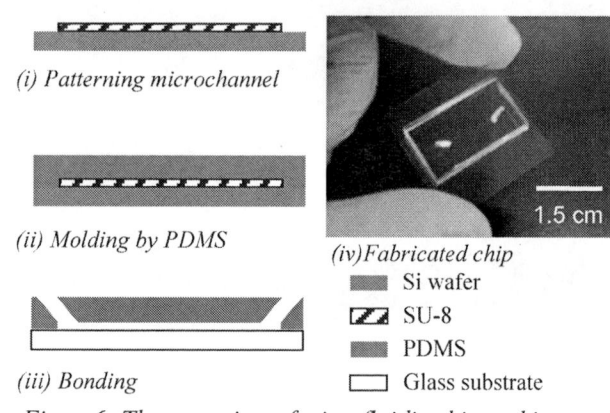

(i) Patterning microchannel

(ii) Molding by PDMS

(iii) Bonding

(iv) Fabricated chip

■ Si wafer
▨ SU-8
▨ PDMS
□ Glass substrate

Figure6: The over view of microfluidic chip and its fabrication process.

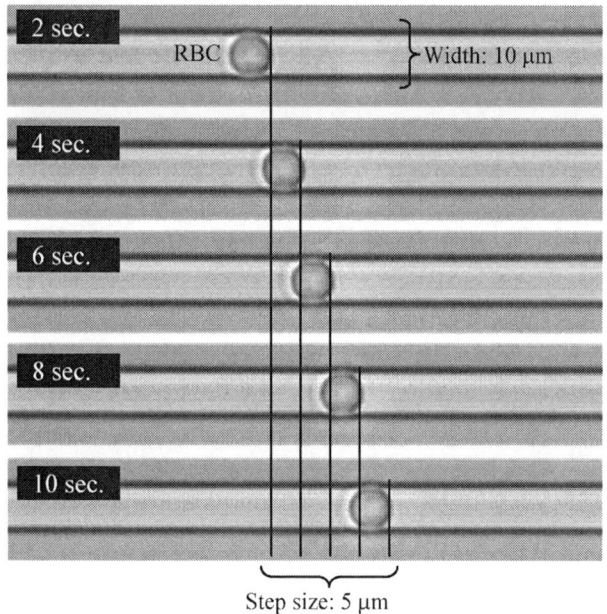

(a) Example of cell position control

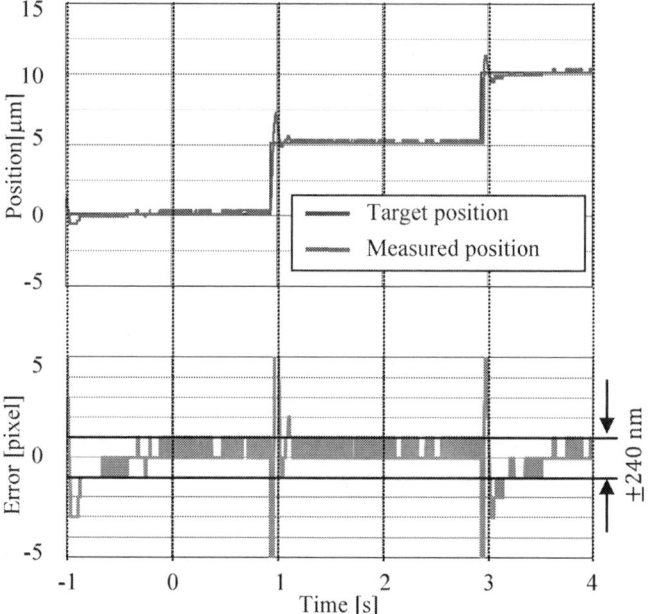

(b) Time response of cell position

Figure7: Demonstration of precise cell positioning using the developed system.

microfluidic chip. Cells injected to the microchannel through the tube, and then the position is detected by an online high speed vision where 1 pixel corresponds to 240 nm. The piezoelectric actuator is controlled based on the position of the cell with the sampling time of 1 ms.

Sample preparation

We use red blood cell (RBC) as the target cells. To achieve single RBC evaluation from massive RBCs in whole blood, we dilute blood by the standard saline solution. RBCs are obtained from a volunteer subject who has read and singed the donor consent for the evaluation. The blood is withdrawn by a licensed medical doctor. The procedures of the preparations are described as follows:
i) The micochannel is filled with standard saline solution.
ii) The blood is diluted by saline with the density of 2 %.
iii) The blood-saline mixture is injected into the microchannel from the sample inlet.

Fabrication of microfluidic chip

Figure 6 shows the fabrication process of the microfluidic chip. The microfluidic chip is fabricated by the soft-lithography using PDMS for the disposable application.
i) Spincoated SU-8 is patterned by the leaser lithography.
ii) The fabricated pattern is transcribed to PDMS.
iii) The molded PDMS is bonded with the glass substrate using O_2 plasma treatment after punched to make the ports.

Since, the height and the diameter of a RBC are typically ranged in 2 through 3 μm and 5 through 8 μm, respectively, the height and width of the microchannel is designed as 4 μm and 10 μm, repetitively.

Cell positioning

Figure 7(a) shows a series of photos when the position of RBC is controlled by 5 μm step input, and the total manipulation distance is 25 μm. Figure 7(b) shows a typical response of position control, where both the target and the

978-1-4799-3510-9/14 $31.00 © 2014 IEEE

measured positions of the RBC are given with respect to time. The blue, red and green curves show the target position, the measured position, and the error between them, respectively. The amplitude of each step input is 21 pixels (5.0 μm). The error of cell positioning is within 2.5 μm including overshoot, and is within ± 240 nm (± 1 pixel) in an equilibrium state. From these results, we could confirm that the elasticity existing in a fluid circuit of the microfluidic chip works as the flow reduction mechanism, and the mechanism can drastically reduce the flow in the microchannel generated by a macro pump system placed outside of a microfluidic chip. Moreover, we can control the cell position with the extremely high resolution of less than 240 nm corresponding of the limitation of pixel of the vision sensor.

CONCLUSIONS

In this paper, we proposed a novel concept, virtual flow reduction mechanism, for high resolution in cell positioning in a microfluidic chip. The mechanism based on the elasticity existing in a fluid circuit of the microfluidic chip, and plays an essential role for the high resolution cell positioning. Focusing on this characteristic, we successfully achieved cell position control with the resolution of 240 nm, corresponding to the limitation of pixel of vision system.

Since the proposed mechanism based on the elasticity of a microfluidic chip without any additional reduction mechanism, the system is kept simple and no additional effort to control multiple units is required. Moreover, there is a great advantage that we can easily construct a high resolution cell manipulation system by combining a macro level actuator and a position sensor, even though the resolution of the actuator is larger than the desired resolution for cell manipulation.

ACKNOWLEDGEMENTS

A part of this work was supported by "Hyper Bio Assembler", "Nanotechnorogy Platform Project (Nanotechnology Open Facilities in Osaka University)" of Ministry of Education, Culture, Sports, Science and Technology, and the Japan Society for the Promotion of Science.

REFERENCES

[1] H. J. Rippon and A. E. Bishop, "Embryonic stem cells", *Cell Proliferation*, Vol. 37, pp.33-34, 2004

[2] K. Takahashi, K. Tanabe, M. Ohnuki, M. Narita, T. Ichisaka, K. Tomoda, and S. Yamanaka, "Induction of pluripotent stem cells from adult human fibroblasts by defend factors", *Cell*, Vol. 131, No.5, pp.861-872, 2007.

[3] T. Wakayama, A. C. F. Perry, M. Zuccotti1, K. R. Johnson and R. Yanagimachi, "Full-term development of mice from enucleated oocytes injected with cumulus cell nuclei", *Nature*, Vol. 394, pp. 369-374, 1998.

[4] Y. Sun, K-T. Wan, K. P. Roberts, J. C. Bischof, and B. J. Nelson, "Mechanical property characterization of mouse zona pellucida", *IEEE Trans. NanoBiosci.*, Vol.

2, pp. 279-286, 2003.

[5] W. S. N. Trimmer, "Microrobots and micromechanical systems". *Sens. Actuators*, Vol. 19, pp. 267-287, 1989.

[6] W. Fukui, M. Kaneko, T. Kawahara, Y. Yamanishi, and F. Arai, "Geometrically-constrained cell manipulation for high speed and fine positioning", *Proc. of the 15th Int. Conf. on Miniaturized Systems for Chemistry and Life Sciences (MicroTAS2011)*, pp. 115-117, 2011.

[7] R. L. Chien and J. W. Parce, "Multiport flow-control system for lab-on-a-chip microfluidic devices", *Fresenius J. Anal. Chem.*, Vol. 371, pp. 106–111, 2001.

[8] C. Fütterer, N. Minc, V. Bormuth, J. H. Codarbox, P. Laval, J. Rossier and J.-L. Viovy, "Injection and flow control system for microchannels", *Lab Chip*, Vol. 4, pp. 351-356, 2004.

[9] N. Pamme and C. Wilhelm, "Continuous sorting of magnetic cells via on-chip free-flow magnetophoresis", *Lab Chip*, Vol. 6, pp. 974–980, 2006.

[10] M. Mahalanabis, H. Al-Muayad, M. D. Kulinski, D. Altman and C. M. Klapperich, "Cell lysis and DNA extraction of gram-positive and gram-negative bacteria from whole blood in a disposable microfluidic chip", *Lab Chip*, Vol. 9, pp. 2811-2817, 2009.

[11] A. Ichikawa, T. Tanikawa, S. Akagi and K. Ohba, "Automatic cell cutting by high-precision microfluidic control", J. Robotics and Mechatronics, Vol.23, No.1, 2011.

[12] S. Sakuma and F. Arai, "Cellular force measurement using a nanometric-probe-integrated microfluidic chip with a displacement reduction mechanism", *J. Robotics and Mechatronics*, Vol.25, No.2, pp.277-284, 2013.

[13] Y. Tanaka, K. Morishima, T. Shimizu, A. Kikuchi, M. Yamato, T. Okano and T. Kitamori, "An actuated pump on-chip powered by cultured cardiomyocytes", *Lab Chip*, Vol. 6, pp.362-368, 2006.

[14] V. Studer, A. Pepin, Y. Chena and A. Ajdari, Analyst, Vol. 129, pp. 944- 949, 2004.

[15] H. T. G. Van Lintel, F. C. M. Van De Pol and S. Bouwstra, "A piezoelectric micropump based on micromachining of silicon", *Sens. Actuators*, Vol. 15, pp.153-167, 1988.

[16] M.A. Burns, B. N. Johnson, S. N. Brahmasandra, K. Handique, J. R. Webster, M. Krishnan, T. S. Sammarco, P. M. Man, D. Jones, D. Heldsinger, C. H. Mastrangelo, D. T. Burke, "An integrated nanoliter DNA analysis device", *Science*, Vol. 282, No. 5388, 1998

[17] N. T. Nguyen and Z. Wu, "Micromixers - a review", J. Micromech. Microeng., Vol. 15, pp. R1-R16, 2005.

[18] S. Maruo and J. T. Fourkas, "Recent progress in multiphoton microfabrication", *Laser and Photon. Rev.*, Vol. 2, No.1-2, pp. 100–111, 2008.

CONTACT

* S. Sakuma, Tel: +81-6-6879-7333; E-mail: sakuma@hh. mech.eng.osaka-u.ac.jp
1998.

SINGLE CELL SEPARATION
BY USING ACCESSIBLE MICROFLUIDIC CHIP

Takeshi Hayakawa[1], Takeshi Fukuhara[2], Keitaro Ito[1], and Fumihito Arai[1]
[1]Nagoya University, Nagoya, JAPAN
[2]Tokyo University of Pharmacy and Life Science, Tokyo, JAPAN

ABSTRACT

Here we proposed a novel device named CACS (Convective Accumulation-based Cell Separator) system that is composed of a cell-trapper unit with accessible (open-structured) microfluidic device and a cell-picker device with high aspect ratio PDMS pillars. Utilizing this systematic instrument, we demonstrated proof-of-concept for cell separation through the process of (1) single cell trap in a well (2) single cell picking by pillar-based tweezers and (3) cell recovery. The optimized system for mammalian cells had enabled us to trap-in and pick-up target cells in 100% efficiency as far as tested. Moreover, picked cells showed 74.5% viability, as a result of a less-invasive cell separation technique. Furthermore, a series of different PDMS pillars indicated their varied efficiency in pick-up and recovery steps. We also found out that the aspect ratio of PDMS pillar is critical for tweezing a cell of interest. To improve on the performance of CACS system, further tuning is necessary, especially on the recovery unit for achieving an efficient and damage-free separation. Though CACS system shows low-throughput at this moment, its simplicity is applicable for any cell type by optimizing each pipeline.

INTRODUCTION

Single cell separation is one of the most important techniques in modern biomedical research such as regenerative medicine and antibody therapy. Especially, monoclonality of materials such as stem cell or hybridoma is critical for quality-controlled mass-production. As for monoclonal antibody, it is essential to adopt the repeated steps of limiting dilution for establishing monoclonal hybridoma. This will result in obtaining the hybridoma clone as single-cell derivative with a uniform function. Currently, limiting dilution of hybridoma requires a labor-intensive manual operation, so fluorescent activated cell sorter (FACS) is used to take the advantage that comes from the multi-color analysis at the single cell level as well as in a high-throughput manner. But it is inevitable for some cell types to get injured by the highly pressurized fluidic controller. Besides, comprehensive system requires skilled operation as well as huge initial cost for purchasing. As such, easy-to-handle and affordable device enabling single cell separation is highly demanded.

Recent progress in microfluidic device with advanced fabrication technology had encouraged the development of cell analyzer and sorter to meet cost effectiveness for mass production and single use requirement at clinical settings to avoid contamination and infection. To note, driving force of microfluidics in most of the published works rely on precise pump control to create laminar flow in the microfluidic chip [1]-[3]. Coherent laminar flow requires a complex pump connection, implying fragile and unstable property by small bubbles or debris in fluidic channel, resulting in malfunctioning. Furthermore, due to the packed chip structure, the system requires inlet or outlet connection in order to load or recover samples for the use of microfluidic

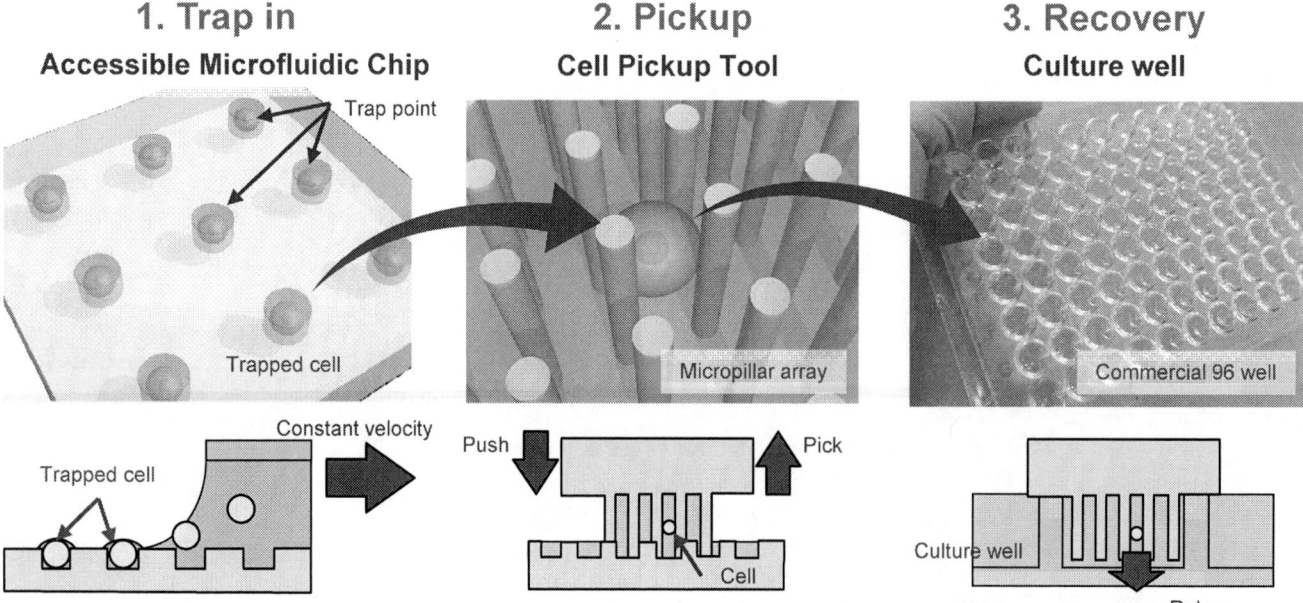

Figure 1: Concept of CACS (Convective accumulation-based Cell Separation) system.

chip. As such, tiny samples such as mammalian cells with 10 µm in size often get tangled and lost in the connection parts. These limitations become a serious problem when we need to sort and recover rare (and precious) cells from the chip for the subsequent experiments.

To overcome these difficulties and limitations, we, hereby proposed a novel, simple, and easy to use single cell separation system by using accessible (open-structured) microfluidic chip and cell pickup tool.

CONCEPT

In the CACS system, cells firstly get loaded onto the accessible microfluidic chip (Trap-in unit), followed by recovering by the cell pickup tool (Pick-up unit). Trap-in unit equips to solve the problems caused by bubbles mixing and the connections in microfluidic channel. Trap-in unit simply works with no connectors to the chip and no precise flow control. With open-top structure, air bubbles are carried away from the top of the chip. In combination with the convective accumulation method, Trap-in unit provides another advantage for the trapped cells to be accessible from the top for the pick-up device. After trapping single cell in a well, our Pick-up tool makes it easier to recover cell of the target. This is different from any current micromanipulator controlled under microscope, as we developed a Pick-up unit that has a soft micropillar array. By tweezing cell with micropillar array, it becomes possible to collect trapped cell in a well on the accessible microfluidic chip. This process can be performed in a quick and accurate manner with visual alignment. With this CACS system, we made cell separation simpler with minimum damage to cells.

CONVECTIVIE ACCUMULATION-BASED SINGLE CELL TRAPPING

To trap single cell on microfluidic chip, we employed convective accumulation method [4]-[6]. Figure 2 shows the principle of the Trap-in unit in CACS system. Firstly, small particles such as cells were accumulated at an air-liquid interface formed in between the cover glass and substrate chip. Secondly, by sliding the cover glass towards one direction, the cell is trapped in a well of the substrate chip as shown in (a). By designing trap wells of defined size in two dimensional space, loaded cells get trapped as single cell per well (b). The advantages of Trap-in unit are the followings:

> Open-structured and easy access to cell of target
> Arbitrary patterning of trap points in size and area
> Sequential trapping by defined parameters
> Less-invasive, free from pressure and rapid flow

Our group previously succeeded in single cell separation of circulating tumor cells using flow-based separation device [7]. Given the advantages in our previous study, we extended to apply in flow-free CACS separation device against typical mammalian cell (10–15 µm), by patterning the well with the size of 20 µm in diameter.

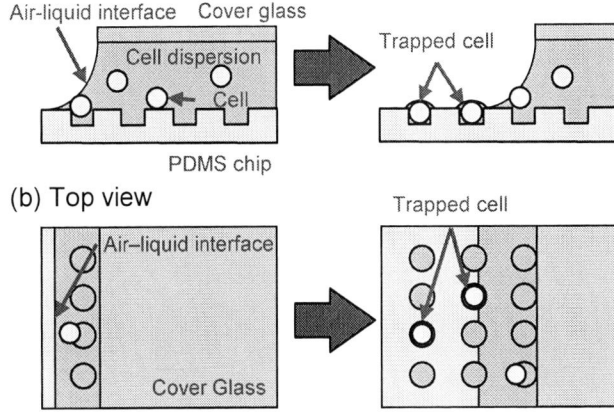

Figure 2: Principle of CACS and design of Trap-in microfluidic unit. (a) side view and (b) top view

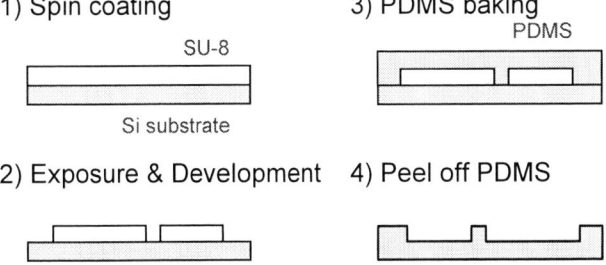

Figure 3: Fabrication process of CACS Trap-in chip.

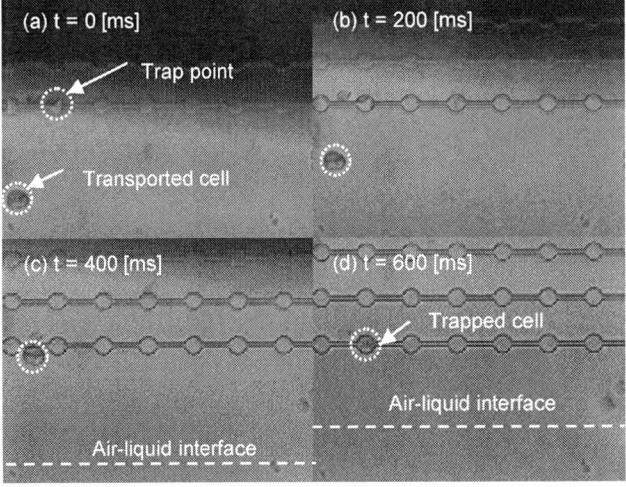

Figure 4: Kinetic process of Trap-in step in CACS system.

Fabrication

To fabricate an accessible Trap-in microfluidic chip, we used polydimethylsiloxane (PDMS) molding (Fig. 3).

Fabrication processes are the followings:

1. SU-8 (Nippon Kayaku Co. Ltd.) was spin coated on silicon substrate.
2. Coated SU-8 was patterned by the mold.
3. PDMS was poured into the fabricated mold and

baked.

4. PDMS was peeled off from the mold.

Experimental result

Figure 4 shows the sequential photographs demonstrating single cell trap by the convective accumulation method. Figure 4 (a) - (c) show that dispersed cell located close to air-liquid interface was transported in a guided direction. After that, the transported cell was successfully trapped in a designed well on the chip (Figure 4 (d)). We observed a significant amount of cells trapped into the designed wells in a sequential manner.

PICKUP TOOL WITH HIGH ASPECT RATIO PDMS PILLARS

To achieve a single cell collection from the accessible microfluidic chip, we developed a cell pickup tool. Conventionally, single cell manipulation or collection has to be performed in open space, such as culture dishes, by micromanipulator with 3 degrees of freedom (DOF) [8][9]. However, CACS system can trap single cells on an accessible microfluidic chip simply by the convective accumulation method as mentioned in the previous section. In brief, we can immobilize single cells two dimensionally and reduce DOF on the motion of cells. As such, there is no need for 3 DOF instead a simple 1 DOF manipulation is sufficient for the collection of trapped single cells. For further simplification of the system, a pickup tool with PDMS pillars was used. We can collect trapped single cells by rough alignment of the tool on trap point of the microfluidic chip by human eyes and followed by pushing the tool on the trapped ce

lls by human hand. Trapped cells are enfolded by PDMS pillars and collected from the accessible microfluidic chip. To reduce the cell damage as much as possible, we ended up using thin and high aspect ratio PDMS pillars.

Fabrication

To fabricate the cell pickup tool with high aspect ratio PDMS pillars, we used PDMS molding and employed deep reactive ion etching (D-RIE) for fabrication of silicon mold (Figure 6). Detailed fabrication processes are the followings:-

1. SU-8 (Nippon Kayaku Co. Ltd.) was spin coated on silicon substrate.
2. Coated SU-8 was patterned as the D-RIE etching mask.
3. Silicon substrate was etched by D-RIE and residual SU-8 was removed by O_2 plasma washing after the D-RIE process.
4. PDMS was poured into the fabricated mold, baked and peeled.

Figure 7 shows the scanning electron microscope (SEM) image of fabricated silicon mold and PDMS pillars with an aspect ratio of 10 (5 μm in diameter and 50 μm in height).

Experimental result

Firstly, to evaluate on the cytotoxic damage at the pickup

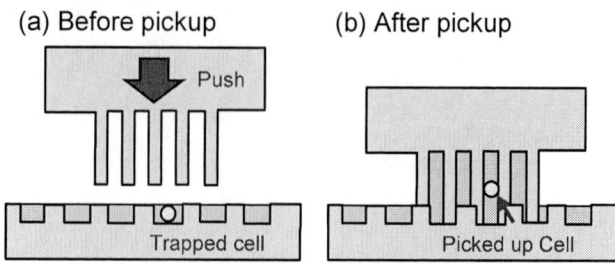

Figure 5: Schematic view of Pick-up unit.

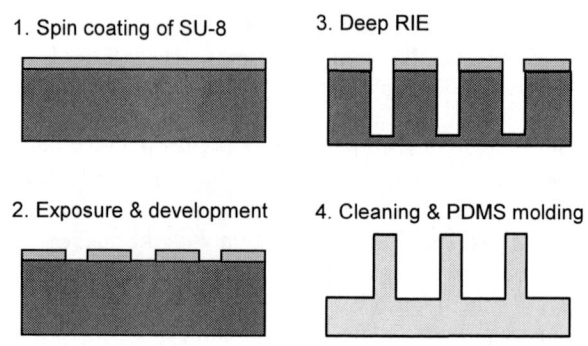

Figure 6: Fabrication process of CACS Pick-up device.

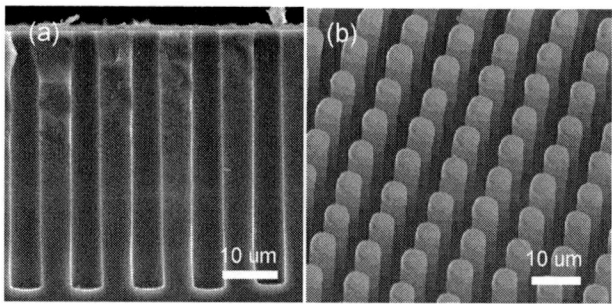

Figure 7: SEM images of Pick-up device with PDMS pillars. (a) Fabricated silicon mold and (b) Molded PDMS pillars

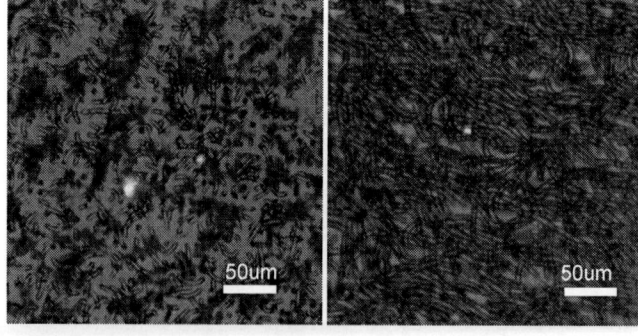

Figure 8: Visualization of cells on Pick-up unit. Merged images illustrated bright field image, live cells (green fluorescence) and dead cells (red fluorescence).

step, we performed cell viability verification after collection. Collected cells were subsequently incubated on pickup tool for 3 hours in cell culture medium. After the incubation, we evaluated viability by staining the cells with Live/Dead viability kit (L-3224, Life Technologies Co. Ltd.), followed by incubation for 40 minutes in 37°C, 5% CO_2 environment. After labeling and incubation, we observed the cells by inverted microscope for bright field and fluorescent images.

In figure 8, representative images show the merged image of bright field and fluorescent image of stained cells. We can segregate live cells in green fluorescence from red fluorescent cells that are dead. Repeated experiments with microscopic observation could capture images of 38 cells with green fluorescence out of a total of 51 cells, which correspond to 74.5 % in viability. We concluded that the pick up process with designed pillars is capable to pick and hold live cells at a practically sufficient level.

WHOLE SYSTEM EVALUATION

Next, we proceeded to evaluate the CACS system through the following three steps:
1. Trap single cell by convective accumulation on the accessible microfluidic chip.
2. Pick up single cell by a tool designed with PDMS pillars.
3. Transfer and release single cell into culture dish.

We conducted the whole process for a total of 28 times, followed by confirming the viability of isolated cells. As a results, 8 wells out of 28 wells contained live cells (28.6 %). We observed single and live cell per well in 4 cases (14.3%) Therefore, current system enabled single cell separation in 14.3 % of success rate.

CONCLUSIONS

We proposed a novel single cell separation system (CACS, Convective Accumulation based Cell Separator) using the advantages of both convective accumulation and tweezing by micropillars. With a different mechanical concept from the typical microfluidic devices, this system utilized accessible microfluidic chip free from controlling the highly pressurized laminar flow that can contribute to the major cause of cellular damage. The following step after trapping was conveyed using the cell pickup tool with PDMS micropillars. Although we need to consider further optimization to increase cellular viability, it is quite noteworthy that PDMS pillars with higher aspect ratio seems to show more efficiency for picking MDCK cells, but not myeloma cells (data not shown). In fact, the kinetics of the pick-up process remains unclear, but it is currently under investigation. The bottom line, however, is that we could manipulate cells by collecting and separating at the single cell level by the convective accumulation method on accessible microfluidic chip and collect by pickup tool from the chip. After which, collected single cells can be transferred to culture wells. Compared with conventional FACS or microfluidic cell sorting devices, it is true that our CACS system shows lower-throughput. Rather than pursuing throughput, CACS system offers quick sorting for preparing intact cells under physiological condition. Given the flexibility to fabricate the size of trapping wells, CACS system opens opportunities for separating larger-sized and damage-sensitive cells (e.g. oocyte, megakaryocyte, oligodendrocyte, embryo etc.). We also believe that CACS system contributes not only to the discovery of unknown cell population but also provides the possibility to track reactive multicellular state *in situ*. Due to the simple configuration of CACS system, it is widely applicable for a new area of research that investigators pursue.

ACKNOWLEDGEMENTS

This work was financially supported by Grant-in-Aid for Japan Society for the Promotion of Science (JSPS) Fellows.

REFERENCES

[1] S. M. Kim, S. H. Lee, and K. Y. Suh, "Cell research with physically modified microfluidic channels: A review", *Lab on a Chip*, 8 pp. 1015-1023, 2008.

[2] M. Yamada, M. Nakashima, and M. Seki, "Pinched Flow Fractionation: Continuous Size Separation of Particles Utilizing a Laminar Flow Profile in a Pinched Microchannel", *Anal. Chem.*, 76, pp. 5465-5471 (2004).

[3] A. J. Mach, J. H. Kim, A. Arshi, S. C. Hur, and D. D. Carlo, "Automated cellular sample preparation using a Centrifuge-on-a-Chip", *Lab on a Chip*, 11 pp. 2827-2834, 2011.

[4] K. Nagayama, "Two-dimensional self-assembly of colloids in this liquid films", *Colloids and surfaces A*, 109, pp. 363-374, 1996.

[5] Y. Yin and Y. Xia, "Self-assembly of Monodispersed spherical colloids into complex aggregates with well-defined sizes, shapes, and structures", Adv. Mater., 13, No. 4, pp. 267-271, 2001.

[6] M. C. Park, J. Y. Hur, K. W. Kwon, S. H Park, and K. Y. Suh, "Pumpless, selective docking of yeast cells inside a microfluidic channel induced by receding meniscus", *Lab on a Chip*, 6, pp. 988-994, 2006.

[7] T.Masuda, Y. Sun, M. Niimi, A. Yusa, H. Nakanishi, and F. Arai, "Cell layoutor: label-free cell isolation and aspiration system of circulating tumor cells", in *Digest Tech. Papers of The 17th International Conference on Miniaturized Systems for Chemistry and Life Sciences (MicroTAS 2013)*, Freiburg, October 27-31, 2013, pp. 1662-1664.

[8] K. Yanagida, H. Katayose, H. Yazawa, Y. Kimura. K. Konnai, and A. Sato, "The usefulness of a piezo-micromanipulator in intracytoplasmic sperm injection in humans," *Human Reproduction*, vol. 14, pp. 448–453, 1998.

[9] A. Ramadan, K. Inoue, and T. A. T. Takubo, "New architecture of a hybrid two-fingered micro-nano manipulator hand: Optimization and design," *Advanced Robotics*, vol. 22, pp. 235–360, 2008.

CONTACT

*Takeshi Hayakawa, Tel: +81-52-789-5026;
E-mail: t-hayakawa@biorobotics.mech.nagoya-u.ac.jp

STUDY OF HOTSPOT COOLING USING ELECTROWETTING ON DIELECTRIC DIGITAL MICROFLUIDIC SYSTEM

Govindraj Bindiganavale[1], Seung Mun You[2] and Hyejin Moon[1]*
[1]University of Texas at Arlington, USA
[2]University of Texas at Dallas, USA

ABSTRACT

This paper presents a novel digital microfluidic (DMF) liquid cooling system using electrowetting on dielectric (EWOD) developed for demonstrating and studying hotspot cooling towards electronics thermal management. The merits of this cooling system lie in the fact that no mechanically moving parts such as valves, pumps and fans are required to achieve cooling pumping, thus having smaller form factor than bulky heat pipes and other conventional cooling systems. This study reveals close profiles of temperature change during coolant drop motion over hotspot as well as importance of phase-change cooling around the meniscus of the droplet.

INTRODUCTION

Recently, there has been a tremendous scaling down in the real estate utilization of an integrated chip (IC) and a subsequent increase in the transistor density. As more transistors are packed due to better lithography resolution, the heat generated also increases [1]. Conventional cooling systems include heat sinks, heat pipes, fans and vents assemblies which are found in majority of the electronic devices today. In order to meet the cooling demands of the latest ICs, these bulky and heavy conventional cooling systems cannot be used. Therefore, there has been abundant research done in developing small and lightweight, yet efficient cooling systems [2-4].

In this paper, a hotspot cooling system using fluids handled drop-wise is presented. In EWOD hotspot cooling, the droplet is made to move across the hotspot directly by

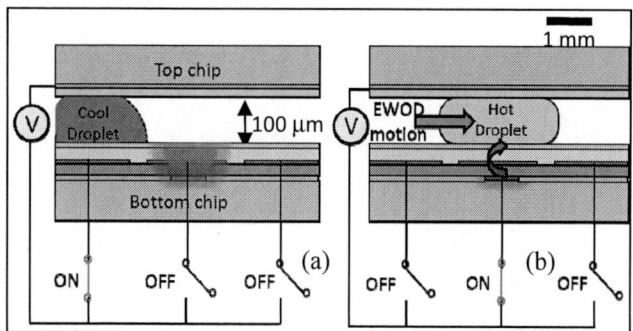

Figure 1: Conceptual demonstration of EWOD cooling of the hotspot. Droplets flow over the hotspot, (a) & (b), reducing its temperature and carry away heat.

sequential electrical actuation of discrete electrodes as shown in Figure 1. This concept of digital microfluidics is based upon the Lippmann's equation [5]. When a potential difference is applied across the liquid drop on an electrode coated with a dielectric and a hydrophobic layer, the surface

tension of the droplet meniscus can be manipulated due to presence of electrostatic charge on the dielectric surface. This charge imposes an electrostatic force which changes the contact angle of the droplet hence making it wet the surface from a non-wetting state. A pressure difference is imposed within the droplet which results in bulk fluid motion in any direction [6]. This novel technique has been exploited to create, move, separate and merge droplets [7] on a chip for different lab-on-chip applications in digital microfluidics, including hotspot cooling [8] for electronics thermal management.

Hotspot cooling using EWOD DMF has many advantageous features. As described above, in EWOD DMF, droplets move on surface without designated fluidic channel, and their motion is controlled by individually addressed electric voltage application. Therefore, by coupling with simple temperature measurement and feedback control, it is readily capable to deliver coolant drops to varying hotspots at the time of use. In addition, EWOD DMF can realize a true near junction cooling by moving coolant droplets directly over the hotspot. A EWOD DMF device only needs a thin electrical insulation layer (typically a few μm), electrodes and dielectric layer (less than 10 μm) over hotspot and does not require any thermal interface materials, therefore, can minimize the thermal resistance from the hotspot to coolant flow. Last, but not the least, EWOD DMF is highly suitable to integrate into 3-Dimensional IC chip stack. As shown in Figure 1, EWOD DMF has a very thin dimension. It only needs two parallel plane chips which can be inserted between the front and back surfaces of the IC chip stack separated by a small space (typically ~ 100 μm) where coolant drops are sandwiched. Therefore, it is easy to integrate between multiple IC chip stacks of the 3-D chip and can efficiently remove heat from the middle of the stack.

EXPERIMENTAL

EWOD Device Fabrication & Setup

As shown in Figure 2, the hotspot device consists of a top chip and a bottom chip, which are placed in a parallel-plate configuration. These chips are fabricated in a clean room using conventional thin film semiconductor fabrication techniques. The bottom chip comprises an ITO (Indium tin oxide) based heater which emulates the hotspot and also doubles as an RTD (Resistance temperature detector). Chromium is used for patterning the EWOD electrodes which are aligned with the heater and define the pathway for the droplet motion. Su-8 (5 μm thick) is used as the dielectric for EWOD and a thin layer of Teflon is coated on the dielectric for hydrophobicity. Similarly, the top chip

978-1-4799-3510-9/14 $31.00 © 2014 IEEE

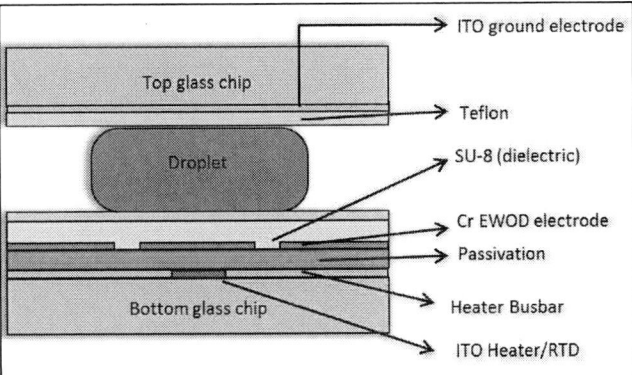

Figure 2: Pictorial cross section of EWOD device.

is completely coated with ITO for EWOD grounding purpose and is also coated with Teflon.

After the top and bottom chips are fabricated, the EWOD device is assembled by placing the top chip on top of the bottom chip separated by spacers. Figure 3(a) shows the bottom chip with the inlet and exit liquid reservoirs. After the EWOD device is interfaced to the electronics using a PCB, liquid is filled through the inlet (Fig. 3(b)) using a syringe pump and drops are generated by EWOD which move from the inlet to the exit over the hotspot. The heater is connected to the DAQ (Digital Acquisition) instrument for measuring the RTD resistance readings.

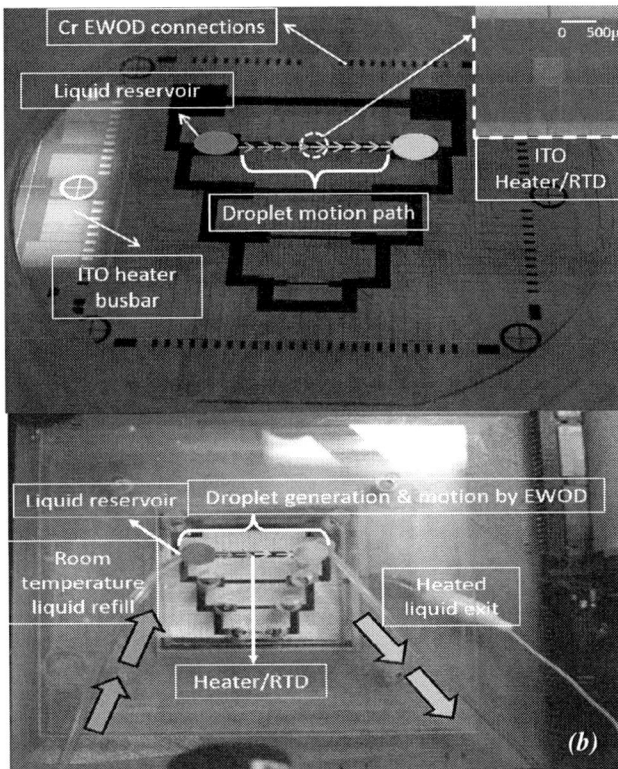

Figure 3: Top view of bottom chip (a). EWOD system showing the liquid inlet and exit piping for filling & removal (b). Drops are generated, moved from the liquid reservoir on chip by EWOD.

ITO Heater & RTD

Indium Tin Oxide is used as the material of choice for the heater/RTD due to its ease of integration with the EWOD device fabrication, optical transparency and good

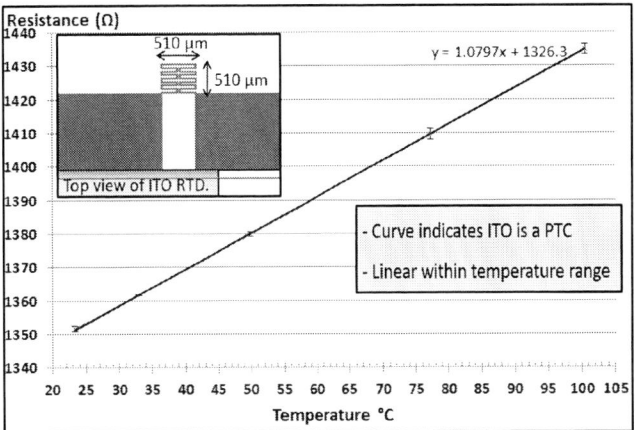

Figure 4: Calibration curve of the ITO thin film RTD showing linear PTC characteristics within the temperature range of 22 °C (R.T.) to 80 °C.

linear PTC (Positive Temperature Coefficient) characteristics (Fig. 4). Wires are directly soldered onto the busbars to establish zero contact resistance. Previous work [9] showed that when the ITO RTD was integrated close to the EWOD electrodes, a lot of noise was created in the RTD readings due to ground bouncing effect. This effect was drastically reduced as higher current was used to power the heater which subdued the noise in the readings.

EWOD Motion and Hotspot Cooling

The experimental procedure starts with turning ON the heater and dispensing liquid in the inlet reservoir electrodes. By using a specific switching pattern, droplets of 600 nL volume are generated from the reservoir and transported over the hotspot to the exit reservoir. As pointed out in the

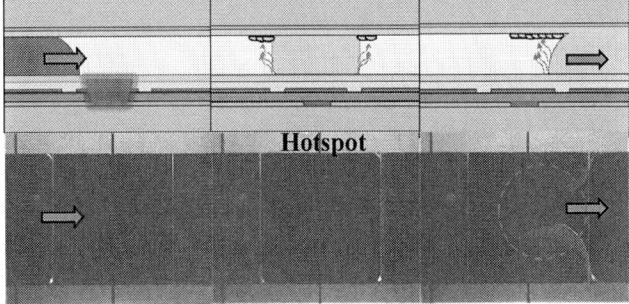

Figure 5: "Crossection" view of EWOD device pictorially demonstrating evaporation at the meniscus (red circle). The vapor then condenses at the top chip (light blue drops in crossection view).

introduction section, as the droplet moves over the hotspot, vapor traces are seen around the advancing and receding menisci of the droplet (Fig. 5). Once a droplet leaves the hotspot, it makes way for another droplet to pass over it. By repeating this process for multiple droplets, the resistance data of the heater over time is collected, analyzed with the calibration data and a temperature plot is made for further analysis.

OBSERVATIONS
Temperature Data Analysis

As the droplets moved over the hotspot, we can see a drop and a subsequent rise in temperature of the hotspot. Each drop and rise corresponds to one cycle of a single liquid droplet cooling. Figure 6(a) shows cooling curves of

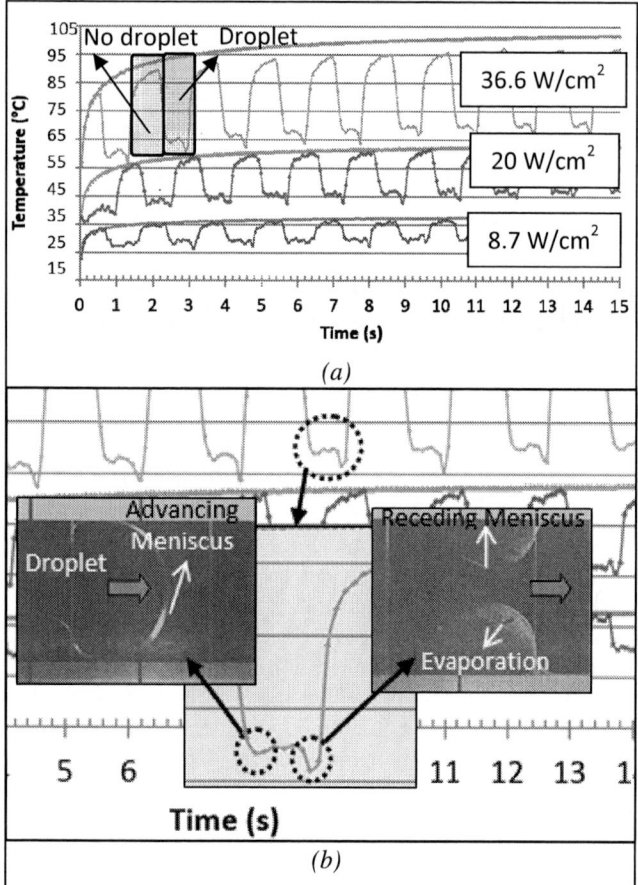

Figure 6: Hotspot cooling result for three heat fluxes (a). Evaporation seen at meniscus of the droplet (b) further reduced temperature of the hotspot during droplet inlet & exit. This effect was more pronounced at exit due to larger dimensions of the thin receding meniscus.

9 droplets for three levels of heat flux provided to the heater. As shown in Figure 6(b), we see some interesting data points when the droplet enters the hotspot.

When the meniscus enters the hotspot, the temperature reduces little more than that when it dwells on the hotspot. Upon exit, this effect is magnified as the temperature drops further. We attribute this repeated effect in all cooling cycles to the enhanced heat transfer due to advancing and receding meniscus of the drop and phase-change at these sites which enhances heat transfer from the hotspot to the drop, hence lowering the temperature. A more in depth analysis will be provided in the next sub-section.

A simple 3-D simulation was performed in COMSOL

Multiphysics to check for any thermal spreading effects. As shown in Figure 7, we see that the area of heat spreading is atleast 4 times the size of the hotspot. This shows us that

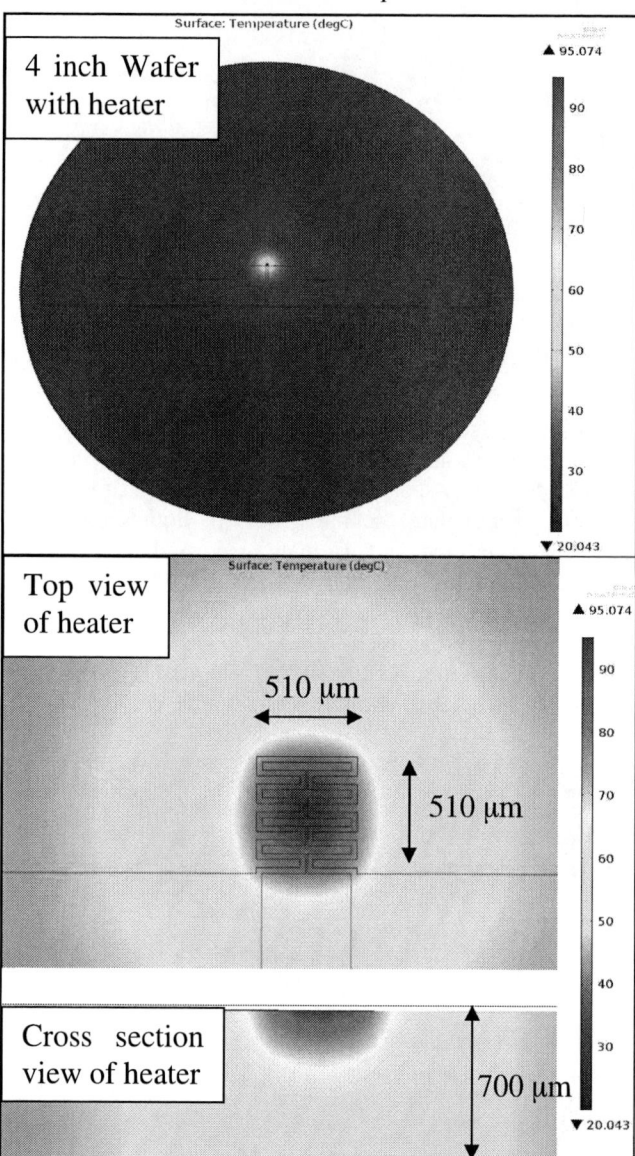

Figure 7: COMSOL Simulation result showing spreading through the top and depth planes. The temperature values for 36.6 W/cm^2 agree to that of experimental ones.

there is considerable spreading through the glass substrate and the temperature readings at steady state matches with that of the experimental values.

Phase-change Cooling Effect

As described in the previous section, it is observed that during the droplet entry and exit on the hotspot, phase-change cooling occurs at the menisci of the droplet. As shown in Figures 6 and 8, during droplet entry, the advancing meniscus appears to be less thick and covering a small area on the hotspot than that of the receding meniscus. As the dimensions of the receding meniscus is larger, more phase-change cooling takes place giving rise to further drop in temperature of the hotspot.

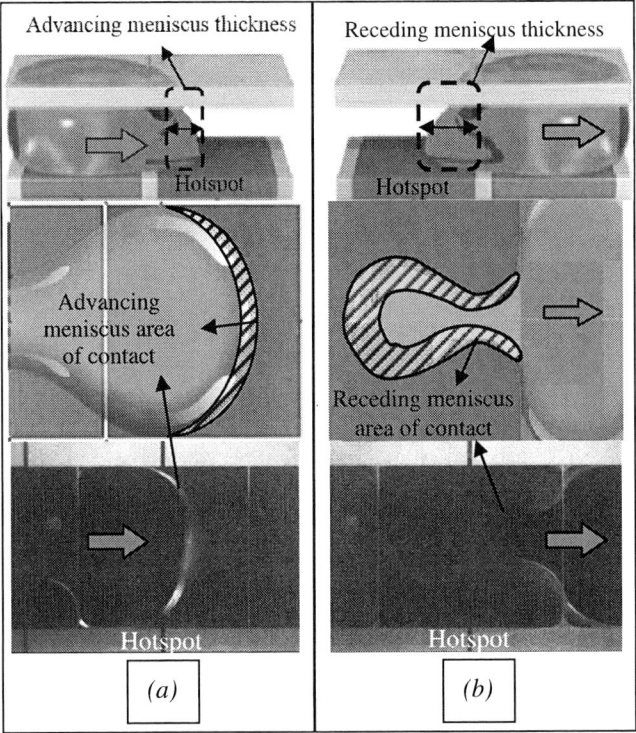

Figure 8: Figures (a) and (b) showing the side and top views of the advancing and receding meniscus and giving a perspective of the dimensions of the meniscus. The receding meniscus is observed to have a larger thickness and surface area in contact with hotspot.

CONCLUSIONS AND FUTURE WORK

This paper presents an innovative method to cool hotspots using EWOD. A EWOD DMF device comprising a top and bottom chip in a parallel-plate configuration was fabricated and tested. Three power levels were supplied to the ITO heater/RTD with a heat flux of 8.7, 20 and 36.6 W/cm^2. Nine droplets were moved across the hotspot by EWOD which lowered the temperature by a maximum change of 30 °C observed for the highest heat flux. For each cooling cycle, it was also observed that the advancing and receding meniscus at the entry and exit of the droplet corresponded to further reduction in temperature due to phase change cooling. Moreover, the phase-change cooling was more pronounced for the receding meniscus due to its larger thickness and more area of contact with the hotspot.

To take advantage of phase-change cooling in future, we propose to use SHNC (Super hydrophilic nanoporous coating), a surface treatment which enables higher spreading area resulting in thin film formation on the hotspot. This will enhance the phase-change heat transfer and reduce the hotspot temperature further.

ACKNOWLEDGEMENTS

This study was supported by the Defense Advanced Research Projects Agency/Microsystems Technology Office (DARPA/MTO) under the supervision of the Program Manager, Dr. Avram Bar-Cohen, with the program grant number: W31P4Q-11-1-0012.

REFERENCES

[1] K. Seshan, *Handbook of Thin Film Deposition*, Elsevier, 2012.

[2] S. -H. Moon and G. Hwang, "Development of the Micro Capillary Pumped Loop for Electronic Cooling", *THERMINIC 2007*, 17th – 19th September 2007, pp.72-76.

[3] E. G. Colgan et al., "A Practical Implementation of Silicon Microchannel Coolers for High Power Chips", *IEEE Transactions on Components and Packaging Technologies*, June 2007, Vol. 30, No. 2, pp. 218-225.

[4] Heffington, S. N., Black, W. Z., Clezer, "Vibration-induced Droplet Atomization Heat Transfer Cell for High Heat-flux Applications", *ITherm 2002, Eighth Intersociety Conf. on Thermal and Thermomechanical Phenomena in Electronic Systems*, vol. 30, pp. 408 – 412, 2002.

[5] M. G. Lippmann, "Relations entre les phénomènes electriques et capillaires," *Ann. Chim. Phys.*, vol. 5, no. 11, pp. 494–549, 1875.

[6] R. Bavière, J. Boutet, Y. Fouillet, "Dynamics of Droplet Transport Induced by Electrowetting Actuation", *Microfluidics and Nanofluidics*, vol. 4, no. 4, pp. 287-294, 2007.

[7] S. K. Cho, H. Moon, C. -J. Kim, "Creating, Transporting, Cutting, and Merging Liquid Droplets by Electrowetting-Based Actuation for Digital Microfluidic Circuits", *Journal of Microelectromechanical Systems*, vol. 12, no. 1, pp. 70-80, 2003.

[8] H. Moon, S. Bindiganavale, Y. Nanayakkara, D. W. Armstrong, "Digital Microfluidic Device using Ionic Liquids for Electronic Hotspot Cooling", *Proc. of the Seventh Intl. ASME conf. on Nanochannels, Microchannels and Minichannels,* Pohang, South Korea, June 22-24, 2009.

[9] G. S. Bindiganavale, H. Moon, S. M. You, M. Amaya, "Digital Microfluidic Device for Hotspot Cooling in ICs using Electrowetting on Dielectric", *Proc. Of the ASME 2012 3rd Micro/Nanoscale Heat & Mass Transfer Intl. Conf. MNHMT2012,* Atlanta, Georgia, March 3-6, 2012, pp. 39-42.

CONTACT

*Hyejin Moon, tel: +1-817-272-2017; hyejin.moon@uta.edu

SURFACE-ACOUSTIC-WAVE DRIVEN POINT SOURCE ATOMIZER INTEGRATED WITH PICOLITER MICRO PUMPS FOR POLYMERIC NANOPARTICLES SYNTHESIS

Shun Sugimoto[1], Motoaki Hara[1], Hiroyuki Oguchi[1], Atsushi Yabe[1] and Hiroki Kuwano[1]

[1]Graguate school of engineering, Tohoku University, Sendai, JAPAN

ABSTRACT

In this study, we report a surface acoustic wave (SAW) driven atomizer for polymeric nanoparticles synthesis based on evaporation-induced self-assembly (EISA). The atomizer consisted of a pair of arc-shaped interdigital transducer (AS-IDT) was designed for converging the SAW into a solvent. The atomizer can generate a point-source narrow mist spray and is very useful to improve the yield of nanoparticle synthesis. As a result of atomizing test with water, we succeeded in an ejection of narrow mist spray from a point on the substrate. The spray was 0.7 mm in width in the vicinity of the substrate, and kept less than 1 mm in width up to 8 mm in height. Also, the atomizer was integrated with SAW driven micro pumps to supply an appropriate amount of liquid at the mist spray ejection point. Using this pump, the discharge of water controlled in picoliter order was achieved.

INTRODUCTION

Polymeric nanoparticle synthesis based on evaporation-induced self-assembly (EISA) is an attractive application of the atomizer in *in vivo* drug and gene delivery, DNA hybridization based biosensors, or immunodiagnostics [1-2]. In these applications, SAW driven atomizers can give an excellent solution [2]. It can be operated with low voltage and miniaturized easily to a chip scale. However, in general SAW driven atomizers have some drawbacks. How to supply a solvent to atomizers is one of them.

It is known in SAW streaming based devices that the ejection mode changes drastically by supplied SAW power and amount of the solvent [3-5]. For efficient atomization with low SAW power, solvent should be supplied in picoliter order. However, because solvent are supplied with a syringe pump manually in literatures, it is difficult to obtain sufficient accuracy and to keep atomization stable for a long time.

Also, it is crucial for efficient EISA to control the direction of atomization. Conventionally, SAW driven atomization is induced in wide area on the substrate [2, 6]. Therefore, a huge reservoir against the atomizer is required to gather the nanoparticles. If a direction of atomization is controlled with narrow spray, nanoparticles can be sorted efficiently using micro arrayed chambers [7-8].

To overcome the above drawbacks, we developed a SAW driven atomizer integrated with picoliter micro pumps. This paper describes the discharge characteristic of the solvent from micro pumps. Also, we present an optimization of SAW convergence using an optical observation of the vibration and the ejection characteristic from the SAW atomizer.

STRUCTURE AND FABRICATION

Figure 1 shows a schematic illustration of our SAW driven atomizer. The atomizer included a pair of AS-IDTs and a channel to obtain point source atomization. The pump which including a reservoir and an IDT pair was located on an edge of the channel. When SAW is radiated from the AS-IDT, SAW was focused on a part of solvent in the channel and atomized it for perpendicular direction against the substrate.

Figure 2 shows a fabrication flow of the atomizer. 200-nm-thick Al film was deposited and patterned on the cleaned 128°Y-X LiNbO₃ substrate (Fig.2 (i)-(iii)). Then, the channel was fabricated by a sand blast technique (Fig.2 (iv)). Depth of the channel was 150 μm. Next, 150-μm-thick resist (SU-8 2100) was patterned as the reservoir (Fig.2 (v)). The fabricated device was, finally, mounted on the printed board for easy handling and wiring as shown in Fig. 3. From this figure, it was confirmed that there are IDTs for the pump,

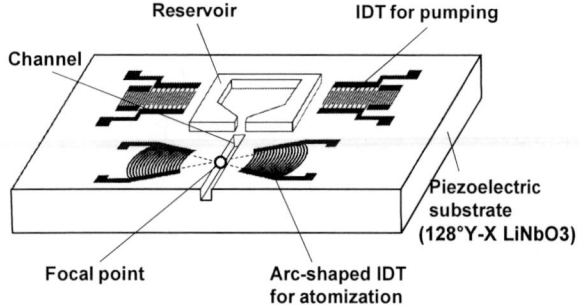

Figure 1: Schematic illustration of the SAW driven atomizer with a picoliter pump and a channel.

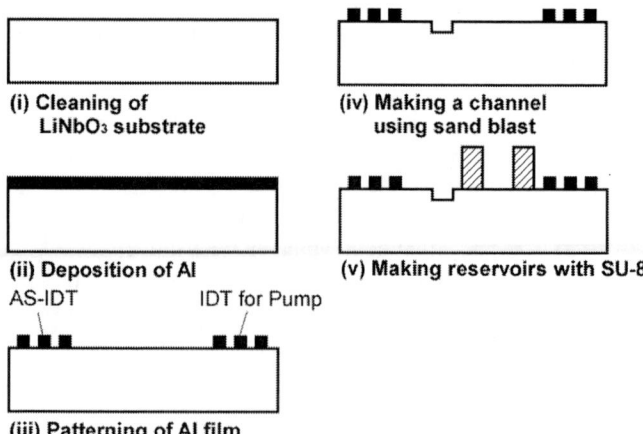

Figure 2: Fabrication flow of the integrated atomizer

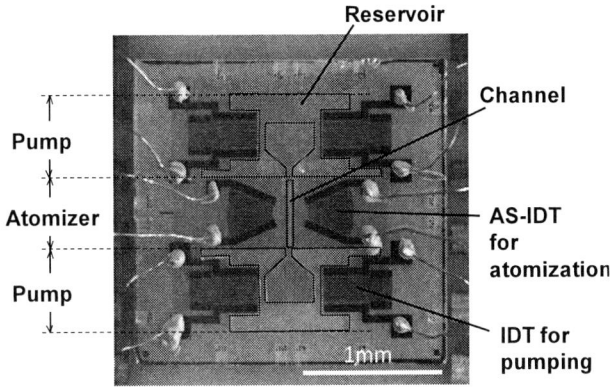

Figure 3: Optical micrograph of the SAW driven atomizer with two picoliter pumps (red colored) and a channel (blue colored).

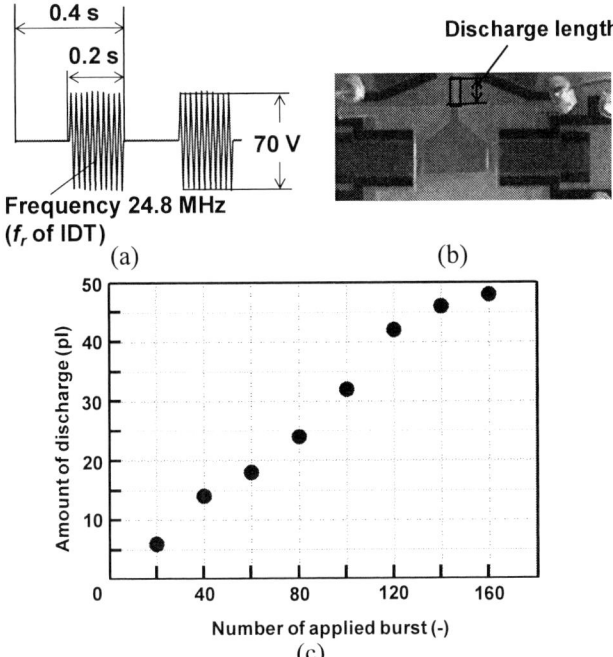

Figure 4: Relationship between the amount of discharge into the cannel and the number of applied burst voltage: (a) applied burst voltage, (b) optical micrograph of the operating pump, (c) discharge characteristics for the burst number

AS-IDTs for the atomizer, a channel and reservoirs on one substrate.

EXPERIMENTAL RESULTS
Testing of the integrated micro pump

The fabricated micro pump was evaluated using water as a working fluid. The pump was driven by the burst voltage to prevent the temperature of the device from rising. Frequency, period, duty and amplitude of applied voltage were 24.8 MHz, 0.4 s, 50 %, and 70 V, respectively, as shown in Fig. 4 (a). Amount of discharge was calculated the cross section of the channel and discharge length shown

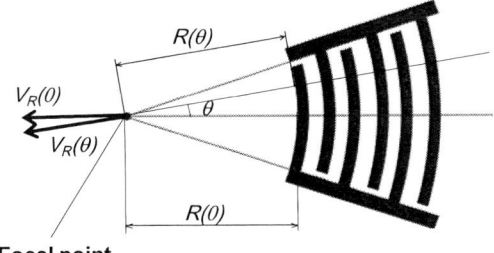

Figure 5: Design of the AS-IDT pattern

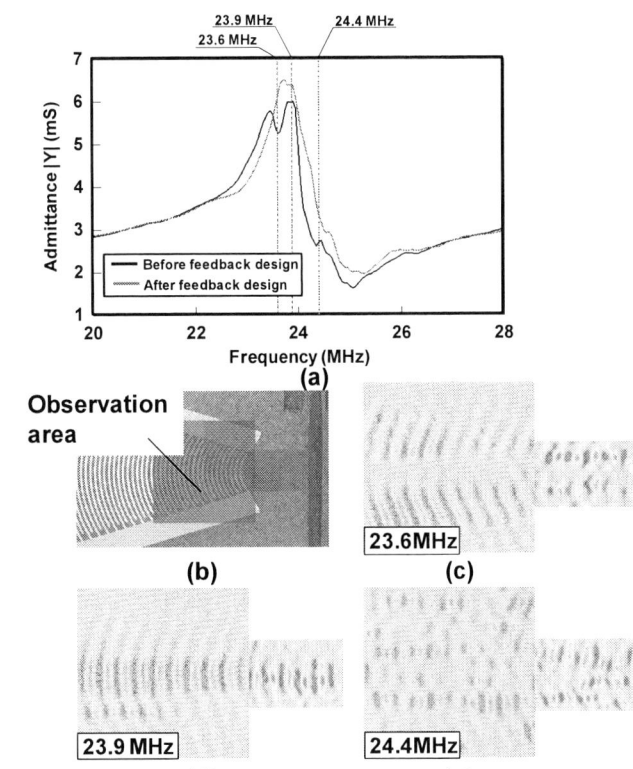

Figure 6: Frequency characteristics of an AS-IDT and vibration distribution of each frequency.

in Fig. 4 (b), in where the cross sectional area of the channel was 200 μm in width × 150 μm in depth.

Figure 4 (c) shows a relationship between the discharge amount and the burst number. It was confirmed that the discharge into the channel was proportional to the burst number. The discharge rate was 0.3 pl a burst.

Evaluation of the AS-IDT with Optical interferometer

The AS-IDT was designed using following equation [9-10]:

$$R(\theta) = R(0) \frac{V_R(\theta)}{V_R(0)} \qquad (1),$$

where $R(\theta)$ and $V(\theta)$ are distance from the focal point and acoustic velocity for the θ as shown in Fig. 5. $V(\theta)$ is depend

978-1-4799-3510-9/14 $31.00 © 2014 IEEE

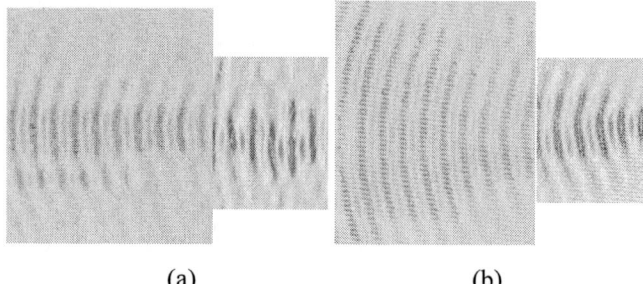

(a)   (b)

Figure 7: Effect of design feedback in the AS-IDT. (a) The first design, (b) The second design improved by the feedback from the optical observation

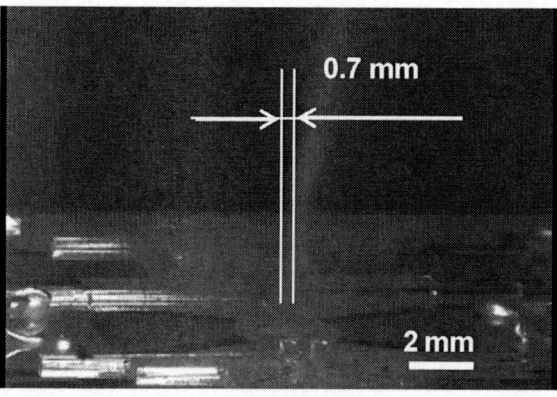

(a)

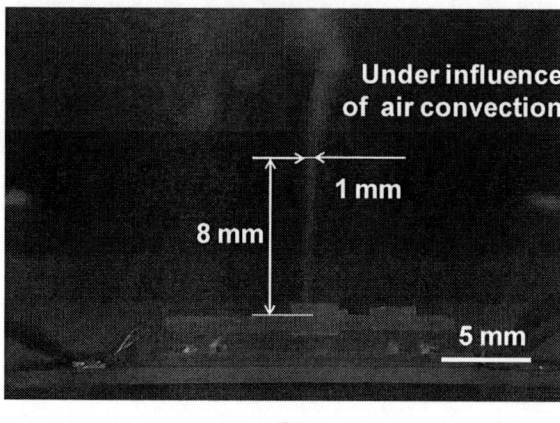

(b)

Figure 8: *Point source atomization using AS-IDT pair: (a) Enlargement of the focal point, (b) Wide view of the atomization to evaluate the ejection height*

on the crystalline of the substrate. At first, we designed the shape of the IDT using the $V(\theta)$ reported in Ref. [9].

Figure 6 (a) and (b) show a frequency characteristic and the optical micrograph of fabricated AS-IDT. To confirm conversing of the SAW, we observed the vibration distribution using Sagnac interferometer (NEOARK: MLD-101-TH) as shown in Fig. 6 (c), (d) and (e) [11]. From this result, it was confirmed that the SAW was concentrated on the center of AS-IDT at resonance (23.9 MHz). However, the both side of AS-IDT strongly vibrated under the resonance (23.6 MHz). We fed back the observation result into the AS-IDT design. Figure 7 shows an improved result. It was confirmed that the AS-IDT vibrated entirely at resonance and fine convergence was obtained. In Figure 6 (a), a frequency characteristic of the improved AS-IDT was shown with red line. From this figure, it was confirmed that split resonance response was improved, and steep response was achieved.

Ejection evaluation of the mist spray

The solvent was atomized from a point source after transferring to the focal point of the AS-IDT. The atomization was observed by a high speed camera (KEYENCE: VW-6000). The working fluid was water in this evaluation. The AS-IDT was also operated by the burst voltage to suppress heating of the device. The burst voltage was same to the signal for the micro pump operation. However, the frequency of burst was adjusted to resonant frequency of AS-IDT.

Figure 8 shows observation results of the point source atomization. From Fig. 8, narrow spray from the focal point could be observed. The spray width was about 0.7 mm near the substrate, and was maintained less than 1 mm in width up to 8 mm in height. In this experiment, since steam pressure of water is too high to evaporate, the mist floated in the air above 8mm in height without evaporation.

CONCLUSION

We reported a surface acoustic wave (SAW) driven atomizer for polymeric nanoparticles synthesis based on evaporation-induced self-assembly (EISA). The atomizer consisted of the arc-shaped interdigital transducer (AS-IDT) and a channel. The pump consisted of the reservoir with a slot and a pair of IDTs. All devices were integrated on the LiNbO$_3$ substrate.

The integrated micro pump was driven by the burst voltage which parameters were frequency during the burst of 24.8 MHz, duty cycle of 50 %, period of 0.4 s and amplitude of 70 V. As an experimental result with water as a working fluid, discharge in which the rate was 0.3 pl a burst was achieved.

Also, vibration distribution of the SAW radiated from the AS-IDT was optically observed using Sagnac interferometer. Feeding back the observed results, fine convergence of SAW was obtained. Using this AS-IDT, we evaluate the atomization of the water in the air. As an experimental result, we succeeded in the observation of the narrow mist spray from the substrate using a high speed camera. The observed spray was 0.7 mm in width in vicinity of the substrate, and maintained less than 1 mm in width up to 8 mm in height.

For the high yield EISA, controlling the direction and the mode of atomization is a key technology. Point source atomizer we developed in this study can give a good solution for the intelligent EISA system.

978-1-4799-3510-9/14 $31.00 © 2014 IEEE

ACKNOWLEDGEMENT

This study was partly performed in R&D Center of Excellence for Integrated Microsystems, Tohoku University under the program "Formation of Innovation Center for Fusion of Advanced Technologies" supported by Special Coordination Funds for Promoting Science and Technology (J120000231). Also, in part of this study was performed under the "Support project to promote the machine industry in 2013 (Ring! Ring! Project)", supported by Public Interest Incorporated Foundation JKA (J130001241).

REFERENCES

[1] C. J. Brinker, Y. F. Lu, A. Sellinger, H. G. Fan, "Evaporation-Induced Self-Assembly: Nanostructures Made Easy", *Advanced materials*, **11** (1999), pp.579-585

[2] J. R. Friend, L. Y. Yeo, D. R. Arifin, A. Mechler. "Evaporative self-assembly assisted synthesis of polymeric nanoparticles by surface acoustic wave atomization", *Nanotechnology*, **19** (2008), 145301

[3] T. Sano, T. Onuki, Y. Hamate, M. Hojo, S. Nagasawa, H. Kuwano, "MICRO BLENDER AND SEPARATOR USING INNER-VORTEX OF DROPLET INDUCED BY SURFACE ACOUSTIC WAVE", *Proc. of Transducers 2009*, Denver, June 21-25, 2009, pp. 370-373.

[4] T. Sano, M. Hojo, Y. Hamate, T. Onuki, S. Nagasawa, H. Kuwano, "A Novel Micro Fluidics Device Using Surface Acoustic Wave", *Trans. The Japan Society of Mechanical Engineers*, **76** (2010), pp.40-42

[5] H. Li, J. R. Friend, L. Y. Yeo. "Microfluidic Colloidal Island Formation and Erasure Induced by Surface Acoustic Wave Radiation", *Phys. Rev. Lett.,* **101** (2008), 084502

[6] M. Kurosawa, T. Watanabe, A. Futami, T. Higuchi. "Surface acoustic wave atomizer", *Sensors and actuators A*, **50** (1995), pp.69-74

[7] R. Frank, "The SPOT-synthesis technique Synthetic peptide arrays on membrane supports—principles and applications", *J. Immunological Methods*, **267**, pp. 13-26

[8] H. Nagai, Y. Murakami, Y. Morita, K. Yokoyama, E. Tamiya, "Development of A Microchamber Array for Picoliter PCR", *Anal. Chem.,* **73** (2001), pp.1043-1047

[9] Y. Nakagawa, K. Yamanouchi, K. Shibayama, "Thirdorder elastic constants of lithium niobate", *J. Appl. Phys.* **44** (1973), pp.3969-3974

[10] T. Sasaki, Y. Katoh, Y. Tanaka, T. Omori, K. Hashimoto, "On Low-Power Droplet Manipulator Employing High-Frequency Surface Acoustic Waves", *IEEJ Trans. Electronics, Information and systems C*, **127** (2007), pp.1186-1191

[11] K. Hashimoto, K. Kashiwa, T. Omori, M. Yamaguchi, O. Takano, S. Meguro, K. Akahane, "A Fast Scanning Laser Probe Based on Sagnac Interferometer for RF Surface and BulkAcoustic Wave Devices", *Dig. of IEEE Int. Microwave Symp.,* (2008), pp.851-854

CONTACT

Shun Sugimoto,
Tel (Fax): +81-22-795-4771;
sugimoto@nanosys.mech.tohoku.ac.jp

TEFLON WETTING AND DEWETTING ON EWOD DEVICE FOR CHEMILUMINESCENCE DETECTOR

Xiangyu Zeng[1], Kaidi Zhang[1], Guowei Tao[1], Shih-Kang Fan[2] and Jia Zhou[1]*
[1]Fudan University, CHINA
[2]National Taiwan University, TAIWAN

ABSTRACT

A hydrophobicity recoverable EWOD (electrowetting-on-dielectric) based chemiluminescence detector with an integrated signal and heater electrode was developed. X-ray-photoelectron-spectroscopy (XPS) was used to reveal the wetting and dewetting mechanism of Teflon on the EWOD device. It was found that the C-O bond formed on the surface of Teflon after the chemiluminescence reaction leaded to the surface permanent wetting. To recover the contact angle of the Teflon surface, the recovery threshold time and heating temperature were proposed experimentally for dewetting to release C-O bond.

INTRODUCTION

Chemiluminescence is one of the most important immuno-detection methods which can be widely used in the fields of food, industry, environment, clinical diagnosis and medicine. It has many advantages, including high detection sensitivity with simple instrument configuration, wide linear range of signal response, and rapid measurement [1-10].

But conventional chemiluminescence analyzers are not suitable for cheap, compact and portable detector as the bulky analysis machines and high volumes of samples and bio-reagents consumption for measurement.

Compared to conventional chemiluminescence analyzers, we reported an EWOD based chemiluminescence detector for rapid and automatic H_2O_2 measurement with high sensitivity, wide linear detection range and low sample consumption, which showed many advantages for highly precise portable diagnosis of blood glucose potentially [11].

However, we found that Teflon on the EWOD turned into hydrophilic after chemiluminescence reaction. A similar phenomenon was reported in Richard Fair's research. They considered that the immuno-detection would contaminate the Teflon surface, and the protein adsorption would render the surface permanently hydrophilic, which was detrimental to the droplet transport [12]. Therefore, they immersed the EWOD chip in the silicone oil to isolate the droplet from the Teflon surfaces to prevent any enzymes contamination and protein adsorption.

Differing from their work, we found that the Teflon surface wetting phenomenon could be dewetted by annealing. It was verified by analysis of XPS. We designed an EWOD device with an electrode for both heating and droplet controlling to make its Teflon surface automatically recoverable.

EXPERIMENT AND ANALYSIS

Material and chemicals

For the device fabrication, the Teflon® AF 2400 was purchased from DuPont. SU-8 2002 with developer was purchased from Microchem, and indium tin oxide (ITO) glass from Weisi Technology Co. Ltd., Guangdong, China.

For the chemiluminescence detection, Horse-radish-peroxidase (HRP), luminol, 4-iodophenol (PIP) were purchased from Yuanye Biological Technology Co. Ltd., Shanghai, China. The H_2O_2 solution was purchased from Pengshen Chemical Reagent Co. Ltd., Jiangsu, China.

System building up and chemiluminescence detection

A diagram of the chemiluminescence detection system is shown in Figure 1(a). The photomultiplier (SenSL, MiniSL-30035- X08, 9 mm^2 active area) below the single planar transparent EWOD device is used for detecting the chemiluminescence signal, and the periphery circuits can provide control signals for the EWOD.

In the field of chemiluminescence detection, the main method is utilizing the luminol-HRP system to measure the concentration of H_2O_2 which is the most common intermediate product in immuno-detection.

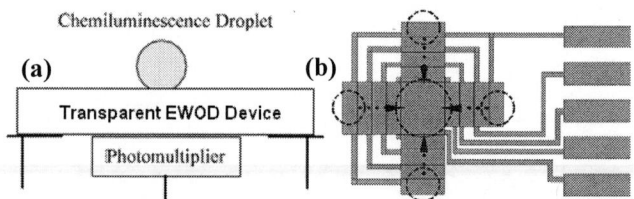

The droplets of four reagents (10μL HRP, 10μL luminol, 10μL PIP(4-iodophenol) and 10μL H_2O_2 were mixed together at the center of the EWOD device, shown in Figure 1(b). After reaction, the mixed droplet will emit blue chemiluminescence signal which can be detected by the photomultiplier [11].

Figure 1: (a) Diagram of chemiluminescence detection system; (b) Diagram of four reagents mixing process and driving electrode pattern.

It should be noticed that the hydrophilic surface after chemiluminescence reaction in our detection not only will be detrimental to the transport of the droplet, causing the permanently failure of the EWOD device, but also will induce a small contact angle and thus lower the detection sensitivity [11], as shown in Fig.2. Therefore, study on hydrophobic recovery and the mechanism of Teflon wetting and dewetting on the EWOD device are carried out experimentally.

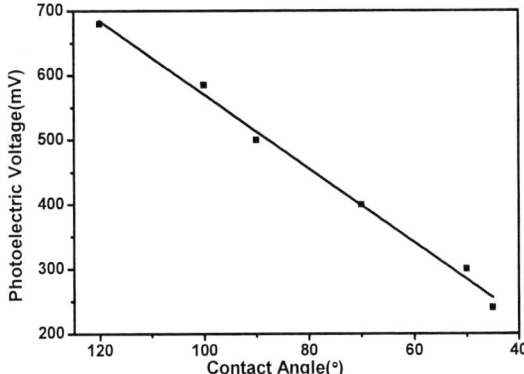

Figure 2: Experimental results of contact angle vs. output voltage of photomultiplier. The original contact angle is 120°. It demonstrates that the more hydrophobic surface, the stronger optical signal the photomultiplier received, the larger photoelectric voltage the photomultiplier output, the higher sensitivity of the chemiluminescence detection. The chemiluminescence reagents were 10μL 100mg/L HRP, 10μL 2mmol/L luminol, 10μL 5mmol/L PIP and 10μL 40mmol/L H_2O_2.

The mechanism of Teflon wetting and dewetting

In Richard Fair's research [12], they considered that the protein adsorption would render the surface permanently hydrophilic. However, we found that the 'permanently hydrophilic' surface could be recovered by annealing, and we considered that the phenomenon of Teflon wetting was just related to the reaction itself.

We tried to put the EWOD device with wetting Teflon surface after chemiluminescence reaction onto a 200°C hot plate. 10min later, the surface of Teflon could dewet to hydrophobic as its initial state, as shown in Figure3.

Figure 3: Photos of DI (de-ionized) water on (a) initial hydrophobic Teflon surface; (b) wetting surface of Teflon after chemiluminescence reaction; (c) dewetting surface of Teflon after annealing.

Since enzymes and proteins cannot be removed on 200°C, we inferred that the wetting of Teflon after chemiluminescence reaction was not caused by the adsorption of those bio-chemicals, but by the bond fractures which was induced by reaction intensity.

In order to verify that the hydrophilic modification was determined by the intensity of chemiluminescence reaction, the contact angle on different devices with different concentration reaction of reagents were tested after the chemiluminescence reactions. We kept the concentration of HRP(100mg/L), luminol (2mmol/L)and PIP (5mmol/L) and mixed them with different concentration solutions of H_2O_2 from 0.25mmol/L to 20mmol/L. Figure 4 shows the results that the stronger the reaction intensity, the smaller the contact angle.

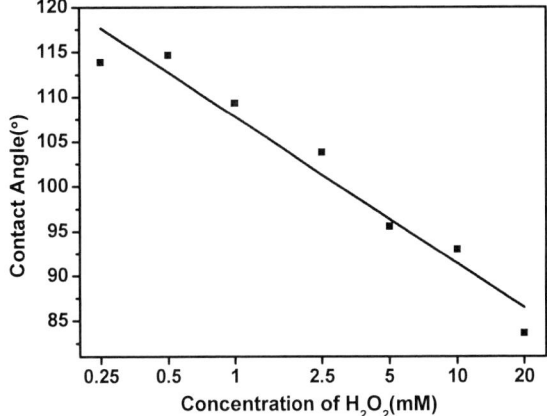

Figure 4: Experimental results of concentration solutions of H_2O_2 vs. contact angle on Teflon after chemiluminescence reaction.

Further surface analysis was used to reveal the essential mechanism of Teflon wetting and dewetting.

Teflon® AF [chemical name: fluorinated (ethylenic-cyclo oxyaliphatic substituted ethylenic) copolymer] is a family of amorphous fluoropolymers based on copolymers of 2,2-bistrifluoromethyl-4,5-difluoro-1,3-dioxole(PDD)[13]. According to its chemical structure which is shown below[14-16], we can get that C, F, O are the three main elements on the surface of our device.

We used XPS to analyze the elements on Teflon surface of our device before reaction, after chemiluminescence reaction (cleaned by DI water and dried in nitrogen airbrush) and after annealing. The XPS spectra are shown in Figure 5. Comparing the three curves in Figure 5, we can see that the peak of element O after reaction is much higher than that before reaction, and it returns to its original state after annealing. The peak of element F after reaction is much lower than that before reaction, and it recovers to the initial high level after annealing as well.

978-1-4799-3510-9/14 $31.00 © 2014 IEEE

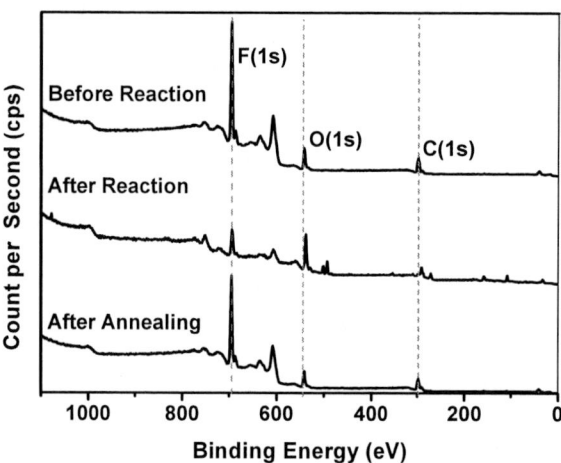

Figure 5: XPS spectra of Teflon surface. The rate of element O vs. element F is much higher after chemiluminescence reaction than that before reaction and after recovery.

We also used XPS spectra to analyze the C bond on the Teflon surface before and after chemiluminescence reaction, as well as after recovery, as shown in Figure 6. We can see that the number of F-C-O bond after reaction is more than that before reaction and after recovery. Comparing the three curves, the two strengthened peaks with 290.5eV and 287.6eV after reaction are all relative with C-O bond [17], which induces the surface of Teflon hydrophilic.

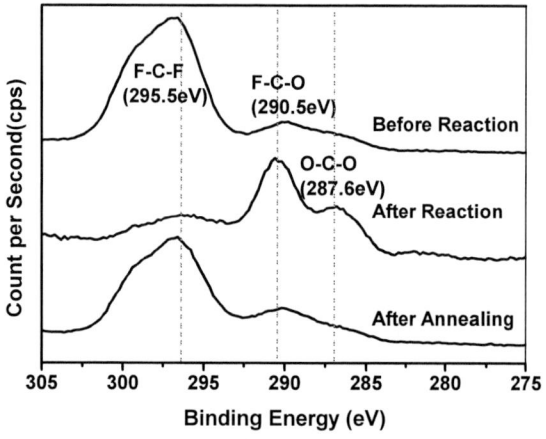

Figure 6: High-solution C1s XPS spectra of Teflon surface. The peak of F-C-O bond is much higher after chemiluminescence reaction than that before reaction and after recovery.

From the analysis of all the XPS spectra, we can conclude that the C-F bond on the surface was broken and the C-O bond was formed after the chemiluminescence reaction, which induced the hydrophobic fluorocarbons surface turning into hydrophilic. After annealing, the element O in C-O bond was released and the surface property recovered to hydrophobic. Therefore, on-chip annealing the device after reaction is necessary for the chemiluminescence detector in practical uses.

The chemiluminescence chip with ITO heater

In order to recover hydrophobic surface of Teflon automatically, an ITO heater was directly integrated at the center of the EWOD chip, as shown in Figure 7. When the two pads of the electrode were connected to the signal (30V, 1kHz) and the ground separately, it would be the heating source; while the two pads were both connected to the signal (30V, 1kHz) or ground at the same time, it would be a typical EWOD controlling electrode.

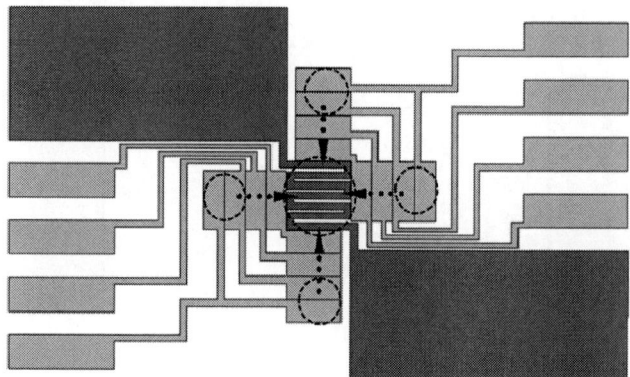

Figure 7: EWOD device for chemiluminescence reaction with heater at center.

When the central heater was used as a typical EWOD controlling electrode, the four reagents could be mixed together and reaction. After chemiluminescence reaction, the Teflon surface underneath the reaction droplet would turn into hydrophilic. By washing the surface with DI water and annealing with the central heater, the hydrophilic surface could be recovered to hydrophobic.

The Figure 8(a) shows the recovered contact angles changing with the annealing time in different annealing temperature after chemiluminescence reaction. Fig.8 (b) illustrates the relationship of the recovered contact angles with different annealing temperature and time after chemiluminescence reaction. It demonstrates that the higher annealing temperature, the better recovery result, and the shorter annealing time.

Since enzyme and protein do not decompose in low temperature such as 100°C, the recovery result also reveals that the essential mechanism of the surface permanent hydrophile after bio-reaction has nothing to do with the protein adsorption, just because of the C-F bond broken in the chemiluminescence reaction.

SUMMARY

In this work, the wetting mechanism of Teflon in an EWOD based chemiluminescence detector was revealed. A series of experiments and XPS analysis demonstrated that the breaking of C-F bond and forming of C-O bond on the surface of Teflon after the chemiluminescence reaction leaded to the surface wetting. By annealing the EWOD chip, the C-O bond could be released, and the Teflon surface could be dewetted.

To recover the hydrophobicity of Teflon, an integrated heater and annealing process on EWOD device were proposed. The recovery relationships between the recovering contact angle, the recovery threshold time and heating temperature were studied.

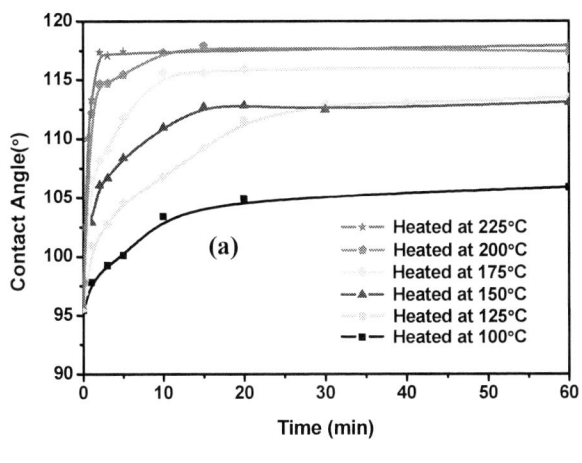

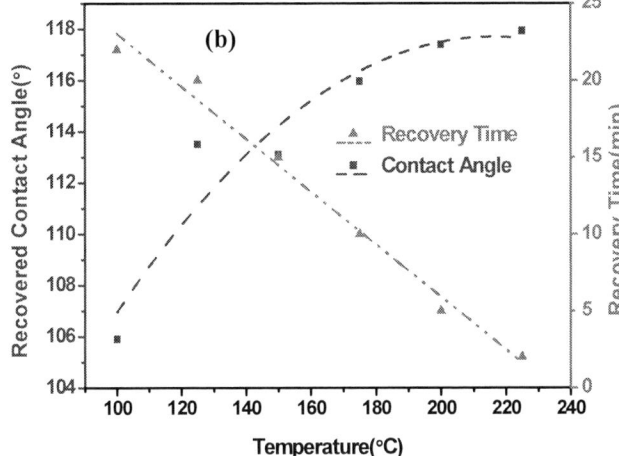

Figure 8: (a) Recovering contact angle of Teflon vs. heating time under different heating temperature after reaction. (b) Relationship between recovering contact angle, recovery threshold time and heating temperature.

ACKNOWLEDGEMENTS

This work was supported by the National Science Foundation of China with Grant No. 61176110.

REFERENCES

[1] S. Okumoto, A. Jones, W.B. Frommer, "Quantitative Imaging with Fluorescent Biosensors", *Annu. Rev. Plant Biol.* 2012, 63, 663.

[2] S.P. Mohanty, "Biosensors: a tutorial review", *IEEE Potentials*, 2006, 25, 35.

[3] R. Monošík, M. Stredanský, E. Šturdík, "Biosensors — classification, characterization and new trends", *Acta Chimica Slovaca*, 2012, 5, 109.

[4] A.C. Matthew, "Optical biosensors in drug discovery", *Nature Reviews Drug Discovery*, 2002, 1, 515.

[5] X. Fan, M.W. Ian, I.S. Siyka, H. Zhu, D.S. Jonathan, Y. Sun, "Sensitive optical biosensors for unlabeled targets: A review", *Analytica Chimica Acta*, 2008, 620, 8.

[6] K. Kaura, B. Singha, A.K. Malika, "Chemiluminescence and Spectrofluorimetric Methods for Determination of Fluoroquinolones: A Review", *Analytical Letters*, 2011, 44, 1602.

[7] D. Christodouleas, C. Fotakis, A. Economou, K. Papadopoulos, M.Timotheou-Potamia, A. Calokerinos, "Flow-Based Methods with chemiluminescence Detection for Food and Environmental Analysis: A Review", *Analytical Letters*, 2011, 44, 176.

[8] F.J. Lara, A.M. García-Campaña, J.J. Aaron, "Analytical applications of photoinduced chemiluminescence in flow systems-A review", *Analytica Chimica Acta*, 2010, 679, 17.

[9] M. Liu, Z. Lin, J-M. Lin, "A review on applications of chemiluminescence detection in food analysis", *Analytica Chimica Acta*, 2010, 670, 1.

[10] L. Gámiz-Gracia, A.M. García-Campaña, J.F. Huertas-Pérez, F.J. Lara, "Chemiluminescence detection in liquid chromatography: Applications to clinical, pharmaceutical, environmental and food analysis-A review", *Analytica Chimica Acta*, 2009, 640, 7.

[11] X. Zeng, K. Zhang, J. Pan, G. Chen, A.Q. Liu, S.K. Fan, J. Zhou, "Chemiluminescence detector based on a single planar transparent digital microfluidic device", *Lab on a Chip*, 2013, 13, 2714-2720.

[12] V. Srinivasan, V. Pamula, M. Pollack, R. Fair. "A digital microfluidic biosensor for multianalyte detection", *Proc. IEEE MEMS Conference*, 2003, pp. 327–330.

[13] P. R. Resnick, W. H. Buck, "TEFLON® AF Amorphous Fluoropolymers", *Modern Fluoropolymers: High performance polymers for diverse applications*, pp. 397–419, Wiley, West Sussex, England (1997)

[14] G. Belanger, P. Sauvageau, C. Sandorfy, "The far-ultraviolet spectra of perfluoro-normal-paraffins", *Chem. Phys. Lett.* 3(8), 649 (1969)

[15] K. Seki, H. Tanaka, T. Ohta, Y. Aoki, A. Imamura, H. Fujimoto, H. Yamamoto, H. Inokuchi, "Electronic structure of poly(tetrafluoroethylene) studied by UPS,VUV absorption, and band calculations", *Physica Scripta,* 41, 167 (1990)

[16] W. Ren, "Fluorinated Resins Teflon AF", *New Chemical Materials*, 1991, 12, 18-22

[17] S. Ding, P. Wang, W. Zhang, J. Wang, Z. Wen, Z. Xia, "Interaction between aluminum deposition by evaporation and Teflon AF film", *ACTA Metallurgica Sinica*, Vol 37, No.3 ,243-246 (2001)

CONTACT

*Jia Zhou, tel: +8621-5566-4601; jia.zhou@fudan.edu.cn

TRANSIENT INERTIAL FLOWS: A NEW DEGREE OF FREEDOM FOR PARTICLE FOCUSING IN MICROFLUIDIC CHANNELS

Michael H. Winer[1], Ali Ahmadi[2], and Karen C. Cheung[1]

[1]The University of British Columbia, Vancouver, CANADA

[2]The University of British Columbia Okanagan, Kelowna, CANADA

ABSTRACT

This paper reports a new inertial particle focusing approach using a rapidly changing flow rate in microfluidic channels. The transient flow rate effects manipulates the relative velocity between the particles and the main flow, therefore changes the balance between the (inertial) hydrodynamic forces acting on the particle. A comparative analysis was conducted with a theoretical basis in inertial flow phenomena for both constant and transient flow rates in straight rectangular channels. Results found using an apparent size correlation (ASC) three-dimensional particle tracking method indicate that particles focus in different equilibrium positions depending on whether a constant or changing flow rate is applied.

INTRODUCTION

Manipulation of cells and particles in microfluidic systems using inertial forces has shown enormous potential for numerous applications including cell/particle filtration, separation and flow cytometry [1, 2]. In essence, the inertial focusing of particles is based on the manipulation of particle positions in the directions perpendicular to the main flow direction by lateral hydrodynamic forces.

In recent years, numerous system designs have been proposed to achieve higher throughput size-based separation using the inertial focusing method [3-6]. In many of these studies, the lateral forces on the particle were resulted from a steady-state flow in the channels, and depending on the geometry and scale of the system, different equilibrium positions and focusing regimes were achieved.

It is known that these lateral forces strongly depend on the relative velocity of the particle and the main flow [7]. Therefore, by manipulating the relative velocity between the particles and flow, new focusing paradigms can be established. This paper demonstrates the use of transient flow for manipulating the relative velocity of the particle and therefore achieving new focusing regimes.

THEORY

The equation of motion of spheres moving in a transient flow is [7]:

$$m \frac{d\vec{V}_p}{dt} = \vec{F}_D + \vec{F}_L + \vec{F}_{AM} + \vec{F}_{SG} + \vec{F}_H + \vec{F}_G \quad (1)$$

where $\vec{V}_p$ and m are the particle velocity and mass respectively, $\vec{F}_D$ is the quasi-steady-state drag force, $\vec{F}_L$ is the lift force due to the fluid shear and particle rotation, $\vec{F}_G$ is the gravitational force, $\vec{F}_{AM}$, $\vec{F}_{SG}$ and $\vec{F}_H$ are the added mass, shear gradient and history forces, respectively, which are due to the transient effects. The equilibrium positions are obtained when the lateral forces balance each other, and velocity of the particles become zero. In addition to the transient and drag forces, the lift force contains a dominant relative velocity term:

$$|\vec{F}_L| = \frac{\pi}{8} \rho_f \, C_L \, \vec{V}_{rel} \cdot \vec{V}_{rel} d_p^2 \quad (2)$$

where ρ_f, C_L, d_p and $\vec{V}_{rel}$ are the fluid density, lift coefficient, particle diameter and the relative velocity, respectively. The lift force, F_L has two components,

$$\vec{F}_{L,\omega} = \vec{\omega} \times \vec{V}_{rel} \quad (3)$$

and

$$\vec{F}_{L,\Omega} = \vec{\Omega} \times \vec{V}_{rel} \quad (4)$$

where $\vec{\omega}$ and $\vec{\Omega}$ are the fluid vorticity and particle angular velocity, repectively. $\vec{F}_{L,\omega}$ and $\vec{F}_{L,\Omega}$ are the vorticity-induced and spin-induced lift forces acting on the particle, respectively.

By changing the flow velocity as a function of time, the balance of forces in equation (1) changes, and therefore the velocity of the particle and consequently the relative velocity will change. This will lead to new balance of forces and different lateral equilibrium positions.

EXPERIMENTAL DESIGN

Experimental Setup

Figure 1 is a schematic of the experimental setup used for all experiments outlined in this paper. To view the beads and microfluidic channel, a Nikon Eclipse TE2000-U microscope was used with a standard 10 X optical lens ($NA = 0.3$, $WD = 16$ mm). A high-speed camera (Miro eX4, Vision Research, USA) with a resolution of 320 x 240 pixels, exposure time of 134 μs and a frame rate of 7,150 fps was used to capture images. Data was acquired using a defocusing apparent size (ASC) technique to correlate the apparent size of the particles to their vertical position in the channel [8]. Images of the particles were taken in bright-field after it was found that the resolution and exposure time settings made it difficult to contrast the fluorescing beads with the background. The applied pressure is related to the Reynolds number (Re) as

$$P_a(mbar) = 4.7088 \times 10^{12} \, Re \, \frac{\mu_f}{\rho_f} \frac{(2H + 2W)^2}{2W}, \quad (5)$$

978-1-4799-3510-9/14 $31.00 © 2014 IEEE

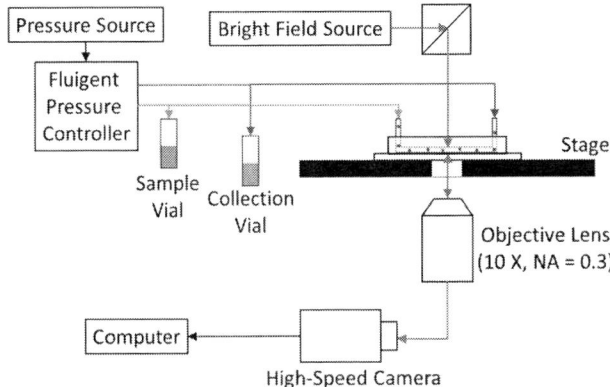

Figure 1: Experimental schematic. The microfluidic chip is placed on the stage, where a bright field source (for imaging PS beads) are able to capture information. Data is collected using Phantom CV 2.2 software (Vision Research, USA). Schematic is not to scale.

A Fluigent pressure control system was used to regulate applied pressure into the channel and through cross-sectional area. Three Reynolds numbers (Re = 16.52, 22.95, 32.12) were chosen to give an overall representation of transient effects with changing flow rate.

Particle Suspensions

Experiments described in this paper used 15.5 μm (± 1.52 μm) green (emission wavelength = 502 nm) fluorescent polystyrene (PS) beads (Bangs Laboratories, CC FS07F, USA) and 2 μm red (emission wavelength = 612 nm) fluorescent PS beads (Thermo Scientific, R0200, USA). The smaller beads were chosen to measure the flow profile and velocity of the fluid. The larger size was chosen to mimic the size of many cells typically analyzed in microfluidic assays including white blood cells or circulating tumor cells. Solutions were made to 0.1 wt% beads using DI water. 1 vol% of Tween-20 surfactant (Sigma-Aldrich, P1379, Canada) was also added to reduce particle aggregation.

Microfluidic Device Fabrication

Devices were fabricated using soft lithography techniques are previously described [9]. Lithography masters were fabricated on 4-inch silicon wafers (University Wafer, USA) with SU-8 3050 (MicroChem, USA) using spin-coating to create a 100 x 85 μm x 4 cm straight rectangular channel design (width x height x length respectively). Microfluidic devices were then fabricated using the masters with poly(dimethylsiloxane) (PDMS) elastomer (Sylgard 184, Dow Corning, USA). The final PDMS chip was bonded using oxygen-enriched plasma to a 75 x 50 x 1 mm No. 1 glass slide (VWR Scientific, PA, USA). Consider Reynolds number,

$$Re = Q \frac{\rho_f}{\mu_f} \frac{D_H}{HW} \qquad (3)$$

where Q is the applied flow rate, ρ_f is the fluid density, μ_f is the fluid viscosity, D_H is the hydraulic diameter of the channel and H and W are the height and width of the channel respectively. The cross-sectional area of the device was chosen to approximate a 1:5 ratio of the particle diameter to the hydraulic diameter of the channel, which in previous work has been shown to produce a wide variety of inertial focusing positions when varying flow rate (and therefore Reynolds number) [5, 10].

Particle Image Velocimetry (PIV) in Transient Flow

To achieve the desired transient effect, control the flow precisely and stop the flow as fast as possible, the Fluigent system was set up to control flow from both the inlet and outlet of the channel. In this way, the flow was quickly stopped by applying a reverse pressure from the outlet against the inlet applied pressure. The inlet pressure was set to the chosen pressure related to Reynolds number (from equation 5). After reaching steady-state (i.e. constant) flow regime, the outlet pressure was increased to match the inlet pressure, effectively stopping the flow in less than 200 ms (this time changes depending on the applied constant pressure). The velocity of the flow during this transient flow period was obtained by tracking small 2 μm beads using the developed particle tracking algorithm. A solution of the beads was fed into the microfluidic channel, and a series of data sets were acquired for the chosen Reynolds numbers. This experimental procedure was repeated at the same pressures (Re) with the larger 15.5 μm beads to obtain the particle velocities.

RESULTS
Relative Velocity Measurements

To highlight the effects of the transient flow in manipulating the relative velocity, the relative velocity between the particle and the transient flow was measured.

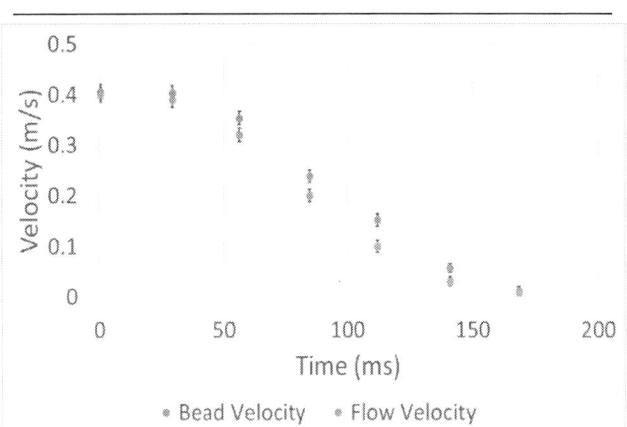

Figure 2: Velocity of fluid and 15.5 μm beads vs. time during application of pressure opposing the inlet driving pressure. Steady flow at Re = 32.12. The maximum difference between fluid velocity and bead velocity occurs at t ≈ 110 ms. Similar plots were derived for all Re tested.

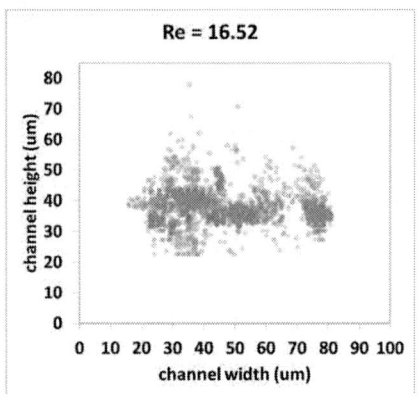 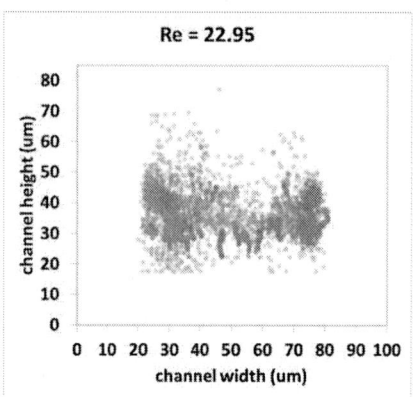 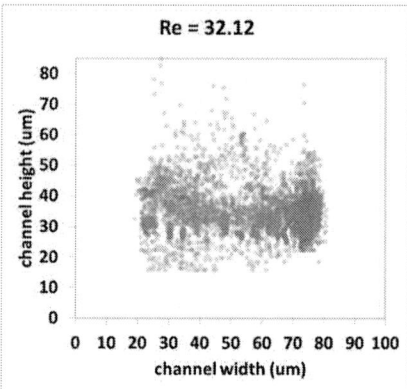

Figure 3: Transient Study Experimental Results. Comparison of constant flow rate data (orange) with transient data (blue) in the channel cross-section for three Reynolds numbers chosen.

The velocity of the fluid and the particles was measured during the transient flow time period. As an example, Figure 2 shows the observed trend for Re = 32.12. The result indicates that by changing the flow rate, a relative velocity is created which can be used to manipulate the focusing positions of the particles. The maximum relative velocity for this Reynolds number is observed at 110 ms with a value of 55 mm/s. The time region around 110 ms is the time period that the highest deviation from the steady-state force balances are expected. Therefore, this information provides insight into the transient flow experiment to find the optimal time period at which the equilibrium focusing positions of the particles are expected to have the greatest difference from the constant flow rate case. This procedure of determining the relative velocity was repeated for all Reynolds numbers.

Transient Flow Focusing Positions

To investigate the effects of the induced relative velocity, the focused positions of the 15.5 μm particles were measured for steady state and transient flows. To observe the largest deviation from the steady state positions, the focused positions in the transient experiment were captured in the time period that had the maximum relative velocity (a time period ranging from 5-10 ms).

Using the ASC three-dimensional particle tracking method, captured images for the transient experiment were translated into a 2D cross-sectional plot of the channel that illustrates the difference between the equilibrium focusing positions as a result of the applied constant and transient flow rates. The results were also normalized to the same number of data points to ensure statistical equivalence.

The focusing positions for steady state and transient flow rates are shown in Figure 3 for three Reynolds numbers (Re= 16.72, 22.95 and 32.12). The difference in the observed focused positions can be discussed from different aspects. In the transient measurements, the velocity of flow during the capturing period is lower than the velocity during the steady state flow. Therefore, based on the results of the previous steady-state flow studies, it was expected that the transient measurement show more scattered behavior in z-direction (channel height) [4, 10]. Interestingly, it appears that the constant flow rate has a larger z-axis scatter than the transient counterparts. To highlight the observed scattered positions, data was summed along the channel width (y-axis) to create a more qualitative density plot of the particle focusing positions (see Figure 4). This is an interesting observation which highlights the potential of using transient flow in achieving higher throughput microfluidic particle focusing. Furthermore, despite the trends observed in previous studies [8], the transient results show some additional horizontal spreading in the y-axis. The use of transient flows results in a short-term increase in the relative velocity of the particles,

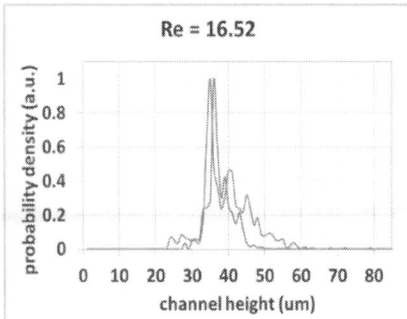 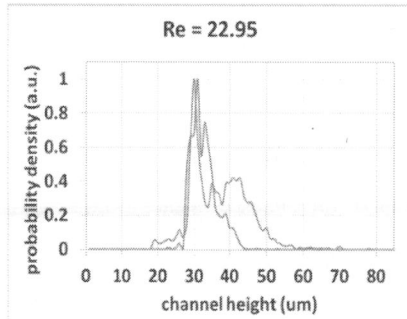 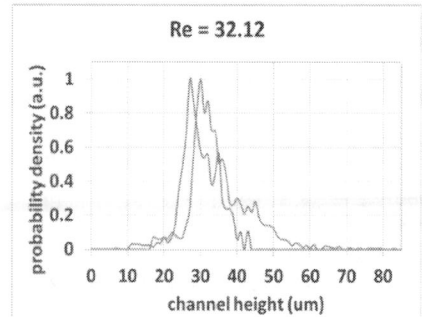

Figure 4: Histogram data for both steady-state (constant, orange) and transient (blue) flows. The data was normalized across the channel width to emphasize the slightly wider scatter in the z-axis (channel height) in the constant flow rate data. Note the y-axis is a probability density, normalized based on the highest number of particles found in each data set.

thereby increasing the lift force as defined in equation (2). We hypothesize that the combination of more spreading in the y-axis and less spreading in the z-axis in the transient data can be attributed to the larger lift force due to the increased relative velocity term. This increased lift force affects the overall balance of the forces as described in equation (1), which in turn explains the new inertial focusing positions.

Overall, transient flow provides a powerful tool for achieving new focusing paradigms and subsequently optimized particle focusing.

CONCLUSIONS

This work is the first experimental effort to examine transient flow effects on inertial particle focusing in microfluidic systems. The results shown in this paper indicate that the use of a transient flow rate can result in alternative equilibrium focusing positions for particles in microfluidic channels. This is theoretically expected due to the effects of the relative velocity term on the lift coefficient, a term related to the lift force, one of two major forces responsible for inertial focusing.

Future work will explore the use of transient flows to create new and useful focusing positions in inertial microfluidic systems.

ACKNOWLEDGEMENTS

This work was funded in part by the Natural Sciences and Engineering Research Council of Canada (NSERC), CMC Microsystems, and the Canada Foundation for Innovation (CFI). Special thanks to Eric Cheng, Jonas Flueckiger and Samantha Grist for expertise in MATLAB, and general experimental advice.

REFERENCES

[1] X. Xuan, *et al.*, "Particle focusing in microfluidic devices," *Microfluidics and Nanofluidics,* vol. 9, pp. 1-16, 2010.

[2] D. Spencer and H. Morgan, "Positional dependence of particles in microfludic impedance cytometry," *Lab on a Chip,* vol. 11, pp. 1234-1239, 2011.

[3] A. A. S. Bhagat, *et al.*, "Enhanced particle filtration in straight microchannels using shear-modulated inertial migration," *Physics of Fluids,* vol. 20, pp. 101702-4, 2008.

[4] D. Di Carlo, *et al.*, "Equilibrium Separation and Filtration of Particles Using Differential Inertial Focusing," *Analytical Chemistry,* vol. 80, pp. 2204-2211, 2008.

[5] D. Di Carlo, *et al.*, "Continuous inertial focusing, ordering, and separation of particles in microchannels," *Proceedings of the National Academy of Sciences,* vol. 104, pp. 18892-18897, November 27, 2007.

[6] S. S. Kuntaegowdanahalli, *et al.*, "Inertial microfluidics for continuous particle separation in spiral microchannels," *Lab on a Chip,* vol. 9, pp. 2973-2980, 2009.

[7] E. Loth and A. J. Dorgan, "An equation of motion for particles of finite Reynolds number and size," *Environmental Fluid Mechanics,* vol. 9, pp. 187-206, 2009.

[8] M. Winer, *et al.*, "Apparent Size Correlation: A Simple Method To Determine Vertical Positions of Particles Using Conventional Microscopy," presented at the IEEE MEMS Conf., San Francisco, USA, 2014.

[9] D. C. Duffy, *et al.*, "Rapid Prototyping of Microfluidic Systems in Poly(dimethylsiloxane)," *Analytical Chemistry,* vol. 70, pp. 4974-4984, 1998.

[10] A. Bhagat, *et al.*, "Inertial microfluidics for continuous particle filtration and extraction," *Microfluidics and Nanofluidics,* vol. 7, pp. 217-226, 2009.

BIAXIAL STRAIN IN SUSPENDED GRAPHENE MEMBRANES FOR PIEZORESISTIVE SENSING

A.D. Smith[1], F. Niklaus[1], S. Vaziri[1], A.C. Fischer[1], M. Sterner[1], F. Forsberg[1], S. Schröder[1], M. Östling[1], and M.C. Lemme[2]

[1]KTH – Royal Institute of Technology, Stockholm, Sweden, [2]University of Siegen, Germany

ABSTRACT

Pressure sensors based on suspended graphene membranes have shown extraordinary sensitivity for uniaxial strains, which originates from graphene's unique electrical and mechanical properties and thinness [1]. This work compares through both theory and experiment the effect of cavity shape and size on the sensitivity of piezoresistive pressure sensors based on suspended graphene membranes. Further, the paper analyzes the effect of both biaxial and uniaxial strain on the membranes. Previous studies examined uniaxial strain through the fabrication of long, rectangular cavities. The present work uses circular cavities of varying sizes in order to obtain data from biaxially strained graphene membranes.

INTRODUCTION

Graphene has a number of interesting properties, which make it suitable for pressure sensor applications. It consists of a monolayer of carbon atoms and its extraordinary thinness gives graphene membranes a high sensitivity to forces applied to it. In addition, it has a high carrier mobility [2, 3]. Further, it is mechanically stable with a Young's modulus of 1 TPa [4, 5] and can be elastically strained by up to 20% [6]. Since graphene is one atom thick, graphene membranes will be deflected by small changes in pressure – resulting in changes to its electronic properties. A further advantage of graphene is its impermeability to gasses: not even helium can permeate a graphene membrane [7, 8].

The piezoresistive effect in graphene has been demonstrated by uniaxially straining monolayer graphene membranes, which are sealing air filled cavities [1]. The uniaxial strain was induced by geometry, i.e. with thin rectangular cavities and suspended graphene membranes sealing the cavities. These devices were then placed into a pressure chamber. As air is pumped out of the chamber, the pressure difference between the air in the chamber and the air trapped underneath the graphene membrane changes. This creates a strain on the membrane as the trapped air presses against it. Typical values of the extracted gauge factors for graphene range from 1 to 4 [1, 9, 10]. Very high sensitivity has been achieved for these devices with values that are tens to hundreds of times more sensitive per membrane area than the sensitivity of conventional silicon pressure sensors. This allows for aggressive scaling of graphene-based sensors [1]. Simulations further corroborate measurement data for both uniaxially and biaxially strained graphene [1]. In uniaxially strained graphene, the graphene is strained in only one direction whereas in biaxial strained graphene, the graphene is strained in two directions

(dimensions) simultaneously. This work for the first time provides measurements on the piezoresistive effect for biaxially strained graphene. The data is compared to previous simulations as well as uniaxial strain data. The data yields an average measured gauge factor of 6.1. Devices with circular membranes of 18, and 24 μm diameter were investigated in order to assess how the size of the cavity affects sensitivity.

FABRICATION

Devices were fabricated beginning with a p-type doped silicon substrate with a 1.5 μm layer of thermally grown SiO_2. Cavities of various sizes were etched 1.5 μm into the oxide to the substrate (Figure 1) using reactive ion etching (RIE). A second set of 603 nm deep cavities were then etched into the oxide layer for the contact metallization. These cavities were filled with Ti/Au (Figure 2) using metal evaporation and lift off. This creates electrical contacts, which are embedded in the SiO_2 layer. The reason for embedding contacts is to eliminate additional process steps after the graphene is transferred. The contacts are made of 150 nm of Ti to act as an adhesion layer to the oxide and 500 nm of gold. Chemically vapor deposited (CVD) graphene was then transferred from copper foil onto the top of the contacts (Figure 1) using a conventional poly(Bisphenol A) carbonate (PC) transfer process and $FeCl_3$ to etch the copper foil. The graphene was then patterned with a photoresist mask using O_2 plasma and placed in a chip package (Figure 1). SEM images of wire bonded devices are shown in Figure 3. The main image shows a 24 μm circular cavity as used for measurements with biaxially strained graphene. The inset shows a rectangular cavity from previous experiments used for measuring uniaxial strain.

EXPERIMENTAL SETUP

In order to measure the devices, a pressure chamber equipped with a reference commercial pressure sensor was used. The graphene devices were connected to a Wheatstone bridge and current was pulsed in order to eliminate temperature related effects caused by Joule heating. Further, the pressure chamber was vented and pumped with inert gas in order to reduce parasitic gas sensing effects.

RESULTS

Graphene membranes with 24 μm cavity diameter were measured in argon (Non-suspended devices have previously been measured in argon, nitrogen, oxygen, CO_2 and air in order to assess the effect of molecular sensing and it was determined that argon was the most inert and stable environment [1]). Figure 4 shows the voltage output versus time (blue) of a 24 μm cavity diameter device in argon gas. The voltage change in the device follows reasonably the commercial sensor measuring pressure variation in the chamber for several pump and vent cycles, where argon is pumped out of the chamber and then purged back into the chamber.

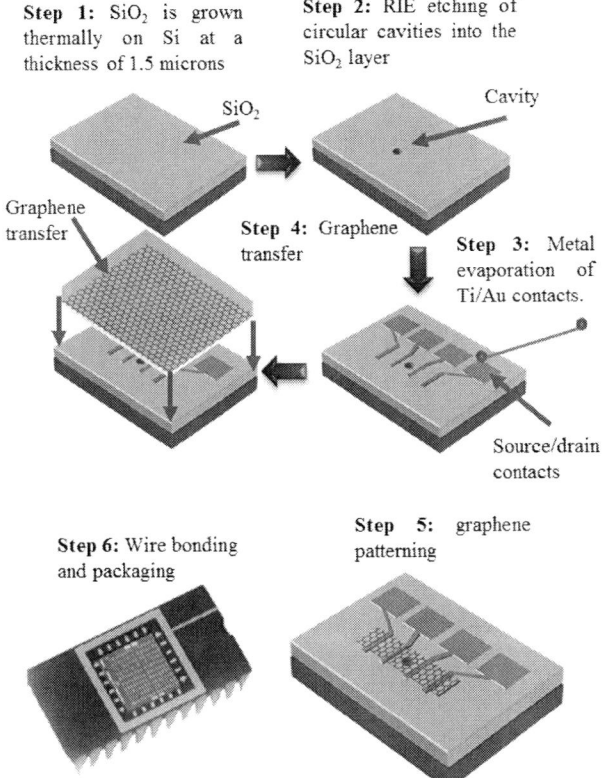

Figure 1: Fabrication of graphene pressure sensors. First, thermal SiO₂ is grown on a Si substrate. Then cavities are etched into the substrate using Reactive Ion Etching (RIE). Ti/Au contacts are then patterned on the devices followed by graphene transfer and etching. Finally, chips are packaged and wire bonded.

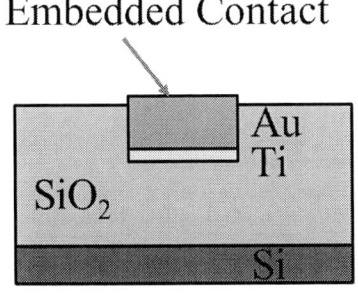

Figure 2: Schematic: contacts made of 150 nm of Ti and 500 nm of Au embedded into the SiO₂.

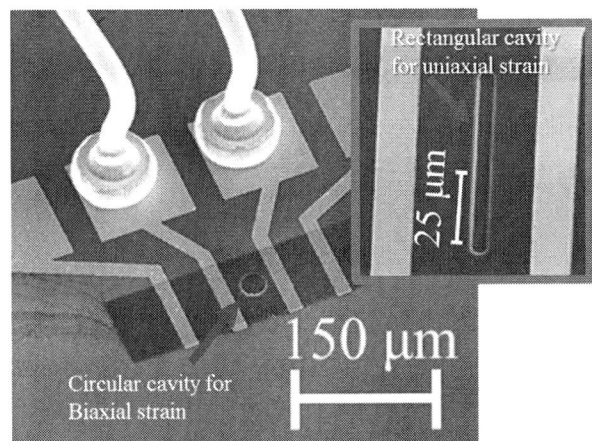

Figure 3: SEM image of a wire bonded device with a circular cavity of 24 μm diameter to induce biaxial strain. The inset shows a rectangular cavity device for inducing uniaxial strain.

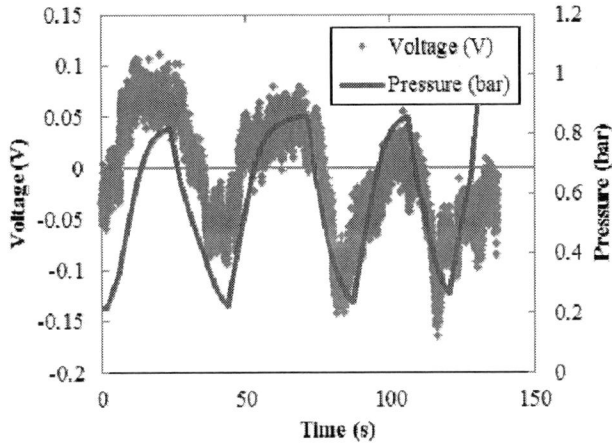

Figure 4: Pressure measurement in argon gas with a 24 μm diameter cavity device. The percent change in resistance was calculated for the cavity region and used with the average strain to extract an average gauge factor of 6.1.

An equivalent resistance model was made in order to estimate the resistance change in the membrane area of the graphene patch. This model was then used in order to calculate the gauge factor of the devices in a similar method to [1]. The model simplifies the circular area into a square of equivalent area as shown in Figure 5, where the resistance R_2 is the resistance of the cavity region. R_2 is calculated by taking the total change in resistance and assuming that the resistance of the non-cavity regions remains constant.

Figure 5: Close-up SEM image of a cavity. Highlighted regions show resistances for respective areas.

In order to estimate the resulting strain in the system, a COMSOL model was used. This model has previously been compared to literature data by Bunch et al. [7] and Koenig et al. [8] and found to be very accurate [1]. Figure 6 shows the simulated deflection of a membrane with a diameter of 24 μm. For a pressure difference of 0.8 bar, the maximum estimated deflection is 0.92 μm. For a given pressure, the average strain is then calculated. Gauge factors are then extracted by taking the percent change in resistance and dividing it by the average change in strain over the membrane. Figure 7 plots gauge factors of circular cavity devices (biaxially strained, red open circles). It further compares them with rectangular cavity devices from [1] (uniaxially strained, blue open squares). An average gauge factor of 6.1 has been derived for biaxial strain, which varies within a reasonable range depending on the pressure difference. Solid squares and circles represent our simulations for uniaxial and biaxial strains, respectively. Finally, we have added simulation data from Huang et al [9], which is in the same range.

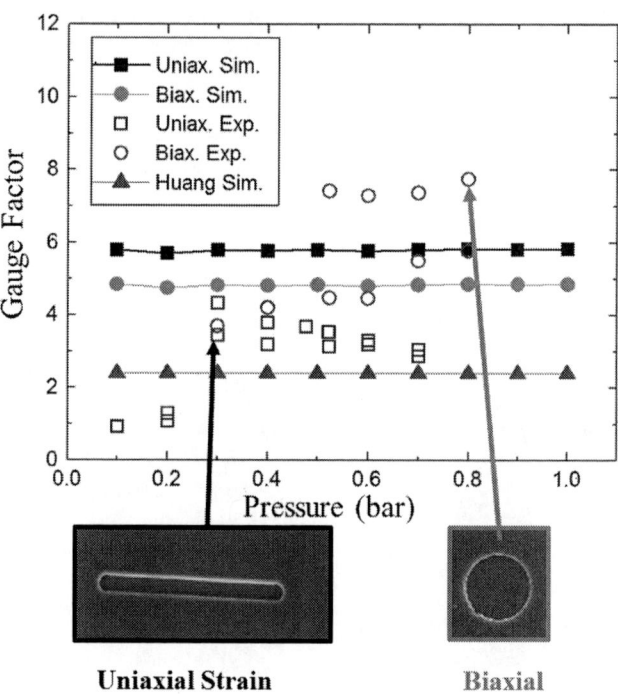

Figure 7: Comparison of gauge factors for circular devices with 24 μm diameter (circular data points) to rectangular devices (square data points) from previous literature [1]. The rectangular cavities result in uniaxial strain and the circular cavities result in biaxial strain. Experimental data (hollow data points) is also compared to various simulations (solid data points) [1, 2].

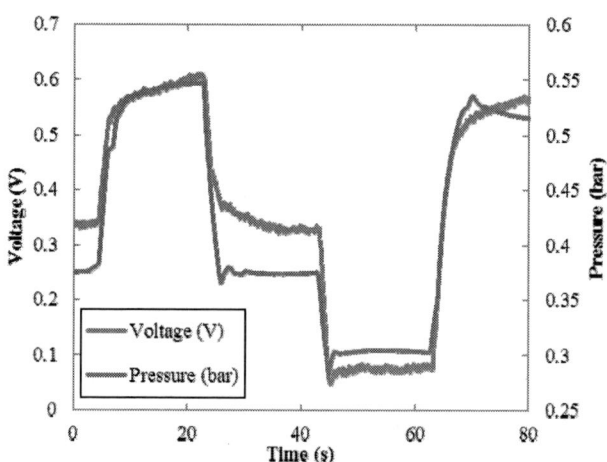

Figure 8: Experimental results of a graphene sensor with 18 μm cavity diameter. The percent change in resistance is calculated for the cavity region and used with the average strain to arrive at a gauge factor of 89.

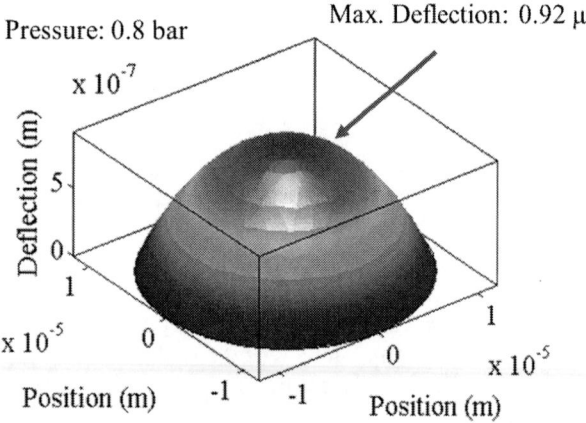

Figure 6: Simulated deflection due to a pressure difference of a suspended membrane over a circular cavity using COMSOL Multiphysics. This example is for 0.8 bar pressure difference. The average strain was also determined from this model and used to calculate a gauge factor for the circular cavity devices.

We note that in contrast to experiments on uniaxial strain, not all circular devices showed similar behavior to the 24 μm cavity case. We found an 18 μm diameter membrane device with a gauge factor of 89. The comparison of the voltage output and commercial pressure sensor output for this device is shown in Figure 8. Several

978-1-4799-3510-9/14 $31.00 © 2014 IEEE

things are noteworthy about this device. First, it has a much higher gauge factor than typical. Second, there is less noise in the data compared to all other devices tested. While this is an encouraging result, we are currently lacking a clear explanation for this discrepancy and further investigations are ongoing.

CONCLUSION

We have experimentally investigated the influence of the cavity shape on suspended graphene membrane pressure sensors for the first time. Our results confirm earlier simulations, which predicted gauge factors in graphene that are nearly independent of crystallographic orientation and strain direction. The experimental comparison yields higher gauge factors for the 24 μm devices as in previously reported data. A smaller cavity size of 18 μm yields a gauge factor of 89.

ACKNOWLEDGEMENTS

The authors thank A. Paussa, D. Esseni, P. Palestri and L. Selmi for fruitful discussions on simulations regarding the piezoresistive effect in graphene. Support from the European Commission through an ERC Advanced Investigator Grant (OSIRIS, No. 228229) and two Starting Grants (M&M' s, No. 277879 and InteGraDe, No. 307311) as well as the German Research Foundation (DFG, LE 2440/1-1) and the Italian MIUR through the Cooperlink project (CII11AVUBF) is gratefully acknowledged.

REFERENCES

[1] A. D. Smith, *et al.*, "Electromechanical Piezoresistive Sensing in Suspended Graphene Membranes,"*Nano Letters,*vol. 13 pp. 3237-3242, 2013.

[2] K. I. Bolotin, *et al.*, "Ultrahigh electron mobility in suspended graphene," *Solid State Communications,* vol. 146, pp. 351-355, 2008.

[3] S. Morozov, *et al.*, "Giant intrinsic carrier mobilities in graphene and its bilayer," *Physical Review Letters,* vol. 100, p. 016602, 2008.

[4] C. Lee, *et al.*, "Measurement of the elastic properties and intrinsic strength of monolayer graphene," *Science,* vol. 321, pp. 385-388, 2008.

[5] G.-H. Lee, *et al.*, "High-Strength Chemical-Vapor– Deposited Graphene and Grain Boundaries," *Science,* vol. 340, pp. 1073-1076, 2013.

[6] H. Tomori, *et al.*, "Introducing nonuniform strain to graphene using dielectric nanopillars," *arXiv preprint arXiv:1106.1507,* 2011.

[7] J. S. Bunch, *et al.*, "Impermeable atomic membranes from graphene sheets," *Nano Letters,* vol. 8, pp. 2458-2462, 2008.

[8] S. P. Koenig, *et al.*, "Ultrastrong adhesion of graphene membranes," *Nature Nanotechnology,* vol. 6, pp. 543-546, 2011.

[9] M. Huang, *et al.*, "Electronic– Mechanical Coupling in Graphene from in situ Nanoindentation Experiments and Multiscale Atomistic Simulations," *Nano Letters,* vol. 11, pp. 1241-1246, 2011.

[10] S.-E. Zhu, *et al.*, "Graphene based piezoresistive pressure sensor," *Applied Physics Letters,* vol. 102, pp. 161904-161904-3, 2013.

[11] A. D. Smith, *et al.*, "Pressure sensors based on suspended graphene membranes," *Solid-State Electronics,* 2013.

CONTACT

M.C. Lemme; max.lemme@uni-siegen.de

FABRICATION OF GOLD NANOPARTICLE-EMBEDDED NANOCHANNELS FOR SURFACE-ENHANCED RAMAN SPECTROSCOPY

Keisuke Suekuni, Toshimitsu Takeshita, Koji Sugano, and Yoshitada Isono
Department of Mechanical Engineering, Kobe University, Kobe, JAPAN

ABSTRACT

A micro/nanofluidic device including linearly-arranged gold nanoparticles embedded into nanochannels was developed toward highly-sensitive Surface-Enhanced Raman Spectroscopy (SERS) of multicomponent analysis with low background signal. The nanochannels array was fabricated by a "photo"-lithography-based process without costly and time-consuming process such as an electron beam lithography and a focused-ion-beam etching. Then particles with diameters of 100 nm are linearly arranged into the nanochannels by a nanotrench-guided self-assembly process followed by microchannel fabrication for introducing solutions. The device was successfully fabricated and it was evaluated for SERS. The fabricated structure was active for SERS analysis with 4,4'-bypiridine as a target molecule.

INTRODUCTION

Raman spectroscopy has proved to be an extremely helpful tool for analytical investigations in many fields such as biology, chemistry, and medicine. Raman spectra include molecule structural information which is shown as several peaks at Raman shifts derived from a molecular structure so that they allow identification of molecules with label-free method. Although Raman scattering light is significant weak due to its low scattering cross section, an enhancement could be achieved by bringing the molecules into contact with metal nanoparticles. It is called Surface-Enhanced Raman Spectroscopy (SERS). The enhancement of Raman signals is caused by an electromagnetic enhancement based on localized plasmonic resonance of metal nanoparticles. Particle-particle contacts have been utilized for highly-sensitive analysis since the contact spot, called "hot spot", generates higher enhancement of the electromagnetic enhancement. Especially the highest electromagnetic enhancement factor of more than 10^5 could be achieved when connection directions of particles are matched to a polarization direction of incident light [1,2]. The enhancement factor of SERS signal has been considered to be a square of the electromagnetic enhancement factor.

One of problems in general nanoparticle-based SERS is that it is difficult to stably fabricate the optimal configuration of particles mentioned above. Furthermore it is crucial to reduce background noise from dispersion liquid in order to pick up significant low signals. The second problem is that a spectrum shows many overlapped peaks when a solution includes a number of species. In order to solve the problems mentioned above we developed the micro/nanofluidic SERS platform with nanoparticle-embedded nanochannels. The straight nanochannels lead straight-line arrangement of nanoparticles with hot spots using a nanotrench-guided self-assembly. The linearly-arranged nanoparticles are

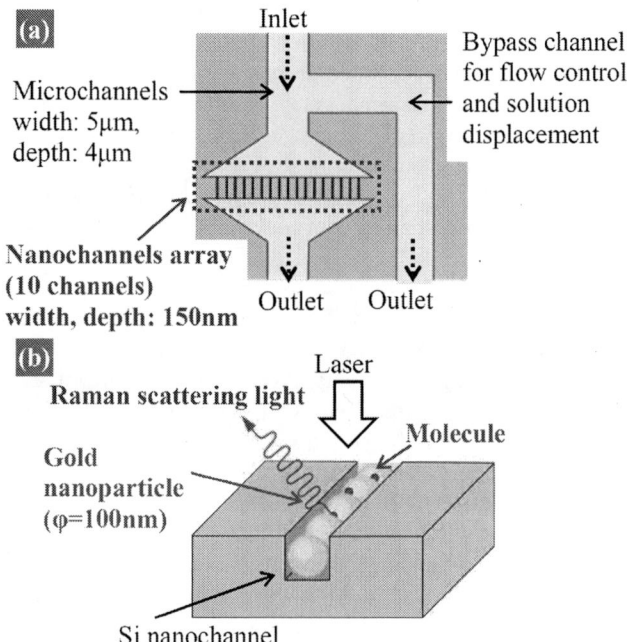

Figure 1: (a) Overview of the proposed device with nanochannels connected to the microchannels. (b) linearly-arranged gold nanoparticles into the nanochannel.

expected to allow us huge SERS enhancement. The second advantage is that solutions are confined into the nanochannels. In this configuration a volume of dispersion liquid is significant small. This leads low background noise from the dispersion. Furthermore analyte molecules are confined in nanochannels decreasing the number of species to a few kinds. This is expected to allow us to analyze the molecules separating them.

In this paper, as the first step for the purpose of this study, we report about the fabrication of the proposed device. The nanochannels were fabricated by "photo"-lithography-based process without costly and time-consuming process. Furthermore the fabricated structure was evaluated for SERS analysis.

DEVICE CONCEPT

Figure 1 shows the proposed device. It includes nanoparticle-embedded nanochannels and microchannels to introduce a solution into nanochannels. The microchannel width and depth are 5 μm and 4 μm, respectively. The nanochannel width and depth are ranged from 150 to 200 nm with the length of 4 μm. The nanochannel volume is calculated to about 0.1 fL. We used gold nanoparticle with the mean diameter of 100 nm.

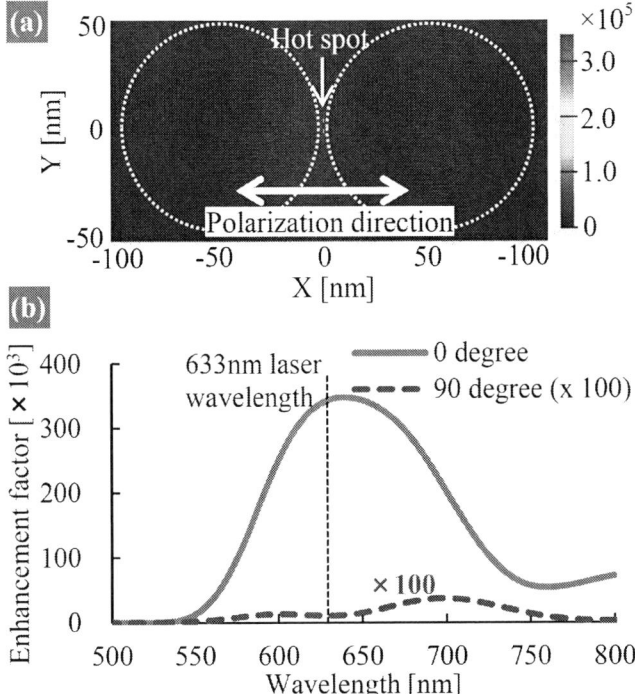

Figure 2: FDTD simulation results for electromagnetic enhancement factor depending on a polarization angle, 0 and 90 degree to polarization direction. (a) Contour plots of enhancement factor at 633 nm of incident light in a particle dimer. (b) Enhancement spectra at a contact spot.

In this study we utilize linearly-arranged particles with hot spots to enhance Raman signals matching the arrangement direction of particles to the polarization direction of incident light. Figure 2 shows the simulation results of electromagnetic field enhancement using Finite-difference time-domain (FDTD) method. The electromagnetic enhancement factor was 3.5×10^5 at the wavelength of 633 nm, which indicates 1.2×10^{11} of SERS enhancement factor.

FABRICATION PROCESS

The device was fabricated according to the following process flow as shown in Fig. 3. In order to achieve the concept, the nanochannel array was fabricated by a "photo"-lithography-based process without costly and time-consuming process such as electron-beam (EB) lithography and focused-ion-beam (FIB) etching.

Nanochannels fabrication

(a) Thermal oxide SiO_2 layer with 200-nm thickness on a silicon wafer was patterned with 5-μm line and space mask pattern by photolithography and a consequent wet etching using buffered hydrofluoric acid (BHF) with an undercutting. The undercut width was controlled from 150 to 200 nm by an etching time. (b) Cr layer with the thickness of 10-20 nm was deposited and the photoresist was lifted off for the nanopattern formation. The width of nanochannel was determined by the undercut width of SiO_2 during the etching.

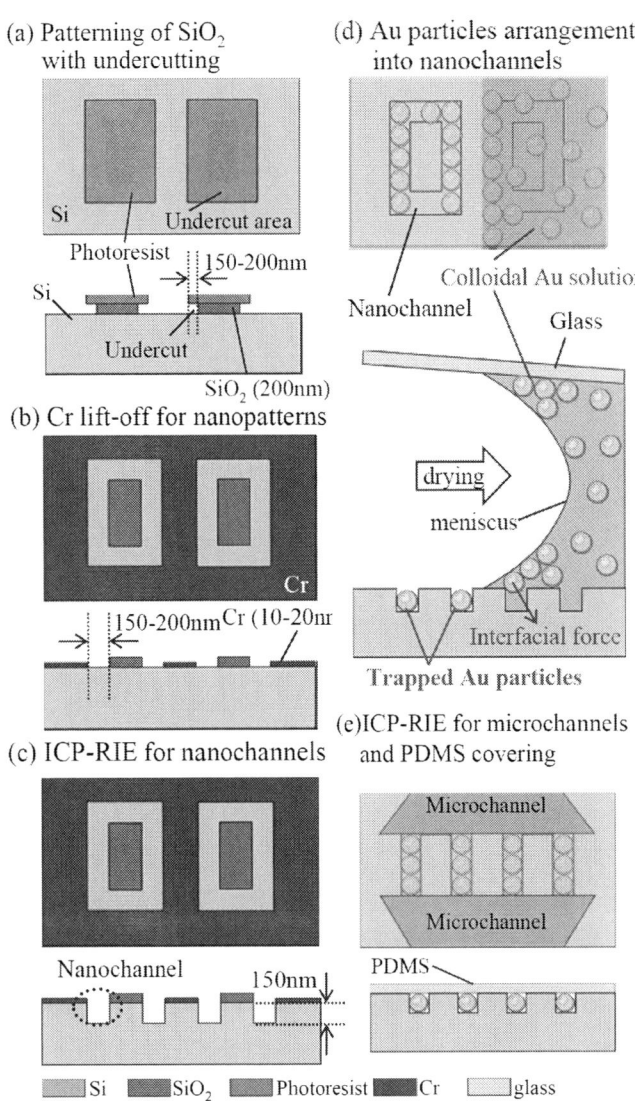

Figure 3: Fabrication process of the micro/nanofluidic SERS device with arranged gold nanoparticles in the nanochannel by photolithography-based process

(c) Si was etched using the nanopatterns of SiO_2 and Cr layers as a mask by an inductively-coupled plasma reactive ion Etching (ICP-RIE) for the nanochannel fabrication. The etching was carried out using SF_6 and C_4F_8 gasses. The etched depth was set to 100-150 nm in order to embed nanoparticles with a diameter of 100 nm into the nanochannels. Then SiO_2 and Cr layers were removed.

Gold nanoparticle arrangement into nanochannels

(d) Gold nanoparticles were arranged into the nanochannels by the nanotrench-guided self-assembly process [3,4]. The gold nanoparticle colloid dispersed in water was purchased from BBI solutions. The gold solution with a concentration of 0.0001 wt.% was prepared and introduced into two substrates, Si substrate with the fabricated nanochannels and a glass substrate as shown in Fig.

a)-c) Nanochannel fabrication

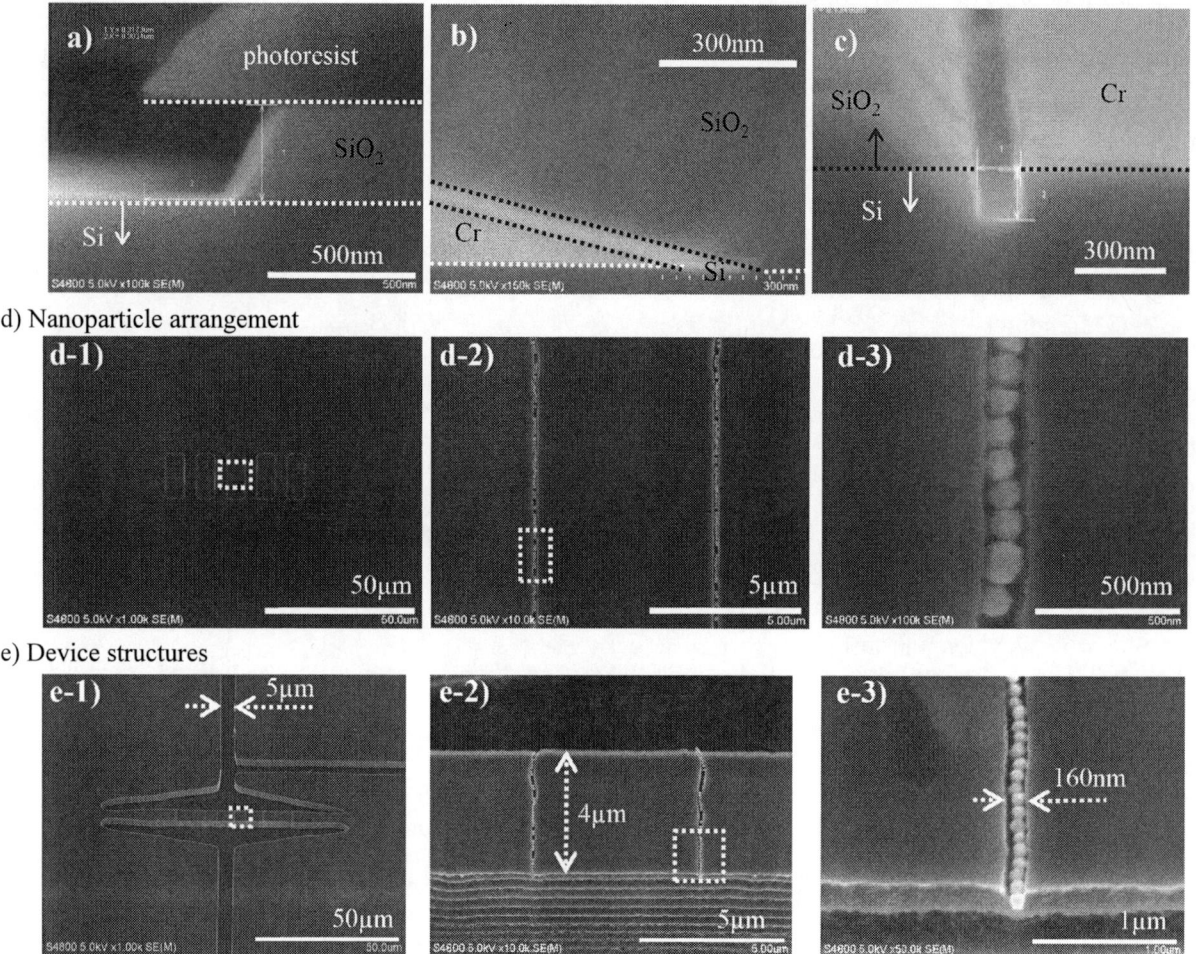

d) Nanoparticle arrangement

e) Device structures

Figure 4: SEM images of the fabricated structures. Alphabets in the figures correspond to the fabrication process shown in Fig. 3. a)-c): Nanochannel fabrication, d): Nanoparticle arrangement, e): Device structures.

3 (d). By drying the aqueous dispersion between the substrates, the water surface line moves backward and the particles are concentrated near the edge of the meniscus. The capillary and the interfacial forces drag and press the particles onto the Si substrate. When the meniscus passes over the nanochannels, the particles are trapped onto them. Then water bridges form between the trapped particles. When drying the water the particles attract each other so that they form particle-particle contacts acting as hot spots.

Microchannel fabrication and covering channels

(e) 2-μm deep microchannels were fabricated by photolithography and ICP-RIE. After removing the photoresist, UV/O$_3$ treatment was conducted for 20 min at 60 °C in order to remove dicarboxy acetone molecules originally attached on the particles. Then the nano/microchannels were covered by a poly-dimethylsiloxane (PDMS) sheet

RESULTS

Fabrication

Figure 4 shows SEM images of the fabricated structures.

Figure 4 (a) shows the structure after the SiO$_2$ undercutting. The SiO$_2$ layer underneath the photoresist was etched controlling the undercut width by the etching time. The nanoparttern after Cr layer deposition and photoresist lift off is shown in Fig. 4 (b). The area which was covered by the photoresist was exposed. The etched nanochannel is shown in Fig. 4 (c). The nanochannel was successfully fabricated around the micron-scale mask pattern of the first photolithography using the proposed process. The nanochannel width can be controlled by the undercut width of SiO$_2$ etching. A batch process for thousands of nanochannels is possible using the proposed process. In addition nanogap electrodes for actuators and sensors can be also fabricated.

Figure 4 (d) shows gold nanoparticles arrangement into the nanochannels. The self-assembly process fabricated linearly-arranged particles with many particle-particle hot spots. The linearly-arranged particles are expected to generate huge electromagnetic enhancements of incident light and Raman scattering light resulting in high SERS intensity according to the simulation results.

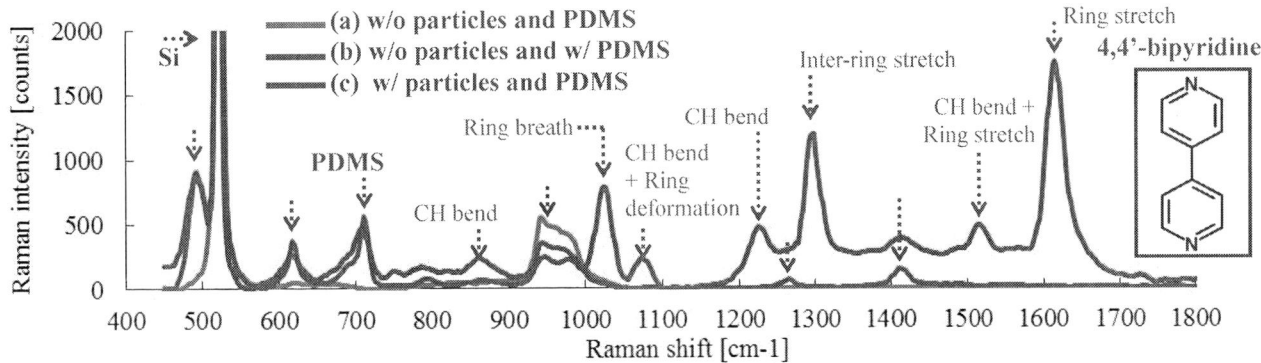

Figure 5: Raman spectra measured at the nanochannels with the 4,4'-bipyridine solution. Blue, green and red arrows indicate Si-, PDMS- and 4,4'-bipyridine-derived Raman peaks, respectively.

The array of 10 nanochannels connected to the microchannels was successfully fabricated as shown in Fig. 4 (e). The nanochannel length was determined by the second mask pattern. In this experiment 4-μm-long nanochannels were fabricated. Then the fabricated channels were successfully covered by the PDMS sheet.

SERS Experiments

We evaluated the fabricated structure whether it is active or not for SERS analysis. 4,4'-bipyridine (Wako Pure Chemical Industries), which is a sort of pesticides was used as a target molecule. The molecule solution with the concentration of 10^{-3} M in water was provided to the nanochannels. Then Raman spectra were acquired at the nanochannels using a micro Raman spectroscope which equips 633-nm wavelength laser with about 2-μm beam spot. The integration time was set to 10 sec.

We acquired the Raman spectra of the target solution with three different cases, (a) only Si nanochannels without the particles and the PDMS cover, (b) nanochannels with the PDMS cover without the particles, and (c) nanochannels with the linearly-arranged particles and the PDMS cover. Figure 5 shows the acquired Raman spectra of the three cases. The spectrum of the case (a) shows only Si-specific peaks at around 520 and 950 cm^{-1}. In the case (b), Si- and PDMS-specific peaks [5] were observed. Although no 4,4'-bipyridine-specific peaks was observed in the above two cases, the molecule-specific peaks [6] were strongly appeared in the case (c) which includes particles. This result means that linearly-arranged particles in the nanochannels strongly enhanced the Raman signal from the molecules.

Furthermore no signal of water molecules around 1636 cm^{-1} was observed in all cases. This indicates that the background noise from the dispersion was sufficiently small for this experiment.

CONCLUSIONS

We developed the micro/nanofluidic SERS device with linearly-arranged gold nanoparticles embedded into nanochannels. The photolithography-based fabrication process for nanochannels was developed. The gold nanoparticles with the diameter of 100 nm were arranged linearly in the nanochannels in order for huge enhancement of Raman signal. The fabricated structure was active for SERS analysis according to the experimental result of 4,4'-bipyridine molecules measurement. The developed device is expected to realize highly-sensitive SERS analysis for multicomponent with low background noise.

REFERENCES

[1] S. Nie and S. Emory, "Probing Single Molecules and Single Nanoparticles by Surface-Enhanced Raman Scattering", *Science*, vol.275, pp. 1102-1106, 1997.

[2] K. Yoshida, T. Itoh, H. Tamura, V. Biju, M. Ishikawa, and Y. Ozaki, "Quantitative evaluation of electromagnetic enhancement in surface-enhanced resonance Raman scattering from plasmonic properties and morphologies of individual Ag nanostructures", *Physical Review B*, vol. 81, pp. 115406-1-9, 2010.

[3] K. Sugano, T. Ozaki, T. Tsuchiya, O. Tabata, "Fabrication of gold nanoparticle pattern using combination of self-assembly and 2-step transfer", *Sensors and Materials*, vol. 23, no. 5, pp.263-275, 2011.

[4] K. Sugano, R. Hiraoka, T. Tsuchiya, O. Tabata, M. Klaumünzer, M. Voigt, and W. Peukert, "Nanogap controllable self-assembly of gold nanoparticles using nanotrench template", *Digest Tech. Transducers'11 Conference*, Beijing, 2011, pp.2570-2573.

[5] T. Park, S. Lee, G. H. Seong, J. Choo, E. K. Lee, Y. S. Kim, W. H. Ji, S. Y. Hwang, D.-G. Gweond and S.Lee, "Highly sensitive signal detection of duplex dye-labelled DNA oligonucleotides in a PDMS microfluidic chip: confocal surface-enhanced Raman spectroscopic study", *Lab Chip*, vol. 5, pp.437-442, 2005.

[6] M. Suzuki, Y. Niidome, S. Yamada, "Adsorption characteristics of 4,4'-bipyridine molecules on gold nanosphere films studied by surface-enhanced Raman scattering", *Thin Solid Films*, 496, pp.740-747, 2006.

CONTACT

*K. Sugano, tel: +81-78-803-6314;
sugano@mech.kobe-u.ac.jp

FINFET WITH FULLY PH-RESPONSIVE HFO$_2$ AS HIGHLY STABLE BIOCHEMICAL SENSOR

S. Rigante[1], M. Wipf[2], A. Bazigos[1], K. Bedner[4], D. Bouvet[1], and A.M. Ionescu[1]

[1]Nanoelectronic Devices Laboratory, EPFL, Lausanne, Switzerland
[2]Department of Physics, University of Basel, Switzerland
[4]Laboratory for Micro- and Nanotechnology, Paul Scherrer Institute, Switzerland

ABSTRACT

In this work, highly scaled FinFETs (Fin Field Effect Transistors) are proposed as both sensing and circuit units of a lab-on-a-chip platform. The FinFET-based sensors with an HfO$_2$ gate oxide demonstrate full pH-response with $\Delta V_{th} \approx 56$ mV/pH. High readout sensitivity $S_{out} = \Delta I_d/I_d \approx 43\%$ is achieved in combination with excellent device electronic properties, i.e. SS = 77 mV/dec and $I_{on}/I_{off} = 1.5 \times 10^6$. High long-term stability is proven over 4.5 days with a drift in time limited at 0.14 mV/h

INTRODUCTION

Challenges of SiNW-based sensors

In the last decade, SiNWs have shown their potential as label-free sensors [1]. With respect to other emerging technologies they provide a real-time and direct signal when a change in charge at the surface occurs. According to the read-out, the variation of the surface potential can be detected as a variation of either the threshold voltage, V_{th}, or drain current, I_d, but no other transduction is needed. Despite their evident efficiency, the integration with CMOS ICs is still a challenge. To achieve such integration the research should focus on the optimization of the device architecture and power supply constraints. Commonly, single SiNWs are designed with rather large width and reduced thickness ($H_{SiNW}/W_{SiNW} < 1$) [2-5] and high voltage are applied, especially at the back-gate [3,4]. The device reliability and its long-term stability are rarely addressed. High-k dielectric materials, such as AlO$_3$ and HfO$_2$, have already been investigated and demonstrated to provide pH response close to the Nernst limit of $\Delta V_{th} \approx 59$ mV/pH [5].

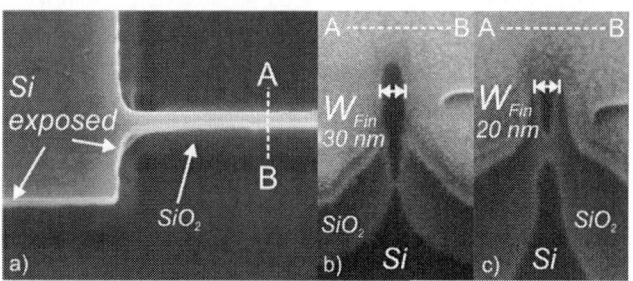

Figure 1: SEM top view of a FinFET detail (a) and wire cross-sections by FIB: Si vertical fin with W_{Fin} = 30 nm, H_{Fin} = 85 nm (b) and W_{Fin} = 20 nm, H_{Fin} = 65 nm (c).

In this work, we propose FETs (Field Effect Transistor) which feature an HfO$_2$ gate oxide and a Double-Gate (DG) architecture for optimized channel control, sensitivity and stability. Such devices are known as FinFETs and their excellent electronic properties are well known in nanoelectronics [6]. Figure 1 shows SEM (Scanning Electron Microscope) images of the FinFET implemented as sensing units: (a) is a top view of the Si-Fin at the anchor point with its contact pad, (b) and (c) are cross-sections obtained by FIB (Focused Ion Beam) showing the vertical fins with W_{Fin} = 30 nm and W_{Fin} = 20 nm, respectively, and aspect ratio $H_{Fin}/W_{Fin} > 3$. In previous works, we published preliminary implementations of the FinFET as ionic sensor [7, 8]. Herein, we present the latest sensing platform after having additionally achieved: (i) full pH response based on HfO$_2$ gate oxide, (ii) highly stable long-term reliability and repeatability.

Device and Microfluid Platform Development

The FinFETs were fabricated using a local SOI technology on Boron doped Si-bulk wafers [7, 8]. Fin widths from 15 to 40 nm and $H_{Fin}/W_{Fin} > 3$ have been achieved. Source and drain contacts were implanted with a Phosphorous dose $\approx 10^{16}$/cm^2 at 30 keV, aimed at reaching a contact doping level $N_a = 10^{20}$-10^{21}/cm^3 and expected penetration depth of 150 nm. The fin surface was covered by 8 nm of HfO$_2$ deposited by ALD (Atomic Layer Deposition). In order to prevent any surface degradation the use of highly concentrated HF acid (Hydrofluoric) solution and high power O$_2$-plasma was avoided after ALD.

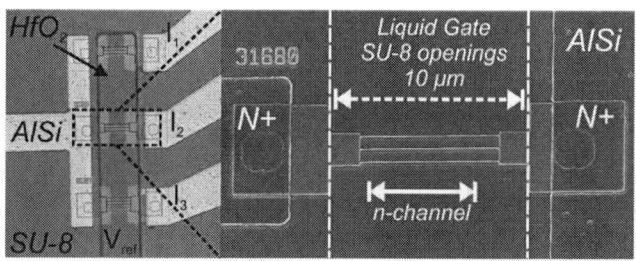

Figure 2: Optical image of the sensors with independent outputs and SU-8 openings over the sensing channels (left); SEM detail of a three SiNW FinFET (right).

The metal connections were fabricated by a lift-off process of AlSi with 1% Silicon to avoid Al diffusion at the junctions. To limit the contact with the liquid at the Si surface, SU-8 openings were patterned next to the sensor channels, as shown in Fig. 2 (left). The correct superposition of the liquid opening and doped regions (Fig.2, right) allow the n-channel to conduct when the solution is biased at the

978-1-4799-3510-9/14 $31.00 © 2014 IEEE

proper potential. Liquid and metal gate FinFETs are both available on the same die, as represented in Fig. 3.

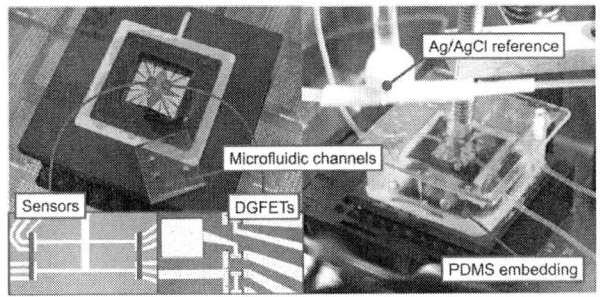

Figure 3: Die-chip carrier assembly with FinFET sensors and transistors; a PDMS cube is patterned with microchannels at its bottom (left); sensing platform with PDMS embedding and Ag/AgCl reference electrode (right).

Once the die was diced out form the wafer, it was connected to the chip carrier by ball-bonding with Au wires. For the assembly of the microfluidic platform, PDMS microchannels were aligned with respect to the SU-8 openings and liquid Epoxy fixed them at chip carrier. The channels were then connected by PTFE tubes and a full PDMS embedding seals the whole system (Fig. 3).

MEASUREMENTS
pH Sensing

To perform pH measurements in a liquid environment the PTFE tubes were connected to a tubing pump and a valve selector system. The liquid potential was controlled by an Ag/AgCl flow through reference electrode included in the tubing. The first pH measurement was meant to evaluate the device threshold voltage shift, ΔV_{th}, at different pH values. The FinFET drain current, I_d, was measured at constant source and drain voltage, $V_{ds} = 80$ mV, and back-gate potential, $V_b = 0$ V. The liquid potential, V_{ref}, was swept from 0.5 to 2.5 V, through the reference electrode at fixed pH = 3. The resulting $I_d(V_{ref})$ transfer characteristic is illustrated in Fig. 4 on the left Y-axis. The liquid gate devices exhibit the same good electrical behavior as for the metal gate devices. Prior to liquid measurements, metal gate FinFETs were characterized, achieving excellent electrical properties of SS = 77 mV/dec and $I_{on}/I_{off} = 1.5 \times 10^6$ for $W_{Fin} = 30$ nm, $H_{Fin} = 80$ nm and $L_{Fin} = 10$ μm. The right Y-axis of Fig. 4 shows the $I_d(V_g)$ corresponding to the same FinFET fabricated with a metal gate. For comparison, we set $V_{ref} = V_g + \Delta V_{sol}$, with $\Delta V_{sol} = 0.75$ V. The liquid gate FinFETs performs a less steep Subthreshold slope, SS = 180 mV/dec. Highly probably, this is due to the higher I_{off} current level, which may depend on different measurement set-up conditions, resulting higher leakage currents. Afterward, the valve selector system was used to exchange the solutions at different pH values. Steady-state measurements were performed between $3 \leq pH \leq 9$ and the rest $I_d(V_{ref})$ transfer characteristics are also reported in the Fig. 4 on the left Y-axis.

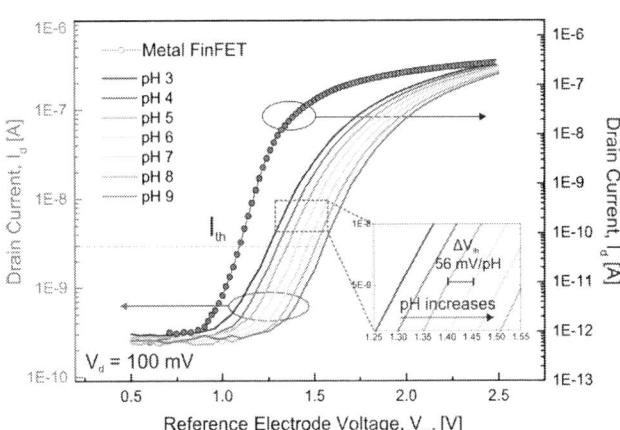

Figure 4: $I_d(V_{ref})$ characteristic for a 3 SiNW metal (right Y-axis) and liquid gate (left Y-axis) FinFET at $V_d = 100$ mV, $V_b = 0$ V. The FinFET sensor characteristics have been obtained for $3 \leq pH \leq 9$ with the inset showing the curve shift $\Delta V_{th}/pH$ due to the surface potential variation at different pH values.

For the data analysis, we can differentiate between the intrinsic sensor sensitivity, $S = V_{th}/pH$, and the readout sensitivity, $S_{out} = \Delta I_d/I_d$. While the former only depends on the oxide surface the second one is linked to the SS of the FETs as well. For all pH transitions, $S \approx 56$ mV/pH is achieved. The fabricated FinFETs finally present a full pH sensitivity and, as a consequence, sensitivity with respect to other chemicals is suppressed [5,8]. Afterwards, V_{ref} was fixed at 1.5 V and I_d measurements vs. time were performed and reported in Fig. 5.

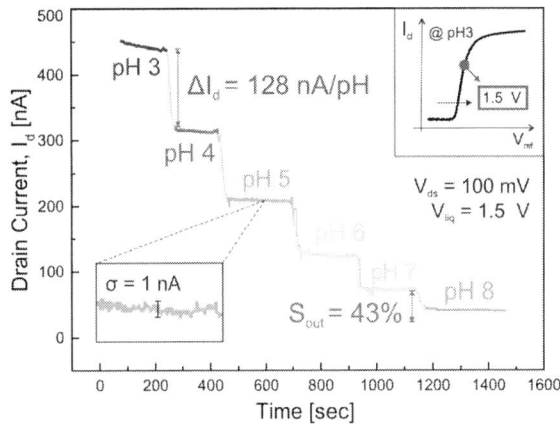

Figure 5: Measured Drain Current I_d, for a 5 SiNW FinFET sensor during a time period of 25 minutes for $3 \leq pH \leq 9$ with reference electrode $V_{ref} = 1.5$ V and $V_{ds} = 100$ mV.

The devices achieved a high current variation with a maximum $\Delta I_d/I_d = 43\%$ for the transition pH 7→8 and averaged $\Delta I_d \approx 80$ nA/pH for $3 \leq pH \leq 8$. At a different bias, $V_{ref} = 2$ V, the sensor provided a maximum $\Delta I_d \approx 271$ nA/pH but the relative current variation does not exceed

$\Delta I_d/I_d = 28\%$. Studying in overall the current transitions we can assume a negligible background noise, with $\Delta I_d/\sigma_{\Delta Id} > 60$. Tables 1 and 2 summarize the data obtained at $V_{ref} = 1.5$ and $V_{ref} = 2$ for each transition. The characterization of time-dependent measurements is important for fast kinetic reactions and small surface potential variation. Steady-state measurements could entail hysteretic effects affecting the detection of small ΔV_{th}. According to the type of measurement the point of biasing can be adjusted for high ΔI_d or $\Delta I_d/I_d$.

TABLE 1: FinFET performances as pH sensor, $V_{ref} = 1.5$ V

	FinFET, $V_{ref1} = 1.5$ $V_{@ph3}$					
pH	3→4	4→5	5→6	6→7	7→8	3-8
S_{out}[%]	29	33	40	40	43	≈ 31%
ΔI_d [nA]	128	105	85	50	32	≈ 80 nA/pH
$\Delta I_d/\sigma_{Id}$	45	88	85	71	80	≈ 61

TABLE 2: FinFET performances as pH sensor, $V_{ref} = 2$ V

	FinFET, $V_{ref1} = 2$ $V_{@ph3}$					
pH	3→4	4→5	5→6	6→7	7→8	3-8
S_{out}[%]	9	7.5	22	15	28	≈ 14%
ΔI_d [nA]	130	100	271	148	232	≈ 176 nA/pH
$\Delta I_d/\sigma_{Id}$	27	31	117	123	300	≈ 99

Long Term Stability

The threshold voltage V_{th} was monitored for 4.5 *days*, keeping the liquid environment at constant pH = 6. Every 30 minutes the system pumping was automatically activated to renew the liquid on top of the sensors. After a stabilization time of several minutes the $I_d(V_{ref})$ characteristic was traced by sweep of the reference electrode. The V_{th} was then extracted at the same $I_d = 2$ nA and plotted, as shown in Figure 6. The FinFETs behaved in an extremely stable way with a drift of $\delta V_{th}/\delta t \approx 0.14$ mV/h for multiple wire FinFETs, with different fin thickness, $T_{Fin} = 30$ nm for D_1 and $T_{Fin} = 20$ nm for D_2.

The presented data also describe the measurement repeatability. The V_{th} measurement was, in fact, acquired more than 200 times. Assuming the intercept of a population of data equal to the mean V_{th} for a specific device, its standard deviation $\sigma_{Vth} \approx 0.8$ mV is an indicator of the V_{th} fluctuation, independent from the time drift.

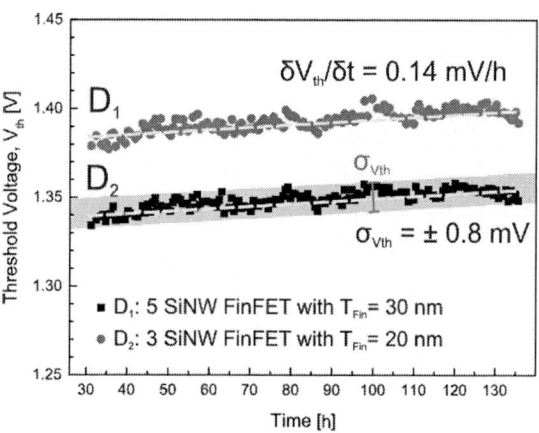

Figure 6: Long-term stability measurement over 4.5 days; the threshold voltage V_{th} is monitored for different FinFET sensors at constant pH = 6.

Moreover, we have compared two identical FinFETs, D3 and D4, located at two different positions of one die, measured through channel A (CH A) and channel B (CH B), respectively. The devices are expected to have the same V_{th}. By subtracting the two V_{th} sets of data, a residual $\Delta V_{th} \approx 0.6$ mV with its corresponding standard deviation $\sigma_{\Delta Vth} \approx 1$ mV is obtained, as shown in Figure 7.

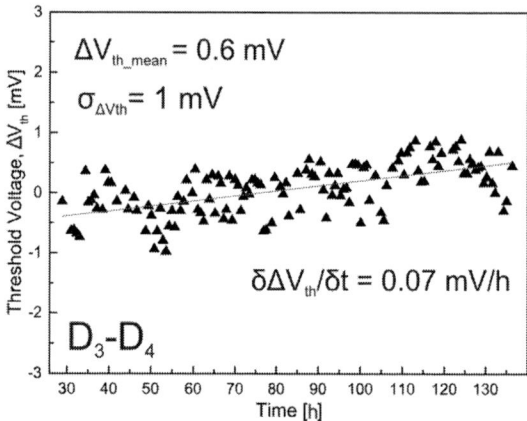

Figure 7: Result of the subtraction of the date set of V_{th} for two identical FinFETs located at two different positions on the same die.

The reliability of the FinFETs and the fabrication process was then demonstrated at the die level. A small drift in time, $\delta\Delta V_{th}/\delta t \approx 0.07$ mV/h was also observed for the relative ΔV_{th}. These data have proven two important sensor properties, stability and reliability at single sensor and wafer/die level.

978-1-4799-3510-9/14 $31.00 © 2014 IEEE

CONCLUSIONS

Herein, we have demonstrated the advantages of implementing FinFETs as sensing units. The same FinFET architecture has been implemented as both sensing and circuit device with excellent outputs. We have optimized the FinFET structures to achieve good electronic proprieties, i.e. high SS and I_{on}/I_{off}. The surface oxide quality for full pH response has been drastically improved [8]. Boosting both electronic and sensing properties have resulted in pH sensors with state-of-the-art features: (i) high current variation with readout sensitivity of $S_{out} \approx 31\%$, $\Delta I_d \approx 176$ nA/pH and signal to noise ratio $S/N > 60$ [2-4, 9], (ii) extremely low power consumption with $P_{sens} < 0.1$ μW and low voltage supply [3,4] and (iii) enhanced long-term stability [2]. Moreover, the FinFETs, thanks to a precise process and their well-defined geometry, have been proved reliable by comparing same entities at different positions.

In conclusion, the presented sensing platform based on high-stability, low-power FinFETs on Si-bulk can be implemented for efficient chemical label-free sensing towards non-invasive simultaneous monitoring of human physiological signals (pH and other biological entities). The use of scalable high-k dielectric FinFETs for both applications is in accordance with the material constraints which come along Moore's Law of scaling, in contrast to other SiNW-based sensors, and paves the way towards sensing integrated circuits.

ACKNOWLEDGEMENTS

The presented work has been financially supported through the Swiss Federal Program Nano-Tera "NanowireSensor" under contract reference 611_61.

REFERENCES

[1] F. Patolsky, C. M. Lieber, "Nanowire Nanosensors", Mater. Today, vol. 8, pp. 20-28, 2005.

[2] I. Park, Z. Li, A. P. Pisano, R. S. Williams, "Top-Down Fabricated Silicon Nanowire for Real-Time Chemical Detection", Nanotechnol., vol. 21, pp. 15501-15510, 2010.

[3] N. Elfstrom, A. E. Karlstrom, J. Linnros, "Silicon Nanoribbons for Electrical Detection of Biomolecules", Nano Lett., vol. 8, pp. 945-949, 2008.

[4] S. K. Yoo, S. Yang, J.-H. Lee, "Hydrogen Ion Sensing Using Schottky Contacted Silicon Nanowire FETs", Trans. On Nanotech., vol. 7, pp. 745748, 2008.

[5] A.Tarasov, M. Wipf, R. L. Stoop, K. Bedner, "Understanding the Electrolyte Background for Biochemical Sensing with Ion-Sensitive Field-Effect Transistors", ACS Nano, vol. 6, pp. 9291-9298, 2012.

[6] J.-P. Colinge, "FinFETs and Other Multi-Gate Transistors", Springer, 2008.

[7] S. Rigante, P. Scarbolo, D. Bouvet, M. Wipf, A. Tarasov, K. Bedner, A. M. Ionescu, "High-k dielectric FinFETs towards sensing integrated circuits," ULIS 2013, vol. 73, pp.19-21, 2013.

[8] S. Rigante, M. Wipf, A. Tarasov, D. Bouvet, K. Bedner, R. L. Stoop, A. M. Ionescu, "Integrated finfet based sensing in a liquid environment", TRANSDUCERS & EUROSENSORS XXVII 2013, vol. 681, pp. 16-20, 2013.

[9] Y. Cui, Q. Wei, H. Park, C. Lieber, "Nanowire Nanosensors for Highly Sensitive and Selective Detection of Biological and Chemical Species", Science, vol. 293, pp. 1289-1292, 2001.

CONTACT

*S.Rigante; sara.rigante@epfl.ch.

FREQUENCY DEPENDENT AC ELECTROOSMOTIC FLOW IN NANOCHANNELS

Wesley T.E. van den Beld[1], Wouter Sparreboom[1], Albert van den Berg[1], and Jan C.T. Eijkel[1]

[1]BIOS Lab-on-a-Chip group, MESA+ Institute for Nanotechnology, Twente University, the Netherlands

ABSTRACT

We report frequency-dependent bidirectional AC electroosmotic flow (AC-EOF) in a nanochannel with double layer overlap. This work follows our report in µTas 2008 of unidirectional AC-EOF in nanochannels [1]. Observed is a bidirectional pumping behavior; simulations of the low frequency pumping confirm a direction opposite to that of AC-EOF in microchannels. By this frequency-dependent bidirectional pumping, nanochannel AC-EOF behaves in fundamentally different way than microchannel AC-EOF. Generally, the results are of importance for the understanding of ion and liquid transport in nanoconfinement.

INTRODUCTION

The pumping mechanism of AC-EOF in microchannels has been investigated in several studies; main advantages of using AC-EOF compared to traditional DC-EOF are the use of low voltages (~1V) and no need for external electrodes. The operational behavior in microchannels has been modeled theoretically and confirmed experimentally in other studies [4,5]. At present, AC-EOF has not been investigated in nanochannels, neither in experiment or simulation apart from our previous report. Here we present novel frequency dependent measurements in combination with simulation results.

EXPERIMENTAL

Glass chips with 50 nm high nanochannels were fabricated using cleanroom technology and sacrificial layer etching of chromium [2,3]. An asymmetric gold electrode array of 100 electrode pairs was integrated along the channel axis with dimensions as shown in Figure 1.

The fabricated device is displayed in Figure 3. Channels were filled with a 100 µM KNO3 solution, leading to a Debye length of 30 nm. This concentration is chosen to have overlapping double layers in the nanochannel. A PDMS layer with microchannels is attached to the chip to allow solution exchange. The measurement setup is schematically shown in Figure 2. Measurements were performed by first filling the nanochannels with 100 µM solution and subsequently filling the connecting microchannels in the PDMS layer on one side with 50 µM and on the other side with 200 µM solution. Then a sinusoidal signal with an amplitude of 400 mV is applied to the electrode pairs using an impedance analyzer, with the large and small electrodes 180 degrees out of phase. The time-dependent increase or decrease of impedance then indicates pumping speed and direction.

SIMULATION

Simulations were performed in COMSOL Multiphysics 4.3a. The simulated 2D-geometry consists of the nanochannel with two pairs of electrodes with dimensions of geometry A as made clear in Figure 1. For the oxide surfaces a capacitance of 1 F/m² and for the metal surfaces 0.2 F/m² is modeled. The boundary condition for the oxide is a constant potential assumption of -0.09 V. At the two ends of this channel two microreservoirs are added to take into account the entrance effects. The behavior is mathematically described by three sets of equations:

- The Nernst-Plank equation is used to describe ionic transport.
- The Poisson equation is included to calculate the electric potential.
- Stokes flow is describing the fluid flow.

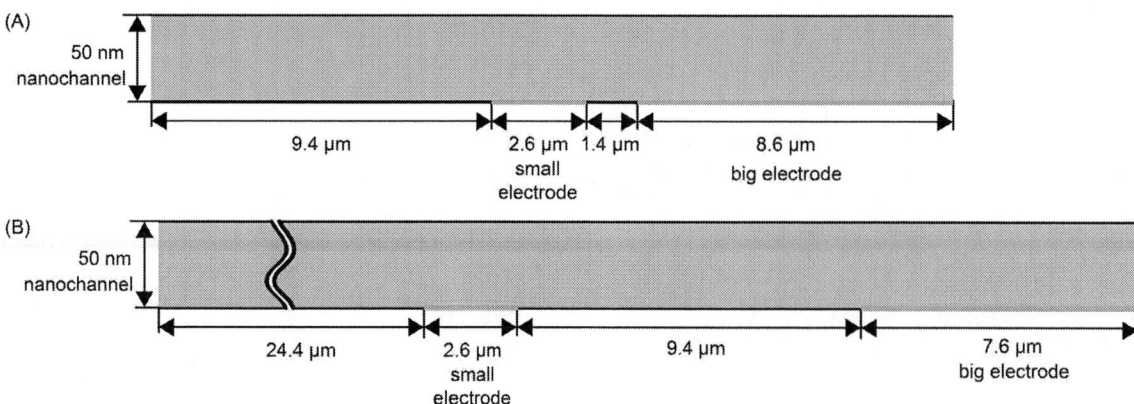

Figure 1: A cross-sectional schematic drawing with both tested geometries showing one repetitive electrode pair of the array.

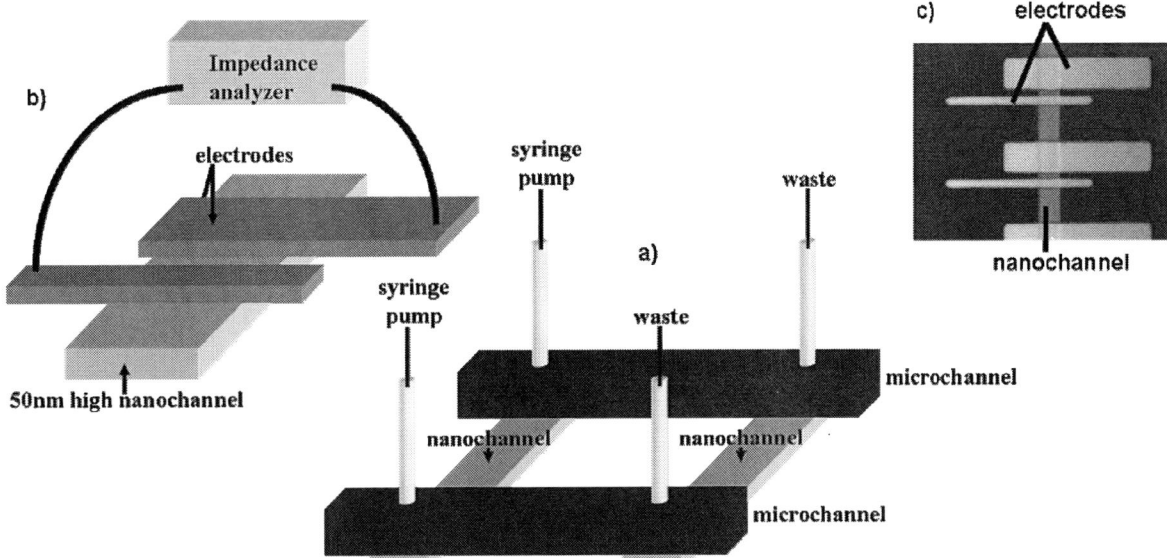

Figure 2: Schematic overview of the measurement setup. The nanochannels are interfaced by two microchannels, each connected to a syringe pump injecting liquids with different conductivities (a). The electrodes are connected to an impedance analyzer for actuation and measuring of the pumping velocity in the nanochannel (b). Photomicrograph of the asymmetric electrodes array situated in the nanochannel (c). [2]

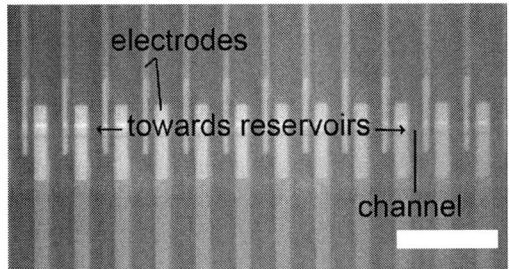

Figure 3: Top-view micrograph of the fabricated device with nanochannel and asymmetric electrode array, the scale bar is 50 μm [1].

RESULTS - EXPERIMENTS

Bidirectional flow is observed in the measurements when changing the applied electrode frequency; results of simulations and measurements are shown in Figure 4. Error bars in the measurements are shown , when multiple experiments were performed at these frequencies. Experiments and simulations show identical flow directions for low frequency actuation, though the magnitude of the simulated flow for geometry A is about threefold lower than the measured flow. In the experiments the flow changes direction at 200 Hz for geometry A and around 30 Hz for geometry B. Lastly, the observed maximum flow speed of 8 μm/s is comparable to the flow speed observed in microchannels at the same applied voltage [4].

Temperature influence on the system can be safely neglected: calculations indicate that the generated power (0.5 μW) after a typical experiment duration of 1000 s only causes a temperature increase of 4 mK in the chip (lumped capacitance 1.2 J/K).

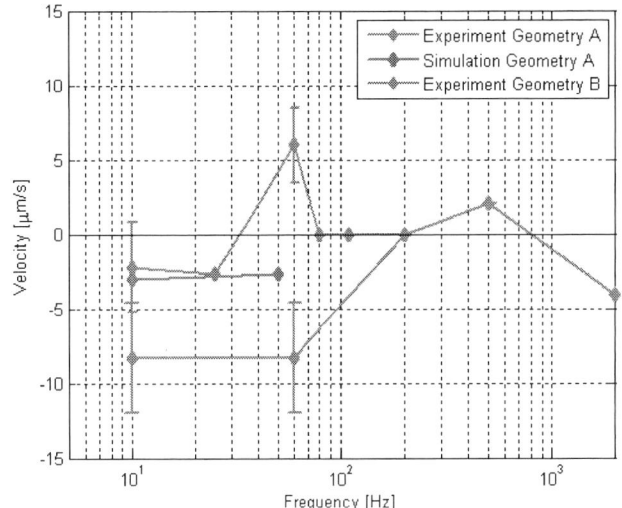

Figure 4: Frequency response of pumping velocity, simulation and measurement [2]. Bidirectional pumping behavior as function of frequency was clearly experimentally observed.

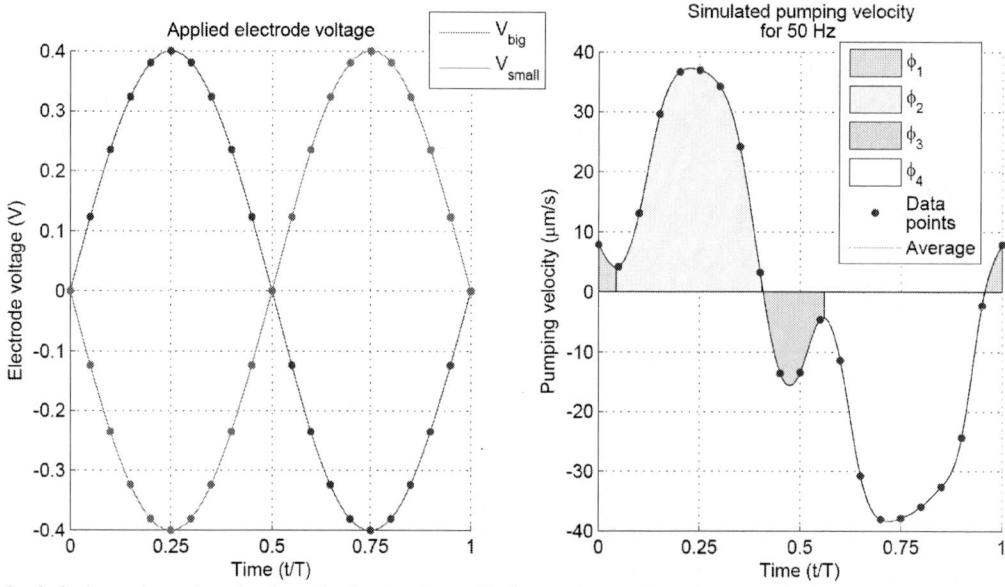

Figure 5. On the left the voltage applied on the big and small electrodes in the array on the bottom of the nanochannel, the simulated pumping velocity on the right for an actuation frequency of 50 Hz. In the pumping behavior four phases ϕ_1-ϕ_4 can be discriminated.

RESULTS - SIMULATIONS

In nanochannels, the operational principle is significantly different from microchannels. As a result of the large surface-to-volume ratio in nanoconfinement, the surface charge on the walls is dominating the flow behavior. Oxide surfaces deprotonate in contact with an aqueous solution as a function of the acidity. When the salt concentration is chosen such that the negative oxide surface charge cannot be screened, double layers will overlap in the nanochannel. Therefore the nanochannel is mostly filled with positive ions and negative ions are mostly excluded.

In a symmetric electrode configuration no net directional flow can be present in the nanochannel due to the balance in forces. However, in an asymmetric configuration a net directional liquid flow can be generated. When a sinusoidal voltage is applied on the electrodes, they attract or repel ions, involving RC-times related to the geometry and the ion concentrations in the channel. Since ions are accumulated or repelled, charge gradients are induced, leading to electric fields and thus generating coulomb forces. When these forces are aligned a directional flow is obtained.

At low frequencies, simulations indicate the flow direction is largely oscillatory and in-phase with the applied electrode potential. The electrodes are screened for the majority of time for these low frequencies, flow is generated by a type of electroosmotic flow driven by the field in the interelectrode spaces. Simulation results of the pumping velocity versus time actuated with 50 Hz are displayed in Figure 5. Four pumping phases can be distinguished, two small (phase 1 and 3) and two large flow contributions (phase 2 and 4). The overall direction of the flow is determined by the dominant flow, which is caused by the large charge gradient in the large interelectrode space. Minor differences in this largely oscillatory flow driven by conventional electroosmotic forces lead to a net flow, as will be treated now.

The abundant positive ions are dominating the pumping behavior, negative ions have a negligible influence. Figure 6 reveals the positive ion distribution in the four different phases. The origin of the minor differences is as follows. In the first phase, the positive ions are repelled from the big electrode and injected in the large interelectrode space (LIES) and the small interelectrode space (SIES), generating charge gradients. The largest charge gradient is present in the LIES, however it is counteracted by the smaller gradient in the SIES, leading to a small net flow in this phase.

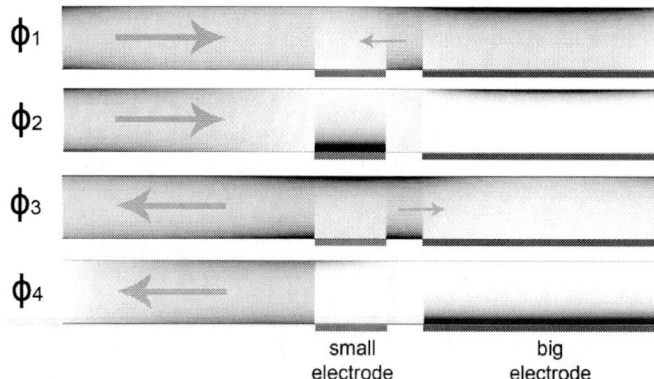

Figure 6. Simulated distribution of the positive ions in the nanochannel in the four phases (dark indicates higher concentration) as well as the flow direction (green arrow). In phase 1 and 3 there is a large gradient, counteracted by a smaller gradient, while in phase 2 and 4 there is a large gradient without a significantly counteracting gradient.

978-1-4799-3510-9/14 $31.00 © 2014 IEEE

This smaller gradient in the SIES disappears when the small electrode starts to attract the positive ions. Then the pumping velocity starts to increase (phase two). Further screening of the small electrode is causing a stronger gradient in the LIES while the gradient in the SIES disappears, generating a strong flow. This flow will continue until the small electrode starts to repel the previously accumulated positive ions, injecting them in the LIES and the SIES. The consequence is that the charge gradient in the LIES switches direction, and therefore the pumping velocity switches sign.

In the third phase, a large gradient is present in the LIES, counteracted by a gradient in the SIES, similar to phase one. However, in phase three the charge is removed faster from the small electrode, hence generating a rapid gradient, having a larger accumulated influence on the net flow than phase one.

This description holds to the point that the big electrode starts to accumulate the positive ions, inducing a large charge gradient in the LIES and removing the gradient from the SIES (phase four). The mechanism is similar to that of phase two. The difference is that the gradient in the LIES is stronger, since the big electrode attracts more positive ions for screening. When the applied voltage on the big electrode approaches the oxide voltage, it repels the positive ions and injects them into the SIES and LIES. The pumping velocity switches and phase one starts again.

Summarizing, in simulation a largely oscillatory flow has been observed with the same direction as in the experiment. In phases one and three the positive ions are repelled from previously charged electrodes, causing opposing gradients and causing a small net force. In phases two and four, the positive ions are attracted, generating a large gradient and thus a large net force. Due to these subtle differences in charge gradients through the several phases a net flow results. The simulations thus provide an explanation for the pumping behavior of AC-EOF in nanochannels. At present we are adjusting our code to also simulate the high frequency behavior.

CONCLUSIONS

The AC-EOF pumping direction at low frequencies in nanochannels is opposite from microchannels[4]. In microchannels, hydrodynamic coupling causes the bulk liquid to move, generated by the induced electroosmotic forces above the electrodes. In nanochannels the force is mainly generated in interelectrode spaces, similar to traditional DC-EOF. Since an alternating signal is applied, an oscillatory flow results. Because of minor differences in the involved charging processes above small and big electrodes, different charge gradients are induced leading to a net flow.

REFERENCES

[1] Sparreboom, W. and Cucu, C.F. and Eijkel, J.C.T. and van den Berg, A. (2008) *Ion pumping in nanochannels using an asymmetric electrode array*. In: 12th International Conference on Miniaturized Systems for Chemistry and Life Sciences, 12-16 Oct 2008, San Diego, California.

[2] Sparreboom, W. (2009). *AC electro-osmosis in nanochannels*. University of Twente. thesis.

[3] Sparreboom, W., Eijkel, J. C., Bomer, J., & van den Berg, A. (2008). Rapid sacrificial layer etching for the fabrication of nanochannels with integrated metal electrodes. *Lab on a Chip, 8*(3), 402-407.

[4] Brown, A. B. D., Smith, C. G., & Rennie, A. R. (2000). Pumping of water with ac electric fields applied to asymmetric pairs of microelectrodes. *Physical review E, 63*(1), 016305.

[5] Ramos, A., Morgan, H., Green, N. G., & Castellanos, A. (1999). AC electric-field-induced fluid flow in microelectrodes. *Journal of colloid and interface science, 217*(2), 420-422.

CONTACT

W.T.E. van den Beld, tel: +31-53-4892154; w.t.e.vandenbeld@utwente.nl

FULLY MONOLITHIC AND ULTRA-COMPACT NEMS-CMOS SELF-OSCILLATOR BASED-ON SINGLE-CRYSTAL SILICON RESONATORS AND LOW-COST CMOS CIRCUITRY

Julien Philippe, Grégory Arndt, Eric Colinet, Mylène Savoye,
Thomas Ernst, Eric Ollier and Julien Arcamone
CEA, LETI, Minatec Campus, Grenoble, FRANCE

ABSTRACT

We report on the first experimental demonstration of a self-oscillator based on a single-crystal silicon NEMS resonator monolithically co-integrated with a CMOS circuitry. The latter, composed only by seven transistors, is manufactured with a very low-cost 0.35μm technology. The NEMS-CMOS self-oscillator pixel is as small as 50x70 μm^2 (pads excluded) and can oscillate near 8MHz. In this paper are described the NEMS-CMOS oscillator characteristics and the implementation method of the self-oscillating loop.

INTRODUCTION

Nano Electro Mechanical Systems (NEMS) resonators constitute a very promising solution for several oscillator-based applications. The techniques used for their fabrication are the same as the CMOS technology making them very attractive because of their low cost production and their possibility of very large scale integration. However, the use of such nano-resonators to implement an oscillator requires a CMOS electronic circuit. A simple approach consists in processing the mechanical and electronic parts on separate dies and to connect them by wire-bonding (Fig. 1). Nevertheless, NEMS-CMOS co-integration appears as the best option to achieve the highest level of performance. Indeed, unlike the off-chip CMOS configuration, the expected benefits are the compactness, an enhanced Signal-to-Noise Ratio (SNR) and a stronger immunity to parasitic feed-through signals [1]. The current state-of-the-art features several demonstrators of M/NEMS-CMOS co-integrated oscillators.

For the oscillator realization, previous demonstrations used various resonator materials (e.g metal [2-4], mono- [5] or polycrystalline silicon [6, 7], silicon alloy [8]) for various applications (pressure sensor [8], time reference [3, 9], communication and data transfer [2, 10], accelerometers [11], gas and mass sensing [4]).

These devices rely on various CMOS technologies (0.35μm bulk CMOS [6, 9], FDSOI [5], 1.5μm bulk CMOS [7]). Two main oscillator architectures have been reported, namely Phase-Locked-Loop (PLL) [12, 13] and self-oscillating loops [6, 14]. This work follows the latter approach with a monolithically integrated NEMS-CMOS self-oscillator using single-crystal Si resonator. Moreover, up to our knowledge, we achieved the most compact NEMS-CMOS pixel (50x70 μm^2). A comparison with the state-of-the-art is provided in Table 1 (the devices areas indicated do not include the pads).

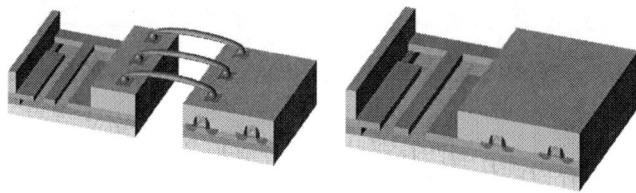

Figure 1: Stand-alone NEMS with its off-chip electronic circuit connected by wire-bonding (left) and a co-integrated NEMS-CMOS system (right).

A description of this hybrid system, more particularly on the mechanical beam and the electronic circuit characteristics, is provided. The strategy to implement this NEMS-CMOS system in a self-oscillating loop without damaging the beam is then explained in detail. Finally, the experimental results of this very compact and low-cost oscillator are exposed.

DESIGN

The mechanical and electronic parts were performed on the top 1μm thick Si layer of SOI wafers with a 1μm thick buried oxide. A pre-CMOS integration approach was followed for this realization.

Table 1: State of the art of NEMS-CMOS co-integrated oscillators.

References	Technology used	Resonator material /shape	Resonance frequency	Device area (μm²)
[10]	0.18μm	SiC / cantilever	8.29 and 11.59 MHz	6250000
[12]	TSMC 0.35μm	Metal / capacitive combs	117.3 kHz	298000
[7]	ON Semiconductor	Si poly / dome	10-100 MHz	63000
[6]	AMS 0.35μm	Metal / cantilever	6.32 MHz	60000
[3]	TSMC 0.35μm	Metal / ring	1.39 and 9.34 MHz	37700
[2]	TSMC 0.35μm	Nickel / disk	10.92 MHz	20600
This work	ST 0.35μm	Si mono / cantilevers	7-8 MHz	3500

978-1-4799-3510-9/14 $31.00 © 2014 IEEE

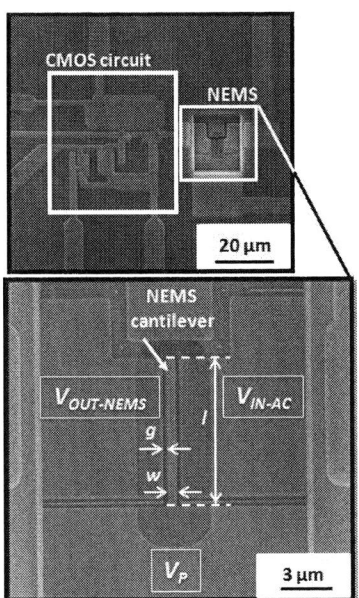

$$G_{MOS} \cdot G_{NEMS} \geq 1 \qquad (1)$$

$$\varphi_{MOS} + \varphi_{NEMS} = 0° \qquad (2)$$

This electronic circuit is monolithically linked to the mechanical part, thereby drastically decreasing the electrical signal losses at the beam output. A very simple design was proposed for the CMOS part: seven transistors are used in two different blocks. An amplification stage provides the right gain and phase shift just after the NEMS output while a unity gain buffer allows an impedance matching with the external apparatus. The schematic of the whole system in a closed-loop configuration is presented in Fig. 3. The same NEMS-CMOS system exists in an open-loop configuration too. The next part will describe the strategy to determine the appropriate conditions such that the system self oscillates.

Figure 2: SEM micrograph of the NEMS-CMOS device (upper) with a zoom on the NEMS cantilever (lower) with its electrodes and dimensions (l=8.3μm, w=0.5μm, g=0.2μm). l, w, g respectively correspond to the length, the width of the beam and the gap between the electrodes and the cantilever.

METHODOLOGY

Open-loop characterization to reach Barkhausen condition

An open-loop characterization of the system was necessary in order to retrieve the suitable NEMS polarizations (V_P) and electronic circuit voltages (V_{GP}, V_B, V_{DD-AMP}, V_{DD-BUF}) for the Barkhausen conditions fulfillment. This experimental study was performed under vacuum (0.05mbar).

The transmission parameter S_{21} of the hybrid system is determined for several supply voltages V_{DD-AMP} of the amplification stage and for several beam polarizations V_P. V_{GP} and V_B are tuned according to V_{DD-AMP} in order to get a maximal gain of the CMOS amplifier. According to the design of the buffer stage, V_{DD-BUF} is fixed at 3.3V for a better impedance matching with external measurement apparatus. An example of S_{21} response of the system is shown on Fig. 4, which is obtained for $V_{DD-AMP} = 3.3$V. As the output of the CMOS amplification stage is directly linked to the NEMS input in closed-loop configuration, the contribution of the buffer part must be suppressed in the S_{21} response (Fig. 4).

After the NEMS patterning, the CMOS circuit is fabricated using a low-cost 0.35μm 3.3V bulk technology from STMicroelectronics without modifying the process. Finally, NEMS resonators are released after protecting the CMOS part with an adequate layer. For further details on the technological process, please refer to [15]. An SEM view is presented in Fig. 2 with the mechanical beam dimensions. The nano-resonator uses an electrostatic actuation and detection scheme. Furthermore, the mechanical beam is designed to resonate around at 8MHz. The CMOS circuit provides both a suitable gain and phase shift such that the Barkhausen conditions, expressed in (1) and (2), are fulfilled by the hybrid NEMS-CMOS system. G_{NEMS}, φ_{NEMS}, G_{MOS}, φ_{MOS} respectively correspond to the gain and the phase of the NEMS and CMOS parts. Some of the parameters are explained in Fig. 3.

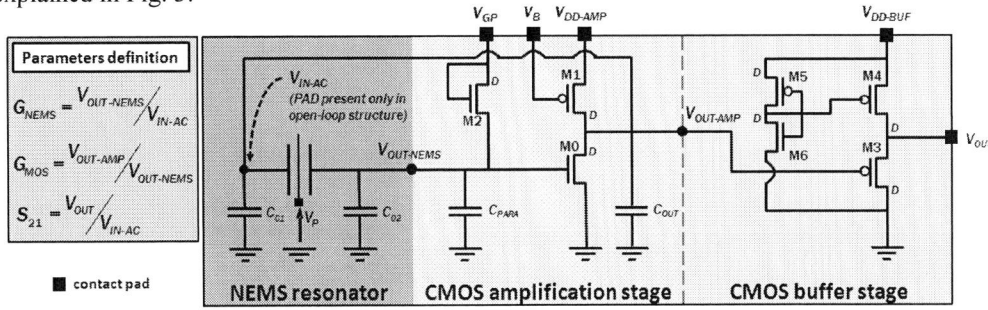

Transistors	M0	M1	M2	M3,6	M4	M5
W (μm)	1.2	6.7 (x10)	1	10	4	0.8
L (μm)	0.5	1	1	0.5	0.625	0.5

Figure 3: Electrical schematic of the NEMS-CMOS self-oscillator and main parameters definition (on the left). The table below gives the CMOS transistors dimensions.

978-1-4799-3510-9/14 $31.00 © 2014 IEEE

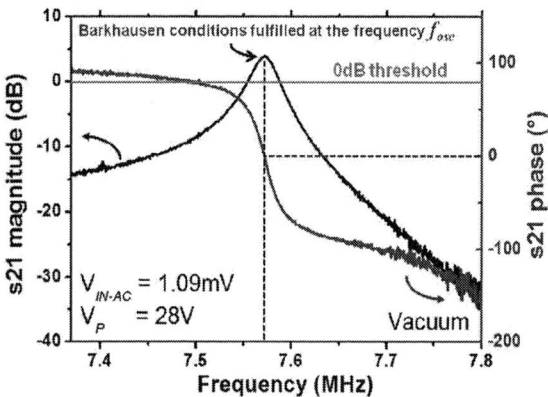

Figure 4: Electrical open-loop response in vacuum (0.05mbar) of the NEMS CMOS system. The Barkhausen criteria are satisfied at the frequency f_{osc}. A pad for external NEMS actuation (not present in the closed-loop configuration) was used for the characterizations of the open-loop structure.

Requested conditions for stable steady-state self-oscillations

Thanks to the previous study, it was possible to determine the minimal NEMS voltage V_P fulfilling the Barkhausen conditions as a function of the CMOS amplifier supply voltage $V_{DD\text{-}AMP}$, and so the theoretical feasibility of the self-oscillator. However, these results are not sufficient to predict the stability of the device. Indeed a saturation mechanism is needed to stabilize the oscillation amplitude at a value smaller that the pull-in distance. A possible solution consists in designing a limiter circuit. Nevertheless this option is potentially area and power consuming. Thus, the CMOS amplification stage was designed such that the oscillation's amplitudes are limited by the saturation of the circuit. This method makes our system very simple and compact. Before the characterization of the closed-loop, an electrical study of the amplifier stage was performed in order to determine the gain variation for different ac input signals and supply voltages $V_{DD\text{-}AMP}$ at the oscillation frequency f_{osc}.

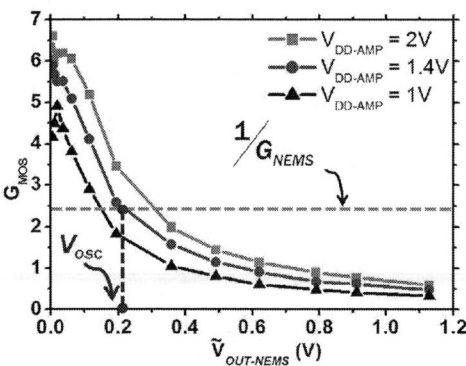

Figure 5: CMOS amplifier response at the oscillation frequency f_{osc} as a function of the circuit ac input voltage $\tilde{v}_{OUT\text{-}NEMS}$ and for different circuit supply voltages. This graph allows extracting the maximum NEMS oscillation amplitude voltage v_{OSC} before pull-in collapse.

The gain of the mechanical system G_{NEMS} at f_{osc} is first obtained by the combination of the hybrid system open-loop characterization and the CMOS frequency response and then superposed to the amplifier stage characteristics (see Fig. 5). According to Fig. 5, the CMOS gain is increasing with $V_{DD\text{-}AMP}$ for a same $\tilde{v}_{OUT\text{-}NEMS}$ (ac term of the NEMS output voltage). This latter is directly proportional to the mechanical amplitude X_{NEMS} of the beam according to (3), where C_{IN} represents the capacitance between the resonator output and the circuit input, S is the electrostatic actuation and detection coupling area, ε_0 the vacuum permittivity, g the gap between the electrodes and the beam, Q the quality factor of the resonator and k_{EFF} the effective stiffness of the mechanical beam.

$$X_{NEMS} = \frac{v_{OSC} \cdot C_{IN} \cdot g^2}{V_P \cdot \varepsilon_0 \cdot S} \qquad (3)$$

$$G_{NEMS} = \frac{\varepsilon_0 \cdot S^2 \cdot Q \cdot V_P^2}{g^4 \cdot C_{IN} \cdot k_{EFF}} \qquad (4)$$

In order to evaluate v_{OSC} and X_{NEMS}, after generation of the self-oscillations, G_{NEMS} and G_{MOS} are plotted on the same graph. The value of v_{OSC} is given by the intersection abscissa point of G_{MOS} and $1/G_{NEMS}$ according to (1). In Fig. 5 is shown the case where v_{OSC} is determined for $V_{DD\text{-}AMP} = 1.4V$. For the preservation of the system integrity during the steady-state oscillations, X_{NEMS} must not exceed a given proportion of g, which means that v_{OSC} must not be larger than a certain value. Furthermore, for a same given value of $1/G_{NEMS}$, v_{OSC} is decreasing with $V_{DD\text{-}AMP}$. In conclusion, the safest and optimized approach for the self-oscillation establishment is first to work with a low value of $V_{DD\text{-}AMP}$, and then finding the V_P threshold value fulfilling the Barkhausen criteria. The selection of this polarization is crucial for the amplitude oscillation limitation because of the dependence between G_{NEMS} and V_P explained in (4).

EXPERIMENTAL RESULTS

Thanks to the previous characterizations and the strategy proposed for the oscillator construction, we were able to make our NEMS-CMOS system oscillate.

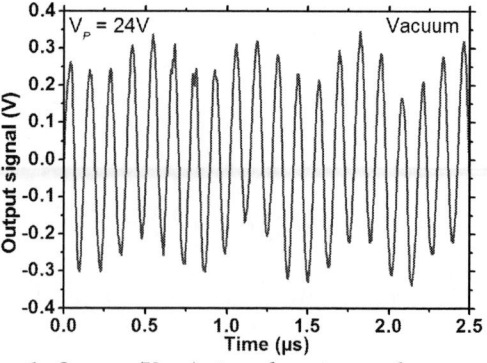

Figure 6: Output (V_{OUT}) time-domain steady-state signal of the NEMS-CMOS closed-loop in vacuum (0.05mbar).

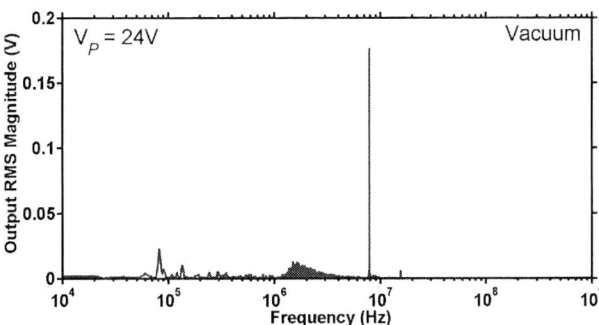

Figure 7: Fast-Fourier transform of the closed-loop output signal (V_{OUT}): the NEMS-CMOS signal is the single peak of the spectrum (no parasitic oscillations). This electrical measurement was performed in vacuum (0.05mbar).

A closed-loop structure was used for this purpose. Electrical characterizations were performed in vacuum environment (0.05mbar). The oscillations were obtained with V_{DD-AMP} = 1.4V and V_P = 24V. Figures 6 and 7 respectively depict the time-domain and frequency-domain steady-state electrical responses of the NEMS-CMOS self-oscillator.

The self-oscillations are stabilized at 7.83MHz with a 0.18V rms amplitude, without any other parasitic oscillations over a wide spectrum. This result shows the strong interest of the co-integrated NEMS-CMOS strategy for oscillators' realization because a better SNR and SBR (Signal-to-Background Ratio) is achieved and the monolithic connection between the electronic and the mechanical parts provides less parasitic oscillations respect to the off-chip configuration.

CONCLUSION

In conclusion, this work demonstrates a 7.83MHz self-oscillator based on a co-integrated single-crystal Si NEMS resonator with a low cost and very compact CMOS circuitry. A detailed study of the open-loop electrical response was necessary to retrieve the Barkhausen conditions while ensuring the system would stabilize at an oscillation amplitude smaller than the pull-in distance. The use of the CMOS saturation constitutes a very simple way to limit the oscillations and to preserve the integrity of the mechanical part when compared to other solutions such as a limiter circuit. As a very compact and functional oscillator is achieved, this work paves the way for ultra-dense arrays of NEMS-CMOS oscillator pixels for emerging applications, such as NEMS-based gas sensors [16] or NEMS-based mass spectrometers [17].

ACKNOWLEDGEMENTS

The authors thank the European Research Council (Delphins project) and EU-FP7 (Minami) for financial support and ST Microelectronics for the CMOS circuits' fabrication.

REFERENCES

[1]   G. Arndt et al, *Sensors and Actuators A*, 172 (2011), pp.293-300.

[2]   W-L. Huang et al, *Proc. MEMS 2008*, pp. 10-13.

[3]   V. Pachkawade, et al, *J. Sensors*, Vol. 13, Number 8 (2013), pp. 2882-2889.

[4]   J. Verd et al, *Lab Chip*, 11, 2670 (2011), pp. 2670-2672.

[5]   E. Ollier et al, *Proc. MEMS 2012*, pp. 1368-1371.

[6]   J. Verd et al, *Electron Device Letters*, Vol. 29, Number 2 (2008), pp. 146-148.

[7]   M.K. Zalalutdinov et al, *J. Microelectromech. Syst*, Vol 19, Number 4 (2010), pp. 807-815.

[8]   K. Allidina et al, *Proc. ICECS 2009*, pp. 583-586.

[9]   W-C. Chen et al, *J Microelectromech. Syst*, Vol 21, Number 3 (2012), pp. 688-701.

[10] F. Nabki et al, *J.Solid-State Circuits*, Vol 44, Number 8 (2009), pp.2154-2168.

[11] S.S. Tan et al, *J. Micromech. Microeng.*, 21 (2011), pp. 329.

[12] H-C. Li et al, *Proc. Sensors 2012*, pp. 1-4.

[13] Mo Li et al, *Nano Letters 2010*, 10, pp.3899-3903.

[14] G. Jourdan et al, *Sensors and Actuators A*, 189 (2013), pp.512-518.

[15] J. Arcamone et al, *Tech. Digest. IEDM 2012*, pp. 15.4.1-15.4.4.

[16] J. Arcamone et al, *IEDM, 2011 IEEE International*, pp. 29.3.1-29.3.4, 5-7 Dec. 2011.

[17] M. S. Hanay et al, *Nature Nanotechnology*, Vol 7, Number 9 (2012), pp. 602-615.

CONTACT

J.Philippe, tel: +33-438789971; Julien.Philippe@cea.fr

A GRAPHENE NANOSENSOR FOR DETECTION OF SMALL MOLECULES

Cheng Wang[1,2], Jinho Kim[1], Jing Zhu[1], Renjun Pei[3], Guohua Liu[2],
James Hone[1], Milan Stojanovic[4], and Qiao Lin[1]

[1]Department of Mechanical Engineering and
[4]Department of Medicine, Columbia University, New York, USA
[2]Department of Electronics and Microelectronics, Nankai University, Tianjin, CHINA
[3]Suzhou Institute of Nano-Tech and Nano-Bionics, Chinese Academy of Sciences, Suzhou, CHINA

ABSTRACT

This paper presents an aptamer-based graphene nanosensor capable of detecting small molecules. To address difficulties in direct detection of small moelcules associated with their low electric charges, we use a competitive sensing approach as demonstrated with dehydroepiandrosterone sulfate (DHEA-S) as a target analyte, which is a small molecular steroid hormone with important applications in clinical diagnostics. A DHEA-S aptamer is captured by a complementary short DNA probe immobilized on the graphene and released upon exposure to DHEA-S in solution due to the binding between DHEA-S and the aptamer. The aptamer release is detected by measuring the change in the conductivity of graphene. Experimental results show that the time rate of aptamer release from the graphene is inversely proportional to DHEA-S concentration in solution. Thus, the nanosensor can potentially enable label-free, specific and quantitative measurement of DHEA-S and other small molecules.

INTRODUCTION

Graphene is a nanomaterial consisting of a monolayer of carbon atoms arranged as a two-dimensional honeycomb lattice [1]. The single-atom thickness offers graphene unique electrical properties, such as very high carrier density and mobility [2]. Owing to these properties, graphene has been intensively investigated for use in electric field effect devices [3]. In particular, biosensors that use graphene field effect transistors (GFET) hold great promise thanks to graphene's strong chemical stability and modifiable surface chemistry [4]. Typically, detection of target analytes in media using such GFET-based biochemical sensors is achieved by measuring changes in the conductivity of graphene upon the binding of targets on the graphene surface.

Existing GFET based biosensors are limited to the detection of relatively large-sized molecules with high electric charges (such as proteins [4] and DNA [5]) that can significantly change the conductivity of graphene. On the other hand, the use of graphene nanosensor for the detection of small molecule analytes is considerably more difficult as graphene interactions with small molecules, which in general are of low charge, often cause insufficient change in graphene conductivity. To our knowledge, such graphene-based small molecule biosensors have not yet been demonstrated. This paper presents an aptamer-functionalized graphene nanobiosensor that is capable of detecting small molecule analytes. For demonstration, we use

dehydroepiandrosterone sulfate (DHEA-S; molecule weight: 426.5), a small molecule (a steroid endogenous hormone) involved in the human metabolism, aging, and pathological changes to organs [6]. In current clinical tests, measurement of DHEA-S traditionally relies on assays such as immunofluorescence assay (IFA), radioimmunoassay (RIA) or enzyme-linked immunosorbent assay (ELISA), which requires labeling of samples and often has limited accuracy. By combining graphene with a target-specific aptamer [7], we enable an accurate and label-free detection of DHEA-S down to the nanomolar concentration level. The competitive binding approach employed in our device, where a charged aptamer is released from the graphene surface upon binding to a target molecule, allows us to measure a significant change in the conductivity of the graphene even for targets with a very low electric charge. Experimental results demonstrate that the graphene nanosensor potentially offers a viable approach for label-free, quantitative, and specific detection of DHEA-S and other small molecules.

PRINCIPLE AND DESIGN

The target molecule detection principle of our device employing a competitive binding approach is illustrated in Figure 1. In the device, short DNA oligomers (used as a probe), which can complementarily hybridize to capture aptamer strands, are immobilized on the graphene surface (Figure 1a). Then the device is exposed to aptamer molecules where one end is complementary to the probe and will partially hybridize onto the probe (Figure 1b).

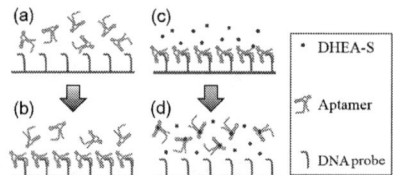

Figure 1: Target detection via competitive binding. Aptamer molecules are (a) introduced to graphene functionalized with a DNA probe, and (b) captured via hybridization. Target molecules (c) are introduced, and (d) competitively release the aptamer from the surface.

Because aptamers are highly charged, the surface-bound aptamers will form an electrical double layer (EDL) just above the graphene surface. The surface charge, which effectively applies a top-gate electric field effect, regulates the electric conductance of the graphene by inducing a change in the charge carrier density in the bulk of the

978-1-4799-3510-9/14 $31.00 © 2014 IEEE   1075

graphene [8, 9]. When a target molecule, which competes with the DNA probe to specifically interact with the aptamer, is introduced into the device (Figure 1c), the aptamer will be released from the graphene surface into the solution (Figure 1d). This results in a decrease of the surface charge density, which in turn results in a change in the graphene's electric conductivity.

Our graphene nanosensor consists of a pair of planar source and drain electrodes (thickness: 5 nm Cr and 45 nm Au; size: 60 μm × 1 mm) on a silicon (Si) substrate with a silicon dioxide (SiO_2) insulation layer (300 nm). Graphene rests on top of the planar electrodes forming a GFET module. A microchannel (length: 10 mm, width: 1 mm, height: 40 μm) fabricated in polydimethylsiloxane (PDMS) is covered on the sensor for fluid handling during sensor functionalization and target solution injection. A platinum (Pt) wire inserted into a designated inlet in the PDMS microchannel provides top-gate voltages to the graphene during measurements. Conductive copper (Cu) tape attached to the bottom surface of the device provides a back-gate ground potential.

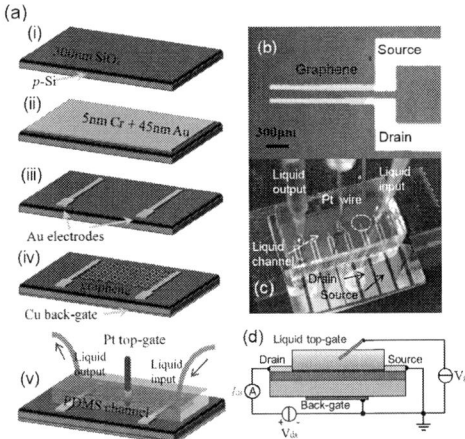

Figure 2: Device fabrication and testing. (a) Fabrication process. (b) Micrograph of a fabricated device. (c) Photograph of a device prepared for measurement. (d) Measurement setup for device testing.

EXPERIMENTAL

Device Fabrication

The device fabrication started with the deposition and patterning of Cr and Au layers on a silicon dioxide-coated silicon substrate by thermal evaporation to form the source and drain electrodes (Figure 2a(i-iii)). Graphene was synthesized via chemical vapor deposition (CVD) by heating an annealed Cu foil (Alfa Aesar) in a methane environment [10], and was transferred onto the electrodes using Cu foil wet chemical etching and polymethylmethacrylate (PMMA)/PDMS protected mechanical rolling press [10] to construct the graphene electric conduction channel (Figure 2a(iv)). Meanwhile, a microfluidic channel was fabricated by casting PDMS via soft lithography using an SU-8 mold. Finally, the PDMS microchannel graphene and electrode bearing substrate, the top-gate Pt wire, and the Cu back-gate

were assembled to complete the nanosensor (Figure 2a(v)). A micrograph of CVD transferred onto the electrodes of a silicon substrate and a photograph of a packaged graphene nanosensor device are shown in Figure 2b and 2c, respectively. The measurement circuitry of our setup is shown as Figure 2d.

Aptamer Functionalization.

The graphene surface of the fabricated device was functionalized with DNA probe and target-specific aptamer molecules by following a room-temperature procedure. The device was incubated for 1 hour in a methanolic solution of 1-pyrenebutanoic acid succinimidyl ester (PASE, 2 mM), (Figure 3a). PASE served as a linker molecule as its pyrenyl group can irreversibly absorb the basal plane of graphene via π-stacking [4]. After rinsing with methanol and phosphate buffered saline (PBS, pH = 7.4), the device was immersed in a PBS solution with 1 μM of DNA probe for 12 hours, whose 5' end was labeled with amino group. This amino group modified probe allows for the immobilization of the probe on the PASE-coated graphene surface (Figure 3b and 3c). Finally, after rinsing with PBS, the device was immersed in a PBS solution with 0.5 μM of DHEA-S aptamer for 2 hours to induce the hybridization of the aptamer onto probe molecules (Figure 3d and 3e).

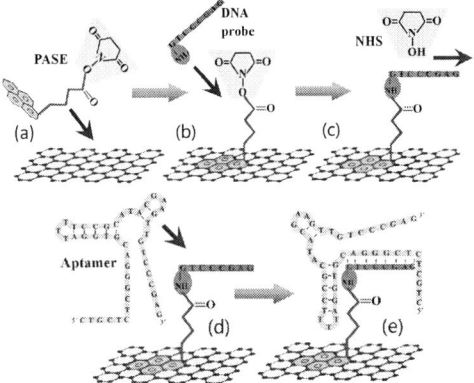

Figure 3: Functionalization of graphene with an aptamer. A PASE linker is (a) absorbed onto graphene. An amino group modified DNA probe (b) is conjugated by PASE, a NHS molecule (c) is released. An aptamer (d) is introduced and complementarily hybridized (e) on the probe.

Experimental Procedure.

In the measurement circuitry (Figure 2h), a DC voltage V_{ds} was applied between the drain and source electrodes. A bias voltage V_{gs} was applied between the top-gate (through a Pt wire) and the source electrodes. A multimeter (Agilent 34410A) was connected in series to the drain-source loop to measure the drain current I_{ds} while the source and Cu back-gate were connected and grounded.

Various concentrations of DEHA-S (0, 2.5, 10, 25, 100, 250 and 1000 nM) were introduced in PBS buffers (pH = 7.4) at a constant flow rate (1 μL/min) with an external injection pump (Syringe NE-1000). For each DHEA-S concentration, the detection was repeated three times. All the experiments were performed at room temperature.

RESULTS AND DISCUSSION

Electrical Characterization

Appropriate electrical conditions were first determined to maximize the output signal while reducing the noise level, and to obtain a high signal-to-noise ratio (SNR). When the drain-source voltage, V_{ds}, is lower than 3 V [8], the drain current output, I_{ds} varies linearly with V_{ds} according to the following equation [11]:

$$I_{ds} = \mu \frac{W}{L} C V_{ds} (V_{gs} - V_{Dirac})$$

where, C is the unit top-gate capacitance, V_{Dirac} is the threshold voltage on the Dirac point, V_{gs} is the top-gate voltage (which directly regulates the graphene's conductance), μ is the carrier mobility, and W and L are the width and length of graphene conductive channel respectively. To characterize, we scanned V_{gs} from -0.8 to 2.2 V to determine the appropriate V_{gs}.

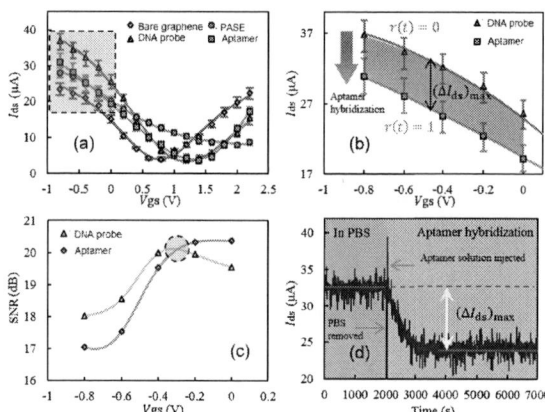

Figure 4: Electrical characterization of the graphene nanosensor. (a) I-V characteristics obtained as graphene was successively functionalized with PASE, DNA probe and aptamer molecules, with a magnified view of the DNA probe and aptamer functionalized surfaces for the indicated region shown in (b). (c) Measurement of the signal-to-noise ratio respectively in the absence and presence of the aptamer. (d) Time-dependent change of the current response (I_{ds}) upon introduction of the aptamer ($V_{gs} = -0.3$ V).

Results obtained from the electrical characterization experiments are shown in Figure 4a. Four different electronic transport characteristic curves (I_{ds} vs. V_{gs}) were obtained during each of the functionalization stages. The shifts in the Dirac point (represented by the value of V_{gs} for the minimum I_{ds}) indicate that the electronic properties of graphene were altered by the surface functionalization. It can also be seen that as V_{gs} varied from -0.8 to 0 V, I_{ds} exhibited a pronounced change, denoted ΔI_{ds}, when the aptamer was captured by the DNA probe on graphene surface (Figure 4b). As the nanosensor is based on measurement of the aptamer

release-induced change in I_{ds}, this range of V_{gs} was chosen for the device operation.

The accuracy of the nanosensor is also dependent on the SNR, defined as the ratio of the maximum value of ΔI_{ds} (($\Delta I_{ds})_{max}$) to the standard deviation of fluctuations in I_{ds}. This ratio was observed to significantly depend on whether the aptamer was present on the surface (Figure 4c), because the aptamer hybridization and dehybridization processes change the physical characteristics of the nanosensor. Thus, we chose $V_{gs} = -0.3$ V, the crossover point of the SNR curves of aptamer-laden and aptamer-free graphene for the operation of the nanosensor. At this top-gate voltage, SNR was consistently at a sufficiently high level (approximately 20.2 dB) for accurate analyte detection.

We then tested the hybridization of the aptamer to the DNA probe (Figure 4d). The graphene nanosensor coated with the DNA probe yielded I_{ds} around 33.2 µA in PBS buffer. When the device was exposed to an aptamer solution (0.5 µM in PBS), the I_{ds} exponentially decreased to approximately 23.6 µA. This suggests that the nanosensor is capable of detecting hybridization, with a maximum drain current change of $(\Delta I_{ds})_{max} \approx 9.6$ µA.

DHEA-S Measurement Results

We then demonstrated measurements of DHEA-S at varying concentrations using our device under a top-gate voltage of $V_{gs} = -0.3$ V. In Figure 5, the current response I_{ds} was normalized by the known $(\Delta I_{ds})_{max}$ to present a relative aptamer hybridization ratio.

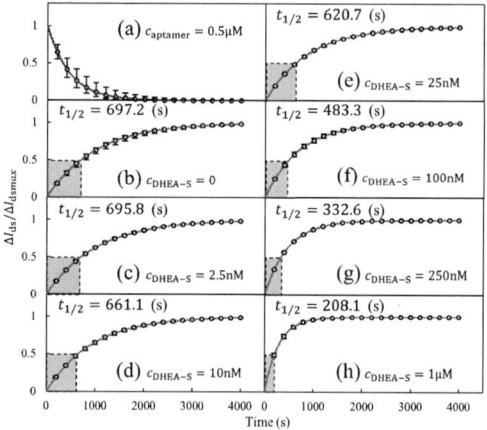

Figure 5: Changes in the drain current I_{ds} of the nanosensor. Aptamer (a) first hybridized to the graphene-attached DNA probe ($c_{aptamer} = 0.5$ µM), and then dehybridized from the probe by DHEA-S solution of varying concentrations: (b) $c_{DHEA-S} = 0$ nM, (c) $c_{DHEA-S} = 2.5$ nM, (d) $c_{DHEA-S} = 10$ nM, (e) $c_{DHEA-S} = 25$ nM, (f) $c_{DHEA-S} = 100$ nM, (g) $c_{DHEA-S} = 250$ nM and (h) $c_{DHEA-S} = 1$ µM.

Considering that the aptamer hybridization causes decreases in I_{ds}, the normalized values can be represented as:

$$\frac{\Delta I_{ds}}{(\Delta I_{ds})_{max}} = 1 - r(t) = 1 - e^{-kt} \quad (1)$$

Thus, it is clear that higher c_{DHEA-S} leads to faster aptamer dehybridization. To quantify the response of our device to different target concentrations, we calculated the time duration of the dehybridization to reach the half maximum of the initial aptamer hybridization level (half time, $t_{1/2}$). As shown in Figure 6, in a range of c_{DHEA-S} from 0 to 1 µM, the $t_{1/2}$ value (average of 3 repeated measurements) decreased from 697.2 s to 208.1 s. The relationship between $t_{1/2}$ and c_{DHEA-S} can be well fitted by an inverse proportional function:

$$t_{1/2} = \frac{185.5 \times 10^3}{c_{DHEA-S} + 269.7} \quad (2)$$

In this case, we can quantitatively calculate c_{DHEA-S} by using the measured $t_{1/2}$ value.

Further, we estimated the resolution of the nanosensor in DHEA-S sensing. By fitting Eq. (2) (Figure 6), the sensitivity of the half time with respect to aptamer dehybridization was determined to be approximately -1.32 s/nM. The largest error in our experiments was approximately 152.3 s. Hence, we estimate the limitation of detection (LOD) of DHEA-S concentration to be approximately 115.38 nM.

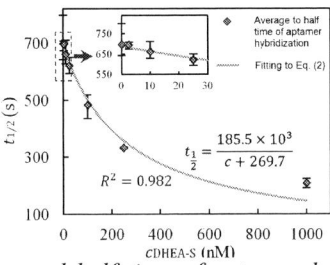

Figure 6: Measured half time of aptamer hybridization at various DHEA-S concentrations. The solid curve represents Eq. (2) fitted to the experimental data.

CONCLUSION

We presented an aptamer-based graphene nanosensor capable of detecting small molecule targets. A target-specific aptamer is captured on the graphene surface via hybridization to a complementary short DNA strand (serving as a capture probe), and then competitively released when the target is introduced. Because the changes in the electrical properties of graphene are induced by the release of the aptamer (which carry a significant charge) from the sensor surface, our device is capable of label-free detection of small molecules with low electrical charges. We demonstrated this principle using DHEA-S as a representative target, whose detection is in general difficult on conventional platforms without using molecular labeling groups. Experimental results have shown that detection of DHEA-S can be achieved in a label-free manner using the nanosensor, in which the time rate of competitive aptamer dehybridization from the graphene-attached DNA probe was correlated to the concentration of target molecules. This indicates the potential of the nanosensor to enable label-free detection and quantification of clinically significant small molecules.

ACKNOWLEDGEMENTS

This work was supported by the National Science Foundation (under grant number CBET-0854030), the National Institutes of Health (under grant numbers 8R21GM104204 and CA147925-01), and the Raymond and Beverly Sackler Program at the Interfaces of Biophysical and Medical Sciences. C. W. also gratefully acknowledges a National Scholarship (award number 201206200032) from the China Scholarship Council.

REFERENCES

[1] K. Novoselov, A. Geim, S. Morozov, D. Jiang, Y. Zhang, S. Dubonos, I. Grigorieva, and A. Firsov, "Electric Field Effect in Atomically Thin Carbon Films", *Science*, vol. 306, pp. 666-669, 2004.

[2] K. Novoselov, D. Jiang, F. Schedin, T. Booth, V. Khotkevich, S. Morozov, and A. Geim, "Two-Dimensional Atomic Crystals", *PNAS*, vol. 102, pp. 10451-10453, 2005.

[3] I. Meric, C. Dean, N. Petrone, L. Wang, J. Hone, P. Kim, and K. Shepard, "Graphene Field-Effect Transistors based on Boron-Nitride Dielectrics", *Proc. IEEE*, vol. 101, pp. 1609-1619, 2013.

[4] Y. Ohno, K. Maehashi, and K. Matsumoto. "Label-Free Biosensors based on Aptamer-Modified Graphene Field-Effect Transistors", *J. Am. Chem. Soc.*, vol. 132, pp. 18012-18013, 2010.

[5] X. Dong, Y. Shi, W. Huang, P. Chen, and L. Li. "Electrical Detection of DNA Hybridization with Single-Base Specificity Using Transistors based on CVD-Grown Graphene Sheets" *Adv. Mater.*, vol. 22, pp. 1649-1653, 2010.

[6] S. Lamberts, A. Van den Beld, and A. van der Lely, "The Endocrinology of Aging". *Science*, vol. 278, pp. 419-424, 1997.

[7] K. Yang, R. Pei, D. Stefanovic, and M. Stojanovic, "Optimizing Cross-Reactivity with Evolutionary Search for Sensors", *J. Am. Chem. Soc.*, vol. 134, pp. 1642-1647, 2012.

[8] I. Meric, M. Han, A. Young, B. Ozyilmaz, P. Kim, and K. Shepard, "Current Saturation in Zero-bandgap, Top-gated Graphene Field-Effect Transistors", *Nat. Nanotechnol.*, vol. 3, pp. 654-659, 2008.

[9] P. Debye, "Dieletric Properties of Pure Liquids", *Chem. Rev.*, vol. 19, pp. 171-182, 1936.

[10] S. Lee, C. Chen, V. Deshpande, G. Lee, I. Lee, M. Lekas, A. Gondarenko, Y. Yu, K. Shepard, P. Kim, and J. Hone, "Electrically Integrated SU-8 Clamped Graphene Drum Resonators for Strain Engineering", *Appl. Phys. Lett.*, vol. 102, pp. 153101-4, 2013.

[11] B. Kim, H. Jang, S. Lee, B. Hong, J. Ahn, and J. Cho, "High-Performance Flexible Graphene Field Effect Transistors with Ion Gel Gate Dielectrics", *Nano Lett.*, vol. 10, pp. 3464-3466, 2010.

CONTACT

*Cheng Wang, +1-212-854-3221, cw2717@columbia.edu

INTERROGATING CONTACT-MODE SILICON CARBIDE (SiC) NANOELECTROMECHANICAL SWITCHING DYNAMICS BY ULTRASENSITIVE LASER INTERFEROMETRY

Tina He[†], Jaesung Lee, Zenghui Wang, and Philip X.-L. Feng[†]
Electrical Engineering, Case Western Reserve University, Cleveland, OH 44106, USA
[†]Email: ting.he@case.edu; philip.feng@case.edu

ABSTRACT

We report on initial experimental demonstration of probing the dynamics of nanoscale contacts in robust nanoelectromechanical switches based on silicon carbide (SiC) nanocantilevers. For the first time, we measure the dynamical behavior of contact-mode SiC nanoscale electromechanical switches by directly probing the vibrating tips of the SiC nanocantilevers, using ultrasensitive laser interferometric techniques. First, we devise a novel '*pump*-and-*probe*'-type optical technique in which we use an RF-modulated 405nm (blue) laser to excite the SiC cantilevers, while using a 633nm (red) laser interferometer to probe the dynamics of their tips. Second, we directly actuate the SiC devices via their electrostatic gates, while monitoring the cantilevers' motions optically. By actuating the SiC cantilever switches near resonance with increasing amplitudes, we reveal new characteristics in motion dynamics when the devices are making contacts periodically. We demonstrate *milli-Volt actuated* SiC NEMS with *cantilever tips tapping on the contact electrodes periodically*, for *>10 billion cycles* (in 'cold' switching mode).

INTRODUCTION

Contact-mode switches based on nanoelectromechanical systems (NEMS) in advanced nanomaterials have emerged as an attractive alternative toward low-voltage logic and ultralow-power applications [1-11]. While deeply scaled NEMS switches offer ideally abrupt switching with minimal leakage, ultrasmall footprints and high speeds (often $\geq 10^3$ better than their MEMS counterparts), their reliability and lifetime are an open challenge, with limited knowledge about the contacts at genuinely nanoscale (often with areas on the order of $\sim 100 nm^2$ or even smaller). It has been shown that most of the recently demonstrated NEMS switches suffer from low operating cycles and very short lifetimes [8].

Among the emerging NEMS switches, SiC devices have exhibited outstanding performance and potential toward robust NEMS logic with long lifetimes [9-11]. Over long cycles of operations (in air), however, degradations in I_{on}/I_{off} are visible in high-precision measurements [9-11]. This demands *quantitative* understandings through investigations of details at nanoscale contacts in individual switching events, and their evolutions over long cycles, which all remain challenging. Lately we have performed precise electrical measurements with devices in quasi-static (DC) and ~1kHz AC operations [9-11]. However, the external circuits for reading out the switching events often limit the operating speed (far below the intrinsic switching speed offered by the

NEMS), and comprise the sensitivity for resolving switch-on events, manifesting the difficulties in dealing with the rapidly changing contact resistance within each switching event [10].

Here we endeavor to measure the fast-moving SiC devices (optically instead of electrically) while their tips are making contacts, and to gain initial insights into the contact dynamics, by driving the cantilevers near resonance. Operating near resonance has several clear advantages over previous slow quasi-static or kHz testing: (i) making it possible to approach the device intrinsic speed at radio frequencies (RF); (ii) enabling the switch to attain much larger displacements to make contact, at much lower actuation voltages; (iii) possible for making contact when the cantilever tip velocity approaches 0, thus a very soft tapping or landing with least impact. Exploiting near-resonance vibrations for devices to approach adjacent electrodes has been explored in much larger MEMS microdisks [12] and nanowire NEMS [13,14], all electrically, but studies on nanocantilever switches, and motions measured by optical or other techniques, have not yet been reported.

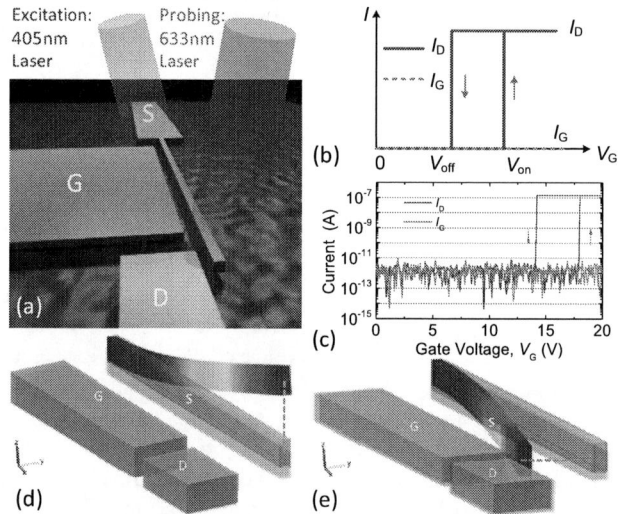

Figure 1: Contact-mode SiC nanoelectromechanical switch for optical probing. (a) 3D illustration of device and basic concept of non-destructive optical probing. (b) Illustration & (c) Measured I-V curves of the basic SiC NEMS switching characteristics in a single quasi-static cycle. (d) & (e) Illustration of out-of-plane and in-plane mode shapes.

DEVICE FABRICATION AND BASICS

The devices are 3-terminal lateral switches as illustrated in Fig. 1a. The cantilevers have lengths in μm, and widths, thicknesses and gaps all of ~200–300nm or smaller. The devices are fabricated by a surface nanomachining process. Starting from a 500nm-thick SiC film deposited by LPCVD on 500nm SiO$_2$ on a 4-inch Si wafer, we define and transfer the designed patterns of devices by wafer-scale electron-beam lithography (EBL) and reactive ion etch (RIE), and finally we release the NEMS after a high-yield vapor hydrofluoric (HF) acid etching of SiO$_2$.

In quasi-static (DC) operation, the gate (G) actuates the cantilever (source, S) electrostatically with increasing gate voltage V_G, causing the cantilever to deflect so that its tip makes contact to the drain (D), forming a conductive path (D to S) and turning on the switch. Figure 1b illustrates the expected *I-V* characteristics with abrupt switch-on and off by sweeping V_G; Fig. 1c shows the typical measured *I-V* characteristics, including I_{on}, I_{off}, V_{on}, V_{off}, and hysteresis. It has been demonstrated that thus abrupt switching can operate in quasi-static mode for $>10^4$ fully-recorded cycles [9] and in ~1kHz AC mode for $>10^7$ cycles [10,11], all in air.

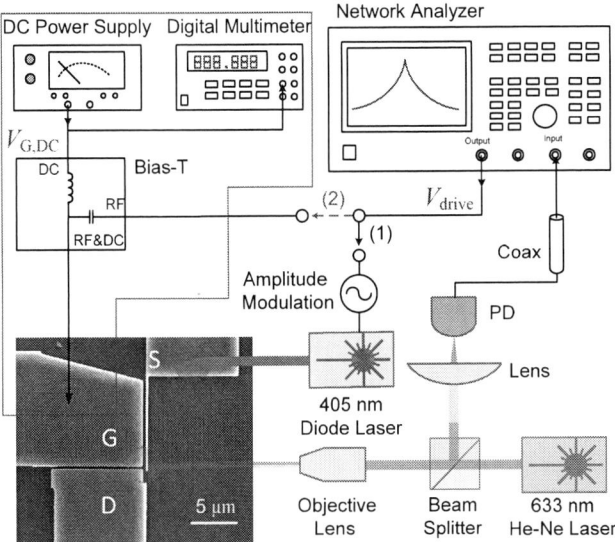

Figure 2: Schematic of the optoelectronic measurement system for interrogating the dynamical behavior of the SiC NEMS contacts. A 633nm laser is focused on the cantilever tip to read out motions interferometrically with an RF network analyzer. In scheme (1) with optical excitation, the RF-modulated 405nm 'pump' laser is focused on the source (S) electrode. In scheme (2) with electrostatic actuation (blue dashed box), the RF drive signal and the DC voltage are applied to the gate (G) electrode through a bias-T.

EXPERIMENTAL TECHNIQUES

For *least destructive yet highly sensitive probing*, we drive the cantilever near its resonance and read out the tip motions with laser interferometry. Figure 2 depicts the system diagram and circuits for probing the switching dynamics in the cantilever tip-contact electrode region. The device operates inside a vacuum chamber with a quartz window (all tests are at ~10mTorr, room temperature).

Excitation: Optical and Electrical

We first employ an RF-modulated laser (405nm) to

photothermally excite the SiC cantilevers, as shown in scheme (1) in Fig. 2. We modulate the 'pump' laser intensity by tuning V_{drive}, the RF drive voltage from the network analyzer. When the drive frequency matches the resonance frequency of the cantilever, the tip displacement is amplified by the resonance quality (Q) factor, as compared to the deflection levels achieved in the previous quasi-static (DC) or ~kHz AC (off resonance) actuation [9-11].

Beyond optical excitation, scheme (2) in Fig. 2 (in the dashed-blue enclosure) illustrates the electrostatic excitation. Herein we drive the device through gate G by combining a DC polarization voltage $V_{G,DC}$ and an RF signal V_{drive}. The DC component $V_{G,DC}$ pulls the cantilever to an equilibrium position where the gap is smaller than the as-fabricated initial gap; tuning the RF signal drives the cantilever near resonance with tunable amplitude. This technique enables enhanced control of electromechanical actuation and switching dynamics, and can also be incorporated into the recently demonstrated multi-gate device designs [15].

Detection: Optical Interferometry

We read out the motions of cantilever tips in a wide frequency range by an ultrasensitive laser interferometer illustrated in Fig. 2. The 633nm He-Ne laser beam first goes through a beam expander and a neutral density (ND) filter, and then a 50× microscope objective lens and focuses on the tip of the cantilever. Light waves are reflected from top and bottom surfaces of the suspended cantilever, and from the Si substrate beneath it. The reflected beam is then redirected by a beam splitter, and focused onto a photodetector (PD), downstream which the optoelectronic signal is analyzed by the RF network analyzer. The probing interferometry is capable of detecting the intrinsic thermomechanical (Brownian-motion) resonance modes of the devices, with sub-pm/√Hz-level displacement sensitivities.

The cantilever vibrates in two flexural modes: out-of-plane (Fig. 1d) and in-plane (Fig. 1e); both can be detected by the ultrasensitive laser interferometry. We calculate the fundamental-mode resonance frequency of a singly-clamped cantilever by using $f_{res} = (1/2\pi)\sqrt{k_{eff}/M_{eff}}$, where k_{eff} is the effective stiffness and M_{eff} the modal mass of the cantilever. The fundamental frequencies are estimated to be 3.25MHz (for in-plane) and 6.32MHz (for out-of-plane) in a cantilever with dimensions $L{\times}w{\times}t \approx$ 8μm×200nm×500nm. We focus on exploiting the in-plane mode to efficiently enhance the motion amplitude for making lateral contacts.

MEASUREMENT RESULTS

Optical Excitation and Optical Detection

Figure 3 presents a typical calibration measurement of resonance modes from a 10μm-long cantilever. The laser spots are positioned as shown in Fig. 2, scheme (1). The resonance at 2.492MHz is in-plane mode and that at 6.802MHz is out-of-plane mode. The peak amplitudes of both modes exhibit linear dependency on V_{drive} (Fig. 3c&e).

We study a family of cantilevers with varying length (L) but same thickness $t\approx$500nm and width $w\approx$200nm. Figure 4

978-1-4799-3510-9/14 $31.00 © 2014 IEEE

shows measured data from an $L\approx4\mu m$ cantilever, with higher resonance frequencies: f_{res}=19.73MHz (in-plane mode) and f_{res}=37.31MHz (out-of-plane mode). The signal amplitude is lower (smaller L). Figure 5 summarizes the effect of varying cantilever length on measured f_{res} and Q values. The dashed lines in Fig. 5a are analytical results of f_{res} as a function of L, showing good agreement between measured data and theoretical analysis. Q factors vary from 1000 to 2000 for both in-plane and out-of-plane modes. Longer cantilevers exhibit higher Qs, suggesting lower clamping losses.

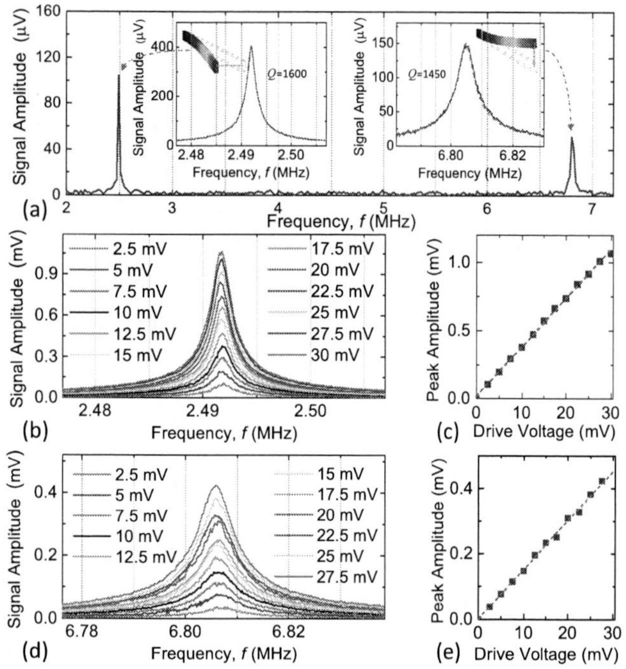

Figure 3: Measured resonances from an $L\approx10\mu m$ SiC cantilever ($w\approx200nm$, $t\approx500nm$). (a) Wide-range frequency spectrum shows resonances from both in-plane (f_{res}=2.492MHz) and out-of-plane (f_{res}=6.802MHz) modes. Insets: zoom-in view of the two modes with illustrations of the mode shapes. (b) Driven response of the in-plane mode. Legends show V_{drive} (rms values). (c) Peak amplitude versus V_{drive} for the in-plane mode. (d) and (e) Data from the same measurements for the out-of-plane mode.

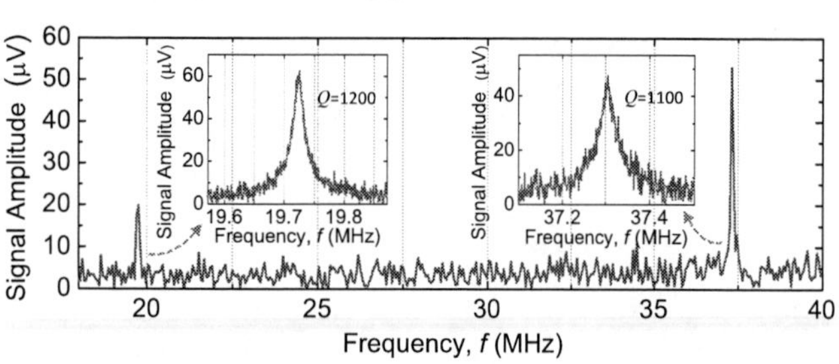

Figure 4: Measured resonances from a short ($L\approx4\mu m$) SiC cantilever ($w\approx200nm$, $t\approx500nm$). The insets show zoom-in scans of both in-plane (left) and out-of-plane (right) modes. Dashed lines show fitting for obtaining Q factors.

Electrical Excitation and Optical Detection

After verifying the both in-plane and out-of-plane resonances, we drive the devices via their electrostatic gates and focus on investigating the in-plane motions. On a cantilever with $L\times w\times t\approx$ 8μm×200nm×500nm, we first perform the two-laser '*pump-&-probe*' technique to find the resonance frequencies at f_{res}=3.706MHz (in-plane mode) and f_{res}=10.42MHz (out-of-plane mode), respectively, as shown in Fig. 6a. Then we switch to scheme (2) in Fig. 2 for electrostatic excitation. Figure 6b shows data from the electrostatic drive for the same device with gate DC voltage $V_{G,DC}$=0.44V and RF drive voltage V_{drive}=50mV (rms value). The measured resonance frequencies shift slightly down (due to electrostatic softening) to f_{res}=3.67MHz (in-plane) and f_{res}=10.32MHz (out-of-plane).

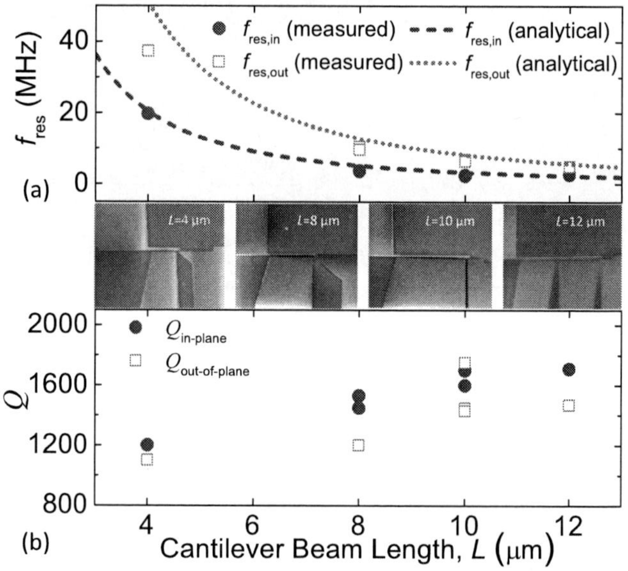

Figure 5: Effect of varying cantilever length on measured (a) f_{res} & (b) Q. The insets show the SEM images of devices with different lengths measured in this study.

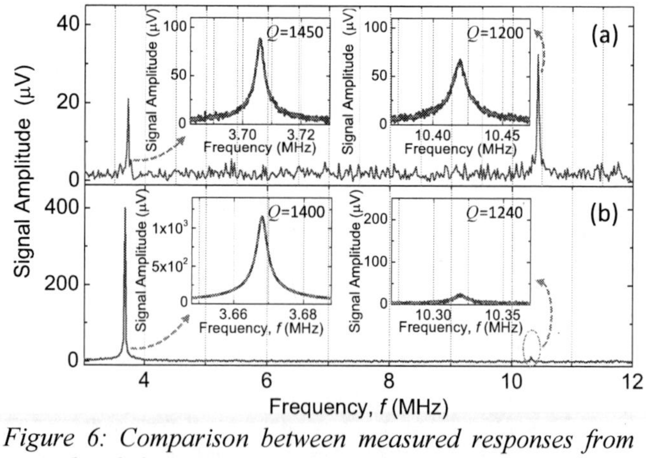

Figure 6: Comparison between measured responses from optical and electrostatic excitations. (a) Resonance spectrum of an $L\approx8\mu m$ device measured with optical drive. (b) Data from the same device with electrical drive. Insets show resonances with fitting for extracting the Qs.

As soon as we implement scheme (2) and compare data in Fig. 6b with those in Fig. 6a, we clearly observe that, while

Qs stay at the same levels, the amplitude of the out-of-plane mode becomes much smaller *relative to* that of the in-plane mode. This indicates that the electrostatic excitation is more selective to the modes, and couples much more strongly and efficiently to the in-plane mode. This is intuitive because the local actuating gate is in lateral configuration, thus coupling to in-plane motions more directly and effectively. This fact plays in favor for studying the in-plane contact-mode switching dynamics. *Experimentally elucidating this highly intuitive fact is the key to achieving high amplitudes with low actuation voltages in the milli-Volt range* as shown in Fig. 7.

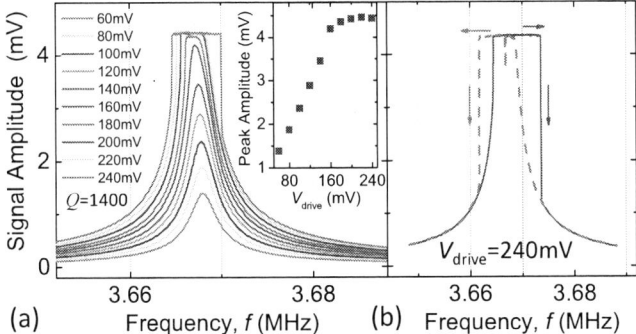

Figure 7: Frequency-domain probing of the nanocontact dynamics – nanocantilever tip approaching and tapping the contact (drain) electrode, during high-amplitude operations of the SiC NEMS in resonance mode, with milli-Volt actuation. (a) Resonance data of an $L \approx 8\mu m$ device at $V_{G,DC}=0.44V$ with varying RF drive voltage. Inset shows the peak amplitude versus drive voltage. Contacting events exemplify as amplitude clipping at high drive voltages. (b) Opposite frequency sweeps (indicated by arrows) measured at RF drive $V_{drive}=240mV$ (rms value), exhibiting hysteresis.

Figure 7a shows the resonance amplitude increases as we increase the RF driving signal V_{drive} while maintaining the DC voltage constant at $V_{G,DC}=0.44V$. The peak amplitude depends on V_{drive} linearly till the V_{drive} reaches ~160mV, where 'amplitude clipping' starts to take place. Data in Fig. 7a clearly demonstrate that the amplitude is eventually clipped at a constant level, limited by the tunable nanogap, indicating the cantilever tip is making contact to the drain (D) electrode. In order to study the intrinsic lifetime of the switches, we then drive the devices at their center frequencies with the clipping amplitudes, with the cantilever tip tapping the drain in each resonant-motion cycle. Operations in this 'tapping' switching mode (~0.27μs per cycle, easily sustain >1 hour and longer) have quickly surpassed 10 billion cycles without failure in our current testing (these are 'cold switching' events as there is not yet a bias current I_{DS} when the contact is being made). With V_{drive} beyond the amplitude clipping level, we sweep the frequency up and down within a 40kHz span centered at the in-plane mode resonance $f_{res}=3.67MHz$. The amplitude *versus* frequency curves in Fig. 7b demonstrate hysteresis, in which the transition from non-contacting to tapping is gradual, while the transition from tapping to non-contacting appears to be abrupt.

CONCLUSIONS

In summary, we have demonstrated in experiments the direct optical probing of dynamic motions of contact-mode SiC nanocantilever-structured NEMS switches, by focusing on monitoring the cantilever tips in the tip-drain contact region, using an ultrasensitive laser interferometry. The dual-laser-beam '*pump-&-probe*' techniques enable efficient excitation and detection of both the in-plane and out-of-plane modes of vibrations. Combining efficient electrostatic drive of the in-plane mode near resonance with optical readout, we have achieved *milli-Volt*-level electrostatic actuation for contact-mode SiC NEMS switching. We have demonstrated switching lifetimes of over 10 billion cycles with this much less destructive contacting scheme (in 'cold switching' mode), without stiction or other device failure observed.

ACKNOWLEDGEMENTS

The authors thank the support from DARPA (NEMS Program, D11AP00292) and NSF (CCF-1116102). We are also grateful to M. Mehregany and S. Rajgopal for providing generous help on material processing.

REFERENCES

[1] T. Rueckes, *et al.*, "Carbon nanotube–based…", *Science*, vol. 289, pp. 94-97, 2000.

[2] M. L. Roukes, "Mechanical computation,…", in *Tech. Digest of IEDM*, San Francisco, CA, Dec. 13-15, 2004, pp. 539-542.

[3] Y. Hayamizu, *et al.*, "Integrated three-dimensional…," *Nature Nanotechnology*, vol. 3, no. 5, pp. 289-294, 2008.

[4] X. L. Feng, *et al.*, "Low voltage…", *Nano Letters*, vol. 10, no. 8, pp. 2891-2896, 2010.

[5] V. Pott, *et al.*, "Mechanical computing redux: relays…", *Proc. IEEE*, vol. 98, no. 12, pp. 2076-2094, 2010.

[6] S. W. Lee, *et al.*, "A fast and low-power…", *Nature Communications*, vol. 2, no. 220, 2011.

[7] S. Chong, *et al.*, "Integration of…," *IEEE Trans. Elec. Dev.*, vol. 59, no. 1, pp. 255-258, 2012.

[8] O. Y. Loh, *et al.*, "Nanoelectromechanical contact…", *Nature Nanotechnology*, vol. 7, pp. 283-295, 2012.

[9] T. He, R. Yang, P. X.-L. Feng, *et al.*, "Robust silicon carbide…", in *Proc. 26th IEEE Int. Conf. on MEMS*, Taipei, Taiwan, Jan. 20-24, 2013, pp. 516-519.

[10] T. He, P. X.-L. Feng, *et al.*, "Time-domain AC…", in *Tech. Digest, 17th Int. Conf. on Solid-State Sensors, Actuators & Microsystems (Transducers'13)*, Barcelona, Spain, June 16-20, 2013, pp. 669-672.

[11] T. He, R. Yang, P. X.-L. Feng, *et al.*, "Silicon carbide (SiC) nanoelectromechanical switches…", in *Tech. Digest of IEDM*, Washington DC, Dec. 9-11, 2013, paper no. 4.6.

[12] L. Yang, C. T.-C. Nguyen, *et al.*, "A resonance dynamical…", in *Proc. IEEE Int. Freq. Contr. Symp.*, Honolulu, Hawaii, May 19-21, 2008 pp. 640-645.

[13] P. X.-L. Feng, *et al.*, "Very low voltage…", US Patent, No. US 8,115,344 B2, 2012 (App. No. US 2010/0140066 A1).

[14] X. L. Feng, *et al.*, "Silicon carbide (SiC) top-down…", in *Tech. Digest, 15th Int. Conf. on Solid-State Sensors, Actuators & Microsystems (Transducers'09)*, Denver, CO, June 21-25, 2009, pp. 2246-2249.

[15] T. He, R. Yang, P. X.-L. Feng, *et al.*, "Dual-gate silicon carbide (SiC)…", in *Proc. 8th IEEE Int. Conf. on NEMS*, Suzhou, China, Apr. 7-10, 2013, pp. 554-557.

978-1-4799-3510-9/14 $31.00 © 2014 IEEE

MATRIX INDEPENDENT LABEL-FREE NANOELECTRONIC BIOSENSOR

Rahim Esfandyarpour[1,2], Mehdi Javanmard[2], Zahra koochak[1], James S. Harris[1], Ronald W. Davis[2]

[1] Department of Electrical Engineering, Stanford University
[2] Stanford Genome Technology Center; 855 California Ave., Palo Alto, CA 94304, USA

ABSTRACT

Previously we presented a novel, label free and real time electrical impedance biosensor, referred to as the nanoneedle biosensor. The nanoneedle is an ultrasensitive and localized device, which has the ability to directly measure biomolecular binding as a function of time. Label-free electronic sensors generally operate in low salt-concentration buffers [1] making them impractical for analyzing complex biological mixtures, which have salt concentrations as high as several hundred millimolar. In this study we have demonstrated detection of a clinically relevant cancer biomarker suspended in high salt buffer (> 100mM) with a detection limit of 2 ng/ml.

INTRODUTION:

Rapid handheld detection of protein biomarkers can open up the potential for point-of-care diagnostics. Proteins are traditionally detected using the sandwich ELISA technique that requires the use of secondary antibodies, several wash steps, fluorescent labels, and expensive optical readout equipment. Thus typically any clinical protein tests must be outsourced to a lab, which can take days to return back with the results. The use of electronic detection allows for minaturizing the readout instrumentation and making it portable with significant reduction of cost. Label-free one step detection using electronic readout can result in translation of protein detection technologies from bench to patient bedside. Over the last thirty years several groups have attempted label-free electronic and mechanical detection of proteins through various methods. One method is based on detection of change in double layer capacitance at the surface of microelectrodes. The main difficulty with this approach is lack of consistency and low sensitivity. The main reason is due to the fact that the debye layer only extends a few nanometers above the electrode surface, whereas probe antibodies extend roughly 10 nm above the surface, thus any antigen binding to the surface will not perturb the debye layer of ions at the electrode surface. Mechanical sensing techniques [2] such as the use of cantilevers and CMUTs have shown promise in label free gas sensing, however because the relatively high viscosity of aqueous solution compared to gas, the quality factor of the resonator significantly drops by several orders of magnitude once placed in fluid. Some groups however have come up with novel approaches to improve quality factor of mechanical resonators in fluid. Field affect transistor based approaches have also shown promise for achieving high sensitivity detection [1]. The main difficulty however is the requirement for operation in low salt concentrations due to the fact that high salt concentration screens any field emitted from charged macromolecules like protein or DNA. As a result this makes FET based devices unsuitable for performing protein assays in complex biological mixtures like serum, which have salt concentrations as high as several hundred millimolar.

Previously we presented, the nanoneedle biosensor, a label-free method for detecting proteins in real time [3]. Here we adapt this assay to make a matrix independent label-free electronic assay capable of detecting minute traces of target protein in aqueous solution independent of the salt concentration with high specificity. We have demonstrated proof of concept for highly sensitive detection of two proteins, streptavidin and vascular endothelial growth factor (VEGF) in buffer. VEGF is a signaling kinase protein that stimulates angiogenesis particularly during embryonic development. Overexpression of VEGF has been implicated in diseases such as cancer and vascular disease. Although VEGF detection is not yet used as a clinical biomarker, VEGF plays an important role in rheumatoid arthritis, diabetic retinopthy, age-related macular degeneration, angiosarcoma, and pre-eclampsia, thus its detection and monitoring can be of significant clinical relevance [4].

NANONEEDLE STRUCTURE:

The nanoneedle consists of two nanoscale electrodes separated by a nanoscale-insulating layer. The tip of the needle is functionalized with antibodies to capture the target protein and act as the active sensing area. Upon capture of the target protein, protein abundance can be quantified based on changes in impedance. Various thicknesses and geometrical designs have been fabricated and tested. The final sensor design used for the shown results consists of 100 nm thick conductive electrodes with a 40 nm middle oxide layer. The top protective oxide layer thickness is 20 nm and the bottom oxide layer thickness is 200 nm (figure 1). The width of the nanoneedle tip is 3 μm[5-8].

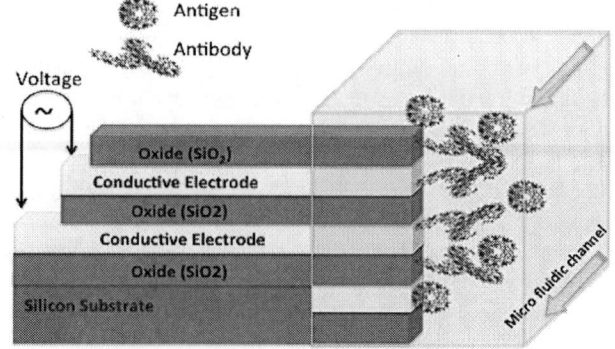

Figure 1. Side view of a Nanoneedle sensor in a

microfluidic channel (not to scale). Thin film electrodes are separated by a 40 nm thin film insulative layer. The top electrode is covered with a protective oxide layer. The surface of the needle is functionalized with antibodies. Capture of antigens on the surface modulates the impedance at the tip of the needle.

SENSOR NANOFABRICATION:

We arrayed 360 nanoneedles (4th generation), 15 bundles each containing 24 nanoneedles (Figure 2) on a four inches silicon wafer. We fabricated the devices using the following protocol: First 200 nm of SiO_2 was thermally grown on a silicon wafer to insulate the substrate from the other layers. This process was done in a high temperature atmospheric furnace, which is made of pure quartz that allows for the introduction and mixing of gases (N_2, Ar, O_2 and H_2) to grow silicon dioxide (SiO_2) on the silicon wafers. Almost three hours was needed to grow 200 nm of SiO_2 on a wafer with the crystal orientation of 100 at the temperature of 1100 C^o. In the next few steps the bottom electrode was patterned and fabricated through an optical lithography process followed by a lift off process. First the wafer was pre-baked on a hot plate at 200C^o for two hours (or it can be done in an oven at 200C^o for 6-8 hours). Then the manual resist spinning was performed by applying 10 drops of Hexamethyldisilizane (HMDS) to the wafer. It was followed by applying MaP 1215 resist to the wafer. This wafer was then transferred to a contact aligner system to perform precision mask-to-wafer alignment followed by near-UV photoresist exposure. The exposer time was set to 3 seconds. Exposed resist was developed in MaD 331 for 35 seconds followed by immediately rinsing in water. The patterned wafer then transferred to an evaporation system to deposit the bottom electrode.

In the evaporation system, first 3 nm of Cr was deposited as an adhesive layer, which was followed by deposition of 100 nm of gold as the bottom conductive electrode. This wafer was then transferred to a Pyrex Petri dish, which was full of Acetone to perform the lift off process. After leaving the wafer in Acetone for about 30 minutes, it was rinsed in water. Only the deposited gold layer in the patterned regions will remain. The next step was deposition of 40 nm of silicon dioxide (sensor area) by using plasma-assisted atomic layer deposition (ALD) technique. The reason for choosing ALD over CVD process was to grow a high quality, conformal, uniform, pinhole and particle free deposited oxide film. For this process, we used BTBAS as our precursor, which is a liquid chemical precursor. The substrate temperature during deposition was set to 300 C^o while the Argon (Ar) was used as our purging gas. The process was then followed by fabrication of the top gold electrode. The same procedure for the bottom electrode was followed to fabricate the top gold electrode. This conductive electrode was then covered with a protective oxide layer to prevent the exposure of top electrode to the solution. This layer was deposited by using a plasma-enhanced chemical vapor deposition (PECVD) system,

which is designed for the low temperature (350°C) deposition of silicon dioxide. Then several etching steps followed a lithography step to form the channels underneath the fabricated sensors. In order to remove the oxide layer from the measuring pads; the patterned wafer went through a wet etching step (6:1 Buffered Oxide Etch) to expose the bonding pads to allow wire bonding. The final step was the forming of sharp edges on the sensor ends of the devices in the channel through a focused ion beam (FIB) etching step. The main ion species used for this physical etching method was Ga^+ and the accelerating voltage was around 25 kV.

Finally a premade PDMS micro channel was bound on top of the fabricated devices. These micro channels were patterned on the soft PDMS bars.

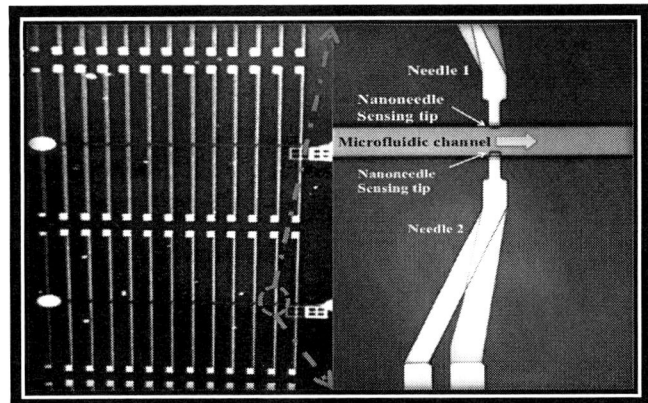

Figure 2: Optical micrograph of a fabricated array of nanoneedles sensors (top view) without a PDMS channel.

SiO_2 SURFACE MODIFICATION:

Mooney et al [9]. fully studied attachment of proteins on the silicon dioxide (SiO_2) surface. In their work, SiO_2 surfaces derivatized with octadecyltrichlorosilane (ODTS) to form a C_{18} surface, which results the adsorption of Biotinylated BSA onto a C_{18}-modified hydrophobic silica surface. We went through a similar procedure to form Si-O bonds on the active sensing region of the nanoneedle biosensors, which is made of silicon dioxide.

First, the chips were cleaned and rinsed several times with Isopropyl alcohol (IPA), Acetone and DI water respectively [10]. All the chips then dried out with a N_2 blowgun. Since hydrated silica surface can hydrolyzes the ODTS and forms a more stable and homogeneous film compare to a nonhydrated surface; the cleaned chips were placed in deionized water overnight to fully hydrate the surface. Then all the chips were dried out with nitrogen to remove the mobile water molecules. Any leftover of mobile water could react with the ODTS and cause it to polymerize rather than react with the surface. After this step, all the chips were transferred into a Pyrex Petri dish, which was full of 0.5% (v/v) solution of ODTS in anhydrous toluene. The chips were left there for 16 hours. The pyrex petri dish was all covered with a parafilm, (Prafilm "M", Laboratory film, American National Can TM), except inlet and outlet holes during the process. Nitrogen gas was purging to the dish through the inlet to minimize exposure of ODTS solution to atmospheric water vapor to avoid their reaction, which could form an undesirable organo-silicon polymer. To remove the residual ODTS, all the sensors were washed three times with toluene. After drying the sensors with blown nitrogen, they were baked in an oven with the set temperature of 120C° for 30 minutes. This step was done to complete the formation of Si-O bonds. All the ODTS-silanized sensors stored in a Petri dish fully covered with a parafilm while the nitrogen was purging into it, until they were used.

MEASUREMNTS METHODS:

In order to perform an Electrochemical Impedance Spectroscopy (EIS) on the sensors, a Versa STAT3 potentiostat (Princeton Instruments, Princeton, New Jersey) was used. A sinusoidal voltage signal was applied to the top electrodes and the current entering the bottom electrode was measured and used to calculate the impedance. As we showed in our previous work [3] the optimal frequency at which maximum change in measured impedance occurs due to biomolecular binding is between 1 and 100 kHz, thus we chose 15 kHz as the operating frequency (250 mV RMS AC signal) for our measurements. For all of our measurements, the sampling rate was one sample per 4 s. In our experiments, a LabSmith uProcess System with programmable syringe pumps is used to have a precise control over the flow rate, flow time and volume of the samples in our microfluidic channel (figure 3).

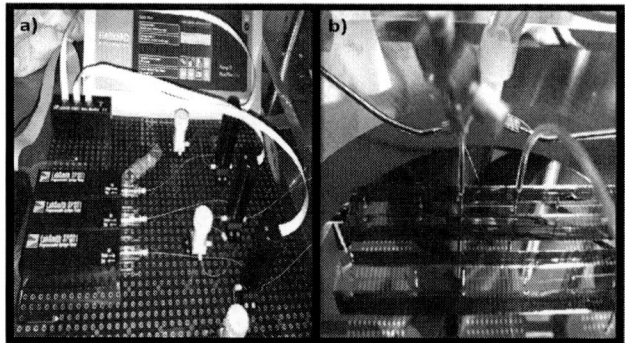

Figure 3: a) LabSmith uProcess System with programmable syringe pumps b) Nanoneedle array with PDMS microchannels on top.

EXPERIMENTAL RESULTS:

As a practical example with clinical relevance, we demonstrated the detection of Vascular Endothelial Growth Factor (VEGF) for cancer diagnosis. Anti-VEGF was immobilized on the sensor surface using ODTS salinization chemistry [10].

In order to immobilize Anti-VEGF antibodies at the surface of our devices, first 50μg /mL of Anti-VEGF which was diluted in 1X phosphate buffered saline (PBS), injected to the channel. Then these spotted chips were incubated overnight in a humidifier. The chips were then rinsed with

washing buffer, for 10 minutes with the flow rate of 5μL/min to wash off all non-bound anti-VEGF molecules. Then impedance of the sensor was measured as shown in figure 4. In step 1, the wash buffer was flowing to the microfluidic channel with a flow rate of 5μL/min to find the base line of impedance; before binding of VEGF proteins to the immobilized Anti-VEGF proteins at the sensor surface occurred. Then VEGF (50 ng/ml, suspended in Phosphate Buffer Saline with 114 mM NaCl) was injected in to the channel with the flow rate of 1 μL/min (step 2) to be bound to the Anti-VEGF proteins. To avoid the mixture of different samples in the microfluidic channel, an air bubble was injected to the channel between each step (steps 3,5 and 7). Afterwards the wash buffer was injected into the channel to wash off all non-bound and loosely bound VEGF proteins (step 4). As seen in figure 3 (step 4), the sensor impedance is increased. The impedance levels after both wash steps (step 1 and 4) (after Anti-VEGF immobilization and after VEGF binding) are compared with each other to quantify the number of VEGF proteins captured by the anti-VEGF molecules at the sensor surface. As a control experiment to verify the specific binding of VEGF proteins to the anti-VEGF proteins, we continued the experiment by injection of 40 μg/ml of Streptavidin into the channel (step 6). Concentration of injected Streptavidin was three orders of magnitude larger than injected VEGF while the injection flow rate was the same. Injection of Streptavidin onto the sensor surface results in a decrease in impedance similar to the positive experiment. The difference in electrical response of the control experiment and positive experiment lies in the final wash step (step 8). As expected no binding occurred between Streptavidin and Anti-VEGF molecules and the measured impedance level came back to its previous level after this wash step (step 8).

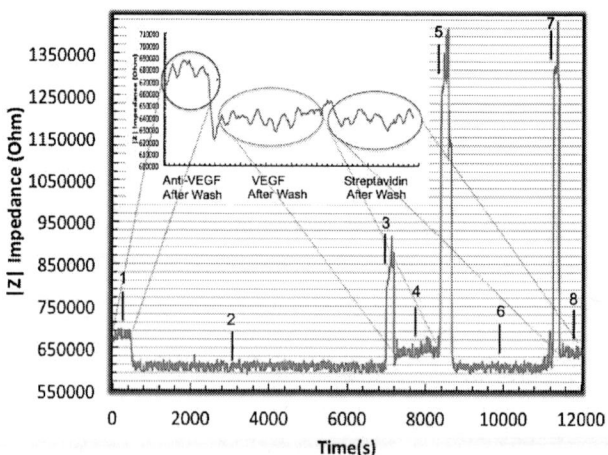

Figure 4: Detection of Vascular Endothelial Growth Factor (VEGF) for cancer diagnosis with a control experiment. Anti-VEGF was immobilized on the ODTS-Silanized sensor surface over the night (1). VEGF with the concentration of 50ng/ml was detected (3). As a control experiment, 40 μg/ml of Streptavidin was injected

to the channel (6). Impedance level comes back to its previous level after the wash step, meaning no binding to Anti-VEGF (8). Step 1, step 4 and step 8 are magnified and fitted on the left top corner of the figure.

DISCUSION:

The presence of protein on the surface of the nanoneedle sensor results in a decrease in the impedance across the two nanoelectrodes. This behavior is contrary to most label-free sensors where the presence of proteins at the surface results in an increase in measured impedance. We discussed the physical mechanisms behind this response in our previous publications [3]. Here we re-describe the three main mechanisms in brief detail. The first mechanism, which dominates the change in the imaginary component of the impedance, is the fringing field capacitance between the two electrodes at the tip of the needle. The binding of protein to the surface results in an increase in the relative permittivity between the two electrodes, thus resulting in an increase in capacitance thus a decrease in impedance. The second mechanism, which dominates the change in the real component of the measured impedance, is an increase in the AC coupling current across the electrodes. Charged biomacromolecules such as protein and nucleic acids contain a sheath of salt ions surrounding the backbone of the molecules. The presence of these salt ions effectively results in an increase in conductivity between the electrodes, thus resulting in a decrease in the real component of the impedance across the electrodes. The third, and least significant cause for decrease in the real component of the impedance is the modulation of the faradaic or tunneling current across the electrodes. The presence of the proteins at the interface creates new paths for electrons to tunnel and pass through the biomolecule, thus decrease the real component of the impedance across the sensor.

CONCLUSION:

In conclusion, we have developed a label-free matrix independent electronic biosensor capable of detecting low concentrations of proteins independent of the salt concentration of the buffer. This matrix-independence will ultimately make this sensor a good candidate for detecting proteins in complex biological samples and even clinical samples. We have also demonstrated the ease of ability to fabricate an array of sensors along a single microfluidic channel, opening up the possibility for multiplexed detection. Our results here encourage us to proceed with testing our sensors on complex biological and clinical samples, and performing multiplexed protein biomarker assays. Although, we are currently using commercial equipment for performing the readout electronics, we envision integrating our sensors with either a lock-in-amplifier on a printed circuit board, and ultimately even a custom designed CMOS solution, which will result in a true handheld system.

ACKNOWLEDGEMENTS:

This work was funded by the National Institutes of Health grant P01HG000205.

REFERENCES

[1]   G. Zheng, F. Patolsky, Y. Cui, W. U. Wang, and C. M. Lieber, "Multiplexed electrical detection of cancer markers with nanowire sensor arrays," Nature biotechnology, vol. 23, no. 10, pp. 1294-1301, 2005.

[2]   D. Lange, C. Hagleitner, A. Hierlemann, O. Brand, and H. Baltes, "Complementary metal oxide semiconductor cantilever arrays on a single chip: mass-sensitive detection of volatile organic compounds," Analytical Chemistry, vol. 74, no. 13, pp. 3084-3095, 2002.

[3]   R. Esfandyarpour, M. Javanmard, Z. Koochak, H. Esfandyarpour, J. S. Harris, and R. W. Davis, "Label-free electronic probing of nucleic acids and proteins at the nanoscale using the nanoneedle biosensor," Biomicrofluidics, vol. 7, pp. 044114, 2013.

[4]   N. Murukesh, C. Dive, and G. C. Jayson, "Biomarkers of angiogenesis and their role in the development of VEGF inhibitors," British journal of cancer, vol. 102, no. 1, pp. 8-18, 2009.

[5]   R. Esfandyarpour, H. Esfandyarpour, J. S. Harris, and R. W. Davis, "Simulation and fabrication of a new novel 3D injectable biosensor for high throughput genomics and proteomics in a lab-on-a-chip device," Nanotechnology, vol. 24, no. 46, pp. 465301, 2013.

[6]   R. Esfandyarpour, M. Javanmard, J. S. Harris, and R. W. Davis, "Thin Film Nanoelectronic Probe for Protein Detection." (Cambridge Univ Press), 2013.

[7]   R. Esfandyarpour, H. Esfandyarpour, M. Javanmard, J. S. Harris, and R. W. Davis, "Electrical Detection of Protein Biomarkers Using Nanoneedle Biosensors." (Cambridge University Press), 2012.

[8]   R. Esfandyarpour, H. Esfandyarpour, M. Javanmard, J. S. Harris, and R. W. Davis, "Microneedle biosensor: A method for direct label-free real time protein detection," Sensors and Actuators B: Chemical, vol. 177, pp. 848-855, 2013.

[9]   J. Mooney, A. Hunt, J. McIntosh, C. Liberko, D. Walba, and C. Rogers, "Patterning of functional antibodies and other proteins by photolithography of silane monolayers," Proceedings of the National Academy of Sciences, vol. 93, no. 22, pp. 12287-12291, 1996.

[10]   T. T. Huang, J. Sturgis, R. Gomez, T. Geng, R. Bashir, A. K. Bhunia, J. P. Robinson, and M. R. Ladisch, "Composite surface for blocking bacterial adsorption on protein biochips," Biotechnology and bioengineering, vol. 81, no. 5, pp. 618-624, 2003.

MECHANICAL PROPERTIES OF FEW LAYER GRAPHENE CANTILEVER

Kazuma Matsui[1], Akira Inaba[1], Yuta Oshidari[2], Yusuke Takei[1], Hidetoshi Takahashi[1],
Tomoyuki Takahata[1], Reo Kometani[2], Kiyoshi Matsumoto[1] and Isao Shimoyama[1]

[1]Graduate School of Information Science and Technology, The University of Tokyo, Tokyo, Japan
[2]Department of Mechanical Engineering, The University of Tokyo, Tokyo, Japan

ABSTRACT

We report the resonant frequency measurement of few-layer (1-, 2-, and 3-layer) graphene (FLG) cantilevers by optical heterodyne interferometry. The micro-sized FLG cantilevers with and without the diamond-like carbon weights were fabricated using focused ion beam. The Young's modulus was able to be calculated from the measured resonant frequency. The calculated Young's modulus was larger than the literature data [1]. This result suggests that the overlapped structure of the FLG cantilever makes the structure rigid.

INTRODUCTION

The reported Young's modulus of MLG graphene is ~1 TPa, and the researchers have studied the mechanical properties of the rigid graphene structures such as doubly clamped beams (DCBs) [1]. However, there has been no report about few-layer graphene (FLG) cantilevers. The purpose of this study is to measure the resonant frequencies of the FLG cantilevers.

Graphene has a planar sheet structure of carbon atoms. The graphene structures, including DCBs and cantilevers, can be suspended in few-layer thickness due to its high stiffness [1]. Taking advantage of the mechanical property, a suspended few-layer graphene (FLG) attracts attention as a resonator with low mass for radiation pressure noise detection [1, 2]. It is desired to understand the mechanical property, such as resonant frequency and spring constant, of the suspended FLG to design the resonator.

The mechanical properties of the multilayer graphene (MLG) cantilever and the FLG DCB have been measured [1-5]. However, the FLG cantilever's property can not be predicted from them. The mechanical properties of the MLG cantilever and DCB follow Timoshenko beam theory unlike FLG beams [2, 3]. The properties of the FLG DCB depend on the tension and the overlap on the edge, and we can not separate the effects of the two factors. The tension of the cantilever is removed because the cantilever has a free end.

In this study, we measured the resonant frequencies of the FLG cantilevers with and without the diamond-like carbon (DLC) weights by optical heterodyne interferometry. As shown in Figure 1, we fabricated the graphene cantilevers with the DLC weights using focused ion beam (FIB) (Figure 1(a)), and measured the resonant frequencies (Figure 1(b)). The Young's moduli were calculated from the measured resonant frequencies using two structures: the FLG cantilevers with and without the DLC weights. Thus, we evaluated the rigidity of the structure.

Although the spring constant measurement of the MLG cantilever using AFM was reported [2, 3], the AFM method

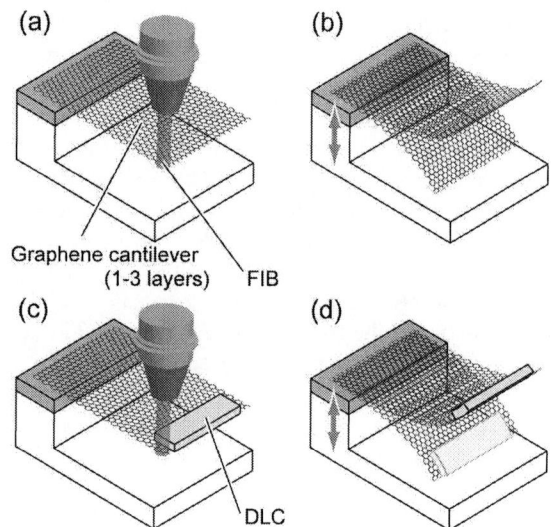

Figure 1: Schematics of the (a,c) fabrication and (b,d) the mechanical property measurement of a FLG cantilever with and without the DLC.

Table 1: Design of the FLG cantilevers **without** the DLC.

	1 layer	2 layers	3 layers
length [μm]	1-3	1-5	1-10
width [μm]	1-4	1-5	1-10

can not be applied to FLG cantilever. It is because the FLG cantilever is thought to have a much smaller spring constant than one of the MLG cantilever.

DESIGN AND FABRICATION

Design of Cantilever without DLC

Table 1 shows the design of the FLG cantilever without DLC. The sizes are different for each layer's cantilever because it is more difficult to fabricate a large-area cantilever when the graphene has fewer layers.

Design of Cantilever with DLC

The sizes of the cantilevers with the DLC are designed to be 1.3 μm in length and 1.8 μm in width. The DLC weight on the FLG cantilever is designed to be 0.5 μm in length, 1.5 μm in width, and 20 nm in thickness. The calculated DLC weight is 5.7×10^{-17} kg, where the mass density is 3.8×10^3 kg/m^3 as reported in Ref.[6]. The mass is large enough compared to the effective mass of a 3-layer graphene cantilever, 4.6×10^{-18} kg (1.5 μm in length and 1.9 μm in width). Therefore, we considered the weight as a concentrated load on the cantilever.

978-1-4799-3510-9/14 $31.00 © 2014 IEEE

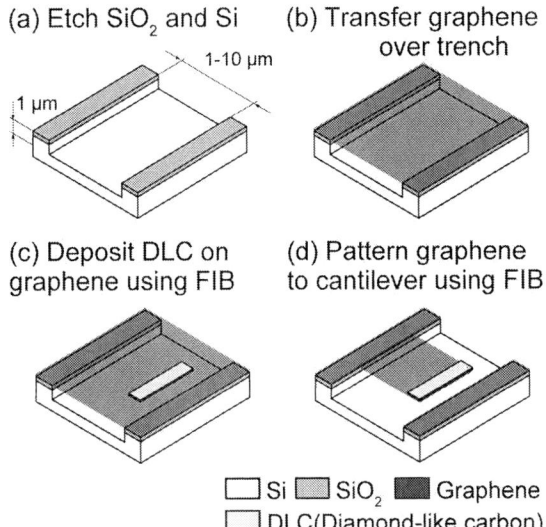

(a) Etch SiO$_2$ and Si (b) Transfer graphene over trench

(c) Deposit DLC on graphene using FIB (d) Pattern graphene to cantilever using FIB

☐ Si ☐ SiO$_2$ ■ Graphene
☐ DLC(Diamond-like carbon)

Figure 2: Fabrication steps of the graphene cantilever with the DLC.

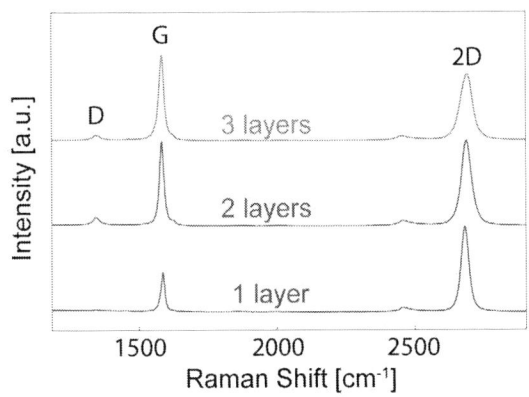

Figure 3: Raman spectra of the FLG.

Fabrication

Figure 2 shows the fabrication of the graphene cantilever with the DLC weight. Firstly, trenches of 1 μm in depth and 1-10 μm in width were etched on a Si wafer with a 300-nm-thick oxidized layer. Secondly, the suspended graphene was formed by transferring the FLG over the trenches (which was ordered to Graphene Platform, Inc., The Woodland, Texas, USA). The number of the graphene layers was confirmed by examining the intensity ratio of 2D-band and G-band (I_{2D}/I_G) in the Raman spectrum (Figure 3) [7]. The ratios I_{2D}/I_G of each graphene were roughly 2, 1, and 2/3, respectively. The ratios indicate that the each graphene has 1, 2, and 3 layers, respectively. Thirdly, a DLC weight of 0.5 μm in length, 1.5 μm in width, and 20 nm in thickness was deposited on the suspended graphene using FIB. At last, the graphene cantilevers were patterned from the suspended graphene using FIB.

FIB (Xvision 200TB, Hitachi High-Technologies Corporation, Tokyo, Japan) cut the suspended graphene in the conditions of 30 keV in acceleration voltage, 11.7 pA in

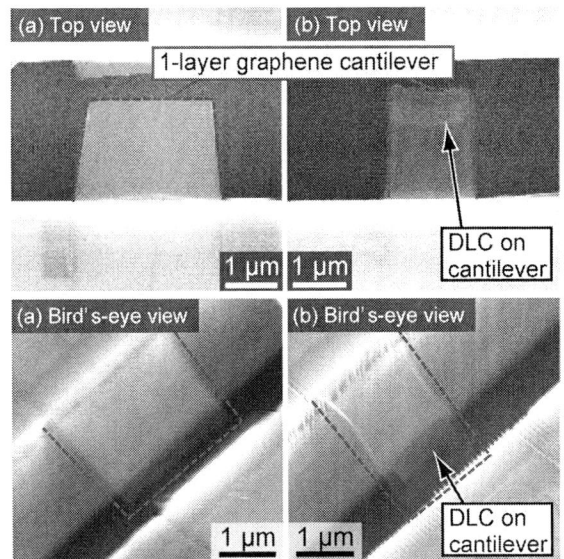

Figure 4: SEM images of the graphene cantilevers. (a) The 1-layer graphene cantilever. (b) Those with the DLC.

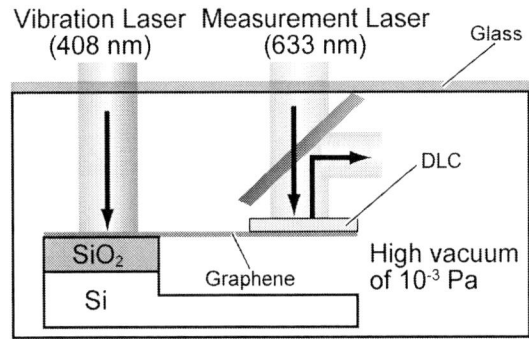

Figure 5: An interferometric optical setup for measuring the resonant frequency of the FLG cantilever.

beam current, 14 nm in beam size, 50 μs in dwell time, and 5-60 in ions/cm^2 dose. Figure 4 shows SEM images of the 1-layer graphene cantilevers with and without the DLC weights.

SETUP AND MEASUREMENT

We measured the resonant frequencies of the graphene cantilevers with and without the DLC weights, and calculated the Young moduli from the resonant frequencies

Optical Setup

Figure 5 shows an optical setup for measuring the resonant frequency of the FLG cantilever using an interferometer (MLD-230-V-100, NEOARK Corporation, Tokyo, Japan). In this system, a thermal energy was produced onto the substrate near the graphene cantilever by an excitation laser [8]. A photothermal excitation of the substrate vibrated the graphene cantilever. The reflectance intensity variation of the measurement laser incident on the cantilever was detected by a photodetector. We performed the experiment under high vacuum of 10^{-3} Pa at room temperature.

978-1-4799-3510-9/14 $31.00 © 2014 IEEE

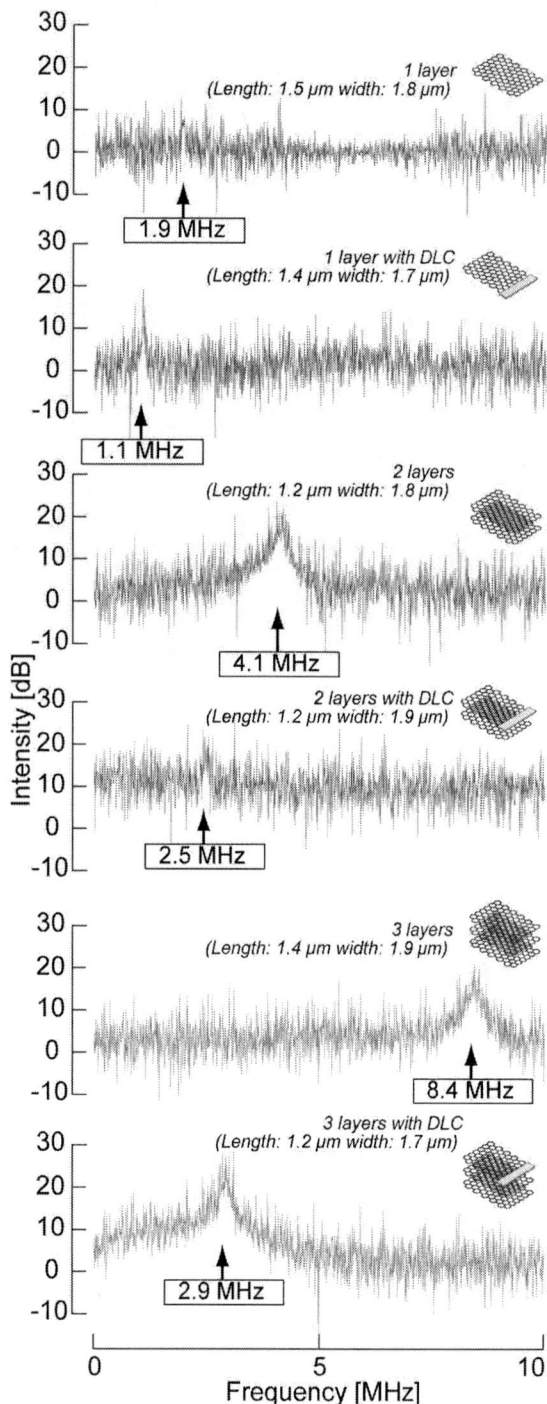

Figure 6: Vibration spectra of the FLG cantilevers with and without the DLC weights.

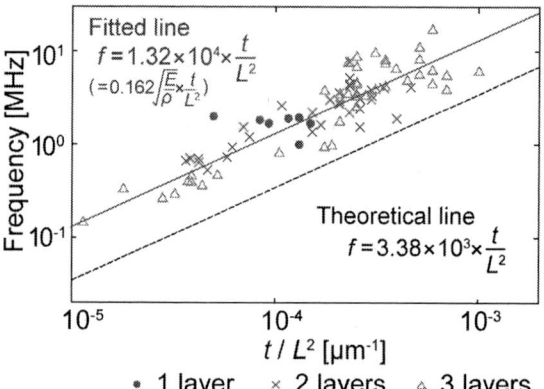

Figure 7: Measurement of resonant frequencies of the FLG cantilevers without the DLC weights.

Vibration Spectra of Cantilever

We measured the resonant frequencies of graphene cantilevers with and without the DLC weights. We confirmed the differences of the resonant frequencies between the each-layer graphene cantilevers with and without the DLC.

Figure 6 shows the vibration spectra around the fundamental resonant frequencies of 1-, 2-, and 3-layer graphene cantilevers with and without the DLC weights. The width and length of the cantilevers were 1.2-1.5 μm and 1.7-1.9 μm, respectively.

The 1-layer graphene has a small reflection intensity, and the amplitude at the resonance of the 1-layer graphene cantilever was smaller than those of the 2- or 3-layer graphene cantilever.

These vibration spectra showed that thinner graphene cantilevers had lower resonant frequencies. These spectra also showed that the resonant frequencies decreased when a DLC weight was deposited, because the mass of the cantilever increased. These results indicated the mechanical properties of the graphene cantilever roughly followed the beam equation.

Resonant Frequency of Cantilever without DLC

We measured the resonant frequencies of 81 graphene cantilevers without the DLC weights, and evaluated the rigidity of the structure.

Figure 7 shows the correlation between the resonant frequency and the graphene cantilever dimension (t/L^2 where t is the thickness and L is the length). The theoretical resonant frequency f is calculated from the following equation [1]:

$$f = 0.162 \times \sqrt{\frac{E}{\rho}} \times \frac{t}{L^2}. \quad (1)$$

This theoretical equation (1) can be described as

$$f = 3.38 \times 10^3 \times \frac{t}{L^2}, \quad (2)$$

where ρ is 2.2×10^3 kg/m³ and E is 1 TPa [1]. Besides, we obtained the following equation from the fitted line of the measured data:

$$f = 13.2 \times 10^3 \times \frac{t}{L^2}. \quad (3)$$

The thicknesses of the graphene cantilevers were calculated by regarding the thickness of the 1-layer graphene as 0.335 nm [9]. According to equation (2) and equation (3), the measured resonant frequency was 3.9 times higher than the resonant frequency calculated from Young's modulus and mass density written in the literature [1]. We calculated the Young's modulus (E^*) to be 15 TPa from the measured

978-1-4799-3510-9/14 $31.00 © 2014 IEEE   1089

Table 2: Resonant frequencies and spring constants of graphene cantilevers with the DLC weights. The spring constant, k, is calculated from the measured frequency, f, using the equation (4).

	1 layer (N=3)	2 layers (N=8)	3 layers (N=13)
f [MHz]			
Mean	1.1	2.7	2.9
SD	0.15	0.56	1.0
k [N/m]			
Mean	2.7×10^{-3}	1.7×10^{-2}	2.1×10^{-2}
SD	0.74×10^{-3}	0.67×10^{-2}	1.5×10^{-2}

resonant frequency using the equation (1) and equation (3), where ρ is the literature data (2.2×10^3 kg/m^3).

E^* is larger than the literature data (1 TPa) [1]. This result suggested that the graphene cantilever's edge was overlapped, and the overlapped structure increased the resonant frequencies of the FLG cantilevers [10]. For example, the resonant frequency of a 1-layer graphene cantilever of 1.5 μm in length and 1.5 μm in width is calculated to be 0.51 MHz. On the other hand, when the cantilever had a 2-layer overlapped edge of 100 nm in width, the moment of inertia increases and the resonant frequency is calculated to be 1.1 MHz. Thus, the overlapped structure of the FLG cantilever has the rigid characteristic.

This finding also suggests that the overlapped structure has an effect on the FLG DCB. Previously, the measured resonant frequencies of the FLG DCBs were higher than the theoretical frequencies [1, 5]. The increase has been thought to be due to two factors: the tension [1] and the overlap at the edge [5]. The result of this study indicates that the overlapped structure increases the resonant frequencies of the doubly clamped graphene beam.

Resonant Frequency of Cantilever with DLC

We measured the resonant frequencies of the 24 graphene cantilevers of 1.3±0.2 μm in length and 1.8±0.1 μm in width with the DLC weights. Table 2 shows the measured resonant frequencies and the spring constants of the graphene cantilevers with the DLC weights. The spring constant, k, is given by following equation:

$$k = m \times (2\pi f)^2 . \quad (4)$$

The Young's moduli of the 1-, 2-, and 3-layer graphene cantilever were calculated to be 520, 400, and 150 TPa using following equation:

$$E = \frac{kL^3}{3I} . \quad (5)$$

The Young's moduli were larger than E^*. The result indicates that the thickness of the overlapped edge of the cantilever with the DLC was larger than the one of the cantilever without the DLC.

CONCLUSIONS

In conclusion, we fabricated the FLG cantilevers with and without the DLC, and measured their resonant frequencies. The calculated Young's modulus of the cantilever was larger than 1 TPa. These results indicate that the overlap at the edge increases the moment of inertia of the FLG cantilevers.

ACKNOWLEDGEMENTS

The photolithography masks were fabricated using an EB lithography apparatus (F5112+VD01) donated by Advantest Corporation at the VLSI Design and Education Center (VDEC) of the University of Tokyo. This work was supported by JSPS KAKENHI Grant Number 25000010. This work was conducted in Research Hub for Advanced Nano Characterization, The University of Tokyo, supported by the Ministry of Education, Culture, Sports, Science and Technology (MEXT), Japan.

REFERENCES

[1] J. Scott Bunch et al., "Electromechanical Resonators from Graphene Sheets," *Science*, vol. 315, pp. 490-493, 2007.

[2] R. Rasuli et al., "Mechanical Properties of Graphene Cantilever from Atomic Force Microscopy and Density Functional Theory," *Nanotechnology,* vol. 21, p. 185503, 2010.

[3] P. Li et al., "Graphene Cantilever Beams for Nano Switches," *Appl. Phys. Lett.*, vol. 101, p. 093111, 2012.

[4] Y. Oshidari et al., "High Quality Factor Graphene Resonator Fabrication Using Resist Shrinkage-Induced Strain," *Appl. Phys. Express.*, vol. 5, p. 117201, 2012.

[5] S. Shivaraman et al., "Free-Standing Epitaxial Graphene," *Nano Lett.*, vol. 9, pp. 3100-3105, 2009.

[6] J. Igaki et al., "Mechanical Characteristics and Applications of Diamondlike-Carbon Cantilevers Fabricated by Focused-ion-beam Chemical Vapor Deposition," *J. Vac. Sci. Technol. B*, vol. 24, pp. 2911-2914, 2006.

[7] C-D. Liao et al., "Chemical Vapor Deposition Synthesis and Raman Spectroscopic Characterization of Large-Area Graphene Sheets," *J. Phys. Chem. A*, vol. 117, pp. 9454-9461, 2013.

[8] B. Ilic et al., "Theoretical and Experimental Investigation of Optically Driven Nanoelectromechanical Oscillators," J. *Appl. Phys.*, vol. 107, p. 034311, 2010.

[9] Z. H. Ni et al., "Graphene Thickness Determination Using Reflection and Contrast Spectroscopy," *Nano Lett.*, vol. 7, pp. 2758-2763, 2007.

[10] J. C. Meyer et al., "The Structure of Suspended Graphene Sheets," *Nature*, vol. 446, pp. 60-63, 2007.

CONTACT

*Kazuma Matsui, tel: +81-3-5841-6318;
k_matsui@leopard.t.u-tokyo.ac.jp

NANO-OPTO-MECHANICAL MEMORY BASED ON OPTICAL GRADIENT FORCE INDUCED BISTABILITY

B. Dong[1,2], J. G. Huang[3], H. Cai[2], P. Kropelnicki[2], A. B. Randles[2], Y. D. Gu[2] and A. Q. Liu[1†]

[1]School of Electrical & Electronic Engineering, Nanyang Technological University, Singapore 639798
[2]Institute of Microelectronics, A*STAR (Agency for Science, Technology and Research),
Singapore 117685
[3]School of Mechanical Engineering, Xi'an Jiaotong University, Xi'an 710049, China

ABSTRACT

A bistable nano-opto-mechanical memory is designed, fabricated and experimentally demonstrated. A doubly-clamped silicon beam is deformed by optical gradient force generated from the ring resonator. The doubly-clamped silicon beam can be bended due to attractive optical gradient force generated by ring resonator. Due to the non-linear behavior of optical gradient force, the silicon beam has two stable positions which can be switched by controlling the light power transmitted inside the ring resonator. The nano-size of the memory enable for large scale integration, high speed operation and low power consumption. It has other potential applications such as optical switch, logic gate and actuator.

INTRODUCTION

Optical memory, as an important element of future all optical computing and all optical network, is attracting wide attention. Most of current optical memory depends on the high nonlinear properties of materials such as silicon, whose optical properties can be changed upon triggered by light [1]. However, such kind of memory normally requires high operating power or complex structure. Due to the ever-increasing speed of fibre-optic-based telecommunications, the high-speed optical memories adapted to densely on-chip integration become critical for buffering of decisions and telecommunication data [2].

In this paper, a bistable opto-mechanical memory is demonstrated, which employs Whisper Gallery Mode ring resonator (RR) to generate optical gradient force over a doubly-clamped silicon beam [3]. The opto-mechanical interaction between the RR and nano-sized silicon beam allows for a high operating speed with low operation power due to the bistability of the opto-mechanical system [4-6]. It has applications in various fields such as all optical communication, computing and opto-mechanical systems.

DESIGN AND THEORY

The NOM memory consists of a ring resonator, a doubly-clamped silicon beam and a bus waveguide. Lights are coupled in and out of the ring resonator through the bus waveguide as shown in Fig. 1(a). Optical gradient force is generated between the ring resonator and the doubly-clamped silicon beam. The ring resonator is supported by rib structure and released from the substrate to reduce loss induced by the releasing process. The outer radius of the ring resonator is 20 μm with a width of 450 nm. The schematic of the 200-nm-wide doubly clamped silicon beam

is shown in Fig. 1(b). The outer radius R is 20.35 μm. The silicon beam is clamped by two anchors, so that it can move along the x direction but restrict the movement in the z and y directions. The red-detuned control light, known as write light, is coupled from the bus waveguide into the ring

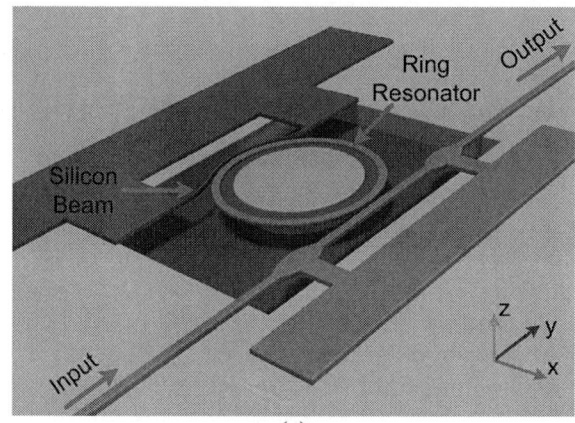

(a)

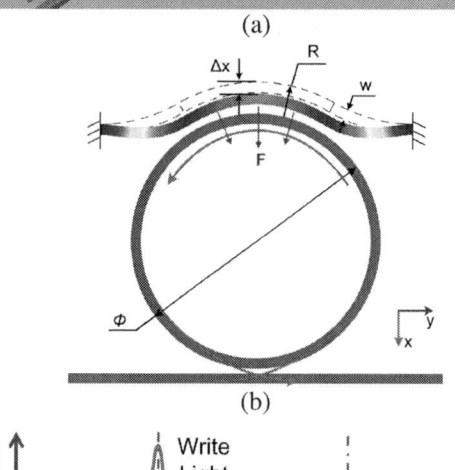

(b)

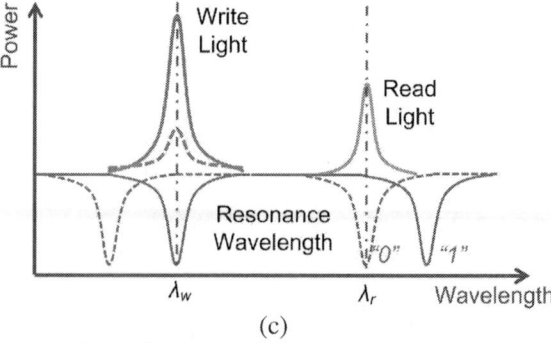

(c)

Figure 1: (a) Schematic of opto-mechanical memory, (b) layout of the optical memory and (c) spectrum illustration of the optical memory.

978-1-4799-3510-9/14 $31.00 © 2014 IEEE

resonator. Thereafter, optical gradient force is conducted between the ring resonator and the silicon beam, which pulls the silicon beam towards the ring. The signal light, known as read light, is also coupled into the ring resonator and sensed at the output port. The wavelength of read light overlaps with the resonance wavelength of ring resonator. When increasing the power of the write light, the optical gradient force is increased and the mechanical arc moves towards the ring resonator along x direction. The displacement x of the silicon beam results in the change of effective refractive index n of the ring resonator, causing a red shift of the ring resonance wavelength $_{r}$, as shown in Fig. 1(c). The read light can transmit through the waveguide since its wavelength is no longer overlapped with the resonance wavelength. Therefore, the displacement of the silicon beam can affect the transmission of read light. Therefore, by pumping a red-detuned write light with different power level, the transmission of read light can be controlled and the status of the doubly-clamped silicon beam, which is known as memory state, can be switched.

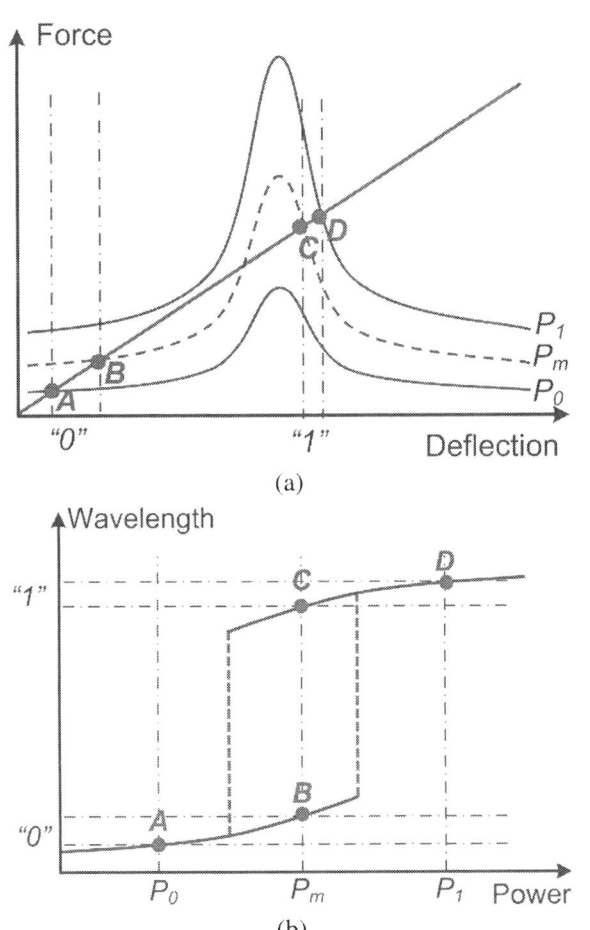

Figure 2: (a) Force analysis of the opto-mechanical memory and (b) working principle of the bi-stable optical memory

The red-detuned single wavelength write light induces optical force over the doubly-clamped silicon beam, and the deformation of the silicon beam can affect the amplitude of the optical force as well. The optical gradient force can be expressed as

$$F_{optical}(x) = -\frac{2\gamma_e P_{optical}}{\omega_w} \frac{g_{om}}{\Delta(x)^2 + \gamma^2} ,$$

where γ_e is the external damping due to waveguide-ring coupling, ω_w is the write light frequency, g_{om} is the optomechanical coupling coefficient, is the total damping coefficient and (x) is the detuning, which is defined as

$$\Delta(x) = \omega(x) - \omega_w .$$

$\omega(x)$ is the resonance frequency of the ring resonator and ω_w is the frequency of the write light. The force as a function of deformation in x direction is shown in Fig. 2(a), which is a Lorenz shape. The maximum force happens at zero detuning, which means that the frequency of the write light and the resonance frequency of the ring resonator overlap. The optical gradient force is balanced by the mechanical spring force of the silicon beam. The two curves have several intersections which indicate stable positions of the silicon beam. When the power of the write light is changed from low level (P_0) to high level (P_1), the stable position is shift from point "A" to "D". It is noticed that the two curves have three intersections and two of which ("B" and "C") are stable positions at moderate power level (P_m). Therefore, when increase from P_0 to P_m, the silicon beam change from "A" to "B", when decrease from P_1 to P_m, the silicon beam change from "D" to "C". The deformation difference between "A" and "B" is quite small, so they are defined as memory state "0", while "C" and "D" are defined as memory state "1". The status can either switch between "0" and "1", or maintain at the status, which depends on the power level of the write light.

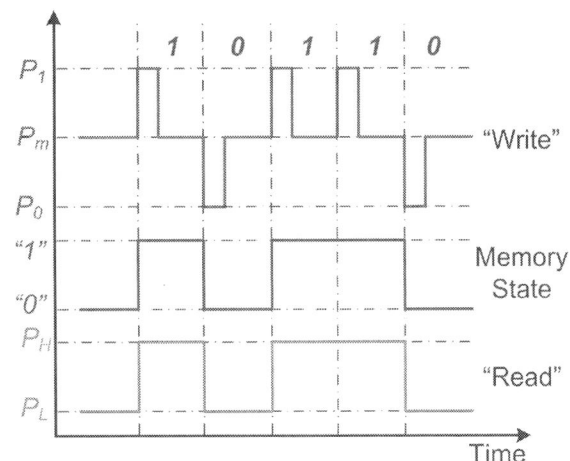

Figure 3: Demonstration of "Write" and "Read" processes.

Figure 3 shows the operation of the optical memory. By pumping in a pulsed high power (P_1) write light, the memory can switch to state "1" while the transmission of

978-1-4799-3510-9/14 $31.00 © 2014 IEEE   1092

read light is at high power level (P_H). The memory is switched to state "0" by further reducing the write light power to P_0. As a result, the transmission of the read light is low (P_L). Due to the bistability of the memory, the memory can maintain its state with a moderate power level (P_m) to save power consumption.

FABRICATION

Figure 4 shows the scanning electron microscope (SEM) image of the doubly-clamped silicon beam. The optical memory is fabricated by nano-photonic fabrication processes using standard silicon-on-insulator wafer, with a 220-nm thick silicon structure layer and a 2-μm buried oxide layer. The waveguide structures have a width of 450 nm and a height of 220 nm for a single mode transmission. The silicon beam is designed to have a width of 200 nm while the coupling gap between the ring resonator and the silicon beam is 150 nm. The waveguides and ring resonators are patterned by deep UV lithography, followed by plasma dry etching to transfer the photo resist pattern into the structure layer. The rib structure of actuation ring has a 70-nm silicon slab layer to support the ring resonator. After etching, a 2-μm SiO$_2$ layer is deposited on the structure layers to ensure a low optical loss. A 40-nm Al$_2$O$_3$ is deposited and patterned, which is used as the protection film to protect those fixed structures and leave the window area open for suspended structures. Finally, HF vapor selectively undercut the buried oxide layer in the window area to release the movable structures.

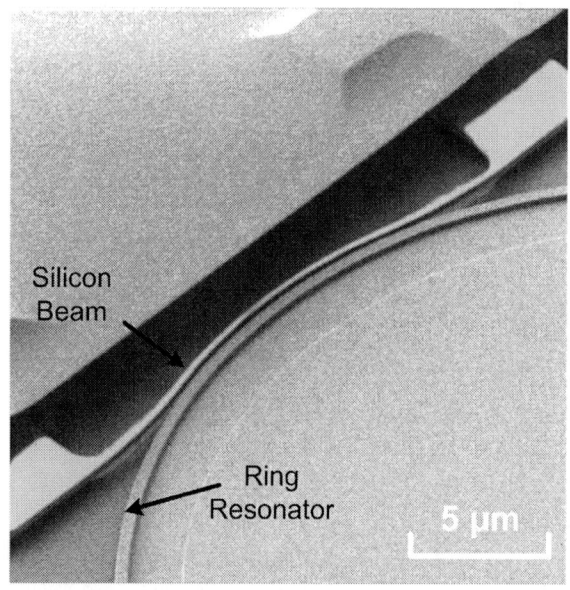

Figure 4: SEM images of the optical memory.

EXPERIMENTS AND DISCUSSIONS

The optical memory is experimentally tested with the setup shown in Fig. 5. The write light go through an electro-optics modulator, to modulate the light power level among P_1, P_m and P_0. The modulated light is then amplified by Erbium Doped Fibre Amplifier (EDFA). The write light and read light are then pumped into the device together through the 2 × 1 coupler. After passing through the device, output light is monitored by the spectrum analyzer and the photo detector. The transmission spectra of the optical memory at various power of the red detuned (0.36 nm) write light is shown in Fig. 6. At high write light power (-6 dBm), the resonance wavelength is red shifted and the memory is at state "1". The resonance wavelength is blue shifted at lower write power level (-10 dBm) and the memory is at state "0". At moderate power level (-8 dBm), the resonance wavelength maintains the same as previous status. Therefore, by modulating the write lights power, the resonance wavelength of the ring resonator are modulated and the state of the optical memory is switched.

Even though the current memory is one unit, it can be expanded to memory array easily while maintaining small footprint. The silicon beam structure and ring resonator can also be optimized to further reduce the operating power.

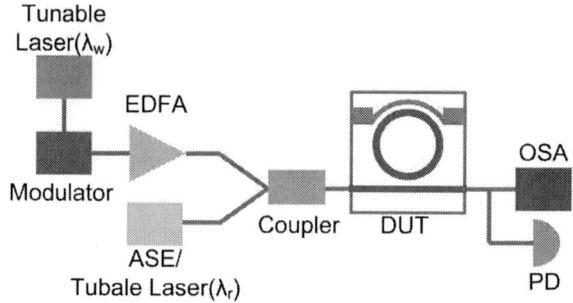

Figure 5: Experimental Setup for memory characterization.

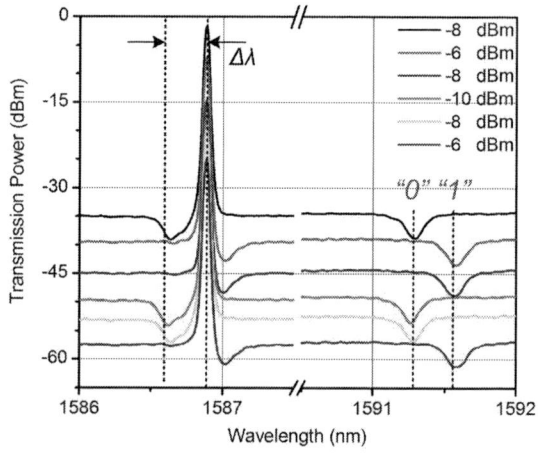

Figure 6: Transmission spectra of optical memory with various input powers.

CONCLUSIONS

In summary, a bistable nano-opto-mechanical memory is experimentally demonstrated. The optical memory is driven by the evanescent wave coupled optical gradient force. Two stable positions of a doubly-clamped silicon beam, named as "0" and "1", can be switched by pumping light with different power level. The proposed NEMS actuator has merits such as small dimensions, low power consumption and easy integration, which offer great

978-1-4799-3510-9/14 $31.00 © 2014 IEEE

potential for a wide span of nano-resonator based applications ranging from all-optical communications to fundamental sciences.

ACKNOWLEDGMENTS

The authors would like to acknowledge the support from the Science and Engineering Research Council of A*STAR, Singapore, under SERC Grant 1021650084.

REFERENCES

[1] T. Tanabe, M. Notomi, S. Mitsugi, A. Shinya, and E. Kuramochi, "Fast bistable all-optical switch and memory on a silicon photonic crystal on-chip," Opt. Lett., vol. 30, pp. 2575-2577, 10/01 2005.

[2] M. T. Hill, H. J. S. Dorren, T. de Vries, X. J. M. Leijtens, J. H. den Besten, B. Smalbrugge, et al., "A fast low-power optical memory based on coupled micro-ring lasers," Nature, vol. 432, pp. 206-209, 2004.

[3] W. M. Zhu, T. Zhong, A. Q. Liu, X. M. Zhang and M. Yu, "Micromachined optical well structure for thermo-optic switching", Appl. Phys. Lett., vol. 91, 261106, 2007.

[4] B. Dong, H. Cai, G. I. Ng, P. Kropelnicki, J. M. Tsai, A. B. Randles, M. Tang, Y. D. Gu, Z. G. Suo and A. Q. Liu, "A nanoelectromechanical systems actuator driven and controlled by Q-factor attenuation of ring resonator," Applied Physics Letters, Vol 103, 181105, 2013.

[5] M. Ren, J. Huang, H. Cai, J. M. Tsai, J. Zhou, Z. Liu, Z. Suo, and A. Q. Liu, "Nano-optomechanical actuator and pull-back instability," ACS Nano, Vol 7, pp.1676–1681, 2013.

[6] H. Cai, B. Dong, J. F. Tao, L. Ding, J. M. Tsai, G. Q. Lo, A. Q. Liu and D. L. Kwong, "A nanoelectromechanical systems optical switch driven by optical gradient force," Applied Physics Letters, Vol 102, 023103, 2013.

CONTACT

*A. Q. Liu, +65-67904336; eaqliu@ntu.edu.sg

SUBMICRON THREE-TERMINAL SIGE-BASED ELECTROMECHANICAL OHMIC RELAY

Maliheh Ramezani[1,2], Stefan Cosemans[1], Jeroen De Coster[1], Xavier Rottenberg[1], Veronique Rochus[1], Haris Osman[1], Harrie A. C. Tilmans[1], Simone Severi[1], and Kristin De Meyer[1,2]

[1] Imec, Leuven, BELGIUM
[2] Katholieke Universiteit Leuven, Leuven, BELGIUM

ABSTRACT

This paper demonstrates functional NEM cantilever relays fabricated in a CMOS-compatible low-T (400°C) CVD SiGe process flow. Devices with a length in the micrometer range (<3μm), a width in the range 0.2-1μm, a thickness and a gap of below 100nm were successfully fabricated and characterized. A high on/off current ratio (of better than 10^8:1), a subthreshold swing (S) better than 150μV/decade and "essentially zero" off-state leakage current were experimentally observed. A life time of minimum 10^3 switching cycles was demonstrated. A maximum current density of around 10 μA/μm^2 without causing stiction due to Joule-heating was found.

INTRODUCTION

Due to some physical phenomenon like subthreshold leakage, short-channel effects and hot carrier injection in miniaturized conventional CMOS transistors, the stand-by power dissipation keeps on growing steadily in the recent technology nodes. These fundamental issues end up with non-ideal switching behavior of the MOS transistors. Therefore to avoid huge static power dissipation in low power applications, new alternatives with steep subthreshold slope are desired. A NEM relay offers a number of interesting features which makes it an attractive solution for low and ultra-low power systems. Electromechanical relays show infinite sub-threshold slope, almost zero off-state leakage current, and it can be manufactured on top of CMOS circuits to add some extra functionalities. In particular the low leakage makes the NEM relay an ideal power gating device [1]. Thanks to the inherent hysteresis behavior, it can also be employed to realize hybrid non-volatile CMOS-MEMS memory array for low power applications [2]. Additionally the NEM relays are radiation hard and robust in harsh environment like high temperature [3].

The importance of scaling simultaneously the motional volume, i.e. the volume of the moving part, and the contact area of a NEM relay is well understood for logic and memory applications [1]. Some attempts at scaling the relay device dimensions below the μm range have recently been shown [4], but either the number of operational cycles or the contact current density were negatively impacted by the device scaling. Several NEM switches based on carbon nanotubes (CNT) and nanowires (NW) have been reported previously however these devices behaves quite dreadfully in terms of reliability. Besides that there are substantial obstacles like controlling the position, size and population

of CNT and NW [5]. Besides the bottom-up approach, a number of top-down designs including lateral and vertical types of MEMS relays were proposed [6-7] which mainly are not efficient in terms of footprint or were realized using E-beam lithography which is not efficient for mass production. One of the main challenges is represented by the need of finding a technology that provides a reliable device down to sub-micrometer size. The investigation presented in this paper on the use of poly-SiGe aims to tackle both of these challenges. Using a reliable SiGe mechanical material [8], directly on top of CMOS Back-End-of-Line (BEOL) gives the possibility to aggressively scale the device with direct access to CMOS interconnections.

In this paper, we propose for the first time a 400°C top-down CMOS compatible vertical process flow that makes use of standard DUV lithography to print minimum feature sizes of 200nm on 200mm silicon wafers and allow for sub-100nm air-gap and beam thickness definition. To study the switching behavior, cantilever relays were fabricated and electrically characterized. The operation of nano cantilever with extremely small motional volume of 0.008μm^3 (L/W/H = 900nm/200nm/40nm) and the contact area (SiGe-SiGe contact) of 0.02μm^3 was experimentally demonstrated which is among smallest fabricated NEMS relays recently [9].

DEVICE STRUCTURE AND FABRCATION PROCESS

A 3D illustration of the 3-terminal (3T) relay, with graphically indicated motional volume and contact area (A_{co}), is depicted in Figure 1. Table 1 summarizes the geometrical and material properties of the fabricated NEM relays.

The schematic diagram of the fabrication process is shown in Figures 2. It starts with processing of buried metal lines which functions as a routing layer between devices and to the bond-pads.

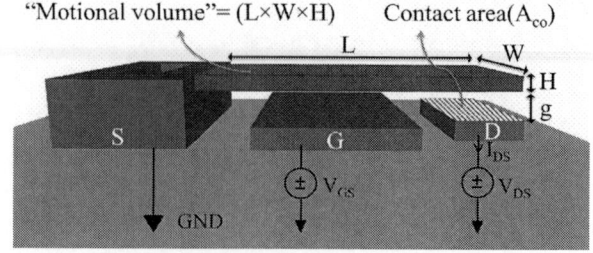

Figure 1: 3D schematic view of 3T NEM relay structure

Table 1. Nano-relay geometrical and material properties

Parameter	Value
Elastic modulus SiGe	115 [GPa]
Resistivity SiGe	9800 [nΩ.m]
Boron doping SiGe	10^{21} [at/cm^3]
Residual stress SiGe	-50 [MPa], Compressive
Beam Length L	0.9 - 3 [μm]
Beam Width W	200, 500, 1000 [nm]
Beam Thickness H	100, 300 [nm]
Actuation gap g	100 [nm]

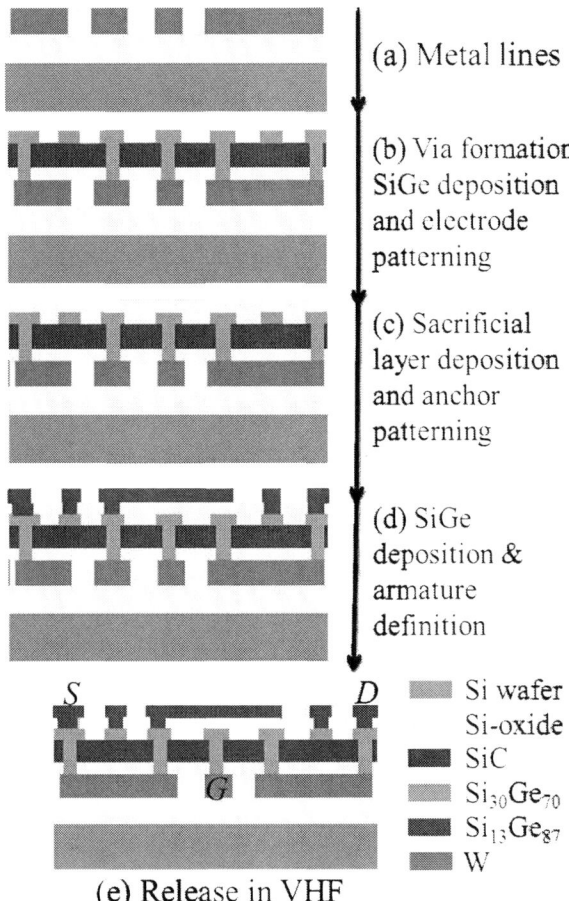

(a) Metal lines

(b) Via formation, SiGe deposition and electrode patterning

(c) Sacrificial layer deposition and anchor patterning

(d) SiGe deposition & armature definition

(e) Release in VHF

Si wafer
Si-oxide
SiC
$Si_{30}Ge_{70}$
$Si_{13}Ge_{87}$
W

Figure 2: Schematic view of SiGe nano-relay fabrication process.

The process begins with forming 400nm deep trenches in 1µm thick High Density Plasma (HDP) Si-oxide followed by a conformal depositing of a 350nm Chemical Vapor Deposition (CVD) tungsten. To complete metal line process, a Chemical Mechanical Polishing (CMP) step is performed to polish away the extra tungsten on top of Si-oxide. Then a passivation layers made of SiC and Si-oxide is deposited on top of the metal lines. The SiC material is used as a protection layer of the CMOS BEOL during the vapor-HF release process [5]. In order to have electrical connection between the buried metal layer and the bottom electrode of the device, some vias need to be opened in the passivation layer. The vias is patterned by making use of a dark-field DUV photolithography process followed by an etch process stopping on the metal layer. Afterwards a 200nm $Si_{40}Ge_{60}$ was deposited with a conformal CVD process so that the vias could be completely filled. This layer was then patterned to form the bottom electrode of the NEMS structure where the contact area and gate of the device are defined. The wafer is next planarized using a SiO_x polishing process (CMP), followed by the deposition of the sub-100nm relay sacrificial layer. After patterning anchors in Si-oxide, a thin SiGe structural layer is defined. The deposition conditions of the ultra-thin SiGe structural layer was optimized to achieve minimum resistivity and low stress gradient [10]. The device is finally released in Vapor HF.

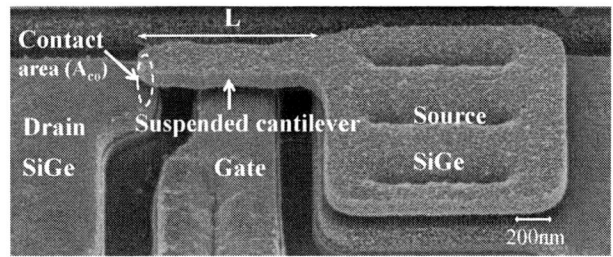

Figure 3: SEM picture of a fabricated 3T cantilever relay where L, W, g and H are 900,200,100 and 100nm respectively.

SEM image of a fabricated 3T nano-relay using the mentioned fabrication technology is exhibited in Figure 3. The minimum achievable device footprint is mainly limited by the minimum feature size that can be printed using the DUV lithography which was around 200nm. The smallest fabricated device had a width of 200nm and length of 900nm, gap and beam thickness of 100nm (Figure 3). Currently we are working on sub-100nm beam thicknesses and actuation gaps to reduce the pull-in voltage of this device without compromising the footprint.

RESULTS AND DISCUSSION

Electrical characterization of fabricated nano-relays confirms that the gate leakage currents and drain off-state current are zero. During the device actuation, the sub-threshold slope is extremely high but the value is not quantified since no AC measurements have been yet performed.

The performance of 10 devices from 10 different dies (from the same wafer) displays a variation in the measured pull-in and pull-out voltages (see Figure 4), something that can be attributed to variations in the actuation gap or in the structural layer thickness.

978-1-4799-3510-9/14 $31.00 © 2014 IEEE

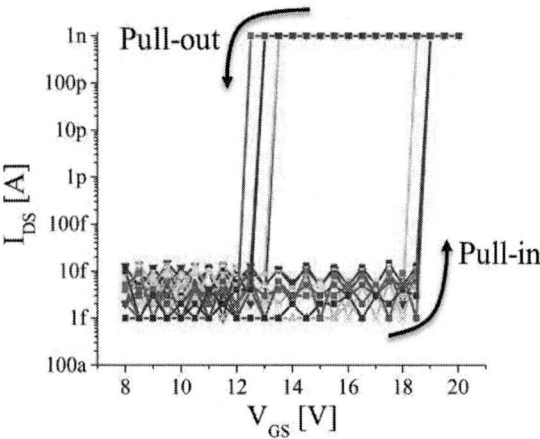

Figure 4: Measured hysteresis I_{DS}-V_{GS} curves for 10 different relays (indicated by a different color) with L/W/H = 3μm/1μm/300nm. The compliance current was set to 1nA and V_{DS} = 2V.

The hysteresis I_{DS}-V_{GS} measurement over 10^3 operating cycles illustrates repetitive operation of the relay and also its stable pull-in and pull-out voltages over the cycling (see Figure 5). The beam pulls-in and -out successfully at around 20V and 14V respectively. Since the cycling is done in air while the device is not packaged yet, the contact surface might get additional oxidation and/or absorb contamination, and as a result, the current degrades gradually during the cycling. Figure 6 highlights that by applying high drain voltage V_{DS}, standard contact performance can be restored after cycling in air. This indicates that the contact reliability is not intrinsically degraded but strongly affected by the environment in which the device operates. With increasing the drain bias the beam exercises higher force on the contact and more asperities come into contact. Larger contact area results into larger hysteresis window due to stronger adhesion forces, as shown in Figure 6. So higher V_{DS} leads to wider hysteresis window and ultimately may result in permanent stiction.

As we expect from theory pull-in voltage is independent of relay width. Ideally the pull-out voltage also should not be affected with the width of the relay where there is no considerable amount of adhesion force at contact. Therefore the hysteresis window should remain unchanged with varying the relay width [11]. It was experimentally observed that the beam width does not have a major impact on pull-in voltage however it can lower the pull-out voltage due to the larger contact area (Figure 8). Lower pull-out voltage might increase the chance of stiction and deteriorates the device reliability. The trend in Figure 6 indicates the major impact of larger adhesion force for wider devices on the pull-out voltage.

In a limited range of voltages and widths where no device stiction occurs, it is possible to control the hysteresis window of the device. For memory application, this in turn will enable the device to switch between volatile and non-volatile regime. However for logic applications, it is desired to decrease the hysteresis window.

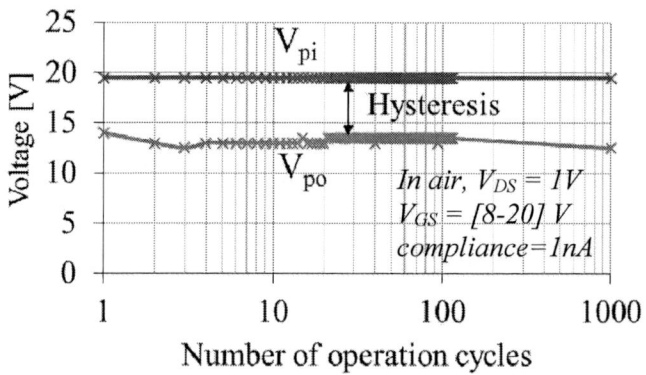

Figure 5: Measured pull-in and pull-out voltages vs. number of operating cycles for L/W/H = 3/1/0.3 μm.

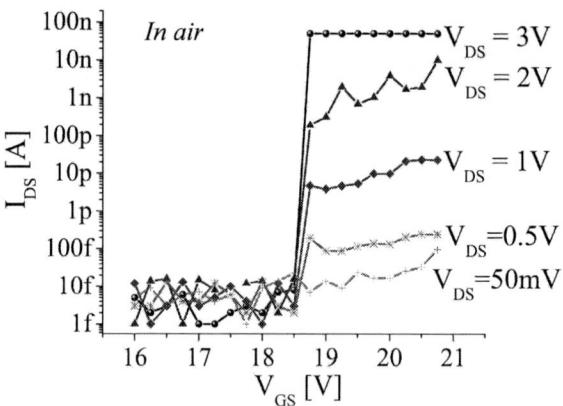

Figure 6: Measured I_{DS}-V_{GS} curves for different V_{DS} after around 100 cycles in air for L/W/H=3/1/0.3μm. The current degraded from 1nA (compliance current) at first cycles to 10pA at the 100th cycle in air. However a higher voltage at contact (V_{DS}) restored the current.

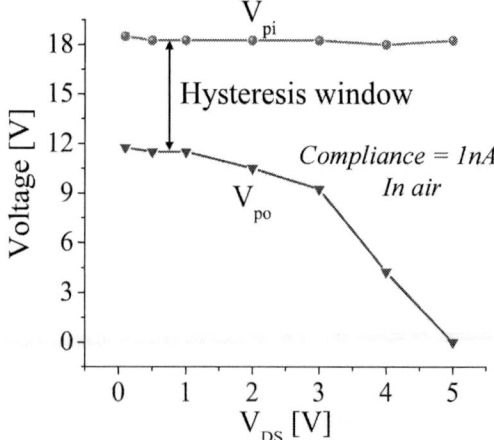

Figure 7: Measured hysteresis window versus V_{DS} for L/W/H = 3μm/1μm/300nm.

978-1-4799-3510-9/14 $31.00 © 2014 IEEE

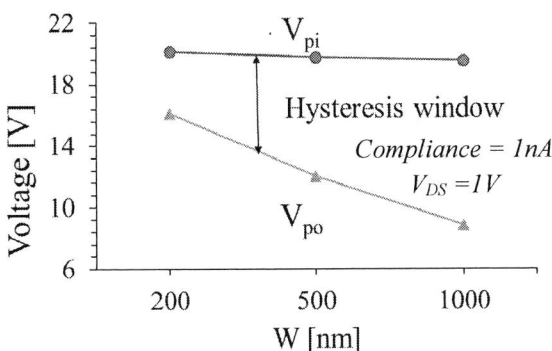

Figure 8: Measured hysteresis window versus beam width for L/H=3μm/300nm.

The maximum current density that can flow through the device without causing permanent stiction (see Figure 9) determines the real value of the contact resistance, estimated to be in the order of 5MΩ after subtracting the parasitic source and drain (series) resistance for contact area of below $0.02\mu m^2$. As Figure 9 shows, devices with smaller contact area have capability to pass higher current density and this is due to smaller contact force. Basically for devices with larger contact area, the maximum current before failure is mostly limited by stiction due to adhesive intermolecular force in contact whereas the current limitation for narrower devices mainly comes from local Joule heating and micro-welding of the SiGe-SiGe bonds [4]. For this reason, stiction for devices with smaller contact area occurs at higher current densities.

Maximum current which can flow from source to drain without failure is a function of the structural dimensions (Figure 9) though typically on-current of 100nA was measured without failure. This result suggests to apply a current compliance of 100nA throughout the measurement to prevent high current related problems such as melting and fuse failure.

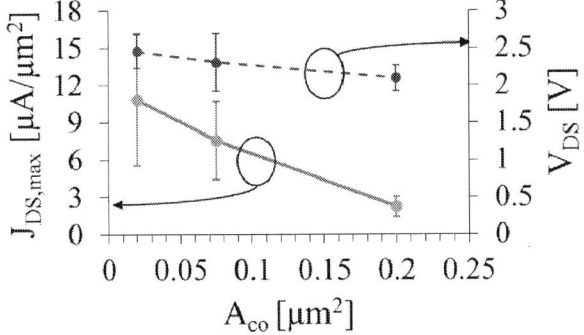

Figure 7: Maximum allowable drain current density measured without causing failure of the relay at V_{GS}=20V ($\approx 1.1V_{pi}$). (L/H =1.5μm/100nm)

CONCLUSION

Nano-relays were fabricated by a proposed CMOS compatible poly-SiGe-based process flow and electrically characterized. However a minimum of 10^3 operational cycles was demonstrated, the device after 10^3 cycles is still working properly and is expected to work much longer in an inert environment rather than air. The amount of contact resistance is expected to be improved from a few MΩ to few tens of KΩ after using an extra coating and packaging the device. It was also observed that the maximum current of around 100nA can flow through the device without causing stiction.

REFERENCES

[1] V. Pott, K. Hei, N. Rhesa, J. Jaeseok, A. Elad, and Tsu-Jae King Liu. "Mechanical computing redux: relays for integrated circuit applications", *Proc. IEEE* 98, no. 12, 2010, pp. 2076-2094.

[2] R. Gaddi,, C. Schepens, Ch. Smith, C. Zambelli, A. Chimenton, and P. Olivo. "Reliability and performance characterization of a MEMS-based non-volatile switch", in *IRPS conference* 2011, pp. 2G-2.

[3] X. Wang, S. Narasimhan, A. Krishna, Francis G. Wolff, S. Rajgopal, T. Lee, M. Mehregany, and S. Bhunia. "High-temperature (> 500° C) reconfigurable computing using silicon carbide NEMS switches", in *DATE conference* 2011, pp. 1-6.

[4] W. W. Jang, J. O. Lee, J. B. Yoon, M. S. Kim, J. M. Lee, & W. S. Lee. "Fabrication and characterization of a nanoelectromechanical switch with 15-nm-thick suspension air gap", *Appl. Phys. Lett.*, 92(10), pp. 103110-103110, 2008.

[6] R. L. Badzey, G. Zolfagharkhani, A. Gaidarzhy, and P. Mohanty. "A controllable nanomechanical memory element", *App. Phys. Lett.* 85, no. 16, pp. 3587-3589, 2004.

[7] R. Parsa, M. Shavezipur, W. S. Lee, S. Chong, D. Lee, H-SP Wong, R. Maboudian, and R. T. Howe, "Nanoelectromechanical relays with decoupled electrode and suspension", in *MEMS conference* 2011, pp. 1361-1364.

[8] K. Akarvardar, D. Elata, R. Parsa, G. C. Wan, K. Yoo, J. Provine, P. Peumans, R. T. Howe, and H-SP Wong. "Design considerations for complementary nanoelectromechanical logic gates", in *Tech. Dig. IEDM* 2007, pp. 299-302.

[9] L. Haspeslagh, J. De Coster, O. V. Pedreira, I. De Wolf, B. Du Bois, A. Verbist, R. Van Hoof et al. "Highly reliable CMOS-integrated 11MPixel SiGe-based micro-mirror arrays for high-end industrial applications", in *Tech. Dig. IEDM* 2008, pp. 1-4.

[10] T. He, R. Yang, S. Rajgopal, S. Bhunia, M. Mehregany, and Ph. X-L. Feng. "Dual-gate silicon carbide (SiC) lateral nanoelectromechanical switches", in *NEMS conference* 2013, pp. 554-557.

[11] T. B. Asafa, N. Tabet, and S. A. M. Said. "Taguchi method–ANN integration for predictive model of intrinsic stress in hydrogenated amorphous silicon film deposited by plasma enhanced chemical vapor deposition", Neurocomputing 2012.

[12] G. M. Rebeiz, *RF MEMS: Theory, Design, and Technology,* John Wiley and Sons, 2003.

PIEZOELECTRIC BUCKLING-BASED NEMS RELAYS FOR MILLIVOLT MECHANICAL LOGIC

Usama Zaghloul[1,2] and Gianluca Piazza[1]

[1]Electrical and Computer Engineering Department, Carnegie Mellon University, Pennsylvania, USA
[2]Microelectronics Department, Electronics Research Institute, Cairo, Egypt

ABSTRACT

We report on the design, fabrication, characterization, and scaling analysis of buckling-based NEMS relays that use, for the first time, piezoelectric actuation. The generated stress from anchored piezoelectric films is employed to buckle a clamped-clamped beam to connect the source and drain, while the residual stress is used to control the actuation voltage. This demonstration is the first of its kind to exploit residual stress to deliver a highly scalable switching mechanism that exhibits low actuation voltage (~1.8 V), and uniquely achieves an equivalent electric body bias via mechanical methods. Analytical and FEA simulation show a linear dependence of the switching voltage on the residual stress, while the voltage vs. stress tuning slope is reduced linearly with scaling the piezoelectric film thickness (-16.6 mV/MPa for 10 nm thick AlN film). A scaling analysis shows that millivolt switching is possible for aggressively miniaturized relays.

INTRODUCTION

The fundamental physical limitations that hinder further scaling of metal-oxide-semiconductor (MOS) transistors have propelled an interest in using NEMS relays for mechanical logic applications. Driven by their extremely steep subthreshold slope [1] and suitability for harsh environments [2] compared to MOS devices, NEMS relays were demonstrated for the implementation of digital logic circuits [3] and memories [4]. Among various reported actuation techniques, electrostatic and piezoelectric methods are the most widely used to drive these relays [3, 4]. Also, piezoelectric relays are generally preferred over their electrostatic counterparts because of the lower actuation voltage, faster switching, and smaller footprint [3]. The piezoelectric NEMS switches demonstrated to date have relied on transferring in-plane strain to a transverse moment by offsetting the neutral axis of a beam with a multilayer stack [1, 3, 5]. Despite the promising demonstrations, such geometries suffer from residual stresses, which make miniaturization harder as it is challenging to maintain a nanoscale switching gaps [1].

We addressed this challenge by introducing a clamped-clamped beam geometry that is formed by a single material and locating the actuators at the anchors (outside the moving structure). This approach also positively exploits residual stress control in a single film to pre-load the beam and tune the actuation voltage.

SWITCHING AND TUNING CONCEPTS

Relay concept and operation

The proposed relay has four terminals (Fig. 1i): source, drain, gate, and body. It consists mainly of a suspended bridge made of AlN which is attached on both sides to anchored piezoelectric actuators. These actuators are composed of a piezoelectric AlN film sandwiched between two metal electrodes, the gate and body, which are used to apply an electric field across the piezoelectric layer. The bridge carries a metal tip at its central section to form the conductive channel between the source and drain, which are located transversely with respect to the bridge.

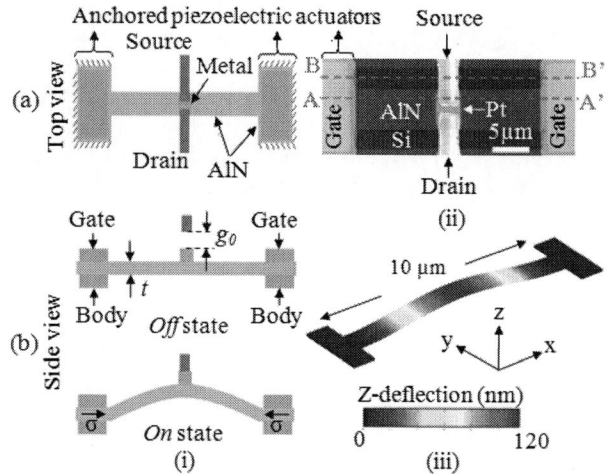

Figure 1: (i) Cartoon schematic describing the concept of switching: (a) top view, (b) side view. (ii) SEM image of a fabricated relay based on 100 nm thick piezoelectric AlN film. (iii) COMSOL simulation for 100 nm thick AlN film buckled under 200 MPa compressive residual stress and 1 V bias.

If the longitudinal stress of the beam is slightly compressive or tensile, the resulting bridge central z-deflection, d, is very small and the switch remains in the *off* state (Fig. 1i). Once the stress exceeds the critical value given by

$$\sigma_c = \frac{E\pi^2}{3}\left(\frac{t}{L}\right)^2 \qquad (1)$$

where E, t, and L are the elastic modulus, thickness, and length of the beam, respectively, the bridge buckles and the deflection increases remarkably. If d is larger than the switching gap, g_0, the metal tip connects the source and drain and switching occurs. The deflection, d, is given by [6]

$$d = \frac{2L}{\pi}\left(\frac{\sigma}{E} - \frac{\pi^2}{3}\left(\frac{t}{L}\right)^2\right)^{1/2} \qquad (2)$$

where σ is the net stress applied to the beam, and is given by

$$\sigma = \sigma_{residual} + \sigma_{piezo} \qquad (3)$$

where $\sigma_{residual}$ is the residual stress resulting from the fabrication process, and $\sigma_{piezo} = e_{31}\frac{V}{t}$ is the stress generated by the anchored piezoelectric actuators, (e_{31} is the transverse piezoelectric coefficient, and V is the applied potential across the piezoelectric film between the gate and body). In view of this analysis, the stress generated from the piezoelectric

978-1-4799-3510-9/14 $31.00 © 2014 IEEE

actuators, which is dependent on V, can be employed to pre-load the bridge and therefore control its buckling and the switching event.

Tuning of switching voltage

The relay design was conducted analytically (Eq. 1-3) and via FEA where COMSOL 3D coupled piezo-solid simulations were performed to validate the analytical model. The beam displacement and the switch actuation voltage were computed as a function of the residual stress, switching gap height, g_0, and beam geometry (Fig. 2, 3).

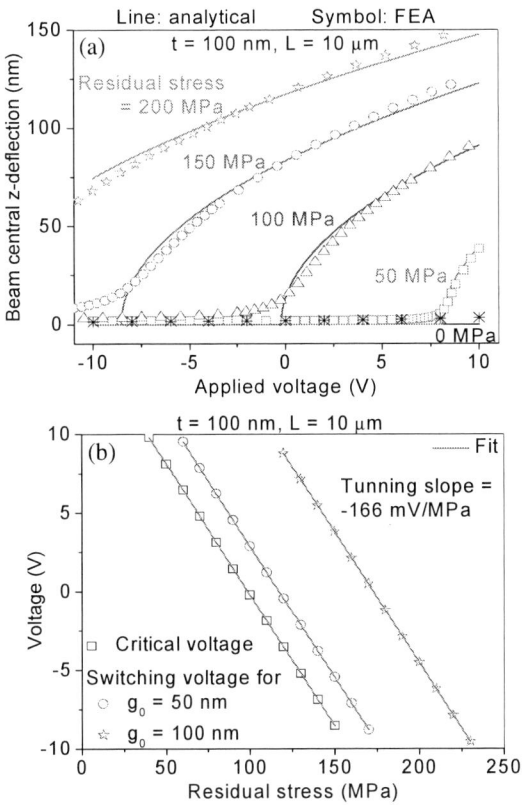

Figure 2: Tuning the actuation voltage via residual stress: (a) FEA and analytical calculation of beam deflection under different voltage and stress levels; (b) the critical and switching voltages are tuned linearly with the residual stress.

Analytical an FEA simulation for the beam central deflection under different applied voltages for a series of residual stress values is plotted in Fig. 2a. The agreement between the results of both methods is obvious, while the relatively small deviation of FEA data may be attributed to the beam width effect, which is not considered in the analytical model [6]. It is evident from the figure that the voltage required to yield a specific deflection, hence the switching voltage, can be tuned by modifying the residual stress, which is a key advantage of the proposed buckling-based design. Moreover, Fig. 2b highlights that the switching voltage is tuned linearly with the film residual stress for different switching gaps, g_0. The tuning slope is very small (-166 mV/MPa), and is independent of g_0. This is a key finding considering the

recently reported data on the precise control of the intrinsic stress in thin AlN films [7]. This level of control enables accurate tuning of the actuation voltage for the proposed buckling-based relays. The figure also shows that the critical voltage (the voltage at which buckling occurs) is linearly dependent on the stress. Furthermore, the proposed switch can be operated as a 'normally open' or 'normally closed' device by modifying the residual stress. In the 'normally closed' operation, the deflection due to residual stress is typically larger than g_0, and the tensile stress generated by the piezoelectric actuators can pull the bridge far from the contact area, switching the relay to the *off* state (see Fig. 1i).

Another important design parameter for the proposed buckling-based relays is the length of the suspended bridge, L. Based on Eq. 1 and 2, the critical load of the bridge as well as its central deflection can be controlled by modifying L (or more precisely the t/L ratio). As a result, for a given technology node (t, g_0, and residual stress), the switching voltage can be tuned by controlling L as plotted in Fig. 3. The figure also shows that both 'normally open' and 'normally closed' relays can be realized simultaneously for a given technology node, just by changing L.

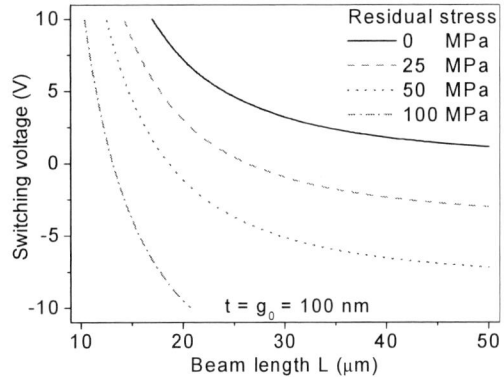

Figure 3: Tuning the actuation voltage via the beam length.

SCALING ANALYSIS

The miniaturization of NEMS relays for digital logic applications is one of the most important challenges to overcome. Reduced device footprint will come with lower energy consumption and faster switching [3, 8]. For the proposed piezoelectric relays, the key design parameters which affect the scaling of the technology are the piezoelectric film thickness, t, and the switching gap height, g_0. Fig. 4a presents the switching voltage calculated for different technology nodes, each identified by a specific t and g_0. Note that the residual stress is also a major specification for each node, but for simplicity we kept it fixed at 50 MPa in this scaling analysis. The results highlight that millivolt switching at aggressively reduced relay footprint (smaller L) is achievable if t and g_0 are appropriately scaled. For example, for the technology node $t = g_0 = 1$ nm, the switching voltage can be reduced to 10 mV for $L = 175$ nm. Considering the recently reported data on the synthesis of ultra-thin (down to 10 nm) AlN piezoelectric films with very high quality [1], and with future progress in thin film deposition techniques, the demonstration of millivolt switching using this technology can be envisioned. Finally, since the generated stress and so the switching voltage is independent of the area of the anchored piezoelectric actuators, the actuators footprint can be drastically scaled to reduce the gate-body capacitance and consequently minimize the switching energy.

The critical and switching voltages computed as a function of the residual stress for different technology nodes are presented in Fig. 4b. It is evident from the figure that the voltage vs. stress tuning slope is reduced linearly with scaling the device (t and g_0). This is another important feature for the proposed switching technology which enables a more precise control of the effect of residual stress on switching voltage with scaling. For the technology node $t = g_0 = 1$ nm, the tuning slope is significantly reduced to -1.6 mV/MPa, which further confirms the possibility of achieving millivolt switching using the proposed buckling-based switching technology.

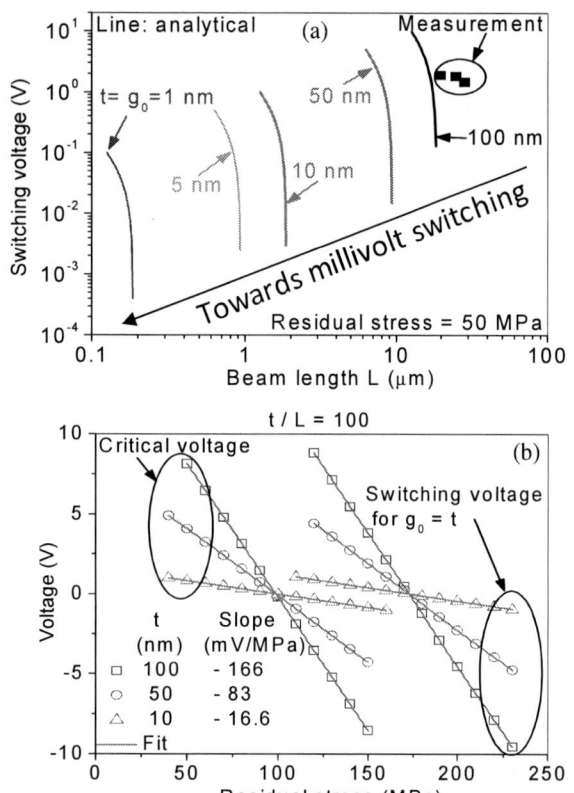

Figure 4: Scaling of piezoelectric buckling-based NEMS relays.

MICROFABRICATION PROCESS FLOW

The NEMS relays fabricated in this work are based on a 100 nm thick piezoelectric AlN film, while the formed switching gap height, g_0, is around 70 nm. The switches were fabricated using seven photolithography masks as explained in Fig. 5i. First, well-textured ultra-thin Pt layers (8 nm) were synthesized over adhesive Ti films (2 nm), both using dc sputtering on (100) high resistivity Si substrate. The Ti/Pt layers were patterned by lift-off to define the bottom electrode. Then, a 100 nm thick highly c-axis oriented piezoelectric AlN film with low compressive stress (~10 MPa) was deposited by reactive sputtering over the Pt layer using the two developed recipes reported in [1]. The full width at half maximum of

x-ray diffraction rocking curve measurements for the sputtered AlN films was 2.4°. Next, vias were opened in the AlN film using hot phosphoric acid at 150 °C to access the bottom electrode. This is followed by patterning the top Pt layer using lift-off, which defines the actuator's top electrode and the relay conductive channel (Fig. 5i-a). The lateral dimensions of the bridge were then defined by dry etching of the AlN film using $Cl_2/BCl_3/Ar$ chemistry (Fig. 5i-b). Next, a sacrificial layer of amorphous Si (~70 nm) was sputtered and patterned by lift-off to define the switching gap, followed by patterning the source and drain from Pt by lift-off. Then, a photoresist mask was used to define small release openings around the bridge (Fig. 5i-c). Next, XeF_2 was used to etch the sacrificial layer and the Si substrate to release the bridge. Finally, the photoresist was removed using plasma ashing and descumming (Fig. 5i-d).

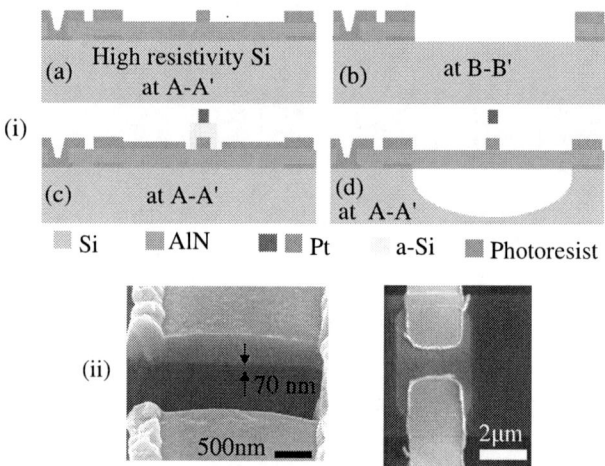

Figure 5: (i) Fabrication process flow described at the cross sections A-A' and B-B' shown in Fig 1-ii. (ii) SEM images showing the nm gap (left) and the contact area after the beam buckling (right).

EXPERIMENTAL RESULTS

The I-V characteristics for a finished NEMS relay based on a 100 nm thick piezoelectric AlN film and a bridge length of 25 μm is presented in Fig. 6. The measurements were performed while the body terminal was biased between -1.3 and -2.3 V. The actuation voltage is reduced to ~1.8 V compared to 3.6 V with no residual stress, clearly showing that the control of residual stress can be used to pre-load (bias) the actuator mechanically. Moreover, the measured actuation voltage is close to the predicted value from the analytical solution at a compressive residual stress of 10 MPa (~2.23 V). Here it is worth mentioning that a comparable actuation voltage has been recently reported for the triple-beam NEMS relays which use a stress compensated geometry, thinner piezoelectric AlN film (25 nm), and much smaller switching gap (28 nm) [1]. This emphasizes the power and simplicity of the proposed buckling-based relays in achieving the switching function at much smaller actuation voltages compared to the NEMS relays that use conventional piezoelectric actuators based on beam bending techniques. Also, the contact resistance for this specific switch is around 8 kΩ, while the *off* state current is extremely low (the noise level of the instrument).

Fig. 6 shows that the threshold voltage, V_{th}, can be further controlled and lowered by applying an electrical body bias, V_B. The average measured V_{th} under different V_B is plotted in Fig. 6b, and highlights the linearity of the actuation mechanism. Although the standard deviation of V_{th} is relatively large (~40 mV), it can be further reduced by scaling t. Also, the little changes in the net actuation voltage, V_G-V_B, for different V_B emphasizes the stability and repeatability of the switching voltage. Finally, hysteresis in the switching process is evident (Fig. 7) and limits the minimum swing voltage to ~250 mV. This can be attributed to the beam buckling hysteresis and/or the influence of the surface forces between the contact pairs.

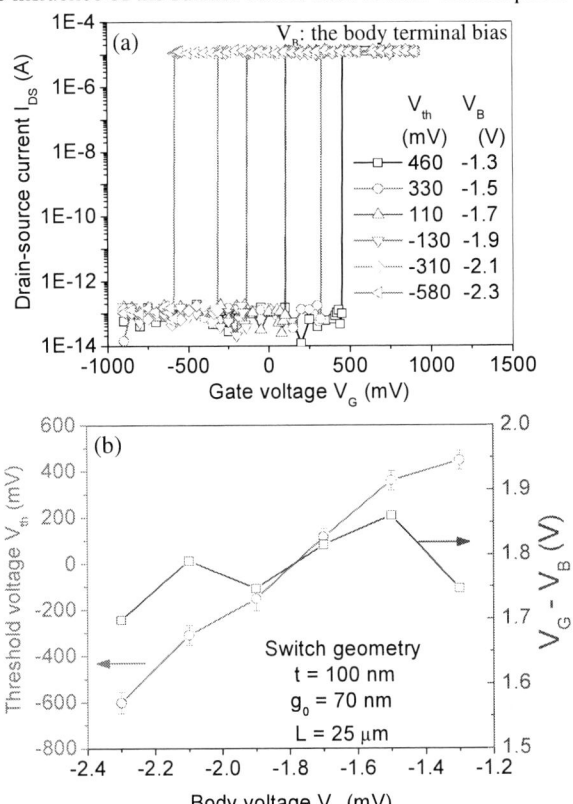

Figure 6: The I-V characteristics for a buckling-based relay.

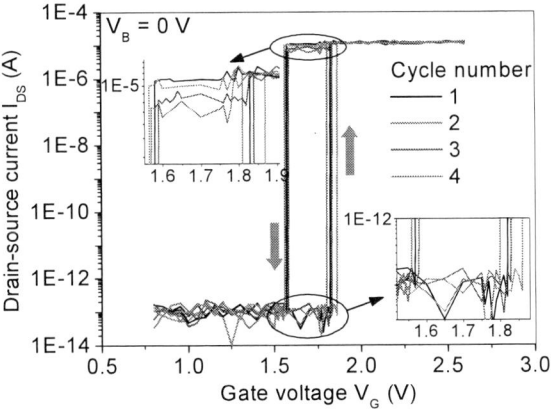

Figure 7: The I-V measurements for four consecutive actuation cycles.

CONCLUSION

The demonstrated buckled-beam relay provides a paradigm shift in the implementation of NEMS switches for ultra-low power computing and memory applications. By introducing a clamped-clamped beam geometry that is formed by a single material (AlN) and employing anchored piezoelectric actuators, the difficulties associated with controlling the stress in the film stacks of conventional piezoelectric actuators is eliminated. Also, the stress control in the beam material is exploited to pre-load it and tune/reduce the actuation voltage precisely (experimental data show a 50% reduction in the actuation voltage at 10 MPa residual stress). Also, the switching voltage is independent of the actuator area, and therefore the relay footprint can be aggressively scaled to minimize energy consumption. Finally, the scaling analysis and the experimental results highlight that millivolt switching at aggressively scaled footprints is feasible thanks to the higher scalability of the proposed relays compared to conventional piezoelectric switches.

ACKNOWLEDGEMENTS

This work was supported by the DARPA NEMS program and the NSF DMREF project (award # CMMI-1334241 and CMMI-1334572). We acknowledge the help from the staff of the nanofabrication facility at Carnegie Mellon University.

REFERENCES

[1] U. Zaghloul, G. Piazza, "0–25 nm piezoelectric nano-actuators and NEMS switches for millivolt computational logic", in *Digest Tech. Papers MEMS'13 Conference*, Taipei, January 20-24, 2013, pp. 233-236.

[2] T. Lee, S. Bhunia, M. Mehregany, "Electromechanical computing at 500°C with silicon carbide", *Science*, vol. 329, pp. 1316-1318, 2010.

[3] N. Sinha, T. Jones, Z. Guo, G. Piazza, "Body-biased complementary logic implemented using AlN piezoelectric MEMS switches", *J. Microelectromech. Syst.*, vol. 21, pp. 484–496, 2012.

[4] W. Kwon, J. Jeon, L. Hutin, and T.-J. K. Liu, "Electromechanical diode cell for cross-point nonvolatile memory arrays," *IEEE Electron Device Letters*, vol. 33, pp. 131-133, 2012.

[5] R. Proie, R. Polcawich, J. Pulskamp, T. Ivanov, M. Zaghloul, "Development of a PZT MEMS switch architecture for low-power digital applications", *J. of Microelectromech. Sys.*, vol. 20, pp. 1032-1042, 2011.

[6] B. Halg, "On a micro-electro-mechanical nonvolatile memory cell", *IEEE Trans. on Electron Devices*, vol. 37, pp. 2230-2236, 1990.

[7] V. Felmetsger, P. Laptev, R. Graham, "Deposition of ultrathin AlN films for high frequency electroacoustic devices", *J. Vac. Sci. Technol. A*, vol. 29, Art. # 021014, 2011.

[8] C. Pawashe, K. Lin, K. Kuhn, "Scaling limits of electrostatic nanorelays", *IEEE Trans. on Electron Devices*, vol. 60, pp. 2936-2942, 2013.

CONTACT

*Gianluca Piazza: piazza@ece.cmu.edu
Usama Zaghloul:uzheiba@andrew.cmu.edu

NANOELECTROMECHANICAL TUNNELING SWITCHES BASED ON SELF-ASSEMBLED MOLECULAR LAYERS

Farnaz Niroui[1], Parag B. Deotare[1], Ellen M. Sletten[1], Annie I. Wang[1],
Eli Yablonovitch[2], Timothy M. Swager[1], Jeffrey H. Lang[1], and Vladimir Bulović[1]
[1]Massachusetts Institute of Technology, Cambridge, Massachusetts, USA
[2]University of California at Berkeley, Berkeley, California, USA

ABSTRACT

We propose nanoelectromechanical (NEM) switches that operate via electromechanical modulation of tunneling current through several-nanometer-thick switching gaps. In such a device, direct contact between electrodes is avoided by utilizing self-assembled molecular layers to define the switching gap. Electrostatic compression of the molecular layer reduces the tunneling gap leading to an exponential increase in the tunneling current, turning on the switch. With removal of an applied voltage, the compressed layer provides the elastic restoring force necessary to overcome the surface adhesive forces, turning off the switch. Thus, the proposed tunneling NEM switch may enable low-voltage operation while simultaneously mitigating device failure due to stiction. This principle is experimentally investigated using a prototype two-terminal tunneling NEM switch with a switching gap formed by a fluorinated decanethiol layer. In this device, the presence of the molecular film promotes repeatable switching. A comparison of the switch operation with a theoretical model indicates electrostatic compression of the molecular switching gap.

INTRODUCTION

As complementary metal oxide semiconductor (CMOS) transistors continue scaling down in size, the accompanying increase in power dissipation [1] drives the need for alternative switching technologies. Nanoelectromechanical (NEM) switches have emerged as promising candidates [2]. They exhibit abrupt switching with large on-off current ratios, near-infinite sub-threshold slopes, and near-zero off-state leakage currents thereby enabling power-efficient operation. However, NEM switches require large actuation voltages and commonly experience irreversible adhesion between their electrodes, referred to as stiction. Reducing the thickness of the switching gap is an effective approach to decrease the operating voltage [3] but it increases the surface adhesive forces which further promotes stiction. Thus, these issues must be simultaneously addressed to enable extensive integration of NEM switches in low-power applications.

To this end, we propose a switching mechanism based on the electromechanical modulation of tunneling currents through nanoscale switching gaps. Contrary to conventional NEM switches, direct contact and the accompanying stiction between electrodes is avoided by introducing a self-assembled molecular layer between the electrodes. The molecular layer enables the definition of nanometer-thick switching gaps and provides nanoscale force control imposed by its deformation during the switching process. Together, these effects facilitate the formation of a low-voltage, low-stiction tunneling NEM switch, experimentally investigated as a two-terminal device in this paper.

OPERATION AND DESIGN PRINCIPLES

In its simplest form, the proposed tunneling NEM switch comprises a nanometer-thick film of molecules sandwiched between two conducting electrodes, as seen in Figure 1. The width of the switching gap is defined by the thickness of the molecular film. To enable control of the thickness with nanometer precision, self-assembled molecules are utilized. Molecules are selected with functional end groups that enable selective adsorption onto the electrode surfaces. Through self-assembly, they then form a uniform film with a thickness dependent on the length of the molecules. Molecular layers thus facilitate the formation of nanometer switching gaps which are otherwise difficult to achieve in conventional NEM switches. This offers a pathway to low-voltage switching operation.

During the switching process a voltage is applied across the metal-molecule-metal structure. The induced electrostatic force reduces the switching gap between the electrodes by compressing the molecular layer, leading to an exponential increase in the tunneling current to turn on the switch. As the voltage is removed, the elastic force of the deformed molecules provides the restoring force necessary to overcome the surface adhesive forces to turn off the switch. This operating principle is shown in Figure 1.

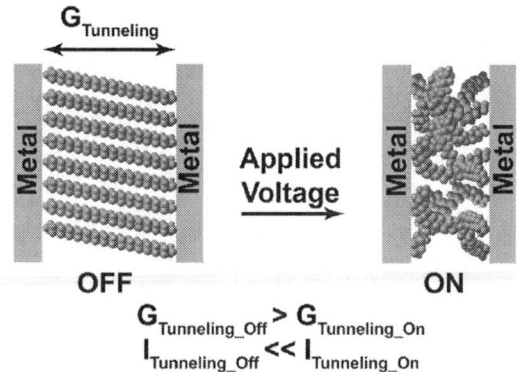

Figure 1: Operating principle of the tunneling NEM switch. An applied voltage results in compression of the molecular layer, reducing the tunneling gap ($G_{Tunneling}$) and exponentially increasing the tunneling current ($I_{Tunneling}$).

FABRICATION

A prototype switch is fabricated in the form of a laterally actuated cantilever as shown in Figure 2. Five layers of poly(methyl-methacrylate) (PMMA), a positive electron beam resist, are spun over a silicon (Si) substrate with 2 µm-thick thermal oxide (SiO_2). Each layer is spun at 2000 rpm for 45 s and baked at 180 °C for 90 s leading to a total thickness of about 1.5 µm. The initial three layers are PMMA with molecular weight of 495 kg/mol (PMMA 495 A6) followed by two layers of 950 kg/mol PMMA (PMMA 950 A4). Device features are then defined by patterning the PMMA film using electron-beam lithography. The resist is developed in 1:3 dilution of methyl isobutyl ketone (MIBK) in isopropanol for 3 min, thoroughly rinsed in isopropanol and dried under nitrogen. Then, 100 nm of gold (Au) is deposited over the substrate using thermal evaporation to form the electrodes.

The lower molecular weight PMMA has a faster dissolution rate in MIBK than the higher molecular weight PMMA. After developing, this leads to a PMMA profile with an undercut, the bottom section being thinner than the top. This prevents sidewall coverage during Au evaporation and ensures electrical isolation between the switch components. This undercut profile along with the high aspect ratio of the PMMA support allows the structure to freely deflect upon application of a force between Electrodes 1 and 2 as defined in Figure 2. The device fabrication is completed by self-assembly of fluorinated decanethiol molecules in vapor phase onto the Au surfaces through the use of thiol chemistry. The molecular film forms over the entire substrate including the switching gap. Results indicate that through the assembly process Electrodes 1 and 2 collapse onto each other with the molecules sandwiched between the electrodes forming a switching gap narrower than originally patterned leading to a metal-molecule-metal junction with a movable electrode. Figure 3 shows a scanning electron micrograph of a fabricated switch prior to deposition of the fluorinated decanethiol film.

RESULTS AND DISCUSSION

The current-voltage (I-V) characteristic of a fabricated switch with an off-state switching gap of about 15 nm in the absence of a molecular layer is shown in Figure 4a. The initial actuation is successful; however, this operation is not subsequently repeatable due to stiction. In this state, the surface adhesive forces overcome the elastic restoring force of the active electrode causing permanent adhesion between Electrodes 1 and 2. A second device of the same structure tested after self-assembly of fluorinated decanethiol molecules exhibits repeatable operation with a current-voltage characteristic resembling that of quantum tunneling as shown in Figure 4b. The roughness of the Au upon evaporation may contribute to the variation in the results over the different measurement runs. The reduced stiction can be attributed to the reduction in surface energy, the introduction of an elastic restoring force provided by the

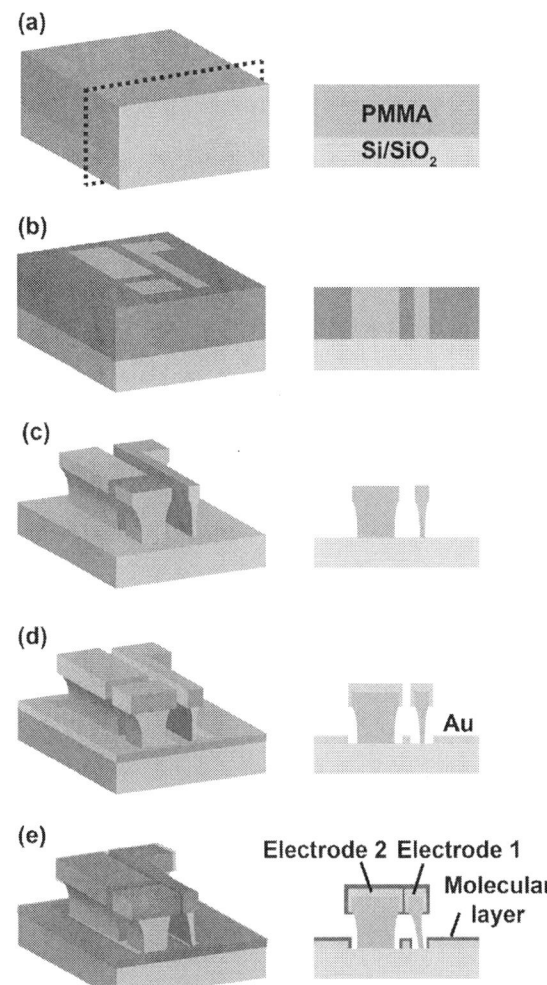

Figure 2: Fabrication process of the tunneling NEM switch; (a) multilayer PMMA electron-beam resist is spun on a Si/SiO_2 substrate and baked after each spin, (b) resist is electron-beam-patterned to define the switch components, (c) resist is developed in 1:3 solution of MIBK in iso-propanol, (d) 100 nm of Au is evaporated over the substrate, (e) fluorinated decanethiol is self-assembled in vapor phase over Au, bridging the gap between Electrodes 1 and 2.

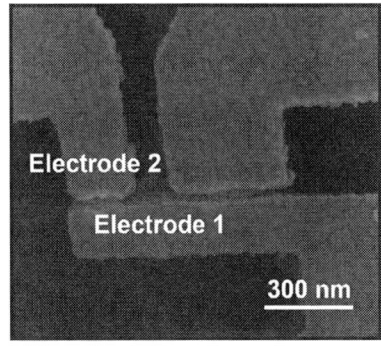

Figure 3: Scanning electron micrograph of a tunneling NEM switch prior to the self-assembly of fluorinated decanethiol.

978-1-4799-3510-9/14 $31.00 © 2014 IEEE   1104

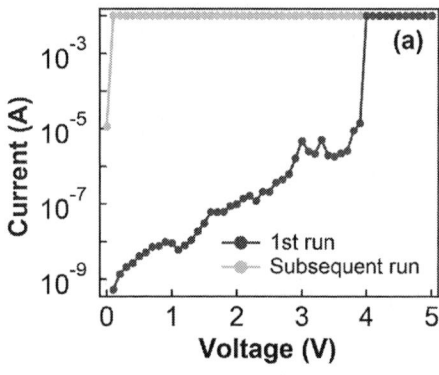

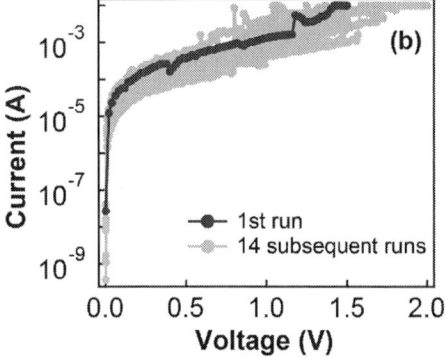

Figure 4: (a) I-V characteristics of a tunneling NEM switch in the absence of fluorinated decanethiol molecular layer showing device failure due to stiction after the first run; (b) I-V characteristics of the switch after self-assembly of fluorinated decanethiol showing repeatable operation.

molecular layer, and the lack of direct contact between the electrodes.

The conduction mechanism through self-assembled monolayers of organic molecules has been previously found to be governed by direct tunneling and explained based on the Simmons model for $V < \Phi/q$ [4-6]:

$$ I = \left(\frac{qA}{4\pi^2\hbar G^2} \right) \left\{ \left(\Phi - \frac{qV}{2} \right) \exp\left[-\frac{2(2m)^{1/2}}{\hbar} \alpha \left(\Phi - \frac{qV}{2} \right)^{1/2} G \right] \right. $$
$$ \left. - \left(\Phi + \frac{qV}{2} \right) \exp\left[-\frac{2(2m)^{1/2}}{\hbar} \alpha \left(\Phi + \frac{qV}{2} \right)^{1/2} G \right] \right\} \quad (1) $$

where m is the electron mass, q is the electron charge, A is the overlapping area between Electrodes 1 and 2, G is the tunneling distance, Φ is the molecular layer tunneling barrier height, and α is an adjustable parameter that accounts for the effects of barrier shape and electron effective mass.

The Simmons model is used to generate a theoretical current-voltage curve expected for the switch at a constant switching gap. To acquire a close fit to the experimental data, Φ of 1.4 eV and α of 0.6 are selected, within the range of values reported in the literature for monolayers of similar molecules [5-6]. It should be noted that the selected values of Φ and α are reasonable approximates but the exact values vary depending on the specifics of the tunneling junction. Based on the experimental data, the initial thickness of the tunneling gap is estimated to be 1.0 nm, which is consistent

with the expected thickness of a monolayer of decanethiol [7]. The switching gap of 1.0 nm deduced through the experimental results suggests collapse of Electrode 1 onto Electrode 2 through the self-assembly process to form a metal-molecule-metal structure with a gap smaller than that present prior to the molecular layer deposition. The theoretical expected current-voltage characteristic of the switch at a constant switching gap is shown in Figure 5. A close fit with the experimental data is obtained at lower voltages but a deviation is observed as the voltage increases.

Applying a voltage across the metal-molecule-metal capacitive structure induces an electrostatic force. If sufficient to overcome the elastic restoring force of the molecular layer, it can deform the film to reduce the gap between Electrodes 1 and 2. The decrease in the electrode-electrode tunneling distance results in a faster increase in the tunneling current than that expected for a constant gap which can account for the observed deviation in Figure 5 between the theoretical (red) and experimental (blue) results. The increase in the tunneling current due to compression of self-assembled monolayers has been previously reported under an applied mechanical force [8-10]. In these studies, the decrease in the tunneling distance due to compression is identified as a cause of increased conduction. The feasibility of this process to occur with an applied electrostatic force within the system under study is explored by modifying the Simmons model to account for the compression of the molecular layer. To do so, the balance of electrostatic, van der Waals and spring restoring forces at equilibrium are used to determine the electrode-electrode distance at each applied voltage. This is implemented in the Simmons model to yield the corresponding tunneling current. The force balance equation is given by

$$ M\frac{d^2z}{dt^2} = \frac{\varepsilon_r\varepsilon_0 AV^2}{2(G_0 - z)^2} + \frac{A_H A}{6\pi(G_0 - z)^3} - k(L - G_0 + z) \quad (2) $$

where M is the mass of Electrode 1, G_0 is the initial distance between the electrodes, z is the displacement of Electrode 1, L is the initial thickness of the molecular layer, A is the overlapping area between the electrodes, A_H is the Hamaker constant, ε_0 is the permittivity of free space, ε_r is the dielectric constant of the molecular layer, and k is the spring constant of the layer related to the Young's modulus Y by $k = YA/L$. The relative permittivity of the molecular layer is assumed to be 2.0, within the range of values reported for alkanethiols [11]. The Hamaker constant is considered to be 3×10^{-19} J based on the values reported for capacitive structures with Au electrodes [12]. Considering the structural profile of Electrode 1, its spring constant is estimated to be orders of magnitude smaller than that expected for the molecular layer; thus, its contribution to the force balance equation is neglected.

The simulated current-voltage characteristic using the modified model with an electromechanical decrease in the tunneling gap is shown in Figure 5. Compared to the simulated result with a constant switching gap, the modified

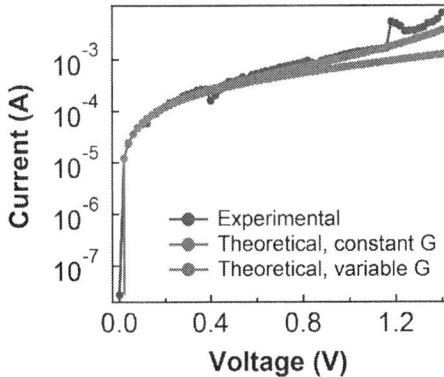

Figure 5: Experimental I-V characteristics of the tunneling NEM switch (blue) compared to the theoretical characteristics based on the Simmons model of tunneling with a constant gap (red) and a variable gap (green).

model yields a closer fit between the experiment and the theory, corresponding to about 26% decrease in the tunneling distance. The theoretical model suggests a Young's modulus of approximately 0.23 GPa for fluorinated decanethiol, in agreement with the measurements reported in the literature for alkanethiols of similar molecular structure [7]. The results imply electrostatically-induced compression of the fluorinated decanethiol layer modulating the tunneling current through the metal-molecule-metal structure. This compression leads to about a 10-fold increase in the current. Considering the exponential dependence of the tunneling current on the tunneling distance, the on-off current ratio of the NEM switch can be optimized by using a thicker molecular switching gap. The mechanical properties of the molecular layer can also be selected to maximize the extent of tunneling gap modulation feasible within the desired applied force considering that a lower Young's modulus enables a lower actuation voltage.

CONCLUSION

Electromechanical modulation of tunneling current through self-assembled molecular layers is proposed as a switching mechanism for low-voltage and low-stiction NEM switches. The use of the molecular layer facilitates formation of nanoscale switching gaps to reduce the actuation voltage. Avoiding direct contact between the electrodes and introducing the restoring force from the deformed molecular layer help to overcome stiction and enhance repeatable switching. The experimental results support the feasibility of the proposed tunneling NEM switch. Further investigation of the off-state tunneling distances and mechanical properties of the molecular film to optimize the metal-molecule-metal junction are needed to develop tunneling NEM switches with optimum performance.

ACKNOWLEDGEMENTS

This work is supported by the National Science Foundation (NSF) Center for Energy Efficient Electronics Science (E³S) Award ECCS-0939514.

REFERENCES

[1] D. J. Frank, R. H. Dennard, E. Nowak, P. M. Solomon, Y. Taur, H. P. Wong, "Device scaling limits of Si MOSFETs and their application dependencies", *Proc. of the IEEE*, vol. 89, no. 3, pp. 259-288, 2001.

[2] O. Y. Loh, H. D. Espinosa, "Nanoelectromechanical contact switches", *Nature Nanotech.*, vol. 7, pp. 283-295, 2012.

[3] J. O. Lee, Y. Song, M. Kim, M. Kang, J. Oh, H. Yang, J. Yoon, "A sub-1-volt nanoelectromechanical switching device," *Nature Nanotech.*, vol. 8, pp. 36-40, 2013.

[4] J. G. Simmons, "Generalized formula for the electric tunnel effect between similar electrodes separated by a thin insulating film," *J. Appl. Phys.*, vol. 34, pp. 1793-1803, 1963.

[5] R. E. Holmlin, R. Haag, M. L. Chabinyc, R. F. Ismagilov, A. E. Cohen, A. Terfort, M. A. Rampi, G. M. Whitesides, "Electron transport through thin organic films in metal-insulator-metal junctions based on self-assembled monolayers," *J. Am. Chem. Soc.*, vol. 123, pp. 5075-5085, 2001.

[6] W. Wang, T. Lee, M. A. Reed, "Electronic transport in self-assembled alkanethiol monolayers," *Physica E.*, vol. 19, pp. 117-125, 2003.

[7] F. W. DelRio, C. Jaye, D. A. Fischer, R. F. Cook, "Elastic and adhesive properties of alkanethiol self-assembled monolayers on gold," *Appl. Phys. Lett.*, vol. 94, pp. 131909, 2009.

[8] V. B. Engelkes, C. D. Frisbie, "Simultaneous nanoindentation and electron tunneling through alkanethiol self-Assembled monolayers," *J. Phys. Chem. B*, vol. 110, pp. 10011-10020, 2006.

[9] J. Zhao, K. Uosaki, "Electron transfer through organic monolayers directly bonded to silicon probed by current sensing atomic force microscopy; Effect of chain length and applied force," *J. Phys. Chem. B*, vol. 108, pp. 17129-17135, 2004.

[10] H. Song, H. Lee, T. Lee, "Intermolecular chain-to-chain tunneling in metal-alkanethiol-metal junctions," *J. Am. Chem. Soc.*, vol. 129, pp. 3806-3807, 2007.

[11] H. B. Akkerman, R. C. G. Naber, B. Jongbloed, P. A. van Hal, P. W. M. Blom, D. M. de Leeuw, B. de Boer, "Electron tunneling through alkanedithiol self-assembled monolayers in large-area molecular junctions", *Proc. Natl. Acad. Sci. U.S.A.*, vol. 104, pp. 11161-11166, 2007.

[12] G. Palasantzas, P. J. van Zwol, J. Th. M. De Hosson, "Transition from Casimir to van der Waals force between macroscopic bodies," *Appl. Phys. Lett.*, vol. 93, pp. 121912, 2008.

CONTACT

*F. Niroui, tel: +1-617-3248110; fniroui@mit.edu

978-1-4799-3510-9/14 $31.00 © 2014 IEEE

TRANSITION OF Q-DOT DISTRIBUTION ON MICROTUBULE ARRAY ENCLOSED BY PDMS SEALING FOR AXONAL TRANSPORT MODEL

Kazuya Fujimoto[1], Hirofumi Shintaku[1], Hidetoshi Kotera[1] and Ryuji Yokokawa[1,2]
[1]Kyoto University, Kyoto, JAPAN
[2]PRESTO, JST, JAPAN

ABSTRACT

In this paper, we show a reconstruction of kinesin driven transport system in three dimensionally (3D) enclosed channels whose scale is similar to axons. Our experimental method enabled successful motility of a large number of kinesin-labeled Q-dots on microtubules (MTs) in enclosed channels. To control the direction of kinesin motility, we prepared a polarity-defined MT array in channels. Due to the directional motility of kinesin, time evolutional accumulation of transported Q-dot at one end of enclosed channels, where corresponds to microtubule (MT) plus end, was observed. This is the first step for an *in vitro* model of motor protein-based active transport with a 3D spatial confinement mimicking intracellular environment, which is applicable to analyze a regulation mechanism of intracellular transport.

INTRODUCTION

Kinesin and dynein are famous and important categories of motor proteins cooperating with MTs, fibrous polymers of tubulin dimers. Active transport in neuron (axonal transport) is a well-studied intracellular transport system driven by kinesin and dynein. Axon is a long and slender projection connecting neurons each other over the distance beyond centimeter scale. MTs are organized in an axon with their minus end on cell body and plus end toward the periphery of the axon, and serving as a rail of bidirectional motility of kinesin and dynein. Various intracellular materials such as proteins, RNAs or organelles are transported by kinesin and dynein. Because most of necessary proteins in axon and synapse need to be transported from cell nuclei, axonal transport is essential for nerve cell growth [1].

Recently, numbers of researchers have found a relationship between particular neurodegenerative diseases and defects of axonal transport [2]. Although many features of transport have been revealed such as relationship between tau protein and mitochondria transport [3], mechanisms of axonal transport is still difficult to fully understand. One reason of the difficulty is its complicated interaction between various factors such as motor proteins, linking proteins with vesicle surface or MT associated proteins. By reconstructing such a system in an artificial environment, with simplified factors, quantitative analyses of axonal transport become expectable. Detailed knowledge about regulation mechanism of the transport will help us to understand neurodegenerative diseases and to develop medical treatment.

In the past decades, *in vitro* utilization of motor protein system has been reported, with aims such as analysis of their mechanical property or application as nanoactuators [4]. Many techniques for controlling MT positions or moving directions have been reported by surface-machined micro pattern on substrates [5]. In this paper, we established a method to reconstruct a transport of kinesin coated Q-dots in 3D enclosed microchannels. In experiments, because kinesin in smaller volume has larger chance to bind to MTs, the ratio of transported Q-dots against diffusing Q-dots in solution is much higher in enclosed micro channel than in conventional assay. As a result, our methods realized an observation of directional transition of Q-dot distribution.

EXPERIMENTS

Device fabrication and flow cell set up

Microtrack was fabricated by photo lithography (Fig. 2a) on a glass surface. Fused silica glass substrate (Corning 7980, Corning) was cleaned with Piranha solution ($H_2SO_4:H_2O_2 =$ 3:1) for 10 minutes at 120°C. Aluminum (Al) was deposited on the substrate and SU-8 2000.5 (MicroChem) photoresist was spin coated. The photoresist and Al thickness are approximately 500 nm and 150 nm, respectively. After resist coating and soft bake, the substrate was exposed to UV light with a photomask having track patterns using a mask aligner (PEM 800, Union Optical). The photoresist was successfully developed by SU-8 developer (MicroChem) for 45 s at the exposure of 40 mJ/cm^2. Al was removed in Al etchant for 5 minutes just before use.

A flow cell was constructed by bonding a PDMS chip (about 0.5 mm of thickness) and grass substrate with a double side tape. Microactuator was set on a microscope stage and was used to push a PDMS top of the flow cell (Fig. 2b).

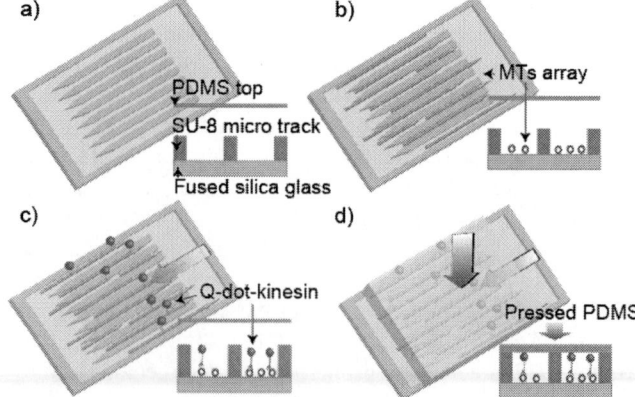

Fig. 1: Overview of experiments. a) Flow cells are constructed by bonding microtrack fabricated fused silica glass and PDMS chip. b) MTs array is prepared by utilizing gliding motility on kinesin coated surface and immobilize MT array. c) Q-dot labeled kinesin (Q-K) was introduced on MT array. d) Q-Ks are transported in microchannels after lowering PDMS top.

978-1-4799-3510-9/14 $31.00 © 2014 IEEE

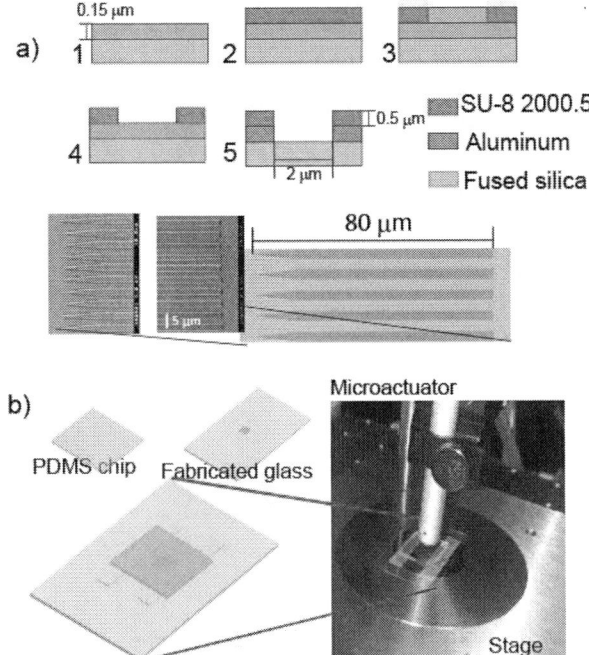

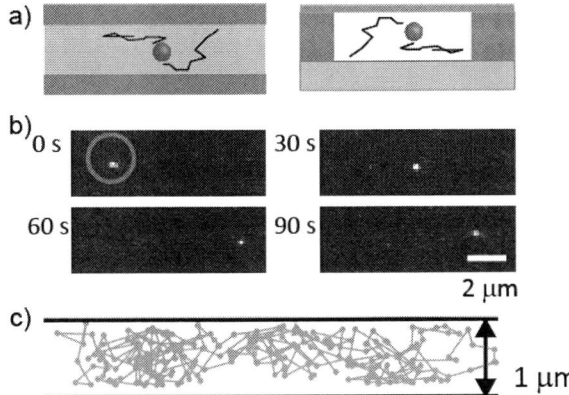

Fig.2: Microtrack fabrication process and experimental set up. a) Fabrication process and SEM images of the microtracks. b) Flow cell construction and microscope setting. Micro actuator is set on a microscope stage to lower the PDMS top.

Fig. 3: Observation of Q-dot confinement in microchannel. a) Concept of Q-dot observation for confirmation of microtrack enclosure confirmation. b) A Q-dots in a microchannel was observed over a minute. c) Trajectory shows Brownian motion of Q-dot confined in the microchannel.

Enclosure of microtracks

To confirm the enclosure of microtracks with a PDMS, we introduced 1 nM Q-dot solution into flow cell and lowered PDMS by the microactuator. Individual Q-dots were continuously observed in enclosed microtrack . Brownian motion confined in the single channel over a minute was shown from the Q-dots trajectories, indicating a successful enclosing of microtracks (Fig. 3).

Proteins preparation

His-tagged *D. melanogaster* kinesin (His-K) and biotinylated *D. melanogaster* kinesin (bio-K) were purified according to R. Yokokawa *et al.* [6]. His-K was diluted in BRB 80 (80 mM PIPES, 2 mM MgCl2, 1 mM EGTA, pH 6.8) including 0.3mg of casein. Bio-K was mixed with Q-dot 655 (Q10121MP, invitrogen) at a molar ratio of 10 : 1 and incubated at 4°C for 15 minutes to prepare kinesin coated Q-dot (Q-K).

Tubulin was purified from porcine brains by two cycles of temperature-dependent polymerizations. Some of the tubulin was labeled with the fluorescent dye TMR (Tetramethyl Rhodamine, C1711, Invitrogen). Non labeled and TMR labeled tubulins were mixed at a ratio of 10 : 1 and polymerized by 40-min incubation at 37°C within 1mM of GTP and $MgSO_4$. Before use, polymerized MTs were diluted ten times with BRB80 containing 20 mM paclitaxel (T1912, Sigma), which stabilizes MTs.

Kinesin assay in microchannel

MT array with defined polarity was prepared in a flow cell as explained above (Fig. 1a). MT solution was introduced into the His-K coated flow cell. After 5-minute incubation to immobilize MTs on the surface, the flow cell was filled with BRB 80 including 36 µg/ml catalase, 140 mM β-mercaptoethanol (β-Me) and 20 mM dithiothreitol (DTT). MTs immobilized nearby one end of microchannels were depolymerized by illuminating mercury light through a fluorescent cube (λ = 420-500 nm). Then, after rinsing, a motility solution, which is BRB 80 including (1 mM ATP, 140 mM β-Me, 20 mM DTT and enzymatic oxygen scavengers, 36 µg/ml Catalase, 216 µg/ml Glucose oxidase and 25 mM Glucose) was introduced [7]. MTs started to glide into a microtrack, from single end of tracks which are not processed by mercury lightcon and formed polarity defined array. After about 10-minute observation and rinsing, arrayed MTs in microtracks were immobilized by 0.1% glutaraldehyde (GA) solution. Excess GA was quenched with 0.1 M glycine (Fig. 1b).

The Q-K solution was diluted with motility solution into 10 nM of Q-K (Fig. 1c). In microchannels, it was concerned that reagents contained in very small volume of solutions will rapidly run out and cause fast deactivation of kinesin or depolymerization of MTs. To keep enclosed MT and kinesin activity in good condition for long time, motility solution for Q-K assay contains 10 mM ATP, 80mM DTT, 864 µg/ml GOD and 100 mM Glucose, in addition, 1 mM $MgCl_2$ and 3 mg/ml casein is also added for stable kinesin motility. Concentrations of other regents were the same as the motility solution used for MT assay. After introducing Q-K solution into a flow cell, PDMS top was carefully lowered by the microactuator (Fig. 1d). Q-dot transport by kinesin motility in microchannel was observed under a fluorescent microscope.

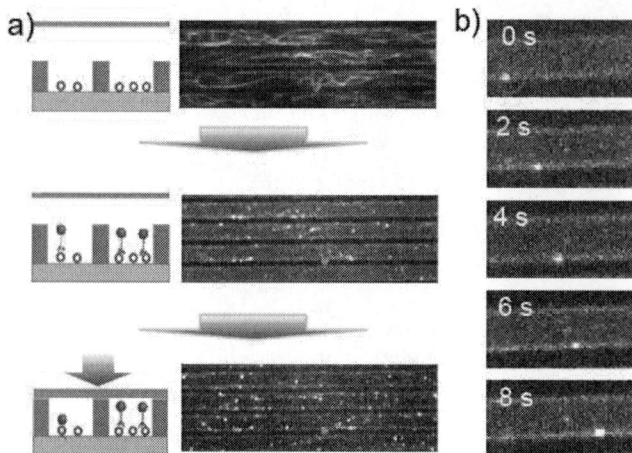

Fig. 4: MT-kinesin assay observation. a) MTs were introduced into microtracks with defined polarity. Introduced Q-K showed motility on MT array in microtrack. Even after lowering PDMS top to enclose the microtracks, Q-K motilities were observed. b) Sequential images of individual Q-K motility.

Table 1 Kinesin motility in microchannels

	Microchannels	Control
Velocity ± S.E. (µm/s)	0.26 ± 0.03	0.27 ± 0.02
Run length ± S.E. (µm)	3.7 ± 0.8	5.2 ± 0.7

N = 20

RESULTS AND DISCUSSION
Kinesin motility in enclosed channels

MT array was prepared in microtracks and Q-dot transport on the MT array was observed before and after the enclosure of the microtracks (Fig. 3a). Q-dot transport was recorded (Fig. 3b) with a CCD camera after enclosing the microtracks. Kinesin motility was evaluated in terms of velocity and run length compared with those obtained in control experiments in a flow cells without microtracks. Table 1 shows that motility was preserved in microchannels although velocity decreased ~4% and run length decreased ~30 %.

Transition of Q-dot distribution by directional transport

Large number of Q-dots was unidirectionally transported by kinesin motility toward the plus end of MTs. At the both ends of microchannel array PDMS layer deforms and the channel height gradually decreases (Fig. 5a, illustration), which creates enclosed reservoirs. Once Q-Ks reached one end of microchannels, they are accumulated at one of reservoirs (Fig. 5a), which can be also observed in enlarged sequential images (Fig. 5b).

To evaluate Q-dots accumulation quantitatively, we plot the intensity profiles of microchannels for every 150 s (Fig. 6). Intensity profiles along microchannels are calculated and plotted by summing the intensity values of pixel along orthogonal direction to the channels. The y axis is normalized intensity

and the x axis is position along microchannel shown in Fig. 5a. The origin of x axis corresponds to the left end of microchannels.

Intensity increase is quantitatively obtained at $x \sim 75$ µm, which corresponds to right end of the channel array. Here, we were able to demonstrate unidirectional transport of Q-Ks in microchannels, resulting in the Q-dot accumulation in the reservoir area. However, intensity profile along the microchannels did not clearly reflect the Q-K transport presumably because of low signal-to-noise ratio caused by nonspecifically immobilized Q-Ks or MTs.

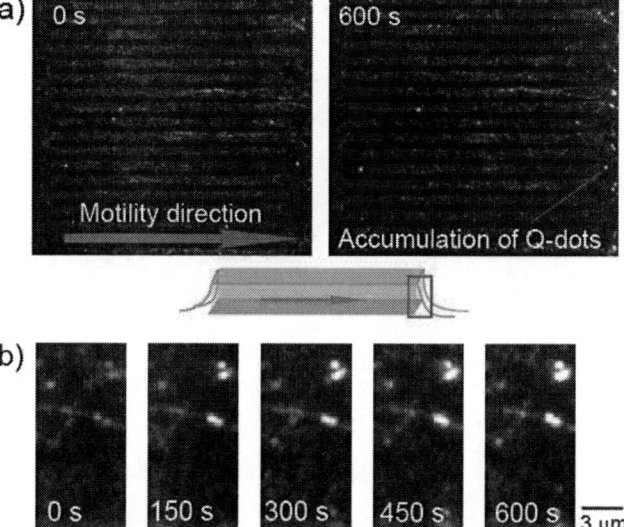

Fig. 5: Q-dot transport in microchannels. a) Overview of the microchannel array. Q-Ks were accumulated at the right reservoir after 600-s assay duration. b) Sequential images focused on the reservoir area.

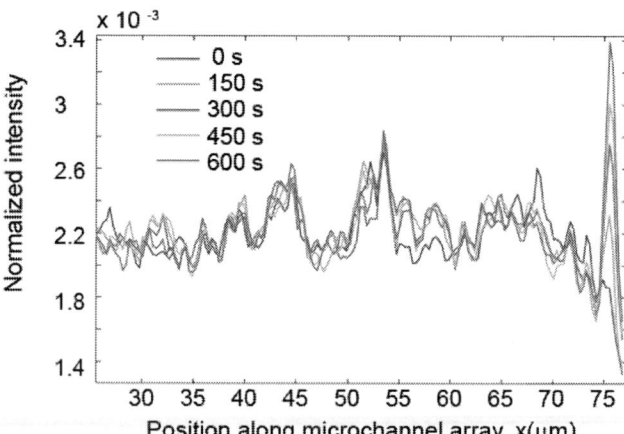

Fig. 6: Intensity profiles along the microchannel array. The overall fluorescent intensities along the channel length at different times is plotted. The fluorescent intensity at x ~ 75 µm increased over time.

Theoretical model for the transition of Q-dot distribution

To simulate active transport in microchannels, we developed a mathematical model referring a study about intracellular vesicle distribution [8]. Our model consists of two partial differential equations (PDEs) about two densities: n_0 is for Q-dots not bound to MTs (free Q-dots in a solution), and n_1 is for Q-dots that are bound to MTs. Equations represent temporal change of n_0 and n_1, which is illustrated in Fig. 7a,

$$\frac{\partial n_0}{\partial t} = D \frac{\partial^2 n_0}{\partial x^2} - k_{on} n_0 + k_{off} n_1 \qquad (1)$$

$$\frac{\partial n_1}{\partial t} = -v \frac{\partial n_1}{\partial x} + k_{on} n_0 - k_{off} n_1, \qquad (2)$$

where D is the diffusion constant for Q-dot and v is velocity. k_{on} and k_{off} are binding and dissociation constants of Q-K to MT, respectively. Time evolution of particle distribution was calculated with finite difference method, where we adopted boundary conditions: Q-dots were uniformly distributed at $t = 0$, the total number of Q-dots in a channel was constant, and Q-dots that reached to the end of the channel showed no further movement resulting in the accumulation of Q-dots. Q-dot distributions on MTs (n_1) for 0–600 s are plotted in Fig. 7b. These concentration profiles qualitatively match to our experimental results shown in Fig. 6, because Q-dots are greatly accumulated at the end of microchannels. To further compare the simulation and experimental results, time course of normalized fluorescent intensity and calculated density at $x \sim 75$ μm are plotted in Fig. 8a and Fig. 8b, respectively. Again, two plots show the similar trend of increase. From the result, possibility to simulate time course evolution in our experiment with the PDEs model is indicated.

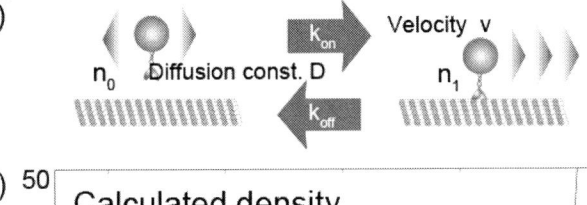

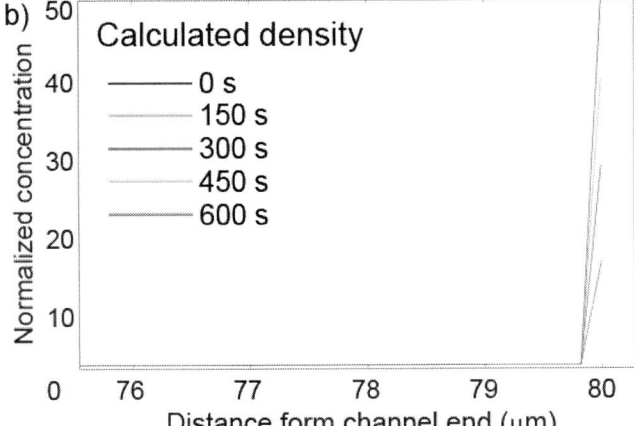

Fig.7: Numerical model of transport. a) Concept of PDEs model assuming stochastic binding/unbinding mechanism. b) Calculated time evolution of Q-dot distribution using the model.

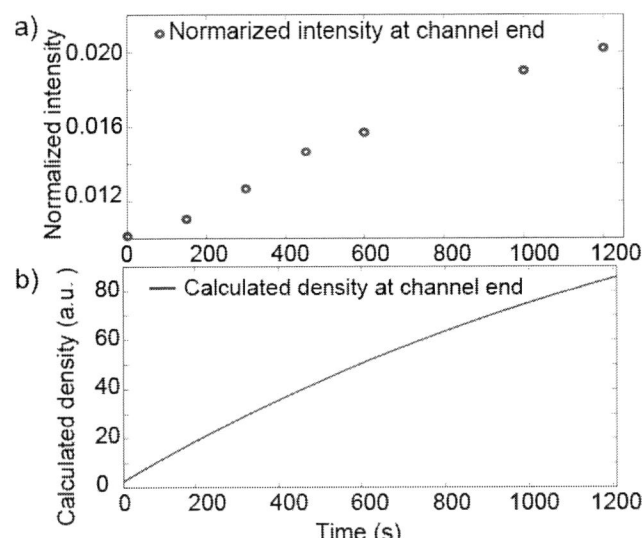

Fig. 8: Comparison between results from experiment and numerical simulation. a) Normalize intensity obtained in the experiment. b) Calculated density. Linear increase in the time course in two graphs indicates the accumulation of Q-dots at the channel end.

CONCLUSION

We proposed a reconstruction of axonal transport *in vitro* in 3D enclosed microchannels. In the experiments, active transport driven by kinesin on polarity defined MT array in microchannels was established. In addition, transition of Q-dot distribution mediated by kinesin motility was observed, which cannot be demonstrated without a confinement of kinesin and MT system in microchannels. This is the first step for understanding the axonal transport using reconstructed intracellular transport.

ACKNOWLEDGEMENTS

This work was supported by PREST, JST, JSPS KAKENHI Grant Number 25709018 to R.Y. and Grant-in-Aid for JSPS Fellows Grant Number 251836 to K.F.

REFERENCES

[1] E. Perlson, S. Maday, M.-M. Fu, A. J. Moughamian, and E. L. F. Holzbaur, *Trends Neurosci.*, vol. 33, no. 7, pp. 335–44, 2010.

[2] S. Roy, B. Zhang, V. M.-Y. Lee, and J. Q. Trojanowski, *Acta neuropathologica*, vol. 109, no. 1. pp. 5–13, 2005.

[3] K. Shahpasand, I. Uemura, T. Saito, T. Asano, K. Hata, K. Shibata, Y. Toyoshima, M. Hasegawa, and S.-I. Hisanaga, *J. Neurosci.*, vol. 32, no. 7, pp. 2430–41, 2012.

[4] M. G. L. van den Heuvel and C. Dekker, *Science*, vol. 317, no. 5836, pp. 333–6, 2007.

[5] T. Korten, A. Månsson, and S. Diez, *Curr. Opin. Biotechnol.*, vol. 21, no. 4, pp. 477–88, 2010.

[6] R. Yokokawa, T. Murakami, T. Sugie, and T. Kon, *Nanotechnology,* vol. 19, pp. 125505, Feb. 2008.

[7] M. Yokokawa, K. Fujimoto, M. Kitamura, R. Yokokawa, and H. Kotera, *Proc. MEMS 2011*, pp. 1349–1352, 2011.

[8] A. Dinh, C. Pangarkar, T. Theofanous, and S. Mitragotri, *Biophys. J.*, vol. 90, pp L67-L69, 2006.

CONTACT

*K. Fujimoto, tel: +81-075-383-3687;
Fujimoto.kazuya.78m@st.kyoto-u.ac.jp

WAFER-SCALE FABRICATION OF SCANNING THERMAL PROBES WITH INTEGRATED METAL NANOWIRE RESISTIVE ELEMENTS FOR SENSING AND HEATING

K. Hatakeyama[1], E. Sarajlic[1,2], M.H. Siekman[1], L. Jalabert[3], H. Fujita[3], N. Tas[1] and L. Abelmann[4]

[1]MESA+ Research Institute, University of Twente, THE NETHERLANDS
[2]SmartTip B.V., THE NETHERLANDS
[3]Institute of Industrial Science, the University of Tokyo, JAPAN
[4]KIST Europe, Saarbrücken, GERMANY

ABSTRACT

Scanning Thermal Microscopy (SThM) and micro-thermal analysis allow the study of thermal phenomena at micro- and nanoscale. We present a novel scanning resistive probe aimed for thermal imaging and localized thermal analysis. The probe features an AFM cantilever with a sharp pyramidal tip. Metal nanowires are integrated at the inner edges of the pyramidal tip forming an electrical cross-junction at the apex. The nanometer- sized cross-junction, addressable through microelectrodes, can be utilized both as a local temperature sensor and a heater. We have fabricated a first prototype of the probe with a 150 μm long, 36 μm wide and 0.5 μm thick silicon nitride cantilever. Platinum nanowires, 300 nm wide and 100 nm thick, are successfully integrated using a wafer-scale fabrication process based on corner lithography. We have experimentally characterized electrical and thermal properties of the probe demonstrating its proper functioning.

INTRODUCTION

Resistive probes are applied in scanning thermal microscopy (SThM) to measure spatial variations in the thermal conductivity and diffusivity of the sample, in addition to its topography [1]. In contrast to thermocouple probes, the resistive probes can both sense temperature and supply heat. Therefore they can also be used in micro-thermal analysis to visualize the spatial distribution of phases, contaminants and components in polymers and biomaterials [2]. At the heart of these techniques are modified AFM probes with an electrical junction integrated at the end of a sharp tip. In the past such probes are realized by multiple-level direct-write EB lithography [3] and direct deposition of metal by focused electron beam (FEB) [4]. Both demonstrated methods are rather expensive, time consuming and unsuitable for high-volume manufacturing. At the MEMS 2010 conference, we have presented a resistive probe with a pyramidal tip composed of freestanding nanowires [5]. Although our probe was fabricated by standard micromachining at the wafer scale, the poor mechanical stability of the nanowire tip severely limited its practical application.

Here we present a novel resistive probe, which has a solid and mechanically robust sharp tip. Directly above the tip apex a nanometer-sized sensing and heating element is integrated. The solid tip is strong enough to perform both contact and tapping mode AFM imaging. During the imaging the nanowire element can be employed as a temperature sensor or serve as a highly localized heat source.

PROBE DESIGN

Our novel resistive scanning probe is schematically illustrated in Figure 1.

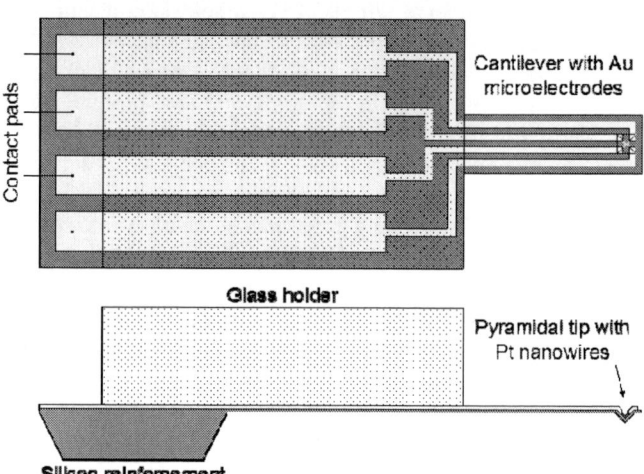

Figure 1: Top and side views of the novel scanning thermal probe. The probe consists of a silicon nitride cantilever with a sharp pyramidal tip. Platinum nanowires integrated at the inner edges of the pyramidal tip form a four-terminal thermal element directly above the tip apex. The thermal element, addressable through the microelectrodes, can be used both as a temperature senor and a heater.

The probe consists of an AFM-type cantilever attached on a glass holder. At the free end of the cantilever a sharp pyramidal tip is located. Platinum nanowires are incorporated at the inner edges of the pyramidal tip forming an electrical cross-junction at its apex. The nanometer-sized cross-junction is addressable through the gold microelectrodes integrated on the cantilever and the probe holder. Underneath the contact pads a part of the bulk silicon is preserved in order to serve as a mechanical support enabling an easy connectivity of the probe with external electronics either by wire bonding or by pin connectors.

978-1-4799-3510-9/14 $31.00 © 2014 IEEE   1111

MICROFABRICATION

The probe is realized in a wafer-scale fabrication process using only standard surface and bulk micromachining and conventional contact lithography. The basic fabrication steps are shown in Figure 2.

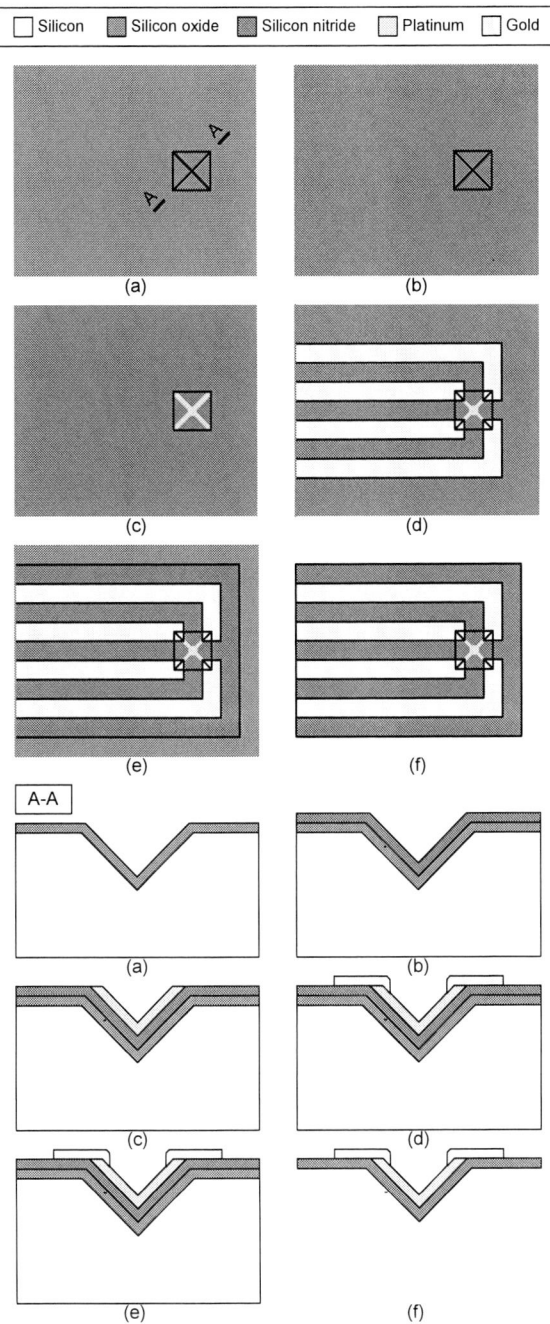

Figure 2: A wafer-scale fabrication process based on standard micromachining and conventional optical lithography: (a) etching of a pyramidal pit followed by oxidation sharpening, (b) silicon nitride deposition, (c) formation of Pt nanowires by corner lithography, (d) sputtering of electrodes, (e) etching of the probe layout and (f) anodic bonding (not shown) and probe release.

The fabrication process starts on a (100)-oriented standard silicon wafer. In the selected wafer a pyramidal cavity is formed by wet anisotropic etching of silicon in KOH (Step a). The pyramidal pit serves as mold for a probe tip. In order to achieve a sharper tip the mold is thermally oxidized at low temperature [6]. After the oxidation sharpening a silicon-rich nitride layer is deposited by LPCVD (Step b). In the sharp corner of the pyramidal pit platinum nanowires are formed by corner lithography (Step c). Corner lithography [7,8] is a 3D nano patterning technique based on the fact that the conformal deposition and subsequent isotropic etching of a layer, leaves a residue of the material in sharp concave corners, as illustrated in Figure 3.

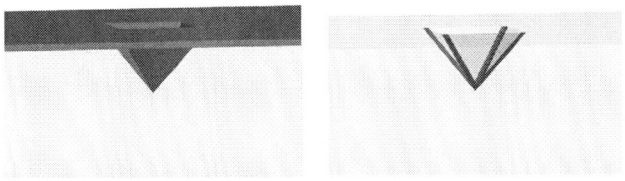

Figure 3: Nanowires inside a pyramidal pit can be formed by corner lithography. A conformally deposited layer is partially removed by time-controlled isotropic etching leaving nanometer-sized features in the sharp corners.

After the nanowire formation, gold microelectrodes and contact pads are formed by a lift-off process (Step d). Subsequently, the probe layout is defined by patterning of the silicon nitride layer by RIE (Step e). In the next fabrication step, which is not illustrated in Figure 2, the patterned silicon wafer is anodically bonded to a pre-diced glass substrate. After the anodic bonding the silicon wafer is partially dissolved in a hot TMAH solution (Step f). The patterned silicon nitride layer serves as a mask guiding the etching process in such a way that silicon is almost completely removed from the backside leaving only a small part underneath the contact pads. The preserved silicon serves as a mechanical reinforcement.

SEM micrographs of a successfully fabricated scanning thermal probe are shown in Figure 4. The entire probe has a dimension of 3400 µm by 1600 µm. The cantilever is 150 µm long, 36 µm wide and 0.5 µm thick. Platinum nanowires integrated at the inner edges of the pyramidal tip are 300 nm wide and 100 nm thick. The microelectrodes and contact pads are made of 100 nm thick gold layer. The nanometer-sized cross-junction formed by platinum nanowires is addressable through the gold microelectrodes.

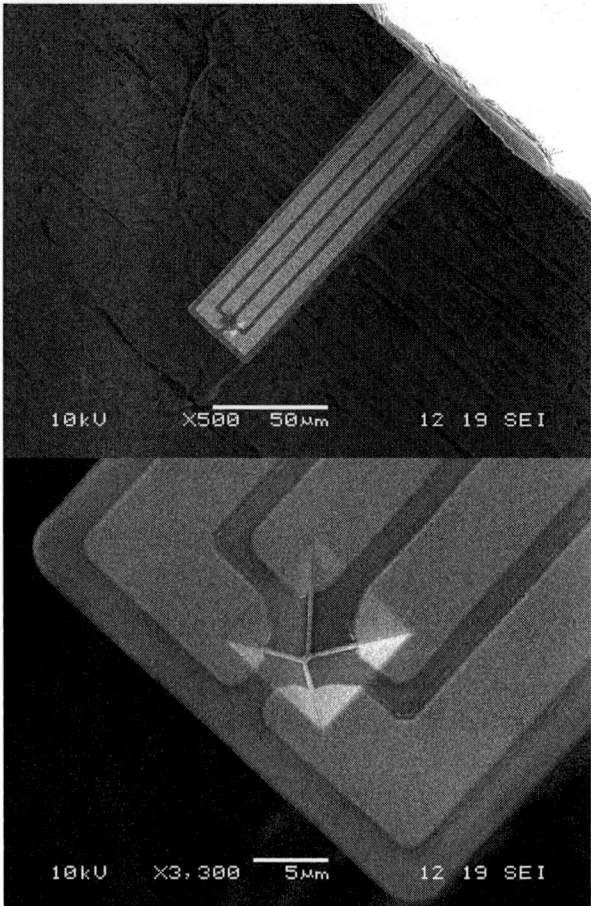

Figure 4: SEM figures of a novel resistive AFM probe. Top side of the cantilever with gold microelectrodes and platinum nanowires integrated at the inner edges of the pyramidal tip is shown. The scanning tip is facing downwards and is not visible on the micrographs.

EXPERIMENTAL CHARACTERIZATION

After fabrication, the probe was mounted on a printed circuit board (PCB) and wire bonds were made between the electrodes on the PCB and the contact pads on the probe base (1, 2, 3 and 4). After the wire-bonding, the electrical resistance between the electrodes is measured by a multimeter. The resistance of the electrical cross-junction at the tip apex and the electrical leads are shown in Table 1.

Table 1: Measured resistance of the tip apex and other electrodes from the contact pad to the apex.

Location	Resistance(Ω)
Cross-junction	5.2
Lead 1	485.4
Lead 2	434.7
Lead 3	452.5
Lead 4	503.5

Resistance of the nanowire cross-junction at the tip apex changes due to the temperature variations. The relation

between the resistance change ΔR and the temperature change ΔT is given by:

$$\Delta R = R_0 \alpha \Delta T \qquad (1)$$

where R_0 is the initial resistance and α is the temperature coefficient.

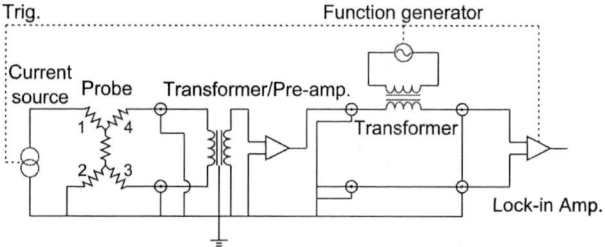

Figure 5: A schematic diagram of the experimental setup.

Figure 5 shows a schematic diagram of the experimental setup. In order to perform a four-point resistivity measurement of the cross-junction, a Keithley Model 6221 AC and DC Current Source was used. The measurement were performed by applying AC current of 30 µA at 377 Hz. The voltage over the cross-junction was amplified by a Stanford Research Systems Model SR554 Transformer Preamplifier. To achieve the low noise measurement, an offset cancelation circuit was introduced. This circuit increases the sensitivity of the lock-in amplifier allowing measurement of the resistance change of the cross-junction element. The resistance change was measured by an Anfatec elockIn 204/2 with a frequency bandwidth of 1 Hz.

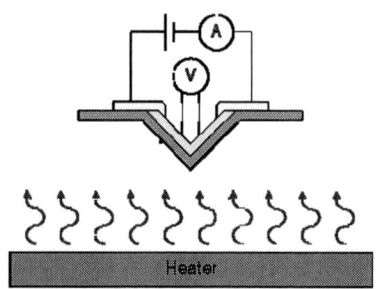

Figure 6: A schematic view of the experimental setup. The probe tip is placed above an external heater. Temperature changes are measured by the four-point resistivity measurements of the nanowire cross-junction located at the tip apex.

For preliminary thermal measurements, the probe is placed above a silicon resistive heater with the angle of a few tenth of degree as illustrated in Figure 6. The distance between the probe and the heater was determined by a piezo scanner placed below the heater. In the first measurement, the heater was turned on and off every 20 seconds by applying a DC current of 100 mA. Electrical resistance of the nanowire cross-junction varied depending on the temperature of the heater (Figure 7). The time response, which was in the order of seconds, was entirely determined by the external heater.

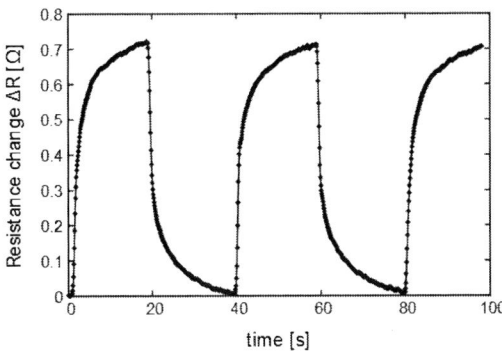

Figure 7: The measured change in electrical resistance of the metal nanowire sensor due to the temperature change induced by the heater.

In the second measurement, the probe was used as both a heater and a sensor. The current on the Si heater was kept off so as to use it as a heat sink. The current applied to the probe and the schematic diagram of the setup is as same as the first measurement (Figure 5). In order to observe the effect of the heat flow from the probe to the sink, the distance between two of them was varied 20 μm. The resistance changes during the retraction(blue) and the approach(red) are shown in Figure 8.

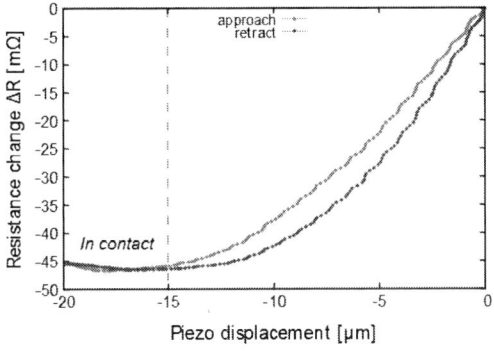

Figure 8: The measured change in electrical resistance of the metal nanowire sensor during the retraction and the approach of the probe.

While decreasing the distance, the resistance decreased and the change became constant around -15 μm since the probe landed on the sink. The curve in the graph can be explained by considering the probe has two parallel heat resistance to the sink. One is between the cross-junction and the sink, the other is between the cantilever and the sink. The latter path most likely is the dominant conduction path. After the landing, the resistance change increased. Possibly, this is because the piezo pushed up the sink even more, the end of the cantilever was moved away due to the cantilever bending, and increased the heat resistance. The loop between approach and the retraction is because of the piezo scanner hysteresis. An important design issue is the heat resistance between the tip and the cantilever. To optimize the spatial resolution of the probe it is important to make this resistance as high as possible to reduce the parasitic path to the sink through the cantilever beam.

CONCLUSION

We have devised and successfully fabricated a novel resistive thermal probe. Electrical and thermal properties of the probe were experimentally determined and its basic functioning is demonstrated. In the future we are planning to employ our probe for thermal imaging in an atomic force microscope.

ACKNOWLEDGEMENTS

The authors would like to thank the staff of the MESA+ Clean Room for their assistance with probe fabrication and P. Linders from Deltamask for the mask processing. The authors gratefully acknowledge J.W. Berenschot from University of Twente for valuable discussions. The authors would like to thank H.B. Waayer from University of Twente for the PCB designing.

REFERENCES

[1] A. Majumdar, "Scanning Thermal Microscopy", *Annu. Rev. Mater. Sci.* 29, pp. 505-585, 1999.

[2] H.M. Pollock, A. Hammiche, "Micro-thermal analysis: techniques and applications", *J. Phys. D: Appl. Phys.* 34, pp. R23-R53, 2001.

[3] H. Zhou, G. Mills, B.K. Chong, A. Midha, L. Donaldson, J.M.R. Weaver, "Recent progress in the functionalization of atomic force microscope probes using electron-beam nanolithography", *J. Vac. Sci. Technol. A* 17 (4), pp. 2233-2239, 1999.

[4] K. Edinger, T. Gotszalk, I.W. Rangelow, "Novel high resolution scanning thermal probe", *J. Vac. Sci. Technol. B* 19 (6), pp. 2856-2860, 2001.

[5] E. Sarajlic, R. Vermeer, M.Y. Delalande, M.H. Siekman, R. Huijink, H. Fujita, L. Abelmann, *Proc. MEMS 2010 (Hong Kong, China)*, pp. 328-331, 2010.

[6] S. Akamine, C.F. Quate, "Low temperature thermal oxidation sharpening of microcast tips", *J. Vac. Sci. Technol. B* 10 (5), pp. 2307-2310, 1992.

[7] E. Sarajlic, J.W. Berenschot, G. Krijnen, M.C. Elwenspoek, " Fabrication of 3D nanowire frames by conventional micromachining technology", *Proc. Transducers '05 (Seoul, South Korea)*, Vol. 1, pp. 27-30, 2005.

[8] J.W. Berenschot, N.R. Tas, H.V. Jansen, M.C. Elwenspoek, "3D-nanomachining using corner lithography", *3rd IEEE Int. Conf. on Nano/Micro Engineered and Molecular Systems (NEMS2008), (Sanya, China)*, pp. 729-732, 2008.

CONTACT

K. Hatakeyama, tel: +31534894354; k.hatakeyama@utwente.nl

CAPILLARY EFFECT BASED TSV FILLING METHOD

Jiebin Gu[1], Xiang Jiang[1], Heng Yang[1] and Xinxin Li[1]
[1] Shanghai Institute of Microsystem and Information Technology, CHINA

ABSTRACT

Via-filling is a critical and costly process of TSV fabrication and is usually realized by Cu-plating, CVD, etc. Liquid metal via-filling made by pressing solder from a melting solder pool into via holes faces problems of wafer breakage caused by pressure differential and cutting off solder-vias from the pool. A liquid bridge pinch-off effect based cutting-off method, together with a 'wafer sandwich' structure, has the feasibility to solve the above problems faced by the pressure assisted liquid metal via-filling approach.

INTRODUCTION

Via-filling is a critical and time-consuming process of TSV fabrication and is usually realized by Cu-plating, CVD, etc. [1]. Several studies have focused on liquid metal via-filling. Our previous work of a solder ball based 'solder-pump' method utilize surface tension only to drive liquid solder for via-filling, which however can only pump liquid solder from small holes to larger holes [2, 3, 4]. Furthermore, electrostatic force make it a troublesome work to handle solder balls less than 200μm. Therefore, it is very difficult to make small footprint TSVs using the 'solder-pump' method.

Pressure assisted liquid metal via-filling has also been studied[5, 6]. However, the fragility of silicon wafer make it cannot stand the high pressure differential, therefore these works are still on die-level. Another problem liquid metal via-filling method encounter is to cut off the filled pillars from liquid metal pool after filling. In this paper, we explore, for the first time, a liquid bridge pinch-off effect based via-filling method, which, together with a 'wafer sandwich' structure, has the feasibility to solve the problems faced by pressure assisted liquid metal via-filling for TSV fabrication.

MECHANISM

Structure

The configuration diagram is illustrated in Figure 1, where wafer for via-filling is sandwiched by a cover and a nozzle wafer. The nozzle wafer and via wafer is vertically aligned. The 'sandwich' is placed on top of a liquid solder pool, which is unwettable to all the three wafers. A chamber, whose pressure is controllable, encloses the 'sandwich'. The gaps between the wafers provide ventilating channels between the via-holes and the chamber. Assume that the chamber and atmospheric pressure is P_i and P_o respectively, and their pressure differential is ΔP.

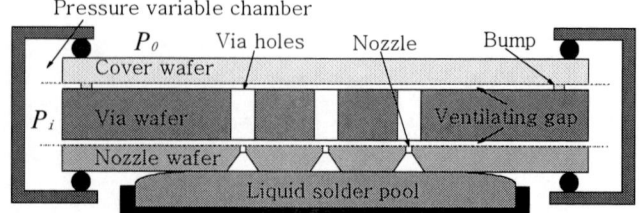

Figure 1: Configuration of the liquid metal via-filling method: a 'wafer sandwich', a pressure variable chamber and a liquid solder pool.

Theory

The via-filling is realized by first decreasing then recovering the chamber pressure (P_i), as shown in Figure 2 and 3. In the first process of decreasing chamber pressure, as soon as ΔP reaches to $\Delta P_{penetrate}$ ($2\gamma/r_{nozzle}$, where γ is surface tension of the liquid solder, and r_{nozzle} is radius of the nozzle), the liquid solder can be sucked into the via-holes by overcoming its surface tension. The cover-wafer and the spacing-controlled ventilating gaps prevent solder from flowing out of the via-holes. The functionality of the sandwich structure is to insure that pressure on the two sides of the via-wafer is always neutralized. The risk of wafer breakage therefore can be eliminated.

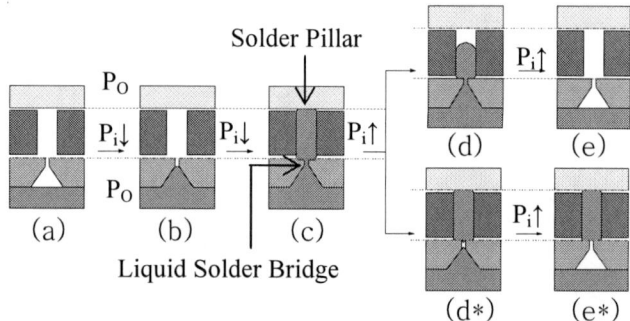

Figure 2: Pinch-off effect based liquid metal via-filling mechanism. (a) initial state, (b-c)solder being sucked into via holes by dropping P_i, (d-e)scenario of solder pillars being pushed back by recovered P_i; (d*-e*) scenario of liquid bridge being pinched off by recovered P_i.

After the sucking-in process, each filled solder pillar in via-hole still connects the solder pool through a liquid solder bridge in nozzle, as shown in Figure 2(c). The chamber pressure (P_i) is then slowly recovered, and accordingly ΔP starts to fall. Both of the solder pillars and the liquid solder bridges need a minimum ΔP to hold or sustain, which are assumed to be $\Delta P_{pillar-hold}$ and $\Delta P_{bridge-sustain}$ respectively. Depending on the relative

978-1-4799-3510-9/14 $31.00 © 2014 IEEE

value of $\Delta P_{pillar-hold}$ and $\Delta P_{bridge-sustain}$, either of two scenarios can occur as ΔP fall. If $\Delta P_{bridge-sustain} < \Delta P_{pillar-hold}$, the increasing P_i can push the solder pillars back to the solder pool, as shown in Figure 2(d-e). However, if $\Delta P_{pillar-hold} < \Delta P_{bridge-sustain}$, as shown in Figure 3, when ΔP drops to the point that liquid bridges are no longer sustainable and to be pinched off by surface tension, ΔP is still strong enough to hold the solder pillars. The second scenario is desired as the solder pillars can be cut off from the solder pool, as shown in Figure 2(d*-e*). The liquid bridge rupture in each nozzle is independent to each other. Failure of one liquid bridge breakage does not affect others. The sucking-in and cutting-off together complete the via-filling process. After solidification, the filled via-wafer can be separated from the 'sandwich'. The cover-wafer and the nozzle-wafer are reusable.

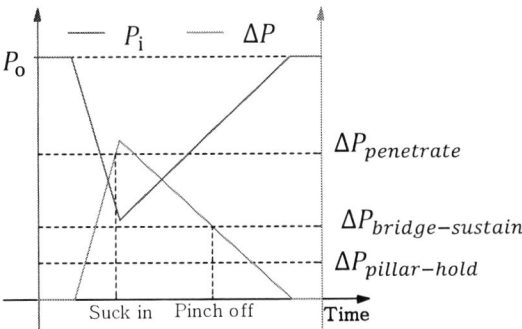

Figure 3: The liquid metal via-filling can be done through a cycle of chamber pressure change. The blue and red line show the change of P_i and ΔP respectively.

Take advantage of the above method, massive TSV via-filling can be realized through a cycle of pressure change, as shown in Figure 3. One consideration for TSV filling is surface smoothness after via filling. Because of surface tension, both ends of the filled pillars will present slight dome-like, which results an uneven surface. This surface roughness can be levelled by polymer coating, such as BCB, thus avoid CMP process.

NOZZLE DESIGN

Nozzle size

The geometry of the nozzles and the minimum via holes fillable can be derived from $\Delta P_{pillar-hold}$, $\Delta P_{bridge-sustain}$ and $\Delta P_{penetrate}$. The $\Delta P_{pillar-hold}$ can be get by the equation [2]:

$$\Delta P_{pillar-hold} = 2\gamma/r_{via},$$

where r_{via} is radius of the via-holes. Liquid bridge can be categorized to several types. In the above case, the liquid bridge formed in the nozzle orifice is constant-pressure liquid bridge, whose minimum sustain pressure

($\Delta P_{bridge-sustain}$, otherwise known as rupture pressure) is a function of its aspect ratio (λ) and dimension [7]. The lower boundary of the dotted-diagram in Figure 4 gives the relationship between the dimensionless rupture pressure ($\kappa = \frac{\Delta P}{\gamma/r_{nozzle}}$) and aspect ratio of constant-pressure liquid bridge, which can be numerically calculated by solving Laplace-Young equation with specific boundary [8]. From Figure 4, it can be found that the less the aspect ratio, the less pressure is needed to rupture the liquid bridge and $\pi/2$ is a dividing point. When aspect ratio is higher than $\pi/2$, the κ is 1, which means $\Delta P_{bridge-sustain} = \gamma/r_{nozzle}$. When aspect ratio is lower than $\pi/2$, its corresponding κ can be obtained through the rupture line. To simplify the problem, for $\lambda \geq \pi/2$, rupture condition $\Delta P_{pillar-hold} < \Delta P_{bridge-sustain}$ gives $r_{nozzle} < r_{via}/2$. Therefore, the diameter of the nozzle should be at least less than half of the via-holes.

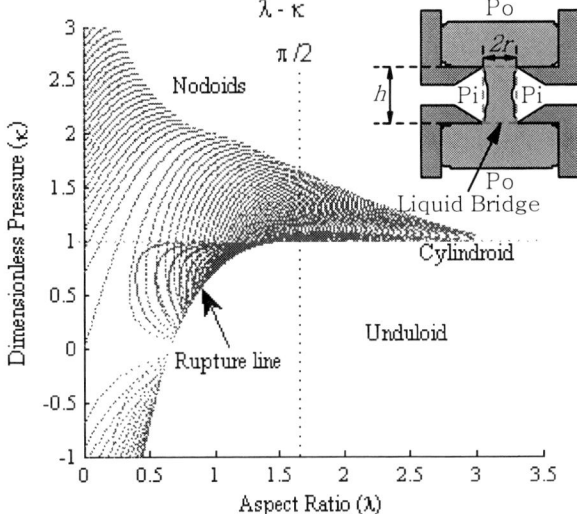

Figure 4: The lower boundary of the dotted-diagram stands for the corresponding relationship between dimensionless rupture pressure ($\kappa = \Delta P/(\gamma/r)$) and aspect ratio ($\lambda = h/(2r)$) of constant-pressure liquid bridge. This rupture line can be used as a reference for nozzle design.

In terms of $\Delta P_{penetrate}$, the minimum r_{nozzle} is a function of maximum ΔP and surface tension γ. For ΔP is one Bar and γ is 0.55N/m [9], the minimum nozzle diameter that can be penetrated is about 20μm, therefore the minimum via hole fillable is about 40μm. However smaller via filling is possible by increasing P_o.

Nozzle structure

Besides dimension, the structure of the nozzle is also critical. For constant-pressure liquid bridge, its surface should be free to atmosphere, which in this case is the chamber atmosphere. Our first nozzle structure is a DRIE etched orifice as shown in Figure 5(a). Experiments show that liquid bridges do not rupture in the orifices. Later it is understood that in this case, the surface of the liquid bridge hermetically touch the sidewall of the nozzle orifices, and it is no longer a genuine constant-pressure liquid bridge. Two others nozzle structures are then proposed. Figure 5(b) shows a nozzle with buried air channel, which can be realized by the 'Micro-holes inner-etch' process [10], but its fabrication is complicated. An alternative is an orange-shape nozzle, as shown in Figure 5(c), which uses the inner edges of the orange slices to form the nozzle while the gaps between the slices provide interface to chamber atmosphere. The orange structure is used in this paper because of its easy fabrication.

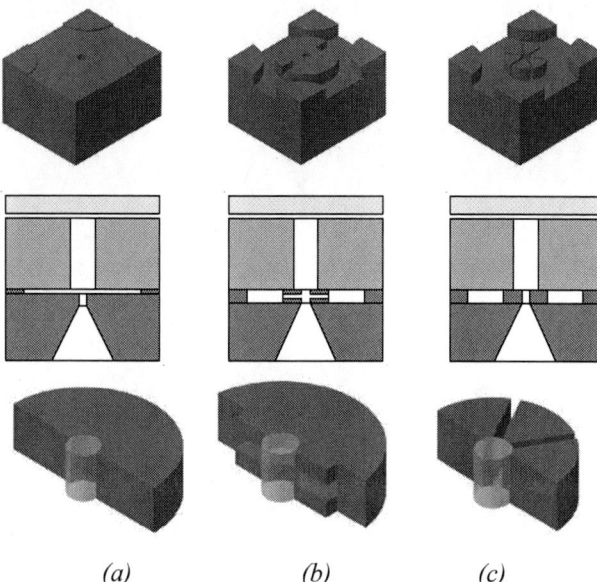

(a)   *(b)*   *(c)*

Figure 5: Three types of nozzle. (a) Orifice nozzle with gap bumps, (b) nozzle with buried air channel, (c) orange nozzle. Due to their dimension, all the nozzle structures only form top layer of a wafer, and therefore all the wafers are etched through from backside. The top array is 3D schematic of nozzles. The middle array is cross-section schematics of via-hole and nozzle assembly. The bottom array is 3D schematic of liquid bridge formed in nozzle.

For the ventilating gap and the orange slice gap, both of them should be small enough to offer a surface tension higher than pressure differential, so that liquid solder cannot be pushed into these gaps during filling process. Depending on the maximum ΔP, simulation in Surface Evolver shows the gap should be in range of a few to tens of microns.

FABRICATION AND EXPERIMENT

The cover-wafer is made of Pyrex glass wafer with etched bumps to provide about 5μm high ventilating gap. Both of the via and nozzle wafer are made of double-side polished 4-inches silicon wafers, with thickness of about 425μm. The via-holes and the orange nozzle heads are

done by DRIE. Nozzle wafer is further etched through from backside by KOH etching (alternatively can be done by DRIE). Figure 6 shows a typical fabricated orange nozzle. All the via and nozzle wafers are finally coated with 2 micron SiO_2 film by thermal oxidation. The wafer alignment is assisted by a pin-dowel approach [2].

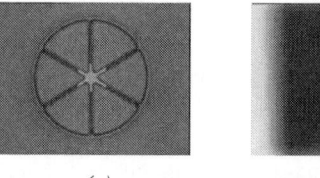

(a)   *(b)*

Figure 6: Optical view of a typical orange nozzle. (a) top view. (b) bottom view. Note the green color is microscope light.

A specific reflow oven is made to accomplish the experiment as shown in Figure 7. The solder used is lead-free, with melting point of ~ 220°C. The solder pool is maintained at 50°C above the melting point during the filling process. Before the filling process, oxygen in the chamber is purified to avoid oxidation of the liquid solder bridge surface, which might severely affect pinch-off effect. The filling process is as following: the chamber pressure is first slowly vacuumed until a sudden filling of via holes is observed through the microscope. The chamber pressure is then stabilized for 1 minute, after which the chamber pressure is slowly recovered with nitrogen gas until atmosphere pressure is reached. The whole process can be finished in several minutes. The wafer sandwich, along with the solder pool is then cooled down before taking out the via-wafer for examination.

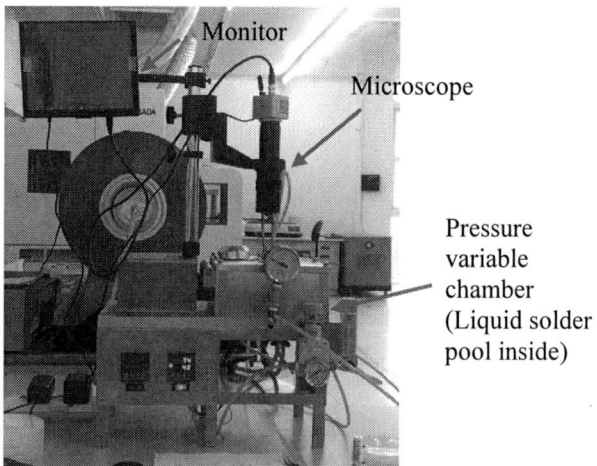

Figure 7: The reflow oven made for the liquid metal TSV via-filling experiment.

RESULTS

In all the experiments, the size of the via holes and the nozzle are φ200×425μm andφ40×63μm respectively, and they are arranged in a matrix of 4×5. After solidification, although with slight adhesion, the via-wafer can be separated from the nozzle-wafer without much effort. 18 of the 20 vias are filled. The results are shown in Figure 8. For all the pillars in via-holes, the top surfaces, that hit the cover-wafer, are dome-like with a flat top, as shown in Figure 8(a). For the bottom surfaces, where the liquid bridge rupture happened, their surfaces are bumpier, as shown in Figure 8(b) and 9(a). Furthermore, a tuber is observed on each bottom surface, as shown in Figure 9(a). These tubers are supposed to be liquid bridge leftover in the nozzles after pinch off.

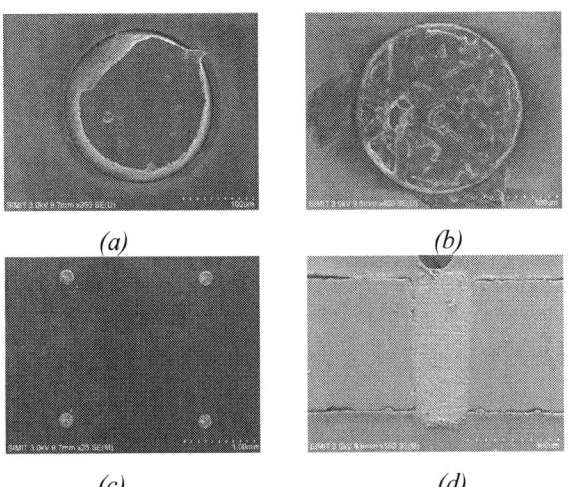

(a)   *(b)*

(c)   *(d)*

Figure 8: SEM view of typical filled solder pillars. (a) Top surface. (b) Bottom surfaces. (c)Birdseye view of bottom surfaces. (d)Cross-section view.

For purpose of comparison, an experiment of pulling the solder bridges off in solid state was carried out. In the comparative experiment, the liquid bridges were solidified without the pinch-off process. The via-wafer was then pulled off from the nozzle wafer by force. The pulling-off results cusps, which are much taller and shaper than the tubers left after pinch-off, as shown in Figure 9. The height of a typical cusp is about 70μm while it is about 30μm (half the height of the nozzle) for a typical tuber. By improving nozzle, it is possible to further reduce the tuber.

In addition, no solder pillars fall off from the via-holes after 48 cycles of temperature cycling test (-25 ~ +125℃).

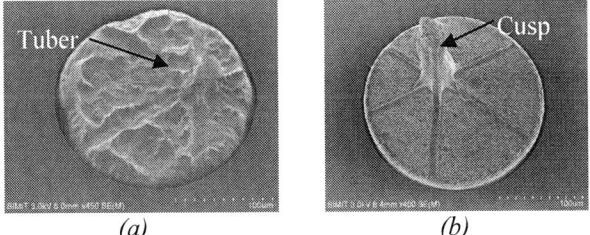

(a)   *(b)*

Figure 9: Comparison of pillar surfaces (a) pinched off in liquid state and (b) pulled off in solid state.

CONCLUSION

In this paper, a surface tension pinch-off effect based TSV via-filling method is presented and verified by experiments. This method is capable to fill TSV with diameter down to 40 microns without aspect ratio restriction. The proposed filling mechanism can realize TSV-filling within a cycle of pressure change in few minutes. This low-cost and fast TSV filling process has a high potential for industry applications, such as MEMS wafer-level packaging.

ACKNOWLEDGEMENT

The research is supported by Chinese 973 program (2011CB309503) and Major National Science and Technology Projects of China (2012ZX02503002).

REFERENCES

[1] A. C. Fischer, S. J. Bleiker, N. Somjit, N. Roxhed, T. Haraldsson, G. Stemme and F. Niklaus. "*High Aspect Ratio Tsvs Fabricated by Magnetic Self-Assembly of Gold-Coated Nickel Wires.*" JMEMS, (2012).

[2] Gu, J., et al. (2009). "*A novel capillary-effect-based solder pump structure and its potential application for through-wafer interconnection.*" Journal of Micromechanics and Microengineering 19(7): 074005.

[3] Jiebin, G., et al. (2010). "*A novel vertical solder pump structure for through-wafer interconnects*". Micro Electro Mechanical Systems (MEMS), 2010 IEEE 23rd International Conference on.

[4] Jiebin, G., et al. (2011). "*Solder Pump Technology for Through-Silicon via Fabrication.*" Microelectromechanical Systems, Journal of 20(3): 561-563.

[5] Young-Ki, K., et al. (2012). *Advanced TSV filling method with Sn alloy and its reliability*. 3D Systems Integration Conference (3DIC), 2011 IEEE International.

[6] Y.K. JEE, J. Y., K. W. PARK, T.S. OH (2009). "*Zinc and Tin-Zinc Via-Filling for the Formation of Through-Silicon Vias in a System-in-Package.*" Journal of Electronic Materials 38(5).

[7] Lowry, B. J. (2000). "*Fixed boundary dual liquid bridges in zero gravity.*" Physics of Fluids 12(5): 1005-1015.

[8] Bowick, M. J. and Z. Yao (2011). "*Crystalline order on catenoidal capillary bridges.*" EPL (Europhysics Letters) 93(3): 36001.

[9] Kaban I, Mhiaoui S, Hoyer W and Gasser J-G 2005 *Surface tension and density of binary lead and lead-free Sn-based solders*, J. Phys.: Condens. Matter.177867–73.

[10] Jiachou, Wang, Xia Xiaoyuan and Li Xinxin. "*Monolithic Integration of Pressure Plus Acceleration Composite Tpms Sensors with a Single-Sided Micromachining Technology.*" Microelectromechanical Systems, Journal of 21, no. 2 (2012): 284-293.

CONTACT

*Jiebin Gu, tel: +86-021-62511070-5585; email: j.gu@mail.sim.ac.cn

CHARACTERISATION AND SIMULATION OF LOW TEMPERATURE Si-Si DIRECT BONDING THROUGH VELCRO-LIKE SURFACES BASED ON POROUS SILICON

Shervin Keshavarzi[1,2], Ulrich Mescheder[1], and Holger Reinecke[2]

[1]Institute of Applied Research (IAF), Furtwangen University, Furtwangen, Germany
[2]Institute of Microsystems Engineering (IMTEK), Freiburg University, Freiburg, Germany

ABSTRACT

Velcro-like (needle like) surfaces based on porous silicon can result in strong permanent bonding at room temperature with capability of multiple bonding and un-bonding similar to Velcro® principle. However, understanding of the interaction mechanisms between such surfaces is crucial for employing this type of bonding for wafer or chip bonding and IC packing applications. In this paper, a simple Si technology to create Velcro-like surfaces is described and characterized, and the interaction mechanisms between such surfaces are modelled based on Van der Waals force approach.

INTRODUCTION

Further miniaturization of chip size needs new bonding techniques to mutate the conventional pick and place methods. Additionally, low temperature wafer bonding or detachable chip-bonding onto a wafer platform provide better process compatibility or more flexibility. Porous Silicon (PS) based bonding allows strong permanent bonds between needle like surfaces in a self-organized process as well as multiple bonding and un-bonding of chips/wafers similar to Velcro® principle [1]. This approach can provide low temperature Si-Si direct bonding, a fully CMOS compatible approach suitable in system integration using the Si-motherboard concept.

With standard structuring techniques brush-like metal structures ("micro-brush") and permanent mechanical and electrical bonds of chip or dies on Si at temperature below soldering temperature are obtained [2]. In this case, the bonding structures are stiff and hardly suitable for Van der Waals (VdW) based bonding technique. Although this limitation can be solved using carbon nanotubes [3], it is very difficult to integrate this concept into standard Si process. This work demonstrates, characterizes and models a modest Si based technology allowing a strong permanent bonds at room temperature with multiple bonding and un-bonding capability of the same chips.

VELCRO-LIKE SURFACES

The Velcro-like surfaces can be simply created through a self-organized fabrication process by anodizing the surfaces of Si wafers in HF/Ethanol electrolyte. The pore morphology of porous Si particularly depends on anodization current density and material properties, such as substrate type, doping concentration, and electrolyte. The properties can be adjusted in such a way that changes the pore shape from sponge type to needle type. Based on the results presented in [4] the proper anodization conditions and electrolyte concentration are defined to create the needle-like surfaces. At transition between porous silicon formation and electropolishing, needles (Fig. 1) are created via anodizing the surfaces of low doped p-Si wafers at 70mA/cm² in 7wt. % HF (in water) electrolyte concentration for 40 minutes, and dried through Ethanol and Pentane for 15 minutes, respectively. As shown, the single needles are clustered through a self-bonding mechanism due to capillary force during rinsing and drying processes. However, the surfaces with single needles can be obtained by some additional treatments [1].

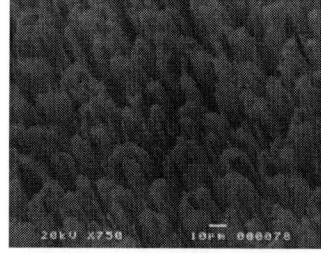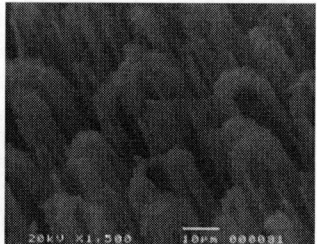

Figure 1: SEM pictures of generated clustered needle like surfaces. A needle like surface in 45° tilted view with 750x (left) and 1500x (right) magnifications.

The needle densities, diameter of curvatures and heights are measured using SEM technique (Table. 1).

Table 1: Surface properties of the specific fabricated needle like surface.

Wafer Type	p^- Si
Clustered needle density (Nr/cm^2)	4.41×10^5
Clustered needle diameter of curvature (µm)	Mean = 8.36 STD = 2.855
Clustered needle height (µm)	Mean = 16.70 STD = 1.17

BONDING FORCE MEASURMENT

The fabricated surfaces are diced to $0.64cm \times 0.64cm$ chips and bonded together by pressing the surfaces together at specified applied load at room temperature. The pull-off force is measured through a special designed pull-off measurement unit, and the results are shown in Figure. 2. The bond strength is defined as the measured pull-off force at which the bonded chips are detached. The pull-off measurement unit consists of a force sensor (Burster 8523-20) and exhibits a high dynamic range from 1 mN to 20 N.

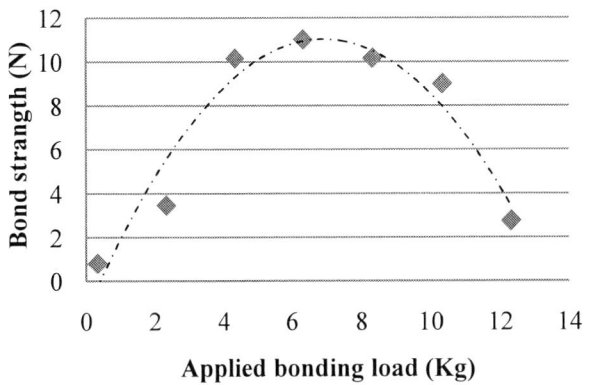

Figure 2: Bond strength at various applied bonding weight for active bonding area of 0.4096 cm².

A maximum bonding force of 11.02N is obtained for a 6.33kg weight load which corresponds to bond strength of 0.27 MPa.

The bonded chips can be detached mechanically and the single chips can be rebonded several times. For rebonding, the optimal bonding weight is restricted to values between 2.33kg and 4.33kg in which up to 40% of the bond strength is available even after 9 rebonds of the same chips (Fig.3).

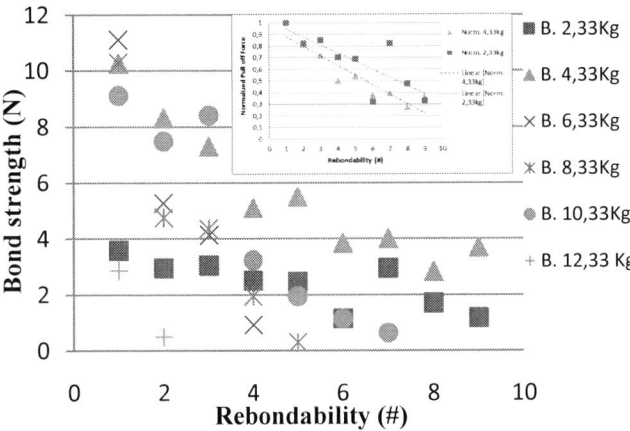

Figure 3: Bond strength rebondability results at constant bonding weight loads for effective bonding area of 0.4096cm², inset: normalized pull-off force for bonding weights of 2.33kg and 4.33kg.

BONDING MECHANISMS

In general, several interactive forces such as Van der Waals (VdW) forces, electrostatic forces, capillary forces, and surface tension can be observed between two solid objects. Among them, capillary and surface tension forces are affected impressively by environment, and can be diminished or eliminated through a dry or a vacuum condition [5]. However, the Van der Waals forces are always presented between two objects, and are dominant when the distance between the involved objects is in the range of few nanometers [6].

The VdW force between atoms or molecules is the combination of three different forces, Keesom, Debey, and London dispersion forces. The interaction energy of all is the same and shows the same dependence on distance r [7]:

$$w_i(r) = \frac{c_i}{r^6} \qquad (1)$$

where C_i is the corresponding energy constant, and r is the separation distance between two molecules or atoms. The VdW force exhibits strong reliance to the distance r ranging from r^{-6} to r^{-2} [1].

BONDING MODEL

The interaction between two Velcro-like surfaces can be modelled as one needle like surface in interaction with a perfectly flat surface for the case of needle-substrate interactions plus side interaction of each needle with needles from opposite surface. Under this hypothesis, the effect of each needle is local and considered separately from other needles. Then the total interaction force can be obtained by summing up all individual contributed needles [8].

A simplified model of the bonding mechanism for needle like surfaces is proposed in Figure 4. A conical shaped needle is represented as combination of a cylindrical body and a half sphere head. Then the total interaction is divided in two parts: i) Interaction between needles and substrate ii) Interaction between parallel or neighbored needles from opposite surfaces

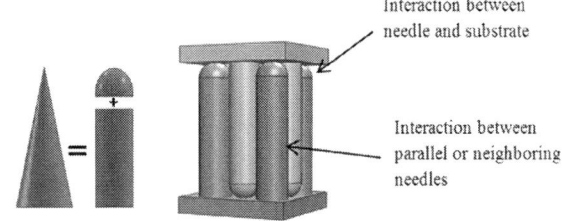

Figure 4: Schematic representation of transformation of a conical needle to a cylindrical body and a half sphere head and their interactions in the bonding mechanism.

Van der Waals force between a needle and a substrate

A needle can be simply represented as a cylinder body with a half sphere head. In this case, the interaction between a needle and a flat surface can be obtained by combining the interaction between the half sphere head and the substrate, and the interaction between the cylinder and the substrate, as shown in Figure 5 [5].

The interaction energy between two objects can be presented in general as [7]:

$$U = -\frac{\pi C \rho}{6d^3} \qquad (2)$$

where ρ is the number of atom per unit volume in two interacting bodies, C is the coefficient for particle-particle pair interaction, and d is the distance between two objects.

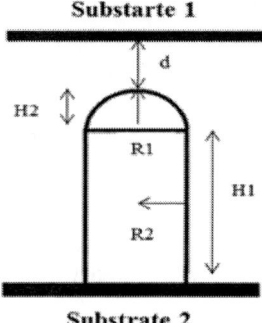

Figure 5: Transformation of a needle to a half sphere head and a cylinder body and its geometry parameters.

The VdW interaction energy U_{hs} between the half sphere head and the substrate is then obtained by summing up all the individual contributions of the half sphere heads as:

$$U_{hs} = -\frac{\pi^2 C \rho_1 \rho_2}{6} \int_0^{H_2} \frac{(2R_1 - z)z}{(d+z)^3} dz \quad (3)$$

where d is the distance between tip of the needle and the flat surface, H_2 is the height of the half sphere head, R_1 is the radius of half sphere, and ρ_1 and ρ_2 are the numbers of atom per unit volume in the needle and the substrate, respectively.

Then, the Van der Waals interaction force between the half sphere head and the substrate becomes:

$$F_{hs} = \frac{\partial U_{hs}}{\partial d} \quad (4)$$

$$F_{hs} = -\frac{\pi^2 C \rho_1 \rho_2}{6} \left[\frac{H_2^2 (H_2(d - R_1) - 3R_1 d)}{d^2 (H_2 + d)^3} \right] \quad (5)$$

The Van der Waals interaction energy U_{cy} and force between the cylinder and the substrate is obtained in the same way and are as follow:

$$U_{cy} = -\frac{\pi^2 C \rho_1 \rho_2}{2} \int_0^{H_1} \frac{R_2^2}{(d+H_2+z)^3} dz \quad (6)$$

$$F_{cy} = \frac{\partial U_{cy}}{\partial d} - \frac{\pi^2 C \rho_1 \rho_2}{2} \left[\frac{1}{(d+H_2+H_1)^3} - \frac{1}{2(d+H_2)^3} \right] \quad (7)$$

where H_1 is height of the cylinder, and R_2 is radius of the cylinder.

Finally, the interaction force between a needle and the flat surface is calculated by summing Eq. (5) and Eq. (7).

$$F_{needle-substrate} = F_{hs} + F_{cy} \quad (8)$$

Van der Waals force between parallel or neighbored needles

The interaction between parallel or neighbored needles from opposite surfaces can be divided in two parts. i) Interaction between two parallel cylinders ii) Interaction between a half sphere and a cylinder. However, according to Yongan and Dongqing, the Van der Waals force between a spherical body and so half a sphere body and a vertical cylinder is almost impossible to calculate analytically [10]. Since the half sphere head is a very small part of a needle, its contribution and generated VdW force is very small and can be neglected. So in this case, the whole needle is considered as a cylinder.

The interaction between parallel or neighbored needles is modeled as an interaction between two parallel cylinders with different radiuses in respect to distance between two surfaces (the penetration depth) as shown in Fig. 6.

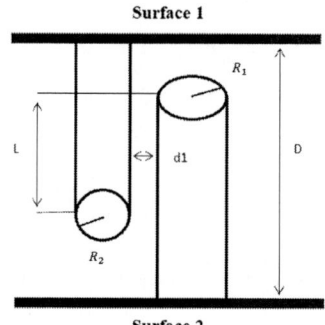

Figure 6: 2D schematic of interaction between two parallel cylinders.

In this case, the interaction energy between two parallel cylinders is represented as [7]:

$$U_{cy-cy} = \frac{-AL}{12\sqrt{2}(d1)^{3/2}} \left[\frac{R_1 R_2}{R_1 + R_2} \right]^{1/2} \quad (9)$$

where $A = C\pi \rho_1 \rho_2$ is the Hamaker Constant, d_1 is the distance between two cylinders, and $L = D - 2(d + H_2)$ is the length of cylindrical interaction.

Then the Van der Waals force between two parallel cylinders is:

$$F_{cy-cy} = \frac{\partial U_{cy-cy}}{\partial d1} \frac{-AL}{8\sqrt{2}(d1)^{5/2}} \left[\frac{R_1 R_2}{R_1 + R_2} \right]^{1/2} \quad (10)$$

At the end, the total Van der Waals interaction force between two Velcro-like surfaces considering same needle densities can be obtained as:

$$F_{total} = 2 \sum_0^m \left(F_{Needle-substrate} \right) + \sum_0^n (F_{cy-cy}) - \sum_0^n (F_{cy-cy}) \quad (11)$$

where m is number of needle in one surface and n is the number of neighboring needles.

Since the surfaces have the same needle density, each needle can have 1 to 6 neighbors depending on the interlacing configuration. However, the interaction of the closest needle is dominant since the VdW strongly depends on the distance between two objects.

SIMULATION

To check the validity and accuracy of the proposed

model, the measured parameters (Table 1) are fed into the model through MATLAB. For an effective bonding area of 0.4096 cm², 180000 needles are generated with Gaussian distributions in height and radius of curvatures. The overall pull-off force in respect to the distance between two surfaces for various distances of closest approach (DCA) considering only one neighbor for each needle is then calculated and shown in Fig. 7. The Hamaker constant of $6.5 \times 10^{-20} J$ is considered. The linear increase of pull-off force at the small distances is a result of more needles involved in bonding and the longer side interaction between interlaced needles in the bonding mechanism. A maximal VdW occurs at a distance corresponding to the mean value of height. For larger distances a linear decrease of pull-off force is resulting from the involvement of less needles and shorter side interaction between them.

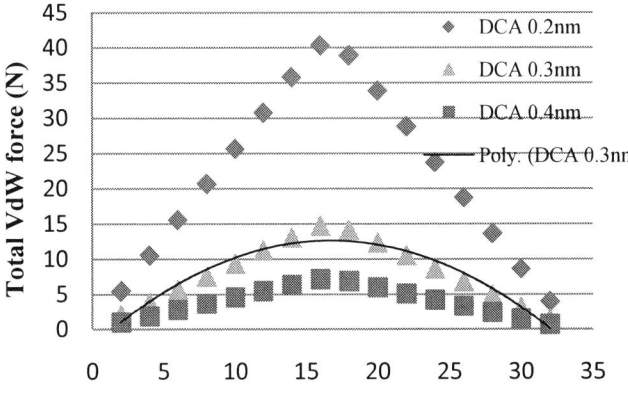

Figure 7: The total simulated VdW force in respect to distance between two surfaces for various distances of closets approach between interacted needles and substrates.

DISCUSION

It is well known that the force of adhesion increases by increasing the contact area or interacting area [11]. But it is important to realize that when two real surfaces are brought into contact, only the asperities on the surfaces are in real contact. In the case of two needle like surfaces in contact and needles made of porous silicon which are deformable, it can be concluded that in the bonding process during interlacing of the needles, the aspirates on the surface of needles are deformed and the real distance between interlaced needles or contributed needles and substrates are shrunk and reaching the closest approaching distance (0.3nm) [11, 12]. By assuming breaking of 10-20% of needles during bonding process (observed from rebondability tests), the DCA of 0.3nm simulation results can be fitted very well to the measurement results.

CONCLUSION

A simple Si based technology resulting in strong permanent bonds at room temperature with multiple bonding and un-bonding capability of the same wafers/chips similar to the Velcro principle is presented and modelled.

The interaction between two Velcro-like surfaces is modelled as one needle like surface interacting with a perfectly flat surface. The effect of each needle is considered locally and the total bonding force is obtained by summing up the all individual contributed needles. The simulation results for closest distances of 0.3nm and the measurement results show very compatible relations for the specific sample, and indicate the validity of the proposed bonding model.

ACKNOWLEDGEMENTS

The authors would like to thank the Ministry of Science, Research and Art at the state of Baden Württemberg, Germany for providing the financial support in the framework program PhD graduate college, "Generation mechanisms for microstructures."

REFERENCES

[1] P. Jonnalagadda, U. Mescheder, and A. Kovacs. "Nanoneedles based on porous silicon for chip bonding with self with self-assembly capability". *J. physica status solidi.* Apr. 2011.

[2] S-H. Lee, J. Chae, N. Yazdani, and K. Najafi, Proc. MEMS, Istanbul, Turkey. 22-26 January, pp 342-345, 2006.

[3] Micro-fastening system and method of manufacture, U.S. Patent 7.181.811. issued February 27, 2007.

[4] H. Föll, M. Christophersen, J. Carstensen, and G. Hasse. *Mater. Sci. Eng.* R 39, 93-141, 2002.

[5] S. lining, W. Lefeng, R. Wibin, and Ch. Liguo. "Considering Van der Waals Force in Micromanipulation Design". Proceeding of the 2007 IEEE International Conference on Mechatronics and Automation, Herbin, China.

[6] S. T. Chow. "Nanoscale surface roughness and particle adhesion on structured substrates". *Nanotechnology* 18(2007) 116713 (4pp).

[7] J. Israelachvili. "Intermolecular and surface force". 2nd edition. Academic press, London.

[8] G. Adams, M. Nosonovsky. "Contact modeling forces". Tribology International 33 (2000) 431-442.

[9] G. Adams, M. Nosonovsky. "Contact modeling forces". Tribology International 33 (2000) 431-442.

[10] G. Yongan, and L. Dongqing. "The Van der Waals interaction between a spherical particle and a cylinder". *J. Colloid and Interface Science* 217, (1999) 60-69.

[11] Q. Li, V. Rudolph, and W. Peukert. "London van der Waals adhesiveness of rough particles". Power Technology 161 (2006) 248-255.

[12] Y. Rabinovich, A. Jishua, A. Ata, S. Rajiv, and M. Brij. "Adhesion between nanoscale rough surfaces". *J. of Colloid and Interface Science* 232, 10-16 (2000).

CONTACT

*Shervin Keshavarzi, tel: +49 7723 9202809 ; kesh@hs-furtwangen.de

FROM CHIPS TO DUST: THE MEMS SHATTER SECURE CHIP

N. Banerjee[1], Y. Xie[1], Md. M. Rahman[1], H. Kim[1,2] and C. H. Mastrangelo[1,2]
[1]Department of Electrical and Computer Engineering, University of Utah, Salt Lake City, USA
[2]Department of Biomedical Engineering, University of Utah, Salt Lake City, USA

ABSTRACT

This paper presents the implementation of a transience mechanism for silicon microchips via low-temperature post-processing steps that transform almost any electronic, optical or MEMS substrate chips into transient ones. Transience is achieved without any hazardous or explosive materials. Triggered chip transience is achieved by the incorporation of a distributed, thermally-activated expanding material on the chip backside. When heated at 160°C the expanding material produces massive chip cleavage mechanically shattering the chip into a heap of silicon dust.

INTRODUCTION

In today's critical military, financial and corporate applications microchips are increasingly used to process and store sensitive information. In every case of theft or loss of equipment there is a possibility of that such information being compromised by malicious subjects. The utilization of software data encryption and access passwords provides a first level of information protection, but encryption can eventually be cracked and the captured hardware can be copied and replicated. In these situations it is therefore increasingly more desirable to utilize microchips which incorporate triggered hardware destruction or disabling transience mechanisms that cannot be cracked.

Several hardware transience methods have been previously reported. These methods include the incorporation of energetic materials that ignite or explode chips [1-4] and chip dissolution by chemically corrosive agents [5-7]. In both of these methods the long term stability, safe handling, and disposal of these substances limits their practical lifetime and range of applications. Therefore it is highly desirable to develop a new transience mechanism that to does not use hazardous corrosive or explosive materials.

In this paper we demonstrate a new safer microchip transience mechanism. Instead of being based on corrosive or explosive materials, the new transience mechanism is based on mechanical stress generation. Mechanical stress is generated within the chip substrate using a distributed high-force actuating material. When the transience is activated, the actuator high level of stress mechanically shatters the chip literally reducing it to a heap of silicon dust as shown in Fig. 1.

SHATTER TRANSIENCE

The actuating material is introduced into the silicon chips through low-temperature post-processing and micro-packaging steps; therefore the new method can transform almost any off-the-shelf silicon chip into a transient device.

First an OTS chip is partially diced on the backside with a series of closely spaced cuts at orthogonal directions. This dicing process produces a series of pillars and grooves that behave as cracks in the silicon substrate. The backside grooves are next refilled with the actuator material. The transience actuator consists of an expanding microscopic material that exerts pressure against the thin silicon walls or pillars left by the dicing process thus inducing fractures and shattering. If the induced stress is sufficiently large the cracks propagate all the way through to the top surface thus fracturing the chip into a large number of pieces as shown in Fig. 2 below.

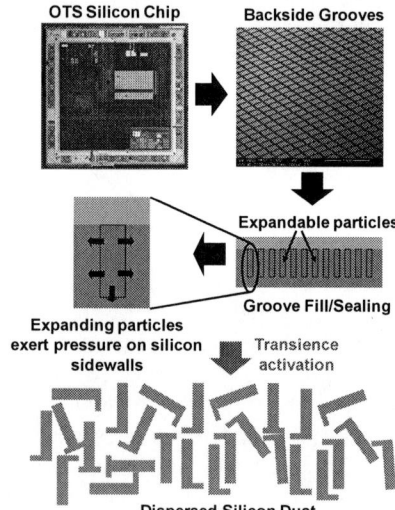

Figure 2: Schematic of the shatter-chip transience mechanism.

Each diced groove can be treated as a partial crack in the silicon substrate. In order to achieve transience the stress must be sufficiently high to propagate the crack to the top surface of the chip. The stress and pressure required to shatter the chip can be thus approximately calculated from fracture mechanics theory.

The Griffith fracture stress required for mode I (uniform stress) crack propagation as shown in Fig. 3 is

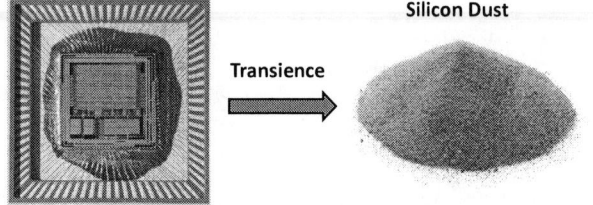

Figure 1: Schematic showing silicon chips being reduced to silicon dust by shatter transience action.

978-1-4799-3510-9/14 $31.00 © 2014 IEEE   1123

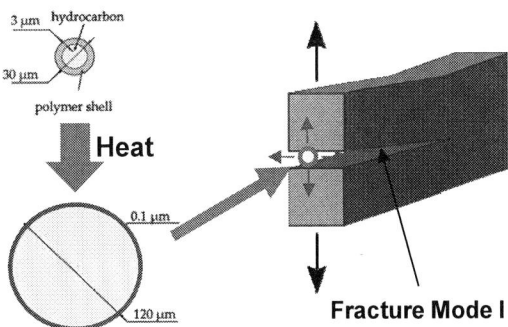

Figure 3: Large expansion of thermally expandable microspheres at elevated temperatures (130-160°C). The expanding microspheres produce a tensile stress that propagates a crack in the substrate causing chip fracture.

$$\sigma_f = (2E\gamma/\pi a)^{\frac{1}{2}} = K_{Ic}/\alpha\sqrt{\pi a} \qquad (1)$$

where E is the substrate Young's modulus, γ is the crystal free energy and a is the depth of the groove. The parameter K_{Ic} is the critical stress intensity factor or fracture toughness and $\alpha = 1$ for edge cracks. For single-crystal silicon, K_{Ic} is typically 0.7-1.3 MPa m$^{0.5}$. A wedging force of 40N generated over a 5 mm long, 300 μm deep grooves corresponding to a uniform wedge pressure of 26 MPa is thus sufficient to cleave the silicon chip under mode I fracture.

3D Device simulation: The fracture stress calculated from Eq. (1) depends on the crack geometry and the energy release rate of the crack propagation. Bursting and cracking of silicon microstructures is well known to depend not only on the shape of the crack but also many unknown factors such as the roughness of the surfaces [8]. All of these factors produce fracture stresses much less than the silicon yield ~7 GPa. Because of these uncertainties simulations are unable to provide quantitative fracture stress information, but they can provide the most likely locations where cracks propagate. We have performed COMSOL device simulations to determine the locations of higher stress under constant internal pressure caused by the expanding material. Fig. 4 shows an example stress distribution at the bottom of the grooves. The highest stress

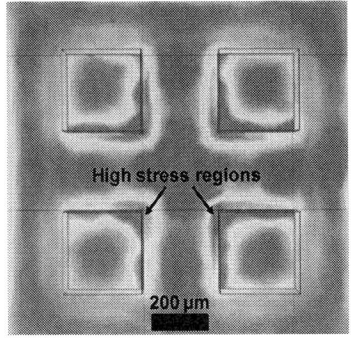

Figure 4: Normalized stress distribution at the bottom of the grooves. The points of highest stress are at the corner and edges of the grooves.

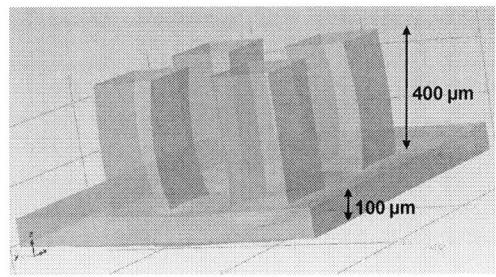

Figure 5: Example deformation of remaining silicon posts under uniform internal pressure and no pressure at the periphery.

occurs at the corners and along the edges. Fig. 5 shows qualitatively the deformation of the post structures and remaining silicon under internal pressure no side pressure. The deformation is consistent with a mode I fracture of a doubly supported beams. The actual type of fracture depends on the specific packaging scheme utilized.

Expanding actuating material: The expanding material used in our experiments consists of thermally expandable polymeric microspheres (Dualite U005-190D, Henkel) [9-17], 40 μm in diameter. These are polymeric core/shell particles in which a volatile hydrocarbon is encapsulated by a thermoplastic shell. When these microspheres are heated beyond a critical temperature (130-160°C) the hydrocarbon boils acting as a blowing agent and the pressure rises beyond the pressure required for plastic deformation of the shell. They thus expand and increase their volume dramatically expanding 60 times of their original volume. The volume increase is retained upon cooling thus leading to a density reduction from around 1100 kg m^{-3} to about 30 kg m^{-3}. The expanding particles thus create internal stresses and the crack wedging action inside the grooves.

Blowing agent pressure: Beyond the critical temperature the blowing agent is completely converted into a vapor; therefore we can estimate the pressure P_b exerted by the blowing agent pressure from its density and molecular weight.

$$P_b \approx \frac{k_B T \cdot N_b}{V_b} = \frac{k_B T \cdot \rho_l \cdot V_b \cdot G}{V_b m_w} = \frac{k_B T \cdot \rho_l \cdot G}{m_w} \qquad (2)$$

where N_b is the number of molecules of the blowing agent enclosed in the microparticle volume V_b and ρ_l and m_w are the liquid density and molecular weight of the blowing agent, respectively. G is Avogadro's number, T is the temperature, and k_B is Boltzmann's constant. Note that the maximum blowing agent pressure occurs when the blowing agent is instantaneously converted into a gas; hence Eq. (2) expression is independent of the volume of the microparticle. For a blowing agent such isopentane [15-16] with liquid density of ρ_l =626 kg/m^3, and molecular weight m_w=72.1 g/mole, at T=160°C, the blowing agent pressure is $P_b \approx 31$ MPa. This is the maximum pressure available for stressing the microchip. The internal pressure varies between 20-30 MPa for common blowing agents. Experimentally it has been determined that stresses as high as 30 MPa are attained consistent with our simple estimate.

DEVICE FABRICATION

In order to convert a conventional silicon chip into a transient chip, first the chip backside is partially diced to form a network of narrow and closely spaced grooves. The dicing was done using Disco DAD 641 dicing machine with 250 μm thick diamond blades. The grooves were 250 μm wide and 400 μm deep spaced 250 μm apart. These grooves were then filled with the thermally expandable microspheres 40 μm in diameter and the excess particles were removed using a razor blade.

The chip backside is next attached to a pyrex glass cap with a room temperature adhesive thus sealing the filler material within the grooves. The seal assures that the filler particles do not force their way out of the grooves during expansion. Some of the chips were next packaged within a harness with the chips sandwiched between two aluminum plates tightly screwed together as shown in the schematic of Fig. 6. The harness ensures that the expanding

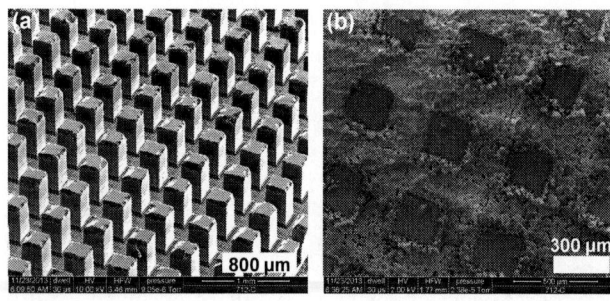

Figure 6: Schematic of packaging scheme and test fixture

microparticles exert pressure in all directions without delamination of the glass sealing cap. Tests were carried out on chips with and without the harness.

EXPERIMENTAL RESULTS

In order to test our concept, we incorporated the transience mechanism on a test off-the-shelf silicon microchip. The backside of the chip was diced into 250 μm wide deep grooves. Fig. 7(a) shows SEM pictures of the backside grooves. After dicing the grooves were filled with expandable microparticles up to the brim as shown in Fig. 7(b). Figs. 8(a) and (b) shows photographs of the backside

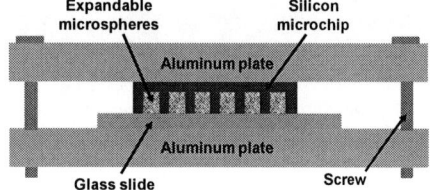

Figure 7: (a) SEM of dicing grooves on the chip backside. (b) SEM of grooves filled with thermally actuated microparticles.

of a chip backside after refill and under unconstrained expansion conditions at 160°C. Note that the microparticle volume increases greatly. A separate chip was next sealed with a glass cap and aluminum harness. The assembly was

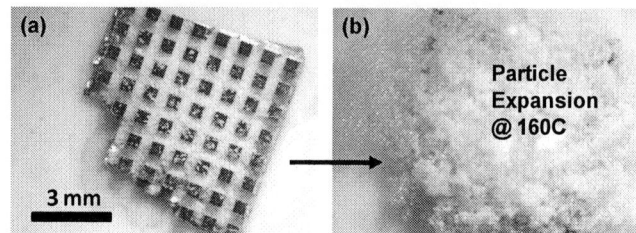

Figure 8: (a) Partially diced silicon chip with microparticle filled grooves. (b) Unconstrained expansion produces a 4-fold diameter expansion of the particles.

next placed on a heater at a temperature of 160°C. Within two minutes the polymeric microspheres start expanding rapidly and immediately fracture the whole microchip into small silicon pieces. Fig. 9 shows the typical results of constrained microparticle expansion for a chip with no harness. The chip is extensively fractured forming a three-dimensional heap of silicon dust and expanded particles. In

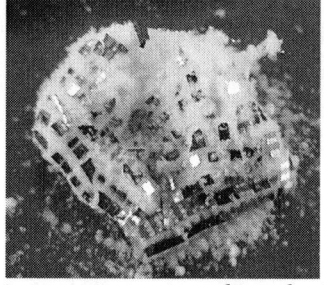

Figure 9: Chip transience is achieved via constrained expansion of the microparticles. The typical result is a heap consisting of a mixture of particles and silicon dust.

general the harness is not needed if the width of the grooves exceeds 200 μm.

Fig. 10 shows optical and SEM photographs of the chip before and after transience. The average distance between cracks is approximately the same as the groove pitch. The three dimensional nature of the heap facilitates the

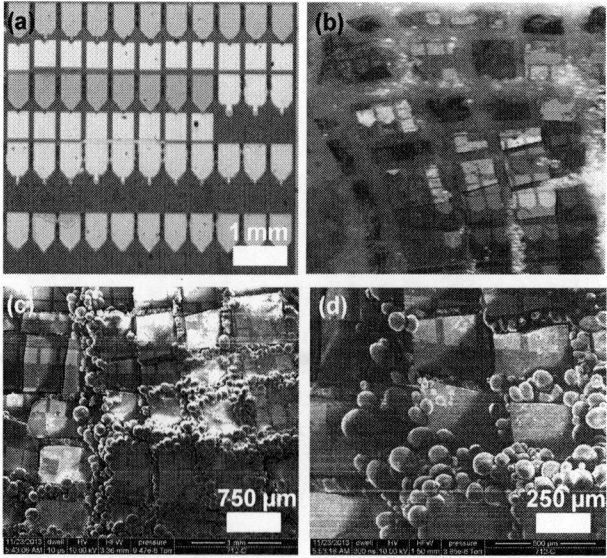

Figure 10: SEM and optical photographs of the test chip before and after transience showing chip shattering.

dispersion of the silicon dust to the surrounding environment. Experimentally we have tested several combinations of groove width and pitch ranging from 45-250 μm in width and 250-1000 μm pitch to produce the best results. In general we found that the shatter action is more effectively produced with wider grooves.

SUMMARY

This paper presents the successful implementation of a transience mechanism that can be incorporated in most microchips through low-temperature post processing. Chip transience is caused via the triggered expansion of thermally expanding microparticles stored within grooves formed on the backside of the chip. This transience mechanism does not utilize any corrosive agents or explosive substances therefore it produces transient chips that can be safely handled without any safety concerns. This transience method is modular and add-on thus it can be used to on-demand disable circuitry/sensors/chips serving diverse applications.

ACKNOWLEDGEMENTS

We would specially like to thank Shashank. S. Pandey for his help and support during development and testing of these devices.

REFERENCES

[1] G. Smolker, "Method and apparatus for protecting sensitive information contained in thin film microelectronic circuitry," United States Patent application number 425405, The United States of America as represented by the Secretary of the Navy, Washington, DC. May 1975.

[2] D. J. Shiel and D. L. Davis, "Method and appparatus for fast self-destruction of a CMOS integrated circuit," United States Patent application number 581436, Intel Corporation. Santa Clara. CA. April 1998

[3] F. Z. Keister and G. S. Smolker, "Single layer self-destruct circuit produced by co-deposition of Tungsten oxide and Aluminium," United States Patent application number 90204, The United States of America as represented by the Secretary of the Navy, Washington, DC. June 1973.

[4] S. Pandey and C. H. Mastrangelo, "An exothermal energy release layer for microchip transience," *Proc. IEEE Sensors Conf.*, Baltimore, Md, Nov. 3-6, 2013, pp. 1759-1762.

[5] S-W. Hwang et al. "A physically transient form of silicon electronics", *Science*, 337, 1640 (2012).

[6] N. Banerjee, Y. Xie, H. Kim and C. H. Mastrangelo, "Microfluidic device for triggered chip transience," *Proc. IEEE Sensors Conf.*, Baltimore, Md, Nov. 3-6, 2013, pp. 920-923.

[7] X. Gu, X. Lou, R. Song, Y. Zhao and L. Zhang, "Simulation research on a novel microfluidic self-destruct device for chips", *IEEE Nano/Micro*

Engineered and Molecular Systems, pp 373-378, Xiamen, China, 2010.

[8] A. K. Henning, S. Patel, M. Selser and B. A. Cozad, "Factors affecting silicon membrane burst strength," in Reliability, Testing, and Characterization of MEMS/MOEMS III, edited by Danelle M. Tanner, Rajeshuni Ramesham, *Proceedings of SPIE* Vol. 5343 (SPIE, Bellingham, WA, 2004), pp. 145-153.

[9] Henkel microspheres home page: www.henkelna.com.

[10] Morehouse, D. S. J.; Tetreault, R. J. U.S. Pat. 3,615,972, 1964.

[11] J. Lundqvist, Eur. Pat. 0 486 080 B1, 1992.

[12] T. Yokomizo, K. Tanaka, K. Niinuma, Jpn. Pat. 9019635, 1997.

[13] M. Jonsson, O. Nordin, A. Larsson Kron, and E. Malmström, "Thermally expandable microspheres with excellent expansion characteristics at high temperature", *J. Applied Polymer Science,* 2010, 117, 384-392.

[14] M. Jonsson, O. Nordin, E. Malmström and C. Hammer, "Suspension polymerization of thermally expandable core/shell particles", *Polymer* 2006, 47, 3315-3324

[15] J. Fredlund, *Synthesis of Thermo Expandable Microspheres*, MSc Thesis, KTH Sundsvall, Sweden, 2011.

[16] J. Magnus, *Thermally Expandable Microspheres Prepared via Suspension Polymerization - Synthesis, Characterization, and Application,* PhD Thesis, KTH Stockholm, Sweden, 2010.

[17] Masayasu Fujino, Takashi Taniguchi, Yasuhiro Kawaguchi, Masahiro Ohshima, "Mathematical models and numerical simulations of a thermally expandable microballoon for plastic foaming," *Chemical Engineering Science*, Volume 104, 18 Dec. 2013, pp. 220-227.

CONTACT

*Niladri Banerjee, tel: +1-801-556-4133; niladribanerjeehere@gmail.com

LONG TERM GLASS-ENCAPSULATED PACKAGING FOR IMPLANT ELECTRONICS

Jay Han-Chieh Chang, Yang Liu and Yu-Chong Tai
California Institute of Technology, USA

ABSTRACT

Hermetic Titanium-alloy packaging (e.g., used in pacemakers and cochlear implants) has been accepted as the industrial standard for decades. However, two remaining issues of this well-known technology are the size and limited number of feedthroughs [1]. On the other hand, the next generation wireless intraocular retinal prosthetic devices do require unprecedented small size and large number of leads to be fitted inside a human eyeball so the traditional metal packaging is difficult to be implemented. Therefore, these new generation of microimplants will need a new packaging scheme. This paper then reports a new long-term packaging method using glass encapsulation featuring a controlled failure mode from fast diffusion to slow undercut. The results is promising that this new packaging scheme could survive more than 10 years by accelerated "active" lifetime soaking test (i.e., with electric field applied) in 0.9 wt.% saline solution. As a whole, this new method provides several advantages including easy employment, controllable long life time, and enhanced heat dissipation.

INTRODUCTION

Smaller biomedical implant electronic devices have drawn a lot of attention recently because of their potential as a platform for healthcare research and *in vivo* treatments. Much effort has been done to develop smaller devices such as pacemakers [2], cochlear implants [3], neural stimulators [4], and retinal prosthesis [5]. Without exception, all implants need reliable packaging and the desirable requirements for the packaging of next generation implants include, but are not limited to, (1) biocompatible surfaces that cells and tissues like; (2) long lifetime in corrosive body fluids; (3) easy deployment, and (4) a high-density and high-lead-count feedthrough capability. The fact is that the sizes of implant devices have been reduced continuously for many decades. However, one major barrier to realizing further downsizing for "micro" implants is still the lack of reliable high-density and high-lead-count packaging technology.

Packaging materials applied to implant devices include metals, glasses, polymers, etc. Among them, for examples, biocompatible metal cans and glass caps combined with laser welding [6] and brazing ring [7] hermetic encapsulation have been demonstrated on some devices with acceptable lifetime. However, these packaging techniques are both expensive and time-consuming, and can only be applied to devices with low-density and small number of electrodes or feedthroughs. In addition, metal can packaging could limit the frequency range of the implant if using wireless design. For an example, the metallic brazing ring could modify the inductivity of the implant antenna to reduce the communication distance between the implant and external control [8]. Polymers, with

the advantages of light-weight, transparency, and flexibility, have also been used as packaging materials for implants [9]. However, even with multi-layer structures, polymer's inherent high water vapor transmission rate (WVTR), which can be several orders of magnitude higher than those of metals and glasses, has always been a concern for long term packaging.

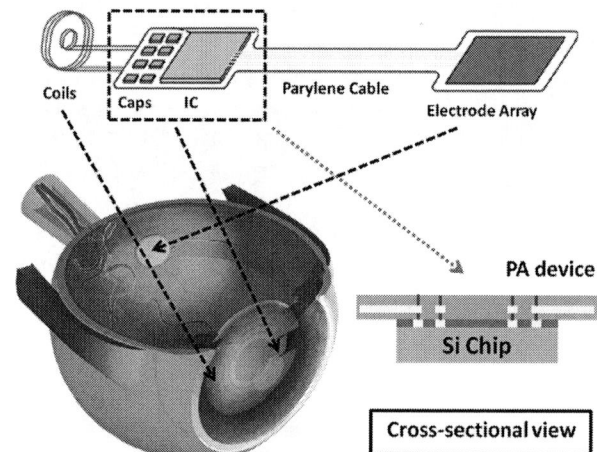

Figure 1. Schematics of the intraocular epiretinal implant and the cross-sectional view of the parylene-supported device integrated with IC chips.

Our lab has been developing microimplant devices for many years. Recently, we have focused on high-resolution retinal prosthesis with 1024 electrodes to partially restore vision for blind people with retinal degenerative diseases [10]. As shown in Figure 1, the device is designed with a flexible and biocompatible parylene-metal-parylene sandwich skin structure [11]. In consideration of the need for 1,024 channels, both the high-density connection and packaging have to be studied. High-yield connection for high-density multi-channel chips has been reported [12-14]. The important remaining work is the long-term reliable packaging of this implant device for its use inside human eyeballs.

DESIGN

Previously we studied a parylene-metal-parylene packaging structure, as shown in Figure 2, which could protect chip samples for up to 7 years [15]. It was found that electrochemical corrosion was the main failure mechanism of this packaging scheme, which is similar to many other implant devices. Therefore, we concluded that water vapor diffusion through the protecting barrier and the following corrosion on conductive epoxy and/or electrode metal pads

978-1-4799-3510-9/14 $31.00 © 2014 IEEE

were the main failure mechanisms. Although a thicker barrier can be adopted to extend the lifetime, the size enlargement, the device stiffness, and the heat dissipation will become the resulting concomitants. Therefore, a new packaging scheme with glass (100 μm used here) encapsulation is studied here as shown in Figure 3. The nature of low WVTR and high thermal conduction of the glass, compared to normal polymers, make this new design very attractive for study. Theoretically, with suitable interface treatments, this approach can change the main failure mechanism from fast diffusion to slow undercut and this is experimentally tested in the following.

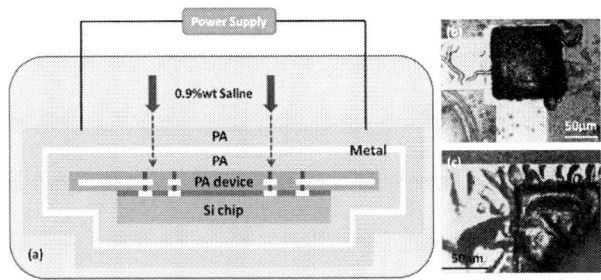

Figure 2. (a) Schematics of the testing samples with parylene-metal-parylene packaging under active soaking test in 0.9% wt. saline solution. Corrosion of conductive epoxy (b) and electrode metal pad (c) are found to be the main failure modes of this packaging approach.

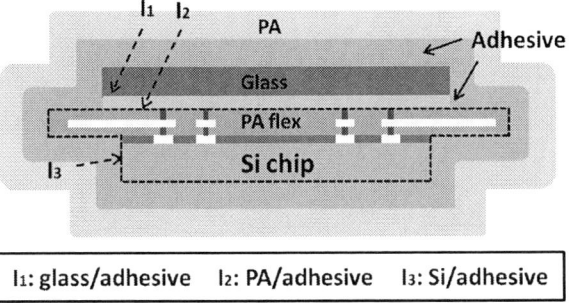

l₁: glass/adhesive   l₂: PA/adhesive   l₃: Si/adhesive

Figure 3. Schematics of the proposed structure of the glass-encapsulated packaging on the testing sample (dash-line).

EXPERIMENTS
Experiments on adhesives

Figure 3 shows that additional biocompatible adhesives and parylene coating are needed to add a piece of protecting glass in our proposed structure, which creates a total of 3 different kinds of interfaces. The first experiment was to determine which adhesive and/or adhesion promoter could create the long-lasting interfaces. In our experiments, biocompatible silicone and epoxy were selected as the first candidates of adhesives. Silicone adhesion promoter (SAP) was first applied by dip coating, where SAP is a formulated silane applied in a solvent and is used to improve the bond strength between adhesives and other substrate materials. After adhesion promoter coating, silicone and epoxy were

applied and cured according to the recipe. Figure 4 shows the soaking experiments with the prepared samples (with 3 different interfaces) in 97°C saline solution. The undercuts along the interfaces between substrates and adhesives were measured as the main parameters to evaluate the adhesion. Accelerated soaking tests show that the biocompatible silicone (MED-6219) with SAP (MED-160) have the lowest undercut rates (as shown in Table 1) so this was chosen as the bonding method for more experiments. Figure 5 also demonstrates the comparison of testing samples with and without SAP.

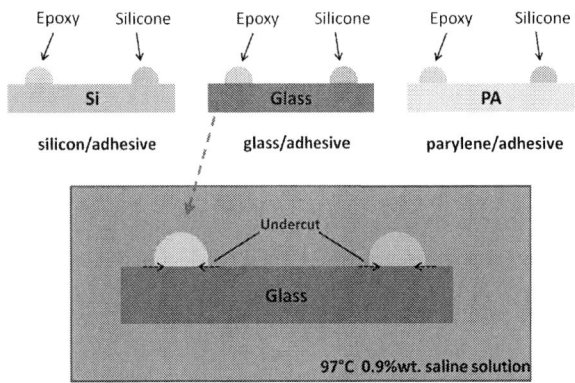

Figure 4. Schematic diagrams to show testing samples with three different interfaces soaked in 97°C saline solution for undercut observation.

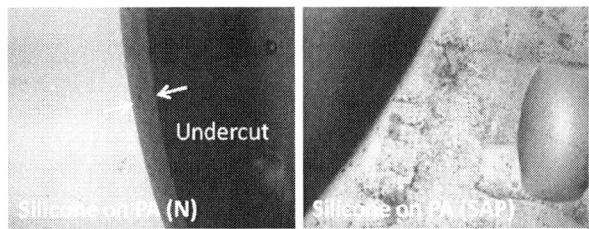

Figure 5. Comparison of undercut rates @97°C saline solution. (Left) Without silicone adhesion promoter (SAP), there was a huge undercut between silicone and parylene substrate. (Right) With SAP, very little undercut was observed.

Table 1. Collected undercut data of two different adhesives, epoxy (M-121HP) and silicone (MED-6219) on three different substrates with and without silicone adhesion promoter (MED-160).

UNDERCUT (μm/day)	Si substrate		Glass substrate		PA substrate	
	N	SAP	N	SAP	N	SAP
Epoxy (M-121HP)	31.2	4.8	17.4	4.9	19.4	3.8
Silicone (MED-6219)	12.5	2.2	13.3	2.9	22.3	3.2
N: no treatment; SAP: silicone adhesion promotor (MED-160)						

Soaking experiments to predict lifetime

We designed high-density, multi-channel resistor chips to simulate the real high lead-count ASIC chips, as shown in Figure 6. These resistor chips were connected to the parylene flex by low-temperature adhesive bonding [13-14] and the conductive epoxy squeegee technique [12]. The connected devices were then finished with the glass-encapsulated packaging scheme (plus biocompatible silicone and parylene coatings as the handling protection), and actively soaked in saline solution until the first failure was detected. In this work, the failure was defined as 50% change in line resistance. The observed failure mode was found to be the very slow undercut along the interface between the glass and silicone adhesive, as shown in Figure 7. Figure 8 also shows the resistance aging curves of the testing samples (E) with various different packaging schemes soaked in high temperature saline solution. Samples without any protection (A), with 40 µm parylene (B), with 40 µm parylene plus 5mm silicone combination (C), and with 20 µm parylene plus 0.5 µm metal (Ti-Au-Ti) plus 20 µm parylene (D) were prepared as comparisons. The data were used to extrapolate the lifetime at body temperature by Arrhenius relationship [16-17], as shown in Table 2. The equation of the relationship at a given system temperature T (in Kelvin) is expressed by:

$$MTTF \sim A\exp(\frac{-E_a}{kT}) \qquad (1)$$

Where A is the pre-exponential constant, E_a is the activation energy of failures (in eV), and k is the Boltzmann's constant (8.62×10^{-5} eV-K^{-1}). For each failure mode, the corresponding activation energy can be calculated at various high temperatures (more than 2).

With this glass-encapsulated packaging scheme, the lifetime can be extended to more than 10 years. Note that the size of the resistor chip is designed as 6 mm × 6 mm, while the glass applied here is 10 mm × 10 mm, so there is a 2 mm of undercut limit on each side, which translates to around 5 years/mm of life time design. Accordingly, longer lifetime can be designed with longer undercut length.

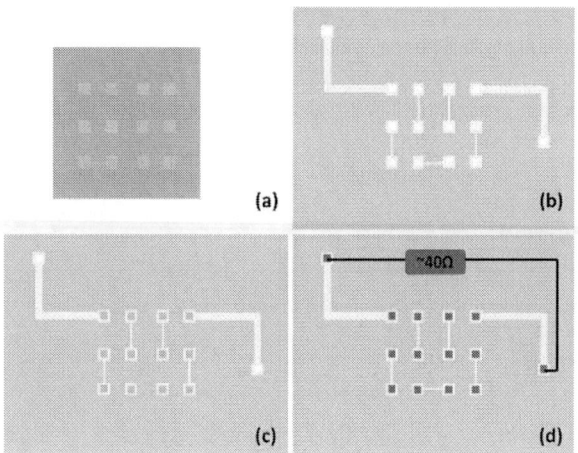

Figure 6. Schematics of (a) high-density multi-channel resistor chip to simulate the real ASIC chip, (b) parylene flex designed to integrate with the chip, (c) alignment of the parylene flex with the chip, and (d) testing sample with around 40 Ω line resistance after squeegee connection.

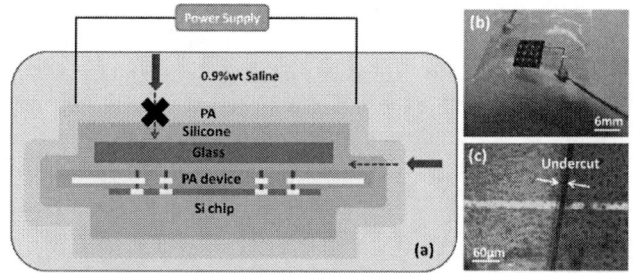

Figure 7. (a) Schematics of the testing samples with glass-encapsulated packaging under active soaking test. Silicone and parylene coating were applied to protect the samples. (b) A real testing sample connected with external wires for testing. (c) The very slow undercut along the interface was the main failure.

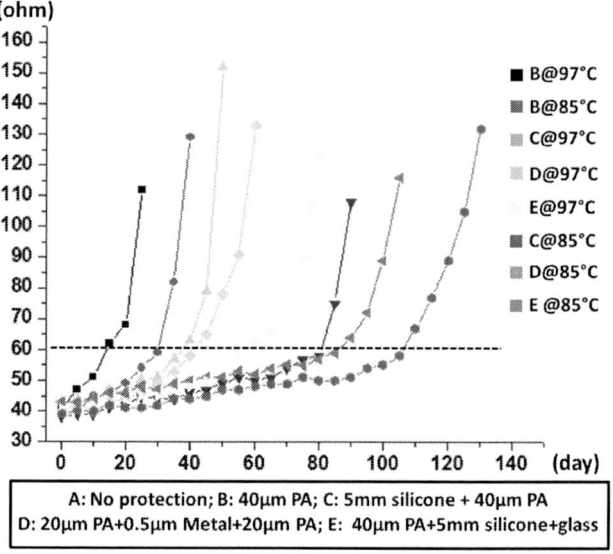

A: No protection; B: 40µm PA; C: 5mm silicone + 40µm PA
D: 20µm PA+0.5µm Metal+20µm PA; E: 40µm PA+5mm silicone+glass

Figure 8. Aging curves of testing samples with various different packaging schemes at 85° and 97°C saline solution. The failure is defined as 50% changes in line resistance (60 Ω in this testing).

Table 2. Extrapolated body-temperature lifetime calculated by Arrhenius relationship. Each failure mode corresponds to a specific activation energy (E_a)

Protection	97°C (day)	85°C (day)	Ea (eV)	37°C
A	1.7	2.6	-0.40	20 days
B	15	31	-0.69	2.7 years
C	40	82	-0.68	6.9 years
D	42	87	-0.69	7.7 years
E	59	107	-0.68	10.3 years

CONCLUSION

A long-term, glass-encapsulated packaging technique for implant devices was proposed. Experiments on adhesives to create the long-lasting interfaces with the proposed glass-encapsulated structure were then studied. It was found that the combined biocompatible silicone (MED-6219) and adhesion promoter (MED-160) gave the best results. Lifetime test samples with resistor chips on parylene flex were prepared and gone through active and accelerated soaking tests. The extrapolated lifetime at body temperature was calculated by Arrhenius relationship to be more than 10 years.

This proposed glass-encapsulated packaging approach changes the failure mode from fast diffusion through barrier layers to slow undercut of the glass. This way, the desirable life time can be designed by with long enough undercut length.

ACKNOWLEDGEMENTS

This work is supported by the NSF ERC center of Biomimetic MicroElectronic Systems (BMES). The authors would like to thank Mr. Trevor Roper for assistance with fabrication and equipment maintenance, and other group members of the Caltech Micromachining Laboratory for the fruitful discussions.

REFERENCES

[1] Second Sight: http://2-sight.eu/en/home-en

[2] D. Difrancesco, and P. Tortora, "Direct activation of cardiac pacemaker channels by intracellular cyclic AMP", *Nature*, vol. 351, pp. 145-147, 1991.

[3] H. McDermott, "An advanced multiple channel cochlear implant", *IEEE Trans. Biomed. Eng.*, vol. 36, pp. 789-797, 1989.

[4] G. Gudnason, and E. Bruun "A chip for an implantable neural stimulator", *Analog Integrated Circuit and Signal Processing*, vol. 22, pp. 81-89, 1999.

[5] J.H. Chang, Y. Liu, D. Kang, M. Monge, Y. Zhao, C.C. Yu, A. Emami, J. Weiland, M. Humayun, and Y.C. Tai, "Packaging study for a 512 channel intraocular epiretinal implant", in *Digest Tech. Papers MEMS'13 Conference*, Paris, Jan 29-Feb 2, 2012, pp. 353-356.

[6] X. Cao, M. Jahazi, J.P. Immarigeon, and W. Wallace, "A review of laser welding techniques for magnesium alloys", *Journal of Materials Processing Technology*, vol. 171, pp. 188-204, 2006.

[7] A. Ginggen, Y. Tardy, R. Crivelli, T. Bork, and P. Renaud, "A telemetric pressure sensor system for biomedical applications", *IEEE Trans. Biomed. Eng.*, vol. 55, pp. 1374-1381, 2008.

[8] T. Bork, A. Hogg, M. Lempen, D. Muller, D. Joss, T. Bardyn, P. Buchler, H. Keppner, S. Braun, Y. Tardy, and J. Burger, "Development and *in-vitro* characterization of an implantable flow sensing transducer for hydrocephalus", *Biomed. Microdevices*, vol. 12, pp. 607-618, 2010.

[9] J.J. Senkevich, and S.B. Desu, "Compositional studies of near-room-temperature thermal CVD poly(chloro-p-xylylene)/SiO_2 nanocomposites", *Appl. Phys. A mater. Sci. Process*, vol. 70, pp. 541-546, 2000.

[10] M. Humayun, R. Propst, E. de Juan, Jr., K. McCormick, and D. Hickingbotham, "Bipolar surface electrical stimulation of the vertebrate retina", *Archives of Ophthalmology*, vol. 112, pp. 110-116, 1994.

[11] J.H. Chang, R. Huang and Y.C. Tai, "High density IC chip integration with parylene pocket", in Digest Tech. Papers NEMS'11 Conference, Kaohsiung, Feb 20-23, 2011, pp. 1067-1070.

[12] J.H. Chang, R. Huang, and Y.C. Tai, "High density 256-channel chip integration with flexible parylene pocket", in *Digest Tech. Papers Transducers'11 Conference*, Peking, June 5-9, 2011, pp. 378-381.

[13] J.H. Chang, D. Kang and Y.C. Tai, "High yield packaging for high-density multi-channel chip integration on flexible parylene substrates", in Digest Tech. Papers MEMS'12 Conference, Paris, Jan 29-Feb 2, 2012, pp. 353-356.

[14] J.H. Chang, Y. Liu, Y.C. Tai, "A low-temperature parylene-C-to-silicon bonding using photo-patternable adhesives and its application", in *Digest Tech. Papers Transducers'13 Conference*, Barcelona, June 16-20, 2013, pp. 2217-2220.

[15] J.H. Chang, Y. Liu, D. Kang and Y.C. Tai, "Reliable packaging for parylene-based flexible retinal implants", in *Digest Tech. Papers Transducers'13 Conference*, Barcelona, June 16-20, 2013, pp. 2612-2615.

[16] J.H. Chang, B. Lu, and Y.C. Tai, "Adhesion enhancing surface treatments for parylene deposition", in *Digest Tech. Papers Transducers'11 Conference*, Peking, June 5-9, 2011, pp. 390-393.

[17] D. Difrancesco, and P. Tortora, "Accelerated aging and lifetime prediction: Review of two non-Arrhenius behaviour due to two competing processes", *J. Polymer Degradation and Stability*, vol. 90, pp. 395-404, 2005.

CONTACT

*Jay Han-Chieh Chang, tel: +1-626-3952267; jaychang@caltech.edu

LOW-TEMPERATURE GOLD-GOLD BONDING USING SELECTIVE FORMATION OF NANOPOROUS POWDERS FOR BUMP INTERCONNECTS

Hayata Mimatsu[1], Jun Mizuno[2], Takashi Kasahara[1], Mikiko Saito[2], Shuichi Shoji[1], and Hiroshi Nishikawa[3]

[1]Major in Nano-science and Nano-engineering, Waseda University, Japan
[2]Institute for Nano-science and Nano-technology, Waseda University, Japan
[3]Joining and Welding Research Institute, Osaka University, Japan

ABSTRACT

We proposed low-temperature Au-Au bump interconnects bonding using nanoporous Au-Ag powders as a connective adhesion. The nanoporous powders were formed by de-alloying Au-Ag alloy in HNO_3 solution. To optimize the pore size, the influence of the annealing temperature on the porous structures was investigated. Selective transfer of the nanoporous powders on bumps was obtained by stamping process. Bonding strength of about 2.4 MPa was achieved at 150 °C by using nanoporous Au-Ag powders. Bonding interface was evaluated by scanning acoustic microscope and scanning electron microscopy. This result indicated that the nanoporous powder is a useful material for low-temperature Au-Au bonding.

INTRODUCTION

Recently, miniaturization, energy saving system and speed-up techniques have been required in field of various devices packaging technology. Two-dimensional (2D) assemblies including a multi-chip module (MCM) are one of the methods to reduce the area of packaging. Flip-chip assemblies are frequently used for three-dimensional (3D) large-scale integration (LSI) [1,2]. Compared with wire bonding, flip-chip bump interconnections have several advantages such as high packaging density, wide bandwidth, and a decrease in signal delay owing to the short connection lengths between chips [3]. High-temperature process is required for the diffusion bonding of metal bumps. Low-temperature bonding has been studied in order to reduce the influence of heat process on MEMS devices caused by the differences of coefficient of thermal expansion of the elements [1]. Plasma treatment or ultrasonic bonding is used for Au-Au low-temperature bonding [4,5]. Recently, surface structure controlling techniques for the material for low-temperature bonding have been reported. The use of nanoporous structures on the surface is one of the efficient methods, due to sponge-like flexible features and highly reactive surface of the nano- or mesoscale structures. Since nanoporous structures have a large surface-to-volume ratio, they have been widely used in various applications in sensors, actuators, and clean-energy devices [6,7]. For example, the electrical connective packaging by using nanoporous gold pads was reported [7]. However, the formation of a nanoporous bump structure on the substrate involves many steps of sputtering, lithography, evaporation, and etching.

In our previous work, low-temperature Au-Au bonding using a nanoporous metal sheet as an adhesive layer was developed [8]. The nanoporous sheet was sandwiched between two Au substrates. The bonding behaviors were evaluated. Au-Au bonding using the sheet was achieved by a simple process without any surface treatments. It was found that a nanoporous sheet is useful to reduce bonding temperature. In this study, we demonstrate low-temperature bonding using Au-Ag nanoporous powders as a joint layer applicable for micro-meter-scale bump interconnects.

EXPERIMENTAL PROCEDURE

Concept of Selective Formation

A selectively formation of nanoporous powders on bumps are the key technology. The concept of selective nanoporous powder formations on the Au bump by stamping and low temperature bonding for bump interconnects are illustrated in Fig. 1 (a) and (b). Nanoporous Au-Ag powder is spreaded on the Si wafer. The Au bump formed on MEMS device pushes down to it, and then release (Fig. 1 (a)). Nanoporous powders on Si were selectively transferred on the Au bumps. The bonding is performed under appropriate pressure and heating (Fig. 1 (b)).

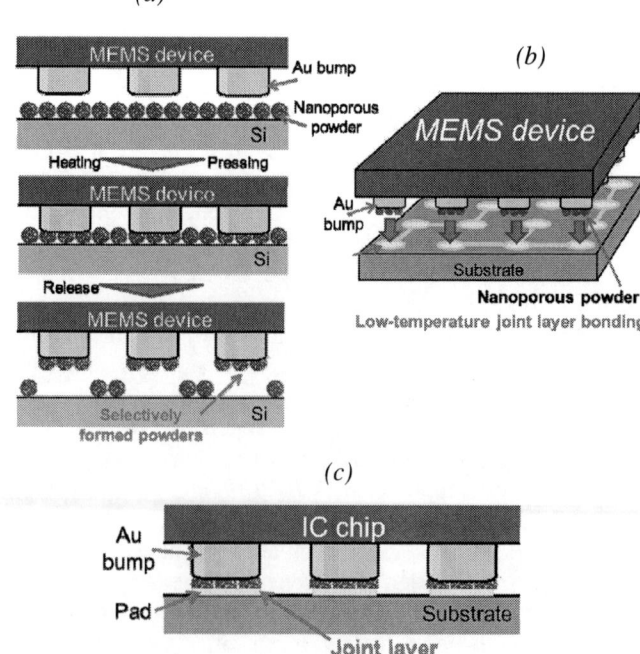

Figure 1: Concept of low-temperature Au-Au bonding using nanoporous powders for bump interconnects

Formation of Nanoporous Powders

Nanoporous structured metals were typically fabricated by extracting one component from a metal alloy [9]. This dealloying process has been demonstrated for various alloy combinations such as Cu–Mg, Pd–Al, Au–Sn, and Au–Ag. We selected the Au–Ag alloy because the resistivities of Ag and Au are lower than those of all other metals except Cu. Figure 2 shows the fabrication process of nanoporous powders. In this study, nanoporous powders were formed by dealloying Ag from Au-Ag alloy as follows. First, we prepared a 100-μm-thick Au-Ag sheet with a Au/Ag mass ratio of 25:75. Ag of the alloy sheet was etched in HNO_3 under ultrasonication for 15 minutes (Fig. 2 (a)). After etching, the shape of the Au-Ag sheet fell apart and turned to powders. The powders were rinsed in water and isopropyl alcohol (IPA) and dried in the atmosphere (Fig. 2 (b) and (c)). Fabricated powders were shown in Fig 3. About several nano-meters pore structures were formed on the powder surface.

Evaluation of Nanoporous Powders

Scanning electron microscopy (SEM) was used to observe the surface of the powders and their changes after the annealing at 50, 100 and 150 °C for 20 min. Au-Au bonding with nanoporous powders as a joint layer was demonstrated. Figure 4 shows a schematic diagram of the Au–Au bonding. Two 1 cm² Si/Ti/Au substrates were prepared by the deposition of a Ti layer on a Si substrate, followed by the deposition of a Au layer by evaporation. The thickness of the Si wafer was 525 μm. A nanoporous Au-Ag powders were then sandwiched between the substrates. The two substrates and the nanoporous sheet were bonded under the pressure of 10 MPa for 20 min. The bonding strengths were measured by a tensile test.

Selective Formation of Powders on the Bump

Nanoporous powders were selectively bonded to Au bumps by stamping (Fig. 1 (a)). Since natural oxide films on Si substrate inhibit local diffusions of Au, the powders are attached to only on the Au bumps. Bump pattern was fabricated on a Si wafer by electron beam (EB) evaporation. The bump number was 100 of 10 × 10 array. The bump diameter was 200 μm, the height was 300 nm, and the pitch between bumps was 500 μm. The selective formation of nanoporous powders on the bumps was as follows. First, nanoporous powders were spread on a Si wafer. The bumps on the substrate were attached to the Si wafer. Then, the substrates were pressed at 50 °C for a few minutes. After this process, the substrate with bumps was rinsed in water in order to remove powders around the Au bumps.

After forming of powders on the bump, a Si substrate with Au membrane and the Au bump with nanoporous powders were bonded (Fig. 5). Bonding was performed under a pressure of 10 MPa for 20 min at 150 °C. Bonding interface was evaluated by scanning acoustic microscope (SAM, from PVA TePla Analytical Systems GmbH). The

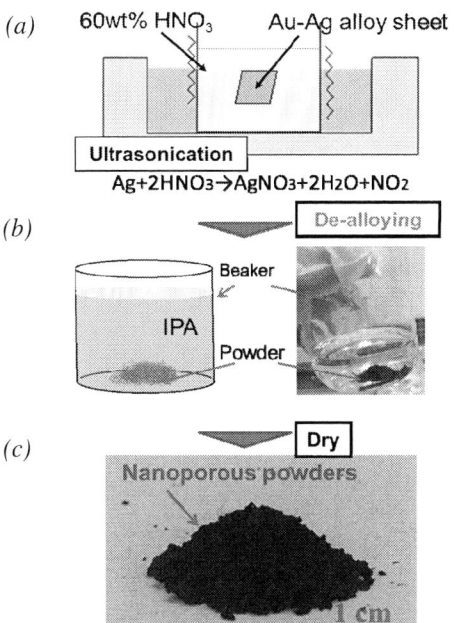

Figure 2: Formation of nanoporous powders

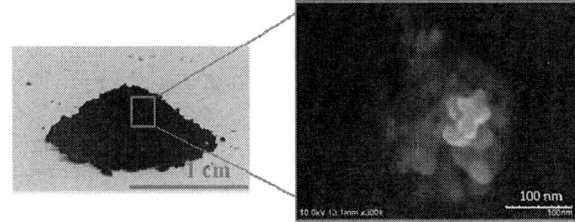

Figure 3: Nanoporous powder formed by de-alloying Au-Ag alloy

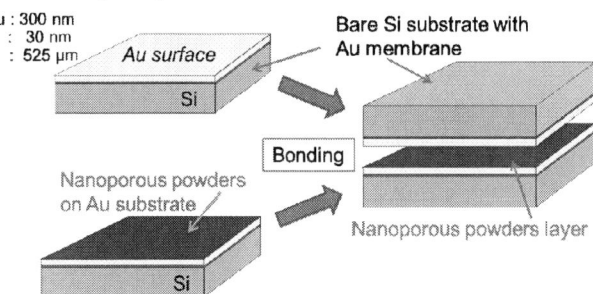

Figure 4: Bare Au sample; Au membrane formed Si substrate and nanoporous powders layer on a bare substrate

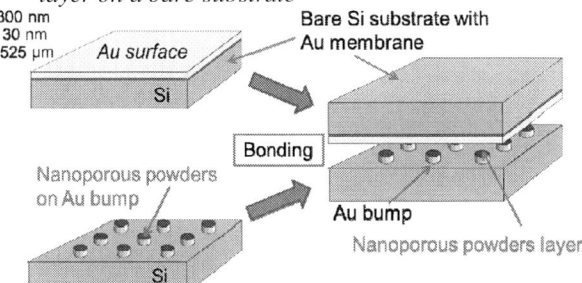

Figure 5: Au bump sample; Au membrane formed Si substrate and nanoporous powders layer on bumps

interface after peeling was observed by SEM.

RESULT AND DISCUSSIONS

Changes in the Nanoporous Structures of Powder

SEM images of the powders after annealing at 50, 100 and 150 °C are shown in Fig. 6. The ligament sizes of the sheet surface increased with increasing temperature. These results indicated that metal diffusions of Au and Ag occurred even at low temperature because of the high reactive surface of the powders [7,8].

Bonding Strengths Measurement

Figure 7 shows the bonding strengths measured by tensile tests. The bonding strengths with and without the powders were measured, at 50, 100, 150, 200, 250 °C. Au-Au substrates pressed at lower than 150 °C were not bonded, probably due to organic contamination of the Au substrates [11]. The bonding strengths were around 0.5 and 1.1 MPa at annealing temperatures of 200 and 250 °C, respectively. On the other hand, bonding strengths with the nanoporous

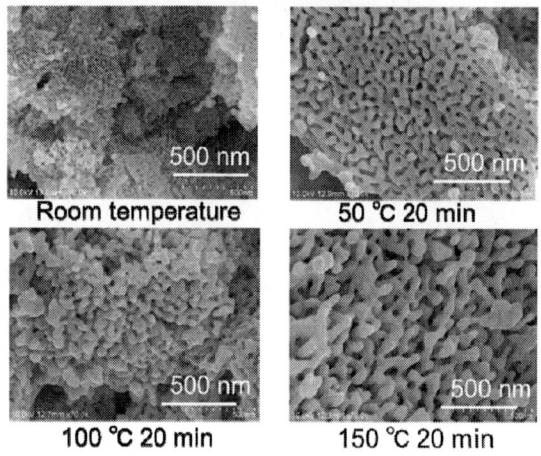

Figure 6: Changes in structure of nanoporous powders with increasing annealing temperature.

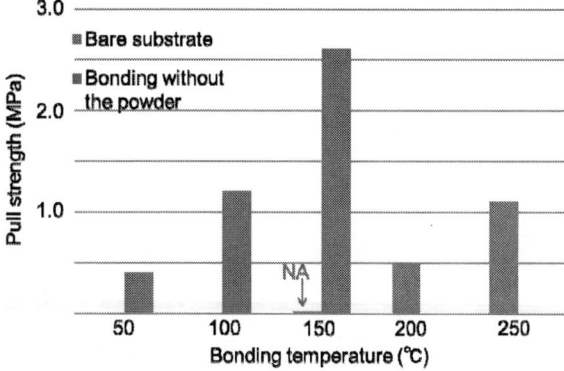

Figure 7: Bond strengths vs. bonding temperature.

powders were about 0.4, 1.2, and 2.6 MPa at annealing temperatures of 50, 100, and 150 °C, respectively. The strengths dramatically increased, compared with those without the nanoporous powders. It is suggested that the increase in the bonding area because of flexibility in nanoporous structure helps Au and Ag diffuse from powders to Au substrate.

Stamp Attachment and Bonding Evaluation

Figure 8 (a) shows that the selective formation of powders on the bumps. Nanoporous powders were bonded to the top of Au bump successfully. The magnified surface structure of nanoporous powder after is shown in Fig. 8 (b). Pore structures and ligament maintained on the powder surface.

The strength of Au-Au bonding with nanoporous powders as a joint layer between Au bump and Au substrate was about 2.4 MPa at annealing temperatures of 150 °C. It indicated that bonding of the Au bump substrate showed comparable pull strength to the case of bare Au substrate.

SAM image of bonding interface is shown in Fig. 9. Figure 10 shows SEM and optical microscope images of peeled surface. White circles represent Au bumps while black parts are the actually bonded nanoporous joint layer. SAM image indicated that nanoporous powders were successfully bonded on bumps. Bulk destruction of Au bumps was observed as shown in Fig. 10. These results indicated that low-temperature bump bonding using nanoporous powders was successfully achieved.

Figure 8: Images of patterned nanoporous powder. (a) The selective formation of nanoporous powders. (b) The magnified surface structure.

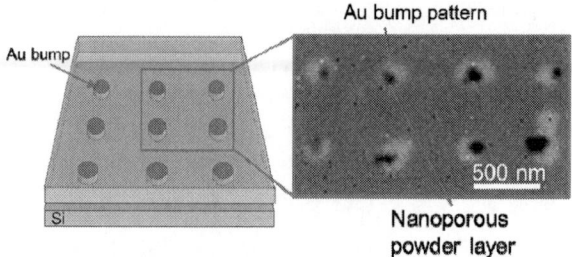

Figure 9: Evaluations of bonded interface of SAM

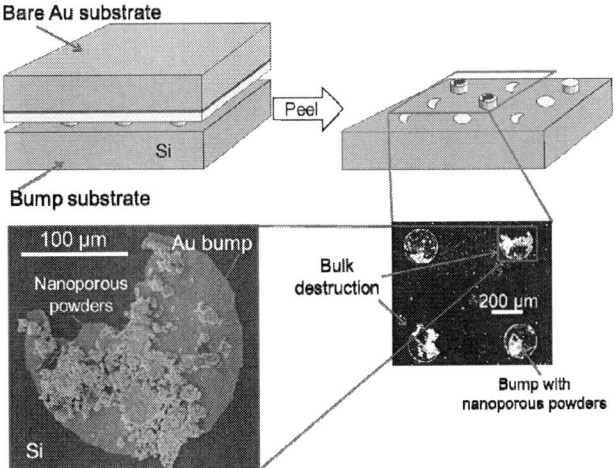

Figure 10: Evaluations of bonded interface of SEM and optical microscope.

CONCLUSIONS AND FUTURE WORK

Conclusion

In this study, low-temperature Au-Au bonding method by using nanoporous Au-Ag powders as a joint layer between bump interconnects was proposed. The nano-porous powders were fabricated by dealloying Au-Ag alloys with Au:Ag mass ratio of 25:75 in HNO_3 under ultrasonication. Changes in the structures of nanoporous powders were observed after annealing them under 50, 100 and 150 °C. Selective formation of powders was demonstrated by stamping process. The Au-Au bonding strength of 2.4 MPa was achieved at at 150 °C by using nanoporous Au-Ag powders as a joint layer. Bonding interface and bulk destruction were evaluated by SAM, SEM, and optical microscope. Low-temperature bonding with nanoporous powders as a joint layer is useful and available for bump interconnects in a field various device or MEMS packaging.

We are going to evaluate electric behavior of nanoporous joint layer, and will apply the proposed Au-Au bonding method for micro bumps, smaller than 100 μm.

ACKNOWLEDGEMENTS

This work is partly supported by the Japan Ministry of Education, Culture, Sports, Science and Technology (MEXT), a Grant-in-Aid for Scientific Basic Research (S) No. 23226010, Specially Promoted Research "Establishment of Electrochemical Device Engineering," and a Grant-in-Aid for Cooperative Research Project Nationwide Joint-Use Research Institute on Advanced Materials Development and Integration of Novel Structured Metallic and Inorganic Materials from MEXT. This work is also partly supported by JSPS KAKENHI Grant Number 25289241. The authors acknowledge the support from the MEXT Nanotechnology Platform Support Project of Waseda University.

REFERENCES

1) K. Sakuma, N. Nagai, M. Saito, J. Mizuno, and S. Shoji, "Effects of Excimer Irradiation Treatment on Thermocompression Au–Au Bonding", IEEJ Trans. Electr. Electron. Eng. **4** (2009) 339.

2) Y. Tsukada, M. Mashimoto, and N. Watanuki" Novel chip replacement method of encapsulated flip-chip bonding ", Proc. IEEE/EIA Electronic Components & Technology, 1993, p. 199.

3) M. M. V. Taklo, P. Storas, K. Schjolberg-Henriksen, H. K. Hasting, and H. Jakobsen, "Strong, high-yield and lowtemperature thermocompression silicon wafer-level bonding with gold". J. Micromech. Microeng. **14** (2004) 884.

4) L. Wood, C. Fairfield, and K. Wang, "Plasma Cleaning of Chip Scale Packages for Improvement of Wire Bond Strength" , Proc. Int. Symp. Electric Materials and Packaging, 2000, p. 406.

5) S. Yamamoto, E. Higurashi, T. Suga, and R. Sawada, "Low-temperature hermetic packaging for microsystems using Au–Au surface-activated bonding at atmospheric pressure" , Transducers, 2011, p. 1384.

6) H. Oppermann and L. Dietrich, "Nanoporous gold bumps for low temperature bonding" , Microelectron. Reliab. **52** (2011) 356.

7) Y. C. Lin, W. S. Wang, L. Y. Chen, T. Gessner, and M. Esashi, "Anodically-bondable LTCC substrates with novel nano-structured electrical interconnection for MEMS packaging", Proc. Transducers'11, 2011, 2351.

8) H. Mimatsu, J. Mizuno, T. Kasahara, M. Saito, H. Nishikawa, and S. Shoji, "Low-Temperature Au–Au Bonding Using Nanoporous Au–Ag Sheets", Jpn. J. Appl. Phys. **52** (2013).

9) M. Hakamada, Y. Chino, and M. Mabuchi, "Deformation Characteristics of Recycled AZ91 Mg Alloy Containing Oxide Contaminants" , Mater. Lett. **64** (2010) 2341..

10) L. H. Qian and M. W. Chen, "Ultrafine nanoporous gold by low-temperature dealloying and kinetics of nanopore formation" , Appl. Phys. Lett. **91** (2007) 083105.

11) N. Unami, K. Sakuma, J. Mizuno, and S. Shoji, "Effects of Vacuum Ultraviolet Surface Treatment on the BondingInterconnections for Flip Chip and 3-D Integration" , Jpn. J. Appl. Phys. **49** (2011).

CONTACT

H. Mimatsu, tel: +81-3-5286-3384;
mimatsu@shoji.comm.waseda.ac.jp
J. Mizuno, tel: +81-3-3205-3181;
mizuno@waseda.jp

MICRO DEVICES INTEGRATION WITH LARGE-AREA 2D CHIP-NETWORK USING STRETCHABLE ELECTROPLATING COPPER SPRING

Wei-Lun Sung[1], Wei-Cheng Lai[1], Chih-Chung Chen[2], Kevin Huang[2] and Weileun Fang[1]
[1] Power Mech. Eng. Dept., National Tsing Hua Univ., Hsinchu, Taiwan
[2] imec Taiwan Inc., Hsinchu, Taiwan

ABSTRACT

This study presents a large-area multi-devices integration scheme using stretchable electroplated copper spring. Each device is located on the silicon-node of a 2D chip-network distributed, which are mechanically and electrically connected to surrounding devices by stretchable copper spring. The springs stretch and expand the functional devices by several orders of magnitude area forming a variable-density network of interconnected devices. Advantages of this approach include: (1) using existing process technologies and materials for semiconductor in large-area applications, compatible with foundry fabrication processes; (2) stretchable electroplated copper springs with large maximum strain act as both mechanical and electrical connections between devices; (3) silicon-nodes act as hubs for device implementation and integration; and (4) the chip-network can be applied to 2D-curved (spherical) surfaces. The proposed expandable network using stretchable springs integrated with multiple devices has been implemented and tested.

INTRODUCTION

Large-area electronics has attracted much research effort from improving conventional approaches such as pick-and-place assembly [1-2] to novel techniques such as printed organic/inorganic technologies [3-4]. Pick-and-place assembly is an obvious way to distribute electronic components over large area. This approach is straightforward and has been practiced for a long time. However, pick-and-place assembly is inefficient as it is a serial process and is difficult to accommodate curved surfaces. Printed organic/inorganic technologies employ repetitive transferring or printing method. It is easier to be applied to flexible substrates but printed technologies are still lacking in device performance, reliability and the ability to integrate with MEMS devices.

The large-area electronics can find various applications such as wearable, flexible, and ambient devices. In these applications, the span of electronics components is larger than the area they occupy during fabrication [5]. Therefore, the cost per unit area is an important factor when evaluating the applicability in large-area electronics. One way to reduce this cost is using stretchable springs to spread out the functionality of devices over a larger surface [6-10]. Electronic devices can be fabricated close together in high density by modern microelectronics fabrication, and spread out at the time deployment to cover a larger area.

Large-area electronics with stretchable springs have been explored for various applications, such as optical system with a curved image surface [6], temperature sensor network [7], structural health monitoring system [8] and Bio-inspired network [9]. They have been achieved using silicon [6, 8, 10] or polymer [7, 9] stretchable spring.

This study further extends the concept to exploit the electroplated copper as stretchable springs since the processes of making these springs are relatively inexpensive. Moreover, copper has higher breaking strain than silicon and can provide good electrical paths between devices. This study demonstrates a network of interconnected devices integration scheme with different functions devices using stretchable electroplated copper springs. The springs mechanically and electrically connect surrounding devices forming a large-area network of devices fabricated on silicon-nodes. Silicon acts as node for device implementation and integration to form 2D chip-network.

DESIGN CONCEPT

The proposed design of large-area 2D chip-network consists of silicon-nodes and stretched springs, as shown in Fig.1. Fig.1a shows a pre-stretched 3×3 chip-network. Each device is fabricated on the silicon-node of the 2D chip-network distributed. The network can be expanded to

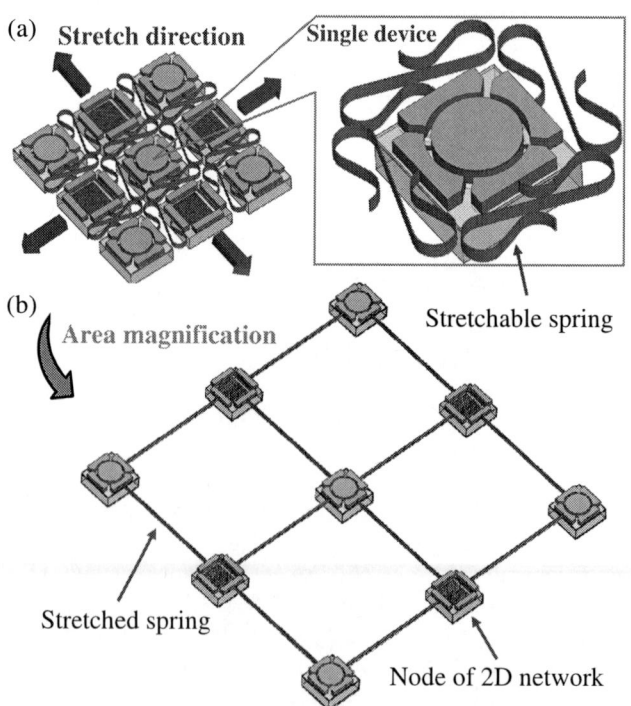

Figure 1: Schematic of 2D network with stretchable spring (a) pre-stretched 3×3 chip-network; (b) area magnification after fully expanding the network.

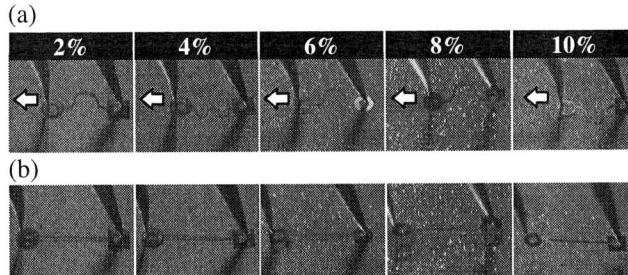

Figure 2: Preliminary tests of the electroplated copper springs (a) designs of 2-10% strain pre-stretched springs; (b) full extension.

Table 1. Five types of design for various stretchable spring.

	Type 1	Type 2	Type 3	Type 4	Type 5
Spring shape					
Stretched Length (mm)	5.35	5.05	8.18	12.17	27.92
Linear expansion factor	7.6	7.2	11.7	17.4	40
Max. strain (%)	3.8	2	4	6	8

increase the area coverage of functional devices by several orders of magnitude. The stretched spring provides mechanical and electrical connection between surrounding devices, as shown in Fig.1b.

Fig.2 shows preliminary stretch tests of five different electroplated copper springs. Micrographs in Fig.2a-b respectively display the springs before and after the stretching experiments using probes. As the springs are stretched to full extension, the maximum strain, ε_{max}, encountered in each section is given by $\varepsilon_{max} = t/D$ [10], where t is the width of the spring defined by lithography process and D is the diameter of the minimum curvature. The maximum strain varies from 2-10% for these five designs. According to the above results, various designs of stretchable springs are shown in Table 1 with <10% strain. The thickness and width of the springs are 20μm and 10μm, respectively. Type1 is a spiral-spring and Type2-5 are flexures-springs with different number of windings.

FABRICATION PROCESS

The fabrication process uses conventional steps and materials to form stretchable springs using electroplated copper. Moreover, integration with semiconductor and MEMS devices manufactured using wafer technology is easily achievable. Fig.3 shows fabrication process steps. In Fig.3a, the 1st Cr/Au (20nm/200nm) films are deposited and patterned on a silicon wafer as the bottom electrode and electrical routing. In Fig.3b, a 5 μm thick Parylene film is deposited by chemical vapor deposition (CVD) and patterned by oxygen plasma. The Parylene film serves as a dielectric layer with relatively high dielectric constant. In Fig.3c, the 2nd Cr/Au (20nm/200nm) film is deposited and patterned as the top electrode and electrical routing. As in Fig.3d,

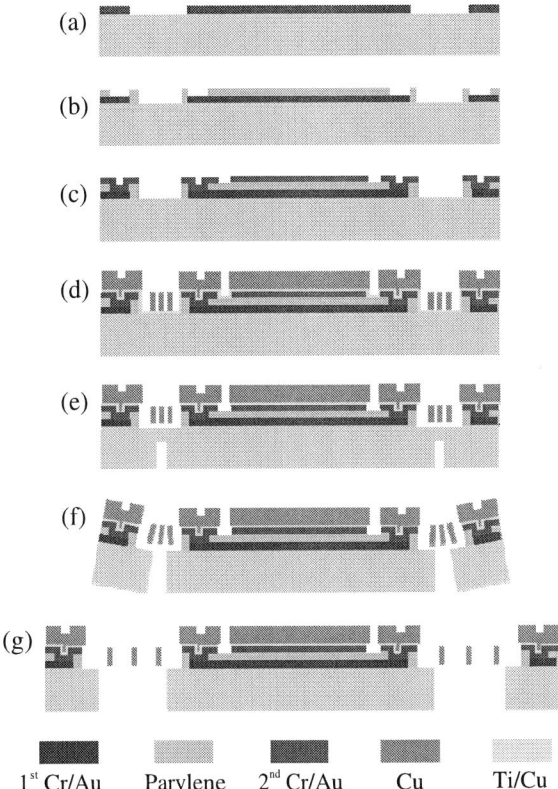

Figure 3: Fabrication process of large-area device network with stretchable spring.

the Ti/Cu (500nm/200nm) is deposited as the seed layer for electroplating and as the sacrificial layer of the spring. Then, a thicker copper (20μm) structure is formed by using photoresist mold and electroplating. After that, dicing saw is used to define the splitting trench on the backside of the wafer. 100μm of silicon is left after the dicing process. Next, the Ti/Cu sacrificial layer under the spring is removed as shown in Fig.3e. The splitting trench is broken in Fig.3f. Finally, in Fig.3g, the chip-network is ready for large-area deployment. Note that the silicon-node can be fabricated using other approaches (such as bonding technology, bulk micromachining, etc.) to implement a variety of micro devices.

RESULTS AND DISCUSSIONS

Typical fabrication results (before device splitting) are shown in Fig.4. Fig.4a shows tactile and proximity sensors integrated with stretchable springs. The sensors are housed on the silicon-nodes. The upper and lower electrodes could serve as electrical routing and sensing element. The electrical signals are transmitted by stretchable copper springs to the scanning circuitry. Fig.4b shows the SEM micrographs of Type1 (spiral-spring) and Type5 (flexures-springs) springs. The electroplated copper springs are suspended on the substrate. The thickness and width of copper spring are 20μm and 10μm, respectively. The results show the copper springs have good yield and low stress.

(a)

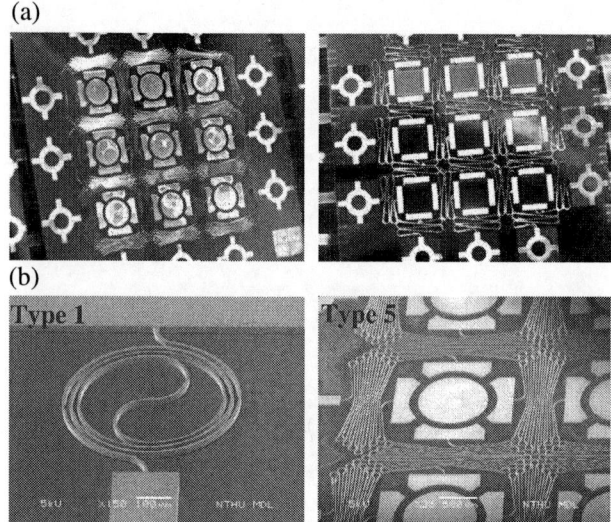

(b)

Figure 4: Typical fabrication results (before device splitting). (a) pre-stretched chip network; (b) SEM pictures of type1 and type5 springs.

Fig.5a shows the setup for stretching tests. The boundaries of chip-network are fixed by probes, and the springs of the network are stretched by X-Y position-stages. This way, the displacement of stretched springs can be precisely controlled. For the array architecture, all four directions should be simultaneously stretched to achieve precise positioning of chips. The micrographs in Fig.5b-c display the test results for different spring designs. Fig.5b shows the chip-network before and after stretching. The chips are split along the yellow dotted line which denotes the back side splitting trench (Fig.5b, top right), and the network is expanded using the setup of stretching tests. After springs are stretched, the nodes are seen distributed in the 2D plane. Fig.5b (bottom right) also shows a zoomed-in silicon-node of chip-network after the springs are stretched from flexures-springs to straight lines. Moreover, the stretched spring still connects to surrounding devices. The 2D chip-network reaches its maximum coverage area when springs are stretched to straight lines. Chip-networks with Type5 springs have maximum areal expansion of 120-fold. As indicated in Fig.5c, the area expansions of chip-network with Type1 and Type4 springs are respectively 7-fold and 30-fold. For large-area electronic applications, the stretchable springs approach can provide large area expansion from an initially small area (pre-stretched) to reduce the cost per unit area and provide low-resistance electrical connection between devices

Micrographs in Fig.6a show the pre-stretched chip-network with 5×5 nodes which is initially fixed on a deflated balloon and then stretched on curved surface by inflating the balloon. While the balloon is being inflated, the stretch of springs is clearly observed. Finally, the springs are fully stretched by the inflated balloon, as shown in Fig.6b. It demonstrates the flexible and stretchable springs allow the 2D chip-network to properly conform to the curved surface.

(a)

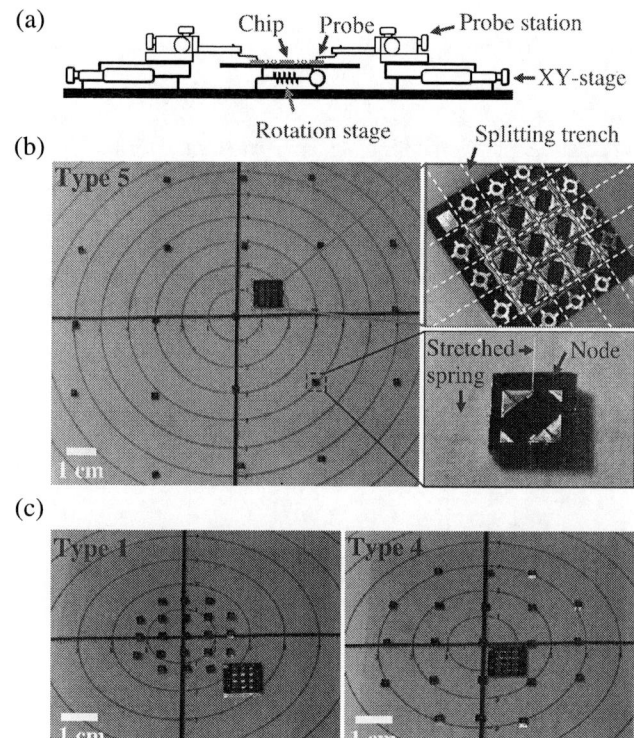

(b)

(c)

Figure 5: Stretching setup and stretch results for different springs. (a) setup for stretching tests; (b) chip-network before and after stretching; (c) different spring type stretch result

(a)

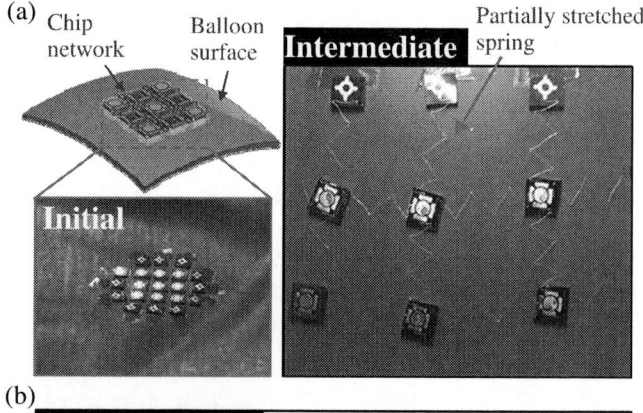

(b)

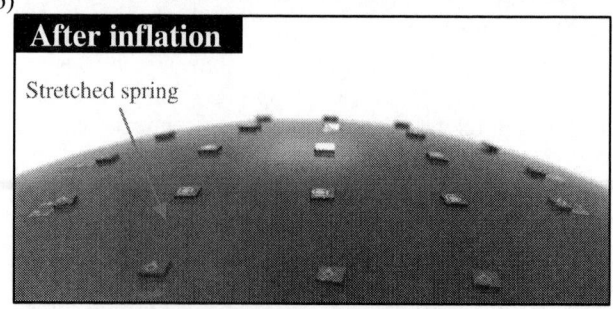

Figure 6: Demonstration of the capability of the 2D chip-network on curved surface by inflating a balloon. (a) initially fixed on a deflated balloon and intermediate step; (b) spring stretched to full length.

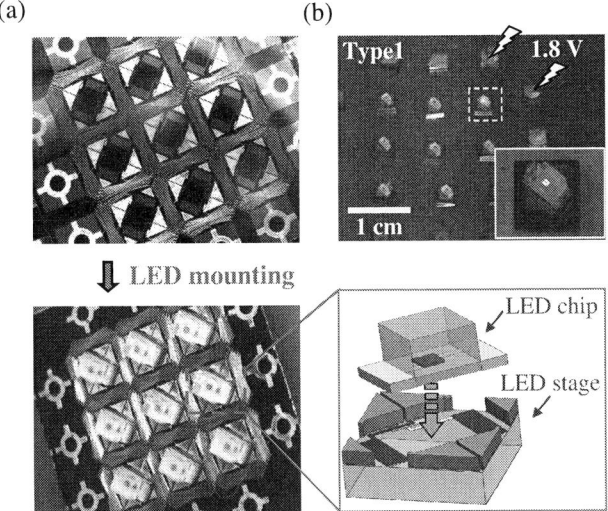

(a)   (b)

⬇ LED mounting

LED chip

LED stage

Figure 7: LED chip integration and result. (a) demonstrates the LED mounting on 2D chip-network.; (b) lighting test result.

Fig.7 shows the integration of 3×3 LED chips with the presented 2D chip-network. The commercially-available LED chip size is 1.6 mm × 0.8 mm. In Fig.7a, the LED chips are mounted on the nodes of the chip-network. Solder paste is coated on the electrodes on the nodes using fluid dispensing systems. After mounting the LED chip, the chip-network is heated to 230°C for 10 minutes. Fig.7b shows the lighting test with stretched Type1 spring by applying 1.8V on two signal I/O nodes. Thus, the LED light can be selectively controlled by applying different I/O nodes.

CONCLUSIONS

In this study, large-area multi-devices integration with stretchable electroplated copper spring has been proposed and demonstrated. The silicon-nodes house the device for 2D chip-network. Stretchable electroplated copper springs serve as mechanical and electrical connection between surrounding devices. The springs can be stretched and the network expands the coverage area of functional devices by several orders of magnitude. The network expansion test is performed by using X-Y position-stages and balloon inflation. Results demonstrate the capability to conform to 2D curved surface and a maximum areal expansion of 120 folds. Other devices (temperature sensor, tactile sensor, and proximity sensor, etc.) have been fabricated on the silicon-nodes with stretchable springs. The performance of these micro-devices will be characterized.

ACKNOWLEDGEMENTS

This research is based on the work supported by National Science Council of Taiwan under grant number NSC 99-2221-E-007-040-MY3 and NSC 99-2221-E-007-039-MY3. The author would like to express his appreciation to CNMM of the National Tsing Hua U. and the Nano Facility Center of National Chiao Tung U. in providing fabrication facilities.

REFERENCES

[1] J. Kahler, N. Heuck, A. Stranz, A. Waag, E. Peiner, "Pick-and-Place silver sintering of small-area chips", *IEEE Trans.s on CPMT*, vol. 2, pp. 199-207, 2012.

[2] T. Fritzsch, R. Jordan, H. Oppermann, O. Ehrmann, M. Topper, T. Baumgartner, K. D. Lang, "Cost effective flip chip assembly and interconnection technologies for large area pixel sensor applications," *Nucl. Instrum. Meth. A*, vol. 650, pp. 189-193. 2011.

[3] R. Parashkov, E. Becker, T. Riedl, H. H. Johannes, W. Kowalsky, "Large area electronics using printing methods," *Proc. IEEE*, vol. 93, pp. 1321-1329, 2005.

[4] G. B. Blanchet, Y. L. Loo, J. A. Rogers, F. Gao, C. R. Fincher," Large area, high resolution, dry printing of conducting polymers for organic electronics," *Appl. Phys. Lett.*, vol. 82, pp. 463-465, 2003.

[5] A. C. Arias, J. D. MacKenzie, I. McCulloch, J. Rivnay, A. Salleo, "Materials and Applications for Large Area Electronics: Solution-Based Approaches," *Chem. Rev*, 110, pp. 3-24, 2010.

[6] S. B. Rim, P. B. Catrysse, R. Dinyari, K. Huang and P. Peumans, "The optical advantages of curved focal plane arrays," *Optics Express*, Vol. 16, pp. 4965-4971, 2008.

[7] Z. Guo, K. Kim, G. Lanzara, N. Salowitz, P. Peumans and F. K. Chang, "Micro-Fabricated, Expandable Temperature Sensor Network for Macro-Scale Deployment in Composite Structures," *Aerospace Conference*, Big Sky, MT, March 5-12, 2011.

[8] K. Huang and P. Peumans, "Stretchable silicon sensor networks for structural health monitoring," *Proc. of SPIE*, vol. 6174, 2006.

[9] N. Salowitz, Z. Guo, Y. H. Li, K. Kim, G. Lanzara and F. K. Chang, "Bio-inspired stretchable network-based intelligent composites," *J. Compos. Mater.*, vol. 47 no. 1, pp. 97-105, 2012.

[10] K. Huang, R. Dinyari, G. Lanzara, J. Y. Kim, J. Feng, C. Vancura, F. K. Chang and P. Peumans, "An Approach to Cost-Effective, Robust, Large-Area Electronics using Monolithic Silicon," *IEDM*, Washington, DC, Dec. 10-12, pp. 217-220, 2007.

CONTACT

* W. Fang, Tel: +886-3-5742923; fang@pme.nthu.edu.tw

SOLID-STATE ISFET FLOW METER FABRICATED WITH A PLANAR PACKAGING PROCESS FOR INTEGRATING MICROFLUIDIC CHANNEL WITH CMOS IC CHIP

Jun-Jie Wang[1], Chen-Fu Lin[2], Ying-Zong Juang[2], Hann-Huei Tsai[2], Hsin-Hao Liao[2], Che-Hsin Lin[1]*

[1] Department of Mechanical Engineering, National Sun Yat-sen University, Kaohsiung, Taiwan
[2] National Chip Implementation Center, National Applied Research Laboratories, Hsinchu, Taiwan

ABSTRACT

This study presents a solid-state ISFET flow meter fabricated with an innovative planar packaging process. It is challenge to integrate a CMOS chip with the microfluidic channel since the physical dimension of the CMOS chip may cause the leakage during packaging the IC chip with microfluidic channels. This study develops a planar packaging process which uses PDMS as the adhesive to embed the IC chip in a cavity over the PCB. Result shows that the interfacial step between the IC chip and the surrounding PDMS substrate is less than 5 μm such that microfluidic channel can be sealed free of leakage with the developed planar package process. The sealed ISFET chip is used for measuring the flow rate of non-ionic solutions including acetone, ethanol and glycerol of slow flow rate. Results show that the flow rate measurement exhibited good reproducibility in the flow rate ranging from 66 to 1700 μm/s. Moreover, the whole packaging process can be achieved in 40 min. The developed method provides a simple yet efficient method to integrate CMOS IC chip with microfluidic systems.

INTRODUCTION

Microfluidic systems have been used various applications for physical and chemical reactions including fluid transportation, mixing, sample separation, droplet formation and the biocatalytic reactions [1, 2]. In this regards, there are in a number of parameters which need to be measured such as temperature, optical intensity, acidity and flow rate. To achieve these measurements in a small device, MEMS-based sensors have play a successful role in these applications. In general, microfabricated sensors are produced directly on the substrates using various materials including metals, silicon or polymers. The produced sensor devices usually exhibit simple function such as current or potential measurements. Therefore, the external electronic equipment is required to read the signals from the microfabricated sensor devices. However, the signals from the sensor devices are usually small and easy to decay during the signal transduction to the external equipment. Alternatively, the rapid development of integrated circuit (IC) technology has brought great advantages on producing electronic circuits and sensor devices on a small chip. For example, the CMOS ISFET is to combine the sensing electrodes, parallel FET structures and signal conditioning circuits on the surface of a single IC chip [3]. The sensing performance is much higher and more stable in compare with

the laboratory produced sensors since these IC devices were fabricated with standard foundry and reliable industrial grade processes. Several researches have integrated CMOS IC chip such as light emission device and optical sensing device into microfluidic system for fluorescence detections. However, these devices were typically assembled off-channel without contacting the liquid sample in the microfluidic channel [4, 5]. Alternatively, a number of the CMOS sensor chips have been reported for a variety of sensing applications including pressure sensors, gas sensors, microphone, motion sensors and even the chemical sensors. However, these CMOS sensors were typically purchased in the form of commercial end-products and were used along for a single purpose detection.

Recently, the integration of microfluidic systems with CMOS ICs has attract researchers who works in microfluidic systems. The high performance CMOS ICs provide many advantages including high detection limit, low noise and easy circuit integration. The detection site of the IC chip can also be modified with functional materials for gas or ion sensing [6] or immobilized with biomolecules for bioanalytical applications [7]. Integrated microfluidic systems and CMOS chip has many advantages, the first one through the CMOS chip can be designed to operate in GHz range for high speed measurements. Although the performance of the sensors could be enhanced with the assistance of the IC chips, the packaging of the IC chip with the microfluidic channels is still challenging. For a successful packaging of the microfluidic channel, the detection site on the IC should be well exposed to the liquid environment for detecting the signals. However, the physical dimension of the chip may usually cause the difficulties while sealing the microfluidic channel. Surface flatness is one of the major problems which causes the failure of the bonding process [8]. In order to integrated IC chip with microfluidic device, a double layer SU-8 photoresist structure was used to compensate the thickness of the IC chip [9]. However the coating thickness of the first SU-8 layer was difficult to controlled and might resulted in a step on the chip edge and SU-8 surface. In addition, the sensing surface of CMOS chip may exposed to thermal processes and organic solvents during the SU-8 lithography procedures, which may have the risks to damage the IC chips. Therefore, there is a need to develop a simple yet rapid method to integrate the CMOS IC chip and microfluidic system. Recently, solid-state IC chips have been integrated with microfluidic systems using elastomeric materials of PDMS.[10] Liquid metal interconnects and the corresponding sample delivery channel were integrated in

978-1-4799-3510-9/14 $31.00 © 2014 IEEE

flexible elastomer substrate. Nevertheless, the liquid based interconnects were not as reliable as the wire bonding interconnects.

This paper aim to develop a simple yet efficient method to integrate CMOS IC with microfluidic channel. Prior to the sealing process of the microfluidic channel, the ISFET IC chip is mounted in the PDMS elastomer and exposed the top surface of the IC chip. A flat surface can be formed with this approach since the PDMS formation and the chip mounting process is done simultaneously. The other PDMS substrate with microfluidic channel can be covered on the IC-mounted PDMS without leakage. The developed microfluidic device is used to measure the flow rate of organic solvent inside the microfluidic channel. The mounting flatness of the IC chip on the is measured. The sensing performances for measuring the flow rate of alcohol, acetone and glycerol are experimentally investigated.

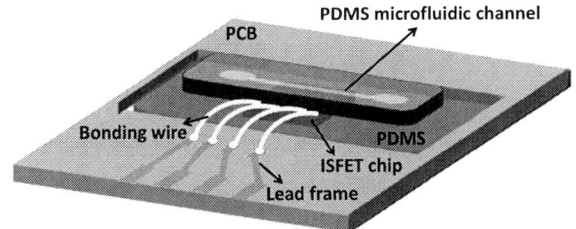

Figure 1: The schematic for the concept for planar packaging ISFET IC with microfluidic channel.

DESIGN CONCEPT

Figure 1 shows the schematic illustration for the concept of the developed planar packaging method. The CMOS IC chip was placed and planarized with PDMS in the cavity of the PCB. The microfluidic channel can be bonded with the bottom PDMS layer with IC chip without leakage. A glass-fiber print circuit board (PCB) was used as the substrate for producing proposed ISFET-based flow sensor. Prior to the fabrication process, the lead frame for connecting signals from the ISFET chip was firstly patterned. A concave cavity with the depth of around 1.0 mm was firstly machined on the PCB using a mini-CNC (Computer Numerical Control) machine. Note that the size and the machining accuracy for producing this cavity is not an important issue. A constant amount of PDMS was the filled into the cavity then the IC chip was then applied on the PDMS using a flat glass substrate. The formed PDMS structure was cured to planar mount the IC chip with the PDMS elastomer. The mounted IC chip was then wire bonded the pads to the lead frame using aluminum wires for signal transduction. Another PDMS substrate patterned with microfluidic channel was finally bonded onto the substrate to seal the microfluidic channel with the ISFET sensing chip.

THE ISFET AND PACKAGING PROCESS

In this study, the sensing elements are produced using a standard semiconductor manufacturing process. Figure 2A presents the photo image of the CMOS ISFET chip for flow

rate measurement. The ISFET was fabricated with a standard 0.35 μm 2P4M (2 layers of polysilicon and 4 layers of metal) foundry process by TSMC (Taiwan Semiconductor Manufacturing Company, Taiwan). The chip size is with 3.6 × 2.7 mm² and there is an active sensing region of the aluminum gate with the area of 200×200 μm². There were four gold electrodes surrounding the active sensing area, which were used as the reference electrodes for the measurement (Fig. 2A). Figure 2B presents the cross-section schematic for the design and the circuit connection for the ISFET flow meter. The thickness of the top aluminum layer for the active sensing area was with 0.925 um. Note that the aluminum at the active area would form a native alumina oxide layer of around 10.0 nm. During electric measurement, an electric potential was applied on the reference electrodes to turn-on the FET device. The polarized molecules in the fluids would attach onto the alumina sensing area and establish a gate potential with the corresponding current from drain to source (I_{DS}). The Due the solution has flow rate changes in the micro-channel, the gate equivalent voltage will also change on the sensing area. Therefore, to detect the flow rate by the change of drain current in the microfluidic channel.

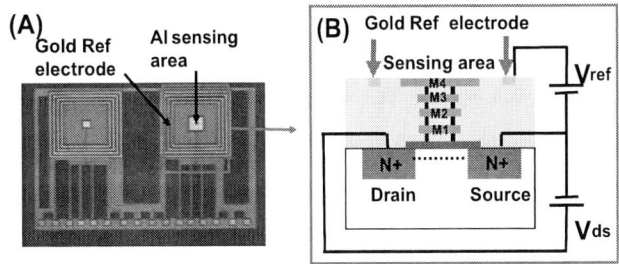

Figure 2: (A) The eagle view for the ISFET chip used in the present study. (B) The schematic showing the cross section view of the ISFET chip.

Figure 3 presents the schematic for the simplified fabrication process for the developed PDMS with IC packaging method. As described above, a PCB was first patterned with the lead frame for wire bonding. In order to achieve the planar mounting of the IC chip in the PDMS elastomer. A 1.5-mm thick PMMA substrate was spin coated with a thin layer of silicon oil with the thickness of around 10 μm. The silicon oil layer was used to stick the ISFET chip and used as the de-molding coating during the mounting process.(Fig. 3A) The PMMA substrate with the IC chip and the surrounding PDMS was placed on a 70°C hotplate for 5 min to slightly curing the PDMS.(Fig. 3B) A concave cavity with the size of 5.0 x 4.0 x 2.0 mm was carved using a micro CNC machine to mount the IC chip where the surrounding cavity was with 1.0 mm in depth.(Fig. 3C) The formed cavity was then again filled with PDMS and placed in a vacuum chamber to remove the air bubble in the PDMS.(Fig. 3D) The PMMA substrate with IC chip was then aligned and covered on the carved PCB substrate.(Fig. 3E) The extra PDMS will be squeezed out of the cavity via the designed flooding cavity. The bonded substrate was then placed in a 80°C oven for 20 min to cure the PDMS. (Fig. 3F) The PMMA substrate could be easily removed from the cured PDMS due to the existence of silicon oil. Finally, PDMS substrate with patterned microfluidic channel was bonded onto the planarized substrate on the PCB with the assistance of O₂ plasma activation. (Fig. 3G) With is approach, microfluidic system and the CMOS IC chip can be successfully without leakage.

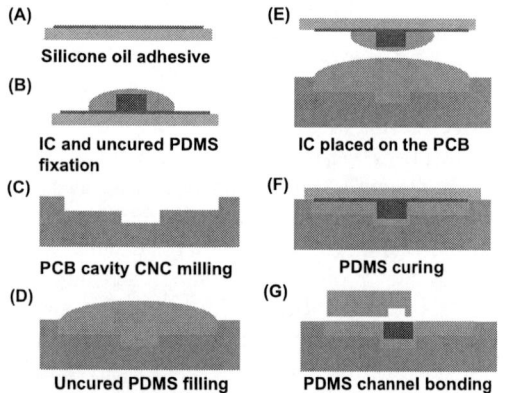

Figure 3: The simplified packaging process for integrating the CMOS IC chip with the microfluidic channel.

EXPERIMENTAL SETUP

The experimental setup for evaluating the sensing performance the proposed microfluidic flow meter is shown in figure 4. The flow rate for the liquid was controlled using a programmed automatic titrator (907 Titrando, Metrohm, Germany). The liquid flow can be stable injected into the microfluidic device with the flow rate of 66, 330, 660, 1060 and 1700 μm/s in the microfluidic channel. The electrical signal from the ISFET chip was collected via a modular DC source/monitor (4142B, Agilent, USA).

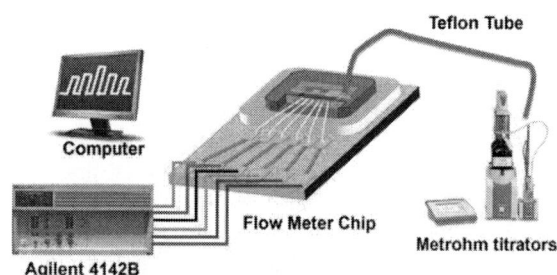

Figure 4: Experimental setup for testing the ISFET-based flow meter packaged with the developed process.

RESULT AND DISCUSSION

Figure 5A presenters the photo image for the device after packaging. The packaging planarity between the PDMS and IC chip was investigated using a surface profilometer (SJ-400, Mitutoyo, Japan). The measurement was performed by scanning from PDMS site to the IC chip using a scan speed of 0.5 mm/s. The measured height variation (R_{max}) between the IC chip and the PDMS surface was less than 5 μm, indicating a nice planarity was achieved using the

developed method.(Fig. 5B) The height variation was majorly caused by the polymerization shrinkage of the PDMS elastomer.

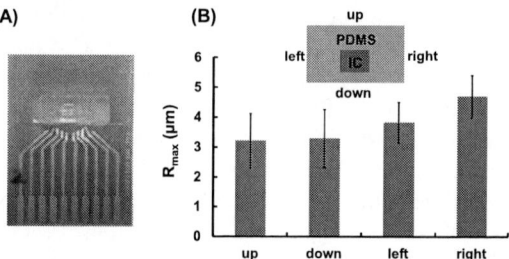

Figure 5: (A)A photo image for the device after packaging (B) Measured step height (R_{max}) between the IC and PDMS.

Figure 6 shows the SEM images for the cross-section view (Fig. 6A) and the top view (Fig. 6B) at the interface of ISFET chip and PDMS layer. It is clear that there was no significant gap between the IC chip and the surrounding PDMS mounting material. The top view SEM also indicate that the PDMS surface was also flat such that the top PDMS layer with microfluidic channels can be sealed without leakage. Planar packaging of the CMOS IC and the microfluidic system can be achieved with this approach.

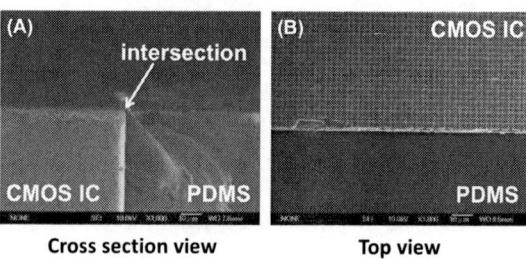

Figure 6: SEM images for the (A) cross-section view and (B) top view at the packaging interface.

Figure 7 presents the measured I_{DS} for detecting the flow rate of ethanol inside the microfluidic channel with an applied reference voltage (V_{Ref}) of 3.0 V and a constant source-drain voltage (V_{SD}) of 1.0 V. Five repeating measurements were performed using the developed ISFET flow sensor. Note that the flow rates were set stepwise at 66, 330, 660, 1060 and 1700 μm/s, respectively. Result shows that the current response increased with the increasing flow rate and the solid-state ISFET flow sensor exhibited a good repeatability for flow rate measurements.

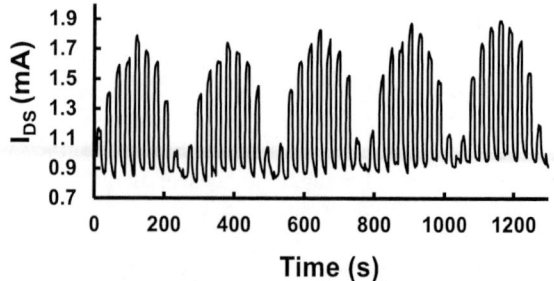

Figure 7: Measured I_{DS} for five repeating measurements on ethanol of the flow rate of 66, 330, 660, 1060 and 1700 μm/s, respectively. (V_{Ref}= 3 V).

The applied reference potential to activate the ISFET was also an important factor for the obtained current response and the sensitivity during flow rate measurement. Therefore, this study also tested the effect on the applied reference voltages ranging from 2.5, 3.0, 3.5, 4.0 V. Figure 8 shows the relationship between the flow velocity and the I_{DS} at different applied V_{Ref} for detecting ethanol. Results showed that the current response exhibited the optimal linearity while applying a reference voltage of 3.0 V. It is also noted that the current responses showed no significant correlation with the flow rate of the fluid. Since the distance between the aluminum gate structure and the gold reference electrode is only 100 μm, the strong gate potential induced by the applied reference voltage dominated the source-drain current. Therefore, the charge variation due to the less dominant fluid flow did not attribute enough current variation.

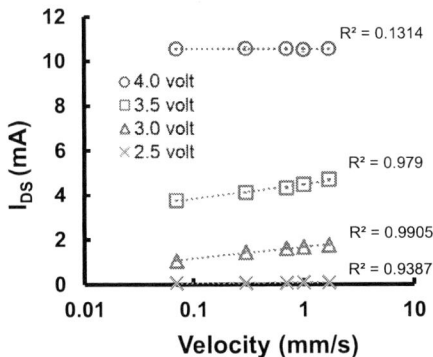

Figure 8: The relationship between the velocity and the measured I_{DS} at different applied V_{Ref}.

The developed ISFET flow sensor was also used to measure the solvents of different electrical properties. Figure 9 shows the flow rate measurements for detecting different solvents of acetone, glycerol and ethanol, respectively. Results show that the sensor exhibited good sensing properties on detecting different solvents. It is also noted that the alcohol showed better response than the acetone, which might be due to the trace water content in the solution. The dissociation of water might play a role on the gate potential.

CONCLUSION

This study developed a simple and low-cost method to plenary packaging the solid-state CMOS chip with microfluidic channel. The IC chip was planarity mounted in a PDMS elastomer then sealed with a PDMS substrate with a microfluidic channel structure. The developed method was demonstrated by packaging a ISFET chip with a microfluidic channel for solvent flow rate measurement. Results showed that the ISFET-based flow sensor exhibited a nice sensing performance on detecting various solvent of alcohol, acetone and glycerol. The method developed in the present study provides a rapid way to integrate IC chip with the microfluidic systems.

ACKNOWLEDGEMENTS

The authors would like to thank the financial supports from National Science Council of Taiwan.

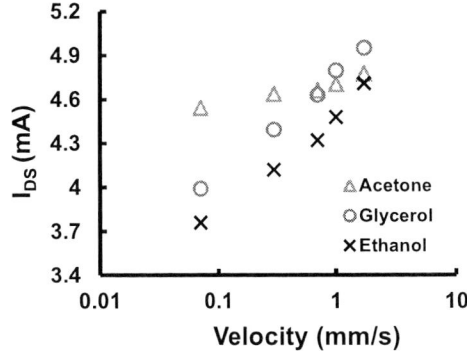

Figure 9: Flow rate measurement for different solvents of acetone, glycerol and ethanol, respectively. (V_{Ref}= 3.5 V)

REFERENCES

[1] W. C. Chang, L. P. Lee, and D. Liepmann, "Biomimetic technique for adhesion-based collection and separation of cells in a microfluidic channel," *Lab on a Chip,* vol. 5, pp. 64-73, 2005.

[2] Y. C. Tan, V. Cristini, and A. P. Lee, "Monodispersed microfluidic droplet generation by shear focusing microfluidic device," *Sensors and Actuators B: Chemical,* vol. 114, pp. 350-356, 2006.

[3] A. Tixier-Mita, T. Takahashi, and H. Toshiyoshi, "Integration of Chemical Sensors with LSI Technology - History and Applications," *IEICE Transactions on Electronics,* vol. E95c, pp. 777-784, May 2012.

[4] M. Behnam, G. V. Kaigala, M. Khorasani, S. Martel, D. G. Elliott, and C. J. Backhouse, "Integrated circuit-based instrumentation for microchip capillary electrophoresis," *IET Nanobiotechnology,* vol. 4, pp. 91-101, Sep 2010.

[5] K. S. Shin, K. K. Pack, J. H. Park, T. S. Kim, B. K. Ju, and J. Y. Kang, "Parasitic bipolar junction transistors in a floating-gate MOSFET for fluorescence detection," *IEEE Electron Device Letters,* vol. 28, pp. 581-583, Jul 2007.

[6] D. Strle, B. Stefane, U. Nahtigal, E. Zupanic, F. Pozgan, I. Kvasic, *et al.,* "Surface-Functionalized COMB Capacitive Sensors and CMOS Electronics for Vapor Trace Detection of Explosives," *IEEE Sensors Journal,* vol. 12, pp. 1048-1057, May 2012.

[7] V. Stadler, M. Beyer, K. Konig, A. Nesterov, G. Torralba, V. Lindenstruth, *et al.,* "Multifunctional CMOS microchip coatings for protein and peptide Arrays," *Journal of Proteome Research,* vol. 6, pp. 3197-3202, Aug 2007.

[8] E. Ghafar-Zadeh, M. Sawan, and D. Therriault, "A microfluidic packaging technique for lab-on-chip applications," *Advanced Packaging, IEEE Transactions on,* vol. 32, pp. 410-416, 2009.

[9] H. Lee, Y. Liu, D. Ham, and R. M. Westervelt, "Integrated cell manipulation system—CMOS/microfluidic hybrid," *Lab on a Chip,* vol. 7, pp. 331-337, 2007.

[10] B. W. Zhang, Q. Dong, C. E. Korman, Z. Y. Li, and M. E. Zaghloul, "Flexible packaging of solid-state integrated circuit chips with elastomeric microfluidics," *Scientific Reports,* vol. 3, Jan 22 2013.

A TUNABLE LASER BASED ON NANO-OPTO-MECHANICAL SYSTEM

M. Ren[1,2], H. Cai[2], Y. D. Gu[2], P. Kropelnicki[2], A. B. Randles[2] and A. Q. Liu[1,2†]

[1]School of Electrical & Electronic Engineering, Nanyang Technological University, SINGAPORE 639798

[2] Institute of Microelectronics, A*STAR, SINGAPORE 117685

ABSTRACT

This paper presents an external cavity tunable laser based on nano-opto-mechanical system by integrating the gain laser diode and the opto-mechanical ring resonators on a silicon chip. An optical force controlled tuning approach is demonstrated whereby the lasing light itself adjusts the lasing wavelength by controlling the mechanical displacement of the silicon ring resonator. In the experiments, a 24-nm wavelength tuning is realized due to a deflection of 14-nm. The optomechanical wavelength tuning coefficient is 214 GHz/nm. The demonstrated device has potential applications for optical communication system, pulse trapping/release, and chemical sensing, with easy on-chip integration on a silicon platform.

INTRODUCTION

Tunable lasers are employed to reduce the amount of spare lasers with specific wavelengths for cost-effective wavelength division multiplexing systems or for passive optical network systems [1-2]. There are several approaches for wavelength tuning, such as by changing the real part of the effective refractive index, n_{eff}, or the length of the laser cavity, which are achievable by electronic control current or voltage. The refractive index control is available either by free-carrier plasma effect or by thermal tuning [3-4]. Both controls usually exhibit the essential disadvantage that the electrical power must be supplied, which heats up the devices and deteriorates other laser parameters such as optical power and cavity efficiency. Micro-electromechanical system (MEMS) technology has been used for external-cavity tunable lasers without power consumption due to the electrostatic force, but they suffer from low external cavity efficiency due to the free-space light travelling [5-7]. By applying nano-waveguide for light guiding in the external cavity may improve the laser performance with lower transmission loss. Nano-electromechanical system technology can be adopted to actuate nano-waveguide without electrical power consumption [8-9]. However, the nano-sized beam increases the impedance, making the devices susceptible to RF noise.

Optical force is a promising approach to actuate free-hanging nano-sized waveguide without electrical power consumption, which is compatible for high density on-chip integration [11-13]. In ring resonator based structures, refractive index can be changed by the movement of the free-standing ring resonator, thus making it possible to use optical force for wavelength tuning [14-15].

In this paper, a gain laser diode and opto-mechanical ring resonators are integrated to construct an external cavity tunable laser on a silicon chip. The lasing wavelength is tuned by controlling the displacement of the free-standing ring resonator. Except for the pumping power applied on the gain chip, no additional electrical power is needed for the tuning of the effective refractive index. The driving force comes from the lasing light in the laser cavity itself.

DESIGN AND SIMULATION

Figure 1 illustrates the schematic of the optical force driven tunable laser. It is constructed by a commercial gain chip and a coupled ring structure (CRS). For the two ring resonators that couples with each other, one ring is fixed on the silicon dioxide (SiO_2) layer, while the other ring is free-standing and can be actuated. The movable ring is supported by an anchor at the center and four nano-waveguide spokes. Both the bus-waveguide and the ring resonators are 450 nm in width and 220 nm in height. An air gap is formed between the free-standing ring resonator and SiO_2 layer. The CRS functions as a wavelength-selective mirror: the broadband light provided by the gain chip is reflected by the CRS, and the reflectivity varies for different wavelengths. The wavelength with maximal reflectivity is called the gain peak wavelength, which obtains the highest gain while suppressing other modes, and is finally lased out from the laser cavity.

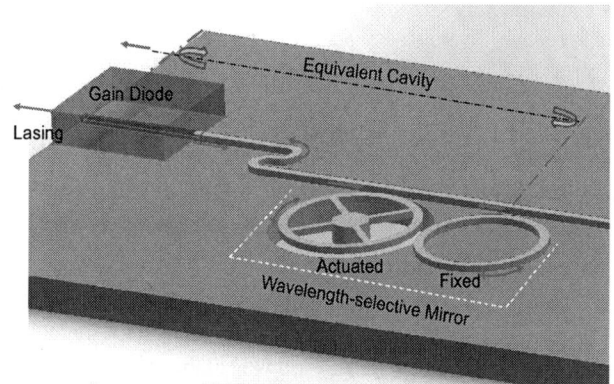

Figure 1: Schematic of tunable laser based on nano-opto-mechanical system.

When the gain provided by the laser diode exceeds the loss in the entire cavity, it is lased at the gain peak wavelength λ_r. At this state, the lasing wavelength λ_r satisfies the resonant condition of both ring resonators. Therefore, the light constrained in the free-standing ring induces a gradient optical force, which deflects the free-standing waveguide with a nano-scale displacement d (Fig. 2(a)). Consequently, the nano displacement changes the effective refractive index of the ring resonator, and thus tunes the gain peak

978-1-4799-3510-9/14 $31.00 © 2014 IEEE

wavelength. The position of λ_r is a function of d. Nano-scale displacement usually causes nano-scale wavelength shift for a single ring resonator [10]. According to Vernier Effect, the wavelength tuning ability is amplified by tens fold for the coupled ring structure. In this paper, the optomechanical wavelength tuning coefficient is $\partial\lambda_r / \partial d = 1.36 \cdot \exp(0.03d)$, as indicated in Fig. 2(b).

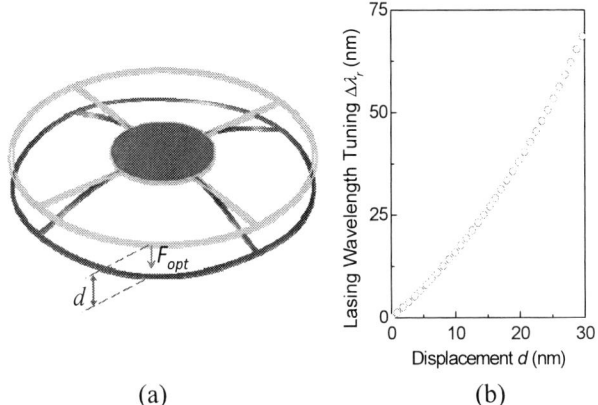

(a)   (b)

Figure 2: (a) Illustration of deflection under optical force; (b) calculation of the optomechanical lasing wavelength tuning coefficient of the coupled ring structure.

The gradient optical force can be expressed as [12]

$$F_{opt} = -\frac{2\tau_e^{-1}P}{\lambda_r((\lambda_0 - \lambda_r)\cdot 2\pi c\lambda_r^{-2})^2 + (\tau_i^{-1} + \tau_e^{-1})^2} \cdot \frac{\partial\lambda_r}{\partial d}, \quad (1)$$

where P is the light power confined in the bus waveguide, λ_r is the lasing wavelength generated in the external cavity tunable laser, λ_0 is the resonance determined by the coupled ring structure. $1/\tau_e$ and $1/\tau_i$ are the extrinsic and intrinsic decay rate respectively.

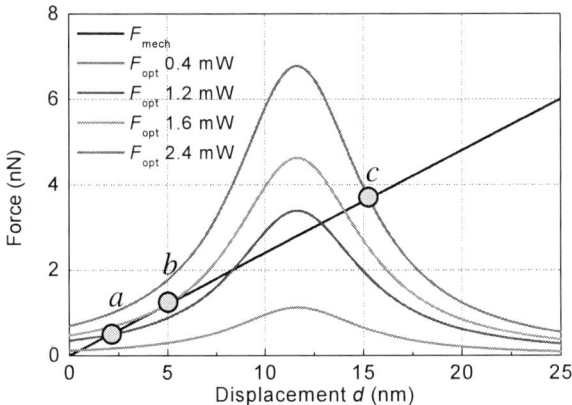

Figure 3: Optical force F_{opt} and mechanical force F_{mech} vs. displacement d, under different optical powers.

The deflection of the free-standing waveguide and the lasing light perform interaction in the laser cavity. The defection can be controlled by the lasing optical power, thus can be controlled by the pumping level of the laser diode. The deflection, on the other side, changes the resonance of the coupled ring structure, and shifts the lasing wavelength. The stable lasing wavelength λ_r is achieved when a balance

between the lasing power and the displacement is built up. The position of the lasing wavelength is determined by the optical power in the laser cavity.

Figure 3 presents the static analysis of the free-standing waveguide, under different optical powers. The static displacement positions are determined with the graphical method. The mechanical force F_{mech} is proportional to the effective stiffness k_{mech}, which is 0.24 N/m for the free-standing ring, while the optical force F_{opt} is a function of both displacement d and the optical power P. The cross points indicate the state whereby force equilibrium is achieved:

$$F_{opt} + k_{mech}d = 0. \quad (2)$$

At optical power level of $P = 1.2$ mW, there are 3 cross points, and the free-standing waveguide is stabled at point a. Therefore, the free-standing waveguide has a 2-nm displacement. This displacement increases gradually with the optical power until it reaches 4.8 nm (point b). By further increasing the optical power, there is only one cross point. For example, at $P = 2.4$ mW, the force-equilibrium condition is achieved at point c.

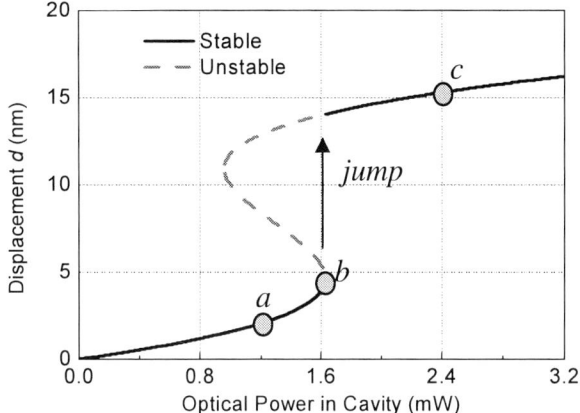

Figure 4: Nonlinear displacement due to different optical powers in the laser cavity.

The nonlinear curve in Fig. 4 shows the achievable displacement of the free-standing waveguide as a function of optical power in the laser cavity. Fig. 4 combines all the force-equilibrium points when the optical power increases from 0 to 3.2 mW. The stable points are plotted in the solid line and the unstable points are plotted in the dashed line. Only stable points are achievable in experiments. The control of displacement is continuous when the optical power is tuned from 0 to 1.6 mW. Due to the existence of the unstable force-equilibrium points, the curve has a step at point b and the displacement jumps from 4.8 nm to 15.2 nm, which shows the discontinuity in lasing wavelength tuning.

FABRICATION AND EXPERIMENTS

The opto-mechanical tunable laser is fabricated on an SOI wafer using nano-silicon photonics fabrication processes and MEMS packaging technology. For the fabrication of the photonic external cavity, the silicon waveguide is patterned

978-1-4799-3510-9/14 $31.00 © 2014 IEEE   1144

by reactive-ion-etching process and the movable structure is formed by controlling the time of the HF vapor etching. The commercial gain laser diode is bonded to the SOI chip by using Au/Sn solder. The misalignment between the waveguides of the laser diode and the photonic external cavity is controlled within 200 nm. Figure 5 is the SEM image of the free-standing ring, which has a radius of 25 μm. An internal disk with a radius of 8 μm works as an anchor. Four 600-nm width spokes connect both the anchor and the free-standing ring. The air gap between the free-standing waveguide and the substrate is controlled to be 150-nm.

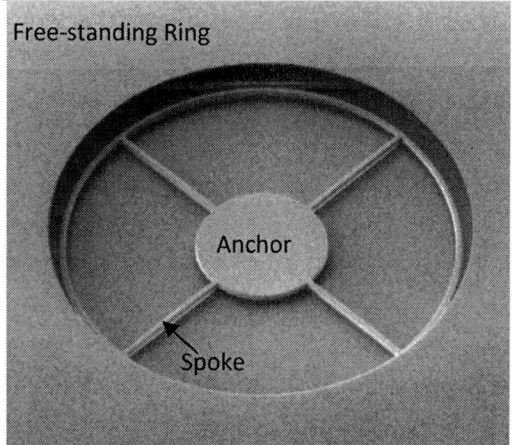

Figure 5: SEM image of the free-standing ring with 25-μm radius supported by an anchor with 8-μm radius and four spokes with 600-nm width.

In the experiment, the transmission spectrum of the couple ring structure is studied as shown in Fig. 6. A broadband light is coupled in and out of the bus waveguide via tapered optical fibers, and the transmission spectrum is detected by an optical spectrum analyzer. The broadband light source covers the range from 1500 to 1610 nm. The absorption dips in Fig. 6 correspond to those reflection peaks in the reflection spectrum. Dip A is the resonance of the free standing ring, while dip B is the resonance of the fixed ring. At the initial state, the merged resonance is observed at λ_1 since the resonance of the two ring resonators match at λ_1. The resonance split is observed at λ_2 and λ_3, and the split distance is 0.13 nm and 0.16 nm, respectively. Once the free-standing waveguide deflects, the dips of the free-standing ring (i.e. dip A) is red-shifted while the dips of the fixed ring (i.e. dip B) is kept static. Therefore, the merged resonance moves from λ_1 to λ_2 and λ_3.

Figure.7 shows the single mode lasing outputs under different gain currents. When the laser diode works at 330 mA, a single mode laser wavelength at $\lambda_1 = 1535.8$ nm is detected. Subsequently, $\lambda_2 = 1538.8$ nm and $\lambda_3 = 1541.8$ nm are lased at 365 mA and 400 mA, respectively. The spacing between two adjacent lasing wavelengths is 3.0 nm and the output power increases from -4dBm to 0 dBm. When the pumping current of the laser diode is increased to 410 mA, the lasing wavelength is tuned to $\lambda_4 = 1524.7$ nm, with a large wavelength jump towards to the blue detuning direction, and the lasing power increases to 1.5 dBm.

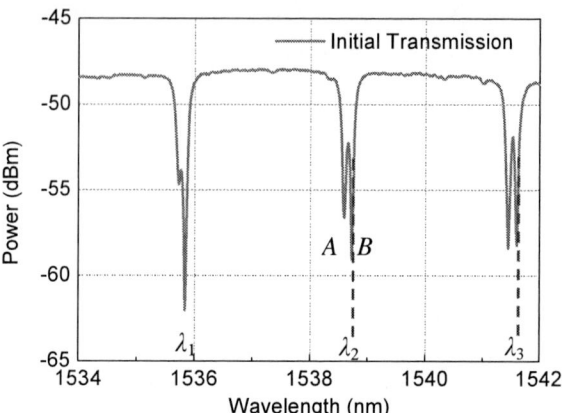

Figure 6: Experiments results of the static transmission spectrum of the coupled ring structure. The resonance of the fixed ring and the free-standing ring merges at λ_1.

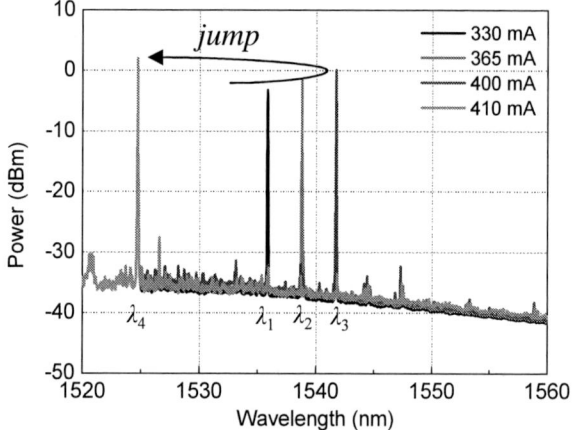

Figure 7: Experimental results of a set of single mode output at different gain currents.

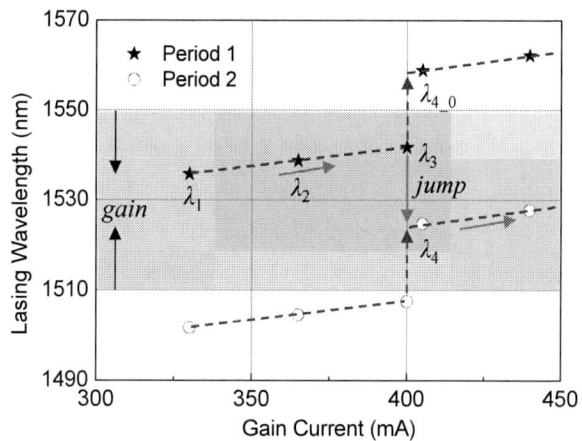

Figure 8: Illustration of lasing wavelength tuning tendency. The wavelength blue-jump corresponds to the discontinuity of the displacement.

The blue jump of lasing wavelength is illustrated in Fig. 8. First, the displacement of the free-standing waveguide has a non-continuous jump at point b as shown in Fig. 4, indicating the discontinuity of the lasing wavelength with the

978-1-4799-3510-9/14 $31.00 © 2014 IEEE   1145

increase of optical power. Second, due to the periodical property of the ring resonator, the coupled ring structure has two wavelength-selective periods within the wavelength range of 1490 to 1570 nm. On the other hand, the 3-dB gain spectrum of the laser diode convers 1510 to 1550 nm, which does not match with any of the two periods. As a result, the tuning of lasing wavelength is non-monotonic. It shifts from λ_3 to λ_4, instead of λ_{4_0}, which is out of the gain spectrum. In the experiment, the acquired tuning range is 24 nm (3000 GHz), which corresponds to a 14-nm displacement of the free-standing waveguide. The optomechanical wavelength tuning coefficient is 214 GHz/nm.

CONCLUSION

In conclusion, a tunable laser based on nano-opto-mechanical system is designed, fabricated and experimented. Optical power, instead of traditional electrical power, is applied for lasing wavelength tuning. A 24-nm tuning range is obtained with a mechanical displacement of 14 nm. The tuning process exhibits nonlinear mechanism due to the interaction between the lasing light and the mechanical movement. It has prospective applications in laser tuning, nonlinear signal processing and on-chip cell-manipulation.

ACKNOWLEDGEMENTS

This work was supported by the Science and Engineering Research Council of A*STAR (Agency for Science, Technology and Research), SERC Grant 1021650084.

REFERENCES

[1] F. J. Duarte, *Tunable lasers handbook*, Academic Press, 1995.

[2] M. C. Amann and J. Buus, *Tunable laser diodes*, Artech House, 1998.

[3] Y. D. Jeong and Y. H. Won, "Tunable single-mode Fabry-Perot laser diode using a built-in external cavity and its modulation characteristics", *Opt. Lett.*, vol. 31, pp. 2586-2588, 2006.

[4] L. Levin, "Mode-hop-free electro-optically tuned diode laser", *Opt. Lett.*, vol. 27, pp. 237-239, 2002.

[5] A. Q. Liu and X. M. Zhang, "A review of MEMS external-cavity tunable lasers", *J. Micromech. Microeng.*, vol.17, pp. R1-R13, 2007.

[6] H. Cai, A. Q. Liu and X. M. Zhang, "A miniature tunable coupled-cavity laser constructed by micromachining technology", *Appl. Phys. Lett.*, vol. 92, 031105, 2008.

[7] W. M. Zhu, T. Zhong, A. Q. Liu, X. M. Zhang and M. Yu, "Micromachined optical well structure for thermo-optic switching", *Appl. Phys. Lett.*, vol. 91, 261106, 2007.

[8] R. R. He, X. L. Feng, M. L. Roukers and P. D. Yang, "self-transducing silicon nanoware electromechanical systems at room tempreture", *Nano Lett.*, vol. 8, pp. 1756-1761, 2008.

[9] H. G. Craighead, "Nanoelectromechanical Systems", *Science*, vol. 290, pp. 1532-1535, 2000.

[10] M. L. Povinelli, M. Lončar and J. D. Joannopoulos, "Evanescent-wave bonding between optical waveguide", *Opt. Lett.*, vol. 30, pp. 3042-3044, 2005.

[11] W. H. P. Pernice, M. Li and H. X. Tang, "Theoretical investigation of the transverse optical force between a silicon nanowire waveguide and a substrate", *Opt. Express*, vol. 17, pp. 1806-1816, 2009.

[12] M. Ren, J. G. Huang, H. Cai, J. M. Tsai, J. X. Zhou, Z. S. Liu, Z. G. Suo, and A. Q. Liu, "Nano-optomechanical actuator and pull-back instability", *ACS Nano*, vol. 7, pp.1676–1681, 2013.

[13] J. Rosenberg, Q. Lin and O. Painter, "Static and dynamic wavelength routing via the gradient optical force", *Nat. Photon.*, vol. 3, pp. 478-483, 2009.

[14] B. Dong, H. Cai, G. I. Ng, P. Kropelnicki, J. M. Tsai, A. B. Randles, M. Tang, Y. D. Gu, Z. G. Suo and A. Q. Liu, "A nanoelectromechanical systems actuator driven and controlled by Q-factor attenuation of ring resonator", *Appl. Phys. Lett.*, vol. 103, 181105, 2013.

[15] W. Bogaerts, P. D. Heyn, T. V. Vaerenbergh, K. D. Vos, S. K. Selvaraja, T. Claes, P. Dumon, P. Bienstman, D. V. Thourhout and R. Baets, "Silicon microring resonators", *Laser & Photon. Rev.*, vol. 6, pp. 47-73, 2012.

CONTACT

[†]A. Q. Liu, tel: +65-67904336; eaqliu@ntu.edu.sg.

A TUNABLE OPTICAL IRIS BASED ON ELECTROMAGNETIC ACTUATION FOR A HIGH-PERFORMANCE MINI/MICRO CAMERA

Hee Won Seo[1], Jeong Byung Chae[1], Sung Jin Hong[1], In Uk Shin[1], Kyehan Rhee[1],
Jong-hyeon Chang[2] and Sang Kug Chung[1]
[1]Myongji University, Yongin, Gyeonggido, South Korea
[2]SAIT—Samsung Advanced Institute of Technology, Yongin, Gyeonggido, South Korea

ABSTRACT

This paper presents a tunable iris based on electromagnetic actuation for a tiny high-performance camera in mobile devices such as smart phones and pads. To investigate the effect of a magnetic field on a ferrofluid, the contact angle modification and transportation of a sessile ferrofluid droplet are tested using a neodymium magnet and electric coil. The variation of the contact angle of the ferrofluid droplet is 21.3° for the neodymium magnet and 18.1° for the electric coil based on electromagnetic induction. And the transportation of the ferrofluid droplet is also demonstrated using the neodymium magnet and electric coil. As the concept proof, the pretest of a tunable iris operated by electromagnetic actuation is conducted by using a hollow cylinder cell. In the initial state, the ferrofluid is in the relax state, so the cylinder cell shows the largest aperture (4.06 mm). When an electrical current is applied to an electric coil wound around the outside of the cylinder cell, the ferrofluid initially placed in the hydrophobic sidewall inside the cylinder cell is actuated and pulled to the center. The aperture under the current is modified from 4.06 mm at 0 A to 3.2 mm at 2 A. Finally, the envisioned tunable iris consisted of two connected circular microchannels is realized using a MEMS technology. The iris size is 9×9×2 mm³, and the variation of the aperture diameter is from 1.72 mm at 0 A to 1.09 mm at 2.6 A.

INTRODUCTION

To meet the increased demand of a tiny high-performance camera for mobile devices such as smart phones and pads, development of a new type of optical element has become important[1, 2]. Liquid based tunable optical elements such as a liquid lens and iris have received substantial attention from optical and MEMS societies for miniaturization and speed[3-9]. In this paper a tunable optical iris based on electromagnetic actuation is presented, along with experimental verification. In optics, an iris, an aperture stop, is placed in the light path of a lens or objective and regulates the amount of light that passes through the lens by controlling the size of the aperture, an opening at it center. The iris not only controls light flux, field of view, depth of field (DOF), but also blocks scattered light and improves image quality by limiting spherical aberration[10-14]. Hence, the iris is an indispensable element in most optical systems. However, the conventional mechanical iris, consisting of movable sliding blades, requires a complicated sliding rotary mechanism that has to be operated by bulky motors and is therefore difficult to miniaturize[15-17].

Figure 1 describes the envisioned tunable mini/micro iris design and its working principle. According to Faraday's law of electromagnetic induction, when an electrical current flows in an electric coil, a magnetic field is generated in its surroundings[18-20]. In this work, the magnetic field is used to actuate or pull an optically opaque ferrofluid initially filled inside the sub-channel of the iris to the center of the main channel, resulting in controlling the diameter of an aperture in Fig. 1(b).

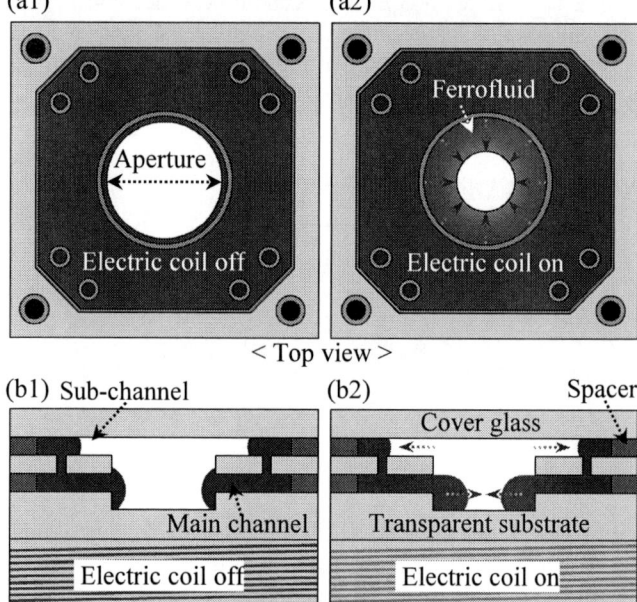

Figure 1: Schematic diagram of a tunable optical iris operated by electromagnetic actuation: When an electrical current is applied to an electric coil, a ferrofluid initially filled inside the sub-channel of the iris is electromagnetically actuated and pulled to the center of the main channel, resulting in changing the aperture of the iris. (a) Top view; (b) Side view.

EXPERIMENT RESULTS

Figure 2 shows the schematic exploded diagram of the envisioned tunable optical iris and the images of a microfabricated iris and test. The iris consists of three plates and two spacers, forming two connected circular microchannels in Fig. 2. Especially, the middle plate has a center hole for an air flow and middle holes for an opaque ferrofluid flow, respectively.

978-1-4799-3510-9/14 $31.00 © 2014 IEEE

(a)

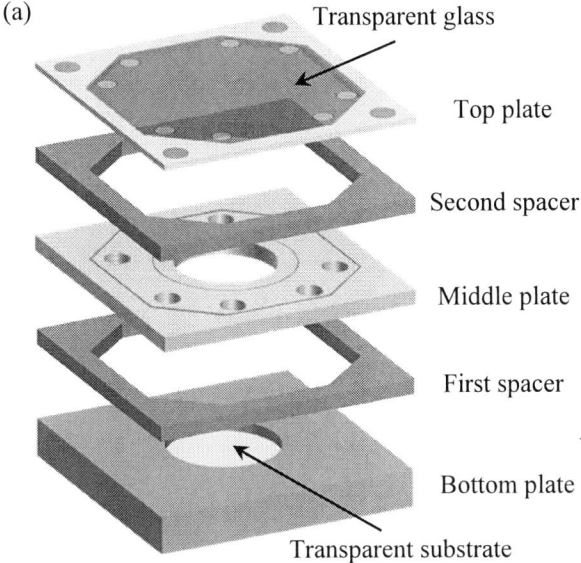

Transparent glass

Top plate

Second spacer

Middle plate

First spacer

Bottom plate

Transparent substrate

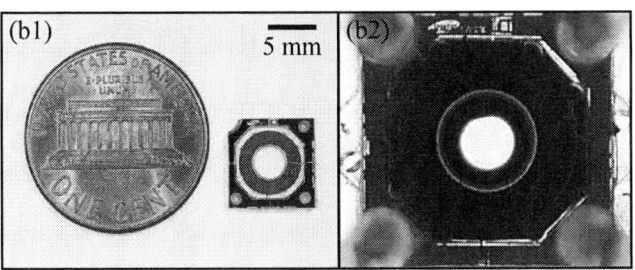

Figure 2: (a) Schematic exploded diagram of the proposed optical iris; (b) A microfabricated iris and its test image.

The schematic diagram of experiment setups is shown in Fig. 3. The setups mainly consisted of electrical and optical systems. A power supply (IT6720, ITECH Electronic Co., Ltd.) is used to actuate a tunable optical iris placed on a 3D traverse system. And test images are obtained by observations using a charge coupled device (CCD, EO-1312C, Edmund Optics) with a zoom lens (VZMTM 450i eo, Edmund Optics) and saved on a personal computer.

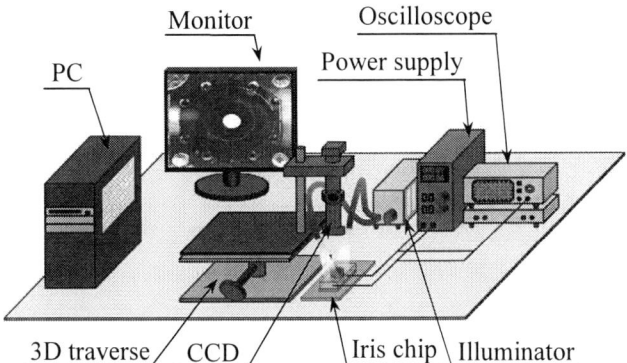

Figure 3: Schematic diagram of experimental setups mainly consisting of electrical and optical systems.

To investigate the effect of a magnetic field on a ferrofluid (EMG 707, Ferro Tec), the contact angle modification of a sessile ferrofluid droplet is firstly tested using a neodymium magnet (1 mm(D) × 1 mm(H)) and electric coil (160 μm diameter with 12.7 Ω), as shown in Fig. 4. When the magnet approaches a ferrofluid droplet (0.3 μl) placed on a hydrophobic substrate from beneath, the droplet spreads on the substrate due to a magnetic field from the magnet in Fig. 4(a). The contact angle of the droplet is modified from 96.41° to 75.15°. The similar test is conducted using the electric coil. When an electric current (2.5 A) is applied to the coil placed beneath a substrate, the initial contact angle (94.72°) of a ferrofluid droplet (0.7 μl) is modified to 76.64° as it spreads on the substrate in Fig. 4(b). The variation of the contact angle of the ferrofluid droplet is 21.3° for the neodymium magnet and 18.1° for the electric coil based on electromagnetic induction.

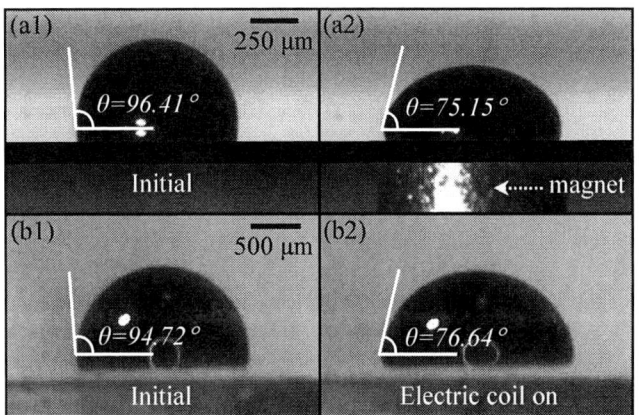

Figure 4: A sessile ferrofluid droplet experiment using a neodymium magnet and electric coil: (a) When the magnet approaches the droplet from beneath, it spreads on the surface (Contact angle modification: 96.41°→75.15°); (b) When a current (2.5 A) is applied to the coil, the droplet actuated by the induced magnetic field spreads on the surface (Contact angle modification: 94.72°→76.64°).

The transportation of a ferrofluid droplet is also conducted using the same permanent magnet and electric coil. When the permanent magnet placed beneath a substrate moves from the left to the right, a ferrofluid droplet placed on the substrate follows the motion of the magnet and is transported to the right in Fig. 5(a). The similar experiment is carried out using the electric coil. When an electric current (2.5 A) is applied to the electric coil placed beneath a substrate, the ferrofluid droplet moves towards the inside of the coil circumference owing to electromagnetic induction, as shown in Fig. 5(b).

As the concept proof, the pretest of a tunable iris operated by electromagnetic actuation is conducted by using a hollow cylinder cell. Figure 6 shows sequential images when a current is applied to the cylinder cell. In the initial state, the ferrofluid is in the relax state, so the cylinder cell

978-1-4799-3510-9/14 $31.00 © 2014 IEEE　　　1148

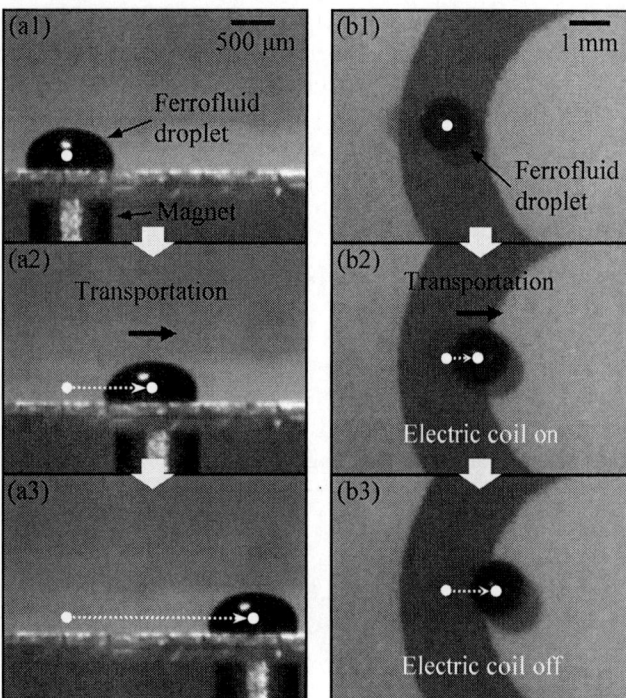

Figure 5: The transportation of a ferrofluid droplet using a neodymium magnet and electric coil.

shows the largest aperture (4.06 mm). When an electrical current is applied to an electric coil wound around the outside of the cylinder cell, the ferrofluid initially placed in the hydrophobic sidewall inside the cylinder cell is actuated and pulled to the center in Fig. 6. The aperture diameter decreases when the current increases. The aperture diameter is modified from 4.06 mm at 0 A to 3.2 mm at 2.0 A.

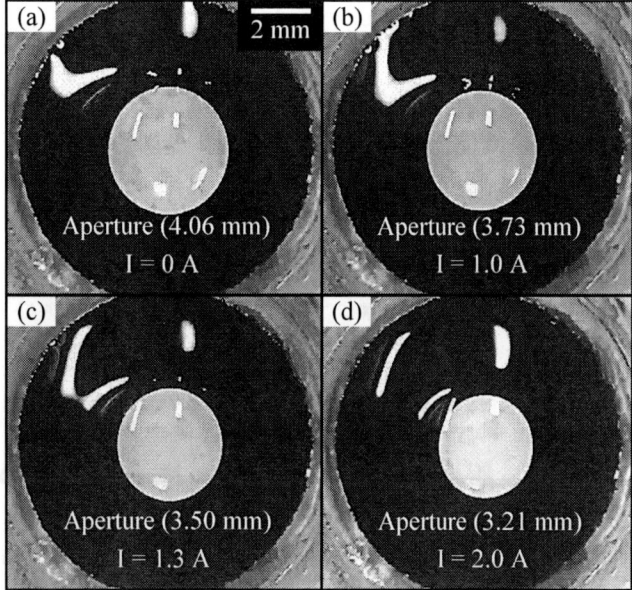

Figure 6: The pretest of a tunable iris operated by electromagnetic actuation using a hollow cylinder cell: (a) Initial; (b-d) When a current is applied to the cell, a ferrofluid initially placed in the hydrophobic sidewall inside the cell is pulled to the center of the iris, changing the aperture from 4.06 mm at 0 A to 3.21 mm at 2.0 A.

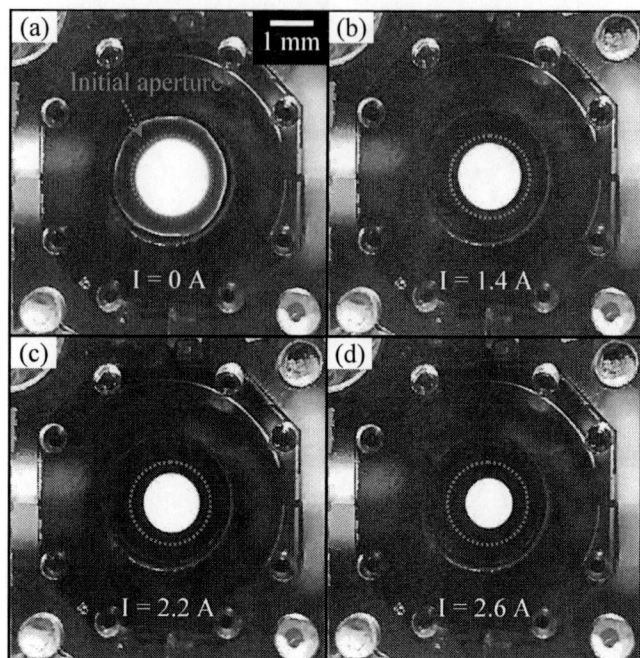

Figure 7: The microfabricated iris: (a) Initial; (b) When a current is applied to a coil attached on the bottom of the iris, the aperture of the iris is continuously changed from 1.72 mm to 1.09 mm.

Finally, the envisioned tunable optical iris is realized using a MEMS technology. The iris has an aqueous diaphragm in two connected circular microchannels. The iris size is $9 \times 9 \times 2$ mm^3, and the aperture diameter is modified from 1.72 mm at 0 A to 1.09 mm at 2.6 A. Figure 8 shows the actuated aperture diameter versus applied current. The aperture diameter is in inverse proportion to the applied current.

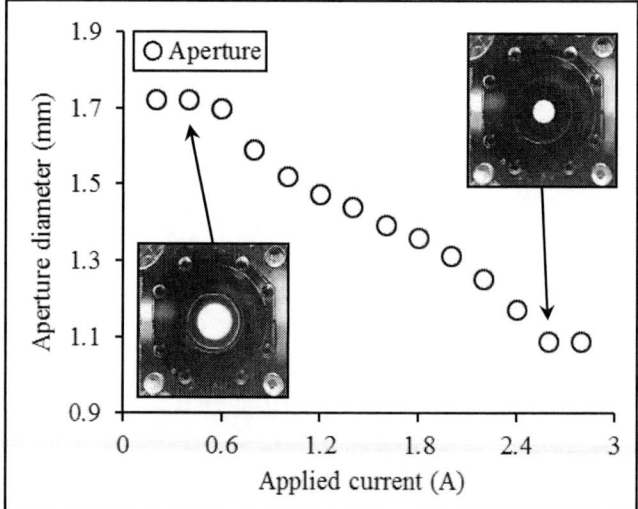

Figure 8: Actuated aperture versus the applied current: When an electric current applied to the microfabricated iris is varied from 0 A to 3 A, the aperture of the iris is modified from 1.72 mm to 1.09 mm.

CONCLUSIONS

An electromagnetically driven tunable optical iris is developed for a tiny high-performance camera in mobile devices. First, the contact angle modification of a sessile ferrofluid droplet is tested using a neodymium magnet and electric coil to study the effect of a magnetic field on the ferrofluid. The amount of contact angle modification is 21.3° for the neodymium magnet (magnitude) and 18.1° for the electric coil (I = 2.5 A). Second, the transportation of a sessile ferrofluid droplet is also conducted using the neodymium magnet and electric coil. Third, the pretest of a tunable iris operated by electromagnetic actuation is conducted by using a hollow cylinder cell. When an electrical current is applied to an electric coil wound around the outside of the cylinder cell, the ferrofluid initially placed in the hydrophobic sidewall inside the cylinder cell is actuated and pulled to the center. The variation of the aperture diameter is from 4.06 mm at 0 A to 3.2 mm at 2 A. Lastly, a MEMS-based tunable optical iris ($9\times9\times2$ mm^3) is realized. The iris consists of two connected circular surface microchannels, which are the passages of a transient air and opaque ferrofluid. The aperture diameter is in inverse proportion to the applied current. The aperture diameter is modified from 1.72 mm at 0 A to 1.09 mm at 2.6 A. This tunable optical iris can be applied to not only portable electronic applications but also biomedical applications such as optical coherence tomography (OCT) and microsurgery.

ACKNOWLEDGEMENTS

This work was supported by the Fundamental Research Supporting Program (2011-0012100) and Basic Science Research Program (2011-0025039) of National Research Foundation of Korea(NRF).

REFERENCES

[1] S. Kuiper and B. H. W. Hendriks, "Variable-focus liquid lens for miniature cameras," *Applied Physics Letters,* vol. 85, pp. 1128-1130, 2004.

[2] J. H. Chang, K. D. Jung, E. Lee, M. Choi, and S. Lee, "Microelectrofluidic iris for variable aperture," in *SPIE MOEMS-MEMS*, vol. 8252, p. 82520, 2012.

[3] B. Berge and J. Peseux, "Variable focal lens controlled by an external voltage: An application of electrowetting," *The European Physical Journal E,* vol. 3, pp. 159-163, 2000.

[4] Y. Hongbin, Z. Guangya, C. F. Siong, and L. Feiwen, "Optofluidic variable aperture," *Optics Letters,* vol. 33, pp. 548-550, 2008.

[5] T. Krupenkin, S. Yang, and P. Mach, "Tunable liquid microlens," *Applied Physics Letters,* vol. 82, pp. 316-318, 2003.

[6] L. Li, C. Liu, H. Ren, and Q. H. Wang, "Adaptive liquid iris based on electrowetting," *Optics Letters,* vol. 38, pp. 2336-2338, 2013.

[7] L. Li, C. Liu, and Q. H. Wang, "Electrowetting based liquid iris," *IEEE Photonics Technology Letters,* vol. 25, pp. 989-991, 2013.

[8] C. G. Tsai and J. A. Yeh, "Circular dielectric liquid iris," *Optics Letters,* vol. 35, pp. 2484-2486, 2010.

[9] Z. Wan, H. Zeng, and A. Feinerman, "Area-tunable micromirror based on electrowetting actuation of liquid-metal droplets," *Applied Physics Letters,* vol. 89, pp. 2011071-2011073, 2006.

[10] J. Draheim, T. Burger, J. G. Korvink, and U. Wallrabe, "Variable aperture stop based on the design of a single chamber silicone membrane lens with integrated actuation," *Optics Letters,* vol. 36, pp. 2032-2034, 2011.

[11] C. Kimmle, U. Schmittat, C. Doering, and H. Fouckhardt, "Compact dynamic microfluidic iris for active optics," *Microelectronic Engineering,* vol. 88, pp. 1772-1774, 2011.

[12] H. Ren, S. Xu, and S. T. Wu, "Optical switch based on variable aperture," *Optics Letters,* vol. 37, pp. 1421-1423, 2012.

[13] L. Li, C. Liu, H. Ren, and Q. H. Wang, "Fluidic optical switch by pneumatic actuation," *IEEE Photonics Technology Letters,* vol. 25, pp. 338-340, 2013.

[14] C. U. Murade, J. M. Oh, D. v. d. Ende, and F. Mugele, "Electrowetting driven optical switch and tunable aperture," *Optics Express,* vol. 19, pp. 15525-15531, 2011.

[15] J. H. Chang, K. D. Jung, E. Lee, M. Choi, S. Lee, and W. Kim, "Variable aperture controlled by microelectrofluidic iris," *Optics Letters,* vol. 38, pp. 2919-2922, 2013.

[16] P. Müller, N. Spengler, H. Zappe, and W. Mo☐nch, "An optofluidic concept for a tunable micro-iris," *Journal of Microelectromechanical Systems,* vol. 19, pp. 1477-1484, 2010.

[17] P. Müller, R. Feuerstein, and H. Zappe, "Integrated optofluidic iris," *Journal of Microelectromechanical Systems,* vol. 21, pp. 1156-1164, 2012.

[18] H. C. Cheng, S. Xu, Y. Liu, S. Levi, and S. T. Wu, "Adaptive mechanical-wetting lens actuated by ferrofluids," *Optics Communications,* vol. 284, pp. 2118-2121, 2011.

[19] J. O. Kwon, J. S. Yang, S. J. Lee, K. Rhee, and S. K. Chung, "Electromagnetically actuated micromanipulator using an acoustically oscillating bubble," *Journal of Micromechanics and Microengineering,* vol. 21, p. 115023, 2011.

[20] B. J. Nelson, I. K. Kaliakatsos, and J. J. Abbott, "Microrobots for minimally invasive medicine," *Annual Review of Biomedical Engineering,* vol. 12, pp. 55-85, 2010.

CONTACT

S. K. Chung, Tel: +82-31-330-6346; skchung@mju.ac.kr

CALORIMETRIC DEVICE FOR NON-DESTRUCTIVE MEASUREMENT OF THE THERMAL DIFFUSIVITY DEPENDENCY BY PHASE DELAY

Takahiro Suzuki, Yasumasa Ichikawa, Tomoyuki Takahata, Kiyoshi Matsumoto and Isao Shimoyama
The University of Tokyo, Tokyo, Japan

ABSTRACT

We developed a calorimetric device for measuring thermal diffusivity non-destructively. The calorimetric device was based on the principle that temperature phase delay between a heater and a resistance temperature detector (RTD) is affected by thermal diffusivity of the region between them. The device consisted of an Au wire, as oscillating heat source and a piezoresistor as an RTD. We exerted the experiment, in which air pressure was changed, and observed that the device measured thermal diffusivity of the air.

INTRODUCTION

Thermal diffusivity is one of the fundamental physical properties in biomedical measurement. Additionally it is a performance index to measure the surface layer property [1,2]. For example, water holding capability of skin and its moisture should be known if the thermal diffusivity of human skin is measured [3].

Thermal diffusivity has been measured in various ways, for example laser flash method, temperature wave analysis method and so on. In these conventional methods, complicated experimental setup and long measurement time have been the issues because in biomedical measurement, physical property changes significantly with time [4-6]. Moreover, in conventional methods the target materials must be cut out into thin film for evaluating. This kind of method is not appropriate especially in biomedical measurement. Therefore, an instant and non-destructive measurement for thermal diffusivity is required.

In this study, we proposed a MEMS calorimetric device for measuring thermal diffusivity non-destructively and improving the measurement time compared with conventional methods. The concept of this research is shown in **Figure 1**. The device can measure thermal diffusivity in small region, which should be able be applied to other MEMS devices. The device had two parts: Gold wire as a heater and a piezoresistor as a resistance temperature detector (RTD). When an alternating current (AC) is applied to the heater, phase delay occurred between the heater temperature and the RTD temperature. Phase delay between these temperatures depends on thermal diffusivity of the medium between the heater and RTD. Therefore the thermal diffusivity can be obtained by measuring this phase delay.

PRINCIPLE AND SIMULATION

Considering a model with a heating point and a detecting point separated in a certain distance (Figure 2 A), when the heating point's temperature is changed periodically in a homogeneous infinite region, phase delay occurs

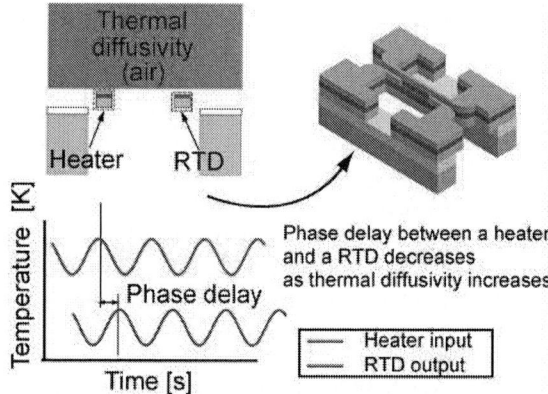

Figure 1: The concept of this study and the design of the device. Phase delay changes by thermal diffusivity of material.

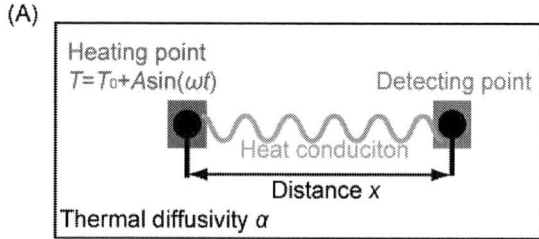

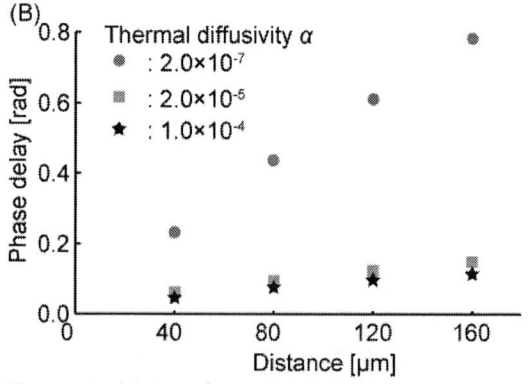

Figure 2: (A) Simulation model: thermal diffusivity α and distance x was changed. (B) Result of Simulation: relation among phase delay, x and α

between these two points. It is generally known that when the distance x, the angular frequency ω of the heating point's temperature change and the thermal diffusivity of the material α are given, the phase delay θ is expressed by the following equation [7].

$$\theta = x\sqrt{\frac{\omega}{2\alpha}} \ . \ (1)$$

978-1-4799-3510-9/14 $31.00 © 2014 IEEE   1151

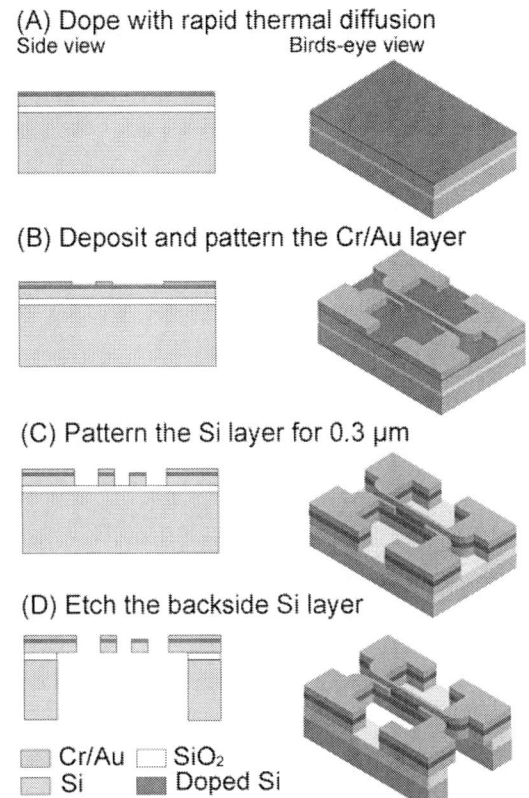

(A) Dope with rapid thermal diffusion
Side view   Birds-eye view

(B) Deposit and pattern the Cr/Au layer

(C) Pattern the Si layer for 0.3 µm

(D) Etch the backside Si layer

Cr/Au   SiO₂
Si   Doped Si

Figure 3: Fabrication process of the sensor. All the sensors are fabricated on the same SOI wafer.

Therefore, α can be expressed by

$$\alpha = \frac{\omega}{2k^2} \quad (k = \frac{\theta}{x}), \quad (2)$$

Our device consisted of a heater and an RTD. When the voltage is applied to the heater, phase delay between the heater and the RTD occurs. It is predicted that the phase delay has relationship to k, ω, and as shown in equation (2). Here, k is calculated by phase delay and ω can be controlled. Thermal diffusivity can be calculated by measuring the phase delay.

To confirm relationship among the phase delay, x and, α we carried out simulations using FEM method (COMSOL Multiphysics). Figure 2(A) shows the model used in the simulations. The values of α was varied to be 2.0×10^{-7}, 2.0×10^{-5}, and 1.0×10^{-4} m²/s. The heating point was designed at the center of the material. The temperature of the heating point was changed periodically. The value of ω was 2π rad/s. The values of the distance x was varied to be 40, 80, 120, and 160 µm. The result of simulations is shown in Figure 2(B). The result indicates that the phase delay increases with the distance for a certain thermal diffusivity. Additionally, phase delay is bigger for smaller thermal diffusivity, considering the same distance The result indicates that relationship between phase delay, distance and thermal diffusivity, is consistent with equation (1).

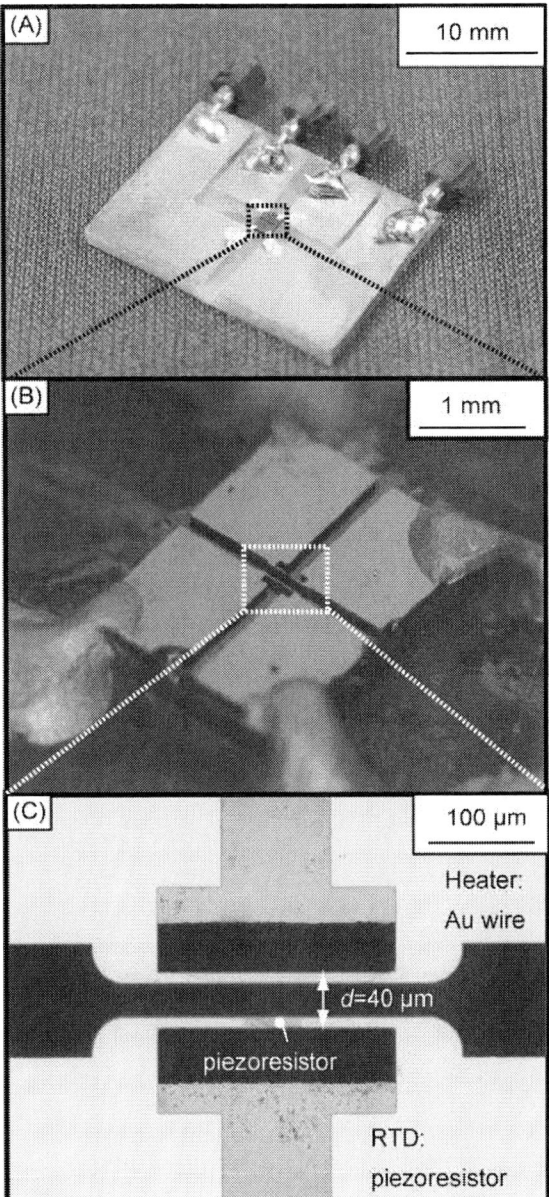

Figure 4: (A) The device with circuit board (B) Overview of the device (C) Close-up view of the device

SENSOR DESIGN AND FABRICATION

The design of the calorimetric device is shown in Figure 1. The device consisted of a gold wire as heat source and a piezoresistor as an RTD. A piezoresistor is supposed to be more sensitive to temperature than a gold wire. To reduce heat transmitted to the backside Si layer, the gold wire and the piezoresistor were suspended. To investigate the effect of the distance x between the heater and the RTD on phase delay, four kinds of the devices were fabricated, in which distance x was varied to be 40, 80, 120, and 160 µm.

The fabrication process is shown in **Figure 3**. We used a 0.3 µm/0.4 µm/300 µm–thick SOI (silicon on insulator) wafer. The process began with doping with rapid thermal

978-1-4799-3510-9/14 $31.00 © 2014 IEEE

diffusion (**Figure 3(A)**). Next, Cr/Au layers(5 nm/35 nm thick) were deposited and patterned (**Figure 3(B)**). Then, top Si layer was etched by DRIE (**Figure 3(C)**). To make a gold wire and a piezoresistor suspended above the substrate, the backside Si layer was etched by DRIE and the glass layer was etched using HF vapor (**Figure 3(D)**). Finally, the device was put on a circuit board and wired by a conductive paste. The photographs of the device are shown in **Figure 4**. The resistance of the Au wires were from 56Ω to 57 Ω. The resistance of the piezoresistors were from 13 kΩ to17 kΩ.

SENSOR CALIBRATION

We measured the temperature coefficient of the RTD using a temperature-controlled bath, where the water temperature was kept and changed form 288 K to 318 K with 3 K step. We put the device in the bath and measured the resistance change of RTD with a source meter applying 1 mA current. This source meter was used for controlling the bath temperature. The result of resistance change of RTDs versus temperature is shown in **Figure 5**. Linear relationship between temperature and resistance change was confirmed. The proportional coefficient was calculated as 1.9×10^3 ppm/K. Hence, the phase delay of temperature signals can be obtained by measuring the resistance change.

EXPERIMENT AND RESULT

Figure 6 shows the experimental system. We used a function generator as an AC source. The center of the AC voltage was 0.7 V and the amplitude was 0.5 V. The angular frequency was 2π rad/s. The voltage was amplified ten times by an amplifier. The amplified voltage was used to applied to the heater. The resistance change of the RTD was measured by a bridge circuit.. The output of the bridge was amplified by a lock-in amplifier. The applied voltage and the RTD signal were measured by an oscilloscope. The peak of the voltage applied to the heater matches with that of the temperature change of the heater [8]. The resistance change of the RTD was proportional to the temperature change of the RTD. Phase delay between the heater and the RTD was able to be calculated from the applied voltage and the RTD signal.

To investigate the effect of the device itself on the phase delay, the device was put in a vacuum desiccator. Air pressure in the desiccator can be changed from atmospheric pressure to 50 Pa. The graphs of relationship between the applied voltage and the RTD signal are shown in **Figure 7(A)** and **Figure 7(B)**. The distance between the heater and the RTD was 40 µm. **Figure 7(A)** shows the result at atmospheric pressure and **Figure 7(B)** shows the result in the vacuum of 50 Pa. Generated heat is in proportion to the square of the applied voltage. Therefore, the RTD signal is in proportion to the square of the applied voltage. Even though the amplitude of applied voltage is the same, the amplitude of the RTD signal at atmospheric pressure was 7.5×10^{-1} V and that of 50 Pa was 1.5×10^{-2} V. The result indicated that the RTD signal is affected highly by the surrounded condition.

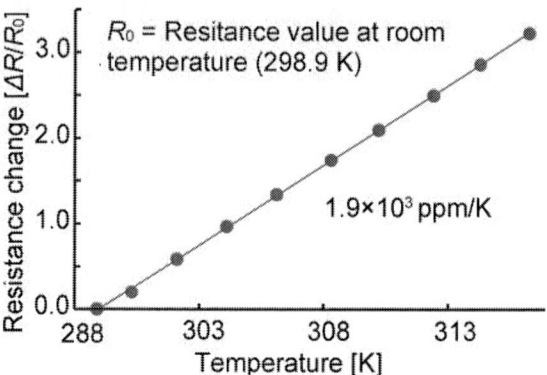

Figure 5: Resistance change of RTDs versus Temperature.

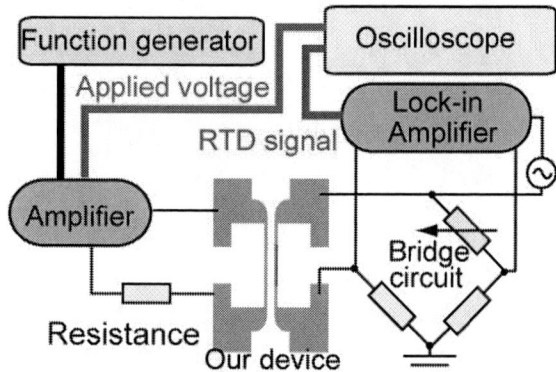

Figure 6: Experimental system to measure the phase delay between a heater and a RTD.

Phase delay was calculated by correlation of the square of the applied voltage and the RTD signal. **Figure 8** shows relationship between the pressure and the phase delay. The distance between the heater and the RTD was 40 µm. The pressure was varied to be 50, 200, 400, and 100000 (atmospheric pressure) Pa. As pressure decreases, phase delay increases. This result indicates that as pressure decreases, thermal diffusivity decreases.

Figure 9 shows the relationship between the phase delay and the distance of the heater and the RTD. The distance was 40, 80, 120, and 160 µm. The distance from the heater and the RTD is in proportion to the phase delay. The fitted line was evaluated by least square method. The value of measured gradient k was 396.7 rad/m. Thermal diffusivity was calculated by the equation (2). The calculated thermal diffusivity was 2.0×10^5 m^2/s. The theoretical thermal diffusivity of the air at 300 K is 2.2×10^5 m^2/s, which is close with the calculated value. This result indicates that our device can measure thermal diffusivity of the air. With this device, when a material was fully contacted with the proposed device, thermal diffusivity of the contacted material could be measured.

CONCLUSION

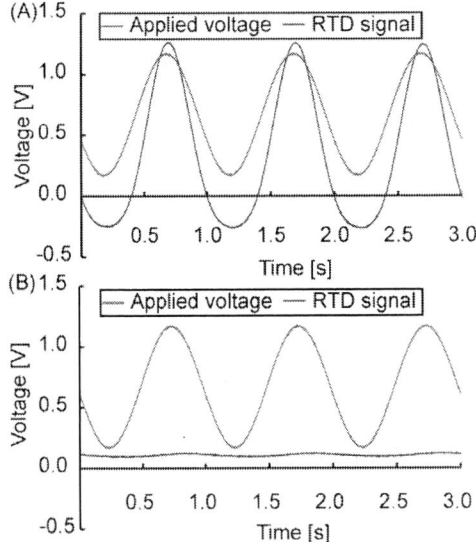

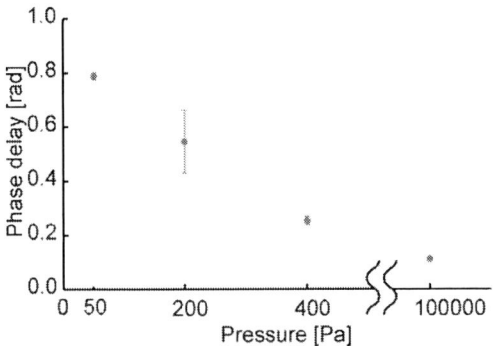

Figure 7: (A): Relation between applied voltage and RTD signal at atmospheric pressure. (B): Relation between applied voltage and RTD signal at 50Pa.

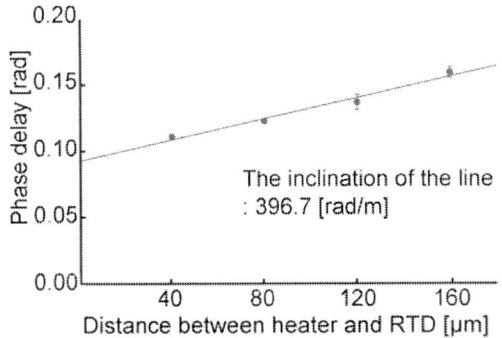

Figure 8: Relation between pressure and phase delay. The distance between the heater and the RTD is 40 μm.

In conclusion, we proposed the calorimetric device, which can non-destructively measure the thermal diffusivity

Figure 9: Phase delay between a heater and a RTD.
Experiment condition is in the air and at vacuum.

by phase delay between heating source and RTD output. We confirmed our device can measure thermal diffusivity of the air and thermal diffusivity of the air changes as air pressure changes.

ACKNOWLEDGEMENT

The photolithography masks were made using the University of Tokyo VLSI Design and Education Center (VDEC)'s 8 inch EB writer F5112 +VD01 donated by ADVANTEST Corporation.

REFERENCES

[1] N. Taketoshi, *et al.*, "Development of a Thermal Diffusivity Measurement System for Metal Thin Films Using a Picosecond Thermoreflectance Technique," *Meas. Sci. Technol.*, 12, pp. 2064-2073, 2001

[2] J.C. Withers, *et al.*, "Thermal Diffusivity/Conductivity of Compacts of C60 Buckminsterfullerene and a C60-C70 Mixture," *J. Am. Ceram. Soc.*, 76, pp. 754-756, 1993

[3] Thune P, *et al.*, "The water barrier function of the skin in relation to the water content of stratum corenum, pH and skin liquids. The effect of alkaline soap and syndet on dry skin in eldery, non-atopic patients," *Acta Dermato-venereologica*, vol. 68, No. 4, pp.277-283, 1988

[4] B. Tetsuya, "Ultra Fast Laser Flash Method for Measuring Thermal Diffusivity of Thin Films," *Transactions of The Japan Society for Precision Engineering*, vol. 73, No. 8, pp. 864-870, 2008

[5] Y. Haruhiko, "AC Calorimetry," *Netsu Sokutei*, vol. 29, No. 1, pp. 5-10, 2001

[6] U. satoru, "Noninvasive Measurement of Thermal Conductivity and Thermal Diffusivity of Biological Materials (Preliminary Examination of Heat Transfer Models and Measurement Accuracy)," *Transactions of the Japan Society of Mechanical Engineers*, Series B, vol. 72, No. 723, pp. 186-191, 2006

[7] S. nobuhiro, *et al.*, "Heat Tranfer Enineering (in Japanese)," Morikita Shuppan Co., Ltd., pp. 30-32, 1988

[8] K. yoshimitsu, "A New Method for Measuring Flow Velocity Using Periodic Heating and the Sensing Characteristics of a Micro Flow Sensor Manufactured by MEMS Technique," *Transactions of the Japan Society of Mechanical Engineers*, Series B, vol. 72, No. 723, pp. 8-15, 2006

CONTACT

Takahiro Suzuki, The University of Tokyo, 7-3-1 Hongo, Bunkyo-ku, Tokyo, 113-8656, Japan
Tel: +81-3-5841-0461, Fax: +81-3-3818-0835
E-mail: suzuki@lepard.t.u-tokyo.ac.jp

CAPACITIVE FEEDBACK CONTROLLED PZT MICRO MIRROR ARRAYS FOR WAVELENGTH SELECTIVE SWITCH

Ryohei Uchino, Tokiko Misaki, Tsuyoshi Fujimura, and Osamu Torayashiki

Sumitomo Precision Products Co., Ltd., Japan

ABSTRACT

We developed a single-axis mechanical micro mirror array used for gridless wavelength selective switch (WSS). The mirrors are driven by lead zirconate titanate (PZT) unimorph actuators, which is adequate for low-voltage actuation and low interference with adjacent mirrors in operation. In addition, the mirror tilt angle is feedback-controlled using comb-shaped capacitance in order to realize high control resolution. We fabricated a prototype mirror array by using simple semiconductor wafer process technologies, and evaluated its basic performance. In conclusion, we realized a high fill factor of over 98%, 5° mechanical angle at 10 V, and high control resolution under 0.01°.

INTRODUCTION

In the next generation networking infrastructure, high-speed and flexible network operation without optical-electrical conversion is an essential technology. Wavelength selective switch (WSS) is a key device that is used in reconfigurable optical add/drop multiplexer (ROADM) networks, as it enables high speed, large capacity, and flexible optical communications. A WSS is based on various type of technologies such as liquid crystals (LC) , liquid crystals on silicon (LCOS) , and micro electro mechanical systems (MEMS) [1, 2]. A gridless controllable WSS has already been realized using LCOS technology [3], but the optics and system of LCOS are relatively complex [4].

In this research, we focused on a MEMS based WSS. In the case of a MEMS mirror, we do not need to care about polarization diversity or a complex control system with heater equipment. To realize a gridless WSS using a MEMS mirror, we chose a piezoelectric actuator to drive the mirror. With a piezoelectric actuator, interference of the mirror driving force can be avoided between adjacent mirrors, which enables a high fill factor array. While an electrostatic-actuated WSS requires a more complicated structure to reduce driving interference [5], a piezo-actuator can be operated without crosstalk. Furthermore, a piezoelectric-actuated mirror is expected to tilt a larger angle than an electrostatic type at same voltage, which enables a large port-count WSS. But, it is not easy to control a piezoelectric actuator precisely because the piezoelectric film has thermal behavior, hysteresis and creep characteristics. To resolve these technical issues, we developed a newly designed micro mirror array combining a piezoelectric actuator with capacitive feedback control, and its basic performance is evaluated.

DESIGN AND PROCESS

Design Concept

Figure 1 shows the schematic concept of the mirror array. The mirror array is composed of lead zirconate titanate (PZT) unimorph actuators, mirrors, and capacitive detectors of the mirror angle. The PZT actuator is suitable for obtaining large vertical displacement and a high mirror tilt angle under low driving voltage. The mirror angle can be controlled by using a capacitive detector that is incorporated into the mirror. The capacitive detector is shaped like a comb, and the changes in capacitance are fed back to control the actuator driving voltage.

The overlapping area between the movable comb and the stationary comb decreases as the mirror angle increases. The area between comb electrodes at the mirror angles in Fig. 2 (a, b, c) is expressed as

$$S = (L_2 - L_1)(T - \frac{L_1 \sin\theta}{2} - \frac{L_2 \sin\theta}{2}) \qquad (1)$$

where θ, T, L_1, and L_2 are mirror tilt angle, thickness of the comb, X-coordinate of the bottom of overlapped area, and X-coordinate of the top of overlapped area respectively. And, the overlapping area between comb electrodes in Fig. 2 (c, d, e) is expressed by Equation (2).

$$S = \frac{1}{2}(\frac{T}{\sin\theta} - L_1)(T - L_1 \sin\theta) \qquad (2)$$

The total capacitance of the mirror angle detector is obtained by calculating the sum of the capacitance of each pair of electrodes. And, the differential signal between the angle detector and the reference capacitance is output to the feedback control system.

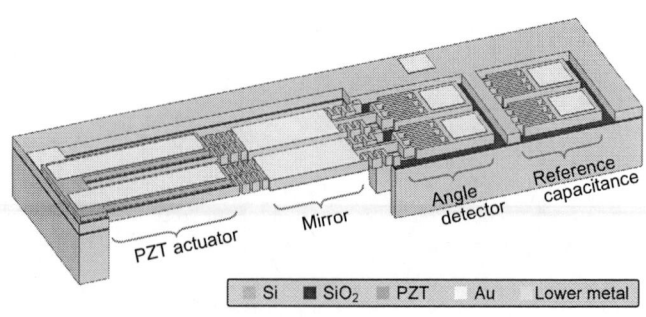

Figure 1: Schematic concept of mirror array composed of PZT unimorph actuators, mirrors, and capacitive detector of the mirror tilt angle.

978-1-4799-3510-9/14 $31.00 © 2014 IEEE

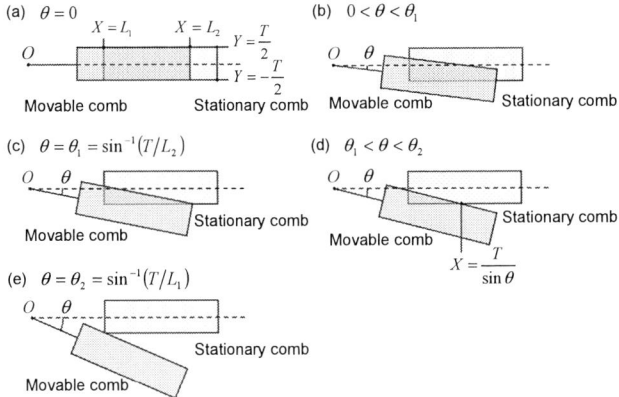

Figure 2: Change in overlapping area between comb electrodes with increase in mirror angle.

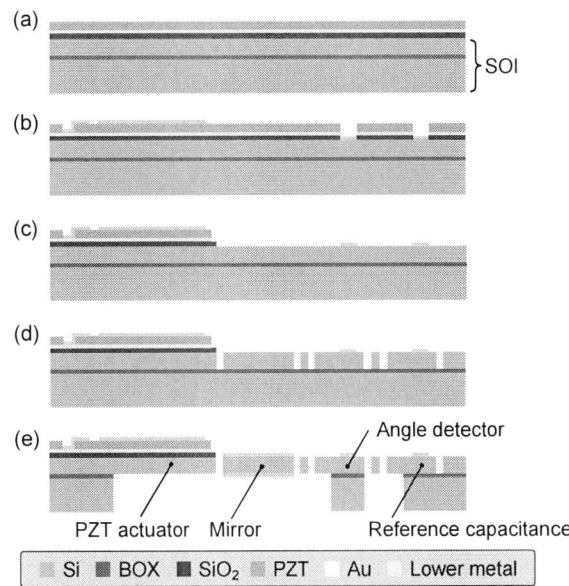

Figure 3: Fabrication process flow of mirror array. (a)PZT thin film is deposited on SOI wafer. (b)Contact electrodes are patterned. (c)PZT thin film is patterned. (d)Mirrors and comb-shaped electrodes are fabricated. (e)PZT actuators are fabricated and reflective films are deposited on both sides of the mirror.

Fabrication Process

Figure 3 explains the fabrication process of the mirror array. A PZT thin film is deposited on a silicon on insulator (SOI) wafer, and PZT unimorph actuators and capacitive detectors are fabricated by ion beam etching (IBE) and deep reactive ion etching (DRIE) process. In step (a), the PZT thin film is sputtered on the device layer of an SOI wafer. The thickness of the device layer is 10 μm. And, the thickness of the PZT film is about 3 μm. In step (b), the contact metal electrodes of the actuator and the comb-shaped detector are patterned. In step (c), the PZT film and SiO$_2$ layer on the actuator are shaped. In step (d), patterns of the actuators, mirrors, and comb electrodes are fabricated by DRIE process. In step (e), the handle layer of an SOI wafer is etched, and the actuators, the mirrors, the comb electrodes are completed. And, reflective films such as Au are deposited on both sides of the mirrors.

As is clear from this process flow, the mirror array switch composed of PZT actuators, mirrors, and capacitive detectors can be fabricated from a single SOI wafer and a PZT thin film. We only used simple wafer process, and any integration process after etching is not needed. It was seen that this process was advantageous in terms of production cost.

A SEM picture of the mirror array and capacitive detectors is shown in Fig.4. The detector has comb-shaped Si electrodes, and capacitance between the comb electrodes changes in accordance with the mirror angle. Movable combs are incorporated into the mirror, and the stationary combs are anchored to the buried oxide (BOX) layer. Typically, the length of a single comb is 50 μm, and the width of the comb is 3 μm. The gap between the two combs is 2 μm.

The array we fabricated includes 50 mirrors. The length of a single mirror is over 1000 μm, and the width of the mirror is about 150 μm. The length of the PZT actuator is designed to be 3000 μm from the aspect of the natural frequency of mirrors. And, the gap between the mirrors is under 2 μm, which means a high fill factor of over 98% is achieved.

Figure 4: Comb-shaped capacitive detectors at mirror tilt angle.

RESULTS

Open-loop Control

After fabricating the mirror array, we evaluated basic mirror performance in open loop control. The mirror angle is measured by using a laser light source and position sensitive detector (PSD). The mirror tilt mechanical angle with the drive voltage under open loop control is shown in Fig.5. The mirror tilt angle reaches almost 5° at under 10 V. And, the polarization inversion was not observed in the voltage range from 5 to -10 V. The maximum mirror angle hysteresis is 0.4°.

Then, we measured the natural frequency of the mirror in atmospheric environment by using a frequency response

978-1-4799-3510-9/14 $31.00 © 2014 IEEE

analyzer (FRA). The natural frequency is about 1055 Hz and increases slightly as mirror angle increases (Fig. 6). The plausible reason for this is that the hinge that is connected to the mirror and the actuator is pulled when the PZT actuator moves a lot. This increases the tensional force and subsequently the frequency of the mirror.

As shown in Fig. 6, the quality factor of the mirror is about 50. The increase in quality factor as the mirror angle increases is attributed to the air damping effect. The overlapping area between the mirrors and the comb electrodes decreases as the mirror tilts more. The gap between the mirrors and combs is 2 μm, so the viscosity resistance of the air cannot be ignored.

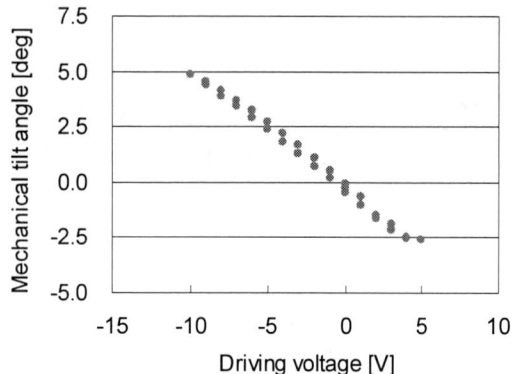

Figure 5: Mirror driving property under open loop control.

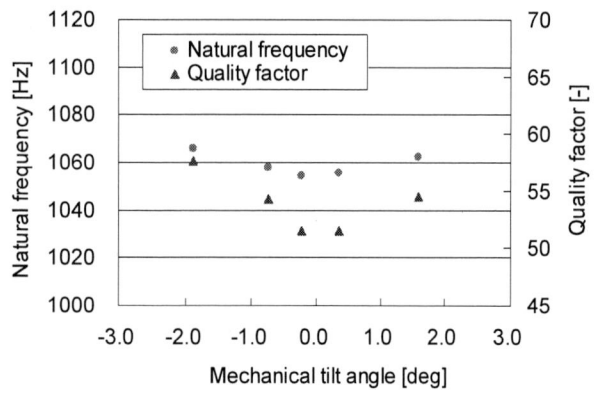

Figure 6: Natural frequency of mirror and quality factor with mechanical tilt angle.

Closed-loop Control

Highly accurate open-loop control of a PZT actuator is difficult because of hysteresis, creep, and thermal expansion of the piezoelectric material. In order to solve these problems, we adopted capacitive feedback control. A prototype for evaluation is composed of MEMS, LSI, and circuit board (Fig. 7). We used ready-made LSI for an accelerometer, and prepared the circuit board for closed-loop control. Capacitive angle detectors of three contiguous mirrors are connected to the 3ch CV converter, and the mirror angle is controlled individually.

The three contiguous mirrors 1, 2, and 3 showed similar control properties (Fig.8). The angular difference of the three mirrors is 0.05° in the range from 3 V to 8 V of control voltage. At both ends of the control range under 3 V and over 8 V, the amount of capacitance change is not large enough for high resolution control, and the angular difference of the three mirrors got larger than 0.05°.

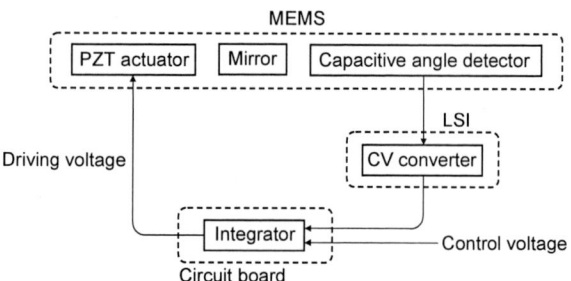

Figure 7: Control composition of prototype for closed-loop evaluation.

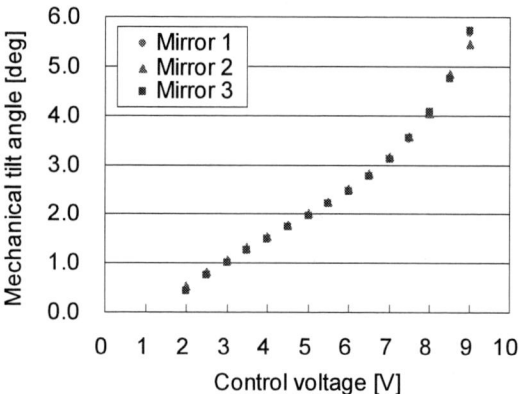

Figure 8: Driving property of three contiguous mirrors under closed loop control using capacitive angle detector.

Mirror Driving Interaction

In order to check driving interaction in closed loop control, we conducted an experiment in which several mirrors were driven at the same time. The mirror 2, which was located in the center of the three adjacent mirrors, was controlled to a constant angle. Then, the mirror 1 and 3, which were located on opposite sides of the mirror 2 were tilted simultaneously and periodically.

The angle stability of the mirror 2 is shown in Fig. 9. While CV output voltages from the mirror 1 and 3 change simultaneously, the output voltage from the mirror 2 seems to keep a constant angle. The change of the actual mirror angle is estimated under 0.01° from PSD output. This result indicates that a mirror driving interference is very small, and the mirror can be controlled with high resolution.

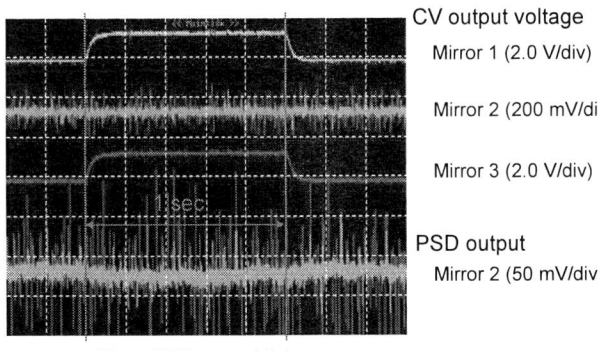

CV output voltage
Mirror 1 (2.0 V/div)

Mirror 2 (200 mV/div)

Mirror 3 (2.0 V/div)

PSD output
Mirror 2 (50 mV/div)

Time (200 msec/div)

Figure 9: Mirror driving interference between three contiguous mirrors.

CONCLUTIONS

This research newly developed a single-axis mechanical micro mirror array used for gridless wavelength selective switch (WSS). The advantages of this device are gridless control and low voltage operation. We realized those advantages by implementing both piezoelectric actuators and capacitive detectors for detecting the mirror angle in a mirror array chip. Consequently, we succeeded in fabricating a mirror array of a high fill factor of over 98%. Our device shows a 5° mechanical mirror angle at 10 V under open loop control. And, a control resolution of under 0.01° is achieved under closed loop control using a capacitive feedback method. We also confirmed that the mirror tilts independently without affecting the adjacent mirrors.

REFERENCES

[1] D. M. Marom, D. T. Neilson, D. S. Greywall, C.-S. Pai, N. R. Basavanhally, V. A. Aksyuk, D. O. López, F. Pardo, M. E. Simon, Y. Low, P. Kolodner, C. A. Bolle, "Wavelength-Selective 1 x K Switches Using Free-Space Optics and MEMS Micromirrors: Theory, Design, and Implementation", *J. Light. Tech.,* vol. 23, Issue 4, pp. 1620-1630, 2005.

[2] S. Maruyama, K. Takahashi, H. Fujita, H. Toshiyoshi, "A MEMS Digital Mirror Array Integrated with High-Voltage Level-Shifter", *Digest Tech. Papers Transducers 2009 Conference,* pp. 2314-2317, Denver, June 21-25, 2009.

[3] S. Frisken, G. Baxter, D. Abakoumov, H. Zhou, I. Clarke, S. Poole, "Flexible and Grid-less Wavelength Selective Switch using LCOS Technology", *Proc. OFC/NFOEC 2011,* pp. 1-3, Los Angeles, March 6-10, 2011.

[4] P. Wall, P. Colbourne, C. Reimer, S. McLaughlin, "WSS Switching Engine Technologies", *Proc. OFC/NFOEC 2008,* pp. 1-5, San Diego, February 24-28, 2008.

[5] M. Usui, S. Uchiyama, E. Hashimoto, K. Hadama, Y. Ishii, N. Matsuura, T. Sakata, N. Shimoyama, Y. Sato, H. Ishii, T. Matsuura, F. Shimokawa, Y. Uenishi, "Electrically separated two-axis MEMS mirror array module for wavelength selective switches", *Proc. OMEMS 2009,* pp. 158-159, Clearwater, August 17-20, 2009.

CONTACT

*R. Uchino, tel: +81(6) 6489-5917; uchino-r@spp.co.jp

CARBON SP^2-SP^3 TECHNOLOGY: GRAPHENE-ON-DIAMOND THIN FILM UV DETECTOR

Kaiyuan Yao[1], Chen Yang[1], Xining Zang,[1] Fei Feng[1,2] and Liwei Lin[1]

[1]Berkeley Sensor and Actuator Center, University of California, Berkeley, USA
[2]Shanghai Institute of Microsystem and Information Technology, Chinese Academy of Sciences, China

ABSTRACT

This work presents a graphene-diamond-metal thin film system as ultraviolet (UV) light sensor on a flexible substrate. New scientific and engineering breakthroughs are: (1) first experimental investigation of electrical and optical properties of carbon-based sp2-sp3 (graphene-diamond) junction; (2) a *peel-and-stick* fabrication process to make flexible diamond thin films from CVD-grown micro crystalline diamond (MCD) wafers; and (3) a sandwich-like vertical sensor structure with graphene as a transparent top electrode, and metal (Ti-Au) as an ohmic bottom contact electrode. As such, the proposed detector/architecture can open up a new class of scheme to build diamond-based, attachable, portable and wearable optoelectronic systems.

INTRODUCTION

Diamond for Ultraviolet Photo Detection

Various wide-bandgap materials have been proposed and prototyped for ultra-violet (UV) photon detection, including diamond [1-3], ZnO [4], SiC [5], GaN [6], AlN [7], *etc*. Among them, diamond is an ideal candidate owing to its exceptional optical, electrical and thermal properties. First of all, its large bandgap (5.5eV) provides wide transparency window from UV to mid-infrared. Most reported diamond photodetectors have a cut-off wavelength at around 260nm. Since solar radiation on earth is longer than 280nm, diamond photodetectors can be completely solar-blind as well as making most use of the solar-free spectrum (<280nm range). The wavelength discrimination from UV to solar spectrum enables the so called Solar-Blind UV (SBUV) technology. SBUV has wide applications in both industry and military, such as corona/UV detection on electrical powerlines, manufacturing facilities, environmental and biological monitoring, muzzle flash, missile plume, and security surveillance. Furthermore, diamond is known to have high thermal stability, radiation hardness and chemical resistance which help the resilience and stability of diamond-based devices without good cooling and protection systems usually required for silicon-based detectors.

Devices based on single crystalline diamond require high cost for mass production such that systems based on various chemical-vapor-deposition (CVD) techniques have been developed to grow micro-crystalline (MCD) and nano-crystalline diamond (NCD) at the wafer scale with relatively low cost [8]. Although MCD and NCD have lower electrical, optical, mechanical and chemical qualities than their single-crystalline counterparts, photodetector devices based on MCD and NCD have been successfully demonstrated with good performances [3].

Graphene Transparent Electrode and Heterojunction

In previous works [1-3], diamond UV detectors are built with an interdigitated metal electrode structure which blocks ~50% of incident UV light. Planar electrodes also cause non-uniform electric field distribution in diamond, which reduce the effectiveness in collecting the photo-generated electron-hole pairs in the diamond films. It has been shown that graphene is 98% transparent in UV range and functions well as a transparent electrode [9]. Therefore, it is expected that the performance of a diamond photodetector could be improved by replacing the lateral, interdigitated metal electrodes with a vertical, sandwich electrode structure: uniformly-covered graphene top electrode and ohmic metal bottom electrode.

A Schottky-like heterojunction is established at the interface between graphene and a semiconductor, such as the graphene-silicon system [10, 11]. The Schottky barrier height depends on the work function of graphene, semiconductor electron affinity energy, and other possible interface states. When the Schottky-junction is reversely biased, the depletion width in the semiconductor will expand with a built-in electric field. Consequently, photo-carriers generated due to band-to-band excitation within the depletion region can be separated and swept to opposite electrodes as the drift current. Minority photo-carriers generated outside the depletion region will be driven by the carrier concentration gradient and diffused into the depletion region as the diffusion current. Furthermore, certain amount of carriers from graphene may jump across the Schottky barrier and become the internal photoemission current. The various contributions of these three components add up to become the overall photocurrent inside the heterojunction.

CONCEPT

Figure 1 shows the schematic structure of the sp2-sp3, graphene-diamond UV detector. The active layer is a 2μm-thick MCD thin film, which is sandwiched by a single-layer graphene as the top electrode to allow maximum light penetration and a bottom metal ohmic contact (Ti/Au 150nm/200nm) is grounded. The single-layer graphene has 98% transmission in UV range, as well as good conductivity (~10^2-10^3 ohm/sq.) as an electrode. The bias voltage is applied through a contact pad (SiO$_2$/Ti/Au 50nm/5nm/50nm) deposited between the diamond and graphene.

Figure 2 illustrates the band diagram of the graphene-diamond heterojunction where E$_g$, E$_c$, E$_v$, qχ are the bandgap (~5.5eV), conduction band edge, valence band edge and electron affinity energy of MCD. Also, qΦ$_G$ represents

978-1-4799-3510-9/14 $31.00 © 2014 IEEE

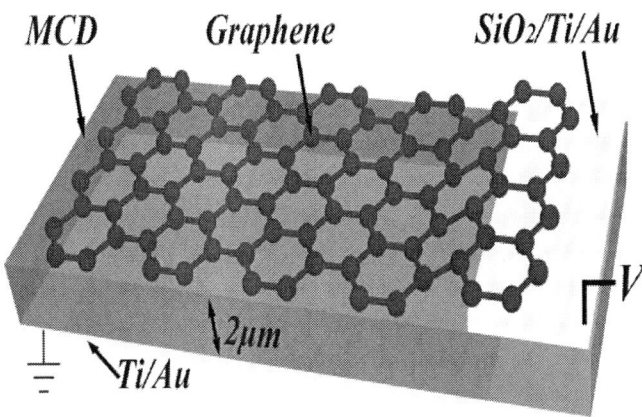

Figure 1: Schematic diagram of the graphene-diamond thin film UV detector. A 2 μm micro-polycrystalline diamond (MCD) is sandwiched by a top electrode (graphene) and bottom electrode (Ti/Au). A metal contact pad is deposited on top of the diamond film below the graphene electrode.

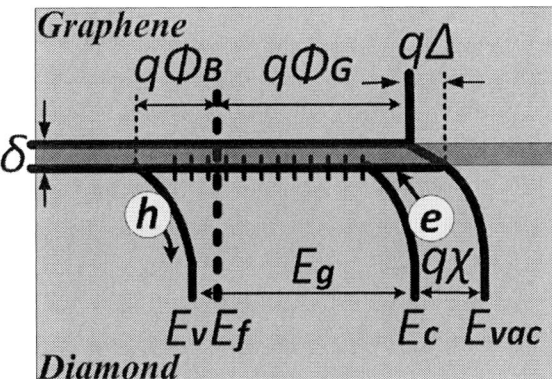

Figure 2: The band diagram of the graphene-diamond heterojunction.

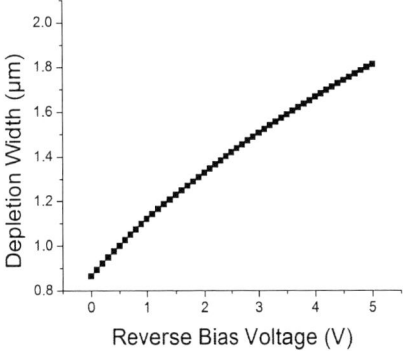

Figure 3: Estimated depletion width of graphene-diamond heterojunction under applied reverse bias from 0V-5V.

the work function of graphene (~4.5eV). Diamond's electron affinity energy depends strongly on post processing conditions, such as annealing, with a magnitude around 0.7eV-1.45eV for annealing temperature from 500°C to 1150°C [12]. Negative electron affinity can also be induced under some conditions. An interface layer of thickness δ and potential drop Δ could be introduced and assuming $q\chi \sim 0.7eV$, the final Schottky barrier height is therefore expressed as:

$$q\Phi_B = E_g - (q\Phi_G - q\chi) - q\Delta = 1.7eV - q\Delta \quad (1)$$

The MCD layer is doped with boron with $N_A \sim 10^{15}$ cm^{-3}. The depletion width (W_D) with an external bias voltage can be calculated by:

$$W_D = \sqrt{\frac{2\varepsilon}{qN_A}\left(V_{bi} - V - \frac{kT}{q}\right)} \quad (2)$$

where ε is permittivity of MCD (~5.5 ε_o), V_{bi} is the built-in potential, V is applied bias voltage. As shown in Fig. 3, the depletion width is estimated to be 0.85μm at equilibrium condition and is expanded to 1.8 μm under a 5V reverse bias. This analysis is the foundation for the 2 μm-thick MCD layer, which is roughly the sum of depletion and diffusion length (~0.5μm in [13]) under a typical magnitude of reverse bias. As such, photo-carriers generated both inside the depletion region and within the distance of the diffusion length outside the depletion region can be effectively collected.

FABRICATION

Device fabrication process is shown in Fig. 4. The MCD is grown by the Hot Filament Chemical Vapor Deposition (HFCVD). The substrate is a silicon wafer with thermally grown oxide. It is heated to 720 °C by an array of tungsten wires when a gas mixture of CH_4, H_2 and TMB ($B(CH_3)_3$) is introduced into the CVD chamber. After the HFCVD process, the MCD (together with the substrate) is annealed at 500°C for 45mins to enhance the crystalline structure and optical property as described in [3].

Ti/Au (150nm/200nm) is deposited by electron beam evaporation and annealed at 450°C for 45mins to form ohmic contact as the bottom electrode.

In Fig. 4(2), a double-sided tape is first is mechanically punched to have small opening holes and adhered to the diamond film. The bonding strength between the MCD and silicon substrate has been weakened due to the large thermal mismatch between the two materials. After several trial and error processes, Nitto Denko 3195 VS is chosen as the best tape as its mechanical properties (stiffness, adhesion force … etc.) match well with MCD to minimize possible damages on the diamond thin film. Experimentally, diamond films from a 4-inch wafer have been successfully peeled off without observable damages.

In Figures 4(3) and (4), the CVD-grown graphene (single layer graphene from Graphene Supermarket) is transferred onto the MCD diamond film. The assembly is then transferred and attached to a flexible polymer substrate and silver paste is used to construct conductive interconnects between the polymer substrate and the tape as illustrated. The electrical contact for the top electrode is established via a pre-patterned metal pad made of SiO2/Ti/Au with the thickness of 50nm/5nm/50nmm respectively. The graphene is originally grown a copper foil and the copper is etched away in saturated $FeCl_3$ solution and rinsed in deionized water, resulting in a single-layer graphene floating on top of the water surface. After dipping the device chip into the deionized water and picking up the graphene sheet on top, the transfer process is accomplished. Drying and annealing at 120°C is conducted next to minimize possible defects at the graphene-diamond interface.

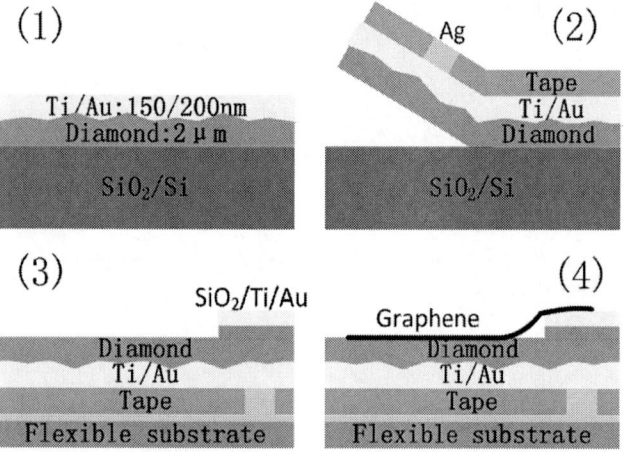

Figure 4: Fabrication process: (1) Hot Filament Chemical Vapor Deposition (HFCVD) of MCD, evaporation and annealing of bottom ohmic contact. (2) The peeling process from the silicon wafer. (3) Sticking to an arbitrary, flexible substrate. (4) Graphene transfer.

RESULTS AND DISCUSSION

The transferred graphene is visible under Scanning Electron Microscopy (SEM). The boundary of the graphene sheet can be identified in Fig. 5 (a). The left part with brighter color is the bare MCD and the right part with a bit dark color is MCD with graphene on top. As expected, bare MCD shows higher brightness under SEM due to stronger electron charging effect, while graphene has lower brightness since electrons can quickly move away. In Fig. 5 (b) we notice that there are graphene micro-ripples formed in a relatively random fashion. This is expected for graphene to release film stress during the post-transfer annealing process, similar to the buckling effect in MEMS micro-bridges and cantilevers. It should be noticed that these ripples could induce interface defects and trapping of incident photons and recombination of photo-generated carriers. Further investigations are needed to improve interface quality for better efficiency.

Raman spectroscopy shown in Fig. 6 validates the successful transfer of graphene on diamond. The sample shows a strong sp^3 peak at ~1332cm^{-1} from the 2μm diamond film, as well as G peak (1582 cm^{-1}) and 2D band (~2700 cm^{-1}), which implies the existence of the graphene. Optoelectronics performance is characterized by measuring current-voltage (I-V) curves (Fig. 7) in dark environment and under illumination. A 300W Xenon lamp is used as the light source. The broadband radiation from Xenon lamp covers the wide spectrum range from UVC to NIR. Single-wavelength light is obtained by filtering the Xenon lamp's broadband radiation using a monochromator installed with a holographic grating. Optical power density is calibrated by a commercial silicon photodetector. The prototype device has an active area of ~0.87cm^2.

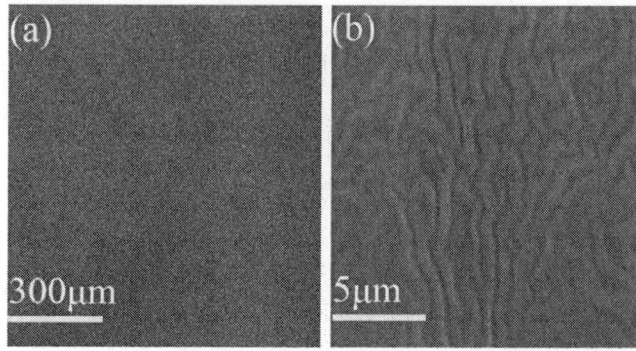

Figure 5: SEM pictures of graphene transferred onto MCD: (a) boundary of the transferred graphene on the right side, and (b) graphene ripples observed at high magnification.

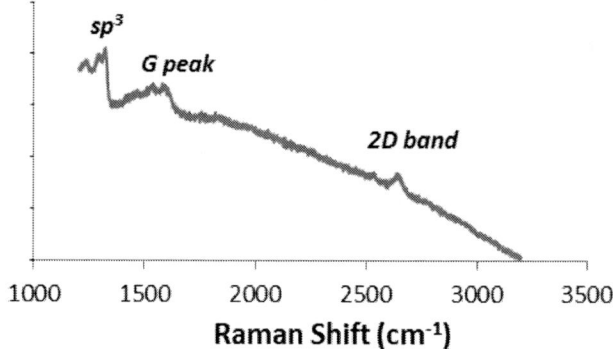

Figure 6: Raman spectroscopy of graphene-on-diamond material system.

Under the dark environment, the photodetector shows a typical rectifying I-V curve (black curve in Fig. 7). When the junction is reversely biased, higher potential is applied at the top graphene electrode. Experimentally, it is observed that under forward bias at V=-5V, the output current is 6μA while under reverse bias at V=5V, the output current is 140nA. The ideality factor and barrier height are two critical parameters in characterizing the Schottky junction and can be extracted from the forward bias I-V data as shown in Fig. 8 by the models proposed in [14]. It is found that the ideality factor remains around unity when the forward bias voltage is small, indicating an ideal thermionic emission and it increases significantly (almost linearly) with respect to the bias voltage, probably due to the increasing quantum tunneling effect through the Schottky barrier. On the other hand, the zero-bias barrier height is 1.57eV which is reasonably close to the estimation (Eq. (1)) based on the band diagram. The barrier height gradually drops under the forward bias due to the image-force lowering effect, as commonly observed in other Schottky diodes.

Under the UV illumination at 220nm (the blue curve in Fig. 7), significant increase in photocurrent can be observed. As expected, photocurrent under the reverse bias is much larger than that under the forward bias because the depletion region where most photo-carriers can be effectively collected is enlarged under the reverse bias.

978-1-4799-3510-9/14 $31.00 © 2014 IEEE   1161

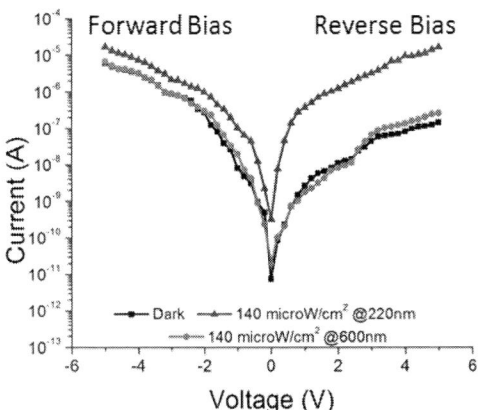

Figure 7: I-V characteristics of the graphene-diamond heterojunction photodetector in dark (black) and under 220nm (blue) and 600nm (red) light illumination.

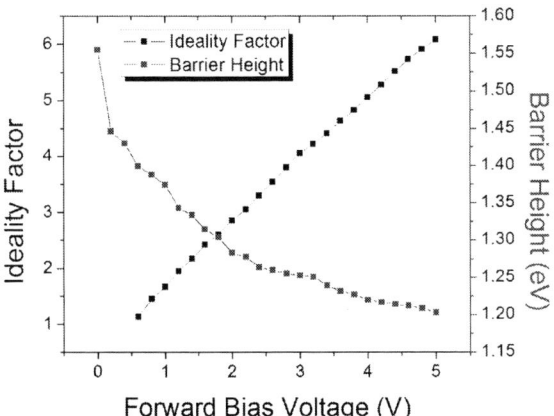

Figure 8: The ideality factor and barrier height calculated from the dark I-V plot of graphene-diamond heterojunction.

The optical responsivity at 5V under the reverse bias and 220nm light illumination is calculated at 0.13A/W, which corresponds to a quantum efficiency of 72%. On the other hand, under visible light illuminations (600nm in Fig. 7), the photodetector only produces negligible responses since the photon energy is smaller than the bandgap of diamond and not high enough to excite the photocurrent.

CONCLUSION

Thin film photodetector based on the graphene-on-diamond (carbon sp^2-sp^3) heterojunction is fabricated and tested for application in solar-blind UV detectors. A novel "peel-and-stick" method is demonstrated in order to make ultrathin microcrystalline diamond films and graphene is transferred and attached to the back and smooth side of the diamond films. I-V curves of the graphene-diamond photodetectors have been characterized under various environments, including dark, UVC illumination and visible light, respectively (Fig. 7).

The demonstrated devices could find promising application in highly-sensitive solar-blind UV detectors. Further improvement on the interface between graphene and diamond could lead to even higher optical responsivity.

ACKNOWLEDGEMENTS

The authors greatly appreciate help from Miss Sarah Brittman and Miss Fan Cui. The device is fabricated in the UC Berkeley Marvel Nanofabrication Laboratory. The project is supported in part by Siemens Inc.

REFERENCES

[1] Meiyong Liao, *et al.* "Thermally Stable Visible-Blind Diamond Photodiode Using Tungsten Carbide Schottky Contact," *Applied Physics Letters*, vol. 87, 022105, 2005

[2] Meiyong Liao, *et al.* "Single Schottky-Barrier Photodiode With Interdigitated-Finger Geometry: Application To Diamond," *Applied Physics Letters*, vol. 90, 123507, 2007

[3] Robert D. McKeag, *et al.* "Diamond UV Photodetectors: Sensitivity And Speed For Visible Blind Applications," *Diamond and Related Materials*, vol. 7, pp 513-518, 1998

[4] Lei Luo, *et al.* "Fabrication and Characterization of ZnO Nanowires Based UV Photodiodes," *Sensors and Actuators - A: Physical*, Vol. 127, pp. 201-206, 2006

[5] Yan, Feng, *et al.* "4H-SiC UV Photo Detectors With Large Area And Very High Specific Detectivity," *Quantum Electronics, IEEE Journal of* 40.9, 1315-1320, 2004

[6] Li, Xiang-yang, *et al.* "GaN Based Ultraviolet Detectors And Its Recent Development," *Infrared and Laser Engineering* 35.3, 276, 2006

[7] Tsai, Dung-Sheng, *et al.* "Solar-Blind Photodetectors for Harsh Electronics," *Scientific Reports,* 4, 2013.

[8] Sussmann, R. S., *et al.* "Properties Of Bulk Polycrystalline CVD Diamond," *Diamond and Related Materials* 3.4: pp. 303-312, 1994

[9] Weber, Constans M., *et al.* "Graphene-Based Optically Transparent Electrodes for Spectroelectrochemistry in the UV–Vis Region." *Small* 6.2 pp.184-189, 2010

[10] Heejun Yang, *et al.* "Graphene Barristor, a Triode Device with a Gate-Controlled Schottky Barrier," *Science,* Vol. 336, pp.1140-1143, 2012

[11] Chun-Chung Chen, *et al.* "Graphene Silicon Schottky Diodes," *Nano Letters,* Vol. 11, pp. 1863–1867, 2012

[12] P. K. Baumann, *et al.* "Electron Affinity And Schottky Barrier Height Of Metal–Diamond (100), „(111), and„(110)…Interfaces," *J. Appl. Phys.,* Vol. 83, No. 4, 1998

[13] T. Malinauskas, *et al.* "Optical Evaluation Of Carrier Lifetime And Diffusion Length In Synthetic Diamonds," *Diamond & Related Materials,* Vol. 17, 1212–1215, 2008

[14] V. Mikhelashvili, *et al.* "Extraction Of Schottky Diode Parameters With A Bias Dependent Barrier Height," *Solid-State Electronics,* Vol. 45 pp. 143-148, 2001

CONTACT

*Kaiyuan Yao: kyyao@berkeley.edu

CLOSE-PACKED LIQUID-FILLED TUNABLE MICROLENS ARRAY

Yoshinobu Iimura[1], Hiroaki Onoe[1, 2] and Shoji Takeuchi[1, 2]
[1]Institute of Industrial Science, The University of Tokyo, JAPAN
[2]ERATO Takeuchi Biohybrid Innovation Project, Tokyo, JAPAN

ABSTRACT

This paper describes close-packed liquid-filled tunable microlens arrays for optical devices such as integral imaging systems. These lenses are simply composed of poly(dimethylsiloxane) (PDMS) microchannels. Applied pressure deforms the top membrane of microchannels to form convex lenses. These lenses have three advantages: (i) Uniform deformation by pressure-driven actuation, (ii) Adjustable optical characteristics without patterned electrode, (iii) High-density integration of tunable microlenses. We fabricated three types of lenses: "Honeycomb", "Ladybird" and "Spiderweb" as close-packed structure and showed that the "Spiderweb" type packing is the most suitable for close packing. We believe that these lenses can be applied to optical applications such as 3D displays based on integral photography.

INTRODUCTION

Tunable lens array is used for many optical devices, for example, optical interconnections, high-density data storage and integral imaging systems for three dimensional displays [1]. Especially in integral imaging systems, integration degree of lenses is very important because it is directly linked to the resolution of displays. To realize highly integrated tunable lens arrays, some researches employed lenses based on liquid crystals (LC) [2-4]. However, the LC lenses suffer from aberrations due to non-uniformities in the electric field and require electrodes, which cause optical distortion, in the lens part. To solve these problems, a promising approach is liquid-filled tunable lenses [5-9]. However, it is difficult to increase integration degree of liquid-filled lenses because hard pillars are needed around lenses. In this study, we propose a design methodology of liquid-filled lenses which have soft pillars for high-density integration. Soft pillars enlarge lens areas by deflecting with membranes on chambers depending on the design of the lenses and pillars. We designed three types of lens arrays. One of them, we call "Honeycomb", is based on honeycomb structure which is closest packing structure using regular hexagon. The others, we call "Ladybird" and "Spiderweb", are based on hexagonal close-packed structure which is closest packing structure for circles. We evaluated which design is suitable for closer packing for the liquid-filled tunable microlens array.

DESIGN AND FABRICATION

Design

Figure 2 (a-c) shows the design of three types of lenses: (a) Honeycomb lens, (b) Ladybird lens and (c) Spiderweb lens. The design of Honeycomb lens is an array of regular

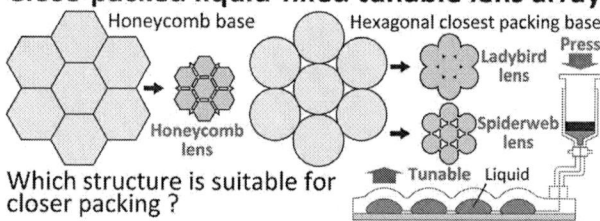

Figure 1: Concept of our work. We propose three types of close-packed liquid-filled tunable lens arrays with different lens geometries. Honeycomb structure is one of the closest packing of hexagonal structures. Hexagonal close-packed structure is the closest packing structure for circles. We evaluated their effective lens area, optical lens effects and durability by comparing these three types of lens arrays.

hexagons, and there are rectangular pillars between hexagons. In Honeycomb lens, the distance between the parallel sides of a hexagon is 433 μm and the length and width of pillars are 200 μm and 100 μm, respectively. The design of Ladybird lens is an array of connected circles. In Ladybird lens, the diameter of chamber is 600 μm. The design of Spiderweb lens has connecting paths between circles. In Spiderweb lens, the diameter of microchannels in the lens part is 500 μm and the length and width of connecting paths are 100 μm and 100 μm, respectively. From these designs, Ladybird lens should have an advantage in effective lens area, and Spiderweb lens should have an advantage in durability because of its large adhesive surface. We fabricated these three types of lenses and then evaluated shapes of deformed membrane of these lenses.

Fabrication

A fabrication process of a liquid-filled tunable microlens array is shown in Figure 2 (d). After constructing SU-8 mold (SU-8 Series; MicroChem Corp., MA), we spin-coated poly(dimethylsiloxane) (PDMS, Sylgard 184 Silicone Elastomer, Dow Corning) on the mold at 500 rpm for 10 sec and 1200 rpm for 30 sec to make replicas. The replica was peeled-off and put onto a Polyethylene terephthalate (PET) film. The replica and a glass substrate (Matsunami glass Co., Ld., Japan) were treated with O_2 plasma. Subsequently, the replica was bonded to the glass substrate, and connected with inlets for tubing. We used syringes (2.5 ml, TERUMO JAPAN) and Teflon tubes to induce solution into microchannels. Figure 2 (e) and (f) show the device dimension, and macroscopic picture of fabricated device,

978-1-4799-3510-9/14 $31.00 © 2014 IEEE   1163

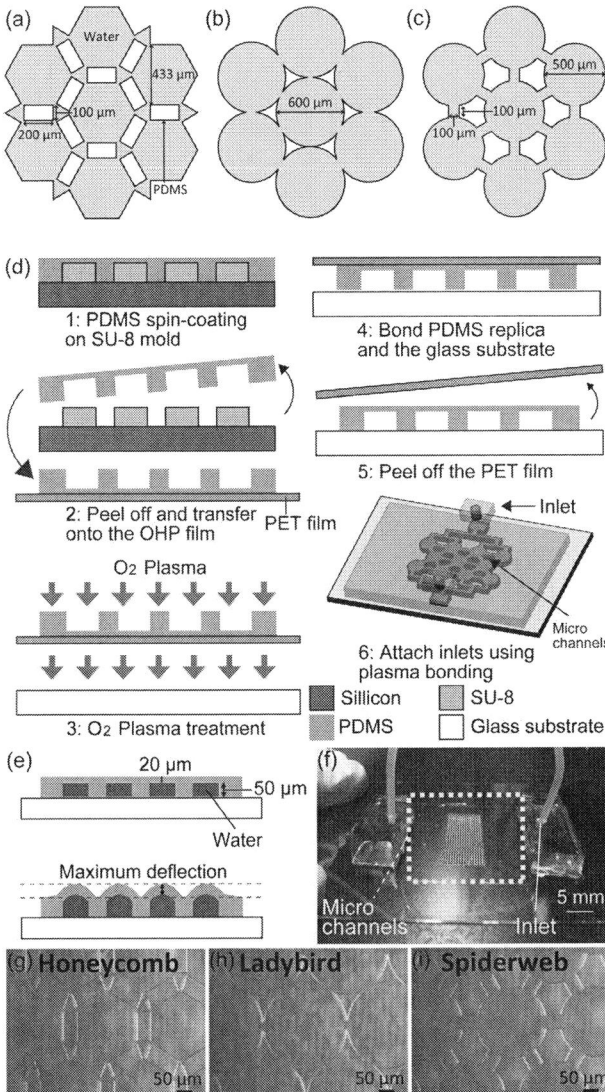

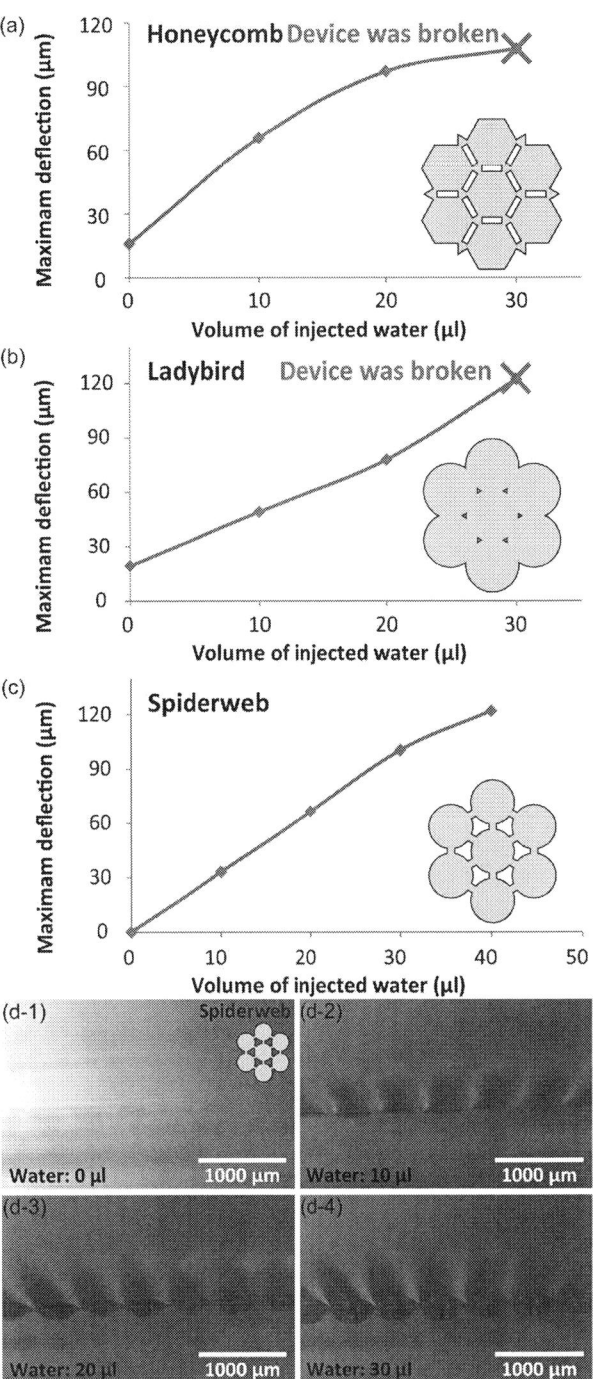

Figure2: (a-c) Designs of lenses, (a) Honeycomb lens, (b) Ladybird lens and (c) Spiderweb lens, respectively. (d) Fabrication process of the liquid-filled tunable lens array. (e) The structure and deflection of the lens array. (f) Macroscopic picture of fabricated devices (Spiderweb lens). Water was injected from inlets using syringe pumps to deform the top membrane of microchannels. (g-i) Microscopic pictures of fabricated devices, (g) Honeycomb lens, (h) Ladybird lens and (i) Spiderweb lens, respectively.

respectively. In all lenses, the thickness of the PDMS membrane was about 70 μm and the height of microchannels is about 50 μm. Therefore, the thickness of the membrane over microchannels is about 20 μm. We defined maximum deflection as shown in Figure 2 (e). Figure 2 (g-i) show microscopic pictures of fabricated devices, (g) Honeycomb lens, (h) Ladybird lens and (i) Spiderweb lens, respectively.

Figure 3: (a-c) Relations between maximum deflection and volume of injected water when the design is (a) Honeycomb lens, (b) Ladybird lens and (c) Spiderweb lens, respectively. When designs are "Honeycomb lens" and "Ladybird lens", lenses were broken. (d) Deformed shapes of Spiderweb lenses.

RESULTS
Measurement of Membrane Deformation

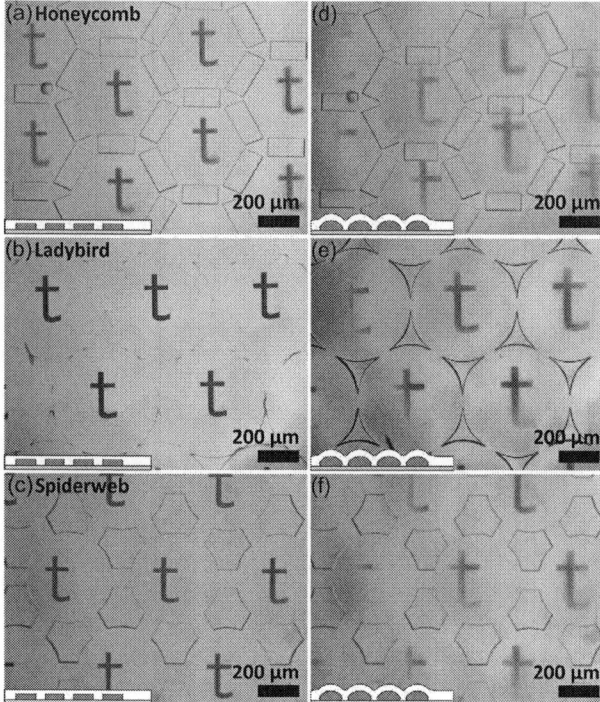

Figure 4: (a-c) Character "t" through lenses when the lens surfaces were flat, (a) Honeycomb lens, (b) Ladybird lens and (c) Spiderweb lens, respectively. (d-f) Character "t" magnified by deformed lenses, (d) Honeycomb lens, (e) Ladybird lens and (f) Spiderweb lens, respectively. These images are composite of lenses and characters "t".

We measured relations between the maximum deflection of the membrane and volumes of injected water to confirm which lens is deformed most. We measured the deflection of the membrane under the microscopic with multi angle lenses (VHX-D 500, Keyence, Co., Ld., Japan). Figure 3 (a-c) shows the relations between the maximum deflection and volumes of injected water for (a) Honeycomb lens, (b) Ladybird lens and (c) Spiderweb lens. Figure 3 (d) shows the deformed membrane of the Spiderweb-type lenses taken from the view angle of 86 degree. In the Honeycomb lens, limit value of the maximum deflection was about 110 μm. In contrast, limit values of maximum deflections were about 120 μm for Ladybird lens and Spiderweb lens. In Honeycomb lens and Ladybird lens, devices were broken when the maximum deflection reached at about 110 μm and about 120 μm because of their small bonding areas. We conclude that the Spiderweb lens is deformed most from these results. Figure 4 (a-c) shows characters "t" through lenses when the lens surfaces were flat. Figure 4 (d-f) shows characters "t" magnified by lenses when the lens surfaces were deformed. These results indicate that these lenses acquire an optical characteristic that magnify images when membrane is deformed.

Measurement of lens profile

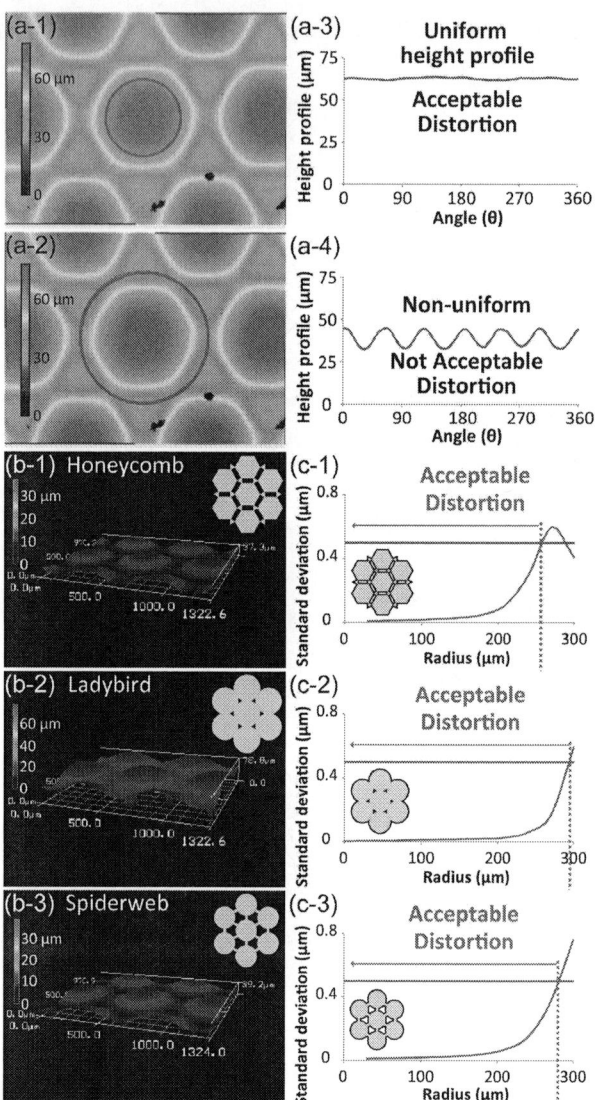

Figure 5: (a-1, a-2) Surface profiles of lenses. (a-3, a-4) Height profiles on red circles. (b) 3D images of lens profiles. (c) Relations between radius of red circles and standard deviations of height profiles.

We measured surface profiles of lenses using a laser interferometer (VK-X 200, Keyence, Co., Ld., Japan) to estimate areas that can be used as lenses. Figure 5 (a-1, a-2) shows surface profiles of lenses. Figure 5 (a-3, a-4) shows height profiles on red circles in Figure 5 (a-1, a-2). If height profiles are uniform to all radiuses, there is no distortion on the red circle. In contrast, if height profiles are non-uniform, there is a distortion on the red circle. Figure 5 (b) shows 3D images of lens profiles and Figure (c) shows relations between radius of red circle and standard deviations of height profiles. The standard deviation correlates to the distortion on a red circle. We defined a threshold value of standard deviation as 0.5 μm: the standard deviation less than 0.5 μm should be regarded as acceptable distortion. Therefore, areas

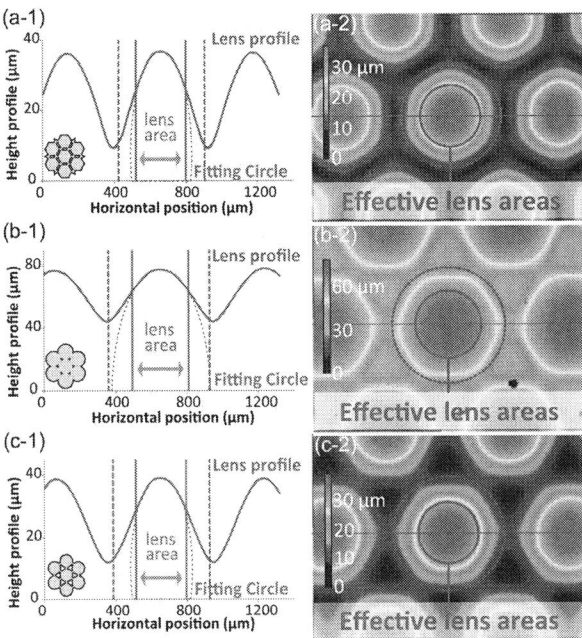

Figure 6: Relation between position and height profile of lenses and area that can be used as lenses. We calculated areas by fitting circle to height profiles of lenses.

Table 1: Relation between designs and areas that can be used as lenses. We decided that the Spiderweb lens array is best from all results.

Design	Diameter of lenses (μm)	Limit value of maximam deflection (μm)	Broken or not
Honeycomb	2.8 x 10²	107.68	Broken
Ladybird	3.1 x 10²	122.31	Broken
Spiderweb	2.8 x 10²	121.92	Not

that can be used as lenses were about 240 μm, 270 μm and 270 μm, when lens designs are Honeycomb lens, Ladybird lens and Spiderweb lens, respectively.

Moreover, we estimated effective lens areas by fitting circle shapes to lens profiles. Common convex lenses have perfect sphere shapes. Therefore, it is expected that areas fitted to circle shapes can be used as lenses. Figure 6 (a-c) shows the relations between horizontal position and height profiles of lenses, and the effective lens areas for (a) Honeycomb lens, (b) Ladybird lens and (c) Spiderweb lens. These results show that diameters of the effective lens areas are (i) 280 μm, (ii) 310 μm and (iii) 280 μm when designs of lenses are (i) Honeycomb lens, (ii) Ladybird lens and (iii) Spiderweb lens, respectively. Table 1. summarize the characteristics of the three types of liquid-filled tunable lens arrays. We concluded that the Spiderweb lens is suitable for closest packing because this lens has advantages in durability and the effective lens areas, compared with other two lenses.

CONCLUSION

We proposed three types of liquid-filled tunable lens arrays for closest packing. We measured profiles of deformed membrane and evaluated their effective lens areas and durability. We concluded that the Spiderweb-type lens is suitable for close-packing because of its durability and the effective lens areas. The deformed shapes around pillars were not ideal sphere shapes, but we believe that the effective lens area can be enlarged by using distortion correction with image processing. We believe that these lenses can be applied to high-resolution 3D displays that can adjust feelings of three-dimensional effects.

ACKNOWLEDGEMENTS

This work was supported by Takeuchi Biohybrid Innovation Project, ERATO, Japan Science and Technology Agency (JST).

REFERENCES

[1] J. Arai, H. Kawai and F. Okano, "Microlens arrays for integral imaging system", *Appl. Opt*, vol. 45, pp. 9066-9078, 2006.

[2] H. T. Dai, Y. J. Liu, X. W. Sun and D. Luo, "A negative-positive tunable liquid-crystal microlens array by printing", *Opt. Express*, vol. 17, pp. 4317-4323, 2009.

[3] Y. J. Liu, X. W. Sun and P. Shum, "Tunable fly's-eye lens made of patterned polymer-dispersed liquid crystal", *Opt. Express*, vol. 14, pp. 5634-5640, 2006.

[4] H. Ren, S. Xu and S.-T. Wu, "Polymer-stabilized liquid crystal microlens array with large dynamic range and fast response time", *Opt. Letters*, vol. 38, pp. 3144-3147, 2013.

[5] N.Chronis, G. L. Liu, K.-H. Jeong and L. P. Lee, "Tunable liquid-filled microlens array integrated with microfluidic network", *Opt. Express*, Vol. 11, pp. 2370-2378, 2003.

[6] J. Chen, W. Wang, J. Fang and K. Varahramyan, "Variable-focusing microlens with microfluidic chip", *J. Micromech. Microengr.*, Vol. 14, pp. 675-680, 2004.

[7] M. Agarwal, R. Gunasekaran, P. Coane and K. Varahramyan, "Polymer-based variable focal length microlens system", *J. Micromech. Microengr.*, Vol. 14, pp. 1665-1673, 2004.

[8] A. Werber and H. Zappe, "Tunable Pneumatic Microoptics", *J. Microelectromech. Syst.*, Vol. 17, pp. 1218-1227, 2008.

[9] A. Pouydebasque, C. Bridoux, F. Jacquet, S. Moreau, E. Sage, D. Saint-Patrice, C. Bouvier, C. Kopp, G. Marchand, S.Bolis, N. Sillon and E. Vigier-Blanc, "Varifocal liquid lenses with integrated actuator, high focusing power and low operating voltage fabricated on 200 mm wafers", *Sensor and Actuator A: Physical*, Vol. 172, pp. 280-286, 2011.

CONTACT

*Yoshinobu Iimura, tel: +81-3-5452-6650; yiimura@iis.u-tokyo.ac.jp

978-1-4799-3510-9/14 $31.00 © 2014 IEEE

COMPACT TUNABLE HYPERSPECTRAL IMAGING SYSTEM

Phuong-Ha Cu-Nguyen[1], Adrian Grewe[2], Csaba Endrödy[3], Stefan Sinzinger[2], Hans Zappe[1], and Andreas Seifert[1]

[1]Department of Microsystems Engineering – IMTEK, University of Freiburg, Germany
[2]Optical Engineering, Ilmenau University of Technology, Germany
[3]Micromechanical Systems, Ilmenau University of Technology, Germany

ABSTRACT

We present a compact hyperspectral imaging system consisting of a combined refractive-diffractive optical element for extended longitudinal chromatic aberration, which provides spectrally resolved information of an object. This fully integrated hybrid device features a diffractive optical element (DOE) acting as a positive lens, a tunable concave liquid-filled membrane lens, and integrated magnetic actuation for hydraulically tuning the focal length of the membrane lens. The lens system is demonstrated to generate a hyperspectral data cube in the visible wavelength range by imaging the light from a pinhole array in a confocal approach onto a CCD camera.

INTRODUCTION

Hyperspectral imaging is similar to spatially resolved spectroscopy of a reflecting object where a complete optical spectrum can be acquired at any position of an object within a short time. Due to its suitability and commercial availability, it has been implemented and established in very diverse applications as a detection method. The traditional approach to generate hyperspectral images is push broom scanning, commonly used in satellite cameras for earth observations from space. A push broom sensor employs a slit aperture scanning along the object. Each column of the image is laterally dispersed, using a prism, and imaged onto a detecting area, where the rows contain the spectral information [1].

An alternative approach is to spread and separate the spectral components of the light beam along the optical axis, for example by increasing the longitudinal chromatic aberration (LCA) of a lens system. By this way, a polychromatic or white light point source is spread into multiple points along the optical axis, and each point contains a spectral component with a certain band width of the light source. By using a pinhole in the image plane in front of the detector, only those wavelengths which focus directly onto the pinhole can pass the aperture and reach the detector area while other wavelengths are blocked. This approach, known as chromatic confocal microscopy, offers the maximum intensity for the wavelength to be detected, and has been introduced and implemented previously [2, 3, 4].

We present here a novel extension of chromatic confocal microscopy with an adaptive tunable optical system for hyperspectral image sensing. The main component is a compact hyperchromatic lens module which is realized by the combination of a DOE and a tunable liquid-filled membrane-lens. The lens system is equipped with a fully

integrated magnetic actuator and was fabricated by a low cost manufacturing method, yielding a flexible and fast operating system.

OPTICAL PRINCIPLE

The core of our presented system is a tunable hyperchromatic lens consisting of a DOE, acting as a positive lens, and a tunable refractive concave lens. This refractive-diffractive combination allows the lens system to have positive refractive power with maximized LCA. As a result, the light originating from an object point is transmitted through the system and the focal points of different wavelengths are distributed along the optical axis in the conjugated wavelength-dependent image planes. The confocal concept is implemented by inserting a pinhole in the optical path of the ray, functioning as a spatial filter, which allows the selected wavelength to cross the pinhole while blocking the others. By changing the focal length of the refractive lens, different wavelengths can traverse the pinhole and reach the detector.

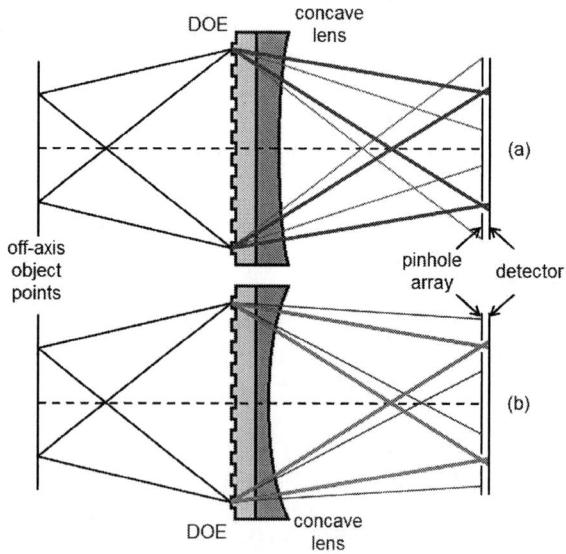

Figure 1: Schematic drawing of a hyperchromatic lens with off-axis multi-point hyperspectral sensing. By changing the curvature of the liquid lens, different spectral components from the object points are focused in the image plane. (a) Blue in focus; (b) red in focus.

We have shown previously that this lens system allows wavelength scanning for an on-axis point source over a large spectral wavelength range from 450 to 900 nm, thereby providing narrow bands with spectral widths below 15 nm [5].

Adopting the same concept, an off-axis point source array is now employed for two-dimensional hyperspectral sensing, as sketched in Figure 1. The idea is realized by generating an array of point sources in the object plane of the lens system and a pinhole array directly before the detector in the image plane. All wavelengths which are out of focus and blurring the image are filtered out by the pinhole array so that a two-dimensional image of narrow spectral bandwidth can be generated. By changing the curvature of the concave lens, images of different wavelengths are formed in the image plane on the detector. For example, in Figure 1(a) the blue light emerging from the object point sources generates the image, whereas in Figure 1(b) the image contains the information of red light from the object.

STRUCTURE OF THE HYBRID LENS

The layout of the integrated hyperchromatic lens, including the electromagnetic actuator, is shown in Figure 2. The lens system consists of two chambers: one is the reservoir volume for the actuator while the other one is the optical cavity of the liquid-filled membrane-lens. The two chambers are connected by a channel, sealed with a polymeric membrane, and filled with DI water with a refractive index of 1.33.

When a current is applied to the coil of the electromagnet, an induced repulsive force is generated to push the magnet upwards, hence pulling the optical liquid from the lens chamber and thereby causing a concave deformation of the lens' membrane and thus a change in focal length. Depending on the amplitude of the current, the focal length can be tuned in a wide range with the resolution defined by the accuracy of the current source.

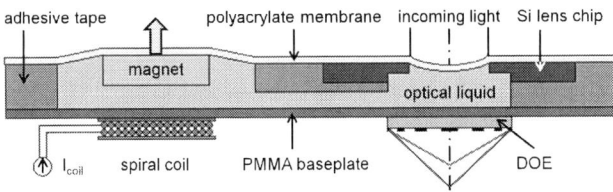

Figure 2: Layout of the hyperchromatic lens with integrated electromagnetic actuation scheme, a positive DOE and a deformable concave lens.

FABRICATION PROCESS

We developed a flexible, low cost fabrication process for the lens system. A summary of the process is given by the assembly steps outlined in Figure 3. The solid supporting structure of the fluidic lens and actuation scheme is made from the high performance adhesive transfer tape 3M™467MP 200MP. The tape was patterned by a CO_2 laser and stacked together to form the frame structure and alignment posts.

The essential elements of the system, including the Si lens chip and DOE, are fabricated by precise photolitho-

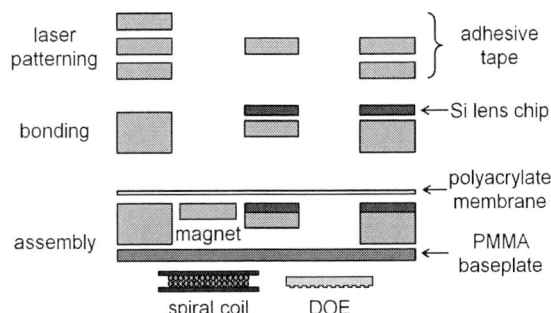

Figure 3: Process chart for assembling the hyperchromatic lens with integrated electromagnetic actuator.

graphic processes. The Si lens chip was fabricated by standard Si bulk micromachining including KOH etching from the back side and DRIE (deep reactive ion etching) from the front side to form a circular aperture of 3 mm.

The DOE was designed as a Fresnel zone plate with a focal length of $f_0 = 20$ mm at a wavelength of 550 nm, diffraction limited on the optical axis, and optimized for low spherical aberration. It is produced on quartz glass, fabricated by 2-step lithography using ICP-RIE etching, thereby forming a 4-level structure.

The actuator consists of a permanent NdFeB magnet, with a residual magnetization of $1.17 - 1.21$ T, attached on the polymeric membrane inside the actuating cavity, and driven by the magnetic field of a spiral coil with 500 windings of $100\,\mu$m copper wire and inductance of 3 mH, which is bonded on the PMMA baseplate outside the actuation cavity.

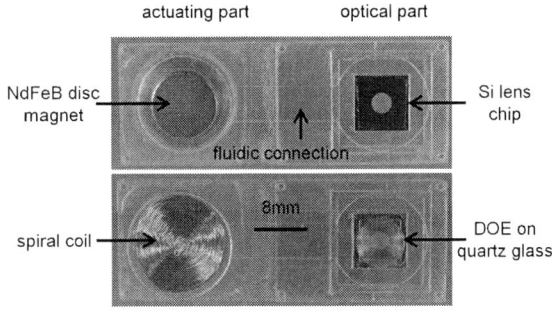

Figure 4: Photo of the back and front of the assembled hyperchromatic lens with integrated electromagnetic actuator.

Stacking and assembling of the lens components was performed with the aid of the integrated alignment structures. On the top and bottom of the solid frame are precise square posts which define the position of the diced Si lens chips and the DOE and allow precisely aligned mounting.

The optical aperture and actuating cavity were covered by a polyacrylate membrane (3M™VHB™4905) [6]. The whole frame was bonded onto a $350\,\mu$m thick PMMA baseplate and the cavity was filled with DI water. A complete prototype of the tunable hyperchromatic lens was assembled and is shown in Figure 4.

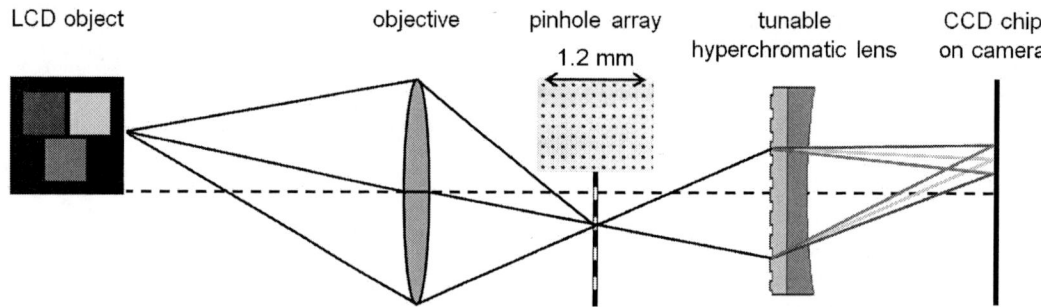

Figure 5: Schematic of the measurement setup for off-axis hyperspectral sensing. The object to be imaged is a colored pattern on an LCD screen. Based on the actuation current of the tunable lens module and the arrangement of the optical components, the spectral information of the object can be extracted on the CCD chip.

MEASUREMENT SETUP

To characterize the hyperspectral module, the measurement setup shown in Figure 5 was employed. To demonstrate the functionality of the system, colored rectangular areas with RGB values of 255 for either red, green or blue on an LCD monitor were taken as the object. This LCD object is first imaged into an intermediate image plane, where we place a pinhole array with hole diameters of 40 μm at a spacing of 120 μm to generate the array of point sources as the object for the hyperchromatic lens.

The intermediate image covers an area of 1.2 by 1.2 mm^2 on the pinhole array and forms an object consisting of 11 by 11 measuring points. From each of these point sources, a ray bundle originates with the spectral information of conjugated points in the original object plane. The array of object points is imaged by the hyperchromatic lens onto the CCD chip of the camera. The pinhole array before the detector is virtually simulated by an appropriate choice of the pixels on the CCD. The hyperchromatic lens is used in a 4f setup with equal object and image distance of 52 mm, yielding an angle of view of 1.32°, hence off-axis aberrations can be neglected.

RESULTS AND DISCUSSION

The change in focal length of the hyperchromatic lens by tuning the current of the actuating coil is the basic principle of the wavelength scanning process. In the image plane, only specific wavelengths with the focal length defined by the object and image distances as well as the actuating current can generate a sharp image of the object points.

The intensity of the wavelength, assigned by the actuation setting, can be determined by measuring the gray scale intensity of the sharply imaged points. For evaluating the intensity of the CCD images, circular areas with the size of the pinhole (40 μm) were taken. The chosen diameter ensures that no overlap between adjacent points appears and out-of-focus wavelengths are filtered out as far as possible. Choosing a smaller diameter would deliver a better spectral resolution but only at the cost of considerably reduced intensity. The spectral bandwidth of the light from a sharply imaged point was previously measured with a value of about 13 nm [5].

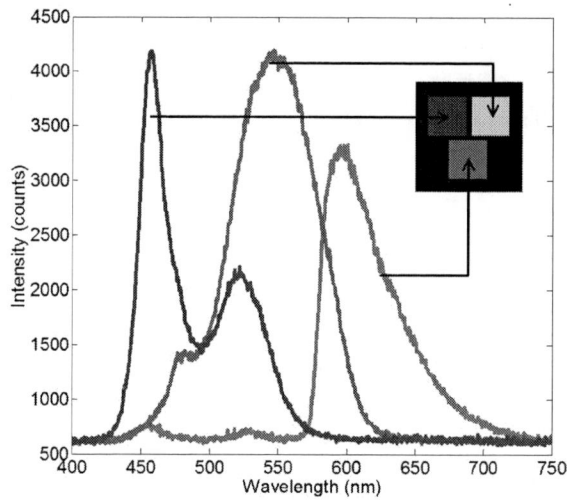

Figure 6: Reference spectra of three RGB areas of the object, recorded by an optical spectrum analyzer.

By changing the current over a wide range from 1 to 140 mA with a resolution of 1 mA, a spectral tuning over the full visible range of 400 – 730 nm is realized, thus forming a three-dimensional $11 \times 11 \times 140$ hyperspectral data cube.

The spectra of the R, G & B squares on the object are shown in Figure 6 and were measured by a commercial optical spectrometer directly in front of the LCD screen as a reference. Figure 7 shows the intensity variations of three arbitrary points of the object indicated by the arrows. The intensities were measured in terms of gray scale values on the CCD camera when the hyperchromatic lens was actuated by increasing the coil current. It was demonstrated earlier [5] that the correlation between wavelength and coil current was highly linear; therefore the intensity is plotted as a function of the coil current and corresponding wavelength.

When we compare the peak positions of the CCD-based measurement with the peaks of the spectrometer-based reference measurement, we see excellent agreement. The amplitude differences can be explained by different spectral sensitivities of the spectrometer and the CCD chip, and by the impact of the hyperspectral lens, i.e. for example the

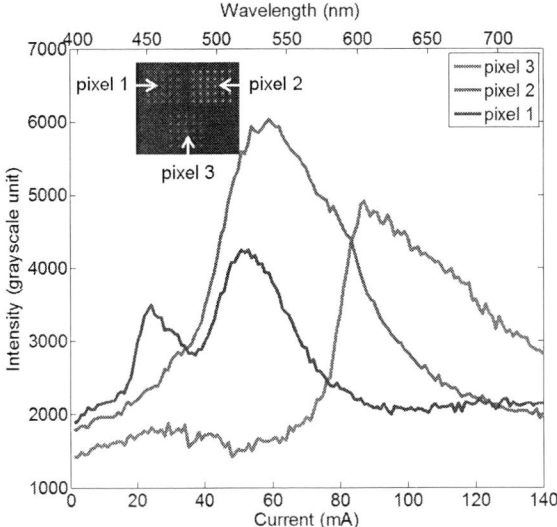

Figure 7: Intensity in terms of gray scale values for the R, G and B "pixels" imaged and measured on the CCD camera as a function of coil current and related wavelength.

wavelength-dependent diffraction efficiency of the DOE, on the CCD-based measurement.

Figure 8 displays eight of the 140 hyperspectral images of the entire object obtained on the CCD chip, corresponding to 8 actuation states. Each image delivers a specific wavelength information of the object. The series of images shows that two-dimensional images of narrow spectral bandwidth can be generated over the entire visible spectral range.

The images reveal that out-of-focus wavelengths contribute to the spectral information of the image as blurred rings around the sharp points, albeit with a strong decay due to the longitudinal chromatic aberration. Accordingly, the spacing of the pinhole array was chosen to avoid cross-talk between adjacent points. As a result, the spatial information of the object is affected and the spatial resolution is reduced. This can be improved by laterally moving the pinhole array in both directions to fill the gaps between the pinholes.

CONCLUSION

A compact tunable hyperchromatic lens for imaging a two-dimensional object with highly resolved spectral information was established. The fully integrated hybrid device was demonstrated to generate a hyperspectral data cube in the visible wavelength range of 400 – 730 nm with a spectral sampling interval of 2.4 nm. The spectra of the image points derived from the hyperspectral images are in good agreement with reference measurements from a spectrometer, measuring the spectral properties of the object directly. Due to the confocal principle of the system, realized by a pinhole array in an intermediate image plane, reduced lateral information and resolution compared to full-field imaging results. An improved system will feature a 2D electrostatically movable pinhole array to improve the spatial resolution. Pinhole diameter and spacing will be optimized

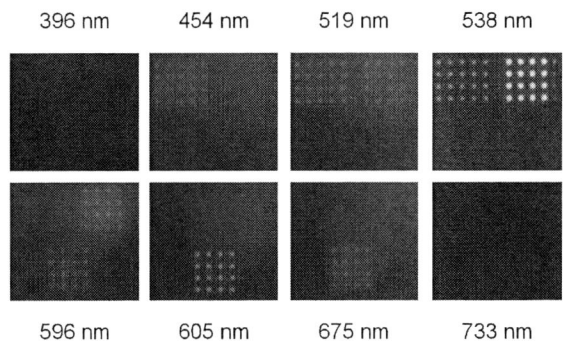

Figure 8: Eight of the 140 hyperspectral images acquired on the CCD camera during the scanning process for coil currents ranging from 1 to 140 mA. The corresponding wavelengths are specified in the figure.

in the same context to enhance the hyperspectral features of the developed system.

ACKNOWLEDGEMENT

This work is supported by the German Federal Ministry of Education and Research BMBF in the project 'Optical Microsystems for Ultra-compact Hyperspectral Sensors' (OpMiSen, FZK 16SV5575K).

REFERENCES

[1] P. Mouroulis and M. McKerns, "Pushbroom imaging spectrometer with high spectroscopic data fidelity: experimental demonstration," *Opt. Eng.*, vol. 39, no. 3, pp. 808–816, 2000.

[2] K. Koerner, C. Kohler, E. Papastathopoulos, A. Ruprecht, T. Wiesendanger, C. Pruss, and W. Osten, "Arrangement for rapid locally resolved flat surface spectroscopic analysis or imaging has flat raster array of pinholes turned about acute angle relative to spectral axis on detector matrix which fills up with elongated su-matrices," Patent DE20 061 007 172, 2013.

[3] M. Hillenbrand, A. Grewe, and S. Sinzinger, "Parallelized chromatic confocal systems enable efficient spectral information coding," *Opt. Des. Eng., SPIE Newsroom*, DOI 10.1117/2.1201301.004642, 2013.

[4] A. Grewe, M. Hillenbrand, and S. Sinzinger, "Hyperspectral imaging sensor systems using tunable lenses," *Photonik 1/2013*, pp. 38–41, 2013.

[5] P.-H. Cu-Nguyen, A. Grewe, M. Hillenbrand, S. Sinzinger, A. Seifert, and H. Zappe, "Tunable hyperchromatic lens system for confocal hyperspectral sensing," *Opt. Express*, vol. 21, no. 23, pp. 27 611–27 621, 2013.

[6] W. Zhang, H. Zappe, and A. Seifert, "Polyacrylate membranes for tunable liquid-filled microlenses," *Opt. Eng.*, vol. 52, no. 4, pp. 046 601–046 601, 2013.

CYLINDRICAL LENS WITH INTEGRATED PIEZO ACTUATION FOR FOCAL LENGTH TUNING AND LATERAL SCANNING

Moritz Stürmer, Andreas Schatz, and Ulrike Wallrabe
University of Freiburg, IMTEK - Institute of Microsystems Engineering,
Laboratory for Microactuators, GERMANY

ABSTRACT

We present a tunable cylindrical lens with an integrated actuation. Our device provides a large usable aperture of ca. 4 x 8 mm^2 and a tuning range of the focal length of more than 20 dpt. Additionally, it provides the possibility to move the lens vertex along one axis. The lens therefore enables scanning of the generated line focus. In this paper we describe the design with an actuator geometry which is optimized for good cylindricity, i.e. ratio of refractive power in the major and minor lens direction. Subsequently, a process for prototype fabrication as well as a characterization of the tuning and scanning behavior are shown.

INTRODUCTION

Lenses with a tunable focal length offer many advantages compared to systems of fixed lenses like, for instance for focusing at high speeds or for very compact device designs. Therefore, many tunable spherical lenses have been presented within the last years. Amongst the various working principles, the fluid-membrane approach has shown to be very promising since it allows for a large tuning range and integration of the actuators, e.g. [1-3]. However, only few approaches [4-6] exist for tunable cylindrical lenses, especially such lenses with fairly large clear apertures of several millimeters as we show in this work. Cylindrical lenses are useful for a broad range of applications like line focusing for laser scanning or beam shaping for edge-emitting lasers. These applications can also benefit from the properties of tunable lenses, i.e. a flexible working distance.

Further, our lens features a translation capability of the lens position in one direction which means that also the generated image can be shifted in this direction. This enables applications where images are generated line by line, e.g. for spectral imaging.

DESIGN

The design is based on a silicone (PDMS) membrane (thickness $t \sim 200\ \mu m$) into which two independently controllable piezo bending actuators are embedded (Figure 1). They are supported on a rigid frame which provides a fixed support at the edges. Further, this frame acts as a spacer to the glass substrate. The integrated actuators allow for a small overall thickness of the device of ca. 1.6 mm. The cavity between substrate and membrane is filled with an incompressible, transparent fluid. When actuated, the bending actuators displace the fluid underneath which generates a pressure inside the lens and therefore the membrane gets deflected.

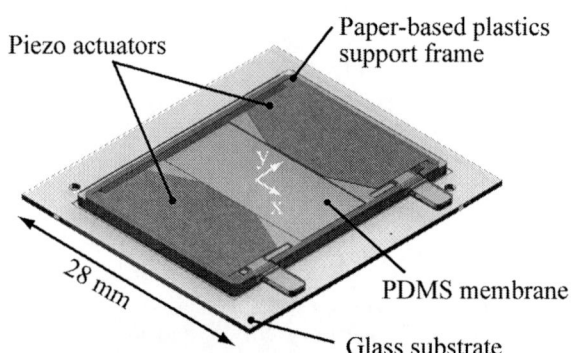

Figure 1: 3D sketch of the device design.

Optimization of the actuator geometry

In a tunable cylindrical lens the curvature in one direction (major direction) of the lens should be tunable while it remains zero, or at least constant, in the other (minor) direction. The geometry of the boundary constriction at the edges of the membrane limits the cylindrical behavior of the lens and results in a spherical membrane deflection. A rectangular membrane can be optimized for cylindrical deflection by widening the areas at the short ends of the rectangle as it was shown with a "dogbone" shaped boundary clamping in [3].

In our device design a second effect makes the membrane not bend cylindrically: the piezo actuators have an isotropic strain in both plane directions which leads to a stronger absolute deflection in the corners than at the long edges of the actuator. Both effects are reducible by cutting away a certain portion of the piezo actuator at the corners, resulting in the trapezoidal actuator shape as shown in Figure 1.

To identify a good choice for the actuator geometry the membrane deflections were analyzed using a finite element simulation, and the shape of the bending actuators was varied by cutting out different sized triangles at the corners (cf. Figure 2). A 3D model of the actual device design was set up in COMSOL Multiphysics with the simplification that the lens fluid was replaced by a constant volume constriction. A quality criteria for the lens cylindricity is the flatness of the profile in the minor (x) lens direction. As a measure for the flatness the mean absolute deviation from a horizontal regression line is used. Only datapoints within the relevant width of 8 mm are considered. This is the distance of the two piezo actuators and therefore the maximum diameter of a circular beam which could be focused with this lens. Figure 2 depicts the result of a parametric sweep of the triangular cutouts c_x and c_y at the corners. The starting size for the piezos is of 8 x 16 mm^2.

978-1-4799-3510-9/14 $31.00 © 2014 IEEE

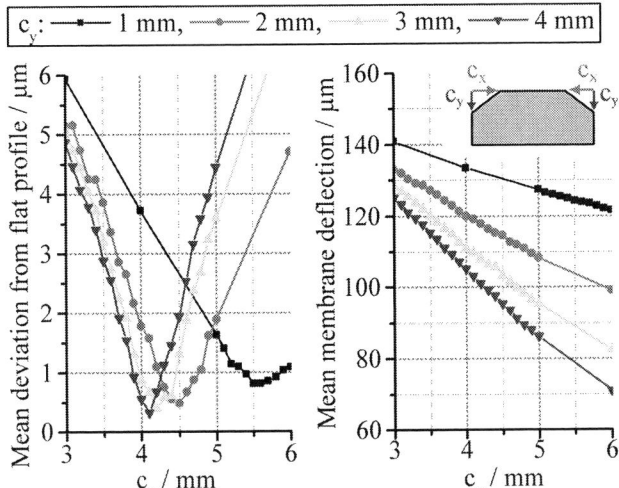

Figure 2: Left: Simulation of the average absolute deviation from a horizontal profile in minor lens direction. Right: Simulation of the mean membrane deflection. Both in dependence of different actuator cutouts and with 150 V applied to left and right actuator.

From the simulated data in Figure 2 becomes clear that the choice of an actuator geometry is a tradeoff between flatness and achievable membrane deflection. We identify the cutout with a size of 4.5 mm in x-direction and 2 mm in y-direction to be a good choice for the actuator which yields less than 0.5 µm average deviation from a purely horizontal membrane profile. For the production of prototypes and further investigation we therefore chose this geometry and compared it to a geometry with a 3 mm cutout in x- and y-direction.

Figure 3 illustrates the difference in the deflected membrane profile for a rectangular and a trapezoidal actuator shape with $c_x = 4.5$ mm and $c_y = 2$ mm. The length over which the membrane is nearly flat in the minor (x) lens direction is increased.

The green lines show as additional feature that the vertex position of the membrane in the major (y) direction can be shifted with different voltages at the left and right actuator.

FABRICATION

For demonstration of the working principle we fabricated the tunable lens using a quick prototyping process which is similar to a procedure we have introduced earlier for tunable spherical lenses [1]. The outer contour of the piezo actuators is laser cut from commercially available bending actuators which consist of a passive steel sheet and an active PZT layer (both ~ 100 µm) with electrodes on both sides. The actuators are mounted onto the support frame using an epoxy glue. Paper based plastics (PCB material FR-2) is chosen for this frame because it provides sufficient mechanical stiffness and can be easily laser structured. The frame with the actuators is cast in PDMS in a machined brass mold, in which the optical surface in the lens area as well as a precisely controlled membrane thickness are achieved by silicon insets. It is important to cure the silicone at low temperature, i.e. room temperature, in order to avoid high shrinkage and therefore a pre-stress in the membrane. In a final step the lens chamber which consists of actuators, support frame, and membrane is bonded onto a glass substrate and filled with the lens fluid at the same time, as depicted in Figure 3. To promote adhesion the glass substrate is equipped with a rough trench which is also obtained by laser ablation. A small amount of PDMS is filled into this trench and cured inside a container filled with the lens liquid. Filling the lens in-situ with its assembly avoids the need for a separate filling step and reduces the risk of trapping air bubbles inside the device. A perfluoropolyether (refractive index $n \sim 1.3$) is used as an optical liquid since it avoids PDMS swelling and does not evaporate through the membrane, which is usually considered to be a limitation of silicone membranes [7].

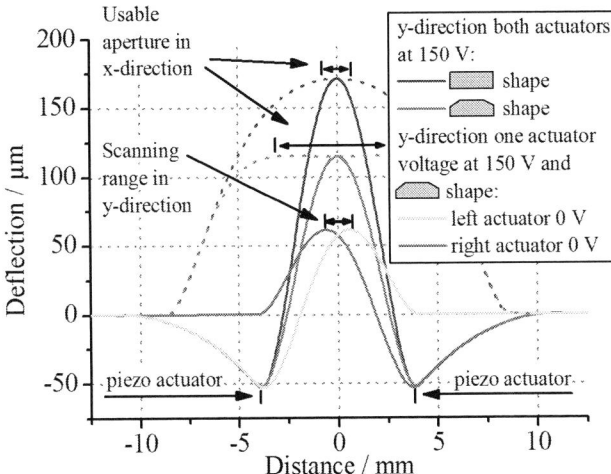

Figure 3: Simulated profiles of the deflected lens along x- and y-axis (cf. Figure 1) for different actuator geometries and actuation voltages.

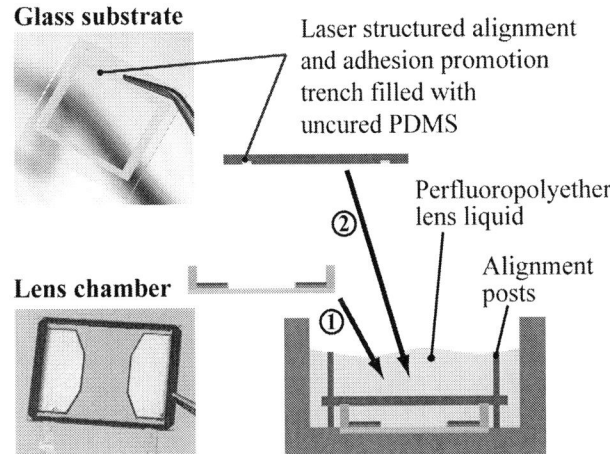

Figure 3: Sketch of the manufacturing procedure with integrated filling and bonding and photographs of the components.

CHARACTERIZATION

Tunability and cylindricity

The properties of the prototypes were characterized by scanning the lens surface with a laser triangulation sensor (Keyence LK-G32) while varying the actuation voltages. In a first step the cylindrical properties of the lens were verified. Therefore the actuation voltages on both actuators were slowly ramped in parallel from − 40 V to 350 V and back. This voltage range is chosen to stay within a safe operation region where the piezo is not damaged neither by depolarization nor by breakthroughs. The refractive powers in major and minor lens direction were calculated from the curvature under the assumption that the refractive effect of the lens is only achieved by the refraction of the lens fluid. The curvature is obtained by a regression of the measured surface profile. For the regression an aperture of 4 x 8 mm was assumed in order to stay within in an area of good agreement with the spherical membrane shape in y-direction and a width in x-direction where the measurement data still shows a flat linear behavior.

In Figure 4 the result of this measurement is shown for both manufactured prototypes. The lens with the actuators that have a cutout of 3 mm in both directions has a larger tuning range in the major lens direction from -2 to +24 dpt compared to the one with a cutout of 4.5 and 2 mm in x- and y-direction which shows a range of -3 to +14 dpt. However, this lens shows a better cylindrical behavior with a refractive power in minor lens direction that remains almost constant at -0.16 ± 0.06 dpt while it varies for the other design from -0.2 to +0.6 dpt.

This is also confirmed by the mean deviation from a perfectly flat profile depicted in the bottom part of Figure 4. Over the whole actuation range it stays for the optimized actuator geometry below the deviation which is shown by the lens with symmetrical cutouts. In general this behavior is already expected from the simulation results even though the absolute values of the deviation are much higher. This can be attributed to tolerances within the fabrication process. It is to be noted that the standard deviation of the measurement with the used setup is ca. 0.15 μm. It is therefore ensured that the measured values are far above the noise level of the measurement setup.

Response time

One advantage of tunable lenses is the high focusing speed they can reach in comparison to conventional lens systems. To demonstrate this feature a voltage step from minimum to maximum driving voltage was applied to both actuators at the same time and the deflection at the center of the membrane was measured, again with the above used triangulation sensor. The sampling rate is 50 kHz to get a sufficient time resolution. The results in Figure 5 show that 99 % of the saturation deflection were already reached within 10 ms after the voltage step. Actuation frequencies of up to 100 Hz are therefore possible. Even higher speeds are achievable with optimized voltage ramps instead of steps and with smaller deflection differences. The high tuning speed results from the low viscosity of the lens liquid (~ 60 mPa s) which only damps the system weakly.

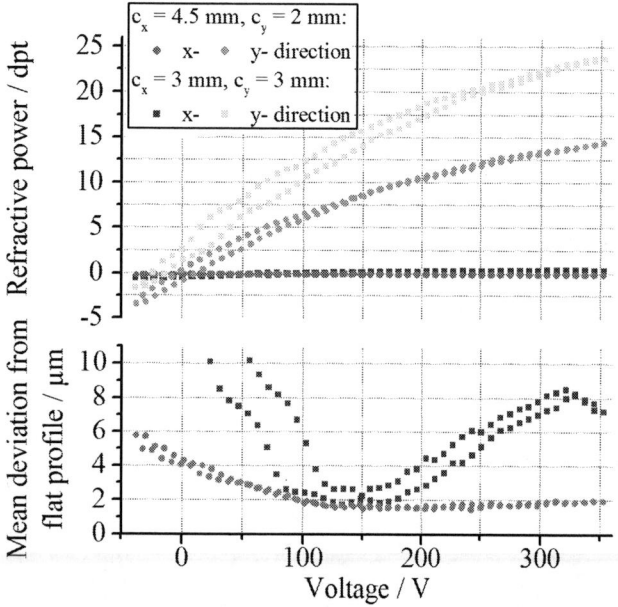

Figure 4: Top: Refractive power in major and minor lens direction obtained from the surface profile in dependence of the applied voltage on both actuators. Bottom: Mean deviation from a flat surface profile in minor lens direction.

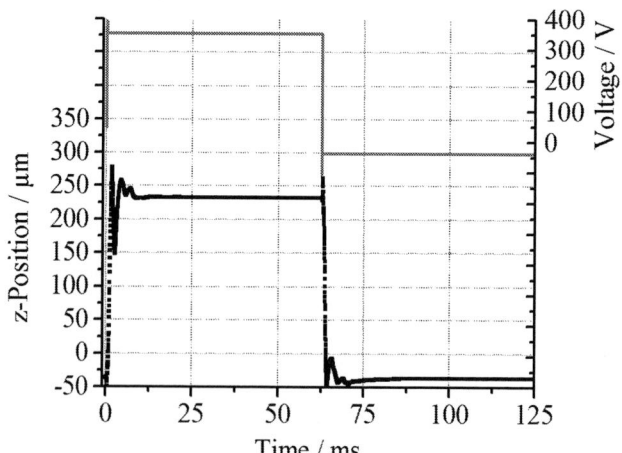

Figure 5: Step response of the deflection in the membrane center to the depicted voltage step on both actuators.

Lateral scanning

In a third experiment the lateral scanning capability was verified by varying the actuation voltage ratio of left and right actuator. The prototype with a symmetrical cutout of 3 mm was chosen for this experiment because it shows higher deflections and therefore a broader lateral translation capability. The voltages depicted in the upper part of Figure 6 were applied. They are chosen with a higher slew rate above 150 V than below. This is done to compensate for the characteristic actuator behavior which becomes flatter at higher voltages (cf. Figure 4). The position of the lens vertex was monitored. The result in the middle part of Figure 6 shows that a scanning range of 1.1 mm could be achieved, while the focal length in the major lens direction could be kept almost constant at 13.2 ± 0.8 dpt.

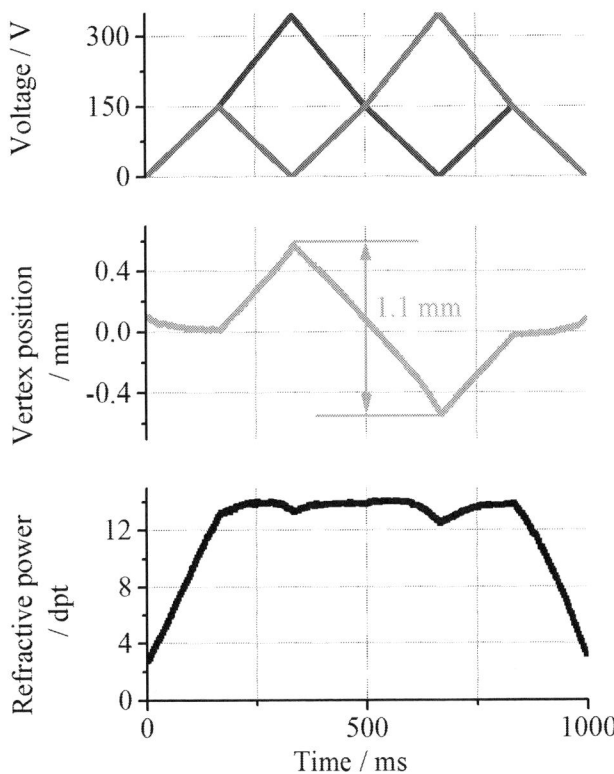

Figure 6: Top: Voltages applied to left and right piezo actuator. Middle: Resulting position of the lens vertex. Bottom: Resulting refractive power in the major lens direction.

CONCLUSIONS

We have shown a cylindrical lens which is tunable with the integrated piezo actuators over a wide range of more than 20 dpt. An almost purely cylindrical behavior could be achieved by optimizing the shape of the actuators. A response time of 10 ms proves that this device is capable for high-speed focusing applications. Because two actuators are controllable independently it is also possible to move the lens vertex by more than 1 mm while keeping the refractive power almost constant. This enables a translation movement of the generated focus.

Further optimization can be done by extending the actuation range with the use of bimorph piezo actuators and by considering the hysteresis behavior of the actuators. Also, integrating an offset lens by pre-shaping the membrane with the profile of a cylindrical lens can increase refractive power while keeping it more constant in lateral scanning operation.

It is to be noted that in principle the manufacturing process is easily extendable for processing of two perpendicularly oriented lenses on the two sides of one substrate. This allows for a fully customized beam shaping and steering in both directions of the imaging plane.

ACKNOWLEDGEMENTS

This research is supported by German Research Foundation (DFG) grant WA 16547/1-2 within the Priority Program "Active Micro-optics".

REFERENCES

[1] F. Schneider, C. Müller, and U. Wallrabe, "A low cost adaptive silicone membrane lens", *J. Opt. A Pure Appl. Opt.*, vol. 10, no. 4, p. 044002, Apr. 2008.

[2] J. Draheim, F. Schneider, T. Burger, R. Kamberger, and U. Wallrabe, "Single chamber adaptive membrane lens with integrated actuation", in *International Conference on Optical MEMS and Nanophotonics*, 2010, pp. 15–16.

[3] A. Pouydebasque, C. Bridoux, F. Jacquet, S. Moreau, E. Sage, D. Saint-Patrice, C. Bouvier, C. Kopp, G. Marchand, S. Bolis, N. Sillon, and E. Vigier-Blanc, "Varifocal liquid lenses with integrated actuator, high focusing power and low operating voltage fabricated on 200 mm wafers", *Sensors Actuators A Phys.*, vol. 172, no. 1, pp. 280–286, Dec. 2011.

[4] S. Leopold, D. Paetz, F. Knoebber, O. Ambacher, S. Sinzinger, and M. Hoffmann, "Tunable cylindrical microlenses based on aluminum nitride membranes", In *Proc. SPIE 8616, MOEMS and Miniaturized Systems XII*, 861611, March 2013.

[5] Y.-H. Lin, H. Ren, K.-H. Fan-Chiang, W.-K. Choi, S. Gauza, X. Zhu, and S.-T. Wu, "Tunable-focus cylindrical liquid crystal lenses", *Japanese Journal of Applied Physics*, vol. 44, pp. 243–244, Jan. 2005.

[6] W. Zhao, D. Liang, J. Zhang, C. Liu, S. Zang, and Q. Wang, "Variable-focus cylindrical liquid lens array", In *Proc. of SPIE Vol. 8769, Int. Conf. on Optics in Precision Engineering and Nanotechnology*, June 2013.

[7] J. N. Lee, C. Park, and G. M. Whitesides, "Solvent compatibility of poly(dimethylsiloxane)-based microfluidic devices", *Analytical Chemistry*, vol. 75, no. 23, pp. 6544–54, Dec. 2003.

CONTACT

U. Wallrabe; wallrabe@imtek.uni-freiburg.de

FRESNEL LENS BASED ON SILICON NANOWIRES

Yen-Sheng Lu, Jayer Fernades, Hewei Liu, and Hongrui Jiang
University of Wisconsin, Madison, USA

ABSTRACT

Silicon nanowire reflective Fresnel lenses are demonstrated and their focusing and lensing properties characterizations are shown. The Fresnel lenses were fabricated using conventional lithography and metal-assisted chemical etching. The lenses have dark zones made of sub-wavelength silicon nanowires and bright zones formed by polished silicon surface. High intensity contrast exists between the bright and the dark zones, leading to an efficient focusing performance in the spectrum of visible light. The reflective Fresnel lens has the potential to reflect and focus light onto photosensitive dye in dye-sensitive solar cells in order to improve their light absorption efficiency and hence, the photocurrent.

INTRODUCTION

Fresnel lenses with their flat surfaces and small volumes are attractive for focusing, collimating and bending light in miniaturized optical systems and integrated optics [1-5]. Unlike curved lenses or mirrors, Fresnel lenses/zone plates use diffraction instead of refraction or reflection. The incident light, encountering the binary Fresnel lens composed of alternative transparent (bright) and opaque (dark) zones, diffracts towards a focus spot. Most of the available Fresnel lenses are the transmission type. Recently, a reflective Fresnel lens based on carbon nanotube forests was proposed, showing high contrast focusing of light owing to the near-perfect optical absorption of carbon nanotubes [6-9].

Carbon nanotubes with their low index of refraction and surface roughness effectively prevent light reflection from the opaque region, leading to a sharp focus. The reflective Fresnel lens can be used for efficiently focusing and collimating light in optical data transfer and communication systems. Besides, two-dimensional source arrays for neural network architectures can be realized using reflective Fresnel lens arrays [8]. However, carbon nanotubes require chemical vapor deposition at a high temperature of 540 °C, making integration with silicon based electronic circuits difficult.

Silicon nanowires with spacing smaller than the wavelength of incident light have been widely used for photovoltaic applications due to their remarkable antireflection behavior [10-14]. Vertically aligned silicon nanowires, embedded in Si substrate and formed by wet chemical etching at room temperature, provide an alternative material with strong light absorption. Silicon nanowires longer than 750 nm have a light reflection of approximately 0.1% over the wide spectral range of 300–800 nm, making it an appropriate candidate for forming the dark zones in Fresnel lenses [15-18]. Here, we demonstrate silicon-based

Fresnel lenses, where the dark zones are composed of silicon nanowires formed directly on the silicon substrate.

DESIGN

Figure 1 illustrates the operating principle of our silicon-based reflective Fresnel lens. It contains alternating dark and bright zones and renders a focused spot by the constructive interference of light reflected from the bright zones. We utilize alternating black silicon nanowires and polished silicon surface as the dark and bright zones. The focal length f of the Fresnel lens is related to the radii of the concentric zones as described in equation 1

$$r_n^2 = nf\lambda, \tag{1}$$

where r_n is the radius of the nth zone, and λ is the wavelength [18,19].

The three Fresnel lenses were designed to have focal lengths of 0.5, 1 and 2 cm, respectively. All of the focal lengths are calculated for a wavelength of 633 nm. The lens parameters such as focal length, innermost radius, number of zones and minimum line width are listed in Table 1. The lens with a focal length at 0.5 cm has an innermost radius of 56.3 µm and a minimum linewidth of 8.3µm. For the lens with a longer focal length of 2 cm, the center zone radius is set at 112.6 µm and the minimum linewidth is 15.3 µm at the 14th zone. The increase in zone number will suppress higher order foci and helps obtain a sharp focus. However, the smaller the minimum linewidth, the more difficult the fabrication becomes. For our design, the geometry of the lens can be easily fabricated using conventional photolithography techniques and metal-assisted chemical etching.

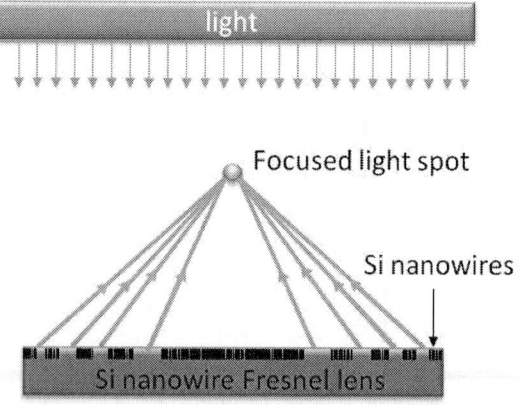

Figure 1: Schematic of reflective silicon nanowire Fresnel lens. The silicon nanowires and polished silicon surface form the dark and bright zones, respectively. The Fresnel lens reflects and focuses the incident light to a light spot above the lens.

Table 1: Parameters of fabricated silicon nanowire Fresnel lenses

Focal length	0.5 cm	1 cm	2 cm
Innermost radius	56.3 μm	79.6 μm	112.6 μm
Number of zones	12	14	14
Minimum line width	8.3 μm	10.8 μm	15.3 μm

FABRICATION PROCESS

The fabrication process is relatively simple and is schematically shown in Figure 2. Single polished p-type silicon wafers with resistivity of 1-100 ohm-cm were used for fabricating the Fresnel lenses. The geometry was defined using conventional photolithography. Photoresist S1813 was spun on the wafer at a speed of 2500 rpm for 30s and then soft baked at 115 ℃ for 1 min. The exposure time of the photoresist was set to 9.5 s with a dose of 10 mW/cm^2. The photoresist was developed using MF-321 for 1min. Silicon nanowires were formed by immersing the sample in metal-assisted chemical etchant at room temperature for 5 min. The etching solution contains 17.68M HF, 2.05M H$_2$O$_2$ and 0.0015M AgNO$_3$ [10]. The patterned photoresist acted as a mask and protected the underlying polished silicon surface from being attacked by the chemical etchant. The non-protected regions were covered with silver (Ag) nanoparticles and formed the silicon nanowires after the wet chemical etching. Following the removal of photoresist using acetone, the binary Fresnel lenses with alternating bright and dark zones were formed (Figure 2c).

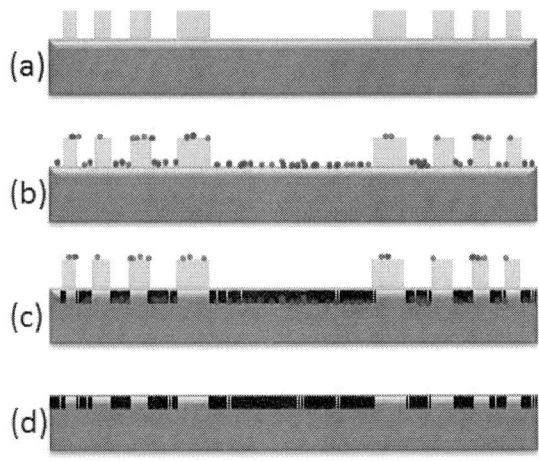

Figure 2: Fabrication process flow of a Fresnel lens based on silicon nanowires. (a) Fresnel photoresist patterns on silicon surface. (b) Silver ions reduced to form nuclei, growing into nanoparticles (c) Silicon nanowires formed by immersing the sample in chemical etchant for 5 min. (d) Removal of photoresist using acetone.

EXPERIEMNT RESULT AND DISCUSSION
Etching of Silicon Nanowires

The SEM images of the Fresnel lens with a focal length of 0.5 cm are shown in Figure 3. From the plan view, the Fresnel lens shows a remarkable difference in the morphology between the opaque zones decorated with silicon nanowires and the bright zones of polished silicon. The innermost zone decorated with silicon nanowires had a radius of 56.3 μm, where the silicon nanowires are uniformly distributed in the dark zones. The cross-sectional view (Figure 3b) shows the silicon nanowires have an average height of 2 μm after etching for 5 min.

The etching process of silicon nanowires comprises of silver (Ag) nanoparticle deposition and electroless chemical deposition. AgNO$_3$ is the chemical precursor for Ag nanoparticles in the chemical etchant where H$_2$O$_2$ reduces Ag$^+$ to Ag nanoparticles on the Si surface (Figure 2b). The chemical reaction can be expressed as

$$H_2O_2 + 2Ag^+ \rightarrow O_2 + 2H^+ + 2Ag \qquad (2)$$

The electroless etching mechanism can be summarized by two half-cell reactions in the galvanic cell to explain the etching behavior at the contact interface of Ag nanoparticles and the silicon substrate [15,20,21]:

Cathode reaction that occurs at the Ag nanoparticle site:
$$H_2O_2 + 2H^+ + 2\,e^- \rightarrow 2\,H_2O \qquad (3)$$

Anode reaction that occurs at the Si substrate:
$$Si + 2H_2O \rightarrow SiO_2 + 4H^+ + 4e^- \qquad (4)$$
$$SiO_2 + 6HF \rightarrow H_2SiF_6 + 2H_2O \qquad (5)$$
The overall reaction:
$$Si + 2H_2O_2 + 6HF \rightarrow H_2SiF_6 + 4H_2O \qquad (6)$$

The galvanic cell is established because the electrical potential of Ag nanoparticles is higher than the Fermi level of the Si substrate. Ag nanoparticles acting as the cathode accumulate the electrons generated from the oxidation of the surrounding Si (anode). The newly formed oxide is immediately attacked and removed away by HF. The Ag nanoparticles sink into the pits generated in Si substrate. A continuous etching process will make the pits deeper. After removing Ag nanoparticles using HNO$_3$ and H$_2$O$_2$, the vertically aligned silicon nanowires are obtained.

The silicon etching process is dependent on the concentration of AgNO$_3$, H$_2$O$_2$ and HF. For example, the concentration of AgNO$_3$ influences the Ag particle size and the density of particles formed on the silicon surface [15]. The randomly distributed silicon nanowires with a spacing of less than the incident wavelength (subwavelength structure) have a lower light reflection compared to well-ordered silicon nanowires. These cone-like silicon nanowires let the incident light scatter randomly inside the nanowires and elongate the travel path, leading to low reflection, as explained by effective medium theory [16].

978-1-4799-3510-9/14 $31.00 © 2014 IEEE

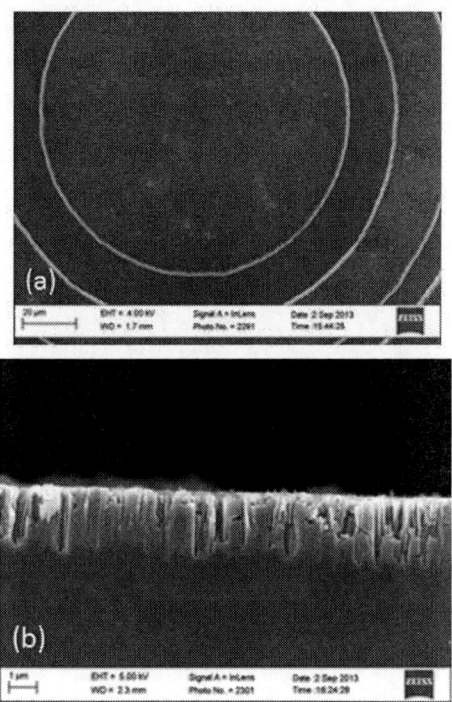

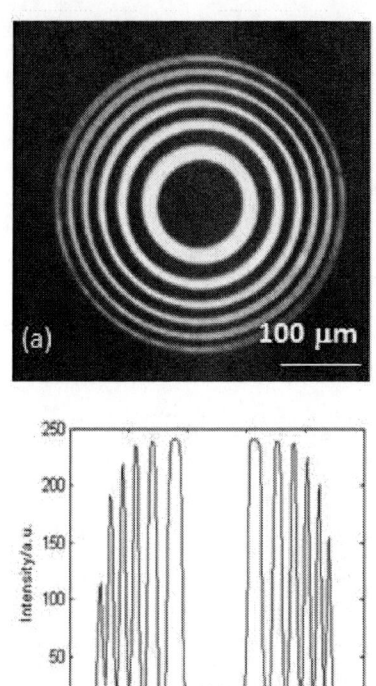

Figure 3: SEM image of the silicon nanowire Fresnel lens. (a) The odd and even regions are composed of silicon nanowires and polished silicon surface, respectively. The innermost zone has a radius of 56.3 μm. (b) The vertically aligned silicon nanowires have the average height of 2μm.

Figure 4: (a) Image of Fresnel lens under an optical microscope. (b) The lens shows a strong contrast in light intensity between bright and dark regions.

Optical Performance of the Fresnel Lens

The optical performance of the Fresnel lens with a focal length of 0.5 cm was characterized under an optical microscope and the images were analyzed using MATLAB as shown in Figure 4 and 5. From the plan view captured using a microscope with an objective lens (5×), the Fresnel lens has 12 zones. The bright zones are composed of polished silicon surface with a light reflectivity of nearly 40% when averaged over all angles of incidence under the light spectrum from 400–1100 nm [21]. The dark silicon nanowires have a low light reflectivity close to 0.1 %, which blocks most of the incident light acting as superior light absorbers [12,13]. The light intensity plot in Figure 4b shows a high-contrast light intensity distribution between the bright and dark zones. For the fabricated Fresnel lens, the intensity ratio between bright and dark zone can achieve 20. In Figure 5, a focused spot was observed at the focal plane, showing good uniformity. The three-dimensional intensity profile is plotted across the focused spot of the lens. The profile is smooth along all directions. The sharpness of the focus spot can be increased by enhancing the depth of the silicon nanowires and the number of zone region [15]. However, the increased number of zone regions and the reduced linewidth of the outmost ring will be a challenge to fabricate using conventional lithography techniques.

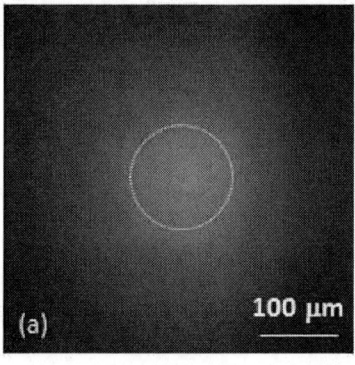

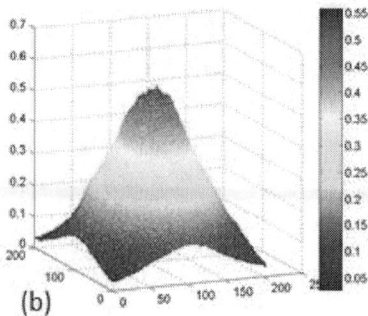

Figure 5: (a) The focused image of reflected light captured at the focal plane. (b) The light intensity across the focus spot.

CONCLUSIONS

Reflective silicon nanowire Fresnel lenses were realized using cost-effective photolithography techniques and metal assisted chemical etching. Because of the reflectivity difference in silicon nanowires and polished silicon surface, these reflective Fresnel lenses demonstrate focused spots above the lens, indicating the potential to be integrated with light energy conversion devices. For example, such Fresnel lenses can reflect and focus light onto the photosensitive dye in dye-sensitive solar cells to improve their light absorption efficiency and thus the photocurrent.

ACKNOWLEDGEMENTS

This research was supported by the National Institute of Health through the program (Grant No. 1DP2OD008678).

REFERENCES

[1] K. Rastani, A. Marrakchi, S. F. Habiby, W. M. Hubbard, H. Gilchrist and R. E. Nahory, "Binary Phase Fresnel Lenses for Generation of Two-dimensional Beam Arrays", *Appl. Opt.*, vol. 30, pp. 1347-1354, 1991.

[2] K. Kodate, E. Tokunaga, Y. Tatuno, J. L. Chen, and T. Kamiya, "Efficient Zone Plate Array Accessor for Optoelectronic Integrated Circuits: Design and Fabrication", *Appl. Opt.*, vol.29, pp. 5115-5119, 1990.

[3] F. Mahmoud, J. K. Keith, P. Scott, O. Nordman, and P. Nasser, "Design and Fabrication of Circular Grating Coupled Distributed Bragg Reflector Lasers", *Opt. Eng.*, vol. 37, pp. 1169-1174, 1998.

[4] B. Morgan, C. M. Waits, J. Krizmanic, and R. Ghodssi, "Development of A Deep Silicon Phase Fresnel Lens Using Gray-Scale Lithography and Deep Reactive Ion Etching", *J. Microelectromech. Syst.*, vol.13, pp. 113-120, 2004.

[5] M. Ferstl and A.-M. Frisch, "Static and Dynamic Fresnel Zone Lenses for Optical Interconnection", *J. Mod. Opt.*, vol. 43, pp. 1451-1462, 1996.

[6] Z. P. Yang , L. Ci , J. A. Bur , S. Y. Lin, and P. M. Ajayan, "Experimental Observation of an Extremely Dark Material Made By a Low-Density Nanotube Array", *Nano Lett.*, vol. 8, pp. 446-451, 2008.

[7] T. de los Arcos, P. Oelhafen, D. Mathys, Nanotechnology, "Optical Characterization of Alignment and Effective Refractive Index in Carbon Nanotube Films", *Nanotech.*, vol.18, pp-265706, 2007.

[8] R. Rajasekharan, H. Butt, Q. Dai, T. D. Wilkinson and G. A. J. Amaratunga, "Can Nanotubes Make a Lens Array ? " *Adv. Mater.*, vol. 24, pp.OP170-OP173, 2012.

[9] H. Butt, R. Rajesekharan, Q. Dai, S. Sarfraz, R. V. Kumar, G. A. J. Amaratunga, and T. D. Wilkinson, "Cylindrical Fresnel Lenses Based on Carbon Nanotube Forests", *Appl. Phys. Lett.*, vol. 101, pp. 243116, 2012.

[10] W. C. Hsu, Y.-S. Lu, J.-Y. Chyan and J. A. Yeh, "High-Efficiency 6 Multicrystalline Black Solar Cells Based on Metal-Nanoparticle-Assisted Chemical Etching", *Int. J. Photoenergy*, vol. 2012, pp.197514, 2012.

[11] K. Peng, Y. Xu, Y. Wu, Y. Yan, S.-T. Lee, J. Zhu, "Aligned Single-Crystalline Si Nanowire Arrays for Photovoltaic applications", *Small*, vol. 1, pp. 1062-1067, 2005.

[12] C. Lin, N. Huang, and M. L. Povinelli, "Effect of Aperiodicity on the Broadband Reflection of Silicon Nanorod Structures for Photovoltaics", *Opt. Express*, vol. 20, pp. A125-A132, 2012.

[13] C. Lin and M. L. Povinelli, "Optical Absorption Enhancement in Silicon Nanowire Arrays with A Large Lattice Constant for Photovoltaic Applications", *Opt. Express*, vol. 17, pp.19371-19381, 2009.

[14] L. Tsakalakos, J. Balch, J. Fronheiser, M.-Y. Shih, S. F. LeBoeuf, M. Pietrzykowski, P. J. Codella, B. A. Korevaar, O. Sulima, J. Rand, A. Davuluru, and U. Rapolc, "Strong Broadband Optical Absorption in Silicon Nanowire Films", *J. Nanophoton.*, vol.1, pp.013552, 2007.

[15] Y.-T. Lu and A. R. Barron, "Nanopore-Type Black Silicon Anti-Reflection Layers Fabricated by A One-Step Silver-Assisted Chemical Etching", *Phys.Chem. Chem. Phys.*, vol. 15, pp. 9862-9870, 2013.

[16] T.-H. Pei, S. Thiyagu, and Z. Pei, "Ultra High-density Silicon Nanowires for Extremely Low Reflection in Visible Regime", *Appl. Phys. Lett.*, vol. 99, pp.153108, 2011.

[17] S. Kato, Y. Kurokawa, Y. Watanabe, Y. Yamada, A. Yamada, Y Ohta, Y Niwa, and M Hirota, "Optical Assessment of Silicon Nanowire Arrays Fabricated by Metal-Assisted Chemical Etching", *Nanoscale Res. Lett.*, vol. 8, pp. 216, 2013.

[18] Y. H. Fan, H. Ren, S.T. Wu, "Switchable Fresnel Lens Using Polymer-stabilized Liquid Crystals", *Opt. Express*, vol. 11 , pp. 3080-3086, 2003

[19] H. C. Kim, H. Ko, and M. Cheng, "Optical Focusing of Plasmonic Fresnel Zone Plate-Based Metallic Structure Covered with A Dielectric Layer", *J. Vac. Sci. Technol. B*, vol. 26, pp.2197-2203, 2008.

[20] Z. Huang, N. Geyer, P. Werner, J. de Boor, U. Gösele, "Metal-Assisted Chemical Etching of Silicon: A Review", *Adv. Mater.*, vol. 23, pp.285-308, 2011.

[21] H.-P. Wang, K.-Y. Lai, Y.-R. Lin, C.-A. Lin and J.-H. He, "Periodic Si Nanopillar Arrays Fabricated by Colloidal Lithography and Catalytic Etching for Broadband and Omnidirectional Elimination of Fresnel Reflection", *Langmuir*, vol. 26, pp.12855 -12858, 2010.

CONTACT

*H. Jiang, tel: +1- 608- 2659418; hongrui@engr.wisc.edu

ENHANCED WAVELENGTH SELECTIVE INFRARED EMISSION USING SURFACE PLASMON POLARITON AND THERMAL ENERGY CONFINED IN MICRO-HEATER

Takahiro Sawada[1], Katsuya Masuno[2], Shinya Kumagai[1],
Makoto Ishii[2], Shouichi Uematsu[2], and Minoru Sasaki[1]

[1]Toyota Technical Institute, Nagoya, Aichi, JAPAN
[2]Yazaki Corporation, Susono, Shizuoka, JAPAN

ABSTRACT

A new surface plasmon polariton (SPP) based wavelength selective IR emitter is combined with micro-heater. IR emitted from the micro-heater is basically confined except SPP propagation on the metal grating carrying IR energy to the outside. The limited condition for SPP excitation realizes the narrow wavelength filtering. SPP related emission is obtained having the peak width similar order compared with the bandwidth of gas absorption. Since the micro-heater can minimize the thermal conduction loss, the high efficiently is expected at SPP related wavelength.

INTRODUCTION

The blackbody having the broad spectrum is still used as the mid-infrared emitter. Mid-infrared (IR) has many applications (e.g., air pollution monitoring, home electronics equipment (fire alarms, heating ventilation, air conditioning), and medical applications), since the molecular vibration can be measured and the identification of the molecule can be possible. The common compact gas sensor is the non-dispersive IR (NDIR) sensor. Figure 1(a) shows the illustration for explaining its principle. The sensor consists of the emitter, the optical path, the optical band pass filter (BPF), and the detector. The blackbody has a broad spectrum as shown in the left curve [1]. The collimated IR propagates the optical path through which the gas flows. The gas absorbs the specific IR wavelength as shown in the middle curve. BPF at the downstream trims the spectrum as shown in the right curve and selects the wavelength region which has the information of absorption due to the gas [2]. The intensity of the transmitted light is measured by the detector. The gas sensing is based on the relation that the higher gas concentration decreases the intensity of the transmitted light following Lambert-Beer law.

Power consumption is the primary concern especially for the battery powered sensors. Miniature incandescent lamp consumes several dozens and hundreds mW, which takes large percentage of entire power consumption. Figure 1(b) shows one example the incandescent light bulb. CO_2 molecule absorbs the wavelength of 4.2-4.3µm. The use of this wavelength region avoids the cross-talk with other gases giving the measurement accuracy. On the other hand, the large percentage of the input power is wasted as the spectral region outside the absorption band of the target gas. In case of Fig. 1(c), only about 3% of the total IR power is used for CO_2 gas sensing. The absorption bandwidth is about 100 nm for many gases. Examples are 125 nm for CO_2, 300 nm for

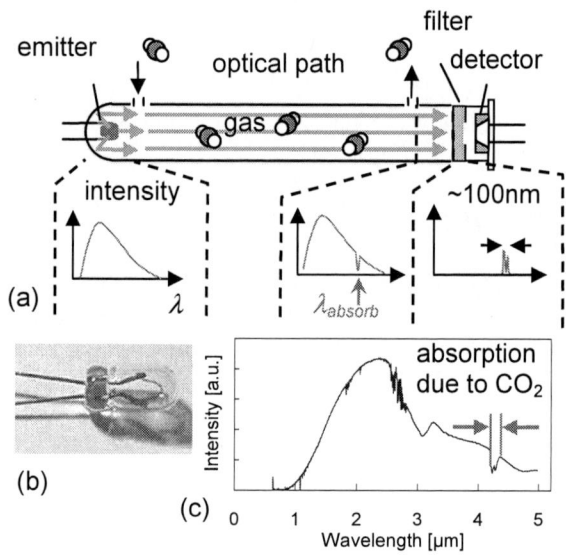

Figure 1:(a) Schematic drawing of NDIR gas sensor. (b) IR emitter (Perkin Elmer IRL715). (c) Typical spectrum observed from (b).

CH_4, and 250 nm for CO [3]. This band is generated by the vibration accompanied by the rotation of molecule. If the emitter has the wavelength selectivity of this band width of the target gas, the higher efficiency is expected. There is no appropriate emitter for the sensor including the light emitting diode [3] or the quantum cascade laser.

Recently, the plasmonic thermal emitter (PTE) emerges as a promising candidate of the wavelength selective emitter [4]. Si and W based heaters with photonic structures perform higher emissivity in the specific IR region [5]. However, typical peak width of emission spectrum is 1.0-1.5 µm, wider than the band width of gases [6]. Previously we have proposed a new principle confirmed using the bulk NiCr heater [7].

In this study, the micro-heater is combined since it is advantageous for not only miniaturization but also for the good thermal isolation of the suspended structure realizing high temperature region with smaller power. The wavelength selective emission is tried to be realized.

PRINCIPLE

Figure 2 shows a schematic drawing of the principle proposed. The grating is placed parallel to the heater. At the left part, the grating faces to the heater. At the right part, the

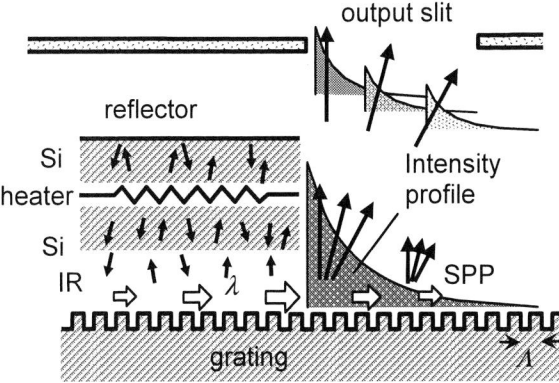

Figure 2: Schematic drawing of the emitter proposed. Λ and λ are grating pitch and IR wavelength, respectively.

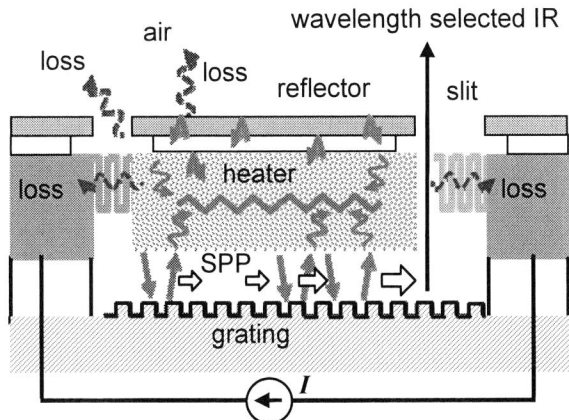

Figure 3: Schematic drawing illustrating the energy flows.

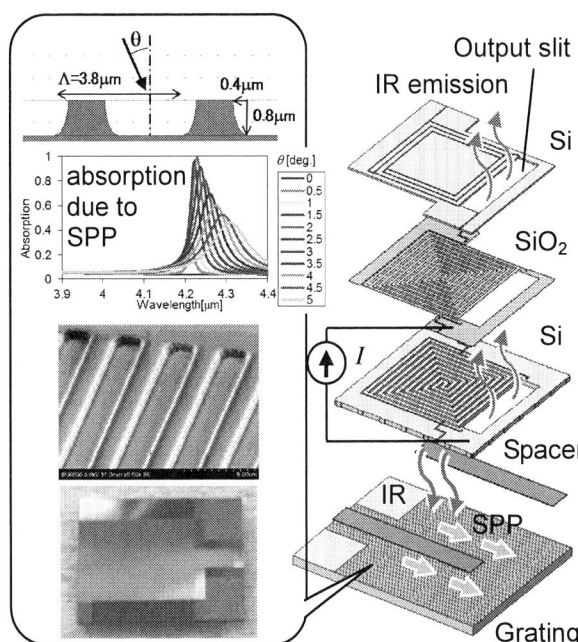

Figure 4: Schematic drawing of IR micro-emitter and its construction. Grating is fabricated from Si wafer, on which 160/10nm-thick Au/Cr layers are deposited.

output end exposes to the open space. IR from the heater is based on blackbody emission and it is incident on the grating at any angle. Basically, IR is reflected back and confined in the cavity between the reflective grating and the reflector preventing energy loss. The exception is the coupling with surface plasmon polariton (SPP) on the grating. SPP carries IR energy along the grating surface to the output end. Then, IR is emitted by the reverse process. The wavelength selection is obtained from the narrow coupling condition with SPP. SPP propagation distance extends to mm level in IR region [8, 9].

MICRO-HEATER FOR EFFICIENCY

Figure 3 shows the schematic drawing illustrating the energy flows. The micro-heater can be combined with the suspension with the good thermal isolation, which is known to be advantageous for making the high temperature region using the small energy. The ability of the blackbody emission is decided by the temperature. Considering the energy flow from high temperature region to IR emission, the other energy flow becomes the loss. (1) The high thermal isolation suppresses the thermal conduction loss. (2) The convection loss via air flow can be reduced using the vacuum package.

(3) The radiation can be confined with the micro-cavity design based on the mechanical structure. So, micro-structures are advantageous for reducing energy losses as well as the small size. This will improve the energy efficiency of the emitter.

ESTIMATION OF SPP EXCITATION

The inset of Fig. 4 shows 2D model of the grating. The grating pitch Λ is 3.8 µm. The depth is 0.8µm with the lateral widening by 0.4µm supposing the elliptical profile. Incident angle θ is the parameter. This will give the theoretical TM-polarized absorption spectra with various incident angles (DiffractMOD, RSoft Design Group). Peaks correspond to the coupling with SPP. A large peak is at the wavelength λ=4.23µm and θ=2°. Since this peak wavelength is near to the absorption wavelength of CO_2, the design value is 3.8 µm pitch and 0.8 µm depth.

DEVICE DESIGN

Figure 4 (at right side) shows the schematic drawing of the micro-emitter and its assembly. The micro-heater is fabricated from SOI wafer etching Si from both sides. The top side is the output slit. The micro-heater is at the bottom side. The current flows inside the suspended spiral bridge generating the high temperature region. This structure is mechanically reinforced by the buried oxide and device Si layers. Elements of grating, spacer, and micro-heater are stacked. Their alignment accuracy is not required.

FABRICATION

Figure 5 shows the fabrication sequence illustrating the

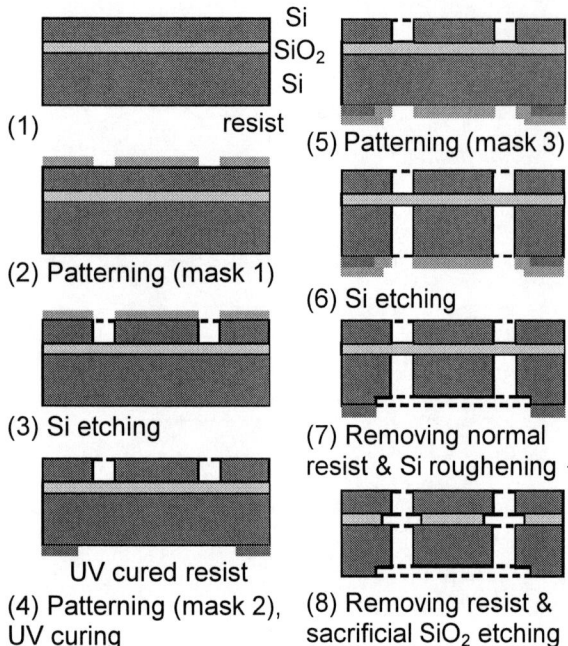

Figure 5: Fabrication sequence of microheater.

(1)   Si / SiO$_2$ / Si / resist

(2) Patterning (mask 1)

(3) Si etching

(4) Patterning (mask 2), UV curing   UV cured resist

(5) Patterning (mask 3)

(6) Si etching

(7) Removing normal resist & Si roughening

(8) Removing resist & sacrificial SiO$_2$ etching

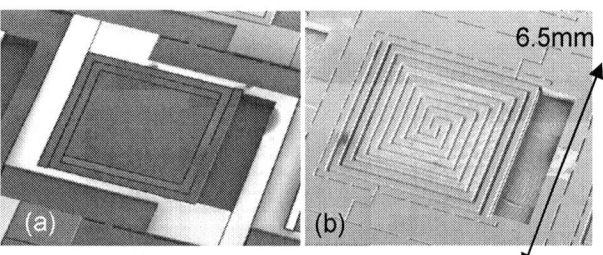

Figure 6: Fabricated micro-heater. (a) Front side for suppressing IR emission. (b) Back side with roughed surface for enhancing IR emission.

Figure 7: Emitter assembled on TO-8 can.

cross-section. SOI wafer is used (step 1). The device, buried oxide, and handle layers are 75, 3 and 200 μm thick. The handle layer is Boron doped <0.02 Ω-cm and becomes the micro-heater. The first mask pattern is transferred on the device layer defining the slit and cover (step 2). The photoresist used is AZ1500, 38 cp, about 3 μm thick. The device Si layer is etched (step 3). The second mask pattern is transferred on the backside handle layer defining a part of the spiral heater (step 4). The center part is open. This mask is UV-cured. Then, the third mask pattern is transferred on the second mask pattern. This resist is not UV-cured keeping as

the normal resist. These two patterns connect each other making the spiral heater structure (step 5). The handle Si layer is etched (step 6). The normal resist layer is removed with the flush exposure and developing (step 7). Using the underlying UV-cured resist mask remained, the additional Si etching is carried out. This is for roughening Si surface at the center area of 3.15x3.15mm^2. IR emission from the micro-heater is enhanced since the surface reflection decreases with the sub-wavelength roughening. The surface roughness increases from 2 to 83 nm Ra (experimentally 1.3-times larger emissivity is observed at the wavelength of 650 nm). After removing the resist layer, the buried SiO$_2$ layer is removed releasing the heater structure (step 8). Figure 6 shows the fabricated micro-heater. The chip size is 6.5x6.5mm^2. The suspended heater is 150 μm wide and about 32 mm long for obtaining the thermal isolation. The whitish region corresponds to the roughened surface.

Grating is fabricated from Si substrate. The cycle etching of the deep RIE is not used for avoiding the scalloping. Plasma etching is carried out using SF$_6$ and C$_4$F$_8$ (56%) gases at the same time. Simultaneous C$_4$F$_8$ gas introduction is for obtaining the anisotropy. 160/10nm-thick Au/Cr layers are deposited on the grating. The photo of the grating is included in the inset of Fig. 4.

RESULTS

Figure 7 shows one micro-emitter assembled on TO-8 can. The grating and the micro-heater are simply stacked. The alignment accuracy is not required. The spacer shown in Fig. 4 is the polyimide tape (thickness: about 55 μm). For the electrical connection between the metal and Si material, wide Cu foil (not wire) is used with Ag paste (Dotite FA-545). This foil is then wired to TO-8 pin with soldering. For activating the micro-heater, the current flows from TO-8 pins. Since the current value is sub-A, the electrical contact condition or the interface resistance becomes important for not to increase the temperature outside of the micro-heater. The temperature is considered to be 300-400 °C.

Figure 8 shows FT-IR (Thermo Scientific, Nicolet 6700) emission spectra. Since the peak at λ=3.5μm is observed only when the micro-heater is combined with the grating, this peak is considered to correspond to SPP. The peak shape is rather trapezoid and the peak width for the single SPP is considered to be narrower. Many peaks are considered to gather together. The difference of the peak wavelength from the designed value is attributed to the different groove profile obtained actually. Since the groove becomes deeper for aiming the higher contrast compared to the previous study [10], its profile becomes important. The peak width is about 190nm and comparable with the bandwidth of CO$_2$ absorption at 4.2-4.3μm. When the input power increases from 0.3 to 1.9W, SPP related peak grows clearly.

Figure 9 shows the intensity at the wavelength of 3.2, 3.5, and 3.6μm as the function of the input power. Basically, the intensity increases against the input power. The increasing ratio of the intensity at the wavelength of 3.2 and 3.6μm indicates the blackbody emission under the

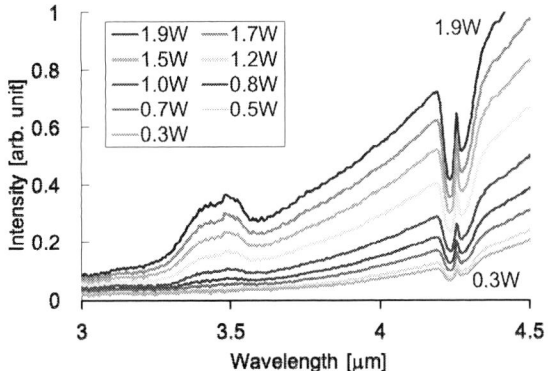

Figure 8: FT-IR spectra around SPP related peak changing the input power to the micro-heater.

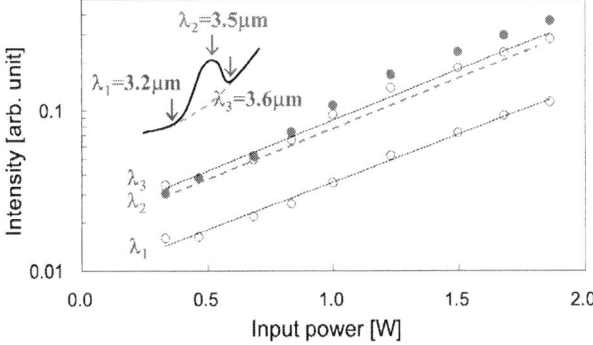

Figure 9: IR intensity at the wavelength of 3.2, 3.5, and 3.6μm as the function of the input power.

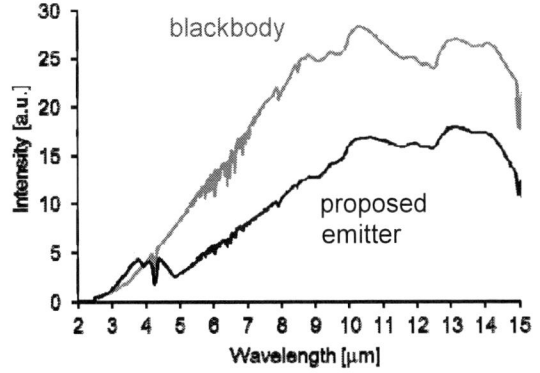

Figure 10: As-measured spectra over the wider wavelength region with (proposed emitter) and without the grating (blackbody).

temperature increase. When the input power is 0.6-1W, the peak intensity (3.5μm) increases steeper than those at the surrounding wavelengths (3.2 and 3.6μm). The increasing ratio over 1W is similar to those of the other wavelength. This is explained by the high contact resistance generating the heat and the blackbody emission from the front side mixing with the controlled emission.

Figures 10 show the as-measured spectra over the wider wavelength region with (proposed emitter) and without the grating (blackbody), respectively. The sample is different from one of Fig. 8. The change of the spectrum shape can be attributed to the grating quality at present. SPP related peak is around 3-5μm. Comparing two spectra at the longer wavelength over 5μm, the intensity of the black curve is suppressed for the proposed emitter. The output intensity at around 3-5μm wavelength is larger than that of the gray curve. As the total setup of the micro-emitter, SPP related wavelength is enhanced and other wavelength is suppressed.

ACKNOWLEDGEMENTS

A part of this work was supported by "Nanotechnology Platform Japan" of the Ministry of Education, Culture, Sports, Science and Technology (MEXT) and MEXT program for Forming Strategic Research Infrastructure (S1101028), Japan.

REFERENCES

[1]   E. Hecht, *Optics*, Chapter 13 (Addison Wesley, 2002).

[2]   P. W. Atkins, *Physical Chemistry* Sixth ed. Chapter 16 (Oxford University Press, 1998).

[3]   S. Alexsandrov, G. Gavrilov, A. Kapralov, S. Karandashov, B. Matveev, G. Sotnikova, N. Stus', "Portable Optoelectronic Gas Sensors Operating in the Mid-IR Spectral Range (λ=3-5 μm)", *Proc. SPIE* 4680 (2002) pp.188-194.

[4]   J.-J. Greffet , R. Carminati , K. Joulain , J.-P. Mulet , S. Mainguy and Y. Chen   "Coherent emission of light by thermal sources", *Nature*, 416 (2002) pp. 61-64.

[5]   H. Sai, Y. Kanamori, and H. Yugami, "Tuning of the thermal radiation spectrum in the near-infrared region by metallic surface microstructures", *J. Micromech. Microeng.* 15, (2005) pp. S243-S249.

[6]   H. T. Miyazaki, K. Ikeda, T. Kasaya, K. Yamamoto, Y. Inoue, K. Fujimura, T. Kanakugi, M. Okada, K. Hatade, and S. Kitagawa, "Thermal emission of two-color polarized infrared waves from integrated plasmon cavities", *Appl. Phys. Lett.* 92 (2008) 141114.

[7]   K. Masuno, S. Kumagai, M. Sasaki, "Reflection-type Wavelength Selective IR Emitter Using Surface Plasmon Polariton", *Optics Letters*, Vol. 36, No.3 (2011) pp.376-378.

[8]   A. Ordal, R. J. Bell, R. W. Alexander, Jr., L. L. Long, and M. R. Querry, "Optical properties of Au, Ni, and Pb at submillimeter wavelengths", *Appl. Opt.* 26 (1987) pp.744-752.

[9]   T. Okamoto, and K. Kajikawa, *Plasmonics*, Chapter 5 (Koudansya Scientific, 2010) (in Japanese).

[10] K. Masuno, T. Sawada, S. Kumagai, M. Sasaki, "Multi-Wavelength Selective IR Emission Using Surface Plasmon Polaritons for Gas Sensing", *IEEE Photon. Technol. Lett.* Vol. 23, No. 22 (2011) 1661-1663.

CONTACT

*M. Sasaki, tel: +81-52-809-1840; mnr-sasaki@toyota-ti.ac.jp

EFFECT OF NEEDLE SHAPE ON PERFORMANCE OF NEEDLE-TYPE ELECTRO TACTILE DISPLAY

Norihide Kitamura[1], Julien Chim[1], and Norihisa Miki[1,2]
[1]Keio University, Kanagawa, Japan
[2]JST PRESTO, Japan

ABSTRACT

This paper describes how the shape of micro-needle electrodes of an electro tactile display affects the tactile sensation. The electro tactile display can display the tactile sensations by stimulating tactile receptors with electric current. Micro-needle electrodes can drastically decrease the electrical impedance because the needles can penetrate through the stratum corneum which has higher impedance than dermis. When the tip radius of the needle is too small, the impedance between finger and micro-needles becomes large due to the small contact area. On the other hand, when the needle tip has large radius, the needle cannot go through the surface of fingers and the impedance does not decrease. In this work, we experimentally investigated the optimum shape of the micro-needle electrodes for electro tactile display applications.

INTRODUCTION

Tactile display, which can display tactile sensations by stimulating the tactile receptors inside our skin, is a promising information communication technology. In prior works, mechanical vibration is mainly utilized to stimulate the tactile receptors [1]. However, since several tens of micro meters deformation of skin is required, they have to use large actuators and thus, they have disadvantages of high electric power consumption, large size, and low-flexibleness of the device. In contrary, electro tactile displays that stimulate the receptors with electric current are advantageous in miniaturization and flexibility [2]. However, they still required several tens of V to stimulate tactile receptors (threshold voltage), although they are still in the low level compare to the mechanical devices, because stratum corneum, the surface layer of the skin, has high impedance of about 100 kΩm. In our previous report, to solve this problem, we changed the electrode shape from flat plates to micro-needles and attempted to stimulate the receptors from the inside of fingers by penetrating the stratum corneum, as shown Figure 1. Micro-needle arrays are mainly researched for the Brain Machine Interface (BMI) or drug deliveries and adjusted their length, which was about the corneum thickness to reduce the impedance without pains [3]. For BMI devices, several fabrication processes of micro-needle arrays were proposed: mechanical processing of silicon [4], using negative photoresists and micro lenses, and using soft lithography [5]. In these works, it was challenging to make needles with enough long needles on the soft substrates. Therefore, we used electrochemical etching for making needles. As a result, we achieved to confirm that the needle-electrode device is about 20 times superior to normal electrotactile devices which use flat electrodes at the threshold voltage of tactile sensations in terms of the threshold voltage [6]. Through our research, we realized that the shape of needle greatly affected the performance of an electro tactile display with a micro-needle electrode array. In this paper, we investigated the relationship between processing conditions and needle shape in the electrochemical etching for controlling the needle tip radii. Then, we revealed the optimum needle shape to stimulate tactile receptors by perception tests.

FABRICATION

The proposed micro-needle array consists of polydimethylsiloxane (PDMS) part and titanium wires and the device was fabricated by 2 steps: arraying titanium wires and processing the wires. Figure 2(A) shows the fabrication process of the wire array. (a) A wire holder was made by mechanical processing. The holder consists of 2 parts. The upper side had holes whose diameters were adjusted to the wire thickness for arrangement of wires. The lower side had a hole that was for adjusting the length of wires (initial length). (b) Titanium wires that have flat cut-end are bent and set to the holder. (c) PDMS was poured on the holder and baked with a hot plate. (d) After PDMS hardened and the wires were fixed, they were peeled off from the holder. Figure 2(B) shows the image of the wire etched to have the needle shape. Needles were etched with electrochemical etching which can etch titanium safely [7]. In the electrochemical etching, the distance between anodic terminal and cathodic terminal (etching distance) is an important factor to control the shape of needle because the etched quantity depends on the distance. Therefore, we controlled this distance by using PMMA spacer and controlled the initial length by using the wire holder.

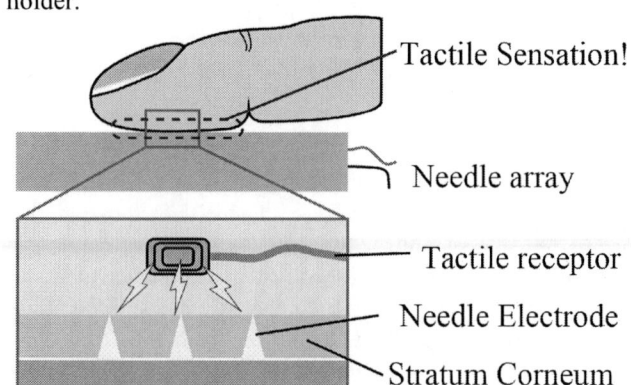

Figure 1: Schematic image of the electro tactile display with micro-needle electrode array. It uses needle type electrodes and stimulates tactile receptors from the inside of the skin.

978-1-4799-3510-9/14 $31.00 © 2014 IEEE

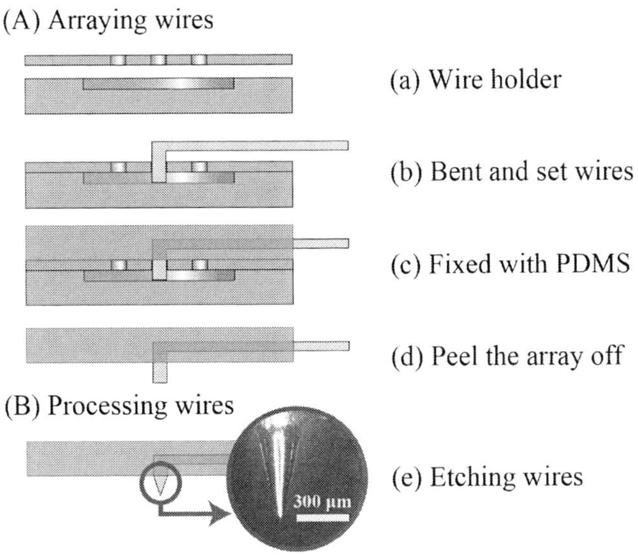

(A) Arraying wires

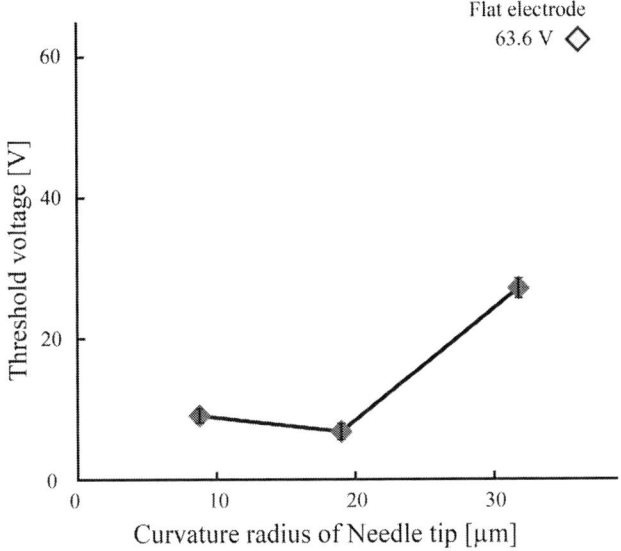

(a) Wire holder

(b) Bent and set wires

(c) Fixed with PDMS

(d) Peel the array off

(B) Processing wires

(e) Etching wires

300 μm

Figure.2 Device fabrication Process. (A)(a) Titanium wire Holder made by 2 parts. (b) Wires are set to the holder. (c) PDMS is poured on the holder. (d) The array peeled off from the holder. (B)(e) Wires are etched to the needles.

Figure.3 Results of Perception test with 600 μm needles which radii were different.

EXPERIMENTAL

We prepared a flat type electrode array and 3 micro-needle arrays 600 μm in length: the thickness of the stratum corneum at the finger tips, and they have different needle tip radii: 10, 20, and 30 μm, and conducted tactile stimulation experiment. Subjects placed and pushed their fingers on the micro-needle array and let the needles penetrate through the stratum corneum. Then, the positive square pulses are supplied by a function generator via a high speed bipolar power supply. Pulse voltages are gradually increased until subjects felt the tactile stimuli and record this voltage as the threshold voltage). Figure 3 shows threshold voltages of each micro-needle electrode array and flat electrode device. We can see the micro-needle electrode devices could drastically reduce the threshold voltages: from 64 V to 7 V. In addition, we revealed that the impedance increases both the needle tip is sharp and blunt, since too sharp needle has small contact areas and too blunt needles are not able to get enough penetration. In the final analysis, 20 μm can stimulate tactile receptors with lowest voltages and this is the optimum radius for the micro-needle electrode array for electrotactile display.

CONCLUSION

Electrochemical etching, one way of the fabrication process of micro-needle electrode array, was demonstrated. The needle shape that can stimulate with the lowest voltage was experimentally determined. In addition, the needles at that shape can stimulate in good stability. Therefore, the most efficient needle shape was presented.

ACKNOWLEDGEMENTS

This work was supported in part by JST PRESTO (Information Environment and Humans)

REFERENCES

[1] J. Watanabe, H. Ishikawa, X. Arouette, Y. Matsumoto, N. Miki, "Demonstration of vibrational braille code display using large displacement micro-electro-mechanical systems actuator", Japanese Journal of Applied Physics, Vol.51, 2012

[2] H. Kajimoto, N. Kawakami, T. Maeda, S. Tachi, "Electrocutaneous display as an interface to a virtual tactile world", *Proceedings IEEE Virtual Reality*, pp.289-290,2001

[3] P. Griss, P. Enoksson, H. Tolvanen-Laakso, P. Merilainen, S. Ollmar, G. Stemme, "Spiked biopotential electrodes", Journal of Microelectromechanical systems 10(9), 323-328, 2000

[4] R. A. Normann, P. K. Cambell, K. E. Jones, "Micromachined, silicon based electrode arrays for electrical stimulation of or recording from cerebral cortex", Proc. 1991 IEEE Int. Conf. on Micro Electro Mechanical Systems. , pp. 247–252, 1991

[5] Y. Ami, N. Miki, H. Tachikawa, N. Takano, "Formation of polymer microneedle arrays using soft lithography", J. Micro/Nanolith. MEMS MOEMS. 10(1), 011503, 2011

[6] N. Kitamura, J. Chim, N. Miki, "Micro-Needle Electrode Array for Electro Tactile Display", *Proc. Transduce*, pp.106-107, 2013

[7] T. Deguchi, "Electrolytic Etching Machining by Ethylene Glycol Solutions", *The journal of the Surface Finishing Society of Japan*,Vol.61, No.4, pp.305-306, 2010

CONTACT

*N. Kitamura, tel: +81-45-563-1141;
norihide.kitamura@z2.keio.jp

INCLINATION-INDEPENDENT TRANSFORMATION OF LIGHT BEAMS USING HIGH-THROUGHPUT UNIQUELY-CURVED MICROMIRRORS

Yasser M. Sabry[1,2], Diaa Khalil[2,3], Bassam Saadany[2] and Tarik Bourouina[1,2]

[1]Université Paris-Est, Laboratoire ESYCOM, ESIEE Paris, Noisy-le-Grand Cedex, France

[2]MEMS Division, Si-Ware Systems, Cairo, Egypt

[3]Electron. and Comm. Eng. Dept., Faculty of Eng., Ain-Shams University, Cairo, Egypt

ABSTRACT

We report a novel class of specifically-designed curved micromirrors enabling phase transformation of light beams independent of their inclination angle. The surface also exhibits a linear relationship between inclination angle and transversal displacement of the beam. The micromirrors were alkaline-free etched to depth levels of more than 300 μm, thus enabling high optical throughput. Measurements at both 675 and 1550 nm wavelengths show stable dimensions for the optical beam spot with less than ± 5% dependence on the inclination angle up to 60 degrees. The presented micromirros have applications in optical beam shaping and scanning, displacement/rotation sensing and imaging.

INTRODUCTION

Deeply-etched micromirros with circular, parabolic or elliptical cross sections were reported in literature for the purpose of making variable optical attenuation [1-2], increasing the coupling efficiency of tunable lasers [3] and improving the quality factor of optical cavities [4]. The reported micromirros have conventional surface profiles and are used to treat the optical beam only for a given inclination angle. Consequently, any variation in the inclination angle will lead to variation in the properties of the reflected beam. Moreover, the height of the etched silicon micromirrors is usually limited to 70-80 μm and, thus, standing as an obstacle for having high optical throughput, which is a critical parameter for high performance optical systems.

Recently, a micro-optical bench comprising acylindrical micromirror with 150-μm etching depth was introduced [5]. The profile of the acylindrical micromirror was generated in a specific way coupled to the motion of a MEMS actuator with a one degree of freedom. The system was optimized to minimize the above mentioned variation but not eliminating it all together, even theoretically. Deeper micromirros were not reported without the use of alkaline etching [6-7]. In the latter case, wet anisotropic etching yields very high surface quality and vertical surface close to ideal but on the expense of: 1) restricting only to flat optical surfaces arrangement and 2) prohibiting the deep etching of in-plane curved surface, which is of particular interest in this report.

The goal of this work is twofold. First, to introduce a new class of curved micromirros that are able to transform (i.e. collimate, focus, guide in free-space...) the incident optical beam in exactly the same way independent of the beam inclination angle on the micromirror surface. The second objective is to achieve a technological advancement by realizing deeper micromirrors height without restriction on their surface profile. This significantly improves the optical throughput of the MEMS micro-optical systems. The proper design of the micromirror curvature and the quality of deep etching are then verified by optical measurements.

UNIQUELY-CURVED ACYLINDRICAL MICROMIRRORS

First consider the transformation of a Gaussian beam (GB) after reflection on a surface, which is curved along one dimension. This allows to evaluate both cylindrical and acylindrical reflectors. For this purpose, consider an incident GB on a curved reflector with an angle of incidence θ_{inc} as shown in Fig. 1. In this case, the incidence plane is referred to as the tangential plane (t-plane) and the plan perpendicular to the tangential plan while containing the beam optical axis is referred to as the sagittal plane (s-plane). Let the reflector surface has a radius of curvature R in the t-plane while the surface cross section is flat in the s-plane. The incident GB undergoes a transformation in the t-plane described by the transformation matrix M given by Eq. (1) [8]:

$$M = \begin{bmatrix} 1 & 0 \\ \dfrac{2}{R\cos\theta_{inc}} & 1 \end{bmatrix} \qquad (1)$$

The corresponding focal length of the transformation is given by Eq. (2):

$$f = -0.5 R \cos\theta_{inc} \qquad (2)$$

The reflected GB waist radius w_{out} is related to the incident GB counterpart by Eq. (3) [9]:

$$w_{out} = \frac{w_{in}}{\sqrt{\left(1 - f^{-1}d_{in}\right)^2 + f^{-2}z_o^2}} \qquad (3)$$

where w_{in} is the waist radius of the incident GB, d_{in} is the distance between the location of the incident GB waist and the point of incidence and z_o is the Rayleigh range of the incident GB. The reflected beam waist radius as given by Eqs. (2) and (3), and consequently the divergence angle and the spot size, varies significantly as the inclination angle is changed. The presented solution is based on the design of a curved surface with locally-varying radius of curvature that is function of the inclination angle [10]. The radius of curvature has the unique proportionality with the reciprocal of the cosine of the inclination angle to counteract the latter in Eq. (2). Only this specific radius of curvature results in a constant focal length independent of the inclination angle.

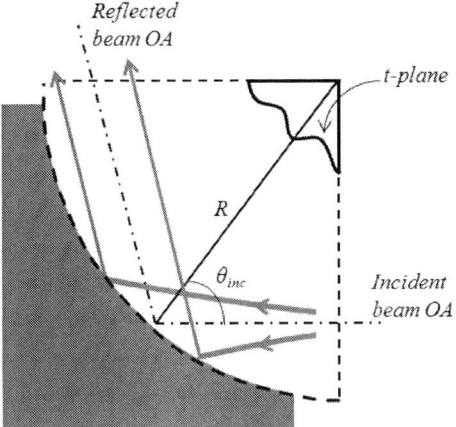

Figure 1: A Gaussian beam is incident on a curved reflector with an inclination angle θ_{inc} where the reflector radius of curvature is R. The tangential plane contains the incident as well as the reflected beam Optical Axis (OA) while the sagittal plane is normal to the tangential one and contains the beam OA.

Keeping the focal length of the curved surface constant and independent of the incidence angle is imposed by the introduction of a new acylindrical surface profile satisfying Eq. (4):

$$-0.5R(\theta_{inc})\cos(\theta_{inc}) = \text{constant} = f_o \qquad (4)$$

where $R = R(\theta_{inc})$ is now a function of the inclination angle, leading to an acylindrical shape for the mirror. Substituting for the radius of curvature of the surface and the cosine of the inclination angle in terms of the surface first and second derivatives, Eq.(4) can be rewritten as a second order non-linear differential equation:

$$2f_o y'' - y'^3 - y' = 0 \qquad (5)$$

The analytical solution of Eq. (5) was found and arranged in the expression in Eq. (6):

$$y = y_v - 2f_o \tan^{-1}\left(\exp\left[\frac{-x_v}{2f_o}\right] \sqrt{\exp\left[\frac{x}{f_o}\right] - \exp\left[\frac{x_v}{f_o}\right]} \right) \qquad (6)$$

where x_v, y_v are the curved surface vertex coordinates, and x and y are the Cartesian coordinate system. A comparison between the presented surface and the conventional cylindrical profile is given in Fig. 2 when $(x_v, y_v) = (0,0)$ for simplicity. The local radius of curvature of the surface has the following properties. It has its smallest value at the vertex of the surface and increases monotonically away from the vertex. It increases gradually with $|y|$ away from the vertex to about $|y| \sim 2f_o$ then it increases rapidly. The cosine of the inclination angle decreases gradually with $|y|$ such that multiplying the radius of curvature by the cosine of the inclination angle remains constant at any portion or point of the surface and is equal to $2f_o$. This perfectly

constant behavior is a unique feature of the invented profile. The inclination angle, and thus the reflection angle, varies linearly with the vertical displacement of the incident beam with respect to the horizontal axis. Indeed, this linear relation between the inclination angle and the displacement is ideal for optical scanning and sensing applications

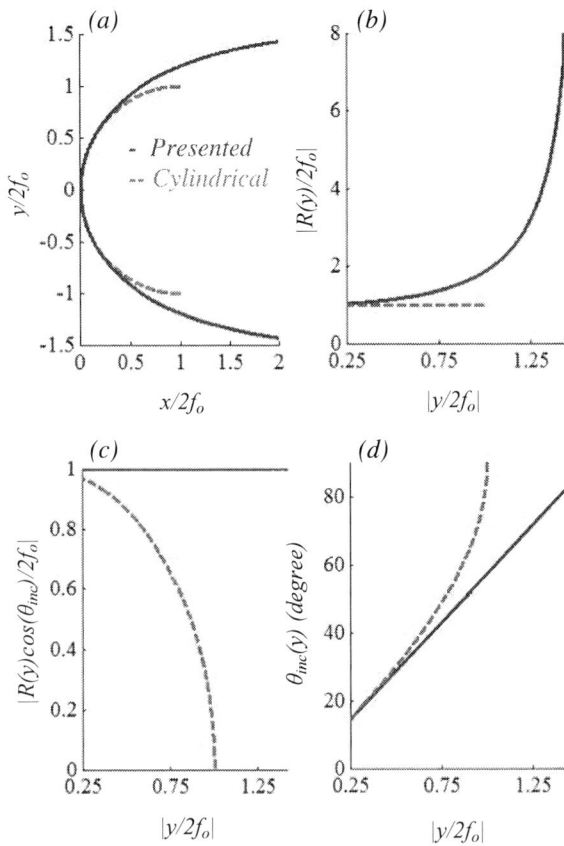

Figure 2: Comparisons between the profiles of the presented mirror surface and the conventional cylindrical one where f_o is the focal length and x/y are the Cartesian coordinate system. (a) Cross sectional mirror profile. (b) Local radius of curvature (c) Multiplication of the radius of curvature by the cosine of the inclination angle. (d) The inclination angle versus the distance between the axis of the beam and the axis of the mirror.

HIGH OPTICAL THROUGHPUT ETCHING

The optical throughput is a quantitative measurement of how much light energy gets through an optical instrument and transferred from a light source to a detector [11]. From an optical system point of view following a light source, the optical throughput is the area of the entrance aperture times the solid angle the source subtends as seen from the aperture. The throughput is a geometrical quantity that is invariant in the system and defined by the least optimized part of it. As a consequence of the limited micromirros size in micro-optical benches, deeply-etched micro-optical benches usually have low optical throughput. The low optical throughput of a micro-optical bench may be not large enough to serve an adequate signal to noise ratio in miniaturized sensors and instruments. In this report we achieved a technological advancement by etching micromirrors with deep reactive ion etching (DRIE) technology with a depth that is larger than 300 μm without the use of alkaline etching. We avoided the use of alkaline etching that can harm the designed curvature of curved micromirrors. Therefore, in addition to the coherent light sources, the micromirror is able to receive light from non-coherent sources that are usually emitting light with relatively large spot size. Moreover, the deeply-etched micromirros were successfully obtained with a sidewall angle whose deviation from the verticality is smaller than 0.1 degree. Followed by a smoothing step to reduce the scallops resulting from the DRIE, peak-to-peak surface roughness in the order of 50 nm was obtained forming a high-quality optical surface. A scanning electron microscope image of the overall structure of a fabricated curved micromirrors with the profile introduced in the previous section is shown in Fig. 3.

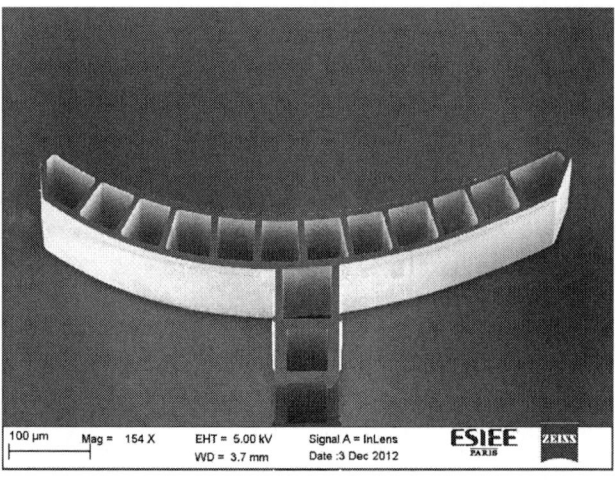

Figure 3: Scanning electron microscope image of the overall structure of a fabricated curved micromirror with the profile introduced in the previous section.

OPTICAL MEASUREMENTS

Optical measurements were carried out in order to characterize the scanned beam spot size and intensity profile versus different inclination angles. A 4/125 OZ Optics optical fiber was fed from a 675 nm laser source to generate a GB in the visible spectrum. The optical fiber was inserted on the substrate with its axis aligned to the principal axis of the curved micromirror. The fiber position was adjusted to have collimated reflected beam. Other positions are also possible, for example for having focused beam. The inclination angle between the optical beam and the surface of the micromirror was controlled by varying the displacement between the optical fiber axis and the principal

axis of the curved micromirror while keeping fixed distance between the optical fiber cleaved face and the micromirror surface. The reflected beam spot size and intensity profile were captured using DataRay Inc. BeamScope™-P8 scanning slit beam profiling system. A comparison between the resulting beam spot radius using the presented micromirror and the conventional cylindrical one is shown in Fig. 4, where the variations in the beam spot radius are minimal for the case of the presented micromirror.

Optical measurements were also carried out using a standard single-mode fiber fed from a laser source operating at 1550 nm. Indeed, the near infrared operation of a wide-angle optical scanner based on the presented micromirror has direct application in biomedical imaging, for example optical coherence tomography. The experiment was conducted for micromirror focal length f_o of 100, 200 and 400 μm and the results are shown in Fig. 5. The optical beam spot variation is less than $\pm$ 5% up to 60 degrees inclination angles.

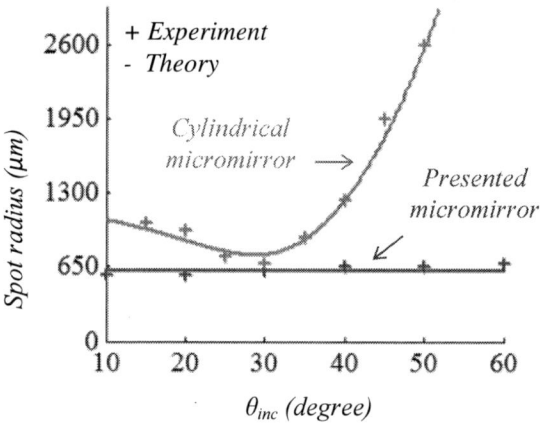

Figure 4: Reflected beam size versus inclination angle θ_{inc} at λ=675 nm using the presented micromirror and the conventional cylindrical profile.

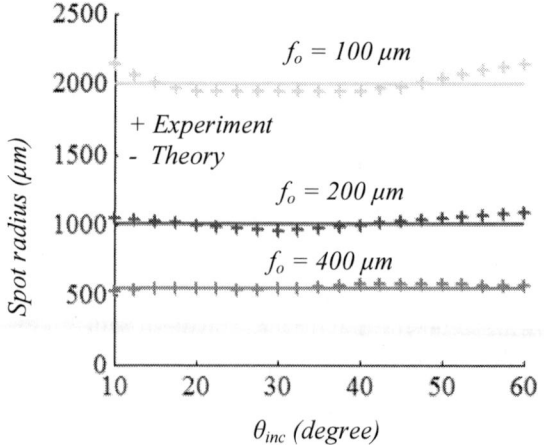

Figure 5: Reflected beam size versus inclination angle θ_{inc} using the presented micromirror at λ=1550 nm for different focal lengths

The measured normalized intensity profile of the beam along the transverse direction is shown in Fig. 6 for inclination angles of 10 and 50 degrees. The profiles show very good agreement with the theoretical Gaussian intensity profile highlighting the high optical quality of the presented curved micromirrors. Gaussian fitting of the data results in a root mean square error that is less than 3 %. The reflected beam spot is of elliptical type because the micromirror is flat in the out-of-plane direction with respect the silicon substrate. In fact, achieving curved profile in the out-of-plane direction is also possible based on plasma etching as recently reported [12-14].

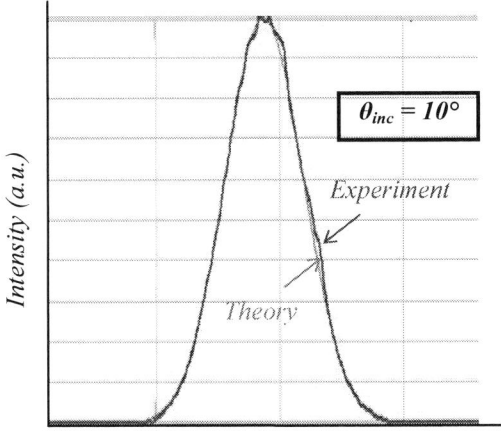

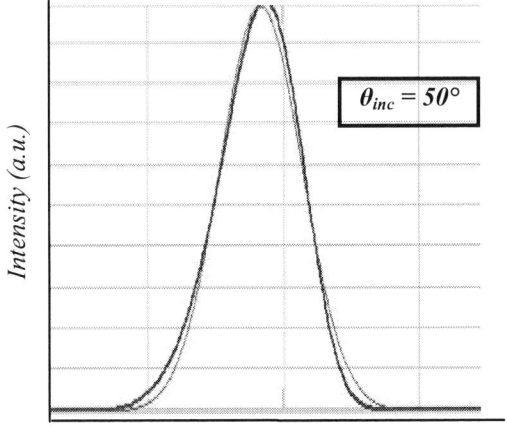

Figure 6: Reflected beam profiles for given inclinations (of 10 and 50 degrees) using the presented micromirror at λ=1550 nm.

REFERENCES

[1] H. Cai, X. M. Zhang, C. Lu and A. Q. Liu, "MEMS variable optical attenuator with linear attenuation using normal fibers", in *Digest Tech. Papers Transducers '05 Conference*, pp. 1171-1174.

[2] X. M. Zhang, A. Q. Liu, H. Cai and A. B. Yu, "Retro-reflection VOA using parabolic mirror for low insertion loss and linear attenuation relationship" in *Proc. MEMS 2007*, pp. 727-730.

[3] J. F. Tao, A. B. Yu, H. Cai, W. M. Zhu, Q. Z. Zhang, J. Wu, J., ... and A. Q. Liu, "Ultra-high coupling efficiency of MEMS tunable laser via 3-dimensional micro-optical coupling system" in *Proc. MEMS* 2011, pp. 13-16.

[4] M. Malak, N. Pavy, F. Marty, Y. Peter, A. Q. Liu and T. Bourouina, "Stable, high-Q Fabry-Perot resonators with long cavity based on curved, all-silicon, high reflectance mirrors" in *Proc. MEMS* 2011, pp. 720-723.

[5] Y. M. Sabry, D. Khalil, B. Saadany and T. Bourouina, "Integrated wide-angle scanner based on translating a curved mirror of acylindrical shape" Optic. Express, 21 (2013), pp. 13906-13916.

[6] R. Agarwal, S. Samson and S. Bhansali, "Fabrication of vertical mirrors using plasma etch and KOH: IPA polishing" *J. Micromech. Microeng.* vol 17, pp. 26-35, 2007.

[7] D. Lee, K. Yu, U. Krishnamoorthy and O. Solgaard, "Vertical mirror fabrication combining KOH etch and DRIE of (110) silicon" *J. Microelectromech. Syst.* vol. 18, pp. 217-227, 2009.

[8] A.E. Siegman, Lasers, University Science Books, Mill Valley, CA, 1986.

[9] P.F. Goldsmith, Quasioptical Systems (Chapman & Hall, 1998).

[10] Y. M. Sabry, D. A. M. Khalil, B. A. Saadany, and, T. E. Bourouina, "Aspherical optical surfaces and optical scanners," *U.S. patent application* 61676336, 2012.

[11] G. Brooker, Modern classical optics, Oxford University Press, vol. 8, 2003.

[12] Y. M. Sabry, B. Saadany, D. Khalil, and T. Bourouina, "Silicon micromirrors with three-dimensional curvature enabling lens-less efficient coupling of free-space light" *Light Sci. Appl.* vol 2, e94, 2013.

[13] Y. M. Sabry, T. E. Bourouina, B. A. Saadany, and D. A. M. Khalil, "Integrated monolithic optical bench containing 3-D curved optical elements and methods of its fabrication" U.S. patent application 20130100424, 2013.

[14] Y. M. Sabry, D. Khalil, B. Saadany and T. Bourouina, "Multi-step etching of three-dimensional sub-millimetre curved silicon microstructures with in-plane principal axis" *Microelectron. Eng.*, vol. 114, 2014.

CONTACT

*Y. M. Sabry, tel: +20-100-1834833; ysabry@ieee.org , yasser.sabry@si-ware.com

MAGNETOSTRICTIVE TYPE TACTILE SENSOR BASED ON METAL EMBEDDED POLYMER ARCHITECTURE

H.-C. Chang[1], W.-L Sung[1], H.-S. Hsieh[1], J.-H. Wen[2], C.-C. Fu[2], S.-C. Liao[3], C.-H. Lai[3], W.-C. Lai[4], C.-H. Chang[4], C.-P. Chang[4], C.-H. Chen[4], and W. Fang[1,2]

[1]Dept. of Power Mechanical Eng., [2]NEMS Inst., and [3]Dept. of Materials Science & Eng.
National Tsing Hua University, Hsinchu, TAIWAN
[4]WinMEMS Technologies Co., Ltd., Taoyuan, TAIWAN

ABSTRACT

This study presents new process scheme to fabricate polymer structure with embedded metal on silicon substrate. The primary merit of presented process scheme is: simple approach for the integration of 3D structures with different materials (e.g. metal, glass, polymer) on substrate. To demonstrate the feasibility of the proposed process scheme, a tactile sensor design consisting of polymer structure with embedded 3D Ni coil winding is implemented. As the polymer diaphragm deformed by tactile force, the magnetostriction effect of the 3D Ni coil inductor will induce the permeability change. Thus, the permeability change as well as the tactile force can be detected by the inductance variation. Preliminary measurements show the sensitivity of magnetostrictive type tactile sensor based on the proposed metal embedded polymer architecture is near 1.33%/N at the sensing range of 0~1N.

INTRODUCTION

In the past decade, various micromachining technologies have been successfully developed and further lead to many MEMS products penetrating into our daily life. Based on these frameworks, the integration of 3D structures with various materials on silicon substrate could increase the variety of MEMS devices and applications. For instance, a flexible tactile sensor could be realized using a planar coil attached to the polymer substrate [1], a micro coil could be embedded in the Si to produce magnetic field [2], and even a solid glass micro probe could be fused with Si via [3]. Moreover, numerous complex 3D metal structures have also been demonstrated, such as a magnetic core surrounded by a solenoid or a toroidal coil [4-6]. However, it remains challenging to integrate metal structures of complicated 3D shape with polymer or glass materials using the existing planar fabrication technologies.

In general, 3D structures could offer the characteristics of high aspect ratio, low impedance, and compact footprint. These characteristics are of useful for the development of electromagnetic components, such as coil winding, magnetic core or circuitry via. Based on the characteristics of 3D structures, a magnetostrictive type sensor could also take benefits from highly integration of coil and magnetic films to further enhance the performance. In the prior literatures, the magnetostriction effect [7] has been exploited for diverse sensor applications with the advantages of high gauge factor, such as strain gauge [8] and pressure sensor [9].

This study presents a novel process scheme to embed the electroplated 3D metal structures with polymer or glass by using the Si molding. The major steps of the proposed process scheme are: (1) implement the complicated 3D metal structures using the existing electroplating foundry; (2) implement a Si mold with multi etching cavities by DRIE; (3) easy assembly of the Si mold and 3D metal structure is achieved by design; (4) integration of the 3D metal structure and polymer housing is easily achieved by molding and de-molding. To demonstrate the feasibility of the proposed process scheme, a magnetostrictive tactile sensor is designed and implemented. The tactile sensor has a 3D Ni coil winding embedded inside a 3D PDMS structure with thin flexible diaphragm and thick rigid supporter.

DESIGN CONCEPT

In light of the motivation described above, a tactile sensor design consisting of the Ni coil winding and PDMS architecture is proposed. The Ni coil winding comprises of a planar coil, interconnection via, and two bonding pads, as shown in Fig.1a. The Ni coil winding has complicated 3D shape, and is embedded inside the PDMS structure except the exposed bonding pads. The PDMS structure, consisting of a thick supporting frame and a thin diaphragm, also has a complicated 3D shape. The Ni coil is arranged near the edge of the PDMS diaphragm. As the flexible diaphragm deformed by tactile force, the permeability change of Ni planar coil will be induced by the magnetostriction effect.

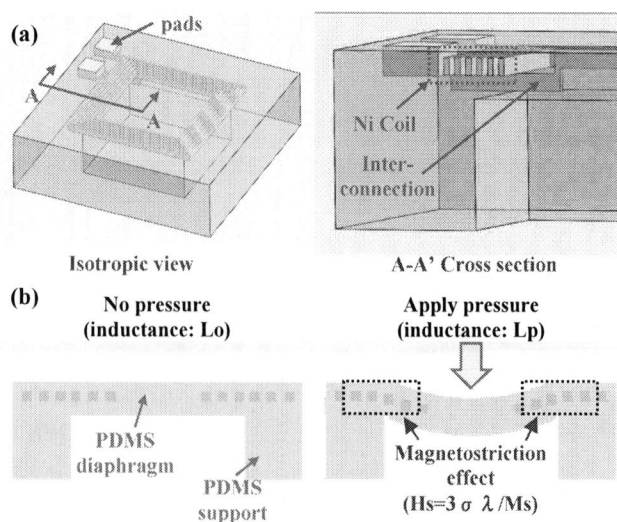

Figure 1: The design concept of this study (a) design schema; (b) sensing principle.

978-1-4799-3510-9/14 $31.00 © 2014 IEEE   1189

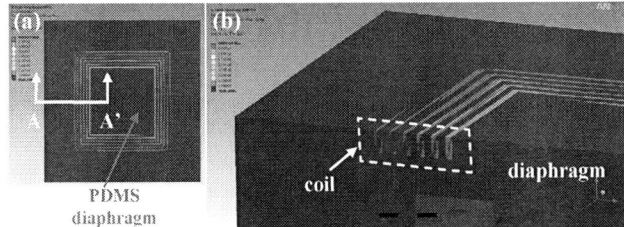

Figure 2: The simulation results of stress distribution on the membrane: (a) top view; (b) cross-section view.

Thus, the diaphragm deformation as well as the tactile force can be detected by the variation of inductance resulted from the permeability change of Ni planar coil, as in Fig.1b. In this regard, the stress distribution on the embedded magnetic Ni coil is also a critical design consideration. In order to enhance the magnetostriction effect, the embedded coil winding is arranged near the PDMS diaphragm edge with larger stress after tactile load. Fig.2 shows a typical FEM simulation results to predict the stress distribution of the coil embedded in PDMS diaphragm.

FABRICATION

Based on the design concept described above, the proposed fabrication process scheme is shown in Fig.3. The photoresist and thermal oxide layers are respectively deposited and patterned as the first and second etching masks for DRIE. As shown in Fig.3a, the Si mold with cavities of different depths could be fabricated after two sequential DRIE processes. The cavities will serve as the mold for the following PDMS molding. The first DRIE process would define the shape of the PDMS supporter, and the second DRIE process would define the thickness and shape of PDMS diaphragm.

Next, the 3D Ni coil supporting by Ni-frame (prepared by electroplating foundry) is assembled in the Si cavity, as shown in Fig.3b. Thickness of Ni structure is smaller than the depth of Si cavity to ensure the Ni structure will be fully embedded inside the PDMS after molding. Moreover, the planar dimensions of Si cavity are also slightly larger than that of Ni structure for the consideration of assembly tolerance. Note the Ni-frame acts as the temporary supporting structure for the Ni magnetic coil with complicated 3D shape. In addition, the supporting frame with a relatively regular shape (such as rectangle, square, or circle) is employed to ease the assembly process despite the complicated shape of 3D magnetic coil.

After that, the liquid PDMS is filled into the Si mold and then cured at 100°C for 90 minutes, as shown in Fig.3c-d. Vacuum chamber is employed to remove the bubbles trapped in PDMS before molding and during curing. As shown in Fig.3e, the etching process is employed to remove the PDMS to expose the bonding pads of the Ni coil winding. Moreover, the thickness of the PDMS diaphragm and the PDMS supporter are also defined in this process. Thus, the PDMS diaphragm with embedded 3D Ni coil winding is realized. As illustrated in Fig.3f, the supporting Ni-frame and the rest Si

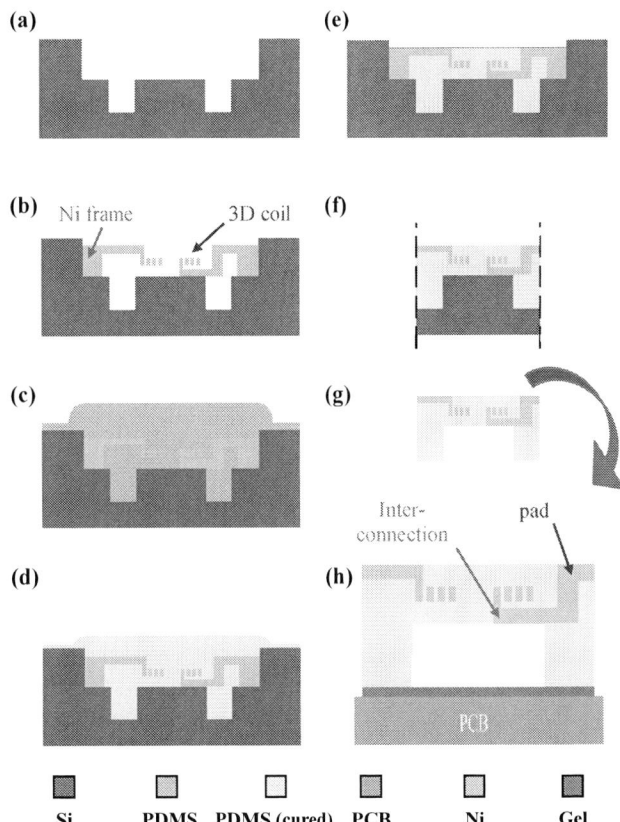

Figure 3: The fabrication process flow of magnetostrictive type tactile sensor based on metal embedded polymer architecture.

mold is cut by the dicing process. Thus, the supporting frame is separated with the magnetic coil, and the foot print of the tactile sensor is also determined. After the dicing process, the entire tactile sensor is released from the residual Si mold by KOH etching, as shown in Fig.3g. The etching selectivity between the PDMS and Si in KOH solution is very high, and thus the influence of KOH etching on PDMS supporter is ignored. Finally, the tactile sensor which comprises of the Ni coil winding embedded in the PDMS architecture will be adhered by the gel and wire bonding on the PCB for the following testing, as illustrated in the Fig.3h.

The SEM micrographs in Fig.4 show the typical electroplated 3D Ni coil winding on supporting frame which prepared by electroplating foundry. As shown in Fig.4a, the supporting frame, coil and pads are clearly observed. The zoom-in micrograph in Fig.4b further shows the magnetic coil anchor at the supporting frame. The micrograph in Fig.4c displays the bonding pads and magnetic coil which will be disconnected from the Ni frame for electrical isolation after dicing process. As indicated in the zoom-in micrograph in Fig.4d, the 3D Ni coil winding and supporting frame prepared by foundry consists of 5 electroplating layers. In short, the first and second layers are respectively the interconnection and via between the pads and coil. The magnetic coil is formed by the third layer. The forth layer is for the connection between the coil and frame, and the fifth layer acts as the bonding pads. Moreover, the supporting frame consists of these five layers. The micrographs in Fig.5 respectively show the Si mold for tactile sensor, and the trenches of micro Golden Gate Bridge on Si mold patterned by DRIE. The Si mold enables the formation of complicated 3D structure by molding. Note that the characteristics of filling materials (eg. viscosity) and molding conditions (eg. temperature) should be considered while designing the pattern of Si mold. Micrograph in Fig.6a displays the 3D Ni coil winding together with its supporting frame has been properly placed inside the Si mold. Fig.6b further shows the result after the PDMS filled into the Si mold to cover the 3D Ni structure. The result also indicates that bubbles have been properly removed from PDMS by vacuum chamber.

Micrographs in Fig.7a-b respectively display the front-side and back-side views of a typical fabricated tactile sensor. The tactile sensor has been removed from the Si mold. The Ni coil embedded in PDMS structure is observed in the front-side micrograph. It is also observed that the Ni frame has been removed by dicing process and thus the electrical isolation of magnetic coil is achieved. Moreover, the thickness variation between the PDMS diaphragm and supporter is also observed from the back-side micrograph. The packed sensor on PCB for the following tests is shown in Fig.7c. Micrograph in Fig.7d further displays another fully released PDMS structure with embedded Ni coil. The PDMS structure implemented using the Si mold in Fig.5b has the Golden Gate Bridge pattern on the flexible diaphragm.

MEASUREMENT

To demonstrate the presented concept, this study establishes the setup consisting of a force gauge, a position stage, and a LCR meter, as in Fig.8. The fabricated sensor is placed on the manually controlled position stage and the load applied on sensor is monitored by a commercial force gauge. As the measurement is performed, the position stage will be moved up step by step to gradually specify the load applied on the tactile sensor. Meanwhile, the inductance change induced by the applied load is detected by the LCR meter. Consequently, the variation between inductance change and tactile load is determined. Typical measurement results in Fig.9 show the variation of inductance change ($(Lp-Lo)/Lo$, in %) with applied load (in N) for the presented tactile sensor. The square dots depict the results for tactile sensor with a sensing diaphragm of $1584\mu m \times 1584\mu m$ and a Ni coil winding of 4 turns. As a result, the sensitivity of presented tactile sensor is 1.33%/N at the sensing range of 0~1N.

CONCLUSIONS AND FUTURE WORK

This study presents a novel fabrication scheme consisted of a 3D metal structure embedded in a 3D polymer architecture. Based on appropriate design of Si mold and 3D metal structure, various applications could be implemented in this process scheme. In application, a tactile sensor consisted of 3D Ni coil winding embedded in the 3D PDMS structure has been fabricated and verified. To demonstrate

the feasibility of the fabricated device, the variation of sensing inductance with the tactile load has been characterized. Measurements show the sensitivity is about 1.33%/N for the sensing range of 0~1N. Moreover, PDMS structure with complicated patter, such as micro Golden Gate Bridge, is also performed to demonstrate the polymer filling ability on Si mold.

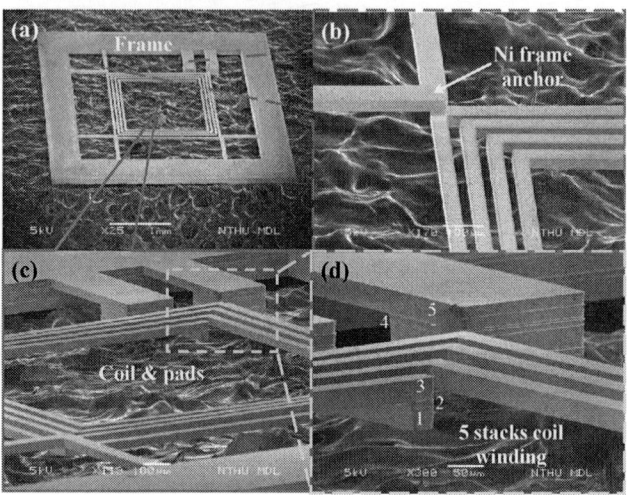

Figure 4: SEM photos of 3D Ni coil winding (a) entire view; (b) frame anchor; (c) coil and pads; (d) multi-stacks.

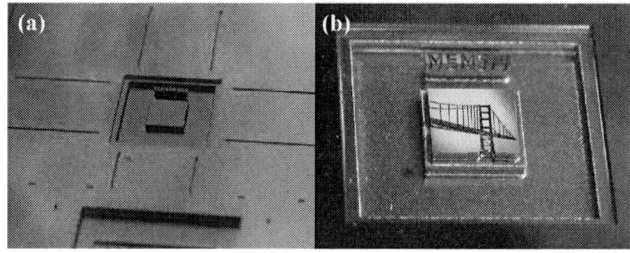

Figure 5: Micrographs of two different silicon molds respectively for tactile sensors consisting (a) flat diaphragm, and (b) diaphragm with micro Golden Gate Bridge pattern.

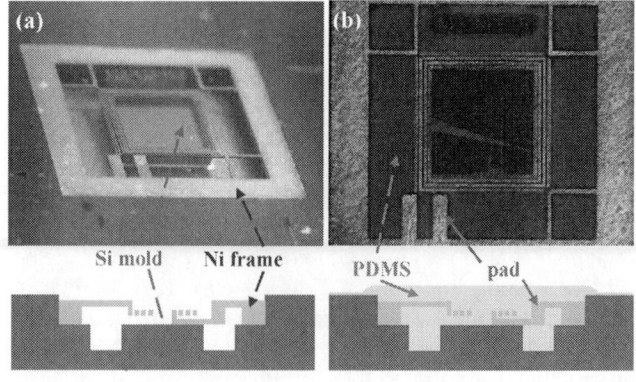

Figure 6: Micrographs of Ni coil winding embedded PDMS architecture (a) tactile sensor; (b) micro Golden Gate Bridge.

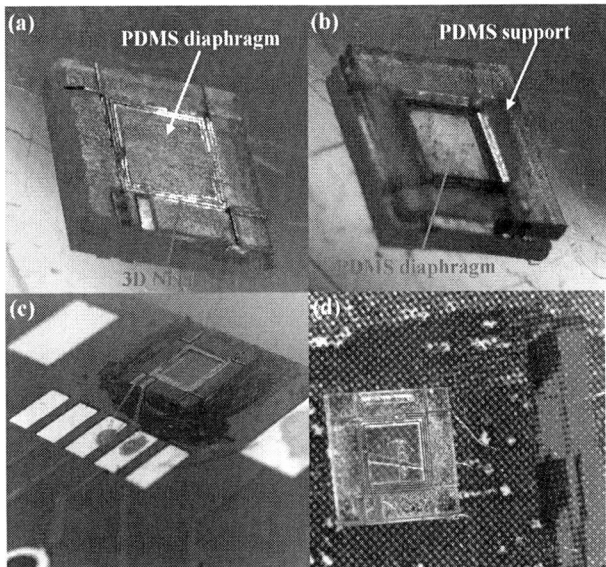

Figure 7: Micrographs of fabricated device (a-b) front & backside view; (c) on test PCB; (d) Micro Golden Gate Bridge.

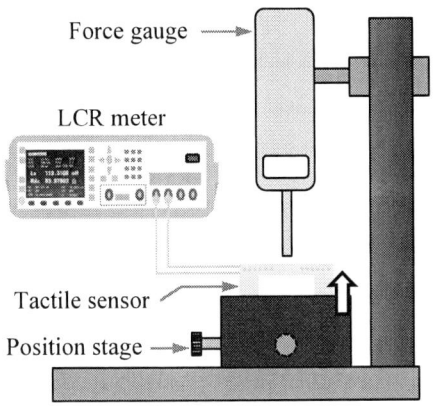

Figure 8: The schema of measurement setup (consists of force gauge, position stage, and LCR meter).

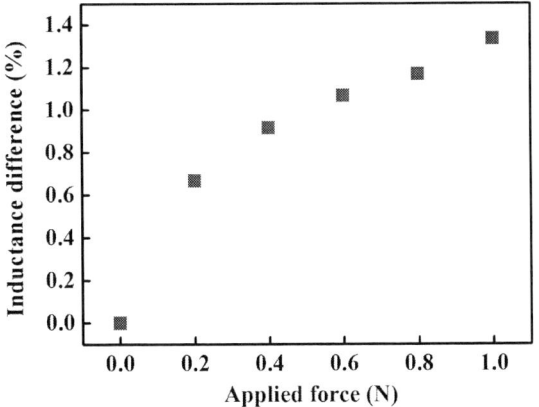

Figure 9: The sensitivity curve of magnetostrictive type tactile sensor based on metal embedded polymer architecture.

In summary, the concept of this study could be further extended to embed numerous micro 3D structures into micro devices of different materials (such as epoxy, glass) by molding on Si mold. Further investigation will be focused on the glass architecture implemented on this fabrication platform due to the higher rigidity of glass after solidification.

ACKNOWLEDGEMENTS

The authors would like to thank WinMEMS Technologies Co., Ltd. for the support of electroplating. This work was supported in part by the National Science Council of Taiwan under the grant of NSC-102-2218-E-007-003, NSC 102-2622-E-007-014-MY3, and the grant for "the Toward World-Class University Project from National Tsing Hua University, Taiwan".

REFERENCES

[1] S. Wattanasarn, K. Noda, K. Matsumoto, I. Shimoyama, "3D flexible tactile sensor using electromagnetic induction coils", *IEEE MEMS 2012*, pp. 488-491, 2012.

[2] Q. Ramadan, V. D. Samper, D. P. Puiu, C. Yu, "Fabrication of three-dimensional magnetic microdevices with embedded microcoils for magnetic potential concentration", *IEEE J. MEMS*, 15, no. 3, pp. 624-638, 2006.

[3] C.-W. Lin, C.-W. Chang, Y.-T. Lee, R. Chen, Y.-C. Chang, W. Fang, "Glass microprobe with embedded silicon vias for 3D integration", *IEEE MEMS 2009*, pp. 200-203, 2009.

[4] C. H. Ahn, M. G. Allen, "Micromachined planar inductors on silicon wafers for MEMS applications", *IEEE Trans. on Industrial Electronics*, 45, no. 6, pp. 866-876, 1998.

[5] X. Yu, M. Kim, F. Herrault, C.-H. Ji, J. Kim, M. G. Allen, "Silicon-embedded 3D toroidal air-core inductor with through-wafer interconnect for on-chip intergration", *IEEE MEMS 2012*, pp. 325-328, 2012.

[6] B. Fernández, J. Kubby, "Design, processing and materials for large-stroke actuators", *Proc. of SPIE*, 6467, 64670T, 2007.

[7] C. S. Schneider, P. Y. Cannell, K. T. Watts, "Magnetoelasticity for large stresses", *IEEE Transactions on Magnetics*, 28, pp. 2626-2631, 1992.

[8] A. B. Amor, T. Budde, H. H. Gatzen, "A magnetoelastic microtransformer-based microstrain gauge", *Sensors and Actuators A*, 129, pp. 41-44, 2006.

[9] H.-C. Chang, S.-C. Liao, H.-S. Hsieh, S.-J. Lin, C.-H. Lai, R. Chen, W. Fang, "A novel inverse-magnetostrictive type pressure sensor with planar sensing inductor," *IEEE MEMS 2013*, pp. 685-688, 2013.

CONTACT

Weileun Fang, Dept. of Power Mechanical Engineering, National Tsing-Hua University, Hsinchu, Taiwan, Tel:+886-3-5742923, E-mail: fang@pme.nthu.edu.tw

A MULTI-MATERIAL Q-BOOSTED LOW PHASE NOISE OPTOMECHANICAL OSCILLATOR

Turker Beyazoglu, Tristan O. Rocheleau, Karen E. Grutter, Alejandro J. Grine,
Ming C. Wu and Clark T.-C. Nguyen
University of California, Berkeley, USA

ABSTRACT

A Radiation Pressure driven Optomechanical Oscillator (RP-OMO) comprised of attached concentric rings of polysilicon and silicon nitride has achieved a first demonstration of a mixed material optomechanical device, posting a mechanical Q_m of 22,300 at 52 MHz, which is more than $2\times$ larger than previous single-material silicon nitride devices [1]. With this Q_m, the RP-OMO exhibits a best-to-date phase noise of -125 dBc/Hz at 5 kHz offset from its 52-MHz carrier—a 12 dB improvement from the previous best by an RP-OMO constructed of silicon nitride alone [1]. The key to achieving this performance is the unique mechanical Q-boosting design where most of the vibrational energy is stored by the high-Q_m polysilicon inner ring which in turn boosts the overall Q_m over that of silicon nitride, all while retaining the high optical Q_o >190,000 of silicon nitride material. Simultaneous high Q_o and Q_m reduces the optical threshold power for oscillation, allowing this multi-material RP-OMO to achieve its low phase noise with an input laser power of only 3.6 mW.

INTRODUCTION

Recent advancements in cavity opto-mechanics have allowed researchers to exploit coupling between the optical field and mechanical motion of an optical cavity to affect cooling [2] or amplification [3] of mechanical motion. Cooling the mechanical motion of micro-scale objects has been of high scientific interest, since it facilitates observation and exploration of certain quantum phenomena, e.g., the standard quantum limit of detection [4]. On the other hand, amplification of the mechanical motion allows realization of micro-scale devices for practical applications, such as light-driven low-phase noise signal generation by radiation pressure driven optomechanical oscillators (RP-OMO's) [3] [5].

Indeed, the ability to achieve self-sustained oscillation with no need for feedback electronics makes an RP-OMO compelling for on-chip applications where directed light energy, e.g., from a laser, is available to fuel the oscillation. In addition to stand-alone oscillator applications, RP-OMO's have been suggested for deployment as combined mixer+oscillators in homodyne receivers and RF sub-carrier links [5], and as reference/microwave oscillators to reduce power consumption in chip scale atomic clocks (CSAC) [1].

To be useful in such applications, the output of an RP-OMO must be sufficiently stable, as gauged over short time spans by its phase noise. To date, the work of [1] achieves the best in class phase noise for such devices of -113 dBc at a 5-kHz offset from a 74-MHz carrier by maximizing the mechanical Q_m of its optomechanical structure—a result of

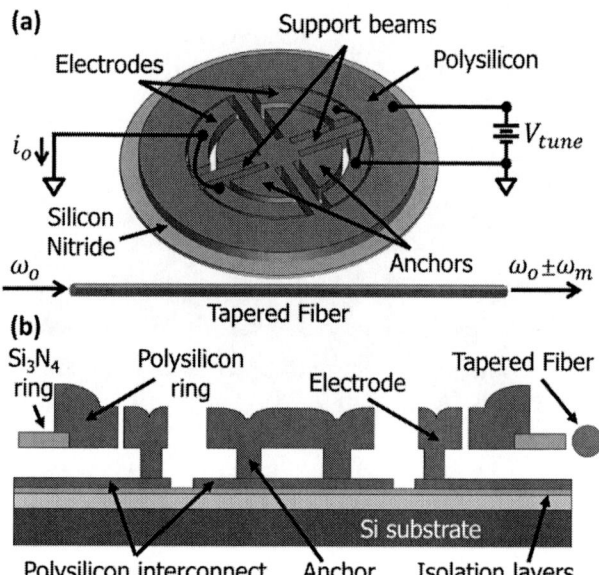

Fig. 1: (a) Perspective-view and (b) cross-sectional schematics of the Q-boosted RP-OMO. Here, the polysilicon inner ring is mechanically coupled at its outer edge to a concentric high optical Q_o silicon nitride ring. A tapered fiber provides optical coupling, while polysilicon electrodes inside the ring enable frequency tuning and electrical input-output.

recognizing that mechanical Q_m has the strongest impact on phase noise, much more than optical Q_o. However, the performance of [1], although good, is still not sufficient, mainly because it uses a single material (silicon nitride) to set both its mechanical and optical Q's.

The RP-OMO described in this work (*cf.* Fig. 1) circumvents this limitation by combining a nitride optical material with a lower mechanical loss polysilicon material that shares its energy to effectively boost the overall mechanical Q_m from 10,400 for a nitride device alone to 22,300. As a result of its high mechanical Q_m, the RP-OMO posts a phase noise of -125 dBc/Hz at 5 kHz offset from its 52-MHz carrier, which is 12 dB better than the previous state-of-the-art RP-OMO constructed of silicon nitride alone [1]. The doped polysilicon structure and electrodes additionally allow tuning of the RP-OMO's oscillation frequency via DC voltage as indicated by V_{tune} in Fig. 1(a), enabling future deployment of the multi-material RP-OMO as a locked oscillator in the targeted CSAC application [1] depicted in Fig. 2.

DEVICE STRUCTURE AND OPERATION

The Q-boosted RP-OMO, summarized in perspective-

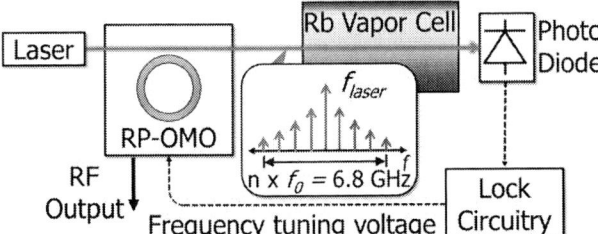

Fig. 2: Targeted CSAC application where an RP-OMO's higher harmonic locks to a Rb vapor cell to borrow its long term stability. Voltage controlled tunability of the RP-OMO provides a simple feedback mechanism for locking where the tuning voltage emanates from locking circuitry.

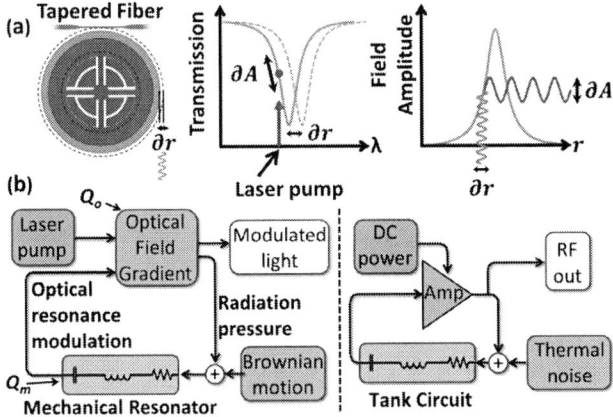

Fig. 3: (a) Optomechanical oscillator dynamics: Radiation pressure from light in the cavity changes the radius which in turn changes the optical field, raising the radiation pressure, and so on, to generate a growing cycle. (b) System block diagrams comparing an RP-OMO with an electronic oscillator. The dynamics of the RP-OMO is analogous to an electronic oscillator where the optical field (with the high Q_o resonance) sets the gain and the mechanical resonator serves as the tank circuit feedback element.

view and cross-section in Fig. 1, comprises a high mechanical Q_m polysilicon inner ring physically attached at its outer edge to a concentric high optical Q_o (but comparatively low mechanical Q_m) silicon nitride ring. Spokes attached to the inner edges of the polysilicon ring extend radially inwards to a common central anchor and serve to support the entire multi-ring device in a completely balanced fashion, where inward forces along the spokes are met with equal and opposite ones, cancelling energy leakage from the spokes to the substrate. Polysilicon electrodes inside the ring overlap its inner edge to form capacitive gaps that then allow electrical interrogation and control (in addition to optical).

To operate the device, an input laser is blue-detuned, i.e., at a wavelength slightly shorter, from the optical resonance of the nitride ring and coupled into the ring via a tapered fiber [6]. Enhanced by the Q_o, the circulating light generates a radiation pressure force that displaces the mechanical resonator which in turn shifts the optical resonance. As depicted in Fig. 3(a), initially Brownian mechanical mo-

tion modulates the optical pump field, which in turn generates a resonant radiation pressure force that modifies the mechanical dynamics. The coupling of the two degrees of freedom is described by the differential equations [5]:

$$\ddot{r}(t) + \Gamma_m \dot{r}(t) + \omega_m^2 r(t) = \frac{F_{rp}(t)}{m_{eff}}$$
$$= \frac{1}{m_{eff}} \frac{2\pi n}{c} |A(t)|^2 \tag{1}$$

$$\dot{A}(t) + A(t)\left[\frac{\omega_o}{2Q_L} - i\Delta\omega + i\frac{\omega_o}{r_o}r(t)\right] = i\sqrt{\frac{\omega_o}{Q_e}}|S|^2 \tag{2}$$

where $r(t)$ is the radial displacement of the mechanical resonator from equilibrium, Γ_m is the mechanical damping rate, ω_m is the mechanical resonance frequency, n is the effective refractive index for the optical mode, c is the speed of light, $A(t)$ the optical field circulating in the optical cavity, $\Delta\omega$ the detuning of laser from optical resonance frequency ω_o, $|S|^2$ the input optical power, Q_L the loaded quality factor of the optical resonance, Q_e the quality factor associated with coupling loss, and m_{eff} the mode dependent effective mass of the mechanical resonator.

Effectively, the resonant radiation pressure force modifies the mechanical dynamics by acting as a negative mechanical damping that completely cancels out the intrinsic mechanical loss when the circulating optical power reaches a threshold value. From a feedback loop perspective, the optical pump power and Q_o sets the gain from the mechanical motion to the radiation pressure which gets positively fed back to the mechanical resonator as depicted in Fig. 3(b), which further shows how the RP-OMO is in fact not so different from a conventional MEMS oscillator. Indeed, in both cases, the MEMS resonator in positive feedback serves as an ultra-high-Q bandpass biquad that accentuates the signal at resonance while suppressing noise off resonance. In this regard, high Q_m is of utmost importance if either system is to exhibit low close-to-carrier phase noise as predicted by the well-known Leeson's equation [7]:

$$L(f) \cong 10\log\left[\frac{2FkT}{P_{sig}}\left(1 + \frac{1}{Q^2}\left(\frac{f_c}{2\Delta f}\right)^2\right)\right] \tag{3}$$

where $L(f)$ is the single side-band phase noise at an offset Δf from the carrier frequency f_c. F is a fitting parameter often termed as effective noise figure, k is the Boltzmann's constant, T is the absolute temperature, and P_{sig} is the output power of the oscillator having a tank-circuit element with quality factor Q, which is the mechanical quality factor for the case of an RP-OMO.

Meanwhile, the Q_o of the structure governs the optical field gradient that in turn sets the loop gain of the system, so must be at least high enough to initiate self-sustained oscillation. Here, the silicon nitride component of the Q-boosted RP-OMO provides a high-Q_o optical cavity that supports a whispering gallery mode resonance in which the optical field propagates along the silicon nitride ring's circumference. For maximum optical Q_o, the optical mode must not overlap with potential sources of optical loss, which dictates

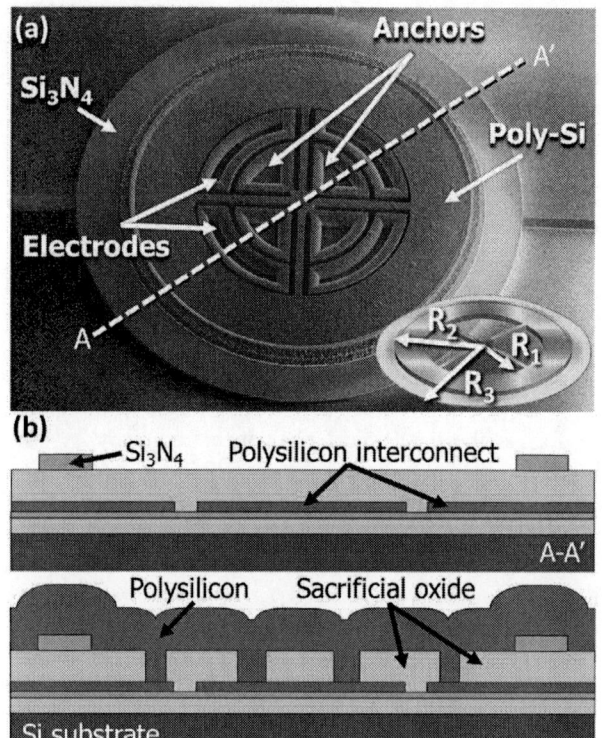

Fig. 4: *(a) Colorized SEM image of the Q-boosted RP-OMO with an inset of mode shape by FEM simulation. R_1 and R_3 are the inner radius of polysilicon and outer radius of Si_3N_4 rings, respectively. R_2 represents the outer radius for polysilicon and inner radius for Si_3N_4 rings where both are coupled. (b) Summary of the fabrication process flow in which LTO and Si_3N_4 layers are deposited for electrical isolation and etch stop followed by polysilicon interconnect deposition and etch. Another LTO layer is deposited and CMP'ed to a final thickness of 2 μm, leaving a planar surface for the 500 nm Si_3N_4 film. After an anchor etch step, 2 μm of polysilicon is deposited and etched stopping on LTO or Si_3N_4. Finally, devices are released in 49% HF, yielding the final cross-section of Fig. 1(b).*

a minimum distance between the scatter-prone polysilicon-nitride attachment interface and the outer edge of the nitride ring. On the other hand, for maximum mechanical Q_m (as will be seen), the width of the nitride ring should be minimized relative to that of the polysilicon one. The nitride ring width thus serves as a design parameter through which RP-OMO performance can be optimized.

MECHANICAL *Q*-BOOSTING

Again, the key to the phase noise performance obtained here is the high mechanical Q_m; and the key to the high Q_m is a concept introduced in [8] dubbed *Q*-boosting. *Q*-boosting is a mechanical circuit-based approach where a high-Q resonator raises the functional Q of a low-Q resonator in a mechanically coupled system by sharing its energy while adding relatively no loss [8]. In the multi-material RP-OMO, a higher Q_m polysilicon ring effectively supplies the added energy to a low Q_m (but high Q_o) nitride ring, where

both vibrate together in the breathing contour mode shape depicted in the inset of Fig. 4(a).

Neglecting the loss at the nitride-polysilicon interface and possible change in the structure's anchor loss due to coupling of two materials, the functional mechanical $Q_{m,tot}$ of the composite structure can be expressed as:

$$Q_{m,tot} = \omega_m \frac{KE_{SiN} + KE_{pSi}}{E_{lost/cycle}} \quad (4)$$

where $E_{lost/cycle}$ is the total mechanical loss per cycle in the polysilicon and silicon nitride rings; and KE_{SiN} and KE_{pSi} are their respecitve kinetic energies, given by:

$$KE_{SiN} = \frac{1}{2} \cdot m_{SiN} \cdot V_{R_2}^2$$
$$KE_{pSi} = \frac{1}{2} \cdot m_{pSi} \cdot V_{R_2}^2 \quad (5)$$

where V_{R_2} denotes the radial velocity at radius R_2, m_{SiN} and m_{pSi} are effective lumped masses of the silicon nitride and polysilicon rings at the coupling location, respectively, given by $m_{eff} = 2U/(\omega_m^2 \cdot \Re^2(R_2))$ with U being total stored energy in the mechanical mode, and $\Re(r)$ being radial displacement amplitude at radius r. Using (5) in (4) the functional $Q_{m,tot}$ simplifies to:

$$Q_{m,tot} = Q_{m,pSi} \frac{1 + \frac{m_{SiN}}{m_{pSi}}}{1 + \frac{m_{SiN}}{m_{pSi}} \cdot \frac{Q_{m,pSi}}{Q_{m,SiN}}} \quad (6)$$

which shows that the total $Q_{m,tot}$ of the RP-OMO structure depends on the Q_m and effective mass of both structures.

EXPERIMENTAL RESULTS

Fig. 4(a) presents a colorized SEM image of a fabricated *Q*-boosted RP-OMO together with the fabrication process. The doped polysilicon mechanical structure and inner capacitive gap electrodes are anchored and electrically connected to a thin layer of conductive polysilicon patterned on the substrate to serve as interconnects that facilitate electrical interrogation and read-out. The electrodes additionally allow tuning of the RP-OMO's oscillation frequency (such as needed for CSAC application [1]) via well-known voltage-controllable electrical stiffness [9].

Fig. 5(a) shows the experimental setup used to characterize the RP-OMO that basically employs the custom-built vacuum probe system of [1]. Measurement of Brownian noise shown in Fig. 5(b) reveals a multi-material RP-OMO boosted Q_m of 22,300, which is more than 2× higher than demonstrated in a previous silicon nitride RP-OMO [1]. To gauge the degree to which (6) matches the measured $Q_{m,tot}$ requires knowledge of the $Q_{m,pSi}$ of a spoke-supported polysilicon ring and the $Q_{m,SiN}$ of an unsupported nitride ring. The former is readily measured to be on the order of 48,000 on actual polysilicon spoke-supported rings operating in their first radial-contour modes [10]. The Q_m of an unsupported nitride ring, on the other hand, is much more elusive, since any real fabricated nitride ring does have supports, so

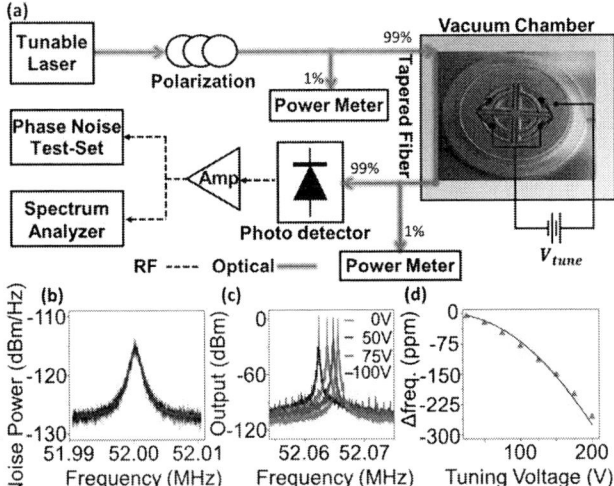

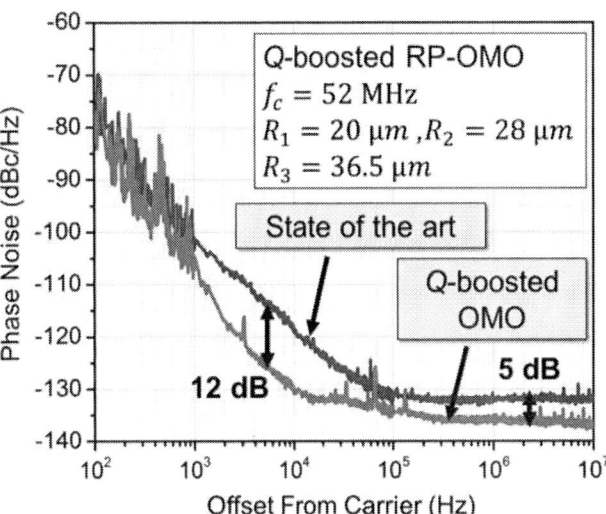

Fig. 5: (a) Schematic description of the experimental measurement setup. The RP-OMO is characterized in a custom-built vacuum chamber as described in [1]. An Agilent E5505A phase noise test system is used for phase noise measurements. (b) Measured Brownian motion of the RP-OMO from which Q_m=22,300 is extracted (c)-(d) demonstrate frequency tuning vs. applied tuning voltage and also (via curve-fitting) indicate a 440 nm resonator-to-electrode gap spacing.

suffers from anchor loss not present in an unsupported (levitated) ring. One reasonable approximation, however, might be the highest Q_m of 10,400 measured among several fabricated spoke supported 1st radial-contour mode nitride rings [1] at the frequency of interest. With the above Q_m values and 2.51 ng nitride and 5.91 ng polysilicon effective masses calculated from the device dimensions given in Fig. 6, Eq. (6) predicts a $Q_{m,tot}$ of 23,100 for the composite RP-OMO which agrees well with the measured value of 22,300.

Fig. 5(c) and (d) present RP-OMO output spectra under several tuning voltages and measured plots gauging oscillating RP-OMO frequency versus tuning voltage, where a relatively large 440 nm electrode-to-resonator gap spacing still allows a 3 ppm/V frequency shift suitable for locking to the Rb vapor cell in a CSAC.

Fig. 6 presents the measured phase noise for the Q-boosted RP-OMO of -125 dBc/Hz at 5 kHz offset from its 52-MHz carrier, which is 12 dB better than the previous state of the art RP-OMO constructed of silicon nitride alone [1], despite the use of an input laser power of only 3.6 mW—more than 2× smaller than that of the previous state-of-the-art [1].

CONCLUSIONS

A multi-material RP-OMO structure has been shown to boost the Q_m of a silicon nitride RP-OMO by more than 2× toward realization of the simultaneous high Q_m >22,000 and Q_o >190,000 needed to maximize RP-OMO performance. The Q-boosted RP-OMO bests the previous state-of-the-art by reducing the phase noise at 5 kHz offset from the carrier by a measured 12 dB that matches the prediction of Eq. (3) with the improved Q_m. The design is shown to have little or

Fig. 6: Phase noise spectra of the Q-boosted RP-OMO compared to the previous best Si_3N_4-only RP-OMO [1]. As expected, the enhanced Q_m lowers the phase noise, achieving a 12 dB improvement at 5 kHz offset.

no effect on the optical properties of the high Q_o silicon nitride, allowing retention of high Q_o despite the introduction of a scatter-prone material interface in the vicinity of the optical resonance. While polysilicon is chosen for its high Q_m in the RP-OMO of this work, the design is applicable to any material of choice as long as it can be integrated with another high Q_o material of choice. The use of high Q_m doped polysilicon as one of the materials further enables electrical interrogation and readout of the RP-OMO, as well as an electrical stiffness-based voltage controlled frequency tuning very much needed for locking in a target low-power CSAC application [1].

Acknowledgement: This work was supported under the DARPA ORCHID program.

REFERENCES

[1] T. O. Rocheleau et al., *Proceedings, IEEE Int. Conf. on MEMS*, 2013, pp. 118-121.

[2] A. Schliesser et al., *Phys. Rev. Lett.*, vol. 97, no. 24, pp. 243905(4), 2006.

[3] H. Rokhsari et al., *Opt. Express*, vol. 13, no. 14, pp. 5293-5301, 2005.

[4] T. J. Kippenberg et al., *Science*, vol. 321, pp. 1172-1176, 2008.

[5] M. Hossein-Zadeh et al., *IEEE J. Sel. Topics Quantum Electron.*, vol. 16, no. 1, pp. 276-287, 2010.

[6] J. C. Knight et al., *Opt. Lett.*, vol. 22, no. 15, pp. 1129-1131, 1997.

[7] D. B. Leeson, *Proc. IEEE*, vol. 54, pp. 329-330, 1966.

[8] Y. W. Lin et al., *in Dig. of Tech. Papers Transducers'07*, pp. 2453-2456.

[9] H. Nathanson et al., *IEEE Trans. Electron Devices*, vol. 14, no. 3, pp. 117-133, 1967.

[10] S.-S. Li et al., *Proceedings, IEEE Int. Conf. on MEMS*, 2004, pp. 821-824.

NOVEL TUNABLE OPTICAL MODULATION LENS USING MAGNETORHEOLOGICAL EFFECT

Fu-Ming Hsu[1], Rongshun Chen[1], and Weileun Fang[1,2]

[1]Power Mechanical Engineering dept., [2]NEMS Inst., National Tsing Hua University, HsinChu, TAIWAN

ABSTRACT

This study extends the fluid dispensing and sealing technology to realize a novel MR fluid lens (MR fluid: liquid polymer with magnetic particles) for light intensity modulation. Merits of the device: Optical transmittance of lens is controlled by (1) weight fraction of magnetic powder, and (2) orientation of columnar particles controlled by magnetic field. In applications, the MR fluid lens is realized on glass substrate and suspended MEMS structures. The light intensity modulation of MR fluid lens (diameter: 2000μm) by magnetic field is demonstrated. Measurements show the NdFeB liquid polymer (10wt%) has a 40% dark area change and 290% laser transmittance difference after applying magnetic field.

INTRODUCTION

According to the design specifications of various MEMS and MOEMS applications, the 3D micromachined structures are frequently required presently. However, the silicon micromachining process is mainly developed to fabricate the planar 2D structures. The available functional mechanical and optical components provided by 3D structures are much more than the 2D counterparts, so as to increase the variety of applications [1]. Presently, many available technologies have been reported to implement MEMS based micro optical systems and different microlens. For example, the photoresist reflow lens formation technology has been extensively investigated [2]. Moreover, the approach to dispense the UV curable polymer on chip for the microlens formation is reported in [3,4]. The electrostatic force has been employed to modulate the shape of polymer microlens [5]. The applications of microlens are widely extended from [6] to [8].

Magnetic particles dispersed in fluid could be redistributed by external magnetic field to form the chain-like configuration. This phenomenon is called magnetorheological effect [9]. The chain-like configuration is due to the magnetic dipole motion and attraction between magnetic particles. This redistribution of magnetic particles could cause the change of material characteristics from isotropic to anisotropic. As a result, the anisotropic mechanical, magnetic, and thermal properties for materials can be achieved through this approach [10]. For instance, this effect has been employed in [11] to change the optical properties of liquid magnetic polymer. The strength of magnetorheological effect depends on some factors, such as size and shape of magnetic particles, viscosity of carrier fluid, volume fraction of particles in MPC, and the intensity of applied magnetic field [10].

This study further exploits the MR fluid (MR fluid: liquid polymer with magnetic particles) dispensing and sealing technology in [12] to realize a novel plano-convex lens. The microlens with polymer sealed MR fluid could be applied for light intensity modulation using the external magnetic field. For instance, the distribution of magnetic powders is modulated by the direction of applied magnetic field; and thus the optical transmission through the microlens could be changed accordingly. The properties of the fabricated MR fluid microlens, such as lens profile, M-H curve, dark area change, and optical transmittance are characterized. The integration of the suspended MEMS structures with MR fluid microlens is also demonstrated to show its potential applications.

DESIGN AND FABRICATION

Fig.1a-b shows the MR fluid lens formed on glass substrate. In Fig.1a, the magnetic particles are mixed with liquid polymer to form MR fluid. The liquid MR fluid is

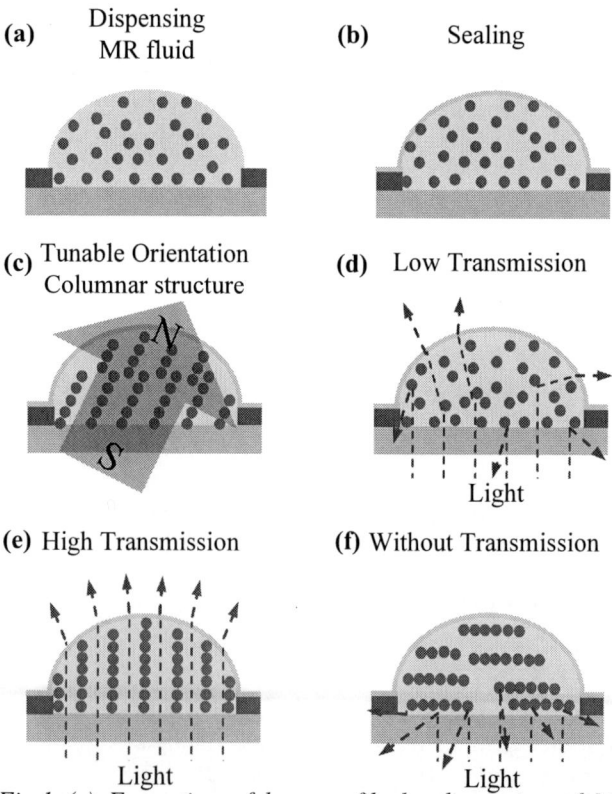

Fig.1 (a) Formation of lens profile by dispensing of MR fluid, (b) MR fluid sealed by polymer film, (c) Formation and orientation tuning of columnar particles by magnetic field, and (d) Changing optical transmission by particle distribution using magnetic field

dispensed by pneumatic system to form plano-convex lens. The lens curvature is controlled by the surface tension and volume of dispensed MR fluid [3,4].The MR fluid lens is sealed by the CVD deposited polymer film (Fig.1b). In Fig.1c, the magnetic particles form the columnar structure and then the orientation of columnar structure is controlled by an applied magnetic field. Thus, as summarize in Fig.1d-f, this approach enables the modulation of light transmission by using the applied magnetic field.

Fig.2 shows the fabrication and integration processes for MR fluid lens. In Fig.2a-b, thermal oxide layer (1μm) was grown on (100) Si substrate (525μm) and the back side cavity pattern was defined by photolithography. The exposed SiO_2 layer was etched by buffered oxide etch (BOE) solution. Then, Si cavity was realized by tetramethylammonium hydroxide (TMAH) solution, and the cavity depth was controlled by etching time. The front side suspended MEMS structures were patterned by DRIE as shown in Fig.2c. The 1μm NdFeB particles of magnetic anisotropy were mixed with polymer for the preparation of MR fluid. After that, the integration of MR fluid with MEMS structure was achieved using the dispensing process, as in Fig.2d. The volume of MR fluid dispended on structure was controlled by pneumatic instruments. The MR fluid lens was implemented after the MR fluid (polymer liquid with NdFeB particles) sealed by CVD parylene (Fig.2e). Note the lens profile is determined by the volume of dispensed MR fluid in Fig.2d. Thus, distribution of magnetic particles sealed inside the polymer liquid can be modulated by the applied magnetic field for the applications of optical modulation.

Micrograph in Fig.3a displays the 6×6 MR fluid lens (2000μm in diameter) array on glass substrate. The zoom in micrograph in Fig.3b shows the MR fluid lens sealed

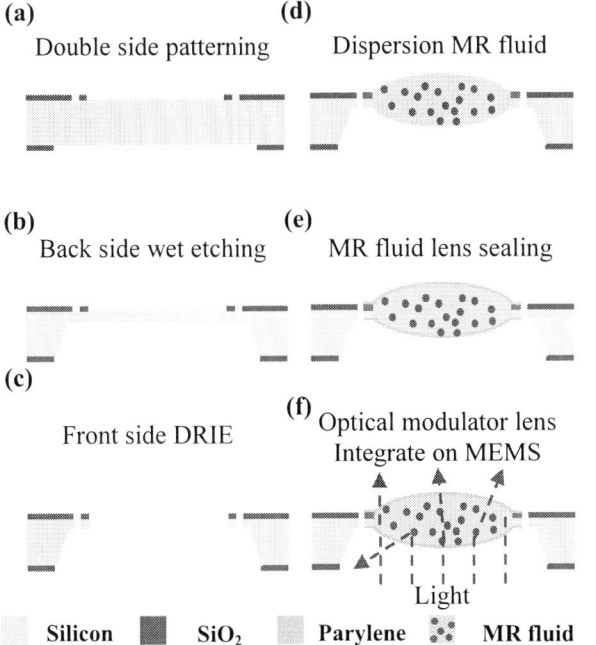

Fig.2 Process steps for the fabrication and integration of MR fluid lens with MEMS structures

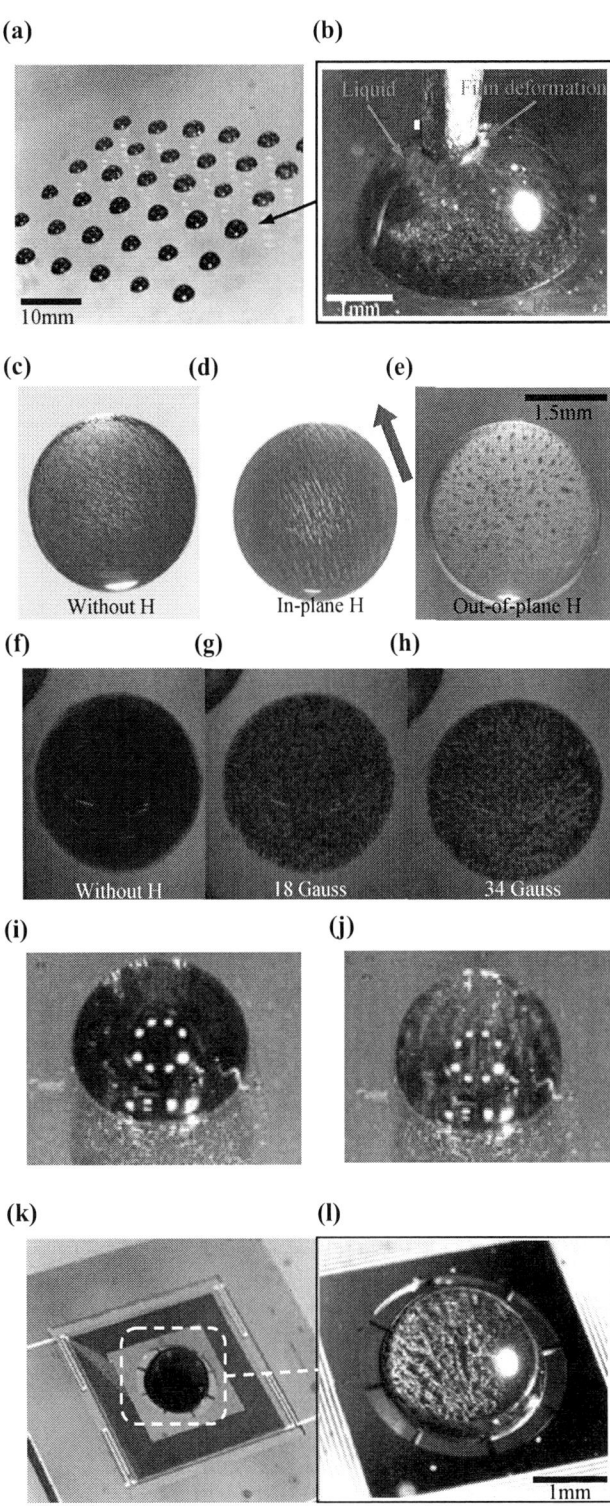

Fig.3 (a) Micrograph fabricated MR-fluid Lens array on glass, (b) Zoom in of a MR-fluid lens with NdFeB particles randomly distributed inside and sealed by a deformable polymer film, (c-e) NdFeB powder inside the lens aligned by magnetic field H, (f-h) NdFeB powder aligned by out-of-plane H of different strength, (i-j) modulation of transmitting laser light by H, and (k-l) integration of MR fluid lens with MEMS structure.

by transparent polymer film. The NdFeB particles random distributed in lens and polymer lens surface deformed by probe are clearly observed. The micrograph in Fig.3c shows the distribution of NdFeB particles in microlens when no magnetic field applied. The micrographs in Figs.3d-e respectively show the distribution of NdFeB particles modulated by magnetic field of in-plane and out-of-plane directions. Micrographs in Fig.3f-h show particle alignment by varying the strength of out-of-plane magnetic field (0~34 Gauss). Fig.3i-j shows the transmitting of laser light modulated by magnetic field. The transmitting light changed from dark to bright after applying out-of-plane magnetic field. Fig.3k shows the integration of MR fluid lens with MEMS structure. Fig.3l shows the zoom in of MR fluid lens on Si support.

MEASUREMENT RESULT

Fig.4a shows the measured lens (of three different MR fluids) radius of curvature ranging 1.08mm~2.15mm. Measured M-H curves in Fig.4b depict magnetic characteristics of MR fluids (10, 20, 30wt%) are: Coercivity:3561~3561Oe, Remanence: 162~450emu/cm^3, and Magnetization: 219~621 emu/cm^3. Fig.5 shows optical transmittance tests of MR fluid lens. Fig.5a indicates the dark area of lens determined by commercial image software (Image J).

Fig.5b-c shows the measured dark areas for lenses of different MR-fluids (10, 20, 30wt% NdFeB) driving at different strength and orientation

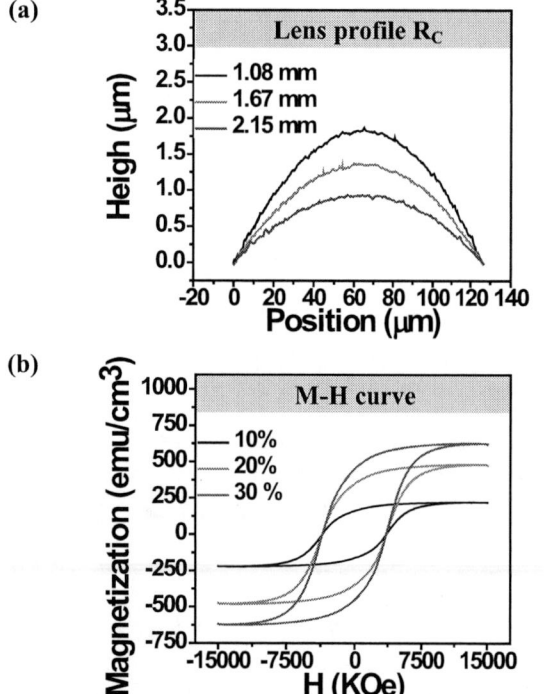

Fig.4 Measurement results of MR fluid lens with different wt% of NdFeB powder, (a) Lens profile measured by optical interferometer, and (b) M-H curve of MR-fluid characterized by VSM

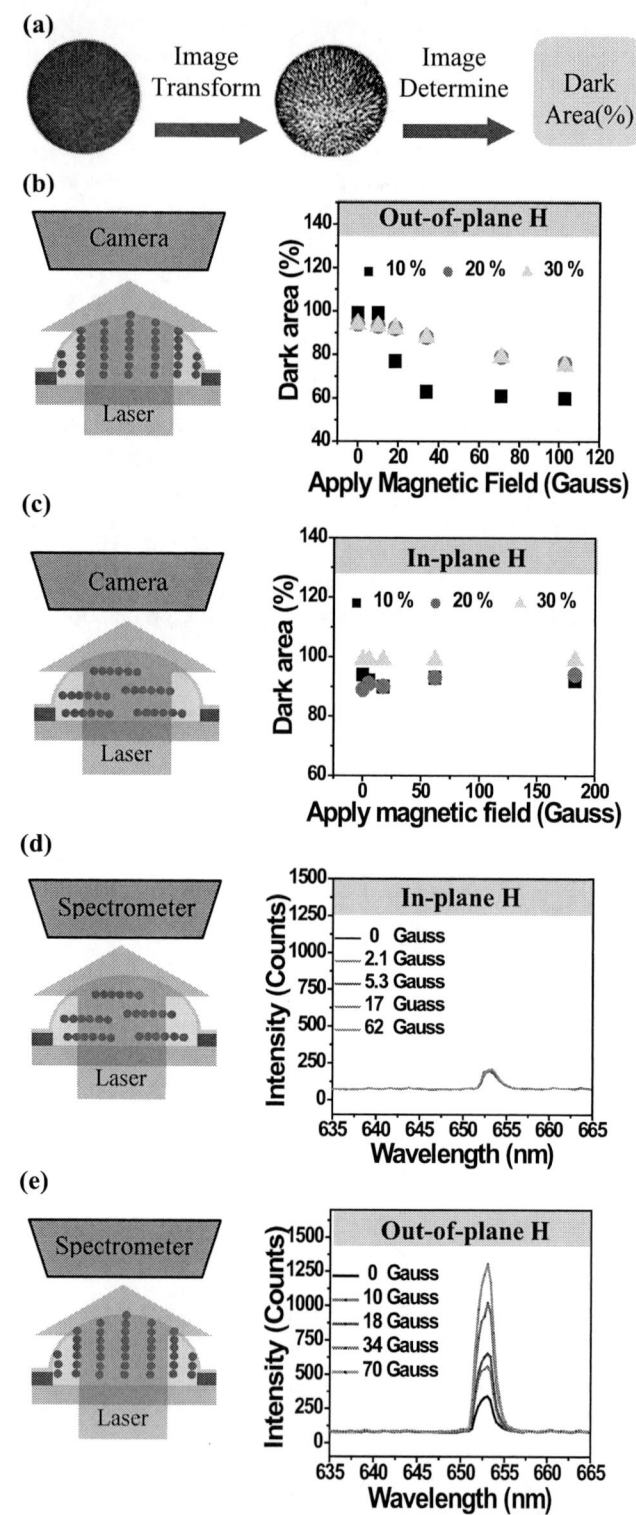

Fig.5 Measurement results of MR fluid lens optical transmission properties , (a) dark area determined by image processing software, (b-c) dark area of MR fluid lenses with different wt% NdFeB driven at out-of-plane and in-plane H of various strength, (d) intensity of laser light after transmitting via MR fluid lens of 10wt% NdFeB at different driving magnetic fields (both strength and orientation)

(in-plane and out-of-plane) of magnetic fields. It indicates the dark-area of 10wt% MR fluid lens has 40% changes as out-of-plane magnetic field exceed 35Gauss (due to the particle-alignment shown in Fig.3e). High wt% MR fluid and small magnetic filed may not change the dark area effectively. Optical spectrometer measurements in Fig.5d further show the 10wt% NdFeB MR fluid lens modulate the intensity of transmitting laser light (654nm) for 290% by 70Gauss out-of-plane magnetic field. However, the intensity of transmitting laser light has no significant change by MR fluid lens using in-plane magnetic field. Fig.6 shows the variation of dark area with view angle (0°~40°) for 10wt% NdFeB MR fluid lens at 220 Gauss out-of-plane magnetic field. The dark area increases from 45% to 60% as view angle exceeding 20°. Table.1 summarizes the characteristics of dark area and light transmittance for 10wt% NdFeB MR fluid lens at different applying magnetic fields.

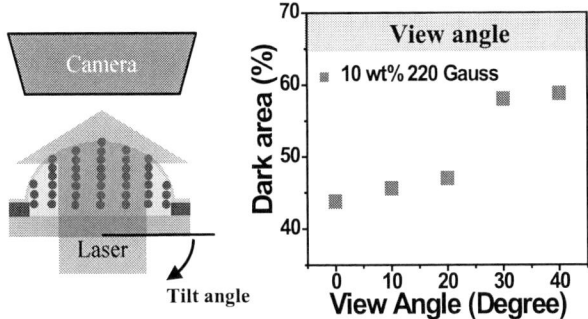

Fig.6 Variation of dark area with view angle for 10wt% NdFeB MR fluid lens driven at 220 Gauss

Table.1 Characteristics of dark area and light transmittance for 10wt% NdFeB MR fluid lens under different magnetic fields

	Dark Area (%)		Light transmission (Counts)	
	In-plane (0-183 Gauss)	Out-of-plane (0-103 Gauss)	In-plane (0-62 Gauss)	Out-of-plane (0-70 Gauss)
10 wt%	99-99	99-60	193-233	340-1306
Variation (±%)	+0%	+40%	+20.7%	+290%

CONCLUSIONS

In summary, this study has successfully implemented microlens with sealed MR fluid for light modulation by magnetic field. Fabrication of MR fluid lens on glass substrate and MEMS structures are demonstrated. Applications of light modulation by magnetic field have also been demonstrated. Various optical transmittance of lens is controlled by weight fraction of magnetic powders, and orientation of columnar particles controlled by magnetic field. This study has characterized the performances enhancement of NdFeB MR fluid (10wt%) by orientation of columnar particles controlled during the proposed strength and direction of magnetic field. The improvement of the dark area change and laser transmittance difference are 40% and 290%, respectively.

ACKNOWLEDGMENTS

This paper was partially supported by National Science Council of Taiwan under grant number NSC 102-2221-E-007-027-MY3 and Brain Research Center at the National Tsing Hua University, Taiwan, under contract 102A0129JA. The authors would like to express his appreciation to the Nano Science and Technology Center of National Tsing Hua University, and Nano Facility Center of National Chiao Tung University in providing the fabrication facilities.

REFERENCES

[1] N.S. Sharr, et al., "Cascaded mechanical alignment for assembling 3d mems", *IEEE Conference MEMS*, pp. 1064-1068,2008.

[2] P. Heremans, et al., " Mushroom microlenses: optimized microlenses by reflow of multiple layers of photoresist" , *IEEE Photonics technology letters*,vol. 9, no. 10, pp.1367-1369, october 1997.

[3] S.-Y. Hsiao, et al., "The implementation of concave micro optical devices using a polymer dispensing", *J. Micromech. Microeng*, vol. 18, pp. 085009, 2000.

[4] C.-C. Lee, et al., "Formation and integration of a ball lens utilizing two phase liquid technology," *IEEE Conference MEMS*, pp. 172-175, 2009.

[5] K. Y. Huang, et al., "Design and fabrication of a copolymer aspheric bi-convex lens utilizing thermal energy and electrostatic force in a dynamic fluidic ", *Optics Express*, vol. 18, no. 6, pp.6014-6023,15 march 2010.

[6] V. K. Singh, et al., "Deposition of thin and uniform photoresist on three-dimension al structures using fast flow in spray coating" , *J. Micromech. Microeng*,vol. 15,pp. 2339-2345,2005.

[7] T. Stone, et al.," , *Applied Optics*, vol. 27, no. 14, pp. 2960-2971, 15 july 1988.

[8] P. Wu, et al.,, " Wavelength-multiplexed submicron holograms for disk-compatible data storage" ,*Optics Express*, vol. 15,no. 26, pp. 17798-17804, 24 december 2007.

[9] A. G. Olabi, et al., "Design and application of magneto-rheological fluid", *Materials and Design,* 28 pp. 2658-2664, 2007.

[10] M. R. Jolly, et al., "A model of the behaviour of magnetorheological materials", *Smart Mater. Struct.,* 5, pp. 607-614. 1996.

[11] Jianping Ge, et al., "Magnetochromatic Microspheres: Rotating Photonic Crystals", *J. Am. Chem. Soc.,* 127, pp. 15687-15694. 2009.

[12] F.-M. Hsu, et al., "MEMS structure with tunable stiffness using the magnetorheological effect", *IEEE Conference MEMS*, pp. 9-12, 2014.

OPTICAL CONTROL AND TUNING OF THERMAL-PIEZORESISTIVE SELF-SUSTAINED OSCILLATORS

Harris J. Hall[1], Luda Wang[1], J. Scott Bunch[1], Siavash Pourkamali[2], and Victor M. Bright[1]
[1]University of Colorado, Boulder, USA
[2]University of Texas, Dallas, USA

ABSTRACT

The ability to frequency tune and provide on/off control of electrically driven thermal-piezoresistive self-sustained oscillators through the application of HeNe (632 nm wavelength) laser illumination to devices is reported in this work. Photoexcitation of charge carriers is presented as the physical mechanism to control the piezoresistive coefficient and electrical resistivity enabling these abilities. The results are significant in that they offer a novel means to directly control the electronic output of RF oscillators through photonics.

INTRODUCTION

Developing effective interconnect mechanisms between radio frequency (RF) and optical signals remains an area of interest to further leverage the benefits of optical communication and control between electronic devices. Micro and nanoscale mechanical resonators have played a central role in providing on-chip methods of integration between these two domains with the goal of realizing photonic integrated circuits in mind. Opto-mechanical resonators have gained considerable attention where suspended structures are deflected with nanoscale precision using the optical gradient force [1], actuated into oscillation with continuous wave laser illumination [2], and electrically actuated to modulate optical waveguides [3].

Thermally-actuated piezoresistively readout resonant devices fabricated from single-crystal silicon have gained interest in the RF MEMS community, particularly for their ability to exhibit self-sustained oscillation under DC bias [4]. Previous work, expanded their range of operation into the VHF regime [5] and applied their use towards a variety of sensing applications. In this work we show how the steady-state performance of these devices operating in self-sustained oscillation changes when under direct illumination of HeNe laser light. Specifically we demonstrate the ability to both frequency tune and provide on/off control of the oscillation by adjusting the laser power.

Device Background

The devices examined in this work are the same as presented in [5] which described their fabrication and performance (unilluminated and at fixed low laser power) as both resonators and self-sustained oscillators. The device dimensions and 2-terminal I-shaped geometry are shown in Table 1. These devices were intended to operate in the in-plane longitudinal structural mode through cyclic Joule heating of the actuator arms, however evidence presented in [5] suggested that the actual mechanical mode of operation

is likely of an alternate shape. The devices were fabricated from n-type (phosphorus doped) single crystal silicon ($\rho \sim 0.1\Omega$-cm, $\sim\sim 8.1 \times 10^{16}$ cm^{-3} at 300K) using an SOI wafer (oxide layer $\sim 3\mu$m) based process. The strong negative piezoresistive coefficient of this material enables the ability for self-sustained oscillation under constant DC bias through internal feedback [4].

Table 1: Top View of Device A3 with average device dimensions[n=7] in μm (h = device thickness)

Device	C6	A3
h	1.8	1.6
a	5.73	6.46
b	5.70	4.17
c	0.94	1.38
L	2.02	1.47
W	0.34	0.36
W_b	1.59	1.54
L_b	2.56	2.56

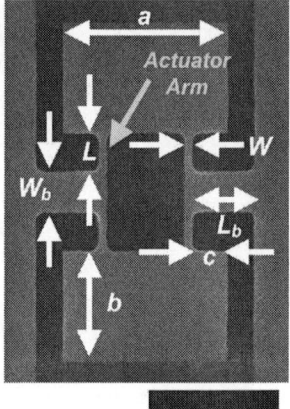

5 μm

THEORY

The performance of these devices under photoillumination can be considered from the perspective of the device's motional conductance at resonance ($\omega=\omega_o$), g_m, defined in [6] as

$$g_m = 2\alpha E \pi_l Q_m \frac{I_{DC}^2}{C_{th}\omega_o} \qquad (1)$$

where Q_m is mechanical quality factor, I_{DC} is the applied DC current, C_{th} is the effective thermal capacitance of the lumped actuator element, E is the Young's modulus, π_l is the longitudinal piezoresistive coefficient, and α is the coefficient of thermal expansion. It is critical to note that (1) is based upon a lumped model for the actuator arms at the center of the device. Thus it is the photon interaction that occurs locally to this area that is of primary interest. Of these parameters shown in (1) it can be expected that three of them may be affected by the HeNe illumination.

First, the resistivity of the silicon should decrease due to photoexcitation of carriers since the photon energy of HeNe laser light (~ 1.96 eV) exceeds both the extrinsic bandgap (~ 0.15 eV at 300K) and the intrinsic bandgap of

978-1-4799-3510-9/14 $31.00 © 2014 IEEE

silicon (~1.12 eV). Thus increasing illumination should increase the I_{DC} and alter the device power dissipation, depending upon how the device is biased. This will in turn alter the Joule heating occurring and change the steady-state temperature. It can also be expected that charge carriers pumped high above the conduction band minimum will relax via phonon scattering mechanisms causing additional heating of the material. Since the Young's modulus is temperature dependent it should thus change with HeNe laser illumination.

Finally, per the many-valleys model [7], the piezoresistive effect in n-type silicon is due to a redistribution of electrons in six potential wells located along [100] equivalent orientations in momentum space when stress is imparted to the material. Kanda in [8] presents the relation between the phenomenological and solid-state descriptions of this effect, as well as theoretical predictions for piezoresistive changes with temperature and doping concentration. These are presented in the form of a piezoresistive factor, $P(N,T)$, that appropriately scales the room temperature piezoresistive coefficient (see Figure 1). While this formulation is less accurate for large doping concentrations, the trends for both carrier concentration and temperature are valid and conceptually extensible to the effects of photoexcitation. The resultant increase in the total carrier concentration from excited carriers and the aforementioned steady-state temperature changes can both be expected to alter the piezoresistive coefficient.

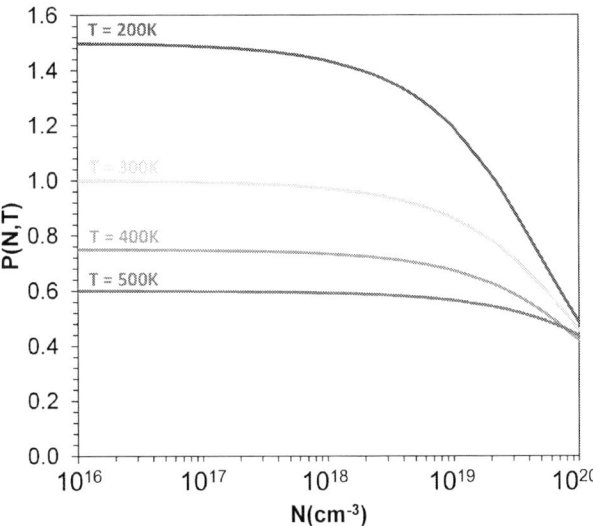

Figure 1: The adjusted piezoresistance factor $P(N,T)$ for n-type silicon as a function of carrier concentration and temperature, after Kanda [8]. The additional carrier concentration from photoexcitation adds to the dopant concentration (assumed fully ionized) in this model.

The impact of any reduction of the piezoresistive coefficient would be a proportional degradation of the magnitude of the motional conductance. Ultimately, this would result in shutoff of the self-sustained oscillation when the $R_A g_m \leq -1$ gain condition required, as described in [4], is not satisfied;

with R_A being the resistance of the actuator arm.

EXPERIMENTAL METHOD

The experimental testbed used for this work is shown in Figure 2 and was identical to that used in previous work [5] for identifying the frequency of mechanical oscillation in ambient air. The applied DC bias voltage to the devices was first fixed to a value above the threshold needed to initiate self-sustained oscillation. This was verified by measuring the AC voltage output in the time-domain using a digital oscilloscope (Agilent DSO7032A, DC coupled, 1 MΩ input impedance). The HeNe laser was then attenuated by a filter wheel to a low power ($\leq$ 1.21 mW) and positioned at the center of one of the outer plate edges. The photodiode captures the light reflected off of the underlying substrate and the device surface. The mechanical motion of the device structure modulates the intensity of this reflected laser return at the frequency of oscillation. A spectrum analyzer (Agilent 9320B) measured and recorded the corresponding electrical output of the photodiode. Once the device was verified to be functioning under low laser illumination, the electrical and mechanical (optical) responses were then measured as the laser power was incrementally increased using a filter wheel. Device operation was presumed to be in steady-state with measurements collected after several minutes for a given condition. The applied laser power for each filter wheel setting was measured after the experiment using a calibrated optical power meter at the output of the final objective lens.

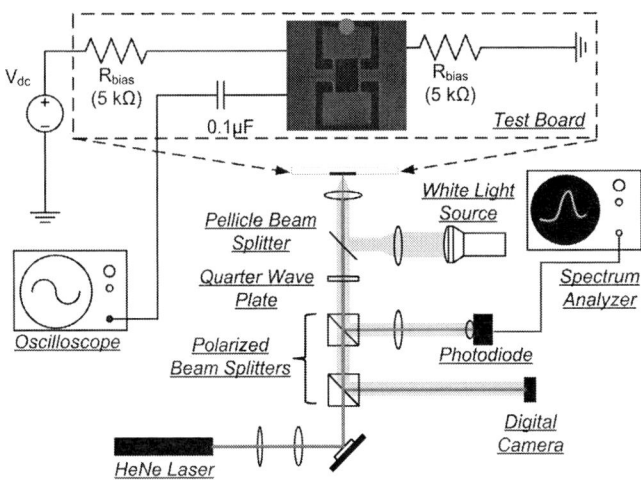

Figure 2: Optical testbed (ambient air, room temperature, ambient lighting) used for sensing the spectral response of the structural mode while measuring electrical output of the device in self-sustained oscillation. The filter wheel used to control laser power (not shown) is positioned prior to the beam splitters. (Position of laser shown on inset SEM image)

RESULTS AND DISCUSSION

Two central effects of increasing the laser power

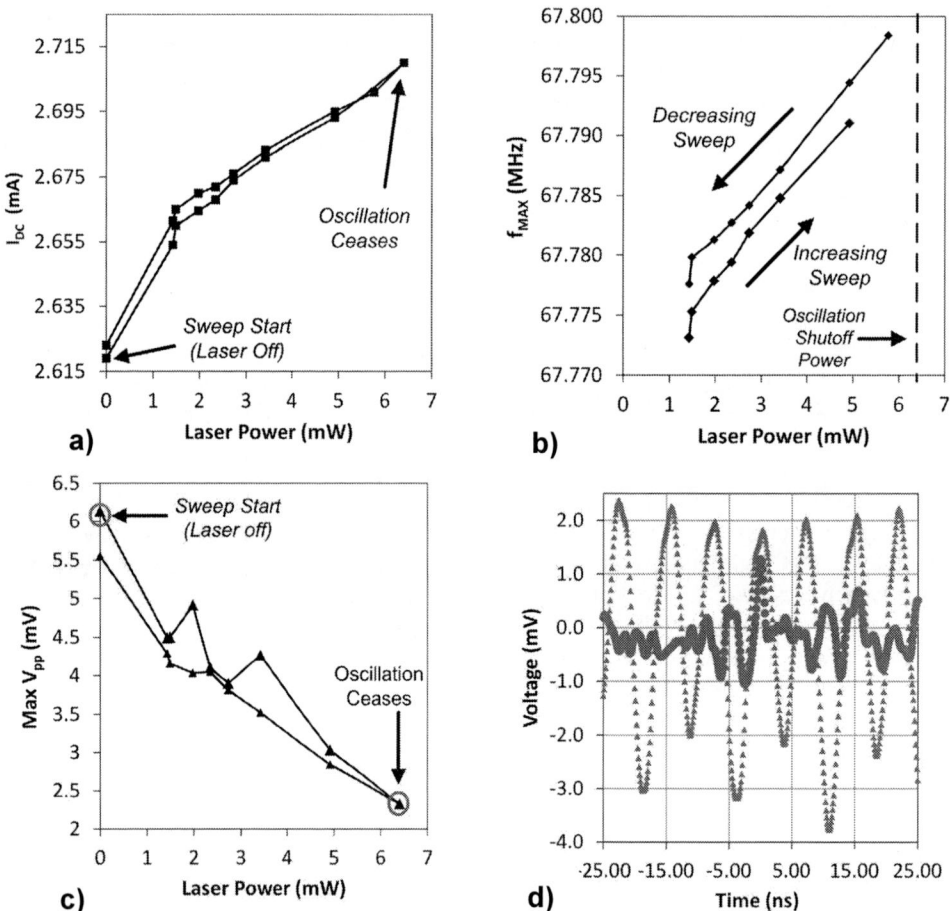

Figure 3: Steady-state experimental results for bi-directional laser power sweep of Device C6 under ambient air test conditions, V_{dc} = 31.61 V (a) DC Current as measured by multimeter (b) Peak frequency of oscillation from 1D Lorentzian fit to photodiode output signal measured by spectrum analyzer (c) Maximum peak voltage difference of AC output signal as measured by digital oscilloscope d) Raw oscilloscope traces corresponding to initial laser off signal, and under shutoff illumination.

with the beam position fixed at a plate edge were observed for Devices C6 and A3 during experiment conduct. First, gradual degradation in the output signal strength and eventually complete quenching of the self-sustained oscillation, both electrically and mechanically occurred. Upon lowering the laser power back below this shutoff power level the self-sustained oscillation was restored. Second, increasing the laser power level caused increases in both the DC current and the frequency of device oscillation although this effect was not consistently linear for both devices at higher laser powers.

Experimental results for a bi-directional sweep of laser power up to a level beyond shutoff and back down to the initial test condition (laser off) are shown for Device C6 in Figures 3a-d). The frequency of oscillation shown in Figure 3b) is from fitting a 1D Lorentzian profile, as shown in [5], to the photodiode signal measured by the spectrum analyzer. The maximum raw peak-peak voltage observed in the electrical signal throughout the sweep is shown in Figure 3c), to depict the degradation in signal strength.

The variation of I_{DC} and peak frequency of oscillation, f_{MAX}, shown in Figure 3a-b) is consistent with considering the device as a self-oscillating photoresistor wired with bias resistors as shown in Figure 2. The photoexcitation of carriers lowers device resistance, which in turn decreases the DC power dissipation in the device, lowering the operating temperature, which stiffens the Young's modulus and ultimately causes an increase in . frequency of oscillation. For comparison, the experimentally measured variation of I_{DC} and f_{MAX} results for Device A3 are shown in Figure 4 with a uni-directional (increasing) laser power sweep up to oscillation shutoff. It is clear that both I_{DC} and f_{MAX} reach a maximum and then decrease before oscillation shutoff. Since the laser was positioned approximately ~1.5um closer to the actuator arms for Device A3 than Device C6 the actual effective laser fluence on the actuator arms is higher. Thus the max peaks observed are likely due to the saturation limit of carrier excitation being reached in the actuator arms allowing scattering loss mechanisms to dominate and cause

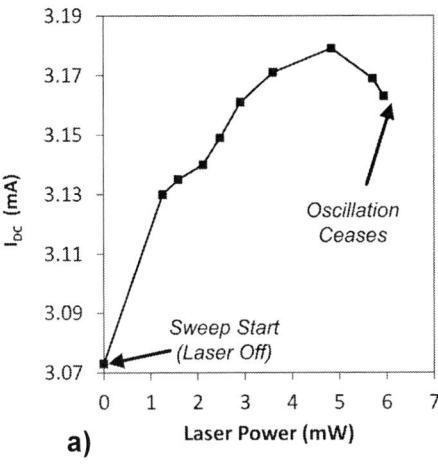
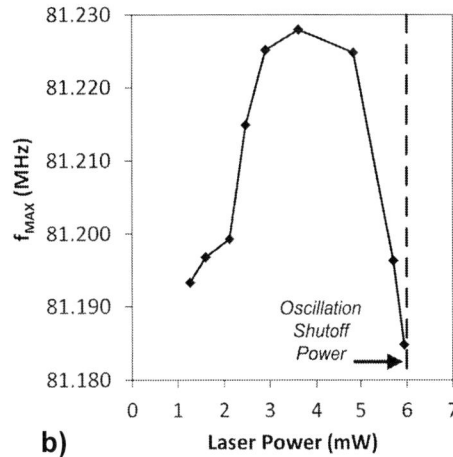

Figure 4: Steady-state experimental results for an increasing laser power sweep of Device A3 under ambient air test conditions, $V_{dc} = 37.96$ V (a) DC Current as measured by multimeter (b) Peak frequency of oscillation from 1D Lorentzian fit to photodiode output signal measured by spectrum analyzer. The b dimension of Device A3 is ~1.5 um smaller than Device C6, thus the actual laser fluence on the actuator arms is higher than that for C6 for the same diameter laser spot.

an increase in electrical resistance.

It was also observed that positioning the laser at a low power setting closer to the actuation arms tended to degrade the electrical output in a similar fashion to that seen with measured data at the outer edge position, with quenching of the electrical output also possible. This observation supports the theory that the oscillation shutoff behavior observed is attributed to the piezoresistive coefficient and not an artifact of thermoelastic effects.

CONCLUSIONS

This work demonstrates thermal-piezoresistive oscillators can be frequency tuned and controllably quenched using continuous wave HeNe illumination of the structure. Such on/off control is believed to be novel to this class of device as it is predicated upon direct influence of the piezoresistivity of the material. This method of control potentially offers a unique way of integrating these devices with on-chip photonic circuitry. In addition, it invites further study of the effects of photoillumination upon the piezoresistivity of materials, which is an area that seems to be largely unexplored even for silicon. Formalizing more refined theory in this regard and further examining the modulation of piezoresistivity remain topics for future work.

ACKNOWLEDGEMENTS

The authors would like to thank Timothy May of the Integrated Teaching and Learning Laboratory and the Bunch Lab at CU Boulder for equipment support. Special thanks goes to Dr. Catalin Badescu of AFRL for discussions regarding the treatment of photoexcitation on piezoresistance. Additional thanks goes to Jason Gray and Joseph Brown of the Bright group for useful discussion and support. The lead author would like to acknowledge the DoD SMART Scholarship program for financial support.

REFERENCES

[1] B. Dong, H. Cai, J. M. Tsai, P. Kropelnicki, A. B. Randles, M. Tang, D. L. Kwong, and A. Q. Liu, "NEMS actuator driven by optical gradient force," in *Transducers '13*, June 16-20 2013, pp. 900–903.

[2] T. Kippenberg and K. Vahala, "Cavity opto-mechanics," *Opt. Express*, vol. 15, no. 25, pp. 74–77, 2007.

[3] S. Sridaran and S. A Bhave, "Electrostatic actuation of silicon optomechanical resonators.," *Optics Express*, vol. 19, no. 10, pp. 9020–6, May 2011.

[4] A. Rahafrooz and S. Pourkamali, "Thermal-Piezoresistive Energy Pumps in Micromechanical Resonant Structures," *IEEE Trans. Electron Devices*, vol. 59, no. 12, pp. 3587–3593, Dec. 2012.

[5] H. J. Hall, D. E. Walker, L. Wang, R. C. Fitch, J. S. Bunch, S. Pourkamali, and V. M. Bright, "Mode selection behavior of VHF thermal-piezoresistive self-sustained oscillators," in *Tranducers '13*, June 16-20 2013, pp. 1392–1395.

[6] A. Rahafrooz and S. Pourkamali, "High-Frequency Thermally Actuated Electromechanical Resonators With Piezoresistive Readout," *IEEE Trans. Electron Devices*, vol. 58, no. 4, pp. 1205–1214, 2011.

[7] W. P. Mason, *Physical Acoustics*. New York: Academic Press, 1964, pp. 183–190.

[8] Y. Kanda, "A graphical representation of the piezoresistance coefficients in silicon," *IEEE Trans. Electron Devices*, vol. 29, no. 1, pp. 64–70, Jan. 1982.

CONTACT

*H.J. Hall, Tel: +1-303-492-7151; Email: Harris.Hall@colorado.edu

PHOTOTHERMAL PROBING OF PLASMONIC HOTSPOTS WITH NANOMECHANICAL RESONATOR

Silvan Schmid, Kaiyu Wu, Tomas Rindzevicius, and Anja Boisen

Department of Micro- and Nanotechnology, Technical University of Denmark, DTU Nanotech, DK-2800 Kgs. Lyngby, Denmark

ABSTRACT

Plasmonic nanostructures (hotspots) are key components e.g. in plasmon-enhanced spectroscopy, plasmonic solar cells, or as nano heat sources. The characterization of single hotspots is still challenging due to a lack of experimental tools. We present the direct photothermal probing and mapping of single plasmonic nanoslits via the thermally induced detuning of nanomechanical string resonators. A maximum relative frequency detuning of 0.5 % was measured for a single plasmonic nanoslit for a perpendicularly polarized laser with a power of 1350 nW. Finally, we show the photothermal scan over a nanoslit array.

INTRODUCTION

Sub-wavelength noble metal structures (nanoparticles or nanovoids) support localized surface plasmon (LSP) resonances that usually occur in the visible and near-infrared spectral region. The incident light can couple to LSP modes producing extremely large field enhancements, so-called hotspots. These strong field confinements are prominently utilized e.g. in surface enhanced Raman scattering (SERS) spectroscopy [1], in plasmonic solar cells [2], or as nano heat sources with a wide range of potential applications [3]. So far there have been a few attempts to investigate the photothermal heating of single hotspots using thermal optical techniques [4] and an AFM tip [5]. However, the desirable direct investigation of the heating mechanisms in plasmonic structures is still a challenge due to the lack of experimental tools [3]. Herein we present a novel technique to probe and image plasmonic structures with nanoscale resolution by measuring the photothermally induced frequency detuning of highly temperature sensitive nanomechanical resonators. We employ the high temperature sensitivity of a nanomechanical string resonator [6] to directly probe the heating pattern produced by a gold nanoslit illuminated by a scanning laser beam. The experimental approach allows a sensitive heat mapping of single LSPs thereby helping to shed light on the underlying thermal effects in plasmonic hotspots.

The principle of the presented photothermal probing of plasmonic nanostructures is based on the temperature induced resonance frequency detuning of a nanomechanical string resonator. The plasmonic hotspot absorbs light, which results in a local heating of the string. The resonance frequency of string resonators is a function of its tensile pre-stress. The heating induced thermal expansion of the string lowers the tensile stress and hence the resonance frequency of the nanomechanical string resonator. We have previously demonstrated that the method is applicable in analyzing single nanoparticles [6] and IR chemical fingerprints of airborne particles accumulated on the nanostring surface [7].

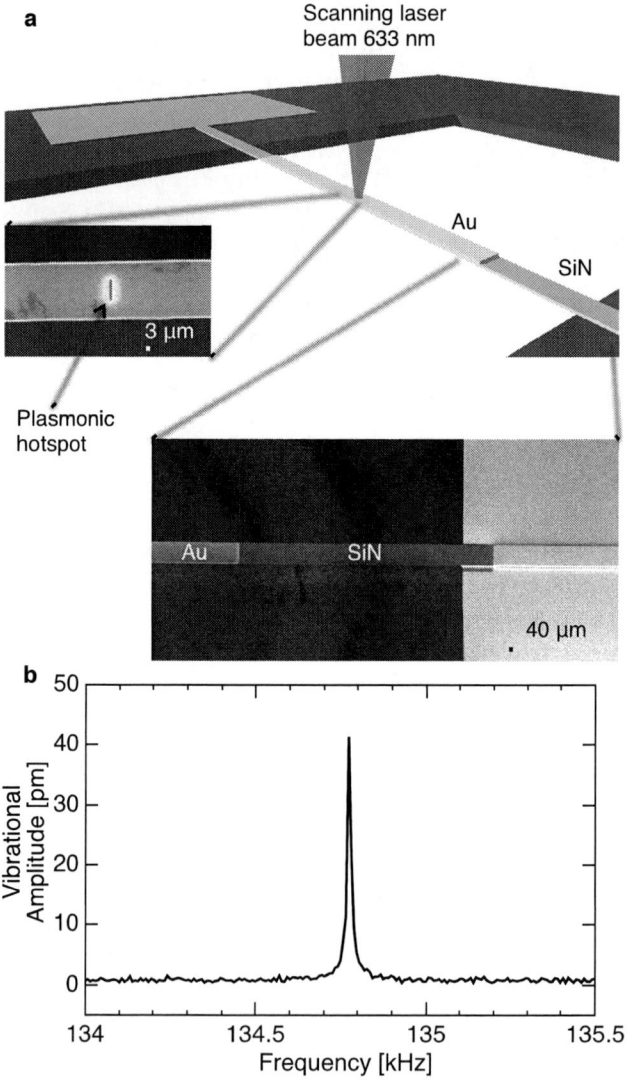

Figure 1: (a) Schematic depiction of experimental setup. A nanomechanical SiN string resonator is partially coated with an Au layer. A plasmonic hotspot (etched with a FIB) is located in the string center and probed with a focused laser beam from a laser-Doppler vibrometer. Insets show scanning electron microscope images of a plasmonic nanoslit (hotspot) and an anchor where the Au layer was removed with FIB milling. (b) Thermal fluctuation spectrum of a 900 nm long SiN string.

978-1-4799-3510-9/14 $31.00 © 2014 IEEE

Fig. 1a schematically shows the measurement arrangement. The focused scanning laser beam of a laser-Doppler vibrometer is used to measure the resonance frequency of the nanomechanical string resonator and to probe the plasmonic nanoslit by means of it photothermal response. The inset shows a typical gold-coated silicon nitride string with a single nanoslit.

The relative resonance frequency detuning is a linear function of the absorbed power P [7]

$$\frac{\Delta f}{f} = - \frac{\alpha E}{\kappa} \frac{L}{16 \sigma A} P, \quad (1)$$

where α is the thermal expansion coefficient, E the Young's modulus, $\varkappa$ the thermal conductivity, L the length, σ the tensile pre-stress, and A the cross section area of the string. The thermal conductivity of Au is more than two orders of magnitude larger than of SiN. To enhance the thermal isolation of the plasmonic nanostructure and hence the sensitivity, the Au layer close to the anchors was removed using a focused ion beam. A SEM image of a string anchor with the removed Au layer is shown in the inset of Fig. 1a. A welcome side effect of the cleared anchors is the positive effect on the quality factor of the string resonator [8,9].

EXPERIMENTAL
Chip fabrication
Stoichiometric silicon nitride (SiN) with a thickness of 157 nm was deposited on a double-side polished silicon wafer in a low-pressure chemical vapor deposition (LPCVD) furnace. First, the string structures were structured by photolithography and subsequently dry-etched into the top SiN layer. The unreleased strings are then covered with a 500 nm thick plasma-enhanced chemical vapor deposition (PECVD) SiN layer. Second, holes for the backside release are defined and dry-etched into the backside SiN layer. A KOH etch from the backside through the entire Si wafer releases the SiN strings on the front side. The protective frontside PECVD SiN layer is removed in buffered HF. The fabrication process is equal to one used for the strings in [6]. The strings were then covered with an 80 nm thick Au layer by physical vapor deposition. Finally, 1 μm long nanoslits with thicknesses between 50 to 150 nm were etched in the center of the strings with a focused ion beam (FIB), using a 30 keV Ga+ beam with a current of 9 pA.

Experimental setup
The resonance frequency is determined with a laser-Doppler vibrometer (MSA from Polytec GmbH) in high vacuum from the thermal fluctuation spectrum of the nanostring. The measurements were done at room temperature. A typical measured thermal spectrum is shown in Fig. 1b.

RESULTS AND DISCUSSION
First, the photothermal heating of single plasmonic nanoslits is simulated using the finite element method

(FEM) for two laser polarization cases. The simulated absorption efficiencies are then compared to the polarization dependent photothermal heating measured with nanomechanical string resonators. Finally, we show a nanomechanical photothermal mapping of single plasmonic nanoslits and a scan over a nanoslit array.

FEM Simulations
Fig. 2 shows the simulated electric field in a string cross-section for s-polarized (perpendicular) and p-polarized (parallel) incident light with respect to the plasmonic nanoslit. In the s-polarized case the incident field is enhanced around the slit due to the generation of a LSP. Additionally, a surface plasmon polariton (SPP) is excited at the metal-insulator interface. In the p-polarized mode, the LSP generation is minimal. In Fig. 3, the resistive loss in the metal of a string with and without a slit is compared for both polarization modes. For p-polarization, the nanoslit does not significantly alter the resistive loss (no additional heating), whereas the resistive loss is increased roughly 5 times for s-polarization (increased heating due to hotspot).

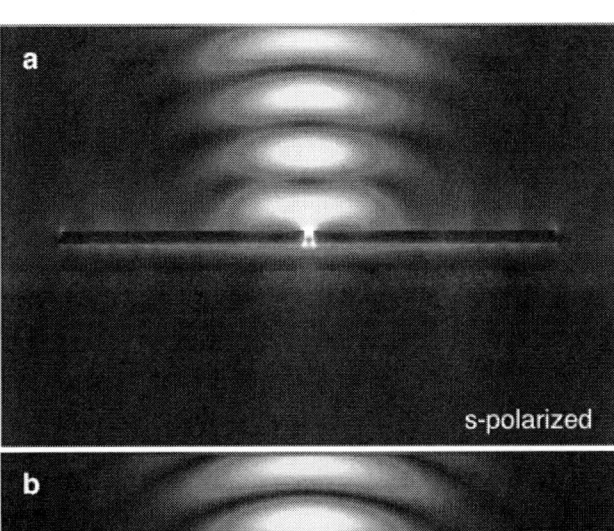

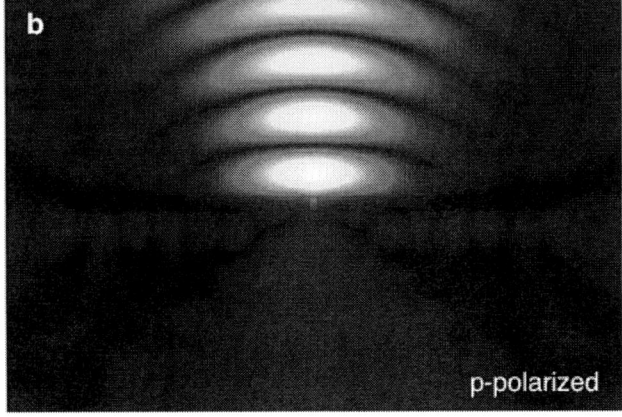

Figure 2: Distribution of the modulus amplitude of the electric field (range from 0V/m - 2V/m) in the case of (a) s-polarization and (b) p-polarization, simulated with FEM (Comsol). Details: SiN string (157 nm thick and 3 μm wide) covered with 80 nm Au, illuminated with a laser (λ=633 nm, laser beam diameter is 800 nm), with 60 nm nanoslit.

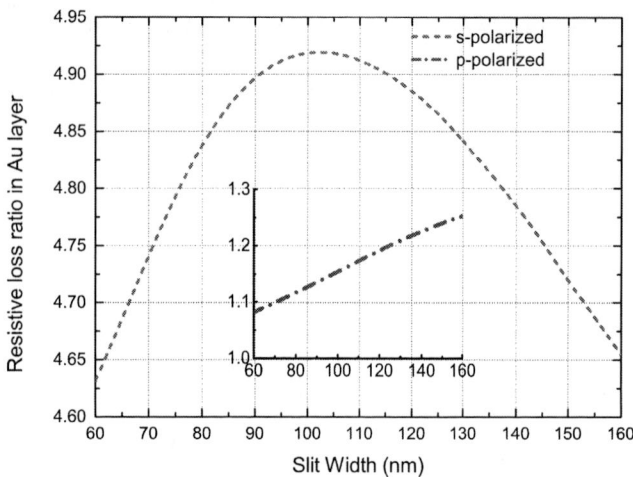

Figure 3: Ratio of the ohmic loss in Au between a gold film with a slit and a continuous gold film under s- and p-polarized excitation simulated using the FEM (Comsol). String details: SiN string (157 nm thick and 3 μm wide) covered with 80 nm Au, illuminated with a laser (λ=633 nm, laser beam diameter – 800 nm), with 60 nm nanoslit.

Polarization dependence

Fig. 4 shows the experimental results measured with a set of three equal slits with varying angle (see Fig. 4a). A photothermal mapping of two hotspots is shown in Fig. 4b with a spatial resolution of ~375 nm. The thermal heating is measured by the relative frequency shift (Δf/f) of the string resonator. A line scan over all three slits is depicted in Fig. 4c. In the s-polarization case the produced frequency shift is roughly double in magnitude compared to the p-polarization one (see Fig. 4b & 4c). The frequency detuning is directly proportional to the absorbed power, and therefore the s-polarization results in double the heating compared to the p-polarization case. The laser beam is only partially polarized with a power ratio of 4:1 in the two directions of polarization. From Fig. 3 we extract the heating ratio ~5:1 of s- vs. p-polarization. Adding up all the contribution from both polarizations (4x5 + 1x1 = 21, 4x1 + 1x5 = 9) a heating ratio (or frequency detuning ratio) of 21:9 can be obtained. This is well in agreement with the measured double frequency detuning of s-polarization compared to p-polarization.

Nanoslit array

Fig. 5 shows the photothermal scan of a plasmonic nanoslit array measured with an s-polarized scanning laser. The array consists of 5, 1 μm long, nanoslits with increasing spacings (600 nm and 1200 nm). The array is shown in Fig. 5a. The photothermal scan of the array is shown in Fig. 5b, with a line resolution of ~500 nm. The thermal heating is measured by the relative frequency shift (Δf/f) of the string resonator. It can be clearly seen, that the photothermal heating is enhanced in the center of the array with a pitch of 600 nm between the nanoslits. The string used in the array mapping is double as wide as the string used for the single nanoslit mapping (Fig. 4b). According to

the analytical model (1), the wider string has half the sensitivity compared to the narrower string. The laser power used for the wider string on the other hand was roughly 2 times larger. Therefore, the measured relative frequency detuning is approximately equal in both measurements (Fig. 4b & Fig. 5b).

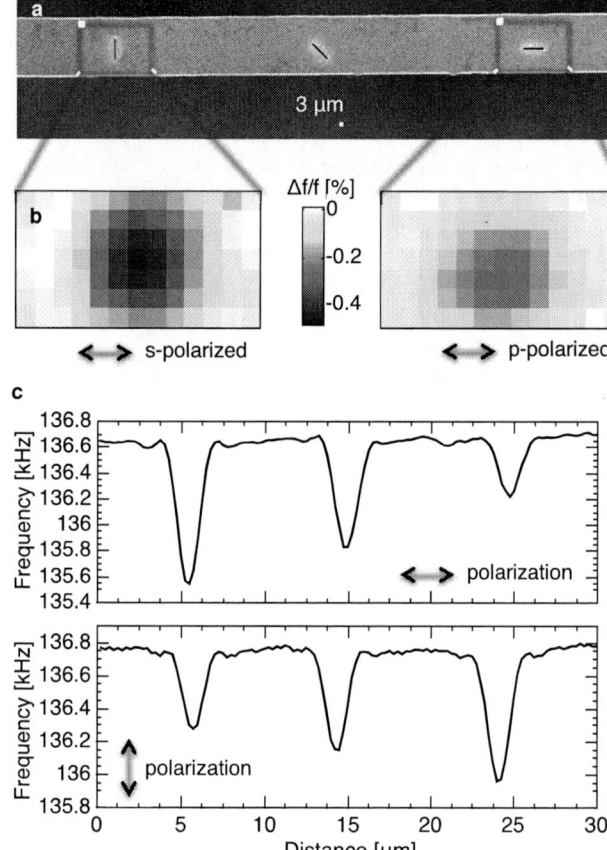

Figure 4: (a) SEM image of an Au coated SiN string (3 μm wide, 157 nm thick and covered with 80 nm Au) featuring plasmonic nanoslits in the center (90 nm wide and 1 μm long). (b) Photothermal mapping of vertical and horizontal plasmonic slits with a partially horizontally polarized laser beam. The photothermal heating is measured by the relative frequency shift (Δf/f) of the string resonator. Measured with 600 nW laser power and a spot size of ~1 μm. (c) Resonance frequency of the string for a scan over all three plasmonic nanoslits for two opposite partial polarizations. Measured with 1350 nW laser power.

CONCLUSION

We present the photothermal probing and mapping of single plasmonic nanostructures (hotspots) using the temperature induced resonance frequency detuning of nanomechanical string resonators. The absorption of the tested plasmonic nanoslits is maximal for a perpendicular polarization (s-polarization). The measured polarization dependent photothermal heating is in agreement with respective FEM simulations. The heating is based on the creation of a LSP inside the nanoslit and SPPs created at the Au/SiN interface. A maximum relative frequency detuning of 0.5 % was measured for a single plasmonic nanoslit for a total laser power of 1350 nW. By scanning the probing laser over the string surface, high-resolution photothermal maps of single plasmonic nanoslits are obtained. We further showed the scanning of a nanoslit array. The presented technique provides a unique way to directly probe and characterize single plasmonic nanostructures. Highly temperature sensitive nanomechanical string resonators can help to improve our understanding of the heat generation mechanisms in single and interacting plasmonic nanostructures.

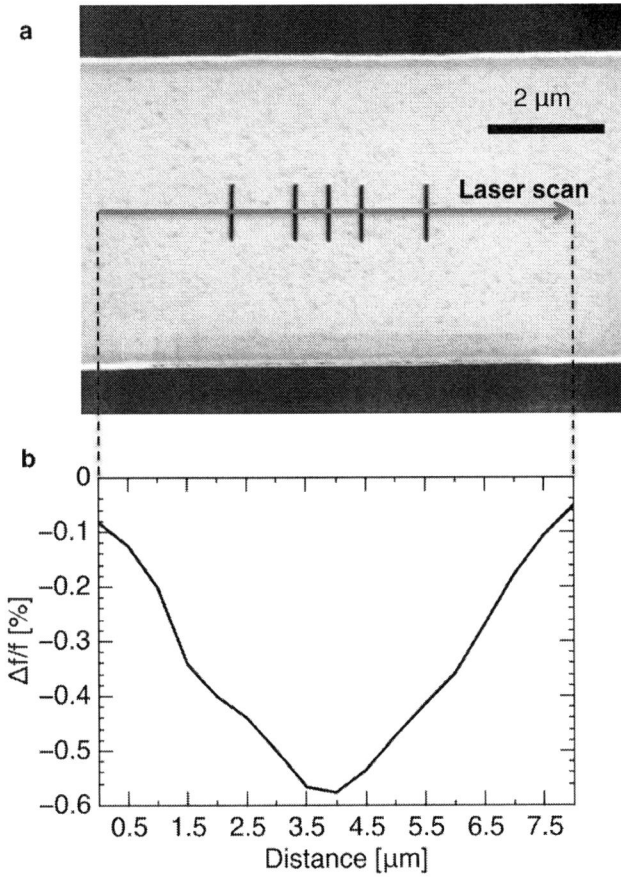

Figure 5: (a) SEM image of nanoslit array on SiN string (157 nm thick and 6 μm wide) covered with 80 nm Au. (b) Photothermal scan of the nanoslit array representing the relative frequency detuning of the string by the backaction of the scanning laser (1.35μW total power). The scan has a lateral resolution of 500 nm.

ACKNOWLEDGEMENTS

This research is supported by the Villum Foundation's Young Investigator Program (Project No. VKR023125). The authors thank Zoltán Imre Balogh and Adam Fuller for the support with the focused ion beam milling, and Jens Q.

Adolphsen for the cleanroom support.

REFERENCES

[1] S. Nie and S. R. Emory, "Probing Single Molecules and Single Nanoparticles by Surface-Enhanced Raman Scattering," *Science*, vol. 275, no. 5303, pp. 1102–1106, 1997.

[2] H. A. Atwater and A. Polman, "Plasmonics for improved photovoltaic devices.," *Nat. Mater.*, vol. 9, no. 3, pp. 205–13, 2010.

[3] G. Baffou and R. Quidant, "Thermo-plasmonics: using metallic nanostructures as nano-sources of heat," *Laser Photon. Rev.*, vol. 7, no. 2, pp. 171–187, 2013.

[4] G. Baffou, C. Girard, and R. Quidant, "Mapping Heat Origin in Plasmonic Structures," *Phys. Rev. Lett.*, vol. 104, no. 13, p. 136805, 2010.

[5] B. Lahiri, G. Holland, V. Aksyuk, and A. Centrone, "Nanoscale Imaging of Plasmonic Hot Spots and Dark Modes with the Photothermal-Induced Resonance Technique.," *Nano Lett.*, vol. 13, pp. 3218–3224, 2013.

[6] T. Larsen, S. Schmid, L. G. Villanueva, and A. Boisen, "Photothermal analysis of individual nanoparticulate samples using micromechanical resonators.," *ACS Nano*, vol. 7, no. 7, pp. 6188–93, 2013.

[7] S. Yamada, S. Schmid, T. Larsen, O. Hansen, and A. Boisen, "Photothermal infrared spectroscopy of airborne samples with mechanical string resonators.," *Anal. Chem.*, vol. 85, no. 21, pp. 10531–5, 2013.

[8] S. Schmid, K. D. Jensen, K. H. Nielsen, and A. Boisen, "Damping mechanisms in high-Q micro and nanomechanical string resonators," *Phys. Rev. B*, vol. 84, no. 16, p. 165307, 2011.

[9] P.-L. Yu, T. Purdy, and C. Regal, "Control of Material Damping in High-Q Membrane Microresonators," *Phys. Rev. Lett.*, vol. 108, no. 8, pp. 1–5, 2012.

CONTACT

*S.S. Public, tel: +45 45 25 57 32; sils@nanotech.dtu.dk

RADIATION-PRESSURE ENHANCED OPTO-ACOUSTIC OSCILLATOR

Matthew J. Storey, Siddharth Tallur, and Sunil A. Bhave
Cornell University, Ithaca, USA

ABSTRACT

This paper presents a driving scheme for silicon opto-acoustic oscillators (OAO) by simultaneously exploiting radiation-pressure (RP) and RF feedback oscillation mechanisms to achieve significantly lower phase noise than could be realized by either phenomenon solely. A theoretical model and experimental results are presented corroborating this scheme, demonstrating a silicon OAO operating at 175 MHz with a phase noise of -128.6 dBc/Hz at 1 MHz offset with 2.77 dBm RF output power, resulting in a 10dB far-from-carrier phase noise improvement.

INTRODUCTION

Reference oscillators are ubiquitous elements used in virtually every communication system in existence. The need for miniaturized, batch manufacturable oscillators as chip scale timing references stems from the need to replace the well-established, high performing, albeit expensive quartz oscillators without compromising on performance. MEMS oscillators have recently found applications in various consumer electronic applications. With numerous advances in fabrication technology and materials processing, these oscillators are being pushed to create a presence in the high performance base-band market and high frequency applications. Scaling MEMS oscillators to high frequencies presents challenges in terms of reduced transduction efficiencies and material limits on quality factors. Opto-mechanical transduction offers higher sensitivity and opens up possibilities to interrogate high frequency mechanical resonances hitherto inaccessible. In the past, our group has demonstrated an opto-mechanically transduced MEMS oscillator designed in silicon nitride with zero flicker noise [1], which greatly simplifies the oscillator design and does away with active noise sources that would otherwise add flicker noise, thereby degrading the oscillator phase noise.

Opto-mechanical resonator based oscillators have been previously demonstrated in both open and closed loop configurations utilizing radiation–pressure (RP) [1,2] and RF feedback (RF) [3], respectively. Simultaneous incorporation of these oscillating mechanisms can be achieved with a 2-coupled-ring opto-mechanical resonator. Figure 1 shows an SEM of our device. The coupled resonator can be transduced through either capacitive electrostatics or evanescent optical coupling. While each ring utilizes a different forcing mechanism, the displacement is conserved and transferred between ring resonators through the $\lambda/2$ coupling beam.

In the following sections, a model is presented for the 2-coupled-ring opto-mechanical cavity dynamics utilizing both RP and RF feedback forces. This model was non-dimensionalized to examine the relative affects between

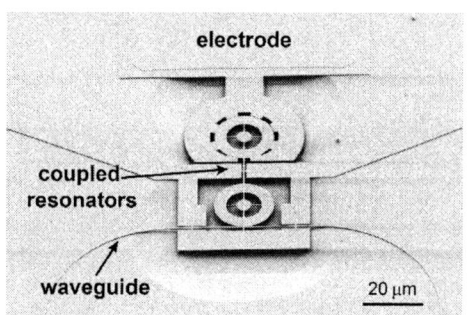

Figure 1: Scanning electron micrograph (SEM) of the 2-coupled-ring resonator. The resonator-waveguide gap is 100nm, each ring has an inner radius of 5.7μm and outer radius of 9.5μm and the resonator-electrode gap is 130nm.

these driving mechanisms and simulations were performed. The 2-coupled-ring resonator was fabricated and tested in vacuum at low temperatures under three operating conditions - RP, RF feedback, and both simultaneously. The phase noise of the oscillator was measured in all three cases and compared at both close-to-carrier and far-from-carrier offset frequencies. We present and discuss these results in subsequent sections.

OAO MODEL

Theoretical Framework

The dynamics of an opto-mechanical cavity have been extensively studied in previous work [4]. The displacement u and optical field $\sqrt{\hbar\Omega_L}a$ inside the cavity are related through the following coupled equations of motion [5]

$$m\ddot{u} + m\gamma_0\dot{u} + m\Omega_0^2 u = F_{rp} + F_{rf}, \qquad (1)$$

$$\dot{a} = i(\Delta_0 + g_{om}u)a - \tfrac{1}{2}\kappa a + \tfrac{1}{2}\kappa n_{max}^{1/2}. \qquad (2)$$

Here, Ω_0 is the mechanical resonance frequency, m is the effective mass, and γ_0 is natural damping of the harmonic resonator. The total optical cavity detuning $\Delta = \Delta_0 + g_{om}u$ is a function of both the laser detuning at zero displacement Δ_0, the dynamic displacement of the cavity u, and the opto-mechanical coupling coefficient g_{om}. The total (loaded) optical linewidth $\kappa = \kappa_i + \kappa_{ex}$ can be expressed as a sum of the intrinsic and extrinsic (coupling) linewidths, respectively. The normalized intra cavity photon number can be expressed as $n = |a|^2$ and the input power P_{in} can be expressed in terms of the maximum intra cavity photon number $n_{max} = 4P_{in}\kappa_{ex}/\kappa^2\hbar\Omega_L$ [5].

The 2-coupled-ring design of our opto-mechanical resonator allows for a unique transduction scheme via simultaneous forcing through both radiation pressure and capacitive electrostatic forces. The force on the cavity generated through radiation pressure is only dependent on the opto-mechanical coupling coefficient and the intra cavity photon number $F_{rp} = \hbar g_{om} n$ [5].

The RF electrostatic force, however, depends on the feedback loop design incorporating the resonator. In our setup, the cavity's output optical power is sent to a photodetector and converted into a photocurrent. The optical field transmitted from the cavity can be expressed in terms of the input optical field and intra cavity field $a_{out} = a_{in} - \sqrt{\kappa_{ex}} a$ where $|a_{out}|^2$ is normalized to the output power and $|a_{in}|^2$ is normalized to the input power (P_{in}) [6]. The total photocurrent is proportional to the output power

$$i_{tot} = |a_{out}|^2 = |a_{in}|^2 + \kappa_{ex} n - 2\,Re\{a_{in}^* a \sqrt{\kappa_{ex}}\}. \quad (3)$$

Since the input power is constant, the current fluctuations are contained in the difference between the output and input power. Therefore, the photocurrent that is fed back to the resonator can be expressed as $i_{rf} = |a_{out}|^2 - |a_{in}|^2$. The photocurrent is then amplified (gain G) with appropriate phase shift and electrostatically applied to the opto-mechanical resonator to close the feedback loop. The general forcing function for the RF feedback is given as $F_{rf} = G i_{rf}$ [7].

Non-dimensionalized Dynamics

To better understand the relative interaction between these two forcing mechanisms, the coupled equations of motion are put into a dimensionless form [5]

$$\tilde{u}'' + \tilde{\gamma}_0 \tilde{u}' + \tilde{u} = c_{om}\tilde{n} + c_{om}\tilde{G}\big(\tilde{\kappa}_{ex}\tilde{n} - 2\sqrt{\tilde{\kappa}_{ex}}Re\{\tilde{a}\}\big), \quad (4)$$

$$\tilde{a}' = i\big(\tilde{\Delta}_0 + \tilde{u}\big)\tilde{a} - \tfrac{1}{2}\tilde{\kappa}\tilde{a} + \tfrac{1}{2}\tilde{\kappa}. \quad (5)$$

Here, the time was scaled by the mechanical resonance frequency $\tau = \Omega_0 t$ and the displacement was scaled as $\tilde{u} = u\, g_{om}/\Omega_0$. All other frequencies were scaled by Ω_0 [5] such that $\tilde{\gamma}_0 = \gamma_0/\Omega_0$, $\tilde{\kappa} = \kappa/\Omega_0$, and $\tilde{\Delta}_0 = \Delta_0/\Omega_0$. The normalized extrinsic linewidth is a fraction of the total normalized linewidth, which depends on the coupling. In general, $\tilde{\kappa}_{ex} = \mu\tilde{\kappa}$ where $0 < \mu < 1$, but for the rest of this analysis the device is assumed to be critically coupled such that $\mu = 0.5$.

The intra cavity optical field was normalized by the maximum photon number such that $\tilde{a} = a/\sqrt{n_{max}}$ and $\tilde{n} = n/n_{max}$. The strength of the radiation pressure force was scaled as $c_{om} = 2n_{max}u_{zpm}^2 g_{om}^2/\Omega_0^2$ and a detailed explanation of the radiation pressure coupling strength can be found in [5]. The strength of the RF feedback force is both a combination of the input optical power contained in c_{om} and the scaled gain $\tilde{G}$ from the amplifier. Therefore, the total dimensionless strength of the closed loop RF feedback force is given by $c_{om}\tilde{G}$.

Phase Noise Improvement

Assuming the harmonic oscillations of the opto-mechanical cavity have an energy that is proportional to the square of the displacement, the oscillation linewidth can be expressed as [8]

$$\delta v = \Delta v \left(\frac{k_B T}{m_{eff}\Omega_0^2}\right)\frac{1}{u^2}. \quad (6)$$

Here, δv is the narrowed linewidth, Δv is the natural linewidth of the resonator, and m_{eff} is the effective mass. The relationship between the oscillator's linewidth and its phase noise $\mathcal{L}$ (dBc/Hz) at a carrier offset frequency Δf in the $1/f^2$ regime is given by [8]

$$\delta v = 2\pi\Delta f^2 10^{\mathcal{L}/10}. \quad (7)$$

By only varying the driving schemes of an OAO, a change in the displacement can vary the degree in which the linewidth narrows. This in turn can change the phase noise at a given carrier offset. Equations (6) and (7) were combined and the phase noise difference at a constant offset was solved for as a function of the displacement ratio

$$\Delta\mathcal{L} = \mathcal{L}_2 - \mathcal{L}_1 = 20\,log_{10}\big(u_1/u_2\big). \quad (8)$$

The equation above shows that the phase noise improvement is proportional to the ratio of oscillation energies. Therefore, if the displacement of second driving scheme is larger than the first driving scheme ($u_2 > u_1$), then there will be an improvement in the phase noise ($\Delta\mathcal{L} < 0$).

Numerical Simulations

The relative displacements of the different driving schemes were compared by numerically integrating equations (4) and (5) for different cases of c_{om} and $\tilde{G}$. Our OAO operates in the unresolved sideband regime (USR) and typically exhibit mechanical quality factors on the order of a couple thousand, so the dimensionless parameters chosen were $\tilde{\gamma}_0 = 0.0005$, $\tilde{\kappa} = 10$, and $\tilde{\Delta}_0 = 3$ [5].

The values for c_{om} and $\tilde{G}$ for the three cases are determined from the threshold behavior of both the RF feedback and radiation pressure induced oscillations. The threshold for radiation pressure oscillations was found to be $c_{om} \approx 0.015$ and the threshold for closed loop RF feedback was $c_{om}\tilde{G} \approx 0.007$.

The first case is just RP oscillations, so the amplifier gain was set to $\tilde{G} = 0$ and the radiation pressure coupling coefficient was set to twice the threshold at $c_{om} = 0.03$. The second case is for just RF feedback, so the radiation pressure force is set well below threshold at $c_{om} = 0.005$ and the amplifier gain was set to just above RF feedback threshold at $\tilde{G} = 2$. In the third case, both forces were placed above threshold. Since the RF feedback force is a function of both gain terms, two comparisons arise for RP enhanced RF feedback oscillations. As c_{om} is raised above

978-1-4799-3510-9/14 $31.00 © 2014 IEEE   1210

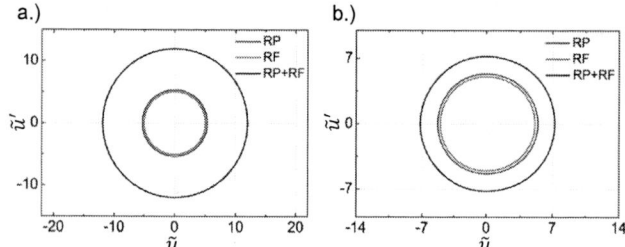

Figure 2: Limit cycle comparisons of the three driving schemes. For RP enhanced RF feedback, plot (a) is when amplifier gain is held constant and plot (b) is for the total closed loop gain held constant.

threshold (0.03), either the amplifier gain $\tilde{G}$ can be held constant at 2 or the total closed loop gain can be held constant at $c_{om}\tilde{G} = 0.01$, which requires reducing the amplifier gain to $\tilde{G} = 1/3$.

Figure 2 shows the simulation results of the three cases. When the driving scheme was only radiation pressure or RF feedback, the oscillator reached a steady limit cycle with scaled displacement amplitude of approximately 5. When both driving schemes were incorporated while holding the amplifier gain constant, the scaled amplitude reached a value of 12 (plot (a) of figure 2). If the total closed loop gain was held constant, the scaled amplitude reached a value of 7.2 (plot (b) of figure 2). Plugging these values into equation (8), the resulting phase noise improvement would be approximately 8 dB and 3 dB, respectively.

EXPERIMENTAL SETUP

The 2-coupled-ring opto-mechanical resonator was fabricated on a silicon-on-insulator (SOI) wafer and the fabrication process was described in detail in [3]. All experiments were performed in a Lakeshore probe station under vacuum (30μTorr) using Liquid Nitrogen to cool the chamber to 80 K. An optical probe was used to send continuous wave (CW) light into the device through a pair of on-chip grating couplers. A GSG probe was used to apply an RF signal to the bond pads, which connected to the

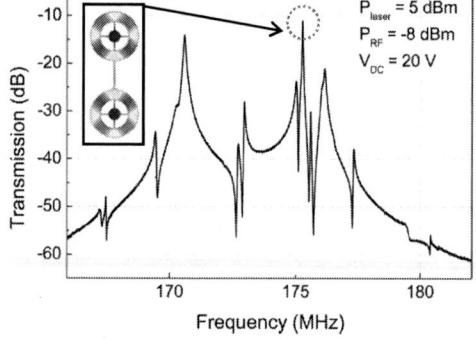

Figure 3: RF Transmission spectrum for the opto-acoustic resonator measured in vacuum (30μTorr) and low temperature (80 K). The mechanical mode at 175.3 MHz corresponds to the fundamental radial expansion mode (inset: mode-shape) with a quality factor of 6,000.

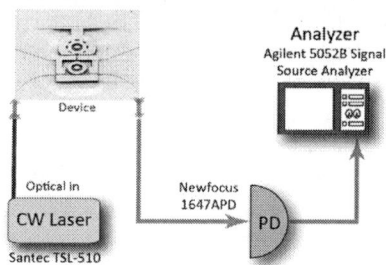

Figure 4: Experimental setup for radiation pressure oscillations

electrodes around the resonator.

Open loop measurements were performed on the opto-acoustic resonator to determine the RF transmission spectrum of the mechanical mode of interest. As the applied RF signal was swept, the transmitted power was sent to a photodetector and the resulting photocurrent was input to a Network Analyzer. Figure 3 shows the electromechanical transmission measurement for the fundamental radial mode. The resonance frequency (175.3 MHz) and mechanical quality factor (6,000) were estimated through a Lorentzian curve fit.

For comparison purposes, the phase noise performance of the opto-mechanical resonator based oscillator was first evaluated for RP and RF feedback separately. Since the RP and RF feedback driving mechanisms differ in their application to the resonator, this experiment required two separate setups. Figure 4 shows the setup for a RP driven opto-mechanical oscillator. A CW diode laser was input to the device and then the output optical power is measured with a Newfocus 1647 photodetector and sent to the phase noise analyzer. RP induced oscillations are achieved by blue detuning the laser within the optical resonance and using an input power above threshold [1,2].

To ensure that the only means of achieving oscillations is by closing the feedback loop, the laser power is reduced below the threshold for RP oscillations. To create a RF feedback oscillator, the output RF signal from the photodetector is amplified and the required phase shift is introduced to overcome the FB oscillation threshold. The signal is sent through a 3dB splitter and one half is applied to the resonator through the GSG probe while the other half

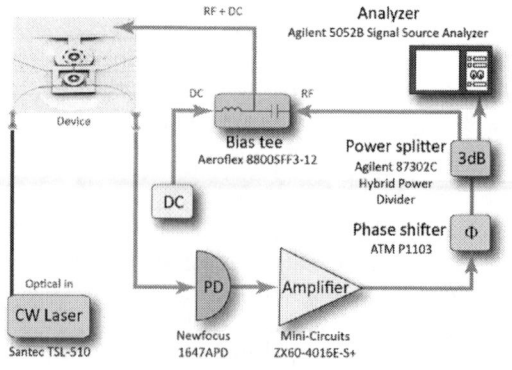

Figure 5: Experimental setup for feedback oscillations

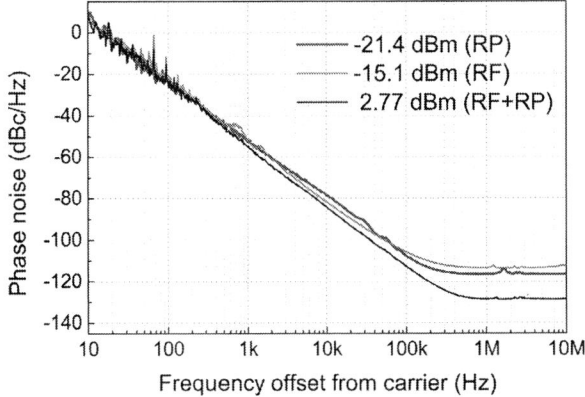

Figure 6: Phase noise data for all three driving mechanisms. For each plot, the RF output signal power and driving scheme is given.

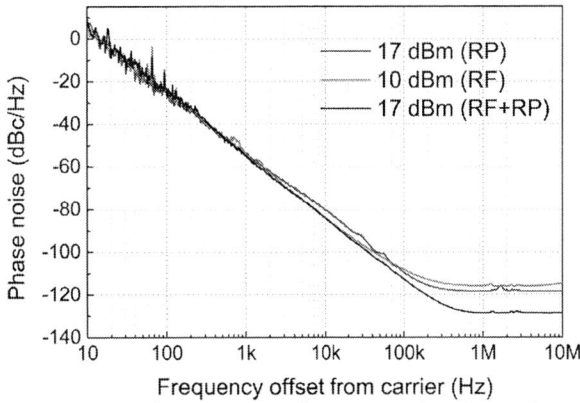

Figure 7: Phase noise data normalized to the power consumption. For each plot, the input optical power and driving scheme is given.

is sent to the phase noise analyzer. Figure 5 shows the experimental setup for the RF feedback oscillator. To achieve oscillations induced simultaneously by both RP and RF feedback, the closed loop setup shown in Figure 5 is used with the laser blue detuned and with the optical power raised above RP threshold conditions.

RESULTS

Using the setup in figure 4, the device was driven into RP oscillations at an input optical power of 17 dBm with a determined RP threshold of 11 dBm. The setup in figure 5 was then used to obtain RF feedback oscillations with an input optical power below RP threshold (10 dBm). While in the same closed loop configuration, the input optical power was increased well above threshold (17 dBm) to observe RP enhanced RF feedback oscillations. Any further increase in laser power would result in thermal nonlinearities and chaotic oscillations. All of the closed loop measurements were carried out with 30 V DC bias. Figure 6 shows the resulting phase noise for the three measurements, along with the carrier powers and driving scheme. The individual RP and RF feedback oscillations had the lowest carrier powers (-21.4 dBm and -15.1 dBm, respectively) with comparable phase noise performance. The simultaneous RP and RF feedback oscillations demonstrated a large increase in carrier power (2.77 dBm) while experiencing an improvement in the far-from-carrier phase noise.

The phase noise measurements were normalized to the oscillator's laser and amplifier power consumption and compared in figure 7 and table 1 (photodetector power consumption was too large and would have diminished the

normalization's effectiveness). The RP enhanced RF feedback oscillations had no $1/f^4$ or higher order noise and the close-to-carrier phase noise is dominated by $1/f^3$ flicker noise from both the amplifier and input laser. A 10dB improvement in far-from-carrier phase noise was observed.

CONCLUSION

The 2-coupled-ring resonator based oscillator with a mechanical quality factor of 6,000 and resonant frequency of 175.3 MHz was tested under vacuum (30μTorr) and low temperature (80 K) for different driving schemes. Simultaneous RP and RF feedback induced oscillations resulted in an increased RF output signal power (2.77 dBm) when compared to only RP (-21.4 dBm) and RF feedback (-15.1 dBm) oscillations. The combined driving scheme resulted in more energy stored in the oscillator and a 10dB improvement in far-from-carrier phase noise (1MHz).

Acknowledgement: This work was supported under the DARPA ORCHID program. We would also like to acknowledge Tanay Gosavi and Professor Clark Nguyen for discussions of phase noise models.

REFERENCES

[1] S. Tallur, S. Sridaran and S. A. Bhave, *MEMS 2012*, pp. 19-22.

[2] T. O. Rocheleau, C.T.-C. Nguyen, et al., *MEMS 2013*, pp. 118-121.

[3] S. Sridaran and S. A. Bhave, *Transducers '11*, pp. 2920-23.

[4] T.J. Kippenberg and K. J. Vahala, *Science*, vol. 321, pp. 1172-1176 (2008)

[5] M. Poot, H. Tang, et al, *Phys. Rev. A*, vol. 86, p. 053826 (2012)

[6] A. Schliesser, T. Kippenberg, et al, *Nature Physics*, vol. 4, pp. 415-19 (2008)

[7] G. I. Harris, U. L. Andersen, J. Knittel, W. P. Bowen, *Phys. Rev. A*, vol. 85, p. 061802 (2012)

[8] H. Rokhsari, M. Hossein-Zadeh, A. Hajimiri, K. Vahala, *Appl. Phys. Lett.*, vol. 89, p. 261109 (2006)

Table 1: Phase noise comparisons at offset frequencies.

Operation mode	(Normalized power consumption) Phase Noise (dBc/Hz)		
	1kHz Offset	100kHz Offset	1MHz Offset
RP	-53.62	-109.6	-118.4
RF	-53.95	-108.2	-115.8
RP+RF	-54.74	-112.6	**-128.6**

THERMOPILE INFRARED ARRAY SENSOR FOR HUMAN DETECTOR APPLICATION

J. Tanaka, M. Shiozaki, F. Aita, T. Seki, and M. Oba,
OMRON Corporation, Kyoto, JAPAN

ABSTRACT

This paper reports the design of thermopile infrared sensor for human detector application. Sensitivity and response time of thermopile infrared sensor element are important for human detector application. In order to fulfill the specification, we developed S-shaped structure for thermopile infrared sensor element and fabrication process of chip scale vacuum package for mass production of the thermopile infrared sensors. As the result, 140V/W sensitivity and 17msec response time of the thermopile infrared sensor element are achieved. The thermopile infrared sensor is all fabricated by CMOS processes.

INTRODUCTION

Recently, uncooled infrared array sensors are used for human detector application. Uncooled infrared array sensors detect the presence of humans by sensing infrared rays emitted from human body. The sensors are utilized for energy savings of air conditioners and lights in the office or home. It is expected that the infrared sensor applications are not limited only for energy-saving fields, but also extended to safety and security fields, including the monitoring for senior citizens. Because the functions are not used only to detect the presence of humans but also to obtain information on the positions and the movements of humans as well as temperature distribution.

For this purpose, infrared sensors must be assembled in arrays so that they can detect line and plane information, rather than mere spot information. Currently, existing infrared array sensors typically used for infrared cameras are very expensive. Therefore, the development of inexpensive infrared array sensors is demanded.

The typical detection methods used for infrared sensors are (i) thermopile, (ii) pyroelectric, (iii) bolometric, and (iv) diode methods. Thermopile has potential of low cost because it can be fabricated by CMOS process. However, the thermopile method has a weak point that thermoelectric effect is lower and less sensitive than the other methods.

In previous work, the method to increase sensitivity has developed by increasing the aperture ratio trough an increase of the number of thermopiles or expanding the size of the absorbing film [1]. However this method goes against the trend of downsizing. And another paper has reported that used vacuum packaging and two pairs of thermocouples of p-n polysilicon [2]. Sensitivity becomes higher because heat insulation of thermopile infrared sensor element becomes higher. But response time of thermopile infrared sensor element becomes late.

On the other hand, chip scale vacuum packaging technology has developed [3]. In this technology, only good dies in a wafer can be simultaneously packaged. But this technology is not suitable for mass production because solder ejection method and vacuum sealing method have less compatibility with CMOS processes.

This paper reports the design of thermopile infrared sensor for human detector application. Sensitivity and response time of thermopile infrared sensor element are important for human detector application. In order to fulfill the specification, we developed S-shaped structure for thermopile infrared sensor element and fabrication process of chip scale vacuum package for mass production of the thermopile infrared sensors.

DESIGN AND PROCESS

Specification of thermopile infrared array sensor

The schematic image of using infrared array sensor is shown in Figure 1. And our target specification of the thermopile infrared sensor package is shown in Table 1. We assume that human detector is used by mounting under the ceiling for detecting the presence of humans. Human stays or moves in the detection area. Height from the floor to the thermopile infrared array sensor is about 3.0m. Height from the floor to the detection area is about 1.0m. Size of human body seen from the ceiling is about 40cm × 80cm. Moving speeds at various cases in the office or room was assumed less than 3.0m/sec. Infrared rays radiated from human body are converged to thermopile infrared sensor elements through silicon lens, and thermopile infrared sensors detect the temperature difference between surface temperature of human body and temperature of the room. The temperature difference between surface of the human body and temperature of the room was about 1 - 4 degrees Celsius.

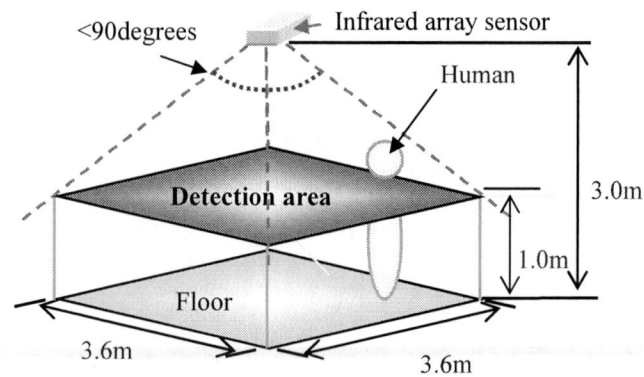

Figure 1: The schematic image of using infrared array sensor

From the schematic image of use, viewing angle of the thermopile infrared array sensor needs over 90 degrees. Minimum sensing area of thermopile infrared element is about 0.2m in the center of the detection area. Maximum sensing area of thermopile infrared sensor element is about 0.4m in the corner of the detection area. The number of infrared sensor element was set at 256 (=16 × 16) in order to detect the temperature of human body in 2-8 pixels. It is sufficient if flame rate of thermopile infrared sensor operates at 10 fps to detect human in the office or room. So response time τ of thermopile infrared sensor element was set at <33msec (=1000msec/10fps/3). Temperature resolution (Netd) has to be less than 0.3 degrees Celsius for detecting human body with a high degree of accuracy. So we specified that the sensitivity of thermopile infrared sensor element is more than 100V/W.

Table 1: Target specification of the thermopile infrared sensor for human detector application

Item	Value	Background
Number of sensor elements	256 (=16 x 16)	-Height from floor to infrared array sensor -Height from floor to detection area -Size of human body
Response time	Faster than 33msec	-Speed of human moving
Sensitivity	More than 100V/W	- Temperature difference between human body and room

Design of the thermopile infrared sensor element

Sensitivity of thermopile infrared sensor element is given by

$$S = n \cdot \alpha \cdot Rth \cdot A_b \cdot A_r \qquad (1)$$

Where n is the number of thermocouple pairs, α the Seebek coefficient, R_{th} the thermal resistance between the hot and cold junctions, A_b the infrared radiation absorpitivity, and A_r is the fill factors. The design concept of a vacuum sealed thermopile infrared sensor element is shown in Figure 2. In case that thermopile infrared sensor element was vacuum-sealed, the thermal resistance between the hot junction and cold junctions almost equals to the thermal resistance of the membrane $Rth_{(membrane)}$ because of the very large thermal resistance of the filling gas $Rth_{(gas)}$. On the other hand, response time τ of the thermopile infrared sensor element is given by

$$\tau \propto R_{th} \qquad (2)$$

To increase sensitivity, small number of thermocouple pairs n is suitable. Small n yields high thermal resistance R_{th} and low electrical resistance of the thermopile infrared sensor element. Low electrical resistance can achieve low electrical noise. Electrical noise of the thermopile infrared sensor element is proportional to the square root of the electrical resistance. In previous report [2], $n=2$ was adapted for

achieving high sensitivity and low electrical noise. But response time of the thermopile infrared sensor element becomes longer because of high thermal resistance.

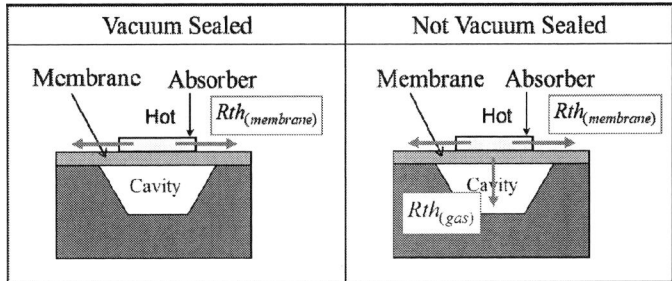

Figure 2: The design concept of a vacuum-sealed sensor

Schematic diagram for calculating sensitivity is shown in Figure 3. Figure 3(a) is schematic diagram of the thermopile infrared sensor element. The thermopile infrared sensor element consists of two beams and a membrane. Length of each beam is L, and width of each beam is W. Figure 3(b) is schematic diagram of the cross section of the beam. Beam consists of multiple thermocouples and two sidewalls. Width of each beam W and length of each beam L are given by

$$W = l \cdot n + 2r \qquad (3)$$

$$L = -2 \cdot l \cdot n + (2A - 4d - 2r) \qquad (4)$$

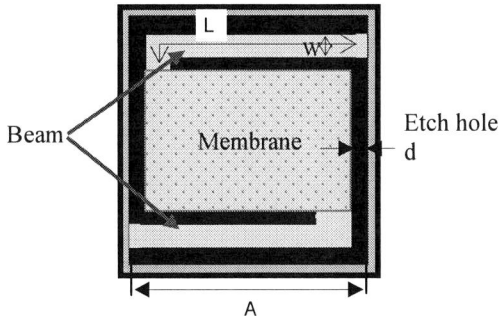

(a) Thermopile infrared sensor element

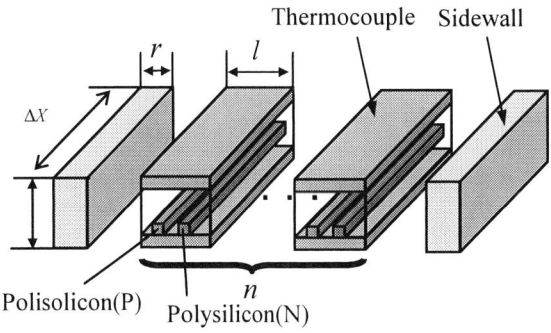

(b) the cross section of the beam

Figure 3: Schematic diagram for calculation

978-1-4799-3510-9/14 $31.00 © 2014 IEEE

The thermal resistance of the thermopile infrared sensor element $Rth_{(membrane)}$ is given by

$$1\Big/ R_{th(membrane)} = 2\left[\sum_{1}^{n}1\Big/R_{th(thrmocouple)} + 2\Big/R_{th(sidewall)}\right] \quad (5)$$

$Rth_{(thermocouple)}$ is the thermal resistance of the thermocouple pairs, $Rth_{(sidewall)}$ is the thermal resistance of the sidewall. The dependence of sensitivity S and electrical resistance on the number of thermocouple pairs n is calculated on that basis. The dependence of response time τ is also calculated by using a finite element method. So the number of thermocouple pairs n is optimized.

Calculation results are shown in Graph 1. Graph 1(a) shows the calculation result of the Sensitivity/$\sqrt{\text{Resistance}}$. Graph 1(b) shows the calculation result of response time. Sensitivity/$\sqrt{\text{Resistance}}$ decreases with an increasing number of thermocouple pairs n. On the other hand, response time becomes fast with an increasing number of thermocouple pairs n. The number of thermocouple pairs n is set at $n=10$ because response time is faster than 33msec and sensitivity is the highest. Sensitivity and electrical resistance are calculated at 200V/W and 120kΩ.

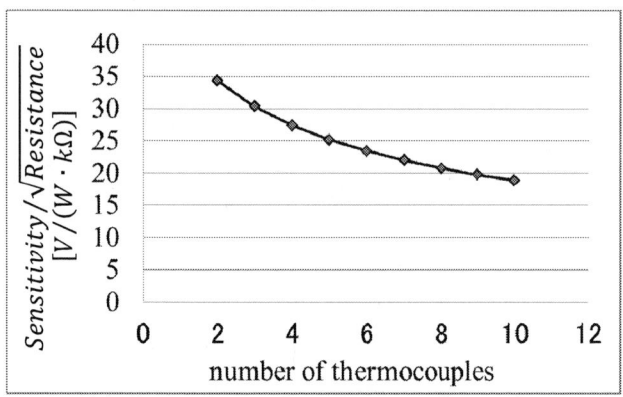

(a) Dependence of sensitivity on the number of thermocouple pairs n

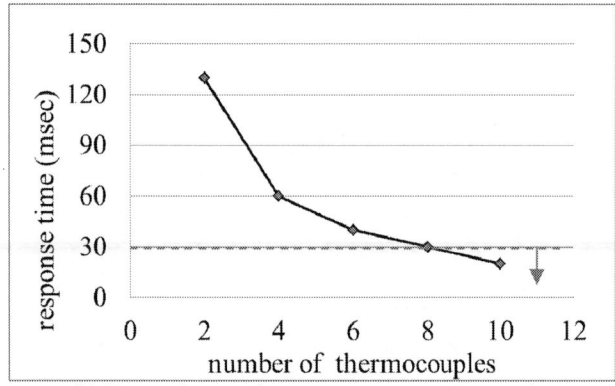

(b) Dependence of response time on the number of thermocouple pairs n

Graph 1: Calculation results

The images of thermopile infrared sensor which was fabricated are shown in Figure 4. Figure 4(a) is the photograph of the thermopile infrared sensor element. The size of the thermopile infrared sensor element is 220um × 220um. As a result, structure of the designed thermopile infrared sensor element becomes S-Shaped. S-Shaped structure of the infrared sensor element is suitable for human detector application. Figure 4(b) is the SEM images of thermopile infrared array sensor chip. The number of thermopile sensor elements are 256(= 16 × 16).

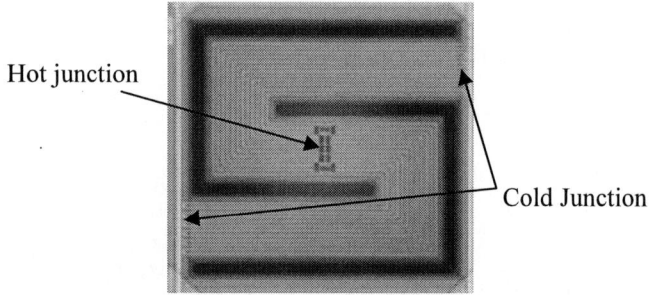

(a)Photograph of the thermopile infrared sensor element

(b) SEM image of the sensor chip (16X16 arrays)

Figure 4: The images of fabricated infrared sensors

Fabrication process of the chip scale packaging

The schematic structure of chip scale package for thermopile infrared sensor is shown in Figure 5. The package consists of sensor chip and lid. A ring shaped metal pattern is formed on both the device chip and the lid.

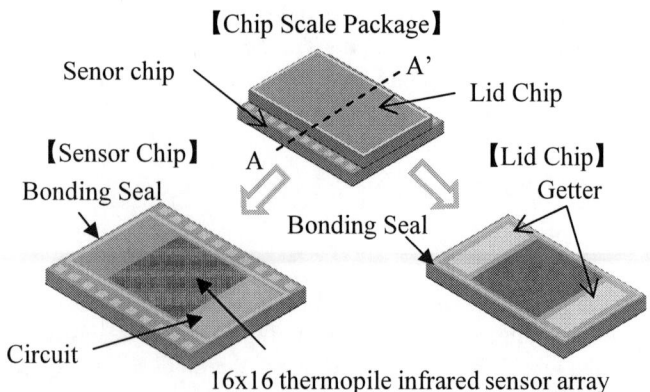

Figure 5: The schematic structure of the vacuum package

978-1-4799-3510-9/14 $31.00 © 2014 IEEE

Fabrication process of the vacuum package is described (1)-(4). (1)Sensor wafer and Lid wafer are loaded in the vacuum camber. Pressure of the vacuum chamber is lower than 1.0*10^(-5) Pa. (2)The two wafers are baked at 200 degrees Celsius for outgassing. (3)Then the two wafers are contacted and bonded at 350 degrees Celsius. (4)After bonding, the wafer is cut to each package size. The vacuum package is all fabricated by CMOS processes. The photograph of fabricated vacuum packaging is shown in Figure 6.

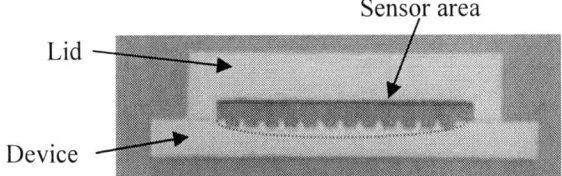

Figure 6: The photograph of chip scale package
(A-A' cross section)

RESULT AND DISCUSSION

The sensitivity and response time were measured by using the optical and electrical setup shown in Figure 7. The sensitivity of the sensor was measured by illuminating it with infrared radiation emitted by a black body at 77 degrees Celsius. The response time was measured by illuminating the sensor by an optical chopper. The output signal of the sensor was first amplified and then introduced into an oscilloscope to obtain a waveform tracing for evaluating the response time.

Figure 7: Schematic of evaluation system

The sensitivity of the thermopile infrared sensor element before bonding sensor wafer and lid wafer is measured in the vacuum chamber. The results were shown in Graph 2. Sensitivity of the thermopile sensor element achieved 180V/W at lower than 5Pa. Electrical resistance of the thermopile infrared sensor element is 110kΩ. Response time of the sensor element was 17msec.

Then, sensitivity of the infrared sensor element after bonding is measured. The sensitivity is 140V/W.As a result, pressure of vacuum chamber after bonding device wafer and lid wafer is estimate about 35Pa. It is sufficient level to provide adequate thermal isolating characteristics.

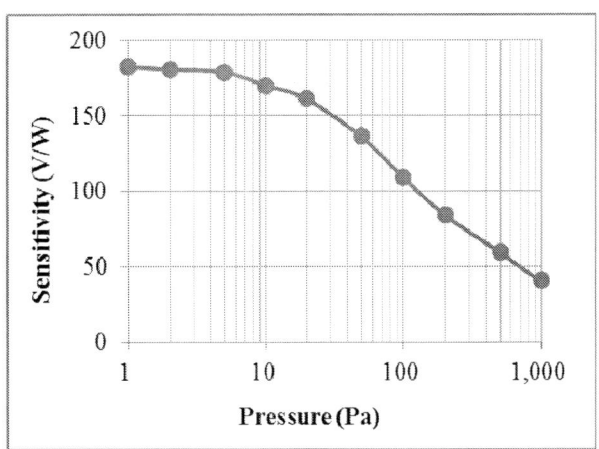

Graph 2: The Graph of sensitivity of thermopile infrared sensor element

CONCLUSION

This paper reports the design of thermopile infrared sensor for human detector application. Sensitivity and response time of thermopile infrared sensor element are important for human detector application. In order to fulfill the specification, we developed S-shaped structure for thermopile infrared sensor element and fabrication process of chip scale vacuum package for mass production of the thermopile infrared sensors. As the result, 140V/W sensitivity and 17msec response time of thermopile sensor are achieved. And thermopile infrared sensors are all fabricated by CMOS processes.

And, in the future, it is possible to increase element number of thermopile infrared arrays by performing the pressure of vacuum below 10Pa with degasifying process before bonding sensor wafer and lid wafer.

ACKNOWLEDGEMENTS

A part of this work was supported by New Energy and Industrial Technology Development Organization (NEDO).

REFERENCES

[1] T.Ishikawa, M.Kimata, Proc. SPIE, Vol.4130, 152, 2000
[2] M. Hirota, "120 x 90 element thermopile infrared focal plane array with precisely patterned Au-black absorber", Sensors and Actuators A135, 2007, pp. 146-151.
[3] M.Takeda, M.Kimata, "Chip scale vacuum packaging for uncooled IRFPA", IEEJ Trans. FM, Vol. 127, No. 7, 2007

TRANSFER-PRINTED COMPOSITE MEMBRANES FOR ELECTRICALLY-TUNABLE ORGANIC OPTICAL MICROCAVITIES

Annie Wang*, Wendi Chang*, Apoorva Murarka*, Jeffrey H. Lang, Vladimir Bulović
Massachusetts Institute of Technology, USA

ABSTRACT

We demonstrate a method for fabricating organic optical microcavities which can be electrostatically actuated to dynamically tune their optical output spectra. Fabrication of an integrated organic micro-opto-electro-mechanical system (MOEMS) cavity is enabled by the solvent-free additive transfer of a composite membrane. Electrical actuation and optical characterization of a completed cavity show resonance tuning greater than 20 nm for membrane deflections of over 200 nm at 50 V.

INTRODUCTION

Tunable-MEMS optical microcavities have been previously demonstrated as reconfigurable filters [1] or inorganic vertical cavity surface-emitting lasers that operate at infrared wavelengths [2]. Organic-based light-emitting materials are of interest due to their broad visible spectrum, but previously reported tunable organic light emitting cavities either required a bulky external mirror stage [3] or their output spectra were determined during fabrication, not dynamically [4]. The availability of a compact, single-system tunable visible light source would benefit a wide range of fields including remote sensing, spectroscopy, and biomedical applications.

The substitution of organic emissive materials for the inorganic emissive materials previously demonstrated in tunable air-gap MEMS structures [5] is complicated. Frequently, the organic semiconductors are incompatible with conventional MEMS and CMOS processes that involve exposure to wet chemistries, solvents, plasma, or elevated process temperatures. In this work, fabrication of integrated tunable organic optical microcavities is enabled by the solvent-free, additive membrane transfer process we have developed. By electrostatically actuating the composite membrane, we observe the cavity resonances shift by greater than 20 nm for estimated electrostatic pressures up to 2.5 kPa. Compared to previously reported all-optical photonic crystal based pressure sensors [6], this optically-pumped device potentially allows single-point contactless-readout for large area pressure sensor arrays. Additionally, the device structure and transfer technique are easily applicable to large area fabrication of electrostatically tuned organic lasers.

DEVICE STRUCTURE / FABRICATION

A schematic cross-section of the device structure and the test setup is shown in Figure 1. The membrane-microcavity structure is formed by the additive transfer of a 700 nm-thick composite membrane onto a pre-patterned bottom substrate. In contrast to the additive transfer-

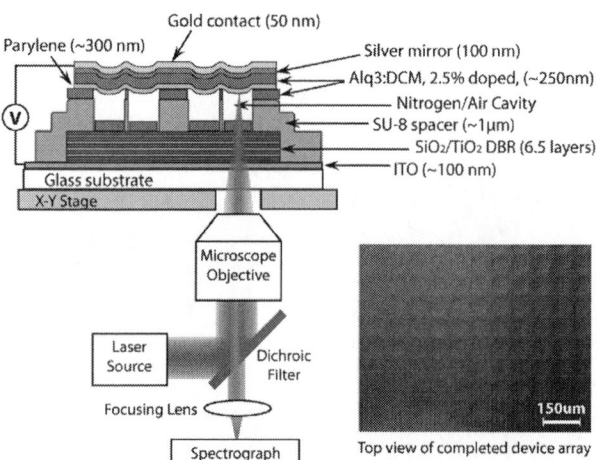

Figure 1: Cross-sectional view of a MEMS tunable cavity device structure with a diagram of the optical testing setup. A microscope objective focuses laser excitation and captures cavity emission. Subsequent filter and lens focus emission into spectrograph. Inset: Fabricated device array from top view.

printing of suspended gold membranes reported at MEMS 2012 [7], the transfer method reported here avoids the use of solvents, which can damage the organic active layer [tris-(8-hydroxyquinoline)aluminum (Alq$_3$) doped with 4-(dicyanomethylene)-2-methyl-6-(4-dimethylaminostyryl)-4H-pyran (DCM)].

Figure 2 details the fabrication process for the membrane-microcavity structure. The bottom substrate consists of a 12.7 mm x 12.7 mm glass substrate coated with indium tin oxide (ITO), which serves as a transparent bottom electrode. Alternating layers of SiO$_2$ and TiO$_2$ are sputtered to form a distributed Bragg reflector (DBR) mirror with a 100 nm stop-band centered at 620 nm, which matches the peak photoluminescence (PL) emission of the organic layer. The DBR passes an excitation wavelength of 400 nm with less than 10% reflectivity, but reflects over 98% of the emission wavelength of 620 nm. The final 1 μm-thick layer of TiO$_2$ acts as an optical cavity spacer on top of the DBR. Circular cavities 50 μm in diameter and ~1μm deep are then patterned in a SU-8 (MicroChem Corp.) layer atop the DBR stack.

The composite membrane is fabricated separately, starting with the chemical vapor deposition of a 300 nm thick layer of a transparent polymer, parylene, onto a glass carrier substrate. A 250 nm-thick layer of Alq$_3$ doped at 2.5% with DCM is thermally co-evaporated through a shadow mask onto both the membrane and the cavity-patterned substrate. Next, a 100 nm-thick silver

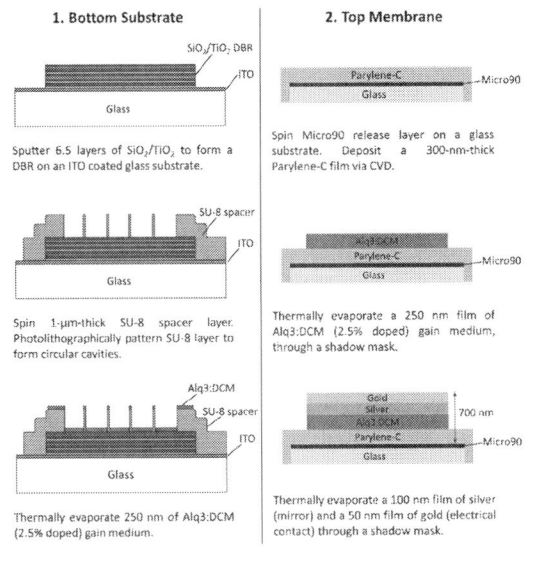

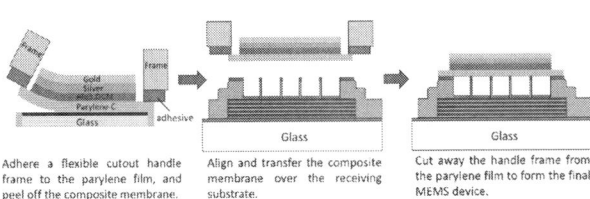

Figure 2: Schematic diagram of the membrane-microcavity fabrication process. The bottom SiO₂/TiO₂ DBR mirror and SU-8 spacer layer are patterned on ITO-coated glass. Separately, a 700 nm-thick composite membrane comprising ultra-thin parylene, Alq₃:DCM gain material, silver mirror, and gold contact layers is prepared on a glass carrier. The membrane is released by peeling off the carrier and is directly transferred to the patterned substrate to complete the device.

mirror and a 50 nm-thick gold contact are deposited in sequence onto the membrane.

As illustrated in Figure 2, the completed 700 nm-thick composite membrane is attached to a flexible cutout handle frame and released from the carrier by peeling. The membrane is then transferred to the cavity-patterned substrate, and it adheres lightly to the substrate due to the bottom parylene layer. The handle is then removed. The use of ultra-thin parylene enables solvent-free transfer of a large area (8 mm x 8 mm) composite membrane that encloses an array of microcavities (Figure 1, inset).

ELECTROSTATIC ACTUATION
Electromechanical Characterization

The composite membrane deflections under electrostatic actuation are characterized using an optical interferometer (Wyko NT9100, Bruker Nano Inc.). Voltage biases from 0 V to 50 V are applied between the top gold membrane contact and the bottom ITO electrode. The electrostatic force of attraction between the flexible top membrane and the rigid bottom electrode causes the

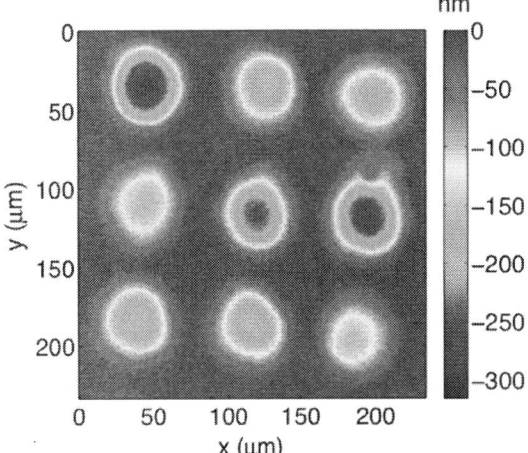

Figure 3: Net membrane deflection over multiple circular cavities. Optical interferometric measurement at 0 V was subtracted from that at 50 V to generate the net deflection contour image. The non-uniform deflection profiles over identical 50 μm-diameter cavities reflect non-uniform tension across the membrane, likely due to the transfer process.

suspended composite membrane to deflect into the circular air cavities (Figure 3). For the single cavity shown in Figure 4, the 700 nm-thick membrane has a maximum deflection of 106 nm when no voltage is applied and it increases to 325 nm at 50 V actuation. The deflection at 0 V is due to the transferred membrane sagging into the cavity because of insufficient tension in the membrane. Membrane tension is not uniform across the membrane, resulting in varying maximum deflections across the multiple cavities shown in Figure 3. At 50 V applied bias, maximum deflections across these cavities range from 140 nm to 290 nm. This non-uniformity likely arises due to the membrane transfer process rather than the patterning of the bottom substrate.

Young's Modulus of the Composite Membrane

The composite membrane undergoes reversible deflections over the range of applied voltages shown in Figure 4. Since the maximum deflection, w_0, is on the order of the thickness, h, of the membrane, the theory of bending of plates that is valid for deflections much smaller than the thickness of the plates is inapplicable. Hence, membrane deflections are modeled using the energy method, as outlined in [7-8]. It is assumed that the deflection profile, $w(r)$, of the membrane, obtained by applying spatially-uniform pressure, q, can be modeled as

$$w(r) = w_0 \left(1 - \frac{r^2}{a^2}\right)^2 \qquad (1),$$

where r is the radial distance from the center of the cavity of radius a, and w is the deflection of the membrane at r. Since the deflection of the membrane is non-zero when no voltage is applied, we fit the above equation to the difference in the membrane deflection at any applied voltage and at 0 V. Figure 5 shows the fit of the deflection

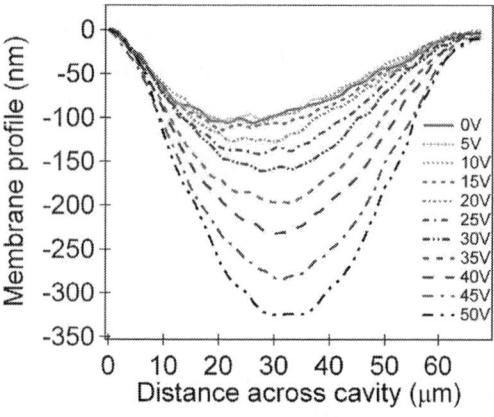

Figure 4: Membrane deflection profile as a function of applied voltage. The single membrane-cavity profiled here corresponds to the center cavity in the contour image in Figure 3. Note that the actual profile, not the net deflection, is shown in this figure.

profile from Equation 1 to the measured *net* deflection profile at 50 V. This close fit implies that the radially-varying electrostatic pressure, *q(r)*, acting on the membrane at any non-zero applied voltage can be approximated as an effective uniform pressure, q_{eff}, using

$$2\pi \int_0^a q(r)\delta w_0 \left(1 - \frac{r^2}{a^2}\right)^2 rdr$$

$$= 2\pi q_{eff} \int_0^a \delta w_0 \left(1 - \frac{r^2}{a^2}\right)^2 rdr \qquad (2)$$

The effective pressure is then used to calculate the effective Young's modulus, *E*, of the composite membrane. At 50 V actuation, q_{eff} is 2.5 kPa for the cavity shown in Figure 4. Assuming a Poisson's ratio of 0.4 for the 700 nm-thick composite membrane, the effective *E* is 4 GPa. This estimated Young's modulus is comparable to that of other organic films [9-10], suggesting that the mechanical deflection characteristics of the composite membrane in the large deflection regime are primarily due to its organic film constituents.

OPTICAL CHARACTERIZATION

Optical measurements are performed using a 400 nm excitation laser at 1 kHz repetition rate with averaged pulse energy of 0.2 nJ. As illustrated in Figure 1, a microscope objective focuses a collimated laser beam through the planar DBR into the cavity. The same objective captures subsequent cavity mode emission. A 405 nm long pass filter diverts any residual laser light, and the emission is focused into a grating spectrograph for spectral analysis. The laser excites the optically active organic Alq3:DCM layer, which has a broad emission spectrum [11]. This organic photoluminescence (PL) is enhanced by the vertical resonator cavity created by the planar DBR and the silver mirror layer within the composite membrane.

Figure 6 shows typical measured spectra of the longitudinal cavity mode of the structure illustrated in

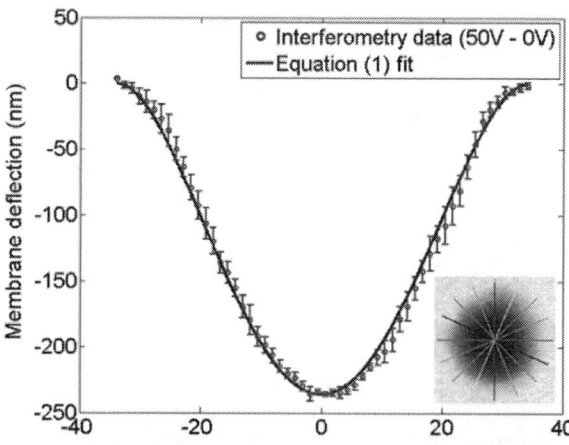

Figure 5: Eight different diametrical net deflection profiles of the composite membrane at 50 V actuation are averaged and the mean deflection profile is plotted. Equation (1) is fitted to this average net deflection profile (inset: difference between interferometry data at 50 V and 0 V for a single cavity, with colored lines indicating the 8 different diametrical profile locations).

Figure 1. The cavity length is tuned to provide the observed change in device emission wavelengths by electrostatically actuating the composite membrane. The emission of the device shows reversible tuning over 20 nm. Each longitudinal mode emission was fitted to a Gaussian function to find mode peak and width.

Figure 7 plots the fitted peak of the measured 637 nm cavity mode, the central peak shown in Figure 6. We observed a reversible blue shift of mode emission up to 23 nm, or a 3.6% change in peak wavelength. The device fabricated had a total optical path length of approximately 6.34 μm as calculated using the measured free spectral range, which is defined by the separation between the spectral mode peaks. This corresponds to approximately 20 longitudinal modes or 10 wavelengths within the vertical cavity resonator. Thus, the total physical deflection of the composite membrane scales by a factor of 10, resulting in an estimated membrane deflection of 230 nm at 50 V. This calculated membrane deflection from the optical data is indicated on the right vertical axis of Figure 7; the same figure also plots, in shaded grey, the maximum and minimum deflection from the interferometry data for the corresponding voltages. The optical resonance shift is in good agreement with the interferometry measurements.

Since the electrostatic force on the membrane is proportional to the square of the applied voltage, a quadratic relationship between the membrane deflection and the applied voltage is observed. Consequently, peak emission shift is proportional to the square of the applied voltage. This relationship does not hold at high voltages as the membranes have been observed to collapse into the patterned cavities due to pull-in, and stiction prevents further electrostatic modulation of membrane deflection.

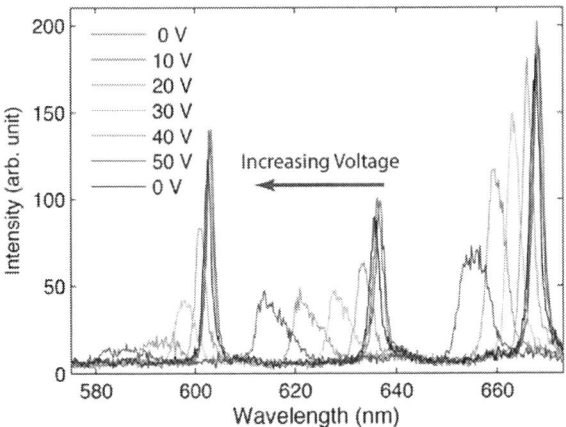

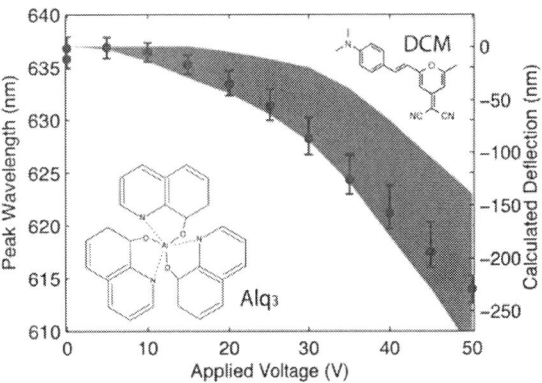

Figure 6: Longitudinal mode emission spectra for a range of applied voltage. The 637 nm center mode shifts to 614 nm under 50 V actuation, indicating a deflection of the composite membrane of over 200 nm. The measurements were taken in order of increasing voltage. The final measurement at 0 V (indicated by the black 0 V spectrum) shows that the tuning is reversible.

Figure 7: Fitted peak wavelength and corresponding calculated membrane deflection as a function of applied voltage to illustrate quadratic voltage dependence and tunable range. The optical measurements (blue) were taken in order of increasing applied voltage; an additional 0 V measurement (lower 0 V data point) illustrates reversibility. This is compared to the interferometric data (grey), which is shaded to show the range of measured membrane deflection from Figure 3. Inset: Chemical structures of emissive organic host Alq_3 and dopant DCM laser dye.

CONCLUSION

We have demonstrated the use of a solvent-free additive membrane transfer technique in an electrostatically tunable organic device. Electrical actuation and optical characterization of a completed device show spectral tuning greater than 20 nm for membrane deflections of over 200 nm at 50 V.

Shorter optical cavities with fewer modes, fabricated using the outlined technique, could yield higher tunable range at a lower applied voltage. With high tunable range and mechanical pressure sensitivity, the demonstrated device structure can be applied towards pressure sensing and tunable organic lasing device applications. Moreover, the composite membrane elements can be varied in both geometry and composition to optimize device performance and to explore other physical phenomenon.

ACKNOWLEDGEMENTS

The authors gratefully acknowledge funding from the National Science Foundation Center for Energy Efficient Electronics Science. We also thank G. Akelsrod and P. Deotare for useful discussions on optical characterization. *These authors contributed equally to this work.

REFERENCES

[1] M. W. Pruessner, T. H. Stievater, W. S. Rabinovich, "Reconfigurable filters using MEMS resonators and integrated optical microcavities," in *Proc. IEEE MEMS*, 2008, pp. 766–769.

[2] T. Ansbaek, I.-S. Chung, E. S. Semenova, O. Hansen, K. Yvind, "Resonant MEMS Tunable VCSEL," *IEEE J. Sel. Top. Quantum Electron.*, vol. 19, no. 4, p. 1702306, 2013.

[3] M. Zavelani-Rossi, G. Lanzani, S. De Silvestri, M. Anni, G. Gigli, R. Cingolani, G. Barbarella, L.

Favaretto, "Single-mode tunable organic laser based on an electroluminescent oligothiophene," *Appl. Phys. Lett.*, vol. 79, no. 25, p. 4082-4084, 2001.

[4] D. Schneider, T. Rabe, T. Riedl, T. Dobbertin, M. Kroger, E. Becker, H.-H. Johannes, W. Kowalsky, T. Weimann, J. Wang, P. Hinze, "Ultrawide tuning range in doped organic solid-state lasers," *Appl. Phys. Lett.*, vol. 85, no. 11, p. 1886-1888, 2004.

[5] C. J. Chang-Hasnain, "Tunable VCSEL," *IEEE J. Sel. Top. Quantum Electron.*, vol 6, no. 6, pp. 978-987, 2000.

[6] Y. Lu, A. Lal, "Photonic crystal based all-optical pressure sensor," in *Proc. IEEE MEMS*, 2011, pp. 621–624.

[7] A. Murarka, S. Paydavosi, T. Andrew, A. Wang, J. Lang, V. Bulovic, "Printed MEMS membranes on silicon," in *Proc. IEEE MEMS*, 2012, pp. 309-312.

[8] S. Timoshenko, *Theory of Plates and Shells*, McGraw-Hill, 1959.

[9] T. A. Harder, T.-J. Yao, Q. He, C.-Y. Shih, Y.-C. Tai, "Residual stress in thin-film parylene-C," in *Proc. IEEE MEMS*, 2002, pp. 435–438.

[10] J. M. Torres, N. Bakken, C. M. Stafford, J. Li, B. D. Vogt, "Thickness dependence of the elastic modulus of tris(8-hydroxyquinolinato) aluminium," *Soft Matter*, vol. 6, no. 22, p. 5783-5788, 2010.

[11] A. Uddin, C. B. Lee, "Exciton behaviours in doped tris (8-hydroxyquinoline) aluminum (Alq_3) films". *Phys. Status Solidi C*, vol. 8, pp 80–83, 2011.

CONTACT

*A. Wang, tel: +1-617-258-9139; aiwang@mit.edu

978-1-4799-3510-9/14 $31.00 © 2014 IEEE

TUNABLE METAMATERIALS BY CONTROLLING SUB-MICRON GAP FOR THE THZ RANGE

Akihiro Isozaki[1], Tetsuo Kan[1], Hidetoshi Takahashi[1], Natsuki Kanda[2,3], Natsuki Nemoto[4], Kuniaki Konishi[2], Makoto Kuwata-Gonokami[2,4], Kiyoshi Matsumoto[5] and Isao Shimoyama[1,5]

[1]Graduate School of Information Science and Technology, The University of Tokyo, Tokyo, Japan
[2]Photon Science Center, The University of Tokyo, Tokyo, Japan
[3]RIKEN, Saitama, Japan
[4]Department of Physics, The University of Tokyo, Tokyo, Japan
[5]Information and Robot Technology Research Initiative, The University of Tokyo, Tokyo, Japan

ABSTRACT

We propose a tunable metamaterial actuated by pneumatic force. The unit of the proposed metamaterial has a pair of sprit-ring-resonators (SRR), and the gap between them is controllable around sub-micron-order in size. One SRR is formed on a thin cantilever, which can be bent due to differential pressure between the upper and lower surfaces of the cantilever, whereas the other is fixed on a thin membrane. We confirmed that controlling the gap ranging from sub-micron-order to a few-micron-order was suitable for tuning a resonant frequency of a terahertz metamaterial.

INTRODUCTION

Metamaterials are artificial materials composed of array of unit cells smaller than the wavelength of interest to provide optical properties unavailable in nature. Researchers have been challenging to actively control metamaterial properties for application in optical modulators [1, 2]. Recently, many tunable metamaterials constructed by MEMS have been reported [3-7] in the terahertz (THz) range because MEMS devices have comparable size to the wavelength of THz wave, and can directly actuate structural configurations of the unit cells, namely "meta-atom". It is reported that the optical properties of the meta-atom largely depends on a distance from a neighbor-most meta-atom[8]. The MEMS metamaterials have a potential to be able to take advantage of the distance dependency of the optical properties by combining an actuation element.

The closer distance between two meta-atoms is, the stronger distance dependency of the optical properties is. Previous MEMS tunable metamaterials used the distance dependency when distance between two meta-atoms is around 10-micron-order, which corresponds to one-tenth of the wavelength of THz light. However, it has been difficult to design further closely packed meta-atoms to obtain the large modulation capability with a sub-micron gap, because a design of movable parts with such narrow gaps usually suffers from a stiction problem. In this study, we propose actuating meta-atoms in a vertical direction to the plane, where the meta-atoms is formed. Our proposed structure can avoid the stiction problem and take advantage of the distance dependency ranging from sub-micron- to a few-micron-order.

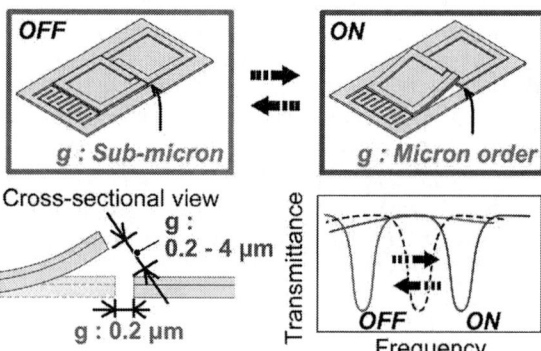

Figure 1: Behavior of a tunable metamaterial unit. By controlling a gap between double SRR, transmission spectrum is tuned.

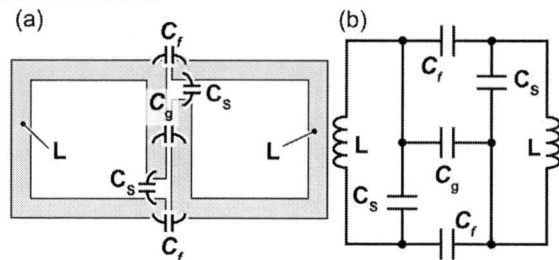

Figure 2: Equivalent circuit mode of the proposed double SRR.

PRINCIPLE AND DESIGN

Principle

We employed a sprit ring resonator (SRR) as a typical meta-atom structure. A configuration of the proposed tunable metamaterial is shown in Fig. 1. It is well known that SRRs work as an LC circuit when the frequency of the incident light matches the resonant frequency of the SRR [9]. Changing a gap size between a double SRR leads to a change of resonant frequency of the metamaterial because a capacitance component C in the double SRR depends on the gap size. In this letter, we assume that an equivalent circuit model of the proposed double SRR is shown in Fig. 2. The capacitance component C_g and C_f strongly depends on the gap size between each SRR. The equivalent circuit model gives relation between the resonant frequency f_R and capacitance component C_g and C_f as,

978-1-4799-3510-9/14 $31.00 © 2014 IEEE

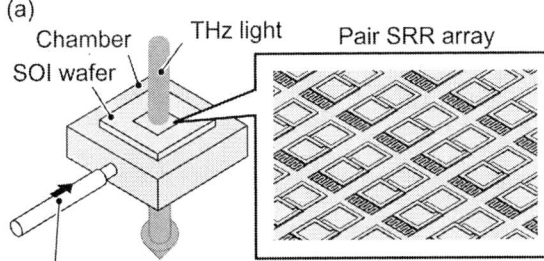

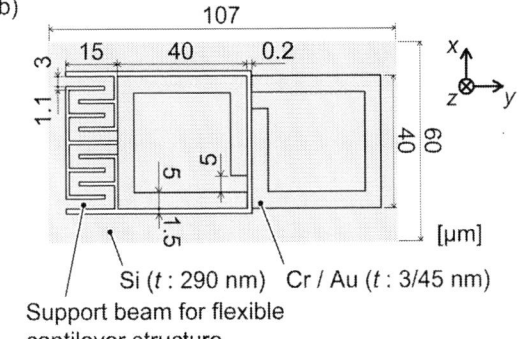

Figure 3: (a) Configuration of the proposed tunable metamaterial which is actuated by pneumatic force. (b) Design of the double SRR formed on 290-nm-thick Si membrane.

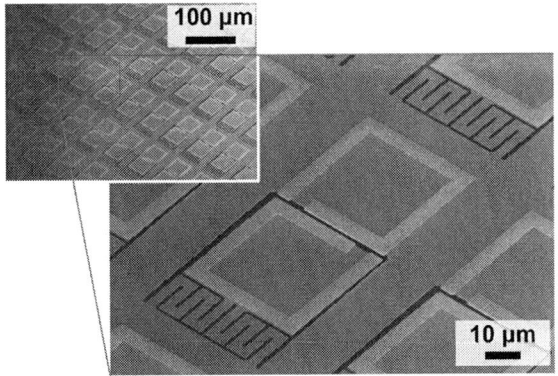

Figure 4: SEM images of the fabricated metamaterial.

$$f_R = \frac{1}{2\pi\sqrt{LC_s}} \frac{1}{\sqrt{1 + \dfrac{C_g C_f}{C_s(2C_f + C_g)}}} \ . \qquad (1)$$

A relation between the gap g and capacitance C of two parallel plates is given as [10, 11],

$$C \propto g^{-0.222} \ . \qquad (2)$$

From equation (1) and (2), the relation between the gap g and resonant frequency f_R is given as,

$$f_R = \frac{1}{2\pi\sqrt{LC_s}} \frac{1}{\sqrt{1 + Ag^{-0.222}}} \ , \qquad (3)$$

where A is constant coefficient. From equation (3), the shift of the resonant frequency by the gap size change is significant as the gap size decreases. In this study, to take

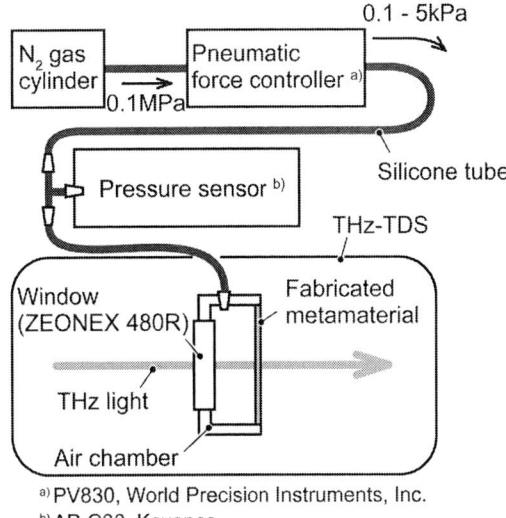

a) PV830, World Precision Instruments, Inc.
b) AP-C30, Keyence

Figure 5: Pneumatic actuation setup for tuning the fabricated metamaterial.

advantage of this dependency, the gap size is designed to be sub-micron-order in the initial state.

Design

Double SRR array is formed on an SOI wafer, which was fixed on an air chamber (Fig. 3a). Pneumatic force was applied through a silicone tube for actuating one of two SRRs. A detailed design of the double SRR is shown in Fig. 3b. The double SRR, which is comprised of an Au layer, is formed on a 290-nm-thick silicon membrane. The typical size, width and thickness of each SRR are 40, 5 and 0.045 μm, respectively. Trench was formed around one of the two SRRs to make it a cantilever structure. Therefore, the SRR on the cantilever is movable only in the vertical direction to the SOI wafer by the pneumatic force. Thus, the cantilever structure avoids the stiction problem in spite of the gap between the double SRR being narrow. A zigzag structure of the support beam made the cantilever flexible. The width of the zigzag beam is 3 μm.

The gap size between the double SRR is designed to be 200 nm, which is roughly one-thousandth of the wavelength of THz light. When pneumatic force was zero, the gap size kept 200 nm. When pneumatic force was applied, the cantilever bent and the gap size increased.

A fabrication process of the proposed metamaterial is basically the same as a process described in our previous research [7]. Scanning electron microscope (SEM) images of the fabricated metamaterial are shown in Fig. 4. The cantilever structures keep flat states when no pneumatic force is applied.

PNEUMATIC ACTUATION

A setup for pneumatic actuation is shown in Fig. 5. A source of pneumatic actuation was nitrogen gas, which supplied gas with 0.1 MPa to a pneumatic force controller. Controlled pneumatic force was applied to the fabricated metamaterial, which was fixed on the air chamber. The

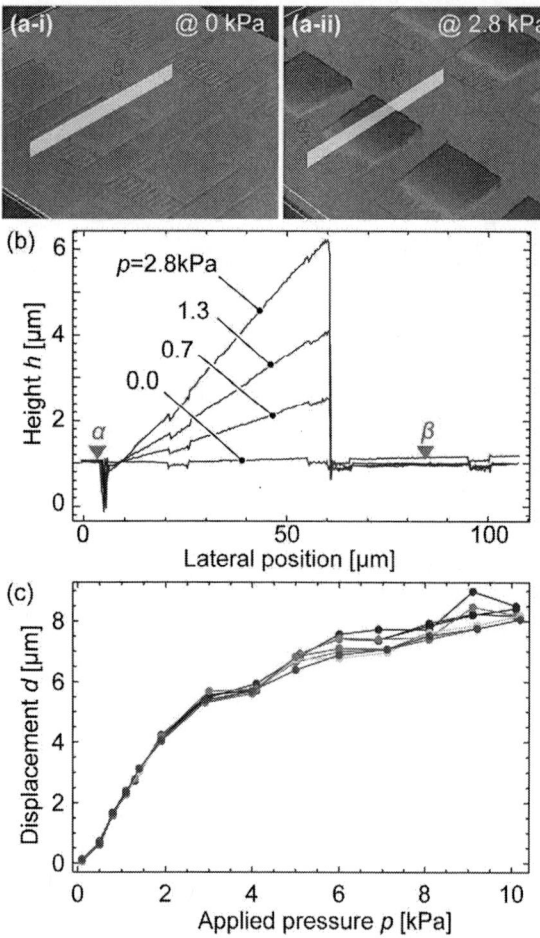

Figure 6: (a) Three-dimensional profile of the fabricated tunable metamaterial obtained by the laser three-dimensional profiler. (b) Cross sectional profiles of four states of a cantilever. (c) Displacement at the tips of eight cantilevers in the vertical direction.

pneumatic force controller and the air chamber were connected with a silicone tube. In the middle of the tube, a pressure sensor was set to monitor pneumatic force. An actuation property of the cantilever structure was evaluated with a laser microscope. Figure 6a shows profile images at 0 kPa (a-i) and 2.8 kPa (a-ii). When pneumatic force increased, the structure went upward and the gap g between the SRRs increased (Fig. 6b). A displacement of the tip of the actuated cantilever was 6 μm at 2.8 kPa. Figure 6c shows the relation between the applied pressure and the displacement d of the tips of eight cantilevers. In the range of 0 to 2 kPa, the relation is roughly linear. Above 2 kPa, the stability of the displacement is decreased. Therefore, in this actuation setup, the stable mechanical deformation of the SRRs is achieved from 0 to 4 μm.

TRANSMITTANCE MEASUREMENT

To evaluate a tunability of the fabricated metamaterial, we measured the transmission spectra of the fabricated metamaterial using the terahertz time domain spectroscopy

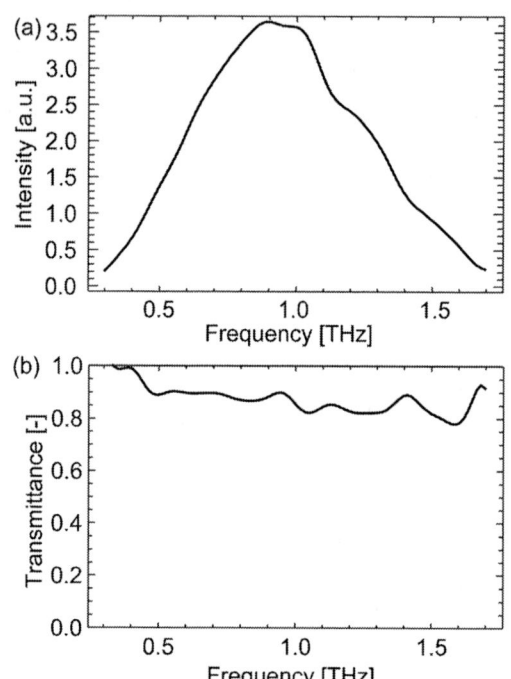

Figure 7: Spectrum of (a) emitter (LiNbO₃) and (b) the window substrate (Zeonex 480R).

(THz-TDS) system [12]. The system allows us to obtain the waveform of a terahertz wave directly in the time domain. In this study, we used an emitter and a detector based on non-linear optical crystals, which is $LiNbO_3$ and $ZnTe$, respectively. A spectrum of the emitter is shown in Fig. 7a. Transmission spectra shown in the following sections were obtained by normalizing with the emitted spectrum.

Figure 7b shows a transmission spectrum of a 3-mm-thick window substrate (Zeonex 480R, Zeon Manufacturing Vietnam co., Ltd), which is used in our pneumatic actuation setup shown in Fig. 5. In the range of 0.3 to 1.7 THz, the measured transmittance is over 0.8.

Transmittances of the fabricated metamaterial were measured to investigate the tunability of the resonant frequency. The measured transmission spectra are shown in Fig. 8 for the incident polarization parallel to x-axis, which is defined in Fig. 2. As shown in Fig. 8a, each spectrum has a signature dip in the vicinity of 0.9 THz. Figure 8b shows the enlarged view of the signature dips. These results indicate that the resonant frequency shifted depending on the gap g, which was changed by the applied pneumatic force. We picked up the resonant frequencies, and showed them in Fig. 9. This graph confirmed that blue shift of the resonant frequency was occurred, and the experimental results are fitted well by equation (3). An obtained equation of fitting curve is

$$f_R = \frac{1.07}{\sqrt{1 + \dfrac{0.490}{g^{0.222}}}} . \quad (4)$$

These results indicate that controlling gap in the range bellow a few-micron is effective for tunable metamaterials

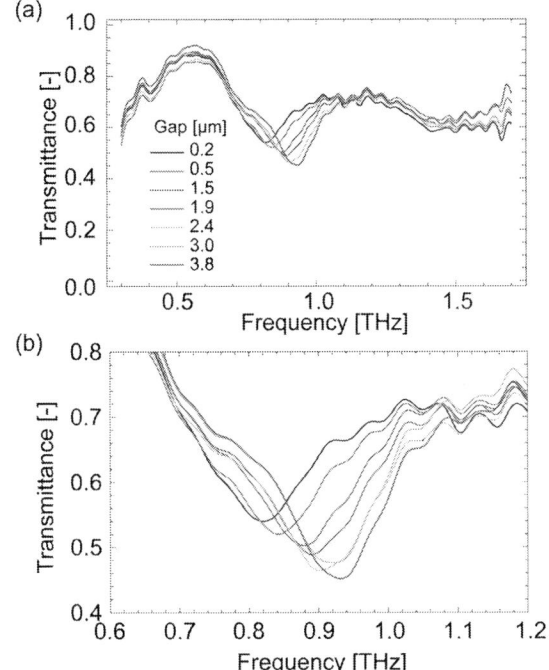

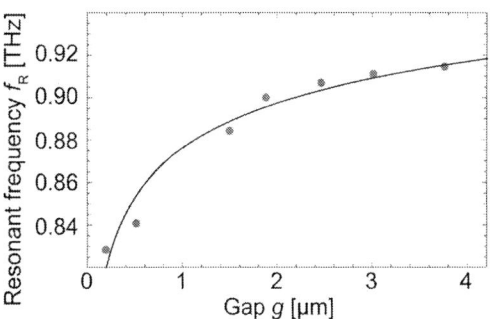

Figure 9: Relation between gap g and resonant frequency f_R. Dots are measurement data and the solid black line is fitting curve.

Figure 8: Spectra of the fabricated metamaterial in the range (a) from 0.2 to 2.0 THz and (b) from 0.6 to 1.2 THz.

compared with the range around 10 micron-order.

CONCLUSIONS

We fabricated the tunable metamaterial that has the 200-nm gap in the initial state. The controllable sub-micron gap made a tunability of the metamaterial property effective below a few-micron deformation. The measurement resonant frequency of the proposed metamaterial was changed from 0.83 to 0.90 THz by only 2 μm deformation.

ACKNOWLEDGMENTS

The photolithography masks were fabricated using an EB lithography apparatus (F5112+VD01) donated by Advantest Corporation at the VLSI Design and Education Center (VDEC) of the University of Tokyo. This work was partially supported by the Murata Science Foundation. This research was partially supported by the Photon Frontier Network Program, the Special Coordination Funds for Promoting Science an d Technology funded by the Ministry of Education, Culture, Sports, Science, and Technology (MEXT), Japan, and the JSPS through its FIRST Program.

REFERENCES

[1] H.-T. Chen, W. J. Padilla, J. M. O. Zide *et al.*, "Active terahertz metamaterial devices," *Nature,* vol. 444, no. 7119, pp. 597-600, 2006.
[2] W. L. Chan, H. T. Chen, A. J. Taylor *et al.*, "A spatial light modulator for terahertz beams," *Applied Physics Letters,* vol. 94, no. 21, pp. 213511, May, 2009.

[3] A. Q. Liu, W. M. Zhu, D. P. Tsai *et al.*, "Micromachined tunable metamaterials: a review," *Journal of Optics,* vol. 14, no. 11, pp. 114009, Nov, 2012.
[4] H. Tao, A. C. Strikwerda, K. Fan *et al.*, "Reconfigurable Terahertz Metamaterials," *Physical Review Letters,* vol. 103, no. 14, pp. 14740, Oct 2, 2009.
[5] W. M. Zhu, A. Q. Liu, X. M. Zhang *et al.*, "Switchable Magnetic Metamaterials Using Micromachining Processes," *Advanced Materials,* vol. 23, no. 15, pp. 1792-1796, Apr 19, 2011.
[6] W. M. Zhu, A. Q. Liu, T. Bourouina *et al.*, "Microelectromechanical Maltese-cross metamaterial with tunable terahertz anisotropy," *Nature Communications,* vol. 3, pp. 1274, 2012.
[7] T. Kan, A. Isozaki, N. Kanda *et al.*, "Spiral metamaterial for active tuning of optical activity," *Applied Physics Letters,* vol. 102, no. 22, pp. 221906, Jun 3, 2013.
[8] N. Liu, H. Liu, S. N. Zhu *et al.*, "Stereometamaterials," *Nature Photonics,* vol. 3, no. 3, pp. 157-162, Mar, 2009.
[9] A. Isozaki, T. Kan, K. Takano *et al.*, "Batch fabrication of a double-layer metamaterial resonator using scalloping structures," *Journal of Micromechanics and Microengineering,* vol. 23, no. 8, pp. 085006, Aug, 2013.
[10] T. Sakurai, and K. Tamaru, "Simple formulas for two- and three-dimensional capacitances," *Electron Devices, IEEE Transactions on,* vol. 30, no. 2, pp. 183-185, 1983.
[11] F. Stellari, and A. L. Lacaita, "New formulas of interconnect capacitances based on results of conformal mapping method," *Ieee Transactions on Electron Devices,* vol. 47, no. 1, pp. 222-231, Jan, 2000.
[12] M. Hangyo, M. Tani, and T. Nagashima, "Terahertz time-domain spectroscopy of solids: A review," *International Journal of Infrared and Millimeter Waves,* vol. 26, no. 12, pp. 1661-1690, Dec, 2005.

CONTACT

Akihiro Isozaki, Tel: +81-3-5841-6318; E-mail: a_isozaki@leopard.t.u-tokyo.ac.jp

978-1-4799-3510-9/14 $31.00 © 2014 IEEE

UNCOOLED MULTI-BAND IR IMAGING USING BIMATERIAL CANTILEVER FPA

Wei Ma[1], Shuyang Wang[1], Yongzheng Wen[1], Yuejin Zhao[2], Liquan Dong[2], Ming Liu[2], Xiaohua Liu[2], and Xiaomei Yu[1]

[1]Institute of Microelectronics, Peking University, Beijing, CHINA
[2]School of optoelectronics, Institute of Beijing Technology, Beijing, CHINA

ABSTRACT

This paper presents the design, fabrication and performance of a 256×256 bimaterial cantilever focal plane array (FPA) which is able to work in the three infrared (IR) atmospheric windows of 1~2.5μm, 3~5μm and 8~14μm simultaneously. The FPA employs a silicon-framed structure by selectively etching away the substrate with Deep Reactive Ion Etching technique, and a stacked layer of chromium and SiN_x serves as the multi-band absorber. The images of short wavelength, middle wavelength and long wavelength infrared were captured successfully with the same FPA by combining the Chromium nano-film with silicon nitride as the multi-band IR absorber. The measured sensitivity of the FPA is 0.18μm/K.

INTRODUCTION

IR imagers have attracted attention for many applications such as night vision, environmental monitoring and security surveillance, among which multi-band detection is a highly favored feature [1]. At present, multi-band IR imaging systems have almost been dominated by cooled photon imagers such as HgCdTe, InGaAsSb or Quantum Well (QW) detectors due to the convenience to control the cut-off wavelength, and by vertically integrating multiple detectors in one pixel, their response can be tailored for multi-band application [2, 3]. These detectors need ponderous cooling equipment to suppress the dark current and thus limited for civil use. As for uncooled systems, multi-band detection was realized either by changing the resonance gap of VO_x bolometer [4] or by separate detection using different sensors with a post image fusion [5]. These two approaches require either a manual adjustment of detection bands or a post-imaging process. Other attempt employs 2D plasmonic crystals as the wavelength selective absorber [6], but requires additional fabrication steps and precise control of the size of the surface structure.

This paper demonstrates a simple way to realize multi-band imaging of bimaterial FPA by introducing a stacked layer of chromium and SiN_x as the multi-band absorber. Bimaterial cantilever based IR detectors, utilizing the thermal expansion coefficient mismatch of two materials, will deform with temperature change due to absorption of IR radiation and the deformation of the FPA pixels can be conveniently readout by an optical system [7].

MULTI-BAND ABSORPTION MECHANISM

Due to the Si-N stretching vibrational mode, silicon nitride exhibit a high loss tangent around 11μm [8].

Therefore, SiN_x with enough thickness is commonly used as the absorber for long wavelength infrared in bimaterial cantilever bolometers, but the absorption drops to almost zero at short and middle wavelength. Thick metal films will reflect all the incident IR radiation while transmission is predominant in very thin metal films. A maximum of 50% absorption would occur when the sheet resistance of the metal films reaches a critical value of 188Ω/□ if the influence of substrate is ignored [9]. Therefore, by stacking SiN_x and appropriate thin metal films, multi-band absorption can be realized. The measured absorption spectrum of 800nm SiN_x is illustrated in Figure 1, showing a maximum of 76% absorption at 11.47μm. The absorption of a stacked layer of 20.8 nm Cr and 800nm SiN_x films was also measured, more than 40% absorption in short wavelength and middle wavelength IR and more than 85% in long wavelength IR are achieved. The broadening of the absorption band accounts for the multi-band response of the fabricated FPA, which will be discussed later.

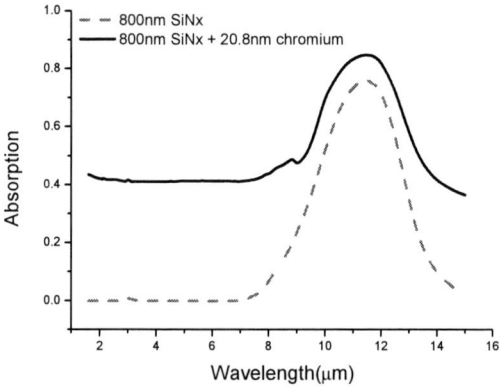

Figure 1: The measured absorption spectra of 800nm SiNx film (red/dashed line) and the stacked layer of 800nm SiNx and 20.8 nm Cr nano-films.

FPA DESIGN

An FPA of 256×256 with the pixel dimension of 100μm×80μm was designed. The schematic view is shown in Figure 2. For the cantilever pixel, the 60μm×40μm IR absorber/mirror is supported by a four-fold leg anchored to the silicon frame on both sides. The four-fold leg consists of two bimaterial deflection legs and two thermal isolation legs separated by each other alternatively. The IR absorber is a stacked layer of SiN_x and thin Cr films which serves as the function of multi-band absorptions as well as the adhesion layer of gold mirror to the SiN$_x$ structure. Note that the IR wave propagates back and forth in the absorber due to the existence of the gold mirror, which is equivalent to a double of the film thickness. The bimaterial deflection leg is composed of a layer of SiN$_x$ and a thick layer of gold, the thermal expansion coefficients of which differ by one order of magnitude to give a bending response when temperature changes. The thermal isolation leg is a layer SiN$_x$ with extremely small thermal conductance to minimize the heat exchange between different pixels.

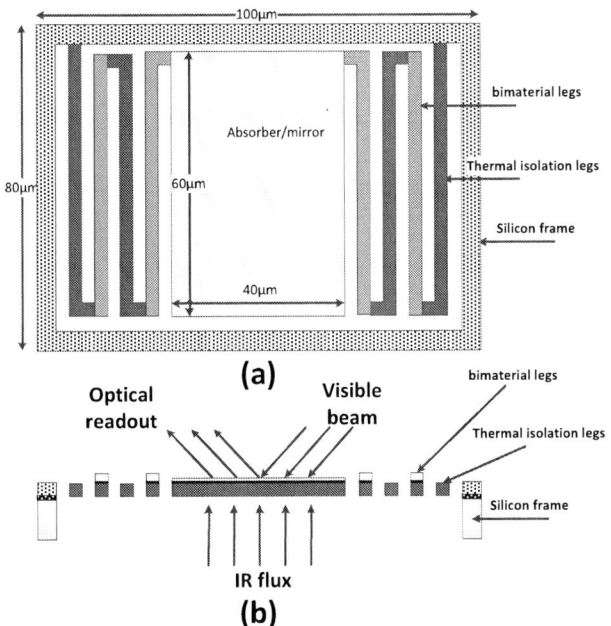

Figure 2: Schematic diagram of the top (a) and cross section (b) view of a cantilever pixel.

To estimate the thermal response of the FPA, the Finite Element Analysis software ANSYS was used to simulate the deformation of the cantilever upon IR flux. The substrate is fixed to a temperature of 25°C and neglecting heat exchange between the cantilever and air. When IR flux of 100pW/μm^2 is cast on the absorber, the temperature distribution and cantilever deformation are simulated (Figure 3). It is evident that the temperature drops mainly on the thermal isolation legs due to its low thermal conductance. The absorber reaches the highest temperature of 30.2°C and the deflection is 0.47μm and 0.50μm on each side. An important figure of merit of bimaterial FPA is its thermo-mechanical sensitivity, defined as the amount of deformation over a unit temperature change, $\Delta z/\Delta T$, where Δz is the maximum deflection of the cantilever and ΔT is the temperature change. Therefore, the sensitivity of the designed FPA is calculated to be 0.19 μm/K.

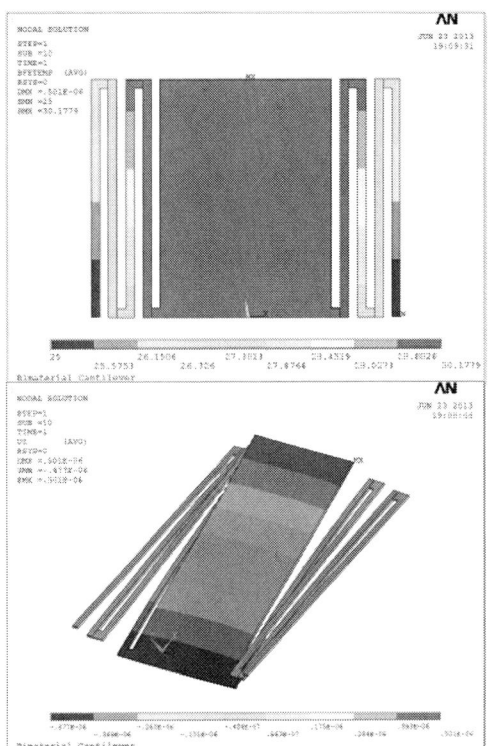

Figure 3: The simulated temperature distribution (top) and deformation (bottom) of the cantilever upon IR flux by ANSYS.

FABRICATION

Typical fabrication of IR FPA employs a surface sacrificial layer micromachining technique. IR radiation has to penetrate through the silicon substrate and suffers a considerable loss of about 48% [10]. This problem can be solved by the removal of silicon substrate under the cantilever, allowing a direct exposure of the absorber to IR radiation.

The fabrication of silicon-framed FPA employs a bulk silicon micromachining process including 4 lithography steps which is schematically shown in Figure 4. Firstly, a SiO$_2$ film of 200nm was deposited on a 400μm double polished wafer to form the etching buffer layer and then a 450nm layer of low stress SiN$_x$ was deposited with low pressure chemical vapor deposition (LPCVD) technique. Next, a thick layer of 15/300nm Cr/Au was sputtered and wet etched to form bimaterial legs. Following that, another thin film of 15/50 nm Cr/Au was sputtered and patterned for the mirror by lift-off technique. Cr functions as the adhesion layer between Au and SiN$_x$, and also as the absorbing material for short and middle wavelength infrared. The thin layer of 50 nm gold was chosen to decrease the surface roughness and guarantee the uniformity of the films as a mirror. The SiN$_x$ was patterned and dry etched to form the pixel structure afterwards. Self-aligned process was applied in the leg area to improve the accuracy of lithography step. At last, the substrate was selectively removed by Deep Reactive Ion Etching (DRIE) in the areas defined by double-side lithography and stopped at the SiO$_2$ buffer layer. The SiO$_2$

buffer layer was finally etched in a BHF solution to release the structure completely. Figure 5 shows the SEM picture of the fabricated FPA, which shows a good uniformity among pixels. The bending of cantilever at room temperature is due to intrinsic stress.

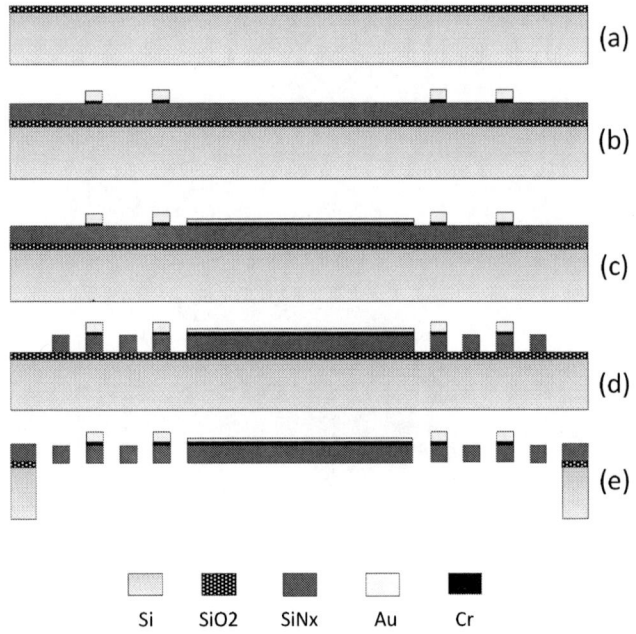

Si   SiO2   SiNx   Au   Cr

Figure 4: The bulk micromachining process of silicon framed FPA: (a) Deposition of SiO2 buffer layer. (b) Deposition of SiN$_x$ and thick layer of gold for bimaterial leg. (c) Deposition and patterning of a thin layer of gold for mirror. (d) Etching of SiN$_x$ to form the pixel. (e) Removal of the substrate.

Figure 5: The SEM photo of PFA, and inset is a zoom-in pixel.

RESULTS AND DISCUSSION

The performance of the FPA was evaluated by both the thermal mechanical response and imaging results. Figure 6 shows the FPA surface topography at different temperatures. Deflection difference divided by temperature difference gives the device sensitivity of 0.18μm/K.

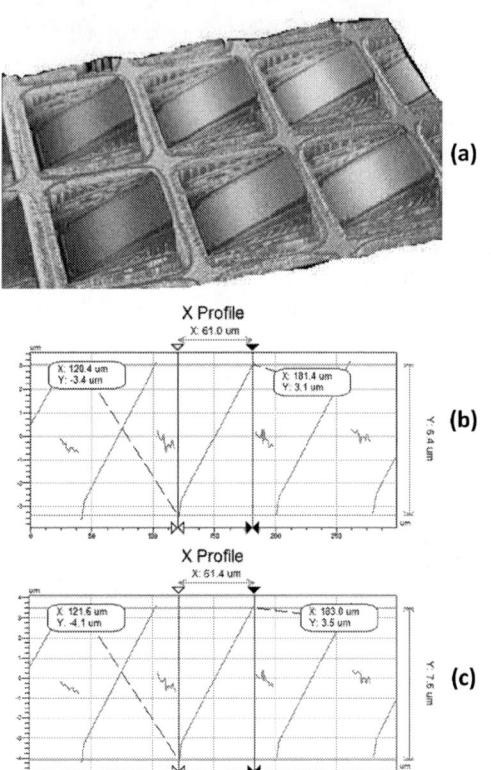

Figure 6: Surface topography (a) and the X profile of reflector at 27.5□ (b) and 34.3□ (c).

Multi-band imaging results were obtained by an optical readout system shown in Figure 7. The FPA was sealed in a vacuum chamber to minimize the heat exchange with the atmosphere. The IR radiation was cast on the back side of the FPA through a wavelength selective IR lens, thus avoiding any interference of responses at different bands. The front side of the FPA was illuminated by a beam of parallel visible light, and the light reflected by the mirrors of the FPA was filtered by a pin hole before reaching a CCD. Therefore, the deflection of the cantilever was converted to a gray scale image.

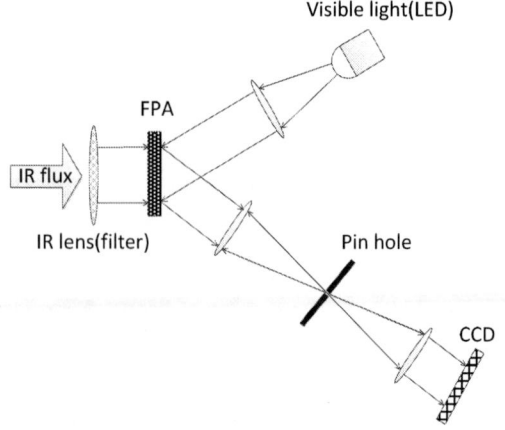

Figure 7: The optical readout system of our FPA.

The human body at room temperature has a radiation peak around 10 μm, and was detected for long wavelength IR with an 8~14μm lens. A heated soldering iron was used to verify the FPA's response at middle wavelength with a 3~5μm lens to eliminate influence of radiation from other bands. The short wavelength response was tested by a 1064 nm laser with the energy of 0.349mJ, directly detected by the FPA due to its good monochromaticity. The captured image of human body, heated soldering iron and laser spot were shown in Figure 8.

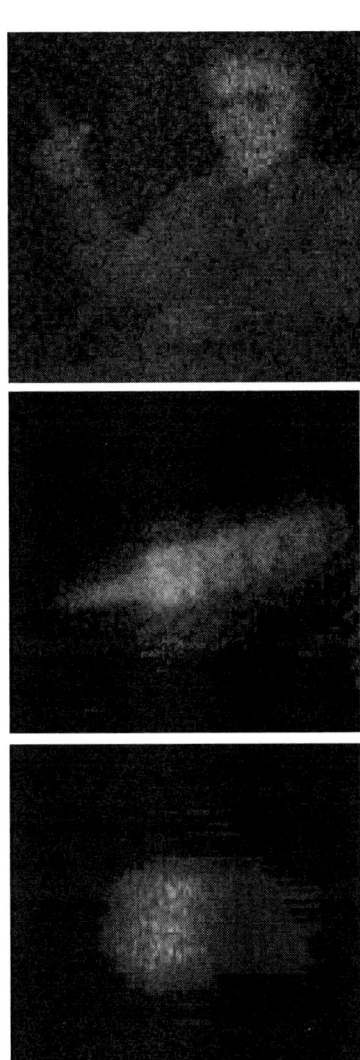

Figure 8: IR image of human body (top), heated soldering iron (middle) and laser spot (bottom) at room temperature.

CONCLUSIONS

In summary, multi-band IR FPA was implemented by a bulk micromachining process of selective removal of the substrate. This silicon-framed structure avoids the energy loss caused by the substrate and thus increases the IR absorption efficiency greatly. The stacked absorber of SiN_x and Cr nano-films contributes to the multi- band response of the device. The sensitivity of the FPA was measured to be 0.18μm/K. Images from three infrared bands were successfully captured.

ACKNOWLEDGEMENTS

The work was funded by the National Natural Science Foundation of China (grants No. 61036006 and61275104).

REFERENCES

[1] A. Rogalski, J. Antoszewski, and L. Faraone, "Third-generation infrared photodetector arrays," *Journal of Applied Physics,* vol. 105, pp. 091101-091101-44, 2009.

[2] S. V. Bandara, S. D. Gunapala, J. K. Liu, S. B. Rafol, D. Z. Ting, J. M. Mumolo, *et al.,* "Four-band quantum well infrared photodetector array," *Infrared Physics & Technology,* vol. 44, pp. 369-375, 2003.

[3] M. N. Abedin, T. F. Refaat, Y. Xiao, and I. Bhat, "Characterization of dual-band infrared detectors for application to remote sensing," in *Proc. SPIE,* 2005, pp. 38-45.

[4] Q. Cheng, S. Paradis, T. Bui, and M. Almasri, "Design of dual-band uncooled infrared microbolometer," *Sensors Journal, IEEE,* vol. 11, pp. 167-175, 2011.

[5] M. Erdtmann, L. Zhang, and G. Jin, "Uncooled dual-band MWIR/LWIR optical readout imager," in *SPIE Defense and Security Symposium,* 2008, pp. 694012-694012-11.

[6] S. Ogawa, K. Okada, N. Fukushima, and M. Kimata, "Wavelength selective uncooled infrared sensor by plasmonics," *Applied Physics Letters,* vol. 100, pp. 021111-021111-4, 2012.

[7] Y. Zhao, M. Mao, R. Horowitz, A. Majumdar, J. Varesi, P. Norton, *et al.,* "Optomechanical uncooled infrared imaging system: design, microfabrication, and performance," *Microelectromechanical Systems, Journal of,* vol. 11, pp. 136-146, 2002.

[8] M. Klanjšek Gunde and M. Ma ek, "Infrared optical constants and dielectric response functions of silicon nitride and oxynitride films," *Physica status solidi (a),* vol. 183, pp. 439-449, 2001.

[9] W. Lang, K. Kühl, and H. Sandmaier, "Absorbing layers for thermal infrared detectors," *Sensors and Actuators A: Physical,* vol. 34, pp. 243-248, 1992.

[10]   Y. Yi, X. Yu, W. Xu, M. Liu, L. Dong, X. Liu, *et al.,* "An uncooled microcantilever IR detector based on bulk silicon technique," in *Solid-State Sensors, Actuators and Microsystems Conference, 2009. TRANSDUCERS 2009. International,* 2009, pp. 485-488.

CONTACT
*Xiaomei Yu, Tel: +86-10-62766592; yuxm@pku.eud.cn

---

[1] The first two authors contributed equally to this work.

VERY LOW POWER CONSUMPTION MEMS SCANNER WITH ALKALI ELECTRET COMB DRIVE

Tatsuhiko Sugiyama[1], Mitsuru Aoyama[1], Keiko Kawai[1] and Gen Hashiguchi[1]
[1]Shizuoka University, Hamamatsu, Shizuoka, JAPAN

ABSTRACT

This paper reports the very low power consumption MEMS scanner that utilizes the electrostatic field generated by alkali-ion electret. The alkali-ion electret formed on comb electrodes of the scanner provides built-in potential for the electro-static actuator so that no bias voltage is necessary. The power consumption of prototype MEMS scanner was 0.57 µW (bias voltage: DC 0 V, driving voltage: AC 9 V_{pp}, deflection angle: 12°, resonance frequency: 1.4 kHz,). It is possible to realize further low power consumption by increasing the force factor of comb-electrodes.

INTRODUCTION

MEMS scanners are developing for mobile projectors and head-up displays of car use. There are several driving methods of MEMS scanner, such as electromagnetic, piezoelectric and electrostatic force [1][2][3]. The electromagnetic and piezoelectric methods are not suitable for miniaturization and mass commercialization because these require complicated assembly. The electrostatic actuators such as comb drive actuator are suitable for miniaturization and mass production because these actuators are fabricated by conventional silicon process. However, the tens of bias or AC voltage need to generate the electrostatic force, the appropriate booster circuit is necessary; resulting in increase of the power consumption of the system. This aspect of electrostatic actuators is the weak point.

Recently, electret films, which stored charges semi-permanently, have been studied for application to vibration energy harvesting. But if the electret fimes can be formed on very narrow and high aspect Si structures in MEMS devices, its application is beyond the scope of energy harvesting. Electret films are corresponding to built-in bias voltage of a MEMS device so that no bias voltage drive of electrostatic MEMS actuator can be achieved by forming the electret on the electrodes. The conventional methods of fabricating electret, such as corona discharge [4], ion implantation [5], electron beam injection [6], or molecule ionization by X-ray radiation [7], are expensive to manufacture and difficult to form on very narrow and high aspect electrodes. We have been developed the completely new electret films (**alkali-ion electret**) that can be formed on very narrow vertical sidewall of comb electrodes [8]. To apply the alkali–ion electret on comb electrodes of MEMS scanner, no bias and low driving voltage MEMS scanner can be fabricated. Figure 1 shows the concept of electret MEMS scanner. Since the electret films are formed on stepped comb electrodes of the scanner and induces the electrical field

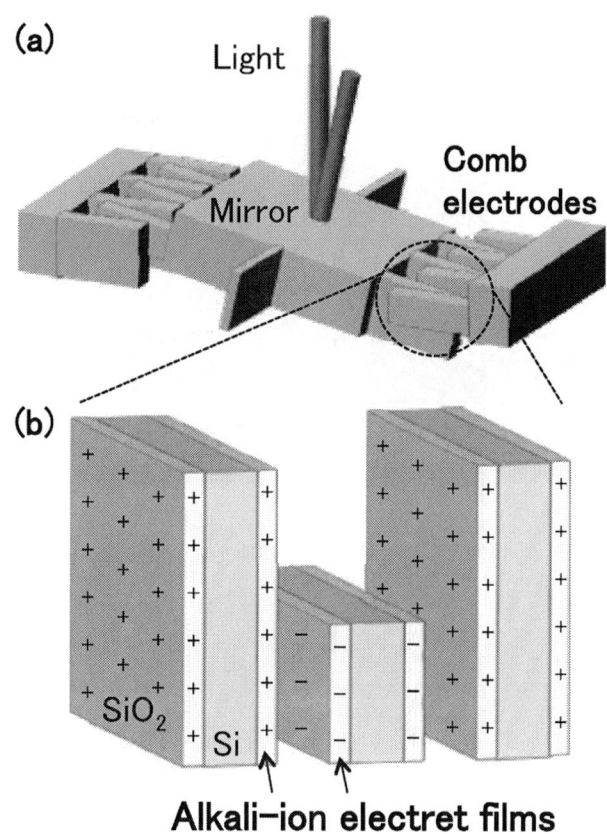

Figure 1: (a) Schematic diagram of MEMS scanner with alkali electret. (b) Cross sectional image of comb electrodes on which the alkali-ion electret film is formed.

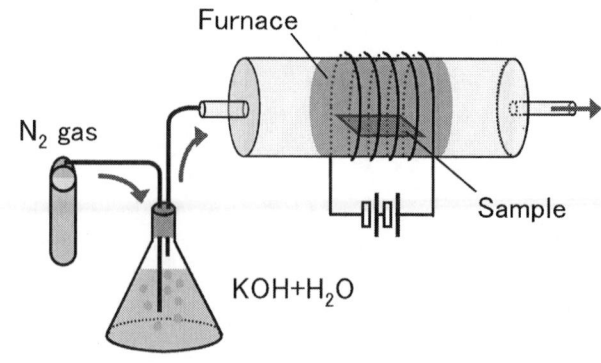

Figure 2: Oxidation setup with bubbling steam of KOH solution.

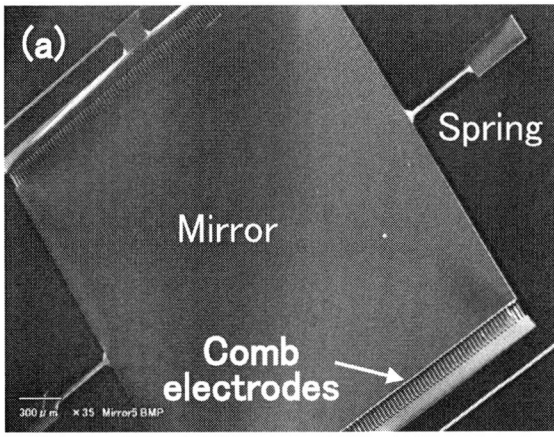

Figure 3: (a) SEM image of fabricated MEMS scanner after alkali-ion oxidation. (b) The magnified view of step comb electrodes.

between the opposing comb-drive electrodes, the bias voltage is not necessary. The potential difference between the comb-electrodes can be increased up to hundreds of volts by the electret, so that the driving AC voltage can be lowered. In this paper, we will experimentally demonstrate the prototype electret MEMS scanner at resonance mode and propose the scanner at non-resonance mode.

EXPERIMENTS
Formation of the alkali-ion electret

The electrostatic MEMS scanner was made of silicon on insulator (SOI) substrate. The device layer thickness of the SOI was 50 µm. The MEMS mirror had stepped comb electrodes. The deep electrodes were 50 µm high and shallow ones were 25 µm. The gap between the electrodes was 6 µm.

The fabricated MEMS scanner was thermally oxidized with bubbling stream of KOH solution to form a silicon oxide film including potassium ions on the etched sidewalls of comb electrodes uniformly. Figure 2 shows an illustration of our oxidation setup. The oxidation furnace

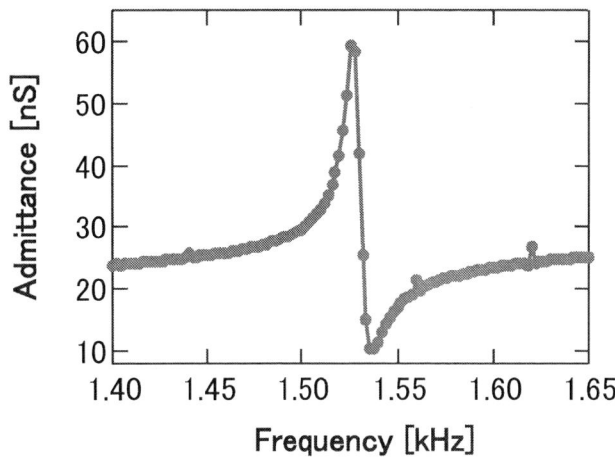

Figure 4: A measured admittance curve as a function of 0.1 V_{pp} AC frequency. DC bias is not applied to the device.

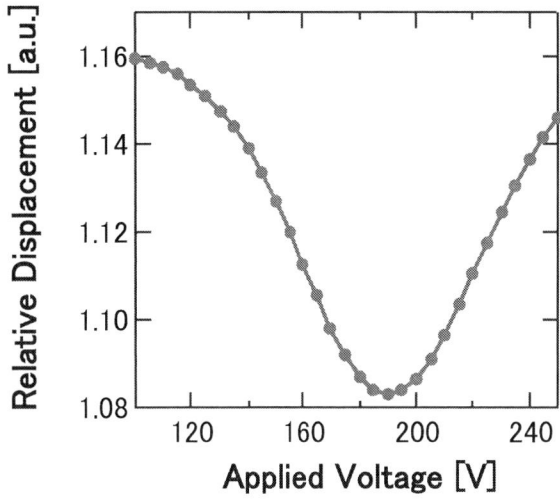

Figure 5: Relative displacement between the opposing comb electrodes when DC voltage is applied to the comb electrode.

temperature was set at 930 °C and 40 wt% KOH solution was maintained at 90 °C in a hot bath and the oxidized time was 30 hours. The carrier gas of the bubbling was pure nitrogen gas with flow rate of 2.5 Little/min. The silicon oxide film thickness was about 1 µm. Figure 3 shows the fabricated MEMS scanner on which the silicon oxide film including potassium ions was formed. After the oxidation, the films included potassium ions on comb-drive actuator were converted into the electret films by bias-temperature (BT) procedure. The temperature was about 750 °C and the bias voltage of 200 V was applied between opposite electrodes during BT procedure. Figure 4 shows the

978-1-4799-3510-9/14 $31.00 © 2014 IEEE   1230

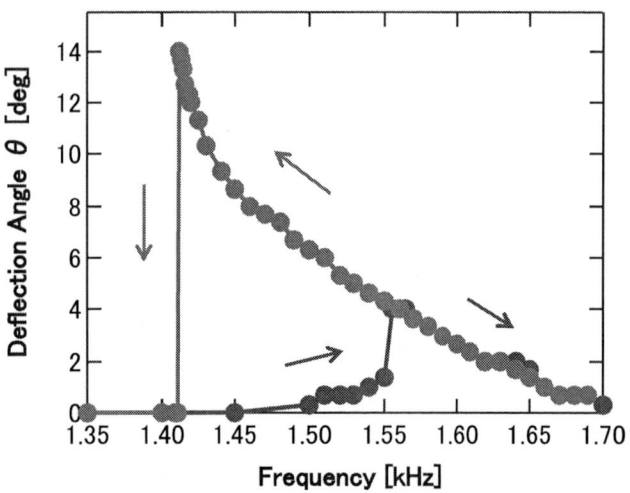

Figure 6: Relationship between the sweep direction of frequency and the deflection angle of the scanner.

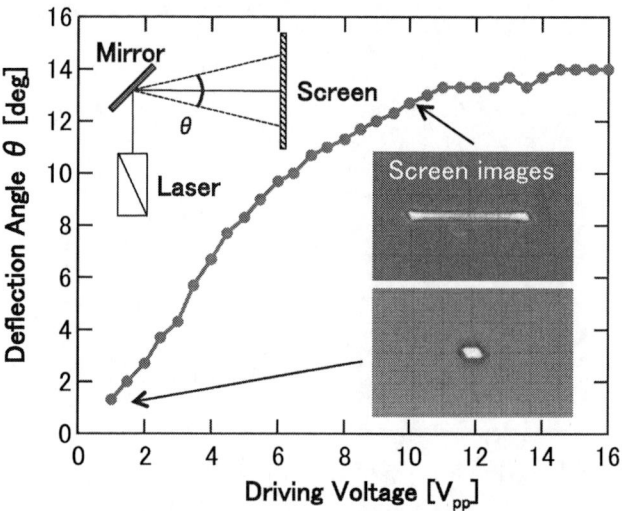

Figure 7: Relationship between driving voltage and deflection angle. Bias voltage is not applied.

frequency dependence of the admittance with the 0.1 V_{pp} AC excitation without DC bias after BT procedure. An obvious resonance peak was observed at 1.53 kHz; this strongly suggests that there is static electric field between the opposing comb electrodes. In order to identify the voltage difference between the comb electrodes, we applied external DC voltage to the movable comb electrodes, to which DC bias was applied during the BT procedure, while its opposed comb electrodes was connected to ground. Figure 5 shows a measured relative displacement as a function of applied DC voltage. The displacement was a like parabola that bottomed on 190 V. Usually, the displacement of electrical static actuators is determined by the balance between electrical static force and mechanical spring force, and from the ideal modeling of comb-drive actuators (no consideration of any parasitic effects such as fringe fields), the displacement is written as [9]

$$x = \frac{n\varepsilon b V_0^2}{kd} = A V_0^2, \quad A = \frac{n\varepsilon b}{kd}, \quad (1)$$

where n is the number of comb electrode pairs, ε is the permittivity, b and d are the thickness and the gap of the comb electrodes, respectively, V_0 is the applied voltage, and k is the spring constant of the device. If the internal voltage V_{in} by the electret is existed, equation (1) is written as follows.

$$x = A(V_0 - V_{in})^2. \quad (2)$$

From figure 5 and equation (2), the 190 V internal voltage between the comb electrodes by the electret was estimated.

Characteristics of the Electret MEMS Scanner

Figure 6 shows the relationship between the sweep direction of frequency and the deflection angle of the scanner. The input AC voltage was 20 V_{pp}. The blue line was the result that input frequency was swept from low frequency to high and the red line was from high to low. The frequency curve was asymmetry near the resonance peak and there was hysteresis that depends on the sweep direction in the frequency. This phenomenon is observed when the structure with vertical comb electrodes is twisted large to out–of–plane [10].

The deflection angle at resonance frequency was measured by triangular method. Figure 7 shows the driving voltage dependence of the deflection angle. The deflection angle 12 ° at the driving voltage 9 V_{pp} was obtained. This driving voltage is very low compared with traditional electrostatic MEMS scanner of resonance mode. The power consumption was 0.57 µW at this condition. The angle was saturated as the driving voltage increased. This reason is thought that the damping effect of air and the hardening of the spring.

The force factor that is barometer of the efficiency is written as [11]

$$K = \frac{2n\varepsilon b V_{in}}{d} \quad (3)$$

The value of the factor in the prototype electret MEMS scanner was 2 x 10^-6. This value is relatively small because the gap of the comb electrodes is large. It is possible to realize further low power consumption by increasing the force factor to be done by small gap comb electrodes.

978-1-4799-3510-9/14 $31.00 © 2014 IEEE   1231

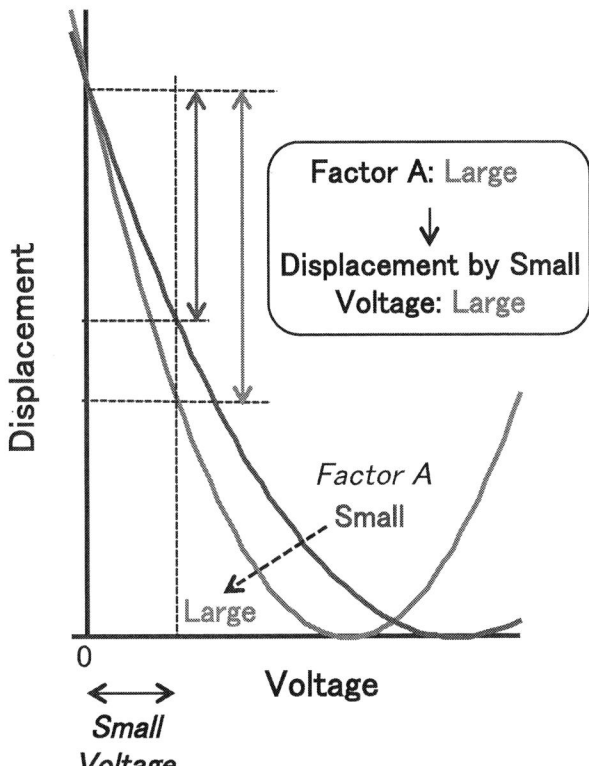

Factor A: Large
↓
Displacement by Small Voltage: Large

Figure 8: Concept of non-resonance mode electret MEMS scanner.

DICUSSION

The prototype electret MEMS scanner was driven at resonance frequency. For versatile scanner, however, the scanner at non-resonance mode is necessary. Figure 8 shows the concept of driving at the non-resonance mode electret MEMS scanner. From equation (2), the relationship between displacement and applied voltage on the electret comb-drive actuator become parabolic carve which bottomed on the internal voltage by the electret. The displacement by small-applied voltage close to zero voltage is large. Therefore, it is possible to realize the electret MEMS scanner of non-resonance mode that is driven by small-applied voltage and has large displacement. In the design, it is preferable that the factor $A = n\varepsilon b / kd$ of equation (1) should be large because the displacement by small-applied voltage is large.

CONCLUSIONS

We have demonstrated the very low power consumption MEMS scanner. The power consumption of prototype scanner was 0.57 μW (bias voltage: DC 0 V, driving voltage: AC 9 V_{pp}, deflection angle: 12°, resonance frequency: 1.4 kHz,). The MEMS scanner was driven by electrostatic force, which was induced by the alkali-ion electret formed on the comb electrodes, so that the no bias voltage is necessary. It is possible to realize further low power consumption by increasing the force factor of comb-electrodes. Furthermore, the electret MEMS scanner of non-resonance mode was proposed. At the non-resonance mode, the large displacement is possible by small AC voltage.

REFERENCES

[1] J. W. Judy and R. S. Muller, "Magnetic microactuation of torsional polysilicon structures", *Sensor and Actuators A,* 53 (1996), pp. 392-397.

[2] K. Yamada and T. Kuriyama, "A novel asymmetric silicon micro-mirror for optical beam scanning display", *Proc. MEMS 1998, pp. 110-115.*

[3] K. E. Petersen, "Silicon Torsional Scanning Mirror", *IBM Journal of Research and Development*, 24 (1980), 5, pp. 631-637.

[4] R. A. C. Altafim, J. A. Giacometti, J. M. Janiszewski, "A novel method for electret production using impulse voltages", *Proc. Symp. (Cat. No.91CH3029-6) th Int Electrets (ISE 7)*, 1991, pp. 267-271.

[5] U. Mescheder, P. Urbanovic, B. Muller, S. Baborie, "Charging of SiO_2 Electret Film by Ion Implantation for MEMS Based Energy Harvesting Systems", *Proceedings of PowerMEMS2008+microEMS2008*, Sendai, Japan, November 9-12, 2008.

[6] P. Gunther, "SiO_2 electrets for electric-field generation in sensors and actuators", *Sensors and Actuators A: Physical*, vol. 32, pp. 357-360, 1992.

[7] K. Hagiwara, M. Honzumi, M. Goto, T. Tajima, Y. Yasuno, H. Kodama, K. Kidokoro, K. Kashiwagi, Y. Suzuki, "Novel through-substrate charging method for electret generator using soft X-ray irradiation", *PowerMEMS2009*, Washington D.C., 2009.

[8] T. Sugiyama, M. Aoyama, Y. Shibata, M. Suzuki, T. Konno, M. Ataka, H. Fujita and G. Hashiguchi, "SiO_2 Electret Generated by Potassium Ions on a Comb-Drive Actuator", *Appl. Phys. Express*, vol.4, 114103, 2011.

[9] W. C. Tang et al.: Proc. An Investigation of Micro Structures, Sensors, Actuators IEEE Machines and Robots Micro Electro Mechanical Systems, p. 53, 1989.

[10] H. Schenk, P. Dürr, T. Hasse, D. Kunze, H. Lakner, H. Kück, "Large Deflection Micromechanical Scanning Mirrors for Linear Scans and Pattern Generation", IEEE J. Selected Topics in Quantum Electron, Vol. 6, No. 5, p. 715-722, 2000.

[11] Y. Nishimori, H. Ooiso, S. Mochizuki, N. Fujiwara, T. Tsuchiya, G. Hashiguchi, "Multi-DOF equivalent circuit for a comb-drive Actuator", Jpn. J. Appl. Phys., vol. 48, 124504, 2009.

CONTACT

T. Sugiyama, e-mail: t.sugiyama@rie.shizuoka.ac.jp

A LOW-LOSS RF MEMS SILICON SWITCH USING REFLOWED GLASS STRUCTURE

Jeongki Hwang, Sung-Hyun Hwang, Yong-Seok Lee, and Yong-Kweon Kim
Seoul National University, South Korea

ABSTRACT

This paper reports a low-loss RF MEMS silicon switch that employs reflowed glass as a structural material beneath a contact metal. The glass structure is inserted into a silicon substrate through the glass reflow process to reduce the insertion loss of the RF MEMS switch. To verify this enhancement in the loss characteristic, two types of RF MEMS switches were fabricated with two different structural materials underneath the contact metal: silicon or reflowed glass. The reference RF MEMS switch is totally made of silicon. In the frequency range of 5 to 30 GHz, S-parameter measurements of the fabricated switches were carried out and the insertion losses were measured as 0.12 ~ 0.33 dB and 0.38 ~ 0.53 dB for the proposed and the reference RF MEMS switch, respectively. The insertion loss was reduced as much as 0.26 dB for the proposed RF MEMS switch in the measured frequency range and it was shown that the insertion loss of the RF MEMS switch is effectively reduced by changing the structural material under the contact metal.

INTRODUCTION

For a couple of decades, RF MEMS switches have been studied as a promising replacement for solid-state RF switches because of their outstanding performances at high frequency: low insertion loss, excellent linearity, and great isolation [1-3]. This expectation is still the same today for high frequency applications such as satellite networks, defense systems, base-station antennas, etc. [3-5]. However, few RF MEMS switches have been reported to be in volume production because the reliable operation of an RF MEMS switch is a hard feature to realize. In view of this, a silicon-structured RF MEMS switch is an excellent candidate for commercialization because the high Young's modulus of silicon can greatly benefit the mechanical reliability of an RF MEMS switch.

Kim et al. [6, 7] have presented a Single Crystalline Silicon (SCS) RF MEMS switch with the intention to realize a mechanically reliable RF MEMS switch. They showed long life-time over 10^9 cycles and high throughput. However, switches reported in [6, 7] had electromagnetic losses brought about by the switch structure because the silicon structure near the transmission line acts as a conductive dielectric substrate and degrades the propagating signal. Considering that the low-loss characteristic is always an appreciable feature for an RF switch, the loss induced by the switch structure needs to be enhanced.

In this paper, we present a low-loss RF MEMS switch that employs reflowed glass as part of the switch structure. The reflowed glass is inserted into the switch structure only under the contact metal to reduce insertion loss. Design of the proposed RF MEMS switch and the fabrication process

are presented. Fabrication results and the scanning electron microscopy (SEM) images are provided. In addition, S-parameter measurements were carried out for the proposed and reference RF MEMS switches. The mechanically stable RF MEMS silicon switch with lowered insertion loss is realized by inserting the glass structure under the contact metal.

DESIGN AND ANALYSIS

To experimentally demonstrate the improvement of the loss characteristic, both reference and the proposed RF MEMS switches are designed. The reference RF MEMS switch is made of silicon totally while the proposed one is made of silicon and reflowed glass. Perspective and cross-sectional view of the proposed RF MEMS switch is presented in Figure 1.

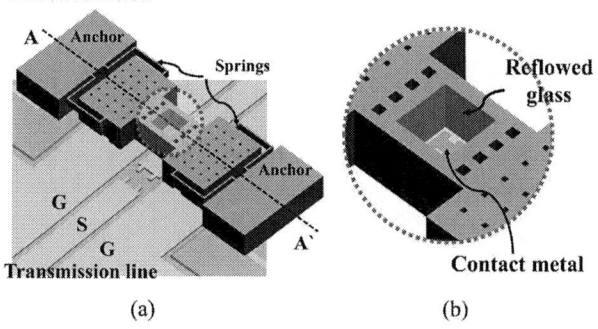

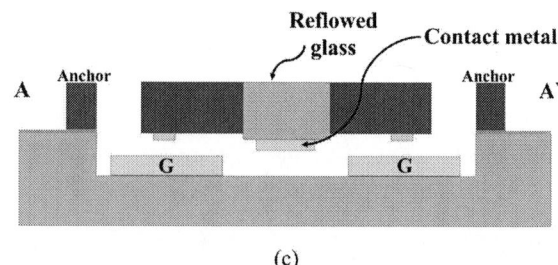

Figure 1: (a) Perspective view of the proposed RF MEMS switch. The switch structure is detached from the bottom substrate to show the separated transmission line. (b) Enlarged view of the reflowed glass structure. The contact metal is visible through the reflowed glass structure. (c) Cross-sectional view of the proposed RF MEMS switch. The signal line of the transmission is separated at the cross-section A-A`.

In this paper, only the region under the contact metal is changed from silicon to glass. This is because the electrode should be made of silicon for an electrostatic operation. In addition, in the proposed switch structure, the

978-1-4799-3510-9/14 $31.00 © 2014 IEEE

electromagnetic wave is concentrated intensely near the contact metal, and not near the electrodes region.

The proposed RF MEMS silicon switch is constructed on the transmission line to perform the switching operation. As shown in Figure 1 (a), the transmission line is physically separated under the contact metal. This disconnection correspond to switch's off-state, in which no electromagnetic wave can propagate along the transmission line. However, when the switch moves down, the contact metal joins two separated transmission lines and the switch's on-state is realized. In this state, the electromagnetic wave can go end to end of the transmission line.

During the switch's on-state, the switch structure comes closer to the transmission line as well as the contact metal. This movement is necessary for the switching operation of RF MEMS silicon switch. In this situation, the electromagnetic wave propagated throughout the transmission line undergoes the electromagnetic wave loss induced by the switch structure. The switch structures acts as a conductive dielectric and this effect gets severe as the switch structure is closely located near the transmission line. The loss is due to electromagnetic wave scattering at the conductive dielectric. As a result, the loss induced by the switch structure is dependent on the conductance of the switch's structural material. The effect of a material conductance on the electromagnetic wave is researched in [8, 9].

Table 1: Design values of the reference and proposed RF MEMS silicon switches.

Items	Reference RF MEMS switch	Proposed RF MEMS switch
Switch structure (L x W x T)	704 x 248 x 70 µm³	
Material under the contact metal (Size)	TEOS and Silicon (N/A)	Reflowed glass (80 x 80 µm²)
Contact metal area	60 x 60 µm²	
Spring constant	770 N/m	
Initial gap	2 µm	
Pull-in voltage	66.4 V	

Design values for the reference and proposed RF MEMS switches are presented in Table 1. The two switches are different only in the structural material under the contact metal. For the proposed switch, the area of reflowed glass structure is set at 80 µm x 80 µm to cover the contact metal entirely, which has a dimension of 60 µm x 60 µm. The other design values are the same for the two models.

For high restoring force and mechanical rigidity of the switch structure, a spring constant was designed as high as 770 N/m. The restoring force provided by the spring components is important factor in realizing the fast operation of the switch and to alleviate the switch stiction problem. An initial gap between the driving electrodes was designed at 2 µm. This value is designed for an isolation characteristic under -30 dB. For the designed spring constant and initial gap, the resulting pull-in voltage was calculated to be 66.4 V.

FABRICATION

A schematic view of the overall fabrication process is presented in Figure 2. The fabrication steps are divided into three substrates: silicon, glass, and bonded. In the silicon and glass substrate fabrication processes, the switch structure and the transmission line is prepared, respectively. The two fabricated wafers are then anodically bonded together and processed further to release the complete switch structure on the transmission line.

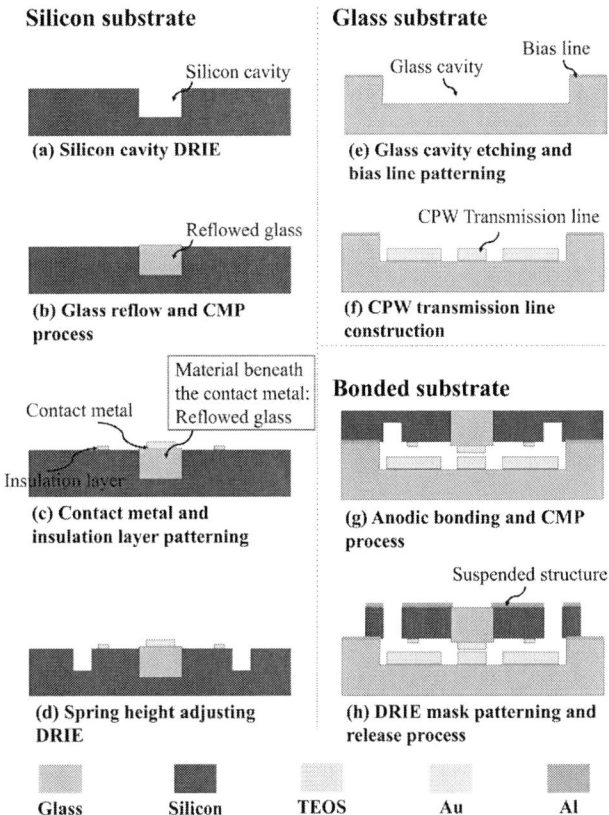

Figure 2: Overall fabrication process. (a) Silicon cavity DRIE. (b) Glass reflow and CMP process. (c) Contact metal and insulation layer patterning. (d) Spring height adjusting DRIE. (e) Glass cavity etching and bias line patterning. (f) CPW transmission line construction. (g) Anodic bonding and CMP process. (h) DRIE mask patterning and release process.

The fabrication process for the silicon substrates starts with the Deep Reactive Ion Etching (DRIE) process for cavity formation. This cavity provides space for reflowed glass to fill in the next step. The following step is the glass reflow process, in which glass is thermally inserted into the cavity. Next, the silicon substrate is polished through the chemical mechanical polishing (CMP) process and the isolation layer is deposited onto the silicon region. The isolation layer prevents the two electrodes from electrically connecting each other in the switching operation. Then, the contact metal is deposited onto the reflowed glass structure. The final step for the silicon substrate is the DRIE process, which etches the silicon substrate partially to adjust the height of the springs.

Meanwhile, glass substrate is prepared by etching the substrate and patterning the bias line. Then, electroplating is carried out to construct a transmission line in the cavity. The roughness of the CPW plane is highly important for the even operation of the switch because the bumpy surface of the transmission line would initially contact the silicon electrodes.

The two wafers are anodically bonded and polished by the CMP process after completing the fabrication of silicon and glass substrates. Next, a mask for the DRIE process is patterned on the silicon side. Finally, a switch structure is released with the DRIE process.

The SEM images of the fabricated RF MEMS silicon switch are shown in Figure 3. The dimension for switch structure is measured to be 704 x 208 x 72 μm^3 and the area of reflowed glass structure is measured to be 80 x 80 μm^2. The measurement results revealed that the structure thickness and the initial gap have fabrication errors. The thickness of the switch structure is determined by the CMP process of the bonded wafer. The targeted thickness of the structure was 70 μm while the fabricated thickness was measured to be 72 μm. A thicker spring will results in more voltage needed for the switch's operation. The measured initial gap was 2.1 μm, whereas the designed value was 2.0 μm. The increase in the initial gap will also result in the rise of the pull-in voltage.

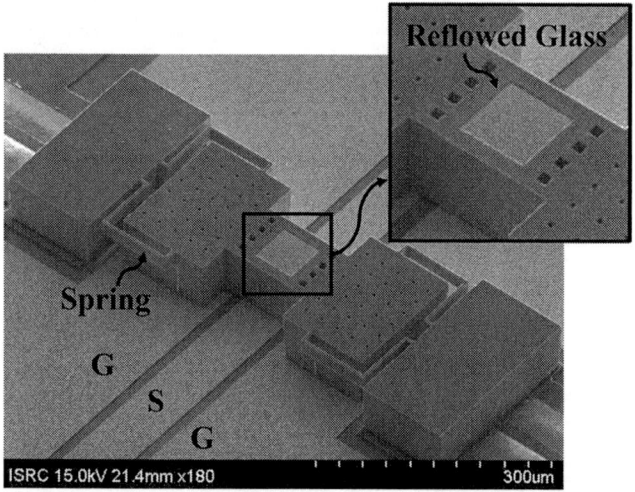

Figure 3: SEM image of the fabricated RF MEMS switch. The reflowed glass structure is inserted in the center of the switch structure.

MEASUREMENT
Pull-in voltage measurement

For the pull-in voltage measurement, a DC voltage supply and digital multi-meter were used to measure DC resistance of the transmission line while applying the driving voltage to the switch structure. Pull-in voltage was determined to be the voltage level where the transmission line resistance changes from infinite to a finite value. This means that the contact metal physically connects the separated transmission lines. The applied voltage is raised slowly to check the pull-in voltage accurately during the measurement. For the ten randomly-selected samples, the average pull-in voltage was measured to be 77.2 V, which is higher than the designed value of 66.4 V. This calculates to a 22 % error, which is due to a fabrication error from a thicker spring and longer initial gap. Using the actual dimensions of the fabricated switch, the pull-in voltage was recalculated to be 81.0 V. This reduced the error in pull-in voltage down to 4 %.

S-parameter measurement

The insertion losses of the reference and proposed models were measured in the switch's on state to investigate the enhancement in the insertion loss of the RF MEMS switch. The measurement was set at a frequency range of 5 to 30 GHz. The resulting S-parameters are shown in Figure 4. The black squares and blue triangles represent the reference and proposed switches, respectively. The proposed switch exhibits superior insertion loss characteristic, which is enhanced about 0.26 dB from the reference switch. In the measurement frequency range, the reference model showed insertion loss of 0.38 ~ 0.54 dB, whereas the proposed model showed 0.12 ~ 0.33 dB.

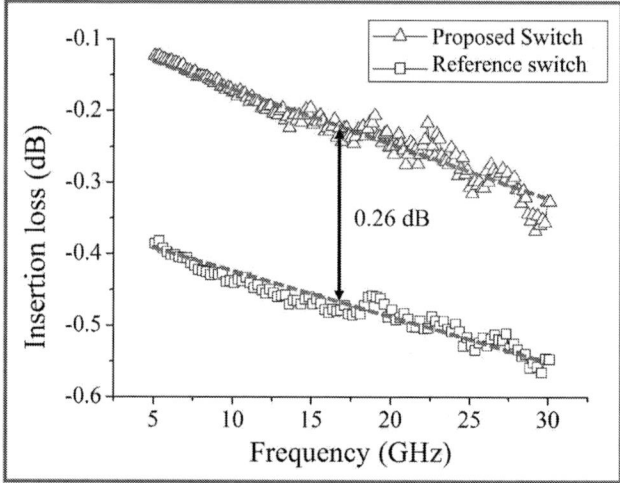

Figure 4: Measured insertion losses of the reference and proposed RF MEMS switch.

The return losses of the two switches are also measured and shown in Figure 5. Both showed return losses under -25 dB.

Isolation characteristic of the reference and proposed switches were measured in the switching-off state. Both switches have isolation characteristic under -29 dB as shown in Figure 6. This means that the switch structure is positioned at its initial position.

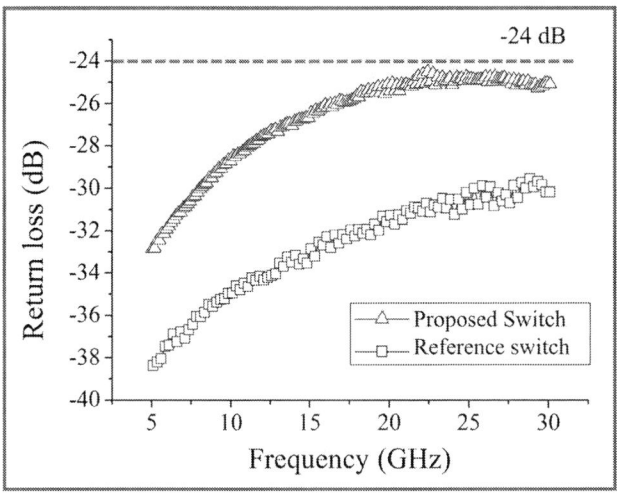

Figure 5: Measured return losses of the reference and proposed RF MEMS switch.

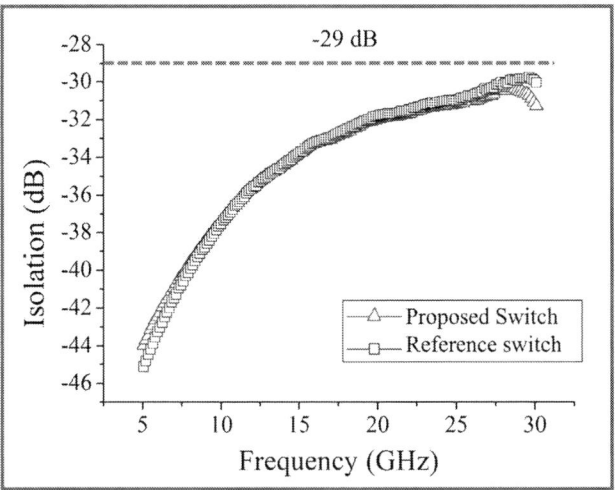

Figure 6: Measured isolations of the reference and proposed RF MEMS switch.

CONCLUSIONS

In this paper, we first proposed and successfully demonstrated a low-loss RF MEMS silicon switch that contains a reflowed glass structure underneath the contact metal. The proposed switch was successfully fabricated with the suggested fabrication method. A DC operation test was carried out and the average pull-in voltage was measured to be 77.2 V. An enhancement in the insertion loss is effectively realized by changing the structural material under the contact metal from silicon to reflowed glass. S-parameter measurement results revealed that the proposed switch shows about 0.26 dB improved insertion loss than the reference model in a frequency range of 5 to 30 GHz.

ACKNOWLEDGEMENTS

This research was supported by the Dual Use Technology Program through the Dual Use Program Cooperation Center (DUPC) (Grant Number 11-DU-EE-02).

REFERENCES

[1]   G. M. Rebeiz, *RF MEMS theory, design and technology*. Hoboken NJ: Wiley, 2003.

[2]   U. L. Rohde and M. Rudolph, *RF/microwave circuit design for wireless applications*. Hoboken NJ: Wiley, 2013.

[3]   G. M. Rebeiz, C. D. Patel, S. K. Han, C. Ko, and K. M. Ho, "The Search for a Reliable MEMS Switch???: Metal-Contact Switches," *Microwave Magazine, IEEE,* vol. 14, pp. 57-67, 2013.

[4]   O. Bayraktar, O. A. Civi, and T. Akin, "Beam switching reflectarray monolithically integrated with RF MEMS switches," *IEEE Transactions on Antennas and Propagation,* vol. 60, pp. 854-862, 2012.

[5]   A. Pourziad, S. Nikmehr, and H. Veladi, "A Novel Multi-State Integrated RF Mems Switch for Reconfigurable Antennas Applications," *Progress In Electromagnetics Research,* vol. 139, pp. 389-406, 2013.

[6]   J.-M. Kim, J.-H. Park, C.-W. Baek, and Y.-K. Kim, "The SiOG-based single-crystalline silicon (SCS) RF MEMS switch with uniform characteristics," *Journal of Microelectromechanical Systems,* vol. 13, pp. 1036-1042, 2004.

[7]   J.-M. Kim, S. Lee, C.-W. Baek, Y. Kwon, and Y.-K. Kim, "Mechanically robust single crystalline silicon (SCS) single-pole-double-throw (SPDT) MEMS switch and its application to dual-band wlan filter," in *IEEE 20th International Conference on Micro Electro Mechanical Systems, 2007. MEMS.,* 2007, pp. 807-810.

[8]   W. H. Haydl, "Conductive substrate losses in coplanar and microstrip transmission lines," in *Microwave Conference, 1997. 27th European,* 1997, pp. 532-537.

[9]   S. Kang, S. Park, H. Kim, and K. Chun, "The RF Characteristics of the RF MEMS switch as the substrate resistivity," in *Global Symposium on Millimeter Waves, 2008. GSMM 2008.,* 2008, pp. 285-287.

CONTACT

*Yong-Kweon Kim, tel: +82-2-880-7440;
fax: +82-2-873-9953; e-mail: yongkkim@snu.ac.kr

A UHF 4TH-ORDER BAND-PASS FILTER BASED ON CONTOUR-MODE PZT-ON-SILICON RESONATORS

Hadi Yagubizade[1], Milad Darvishi[2], Miko C. Elwenspoek[1] and Niels R. Tas[1]
[1]MESA$^+$ Institute for Nanotechnology, [2]CTIT Institute,
University of Twente, Enschede, THE NETHERLANDS

ABSTRACT

A UHF 4th-order band-pass filter (BPF) based on the subtraction of two 2nd-order contour-mode resonators with slightly different resonance frequencies is presented. The resonators consists of a 1 μm pulsed-laser deposited (PLD) lead zirconate titanate (PZT) thin-film on top of a 3 μm silicon (PZT-on-Si). The resonators are actuated in-phase and their outputs are subtracted. Utilizing this technique, the outputs of the resonators are added up constructively while the feed-through signals are eliminated. The BPF presented a bandwidth of approximately 28.6 MHz and more than 30 dB stopband rejection at around 700 MHz.

INTRODUCTION

RF-MEMS filters are providing new opportunities for the next generation of wireless communication systems enabling low power consumption and high level of integration. FBAR [1] is most successful among available RF-MEMS filters and is already commercialized. The current demand requires the integrability of filters at different resonance frequencies compacted in the same die fabricated in a single process. Due to the lateral dependency of Lamb wave resonators as well as their high quality factor performance, they are the promising solution for future single-chip multi-frequency BPFs. Available Lamb wave BPF are based on mechanical and/or electrical coupling techniques [2, 3, 4] and they are still in the research stage.

In this paper, we propose a new technique of differentially readout of two in-phase actuated resonators, which is not based on the traditional methods such as mechanical and/or electrical couplings. The presented method has been presented recently operating at low frequencies [5], however this method requires to be developed further at UHF and SHF bands. In this paper, the technique is developed further at UHF band.

This method is effective for improving the stopband rejection by canceling the feed-through signal, which is more crucial in high-dielectric materials such as PZT especially at UHF and SHF [6].

THEORY AND MODELLING

Fig. 1 shows a simplified equivalent circuit of the filter concept. The transfer function of each resonator can be expressed as:

$$H_i(s) = \frac{V_{\text{out},\,i}}{V_{\text{in}}}$$
$$= \frac{R_s}{L_{\text{m},\,i}} \times \frac{s}{s^2 + \omega_{\text{-3dB},\,i} \cdot s + \omega_{\text{res},\,i}^2}, \quad i = 1, 2 \quad (1)$$

where $\omega_{\text{-3dB},\,i}$ and $\omega_{\text{res},\,i}$ are respectively the bandwidth $(2R_s + R_{\text{m},\,i})/L_{\text{m}.i}$ and the resonance frequency $1/\sqrt{L_{\text{m},\,i}C_{\text{m},\,i}}$ of the ith path.

Assuming $L_{\text{m},\,1} = L_{\text{m},\,2} = L_{\text{m}}$ and $R_{\text{m},\,1} = R_{\text{m},\,2} = R_{\text{m}}$, the total transfer function of the filter after subtraction can be expressed as

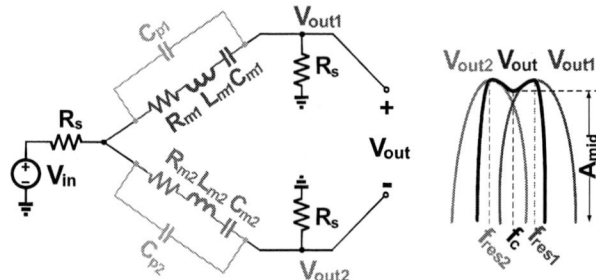

Figure 1: Equivalent circuit of 4th-order filter with differential readout of two in-phase actuated resonators.

$$H(s) = \frac{V_{\text{out1}} - V_{\text{out2}}}{V_{\text{in}}} = \frac{R_s}{L_{\text{m}}}$$
$$\times \frac{2\omega_c \cdot \Delta\omega_{\text{res}} \cdot s}{\left(s^2 + \omega_{\text{-3dB}} \cdot s + \omega_{\text{res},\,1}^2\right)\left(s^2 + \omega_{\text{-3dB}} \cdot s + \omega_{\text{res},\,2}^2\right)}, \quad (2)$$

where $\omega_c = (\omega_{\text{res},\,1} + \omega_{\text{res},\,2})/2$ and $\Delta\omega_{\text{res}} = \omega_{\text{res},\,1} - \omega_{\text{res},\,2}$.

We assume two case studies for representation, where two resonators at resonance frequencies of f_{res1}=690 MHz and f_{res2}=710 MHz and $L_{\text{m},1}$=$L_{\text{m},2}$=6 μH are considered. For the first case, a bandwidth of $\omega_{\text{-3dB},\,i}$=8 MHz ($\omega_{\text{-3dB},\,i}$<$\Delta\omega_{\text{res}}$), and for the second case, a bandwidth of $\omega_{\text{-3dB},\,i}$=30 MHz ($\omega_{\text{-3dB},\,i}$>$\Delta\omega_{\text{res}}$) are considered. For the first case ($\omega_{\text{-3dB},\,i}$<$\Delta\omega_{\text{res}}$), the transmission of each resonator is presented in Fig. 2. By subtracting the output of two 2nd-order resonators, a 4th-order BPF centered at f_{res1} =700 MHz (Fig. 2) is obtained. As seen, the filter response shows the same insertion loss as each resonator with a ripple of 2.23 dB. For the second case ($\omega_{\text{-3dB},\,i}$>$\Delta\omega_{\text{res}}$), the transmission of each resonator as well as the subtracted filter response is presented in Fig. 3. In this case, the filter has zero ripple, however the insertion loss is lower than that of each resonator. Considering $\omega_{\text{-3dB},\,i}$=$(2R_s + R_{\text{m},\,i})/L_{\text{m}.i}$, in both cases, $L_{\text{m},1}$ has been kept constant. Therefore by increasing the bandwidth of each resonator ($\omega_{\text{-3dB},\,i}$) in the second case, it is assumed that the motional impedance of the resonators is increased which will lead to increasing the insertion loss of the second case compared to the first case. For designing a BPF with zero ripple performance, there should be an optimum condition between the cases of $\omega_{\text{-3dB},\,i}$>$\Delta\omega_{\text{res}}$ and $\omega_{\text{-3dB},\,i}$<$\Delta\omega_{\text{res}}$. To investigate further on this issue, we will have a closer look at the operation of the filter at the center frequency of the filter, $\omega_c = (\omega_{\text{res},\,1} + \omega_{\text{res},\,2})/2$.

Assuming ω around the center frequency ($\omega = \omega_c + \delta\omega$) with a small frequency variation ($\delta\omega$), $\delta\omega \ll \omega_c$, allows (1) to be simplified as:

978-1-4799-3510-9/14 $31.00 © 2014 IEEE

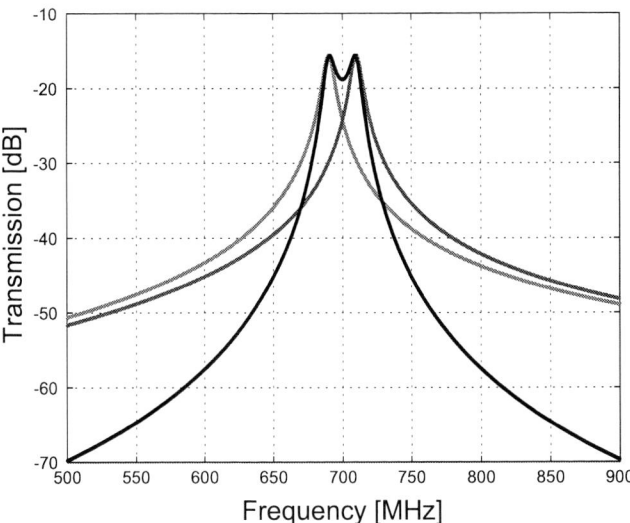

Figure 2: The simulation results of two 2nd-order resonators with $\omega_{\text{-3dB}, i} = 8\,\text{MHz}$ *and* $\Delta\omega_{res} = 20\,\text{MHz}$ ($\omega_{\text{-3dB}, i} < \Delta\omega_{res}$).

$$H_i(j\omega) \cong \frac{A_0}{1 + \frac{j}{\omega_{\text{-3dB}}/2} \times (\delta\omega \pm \Delta\omega_{res}/2)}, \qquad (3)$$

where $A_0 = R_s / (2R_s + R_m)$.

Using (3), the amplitude and the phase of each resonator, can be expressed as:

$$A_{\text{mid.}} = |H_i(j\omega)| = \frac{A_0}{\sqrt{1 + \frac{\Delta\omega_{res}^2}{\omega_{\text{-3dB}}^2}}}, \qquad (4)$$

$$\varphi = \angle H_i(j\omega) = \pm\, tan^{-1}\frac{\Delta\omega_{res}}{\omega_{\text{-3dB}}}. \qquad (5)$$

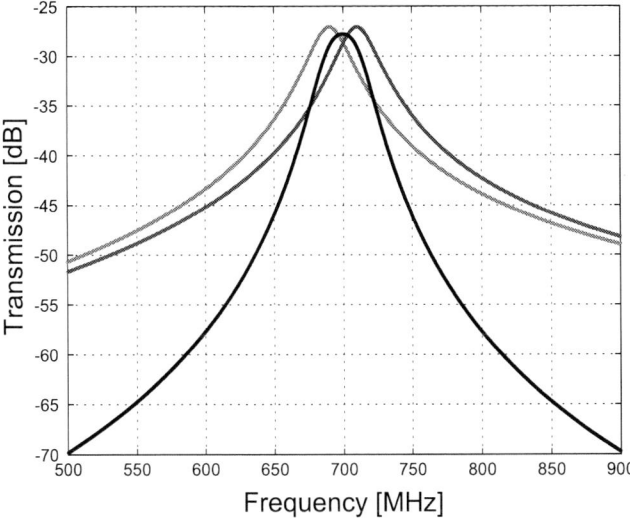

Figure 3: The simulation results of two 2nd-order resonators with $\omega_{\text{-3dB}, i} = 30\,\text{MHz}$ *and* $\Delta\omega_{res} = 20\,\text{MHz}$ ($\omega_{\text{-3dB}, i} > \Delta\omega_{res}$).

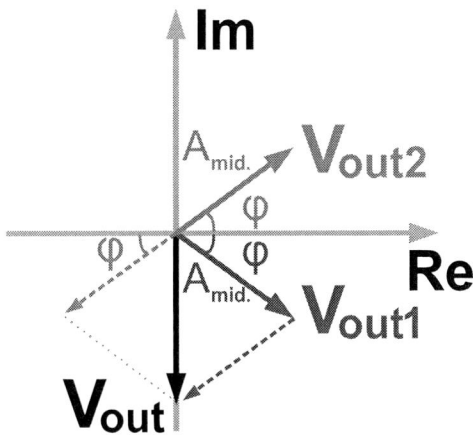

Figure 4: Complex plane representation of two in-phase actuated resonators (V_{out1}* and *V_{out2}*) and subtracted* V_{out} *at center frequency* (f_c).

The amplitude and the phase of the total output of the filter at ω_c are shown in Fig. 4. By increasing $\Delta\omega_{res}$, φ increases and the output vectors of each resonator are adding up more constructively. However, at the same time, the magnitude of each resonator decreases. Therefore, there must be an optimum $\Delta\omega_{res}$ to maximize the total output of the filter at ω_c ($A_{\text{mid.}}$) and hence reducing the ripple of the filter. It can be shown that the optimum $\Delta\omega_{res}$ equal the bandwidth of each resonator ($\Delta\omega_{res} = \omega_{\text{-3dB}, i}$), which leads to an optimum value of $A_{\text{mid.}} = A_0$ and $\Delta\varphi = \pi/2$. The optimum condition has been shown in Fig. 5 assuming $\Delta\omega_{res} = \omega_{\text{-3dB}, i} = 20\,\text{MHz}$. If $\Delta\omega_{res} > \omega_{\text{-3dB}, i}$, the filter response will contain a certain ripple which increases with increasing $\Delta\omega_{res}$. If $\Delta\omega_{res} < \omega_{\text{-3dB}, i}$, the filter response will have zero ripple but the insertion loss of the filter will increase.

By assuming the same equal quality-factors ($Q_1 = Q_2$) for the resonators, they will have about the same bandwidths ($\omega_{\text{-3dB}, 1} \approx \omega_{\text{-3dB}, 2}$). Based on the formula presented in (4) and (5), this leads to $\varphi_1 \approx -\varphi_2$. However, if the fabrication variation of the resonators is leading to a considerable quality-factor differences, then the bandwidths will be different as well as the insertion losses of the resonators. This can lead to some distortion on the filter performance. To investigate further on this issue, a mismatching between the resonators has been applied to the ideal case presented in Fig. 5. By assuming an exaggerated value of 20% mismatch between the amplitude of the resonators ($V_{out, 2} = 0.8 \times V_{out, 1}$), there will be a certain distortion in the filter response as illustrated in Fig. 6. As seen, the mismatch between the resonators is leading to a distortion in the filter behavior and still shows an acceptable performance as a BPF. As seen, in Fig. 6, if there is an intersection between the responses in the region where the resonators are in phase, there will be a notch in the resulted filter response.

EXPERIMENTS

The proposed technique is realized using two contour-mode resonators, Fig. 7, with slightly different resonance frequencies approximately at 700 MHz. The configuration and the SEM image of the device are shown in Fig. 7(a) and (b). To cancel the feed-through signal of each resonator, the bottom-electrode patterning method [7] has been utilized for both resonators. Each resonator consists of 6 fingers with 3 fingers for the input and output ports. Each finger has

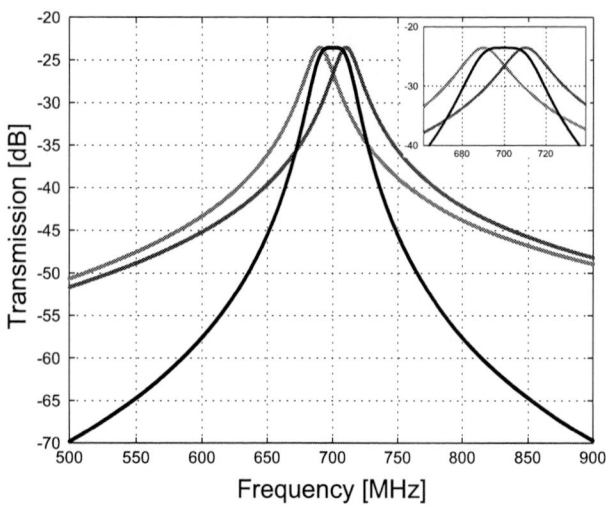

Figure 5: The simulation results of two 2nd-order resonators with $\omega_{-3dB,\,i} = \Delta\omega_{res} = 20\,\mathrm{MHz}$.

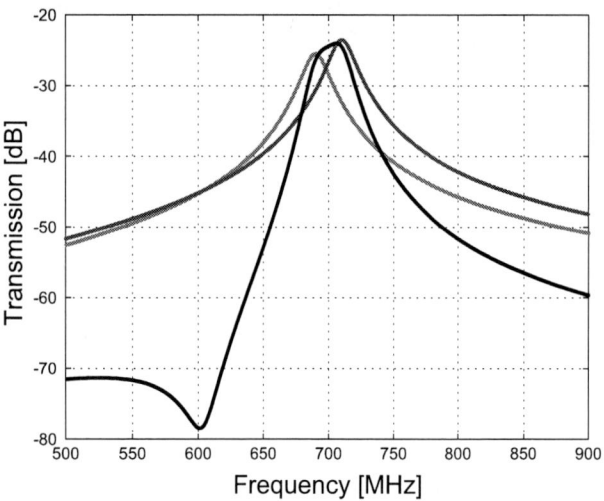

Figure 6: The result of 20% mismatch between the output of the resonators ($V_{out2} = 0.8 \times V_{out1}$)

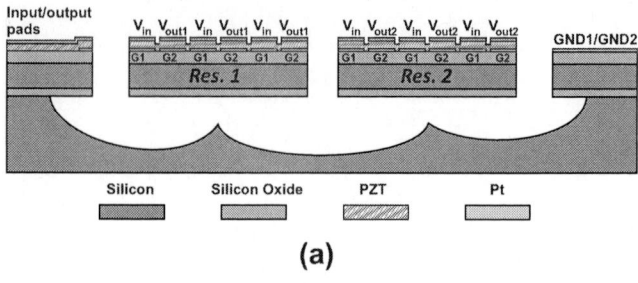

(a)

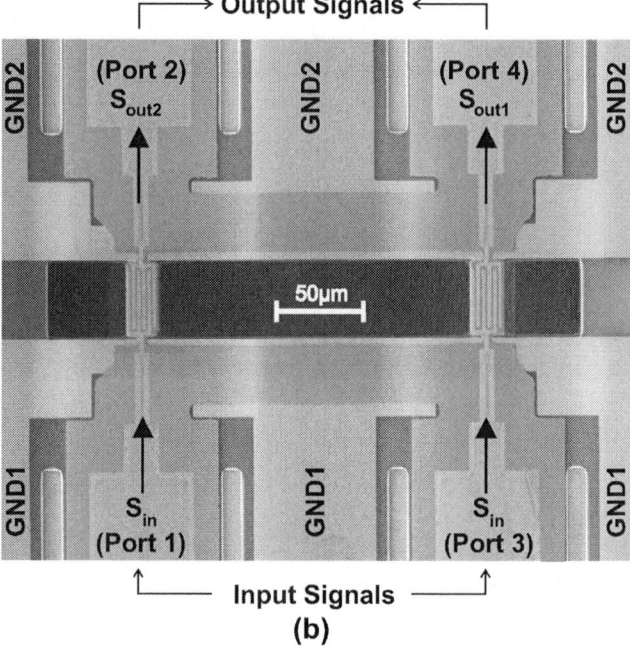

(b)

Figure 7: (a) The cross-section schematic of two bottom-electrode patterned resonators, (b) Scanning electron micrograph (SEM) of two 2nd-order PZT-on-silicon resonators with slightly different resonance frequencies around 700 MHz.

around 2.2 μm width with 1.8 μm spacing in between. The resonators consist of a composite stack of (bottom to top) SiO$_2$/Si/SiO$_2$/Ti-Pt/PZT/Pt. The fabrication process is similar to the earlier work presented in [7]. A 1 μm PLD-based PZT thin-film is utilized on top of an around 3 μm silicon layer using a highly resistive silicon-on-insulator (SOI) wafer with 0.5 μm buried oxide (BOX) layer.

The resonators were characterized using GSGSG probes and a 4-port Agilent N5244A network analyzer. A 4-port SOLT calibration has been performed using Agilent Electronic Calibration Module up to the probes. All the measurements have been done by applying 0 dBm input power. The frequency response of each fabricated resonator (S_{21} and S_{43}) and the differential readout of the filter (S_{dc21}) at DC-bias voltage of 0 V are shown in Fig. 9(a). The phase change of each resonators is presented in Fig. 9(b). The phase of the resonators at f_c is around $\varphi_i = \pm120°$. The BPF is characterized using 50 Ω termination and shows the bandwidth of approximately 28.6 MHz with 30 dB stopband rejection. The transmission of the filter (S_{dc21}) at different DC-bias voltages, 0-4 V,

with 1 V increments are presented in Fig. 9.

As presented in Fig. 8(a), by using the differential readout technique, the feed-through signals are canceled out and the stopband rejection of the filter stays quite constant compared to each single resonator response. Utilizing the presented method, there is more 20 dB improvement in the stopband rejection. As seen in Fig. 8(b), the phases ($\varphi_i = \pm120°$) are not in the optimum condition ($\Delta\varphi = \pi/2$), therefore it leads to a ripple of 3.52 dB at 0 V DC-bias voltage. By applying DC-bias voltage, the bandwidth of each resonator is increasing which causes the reduction of the ripple to 1.55 dB at 4 V.

By applying the DC-bias voltages, the transverse piezoelectric coefficient (e_{31}) of PZT layer increases and subsequently causes the reduction of motional impedance and insertion loss of the resonators. As seen in Fig. 9, the insertion loss of the filter is reduced from -23.75 dB up to -14.18 dB with DC-bias voltage of 0 V till 4 V. In-band IIP3 measurement [3] of the filter is presented in Fig. 10 using the DC-bias of 4 volt. The IIP3 at different DC-bias voltages is presented as a subfigure of Fig. 10.

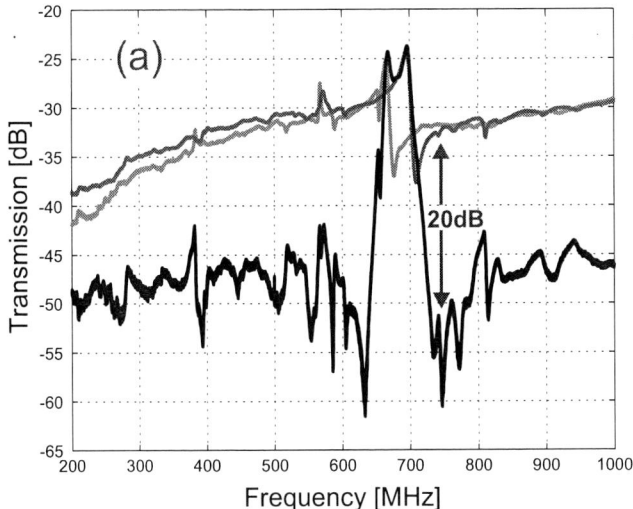

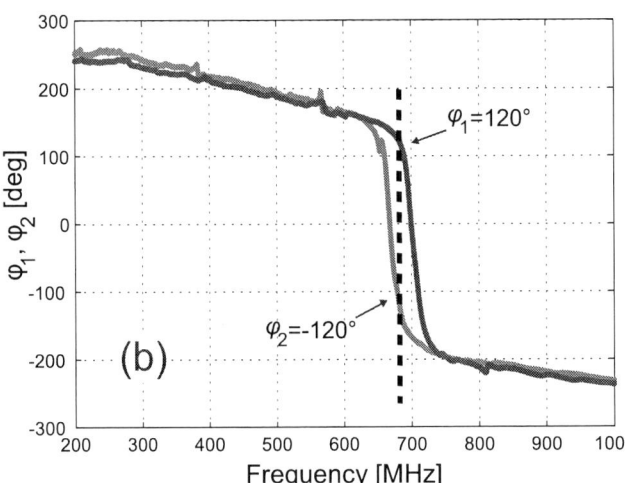

Figure 8: (a) Measured transmission gain of individual resonators and the filter (Sdc21) with 50 Ω termination at DC-bias of 0 V, (b) The measured phase change of the resonators.

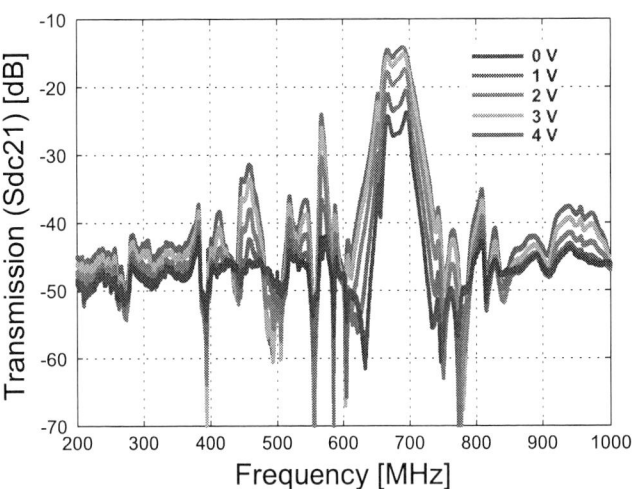

Figure 9: Measured transmission gain of the filter at different DC-bias voltages from 0-4 V with 1 V increments.

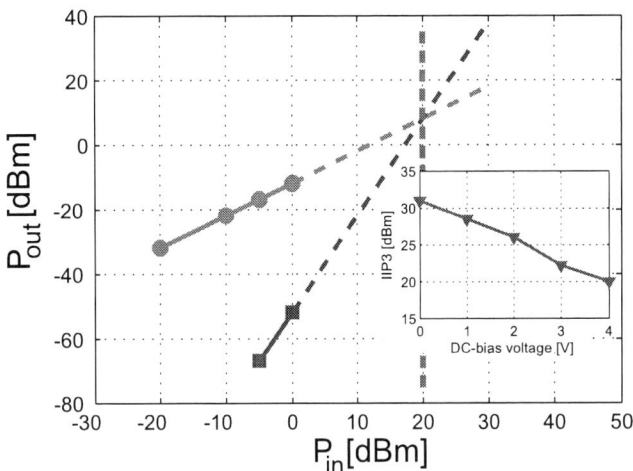

Figure 10: Measured IIP3 for the 700 MHz 4th-order filter at DC-bias of 4 V.

CONCLUSIONS

A new 4th-order BPF technique was demonstrated using the subtraction of two in-phase actuated 2nd-order contour-mode Lamb-wave resonators with slightly different resonance frequencies around 700 MHz. Utilizing this technique, the output of the resonators were added up constructively as the phase of the resonators was $\varphi_1 = -\varphi_2$, while the feed-through signals were eliminated. The BPF was presented with a bandwidth of approximately 28.6 MHz and 30 dB stopband rejection. This technique is a powerful approach for RF-MEMS filter design as well as resolving design issues associated with feed-through at high frequencies for materials with high-dielectric constant such as PZT.

ACKNOWLEDGMENTS

The authors would like to thank SolMateS B.V. for providing the PLD-based PZT thin-film.

REFERENCES

[1] R. Ruby et al., "Ultra-Miniature High-Q Filters and Duplexers Using FBAR Technology," in *Solid-State Circuits Conference*, 2001.

[2] H. Chandrahalim et al., "A $Pb(Zr_{0.52}Ti_{0.48})O_3$ transduced fully-differential mechanical-coupled frequency-agile filter," vol. 30, pp. 1296–1298, 2009.

[3] C. Zuo et al., "Very high frequency channel-select MEMS filters based on self-coupled piezoelectric AlN contour-mode resonators," *Sens. Actuators, A*, vol. 160, pp. 132–140, 2010.

[4] C. Zuo and G. Piazza, "Single-ended-to-differential and differential-to-differential channel-select filters based on piezoelectric AlN contour-mode MEMS resonators," in *Frequency Control Symposium (FCS)*, 2010.

[5] H. Yagubizade et al., "A 4th-order band-pass filter using differential readout of two in-phase actuated contour-mode resonators," *Appl. Phys. Lett.*, vol. 103, p. 173517, 2013.

[6] H. Chandrahalim et al., "PZT-transduced high-overtone width-extensional resonators above 1GHz," in *IEEE International Ultrasonics Symposium*, 2009.

[7] H. Yagubizade et al., "Pulsed-Laser Deposited $Pb(Zr_{0.52}, Ti_{0.48})O_3$-on-Silicon Resonators With High-Stopband Rejection Using Feed-Through Cancellation," *Appl. Phys. Lett.*, vol. 102, p. 063509, 2013.

AN 880 MHZ LADDER FILTER FORMED BY ARRAYS OF LATERALLY VIBRATING THIN FILM LITHIUM NIOBATE RESONATORS

Songbin Gong[1] and Gianluca Piazza[2]
[1]University of Illinois at Urbana Champaign, Urbana, IL, USA
[2]Carnegie Mellon University, Pittsburgh, PA, USA

ABSTRACT

This paper reports on the first implementation of a ladder filter using Lithium Niobate based laterally vibrating resonator arrays. This demonstration is made possible by optimizing the device design parameters and using a distributed configuration of resonator arrays to simultaneously reduce spurious vibrations and insertion loss in a low impedance RF system. The designed distributed ladder filter, with a chip footprint of 0.5 mm^2, consists of 37 resonators that are divided into 3 series and 2 shunt resonator arrays. The fabricated ladder filter based on LN LVRs showed an insertion loss of 3.5 dB and fractional bandwidth of 2.2%.

INTRODUCTION

In the recent years, the fast growth in the mobile marketplace has sparked great interest in engineering the next generation RF front ends that can meet the increasing demand for higher bandwidth and additional functionalities. New radio architectures that utilize the crowded radio frequency (RF) spectrum efficiently and adaptively are promising solutions for overcoming the data rate limits imposed by the existing static spectrum utilization methods [1]. However, the key enabling technologies for implementing these adaptive RF architectures have not yet been developed. One of the missing components in such architectures is a high performance frequency agile RF filter that can provide dynamic multiplexing at the chip scale. Surface and bulk acoustic wave devices have been the conventional multiplexing solutions for telecommunications [2, 3]. However, the limitations in their capability of monolithically integrating various frequencies on a single substrate have hindered their deployment as the building blocks for frequency agile RF filters. Piezoelectric microelectromechanical (MEMS) resonators that vibrate in the lateral direction have demonstrated multi-frequency capabilities using several widely available piezoelectric materials [4-6]. Nevertheless, none of the devices developed to date is capable of concurrently demonstrating high k_t^2 and Q. The authors have recently demonstrated a new class of laterally vibrating MEMS resonators that harness the high electromechanical coupling coefficient and low acoustic loss of transferred lithium niobate (LN) thin films [7-9]. These devices hold strong potential in providing a frequency-agile RF filtering platform owing to their record high figure of merit (FoM) and lithographically definable center frequencies. However, previous device demonstrations have shown that conventional designs of laterally vibrating resonators in thin film lithium niobate exhibit large spurious responses that appear in the filter pass band. In this work, the dependence of spurious mode excitation on design parameters including device orientation, size

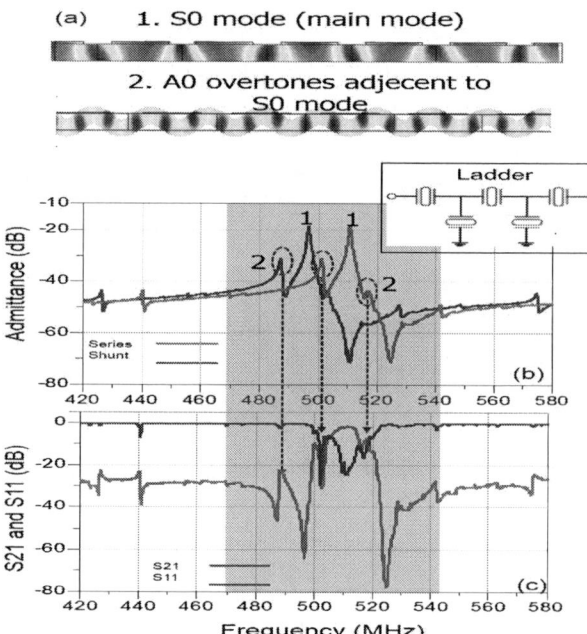

Figure 1: (a) The displacement mode shape for S0 and A0 mode resonances observed in X-cut LN resonators. (b) The measured admittance response of two LN LVRs oriented 30° to +Y axis, showing the main S0 (labeled as 1) and A0 (labeled as 2) resonances. (c) The simulated response of a ladder filter comprised by series (red) and shunt (blue) LVRs. Note that the series and shunt branches are formed by resonators of different frequencies. The A0 mode overtones introduce ripples in the overall filter response (adjacent to or in the pass band). For synthesizing a filter response with flat pass-band and clean out of band rejection, a spurious-free spectral response is desired in the frequency region highlighted in green.

and electrode thickness was investigated. The findings were leveraged to optimize the layout of LN laterally vibrating resonator (LVRs) and demonstrate devices with high FoM and spurious free response simultaneously. The same designs were subsequently used to synthesize a pass band response centered at 880 MHz formed by arraying 37 resonators in a ladder configuration. The effective spurious mode suppression techniques developed in this work enabled the fabricated ladder filter to show an insertion loss (IL) of 3.5 dB and fractional bandwidth (FBW) of 2.2% with minimum ripple.

SPURIOUS MODE SUPPRESSION

The X-Cut LN LVRs previously demonstrated by the authors were oriented 30° to +Y-axis to exploit the high electromechanical coupling for S0/lateral mode vibrations. A record high k_t^2 of 21.7% was achieved, but spurious

978-1-4799-3510-9/14 $31.00 © 2014 IEEE

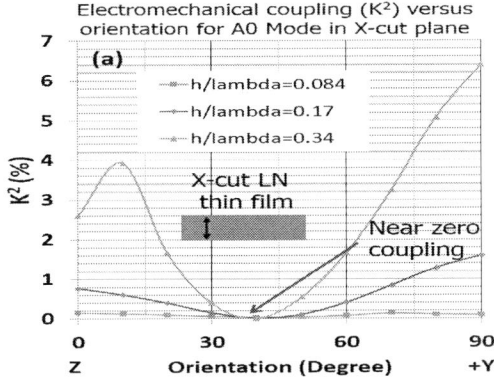

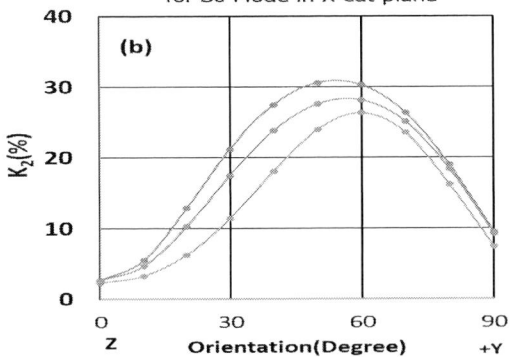

Figure 2: The simulated electro-mechanical coupling (K^2) versus device orientation for (a) A0 mode and (b) S0 mode vibrations. An orientation of 50° to $+Y$ is chosen for this work to subdue the A0 spurious response near the intended S0 resonance.

vibrations were present [9]. We identified through COMSOL simulations (Fig.1 (a)) that these spurious modes are A0 mode vibrations and are major sources of pass-band ripples in the comprised RF filters. This is clearly shown in the simulated filter response of Fig. 1. A spurious-free range of frequencies adjacent to the intended S0 resonance is required to attain an optimally flat pass-band. This presents a new challenge as the suppression of A0 modes in LN LVRs has never been investigated.

In this work, a three-prong approach was pursued to eliminate or appreciably subdue the A0 mode spurious response with the objective of substantially extending the spurious-free spectral range. These methods to subdue spurious resonances are based on understanding the influence of the device orientation, electrode thickness, and size on its performance. The rationale behind each method is explained in the following subsections.

Device Orientation

We have previously shown that the electromechanical coupling (k_t^2) is a function of device orientation for MEMS resonators enabled by transferred lithium niobate thin film. Such dependence is caused by the anisotropic crystal structure of LN. We have theoretically investigated and experimentally proven that the S0 mode of vibration in X-Cut LN has a relationship between the electromechanical cou

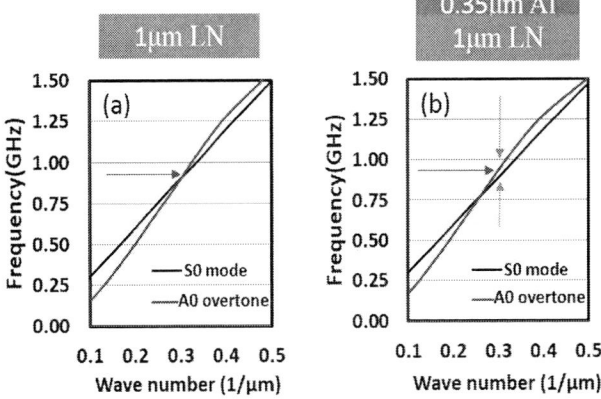

Figure 3: Comparison of the calculated dispersion relationship for S0 and A0 modes of vibration in (a) 1 μm electrode-less LN thin film (b) a film stack consisted of 0.35 μm on top of 1μm LN thin film. Both calculations were done for wave propagation along 50° to $+Y$-axis.

pling (K^2) and device in-plane orientation as shown in Fig. 2(b). The A0 modes, on the other hand, have not yet been studied since they have a lower phase velocity and are not as effective at implementing monolithic multi-frequency devices. We calculated the A0 mode electromechanical coupling as a function of the orientation in the X-cut plane using the method described in [7]. The relationship, as seen in Fig. 2(a), suggests that the electromechanical coupling for A0 mode vibrations reduces to near-zero at an orientation of 50° to $+Y$-axis, while the k_t^2 for the main S0 mode remains at a value of 10% at this orientation. For the purpose of eliminating spurious vibrations, we select this particular orientation so as to reduce the energy coupled into A0 modes at the expenses of a reduced (although acceptable and high) coupling coefficient for the S0 mode.

Electrode Thickness

The investigation of A0 mode propagation in transferred LN thin film has also revealed a dispersion relationship as seen in Fig. 3. The dispersion curve for the electrode-less LN thin film indicates that the wave numbers of an A0 overtone and the intended S0 mode will coincide near the designed center frequency (880MHz). As a result, the unwanted A0 spurious response is expected to appear near the S0 mode for the 880MHz resonators. It was shown for AlN-based resonators that the composition of the device film stack has an influence on the dispersion of lamb waves [10]. This presents an opportunity to place the A0 overtone far apart from the main S0 resonance by engineering the film stack forming the LN LVRs and hence manipulating the dispersion curves. Since only transferred LN thin film and aluminum electrodes are available in the device stack, the electrode thickness was varied to shift the intersection point of the A0 and S0 dispersion apart from the designed center frequency. An electrode thickness of 350 nm was chosen based on a tradeoff between the maximum achievable separation between A0 and S0 modes and additional fabrication complexity introduced by the use of thicker electrodes.

978-1-4799-3510-9/14 $31.00 © 2014 IEEE

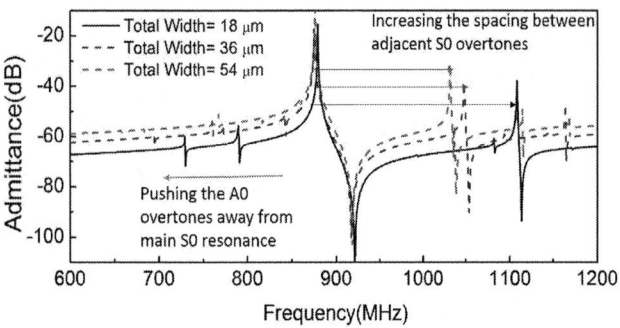

Figure 4: COMSOL simulated admittance responses for resonators of 3 different total widths. Decreasing the device total width effectively creates larger spectral separation between adjacent harmonics for both A0 (phase velocity=3000m/s) and S0 (6000m/s) modes. A spurious free spectral region (as highlighted in green in Fig. 2) is attained by using smaller width devices at the cost of lower device static capacitance and higher termination impedance.

Table 1 The summary of the key design parameters*.

Para.	Description	Value
w_p	Electrode pitch for the shunt resonators	3.2μm
w_s	Electrode pitch for the series resonators	3.15μm
$w_{p\text{-}m}$	Aluminum electrode width for the shunt resonators	2.2μm
$w_{s\text{-}m}$	Aluminum electrode width for the series resonators	2μm
t_{Al}	Aluminum electrode thickness for the series resonators	0.35μm
N_p, N_s	Number of total unweighted electrodes for resonators	5
N_w	Number of total weighted electrodes for resonators	2
W_p	Total device width for shunt resonators	19.2μm
W_s	Total device width for the series resonators	18.9μm

*The weighted electrode pitch and width are half of unweighted pitch and width respectively. The two weighted electrodes are placed at the edges of the resonator, similarly to the designs in [8].

high frequencies. This overmoded operation can be understood by treating the suspended LN thin film as an acoustic waveguide in the lateral direction with reflective boundaries at the device edges. The overtones that have their period matched to the inter-digitated electrodes are more intensively excited while other overtones remain as moderately coupled spurious modes in the device response. It was shown that the spurious S0 overtones could be suppressed using optimal electrode patterns [11, 12]. Nevertheless, the A0 overtone spurious vibrations remain more difficult to overcome due to the lower phase velocity of the A0 vibration (~3000m/S). The lower phase velocity results in smaller spacing between adjacent overtones. This is problematic for implementing a wideband filter as one or more of the closely spaced overtones may fall into the designed flat pass-band or nearby it. The spacing between adjacent overtones is inversely related the total device width. Therefore, the spurious free spectral range can be extended by simultaneously reducing the total device width (Fig. 4) and engineering the dispersion as described in the last subsection. It is important to note that this A0 suppression technique results in a lower device static capacitance as fewer inter-digitated electrodes are incorporated in the body of the resonator. This tradeoff, however, can be overcome by using arrays of resonators.

LADDER FITLER

The described spurious mode suppression techniques were all simultaneously implemented in the resonators designed to form an 880MHz ladder filter. The ladder filter relies on the use of series and shunt resonators with precise frequency spacing in between their resonances. In this work, a 30MHz resonance difference was attained by assigning different pitch width (3.15 and 3.2 μm) to the series and shunt resonators (Fig. 5(a)). The pitch width was selected based on COMSOL simulation results as shown in Fig. 6(a). The other key design device parameters are shown in in Table 1. In order to simultaneously use small size resonators and attain a high static capacitance for each filter element, arrays of resonators were chosen as the building blocks to synthesize the ladder filter. As seen in Fig. 5(b), a total number of 37

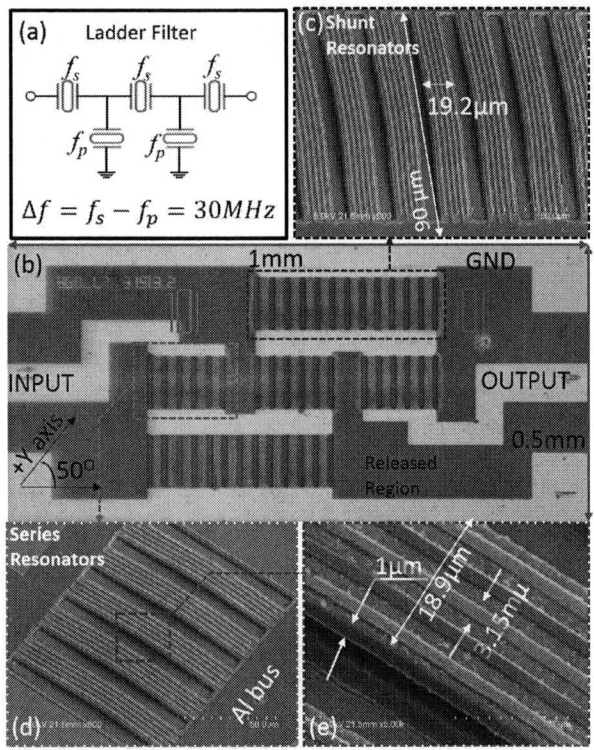

Figure 5: (a) The ladder filter comprised by 3 series (red) and 2 shunt (blue) resonator arrays. (b) The optical microscope and SEM image of the fabricated ladder filter. (c) The SEM of shunt resonator consisted of 12 distributed sub-resonators with a pitch of 3.2 μm, and a length of 90 μm. (d) The series resonator comprised by 5 parallel sub-resonators with a pitch of 3.15 μm. The resonance frequency difference between series and shunt resonators is implemented through the definition of different pitches. (e) The zoomed-in view of the resonator boundary and sidewall definition for the transferred LN thin film.

Device Size

As seen in Fig. 1, multiple A0 overtones appear in the measured response due to the overmoded nature of LVRs at

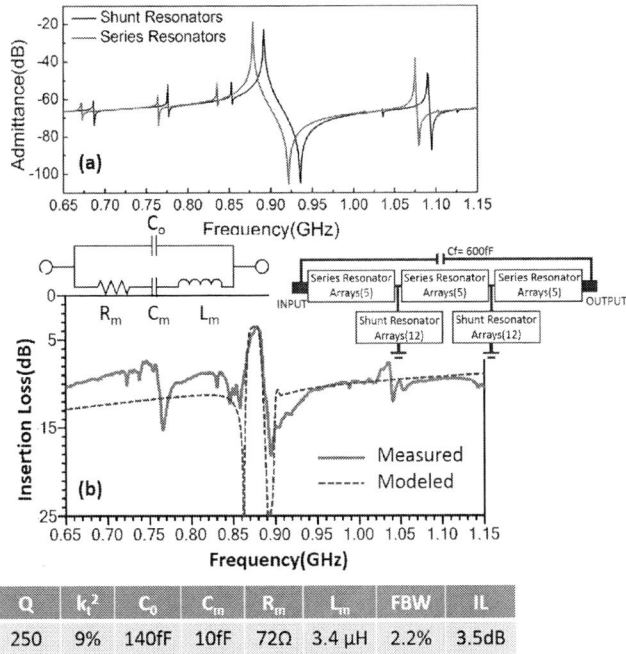

Q	k_t^2	C_0	C_m	R_m	L_m	FBW	IL
250	9%	140fF	10fF	72Ω	3.4 µH	2.2%	3.5dB

Figure 6: (a) COMSOL simulated frequency shift for the series and shunt resonance. Resonators with two frequencies are required by the design of ladder filters and implemented by changing device pitch width from 3.15 to 3.2 µm. (b) The measured and simulated distributed ladder filter response. The simulation is done using the equivalent circuit model for the individual resonator (shown in the inset) to construct the distributed ladder filter. The parameters used in the simulation are shown in the table. The measured filter response showed an IL of 3.5dB with a FBW of 2.2%. The discrepancy between the simulation and measurement below 850 MHz is not yet clear and still under analysis.

resonators were employed to realize the 880 MHz ladder filter. The resonators are divided into 3 series arrays, each formed by 5 resonators, and 2 shunt arrays, each formed by 12 resonators. The ladder filter (Fig. 5) was fabricated with a process that leverages multiple micro-machining techniques, including crystal ion slicing, LN RIE, and wet etching for release [7-9]. The fabricated filter occupies a chip area of 0.5mm by 1mm. This first demonstration (Fig. 5(b)) of a ladder filter using LN LVRs has shown an IL of 3.5 dB and FBW of 2.2%. The current out-of-band rejection is affected by a large feed through capacitance. The source of such capacitance is still under investigation.

CONCLUSION

A ladder filter based on LN LVRs was demonstrated by first optimizing the resonator layout to subdue spurious responses and then using 37 of these devices in arrays to form the ladder-connected shunt and series branches. The fabricated ladder filter showed an insertion loss of 3.5 dB and fractional bandwidth of 2.2%, further validating the LN LVRs' potential as the key enabler for adaptive RF filtering.

ACKNOWLEDGEMENTS

The authors would like to thank the DARPA RF-FPGA program for the funding support.

REFERENCES

[1] C. T. -C. Nguyen, "MEMS-based RF channel selection for true software-defined cognitive radio and low-power sensor communications." *Communications Magazine, IEEE* 51, no. 4 (2013): 110-119.

[2] T. Kimura, M. Kadota, and Y. Ida, "High Q SAW resonator using upper-electrodes on grooved-electrodes in LiTaO₃," *International Microwave Symposium Digest (MTT), 2010 IEEE MTT-S International*, pp.1740-1743, 23-28 May 2010.

[3] R. Ruby, "Review and comparison of bulk acoustic wave FBAR, SMR technology," In *2007 IEEE Ultrasonics Symposium.*, pp. 1029-40, 2007.

[4] G. Piazza, P.J. Stephanou, and A. P. Pisano, "Piezoelectric Aluminum Nitride Vibrating Contour-Mode MEMS Resonators," *Microelectromechanical Systems, Journal of*, vol.15, no.6, pp.1406,1418, Dec. 2006.

[5] C-M. Lin, Y-J. Lai, J-C. Hsu, Y-Y. Chen, D. G. Senesky, and A. P. Pisano, "High-Q aluminum nitride Lamb wave resonators with biconvex edges." *Applied Physics Letters* 99 (2011): 143501.

[6] S. Gong, N-K. Kuo, and G. Piazza, "A 1.75 GHz piezoelectrically-transduced SiC lateral overmoded bulk acoustic-wave resonator," *Solid-State Sensors, Actuators and Microsystems Conference (TRANSDUCERS), 2011 16th International*, pp.922-925, 5-9 June 2011.

[7] S. Gong, and G. Piazza, "Design and Analysis of Lithium Niobate based High Electromechanical Coupling RF-MEMS Resonators for Wideband Filtering," *Microwave Theory and Techniques, IEEE Transaction on*, vol. 61, no.1, pp.403-414, Jan. 2013

[8] S. Gong, and G. Piazza, "Figure of Merit Enhancement for Laterally Vibrating Lithium Niobate MEMS Resonators," *Electron Devices, IEEE Transaction on*, vol. 60, no. 11, pp. 3888-3894, Nov. 2013.

[9] S. Gong and G. Piazza, "Multi-frequency wideband RF filters using high electromechanical coupling laterally vibrating lithium niobate MEMS resonators," *Micro Electro Mechanical Systems (MEMS), 2013 IEEE 26th International Conference on*, pp.785-788, 20-24 Jan. 2013

[10] C-M. Lin, Y-Y. Chen, and A. P. Pisano. "Theoretical investigation of Lamb wave characteristics in AlN/3C–SiC composite membranes." *Applied Physics Letters* 97, no. 19 (2010): 193506-193506.

[11] R. Wang, S. A. Bhave, and K. Bhattacharjee, "Thin-film high k_t^2Q, multi-frequency lithium niobate resonators," *Micro Electro Mechanical Systems (MEMS), 2013 IEEE 26th International Conference on*, pp. 165-168, Jan. 2013.

[12] S. Gong, and G. Piazza, "Weighted electrode configuration for electromechanical coupling enhancement in a new class of micromachined Lithium Niobate laterally vibrating resonators," *Electron Devices Meeting (IEDM), 2012 IEEE International*, pp.15.6.1,15.6.4, 10-13 Dec. 2012

CONTACT

*S. Gong, songbin@illinois.edu;
 G. Piazza, piazza@ece.cmu.edu

CAPACITIVE SILICON RESONATOR STRUCTURE WITH MOVABLE ELECTRODES TO REDUCE CAPACITIVE GAP WIDTHS BASED ON ELECTROSTATIC PARALLEL PLATE ACTUATION

Nguyen Van Toan[1,2] and Takahito Ono[1]

[1]Graduate School of Engineering, Tohoku University, Sendai City, JAPAN
[2]Microsystem Integration Center (μSIC), Tohoku University, Sendai City, JAPAN

ABSTRACT

This paper presents the design and fabrication of a capacitive silicon resonator with movable electrodes to obtain smaller capacitive gap widths, which results in smaller motional resistance and lower insertion loss. It also helps to increase the tuning frequency range for the compensation of temperature drift of the silicon resonator. The resonant frequency of the fabricated device with a length of 500 μm, width of 440 μm and thickness of 5 μm is observed at 9.65 MHz, and the quality factor is 49,000. Using electrostatically-driven movable electrode structure, it is shown that the motional resistance is reduced by 200 times, the output signal (insertion loss) is improved by 21 dB and the tuning characteristic of the frequency is 7 times larger than those of the structures without movable electrode structures.

INTRODUCTION

A capacitive silicon resonator is one of the most promising options to replace quartz crystal resonators due to the capability of further miniaturization [1-2]. However, its main difficulty is achievement of a small motional resistance and hence, resulting low insertion loss to satisfy the condition for the oscillation. Also, a smaller motional resistance of the silicon resonator results in lower phase noise in the oscillators; therefore, the motional resistance R_m should be as small as possible. There are many ways for reducing the motional resistance such as increasing the polarization voltage V_{DC}, increasing the overlap area of capacitive, increasing the Q factor, and decreasing the gap widths. The gap reduction is considered as one of the best ways since the motional resistance is proportional to the fourth order of the gap width.

Some researches have been reported to obtain a small motional resistance [3-4]. The motional resistance of resonant structures with a solid-filled gap [3] shows a small value. However, the resonator structure suffers from material mismatch which decreases the mechanical quality factor (Q) due to the high interface loss [1]. Fabrication method of resonators with small capacitive gap using a thin oxide film as a sacrificial layer has been presented in [4]. However, this approach is slightly complex and high temperature process is required.

A fabrication and packaging process of silicon resonators capable of the integration of LSI towards application of timing device were reported in our previous research [2, 5]. However, the motional resistance has a big value due to the large capacitive gap width between the electrode and the resonant body. In this research, a simple fabrication process for making capacitive silicon resonators with narrow gap using direct etching is proposed. A capacitive silicon resonator with movable electrodes for reducing the electrodes to resonant body gap widths based on electrostatic parallel plate actuation is fabricated and evaluated.

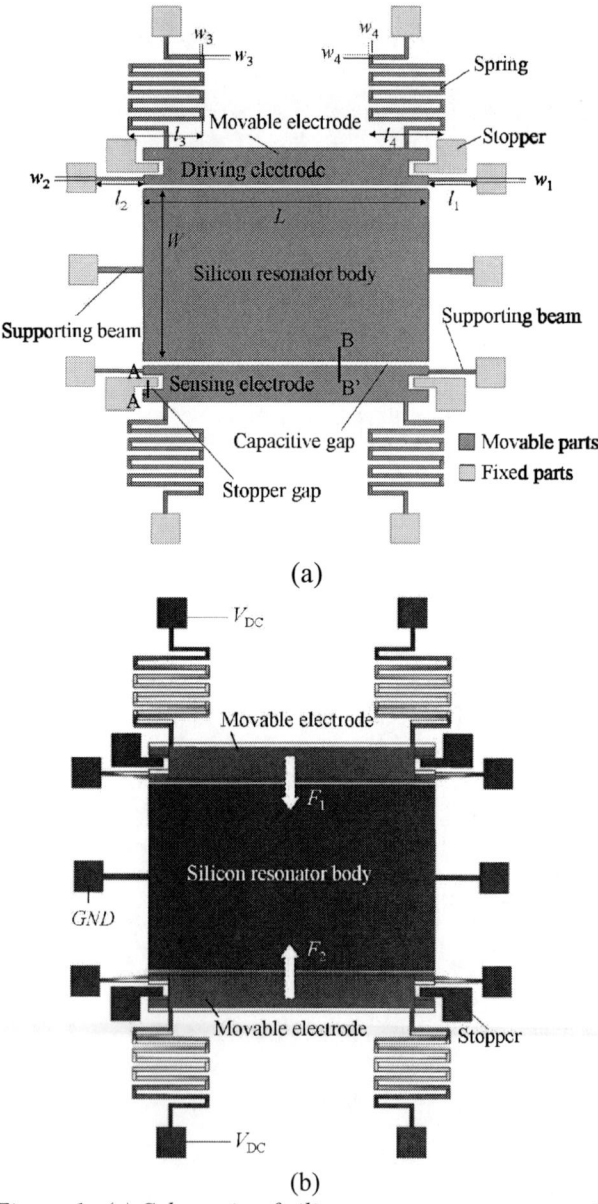

Figure 1: (a) Schematic of silicon resonator structure with movable electrodes. (b) FEM simulation results

RESONATOR STRUCTURE AND WORKING PRINCIBLE

The resonator structure in this research is shown in Fig. 1(a). It basically consists of a silicon resonant body, movable electrodes and stoppers. The resonant body is supported by two supporting beams on the sides, and also placed between two electrodes (driving and sensing electrodes) with narrow capacitive gap widths. Each electrode is supported by two springs and two beams, and can be electrostatically actuated. The stoppers are formed on a substrate, which make narrow gap between the resonant body and movable electrodes. The parameters of the resonator are summarized in Table 1.

The DC polarization voltage V_{DC} is applied to driving/sensing movable electrodes while the resonant body is connected to ground. Thus, same values of electrostatic forces ($F_1 = -F_2$, as shown in Fig. 1(b)) are generated on the both sides of the resonant body. When V_{DC} increased, the movable electrodes will move toward the resonant body due to electrostatic force between the movable electrode and the resonant body. The electrodes will be attracted to the resonant body and they touch the stoppers when V_{DC} is increased; thus, the capacitive gaps can be reduced. Figure 1(b) shows the schematic of the actuation of silicon resonator structure with movable electrodes simulated by FEM simulation using COMSOL software. Both of the movable electrodes move toward the resonant body when a DC voltage is applied. In this design, the reduced gap size ($g_{reduced}$) of the silicon resonator with the movable electrodes can be given by

$$g_{reduced} = g_{B-B'} - g_{A-A'}, \qquad (1)$$

where $g_{B-B'}$ is the capacitive gap width and $g_{A-A'}$ is the stopper gap width, as shown in Fig. 1(a). The capacitive gap width is designed to be larger than the stopper gap width to make smaller gaps between the resonant body and electrodes when electrodes are actuated.

When an AC voltage V_{AC} is applied to the driving electrode, resulting electrostatic force causes bulk acoustic wave in the resonant body. The resonator is electrostatically exited and vibrated in the horizontal width extensional mode at the resonant frequency by AC voltage in addition to DC voltage. The output voltage of the resonator is obtained by the capacitive detection between a sensing electrode and the resonant body. Small changes in the size of the capacitive gap width generate a voltage on the sensing electrode.

EXPERIMENTAL METHODOLOGY

The silicon resonator structures are formed by a combination of electron beam lithography (EBL) and photolithography, and deep reactive ion etching (DRIE) process on a silicon on insulator (SOI) wafer, and then transferred to an LTCC (low temperature co-fired ceramic) substrate by using an anodic bonding technique. The handling silicon layer of the SOI wafer was etched out by plasma etching, and the buried SiO_2 layer was removed. The fabrication process of the silicon resonator has been presented in the details in our previous research [2, 5]. The resonator structures with and without movable electrodes are

successfully fabricated as shown in Figs. 2(a) and 2(c). The stopper gap width with A-A' cross section and the capacitive gap width with B-B' cross section of the structure with the movable electrodes are 400 nm and 500 nm, respectively, as shown in Figs. 2(b) and 2(d). Therefore, the final gap widths can reach at 100 nm in this design. The silicon resonator structure without the movable electrode with 400 nm of the capacitive gap size is fabricated as A-A' cross section is shown in Figs. 2(c) and 2(d).

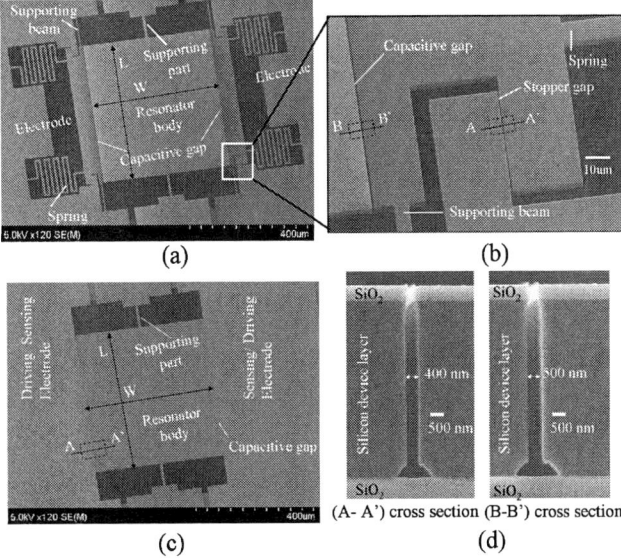

(a) (b)

(c) (d)

(A- A') cross section (B-B') cross section

Figure 2. (a) SEM view of the fabricated silicon resonator structure with movable electrodes. (b) A close-up of a corner of the resonator structure. (c) SEM photograph of the fabricated silicon resonator without movable electrodes. (d) Profile shape of (A-A') and (B-B') cross section.

EXPERIMENTAL RESULTS

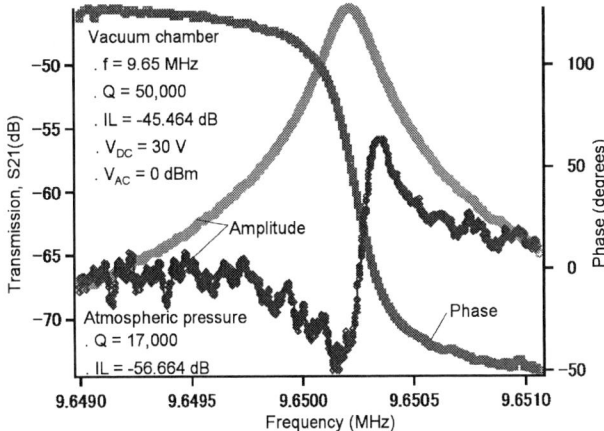

Figure 3. Transmission and phase response of resonator #1 and frequency response of resonator #1 both under vacuum chamber pressure and atmospheric pressure.

The resonators are placed inside a vacuum chamber with the coaxial feed-through. The resonance characteristics of silicon resonator structures with (resonator #1) and without (resonator #2) the movable electrodes are observed. The specifications are summarized in Table 1. As shown in Fig. 3, the measurement results of the transmission S_{21} and phase

978-1-4799-3510-9/14 $31.00 © 2014 IEEE   1246

response are indicated for the resonator #1 with a length of 500 μm, width of 440 μm, thickness of 5 μm and the reduced capacitive gap size of 100 nm. A resonant peak, which is observed under measurement conditions V_{DC} of 30 V and V_{AC} of 0 dBm, is found at 9.65 MHz with a high Q factor of 50,000. Besides, figure 3 also presents a comparison of measured frequency characteristics in the vacuum chamber with pressures of 0.01 Pa and ambient atmosphere pressure of 10^5 Pa. The Q factor of 50,000 at 0.01 Pa is larger than that of 17,000 in ambient atmosphere due to absence of air viscous damping.

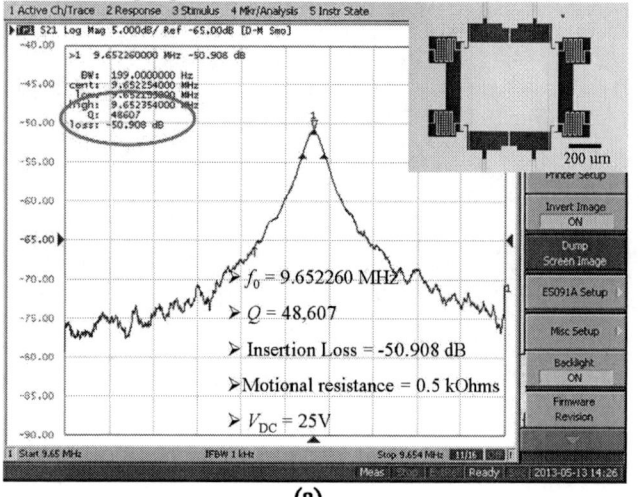

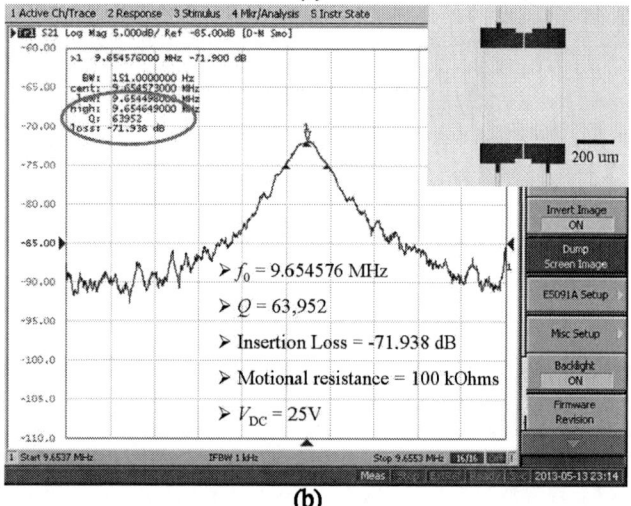

Figure 4. (a) Frequency response of resonator #1. (b) Frequency response of resonator #2.

Although the resonators #1 and #2 have been designed with the same dimensions, and fabricated on the same SOI wafer, their resonance frequencies are slightly different as can be seen in Figs. 4(a) and 4(b) (9.652260 MHz and 9.654576 MHz). This is due to the fabrication tolerances such as scalloping, notching and over etching that makes the effective spring constant and effective mass slightly changed. A comparison of the performance of the resonators #1 and #2 is shown in Table 1.

The frequency responses of the resonators #1 and #2, which are performed under measurement conditions V_{DC} of

25 V and V_{AC} of 0 dBm, are shown in Table 1 and Fig. 4. The measured transmission results in Figs. 4(a) and 4(b) clearly show that the output signal of the structure with movable electrodes is 21 dB larger than that of structure without movable electrodes at resonant frequency. The calculation of equivalent circuit model of the resonator #1 based on measured data shows the motional resistance $R_m = 0.5$ kΩ, the motional capacitance $C_m = 6.50 \times 10^{-16}$ F and the motional inductance $L_m = 4.18 \times 10^{-1}$ H in the reduced capacitive gap size of 100 nm while that of the resonator #2 shows motional resistance $R_m = 100$ kΩ, motional capacitance $C_m = 2.54 \times 10^{-18}$ F and motional inductance $L_m = 1.07 \times 10^2$ H in the capacitive gap size of 400 nm. The motional resistance value of resonator #1 is 200 times smaller than that of resonator #2 because of the reduction of the capacitive gap size. However, the Q factor decreases from 64,000 to 49,000 in comparison of the resonators #1 and #2.

The drop in the measured Q factor is found to be attributed to the various loading effects such as any series resistance in the resonator path, including termination resistors, as well as other resistances in the silicon structure of the resonator [4]. All the loading effects can be collected in the equivalent resistor R_{loaded}, connected in series with the resonator. The definitions of the unloaded Q factor and loaded Q factor are presented in [1, 4, 6] and as shown below.

$$Q_{unloaded} = \frac{2\pi f_0 L_m}{R_m}, \qquad (2)$$

$$Q_{loaded} = \frac{2\pi f_0 L_m}{R_m + R_{loaded}}, \qquad (3)$$

The $Q_{unloaded}$ factor represents for the intrinsic properties while Q_{loaded} factor shows the measured Q factor including the loading effect of R_{loaded}. The relationship between $Q_{unloaded}$ factor and Q_{loaded} factor can be found by the combination of Eq. 2 and Eq. 3.

$$Q_{loaded} = Q_{unloaded} \frac{R_m}{R_m + R_{loaded}}. \qquad (4)$$

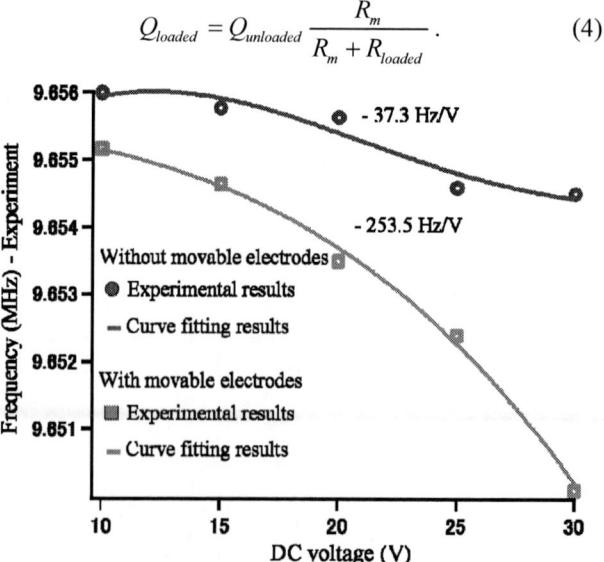

Figure 5. Electrostatic tuning characteristic of resonators #1 and #2.

As discussed above, when the capacitive gap width

reduces from 400 nm (resonator #2) to 100 nm (resonator #1), the motional resistance value is dramatically decreased by 200 times from 100 kΩ to 0.5 kΩ. This causes the drop in the measured Q_{loaded} value, as presented in Eq. 4. Therefore, the Q factor of structure with movable electrodes is smaller than that of the structure without movable electrode

The resonant peak shifts to lower frequency if the polarization voltage is increased due to the effect of electrical stiffness. Figure 5 shows comparisons of the tuning slope between the structures with and without the movable electrodes. The measured electrostatic tuning characteristic for resonator #2 with 400 nm capacitive gap is around - 37.0 Hz/V while that of resonator #1 is around - 253.0 Hz/V. There is 7 time difference of tuning characteristic between resonators #1 and #2.

CONCLUSION

In this paper, we designed, fabricated and evaluated capacitive silicon resonators with movable electrode structures for lower motional resistance, lower insertion loss and wider tuning frequency range for the compensation of the temperature drifts. Frequency characteristics of the silicon resonator of resonant frequency 9.65 MHz with a length of 500 µm, width of 440 µm and thickness of 5 µm were evaluated, and a high Q factor of 49,000 is achieved at a polarization voltage of 25 V. The measurement results have shown that the motional resistance is reduced by 200 times, the output signal is increased by 21 dB and the tuning characteristic of the frequency is also increased by 7 times than that without movable electrode structures.

REFERENCES

[1] J.T.M.V Beek and R. Puers, "A review of MEMS oscillators for frequency reference and timing applications" *J. Micromech. Microeng* 22 013001, 2012.

[2] N.V. Toan, H. Miyashita, M. Toda, Y. Kawai and T. Ono, "Fabrication of an hermetically packaged silicon resonator on LTCC substrate" *J. Microsystem Techonlogies*, 19, pp. 1165-1175, 2012.

[3] Y.W. Lin, S.S. Li, Y. Xie, Z. Ren, and C.T.C. Nguyen, "Vibrating micromechanical resonators with solid dielectric capacitive transducer gaps" *In frequency control symposium and exposition*, pp. 128-34, 2005.

[4] S. Pourkamali, G.K. Ho and F. Ayazi, "Low impedance VHF and UHF capacitive silicon bulk acoustic wave resonators", *In IEEE transactions on electron device* 54, pp. 2017-2023, 2007.

[5] N.V. Toan, H. Miyashita, M. Toda, Y. Kawai and T. Ono, "Fabrication and packaging process of silicon resonator capable of the integration of LSI for application of timing device", *The 26th IEEE international conference on Micro Electro Mechanical Systems*, pp. 377-380, 2013.

[6] G.K. Ho, K. Sundaresan, S. Pourkamali and F. Ayazi, "Micromechincal IBARs: tunable high Q resonator fro temperature compensated reference oscillators", *Journal of micromelectromechanical systems*, 19, pp. 503-15, 2010

CONTACT

*N.V. Toan, tel: +81-22-795-5806;
E-mail: nvtoan@nme.mech.tohoku.ac.jp

Table 1: Summarized parameters of the silicon resonators with and without movable electrode structures

	With movable electrodes (Resonator #1)	Without movable electrodes (Resonator #2)
Parameters:		
. Length of resonant body - L	. 500 µm	. 500 µm
. Width of resonant body - W	. 440 µm	. 440 µm
. Thickness of SOI device layer- t	. 5 µm	. 5 µm
. Capacitive gap - g	. 100 nm	. 400 nm
. Length of the supporting beams - l_1, l_2	. 100 µm	. none
. Width of the supporting beams - w_1, w_2	. 3 µm	. none
. Length of the springs - l_3, l_4	. 120 µm	. none
. Width of the springs - w_3, w_4	. 6 µm	. none
Applied conditions:		
. V_{DC}	. 25 V	. 25 V
. V_{AC}	. 0 dBm	. 0 dBm
. Pressure level of vacuum chamber	. 0.01 Pa	. 0.01 Pa
Measurement results:		
. Measured frequency - f_0	. 9.652260 MHz	. 9.654576 MHz
. Quality factor - Q	. 48, 607	. 63,952
. Insertion loss - IL	. - 50.908 dB	. - 71.938 dB
. Motional resistance - R_m	. 0.5 kΩ	. 100 kΩ
. Motional capacitance - C_m	. 6.50×10^{-16} F	. 2.54×10^{-18} F
. Motional inductance - L_m	. 4.18×10^{-1} H	. 1.07×10^{2} H
. Feed-through capacitance - C_f	. 2.21×10^{-13} F	. 5.53×10^{-14} F
. Tuning slope frequency	. - 253.5 Hz/V	. - 37.3 Hz/V

978-1-4799-3510-9/14 $31.00 © 2014 IEEE

COMBINED ELECTRICAL AND MECHANICAL COUPLING FOR MODE-RECONFIGURABLE CMOS-MEMS FILTERS

Chao-Yu Chen[1], Ming-Huang Li[1], Chi-Hang Chin[1], Cheng-Syun Li[1], and Sheng-Shian Li[1,2]

[1]Institute of NanoEngineering and MicroSystems, [2]Department of Power Mechanical Engineering
National Tsing Hua University, Hsinchu, Taiwan

ABSTRACT

This work presents a novel filter scheme which combines both electrical and mechanical coupling mechanisms implemented in a CMOS-MEMS filter to simultaneously attain small percent bandwidth through weakly mechanical link and decent stopband rejection via differentially electrical configuration. As compared to the traditional parallel-class (i.e., electrically-coupled) filters and mechanically-coupled filters, the proposed oxide-rich filter structure features flexible electrical routing and non-conductive mechanical filter couplers, hence enabling common-mode to differential (CIDO) and differential to common-mode (DICO) reconfigurable modes all within a single device. The proposed 8.6-MHz CMOS-MEMS filter has been successfully demonstrated with a narrow passband of 35 kHz (0.41% bandwidth) and stopband rejection more than 20 dB under proper termination.

INTRODUCTION

To meet the demand of miniaturization for modern wireless communication systems and sensor networks, the low-power and compact transceivers play an important role from the Internet of Things (IoT) to smart life for human beings. However, a few discrete mechanical transducers, such as reference oscillators (e.g., quartz) and band-select filters (e.g., SAW), in current front-end communication systems are still necessitated, thus impeding the trend toward on-chip integrated transceivers. To address this issue, the MEMS oscillators [1][2] and filters [3]-[6] have been intensively studied in the past decade, so far realizing small form factor and bringing cost-effective and robust features to replace their bulky counterparts abovementioned. Although the MEMS-based devices greatly reduce the RF front-end chip area as in [7], the additional wire-bonding or packaging technology (e.g., TSV) is still necessary to form electrical connections between the MEMS devices and CMOS circuits (i.e., system in package, SiP), therefore hindering the single chip solution in the future.

On the other hand, CMOS-MEMS platforms [8] take advantage of the existing layers in CMOS back-end-of-line (BEOL) to fabricate the mechanical devices. Through the standard CMOS foundry processes, the MEMS resonators feature not only high Q and small footprint, but integration with ICs. The CMOS-MEMS filters to date are designed into two main categories, including electrically-coupled [9][10] and mechanically-coupled [11][12] filter topologies. The electrically-coupled design can easily achieve a narrow filter passband using two or more resonator output signals coupled by an additional readout circuit (e.g., TIA or Buffer).

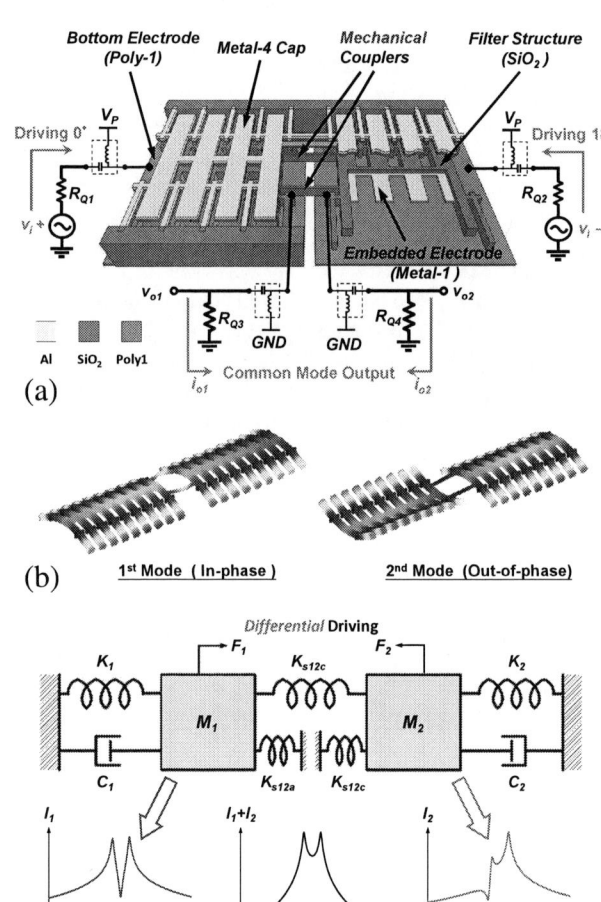

Fig. 1: (a) Perspective-view schematic, (b) finite element simulation of the proposed filter structure and (c) qualitative description of the equivalent model for the proposed filter in a differential to common-mode (DICO) configuration.

However, the use of extra circuits may lead to more power consumption; in addition, the process variation of resonators would cause frequency discrepancy, resulting in large in-band ripples of the filter. On the contrary, the filter bandwidth of mechanically-coupled filters is well defined through the physical filter coupler but very difficult to suppress the background feedthrough signals from the substrate under a typical two-port configuration.

In this work, we propose a novel coupling concept which combines electrically and mechanically-coupled mechanisms together for narrow-bandwidth filter implementation. Based on the thorough modeling of capacitive transducers [13], the

978-1-4799-3510-9/14 $31.00 © 2014 IEEE

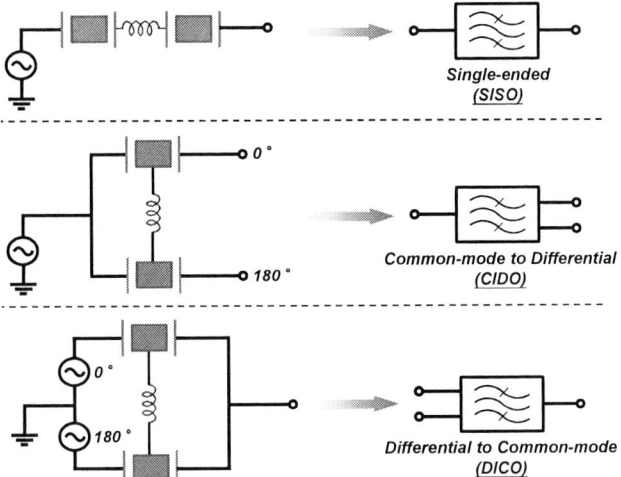

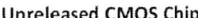

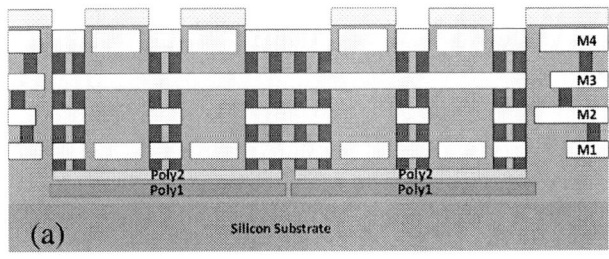

Fig. 2: Concept of the mode-reconfigurable CMOS-MEMS filter.

arraying design [12] and deep submicron gap process [14] are adopted in the proposed filter to take advantage of the low motional impedance (R_m) resonant tanks. Moreover, two weak filter couplers close to nodal locations of the constituent resonators (i.e., low-velocity coupling) together with a differentially operating configuration facilitate the filter termination, thus resulting in flat passband and decent stopband rejection. By using the proposed concept, the CMOS-MEMS filter structure enables the operation of (i) single-ended (SISO), (ii) common-mode to differential (CIDO), and (iii) differential to common-mode (DICO) configurations in a single device, paving a way for the future filter implementation.

CMOS-MEMS FILTER DESIGN

Fig. 1(a) presents the perspective view schematic of the proposed filter operated in a differential to common-mode (DICO) configuration (one of the implementations). The oxide-rich filter structure is composed of two free-free beam (FFB) resonator arrays mechanically linked by two flexural filter couplers close to the nodal locations of the vibrating FFBs to enable narrow passband (i.e., low velocity coupling scheme [13]). Notably, the oxide-rich structure [15] and mechanically arrayed design would provide higher quality factor (Q) than its metal-alloy counterpart [12] and simultaneously increase the effective transduction areas, thus evidently reducing the insertion loss (*I.L.*) of the filter.

Unlike the traditional mechanically-coupled filters operating in a two-port biasing and excitation configuration (SISO), a differentially driving signals (v_i+ and v_i-) together with a dc bias (V_P) are applied onto the bottom poly-Si electrodes (Poly-1) to concurrently drive the filter structure into vibration with two physical mode shapes shown in the Fig. 1(b). The common-mode output current can then be sensed by the embedded electrodes (Metal-1) inside the SiO_2 filter structure. To understand the working principle, Fig. 1(c) illustrates the transfer function plots of the proposed filter based on a DICO configuration. In this system, the

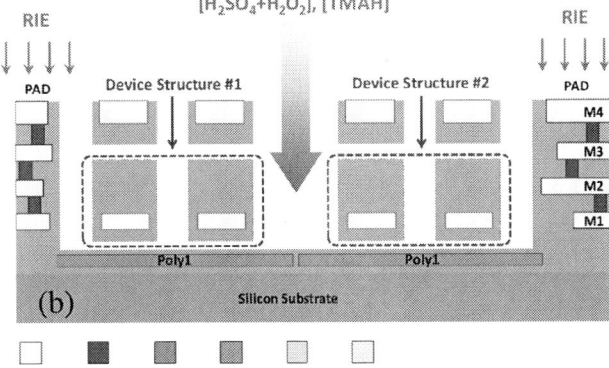

Fig. 3: Cross-sectional view of the CMOS-MEMS filter (a) before and (b) after post-CMOS processing. Note that the device structure #1 and #2 are electrically isolated, due to non-conductive couplers.

motional current (I_i) can be individually detected from a single resonant unit, given by

$$I_i = \eta_e \left(j\omega X_i \right) \tag{1}$$

where X_i is the displacement of a single resonant tank; i numbers each resonant tank of the filter (i= 1, 2); and η_e is the electromechanical coupling factor of the capacitive transducer. The desirable filter spectrum (I_1+I_2) is formed by summing the motional signals I_1 and I_2 together. By the use of the differential operation, the undesired parasitic feedthrough signals can be effectively eliminated. Therefore, this topology features evident improvement on noise floor without the use of any de-embedding technique as compared to our previous work [12].

Moreover, the proposed filter is also capable of realizing various configurations as different electrical setups are carried out, such as (i) a traditional mechanically-coupled filter (single-ended, SISO) operated in a typical two-port configuration and (ii) a common-mode to differential filter (CIDO), respectively, as depicted in Fig. 2. With this flexibility, the designer can freely chose the operation conditions dependent on the system requirements. Furthermore, the feature of the unique port reconfigurable capability (i.e., CIDO and DICO) physically realizes an on-chip single-to-balance converter without any discrete balun transformer, facilitating the implementation of a miniaturized, integrated transceiver.

FABRICATION PROCESS

978-1-4799-3510-9/14 $31.00 © 2014 IEEE

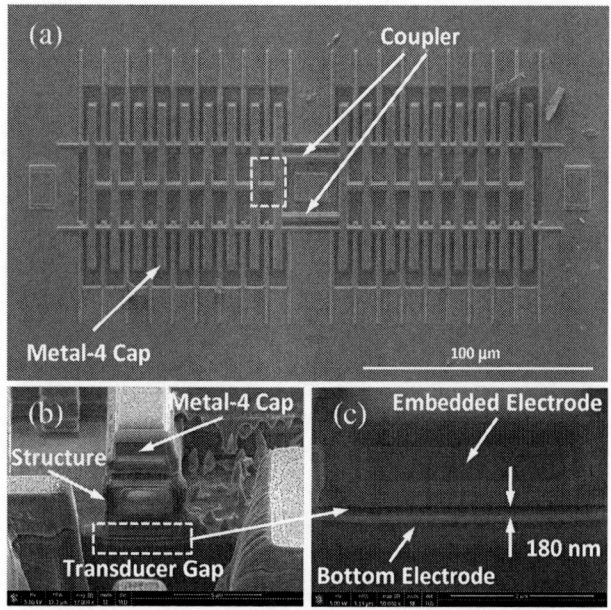

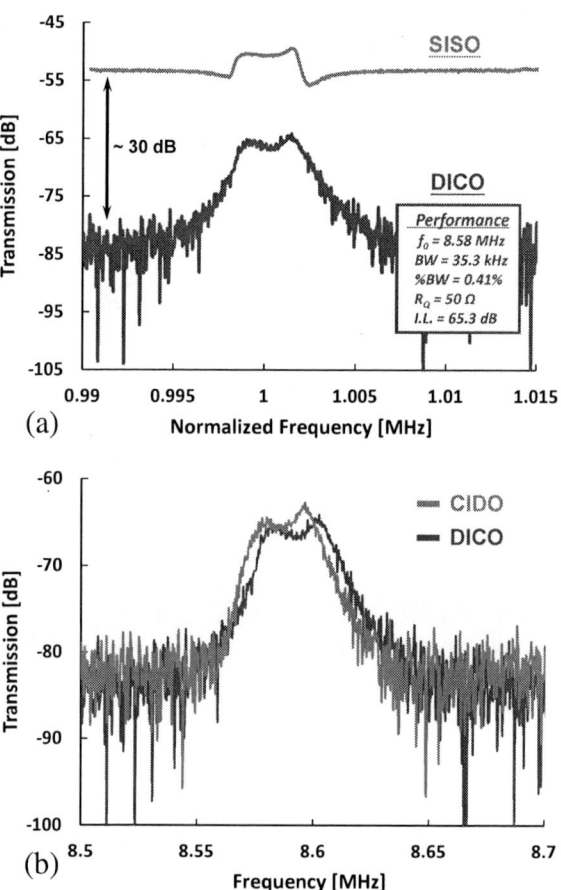

Fig. 4: SEM pictures of the proposed CMOS-MEMS filter: (a) the global view, (b) cross-section of a free-free beam structure, and (c) transducer's gap-zoomed view.

As aforementioned, the oxide-rich filter is fabricated using a standard foundry-oriented CMOS-MEMS platform [14]. Fig. 3 indicates the cross-sectional view and post-CMOS process flow for the proposed CMOS-MEMS filter. To well define the structure profile, a two-step wet etching process [14] is adopted in this work for achieving the tiny transducer's gap without the special structure design (i.e., pull-in frame designed in [12]). First, a metal etchant composed of sulfuric acid (H_2SO_4) and hydrogen peroxide (H_2O_2) is utilized to remove the sacrificial metal layer under the passivation-free region due to the high selectivity to SiO_2. Then a TMAH wet etching is used to remove the exposed polysilicon layer (Poly-2) for realizing deep submicron gap spacing. Note that the top Metal-4 cap serves to protect the device structure during the final RIE pad-opening process.

After the maskless release process, Fig. 4(a) presents a

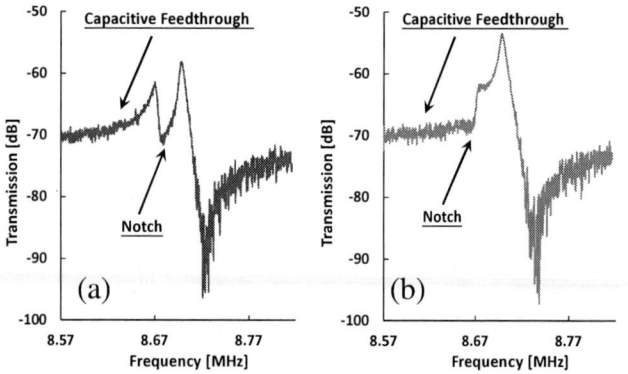

Fig. 5: CMOS-MEMS filter under a differential driving scheme while the output signals are individually accessed from (a) device structure #1 and (b) device structure #2 (denoted in Fig. 3(b)). The measured frequency characteristics match quite well with the theoretical prediction (cf. Fig. 1(c)).

Fig. 6: Comparison of the transmission spectra: (a) SISO vs. DICO operation, and (b) CIDO vs. DICO operation.

global SEM view of the fabricated CMOS-MEMS filter on which the metal protection layer (Metal-4 Cap) is clearly seen. Fig. 4(b) further shows the cross-sectional view from a focused ion beam (FIB) cut area indicated in Fig. 4(a) while a zoom-in of the transducer's gap (around 180 nm) can be easily seen in Fig. 4(c).

MEASUREMENT RESULTS

To verify the performance of the DICO configuration illustrated in Fig. 1(c), the fabricated filters are tested under a 1 mTorr environment using a cryogenic vacuum chamber with a proper bias voltage (around 60 V). Since the non-conductive filter couplers are used to connect two distinct resonant tanks, we can drive this filter into resonance through differential signals while individually accessing output signals from different parts of the filter structure (i.e., #1 and #2 denoted in Fig. 3(b)) with the measured spectra shown in Fig. 5(a) and (b), respectively. However, the measured transmission exhibits some discrepancy in background feedthrough level as compared to an ideal simulation (cf. Fig. 1(c)) due to the existence of undesired feedthrough capacitance (C_f) coupling from the substrate, but still in good agreement with our theoretical prediction in (1).

As abovementioned, the proposed filter is capable of operating in various configurations. Fig. 6(a) and (b) successfully demonstrate the mode-reconfigurable capability

978-1-4799-3510-9/14 $31.00 © 2014 IEEE   1251

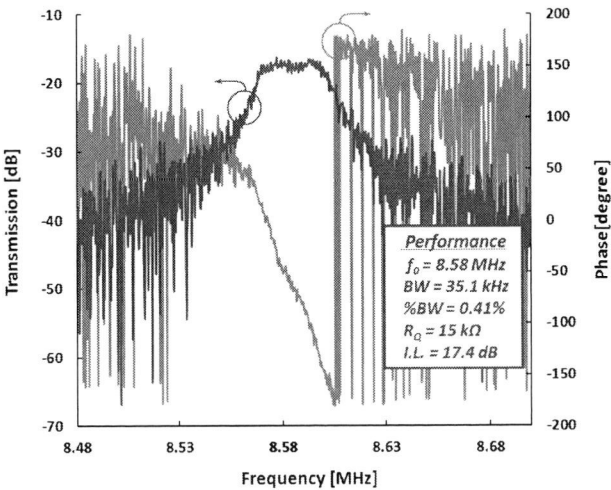

Fig. 7: Terminated frequency spectrum of the CMOS-MEMS filter in a DISO configuration.

in a single device. Notably, as compared to the conventional mechanically-coupled filter operated in SISO setup, the differential configuration provides feedthrough cancellation that creates a 30-dB improvement on the feedthrough floor as shown in Fig. 6(a). In addition, there is no difference on bandwidth between SISO and DICO operation, implying that the filter bandwidth is preserved and determined by the geometry design and location of the filter couplers. Also, the comparison of CIDO and DICO configurations is provided in Fig. 6(b) where the filter spectra show very similar performance in terms of transmission magnitude and bandwidth. Note the deviation of the filter center frequency is mainly caused by different diving conditions.

Fig. 7 finally presents a terminated spectrum and phase transition for the 8.6-MHz CMOS-MEMS filter under a DICO configuration with the stopband rejection greater than 20 dB thanks to the differential operation. As a result, the flatten passband is attained with a narrow 0.41% bandwidth and insertion loss of 17.4 dB when the port resistance (R_Q) of 15 kΩ is adopted in the termination network.

CONCLUSIONS

In this work, we demonstrate the micromechanical filter fabricated by a foundry-orientation CMOS-MEMS platform. To attain low-loss and channel-select features, the high-Q structural material (i.e., silicon dioxide) and beam-arrayed resonator design are implemented in the proposed filter. Moreover, the combination of metal and polysilicon etching processes is adopted for the post-CMOS release to create deep submicron transducer's gap to further reduce the motion impedance. Finally, the novel concept combining electrical and mechanical coupling for future multi-mode systems is reported in this work. By employing this topology, various operating configurations can easily be realized in a single device. As a result, the proposed filter demonstrates a narrow percent bandwidth (0.41%) and acceptable stopband rejection benefiting from the precise coupler design and differential operation.

ACKNOWLEDGEMENTS

The author would like to thank the supported by the National Science Council (NSC) of Taiwan under grant of NSC-101-2628-E-007-008-MY2, also appreciate the TSMC and National Chip Implementation Center (CIC), Taiwan, for the device manufacturing and CAD tools providing.

REFERENCES

[1] Y.-W. Lin, *et al.*, "Series-resonant VHF micromechanical resonator reference oscillators," *IEEE J. Solid-State Circuits*, vol. 39, no. 12, pp. 2477-2491, 2004.

[2] C. T.-C. Nguyen, "MEMS technology for timing and frequency control," *IEEE Trans. Ultrason., Ferroelect., Freq. Contr.*, vol. 54, no. 2, pp. 251-270, 2007.

[3] S. Pourkamali, *et al.*, "Electrically coupled MEMS bandpass filters part II. Without coupling elements," *Sensors and Actuators A*, vol. 122, pp. 317-325, 2005.

[4] D. Weinstein, *et al.*, "Dielectrically transduced single-ended to differential MEMS filter," *Dig. ISSCC'06*, pp. 318-319, 2006.

[5] S.-S. Li, *et al.*, "An MSI micromechanical differential disk-array filter," *Dig. Transducers'07*, pp. 307-311, 2007.

[6] S. Wang, *et al.*, "Encapsulated mechanically coupled fully-differential breathe-mode ring filters with ultra-narrow bandwidth," *Dig. Transducers'11*, pp. 942-945, 2011.

[7] A. Heragu, *et al.*, "A 2.4 GHz MEMS based sub-sampling receiver front-end with low power channel selection filtering at RF," *IEEE RFIC'12*, pp. 257-260, 2012.

[8] W.-C. Chen, *et al.*, "A generalized CMOS-MEMS platform for micromechanical resonators monolithically integrated with circuits," *J. Micromech. Microeng.*, vol. 21, no. 6, pp. 065012, May 2011.

[9] J. Lopez, *et al.*, "A CMOS–MEMS RF-tunable bandpass flter based on two high-Q 22-MHz poly-silicon clamped-clamped beam resonators," *IEEE Electron Device Lett.*, vol. 30, no. 7, July 2009.

[10] J Giner, *et al.*, "A fully integrated programmable dual-band RF filter based on electrically and mechanically coupled CMOS-MEMS resonators," *J. Micromech. Microeng.*, vol. 22, no. 5, 055020 (6pp), April 2012.

[11] J. Giner, *et al.*, "A CMOS-MEMS filter using a v-coupler and electrical phase inversion," *IEEE IFCS10*, pp. 344-348, 2010.

[12] C.-Y. Chen, *et al.*, "Design and characterization of mechanically-coupled CMOS-MEMS filters," *Dig. Transducers'13*, pp. 2288-2291, 2013.

[13] F. D. Bannon III, *et al.*, "High-Q HF microelectromechanical filters," *IEEE J. Solid-State Circuits*, vol. 35, no. 4, pp. 512-526, 2000.

[14] C.-H. Chin, *et al.*, "A CMOS-MEMS resonant gate field effect transistor," *Dig. Transducers'13*, pp. 2284-2287, 2013.

[15] W.-C. Chen, *et al.*, "VHF CMOS-MEMS oxide resonators with $Q > 10,000$," *IEEE IFCS12*, pp. 1-4, 2012.

CONTACT

* S.-S. Li, Tel: +886-3-516-2401; ssli@mx.nthu.edu.tw

DYNAMIC CHARACTERIZATION OF TUNABLE RF MEMS PRODUCTS

Dana DeReus, Shawn Cunningham, Saravana Natarajan, Art Morris, Jeff Hilbert
Wispry, Inc., Irvine CA 92706

ABSTRACT

The purpose of this paper is to present results of our dynamic characterization of tunable RF MEMS capacitor products that will provide tunable RF solutions for wireless communications. We will present measurement and simulation results for switching time and frequency response of the OPEN and CLOSED MEMS tunable capacitor. The dynamic behavior measurements and models will be used to optimize RF performance.

INTRODUCTION

Background

Tunable RF MEMS products are often based on tunable MEMS capacitors that are actuated by electrostatics, because they are known for their extremely low loss, high Q, high linearity, and low power consumption. The MEMS performance superiority is in contrast to other tunable technologies such as ferroelectric capacitors (BST), varactor diodes, and semiconductor switches. The RF MEMS tunable capacitors exemplified in this paper have been integrated with CMOS for high volume, low cost manufacturing, BEOL packaging, and testing. The CMOS-MEMS integration is not without challenges that have witnessed several attempts. Incomplete integration leads to multi-chip (CMOS chip and MEMS chip, stacked or side-by-side) approaches or partial integrations (CMOS wafer + post-processing in MEMS-only fabrication process).

Many RF applications for wireless communications [1] have been demonstrated with tunability provided by MEMS tunable capacitors. These product applications have included tunable filters [2-3], phase shifters [4], antenna tuning [5-6], and amplifiers. The MEMS devices described in this paper have been designed into cell phone product families for antenna tuning. The product families are based on the size of the tunable capacitor array.

The MEMS tunable capacitor mechanical characteristics are critical to the understanding of the nonlinear behavior and the impact on intermodulation distortion and power handling. These capacitor mechanical characteristics include the mass, spring, and damping properties that are determined by modeling [7] and measurement [8-9]. The linearity and stability have been characterized to understand the effects of the tunable capacitor behavior [10-11].

Design and Fabrication

The tunable RF MEMS capacitor is fabricated in a CMOS-integrated, wafer level encapsulated process flow described by Stamper, et al. [12]. The integrated CMOS-MEMS process is a CMOS-first process that is based on a 0.18 μm, 50V-LDMOS/5V CMOS process on 200 mm wafers. The tunable MEMS capacitor is based on a unique sprung cantilever design described by DeReus, et al. [13] and RF performance described by Natarajan, et al. [14].

The tunable MEMS capacitor fabrication x-section is shown in Figure 1, which shows the MEMS capacitor fabricated on the high voltage RF CMOS. The MEMS capacitors are fabricated with a planarized silicon sacrificial layer above and below the MEMS capacitor. The silicon sacrificial layer is removed by XeF_2 through release holes in a first CVD oxide layer. The release holes are sealed by a second CVD oxide and nitride depositions to achieve a cavity pressure of 1 Torr. The MEMS capacitor is an AlCu-Oxide-AlCu composite. The capacitor is defined by 80 nm oxide and 15 nm ALD Al_2O_3 layers on the fixed electrode and 80nm oxide on the moving electrode.

Figure 2 shows a product floor plan with CMOS partition above a 4x4 array of MEMS tunable capacitor unit cells. The CMOS functionality includes a serial interface (SPI and RFFE) and charge pump for providing the 40V operating voltage. Each MEMS device is encapsulated in an individual cavity that is transparent. The transparency of the encapsulation enables the measurement of frequency response and switching time by a laser Doppler vibrometer.

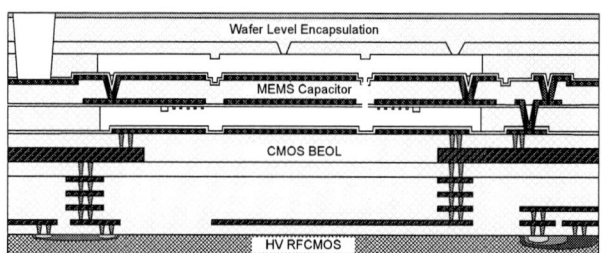

Figure 1: HV RF CMOS-MEMS integrated process cross-section including hermetic wafer level encapsulation.

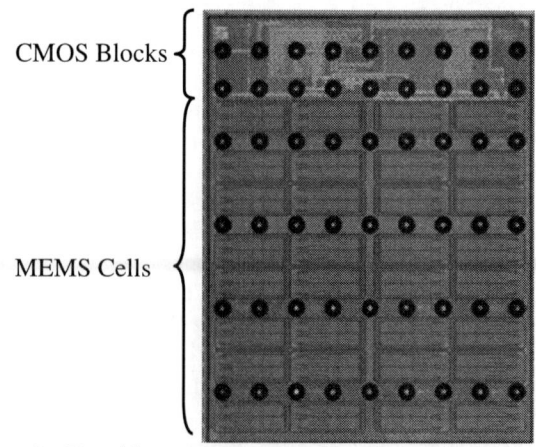

Figure 2: Tunable capacitor product floor plan showing CMOS partition and 4x4 array of tunable capacitor cells.

DYNAMIC CHARACTERIZATION

The dynamic characterization is aimed at measuring the properties of the tunable MEMS capacitor in the OPEN and CLOSED state as shown in Figure 3. Figure 3 is produced by making white light interferometer measurements on a Wyko NT8000 at different voltages and represented by the fully CLOSED state shown. The OPEN state constitutes the unactuated MEMS device that is characterized by its frequency response, closing switching time, and opening switching time. The CLOSED state constitutes the actuated MEMS device that is characterized by the frequency response of the closed capacitor plate; a more constrained feature.

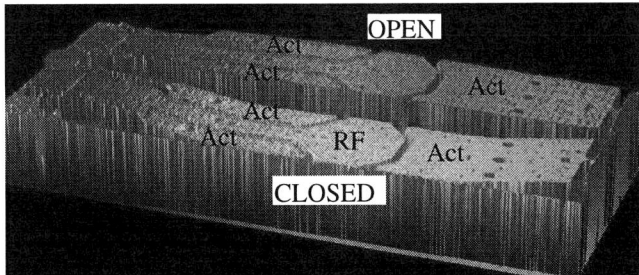

Figure 3: White light interferometer measurement of an OPEN and CLOSED tunable capacitor.

The dynamic characterization has employed various measurement techniques. Laser Doppler Vibrometer measurements (LDV) have been used to measure OPEN and CLOSED state frequency response and switching times. In addition, the switching time has been measured using methodologies of the power detector and the RC time constant. The Polytec Model MSA-500LDV is a scanning system that has been used to measure the frequency response and the mode shapes for the entire length of the tunable capacitor. The measurement is made through the transparent thin film encapsulation using a 10V (< pull-in voltage) swept sinusoidal signal that is applied to the actuator plates (Act) and no signal applied to the RF capacitor (RF). The measurement through the lid is challenging because of reflections from the top and bottom surface of the lid that can interfere with the primary reflection from the top surface of the MEMS device, the thickness of the lid, and the air gap between the bottom of the encapsulation and the beam.

The frequency response spectrum is shown in Figure 4, which shows the scanned output from the RF capacitor. The frequency was swept from 10 kHz to 2.5 MHz that shows the primary modal frequency to be 64 kHz. The frequency response spectrum simulated in MEMS+ is shown in Figure 5. The excited frequency modes up to 2 MHz are listed in Table 1, where they are compared with simulation results generated by FEA and by Coventor's MEMS+ high-order, FEA methodology. Two simulated mode shapes, from FEA, are shown in Figure 6. The primary mode, Mode 1, has a resonance at 64 kHz, which matches the frequency predicted by FEA and MEMS+. A higher order mode, Mode 6, is seen in Figure 6(b) at a frequency of 601 kHz. Mode 1 exhibits significant vertical motion of the RF capacitor plate leading to a variable Cmin, if this resonance mode is excited. Mode 6 has a node line located on the capacitor so the vertical displacement is minimal but tilting is exhibited.

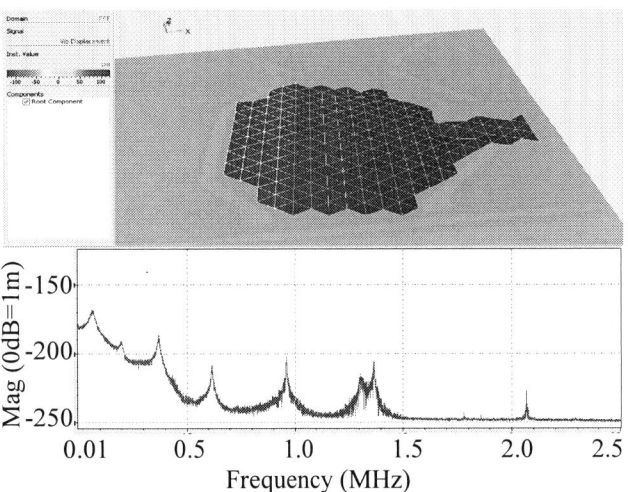

Figure 4: LDV output for the frequency response spectrum of an electrostatic actuated MEMS tunable capacitor.

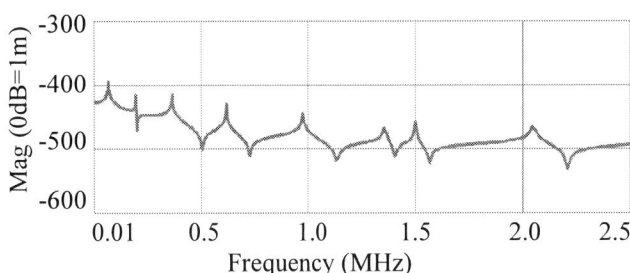

Figure 5: MEMS+ frequency response spectrum of an electrostatic actuated MEMS tunable capacitor.

Table 1: Measured and simulated modal frequencies for RF MEMS tunable capacitor.

Mode	Measured (LDV, kHz)	FEA (kHz)	MEMS+ (kHz)
1	64	63	67
2	199	192	194
4	369	363	362
6	616	601	619
9	959	936	973
11	1368	1354	1357
15	2068	1996	2043

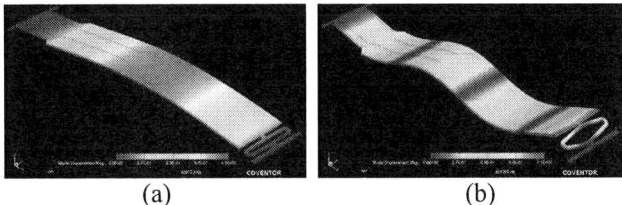

Figure 6: OPEN state modal analysid showing (a) Mode 1 (63 kHz) and (b) Mode 6 (601 kHz).

The switching time has been measured using 3 different approaches: laser Doppler vibrometer, RF power detector, and an RC detector method. The laser Doppler vibrometer is applied only at the wafer level, where the tunable capacitor is visible through the wafer level encapsulation and the actuator electrodes are driven through electrical probes. The LDV measurements are critical for estimating the damping coefficients. Once the wafers complete the solder bumping process and flip chip assembly to a substrate, the MEMS device is no longer visible for LDV. The RF power detector method has been applied at the wafer level to verify LDV measurements and at the package level. The RC detector method was developed for final product test (FT), so switching time (ON/OFF) is measured for every device. A 40V square wave was used to generate the closing and opening time response measured by LDV in Figure 7. During the closing state, we do not observe significant bounce at this scale. During the opening state, overshoot and settling are observed that are characteristic of the underdamped system. A detailed image of the overshoot and oscillations during opening are shown in Figure 8, which shows the measured response and the MEMS+ simulation using Rayleigh damping.

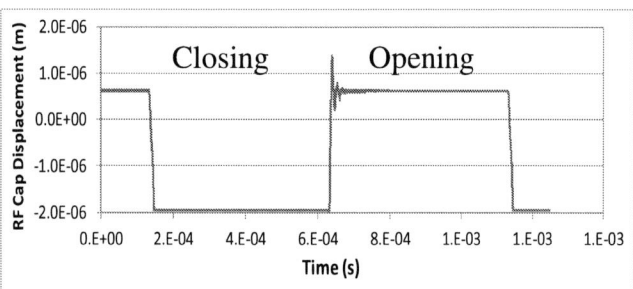

Figure 7: Tunable MEMS Capacitor Closing and Opening response measured by LDV.

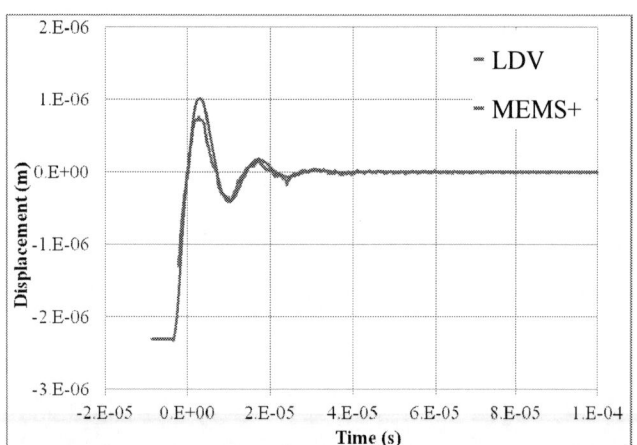

Figure 8: Tunable capacitor opening oscillations showing match between LDV measurements and MEMS+ dynamic model with a Rayleigh damping mode.

The dynamic characterization of the CLOSED state is focused on the frequency response of the closed RF capacitor. The tunable capacitor is closed by applying 40V

between the actuator plates that are driven by the integrated charge pump. With the capacitor closed, a high frequency generator provides a swept sine wave signal across the RF port at a power level of 21 dBm. The high frequency response is measured by the Polytec UHF 120 LDV system, which is a non-scanning system that monitored a single point in the center of the RF capacitor. Because a single point was imaged, some of the vibration modes are not detected or shown in the frequency response.

The measured frequency spectrum is shown in Figure 9, which shows a primary resonance of 4.4 MHz that has been compared to simulation results in Table 2. The measured resonant frequency was 4.4 MHz compared to the simulations of 3.3 MHz and 2.7 MHz. The primary mode of the closed capacitor is the 14th mode of the partially constrained beam. The simulated results are achieved by two different methods. In the MEMS+ environment, the device is simulated in two steps. The 1st step applies 40V between the actuator plates to close the beam and constrain the device. The 2nd step performs a modal analysis on the closed beam. For the finite element model, a mechanical constraint is applied to the actuator plates to partially constrain the overall beam. This is not as realistic as the electrostatic loading, but it allows us to study the effect of the constraint on the capacitor frequency. With a broad range of the capacitor area fixed, the resonance was 3.3 MHz, which is higher than the electrostatic constraint. The least rigid mechanical constraint estimated a resonance of only 2.3 MHz. This shows the stiffness of the constraint does not explain the difference between measured and simulated frequencies. When the capacitor motion is limited by contact elements, the estimated resonance is 3.8 MHz for partial contact. This suggests the difference between simulated and measured is due to variable contact in the capacitor interface.

Figure 10 shows the primary resonant mode of the closed capacitor. The entire beam length is not fixed, but the primary mode displacements are located around the capacitor. The measured mode shape was not captured because the UHF measurement was non-scanning.

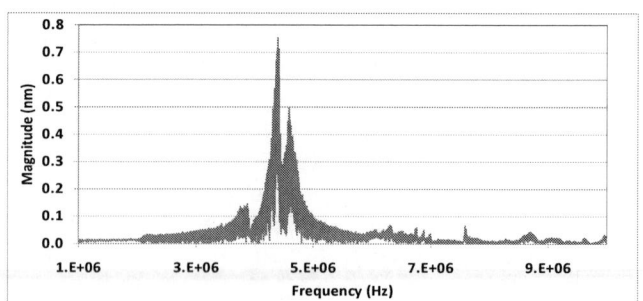

Figure 9: Frequency response of the CLOSED RF capacitor measured by the UHF LDV and simulated by MEMS+.

Table2: Measured and simulated primary modal frequency of tunable capacitor in the CLOSED state.

Mode	Measured	FEA	MEMS+
14	4.4MHz	3.3 MHz	2.7 MHz

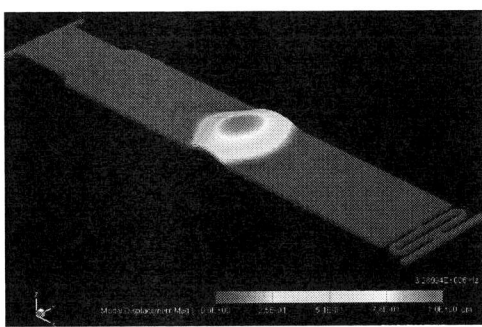

Figure 10: Primary frequency mode of CLOSED capacitor (Mode 14) estimated at 2.7 MHz with electrostatic constraint (MEMS+) and 3.3 MHz with mechanical constraint (FEA).

At the primary resonant frequency of 4.4 MHz, the effect of input power was measured by varying the input power from 6 dBm to 21 dBm to measure the displacement amplitude of the capacitor for various power levels. Figure 11 shows the capacitor displacement at 4.4 MHz for different power levels. At 21 dBm, the capacitor displacement is 7 nm with negligible displacement at +/- 50kHz.

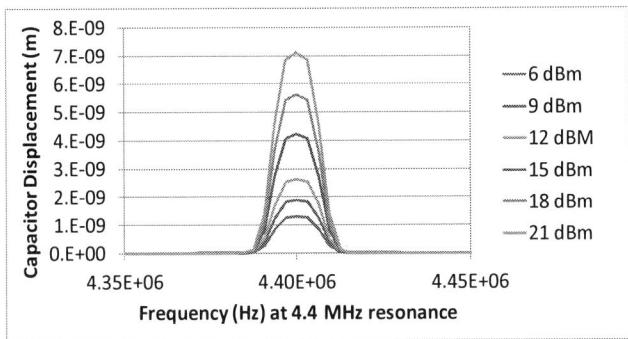

Figure 11: Capacitor displacement as a function of input power level at the 4.4 MHz resonance.

CONCLUSIONS

The frequency response of the OPEN and CLOSED capacitor has been measured using a laser Doppler vibrometer. The OPEN and CLOSED state resonant frequencies are 64 kHz and 4.4 MHz, respectively, which agree with simulation results. The switching times to open and close the tunable capacitor were measured by LDV and correlated with MEMS+ simulations. The measured and simulated switching times are in excellent agreement. Power detector and RC detector methods have been implemented to measure switching time of packages, which is in general agreement with the LDV results.

ACKNOWLEDGEMENTS

The authors would like to acknowledge Sandipan Maity of Coventor for MEMS+ support and Toby Bahn and Jerome Eichenberg of Polytec for LDV support.

REFERENCES

[1] J. Hilbert, "RF-MEMS for Wireless Communications", *IEEE Comm. Magazine*, vol. 46, pp. 68-74, 2008.

[2] J. R. De Luis, et al., "A Novel Frequency Control Loop for Tunable Notch Filters", *IEEE Trans. Microwave Theory and Tech.*, vol. 59, pp. 2265-2274, 2011.

[3] J. R. De Luis, et al., "A Tunable Asymmetric Notch Filter using RFMEMS", *in Digest of IEEE MTT-S Microwave Symp.*, Anaheim, CA, 23-28 May, 2010, pp. 1146-1149.

[4] J. Hung, L. Dussopt, and G. Rebeiz, "Distributed 2- and 3-bit W-band MEMS Phase Shifters on Glass Substrates", *IEEE Trans. Microw. Theory Tech.*, vol. 52, pp. 600-606, 2004.

[5] J. R. De Luis, et al., "Tunable Antenna Systems for Wireless Transceivers", *Intl. Symp. Antennas and Propagation (APSURSI)*, Spokane, WA, 3-8 July, 2011, pp. 730-733.

[6] A. S. Morris, "High Performance Tuners for Handsets", *in Digest of IEEE MTT-S Microwave Symp.*, Baltimore, MD, 5-10 June, 2011, pp. 1-4.

[7] M. Innocent, et al., "Analysis of the Nonlinear Behavior of a MEMS Variable Capacitor", *Proc. Modeling and Simulation of Microsystems,* San Juan, PR, 2002, pp. 234-237.

[8] M. Innocent, et al., "Measurement of the Nonlinear Behavior of a MEMS Variable Capacitor", *Proc. 17th IEEE Intl. Conf. on MEMS*, Maastricht, The Netherlands, 25-29 Jan., 2004, pp. 773-776.

[9] D. Girbau, A. Lazaro, and L. Pradell, "Characterization of Dynamics in On-Wafer RF MEMS Variable Capacitors using RF Measurement Techniques", *IEEE Trans. on Microwave Theory and Tech.*, vol. 52, pp. 2627-2633, 2004.

[10] L. Dussopt and G. M. Rebeiz, "Intermodulation Distortion and Power Handling in RF MEMS Switches, Varactors, and Tunable Filters", *IEEE Trans. Microw. Theory and Tech.*, vol. 51, pp. 1247-1256, 2003.

[11] K. Chen, et al., "Antibiased Electrostatic RF MEMS Varactors and Tunable Filters", *IEEE Trans. on Microwave Theory and Tech.*, vol. 58, pp. 3971-3980, 2010.

[12] A. K. Stamper, et al., "Planar MEMS RF Capacitor Integration", in *Digest Tech. Papers Transducers 2011 Conference*, Beijing, 5-9 June, 2011, pp. 1803-1806.

[13] D. R. DeReus, et al., "Tunable Capacitor Series/Shunt Design for Integrated Tunable Wireless Front End Applications", *MEMS 2011*, Cancun, MX, 23-27 June, 2011, pp. 805-808.

[14] S. P. Natarajan, et al., "CMOS Integrated Digital RF MEMS Capacitors", *Si Mono. Int. Circuits in RF Sys. (SiRF)*, Phoenix, AZ, 16-19 January, 2011, pp. 173-176.

CONTACT

*S. J. Cunningham, +1-949-458-9477 x205, shawn.cunningham@wispry.com.

ETCH-HOLE-ASSISTED ENERGY DISPERSION FOR ENHANCING QUALITY FACTOR IN SILICON BULK ACOUSTIC RESONATORS

Cheng Tu and Joshua E.-Y Lee

[1] Department of Electronic Engineering, City University of Hong Kong, Hong Kong
[2] State Key Laboratory of Millimeter Waves, City University of Hong Kong, Hong Kong

ABSTRACT

This paper empirically demonstrates how the quality factor (Q) of a width-extensional mode single-crystal silicon bulk-acoustic-resonator (SiBAR) can be enhanced by three times by strategic placement of holes on the structure. The holes serve to disperse the strain energy field concentrated primarily around the nodal lines, ultimately re-distributing strain energy away from the anchors. This in turn reduces anchor loss and thus enhances Q. These results agree well with our finite-element (FE) simulations. We envisage that the concepts reported herein can be extended to even higher performance resonators like piezoelectric aluminum nitride contour mode resonators.

INTRODUCTION

Micromechanical resonators hold an important role for frequency references, filters and resonant sensors. In all these cases, enhancing Q is of great importance since higher Q leads to lower phase noise level in frequency references, higher selectivity in filters and better resolution in resonant sensors. On this note, extensional bulk mode resonators operating in VHF band typically yield very high frequency-Q products in the order of 10^{12} under vacuum [1-2]. In such resonators, anchor loss is seen as the dominant energy loss source that limits Q [2-5]. This is still true even when the anchors are located at the nodal points of the resonators since the anchors are inevitably finite in size, while the ideal nodal point is infinitesimal. To date, many resonator designs have been proposed to minimize the anchor loss. Harrington et al [2] demonstrated that Q could be improved by introducing in-plane reflectors placed at a certain distance away from the anchors. These reflectors serve to reflect any incident elastic wave back into the resonator body and thus increase the Q. Li et al [3] alternatively used a quarter-wavelength support beam to deliberately create an acoustic impedance mismatch between the resonator body and the substrate, thus preventing the elastic energy from going out and minimizing energy loss of the resonant system. This idea was further developed by Taş et al [4], who used quarter-wavelength beams with alternating low and high impedances as the anchor to transform the impedance of the substrate to a very small value. Alternatively, phononic-crystal strips with band-stop characteristic were employed by Hsu et al [5] to reduce the anchor loss. All the above-mentioned methods are related to geometry design outside the resonator which increases the real estate the whole device occupies. This work explores a more compact solution to boost Q via strategic placement of etch holes on the resonator body. This approach is compliant with fabrication methods in light of the need for etch holes for top side release.

These etch holes are common perforation features when fabricating silicon-on-insulator micromechanical resonators to realize free standing structures. It has been found that perforating flexural beam resonators with slots can suppress thermoelastic damping (TED) to engineer high Q [6]. With bulk modes, we previously found that uniformly distributed etch holes have the opposite effect, drastically reducing Q by increasing TED instead [7-8]. In this work, we show that when placed strategically, holes can be used to significantly enhance Q particularly where in the cases where Q is limited by anchor loss like in the width-extensional (WE) mode.

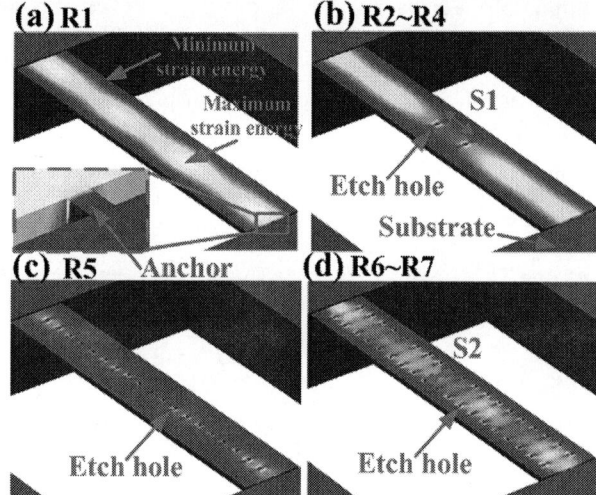

Figure 1: FE simulation of strain energy distribution for SiBARs with and without etch holes. Designs R2~R4 contain only 2 holes, differing in their separation as denoted by S1. Device R5 contains a line of holes down the center of the bar. Devices R6~R7 have two lines of holes placed symmetrically on each side of the center line, differing in their separation as denoted by S2.

Table 1. Summary for hole spacing between devices R1 to R7 in terms of S1 and S2 (All dimensions in µm).

	R1	R2	R3	R4	R5	R6	R7
S1	NA	180	260	220	90	90	90
S2	NA	NA	NA	NA	0	110	160

DESIGN AND SIMULATION

We designed a series of seven SiBARs of different hole patterns (denoted by R1~R7) of the same length (1800µm) and width (300µm). Dimensions of the anchors (10µm×8µm) were all the same for R1~R7. The FE simulations of strain energy distribution across these resonators are shown in Fig 1.

978-1-4799-3510-9/14 $31.00 © 2014 IEEE

R1 is a reference SiBAR with no holes. Since the alternating electrostatic force is applied in the width direction of resonator (electrodes are not shown in Fig 1 for clear visualization), the elastic wave in the width direction should be dominant. At the same time, elastic waves in the length direction are also generated due to a non-zero Poisson's ratio of the structural material. This is verified by the standing wave formed in the length direction as seen in Fig 1(a). As a result, part of the elastic energy propagating in the length direction escapes to the substrate from the anchor as shown in the inset of Fig 1(a). This contributes to anchor loss.

Devices R2~R4 contain a pair of $20 \times 20 \mu m^2$ holes placed along the nodal line along the length of the bar resonator, concentrated around its center. The spacing between the pair of holes was designed to be slightly different among devices R2~R4 for comparison. As seen from Fig 1(b), the insertion of two holes disperses the strain energy field concentrated in between and makes the energy more uniformly distributed along the nodal line near the anchors. This can be seen when contrasted against the strain energy distribution in the plain SiBAR shown in Fig 1(a). Device R5 has a series of holes along the nodal line, which are spaced 90μm apart. Devices R6~R7 contain two rows of holes that run symmetrically along each side of the nodal line. The spacing between the holes in each row is also 90μm. Between R6 and R7, the distance of the holes from the nodal line was designed to be slightly different. In Fig 1, the center-to-center separation between the holes in the directions along the length and width are denoted by parameters S1 and S2 respectively. Details of the respective hole-patterns are summarized in Table 1.

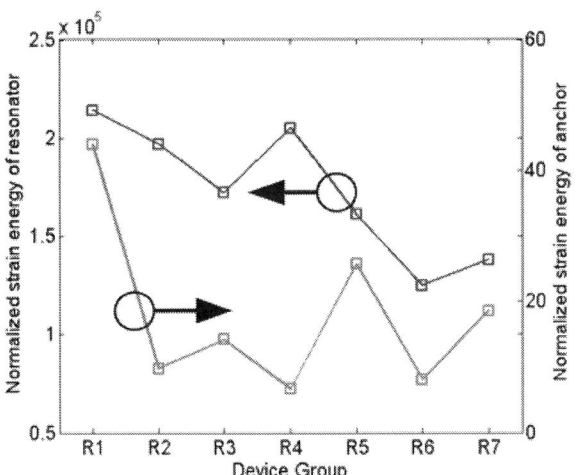

Figure 2: Comparison of FE-computed normalized strain energies for resonator body (E_r) and anchors (E_a) among different device groups of the same thickness (25μm).

The anchor-limited Q (Q_{anchor}) can be computed by taking the ratio of the relative strain energies for the resonator body (E_r) and anchors (E_a), as illustrated in the following equation [9]:

$$Q_{anchor} = \eta \frac{E_r}{E_a}, \qquad (1)$$

where η is a scale factor which we determine experimentally. E_r and E_a can be computed through FE simulation. For fair comparison among different devices, E_r and E_a were both normalized against the maximum displacement on the sides of each resonator body. Fig 2 compares the normalized E_r and E_a for devices that were 25μm-thick. A similar trend for normalized E_r and E_a can be found for thinner devices that were 10μm-thick. It can be seen from Fig 2 that different patterns of holes yield different energy distributions between the resonator body and anchors. The strain energies in the resonator body generally follow a decreasing trend as the number of the holes is increased. Moreover, it can be seen that devices with holes (R2~R7) all have less anchor energy compared to the plain one (R1). In particular, Fig 2 shows that, although devices R1 and R4 have almost the same values for E_r (difference is less than 5%), E_a for R4 is about six times smaller than that for R1. This translates to about six times increase in Q_{anchor} according to equation (1).

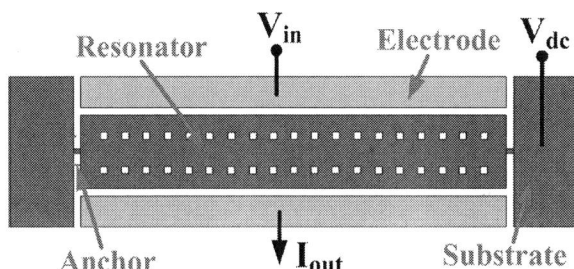

Figure 3: Circuit schematic of the bias configuration used for electrical characterization of the SIBAR devices.

MEASUREMENT AND DISCUSSION

A series of seven SiBARs following the above designs (i.e. R1~R7) were fabricated using a standard SOI MEMS process. We electrically characterized six dies of the seven SiBARs (i.e. total of 42 devices) in mTorr levels of vacuum in a probe station (Janis Research ST-100) using a network analyzer (Agilent E5061A). Short-Open-Load-Through (SOLT) calibration was performed prior to measuring the devices. All measured results were acquired using 50Ω termination impedance. Half of the dies we characterized had device thicknesses of 25μm while the other half had thicknesses of 10μm. The devices were all actuated and sensed using capacitive electrodes based on the biasing configuration shown in Fig 3. An actuation power level was kept at 0dBm together with a bias voltage (V_{dc}) of 50V.

Fig 4 plots the measured resonant frequencies of all 42 devices, which agree well with the FE simulation predictions. It can be observed that devices with two holes (R2~R4) have resonant frequencies about just 0.4% lower than that for plain device (R1). However, this disparity increases up to about 5% for devices R5~R7 wherein the number of holes is larger. Fig 5 compares the measured S21 electrical transmission magnitudes for R4 (hole-pair separation of 220μm) against

978-1-4799-3510-9/14 $31.00 © 2014 IEEE   1258

R1 (reference SiBAR with no holes). These two sets of devices were of the same thickness (25μm). It can be found that R4 devices had extracted Qs that were on average three times that of R1 devices. This significant boost in Q agrees with the prediction from FE simulation. In comparison, Fig 6 shows the measured transmission magnitudes for the same pair of devices with device thickness of 10μm. It can be seen that the increase in average values of Qs is just 40% which is much less than that for its thicker counterparts.

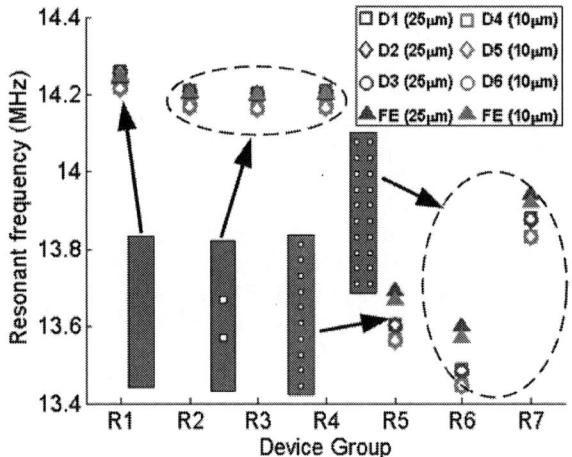

Figure 4: Comparison of measured and FE simulated resonant frequencies among different device groups.

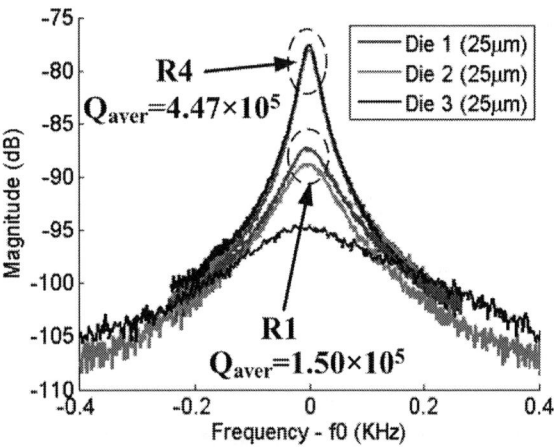

Figure 5: Extracted S21 electrical transmission (magnitude) for devices R1 and R4 for thickness of 25μm. Each device group was measured from the same three die samples.

Fig 7 and 8 show the extracted Qs for devices that were 25μm-thick and those that were 10μm-thick as well. Both figures show that the trends in the variation of Q among devices R1~R7 are reasonably repeatable among the six dies measured. It can be found that R4 devices always had the highest Q for different dies and different device thickness while R5 devices had the lowest. Comparison between R5 and R6~R7 shows that perforating the nodal line, where the maximum strain, leads to significant drop in Q compared to putting the holes away from the nodal line.

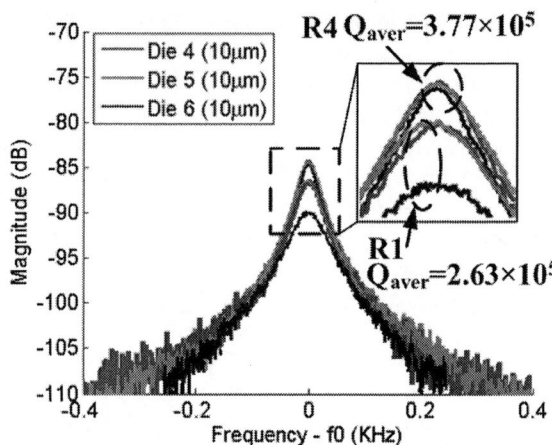

Figure 6: Extracted electrical transmissions (magnitude) for R1 and R4 with thickness of 10μm. Each device group was measured from the same three die samples.

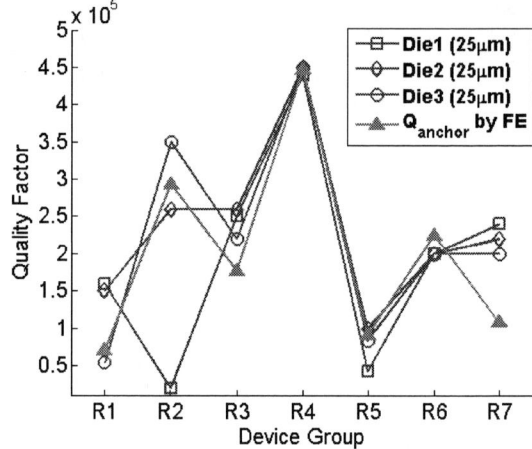

Figure 7: Comparison of measured Q and FE-computed Q_{anchor} among three device groups with the same device thickness of 25μm.

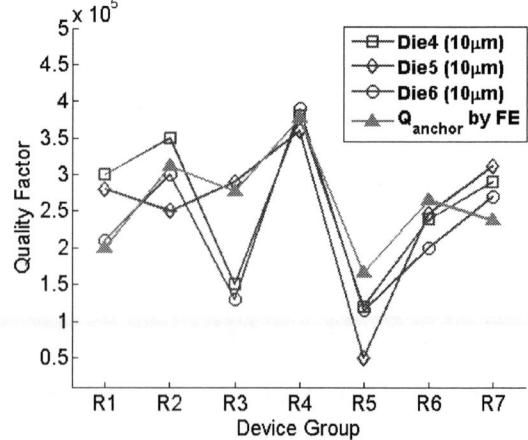

Figure 8: Comparison of measured Q and FE-computed Q_{anchor} among three device groups with the same device thickness of 10μm.

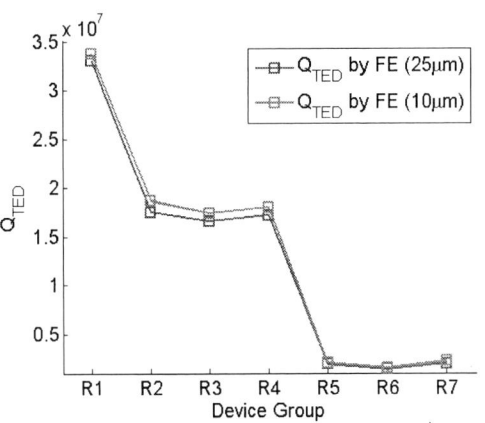

Figure 9: Comparison of FE simulated Q_{TED} among different device groups.

By scaling the ratio of E_r and E_a to the average value of measured Q for R4, the scale factor η was obtained and found to be 14.6. The values of Q_{anchor} for each device design among R1~R7 were then computed based on Equation (1). The results are shown in Fig 7 and 8. The variation trends of computed Q_{anchor} closely track the measured Q for both device thicknesses. Using FE simulation, we also computed the TED-limited Q (Q_{TED}) [8], which was found to be an order of magnitude higher than the measured Q. It is thus unlikely that TED plays a significant role in determining Q. Rather, the currently measured Q appears instead to be limited by anchor loss.

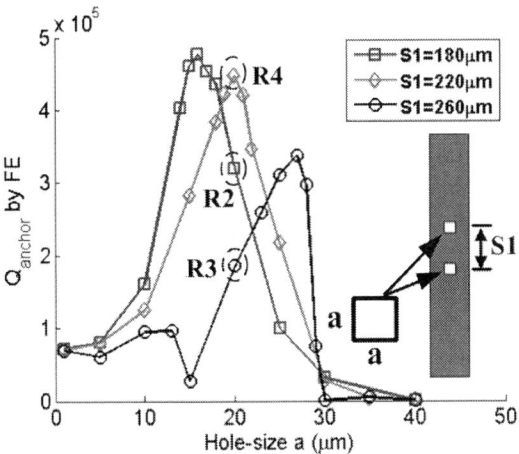

Figure 10: Comparison of FE-computed Q_{anchor} for devices with different hole-spacing (S1) and different hole-size (a). Device thickness of the model was 25μm.

The holes in all the above mentioned devices (R2~R7) all had the same physical dimensions of 20 μm by 20μm. On this note, we have computed Q_{anchor} by FE based for different hole-spacing and hole-size based on the dual-hole design that is depicted in Fig 1(b). The simulations were done for a device thickness of 25μm. The results are shown in Fig 10. It can be seen that, for a given hole-spacing, there exists an optimal value of hole-size which yields highest Q_{anchor}. However, it can also be seen that the highest achievable Q_{anchor} increases as hole-spacing is decreased from 260μm to

180μm. Our FE simulation predicts that further reducing the hole spacing (smaller than 100μm) will lead to a maximum of achievable Q_{anchor} at about 5×10^5 that does not further increase when the hole-size is further reduced.

CONCLUSION

This paper has demonstrated that significant increase in Q can be achieved in width-extensional mode resonators by re-distributing strain energy away from the anchors. By placing two holes around the center of the resonator yields the highest Q measured so far compared to inserting a line of holes. Furthermore, we have shown that anchor loss could be flexibly engineered by tuning the hole-spacing and hole-size. If the hole-spacing is fixed, there exists an optimal hole-size for which a minimum on the anchor loss can be found.

ACKNOWLEDGEMENTS

This work was supported by the Early Career Scheme of the Research Grants Council of Hong Kong (project number CityU 124312).

REFERENCES

[1] S. Pourkamali, et al., "Vertical capacitive SiBARs," *Proc. MEMS*, 2005, pp. 211–214.

[2] B. P. Harrington and R. Abdolvand, "In-plane acoustic reflectors for reducing effective anchor loss in lateral-extensional MEMS resonators," *J. Micromech. Microeng.*, vol. 21, no. 8, p. 085021, July. 2011.

[3] S.-S. Li, et al., "Micromechanical 'hollow-disk' ring resonators," *Proc. MEMS*, 2004, pp. 821–824.

[4] V. Taş, et al., "Reducing anchor loss in micromechanical extensional mode resonators," *IEEE Trans. Ultrason. Ferroelectr. Freq. Control*, vol. 57, no. 2, pp. 448-454, Feb. 2010.

[5] F.-C. Hsu, et al., "Reducing support loss in micromechanical ring resonators using phononic band-gap structures," *J. Phys. D: Appl. Phys.* vol. 44, no. 37, p. 375101, Aug. 2011.

[6] R. N. Candler, et al., "Impact of geometry on thermoelastic dissipation in micromechanical resonant beams," *J. Microelectromech. Syst.*, vol. 15, no. 4, pp. 927–934, Aug. 2006.

[7] C. Tu and J. E.-Y. Lee, "Thermoelastic dissipation in etch-hole filled Lamé bulk-mode silicon microresonators," *IEEE Electron Device Lett.*, vol. 33, no. 3, pp. 450-452, Feb 2012.

[8] C. Tu and J. E.-Y. Lee, "Increased dissipation from distributed etch holes in a lateral breathing mode silicon micromechanical resonator," *Appl. Phys. Lett.*, vol. 101, no. 2, pp. 023504-1–023504-4, Jul. 2012.

[9] J. E.-Y. Lee, et al., "Study of lateral mode SOI-MEMS resonators for reduced anchor loss," *J. Micromech. Microeng.*, vol. 21, no. 4, p. 045010, Apr. 2011.

CONTACT

*Cheng Tu, tel: +852-34422656; mems305@gmail.com

NANO-OPTO-ELECTRO-MECHANICAL (NOEM) OSCILLATOR WITH CONTROLLABLE NON-LINEAR DYNAMICS

B. Dong[1,2], J. G. Huang[3], H. Cai[2], P. Kropelnicki[2], A. B. Randles[2], Y. D. Gu[2] and A. Q. Liu[1†]

[1]School of Electrical & Electronic Engineering, Nanyang Technological University, Singapore 639798
[2]Institute of Microelectronics, A*STAR (Agency for Science, Technology and Research), Singapore 117685
[3]School of Mechanical Engineering, Xi'an Jiaotong University, Xi'an 710049, China

ABSTRACT

In this paper, a nano-opto-electro-mechanical (NOEM) oscillator with controllable non-linear dynamics is demonstrated. An optical gradient force driven cantilever shows high non-linearity due to non-linear behavior of the optical force. The non-linear dynamics of the oscillator which utilizes opto-mechanical interaction is well studied and controlled by varying the optical signal, including the power and wavelength. The resonance frequency of the oscillator can be tuned by up to 50 KHz by changing the detuning condition of the opto-mechanical system. It has potential applications such as optical navigation sensors, actuators and optical switches.

INTRODUCTION

Microelectromechanical Systems (MEMS) based oscillator has been intensively developed over the past decade as a promising alternative for various applications like timing and frequency control. [1-2] However, micro-scale dimension of MEMS oscillators restrict its applications in high frequency field. Therefore, Nanoelectromechanical Systems (NEMS) based oscillator provides an alternative of current MEMS oscillator to achieve high frequency and high Q-factor.

Opto-mechanics has been developed recently which utilize near-field cavity optomechanics to achieve effective actuation of nano-scale devices [3]. Optomechanical phenomena in photonics devices provide a new means of light-light interaction mediated by optical force actuated mechanical motion, enabling all-optical signal processing without resorting to electro-optical conversion or non-linear materials [4-6]. In this paper, a nano-opto-electro-mechanical (NOEM) oscillator with controllable non-linear dynamics is demonstrated. The non-linear dynamics of the oscillator which utilizes opto-mechanical interaction is well studied and controlled by varying the optical signal, such as power and wavelength. It has potential applications such as optical navigation sensors, actuators and optical switches.

DESIGN AND THEORY

The oscillator consists of a cantilever beam, a tunable ring resonator, two bus waveguides and electrical pads, as shown in Figure 1(a). High power control light is pumped into the bus waveguide through control input port, and coupled into the tunable ring resonator. The tunable ring resonator has a diameter of 40 μm, and it can generate an attractive optical gradient force over the cantilever beam due to evanescent wave interaction. Signal light is pumped

from the input port of the signal waveguide, and transmit through the slit to the output port. Part of the signal waveguide is cantilever beam, which has a width of 450 nm and length of 15 μm. There is also an electrical pad close to the cantilever, which provides an attractive electrostatic force over the cantilever beam. The cantilever beam is "pulled" by optical force generated from the ring resonator, and balanced by the mechanical spring force and electrostatic force, as shown in Figure 1(b). When the optical gradient force is modulated with frequency close to the resonance frequency of the cantilever beam, the cantilever behaves as an oscillator. The waveguide of ring resonator is sandwiched by two silicon pads, which is P-doped and N-doped respectively. The resonance wavelength of the ring resonator can be controlled by varying the injection current through the p-i-n junction. The control light can therefore be modulated by modulating the injection current. The motion of the cantilever can be monitored by monitoring the transmitted signal light's power.

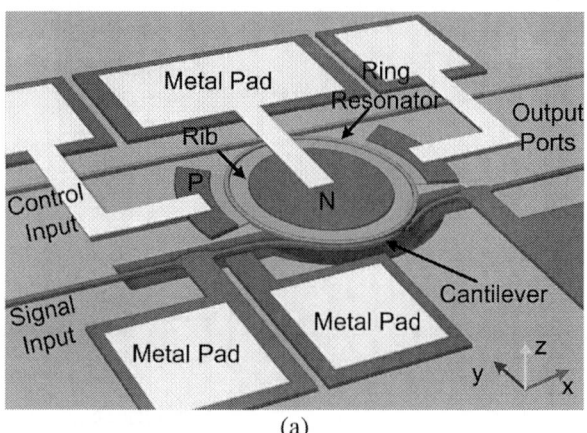

(a)

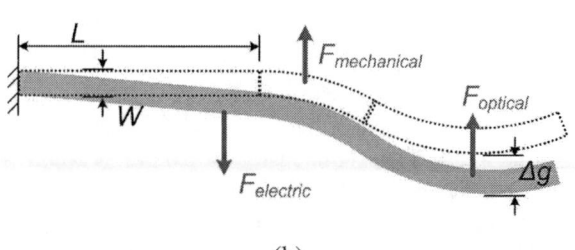

(b)

Figure 1: (a) Schematic of opto-mechanical oscillator and (b) force analysis of the cantilever beam.

The optical force applied on the cantilever can be expressed as [7]

978-1-4799-3510-9/14 $31.00 © 2014 IEEE

$$F_{optical}(x) = -\frac{2\gamma_e P_{optical}}{\omega_c} \frac{g_{om}(x)}{\Delta(x)^2 + \gamma^2} \qquad (1)$$

where γ_e is the external damping due to waveguide-ring coupling, ω_c is the control light frequency, $g_{om}(x)$ is the optomechanical coupling coefficient, γ is the total optical damping coefficient and (x) is the light detuning which is defined as $\Delta(x) = \omega_c - \omega_r(x)$, and ω_r is the resonance frequency of the ring resonator., The optical force can be controlled by the light detuning and optical power as shown in Eq. (1). Due to non-linearly of the optical gradient force, it can cause non-linear dynamic behavior of the cantilever, which typically happen at large deformation. To investigate the optical force induced nonlinear behavior of the oscillator, the optical force is expended at the equilibrium point to third order as [6]

$$F_{optical}(x) = F(x_0) + k_1 \delta x + k_2 \delta x^2 + k_3 \delta x^3 + o(\delta x^4) \qquad (2)$$

where the coefficient k_1 is the force constant corresponding to the optical spring effect, k_2 and k_3 are the quadratic and cubic nonlinear coefficients, respectively. In fact, the quadratic term can be ignored for the primary resonance. Therefore, the cubic factor has dominant effects on the resonance of cantilever beam.

Due to existence of nonlinear coefficients, the cantilever waveguide can be modeled as a nonlinear oscillator driven by the optical force with quadratic and cubic nonlinearities, which can be expressed as a typical Duffing oscillator [8]

$$\ddot{x} + 2\xi\dot{x} + x + \eta x^3 = F\cos\Omega t \qquad (3)$$

where ξ is the damping ratio, η is cubic stiffness parameter, F is the excitation amplitude and Ω is the excitation frequency. The solution to the above equation is given by [8]

$$\overline{F}^2 = 4a^2(\overline{\xi}^2 + (\sigma - \frac{3}{8}\alpha_3 a^2)^2) \qquad (4)$$

where $F = \frac{1}{\omega_m^2}\overline{F}$, $\xi = \frac{1}{\omega_m^2}\overline{\xi}$, a is the resonance amplitude and σ is the detuning of driving force frequency defined as $\omega - \omega_m$ and α_3 is the cubic coefficient defined as $\alpha_3 = -k_3/m$. The cubic factor α_3 as a function of laser detuning is shown in Figure 2. It shows that the cubic term can either be positive or negative, which is depend on the later detuning. Therefore, by changing the laser detuning condition, the cubic factor can be modified.

When the driven force is fixed, the different cubic coefficient will result different behavior, as shown in Figure 3. Unlike the amplitude response in the linear case, the amplitude response in the nonlinear case can be multiple-valued. The response curves lean toward the lower frequencies for negative values, resulting in a softening response; the resonance curves lean toward the high frequencies for positive values, resulting in a hardening response. The absolute value of α_3 determines the frequency shift for both softening and hardening response.

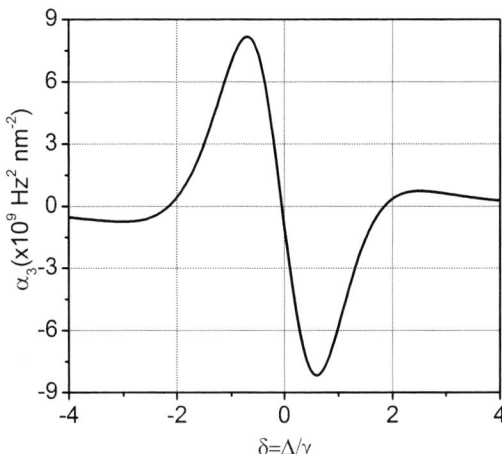

Figure 2: Duffing coefficients at different wavelength detuning condition.

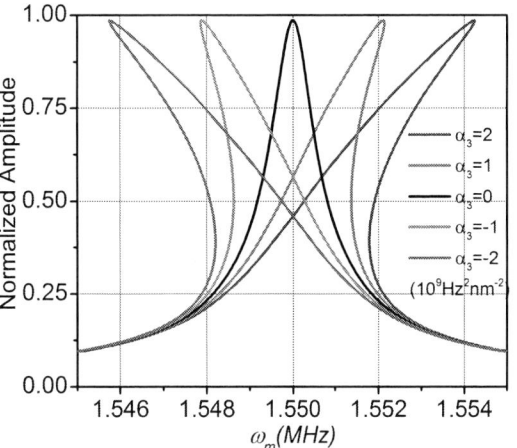

Figure 3: Mechanical response of oscillator at different cubic factor.

The cubic coefficient can be easily tuned by the input power, wavelength detuning and the equilibrium point. It can be seen that the cubic coefficient is different at different deformation. When the input light is red detuned, the cubic coefficient is positive, which means the oscillator shows the hardening nonlinearity. When the input light is blue detuned, the cubic coefficient is negative, which means the oscillator shows the hardening nonlinearity.

To effectively control the cubic factor via light detuning, a p-i-n junction is used to modulate the resonance wavelength of the ring resonator, which is shown in Figure 4. When applying a forward bias current, the free electrons is injected into the waveguide, changing the refractive index of the waveguide. Therefore, the resonance wavelength can be well tuned by controlling the injection current across the p-i-n junction. At fixed-wavelength pumping light, the resonance wavelength of the ring resonator can be controlled to be either blue or red detuned. Furthermore, due to the fast modulation speed, which is in nano second level, the resonance behavior of the cantilever beam can be rapidly

978-1-4799-3510-9/14 $31.00 © 2014 IEEE   1262

changed to realize in-time control by changing the detuning properties.

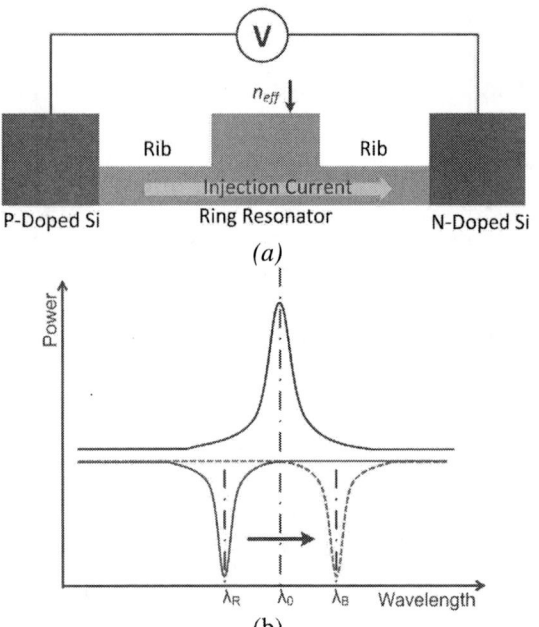

Figure 4: (a) Schematic of the p-i-n tunable ring resonator and (b) spectrum demonstration of control light.

FABRICATION AND DISCUSSIONS

The nonlinear oscillator is fabricated by nano-photonic fabrication processes using standard silicon-on-insulator wafer, with a 220 nm thick silicon structure layer and a 2 μm buried oxide layer. The waveguide structures have a width of 450 nm and a height of 220 nm for a single mode transmission. The silicon bean is designed to have a width of 200 nm while the coupling gap between the ring resonator and the silicon beam is 150 nm. The waveguides and ring resonators are patterned by deep UV lithography, followed by plasma dry etching to transfer the photo resist pattern into the structure layer. The p-i-n junction of the ring resonator has a 70-nm silicon slab layer for doping and supporting. After etching, a 2-μm SiO$_2$ layer is deposited on the structure layers to ensure a low optical loss. A 40-nm Al$_2$O$_3$ is deposited and patterned, which is used as the protection film to protect those fixed structures and leave the window area for suspended structures open. Finally, the cantilever beam is released from the substrate via HF vapor release processes. Figure 5 shows the scan electron microscopy (SEM) image of silicon cantilever beam and signal waveguide.

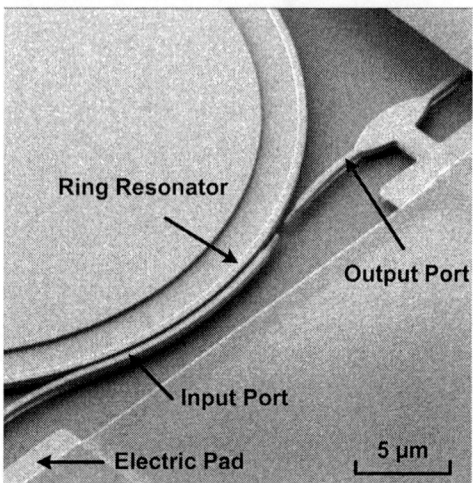

Figure 5: SEM images of the silicon cantilever beam and signal waveguide.

The oscillator is tested at static control light. Figure 6 shows the transmission spectra of the oscillator at red detuned pumping light. By change the detuning of the control light, the resonance wavelength shifts about 0.48 nm, which correspond a 16-nm displacement of the cantilever.

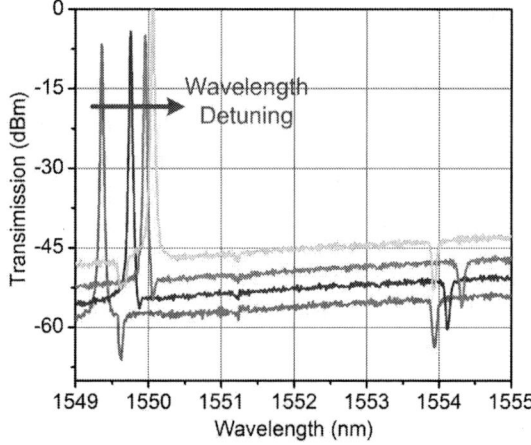

Figure 6: Transmission spectra of the tunable oscillator at red detuned control light.

CONCLUSIONS

In conclusion, an opto-mechanical oscillator with controllable non-linear dynamics is designed, fabricated and experimentally demonstrated. The non-linear behavior of optical force driven oscillator shows great potential for tunable all optical oscillator. Fabricated with CMOS compatible process, this opto-mechanical oscillator can be easily packaged and integrated with other photonic devices. It has potential applications such as optical resonator type gyroscope, accelerometer and optical communication devices.

ACKNOWLEDGMENS

The authors would like to acknowledge the support from the Science and Engineering Research Council of A*STAR, Singapore, under SERC Grant 1021650084.

REFERENCES

[1] D. K. Agrawal, J. Woodhouse, and A. A. Seshia, "Modeling nonlinearities in MEMS oscillators," Ultrasonics, Ferroelectrics and Frequency Control, IEEE Transactions on, vol. 60, pp. 1646-1659, 2013.

[2] W. M. Zhu, T. Zhong, A. Q. Liu, X. M. Zhang and M. Yu, "Micromachined optical well structure for thermo-optic switching", Appl. Phys. Lett., vol. 91, 261106, 2007.

[3] G. Anetsberger, O. Arcizet, Q. P. Unterreithmeier, R. Riviere, A. Schliesser, E. M. Weig, et al., "Near-field cavity optomechanics with nanomechanical oscillators," Nat Phys, vol. 5, pp. 909-914, 2009.

[4] B. Dong, H. Cai, G. I. Ng, P. Kropelnicki, J. M. Tsai, A. B. Randles, M. Tang, Y. D. Gu, Z. G. Suo and A. Q. Liu, "A nanoelectromechanical systems actuator driven and controlled by Q-factor attenuation of ring resonator," Applied Physics Letters, Vol 103, 181105, 2013.

[5] H. Cai, B. Dong, J. F. Tao, L. Ding, J. M. Tsai, G. Q. Lo, A. Q. Liu and D. L. Kwong, "A nanoelectromechanical systems optical switch driven by optical gradient force," Applied Physics Letters, Vol 102, 023103, 2013.

[6] H. Li, Y. Chen, J. Noh, S. Tadesse, and M. Li, "Multichannel cavity optomechanics for all-optical amplification of radio frequency signals," Nat Commun, vol. 3, p. 1091, 10/02/online 2012.

[7] M. Ren, J. Huang, H. Cai, J. M. Tsai, J. Zhou, Z. Liu, Z. Suo, and A. Q. Liu, "Nano-optomechanical actuator and pull-back instability," ACS Nano, Vol 7, pp.1676–1681, 2013.

[8] I. Kovacic, M. J. Brennan, *The Duffing Equation: Nonlinear Oscillators and their Behaviour*, Wiley, 2011.

CONTACT

*A. Q. Liu, +65-67904336; eaqliu@ntu.edu.sg

ON/OFF SWITCHABLE HIGH-Q CAPACITIVE-PIEZOELECTRIC ALN RESONATORS

Robert A. Schneider and Clark T.-C. Nguyen

Berkeley Sensor & Actuator Center, University of California, Berkeley, CA, USA

ABSTRACT

Voltage-controlled on/off switching based on electrode collapse around a 301-MHz capacitive-piezoelectric AlN contour-mode disk resonator has been demonstrated while still allowing on-state Q's as high as 8,800—the highest yet demonstrated around 300MHz in sputtered AlN. The key to on/off switching of this device is the structure of its capacitive-piezoelectric transducer, which provides a suspended electrode atop the suspended AlN resonator that can be pulled electrostatically towards the substrate, pinning the combined resonator-electrode structure to the substrate, thereby opening new conduits for energy loss that suppress signal transmission. This on/off switchability obviates the signal path micromechanical switches needed by conventional attached-electrode AlN resonator counterparts. The Q approaching 9,000, together with the demonstrated on/off switchability and electromechanical coupling k_{eff}^2 up to 0.78%, make this capacitive-piezoelectric resonator a strong contender among resonator technologies targeting RF channel-selecting communication front-ends.

INTRODUCTION

RF channel-selection promises to reduce noise and suppress interferers in wireless transceivers [1], yet research efforts confront a daunting challenge: the need for a switchable bank of small percent bandwidth filters. So far, neither piezoelectric nor capacitively transduced resonators can provide the simultaneous high-Q, high electromechanical coupling (k_{eff}^2), low motional resistance (R_x), self-switchability, and power handling needed to select individual channels over a wide RF range. Indeed, each technology provides only a subset of the desired properties.

Channel-selecting filters require higher Q's than the 2,100's conventional AlN resonators can currently deliver [2]. To illustrate, Figure 1 presents low-Q and high-Q simulations for 0.25% bandwidth 2nd, 3rd, and 4th order filters, for which insertion loss improves dramatically with Q's of 8,800. Since insertion loss (which equates to noise figure), increases as bandwidth drops, the minimum achievable bandwidth is inversely dependent on the Q's of a filter's constituent resonators. High-Q is thus essential: it allows for higher-order filters, reduced noise, and freedom to reduce a filter's bandwidth.

Capacitive gap transduced resonators are of great interest because they achieve very high Q's at UHF frequencies, e.g., Q's over 40,000 at 3GHz, that satisfy the needs of RF channel-selection [3]. They are also self-switchable via a DC bias—helpful for implementing filter banks without the signal path switches of Figure 2(a), which add 0.2-0.5dB of insertion loss and require substantial on-chip real estate [4], so increase cost. Self-switching filters, shown in the architecture

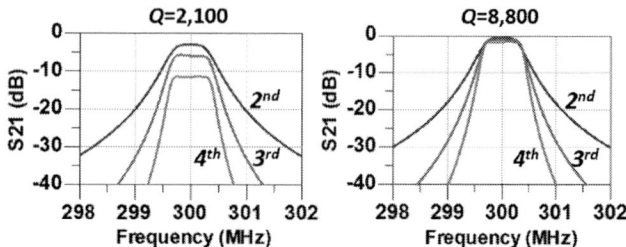

Figure 1: Simulated 0.25%-bandwidth 2nd, 3rd, and 4th order Chebyshev micromechanical filter responses using low-Q and high-Q constituent resonators.

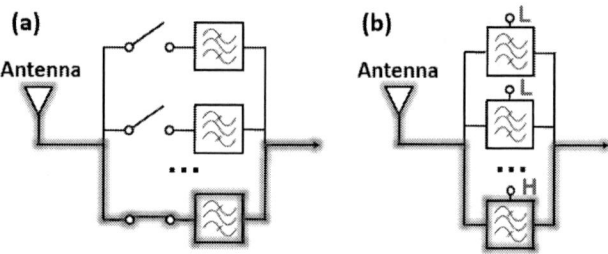

Figure 2: (a) Filter bank architecture requiring series switches. (b) Improved design with self-switched devices.

of Figure 2(b), are preferable, as they dispense with the need for such switches. Unfortunately, capacitive resonators are often plagued by very high R_x and insufficient k_{eff}^2, precluding their use in many applications.

On the other hand, piezoelectric contour-mode AlN resonators offer very low R_x, high k_{eff}^2, and strong power handling, excelling where capacitive resonators struggle. However, in addition to their much lower Q's, traditional AlN resonators currently lack a self-switching capability, meaning they can only be used to implement the undesirable filter bank architecture of Figure 1(a). Pursuant to solving these problems, this paper introduces methods for improving the Q's of piezoelectric AlN contour-mode resonators while simultaneously making them self-switchable.

SWITCHABLE RESONATOR DESIGN

Figure 3(a) presents a perspective-view schematic of the demonstrated on/off switchable capacitive-piezoelectric AlN resonator. Like the device of [5], separating the electrodes from the resonant structure and suspending them at close distance eliminates energy loss from lossy metal electrodes and from the metal-to-piezoelectric interface. The resultant increase in Q does come at the cost of marginally reduced k_{eff}^2 due to reduction in electric field strength in the AlN when electrode-to-resonator gaps are introduced. Design parameters for the 300-MHz resonator, including the frequency-set-

978-1-4799-3510-9/14 $31.00 © 2014 IEEE

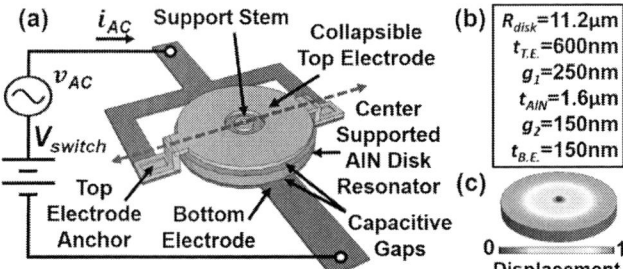

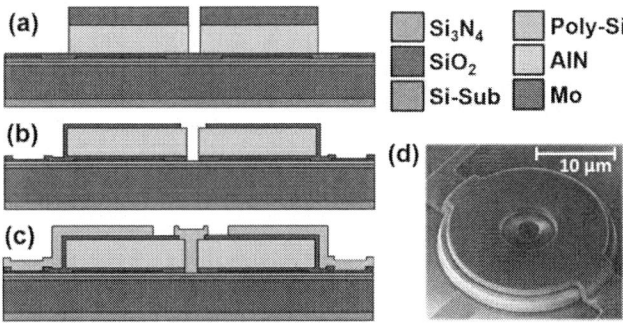

Figure 3: (a) Perspective-view schematic of the 300-MHz switchable capacitive piezoelectric disk resonator of this work. (b) Device dimensions. (c) FEM-simulated radial-contour mode shape at 300 MHz.

Figure 5: Fabrication process illustration. (a) Post AlN etch. (b) Post anchor etch. (c) Post polySi etch, pre-release. (d) SEM of the completed device.

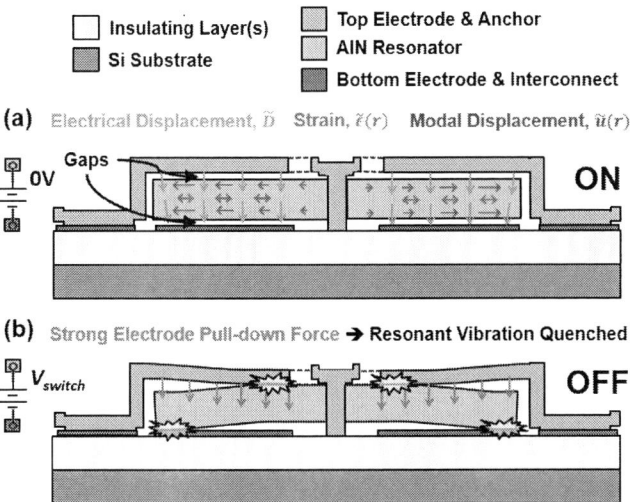

Figure 4: Disk resonator cross-sections corresponding to the vertical plane of the purple dashed line in Figure 3. (a) ON State: $V_{switch}=0V$ and the device resonates freely. (b) OFF State: V_{switch} incites pull-down of the top plate and collapse of the structure to impede resonance.

ting disk radius and vertical layer dimensions in top-down order, are given in Figure 3(b). Figure 3(c) presents the radial-contour modal displacement field at the predicted eigenfrequency.

The key to on/off switching of this device is the structure of its capacitive-piezoelectric transducer, which provides a suspended electrode atop the suspended AlN resonator that can be pulled electrostatically towards the substrate, pinning the combined resonator-electrode structure to the substrate, thereby opening new conduits for energy loss that compromise signal transmission. Figure 4 presents cross sections of the switchable disk for both the ON and OFF states. In the ON state, cf. Figure 4(a), an input voltage v_{AC} between the top and bottom electrodes drives the device into resonant vibration through the reverse piezoelectric effect, in the absence of a DC switching voltage.

To shut the device OFF, one merely applies a switching bias voltage V_{switch} between the top and bottom electrodes until the electrostatic force acting downward on the top electrode is large enough to pull the top plate down, taking with it the AlN resonator structure so that both end up pinned to

the substrate. With top and bottom electrodes abutted against, but not attached, to the AlN structure, frictional loss becomes quite large. In addition, direct contact of the resonator with the substrate, without the Bragg reflectors used by solidly-mounted resonators (SMR's) [6], also steals considerable energy. All told, the energy loss inflicted by voltage-induced collapse is large enough to completely remove the resonant peak. Of course, the real magic of such a device is the ability to turn ON again. Upon removal of the switching voltage, the top plate springs back up, resuming normal operation.

DEVICE FABRICATION

The fabrication process of this work includes four lithography steps and allows for versatile top electrode configurations through the use of a conformal CVD doped polysilicon deposition. The process begins with a 150mm wafer upon which 2μm of SiO_2 is grown, for electrical isolation, followed by a 250nm low-stress silicon nitride deposition to protect the underlying oxide during a future release step. Since the release will be done via vapor HF, a 100-nm-thick AlN layer is sputter deposited above the silicon nitride to act as a barrier layer to mitigate silicon nitride's incompatibly with vapor HF etching. In retrospect, this AlN layer should be kept as thin as possible, or eliminated altogether, to minimize piezoelectric coupling to the substrate.

150nm of molybdenum is then sputter deposited and dry-etched in an SF_6 plasma to form the bottom interconnect layer. Next, 300nm of low temperature oxide (LTO) is deposited, followed by a quick CMP (~150nm) to eliminate surface topography whilst forming a thin sacrificial layer. 1.6μm of AlN is then sputter deposited in a Tegal Endeavor AT system to form the resonator, followed by LPCVD deposition of a 1-μm-thick LTO oxide to serve as a hard mask during AlN etching. After patterning the hard mask to delineate resonator features and expose AlN, the AlN is dry etched using a 90/35/10 sccm $Cl_2/BCl_3/Ar$ etch chemistry, resulting in the cross section depicted in Figure 5(a), followed by a final oxide etch to remove the hard mask above the resonators. A final 250-nm-thick LTO deposition then follows to define the top sacrificial electrode-to-resonator gap spacer, which is patterned in the third lithography step to define anchor openings to the substrate, cf. Figure 5(b).

978-1-4799-3510-9/14 $31.00 © 2014 IEEE

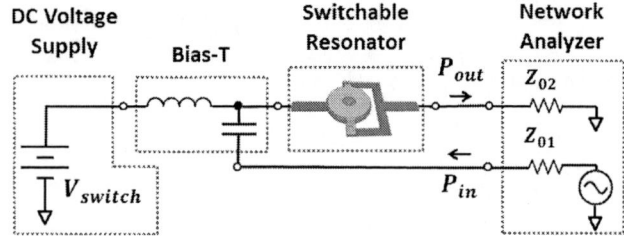

Figure 6: Experimental setup to measure impedance and apply a DC switching voltage to a switchable resonator.

Finally, a dual purpose 600-nm-thick LPCVD doped polysilicon deposition forms the supporting stem for the disk as well as the top collapsible electrode. After a final lithography step and plasma etch to pattern the polysilicon electrode, cf. Figure 5(c), the wafer is cleaved. Devices are cleaned and put on a hot plate at 200°C prior to release in a Primaxx vapor HF etch system. Figure 5(d) presents the SEM of a completed device.

EXPERIMENTAL RESULTS

Figure 6 presents a schematic of the test setup used, where a DC switching voltage V_{switch} is added to the input RF signal using a bias-T. On-state operation of the switchable resonator entails applying an AC input voltage and sensing the output current while the DC switch-control voltage V_{switch} is turned off. At 300 MHz, this is best implemented using the microwave test setup shown, where the passive switchable resonator, held under vacuum for maximum Q, has a highly frequency-selective impedance to be characterized. Here, a two-port S-parameter measurement is performed, despite being a one-port device, because the parameter of greatest interest for a filter is its signal transmission. Although a single resonator is measured in this experiment, the same measurement method can be used for a higher-order filter of multiple coupled resonators. Also, since the impedance of a single device is on the order of 1-10 kΩ, a two-port measurement provides a more accurate estimate of impedance than does a one-port measurement, for which a small calibration error results in a large error in measured admittance. The resonator parameters R_x, Q, and k_{eff}^2 for a device are extracted directly from the S_{21} frequency characteristic measured by the network analyzer as follows:

$$R_x = 2Z_0\left(-1 + 10^{-S_{21,dB}/20}\right) \quad (1)$$

$$Q = (f_{peak}/\Delta f_{3dB})(R_x + 2Z_0)/R_x \quad (2)$$

$$k_{eff}^2 = (\pi^2/4)(f_p - f_s)/f_s \quad (3)$$

where $Z_0 = Z_{01} = Z_{02} = 50\Omega$ represent the characteristic impedances of the measurement system, $f_s = f_{peak}$ is the series resonance frequency at which maximum transmission occurs, Δf_{3dB} is the 3-dB bandwidth around the f_s peak, and f_p is the parallel resonance frequency.

On/Off Switchability

To characterize the on/off switchability of the device, S_{21} measurements are made in between repeated transitions between the ON and OFF states of V_{switch}=0V and V_{switch}=220V,

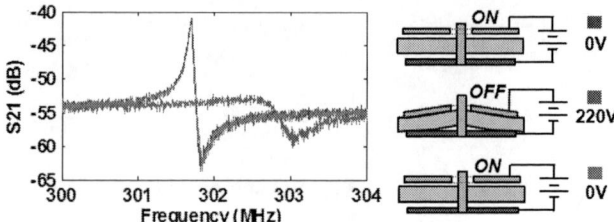

Figure 7: Demonstration of switching capability. The resonator is ON when no switching voltage is applied. Next, a DC voltage of 220V is applied, causing the top electrode to collapse, effectively turning the device OFF. When the switching bias is removed, the resonator turns back ON, with no degradation in performance.

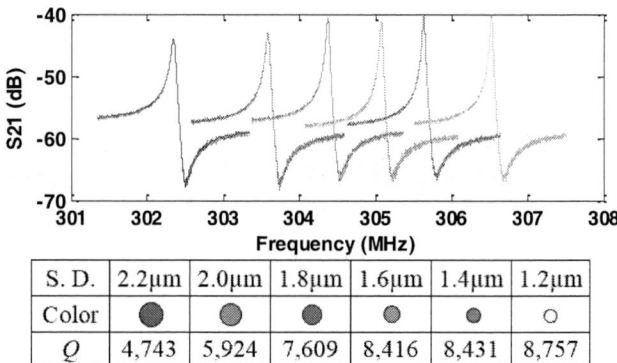

S. D.	2.2μm	2.0μm	1.8μm	1.6μm	1.4μm	1.2μm
Color	●	●	●	●	●	○
Q	4,743	5,924	7,609	8,416	8,431	8,757

Figure 8: Q and resonance frequency dependence on stem diameter for six adjacent resonators.

respectively. Figure 7 presents three superposed S_{21} measurements made before, while, and after applying the needed switching voltage, as well as device cross sections. The measurements made before and after turning the device off are virtually identical and the measurement was cycled 20 times with no failure or initial frequency or Q changes over the duration of the experiment.

Here, the needed pull-down voltage of 220V is high, but not dissimilar from voltages normally needed to actuate RF MEMS switches. The good news is smaller gaps and proper device suspension design can make the needed voltages much smaller. Numerous approaches toward achieving this goal warrant further investigation, including single point electrode anchoring and folded beam electrode suspensions, as are used for some MEM relays, as well as their fabrication techniques [7].

High-Q Operation

While capacitive piezoelectric transduction is essential to maximizing Q, anchor minimization and post-release cleaning are equally important. Figure 8, which plots S_{21} as a function of stem diameter, shows how the Q's of these devices also depend strongly on the widths of their stem anchors. Here, the smallest anchor diameter of 1.2 μm yields a measured Q of 8,757, which is almost twice that of a 2.2 μm-stem version, and 4.2× higher than the ~2,100 typically reported for conventional AlN resonators in this frequency range [2]. The resonance frequency changes seen in Figure 8

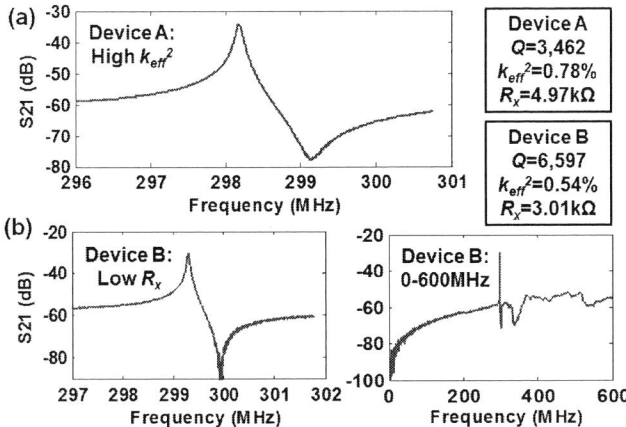

Figure 9: (a) Frequency response for Device A with a high measured k_{eff}^2 of 0.78%. (b) Frequency response for Device B with a k_{eff}^2-Q product of 35.6 and an R_x of 3.01kΩ. The wideband response for Device B shows no spurious mechanical modes from DC to 600 MHz.

agree with predictions that estimate resonance frequency as a function of polysilicon anchor size.

After release, the devices have poor Q's of only several hundred. One probable reason revealed upon SEM inspection is that a thin layer of etch residue remains on the surface of the disks right after release. A post-release procedure comprised of dipping the devices in DI water for several minutes, drying with an N_2 gun, and placing on a hotplate at 200°C in air for several minutes, reliably raises Q's up to 6,000. A 5 minute anneal at 500°C in an N_2 environment yields further Q-improvement, as was done in [5], and was necessary to achieve Q-values above 8,000.

Strong Electromechanical Coupling

Pursuant to increasing the k_{eff}^2's and lowering the R_x's of these resonators, the process was rerun on a second wafer with more aggressive gap spacings of 120-150 nm. Through a combination of smaller gaps and better AlN quality, a 4× reduction in R_x (for devices of comparable Q's) and a 2× increase in k_{eff}^2 were realized. However, due to a problem at the anchor etch step arising from changing to a 1.5μm-thick polySi interconnect, only a small subset of devices survived the release that unfortunately did not include the devices with very small anchors for maximum Q. Stem diameters are 1.8μm for the measured disks.

As expected, the k_{eff}^2's on the order of the 0.78% measured in Figure 9(a) are smaller for these capacitive-piezoelectric resonators than for conventional contour-mode AlN resonator counterparts (with attached electrodes), but they are nonetheless much larger than achieved by capacitive-gap transduced devices at this frequency, and more importantly, are plenty large enough to satisfy the needs of channel-selecting RF front-ends.

Strong evidence of the efficacy of capacitive piezoelectric transduction is presented in Figure 9(b), which demos a k_{eff}^2-Q product of 35.6 on par with the 31.5 and 39.9 exhibited by the conventional attached electrode counterparts of [2] at the same frequency. Additionally, Figure 9(c) presents a

clean, spur-free wideband response for the same resonator over a 600MHz range, indicating the device's superior ability to excite a single desired mode shape.

Looking forward, based on measurements of devices on each of the two runs, k_{eff}^2-Q's on the order of 70 are not unreasonable to expect for this device, as are single disk motional impedances of 1-2kΩ. If achievable, such individual disks, when mechanically combined to form array resonators, could realize sub-100Ω R_x's, making them amenable to low impedance termination in practical RF systems.

CONCLUSIONS

With this work, capacitive-piezoelectric resonators have demonstrated not only Q's approaching 9,000 at 300MHz, but also a new on/off switching capability, both of which represent important first steps toward implementing switchable AlN channelizing filter banks for RF front ends. Although more extensive study is needed on the reliability of the collapse-based switching mechanism, this technology certainly raises eyebrows and encourages work towards developing higher-order filters that harness the high Q and self-switching. Such work is ongoing.

ACKNOWLEDGEMENTS

The authors would like to thank the staff of the Berkeley Nanolab and Zeying Ren for valuable assistance. This work was supported by the DARPA CSSA program.

CONTACT

*Robert Schneider, schneid@berkeley.edu

REFERENCES

[1] C. T.-C. Nguyen, "Integrated micromechanical circuits for RF front ends," in *Solid-State Device Research Conference, 2006. ESSDERC 2006. Proceeding of the 36th European*, 2006, pp. 7–16.

[2] C. Zuo, N. Sinha, and G. Piazza, "Very high frequency channel-select MEMS filters based on self-coupled piezoelectric AlN contour-mode resonators," *Sens. Actuators Phys.*, vol. 160, no. 1, pp. 132–140, 2010.

[3] T. L. Naing, T. Beyazoglu, L. Wu, M. Akgul, Z. Ren, T. O. Rocheleau, and C. T.-C. Nguyen, "2.97-GHz CVD diamond ring resonator with Q> 40,000," in *Frequency Control Symposium (FCS), 2012 IEEE International*, 2012, pp. 1–6.

[4] S. S. Li, Y. W. Lin, Z. Ren, and C. T.-C. Nguyen, "Self-switching vibrating micromechanical filter bank," in *Frequency Control Symposium and Exposition, 2005. Proceedings of the 2005 IEEE International*, pp. 135–141.

[5] L. W. Hung and C. T.-C. Nguyen, "Capacitive-piezoelectric AlN resonators with Q> 12,000," in *Micro Electro Mechanical Systems (MEMS), 2011 IEEE 24th International Conference on*, 2011, pp. 173–176.

[6] K. M. Lakin, K. T. McCarron, and R. E. Rose, "Solidly mounted resonators and filters," in *Ultrasonics Symposium, 1995. Proceedings., 1995 IEEE*, 1995, vol. 2, pp. 905–908.

[7] T.-J. K. Liu, L. Hutin, I.-R. Chen, R. Nathanael, Y. Chen, M. Spencer, and E. Alon, "Recent progress and challenges for relay logic switch technology," in *VLSI Technology (VLSIT), 2012 Symposium on*, 2012, pp. 43–44.

PARAMETRIC FILTERING SURPASSES RESONATOR NOISE IN ALN CONTOUR-MODE OSCILLATORS

Cristian Cassella, Nicholas Miller, Jeronimo Segovia-Fernandez, Gianluca Piazza
Carnegie Mellon University

ABSTRACT

In this work we present a new method to lower the phase noise in acoustic resonator based oscillators. This method uses the nonlinear dynamics of a parametric divider made to work close to the bifurcation region. We call this technique "*parametric filtering*". This approach was applied to an oscillator based on an aluminum nitride contour-mode resonator vibrating around 227 MHz. This class of resonators exhibits frequency flicker noise that limits the oscillator phase noise close to the carrier. By means of parametric filtering we obtained an improvement in the phase noise of more than 14 and 19 dB, respectively at 1 and 10 kHz offsets. This technique can be applied to any MEMS oscillator and represents the first demonstration of open loop phase noise filtering.

INTRODUCTION

Microelectromechanical (MEMS) resonators represent good candidates for replacing Quartz and Surface Acoustic Wave (SAW) devices as frequency selective elements in oscillator circuits. The possibility to build both the resonant element and the controlling electronics by using the same fabrication techniques allows building compact and fully integrated high frequency oscillators. However, MEMS oscillators in the high frequency range exhibit worse phase noise than oscillators based on SAW and Quartz crystals. This issue is mainly due to two reasons: i) MEMS resonators have a lower quality factor, Q, and consequently are worse at filtering the fast frequency fluctuations due to noise coming from the circuit; ii) some MEMS resonators exhibit high intrinsic frequency fluctuations [3].

The approach commonly used to improve the phase noise of oscillators consists of increasing the quality factor of the resonant elements. However, although having high Q is desirable to limit the effect of the noise coming from the circuit, it leads to lower power handling for a given device size and hence a worst oscillator noise floor. Simultaneously, nonlinearities would emerge at lower power levels with a high Q. In general, nonlinearities cause coupling between amplitude and phase noise and therefore further degrade the phase noise of the oscillator.

Some recent demonstrations have shown that the oscillator phase noise can be lowered by using nonlinear resonators made to work in critical operational states (COS) [1-2]. By making the resonator working close to a bifurcation condition, it was shown that it is possible to slow down the resonator dynamics and hence lower the phase noise of the oscillator. This technique, as well as the increase of Q, are not effective, though, when the resonator noise is dominant over the circuit noise. In fact, in the case in which the frequency noise in the MEMS resonator is much higher than the noise coming from

the circuit, the resulting oscillator phase noise is limited by the acoustic device and it is independent of the resonator Q. In this work a new methodology to surpass the limitations introduced by noisy MEMS resonators is presented. We have called this technique "*parametric filtering*", since it uses the nonlinearities of a parametric divider made to work close to the division bifurcation. This method has been applied to a MEMS oscillator based on an AlN Contour-Mode Resonator (CMR) [4].

PARAMETRIC FILTER

The technique we are presenting in this work allows reducing the phase noise of the output signal of a MEMS oscillator by connecting at its output port an external circuit. The core of this external circuit includes a varactor parametric divider [5,6] (Fig.1). The divider is biased by a dc-voltage, V_b, which sets the varactor working point. An input signal, labeled here as *pump,* is superimposed to the varactor dc-bias to modulate the varactor capacitance around its dc-value. As we will show later, in the case of 2:1 parametric dividers, , the parametric circuit starts behaving as a frequency divider for a given peak-to-peak value of the pump voltage. The activation of the frequency division happens through a system bifurcation phenomenon, which changes the frequency response of the parametric element. In fact, after reaching bifurcation, the circuit starts absorbing energy from the pump and converting it to a new signal having frequency approximately equal to half of the pump frequency. As we will prove in this work, parametric dividers are capable of filtering the phase noise fluctuations of the pump signal, and, as a consequence, they can generate sub-harmonic signals that have much lower phase noise than what expected by looking just at linear division mechanisms. It is important to note that this phase noise reduction happens just when the parametric circuit is made to work close to the division bifurcation. Therefore, the external circuit immediately following the output of the oscillator circuit has two main purposes. The first is to allow frequency division by regulating the pump voltage at the input of the parametric divider. The second function is ensuring that the parametric divider working point is in the vicinity of the division bifurcation.

The overall circuit that was used in this work to accomplish these functions is shown in Fig.1. It includes a low pass filter (Minicircuits SLP-300+) connected at the oscillator output port to cancel the oscillator harmonics. A frequency doubler (Minicircuits FK-3000+), with 11 dB of conversion loss at the frequency of interest, is used to generate a signal at two times the oscillator main output frequency. The doubler output is sent to the input port of a resistive power combiner (Minicircuits ZFRSC-42-S+) which causes 6 dB of insertion loss for each input signal and ensures 6 dB of isolation between the two input ports. The output of the power combiner is properly amplified to reach the pump threshold voltage needed for

978-1-4799-3510-9/14 $31.00 © 2014 IEEE

activating the division. This was done by using an external amplifier (Minicircuits ZKL-2+) biased to provide a gain of 40 dB. The output of the amplifier is filtered by a high pass filter (Minicircuits SHP 400+) having insertion loss < 0.5 dB at the frequency of interest. The output of the filter is sent to a parametric divider, which was built on an FR4 printed circuit board (PCB). The divider includes a Skywork SMV1248 as the variable capacitor. The varactor dc-bias is provided through the use of a bias-T. The output of the parametric divider is split in two equal paths by using a resistive power splitter (Minicircuits ZFRSC-42-S+). One path is sent to a Signal Source Analyzer for measurement, whilst the other is sent back to the power combiner. This connection is made to use a portion of the output signal to increase the level of the pump signal and, consequently to reduce the amplifier power consumption needed to activate division. We observed that having a portion of the pump strictly correlated to the sub-harmonic signal is crucial in ensuring proper operation of the circuit.

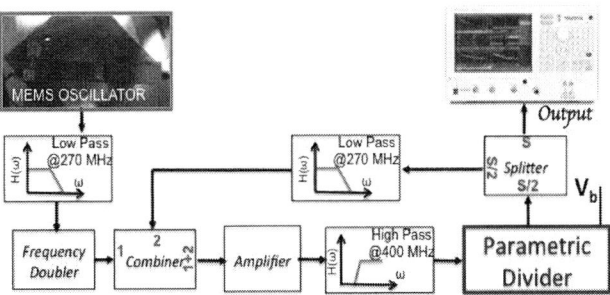

Figure 1: Filtering topology investigated in this paper. The output at ω_{out} of the feedback-loop oscillator, is filtered around the oscillation frequency and sent to a frequency doubler (1). The output of the frequency doubler is amplified (2), filtered (3) and sent to the parametric divider input (4). The varactor dc-bias, V_b, sets the divider working point.

ALN CONTOUR-MODE OSCILLATORS

AlN CMRs [4] represent an emerging technology for high-frequency references. In particular, the moderately high quality factors, Q, and electromechanical coupling, k_t^2, make this class of devices intrinsically capable of low power consumption and low phase noise. However, it was experimentally shown that the phase-noise achievable in oscillators using this class of resonators is degraded by the high level of flicker frequency noise intrinsic to the resonator. We further demonstrate this in here. In this work we used a 227 MHz AlN CMR with loaded Q and k_t^2 respectively equal to 2000 and 0.12% (Fig.2). We measured the frequency fluctuations of the resonators in an open loop measurement [3]. Then we built a feedback loop oscillator formed by the resonator, mounted in a two-port configuration on a FR4 circuit board (Fig.2), an amplifier (Minicircuits ZKL-1R5) having maximum gain equal to 35 dB, a phase shifter (ATM P1213) and a resistive power splitter (Minicircuits ZFRSC-42-S+) that allowed probing the oscillator output (Fig.3). By measuring the phase-noise of the output signal, we confirmed that the measured closed-loop phase-noise matched closely the predicted value calculated by assuming the resonator frequency noise to be the only noise source in the oscillator (Fig.3).

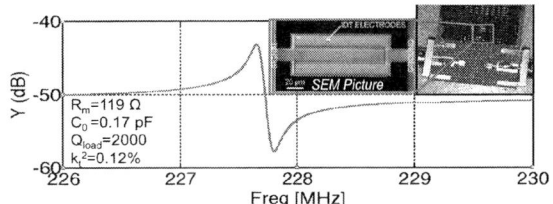

Figure 2: Admittance response of the 227 MHz AlN CMR, mounted on a FR4 circuit board

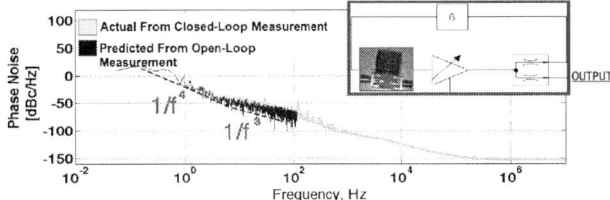

Figure 3: In black: prediction of the oscillator phase-noise at low frequency offset obtained by open-loop measurement of the resonator frequency fluctuations. In green: measured phase noise of the feedback-loop oscillator shown in the inset, and based on the same resonator.

PARAMETRIC DIVIDERS

Parametric amplifiers and dividers have represented an alternative low noise approach to implement signal amplification and frequency division in the microwave frequency range. Their key element consists of a variable reactance whose value is modulated by a voltage at a frequency related to the output signal frequency. A reactive load is placed in series to the variable reactance to have an average resonance frequency at the desired output frequency, f_{out}. In 2:1 capacitive parametric amplifiers and dividers the variable capacitance is modulated by using an external signal, the *pump signal*, having a frequency equal to $2f_{out}$. This modulation permits to increase the net charge flowing out of the capacitor. For this reason parametric amplifiers and dividers are analyzed as equivalent negative conductances placed in parallel to the output load seen at the output port [5]. The value of this negative conductance is mostly set by the amplitude of the modulation voltage, V_{in}, as well as by the dc-bias, V_b, used to set the varactor working point.

Capacitive parametric amplifiers can become unstable for given values of V_{in} and V_b. When this happens at least one system eigenvalue shows positive real part and null imaginary part and parametric amplifiers start behaving as parametric dividers. The value of the negative conductance is a function of frequency, f, and in the case of degenerate amplifiers [5] presents a maximum at f equal to $f_{pump}/2$. In order to apply the concept of parametric filtering to a 227 MHz AlN CMR, we have designed a parametric divider that enabled division from 454 to 227 MHz. The parametric divider design was conducted analytically and numerically by using ADS. Analytically, we used a simplified RLC circuit to describe the system and derived the value of the peak-to-peak modulation voltage needed to create the instability at $f_{pump}/2$ (Fig. 4-a). This simplification can be used to model the threshold voltage

978-1-4799-3510-9/14 $31.00 © 2014 IEEE   1270

even in those cases in which the divider topology includes more lumped elements. In fact, the equivalent impedance seen by the variable capacitor has to be inductive at the frequency of interest for maximizing the energy transmitted to the resistive load with respect to the energy stored by the reactive components.

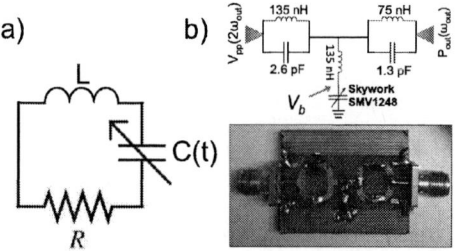

Figure 4:a) Simplified RLC circuit for symbolic analysis of sub-harmonic oscillations. b) Electric schematic of the parametric divider (top) with view of the PCB built on an FR4 substrate (bottom). The capacitance value of each capacitor in the two tanks is nominal and does not take into account the circuit parasitics.

The differential equation that describes the simplified system of Fig.4-a can be written as:

$$L[C(t)V''(t) + 2C'(t)V'(t) + V(t)C''(t)] + RC'(t)V(t) +$$

$$+ RC(t)V'(t) + V(t) = 0 \qquad [Eq.1]$$

where $V(t)$, $V'(t)$ and $V''(t)$ are respectively the voltage across the varactor and its first and second derivative with respect to time. $C(t)$, $C'(t)$ and $C''(t)$ are respectively the capacitance value and its first and second derivative with respect to time. L is the circuit equivalent inductance value, and R is the output load. Assuming a linear approximation for the capacitance, we can express $C(t)$ as

$$C(t) = C_0(V_b) + C_1 V_{pp} \sin(2\omega_{out} t) \qquad [Eq.2]$$

where $C_0(V_b)$ is the *static* capacitance, defined as the average capacitance in a modulation period. V_{pp} is the peak-to-peak modulation voltage and C_1 is the linear approximation of $C(V)$ around the dc-bias, V_b. We found that one eigenvalue of the differential equation becomes unstable when

$$V_{pp}^a = \frac{2C_0 \sqrt{[(\frac{\omega_{out}}{\omega_{LC}})^2 - 1]^2 + (C_0 \omega_{out} R)^2}}{C_1 \sqrt{1 + 4C_0^2 \omega_{out}^2 R^2}} \qquad [Eq.3]$$

By considering $C_0(V_b)$ in the pF range, and assuming the load resistance to be close to 50 Ω, (Eq.3) can be approximated at the frequency of interest to:

$$V_{pp}^a = \frac{2C_0 \sqrt{[(\frac{\omega_{out}}{\omega_{LC}})^2 - 1]^2 + (C_0 \omega_{out} R)^2}}{C_1} \qquad [Eq.4]$$

where ω_{LC} is the average resonance of the varactor calculated as $1/\sqrt{LC_0(V_b)}$. (Eq.4) shows than that a high modulation ratio (*i.e.* $C_1/C_0(V_b)$) and an average resonance frequency close to the desired output frequency are needed to reduce the

activation voltage, and consequently the threshold power at which division happens.

To physically implement the parametric divider, the simplified circuit described by Eq. 1 was modified to the one of (Fig.4-b). This modification was required in order to ensure proper power flow from the pump to the output and vice versa isolate the output from the pump. An output notch filter was introduced to prevent the pump energy from flowing to the output port instead of being used to modulate the variable capacitance. An input notch filter was used to reduce the output power leaking back to the pumping port. An inductor was also placed in series to the variable capacitor to move the system eigenfrequency close to 227 MHz. Skywork SMV1248 was chosen for the variable capacitor because of its high $C_1/C_0(V_b)$ ratio. The values of C_1 and $C_0(V_b)$ were also extracted experimentally at the frequency of interest, and for a reverse dc-bias approximately equal to 2 V. Transient simulations in Agilent ADS were performed to detect the threshold power needed for activating the sub-harmonic generation. The simulated threshold power shows good agreement with the measured value (Fig.5).

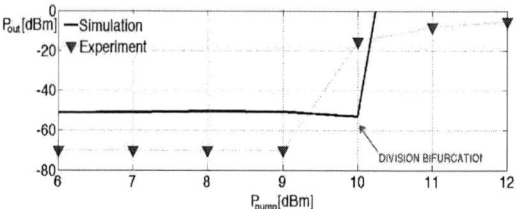

Figure 5: In black – variation of the output power, P_{out} at f_{out}, with respect to the pump power, P_{pump}, at $2f_{out}$ simulated by Fast Fourier Transform in transient analysis. Blue triangles – output power measured at the output port of the circuit shown in (Fig. 4-b) as a function of P_{pump}.

PARAMETRIC DIVIDER'S PHASE NOISE

The method of phase noise reduction that we are presenting in this work is based on the usage of a parametric divider as a phase noise filter. A parametric divider made to work close to the division bifurcation is capable of slowing its dynamics. As a consequence, it cannot follow the fast phase fluctuations of the pump signal. As described in [6], an admittance function, $Y_T(\omega_{out})$, can be used to analytically model the impact of the divider working point on the phase noise spectral density of the sub-harmonic signal. $Y_T(\omega_{out})$ is defined here as the linearized admittance seen across the varactor terminal for different dc-biases and pump power levels. The phase noise of the sub-harmonic signal at a given offset frequency, Ω, can be written as

$$|\Delta\varphi_T(\Omega)|^2 \propto \frac{\left|\frac{dY_T}{d\varphi}\right|^2 \cdot \left|\frac{\varphi_s(\Omega)}{2}\right|^2}{\left|\frac{dY_T}{d\varphi}\right|^2 + \left|\frac{dY_T}{d\omega}\right|^2 \cdot \Omega^2} \qquad [Eq.5]$$

where $dY_T/d\varphi$ is the sensitivity of the admittance function to phase variation in the pump signal. (Eq.5) shows that for small Ω the PN of the divider tends to the linear prediction

$\varphi_s^2(\Omega)/4$, where $\varphi_s^2(\Omega)$ is the square of the phase noise of the pump signal at the frequency offset, Ω. For larger Ω and especially for a large derivative of the admittance with respect to frequency, the second term at the denominator starts playing a more important role. Before reaching the division bifurcation the divider stores energy and uses it to modulate the varactor capacitance. However, after bifurcation the circuit absorbs real power and the input reflection coefficient seen at the pump port greatly moves towards much lower VSWR values. The almost infinte variation of the admittance function $Y_T(\omega_{out})$ when bifurcation is approached reflects the fast transition from the state in which the circuit stores energy and the state in which it absorbes real power and releases it at half of the input frequency. Therefore, in the attempt of reducing the original phase noise of the pump signal by more than the 6 dB suggested by the linear theory, we controlled the system working point to reach operational states that are close to the division bifurcation.

EXPERIMENTAL RESULTS

The phase noise of the parametric filter output signal (Fig.1) was experimentally analyzed for different varactor dc-biases. By controlling the static capacitance value it is possible to change both the $\frac{\omega_{out}}{\omega_{LC}}$ ratio defined in (Eq.4) and $C_1/C_0(V_b)$. This consequently changes the divider working point and leads it to operate close to the bifurcation condition.

By using this approach the oscillator phase noise was drastically lowered with respect to the original, unfiltered response. As expected, the phase noise performance was affected by the divider working point, and hence by V_b. Values of the phase noise at 1 and 10 kHz offsets, as we approach the division bifurcation by sweeping V_b are shown in Fig.6. An improvement of 14 and 19 dB, respectiveley at 1 and 10 kHz offsets, could be simultaneously achieved. Moreover an improvement of more than 5 dB was also measured in the floor noise, although a 15 dB reduction of the output power was recorded at the output of the parametric filter. These results confirm that the parametric filter does not perturbe the closed-loop fixed point, but acts, instead, as an external phase noise filter at the oscillator output. The comparison of the best measured phase noise with respect to the unfiltered response is shown in Fig.7.

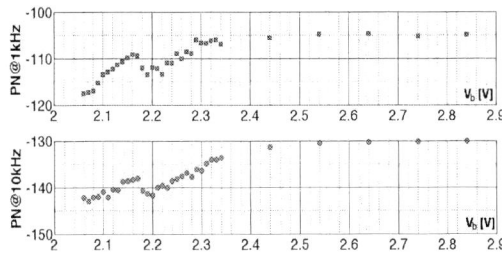

Figure 6: _Squares_: phase noise at 1kHz offset for different varactor DC-biases. _Circles_: phase noise at 10 kHz offset at different reverse varactor dc-biases, V_b (Fig.1). When no parametric filtering is used the phase noise is -104 dBc/Hz at 1 kHz offset and -125 dBc/Hz at 10 kHz offset. The use of

parametric filtering enables 14 dB improvement in the phase noise at Ω=1 kHz and 19 dB at Ω=10 kHz.

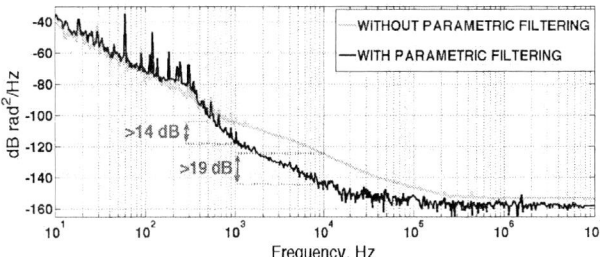

Figure 7: _In green_: phase noise, PN, for the feedback loop oscillator (Fig.3) without parametric filtering. The output power is 3 dBm. _In black_: phase noise at the output of the parametric filter when V_b is 2.07V and the system is working close to the division bifurcation. The output power is -9dBm and is mostly set by the quadratic nonlinearities in the parametric divider.

CONCLUSIONS

A new technique to reduce the phase noise in MEMS oscillators was introduced. We called this technique *parametric filtering* since it uses a parametric divider working close to the division bifurcation to achieve noise canceling. We applied this technique to an oscillator based on an AlN CMR operating at 227 MHz. The phase noise of this oscillator is limited by the resonator frequency fluctuations. Improvement of 14 and 19 dB in the phase noise respectively at 1 and 10 kHz offsets were demonstrated. This technique is broadly applicable to any MEMS based oscillator.

REFERENCES

[1] Yurke, B., Greywall, D. S., Pargellis, A. N., & Busch, P. A., Phys. Rev. A, vol. 51(5), 4211 (1995).

[2] J. Segovia, C. Cassella, G. Piazza, "Close-in Phase Noise Reduction in an Oscillator based on 222 MHz Non-Linear Contour Mode AlN Resonators", accepted for publication at IEEE IFCS 2013, July 2013.

[3] N. Miller and G. Piazza, "Vector Network Analyzer Measurements of Frequency Fluctuations in Aluminum Nitride Contour-Mode Resonators", accepted for publication at IEEE IFCS 2013, July 2013.

[4] G. Piazza, P.J. Stephanou, A.P. Pisano, Journal of MicroElectroMechanical Systems, vol. 15, no.6, pp. 1406-1418 (2006).

[5] J.C. Decroly, L. Laurent and J.C. Lienard, Parametric Amplifiers (PHILIPS Technical Library, 1973).

[6] A. Suarez, Analysis and Design of Autonomous Microwave Circuits, (Wiley, 2009), pp.244-245.

[7] Guckenheimer, J., and Holmes, P., _Nonlinear Oscillations, Dynamical Systems, and Bifurcations of Vector Fields_ (Applied Mathematical Sciences, Book 42, _Springer_, 2002).

POLYCIDE CONTACT INTERFACE TO SUPPRESS SQUEGGING IN MICROMECHANICAL RESOSWITCHES

Yang Lin, Ruonan Liu, Wei-Chang Li, and Clark T.-C. Nguyen
Berkeley Sensor and Actuator Center
Department of Electrical Engineering and Computer Sciences
University of California, Berkeley, CA USA

ABSTRACT

The use of a Pt-silicide-based contact interface has greatly reduced impact-induced energy loss in comb-driven resonant micromechanical switches (a.k.a., resoswitches) to the point where squegging phenomena (whereby impacts do not occur on every cycle) are eliminated, so no longer constrain the clock frequency of recently demonstrated mechanical charge pumps [1]. This opens the application range of such charge pumps to power converters capable of delivering currents much higher than the low current-draw MEMS dc-biasing applications targeted by [1]. The key to eliminating squegging in the present work is contact engineering, where softer contact materials, including Au, Ag, and Ni, steal too much energy on each impact; but harder contact materials, like Pt-silicide, allow more elastic impact while still maintaining low contact resistance due to the large impulsive force generated by impact—a distinct advantage of the resoswitch over conventional non-resonant counterparts.

INTRODUCTION

If high bias voltages on the order of 50-200V were available on-chip, the corresponding increase in the electromechanical coupling strengths of many capacitively transduced MEMS devices, from gyroscopes, to microphones, to timing oscillators, would bring about substantial performance and cost benefits. Indeed, the size of a mechanical element can often be made substantially smaller when capacitive coupling becomes stronger, since less overlap is required to achieve a given coupling strength. To date, voltages exceeding the supply voltages are generally provided by semiconductor-based charge pumps that use unconventional, expensive transistors designed to support higher voltages. Even so, such devices are still limited by pn-junction and dielectric breakdown limits, so constrain the voltages ultimately achievable by charge pumps using them.

To remedy this, the MEMS-based Dickson charge pump demonstrated in [1] use micromechanical resoswitches [2] to remove the diode voltage drop and junction breakdown issues that plague conventional transistor versions, allowing them to transfer charge with any input voltage level and achieve much higher voltages, perhaps eventually as high as 200V. The resoswitch device is simple enough that its fabrication steps are often already present in the fabrication process flows of many existing MEMS applications, so high voltage and its benefits might come almost for free.

But there were imperfections. In particular, the work of [1] used gated-sinusoid excitation signals to synchronize the movement of charge through the pump topology and to

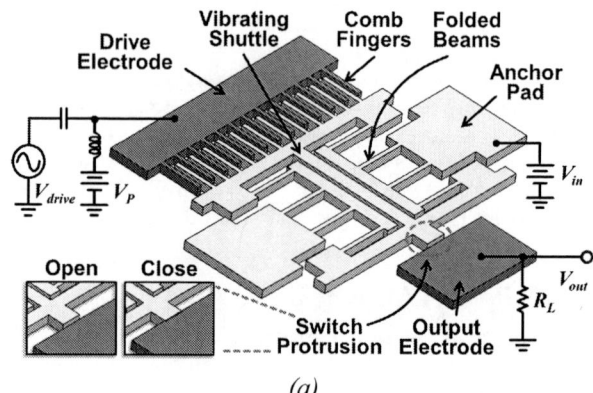

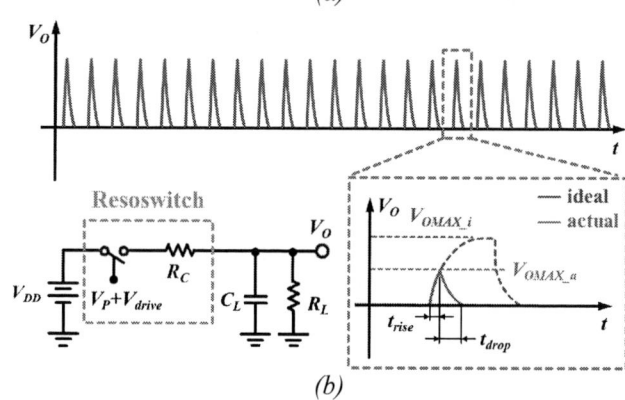

Fig. 1: (a) Detailed schematic of a comb-driven resoswitch with flat contact point on the opposite side of the drive fingers. (b) Simplified equivalent circuit and expected output waveform with zoom in on the shape of each spike.

overcome difficulty with squegging—a phenomenon where an oscillation amplitude is not constant, but rather grows and shrinks with a certain period [3], as shown (later) in Fig. 6.

This work explores contact engineering to overcome squegging and ultimately uses a Pt-silicide-based contact interface to greatly reduce impact-induced energy loss in resoswitches to the point where squegging phenomena are eliminated, so no longer constrain the clock frequency of recently demonstrated mechanical charge pumps. This opens the application range of such charge pumps to power converters capable of delivering currents much higher than the low current-draw MEMS dc-biasing applications previously targeted by [1].

RESOSWITCH SQUEGGING

To explain squegging in the subject resoswitch, some description of device structure and operation is in order.

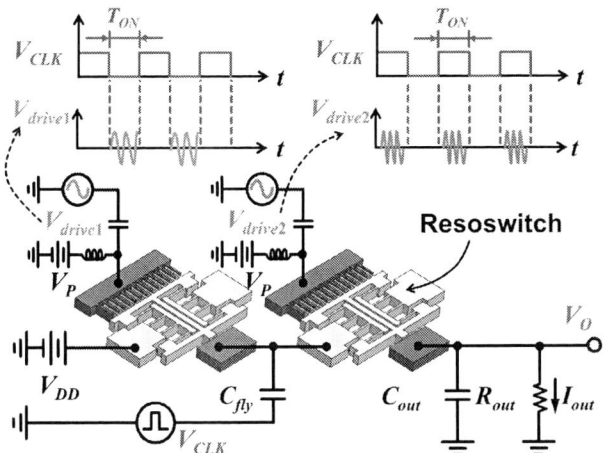

Fig. 2: Schematic of a single-stage MEMS Dickson's charge pump and gated-sinusoid waveforms needed to affect synchronized pumping.

Comb-Driven Resoswitch Structure and Operation

Fig. 1(a) presents the perspective-view schematic of one resoswitch variant, with others shown later in Fig. 8. Like the device of [1], these devices feature folded-beam suspensions and comb-finger transducers that can drive their conductive shuttle protrusions to impact with a switch (or output) electrode. They differ, however, in the placement and type of impact points, which now include flat or sharp points placed either on the opposite or same side as the fingers. Same side placement allows the fingers to apply additional impact force over and above the momentum-based force when fingers and contact point are on opposite sides.

The operation of any of these devices is simple and shown for the device of Fig. 1(a): Drive the capacitive combs with a combined (dc-bias + resonance ac) voltage hard enough to affect impacting, in turn closing a mechanical switch that then periodically transfers charge from the supply V_{in} to the awaiting load (R_L or C_L) at the output. Fig. 1(b) presents the equivalent circuit for the hookup of Fig. 1(a), where R_C is a combination of interconnect resistance and switch contact resistance (but dominated by the latter). The expected normal output waveform from this comb-driven resoswitch is also shown to be a series of spikes shaped by the RC circuit formed between R_C, R_L and C_L. The governing expression from which the contact resistance R_C can be extracted takes the form

$$\frac{V_{DD} \cdot R_L}{R_C + R_L} \cdot \left(1 - exp\left(-\frac{t_{rise}}{(R_C || R_L)C_L} \right) \right) = V_{OMAX\_a} \quad (1)$$

where t_{rise} is the charging time of the spike, i.e., the switch contact time; and V_{OMAX_a} is the measured spike amplitude.

Charge Pump Operation

In the actual charge pump application, shown in Fig. 2, the waveforms required to actuate the devices are actually gated sinusoids rather than pure ones. This is a consequence of finite tolerances achievable via planar microfabrication

that produce resoswitches with slightly different resonance frequencies. Because of this, the devices cannot be actuated simply by a single signal at one frequency phase shifted to service different pumping phases. Rather, each device requires a different frequency to affect resonant switching. To synchronize impact-based charge transfer events, gated sinusoids tailored to the resonance frequency of each individual device are needed, as illustrated in Fig. 2. Here, impacting charge transfer occurs only during periods when the gate is "on". The amount of charge transferred is a function of the total time of impact, i.e., during which the shuttle and electrode are in contact, which for the low frequency design of the present discussion, is generally only a small fraction of the resonance period, as shown in Fig. 1(b). Indeed, one of the factors that sets the needed "gate-on" period in a charge pump is this impact residence time.

Ultimately, the other factor governing the "gate-on" time is squegging. As mentioned, squegging refers to a phenomenon where an oscillation amplitude is not constant, but rather grows and shrinks with a period [3] governed by the degree to which losses vary with time. For the resoswitch device, the loss in the system rises abruptly from its free vibration value once impacting occurs, during which each impact steals an amount of energy from the system governed by the elasticity of the impact. Squegging occurs when the energy stolen on impact is large enough to reduce the amplitude of motion so that no impact occurs on the next cycle(s). Rather, energy must build up towards another impact, after which energy is lost again, and the cycle continues with a period essentially governed by impact loss.

Of course, a squegged waveform, generates fewer impacts per "gate-on" period T_{on}, thereby requiring a longer T_{on} for a given amount of charge transfer and a smaller pumping frequency. This in turn means less pumping ability, so smaller current (or power) delivery to a load, i.e., squegging compromises the ultimate power delivery of a power converter. Thus, the power delivery capability of a MEMS-based charge pump (or other power converter type) cannot be maximized unless squegging is eliminated.

Squegging Model

The energy loss per impact can be modeled by the restitution law governing the relationship between velocity before and after impact: [3]

$$dx/dt \,|_{after\ impact} = -r \cdot dx/dt \,|_{before\ impact} \quad (2)$$

where x is displacement, and $r<1$ is the coefficient of restitution, governed largely by the contact material interface. Loss of velocity after impact, of course, means loss of kinetic energy, and the more energy lost per impact, the longer it takes to recover, and the larger the number of non-impact cycles during squegging. From (2), squegging is minimized via use of materials with higher hardness, i.e., less plastic deformation when impacting, for which r is closer to 1. In this regard, polysilicon is preferred over most metals. Polysilicon alone, however, is too resistive; but a polycide that combines polysilicon and a metal makes good sense.

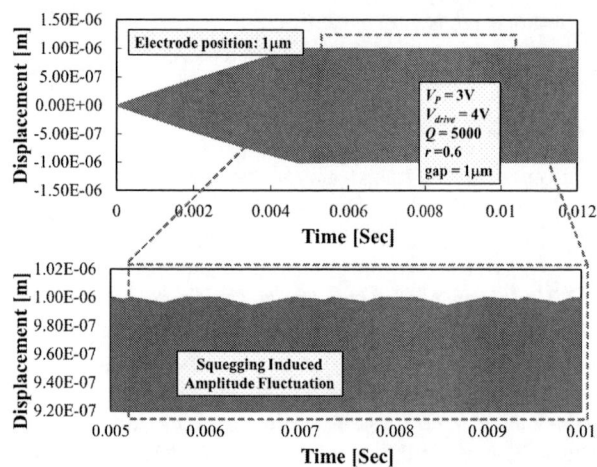

Fig. 3: Theoretically simulated resoswitch shuttle displacement by solving (2) and equation of motion numerically in MATLAB.

The squegging behavior of the resoswitch can be captured theoretically by solving a group of ODE's consisting of (2) and the equation of motion. Fig. 3 shows the simulated shuttle displacement with the parameters summarized in the inset table, where a fluctuation of displacement amplitude is clearly observed.

Simulation also reveals that besides choosing harder structural materials, careful mechanical design can also suppress squegging. For instance, reducing switch gap constrains the system to smaller displacements, so smaller energy deficits to recover after lossy impact, hence, less squegging. In addition, locations of electrodes relative to the contacts can affect squegging. As mentioned, placement of comb electrodes on the same side as the contact allows the electrode-generated force to drive the shuttle into the contact point, so adds to the impulsive impact force, making for more efficient energy recovery. In contrast, having drive electrodes and contacts on the opposite sides relies on less efficiently generated moment forces when contacting, so displays much more simulated squegging.

FABRICATION PROCESSES

Pursuant to gauging the influence of contact interface design on squegging, comb-transduced resoswitches were realized in various materials with various contact interfaces. These include electroplated nickel devices coated with Ruthenium to serve as soft contact resoswitches; and polycide devices to serve as hard contact ones. All devices utilized simple one-mask surface-micromachining fabrication process flows, such as used in [1], where oxide mesas that remain after a timed release etch serve as anchors for suspended structures. Ni/Ru device fabrication is rather straightforward, comprising a molded Ni electroplating above an oxide spacer, followed by subsequent release and coating with 10-20nm of Ru via sputtering, which was found to be sufficiently conformal to coat the sidewalls of the structures. Fig. 4 presents SEMs of a Ni/Ru device.

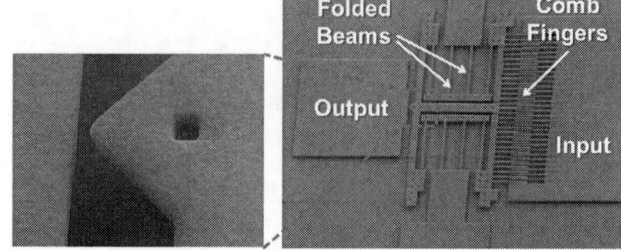

Fig. 4: SEM photos of the comb-driven resoswitch using electroplated Ni as structural material with sputtered Ru covering contact interfaces.

Fig. 5: SEM photos of the poly silicon resoswitch with Pt-silicide covering the entire structure.

Polycide devices were achieved by first fabricating one-mask polysilicon device, coating them with 25nm of Pt via atomic layer deposition (ALD), then RTA annealing for 3 mins. to form platinum polycide. The polycide forms only over polysilicon, allowing simple removal of unreacted Pt over oxide mesa sidewalls via liquid regia. The end result is a polycided resoswitch that actually sports a lower contact resistance than the Ru-coated polysilicon devices of [1], since the polycide layer actually ends up being thicker than [1]'s 2-5nm Ru coating. Fig. 5 presents SEMs of the polycide device. Here, somewhat non-uniform silicidation [4] over the polysilicon surface does increase the contact resistance over what could have been, but still delivers sufficient conductivity for charge pumping of MEMS devices.

MEASUREMENT RESULTS

Fig. 6(a) and (b) present voltage output waveforms for the fabricated Ni/Ru and PtSi resoswitches, respectively, when driven hard enough to impact across 1000nm switch gaps. Clearly, the softer Ni/Ru contact interface induces considerable squegging, where periods of no contact are clearly visible. On the other hand, the polycide device displays much less squegging and appears to make contact at all times, although some impacts are less forceful than others, so achieve larger contact resistances, hence, slightly lower output voltages. Using (1) with the parameters summarized in Fig. 6, the contact resistances are found to be 49Ω and 870Ω for the Ni/Ru and PtSi devices, respectively. These are larger than desired for high power converters, but are comfortably sufficient for low power, high voltage charge pump applications, like MEMS dc biasing.

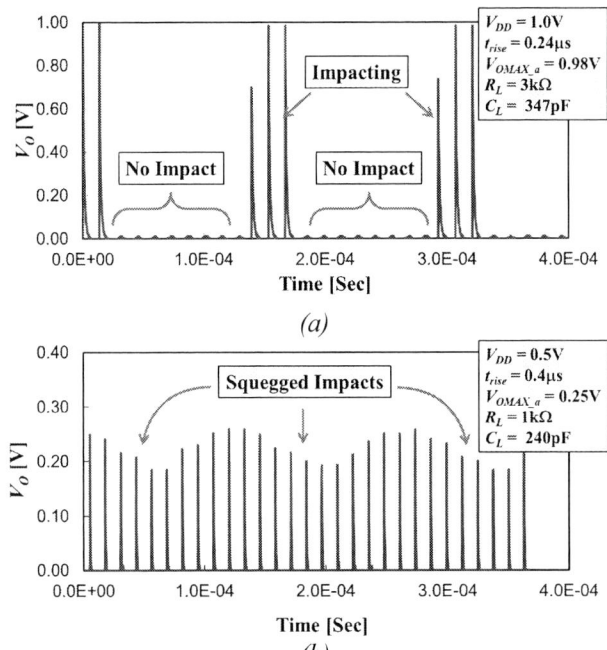

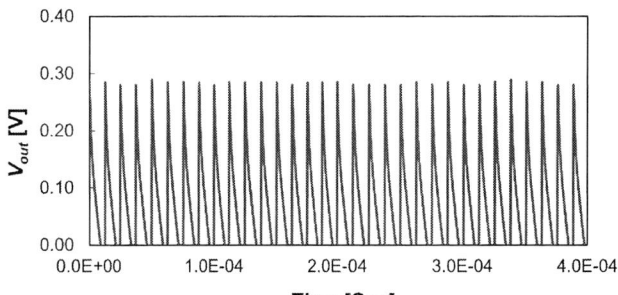

Fig. 7: Output waveform of the PtSi resoswitch with a 500nm lateral switch gap showing much less squegging induced amplitude fluctuation.

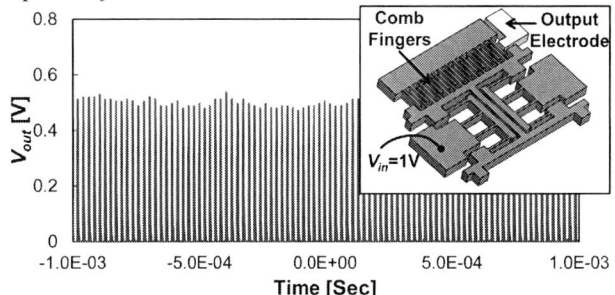

(a)

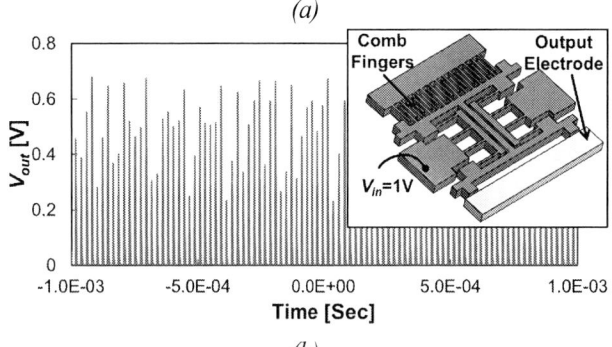

(b)

Fig. 6: Squegged output waveforms of the (a) Ni/Ru and (b) PtSi resoswitches tested using the circuit of Figure 1. Here, both resoswitches had switch axis gaps of 1000nm.

Fig. 7 presents the measured output waveform from a resoswitch identical to that of Fig. 1(a), but with a smaller switch gap of only 500nm. Here, the smaller gapped device exhibits much less squegging, in agreement with simulation. Fig. 8 further shows that placement of actuating comb fingers on the same side as output electrode suffers less squegging than configuring them on opposite sides.

CONCLUSIONS

The reduction of squegging demonstrated here not only solves an important issue with resoswitches that previously constrained the operation frequency of charge pumps using them, but also identifies polycides or silicides as compelling contact interfaces. Although the use of a polycide contact yielded a rather high contact resistance for the low frequency switches demonstrated here, higher frequency disk-based resoswitches, such as demonstrated in [2], should exhibit much smaller resistances, since their impact force is much larger. With high frequency, no squegging, and potentially good contact resistance, disk resoswitches using polycide material might just fit the bill for their targeted power amplifier and converter applications.

Acknowledgement: This work was funded by DARPA under the NEMS Program.

Fig. 8: Output waveforms of the PtSi resoswitch with output electrode placed on (a) the same side and (b) opposite side of the actuating comb fingers.

("Resoswitch")," in *Technical Digest, 2008 Solid-State Sensor, Actuator, and Microsystems Workshop*, Hiltion Head, SC, Jun. 2008, pp. 40-43.

[3] C. Budd and F. Dux, "Chattering and related behaviour in impact oscillators," *Philosophical Transactions of the Royal Society of London. Series A: Physical And Engieering Sciences*, vol. 347, no. 1683, pp. 365-389, May 1994.

[4] P. Shrestha, *et al.*, "Investigation of Volmer-Weber growth during the nucleation phase of ALD platinum thin films and template based platinum nanotubes," *ECS Transactions*, vol. 33, no. 2, pp. 127-134, 2010.

REFERENCES

[1] Y. Lin, *et al.*, "A micromechanical resonant charge pump," in *Proc., Transduers 2013*, Barcelona, Spain, Jun. 2013, pp. 1727-1730.

[2] Y. Lin, *et al.*, "The micromechanical resonant switch

CONTACT

*Yang Lin, tel: +1-510-5173931; linyang@berkeley.edu

STABLE CHARGE-BIASED CAPACITIVE RESONATORS WITH ENCAPSULATED SWITCHES

Eldwin J. Ng, Kimberly L. Harrison, Camille L. Everhart, Vu A. Hong, Yushi Yang,
Chae Hyuck Ahn, David B. Heinz, Roger T. Howe and Thomas W. Kenny
Stanford University, Stanford, CA, USA

ABSTRACT

A large dc bias voltage (tens of volts) is often useful for the operation of capacitive MEMS devices. Charge-biasing techniques have been demonstrated to be able to replace a bias voltage source by trapping an equivalent charge on an electrically floating electrode. This work presents a charge-biased resonator that uses an electrostatically actuated mechanical switch with a pull-in voltage of 36 V to introduce a voltage-equivalent charge of 10 – 15V onto a resonant body. The switch and resonator are hermetically sealed in a clean vacuum cavity, within an epitaxial polysilicon encapsulation process (*epi-seal*). No charge leakage has been observed, even at an elevated temperature of 125°C for weeks.

INTRODUCTION

Many capacitively transduced MEMS devices are operated with a dc bias voltage to boost transduction. The bias voltage for typical resonators is usually in the range of tens of volts, and even higher voltages (hundreds of volts) are not uncommon, especially when large transduction gap sizes (> 1 μm) are involved [1]. To generate such high voltages in modern day CMOS requires charge pump circuitry and a substantial amount of power. The stability of the generated bias voltage is also of concern for ppb-stability MEMS resonators, as the frequency stability could be affected through the electrical softening effect.

Charge-biasing techniques have previously been demonstrated [2-5] and show great potential for low power timing references. By breaking off the electrical connection to a voltage-biased resonant body, a charge remains on the resonant body and performs the same function as a bias voltage. Once the charge has been applied, the bias voltage source is ideally no longer required, obviating the need for charge pumping circuits and hence conserving power. However, one caveat that has plagued the concept of charge biasing is that the charge can leak through parasitic resistances. As such, it is desired to increase the charge leakage time constant $\tau = RC$, and some demonstrated methods include: 1) increasing the stored charge by inserting a large capacitor [2]; 2) increasing the leakage resistance by controlling the environment, e.g. decreasing the temperature or humidity [6]; 3) trapping charge at the surface with an oxide-nitride layer [3, 5] or within an electrically floating electrode surrounded by oxide [4]. Of these methods, charge trapping is seen to work the most reliably, with 19 months of no observed charge leakage on an electrically floating electrode surrounded by a thick 3 μm layer of oxide for a capacitive micro-machined ultrasonic

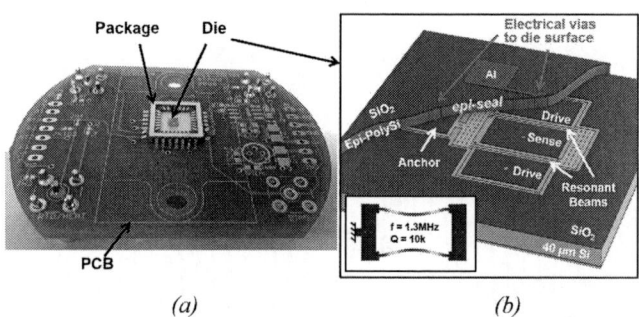

Figure 1. (a) Printed circuit board with a gold package and an epi-sealed resonator die. (b) Cutaway schematic of an epi-seal encapsulated DETF resonator.

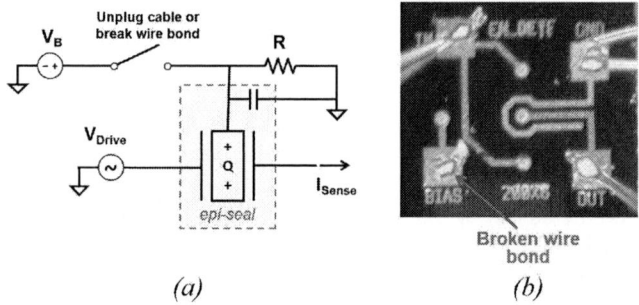

Figure 2. (a) Circuit schematic for charge-biasing the resonator. The bias voltage supply is disconnected from the resonator, leaving a charge on the resonant body. The charge can then leak out through parasitic resistances. (b) A broken bias voltage wire bond to the die.

transducer in [4].

With the *epi-seal* encapsulation process demonstrating excellent long term stability for resonators [7], it is desired to investigate if the stability of such charge-biasing techniques extend to *epi-seal*ed devices with silicon surfaces free of native oxide and similarly anchored on a thick buried oxide layer.

CHARGE LEAKAGE

To investigate charge leakage from the resonant body, initial experiments were performed involving two different methods of electrically isolating the resonant body. A well-characterized double-ended tuning fork (DETF) resonator [8] fabricated within the *epi-seal* process was first mounted on a gold package, which was then soldered to a printed circuit board (PCB) as shown in Fig. 1. Frequency sweeps were performed on the device using an Agilent 8753ES network analyzer. A bias voltage was applied to the resonant body, and thereafter disconnected abruptly to leave

978-1-4799-3510-9/14 $31.00 © 2014 IEEE

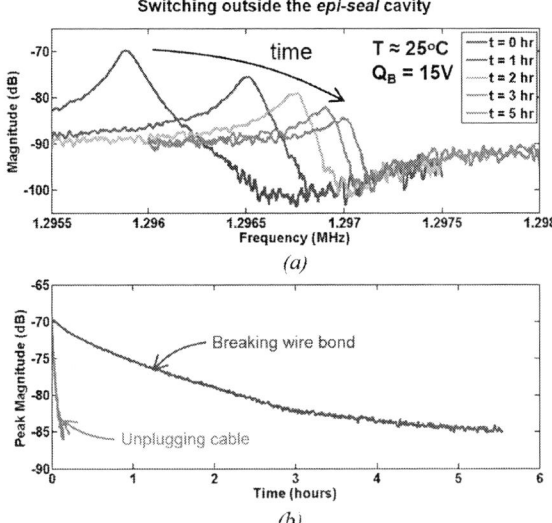

Figure 3. (a) Resonant peak decay when charge-biased by breaking the bias voltage wire bond to the die. (b) The transient peak decay after breaking the wire bond and unplugging the cable from the PCB.

a charge on the resonator. The circuit diagram is shown in Fig. 2a and the bias voltage can be disconnected by either unplugging the bias voltage cable to the PCB, or by breaking the wire bond (Fig. 2b).

Measured results at room conditions show that charge leakage ensues once the bias voltage is disconnected, as observable from the resonant peak decay (Fig. 3a). Plotting the transient peak magnitudes (Fig. 3b), a much faster leakage is noted for the case of unplugging the bias voltage cable, compared to breaking the wire bond directly off the die. This could be due to parasitic leakage paths present on the PCB, package, and connectors. To achieve stable resonators, the question then becomes: Can the parasitic resistances be removed almost entirely such that no noticeable leakage occurs?

EPI-SEALED CHARGES

Flash memory today is an example of extreme charge retention. Fowler-Nordheim tunneling is used to introduce tens of electrons onto a floating gate electrode sandwiched between dielectric layers, and these electrons are reliably trapped for years.

For reliable charge-biasing of capacitive MEMS devices, there similarly needs to be: 1) an environment through which charges cannot dissipate, and 2) a method for introducing charges with zero leakage through the charge-introduction mechanism. It is noted that long term charge trapping is achieved in [4] by having the electrically floating electrode entirely covered with a thick oxide (3 µm) and using an extremely high electric field (a few MV/cm) to inject charge into the floating electrode.

In this work, a similar approach is taken, but instead of having the electrically floating electrode entirely covered with a thick oxide layer, only the resonator anchor sits on the thick buried oxide layer; elsewhere, the silicon surfaces

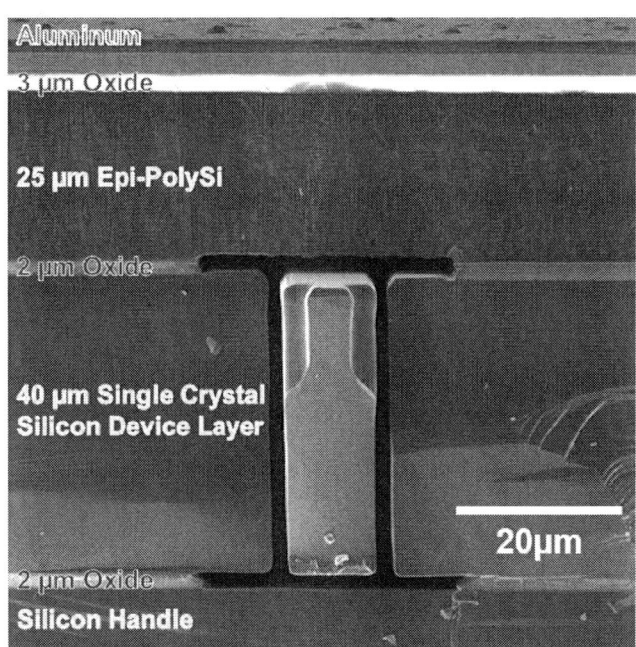

Figure 4. Cross-section SEM of an epi-seal encapsulated representative beam.

are free of native oxide and within the *epi-seal* cavity, which provides the closest-to-ideal packaging environment for MEMS demonstrated thus far, with a clean, hermetic vacuum encapsulation [9]. Charges are then introduced with a mechanical contact switch located within the *epi-seal*ed cavity to reduce the charge leakage pathways. This has the added benefit of leveraging the existing *epi-seal* fabrication process without any modifications – in other words, these switches can be included for free.

Fabrication Process

The *epi-seal* process is a high vacuum (sub-Pa), ultra-clean (high temperature seal, native oxide-free), high yield, wafer-scale commercial process. For this fabrication run, a 40 µm-thick, n-doped (antimony doped, ~1e18 cm^{-3}) device layer was used. The cross-section of a beam representing the final encapsulated resonating and switching beams is shown in Fig. 4. The device is surrounded by inert gas in the hermetically *epi-seal* cavity, and the electrically floating anchor of the resonant body sits only on the buried (lower) silicon oxide layer, which at 2 µm-thick is a near-ideal electrical insulator.

Electrical Isolation by Switching within *epi-seal* Cavity

A mechanical contact switch can be realized by means of a clamped-clamped beam amongst many possibilities. In this work, a 300 µm long by 3 µm wide clamped-clamped beam was designed to be electrostatically pulled towards a pair of gate electrodes (Fig. 5a). The gap varies linearly from 0.7 µm at the end to 1.5 µm at the middle (Fig. 5b), designed to help reduce the pull-in voltage but also to prevent collapse of the beam into the gate. At the center, a 6 µm wide contact post, which is connected to the resonant

978-1-4799-3510-9/14 $31.00 © 2014 IEEE

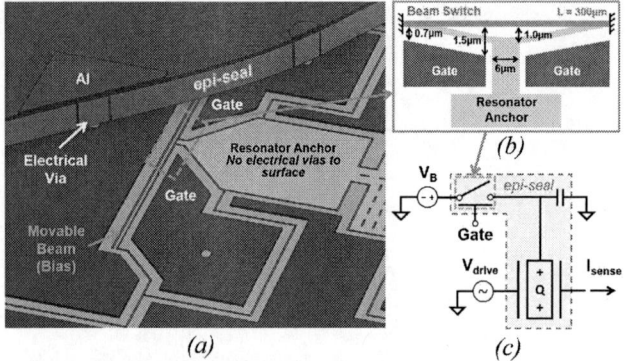

Figure 5. (a) Design of contact switch within the epi-seal cavity. (b) The beam can be pulled in to make contact with the resonator anchor for charge introduction. (c) The switch is now placed within the epi-seal cavity, compared to Fig. 2.

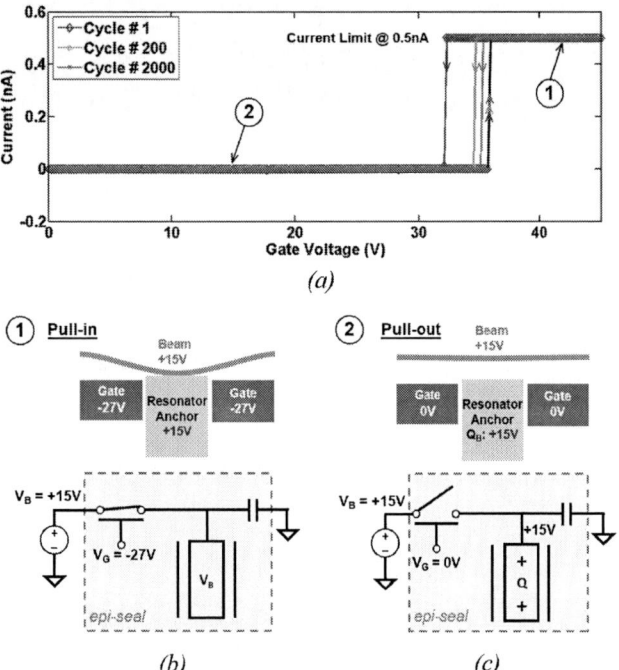

Figure 6. (a) Measured pull-in / pull-out characteristics of a mechanical beam switch. (b) The desired voltage is applied to the beam and the remaining voltage difference required for pull-in is applied to the gate. (c) The gate voltage is then zeroed, allowing the beam to pull out. The bias and gate voltages can subsequently be released.

body's anchor, protrudes out by 0.5 μm to physically contact the center of the pulled in beam and prevent the beam from collapsing into the gate. The circuit diagram is shown in Fig 5c, and it is worth emphasizing that the switch is placed within the *epi-seal* cavity. The resonant body's anchor, unlike that in the previous test (Fig. 1), has no electrical via and is hence isolated from the die surface. Charges are introduced to the resonant body when the beam is pulled in, and once the beam has pulled out, there is effectively no electrical pathway for leakage.

In an effort to first experimentally characterize the pull-

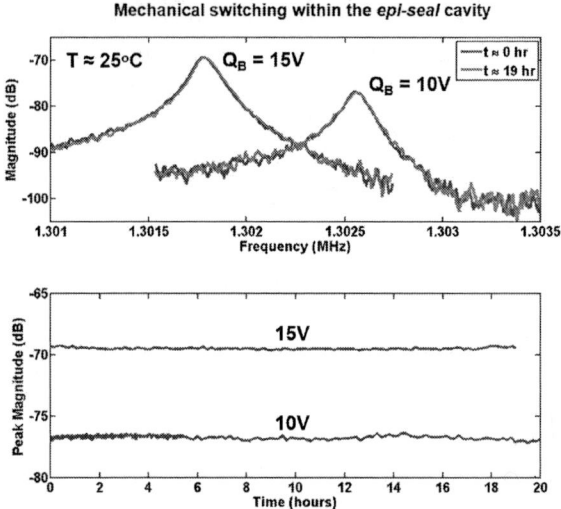

Figure 7. Voltage-equivalent charges can be applied with no noticeable leakage. The charge can be reprogrammed by repeating the same pull-in / pull-out technique.

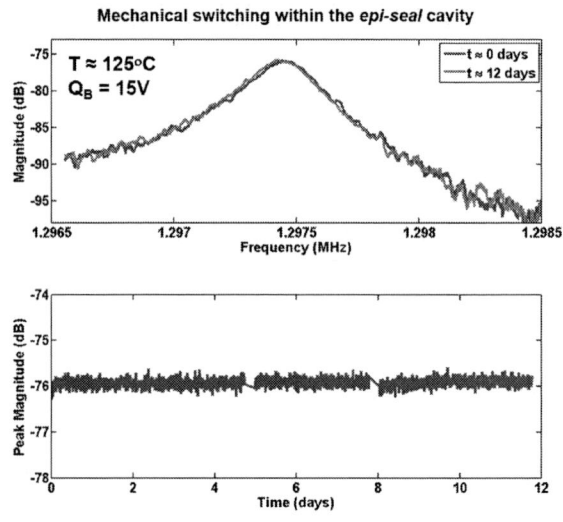

Figure 8. No charge leakage was observed, even at elevated temperatures of 125°C.

in switch mechanism to the isolated resonant body's anchor, an identical switch mechanism was designed, but this time with an electrical connection to the contact post such that the beam contact can be sensed via a resistance change between the beam and the contact post. Pull-in and pull-out characteristics are seen (Fig. 6a), with the current limit set to 0.5 nA for this characterization to prevent burning out the switch. The pull-in voltage is seen to be about 36 V and the pull-out voltage ranges from 32 to 35 V. It is also seen that thousands of pull-in/pull-out cycles can be performed, allowing for the introduced charge to be reprogrammed multiple times.

When charging an actual resonator, it is difficult to detect pull-in as no electrical vias are present on the resonant body's anchor. Thus, it is assumed that the pull-in voltage remains within the same range as the characterized beam, and to be certain that the beam is pulled in, a slightly higher voltage difference (in this case 42 V, compared to the measured 36 V) is assumed to be required for pull-in. The desired voltage for the resonator (e.g. +15V) is applied to the beam, and the remaining voltage difference for pull-in (e.g. -27 V) is applied to the gate (Fig. 6b). The surrounding field potential is held at ground.

Reducing the gate voltage to ground (Fig. 6c) then causes the beam to pull-out, breaking the electrical contact with the resonant body while biased at the desired voltage (e.g. +15V). The applied voltages on the beam and gate can then be removed. The electrically floating resonant body is now anchored on the buried oxide layer and is entirely within the *epi-seal* cavity with no electrical connection to the outside of the cavity.

Fig. 7 shows the frequency sweeps of a resonator that has been charged with bias voltages of 10 V then reprogrammed to 15 V, with a difference in the peak magnitude as expected. No leakage is noticed even after 19 hours at room conditions. To further confirm the stability, the device was heated to 125°C and held for 12 days – again no charge leakage was noticed (Fig. 8), attesting to the inert environment of the *epi-seal* cavity.

CONCLUSION

Demonstrated in this work is a resonator that is charge-biased by mechanical contact switching within an *epi-seal* cavity – a well-controlled inert environment. A voltage-equivalent charge can be programmed multiple times as desired with no charge loss detected, even at an elevated temperature of 125°C. This may obviate the need for circuitry to maintain a stable dc bias voltage and reduce the required power for capacitive MEMS resonators. Further long-term stability testing is underway, but the experimental confirmation thus far of the absence of charge leakage pathways also provides further evidence for the ultra-clean nature of the *epi-seal* environment.

ACKNOWLEDGEMENTS

This work was supported by DARPA grant N66001-12-1-4260, "Precision Navigation and Timing program (PNT)," managed by Dr. Robert Lutwak, and DARPA grant FA8650-13-1-7301, "Mesodynamic Architectures (MESO)," managed by Dr. Jeff Rogers. The fabrication work was performed at the Stanford Nanofabrication Facility (SNF) which is supported by National Science Foundation through the NNIN under Grant ECS-9731293.

REFERENCES

[1] V. Thakar and M. Rais-Zadeh, "Optimization of tether geometry to achieve low anchor loss in Lamé mode resonators," *IEEE International Frequency Control Symposium 2013,* July 2013.

[2] S. S. Li, Y. W. Lin, Y. Xie, Z. Y. Ren, and C. T. C. Nguyen, "Charge-biased vibrating micromechanical resonators," *IEEE Ultrasonics Symposium, 2005,* pp. 1596-1599, 2005.

[3] A. K. Samarao and F. Ayazi, "Self-Polarized Capacitive Silicon Micromechanical Resonators via Charge Trapping," *IEEE International Electron Devices Meeting 2010,* 2010.

[4] M.-C. Ho, M. Kupnik, K. K. Park, and B. T. Khuri-Yakub, "Long-term measurement results of pre-charged CMUTs with zero external bias operation," in *IEEE International Ultrasonics Symposium 2012* 2012, pp. 89-92.

[5] K. K. Park, M. Kupnik, H. J. Lee, O. Oralkan, and B. T. Khuri-Yakub, "Zero-Bias Resonant Sensor with an Oxide-Nitride Layer as Charge Trap," *IEEE Sensors Conference 2010,* pp. 1024-1028, 2010.

[6] R. G. Hennessy, M. M. Shulaker, M. Messana, A. B. Graham, N. Klejwa, J. Provine, *et al.*, "Vacuum Encapsulated Resonators for Humidity Measurement," *Sensors and Actuators B-Chemical,* vol. 185, pp. 575-581, Aug 2013.

[7] B. Kim, R. N. Candler, M. A. Hopcroft, M. Agarwal, W. T. Park, and T. W. Kenny, "Frequency Stability of Wafer-Scale Film Encapsulated Silicon Based MEMS Resonators," *Sensors and Actuators A,* vol. 136, pp. 125-131, May 2007.

[8] M. A. Hopcroft, B. Kim, S. Chandorkar, R. Melamud, M. Agarwal, C. M. Jha, *et al.*, "Using the Temperature Dependence of Resonator Quality Factor as a Thermometer," *Applied Physics Letters,* vol. 91, Jul 2007.

[9] R. N. Candler, M. A. Hopcroft, B. Kim, W. T. Park, R. Melamud, M. Agarwal, *et al.*, "Long-Term and Accelerated Life Testing of a Novel Single-Wafer Vacuum Encapsulation for MEMS Resonators," *Journal of Microelectromechanical Systems,* vol. 15, pp. 1446-1456, Dec 2006.

CONTACT

*E. J. Ng, email: eldwin@mems.stanford.edu

STABLE PULL-IN ELECTRODES FOR NARROW GAP ACTUATION

Eldwin J. Ng[1], Yushi Yang[1], Vu A. Hong[1], Chae Hyuck Ahn[1], David L. Christensen[1],
Brian A. Gibson[2], Kamala R. Qalandar[2], Kimberly L. Turner[2], Thomas W. Kenny[1]
[1]Stanford University, USA
[2]University of California, Santa Barbara, USA

ABSTRACT

This paper reports the use of movable electrodes that can be electrostatically pulled in to achieve narrow gaps beyond lithography / etch capabilities. Width-extensional resonators with frequencies of 50 MHz and quality factors of 150k are demonstrated with such movable electrodes to have a significantly lower motional impedances when pull-in occurs. Sub-ppm stability over 10^5 pull-in/pull-out cycles is measured using temperature-compensated resonators within the *epi-seal* epitaxial polysilicon encapsulation process. The pull-in phenomena is reversible, but can be made permanent by electrically welding the pulled-in electrode to a stop.

INTRODUCTION

The performance of most capacitive MEMS devices has a significant dependence on the dimensions of the capacitive transduction gap. In particular, the motional impedance of a resonator is lowered by having narrower gaps, contributing to a reduction in the phase noise of resonator-based oscillators. By shrinking the gap size, bias voltages can also be reduced considerably.

A number of methods have been presented to create deep submicron gaps in silicon. Deep reactive ion etching (DRIE) forms the basis for most of these methods, creating trench gap widths with aspect ratios of up to about 100:1 [1, 2]. To further decrease the gap size, one could employ a sacrificial oxide / polysilicon refill process (e.g. the High Aspect-Ratio Combined Poly and Single-Crystal Silicon (HARPSS) [3]). However, in the effort to produce a device compatible with the *epi-seal* epitaxial polysilicon encapsulation process (a high vacuum, ultra-clean, high yield, wafer-scale commercial process that has been used for stable resonators) [4, 5], such a sacrificial oxide / polysilicon refill process cannot be used, as the high temperature (>1000°C) bake step in an epitaxial reactor removes the native oxide and the polysilicon surface roughens due to silicon migration [6]. This roughening of the polysilicon surface on the order of hundreds of nanometers precludes the use of the material for submicron narrow gaps within the *epi-seal* process. Another method of reducing the gap dimension is to use an epitaxial gap tuning process [7] that produces smooth monocrystalline silicon on both sides of the gap. However, the selective epitaxial deposition step is dependent on the crystal growth direction, and could be undesirable for curved structures.

Alternatively, instead of defining the gap dimension during the fabrication process, another possibility is to use a movable electrode, where the gaps are defined with regular DRIE, and the electrode-to-device gap can be narrowed

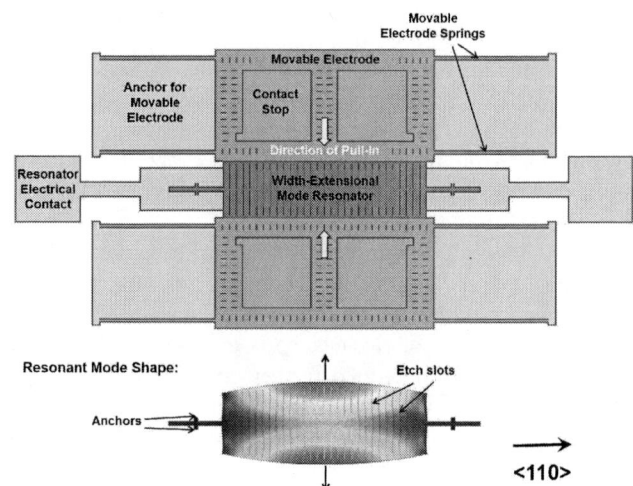

Figure 1: Width-extensional mode resonator with movable electrodes that can be pulled in to a stop.

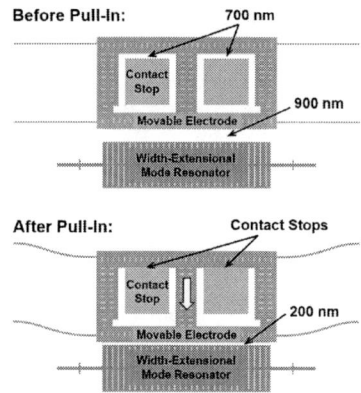

Figure 2: Gap reduction by pulling in a movable electrode.

post-process by electrostatically pulling in the electrode. A significant reduction in the motional impedance of a resonator has thus been demonstrated previously [8], but the stability of such a device is in question especially when repeated pull-in cycles are performed. This work reports on a width-extensional mode resonator with pull-in electrodes that is fabricated in the *epi-seal* process. Sub-ppm stability is demonstrated over extended on-off bias voltage cycle testing.

DEVICE CONCEPT AND DESIGN

To demonstrate the pull-in concept within the *epi-seal* platform, a 50 MHz width-extensional mode resonator [9] is designed with movable electrodes suspended on springs of dimensions 3 μm wide by 160 μm long (Fig. 1).

978-1-4799-3510-9/14 $31.00 © 2014 IEEE

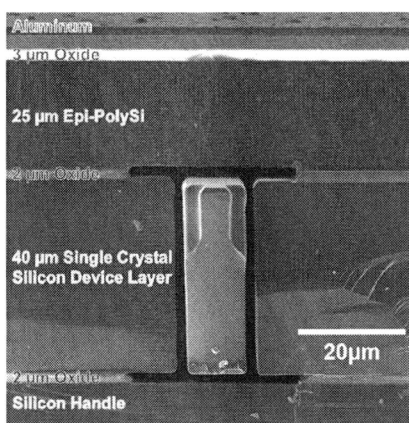

Figure 3: Cross-section SEM of an epi-seal encapsulated beam.

To operate the resonator, a bias voltage is applied to the resonator body to amplify the ac actuation force on the resonator body. In this case, it also serves a second purpose, which is to electrostatically pull in the movable electrodes (Fig. 2). To prevent the movable electrode from physically contacting and shorting to the resonator, a stop is placed to limit further movement of the movable electrode towards the resonator. The designed gap is hence the difference between the resonator-electrode gap and the electrode-stop gap: with a resonator-electrode gap of 900 nm and an electrode-stop gap of 700 nm, the designed gap is 200 nm.

FABRICATION PROCESS

The devices were fabricated in an *epi-seal* fabrication process [4]. *Epi-seal* is a high vacuum (sub-Pa), ultra-clean (high temperature seal, native oxide-free), high yield, wafer-scale process that is similar to the one used commercially at SiTime. For this fabrication run, a 40 μm-thick, highly n-doped (phosphorus doped, 6e19 cm^{-3}) device layer was used. The cross-section of a beam representing a final encapsulated device is shown in Fig. 3. It is useful to note here that although the walls of the trench are smoothed by the hydrogen anneal to the nanometer level, they are not flat, and some residual unevenness is observed. This has a direct impact on the contact, as it means that the contact between two surfaces is likely to be asperity dominated.

RESULTS
Pull-In and Pull-Out

Frequency sweeps were performed using an Agilent 8753ES network analyzer connected to the movable electrodes and a dc bias voltage applied on the resonator. The bias voltage was ramped up in steps, and for each voltage step, a frequency sweep was performed.

The peak magnitude of the frequency response depends on the bias voltage as well as the gap size, with the motional impedance R_m given by [9]:

$$R_m \propto \frac{g^4}{V_{Bias}^2} \qquad (1)$$

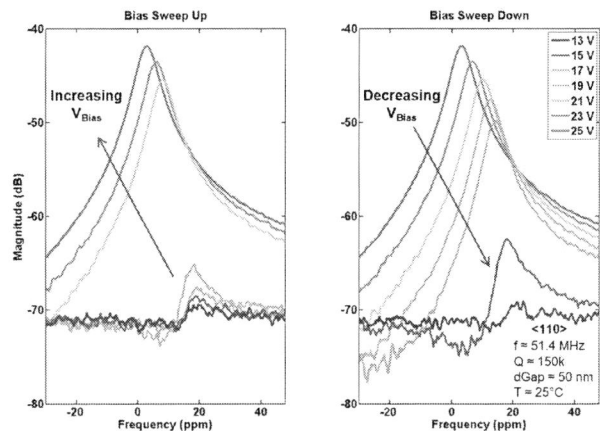

Figure 4: Frequency response of a resonator with pull-in electrodes at various bias voltages.

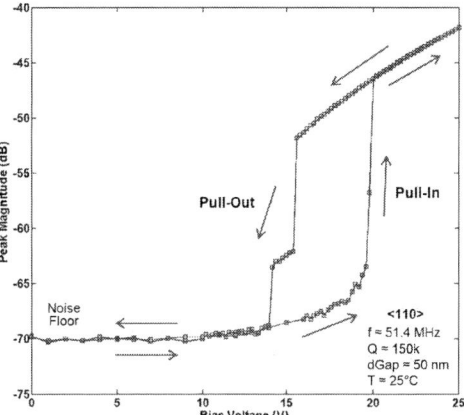

Figure 5: Peak magnitude of the resonator with pull-in electrodes as a function of the bias voltage. Pull-in / pull-out characteristics are observed.

where g is the gap size and V_{Bias} is the bias voltage. With a decrease in gap or an increase in the bias voltage, a motional impedance decreases, or equivalently, the peak magnitude increases. Frequency response plots for a width-extensional resonator with a designed gap of 200 nm are shown for several bias voltages in Fig. 4, while in Fig. 5, the peak value of the frequency response is shown as a function of the bias voltage. A sudden increase in the peak magnitude is seen at the pull-in voltage, corresponding to a sharp decrease in gap size. Further increasing the bias voltage causes the peak magnitude to increase even further as expected from (1). On the other hand, when the bias voltage is swept downwards after pull-in has occurred, a hysteresis curve is observed due to the pull-out voltage being lower than the pull-in voltage.

In addition, a double step is seen in the pull-in and pull-out curves (Fig. 5), because although the two electrodes on either side were designed to have identical pull-in voltages, the actual fabricated parts are not exactly identical, and the observed steps are due to one of the electrodes pulling in / out before the other.

978-1-4799-3510-9/14 $31.00 © 2014 IEEE

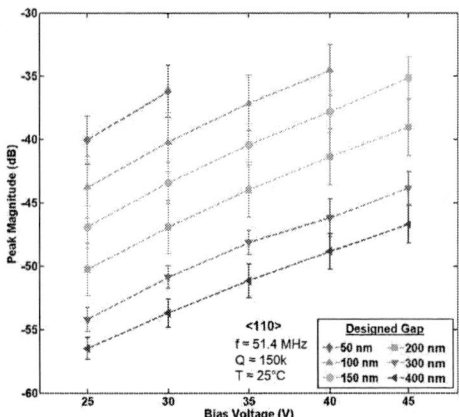

Figure 6: Peak magnitude as a function of bias voltage for various designed gaps.

Various Designed Gap Sizes

Devices with different designed gap sizes were tested, and Fig. 6 shows the variation of the peak magnitude with bias voltage for several designed gaps. About 5 devices of each gap size from around the wafer were tested, and the spread is reflected in the error bars. These devices were tested to the maximum bias voltage until failure (shorting of electrode to resonator), and it is observed that there is a region at the top of the plot beyond which a larger bias voltage is unable to achieve a higher peak magnitude (or a lower motional impedance) without device failure. It is also noted that the peaks are relatively lower than reported in [9]. This could be due to the non-uniform shape of the extensional mode, or the uneven trench profile, causing the average actuation gap to be larger than the designed gap.

Stability

A resonator's frequency stability is crucial for timekeeping applications, and the pull-in electrodes could adversely affect the stability. To investigate the stability at the sub-ppm level, it is important to first remove the effect of temperature fluctuations [10]. Heavily n-type doped silicon is known to change the frequency-temperature dependency [11], and frequency-temperature curves for width-extensional resonators aligned with <110> and <100> direction were measured for a doping level of 6e19 cm^{-3} (Fig. 7). Particularly useful is the frequency-temperature turnover point at 105°C for the <100> orientation, as operating at such a point minimizes the effect of temperature variation on frequency. <100> width-extensional resonators were thus chosen to investigate the stability of the pull-in electrodes. These resonators were placed in a Thermotron S1.2C temperature chamber and maintained at the turnover point of 105°C. Repeated bias voltage on-off cycles (from 0 to 25 V) were performed. The frequency stability and peak magnitude during the cycling is shown in Fig. 8. The first device was operated at an ac drive power of -15 dBm, and the same experiment was repeated on a second device with a higher ac drive power of -5 dBm to reduce noise. Small changes in the peak magnitude over

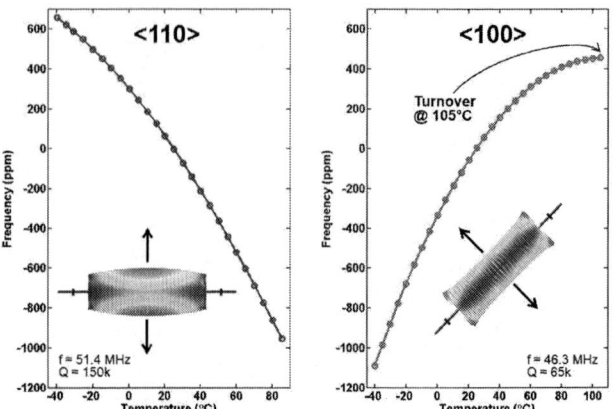

Figure 7: Temperature dependence of the resonant frequency for width-extensional resonators aligned to the <110> and <100> directions for 6e19cm^{-3} phosphorus-doped monocrystalline silicon.

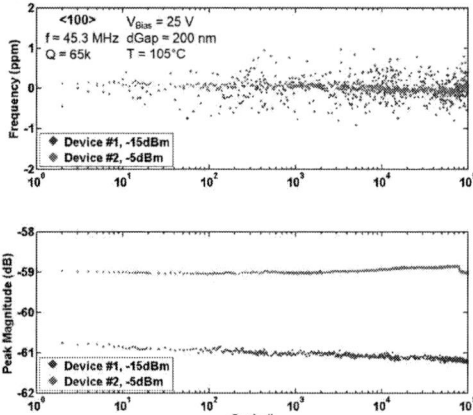

Figure 8: Frequency and peak magnitude stability of temperature-compensated width-extensional resonators with pull-in electrodes for 10^5 bias voltage on-off cycles.

10^5 cycles were observed for both devices, but the frequency remains largely stable at well below ±1 ppm.

Frequency sweeps were also performed periodically during the bias voltage cycling to investigate the contact and the variation of the pull-in and pull-out voltages (Fig. 9). Some variation in the pull-out voltage is observed (between 16 and 19 V), but the pull-in voltage remains stable, varying by no more than 0.2 V (the voltage sweep step size) throughout the 10^5 cycles.

Welded Electrodes

Furthermore, the electrodes can be welded to the stop to eliminate the pull-in / pull-out effect entirely. When the electrode is pulled into in contact with the stop, a current (of about 10mA) can be used to heat and permanently fuse the contacting silicon surfaces together, resulting in permanently pulled-in electrodes that behave like regularly anchored electrodes. Fig. 10 shows a peak magnitude vs. bias voltage plot of a device with welded pull-in electrodes.

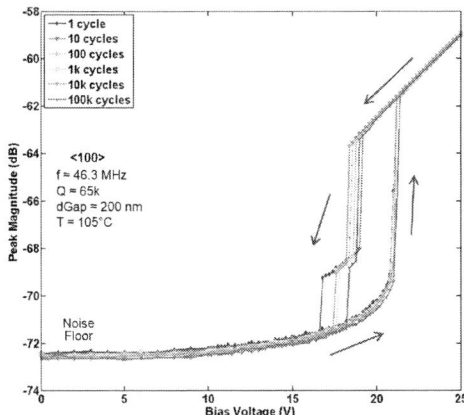

Figure 9: Peak magnitude as a function of the bias voltage as measured over 10^5 bias voltage on-off cycles.

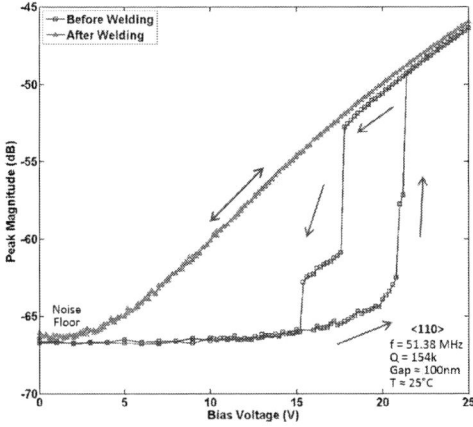

Figure 10: Comparison of a resonator before and after permanently welding the pull-in electrodes to the stop.

CONCLUSION

Demonstrated in this paper is a technique for achieving narrower gaps than lithography / etch capabilities allow within the *epi-seal* process. Lower motional impedances are observed for resonators when the electrodes are pulled in, and sub-ppm stability over 10^5 pull-in / pull-out cycles is achieved. In addition, the movable electrodes can be electrically welded to a stop to eliminate the pull-in / pull-out characteristics.

ACKNOWLEDGEMENTS

This work was supported by DARPA grant N66001-12-1-4260, "Precision Navigation and Timing program (PNT)," managed by Dr. Robert Lutwak, and DARPA grant FA8650-13-1-7301, "Mesodynamic Architectures (MESO)," managed by Dr. Jeff Rogers. The fabrication work was performed at the Stanford Nanofabrication Facility (SNF) which is supported by National Science Foundation through the NNIN under Grant ECS-9731293.

REFERENCES

[1] F. Marty, L. Rousseau, B. Saadany, B. Mercier, O. Francais, Y. Mita, *et al.*, "Advanced etching of silicon based on deep reactive ion etching for silicon high aspect ratio microstructures and three-dimensional micro- and nanostructures," *Microelectronics Journal,* vol. 36, pp. 673-677, Jul 2005.

[2] K. J. Owen, B. VanDerElzen, R. L. Peterson, and K. Najafi, "High Aspect Ratio Deep Silicon Etching," *IEEE International Conference on Micro Electro Mechanical Systems (MEMS) 2012,* 2012.

[3] F. Ayazi and K. Najafi, "High aspect-ratio combined poly and single-crystal silicon (HARPSS) MEMS technology," *Journal of Microelectromechanical Systems,* vol. 9, pp. 288-294, Sep 2000.

[4] R. N. Candler, M. A. Hopcroft, B. Kim, W. T. Park, R. Melamud, M. Agarwal, *et al.*, "Long-Term and Accelerated Life Testing of a Novel Single-Wafer Vacuum Encapsulation for MEMS Resonators," *Journal of Microelectromechanical Systems,* vol. 15, pp. 1446-1456, Dec 2006.

[5] A. Partridge and M. Lutz, "Episeal pressure sensor and method for making an episeal pressure sensor," U. S. Patent 6928879, 2005.

[6] E. J. Ng, S. Wang, D. Buchman, C.-F. Chiang, T. W. Kenny, H. Muenzel, *et al.*, "Ultra-Stable Epitaxial Polysilicon Resonators," *Solid-State Sensors, Actuators, and Microsystems Workshop, Hilton Head 2012,* pp. 271-274, 2012.

[7] E. J. Ng, C. F. Chiang, Y. Yang, V. A. Hong, C. H. Ahn, and T. W. Kenny, "Ultra-high aspect ratio trenches in single crystal silicon with epitaxial gap tuning," *Solid-State Sensors, Actuators and Microsystems Conference (Transducers) 2013,* pp. 182-185, Jun 2013.

[8] G. Dimitri, K. Andreas, B. Lionel, L. Bernard, C. Dominique, and C. Chantal, "Design, realization and testing of micro-mechanical resonators in thick-film silicon technology with postprocess electrode-to-resonator gap reduction," *Journal of Micromechanics and Microengineering,* vol. 13, p. 134, 2003.

[9] S. Pourkamali, G. K. Ho, and F. Ayazi, "Low-impedance VHF and UHF capacitive silicon bulk acoustic wave resonators," *IEEE Transactions on Electron Devices,* vol. 54, pp. 2017-2030, Aug 2007.

[10] E. J. Ng, H. K. Lee, C. H. Ahn, R. Melamud, and T. W. Kenny, "Stability of Silicon Microelectromechanical Systems Resonant Thermometers," *IEEE Sensors Journal,* vol. 13, Mar 2013.

[11] M. Shahmohammadi, B. P. Harrington, and R. Abdolvand, "Turnover Temperature Point in Extensional-Mode Highly Doped Silicon Microresonators," *IEEE Transactions on Electron Devices,* vol. 60, pp. 1213-1220, Mar 2013.

CONTACT

*E.J. Ng, email: eldwin@mems.stanford.edu

ULTRA-STABLE NONLINEAR THIN-FILM PIEZOELECTRIC-ON-SUBSTRATE OSCILLATORS OPERATING AT BIFURCATION

H. Fatemi, M. Shahmohammadi, and R. Abdolvand
Oklahoma State University, Tulsa, Oklahoma, USA

ABSTRACT

Presented is a ~27MHz oscillator incorporating a two-port thin-film piezoelectric-on-silicon resonator with a phase noise (PN) of -139 dBc/Hz at 1kHz and -157 dBc/Hz at 1MHz from the carrier. The close-to-carrier PN is equivalent to -148 dBc/Hz when normalized to 10MHz and is the lowest reported to date for MEMS oscillators [1]. The measured RMS jitter for this oscillator is $290psec$ from 12kHz to 5MHz offset from the carrier. In addition, it is experimentally proven that the oscillator PN settles to a minimum value once the resonator is driven at the bifurcation point in the closed-loop oscillator circuit.

INTRODUCTION

Reference oscillators are demanded in almost every electronic system for various applications. Some of these applications such as high speed data communication have very tight phase noise (PN) requirements which cannot be achieved using passive electrical components due to the low quality factor (Q) of the on-chip inductors [2]. Contrarily, quartz crystal resonators exhibit very high Q and have been utilized in oscillators with very low PN values and excellent temperature stability for high performance applications [3, and for a brief history refer to 4]. The drawback however is their relatively large size and the fact that their fabrication process is incompatible with the mature silicon micro-fabrication technology. MEMS resonators, on the other hand can potentially be integrated with CMOS and offer a competing alternative for the bulky quartz resonators in a wide variety of consumer electronics where size and cost are critical factors. In recent years, both piezoelectric and capacitive MEMS oscillators have been reported with excellent PN performance [5, 6].

The far-from-carrier phase noise in an oscillator is inversely proportional to the carrier power [7]. Thus, in applications with stringent requirement on stability, it is desired to operate the oscillator at high power levels or in other words, at large ac voltage amplitudes. However, the nonlinear terms in the resonator's equation of motion gradually become dominant and shift the resonance frequency at large applied powers (known as amplitude-frequency dependency). Further increase in the applied power will push the resonator to the bifurcation point at which the transmission plot exhibits hysteresis [8]. Previously, it was presumed that at or beyond the bifurcation, the oscillator becomes unstable and its performance degrades. Recently, stable oscillators have been reported that are believed to drive the resonator into the nonlinear operation region and preliminary results show that the phase noise continues to improve even in this range [9, 10].

Piezoelectrically-transduced resonators offer larger electromechanical coupling coefficient relative to that of electrostatic resonators, which results in achieving lower motional resistance [11, 12]. Furthermore, the need for large polarization voltages is eliminated due to the self-generating property of piezoelectric transduction. Thin-film piezoelectric-on-silicon (TPoS) resonators are a subcategory of piezoelectric resonators with superior power handling and high quality factors [13]. This makes the TPoS technology a suitable candidate for implementing very stable oscillators operating at high vibration amplitudes.

The oscillator of this work comprises a two-port ~27MHz TPoS resonator with high quality factor in air and a commercially available wide bandwidth amplifier in an inverting configuration. The oscillation power is simply increased by incrementing the supply voltage of the circuit. It will be shown that the oscillator is most stable (exhibits lowest PN) when the resonator is operated at the bifurcation point. The reported close-to-carrier PN (@1kHz offset) is the lowest value reported to date in literature for MEMS oscillators.

Among works on nonlinear MEMS oscillators, a few have experimentally confirmed the nonlinear operation of the resonator in closed-loop [9, 14, 15]. To examine the nonlinear operation in closed-loop, the current through and the voltage across the port(s) of the device have to be measured simultaneously to collect the phase data at each operating frequency. Next, the admittance curve versus the oscillation frequency is generated. While previous works show the nonlinear operation of a one-port resonator in closed-loop, the effect of operation at or beyond bifurcation on the phase noise is not explicitly discussed. In this work, a simple practical method is presented to verify the operation at bifurcation in closed-loop and the corresponding effect on the PN is highlighted.

RESONATOR CHARACTERIZATION AND OSCILLATOR DESIGN

The S-parameters were measured using an Agilent E8358A PNA network analyzer, a pair of GSG probes, and an RF amplifier (Mini-Circuits® ZFL-2500VH). Prior to the measurements, short-open-load-thru (SOLT) calibration was carried out on a reference substrate to calibrate the measurement setup including the amplifier. All the measurements are performed at atmospheric pressure and ambient temperature.

978-1-4799-3510-9/14 $31.00 © 2014 IEEE

The TPoS resonator of this work is a 5th-order harmonic two-port design with unloaded Q >10,000 in air and a motional resistance of ~200 at 27.7MHz. This resonator exhibits frequency softening as the applied power is increased. The transmission (S21) for different input powers is plotted in Figure 1.

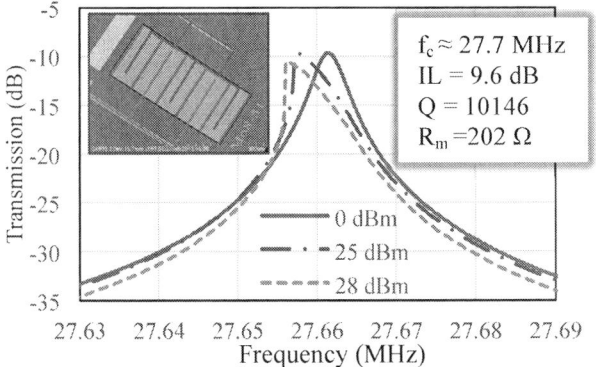

Figure 1: Frequency response of the TPoS resonator of this work at different applied powers. Inset: SEM of the resonator signifying the ports in colors.

To study the effect of self-heating on the nonlinear behavior of the resonator, the temperature increase in the resonator was estimated at bifurcation. First, the dissipated power (P_d) in the resonator is determined at bifurcation to be 76mW [16]. Neglecting the convection and radiation, the only source of heat loss in the body of the resonator is through the four tethers holding the device in place. Thus, the temperature increase in the resonator can be calculated using equation (1) where R_{th} is the thermal resistance of one tether. R_{th} is approximated by equation (2) where l, w, and t are the length, width and the thickness respectively and k is the material thermal conductivity.

$$\Delta T = P_d * R_{th} * \frac{1}{4} \qquad (1)$$

$$R_{th} = \frac{l}{w \cdot (t_{AlN} k_{AlN} + t_{Si} k_{Si})} \qquad (2)$$

After substituting the values, the maximum temperature increase at bifurcation is slightly larger than 2°C which would be equivalent to less than $-60ppm$ shift in the resonance frequency. Considering that the calculated temperature change is an overestimation when ignoring the convection, the measured $-105ppm$ frequency shift at bifurcation indicates that the self-heating is only partially responsible for the frequency shift at high powers and the elastic nonlinearities are the primary contributor.

The oscillator as pictured in Figure 2 incorporates a wide bandwidth amplifier (TI THS4211) in the inverting configuration. This amplifier can operate with supply voltages up to 15V (±7.5V). The loop gain is set to 10 by R_2 and R_3 and the resonator is placed in the feedback loop through its two ports. The three-fingered port (marked in red on the inset of Figure 1) is connected to the oscillator

output and the two-fingered port (marked in blue) is connected to the input of the oscillator while the bottom metal plate is connected to the circuit ground.

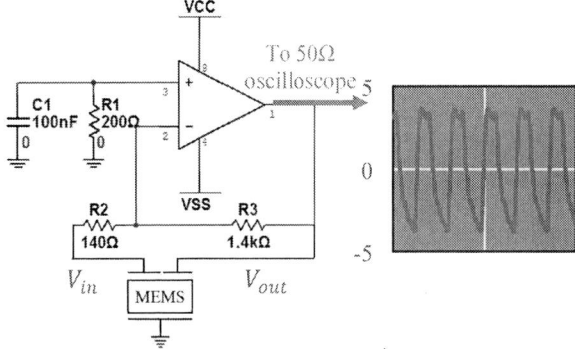

Figure 2: Schematic of the oscillator circuit. The inset on the right shows the output waveform with ±5.5 supply voltage.

OSCILLATOR PHASE NOISE

In order to drive the resonator into the nonlinear regime the supply voltage of the circuit and hence the oscillation voltage amplitude (V_{out}) are increased gradually (Figure 3). Due to the asymmetry of the resonators' input and output ports and the impedances connected to them, this 5th-order resonator acts as a voltage step-up transformer and thus, the voltage on the two-fingered port (V_{in}) is larger [17]. The oscillation frequency reduces as expected considering the observed frequency softening in the frequency response of the resonator (Figure 1). Both the oscillation frequency and PN are plotted versus the oscillation amplitude in Figure 4.

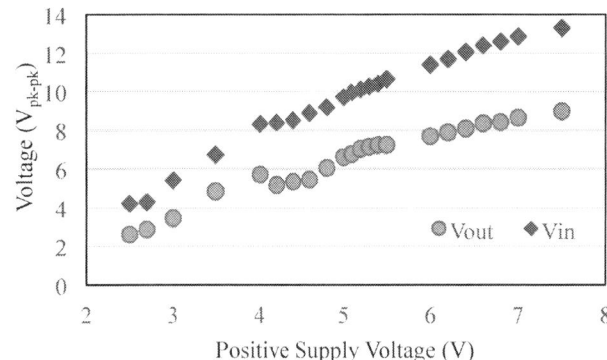

Figure 3: Voltages on the input and output ports of the resonator in closed-loop versus the positive supply voltage, showing a steady increase as the supply voltage is increased.

The impaired PN below $6V_{pk-pk}$ ($|V_{supply}| < 5V$) is believed to be due to the existence of subharmonic oscillation of the order of $1/2$, which is the result of unsymmetrical nonlinearities of the system [18]. The subharmonic oscillation, disappears when the V_{supply} is increased beyond $\pm 5V$. After this point, the close- and far-from-carrier PN values improve until they reach their lowest values when the oscillation voltage is around $7.2V_{pk-pk}$. PN is plotted in Figure 5 for offset frequencies from 10Hz to 5MHz for $|V_{supply}|$ greater than $4V$. The increase in the PN at small offsets from the carrier (@10Hz) for oscillation amplitudes larger than ~7V_{pk-pk} is not well understood at this point and requires further investigation. The authors speculate that it is either the result of external factors [15] with the most dominant one being the temperature fluctuations since the uncompensated TPoS resonators such as the one used in this paper have relatively large temperature coefficient of frequency [19], or it is due to the resonator nonlinearities as is discussed for quartz resonators [20]. The lowest measured PN plot is illustrated in Figure 6 along with the equivalent PN normalized to 10MHz. RMS jitter was measured from 12kHz to 5MHz to be 290 $fsec$.

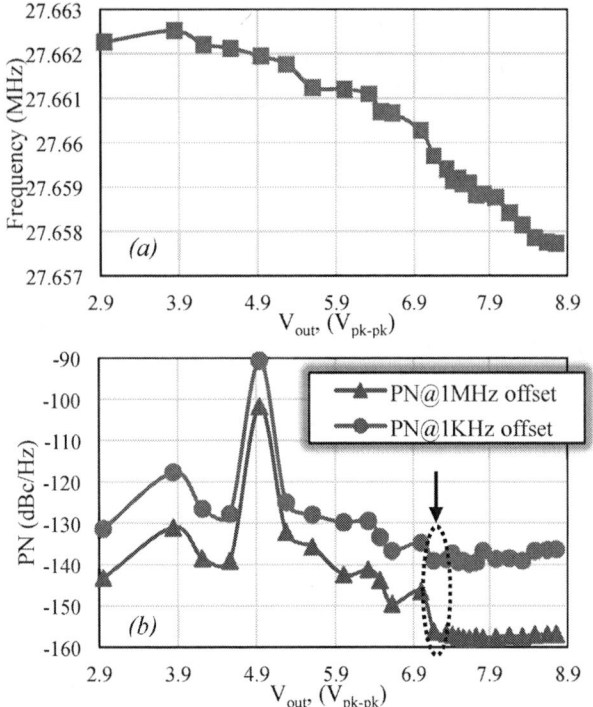

Figure 4: (a) Oscillation frequency and (b) Phase noise at 1KHz and 1MHz offset frequencies from the carrier versus the oscillator output voltage showing a noticeable drop at $7.2V_{pk-pk}$.

VERIFICATION OF THE NONLINEAR OPERATION IN CLOSED-LOOP

In order to understand the closed-loop behavior of the resonator, a method is developed based on the correlation between the frequency shift and the applied power. First, the resonance frequency and the output voltage of the resonator were recorded for a wide range of applied power in an open loop configuration. Power was applied to the three-fingered port by the network analyzer and the output voltage was measured with a high frequency probe at the two-fingered port once with 50Ω and again with an

open termination (Figure 7(a)). The resonator was driven to the edge of bifurcation by increasing the applied power. The measured S_{11} data for both terminations is plotted in Figure 7(b). The slight frequency difference as indicated on the plot, is the result of altered stiffness of the piezoelectric film when terminated to different impedances [21]. The frequency softening starts at lower applied power for a larger termination due to the stronger induced electric field on the output port and thus the larger energy density stored in the resonator.

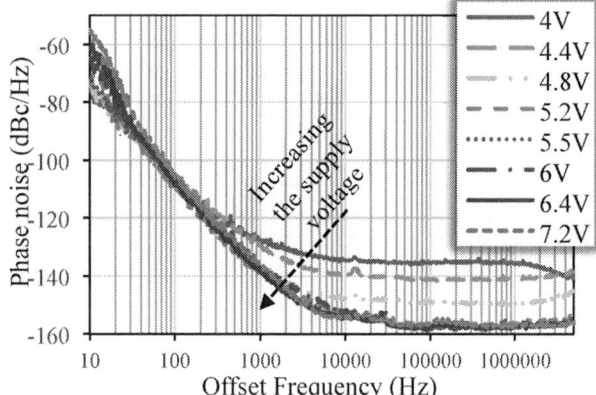

Figure 5: Phase noise plots for supply voltages greater than 4V showing a steady decrease as the voltage is increased.

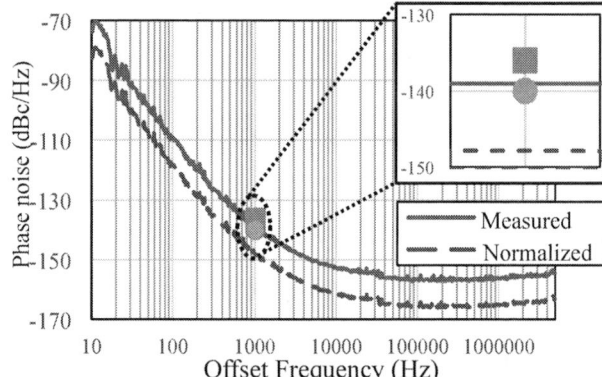

Figure 6: Measured and normalized (to 10MHz) phase noise plots of the oscillator with oscillation voltage of $7.2V_{pk-pk}$. The square and circle represent the best normalized phase noise values reported in literature for MEMS piezoelectric [5] and capacitive [6] oscillators respectively.

Next, to make a fair comparison between the open- and closed-loop measurements the normalized frequency shift ($\frac{\Delta f}{f_0}$) is plotted versus the voltage on the two-fingered port (Figure 8). Just before the bifurcation point, the resonator exhibits $\frac{\Delta f}{f_0}$ of -9×10^{-5} with both 50Ω and open terminations. This value of the normalized frequency shift corresponds to a peak-to-peak oscillation voltage around $7.2V$ for the closed-loop measurement,

978-1-4799-3510-9/14 $31.00 © 2014 IEEE   1287

and that coincides with the PN reaching its minimum value after a significant drop as pointed out in Figure 4 (b). This proves that the oscillator exhibits the lowest close-to-carrier phase noise at and beyond the bifurcation of the TPoS resonator.

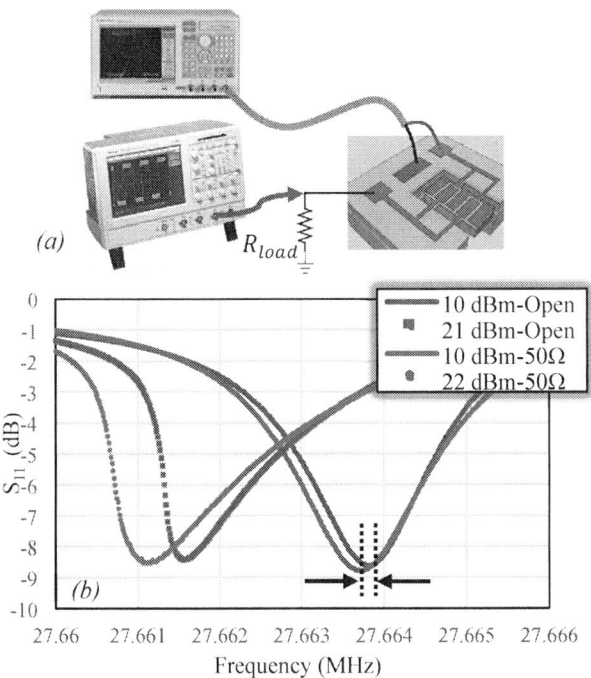

(a)

(b)

Figure 7: (a) Schematic of the one-port measurement setup. (b) S_{11} plots in linear region and at the edge of bifurcation measured with 50Ω and open termination. Note the slight frequency difference between the two cases at low input powers.

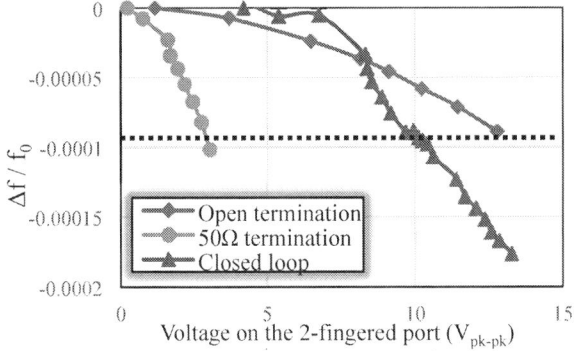

Figure 8: Normalized frequency shift ($\frac{\Delta f}{f_0}$) vs. the output voltage (the voltage on the two-fingered port) of the resonator.

CONCLUSION

A 27MHz oscillator is built incorporating a thin-film piezoelectric-on-substrate MEMS resonator. The resonator is driven into the nonlinear region while in closed-loop by increasing the supply voltage of the circuit. It is shown that the lowest phase noise is achieved when the resonator is operating at or beyond its

bifurcation point. A phase noise of -139 dBc/Hz of at 1kHz offset from the carrier is reported that is equivalent to -148 dBc/Hz when normalized to 10MHz and is the lowest value reported to date for MEMS resonators. The far-from-carrier phase noise is -157 dBc/Hz and the jitter is 290 $fsec$ from 12KHz to 5MHz.

ACKNOWLEDGEMENT

Authors wish to thank Brandon Harrington for his contribution in the fabrication of the devices.

REFERENCES

[1] J. Beek and R. Puers, *J. Micromechechanical Microeng.* vol. 22, no. 1, 2012.
[2] C.T. Nguyen and R. T. Howe, *J. Solid-State Circuits,* vol. 34, no. 4, pp. 440-455, Apr. 1999.
[3] T. McClelland, *et al., Int'l Frequency Control Symp. and the European Frequency and Time Forum,* pp.1-5, May 2011.
[4] M. E. Frerking, *Frequency Control Symp.,* pp.33-46, Jun. 1996.
[5] J. Hu, *et al., Radio Frequency Integrated Circuits Symp.,* pp.325-328, May 2010.
[6] V. Kaajakari, *et al., Electron Device Letters,* vol.25, no.4, pp.173-175, April 2004.
[7] A. Hajimiri and T. Lee, *J. Solid-State Circuits,* vol. 33, no. 2, pp. 179-194, Feb. 1998.
[8] V. Kaajakari, *et al., J. Microelectromechechanical Systems,* vol. 13, no. 5, pp. 715- 724, Oct. 2004.
[9] H. K. Lee, *et al., Tranducers,* pp. 510-513, 2011.
[10] M. Shahmohammadi, *et al., Frequency Control Symp.,* pp. 613-617, Jun. 2010.
[11] R. Abdolvand, *et al., Trans. on Ultrasonics, Ferroelectrics and Frequency Control,* vol.55, no.12, pp. 2596-2606, 2008.
[12] G. Piazza, *et al., European Solid-State Device,* pp.182-185, Sep. 2006.
[13] M. Shahmohammadi, *et al., Microwave Symp.,* pp. 1452-1455, May 2010.
[14] D.T. Chang, *et al., Int'l Conference on MicroElectroMechanical Systems,* pp.781-784, Jan. 2013.
[15] H.K. Lee, *et al., J. Microelectromechanical Systems,* vol. 20, no. 6, pp. 1228-1230, Dec. 2011.
[16] H. Fatemi and Reza Abdolvand, *Int'l Conference on MicroElectroMechanical Systems,* pp.461-464, Jan. 2013.
[17] H. Fatemi and Reza Abdolvand, *Int'l Frequency Control Symp.,* pp.1-4, May 2012.
[18] C. Hayashi, *J. Applied Physics,* vol. 24, no. 5, pp. 521-529, May 1953.
[19] M. Shahmohammadi, *et al., Trans. On Electron Devices,* vol. 60, no. 3, pp. 1213-1220, Mar. 2013.
[20] J. J. Gagnepain, *et al., Frequency Control Symp.,* pp. 218–225, 1983.
[21] M. Shahmohammadi, *et al., Int'l Frequency Control Symp. and European Frequency and Time Forum,* pp. 1-5, May 2011.

VARIABLE CAPACITOR WITH SWITCHING MECHANISM FOR WIDE TUNING RANGE

Dae-Hyun Baek, Youngkee Eun, Dae-Sung Kwon, Min-Ook Kim, Taeyoung Chung, and Jongbaeg Kim
School of Mechanical Engineering, Yonsei University, Korea

ABSTRACT

We developed a variable capacitor with mechanical switching mechanism and reversible mechanical latching component to enhance tuning ratio. The switching mechanism could connect four sets of capacitors arranged in parallel sequentially by controlling the displacement of a microactuator for abrupt and coarse tuning of total capacitance. Continuous and fine tuning was also achieved by gap-closing mode of interdigitated capacitors. The resultant maximum tuning ratio was 5.71 by combining coarse and fine tuning.

INTRODUCTION

Frequency selection to remove noise and interference is very important for the operation of radio frequency (RF) systems composed of electrical components such as capacitors, switches, and inductors. In particular, since the phase noise occurring in voltage controlled oscillator (VCO) is inversely proportional to the selectivity of LC tanks, it is required to employ frequency-tunable device with high selectivity [1]. The variable capacitor, one of the frequency-tunable devices, have been researched and integrated in various types of electrical devices such as tunable filter [2], phase shifter [3, 4], antenna [5], impedance tuner [6], and VCO [1, 7].

Since the devices manufactured by micromachining technology do not respond to a noise signal with frequencies above their resonant frequency, a number of micro electro mechanical system (MEMS) variable capacitors have been developed. Parallel plate type variable capacitors are the most commonly used MEMS tunable capacitors because they exhibit relatively high quality factor, simplicity of fabrication and low power consumptions [8-13]. They are designed as a form of stacked layers with a gap between the upper and the lower conductive plates which form a capacitive structure, where the capacitance is tuned by adjusting the gap between the conductive layers. However, the tuning ranges of parallel plate type are limited due to the pull-in phenomenon, which restricts the continuous tuning range down to 50%. Moreover the nonlinear relationship among the control voltage, the gap between the plates, and capacitance produces a nonlinear capacitance-voltage response. To overcome the limitation of parallel plate type variable capacitors, a number of researches have been developed. Capacitors with higher tuning ratio, from 100% to over 1000%, were reported in the previous researches [14-16]. However, these devices were also affected by pull-in effect, and thus their tuning behavior was very sensitive to the tuning voltage in the range of operation. The behavior of these capacitors in the vicinity of pull-in was similar to that of on-off switch, where a small change in voltage results in a large jump in capacitance, making fine

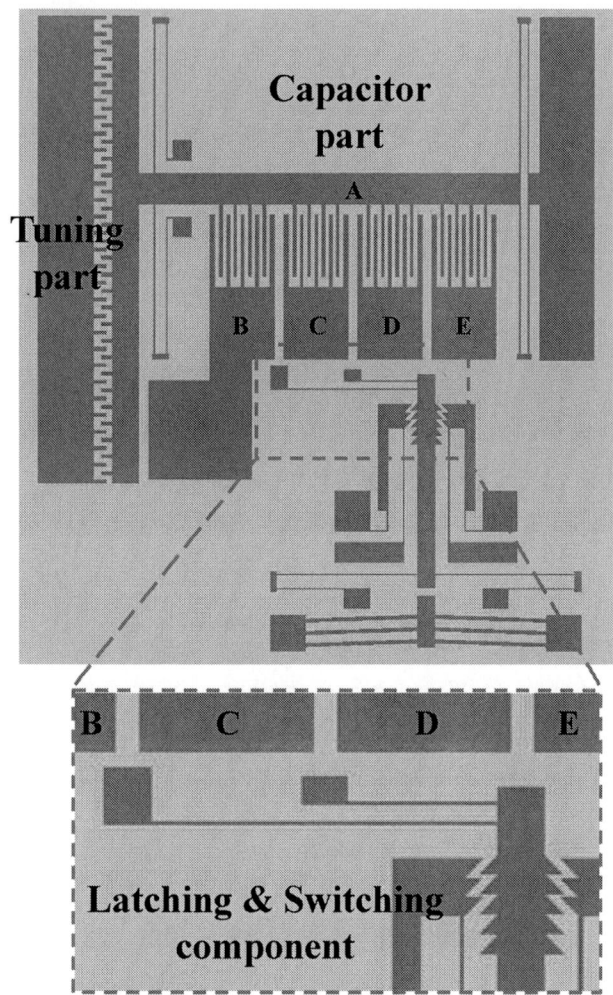

Figure 1: Schematic view of the variable capacitor with switching components. The device consists of four fixed interdigitated capacitor electrodes, movable comb electrode, and latchable switches. Since the gaps between the fixed electrodes and three switches are differently designed, switches are tuned on sequentially from left to right by a single actuator.

tuning very difficult.

In this paper, we suggest a novel MEMS variable capacitor with switching mechanism and reversible mechanical latching component, which enables the fine and continuous tuning of capacitance in the whole tuning range over 500%. The switches allow multiple sets of capacitive structures to be sequentially connected in parallel for a wide tuning range, and the reversible latching component is employed to maintain the connected state of the additional sets of capacitors without additional power consumption for relatively low power operation.

978-1-4799-3510-9/14 $31.00 © 2014 IEEE

a. Switching mechanism

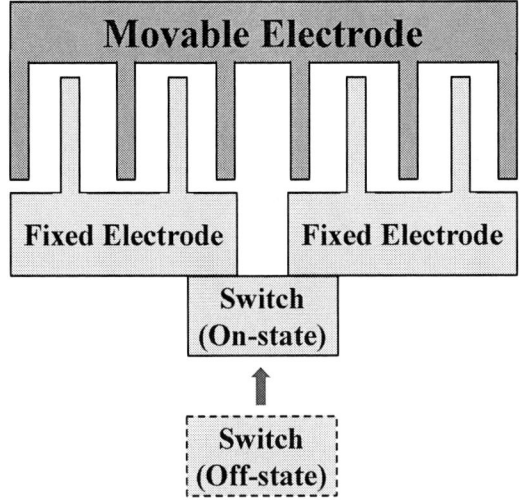

b. Tuning mechanism

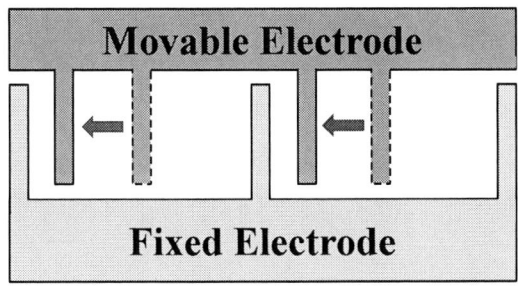

Figure 2: (a) An abrupt switching-on action to turn the electrode from inactive to active state. Once each of the inactive electrodes is turned on, the on-state is maintained by the latch mechanism even after the actuation power is off. (b) The gap-closing motion allows continuous tuning of the total capacitance.

DESIGN AND WORKING PRINCIPLE

The device consists of four sets of interdigitated capacitor electrodes, a movable comb electrode for actuation of tuning part, and three latchable switches driven by a single chevron actuator as shown in Figure 1. Gap-closing motion between interdigitated capacitor electrodes is driven by electrostatic comb actuator, while the on/off motion of switches and release of latch are performed by thermal actuators. Three switches are connected to a single actuator, and the gap distance between each of three switches and fixed electrodes is differently designed. When the actuator is moved, the switch with smallest gap will be closed first, and as the actuator motion is continued, the second and the third switches are closed sequentially. During this procedure, the initial capacitance, A-B, between the movable (A) and fixed (B) electrodes, increases to the larger capacitances, A-(B+C), A-(B+C+D), and A-(B+C+D+E). This abrupt switching-on

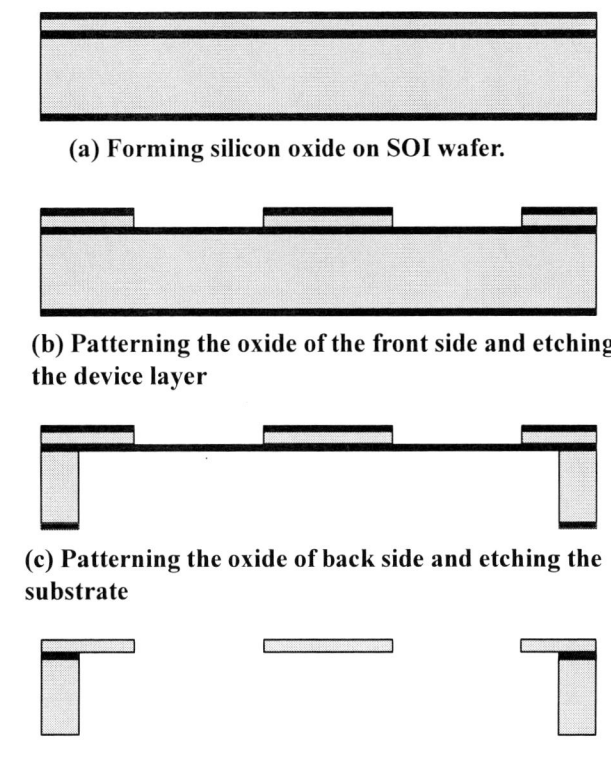

Figure 3: Fabrication process for the variable capacitor. The device is fabricated on SOI wafer by patterning and deep-etching both front- and backsides.

action to turn the electrode from inactivate to activate state is also described in Figure 2(a). Once each of the inactive electrodes is turned on, the on-state is maintained by the latch mechanism even after the actuation power is off, and the latch can be released by a single action of thermal actuator, enabling reversible switching. The mechanism for tuning the capacitance is shown in Figure 2(b). The movable electrode is displaced by electrostatic actuator, resulting in capacitance change by gap-closing motion. This gap-closing allows continuous tuning of the total capacitance.

FABRICATION

Figure 3 shows the fabrication process for the proposed variable capacitor with switching mechanism. The device is fabricated on silicon-on-insulator (SOI) wafer. The thickness of silicon device layer and buried silicon dioxide are 20 μm and 1 μm, respectively. The process starts with forming 1 μm-thick silicon oxide on both sides of the SOI wafer as shown in Figure 3(a). The oxide layer is then patterned by photolithography, and the capacitors, actuators, switches and latch mechanisms are formed by deep reactive ion etching (DRIE) of silicon using the oxide layer as etch-mask (Figure 3(b)). The similar process is done on the backside of the wafer to open large holes on the substrate (Figure 3(c)). Finally, the oxide etch-mask and buried oxide layers are etched to release the movable structures (Figure 3(d)).

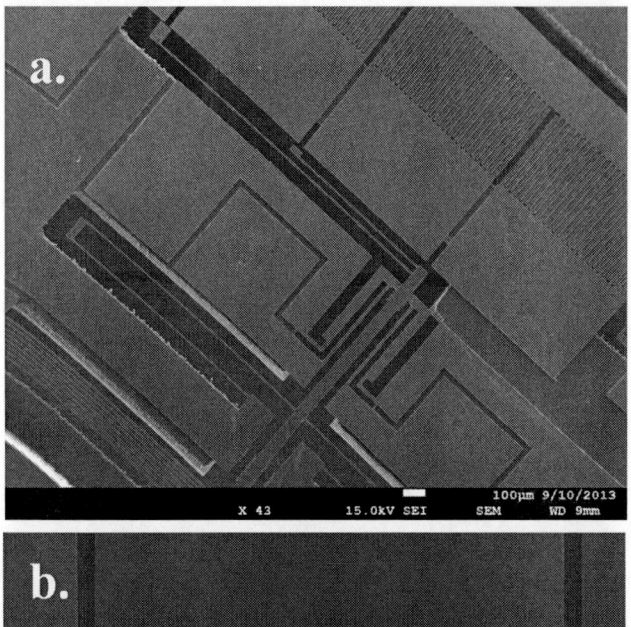

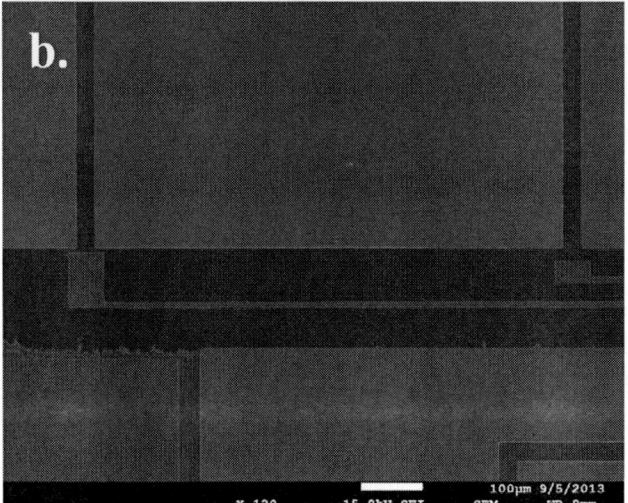

Figure 4: SEM images of the fabricated device. (a) An overview of the capacitor and components to switch and latch. (b) Close-up views of the switches for sequential switching by single actuator.

RESULTS AND DISCUSSION

Figure 4 shows the scanning electron microscope (SEM) images of the fabricated device. Top view is shown in Figure 4(a), and close-up view of switches for sequential switching is shown in Figure 4(b).

Figure 5(a) shows the change in capacitance tuned by gap-closing mode for four switching steps with respect to the tuning voltage. The black, red, blue and green line is the tuned capacitance at each mode, when the number of activated switch is 0, 1, 2 and 3, respectively. As the number of the "on-stated" switch increases, it is easily shown that the maximum capacitance increases, and the tuning range is getting wider. Using both switching mode and continuous tuning mode, the maximum capacitance of 24.11 pF was achieved by activating all of three switches, giving the tuning ratio of 5.71. By adjusting the switches and movable electrode simultaneously, any of the desired capacitance within the shaded area can be achieved. It is also possible to

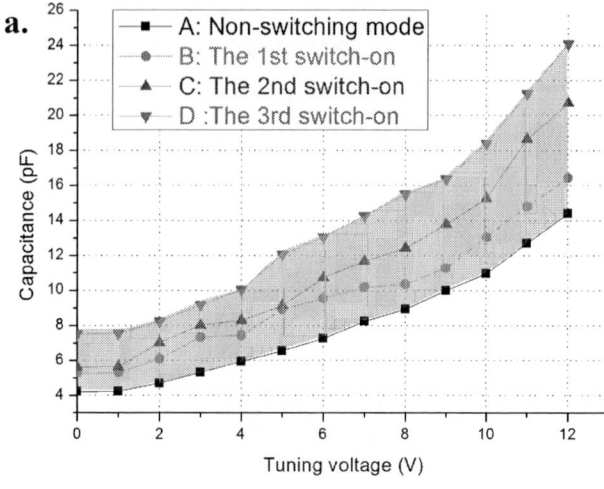

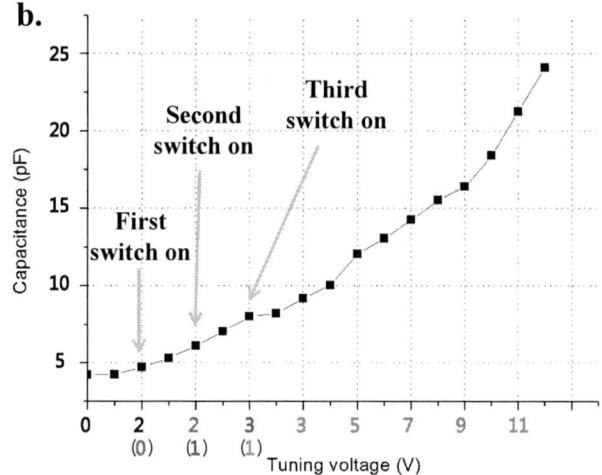

Figure 5: (a) The change in capacitance tuned by gap-closing mode for four switching steps with respect to the tuning voltage. (b) A continuous increment of capacitance from the minimum to the maximum value with relatively low tuning voltage. The black, red, blue, and green colors represent the tuning voltages at each state of switch-off, the first switch-on, the second switch-on, and the third switch-on, respectively.

obtain the continuous increment of capacitance from the minimum value without switching action to the maximum value with all the switches turned on, with relatively low tuning voltage, as plotted in Figure 5(b). The black, red, blue and green colored numbers on the x-axis represent the tuning voltages applied to generate this C-V plot at each state of switch-off, the first switch-on, the second switch-on, and the third switch-on, respectively.

CONCLUSIONS

In summary, large tuning ratio of a variable capacitor was achieved using sequential switching by single actuator and reversible latching component. The switching mechanism integrated in the variable capacitor has shown improved capacitance tuning ratio by activating the inactive capacitors to form multiple sets of capacitors connected in

978-1-4799-3510-9/14 $31.00 © 2014 IEEE   1291

parallel. The capacitance was increased from 4.22 pF to 24.11 pF, showing the maximum capacitance tuning ratio of 5.71. In addition, it was experimentally confirmed that the continuous and fine tuning of capacitance was possible by properly adjusting the switching action and tuning voltage. The proposed variable capacitor successfully demonstrated a novel mechanism for enhanced tunable range with fine tuning.

ACKNOWLEDGEMENTS

This research was supported by the Converging Research Center Program through the Ministry of Science, ICT and Future Planning, Korea (2013K000388) and the Pioneer Research Center Program through the National Research Foundation of Korea funded by the Ministry of Science, ICT & Future Planning (2010-0019313).

REFERENCES

[1] D. J. Young, "A Low-Noise RF Voltage-Controlled Oscillator Using On-Chip High-Q Three-Dimensional Coil Inductor and Micromachined Variable Capacitor", *Solid-state Sensor and Actuators Workshop*, South California, June 8-11, 1998, pp. 128-131.

[2] C. L. Goldsmith, A. Malczewski, Z. J. Yao, S. Chen, J. Ehmke, D. H. Hinzel, "RF MEMs Variable Capacitors for Tunable Filters", *Int. J. RF and Microwave CAE.*, vol. 9, pp. 362-374, 1999.

[3] G. M. Rebeiz, G. Tan, J. S. Hayden, "RF MEMS Phase Shifter: Design and Applications", *IEEE Microwave Magazine*, vol. 3, pp. 72-81, 2002.

[4] N. S. Barker, G. M. Rebeiz, "Distributed MEMS True-Time Delay Phase Shifters and Wide-Band Switches", *IEEE Trans. Microw. Techn.*, vol. 46, pp. 1881-1890, 1998.

[5] J. Schoebel, T. Buck, M. Reimann, M. Ulm, M Schneider, "Design Considerations and Technology Assessment of Phased-Array Antenna Systems With RF MEMS for Automotive Radar Applications", *IEEE Trans. Microw. Tech.*, vol. 53, pp. 1968-1975, 2005.

[6] Y. Lu, "A Novel MEMS Impedance Tuner Simultaneously Optimized for Maximum Impedance Range and Power Handling", *Microwave Symposium Digest*, June 12-17, 2005, pp. 927-930.

[7] M. Behera, V. Kratyuk, S. K. De, N. R. Aluru, Y. Hu, K. Mayaram, "Accurate Simulation of RF MEMS VCO Performance Including Phase Noise", *J. Microelectromech. Syst.*, vol. 14, pp. 313-324, 2005.

[8] M. Barki-Kassem, R. R. Mansour, "A High-Tuning-Range MEMS Variable Capacitor Using Carrier Beams", *Can. J. Elect. Comput. Eng.*, vol. 31, pp. 89-95, 2006.

[9] Z. Xiao, W. Peng, R. F. Wolffenbuttel, K. R. Farmer, "Micromachined Variable Capacitors with Wide Tuning Range", *Sens. Act. A: Phys.*, vol. 104, pp. 299-305, 2003.

[10] S. C. Saha, "Modeling, Design and Simulation of Tunable Band Pass Filter Using RF MEMS Capacitance and Transmission Line", *Proc. of SPIE*, June 05, 2006, 60350C.

[11] X. Chen, C. H. J. Fox, S. McWilliam, "Modelling of a Tunable Capacitor with Piezoelectric Acutation", *J. Micromech. Microsyst.*, vol. 14, pp. S102-107, 2004.

[12] G. V. Ionis, "A Zipper-Action Differential Micro-Mechanical Tunable Capacitor", *Microelectromech. Syst. Conference*, Aug 24-26, 2001, pp. 29-32.

[13] R. L. Borwick III, P. A. Stupar, J. DeNatale, R. Anderson, C. Tsai, K. Garrett, R. Erlandson, "A High Q, Large Tuning Range MEMS Capacitor for RF Filter Systems", *Sens. Act. A: Phys.*, vol. 103, pp. 33-41, 2003.

[14] A. J. Gallant, D. Wood, "The Modeling and Fabrication of Widely Tunable Capacitors", *J. Micromech. Microeng.*, vol. 13, pp. S178-S182, 2003.

[15] J. Chen, J. Zou, C. Liu, J. E. Schutt-Ainé, S. Kang, "Design and Modeling of a Micromachined High-Q Tunable Capacitor With Large Tuning Range and a Vertical Planar Spiral Inductor", *IEEE Trans. Electr. Dev.*, vol. 50, pp. 730-739, 2003.

[16] T. G. S. M. Rijks, P. G. Steenken, J. T. M. V. Beek, M. J. E. Ulenaers, A. Jourdain, H. A. C. Tilmans, J. D. Coster, R. Puers, "Microelectromechanical Tunable Capacitors for Reconfigurable RF Architectures", *J. Micromech. Microeng.*, vol. 16, pp. 601-611, 2006.

CONTACT

*J. Kim, tel: +82-2-2123-2812; kimjb@yonsei.ac.kr

MEMS 2014 KEYWORD INDEX

Scroll to the keyword and select a Blue link to open a paper. After viewing the paper, use the bookmarks to the left to return to the beginning of the Keyword Index.

3D...177, 366, 425, 1189
3D Fabrication 660
3D Ice Printing...52
3D Nanofabrication..437
3D Polydimethylsiloxane Foil 927
3D Silicon Microelectrodes 951

A

Accelerometer (s)..28, 761
Accelerometer Gyrometer Co-Integration 725
Acoustic Impedance Matching 845
Acoustic Sensor 849
Acoustics..897
Activated Carbon-Graphite Configuration 401
Active Scaffold 917
Actuator (s) 192, 909, 913
Adaptive Focusing 909
Additive Manufacturing...510
Adhesion Forces...596
Adjustable Sensing Ranges 713
Airflow Sensing 777
AlGaN/GaN HEMT-Based FET...250
Aligned..159
Alumina..449, 1011
Aluminum Nitride(AlN) 124, 636, 688, 729, 1265
Alzheimer Disease 314
Amorphous Carbon...143
Amplification 242
Anchor Loss..1257
Angular Acceleration Sensor 696
Annealing...825
Anodization...1011
Apparent Size Correlation..939
Aptamer..250
Arterial Diameter Monitoring 869

Arterial Dysfunction 869

Artificial Cell 17

Artificial Cilia 927

Artificial Finger 785

Assembly 433, 1015

Assist-Free 433

Asthma 935

Atomic Clock 552

Atomic Force Microscope...100, 128, 733, 789

Atomic Layer Deposition 167, 342, 449, 584

Atomization...1043

Autonomous 218

B

Back Action 725

Bacteria 246

Bacterial Cellulose 518

Band-Pass Filter (BPF) 1237

Battery 358

Bearings 200

Bent Cantilevers..177

Bi-Directional Gas Pump 294

Bioactuator...196

Biodegradable 358

Biofuel Cell 163

Bioimaging 322

Bioinspired 696, 741

Biomedical Application 901

BioMEMS 177, 238, 358

Biomimetics 741

Biomolecular Interactions...272

Biosensor (s) 1075, 1083

Birdbath Resonator Gyro 20

Bistable Micro Valve 1023

Blood Coagulation 330

Blood Storage..833

Boron-Doped Diamond...322

Bosch Process 459

Breath Diagnostics...935

BTX...294

Buckled Beam 692

Buckling...1099

Buckypaper 441

Bulk PZT 524

Bulk Tungsten 502

Bundles 48

C

C-Reactive Protein 250

Cancer Diagnosis Devices 889

Cantilever 769

Capacitance Measurement 222

Capacitive Gap Width 1245

Capacitive Micromachined Ultrasonic Transducers 584

Capacitive Sensing 680

Carbon Nanotubes 48, 167, 342

Cardiomyocyte 159

Cell 290

Cell Cell Interaction 943

Cell Characterization 901

Cell Culture 280, 518

Cell Culturing Device 181, 264

Cell Manipulation 1031

Cell Pairing 185

Cell Separation 951

Cell Surgery 885

Cell Targeting Aptamer 242

Cell Trapping 947

Cell Viability 865

Cellular Aggregate 925

Cellular Materials 510

Cellular Parylene 374

Cemented Carbide 652

Ceramic Micromachining 494

Chain-Like Structure 1197

Charge Bias 1277

Charge Pump 1273

Chemical Sensor 302, 318

Chemical Sensors & Systems 226

Chemical Synthesis 92

Chemiluminescence 1047

Chip Destruction 1123

Chip Scale Package 1213

Chip Security 1123

Circulatory 1007
Closed-Loop 761
Closest Packing 1163
Cluster 729
CMOS-Based Amperometric Biosensor Array 322
CMOS-IC 1139
CMOS-MEMS 676, 757, 1249
CMOS-MEMS Microphone 136
Co-Culture 943
Co-Integration 1071
Column 1007
Comb Drive 1229
Combinatorial 334
Combined Electrodes 409
Complementary 676
Complementary Metal Oxide Semiconductor 1071
Composite Membrane 1217
Confocal Hyperchromatic System 1167
Confocal Microscopy 881
Contact Lens Tonometer 624
Contact Material 143
Continuous Medium Exchange 234
Controllable Delivery 865
Convective Accumulation 1035
Cooling 1039
Copper On Polyimide 528
Coriolis Flow Sensor 975
Corner Lithography 1111
Coronary Artery Disease 893
Creep 660
Critical Bifurcation 749
Crosslinking 471
Cryopreservation 829
CTE 40
Cu Etching 528
Cubic 425
Cuff Electrode 9
Curve Fit 580
Curved Micromirror 1185
Curved Piezoelectric Micromachined Ultrasonic Transducers 124
CVD 486
Cylindrical Lens 1171

D

Deep Etching	502, 1185
Deformability	833
Degradable Scaffold	518
Degradation	616
Device Tuning	660
Diamond	1159
Diaphragm with Electrode	544
Dielectric	1011
Dielectric Liquid	713
Dielectrophoresis	268, 837, 951
Differential	1249
Differential Readout	1237
Differential Scanning Calorimeter	306
Diffraction Optics	608
Digital Microfluidics	272, 334, 829, 955
Direct Bonding	1119
Direct Laser Writing	471
Direct Prototyping	298
Disk Resonator Gyroscope	749
Displacement Sensor	1185
Dissolvable Material	604
Dither	608
DRIE	36, 459
Droplet (s)	92, 486, 664, 983
Droplet Merging	1003
Droplet Micro-Array	96
Drosophila Melanogaster	196
Drug Delivering	52
Dual Axis	797
Dual Axis Confocal Endoscope	805
Dual Mode	120
Duffing Nonlinearity	749
Dynamic Mask Lithography	733
Dynamics	1253

E

Eccentric Mass	370
Electret	366, 417, 704, 717, 1229
Electro Tactile Display	1183
Electrocaloric cooling	544

Electrocardiography 841

Electrochemical Etching 1183

Electrochemical Impedance 104

Electrochemiluminescence-Induced Fluorescence 108

Electrohydrodynamic 963

Electrolysis 809

Electromagnetic 378, 425, 564

Electromagnetic Actuation 1147

Electromechanic 218

Electron Beam Lithography 437

Electron Emitter 467

Electronic Article Surveillance 76

Electroosmotic Flow 1067

Electrophoresis 959

Electroplated Iron Cobalt 520

Electroplating 40, 64, 453, 1135

Electroporation 234, 268, 817

Electrospray 17, 100

Electrostatic 366

Electrostatic Actuation 805

Electrostatic Force 913

Electrostatic Generators 385

Electrothermal 789

Electrowetting 72, 222, 284, 1011, 1039

Electrowetting-On-Dielectric(EWOD) 338, 955

Electrowetting-On-Dielectric (EWOD), Recoverable 1047

Embryo Vitrification 829

Embryonic Body 181, 280

Emitter Array 463

Encapsulation 56, 588, 1281

Encased Cantilevers 128

Endoscopy 200

Endothelial Cells 238

Endpoint 459

Energy Harvest.............159, 350, 366, 370, 374, 378, 382, 397, 413, 417, 421, 425, 429, 568

Energy Scavenging 897

Energy Storage Device 401

Enzyme 959

Epi-Seal 588, 1277

Escherichia Coli 175

Etch Hole 1257

Etching 656

Eutectic Bonding..64
EUV...482
Excitation...108
Extracellular Electron Transport (EET)...362
Extraction...995

F

Fabrication...478, 620
Fatigue...640
FENO..935
Ferrofluid (s)..200, 350
Fiber MEMS...60
Film Stress...482
Filter Termination..1249
Fine Line..478
FinFET...1063
Flapping Wing...648
Flexible..84, 306, 389, 556, 817
Flexible Device..737
Flexible Electronics...230, 548
Flexible Printed Circuit..346
Flexible Sensor...528
Flow Control...1031
Flow Meter...1139
Flow Regulator..935
Fluidics..56
Force Map..151
Force Sensor...680, 769
Frequency Divider...210
Frequency Split..612
Fresnel Lens...1175
Fuel Cell..393
Functional Electrical Stimulation...616
Fused Silica...20

G

G-Sensitivity...32
Galinstan..540, 967
Gas Analysis...171
Gas Chromatography..1007
Gas Sensing..230
Gas Sensor (s)...326, 506, 1179

Glancing Angle Deposition..437

Glass Reflow..1233

Gradient Force 721

Graphehe FET..230

Graphene 326, 362, 393, 486, 1055, 1075, 1159

Graphene Cantilever 1087

Graphene Loudspeaker 556

Graphene Woven Fabric 624

Gratings..490

Grayscale Mask Fabrication 1027

Grid Filter...482

Gyroscope 32, 612, 721

Gyroscope Quadrature 36

H

Hard Magnetic Layer 520

Heat Effects...825

Heat-Transfer Resistance 272

Heater Electrode...1047

Helicobacter Pylori 821

Hemispherical 672

Hemispherical Shell Resonator..40

Hemispherical Shells 494

HfO$_2$ 1063

High Aspect Ratio..374

High Flow Knudsen Pump...112

High Frequency...955

High K Dielectric...584

High Quality Factor 24, 628

High Resolution..624

High Resolution Trimming..494

High Throughput Screening..334

HIV/AIDS Diagnostics...256

Honeycomb..449

Hot-Wire 777

Hotspot...1039

Human Detector...1213

Humidity Sensor 532

Hybrid 354

Hybrid Valves 260

Hydrodynamic Trapping..246

Hydrogel 733

HydrophilicityControl 925

Hydrophobic 979

Hyperspectral Imaging 1167

Hyperthermia Treatments 877

Hysteresis Window 1095

I

Ice Mould Microfluidics 52

Immersion 498

Immunosensor 809

Impedance Sensing 338

Implants 1127

In-Plane Capacitive Flow Sensors 971

In-Plane Gap Closing 136

Inclinometer 797

Induced Pluripotent Stem Cell 181

Inductively Coupled Plasma 502

Inertial Focusing 939, 1051

Inertial Measurement Units 28

Inertial MEMS 725

Infrared Absorptance 644

Infrared Focal Plane Array 1225

Injection Molding 560

Inkjet 963, 967

Inkjet Printing 506, 536

Insect 163

In-situ Gyroscope Calibration 608

Integrated 729

Integrated Heater 905

Integration 1189

Internet of Things 773

Intracellular Recording 155

Intrinsic Losses 632

Intrinsic Stress 177

Invar 40

Ion Assisted Breakdown 171

Ion Concentration Polarization 276

Ion Selective Membrane 421

Ion Source 463

Ionic Liquid 326, 463

iPS Cell 280

Iridium Oxide 616

J

Jet Flow 288

K

Kinesin 1107
Kinesin Motor Protein 314

L

Lab-on-a-Chip 272, 330
Lab-on-a-Disc 260
Label Free 1083
Laplace Trap 1003
Large Area Electronics 1135
Large Area Microfabrication 1217
Laser Micro Machining 524
Laser Scribing 556
Laterally Vibrating Resonators 1241
Lead Free Piezoelectric Thin Film 397
Lead ZirconateTitanate (PZT) 1155, 1237
Leukemia 837
Linear Acceleration Sensitivity 32
Lipid Bilayer 17, 457
Liposome 17, 288, 457
Liquid Based Transmitter 765
Liquid Bridge 769, 1115
Liquid Crystal Elastomer 905
Liquid Deposition 100
Liquid Filled Lens 1163
Liquid Lens 909
Liquid Medium 765
Liquid Metal 540, 664, 967
Liquid Sensor 132
Lithium Niobate 1241
Lithographic Microfabrication 112
Lithography 498
Live Bacteria Detection 5
Living Battery 163
Local Stimulation at Single Cell Level 264
Localized Nano-Electroporation 865
Lorentz Force 80, 700
Low Cost Sensor 536

Low Frequency 346, 429
Low Loss Radio Frequency Switch 1233
Low Phase Noise..1209
Low Power 717
Low Temperature Bonding 1131

M

Magnesium...358
Magnetic 564
Magnetic Bead 809
Magnetic Gradient 44
Magnetic Nanoparticle 917
Magnetic Resonance Force Microscopy 151
Magnetic Sensor..80
Magnetometer..80, 700
Magnetophoresis 256
Magnetorheological Effect..1197
Magnetostriction 1189
Manganese Dioxide 405
Mass Sensing 128
Mechanical Filter 1249
Mechanical Latch 1289
Mechanical Property 652
Mechanical Stress 917
Mechanomyogram 845
Membrane Protein..288, 457
Membranes...632
MEMS Aircraft 648
MEMS Force Sensor..979
MEMS Gyroscope 32, 749, 801
MEMS Infrared Devices 644
MEMS Oscillator 1285
MEMS Pump 564
MEMS Resonator..120, 1257
Meniscus Visualization...668
Mesoporous Carbon 298
Mesoporous Silica..995
Metamaterial (s) 664, 1221
Metamaterial Perfect Absorber 84
Metglas..76
Micro Beads 592
Micro Bearings...983

Micro Electrode Array 841
Micro Electro Mechanical System (MEMS) 36, 132, 330, 362, 467, 506, 568, 632, 640,
....656, 741, 753, 777, 931, 971, 1055, 1241, 1289
Micro Gas Chromatography 294
Micro Heater(s)..514, 1179
Micro Knife...885
Micro Medical Robot..188
Micro Plasma...171
Micro Scale..393, 721
Micro Shell..721
Micro Supercapacitor..405
Micro Thruster...999
Micro Total Analysis System...5
Micro Trace Heavy Metal Sensor...298
Micro Variable Lens..72
Micro Variable Prism...72
Microactuator...905
Microarray..254
Microassembly..901
Microbial Fuel Cell..393
Microbial Supercapacitor...362
Microbolometer...572
Microbubble..104
Microcantilever...330
Microchamber..175
Microchannels...1107
Microelectrode..857
Microelectrode Array..254
Microfabricated Magnetizing Mask..520
Microfabrication 56, 254, 490, 518
Microfluidic (s)...............................5, 92, 185, 238, 246, 250, 256, 284, 809, 821, 939, 987, 995, 1027, 1051
Microfluidic Channel...310, 947, 1139
Microfluidic Chip...1035
Microfluidic System...88
Microhotplate...48
Microlens...1015
Microlens Array...1163
Micromachined...753
Micromachining Compatible..644
Micromanipulator..901
Microneedle...813, 817, 1183
Microscale Magnetic Patterning..520

Microscanner...805, 1185

Microsurgery...1147

Microtubule (s)...314, 1107

Microvalve..987

Millimeter-Waves..206

Millivolt Switching..1099

Minimal Invasive..885

Mirror...1229

Mirror Array...1155

Molecular Sensor..222

Monolithic..112

Motion Control...310

Motion Harvesting...350

Motor Unit EMG Recording..857

Movable Electrode..1245

MR Fluid Lens..1197

Multi-Band..1225

Multi-Channel Recording...857

Multi-Degree-of-Freedom Sensors..28

Multi-Parameter Measurement System..975

Multi-Physics Modeling...700

Multidrug Resistance Degree..837

Multilayer Magnetic anisotropy...44

N

Nano Channels...995

Nano Fountain Pen..100

Nano Opto Electro Mechanical..1261

Nano Opto Mechanical Systems..1091

Nano Pillar...147

Nano Technology...1083

Nanochannel..1019, 1059, 1067

Nanocomposite...354

Nanocrystalline Silicon..467

Nano Electro Mechanical (NEM)..1095

Nanoelectromechanical Relay(s)...143, 1099

Nanoelectromechanical Switch...143, 1103

Nano Electro Mechanical Systems (NEMS)...1055, 1071, 1079

Nanofiber...498, 917

Nanofiber Forests..644

Nanofluidic...1059

Nanofluidic Crystal Sensing...276

Nanofluidic Device 1019
Nanomanufacturing...474
Nanomechanical Resonators 1205
Nanoparticle 421, 1059
Nanoparticle Characterization 116
Nanoparticles Sorting..1015
Nanoporous Powder 1131
Nanoscale..471
Nanoscale Contact 1079
Nanostructure (s)...474, 648
Nanowire 155, 246, 652
Narrow Gaps 1281
Nd-Fe-B Magnet 151
NDT 490
NEMS Oscillator Stability 116
Neural Cell 254
Neural Interface 873
Neural Microelectrodes...616
Neural Probe 155, 853
Neural Recording 9
Neural Signal 853
Neural Stimulation 9
No Back-Plate 136
Non-Contact Pairing 1003
Non-Destructive Measurement 1151
Non-Equilibrium 310
Non-Linear Dynamics...1261
Nonlinear Harvester 397
Nonlinearity Tuning..801
Normal Physiological Condition...276

O

On-Chip Chemical Reaction Monitoring...338
On-Chip Control 1023
On-Chip Li-Ion Capacitor...401
On-Chip Mixing 829
Optical Coherence Tomography (OCT) 1147
Optical Control..1201
Optical Emission Spectroscopy 459
Optical Gradient Force 1091
Optical Measurement 580
Optical Memory 1091

Optical MEMS 433
Optical Resonators 1193
Optical Stimulation 853
Optical Switch..1155
Optical Tweezers 921
Optically Tunable..84
Optically-Induced Electroporation..234
Optics Array 72
Opto-Acoustic Oscillator 1209
Opto-Mechanical..1143
Opto-Mechanical System 869
Optofluidic (s) 797, 909, 1015
Optogenetics 196, 813
Optomechanics...1209
Organic..1217
Oscillating Droplet...897
Oscillator (s)..210, 214, 218, 1193, 1201, 1261
OSTE..96
OSTE+..987

P

Package 1139
Packaging...1127
Paper 620
Paper Based Electronics..781
Paper Sensor..536
Parameter Optimization 700
Parametric Amplification...210
Parametric Resonance Excitation 805
Particle Tracking..939, 1051
Particle Transport...927
Particle Trapping..1019
Parylene..9, 132, 185, 841, 1127
Parylene C 104, 825
Pattern 413
Peel-Off..514
Periodic Heating Method 1151
Periprosthetic Joint Infection 5
pH Sensor..1063
Phase Change 1039
Phase Noise...210, 1193
Photo Detector 147, 1159

Photo Sensor 486

Photolithography 540

Photonic Crystal 576

Photonic Sintering 532

Photoresist 389

Photothermal Probing 1205

Pierce Typed Electron Emitter 467

Piezoelectret 374

Piezoelectric 159, 370, 429, 704, 745

Piezoelectric Actuator 1099, 1155, 1171

Piezoelectric Coefficient 620

Piezoelectric Effect 897

Piezoelectric Energy Harvester 397

Piezoelectric Material 620

Piezoelectric Resonator (s) 214, 688

Piezoelectronics 781

Piezoresistive 785, 789, 1201

Piezoresistive Cantilever 290

Piezoresistive Detection 640

Piezoresistor 1151

Pirani Gauge 676

Piston-Cylinder Actuator 592

Plasma FET 171

Plasmonic Nanostructures 1205

Plastic Deformation 660

PMUT 745, 753

Pneumatic Balloon Actuator 925

Pneumatic Force 913

Pneumatic Switch 1023

Point of Care 256

Poly-SiGe 1095

Polyacrylamide Gel Electrophoresis 1027

Polycide 1273

Polycrystalline Diamond 628

Polydimethylsiloxane(PDMS) 192, 514, 873, 987

Polydimethylsiloxane (PDMS) Transfer 544

Polylactic Acid 532

Polymer Dielectrics 230

Polymer MEMS 560

Polymer Micromachining 528

Polypyrrole 354

Porous Carbon 389, 405

Porous Silicon Dioxide 417

Positrons..453

Prandtl-Tube 999

Pre-Sealed Reagents..52

Precision..797

Pressure Sensor 104, 120, 717, 757, 845, 849, 893, 975

Print..560

Printed Electronics 506, 532

Printing..963, 967

Probe 861

Process Development..502

Protein Detection 1083

Protein Electrophoresis 1027

Protein Measurement 921

PTFE Electret...413, 684

Pull-In 1281

Pumping 1067

Q

Q-Enhancement..789

Quality Factor 612, 632, 725, 801, 1257

Quantitative Analyses 294

Quantum Tunneling 1103

Quinolones Resistance 821

R

Radiation Detection 793

Radiation Pressure 1209

Radio Frequency Microelectromechanical Systems.............................206, 1193

Radio Frequency Microelectromechanical Systems Switch.......................1233

Radiotracer..284

Rarefied Plume...999

Rate Integrating Gyroscope 20, 24

Reactive Ion Etching...463

Reconfigurable...540

Reconfigurable Waveguide Iris 206

Red Blood Cell...1031

Reduction Mechanism 1031

Relay 1095

Reliability..568, 596

Replication 560

Resonant..132

Resonant Heating 877

Resonant Microcantilever 302

Resonant Pressure Sensor 120

Resonant Radiation Sensor 704

Resonant Switch..1273

Resonator 80, 612, 628, 672, 717, 1265, 1277

Restenosis 877, 893

Reverse Electrodialysis 421

RF Filters 1241

RF MEMS 1237, 1253

Ring 951

Ring Gyroscope 24

RNA Purification 260

Rotary Motors 200

Rotating Gear 429

Rotational 684

Rubidium 552

Rubidium Azide 552

Rubrene Solution 108

Ruthenium Oxide 167

S

Sacrificial & Biodegradable..656

Sacrificial Layer...548

SAM..514

Sample Comparison Method..931

Scanner..1229

Scanning Probe Microscopy 128

Scanning Thermal Probe Microscopy...1111

Sciatic Nerve 873

Sealing Technique...592

Selection..242

Selective Formation 1131

SELEX 242

Self-Assembled 147

Self-Assembled Molecules 1103

Self-Assembly...56

Self-Oscillator..1071

Self-Propelled Droplets...310

Semicircular Channels 696

Sensing Material 302

Sensor (s) 729, 773, 1055

Sensor Characterization 825

Sensor Device 536

SERS 1059

Shape Regulation 175

Shear Piezoresistance 600

Shear Stress 238

Shock Sensor 692

Sidewall 36

Sigma-Delta Modulator 761

Silica 449

Silicon 648

Silicon Carbide (SiC) 793, 1079

Silicon Microwire 604

Silicon Migration Technology 445

Silicon Nanowire (s) 600, 1175

Silicon Nanowire Probe 151

Silicon Optical Micro-Bench 881

Silicon Resonator 1245

Silicon Tweezer 13

Silk Fibroin 604

Silver Nanowire 785

Simulation 927

Single Cell 943

Single Cell Manipulation 1035

Single Cell Separation 1035

Single Layer Silicon-On-Insulator Process 409

Single Nucleotide Polymorphism Polymerase Chain Reaction 821

Sliding 979

Small Molecules 1075

Smart Cut 793

Soft X-Ray Charging 409

Solar 576

Solid Phase Extraction 284

Solid State Supercapacitor 342

Somatic Muscle 196

Sorting 92

Speed Sensor 684

SPF 482

Spheroplast 175

SPIH 350

Spiral 913

Spring Forces 596

Sputtering 453
Squeeze-Film Damping 28
Squegging 1273
Stability 1063
Stable..1277, 1281
Stamp Bonding...1131
Stencil Printing..478
Stent 877
Stiction 588
Stimulation...861
Stored Red Blood Cells..833
Strain Sensor 737, 785
Structural Layer 548
Subpixel Resolution...580
Supercapacitor...167, 354, 389
Superhydrophobic Surface 668
Surface Acoustic Wave..1043
Surface Energy Patterning 96
Surface Plasmon Polariton..1179
Surface Plasmon Resonance 147
Surface Ratchets...983
Surface Tension 222, 991
Suspended Nanochannel Resonators 116
Swallowing 849
Switch 1079, 1265, 1289
Switchable Cavity Resonator 206
Switched...218

T

Tactile Display 931
Tactile Sensor..441, 680, 709, 713, 1189
TASCL 280
Tau Protein...314
Temperature Compensation 214
Tensile Test...652
Terahertz 1221
Tetramethylammonium Hydroxide...857
Thermal Conductivity 628
Thermal Diffusivity 1151
Thermal Flexure Actuator 548
Thermal Flow Sensor..975
Thermal Imaging..572

Thermal Performance...48

Thermal Read-Out...696

Thermal Sensor...1213

Thermally Responsive Solution...592

Thermo-Acoustic Transmitter...765

Thermodynamic Parameters...302

Thermophotovoltaic...576

Thermopile Sensor...1213

Thermotropic...905

Thin-Film Magnesium (Mg)...656

Thiol-Ene Polymer...96

Thread...959

Through-Glass Vias (TGVs)...64

Through-Silicon Via...1115

THz...664

THz Polarization Rotation...88

Time Constant...833

Time-Domain...596

Timing References...214

Tissue Assay...13

Tongue Motion...849

Torque Sensor...680

Torsional Resonator...793

Touch Sensing...781

Traction Force...290

Transducer...745

Transfer Technology...322

Transience...1123

Transient Flow...1051

Transport...1107

Trehalose...163

Triboelectric Generator...346

Tunable...1163, 1221

Tunable Capacitor...1253

Tunable Laser...1143

Tunable Lens...881, 1171

Tunable Magnetic Anisotropy...44

Tunable Optical Liquid Lens...1147

Tunable Optical Microcavity...1217

Tunable Optical Modulation...1197

Tunable Spectral Filter...1167

Tuning...1201

Two-Phase...943
Two-Photon..471
Two-Photon Stereolithography..............................921
Two-Stage Solidification...................................44

U

Ultra-Thin Membranes......................................889
Ultrasonic..745, 769
Ultrasonic-Micromachining.................................494
Ultrasound..753, 885
Ultrasound Transducer.....................................765
Ultraviolet..1159
Uncooled...1225
Uncooled Infrared Detector................................688

V

Vacuum..761
Vacuum Filtration...441
Van der Waals Force......................................1119
Vapor Cell..552
Vapor-Liquid-Solid..604
Vapor-Liquid-Solid Growth.................................155
Variable Capacitor.......................................1289
Velcro Principle...1119
Vertical Electrets..409
Vibration Energy Harvesting...............................385
Virtual Surface...931
Volume Precision..955

W

Wafer Bonding...453
Wafer Level Process.......................................564
Wafer Level Sealing..24
Wafer-Scale...556
Water Droplet...979
Waveguide...813
Wavelength Selective Device..............................1167
Wavelength Selective IR Emission.........................1179
Wearable..370
Wetting Dynamics..668
Whole Angle Mode...20
Wineglass...672

Wireless Optrode 813
Wireless Sensor Node 773
Wrinkling 474

X

X-Ray 893
X-Ray Imaging 490

Z

Zebrafish ECG 841
Zinc Oxide Nanowires 781

MEMS 2014 Author Index

Scroll to the author and select a **Blue** link to open a paper. After viewing the paper, use the bookmarks to the left to return to the beginning of the Author Index.

A

Author	Pages
Abdelgawad, M.	829
Abdolvand, R.	132, 1285
Abe, Y.	185
Abelmann, L.	100, 1111
Adachi, C.	108, 147
Adachi, J.	147
Ahamed, M.J.	24
Ahmadi, A.	939, 1051
Ahmed, S.M.	548
Ahn, C.H.	24, 80, 588, 749, 773, 1277, 1281
Aita, F.	1213
Ajiki, Y.	147
Akashi, T.	709
Akhbari, S.	124
Akin, T.	32
Akita, S.	737
Akiyama, Y.	163, 196
Allen, M.G.	1, 358
Alper, S.E.	32
Alveringh, D.	680
Amato, B.	632
Amjadi, M.	785
An, S.	112
Anand, S.V.	192
Ando, Y.	155
Ang, W.C.	688
Angelescu, D.E.	797
Ansell, O.	459
Aoki, R.	845
Aoyagi, S.	417
Aoyama, M.	1229
Appel, J.H.	889
Arai, F.	1031, 1035
Arcamone, J.	1071

Ardanuç, S. **524, 608**
Arie, T. **737**
Armutlulu, A. **358**
Arndt, G. **1071**
Arnold, D.P. **520**
Asadnia, M. **741**
Asano, T. **196**
Ashby, P.D. **128**
Ashizawa, H. **717**
Askari, S. **24**
Assadsangabi, B. **200**
Ataka, M. **717**
Averitt, R.D. **84**
Ayala, C.L. **143**
Ayazi, F. **28, 40, 120, 612**
Azgin, K. **32**

B

Baborowski, J.J. **490**
Bae, J. **72**
Baek, D. **1289**
Baghchehsaraei, Z. **206**
Bagherinia, M. **700**
Bahri, D. **128**
Banerjee, N. **1123**
Bargatin, I. **449**
Barnett, R. **459**
Barutçu, D. **350**
Basset, P. **385**
Bazigos, A. **1063**
Bedener, K. **1063**
Beh, S.P. **196**
Beld van den, W.T.E. **1067**
Bellamkonda, R.V. **873**
Benthem van, K. **628**
Berenschot, J.W. **100**
Berg van den, A. **1067**
Bergonzi, G. **552**
Berthelot, A. **640**
Beyazoglu, T. **1193**
Bhave, S.A. **1209**
Bhugra, H. **204**

Bian, W.	413, 684
Bidstrup, S.A.	358
Bienstman, J.	218
Bierman, D.M.	576
Bindiganavale, G.S.	1039
Binh-Khiem, N.	765, 769
Bleiker, S.J.	143
Boden, T.J.	721
Boer de, M.J.	696, 999
Boero, G.	177
Bohringer, K.F.	983
Boisen, A.	1205
Booth, R.	238
Borana, J.	865
Borowsky, D.	572
Boser, B.	753
Bourouina, T.	385, 725, 1185
Bouvet, D.	1063
Boyd, C.W.	20
Boyle, D.	260
Briand, D.	370, 429, 506, 532
Bright, V.M.	1201
Brookhuis, R.A.	680
Brücker, C.	927
Brugger, J.	56, 177
Buja, F.	580
Bulbul, A.	226
Bulovic, V.	1103, 1217
Bunch, J.S.	1201
Burtghartz, J.	572
Büttgenbach, S.	869
Byun, I.	514

C

Cai, H.	688, 692, 1091, 1143, 1261
Cakmak, O.	330
Campanella, H.	688
Cardot, F.	490
Carlborg, C.F.	987
Carraro, C.	389
Cassella, C.	1269
Celanovic, I.	576

Cermak, N. ... 116
Chae, J. ... **362, 889**
Chae, J.B. ... **1147**
Chang, C.-H. ... **1189**
Chang, C.-I. ... **136, 757**
Chang, C.-J. ... **234**
Chang, C.-P. ... **1189**
Chang, D.C. ... **36**
Chang, D.T. ... **721**
Chang, H.-C. ... **757, 1189**
Chang, H.L. ... **761**
Chang, J. ... **230**
Chang, J.-H. ... **1147**
Chang, J.H.-C. ... **656, 1127**
Chang, K.W. ... **250**
Chang, W. ... **1217**
Chang, W.H. ... **5, 250**
Chao, C.Y. ... **821**
Chatziioannou, A.F. ... **284**
Che, Y.J. ... **821**
Chen, A. ... **1023**
Chen, B.-A. ... **861**
Chen, C.C. ... **1135**
Chen, C.-H. ... **1189**
Chen, C.-N. ... **648**
Chen, C.-Y. ... **1249**
Chen, C.Z. ... **995**
Chen, D.P. ... **644**
Chen, F. ... **761**
Chen, J. ... **502**
Chen, J. ... **833**
Chen, P.-C. ... **13**
Chen, P.H. ... **478**
Chen, R. ... **1197**
Chen, S. ... **284, 338, 1011**
Chen, S. ... **502**
Chen, S.J. ... **382**
Chen, W.-C. ... **44**
Chen, X. ... **877**
Chen, Y.-C. ... **943**
Chen, Y.J. ... **644**
Chen, Y.-Y. ... **861**

Chen, Z.J.	644
Cheng, C.-L.	676, 757
Cheng, Y.	943
Cheung, K.C.	939, 1051
Chia, E.M.	664
Chim, J.	1183
Chin, C.-H.	1249
Chin, L.K.	1015
Chiquet, M.	177
Chishiro, T.	592
Chmielewski, D.	128
Cho, I.-J.	853
Cho, J.W.	963
Cho, J.Y.	20
Cho, K.	498, 917
Cho, S.K.	188, 991
Choi, E.	421
Choi, K.	72
Choi, N.	853
Choi, S.	222
Choi, S.	72
Choi, W.	540, 967
Choi, Y.	873
Choi, Y.	72
Christensen, D.L.	1281
Chronis, N.	893
Chung, D.	256
Chung, S.	246
Chung, S.K.	897, 901, 1147
Chung, T.	1289
Chung, Y.	873
Clark, B.	608
Clerc, P.-A.	552
Colinet, E.	1071
Corigliano, A.	700
Cosemans, S.	1095
Coster De, J.	1095
Cottone, F.	385
Cu-Nguyen, P.-H.	1167
Cui, T.	833
Cui, Y.L.	556
Cunningham, S.	1253

D

Dahmardeh, M.	877
Dam van, R.M.	284, 338
Darvishi, M.	1237
Davami, K.	449
Davis, R.W.	1083
Decrop, D.	272
Degertekin, F.L.	584
Deimerly, Y.	725
Deléglise, S.	140
Dellea, S.	640
Demircan, Y.	837
Dempsey, N.M.	520
Deotare, P.B.	1103
Deotte, J.	510
DeReus, D.	1253
Despont, M.	143, 490, 552
Dickens, A.	36
Dijkstra, M.	999
Dimov, N.	260
Domicone, N.	909
Dommann, A.	490
Dong, B.	88, 692, 1091, 1261
Dong, L.	1225
Dooraghi, A.A.	284
Drechsler, U.	143
Droogendijk, H.	696
Du, J.-C.	616
Duan, X.	805
Ducrée, J.	256, 260
Duerig, U.T.	143
Duncombe, T.A.	1027

E

Edura, T.	108
Eijkel, J.C.T.	1067
Elbuken, C.	330
Elfrink, R.	568
Elwenspoek, M.	1237
Endrödy, C.	1167
English, A.W.	873

Erdem, M. 837
Ermek, E. 330
Ernst, T. 1071
Esashi, M. 322, 467, 482, 709
Esfandyarpour, R. 1083
Etter, D. 572
Eun, C.K. 64
Eun, Y. 1289
Everhart, C.L. 1277

F

Fan, K. 84
Fan, S.K. 1047
Fang, N. 510
Fang, S.P. 498, 917
Fang, W. 44, 136, 676, 757, 1135, 1189, 1197
Fatemi, H. 1285
Fedder, G.K. 544, 801
Feng, F. 1159
Feng, J. 188
Feng, P.X.-L. 793, 1079
Feng, Y. 374
Fernandes, J. 1175
Fernandez-Bolanos, M. 143
Fiorentino, G. 48, 342
Fischer, A.C. 1055
Fischer, P. 437
Fitzgerald, P. 596
Forsberg, F. 1055
Foster, J.S. 564
Franca, E. 498
Frolet, N. 532
Fu, C.-C. 1189
Fu, Q.Y. 409
Fujii, T. 652
Fujimoto, K. 1107
Fujimura, T. 1155
Fujita, H. 314, 713, 717, 1111
Fujiyoshi, M. 709
Fukuhara, T. 1035
Fukushi, H. 709
Funabashi, H. 709

Funahashi, A.	264
Fung, S.	628

G

Gadhari, N.	177
Galayko, D.	385
Galchev, T.	350
Gammaitoni, L.	385
Gao, J.	544
Garraud, A.	520
Gaughran, J.	260
Gavcar, H.D.	32
Geerlings, J.	100
Geisberger, A.	36
Ghannad-Rezaie, M.	893
Ghovanloo, M.	813
Gianchandani, Y.B.	64, 76, 112, 294, 494, 885
Gibbs, J.G.	437
Gibson, B.A.	1281
Glynn, M.	256
Goedbloed, M.	568
Goericke, F.	729
Gong, S.	1241
Gonzales, J.M.	132
Gore, R.K.	873
Goryu, A.	155
Green, S.R.	76
Grewe, A.	1167
Grine, A.J.	1193
Grinsven van, B.	272
Groenesteijn, J.	696, 975, 999
Grogg, D.	143
Grutter, K.E.	1193
Gu, J.	1115
Gu, Y.A.	688
Gu, Y.D.	692, 1091, 1143, 1261
Gudeman, C.S.	564
Guedes, A.	745, 753
Gulari, M.N.	893
Gullo, M.R.	56
Gündüz, U.	837
Guo, D.	544

H

Haase, T. 459
Haesler, J. 552
Hagleitner, C. 143
Hall, H.J. 1201
Hamada, A. 147
Hamano, H. 17
Han, C.H. 660
Han, M.D. 346, 425
Hans, H. 741
Hansson, J. 987
Hara, M 1043
Hara, M. 397, 463, 636
Hara, Y. 310
Harada, S. 737
Haraldsson, T. 96, 987
Harris, J.S. 1083
Harrison, K.L. 1277
Hasegawa, S. 931
Hashiguchi, G. 717, 1229
Hashimoto, S. 310
Hata, Y. 709
Hatakeyama, K. 1111
Hayakawa, T. 1035
Hayasaka, T. 322
Hayase, M. 60
Hayashi, M. 60
Hayashi, S. 280
He, T. 1079
Heidari, A. 745
Heinz, D.B. 588, 1277
Heo, Y.J. 175
Herr, A.E. 1027
Herrault, F. 358
Herrera, M.D. 869
Herrmann, I. 572
Hilbert, J. 1253
Hill, B.E. 773
Hill, M. 596
Hiraiwa, T. 264
Hirayama, K. 175, 518

Hiroi, N. 264

Hirooka, M. 196

Hoang-Giang, D. 765

Hoang-Phuong, P. 765

Honda, S. 592

Honda, W. 737

Hone, J. 1075

Hong, J. 947

Hong, S.J. 1147

Hong, V.A. 24, 80, 588, 749, 773, 1277, 1281

Hong, Y. 963

Hoof Van, C. 218

Hopcroft, M.A. 773

Horsley, D.A. 80, 124, 628, 700, 745, 749, 753

Hoshino, K. 196

Hoshino, T. 196

Hou, M.T. 648, 713

Howe, R.T. 1277

Hsia, B. 389

Hsieh, H.-C. 226, 1007

Hsieh, H.-S. 1189

Hsu, F.-M. 44, 1197

Hsu, L.-S. 441

Hsueh, H.-T. 809

Hu, J. 502

Huang, C.-C. 861

Huang, C.W. 478

Huang, J.G. 692, 1091, 1261

Huang, K. 1135

Huang, L.X. 721

Huang, P.-C. 861

Huang, R.F. 88

Hussain, A.M. 548

Hussain, M.M. 548

Hutter, F. 572

Hwang, J. 1233

Hwang, S.-H. 1233

I

Ichikawa, Y. 1151

Iimura, Y. 1163

Ikegami, N. 467

Ikeuchi, M. 181, 280, 921
Ikuta, K. 181, 280, 921
Imashioya, T. 604
Imato, T. 108
Inaba, A. 326, 1087
Inoue, K.Y. 322
Inoue, S. 652
Ionescu, A.M. 1063
Ischer, S. 552
Ishida, M. 155, 604, 857
Ishihara, R. 342
Ishii, M. 1179
Ishikawa, H. 931
Ishikawa, T. 322
Ishimatsu, R. 108
Isono, Y. 600, 1059
Isozaki, A. 1221
Isozaki, A.I. 913
Ito, K. 1035
Itoh, T. 60
Iwabuchi, K. 196
Iwai, K. 393
Iyer, S.S. 721

J

Jacot-Descombes, L. 56
Jalabert, L. 1111
Jambunathan, M. 568
Jamsaid, A. 92
Janphuang, P. 370, 429
Jao, P.F. 498, 917
Javanmard, M. 1083
Jayaprakash, V. 393
Jennings, J. 453
Jeon, E. 222
Jeong, H. 885
Jeong, H.-H. 437
Jeong, Y. 28
Jia, K. 36
Jia, Y. 306
Jiang, H. 1175
Jiang, X. 1115

Johansson, S.B. 935
Jourdan, G. 725
Joyce, R.J. 721
Juang, Y.Z. 1139
Jung, U.G. 290

K

Kamiya, K. 185, 288, 457
Kan, T. 147, 290, 1221
Kan, T.K. 913
Kanda, N. 1221
Kaneko, M. 1031
Kaneko, T. 845, 849
Kang, H.M. 1019
Kang, J.Y. 853
Kang, X.-Y. 616
Kang, Y.W. 250
Kao, C.Y. 821
Karita, Y. 518
Karlsson, J.M. 987
Karsten, S.L. 314
Kasahara, T. 108, 1131
Kashyap, K. 648
Katz, A.J. 917
Kaufman, R. 490
Kaufmann, J. 552
Kawai, K. 1229
Kawai, Y. 151
Kawano, R. 185, 288, 457
Kawano, T. 155, 604, 857
Keesara, V. 28
Kenny, T.W. 24, 80, 588, 749, 1277, 1281
Keshavarzi, S. 1119
Khalil, D. 1185
Khatri, B. 905
Kilinc, N. 330
Kim, B.J. 514
Kim, B.J. 825
Kim, C.J. 284, 338, 668, 1011
Kim, D. 421
Kim, D. 540, 967
Kim, D. 853

Kim, G.J.	498, 917
Kim, H.	226, 238, 672, 1007, 1123
Kim, H.	1003
Kim, J.	246
Kim, J.	1003
Kim, J.	1075
Kim, J.	1289
Kim, J.H.	1019
Kim, K.T.	917
Kim, M.-S.	389
Kim, M.O.	1289
Kim, S.	222
Kim, S.H.	963
Kim, T.G.	853
Kim, Y.	72
Kim, Y.K.	853
Kim, Y.-K.	1019, 1233
Kimura, H.	264
Kimura, T.	264, 600
Kinahan, D.J.	256, 260
Kippenberg, T.J.	140
Kirby, D.J.	721
Kitamura, N.	1183
Klaassen, A.	128
Koelmans, W.W.	143
Koh, Y.	1019
Kojima, A.	467
Kokorian, J.	580
Kometani, R.	1087
Konishi, K.	1221
Konishi, S.	592, 925
Koochak, Z.	1083
Kosemura, Y.	931
Koshida, N.	467
Kosla, C.	596
Kotera, H.	314, 1107
Kotler, C.	490
Kottapalli, A.G.P.	741
Kozinda, A.	167, 354
Kraft, M.	761
Krijnen, G.J.M.	680, 696
Kropelnicki, P.	688, 692, 1091, 1143, 1261

Kubota, Y. 155, 604
Külah, H. 837
Kumagai, S. 1179
Kumar, A. 648
Kumemura, M. 717
Kunisawa, E. 108
Kuo, J.-C. 441, 809
Kuo, P.-H. 809
Kuo, W.-C. 959
Kurabayashi, K. 805
Kurihara, K. 560
Kuroda, K. 652, 1031
Kuwano, H. 397, 463, 636, 1043
Kuwata-Gonokami, M. 1221
Kwon, D. 246
Kwon, D.S. 1289
Kwon, K. 421, 813
Kwon, Y. 222
Kwon, Y. 72

L

Ladhane, K. 793
Lai, C.-H. 1189
Lai, W.-C. 1189
Lai, W.-M. 44
Lai, W.C. 1135
Lake, J.J. 721
Lal, A. 13, 524, 608
Lammerink, T.S. 975
Lammertyn, J. 272
Lang, J. 1217
Lang, J.H. 1103
Langfelder, G. 640
Larsen, T. 632
Lasalandra, E. 700
Lazari, M. 284
Le Roy, D. 520
Lee, C.S. 963
Lee, G.B. 5, 250, 821
Lee, G.-B. 234
Lee, H. 510, 813, 853
Lee, I. 733

Lee, J.	116, 222, 733, 793, 947, 1079
Lee, J.B.	540, 967
Lee, J.E.-Y.	1257
Lee, J.	72
Lee, K.I.	963
Lee, M.S.	5
Lee, S.	204
Lee, S.	947
Lee, S.	1003
Lee, S.S.	704
Lee, Y.	967
Lee, Y.S.	1019
Lee, Y.-S.	1233
Lee, Y.R.	897, 901
Lei, C.	644
Lei, D.	204
Lemme, M.C.	1055
Lenert, A.	576
Leong, K.C.	688
Li, C.-S.	1249
Li, H.	52
Li, H.	805
Li, M.	80, 700
Li, M.-H.	1249
Li, N.	502
Li, S.	159
Li, S.	401, 405
Li, S.-S.	1249
Li, T.	494, 885
Li, W.	813
Li, W.	1273
Li, X.	624, 781
Li, X.X.	302, 995, 1115
Li, Z.H.	9, 52, 276, 624, 817
Liang, K.-C.	676
Liang, Z.	817
Liao, H.H.	1139
Liao, J.C.	889
Liao, K.W.	713
Liao, S.-C.	1189
Lim, B.	963
Lim, H.B.	777

Lin, C.F.	1139
Lin, C.H.	478, 959, 1139
Lin, C.L.	250
Lin, C.-T.	809
Lin, C.-W.	861
Lin, C.-Y.	861
Lin, L.	124, 159, 167, 230, 354, 393, 486, 628, 1159
Lin, N.	777
Lin, Q.	242, 306, 1075
Lin, Y.	1273
Lin, Y.C.	5
Liu, A.Q.	88, 664, 692, 1015, 1091, 1143, 1261
Liu, C.-S.	612
Liu, G.	1075
Liu, H.	1175
Liu, H.B.	777
Liu, J.	829
Liu, J.-Q.	616
Liu, L.	36
Liu, M.	1225
Liu, R.	210, 1273
Liu, S.W.	366, 777
Liu, S.Y.	382
Liu, W.	346, 425
Liu, X.	159, 1225
Liu, X.Y.	781
Liu, Y.	230
Liu, Y.	656, 1127
Llobera, A.	869
Lo, S.-C.	136
Lockhart, R.	370, 429
Longoni, A.F.	640
Lötters, J.C.	975, 999
Lu, M.-Y.	234
Lu, S.-S.	809
Lu, Y.	745
Lu, Y.-S.	1175
Luo, J.	502
Luo, X.	64
Luo, Y.	877
Lye, S.W.	366
Lynn, K.G.	453

M

Ma, C.-W.	441, 809
Ma, K.	999
Ma, P.	334
Ma, W.	1225
Ma, X.	338
Maboudian, R.	389
Maeda, R.	60, 560
Maeda, S.	310
Mahadeva, S.K.	620
Makihata, M.	709
Man, T.	624
Manalis, S.R.	116
Mansour, R.R.	789
Mansukhani, K.	729
Mao, H.Y.	644
Marelli, M.	177
Marentis, T.C.	893
Marette, A.	532
Mariani, S.	700
Mark, A.G.	437
Märki, D.	532
Martinez, A.	358
Marty, F.	385, 797
Mastrangeli, M.	56
Mastrangelo, C.H.	672, 1123
Masuno, K.	1179
Matchkov, V.	869
Matova, S.	568
Matsue, T.	322
Matsui, K.	1087
Matsumoto, K.M.	147, 290, 326, 765, 769, 845, 849, 913, 979, 1087, 1151, 1221
Matsumoto, Y.	264
Matsunami, S.	108
Mattana, G.	532
Maurer, K.	1027
McAuley, D.	260
McGaughey, A.J.H.	544
McNeil, A.	36
Mehanathan, N.	40
Meinert, T.	881

Meiss, T. 536
Meng, B. 346, 425
Meng, E. 104, 825
Mercier, B. 797
Mescheder, U. 1119
Metcalfe, G.D. 84
Meyer De, K. 1095
Miao, J.M. 366, 741, 777
Michelassi, F. 13
Miki, N. 185, 264, 931, 1183
Miller, N. 1269
Mimatsu, H. 1131
Ming, A.J. 644
Minh, L.V. 397
Minh-Dung, N. 769, 845
Misaki, T. 1155
Misawa, N. 318
Mitsuya, H. 717
Mizuno, J. 108, 1131
Molina-Lopez, F. 506
Moon, H. 955, 1039
Morana, B. 48
Morikaku, T. 652
Morishima, K. 163, 196
Moriyama, M. 482
Morris, A. 1253
Mostafazadeh, A. 330
Mugele, F. 128
Mukherjee, T. 801
Munoz-Berbel, X. 869
Murakami, Y. 318
Muramatsu, Y. 925
Murarka, A. 1217
Muroyama, M. 709

N

Nadig, S. 524, 608
Naghsh Nilchi, J. 210
Nagumo, O. 560
Naing, T.L. 210
Najafi, K. 20
Najar, H. 628

Nakagawa, T.	378
Nakamura, N.	163
Nakano, M.	322
Nakayama, T.	709
Nam, Y.	576
Namazu, T.	652
Narimannezhad, A.	453
Natarajan, S.	1253
Neels, A.	490
Nemoto, N.	1221
Ng, E.J.	24, 80, 588, 749, 773, 1277, 1281
Nguyen, C.T.-C.	210, 1193, 1265, 1273
Nguyen, H.D.	721
Nguyen, S.D.	971
Niedermann, P.	490
Nievergelt, A.	128
Niimi, Y.	528
Nikapitiya, N.Y.J.B.	955
Niklaus, F.	1055
Niroui, F.	1103
Nishida, T.	417
Nishijima, T.	181
Nishikawa, H.	1131
Nishino, H.	467
Nitzan, S.H.	749
Noda, K.	765, 849
Noh, S.	238
Nomura, K.	378
Nonomura, Y.	709
Nooijer de, C.	568
Norford, L.K.	777
Novelli, P.	893
NuLi, Y.	616
Numakunai, S.	92
Numano, R.	155

O

Oba, M.	1213
Oberhammer, J.	206
Oguchi, H.	463, 636, 1043
Oh, S.W.	963
Ohno, H.	163

Oi, H.	155, 604
Oka, K.	264
Oka, Y.	433
Olcum, S.	116
Oldham, K.	805
Ollier, E.	1071
Olsen, T.	242
Omura, Y.	709
Oniku, O.D.	520
Ono, T.	151, 1245
Onoe, H.	518, 1163
Oohira, F.	433
Osaki, T.	17, 185, 288, 457
Oshidari, Y.	1087
Osman, H.	1095
Osonoi, M.	280
Östling, M.	1055
Ota, H.	264
Ou, W.	644
Ouyang, W.	276
Overstolz, T.	552
Özgür, E.	837

P

Pai, M.	204
Pai, P.	171
Pan, S.S.	777
Pan, T.	1023
Pan, W.	204
Pandraud, G.	48
Paprotny, I.	971
Pardon, G.	96
Park, I.	246, 785
Park, I.S.	897, 901
Park, J.	421
Park, J.	656, 841
Park, J.	885
Park, S.	268
Park, S.	873
Patterson, W.C.	520
Paul, O.	350
Pavy, N.	797

Payer, K. 116
Pei, R. 242, 1075
Perahia, R. 721
Perez-Ruiz, E. 272
Petsch, S. 905
Pezous, A. 552
Phan, H.P. 769
Philippe, J. 1071
Piazza, G. 1099, 1241, 1269
Pichitpajongkit, A. 785
Pisano, A. 729
Plaza, J.A. 869
Pourkamali, S. 1201
Pruitt, B.L. 773
Przybyla, R. 753
Puers, R. 218
Purwar, P. 947
Pyne, D.G. 829

Q

Qalandar, K.R. 1281
Qin, Y. 112, 294
Qiu, Z. 805
Quang-Khang, P. 769

R

Rahman, M.M. 672, 1123
Rais-Zadeh, M. 214
Rajagopalan, J. 192
Ramezani, M. 1095
Randles, A.B. 688, 692, 1091, 1143, 1261
Rane, T.D. 334
Reindl, T. 437
Reinecke, H. 1119
Reith, P. 905
Ren, H. 362
Ren, M. 664, 1015, 1143
Ren, T.L. 362, 556
Renaud, M. 568
Revol, V. 490
Rey, P. 640, 725
Rhee, K. 1147

Rigante, S. 1063
Rindzevicius, T. 1205
Rinnerbauer, V. 576
Rix, R. 905
Robert, P. 725
Rocheleau, T.O. 210, 1193
Rochus, V. 1095
Rockenbach, A. 927
Rodríguez-Rodríguez, R. 869
Rojas, J.P. 548
Rooij de, N.F. 370, 429, 506, 532
Roozeboom, C.L. 773
Rottenberg, X. 218, 1095
Rovers, M. 568
Roxhed, N. 935
Rozen, O. 753
Rubel, P.J. 564
Ruh, D. 905
Ryu, S. 785

S

Saadany, B. 1185
Sabry, Y.M. 1185
Sader, J.E. 128
Saif, M.T.A. 192
Saito, M. 1131
Saito, N. 600
Sakuma, S. 1031
Sammoura, F. 124, 167, 354
Sanders, R.G.P. 100, 696
Santagata, F. 48
Santhanam, S. 544
Santra, T.S. 865
Sarajlic, E. 100, 1111
Sarkar, N. 789
Sarro, P.M. 48, 342
Sasaki, M. 1179
Sato, K. 254
Savoye, M. 1071
Sawada, T. 1179
Sawahata, H. 155, 857
Schaijk van, R. 568

Schatz, A.	1171
Schelling, C.	572
Schliesser, A.	140
Schmid, S.	632, 1205
Schnakenberg, U.	927
Schneider, R.A.	1265
Schröder, S.	1055
Schuhladen, S.	905
Sebastian, A.	143
Segovia-Fernandez, J.	1269
Seifert, A.	881, 1167
Seki, T.	1213
Sekiguchi, T.	92
Sekitani, T.	68
Senior, D.E.	498
Senkal, D.	24
Seo, H.W.	1147
Seo, Y.J.	151
Seren, H.R.	84
Serien, D.	471
Serrano, D.E.	28
Severi, S.	1095
Shahmohammadi, M.	1285
Shang, H.	917
Shao, P.	40, 612
Sheen, H.-J.	861
Shehata, N.	833
Shelton, S.	124, 745, 753
Shen, C.	405
Shen, C.W.	298
Shen, W.	116
Shi, Y.Z.	1015
Shibata, S.	528
Shih, W.-P.	861
Shikida, M.	528
Shimada, N.	921
Shimokawa, F.	433
Shimomura, S.	925
Shimoyama, I.S.	147, 290, 326, 765, 769, 845, 849, 913, 979, 1087, 1151, 1221
Shin, I.U.	1147
Shin, J.H.	897, 901
Shinozaki, R.	433

Shintaku, H.	314, 1107
Shiozaki, M.	1213
Shirai, T.	378
Shkel, A.M.	24
Shoji, K.	163
Shoji, S.	92, 108, 1131
Shorman, E.	520
Shusteff, M.	510
Siekman, M.H.	1111
Silvestri, C.	48
Simonsen, U.	869
Sin, L.Y.	889
Sinzinger, S.	1167
Sletten, E.M.	1103
Smith, A.D.	1055
Sochol, R.D.	393
Soljacic, M.	576
Someya, T.	68
Son, Y.	853
Song, L.	502
Song, Q.H.	88, 664
Song, S.H.	704
Sonmezoglu, S.	32
Sorenson, L.D.	40, 612, 721
Spadaccini, C.M.	510
Sparreboom, W.	975, 1067
Spasic, D.	272
Spengen van, W.M.	580
Staufer, U.	580
Steimle, R.	36
Stemme, G.	935
Sterner, M.	1055
Stoeber, B.	536, 620
Stojanovic, M.	242, 1075
Storey, M.J.	1209
Stuermer, M.	1171
Su, T.-H.	749
Subramaniyan, S.P.	314
Suekuni, K.	1059
Sugano, K.	600, 1059
Sugimoto, S.	1043
Sugiyama, T.	717, 1229

Sun, G. 668
Sun, S.H. 572
Sun, X.M. 346, 425
Sun, Y. 829, 833
Sun, Y.-C. 44, 136, 676
Sung, W.K. 28
Sung, W.-L. 1135, 1189
Suzuki, A. 310
Suzuki, M. 163, 417
Suzuki, T. 433
Suzuki, T. 463
Suzuki, T. 1151
Suzuki, Y. 374, 409
Suzuki, Y. 482
Swager, T.M. 1103

T

Tabata, Y. 925
Tabib-Azar, M. 171
Tabrizian, R. 120
Taguchi, Y. 264
Tai, J. 841
Tai, Y.-C. 656, 841, 1127
Tajima, S. 925
Takagi, H. 560
Takahashi, H.T. 290, 913, 979, 1087, 1221
Takahashi, T. 417
Takahata, K. 200, 877
Takahata, T. 765, 845, 1087, 1151
Takami, Y. 652
Takao, H. 433
Takei, K. 737
Takei, Y. 326, 849, 1087
Takenaka, Y. 264
Takeshita, T. 600, 1059
Takeuchi, S. 17, 175, 185, 254, 288, 457, 471, 518, 1163
Tallur, S. 1209
Tan, C.S. 688
Tanaka, J. 1213
Tanaka, S. 322, 467, 482, 709
Tanamoto, R. 264
Tang, J. 76

Tang, W. 346
Tao, G.W. 1047
Tao, K. 366
Tarhan, M.C. 314
Tas, N.R. 100, 1111, 1237
Tatar, E. 801
Tavassoli, V. 40, 612
Tee, M.H. 200
Teh, K.S. 354
Tekes, C. 584
Teng, F. 298
Teramachi, Y. 925
Terao, K. 433
Thakar, V.A. 214
Thanh-Vinh, N. 765, 979
Thomas, D. 459
Thron, A. 628
Tian, H. 362, 556
Tian, H.-C. 616
Tichelaar, F.D. 342
Tilmans, H.A.C. 218, 1095
Ting, S.K. 88
Toan, N.V. 1245
Toda, M. 151
Tomida, M. 318
Tonooka, T. 17
Torayashiki, O. 1155
Torri, G.B. 218
Totsu, K. 482
Triantafyllou, M. 741
Tsai, D. 861
Tsai, H.H. 1139
Tsang, M. 358
Tseng, F.-G. 865
Tsujimura, H. 196
Tsukagoshi, T. 290
Tsuwaki, M. 108
Tu, C. 1257
Tuan, Y.-T. 757
Turner, K.L. 1281

U

Uchino, R. 1155
Uematsu, S. 1179
Ueno, R. 514
Urey, H. 330
Utermöhlen, F. 572

V

Vandenryt, T. 272
Varel, C. 983
Vargo, S. 459
Vásquez Quintero, A. 532
Vaziri, S. 1055
Verhagen, E. 140
Vericella, J. 510
Verma, N. 947
Villanueva, L.G. 632
Viswanath, A. 494
Vollebregt, S. 48, 342
Vullers, R.J.M. 568

W

Wada, T. 417
Wagner, P. 272
Wahl, S. 797
Waizmann, U. 437
Wakasa, Y. 378
Wallrabe, U. 1171
Walus, K. 620
Wang, A.I. 1103, 1217
Wang, B. 306
Wang, C. 136
Wang, C. 833
Wang, C. 1075
Wang, C.H. 5, 821
Wang, E.N. 576
Wang, J.J. 1139
Wang, L. 1201
Wang, P.-C. 865
Wang, Q. 636
Wang, R. 817

Wang, S.	1225
Wang, T.D.	805
Wang, T.H.	334
Wang, W.	9, 52, 276
Wang, X.H.	159, 298, 401, 405, 413, 684
Wang, Y.-H.	781, 861
Wang, Y.L.	250
Wang, Z.	568
Wang, Z.	793, 1079
Warren, R.	167, 393
Washabaugh IV, E.P.	825
Wasserman, S.C.	116
Watanabe, J.	931
Weber, A.	813
Weber, M.H.	453
Weber, N.	881
Wei, K.	474, 909
Wei, Z.	817
Weis, J.	437
Weis, S.	140
Weisgraber, T.	510
Wen, J.-H.	1189
Wen, Y.	1225
Werthschützky, R.	536
Wheeler, B.	498
White, R.M.	971
Wiegerink, R.J.	680, 696, 975, 999
Wijngaart van der, W.	96, 987
Williams, B.J.	192
Winebarger, P.	36
Winer, M.H.	939, 1051
Wipf, M.	1063
Witters, D.	272
Wong, M.	445
Woo, J.-K.	20
Wouden van der, E.	975
Wraback, M.	84
Wright, P.K.	971
Wu, C.	13
Wu, J.J.	5, 821
Wu, J.Y.	382
Wu, K.	1205

Wu, M.C. 1193

Wu, M.X. 52

Wu, S. 200

Wu, W. 276

Wu, W.G. 644

Wu, X. 413, 684

X

Xie, D. 556

Xie, L. 459

Xie, Y. 672, 1123

Xing, X. 951

Xiong, S. 1015

Xiong, W.J. 9

Xu, M. 668

Xu, P.C. 302, 995

Xu, T. 584

Xue, F. 777

Y

Yabe, A. 1043

Yablonovitch, E. 1103

Yagi, S. 604

Yagubizade, H. 1237

Yagyu, H. 378

Yahiro, M. 147

Yamada, H. 709

Yamagiwa, S. 604, 857

Yamaguchi, U. 709

Yamamoto, K. 729

Yang, B. 616

Yang, C. 124, 628, 1159

Yang, C.-S. 616

Yang, H. 1115

Yang, R. 793

Yang, S.Y. 5

Yang, Y. 24, 80, 588, 749, 773, 1277, 1281

Yang, Y. 556

Yang, Y.-A. 959

Yang, Y.-J. 441, 809

Yao, K. 1159

Yao, S.-C. 544

Yaralioglu, G.G. 330
Yasukawa, A. 181
Yeh, J.A. 648, 713
Yobas, L. 951
Yokokawa, R. 314, 1107
Yoo, J.H. 540, 967
Yoo, K. 540, 967
Yoon, C.K. 704
Yoon, D.H. 92
Yoon, E. 943
Yoon, E.-S. 853
Yoon, J.-B. 660
Yoon, J.H. 498, 917
Yoon, Y.H. 660
Yoon, Y.K. 498, 917
Yoshida, R. 463
Yoshida, S. 254, 322, 467
Yoshikawa, Y. 417
Yossifon, G. 268
You, S.M. 955, 1039
Young, D. 793
Yu, H. 995
Yu, H.Q. 9, 52
Yu, J. 861
Yu, L. 104
Yu, X. 1225
Yu, Y.J. 544
Yuan, J. 991
Yuan, W.Z. 761

Z

Zaghloul, U. 1099
Zang, X.N. 354, 486, 1159
Zappe, H. 881, 905, 1167
Zec, H.C. 334
Zeimpekis, I. 761
Zeng, F. 445
Zeng, X.Y. 1047
Zentel, R. 905
Zhang, G.Q. 48
Zhang, H. 276
Zhang, H.X. 346, 425

Zhang, H.Z.	9, 52
Zhang, J.	84
Zhang, K.D.	1047
Zhang, S.	502
Zhang, W.	88, 664
Zhang, X.	84
Zhang, X.	841
Zhang, X.S.	346, 425
Zhang, Y.	60
Zhang, Y.	502
Zhang, Y.	624
Zhao, C.	781
Zhao, L.	449
Zhao, X.	84
Zhao, Y.	474, 909
Zhao, Y.	1225
Zheng, S.	817
Zheng, X.	510
Zheng, Y.	833
Zhou, J.	1047
Zhu, H.	624
Zhu, H.-Y.	616
Zhu, J.	242, 306, 1075
Zhu, W.M.	88, 664
Zhu, Y.	688
Ziaie, B.	704
Ziegler, D.	128